WITHDRAWN

**BUSINESS REPLY MAIL**

FIRST CLASS MAIL PERMIT NO. 26 NEW YORK, NY

*POSTAGE WILL BE PAID BY ADDRESSEE*

**McGraw-Hill Book Company**
E. T. Matthews — 26th Floor
1221 Avenue of the Americas
New York, NY 10124-0025

# STANDARD HANDBOOK FOR ELECTRICAL ENGINEERS

## Other McGraw-Hill Reference Books of Interest

### Handbooks

*Baumeister* • Marks' Standard Handbook for Mechanical Engineers
*Bovay* • Handbook of Mechanical and Electrical Systems for Buildings
*Brady and Clauser* • Materials Handbook
*Brater and King* • Handbook of Hydraulics
*Croft, Carr, and Watt* • American Electricians' Handbook
*Harris* • Shock and Vibration Handbook
*Hicks* • Standard Handbook of Engineering Calculations
*Hicks and Mueller* • Standard Handbook of Professional Consulting Engineering
*Juran* • Quality Control Handbook
*Kurtz* • Handbook of Engineering Economics
*Maynard* • Industrial Engineering Handbook
*Pachner* • Handbook of Numerical Analysis Applications
*Parmley* • Mechanical Components Handbook
*Parmley* • Standard Handbook of Fastening and Joining
*Perry and Green* • Perry's Chemical Engineers' Handbook
*Raznjevic* • Handbook of Thermodynamic Tables and Charts
*Rohsenow, Hartnett, and Ganic* • Handbook of Heat Transfer Applications
*Rohsenow, Hartnett, and Ganic* • Handbook of Heat Transfer Fundamentals
*Rothbart* • Mechanical Design and Systems Handbook
*Seidman and Mahrous* • Handbook of Electric Power Calculations
*Smeaton* • Motor Application and Control Handbook
*Smeaton* • Switchgear and Control Handbook
*Tuma* • Engineering Mathematics Handbook
*Tuma* • Handbook of Physical Calculations
*Tuma* • Technology Mathematics Handbook

### Encyclopedias

Concise Encyclopedia of Science and Technology
Encyclopedia of Electronics and Computers
Encyclopedia of Energy
Encyclopedia of Engineering

### Dictionaries

Dictionary of Electrical and Electronic Engineering
Dictionary of Electronics and Computer Science
Dictionary of Mechanical and Design Engineering
Dictionary of Scientific and Technical Terms

# STANDARD HANDBOOK FOR ELECTRICAL ENGINEERS

**Donald G. Fink** *Editor*
*Director Emeritus, Institute of Electrical and Electronics Engineers;*
*formerly Vice President—Research, Philco Corporation;*
*President of the Institute of Radio Engineers;*
*Editor of the Proceedings of the IRE; Fellow of the IEEE;*
*Fellow of the IEE (London); Eminent Member, Eta Kappa Nu;*
*Member of the National Academy of Engineering*

**H. Wayne Beaty** *Editor*
*Vice President, Loadmaster Systems, Inc.;*
*Former Senior Editor, Electrical World;*
*Senior Member of the Institute of Electrical and Electronics Engineers*

**Twelfth Edition**

**McGRAW-HILL BOOK COMPANY**

New York St. Louis San Francisco Auckland Bogotá
Hamburg Johannesburg London Madrid Mexico
Milan Montreal New Delhi Panama
Paris São Paulo Singapore
Sydney Tokyo Toronto

*The Library of Congress cataloged the First Issue as follows:*

Standard handbook for electrical engineers. 1st- ed.
v, diagrs., tables. 18–24 cm.
Editors: 4th–6th, F. F. Fowle.—7th- A. E. Knowlton
1. Electric engineering—Handbooks, manuals, etc.
I. Fowle, Frank Fuller, date, ed. II. Knowlton, Archer Eben, date, ed.
TK151.S8 56-6964
Library of Congress [r58n[3]19]
ISBN 0-07-020975-8



1234567890 KPH KPH 8932109876

ISBN 0-07-020975-8

The editors for this book were Harold B. Crawford and Jim Halston, the designer was Mark E. Safran, and the production supervisor was Thomas G. Kowalczyk. It was set in Times Roman by Progressive Typographers and University Graphics, Inc.

Printed and bound by Arcata Graphics/Kingsport

# CONTENTS

**Section 9. Hydroelectric Power Generation** . . . . . 9-1

Hydraulic turbines; elements of a hydroelectric plant; power plant settings; design factors; efficiency performance; reversible pump turbines; model tests; speed control; cavitation

**Section 10. Power-System Components** . . . . . 10-1

Transformer theory, connections, design, insulation, testing, loading practices; circuit breaker principles, design, and operation; metal-clad switchgear; voltage regulators; methods of regulation; power capacitor application and principles; line drop compensation; fuses, switches, molded-case breakers, and buses

**Section 11. Alternate Sources and Converters of Power** . . . . . 11-1

Solar power; geothermal power; wind power; energy storage methods; magnetohydrodynamics; nuclear fusion; thermoelectric and thermionic conversion; batteries and fuel cells

**Section 12. Economics of Bulk Electric Power Supply** . . . . . 12-1

Primary sources of power; energy storage systems; by-product power; fuels; development of overall costs; reserve capacity; reliability; availability; capacity factor; load factor; generating capacity mix; interconnection and pooling

**Section 13. Project Economics** . . . . . 13-1

Bottom-line economic measurements; value of money; decision criteria; after-tax cash flows; financing effects; leasing; rate-of-return requirements; investments; risk and reward

**Section 14. Transmission Systems** . . . . . 14-1

Overhead ac power transmission; economics; electrical properties of conductors; system stability; corona, audible noise, and ozone; line insulation; mechanical designs; operation and maintenance; underground transmission systems; cable systems; electrical characteristics; installation methods; accessories; future developments

**Section 15. Direct-Current Power Transmission** . . . . . 15-1

Converter behavior; terminal design; filters; system protection; switching; subsynchronous oscillations; overhead lines and cables; valves; transformers and reactors; breakers; testing; operation and maintenance

**Section 16. Power-System Interconnections** . . . . . 16-1

Control of generation and power flow; schedule deviations; economic dispatch; control application; power system security; relaying and protection; power-line carrier; carrier current; carrier communication; carrier relaying

# CONTRIBUTORS

**Michael M. Adibi** *President, IRD Corporation* (SEC. 25)

**P. F. Albrecht** *Senior Applications Engineer, Electric Utility System Engineering Department, General Electric Company* (SEC. 15)

**Roy P. Allen** *Manager, Advanced Development Engineering, General Electric Company* (SEC. 6)

**Charles L. Amick** *Lighting Consultant* (SEC. 26)

**R. E. Appleyard** *Engineering Consultant* (SEC. 7)

**Mark A. Baker** *Vice President, Technology, Alcoa Conductor Products Company* (SEC. 4)

**Lionel O. Barthold** *Chairman, Power Technologies, Inc.* (SEC. 14)

**Joseph Basilesco** *President, Basilesco Consultants* (SEC. 17)

**Norman B. Belecki** *Group Leader, Electricity Division, National Bureau of Standards* (SEC. 3)

**James R. Birk** *Director, Advanced Conversion and Storage Department, Electric Power Research Institute* (SEC. 11)

**Keith F. Blurton** *Vice President, PCK Technology Division, Kollmorgen Corporation* (SEC. 24)

**Warren B. Boast** *Anson Marston Distinguished Professor Emeritus of Electrical Engineering, Iowa State University* (SEC. 2)

**B. P. Boehle** *Switchgear Division, Brown Boveri Company* (SEC. 10)

**Robert W. Bohl** *Professor, Metallurgy and Nuclear Engineering, University of Illinois* (SEC. 4)

**G. D. Breuer** *Consultant, Electric Utility System Engineering Department, General Electric Company* (SEC. 15)

**Jerome B. Brewster** *Scientist, Westinghouse Electric Corporation* (SEC. 22)

**Harold E. Campbell** *Retired, formerly Senior Engineer, General Electric Company* (SEC. 18)

**Allen L. Clapp** *Managing Director, Clapp Research Associates* (SECS. 5, 11, 13, 18, 19)

**Kara Clark** *Applications Engineer, General Electric Company* (SEC. 15)

**Nathan Cohn** *General Partner and Senior Associate, Network Systems Development Associates* (SEC. 16)

**Arthur D. Crino** *Manager of Engineering, Siemens-Allis, Inc.* (SEC. 10)

**Edward J. Croop** *Consulting Scientist, Westinghouse Electric Corporation* (SEC. 4)

**John Cutting** *Director of Power Systems, Autodynamics, Inc.* (SEC. 11)

**J. D. Cypert** *Manager, Energy Utilities Engineering Support Center, IBM Corporation* (SEC. 25)

**Joseph D'Auria** *Project Manager, Resource Dynamics Corporation* (SEC. 21)

**Thomas W. Dakin** *Consultant* (SEC. 4)

**Robert C. Degeneff** *Manager, HVDC Transmission Engineering, General Electric Company* (SEC. 15)

**Anthony M. DiGioia, Jr.** *President, GAI Consultants, Inc.* (SEC. 14)

**Dale A. Douglas** *Senior Engineer, Power Technologies, Inc.* (SEC. 14)

**Jack H. Easley** *Deceased, formerly Senior Engineer, General Electric Company* (SEC. 18)

**Edward F. Eaton** *Manager, Marine Applications Engineering, General Electric Company* (SEC. 23)

**Walter A. Elmore** *Consulting Engineer and Section Manager, Relay Instrument Division, Westinghouse Electric Corporation* (SEC. 16)

**Arnold P. Fickett** *Director, Energy Utilization Department, Electric Power Research Institute* (SEC. 11)

**Harold J. Fielder** *Senior Applications Engineer, General Electric Company* (SEC. 15)

**Donald G. Fink** *Director Emeritus, Institute of Electrical and Electronics Engineers* (SEC. 1)

**Clarence W. Flairty** *Manager, HVDC Control Engineering, General Electric Company* (SEC. 15)

**R. X. French** *Manager, Electrical Department, Sargent & Lundy Engineers* (SEC. 12)

**N. Richard Friedman** *Chairman, Resource Dynamics Corporation* (SEC. 21)

**Albert L. Gaines** *President, A. Gaines Company* (SEC. 5)

**Donald Galler** *Senior Engineer, Alexander Kusko, Inc.* (SEC. 22)

**Jalal Gohari** *Principal Electrical Engineer, Gibbs & Hill, Inc.* (SEC. 16)

**Ian S. Grant** *Manager, Software Products Department, Power Technologies, Inc.* (SEC. 14)

**L. Gyugyi** *Manager, Power Electronics Department, Westinghouse Electric Corporation* (SEC. 22)

**Harry W. Hale** *Professor of Electrical Engineering, Iowa State University* (SEC. 2)

**Donald H. Hall** *Senior Engineer, General Electric Company* (SEC. 6)

**William F. Hamilton** *Consultant* (SEC. 23)

**James H. Harlow** *Manager, Development Engineering, Siemens-Allis, Inc.* (SEC. 10)

**Charles A. Harper** *Manager, Materials Engineering and Technologies, Westinghouse Electric Corporation* (SEC. 4)

**Forest K. Harris** *Retired, Physicist, Electricity Division, National Bureau of Standards* (SEC. 3)

**T. M. Heinrich** *Manager, Motor Drive Systems, Westinghouse Electric Corporation* (SEC. 22)

**David W. Houghtaling** *Senior Applications Engineer, General Electric Company* (SEC. 15)

**Frederick N. Houser** *Consultant, National Research Council* (SEC. 23)

**Fred N. Huffman** *Manager, Thermionic Conversion Department, Thermo Electron Corporation* (SEC. 11)

**Daniel Jackson, Jr.** *Associate Editor, Coal Age* (SEC. 21)

**Erwin T. Jauch** *Senior Applications Engineer, General Electric Company* (SEC. 15)

**Loering M. Johnson** *President, Johnson Consulting Engineers* (SEC. 5)

**Edwin C. Jones, Jr.** *Professor of Electrical Engineering, Iowa State University* (SEC. 2)

**Alexander Kusko** *President, Alexander Kusko, Inc.* (SEC. 20)

**James J. LaForest** *Senior Applications Engineer, General Electric Company* (SEC. 15)

**John Lapp** *Manager, Advanced Development, McGraw-Edison Company* (SEC. 10)

**Einar V. Larsen** *Senior Applications Engineer, General Electric Company* (SEC. 15)

**Kenneth L. Latimer** *Senior Engineer, Arnold Engineering Company* (SEC. 4)

**Wendell G. Leisinger** *Manager, Circuit Breaker Marketing Section, Square D Company* (SEC. 19)

**Enrico Levi** *Professor, Polytechnic Institute of New York* (SEC. 23)

**David Linden** *Consultant* (SEC. 11)

**Donald L. Lott** *Development Engineer, Siemens-Allis, Inc.* (SEC. 10)

**John Lusti** *Vice President, Corporate Technical Development, Otis Elevator Company* (SEC. 21)

**Duane E. Lyon** *Professor of Wood Science, Mississippi State University* (SEC. 4)

**John A. Makuch** *Director, Product Design, Fossil Power Systems, Combustion Engineering* (SEC. 5)

**James C. McIver** *Applications Engineer, General Electric Company* (SEC. 15)

**Robert W. McKnight** *Manager, Communication and Signal Engineering, Association of American Railroads* (SEC. 23)

**Michael F. McNitt-Gray** *Instructor, Pennsylvania State University* (SEC. 11)

**William J. McNutt** *Manager, Advance Development Engineering, General Electric Company* (SEC. 10)

**George H. Miley** *Chairman, Nuclear Engineering Program, University of Illinois* (SEC. 11)

**Richard C. Miller** *Section Manager, Water Resources and Hydraulic Engineering Group, Stone & Webster Engineering Corporation* (SEC. 5)

**Ronald E. Mittelstaedt** *Marketing Supervisor, Siemens-Allis, Inc.* (SEC. 10)

**Larry Mizen** *Senior Engineer, General Electric Company* (SEC. 6)

**James A. Moran** *U.S. Technical Representative, Nokia Engineering* (SEC. 14)

**John D. Mozer** *Technical Supervisor, GAI Consultants, Inc.* (SEC. 14)

**William E. Murray** *Senior Staff Engineer, Douglas Aircraft Company* (SEC. 23)

**Earl H. Myers** *Consultant, Large Motor Engineering* (SEC. 8)

**Wendell Neugebauer** *Project Manager, General Electric Company* (SEC. 27)

**Farhad Nozari** *Senior Applications Engineer, General Electric Company* (SEC. 15)

**R. M. Oates** *Manager, Advanced Systems Laboratory, Westinghouse Electric Corporation* (SEC. 22)

**Jerald D. Parker** *Professor of Mechanical Engineering, Oklahoma State University* (SEC. 11)

**Syed M. Peeran** *Senior Engineer, Alexander Kusko, Inc.* (SEC. 20)

**B. R. Pelley** *Vice President, Worldwide High Power Products, International Rectifier* (SEC. 22)

**Charles A. Popeck** *Manager, Apparatus Engineering, A. B. Chance Company* (SEC. 10)

**R. Ramakumar** *Professor of Electrical Engineering, Oklahoma State University* (SEC. 11)

**Robert W. Rex** *Chairman, Republic Geothermal Company* (SEC. 11)

**Raymond L. Rofini** *Senior Valve Design Engineer, General Electric Company* (SEC. 15)

**Walter J. Ros** *Manager, Transmission and Distribution Product Application Engineering, General Electric Company* (SEC. 18)

**L. T. Rosenberg** *Turbogenerator Consultant* (SEC. 7)

**Daniel Rossiter** *Vice President, Macro Corporation* (SEC. 16)

**W. J. Schmitt** *Switchgear Division, Brown Boveri Company* (SEC. 10)

**H. M. Schneider** *HV Transmission Research Facility, General Electric Company* (SEC. 15)

**Norman R. Schultz** *Deceased, formerly Manager, Power Distribution Systems Engineering Operations, General Electric Company* (SEC. 18)

**Kelly A. Shaw** *Manager of Marketing, Small Motor Division, Siemens-Allis, Inc.* (SEC. 10)

**Sava I. Sherr** *Staff Director of Standards, Institute of Electrical and Electronics Engineers* (SEC. 28)

**Robert J. Soulen, Jr.** *Chief, Temperature and Pressure Division, National Bureau of Standards* (SEC. 3)

**James R. Stewart** *Senior Consultant, Power Technologies, Inc.* (SEC. 14)

**J. D. Stickler** *Project Manager, General Electric Company* (SEC. 15)

**Joseph Urbanek** *Manager, HVDC Valve and Auxiliary Engineering, General Electric Company* (SEC. 15)

**Roland W. Ure, Jr.** *Former Professor of Electrical Engineering and Materials Science, University of Utah* (SEC. 11)

**Anthony L. Von Holle** *Principal Research Engineer, Armco, Inc.* (SEC. 4)

**Grover F. Wachter** *Manager, Repair and Rehabilitation Services, Allis-Chalmers Hydro, Inc.* (SEC. 9)

**Robert D. Weaver** *Project Manager, Electric Power Research Institute* (SEC. 11)

**Jay A. Williams** *Manager, Power Technologies, Inc.* (SEC. 14)

**J. R. Willy** *Vice President, Engineering, Applied Systems, Inc.* (SEC. 10)

**Delano D. Wilson** *President, Power Technologies, Inc.* (SEC. 14)

**Bruce F. Wollenberg** *Principal Consultant, Power System Security, Control Data Corporation* (SEC. 16)

**Peter Wood** *Consulting Engineer, Westinghouse Electric Corporation* (SEC. 22)

**Robert H. Wood, Jr.** *Manager, Busway Marketing, Square D Company* (SEC. 19)

**J. Sam Young** *Retired, McDonnell Douglas Astronautics, Inc.* (SEC. 23)

**L. E. Zaffanella** *Manager, HVT Research Facility, General Electric Company* (SEC. 15)

# PREFACE TO THE TWELFTH EDITION

Since the publication of the preceding edition of this handbook, several changes have taken place which impact on the science and technology of electrical engineering. The accident at Three Mile Island Nuclear Plant in Pennsylvania has caused many changes in the technology associated with generating nuclear energy. The more recent accident at the Soviet Union's Chernobyl nuclear plant will undoubtedly cause an even closer scrutiny of the use of nuclear power.

Industry plans for the use of energy sources included a major sector for nuclear but now a new scenario has been produced for the fuel requirements over the next decade that redistributes the amount planned for nuclear systems.

The need for conservation of energy resources has continued to affect the design of end-use products and associated technologies of monitoring and controlling energy use to reduce both consumption and demand.

Alternative energy sources such as wind power, solar power, geothermal, magnetohydrodynamics, nuclear fusion, storage systems, and fuel cells have taken on more importance in the face of dwindling nonrenewable energy sources.

All of this new technology is covered in this edition of the *Standard Handbook.* An entirely new section on Project Economics has been included which guides the engineer to the important financial considerations of projects. All other sections have been updated and some were substantially rewritten to include new material and to better organize the material for better use of the handbook.

Major new material and revised sections include the new advances in computer technologies in the electrical engineering discipline for power industry management, system planning, operation, plant monitoring and control, design, construction, environmental considerations, and future computer applications.

Important new material on high-voltage transmission systems includes high-voltage direct-current systems that solve many of the interconnection problems associated with large power pools. New design considerations for power-system interconnections are also included.

Exciting new technologies in the consumer end of electrical energy are included. Advances in residential, commercial, and industrial applications of electrical power have resulted in more efficient motors, lighting, and transportation.

A complete listing of standards governing the electrical engineering field is included. The section includes a history of standards, those relating to law, government regulatory standards bodies, and a complete alphabetical list of standards by subject matter.

The aim of the Twelfth Edition is consistent with that of previous editions of the handbook which is to contain in a single volume all pertinent data within its scope, to be accurate and comprehensive in technical treatment, to be of use in engineering practice (as well as in study and in preparation for such practice) and, above all, to be oriented toward practical application, including the impact of economic considerations. The scope of the handbook includes the generation, transmission, distribution, control, conservation, and application of electrical power.

Many phases of the electronic sciences and technology are included but this is not a comprehensive treatise on the subject of electronics. These detailed subjects are treated in the *Electronics Engineers' Handbook*. These two volumes cover the whole field of electrical and electronics engineering.

The Twelfth Edition includes much new and updated material and continues the tradition of having one complete volume of pertinent material for engineering reference and practice.

H. WAYNE BEATY
DONALD G. FINK

*Editors*

# STANDARD HANDBOOK FOR ELECTRICAL ENGINEERS

# SECTION 1

# UNITS, SYMBOLS, CONSTANTS, DEFINITIONS, AND CONVERSION FACTORS

**Donald G. Fink**

*Director Emeritus, Institute of Electrical and Electronics Engineers; Editor, Standard Handbook for Electrical Engineers and the Electronics Engineers' Handbook; Fellow, IEEE*

## CONTENTS

*Numbers refer to paragraphs*

***1. The SI Units.*** The units of the quantities most commonly used in electrical engineering (volts, amperes, watts, ohms, etc.) are those of the metric system. They are embodied in the International System of Units (Système International d'Unités, abbreviated "SI"). The SI units are used throughout this handbook, in accordance with the established practice of electrical engineering publications throughout the world. Other units, notably the cgs (centimeter-gram-second) units, may be used in citations in the earlier literature. The cgs electrical units are listed in Table 1-9 with conversion factors to the SI units.

The SI electrical units are based on the mksa (meter-kilogram-second-ampere) system. They have been adopted by the standardization bodies of the world, including the International Electrotechnical Commission (IEC), the American National Standards Institute (ANSI), and the Standards Board of the Institute of Electrical and Electronics Engineers (IEEE).

***2. CGPM Base Quantities.*** Seven quantities have been adopted by the General Conference on Weights and Measures (CGPM) as *base quantities,* that is, quantities that are not derived from other quantities. The base quantities are length, mass, time, electric current, thermodynamic temperature, amount of substance, and luminous intensity. Table 1-1 lists these quantities, the name of the SI unit for each, and the standard letter symbol by which each is expressed in the International System (SI).

**TABLE 1-1** SI Base Units

| Quantity | Unit | Symbol |
|---|---|---|
| Length | meter | m |
| Mass | kilogram | kg |
| Time | second | s |
| Electric current | ampere | A |
| Thermodynamic temperature* | kelvin | K |
| Amount of substance | mole | mol |
| Luminous intensity | candela | cd |

* Celsius temperature is in general expressed in degrees Celsius (symbol °C). See Par. **6**. Reproduced from ANSI/IEEE Standard 268-1982.

The units of the base quantities have been defined by the CGPM as follows:

*meter.* The length equal to 1 650 763.73 wavelengths in vacuum of the radiation corresponding to the transition between the levels $^2p_{10}$ and $^5d_5$ of the krypton-86 atom (CGPM, 1960).

*kilogram.* The unit of mass; it is equal to the mass of the international prototype of the kilogram (CGPM, 1901).

EDITOR'S NOTE: The prototype is a platinum-iridium cylinder maintained at the International Bureau of Weights and Measures, near Paris. The kilogram is approximately equal to the mass of 1000 cubic centimeters of water at its temperature of maximum density.

*second.* The duration of 9 192 631 770 periods of the radiation corresponding to the transition between the two hyperfine levels of the ground state of the cesium 133 atom (CGPM, 1967).

*ampere.* The constant current that, if maintained in two straight parallel conductors of infinite length, of negligible circular cross section, and placed 1 meter apart in vacuum, would produce between these conductors a force equal to $2 \times 10^{-7}$ newton per meter of length (CGPM, 1948).

*kelvin.* The unit of thermodynamic temperature is the fraction 1/273.16 of the thermodynamic temperature of the triple point of water (CGPM, 1967).

EDITOR'S NOTE: The zero of the Celsius scale (the freezing point of water) is defined as 0.01 K below the triple point, that is, 273.15 K. See Table 1-26.

*mole.* That amount of substance of a system that contains as many elementary entities as there are atoms in 0.012 kilogram of carbon-12 (CGPM, 1971).

NOTE: When the mole is used, the elementary entities must be specified. They may be atoms, molecules, ions, electrons, other particles, or specified groups of such particles.

*candela.* The luminous intensity, in a given direction, of a source that emits monochromatic radiation of frequency 540 × 10[12] Hz and that has a radiant intensity in that direction of 1/683 watt per steradian (CGPM, 1979).

EDITOR'S NOTE: Until Jan. 1, 1948, the generally accepted unit of luminous intensity was the *international candle.* The difference between the candela and the international candle is so small that only measurements of high precision are affected. The use of the term candle is deprecated.

***3. Supplementary SI Units.*** Two additional SI units, numerics which are considered as dimensionless derived units (see Par. **4**), are the radian and the steradian, for the quantities plane angle and solid angle, respectively. Table 1-2 lists these quantities, their units and symbols. The supplementary units are defined as:

*radian.* The plane angle between two radii of a circle which cut off on the circumference an arc equal in length to the radius (CGPM, 1960).

*steradian.* The solid angle which, having its vertex in the center of a sphere, cuts off an area of the surface of the sphere equal to that of a square with sides equal to the radius of the sphere (CGPM, 1960).

**TABLE 1-2** SI Supplementary Units

| Quantity | Unit | Symbol |
|---|---|---|
| Plane angle | radian | rad |
| Solid angle | steradian | sr |

Reproduced from ANSI/IEEE Standard 268-1982.

***4. Derived SI Units.*** Most of the quantities and units used in electrical engineering fall in the category of *SI derived units,* that is, units which can be completely defined in terms of the base and supplementary quantities described above. Table 1-3 lists the principal electrical quantities in the SI system, and shows their equivalents in terms of the base and supplementary units. The definitions of these quantities, as they appear in the *IEEE Standard Dictionary of Electrical and Electronics Terms*[19*] are:

*hertz.* The unit of frequency 1 cycle per second.

*newton.* The force that will impart an acceleration of 1 meter per second per second to a mass of 1 kilogram.

*pascal.* The pressure exerted by a force of 1 newton uniformly distributed on a surface of 1 square meter.

*joule.* The work done by a force of 1 newton acting through a distance of 1 meter.

*watt.* The power required to do work at the rate of 1 joule per second.

*coulomb.* The quantity of electric charge that passes any cross section of a conductor in 1 second when the current is maintained constant at 1 ampere.

*volt.* The potential difference between two points of a conducting wire carrying a constant current of 1 ampere, when the power dissipated between these points is 1 watt.

*farad.* The capacitance of a capacitor in which a charge of 1 coulomb produces 1 volt potential difference between its terminals.

*ohm.* The resistance of a conductor such that a constant current of 1 ampere in it produces a voltage of 1 volt between its ends.

*siemens (mho).* The conductance of a conductor such that a constant voltage of 1 volt between its ends produces a current of 1 ampere in it.

*weber.* The magnetic flux whose decrease to zero when linked with a single turn induces in the turn a voltage whose time integral is 1 volt-second.

*tesla.* The magnetic induction equal to 1 weber per square meter.

---

* Superior numbers refer to the Bibliography, Par. **17**.

**TABLE 1-3** SI Derived Units Used in Electrical Engineering

| Quantity | SI unit | | | |
|---|---|---|---|---|
| | Name | Symbol | Expression in terms of other units | Expression in terms of SI base units |
| Frequency (of a periodic phenomenon) | hertz | Hz | 1/s | $s^{-1}$ |
| Force | newton | N | | $m \cdot kg \cdot s^{-2}$ |
| Pressure, stress | pascal | Pa | $N/m^2$ | $m^{-1} \cdot kg \cdot s^{-2}$ |
| Energy, work, quantity of heat | joule | J | $N \cdot m$ | $m^2 \cdot kg \cdot s^{-2}$ |
| Power, radiant flux | watt | W | J/s | $m^2 \cdot kg \cdot s^{-3}$ |
| Quantity of electricity, electric charge | coulomb | C | $A \cdot s$ | $s \cdot A$ |
| Potential difference, electric potential, electromotive force | volt | V | W/A | $m^2 \cdot kg \cdot s^{-3} \cdot A^{-1}$ |
| Electric capacitance | farad | F | C/V | $m^{-2} \cdot kg^{-1} \cdot s^4 \cdot A^2$ |
| Electric resistance | ohm | Ω | V/A | $m^2 \cdot kg \cdot s^{-3} \cdot A^{-2}$ |
| Conductance | siemens | S | A/V | $m^{-2} \cdot kg^{-1} \cdot s^3 \cdot A^2$ |
| Magnetic flux | weber | Wb | $V \cdot s$ | $m^2 \cdot kg \cdot s^{-2} \cdot A^{-1}$ |
| Magnetic flux density | tesla | T | $Wb/m^2$ | $kg \cdot s^{-2} \cdot A^{-1}$ |
| Celsius temperature | degree Celsius | °C | K | |
| Inductance | henry | H | Wb/A | $m^2 \cdot kg \cdot s^{-2} \cdot A^{-2}$ |
| Luminous flux | lumen | lm | | $cd \cdot sr$* |
| Illuminance | lux | lx | $lm/m^2$ | $m^{-2} \cdot cd \cdot sr$* |
| Activity (of radionuclides) | becquerel | Bq | l/s | $s^{-1}$ |
| Absorbed dose | gray | Gy | J/kg | $m^2 \cdot s^{-2}$ |
| Dose equivalent | sievert | Sv | J/kg | $m^2 \cdot s^{-2}$ |

* In this expression the steradian (sr) is treated as a base unit. See Table 1-2.
Adapted from ANSI/IEEE Standards 268-1973, 268-1976 and 268-1982.

*henry.* The inductance for which the induced voltage in volts is numerically equal to the rate of change of current in amperes per second.

*lumen.* The flux through a unit solid angle (steradian) from a uniform point source of 1 candela; the flux on a unit surface all points of which are at a unit distance from a uniform point source of 1 candela.

*lux.* The illumination on a surface of 1 square meter on which there is uniformly distributed a flux of 1 lumen; the illumination produced at a surface all points of which are 1 meter from a uniform point source of 1 candela.

Table 1-4 lists other quantities and the SI derived unit names and symbols useful in engineering applications. Table 1-5 lists additional quantities and the SI derived units and symbols used in mechanics, heat, and electricity.

***5. SI Decimal Prefixes.*** All SI units may have affixed to them standard prefixes which multiply the indicated quantity by a power of 10. Table 1-6 lists the standard prefixes and their symbols. A substantial part of the extensive range ($10^{36}$) covered by these prefixes is in common use in electrical engineering (e.g., gigawatt, gigahertz, nanosecond, and picofarad). The practice of compounding a prefix (e.g., micromicrofarad) is deprecated (the correct term is picofarad).

***6. Usage of SI Units, Symbols, and Prefixes.*** Care must be exercised in using the SI symbols and prefixes to follow exactly the capital-letter and lowercase-letter usage prescribed in Tables 1-1 through 1-8, inclusive. Otherwise serious confusion may occur. For example, pA is

**TABLE 1-4** Examples of SI Derived Units of General Application in Engineering

| Quantity | SI unit Name | Symbol |
|---|---|---|
| Angular velocity | radian per second | rad/s |
| Angular acceleration | radian per second squared | $rad/s^2$ |
| Radiant intensity | watt per steradian | W/sr |
| Radiance | watt per square meter steradian | $W\cdot m^{-2}\cdot sr^{-1}$ |
| Area | square meter | $m^2$ |
| Volume | cubic meter | $m^3$ |
| Velocity | meter per second | m/s |
| Acceleration | meter per second squared | $m/s^2$ |
| Wavenumber | 1 per meter | $m^{-1}$ |
| Density, mass | kilogram per cubic meter | $kg/m^3$ |
| Concentration (of amount of substance) | mole per cubic meter | $mol/m^3$ |
| Specific volume | cubic meter per kilogram | $m^3/kg$ |
| Luminance | candela per square meter | $cd/m^2$ |

Reproduced from ANSI/IEEE Standard 268-1982.

the SI symbol for $10^{-12}$ of the SI unit for electric current (picoampere) while Pa is the SI symbol for pressure (the pascal).

The spelled-out names of the SI units (e.g., volt, ampere, watt) are not capitalized. The SI letter symbols are capitalized only when the name of the unit stands for, or is directly derived from, the name of a person. Examples are V for volt, after the Italian physicist Alessandro Volta (1745–1827); A for ampere, after the French physicist André Ampère (1775–1836); and W for watt, after the Scottish engineer James Watt (1736–1819). The letter symbols serve the function of abbreviations, but they are used without periods.

**TABLE 1-5** Examples of SI Derived Units Used in Mechanics, Heat, and Electricity

| Quantity | SI unit Name | Symbol | Expression in terms of SI base units |
|---|---|---|---|
| Viscosity, dynamic | pascal second | Pa·s | $m^{-1}\cdot kg\cdot s^{-1}$ |
| Moment of force | newton meter | N·m | $m^2\cdot kg\cdot s^{-2}$ |
| Surface tension | newton per meter | N/m | $kg\cdot s^{-2}$ |
| Heat flux density, irradiance | watt per square meter | $W/m^2$ | $kg\cdot s^{-3}$ |
| Heat capacity | joule per kelvin | J/K | $m^2\cdot kg\cdot s^{-2}\cdot K^{-1}$ |
| Specific heat capacity, specific entropy | joule per kilogram kelvin | J/(kg·K) | $m^2\cdot s^{-2}\cdot K^{-1}$ |
| Specific energy | joule per kilogram | J/kg | $m^2\cdot s^{-2}$ |
| Thermal conductivity | watt per meter kelvin | W/(m·K) | $m\cdot kg\cdot s^{-3}\cdot K^{-1}$ |
| Energy density | joule per cubic meter | $J/m^3$ | $m^{-1}\cdot kg\cdot s^{-2}$ |
| Electric field strength | volt per meter | V/m | $m\cdot kg\cdot s^{-3}\cdot A^{-1}$ |
| Electric charge density | coulomb per cubic meter | $C/m^3$ | $m^{-3}\cdot s\cdot A$ |
| Electric flux density | coulomb per square meter | $C/m^2$ | $m^{-2}\cdot s\cdot A$ |
| Permittivity | farad per meter | F/m | $m^{-3}\cdot kg^{-1}\cdot s^4\cdot A^2$ |
| Current density | ampere per square meter | $A/m^2$ | |
| Magnetic field strength | ampere per meter | A/m | |
| Permeability | henry per meter | H/m | $m\cdot kg\cdot s^{-2}\cdot A^{-2}$ |
| Molar energy | joule per mole | J/mol | $m^2\cdot kg\cdot s^{-2}\cdot mol^{-1}$ |
| Molar entropy, molar heat capacity | joule per mole kelvin | J/(mol·K) | $m^2\cdot kg\cdot s^{-2}\cdot K^{-1}mol^{-1}$ |

Adapted from ANSI/IEEE Standards 268-1973 and 268-1982.

**TABLE 1-6** SI Prefixes Expressing Decimal Factors

| Factor | Prefix | Symbol | Factor | Prefix | Symbol |
|---|---|---|---|---|---|
| $10^{18}$ | exa | E | $10^{-1}$ | deci | d |
| $10^{15}$ | peta | P | $10^{-2}$ | centi | c |
| $10^{12}$ | tera | T | $10^{-3}$ | milli | m |
| $10^{9}$ | giga | G | $10^{-6}$ | micro | μ |
| $10^{6}$ | mega | M | $10^{-9}$ | nano | n |
| $10^{3}$ | kilo | k | $10^{-12}$ | pico | p |
| $10^{2}$ | hecto | h | $10^{-15}$ | femto | f |
| $10^{1}$ | deka | da | $10^{-18}$ | atto | a |

From ANSI/IEEE Standard 268-1982.

It will be noted from Tables 1-1, 1-3, and 1-5 that, with the exception of the ampere, all the SI electrical quantities and units are derived from the SI base and supplementary units, or from other SI derived units. Thus, many of the short names of SI units may be expressed in compound form embracing the SI units from which they are derived. Examples are the volt per ampere for the ohm, the joule per second for the watt, the ampere-second for the coulomb, and the watt-second for the joule. Such compound usage is permissible, but in engineering publications the short names are customarily used.

The use of the SI prefixes with non-SI units is not recommended; the only exception stated in IEEE Standard 268-1976 is the microinch. Non-SI units which are related to the metric system but are not decimal multiples of the SI units (such as the calorie, torr, and kilogram-force) are specially to be avoided.

A particular problem arises with the universally used units of time (minute, hour, day, year, etc.) that are nondecimal multiples of the second. Table 1-7 lists these and their equivalents in

**TABLE 1-7** Time and Angle Units Used in the SI System (Not Decimally Related to the SI Units)

| Name | Symbol | Value in SI unit |
|---|---|---|
| minute | min | 1 min = 60 s |
| hour | h | 1 h = 60 min = 3 600 s |
| day | d | 1 d = 24 h = 86 400 s |
| degree | ° | 1° = (π/180) rad |
| minute | ′ | 1′ = (1/60)° = (π/10 800) rad |
| second | ″ | 1″ = (1/60)′ = (π/648 000) rad |

ANSI/IEEE Standard 268-1982.

seconds, as well as their standard symbols (see also Table 1-18). The watthour (Wh) is a case in point; it is equal to 3 600 joules. The kilowatthour (kWh) is equal to 3 600 000 joules or 3.6 megajoules (MJ). In the mid-1980s the use of the kilowatthour persisted widely, although eventually it was expected to be replaced by the megajoule, with the conversion factor 3.6 megajoules per kilowatthour.

Other aspects in the usage of the SI system are the subject of the following recommendations published by the IEEE:

*Frequency.* The CGPM has adopted the name hertz for the unit of frequency, but cycle per second is widely used. Although cycle per second is technically correct, the name hertz is preferred because of the widespread use of cycle alone as a unit of frequency. Use of cycle in place of cycle per second, or kilocycle in place of kilocycle per second, etc., is incorrect.

*Magnetic Flux Density.* The CGPM has adopted the name tesla for the SI unit of magnetic flux density. The name gamma shall not be used for the unit nanotesla.

*Temperature Scale.* In 1948, the CGPM abandoned centigrade as the name of the temperature scale. The corresponding scale is now properly named the Celsius scale, and further use of centigrade for this purpose is deprecated.

*Luminous Intensity.* The SI unit of luminous intensity has been given the name candela, and further use of the old name candle is deprecated. Use of the term candle-power, either as the name of a quantity or as the name of a unit, is deprecated.

*Luminous Flux Density.* The common British-American unit of luminous flux density is the lumen per square foot. The name footcandle, which has been used for this unit in the United States, is deprecated.

*micrometer and micron.* The names micron for micrometer and millimicron for nanometer are deprecated.

*gigaelectronvolt (GeV).* Because billion means a thousand million in the United States but a million million in most other countries, its use should be avoided in technical writing. The term billion electronvolts is deprecated; use gigaelectronvolts instead.

*British-American Units.* In principle, the number of British-American units in use should be reduced as rapidly as possible. Quantities are not to be expressed in mixed units. For example, a mass should be expressed as 12.75 lb, rather than as 12 lb, 12 oz. As a start toward implementing this recommendation, the following should be abandoned:

1. British thermal unit (for conversion factors, see Table 1-24).
2. horsepower (Table 1-25).
3. Rankine temperature scale (Table 1-26).
4. U.S. dry quart, U.S. liquid quart, and U.K. (Imperial) quart, together with their various multiples and subdivisions. If it is absolutely necessary to express volume in British-American units, the cubic inch or cubic foot should be used (for conversion factors, see Table 1-16).
5. footlambert. If it is absolutely necessary to express luminance in British-American units, the candela per square foot or lumen per steradian square foot should be used (Table 1-27A).
6. inch of mercury (Table 1-22C).

*7. **Other SI Units.*** Table 1-8 lists units used in the SI system whose values are not derived

**TABLE 1-8** Units Used with the SI System, Whose Values Are Obtained Experimentally

| Name | Symbol |
|---|---|
| electronvolt | eV |
| unified atomic mass unit | u |
| astronomical unit* | |
| parsec | pc |

* The astronomical unit does not have an international symbol. AU is customarily used in English, UA in French.

Adapted from IEEE Standard 268-1973, by permission.

from the base quantities, but from experiment. The definitions of these units, given in the *IEEE Standard Dictionary*[19] are:

*electronvolt.* The kinetic energy acquired by an electron in passing through a potential difference of 1 volt in vacuum.

NOTE: The electronvolt is equal to $1.60219 \times 10^{-19}$ joule, approximately (seeTable 1-24B).

*unified atomic mass unit.* The fraction 1/12 of the mass of an atom of the nuclide $^{12}C$.

NOTE: u is equal to $1.660\,53 \times 10^{-27}$ kg, approximately.

*astronomical unit.* The length of the radius of the unperturbed circular orbit of a body of negligible mass moving around the sun with a sidereal angular velocity of 0.017 202 098 950 radian per day of 86 400 ephemeris seconds.

NOTE: The International Astronomical Union has adopted a value for 1 AU equal to $1.496 \times 10^{11}$ meters (see Table 1-14C).

*parsec.* The distance at which 1 astronomical unit subtends an angle of 1 second of arc. 1 pc = 206 264.8 AU = $30\ 857 \times 10^{12}$ m, approximately (see Table 1-14C).

***8. CGS Systems of Units.*** The units most commonly used in physics and electrical science from their establishment in 1873 until their virtual abandonment in 1948 are based on the centimeter-gram-second (cgs) electromagnetic and electrostatic systems. They have been used primarily in theoretical work, as contrasted with the SI units (and their "practical unit" predecessors, see Par. **9**) used in engineering. Table 1-9 lists the principal cgs electrical quantities, their units, symbols, and equivalent values in SI units. The use of these units in electrical engineering publications has been officially deprecated by the IEEE since 1966.

The cgs units have not been used to any great extent in electrical engineering, since many of the units are of inconvenient size compared with quantities used in practice. For example, the cgs electromagnetic unit of capacitance is the gigafarad.

***9. Practical Units (ISU).*** The shortcomings of the cgs systems were overcome by adopting the volt, ampere, ohm, farad, coulomb, henry, joule, and watt as "practical units," each being an exact decimal multiple of the corresponding electromagnetic cgs unit (see Table 1-9). From 1908 to 1948, the practical electrical units were embodied in the International System Units (ISU, not to be confused with the SI units). During these years, precise formulation of the units in terms of mass, length, and time was impractical because of imprecision in the measurements of the three basic quantities. As an alternative, the units were standardized by comparison with apparatus, called prototype standards. By 1948, advances in the measurement of the basic quantities permitted precise standardization by reference to the definitions of the basic units,

**TABLE 1-9** CGS Units and Equivalents

| Quantity | Name | Symbol | Correspondence with SI unit |
|---|---|---|---|
| | | Electromagnetic system | |
| Current | abampere | abA | = 10 amperes (exactly) |
| Voltage | abvolt | abV | = $10^{-8}$ volt (exactly) |
| Capacitance | abfarad | abF | = $10^{9}$ farads (exactly) |
| Inductance | abhenry | abH | = $10^{-9}$ henry (exactly) |
| Resistance | abohm | abΩ | = $10^{-9}$ ohm (exactly) |
| Magnetic flux | maxwell | Mx | = $10^{-8}$ weber (exactly) |
| Magnetic field strength | oersted | Oe | = 79.577 4 amperes per meter |
| Magnetic flux density | gauss | G | = $10^{-4}$ tesla (exactly) |
| Magnetomotive force | gilbert | Gb | = 0.795 774 ampere |
| | | Electrostatic system | |
| Current | statampere | statA | = $3.335\ 641 \times 10^{-10}$ ampere |
| Voltage | statvolt | statV | = 299.792 46 volts |
| Capacitance | statfarad | statF | = $1.112\ 650 \times 10^{-12}$ farad |
| Inductance | stathenry | statH | = $8.987\ 554 \times 10^{11}$ henrys |
| Resistance | statohm | statΩ | = $8.987\ 554 \times 10^{11}$ ohms |
| | | Mechanical units | |
| | (equally applicable to the electrostatic and electromagnetic systems) | | |
| Work/energy | erg | erg | = $10^{-7}$ joule (exactly) |
| Force | dyne | dyn | = $10^{-5}$ newton (exactly) |

and the International System Units were officially abandoned in favor of the absolute units. These in turn were supplanted by the SI units, beginning in 1950.

***10. Definitions of Electrical Quantities.*** The following definitions are based on the principal meanings listed in the *IEEE Standard Dictionary,*[19] which should be consulted for extended meanings, compound terms, and related definitions. The United States Standard Symbols[2,3] (ANSI/IEEE Standard 260-1978 and IEEE Standard 280) for these quantities are shown in parentheses (see also Tables 1-10 and 1-11). Electrical units used in the United States prior to 1969, with SI equivalents, are listed in Table 1-28.

*Admittance (Y).* An admittance of a linear constant-parameter system is the ratio of the phasor equivalent of the steady-state sine-wave current or current-like quantity (response) to the phasor equivalent of the corresponding voltage or voltage-like quantity (driving force).

*Capacitance (C).* Capacitance is that property of a system of conductors and dielectrics

**TABLE 1-10** Standard Symbols for Quantities

| Quantity | Quantity symbol | Unit based on International System | Remarks |
|---|---|---|---|
| Space and time: | | | |
| Angle, plane | $\alpha, \beta, \gamma, \theta, \phi, \psi$ | radian | Other Greek letters are permitted where no conflict results. |
| Angle, solid | $\Omega \cdots \omega$ | steradian | |
| Length | $l$ | meter | |
| Breadth, width | $b$ | meter | |
| Height | $h$ | meter | |
| Thickness | $d, \delta$ | meter | |
| Radius | $r$ | meter | |
| Diameter | $d$ | meter | |
| Length of path line segment | $s$ | meter | |
| Wavelength | $\lambda$ | meter | |
| Wave number | $\sigma \cdots \tilde{\nu}$ | reciprocal meter | $\sigma = 1/\lambda$. The symbol $\tilde{\nu}$ is used in spectroscopy. |
| Circular wave number<br>Angular wave number | $k$ | radian per meter | $k = 2\pi/\lambda$ |
| Area | $A \cdots S$ | square meter | |
| Volume | $V, v$ | cubic meter | |
| Time | $t$ | second | |
| Period | $T$ | second | |
| Time constant | $\tau \cdots T$ | second | |
| Frequency | $f \cdots \nu$ | hertz | |
| Speed of rotation<br>Rotational frequency | $n$ | revolution per second | |
| Angular frequency | $\omega$ | radian per second | $\omega = 2\pi f$ |
| Angular velocity | $\omega$ | radian per second | |
| Complex (angular) frequency<br>Oscillation constant | $p \cdots s$ | reciprocal second | $p = -\delta + j\omega$ |
| Angular acceleration | $\alpha$ | radian per second squared | |
| Velocity | $v$ | meter per second | |
| Speed of propagation of electromagnetic waves | $c$ | meter per second | In vacuum, $c_0$ |
| Acceleration (linear) | $a$ | meter per second squared | |
| Acceleration of free fall<br>Gravitational acceleration | $g$ | meter per second squared | |
| Damping coefficient | $\delta$ | neper per second | |
| Logarithmic decrement | $\Lambda$ | (numeric) | |
| Attenuation coefficient | $\alpha$ | neper per meter | |
| Phase coefficient | $\beta$ | radian per meter | |
| Propagation coefficient | $\gamma$ | reciprocal meter | $\gamma = \alpha + j\beta$ |
| Mechanics: | | | |
| Mass | $m$ | kilogram | |
| (Mass) density | $\rho$ | kilogram per cubic meter | Mass divided by volume |
| Momentum | $p$ | kilogram meter per second | |
| Moment of inertia | $I, J$ | kilogram meter squared | |
| Force | $F$ | newton | |
| Weight | $W$ | newton | Varies with acceleration of free fall |
| Weight density | $\gamma$ | newton per cubic meter | Weight divided by volume |

## TABLE 1-10 Standard Symbols for Quantities *(Continued)*

| Quantity | Quantity symbol | Unit based on International System | Remarks |
|---|---|---|---|
| Moment of force | $M$ | newton meter | |
| Torque | $T \cdots M$ | newton meter | |
| Pressure | $p$ | newton per square meter | The SI name *pascal* has been adopted for this unit. |
| Normal stress | $\sigma$ | newton per square meter | |
| Shear stress | $\tau$ | newton per square meter | |
| Stress tensor | $\sigma$ | newton per square meter | |
| Linear strain | $\epsilon$ | (numeric) | |
| Shear strain | $\gamma$ | (numeric) | |
| Strain tensor | $\epsilon$ | (numeric) | |
| Volume strain | $\theta$ | (numeric) | |
| Poisson's ratio | $\mu$, $\nu$ | (numeric) | Lateral contraction divided by elongation |
| Young's modulus<br>Modulus of elasticity | $E$ | newton per square meter | $E = \sigma/\epsilon$ |
| Shear modulus<br>Modulus of rigidity | $G$ | newton per square meter | $G = \tau/\gamma$ |
| Bulk modulus | $K$ | newton per square meter | $K = -p/\theta$ |
| Work | $W$ | joule | |
| Energy | $E$, $W$ | joule | $U$ is recommended in thermodynamics for internal energy and for blackbody radiation. |
| Energy (volume) density | $w$ | joule per cubic meter | |
| Power | $P$ | watt | |
| Efficiency | $\eta$ | (numeric) | |
| Heat: | | | |
| Thermodynamic temperature | $T \cdots \Theta$ | kelvin | |
| Temperature<br>Customary temperature | $t \cdots \theta$ | degree Celsius | The word *centigrade* has been abandoned as the name of a temperature scale |
| Heat | $Q$ | joule | |
| Internal energy | $U$ | joule | |
| Heat flow rate | $\Phi \cdots q$ | watt | Heat crossing a surface divided by time |
| Temperature coefficient | $\alpha$ | reciprocal kelvin | |
| Thermal diffusivity | $\alpha$ | square meter per second | |
| Thermal conductivity | $\lambda \cdots k$ | watt per meter kelvin | |
| Thermal conductance | $G_\theta$ | watt per kelvin | |
| Thermal resistivity | $\rho_\theta$ | meter kelvin per watt | |
| Thermal resistance | $R_\theta$ | kelvin per watt | |
| Thermal capacitance<br>Heat capacity | $C_\theta$ | joule per kelvin | |
| Thermal impedance | $Z_\theta$ | kelvin per watt | |
| Specific heat capacity | $c$ | joule per kelvin kilogram | Heat capacity divided by mass |
| Entropy | $S$ | joule per kelvin | |
| Specific entropy | $s$ | joule per kelvin kilogram | Entropy divided by mass |
| Enthalpy | $H$ | joule | |
| Radiation and light: | | | |
| Radiant intensity | $I \cdots I_e$ | watt per steradian | |
| Radiant power<br>Radiant flux | $P, \Phi \cdots \Phi_e$ | watt | |
| Radiant energy | $W, Q \cdots Q_e$ | joule | The symbol $U$ is used for the special case of blackbody radiant energy |
| Radiance | $L \cdots L_e$ | watt per steradian square meter | |
| Radiant exitance | $M \cdots M_e$ | watt per square meter | |
| Irradiance | $E \cdots E_e$ | watt per square meter | |
| Luminous intensity | $I \cdots I_v$ | candela | |
| Luminous flux | $\Phi \cdots \Phi_v$ | lumen | |
| Quantity of light | $Q \cdots Q_v$ | lumen second | |
| Luminance | $L \cdots L_v$ | candela per square meter | |
| Luminous exitance | $M \cdots M_v$ | lumen per square meter | |
| Illuminance<br>Illumination | $E \cdots E_v$ | lux | |
| Luminous efficacy† | $K(\lambda)$ | lumen per watt | |
| Total luminous efficacy | $K$, $K_t$ | lumen per watt | |
| Refractive index<br>Index of refraction | $n$ | (numeric) | |

**TABLE 1-10** Standard Symbols for Quantities *(Continued)*

| Quantity | Quantity symbol | Unit based on International System | Remarks |
|---|---|---|---|
| Emissivity† | $\epsilon(\lambda)$ | (numeric) | |
| Total emissivity | $\epsilon$, $\epsilon_t$ | (numeric) | |
| Absorptance† | $\alpha(\lambda)$ | (numeric) | |
| Transmittance† | $\tau(\lambda)$ | (numeric) | |
| Reflectance† | $\rho(\lambda)$ | (numeric) | |
| Fields and circuits: | | | |
| Electric charge<br>Quantity of electricity | $Q$ | coulomb | |
| Linear density of charge | $\lambda$ | coulomb per meter | |
| Surface density of charge | $\sigma$ | coulomb per square meter | |
| Volume density of charge | $\rho$ | coulomb per cubic meter | |
| Electric field strength | $E \cdots K$ | volt per meter | |
| Electrostatic potential<br>Potential difference | $V \cdots \phi$ | volt | |
| Retarded scalar potential | $V_r$ | volt | |
| Voltage<br>Electromotive force | $V, E \cdots U$ | volt | |
| Electric flux | $\Psi$ | coulomb | |
| Electric flux density<br>(Electric) displacement | $D$ | coulomb per square meter | |
| Capacitivity<br>Permittivity<br>Absolute permittivity | $\epsilon$ | farad per meter | Of vacuum, $\epsilon_v$ |
| Relative capacitivity<br>Relative permittivity<br>Dielectric constant | $\epsilon_r$, $\kappa$ | (numeric) | |
| Complex relative capacitivity<br>Complex relative permittivity<br>Complex dielectric constant | $\epsilon_r^*$, $\kappa^*$ | (numeric) | $\epsilon_r^* = \epsilon_r' - j\epsilon_r''$<br>$\epsilon_r''$ is positive for lossy materials. The complex absolute permittivity $\epsilon^*$ is defined in analogous fashion. |
| Electric susceptibility | $\chi_e \cdots \epsilon_i$ | (numeric) | $\chi_e = \epsilon_r - 1$ MKSA |
| Electrization | $E_i \cdots K_i$ | volt per meter | $E_i = (D/\Gamma_e) - E$ MKSA |
| Electric polarization | $P$ | coulomb per square meter | $P = D - \Gamma_e E$ MKSA |
| Electric dipole moment | $p$ | coulomb meter | |
| (Electric) current | $I$ | ampere | |
| Current density | $J \cdots S$ | ampere per square meter | |
| Linear current density | $A \cdots \alpha$ | ampere per meter | Current divided by the breadth of the conducting sheet |
| Magnetic field strength | $H$ | ampere per meter | |
| Magnetic (scalar) potential<br>Magnetic potential difference | $U$, $U_m$ | ampere | |
| Magnetomotive force | $F, F_m \cdots \mathfrak{F}$ | ampere | |
| Magnetic flux | $\Phi$ | weber | |
| Magnetic flux density<br>Magnetic induction | $B$ | tesla | |
| Magnetic flux linkage | $\Lambda$ | weber | |
| (Magnetic) vector potential | $A$ | weber per meter | |
| Retarded (magnetic) vector potential | $A_r$ | weber per meter | |
| Permeability<br>Absolute permeability | $\mu$ | henry per meter | Of vacuum, $\mu_v$ |
| Relative permeability | $\mu_r$ | (numeric) | |
| Initial (relative) permeability | $\mu_o$ | (numeric) | |
| Complex relative permeability | $\mu_r^*$ | (numeric) | $\mu_r^* = \mu_r' - j\mu_r''$<br>$\mu_r''$ is positive for lossy materials. The complex absolute permeability $\mu^*$ is defined in analogous fashion. |
| Magnetic susceptibility | $\chi_m \cdots \mu_i$ | (numeric) | $\chi_m = \mu_r - 1$ MKSA |
| Reluctivity | $\nu$ | meter per henry | $\nu = 1/\mu$ |
| Magnetization | $H_i$, $M$ | ampere per meter | $H_i = (B/\Gamma_m) - H$ MKSA |
| Magnetic polarization<br>Intrinsic magnetic flux density | $J$, $B_i$ | tesla | $J = B - \Gamma_m H$ MKSA |
| Magnetic (area) moment | $m$ | ampere meter squared | The vector product $\boldsymbol{m} \times \boldsymbol{B}$ is equal to the torque |
| Capacitance | $C$ | farad | |
| Elastance | $S$ | reciprocal farad | $S = 1/C$ |
| (Self-) inductance | $L$ | henry | |
| Reciprocal inductance | $\Gamma$ | reciprocal henry | |
| Mutual inductance | $L_{ij}$, $M_{ij}$ | henry | If only a single mutual inductance is involved, $M$ may be used without subscripts |
| Coupling coefficient | $k \cdots \kappa$ | (numeric) | $k = L_{ij}(L_i L_j)^{-1/2}$ |
| Leakage coefficient | $\sigma$ | (numeric) | $\sigma = 1 - k^2$ |
| Number of turns (in a winding) | $N$, $n$ | (numeric) | |
| Number of phases | $m$ | (numeric) | |
| Turns ratio | $n \cdots n_*$ | (numeric) | |

**TABLE 1-10** Standard Symbols for Quantities *(Continued)*

| Quantity | Quantity symbol | Unit based on International System | Remarks |
|---|---|---|---|
| Transformer ratio | $a$ | (numeric) | Square root of the ratio of secondary to primary self-inductance. Where the coefficient of coupling is high, $a \approx n_*$. |
| Resistance | $R$ | ohm | |
| Resistivity | $\rho$ | ohm meter | |
| Volume resistivity | | | |
| Conductance | $G$ | siemens | $G = \text{Re } Y$ |
| Conductivity | $\gamma, \sigma$ | siemens per meter | $\gamma = 1/\rho$<br>The symbol $\sigma$ is used in field theory, as $\gamma$ is there used for the propagation coefficient. |
| Reluctance | $R, R_m \cdots \mathcal{R}$ | reciprocal henry | Magnetic potential difference divided by magnetic flux |
| Permeance | $P, P_m \cdots \mathcal{P}$ | henry | $P_m = 1/R_m$ |
| Impedance | $Z$ | ohm | |
| Reactance | $X$ | ohm | |
| Capacitive reactance | $X_C$ | ohm | For a pure capacitance, $X_C = -1/\omega C$ |
| Inductive reactance | $X_L$ | ohm | For a pure inductance $X_L = \omega L$ |
| Quality factor | $Q$ | (numeric) | See $Q$ in Par. 10 |
| Admittance | $Y$ | siemens | $Y = 1/Z = G + jB$ |
| Susceptance | $B$ | siemens | $B = \text{Im } Y$ |
| Loss angle | $\delta$ | radian | $\delta = (R/\lvert X \rvert)$ |
| Active power | $P$ | watt | |
| Reactive power | $Q \cdots P_q$ | var | |
| Apparent power | $S \cdots P_s$ | voltampere | |
| Power factor | $\cos \phi \cdots F_p$ | (numeric) | |
| Reactive factor | $\sin \phi \cdots F_q$ | (numeric) | |
| Input power | $P_i$ | watt | |
| Output power | $P_o$ | watt | |
| Poynting vector | $S$ | watt per square meter | |
| Characteristic impedance | $Z_0$ | ohm | |
| Surge impedance | | | |
| Intrinsic impedance of a medium | $\eta$ | ohm | |
| Voltage standing-wave ratio | $S$ | (numeric) | |
| Resonance frequency | $f_r$ | hertz | |
| Critical frequency | $f_c$ | hertz | |
| Cutoff frequency | | | |
| Resonance angular frequency | $\omega_r$ | radian per second | |
| Critical angular frequency | $\omega_c$ | radian per second | |
| Cutoff angular frequency | | | |
| Resonance wavelength | $\lambda_r$ | meter | |
| Critical wavelength | $\lambda_c$ | meter | |
| Cutoff wavelength | | | |
| Wavelength in a guide | $\lambda_g$ | meter | |
| Hysteresis coefficient | $k_h$ | (numeric) | |
| Eddy-current coefficient | $k_e$ | (numeric) | |
| Phase angle | $\phi, \theta$ | radian | |
| Phase difference | | | |

† ($\lambda$) is not part of the basic symbol but indicates that the quantity is a function of wavelength.

which permits the storage of electrically separated charges when potential differences exist between the conductors. Its value is expressed as the ratio of an electric charge to a potential difference.

*Coupling Coefficient (k).* Coefficient of coupling (used only in the case of resistive, capacitive, and inductive coupling) is the ratio of the mutual impedance of the coupling to the square root of the product of the self-impedances of similar elements in the two circuit loops considered. Unless otherwise specified, coefficient of coupling refers to inductive coupling, in which case $k = M/(L_1L_2)^{1/2}$, where $M$ is the mutual inductance, $L_1$ the self-inductance of one loop, and $L_2$ the self-inductance of the other.

*Conductance (G)*

1. The conductance of an element, device, branch, network, or system is the factor by which the mean-square voltage must be multiplied to give the corresponding power lost by dissipation as heat or as other permanent radiation or loss of electromagnetic energy from the circuit.

**TABLE 1-11** Standard Symbols for Units

| Unit | Symbol | Notes |
|---|---|---|
| ampere | A | SI unit of electric current |
| ampere (turn) | A | SI unit of magnetomotive force |
| ampere-hour | Ah | Also A·h |
| ampere per meter | A/m | SI unit of magnetic field strength |
| angstrom | Å | 1 Å = $10^{-10}$ m. Deprecated |
| atmosphere, standard | atm | 1 atm = 101 325 Pa. Deprecated. |
| atmosphere, technical | at | 1 at = 1 kgf/cm$^2$. Deprecated. |
| atomic mass unit (unified) | u | The (unified) atomic mass unit is defined as one-twelfth of the mass of an atom of the $^{12}$C nuclide. Use of the old atomic mass (amu), defined by reference to oxygen, is deprecated. |
| atto | a | SI prefix for $10^{-18}$ |
| attoampere | aA | |
| bar | bar | 1 bar = 100 kPa. Use of the bar is strongly discouraged, except for limited use in meteorology. |
| barn | b | 1 b = $10^{-28}$ m$^2$ |
| barrel | bbl | 1 bbl = 42 gal$_{US}$ = 158.99 L |
| barrel per day | bbl/d | This is the standard barrel used for petroleum, etc. A different standard barrel is used for fruits, vegetables, and dry commodities. |
| baud | Bd | In telecommunications, a unit of signaling speed equal to one element per second. The signaling speed in bauds is equal to the reciprocal of the signal element length in seconds. |
| bel | B | |
| becquerel | Bq | SI unit of activity of a radionuclide. |
| billion electronvolts | GeV | The name *gigaelectronvolt* is preferred for this unit. |
| bit | b | In information theory the bit is a unit of information content equal to the information content of a message the *a priori* probability of which is one-half. In computer science the bit is a unit of storage capacity. The capacity, in bits, of a storage device is the logarithm to the base two of the number of possible states of the device. |
| bit per second | b/s | |
| British thermal unit | Btu | |
| calorie (International Table calorie) | cal$_{IT}$ | 1 cal$_{IT}$ = 4.1868 J. Deprecated. |
| calorie (thermochemical calorie) | cal | 1 cal = 4.1840 J. Deprecated. |
| candela | cd | SI unit of luminous intensity |
| candela per square inch | cd/in$^2$ | Use of the SI unit, cd/m$^2$, is preferred. |
| candela per square meter | cd/m$^2$ | SI unit of luminance. The name *nit* is sometimes used for this unit. |
| candle | cd | The unit of luminous intensity has been given the name *candela;* use of the name *candle* for this unit is deprecated. |
| centi | c | SI prefix for $10^{-2}$ |

**TABLE 1-11** Standard Symbols for Units *(Continued)*

| Unit | Symbol | Notes |
|---|---|---|
| centimeter | cm | |
| centipoise | cP | 1 cP = 1 mPa·s. The name is deprecated. |
| centistokes | cSt | 1 cSt = 1 $mm^2/s$. The name centistokes is deprecated. |
| circular mil | cmil | 1 cmil = $(\pi/4)\cdot 10^{-6}$ $in^2$ |
| coulomb | C | SI unit of electric charge |
| cubic centimeter | $cm^3$ | |
| cubic foot | $ft^3$ | |
| cubic foot per minute | $ft^3/min$ | |
| cubic foot per second | $ft^3/s$ | |
| cubic inch | $in^3$ | |
| cubic meter | $m^3$ | |
| cubic meter per second | $m^3/s$ | |
| cubic yard | $yd^3$ | |
| curie | Ci | A unit of activity of radionuclide. Use of the SI unit, the becquerel, is preferred, 1 Ci = $3.7 \times 10^{10}$ Bq |
| cycle | c | |
| cycle per second | Hz, c/s | See hertz. The name *hertz* is internationally accepted for this unit; the symbol Hz is preferred to c/s. |
| darcy | D | 1 D = 1 cP (cm/s) (cm/atm) = 0.986 923 $\mu m^2$. A unit of permeability of a porous medium. By traditional definition, a permeability of one darcy will permit a flow of 1 $cm^3/s$ of fluid of 1 cP viscosity through an area of 1 $cm^2$ under a pressure gradient of 1 atm/cm. For nonprecision work 1 D may be taken equal to 1 $\mu m^2$ and 1 mD equal to 0.001 $\mu m^2$. Deprecated. |
| day | d | |
| deci | d | SI prefix for $10^{-1}$ |
| decibel | dB | |
| degree (plane angle) | . . .° | |
| degree (temperature): | | |
| degree Celsius | °C | SI unit of Celsius temperature. The degree Celsius is a special name for the kelvin, for use in expressing Celsius temperatures or temperature intervals. |
| degree Fahrenheit | °F | Note that the symbols for °C, °F, and °R comprise two elements, written with no space between the ° and the letter that follows. The two elements that make the complete symbol are not to be separated. |
| degree Kelvin | | See kelvin |
| degree Rankine | °R | |
| deka | da | SI prefix for 10 |
| dyne | dyn | Deprecated. |
| electronvolt | eV | |
| erg | erg | Deprecated. |
| exa | E | SI prefix for $10^{18}$ |
| farad | F | SI unit of capacitance |

**TABLE 1-11** Standard Symbols for Units *(Continued)*

| Unit | Symbol | Notes |
|---|---|---|
| femto | f | SI prefix for $10^{-15}$ |
| femtometer | fm | |
| foot | ft | |
| conventional foot of water | $ftH_2O$ | 1 $ftH_2O$ = 2989.1 Pa (ISO) |
| foot per minute | ft/min | |
| foot per second | ft/s | |
| foot per second squared | $ft/s^2$ | |
| foot pound-force | ft·lbf | |
| footcandle | fc | 1 fc = 1 $lm/ft^2$. The name *lumen per square foot* is also used for this unit. Use of the SI unit of illuminance, the lux (lumen per square meter), is preferred. |
| footlambert | fL | 1 fL = $(1/\pi)$ $cd/ft^2$. A unit of luminance. One lumen per square foot leaves a surface whose luminance is one footlambert in all directions within a hemisphere. Use of the SI unit, the candela per square meter, is preferred. |
| gal | Gal | 1 Gal = 1 $cm/s^2$. Deprecated. |
| gallon | gal | 1 $gal_{UK}$ = 4.5461 L<br>1 $gal_{US}$ = 231 $in^3$ = 3.7854 L |
| gauss | G | The gauss is the electromagnetic CGS unit of magnetic flux density. Deprecated. |
| giga | G | SI prefix for $10^9$ |
| gigaelectronvolt | GeV | |
| gigahertz | GHz | |
| gilbert | Gb | The gilbert is the electromagnetic CGS unit of magnetomotive force. Deprecated. |
| grain | gr | |
| gram | g | |
| gram per cubic centimeter | $g/cm^3$ | |
| gray | Gy | SI unit of absorbed dose in the field of radiation dosimetry. |
| hecto | h | SI prefix for $10^2$ |
| henry | H | SI unit of inductance |
| hertz | Hz | SI unit of frequency |
| horsepower | hp | The horsepower is an anachronism in science and technology. Use of the SI unit of power, the watt, is preferred. |
| hour | h | |
| inch | in | |
| conventional inch of mercury | inHg | 1 inHg = 3386.4 Pa (ISO) |
| conventional inch of water | $inH_2O$ | 1 $inH_2O$ = 249.09 Pa (ISO) |
| inch per second | in/s | |
| joule | J | SI unit of energy, work, quantity of heat |
| joule per kelvin | J/K | SI unit of heat capacity and entropy |
| kelvin | K | In 1967 the CGPM gave the name *kelvin* to the SI unit of temperature which had formerly been called *degree kelvin* and assigned it the symbol K (without the symbol °). |
| kilo | k | SI prefix for $10^3$ |
| kilogauss | kG | Deprecated. |

**TABLE 1-11** Standard Symbols for Units *(Continued)*

| Unit | Symbol | Notes |
|---|---|---|
| kilogram | kg | SI unit of mass |
| kilogram-force | kgf | Deprecated. In some countries the name kilopond (kp) has been used for this unit. |
| kilohertz | kHz | |
| kilohm | kΩ | |
| kilometer | km | |
| kilometer per hour | km/h | |
| kilopound-force | klbf | Kilopound-force should not be misinterpreted as kilopond (see kilogram-force). |
| kilovar | kvar | |
| kilovolt | kV | |
| kilovoltampere | kVA | |
| kilowatt | kW | |
| kilowatthour | kWh | Also kW·h |
| knot | kn | 1kn = 1 nmi/h |
| lambert | L | 1 L = $(1/\pi)$ cd/cm². A CGS unit of luminance. One lumen per square centimeter leaves a surface whose luminance is one lambert in all directions within a hemisphere. Deprecated. |
| liter | L | 1 L = $10^{-3}$ m³. The letter symbol l has been adopted for *liter* by the CGPM, and it is recommended in a number of international standards. In 1978 the CIPM accepted L as an alternative symbol. Because of frequent confusion with the numeral 1 the letter symbol l is no longer recommended for USA use. The script letter $\ell$, which had been proposed, is not recommended as a symbol for liter. |
| liter per second | L/s | |
| lumen | lm | SI unit of luminous flux |
| lumen per square foot | lm/ft² | A unit of illuminance and also a unit of luminous exitance. Use of the SI unit, lumen per square meter, is preferred. |
| lumen per square meter | lm/m² | SI unit of luminous exitance |
| lumen per watt | lm/W | SI unit of luminous efficacy |
| lumen second | lm·s | SI unit of quantity of light |
| lux | lx | 1 lx = 1 lm/m². SI unit of illuminance |
| maxwell | Mx | The maxwell is the electromagnetic CGS unit of magnetic flux. Deprecated. |
| mega | M | SI prefix for $10^6$ |
| megaelectronvolt | MeV | |
| megahertz | MHz | |
| megohm | MΩ | |
| meter | m | SI unit of length |
| metric ton | t | 1 t = 1000 kg. The name *tonne* is used in some countries for this unit, but use of this name in the USA is deprecated. |
| mho | mho | Formerly used as the name of the siemens (S). |
| micro | $\mu$ | SI prefix for $10^{-6}$ |

**TABLE 1-11** Standard Symbols for Units *(Continued)*

| Unit | Symbol | Notes |
|---|---|---|
| microampere | μA | |
| microfarad | μF | |
| microgram | μg | |
| microhenry | μH | |
| microinch | μin | |
| microliter | μL | See note for *liter* |
| micrometer | μm | |
| micron | μm | Deprecated. Use micrometer. |
| microsecond | μs | |
| microwatt | μW | |
| mil | mil | 1 mil = 0.001 in |
| mile (statute) | mi | 1 mi = 5280 ft |
| mile per hour | mi/h | Although use of mph as an abbreviation is common, it should not be used as a symbol. |
| milli | m | SI prefix for $10^{-3}$ |
| milliampere | mA | |
| millibar | mbar | Use of the bar is strongly discouraged, except for limited use in meteorology. |
| milligram | mg | |
| millihenry | mH | |
| milliliter | mL | See note for *liter*. |
| millimeter | mm | |
| conventional millimeter of mercury | mmHg | 1 mmHg = 133.322 Pa. Deprecated. |
| millimicron | nm | Use of the name *millimicron* for the nanometer is deprecated. |
| millipascal second | mPa·s | SI unit-multiple of dynamic viscosity |
| millisecond | ms | |
| millivolt | mV | |
| milliwatt | mW | |
| minute (plane angle) | ′ | |
| minute (time) | min | Time may also be designated by means of superscripts as in the following example: $9^h46^m30^s$. |
| mole | mol | SI unit of amount of substance. |
| month | mo | |
| nano | n | SI prefix for $10^{-9}$ |
| nanoampere | nA | |
| nanofarad | nF | |
| nanometer | nm | |
| nanosecond | ns | |
| nautical mile | nmi | 1 nmi = 1852 m |
| neper | Np | |
| newton | N | SI unit of force |
| newton meter | N·m | |
| newton per square meter | N/m² | SI unit of pressure or stress, see pascal. |
| nit | nt | 1 nt = 1 cd/m². The name *nit* is sometimes given to the SI unit of luminance, the candela per square meter. |
| oersted | Oe | The oersted is the electromagnetic CGS unit |

**TABLE 1-11** Standard Symbols for Units *(Continued)*

| Unit | Symbol | Notes |
|---|---|---|
| | | of magnetic field strength. Deprecated. |
| ohm | Ω | SI unit of resistance |
| ounce (avoirdupois) | oz | |
| pascal | Pa | 1 Pa = 1 N/m$^2$<br>SI unit of pressure or stress |
| pascal second | Pa·s | SI unit of dynamic viscosity |
| peta | P | SI prefix for $10^{15}$ |
| phot | ph | 1 ph = lm/cm$^2$<br>CGS unit of illuminance. Deprecated. |
| pico | p | SI prefix for $10^{-12}$ |
| picofarad | pF | |
| picowatt | pW | |
| pint | pt | 1 pt (UK) = 0.568 26 L<br>1 pt (US dry) = 0.550 61 L<br>1 pt (US liquid) = 0.473 18 L |
| poise | P | Deprecated. |
| pound | lb | |
| pound per cubic foot | lb/ft$^3$ | |
| pound-force | lbf | |
| pound-force foot | lbf·ft | |
| pound-force per square foot | lbf/ft$^2$ | |
| pound-force per square inch | lbf/in$^2$ | Although use of the abbreviation psi is common, it should not be used as a symbol. |
| poundal | pdl | |
| quart | qt | 1 qt (UK) = 1.136 5 L<br>1 qt (US dry) = 1.101 2 L<br>1 qt (US liquid) = 0.946 35 L |
| rad | rd | A unit of absorbed dose in the field of radiation dosimetry. Use of the SI unit, the gray, is preferred. 1 rd = 0.01 Gy. |
| radian | rad | SI unit of plane angle |
| rem | rem | A unit of dose equivalent in the field of radiation dosimetry (Use of the SI unit, the sievert, is preferred. 1 rem = 0.01 Sv.) |
| revolution per minute | r/min | Although use of rpm as an abbreviation is common, it should not be used as a symbol. |
| revolution per second | r/s | |
| roentgen | R | A unit of exposure in the field of radiation dosimetry |
| second (plane angle) | . . ." | |
| second (time) | s | SI unit of time |
| siemens | S | 1 S = 1 Ω$^{-1}$<br>SI unit of conductance.<br>The name mho has been used for this unit in the USA. |
| sievert | Sv | SI unit of dose equivalent in the field of radiation dosimetry. Name adopted by the CIPM in 1978. |
| slug | slug | 1 slug = 14.5939 kg |
| square foot | ft$^2$ | |
| square inch | in$^2$ | |

**TABLE 1-11** Standard Symbols for Units *(Continued)*

| Unit | Symbol | Notes |
|---|---|---|
| square meter | $m^2$ | |
| square meter per second | $m^2/s$ | SI unit of kinematic viscosity |
| square millimeter per second | $mm^2/s$ | SI unit-multiple of kinematic viscosity |
| square yard | $yd^2$ | |
| steradian | sr | SI unit of solid angle |
| stilb | sb | 1 sb = 1 $cd/cm^2$<br>A CGS unit of luminance. Deprecated. |
| stokes | St | Deprecated. |
| tera | T | SI prefix for $10^{12}$ |
| tesla | T | 1 T = 1 N/(A·m) = 1 $Wb/m^2$. SI unit of magnetic flux density (magnetic induction) |
| therm | thm | 1 thm = 100 000 Btu |
| ton (short) | ton | 1 ton = 2000 lb |
| ton, metric | t | 1 t = 1000 kg. The name *tonne* is used in some countries for this unit, but use of this name in the USA is deprecated. |
| (unified) atomic mass unit | u | The (unified) atomic mass unit is defined as one-twelfth of the mass of an atom of the $^{12}C$ nuclide. Use of the old atomic mass unit (amu), defined by reference to oxygen, is deprecated. |
| var | var | IEC name and symbol for the SI unit of reactive power |
| volt | V | SI unit of voltage |
| volt per meter | V/m | SI unit of electric field strength |
| voltampere | VA | IEC name and symbol for the SI unit of apparent power |
| watt | W | SI unit of power |
| watt per meter kelvin | W/(m·K) | SI unit of thermal conductivity |
| watt per steradian | W/sr | SI unit of radiant intensity |
| watt per steradian square meter | W/(sr·$m^2$) | SI unit of radiance |
| watthour | Wh | |
| weber | Wb | Wb = V·s<br>SI unit of magnetic flux |
| yard | yd | |
| year | a | In the English language, generally yr |

Adapted from ANSI/IEEE Standard 260-1982.

2. Conductance is the real part of admittance.

*Conductivity (γ).* The conductivity of a material is a factor such that the conduction current density is equal to the electric field strength in the material multiplied by the conductivity.

*Current (I).* Current is a generic term used when there is no danger of ambiguity to refer to any one or more of the currents described below. (For example, in the expression "the current in a simple series circuit," the word current refers to the conduction current in the wire of the inductor and to the displacement current between the plates of the capacitor.)

*Conduction Current.* The conduction current through any surface is the integral of the normal component of the conduction current density over that surface.

*Displacement Current.* The displacement current through any surface is the integral of the normal component of the displacement current density over that surface.

*Current Density (J).* Current density is a generic term used when there is no danger of ambiguity to refer either to conduction current density or to displacement current density or to both.

*Displacement Current Density.* The displacement current density at any point in an electric field is (in the International System) the time rate of change of the electric-flux-density vector at that point.

*Conduction Current Density.* The electric conduction current density at any point at which there is a motion of electric charge is a vector quantity whose direction is that of the flow of positive charge at this point, and whose magnitude is the limit of the time rate of flow of net (positive) charge across a small plane area perpendicular to the motion, divided by this area, as the area taken approaches zero in a macroscopic sense, so as to always include this point. The flow of charge may result from the movement of free electrons or ions but is not in general, except in microscopic studies, taken to include motions of charges resulting from the polarization of the dielectric.

*Damping Coefficient (δ).* If $F$ is a function of time given by

$$F = A \exp(-\delta t) \sin(2\pi t/T)$$

then $\delta$ is the damping coefficient.

*Elastance (S).* Elastance is the reciprocal of capacitance.

*Electric Charge, Quantity of Electricity (Q).* Electric charge is a fundamentally assumed concept, required by the existence of forces measurable experimentally. It has two forms, known as positive and negative.

The electric charge on (or in) a body or within a closed surface is the excess of one form of electricity over the other.

*Electric Constant, Permittivity of Vacuum ($\Gamma_e$).* The electric constant pertinent to any system of units is the scalar which in that system relates the electric flux density $D$, in vacuum, to $E$, the electric field strength ($D = \Gamma_e E$). It also relates the mechanical force between two charges in vacuum to their magnitudes and separation. Thus in the equation $F = \Gamma_r Q_1 Q_2 / 4\pi\Gamma_e r^2$ for the force $F$ between charges $Q_1$ and $Q_2$ separated by a distance $r$, $\Gamma_e$ is the electric constant and $\Gamma_r$ is a dimensionless factor which is unity in a rationalized system and $4\pi$ in an unrationalized system.

NOTE: In the cgs electrostatic system $\Gamma_e$ is assigned measure unity and the dimension "numeric."

In the cgs electromagnetic system the measure of $\Gamma_e$ is that of $1/c^2$ and the dimension is $[L^{-2}T^2]$.

In the International System the measure of $\Gamma_e$ is $10^7/4\pi c^2$, and the dimension is $[L^{-3}M^{-1}T^4I^2]$. Here $c$ is the speed of light expressed in the appropriate system of units (see Table 1-12).

*Electric Field Strength (**E**).* The electric field strength at a given point in an electric field is the vector limit of the quotient of the force that a small stationary charge at that point will experience, by virtue of its charge, to the charge as the charge approaches zero.

*Electric Flux (Ψ).* The electric flux through a surface is the surface integral of the normal component of the electric flux density over the surface.

*Electric Flux Density, Electric Displacement (**D**).* The electric flux density is a quantity related to the charge displaced within a dielectric by application of an electric field. Electric flux density at any point in an isotropic dielectric is a vector which has the same direction as the electric field strength and a magnitude equal to the product of the electric field strength and the permittivity, $\epsilon$. In a nonisotropic medium, $\epsilon$ may be represented by a tensor, and $\boldsymbol{D}$ is not necessarily parallel to $\boldsymbol{E}$.

*Electric Polarization (**P**).* The electric polarization is the vector quantity defined by the equation $\boldsymbol{P} = (\boldsymbol{D} - \Gamma_e \boldsymbol{E})/\Gamma_r$, where $\boldsymbol{D}$ is the electric flux density, $\Gamma_e$ is the electric constant, $\boldsymbol{E}$ is the electric field strength, and $\Gamma_r$ is a coefficient that is set equal to unity in a rationalized system and to $4\pi$ in an unrationalized system.

*Electric Susceptibility ($\chi_e$).* Electric susceptibility is the quantity defined by $\chi_e = (\epsilon_r - 1)/$

$\Gamma_r$, where $\epsilon_r$ is the relative permittivity and $\Gamma_r$ is a coefficient that is set equal to unity in a rationalized system and to $4\pi$ in an unrationalized system.

*Electrization ($E_i$).* The electrization is the electric polarization divided by the electric constant of the system of units used.

*Electrostatic Potential (V).* The electrostatic potential at any point is the potential difference between that point and an agreed-upon reference point, usually the point at infinity.

*Electrostatic Potential Difference (V).* The electrostatic potential difference between two points is the scalar-product line integral of the electric field strength along any path from one point to the other in an electric field resulting from a static distribution of electric charge.

*Impedance (Z).* An impedance of a linear constant-parameter system is the ratio of the phasor equivalent of a steady-state sine-wave voltage or voltage-like quantity (driving force) to the phasor equivalent of a steady-state sine-wave current or current-like quantity (response). In electromagnetic radiation, electric field strength is considered the driving force and magnetic field strength the response. In mechanical systems mechanical force is always considered as a driving force and velocity as a response. In a general sense the dimension (and unit) of impedance in a given application may be whatever results from the ratio of the dimensions of the quantity chosen as the driving force to the dimensions of the quantity chosen as the response. However, in the types of systems cited above any deviation from the usual convention should be noted.

*Mutual Impedance.* Mutual impedance between two loops (meshes) is the factor by which the phasor equivalent of the steady-state sine-wave current in one loop must be multiplied to give the phasor equivalent of the steady-state sine-wave voltage in the other loop caused by the current in the first loop.

*Self-impedance.* Self-impedance of a loop (mesh) is the impedance of a passive loop with all other loops of the network open-circuited.

*Transfer Impedance.* A transfer impedance is the impedance obtained when the response is determined at a point other than that at which the driving force is applied.

NOTE: In the case of an electric circuit the response may be determined in any branch except that which contains the driving force.

*Logarithmic Decrement (Λ).* If $F$ is a function of time given by

$$F = A \exp(-\delta t) \sin(2\pi t/T)$$

then the logarithmic decrement $\Lambda = T\delta$.

*Magnetic Constant, Permeability of Vacuum ($\Gamma_m$).* The magnetic constant pertinent to any system of units is the scalar which in that system relates the mechanical force between two currents in vacuum to their magnitudes and geometrical configurations. For example, the equation for the force $F$ on a length $l$ of two parallel straight conductors of infinite length and negligible circular cross section, carrying constant currents $I_1$ and $I_2$ and separated by a distance $r$ in vacuum, is $F = \Gamma_m \Gamma_r I_1 I_2 l / 2\pi r$, where $\Gamma_m$ is the magnetic constant and $\Gamma_r$ is a coefficient set equal to unity in a rationalized system and to $4\pi$ in an unrationalized system.

NOTE: In the cgs electromagnetic system $\Gamma_m$ is assigned the magnitude unity and the dimension "numeric."

In the cgs electrostatic system the magnitude of $\Gamma_m$ is that of $1/c^2$, and the dimension is $[L^{-2}T^2]$.

In the International System $\Gamma_m$ is assigned the magnitude $4\pi \times 10^{-7}$ and has the dimension $[LMT^{-2}I^{-2}]$.

*Magnetic Field Strength (**H**).* Magnetic field strength is that vector point function whose curl is the current density, and which is proportional to magnetic flux density in regions free of magnetized matter.

*Magnetic Flux (Φ).* The magnetic flux through a surface is the surface integral of the normal component of the magnetic flux density over the surface.

*Magnetic Flux Density, Magnetic Induction (**B**).* Magnetic flux density is that vector quantity producing a torque on a plane current loop in accordance with the relation $\boldsymbol{T} = IA\boldsymbol{n} \times \boldsymbol{B}$, where $\boldsymbol{n}$ is the positive normal to the loop and $A$ is its area.

The concept of flux density is extended to a point inside a solid body by defining the flux density at such a point as that which would be measured in a thin disk-shaped cavity in the body centered at that point, the axis of the cavity being in the direction of the flux density.

*Magnetic Moment ($\boldsymbol{m}$).* The magnetic moment of a magnetized body is the volume integral of the magnetization. The magnetic moment of a loop carrying current $I$ is $\boldsymbol{m} = (1/2)\int \boldsymbol{r} \times d\boldsymbol{r}$ where $\boldsymbol{r}$ is the radius vector from an arbitrary origin to a point on the loop and where the path of integration is taken around the entire loop.

NOTE: The magnitude of the moment of a plane current loop is $IA$, where $A$ is the area of the loop.

The reference direction for the current in the loop indicates a clockwise rotation when the observer is looking through the loop in the direction of the positive normal.

*Magnetic Polarization, Intrinsic Magnetic Flux Density ($\boldsymbol{J}$, $\boldsymbol{B}_i$).* The magnetic polarization is the vector quantity defined by the equation $\boldsymbol{J} = (\boldsymbol{B} - \Gamma_m \boldsymbol{H})/\Gamma_r$, where $\boldsymbol{B}$ is the magnetic flux density, $\Gamma_m$ is the magnetic constant, **H** is the magnetic field strength, and $\Gamma_r$ is a coefficient that is set equal to unity in a rationalized system and to $4\pi$ in an unrationalized system.

*Magnetic Susceptibility ($\chi_m$).* Magnetic susceptibility is the quantity defined by $\chi_m = (\mu_r - 1)/\Gamma_r$, where $\mu_r$ is the relative permeability and $\Gamma_r$ is a coefficient that is set equal to unity in a rationalized system and to $4\pi$ in an unrationalized system.

*Magnetic Vector Potential ($\boldsymbol{A}$).* The magnetic vector potential is a vector point function characterized by the relation that its curl is equal to the magnetic flux density and its divergence vanishes.

*Magnetization ($\boldsymbol{M}$, $\boldsymbol{H}_i$).* The magnetization is the magnetic polarization divided by the magnetic constant of the system of units used.

*Magnetomotive Force ($F_m$).* The magnetomotive force acting in any closed path in a magnetic field is the line integral of the magnetic field strength around the path.

*Mutual Inductance (M).* The mutual inductance between two loops (meshes) in a circuit is the quotient of the flux linkage produced in one loop divided by the current, in another loop, which induces the flux linkage.

*Permeability.* Permeability is a general term used to express various relationships between magnetic flux density and magnetic field strength. These relationships are either (1) *absolute permeability* ($\mu$), which in general is the quotient of a change in magnetic flux density divided by the corresponding change in magnetic field strength; or (2) *relative permeability* ($\mu_r$), which is the ratio of the absolute permeability to the magnetic constant.

*Permeance ($P_m$).* Permeance is the reciprocal of reluctance.

*Permittivity, Capacitivity ($\epsilon$).* The permittivity of a homogeneous, isotropic dielectric, in any system of units, is the product of its relative permittivity and the electric constant appropriate to that system of units.

*Relative Permittivity, Relative Capacitivity, Dielectric Constant ($\epsilon_r$).* The relative permittivity of any homogeneous isotropic material is the ratio of the capacitance of a given configuration of electrodes with the material as a dielectric to the capacitance of the same electrode configuration with a vacuum as the dielectric. Experimentally, vacuum must be replaced by the material at all points where it makes a significant change in the capacitance.

*Power (P).* Power is the time rate of transferring or transforming energy. *Electrical power* is the time rate of flow of electrical energy. The *instantaneous electrical power* at a single terminal pair is equal to the product of the instantaneous voltage multiplied by the instantaneous current. If both voltage and current are periodic in time, the time average of the instantaneous power, taken over an integral number of periods, is the *active power,* usually called simply the *power* when there is no danger of confusion.

If the voltage and current are sinusoidal functions of time, the product of the rms value of the voltage and the rms value of the current is called the *apparent power;* the product of the rms value of the voltage and the rms value of the in-phase component of the current is the *active power;* and the product of the rms value of the voltage and the rms value of the quadrature component of the current is called the *reactive power.*

The SI unit of instantaneous power and of active power is the watt. The germane unit for apparent power is the voltampere, and for reactive power the var.

*Power Factor ($F_p$).* Power factor is the ratio of active power to apparent power.

*Q.* Q, sometimes called *quality factor,* is that measure of the quality of a component, network, system, or medium considered as an energy storage unit in the steady state with sinusoidal driving force which is given by

$$Q = \frac{2\pi \times (\text{maximum energy in storage})}{\text{energy dissipated per cycle of the driving force}}$$

NOTE: For single components, such as inductors and capacitors, the $Q$ at any frequency is the ratio of the equivalent series reactance to resistance, or of the equivalent shunt susceptance to conductance.

For networks that contain several elements and for distributed parameter systems the $Q$ is generally evaluated at a frequency of resonance.

The *nonloaded* $Q$ of a system is the value of $Q$ obtained when only the incidental dissipation of the system elements is present. The *loaded* $Q$ of a system is the value of $Q$ obtained when the system is coupled to a device that dissipates energy.

The "period" in the expression for $Q$ is that of the driving force, not that of energy storage, which is usually half that of the driving force.

*Reactance (X).* Reactance is the imaginary part of impedance.

*Reluctance ($R_m$).* Reluctance is the ratio of the magnetomotive force in a magnetic circuit to the magnetic flux through any cross section of the magnetic circuit.

*Reluctivity (ν).* Reluctivity is the reciprocal of permeability.

*Resistance (R)*

1. The resistance of an element, device, branch, network, or system is the factor by which the mean-square conduction current must be multiplied to give the corresponding power lost by dissipation as heat or as other permanent radiation or loss of electromagnetic energy from the circuit.
2. Resistance is the real part of impedance.

*Resistivity (ρ).* The resistivity of a material is a factor such that the conduction current density is equal to the electric field strength in the material divided by the resistivity.

*Self-inductance (L)*

1. Self-inductance is the quotient of the flux linkage of a circuit divided by the current in that same circuit which induces the flux linkage. If $v$ = voltage induced, $v = d(Li)/dt$.
2. Self-inductance is the factor $L$ in the $\frac{1}{2}Li^2$ if the latter gives the energy stored in the magnetic field as a result of the current $i$.

NOTE: Definitions 1 and 2 are not equivalent except when $L$ is constant. In all other cases the definition being used must be specified.

The two definitions are restricted to relatively slow changes in $i$, that is, to low frequencies, but by analogy with the definitions equivalent inductances may often be evolved in high-frequency applications such as resonators and waveguide equivalent circuits. Such "inductances," when used, must be specified.

The two definitions are restricted to cases in which the branches are small in physical size compared with a wavelength, whatever the frequency. Thus in the case of a uniform 2-wire transmission line it may be necessary even at low frequencies to consider the parameters as "distributed" rather than to have one inductance for the entire line.

*Susceptance (B).* Susceptance is the imaginary part of admittance.

*Transfer Function (H).* A transfer function is that function of frequency which is the ratio of a phasor output to a phasor input in a linear system.

*Transfer Ratio (H).* A transfer ratio is a dimensionless transfer function.

*Voltage, Electromotive Force (V).* The voltage along a specified path in an electric field is the dot product line integral of the electric field strength along this path. As here defined, voltage is synonymous with potential difference only in an electrostatic field.

***11. Definitions of Quantities of Radiation and Light.*** The following definitions are based on the principal meanings listed in the *IEEE Standard Dictionary,*[19] which should be consulted for extended meanings, compound terms, and related definitions. The symbols shown in parentheses are from Table 1-10.

*Candlepower.* Candlepower is luminous intensity expressed in candelas. (Term deprecated by IEEE.)

*Emissivity. Total Emissivity, (ϵ).* The total emissivity of an element of surface of a temperature radiator is the ratio of its radiant flux density (radiant exitance) to that of a blackbody at the same temperature.

*Spectral Emissivity, ϵ(λ).* The spectral emissivity of an element of surface of a temperature radiator at any wavelength is the ratio of its radiant flux density per unit wavelength interval (spectral radiant exitance) at that wavelength to that of a blackbody at the same temperature.

*Light.* For the purposes of illuminating engineering, light is visually evaluated radiant energy.

NOTE 1: Light is psychophysical, neither purely physical nor purely psychological. Light is not synonymous with radiant energy, however restricted, nor is it merely sensation. In a general nonspecialized sense, light is the aspect of radiant energy of which a human observer is aware through the stimulation of the retina of the eye.

NOTE 2: Radiant energy outside the visible portion of the spectrum must not be discussed using the quantities and units of light; it is nonsense to refer to "ultraviolet light," or to express infrared flux in lumens.

*Luminance (Photometric Brightness) (L).* Luminance in a direction, at a point on the surface of a source or of a receiver, or on any other real or virtual surface, is the quotient of the luminous flux (Φ) leaving, passing through, or arriving at a surface element surrounding the point, propagated in directions defined by an elementary cone containing the given direction, divided by the product of the solid angle of the cone ($d\omega$) and the area of the orthogonal projection of the surface element on a plane perpendicular to the given direction ($dA \cos \theta$). $L = d^2\Phi/[d\omega(da \cos \theta)] = dI/(dA \cos \theta)$. In the defining equation $\theta$ is the angle between the direction of observation and the normal to the surface.

In common usage the term *brightness* usually refers to the intensity of sensation which results from viewing surfaces or spaces from which light comes to the eye. This sensation is determined in part by the definitely measurable luminance defined above and in part by conditions of observation such as the state of adaptation of the eye. In much of the literature the term brightness, used alone, refers to both luminance and sensation. The context usually indicates which meaning is intended.

*Luminous Efficacy of Radiant Flux.* The luminous efficacy of radiant flux is the quotient of the total luminous flux divided by the total radiant flux. It is expressed in lumens per watt.

*Spectral Luminous Efficacy of Radiant Flux, K(λ).* Spectral luminous efficacy of radiant flux is the quotient of the luminous flux at a given wavelength divided by the radiant flux at the wavelength. It is expressed in lumens per watt.

*Spectral Luminous Efficiency of Radiant Flux.* Spectral luminous efficiency of radiant flux is the ratio of the luminous efficacy for a given wavelength to the value at the wavelength of maximum luminous efficacy. It is a numeric.

NOTE: The term *spectral luminous efficiency* replaces the previously used terms *relative luminosity* and *relative luminosity factor.*

*Luminous Flux (Φ).* Luminous flux is the time rate of flow of light.

*Luminous Flux Density at a Surface.* Luminous flux density at a surface is luminous flux per unit area of the surface. In referring to flux incident on a surface, this is called *illumination (E)*. The preferred term for luminous flux *leaving* a surface is *luminous exitance (M)*, which has been called *luminous emittance.*

*Luminous Intensity (I).* The luminous intensity of a source of light in a given direction is the luminous flux proceeding from the source per unit solid angle in the direction considered. ($I = d\Phi/d\omega$)

*Quantity of Light (Q).* Quantity of light (luminous energy) is the product of the luminous flux by the time it is maintained; that is, it is the time integral of luminous flux.

*Radiance (L).* Radiance in a direction, at a point on the surface of a source or of a receiver, or on any other real or virtual surface, is the quotient of the radiant flux (*P*) leaving, passing through, or arriving at a surface element surrounding the point, and propagated in directions defined by an elementary cone containing the given direction, divided by the product of the solid angle of the cone ($d\omega$) and the area of the orthogonal projection of the surface element on a plane perpendicular to the given direction ($dA \cos \theta$). $L = d^2P/d\omega(dA \cos \theta) = dI/(dA \cos \theta)$. In the defining equation $\theta$ is the angle between the normal to the element of the source and the direction of observation.

*Radiant Density (w).* Radiant density is radiant energy per unit volume.

*Radiant Energy (W).* Radiant energy is energy traveling in the form of electromagnetic waves.

*Radiant Flux Density at a Surface.* Radiant flux density at a surface is radiant flux per unit area of the surface. When referring to radiant flux incident on a surface, this is called *irradiance (E).* The preferred term for radiant flux *leaving* a surface is *radiant exitance (M),* which has been called *radiant emittance.*

*Radiant Intensity (I).* The radiant intensity of a source in a given direction is the radiant flux proceeding from the source per unit solid angle in the direction considered. ($I = dP/d\omega$)

*Radiant Power, Radiant Flux (P).* Radiant flux is the time rate of flow of radiant energy.

***12. Letter Symbols.***[2,3] Tables 1-10 and 1-11 list the United States Standard letter symbols for quantities and units.

A *quantity symbol* is a single letter (for example, *I* for electric current), specified as to general form of type, and modified when appropriate by one or more subscripts or superscripts. A *unit symbol* is a letter or group of letters (for example, cm for centimeter), or in a few cases a special sign, that may be used in the place of the name of the unit.

Symbols for quantities are printed in italic type, while symbols for units are printed in roman type. Subscripts and superscripts that are letter symbols for quantities or for indices are printed in roman type:

| | |
|---|---|
| $C_p$ | heat capacity at constant pressure $p$ |
| $a_{ij}$, $a_{45}$ | matrix elements |
| $I_i$, $I_o$ | input current, output current |

For indicating the vector character of a quantity, boldface italic type is used, for example, ***F*** for force. Ordinary italic type is used to represent the magnitude of a vector quantity.

The product of two quantities is indicated by writing *ab.* The quotient may be indicated by writing

$$\frac{a}{b} \qquad a/b \qquad \text{or} \qquad ab^{-1}$$

If more than one solidus (/) is required in any algebraic term, parentheses must be inserted to remove any ambiguity. Thus one may write $(a/b)/c$ or $a/bc$, but not $a/b/c$.

Unit symbols are written in lowercase letters, except for the first letter when the name of the unit is derived from a proper name, and except for a very few that are not formed from letters. When a compound unit is formed by multiplication of two or more other units, its symbol consists of the symbols for the separate units joined by a raised dot (for example, N·m for newton meter). The dot may be omitted in the case of familiar compounds such as watthour (symbol Wh) if no confusion would result. Hyphens should not be used in symbols for compound units. Positive and negative exponents may be used with the symbols for units.

When a symbol representing a unit that has a prefix (see Par. **5**) carries an exponent, this indicates that the multiple (or submultiple) unit is raised to the power expressed by the exponent.

*Examples:*

$$2\ \mathrm{cm}^3 = 2(\mathrm{cm})^3 = 2(10^{-2}\ \mathrm{m})^3 = 2\cdot 10^{-6}\ \mathrm{m}^3$$
$$1\ \mathrm{ms}^{-1} = 1(\mathrm{ms})^{-1} = 1(10^{-3}\ \mathrm{s})^{-1} = 10^3\ \mathrm{s}^{-1}$$

*Phasor quantities,* represented by complex numbers or complex time-varying functions, are extensively used in certain branches of electrical engineering. The following notation and typography are standard:

| | Notation | Remarks |
|---|---|---|
| Complex quantity | $Z$ | $Z = \|Z\| \exp(j\phi)$<br>$Z = \mathrm{Re}\, Z + j\, \mathrm{Im}\, Z$ |
| Real part | Re $Z$, $Z'$ | |
| Imaginary part | Im $Z$, $Z''$ | |
| Conjugate complex quantity | $Z^*$ | $Z^* = \mathrm{Re}\, Z - j\, \mathrm{Im}\, Z$ |
| Modulus of $Z$ | $\|Z\|$ | |
| Phase of $Z$, Argument of $Z$ | arg $Z$ | arg $Z = \phi$ |

## TABLE 1-12 Values of Fundamental Physical Constants

| Quantity | Symbol | Value[1] | Uncertainty, ppm |
|---|---|---|---|
| Permeability of vacuum | $\mu_0$ | $4\pi \times 10^{-7}$ H m$^{-1}$ <br> $= 12.566\ 370\ 614\ 4 \times 10^{-7}$ H m$^{-1}$ | |
| Speed of light in vacuum | $c$ | 299 792 458(1.2) m s$^{-1}$ | 0.004 |
| Permittivity of vacuum | $\epsilon_0 = (\mu_0 c^2)^{-1}$ | $8.854\ 187\ 82(7) \times 10^{-12}$ F m$^{-1}$ | 0.008 |
| Fine structure constant, | $\alpha$ | 0.007 297 350 6(60) | 0.82 |
| $\mu_0 ce^2/2h$ | $\alpha^{-1}$ | 137.036 04(11) | 0.82 |
| Elementary charge | $e$ | $1.602\ 189\ 2(46) \times 10^{-19}$ C | 2.9 |
| Planck constant | $h$ | $6.626\ 176(36) \times 10^{-34}$ J Hz$^{-1}$ | 5.4 |
| | $\hbar = h/2\pi$ | $1.054\ 588\ 7(57) \times 10^{-34}$ J s | 5.4 |
| Avogadro constant | $N_A$ | $6.022\ 045(31) \times 10^{23}$ mol$^{-1}$ | 5.1 |
| Atomic mass unit | $1u = (10^{-3}$ kg mol$^{-1})/N_A$ | $1.660\ 565\ 5(86) \times 10^{-27}$ kg | 5.1 |
| Electron rest mass | $m_e$ | $0.910\ 953\ 4(47) \times 10^{-30}$ kg | 5.1 |
| | | $5.485\ 802\ 6(21) \times 10^{-4}$ u | 0.38 |
| Muon rest mass | $m_\mu$ | $1.883\ 566(11) \times 10^{-28}$ kg | 5.6 |
| | | 0.113 429 20(26) u | 2.3 |
| Proton rest mass | $m_p$ | $1.672\ 648\ 5(86) \times 10^{-27}$ kg | 5.1 |
| | | 1.007 276 470(11) u | 0.011 |
| Neutron rest mass | $m_n$ | $1.674\ 954\ 3(86) \times 10^{-27}$ kg | 5.1 |
| | | 1.008 665 012(37) u | 0.037 |
| Ratio, proton mass to electron mass | $m_p/m_e$ | 1836.151 52(70) | 0.38 |
| Ratio, muon mass to electron mass | $m_\mu/m_e$ | 206.768 65(47) | 2.3 |
| Specific electron charge | $e/m_e$ | $1.758\ 804\ 7(49) \times 10^{11}$ C kg$^{-1}$ | 2.8 |
| Faraday constant | $F = N_A e$ | $9.648\ 456(27) \times 10^{4}$ C mol$^{-1}$ | 2.8 |
| Magnetic flux quantum | $\Phi_0 = h/2e$ | $2.067\ 850\ 6(54) \times 10^{-15}$ Wb | 2.6 |
| | $h/e$ | $4.135\ 701(11) \times 10^{-15}$ J Hz$^{-1}$ C$^{-1}$ | 2.6 |
| Josephson frequency-voltage ratio | $2e/h$ | 483.593 9(13) THz V$^{-1}$ | 2.6 |
| Quantum of circulation | $h/2m_e$ | $3.636\ 945\ 5(60) \times 10^{-4}$ J Hz$^{-1}$ kg$^{-1}$ | 1.6 |
| | $h/m_e$ | $7.273\ 891(12) \times 10^{-4}$ J Hz$^{-1}$ kg$^{-1}$ | 1.6 |
| Rydberg constant | $R_\infty$ | $1.097\ 373\ 177(83) \times 10^{7}$ m$^{-1}$ | 0.075 |
| Bohr radius | $a_0 = \alpha/4\pi R_\infty$ | $0.529\ 177\ 06(44) \times 10^{-10}$ m | 0.82 |
| Electron Compton wavelength | $\lambda_C = \alpha^2/2R_\infty$ | $2.426\ 308\ 9(40) \times 10^{-12}$ m | 1.6 |
| | $\bar{\lambda}_C = \lambda_C/2\pi = \alpha a_0$ | $3.861\ 590\ 5(64) \times 10^{-13}$ m | 1.6 |
| Classical electron radius | $r_e = \mu_0 e^2/4\pi m_e = \alpha \bar{\lambda}_C$ | $2.817\ 938\ 0(70) \times 10^{-15}$ m | 2.5 |
| Electron $g$ factor | $\frac{1}{2}g_e = \mu_e/\mu_B$ | 1.001 159 656 7(35) | 0.0035 |
| Muon $g$ factor | $\frac{1}{2}g_\mu$ | 1.001 166 16(31) | 0.31 |
| Proton moment in nuclear magnetons | $\mu_p/\mu_N$ | 2.792 845 6(11) | 0.38 |
| Bohr magneton | $\mu_B = e\hbar/2m_e$ | $9.274\ 078(36) \times 10^{-24}$ J T$^{-1}$ | 3.9 |
| Nuclear magneton | $\mu_N = e\hbar/2m_p$ | $5.050\ 824(20) \times 10^{-27}$ J T$^{-1}$ | 3.9 |
| Electron magnetic moment | $\mu_e$ | $9.284\ 832(36) \times 10^{-24}$ J T$^{-1}$ | 3.9 |

| | | | |
|---|---|---|---|
| Proton magnetic moment | $\mu_p$ | $1.410\,617\,1(55) \times 10^{-26}$ J T$^{-1}$ | 3.9 |
| Proton magnetic moment in Bohr magnetons | $\mu_p/\mu_B$ | $1.521\,032\,209(16) \times 10^{-3}$ | 0.011 |
| Ratio, electron to proton magnetic moments | $\mu_e/\mu_p$ | 658.210 688 0(66) | 0.010 |
| Ratio, muon moment to proton moment | $\mu_\mu/\mu_p$ | 3.183 340 2(72) | 2.3 |
| Muon magnetic moment | $\mu_\mu$ | $4.490\,474(18) \times 10^{-26}$ J T$^{-1}$ | 3.9 |
| Proton gyromagnetic ratio | $\gamma_p$ | $2.675\,198\,7(75) \times 10^8$ s$^{-1}$ T$^{-1}$ | 2.8 |
| Diamagnetic shielding factor, spherical $H_2O$ sample | $1 + \sigma(H_2O)$ | 1.000 025 637(67) | 0.067 |
| Proton gyromagnetic ratio (uncorrected) | $\gamma'_p$ | $2.675\,130\,1(75) \times 10^8$ s$^{-1}$ T$^{-1}$ | 2.8 |
| | $\gamma'_p/2\pi$ | 42.576 02(12) MHz T$^{-1}$ | 2.8 |
| Proton moment in nuclear magnetons (uncorrected) | $\mu_p/\mu_N$ | 2.792 774 0(11) | 0.38 |
| Proton Compton wavelength | $\lambda_{C,p} = h/m_p c$ | $1.321\,409\,9(22) \times 10^{-15}$ m | 1.7 |
| | $\bar{\lambda}_{C,p} = \lambda_{C,p}/2\pi$ | $2.103\,089\,2(36) \times 10^{-16}$ m | 1.7 |
| Neutron Compton wavelength | $\lambda_{C,n} = h/m_n c$ | $1.319\,590\,9(22) \times 10^{-15}$ m | 1.7 |
| | $\bar{\lambda}_{C,n} = \lambda_{C,n}/2\pi$ | $2.100\,194\,1(35) \times 10^{-16}$ m | 1.7 |
| Molar gas constant | $R$ | 8.314 41(26) J mol$^{-1}$ K$^{-1}$ | 31 |
| Molar Volume, Ideal Gas ($T_0 = 273.15$ K, $p_0 = 1$ atm) | $V_m = RT_0/p_0$ | 0.022 413 83(70) m$^3$ mol$^{-1}$ | 31 |
| Boltzmann constant | $k = R/N_A$ | $1.380\,662(44) \times 10^{-23}$ J K$^{-1}$ | 32 |
| Stefan-Boltzmann constant | $\sigma = (\pi^2/60)k^4/\hbar^3 c^2$ | $5.670\,32(71) \times 10^{-8}$ W m$^{-2}$ K$^{-4}$ | 125 |
| First radiation constant | $c_1 = 2\pi hc^2$ | $3.741\,832(20) \times 10^{-16}$ W m$^2$ | 5.4 |
| Second radiation constant | $c_2 = hc/k$ | 0.014 387 86(45) m K | 31 |
| Gravitational constant | $G$ | $6.672\,0(41) \times 10^{-11}$ N m$^2$ kg$^{-2}$ | 615 |
| Ratio, BIPM ampere to SI ampere | $K \equiv A_{B169}/A$ | 1.000 000 7(26) | 2.6 |
| Ratio, BIPM ohm to SI ohm | $R \equiv \Omega_{B169}/\Omega$ | 0.999 999 47(19) | 0.19 |
| Ratio, BIPM volt to SI volt | $V_{B169}/V$ | 1.000 000 2(26) | 2.6 |
| Ratio, kxu to ångström, $\lambda(CuK\alpha_1) \equiv 1.537400$ kxu | $\Lambda$ | 1.002 077 2(54) | 5.3 |
| Ratio, Å* to ångström, $\lambda(WK\alpha_1) \equiv 0.2090100$Å* | $\Lambda^*$ | 1.000 020 5(56) | 5.6 |
| Energy equivalents: | | | |
| 1 u | | 931.501 6(26) MeV | 2.8 |
| 1 proton mass | | 938.279 6(27) MeV | 2.8 |
| 1 neutron mass | | 939.573 1(27) MeV | 2.8 |
| 1 muon mass | | 105.659 48(35) MeV | 3.3 |
| 1 electron mass | | 0.511 003 4(14) MeV | 2.8 |
| 1 electronvolt | 1 eV/$k$ | 11 604.50(36) K | 31 |
| | 1 eV/$hc$ | 8065.479(21) cm$^{-1}$ | 2.6 |
| | 1 eV/$h$ | $2.417\,969\,6(63) \times 10^{14}$ Hz | 2.6 |
| Voltage-wavelength product | $V\lambda$ | 12 398.520(32) eV Å | 2.6 |
| Rydberg constant | $R_\infty hc$ | 13.605 804(36) eV | 2.6 |
| Gas constant | $R$ | 82.056 8(26) cm$^3$ atm mol$^{-1}$ K$^1$ | 31 |
| | | 1.987 19(6) cal mol$^{-1}$ K$^{-1}$ | 31 |

* The digits in parentheses following a numerical value represent the standard deviation of that value, in terms of the final listed digits. Reproduced by permission from Recommended Consistent Values of the Fundamental Constants, *CODATA Bulletin* 11, December 1973 (Ref. 10, Par. 17).

**TABLE 1-13** Numerical Values Used in Electrical Engineering

Functions of $\pi$:

$\pi = 3.141\,592\,654$
$1/\pi = 0.318\,309\,886$
$\pi^2 = 9.869\,604\,404$
$\sqrt{\pi} = 1.772\,453\,851$
$\pi/180^\circ = 0.017\,453\,293$ (= radians per degree)
$180^\circ/\pi = 57.295\,779\,51$ (= degrees per radian)

Functions of $\epsilon$:

$\epsilon = 2.718\,281\,828$
$1/\epsilon = 0.367\,879\,441$
$1 - 1/\epsilon = 0.632\,120\,559$
$\epsilon^2 = 7.389\,056\,096$
$\sqrt{\epsilon} = 1.648\,721\,271$

Logarithms to the base 10:

$\log_{10} \pi = 0.497\,149\,873$
$\log_{10} \epsilon = 0.434\,294\,482$
$\log_{10} 2 = 0.301\,029\,996$
$\log_{10} x = (\ln x)(0.434\,294\,482) = (\log_2 x)(0.301\,029\,996)$

Natural logarithms (to the base $\epsilon$):

$\ln \pi = 1.144\,729\,886$
$\ln 2 = 0.693\,147\,181$
$\ln 10 = 2.302\,585\,093$
$\ln x = (\log_{10} x)(2.302\,585\,093) = (\log_2 x)(0.693\,147\,181)$

Logarithms to the base 2:

$\log_2 \pi = 1.651\,496\,130$
$\log_2 \epsilon = 1.442\,695\,042$
$\log_2 10 = 3.321\,928\,096$
$\log_2 x = (\log_{10} x)(3.321\,928\,096) = (\ln x)(1.442\,695\,042)$

Powers of 2:

$2^5 = 32$
$2^{10} = 1024$
$2^{15} = 32{,}768$
$2^{20} = 1{,}048{,}576$
$2^{25} = 33{,}554{,}432$
$2^{30} = 1{,}073{,}741{,}824$
$2^{40} = 1.099\,511\,628 \times 10^{12}$
$2^{50} = 1.125\,899\,907 \times 10^{15}$
$2^{100} = 1.267\,650\,601 \times 10^{30}$

Logarithmic units:

| Power ratio | Current or voltage ratio | Decibels* | Nepers⁺ |
|---|---|---|---|
| 1 | 1 | 0 | 0 |
| 2 | 1.414 214 | 3.010 300 | 0.346 574 |
| 3 | 1.732 051 | 4.771 213 | 0.549 306 |
| 4 | 2 | 6.020 600 | 0.693 147 |
| 5 | 2.236 068 | 6.989 700 | 0.804 719 |
| 10 | 3.162 278 | 10 | 1.151 293 |
| 15 | 3.872 983 | 11.760 913 | 1.354 025 |

*The decibel is defined for power ratios only. It may be applied to current or voltage ratios only when the resistances through which the currents flow or across which the voltages are applied are equal.

⁺The neper is defined for current and voltage ratios only. It may be applied to power ratios only when the respective resistances are equal.

**TABLE 1-13** Numerical Values Used in Electrical Engineering *(Continued)*

Values of $2^{(2^N)}$:

| Value of $N$ | Value of $2^{(2^N)}$ |
|---|---|
| 1 | 4 |
| 2 | 16 |
| 3 | 256 |
| 4 | 65,536 |
| 5 | 4,294,967,296 |
| 6 | $1.844\,674\,407 \times 10^{19}$ |
| 7 | $3.402\,823\,668 \times 10^{38}$ |
| 8 | $1.157\,920\,892 \times 10^{77}$ |
| 9 | $1.340\,780\,792 \times 10^{154}$ |
| 10 | $1.797\,693\,132 \times 10^{308}$ |

***13. Graphic Symbols.***[5] An extensive list of standard graphic symbols for electrical engineering has been compiled in IEEE Std 315-75 (ANSI Y32.2-1975). Since this standard comprises 110 pages, including 78 pages of diagrams, it is impractical to reproduce it here. Those concerned with the preparation of circuit diagrams and graphic layouts should conform to these standard symbols to avoid confusion with earlier, nonstandard forms. See also Sec. **28**.

***14. Physical Constants.*** Table 1-12 lists the values of the fundamental physical constants, compiled by E. R. Cohen and his coworkers on the Task Group on Fundamental Constants of the Committee on Data for Science and Technology (CODATA), sponsored by the International Council of Scientific Unions. Further details on the methods used to adjust these values to form a consistent set are contained in Ref. 10, Par. 1-17.

***15. Numerical Values.*** Extensive use is made in electrical engineering of the constants $\pi$ and $\epsilon$ and of the numbers 2 and 10, the latter in logarithmic units and number systems. Table 1-13 lists functions of these numbers to 9 or 10 significant digits. In most engineering applications (except those involving the difference of large, nearly equal numbers), five significant digits suffice. The use of the listed values in computations with electronic hand calculators will suffice in most cases to produce results more than adequate for engineering work.

***16. Conversion Factors.*** The increasing use of the metric system in British and American practice has generated a need for extensive tables of multiplying factors to facilitate conversions from and to the SI units. Tables 1-14 through 1-27 have been compiled by the author for this purpose, arranged as follows:

| *Table* | *Quantity* | *SI unit* | *Subtabulation* | *Basis of grouping* |
|---|---|---|---|---|
| 1-14 | Length | meter | 1-14A | Units decimally related to one meter |
| | | | 1-14B | Units less than one meter |
| | | | 1-14C | Units greater than one meter |
| | | | 1-14D | Other length units |
| 1-15 | Area | square meter | 1-15A | Units decimally related to one square meter |
| | | | 1-15B | Nonmetric area units |
| | | | 1-15C | Other area units |
| 1-16 | Volume/capacity | cubic meter | 1-16A | Units decimally related to one cubic meter |
| | | | 1-16B | Nonmetric volume units |
| | | | 1-16C | U.S. liquid capacity measures |
| | | | 1-16D | British liquid capacity measures |
| | | | 1-16E | U.S. and U.K. dry capacity measures |
| | | | 1-16F | Other volume and capacity units |
| 1-17 | Mass | kilogram | 1-17A | Units decimally related to one kilogram |

*(Continued on p. 1-32)*

## TABLE 1-14 Length Conversion Factors

(Exact conversions are shown in **boldface** type. Repeating decimals are underlined.) The SI unit of length is the meter

A. Length units decimally related to one meter

| | Meters (m) | Kilometers (km) | Decimeters (dm) | Centimeters (cm) | Millimeters (mm) | Micrometers ($\mu$m) | Nanometers (nm) | Ångströms (Å) |
|---|---|---|---|---|---|---|---|---|
| 1 meter = | **1** | **0.001** | **10** | **100** | **1 000** | **1 000 000** | **$10^9$** | **$10^{10}$** |
| 1 kilometer = | **1 000** | **1** | **10 000** | **100 000** | **1 000 000** | **$10^9$** | **$10^{12}$** | **$10^{13}$** |
| 1 decimeter = | **0.1** | **0.000 1** | **1** | **10** | **100** | **100 000** | **$10^8$** | **$10^9$** |
| 1 centimeter = | **0.01** | **0.000 01** | **0.1** | **1** | **10** | **10 000** | **$10^7$** | **$10^8$** |
| 1 millimeter = | **0.001** | **$10^{-6}$** | **0.01** | **0.1** | **1** | **1 000** | **1 000 000** | **$10^7$** |
| 1 micrometer (micron) = | **$10^{-6}$** | **$10^{-9}$** | **0.000 01** | **0.000 1** | **0.001** | **1** | **1 000** | **10 000** |
| 1 nanometer = | **$10^{-9}$** | **$10^{-12}$** | **$10^{-8}$** | **$10^{-7}$** | **$10^{-6}$** | **0.001** | **1** | **10** |
| 1 ångström = | **$10^{-10}$** | **$10^{-13}$** | **$10^{-9}$** | **$10^{-8}$** | **$10^{-7}$** | **0.000 1** | **0.1** | **1** |

B. Nonmetric length units less than one meter

| | Meters (m) | Yards (yd) | Feet (ft) | Inches (in) | Mils (mil) | Microinches ($\mu$in) |
|---|---|---|---|---|---|---|
| 1 meter = | **1** | 1.093 613 30 | 3.280 839 89 | 39.370 078 7 | $3.937\,007\,87 \times 10^4$ | $3.937\,007\,87 \times 10^7$ |
| 1 yard = | **0.914 4** | **1** | **3** | **36** | **36 000** | **$3.6 \times 10^7$** |
| 1 foot = | **0.304 8** | **1/3** = 0.333 3 | **1** | **12** | **12 000** | **$1.2 \times 10^7$** |
| 1 inch = | **0.025 4** | **1/36** = 0.027 7 | **1/12** = 0.083 3 | **1** | **1 000** | **1 000 000** |
| 1 mil = | **$2.54 \times 10^{-5}$** | $2.777 \times 10^{-5}$ | $8.333 \times 10^{-5}$ | **0.001** | **1** | **1 000** |
| 1 microinch = | **$2.54 \times 10^{-8}$** | $2.777 \times 10^{-8}$ | $8.333 \times 10^{-8}$ | **$10^{-6}$** | **0.001** | **1** |

C. Nonmetric length units greater than one meter (with equivalents in feet)

| | Meters (m) | Rods (rd) | Statute miles (mi) | Nautical miles (nmi) | Astronomical units (AU) | Parsecs (pc) | Feet (ft) |
|---|---|---|---|---|---|---|---|
| 1 meter = | **1** | 0.198 838 78 | $6.213\,711\,92 \times 10^{-4}$ | $5.399\,568\,04 \times 10^{-4}$ | $6.684\,491\,98 \times 10^{-12}$ | $3.240\,733\,17 \times 10^{-17}$ | 3.280 839 89 |
| 1 rod = | **5.029 2** | **1** | **0.003 125** | $2.715\,550\,76 \times 10^{-3}$ | $3.361\,764\,71 \times 10^{-11}$ | $1.629\,829\,53 \times 10^{-16}$ | **16.5** |
| 1 statute mile = | **1 609.344** | **320** | **1** | 0.868 976 24 | $1.075\,764\,71 \times 10^{-8}$ | $5.215\,454\,50 \times 10^{-14}$ | **5 280** |
| 1 nautical mile = | **1 852** | 368.249 423 | 1.150 779 45 | **1** | $1.237\,967\,91 \times 10^{-8}$ | $6.001\,837\,80 \times 10^{-14}$ | 6 076.115 48 |
| 1 astronomical unit* = | **$1.496 \times 10^{11}$** | $2.974\,628\,17 \times 10^{10}$ | 92 957 130.3 | 80 777 537.8 | **1** | $4.848\,136\,82 \times 10^{-6}$ | $4.908\,136\,48 \times 10^{11}$ |
| 1 parsec = | $3.085\,721\,50 \times 10^{16}$ | $6.135\,611\,02 \times 10^{15}$ | $1.917\,378\,44 \times 10^{13}$ | $1.666\,156\,32 \times 10^{13}$ | 206 264.806 | **1** | $1.012\,375\,82 \times 10^{17}$ |
| 1 foot = | **0.304 8** | 0.060 606 | $1.893\,939 \times 10^{-4}$ | $1.645\,788\,33 \times 10^{-4}$ | $2.037\,433\,16 \times 10^{-12}$ | $9.877\,754\,72 \times 10^{-18}$ | **1** |

D. Other length units

1 cable = **720** feet = **219.456** meters
1 cable (U.K.) = **608** feet = **185.318 4** meters
1 chain (engineers') = **100** feet = **30.48** meters
1 chain (surveyors') = **66** feet = **20.116 8** meters
1 fathom = **6** feet = **1.828 8** meters
1 fermi = **1** femtometer = $\mathbf{10^{-15}}$ meter
1 foot (U.S. Survey) = 0.304 800 6 meter
1 furlong = **660** feet = **201.168** meters
1 hand = **4** inches = **0.101 6** meter
1 league (international nautical) = **3** nautical miles = **5 556** meters
1 league (statute) = **3** statute miles = **4 828.032** meters
1 league (U.K. nautical) = 5 559.552 meters
1 light-year = 9.460 895 2 × $10^{15}$ meters (= distance traveled by light in vacuum in one sidereal year)
1 link (engineers') = **1** foot = **0.304 8** meter
1 link (surveyors') = **7.92** inches = **0.201 168** meter
1 micron = **1** micrometer = $\mathbf{10^{-6}}$ meter
1 millimicron = **1** nanometer = $\mathbf{10^{-9}}$ meter
1 myriameter = **10 000** meters
1 nautical mile (U.K.) = 1 853.184 meters
1 pale = **1** rod = **5.029 2** meters
1 perch (linear) = **1** rod = **5.029 2** meters
1 pica = **1/6** inch (approx.) = 4.217 518 × $10^{-3}$ meter
1 point = **1/72** inch (approx.) = 3.514 598 × $10^{-4}$ meter
1 span = **9** inches = **0.228 6** meter

*** As defined by the International Astronomical Union, 1964. See Ref. 11, Par. 17.**

*(Continued from p. 1-29)*

| *Table* | *Quantity* | *SI unit* | *Subtabulation* | *Basis of grouping* |
|---|---|---|---|---|
| | | | 1-17B | Less than one pound-mass |
| | | | 1-17C | One pound-mass and greater |
| | | | 1-17D | Other mass units |
| 1-18 | Time | second | 1-18A | One second and less |
| | | | 1-18B | One second and greater |
| | | | 1-18C | Other time units |
| 1-19 | Velocity | meter per second | | |
| 1-20 | Density | kilogram per cubic meter | 1-20A | Units decimally related to one kilogram per cubic meter |
| | | | 1-20B | Nonmetric density units |
| | | | 1-20C | Other density units |
| 1-21 | Force | newton | | |
| 1-22 | Pressure | pascal | 1-22A | Units decimally related to one pascal |
| | | | 1-22B | Units decimally related to one kilogram-force per square meter |
| | | | 1-22C | Units expressed as heights of liquid |
| | | | 1-22D | Nonmetric pressure units |
| 1-23 | Torque/bending moment | newton meter | | |
| 1-24 | Energy/work | joule | 1-24A | Units decimally related to one joule |
| | | | 1-24B | Units less than 10 joules |
| | | | 1-24C | Units greater than 10 joules |
| 1-25 | Power | watt | 1-25A | Units decimally related to one watt |
| | | | 1-25B | Nonmetric power units |
| 1-26 | Temperature | kelvin | | |
| 1-27 | Light | candela per square meter | 1-27A | Luminance units |
| | | lux | 1-27B | Illuminance units |

*Statements of Equivalence.* To avoid ambiguity, the conversion tables have been arranged in the form of statements of equivalence; that is, each unit listed at the left-hand edge of each table is stated to be equivalent to a multiple or fraction of each of the units to the right in the table. For example, the uppermost line of Table 1-14B represents the following statements:

1 meter is equal to 1.093 613 30 yards

1 meter is equal to 3.280 839 89 feet

1 meter is equal to 39.370 078 7 inches

1 meter is equal to $3.937\ 007\ 87 \times 10^4$ mils

1 meter is equal to $3.937\ 007\ 87 \times 10^7$ microinches

This table contains similar statements relating the meter, yard, foot, inch, mil, and microinch to each other; that is, conversion factors between the non-SI units, as well as to and from the SI unit, are given. In all, these tables contain over 1700 such statements. Exact conversion factors are indicated in **boldface** type.

*Tabulation Groups.* To produce tables that can be contained on individual pages of the handbook, units of a given quantity have been arranged in separate subtabulations identified by capital letters. Each such subtabulation represents a group of units related to each other deci-

## TABLE 1-15 Area Conversion Factors

**(Exact conversions are shown in boldface type. Repeating decimals are underlined.) The SI unit of area is the square meter**

A. Area units decimally related to one square meter

| | Square meters $(m)^2$ | Square kilometers $(km)^2$ | Hectares **(square hectometers) $(hm)^2$** | Square centimeters $(cm)^2$ | Square millimeters $(mm)^2$ | Square micrometers $(\mu m)^2$ | Barns (b) |
|---|---|---|---|---|---|---|---|
| 1 square meter = | **1** | **$10^{-6}$** | **0.000 1** | **10 000** | **1 000 000** | **$10^{12}$** | **$10^{28}$** |
| 1 square kilometer = | **1 000 000** | **1** | **100** | **$10^{10}$** | **$10^{12}$** | **$10^{18}$** | **$10^{34}$** |
| 1 hectare = | **10 000** | **0.01** | **1** | **$10^{8}$** | **$10^{10}$** | **$10^{16}$** | **$10^{32}$** |
| 1 square centimeter = | **0.000 1** | **$10^{-10}$** | **$10^{-8}$** | **1** | **100** | **$10^{8}$** | **$10^{24}$** |
| 1 square millimeter = | **$10^{-6}$** | **$10^{-12}$** | **$10^{-10}$** | **0.01** | **1** | **$10^{6}$** | **$10^{22}$** |
| 1 square micrometer = | **$10^{-12}$** | **$10^{-18}$** | **$10^{-16}$** | **$10^{-8}$** | **$10^{-6}$** | **1** | **$10^{16}$** |
| 1 barn = | **$10^{-28}$** | **$10^{-34}$** | **$10^{-32}$** | **$10^{-24}$** | **$10^{-22}$** | **$10^{-16}$** | **1** |

B. Nonmetric area units (with square meter equivalents)

| | Square meters $(m)^2$ | Square statute miles $(mi)^2$ | Acres (acre) | Square rods $(rd)^2$ | Square yards $(yd)^2$ | Square feet $(ft)^2$ | Square inches $(in)^2$ | Circular mils (cmil) |
|---|---|---|---|---|---|---|---|---|
| 1 square meter = | 1 | $3.861\,021\,59 \times 10^{-7}$ | $2.471\,053\,82 \times 10^{-4}$ | $3.953\,686\,10 \times 10^{-2}$ | 1.195 990 05 | 10.763 910 4 | 1 550.003 10 | $1.973\,525\,24 \times 10^{9}$ |
| 1 square statute mile = | 2 589 988.1 | **1** | **640** | **102 400** | **3 097 600** | **27 878 400** | $4.014\,489\,60 \times 10^{9}$ | $5.111\,406\,91 \times 10^{15}$ |
| 1 acre = | 4 046.856 41 | **1/640 = 0.001 562 5** | **1** | **160** | **4 840** | **43 560** | **6 272 640** | $7.986\,573\,30 \times 10^{12}$ |
| 1 square rod = | 25.292 852 6 | **$9.765\,625 \times 10^{-6}$** | **1/160 = 0.006 25** | **1** | **30.25** | **272.25** | **39 204** | $4.991\,608\,31 \times 10^{10}$ |
| 1 square yard = | **0.836 127 36** | $3.228\,305\,79 \times 10^{-7}$ | $2.066\,115\,70 \times 10^{-4}$ | $3.305\,785\,12 \times 10^{-2}$ | **1** | **9** | **1 296** | $1.650\,118\,45 \times 10^{9}$ |
| 1 square foot = | **0.092 903 04** | $3.587\,006\,43 \times 10^{-8}$ | $2.295\,684\,11 \times 10^{-5}$ | $3.673\,094\,58 \times 10^{-3}$ | **1/9** = 0.111 111 | **1** | **144** | $1.833\,464\,95 \times 10^{8}$ |
| 1 square inch = | **$6.451\,6 \times 10^{-4}$** | $2.490\,976\,69 \times 10^{-10}$ | $1.594\,225\,08 \times 10^{-7}$ | $2.550\,760\,13 \times 10^{-5}$ | $7.716\,049\,38 \times 10^{-4}$ | **1/144** = 0.006 944 44 | **1** | $1.273\,239\,55 \times 10^{6}$ |
| 1 circular mil = | $5.067\,074\,79 \times 10^{-10}$ | $1.956\,408\,51 \times 10^{-16}$ | $1.252\,101\,45 \times 10^{-13}$ | $2.003\,362\,32 \times 10^{-11}$ | $6.060\,171\,01 \times 10^{-10}$ | $5.454\,153\,91 \times 10^{-9}$ | $7.853\,981\,63 \times 10^{-7}$ | 1 |

Exact conversions are:

1 acre = **4 046.856 422 4** square meters

1 square mile = **2 589 988.110 336** square meters

C. Other area units

1 are = **100** square meters
1 centiare (centare) = **1** square meter
1 perch (area) = **1** square rod = **30.25** square yards = 25.292 852 6 square meters
1 rood = **40** square rods = 1 011.714 11 square meters
1 section = **1** square statute mile = 2 589 988.1 square meters
1 township = **36** square statute miles = 93 239 572 square meters

**TABLE 1-16** Volume and Capacity Conversion Factors

(Exact conversions are shown in **boldface type**. Repeating decimals are underlined.) The SI unit of volume is the cubic meter

A. Volume units decimally related to one cubic meter

| | Cubic meters (steres) (m)$^3$ | Cubic decimeters (dm)$^3$ | Cubic centimeters (cm)$^3$ | Liters (L) | Centiliters (cL) | Milliliters (mL) | Microliters ($\mu$L) |
|---|---|---|---|---|---|---|---|
| 1 cubic meter = | **1** | **1 000** | **1 000 000** | **1 000** | **100 000** | **1 000 000** | **$10^{9}$** |
| 1 cubic decimeter = | **0.001** | **1** | **1 000** | **1** | **100** | **1 000** | **1 000 000** |
| 1 cubic centimeter = | **0.000 001** | **0.001** | **1** | **0.001** | **0.1** | **1** | **1 000** |
| 1 liter = | **0.001** | **1** | **1 000** | **1** | **100** | **1 000** | **1 000 000** |
| 1 centiliter = | **0.000 01** | **0.01** | **10** | **0.01** | **1** | **10** | **10 000** |
| 1 milliliter = | **0.000 001** | **0.001** | **1** | **0.001** | **0.1** | **1** | **1 000** |
| 1 microliter = | **$10^{-9}$** | **0.000 001** | **0.001** | **0.000 001** | **0.000 1** | **0.001** | **1** |

B. Nonmetric volume units (with cubic meter and liter equivalents)

| | Cubic meters (steres) (m)$^3$ | Liters (L) | Cubic inches (in)$^3$ | Cubic feet (ft)$^3$ | Cubic yards (yd)$^3$ | Barrels (U.S.A.) (bbl) | Acre-Feet (acre-ft) | Cubic miles (mi)$^3$ |
|---|---|---|---|---|---|---|---|---|
| 1 cubic meter = | **1** | **1 000** | 6.102 374 41 × $10^{4}$ | 35.314 666 | 1.307 950 62 | 6.289 810 97 | 8.107 131 94 × $10^{-4}$ | 2.399 127 59 × $10^{-10}$ |
| 1 liter = | **0.001** | **1** | 61.023 744 1 | 0.035 314 66 | 1.307 950 62 × $10^{-3}$ | 6.289 810 97 × $10^{-3}$ | 8.107 131 93 × $10^{-7}$ | 2.399 127 59 × $10^{-13}$ |
| 1 cubic inch = | **1.638 706 4 × $10^{-5}$** | **1.638 706 4 × $10^{-2}$** | **1** | **1/1 728** = 5.787 037 03 × $10^{-4}$ | **1/46 656** = 2.143 347 05 × $10^{-5}$ | 1.030 715 32 × $10^{-4}$ | 1.328 520 90 × $10^{-8}$ | 3.931 465 73 × $10^{-15}$ |
| 1 cubic foot = | 2.831 684 66 × $10^{-2}$ | **28.316 846 592** | **1 728** | **1** | **1/27** = 0.037 037 | 0.178 107 61 | **1/43 560** = 2.295 684 11 × $10^{-5}$ | 6.793 572 78 × $10^{-12}$ |
| 1 cubic yard = | 0.764 554 86 | 764.55 485 8 | **46 656** | **27** | **1** | 4.808 905 38 | 6.198 347 11 × $10^{-4}$ | 1.834 264 65 × $10^{-10}$ |
| 1 barrel (U.S.A.) = | 0.158 987 29 | 158.987 294 | **9 702** | 5.614 583 33 | 0.207 947 53 | **1** | 1.288 930 98 × $10^{-4}$ | 3.814 308 05 × $10^{-11}$ |
| 1 acre-foot = | 1 233.481 84 | 1.233 481 84 × $10^{6}$ | 7.527 168 00 × $10^{7}$ | **43 560** | 1 613.333 33 | 7 758.367 34 | **1** | 2.959 280 30 × $10^{-7}$ |
| 1 cubic mile = | 4.168 181 83 × $10^{9}$ | 4.168 181 83 × $10^{12}$ | 2.543 580 61 × $10^{14}$ | **1.471 979 52 × $10^{11}$** | **5.451 776 × $10^{9}$** | 26.217 074 9 × $10^{9}$ | **3 379 200** | **1** |

C. United States liquid capacity measures (with liter equivalents)

| | Liters (L) | Gallons (U.S. gal) | Quarts (U.S. qt) | Pints (U.S. pt) | Gills (U.S. gi) | Fluid ounces (U.S. floz) | Fluidrams (U.S. fldr) | Minims (U.S. minim) |
|---|---|---|---|---|---|---|---|---|
| 1 liter = | **1** | 0.264 172 05 | 1.056 688 | 2.113 376 | 8.453 506 | 33.814 023 | 270.512 18 | 16 230.73 |
| 1 gallon, U.S. = | 3.785 411 8 | **1** | **4** | **8** | **32** | **128** | **1 024** | **61 440** |
| 1 quart, U.S. = | **0.946 352 946** | **1/4 = 0.25** | **1** | **2** | **8** | **32** | **256** | **15 360** |
| 1 pint, U.S. = | 0.473 176 5 | **1/8 = 0.125** | **1/2 = 0.5** | **1** | **4** | **16** | **128** | **7 680** |
| 1 gill, U.S. = | 0.118 294 1 | **1/32 = 0.031 25** | **1/8 = 0.125** | **1/4 = 0.25** | **1** | **4** | **32** | **1 920** |
| 1 fluid ounce, U.S. = | 2.957 353 × $10^{-2}$ | **1/128 = 0.007 812 5** | **1/32 = 0.031 25** | 1/16 = 0.062 5 | **1/4 = 0.25** | **1** | **8** | **480** |
| 1 fluidram, U.S. = | 3.696 691 2 × $10^{-3}$ | **1/102 4 = 9.765 625 × $10^{-4}$** | **1/256 = 3.906 25 × $10^{-3}$** | **1/128 = 0.007 812 5** | **1/32 = 0.031 25** | **1/8 = 0.125** | **1** | **60** |
| 1 minim, U.S. = | 6.161 152 × $10^{-5}$ | **1/61 440** = 1.627 604 16 × $10^{-5}$ | **1/15 360** = 6.510 416 66 × $10^{-5}$ | **1/7 680** = 1.302 083 33 × $10^{-4}$ | **1/1 920** = 5.208 333 3 × $10^{-4}$ | **1/480** = 2.083 333 3 × $10^{-3}$ | **1/60** = 0.016 666 6 | **1** |

D. British Imperial liquid capacity measures (with liter equivalents)

| | Liters (L) | Gallons (U.K. gal) | Quarts (U.K. qt) | Pints (U.K. pt) | Gills (U.K. gi) | Fluid ounces (U.K. floz) | Fluidrams (U.K. fldr) | Minims (U.K. minim) |
|---|---|---|---|---|---|---|---|---|
| 1 liter = | **1** | 0.219 969 2 | 0.879 876 6 | 1.759 753 | 7.039 018 | 35.195 06 | 281.560 5 | 16 893.63 |
| 1 gallon, U.K. = | 4.546 092 | **1** | **4** | **8** | **32** | **160** | **1 280** | **76 800** |
| 1 quart, U.K. = | 1.136 523 | **1/4 = 0.25** | **1** | **2** | **8** | **40** | **320** | **19 200** |
| 1 pint, U.K. = | 0.568 261 5 | **1/8 = 0.125** | **1/2 = 0.5** | **1** | **4** | **20** | **160** | **9 600** |
| 1 gill, U.K. = | 0.142 065 4 | **1/32 = 0.031 25** | **1/8 = 0.125** | **1/4 = 0.25** | **1** | **5** | **40** | **2 400** |
| 1 fluid ounce, U.K. = | 2.841 307 $\times 10^{-2}$ | **1/160 = 0.006 25** | **1/40 = 0.025** | **1/20 = 0.05** | **1/5 = 0.2** | **1** | **8** | **480** |
| 1 fluidram, U.K. = | 3.551 634 $\times 10^{-3}$ | **1/1280 = 7.812 5 $\times 10^{-4}$** | **1/320 = 0.003 125** | **1/160 = 0.006 25** | **1/40 = 0.025** | **1/8 = 0.125** | **1** | **60** |
| 1 minim, U.K. = | 5.919 391 $\times 10^{-5}$ | **1/76 800 =** 1.302 083 33 $\times 10^{-5}$ | **1/19 200 =** 5.208 333 33 $\times 10^{-5}$ | **1/9 600 =** 1.041 666 66 $\times 10^{-4}$ | **1/2 400 =** 4.166 666 66 $\times 10^{-4}$ | **1/480 =** 2.083 333 33 $\times 10^{-3}$ | **1/60 =** 0.016 666 66 | **1** |

E. United States and British dry capacity measures (with liter equivalents)

| | Liters (L) | U.S. dry measures: Bushels (U.S. bu) | U.S. dry measures: Pecks (U.S. peck) | U.S. dry measures: Quarts (U.S. qt) | U.S. dry measures: Pints (U.S. pt) | British dry measures: Bushels (U.K. bu) | British dry measures: Pecks (U.K. peck) | British dry measures: Quarts (U.K. qt) | British dry measures: Pints (U.K. pt) |
|---|---|---|---|---|---|---|---|---|---|
| 1 liter = | **1** | 0.028 377 59 | 0.113 510 37 | 0.908 082 99 | 1.816 165 98 | 0.027 496 1 | 0.109 984 6 | 0.879 876 6 | 1.759 753 4 |
| 1 bushel, U.S. = | 35.239 070 | **1** | **4** | **32** | **64** | 0.968 938 7 | 3.875 754 9 | 31.006 04 | 62.012 08 |
| 1 peck, U.S. = | 8.809 767 5 | **1/4 = 0.25** | **1** | **8** | **16** | 0.242 234 7 | 0.968 938 7 | 7.751 509 | 15.503 02 |
| 1 quart, U.S. = | 1.101 220 9 | **1/32 = 0.031 25** | **1/8 = 0.125** | **1** | **2** | 0.030 279 34 | 0.121 117 3 | 0.968 938 7 | 1.937 878 |
| 1 pint, U.S. = | 0.550 610 5 | **1/64 = 0.015 625** | **1/16 = 0.062 5** | **1/2 = 0.5** | **1** | 0.015 139 67 | 0.060 558 67 | 0.484 469 3 | 0.968 938 7 |
| 1 bushel, U.K. = | 36.368 73 | 1.032 057 | 4.128 228 | 33.025 82 | 66.051 65 | **1** | **4** | **32** | **64** |
| 1 peck, U.K. = | 9.092 182 | 0.258 014 3 | 1.032 057 | 8.256 456 | 16.512 91 | **1/4 = 0.25** | **1** | **8** | **16** |
| 1 quart, U.K. = | 1.136 523 | 0.032 251 78 | 0.129 007 1 | 1.032 057 | 2.064 114 2 | **1/32 = 0.031 25** | **1/8 = 0.125** | **1** | **2** |
| 1 pint, U.K. = | 0.568 261 4 | 0.016 125 89 | 0.064 503 6 | 0.516 028 4 | 1.032 057 | **1/64 = 0.015 625** | **1/16 = 0.062 5** | **1/2 = 0.5** | **1** |

Exact conversion: 1 dry pint, U.S. = **33.600 312 5** cubic inches

F. Other volume and capacity units

1 barrel, U.S. (used for petroleum, etc.) = **42** gallons = 0.158 987 296 cubic meter
1 barrel ("old barrel") = **31.5** gallons = 0.119 240 cubic meter
1 board foot = **144** cubic inches = 2.359 737 $\times 10^{-3}$ cubic meter
1 cord = **128** cubic feet = 3.624 556 cubic meters
1 cord foot = **16** cubic feet = 0.453 069 5 cubic meter
1 cup = **8** fluid ounces, U.S. = 2.365 882 $\times 10^{-4}$ cubic meter
1 gallon (Canadian, liquid) = 4.546 090 $\times 10^{-3}$ cubic meter
1 perch (volume) = 24.75 cubic feet = 0.700 842 cubic meter
1 stere = **1** cubic meter
1 tablespoon = **0.5** fluid ounce, U.S. = 1.478 677 $\times 10^{-5}$ cubic meter
1 teaspoon = **1/6** fluid ounce, U.S. = 4.928 922 $\times 10^{-6}$ cubic meter
1 ton (register ton) = **100** cubic feet = 2.831 684 66 cubic meters

**TABLE 1-17** Mass Conversion Factors

(Exact conversions are shown in **boldface** type. Repeating decimals are underlined.) The SI unit of mass is the kilogram.

A. Mass units decimally related to one kilogram

| | Kilograms (kg) | Tonnes (metric tons) (t) | Grams (g) | Decigrams (dg) | Centigrams (cg) | Milligrams (mg) | Micrograms (μg) |
|---|---|---|---|---|---|---|---|
| 1 kilogram = | **1** | **0.001** | **1 000** | **10 000** | **100 000** | **1 000 000** | **$10^9$** |
| 1 tonne = | **1 000** | **1** | **1 000 000** | **$10^7$** | **$10^8$** | **$10^9$** | **$10^{12}$** |
| 1 gram = | **0.001** | **0.000 001** | **1** | **10** | **100** | **1 000** | **1 000 000** |
| 1 decigram = | **0.000 1** | **$10^{-7}$** | **0.1** | **1** | **10** | **100** | **100 000** |
| 1 centigram = | **0.000 01** | **$10^{-8}$** | **0.01** | **0.1** | **1** | **10** | **10 000** |
| 1 milligram = | **0.000 001** | **$10^{-9}$** | **0.001** | **0.01** | **0.1** | **1** | **1 000** |
| 1 microgram = | **$10^{-9}$** | **$10^{-12}$** | **0.000 001** | **0.000 01** | **0.000 1** | **0.001** | **1** |

B. Nonmetric mass units less than one pound-mass (with gram equivalents)

| | Grams (g) | Avoirdupois ounces-mass ($oz_m$, avdp) | Troy ounces-mass ($oz_m$, troy) | Avoirdupois drams (dr avdp) | Apothecary drams (dr apoth) | Pennyweights (dwt) | Grains (grain) | Scruples (scruple) |
|---|---|---|---|---|---|---|---|---|
| 1 gram = | **1** | 0.035 273 962 | 0.032 150 747 | 0.564 383 39 | 0.257 205 97 | 0.643 014 93 | 15.432 358 4 | 0.771 617 92 |
| 1 avdp ounce-mass = | 28.349 523 1 | **1** | 0.911 458 33 | **16** | 7.291 666 66 | 18.227 166 7 | **437.5** | **21.875** |
| 1 troy ounce-mass = | 31.103 476 8 | 1.097 142 86 | **1** | 17.554 285 7 | **8** | **20** | **480** | **24** |
| 1 avdp dram = | 1.771 845 20 | **1/16 = 0.062 5** | 0.056 966 15 | **1** | 0.455 729 17 | 1.139 322 92 | **27.343 75** | **1.367 187 5** |
| 1 apothecary dram = | 3.887 934 58 | 0.137 142 857 | **1/8 = 0.125** | 2.194 285 70 | **1** | **2.5** | **60** | **3** |
| 1 pennyweight = | 1.555 173 83 | 0.054 863 162 | **1/20 = 0.05** | 0.877 714 28 | **1/2.5 = 0.4** | **1** | **24** | **1.2** |
| 1 grain = | **0.064 798 91** | **1/437.5 =** 2.285 714 29 × $10^{-3}$ | **1/480 =** 0.002 083 33 | 3.657 142 85 × $10^{-2}$ | **1/60** = 0.016 666 66 | **1/24 =** 0.041 666 66 | **1** | **0.05** |
| 1 scruple = | 1.295 978 20 | 4.571 428 58 × $10^{-2}$ | **1/24** = 0.041 666 66 | 0.731 428 57 | **1/3** = 0.333 333 33 | **5/6** = 0.833 333 33 | **20** | **1** |

C. Nonmetric mass units of one pound-mass and greater (with kilogram equivalents)

| | Kilograms (kg) | Long tons (long ton) | Short tons (short ton) | Long hundredweights (long cwt) | Short hundredweights (short cwt) | Slugs (slug) | Avoirdupois pounds-mass ($lb_m$, avdp) | Troy pounds-mass ($lb_m$, troy) |
|---|---|---|---|---|---|---|---|---|
| 1 kilogram = | **1** | 9.842 065 28 × $10^{-4}$ | 1.102 311 31 × $10^{-3}$ | 1.968 411 31 × $10^{-2}$ | 2.204 622 62 × $10^{-2}$ | 0.068 521 77 | 2.204 622 62 | 2.679 228 89 |
| 1 long ton = | 1 016.046 9 | **1** | **1.12** | **20** | **22.4** | 69.621 329 | **2 240** | 2 722.222 22 |
| 1 short ton = | **907.184 74** | **200/224 =** 0.892 857 14 | **1** | **4 000/224 =** 17.857 142 9 | **20** | 62.161 901 | **2 000** | 2 430.555 55 |
| 1 long hundredweight = | 50.802 345 4 | **0.05** | **0.056** | **1** | **1.12** | 3.481 066 4 | **112** | 136.111 111 |
| 1 short hundred weight = | **45.359 237** | **10/224** = 0.044 642 86 | **0.05** | **100/112 =** 0.892 857 14 | **1** | 3.108 095 0 | **100** | 121.527 777 |
| 1 slug = | 14.593 903 | 0.014 363 41 | 0.016 087 02 | 0.287 268 3 | 0.321 740 5 | **1** | 32.174 05 | 39.100 406 |
| 1 avdp pound-mass = | **0.453 592 37** | **1/2 240 =** 4.464 285 71 × $10^{-4}$ | **0.000 5** | **1/112 =** 8.928 571 43 × $10^{-3}$ | **0.01** | 3.108 095 0 × $10^{-2}$ | **1** | 1.215 277 777 |
| 1 troy pound-mass = | 0.373 241 72 | 3.673 469 37 × $10^{-4}$ | 4.114 285 70 × $10^{-4}$ | 7.346 938 79 × $10^{-3}$ | 8.228 571 45 × $10^{-3}$ | 0.025 575 18 | 0.822 857 14 | **1** |

Exact conversions: 1 long ton = **1 016.046 908 8** kilograms
1 troy pound-mass = **0.373 241 721 6** kilogram

D. Other mass units

1 assay ton = 29.166 667 grams
1 carat (metric) = **200** milligrams
1 carat (troy weight) = **3⅙** grains = 205.196 55 milligrams
1 myriagram = **10** kilograms
1 quintal = **100** kilograms
1 stone = **14** pounds, avdp = **6.350 293 18** kilograms

## TABLE 1-18 Time Conversion Factors

(Exact conversions are shown in **boldface** type. Repeating decimals are underlined.) The SI unit of time is the second.

A. Time units of one second and less

| | seconds (s) | milliseconds (ms) | microseconds ($\mu s$) | | picoseconds (ps) |
|---|---|---|---|---|---|
| 1 second = | **1** | **1 000** | **1 000 000** | $\mathbf{10^9}$ | $\mathbf{10^{12}}$ |
| 1 millisecond = | **0.001** | **1** | **1 000** | **1 000 000** | $\mathbf{10^9}$ |
| 1 microsecond = | **0.000 001** | **0.001** | **1** | **1 000** | **1 000 000** |
| 1 nanosecond = | $\mathbf{10^{-9}}$ | **0.000 001** | **0.001** | **1** | **1 000** |
| 1 picosecond = | $\mathbf{10^{-12}}$ | $\mathbf{10^{-9}}$ | **0.000 001** | **0.001** | **1** |

B. Time units of one second and greater

| | mean solar seconds (s) | mean solar minutes (min) | mean solar hours (h) | mean solar days (d) | mean solar weeks (w) | calendar (Gregorian) year (yr) |
|---|---|---|---|---|---|---|
| 1 second = | **1** | **1/60** = 0.016 666 $\underline{6}$ | **1/3 600** = 0.000 277 $\underline{7}$ | **1/86 400** = 1.157 407 $\underline{407}$ × $10^{-5}$ | **1/604 800** = 1.653 439 15 × $10^{-6}$ | 3.168 873 85 × $10^{-8}$ |
| 1 minute = | **60** | **1** | **1/60** = 0.016 666 $\underline{6}$ | **1/1 440** = 0.000 694 4$\underline{4}$ | **1/10 080** = 9.920 634 92 × $10^{-5}$ | 1.901 324 31 × $10^{-6}$ |
| 1 hour = | **3 600** | **60** | **1** | **1/24** = 0.041 666 $\underline{6}$ | **1/168** = 5.952 380 95 × $10^{-3}$ | 1.140 794 50 × $10^{-4}$ |
| 1 day = | **86 400** | **1 440** | **24** | **1** | **1/7** = 0.142 857 14 | 2.737 907 00 × $10^{-3}$ |
| 1 week = | **604 800** | **10 080** | **168** | **7** | **1** | 1.916 534 90 × $10^{-2}$ |
| **1 calendar year = (Gregorian)** | 31 556 952 | 525 949.2 | 8 765.82 | 365.242 5 | 52.117 5 | **1** |

NOTES: The conventional calendar year of 365 days can be used in rough calculations only; the modern calendar is based on the Gregorian year of 365.2425 mean solar days, the value chosen by Pope Gregory XIII in 1582. This value requires that a leap-year day be introduced every four years as February 29, except that centennial years (1900, 2000, etc.) are leap years only when divisible by 400. The remaining difference between the Gregorian year and the tropical year (see below) introduces an error of 1 day in 3300 years.

The tropical year is the interval between successive vernal equinoxes and has been defined by the International Astronomical Union for noon of January 1, 1900 as 31 556 925.974 7 seconds = 365.242 198 79 mean solar days. The tropical year decreases by approximately 5.3 milliseconds per year.

The sidereal year is the interval between successive returns of the sun to the direction of the same star. Sidereal time units, given in Table 1-18C, are used primarily in astronomy.

The SI second, defined by the atomic process of the cesium atom, is equal to the mean solar second within the limits of their definition.

C. Other time units

1 decade = **10** Gregorian years
1 fortnight = **14** days = **1 209 600** seconds
1 century = **100** Gregorian years
1 millennium = **1000** Gregorian years
1 sidereal year = 366.256 4 sidereal days = 31 558 149.8 seconds
1 sidereal day = 86 164.091 seconds
1 sidereal hour = 3 590.170 seconds
1 sidereal minute = 59.836 17 seconds
1 sidereal second = 0.997 269 6 second
1 shake = $\mathbf{10^{-8}}$ seconds

**TABLE 1-19** Velocity Conversion Factors

The SI unit of velocity is the meter per second.

| | Meters per second (m/s) | Kilometers per hour (km/h) | Statute miles per hour (mi/h) | Knots (kn) | Feet per minute (ft/min) | Feet per second (ft/s) | Inches per second (in/s) |
|---|---|---|---|---|---|---|---|
| 1 meter per second = | 1 | **3.6** | 2.236 936 29 | 1.943 844 49 | 196.850 394 | 3.280 839 89 | 39.370 078 7 |
| 1 kilometer per hour = | **1/3.6** = 0.277 777 | 1 | 0.621 371 19 | 0.539 956 80 | 54.680 664 9 | 0.911 344 42 | 10.936 133 0 |
| 1 statute mile per hour = | **0.447 04** | **1.609 344** | 1 | 0.868 976 24 | **88** | **88/60** = 1.466 666 | **88/5 = 17.6** |
| 1 knot = | 0.514 444 | **1.852** | 1.150 779 45 | 1 | 101.268 592 | 1.687 780 99 | 20.253 718 4 |
| 1 foot per minute = | **0.005 08** | **0.018 288** | 0.011 363 | $9.874\ 730\ 01 \times 10^{-3}$ | 1 | **1/60** = 0.016 666 | **1/5 = 0.2** |
| 1 foot per second = | **0.304 8** | **1.097 28** | 0.681 818 | 0.592 483 80 | **60** | 1 | **12** |
| 1 inch per second = | **0.025 4** | **0.091 44** | 0.056 818 | 0.049 373 65 | **5** | **1/12** = 0.083 333 | 1 |

NOTE: The velocity of light in vacuum, $c$ = 299 792 458 meters per second = 670 616 629 statute miles per hour
= 186 282.397 statute miles per second
= 0.983 571 056 feet per nanosecond

Other velocity units

1 foot per hour = $8.466\ 667 \times 10^{-5}$ meter per second
1 statute mile per minute = **26.822 4** meters per second
1 statute mile per second = **1 609.344** meters per second

## TABLE 1-20 Density Conversion Factors

(Exact conversions are shown in **boldface** type. Repeating decimals are underlined.) The SI unit of density is the kilogram per cubic meter.

A. Density units decimally related to one kilogram per cubic meter

| | Kilograms per cubic meter (kg/m³) | Tonnes per cubic meter (t/m³) | Grams per cubic meter (g/m³) | Grams per liter (g/L) | Milligrams per liter (mg/L) | Micrograms per milliliter (μg/mL) |
|---|---|---|---|---|---|---|
| 1 kilogram per cubic meter = | **1** | **0.001** | **1 000** | **1** | **1 000** | **1 000** |
| 1 tonne per cubic meter = | **1 000** | **1** | **1 000 000** | **1 000** | **1 000 000** | **1 000 000** |
| 1 gram per cubic meter = | **0.001** | **0.000 001** | **1** | **0.001** | **1** | **1** |
| 1 gram per liter = | **1** | **0.001** | **1 000** | **1** | **1 000** | **1 000** |
| 1 milligram per liter = | **0.001** | **0.000 001** | **1** | **0.001** | **1** | **1** |
| 1 microgram per milliliter = | **0.001** | **0.000 001** | **1** | **0.001** | **1** | **1** |

B. Nonmetric density units (with kilogram per cubic meter equivalents)

| | Kilograms per cubic meter (kg/m³) | Short tons per cubic mile (short tons/mi³) | Avoirdupois pounds per acre-foot (lb avdp/acre-ft) | Avoirdupois pounds per cubic foot (lb avdp/ft³) | Avoirdupois pounds per cubic inch (lb avdp/in³) | Avoirdupois ounces per U.S. quart (oz advp/U.S. qt) | Avoirdupois drams per U.S. fluid ounce (dr advp/U.S. floz) | Grains per U.S. fluid ounce (grain/U.S. floz) |
|---|---|---|---|---|---|---|---|---|
| 1 kilogram per cubic meter = | **1** | 4 594 934 | 2 719.362 0 | $6.242\ 796\ 1 \times 10^{-2}$ | $3.612\ 729\ 20 \times 10^{-5}$ | $3.338\ 161\ 6 \times 10^{-2}$ | $1.669\ 080\ 82 \times 10^{-2}$ | 0.456 389 28 |
| 1 short ton per cubic mile = | $2.176\ 451\ 9 \times 10^{-7}$ | **1** | $5.918\ 560\ 5 \times 10^{-4}$ | $1.358\ 714\ 5 \times 10^{-8}$ | $7.862\ 931\ 3 \times 10^{-12}$ | $7.265\ 348\ 2 \times 10^{-9}$ | $3.632\ 674\ 1 \times 10^{-9}$ | $9.933\ 093\ 1 \times 10^{-8}$ |
| 1 avdp pound per acre-foot = | $3.677\ 333\ 2 \times 10^{-4}$ | 1 689.600 0 | **1** | $2.295\ 684\ 1 \times 10^{-5}$ | $1.328\ 520\ 9 \times 10^{-8}$ | $1.227\ 553\ 2 \times 10^{-5}$ | $6.137\ 766\ 2 \times 10^{-6}$ | $1.678\ 295\ 5 \times 10^{-4}$ |
| 1 avdp pound per cubic foot = | 16.018 463 4 | 73 598 976 | **43 560** | **1** | **1/1 728 =** $5.787\ 037\ 03 \times 10^{-4}$ | 0.534 722 2 | 0.267 361 1 | 7.310 655 0 |
| 1 avdp pound per cubic inch = | 27 679.905 | $1.271\ 790\ 4 \times 10^{11}$ | **75 271 680** | **1 728** | **1** | **924** | **462** | 12 632.812 |
| 1 avdp ounce per U.S. quart = | 29.956 608 | $1.376\ 396\ 5 \times 10^{8}$ | 81 462.86 | 1.870 130 0 | $1.082\ 251\ 1 \times 10^{-3}$ | **1** | **0.5** | 13.671 874 |
| 1 avdp dram per U.S. fluid ounce = | 59.913 216 | $2.752\ 793\ 0 \times 10^{8}$ | 162 925.72 | 3.740 259 8 | $2.164\ 502\ 3 \times 10^{-3}$ | **2** | **1** | 27.343 748 |
| 1 grain per U.S. fluid ounce = | 2.191 111 9 | 10 067 357 | 5 958.426 3 | 0.136 786 65 | $7.915\ 894\ 0 \times 10^{-5}$ | 0.073 142 86 | 0.036 571 43 | **1** |

C. Other density units

1 grain per gallon, U.S. = 17.118 06 grams per cubic meter
1 gram per cubic centimeter = **1 000** kilograms per cubic meter
1 avdp ounce per gallon, U.S. = 7.489 152 kilograms per cubic meter
1 avdp ounce per cubic inch = 1 729.994 kilograms per cubic meter
1 avdp pound per gallon, U.S. = 119.826 4 kilograms per cubic meter
1 slug per cubic foot = 515.379 kilograms per cubic meter
1 long ton per cubic yard = 1 328.939 kilograms per cubic meter

**TABLE 1-21** Force Conversion Factors

(Exact conversions are shown in **boldface** type. Repeating decimals are underlined.) The SI unit of force is the newton (N).

| | Newtons (N) | Kips (kip) | Slugs-force ($slug_f$) | **Kilograms-force (kilopond) ($kg_f$)** | Avoirdupois pounds-force ($lb_f$ avdp) | Avoirdupois ounces-force ($oz_f$ advp) | Poundals (pdl) | Dynes (dyn) |
|---|---|---|---|---|---|---|---|---|
| 1 newton = | **1** | $2.248\,089\,43 \times 10^{-4}$ | $6.987\,275\,24 \times 10^{-3}$ | 0.101 971 62 | 0.224 808 94 | 3.596 943 09 | 7.233 014 2 | **100 000** |
| 1 kip = | 444 8.221 62 | **1** | 31.080 949 | 453.592 370 | **1 000** | **16 000** | 32 174.05 | 444 822 162 |
| 1 slug-force = | 143.117 305 | 0.032 174 05 | **1** | 14.593 903 | 32.174 05 | 514.784 80 | 1 035.169 5 | 14 311 730 |
| 1 kilogram-**force (kilo-pond) =** | **9.806 650** | $2.204\,622\,62 \times 10^{-3}$ | $6.852\,176\,3 \times 10^{-2}$ | **1** | 2.204 622 62 | 35.273 961 9 | 70.931 638 4 | **980 665** |
| 1 avdp pound-force = | 4.448 221 62 | **0.001** | $3.108\,094\,88 \times 10^{-2}$ | 0.453 592 37 | **1** | **16** | 32.174 05 | 444 822.162 |
| 1 avdp ounce-force = | 0.278 013 85 | **1/16 000 = 0.000 062 5** | $1.942\,559\,30 \times 10^{-3}$ | $2.834\,952\,3 \times 10^{-2}$ | **1/16 = 0.062 5** | **1** | 2.010 878 03 | 27 801.385 |
| 1 poundal = | 0.138 254 95 | $3.108\,094\,9 \times 10^{-5}$ | $9.660\,253\,9 \times 10^{-4}$ | 0.140 980 81 | 0.031 080 95 | 0.497 295 18 | **1** | 13 825.495 |
| 1 dyne = | **0.000 01** | $2.248\,089\,43 \times 10^{-9}$ | $6.987\,275\,24 \times 10^{-8}$ | $1.019\,716\,21 \times 10^{-6}$ | $2.248\,089\,43 \times 10^{-6}$ | $3.596\,943\,10 \times 10^{-5}$ | $7.233\,014\,2 \times 10^{-5}$ | **1** |

The exact conversion is: 1 avdp pound-force = **4.448 221 615 260 5** newtons mally, by magnitude or by usage. Each subtabulation contains the SI unit,* so that equivalent values can be found between units that are tabulated in separate tables. For example, to obtain the equivalence between pounds per cubic foot and tonnes per cubic meter, we read from the fourth line of Table 1-20B:

1 pound per cubic foot is equal to 16.018 463 4 kilograms per cubic meter

From the first line of Table 1-20A we find:

1 kilogram per cubic meter is equal to 0.001 metric ton per cubic meter

Hence:

1 pound per cubic foot is equal to 16.018 463 4 kilograms per cubic meter

= 0.016 018 463 4 metric ton per cubic meter

*Use of Conversion Factors.* Conversion factors are multipliers used to convert a quantity expressed in a particular unit *(given unit)* to the same quantity expressed in another unit *(desired unit).* To perform such conversions, the *given unit* is found at the left-hand edge of the conversion table and the *desired unit* is found at the top of the same table. Suppose, for example, that the quantity 1000 feet is to be converted to meters. The given unit (foot) is found in the left-hand edge of the third line of Table 1-14B. The desired unit, meter, is found at the top of the first column in that table. The conversion factor (**0.304 8,** exactly) is located to the right of the given unit and below the desired unit. The given quantity, 1000 feet, is multiplied by the conversion factor to obtain the equivalent length in meters; that is, 1000 feet is $1000 \times 0.304\ 8 = 304.8$ meters.

The general rule is: Find the given unit at the left side of the table in which it appears, and the desired unit at the top of the same table; note the conversion factor to the right of the given unit and below the desired unit. Multiply the quantity expressed in the given unit by the conversion factor to find the quantity expressed in the desired unit.

Listings of conversion factors (see Refs. 1 and 7) are often arranged as follows:

| *To convert from* | *To* | *Multiply by* |
|---|---|---|
| (Given unit) | (Desired unit) | (Conversion factor) |

The equivalences listed in the accompanying conversion tables can be cast in this form by placing the given unit (at the left of each table) under "To convert from," the desired units (at the top of the table) under "To," and the conversion factor, found to the right and below these units, under "Multiply by."

*Use of Two Tables to Find Conversion Factors.* When the given and desired units do not appear in the same table, the conversion factor between them is found in two steps. The *given unit* is selected at the left-hand edge of the table in which it appears, and an *intermediate conversion factor,* applicable to the SI unit shown at the top of the same table, is recorded. The *desired unit* is then found at the top of another table, in which it appears, and another *intermediate conversion factor,* applicable to the SI unit at the left-hand edge of that table, is recorded. The conversion factor between the given and desired units is the product of these two intermediate conversion factors.

For example, it is required to convert 100 cubic feet to the equivalent quantity in cubic centimeters. The given quantity (cubic feet) is found in the fourth line at the left of Table 1-16B. Its intermediate conversion factor with respect to the SI unit is found (below the cubic meter) to be $2.831\ 684\ 66 \times 10^{-2}$. The desired quantity (cubic centimeters) is found at the top of the third column in Table 1-16A. Its intermediate conversion factor with respect to the SI unit, found under the cubic centimeter and to the right of the cubic meter, is 1 000 000. The conversion factor between cubic feet and cubic centimeters is the product of these two intermediate conversion factors; that is, 1 cubic foot is equal to $2.831\ 684\ 66 \times 10^{-2} \times 1\ 000\ 000 =$

* In Tables 1-16C, 1-16D, 1-16E, and 1-17B, a decimal submultiple of the SI unit (the liter and gram, respectively) is listed since it is most commonly used in conjunction with the other units in the respective tables. The procedure for linking the subtables is unchanged.

**TABLE 1-22** Pressure/Stress Conversion Factors

(Exact conversions are shown in **boldface** type. Repeating decimals are underlined.) The SI unit of pressure or stress is the pascal (Pa).

A. Pressure units decimally related to one pascal

| | Pascals (Pa) | Bars (bar) | Decibars (dbar) | Millibars (mbar) | Dynes per square centimeter (dyn/cm²) |
|---|---|---|---|---|---|
| 1 pascal = | **1** | **0.000 01** | **0.000 1** | **0.01** | **10** |
| 1 bar = | **100 000** | **1** | **10** | **1 000** | **1 000 000** |
| 1 decibar = | **10 000** | **0.1** | **1** | **100** | **100 000** |
| 1 millibar = | **100** | **0.001** | **0.01** | **1** | **1 000** |
| 1 dyne per square centimeter = | **0.1** | **0.000 001** | **0.000 01** | **0.001** | **1** |

B. Pressure units decimally related to one kilogram-force per square meter (with pascal equivalents)

| | Kilograms-force per square meter ($kg_f/m^2$) | Kilograms-force per square centimeter ($kg_f/cm^2$) | Kilograms-force per square millimeter ($kg_f/mm^2$) | Grams-force per square centimeter ($g_f/cm^2$) | Pascals (Pa) |
|---|---|---|---|---|---|
| 1 kilogram-force per square meter = | **1** | **0.000 1** | **0.000 001** | **0.1** | **9.806 65** |
| 1 kilogram-force per square centimeter = | **10 000** | **1** | **0.01** | **1 000** | **98 066.5** |
| 1 kilogram-force per square millimeter = | **1 000 000** | **100** | **1** | **100 000** | **9 806 650** |
| 1 gram-force per square centimeter = | **10** | **0.001** | **0.000 01** | **1** | **98.066 5** |
| 1 pascal = | 0.101 971 62 | $1.019\,716\,2 \times 10^{-5}$ | $1.019\,716\,2 \times 10^{-7}$ | $1.019\,716\,2 \times 10^{-2}$ | **1** |

NOTE: 1 atmosphere (technical) = 1 kilogram-force per square centimeter = **98 066.5** pascals.

C. Pressure units expressed as heights of liquid (with pascal equivalents)

| | Millimeters of mercury at 0°C (mmHg, 0°C) | Centimeters of mercury at 60°C (cmHg, 60°C) | Inches of mercury at 32°F (inHg, 32°F) | Inches of mercury at 60°F (inHg, 60°F) | Centimeters of water at 4°C ($cmH_2O$, 4°C) | Inches of water at 60°F ($inH_2O$, 60°F) | Feet of water at 39.2°F ($ftH_2O$, 39.2°F) | Pascals (Pa) |
|---|---|---|---|---|---|---|---|---|
| 1 millimeter of mercury, 0°C = | **1** | 0.100 282 | 0.039 370 1 | 0.039 481 3 | 1.359 548 | 0.535 775 6 | 0.044 604 6 | 133.322 4 |
| 1 centimeter of mercury, 60°C = | 9.971 830 | **1** | 0.392 591 9 | 0.393 700 8 | 13.557 18 | 5.342 664 | 0.444 789 5 | 1 329.468 |
| 1 inch of mercury, 32°F = | **25.4** | 2.547 175 | **1** | 1.002 824 8 | 34.532 52 | 13.608 70 | 1.132 957 | 3 386.389 |
| 1 inch of mercury, 60°C = | 25.328 45 | **2.54** | 0.997 183 1 | **1** | 34.435 25 | 13.570 37 | 1.129 765 | 3 376.85 |
| 1 centimeter of water, 4°C = | 0.735 539 | 0.073 762 | 0.028 958 | 0.029 040 0 | **1** | 0.394 083 8 | 0.032 808 4 | 98.063 8 |
| 1 inch of water, 60°F = | 1.866 453 | 0.187 173 | 0.073 482 | 0.073 690 0 | 2.537 531 | **1** | 0.083 252 4 | 248.840 |
| 1 foot of water, 39.2°F = | 22.419 2 | 2.248 254 | 0.882 646 | 0.885 139 | 30.479 98 | 12.011 67 | **1** | 2 988.98 |
| 1 pascal = | $7.500\ 615 \times 10^{-3}$ | $7.521\ 806 \times 10^{-4}$ | $2.952\ 998 \times 10^{-4}$ | $2.961\ 34 \times 10^{-4}$ | $1.019\ 74 \times 10^{-2}$ | $4.018\ 65 \times 10^{-3}$ | $3.345\ 62 \times 10^{-4}$ | **1** |

NOTE: 1 torr = 1 millimeter of mercury at 0°C = 133.322 4 pascals.

D. Nonmetric pressure units (with pascal equivalents)

| | Atmospheres (atm) | Avoirdupois pounds-force per square inch (psi) | Avoirdupois pounds-force per square foot ($lb_f/ft^2$, avdp) | Poundals per square foot ($pdl/ft^2$) | Pascals (Pa) |
|---|---|---|---|---|---|
| 1 atmosphere = | **1** | 14.695 95 | 2 116.217 | 68 087.24 | **101 325** |
| 1 avdp pound-force per square inch = | $6.804\ 60 \times 10^{-2}$ | **1** | **144** | 4 633.063 | 6 894.757 |
| 1 avdp pound-force per square foot = | $4.725\ 414 \times 10^{-4}$ | **1/144** = 0.006 94$\underline{4}$ | **1** | 32.174 05 | 47.880 26 |
| 1 poundal per square foot = | $1.468\ 704 \times 10^{-5}$ | $2.158\ 399 \times 10^{-4}$ | 0.031 080 9 | **1** | 1.488 164 |
| 1 pascal = | $9.869\ 233 \times 10^{-6}$ | $1.450\ 377 \times 10^{-4}$ | 0.020 885 4 | 0.671 968 9 | **1** |

NOTE: 1 normal atmosphere = 760 torr = **101 325** pascals.

**TABLE 1-23** Torque/Bending Moment Conversion Factors

(Exact conversions are shown in **boldface** type. Repeating decimals are underlined.) The SI unit of torque is the newton-meter (N·m).

| | Newton-meters (N·m) | Kilogram-force-meters ($kg_f$·m) | Avoirdupois pound-force-feet ($lb_f$·ft, avdp) | Avoirdupois pound-force-inches ($lb_f$·in, avdp) | Avoirdupois ounce-force-inches ($oz_f$·in, avdp) | Dyne-centimeters (dyne·cm) |
|---|---|---|---|---|---|---|
| 1 newton-meter = | **1** | 0.101 971 6 | 0.737 562 1 | 8.850 748 1 | 141.611 9 | **10 000 000** |
| 1 kilogram-force-meter = | **9.806 65** | **1** | 7.233 013 | 86.796 16 | 1 388.739 | **98 066 500** |
| 1 avdp pound-force-foot = | 1.355 818 | 0.138 255 0 | **1** | **12** | **192** | 13 558 180 |
| 1 avdp pound-force-inch = | 0.112 984 8 | $1.152\ 124 \times 10^{-2}$ | **1/12** = 0.083 333 | **1** | **16** | 1 129 848 |
| 1 avdp ounce-force-inch = | $7.061\ 552 \times 10^{-3}$ | $7.200\ 779 \times 10^{-4}$ | **1/192** = 0.005 208 3 | **1/16 = 0.062 5** | **1** | 70 615.52 |
| 1 dyne-centimeter = | $\mathbf{10^{-7}}$ | $1.017\ 716 \times 10^{-8}$ | $7.375\ 621 \times 10^{-8}$ | $8.850\ 748 \times 10^{-7}$ | $1.416\ 119 \times 10^{-5}$ | **1** |

**TABLE 1-24** Energy/Work Conversion Factors

(Exact conversions are shown in **boldface** type. Repeating decimals are underlined.) The SI unit of energy and work is the joule (J).

A. Energy/work units decimally related to one joule

| | Joules (J) | Megajoules (MJ) | Kilojoules (kJ) | Millijoules (mJ) | Microjoules ($\mu$J) | Ergs (erg) |
|---|---|---|---|---|---|---|
| 1 joule = | **1** | **0.000 001** | **0.001** | **1 000** | **1 000 000** | $\mathbf{10^{7}}$ |
| 1 megajoule = | **1 000 000** | **1** | **1 000** | $\mathbf{10^{9}}$ | $\mathbf{10^{12}}$ | $\mathbf{10^{13}}$ |
| 1 kilojoule = | **1 000** | **0.001** | **1** | **1 000 000** | $\mathbf{10^{9}}$ | $\mathbf{10^{10}}$ |
| 1 millijoule = | **0.001** | $\mathbf{10^{-9}}$ | $\mathbf{10^{-6}}$ | **1** | **1 000** | **10 000** |
| 1 microjoule = | **0.000 001** | $\mathbf{10^{-12}}$ | $\mathbf{10^{-9}}$ | **0.001** | **1** | **10** |
| 1 erg = | $\mathbf{10^{-7}}$ | $\mathbf{10^{-13}}$ | $\mathbf{10^{-10}}$ | **0.000 1** | **0.1** | **1** |

NOTE: 1 watt-second = 1 joule.

B. Energy/work units less than ten joules (with joule equivalents)

| | Joules (J) | Foot-poundals (ft·pdl) | Foot-pounds-force (ft·$lb_f$) | Calories (International Table) (cal, IT) | Calories (thermochemical) (cal, thermo) | Electronvolts (eV) |
|---|---|---|---|---|---|---|
| 1 joule = | **1** | 23.730 36 | 0.737 562 1 | 0.238 845 9 | 0.239 005 7 | $6.241\,46 \times 10^{18}$ |
| 1 foot-poundal = | $4.214\,011 \times 10^{-2}$ | **1** | $3.108\,095 \times 10^{-2}$ | $1.006\,499 \times 10^{-2}$ | $1.007\,173 \times 10^{-2}$ | $2.630\,16 \times 10^{17}$ |
| 1 foot-pound-force = | 1.355 818 | 32.174 05 | **1** | 0.323 831 6 | 0.324 048 3 | $8.462\,28 \times 10^{18}$ |
| 1 calorie (Int. Tab.) = | **4.186 8** | 99.354 27 | 3.088 025 | **1** | 1.000 669 | $2.613\,17 \times 10^{19}$ |
| 1 calorie (thermo) = | **4.184** | 99.287 83 | 3.085 960 | 0.999 331 2 | **1** | $2.611\,43 \times 10^{19}$ |
| 1 electronvolt = | $1.602\,19 \times 10^{-19}$ | $3.802\,05 \times 10^{-18}$ | $1.181\,71 \times 10^{-19}$ | $3.826\,77 \times 10^{-20}$ | $3.829\,33 \times 10^{-20}$ | **1** |

*(continued on p. 1–46)*

**TABLE 1-24** Energy/Work Conversion Factors *(Continued)*

C. Energy/work units greater than ten joules (with joule equivalents)

| | Joules (J) | British thermal units, International Table (Btu, IT) | British thermal units, thermochemical (Btu, thermo) | Kilowatthours (kWh) | Horsepower-hours, electrical (hp·h, elec) | Kilocalories, International Table (kcal, IT) | **Kilocalories, thermochemical (kcal,thermo)** |
|---|---|---|---|---|---|---|---|
| 1 joule = | **1** | $9.478\,170 \times 10^{-4}$ | $9.484\,516\,5 \times 10^{-4}$ | $\mathbf{1/(3.6 \times 10^6)}$ $\mathbf{2.77\underline{7} \times 10^{-7}}$ | $3.723\,562 \times 10^{-7}$ | $2.388\,459 \times 10^{-4}$ | $2.390\,057\,4 \times 10^{-4}$ |
| 1 British thermal unit, Int. Tab. = | 1 055.056 | **1** | 1.000 669 | $2.930\,711\,1 \times 10^{-4}$ | $3.928\,567 \times 10^{-4}$ | 0.251 995 8 | 0.252 164 4 |
| 1 British thermal unit (thermo) = | 1 054.35 | 0.999 331 | **1** | $2.928\,745 \times 10^{-4}$ | $03.925\,938 \times 10^{-4}$ | 0.251 827 2 | 0.251 995 7 |
| 1 kilowatthour = | **3 600 000** | 3 412.141 | 3 414.426 | **1** | **1/0.746** = 1.340 482 6 | 859.845 2 | 860.420 7 |
| 1 horsepower hour, electrical = | **2 685 600** | 2 545.457 | 2 547.162 | **0.746** | **1** | 641.444 5 | 641.873 8 |
| 1 kilocalorie, Int. Tab. = | **4 186.8** | 3.968 320 | 3.970 977 | **0.001 163** | $1.558\,981 \times 10^{-3}$ | **1** | 1.000 669 |
| 1 kilocalorie, thermochemical = | **4 184** | 3.965 666 | 3.968 322 | $0.001\,162\,\underline{2}$ | $1.557\,938\,6 \times 10^{-3}$ | 0.999 331 | **1** |

The exact conversion is 1 British thermal unit, International Table = **1 055.055 852 62** joules.

## TABLE 1-25 Power Conversion Factors

(Exact conversions are shown in **boldface** type. Repeating decimals are underlined.) The SI unit of power is the watt (W).

A. Power units decimally related to one watt

| | Watts (W) | Megawatts (MW) | Kilowatts (kW) | Milliwatts (mW) | Microwatts ($\mu$W) | Picowatts (pW) | Ergs per second (ergs/s) |
|---|---|---|---|---|---|---|---|
| 1 watt = | **1** | **0.000 001** | **0.001** | **1 000** | **1 000 000** | **$10^{9}$** | **$10^{7}$** |
| 1 megawatt = | **1 000 000** | **1** | **1 000** | **$10^{9}$** | **$10^{12}$** | **$10^{15}$** | **$10^{13}$** |
| 1 kilowatt = | **1 000** | **0.001** | **1** | **1 000 000** | **$10^{9}$** | **$10^{12}$** | **$10^{10}$** |
| 1 milliwatt = | **0.001** | **$10^{-9}$** | **0.000 001** | **1** | **1 000** | **1 000 000** | **10 000** |
| 1 microwatt = | **0.000 001** | **$10^{-12}$** | **$10^{-9}$** | **0.001** | **1** | **1 000** | **10** |
| 1 picowatt = | **$10^{-9}$** | **$10^{-15}$** | **$10^{-12}$** | **0.000 001** | **0.001** | **1** | **0.01** |
| 1 erg per second = | **$10^{-7}$** | **$10^{-13}$** | **$10^{-10}$** | **0.000 1** | **0.1** | **100** | **1** |

NOTE: 1 watt = 1 joule per second (J/s).

B. Nonmetric power units (with watt equivalents)

| | British thermal units (International Table) per hour (Btu/hr, IT) | British thermal units (thermochemical per minute (Btu/min, thermo) | Avoirdupois foot-pounds-force per second ($ft \cdot lb_f/s$ avdp) | Kilocalories per minute (thermochemical) (kcal/min, thermo) | Kilocalories per second (International Table) (kcal/s, IT) | Horsepower (electrical) (hp, elec) | Horsepower (mechanical) (hp, mech) | Watts (W) |
|---|---|---|---|---|---|---|---|---|
| 1 British thermal unit (Int. Tab.)-per hour = | **1** | 0.016 677 8 | 0.216 158 1 | $4.202\ 740\ 5 \times 10^{-3}$ | $6.999\ 883\ 1 \times 10^{-5}$ | $3.928\ 567\ 0 \times 10^{-4}$ | $3.930\ 148\ 0 \times 10^{-4}$ | 0.293 071 1 |
| 1 British thermal unit (thermo) per minute = | 59.959 853 | **1** | 12.960 810 | 0.251 995 7 | $4.197\ 119\ 5 \times 10^{-3}$ | 0.023 555 6 | 0.023 565 1 | 17.572 50 |
| 1 foot-pound-force per second = | 4.626 242 6 | 0.077 155 7 | **1** | 0.019 442 9 | $3.238\ 315\ 7 \times 10^{-4}$ | $1.817\ 450\ 4 \times 10^{-3}$ | **1/550** = $1.818\ 18\underline{1\ 8} \times 10^{-3}$ | 1.355 818 |
| 1 kilocalorie per minute (thermo) = | 237.939 98 | 3.968 321 7 | 51.432 665 | **1** | 0.016 655 5 | 0.093 476 3 | 0.093 513 9 | 69.733 333 |
| 1 kilocalorie per second (Int. Tab.) = | 14 285.953 | 238.258 64 | 3 088.025 1 | 60.040 153 | **1** | 5.612 332 4 | 5.614 591 1 | **4 186.800** |
| 1 horsepower (electrical) = | 2 545.457 4 | 42.452 696 | 550.221 34 | 10.697 898 | 0.178 179 0 | **1** | 1.000 402 4 | **746** |
| 1 horsepower (mechanical) = | 2 544.433 4 | 42.435 618 | **550** | 10.693 593 | 0.178 107 4 | 0.999 597 7 | **1** | 745.699 9 |
| 1 watt = | 3.412 141 3 | 0.056 907 1 | 0.737.562 1 | 0.014 340 3 | $2.388\ 459\ 0 \times 10^{-4}$ | **1/746** = $1.340\ 482\ 6 \times 10^{-3}$ | $1.341\ 022\ 0 \times 10^{-3}$ | **1** |

NOTE: The horsepower (mechanical) is defined as a power equal to **550** foot-pounds-force per second.
Other units of horsepower are:
1 horsepower (boiler) = 9 809.50 watts
1 horsepower (metric) = 735.499 watts
1 horsepower (water) = 746.043 watts
1 horsepower (U.K.) = 745.70 watts
1 ton (refrigeration) = 3 516.8 watts

**TABLE 1-26** Temperature Conversions

(Conversions in **boldface** type are exact. Continuing decimals are underlined.)

| Celsius (°C)<br>°C = 5(°F − 32)/9 | Fahrenheit (°F)<br>°F = [9(C°)/5] + 32 | Absolute (K)<br>K = °C + 273.15 |
|---|---|---|
| **−273.15** | **−459.67** | **0** |
| **−200** | **−328** | **73.15** |
| **−180** | **−292** | **93.15** |
| **−160** | **−256** | **113.15** |
| **−140** | **−220** | **133.15** |
| **−120** | **−184** | **153.15** |
| **−100** | **−148** | **173.15** |
| **−80** | **−112** | **193.15** |
| **−60** | **−76** | **213.15** |
| **−40** | **−40** | **233.15** |
| **−20** | **−4** | **253.15** |
| −17.77 | 0 | 255.372 |
| **0** | **32** | **273.15** |
| **5** | **41** | **278.15** |
| **10** | **50** | **283.15** |
| **15** | **59** | **288.15** |
| **20** | **68** | **293.15** |
| **25** | **77** | **298.15** |
| **30** | **86** | **303.15** |
| **35** | **95** | **308.15** |
| **40** | **104** | **313.15** |
| **45** | **113** | **318.15** |
| **50** | **122** | **323.15** |
| **55** | **131** | **328.15** |
| **60** | **140** | **333.15** |
| **65** | **149** | **338.15** |
| **70** | **158** | **343.15** |
| **75** | **167** | **348.15** |
| **80** | **176** | **353.15** |
| **85** | **185** | **358.15** |
| **90** | **194** | **363.15** |
| **95** | **203** | **368.15** |
| **100** | **212** | **373.15** |
| **105** | **221** | **378.15** |
| **110** | **230** | **383.15** |
| **115** | **239** | **378.15** |
| **120** | **248** | **393.15** |
| **140** | **284** | **413.15** |
| **160** | **320** | **433.15** |
| **180** | **356** | **453.15** |
| **200** | **392** | **473.15** |
| **250** | **482** | **523.15** |
| **300** | **572** | **573.15** |
| **350** | **662** | **623.15** |
| **400** | **752** | **673.15** |
| **450** | **842** | **723.15** |
| **500** | **932** | **773.15** |
| **1 000** | **1 832** | **1 273.15** |
| **5 000** | **9 032** | **5 273.15** |
| **10 000** | **18 032** | **10 273.15** |

NOTE: Temperature in kelvins equals temperature in degrees Rankine divided by 1.8. [K = °R/**1.8**].

**TABLE 1-27** Light Conversion Factors

(Exact conversions are shown in **boldface** type. Repeating decimals are underlined.)

A. Luminance units. The SI unit of luminance is the candela per square meter (cd/m²)

| | Candelas per square meter (cd/ m²) | Candelas per square foot (cd/ft²) | Candelas per square inch (cd/in²) | Apostilbs (asb) | Stilbs (sb) | Lamberts (L) | Footlamberts (fL) |
|---|---|---|---|---|---|---|---|
| 1 candela per square meter = | **1** | **0.092 903 04** | **6.451 6 × 10⁻⁴** | π = 3.141 592 65 | **0.000 1** | **(0.000 1) π** = 3.141 592 65 × 10⁻⁴ | 0.291 863 51 |
| 1 candela per square foot = | 10.763 910 4 | **1** | **1/144** = 0.006 944 44 | 33.815 821 8 | 1.076 391 04 × 10⁻³ | 3.381 582 18 × 10⁻³ | **π** = 3.141 592 65 |
| 1 candela per square inch = | 1 550.003 1 | **144** | **1** | 4 869.478 4 | 0.155 000 31 | 0.486 947 84 | 452.389 342 |
| 1 apostilb = | **1/π** = 0.318 309 89 | 0.029 571 96 | 2.053 608 06 × 10⁻⁴ | **1** | 3.183 098 86 × 10⁻⁵ | **0.000 1** | **0.092 903 04** |
| 1 stilb = | **10 000** | **929.030 4** | **6.451 6** | 31 415.926 5 | **1** | **π** = 3.141 592 65 | 2 918.635 |
| 1 lambert = | **10 000/π** = 3 183.098 86 | 295.719 561 | 2.053 608 06 | **10 000** | **1/π** = 0.318 309 89 | **1** | **929.030 4** |
| 1 footlambert = | 3.426 259 1 | **1/π** = 0.318 309 89 | 2.210 485 32 × 10⁻³ | 10.763 910 4 | 3.426 259 1 × 10⁻⁴ | 1.076 391 03 × 10⁻³ | **1** |

NOTE: 1 nit (nt) = 1 candela per square meter (cd/m²).
1 stilb (sb) = 1 candela per square centimeter (cd/cm²).

B. Illuminance units. The SI unit of illuminance is the lux (lux)

| | Luxes (lx) | Phots (ph) | Footcandles (fc) | Lumens per square inch (lm/in²) |
|---|---|---|---|---|
| 1 lux = | **1** | **0.000 1** | **0.092 903 04** | **6.451 6 × 10⁻⁴** |
| 1 phot = | **10 000** | **1** | **929.030 4** | **6.451 6** |
| 1 footcandle = | 10.763 910 4 | 1.076 391 04 × 10⁻³ | **1** | **1/144** = 0.006 944 44 |
| 1 lumen per square inch = | 1 550.003 1 | 0.155 000 31 | **144** | **1** |

NOTE: **1 lux (lux) = 1 lumen per square meter (lm/m²)**
**1 phot (ph) = 1 lumen per square centimeter (lm/cm²)**
1 footcandle (fc) = 1 lumen per square foot (lm/ft²)

28 316.846 6 cubic centimeters. The conversion from 100 cubic feet to cubic centimeters then yields 100 × 28 316.846 6 = 2 831 684.66 cubic centimeters.

*Conversion of Electrical Units.* Since the electrical units in current use are confined to the International System, conversions to or from non-SI units in modern practice are fortunately not required. Conversions to and from the older cgs units, when required, can be performed using the conversions shown in Table 1-9. Slight differences from the SI units occur in the electrical units legally recognized in the United States prior to 1969. These differences involve amounts smaller than are customarily significant in engineering; they are listed in Table 1-28.

**TABLE 1-28** U.S.A. Electrical Units Used Prior to 1969, with SI Equivalents

| A. Legal units in the U.S.A. prior to January 1948 | |
|---|---|
| 1 ampere (US-INT) | = 0.999 843 ampere (SI) |
| 1 coulomb (US-INT) | = 0.999 843 coulomb (SI) |
| 1 farad (US-INT) | = 0.999 505 farad (SI) |
| 1 henry (US-INT) | = 1.000 495 henry (SI) |
| 1 joule (US-INT) | = 1.000 182 joule (SI) |
| 1 ohm (US-INT) | = 1.000 495 ohm (SI) |
| 1 volt (US-INT) | = 1.000 338 volt (SI) |
| 1 watt (US-INT) | = 1.000 182 watt (SI) |
| **B. Legal units in the U.S.A. from January 1948 to January 1969** | |
| 1 ampere (US-48) | = 1.000 008 ampere (SI) |
| 1 coulomb (US-48) | = 1.000 008 coulomb (SI) |
| 1 farad (US-48) | = 0.999 505 farad (SI) |
| 1 henry (US-48) | = 1.000 495 henry (SI) |
| 1 joule (US-48) | = 1.000 017 joule (SI) |
| 1 ohm (US-48) | = 1.000 495 ohm (SI) |
| 1 volt (US-48) | = 1.000 008 volt (SI) |
| 1 watt (US-48) | = 1.000 017 watt (SI) |

### *17. Bibliography*

*Standards*

1. ANSI/IEEE Std 268-1982; *Metric Practice.* New York, Institute of Electrical and Electronics Engineers, 1982.
2. Letter Symbols for Quantities Used in Electrical Science and Electrical Engineering; ANSI Standard Y10.5-1968. Also published as IEEE Std 280; New York, Institute of Electrical and Electronics Engineers, 1968.
3. IEEE Standard Letter Symbols for Units of Measurement, ANSI/IEEE Std 260-1978. New York, Institute of Electrical and Electronics Engineers, 1978.
4. IEEE Recommended Practice for Units in Published Scientific and Technical Work, IEEE Standard 268-1973; New York, Institute of Electrical and Electronics Engineers, 1973.
5. Graphic Symbols for Electrical and Electronics Diagrams, IEEE Std 315-1975 (also published as ANSI Standard Y32.2-1975); New York, Institute of Electrical and Electronics Engineers, 1975.
6. SI Units and Recommendations for the Use of Their Multiples and of Certain Other Units; International Standards ISO-1000-1973 (E). Available in the United States from ANSI. New York, American National Standards Institute, 1973. Also identified as IEEE Std 322-1971 and ANSI Z210.1-1972.

*Collections of Units and Conversion Factors*

7. National Bureau of Standards Units of Weight and Measure—International (Metric) and U.S. Customary; *Natl. Bur. Stand. Misc. Publ.* 286; Washington, Government Printing Office, 1967.
8. World Weights and Measures, Handbook for Statisticians, Statistical Papers, Series M, No. 21, Publication Sales Number 66, XVII, 3; New York, United Nations Publishing Service, 1966.

9. The Use of SI Units (The Metric System in the United Kingdom), PD 5686; London, British Standards Institution, 1972. See also British Standard 350, Part 2, 1962, and PD 6203 Supplement 1 (1967).

10. Cohen, E. R., et al.: Recommended Consistent Values of the Fundamental Physical Constants, 1973; *CODATA Bulletin* 11, December 1973; Frankfurt/Main, ICSU CODATA Central Office, 1973.

11. The Introduction of the IAU System of Astronomical Constants into the Astronomical Ephemeris and into the American Ephemeris and Nautical Almanac (Supplement to the American Ephemeris 1968); Washington, United States Naval Observatory, 1966.

12. *Encyclopaedia Britannica* (see under "Weights and Measures"); Chicago, Encyclopaedia Britannica, Inc., 1974.

13. *McGraw-Hill Encyclopedia of Science and Technology* (see entries by name of quantity or unit and vol. 15 under "Scientific Notation"; New York, McGraw-Hill Book Company, 1982.

14. *The World Book Encyclopedia* (see under "Weights and Measures"); Chicago, Field Enterprises Educational Corporation, 1974.

*Books and Papers*

15. Cornelius, P., de Groot, W., and Vermeulen, R.: Quantity Equations, Rationalization and Change of Number of Fundamental Quantities (in three parts); *Appl. Sci. Res.,* 1965, vol. B12, pp. 1, 235, 248.

16. Page, C. H.: Physical Entities and Mathematical Representation; *J. Res. Natl. Bur. Standards,* October–December 1961, vol. 65B, pp. 227–235.

17. Silsbee, F. B. Systems of Electrical Units; *J. Res. Natl. Bur. Standards,* April–June 1962, vol. 66C, pp. 137–178.

18. Young, L.: *Systems of Units in Electricity and Magnetism;* Edinburgh, Oliver & Boyd Ltd., 1969.

19. *IEEE Standard Dictionary of Electrical and Electronics Terms,* IEEE Standard 100-1984; New York, Institute of Electrical and Electronics Engineers, 1984.

# SECTION 2

# ELECTRIC AND MAGNETIC CIRCUITS

**Warren B. Boast**
*Anson Marston Distinguished Professor Emeritus of Electrical Engineering, Iowa State University; Fellow, Institute of Electrical and Electronics Engineers*

**Harry W. Hale**
*Professor of Electrical Engineering, Iowa State University; Senior Member, Institute of Electrical and Electronics Engineers*

**Edwin C. Jones, Jr.**
*Professor of Electrical Engineering, Iowa State University; Senior Member, Institute of Electrical and Electronics Engineers*

## CONTENTS

*Numbers refer to paragraphs*

## *ELECTRIC AND MAGNETIC CIRCUITS*

*By WARREN B. BOAST*

### Electric Charge

***1. Electric Charge.*** According to the electronic theory of electricity, there is an indivisible particle of negative electricity, called the *electron*. An atom of matter consists of one or more electrons and a nucleus which carries a charge of positive electricity. A negatively charged particle of matter is one in which there are more electrons than necessary to neutralize its positive electricity. A positively electrified body is one which has lost some of the electrons it had in the neutral state. Atoms which have gained or lost electrons are called *ions*.

The unit of electric charge in the International System of Units (SI) is the *coulomb*. The magnitude of the charge associated with an electron is approximately $1.602 \times 10^{-19}$ C.

## Electric Potential

*2. **Sources of Electric-Potential Difference.*** A difference of electric potential, sometimes called *electromotive force,* and abbreviated *emf,* is caused by the separation of opposite charges of electricity. This separation may be forced by physical motion, or it may be initiated or complemented by thermal, chemical, or magnetic causes or by radiation. The various classifications of these causes are called:

**a.** Electromagnetic induction (see Par. **33**).

**b.** Contact of dissimilar substances (see Par. **6**).

**c.** Thermoelectric action (see Pars. **3**, **4**, and **5**).

**d.** Chemical action (Sec. **24**).

**e.** Friction between dissimilar substances.

**f.** Photoelectric effect (Sec. **22**).

*3. **Thomson Effect.*** A temperature gradient in a metallic conductor is accompanied by a small voltage gradient whose magnitude and direction depend upon the particular metal. When an electric current flows, there is an evolution or absorption of heat due to the presence of the thermoelectric gradient, with the net result that the heat evolved in a volume interval bounded by different temperatures is slightly greater or less than that accounted for by the resistance of the conductor. In copper, the evolution of heat is greater when the current flows from hot to cold parts and less when the current flows from cold to hot. In iron, the effect is the reverse. Discovery of this phenomenon in 1854 is credited to Sir William Thomson (Lord Kelvin), the English physicist.

*4. **Peltier Effect.*** When a current is passed across the junction between two different metals, an evolution or an absorption of heat takes place. This effect is different from the evolution of heat $i^2r$, owing to the resistance of the junction, and is reversible, heat being evolved when the current passes one way across the junction and absorbed when the current passes in the other direction. The junction is the source of a Peltier electromotive force. When current is forced across the junction against the direction of the emf, a heating action occurs. If the current is forced in the direction of the Peltier emf, the junction is cooled. Refrigerators can be constructed using this principle. Since the Joule effect (see Par. **21**) produces heat in the conductors leading to the junction, the Peltier cooling must be greater than the Joule effect in that region for refrigeration to be successful. The phenomenon was discovered by Jean Peltier, the French physicist, in 1834.

*5. **Seebeck Effect.*** In a closed electric circuit consisting of two different metals, if the two junctions are maintained at different temperatures, within certain temperature ranges, an electric current flows. Thus, if one junction of a copper-iron circuit is kept in melting ice and the other in boiling water, current passes from copper to iron across the hot junction. The resulting device is usually called a *thermocouple.* The phenomenon was discovered in 1821 by Thomas Johann Seebeck.

*6. **Volta Effect, or Contact Potential.*** When pieces of various materials are brought in contact, an emf is developed between them. Thus, in the case of zinc and copper, zinc becomes charged positively and copper negatively. According to the electron theory, different substances possess different tendencies to give up their negatively charged particles. Zinc gives them up very easily; therefore, a number of negatively charged particles pass from it to copper. Measurable emfs are observed even between two pieces of the same substance having different structures, for instance, between a piece of cast copper and electrolytic copper. Frictional electricity is explained in a similar way, except that more intimate contact is necessary where the conductivities of the substances are small.

*7. **Thermocouples and Batteries.*** To utilize contact emfs, means must be devised to supply energy to the system continuously, for example, by heat or by chemical reaction. The thermocouple is an example of the former and the battery of the latter. For further discussion of the use of thermocouples in pyrometry and telemetry, see Sec. **3**. Batteries are treated in Sec. **11**.

*8. **Literature References***

Foecke, Harold A. "Introduction to Electrical Engineering Science"; Englewood Cliffs, N.J., Prentice-Hall, Inc., 1961, Chap. 7.

McGraw-Hill Encyclopedia of Science and Technology, "Thermoelectricity," 5th ed.; New York, McGraw-Hill Book Company, 1982, Vol. 13, pp. 656–665.

Encyclopedia of Physics, "Thermoelectricity," 2d ed.; New York, Van Nostrand Reinhold Co., 1974, pp. 944–946.

## Conductors

***9. Conductors and Insulators.*** Two principal materials used in electrical engineering are *conductors* and *insulators.* A conducting material allows a continuous current to pass through it under the action of a continuous emf. An ideal insulator (more correctly called a *dielectric*) allows only a brief transient current which charges it electrostatically. This charge, or *displacement,* of electricity produces a counter emf equal and opposite to the applied emf, and the flow of current ceases. The division into conductors and dielectrics is not strictly correct but is convenient for practical purposes. A substance may have practically no current when the applied voltage is sufficiently low but may be unsuitable as an insulator at high voltages. Some materials which are practically nonconducting at ordinary temperatures become good conductors when sufficiently heated. For numerical data and tables of conducting and insulating properties of the principal materials used in practice, see Sec. **4**.

Metals and other solid conductors possess free (unbound) electrons, in addition to those associated with the molecules. These free electrons move as if they were the particles of a gas dissolved in the metal. When a positive emf is applied, each electron gains a component velocity which, on account of its negative charge, is in the direction opposite to the emf. The drift of electrons constitutes a current. In their motion the electrons collide with the molecules and give up part of their momentum. This loss is supposed to account for the joulean ($i^2r$) heat set free in the conductor.

***10. Gaseous Conduction.*** A gas may be put in the conducting state by such means as raising its temperature; placing it in the neighborhood of flames, arcs, or glowing metals; or passing an electric discharge through it. The conductivity is due to free electrons. The process by which a gas is made conducting is called *ionization.* The movement of free electrons constitutes the current through the gas.

***11. Electrolytes.*** In liquid chemical compounds (electrolytes), the passage of an electric current is accompanied by chemical decomposition. Atoms of metals and hydrogen travel through the liquid in the direction of the positive current, while oxygen and acid radicals travel against the positive current. Thus, while in solid conductors electricity travels "across" the matter, in electrolytes it travels "with" the matter. For details of electrolytic conduction, see Sec. **24**.

***12. Semiconductors.*** These are a class of electrical materials that are intermediate between conductors and insulators. Their electrical behavior is often sensitive to temperature and to the degrees of impurities in certain regions. They are the basis for transistors and diodes. See Sec. **4**.

## Continuous-Current Circuits

***13. Ohm's Law.*** When the current in a conductor is steady and there are no emfs within the conductor, the value of the voltage $e$ between the terminals of the conductor is proportional to the current $i$, or

$$e = ri \tag{2-1}$$

where the coefficient of proportionality $r$ is called the *resistance* of the conductor. The same law may be written in the form

$$i = ge \tag{2-2}$$

where the coefficient of proportionality $g = 1/r$ is called the *conductance* of the conductor. When the current is measured in amperes and the emf in volts, the resistance $r$ is in ohms and $g$ is in siemens (often called mhos for reciprocal ohms).

When there is a counter emf $e_c$ within the conductor, Ohm's law becomes

$$e_t - e_c = ri \tag{2-3}$$

or

$$i = g(e_t - e_c) \tag{2-4}$$

where $e_t$ is the voltage between the terminals of the conductor.

***14. Cylindrical Conductors.*** For current directed along the axis of the cylinder, the resistance $r$ is proportional to the length $l$ and inversely proportional to the cross section $A$, or

$$r = \rho \frac{l}{A} \tag{2-5}$$

where the coefficient of proportionality $\rho$ (rho) is called the *resistivity* (or *specific resistance*) of the material. For numerical values of $\rho$ for various materials, see Sec. **4**.

The conductance of a cylindrical conductor is

$$g = \sigma \frac{A}{l} \tag{2-6}$$

where $\sigma$ (sigma) is called the *conductivity* of the material. Since $g = 1/r$, the relation also holds that

$$\sigma = \frac{1}{\rho} \tag{2-7}$$

***15. Changes of Resistance with Temperature.*** The resistance of a conductor varies with the temperature. The resistance of metals and most alloys increases with the temperature, while the resistance of carbon and electrolytes decreases with the temperature.

For usual conditions, as for about 100°C change in temperature, the resistance at a temperature $t_2$ is given by

$$R_{t2} = R_{t1}[1 + \alpha_{t1}(t_2 - t_1)] \tag{2-8}$$

where $R_{t1}$ is the resistance at an initial temperature $t_1$ and $\alpha_{t1}$ is called the temperature coefficient of resistance of the material for the initial temperature $t_1$. For copper having a conductivity of 100% of the International Annealed Copper Standard, $\alpha_{20} = 0.00393$, where temperatures are in degrees Celsius (see Sec. **4**).

An equation giving the same results as Eq. (2-8), for copper of 100% conductivity, is

$$\frac{R_{t2}}{R_{t1}} = \frac{234.4 + t_2}{234.4 + t_1} \tag{2-9}$$

where $-234.4$ is called the "inferred absolute zero" because if the relation held (which it does not over such a large range) the resistance at that temperature would be zero. For hard-drawn copper of 97.3% conductivity, the numerical constant in Eq. (2-9) is changed to 241.5.

See Sec. **4** for values of these numerical constants for copper; and for other metals see Sec. **4** under the metal being considered.

For 100% conductivity copper,

$$\alpha_{t1} = \frac{1}{234.4 + t_1} \tag{2-10}$$

When $R_{t1}$ and $R_{t2}$ have been measured, as at the beginning and end of a heat run, the "temperature rise by resistance" for 100% conductivity copper is given by

$$t_2 - t_1 = \frac{R_{t2} - R_{t1}}{R_{t1}}(234.4 + t_1) \tag{2-11}$$

***16. Resistances and Conductances in Series.*** When two or more resistances are connected in series, the equivalent resistance of the combination is equal to the sum of the resistances of the individual resistors, or

$$r_{eq} = r_1 + r_2 + \cdots \tag{2-12}$$

When conductances are connected in series, the equivalent conductance $g_{eq}$ is determined from the relation

$$\frac{1}{g_{eq}} = \frac{1}{g_1} + \frac{1}{g_2} + \cdots \tag{2-13}$$

that is, the reciprocal of the equivalent conductance is equal to the sum of the reciprocals of the individual conductances.

***17. Resistances and Conductances Connected in Parallel.*** The equivalent resistance $r_{eq}$ of a parallel combination of resistors is determined from the relation

$$\frac{1}{r_{eq}} = \frac{1}{r_1} + \frac{1}{r_2} + \cdots \tag{2-14}$$

or in conductance notation

$$g_{eq} = g_1 + g_2 + \cdots \tag{2-15}$$

When two resistances are connected in parallel, Eq. (2-14) reduces to

$$r_{eq} = \frac{r_1 r_2}{r_1 + r_2} \tag{2-16}$$

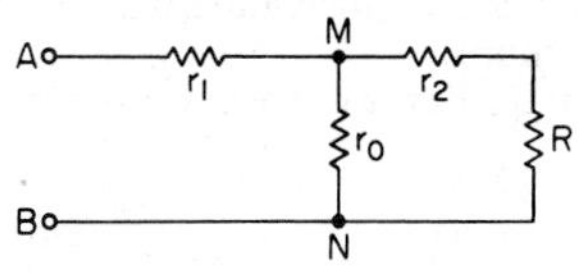

**FIG. 2-1** Series-parallel circuit.

***18. Series-Parallel Circuits.*** In a combination like the one shown in Fig. 2-1, where some of the resistances are in series, some in parallel, and it is required to find the equivalent resistance between $A$ and $B$, the problem is solved step by step, by combining the resistances in series, converting them into conductances, and adding them with other conductances in parallel. For instance, in the case shown in Fig. 2-1, begin by combining the resistance $r_2$ and $R$ into one, and determine the corresponding conductance

$$\frac{1}{R + r_2} \tag{2-17}$$

Then add this conductance to the conductance $1/r_0$. This gives the total conductance between the points $M$ and $N$, and its reciprocal gives the equivalent resistance between these points. The total resistance between the points $A$ and $B$ is found by adding $r_1$ to this resistance. When a network of conductors cannot be reduced to a series-parallel combination, the problem is solved by methods as shown in Pars. **22** and **24** to **28**.

***19. Power and Energy.*** When a steady current $i$ exists in a conductor and the voltage across the terminals of the conductor is $e$, the power, that is, the energy delivered to the conductor per unit time, is

$$P = ei \tag{2-18}$$

If the current is expressed in amperes and the potential difference in volts, the power $P$ is in watts (joules per second). When the voltage and the current are variable, their instantaneous values being represented by $e$ and $i$, respectively, the preceding equation gives the instantaneous power, that is, the instantaneous rate at which energy is being delivered to the conductor.

The total energy delivered to the conductor during a time $t$ is

$$W = eit = eQ \tag{2-19}$$

where $Q$ is the total quantity of electricity (coulombs) which passed through the conductor. If $Q$ is in coulombs (ampere-seconds), then $W$ is in joules (watt-seconds). When the voltage and the current are variable, the toal energy is expressed by

$$W = \int_{t_1}^{t_2} ei\,dt \tag{2-20}$$

where the time interval is $t_2 - t_1$.

***20. Poynting's Law.*** The power density can be determined from the product of the values of the magnetic field and the component of the electric field which is perpendicular to the magnetic field. The flow of energy at any point is in a direction perpendicular to both the fields. The total power is obtained by integrating the expression for the power density over the appropriate surface.

***21. Joule's Law.*** When the conductor contains ohmic resistance only and no counter emfs, we have $e = ri = i/g$, so that the power

$$P = i^2r = \frac{i^2}{g} = \frac{e^2}{r} = e^2g \tag{2-21}$$

This expression is known as Joule's law.

If the conductor contains a counter emf $e_c$, for instance, that developed by a motor or a battery, the power is given by

$$P = e_t i = e_c i + i^2 r \tag{2-22}$$

where $e_t$ is the voltage across the terminals of the conductor. In this expression, $e_c i$ is useful power, and $i^2r$ is the heat loss in the conductor (see Par. **13**).

***22. Kirchhoff's Laws.*** In an arbitrary network of conductors, such as Fig. 2-2, with sources of emf connected in one or more places, the distribution of currents is such that two conditions are satisfied, namely:

**1.** The algebraic sum of the currents toward any junction point is zero.

**2.** The algebraic sum of the voltages around any closed path in the network is zero.

The unknown currents in the branches may be denoted by letters, and a direction for each current may be assumed by marking an arrow along the branch. If the direction of the current is not in the assumed direction of the arrow, the computation will show the value of the corresponding current to be negative. The total number of independent equations is equal to the number of unknown quantities, which can therefore be determined by solving the simultaneous equations. For an example of such equations see Par. **24**.

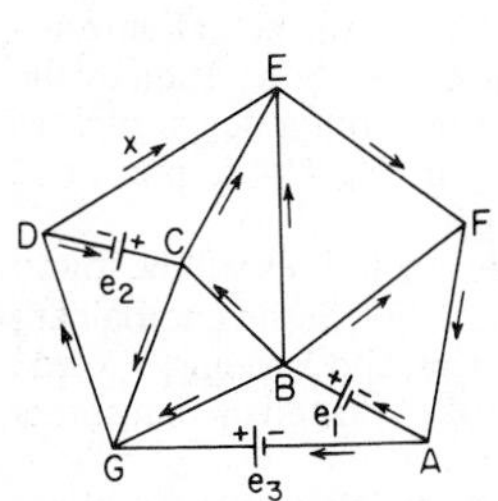


**FIG. 2-2** Network of conductors.

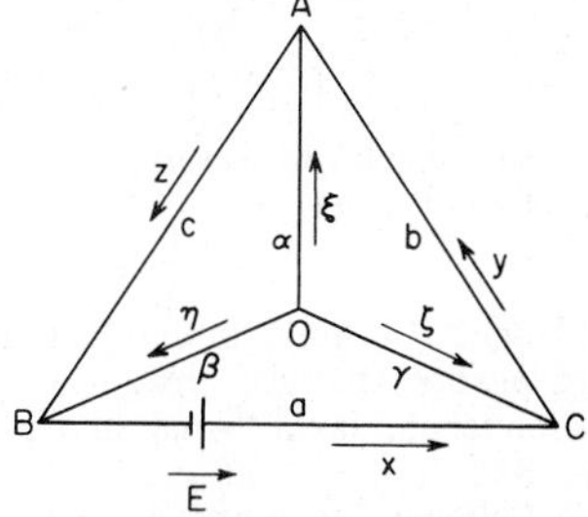


**FIG. 2-3** Wheatstone bridge.

***23. Wheatstone Bridge.*** The combination of six resistances shown in Fig. 2-3 is called the Wheatstone bridge. The resistances are denoted by $a, b, c, \alpha, \beta, \gamma$; the currents, by $x, y, z, \xi, \eta, \zeta$. An electric battery of emf $E$ is connected in the branch $BC$, and the value of $a$ includes the internal resistance of the battery. In practice, a galvanometer is usually connected in the branch

$OA$, and $\alpha$ includes its resistance. When the four resistances $b$, $c$, $\beta$, $\gamma$ are so adjusted that no current flows through $OA$, the bridge is said to be balanced and the condition holds that

$$b\beta = c\gamma \tag{2-23}$$

***24. Unbalanced Bridge.*** When the Wheatstone bridge is not balanced, Ohm's law and Kirchhoff's laws give the following equations:

$$\begin{array}{lll} ax = C - B + E & \alpha\xi = A & \xi + y - z = 0 \\ by = A - C & \beta\eta = B & \eta + z - x = 0 \\ cz = B - A & \gamma\zeta = C & \zeta + x - y = 0 \end{array}$$

Here $E$ is the battery emf, and $A$, $B$, $C$ denote the potentials of these points below that at $O$. These nine equations contain nine unknown quantities, namely, six currents and three potentials. If these are solved as simultaneous equations, any of the unknown quantities may be determined. For instance, the current in the galvanometer circuit is

$$\xi = \frac{E}{D}(b\beta - c\gamma) \tag{2-24}$$

where the "determinant" $D$ is given by[1]

$$D = abc + bc(\beta + \gamma) + ca(\gamma + \alpha) + ab(\alpha + \beta) + (a + b + c)(\beta\gamma + \gamma\alpha + \alpha\beta)$$

For practical forms of the Wheatstone bridge and its application to the measurement of resistance see Sec. **3**.

***25. Networks of Conductors.*** In a general case (Fig. 2-2), as many Kirchhoff equations (Par. **22**) may be written as there are branches in the network; the unknown quantities may be the currents, the resistances, or the voltages, also any combination of these, provided that the total number of unknown quantities is equal to the number of equations and that at least one known quantity exists in each branch of the network.

***26. Loop Currents.*** In some cases it is convenient to consider, instead of the actual currents, fictitious currents in each loop or mesh of the network. The method was originated by Maxwell (*ibid.*, Art. 282*b*). The actual current in each conductor is equal to the algebraic sum of the fictitious currents. For instance, in Fig. 2-4, the current in conductor $f$ is the difference of the fictitious currents $X$ and $Y$. The Kirchhoff equations of voltages are written using the loop currents. No current equations enter into the necessary simultaneous equations. Reconstruction of the actual currents then is accomplished after the solution by the proper addition of the components, such as $f = X - Y$, where $f$ in Fig. 2-4 is presumed measured to the right.

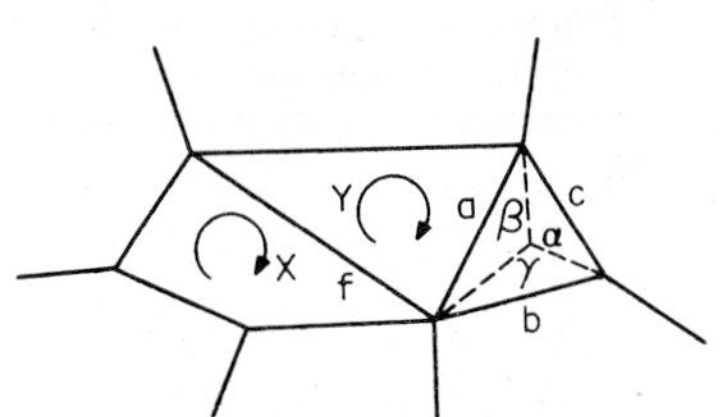

**FIG. 2-4** Method of simplifying networks by loop currents or by transformations.

Examples of such solutions will be found in Hayt and Hughes, "Introduction to Electrical Engineering," pp. 74–79; Seshu and Balbanian, "Linear Network Analysis," pp. 25–31; Smith, "Circuits, Devices, and Systems," 2d ed., pp. 35–36; Brenner and Javid, "Analysis of Electric Circuits," pp. 414–418; and in many other texts (see Par. **32**).

***27. Node Voltages.*** In a similar manner it is sometimes convenient to use a voltage notation in the simultaneous-equation solution. The voltages $A$, $B$, and $C$ of Fig. 2-3 constitute three voltages with respect to the reference node $O$, and these voltages are called *node voltages*. Mathematical procedures using the node voltages as variables of the problem result in establishment of current equations written at the junction nodes. No voltage equations enter into these necessary simultaneous equations. After the results for the node voltages are established, the

[1] Maxwell, J. C. "A Treatise on Electricity and Magnetism"; Vol. I, Art. 347.

unknown conditions within each branch of the network can be determined by equations such as $ax = C - B + E$ for the branch between nodes $B$ and $C$ of Fig. 2-3.

Examples of such solutions will be found in the texts referred to above on the following pages: Hayt and Hughes, pp. 71–73; Seshu and Balbanian, pp. 31–35; Smith, 2d ed., pp. 36–39; Brenner and Javid, pp. 408–414 (see Par. **32**).

***28. Equivalent Star and Delta.*** The number of equations required in the solution of a network may be reduced by replacing a triangular connection of branches of the network (called a "delta" or a "mesh"), such as *abc* (Fig. 2-4), by a star connection $\alpha\beta\gamma$ which is externally equivalent to the delta. It is sometimes advantageous to replace a star by the equivalent delta, and sometimes such successive replacements can reduce a network to a single impedance. This is useful in the calculation of short-circuit currents in power networks, in telephone-cable calculations, and elsewhere.

To replace a delta by a star, put

$$\alpha = \frac{bc}{a+b+c} \qquad \beta = \frac{ca}{a+b+c} \qquad \gamma = \frac{ab}{a+b+c} \tag{2-25}$$

To replace a star by a delta, put

$$a = \beta + \gamma + \frac{\beta\gamma}{\alpha} \qquad b = \gamma + \alpha + \frac{\gamma\alpha}{\beta} \qquad c = \alpha + \beta + \frac{\alpha\beta}{\gamma} \tag{2-26}$$

The six quantities $a, b, c, \alpha, \beta, \gamma$ may be complex quantities, representing impedances (see Par. **148**); or they may be the values of resistance when no reactance is involved or the values of reactance when the resistance is negligible.

To replace a star having $n$ branches by a complete mesh joining every pair of the points $A, B, C, \ldots$ of the star (thus reducing the junction points by one), put

$$\begin{aligned} AB &= \alpha\beta\left(\frac{1}{\alpha} + \frac{1}{\beta} + \frac{1}{\gamma} + \frac{1}{\delta} + \cdots\right) \\ BC &= \beta\gamma\left(\frac{1}{\alpha} + \frac{1}{\beta} + \frac{1}{\gamma} + \frac{1}{\delta} + \cdots\right) \end{aligned} \tag{2-27}$$

and so on, where $AB$ is the impedance of the branch of the mesh between $A$ and $B$, $\alpha$ is the impedance of the branch of the star leading to $A$, etc.

No general formula for replacing a mesh of more than three sides by a star exists.

***29. Superposition Principle.*** The response of a network of constant resistances (or impedances for the ac case) to a number of simultaneously applied excitations is equal to the sum of the responses to those excitations as though taken one at a time.

If in two different locations in a network of constant resistances (impedances), emfs $E_1$ and $E_2$ are applied, the currents that will result in the network may be calculated as the result of superposing one set of currents on another as follows: The first set consists of the currents that $E_1$ acting alone would drive through the branches of the network when $E_2$ is reduced to zero, but with the connection along the branch originally containing $E_2$ not broken. The second set consists of currents that $E_2$ would cause to exist in the various branches of the network when $E_1$ is similarly reduced to zero.

The principle applies also in networks containing sources that supply fixed currents as well as known voltages. In removing such current sources while the component currents caused by other sources are being investigated, the branch containing the current source must be opened.

***30. Thévenin's Theorem.*** If two terminals exist in a network containing voltage and/or current sources and composed of passive constant linear resistances (impedances for the ac case), the network may be considered insofar as all external circuit conditions are concerned as composed of a simple series circuit consisting of an emf equal to the open-circuit voltage of the network across the terminals and of an equivalent resistance (impedance) equal to that observed in looking back into the pair of terminals, with all energy sources (voltages and currents) properly removed. For the evaluation of the equivalent resistance (impedance), the removal of voltage sources must be with the branches in which they existed not broken, and the removal of current sources must be made with the branches open.

***31. Norton's Theorem.*** Analogous to Thévenin's theorem is Norton's theorem, which may be stated as follows: If any two terminals emerge from a network containing voltage and/or current sources and with passive elements composed of constant linear resistances (impedances for the ac case), the network may be considered insofar as all external circuit conditions are concerned as composed of a simple parallel circuit consisting of an ideal current source equal to the short-circuit current of the network and of a shunt conductance (admittance for the ac case) equal to that observed in looking back into the pair of terminals with all energy sources properly removed. The removal of such sources must be done according to the same principle stated in Par. **30**, Thévenin's Theorem.

***32. Literature References***

Seshu, Sundaram, and Balbanian, Norman. "Linear Network Analysis"; New York, John Wiley & Sons, Inc., 1959.

Brenner, Egon, and Javid, Mansour. "Analysis of Electric Circuits," 2d ed.; New York, McGraw-Hill Book Company, 1967.

Foecke, Harold A. "Introduction to Electrical Engineering Science"; Englewood Cliffs, N.J., Prentice-Hall, Inc., 1961.

Smith, Ralph. "Circuits, Devices and Systems," 3d ed.; New York, John Wiley & Sons, Inc., 1976.

Sabbagh, Elias M. "Circuit Analysis"; New York, The Ronald Press Company, 1961.

Hayt, William H., Jr., and Hughes, George W. "Introduction to Electrical Engineering"; New York, McGraw-Hill Book Company, 1968.

## Electromagnetic Induction of EMF

***33. Faraday's Law of Induction.*** According to Faraday's law, in any closed linear path in space, when the magnetic flux $\phi$ (see Par. **47**) surrounded by the path varies with the time, an emf is induced around the path equal to the negative rate of change of the flux in webers per second,

$$e = -\frac{\partial \phi}{\partial t} \qquad \text{volts} \tag{2-28}$$

The minus sign denotes that the direction of the induced emf is such as to produce a current opposing the flux. See the right-handed-screw rule (Par. **62** and Fig. 2-8). If the flux is changing at a constant rate, the emf is numerically equal to the increase or decrease in webers in 1 s.

The closed linear path (or circuit) is the boundary of a surface and is a geometrical line, having length but infinitesimal thickness and not having branches in parallel. It can be changing in shape or position.

If a loop of wire of negligible cross section occupies the same place and has the same motion as the path just considered, the emf $e$ will tend to drive a current of electricity around the wire, and this emf can be measured by a galvanometer or voltmeter connected in the loop of wire. As with the path, the loop of wire is not to have branches in parallel; if it has, the problem of calculating the emf shown by an instrument is more complicated and involves the resistances of the branches.

For accurate results, the simple equation (2-28) cannot be applied to metallic circuits having finite cross section. In some cases, the finite conductor can be considered as being divided into a large number of filaments connected in parallel, each having its own induced emf and its own resistance. In other cases, such as the common ones of dc generators and motors and homopolar generators, where there are sliding and moving contacts between conductors of finite cross section, the induced emf between neighboring points is to be calculated for various parts of the conductors. These can then be summed up or integrated. For methods of computing the induced emf between two points, see Par. **38** and texts on electromagnetic theory (Par. **180**).

In cases such as a dc machine or a homopolar generator, there may at all times be a conducting path for current to flow, and this may be called a "circuit," but it is not a closed linear circuit without parallel branches and of infinitesimal cross section, and therefore Eq. (2-28) does not strictly apply to such a circuit in its entirety, even though approximately correct numerical results can sometimes be obtained.

If such a practical circuit or current path is made to enclose more magnetic flux by a process of connecting in one parallel branch conductor in place of another, then such a change in enclosed flux does not correspond to an emf according to Eq. (2-28). Although it is possible in some cases to describe a loop of wire having infinitesimal cross section and sliding contacts for which Eq. (2-28) gives correct numerical results, the equation is not reliable, without qualification, for cases of finite cross section and sliding contacts. It is advisable not to use equations involving $\delta\phi/\delta t$ directly on complete circuits where there are sliding or moving contacts.

Where there are no sliding or moving contacts, if a coil has $N$ turns of wire in series closely wound together so that the cross section of the coil is negligible compared with the area enclosed by the coil, or if the flux is so confined within an iron core that it is enclosed by all $N$ turns alike, the emf induced in the coil is

$$e = -N\frac{\partial\phi}{\partial t} \qquad \text{volts} \tag{2-29}$$

In such a case, $N\phi$ is called the number of interlinkages of lines of magnetic flux with the coil, or simply the *flux linkage.*

For the above equations, the change in flux may be due to relative motion between the coil and the magnetomotive force (mmf, the agent producing the flux), as in a rotating-field generator; it may be due to change in the reluctance of the magnetic circuit, as in an inductor-type alternator or microphone; it may be due to variations in the primary current producing the flux, as in a transformer; it may be due to variations in the current in the secondary coil itself, as in Eq. (2-30); or it may be due to change in shape or orientation of the loop of coil.

***34. Self-Induction.*** When the flux $\phi$ is produced by a changing current $i$ in the loop or coil itself, the potential difference caused by the change in the flux is called the emf of self-induction. Its direction is related to the current and the emf of resistance by the equation

$$e = ri + L\frac{di}{dt} \tag{2-30}$$

where

$$L = \frac{\partial(N\phi)}{\partial i} = \text{coefficient of self-inductance} \tag{2-31}$$

(see Par. 77).

The resistance voltage $ri$ tends to oppose the current. For a decreasing current, $di/dt$ is negative, and the inductance voltage $L\ di/dt$ tends to maintain the current. This is the "extra current" flowing in the arc caused by interrupting a current with a switch. The direction assumed positive for $e$ in Eq. (2-30) is not the same as the direction assumed positive in Eq. (2-28).

***35. Lenz's Law.*** When the flux through a secondary circuit is changed because of the relative motion of primary and secondary circuits, the direction of the induced current in the secondary is related to the mechanical force between the circuits according to Lenz's law, which is stated by Maxwell as follows: "If a constant current flows in the primary circuit, $A$, and if, by the motion of $A$ or of the secondary circuit $B$, a current is induced in $B$, the direction of this induced current will be such that, by its electromagnetic action on $A$, it tends to oppose the relative motion of the circuits" ("A Treatise on Electricity and Magnetism," by J. C. Maxwell, Vol. II, Art. 542). A generator is an example of this law; the currents induced by the relative motion of the field and armature tend to oppose the motion; that is, it requires mechanical power to maintain the rotation of the generator.

***36. Sinusoidal Flux Variation.*** When the flux through a coil varies sinusoidally with the time, that is, according to sine law, at a frequency of $f$ Hz, the maximum induced emf is

$$E_m = 2\pi f N\phi_m \qquad \text{volts} \tag{2-32}$$

where $\phi_m$ is the maximum instantaneous value of the flux, in webers. The effective value of the induced emf is

$$E = 0.707E_m = 4.44fN\phi_m \qquad \text{volts} \tag{2-33}$$

***37. Average Induced EMF.*** For an interval of time $t_2 - t_1$, no matter what the law of variation of the flux with the time, the average emf is

$$E_{avg} = N\frac{\phi_1 - \phi_2}{t_2 - t_1} \quad \text{volts} \tag{2-34}$$

***38. EMF Induced in a Short Length of Straight Conductor.*** It is sometimes convenient to calculate the emf induced in a short section of the conductor or path. Examples are antennas, skin effect, the path of electrons in a vacuum tube, circuits containing moving contacts, and circuits or parts of circuits made up of straight conductors. A particular solution for the emf is given by the rule for motional emf,

$$e = Blv \quad \text{volts} \tag{2-35}$$

where $B$ is the flux density in teslas (webers per square meter) at the location of the conductor, $l$ is the length of the conductor in meters, and $v$ is the relative velocity in meters per second between the flux and the conductor. The directions of $B$, $l$, and $v$ are assumed mutually perpendicular; if they are not, their projections at right angles to each other are used in Eq. (2-35). The product $Blv$ is equal to the number of magnetic lines of force which cut the conductor per second.

The conductor in Eq. (2-35) is assumed to be of infinitesimal cross section and so short that it can be considered straight. The magnetic field adjacent to it is considered uniform. In computing the value of $v$, it is customary to consider the magnetic lines of force produced by an electromagnetic coil or magnet as moving with the coil or magnet, as, for example, in the case of the rotating field of an ac generator. The direction of the emf is given by Fleming's right-handed rule (Par. **63** and Fig. 2-9).

The voltage induced in a short length of conductor by a short element of alternating current (the conductor and the element of current being of infinitesimal cross section) may be determined by calculating the rate at which lines of force cut the conductor. As stated in Par. **42,** the lines of the magnetic field in air due to a short element of current are circles concentric on the straight line in which the short element lies. The rule can be used that, when the current dies to zero, these circular lines of force collapse, each in its own plane, and that in doing so each line cuts any conductor extending through that plane and induces voltages in them.

If the current varies as a sine wave, the total number of lines $\phi_m$ which cut the conductor in a quarter cycle is computed by integrating the maximum flux densities from the conductor outward. The maximum voltage is given by Eq. (2-32). If effective flux is calculated from effective current, effective volts will be obtained. These rules for calculating induced voltages in metallic conductors and the restrictions to be applied to them are subject to modification in unusual and complicated situations.

The discussion of induced voltages given in the preceding paragraphs does not include voltages produced by radiated fields. For the latter, see texts on modern electromagnetic theory (Par. **180**).

Practical formulas giving the emf induced in ac and dc machines are given in Secs. **7** and **8.** A special method of computation is sometimes employed when coils are moving through a pulsating field. The induced emf at any instant is taken equal to the sum of the emf induced by a constant flux in a moving coil and that induced by a pulsating flux in a stationary coil.

***39. The Fundamental Circuital Laws.***[1] The following summary based on electromagnetic theory has been contributed by Dr. M. S. Vallarta: "Electromagnetic theory is based on two fundamental laws, called the 'circuital laws.' First law (Ampère's law): 'The line integral of the magnetic field strength or intensity taken around any closed path is proportional to the total current flowing across any area bounded by that path.' Second law (Faraday's law): 'The line integral of the total electric field strength taken around any closed path is proportional to the negative rate of change, with respect to time, of the magnetic flux across any area bounded by that path.' In symbols, these laws are

$$\oint H\,ds = I \tag{2-36a}$$

[1] Units as originally stated by Dr. Vallarta have been converted to SI units.

$$\oint \mathscr{E}\, ds = -\frac{d}{dt}\int B_n\, da \qquad (2\text{-}36b)$$

where $B_n$ is the normal magnetic induction. The Maxwell theory, in addition to the foregoing laws, includes the analytical statements of the following two facts: (1) The nonexistence of separate north and south magnetic poles (the total magnetic flux across a closed surface appropriately chosen in a magnetic field is zero; in symbols $\int B_n da = 0$). (2) The electric charge is the source of the electric field (the total dielectric flux across any closed surface enclosing a charge is proportional to that charge; in symbols $\int D_n da = Q$, the enclosed charge in coulombs). These four laws constitute the Maxwell theory.

"Equation (2-36*b*) holds without restriction for constant direct currents. For alternating currents, it holds only for those points the distance of which to the element of current is small compared with the wave length. See Abraham's 'Theorie der Elektrizitat,' Vol. 1, p. 325."

***40. Electromagnetic Wave Propagation Phenomona.***[1] The following was contributed by Dr. E. A. Guillemin: "When the circuital laws mentioned in the foregoing paragraph are applied to nonconducting regions, one of them, namely, Ampère's circuital law [Eq. (2-36*a*)], must be modified to the extent of adding to the total conduction current $I$ the total displacement current $d\psi/dt$ enclosed by the path in question. $\psi$ is the electric flux or displacement and is given by an appropriate surface integral of the flux density $D = \epsilon\mathscr{E}$.

"Differential forms for these circuital-law equations are obtained through dividing both sides by the area which the path encloses and then allowing the path to become smaller and smaller. The limiting forms thus arrived at are written; *(a)* curl $\mathscr{E} = -\partial B/\partial t$ and *(b)* curl $H = J + \partial D/\partial t$, in which $J$, the conduction current density, is zero if the medium is a nonconductor (dielectric). The two additional relationships $\int D_n da = Q$ (Gauss's theorem) and $\int B_n da = 0$ may similarly be expressed in equivalent differential forms through dividing both sides of these equations by the enclosed volume and considering the limits of the resulting ratios as the volume becomes smaller and smaller. The limiting forms are written: *(c)* div $D = \rho$ and *(d)* div $B = 0$, in which $\rho$ is the charge density. Equations *(a)*, *(b)*, *(c)*, and *(d)* are Maxwell's equations in differential form.

"Physically, the operation denoted by curl, when applied to a vector function, yields a measure of the rotational or swirling intensity (also called 'vortex density') of the field represented by that function; the operation divergence (abbreviated div) yields a measure of the flux produced per unit volume at a point. The former is an axial vector (similar to a mechanical torque); the latter is a scalar. In rectangular coordinates, the $x$ component of curl $\mathscr{E}$ is $\partial\mathscr{E}_z/\partial y - \partial\mathscr{E}_y/\partial z$, from which the $y$ and $z$ components are had through simultaneously advancing, by one and by two letters, each coordinate symbol in the cyclic order $xyzxy$ . . . , while

$$\text{div } D = \partial D_x/\partial x + \partial D_y/\partial y + \partial D_z/\partial z$$

"In a charge-free dielectric, for example in free space, the field equations yield the so-called 'wave' equations governing the behavior of the electric and magnetic field intensities. It is thus found that these fields propagate with a finite velocity

$$v = 1/\sqrt{\epsilon\mu}$$

which, for free space, is the velocity of light. In the case of sinusoidally time-varying fields of frequency $f$, the distance $\lambda$ traveled in one period $\tau$ is called a 'wavelength.' Thus, this wavelength $\lambda = v\tau = v/f$. The propagational behavior of the fields is, however, not dependent upon the frequency and proceeds essentially unchanged even at zero frequency or for the dc case.

"The energies stored in the fields travel with them, and this phenomenon is the basic and sole mechanism whereby electric power transmission takes place. Thus, the electrical energy transmitted by means of transmission lines flows through the space surrounding the conductors, the latter acting merely as guides. Hollow pipes, or so-called 'waveguides,' are less conventional forms of transmission lines in which the boundary conditions permitting the propagation of the fields are fulfilled only if $\lambda$ is less than or of the same order of magnitude as the inside diameter. The energy radiated from a broadcast antenna propagates in the same manner along the earth's

[1] Units as originally stated by Dr. Guillemin have been converted to SI units.

surface or into space. The fundamental mechanism of energy propagation is the same in all these applications.

"A significant point about this phenomenon is the fact that electromagnetic energy flows predominantly well through dielectrics (nonconductors). Metals are conductors for current but nonconductors for the flow of energy, while dielectrics are good conductors for the flow of energy.

"Near the surface of a transmission-line conductor, the vector representing energy flow (Poynting's vector) is slightly inclined toward the conductor surface, thus giving rise to a small component of energy flow into the conductor. This component electromagnetic wave causes the conductor current, which in turn causes a loss but does not contribute usefully to the power transmission. Since this component electromagnetic wave is directed normally to the conductor surface and at the higher frequencies attenuates rapidly as it propagates into the conductor, the associated current density is largest at the surface and drops off more or less rapidly with increase of depth—a phenomenon known as the 'skin effect.'

"The usually accepted view that the conductor current produces the magnetic field surrounding it must be displaced by the more appropriate one that the electromagnetic field surrounding the conductor produces, through a small drain on its energy supply, the current in the conductor. Although the value of the latter may be used in computing the transmitted energy, one should clearly recognize that physically this current produces only a loss and in no way has a direct part in the phenomenon of power transmission."

## Magnetic Fields

***41. Early Concepts of Magnetic Poles.*** Substances now called magnetic, such as iron, were observed centuries ago as exhibiting forces upon one another. From this beginning the concept of magnetic poles evolved, and a quantitative theory built upon the concept of these poles, or small regions of magnetic influence, was developed. André Ampère observed forces of a similar nature between conductors carrying currents. Further developments have shown that all theories of magnetic materials can be developed and explained through the magnetic effects produced by electric charge motions.

***42. Ampère's Formula.*** The magnetic field intensity $dH$ produced at a point $A$ by an element of a conductor $ds$ (in meters) through which there is a current of $i$ $A$ is

$$dH = i\,ds\,(\sin\alpha/4\pi r^2) \qquad \text{A/m} \tag{2-37}$$

where $r$ is the distance between the element $ds$ and the point $A$, in meters, and $\alpha$ is the angle between the directions of $ds$ and $r$. The intensity $dH$ is perpendicular to the plane containing $ds$ and $r$, and its direction is determined by the right-handed-screw rule given in Par. **62** and Fig. 2-8.

The magnetic lines of force due to $ds$ are concentric circles about the straight line in which $ds$ lies. The field intensity produced at $A$ by a closed circuit is obtained by integrating the expression for $dH$ over the whole circuit.

***43. The magnetic field due to an indefinitely long, straight conductor*** carrying a current of $i$ $A$ consists of concentric circles which lie in planes perpendicular to the axis of the conductor and have their centers on this axis. The magnetic field intensity at a distance of $r$ m from the axis of the conductor is

$$H = \frac{i}{2\pi r} \qquad \text{A/m} \tag{2-38}$$

its direction being determined by the right-handed-screw rule (Par. **62**).

***44. Magnetic Field in Air Due to a Closed Circular Conductor.*** If the conductor carrying a current of $i$ A is bent in the form of a ring of radius $r$ m (Fig. 2-5), the magnetic field intensity at a point along the axis at a distance $b$ m from the ring is

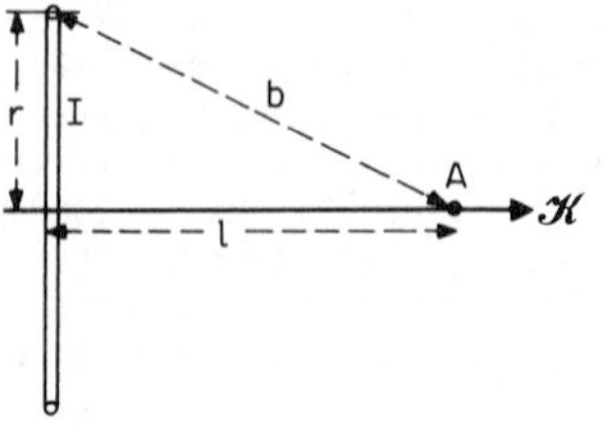


FIG. 2-5 Magnetic field along the axis of a circular conductor.

$$H = \frac{r^2 i}{2b^3} = \frac{r^2 i}{2(r^2 + l^2)^{3/2}} \qquad \text{A/m} \tag{2-39}$$

When $l = 0$,

$$H = \frac{i}{2r} \tag{2-40}$$

and when $l$ is very great in comparison with $r$,

$$H = \frac{r^2 i}{2l^3} \tag{2-41}$$

***45. The magnetic field intensity within a solenoid*** made in the form of a *torus ring,* and also in the middle part of a *long, straight solenoid,* is approximately

$$H = n_1 i \qquad \text{A/m} \tag{2-42}$$

where $i$ is the current in amperes and $n_1$ is the number of turns per meter length.

***46. Magnetic Flux Density.*** The magnetic flux density resulting in free space, or in substances not possessing magnetic behaviors differing from those in free space, is

$$B = \mu_0 H = 4\pi \times 10^{-7} H \tag{2-43}$$

where $B$ is in teslas (or webers per square meter), $H$ in amperes per meter, and the constant $\mu_0 = 4\pi \times 10^{-7}$ is the *permeability* of free space and has units of henrys per meter. In the so-called practical system of units the flux density is frequently expressed in *lines* or *maxwells per square inch* (see Par. **47**). The maxwell per square centimeter is called the *gauss.*

For substances such as iron, and other materials possessing magnetic density effects greater than those of free space, a term $\mu_r$ is added to the relationship as

$$B = 4\pi \times 10^{-7} \mu_r H \tag{2-44}$$

where $\mu_r$ is the relative permeability of that substance under the conditions existing in it compared with that which would result in free space under the same magnetic-field-intensity condition. $\mu_r$ is a dimensionless quantity.

***47. Magnetic Flux.*** The magnetic flux in any cross section of magnetic field is

$$\phi = \int B \cos \alpha \, dA \qquad \text{webers} \tag{2-45}$$

where $\alpha$ is the angle between the direction of the magnetic flux density $B$ and the normal at each point to the surface over which $A$ is measured. In the so-called practical system of units the magnetic *line* (or *maxwell*) is frequently used, where $10^8$ lines is equivalent to 1 Wb.

***48. The density of magnetic energy,*** or the magnetic energy stored per cubic meter of a magnetic field in free space, is

$$\begin{aligned} dW/dv &= \tfrac{1}{2}\mu_0 H^2 = 2\pi \times 10^{-7} H^2 \\ &= \frac{B^2}{2\mu_0} = \frac{B^2}{8\pi \times 10^{-7}} \qquad \text{J/m}^3 \end{aligned} \tag{2-46}$$

In magnetic materials the energy density stored in a magnetic field as a result of a change from a condition of flux density $B_1$ to that of $B_2$ can be expressed as

$$dW/dv = \int_{B_2}^{B_2} H \, dB \tag{2-47}$$

***49. Flux plotting*** by a graphical process is useful for determining the properties of magnetic and other fields in air. The field of flux required is usually uniform along one dimension, and a cross section of it is drawn. The field is usually required between two essentially equal magnetic potential lines, such as two iron surfaces. The field map consists of lines of force and equipotential lines, which must intersect at right angles. For the graphical method, a field map of curvilinear squares is recommended when the problem is two-dimensional. The squares are of different sizes, but the number of lines of force crossing every square is the same.

In sketching the field map, first draw those lines which can be drawn by symmetry. If parts of the two equipotential lines are straight and parallel to each other, the field map in the space between them will consist of lines which are practically straight, parallel, and equidistant. These can be drawn in. Then extend the series of curvilinear squares into other parts of the field, making sure, first, that all the angles are right angles and, second, that in each square the two diameters are equal, except in regions where the squares are evidently distorted, as near sharp corners of iron or regions occupied by current-carrying conductors. The diameters of a curvilinear square may be taken to be the distances between midpoints of opposite sides.

The magnetic field map near an iron corner is drawn as if the iron had a small fillet; that is, a line issues from an iron angle of 90° at 45° to the surfaces.

Inside a conductor which carries current, the magnetic field map is not made up of curvilinear squares, as in free space on air. In such cases special rules for the spacing of the lines must be used. The equipotential lines converge to a point called the "kernel."

The graphical method described can be used for determining other kinds of flow. Solutions of problems involving current flow and electrostatic flux can be obtained by this method.[1]

## Force Acting on Conductors

***50. Force on a Conductor Carrying a Current in a Magnetic Field.*** Let a conductor of length $l$ m carrying a current of $i$ A be placed in a magnetic field the density of which is $B$ in teslas. The force tending to move the conductor across the field is

$$F = Bli \qquad \text{newtons} \tag{2-48}$$

This formula presupposes that the direction of the axis of the conductor is at right angles to the direction of the field. If the directions of $i$ and $B$ form an angle $\alpha$, the expression must be multiplied by $\sin \alpha$.

The force $F$ is perpendicular to both $i$ and $B$, and its direction is determined by the right-handed-screw rule (Par. **62**). The effect of the magnetic field produced by the conductor itself is to increase the original flux density $(B)$ on one side of the conductor and to reduce it on the other side. The conductor tends to move away from the denser field. A closed metallic circuit carrying current tends to move so as to enclose the greatest possible number of lines of magnetic force.

***51. Force between Two Long, Straight Lines of Current.*** The force upon a unit length of either of two long, straight, parallel conductors carrying currents of $i_1$ and $i_2$ A and placed in a nonmagnetic medium (that is, not near masses of iron) is

$$\frac{F}{l} = \frac{2 \times 10^{-7} i_1 i_2}{b} \tag{2-49}$$

where $F$ is in newtons and $l$ (length of the long wires) and $b$ (the spacing between them) are in the same units, such as meters. The force is an attraction or a repulsion according to whether the two currents are flowing in the same or in opposite directions. If the currents are alternating, the force is pulsating. If $i_1$ and $i_2$ are effective values, as measured by ac ammeters, the maximum momentary value of the force may be as much as 100% greater than given by Eq. (2-49). The natural frequency (resonance) of mechanical vibration of the conductors may add still further to the maximum force, so that a factor of safety should be used in connection with Eq. (2-49) for calculating stresses on busbars.

If the conductors are straps, as is usual in busbars, the following form of equation results for thin straps placed parallel to each other, $b$ m apart,

$$\frac{F}{l} = \frac{2 \times 10^{-7} i_1 i_2}{s^2} \left( 2s \tan^{-1} \frac{s}{b} - b \log_\epsilon \frac{s^2 + b^2}{b^2} \right) \qquad \text{N/m} \tag{2-50}$$

[1] See Boast, W. B. "Vector Fields"; New York, Harper & Row Publishers, Inc., 1964, Chap. 23.

where $s$ is the dimension of the strap width in meters and the thickness of the straps placed side by side is presumed small with respect to the distance $b$ between them.

***52. Pinch Effect.*** Mechanical force exerted between the magnetic flux and a current-carrying conductor is also present within the conductor itself and is called *pinch effect.* The force between the infinitesimal filaments of the conductor is an attraction, so that a current in a conductor tends to contract the conductor. This effect is of importance in some types of electric furnaces where it limits the current that can be carried by a molten conductor. This stress also tends to elongate a liquid conductor.

## The Magnetic Circuit

***53. The Simple Magnetic Circuit.*** A simple magnetic circuit is a uniformly wound torus ring (Fig. 2-6). The relation between the mmf $\mathcal{F}$ and the flux $\phi$ is similar to Ohm's law (Par. **13**), namely,

$$\mathcal{F} = \mathcal{R}\phi \qquad \text{At} \tag{2-51}$$

where $\mathcal{R}$ is called the *reluctance* of the magnetic circuit. The relation is sometimes written in the form

$$\phi = \mathcal{P}\mathcal{F} \qquad \text{Wb} \tag{2-52}$$

where $\mathcal{P} = 1/\mathcal{R}$ is called the *permeance* of the magnetic circuit. Reluctance is analogous to resistance, and permeance is analogous to conductance of an electric circuit.

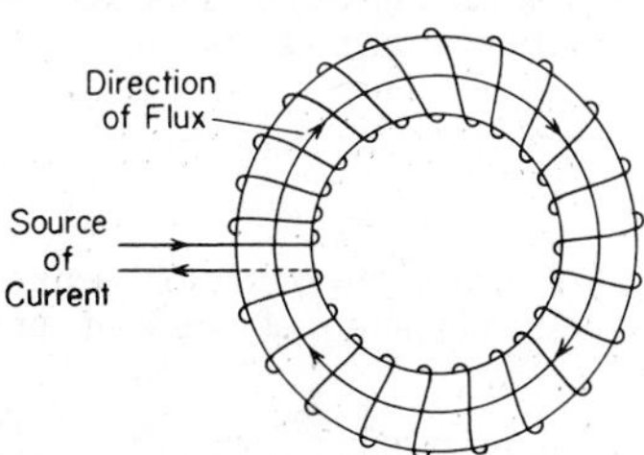


**FIG. 2-6** Closed magnetic circuit.

$$\mathcal{F} = NI \qquad \text{At} \tag{2-53}$$

where $N$ is the number of turns of conductor around the magnetic circuit, as in Fig. 2-6, and $I$ is the current in the conductor, in amperes.

***54. Permeability and Reluctivity.*** The reluctance of a uniform magnetic path (Fig. 2-6) is proportional to its length $l$ and inversely proportional to its cross section $A$,

$$\mathcal{R} = \nu \frac{l}{A} \tag{2-54}$$

and

$$\mathcal{P} = \mu \frac{A}{l} \tag{2-55}$$

In these expressions, $\nu$ is called the "reluctivity" and $\mu$ the "permeability" of the material of the magnetic path, it being assumed that there is no residual magnetism. The dimensions $l$ and $A$ are in metric units. For a vacuum, air, or other nonmagnetic substance, the reluctivity and permeability are usually written $\nu_0$ and $\mu_0$, and their values are $1/(4\pi \times 10^{-7})$ and $4\pi \times 10^{-7}$, respectively.

***55. Magnetic field intensity*** $H$ is defined as the mmf per unit length of path of the magnetic flux. It is known also as the magnetizing force or the magnetic potential gradient.

In a uniform field,

$$H = \frac{\mathcal{F}}{l} \qquad \text{At/m} \tag{2-56}$$

In a nonuniform magnetic circuit,

$$H = \frac{\partial \mathcal{F}}{\partial l} \tag{2-57}$$

Inversely, for a uniform field,

$$\mathcal{F} = Hl \tag{2-58}$$

and for a nonuniform field,

$$\mathcal{F} = \int H\,dl \tag{2-59}$$

By Ampère's law (Pars. **39** and **42**), when this integral is taken around a complete magnetic circuit,

$$\mathcal{F} = \oint H\,dl = I \tag{2-60}$$

where $I$ is the total current, in amperes, surrounded by the magnetic circuit. The circle on the integral sign indicates integration around the complete circuit. In Eqs. (2-58) to (2-60) it is presumed that $H$ is directed along the length of $l$; otherwise, the factor $\cos\theta$ must be added to the product of $H\,dl$ where $\theta$ is the angle between $H$ and $dl$.

***56. Flux density*** $B$ is the magnetic flux per unit area, the area being perpendicular to the direction of the magnetic lines of force. In a uniform field,

$$B = \frac{\phi}{A} \qquad T \text{ (or Wb/m}^2) \tag{2-61}$$

Flux density is also commonly expressed in lines or maxwells per square inch (see Pars. **46** and **47**). Combinations of metric and practical units are also used as demonstrated in Fig. 2-7.

***57. Reluctances and Permeances in Series and in Parallel.*** Reluctances and permeances are added like resistances and conductances (Pars. **16** to **18**), respectively. That is, *reluctances are added when in series, and permeances are added when in parallel.* If several permeances are given connected in series, they are converted into reluctances by taking the reciprocal of each. If reluctances are given in a parallel combination, they are similarly converted into permeances.

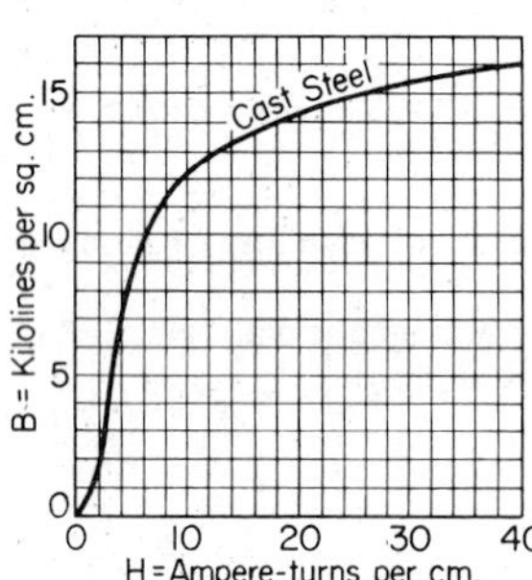


**FIG. 2-7** Typical $B$-$H$ curve.

***58. Magnetization Characteristic or Saturation Curve.*** The magnetic properties of steel or iron are represented by a saturation or magnetization curve (Fig. 2-7). Magnetic field intensities $H$ in ampere-turns per meter or in ampere-turns per centimeter, per meter, or per inch are plotted as abscissas, and the corresponding flux densities $B$ in teslas (webers per square meter) or in kilolines per square centimeter or per square inch as ordinates.

The practical use of a magnetization curve may be best illustrated by an example. Let it be required to find the number of exciting ampere-turns for magnetizing a steel ring so as to produce in it a flux of 168 kilolines. Let the cross section of the ring be 3 by 4 cm and the mean diameter 46 cm. Let the quality of the material be represented by the curve in Fig. 2-7. The flux density is $168/(3 \times 4) = 14$ kilolines/cm². For this flux density, the corresponding abscissa from the curve is about 18 At/cm. The total required number of ampere-turns is then $18 \times \pi \times 46 = 2600$.

For curves of various grades of steel and iron, see Sec. **4**. The principal methods for experimentally obtaining magnetization curves will be found in Sec. **3**.

***59. Ampere-Turns for an Air Gap.*** In a magnetic circuit consisting of iron with one or more small air gaps in series with the iron, the magnetic flux density in each of the air gaps may be considered approximately uniform. If the length across a given air gap in the direction of the flux is $l$ m, the ampere-turns required for that air gap is given by the equation

$$\text{At/m} = \frac{B(T)}{4\pi \times 10^{-7}} = 7.958 \times 10^5\, B(T) \tag{2-62}$$

or

$$\text{At/cm} = 0.7958\, B \qquad \text{lines/cm}^3 \tag{2-63}$$

or

$$\text{At/in} = 0.3133\,B \qquad \text{lines/in}^3 \tag{2-64}$$

The ampere-turns for each portion of iron, computed from iron magnetization curves such as Fig. 2-7, and the ampere-turns for the air gaps are added together to give the ampere-turns for the complete magnetic circuit.

***60. Analysis of Magnetization Curve.*** Three parts are distinguished in a magnetization curve (Fig. 2-7): the lower, or nearly straight, part; the middle part, called the knee of the curve; and the upper part, which is nearly a straight line. As the magnetic intensity increases, the corresponding flux density increases more and more slowly, and the iron is said to approach saturation (see Sec. **4**).

***61. Magnetization per Unit Volume and Susceptibility.*** If a portion of ferromagnetic material is magnetized by an mmf, $H$ At/m, the resulting flux density in teslas may be written as

$$B = \mu_0(H + M) \tag{2-65}$$

where $M$ is the magnetization per unit volume of the material (see Sec. **4**).

The ratio of $M/H$ is symbolized by $\chi$ and is called the *magnetic susceptibility*. It is the excess of the ratio of $B/\mu_0 H$ above unity, that is,

$$\chi = \frac{B}{\mu_0 H} - 1 \tag{2-66}$$

This is a dimensionless quantity. See Sec. **1**.

***62. The Right-Handed-Screw Rule.*** The direction of the flux produced by a given current is determined as shown in Fig. 2-8 (see also Fig. 2-6). If the current is established in the direction of rotation of a right-handed screw, the flux is in the direction of the progressive movement of the screw. If the current in a straight conductor is in the direction of the progressive motion of a right-handed screw, then the flux encircles this conductor in the direction in which the screw must be rotated in order to produce this motion. The dots in the figure indicate the direction of flux or current toward the reader; the crosses, that away from him.

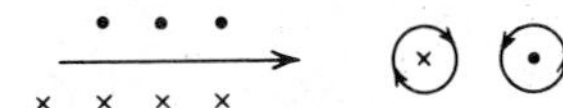

**FIG. 2-8** Relation between directions of current and flux.

***63. Fleming's Rules.*** The relative direction of flux, emf, and motion in a revolving-armature generator may be determined with the right hand by placing the thumb, index, and middle fingers so as to form the three axes of a coordinate system and pointing the index finger in the direction of the flux (north to south) and the thumb in the direction of motion; the middle finger will give the direction of the generated emf (Fig. 2-9). In the same way, in a revolving-armature motor, by using the left hand and pointing the index finger in the direction of the flux and the middle finger in the direction of the current in the armature conductor, the thumb will indicate the direction of the force and, therefore, the resulting motion. These two rules, indicated in Fig. 2-9, are known as Fleming's rules.

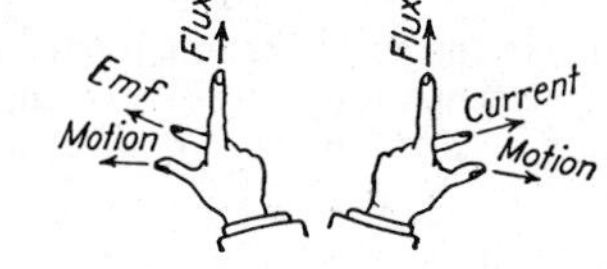


**FIG. 2-9** Fleming's generator and motor rules.

***64. Magnetic Tractive Force.*** The attracting force of a magnet is

$$F = \frac{1}{2}\frac{AB^2}{\mu_0} = \frac{AB^2}{8\pi \times 10^{-7}} \qquad \text{newtons} \tag{2-67}$$

where $B$ is the flux density in the air gap expressed in teslas (webers per square meter) and $A$ is the total area of the contact between the armature and the core, in square meters. The mass that can be supported is dependent upon the gravity field in which the mass and magnet are located.

***65. Magnetic Force, or Torque.*** The mechanical force, or the torque, between two parts of a magnetic or electric circuit may in some cases be conveniently calculated by making use of the principle of *virtual displacements*. An infinitesimal displacement between the two parts is assumed. The energy supplied from the source of current is then equal to the mechanical energy for producing the motion, plus the change in the stored magnetic energy, plus the energy for resistance loss.

When the differential motion *ds* m of a part of a circuit carrying a current *I* A changes its self-inductance by a differential *dL* H, the mechanical force on that part of the circuit, in the direction of the motion, is

$$F = \frac{1}{2} I^2 \frac{dL}{ds} \quad \text{newtons} \tag{2-68}$$

When the motion of one coil or circuit carrying a current $I_1$ A changes its mutual inductance by a differential *dM* H with respect to another coil or circuit carrying a current $I_2$ A, the mechanical force on each coil or circuit, in the direction of the motion, is

$$F = I_1 I_2 \frac{dM}{ds} \quad \text{newtons} \tag{2-69}$$

where *ds* represents the differential of distance in meters. For a discussion of self- and mutual inductance *L* and *M* see Pars. **77** and **85.**

## Hysteresis and Eddy Currents in Iron

***66. The Hysteresis Loop.*** When a sample of iron or steel is subjected to an alternating magnetization, the relation between *B* and *H* is different for increasing and decreasing values of the magnetic intensity (Fig. 2-10). This phenomenon is due to irreversible processes which

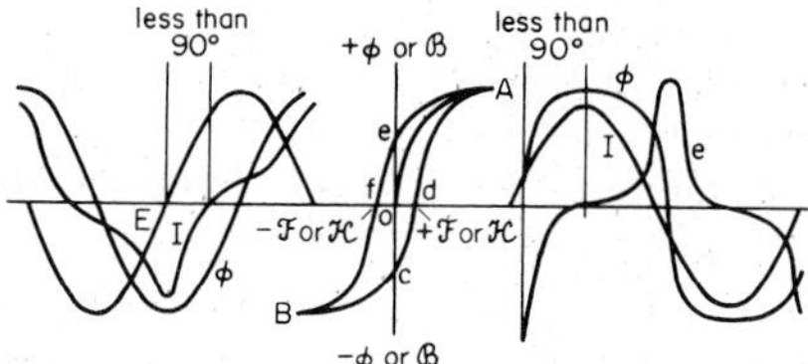


**FIG. 2-10** Periodic waves of current, flux, and emf; hysteresis loop.

result in energy dissipation, producing heat. Each time the current wave completes a cycle, the magnetic flux wave must also complete a cycle, and the elementary magnets are turned. The figure *AefBcdA* in Fig. 2-10 is called the *hysteresis loop.*

***67. Retentivity.*** If the coil shown in Fig. 2-6 is excited with alternating current, the ampere-turns and consequently the mmf will, at any instant, be proportional to the instantaneous value of the exciting current. Plotting a *B-H* (or $\phi$-$\mathcal{F}$) curve (Fig. 2-10) for one cycle, the closed loop *AefBcdA* is obtained. The first time the iron is magnetized, the *virgin,* or *neutral, curve OA* will be produced; but it cannot be produced in the reverse direction *AO,* because when the mmf drops to zero there will always be some remaining magnetism (+*Oe* or −*Oc*). This is called *residual magnetism;* to reduce this to zero, an mmf (−*Of* or +*Od*) of opposite polarity must be applied. This mmf is called the *coercive force.*

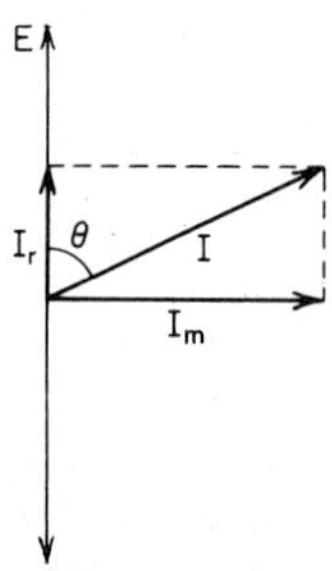


**FIG. 2-11** Components of exciting current; hysteretic angle.

***68. Wave Distortion.*** In Fig. 2-10, the instantaneous values of the exciting current $I$ (which is directly proportional to the mmf) and the corresponding values of the flux $\phi$ and voltage $E$ (or $e$) are plotted against time as abscissas, beside the hysteresis loop. *(a)* If the voltage applied to the coil is *sinusoidal* ($E$, to the left), the current wave is distorted and displaced from the corresponding sinusoidal flux wave. The latter wave is in quadrature with the voltage wave. *(b)* If the *current* through the coil is *sinusoidal* ($I$, to the right), the flux is distorted into a flat-top wave and the induced voltage $e$ is peaked.

***69. Components of Exciting Current.*** The alternating current that flows in the exciting coil (Fig. 2-10) may be considered to consist of two components, one exciting magnetism in the iron, and the other supplying the iron loss. For practical purposes, both components may be replaced by equivalent sine waves and phasors (Fig. 2-11) (see Pars. **143** to **147**). We have

$$I_r = I \cos \theta = \text{power component of current}$$

$$P_h = IE \cos \theta = I_r E = \text{iron loss in watts} \qquad (2\text{-}70)$$

$$I_m = I \sin \theta = \text{magnetizing current}$$

where $I$ is the total exciting current and $\theta$ the angle of time-phase displacement between current and voltage.

***70. Hysteretic Angle.*** Without iron loss, the current $I$ would be in phase quadrature with $E$. For this reason, the angle $\alpha = 90 - \theta$ is called the *angle of hysteretic advance of phase.*

$$\sin \alpha = \frac{I_r}{I} = \frac{I_r E}{IE} = \frac{\text{W loss}}{\text{VA}} \qquad (2\text{-}71)$$

In practice, the measured loss usually includes eddy currents (Par. **73**), so that the name "hysteretic" is somewhat of a misnomer.

***71. The energy lost per cycle from hysteresis*** is proportional to the area of the hysteresis loop (Fig. 2-10). This is a consequence of the evaluation over a cycle of Eq. (2-47).

***72. Steinmetz's Formula.*** According to experiments by C. P. Steinmetz, the heat energy due to hysteresis released per cycle per unit volume of iron is approximately

$$W_h = \eta B_{\max}^{1.6} \qquad (2\text{-}72)$$

The exponent of $B_{\max}$ varies between 1.4 and 1.8 but is generally taken as 1.6. Values of the hysteresis coefficient $\eta$ are given in Sec. **4.**

***73. Eddy-current losses*** are $I^2R$ losses due to secondary currents (Foucault currents) established in those parts of the circuit which are interlinked with alternating or pulsating flux.

Referring to Fig. 2-12, which shows a cross section of a transformer core, the primary current $I$ produces the alternating flux $\phi$, which by its change generates an emf $e$ in the core; this emf then sets up the secondary current $i$. Now, if the core is divided into two *(b)*, four *(c)*, or $n$ parts, the emf in each circuit is $e/2$, $e/4$, $e/n$; and the conductance, $g/2$, $g/4$, $g/n$, respectively. Thus, the loss per lamination will be $1/n^3$ times the loss in the solid core, and the total loss is $1/n^2$ times the loss in the solid core.

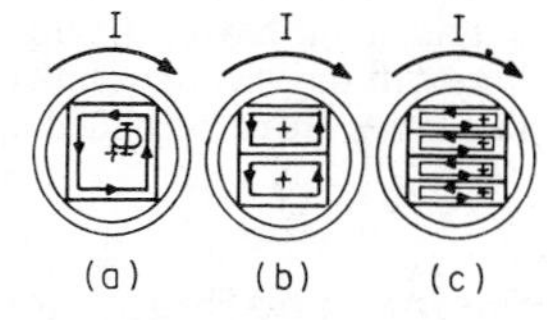


**FIG. 2-12** Section of a transformer core.

Eddy currents can be greatly reduced by laminating the circuit, that is, by making it up of thin sheets each electrically insulated from the others. The same purpose is accomplished by using separately insulated strands of conductors or bundles of wires.

A formula for the eddy loss in conductors of circular section, such as wire, is

$$P_e = \frac{(\pi r f B_{\max})^2}{4\rho \times 10^{16}} \qquad \text{W/cm}^3 \qquad (2\text{-}73)$$

where $r$ is the radius of the wire in centimeters; $f$, the frequency in hertz (cycles per second); $B_{\max}$, the maximum flux density in lines per square centimeter; and $\rho$, the specific resistance in ohm-centimeters.

A formula for the loss in sheets is

$$P_e = \frac{(\pi t f B_{max})^2}{6\rho \times 10^{16}} \quad \text{W/cm}^3 \tag{2-74}$$

where $t$ is the thickness in centimeters; $f$, the frequency in hertz (cycles per second); $B_{max}$, the maximum flux density in lines per square centimeter; and $\rho$, the specific resistance.

The specific resistance of various materials is given in Sec. **4.**

***74. Effective Resistance and Reactance.*** When an ac circuit has appreciable hysteresis, eddy currents, and skin effect (Par. **88**), it can be replaced by a circuit of equivalent resistances and equivalent reactances (Par. **148**) in place of the actual ones. These effective quantities are so chosen that the energy relations are the same in the equivalent circuit as in the actual one. In a series circuit, let the true power lost in ohmic resistance, hysteresis, and eddy currents be $P$; and the reactive (wattless) voltamperes, $Q$. Then the effective resistance and reactance are determined from the relations

$$i^2 r_{eff} = P \qquad i^2 x_{eff} = Q \tag{2-75}$$

In a parallel circuit, with a given voltage, the equivalent conductances and susceptance (Par. **149**) are calculated from the relations

$$e^2 g_{eff} = P \qquad e^2 b_{eff} = Q \tag{2-76}$$

Such equivalent electric quantities, which replace the core loss, are used in the analytical theory of transformers and induction motors.

***75. Core Loss.*** In practical calculations of electrical machinery, the total core loss is of interest rather than the hysteresis and the eddy currents separately. For such computations, empirical curves are used, obtained from tests on various grades of steel and iron (see Sec. **4**).

***76. Separation of Hysteresis Losses from Eddy-Current Losses.*** For a given sample of laminations, the total core loss $P$, at a constant flux density and at variable frequency $f$, can be represented in the form

$$P = af + bf^2 \tag{2-77}$$

where $af$ represents the hysteresis loss and $bf^2$ the eddy, or Foucault-current, loss, $a$ and $b$ being constants. The voltage waveform should be very close to a sine wave. If we write this equation for two known frequencies, two simultaneous equations are obtained from which $a$ and $b$ are determined.

It is convenient to divide the foregoing equation by $f$, because the form

$$P/f = a + bf \tag{2-78}$$

represents a straight line relating $P/f$ and $f$. Known values of $P/f$ are plotted against $f$ as abscissas, and a straight line having the closest approximation to the points is drawn. The intersection of this line with the axis of ordinates gives $a$; $b$ is calculated from the preceding equation. The separate losses are calculated at any desired frequency from $af$ and $bf^2$, respectively.

## Inductance and Reactance

***77. The electromagnetic inductance,*** or the ***coefficient of self-induction,*** $L$ is defined from either of the following fundamental equations:

$$e = L\frac{di}{dt} \tag{2-79}$$

$$W = \tfrac{1}{2} i^2 L \tag{2-80}$$

The first equation expresses the fact that the self-induced voltage $e$ is proportional to the time rate of change of the current $i$ in the circuit, and $L$, the coefficient of proportionality, is the self-inductance. In the second equation, the magnetic energy $W$ stored in the circuit is proportional to the square of the current $i$, where $L$ is the coefficient of proportionality. The units in the above equations are $e$ volts, $i$ amperes, $t$ seconds, $W$ joules, and $L$ henrys.

***78. Torus Ring or Toroidal Coil of Rectangular Section with Nonmagnetic Core*** (Fig. 2-6). The inductance of a rectangular toroidal coil, uniformly wound with a single layer of fine wire, is

$$L = 2 \times 10^{-7} N^2 b \left( \ln \frac{r_2}{r_1} \right) \qquad \text{henrys} \tag{2-81}$$

where $N$ equals the number of turns of wire on the coil; $b$, the axial length of the coil, in meters; and $r_2$ and $r_1$ are the outer and inner radial distances in meters.

***79. Torus Ring or Toroidal Coil of Circular Section with Nonmagnetic Core*** (Fig. 2-6). A toroidal coil of circular section, uniformly wound with a single layer of fine wire of $N$ turns, has an inductance of

$$L = 4\pi \times 10^{-7} N^2 (g - \sqrt{g^2 - a^2}) \qquad \text{henrys} \tag{2-82}$$

where $g$ is the mean radius of the toroidal ring and $a$ is the radius of the circular cross section of the core, both measured in meters.

***80. Inductance of a Very Long Solenoid.*** A solenoid uniformly wound in a single layer of fine wire possesses an inductance of

$$L = \frac{4\pi^2 \times 10^{-7} N^2 R^2}{S} \qquad \text{henrys} \tag{2-83}$$

where $R$ and $S$ are the radius and length of the solenoid in meters, as illustrated in Fig. 2-13. The assumption is made that $S$ is very large with respect to $R$.

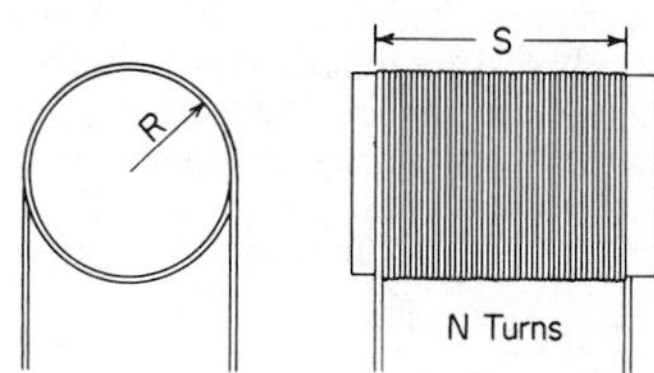


**FIG. 2-13** Cylindrical solenoid.

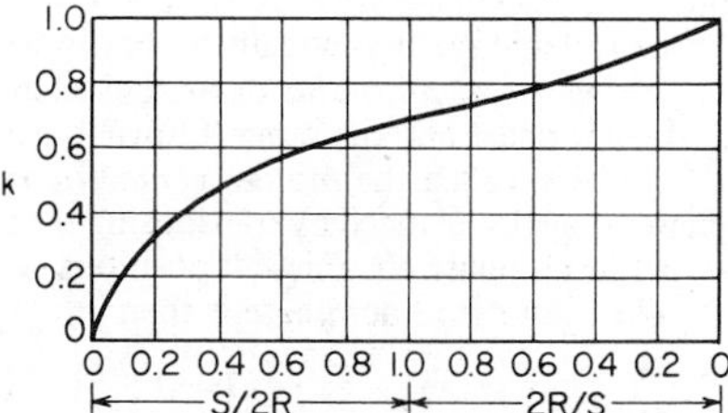


**FIG. 2-14** Factor $k$ of Eq. (2-84). *(From W. B. Boast, "Vector Fields," New York, Harper & Row, Publishers, Inc. 1964.)*

***81. Inductance of the Finite Solenoid.*** The inductance of a short solenoid is less than that given by Eq. (2-84), by a factor $k$ (a dimensionless quantity). The inductance relation then is

$$L = k \frac{4\pi^2 \times 10^{-7} N^2 R^2}{S} \qquad \text{henrys} \tag{2-84}$$

where the values of $k$ for various ratios of $R$ and $S$ are given in Fig. 2-14.

Inductance relations for other configurations of coils are given in W. B. Boast, "Vector Fields," New York, Harper & Row, Publishers, Incorporated, 1964, Chaps. 17 and 18.

***82. Inductance per Unit Length of a Coaxial Cable.*** For low-frequency applications, where skin effect is not predominant (uniform current density over nonmagnetic current-carrying cross sections), the inductance per unit length of a coaxial cable is

$$l = \frac{10^{-7}}{2} \left[ 1 + 4 \ln \frac{R^2}{R_1} + \frac{4R_3^4}{(R_3^2 - R_2^2)^2} \ln \frac{R_3}{R_2} - \frac{3R_3^2 - R_2^2}{R_3^2 - R_2^2} \right] \qquad \text{H/m} \tag{2-85}$$

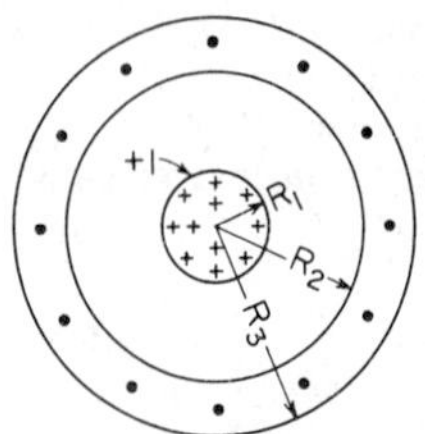


FIG. 2-15 Coaxial cable.

where $R_1$, $R_2$, and $R_3$ are the radii of the inner conductor, the inner radius of the outer conductor, and the outer radius of the outer conductor, in meters, respectively, as shown in Fig. 2-15. For very thin outer shells the last two terms drop out of the equation, and for very small inner conductors the first term becomes less important. For high-frequency applications the first, third, and fourth terms are all suppressed, and for the extreme situation where all the current is essentially at the boundaries formed by $R_1$ and $R_2$, respectively, the inductance per unit length becomes

$$l = 2 \times 10^{-7} \ln (R_2/R_1) \qquad \text{H/m} \tag{2-86}$$

***83. Inductance of Two Long, Cylindrical Conductors, Parallel and External to Each Other.*** The inductance per unit length of two separate parallel conductors is

$$l = 10^{-7} \left(1 + 4 \ln \frac{D}{\sqrt{R_1 R_2}}\right) \qquad \text{H/m} \tag{2-87}$$

where $D$ is the distance between centers of the two cylinders and $R_1$ and $R_2$ are the radii of the conductor cross sections.

If $R_1 = R_2 = R$ and the skin-effect phenomenon applies as at very high frequencies, the inductance per unit length becomes

$$l = 4 \times 10^{-7} \ln (D/R) \qquad \text{H/m} \tag{2-88}$$

***84. Inductance of Transmission Lines.*** The inductance relationships used in predicting the performance of power-transmission systems often involve the effects of stranded and bundled conductors operating in parallel, as well as configurations of these groups of current-carrying elements of one phase group of the system coordinated with similar groups constituting other phases, in polyphase systems. In such systems the several current-carrying elements of a phase are considered mathematically as a cylindrical shell of current of radius $D_s$ (meters) called the *self-geometric mean radius* of the phase; and the mutual distances [between the current in a particular phase and the other (return) currents in the other phases] are replaced by a distance $D_m$ (meters) called the *mutual geometric mean distance* to the return. The inductance of all phases may be balanced by transposing the conductors over the length of the transmission line, so that each phase occupies all positions equally in the length of the line.

The inductance per phase is then one-half as large as that of Eq. (2-88), that is,

$$l = 2 \times 10^{-7} \ln (D_m/D_s) \qquad \text{H/m} \tag{2-89}$$

The following references provide methods for computing the geometric mean distances $D_m$ and $D_s$:

Anderson, P. M. "Analysis of Faulted Power Systems"; Ames, Iowa State University Press, 1973, Chap. 4.

Boast, W. B. "Vector Fields"; New York, Harper & Row, Publishers, Incorporated, 1964, Chap 19.

Stevenson, W. D., Jr. "Elements of Power System Analysis," 3d ed.; New York, McGraw-Hill Book Company, 1975, Chap 2.

Clarke, E. "Circuit Analysis of A-C Power Systems"; New York, John Wiley & Sons, Inc., 1943, Vol. I.

"Electrical Transmission and Distribution Reference Book," 4th ed.; Westinghouse Electric Corporation, 1950.

***85. Mutual Inductance.*** When two independent circuits or coils 1 and 2 are in proximity to each other, a change in current in one is accompanied by a change in its magnetic field, which induces an emf in the other. Thus

$$e_1 = M \frac{di_2}{dt} \qquad \text{and} \qquad e_2 = M \frac{di_1}{dt} \tag{2-90}$$

$M$ is measured in henrys when $i$ is in amperes and $e$ in volts.

When there is no magnetic material, the total energy of the system, stored in the magnetic field, is

$$W = \tfrac{1}{2}L_1 i_1^2 + \tfrac{1}{2}L_2 i_2^2 + M i_1 i_2 \qquad (2\text{-}91)$$

where $L_1$ and $L_2$ are the self-inductances in henrys.

For interference between transmission and telephone lines caused by the mutual-inductance coupling between them, and for transpositions of telephone lines, see Sec. **18.**

For effects of mutual coupling between circuits in power-transmission systems, see Sec. **14.**

***86. Leakage Inductance.*** In electrical apparatus, such as transformers, generators, and motors, in which the greater part of the flux is carried by an iron core, the difference between self-inductance and mutual inductance of the primary and secondary windings is small. This small difference is called "leakage inductance." It is of great importance in the characteristics and operation of the apparatus and is usually calculated or measured separately. The loss in voltage in such apparatus, due to inductance, is associated with the leakage.

***87. Magnetizing Current.*** The mutual inductance of the windings of apparatus with iron cores is not usually stated in henrys, but the effective alternating current required to produce the flux is stated in amperes and is called the *exciting current.* One component of this current supplies the energy corresponding to the core loss (Pars. **69** and **75**). The remaining component is called *magnetizing current.* Solenoids and other coils with only one winding are usually treated in a similar manner when they have iron cores. The exciting current usually does not have a sine-wave form. See Par. **68** and Fig. 2-10.

## Skin Effect

***88. Real, or ohmic, resistance*** is the resistance offered by the conductor to the passage of electricity. Although the specific resistance is the same for either alternating or continuous current, the total resistance of a wire is greater for alternating than for continuous current. This is due to the fact that there are induced emfs in a conductor in which there is alternating flux. These emfs are greater at the center than at the circumference, so that the potential difference tends to establish currents that oppose the current at the center and assist it at the circumference. The current is thus forced to the outside of the conductor, thus reducing the effective area of the conductor. This phenomenon is called *skin effect.*

***89. Skin-Effect Resistance Ratio.*** The ratio of the ac resistance to the dc resistance is a function of the cross-sectional shape of the conductor and its magnetic and electrical properties as well as of the frequency. For cylindrical cross sections with presumed constant values of relative permeability $\mu_r$ and resistivity $\rho$, the function that determines the skin-effect ratio is

$$mr = \sqrt{\frac{8\pi^2 \times 10^{-7} f \mu_r r}{\rho}} \qquad (2\text{-}92)$$

where $r$ is the radius of the conductor and $f$ is the frequency of the alternating current. The ratio of $R$, the ac resistance, to $R_0$, the dc resistance, is shown as a function of $mr$ in Fig. 2-16.

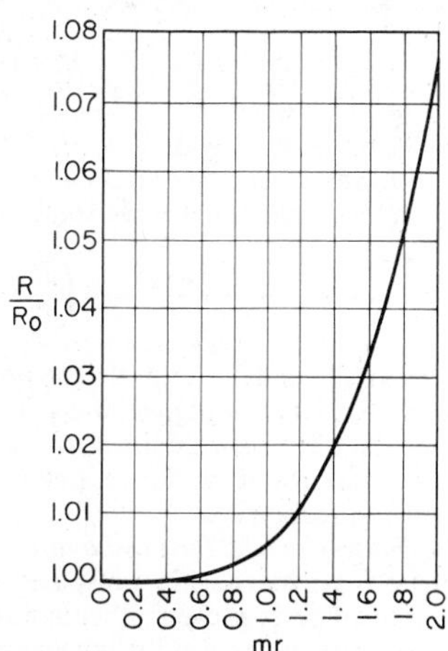


**FIG. 2-16** Ratio of ac to dc resistance of a cylindrical conductor. *(From W. D. Stevenson, Jr., "Elements of Power Systems Analysis," New York, McGraw-Hill Book Company, 1962.)*

***90. The skin effect of steel wires and cables*** cannot be accurately calculated by assuming a constant value of the permeability, which varies throughout a large range during every cycle. Therefore, curves of measured characteristics should be used. See "Electrical Transmission and Distribution Reference Book," 4th ed., 1950.

***91. Skin Effect of Tubular Conductors.*** Cables of large size are often made so as to be, in effect, round, tubular conductors. Their effective resistance due to skin effect may be taken from the curves of Sec. **4**; or from Chap. 25, "Electrical Coils and Conductors," by H. B. Dwight; or from the paper by A. W. Ewan, *Gen. Elec. Rev.*, April 1930, p. 250. The effective resistance may be calculated as described in Chap. 20 of the foregoing book.

The skin-effect ratio of square, tubular busbars may be obtained from semiempirical formulas in the paper A-C Resistance of Hollow, Square Conductors, by A. H. M. Arnold, *Jour. IEE (London),* 1938, Vol. 82, p. 537. These formulas have been compared with tests. The resistance ratio of square tubes is somewhat larger than that of round tubes. Values may be read from the curves of Fig. 4, Chap. 25, of "Electrical Coils and Conductors."

***92. Penetration Formula.*** For wires and tubes (and approximately for other compact shapes) where the resistance ratio is comparatively large, the conductor can be approximately considered to be replaced by its outer shell, of thickness equal to the "penetration depth," given by

$$\delta = \frac{1}{2\pi}\sqrt{\frac{10^7 \rho}{f\mu_r}} \quad \text{meters} \tag{2-93}$$

where $\rho$ is the resistivity in ohm-meters and $\mu_r$ is a presumed constant value of relative permeability. The resistance of the shell is then the effective resistance of the conductor. For Eq. (2-93) to be applicable, $\delta$ should be small compared with the dimensions of the cross section. In the case of tubes, $\delta$ is, evidently, less than the thickness of the tube. See Eq. (30), Chap. 19, of the book "Electrical Coils and Conductors."

## Eddy-Current Loss in Transformer Windings

***93. Eddy-current loss in a winding of rectangular wires*** in series, at power frequencies. The eddy-current loss in the primary winding of a transformer is computed separately from that in the secondary. The main eddy-current loss is that produced by the leakage magnetic flux in the copper winding due to the load current. This is determined from the quantity $D$, given in Eq. (2-94). The loss thus depends on the length of the leakage flux path $s$, in inches, the net extent of copper in the direction of $s$, $w$, in inches, the number $q$ of the rectangular wires counted across the leakage flux (from zero to maximum flux density), the height of the wires $h$, in inches, measured in the direction across the leakage flux, and the frequency $f$, in hertz.

$$D = \left(\frac{h^2 f w}{8.23 s}\right)^2 \quad \text{at } 75°\text{C} \tag{2-94}$$

See paper by H. W. Taylor, *Jour. IEE (London),* 1920, Vol. 58, p. 279. The numerical factor 8.23 varies in proportion to (234.4° + $T$°) for copper of 100% conductivity, at $T$°C.

**Extra loss ratio** or **extra resistance ratio**

$$\frac{R_{ac} - R_{dc}}{R_{dc}} = q^2\left(\frac{D}{9} - \frac{17}{3{,}780}D^2 \cdots\right) - \frac{D}{45} + \frac{D^2}{900} \cdots \tag{2-95}$$

The terms in $D^2$ and the terms that do not involve $q$ are usually negligible in practical cases. If the terms in $D^2$ are not negligible, the series does not converge rapidly and should not be used.

Note that the loss with direct current is $I^2R_{dc}$ and the loss with alternating current is $I^2R_{ac}$, where $I$ is the current in amperes in the winding. The ratio of change in loss is equal to the ratio of change in resistance.

***94. Eddy-Current Loss in Laminated Conductors, Thoroughly Transposed.*** If the copper laminations are thoroughly transposed so that each lamination can be taken to have the same number of amperes as every other lamination, then the magnetic field and current-distribution conditions are the same as if all the laminations were in series. Equations (2-94) and (2-95) are used, $q$ being the total number of laminations across the direction of leakage magnetic flux, from zero to maximum flux density, in either the primary or the secondary, whichever is being computed. The dimension $h$ is in this case the height of one copper lamination, in the direction across the leakage flux.

## Eddy-Current Loss in Armature Coils

***95. Eddy-Current Loss in Coils of Rectangular Wires in Series, at Power Frequencies.*** The phase angle between the currents of the upper half and the lower half of a slot in a 3-phase machine (except in mesh-wound machines) is usually either 0° or 60°. In a certain proportion of the slots, $\phi$ is 0°; and in the remainder, 60°. The value of extra loss ratio is to be computed for each case.

Let the length of leakage flux path (that is, the width of the slot) be $s$ in; the net extent of copper in the direction of $s$ be $w$ in; $q$ be the number of rectangular wires one above the other in the half slot, connected in series, each of height $h$ in; and the frequency be $f$ hertz.

Find $D$ by Eq. (2-94) adjusted for the operating temperature. The extra loss ratio for the lower half of the slot-embedded part of coil is then

$$\frac{R_{ac} - R_{dc}}{R_{dc}} = D\left(\frac{q^2}{9} - \frac{1}{45}\right) \tag{2-96}$$

The extra loss ratio for the upper half of the slot-embedded part of coil is

$$\frac{R_{ac} - R_{dc}}{R_{dc}} = D\left(\frac{7}{9}q^2 - \frac{1}{45} - \frac{2}{3}q^2 \sin^2\frac{\phi}{2}\right) \tag{2-97}$$

where $\phi$ is 0° or 60°, as previously described.

The average of the foregoing gives the extra loss ratio for the complete slot-embedded part of the coil,

$$\frac{R_{ac} - R_{dc}}{R_{dc}} = D\left(\frac{4q^2}{9} - \frac{1}{45} - \frac{q^2}{3}\sin^2\frac{\phi}{2}\right) \tag{2-98}$$

where $q$ is the number of rectangular wires in the *half* slot, one above the other.

Terms in $D^2$ have been omitted. It is to be noted that the amount of insulation between the wires does not affect the loss ratio, in this case. For wires in series, it is usually assumed that the eddy-current loss is negligible in the part of the coil not embedded in the slot.

***96. Eddy-Current Loss in Laminated Conductors, Thoroughly Transposed.*** If the copper laminations are thoroughly transposed as, for instance, in the Punga and Roebel types of transposed conductors illustrated in Sec. 7, so that each lamination can be taken to have the same number of amperes as every other lamination, then (as stated in Par. **94**) the current-distribution conditions are the same as if all the laminations were in series. Equations (2-96) to (2-98) are used, $q$ being the number of copper laminations, one above the other, in the half slot and $h$ being the height of one lamination. The loss in the nonembedded part is taken to be negligible.

***97. Eddy-Current Loss in Laminated Conductors, Not Thoroughly Transposed, Laminations Being Soldered Together at Beginning and End of Coil.*** In this case, circulating currents flow along the upper laminations and return along the lower ones, thus producing extra copper loss in the entire coil, in both the embedded and the nonembedded parts. This extra loss, the *long-path eddy-current loss,* is in addition to the *short-path eddy-current loss* [Eqs. (2-96) to (2-98)], which occurs in the embedded part of the coil.

In calculating the long-path eddy-current loss, any laminated conductor may be called right side up, denoted by $D$ (direct).

Note that, in this discussion, "laminations" are connected in parallel but "conductors" are connected in series in the coil.

For a coil denoted by $D$, let

$$I_0 = +I_b \tag{2-99}$$

where $I_b$ is the vector sum of all the current below it in the same slot.

For conductors wrong side up, denoted by $R$ (reversed),

$$I_0 = -I - I_b \tag{2-100}$$

that is, (− 1) times all the current in the slot below the double line of the conductor considered (see Fig. 2-17), added vectorially, where $I$ is the current in the conductor considered and $I_b$ is the vector sum of all the current below it.

The twisting of the laminated conductors, by which $D$ can be changed to $R$ as desired (Fig. 2-17), is illustrated in Reduction of Armature Copper Losses, by I. H. Summers, *Trans. AIEE,* 1927, p. 102.

The extra loss ratio due to long-path eddy current in a coil with any arrangement of direct and reversed laminated conductors is

$$\frac{R_{ac}-R_{dc}}{R_{dc}} = b^2c^2\left[M_r - 1 + \left(\left|\frac{I_0}{I}\right|^2 + \left|\frac{I_0}{I}\right|\cos\delta\right)N_r\right] \tag{2-101}$$

where the average value of $I_0/I$ is used. The angle $\delta$ is the phase angle between $I_0$ and $I$. The ratio of the embedded part to the total is $b$; that is, $b$ equals the core length divided by one-half the mean length of the coil turns. The ratio of gross height of the laminated conductor to the net copper height of the same conductor is $c$. The ratio $b$ is less than 1; $c$ is greater than 1.

In Eq. (2-101), $M_r$ and $N_r$ are given by

$$M_r - 1 = \frac{4}{45}Dn^4 - \frac{16}{4{,}725}D^2n^8 \cdots \tag{2-102}$$

$$N_r = \frac{D_n^4}{3} - \frac{17}{1{,}260}D^2n^8 \cdots \tag{2-103}$$

where $D$ is computed from one lamination by Eq. (2-94) and $n$ is the number of laminations per conductor. If the terms in $D^2$ are not negligible, the series does not converge rapidly and should not be used.

See Heat Losses in the Conductors of A-C Machines, by Waldo V. Lyon, *Trans. AIEE,* 1921, p. 1378.

***98. Normal Diamond Type of Coil without Special Twisting and with Upper and Lower Currents in Phase.***

Arrangement $DD \cdots RR \cdots$ (Fig. 2-17).

Extra loss ratio due to long-path eddy current

$$\frac{R_{ac}-R_{dc}}{R_{dc}} = b^2c^2\left[\frac{4}{45}Dn^4 - \frac{16}{4{,}725}D^2n^8 \cdots + \frac{q^2-1}{4}\left(\frac{Dn^4}{3} - \frac{17}{1{,}260}D^2n^8 \cdots\right)\right] \tag{2-104}$$

Above the coil, with upper and lower currents 60° out of phase, use expression (2-104).

***99. Diamond Coil with Twisted Conductors, with an Even Number of Turns per Coil (q Even).*** Use the arrangement $DRDR \cdots RDRD \cdots$ (Fig. 2-17). This can be obtained by twisting each conductor, turning it upside down, as it passes the end of the coil away from the terminals. The extra loss ratio due to long-path eddy current is the same whether $\phi$ is 0° or 60° and is

$$\frac{R_{ac}-R_{dc}}{R_{dc}} = b^2c^2\left(\frac{Dn^4}{180} - \frac{D^2n^8}{75{,}600} \cdots\right) \tag{2-105}$$

This is usually very much less than (2-104).

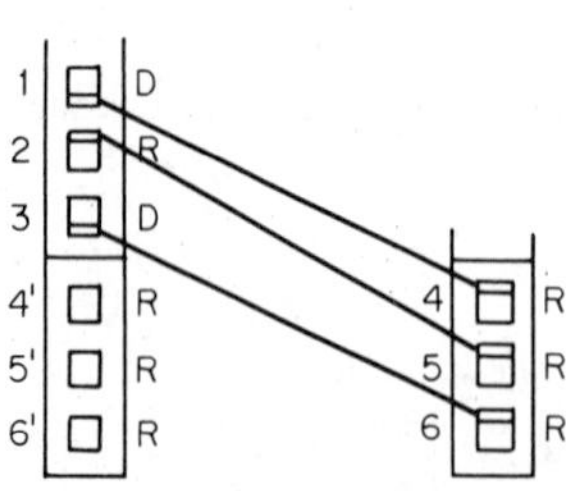


**FIG. 2-17** Armature coils with laminated conductors.

***100. Diamond Coil with Twisted Conductors, with an Odd Number of Turns per Coil (q Odd).*** To obtain the lowest loss, for three turns per coil, use *DRD, RRR,* starting at the top or open end of the slot, as in Fig. 2-17.

For five turns per coil, use *DRDRR, RDDDD,* starting at the top.

For seven turns per coil, use *DRDRDRR, RDDRDDD,* starting at the top.

For all these coils, when $\phi = 0$, use Eq. (2-105). When $\phi = 60°$, the extra loss ratio due to long-path eddy current is not given by Eq. (2-105) but may be determined by Eq. (2-101).

***101. In the embedded part of the coil,*** as stated in Par. **97**, the total extra loss ratio is the sum of the short-path extra loss ratio, given by Eqs. (2-96) to (2-98), computed as if the laminations were thoroughly transposed, and the long-path extra loss ratio, computed as in Pars. **97** to **100.** In the nonembedded part of the coil, that is, the coil end, the total ratio is equal to the long-path extra loss ratio. For an entire winding, the loss may first be computed for direct current; then the watts of extra loss may be determined for the embedded parts and for the coil ends, where $\phi = 0$ and $\phi = 60°$; and so the total watts of loss with alternating current may be found. See also Heat Losses in Stranded Armature Conductors, by W. V. Lyon, *Trans. AIEE,* 1922, p. 199.

***102. In designing a coil,*** if it is desired to reduce the short-path extra loss ratio, a larger number of thinner laminations may be specified. If it is desired to reduce the long-path extra loss ratio, the conductors may be twisted, or a thoroughly transposed type of coil may be specified (see Par. **96**).

## Electrostatics

***103. Electrostatic Force.*** Electrically charged bodies exert forces upon one another according to the following principles:

1. Like charged bodies repel; unlike charged bodies attract one another.
2. The force is proportional to the product of the magnitudes of the charges upon the bodies.
3. The force is inversely proportional to the square of the distance between charges if the material in which the charges are immersed is extensive and possesses the same uniform properties in all directions.
4. The force acts along the line joining the centers of the charges.

Two concentrated charges $Q_1$ and $Q_2$ coulombs located $R$ m apart experience a force between them of

$$F = \frac{Q_1 Q_2}{4\pi\epsilon_0 R^2} \qquad \text{newtons} \tag{2-106}$$

where $\epsilon_0 = 8.85 \times 10^{-12}$ F/m and is the permittivity of free space.

***104. Electrostatic Potential.*** The electric potential resulting from the location of charged bodies in the vicinity is called electrostatic potential. The potential at $R$ m from a concentrated charge $Q$ C is

$$\Phi = \frac{Q}{4\pi\epsilon_0 R} \qquad \text{volts} \tag{2-107}$$

where $\epsilon_0 = 8.85 \times 10^{-12}$ F/m. This potential is a scalar quantity. See also Boast, W. B., "Potential" in *The Encyclopedia of Physics,* 2d ed., pp. 729–731 for surface charge distributions.

***105. Electric Field Intensity.*** The electric field intensity is the force per unit charge that would act at a point in the field on a very small test charge placed at that location.

The electric field intensity $\mathscr{E}$ at a distance $R$ m from a concentrated charge $Q$ C is

$$\mathscr{E} = \frac{Q}{4\pi\epsilon_0 R^2} \qquad \text{N/C} \tag{2-108}$$

where $\epsilon_0$ is $8.85 \times 10^{-12}$ F/m.

***106. Electric Potential Gradient in Electrostatic Fields.*** The space rate of change of the electric potential is the electric potential gradient of the field, symbolized by $\nabla\Phi$. The general relationship between the gradient of the electric potential and the electric field intensity is

$$\mathscr{E} = -\nabla\Phi \qquad \text{V/m} \tag{2-109}$$

The units for the electric potential gradient, volts per meter, are frequently also used for the electric field intensity since their magnitudes are the same.

***107. Electric Flux Density.*** The density of electric flux (symbol $D$) in a region where simple dielectric materials exist is determined from the electric field intensity from

$$D = \epsilon\mathscr{E} = \epsilon_0 K\mathscr{E} \qquad \text{C/m}^2 \tag{2-110}$$

where $\epsilon_0 = 8.85 \times 10^{-12}$ F/m and $K$ is a dimensionless number called the *dielectric constant.* In free space $K$ is unity. For numerical values of dielectric constant of various dielectrics, see Sec. **4**.

***108. Polarization.*** The polarization is the excess of electric flux density that results in dielectric materials over that which would result at the same electric field intensity if the space were free of material substance. Thus

$$P = D - \epsilon_0\mathscr{E} \qquad \text{C/m}^2 \tag{2-111}$$

***109. Crystalline Atomic Materials.*** In simple isotropic materials the directions of the vectors $P$, $D$, and $\mathscr{E}$ are the same. For crystalline atomic structures that are not isotropic, Eq. (2-111) is the only relationship which is meaningful and Eq. (2-110) should not be used.

***110. Electric Flux.*** Electric flux and its density are related by

$$\psi = \int D \cos\alpha \, dA \qquad \text{coulombs} \tag{2-112}$$

where $\alpha$ is the angle between the direction of the electric flux density $D$ and the normal at each differential surface area $dA$.

***111. Capacitance.*** The capacitance between two oppositely charged bodies is the ratio of the magnitude of charge on either body to the difference of electric potential between them. Thus

$$C = \frac{Q}{V} \qquad \text{farads} \tag{2-113}$$

where $Q$ is in coulombs and $V$ is the voltage between the two equally but oppositely charged bodies, in volts. The units of microfarad ($\mu$F), equal to $10^{-6}$ farad, and picofarad (pF), equal to $10^{-12}$ farad, are frequently used.

***112. Elastance.*** The reciprocal of capacitance, called *elastance,* is

$$S = V/Q \qquad \text{farads} \tag{2-114}$$

***113. Electric Field Outside an Isolated Sphere in Free Space.*** The electric field intensity at a distance $r$ m from the center of an isolated charged sphere located in free space is

$$\mathscr{E} = \frac{Q}{4\pi\epsilon_0 r^2} \qquad \text{V/m} \tag{2-115}$$

where $Q$ is the total charge (which is distributed uniformly) on the sphere and $\epsilon_0 = 8.85 \times 10^{-12}$ F/m.

***114. Spherical Capacitor.*** The capacitance between two concentric charged spheres is

$$C = \frac{4\pi\epsilon_0 K}{1/R_1 - 1/R_2} \qquad \text{farads} \tag{2-116}$$

where $R_1$ is the outside radius of the inner sphere, $R_2$ is the inside radius of the outer sphere, $K$ is the dielectric constant of the space between them, and $\epsilon_0 = 8.85 \times 10^{-12}$ F/m.

***115. Electric Field Intensity Created by an Isolated, Charged, Long Cylindrical Wire in Free Space.*** The electric field intensity in the vicinity of a long, charged cylinder is

$$\mathscr{E} = \frac{\Lambda}{2\pi\epsilon_0 r} \qquad \text{V/m} \tag{2-117}$$

where $\Lambda$ is the charge per unit of length in coulombs per meter (distributed uniformly over the surface of the isolated cylinder), $r$ is the distance in meters from the center of the cylinder to the point at which the electric field intensity is evaluated, and $\epsilon_0 = 8.85 \times 10^{-12}$ F/m.

***116. Coaxial Cable.*** The capacitance per unit length of a coaxial cable composed of two concentric cylinders is

$$c = \frac{2\pi\epsilon_0 K}{\ln (R_2/R_1)} \quad \text{F/m} \tag{2-118}$$

where $R_1$ is the outside radius of the inner cylinder, $R_2$ is the inside radius of the outer cylinder, $K$ is the dielectric constant of the space between the cylinders, and $\epsilon_0 = 8.85 \times 10^{-12}$ F/m.

***117. Two-Wire Line.*** The capacitance per unit length between two long, oppositely charged, cylindrical conductors of equal radii, parallel and external to each other, is

$$c = \frac{\pi\epsilon_0 K}{\ln \left[ \dfrac{D}{2R} + \sqrt{\left(\dfrac{D}{2R}\right)^2 - 1} \right]} \quad \text{F/m} \tag{2-119}$$

where $D$ is the distance in meters between centers of the two cylindrical wires each with radius $R$, $K$ is the uniform dielectric constant of all space external to the wires, and $\epsilon_0 = 8.85 \times 10^{-12}$ F/m.

***118. Capacitance of Two Flat, Parallel Conductors Separated by a Thin Dielectric.*** The capacitance is approximately

$$C = \frac{\epsilon_0 KA}{t} \quad \text{farads} \tag{2-120}$$

where $A$ is the area of either of the two conductors, $t$ is the spacing between them, $K$ is the dielectric constant of the space between the conductors, and $\epsilon_0 = 8.85 \times 10^{-12}$ F/m. Strictly, the linear dimensions of the flat conductors should be very large compared with the spacing between them. Good results are obtained from Eq. (2-120) even though the conductors are curved provided that the spacing $t$ is small with respect to the radius of curvature.

***119. Induced Charges.*** The surface of a conducting body, near a charge $Q$, through which no currents are flowing is an equipotential surface, a condition maintained by the motion of positive and negative charges to the parts of the conductor near $Q$ and distant from it. Hence the potential at any point on the conductor, due to all the charges of the system, is a constant. The charges on the conductors are said to be induced by $Q$, and the conductor is said to be electrified by induction.

***120. Electrostatic Induction on Parallel Wires.*** Two insulated wires running parallel to a wire carrying a charge $\Lambda$ C/m display a potential difference (provided that the two wires are not connected to each other or to other conductors) of

$$\phi = \frac{\Lambda}{2\pi\epsilon_0} \ln \frac{b}{a} \quad \text{volts} \tag{2-121}$$

where $b$ and $a$ are the distances of the two insulated wires from the charged wire.

If the two wires are connected together, as, for example, through telephone instruments, the current flowing from one wire to the other is that required to equalize their potential difference.

## The Dielectric Circuit

***121. Circuit Concepts with Capacitive Elements.*** When a continuous voltage is applied to the terminals of a capacitor ($AB$, Fig. 2-18), a positive charge of electricity $+Q$ appears on one plate and a negative charge $-Q$ on the other. A quantity of electricity $Q$ flows through the connecting wires, and this quantity of electricity is said to be displaced through the dielectric. An electrostatic field, as has been described in Par. **106**, then exists between the two charged plates.

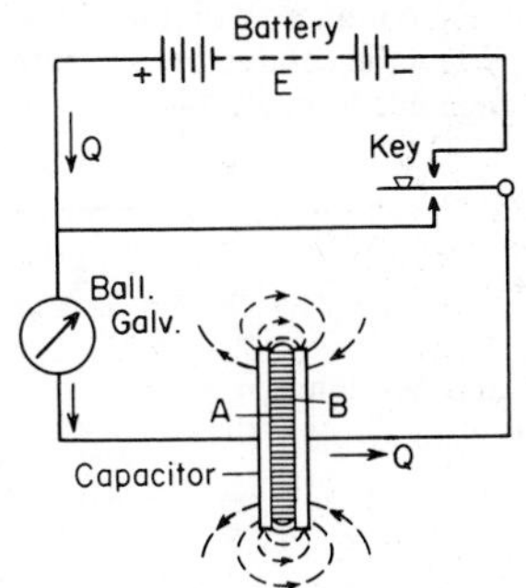


**FIG. 2-18** Circuit containing a capacitor.

When an applied voltage $e$ is changing with time the current $i$ through the capacitive circuit is the time derivative of $Q$ in Eq. (2-113),

$$i = \frac{dQ}{dt} = C\frac{de}{dt} \quad \text{amperes} \tag{2-122}$$

***122. Electrostatic Flux.*** The space between the plates of a capacitor can be treated as a dielectric circuit through which passes a dielectric flux $\psi$, in coulombs. In any dielectric circuit, one coulomb of electrostatic flux passes from each coulomb of positive charge to each coulomb of negative charge, and this is true with any insulating substance or group of substances. That is, electrostatic flux lines end only on charges of electricity. Their number is not affected when they pass from one dielectric to another, unless there is a charge of electricity on the surface of separation (see Par. **40**). Electrostatic flux lines are also called "lines of electrostatic induction."

***123. Capacitors in Series and in Parallel.*** When capacitors are connected in parallel, the equivalent capacitance is equal to the sum of all the capacitances of the component capacitors, or

$$C_{eq} = \Sigma C \tag{2-123}$$

When two or more capacitors are connected in series, the equivalent capacitance is determined from the relation

$$\frac{1}{C_{eq}} = \Sigma \frac{1}{C} \tag{2-124}$$

Analogously, for a series connection of elastances (Par. **112**),

$$S_{eq} = \Sigma S \tag{2-125}$$

and for parallel connection of elastances,

$$\frac{1}{S_{eq}} = \Sigma \frac{1}{S} \tag{2-126}$$

***124. As an example,*** let two capacitances $C_1 = 0.2\ \mu$F and $C_2 = 0.3\ \mu$F be connected in parallel with each other and in series with a third capacitor for which $C_3 = 0.4\ \mu$F. The capacitance of the combination is required. The capacitance of the two capacitors in parallel is $C_1 + C_2 = 0.5\ \mu$F, and the elastance of the combination is 2 Mdarafs. The elastance of the third capacitor $1/C_3 = 2.5$ Mdarafs, and the total elastance of the combination is $2 + 2.5 = 4.5$ Mdarafs. The equivalent capacitance is $1/4.5 = 0.222\ \mu$F.

***125. The capacitance to neutral of a conductor in an ac line*** is defined as the capacitance that, when multiplied by $2\pi f$ and by the voltage to neutral, gives the charging current of the conductor, $f$ being the frequency. This is not the same as the capacitance to a neutral wire measured electrostatically. The voltage to neutral of a single-phase line is one-half the voltage between conductors. The voltage to neutral of a balanced 3-phase line is equal to the voltage between conductors divided by 1.732.

When the conductors are round wires, for either single-phase or 3-phase overhead lines, the capacitance to neutral is

$$c = \frac{2\pi\epsilon_0 K}{\ln \sqrt{\dfrac{s}{d} + \left[\left(\dfrac{s}{d}\right)^2 - 1\right]^{1/2}}} \quad \text{F/m, to neutral} \tag{2-127}$$

or, approximately,

$$c = \frac{0.0388}{\ln (2s/d)} \quad \mu\text{F/mi, to neutral} \tag{2-128}$$

where $s$ is the axial spacing and $d$ is the diameter of the conductors, in the same units.

Values of charging kVA for transmission lines are tabulated in Sec. **14**.

The capacitance of a complete single-phase line is one-half the capacitance to neutral of one conductor. The capacitance of stranded conductors may be approximately calculated by using the outside diameter of the conductors. The capacitance of iron or steel conductors is calculated by the same formulas as that of copper conductors.

The above relations assume equilateral spacing for 3-phase systems. If unbalanced spacings are present and the phases are balanced by transposing the conductors over the length of the line, the approximate capacitance per phase can be obtained from the concepts of geometric mean distances $D_m$ and $D_s$ (Par. **84**). Then, approximately,

$$c = \frac{2\pi\epsilon_0 K}{\ln (D_m/D_s')} \quad \text{F/m} \qquad (2\text{-}129)$$

The self-geometric mean distance $D_s'$ in Eq. (2-129) differs slightly from that for $D_s$ in Eq. (2-89) in that for good conductors the transverse gradient of the electric field is confined principally to the air space about the conductors and $D_s'$ is slightly larger than $D_s$ of Eq. (2-89). In the latter equation, internal flux linkages in the conductor contribute to the meaning of $D_s$.

The following references may be consulted for the many details needed in computing the geometric mean distances of Eq. (2-129):

Anderson, P. M. "Analysis of Faulted Power Systems"; Ames, The Iowa State University Press, 1973, Chap. 5.

Boast, W. B. "Vector Fields"; New York, Harper & Row, Publishers, Inc., 1964, Chaps. 8, 10, and 11.

Stevenson, W. D., Jr. "Elements of Power System Analysis," 3d ed.; New York, McGraw-Hill Book Company, 1975, Chap. 3.

Clarke, E. "Circuit Analysis of A-C Power Systems"; New York, John Wiley & Sons, Inc., 1943, Vol. I.

"Electrical Transmission and Distribution Reference Book," 4th ed.; Westinghouse Electric Corporation, 1950.

***126. Velocity of Propagation on Long Transmission Lines.*** The inductance and capacitive parameters per unit length of a transmission line determine the velocity with which such effects as switching surges are propagated along the line. The velocity of propagation is

$$v = \frac{1}{\sqrt{lc}} \quad \text{m/s} \qquad (2\text{-}130)$$

where $l$ is the inductance per unit length from Eq. (2-89) and $c$ is the capacitance per unit length from Eq. (2-129). Substituting these values gives

$$v = 3 \times 10^8 \sqrt{\frac{\ln (D_m/D_s')}{K \ln (D_m/D_s)}} \quad \text{m/s} \qquad (2\text{-}131)$$

The fact that $D_s'$ is slightly larger than $D_s$ produces a velocity of propagation along the transmission line which is slightly less than the velocity of propagation of electromagnetic radiation in free space ($3 \times 10^8$ m/s). Since the dielectric constant $K$ of the atmosphere surrounding the transmission line may be somewhat greater than unity, the velocity of propagation may be reduced slightly more. Magnetic materials in the conductors tend to increase the inductance in the denominator of Eq. (2-130) and reduce further the velocity of propagation by a small amount.

***127. The Energy Stored in a Capacitor.*** The energy stored in a capacitor is

$$W = \frac{CV^2}{2} = \frac{V^2}{2S} = \frac{VQ}{2} \quad \text{J (W-s)} \qquad (2\text{-}132)$$

where the voltage $V$ is in volts, the charge $Q$ in coulombs, and the capacitance $C$ in farads. Equivalently if $C$ is in microfarads and $Q$ in microcoulombs, $W$ is in microjoules.

The energy stored per unit volume of the dielectric is

$$W' = \tfrac{1}{2}\epsilon \mathscr{E}^2 \quad \text{J/m}^3 \qquad (2\text{-}133)$$

where $\epsilon$ is the permittivity ($8.85 \times 10^{-12}$ F/m multiplied by the dielectric constant) and $\mathscr{E}$ is the voltage gradient (electric field intensity) in volts per meter.

***128. The dielectric strength of insulating materials (rupturing voltage gradient)*** is the maximum voltage per unit thickness that a dielectric can withstand in a uniform field before it breaks down electrically. The dielectric strength is usually measured in kilovolts per millimeter or per inch. It is necessary to define the dielectric strength in terms of a uniform field, for instance, between large parallel plates a short distance apart. If the striking voltage is determined between two spheres or electrodes of other defined shape, this fact must be stated. In designing insulation, a factor of safety is assumed depending upon conditions of operation. For numerical values of rupturing voltage gradients of various insulating materials, see Sec. **4**.

## Dielectric Loss and Corona

***129. Dielectric Hysteresis and Conductance.*** When an alternating voltage is applied to the terminals of a capacitor, the dielectric is subjected to periodic stresses and displacements. If the material were perfectly elastic, no energy would be lost during any cycle, because the energy stored during the periods of increased voltage would be given up to the circuit when the voltage is decreased. However, since the electric elasticity of dielectrics is not perfect, the applied voltage has to overcome molecular friction or viscosity, in addition to the elastic forces. The work done against friction is converted into heat and is lost. This phenomenon resembles magnetic hysteresis (Par. **66**) in some respects but differs in others. It has commonly been called "dielectric hysteresis" but is now often called "dielectric loss." The energy lost per cycle is proportional to the square of the applied voltage. Methods of measuring dielectric loss are described in Sec. **3** (see J. B. Whitehead, "Lectures on Dielectric Theory and Insulation," New York, McGraw-Hill Book Company, 1927, pp. 57–59).

An imperfect capacitor does not return on discharge the full amount of energy put into it. Some time after the discharge an additional discharge may be obtained. This phenomenon is known as *dielectric absorption.*

A capacitor that shows such a loss of power can be replaced for purposes of calculation by a perfect capacitor with an ohmic conductance shunted around it. This conductance (or "leakance") is of such value that its $I^2R$ loss is equal to the loss of power from all causes in the imperfect capacitor. The actual current through the capacitor is then considered as consisting of two components—the leading reactive component through the ideal capacitor and the loss component, in phase with the voltage, through the shunted conductance.

***130. Electrostatic Corona.*** When the electrostatic flux density in the air exceeds a certain value, a discharge of pale violet color appears near the adjacent metal surfaces. This discharge is called *electrostatic corona.* In the regions where the corona appears, the air is electrically ionized and is a conductor of electricity. When the voltage is raised further, a brush discharge takes place, until the whole thickness of the dielectric is broken down and a disruptive discharge, or spark, jumps from one electrode to the other.

Corona involves power loss, which may be serious in some cases, as on transmission lines (Secs. **14** and **15**). Corona can form at sharp corners of high-voltage switches, busbars, etc.; so the radii of such parts are made large enough to prevent this. A voltage of 12 to 25 kV between conductors separated by a fraction of an inch, as between the winding and core of a generator or between sections of the winding of an air-blast transformer, can produce a voltage gradient sufficient to cause corona. A voltage of 100 to 200 kV may be required to produce corona on transmission-line conductors that are separated by several feet. Corona can have an injurious effect on fibrous insulation. For numerical data in application to transmission lines see Secs. **14** and **15**.

## Transient Currents and Voltages

***131. Transient electric phenomena*** occur, for instance, when a load is suddenly changed and an appreciable time elapses before the generators and the line adapt themselves to the new conditions. The currents and the voltages during the intermediate time are known as transient.

***132. Closing a Circuit Containing a Resistance R (Ohms) and an Inductance L (Henrys) in Series with a Continuous EMF.*** When a deenergized circuit is suddenly connected to a source of continuous voltage $E$, the current gradually rises to the final value $i_0 = E/R$ according to the law

$$i = i_0(1 - \epsilon^{-tR/L}) \tag{2-134}$$

where $t$ is time in seconds and $\epsilon$ is the base of natural logarithms. This expression is known as *Helmholtz's law.*

When the source of emf is short-circuited, the current in the remaining circuit decreases to zero according to a similar law

$$i = i_0\epsilon^{-tR/L} \tag{2-135}$$

***133. Alternating EMF,* RL *Circuit.*** When a deenergized circuit containing $R$ and $L$ is suddenly connected at time $t = 0$ to a source of alternating voltage

$$e = E_m \sin(2\pi ft + \alpha)$$

the current in the circuit varies according to the law

$$i = \frac{E_m}{Z}\sin(2\pi ft + \alpha - \phi) - \frac{E_m}{Z}\sin(\alpha - \phi)\epsilon^{-tR/L} \tag{2-136}$$

In this equation, $Z = \sqrt{R^2 + (2\pi fL)^2}$ is the impedance of the circuit and $\phi$ is the phase displacement between the current and the voltage (determined by $\tan\phi = 2\pi fL/R$). The angle $\alpha$ is the phase displacement between the voltage $e$ and the reference wave which passes through zero at the time $t = 0$, and $f$ is the frequency in hertz (cycles per second). The first term in Eq. (2-136) is the current corresponding to the steady-state condition. The second term is the transient, which approaches zero with the time.

***134. Closing a Circuit Containing a Resistance R (Ohms) and a Capacitance C (Farads) in Series.*** When the capacitor is initially uncharged, the current produced by a constant applied emf $E$ is

$$i = i_0\epsilon^{-t/RC} \tag{2-137}$$

where $i_0 = E/R$ is the initial current. In practice some inductance is always present in the circuit, and this smooths down the initial change in current.

The charge on the capacitor, initially uncharged, is

$$q = EC(1 - \epsilon^{-t/RC}) \tag{2-138}$$

***135. Discharging Capacitor.*** When a capacitor, initially charged to a voltage $E_0$, is discharged through resistance $R$, the discharge current at the first instant is $i_0 = E_0/R$. Thereafter, the current varies according to

$$i = i_0\epsilon^{-t/RC} \tag{2-139}$$

The charge on the capacitor decreases according to

$$q = Q_0\epsilon^{-t/RC} \tag{2-140}$$

where $Q_0$ is the initial charge.

***136. Alternating EMF,* RC *Circuit.*** When a deenergized circuit containing $R\Omega$ and $CF$ is suddenly connected at $t = 0$ to a source of alternating voltage $e = E_m \sin(2\pi ft + \alpha)$, the current in the circuit is

$$i = \frac{E_m}{Z}\sin(2\pi ft + \alpha + \phi) - \frac{E_m\cos(\alpha + \phi)}{2\pi fRCZ}\epsilon^{-t/RC} \tag{2-141}$$

In this equation $Z = \sqrt{R^2 + (1/2\pi fC)^2}$ is the impedance of the circuit, and $\phi$ is the phase displacement between the current and the voltage, determined by

$$\cot\phi = 2\pi fCR$$

The angle $\alpha$ is the phase displacement between the voltage $E$ and the reference wave which passes through zero at the time $t = 0$; $f$ is the frequency in hertz. The first term in Eq. (2-141) is the current corresponding to the steady-state condition; the second term is a transient, which approaches zero with the time (compare Par. **133**).

For the case described for Eq. (2-141),

$$q = \frac{E_m}{2\pi f Z} \cos(2\pi f t + \alpha + \phi) + \frac{E_m}{2\pi f Z} \cos(\alpha + \phi)\epsilon^{-t/RC} \qquad (2\text{-}142)$$

***137. Single-Energy and Double-Energy Transients.*** The two preceding cases are examples of single-energy transients; that is, the energy is stored in one form only (electromagnetic or electrostatic), and the energy change consists in an increase or a decrease of the stored energy.

When both inductance and capacitance are present, the energy of the circuit is stored in two forms, and there is a periodic transfer of energy from magnetic to dielectric form, and vice versa, producing electric oscillations.

A *triple-energy transient* occurs, for instance, when a synchronous motor is hunting at the end of a long transmission line which possesses inductance and capacitance. In the latter case, the energy of the system is stored in magnetic, dielectric, and mechanical forms.

***138.* RLC *Circuit.*** For a series circuit of resistance $R$ (ohms), inductance $L$ (henrys), capacitance $C$ (farads), and constant emf $E$ (volts), there are three types of current $i$ (amperes).

Nonoscillatory case (overdamped), when $R^2 > 4L/C$.

Oscillatory case (underdamped), when $R^2 < 4L/C$.

Critical case, when $R^2 = 4L/C$.

***139. Nonoscillatory Cases.*** When the initial charge on the capacitor and the current are zero,

$$i = \frac{E}{2Lb}\left(\epsilon^{(-a+b)t} - \epsilon^{(-a-b)t}\right) \quad \text{amperes} \qquad (2\text{-}143)$$

where

$$a = \frac{R}{2L} \quad \text{and} \quad b = \sqrt{\frac{R^2}{4L^2} - \frac{1}{LC}}$$

Note that $-a + b$ is negative. The charge on the capacitor is

$$q = EC\left(1 - \frac{a+b}{2b}\epsilon^{(-a+b)t} + \frac{a-b}{2b}\epsilon^{(-a-b)t}\right) \quad \text{coulombs} \qquad (2\text{-}144)$$

When the initial charge on the capacitor is $Q_0$ and the initial current is zero,

$$i = \frac{Q_0}{2LCb}\left(\epsilon^{(-a+b)t} - \epsilon^{(-a-b)t}\right) \quad \text{amperes} \qquad (2\text{-}145)$$

$$q = Q_0\left(\frac{a+b}{2b}\epsilon^{(-a+b)t} - \frac{a-b}{2b}\epsilon^{(-a-b)t}\right) \quad \text{coulombs} \qquad (2\text{-}146)$$

***140. Oscillatory Case.*** $R^2 < L/C$. When the initial charge and current are zero,

$$i = \frac{E}{Lg}\epsilon^{-at} \sin gt \quad \text{amperes} \qquad (2\text{-}147)$$

where

$$a = \frac{R}{2L}$$

$$g = \sqrt{\frac{1}{LC} - \frac{R^2}{4L^2}}$$

and $E$ is the applied direct voltage.

$$q = EC\left[1 - \epsilon^{-at}\left(\cos gt + \frac{a}{g}\sin gt\right)\right] \qquad \text{coulombs} \tag{2-148}$$

When the initial charge is $Q_0$ and the initial current is zero,

$$i = -\frac{Q_0}{LCg}\epsilon^{-at}\sin gt \qquad \text{amperes} \tag{2-149}$$

$$q = Q_0\epsilon^{-at}\left(\cos gt + \frac{a}{g}\sin gt\right) \qquad \text{coulombs} \tag{2-150}$$

The frequency of the oscillations in the oscillatory case of the $RLC$ series circuit is

$$f = \frac{g}{2\pi} = \frac{1}{2\pi}\sqrt{\frac{1}{LC} - \frac{R^2}{4L^2}} \qquad \text{Hz} \tag{2-151}$$

When the resistance is negligible, that is, when $R^2$ is very small compared with $4L/C$, the frequency of oscillations is very close to

$$f_0 = \frac{1}{2\pi\sqrt{LC}} \qquad \text{Hz} \tag{2-152}$$

***141. Critical Case.*** $R^2 = 4L/C$. When the initial charge and current are zero,

$$i = \frac{Et}{L}\epsilon^{-tR/2L} \qquad \text{amperes} \tag{2-153}$$

$$q = EC - EC\left(1 + \frac{tR}{2L}\right)\epsilon^{-tR/2L} \qquad \text{coulombs} \tag{2-154}$$

where $E$ is the applied direct voltage.

When the initial charge is $Q_0$ and the initial current is zero,

$$i = -\frac{Q_0 t}{LC}\epsilon^{-tR/2L} \qquad \text{amperes} \tag{2-155}$$

$$q = Q_0\left(1 + \frac{tR}{2L}\right)\epsilon^{-tR/2L} \qquad \text{coulombs} \tag{2-156}$$

***142. Stored Energy.*** When energy is suddenly changed at some point on a transmission line, for instance, because of an indirect lightning stroke, a wave travels along the line, carrying the energy change to the ends of the line. Part of the wave enters the apparatus at the ends, part is reflected, and the rest is converted into heat.

The total energy stored in a transmission at any instant is

$$W = \tfrac{1}{2}Li^2 + \tfrac{1}{2}Ce^2 \qquad \text{joules} \tag{2-157}$$

where $L$ is the inductance of the line in henrys, $i$ instantaneous current, $C$ the capacitance of the line in farads, and $e$ the corresponding instantaneous voltage. The term $\tfrac{1}{2}Li^2$ represents the electromagnetic energy; the term $\tfrac{1}{2}Ce^2$, the electrostatic energy. At certain instants, the current is equal to zero; at others, the voltage is zero. Since energy remains constant (no losses are assumed), maximum values of the two energies must be equal and

$$\frac{e_{max}}{i_{max}} = \sqrt{\frac{L}{C}} \qquad \text{ohms} \tag{2-158}$$

Thus the maximum instantaneous current $i_{max}$ can be calculated from the maximum voltage $e_{max}$, and vice versa.

In the case of a lightning stroke, for example, the maximum voltage is limited by the disruptive strength of the insulation, and the maximum current disturbance may be calculated from the preceding equation.

The quantity $\sqrt{L/C}$ is the *surge impedance* of the line; its reciprocal is the *surge admittance.*

## Alternating-Current Circuits

***143. Phasor Quantities.*** If $e$ and $e'$ (Fig. 2-19) are the components of a phasor voltage $E$ along two perpendicular axes, $E$ may be represented symbolically as

$$\dot{E} = e + je' \tag{2-159}$$

where

$$j = \sqrt{-1} \tag{2-160}$$

The dot over $E$ signifies that this quantity has a direction as well as a magnitude. (Note that here the dot does not represent the first derivative. The dot over the letter is very frequently omitted, as it is usually obvious when a symbol denotes a complex quantity.)

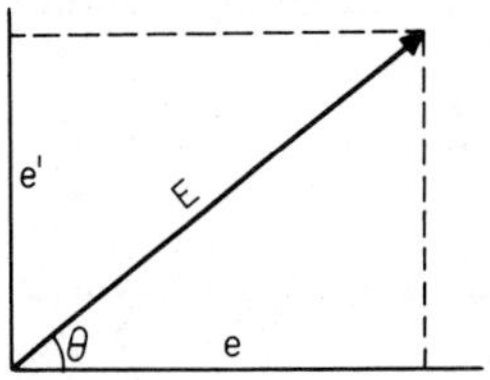

FIG. 2-19 Phasor quantities; axes of reals and imaginaries.

The quantity $E$ in Eq. (2-159) may denote an alternating voltage of sine-wave form of frequency $f$ Hz. The effective or root-mean-square (rms) value of the voltage is

$$E_{\text{eff}} = \frac{E_{\max}}{\sqrt{2}} = 0.707E_{\max} \tag{2-161}$$

The instantaneous value of the voltage is $E_{\max} \sin 2\pi ft$, where $t$ is the time in seconds and the angle $2\pi ft$ is in radians.

The effective value of any alternating voltage or current (not necessarily a sine wave) is defined in Par. **166.**

***144. Addition and Subtraction of Phasors.*** When two phasor voltages are represented as

$$\dot{E}_1 = e_1 + je'_1$$

$$\dot{E}_2 = e_2 + je'_2$$

the sum or the difference of these two phasors is

$$\dot{E}_3 = \dot{E}_1 \pm \dot{E}_2 = (e_1 \pm e_2) + j(e'_1 \pm e'_2) \tag{2-162}$$

***145. Rotation of a Phasor.*** Multiplying a phasor by $j$ rotates it by 90° in the positive direction (counterclockwise). Thus,

$$j\dot{E} = j(e + je') = -e' + je \tag{2-163}$$

Note that $j^2 = -1$. Conversely, multiplying a phasor by $-j$ rotates the phasor by 90° in the negative direction, that is, clockwise.

The phasor $\dot{E}$ may be also represented symbolically (Fig. 2-19) in the polar form as

$$\dot{E} = E\underline{/\theta} = E(\cos\theta + j\sin\theta) \tag{2-164}$$

where $E$ without the dot stands for the magnitude only.

The operator $\underline{/\phi}$ is

$$\underline{/\phi} = \epsilon^{j\phi} = \cos\phi + j\sin\phi \tag{2-165}$$

where $\epsilon$ is the base of the natural (naperian) logarithms. It turns a phasor by the angle $\phi$ in the positive direction. Thus,

$$\begin{aligned}\dot{E}(\cos\phi + j\sin\phi) &= E(\cos\theta + j\sin\theta)(\cos\phi + j\sin\phi)\\ &= E[\cos(\theta+\phi) + j\sin(\theta+\phi)]\end{aligned} \tag{2-166}$$

The operator $\epsilon^{-j\phi} = \cos\phi - j\sin\phi$ turns a phasor by an angle $\phi$ in the negative direction, that is, clockwise.

The angle $\phi$ between two alternating quantities (phasors) represents a time-phase difference, such that the period of one complete cycle ($1/f$) is equal to $2\pi$ rad. The phase displacement between alternating quantities is commonly measured in electrical degrees. One electrical degree (1/360th part of a complete cycle) is $2\pi/360$ electrical rad.

The horizontal line at the bottom of Fig. 2-19 is called the reference phase. The phase angle $\theta$ of $E$ is measured from it.

***146. The absolute, or numerical, value of a phasor*** expressed in the rectangular form $\dot{E} = e + je'$ is equal to

$$|\dot{E}| = \sqrt{e^2 + e'^2} \tag{2-167}$$

The dot and the vertical lines are often omitted.

The rectangular form $e + je'$ may be changed to the polar form $E\underline{/\theta}$ by means of the equations

$$E = |\dot{E}| = \sqrt{e^2 + e'^2}$$

$$\cos\theta = \frac{e}{\sqrt{e^2 + e'^2}} \tag{2-168}$$

$$\sin\theta = \frac{e'}{\sqrt{e^2 + e'^2}}$$

The expression $\theta = \tan^{-1}(e'/e)$ indicates two possible angular positions 180° apart, only one of which is the correct angle for a given phasor. For instance, if $e'$ and $e$ are both negative, $\theta$ is an angle greater than 180° and the appropriate value of $\tan^{-1}(e'/e)$ must be selected. (See "Tables of Integrals and Other Mathematical Data," by H. B. Dwight, No. 401.2.)

The numerical values of two or more phasors *cannot* be added unless the phasors are in phase. Otherwise the phasors must be added according to Eq. (2-162). Similarly, impedances and admittances must be added as in Eqs. (2-175) and (2-181).

***147. Square Root of a Complex Quantity.*** If a complex quantity is expressed in the polar form $r\underline{/\theta}$, its two square roots are $\sqrt{r}\underline{/\pm\theta/2}$. If the quantity is in the rectangular form $a + jb$ or $a - jb$ (where $b$ is positive),

$$\sqrt{a+jb} = \pm\left(\sqrt{\frac{r+a}{2}} + j\sqrt{\frac{r-a}{2}}\right)$$
$$\sqrt{a-jb} = \pm\left(\sqrt{\frac{r+a}{2}} - j\sqrt{\frac{r-a}{2}}\right) \tag{2-169}$$

where $r = \sqrt{a^2 + b^2}$. The positive square roots of $(r+a)/2$ and $(r-a)/2$ are used.

If desired, Eqs. (2-168) may be used by changing the rectangular form to the polar form. In spite of the simplicity of the polar expression, it is sometimes less work to use Eqs. (2-169) than to change to the polar form and back.

***148. Impedance.*** An impedance consisting of a resistance $r$ in series with a reactance $x$ is represented by the impedance operator

$$Z = r + jx \qquad \text{ohms} \tag{2-170}$$

This is a complex quantity but not a rotating phasor.

The letter $x$ has a positive numerical value, equal to $2\pi fL$, for inductive reactance, where $f$ is the frequency in hertz and $L$ is in henrys. It has a negative value $-1/2\pi fC$ for capacitive reactance, where $C$ is in farads.

When inductive and capacitive reactances are in series,

$$x = 2\pi fL - 1/2\pi fC \qquad \text{ohms} \tag{2-171}$$

A sine-wave current in an impedance $Z$ is represented by a current phasor $I$. The voltage drop across the impedance is

$$E = IZ = Ir + jIx \qquad \text{volts} \tag{2-172}$$

$E$ is a phasor such that its horizontal component (Fig. 2-19) is equal to $Ir$ and its vertical component is $Ix$.

The angle $\theta$ between the voltage $E$ and the current $I$ is determined by

$$\tan\theta = x/r \tag{2-173}$$

When $x$ is positive, as for inductive reactance, the current comes to a maximum after the voltage and so lags behind the voltage. This is in agreement with the positions of the phasors for voltage and current. When $x$ is negative, as for capacitive reactance, the current leads the voltage.

When inductive and capacitive reactances are in series, the current is lagging or leading with respect to the voltage according to whether $x$, given by Eq. (2-171), is positive or negative, that is, according to whether the effect of the inductance or capacitance predominates, respectively.

When a current $I = p + jq$ exists in an impedance $Z = r + jx$, the voltage across the impedance is

$$\begin{aligned} E &= IZ = (p + jq)(r + jx) \\ &= (pr - qx) + j(px + qr) \qquad \text{volts} \end{aligned} \tag{2-174}$$

When impedances are connected in series, the complex quantities are added as

$$r + jx = r_1 + jx_1 + r_2 + jx_2 \tag{2-175}$$

If it is not necessary to keep account of phase angles, the numerical value of impedance drop is

$$|E| = |IZ| = |I|\sqrt{r^2 + x^2} \qquad \text{volts} \tag{2-176}$$

Similarly, the numerical value of current is

$$|I| = \frac{|E|}{|Z|} = \frac{|E|}{\sqrt{r^2 + x^2}} \qquad \text{amperes} \tag{2-177}$$

***149. Admittance.*** An admittance consisting of a conductance $g$ in parallel with a susceptance $b$ is represented by the admittance operator

$$Y = g - jb \qquad \text{siemens} \tag{2-178}$$

The quantity $b$ has a positive numerical value $1/2\pi fL$ for inductance, where $L$ is in henrys. It has a negative value $-2\pi fC$ for capacitance, where $C$ is in farads.

When inductive and capacitive susceptances are in parallel,

$$b = 1/2\pi fL - 2\pi fC \qquad \text{siemens} \tag{2-179}$$

If the voltage $e$ is represented by a horizontal phasor $E$, the total current $I$ is obtained by multiplying the expressions for voltage and admittance and is

$$I = EY = Eg - jEb \qquad \text{amperes} \tag{2-180}$$

When $b$ is positive, as for inductance, the current lags behind the voltage; and when $b$ is negative, as for capacitance, the current is leading. When inductive and capacitive susceptances are in parallel, the total current lags or leads the voltage according to whether $b$, given by Eq. (2-179), is positive or negative, that is, according to whether the effect of the inductance or the capacitance predominates, respectively.

When branches of a network are connected in parallel, the complex quantities representing the admittances are added, as

$$g - jb = g_1 - jb_1 + g_2 - jb_2 \tag{2-181}$$

If it is not desired to keep account of phase angles, the numerical value of current is

$$|I| = |EY| = |E|\sqrt{g^2 + b^2} \qquad \text{amperes} \tag{2-182}$$

Similarly the numerical value of voltage is

$$|E| = \frac{|I|}{|Y|} = \frac{|I|}{\sqrt{g^2 + b^2}} \qquad \text{volts} \tag{2-183}$$

***150. The admittance of a circuit or network*** whose impedance is known can be obtained directly from the equation

$$Y = 1/Z \tag{2-184}$$

where $Y$ and $Z$ are complex quantities. From this,

$$Y = g - jb = \frac{1}{r + jx} = \frac{r}{r^2 + x^2} - \frac{jx}{r^2 + x^2} \tag{2-185}$$

If the circuit consists of a resistance $r$ in series with a reactance $x$, it is possible to complete, by Eq. (2-185), the equivalent conductance $g$ and susceptance $b$ of the circuit. These are equivalent to the resistance and reactance, connected in parallel, which together take the same current and power as the actual series circuit, at the same voltage.

In a series-parallel circuit if $r$ and $x$ are the resistance and reactance of the equivalent series circuit, $g$ and $b$ of Eq. (2-185) give the conductance and susceptance of the equivalent parallel circuit.

***151. The impedance of a circuit or a network*** whose admittance is known can be obtained from the equation

$$Z = 1/Y \tag{2-186}$$

where $Y$ and $Z$ are complex quantities. Then

$$Z = r + jx = \frac{1}{g - jb} = \frac{g}{g^2 + b^2} + \frac{jb}{g^2 + b^2} \tag{2-187}$$

The values of $r$ and $x$ are the resistance and reactance in a simple series circuit which is equivalent to the circuit of admittance $g - jb$.

***152. Impedance of Circuit Combinations.*** See Tables 2-1 and 2-2.

***153. Power, Reactive Voltamperes, and Power Factor.*** When the alternating current in a circuit is $I\underline{/\alpha}$ and the voltage is $E\underline{/\beta}$, the power is equal to the product of the effective values of $E$ and $I$ multiplied by the cosine of the phase angle between them and is

$$P = E_{\text{eff}} I_{\text{eff}} \cos(\beta - \alpha) \qquad \text{watts} \tag{2-188}$$

This is sometimes called the *average power.*

The reactive voltamperes, or vars, are

$$Q = E_{\text{eff}} I_{\text{eff}} \sin(\beta - a) \qquad \text{vars} \tag{2-189}$$

When $\alpha$ is less than $\beta$, the current is lagging and expression (2-189) is positive. If, as is customary in the case of power systems, the voltage is taken as a reference, the angle $\beta = 0$ and the relation of the current and voltage is shown in the phasor diagram (Fig. 2-20) for lagging current. The term *power factor* is used for the factor $\cos(\beta - \alpha)$ in Eq. (2-188).

Power and reactive voltamperes can be shown in a complex plane diagram, as in Fig. 2-21. Voltamperes are plotted as

$$E\hat{I}(\text{VA}) = P(W) + jQ \qquad \text{vars} \tag{2-190}$$

where it is understood that $E$ and $I$ are effective values and the vars are positive when current lags the voltage. $\hat{I}$ is the conjugate of the vector $I\underline{/\alpha}$, that is, it has the same magnitude but its angle is $-\alpha$. When a complex quantity is expressed in the rectangular form, the conjugate has the same form except that $j$ is changed to $-j$, or vice versa.

**TABLE 2-1** Impedance of Series-Connected Circuit Elements*

| No. | Circuit | Impedance $Z = R + jX$ (ohms) | Magnitude of impedance $\lvert Z\rvert = [R^2 + X^2]^{1/2}$ (ohms) | Phase angle $\theta = \tan^{-1}\frac{X}{R}$ (radians) | Admittance $Y = 1/Z$ (siemens) |
|---|---|---|---|---|---|
| 1 | R | $R$ | $R$ | $0$ | $\frac{1}{R}$ |
| 2 | L | $j\omega L$ | $\omega L$ | $+\frac{\pi}{2}$ | $-j\frac{1}{\omega L}$ |
| 3 | C | $-j\frac{1}{\omega C}$ | $\frac{1}{\omega C}$ | $-\frac{\pi}{2}$ | $j\omega C$ |
| 4 | $R_1$ $R_2$ | $R_1 + R_2$ | $R_1 + R_2$ | $0$ | $\frac{1}{R_1 + R_2}$ |
| 5 | M, $L_1$ $L_2$ | $j\omega(L_1 + L_2 \pm 2M)$ | $\omega(L_1 + L_2 \pm 2M)$ | $+\frac{\pi}{2}$ | $-j\frac{1}{\omega(L_1 + L_2 \pm 2M)}$ |
| 6 | $C_1$ $C_2$ | $-j\frac{1}{\omega}\left(\frac{C_1 + C_2}{C_1 C_2}\right)$ | $\frac{1}{\omega}\left(\frac{C_1 + C_2}{C_1 C_2}\right)$ | $-\frac{\pi}{2}$ | $j\omega\left(\frac{C_1 C_2}{C_1 + C_2}\right)$ |
| 7 | R L | $R + j\omega L$ | $\left[R^2 + \omega^2 L^2\right]^{1/2}$ | $\tan^{-1}\frac{\omega L}{R}$ | $\frac{R - j\omega L}{R^2 + \omega^2 L^2}$ |
| 8 | R C | $R - j\frac{1}{\omega C}$ | $\left[\frac{\omega^2 C^2 R^2 + 1}{\omega^2 C^2}\right]^{1/2}$ | $\tan^{-1} - \frac{1}{\omega RC}$ | $\frac{\omega^2 C^2 R + j\omega C}{\omega^2 C^2 R^2 + 1}$ |
| 9 | L C | $j\left(\omega L - \frac{1}{\omega C}\right)$ | $\left(\omega L - \frac{1}{\omega C}\right)$ | $\pm\frac{\pi}{2}$ | $-j\frac{\omega C}{\omega^2 LC - 1}$ |
| 10 | R L C | $R + j\left(\omega L - \frac{1}{\omega C}\right)$ | $\left[R^2 + \left(\omega L - \frac{1}{\omega C}\right)^2\right]^{1/2}$ | $\tan^{-1}\frac{\left(\omega L - \frac{1}{\omega C}\right)}{R}$ | $\frac{R - j\left(\omega L - \frac{1}{\omega C}\right)}{R^2 + \left(\omega L - \frac{1}{\omega C}\right)^2}$ |

* From B. Dudley, *Electronics,* December 1942.

The voltampere diagram (Fig. 2-21) has the same shape as the current diagram (Fig. 2-20) except that its vertical components are reversed.

The above relations can be expressed in rectangular form (all quantities are effective values),

$$E = A + jB \tag{2-191}$$

$$I = C + jD \tag{2-192}$$

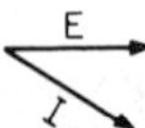


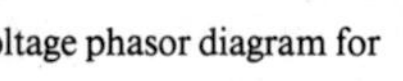

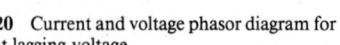

**FIG. 2-20** Current and voltage phasor diagram for a current lagging voltage.

**FIG. 2-21** Complex plane diagram for volt-amperes for current lagging voltage.

**TABLE 2-2** Impedance of Parallel-Connected Circuit Elements*

| | CIRCUIT | IMPEDANCE $Z = R + jX$ (ohms) | MAGNITUDE OF IMPEDANCE $\lvert Z\rvert = \left[R^2 + X^2\right]^{1/2}$ (ohms) | PHASE ANGLE $\theta = \tan^{-1}\frac{X}{R}$ (radians) | ADMITTANCE $Y = 1/Z$ (mhos) |
|---|---|---|---|---|---|
| 1 | $R_1$, $R_2$ | $\frac{R_1R_2}{R_1+R_2}$ | $\frac{R_1R_2}{R_1+R_2}$ | 0 | $\frac{R_1+R_2}{R_1R_2}$ |
| 2 | $L_1$, $L_2$, $M$ | $+j\omega\left[\frac{L_1L_2-M^2}{L_1+L_2\mp 2M}\right]$ | $\omega\left[\frac{L_1L_2-M^2}{L_1+L_2\mp 2M}\right]$ | $+\frac{\pi}{2}$ | $-j\frac{1}{\omega}\left[\frac{L_1+L_1\mp 2M}{L_1L_2-M^2}\right]$ |
| 3 | $C_1$, $C_2$ | $-j\frac{1}{\omega(C_1+C_2)}$ | $\frac{1}{\omega(C_1+C_2)}$ | $-\frac{\pi}{2}$ | $+j\omega(C_1+C_2)$ |
| 4 | $R$, $L$ | $\frac{\omega^2L^2R+j\omega LR^2}{\omega^2L^2+R^2}$ | $\frac{\omega LR}{\left[\omega^2L^2+R^2\right]^{1/2}}$ | $\tan^{-1}\frac{R}{\omega L}$ | $\frac{\omega L-jR}{\omega LR}$ |
| 5 | $R$, $C$ | $\frac{R-j\omega R^2C}{1+\omega^2R^2C^2}$ | $\frac{R}{\left[1+\omega^2R^2C^2\right]^{1/2}}$ | $\tan^{-1}-\omega RC$ | $\frac{1}{R}+j\omega C$ |
| 6 | $L$, $C$ | $j\frac{\omega L}{1-\omega^2LC}$ | $\frac{\omega L}{1-\omega^2LC}$ | $\pm\frac{\pi}{2}$ | $-j\left(\frac{1-\omega^2LC}{\omega L}\right)$ |
| 7 | $R$, $L$, $C$ | $\frac{\frac{1}{R}-j(\omega C-\frac{1}{\omega L})}{\left[(\frac{1}{R})^2+(\omega C-\frac{1}{\omega L})^2\right]}$ | $\frac{1}{\left[(\frac{1}{R})^2+(\omega C-\frac{1}{\omega L})^2\right]^{1/2}}$ | $\tan^{-1}R(\frac{1}{\omega L}-\omega C)$ | $\frac{1}{R}+j(\omega C-\frac{1}{\omega L})$ |
| 8 | $R$, $L$, $C$ | $\frac{\frac{R}{\omega^2C^2}-j\left[\frac{R^2}{\omega C}+\frac{L}{C}(\omega L-\frac{1}{\omega C})\right]}{R^2+(\omega L-\frac{1}{\omega C})^2}$ | $\frac{\left\{(\frac{R}{\omega^2C^2})^2+\left[\frac{R^2}{\omega C}+\frac{L}{C}(\omega L-\frac{1}{\omega C})\right]^2\right\}^{\frac{1}{2}}}{R^2+(\omega L-\frac{1}{\omega C})^2}$ | $\tan^{-1}-\frac{\left[\frac{R^2}{\omega C}+\frac{L}{C}(\omega L-\frac{1}{\omega C})\right]}{(\frac{R}{\omega^2C^2})}$ | $\frac{R+j\omega\left[R^2C-L+\omega^2L^2C\right]}{R^2+\omega^2L^2}$ |
| 9 | $R_1$, $L$, $R_2$, $C$ | $\frac{R_1R_2(R_1+R_2)+\omega^2L^2R_2+\frac{R_1}{\omega^2C^2}}{(R_1+R_2)^2+(\omega L-\frac{1}{\omega C})^2}+j\frac{\omega R_2^2L-\frac{R_1^2}{\omega C}-\frac{L}{C}(\omega L-\frac{1}{\omega C})}{(R_1+R_2)^2+(\omega L-\frac{1}{\omega C})^2}$ | $\left\{\left[\frac{R_1R_2(R_1+R_2)+\omega^2L^2R_2+\frac{R_1}{\omega^2C^2}}{(R_1+R_2)^2+(\omega L-\frac{1}{\omega C})^2}\right]^2+\left[\frac{\omega LR_2^2-\frac{R_1^2}{\omega C}-\frac{L}{C}(\omega L-\frac{1}{\omega C})}{(R_1+R_2)^2+(\omega L-\frac{1}{\omega C})^2}\right]^2\right\}^{\frac{1}{2}}$ | $\tan^{-1}\frac{\left[\omega LR^2-\frac{R_1^2}{\omega C}-\frac{L}{C}(\omega L-\frac{1}{\omega C})\right]}{\left[R_1R_2(R_1+R_2)+\omega^2L^2R_2+\frac{R_1}{\omega^2C^2}\right]}$ | $\frac{R_1+\omega^2R_1R_2C^2(R_1+R_2)+\omega^4L^2C^2R_2}{(R_1^2+\omega^2L^2)(\omega^2R_2^2C^2+1)}+j\frac{\omega\left[R_1^2C-L+\omega^2LC(L-R_2^2C)\right]}{(R_1^2+\omega^2L^2)(\omega^2R_2^2C^2+1)}$ |

* From B. Dudley, *Electronics,* December 1942.

Then the power is

$$P = \text{Re}\,(E\hat{I}) = \text{Re}\,[(A + jB)(C - jD)] = AC + BD \qquad \text{watts} \qquad (2\text{-}193)$$

where Re denotes "real part of." The reactive voltamperes, or vars, can be expressed as

$$Q = \text{Im}\,(E\hat{I}) = \text{Im}\,[(A + jB)(C - jD)] = BC - AD \qquad \text{vars} \qquad (2\text{-}194)$$

where Im denotes "imaginary part of." $Q$ is sometimes called reactive power.

***154. Leading current through an inductive line*** can raise the voltage at the receiving end of the circuit. Referring to Fig. 2-22, let $E$ be the voltage at the generator end of a circuit; $e$, the voltage at the receiver end; and $i$, the line current. Let the load be of such a nature that the current is leading with respect to the voltage $e$. By adding to $e$ the ohmic drop $ir$ in the line (Fig. 2-23) in phase with $i$ and the reactive drop $ix$ in leading quadrature with $i$, the impressed voltage $E$ is obtained. It will be seen that $E$ may be less than $e$ with a leading current. With a lagging current, $E$ is greater than $e$ (Fig. 2-24).

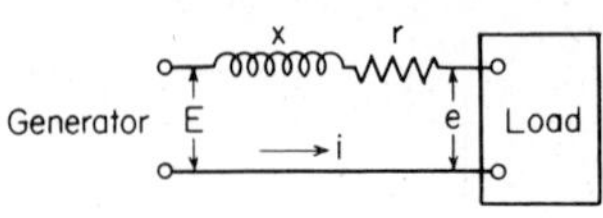


FIG. 2-22 Load connected to inductive line.

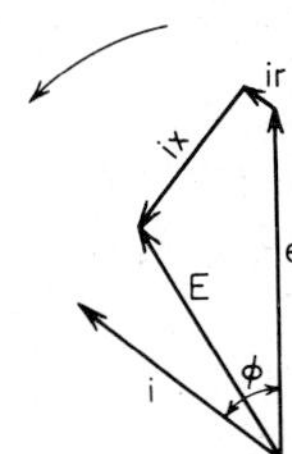


FIG. 2-23 Effect of inductive line with leading current.

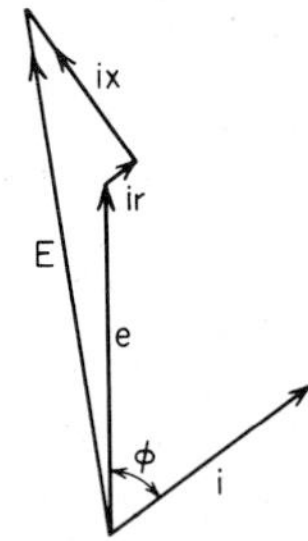


FIG. 2-24 Effect of inductive line with lagging current.

Leading current, as described in this paragraph, is usually obtained in practice from static capacitors or overexcited synchronous machines connected in parallel with the load. Leading current from synchronous machines can be controlled by automatic voltage regulators so as to give constant voltage (or rising voltage) from no load to full load.

A static capacitor may be connected in series with the line. This has the effect of reducing the line reactance or even making it negative, according to Eq. (2-171). It does not reduce the line current as parallel capacitance does. Lagging current through a capacitance will tend to raise the voltage. Series capacitance can improve the division of current between unlike lines in parallel.

***155. Series Resonance.*** In a circuit which contains inductive reactance and capacitive reactance in series, it is possible to obtain a very great rise in voltage across the reactances by adjusting them or the frequency. Thus, according to Par. **148,**

$$E = IZ = I\sqrt{r^2 + (x_l - x_c)^2} \qquad \text{volts} \qquad (2\text{-}195)$$

The voltage across the capacitive reactance is $e = Ix_c$, so that

$$\frac{e}{E} = \frac{x_x}{\sqrt{r^2 + (x_l - x_c)^2}} = \frac{x_c}{Z} \qquad (2\text{-}196)$$

$Z$ (the total impedance) may be less than $x_c$, and in this case $e$ (the drop across the capacitor's terminals) will be greater than $E$ (the total impressed emf). When the frequency is

$$f = \frac{1}{2\pi\sqrt{LC}} \qquad \text{Hz} \qquad (2\text{-}197)$$

the reactances are equal,

$$x_l = x_c \tag{2-198}$$

This condition gives the highest current for voltage $E$. Moreover, if $r$ (the resistance of the circuit) is assumed to be zero,

$$\frac{e}{E} = \infty \tag{2-199}$$

and we have an extreme case of voltage resonance. Actually, some resistance $r$ is always present. Then

$$\frac{e}{E} = \frac{x_c}{r} \quad \text{when } f = \frac{1}{2\pi\sqrt{LC}} \tag{2-200}$$

where $L$ is the coefficient of self-induction in henrys and $C$ the capacitance in farads of the apparatus connected in series. Resonance at one of the higher harmonics of the applied voltage may also take place when Eq. (2-200) is satisfied by the frequency of the harmonic.

***156. Parallel Resonance.*** When an inductive reactance and a capacitive reactance are in parallel, they can be so adjusted or the frequency can be so chosen that current resonance takes place.

Let the total conductance of the combination be $g$, the inductive susceptance $b_l$, and the capacitive susceptance $b_c$. Then the total current is

$$I = Ey = E\sqrt{g^2 + (b_l - b_c)^2} \tag{2-201}$$

The current through the capacitive susceptance is

$$i_c = Eb_c \tag{2-202}$$

Hence

$$\frac{i_c}{I} = \frac{b_c}{\sqrt{g^2 + (b_l - b_c)^2}} = \frac{b_c}{y} \tag{2-203}$$

When the total admittance $y$ is smaller than $b_c$, the total line current $I$ is less than one of its components $i_c$. A similar relation may be proved for $i_l$. When the frequency is

$$f = \frac{1}{2\pi\sqrt{LC}} \quad \text{Hz} \tag{2-204}$$

it follows that

$$b_l = b_c \tag{2-205}$$

and

$$I = Eg \qquad i_l = i_c \tag{2-206}$$

The line current is comparatively small, but there is a large interchange of current between the inductance and the capacitance in parallel.

## Complex (Nonsinusoidal) Waveforms

***157. Examples of Complex Waveforms.*** Figure 2-25 illustrates the effect of the inductance and capacitance in a circuit to which is applied an alternating emf differing from the simple sine wave. The curves were taken simultaneously with an oscillograph. $E$ is the impressed emf; $I_l$, the current taken by an inductance coil; and $I_c$, that taken by a capacitor. Figure 2-26 shows the circuit where the inductance and capacitance are in parallel.

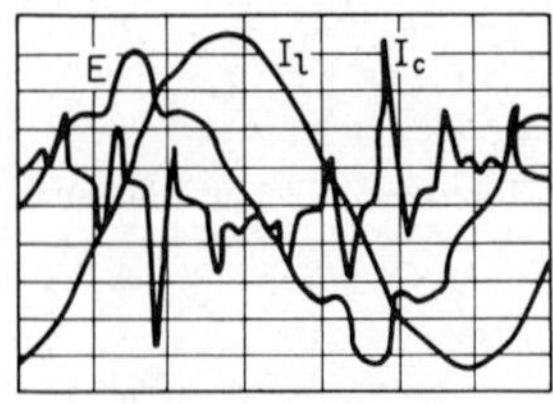


FIG. 2-25 Complex ac waveforms.

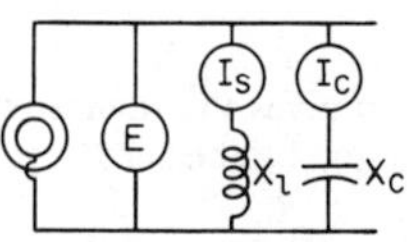


FIG. 2-26 Circuit in which the waveforms of Fig. 2-25 were observed.

***158. Waveform of Reactive EMF Due to Inductive Reactance.*** On the assumption that the reluctance of the iron core in the inductance coil is constant, which is approximately true below the saturation point, the value of the flux is proportional to the current $I_l$. The instantaneous value of the emf (see Par. 77) is

$$e = n\frac{d\phi}{dt} = L\frac{di}{dt} \tag{2-207}$$

That is, the curve $E$ will have its maximum amplitude when the curve $I$ passes through zero. This is not precisely true, because the current needed to supply losses in the resistance and in the iron is in phase with the emf $E$.

***159. Waveform of Current through Capacitive Reactance.*** The capacitive current is proportional to the rate of change of the emf (see Par. **121**); the instantaneous value is

$$i_c = C\frac{de}{dt} \tag{2-208}$$

That is, the curve $I_c$ has its maximum when the rate of change of the curve $E$ is a maximum. When $E$ is a sine curve, $I_c$ is also a sine curve, in quadrature with $E$. When the curve of emf is not a sine curve, as in Fig. 2-25, the maximum amplitude of the current $I_c$ occurs at the point where the slope of the emf curve is a maximum.

***160. Effects of Inductive and Capacitive Reactance on Waveform.*** The curves in Fig. 2-25 show the effect upon the current waveform of inductive reactance and capacitive reactance where the reactances are in parallel. The waveform $E$ produced by the generator contains several harmonics (see Par. **171**). The inductive reactance tends to damp out the higher harmonics, while the capacitive reactance emphasizes them.

***161. Determination of Total Complex Current Waveform.*** When the applied voltage contains higher harmonics (Par. **171**), the total current through an impedance is found by summing the harmonic currents due to each harmonic of the voltage acting alone. Thus, for an inductance, the reactance at the fundamental frequency $f$ is $x_1 = 2\pi fL$; the reactance to the $n$th harmonic is $x_n = 2\pi nfL$; and the impedance to the $n$th harmonic is

$$Z_n = \sqrt{r^2 + (2\pi nfL)^2} \quad \text{ohms} \tag{2-209}$$

Similar procedures can be used for capacitive elements.

***162. Power and Energy.*** The general expression for the energy delivered to an ac circuit with any waveform of current and voltage is

$$W = \int_{t_1}^{t_2} ei\,dt \quad \text{joules} \tag{2-210}$$

where $e$ is the instantaneous value of the voltage in volts; $i$ is the corresponding instantaneous current in amperes; and $t_2 - t_1 = T$ is the interval of time, in seconds, for which the energy is to be determined. The average power delivered during the interval $T$ is

$$P = \frac{1}{t_2 - t_1}\int_{t_1}^{t_2} ei\,dt \quad \text{watts} \tag{2-211}$$

***163. Power and Power Factor, Sine Waveforms.*** When the current and the voltage vary according to the sine law, the power $P = EI \cos \phi$ where $E$ and $I$ are the effective values of the voltage and the current, respectively, and $\phi$ is the phase angle between the two, $\cos \phi$ being known as the power factor of the circuit. See also Par. **153.**

***164. Power, Complex Waveforms.*** When $e$ and $i$ are irregular curves, the average power is found as the average ordinate of a curve the ordinates of which are proportional to the product $ei$. If $e$ and $i$ are resolved into their harmonics, each harmonic contributes its own share of power as if it were acting alone, so that the average power for a large number of cycles is

$$P = E_1 I_1 \cos \phi_1 + E_3 I_3 \cos \phi_3 \cdots \tag{2-212}$$

where $I_1, I_3, \ldots$ and $E_1, E_3, \ldots$ are the effective values of the harmonic currents and voltages, respectively, and the angles $\phi$ are the respective phase displacements.

***165. Energy Component and Reactive Component of Voltage or Current.*** In a simple harmonic circuit with the voltage $E$, current $I$, and phase displacement $\phi$ between the two, $E \cos \phi$ is called the energy component of the voltage; and $E \sin \phi$, the reactive component of the voltage. Analogously, $I \cos \phi$ is the energy component of the current, and $I \sin \phi$ is the reactive component of the current. Similar components are used in circuits with nonsinusoidal currents and voltages, provided that these are first replaced by equivalent sine waves.

***166. Effective Value of Any Wave.*** The effective value of a variable current or voltage is defined as that continuous value which gives the same total $i^2r$ loss. That is, if $I$ is the effective value of a periodic current $i$ and $T$ is the time of one cycle,

$$I^2 rT = \int_0^T i^2 r \, dt \qquad \text{joules}$$

from which

$$I = \sqrt{\frac{1}{T} \int_0^T i^2 \, dt} \qquad \text{amperes} \tag{2-213}$$

This may be expressed by saying that the *effective value of a current or voltage is equal to the square root of the mean square (rms) of the variable values taken throughout one cycle.* Hot-wire instruments and electrodynamometer-type instruments indicate directly the effective values of alternating currents and voltages.

***167. Effective Value of a Sine Wave.*** For sine waves, the effective value is given in Par. **143.** In terms of the maximum value, the effective value is

$$E_{\text{eff}} = 0.707\, E_{\text{max}} \tag{2-214}$$

***168. The crest factor*** is the ratio of the maximum or crest value (peak value) to the effective value; thus,

$$\frac{y_{\text{max}}}{y_{\text{eff}}} = \text{crest factor} \tag{2-215}$$

The crest factor for a sine wave is $\sqrt{2} = 1.414$.

***169. The form factor*** is the ratio of the effective value to the half-period mean value; thus,

$$\frac{y_{\text{eff}}}{y_{\text{mean}}} = \text{form factor} \tag{2-216}$$

The form factor for a sine wave is $\dfrac{\pi}{2\sqrt{2}} = 1.111$.

***170. Waveform Analysis; Fourier's Series.*** In the mathematical treatment of alternating waves, it is most convenient to work with those having sine form. Waves differing from the sine form may be resolved into a fundamental sine wave and its harmonics. The general equation of any alternating wave, as given by Fourier's series, is

$$y = Y_1 \sin (\omega t + \theta_1) + Y_2 \sin (2\omega t + \theta_2) + \cdots + Y_n \sin (n\omega t + \theta_n) \tag{2-217}$$

wherein $y$ is the ordinate of the resultant wave at time $t$; $Y_1, Y_3, \ldots Y_n, \ldots$ are the maximum ordinates or amplitudes of the first, second, . . . , $n$th, . . . harmonics; $\theta_1, \theta_2, \ldots, \theta_n$, the constant angles which determine the relative time-phase position of the corresponding harmonics; and $\omega = 2\pi f$, the angular velocity of the generating vector of the fundamental wave, corresponding to the fundamental frequency $f$ in hertz.

***171. Waveform Analysis; Fischer-Hinnen's Method.*** Waves having like loops above and below the time axis contain only odd harmonics, whereas waves having unlike loops above and below the axis contain both even and odd harmonics. A direct method of wave analysis given by J. Fischer-Hinnen[1] is based on the following equations:

$$A_n = \frac{1}{n}(y_4 + y_8 + y_{12} + \cdots + y_{2n-2} - y_2 - y_6 - y_{10} - \cdots - y_{2n-4}) \tag{2-218}$$

and

$$B_n = \frac{1}{n}(y_1 + y_5 + y_9 + \cdots + y_{2n-1} - y_3 - y_7 - y_{11} - \cdots - y_{2n-3}) \tag{2-219}$$

where $y_1, y_2, \ldots, y_n$ are ordinates at points along the base of the half wave, which is divided into $2n$ equal parts; and $A_n$ and $B_n$ are the ordinates of two quadrature components of the $n$th harmonic. The maximum ordinate of the $n$th harmonic is $\sqrt{A_n^2 + B_n^2}$, and its time-phase displacement from the resultant wave is

$$\theta_n = \tan^{-1}\frac{A_n}{-B_n} \tag{2-220}$$

$\theta_n$ being measured in terms of the $n$th harmonic.

The above equations for the $n$th harmonic do not take into account the harmonics of the harmonics $2n$, $3n$, etc. This correction is practically negligible for all harmonics, except the first or fundamental, and a correction rarely needs to be carried beyond the ninth harmonic.

Since waveforms generated by electric machinery almost never contain even harmonics, they do not enter into the correction. Denoting the corrected values by prime, we have

$$A'_n = A_n - A'_{3n} - A'_{5n} - A'_{7n} - \cdots \tag{2-221}$$

and

$$B'_n = B_n + B'_{3n} - B'_{5n} + B'_{7n} - \cdots \tag{2-222}$$

In applying this to the first harmonic, $A_n$ is the ordinate of the resultant wave at $y_0$ (Fig. 2-28), and $B_n$ is the ordinate displaced 90 time deg therefrom, at $y_3$.

***172. Example of Wave Analysis.*** As an example,[2] assume the wave given in Fig. 2-27, which is split into three harmonics: the first, or fundamental; the third; and the fifth. Figure 2-28 shows the method of determining a given harmonic, in this case the third. The base of the wave is divided into $2n$ or six equal parts, and ordinates are erected. The ordinates are:

$$y_1 = 676 \qquad y_2 = 660 \qquad y_3 = 940 \qquad y_4 = 1004 \qquad y_5 = 554 \qquad y_6 = 0 \tag{2-223}$$

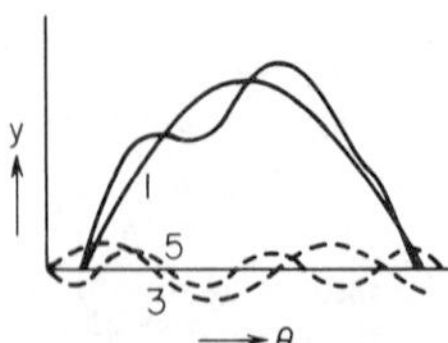


**FIG. 2-27** Waveform analysis, Par. **172.**

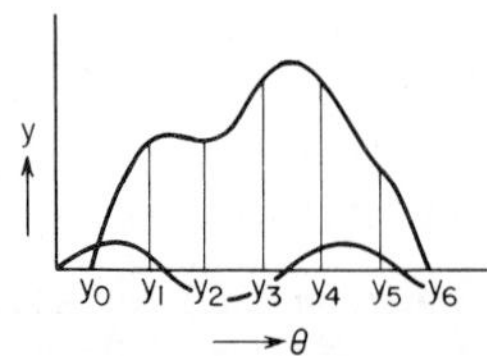


**FIG. 2-28** Waveform analysis, Par. **172.**

[1] Fischer-Hinnen, J. *Elek. Zt.*, 1901, Vol. 22, p. 396. Also Lincoln, P. M. *Elec. J.*, 1908, Vol. 5, 386.
[2] *Elec. J.*, 1908, Vol. 5, p. 386.

Then

$$A_3 = \frac{1}{3}(y_4 - y_2) = \frac{1004 - 660}{3} = 114.7 \tag{2-224}$$

and

$$B_3 - B_2 = \frac{1}{3}(y_1 + y_5 - y_3) = \frac{676 + 554 - 940}{3} = 96.7 \tag{2-225}$$

The maximum ordinate is

$$\sqrt{(114.7)^2 + (96.7)^2} = 150 \tag{2-226}$$

and the phase angle is

$$\theta_3 = \tan^{-1}\frac{-114.7}{96.7} = -50°* \tag{2-227}$$

In a similar manner, it is found that $A_5 = -92.8$ and $B_5 = 37.4$.

In this example, the wave contains only the third and the fifth harmonics; therefore, the fundamental is determined as follows:

$$A_1 = y_0 - A'_3 - A'_5 = 0 - 114.7 + 92.8 = -21.9$$

$$B_1 = y_3 + B'_3 - B'_5 = 940 + 96.7 - 37.4 = 999.3$$

$$\theta_1 = \tan^{-1}\frac{21.9}{999.3} = 1°15' \text{ (approx.)}$$

## Three-Phase Systems

***173. Three-phase* Y *and* Δ *Connections.*** In a balanced 3-phase system, the star connection is also called the *Y* connection (Fig. 2-29). The relations of the currents and the voltages are (Fig. 2-31)

$$E_\Delta = E_Y\sqrt{3} = 1.732E_Y \quad \text{and} \quad I_\Delta = \frac{I_Y}{\sqrt{3}} = \frac{I_Y}{1.732} \tag{2-228}$$

***174. Three-Phase Power.*** In a balanced 3-phase system (Figs. 2-29 and 2-30), the power is

$$P = 3I_YE_Y\cos\phi = 3I_\Delta E_\Delta\cos\phi = I_YE_\Delta\sqrt{3}\cos\phi \quad \text{watts} \tag{2-229}$$

when the currents are in amperes and the voltages in volts.

***175. Voltage Drop in Unsymmetrical Circuits.*** The voltage drop due to resistance and self-

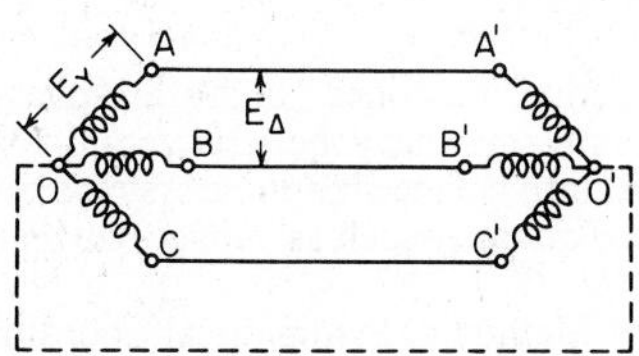

**FIG. 2-29** Three-phase wye connection.

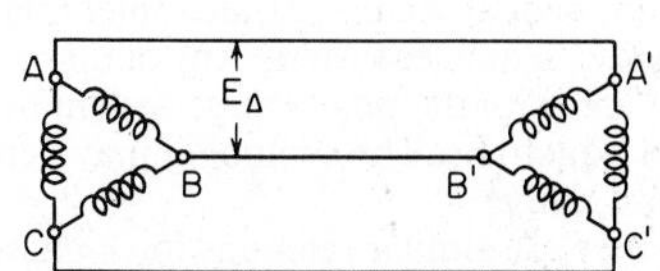

**FIG. 2-30** Three-phase delta connection.

* Fifty degrees in terms of the third harmonic 3*f*, or 50/3° in terms of the fundamental frequency *f*.

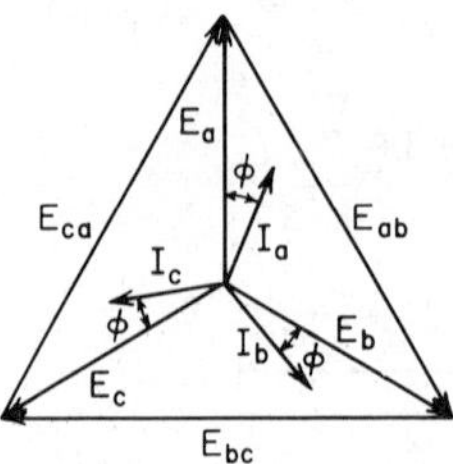


**FIG. 2-31** Three-phase currents and voltages.

and mutual inductance in any conductor of a group of long, parallel, round, nonmagnetic conductors forming a single-phase or polyphase circuit, and with one or more conductors connected electrically in parallel, may be calculated by summing the flux due to each conductor up to a certain large distance $u$. The vectorial sum of all the currents is zero in a complete system of currents in the steady state, and the quantity $u$ cancels out, so that the result is the same, no matter how large $u$ may be. The currents may be unbalanced, and in addition, the arrangement of the conductors may be unsymmetrical.

The voltage drop in any conductor $a$ of a group of round conductors $a, b, c, \ldots$ is

$$I_a R_a = j0.2794(I_a \log_{10} G_a + I_b \log_{10} S_{ab} + I_c \log_{10} S_{ac} + \cdots) \quad \text{V/mi at 60 Hz} \quad (2\text{-}230)$$

where $I_a + I_b + I_c + \cdots = 0$, the values of the currents being complex quantities; $R_a$ = resistance of conductor $a$, per mile; $G_a$ = self-geometric mean distance of conductor $a$; $S_{ab}$ = axial spacing between conductors $a$ and $b$, etc. The values of $G$ and $S$ should be in the same units (see Par. **84**). For further reference on the evaluation of the self- and mutual geometric mean distances see reference given in Par. **84**.

***176. Symmetrical Components—Resolution of an Unbalanced Three-Phase System into Balanced Systems.*** Let the three cube roots of unity, 1, $\epsilon^{j(2\pi/3)}$, $\epsilon^{j(4\pi/3)}$, be 1, $a$, $a^2$, where $j = \sqrt{-1}$,

$$a = 1\underline{/120^\circ} = -0.5 + j0.866$$

and

$$a^2 = 1\underline{/240^\circ} = -0.5 - j0.866$$

Any three vectors $Q_a, Q_b, Q_c$ (which may be unsymmetrical or unbalanced, that is, with unequal magnitudes or with phase differences not equal to 120°) can be resolved into a system of three equal vectors $Q_{a0}, Q_{a0}, Q_{a0}$ and two symmetrical (balanced) 3-phase systems $Q_{a1}$, $a^2Q_{a1}$, $aQ_{a1}$ and $Q_{a2}$, $aQ_{a2}$, $a^2Q_{a2}$, the first of which is of positive phase sequence and the second of negative phase sequence. Thus

$$\begin{aligned} Q_a &= Q_{a0} + Q_{a1} + Q_{a2} \\ Q_b &= Q_{a0} + a^2Q_{a1} + aQ_{a2} \\ Q_c &= Q_{a0} + aQ_{a1} + a^2Q_{a2} \end{aligned} \quad (2\text{-}231)$$

The values of the component vectors are

$$\begin{aligned} Q_{a0} &= \tfrac{1}{3}(Q_a + Q_b + Q_c) \\ Q_{a1} &= \tfrac{1}{3}(Q_a + aQ_b + a^2Q_c) \\ Q_{a2} &= \tfrac{1}{3}(Q_a + a^2Q_b + aQ_c) \end{aligned} \quad (2\text{-}232)$$

The three equal vectors $Q_{a0}$ are sometimes called the "residual quantities," or the zero-phase, or uniphase, sequence system. Any of the vectors $Q_a$, $Q_b$, or $Q_c$ may have the value zero. If two of them are zero, the single-phase system may be resolved into balanced 3-phase systems by the above equations. The symbol $Q$ may denote any vector quantity such as voltage, current, or electric charge.

There are similar relations for $n$-phase systems. See Method of Symmetrical Coordinates Applied to the Solution of Polyphase Networks, by C. L. Fortescue, *Trans. AIEE,* 1918, p. 1027.

***177. The calculation of short-circuit currents*** in 3-phase power networks is a common application of the method of symmetrical components.

The location of a probable short circuit of fault having been selected, three networks are computed in detail from the neutrals to the fault, one for positive, one for negative, and one for zero-phase sequence currents. The three phases are assumed to be identical, in ohms and in mutual effects, except in the connection of the fault itself. Let $Z_1$, $Z_2$, and $Z_0$ be the ohms per phase between the neutrals and the fault in each of the networks, including the impedance of the generators.

Then, for a line-to-ground fault,

$$I_{a1} = I_{a2} = I_{a0} = \frac{I_c}{3} = \frac{E_a}{Z_1 + Z_2 + Z_0} \tag{2-233}$$

where $E_a$ is the line-to-neutral voltage and $I_{a1}$ is the positive-phase-sequence current flowing to the fault in phase $a$ and similarly for $I_{a2}$ and $I_{a0}$. $I_a$ is the total current flowing to the fault in phase $a$.

The component currents in phases $b$ and $c$ are derived from those in phase $a$, by means of the relations $I_{b1} = a^2 I_{a1}$, $I_{b2} = aI_{a2}$, $I_{b0} = I_{a0}$, $I_{c1} = aI_{a1}$, $I_{c2} = a^2 I_{a2}$, and $I_{c0} = I_{a0}$.

Each of the component currents divides in the branches of its own network according to the impedance of that network. Thus, each of the component currents, and therefore the total current, at any part of the power system can be determined.

For a line-to-line fault between phases $b$ and $c$,

$$I_{a1} = -I_{a2} = \frac{E_a}{Z_1 + Z_2} \tag{2-234}$$

and

$$I_{a0} = 0 \tag{2-235}$$

For a double line-to-ground fault between phases $b$ and $c$ and ground,

$$I_{a1} = \frac{E_a}{Z_1 + \dfrac{Z_2 Z_0}{Z_2 + Z_0}} \tag{2-236}$$

$$I_{a0} = -\frac{I_{a1} Z_2}{Z_2 + Z_0} \tag{2-237}$$

and

$$I_{a2} = -I_{a1} - I_{a0} \tag{2-238}$$

See C. F. Wagner and R. D. Evans, “Symmetrical Components,” McGraw-Hill Book Company; P. M. Anderson, “Analysis of Faulted Power Systems,” Iowa State University Press, 1973; W. D. Stevenson, Jr., “Elements of Power System Analysis,” 3d ed., McGraw-Hill Book Company, 1975; and “Electrical Transmission and Distribution Reference Book,” 4th ed., Westinghouse Electric Corporation, 1950.

If there is no current in the power system before the fault occurs, the voltage $E_a$ of every generator is the same in magnitude and phase. Such a condition often is assumed in calculated circuit-breaker duty and relay currents, although the effects of loads on the system can be included in the analysis.

In calculating power-system stability, however, it must be assumed that current exists in the lines before the fault occurs. The voltage $E_a$ becomes the positive-sequence voltage at the point of fault before the fault occurs. A practical method of computing the positive-sequence current under fault conditions is to leave the positive-sequence network unchanged, with each generator at its own voltage and phase angle. The equivalent $Z_1$ of the network need not be computed. Certain 3-phase impedances are inserted between line and neutral at the location of the fault. For a single line-to-line ground fault $Z_2 + Z_0$ is inserted, for a line-to-line fault $Z_2$ is inserted, and for a double line-to-ground fault $Z_2 Z_0/(Z_2 + Z_0)$ is inserted. This gives one phase of an equivalent balanced 3-phase circuit for which the positive-sequence currents driven by all the generators in all the branches under fault conditions can be found by means of a network analyzer or computed on a digital computer. The power transmitted after the fault occurs can be determined from these positive-sequence currents.

If it is desired to find the negative-sequence and zero-sequence currents (some relays are operated by the latter), they can be computed from Eqs. (2-233) to (2-238) that do not involve $E_a$, after finding $I_{a1}$ to the fault.

The impedance $Z_f$ of each arc is mainly resistance. It may be brought into the computation. For single line-to-ground and double line-to-ground faults, $Z_f$ is added to each of $Z_1$, $Z_2$, and $Z_0$. For line-to-line faults, $Z_f$ is added to $Z_2$ only.

***178. Load Studies.*** In calculations relating to the steady-state operation of power systems, in which it is desired to determine the voltage, power, reactive power, etc., at various points, the loads may be designated by kilowatts and kilovars, rather than by impedances. The effect of the impedance of the transmission and distribution lines, transformers, etc., of the network can be computed. The modern method is a process of iterations using a digital computer for the calculations. The division of current in branches, the voltage at various points, and the required ratings of synchronous capacitors can be determined.

Conditions can be estimated at one or two points and a solution for the rest of the network calculated on the basis of these assumptions. If the assumptions are not correct, discrepancies will appear at the end of the work. For instance, two different voltages may be obtained for the same point, one calculated before, and one after, going around a loop of the network. The necessary correction to the first estimates may be based on the discrepancies, thus giving successive approximations which are improvements on the preceding ones.

## Three-Phase Armature Windings

***179. The armature winding of a 3-phase generator or motor*** is an important type of electric circuit. Windings consisting of diamond-shaped coils, with two coil sides per slot, are connected in groups of coils, three groups or phase belts being opposite each pole. In general, the number of slots per pole per phase is a fraction equal to the average number of coils per phase belt. There are a larger number and a smaller number of coils per phase belt, differing by 1. The winding is usually found to be divided into repeatable sections of several poles each, the sections being duplicates of each other.

The number of poles in a section is found by writing the fraction equal to the number of slots divided by the number of poles and canceling factors to the extent possible. The denominator is the number of poles per section, and the numerator is the number of slots per section.

If the final value of the numerator is not divisible by 3, a balanced 3-phase winding cannot be made, since the windings for phases *a, b,* and *c* in a section each require the same number of slots and they must be duplicates except for the phase shift of 120°. This gives rise to the rule for balanced 3-phase windings that the factor 3 must occur at least one more time in the number of slots than in the number of poles.

It can be shown that the slots of a repeatable section have phase angles which, when suitably drawn, are all different and equidistant. They fill the space of 180 elec deg like the blades of a Japanese fan (see "Electrical Coils and Conductors," by H. B. Dwight, Chap. 8, pp. 43 and 44). The angle between the vectors in this fan is

$$\beta = \frac{180}{\text{slots per section}} \quad \text{deg} \tag{2-239}$$

The vectors lying from 0 to 59° may be assigned to phase *a* or $-a$, those from 60 to 119° to phase $-c$ or *c*, and those from 120 to 179° to phase *b* or $-b$.

The phase angles for the upper coil sides of the slots should be tabulated to indicate the proper connections of the winding. Since the diamond coils are all alike, the total resultant voltage developed in the lower coil sides of a phase is a duplicate of that developed in the upper coil sides and can be added on by means of the pitch factor.

The phase angle between two adjacent slots is

$$\frac{\text{Poles per section} \times 180}{\text{Slots per section}} = q\beta \quad \text{deg} \tag{2-240}$$

where $q$ is the number of poles in the repeatable section. From this, the phase angle for every slot in the section can be written. To save numerical work, especially where $\beta$ is a fractional number of degrees, the angles may be expressed in terms of the angle $\beta$, as given in the example below. They may be expressed in degrees and fractions of a degree, but decimal values of degrees should not be used in this part of the work. The required accuracy is obtained by using fractions instead of decimals. Appropriate multiples of 180° should be subtracted to keep the angles less than 180°, thus indicating the relative position of each coil side with respect to the nearest pole. When an odd number times 180° has been subtracted, the coil side is tabulated as $-a$ instead of $a$, etc., since it will be opposite a south pole when $a$ is opposite a north pole. The terminals of a coil marked $-a$ are reversed with respect to the terminals of a coil marked $a$ with which it is in series.

*Example.* 21 slots per repeatable section; 5 poles per section; 1⅖ slots per pole per phase.

$$\beta = \frac{180}{21} = 8\frac{4}{7}\text{ deg} \qquad \text{[by Eq. (2-239)]}$$

It is more convenient in this case to express the angles in terms of $\beta$ rather than by fractions of degrees. Note that $21\beta = 180°$ and $7\beta = 60°$. The range for coils to be marked $\pm a$ is from 0 to $6\beta$, inclusive; coils marked $\mp c$ from $7\beta$ to $13\beta$; and coils marked $\pm b$ from $14\beta$ to $20\beta$. Subtract multiples of $21\beta = 180°$. The angle between two adjacent slots is $q\beta = 5\beta$.

*Tabulation of Phase Angles:*

| 1 | 2 | 3 | 4 | 5 | 6 | 7 | 8 | 9 | 10 | 11 | 12 | 13 | 14 |
|---|---|---|---|---|---|---|---|---|---|---|---|---|---|
| 0 | $5\beta$ | $10\beta$ | $15\beta$ | $20\beta$ | $(25\beta)4\beta$ | $9\beta$ | $14\beta$ | $19\beta$ | $(24\beta)3\beta$ | $8\beta$ | $13\beta$ | $18\beta$ | $(23\beta)2\beta$ |
| $a$ | $a$ | $-c$ | $b$ | $b$ | $-a$ | $c$ | $-b$ | $-b$ | $a$ | $-c$ | $-c$ | $b$ | $-a$ |

| 15 | 16 | 17 | 18 | 19 | 20 | 21 | [22] |
|---|---|---|---|---|---|---|---|
| $7\beta$ | $12\beta$ | $17\beta$ | $(22\beta)\beta$ | $6\beta$ | $11\beta$ | $16\beta$ | $[(21\beta)0]$ |
| $c$ | $c$ | $-b$ | $a$ | $a$ | $-c$ | $b$ | $[-a]$ |

The seven vectors of phase $a$ make a regular fan covering $7\beta = 60°$.

The resultant terminal voltage produced by the coils of phase $a$ is equal to the numerical sum of the voltages in those coils multiplied by the "distribution factor"

$$\frac{\sin(n\beta/2)}{n\sin(\beta/2)} \tag{2-241}$$

where $n$ is the number of vectors in the regular fan covering 60° and $\beta$ is the angle between adjacent vectors, given by Eq. (2-238). The number $n$ is large, and the perimeter approaches the arc of a circle. Expression (2-240) is of the same form as the formula for breadth factor, which also is based on a vector diagram that is a regular fan.

The distribution factor for the winding of the foregoing example is

$$\frac{\sin\dfrac{7\times 60°}{2\times 7}}{7\sin\dfrac{60°}{2\times 7}} = \frac{\sin 30°}{7\sin 4\dfrac{2°}{7}} = \frac{0.5}{7\times 0.0746} = 0.956$$

Other possible balanced 3-phase windings for this example could be specified by having some of the vectors of phase $a$ lie outside the 60° range. This would result in a lower distribution factor. The voltage, and hence the rating of the machine, would be lower by 2% or more than in the case described. The canceling of the harmonic voltages would apparently not be improved, and there would be no advantage to compensate for the reduction in kVA rating, which would correspond to a loss or waste of 2% or more of the cost of the machine.

## General Bibliography

***180. General Reference Literature***


Abraham, M. "The Classical Theory of Electricity and Magnetism," revised by R. Becker, translated into English by J. Dougall; Glasgow, Blackie & Son, Ltd., 1932.

Anderson, P. M. "Analysis of Faulted Power Systems"; Ames, The Iowa State University Press, 1973.

Bitter, F. "Introduction to Ferromagnetism"; New York, McGraw-Hill Book Company, 1937.

Boast, W. B. "Vector Fields"; New York, Harper & Row, Publishers, Incorporated, 1964.

Booker, H. G. "An Approach to Electrical Science"; New York, McGraw-Hill Book Company, 1959.

Brenner, E., and Javid, M. "Analysis of Electric Circuits," 2d ed.; New York, McGraw-Hill Book Company, 1967.

Christie, C. V. "Electrical Engineering"; New York, McGraw-Hill Book Company, 1952.

Clarke, E. "Circuit Analysis of A-C Power Systems"; New York, John Wiley & Sons, Inc., 1943, Vol. I, and 1950, Vol. II.

Dwight, H. B. "Electrical Coils and Conductors"; New York, McGraw-Hill Book Company, 1945.

Dwight, H. B. "Electrical Elements of Power Transmission Lines"; New York, The Macmillan Company, 1954.

"Electrical Transmission and Distribution Reference Book," 4th ed.; Westinghouse Electric Corporation, 1950.

"Encyclopedia of Physics," 3d ed., New York, Van Nostrand Reinhold Co., 1985.

Faraday, M. "Experimental Researches in Electricity," 3 vols.; London, B. Quaritch, 1839–1855.

Fitzgerald, A. E., and Kingsley, C., Jr. "Electric Machinery," 2d ed.; New York, McGraw-Hill Book Company, 1961.

Foecke, H. A. "Introduction to Electrical Engineering Science"; Englewood Cliffs, N.J., Prentice-Hall, Inc., 1961.

Frank, N. H. "Introduction to Electricity and Optics," 2d ed.; New York, McGraw-Hill Book Company, 1950.

Hammond, S. B. "Electrical Engineering"; New York, McGraw-Hill Book Company, 1961.

Harnwell, G. P. "Principles of Electricity and Electromagnetism," 2d ed.; New York, McGraw-Hill Book Company, 1949.

Hayt, W. H., Jr. "Engineering Electromagnetics"; New York, McGraw-Hill Book Company, 1958.

Hayt, W. H., Jr., and Hughes, G. W. "Introduction to Electrical Engineering"; New York, McGraw-Hill Book Company, 1968.

Hayt, W. H., Jr., and Kemmerly, J. E. "Engineering Circuit Analysis"; New York, McGraw-Hill Book Company, 1962.

Jeans, J. H. "Mathematical Theory of Electricity and Magnetism"; New York, Cambridge University Press, 1908.

McGraw-Hill Encyclopedia of Science and Technology, 3d ed.; 1971.

Maxwell, J. C. "A Treatise on Electricity and Magnetism," 2 vols.; New York, Oxford University Press, 1904.

Page, L., and Adams, N. I. "Principles of Electricity"; Princeton, N.J., D. Van Nostrand Company, Inc., 1934.

Peek, F. W., Jr. "Dielectric Phenomena in High-Voltage Engineering"; New York, McGraw-Hill Book Company, 1929.

Rosa, E. B., and Grover, F. W. Formulas and Tables for the Calculation of Mutual and Self-Inductance, *NBS Sci. Paper* 169, 1916. Published also as Pt. 1 of Vol. 8, *NBS Bull.* Contains also skin-effect tables.

Sabbagh, E. M. "Circuit Analysis"; New York, The Ronald Press Company, 1961.

Sears, F. W. "Principles of Physics"; Reading, Mass., Addison-Wesley Publishing Company, Inc., 1946, Vol. 2, Electricity and Magnetism.

Seshu, S., and Balabanian, N. "Linear Network Analysis"; New York, John Wiley & Sons, Inc., 1959.

Skilling, H. H. "Electromechanics"; New York, John Wiley & Sons, Inc., 1962.

Smith, R. J. "Circuits, Devices, and Systems, 3d ed.; New York, John Wiley & Sons, Inc., 1976.

Smythe, W. R. "Static and Dynamic Electricity," 3d ed.; New York, McGraw-Hill Book Company, 1967.

Stevenson, W. D., Jr. "Elements of Power System Analysis," 3d ed.; New York, McGraw-Hill Book Company, 1975.

Wildi, T. "Electric Power Technology"; New York, John Wiley & Sons, Inc., 1981.

Woodruff, L. F. "Principles of Electric Power Transmission"; New York, John Wiley & Sons, Inc., 1938.

## FILTERS

*By HARRY W. HALE AND EDWIN C. JONES, JR.*

***181. Historical Note.*** The basic concept of the electric wave filter was originated by Campbell and Wagner, working independently, in 1915. The continuation of this work has proceeded along two paths. The first of these has been called *image parameter filter design* and the second is often referred to as *insertion loss filter design.*

The design of *image parameter filters* is based on the concept of a cascade of fundamental sections with matched image impedances. This type of filter design was given a marked impetus by the work of Zobel in 1923 on *m-derived filters* and was dominant for more than 30 years.

*Insertion loss filter design* is based on the specification of an appropriate network response in the form of a network function part such as transfer function magnitude or phase and the synthesis of a lumped network to achieve the specified response. This type of filter design had its origins in the work of Norton, Foster, Cauer, Bode, Darlington, and others and was well developed by 1940. It did not gain immediate acceptance, however, because of the requirement for precise and lengthy computations in the design process. The advent of digital computers in the 1950s eliminated the computational burden, and this type of filter design gained rapid acceptance and has largely supplanted image parameter design. Today, design of this type of filter is largely a matter of looking up element values in computer-determined tables for various classes of filters, and modifying these in a routine fashion.

*Image parameter filters* will not be discussed here. Detailed descriptions are available elsewhere.[1*]

***182. Insertion Loss Filter Design.*** The design process is carried out in terms of one of the two general networks shown in Fig. 2-32 according to the type of source. The procedure consists

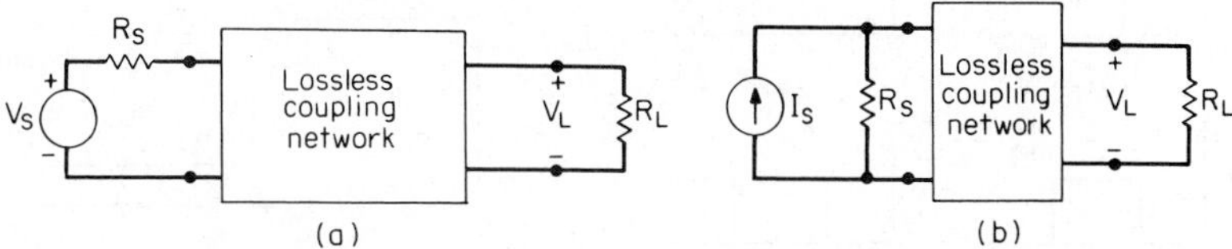


**FIG. 2-32** General form of network. *(a)* With voltage source; *(b)* with current source.

of the following general steps:

1. The filter characteristics are specified.
2. The filter characteristics are approximated in some sense by a realizable network function part. This is often the magnitude of the network transfer function

$$|G_{SL}(j\omega)| = \left|\frac{V_L}{V_S}(j\omega)\right| \tag{2-242}$$

or

$$|Z_{SL}(j\omega)| = \left|\frac{V_L}{I_S}(j\omega)\right| \tag{2-243}$$

for the networks of Fig. 2-32*a* and *b*, respectively. (For convenience, subsequent references to $G_{SL}$ will omit subscripts and also apply to $Z_{SL}$ except where a specific distinction is made.) The phase function

$$\theta(\omega) = \arg G(j\omega) \tag{2-244}$$

is also used in some applications.

3. The coupling network of Fig. 2-32*a* or *b* is synthesized to realize the network function part of step 2. In 1939, Darlington[2] showed that this coupling network can be lossless. In addition, it is a ladder network for many common cases. This is an attractive feature.

The emphasis in the design procedure is on steps 1 and 2, with the result of these being parameters that can be used to enter computer-determined tables of element values for normalized prototypes to which scaling and frequency transformation techniques can then be applied.

---

* Superior numbers refer to bibliography, Par. **209**.

***183. The Classification of Ideal Filters.*** An ideal filter is said to pass (with no attenuation) all frequencies within its passbands and to stop (with infinite attenuation) all frequencies in its stopbands. The following five basic types of filters are commonly used:

1. The low-pass filter passes frequencies from zero to its cutoff frequency and stops all frequencies higher than the cutoff frequency.
2. The high-pass filter stops all frequencies below its cutoff frequency, passes all above it.
3. The bandpass filter passes all frequencies between its lower and upper cutoff frequencies and stops all frequencies outside this range.
4. The band-elimination filter stops all frequencies between its lower and upper cutoff frequencies and passes all frequencies outside this range.
5. The all-pass filter passes all frequencies. Its purpose is to produce a predictable phase shift, and it is used to produce a constant time delay for all frequencies.

These ideal filter characteristics are shown in Fig. 2-33.

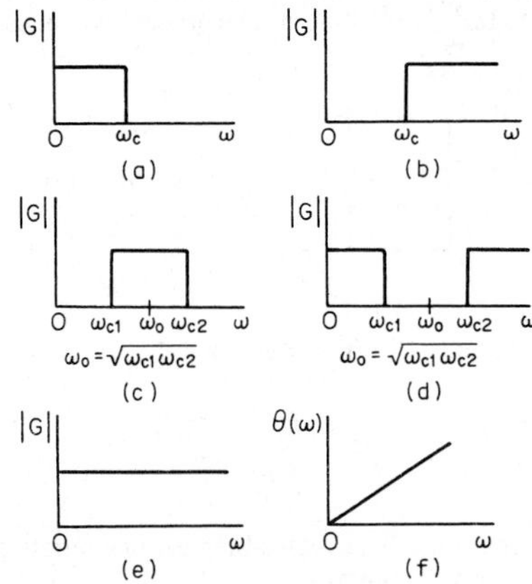


**FIG. 2-33** Ideal-filter characteristics. *(a)* Low-pass; *(b)* high-pass; *(c)* bandpass; *(d)* band-elimination; *(e)* all-pass (magnitude); *(f)* all-pass (phase).

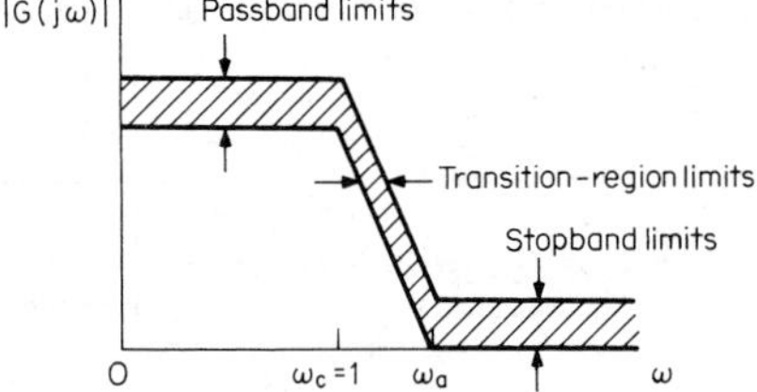


**FIG. 2-34** Limits on frequency response.

***184. The Low-Pass Prototype*** The low-pass, high-pass, bandpass, and band-elimination filters can be designed by a direct process. An alternative procedure consists of first determining a low-pass prototype normalized to make $R_L = 1$ and $\omega_c = 1$, and then using scaling and frequency transformations to arrive at the desired filter. The latter procedure, used here, makes it possible to use a reasonable number of computer-determined tables of element values for low-pass prototypes with various characteristics. This is particularly significant since this type of filter design requires precision in prototype computation to achieve satisfactory results. The following sections are organized around this concept.

***185. The Approximation Problem.*** The ideal low-pass filter characteristics can only be approximated. A basic problem is to select a transfer function magnitude $|G(j\omega)|$ that approximates the ideal characteristic *and* results in a realizable coupling network. This is indicated in Fig. 2-34, where limits on the frequency response are suggested. The approximation problem consists of finding an appropriate $|G(j\omega)|$ that will lie within the shaded region.

Generally, the magnitude-squared transfer function can be of the form

$$|G(j\omega)|^2 = K^2 \frac{A(\omega^2)}{D(\omega^2)} \tag{2-245}$$

It is noted that the numerator and denominator must be even, nonnegative polynomials. There are, of course, other restrictions. Equation (2-245) can be rewritten as

$$\begin{aligned} |G(j\omega)|^2 &= K^2 \frac{N(\omega^2)}{N(\omega^2) + M(\omega^2)} \\ &= K^2 \frac{1}{1 + M(\omega^2)/N(\omega^2)} \end{aligned} \tag{2-246}$$

One approach to the approximation problem is to let $N(\omega^2) = 1$, in which case the magnitude-squared transfer function becomes

$$|G(j\omega)|^2 = \frac{K^2}{1 + M(\omega^2)} \tag{2-247}$$

and the problem is to select a polynomial $M(\omega^2)$ such that

$$|M(\omega^2)| \ll 1 \qquad \omega \ll 1$$

$$|M(\omega^2)| \gg 1 \qquad \omega \gg 1$$

and which will lead to a realizable network.

Low-pass filters based on Eq. (2-247) are ladder structures with inductors as the series elements and capacitors as the shunt elements. If the more general form of Eq. (2-246) is used, the resulting filter will again be a ladder but the series elements may be parallel combinations of inductance and capacitance and the shunt elements may be series combinations of inductance and capacitance.

In order that the coupling network will not contain ideal transformers, the constant $K$ of Eq. (2-245) must have a specific value which is dependent on the terminating resistances and the type of source. In terms of Fig. 2-32, $K$ must be such that

$$|G(0)| = \frac{R_L}{R_S + R_L} \tag{2-248}$$

for a voltage source, and

$$|Z(0)| = \frac{R_S R_L}{R_S + R_L} \tag{2-249}$$

for a current source.

***186. Time-Domain Considerations.*** Filter specifications are ordinarily given in terms of frequency-domain response. Time-domain response is often an important factor also.

The unit step response can be used as a measure of time-domain response. In order to determine this, the function $G(s)$ is needed. This can be determined from the magnitude-squared function of Eq. (2-247) by use of analytic continuation to obtain

$$\begin{aligned} |G(j\omega)|^2_{\omega^2 = -s^2} &= G(s)\,G(-s) \\ &= K^2 \frac{1}{1 + M(-s^2)} \end{aligned} \tag{2-250}$$

The left-half-plane and right-half-plane poles of $G(s)\,G(-s)$ must be assigned to $G(s)$ and $G(-s)$, respectively, if $G(s)$ is to be realizable. The result is

$$G(s) = \frac{k}{\prod_{j=1}^{n} (s - s_j)} \tag{2-251}$$

where the $s_j$ are the left-half-plane poles. With $G(s)$ known, the unit step response can be determined.

The following sections describe five different approximations, four to the ideal low-pass frequency response and one to the ideal all-pass (time-delay) response. In each case the pole locations are tabulated as linear and quadratic factors. The unit-step response for each of the five approximations is described in Par. **193** in terms of tabulated figures of merit.

***187. The Butterworth Approximation.*** The transfer function magnitude

$$|G(j\omega)| = \frac{K}{(1 + \omega^{2n})^{1/2}} \tag{2-252}$$

is known as the $n$th-order Butterworth low-pass approximation. The general form of the approximation is shown in Fig. 2-35.

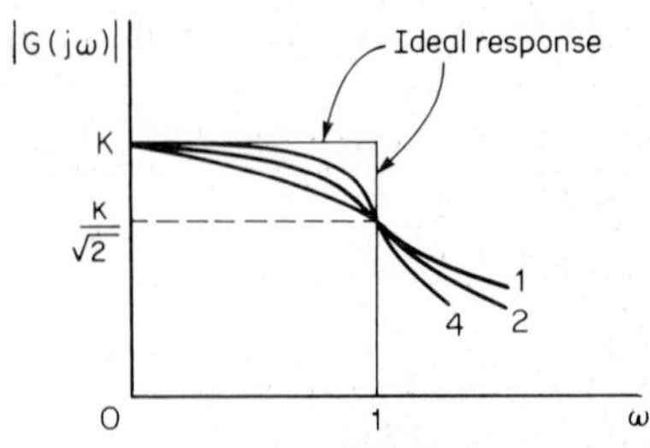


FIG. 2-35 The Butterworth approximation.

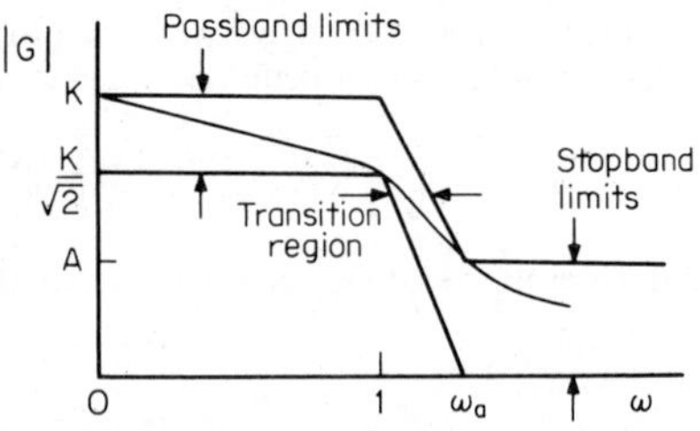


FIG. 2-36 Illustration of determination of the minimum value of $n$ for the Butterworth approximation.

It is clear that increasing $n$ improves the approximation in both passband and stopband. This also has adverse effects on the time-domain response and results in a larger number of elements in the filter. Thus the choice of $n$ is the smallest value that will satisfy some specific specification on the frequency-domain response.

The typical problem of determining $n$ is illustrated in Fig. 2-36. If the specifications require that

$$|G(j\omega)| \leq A \qquad \omega_a \leq \omega < \infty \tag{2-253}$$

then the value of $n$ required is the smallest integer value satisfying the inequality

$$A \geq \frac{K}{(1+\omega_a^{2n})^{1/2}} \tag{2-254}$$

It can be shown that the first $2n-1$ derivatives of the $|G(j\omega)|$ of Eq. (2-252) are zero at $\omega = 0$. This approximation is often called a maximally flat approximation for this reason. It is also noted that the derivative at the cutoff frequency is

$$\frac{d}{d\omega}|G(j\omega)|_{\omega=1} = \frac{-nK}{2^{3/2}} \tag{2-255}$$

The frequency response is often described in terms of the attenuation

$$\alpha = 20 \log \frac{|G(j\omega)|}{|G(0)|} = 10 \log (1+\omega^{2n}) \qquad \text{dB} \tag{2-256}$$

It is noted that at the cutoff frequency

$$\alpha_{\omega=1} = 10 \log 2 = 3.0103 \text{ dB} \tag{2-257}$$

This is the usual interpretation of a cutoff frequency, also referred to as the half-power or "3-dB" frequency. It is also noted that for $\omega \gg 1$

$$\alpha \cong 20\, n \log \omega \qquad \text{dB} \tag{2-258}$$

A specification of minimum attenuation $\alpha_{\min}$ for $\omega \geq \omega_a$ results in

$$\alpha_{\min} \geq 10 \log (1+\omega^{2n}) \tag{2-259}$$

and a solution for $n$ with the equality sign gives

$$n = \frac{1}{2}\,\frac{\log (10^{\alpha_{\min}/10} - 1)}{\log \omega_a} \tag{2-260}$$

which, for large attenuation, is reasonably approximated by

$$n \cong \frac{\alpha_{\min}}{20 \log \omega_a} \tag{2-261}$$

with the next larger integer value being used.

As indicated in Par. **186,** the poles of $G(s)$ can be determined from the magnitude-squared function

$$|G(j\omega)|^2 = \frac{K^2}{1+\omega^{2n}} \tag{2-262}$$

by the use of analytic continuation. The specific pole locations are given in Table 2-3 for $n$ ranging from 1 to 10, where they are given as quadratic and linear factors.

**TABLE 2-3** Linear and Quadratic Factors of Butterworth Polynomials—Poles of Butterworth Transfer Functions

| $n$ | $[G(s)]^{-1}$ | $n$ | $[G(s)]^{-1}$ |
|---|---|---|---|
| 2 | $s^2 + 1.4142136s + 1$ | 8 | $s^2 + 0.3901806s + 1$<br>$s^2 + 1.1111405s + 1$<br>$s^2 + 1.6629392s + 1$<br>$s^2 + 1.9615706s + 1$ |
| 3 | $s + 1$<br>$s^2 + s + 1$ | | |
| 4 | $s^2 + 0.7653669s + 1$<br>$s^2 + 1.8477591s + 1$ | 9 | $s + 1$<br>$s^2 + 0.3472964s + 1$<br>$s^2 + s + 1$<br>$s^2 + 1.5320889s + 1$<br>$s^2 + 1.8793852s + 1$ |
| 5 | $s + 1$<br>$s^2 + 0.6180340s + 1$<br>$s^2 + 1.6180340s + 1$ | | |
| 6 | $s^2 + 0.5176381s + 1$<br>$s^2 + 1.4142136s + 1$<br>$s^2 + 1.9318517s + 1$ | 10 | $s^2 + 0.3128689s + 1$<br>$s^2 + 0.9079810s + 1$<br>$s^2 + 1.4142136s + 1$<br>$s^2 + 1.7820130s + 1$<br>$s^2 + 1.9753767s + 1$ |
| 7 | $s + 1$<br>$s^2 + 0.4450419s + 1$<br>$s^2 + 1.2469796s + 1$<br>$s^2 + 1.8019377s + 1$ | | |

***188. The Chebyshev Approximation.*** The Chebyshev approximation is

$$|G(j\omega)| = \frac{K}{[1+\epsilon^2 C_n^2(\omega)]^{1/2}} \tag{2-263}$$

where $C_n(\omega)$ is the $n$th-order Chebyshev polynomial and $\epsilon$ is a real constant less than one. Specifically,

$$C_n(\omega) = \cos(n \cos^{-1} \omega) \qquad \text{for } 0 < \omega \leqslant 1 \tag{2-264}$$

or

$$C_n(\omega) = \cosh(n \cosh^{-1} \omega) \qquad \text{for } \omega \geqslant 1 \tag{2-265}$$

It is apparent that

$$C_0(\omega) = 1 \tag{2-266}$$

and

$$C_1(\omega) = \omega \tag{2-267}$$

These can be used with the recursion formula

$$C_{n+1}(\omega) = 2\omega C_n(\omega) - C_{n-1}(\omega) \tag{2-268}$$

to develop higher-order polynomials. These are given in Table 2-4.

**TABLE 2-4** Chebyshev Polynomials for $n = 0$ to 10

| $n$ | Chebyshev polynomial |
|---|---|
| 0 | $1$ |
| 1 | $\omega$ |
| 2 | $2\omega^2 - 1$ |
| 3 | $4\omega^3 - 3\omega$ |
| 4 | $8\omega^4 - 8\omega^2 + 1$ |
| 5 | $16\omega^5 - 20\omega^3 + 5\omega$ |
| 6 | $32\omega^6 - 48\omega^4 + 18\omega^2 - 1$ |
| 7 | $64\omega^7 - 112\omega^5 + 56\omega^3 - 7\omega$ |
| 8 | $128\omega^8 - 256\omega^6 + 160\omega^4 - 32\omega^2 + 1$ |
| 9 | $256\omega^9 - 576\omega^7 + 432\omega^5 - 120\omega^3 + 9\omega$ |
| 10 | $512\omega^{10} - 1280\omega^8 + 1120\omega^6 - 400\omega^4 + 50\omega^2 - 1$ |

Each Chebyshev polynomial has real zeros that are within the interval $-1 \leq \omega \leq 1$ and its maximum and minimum values are $+1$ and $-1$, respectively, within the interval. Thus

$$C_n^2(\omega) \leq 1 \qquad \text{for } 0 \leq \omega \leq 1 \tag{2-269}$$

$$C_n^2(0) = 0 \qquad \text{for } n \text{ odd} \tag{2-270}$$

$$C_n^2(0) = 1 \qquad \text{for } n \text{ even} \tag{2-271}$$

$$C_n^2(1) = 1 \qquad \text{for all } n \tag{2-272}$$

$$C_n^2(\omega) > 1 \qquad \text{for } \omega > 1 \tag{2-273}$$

Figure 2-37 shows the frequency response for the Chebyshev approximation for $n$ even and odd. In either case there are $n$ half cycles (from maximum to minimum and the reverse) in the interval $0 \leq \omega \leq 1$.

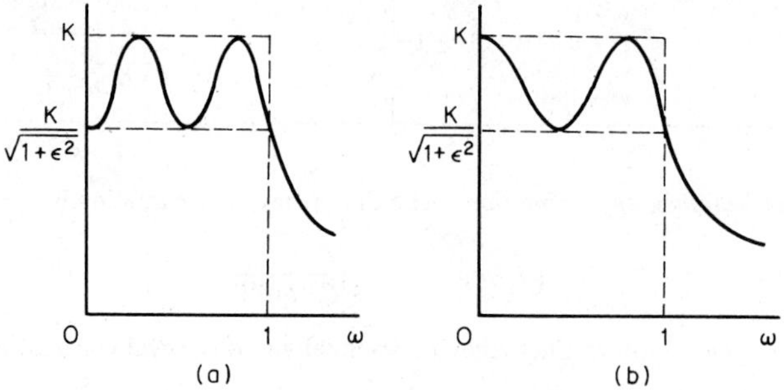


**FIG. 2-37** The Chebyshev approximation. *(a)* $n$ even; *(b)* $n$ odd.

It is also observed that for the special case of $\epsilon = 1$

$$\left.\frac{d|G(j\omega)|}{d\omega}\right|_{\omega=1} = -\frac{Kn^2}{2^{3/2}} \tag{2-274}$$

$$|G(j\omega)| \leq K = G(0) \qquad \text{for all } \omega \tag{2-275}$$

and for $n$ even

$$|G(j\omega)| \leq K = G(0)\sqrt{1+\epsilon^2} \qquad \text{for all } \omega \tag{2-276}$$

The consequence of these conditions is that there are combinations of values of $R_S$, $R_L$, and $\epsilon$

that cannot be realized when $n$ is even. In particular, the even-order Chebyshev approximation cannot be realized for any value of $\epsilon$ when $R_S = R_L$. It will be shown subsequently that the even-order Chebyshev polynomial can be modified to result in a realizable approximation with some sacrifice in stopband performance.

Two parameters, $\epsilon$ and $n$, are necessary to define a particular Chebyshev approximation. Figure 2-38 illustrates the problem for $n$ odd (the results also apply to $n$ even). It is apparent that $\epsilon$ controls the passband limits. The ratio of upper to lower passband limit is $(1 + \epsilon^2)^{1/2}$, and the value of this ratio is sufficient to determine $\epsilon$. The logarithmic function

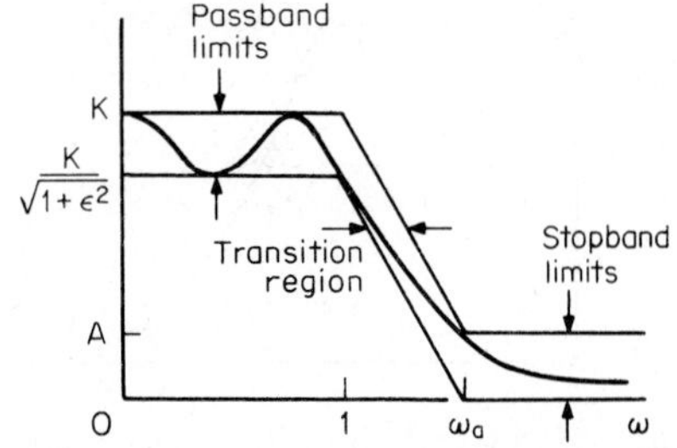


**FIG. 2-38** Pertaining to the identification of $\epsilon$ and $n$.

$$10 \log (1 + \epsilon^2) \tag{2-277}$$

is often used to describe this ratio. Thus a Chebyshev approximation with $\epsilon = 0.7648$ is said to have a 2-dB ripple. Both $\epsilon$ and $n$ affect the stopband response. With $\epsilon$ known and with the specifications requiring that

$$G(j\omega) \leqslant A \qquad \text{for } \omega_a \leqslant \omega < \infty \tag{2-278}$$

the inequality

$$A \geqslant \frac{K}{[1 + \epsilon^2 C_n^2(\omega_a)]^{1/2}} \tag{2-279}$$

must be satisfied. Equation (2-279) can be used with the known value of $\epsilon$ to determine that

$$C_n(\omega_a) \geqslant \frac{(K^2 - A^2)^{1/2}}{A\epsilon} \tag{2-280}$$

Thus $n$ can be determined by using $C_n(\omega_a)$ resulting from the equality sign in Eq. (2-280) in conjunction with Eq. (2-265), with the result that

$$n = \frac{\cosh^{-1} C_n(\omega_a)}{\cosh^{-1} \omega_a} \tag{2-281}$$

with the next larger integer value being used.

It is apparent that decreasing $\epsilon$ to improve passband response either degrades the stopband response or requires a larger $n$.

The attenuation for the Chebyshev approximation is

$$\alpha = 10 \log [1 + \epsilon^2 C_n^2(\omega)] \qquad \text{dB} \tag{2-282}$$

which is approximated by

$$\alpha \cong 10 \log \epsilon^2 C_n^2(\omega) \qquad \text{dB} \tag{2-283}$$

for $\omega \gg 1$. Recognizing that the highest-powered term in $C_n(\omega)$ is $2^{n-1}\omega^n$, Eq. (2-283) becomes

$$\alpha \cong 20 \log \epsilon + 6(n-1) + 20n \log \omega \tag{2-284}$$

as a further approximation.

It was observed for the Butterworth approximation that $\omega = 1$ is the half-power frequency. This is not the case with the Chebyshev approximation except for the special case of $\epsilon = 1$. In some applications it is desirable to normalize the Chebyshev approximation so that $\omega = 1$ is the half-power frequency. In this case, the frequency response appears as in Fig. 2-39. This can be done by determining the frequency at which the magnitude of $G(j\omega)$ becomes $K/(2)^{1/2}$, implying that

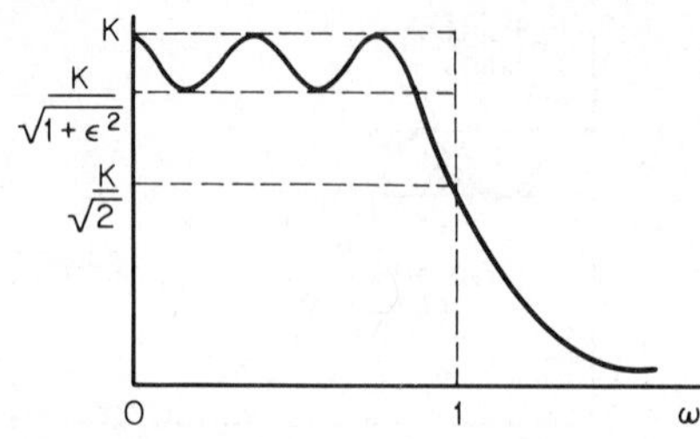


**FIG. 2-39** The Chebyshev approximation ($n$ odd) normalized to make $\omega = 1$ the half-power frequency.

$$\epsilon^2 C_n^2(\omega_3) = 1 \qquad (2\text{-}285)$$

or

$$C_n^2(\omega_3) = \frac{1}{\epsilon} \qquad (2\text{-}286)$$

There is an explicit expression for the resulting value of $\omega$

$$\omega_{hp} = \cosh\left(\frac{1}{n}\cosh^{-1}\frac{1}{\epsilon}\right) \qquad (2\text{-}287)$$

The poles of $G(s)$ can be determined from the magnitude-squared transfer function. Table 2-5 gives the quadratic factors for $n$ ranging from 1 to 10 and ripples of 3, 1, 0.5, 0.1, and 0.01 dB rather than the explicit complex pole locations.

The pole locations and the element values in Table 2-15 in a later section are based on $\omega = 1$ being the end of the ripple as in Fig. 2-38. The half-power frequencies $\omega_{hp}$ of Eq. (2-287) are given in Table 2-16 in a later section for each value of $n$ and $\epsilon$ for ready reference and use in scaling if desired.

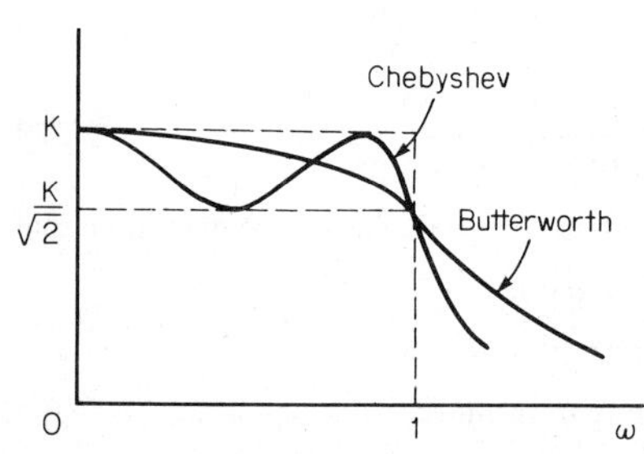


**FIG. 2-40** Butterworth and Chebyshev approximations for $n = 3$ and $\epsilon = 1$.

***189. Comparison of Butterworth and Chebyshev Approximations.*** An equitable comparison of the Butterworth and Chebyshev approximations can be made only when they have the same half-power frequency. In general, this involves normalizing the Chebyshev approximation to a 3-dB bandwidth before comparison. This occurs naturally when $\epsilon = 1$, and considerable insight into the comparative merits of the two approximations can be obtained from this special situation.

Figure 2-40 shows the Butterworth and Chebyshev ($\epsilon = 1$) frequency responses for $n = 3$. The general relationships shown hold for all values of $n$, although the details differ. The two approximations are compared as follows:

1. The Butterworth approximation is superior at and near $\omega = 0$. In fact, it is the optimum low-pass approximation in the sense that of all polynomial approximations it has the greatest number of zero derivatives at $\omega = 0$.
2. The Chebyshev approximation is superior at and near the cutoff frequency. It has a derivative at $\omega = 1$ which is $n$ times that of the Butterworth approximation.
3. The Chebyshev approximation is superior in the stopband. An examination of Eqs. (2-258) and (2-284) for $\epsilon = 1$ shows that the attenuation at high frequencies for the Chebyshev approximation is greater than that for the Butterworth approximation by a constant $6(n - 1)$ dB.
4. The Chebyshev approximation sacrifices smoothness in the passband.

***190. The Modified Chebyshev Approximation.*** It was noted earlier that the even-order Chebyshev approximation was not realizable for equal load and source terminations. The reason for this is that the even-ordered Chebyshev polynomial has a maximum at $\omega = 0$ rather than a zero, and thus the transfer function has a relative minimum at $\omega = 0$. Saal[3] has suggested a modification of the even-ordered Chebyshev polynomial to obtain a new even-ordered polynomial that has a zero at $\omega = 0$, still has maxima and minima of $+1$ and $-1$ within the interval $0 < \omega \leq 1$, and increases monotonically for $\omega > 1$. This procedure yields a set of polynomials $\overline{C}_n(\omega)$ which are given in Table 2-6.

**TABLE 2-5** Linear and Quadratic Factors of Poles of Chebyshev Transfer Functions

| $n$ | Ripple, dB | $[G(s)]^{-1}$ |
|---|---|---|
| 2 | 0.01 | $s^2 + 4.4555282s + 10.4258656$ |
| | 0.10 | $s^2 + 2.3723563s + 3.3140371$ |
| | 0.50 | $s^2 + 1.4256245s + 1.5162026$ |
| | 1.00 | $s^2 + 1.0977343s + 1.1025103$ |
| | 3.00 | $s^2 + 0.6448997s + 0.7079478$ |
| 3 | 0.01 | $s + 1.5893705$<br>$s^2 + 1.5893705s + 3.2760986$ |
| | 0.10 | $s + 0.9694057$<br>$s^2 + 0.9694057s + 1.6897474$ |
| | 0.50 | $s + 0.6264565$<br>$s^2 + 0.6264565s + 1.1424477$ |
| | 1.00 | $s + 0.4941706$<br>$s^2 + 0.4941706s + 0.9942046$ |
| | 3.00 | $s + 0.2986202$<br>$s^2 + 0.2986202s + 0.8391740$ |
| 4 | 0.01 | $s^2 + 0.8217328s + 2.0062682$<br>$s^2 + 1.9838384s + 1.2991615$ |
| | 0.10 | $s^2 + 0.5283127s + 1.3300314$<br>$s^2 + 1.2754598s + 0.6229246$ |
| | 0.50 | $s^2 + 0.3507061s + 1.0635186$<br>$s^2 + 0.8466795s + 0.3564119$ |
| | 1.00 | $s^2 + 0.2790720s + 0.9865049$<br>$s^2 + 0.6737394s + 0.2793981$ |
| | 3.00 | $s^2 + 0.1703408s + 0.9030868$<br>$s^2 + 0.4112391s + 0.1959800$ |
| 5 | 0.01 | $s + 0.8171468$<br>$s^2 + 1.3221713s + 1.0132204$<br>$s^2 + 0.5050245s + 1.5722374$ |
| | 0.10 | $s + 0.5389143$<br>$s^2 + 0.8719817s + 0.6359202$<br>$s^2 + 0.3330674s + 1.1949371$ |
| | 0.50 | $s + 0.3623196$<br>$s^2 + 0.5862455s + 0.4767670$<br>$s^2 + 0.2239258s + 1.0357840$ |
| | 1.00 | $s + 0.2894933$<br>$s^2 + 0.4684101s + 0.4292979$<br>$s^2 + 0.1789167s + 0.9883149$ |
| | 3.00 | $s + 0.1775303$<br>$s^2 + 0.2872500s + 0.3770085$<br>$s^2 + 0.1097197s + 0.9360255$ |
| 6 | 0.01 | $s^2 + 0.3429310s + 1.3719082$<br>$s^2 + 0.9369050s + 0.9388954$<br>$s^2 + 1.2798360s + 0.5058827$ |
| | 0.10 | $s^2 + 0.2293867s + 1.1293868$<br>$s^2 + 0.6266962s + 0.6963741$<br>$s^2 + 0.8560830s + 0.2633614$ |

**TABLE 2-5** *(Continued)*

| $n$ | Ripple, dB | $[G(s)]^{-1}$ |
|---|---|---|
| 6 | 0.50 | $s^2 + 0.1553002s + 1.0230228$<br>$s^2 + 0.4242879s + 0.5900101$<br>$s^2 + 0.5795881s + 0.1569974$ |
| | 1.00 | $s^2 + 0.1243620s + 0.9907323$<br>$s^2 + 0.3397634s + 0.5577196$<br>$s^2 + 0.4641255s + 0.1247069$ |
| | 3.00 | $s^2 + 0.0764590s + 0.9548302$<br>$s^2 + 0.2088899s + 0.5218175$<br>$s^2 + 0.2853490s + 0.0888048$ |
| 7 | 0.01 | $s + 0.5584353$<br>$s^2 + 1.0062656s + 0.5001050$<br>$s^2 + 0.6963574s + 0.9231104$<br>$s^2 + 0.2485271s + 1.2623344$ |
| | 0.10 | $s + 0.3767779$<br>$s^2 + 0.6789303s + 0.3302167$<br>$s^2 + 0.4698343s + 0.7532220$<br>$s^2 + 0.1676819s + 1.0924460$ |
| | 0.50 | $s + 0.2561700$<br>$s^2 + 0.4616024s + 0.2538782$<br>$s^2 + 0.3194388s + 0.6768835$<br>$s^2 + 0.1140064s + 1.0161075$ |
| | 1.00 | $s + 0.2054143$<br>$s^2 + 0.3701438s + 0.2304501$<br>$s^2 + 0.2561474s + 0.6534555$<br>$s^2 + 0.0914180s + 0.9926795$ |
| | 3.00 | $s + 0.1264854$<br>$s^2 + 0.2279188s + 0.2042536$<br>$s^2 + 0.1577247s + 0.6272590$<br>$s^2 + 0.0562913s + 0.9664830$ |
| 8 | 0.01 | $s^2 + 0.9480834s + 0.2716669$<br>$s^2 + 0.8037463s + 0.5422649$<br>$s^2 + 0.5370461s + 0.9249484$<br>$s^2 + 0.1885855s + 1.1955464$ |
| | 0.10 | $s^2 + 0.6432996s + 0.1456123$<br>$s^2 + 0.5453631s + 0.4162103$<br>$s^2 + 0.3644000s + 0.7988938$<br>$s^2 + 0.1279602s + 1.0694918$ |
| | 0.50 | $s^2 + 0.4385859s + 0.0880523$<br>$s^2 + 0.3718151s + 0.3586504$<br>$s^2 + 0.2484389s + 0.7413338$<br>$s^2 + 0.0872402s + 1.0119319$ |
| | 1.00 | $s^2 + 0.3519965s + 0.0702612$<br>$s^2 + 0.2984083s + 0.3408593$<br>$s^2 + 0.1993900s + 0.7235427$<br>$s^2 + 0.0700165s + 0.9941407$ |
| | 3.00 | $s^2 + 0.2169614s + 0.0502939$<br>$s^2 + 0.1839310s + 0.3208920$<br>$s^2 + 0.1228988s + 0.7035754$<br>$s^2 + 0.0431563s + 0.9741735$ |

**TABLE 2-5** *(Continued)*

| $n$ | Ripple, dB | $[G(s)]^{-1}$ |
|---|---|---|
| 9 | 0.01 | $s + 0.4264123$<br>$s^2 + 0.8013929s + 0.2988052$<br>$s^2 + 0.6533015s + 0.5950033$<br>$s^2 + 0.4264123s + 0.9318274$<br>$s^2 + 0.1480914s + 1.1516737$ |
| | 0.10 | $s + 0.2904612$<br>$s^2 + 0.5458885s + 0.2013455$<br>$s^2 + 0.4450123s + 0.4975436$<br>$s^2 + 0.2904612s + 0.8343677$<br>$s^2 + 0.1008761s + 1.0542140$ |
| | 0.50 | $s + 0.1984053$<br>$s^2 + 0.3728800s + 0.1563424$<br>$s^2 + 0.3039745s + 0.4525406$<br>$s^2 + 0.1984053s + 0.7893647$<br>$s^2 + 0.0689054s + 1.0092110$ |
| | 1.00 | $s + 0.1593305$<br>$s^2 + 0.2994433s + 0.1423640$<br>$s^2 + 0.2441084s + 0.4385621$<br>$s^2 + 0.1593305s + 0.7753862$<br>$s^2 + 0.0553349s + 0.9952325$ |
| | 3.00 | $s + 0.0982746$<br>$s^2 + 0.1846958s + 0.1266357$<br>$s^2 + 0.1505654s + 0.4228338$<br>$s^2 + 0.0982746s + 0.7596579$<br>$s^2 + 0.0341304s + 0.9795042$ |
| 10 | 0.01 | $s^2 + 0.7540220s + 0.1701746$<br>$s^2 + 0.6802131s + 0.3518103$<br>$s^2 + 0.5398201s + 0.6457029$<br>$s^2 + 0.3465859s + 0.9395955$<br>$s^2 + 0.1194254s + 1.1212312$ |
| | 0.10 | $s^2 + 0.5150591s + 0.0924569$<br>$s^2 + 0.4646415s + 0.2740926$<br>$s^2 + 0.3687416s + 0.5679852$<br>$s^2 + 0.2367467s + 0.8618778$<br>$s^2 + 0.0815773s + 1.0435134$ |
| | 0.50 | $s^2 + 0.3522999s + 0.0562789$<br>$s^2 + 0.3178143s + 0.2379146$<br>$s^2 + 0.2522189s + 0.5318072$<br>$s^2 + 0.1619345s + 0.8256998$<br>$s^2 + 0.0557988s + 1.0073354$ |
| | 1.00 | $s^2 + 0.2830386s + 0.0450019$<br>$s^2 + 0.2553328s + 0.2266375$<br>$s^2 + 0.2026332s + 0.5205301$<br>$s^2 + 0.1300985s + 0.8144227$<br>$s^2 + 0.0448289s + 0.9960584$ |
| | 3.00 | $s^2 + 0.1746631s + 0.0322899$<br>$s^2 + 0.1575659s + 0.2139255$<br>$s^2 + 0.1250450s + 0.5078181$<br>$s^2 + 0.0802838s + 0.8017108$<br>$s^2 + 0.0276639s + 0.9833464$ |

**TABLE 2-6** Modified Chebyshev Polynomials of Even Order

| $n$ | $\bar{C}_n(\omega)$ |
|---|---|
| 2 | $\omega^2$ |
| 4 | $5.82842713\omega^4 - 4.82842713\omega^2$ |
| 6 | $25.9903811\omega^6 - 36.1865335\omega^4 + 11.1961524\omega^2$ |
| 8 | $109.597711\omega^8 - 210.522708\omega^6$ |
| | $+ 122.034355\omega^4 - 20.109358\omega^2$ |
| 10 | $452.344415\omega^{10} - 1102.492675\omega^8 + 926.297276\omega^6$ |
| | $- 306.717773\omega^4 + 31.568758\omega^2$ |

For a given value of $n$, the modified Chebyshev polynomials have the property that

$$C_n(\omega) > \bar{C}_n(\omega) > C_{n-1}(\omega) \tag{2-288}$$

for $\omega > 1$. Thus a low-pass approximation based on these polynomials for an even value of $n$ will be better in the stopband than that using the Chebyshev polynomial of order $n - 1$ but will not be as good as that using the regular Chebyshev polynomial of order $n$. Thus this modified Chebyshev approximation is a possible alternative in some situations to the increase of $n$ to the next higher odd value to achieve realizability, especially in the case when equal load and source resistances are required.

The extent of the difference between $\bar{C}_n(\omega)$ and $C_n(\omega)$ is shown in Fig. 2-41, where the ratio $\bar{C}_n(\omega)/C_n(\omega)$ for $\omega \geqslant 1$ is shown for even values of $n$.

Table 2-7 gives the poles of $G(s)$, in terms of quadratic factors, for $n = 2, 4, 6, 8, 10$ and ripples of 0.01, 0.1, 0.5, 1.0, and 3.0 dB.

As for the regular Chebyshev approximation, the pole locations and the element values in Table 2-15 in a later section are based on $\omega = 1$ being the end of the ripple. The half-power frequencies are given in Table 2-16 in a later section.

***191. The Legendre-Papoulis Approximation.*** A. Papoulis[4] has described an approximation

$$|G(j\omega)| = \frac{K}{[1 + L_n(\omega^2)]^{1/2}} \tag{2-289}$$

where the polynomial $L_n(\omega^2)$ is derived, using Legendre polynomials of the first kind, to satisfy

**TABLE 2-7** Linear and Quadratic Factors Giving Poles of Modified Chebyshev Transfer Functions

| $n$ | Ripple, dB | $[G(s)]^{-1}$ |
|---|---|---|
| 2 | 0.01 | $s^2 + 6.4541054s + 20.8277385$ |
| | 0.10 | $s^2 + 3.6200009s + 6.5522033$ |
| | 0.50 | $s^2 + 2.3928122s + 2.8627752$ |
| | 1.00 | $s^2 + 1.9825371s + 1.9652267$ |
| | 3.00 | $s^2 + 1.4158936s + 1.0023773$ |
| 4 | 0.01 | $s^2 + 2.2680440s + 1.6170715$ |
| | | $s^2 + 0.9235542s + 2.2098435$ |
| | 0.10 | $s^2 + 1.5452285s + 0.7991383$ |
| | | $s^2 + 0.6059475s + 1.4067406$ |
| | 0.50 | $s^2 + 1.1191759s + 0.4523364$ |
| | | $s^2 + 0.4086395s + 1.0858612$ |
| | 1.00 | $s^2 + 0.9486848s + 0.3398608$ |
| | | $s^2 + 0.3272401s + 0.9921108$ |
| | 3.00 | $s^2 + 0.6845838 + 0.1932927$ |
| | | $s^2 + 0.2013994s + 0.8897428$ |

**TABLE 2-7** *(Continued)*

| $n$ | Ripple, dB | $[G(s)]^{-1}$ |
|---|---|---|
| 6 | 0.01 | $s^2 + 1.4118889s + 0.5893450$ |
| | | $s^2 + 1.0058571s + 0.9699759$ |
| | | $s^2 + 0.3640166s + 1.4018414$ |
| | 0.10 | $s^2 + 0.9995998s + 0.3173248$ |
| | | $s^2 + 0.6819299s + 0.6966158$ |
| | | $s^2 + 0.2448454s + 1.1404529$ |
| | 0.50 | $s^2 + 0.7362729s + 0.1875012$ |
| | | $s^2 + 0.4662515s + 0.5727968$ |
| | | $s^2 + 0.1663135s + 1.0255810$ |
| | 1.00 | $s^2 + 0.6267633s + 0.1428406$ |
| | | $s^2 + 0.3748110s + 0.5343433$ |
| | | $s^2 + 0.1333343s + 0.9906678$ |
| | 3.00 | $s^2 + 0.4534072s + 0.0825380$ |
| | | $s^2 + 0.2315851s + 0.4909174$ |
| | | $s^2 + 0.0820877s + 0.9518235$ |
| 8 | 0.01 | $s^2 + 1.0335548s + 0.3097540$ |
| | | $s^2 + 0.8491714s + 0.5489161$ |
| | | $s^2 + 0.5591534s + 0.9283901$ |
| | | $s^2 + 0.1954150s + 1.2038909$ |
| | 0.10 | $s^2 + 0.7417790s + 0.1718214$ |
| | | $s^2 + 0.5828511s + 0.4083754$ |
| | | $s^2 + 0.3806989s + 0.7943820$ |
| | | $s^2 + 0.1328521s + 1.0725554$ |
| | 0.50 | $s^2 + 0.5496051s + 0.1030188$ |
| | | $s^2 + 0.4007060s + 0.3416991$ |
| | | $s^2 + 0.2600622s + 0.7328336$ |
| | | $s^2 + 0.0906706s + 1.0125586$ |
| | 1.00 | $s^2 + 0.4685593s + 0.0788470$ |
| | | $s^2 + 0.3226466s + 0.3205446$ |
| | | $s^2 + 0.2088595s + 0.7137508$ |
| | | $s^2 + 0.0727952s + 0.9940103$ |
| | 3.00 | $s^2 + 0.3392712s + 0.0458028$ |
| | | $s^2 + 0.1997071s + 0.2963792$ |
| | | $s^2 + 0.1288387s + 0.6922959$ |
| | | $s^2 + 0.0448874s + 0.9731910$ |
| 10 | 0.01 | $s^2 + 0.8175750s + 0.1921666$ |
| | | $s^2 + 0.7133475s + 0.3528340$ |
| | | $s^2 + 0.5571617s + 0.6426718$ |
| | | $s^2 + 0.3555630s + 0.9397252$ |
| | | $s^2 + 0.1222554s + 1.1244354$ |
| | 0.10 | $s^2 + 0.5905036s + 0.1080675$ |
| | | $s^2 + 0.4925060s + 0.2665100$ |
| | | $s^2 + 0.3816272s + 0.5602769$ |
| | | $s^2 + 0.2431732s + 0.8592520$ |
| | | $s^2 + 0.0835800s + 1.0446869$ |
| | 0.50 | $s^2 + 0.4387194s + 0.0652281$ |
| | | $s^2 + 0.3394908s + 0.2246543$ |
| | | $s^2 + 0.2614400s + 0.5216326$ |
| | | $s^2 + 0.1664374s + 0.8217378$ |
| | | $s^2 + 0.0571933s + 1.0075592$ |
| | 1.00 | $s^2 + 0.3742846s + 0.0500294$ |
| | | $s^2 + 0.2735736s + 0.2112433$ |
| | | $s^2 + 0.2101536s + 0.5095395$ |
| | | $s^2 + 0.1337446s + 0.8100361$ |
| | | $s^2 + 0.0459557s + 0.9959854$ |
| | 3.00 | $s^2 + 0.2711265s + 0.0291328$ |
| | | $s^2 + 0.1694797s + 0.1958431$ |
| | | $s^2 + 0.1297678s + 0.4958770$ |
| | | $s^2 + 0.0825542s + 0.7968406$ |
| | | $s^2 + 0.0283639s + 0.9829386$ |

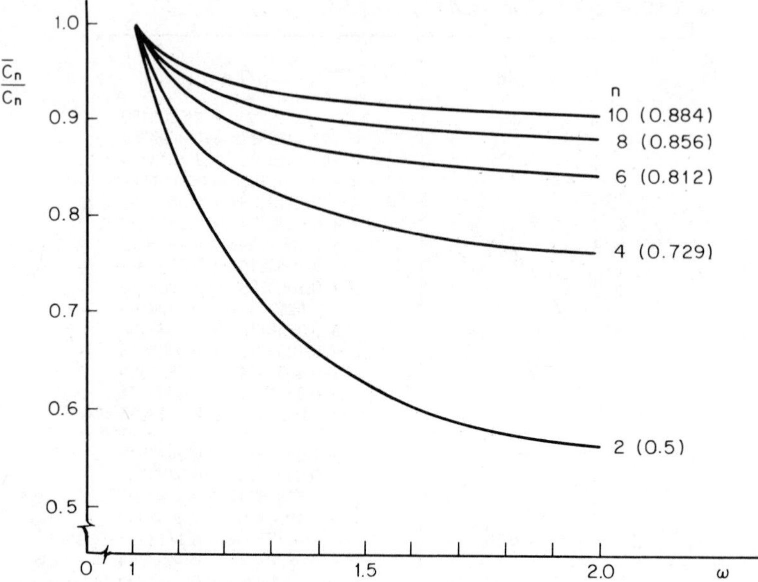


**FIG. 2-41** Ratio of modified Chebyshev polynomial to Chebyshev polynomial. Limiting value as $\omega \to \infty$ is in parentheses for each value of $n$.

the following requirements:

1. $L_n(\omega^2)$ increases monotonically; thus $|G(j\omega)|$ decreases monotonically.
2. $L_n(0) = 0$ and $L_n(1) = 1$.
3. Of all the polynomials satisfying 1 and 2, $L_n(\omega^2)$ has the largest derivative at $\omega = 1$.

The resulting polynomials are tabulated in Table 2-8 for $n = 2$ to 10. The pole locations for $G(s)$ are described in Table 2-9 in terms of the quadratic factors.

Figure 2-42 compares the Legendre-Papoulis approximation to the Butterworth and Chebyshev ($\epsilon = 1$) approximation for $n = 3$. The Legendre-Papoulis approximation is smoother than the Chebyshev approximation but is not as smooth as the Butterworth approximation in the passband. The stopband characteristics of the Legendre-Papoulis approximation are better than those of the Butterworth approximation but are not as good as those of the Chebyshev approximation. It is apparent that the Legendre-Papoulis approximation is in many respects a good compromise between the Butterworth and Chebyshev approximations.

**TABLE 2-8** The Polynomials $L_n(\omega^2)$

| $n$ | |
|---|---|
| 2 | $\omega^4$ |
| 3 | $3\omega^6 - 3\omega^4 + \omega^2$ |
| 4 | $6\omega^8 - 8\omega^6 + 3\omega^4$ |
| 5 | $20\omega^{10} - 40\omega^8 + 28\omega^6 - 8\omega^4 + \omega^2$ |
| 6 | $50\omega^{12} - 120\omega^{10} + 105\omega^8 - 40\omega^6 + 6\omega^4$ |
| 7 | $175\omega^{14} - 525\omega^{12} + 615\omega^{10} - 355\omega^8 + 105\omega^6 - 15\omega^4 + \omega^2$ |
| 8 | $490\omega^{16} - 1680\omega^{14} + 2310\omega^{12} - 1624\omega^{10} + 615\omega^8 - 120\omega^6 + 10\omega^4$ |
| 9 | $1764\omega^{18} - 7056\omega^{16} + 11704\omega^{14} - 10416\omega^{12} + 5376\omega^{10} - 1624\omega^8 + 276\omega^6 - 24\omega^4 + \omega^2$ |
| 10 | $5292\omega^{20} - 23520\omega^{18} + 44100\omega^{16} - 45360\omega^{14} + 27860\omega^{12} - 10416\omega^{10} + 2310\omega^8 - 280\omega^6 + 15\omega^4$ |

**TABLE 2-9** Linear and Quadratic Factors for Poles of Legendre-Papoulis Approximation

| $n$ | $[G(s)]^{-1}$ |
|---|---|
| 2 | $s^2 + 1.4142136s + 1.0000000$ |
| 3 | $s + 0.6203318$ |
| | $s^2 + 0.6903712s + 0.9307119$ |
| 4 | $s^2 + 0.4633774s + 0.9476701$ |
| | $s^2 + 1.0994868s + 0.4307915$ |
| 5 | $s + 0.4680899$ |
| | $s^2 + 0.3071734s + 0.9608963$ |
| | $s^2 + 0.7762796s + 0.4971406$ |
| 6 | $s^2 + 0.2303854s + 0.9696012$ |
| | $s^2 + 0.6179218s + 0.5828947$ |
| | $s^2 + 0.8778030s + 0.2502256$ |
| 7 | $s + 0.3821033$ |
| | $s^2 + 0.1724170s + 0.9764158$ |
| | $s^2 + 0.4748794s + 0.6621299$ |
| | $s^2 + 0.6984636s + 0.3060005$ |
| 8 | $s^2 + 0.1378844s + 0.9808397$ |
| | $s^2 + 0.3885518s + 0.7179832$ |
| | $s^2 + 0.6005680s + 0.3828971$ |
| | $s^2 + 0.7343526s + 0.1675357$ |
| 9 | $s + 0.3256878$ |
| | $s^2 + 0.1101944s + 0.9844435$ |
| | $s^2 + 0.3145676s + 0.7666498$ |
| | $s^2 + 0.4971058s + 0.4635058$ |
| | $s^2 + 0.6187708s + 0.2089807$ |
| 10 | $s^2 + 0.0918020s + 0.9869313$ |
| | $s^2 + 0.2650376s + 0.8012497$ |
| | $s^2 + 0.4283460s + 0.5282527$ |
| | $s^2 + 0.5548108s + 0.2702425$ |
| | $s^2 + 0.6344130s + 0.1217699$ |

The selection of $n$ follows the same general procedure as for the Butterworth approximation. If the specification

$$A \geqslant |G(j\omega_a)| = \frac{K}{(1 + L_n(\omega_a^2))^{1/2}} \tag{2-290}$$

is to be satisfied, then

$$L_n(\omega_a^2) \geqslant \frac{K^2}{A^2} - 1 \tag{2-291}$$

and successively higher-ordered polynomials of Table 2-8 can be evaluated at $\omega_a$ until Eq. (2-291) is satisfied.

***192. The Bessel Approximation.*** The ideal all-pass characteristic of Fig. 2-33

$$|G(j\omega)| = K \tag{2-292}$$

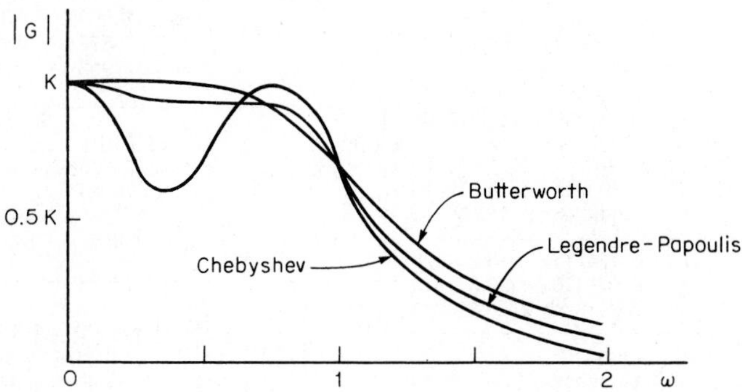


**FIG. 2-42** Comparison of Butterworth, Legendre-Papoulis, and Chebyshev ($\epsilon = 1$) approximations for $n = 3$.

and

$$\theta(j\omega) = T\omega \tag{2-293}$$

implies that the output is a replica (scaled by the constant $K$) of the input but delayed in time by $T$ seconds. The negative derivative of the phase function

$$-\frac{d\theta(j\omega)}{d\omega} = \tau_d(\omega) \tag{2-294}$$

is called the group delay and in the ideal situation is a constant $T$.

One approach to the approximation of the ideal delay characteristic, first described by Thomson,[5] is to make the group delay maximally flat at $\omega = 0$. This leads to a transfer function of the form

$$G(s) = \frac{b_0 K}{b_0 + b_1 s + \cdots + b_{n-1}s^{n-1} + s^n} \tag{2-295}$$

where the denominator polynomial is related to a class of Bessel polynomials. The coefficients for these polynomials are given in Table 2-10 for the approximation to a group delay of 1 second at $\omega = 0$. These polynomials are related by the recursion formula

$$B_n = (2n - 1)B_{n-1} + s^2 B_{n-2} \tag{2-296}$$

The pole locations for $G(s)$ are described in Table 2-11 in terms of the quadratic factors.

**TABLE 2-10** The Coefficients of the Bessel Polynomials

| $n$ | $b_0$ | $b_1$ | $b_2$ | $b_3$ | $b_4$ | $b_5$ | $b_6$ | $b_7$ |
|---|---|---|---|---|---|---|---|---|
| 1 | 1 | | | | | | | |
| 2 | 3 | 3 | | | | | | |
| 3 | 15 | 15 | 6 | | | | | |
| 4 | 105 | 105 | 45 | 10 | | | | |
| 5 | 945 | 945 | 420 | 105 | 15 | | | |
| 6 | 10,395 | 10,395 | 4,725 | 1,260 | 210 | 21 | | |
| 7 | 135,135 | 135,135 | 62,370 | 17,325 | 3,150 | 378 | 28 | |
| 8 | 2,027,025 | 2,027,025 | 945,945 | 270,270 | 51,975 | 6,930 | 630 | 36 |

**TABLE 2-11** Linear and Quadratic Factors for Poles of Bessel Approximation

| $n$ | $[G(s)]^{-1}$ |
|---|---|
| 2 | $s^2 + 3s + 3$ |
| 3 | $s + 2.322185$<br>$s^2 + 3.677814s + 6.459432$ |
| 4 | $s^2 + 5.792422s + 9.140133$<br>$s^2 + 4.207578 + 11.487799$ |
| 5 | $s + 3.646739$<br>$s^2 + 6.703912s + 14.272476$<br>$s^2 + 4.649348s + 18.156314$ |
| 6 | $s^2 + 8.496718s + 18.801128$<br>$s^2 + 7.471416s + 20.852819$<br>$s^2 + 5.031864s + 26.514025$ |
| 7 | $s + 4.971787$<br>$s^2 + 9.516582s + 25.666449$<br>$s^2 + 8.140278s + 28.936544$<br>$s^2 + 5.371354s + 36.596784$ |
| 8 | $s^2 + 11.175772s + 31.977224$<br>$s^2 + 10.409682s + 33.934741$<br>$s^2 + 8.736578s + 38.569256$<br>$s^2 + 5.677968s + 48.432015$ |
| 9 | $s + 6.297019$<br>$s^2 + 12.258736s + 40.589268$<br>$s^2 + 11.208844s + 43.646648$<br>$s^2 + 9.276880s + 49.788507$<br>$s^2 + 5.958522s + 62.041443$ |
| 10 | $s^2 + 13.844090s + 48.667550$<br>$s^2 + 13.230582s + 50.582362$<br>$s^2 + 11.935056s + 54.839151$<br>$s^2 + 9.772440s + 62.625584$<br>$s^2 + 6.217832s + 77.442692$ |

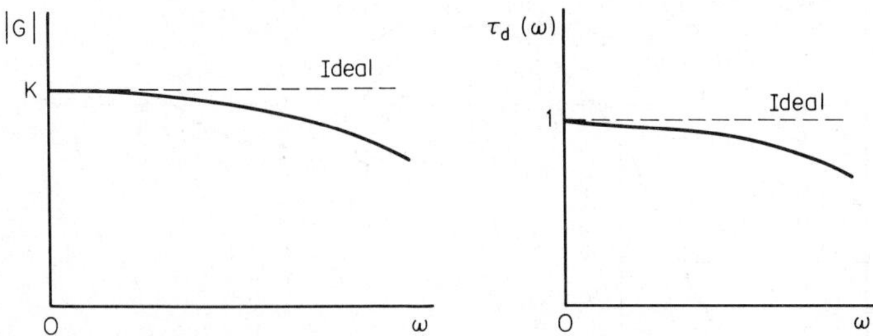


**FIG. 2-43** Magnitude and group delay for maximally flat group-delay approximation.

The general forms of the magnitude and phase functions for this approximation are indicated in Fig. 2-43. It is noted that $|G(j\omega)|$ exhibits a low-pass characteristic. The approximation can be used as a low-pass approximation in situations where linearity of phase in the passband is of concern. In this application, it would be appropriate to normalize so that $\omega = 1$ is the half-power frequency. Here, however, the normalization makes $\tau_d(0) = 1$. The resulting half-power frequencies and the group delays at those frequencies are tabulated in Table 2-16 in a later section.

The parameter $n$ is selected to satisfy either a minimum group delay requirement or a minimum magnitude requirement at some frequency $\omega_n$ that is,

$$\tau_d(\omega_a) \geqslant T_a \tag{2-297}$$

or

$$|G(j\omega_a)| \geqslant A \tag{2-298}$$

The condition of Eq. (2-298) can be stated equivalently in terms of the maximum attenuation at $\omega_a$. Figure 2-44 gives the attenuation versus $\omega$ and the group delay versus $\omega$ for $n = 1$ to 8 and can be used for the selection of $n$.

***193. The Step Response.*** The step response can be described in terms of various figures of merit. Figure 2-45 is used to define three of these, overshoot, rise time, and delay time, that are used here. Tables 2-12, 2-13, and 2-14 give the percent overshoot, rise time, and delay time, respectively, for each of the five approximations that have been described.

The general form of response illustrated in Fig. 2-45 is typical except for the odd-order Chebyshev approximation and the modified Chebyshev approximations when the ripple is large. Figure 2-46 shows the response for the third-order Chebyshev approximation with a 3-dB ripple. Attention is drawn to the fact that the first relative maximum is less than the final value. This characteristic becomes even more pronounced with increasingly large values of $n$, with several relative maxima occurring prior to the peak value, as is illustrated in Fig. 2-47 for $n = 5$.

***194. The Elliptic Filter.*** The five approximations that have been described are all of a class known as all-pole approximations, and the resulting filters are known as all-pole filters because the magnitude-squared function of Eq. (2-247) has no finite zeros. If the ratio of $M(\omega^2)/N\omega^2)$ of Eq. (2-246) is permitted to be a Chebyshev rational function, the

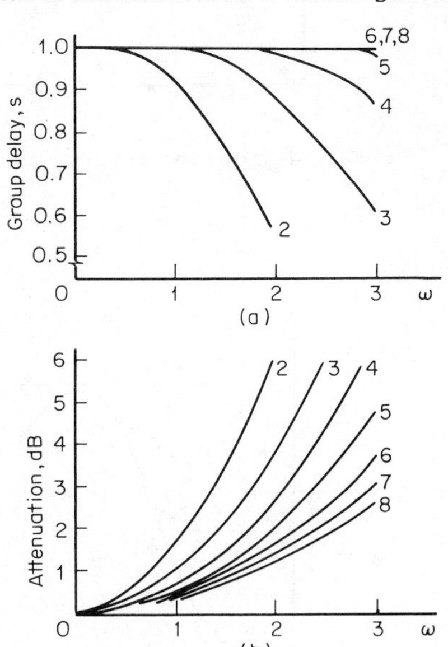


**FIG. 2-44** Group delay and attenuation for the Bessel approximation.

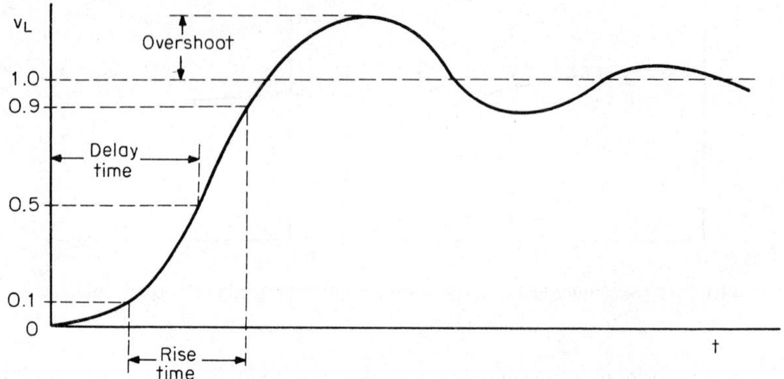


**FIG. 2-45** General form of the step response and definitions of figures of merit.

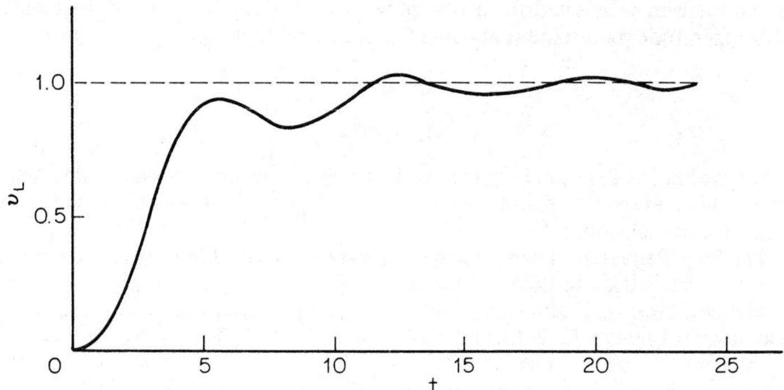


**FIG. 2-46** Step response of third-order Chebyshev approximation with a 3-dB ripple.

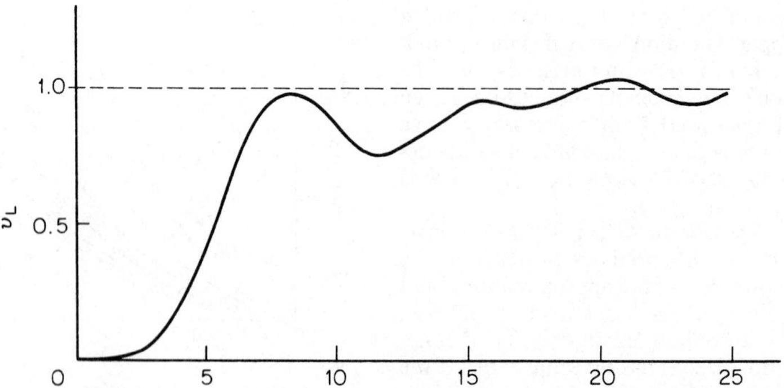


**FIG. 2-47** Step response for fifth-order Chebyshev approximation with a 3-dB ripple.

**TABLE 2-12** Percent Overshoot

| Order | Overshoot Bessel, % | Overshoot Butterworth, % | Overshoot Legendre-Papoulis, % | Overshoot Chebyshev, % | | | | Overshoot Modified Chebyshev, % | | | |
|---|---|---|---|---|---|---|---|---|---|---|---|
| | | | | 3.0 dB | 1.0 dB | 0.5 dB | 0.1 dB | 3.0 dB | 1.0 dB | 0.5 dB | 0.1 dB |
| 2 | 0.4 | 4.3 | 4.3 | 27.2 | 14.6 | 10.7 | 6.7 | 4.3 | 4.3 | 4.3 | 4.3 |
| 3 | 0.7 | 8.1 | 7.5 | 2.7 | 6.4 | 8.9 | 10.2 | | | | |
| 4 | 0.8 | 10.8 | 11.2 | 35.7 | 21.9 | 18.1 | 14.5 | 8.0 | 10.5 | 12.1 | 12.9 |
| 5 | 0.8 | 12.8 | 13.3 | 1.7 | 10.2 | 13.2 | 15.2 | | | | |
| 6 | 0.6 | 14.3 | 15.2 | 38.7 | 24.9 | 21.2 | 18.0 | 8.5 | 12.8 | 15.1 | 16.5 |
| 7 | 0.5 | 15.4 | 16.4 | 2.0 | 12.1 | 15.3 | 17.7 | | | | |
| 8 | 0.3 | 16.3 | 17.6 | 40.4 | 26.5 | 23.0 | 19.8 | 8.7 | 14.0 | 16.6 | 18.5 |
| 9 | 0.2 | 17.1 | 18.3 | 3.1 | 13.3 | 16.7 | 19.1 | | | | |
| 10 | 0.1 | 17.8 | 19.1 | 41.5 | 27.6 | 24.1 | 21.0 | 9.5 | 14.8 | 17.6 | 19.7 |

**TABLE 2-13** Rise Time for Polynomial Filters

| Order | Rise time Bessel | Rise time Butterworth | Rise time Legendre-Papoulis | Rise time Chebyshev | | | | Rise time modified Chebyshev | | | |
|---|---|---|---|---|---|---|---|---|---|---|---|
| | | | | 3.0 dB | 1.0 dB | 0.5 dB | 0.1 dB | 3.0 dB | 1.0 dB | 0.5 dB | 0.1 dB |
| 2 | 1.58 | 2.2 | 2.2 | 1.7 | 1.6 | 1.5 | 1.1 | 2.2 | 1.5 | 1.3 | 0.8 |
| 3 | 1.25 | 2.3 | 2.5 | 3.2 | 2.4 | 2.2 | 1.7 | | | | |
| 4 | 1.04 | 2.4 | 2.7 | 2.5 | 2.5 | 2.4 | 2.2 | 3.2 | 2.6 | 2.4 | 2.1 |
| 5 | 0.91 | 2.6 | 2.9 | 3.7 | 3.0 | 2.8 | 2.5 | | | | |
| 6 | 0.82 | 2.7 | 3.0 | 2.9 | 3.0 | 2.9 | 2.8 | 3.7 | 3.2 | 3.0 | 2.7 |
| 7 | 0.75 | 2.8 | 3.2 | 4.0 | 3.4 | 3.3 | 3.0 | | | | |
| 8 | 0.69 | 2.9 | 3.4 | 3.2 | 3.3 | 3.3 | 3.2 | 4.1 | 3.6 | 3.4 | 3.2 |
| 9 | 0.64 | 3.0 | 3.5 | 4.3 | 3.8 | 3.6 | 3.4 | | | | |
| 10 | 0.60 | 3.1 | 3.6 | 3.5 | 3.6 | 3.6 | 3.5 | 4.4 | 3.9 | 3.7 | 3.5 |

**TABLE 2-14** Delay Times for Polynomial Filters

| Order | Delay time Bessel | Delay time Butterworth | Delay time Legendre-Papoulis | Delay Chebyshev | | | | Delay modified Chebyshev | | | |
|---|---|---|---|---|---|---|---|---|---|---|---|
| | | | | 3.0 dB | 1.0 dB | 0.5 dB | 0.1 dB | 3.0 dB | 1.0 dB | 0.5 dB | 0.1 dB |
| 2 | 0.90 | 1.4 | 1.4 | 1.5 | 1.2 | 1.1 | 0.8 | 1.4 | 1.0 | 0.8 | 0.6 |
| 3 | 0.96 | 2.1 | 2.4 | 3.0 | 2.4 | 2.2 | 1.7 | | | | |
| 4 | 0.98 | 2.8 | 3.3 | 3.6 | 3.3 | 3.2 | 2.7 | 3.9 | 3.3 | 3.1 | 2.6 |
| 5 | 0.99 | 3.5 | 4.2 | 5.2 | 4.6 | 4.3 | 3.8 | | | | |
| 6 | 0.99 | 4.2 | 5.2 | 5.6 | 5.4 | 5.2 | 4.8 | 6.1 | 5.5 | 5.2 | 4.7 |
| 7 | 1.00 | 4.8 | 6.2 | 7.3 | 6.7 | 6.4 | 5.8 | | | | |
| 8 | 1.00 | 5.5 | 7.1 | 7.7 | 7.5 | 7.3 | 6.8 | 8.2 | 7.6 | 7.3 | 6.8 |
| 9 | 1.00 | 6.2 | 8.1 | 9.4 | 8.8 | 8.5 | 7.9 | | | | |
| 10 | 1.00 | 6.8 | 9.0 | 9.7 | 9.6 | 9.4 | 8.9 | 10.3 | 9.7 | 9.4 | 6.9 |

result is an equiripple characteristic in the stopband as well as in the passband. Filters based on such approximations are called elliptic filters because their design often employs elliptic functions. Such filters have three parameters as opposed to only two in the Chebyshev approximations. As a consequence, the design procedures become quite involved and extensive tables of element values are required. Such tables can be found in Zverev[6] and Hansell.[7]

***195. The Prototype Filter Networks.*** The all-pole approximations that have been described can be realized by lossless ladder networks terminated in resistance at both ends. The applicable synthesis techniques are described in the literature,[8,9,10] and only the results are given here. The resulting ladder networks are shown in Figs. 2-48 and 2-49.

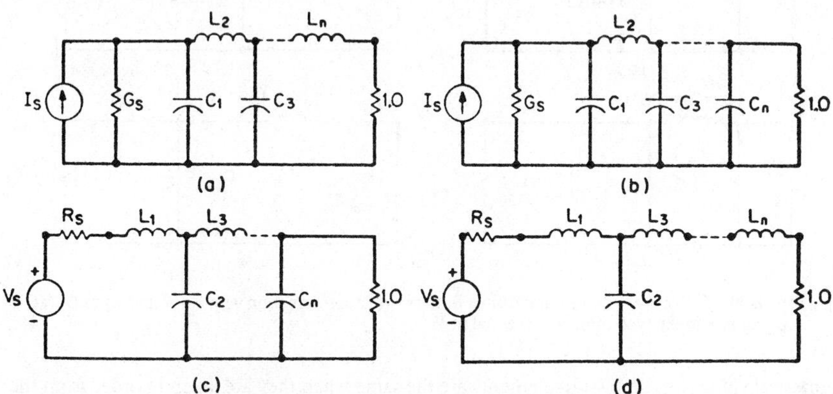


**FIG. 2-48** Four filter networks classified according to type of transfer function and $n$ even or odd: *(a)* $Z_{SL}$, $n$ even; *(b)* $Z_{SL}$, $n$ odd; *(c)* $G_{SL}$, $n$ even; *(d)* $G_{SL}$, $n$ odd.

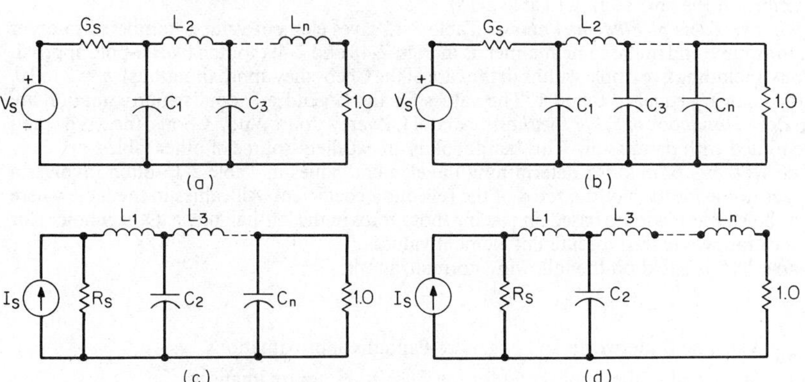


**FIG. 2-49** Filter networks derived from those in Fig. 2-48 by source transformations: *(a)* $G_{SL}$, $n$ even; *(b)* $G_{SL}$, $n$ odd; *(c)* $Z_{SL}$, $n$ even; *(d)* $Z_{SL}$, $n$ odd.

Figure 2-48 shows four networks classified according to the type of transfer function, $Z_{SL}$ or $G_{SL}$, being realized and whether $n$ is even or odd. It is noted that the networks of Fig. 2-48*c* and *d* are duals of those in Fig. 2-48*a* and *b*, respectively. The four networks of Fig. 2-49 can be regarded as being derived from those of Fig. 2-48 by source transformation.

For a given type of source and value of $n$ there are thus two networks that can be used; for example, with a current source input and with $n$ odd, either the network of Fig. 2-48*b* or that of Fig. 2-49*d* can be used. A significant difference is that the first and last lossless elements in the first case are shunt capacitors and in the second case are series inductors.

The notation and organization in Figs. 2-48 and 2-49 are based on the fact that for a given type of approximation and value of $n$ the same numerical element values apply to four different networks. For example, a Butterworth approximation with $n = 2$ and $R_S$ or $G_S$, according to the specific network being used, equal to ½ results in the four networks of Fig. 2-50. Here, the

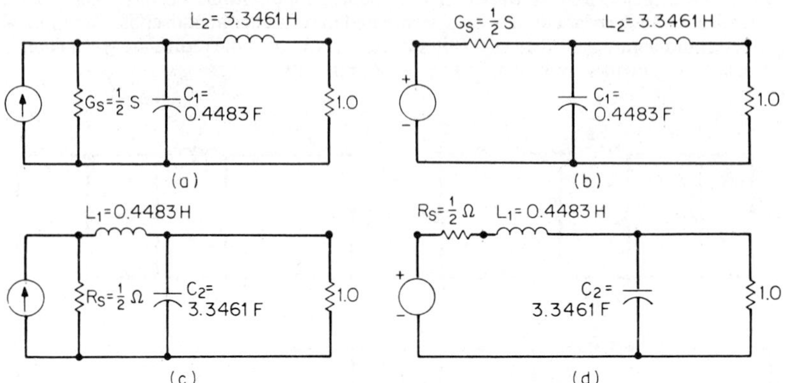


**FIG. 2-50** Four networks resulting from a Butterworth approximation with $n = 2$ and $R_S$ or $G_S$ (as approximate to the type of network) equal to ½.

numerical values for the lossless elements are the same when they are taken in order from the source end. This fact makes it possible to construct a table of element values which uses as parameters for entry (1) the type of approximation, including the ripple width in dB, Eq. (2-277), in the case of the Chebyshev approximations; (2) the value of $n$; and (3) the value of $R_S$ or $G_S$, according to the particular network in Figs. 2-48 and 2-49 that is being used. Such a table is described in the next section (Table 2-15).

***196. The Table of Element Values.*** Table 2-15 gives element values, numbered in order from the source end in the same manner as in Figs. 2-48 and 2-49, for the five all-pole approximations (including five ripple widths in the case of the Chebyshev approximations), $n = 2$ to 10, and for $R_S$ or $G_S$ equal to 0 and 1. The values for the Legendre-Papoulis approximation are taken from *Handbook of Filter Synthesis,* Anatol I. Zverev, John Wiley & Sons, Inc., 1967, and are reprinted with permission. This handbook is an excellent source of other tables.

The synthesis process for determining the element values in Table 2-15 often involves a choice as to the location of the zeros of the reflection coefficient. All values in the table where such a choice was made are based on placing those zeros in the left-half plane. Other choices for their location would lead to different element values.

Table 2-15 is based on the following normalizations:

1. $R_L = 1$.
2. $\omega_{hp} = 1$ for the Butterworth and Legendre-Papoulis approximations.
3. $\omega = 1$ is the end of the ripple band for the Chebyshev approximations.
4. The group delay at $\omega = 0$ is 1 second for the Bessel approximation.

The half-power frequencies for the Chebyshev and Bessel approximations and the group delay of the half-power frequency for the Bessel approximation have been calculated and are tabulated in Table 2-16.

Only two values of $G_S$ or $R_S$, 0 and 1, are used in Table 2-15, since these represent the two most commonly encountered situations. Tables for other values of $G_S$ and $R_S$ can be found in Zverev[6] and Weinberg.[10]

In the case of $n$ odd and $G_S = R_S = 1$ for the Butterworth and regular Chebyshev approximations, the resulting networks are symmetrical; that is, $C_1 = C_n$, $L_2 = L_{n-1}$, etc. In these particular cases, the element values can be modified easily to realize other values of $G_S$ or $R_S$.

**TABLE 2-15** Low-Pass Filter Circuit Element Values

(Table 2-16 gives half-power frequencies for Bessel and Chebyshev circuits. The half-power frequency is 1.000 radians per second for Butterworth and Legendre-Papoulis. The empty spaces indicate unrealizable networks.)

| Filter type | $R_s$ or $G_s$ \ Element number | Second-order network $n = 2$ | | Third-order network $n = 3$ | | | Fourth-order network $n = 4$ | | | | Fifth-order network $n = 5$ | | | | |
|---|---|---|---|---|---|---|---|---|---|---|---|---|---|---|---|
| | | 1 | 2 | 1 | 2 | 3 | 1 | 2 | 3 | 4 | 1 | 2 | 3 | 4 | 5 |
| Bessel | 0 | 1.0000 | 0.3333 | 0.8333 | 0.4800 | 0.1667 | 0.7101 | 0.4627 | 0.2899 | 0.1000 | 0.6231 | 0.4215 | 0.3103 | 0.1948 | 0.0667 |
| | 1 | 1.5774 | 0.4227 | 1.2550 | 0.5528 | 0.1922 | 1.0598 | 0.5116 | 0.3181 | 0.1104 | 0.9303 | 0.4577 | 0.3312 | 0.2089 | 0.0718 |
| Butterworth | 0 | 1.4142 | 0.7071 | 1.5000 | 1.3333 | 0.5000 | 1.5307 | 1.5772 | 1.0824 | 0.3827 | 1.5451 | 1.6944 | 1.3820 | 0.8944 | 0.3090 |
| | 1 | 1.4142 | 1.4142 | 1.0000 | 2.0000 | 1.0000 | 0.7654 | 1.8478 | 1.8478 | 0.7654 | 0.6180 | 1.6180 | 2.0000 | 1.6180 | 0.6180 |
| Legendre-Papoulis | 0 | 1.4142 | 0.7071 | 1.5909 | 1.4270 | 0.7629 | 1.6120 | 1.6616 | 1.4292 | 0.6399 | 1.6372 | 1.7509 | 1.7358 | 1.3945 | 0.6445 |
| | 1 | 1.4142 | 1.4142 | 2.1801 | 1.3538 | 1.1737 | 1.5645 | 1.9584 | 1.4769 | 1.0826 | 1.9990 | 1.5395 | 2.0673 | 1.4780 | 0.9512 |
| Chebyshev 0.01 dB | 0 | 0.4274 | 0.2244 | 0.7997 | 0.7634 | 0.3146 | 1.0421 | 1.1547 | 0.8945 | 0.3564 | 1.1978 | 1.3902 | 1.2739 | 0.9576 | 0.3782 |
| | 1 | | | 0.6292 | 0.9703 | 0.6292 | | | | | 0.7563 | 1.3049 | 1.5773 | 1.3049 | 0.7563 |
| Chebyshev 0.10 dB | 0 | 0.7159 | 0.4215 | 1.0895 | 1.0864 | 0.5158 | 1.2453 | 1.4576 | 1.1994 | 0.5544 | 1.3759 | 1.5924 | 1.5562 | 1.2490 | 0.5734 |
| | 1 | | | 1.0316 | 1.1474 | 1.0316 | | | | | 1.1468 | 1.3712 | 1.9750 | 1.3712 | 1.1468 |
| Chebyshev 0.50 dB | 0 | 0.9403 | 0.7014 | 1.3465 | 1.3001 | 0.7981 | 1.3138 | 1.7279 | 1.3916 | 0.8352 | 1.5388 | 1.6426 | 1.8142 | 1.4291 | 0.8529 |
| | 1 | | | 1.5963 | 1.0967 | 1.5963 | | | | | 1.7058 | 1.2296 | 2.5408 | 1.2296 | 1.7058 |
| Chebyshev 1.0 dB | 0 | 0.9957 | 0.9110 | 1.5088 | 1.3332 | 1.0118 | 1.2817 | 1.9093 | 1.4126 | 1.0495 | 1.6652 | 1.5908 | 1.9938 | 1.4441 | 1.0674 |
| | 1 | | | 2.0236 | 0.9941 | 2.0236 | | | | | 2.1349 | 1.0911 | 3.0009 | 1.0911 | 2.1349 |
| Chebyshev 3.0 dB | 0 | 0.9109 | 1.5506 | 2.0302 | 1.1739 | 1.6744 | 1.0578 | 2.5272 | 1.2292 | 1.7195 | 2.1489 | 1.3016 | 2.6224 | 1.2502 | 1.7406 |
| | 1 | | | 3.3487 | 0.7117 | 3.3487 | | | | | 3.4813 | 0.7619 | 4.5375 | 0.7619 | 3.4813 |
| Modified Chebyshev 0.01 dB | 1 | 0.3099 | 0.3099 | | | | 0.6266 | 1.1938 | 1.1938 | 0.6266 | | | | | |
| Modified Chebyshev 0.10 dB | 1 | 0.5525 | 0.5525 | | | | 0.9297 | 1.4346 | 1.4346 | 0.9297 | | | | | |
| 0.50 dB | 1 | 0.8358 | 0.8358 | | | | 1.3091 | 1.5415 | 1.5415 | 1.3091 | | | | | |
| 1.0 dB | 1 | 1.0088 | 1.0088 | | | | 1.5675 | 1.5537 | 1.5537 | 1.5675 | | | | | |
| 3.0 dB | 1 | 1.4125 | 1.4125 | | | | 2.2574 | 1.5107 | 1.5107 | 2.2574 | | | | | |

**TABLE 2-15** *(Continued)*

| Filter type | | Sixth-order network $n = 6$ | | | | | | Seventh-order network $n = 7$ | | | | | | |
|---|---|---|---|---|---|---|---|---|---|---|---|---|---|---|
| | Element number / $R_s$ or $G_s$ | 1 | 2 | 3 | 4 | 5 | 6 | 1 | 2 | 3 | 4 | 5 | 6 | 7 |
| Bessel | 0 | 0.5595 | 0.3821 | 0.3005 | 0.2246 | 0.1400 | 0.0476 | 0.5111 | 0.3487 | 0.2827 | 0.2288 | 0.1704 | 0.1055 | 0.0357 |
| | 1 | 0.8376 | 0.4116 | 0.3158 | 0.2364 | 0.1480 | 0.0505 | 0.7677 | 0.3744 | 0.2944 | 0.2378 | 0.1778 | 0.1104 | 0.0375 |
| Butterworth | 0 | 1.5529 | 1.7593 | 1.5529 | 1.2016 | 0.7579 | 0.2588 | 1.5576 | 1.7988 | 1.6588 | 1.3972 | 1.0550 | 0.6560 | 0.2225 |
| | 1 | 0.5176 | 1.4142 | 1.9319 | 1.9319 | 1.4142 | 0.5176 | 0.4450 | 1.2470 | 1.8019 | 2.0000 | 1.8019 | 1.2470 | 0.4450 |
| Legendre-Papoulis | 0 | 1.6348 | 1.8088 | 1.8223 | 1.6795 | 1.3486 | 0.5793 | 1.6391 | 1.8312 | 1.8911 | 1.7988 | 1.6845 | 1.3290 | 0.5787 |
| | 1 | 1.5763 | 1.9040 | 1.7442 | 1.9857 | 1.4852 | 0.9160 | 1.8640 | 1.5895 | 2.1506 | 1.7270 | 1.9394 | 1.4770 | 0.8394 |
| Chebyshev 0.01 dB | 0 | 1.2931 | 1.5400 | 1.4922 | 1.3319 | 0.9925 | 0.3907 | 1.3615 | 1.6303 | 1.6290 | 1.5412 | 1.3650 | 1.0137 | 0.3985 |
| | 1 | | | | | | | 0.7969 | 1.3924 | 1.7481 | 1.6331 | 1.7481 | 1.3924 | 0.7969 |
| Chebyshev 0.10 dB | 0 | 1.4035 | 1.7236 | 1.6749 | 1.5999 | 1.2752 | 0.5841 | 1.4745 | 1.7395 | 1.7987 | 1.7107 | 1.6236 | 1.2908 | 0.5906 |
| | 1 | | | | | | | 1.1812 | 1.4228 | 2.0967 | 1.5734 | 2.0967 | 1.4228 | 1.1812 |
| Chebyshev 0.50 dB | 0 | 1.4042 | 1.9018 | 1.7101 | 1.8494 | 1.4483 | 0.8627 | 1.5983 | 1.7252 | 1.9713 | 1.7369 | 1.8677 | 1.4595 | 0.8686 |
| | 1 | | | | | | | 1.7373 | 1.2582 | 2.6383 | 1.3443 | 2.6383 | 1.2582 | 1.7373 |
| Chebyshev 1.0 dB | 0 | 1.3457 | 2.0491 | 1.6507 | 2.0270 | 1.4601 | 1.0773 | 1.7120 | 1.6488 | 2.1194 | 1.6735 | 2.0438 | 1.4692 | 1.0833 |
| | 1 | | | | | | | 2.1666 | 1.1115 | 3.0936 | 1.1735 | 3.0936 | 1.1115 | 2.1666 |
| Chebyshev 3.0 dB | 0 | 1.0876 | 2.6309 | 1.3455 | 2.6578 | 1.2606 | 1.7522 | 2.1828 | 1.3281 | 2.7143 | 1.3613 | 2.6752 | 1.2665 | 1.7593 |
| | 1 | | | | | | | 3.5185 | 0.7722 | 4.6390 | 0.8038 | 4.6390 | 0.7722 | 3.5185 |
| Modified Chebyshev 0.01 dB | 1 | 0.7190 | 1.3728 | 1.6006 | 1.6006 | 1.3728 | 0.7190 | | | | | | | |
| Modified Chebyshev 0.10 dB | 1 | 1.0382 | 1.5163 | 1.7892 | 1.7892 | 1.5163 | 1.0382 | | | | | | | |
| Modified Chebyshev 0.50 dB | 1 | 1.4611 | 1.5085 | 1.9333 | 1.9333 | 1.5085 | 1.4611 | | | | | | | |
| Modified Chebyshev 1.0 dB | 1 | 1.7623 | 1.4528 | 2.0089 | 2.0089 | 1.4528 | 1.7623 | | | | | | | |
| Modified Chebyshev 3.0 dB | 1 | 2.6073 | 1.2711 | 2.1729 | 2.1729 | 1.2711 | 2.6073 | | | | | | | |

**TABLE 2-15** *(Continued)*

| Filter type | $R_s$ or $G_s$ \ Element number | Eighth-order network $n = 8$ | | | | | | | | Ninth-order network $n = 9$ | | | | | | | | |
|---|---|---|---|---|---|---|---|---|---|---|---|---|---|---|---|---|---|---|
| | | 1 | 2 | 3 | 4 | 5 | 6 | 7 | 8 | 1 | 2 | 3 | 4 | 5 | 6 | 7 | 8 | 9 |
| Bessel | 0 | 0.4732 | 0.3212 | 0.2639 | 0.2227 | 0.1806 | 0.1338 | 0.0823 | 0.0278 | 0.4424 | 0.2986 | 0.2465 | 0.2129 | 0.1811 | 0.1463 | 0.1077 | 0.0660 | 0.0222 |
| | 1 | 0.7125 | 0.3446 | 0.2735 | 0.2297 | 0.1867 | 0.1387 | 0.0855 | 0.0289 | 0.6678 | 0.3203 | 0.2547 | 0.2184 | 0.1859 | 0.1506 | 0.1112 | 0.0682 | 0.0230 |
| Butterworth | 0 | 1.5607 | 1.8246 | 1.7287 | 1.5283 | 1.2588 | 0.9371 | 0.5776 | 0.1951 | 1.5628 | 1.8424 | 1.7772 | 1.6202 | 1.4037 | 1.1408 | 0.8414 | 0.5155 | 0.1736 |
| | 1 | 0.3902 | 1.1111 | 1.6629 | 1.9616 | 1.9616 | 1.6629 | 1.1111 | 0.3902 | 0.3473 | 1.0000 | 1.5321 | 1.8794 | 2.0000 | 1.8794 | 1.5321 | 1.0000 | 0.3473 |
| Legendre-Papoulis | 0 | 1.6345 | 1.8542 | 1.9102 | 1.8673 | 1.8019 | 1.6437 | 1.2869 | 0.5372 | 1.6341 | 1.8625 | 1.9349 | 1.8961 | 1.8815 | 1.7860 | 1.6397 | 1.2740 | 0.5358 |
| | 1 | 1.5564 | 1.8501 | 1.8411 | 2.0515 | 1.7672 | 1.9115 | 1.4688 | 0.8205 | 1.7645 | 1.6134 | 2.1585 | 1.7816 | 2.0662 | 1.7755 | 1.8674 | 1.4555 | 0.7695 |
| Chebyshev 0.01 dB | 0 | 1.4036 | 1.6971 | 1.7097 | 1.6708 | 1.5697 | 1.3858 | 1.0275 | 0.4036 | 1.4392 | 1.7371 | 1.7701 | 1.7457 | 1.6949 | 1.5878 | 1.3998 | 1.0370 | 0.4072 |
| | 1 | | | | | | | | | 0.8145 | 1.4271 | 1.8044 | 1.7125 | 1.9058 | 1.7125 | 1.8044 | 1.4271 | 0.8145 |
| Chebyshev 0.10 dB | 0 | 1.4660 | 1.8163 | 1.8070 | 1.8302 | 1.7302 | 1.6380 | 1.3008 | 0.5949 | 1.5182 | 1.7991 | 1.8814 | 1.8343 | 1.8473 | 1.7423 | 1.6476 | 1.3076 | 0.5978 |
| | 1 | | | | | | | | | 1.1957 | 1.4426 | 2.1346 | 1.6167 | 2.2054 | 1.6167 | 2.1346 | 1.4426 | 1.1957 |
| Chebyshev 0.50 dB | 0 | 1.4379 | 1.9571 | 1.7838 | 1.9980 | 1.7508 | 1.8786 | 1.4666 | 0.8725 | 1.6238 | 1.7571 | 2.0203 | 1.8055 | 2.0116 | 1.7591 | 1.8856 | 1.4714 | 0.8752 |
| | 1 | | | | | | | | | 1.7504 | 1.2690 | 2.6678 | 1.3673 | 2.7239 | 1.3673 | 2.6678 | 1.2690 | 1.7504 |
| Chebyshev 1.0 dB | 0 | 1.3691 | 2.0922 | 1.7021 | 2.1453 | 1.6850 | 2.0537 | 1.4751 | 1.0872 | 1.7317 | 1.6707 | 2.1574 | 1.7213 | 2.1582 | 1.6918 | 2.0601 | 1.4790 | 1.0899 |
| | 1 | | | | | | | | | 2.1797 | 1.1192 | 3.1214 | 1.1897 | 3.1746 | 1.1897 | 3.1214 | 1.1192 | 2.1797 |
| Chebyshev 3.0 dB | 0 | 1.0982 | 2.6618 | 1.3687 | 2.7436 | 1.3690 | 2.6852 | 1.2701 | 1.7638 | 2.1970 | 1.3380 | 2.7413 | 1.3827 | 2.7576 | 1.3733 | 2.6915 | 1.2726 | 1.7670 |
| | 1 | | | | | | | | | 3.5339 | 0.7760 | 4.6691 | 0.8118 | 4.7270 | 0.8118 | 4.6691 | 0.7760 | 3.5339 |
| Modified Chebyshev 0.01 dB | 1 | 0.7584 | 1.4274 | 1.7122 | 1.7503 | 1.7503 | 1.7122 | 1.4274 | 0.7584 | | | | | | | | | |
| Modified Chebyshev 0.10 dB | 1 | 1.0880 | 1.5265 | 1.9029 | 1.8301 | 1.8301 | 1.9029 | 1.5265 | 1.0880 | | | | | | | | | |
| Modified Chebyshev 0.50 dB | 1 | 1.5372 | 1.4679 | 2.1033 | 1.8436 | 1.8436 | 2.1033 | 1.4679 | 1.5372 | | | | | | | | | |
| Modified Chebyshev 1.0 dB | 1 | 1.8642 | 1.3833 | 2.2357 | 1.8319 | 1.8319 | 2.2357 | 1.3833 | 1.8642 | | | | | | | | | |
| Modified Chebyshev 3.0 dB | 1 | 2.8062 | 1.1473 | 2.5784 | 1.7813 | 1.7813 | 2.5784 | 1.1473 | 2.8062 | | | | | | | | | |

**TABLE 2-15** *(Continued)*

| Filter type | $R_s$ or $G_s$ \ Element number | Tenth-order network $n = 10$ | | | | | | | | | |
|---|---|---|---|---|---|---|---|---|---|---|---|
| | | 1 | 2 | 3 | 4 | 5 | 6 | 7 | 8 | 9 | 10 |
| Bessel | 0 | 0.4170 | 0.2797 | 0.2311 | 0.2021 | 0.1770 | 0.1504 | 0.1209 | 0.0886 | 0.0541 | 0.0182 |
| | 1 | 0.6305 | 0.3002 | 0.2384 | 0.2066 | 0.1808 | 0.1539 | 0.1240 | 0.0911 | 0.0556 | 0.0187 |
| Butterworth | 0 | 1.5643 | 1.8552 | 1.8121 | 1.6869 | 1.5100 | 1.2921 | 1.0406 | 0.7626 | 0.4654 | 0.1564 |
| | 1 | 0.3129 | 0.9080 | 1.4142 | 1.7820 | 1.9754 | 1.9754 | 1.7820 | 1.4142 | 0.9080 | 0.3129 |
| Legendre-Papoulis | 0 | 1.6298 | 1.8741 | 1.9405 | 1.9223 | 1.9102 | 1.8629 | 1.7785 | 1.6082 | 1.2386 | 0.5065 |
| | 1 | 1.5286 | 1.8122 | 1.8953 | 2.0409 | 1.8453 | 2.0327 | 1.7839 | 1.8537 | 1.4454 | 0.7575 |
| Chebyshev 0.01 dB | 0 | 1.4598 | 1.7731 | 1.8054 | 1.8020 | 1.7664 | 1.7106 | 1.6003 | 1.4096 | 1.0439 | 0.4098 |
| | 1 | | | | | | | | | | |
| Chebyshev 0.10 dB | 0 | 1.4964 | 1.8585 | 1.8600 | 1.9068 | 1.8489 | 1.8579 | 1.7503 | 1.6542 | 1.3124 | 0.6000 |
| | 1 | | | | | | | | | | |
| Chebyshev 0.50 dB | 0 | 1.4539 | 1.9816 | 1.8119 | 2.0432 | 1.8165 | 2.0197 | 1.7645 | 1.8905 | 1.4748 | 0.8771 |
| | 1 | | | | | | | | | | |
| Chebyshev 1.0 dB | 0 | 1.3801 | 2.1111 | 1.7215 | 2.1803 | 1.7307 | 2.1658 | 1.6962 | 2.0645 | 1.4817 | 1.0918 |
| | 1 | | | | | | | | | | |
| Chebyshev 3.0 dB | 0 | 1.1032 | 2.6753 | 1.3774 | 2.7682 | 1.3893 | 2.7655 | 1.3761 | 2.6958 | 1.2744 | 1.7692 |
| | 1 | | | | | | | | | | |
| Modified Chebyshev 0.01 dB | 1 | 0.7795 | 1.4503 | 1.7608 | 1.7903 | 1.8495 | 1.8495 | 1.7903 | 1.7608 | 1.4503 | 0.7795 |
| 0.10 dB | 1 | 1.1165 | 1.5249 | 1.9621 | 1.8183 | 1.9346 | 1.9346 | 1.8183 | 1.9621 | 1.5249 | 1.1165 |
| 0.50 dB | 1 | 1.5832 | 1.4380 | 2.2072 | 1.7663 | 2.0029 | 2.0029 | 1.7663 | 2.2072 | 1.4380 | 1.5832 |
| 1.0 dB | 1 | 1.9273 | 1.3374 | 2.3826 | 1.7118 | 2.0410 | 2.0410 | 1.7118 | 2.3826 | 1.3374 | 1.9273 |
| 3.0 dB | 1 | 2.9356 | 1.0731 | 2.8661 | 1.5644 | 2.1269 | 2.1269 | 1.5644 | 2.8661 | 1.0731 | 2.9356 |

**TABLE 2-16** Half-Power Frequencies for Various Chebyshev and Bessel Filters

| Type of filter | Order of filter | | | | | | | | |
|---|---|---|---|---|---|---|---|---|---|
| | $n = 2$ | $n = 3$ | $n = 4$ | $n = 5$ | $n = 6$ | $n = 7$ | $n = 8$ | $n = 9$ | $n = 10$ |
| Chebyshev 0.01 dB | 3.303615 | 1.877180 | 1.466904 | 1.291217 | 1.199412 | 1.145268 | 1.110609 | 1.087064 | 1.070331 |
| Chebyshev 0.10 dB | 1.943219 | 1.388995 | 1.213099 | 1.134718 | 1.092931 | 1.068001 | 1.051927 | 1.040955 | 1.033131 |
| Chebyshev 0.50 dB | 1.389744 | 1.167485 | 1.093102 | 1.059259 | 1.041030 | 1.030090 | 1.023011 | 1.018167 | 1.014707 |
| Chebyshev 1.00 dB | 1.217626 | 1.094868 | 1.053002 | 1.033815 | 1.023442 | 1.017205 | 1.013164 | 1.010396 | 1.008418 |
| Chebyshev 3.00 dB | 1.000594 | 1.000264 | 1.000149 | 1.000095 | 1.000066 | 1.000048 | 1.000037 | 1.000029 | 1.000024 |
| Modified Chebyshev 0.01 dB | 4.563742 | | 1.532784 | | 1.212468 | | 1.114760 | | 1.072036 |
| Modified Chebyshev 0.10 dB | 2.559727 | | 1.246004 | | 1.099300 | | 1.053929 | | 1.033948 |
| Modified Chebyshev 0.50 dB | 1.691974 | | 1.108290 | | 1.043913 | | 1.023911 | | 1.015073 |
| Modified Chebyshev 1.00 dB | 1.401865 | | 1.061830 | | 1.025105 | | 1.013681 | | 1.008628 |
| Modified Chebyshev 3.00 dB | 1.001188 | | 1.000174 | | 1.000071 | | 1.000039 | | 1.000024 |
| Bessel half-power frequency | 1.3617 | 1.7557 | 2.1140 | 2.4274 | 2.7034 | 2.9517 | 3.1797 | 3.3917 | 3.5910 |
| Delay at half-power frequency | 0.8090 | 0.9349 | 0.9819 | 0.9960 | 0.9993 | 0.9999 | 0.9999 | 1.0000 | 1.0000 |

The Chebyshev filter prototypes have the ripple specified at 1 radian per second. The Bessel filter has a delay of 1 second at very low frequency. The second figure given is the delay at the half-power frequency.

Consider the case of Fig. 2-48*b* with $n = 5$, redrawn as Fig. 2-51*a* to emphasize the resulting symmetry. $G_S$ can be changed to some other value by magnitude scaling the element values in the left half of the network. Thus making $G_S = ½$ would result in the network of Fig. 2-51*b*, where the two capacitors of value $C_3/2$ and $C_3/4$ can be recombined into a single capacitor. Similar operations can be applied to other symmetrical networks. After the application of this procedure the zeros of the reflection coefficient do not remain in the left-half plane.

*197. Practical Filter Design.* Table 2-15 gives element values for normalized networks. Practical filters can be designed from these normalized networks by a process consisting of three general steps.

1. Statement of the specifications for the filter.
2. Translation of the specifications to equivalent statements for the normalized prototype and their use to determine the parameters necessary to enter Table 2-15.

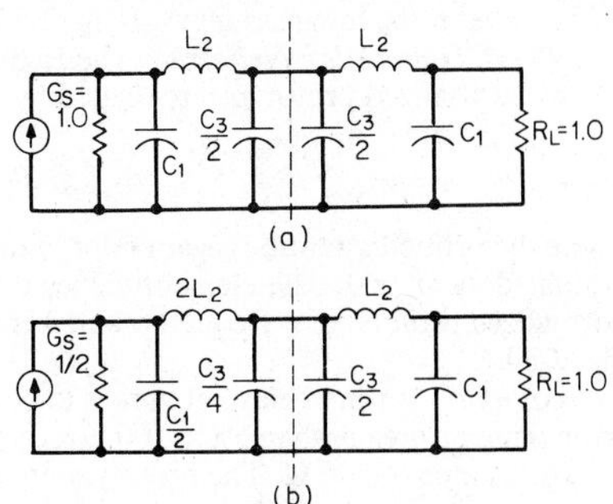


**FIG. 2-51** Illustrating the scaling of symmetrical networks to change the value of $G_S$.

3. Application of frequency transformations and impedance-level scaling to the normalized prototype to determine the filter network.

This process is described in the following sections for the various types of filters, accompanied by examples.

***198. The Low-Pass Filter.*** The low-pass filter is related to the normalized prototype by frequency and impedance-level scaling. The frequency in the prototype is related to the frequency in the low-pass filter by

$$\omega = \frac{\omega'}{\omega_c'} \tag{2-299}$$

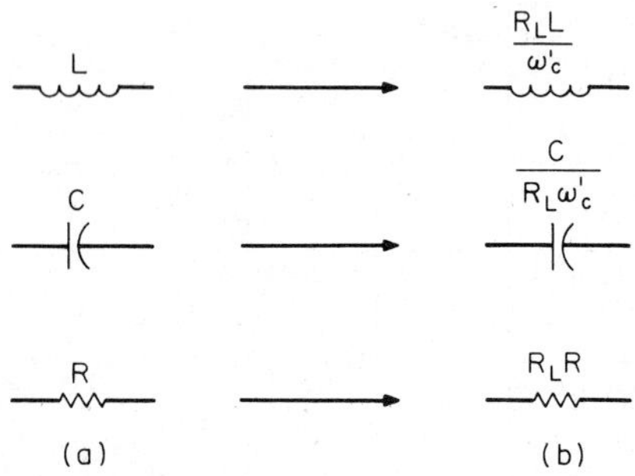

where the primed quantities refer to the practical low-pass filter, with $\omega_c'$ being its cutoff frequency. The application of the frequency scaling implied by Eq. (2-299) and the impedance-level scaling required to change the load resistance from 1.0 to $R_L$ results in the elements of the low-pass filter becoming those indicated in Fig. 2-52*b*.

| (a) | | (b) |
|---|---|---|
| $L$ | → | $\frac{R_L L}{\omega'_c}$ |
| $C$ | → | $\frac{C}{R_L \omega'_c}$ |
| $R$ | → | $R_L R$ |

**FIG. 2-52** Relations between elements of *(a)* a normalized prototype and those of *(b)* a practical low-pass filter.

*Example.* A low-pass filter is to be designed to have *(a)* a maximally flat amplitude versus frequency characteristic at $\omega' = 0$, *(b)* a cutoff (half-power) frequency of 2 kHz, *(c)* a load resistance of 200 Ω, *(d)* a voltage source input with zero resistance, and *(e)* an attenuation of not less than 20 dB at frequencies greater than 3.2 kHz. A Butterworth approximation is indicated. The frequency in the prototype equivalent to 3.2 kHz is

$$\omega = \frac{(3.2 \times 10^3)(2\pi)}{(2 \times 10^3)(3\pi)} = 1.6 \tag{2-300}$$

and the equivalent design specifications for the prototype require that the attenuation be no less than 20 dB for $\omega \geqslant 1.6$. This, using Eq. (2-261), results in

$$n \geqslant \frac{20}{20 \log 1.6} = 4.899 \tag{2-301}$$

and the next larger integer value of 5 is used. The approximate prototype appears in Fig. 2-53*a* with element values taken from Table 2-15 for the Butterworth approximation with $n = 5$ and $R_S = 0$. The application of frequency and magnitude scaling by means of the relations in Fig. 2-52 results in the low-pass filter of Fig. 2-53*b*.

***199. A Time-Delay Network.*** The frequencies in the time-delay network (all-pass filter) and its normalized prototype are related by

$$\omega = \tau_d' \, \omega' \tag{2-302}$$

where the primed quantities again refer to the practical network, with $\tau_d'$ being the group delay evaluated at $\omega' = 0$. The element values of the practical network are related to those of the normalized prototype by frequency and impedance-level scaling. These relations are shown in Fig. 2-54.

*Example.* A time-delay network is to be designed to have *(a)* a time delay of 1 ms, with the error being no greater than 5% at 500 Hz; *(b)* a voltage input with a resistance of 500 Ω; and *(c)* a load resistance of 500 Ω. The prototype frequency equivalent to 500 Hz is

$$\omega = 10^{-3}(2\pi \times 500) = 3.142 \tag{2-303}$$

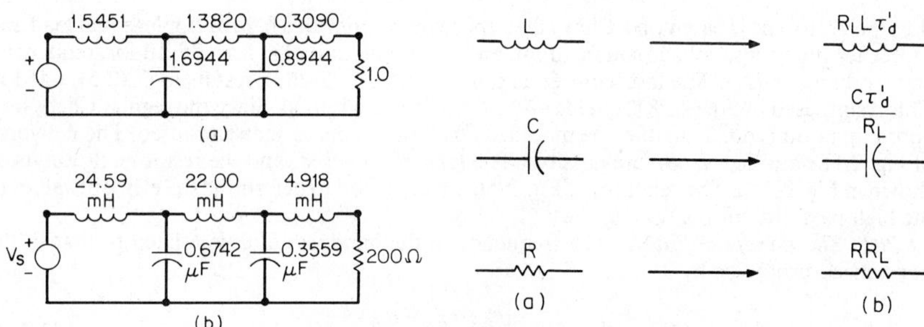


**FIG. 2-53** *(a)* A normalized prototype and *(b)* resulting low-pass filter.

**FIG. 2-54** Relations between elements for *(a)* a normalized prototype and those of *(b)* time-delay network.

and the equivalent specifications for the prototype require a time delay of 1 s, with the error being no greater than 5% at $\omega = 3.142$. Figure 2-44*b* is used to determine that $n = 5$ is required to satisfy this requirement. The prototype can be of the form of either Fig. 2-48*d* or Fig. 2-49*b*. The latter is chosen with the element values taken from Table 2-15 for the Bessel approximation with $n = 5$ and $G_S = 1$. The prototype is shown in Fig. 2-55*a*, and the application of the relations in Fig. 2-54 results in the time-delay network of Fig. 2-55*b*.

***200. The High-Pass Filter.*** The frequency in the high-pass filter is related to that in the normalized prototype by

$$\omega = \frac{\omega_c'}{\omega'} \tag{2-304}$$

where the primed quantities refer to the high-pass filter, with $\omega_c'$ being the cutoff frequency. The relations between the elements of the normalized prototype and those of the high-pass filter are given in Fig. 2-56.

*Example.* A high-pass filter is to be designed to have *(a)* a cutoff frequency of 7.5 kHz, *(b)* an equiripple characteristic in the passband with a 0.5-dB ripple ($\epsilon = 0.3493$), *(c)* an attenuation of at least 35 dB at frequencies below 3 kHz, *(d)* equal load and source resistance of 2000 Ω, and *(e)* a voltage source input. The prototype frequency equivalent to 3 kHz is

$$\omega = \frac{2\pi(7.5 \times 10^3)}{2\pi(3 \times 10^3)} = 2.5 \tag{2-305}$$

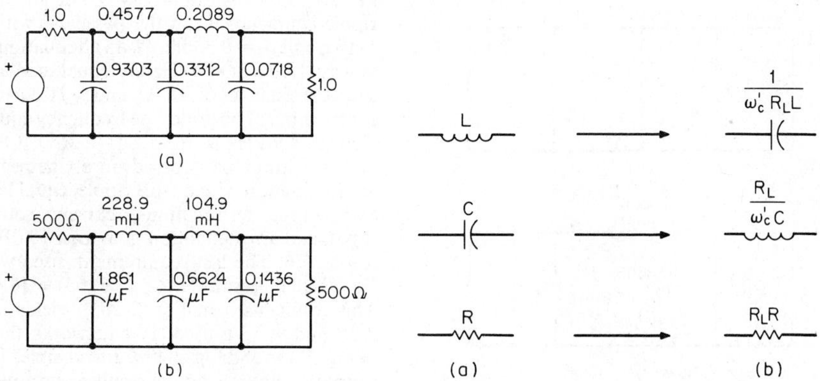


**FIG. 2-55** *(a)* A normalized prototype and *(b)* resulting time-delay filter.

**FIG. 2-56** Relations between elements of *(a)* a normalized prototype and those of *(b)* a high-pass filter.

thus the prototype is based on a Chebyshev approximation with a 0.5-dB ripple, equal load and source terminations, voltage source input, and an attenuation of at least 35 dB for frequencies greater than $\omega = 2.5$. The last requirement, used with Eq. (2-283), results in $C_n(2.5) \cong 161.0$. This result, used in Eq. (2-281), yields $n = 3.69$, and $n = 4$ is used. Since the regular Chebyshev approximation is not available, the modified Chebyshev approximation is used. The networks of Fig. 2-48*c* and Fig. 2-49*a* are possible. The former is chosen, and the resulting prototype is shown in Fig. 2-57*a*. The relations of Fig. 2-56 are then used to determine the element values of the high-pass filter of Fig. 2-57*b*.

***201. The Bandpass Filter.*** The frequency in the bandpass filter is related to that in the normalized prototype by

$$\omega = \frac{\omega_0'}{\beta'}\left(\frac{\omega'}{\omega_0'} - \frac{\omega_0'}{\omega'}\right) \tag{2-306}$$

where the primed quantities refer to the bandpass filter, with $\omega_0'$ being the center frequency and $\beta'$ the bandwidth, $\omega_{c2}' - \omega_{c1}'$, as defined in Fig. 2-33*c*. The elements of the normalized prototype and those of the bandpass filter are related as shown in Fig. 2-58.

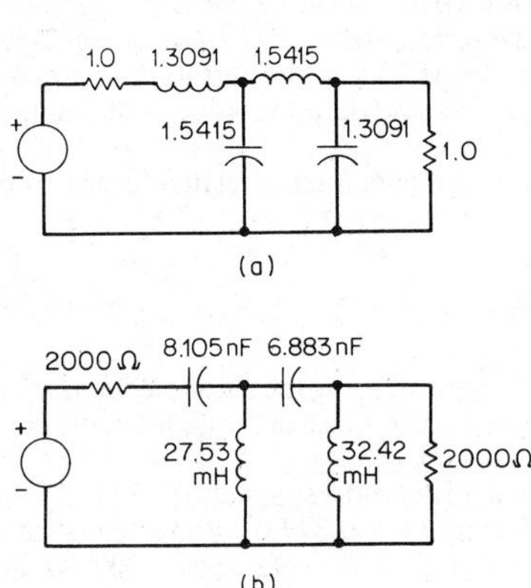


FIG. 2-57 *(a)* A normalized prototype and *(b)* resulting high-pass filter.

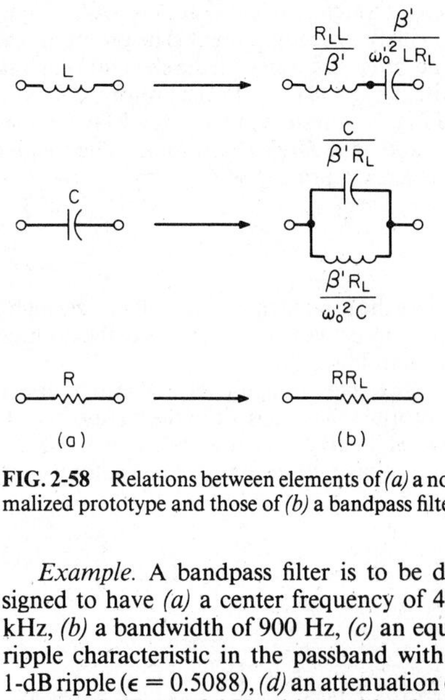

FIG. 2-58 Relations between elements of *(a)* a normalized prototype and those of *(b)* a bandpass filter.

*Example.* A bandpass filter is to be designed to have *(a)* a center frequency of 4.0 kHz, *(b)* a bandwidth of 900 Hz, *(c)* an equiripple characteristic in the passband with a 1-dB ripple ($\epsilon = 0.5088$), *(d)* an attenuation of at least 18 dB at 4.9 kHz, *(e)* equal load and source resistances of 200 Ω, and *(f)* a current source input. The prototype frequency equivalent to 4.9 kHz is, from Eq. (2-306), 1.816 and the prototype is based on a Chebyshev approximation with a 1-dB ripple, equal load and source terminations, current source input, and an attenuation of at least 18 dB at $\omega = 1.816$. The last requirement, used with Eq. (2-283), results in $C_n\ (1.8163) \cong 19.65$. This result, used in Eq. (2-281), yields $n = 2.86$, and $n = 3$ is used. Two networks, those of Fig. 2-48*a* and Fig. 2-49*d*, are possible. The former is chosen, and the resulting prototype is shown in Fig. 2-59*a*. The relations of Fig. 2-58 are then used to determine the element values of the bandpass filter of Fig. 2-59*b*.

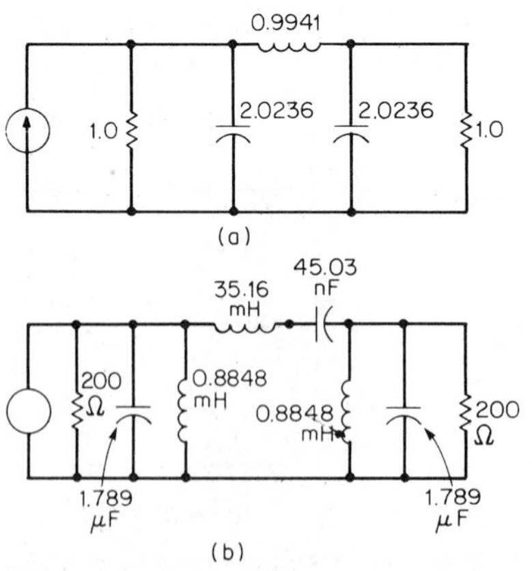


FIG. 2-59 *(a)* A normalized prototype and *(b)* resulting bandpass filter.

The results of the preceding example illustrate a problem in the prototype to bandpass filter transformation. For even the moderate ratio of center frequency to bandwidth of the example, the ratio of series-arm to shunt-arm inductances in the bandpass filter is large. This creates some practical problems arising from the stray capacitances associated with large inductances. The reader is referred to Humpherys[8] for an excellent discussion of this problem and possible ways of overcoming the difficulty. A similarly large ratio of shunt-arm to series-arm capacitances is also present, although the practical problems are not so severe.

***202. The Band-Elimination Filter.*** The frequency in the band-elimination filter is related to that in the normalized prototype by

$$\omega = \frac{1}{\dfrac{\omega_0'}{\beta'}\left(\dfrac{\omega_0'}{\omega'} - \dfrac{\omega'}{\omega_0'}\right)} \tag{2-307}$$

where the primed quantities refer to the band-elimination filter, with $\omega_0'$ being the center frequency and $\beta'$ the bandwidth, $\omega_{c2}' - \omega_{c1}'$, as defined in Fig. 2-33*d*. The elements of the normalized prototype and those of the band-elimination filter are related as shown in Fig. 2-60.

*Example.* A band-elimination filter is to be designed to have *(a)* a center frequency of 12 kHz, *(b)* a bandwith (half-power) of 1.5 kHz, *(c)* a maximally flat characteristic at $\omega' = 0$, *(d)* an attenuation of at least 16 dB at 11.6 kHz, *(e)* a current source input, and *(f)* $R_L = 800\ \Omega$ and $G_S = 0$. Since the prototype frequency equivalent to 11.6 kHz is, from Eq. (2-307), 1.8432, the prototype is based on a Butterworth approximation with an attenuation of at least 16 dB at this frequency. This requirement, used with Eq. (2-260), results in $n = 2.99$, and $n = 3$ is used. Since $G_S = 0$, the only network available is that of Fig. 2-48*b*, and the resulting prototype is shown in Fig. 2-61*a*. The relations of Fig. 2-60 are then used to determine the element values of the band-elimination filter of Fig. 2-61*b*.

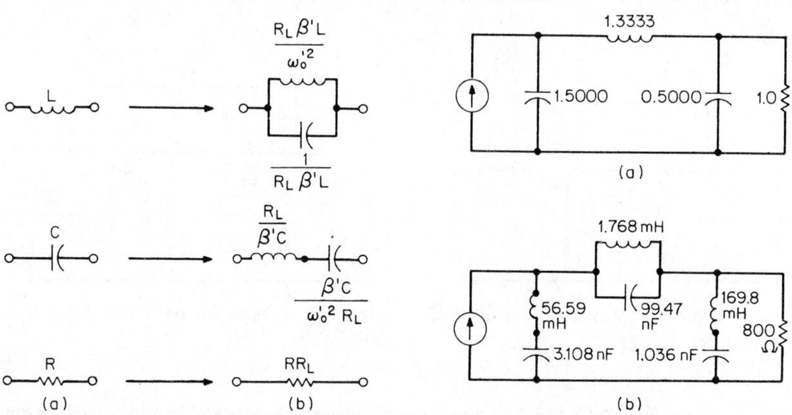


**FIG. 2-60** Relations between elements of *(a)* a normalized prototype and those of *(b)* a band-elimination filter.

**FIG. 2-61** *(a)* A normalized prototype and *(b)* resulting band-elimination filter.

This example illustrates the large ratios of shunt-arm to series-arm inductances and of series-arm to shunt-arm capacitances that result with even a moderate ratio of center frequency to bandwidth. The reader is again referred to Humpherys[8] for a discussion of this difficulty and of steps that can be taken to overcome it.

***203. Active Network Filters.*** Advances in recent years in the development of low-cost integrated circuits have made it possible to incorporate these modules into filters. These are attractive in the lower frequency ranges, though many practical circuits now operate to above 100 kHz. A primary reason for choice of an active filter is to eliminate the inductors; a second advantage is that a voltage or current gain is often possible. Huelsman[11,12] describes three general techniques, of which a few examples are given here. It is done in a way designed to make maximum use of the tables in preceding paragraphs. The operational amplifier is the principal active element employed.

The quadratic factors given in Tables 2-3, 2-5, 2-7, 2-9, and 2-11 are all for low-pass filters. In every case these are denominator polynomials, while the numerator is a constant. These transfer functions may be changed to high-pass functions by the change of variable

$$s = 1/p \tag{2-308}$$

and to bandpass functions by

$$s = \frac{p^2 + \omega_0^2}{Bp} \tag{2-309}$$

where, in both cases, $p$ is the new frequency variable. Also, $B$ is the bandwidth, $\omega_{c2} - \omega_{c1}$, while $\omega_0$ is the center frequency. Both of these are illustrated in Fig. 2-33*c*.

***204. Controlled-Source Realizations.*** The circuit of Fig. 2-62 is a second-order, low-pass active filter having a gain $K$ that is greater than zero. For this network,

$$\frac{V_2}{V_1}(s) = \frac{K\left(\dfrac{1}{R_1R_2C_1C_2}\right)}{s^2 + \left[(1-K)\dfrac{1}{R_2C_2} + \dfrac{1}{R_1C_1} + \dfrac{1}{R_2C_1}\right]s + \dfrac{1}{R_1R_2C_1C_2}} \tag{2-310}$$

Design of this circuit requires choice of a suitable quadratic factor from Tables 2-3, 2-5, 2-7, 2-9, or 2-11, matching coefficients so that the four elements are chosen (two of these may be chosen arbitrarily) and, finally, frequency scaling. When an odd-order circuit is required, a single *RC* section may be added to the output terminals, with care to avoid changes in *K*. When higher-order networks are required, a cascade of these sections is possible.

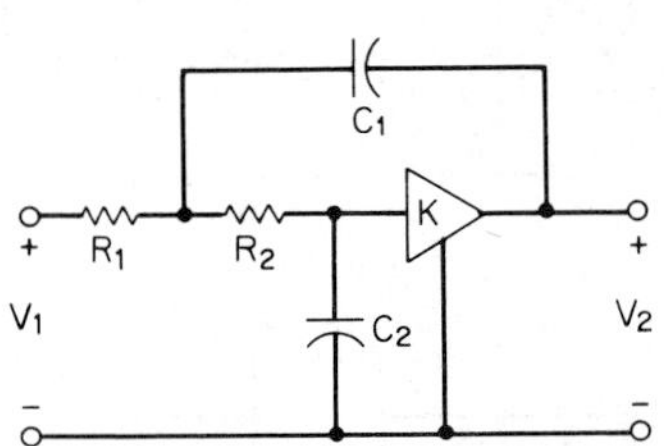


**FIG. 2-62** A low-pass active filter network with gain $K > 0$.

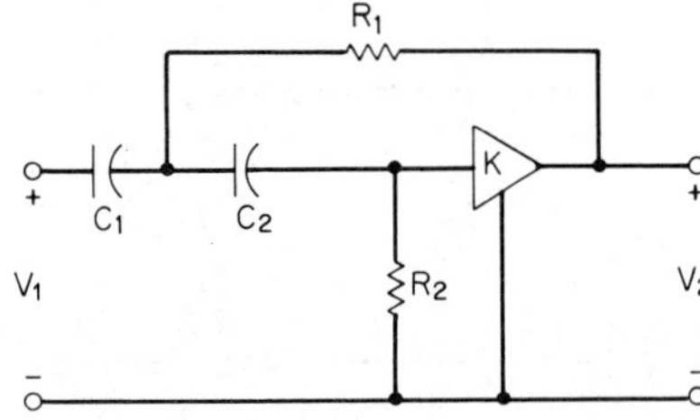


**FIG. 2-63** A high-pass active filter network with gain $K > 0$.

The circuits of Figs. 2-63 and 2-64 give corresponding high-pass and bandpass realizations. Their transfer functions are Eqs. (2-311) and (2-312), respectively. Design techniques are similar to those of the low-pass network.

$$\frac{V_2}{V_1}(s) = \frac{Ks^2}{s^2 + s\left[(1-K)\dfrac{1}{R_1C_1} + \dfrac{1}{R_2C_2} + \dfrac{1}{R_2C_1}\right] + \dfrac{1}{R_1R_2C_1C_2}} \tag{2-311}$$

$$\frac{V_2}{V_1}(s) = \frac{K\left(\dfrac{1}{R_1C_2}\right)s}{s^2 + s\left[(1-K)\dfrac{1}{R_2C_2} + \dfrac{1}{R_3C_2} + \dfrac{1}{R_1C_1} + \dfrac{1}{R_2C_1} + \dfrac{1}{R_1C_2}\right] + \dfrac{1}{R_3C_1C_2}\left(\dfrac{1}{R_1} + \dfrac{1}{R_2}\right)} \tag{2-312}$$

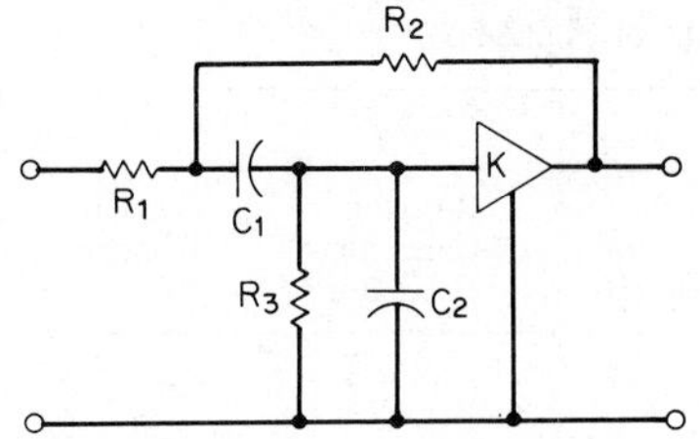


FIG. 2-64 A bandpass active filter network with gain $K > 0$.

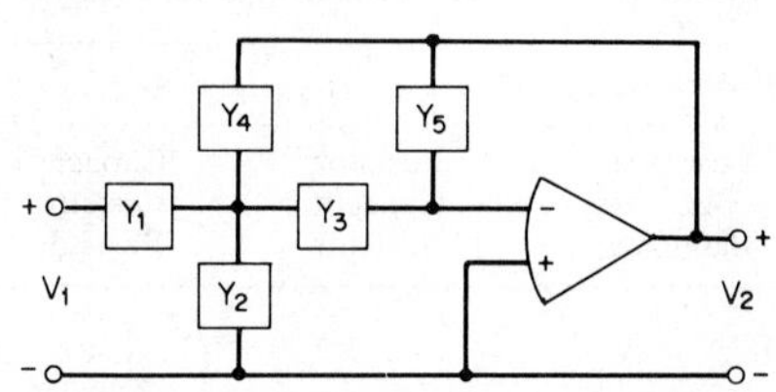


FIG. 2-66 A multiple feedback, infinite-gain active realization for low-pass, high-pass, and bandpass filters. Table 2-17 indicates the choice of the five passive elements for each case.

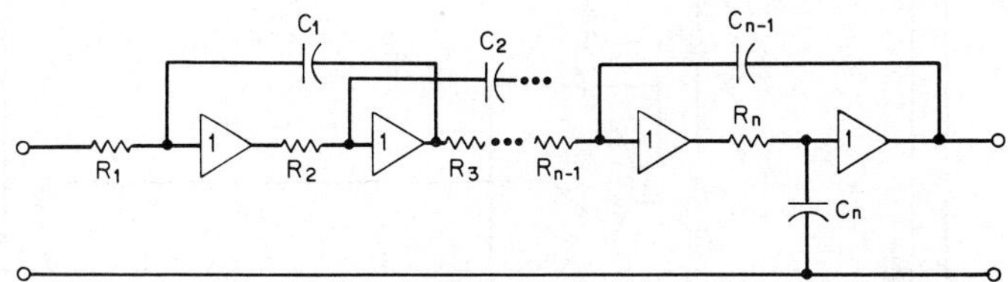


FIG. 2-65 An *RC*-unity-gain amplifier realization of an active low-pass filter.

Mitra[13] and Huelsman[11] both describe a chain network that may be used for low-pass circuits. It is given in Fig. 2-65, and the transfer function is

$$\frac{V_2}{V_1}(s) = \frac{\omega_1\omega_2\omega_3 \cdots \omega_n}{s^n + \omega_1 s^{n-1} + \omega_1\omega_2 s^{n-2} + \cdots + \omega_1\omega_2\omega_3 \cdots \omega_n} \tag{2-313}$$

where

$$\omega_i = \frac{1}{R_i C_i} \tag{2-314}$$

As an example, consider a third-order Bessel filter, for which

$$\frac{V_2}{V_1}(s) = \frac{15}{s^3 + 6s^2 + 15s + 15} \tag{2-315}$$

choose $1/R_1C_1 = 6$, $6(1/R_2C_2) = 15$, and $15(1/R_3C_3) = 15$. If $C_1 = C_2 = C_3 = 1.0$, then

$$R_1 = 1/6 \qquad R_2 = 2/5 \qquad R_3 = 1$$

Frequency and impedance scaling can be applied as required.

***205. Infinite-Gain, Multiple Feedback Realization.***[12,13] The circuit of Fig. 2-66 shows an operational amplifier with five passive elements, which are either resistors or capacitors. The general voltage transfer function is

$$\frac{V_2}{V_1} = \frac{-Y_1Y_3}{Y_5(Y_1 + Y_2 + Y_3 + Y_4) + Y_3Y_4} \tag{2-316}$$

Table 2-17 describes how the five passive elements may be chosen to implement a low-pass, high-pass, or bandpass network.

***206. State-Variable Realization.***[12] This network is a special but important type of infinite gain realization. It has the advantage that low-pass, high-pass, and bandpass configurations can be realized simultaneously, and it is also easy to adjust and to produce in quantity. The network

**TABLE 2-17** Element Choice for Active Filter Circuit of Fig. 2-66

| Filter desired | $Y_1$ | $Y_2$ | $Y_3$ | $Y_4$ | $Y_5$ |
|---|---|---|---|---|---|
| Low-pass | Resistor | Capacitor | Resistor | Resistor | Capacitor |
| High-pass | Capacitor | Resistor | Capacitor | Capacitor | Resistor |
| Bandpass | Resistor | Resistor | Capacitor | Capacitor | Resistor |

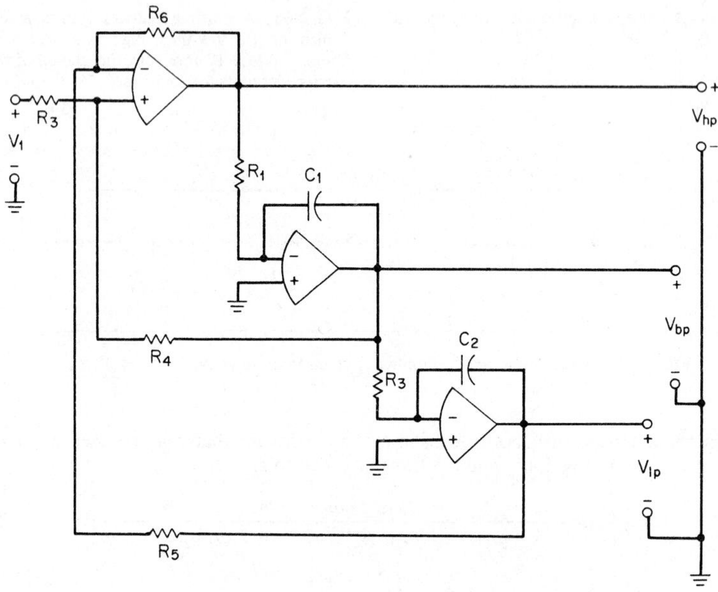


**FIG. 2-67** State-variable realization of second-order active filters.

is shown in Fig. 2-67, and the three possible transfer functions are

Low-pass:
$$\frac{V_{lp}}{V_1} = \frac{\dfrac{R_4(R_5 + R_6)}{R_1R_2R_5C_1C_2(R_3 + R_4)}}{s^2 + s\left[\dfrac{R_3(R_5 + R_6)}{R_1C_1(R_3 + R_4)}\right] + \dfrac{R_6}{R_1R_2R_5C_1C_2}} \tag{2-317}$$

High-pass:
$$\frac{V_{hp}}{V_1} = \frac{\dfrac{s^2 \cdot R_4(R_5 + R_6)}{R_5(R_3 + R_4)}}{s^2 + s\left[\dfrac{R_3(R_5 + R_6)}{R_1C_1(R_3 + R_4)}\right] + \dfrac{R_6}{R_1R_2R_5C_1C_2}} \tag{2-318}$$

Bandpass:
$$\frac{V_{bp}}{V_1} = \frac{\dfrac{-s \cdot R_4(R_5 + R_6)}{R_1R_5C_1(R_3 + R_4)}}{s^2 + s\left[\dfrac{R_3(R_5 + R_6)}{R_1C_1(R_3 + R_4)}\right] + \dfrac{R_6}{R_1R_2R_5C_1C_2}} \tag{2-319}$$

The network has a low output impedance at each terminal, so that *RC* sections can be added for odd-ordered networks. These sections can be cascaded so that networks of order four and higher can be built. They have the property that performance variations with parameter changes are comparable with those of strictly passive networks, but the disadvantage of requiring three operational amplifiers.

As with the other networks, design is a matter of choosing the appropriate quadratic factors from Tables 2-3, 2-5, 2-7, 2-9, or 2-11, matching coefficients with other constraints that arise from technological and economic considerations, and finally, impedance and frequency scaling.

***207. Synthetic Inductance Realizations.***[11] In some cases, it is possible to replace the inductors required in the circuits of Table 2-15 with simulated or synthetic inductors, that is, combinations of resistors, capacitors, and active elements that replace the inductors. The most successful ones are based on the gyrator used in high-pass networks, where one terminal of the synthetic inductor is grounded. This is a subject in which rapid technological changes are being made, and it seems appropriate here to suggest that the potential user study manufacturers' current literature carefully.

***208. Power-Line Filters.*** Power-line filters[14] are passive filters placed in the lines that supply electric power to a device or system. A concern with many electrical systems is high-frequency electrical signal interference propagated along power lines, and power-line filters are designed to reduce or eliminate this problem.

Two types of interference may be identified. The first is the undesired signals that may be present on a power line and that could adversely affect the system being used. The second is the undesired signals that may be returned to the power line from the system being designed, leading to interference signals being sent to other devices or systems, and also leading to noncompliance with various standards for such equipment. The same filter is used to solve both problems. In various countries, the standards used for this problem may vary significantly, requiring the designer to understand the applications well.

A power-line filter is a low-pass filter that is placed as closely as possible to the power-line connections to a device. Typically, they are first-order (a shunt capacitor or a series inductor), second-order (an "L-section"), or third-order (a "$\pi$-section") LC circuits. The basic sections are extended for systems that have single-phase, 3-wire power and for multiphase systems. A typical attenuation characteristic is shown in Fig. 2-68. Other specifications that may be stated include

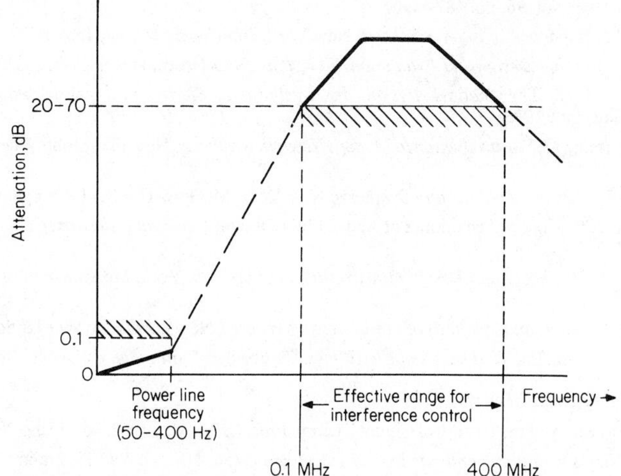


**FIG. 2-68** Power-line filter attenuation characteristic. Numerical values shown are representative values.

the voltage and current ratings of the filter, overvoltage limits, leakage current limits, and environmental and mechanical considerations. Many manufacturers produce these filters.

Standards that must be met by power-line filters may be obtained from the following testing laboratories, and from similar organizations in other countries:

Underwriters Laboratories Inc. (UL)
207 East Ohio St.
Chicago, Ill.

Canadian Standards Association (CSA)
178 Rexdale Blvd.
Rexdale, Ontario, Canada

Svenska Elektriska Materielkontrollanstalten AB (SEMKO)
Franzengatan 5
S-10425 Stockholm, Sweden

Schweizerischer Elektrotechnischer Verein (SEV)
Seefeldstrasse 301
OH-8008 Zurich, Switzerland

Verband Deutscher Elektrotechniker (VDE)
VDE-Prufstelle
Merianstrasse 28
D-6050 Offenbach (Main), West Germany

### *209. Bibliography on Filters*

*Numbered References*

1. Ruston, H., and Bordogna, J. *Electric Networks: Functions, Filters, Analysis;* New York, McGraw-Hill Book Company, 1966.
2. Darlington, S. Synthesis of Reactance 4-Poles Which Produce Prescribed Insertion Loss Characteristics, Including Special Applications to Filter Design; *J. Math. Phys.*, 1939, vol. 18, pp. 257–353.
3. Saal, R., and Ulbrich, E. On the Design of Filters by Synthesis; *IRE Trans. Circuit Theory,* 1958, vol. CT-5, pp. 284–327.
4. Papoulis, A. A New Class of Filters; *Proc. IRE,* 1959, vol. CT-6, pp. 277–281.
5. Thomson, W. E. Delay Networks Having Maximally Flat Frequency Characteristics; *Proc. IEE,* Pt. 3, November 1949, vol. 96, pp. 487–490.
6. Zverev, A. I. *Handbook of Filter Synthesis;* New York, John Wiley & Sons, Inc., 1967.
7. Hansell, G. E. *Filter Design and Evaluation;* New York, Van Nostrand Reinhold Co., 1969.
8. Humpherys, D. S. *The Analysis, Design, and Synthesis of Electric Filters;* Englewood Cliffs, N.J., Prentice-Hall, Inc., 1970.
9. Van Valkenburg, M. E. *Introduction to Modern Network Synthesis;* New York, John Wiley & Sons, Inc., 1960.
10. Weinberg, L. *Network Analysis and Synthesis;* New York, McGraw-Hill Book Company, 1962.
11. Huelsman, L. P. Modern Techniques of Active Filter Design; *Proc. Natl. Electron. Conf.,* 1974, vol. 29, pp. 449–453.
12. Huelsman, L. P. *Theory and Design of Active RC Circuits;* New York, McGraw-Hill Book Company, 1968.
13. Mitra, S. K. *Analysis and Synthesis of Linear Active Networks;* New York, John Wiley & Sons, Inc., 1969.
14. Horowitz, P., and Hill, W. *The Art of Electronics;* Cambridge, Cambridge University Press, 1980.

*Additional General References*

Belevitch, V. Summary of the History of Circuit Theory; *Proc. IEEE,* May 1962, vol. 50, no. 5, pp. 848–855.

Bode, H. W. *Network Analysis and Feedback Design;* Princeton, N.J., D. Van Nostrand Company, Inc., 1945.

Bruton, L. F. *RC-Active Circuits:* Englewood Cliffs, N.J., Prentice-Hall, Inc., 1980.

Cauer, W. "New Theory and Design of Wave Filters"; *Physics,* 1932, vol. 2, pp. 242–268.

Craig, J. W. *Design of Lossy Filters;* Cambridge, Mass., M.I.T. Press, 1970.

Daryanani, G. *Principles of Active Network Synthesis and Design;* New York, John Wiley & Sons, Inc., 1976.

Guillemin, E. A. *Communication Networks;* New York, John Wiley & Sons, Inc., 1935, vol. 2.

Huelsman, L. P. *Active Filters: Lumped, Distributed, Digital, and Parametric;* New York, McGraw-Hill Book Company, 1970.

Johnson, D. E. *Introduction to Filter Theory;* Englewood Cliffs, N.J., Prentice-Hall, Inc., 1976.

Johnson, D. E., Johnson, J. R., and Moore, H. P. *A Handbook of Active Filters;* Englewood Cliffs, N.J., Prentice-Hall, Inc., 1980.

Lam, Harry Y-F. *Analog and Digital Filters;* Englewood Cliffs, N.J., Prentice-Hall, Inc., 1979.

Lindquist, C. S. *Active Network Design;* Long Beach, California, Steward and Sons, 1977.

Mitra, S. K. (ed). *Active Inductorless Filters;* New York, IEEE Press, 1971.

Moschytz, G. S., and Horn, P. *Active Filter Design Handbook;* New York, John Wiley & Sons, Inc., 1981.

Orchard, H. J. The Roots of the Maximally Flat-Delay Polynomials; *IEEE Trans. Circuit Theory,* 1965, vol. CT-12, pp. 452–454.

Szentirmai, G. (ed). *Computer-Aided Filter Design;* New York, IEEE Press, 1973.

Temes, G. C., and Mitra, S. K. *Modern Filter Theory and Design;* New York, John Wiley & Sons, Inc., 1973.

Temes, G. C., and LaPatra, J. W. *Introduction to Circuit Synthesis and Design;* New York, McGraw-Hill Book Company, Inc., 1977.

Van Valkenburg, M. E. (ed). *Circuit Theory: Foundations and Classical Contributions;* Stroudsburg, Pa., Dowden, Hutchinson, and Ross, Inc., 1974.

Van Valkenburg, M. E. *Analog Filter Design;* New York, Holt, Rinehart, and Winston, 1982.

# SECTION 3
# MEASUREMENTS AND INSTRUMENTS

**Forest K. Harris**
*Retired, Physicist, Electricity Division, National Bureau of Standards; Fellow, IEEE*

**Norman B. Belecki**
*Group Leader, Electricity Division, National Bureau of Standards*

**Robert J. Soulen, Jr.**
*Chief, Temperature & Pressure Division, National Bureau of Standards*

## CONTENTS

*Numbers refer to paragraphs*

## *ELECTRIC AND MAGNETIC MEASUREMENTS*

### General

*1.* ***Measurement*** of a quantity consists either in its comparison with a unit quantity of the same kind or in its determination as a function of quantities of different kinds whose units are related to it by known physical laws. An example of the first kind of measurement is the evaluation of a resistance (in ohms) with a Wheatstone bridge, in terms of a calibrated resistance and a ratio. An example of the second kind is the calibration of the scale of a wattmeter (in watts) as the product of current (in amperes) in its field coils and the potential difference (in volts) impressed on its potential circuit.

*2.* ***Units.*** The units used in electrical measurements are related to the metric system of mechanical units in such a way that the electrical units of power and energy are identical with the corresponding mechanical units. In 1960 the name Système International (abbreviated SI),[1] now in use throughout the world, was assigned to the system based on the meter-kilogram-second-ampere (abbreviated mksa).[2] The mksa units are identical in value with the practical units—volt, ampere, ohm, coulomb, farad, henry—used by engineers. Certain prefixes have been adopted internationally to indicate decimal multiples and fractions of the basic units. These are given in Sec. **1**, Par. 5.

*3.* ***A reference standard*** is a concrete representation of a unit or of some fraction or multiple of it, having an assigned value which serves as a measurement base. Its assignment should be traceable through a chain of measurements to the *National Reference Standard* maintained by the National Bureau of Standards (NBS). Standard cells and certain fixed resistors, capacitors, and inductors of high quality are used as reference standards.

*4.* ***The National Reference Standards*** maintained by the NBS comprise the legal base for measurements in the United States. Other nations have similar laboratories to maintain the standards which serve as their measurement base. An international bureau—Bureau International des Poids et Mesures (abbreviated BIPM) in Sèvres, France—also maintains reference standards and compares standards from the various national laboratories to detect and reconcile any differences that might develop between the as-maintained units of different countries.

At NBS the reference standard of resistance is a group of 1-Ω resistors,[3] fully annealed and mounted strain-free out of contact with the air, in sealed containers. The reference standard of

[1] The absolute units of the Système International must not be confused with the old International Units in use before 1948, based on the "mercury" ohm and "silver" ampere.

[2] The SI definition of the *ampere*—the current which, in two infinite parallel conductors spaced 1 m apart, would produce a force of $10^{-7}$ N/m of length—is based on an assigned value of $4\pi \times 10^{-7}$ as the permeability of free space.

[3] Thomas, J. L. *Natl. Bur. Stand. J. Res.*, 1946, vol. 36, p. 107.

capacitance is a group of 10-pF fused-silica-dielectric capacitors whose values are assigned in terms of the calculable capacitor used in the ohm determination. The reference standard of voltage is a group of standard cells continuously maintained at a constant temperature.

The "absolute" experiments from which the value of an electrical unit is derived are measurements in which the electrical unit is related directly to appropriate mechanical units. In recent *ohm* determinations, the value of a capacitor of special design was calculated from its measured dimensions, and its impedance at a known frequency was compared to the resistance of a special resistor. Thus the ohm was assigned in terms of *length* and *time*. The as-maintained ohm is believed to be within 1 ppm of the defined SI unit.[1] Recent *ampere* determinations, used to assign the *volt* in terms of current and resistance, derived the ampere by measuring the force between current-carrying coils of a mutual inductor of special construction whose value was calculated from its measured dimensions. The voltage drop of this current in a known resistor was used to assign the emf of the standard cells which maintain the volt. The stated uncertainty of these ampere determinations ranges from 4 to 7 ppm; and the departure of value of the "legal" volt from the defined SI unit carries the same uncertainty. Since 1972 the assigned emf of the standard cells in the reference group which maintains the legal volt is monitored (and reassigned as necessary) in terms of atomic constants (the ratio of Planck's constant to electron charge) and a microwave frequency, by an ac Josephson experiment in which their voltage is measured with respect to the voltage developed across the barrier junction between two superconductors irradiated by microwave energy and biased with a direct current.[2] This experiment appears to be repeatable within 0.1 ppm. It should be noted that, while the Josephson experiment may be used to maintain the legal volt at a constant level, it is not used to define the SI unit.

*5. **Precision***—a measure of the spread of repeated determinations of a particular quantity—depends on various factors. Among these are the resolution of the method used, variations in ambient conditions (such as temperature and humidity) that may influence the value of the quantity or of the reference standard, instability of some element of the measuring system, and many others. In the National Laboratory, where every precaution is taken to obtain the best possible value, intercomparisons may have a precision of a few parts in $10^7$. In commercial laboratories, where the objective is to obtain results that are reliable but only to the extent justified by engineering or other requirements, precision ranges from this figure to a part in $10^3$ or more, depending on circumstances. For commercial measurements such as the sale of electrical energy, where the cost of measurement is a critical factor, a precision of 1 or 2% is considered acceptable in some jurisdictions.

The use of digital instruments occasionally creates a problem in the evaluation of precision; i.e., all results of a repeated measurement may be identical due to the combination of limited resolution and quantized nature of the data. In these cases the least-count and sensitivity of the instrumentation must be taken into account in determining precision.

*6. **Accuracy***—a statement of the limits which bound the departure of a measured value from the true value of a quantity—includes the imprecision of the measurement, together with all the accumulated errors in the measurement chain extending downward from the basic reference standards to the specific measurement in question. In engineering measurement practice, accuracies are generally stated in terms of the values assigned to the National Reference Standards—the *legal* units. It is only rarely that one needs also to state accuracy in terms of the *defined* SI unit by taking into account the uncertainty in the assignment of the National Reference Standard.

*7. **General precautions*** should be observed in electrical measurements, and sources of error should be avoided, as detailed below:

**a.** The accuracy limits of the instruments, standards, and methods used should be known, so that appropriate choice of these measuring elements may be made. It should be noted that instrument *accuracy classes* state the "initial" accuracy. Operation of an instrument, with energy applied over a prolonged period, may cause errors due to elastic fatigue of control springs or resistance changes in instrument elements because of heating under load. ANSI C39.1 specifies permissible limits of error of portable instruments because of sustained operation.

[1] R. D. Cutkosky. *Trans. IEEE, IM-23*, 1974, pp. 305–309.
[2] B. Field et al. *Metrologia* 1973, vol. 9, p. i55.

**b.** In any other than rough determinations, the *average of several readings* is better than one. Moreover, the alteration of measurement conditions or techniques, where feasible, may help to avoid or minimize the effects of accidental and systematic errors.

**c.** The *range* of the measuring instrument should be such that the measured quantity produces a reading large enough to yield the desired precision. The deflection of a measuring instrument should preferably exceed half scale. Voltage transformers, wattmeters, and watthour meters should be operated near to rated voltage for best performance. Care should be taken to avoid either momentary or sustained overloads.

**d.** *Magnetic fields,* produced by currents in conductors or by various classes of electrical machinery or apparatus, may combine with the fields of portable instruments to produce errors. Alternating or time-varying fields may induce emfs in loops formed in connections or the internal wiring of bridges, potentiometers, etc., to produce an error signal or even "electrical noise" that may obscure the desired reading. The effects of stray alternating fields on ac indicating instruments may be eliminated generally by using the average of readings taken with direct and reversed connections; with direct fields and dc instruments the second reading (to be averaged with the first) may be taken after rotating the instrument through 180°. If instruments are to be mounted in magnetic panels, they should be calibrated in a panel of the same material and thickness. It should also be noted that Zener-diode-based references are affected by magnetic fields. This may alter the performance of digital meters.

**e.** In measurements involving high resistances and small currents, *leakage paths* across insulating components of the measuring arrangement should be eliminated if they shunt portions of the measuring circuit. This is done by providing a guard circuit to intercept current in such shunt paths or to keep points at the same potential between which there might otherwise be improper currents.

**f.** Variations in *ambient temperature* or internal temperature rise from self-heating under load may cause errors in instrument indications. If the temperature coefficient and the instrument temperature are known, readings can be corrected where precision requirements justify it. Where measurements involve extremely small potential differences, thermal emfs resulting from temperature differences between junctions of dissimilar metals may produce errors; heat from the observer's hand or heat generated by the friction of a sliding contact may cause such effects.

**g.** *Phase-defect angles* in resistors, inductors, or capacitors and in instruments and instrument transformers must be taken into account in many ac measurements.

**h.** Large *potential differences* are to be avoided between the windings of an instrument or between its windings and frame. Electrostatic forces may produce reading errors, and very large potential difference may result in insulating breakdown. Instruments should be connected in the ground leg of a circuit where feasible. The moving-coil end of the voltage circuit of a wattmeter should be connected to the same line as the current coil. When an instrument must be at a high potential, its case must be adequately insulated from ground and connected to the line in which the instrument circuit is connected, or the instrument should be enclosed in a screen that is connected to the line. Such an arrangement may involve shock hazard to the operator, and proper safety precautions must be taken.

**i.** *Electrostatic charges* and consequent disturbance to readings may result from rubbing the insulating case or window of an instrument with a dry dustcloth; such charges can generally be dissipated by breathing on the case or window. Low-level measurements in very dry weather may be seriously affected by charges on the clothing of the observer; some of the synthetic textile fibers—such as nylon and Dacron—are particularly strong sources of charge; the only effective remedy is the complete screening of the instrument on which charges are induced.

**j.** *Position influence* (resulting from mechanical unbalance) may affect the reading of an analog-type indicating instrument if it is used in a position other than that in which it was calibrated. Portable instruments of the better accuracy classes (with antiparallax mirrors) are normally intended to be used with the axis of the moving system vertical, and the calibration is generally made with the instrument in this position.

## Detectors and Galvanometers

***8. Detectors*** are used to indicate approach to balance in bridge or potentiometer networks. They are generally responsive to small currents or voltages, and their sensitivity—the value of current or voltage that will produce an observable indication—ultimately limits the resolution of the network as a means for measuring some electrical quantity.

***9. Galvanometers*** are deflecting instruments which are used, mainly, to *detect* the presence of a small electrical quantity—current, voltage, or charge—but which are also used in some instances to measure the quantity through the magnitude of the deflection.

***10. The D'Arsonval (moving-coil) galvanometer*** consists of a coil of fine wire suspended between the poles of a permanent magnet. The coil is usually suspended from a flat metal strip which both conducts current to it and provides control torque directed toward its neutral (zero-current) position. Current may be conducted from the coil by a helix of fine wire which contributes very little to the control torque (pendulous suspension) or by a second flat metal strip which contributes significantly to the control torque (taut-band suspension). An iron core is usually mounted in the central space enclosed by the coil; and the pole pieces of the magnet are shaped to produce a uniform radial field throughout the space in which the coil moves. A mirror attached to the coil is used in conjunction with a lamp and scale or a telescope and scale to indicate coil position.

The *pendulous-suspension* type of galvanometer has the advantage of higher sensitivity (weaker control torque) for a suspension of given dimensions and material and the disadvantage of responsiveness to mechanical disturbances to its supporting platform, which produce anomalous motions of the coil. The *taut-suspension* type is generally less sensitive (stiffer control torque) but may be made much less responsive to mechanical disturbances if it is properly balanced, that is, if the center of mass of the moving system is in the axis of rotation determined by the taut upper and lower suspensions.

***11. Galvanometer sensitivity*** can be expressed in a number of ways, depending on application:

**a.** The *current* constant is the current in microamperes that will produce unit deflection on the scale—usually a deflection of 1 mm on a scale 1 m distant from the galvanometer mirror.

**b.** The *megohm* constant is the number of megohms in series with the galvanometer through which 1 V will produce unit deflection. It is the reciprocal of the current constant.

**c.** The *voltage* constant is the number of microvolts which, in a critically damped circuit (or another specified damping), will produce unit deflection.

**d.** The *coulomb* constant is the charge in microcoulombs which, at a specified damping, will produce unit ballistic throw.

**e.** The *flux-linkage* constant is the product of change of induction and turns of the linking search coil which will produce unit ballistic throw.

All these sensitivities (galvanometer response characteristics) can be expressed in terms of current sensitivity, circuit resistance in which the galvanometer operates, relative damping (see Par. **14**), and period. If we define *current* sensitivity $S_i$ as deflection per unit current, then—in appropriate units—the *voltage* sensitivity (the deflection per unit voltage) is

$$S_e = S_i/R$$

where $R$ is the resistance of the circuit, including the resistance of the galvanometer coil. The *coulomb* sensitivity is

$$\frac{\theta}{Q} = \frac{2\pi}{T_o} S_i \exp\left(\frac{-\gamma}{\sqrt{1-\gamma^2}} \tan^{-1} \frac{\sqrt{1-\gamma^2}}{\gamma}\right)$$

where $T_o$ is the undamped period and $\gamma$ is the relative damping in the operating circuit. The *flux-linkage* sensitivity is

$$\frac{\theta}{\int e\,dt} \approx S_i \frac{2\pi}{T_o} \frac{1}{2R_c} \frac{1}{1-\gamma_0}$$

for the case of greatest interest—maximum ballistic response—where the galvanometer is heavily overdamped, $\gamma_0$ being the open-circuit relative damping, $\int e\,dt$ the time integral of induced voltage or the change in flux linkages in the circuit, and $R_c$ the circuit resistance (including that of the galvanometer) for which the galvanometer is critically damped.[1]

**12. *Galvanometer motion*** is described by the differential equation

$$P\ddot{\theta} + \left(K + \frac{G^2}{R}\right)\dot{\theta} + U\theta = \frac{GE}{R}$$

where $\theta$ is the angle of deflection in radians, $P$ is the moment of inertia, $K$ is the mechanical damping coefficient, $G$ is the motor constant ($G$ = coil area turns × air-gap field), $R$ is total circuit resistance (including the galvanometer), and $U$ is the suspension stiffness. If the viscous and circuital damping are combined,

$$K + G^2/R = A$$

the roots of the auxiliary equation are

$$m = \frac{A}{2P} \pm \sqrt{\frac{A^2}{4P^2} - \frac{U}{P}}$$

Three types of motion can be distinguished.

**a.** *Critically damped* motion occurs when $A^2/4P^2 = U/P$. It is an aperiodic, or deadbeat, motion in which the moving system approaches its equilibrium position without passing through it in the shortest time of any possible aperiodic motion. This motion is described by the equation

$$y = 1 - \left(1 + \frac{2\pi t}{T_o}\right)\exp\left(\frac{-2\pi t}{T_o}\right)$$

where $y$ is the fraction of equilibrium deflection at time $t$ and $T_o$ is the undamped period of the galvanometer—the period that the galvanometer would have if $A = 0$. If the total damping coefficient at critical damping is $A_c$, we can define relative damping as the ratio of the damping coefficient $A$ for a specific circuit resistance to the value $A_c$ it has for critical damping—$\gamma = A/A_c$, which is unity for critically damped motion.

**b.** In *overdamped* motion, the moving system approaches its equilibrium position without overshoot and more slowly than in critically damped motion. This occurs when

$$\frac{A^2}{4P^2} > \frac{U}{P}$$

and $\gamma > 1$. For this case, the motion is described by the equation

$$y = 1 - \left(\frac{\gamma}{\sqrt{\gamma^2 - 1}}\sinh\frac{2\pi t}{T_o}\sqrt{\gamma^2 - 1} + \cosh\frac{2\pi t}{T_o}\sqrt{\gamma^2 - 1}\right)\exp\left(\frac{-2\pi t}{T_o}\gamma\right)$$

**c.** In *underdamped* motion, the equilibrium position is approached through a series of diminishing oscillations, their decay being exponential. This occurs when

$$\frac{A^2}{4P^2} > \frac{U}{P}$$

and $\gamma < 1$. For this case the motion is described by the equation

$$y = 1 - \frac{1}{\sqrt{1 - \gamma^2}}\left[\sin\left(\frac{2\pi t}{T_o}\sqrt{1 - \gamma^2} + \sin^{-1}\sqrt{1 - \gamma^2}\right)\right]\exp\left(\frac{-2\pi t}{T_o}\gamma\right)$$

[1] See Harris, F. K., "Electrical Measurements," p. 316, for the general expression of flux-linkage sensitivity.

*13. **Damping factor*** is the ratio of deviations of the moving system from its equilibrium position in successive swings. More conveniently, it is the ratio of the equilibrium deflection to the "overshoot" of the first swing past the equilibrium position, or

$$F = \frac{\theta_1 - \theta_F}{\theta_F - \theta_2} = \frac{\theta_F}{\theta_1 - \theta_F}$$

where $\theta_F$ is the equilibrium deflection and $\theta_1$ and $\theta_2$ are the first maximum and minimum deflections of the damped system. It can be shown that damping factor is connected to relative damping by the equation

$$F = \exp\left(\frac{\pi\gamma}{\sqrt{1-\gamma^2}}\right)$$

*14. **The logarithmic decrement*** of a damped harmonic motion is the naperian logarithm of the ratio of successive swings of the oscillating system. It is expressed by the equation

$$\ln\frac{\theta_1 - \theta_F}{\theta_F - \theta_2} = \ln\frac{\theta_F}{\theta_1 - \theta_F} = \lambda$$

and in terms of relative damping

$$\lambda = \frac{\pi\gamma}{\sqrt{1-\gamma^2}}$$

*15. **The period*** of a galvanometer (and, generally, of any damped harmonic oscillator) can be stated in terms of its undamped period $T_o$ and its relative damping $\gamma$ as $T = T_o/\sqrt{1-\gamma^2}$.

*16. **Reading time*** is the time required, after a change in the quantity measured, for the indication to come and remain within a specified percentage of its final value. Minimum reading time depends on the relative damping and on the required accuracy (see Table 3-1).

**TABLE 3-1** Minimum Reading Time for Various Accuracies

| Accuracy, percent | Relative damping | Reading time/free period |
|---|---|---|
| 10 | 0.6 | 0.37 |
| 1 | 0.83 | 0.67 |
| 0.1 | 0.91 | 1.0 |

Thus, for a reading within 1% of equilibrium value, minimum time will be required at a relative damping of $\gamma = 0.83$. Generally in indicating instruments, this is known as *response time* when the specified accuracy is the stated accuracy limit of the instrument.

*17. **External critical damping resistance (CDRX)*** is the external resistance connected across the galvanometer terminals that produces critical damping ($\gamma = 1$).

*18. **Measurement of damping*** and its relation to circuit resistance can be accomplished by a simple procedure in the circuit of Fig. 3-1. Let $R_a$ be very large (say, 150 kΩ) and $R_b$ small (say, 1 Ω), so that, when $E$ is a 1.5-V dry cell, the driving voltage in the local galvanometer loop is a few microvolts (say, 10 μV). Since circuital damping is related to *total* circuit resistance ($R_c + R_b + R_g$), the galvanometer resistance $R_g$ must be determined first. If $R_c$ is adjusted to a value that gives a convenient deflection and then to a new value $R_c'$ for which the deflection is cut in half, we have $R_g = R_c' - 2R_c - R_b$. Now let $R_c$ be

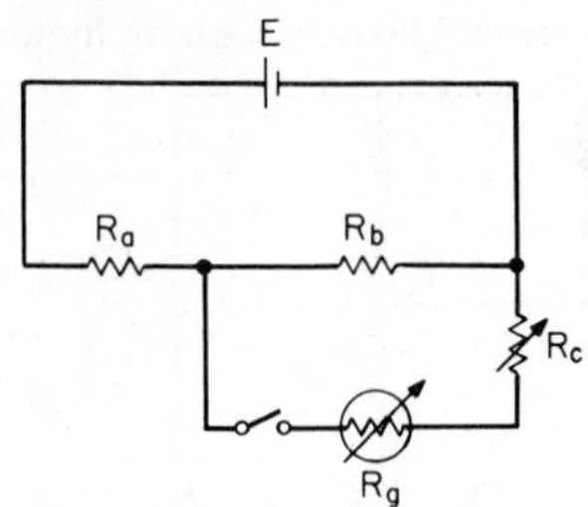


**FIG. 3-1** Determination of relative damping.

set at such a value that, when the switch is closed, the overshoot is readily observed. After noting the open-circuit deflection $\theta_o$, the switch is closed and the peak value $\theta$, of the first overswing, and the final deflection $\theta_F$ are noted. Then

$$\ln \frac{\theta_F - \theta_o}{\theta_1 - \theta_F} = \frac{\pi \gamma_1}{\sqrt{1-\gamma_1^2}}$$

$\gamma_1$ being the relative damping corresponding to the circuit resistance $R_1 = R_g + R_b + R_c$. The switch is now opened, and the first overswing $\theta_2$ past the open-circuit equilibrium position $\theta_o$ is noted. Then

$$\ln \frac{\theta_F - \theta_o}{\theta_2 - \theta_o} = \frac{\pi \gamma_o}{\sqrt{1-\gamma_0^2}}$$

$\gamma_o$ being the open-circuit relative damping. The relative damping $\gamma_x$ for any circuit resistance $R_x$ is given by the relation

$$\frac{R_x}{R_1} = \frac{\gamma_1 - \gamma_o}{\gamma_x - \gamma_o}$$

where it should be noted that the galvanometer resistance $R_g$ is included in both $R_x$ and $R_1$. For critical damping $R_d$ can be computed by setting $\gamma_x = 1$, and the external critical damping resistance $CDRX = R_d - R_g$.

***19. Galvanometer shunts*** are used to reduce the response of the galvanometer to a signal. However, in any sensitivity-reduction network, it is important that relative damping be preserved for proper operation. This can always be achieved by a suitable combination of series and parallel resistance. In Fig. 3-2, let $r$ be the external circuit resistance and $R_g$ the galvanometer resistance such that $r + R_g$ gives an acceptable damping (for example, $\gamma = 0.8$) at maximum sensitivity. This damping will be preserved when the sensitivity-reduction network ($S$, $P$) is inserted, if $S = (n-1)r$ and $P = nr/(n-1)$, $n$ being the factor by which response is to be reduced. The Ayrton-Mather shunt, shown in Fig. 3-3, may be used where the circuit resistance $r$ is so high that it exerts no appreciable damping on the galvanometer. $R_{ab}$ should be such that correct damping is achieved by $R_{ab} + R_g$. In this network, sensitivity reduction is

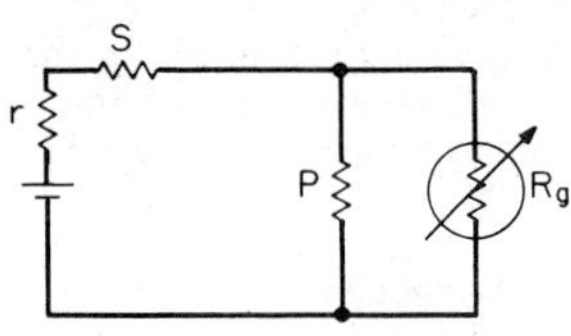

**FIG. 3-2** Galvanometer shunt.

$$n = R_{ac}/R_{ab}$$

and the ratio of galvanometer current $I_g$ to line current $I$ is

$$\frac{I_g}{I} = \frac{R_{ab}}{n(R_g + R_{ab})}$$

***20. The ultimate resolution*** of a detection system is the magnitude of the signal it can discriminate against the noise background present. In the absence of other noise sources, this limit is set by the *Johnson noise* generated by electron thermal agitation in the resistance of the circuit. This is expressed by the formula $e = \sqrt{4k\theta Rf}$, where $e$ is the rms noise voltage developed across the resistance $R$, $k$ is Boltzmann's constant $1.4 \times 10^{-23}$ J/K, $\theta$ is the absolute temperature of the resistor in kelvin, and $f$ is the bandwidth over which the noise voltage is observed. At room temperature (300 K), and with the assumption that the peak-to-peak voltage is $5 \times$ rms value, the peak-to-peak Johnson noise voltage is $6.5 \times 10^{-10} \sqrt{Rf}$ V. If, in a dc system, we use the approximation that $f = 1/3t$, where $t$ is the system's response time, the Johnson voltage is $4 \times 10^{-10} \sqrt{R/t}$ V (peak to peak).

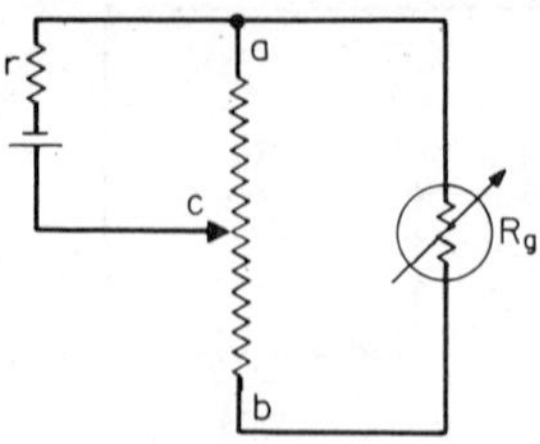

**FIG. 3-3** Ayrton-Mather universal shunt.

By using reasonable approximations, it can be shown that the random brownian-motion deflections of the moving system of a galvanometer, arising from impulses by the molecules in the air around it, are equivalent to a voltage indication $e = 5 \times 10^{-10} \sqrt{R/T}$ V (peak to peak), where $R$ is circuit resistance and $T$ is the galvanometer period in seconds. If the galvanometer damping is such that its response time is $t = 2T/3$ (for $\gamma \approx 0.8$), the Johnson noise voltage to which it responds is about $5 \times 10^{-10} \sqrt{R/t}$ V (peak to peak). This value represents the limiting resolution of a galvanometer, since its response to smaller signals would be obscured by the random excursions of its moving system. Thus, a galvanometer with a 4-s period would have a limiting resolution of about 2 nV in a 100-Ω circuit and 1 nV in a 25-Ω circuit.

It is not surprising that one arrives at the same value from considerations either of random electron motions in the conductors of the measuring circuit or of molecular motions in the fluid that surrounds the system. The resulting figure rests on the premise that the law of equipartition of energy applies to the measuring system and that the galvanometer coil—a body with one degree of freedom—is statically in thermal equilibrium with its surroundings.

*21. **Optical systems*** used with galvanometers and other indicating instruments avoid the necessity for a mechanical pointer and thus permit smaller, simpler balancing arrangements since the mirror attached to the moving system can be symmetrically disposed close to the axis of rotation. In portable instruments, the entire system—source, lenses, mirror, scale—is generally integral with the instrument; and the optical "pointer" may be folded one or more times by fixed mirrors so that it is actually much longer than the mechanical dimensions of the instrument case. In some instances, the angular displacement may be magnified by use of a cylindrical lens or mirror. For a wall- or bracket-mounted galvanometer, the lamp and scale arrangement is external, and the length of the light-beam pointer can be controlled. Whatever the arrangement, the pointer length cannot be indefinitely extended with consequent increase in resolution at the scale. The optical resolution of such a system is, in any event, limited by image diffraction, and this limit—for a system limited by a circular aperture—is $\alpha \approx 1.2\lambda/nd$ where $\alpha$ is the angle subtended by resolvable points, $\lambda$ is the wavelength of the light, $n$ is the index of refraction of the image space, and $d$ is the aperture diameter. In this case, $d$ is the diameter of the moving-system mirror, and $n = 1$ for air. If we assume that points 0.1 mm apart can just be resolved by the eye at normal reading distance, the resolution limit is reached at a scale distance of about 2 m in a system with a 1-cm mirror, which uses no optical magnification. Thus, for the usual galvanometer, there is no profit in using a mirror-scale separation greater than 2 m. Since resolution is a matter of subtended angle, the corresponding scale distance is proportionately less for systems that make use of magnification.

*22. **The photoelectric galvanometer amplifier*** is a detector system in which the light beam from the moving-system mirror is split between two photovoltaic cells connected in opposition as shown in Fig. 3-4. As the mirror of the primary galvanometer turns in response to an input

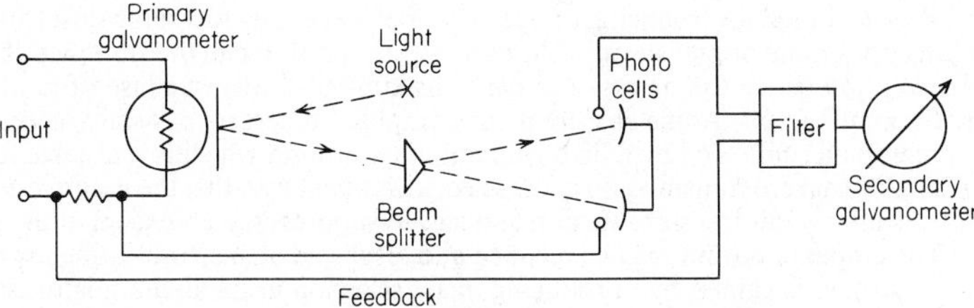


**FIG. 3-4** Photoelectric galvanometer amplifier.

signal, the light flux is increased on one of the photocells and decreased on the other, resulting in a current and thence an enhanced signal in the circuit of the secondary (reading) galvanometer. Since the photocells respond to the total light flux on their sensitive elements, the system is not subject to resolution limitation by diffraction as is the human eye; and the ultimate resolution of the primary instrument—limited only by its brownian motion and the Johnson noise of the input circuit—may be realized.

A feedback loop, energized by the photocell output, is frequently used to stabilize the primary galvanometer and to control its sensitivity. A low-pass filter is sometimes inserted before the secondary galvanometer to increase the response time of the output circuit and thus to enhance the ultimate resolution of the system. Such a detector system must be isolated from mechanical disturbances or made unresponsive to them. The moving system of the primary galvanometer should be very carefully balanced about its axis of rotation. As a further precaution, the moving system is, in some instances, floated with zero buoyancy in a closed, liquid-filled compartment. Since the moving system of the primary galvanometer is usually not perfectly balanced, isolation is required in very low level operation. A simple, effective isolation platform which can support a number of galvanometers is a 24 × 36 in wooden tray carrying a 2-in bed of sand on which is supported a 2 × 18 × 30 in concrete slab. The whole system is supported on four interconnected ¾ × 4¼ in inner tubes at the corners of the tray. It is essential that these inflated cushions be interconnected to provide damping of mechanical disturbances and that the system be enclosed to protect it from air-borne disturbances.

*23. **Electronic*** instruments for low-level dc signal detection are more convenient, more rugged, and less susceptible to mechanical disturbances than is a galvanometer. However, considerable filtering, shielding, and guarding must be used to minimize electrical interference and noise. On the other hand, a galvanometer is an extremely efficient low-pass filter and, when operated to make optimum use of its design characteristics, it is still the most sensitive low-level dc detector. Electronic detectors generally make use of either a mechanical or a transistor chopper driven by an oscillator whose frequency is chosen to avoid the local power frequency and its harmonics. This modulator converts the dc input signal to ac, which is then amplified, demodulated, and displayed on an analog-type indicating instrument or fed to a recording device or a signal processor.

*24. **Magnetic modulator amplifier*** systems use saturable cores to convert the dc input signals to alternating current, which is then amplified, demodulated, and presented on a center-zero dc indicating instrument. In one such system, a 2.5-kHz oscillator provides excitation to two identical but oppositely wound cores, which are carried to saturation in each ac half cycle. Since the two cores carry identical exciting windings, there is no net flux in the input winding which surrounds both. A dc signal in this input winding results in earlier saturation in one core section and later saturation in the other during each excitation frequency in a winding that surrounds both core sections. The double-frequency voltage induced in this winding is proportional to the dc input signal and reversed in phase if the polarity of the input signal is reversed. The ac signal goes to a sharply tuned ac amplifier and thence to a ring demodulator whose phase discrimination is provided by a signal from the oscillator via the frequency doubler. Thus, the zero-center dc instrument receiving the demodulated signal indicates the polarity as well as the magnitude of the input signal. The sensitivity of the device is selected by means of an amplifier gain-control network. The instrument described is essentially a current-sensitive device with a resolution of 1 to 2 nA on its most sensitive range; voltage resolution is also good in a low-resistance circuit.

*25. **AC detectors*** used for balancing bridge networks are usually tuned low-level amplifiers coupled to an appropriate display device. The narrower the passband of the amplifier, the better the signal resolution, since the narrow passband discriminates against noise of random frequency in the input circuit. Adjustable-frequency amplifier-detectors basically incorporate a low-noise preamplifier followed by a high-gain amplifier around which is a tunable feedback loop whose circuit has zero transmission at the selected frequency, so that the negative-feedback circuit controls the overall transfer function and acts to suppress signals except at the selected frequency. The amplifier output may be rectified and displayed on a dc indicating instrument, and added resolution is gained by introducing phase selection at the demodulator, since the wanted signal is regular in phase, while interfering noise is generally random. In detectors of this type, inphase and quadrature signals can be displayed separately, permitting independent balancing of bridge components. Further improvement can result from the use of a low-pass filter between the demodulator and the dc indicator such that the signal of selected phase is integrated over an appreciable time interval up to a second or more.

An alternative type of display is the screen of a cathode-ray oscillograph where the ac signal to be observed is connected to the $y$ terminals of the oscillograph and a synchronous signal of adjustable phase and magnitude is connected to the $x$ terminals. The resulting Lissajous figure on the screen is an ellipse; and, by proper phase adjustment of the $x$ signal, the inphase and quadrature components of an unbalance signal from a bridge correspond, respectively, to the opening of the ellipse and the angle which its major axis makes with the horizontal. In this way, the two components of the bridge balance can be independently adjusted so that, at balance, the ellipse collapses into a horizontal line.

## Continuous EMF Measurements

*26. **A standard of emf*** may be either an electrochemical system or a Zener-diode-controlled circuit operated under precisely specified conditions.

*27. **The Weston standard cell*** has a positive electrode of metallic mercury and a negative electrode of cadmium-mercury amalgam (usually about 10% Cd). The electrolyte is a saturated solution of cadmium sulfate with an excess of $Cd \cdot SO_4 \cdot \frac{8}{3}H_2O$ crystals, usually acidified with sulfuric acid (0.04 to 0.08 N). A paste of mercurous sulfate and cadmium sulfate crystals over the mercury electrode is used as a depolarizer. The saturated cell has a substantial temperature coefficient of emf. Vigoureux and Watts[1] of the National Physical Laboratory have given the following formula, applicable to cells with a 10% amalgam:

$$E_t = E_{20} - 39.39 \times 10^{-6}(t-20) - 0.903 \times 10^{-6}(t-20)^2 + 0.00660 \times 10^{-6}(t-20)^3 - 0.000150 \times 10^{-6}(t-20)^4$$

where $t$ is the temperature in degrees Celsius. Since cells are frequently maintained at 28°C, the following equivalent formula by Hamer[2] is useful:

$$E_t = E_{28} - 52.899 \times 10^{-6}(t-28) - 0.80265 \times 10^{-6}(t-28)^2 + 0.001813 \times 10^{-6}(t-28)^3 - 0.0001497 \times 10^{-6}(t-28)^4$$

These equations are general and are normally used only to correct cell emfs for small temperature changes; i.e., 0.05 K or less. For changes at that level, negligible errors are introduced by making corrections. Standard cells should always be calibrated at their temperature of use (within 0.05 K) if they are to be used at an accuracy of 5 ppm or better.

A group of saturated Weston cells, maintained at a constant temperature in an air bath or a stirred oil bath, is quite generally used as a laboratory reference standard of emf. The bath temperature must be constant within a few thousandths of a degree if the reference emf is to be reliable to a microvolt. It is even more important that temperature gradients in the bath be avoided, since the individual limbs of the cell have very large temperature coefficients (about $+315\ \mu V/°C$ for the positive limb and $-379\ \mu V/°C$ for the negative limb—more than $-50\ \mu V/°C$ for the complete cell—at 28°C). Frequently, two or three groups of cells are used, one as a reference standard which never leaves the laboratory, the others as transport groups which are used for interlaboratory comparisons and for assignment by a standards laboratory.

Recently inert mechanical retainers have been used to hold the electrode materials in place, facilitating interlaboratory transport; such cells need not be hand-carried to avoid injury. Thus a group of cells in a constant-temperature bath (powered by a battery) can be shipped by air to a laboratory almost anywhere in the world for comparison with the local standard.

*28. **The Weston unsaturated cell*** uses the same electrode system as the saturated cell, but its electrolyte is not saturated at ordinary room temperatures. Its concentration is such that it reaches saturation at 3 to 4°C. Its emf at room temperature is about 0.05% higher than that of the saturated cell, and its temperature coefficient of emf is much lower (less than $10\ \mu V/°C$). Here, also, the temperature coefficients of the individual limbs are large, and a copper case is used to ensure their temperature equality. Unsaturated cells are widely used for the reference emf on potentiometers or wherever the accuracy of the voltage reference need not be better than 0.005%. To ensure this accuracy, the cell should be checked annually against the reference voltage from a properly maintained bank of saturated cells, since the emf of the unsaturated cell is expected to decrease slowly with time. The unsaturated cell is usually built with retaining barriers over the electrodes to hold them in place, and they can be shipped by express or even by parcel post. The use of inert retainers in place of the cork barriers formerly employed has reduced voltage drift rate and increased useful life substantially.

---

[1] Vigoureux and Watts *Proc. Phys. Soc. (London),* 1933, vol. 45, p. 172.
[2] Hamer, W. J. *NBS Mono.* 84, 1965.

***29. Precautions in Using Standard Cells***

**a.** The cell should not be exposed to extreme temperatures—below 4°C or above 40°C.

**b.** Temperature gradients (differences between the cell limbs) should be avoided.

**c.** Abrupt temperature changes should be avoided—the recovery period after a sudden temperature change may be quite extended; recovery is usually much quicker in an unsaturated than in a saturated cell. Full recovery of saturated cells from a gross temperature change (for example, from room temperature to a 35°C maintenance temperature) can take up to 3 months. More significantly, some cell emfs have been seen to exhibit a plateau in their response over a 2- to 3-week period within a week or two after the temperature shock is sustained. This plateau can be as much as 5 ppm higher than the final stable value.

**d.** Current in excess of 100 nA should never be passed through the cell in either direction; actually, one should limit current to 10 nA or less for as short a time as feasible in using the cell as a reference. Cells that have been short-circuited or subjected to excessive charging current drift until chemical equilibrium in the cell is regained over an extended time period—as long as 9 months, depending on the amount of charge involved.

***30. Zener diodes*** or diode-based devices have replaced chemical cells as voltage references in commercial instruments, such as digital voltmeters and voltage calibrators. Some of these instruments have uncertainties below 10 ppm, instabilities below 5 ppm per month (including drift and random uncertainties); and temperature coefficient of output as low as 2 ppm/°C.

The best devices, as identified in a testing in selection process, are available as solid-state voltage reference or transport standards. Such instruments generally have at least two outputs, one in the range 1.018 to 1.02 V for use as a standard cell replacement, and one in the range 6.4 to 10 V, the output voltage of the reference device itself. The lower voltage is usually obtained via a resistive divider.

Other features sometimes include a vernier adjustment for the lower voltage for adjusting to equal the output of a given standard cell, and internal batteries for complete isolation. Such devices have performance approaching that of standard cells and can be used in many of the same applications. Some have stabilities (drift rate and random fluctuations) as low as 2 to 3 ppm per year and temperature coefficient of 0.1 ppm/°C.

The current through the reverse-biased junction of a silicon diode remains very small until the bias voltage exceeds a characteristic $V_z$ in magnitude, at which point its resistance becomes abruptly very low so that the voltage across the junction is little affected by the junction current. Since the voltage-current relationship is repeatable, the diode may be used as a standard of voltage as long as its rated power is not exceeded.

However, since $V_z$ is a function of temperature, single junctions are rarely used as voltage references in precise applications. Since a change in temperature shifts the *I-V* curve of a junction, the use of a forward-biased junction in series with the Zener diode permits a current level to be found at which changes in Zener voltage from temperature changes are compensated by changes in the voltage drop across the forward-biased junction.

Devices using this principle fall into two categories: the temperature-compensated Zener diode, in which two diodes are in series opposition, and the reference amplifier in which the Zener diode is in series with the base-emitter junction of an appropriate *npn* silicon transistor. In each case the two elements may be on the same substrate for temperature uniformity. In some precision devices, the reference element is in a temperature-controlled oven to permit even greater immunity to temperature fluctuations.[1]

***31. Potentiometers*** are used for the precise measurement of emf in the range below 1.5 V. This is accomplished by opposing to the unknown emf an equal *IR* drop. There are two possibilities: either the current is held constant while the resistance across which the *IR* drop is opposed to the unknown is varied, or current is varied in a fixed resistance to achieve the desired *IR* drop.

Figure 3-5 shows schematically most of the essential features of a general-purpose constant-current instrument. With the standard-cell dial set to read the emf of the reference standard cell,

[1] Eicke, W. G. *Electron. Instr. Dig.*, May 1970, p. 50. Miller, W. D., and DeFreitas, R. E. *Electronics*, 1975, vol. 48, no. 4, p. 101. Spreadbury, P. J. and Everhart, T. E. Ultrastable Portable Voltage Sources; *IEE Conference Publication* 1979, vol. 174, p. 117.

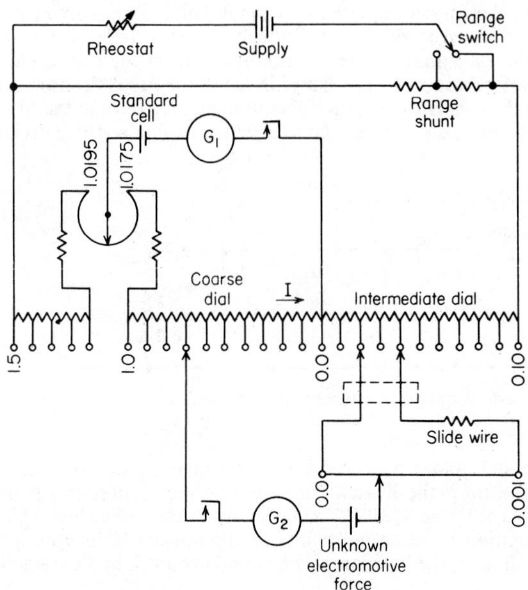


**FIG. 3-5** General-purpose constant-current potentiometer.

the potentiometer current $I$ is adjusted until the $IR$ drop across 10 of the coarse-dial steps plus the drop to the set point on the standard-cell dial balances the emf of the reference cell. The correct value of current is indicated by a null reading of the galvanometer in position $G_1$. This adjustment permits the potentiometer to be read directly in volts. With the galvanometer in position $G_2$, the unknown emf is balanced by varying the opposing $IR$ drop. Resistances used from the *coarse* and *intermediate* dials and the *slide wire* are adjusted until the galvanometer again reads null, and the unknown emf can be read directly from the dial settings. The ratio of the unknown and reference emfs is precisely the ratio of the resistances for the two null adjustments, provided that the current is the same.

The switching arrangement is usually such that the galvanometer can be shifted quickly between the $G_2$ and $G_1$ positions to check that the current has not drifted from the value at which it was standardized. It will be noted that the contacts of the coarse-dial switch and slide wire are in the galvanometer branch of the circuit. At balance, they carry no current, and their contact resistance does not contribute to the measurement. However, there can be only two noncontributing contact resistances in the network shown; the switch contacts for adjusting the intermediate-dial position do carry current, and their resistance does enter the measurement. Care is taken in construction that the resistances of such current-carrying contacts are low and repeatable; and frequently, as in the example illustrated, the circuit is arranged so that these contributing contacts carry only a fraction of the reference current, and the contribution of their $IR$ drop to the measurement is correspondingly reduced.

Another feature of many general-purpose potentiometers, illustrated in the diagram, is the availability of a reduced range. The resistances of the range shunts have such values that, at the 0.1 position of the range-selection switch, only a tenth of the reference current goes through the measuring branch of the circuit, and the range of the potentiometer is correspondingly reduced. Frequently, a $\times 0.01$ range is also available.

In addition to the effect of $IR$ drops at contacts in the measuring circuit, accuracy limits are also imposed by thermal emfs generated at circuit junctions. These limiting factors are increasingly important as potentiometer range is reduced. Thus, in low-range or microvolt potentiometers, special care is taken to keep circuit junctions and contact resistances out of the direct measuring circuit as much as possible, to use thermal shielding, and to arrange the circuit and galvanometer keys so that temperature differences will be minimized between junction points that are directly in the measuring circuit. Generally also, in microvolt potentiometers, the galvanometer is connected to the circuit through a special *thermofree* reversing key so that thermal emfs in the galvanometer can be eliminated from the measurement—the balance point being that which produces zero change in galvanometer deflections on reversal.

An example of the constant-resistance potentiometer is shown in the simplified diagram in Fig. 3-6.[1] It consists basically of a constant-current source, a resistive divider *D* (used in the

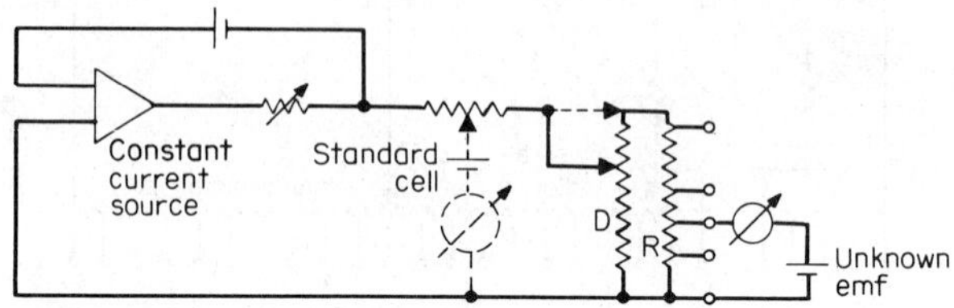


**FIG. 3-6** Constant-resistance potentiometer.

current-divider mode), and a fixed resistor *R* in which the current (and the *IR* drop) are determined by the setting of the divider. The output of the current source is adjusted by equating the emf of a standard cell to an equal *IR* drop as shown by the dashed line. This design lends itself to multirange operation by using tap points on the resistor *R*. Its accuracy depends on the uniformity of the divider, the location of the tap points on *R*, and the stability of the current source.

Another type of constant-resistance potentiometer, operating from a current comparator which senses and corrects for inequality of ampere-turns in two windings threading a magnetic core, is shown in Fig. 3-7.[2] Two matched toroidal cores wound with an identical number of

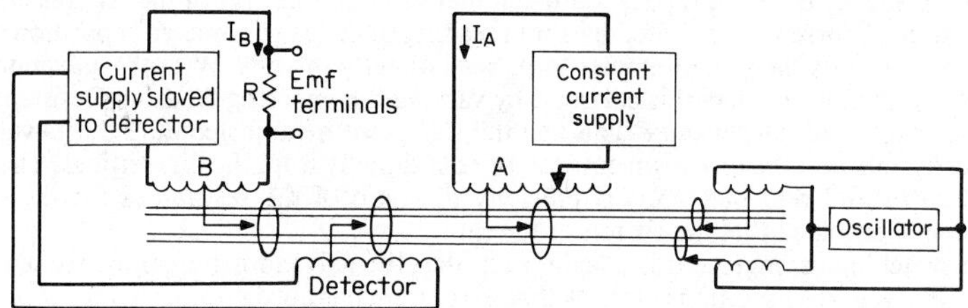


**FIG. 3-7** Current-comparator potentiometer.

turns are excited by a fixed-frequency oscillator. The fluxes induced in the cores are equal and oppositely directed, so that they cancel with respect to a winding that encloses both. In the absence of additional mmf[3] the detector winding enclosing both cores receives no signal.

If, in another winding *A* enclosing both cores, we inject a direct current, its mmf reinforces the flux in one core and opposes the other. The net flux in the detector winding induces a voltage in it. This signal is used to control current in another winding *B* which also threads both cores. When the mmf of *B* is equal and opposite that of *A*, the detector signal is zero and the ampere-turns of *A* and *B* are equal. Thus a constant current in an adjustable number of turns is matched to a variable current in a fixed number of turns, and the voltage drop $I_B R$ is used to oppose the emf to be measured.

The system is made direct-reading in voltage units (in terms of the turns ratio *B*/*A*) by adjusting the constant-current source with the aid of a standard cell circuit (not shown in the

[1] Julie, L. *Trans. IEEE,* 1967, vol. IM-16, p. 187; Luppold, D. S. *Precision DC Measurement,* Chap. 5, Addison-Wesley, 1969.

[2] Kusters, N. L., and McMartin, M. P. *Trans. IEEE,* 1966, vol. IM-15, p. 212.

[3] Magnetomotive force, or magnetic potential difference.

figure). This type of potentiometer has an advantage over those whose continuing accuracy depends on the stability of a resistance ratio; the ratio here is the turns ratio of windings on a common core, dependent solely on conductor position and hence not subject to drift with time.

*32.* ***Voltage dividers*** are used to reduce voltages by a known factor. *Volt boxes,* used to extend the range of potentiometers, are resistive voltage dividers with a limited number of fixed-position taps, as shown in Fig. 3-8. The unknown voltage is connected between appropriately marked input terminals (say, 0 and 150), and the potentiometer is connected between the 0 and 1.5 output terminals. When balance is achieved on the potentiometer in the usual way, the input voltage is obtained by multiplying the potentiometer reading by the indicated factor (×100). While no current is taken by the potentiometer at balance, current is supplied by the source to the divider through its input terminals. Hence, the voltage drop between the input terminals is measured rather than the open-circuit emf of the source. The resistances of working volt boxes range from about 200 to 750 Ω/V, and even higher—to 1 kΩ/V—in reference-standard types.

Since changes in ratio as a result of self-heating are much larger in low- than in high-resistance circuits, particularly in the higher voltage ranges, the present trend is toward the higher ohms per volt figure. However, the quality of insulation becomes more critical in high-resistance boxes, since insulating structures which support the resistance elements are themselves high-resistance leakage paths. Thus, the high-accuracy reference-standard boxes of 1000 Ω/V resistance are provided with a guard circuit which maintains shields at appropriate potentials along the insulating structure of the box.

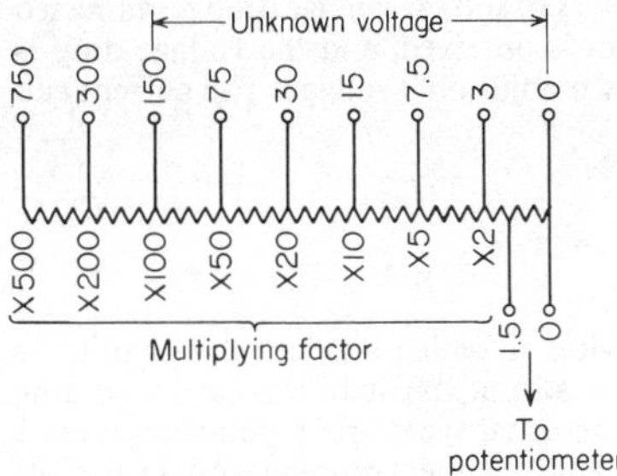


**FIG. 3-8** Volt box.

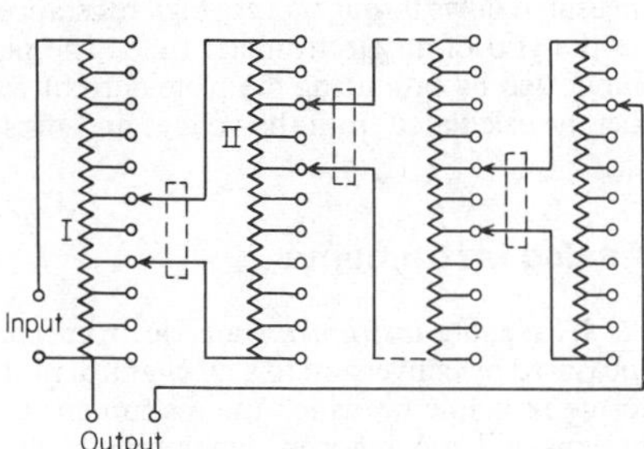


**FIG. 3-9** Decade voltage divider.

*Decade voltage dividers* generally use the Kelvin-Varley circuit arrangement shown in Fig. 3-9. It will be seen that two elements of the first decade are shunted by the entire second decade, whose total resistance equals the combined resistance of the shunted steps of decade I. The two sliders of decade I are mechanically coupled and move together, keeping the shunted resistance constant regardless of switch position. Thus, the current divides equally between decade II and the shunted elements of decade I, and the voltage drop in decade II equals the drop in one unshunted step of decade I. The effect of contact resistance at the switch points is somewhat diminished because of the division of current. The Kelvin-Varley principle is used in succeeding decades except the final one, which has only a single switch contact. Such voltage dividers may have as many as eight decades and have ratio accuracies approaching 1 part in $10^6$ of input.

*33.* ***Spark gaps*** provide a means of measuring high voltages. The maximum gap which a given voltage will break down depends on air density, gap geometry, crest value of the voltage, and other factors (see Sec. 27). Sphere gaps constitute a recognized means for measuring crest values of alternating voltages and of impulse voltages. IEEE Standard 4 has tables of sparkover voltages for spheres ranging from 6.25 to 200 cm diameter and for voltages from 17 to 2500 kV. Sphere gap voltage tables are also available in ANSI Standard 68.1 and in IEC Publication 52.

## Continuous Current Measurements

*34.* ***Absolute current measurement*** relates the value of the current unit—the ampere—to the prototype mechanical units of length, mass, and time—the meter, the kilogram, and the second—through force measurements in an instrument called a *current balance.* Such instruments are to be found generally only in national standards laboratories, which have the responsibility of establishing and maintaining the electrical units. In a current balance, the force between fixed and movable coils is opposed by the gravitational force on a known mass, the balance equation being $I^2(\partial M/\partial X) = mg$. The construction of the coil system is such that the rate of change with displacement of mutual inductance between fixed and moving coils can be computed from measured coil dimensions. Absolute current determinations are used to assign the emf of reference standard cells. A 1-Ω resistance standard is connected in series with the fixed- and moving-coil system, and its drop is compared with the emf of a cell during the force measurement. Thus the National Reference Standard of voltage is derived from absolute ampere and ohm determinations.

*35.* ***The potentiometer method*** of measuring continuous currents is commonly used where a value must be more accurate than can be obtained from the reading of an indicating instrument. The current to be measured is passed through a four-terminal resistor (shunt) of known value, and the voltage developed between its potential terminals is measured with a potentiometer. If the current is small, so that there is no significant temperature rise in the shunt, the measurement accuracy can be 0.01% or better. In general, the accuracy of potentiometer measurements of continuous currents is limited by how well the shunt resistance is known under operating conditions.

*36.* ***Measurement of very small continuous currents,*** down to $10^{-17}$ A, have been accomplished by means of "electrometer" tubes—vacuum tubes designed so that the grid has practically no leakage current either over its insulating supports or to the cathode. The current to be measured flows through a very high resistance (up to $10^{12}$ Ω), and the voltage drop is impressed on the grid of an electrometer tube. The plate current is observed, and the voltage drop is duplicated by producing the plate current with a known adjustable voltage. The current can then be calculated from the voltage and resistance.

## Analog Instruments

*37.* ***Analog instruments*** are electromechanical devices in which an electrical quantity is measured by conversion to a mechanical motion. Such instruments can be classified according to the principle on which the instrument operates. The usual types are: permanent-magnet moving-coil, moving-iron, dynamometer, and electrostatic. Another grouping is on the basis of use: panel, switchboard, portable, and laboratory-standard. Accuracy can also be the basis of classification. Details concerning performance and other specifications are to be found in ANSI Standard C39.1, "Requirements for Electrical Analog Indicating Instruments."

*38.* ***Permanent-magnet moving-coil instruments*** are the most common type in general use. The operating mechanism consists of a coil of fine wire suspended in such a manner that it can rotate in an annular gap which has a radial magnetic field. The torque, generated by the current

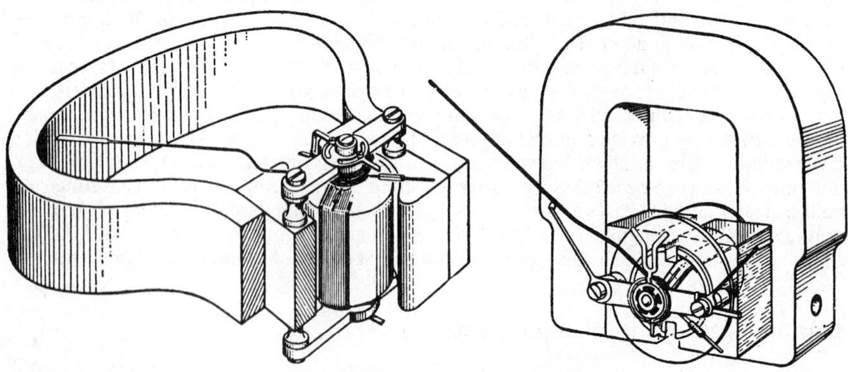

**FIG. 3-10** Permanent-magnet moving coil mechanism (external magnet).

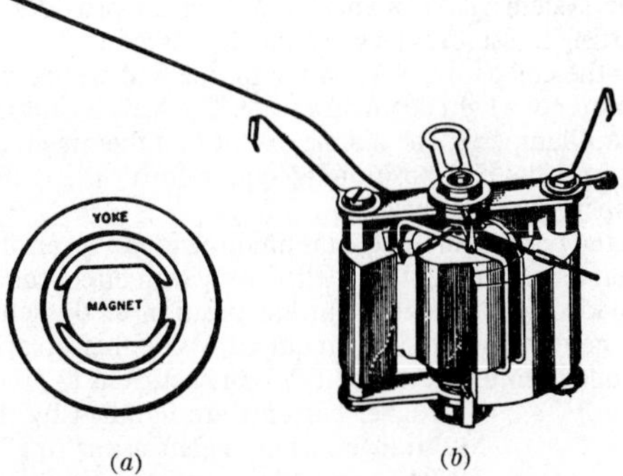


**FIG. 3-11** *(a)* Core magnet construction; *(b)* internal magnet design.

in the moving coil reacting to the magnetic field of the gap, is opposed by some form of spring restraint. The restraint may be a helical spring, in which case the coil is supported by a pivot and jewel, or both the support and the angular restraint is by means of a taut-band suspension (see Par. **10**). Figure 3-10 shows a mechanism using an external magnet with pivots and helical springs. Figure 3-11 shows a similar instrument with an internal magnet. Figure 3-12 shows the

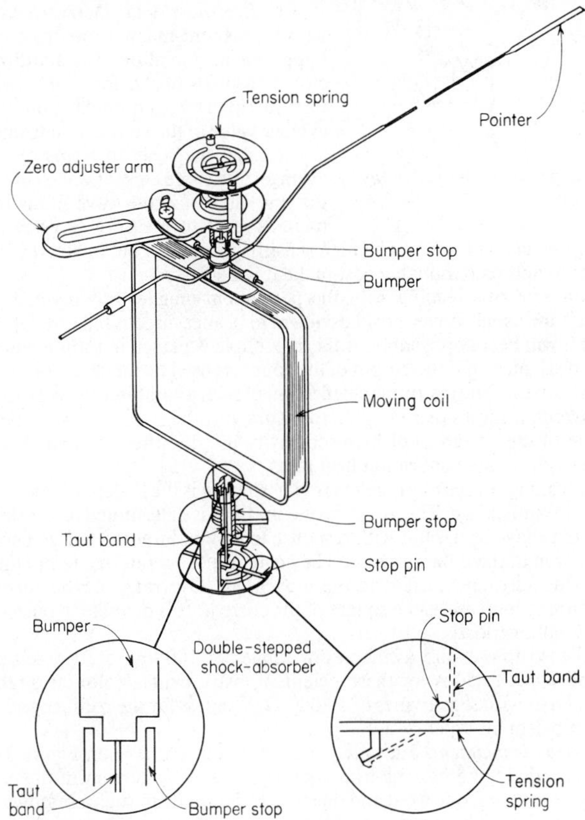


**FIG. 3-12** Construction of a taut-band suspension.

taut-band system. In this system a flat band held under tension provides the support for the coil. By its resistance to twisting it also provides the angular restraint.

The position which the coil assumes when the torque and spring restraint are balanced is indicated by either a pointer or a light beam on a scale. The scale is calibrated in units suitable to the application: volts, milliamperes, etc. To the extent that the magnetic field is uniform, the spring restraint linear, and the coil positioning symmetrical, the deflection will be linearly proportional to the ampere-turns in the coil.

Because the field of the permanent magnet is unidirectional, reversal of the coil current will reverse the torque so that the instrument will deflect only with direct current in the moving coil. Scales are usually provided with the zero current position at the left to allow a full-range deflection. However, where measurement is required with either polarity, a zero center scale position is used. The coil is limited in its ability to carry current to 50 or 100 mA.

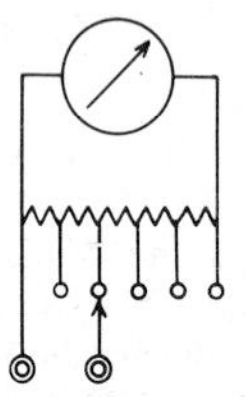

**FIG. 3-13** Multirange ring shunt.

Higher currents are handled by the addition of shunts. Multirange current meters use a "ring" shunt as shown in Fig. 3-13. Figure 3-14 shows the addition of series resistance to the coil to provide multirange voltage measurement. Self-contained ranges of current are available from 1 μA to 50 A while voltage ranges are available from 2 mV to 1 kV self-contained. External shunts to several thousand amperes and multipliers to several hundred volts are available. Voltmeter sensitivity is usually stated in ohms per volt, the reciprocal of full-scale current in amperes. Commercial sensitivities range from about 50 Ω/V to 1 MΩ/V.

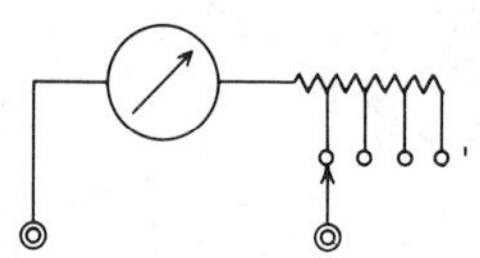

**FIG. 3-14** Multirange circuit for voltmeter.

***39. Rectifiers and thermoelements*** are used with permanent-magnet moving-coil instruments to provide ac operation. The addition of a rectifier circuit, usually in the form of a bridge, gives an instrument in which the deflection is in terms of the average value of the voltage or current. It is customary to label the scale in terms of 1.11 times the average; this is the correct waveform factor to read the rms value of a sine wave. If the rectifier instrument is used to measure severely nonsinusoidal waveforms, large errors will result. The high sensitivity that can be obtained with the rectifier type of instrument and its reasonable cost make it widely used.

To provide a true rms reading with the permanent-magnet moving-coil instrument, a thermoelement is the usual converter. The current to be measured is fed through a resistance of such value that it will heat appreciably. A thermocouple is placed in intimate thermal contact with the heater resistance, and the output of the couple is used to energize a permanent-magnet moving-coil instrument. The instrument deflection of such a combination is proportional to the square of the current; using a square-root factor in drawing the scale allows it to be read in terms of the rms value of the current. For high-sensitivity use, the thermoelement is placed in an evacuated bulb to eliminate convection heat loss.

The prime advantage of the thermoelement instrument is the high frequency at which it will operate and the rms indication. The upper frequency limit is determined by the skin effect in the heater. Instruments have been built with response to several hundred MHz. There is one very important limitation to these instruments. The heater must operate at a temperature of 100°C or more to provide adequate current to the movement. Overrange of the current will cause heater temperature to increase as the square of the current. It is possible to burn out the heater with relatively small overloads.

Commercial instruments are available for ranges of 100 μA to 50 A self-contained for rectifier type and 1 mA to 50 A for thermoelement instruments. Voltmeters range from 1 to 1000 V for both types with sensitivities of 10,000 Ω/V or less for the rectifier and 1000 Ω/V or less for the thermoelement type.

***40. Moving-iron instruments*** are widely used at power frequencies. Figure 3-15 shows the construction of a radial-vane moving-iron type. It operates by current in the coil which surrounds two magnetic vanes, one fixed and one which can rotate in such a manner as to increase the spacing between them. Current in the coil causes the vanes to be similarly magnetized and so to repel each other. The torque produced by the moving vane is proportional to the square of the current and is independent of its polarity. This provides the basis for ac operation and for rms indication by means of a suitable scale.

Many different configurations of the iron vanes are available. In some designs, the scale can be made to show a linear characteristic over 80 to 90% of the range. Instrument sensitivity is in the order of 10 mA at best, with self-contained ranges as high as 200 A. The frequency range is limited by eddy currents and skin effect in the vanes and distributed capacitance in the coil. Most commercial instruments have an upper limit of about 500 Hz, although with special compensation instruments have been made for use to 10 kHz. DC operation will be satisfactory for most moving-iron instruments; if the mean of reversed dc readings is used, most instruments will be within rated ac accuracy. Range extension is accomplished with current and voltage transformers.

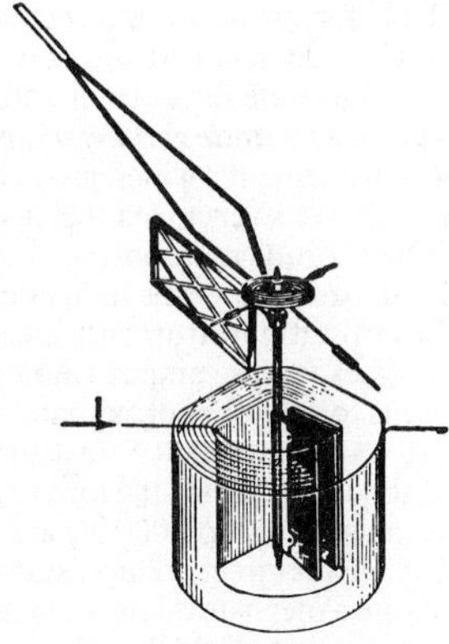

**FIG. 3-15** Radial-vane mechanism.

***41. Dynamometer instrument*** construction is shown in Fig. 3-16. The moving coil rotates in a field set up by the current in a pair of surrounding field coils. The torque is proportional to the product of the moving-coil ampere-turns and the field-coil ampere-turns. Connection of field coils and moving coil in series gives response in terms of the current squared, making the dynamometer an rms reading instrument for current or voltage.

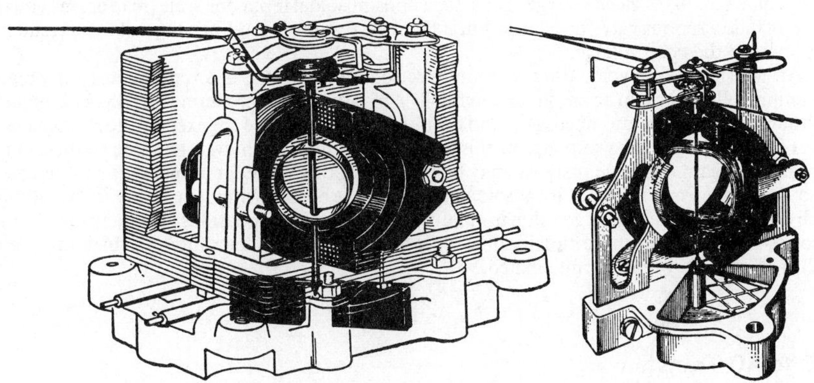

**FIG. 3-16** Weston dynamometer instrument.

While voltmeters and ammeters are available, the common use of the dynamometer is for power measurement. The moving coil (with a suitable series resistance) is connected in parallel with a load to be measured, and the field coils are connected in series with the load. The instantaneous torque is proportional to the product of voltage and current; the mechanical time constant of the movement will integrate this product, and pointer indication is in terms of average power.

Figure 3-17 shows two ways in which a wattmeter may be connected to measure power in a load. With the moving coil connected at A the instrument will read high by the amount of power used by the moving-coil circuit. If connection is made at B, the wattmeter will read high by the power dissipated in the field coils. When using sensitive, low-range meters, it is necessary to correct for this error. Commercial instruments are

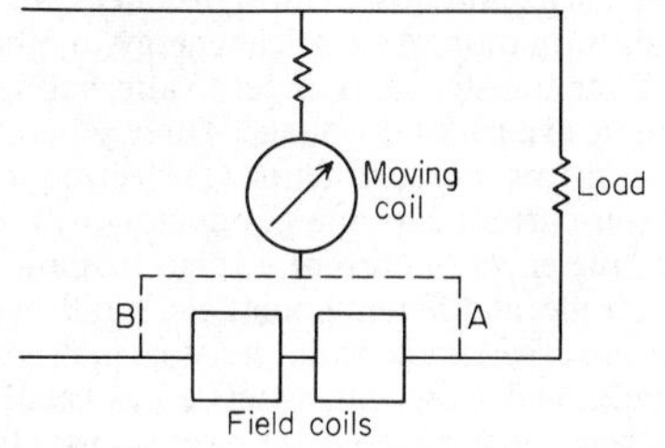


**FIG. 3-17** Alternative wattmeter connections.

available for ranges from a fraction of a watt to several hundred watts self-contained. Range extensions are obtained with current and voltage transformers. In specifying wattmeters, it is necessary to state the current and voltage ranges as well as the watt range.

***42. Electrostatic voltmeters*** are actually voltage-operated in contrast to all the other types of analog instruments, which are current-operated. In an electrostatic voltmeter, fixed and movable vanes are so arranged that a voltage between them causes attraction to rotate the movable vane. The torque is proportional to the energy stored in the capacitance, and thus to the voltage squared, permitting rms indication.

Electrostatic instruments are used for voltage measurements where the current drain of other types of instrument cannot be tolerated. Input resistance (due to insulation leakage) amounts to $10^{13}\ \Omega$ approximately for a range of 100 V (the lowest commercially available) to $3 \times 10^{15}\ \Omega$ for 100,000-V instruments (the highest commonly available). Capacitance ranges from about 300 pF for the lower ranges to 10 pF for the highest. Multirange instruments in the lower ranges (100 to 5000 V) are frequently made with capacitive dividers which make them inoperable on direct voltage, since the series capacitor blocks out dc. Other multirange instruments use a mechanical movement of the fixed electrode to change ranges. These can be used on dc or ac, as can all single-range voltmeters.

***43. Electronic voltmeters*** vary widely in performance characteristics and frequency range covered, depending on the circuitry used. A common type uses an initial diode to charge a capacitor. This may be followed by a stabilized amplifier with a microammeter as indicator. Range may be selected by appropriate cathode resistors in the amplifier section. Such instruments normally have very high input impedance (a few picofarads), respond to peak voltage, and are suitable for use to very high frequencies (100 MHz or more). While the response is to *peak* voltage, the scale of the indicating element may be marked in terms of rms for a sine-wave input, that is, $0.707 \times$ *peak* voltage. Thus, for a nonsinusoidal input, the scale (read as rms volts) may include a serious waveform error; but if the scale reading is multiplied by 1.41, the result is the value of the *peak* voltage.

An alternative network, used in some electronic voltmeters, is an attenuator for range selection, followed by an ac amplifier and finally a rectifier and microammeter. This system has substantially lower input impedance, and limits of frequency range are fixed by the characteristics of the amplifier. The response in this arrangement may be to *average* value of the input signal, but, again, the scale marking may be in terms of rms value for a sine wave. In this case also, the waveform error for nonsinusoidal input must be borne in mind; but if the scale reading is divided by 1.11, the *average* value is obtained. Within these limitations, accuracy may be as good as 1% of full-scale indication in some types of electronic voltmeter, although in many cases a 2 to 5% accuracy may be anticipated.

## DC to AC Transfer

***44. General transfer capability*** is essential to the measurement of voltage, current, power, and energy. The standard cell, the unit of voltage which it preserves, and the unit of current derived from it in combination with a standard of resistance are applicable only to the measurement of dc quantities, while the problems of measurement in the power and communication fields involve alternating voltages and currents. It is only by means of transfer devices that one can assign the values of ac quantities or calibrate ac instruments in terms of the basic dc reference standards. In most instances, the rms value of a voltage or current is required, since the transformation of electrical energy to other forms involves the square of voltages or currents; and the transfer from direct to alternating quantities is made with devices that respond to the square of current or voltage. Three general types of transfer instruments are capable of high-accuracy rms measurements: (1) electrodynamic instruments—which depend on the force between current-carrying conductors, (2) electrothermic instruments—which depend on the heating effect of current, (3) electrostatic instruments—which depend on the force between electrodes at different potentials. While two of these depend on current and the third on voltage, the use of series and shunt resistors makes all three types available for current or voltage transfer. Traditional American practice has been to use electrodynamic instruments for current and voltage transfer as well as power transfer from direct to alternating current; but recent developments in thermoelements[1] have improved their transfer characteristics until they are now the preferred means for *current* and *voltage* transfer, although the electrodynamic wattmeter is still the instrument of choice for *power* transfer up to 1 kHz.

***45. Electrothermic transfer standards*** for current and voltage use a thermoelement consisting of a heater and a thermocouple. In its usual form, the heater is a short, straight wire suspended by two supporting lead-in wires in an evacuated glass bulb. One junction of a thermocouple is fastened to its midpoint and is electrically insulated from it with a small bead. The thermal emf—5 to 10 mV at rated current in a conventional element—is a measure of heater current. Multijunction thermoelements having a number of couples in series along the heater have also been used in transfer measurements. Typical output is 100 mV for an input power of 30 mW.[2]

Monolithic integrated circuit thermoelements consist of two matched heater-element-transistor pairs on a common substrate. The transistor base-emitter junction characteristics are employed to sense the temperature of the heaters. The ac and dc signals to be compared are applied to one of the heaters and the resulting differential transistor output is amplified to produce a dc current through the second heater. The current level is monitored by measuring the voltage drop on a stable resistor in the feedback circuit and used to sense rms equality between ac and dc input signals.[3]

For voltage measurement, a resistor is connected in series with the heater so that output emf is a measure of input voltage. For high-frequency applications, the series element consists of one or more metal-film resistors in a coaxial metal guard cylinder with the thermoelement mounted in a separate cylinder. The guard cylinders are so arranged and connected to the circuit as to minimize the phase-defect angle of the combination. These elements are useful into the radio-frequency range—to 100 MHz or more—and at audio frequencies, their transfer accuracy may be a few parts in $10^6$.

For current measurements, an ac shunt (preferably of coaxial design to minimize its time constant) is used with a low-range thermoelement connected across its potential terminals. Here, also, transfer accuracies of a few parts per million are possible in the audio-frequency range. Equal direct and alternating quantities should give the same thermocouple output (since their heating effect is the same), and the magnitude of the dc quantity can be measured by potentiometer techniques with resistive voltage dividers and standard resistors. Thus, the magnitude of the ac quantity is referred to the basic dc standards of voltage and resistance through the transfer device, avoiding the limitations of ordinary indicating instruments. It is generally the ac-dc differences of the shunt or series resistor, rather than the characteristics of the thermoelement, that determine the upper frequency limit of applicability of a transfer standard.[4]

## Digital Instruments

***46. Digital voltmeters*** (DVMs), displaying the measured voltage as a set of numerals, are analog-to-digital converters in which an unknown dc voltage is compared with a stable reference voltage.[5] Internal fixed dividers or amplifiers extend the voltage ranges. For ac measurements dc DVMs are preceded by ac-to-dc converters. DVMs are widely used as laboratory, portable, and panel instruments because of their convenience, accuracy, and speed. Automatic range changing and polarity indication, freedom from reading errors, and the availability of outputs for data acquisition or control are added advantages. Integrated circuits and modern techniques have greatly increased their reliability and reduced their cost. Full-scale accuracies

---

[1] Hermach, F. L.: *Trans. IEEE,* December 1965, vol. 1, p. 14.

[2] Wilkins, F. J.: *Trans. IEEE,* 1972, vol. IM-21, p. 334.

[3] Ott, W. E. "A New Technique of Thermal RMS Measurement." *IEEE Journal of Solid State Circuits,* 1974, vol. SC9, no. 6, pp. 374–380.

[4] Williams, E. E. "The Practical Uses of AC-DC Transfer Instruments" *NBS Tech. Note* 1166, Oct. 1982; Hermach, F. L. "AC-DC Comparators for Audio-Frequency Current and Voltage Measurements of High Accuracy." *Trans. IEEE,* 1976, vol. IM-25, pp. 489–494, 1976. (36 references.)

[5] Oliver and Cage. *Electronic Measurements and Instruments,* McGraw-Hill, 1971; Stansbury, C. *Trans. AIEE,* 1961, vol. 80, p. 465; Turgel. *Methods Exp. Phys.* 1975, vol. 2, Sec. 9.5.

range from about 0.5% for three-digit panel instruments to 1 ppm for eight-digit laboratory dc voltmeters and 0.016% for ac voltmeters.

***47. Successive-approximation DVMs*** are automatically operated dc potentiometers. These may be based on resistive voltage or current divider techniques or on dc current comparators (Par. **31**). A comparator in a series of steps adjusts a discrete fraction of the reference voltage (by current or voltage division in a resistance network) until it equals the unknown. Various "logic schemes" have been used to accomplish this, and the stepping relays of earlier models have been replaced by electronic or reed switches. Filters reduce input noise (which could prevent a final display) but generally increase the response time. Accuracy depends chiefly on the reference voltage and the ratios of the resistance network.

***48. Voltage-to-frequency-converter (V/f) DVMs*** generate a ramp voltage at a rate proportional to the input until it equals a fixed voltage, returns the ramp to the starting point, and repeats. The number of pulses (ramps) generated in a fixed time is proportional to the input and is counted and displayed. Since it integrates over the counting time, a V/f DVM has excellent input-noise rejection. The ramp is usually generated by an operational integrator (a high-gain operational amplifier with a capacitor in the feedback loop so that its output is proportional to the integral of the input voltage). The capacitor is discharged each time by a pulse of constant and opposite charge, and the time interval of the counter is chosen so that the number of pulses makes the DVM direct-reading. Accuracy depends on the integrator and on the charge of the pulse generator, which contains the reference voltage.

***49. Dual-slope DVMs*** generate a voltage ramp at a rate proportional to the input voltage $V_i$ for a fixed time $t_1$. The ramp input is then switched to a reference voltage $V_r$ of the opposite polarity for a time $t_2$ until the starting level is reached. Pulses with a fixed frequency $f$ are accumulated in a counter, with $N_1$ counts during $t_1$. The counter resets to zero and accumulates $N_2$ counts during $t_2$. Thus $t_1 = N_1 f$ and $t_2 = N_2 f$.

If the slope of the linear ramp is $m = kV$, the ramp voltage is $V_o = mt = kVt$. Thus $V_i t_1 = V_r t_2$, so that $V_i = V_r N_2/N_1$. The time $t_1$ is controlled by the counter to make $N_2$ direct-reading in appropriate units. In principle, the accuracy is not dependent on the constants of the ramp generator or the frequency of the pulses. A single operational integrator, switched to either input or reference voltage, generates the ramps. Since there are few critical components, integrated circuits are feasible, leading to simplicity and reliability as well as high accuracy. Because this is an integrating DVM, noise rejection is excellent.

In pulse-width conversion meters, an integrating circuit and matched comparators are used to produce trains of positive and negative pulses whose relative widths are a linear function of any dc input. The difference in positive and negative pulse widths can be measured using counting techniques, and very high resolution and accuracy (up to 1 ppm, relative to an internal voltage reference) can be achieved by integrating the counting over a suitable time period.[1]

***50. Average ac-to-dc converters*** contain an operational rectifier (an operational amplifier with a rectifier in the feedback circuit), followed by a filter, to obtain the rectified average value of the ac voltage. The operational amplifier greatly reduces errors of nonlinearity and forward voltage drop of the rectifier. For convenience the output voltage is scaled so that the dc DVM connected to it indicates the rms value of a sine wave. Large errors can result for other waveforms; up to $h/n$%, with $h$% of the $n$th harmonic in the wave, if $n$ is an odd number. For example, with 3% of third harmonic the error can be as much as 1%, depending on the phase of the harmonic.

***51. Electronic multipliers*** and other forms of rms-responding ac-to-dc converters eliminate this waveform error but are generally more complex and expensive.[2] In one version, the feedback rms circuit shown in Fig. 3-18, the two inputs of the multiplier $M_1$ are connected together, so that the instantaneous output of $M$ is $v_i^2/V_o$. The operational filter $F$ ($RC$ circuit and operational amplifier) makes $V_o = V_i^2/V_o$, where $V_i^2$ is the square of the rms value. Thus $V_o = V_i$. The conversion accuracy approaches 0.1% up to 20 kHz in transconductance or logarithmic mul-

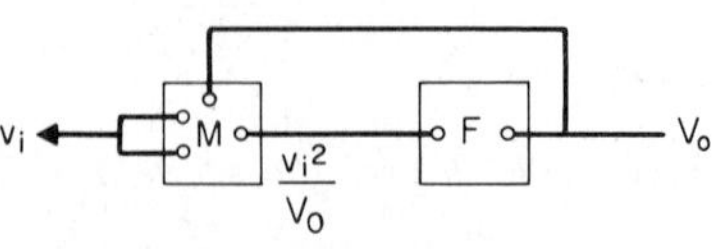

**FIG. 3-18** Electronic rms ac-to-dc converter.

[1] Anderson, A. and Nemeroff, E. "A 7½ Digit Precision Digital Voltmeter Employing Pulse Width Conversion." *Proceedings of the Measurement Science Conference,* Session 2B, 1978.

[2] *Non-Linear Circuits Handbook,* Analog Devices, Inc., 1974.

tipliers, without requiring a wide dynamic range in the instrument, because of the internal feedback. A series of diodes, biased to conduct at different voltage levels, can provide an excellent approximation to a square-law function in a feedback circuit like that of Fig. 3-18.

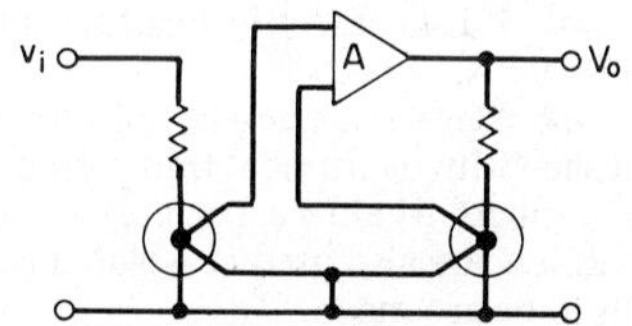


**FIG. 3-19** Dual-thermoelement ac-to-dc converter.

*52. Electronic Multipliers.* Dual thermoelements (see Par. **45**) provide a dc voltage proportional to the true rms value of the input, in the feedback circuit of Fig. 3-19. The input-output curves of the two thermoelements must agree to the accuracy required, but they need not be perfectly square-law. In another type, operation at a fixed current level avoids the tracking error. The dc thermoelement is energized through a Zener diode, and the amplifier *A* controls the gain of input amplifiers (not shown in the figure). A single thermoelement or multijunction thermal converter can be switched alternately between the input and a dc voltage to avoid the tracking error. The dc level is controlled by feedback sensed by the ac component of the thermocouple emf at the switching frequency. A similar scheme, using integrated-circuit thermoelements described in Par. **45**, is employed in most high-accuracy digital meters today. When properly compensated for drift, such a technique can be used to measure ac voltages directly with accuracies better than 200 ppm and indirectly (by substitution of nearly equal ac and dc voltages) with accuracies better than 100 ppm.[1]

*53. Specifications for DVMs* should follow the recommendations of ANSI Standard C39.7, "Requirements for Digital Voltmeters." Accuracy should be stated as the overall limit of error for a specified range of operating conditions. It should be in percent of reading plus percent of full scale and may be different for different frequency and voltage ranges. Accuracy at a narrow range of reference conditions is also often specified for laboratory use. The input configuration (two-terminal, three-terminal unguarded, three- or four-terminal guarded) is important. Number of digits and "overrange" should also be stated.

*54. Errors and precautions.* Because of the sensitivity of DVMs a number of precautions should be taken to avoid in-circuit errors from ground loops, input noise, etc. The high input impedance of most types makes input loading errors negligible, but this should always be checked. On dc millivolt ranges unwanted thermal emfs should be checked as well as the normal-mode rejection of ac line-frequency voltage across the input terminals. Two-terminal DVMs (chassis connected to one input as well as to line ground) may measure unwanted voltages from ground currents in the common line.

Errors are greatly reduced in three-terminal DVMs (chassis connected to line ground only) and are generally negligible with guarded four-terminal DVMs (separate guard chassis surrounding the measuring circuit). Such DVMs have very high common-mode rejection. Some types of DVMs introduce small voltage spikes or currents to the measuring circuit, often from internal switching transients, which may cause errors in low-level circuits.

*55. Digital multimeters* are DVMs with added circuitry to measure quantities such as dc voltage ratio, dc and ac current, and resistance. Voltage ratio is measured by replacing the reference voltage with one of the unknowns. For current, the voltage across an internal resistor carrying the current is measured by the DVM. For resistance a fixed reference current is generated and applied to the unknown resistor. The voltage across the resistor is measured by the DVM. Several ranges are provided in each case.

## Instrument Transformers

The material that follows is a brief summary of information on instrument transformers as measurement elements. For more extensive information, consult American National Standard C57.13, "Requirements for Instrument Transformers," American National Standards Institute; American National Standard C12, "Code for Electricity Metering"; *Electrical Meterman's*

[1] Brodie, B. "A 160 ppm RMS Digital Voltmeter for Use in AC Calibration." *Proceedings of the Measurement Science Conference,* 1982, p. 91.

*Handbook,* Edison Electric Institute; manufacturer's literature; and textbooks on electrical measurements.

***56. AC range extension*** beyond the reasonable capability of indicating instruments is accomplished with instrument transformers, since the use of heavy-current shunts and high-voltage multipliers would be prohibitive both in cost and in power consumption. Instrument transformers are also used to isolate instruments from power lines and to permit instrument circuits to be grounded.

The current circuits of instruments and meters normally have very low impedance, and current transformers must be designed for operation into such a low-impedance secondary burden. The insulation from the primary to secondary of the transformer must be adequate to withstand line-to-ground voltage, since the connected instruments are usually at ground potential. Normal design is for operation with a rated secondary current of 5 A, and the input current may range upward to many thousand amperes. The potential circuits of instruments are of high impedance, and voltage transformers are designed for operation into a high-impedance secondary burden. In the usual design, the rated secondary voltage is 120 V, and instrument transformers have been built for rated primary voltages up to 350 kV.

With the development of higher transmission-line voltages (350 to 765 kV) and intersystem ties at these levels, the coupling-capacitor voltage transformer (CCVT) has come into use for metering purposes to replace the conventional voltage transformer which, at these voltages, is bulkier and more costly. The metering CCVT, shown in Fig. 3-20, consists of a modular

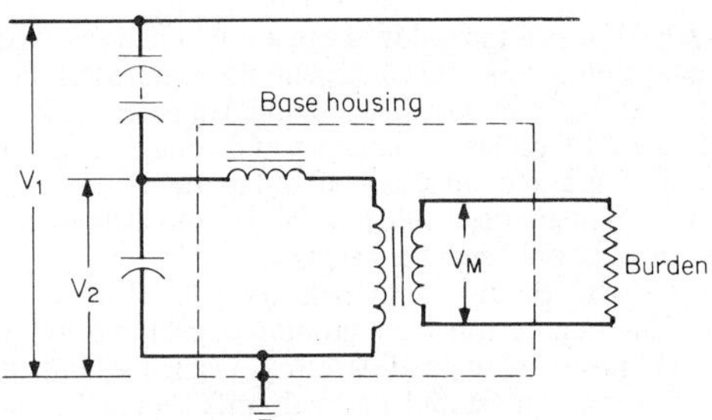


**FIG. 3-20** CCTV metering arrangement.

capacitive divider which reduces the line voltage $V_1$ to a voltage $V_2$ (10–20 kV), with a series-resonant inductor to tune out the high impedance and make available energy transfer across the divider to operate the voltage transformer which further reduces the voltage to $V_M$, the metering level. Required metering accuracy may be 0.3% or better.[1]

Instrument transformers are broadly classified in two general types: (1) dry type, having molded insulation (sometimes only varnish-impregnated paper or cloth) usually intended for indoor installation although large numbers of modern transformers have molded insulation suitable for outdoor operation on circuits up to 15 kV to ground; (2) liquid-filled types in steel tanks with high-voltage primary terminals, intended for installation on circuits above 15 kV.

They are further classified according to accuracy: *(a)* metering transformers having highest accuracy, usually at relatively low burdens; *(b)* relaying and control transformers which in general have higher burden capacity and lower accuracy, particularly at heavy overloads. This accuracy classification is not rigid, since many transformers, often in the larger sizes and higher voltage ratings, are suitable for both metering and control purposes.

Another classification differentiates between single and multiple ratios. Multiple primary windings, sometimes arranged for series-parallel connection, tapped primary windings, or tapped secondary windings are employed to provide multiple ratios in a single piece of equipment. Current transformers are further classified according to their mechanical structure: *(a)* wound primary, having more than one turn through the core window; *(b)* through type, wherein the circuit conductor (cable or busbar) is passed through the window; *(c)* bar type, having a bar,

[1] The field calibration of CCVT's is discussed in *NBS Tech. Note* 1155, 1982.

rod, or tube mounted in the window; *(d)* bushing type, that is, through types intended for mounting on the insulating bushing of a power transformer or circuit breaker.

*57. **Current transformers,*** whose primary winding is series connected in the line, serve the double purpose of (1) convenient measurement of large currents and (2) insulation of instruments, meters, and relays from high-voltage circuits. Such a transformer has a high-permeability core of relatively small cross section operated normally at a very low flux density. The secondary winding is usually in excess of 100 turns (except for certain small low-burden through-type current transformers used for metering, where the secondary turns may be as low as 40) and the primary is of few turns and may even be a single turn or a section of a busbar threading the core. The nominal current ratio of such a transformer is the inverse of the turns ratio, but for accurate current measurement, the actual ratio must be determined under loading corresponding to use conditions. For accurate power and energy measurement, the phase angle between the secondary and reversed primary phasor must also be known for the use condition. Insulation of primary from secondary and core must be sufficient to withstand, with a reasonable safety factor, the voltage to ground of the circuit into which it is connected; secondary insulation is much less, as the connected instrument burden is at ground potential or nearly so.

The overload capacity of station-type current transformers and the mechanical strength of the winding and core structure must be high to withstand possible short circuits on the line. Various compensation schemes are used in many transformers to retain ratio accuracy up to several times rated current. The secondary circuit—the current elements of connected instruments or relays—*must never be opened* while the transformer is excited by primary current, because high voltages are induced which may be hazardous to insulation and to personnel; and because the accuracy of the transformer may be adversely affected.

*58. **Voltage transformers*** (potential transformers) are connected between the lines whose potential difference is to be determined and are used to step the voltage down (usually to 120 V) and to supply the voltage circuits of the connected instrument burden. Their basic construction is similar to that of a power transformer operating at the same input voltage, except that they are designed for optimum performance with the high-impedance secondary loads of the connected instruments. The core is operated at high flux density, and the insulation must be appropriate to the line-to-ground voltage.

*59. **Standard burdens*** and standard accuracy requirements for instrument transformers are given in American National Standard C57.13 (Sec. **28** of this handbook).

*60. **Accuracy.*** Most well-designed instrument transformers (provided they have not been damaged or incorrectly used) have sufficient accuracy for metering purposes. See Sec. **10** for typical accuracy curves. Where higher accuracy is required, see Appendix D of ANSI C12, "The Code for Electricity Metering."

*61. **Testing.*** The insulation of instrument transformers is tested in the same manner as that of power transformers and other similar equipment. For details see ANSI Standard C57.13. The accuracy of instrument transformers can be determined in several ways. A rough check to verify nameplate ratings or check for damage may be made by suitable ammeters and voltmeters. Precise calibration is accomplished in a potentiometer arrangement to compare quantities representing the primary and secondary phasors and using a combination of resistance and inductance (or capacitance) such that inphase and quadrature components are separately balanced to determine transformer ratio and phase angle. Examples are shown in Figs. 3-21 and 3-22.

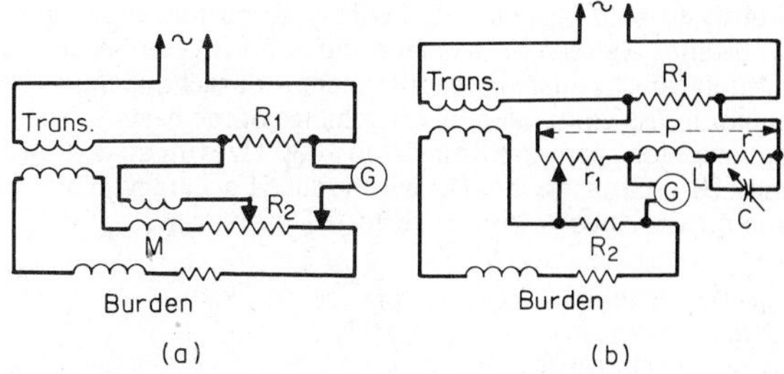


**FIG. 3-21** Ratio and phase angle of current transformer by resistance method.

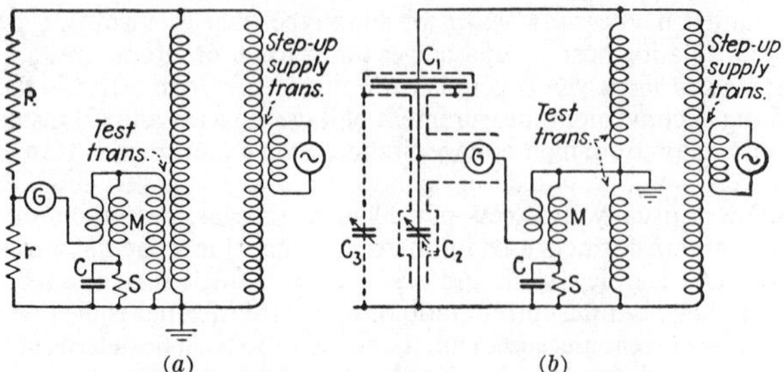


**FIG. 3-22** Resistance and capacitance methods for determining ratio of potential transformers.

Another comparison method uses a "standard" transformer of the same nominal rating as the one being tested. Accuracies of 0.01% are attainable. Commercial test sets are available for this work and are widely used in laboratory and field tests. Commercial test sets based on the current-comparator method[1] and capable of 0.001% accuracy are also available. For further details see ANSI Standard C57.13.

## Power Measurement

***62. Electronic wattmeters*** of 0.1% or better accuracy may be based on a pulse-area principle.[2] Voltages proportional to the applied voltage and to the current (derived from resistors or transformers), govern the height and width of a rectangular pulse, so that the area is proportional to the instantaneous power. This is repeated many times during a cycle, and its average represents active power. Average power can also be measured by a system which samples instantaneous voltage and current repeatedly, at predetermined intervals within a cycle. The sampled signals are digitized and the result is computed by numerical integration. The response of such a system has been found to agree with that of a standard electrodynamic wattmeter within 0.02% from dc to 1 kHz.[3] Depending on sampling speed, measurements can be made to higher frequencies with somewhat reduced accuracy. In the digital instrument, the multiplication involves discrete numbers and thus has no experimental error except for rounding. Such an arrangement is well adapted to the measurement of power in situations where current or voltage waveforms are badly distorted.

In the thermal wattmeter of Cox and Kusters[4] the arrangement is such that if one current $v$ is proportional to instantaneous load voltage and another $i$ is proportional to load current, their sum is applied to one thermal converter and their difference to another. Assuming identical quadratic response of the converters, their differential output may be represented as

$$V_{dc} = \frac{k}{T}\int_0^T [(v+i)^2 - (v-i)^2]\,dt = \frac{4k}{T}\int_0^T vi\,dt$$

which is by definition average power. Multijunction thermal converters with outputs connected differentially are used for the ac-dc transfer of power, with ac and dc current and voltage signals applied simultaneously to both heaters. DC feedback to current input speeds response and maintains thermal balance between heaters, and the output meter becomes a null indicator. This mode of operation can eliminate the requirement for exact quadrature response; and the matching requirement is also eliminated by interchange of the heaters. The Cox and Kusters instrument was designed for operation from 50 to 1000 Hz with ac-dc transfer errors within 30 ppm; and it may be used up to 20 kHz with reduced accuracy. This instrument also is capable of precision measurement with very distorted waveforms.

[1] Kusters, N. *Trans. IEEE,* 1964, vol. IM-13, p. 197.
[2] Sternberg. *RCA Rev.,* 1955, p. 618.
[3] Turgel. *Trans. IEEE,* 1974, vol. IM-23, p. 337.
[4] Cox and Kusters. *Trans. IEEE,* 1976, vol. IM-25, p. 553; 1980, vol. IM-29, p. 426.

In Schuster's[1] thermal wattmeter, the heating powers of the sum $(i_1 + i_2)$ and difference $(i_1 - i_2)$ currents are equalized by a dc current $2I$ added to the difference element, thus $(i_1 + i_2)^2 = (i - i_2 + 2I)^2$, which on time average results in $I^2 = i_1 i_2$. Automated operation is accomplished where the current $I$ is added by a control circuit to equalize the rms value when the system is switched back and forth between sum and difference. An analog-to-digital converter makes it possible to process the dc value with a digital arithmetic unit which calculates time-average quadratic values for power or time-integral quantities for energy. Tests of a portable instrument of this type at three National Laboratories agreed within 15 ppm at unity and 0.5 power factor (lead and lag) at 120 V, 5 A, 60 Hz. In measurement of low-frequency power, resolution of the instrument is 1 ppm.

***63. Laboratory-standard wattmeters*** use an electrodynamic mechanism (see Par. **41**) and are in the 0.1% accuracy class for dc and for ac up to 133 Hz. This accuracy can be maintained up to 1 kHz or more. Such instruments are shielded from the effects of external magnetic fields by enclosing the coil system in a laminated iron cylinder. Instruments having current ranges to 10 A and voltage ranges to 300 V are generally self-contained. Higher ranges are realized with the aid of precision instrument transformers.

***64. Portable wattmeters*** are generally of the electrodynamic type. The current element consists of two fixed coils connected in series with the load to be measured. The voltage element is a moving coil supported on jewel bearings or suspended by taut bands between the fixed field coils. The moving coil is connected in series with a relatively large noninductive resistor across the load circuit. The coils are mounted in a laminated iron shield to minimize coupling with external magnetic fields. Switchboard wattmeters have the same coil structure but are of broader accuracy class and do not have the temperature compensation, knife-edge pointers, and antiparallax mirrors required for the better-class portable instruments.

***65. Line connections*** should be such that the moving-coil end of the voltage circuit and the current coils are on the same side of the circuit being measured to minimize potential differences between the fixed and moving coils. When used with instrument transformers, the moving-coil end of the voltage circuit should be connected to the ground terminal of the voltage transformer, and an electrostatic tie (a resistance of a few thousand ohms) should be connected between this terminal and one of the current terminals. Otherwise, there may be sufficient electrostatic forces between the fixed and moving coils to cause an error, or if their voltage difference is large, insulation between the windings may be broken down.

***66. Correction for wattmeter power consumption*** may be important when the power measured is small. When the wattmeter is connected directly to the circuit (without the interposition of instrument transformers), the instrument reading will include the power consumed in the element connected next to the load being measured. If the instrument loss cannot be neglected, it is better to connect the voltage circuit next to the load and include its power consumption rather than that of the current circuit, since it is generally more nearly constant and is more easily calculated. In some low-range wattmeters, designed for use at low power factors, the loss in the voltage circuit is automatically compensated by carrying the current of the voltage circuit through compensating coils wound over the field coils of the current circuit. In this case, the voltage circuit must be connected next to the load to obtain compensation.

***67. The inductance error*** of a wattmeter may be important at low power factor. At power factors near unity, the noninductive series resistance in the voltage circuit is large enough to make the effect of the moving-coil inductance negligible at power frequencies; but with low power factor, the phase angle of the voltage circuit may have to be considered. This may be computed as $\alpha = 21.6fL/R$, where $\alpha$ is the phase angle in minutes, $f$ is the frequency in hertz, $L$ is the moving-coil inductance in millihenrys, and $R$ is the total resistance of the voltage circuit in ohms (see Par. **76** for correcting the effect of inductance).

***68. Characteristics of wattmeters,*** as stated in makers' literature, are given in Table 3-2.

***69. Wattmeter calibration*** can best be checked on direct current by using normal potentiometer techniques to measure current supplied to the field coils and voltage supplied to the voltage circuit from independent sources, but with an electrostatic tie (a high resistance) between one current terminal and the terminal at the moving-coil end of the voltage circuit to avoid errors from electrostatic forces between fixed and moving coils. The product of the

[1] Schuster. *Trans. IEEE,* 1976, vol. IM-25, p. 529; see also McAuliff, Lentner, Moore, Schuster. *Trans. IEEE,* 1978, vol. IM-27, p. 445.

**TABLE 3-2 Characteristics of Wattmeters**

| Manufacturer and model number of wattmeters | Full-scale accuracy, % | Scale length, in | Resistance of potential circuit, Ω* | Impedance of current circuit, Ω† | Type of damping | Magnetic shield | Weight, lb |
|---|---|---|---|---|---|---|---|
| | | | Portable wattmeters | | | | |
| Weston, Model 310 | 0.25 | 5.25 | 4,500 | 0.04 | Air | Yes | 12 |
| Weston, Model 310, pf = 0.20 | 0.25 | 5.25 | 3,000 | 0.16 | Air | Yes | 12 |
| Sensitive Research, Model DW | 0.25 | 6.4 | 4,500 | 0.20 | Air | Yes | 6 |
| Sens. Res., Model DLW, pf = 0.20 | 0.5 | 6.4 | 68,000 | 1.80 | Air | Yes | 6 |
| Weston, Model 432 | 0.5 | 4.0 | 11,000 | 0.04 | Air | Yes | 3.3 |
| Westinghouse, Type PY-5 | 0.5 | 5.3 | 6,500 | 0.16 | Air | Yes | 4 |
| | | | Switchboard wattmeters | | | | |
| Weston, Models 343 and 498 | 1.0 | 5.1 | 3,500 | 0.03 | Air | Case | 7 |
| General Electric, Type AB-16 | 1.0 | 13.8 | 194,000 | 0.013 | E.M. | Case | 5 |
| General Electric, Type AB-30 | 1.0 | 5.1 | 194,000 | 0.013 | E.M. | Case | 3 |
| General Electric Type AB-40 | 1.0 | 6.9 | 194,000 | 0.013 | E.M. | Case | 3 |
| Westinghouse, Type KY-221 | 1.0 | 4.0 | 4,000 | 0.144 | Air | Yes | 3 |
| Westinghouse, Type KP-241 | 1.0 | 7.1 | 5,700 | 0.08 | E.M. | Yes | 5 |
| Weston, Model 610 | 1.0 | 3.5 | 3,500 | 0.03 | Air | Case | 2 |

* For 150-V (max) range.
† For 5-A coil at 60 Hz.
NOTE: 1 in = 25.4 mm; 1 lb = 0.4536 kg.

measured current and voltage should equal the indicated power without correction for power consumption in either voltage or current circuit. For this test, a "compensated" wattmeter should be connected with its selector switch in the "uncompensated" position. A polyphase meter, with two independent measuring circuits connected by a common moving-element shaft, can be checked as though it were a single-phase meter, by connecting the voltage circuits in parallel to the voltage supply and the current circuits in series to the current supply. However, the following two tests should first be made: (1) independence can be checked by exciting only the current circuit of one element and the voltage circuit of the other—zero scale indication verifies that there is no interaction between the systems; (2) electrodynamic balance between the systems can be checked by connecting both voltage circuits in parallel to the voltage source and both current circuits in series opposition to the current source—here again at rated current and voltage, there should be no deflection of the pointer.

*70. 2-phase 4-wire circuit* (not interconnected) may be treated as equivalent to two single-phase circuits. Two wattmeters are connected as shown in Fig. 3-23; total power is the arithmetical sum of the two instrument readings.

*71. A 2-phase 3-wire circuit* requires two wattmeters connected as shown in Fig. 3-24; total power is the algebraic sum of the two readings. This connection is correct for any condition of load and power factor. One wattmeter may be used, connected as shown in Fig. 3-25, if there is no load across the outer conductors and the phases are balanced as to load and power factor; readings are summed for the two switch positions.

*72. A 2-phase 4-wire interconnected circuit* requires three wattmeters, connected as

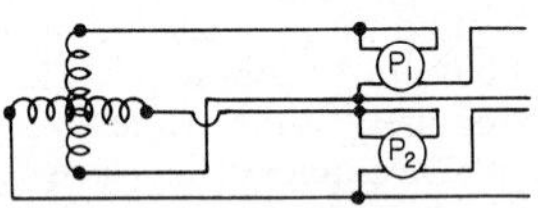

FIG. 3-23 Power in 2-phase 4-wire circuit (not interconnected).

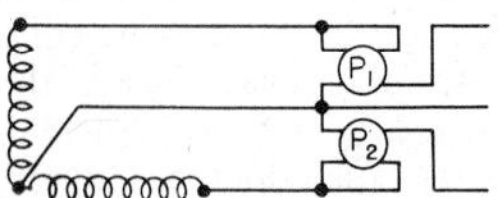

FIG. 3-24 Power in 2-phase 3-wire circuit.

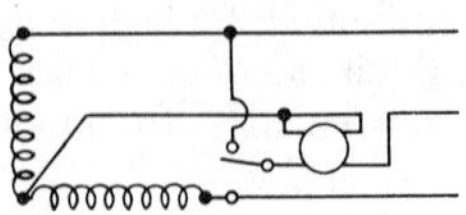
FIG. 3-25 Power in 2-phase 3-wire circuit, one wattmeter.

shown in Fig. 3-26; total power is the algebraic sum of the three readings. This connection is correct under all conditions of load and power factor. It will be noted that the voltage impressed on $P_3$ is 1.414 times the voltage on $P_1$ and $P_2$. Two wattmeters, one in each phase, will give the power only when the load is balanced in all four legs.

*73. **A 3-phase 3-wire circuit*** requires two wattmeters connected as shown in Fig. 3-27, total power is the algebraic sum of the two readings under all conditions of load and power factor. If the load is balanced, at unity power factor, each instrument will read half the load; at 50% power factor one instrument reads all the load and the other reading is zero; at less than 50% power factor one reading will be negative. When the load is balanced, power may be measured with one wattmeter, using a Y box as shown in Fig. 3-28. This arrangement, which creates an artificial neutral, has two branches which have the same impedance and power factor as the wattmeter's voltage circuit, which is the third branch of the Y. Total power is three times the reading of the wattmeter.

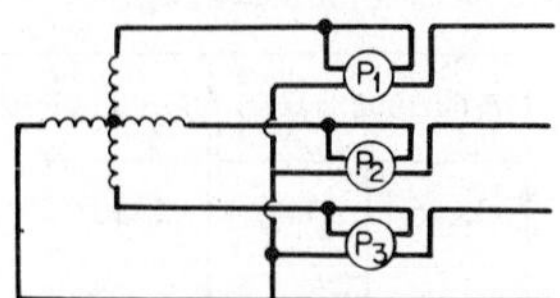


**FIG. 3-26** Power in 2-phase 4-wire interconnected circuit.

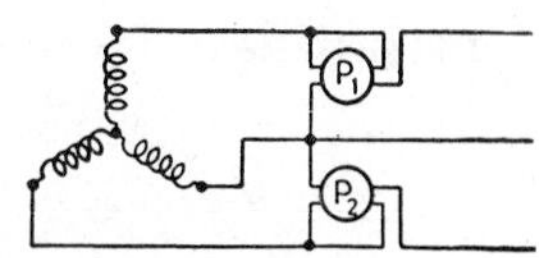


**FIG. 3-27** Power in 3-phase 3-wire circuit, two wattmeters.

*74. **The 3-phase 4-wire circuits*** require three wattmeters as shown in Fig. 3-29. Total power is the algebraic sum of the three readings under all conditions of load and power factor. A 3-phase Y system with a grounded neutral is the equivalent of a 4-wire system and requires the use of three wattmeters. If the load is balanced, one wattmeter can be used with its current coil in series with one conductor and the voltage circuit connected between that conductor and the neutral. Total power is three times the wattmeter reading in this instance.

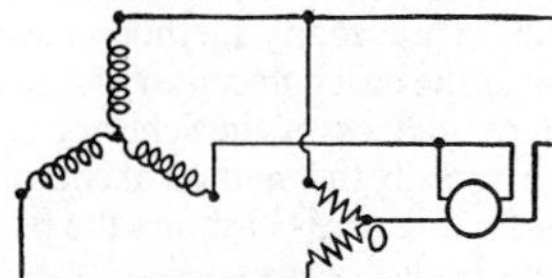

**FIG. 3-28** Power in balanced 3-phase 3-wire circuit, one wattmeter with Y box.

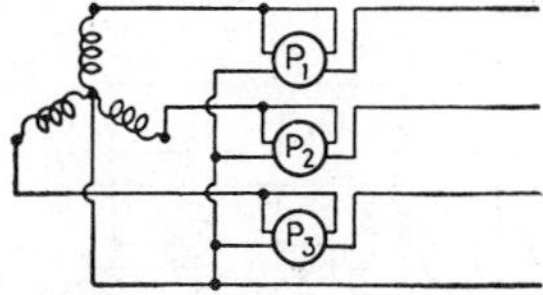


**FIG. 3-29** Power in 3-phase 4-wire circuit using three wattmeters.

*75. **Reactive power*** (reactive voltamperes, or vars) is measured by a wattmeter with its current coils in series with the circuit and the current in its voltage element in quadrature with the circuit voltage.

*76. **Corrections for instrument transformers*** are of two kinds. *Ratio* errors, resulting from deviations of the actual ratio from its nominal, may be obtained from a calibration curve showing true ratio at the instrument burden imposed on the transformer and for the current or voltage of the measurement. The effect of *phase-angle* changes introduced by instrument transformers is to modify the angle between the current in the field coils and that in the moving coil of the wattmeter; the resulting error depends on the power factor of the circuit and may be positive or negative depending on phase relations as shown in the following table (page 3-30). If $\cos\theta$ is the *true* power factor in the circuit and $\cos\theta_2$ is the *apparent* power factor (that is, as determined from the wattmeter reading and the secondary voltamperes), and if $K_c$ and $K_v$ are the true ratios of the current and voltage transformers, respectively, then

$$\text{Main-circuit watts} = K_c K_v \frac{\cos\theta}{\cos\theta_2} \times \text{wattmeter watts}$$

The line power factor $\cos\theta = \cos(\theta_2 \pm \alpha \pm \beta \pm \gamma)$, where $\theta_2$ is the phase angle of the secondary circuit, $\alpha$ is the angle of the wattmeter's voltage circuit (see Par. 67), $\beta$ is the phase angle of the current transformer, and $\gamma$ is the phase angle of the voltage transformer. These angles—$\alpha$, $\beta$,

| Line power factor | Sign to be used for phase angle | | | | | |
|---|---|---|---|---|---|---|
| | $\alpha$ wattmeter | | $\beta$ current transf. | | $\gamma$ voltage transf. | |
| | Lead[1] | Lag | Lead | Lag | Lead | Lag |
| Lead.................. .............. | + | − | − | + | + | − |
| Lag.................................... | − | + | + | − | − | + |

[1] In general, $\alpha$ will be leading only when the inductance of the potential coil has been overcompensated with capacitance.

and $\gamma$—are given positive signs when they act to decrease and negative when they act to increase the phase angle between instrument current and voltage with respect to that of the circuit. This is so because a decreased phase angle gives too large a reading and requires a negative correction (and vice versa), as shown in the preceding table of signs.

*77. **Dielectric loss,*** which occurs in cables and insulating bushings used at high voltages, represents an undesirable absorption of available energy and, more important, a restriction on the capacity of cables and insulating structures used in high-voltage power transmission. The problem of measuring the power consumed in these insulators is quite special, since their power factor is extremely low and the usual wattmeter techniques of power measurement are not applicable. While many methods have been devised over the past half century for the measurement of such losses, the Schering bridge is almost universally the method of choice at the present time. Figure 3-30 shows the basic circuit of the bridge, as described by Schering and Semm in 1920. The balance equations are $C_x = C_s R_1/R_2$ and $\tan\delta_x = \omega R_1 P$, where $C_x$ is the cable or bushing whose losses are to be determined, $C_s$ is a loss-free high-voltage air-dielectric capacitor, $R_1$ and $R_2$ are noninductive resistors, and $P$ is an adjustable low-voltage capacitor having negligible loss. For details of shielding, screening, and operation, one should refer to a text on ac bridge techniques such as Hague, B. F.: *AC Bridge Methods,* 6th ed., Pitman, 1971.

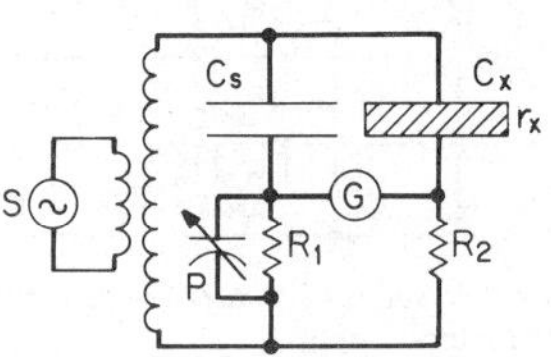


**FIG. 3-30** Schering and Semm's bridge for measuring dielectric loss.

## Power-Factor Measurement

*78. **The power factor of a single-phase circuit*** is the ratio of the true power in watts, as measured with a wattmeter, to the apparent power in voltamperes, obtained as the product of the voltage and current. When the waveform is sinusoidal (and only then), the power factor is also equal to the cosine of the phase angle.

*79. **The power factor of a polyphase circuit*** which is balanced is the same as that of the individual phases. When the phases are not balanced, the true power factor is indeterminate. In the wattmeter-voltmeter-ammeter method, the power factor for a balanced 2-phase 3-wire circuit is $P/(\sqrt{2}EI)$, where $P$ is total power in watts, $E$ is voltage between outside conductors, and $I$ is current in an outside conductor; for a balanced 3-phase 3-wire circuit, the power factor is $P/(\sqrt{3}EI)$, where $P$ is total watts, $E$ is volts between conductors, and $I$ is amperes in a conductor. In the two-wattmeter method, the power factor of a 2-phase 3-wire circuit is obtained from the relation $W_2/W_1 = \tan\theta$, where $W_1$ is the reading of a wattmeter connected in one phase as in a single-phase circuit and $W_2$ is the reading of a wattmeter connected with its current coil in series with that of $W_1$ and its voltage coil across the second phase. At unity power factor $W_2 = 0$; at 0.707 power factor $W_2 = W_1$; at lower power factors $W_2 > W_1$. In a 3-phase 3-wire circuit, power factor can be calculated from the reading of two wattmeters connected in the standard way for measuring power, by using the relation

$$\tan\theta = \frac{\sqrt{3}(W_1 - W_2)}{W_1 + W_2}$$

where $W_1$ is the larger reading (always positive) and $W_2$ the smaller.

***80. Power-factor meters,*** which indicate the power factor of a circuit directly, are made both as portable and as switchboard types. The mechanism of a single-phase electrodynamic meter resembles that of a wattmeter except that the moving system has two coils $M$, $M'$. One coil, $M$, is connected across the line in series with a resistor, whereas $M'$ is connected in series with an inductance. Their currents will be nearly in quadrature. At unity power factor, the reaction with the current-coil field results in maximum torque on $M$, moving the indicator to the 100 mark on the scale, where torque on $M$ is zero. At zero power factor, $M'$ exerts all the torque and causes the moving system to take a position where the plane of $M'$ is parallel to that of the field coils and the scale indication is zero. At intermediate power factors, both $M$ and $M'$ contribute torque, and the indication is at an intermediate scale position. In a 2-phase meter the inductance is not required, coil $M$ being connected through a resistance to one phase, while $M'$ with a resistance is connected to the other phase; the current coil may go in the middle conductor of a 3-wire system. Readings are correct only on a balanced load. In one form of polyphase meter, for balanced circuits, there are three coils in the moving system, connected one across each phase. The moving system takes a position where the resultant of the three torques is minimum, and this is dependent on power factor. In another form, three stationary coils produce a field which reacts on a moving voltage coil. When the load is unbalanced, neither form is correct.

## Energy Measurements

***81. General.*** The subject of metering electric power and energy is extensively covered in the American National Standard C12, "Code for Electricity Metering," American National Standards Institute. It covers definitions, circuit theory, performance standards for new meters, test methods, and installation standards for watthour meters, demand meters, pulse recorders, instrument transformers, and auxiliary devices. Further detailed information may be found in the *Handbook for Electric Metermen,* Edison Electric Institute.

***82. The practical unit of electrical energy*** is the watthour, which is the energy expended in 1 h when the power (or rate of expenditure) is 1 W.

***83. Energy is measured*** in watthours (or kilowatthours) by means of a watthour meter. A watthour meter is a motor mechanism in which a rotor element revolves at a speed proportional to power flow and drives a registering device on which energy consumption is integrated. Meters for continuous current are usually of the mercury-motor type, whereas those for alternating current utilize the principle of the induction motor.

***84. Continuous-current meters*** have been historically of two types: the commutator type, which has not been built since the early 1920s; and the mercury-motor type, now used in dc reduction processing, etc. The dc meter made by the Sangamo Electric Company is representative of the mercury-motor type. Figure 3-31 shows the circuits and scheme of operation. $D$ is a copper disk, having a number of radial slots and floating in a pool of mercury; $F$ is a float which supports the shaft, producing an upward thrust which is taken by a jewel bearing at the top of the shaft; $H$ is a laminated iron core; and $C$ is a chamber filled with mercury. A ring-jewel guide bearing at the bottom of the shaft holds the rotor on a fixed axis. The flux, produced in the core $H$ by the shunt (voltage) coil, cuts the disk at two diametric-

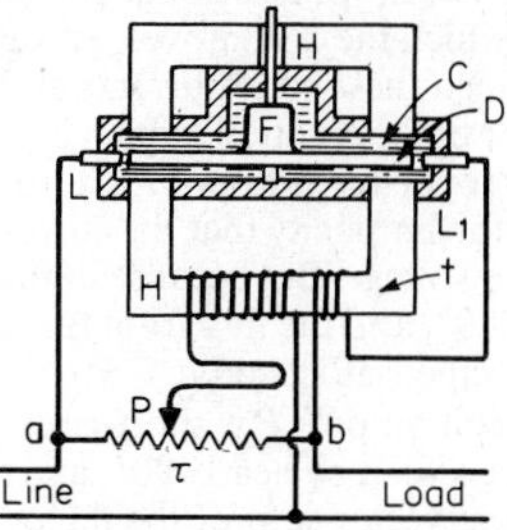


FIG. 3-31 Diagram of Sangamo mercury-motor-type watthour meter.

ally opposite points. The line current passes from $L$ to $L_1$ into the mercury, entering and leaving the copper disk in a flow pattern that is directed along its diameter by the slots. Since this disk is cut by magnetic flux, a torque is produced that is proportional to the product of current and voltage. The usual drag (braking) disk and magnets, gear train, and register are provided at the top of the shaft. The friction of the disk rotating in the mercury may be compensated with the Ayrton shunt *ab* of Fig. 3-31 across the current circuit, allowing a small adjustable current to be supplied by the voltage circuit.

In an alternative compensation method, a small current is introduced at $L$, $L_1$ from a thermocouple whose junction is heated by a coil in series with the voltage circuit. This current is adjusted by a slide-wire resistor in the thermocouple circuit. The increase, with speed, of friction between the mercury and disk is compensated by a series turn $t$ on the core $H$. The rotor speed, proportional to load, is adjusted by shifting the drag magnets radially with respect to the meter shaft, thus altering the retarding torque.

Figure 3-32 shows typical load accuracy curves both for older types of meters and for the more recent types, which are temperature-compensated for large sustained overloads.

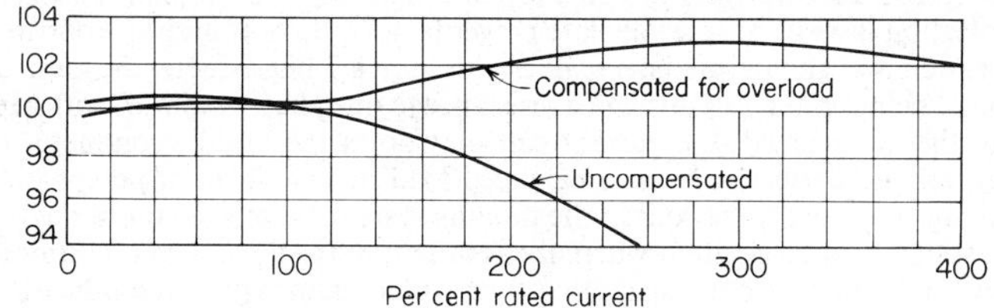


FIG. 3-32 Typical accuracy curves for dc watthour meters.

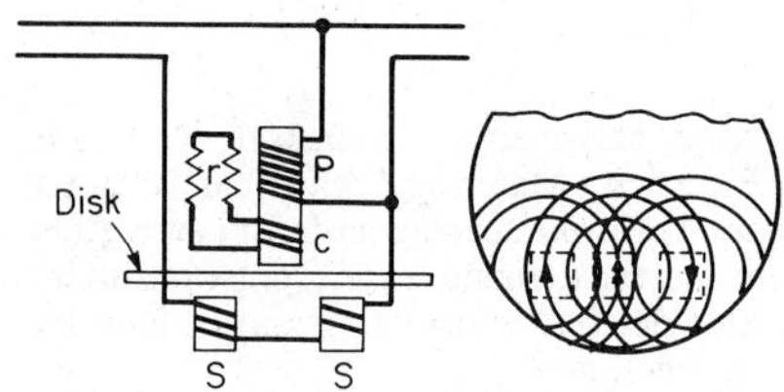


FIG. 3-33 Diagram of induction-type watthour meter.

***85. The ac watthour meters*** measure energy by using the principle of the induction motor. The essential features are shown in Fig. 3-33. $P$ is the voltage coil, $S$ the series (current) coils, and $c$ a compensating coil. An aluminum disk is free to revolve between the poles. The alternating fluxes from these poles will establish currents in the disk as indicated by the arrows in the sketch Fig. 3-33, which shows the poles and a portion of the disk. The voltage winding $P$ has many turns and is highly inductive so that the flux from its pole tip lags the applied voltage by nearly 90°. The fluxes in the current poles, set up by the line current in the series coils $S$, are in phase with that current.

A torque will be produced which is proportional to power in the load circuit if the fluxes from poles $P$ and $S$ are exactly in quadrature at unity power factor in the load. To provide a rotor speed proportional to this driving torque, a retarding (braking) torque must be exerted on the disk that is also proportional to speed. This is accomplished by a permanent-magnet field through which the disk moves, producing eddy currents proportional to its speed. The interaction between these eddy currents and the magnet flux that produced them results in a countertorque (or braking action) which is proportional to rotor speed. Consequently, the speed of the rotor is proportional to driving torque and thus to power in the circuit. The above argument assumes for simplicity that the current in $P$ and the flux from its pole are exactly in quadrature in the line voltage. But because of the ohmic resistance of this coil, the current lag is somewhat less than 90°, and the flux from this pole tip must be brought into quadrature with line voltage by the compensating coil $c$.

This coil on pole $P$ is short-circuited through the resistor $r$. The current in the coil lags the emf induced in it by nearly 90°, and its magnitude can be controlled by adjusting the resistance $r$. Thus the vector sum of the fluxes can be brought into quadrature with the line voltage. In modern watthour meters this coil is usually a single turn stamped from a flat sheet of copper alloy, and the resistance is adjusted by cutting away a portion of the plate provided for this purpose. Another method is to adjust the current in the plate by an adjustable inductance connected to it. If this adjustment is correct, the meter registration will be correct at any load power factor.

With no load current in the current coils, any lack of symmetry in the flux from the voltage pole could produce a forward or reverse torque. Also, there is necessarily friction in the rotor bearings, gear train, and register of the meter which tends to make the disk rotation slower than it should be at very small load currents. To compensate for these tendencies, a controlled driving torque is added by means of a shading-pole loop or plate in the disk gap, to react with the voltage flux. The position of this plate can be adjusted tangentially so that the driving torque it produces by its shading action is constant as long as the line voltage is constant. Its effect on meter registration is inversely proportional to line current and is negligible at large loads, whereas it counteracts the effect of friction and brings the registration to its correct value for light loads.

Even with correct calibration with this adjustment at light load, there may still be a slight "creep" torque (generally forward) on voltage alone. To prevent continuous creep, two anti-creep holes or slots are cut into the disk diametrically opposite. The holes restrict the eddy currents induced in the disk by the voltage flux and cause the disk to stop at a position of least coupling between the conductive path in the disk and the flux field.

***86. Adjustments of induction watthour meters*** are normally made only at *light* and *full* load. The position of the light-load compensation plate can be changed with conveniently placed screws and the light-load speed thus adjusted to be correct. Speed adjustment at full load (but affecting the speed at all loads) is made by shifting the drag magnets relative to the disk axis or by shunting flux by means of a movable soft-iron strip. The lag or power-factor adjustment is made at the meter factory and, if properly made, should never require correction. Temperature compensation for variations of ambient temperature has been secured by using a piece of temperature-sensitive iron-nickel alloy[1] as a shunt across the air gap of the drag magnets. Figure 3-34 shows typical performance characteristics of modern watthour meters.

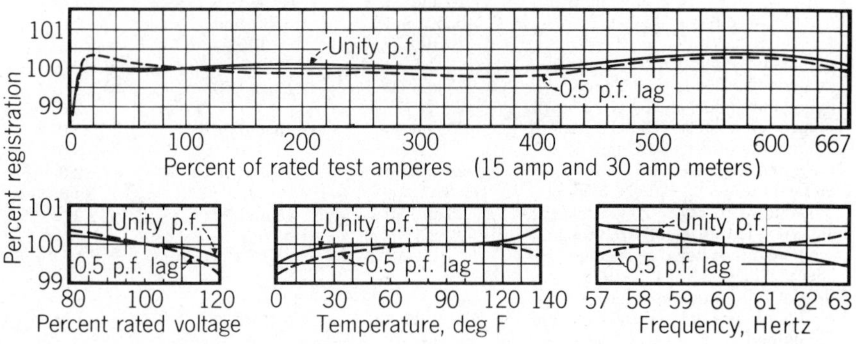


**FIG. 3-34** Characteristic performance curves of watthour meters. Latest meters have flat curves to 600% load.

***87. Meter Ratings.*** Watthour meters were formerly rated only at full or rated load. Maximum load capacities, however, had little relationship to the rated loads as the overload ranges were extended to higher and higher loads. Consequently, in 1960 the industry rerated watthour meters into classes on the basis of their maximum capacities. The class designation of a watthour meter denotes the maximum of the load range in amperes. Generally used classes are 10 and 20 (for use with current transformers), 100, and 200. The previous full-load rating, now known as TA or test ampere rating, is retained simply for purposes of calculating test constants and determining the percent registration of a watthour meter at heavy load (usually the TA value marked on the nameplate) and light load (10% of the TA value).

***88. Typical characteristics*** of the generally used single-phase Class 200, 240-V, 3-wire, 60-Hz watthour meter are shown in Table 3-3.

[1] Kinnard and Faus. *Trans. AIEE,* 1925, p. 275.

**TABLE 3-3** Induction Watthour-Meter Data*

| Item | Duncan MS | General Electric I70S | Sangamo J4S | Westinghouse D4S |
|---|---|---|---|---|
| Speed at rated TA load | 16⅔ | 16⅔ | 16⅔ | 16⅔ |
| Torque at rated load, mmg | 37 | 40 | 40 | 33 |
| Weight of rotor, g | 20 | 22 | 25 | 21 |
| Loss in potential circuit, at rated voltage, W | 0.85 | 0.8 | 0.78 | 0.95 |
| Power factor of potential circuits, % | 19 | 22 | 22 | 13 |
| Loss in current circuit at TA current, W | 0.35 | 0.24 | 0.26 | 0.26 |
| Starting watts | 24 | 22 | 24 | 24 |

* Data supplied by manufacturers.

***89. Acceptance Accuracy Limits*** (from the Code for Electricity Metering, ANSI C12-1975)

| Condition at 23°C ± 5°C | % of class designation: Class 10 Class 20 | Class 100 Class 200 | Max. deviation in % from normal accuracy |
|---|---|---|---|
| Initial performance at unity pf | 1.5 | 1.0 | ±2.0 |
| | 2.5–25 | 1.5–30 | ±1.0 |
| | 50 | 50 | ±1.5 |
| | 75 to 100 | 75 to 100 | ±2.0 |
| Voltage effect, 10% change unity pf | 2.5 and 25 | 1.5 and 15 | ±1.0 |
| Frequency effect, 5% change unity pf | 2.5 and 25 | 1.5 and 15 | ±1.0 |
| Power-factor effect at 0.5 lag | 5, 50, and 100 | 3, 50, and 100 | ±2.0 |
| Temperature effect at +50°C unity pf | 2.5 | 1.5 | ±2.0 |
| Temperature effect at +50°C unity pf | 25 and 50 | 15 and 50 | ±1.0 |
| Temperature effect at +50°C 0.5 lag pf | 5 | 3 | ±3.0 |
| Temperature effect at +50°C 0.5 lag pf | 25 and 50 | 15 and 50 | ±2.0 |
| Temperature effect −20°C unit pf | 2.5 | 1.5 | ±3.0 |
| Temperature effect −20°C unity pf | 25 and 50 | 15 and 50 | ±2.0 |
| Temperature effect −20°C 0.5 lag pf | 5 | 3 | ±4.0 |
| Temperature effect at −20°C 0.5 lag pf | 25 and 50 | 15 and 50 | ±3.0 |
| External field effect (100 At) through conductors arranged as specified in Sec. 5.1.8.10, ANSI C12-1975 | 2.5 | 1.5 | ±1.0 |
| Effect of 7000-A overload for 0.1 s | | 1.5 and 15 | ±1.5 |
| Equality of current circuits, relative to combined | 2.5, 5, 25, 50 | 1.5, 3, 15, 30 | ±1.0 |
| Effect of tilt at 4° from vertical: | | | |
| Forward, backward | 2.5 | 1.5 | ±1.0 |
| Left and right | 25 | 15 | ±0.5 |
| Independence of stators on multistator meter | 5, 10, 15, 25, 50, 75 | 3, 6, 9, 15, 30, 45 | ±1.0 |
| 20,000 A (20 × 50 μs) surge through vertical conductor 1.5 in from meter | 15 | 15 | ±1.0 |

***90. Measurement of energy*** in ac circuits is effected with watthour meters connected in exactly the same manner as are wattmeters for the measurement of power (see Figs. 3-23 to 3-29). In 3-wire 2-phase or 3-phase systems, polyphase meters may be used. Such meters comprise merely two single-phase meters in one case, with a common shaft, and connected to the main circuit in the same manner as two single-phase meters.

***91. Total energy*** in a 3-phase 3-wire circuit is the algebraic sum of the indications of two single-phase meters, just as the total power is the algebraic sum of the readings of two wattmeters. If a polyphase meter is used, the summation is automatically performed, and when one element tends to run backward (power factor less than 50%), it simply reduces the torque of the other one, so that the actual speed is still proportional to the total power in the circuit.

Four-wire systems, unless balanced, require three single-phase meters. A 3-phase system with grounded neutral should be considered a 4-wire system requiring three meters, unless it is completely and continually balanced. Three-element meters, made with three driving-braking disks on one shaft, are generally used instead of the three single-phase meters.

***92. Polyphase Meter Connections.*** Obviously, it is extremely important that the various circuits of a polyphase meter be properly connected. If, for example, the current-coil connections of Fig. 3-27 are interchanged and the line power factor is 50%, the meter will run at the normal 100% power-factor speed, thus giving an error of 100%.

A test for correct connections of the arrangement of Fig. 3-27 is as follows: If the line power factor is over 50%, rotation will always be forward when the potential or the current circuit of either element is disconnected, but in one case the speed will be less than in the other. If the power factor is less than 50%, the rotation in one case will be backward.

When it is not known whether the power factor is less or greater than 50%, this may be determined by disconnecting one element and noting the speed produced by the remaining element. Then change the voltage connection of the remaining element from the middle wire to the other outside wire, and again note the speed. If the power factor is over 50%, the speed will be different in the two cases, but in the same direction. If the power factor is less than 50%, the rotation will be in opposite directions in the two cases.

When instrument transformers are used, care must be exercised in determining correct connections; if terminals of similar instantaneous polarity have been marked on both current and voltage transformers, these connections can be verified and the usual test made to determine power factor. If the polarities have not been marked, or if the identities of instrument transformer leads have been lost in a conduit, the correct connections can still be established, but the procedure is more lengthy.

***93. Use of Instrument Transformers with Watthour Meters.*** When the capacity of the circuit is over 200 A, instrument current transformers are generally used to step down the current to 5 A. If the voltage is over 480 V, current transformers are almost invariably employed, irrespective of the magnitude of the current, in order to insulate the meter from the line; in such cases, voltage transformers are also used to reduce the voltage to 120 V. Transformer polarity markings must be observed for correct registration. The ratio and phase-angle errors of these transformers must be taken into account where high accuracy is important, as in the case of a large installation. These errors can be largely compensated for by adjusting the meter speed.

***94. Reactive voltampere-hour (var-hour) meters*** are generally ordinary watthour meters in which the current coil is inserted in series with the load in the usual manner while the voltage coil is arranged to receive a voltage in quadrature with the load voltage. In 2-phase circuits, this is easily accomplished by using two meters as in power measurements, with the current coils connected directly in series with those of the "active" meters but with the voltage coils connected across the quadrature phases. Evidently, if the meters are connected to rotate forward for an inductive load, they will rotate backward for capacitive loads. For 3-phase 3-wire circuits, and 3-phase 4-wire circuits, phase-shifting transformers are used normally and complex connections result. For detailed information consult the references given in Par. **81**.

***95. Errors of Var-Hour Meters.*** The 2- and 3-phase arrangements described above give correct values of reactive energy when the voltages and currents are balanced. The 2-phase arrangement still gives correct values for unbalanced currents but will be in error if the voltages are unbalanced. Both 3-phase arrangements give erroneous readings for unbalanced currents or voltages; an autotransformer arrangement will usually show less error for a given condition of unbalance than the simple arrangement with interchanged potential coils.

***96. Total var-hours,*** or "apparent energy" expended in a load, is of interest to engineers because it determines the heating of generating, transmitting, and distributing equipment, and hence their rating and investment cost. The apparent energy may be computed if the power factor is constant, from the observed watthours $P$ and the observed reactive var-hours $Q$; thus, var-hours $= \sqrt{P^2 + Q^2}$. This method may be greatly in error when the power factor is not constant; the computed value is always too small.

A number of devices have been offered for the direct measurement of the apparent energy. In one class *(a)* are those in which the meter power factor is made more or less equal to the line power factor. This is accomplished automatically (in the Angus meter) by inserting a movable member in the voltage-coil pole structure which shifts the resulting flux as line power factor changes. In others, autotransformers are used with the voltage elements to give a power factor in the meter close to expected line power factor. By using three such pairs of autotransformers and three complete polyphase watthour-meter elements operating on a single register, with the record determined by the meter running at the highest speed, an accuracy of about 1% is achieved, with power factors ranging from unity down to 40%. In the other class *(b)*, vector addition of active and reactive energies is accomplished either by electromagnetic means or by electromechanical means, many of them very ingenious. But the results obtained with the use of modern watthour and var-hour meters are generally adequate for most purposes.

***97. The accuracy of a watthour meter*** is the percentage of the total energy passed through a meter which is registered by the dials. The watthours indicated by the meter in a given time are noted, while the actual watts are simultaneously measured with standard instruments. Because of the time required to get an accurate reading from the register, it is customary to count revolutions of the rotating element instead of the register. The accuracy of the gear-train ratio between the rotating element and the first dial of the register can be determined by count. Since the energy represented by one revolution, or the watthour constant, has been assigned by the manufacturer and marked on the meter, the indicated watthours will be $K_h \times R$, where $K_h$ is the watthour constant and $R$ the number of revolutions.

***98. Reference standards*** for dc meter tests in the laboratory may be ammeters and voltmeters, in portable or laboratory-standard types, or potentiometers; in ac meter tests, use is made of indicating wattmeters and a time reference standard such as a stopwatch, clock, or tuning-fork or crystal-controlled oscillator together with an electronic digital counter. A more common reference is a standard watthour meter, which is started and stopped automatically by light pulsing through the anticreep holes of the meter under test.

***99. The portable standard*** watthour meter (often called rotating standard) method of watthour-meter testing is that most often used, because only one observer is required and it is more accurate with fluctuating loads. Rotating standards are watthour meters similar to regular meters, except that they are made with extra care, are usually provided with more than one current and one voltage range, and are portable. A pointer, attached directly to the shaft, moves over a dial divided into 100 parts, so that fractions of a revolution are easily read. Such a standard meter is used by connecting it to measure the same energy as is being measured by the meter to be tested; the comparison is made by the "switch" method, in which the register only (in dc standards) or the entire moving element (in ac standards) is started at the beginning of a revolution of the meter under test, by means of a suitable switch, and stopped at the end of a given number of revolutions. The accuracy is determined by direct comparison of the number of whole revolutions of the meter under test with the revolutions (whole and fractional) of the standard. Another method of measuring speed of rotation in the laboratory is to use a tiny mirror on the rotating member which reflects a beam of light into a photoelectric cell; the resulting impulses may be recorded on a chronograph or used to define the period of operation of a synchronous electric clock, etc.

***100. Watthour meters used with instrument transformers*** are usually checked as secondary meters; that is, the meter is removed from the transformer secondary circuits (current transformers must first be short-circuited) and checked as a 5-A 120-V meter in the usual manner. The meter accuracy is adjusted so that, when the known corrections for ratio and phase-angle errors of the current and potential transformers have been applied, the combined accuracy will be as close to 100% as possible, at all load currents and power factors. An overall check is seldom required, both because of the difficulty and because of the decreased accuracy as compared with the secondary check.

***101. General precautions*** to be observed in testing watthour meters are as follows: *(a)* The test period should always be sufficiently long and a sufficiently large number of independent readings should be taken to ensure the desired accuracy. *(b)* Capacity of the standards should be so chosen that readings will be taken at reasonably high percentages of their capacity in order to make observational or scale errors as small as possible. *(c)* Where indicating instruments are used on a fluctuating load, their average deflections should be estimated in such a manner as to include the time of duration of each deflection as well as the magnitude. *(d)* Instruments should be so connected that neither the standards nor the meter being tested are measuring the voltage-circuit loss of the other, that the same voltage is impressed on both, and that the same load current passes through both. *(e)* When the meter under test has not been previously in circuit, sufficient time should be allowed for the temperature of the voltage circuit to become constant. *(f)* Guard against the effect of stray fields by locating the standards and arranging the temporary test wiring in a judicious manner.

***102. Meter Constants.*** The following definitions of various meter constants are taken from the Code for Electricity Metering, 6th edition, ANSI C12-1975.

***103. Register constant*** $K_r$ is the factor by which the register reading must be multiplied in order to provide proper consideration of the register or gear ratio and of the instrument-transformer ratios to obtain the registration in the desired units.

***104. Register ratio*** $R_r$ is the number of revolutions of the first gear of the register, for one revolution of the first dial pointer.

***105. Watthour constant*** $K_h$ is the registration expressed in watthours corresponding to one revolution of the rotor. (When a meter is used with instrument transformers, the watthour constant is expressed in terms of primary watthours. For a secondary test of such a meter, the constant is the primary watthour constant, divided by the product of the nominal ratios of transformation.)

***106. Test current*** of a watthour meter is the current marked on the nameplate by the manufacturer (identified as TA on meters manufactured since 1960) and is the current in amperes which is used as the basis for adjusting and determining the percent registration of a watthour meter at heavy and light loads.

***107. Percentage registration*** of a meter is the ratio of the actual registration of the meter to the true value of the quantity measured in a given time, expressed as a percentage. Percentage registration is also sometimes referred to as the accuracy or percentage accuracy of a meter. The value of one revolution having been established by the manufacturer in the design of the meter, meter watthours $= K_h \times R$, where $K_h$ = watthour constant and $R$ = number of revolutions of rotor in $S$ seconds. The corresponding power in meter watts is $P_m = (3600 \times R \times K_h)/S$. Hence, multiplying by 100 to convert to terms of percentage registration (accuracy):

$$\text{Percentage registration} = \frac{K_h \times R \times 3600 \times 100}{PS}$$

where $P$ = true watts. This is the basic formula for watthour meters in terms of true watt reference.

***108. Average Percentage Registration (Accuracy) of Watthour Meters.*** The Code for Electricity Metering makes the following statement under the heading, Methods of Determination:

The percentage registration of a watthour meter is, in general, different at light load than at heavy load, and may have still other values at intermediate loads. The determination of the average percentage registration of a watthour meter is not a simple matter as it involves the characteristics of the meter and the loading. Various methods are used to determine one figure which represents the average percentage registration, the method being prescribed by commissions in many cases. Two methods of determining the average percentage registration (commonly called "average accuracy" or "final average accuracy") are in common use:

*Method 1.* Average percentage registration is the weighted average of the percentage registration at light load *(LL)* and at heavy load *(HL)*, giving the heavy-load registration a weight of 4. By this method:

$$\text{Weighted average percentage registration} = \frac{LL + 4HL}{5}$$

*Method 2.* Average percentage registration is the average of the percentage registration at light load *(LL)* and at heavy load *(HL)*. By this method:

$$\text{Average percentage registration} = \frac{LL + HL}{2}$$

***109. In-service performance tests,*** as specified in the "Code for Electricity Metering," ANSI C12-1975, shall be made in accordance with a *periodic test schedule,* except that self-contained single-phase meters, self-contained polyphase meters, and 3-wire network meters may also be tested under either of two other systems, provided that all meters are tested under the same system. These systems are the *variable interval plan* and the *statistical sampling plan.*

***110. Periodic Interval System.*** The chief characteristic of this system is that a fixed percentage of the meters in service shall be tested annually. In the test intervals specified below, the word "years" means calendar years. The periods stated are recommended test intervals. There may be situations in which individual meters, groups of meters, or types of meters should be tested more frequently. In addition, because of the complexity of installations using instrument transformers, and the importance of large loads, more frequent inspection and test of such installations may be desirable. In general, periodic test schedules should be as follows:

1. Meters with surgeproof magnets and without demand registers or pulse initiators—16 years.
2. Meters without surgeproof magnets and without demand registers or pulse initiators—8 years.

The chief weaknesses of the above periodic test schedule are that it fails to recognize the differences in accuracy characteristics of various types of meters as a result of technical advance in meter design and construction and fails to provide incentives for maintenance and modernization programs.

***111. The variable-interval plan*** provides for the division of meters into homogeneous groups and the establishment of a testing rate for each group based on the results of in-service performance tests made on meters longest in service without test. The maximum test rate recommended is 25% per year. The minimum test rate recommended provides for the testing of a sufficient number of meters to provide adequate data to determine the test rate for the succeeding year. The provisions of the variable-interval plan recognize the difference between various meter types and encourage adequate meter maintenance and replacement programs. See Section 8.1.8.5 of ANSI C12-1975 for details of operation of this plan.

***112. The statistical sampling program*** included is purposely not limited to a specific method, since it is recognized that there are many acceptable ways of achieving good results. The general provisions of the statistical sampling program provide for the division of meters into homogeneous groups, the annual selection and testing of a random sample of meters of each group, and the evaluation of the test results. The program provides for accelerated testing, maintenance, or replacement if the analysis of the sample test data indicates that a group of meters does not meet the performance criteria. See Section 8.1.8.6 of ANSI C12-1975 for details of the operation of this program.

***113. Ampere-hour meters*** measure only electrical quantity, that is, coulombs or ampere-hours; and, therefore, where they are used in the measurement of electrical energy, the potential is assumed to remain constant at a "declared" value and the meter is calibrated or adjusted accordingly. The outstanding example in the United States of ampere-hour meters is the Sangamo mercury meter, which is practically the same as the Sangamo dc watthour meter (Par. **84**) except that the electromagnet is replaced by permanent magnets. Models are available especially designed for use with storage batteries, electric railways, etc.

***114. Ampere-hour or volt-hour meters for alternating current*** are not practical but ampere-squared-hour or volt-squared-hour meters are readily built in the form of the induction watthour meter. Ampere-hours or volt-hours are then obtainable by extracting the square root of the registered quantities.

***115. Maximum-Demand Meters.*** Some methods of selling energy involve the maximum amount which is taken by the customer in any period of a prescribed length, that is, the maximum demand. Many types of meters for measuring this demand have been developed, but space permits only a brief description of a few. There are two general classes of demand meters in common use: (1) integrated-demand meters and (2) thermal, logarithmic, or lagged-demand meters. Both have the same function, which is to meter energy in such a way that the registered value is a measure of the load as it affects the heating (and therefore the load-carrying capacity) of the electrical equipment.

***116. Integrated-demand meters*** consist of an integrating meter element (kWh or kvarh) driving a mechanism in which a timing device returns the demand actuator to zero at the end of each timing interval, leaving the maximum demand indicated on a passive pointer, display, or chart, which in turn is manually reset to zero at each reading period, generally 1 month. Such demand mechanisms operate on what is known as the block-interval principle. There are three types of block-interval demand registers: (1) the indicating type, in which the maximum demand obtained between each reading period is indicated on a scale or numeric display; (2) the cumulative type, in which the accumulated total of maximum demand during the preceding periods is indicated during the period after the device has been reset and before it is again reset, that is, the maximum demand for any one period is equal or proportional to the difference between the accumulated readings before and after reset; (3) the multiple-pointer form, in which the demand is obtained by reading the position of the multiple pointers relative to their scale markings. The multiple pointers are resettable to zero.

Another form of demand meter, usually in a separate housing from its associated watthour meter, is the recording type in which the demand is transferred as a permanent record onto a tape by printing, punching, or magnetic means or onto a circular or strip chart. A special form of tape recording for demand metering that has come into wide use in recent years is the pulse recorder, in which pulses from a pulse initiator in the watthour meter are recorded on magnetic tape or punched paper tape in a form usable for machine translation by digital-data-processing techniques. Advantages of this system are its great flexibility, freedom from the operating difficulties inherent in inked charts, and freedom from many of the personal errors of manual reading and interpretation of charts.

***117. Thermal, logarithmic, or lagged-demand meters*** are devices in which the indication of the maximum demand is subject to a characteristic time lag by either mechanical or thermal means. The indication is often designed to follow the exponential heating curve of electrical equipment. Such a response, inherent in thermal meters, averages on a logarithmic and continuous basis, which means that more recent loads are heavily weighted but that, as time passes, their effect decreases. The time characteristics for the lagged meter are defined as the nominal time required for 90% of the final indication with a constant load suddenly applied.

***118. Concordance of Demand Meters and Registers.*** The measurement of demand may be obtained with meters and registers having various operating principles and employing various means of recording or indicating the demand. On a constant load of sufficient duration, accurate demand meters and registers of both classifications will give the same value of maximum demand, within the limits of tolerance specified. On varying loads, the values given by accurate meters and registers of different classifications may differ because of the different underlying principles of the meters themselves. In commercial practice, the demand of an installation or a system is given with acceptable accuracy by the record or indication of any accurate demand meter or register of acceptable type.

## Electrical Recording Instruments

***119. Recording instruments*** are, in many instances, essentially high-torque indicating instruments arranged so that a permanent, continuous record of the indication is made on a chart. They are made for recording all electrical quantities that can be measured with indicating instruments—current, voltage, power, frequency, etc. In general, the same type of electrical mechanism is used—permanent-magnet moving-coil for direct current and moving-iron or dynamometer for alternating current. The indicator is an inking pen or stylus that makes a record on a chart moving under it at constant speed. This requires a higher torque to overcome friction, so that the operating power required for a recording instrument is greater than for a simple indicating instrument. Overshoot is generally undesirable, and recording instruments are slightly overdamped, whereas indicating instruments are usually somewhat underdamped. Some recorders use strip charts; graduations along the length of the chart are usually of time intervals, and the graduations across the chart represent the instrument scale. Alternatively, the chart may be circular, with radial graduations for the instrument scale and time markers around the circumference. The chart paper should be well made and glazed to minimize dimensional changes from temperature and humidity. The ink should be in accordance with the maker's specification for the particular paper used so that it is accepted readily and does not run or blot the paper. Chart drives may be electrical or clockwork. In strip charts, perforations along the edges of the paper are engaged by a drive pinion; circular charts are rotated from a central hub.

***120. Potentiometric self-balancing recorders*** are systems incorporating dc potentiometers, used either alone or with a transducer to measure various quantities.

***121. Transducers*** include those for voltage, current, power, power factor, frequency, temperature, humidity, steam or water flow, gas velocity, neutron density, and many other applications.

***122. Types of systems*** are classified according to the means of detecting and correcting electrical unbalance in the potentiometer circuit.

***123. The Leeds & Northrup Micromax recorder*** senses unbalance of a D'Arsonval galvanometer every 2.4 s. When unbalance occurs, a motor-driven cam clamps the galvanometer in its unbalanced position and metal feelers indicate this out-of-balance condition and cause a clutch arm to engage the slide-wire disk onto a constantly running motor so that the slide wires rotate until a balance condition is reached. The sensing time of 2.4 s gives a standard across the chart time of 24 s.

***124. The Tag Celectray recorder*** (see Fig. 3-35), also typical of the galvanometer-operated type, contains the light-beam and phototube combination applied to the D'Arsonval galvanometer. Deflection of the galvanometer changes the intensity of light falling on the vacuum tube, and the output of this phototube is electrically amplified to operate two relays which control the motion of the reversible balancing motor connected to the slide wire.

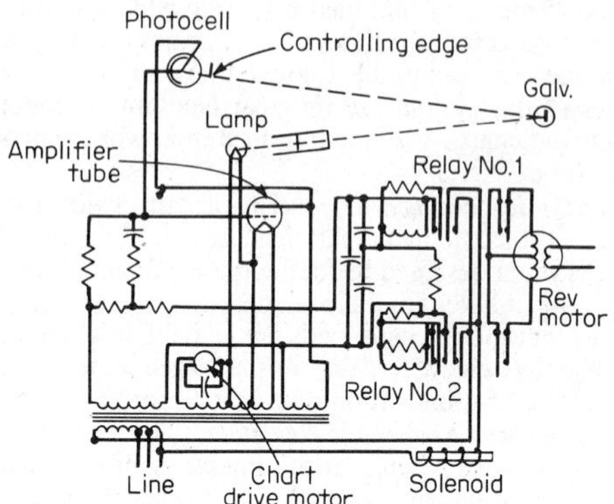


**FIG. 3-35** Tagliabue Celectray recorder.

***125. Electrical self-balancing systems*** employ an amplifier to amplify the dc unbalance in the potentiometer circuit and a reversible motor to produce the required rebalancing action. The *drift* which occurs in dc amplifiers is equivalent to a spurious input signal, so that it has become general to employ dc-to-ac conversion to obtain a stable amplifier, free from zero drift. The ac control signal is amplified by the ac-coupled amplifier and applied to the control winding of a 2-phase induction motor, whose power phase is fed from the power line. Polarity reversal of the dc unbalance signal causes phase reversal of the ac signal into the control winding of the motor, so that the resulting unbalance drives the motor in the correct direction to rebalance the potentiometer.

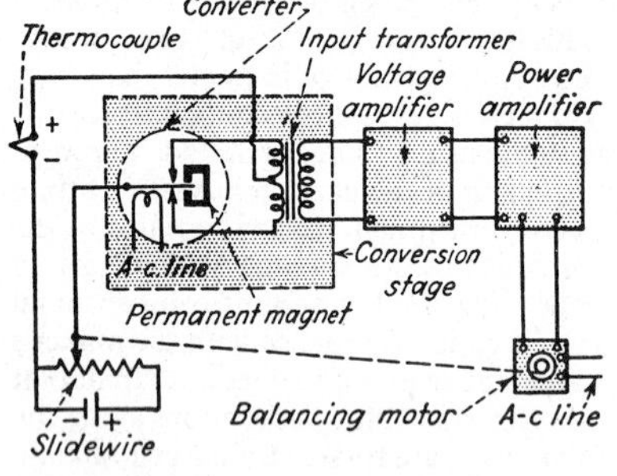


**FIG. 3-36** Brown Electronik recorder.

***126. The Brown Electronik,*** shown schematically in Fig. 3-36, uses a vibrating reed to change the dc unbalance signal to alternating current, which is then amplified and passed to the reversible induction motor. The reed converter is a metal reed driven in synchronism with the power frequency and oscillating between two contacts connected to opposite ends of the primary winding of the input transformer of the ac amplifier. The dc signal is thus applied in turn to alternate sides of the transformer, inducing in it an alternating voltage of power frequency whose magnitude and phase depend on the magnitude and polarity of the dc signal.

***127. The Bailey Pyrotron*** (see Fig. 3-37) employs a saturable core reactor in place of the vibrating reed to give dc-to-ac conversion, amplifiers and the 2-phase motor operation being as in the figure.

***128. Leeds & Northrup Speedomax G,*** whose circuits are shown in Fig. 3-38 in simplified form, illustrates the complete system. The amplifiers are prone to pick up stray electrical signals that could give false indications. To avoid this, a filter network connected to the input serves the double purpose of acting as a damping network to prevent excessive pointer overshoot and, at the same time, screening out electrical strays that would otherwise be picked up and transmitted through the amplifier.

*129. Accuracy* of the order of ¼% may be expected from the potentiometer recorders described above. To maintain this accuracy, the potentiometer is referenced against a standard cell or a reference voltage provided by a Zener diode. This may be performed by the operator pressing a button to give manual standardization whenever desired. A further refinement is to have automatic standardization, in which the operation is initiated by the chart-drive motor at specified intervals.

*130. Range extension* of potentiometric recorders upward is by means of shunt or series resistors. Extension below the basic range of the recorder requires preamplifiers.

*131. Measurement of ac quantities* requires the use of ac-to-dc transducers, for example, thermocouples, rectifiers, etc.

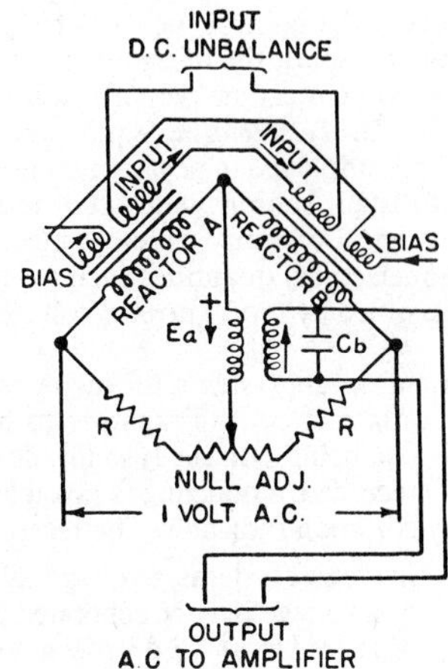


FIG. 3-37 Bailey Pyrotron recorder.

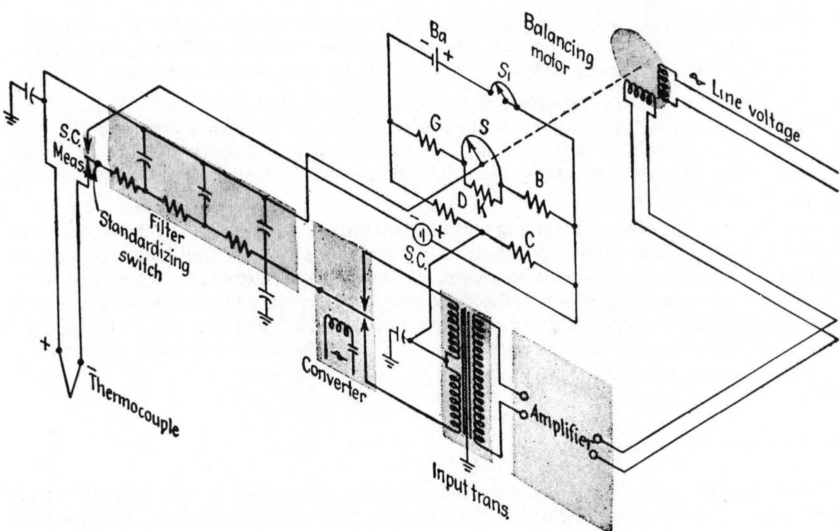


FIG. 3-38 Leeds & Northrup Speedomax G recorder.

*132. Alternating-current potentiometer recorders* are simpler than the dc types, as they require no standardization against a standard cell or Zener reference voltage, and ac-to-dc conversion is not required, eliminating the requirement for a vibrator or saturable reactor. The amplifier and motor-control circuits can be the same as in the dc recorder. By far the greatest application is with ac bridges, where the ac amplifier acts as an unbalance detector. Strain-gage bridges and bridges which employ platinum or nickel resistive elements for narrow-range temperature measurements frequently employ recorders of this type.

*133. Proximity-type recorders* use a high-frequency oscillator whose operation is started or stopped by the insertion of a metal vane into a pair of coils. If the vane is mounted on the pointer of an indicating instrument, the oscillator can sense movement between the pointer and a pair of coils fitted to the oscillator. Servo motion of the coils on displacement of the instrument pointer is accomplished by coupling the oscillator output to the input of a servo amplifier which drives the control motor. This gives a graphic record that follows but does not constrain movement of the instrument pointer. In this way, quantities which can operate an indicating instrument can be recorded without using a transducer.

***134. Telemetering*** is the indicating or recording of a quantity at a distant point. Telemetering is employed in power measurements to show at a central point the power loads at a number of distant stations, and often to indicate total power on a single meter, but practically any electrical quantity which is measured can be transmitted, together with a large number of nonelectrical quantities such as levels, positions, and pressures. Telemetering systems may be classified by type: current, voltage, frequency, position, and impulse.

**a.** In *current* systems, the movement of the primary measuring element calls for a current in the attached control member to balance the torque created by the quantity measured. This balancing current (usually dc) is sent over the transmitting circuit to be indicated and recorded. Totalizing is possible by the addition of such currents from several sources in a common indicator. The receiver may be as much as 50 mi from the transmitter.

**b.** In *voltage* systems, a voltage balance may be produced through a control-member voltmeter, or a voltage may be generated by thermocouples heated by the quantity to be measured, or produced as an *IR* drop as a result of a current torque balance, or generated by a generator driven at a speed proportional to the measured quantity. These voltages, however produced, are recorded at a distance by a potentiometer recorder. Here, also, the recorder may be 50 mi from the transmitter.

**c.** A *variable frequency* may be produced for telemetering by causing the primary element to move a capacitor plate in an rf oscillator or to change the speed of a small dc motor driving an alternator. High-frequency systems cannot be used for transmission over many miles.

**d.** In *position* systems, the movement of the primary element or of a pilot controlled by the primary element is duplicated at a distance. The pilot may be a bridge balancing resistance or reactance, a variable mutual inductance, or a selsyn motor where the position of a rotor relative to a 3-phase stator is reproduced at the receiver end. Satisfactory operation is usually limited to a few miles.

**e.** The *impulse* type of transmission of measured quantities is represented by the largest number of devices. The number of impulses transmitted in a given time may represent the magnitude of the quantity being measured, and these may be integrated by a notching device or by a clutch, or the duration of the pulse may be governed by the primary element and interpreted at the receiver. If the impulses are transmitted at high frequency, inductance and capacitance effects in the transmitting line limit the distance of satisfactory transmission; systems using dc impulses operate over 50 to 250 mi.

## Resistance Measurements

***135. The SI unit of resistance,*** the ohm, has been determined directly in terms of the mechanical units by *absolute-ohm* experiments performed at the National Bureau of Standards and at national laboratories in other countries. The reactance of an inductor or capacitor of special construction whose value can be computed from its dimensional properties is compared with a resistance at a known frequency. The value of this resistance can then be assigned in absolute (or SI) units, in terms of length and time—the dimensions of the inductor or capacitor and the time interval corresponding to the comparison frequency. These measurements are made with high precision, and it is believed that the assigned value of the National Reference Standards of resistance, maintained at the National Bureau of Standards, differs from its intended *absolute* value by not more than 1 part in $10^6$.

***136. The National Reference Standard*** of resistance is a group of five 1-Ω resistors of special construction, sealed in double-walled enclosures containing dry nitrogen and kept in a constant temperature bath of mineral oil at 25°C at the National Bureau of Standards.[1] To ensure that their values are constant, they are intercompared at least weekly, compared with other standards of differing construction quarterly, and compared with similar groups in other major

[1] Thomas, J. L. *NBSJRes*, 1946, vol. 36, p. 107.

national laboratories frequently. Absolute experiments to determine their SI values are performed at rather longer intervals because of the complexity of such experiments—a new experiment of this type may require 5 years or more to complete. This reference group serves as the basis for all resistance measurements made in the country.

***137. Resistance standards,*** used in precise measurements, are made with high-resistivity metal, in the form of wire or strip. Manganin—a copper-nickel-manganese alloy—is generally used in resistance standards because, when properly treated and protected from air and moisture, it has a number of desirable characteristics, including stable value, low temperature coefficient, low thermal emf at junctions with copper, and relatively high resistivity. A copper-nickel-chromium-aluminum alloy, Evanohm, has recently been used for high-resistance standards, since it has the same desirable characteristics as manganin and a much higher resistivity.

Standards with nominal values exceeding a megohm (a million ohms) are generally of films of metals such as Nichrome, a nickel-chromium alloy, deposited on a glass substrate. Four forms of standard are in general use. The Thomas-type 1-Ω standard is widely used as a primary standard. The Reichsanstalt form developed in the German national laboratory is illustrated in Fig. 3-39; the NBS form developed at the National Bureau of Standards is shown in Fig. 3-40.

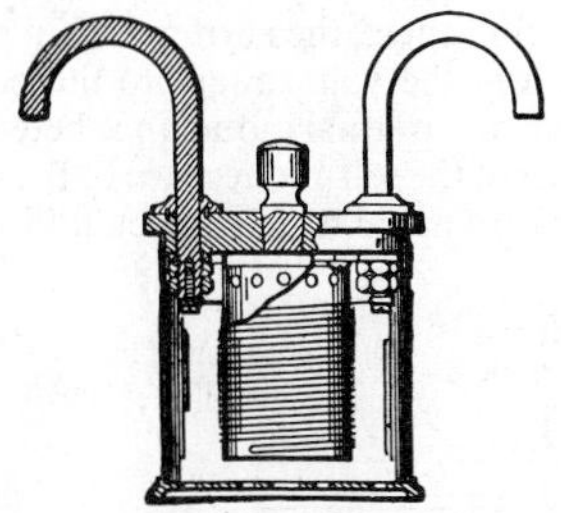

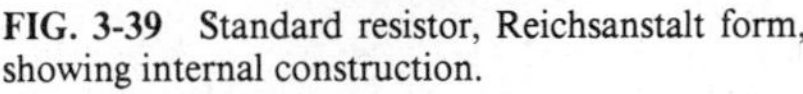

**FIG. 3-39** Standard resistor, Reichsanstalt form, showing internal construction.

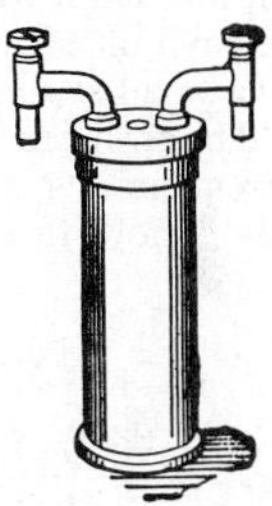

**FIG. 3-40** Standard resistor, NBS form.

All three are designed to be used with their current-terminal lugs in mercury cups and are generally suspended in an oil bath to dissipate heat and to hold the temperature constant at a known value during measurements.

The fourth type, in widespread use for secondary references and as a primary standard at the 10,000-Ω level, consists of one or more coils of Evanohm wound on mica cards or cylindrical formers and terminated in binding posts for use on benchtops. The primary standard version of this type of resistor generally has the resistance elements hermetically sealed in an oil-filled container which also contains some type of resistive temperature sensor.

For highest precision, power dissipation must be kept below 0.1 W, (calibrations at the National Bureau of Standards are generally performed at 0.01 W), although as much as 1 W can be dissipated in stirred oil with very small changes in value. The maker's recommendations should be followed regarding safe operating current levels. High-resistance and low-resistance standards use different terminal arrangements. In all standards of 1 Ω or lower value, and standards up to 10,000 Ω intended for use at the part-per-million (ppm) level of accuracy or better, the current and voltage terminals are separated, whereas in other standards they may not be. The four-terminal construction is required to define the resistance to be measured. Connections to the current-carrying circuit range from a few microhms upward and, in a two-terminal construction, would make the resistance value uncertain to the extent that the connection resistance varies. With four-terminal construction, the resistance of the standard can be *exactly* defined as the voltage drop between the voltage terminals for unit current in and out at the current terminals.

***138. Current standards*** are precision four-terminal resistors used to measure current by measuring the voltage drop between the voltage terminals with the current introduced at the current terminals. These standards, designed for use with potentiometers for precision current measurement, correspond in structure to the shunts used with millivoltmeters for current measurement with indicating instruments. Current standards must be designed to dissipate the

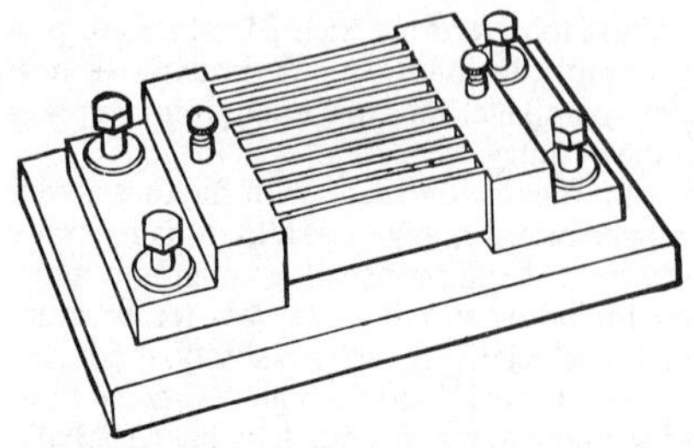

FIG. 3-41 Leeds & Northrup air-cooled resistor or shunt.

heat they develop at rated current, with only a small temperature rise. They may be oil-cooled or air-cooled, the latter design having much greater surface, since heat transfer to still air is much less efficient than to oil. Figure 3-41 shows an air-cooled current standard of 20 $\mu\Omega$ resistance and 2000 A capacity, with an accuracy of 0.04%. Very low resistance oil-cooled standards are mounted in individual oil-filled containers provided with copper coils through which cooling water is circulated and with propellers to provide continuous oil motion.

***139. Alternating-current resistors for current measurement*** require further design consideration. For example, if the resistor is to be used for current-transformer calibration, its ac resistance must be identical with its dc resistance within 1/100% or better, and the voltage difference between its voltage terminals must be in phase with the current through it within a few tenths of a minute. Thin strips or tubes of resistance material are used to limit eddy currents and minimize "skin" effect, the current circuit must be arranged to have small self-inductance, and the leads from the voltage taps to the potential terminals should be arranged so that, as nearly as possible, the mutual inductance between the voltage and current circuits opposes and cancels the effect of the self-inductance of the current circuit. Figure 3-42 shows three types of construction. In (*a*) a metal strip has been folded into a very narrow U; in (*b*) the current circuit consists of coaxial tubes soldered together at one end and to terminal blocks at the other end; in (*c*) a straight tube is used as the current circuit, and the potential leads are snugly fitting coaxial tubes soldered to the resistor tube at the desired separation and terminating at the center.

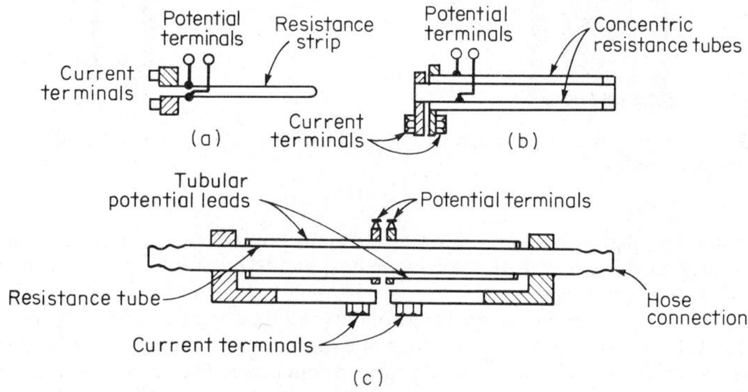


FIG. 3-42 Types of low-inductance standard resistors.

***140. Resistance coils*** consist of insulated resistance wire wound on a bobbin or winding form, hard-soldered at the ends to copper terminal wires. Metal tubes are widely used as winding form for dc resistors, as they dissipate heat more readily than insulating bobbins; but if the resistor is to be used in ac measurements, a ceramic winding form is greatly to be preferred, as it contributes less to the phase-defect angle of the resistor. The resistance wire ordinarily is folded into a narrow loop and wound bifilar onto the form, to minimize inductance. This construction results in considerable associated capacitance of high-resistance coils, for which the wire is quite long; and an alternative construction is to wind the coil inductively on a thin mica or plastic card. The capacitive effect is greatly reduced, and the inductance is still quite small if the card is thin.

Resistors in which the wire forms the warp of a woven ribbon have lower time constants than either the simple bifilar- or card-wound types. Manganin is the resistance material most generally employed, but Evanohm and similar alloys are beginning to be extensively used for very high resistance coils. Enamel or silk is used to insulate the wire; and the finished coil is ordinarily coated with shellac or varnish to protect the wire from the atmosphere. Such coatings do not completely exclude moisture, and dimensional changes of insulation with humidity will result in small resistance changes, particularly in high resistances where fine wire is used.

***141. Resistance boxes*** usually have two to four decades of resistance so that with reasonable precision they cover a considerable range of resistance, adjustable in small steps. For convenience of connection, terminals of the individual resistors are brought to copper blocks or studs, which are connected into the circuit by means of plugs or of dial switches using rotary laminated brushes; clean, well-fitted plugs probably have lower resistance than dial switches but are much less convenient to use. The residual inductance of decade groups of coils due to switch wiring, and the capacitance of connected but inactive coils, will probably exceed the residuals of the coils themselves, and it is to be expected that the time constant of an assembly of coils in a decade box will be considerably greater than that of the individual coils.

***142. Measurement of resistance*** is accomplished by a variety of methods, depending on the magnitude of the resistor and the accuracy required. Over the range from a few ohms to a megohm or more, an ohmmeter may be used for an accuracy of a few percent. A simple ohmmeter may consist of a milliammeter, dry cell, and resistor in a series circuit, the instrument scale being marked in resistance units. For a better value, the voltage drop is measured across the resistor for a measured or known current through it. Here accuracy is limited by the instrument scales unless a potentiometer is used for the current and voltage measurements. This approach is also taken in the wide variety of digital multimeters now in common use. Their manufacturers' specifications indicate a range of accuracies from a few percent to 10 ppm (0.001%) or better from the simplest to the most precise meters. Bridge methods can have the highest accuracy, both because they are null methods in which two or more ratios can be brought to equality, and because the measurements can be made by comparison with accurately known standards. For two-terminal resistors, a Wheatstone bridge can be used; for four-terminal measurements, a Kelvin bridge or a current comparator bridge can be used. Bridges for either two- or four-terminal measurements may also be based on resistive dividers. Because of their extremely high input impedance, digital voltmeters may be used with standard resistors in unbalanced bridge circuits of high accuracy.

***142a. Digital multimeters*** are frequently used to make low-power measurements of resistors in the range between a few ohms and a hundred megohms or so. Resolution of such instruments varies from 1% of full scale to a part per million of full scale. These meters generally use a constant-current source with a known current controlled by comparing the voltage drop on an internal "standard" resistor to the emf produced by a Zener diode. The current is set at such a level as to make the meter direct reading in terms of the displayed voltage, i.e., the number displayed by the meter reflects the voltage drop across the resistor, but the decimal point is moved and the scale descriptor is displayed as appropriate. Multimeters typically use three or more fixed currents and several voltage ranges to produce seven or more decade ranges with the full-scale reading from 1.4 to 3.9 times the range. For example, on the 1000-Ω range, full scale may be 3,999.999 Ω. Power dissipated in the measured resistor generally does not exceed 30 mW, and reaches that level only in the lowest ranges where resistors are usually designed to handle many times that power. The most accurate multimeters have a resolution of 1 to 10 ppm of range on all ranges above the 10-Ω range. Their sensitivity, linearity, and short-term stability make it possible to compare nominally equal resistors by substitution with an uncertainty 2 to 3 times the least count of the meter. This permits their use in making very accurate measurements, up to 10 ppm, of resistors whose values are close to those of standards at hand. Many less expensive multimeters have only two leads or terminals to use to make measurements. In those cases, the leads from the meter to the resistor to be measured become part of the measured resistance. For low resistances, the lead resistance must be measured and subtracted out, or zeroed out.

***143. The Wheatstone bridge*** is generally used for two-terminal resistors. In the low-resistance range where four-terminal construction is normal, the resistance of connections into the network may be a significant fraction of the total resistance to be measured, and the Wheatstone network is not applicable. Figure 3-43 shows the arrangement of a Wheatstone bridge, where $A$, $B$, and $C$ are known resistances and $D$ is the resistance to be measured. One or more of the known arms is adjusted until the galvanometer $G$ indicates a null; then $D = B(C/A)$. In case $D$ is inductive, the battery switch $S_1$ should be closed before the galvanometer key $S_2$ to protect the galvanometer from the initial transient current. In a common form of bridge, $B$ is a decade

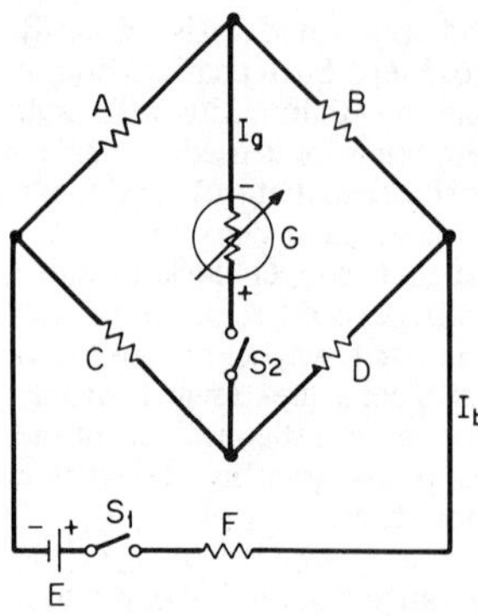


**FIG. 3-43** Wheatstone bridge.

resistance, adjustable in small steps, while $C$ and $A$ (the ratio arms of the bridge) can be altered to select ratios in powers of 10 from $C/A = 10^{-3}$ to $10^3$. If the value of the unknown resistors is not very different from that of a known resistor, accuracy may be improved by substituting the known and unknown in turn into arm $D$ and noting the difference in balance readings of the adjustable arm $B$. Since there has been no change in the ratio arms, any errors they may have do not affect the difference measurement, and only those errors in arm $B$ which were involved in the difference between the settings affect the difference value; in effect, the unknown is measured in terms of a known resistor by a substitution procedure. An alternative form of Wheatstone bridge is frequently assembled from standards and a ratio box of limited range called a "direct-reading ratio set." This latter has a nominal ratio of unity, with ratio adjustments ranging from 1.005000 to 0.995000, that is, four decades of adjustment of which the largest has steps of 0.1%. If a balance is made with the two standards in arms $B$ and $D$ and a second balance with the standards interchanged, their difference is half the difference between the balance readings. A similar technique can be used wherever small resistance differences are involved, for example, in the determination of temperature coefficients.

Bridge sensitivity can be determined in the following way. The voltage that would appear in the galvanometer branch of the circuit with switch $S_2$ open is

$$e = \frac{EBD\,\Delta B}{(B+D)^2}$$

where $E$ is the supply voltage and $\Delta B$ is the amount in proportional parts by which $B$ departs from balance. If, now, the voltage sensitivity of the galvanometer is known for operation in a circuit whose external resistance is that of the bridge as seen from the galvanometer terminals, its response for the unbalance $\Delta B$ can be computed. The current in the galvanometer with $S_2$ closed is

$$I_g = \frac{e}{G + BD/(B+D) + AC/(A+C)}$$

where $G$ is the resistance of the galvanometer. If there is a large current-limiting resistance $F$ in the battery branch of the bridge, the terminal voltage at the $AC$ and $BD$ junction points should be used rather than the supply voltage $E$ in computing $e$. In connecting and operating a bridge, the allowable power dissipation of its components should first be checked to ensure that these limits are not exceeded, either in any element of the bridge itself or in the resistance to be measured.

Resistive voltage dividers can be used to form bridges for either two- or four-terminal resistance measurements. There are two common forms of resistive voltage divider—the Kelvin-Varley divider (see Par. **32**) and the universal ratio set (URS)—with the former being the most commonly encountered. Each behaves as a potential divider with nearly constant input resistance and an open-circuit output potential of some rational fraction of the input, that fraction being given by the dial settings with calibration corrections applied. In the case of the Kelvin-Varley divider the maximum ratio is 0.99999 . . . X and outputs may be selected with a resolution as great as a part in 100 million of the input. Most Kelvin-Varley dividers have input resistances of 10,000 or 100,000 Ω. The URS was specifically designed to calibrate precision potentiometers. Its nominal input resistance is 2111.11 . . . 0 Ω and that is also its full-scale dial designation. Its resolution, or one step of its least-significant dial, is either 1 mΩ or 0.1 mΩ. For bridge applications, either divider type appears as two adjacent (series-connected) bridge elements with a ratio of $r/(R-r)$, where $r$ is the dial setting and $R$ is the full-scale dial setting. In a Wheatstone or two-terminal type of bridge as shown in Fig. 3-43, the divider appears as resistors $A$ and $B$, with $C$ being the known resistor, or standard, and $D$ being the unknown. In that case the balance equation is

$$D/C = (R-r)/r$$

assuming that the low input of the divider is connected to the node between resistors *A* and *C* and its high input to the node between *B* and *D*.

Four-terminal applications are more complex, as four separate balances must be made to obtain the ratio between two resistors. The schematic is given in Fig. 3-44. To measure *B* in

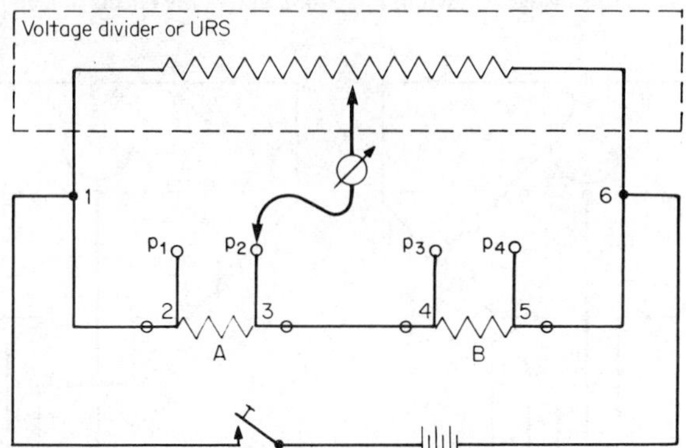


**FIG. 3-44** Four-terminal resistance measurements.

terms of *A*, the lead resistances between node pairs 1–2, 3–4, and 5–6, which we will call *x, y,* and *z*, respectively, must be eliminated. This is done by balancing the circuit with the resistor-side detector lead tied to each of the resistor potential leads at the terminals marked $p_1$, $p_2$, $p_3$, and $p_4$. The result is

$$A/B = (r_2 - r_1)/(r_4 - r_3)$$

where the *r*'s are readings obtained by balancing the divider at each of the potential terminals.

Both types of divider must be calibrated. This can be done by comparison with a more accurate divider, dial by dial. Such a divider can readily be tormed by using a number of nominally equal resistors in series. Each resistor is measured relative to the same standard and the results used to calculate the various ratios in the string of resistors. The string is then used to calibrate each setting of each dial in the voltage divider. In the case of the Kelvin-Varley divider, the dial corrections are interdependent; the correction for the steps in a particular dial depend on the settings of the less-significant dials. A straightforward method of coping with this problem is covered in the literature.[1]

Unbalanced bridge techniques have been made practical by the very high input resistances of modern digital instrumentation and are a satisfactory approach to resistance measurements when the values of the resistors being measured do not differ significantly from one another. They are particularly useful in cases where a process, not expected to change significantly, is being monitored using resistive sensors, such as thermistors or copper or nickel resistors. The simplest case is that of a Wheatstone bridge, such as that shown in Fig. 3-43. In it, the galvanometer, *G*, would be replaced by a digitial meter of suitable sensitivity and sufficiently high input impedance to make bridge loading errors insignificant. The bridge relationship then becomes

$$\frac{B}{A+B} + \frac{v}{V} = C + D$$

where *V* is the voltage applied to the bridge, or $(E - I_b\,F)$, and *v* is the reading of the digital meter. In practice, the meter is generally used to measure *V*, as well as *v*. If the individual elements of the resistor pairs *A, B* and *C, D* are nearly equal, the bridge is nearly at balance, *v* is

[1] Dunn, A. F. "Calibration of a Kelvin-Varley voltage divider," *Trans. IEEE,* 1964, IM-13, pp. 129–139.

small, and measurements of $v$ and $V$ need not be made at high accuracies. Resolution is not generally a problem for resistance element values of 100 Ω and higher as digital meters with least counts of 0.1 and 1 μV microvolts are commonly available.

***144. Mueller Bridge.*** A special form of Wheatstone bridge, known as a Mueller bridge, is commonly used for four-terminal measurements of the resistance of platinum resistance thermometers (PRTs). In this bridge, shown in Fig. 3-45*a* and *b*, the effects of lead resistance of the

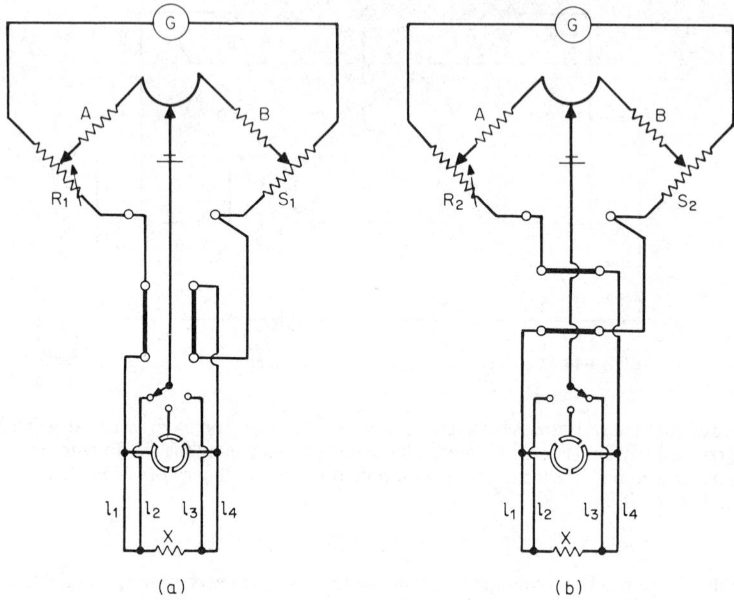


**FIG. 3-45** Mueller bridge.

PRT are eliminated by including two of the leads in adjacent bridge arms and making a second measurement after transposing the leads. The equations are

$$R_1 + l_1 = R_x + l_4 + S_1 \text{ (a)}$$

$$R_2 + l_4 = R_x + l_1 + S_2 \text{ (b)}$$

as the bridge is always used with the ratio arms $A$ and $B$ adjusted to be equal. These two equations may be added to eliminate lead resistances and result in the equation

$$R_x = 0.5(R_1 - S_1 + R_2 - S_2)$$

where $R_1$, $S_1$, $R_2$, $S_2$ are the dial readings (with corrections applied) for conditions $A$, $B$.[1]

There is increasing use of low-frequency square and sinusoidal wave bridges for PRT measurements. These bridges rely on the inherent ratio stability and accuracy of specially designed transformers and the increased sensitivity available with ac amplifiers to provide accuracies rivaling or surpassing those of the best dc bridges while requiring a minimum amount of upkeep. Both manual and automatic balancing types are available. Many contain one or more resistance reference standards kept at constant temperature in ovens. The transfer accuracies (i.e., accuracy available immediately after the bridge reference resistor has been calibrated) are very nearly equal to their least count, generally 0.1 ppm or better. Such bridges

[1] Riddle, J. L., Furukawa, G. T., and Plumb, H. H. Platinum Resistance Thermometry, NBS Monograph 126, 1973.

operate at 400 Hz or less to reduce problems with quadrature balances in the resistance being measured and its leads. They usually cover the range below 100 Ω.[1]

***145. The Kelvin double bridge*** is used for the measurement of low resistances of four-terminal construction, that is, whose current and voltage terminals are separate. Figure 3-46 shows the network. The balance equation is

$$\frac{X}{S}=\frac{A}{B}+\frac{l}{S}\frac{\beta}{\alpha+\beta+l}\left(\frac{A}{B}-\frac{\alpha}{\beta}\right)$$

If the resistances $X$ and $S$ being compared are small, so that the resistance of the link $l$ connecting them is comparable, the term of the balance equation involving $l$ could be significant; but if the ratio $A/B$ is equal to the ratio $\alpha/\beta$, the correction term vanishes. This equality can be demonstrated, after the bridge is balanced, by opening the link $l$; if the inner and outer ratios are equal, the bridge will remain balanced. It should be noted that the resistance of the leads $r_1, r_2, r_3, r_4$ between the bridge terminals and the voltage terminals of the resistors may contribute to a ratio unbalance; these lead resistances should be in the same ratio as the arms to which they are connected. In some Kelvin bridges, small adjustable resistors are provided for balancing leads; another technique is to shunt the $\alpha$ or $\beta$ arm with a high resistance until $A/B=\alpha/\beta$ with the link removed. When this balance is achieved, the link $l$ is replaced and the main bridge balance is readjusted. In some bridges, the outer and inner ratio arms are adjustable only in decimal steps, and the main balance is secured by means of an adjustable standard resistor consisting of a Manganin strip with nine voltage taps of 0.01 or 0.001 Ω each and a Manganin slide wire. Portable bridges may use slide-wire arms and reference resistors to cover a range from 10 μΩ to 10 Ω.

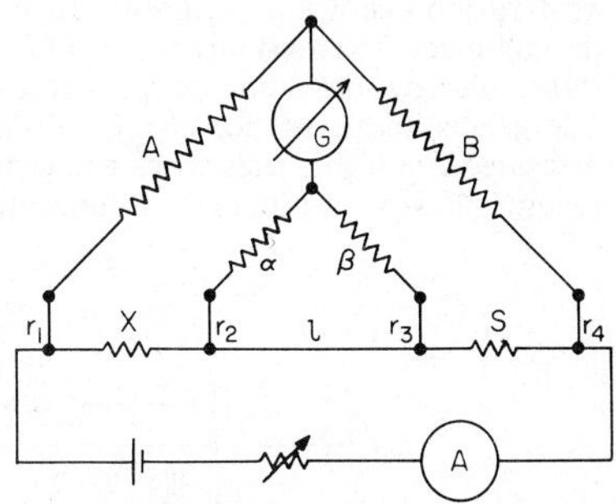

**FIG. 3-46** Kelvin double bridge.

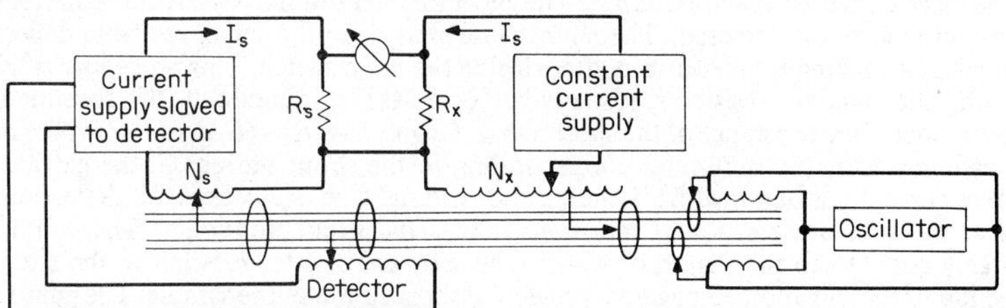

**FIG. 3-47** Current-comparator bridge.

***146. In the current-comparator bridge,***[2] shown schematically in Fig. 3-47, the ratio of resistor currents is evaluated in the comparator (see Par. **31** for details on operation of the dc current comparator) as a balance of ampere-turns in the two circuits. $I_xN_x=I_sN_s$, so that $R_x/R_s=N_x/N_s$ when $I_sR_s$ when $I_sR_s=I_xR_x$. A resistance determination that depends on the evaluation of a ratio is limited by the stability of that ratio. In Wheatstone and Kelvin bridges, the stability of individual resistors sets that limit; in the current-comparator bridge, the ratio is that of windings on a common magnetic core, and therefore stable.

Since this bridge operates in terms of a ratio of currents for equal voltage drops, it can be used to determine power coefficients of low-value resistors.[3] In a Kelvin bridge the ratio of power

[1] Five such bridges are described in "Temperature, its Measurement and Control in Science and Industry," 1982, vol. 5, American Institute of Physics, New York, N.Y.

[2] McMartin, M. P., and Kusters, N. L. *Trans. IEEE,* 1966, vol. IM-15, p. 212.

[3] Dunn, A. F. *Trans. IEEE,* 1966, vol. IM-15, p. 200.

dissipated is $P_s/P_x = R_s/R_x$ in the resistors compared; in the comparator bridge, this ratio is $P_s/P_x = R_x/R_s$. Thus a low-value resistor operated at a substantial power level can be compared directly with a standard of higher resistance operated at a low power level.

***147. Insulation resistance*** is generally measured by deflection methods. In the case of resistances of the order of a few megohms, a Wheatstone bridge may be used with low to moderate accuracy. A portable megohm bridge is made by General Radio Company. It operates as a Wheatstone bridge with an amplifier and dc indicating instrument as the detector system. A choice of high-resistance ratio arms gives ranges of 0.1 to $10^4$ MΩ. On the highest range, the resolution limit is about $10^6$ MΩ. The deflection methods fall in two general classes: (1) direct-deflection methods and (2) loss-of-charge methods.

***148. Direct-deflection methods (insulation resistance)*** involve the simple application of Ohm's law. When the resistance is of the order of 1 MΩ, an ordinary voltmeter can give results that are good enough for most purposes. Two readings are taken, one with the voltmeter directly across the source of voltage, the other with the resistance to be measured connected in series with the voltmeter. The resistance is $R = V(d_1 - d_2)/d_2$, where $V$ is the resistance of the voltmeter, $d_1$ is the voltmeter deflection on the first reading, and $d_2$ is the deflection on the second reading. The greater the resistance of the voltmeter per volt, the higher the resistance that can be measured. For higher resistances, a reflecting galvanometer with high current sensitivity is used. Figure 3-48 is a diagram of the arrangement for measuring the insulation resistance of a cable.

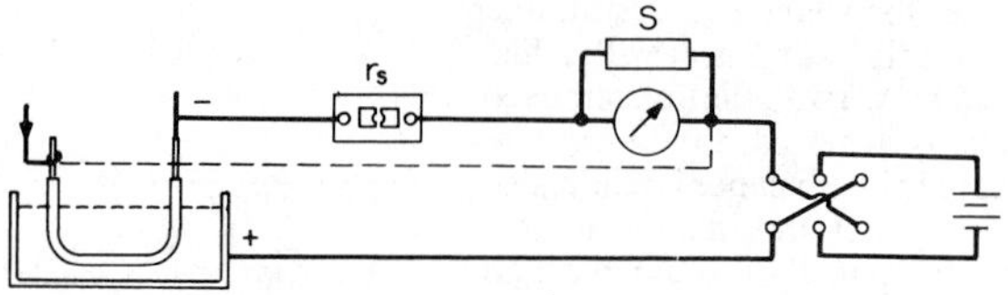


**FIG. 3-48** Diagram for measurement of insulation resistance of cable.

The measurement is made as follows: The galvanometer shunt $S$ is set at the highest shunting value, and the circuit is closed. The shunt is decreased until a large, readable deflection is obtained. The reading is taken 1 min after closing the main switch. This procedure is repeated with only the standard resistor $r_s$ (usually 0.1 or 1 MΩ) in the circuit, the specimen being short-circuited. The resistance of the specimen in megohms is $R = (G/d_1s_1) - r_s$ where $d_1$ is the first reading and $s_1$ the multiplier corresponding to the shunt setting. $G$, the galvanometer megohm constant, is obtained from the second reading, $G = dr_ss$, where $d$ is deflection, $r_s$ the value of the standard resistor in megohms, and $s$ the shunt multiplier. The conductor is preferably negative to the sheath or water. The standard resistor $r_s$ is left in the circuit as a protection to the galvanometer against accidental short circuit in the sample. The guard for the cable ends is shown by the broken line. Removing braid for several inches at the ends of the sample and dipping the ends in hot paraffin tend to reduce leakage across the face of the insulation from sheath to central conductor, especially in damp weather.

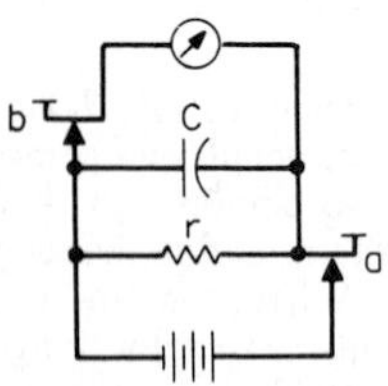


**FIG. 3-49** Leakage method of measuring insulation resistance.

***149. The loss-of-charge method*** of measuring insulation resistance may be used when the resistance is very high, such as the resistance of porcelain and glass, and the surface leakage resistance of line insulators. The principle is shown in Fig. 3-49 where the resistance $r$ to be measured is connected in parallel with a capacitor $C$. Key $a$ is closed and immediately opened, charging the capacitor. Key $b$ is closed immediately after $a$ is opened and the ballistic throw $d_1$ of the galvanometer noted. The process is repeated, but now a time $t$ s is allowed to pass from the instant of charging before key $b$ is closed and a deflection $d_2$ observed. The resistance is

$$r = \frac{t}{2.303C \log_{10}(d_1/d_2)} \quad \text{M}\Omega$$

where $C$ is the capacitance in microfarads.

The insulation resistance of the capacitor is not infinite and should be measured in a similar manner with $r$ removed. The two resistances are in parallel, and the corrected value is

$$r = \frac{r_1 r_2}{r_2 - r_1}$$

Where $r_1$ is the resistance value obtained in the first measurement and $r_2$ is the resistance of the capacitor. For even higher resistance, a growth-of-charge method may be used. In this case, the resistance to be measured is connected in series with a capacitor (preferably an air capacitor), and a known voltage $E$ is applied for $t$ s, the voltage on the capacitor being $e$ at the end of this time. This value, $e$, is best measured with an electrostatic voltmeter connected continuously across $C$. The resistance is

$$r = \frac{t}{2.303C \log_{10}[E/(E - e)]} \quad \text{M}\Omega$$

***150. The Evershed Megger***[1] is an ohmmeter that uses the principle of the ratio meter, as shown in Fig. 3-50, where $A$ is a coil in series with the resistance to be measured and $B$, $B_1$ are coils which, with the resistance $R$, are connected to a hand-driven generator $D$. All three coils are rigidly coupled together, mounted for rotation about an axis $O$, and connected to the circuit by fine copper strips that exert no control force. If the external circuit is open, $B$ and $B_1$ are deflected to the position where they will intercept the least flux from the permanent magnets $M$, $M_1$, that is, opposite the gap in the C-shaped iron piece about which coils $A$ and $B_1$ move. The pointer then stands at "infinity" on the scale. If a finite resistance is connected across the terminals, the current in $A$ will produce a torque and, as the system moves, the coils $B$, $B_1$ exert an opposing torque whose magnitude increases with displacement from the infinity position. The system will come to rest at a position where the torques are equal, depending on the external resistance. This position is practically independent of the voltage generated at $D$. In some Meggers, there is a slip clutch which enables the generator to turn at constant speed when hand-driven. This applies a constant voltage, 500 or 1000 V, to the external circuit and permits a voltage-withstand test of the insulation connected to the external circuit as well as a measure of insulation resistance.

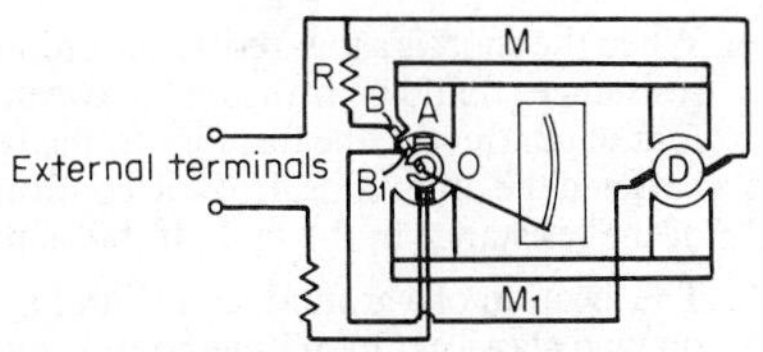


**FIG. 3-50** Evershed Megger.

***151. The resistance of earth connections*** may be measured by a three-electrode method. In Fig. 3-51, $A$ is the connection whose resistance to earth is to be measured; it is temporarily disconnected from the distribution system while ground connection is preserved through a connection at $D$, either temporary or permanent. Two additional "grounds," $B$ and $C$, are established, separated from each other and from $A$ by not less than 15 ft. These auxiliary grounds may be pieces of metal buried in the earth, such as a guy wire or a steel pole, making sufficient contact with ground for a good current reading. Resistances between the three electrodes taken in pairs are measured by a volt-

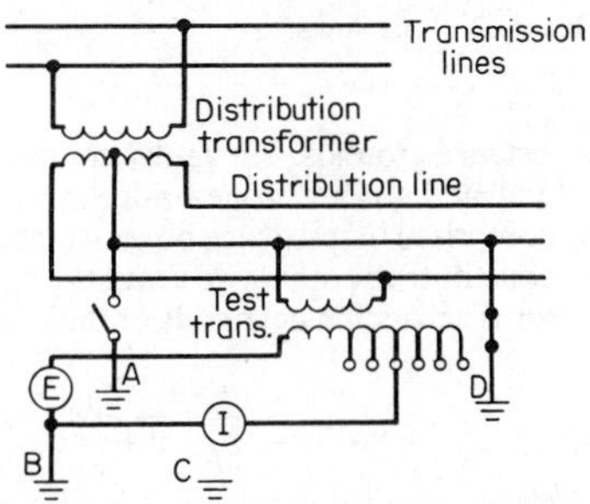


**FIG. 3-51** Resistance of earth connections.

[1] A registered trademark of Evershed and Vignoles, Ltd.

meter-ammeter method. These resistances are $r_{ab}$, $r_{bc}$, $r_{ac}$. Then the resistances are as follows:

At *A*: $$R_a = \frac{r_{ab} - r_{bc} + r_{ac}}{2}$$

At *B*: $$R_b = \frac{r_{ab} - r_{bc} - r_{ac}}{2}$$

At *C*: $$R_c = \frac{r_{bc} - r_{ab} + r_{ac}}{2}$$

The measurement should be made with alternating current, which can be taken from the distribution system through an isolating transformer with secondary taps as shown. A low-range voltmeter is usually required. An Evershed ratio instrument similar to that described in the previous paragraph, is used for the measurement of ground resistance. One of the moving coils is traversed by the current sent through the ground from the attached hand-operated generator; the other is energized by the voltage drop to an auxiliary, driven electrode.

***152. Location of Line Faults.***[1] Faults in electric lines for the transmission and distribution of power, speech, etc., may be divided into two classes, closed-circuit faults and open-circuit faults. *Closed-circuit faults* consist of *shorts,* where the insulation between conductors becomes faulty, and *grounds,* where the faulty insulation permits the conductor to make more or less perfect contact with the earth. *Open-circuit faults,* or *opens,* are produced by breaks in the conductors.

**a.** When the short is a low-resistance union of the two conductors, such as at *M* in Fig. 3-52, the resistance should be measured between the ends *AB;* from this value and the resistance per foot of conductor, the distance to the fault can be computed. A measurement of resistance between the other ends *A′B′* will confirm the first computation or will permit the elimination of the resistance in the fault, if this is not negligible.

**b.** The location of a ground, as at *N* in Fig. 3-52, or of a high resistance short is made by either the two classical "loop" methods, provided that a good conductor remains.

Figure 3-53 shows the arrangement of the *Murray loop test,* which is suitable for low-resistance grounds. The faulty conductor and a good conductor are joined together at the far end, and a Wheatstone-bridge arrangement is set up at the near ends with two arms *a* and *b* comprised in resistance boxes which can be varied at will; the two segments of line *x* and $y + l$ constitute the other two arms; the battery current flows through the ground; the galvanometer is across the near ends of the conductors. At balance,

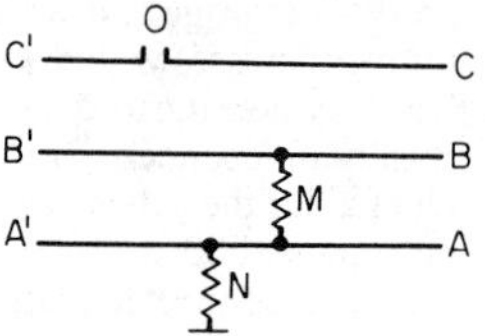

**FIG. 3-52** Line faults.

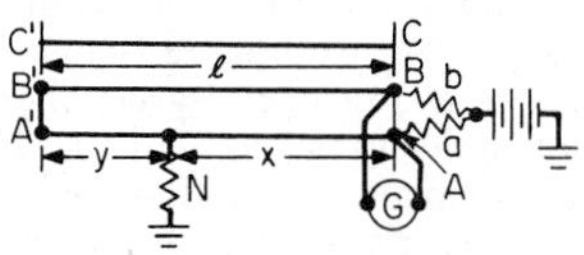

**FIG. 3-53** Murray loop.

$$\frac{a}{b} = \frac{x}{y + l} \quad \text{or} \quad \frac{a + b}{b} = \frac{x + y + l}{y + l} \quad \text{ohms}$$

The sum $x + y + l$ may be measured or known. If the conductors are uniform and alike and

[1] Henneberger, T. C., and Edwards, P. G. Bridge Methods for Locating Resistance Faults in Cable Wires; *Bell System Tech. J.,* 1931, vol. 10, p. 382.

$x$ and $l$ are expressed as lengths, say, in feet,

$$x = \frac{2al}{a+b} \quad \text{ft}$$

If the ground is of high resistance, very little current will flow through the bridge with the arrangement of Fig. 3-53. In that case, battery and galvanometer should be interchanged, and the galvanometer used should have a high resistance. If ratio arms $a$ and $b$ consist of a slide wire (preferably with extension coils), the sum $a + b$ is constant and the computation is facilitated. Observations should be taken with direct and reversed currents, especially in work with underground cables.

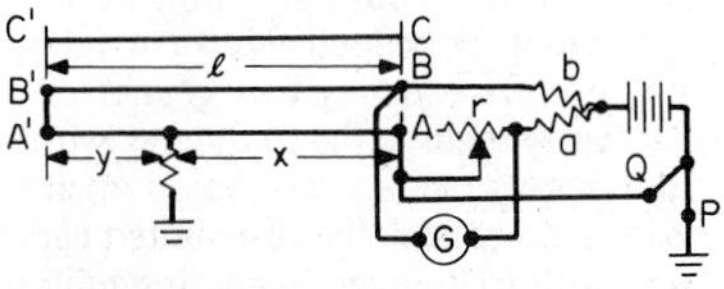

**FIG. 3-54** Varley loop.

In the *Varley loop,* shown in Fig. 3-54, fixed-ratio coils, equal in value, are employed, and the bridge is balanced by adding a resistance $r$ to the near leg of the faulty conductor.

$$\frac{a}{b} = \frac{r+x}{y+l} \quad \text{or} \quad \frac{a+b}{b} = \frac{x+y+l+r}{y+l} \quad \text{ohms}$$

If $a = b$,

$$x = y + l - r \quad \text{or} \quad x = \frac{1}{2(x+y+l-r)} \quad \text{ohms}$$

The total line resistance $x + y + l$ is conveniently determined by shifting the battery connection from $P$ to $Q$ and making a new balance, $r'$. The equation then becomes $x = 1/2(r' - r)$. When $a$ and $b$ are slightly unequal, a second set of readings should be taken with $a$ and $b$ interchanged and the average values of $r$ and $r'$ substituted in the foregoing equations.

**c.** *Opens,* such as $O$ in Fig. 3-52, are located by measuring the electrostatic capacitance to ground (or to a good conductor) of the faulty conductor and of an identical good conductor; the position of the fault is determined from the ratio of the capacitances.

**d.** *Shorts* and *grounds* may be detected by sending through the defective conductor an alternating current of audible frequency, say 1000 Hz. A pickup coil connected to a telephone receiver worn on the head of the tester is then carried along the line; the note in the receiver will cease when the fault has been passed.

## Inductance Measurements

***153. General.*** *The self-inductance, or coefficient of self-induction,* of a circuit is the constant by which the time rate of change of the current in the circuit must be multiplied to give the self-induced counter emf. Similarly, the *mutual inductance* between two circuits is the constant by which the time rate of change of current in either circuit must be multiplied to give the emf thereby induced in the other circuit. Self-inductance and mutual inductance depend upon the shape and dimensions of the circuits, the number of turns, and the nature of the surrounding medium.

***154. Computable standards*** of self- or mutual inductance have been used traditionally in *absolute-ohm* determinations; but they are not suitable for use in assigning the values of other inductors—they are bulky and have relatively large capacitance to ground and considerable coupling to other circuits, their ratio of inductance to resistance is relatively low, and they exhibit appreciable skin effect even at moderately high frequencies, since they must be wound with rather heavy wire. All these undesirable features inevitably follow from the requirement that their values be computable from measured dimensions. Computable self-inductors and the primaries of computable mutual inductors are wound as single-layer solenoids on a dimensionally stable nonmagnetic form. The best of them are on cyclinders of fused silica, and the winding is laid in a groove lapped into the form to ensure uniform winding pitch. The primary winding of a computable mutual inductor is in two or three sections spaced at such intervals that there is a region outside and in its central plane in which its field gradients are very small. The secondary —a multilayer winding—is located in this position so that its position and dimensions will be less critical.

***155. Working standards of inductance*** are usually multilayer coils wound on nonmagnetic forms of Bakelite, marble, or ceramic to ensure reasonable dimensional stability. A toroidal core gives a coil that is practically immune to external magnetic fields. Approximate astaticism is also achieved by using two equal coils, connected in series and so located with respect to each other that their coupling with external fields tends to cancel each other. Since there is always capacitance associated with a winding, the effective value of an inductor will always be a function of frequency to a greater or lesser extent and an accurate statement of value must necessarily include the frequency with which the value is associated. Inductance standards for radio frequencies are wound on open frames. Single-layer winding or "loose basket weave" is essential to reduce the distributed capacitance and the consequent change of effective inductance with frequency. Insulating material is kept to a minimum to reduce dielectric loss.

***156. Inductometers*** are continuously adjustable inductance standards. The Ayrton-Perry inductometer uses pairs of coaxial coils wound on zones of spheres; the outer pair is fixed, and the inner pair can be rotated about a vertical axis. The coils are so proportioned that the scale is uniform over most of its length. This inductometer is not astatic, and its coupling with external fields can cause significant measurement errors. The Brooks inductometer, a better design from several viewpoints, consists of six link-shaped coils. The four stator coils are mounted in pairs above and below the rotor coils, which are located diametrically opposite one another in a flat disk. These two fixed- and moving-coil combinations are so connected that their coupling with external fields tends to cancel. The shape of the link coils gives a scale that is completely uniform except at its extreme ends, and the time constant of the inductometer is much higher than in the Ayrton-Perry arrangement. Ratio of maximum to minimum inductance is about 8:1; and change of calibration with wear in the bearings is negligibly small. Terminals of the fixed and movable coils are usually brought out separately so that inductometers can be used as either adjustable self-inductors or adjustable mutual inductors.

***157. Measurement methods*** at power and audio frequency are (1) null methods employing bridges if accurate values are required or (2) deflection methods in which the inductance is computed from measured values of impedance and power factor, the measurements being made with indicating instruments—ammeter, voltmeter, wattmeter. At radio frequencies, resonance methods are used.

***158. Bridges for inductance measurements*** can assume a variety of forms, depending on available components and reference standards, magnitude and time constant of the inductance to be measured, and a variety of other factors. In a four-arm bridge similar to the Wheatstone network, an inductance can be (1) compared with another inductance in an adjacent arm with two resistors forming the "ratio" arms or (2) measured in terms of a combination of resistance and capacitance in the opposite arm with two resistors as the "product" arms. It is generally better, where possible, to measure inductance in terms of capacitance and resistance rather than by comparison with another inductance, because the problems of stray fields and coupling between bridge components are more easily avoided. The basic circuits will be described for a few bridges which can be used to measure inductance, but the reader must refer to a text on bridges (e.g., Hague, "A-C Bridge Methods," 6th edition, Pitman, 1971) for discussion of shielding, physical arrangement of components, effects of residuals, etc. In the balance equations which will be stated below, the inductance, $L$ or $M$, will be expressed in henrys, the resistance $R$ in ohms, capacitance in farads, and $\omega$ is $2\pi \times$ frequency in hertz. The time constant of an inductor is $L/R$; its storage factor $Q$ is $\omega L/R$.

***159. Inductance comparison*** is accomplished in the simple Wheatstone network shown in Fig. 3-55, in which $A$ and $B$ are resistive ratio arms, $L_x$ and $r_x$ represent the inductor being measured, and $L_s$ and $r_s$ are the reference inductor and the associated resistance (including that of the inductor itself) required to make the time constants of the two inductive arms equal. At balance,

$$\frac{A}{B} = \frac{L_x}{L_s} = \frac{r_x}{r_s}$$

An inductometer may be used to achieve balance, together with an adjustable resistance in the same bridge arm, as indicated in the diagram. If only a fixed-value standard inductor is avail-

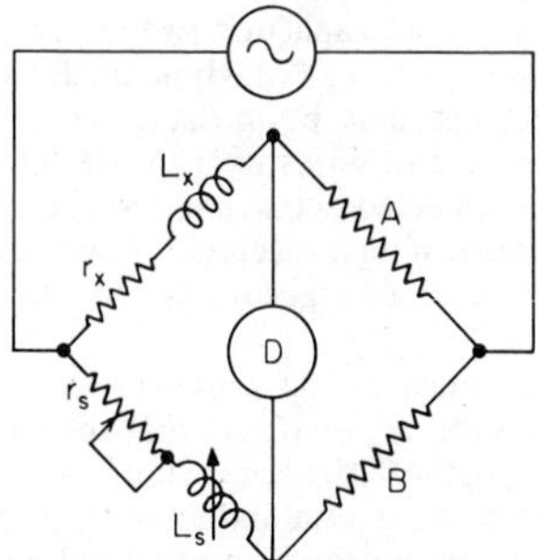


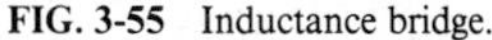

**FIG. 3-55** Inductance bridge.

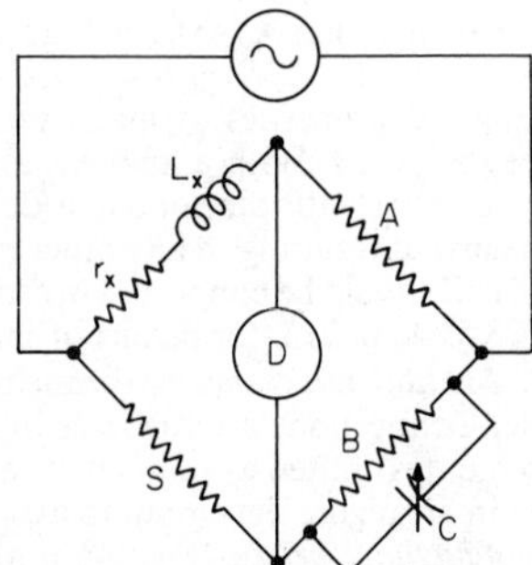


**FIG. 3-56** Maxwell-Wien inductance-capacitance bridge.

able, balance can be secured by varying one of the ratio arms but there must also be an adjustable resistance in series with $L_x$ or $L_s$ to balance the time constants of the inductive arms. Care must be taken to ensure that there is no inductive coupling between $L_s$ and $L_x$, as this would lead to a measurement error.

***160. The Maxwell-Wien bridge*** for the determination of inductance in terms of capacitance and resistance is shown in Fig. 3-56. The balance equations are $L_x = ASC$ and $r_x = AS/B$. This bridge is widely used for accurate inductance measurements. It is most easily balanced by adjustments of capacitor $C$ and resistor $B$; these elements are in quadrature, and therefore their adjustments do not interact.

***161. Anderson's bridge,*** shown in Fig. 3-57, can be used for measurement over a wide range of inductances with reasonable values of $R$ and $C$. Its balance equations are $L_x = CAS(1 + R/S + R/B)$ and $r_x = AS/B$. Balance adjustments are best made with $R$ and $r_x$. This bridge has also been used to measure the residuals of resistors, a substitution method being employed in which the unknown and a loop of resistance wire with calculable residuals are substituted in turn into the $L$ arm. If $A$ and $B$ are equal and if the resistances of the unknown and the calculable loop are matched, the residuals in the various bridge arms do not enter the final calculation, except the residual of $\Delta r_x$, the change in $r_x$ between balances. The elimination of bridge-arm residuals from the *exact* balance equations is characteristic of substitution methods; and quite generally, residuals or corrections to the arms that are unchanged between the balances do not have to be taken into account in the final calculation when the difference is small between the substituted quantities.

***162. Owen's bridge,*** shown in Fig. 3-58, can be used to measure a wide range of inductance

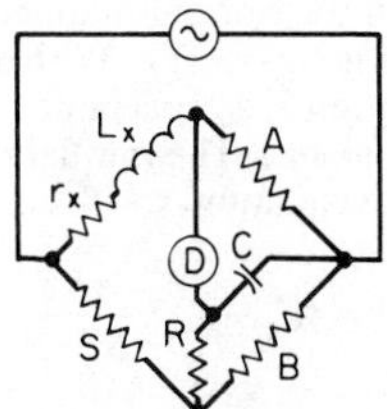


**FIG. 3-57** Anderson's bridge.

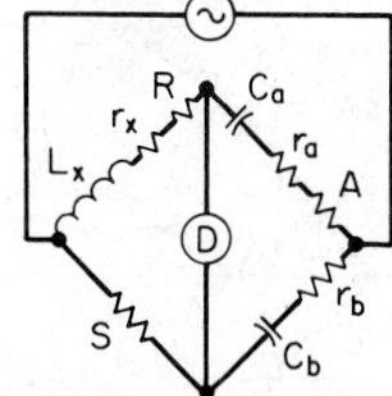


**FIG. 3-58** Owen's inductance-capacitance bridge.

with a standard capacitor $C_b$ of fixed value, by varying the resistance arms $S$ and $A$. In operation, the resistance $S$ and capacitor $C_b(r_b)$ are usually fixed, balance being secured by successive adjustments of $A$ and $R$. At balance

$$r_x + R = (C_b/C_a)S + \omega L_x \omega C_b r_b$$

and $L_x(1 + \tan \delta_b \tan \delta_x) = C_b S(A + r_a)$. If $C_b(r_b)$ is a loss-free air capacitor, so that $r_b = 0$ and $\tan \delta_b = 0$, $r_x = (C_b/C_z)S' - R$ and $L_x = C_b S(A + r_a)$. This is a bridge which is much used for examining the properties of magnetic materials; inductance may be measured with direct current superposed. With a low-reactance blocking capacitor in series with the detector and another in series with the source, a dc supply may be connected across the test inductance without current resulting in any other branch of the network; a high-reactance, low-resistance "choke" coil should be connected in series with the dc source (see Ferguson, *Bell System Tech. J.*, 1927, vol. 6, p. 375, for details of shielding, etc.).

***163. Mutual inductance*** can be measured readily if an adjustable standard of proper range is available. Connections are made as in Fig. 3-59. At balance, $M_x = M_s$, so that the range of measurement is limited to values that can be read on $M_s$ with the desired precision. Care should be taken in arranging the circuit to avoid coupling between the mutual inductors. Campbell's *mutual-inductance bridge* (Fig. 3-60) makes possible the comparison of mutual inductors of

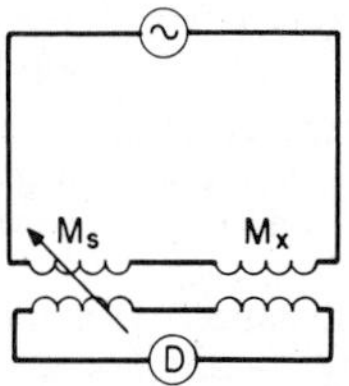


**FIG. 3-59** Comparison of mutual inductometers.

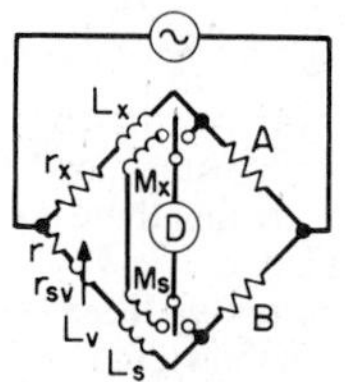


**FIG. 3-60** Campbell's bridge for comparing mutual inductances.

quite different value. The resistances and self-inductances of their primaries must also be balanced. This is accomplished by adjusting $L_v$ and $r$ in the self-inductive bridge with the switches in the detector branch to the right. The mutual-inductance balance is then made with the switches to the left. At balance,

$$\frac{A}{B} = \frac{M_x}{M_s} = \frac{L_x}{L_v + L_s} = \frac{r_x}{r + r_{sv}}$$

The *Maxwell-Wien bridge* can also be adapted to the measurement of mutual inductance as shown in Fig. 3-61. Balancing adjustments are to be made with $r_x$ and $C$. With the selector switch of the supply branch in position 1, the self-inductance $L_x$ of the primary winding is determined in the normal manner. The balance equation is $L_x = ASC_1$. With the selector switch in position 2, a capacitance value $C_2$ is required for balance. The mutual inductance is given by the equation

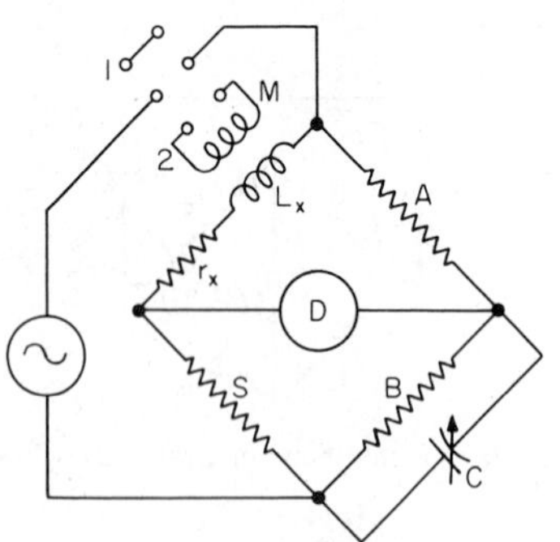


**FIG. 3-61** Maxwell-Wien bridge for mutual inductance.

$$M = (C_1 - C_2)\frac{ABS}{B + S}$$

The values of resistors $A$, $B$, and $S$ must not be changed between the two balances, and the connections should be such that the emf induced in $L_x$ from the winding of $M$ which is in the supply circuit will oppose the emf of self-inductance in $L_x$ (see Harris, *Electrical Measurements,* John Wiley & Sons, Inc., 1952, p. 721, for an analysis of the effect of bridge residuals in this measurement). Mutual inductance may also be measured in any self-inductance bridge by the method illustrated in Fig. 3-62. The coils $L_1$ and $L_2$ are connected in series *(a)* so that the emf of mutual inductance aids the emf of self-inductance. The measured value of total inductance is $\lambda_a = L_1 + L_2 + 2M$. A second measurement is made with one coil reversed in the series connection of *(b)*, so that the emfs of self- and mutual inductance are opposed. The measured value is now $\lambda_b = L_1 + L_2 - 2M$. The mutual inductance is calculated from the expression

$$M = (\lambda_a - \lambda_b)/4$$

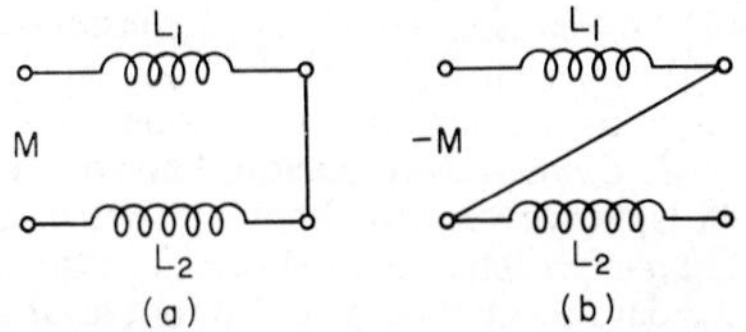

**FIG. 3-62** Mutual inductance connected for measurement of self-inductance.

***164. Iron-cored inductors*** vary in value with frequency and with current, so that measurements must be made at known current and frequency; bridge methods can, of course, be adapted to this measurement, care being exercised to ensure that the current capacities of the various bridge components are not exceeded. In such a case, the waveform of the voltage drop across the circuit branch containing the inductor may not be sinusoidal, whereas that across the other side of the bridge, containing linear resistances and reactances, may be undistorted. Generally, a tuned detector should be used.

***165. Resonance methods*** can be used to measure inductance at radio frequencies. A suitable source is used to establish an rf field whose wavelength is $\lambda$ m. The inductance $L_x$ (microhenrys) to be measured is placed in this field and connected to a calibrated variable capacitor through a thermocouple ammeter (i.e., a current-indicating instrument without reactance). The capacitor is adjusted to resonance at a value of $C$ (picofarads). Then $L_x = 0.2815\lambda^2/C$. If a calibrated inductor $L_s$ of the same order as $L_x$ is available, the wavelength need not be known and a substitution method can be used. The resonance settings are $C_s$ and $C_x$, with $L_s$ and $L_x$, respectively, in the circuit. Then, $L_x = L_sC_s/C_x$. The value of $L_x$ is the effective inductance at the frequency of measurement and includes the effect of associated coil capacitance. The frequency of the source must not be affected by the substitution of $L_x$ for $L_s$.

A *resonance-impedance method,* suitable for high frequencies, is indicated in Fig. 3-63. The capacitor $C$ is adjusted until the same current is indicated by the ammeter with switch $K$ open or closed. (The applied voltage must be constant.) Then $L_x = (1/2\omega^2C)$ H, if $C$ is in farads and the frequency is $f = \omega/2\pi$. The waveform must be practically sinusoidal and the ammeter of negligible impedance. This method may be used to measure the effective inductance of choke coils with superposed direct current (see Turner, *Proc. IRE,* 1928, vol. 16, p. 1559, for details).

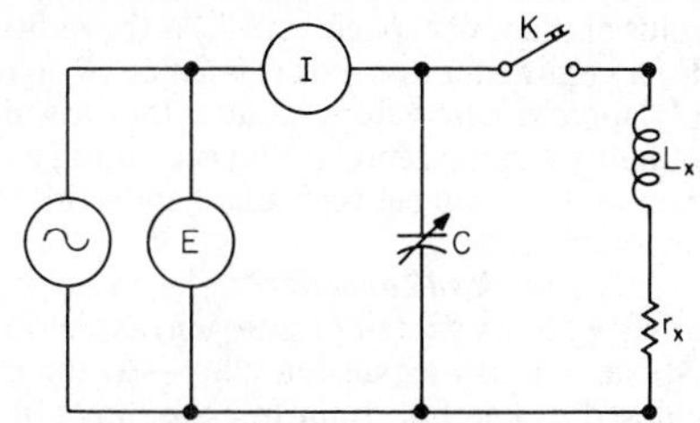

**FIG. 3-63** Resonance-impedance method of measuring inductance.

***166. The residual inductance*** of a resistor or a length of cable at high frequency can often be determined by connecting the resistor in series with a fixed air capacitor and measuring its effective capacitance in an appropriate bridge with and without the series resistor $S$. If $C_1$ and $C_2$ are the measured capacitances in farads, without and with the series resistor, then

$$L = \frac{C_2 - C_1}{\omega^2 C_1 C_2} \quad \text{and} \quad S = \frac{1 - \omega^2 LC_1}{\omega C_1} \tan\delta$$

$S$ is the effective resistance in ohms. $L$ is the residual inductance in henrys, and $\delta$ is the loss angle of the capacitor-resistor combination computed from the second bridge balance.

## Capacitance Measurements

***167. General.*** The *capacitance* between two electrodes may be defined for measurement purposes as the charge stored per unit potential difference between them. It depends on their area, spacing, and the character of the dielectric material or materials, which is affected by the electric field between them. The value of a capacitor, measured in farads or a convenient submultiple of this unit, will be influenced quite generally by temperature, pressure, or any ambient condition that changes the dimensions or spacing of the electrodes or the characteristics of the dielectric. The *dielectric constant* of a material is defined as the ratio of the capacitance of a pair of electrodes, with the material occupying all the space affected by the field between them, to the capacitance of the same electrode configuration in vacuum.

***168. Computable capacitors*** known to a part in $10^6$ or better have recently been constructed at the National Bureau of Standards and at certain other national laboratories as a basis for their *absolute-ohm* determinations. Such capacitors now serve as the "base" unit in assigning values to standard capacitors.[1] The electrode arrangement of these computable capacitors conforms to the geometry prescribed in the recently discovered *Thompson-Lampard theorem:* If four cylindrical conductors of arbitrary sections are assembled with their generators parallel, to form a completely enclosed cylinder in such a way that the internal cross capacitances per unit length are equal, then in vacuum these cross capacitances are each

$$\frac{\ln 2}{4\pi^2 \mu_0 V^2}$$

In the mksa system of electrical units, where $\mu_0$ has the assigned value $4\pi \times 10^{-7}$ and $V$ is the speed of light in vacuum in meters per second, this capacitance is in farads per meter. The capacitance of such a cross capacitor is about 2 pF/m. A practical realization of such a capacitor consists of four equal closely spaced cylindrical rods with their axes parallel and at the corners of a square. Arranged as a three-terminal capacitor and with end effect eliminated, its value can be computed as accurately as its effective length can be measured.

The capacitance of vacuum capacitors with electrodes of simple geometry can be computed approximately in a few cases: (1) Flat, parallel plates with guard ring, $C = 0.08854A/t$ pF, where $A$ is area of the guarded plate in square centimeters and $t$ is spacing in centimeters between electrodes; if dimensions are in inch units, $C = 0.2249A/t$. (2) Coaxial cylinders with guard cylinders at both ends, $C = 0.24161L/\log(R_2/R_1)$ pF for centimeter units, or $C = 0.6137L/\log(R_2/R_1)$ pF for inch units, where $L$ is the length of the guarded cylinder, $R_1$ is the radius of the inner cylinder, and $R_2$ the radius of the outer cylinder. (3) Concentric spheres, where $R_1$ is the radius of the inner sphere and $R_2$ is the radius of the outer sphere, $C = 1.1127R_1R_2/(R_2 - R_1)$ pF for centimeter units, or $C = 2.8262R_1R_2/(R_2 - R_1)$ pF for inch units. These formulas give only approximate values because they assume no contributing field beyond the edges of the bounding surfaces, and take no account of possible eccentricity, lack of parallelism of surfaces, finite width of gap between guard and working electrode, etc., all of which would require small correction terms.

***169. Standard capacitors*** at levels up to $10^3$ pF are generally of a multiple-parallel-plate variety with dry gas (air or nitrogen) as dielectric. Low temperature coefficient is secured by use of Invar—a low-expansion alloy—as the electrode material and a good degree of stability achieved by careful, strain-free mounting of fully annealed components and by hermetically sealing the unit. A very high degree of stability has been achieved in a solid-dielectric construction at the 10-pF level in which a disk of fused silica is provided with fired-on silver electrodes. Direct capacitance is through the interior of the disk between its parallel faces, and a silver coating on the cylindrical face acts as guard electrode and confines the field. Very narrow gaps at the edges of the disk between the guard and active electrodes, together with continuation of the shielding in the mounting arrangement, eliminate the possibility of any portion of the measured capacitance being through an outside path between the parallel-plate electrodes. The assembly is hermetically sealed in dry nitrogen, in a shock-resistant, resilient mounting together with a resistance thermometer so that temperature corrections can be accurately applied. Standards of this type have shown variations less than 1 part in $10^7$ over a year interval. From $10^3$ pF to 1 $\mu$F, standard capacitors generally have clear mica as dielectric. The electrodes may be metal foils laid out between the mica sheets, the assembly impregnated with paraffin, and the excess wax squeezed out under high pressure. In an alternative construction, the mica sheets are silvered, assembled under pressure, and the assembly hermetically sealed. Neither construction is as stable with time as the lower-value air-dielectric units and the mica units are characterized by low but appreciable loss angles, whereas the loss angle of the air-dielectric standards is negligible in almost all applications. Continuously adjustable air capacitors have two stacks of interleaved parallel metal plates, one stack being mounted to rotate on an axis. The maximum capacitance

[1] Cutkosky, R. D. New NBS Measurements of the Absolute Farad and Ohm; *Trans. IEEE,* 1974, vol. IM-23, pp. 305–309.

occurs when the fixed and movable plates completely overlap; the minimum, a small value but not zero, occurs 180° from this position.

*170. **A three-terminal construction*** is required if the value of the capacitor is to be definite and independent of its proximity to other objects. In a nominally two-terminal arrangement, each of the electrodes has some capacitance to surrounding objects or to ground which may depend on spacing and which actually forms a second capacitance circuit in parallel with the capacitor of interest, as will be seen from Fig. 3-64*a* and *b*. It is only in case *c*, where there is an

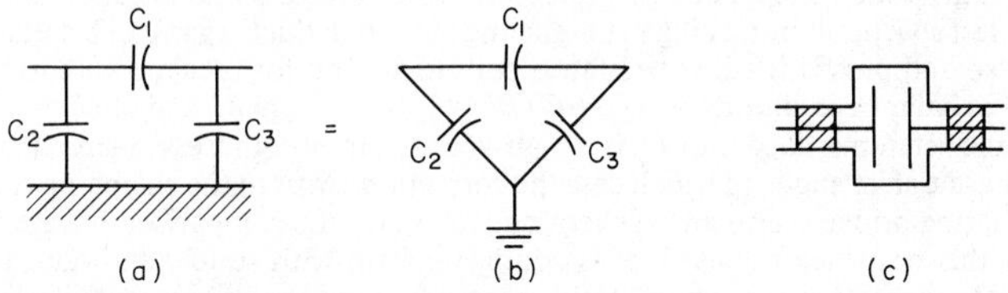


**FIG. 3-64** Two-terminal and three-terminal capacitors.

actual third electrode which completely encloses the other two, that the value can be made definite and completely independent of any object or field outside the assembly. A second advantage of the three-terminal construction is that the direct capacitance between the two enclosed electrodes can be made loss-free, since the solid insulation required to support them mechanically can be in the auxiliary capacitances between the enclosing shield electrode and the shielded electrodes.

*171. **Methods of measuring capacitance*** can be classified as *null* methods, which quite generally involve the use of bridges, and *deflection* methods, in which some characteristic, usually impedance, is measured with the aid of indicating instruments. In the equations that follow, the capacitance $C$ will be expressed in farads and resistance $A$, $B$, $S$ in ohms. $\delta$ will be the *loss angle*, the amount which the current lacks of a true quadrature relation with voltage. The *power factor* of a capacitor is then $\cos(\pi/2 - \delta) = \sin \delta$. The dissipation factor $D$ is the name given to $\tan \delta$. It is convenient to represent a capacitor as consisting of a capacitance $C$ (farads) in series with a resistance $r$ (ohms), such that $\tan \delta = 2\pi f C r$ at a frequency $f$. The *power loss*, for an impressed voltage $E$ (volts), is $P = 2\pi f C E^2 \sin \delta$. Since most bridges yield $\tan \delta$, the power loss can be expressed conveniently as $P = 2\pi f C E^2 \tan \delta$, where $\delta$ is small, or $P = \omega C E^2 D$.

*172. **Bridge methods*** for the comparison of capacitors are to be preferred over methods in which capacitance is determined in terms of inductance, since it is simple to shield capacitors so that their values are completely independent of neighboring objects and their electric fields are completely confined, whereas the magnetic fields of inductors cannot be so confined. Error voltages can enter bridges through coupling of an inductor with an external field, through mutual coupling with eddy-current circuits induced by the inductor in neighboring metal objects, etc. The reader should refer to a text on bridges (e.g., Hague, *A-C Bridge Methods*, 5th ed., Pitman, 1957) for discussion of shielding, physical arrangements of components, effects of residuals, etc.

*DeSauty's bridge*, shown in Fig. 3-65, is a simple Wheatstone network in which capacitors may be compared in terms of a resistance ratio. It should be noted with the loss angles of the two

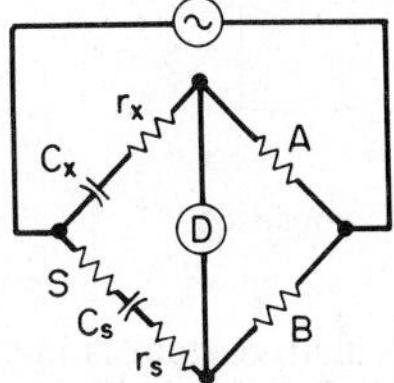


**FIG. 3-65** DeSauty bridge.

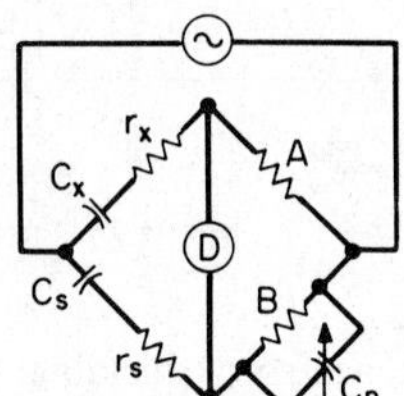


**FIG. 3-66** Schering's bridge.

capacitance arms must be equal, so that a series resistor is inserted in the branch with the smaller loss angle. In the case illustrated, the resistance $S$ is in series with the reference capacitor $C_s$. At balance $C_x = C_s(B/A)$, and $\tan \delta_x = \omega C_x r_x = \omega C_s(r_s + S) = \tan \delta_s + \omega C_s S$.

*Schering's bridge,* shown in Fig. 3-66, has found wide application in measuring the loss angles of high-voltage power cables and high-voltage insulators. For this purpose, the supply voltage is connected as shown, and a ground connection is made at the junction of branches $A$ and $B$, so that the balance adjustments may be made close to ground potential. The adjustable components are generally $A$ and $C_p$. It is also customary to enclose the $A$, $B$, and detector branches in a grounded screen and to protect this low-voltage section against possible breakdown of the test specimen by an air gap paralleling branch $A$. Such a gap can be set to spark over at 100 V or so and provides a low-resistance path to ground for breakdown current from the specimen. The balance equations are $C_x = C_s(B/A)(1 + \tan \delta_s \tan \delta_p)$ and $\tan \delta_x = \omega C_p B + \tan \delta_s$. Usually the reference capacitor $C_s$ is a high-voltage air or compressed-gas capacitor with a negligible phase-defect angle, in which case the correction terms to the balancing equation drop out. The Schering bridge is also an excellent one to use for the comparison of capacitors at low voltage. For this purpose, it is used in its conjugate form with supply and detector branches interchanged to increase sensitivity. $C_p$ must, of course, be connected across branch $A$ instead of $B$ if the loss angle of $C_s$ is greater than that of $C_x$, with a corresponding modification of the balance equations. When the loss angles of $C_s$ and $C_x$ are both very small, adjustable capacitors must be connected across both $A$ and $B$ arms and the difference in the phase-defect angles they introduce into the bridge must equal the difference in loss angles of $C_s$ and $C_x$. This modification of the bridge is made necessary by the fact that the capacitance of an adjustable capacitor cannot be reduced to zero in the usual construction.

The *transformer bridge* has been developed recently into the most precise tool available for the comparison of capacitors, especially for three-terminal capacitors with complete shielding. A three-winding transformer is used so that the bridge ratio is the ratio of the two secondary windings of the transformer which are of low resistance and uniformly distributed around a toroidal core to minimize leakage reactance. A stable ratio, known to better than 1 part in $10^7$, can be achieved in this way. For details of such a transformer construction and the construction and shielding of bridge components, the reader should consult the following papers:

Thompson, A. M. Precise Measurement of Small Capacitances; *Trans. IRE,* Instrumentation, December 1958, vol. I-7.

McGregor, M. C., et al. New Apparatus at NBS for Absolute Capacitance Measurement; *Trans. IRE,* Instrumentation, December 1958, vol. I-7.

Cutkosky, R. D., and Shields, J. Q. Precise Measurement of Transformer Ratios; *Trans. IRE,* December 1960, vol. I-9.

A variety of schemes for balancing adjustment have been used successfully. One of these, employing *inductive voltage dividers,* is shown schematically in Fig. 3-67, but simplified by

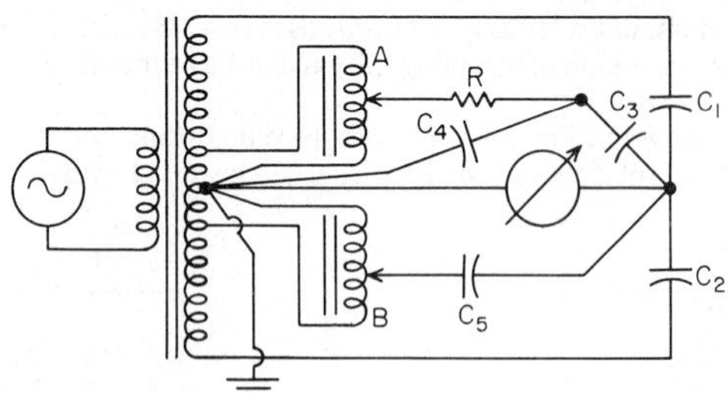


**FIG. 3-67** Transformer bridge for capacitor comparison.

omitting the necessary shielding. Current in phase with the main current is injected at the junction between the capacitors being compared, $C_1$ and $C_2$, to balance their inequality in magnitude. This current, through capacitor $C_5$, is controlled by adjusting the tap position on inductive voltage divider $B$, supplied from an appropriate tap point on the main transformer-ratio arm. Quadrature current, to balance the phase difference between $C_1$ and $C_2$, is similarly injected through $R$ and the current divider $C_3/(C_3 + C_4)$, controlled by adjusting the tap point on divider $A$. The current divider is used so that $R$ may have a reasonable value, a few megohms at most. In the illustrated network, it is assumed that $C_1 > C_2$ and that $\delta_1 < \delta_2$. The balance equations are $C_2 = C_1 + N_B C_5$ and $\delta_2 = \delta_1 + \omega R C_1 \cdot N_A \cdot C_3/(C_3 + C_4)$, where $N_B$ is the fraction of the voltage across $C_2$ which is impressed on $C_5$, that is, the product of the tap-point ratios of the main transformer and divider $B_5$ and $N_A$ is the corresponding fraction of the voltage across $C_1$ which is impressed on $R$. The reactance of $C_3$ and $C_4$ in parallel must be small compared with the resistance of $R$.

New automated impedance measurement instruments have come into being because of the ready availability of microprocessors. Some of these make use of the transformer techniques mentioned above, using relays to balance them by selecting ratios computed by the microprocessor from detector output voltages. Many have purely analog quadrature balance features. At least one measures by passing the same current through the admittance to be measured and a reference resistor and computing the vector impedance of the unknown from the vector ratio of the voltage drops across it and the reference resistor. This is done using a 90° phase reference generated internally using digital synthesis techniques.

Many automated bridges are intended for testing of precision components over a broad range of frequencies and with programmable direct current or voltage biases. Their accuracies range from a few percent at high frequencies to 0.01% or better at audio frequencies. Their calibration is generally done using fixed-value two- or three-terminal or four-pair-terminal standards.[1]

***173. Detectors*** used in bridge measurements are selected with regard to frequency and impedance. *Vibration galvanometers* can be used at power frequencies in low-impedance circuits; they discriminate well against harmonics and have high sensitivity, but they must be tuned sharply to the use frequency.

*Wave analyzers,* which are commercially available with internal crystal control, also have a narrow passband and a high rejection of frequencies on either side. They can be used with a preamplifier when maximum sensitivity is required, and it is desirable that the preamplifier itself be sharply tuned in its first stage to improve noise rejection. This system can be used at any frequency throughout the audio region.

*Cathode-ray oscilloscopes* of adequate sensitivity (or used with tuned preamplifiers) make particularly good null detectors. If a phase-adjustable voltage from the bridge supply is impressed on the horizontal plates and the unbalance signal in the detector branch impressed on the vertical plates, the resulting Lissajous figure is an ellipse which, with proper phase adjustment, will change its opening with magnitude adjustment and the slope of its major axis with quadrature adjustment in the bridge. Balance is indicated by a straight horizontal trace on the screen. It is essential in this system that the initial stages of amplification be sharply tuned or that the bridge input be sinusoidal, for otherwise the pattern on the screen is confused and difficult to interpret. Phase discrimination of this type in the null detector is of considerable value in achieving balance, as it informs the operator of the individual magnitudes of inphase and quadrature unbalance.

*Telephone receivers* may be used at audio frequencies (maximum sensitivity being at about 1 kHz), but their response is usually quite broad, and the balance point may be masked by the presence of harmonics.

In *resonance methods* at radio frequencies, a thermocouple ammeter can be employed to show the current maximum. A crystal rectifier with an electronic voltmeter is used at ultrahigh frequencies.

---

[1] Tomio, W., Takahashi, K., and Toshio, T. "New Multi-frequency LCZ Meters Offer Higher-Speed Impedance Measurements," July 1983, *Hewlett-Packard J.* pp.32–37; Huntley, L. E. and Jones, R. N. "Lumped Parameter Impedance Measurements" *Proc. IEEE,* 1966, vol. 55, no. 6, pp. 900–911, (42 references); Hall, H. P. "Digital Impedance Comparator with High Resolution" *Trans. IEEE* 1980, vol. IM-29, pp. 337–341; Cutkosky, R. D. "A Programmable Phase-Sensitive Detector for Automatic Bridge Applications," *Trans. IEEE* 1978, vol. IM-27, pp. 401–402; Jones, R. N. "A Technique for Extrapolating 1-kC Values of Secondary Capacitance Standards to Higher Frequencies," *NBS Tech. Note* 1963, p. 201; Shields, J. Q. "Measurement of 4-Pair Admittances with 2-Pair Bridges" *Trans. IEEE* 1974, vol. IM-23, pp. 345–352.

***174. Precautions in Bridge Measurements.*** The effect of stray magnetic fields can be minimized by using twisted-pair or coaxial leads and by avoiding loops in which an emf could be induced. Inductive coupling between bridge components should be avoided. Capacitive coupling existing between parts of the bridge which are not at the same potential will impress shunt capacitance across one or more of the bridge arms and modify the balance condition. Shielding must be used to minimize these effects. Figure 3-68 shows a completely shielded bridge suitable

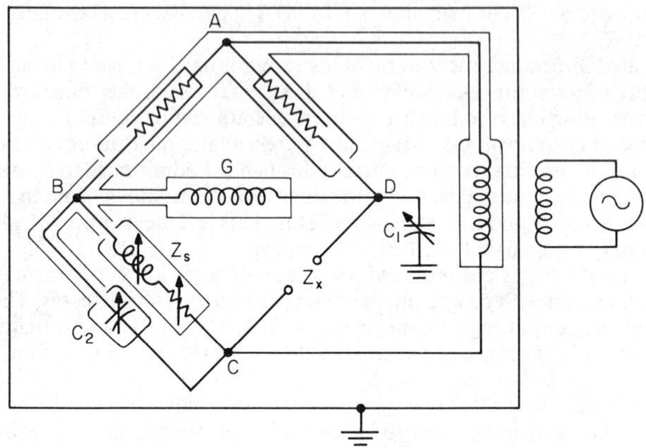


**FIG. 3-68** Completely shielded bridge.

for comparing inductive impedances. The ratio arms *AB* and *AD* are enclosed in metal shields connected to their junction with the source. The balancing inductance and resistance of $Z_s$ are enclosed in a shield which is connected to the source at terminal *C*. The detector *G* is shielded, and its shield is connected to *D*. Definite fixed shunting capacitances are thereby established in these arms. The capacitance between ratio-arm shield and reference-standard shield is eliminated by enclosing the former in a second shield which is extended over the secondary winding of the supply transformer and connected to *C*. Thus, the entire bridge is encased in a grounded shield; *D* is connected to ground through an adjustable capacitor $C_1$, and a shielded balancing capacitor is connected across arm *BC*. It is often required that the terminals *B* and *D* shall not be connected directly to ground, but that they must be maintained at ground potential. *Auxiliary Wagner arms* are used for this purpose. In Fig. 3-69 impedances *M* and *N* are connected in series across the source, with their junction point *a* grounded. If their ratio is such that $M/N = A/B$, the junction points *b* and *c* will also be at ground potential and admittances to ground from the other two bridge corners will shunt the auxiliary Wagner arms rather than the working arms of the bridge. The ratio *M*/*N* must be balanced for phase as well as magnitude with the main bridge arms. This balance can best be achieved by connecting the detector temporarily between *a* and *b* or *c*.

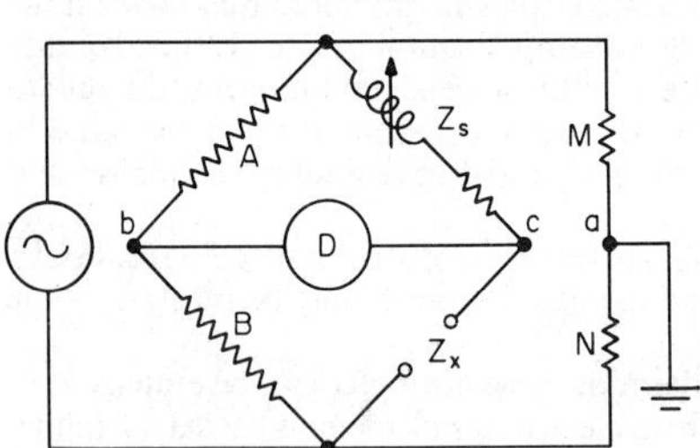


**FIG. 3-69** Illustrating the Wagner arms.

*Substitution methods* should be used where possible to eliminate systematic errors in ratio arms and adjustable balancing components, as well as the effects of stray admittances and the residuals in arms that are unchanged by the substitution. The technique used is essentially a difference measurement, and substitution methods are particularly effective when the known standard that is substituted has very nearly the same value as the unknown.

***175. Resonance methods*** are used for capacitance measurements at radio frequencies, a coil of known inductance $L$ (microhenrys) at a known wavelength $\lambda$ (m) being employed. Resonance is produced by varying $\lambda$ and is detected with a thermocouple ammeter. At resonance $C_x = 0.2815\lambda^2/L$ in picofarads. $\lambda$ and $L$ need not be known if a substitution method is used in which an adjustable capacitor with a range that includes $C_x$ is connected in place of the unknown capacitor and adjusted to resonance without altering the frequency, so that $C_x = C_s$. The leads used to connect the capacitors into the circuit must not be changed in length or position in making the substitution.

***176. A cavity resonator*** can be used at frequencies of the order of 200 to 1000 MHz for measuring the characteristic of insulating material placed between electrodes within the cavity. The arrangement is shown in Fig. 3-70. Resonance is established with excitation of a small loop

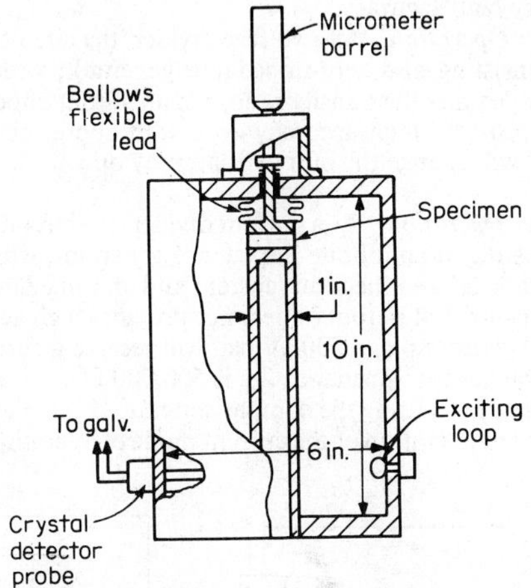


**FIG. 3-70** Diagram of a cavity resonator for about 200 MHz.

of wire within the cavity by connection to an oscillator, and resonance is shown by a crystal-rectifier probe connected to an electronic voltmeter.

## Inductive Dividers

***177. General.*** Inductive dividers are employed in precise voltage- and current-ratio applications. The ratios are used for comparing impedances and for calibrating devices with known nominal ratios such as other dividers, synchros, and resolvers. A divider usually consists of an autotransformer adjustable in decade steps. Such transformers, with high ratio accuracy for voltage or current comparison, have been made by using high-permeability magnetic-core materials and ingenious winding and connection techniques. Such a transformer can be represented electrically by the equivalent circuit of Fig. 3-71. The components of this circuit can be

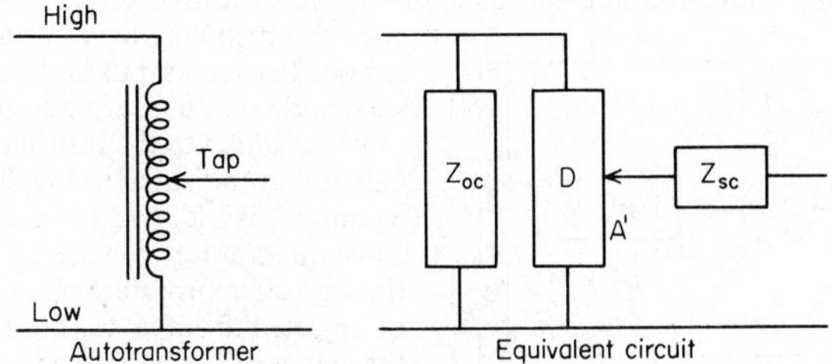


**FIG. 3-71** Autotransformer and equivalent circuit.

measured directly and will predict the performance of the divider. *D* is a perfect divider with infinite input impedance and zero output impedance. $A'$ is the transfer ratio of *D*, the ratio of the voltage between the open-circuited *tap point* and the *low* end to the voltage between the *high* and *low* ends. $A'$ is also the ratio of the short-circuit current between the high and low ends to the current into the tap point and out of the low point. $Z_{oc}$ is impedance between high and low points, with the tap point open-circuited. This impedance is quite high and is a function of input voltage and frequency primarily. Its major components are the winding capacitance, charging inductance, and leakage reactance in parallel. $Z_{sc}$ is the impedance between tap and low points, with the high and low points short-circuited together. This impedance is quite low and is a function of frequency and setting. Its major components are winding and contact resistances and the leakage inductance in series. The autotransformer configuration can produce voltage and current ratios of very high accuracy.

***178. Voltage-Divider Operation.*** As a voltage divider, the circuit can be represented by a Thévenin equivalent consisting of a zero-impedance generator, with a voltage which is the product of the input voltage times the transfer ratio, and an output impedance equal to $Z_{sc}$. This low-output impedance provides high accuracy even with appreciable load admittance. For example, a 5000-Ω load will change the output voltage by only 0.1% if the output impedance is 5 Ω.

***179. Current-Divider Operation.*** As a current divider, the circuit can be represented by a Norton equivalent consisting of an infinite-impedance generator, with a current which is the product of the transfer ratio $A'$ and the input current and an impedance equal to $Z_{oc}$ in shunt across the output. This high-shunt output impedance provides high accuracy even with appreciable load impedance. For example, a 500-Ω load will receive a current only 0.1% less than short-circuit current if the output impedance $Z_{oc}$ is 500,000 Ω.

***180. Impedance comparison,*** using the comparison ratio $A'/(1 - A')$, can be accomplished in either the voltage mode of operation or the current mode of operation, as shown in Fig. 3-72.

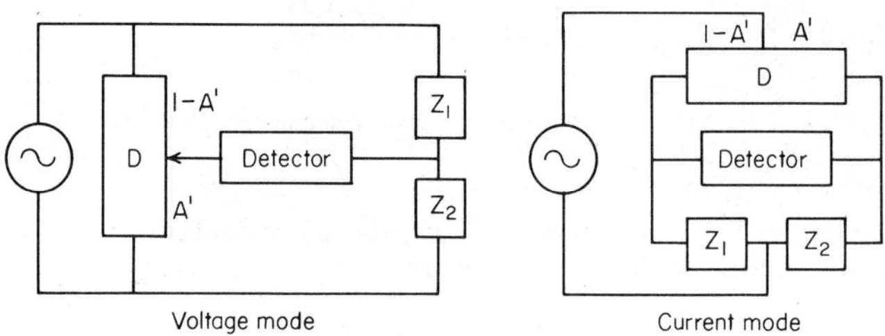


**FIG. 3-72** Voltage mode (left) and current mode for impedance comparison.

For impedance comparison, the divider impedances $Z_{oc}$ and $Z_{sc}$ are of no consequence. In the voltage-ratio mode, $Z_{oc}$ is outside the bridge circuit, and at null no current is drawn through $Z_{sc}$. In the current-ratio mode, $Z_{sc}$ is outside the bridge circuit, and at null there is no voltage across $Z_{oc}$. In either mode, the balance equation is $Z_2 = Z_1 A'/(1 - A')$.

***181. Quadrature Transfer Ratio.*** A well-designed autotransformer type of inductive divider has very little phase difference between its input and output signals in most applications. Impedances being compared and other circuits being measured do, however, form divider circuits which may have appreciable phase difference between input and output signals. For such bridge circuits, a quadrature component *B* must be added to the transfer ratio to obtain a balance, as indicated in Fig. 3-73. The balance equation is $Z_2 = Z_1(A' + jB)/(1 - A' - jB)$. This quadrature voltage may be developed through an appropriate phase-shifting network or injected from a calibrated-divider source through a suitably shielded transformer.

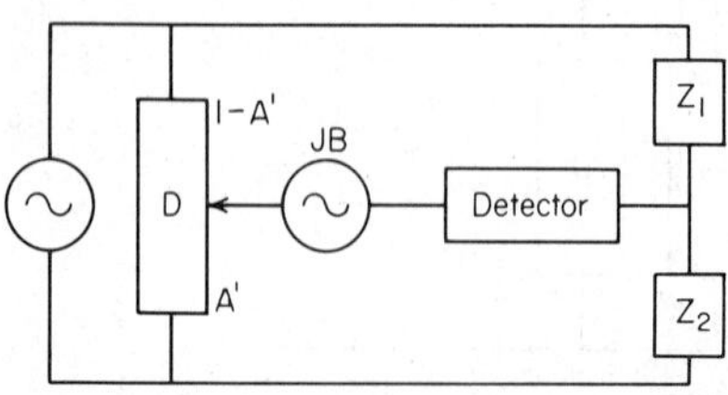


**FIG. 3-73** Quadrature transfer ratio.

***182. Synchro and resolver calibration*** can

be accomplished by supplying a signal to the rotor and measuring the ratio of the resulting stator voltages. The synchro test circuit is shown in Fig. 3-74*a* and of the resolver in Fig. 3-74*b*. For these applications, the divider impedances can usually be ignored, but for measurements of the highest accuracy and at high frequencies, $Z_{oc}$ may cause measurable loading on the stator circuit supplying it.

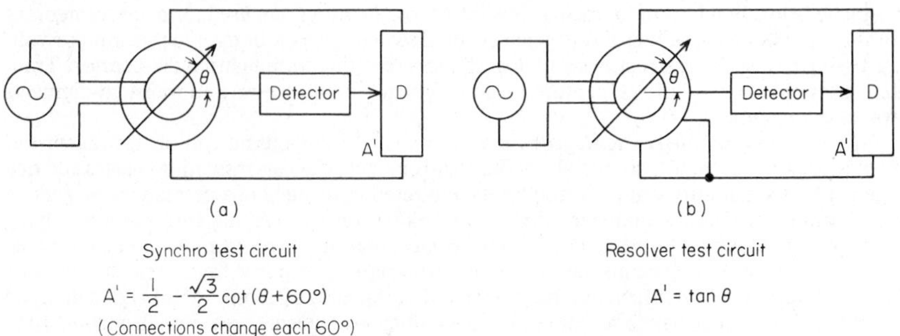


FIG. 3-74 Synchro and resolver calibration.

***183. Calibration*** of other dividers may be checked in terms of a reference divider by using the circuit of Fig. 3-75. A calibrated inductive divider *D* and a calibrated source of quadrature voltage *B* are required to give a null for the desired setting on the unknown divider. The balance equation is $A_u = A + a + j(b + B)$. Corrections for the inductive divider are given in terms of the linearity deviation *a*, the quadrature deviation *b*, and the nominal ratio *A* (usually the dial setting); thus, $A' = A + a + jb$.[1]

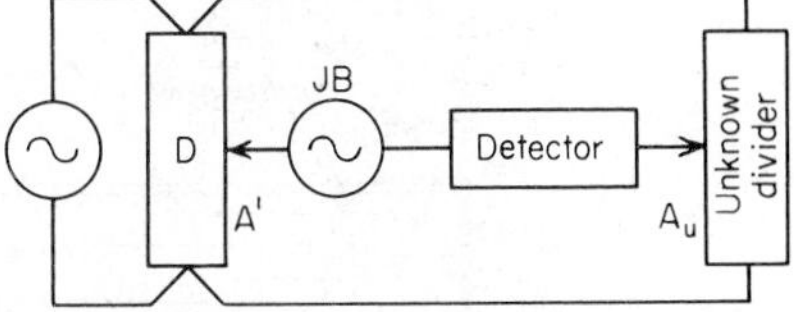


FIG. 3-75 Divider calibration.

***184. Specifications*** of the performance of an inductive voltage divider are stated in terms of the characteristics of its equivalent circuit.

**a.** *Ratio accuracy* is $A' = A + a + jb$, where *A* is the nominal ratio given by the dial setting, usually corresponding to the turns ratio of the autotransformer.

**b.** *Voltage limits* are imposed by (1) the input voltage at which the core saturates and the divider output becomes distorted and (2) the ability of the winding insulation to withstand breakdown.

**c.** *Operating Frequency.* The voltage-frequency limit and the $Z_{oc}$ and $Z_{sc}$ impedances can be varied by core and winding choices so that dividers can be designed for optimum operation in different frequency bands and at different excitation levels. Typical values of the various characteristics are given in the table for two designs which give their best performance at the frequency stated.

| Frequency | Accuracy | | $Z_{oc}$ | | | $Z_{sc}$ | | Volt-frequency limit |
|---|---|---|---|---|---|---|---|---|
| | $a_{max}$ | $b_{max}$ | $R$ | $L$ | $C$ | $R_{max}$ | $L_{max}$ | |
| 60 Hz | $10^{-6}$ | ...... | 1 MΩ | 2 kH | 10 nF | 5 Ω | 500 mH | 2.5 × *f* |
| 1 kHz | $10^{-6}$ | $10^{-5}$ | 500 kΩ | 200 H | 1 nF | 3 Ω | 100 mH | 0.35 × *f* |

[1] Hill and Deacon discuss design and calibration of inductive voltage dividers in *Proc. IEE* 1968, vol. 115, p. 727; Sze describes a method for self-calibration of a voltage divider in *NBS J. Res.* 1968, vol. 72C, p. 49.

## Waveform Measurements

***185. Methods.*** The instantaneous variations of current and voltage in a circuit can be measured by oscillographs, whose basic operating principle may be either that of a D'Arsonval galvanometer whose inertia is low enough to permit it to follow the variations or that of an electron beam which has no sensible inertia and whose deflection is governed by electric or magnetic fields. In addition to tracing waveforms, oscillographs are used for measurements of transient phenomena, such as those which occur in switching operations or in the impulse-voltage testing of insulating structures and disturbances resulting from lightning discharges. Transient phenomena may also be captured using digitizing oscilloscopes and transient digitizers (waveform recorders).

***186. The galvanometer oscillograph*** may have a light low-inertia coil or, for higher-frequency response, a pair of thin metal ribbons tightly stretched across insulating bridges and tied together by a small mirror at their midpoints, mounted in the field of a permanent magnet. A light beam from the galvanometer mirror traces its response to varying current on a moving photographic film or, by means of an intermediate rotating mirror, on a stationary viewing screen. Galvanometer elements have been built with natural response frequencies as high as 8 kHz (a more common construction has a resonance frequency of about 3 kHz) and, if damped at about 0.7 of critical, have a response to signals which is practically free from distortion up to about half their resonant frequency; at resonant frequency, the deflection sensitivity has decreased to about 70% of their dc sensitivity for this damping.

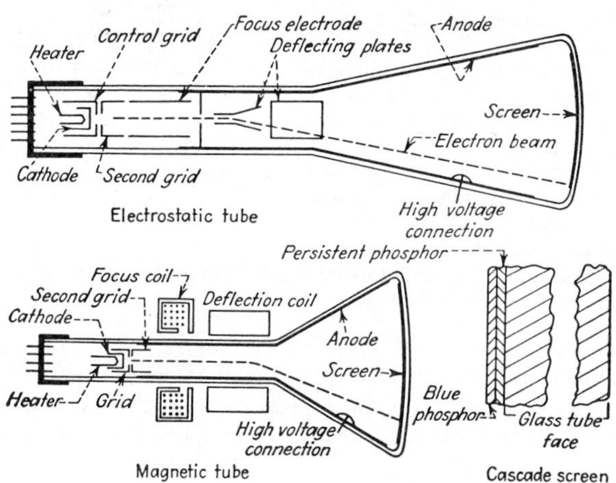


FIG. 3-76 Elements of a cathode-ray oscilloscope.

***187. Cathode-ray-oscillograph*** tube elements are shown in Fig. 3-76. Electrons emitted by a heated cathode are accelerated toward an anode which has a small central aperture. A grid system serves to concentrate the electrons into a narrow beam and focus it into a small spot on a fluorescent screen at the far end of the tube. The grid nearest the cathode controls the beam current and the brightness of the spot on the fluorescent screen. After going through the focusing electrode system, the beam passes between two plate pairs set at right angles to one another and is deflected by the electric fields impressed on the plates. If a sawtooth voltage is impressed on the plate pair controlling horizontal deflection and is synchronized with an ac signal impressed on the vertical-deflection plate pair, the waveshape of the signal is traced out on the fluorescent screen. Since the deflection sensitivity of this system is relatively low, the plate pairs are preceded by amplifiers which control the actual magnitude of deflections on the screen. A variety of circuits is required to control the screen displays for repetitive or nonrepetitive signals, to initiate, form, and synchronize a variety of sweep-signal voltages, etc. Modern oscilloscopes are very flexible tools for studying the nature of both continuous and transient signals and provide uniform response well into the high-frequency region, capable of reliable calibration in both voltage and frequency or time interval.

Digitizing oscilloscopes and transient digitizers (waveform recorders) represent a new class of instrument responding to the need for measurements fast enough to characterize integrated circuit parameters (where speeds are high because of extremely small dimensions of circuit paths, i.e., involvement of small capacitances and inductances) at high accuracy for quality-control purposes. The waveform recorder typically consists of a wideband amplifier; high-speed sample-and-hold and analog-to-digital conversion circuits; temporary (or buffer) memory; display; and digital outputs, all under the control of internal digital logic circuits. The display is generally a CRO fed with an analog signal reconstructed from the digital information, using operator-selected time scales. Better waveform recorders are stated by their manufacturers to have 20 to 100-MHz sampling rates and 0.25 to 0.1% amplitude resolution. As of 1983, neither performance standards nor means of verifying performance (physical standards, measurement techniques, etc.) have been formulated for this class of instrument.

## Frequency Measurements

***188. Reed-type frequency meters*** have a number of steel strips rigidly fastened to a bar at one end and free to vibrate at the other. These strips are located in the field of an electromagnet which is energized from the circuit whose frequency is to be measured, as shown in Fig. 3-77. The strips have been accurately adjusted by solder weights to resonant vibration frequencies that differ by ¼ or ½ Hz, and the one with a period corresponding to the alternations of the voltage will be set into vibration. The free ends of the strips or reeds are turned up and painted white so that the reed which is vibrating will be indicated by an extended white band or blur.

***189. The Weston frequency meter*** is shown in Fig. 3-78, where 1, 1 and 2, 2 are fixed coils, 90° apart, and *c, c* is the movable element consisting of a simple, soft-iron core mounted on a shaft, with no control of any kind. Coil 2, 2 is connected in series with a noninductive resistance $R_2$ and coil 1, 1 in series with an inductance $X_1$. A second noninductive resistance $R_1$ is connected in parallel with 1, 1 and $X_1$. A second inductance $X_2$ is connected in parallel with 2, 2 and $R_2$. The soft-iron core takes up the position of the resultant field produced by the two coils. When the frequency increases, the current decreases in 1, 1 and increases in 2, 2 thus shifting the direction of the resultant field and the position of *c, c* to which the pointer is attached. The opposite effect takes place when the frequency is decreased. The series inductance *X* serves merely to damp the higher harmonics which are present if the voltage waveshape is distorted.

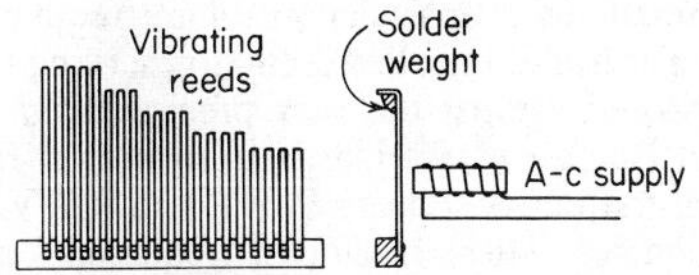


FIG. 3-77 Frequency meter, reed type.

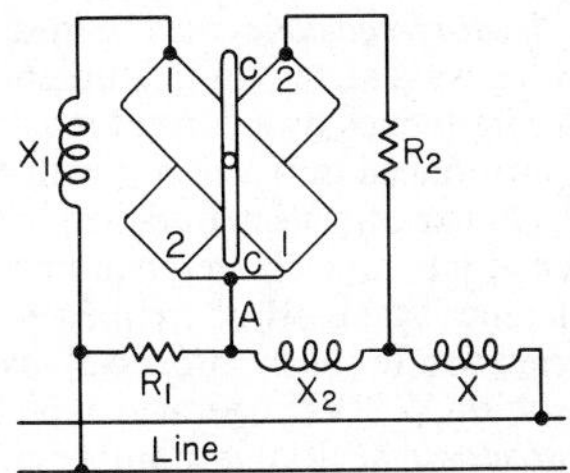


FIG. 3-78 Circuits of the Weston frequency meter.

***190. Resonant circuit meters,*** operating from circuits containing inductance and capacitance, can be made sensitive enough to indicate frequency variations of 0.01 Hz or less. One form, used by the General Electric Company, is shown in Fig. 3-79. In a 60-Hz instrument, one circuit is adjusted for resonance at about 70, another at about 58, and a third at about 36 Hz. The latter two are connected in parallel and then to coil *A*; the first circuit is in series with coil *A'*, both coils being in series with the field *F*. With the center of a 6-in scale marked for 60 Hz, end-scale deflection is obtained for variations of 5 Hz from the central value.

***191. The Leeds & Northrup frequency recorder*** uses a Wien-bridge network with a slide wire between the resistance ratio arms *A* and *B*, as shown in Fig. 3-80. The other two arms have capacitors $C_1$ and $C_2$, one with series and the other with parallel resistance. With a change in frequency, the reactance of one arm is increased and the other is decreased; the movement of the

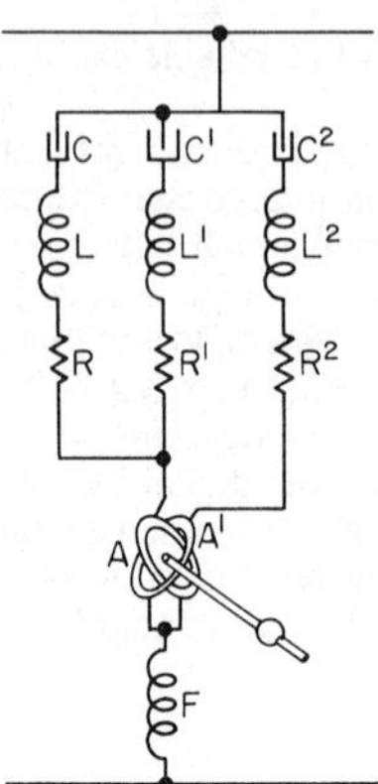


FIG. 3-79 General Electric resonating frequency meter.

detector $G$ along the slide wires $S_1$ and $S_2$ required to restore balance brings the recording pen to a new position corresponding to the changed frequency.

***192. Precise frequency control*** is also accomplished with resonance techniques.

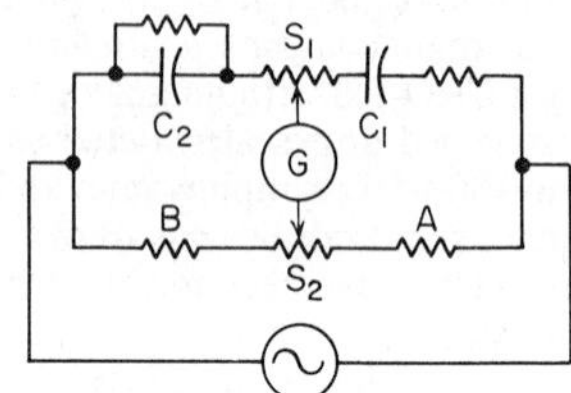


FIG. 3-80 Circuit of Leeds & Northrup frequency recorder.

Small-range indicators or recorders can be built as relays to monitor the frequency of a power system or generator, injecting an appropriate signal into a control system to restore frequency to a particular value. Such control may be made precise enough for use of the system frequency for electric-shock operation. Any tendency to frequency drift may be detected and corrected at the source by comparing an electric clock with a precise pendulum clock or one driven by a quartz-crystal oscillator.

***193. Radio frequencies*** may be measured directly or indirectly. Direct measurement may be made with a wavemeter, an instrument with a tunable circuit and an ammeter to indicate the resonance frequency by a current maximum. In the indirect method, the unknown-frequency signal is introduced into a circuit with a precisely known frequency, and the beat frequency is counted. Quartz crystals maintained in temperature-regulated ovens will control the frequency of an oscillator to much better than 1 part in $10^6$. Such a crystal-controlled oscillator, serving as a local reference standard of frequency, can be monitored against the very precise *standard* frequencies continuously broadcast by the National Bureau of Standards from its low-frequency station WWVL, operated at 60 kHz or its high-frequency stations WWV and WWVH, which broadcast at a large number of higher frequencies. These broadcast frequencies are controlled by crystals operating under conditions that are most favorable to stability and are, in turn, monitored against the frequency of an atomic-beam resonator. The transmitted frequencies, as sent from the bureau stations, are accurate to about 1 part in $10^{12}$. Frequencies from these broadcasts are modified somewhat in transmission by diurnal and moment-to-moment variations in the ionosphere, and their accuracy as received may be reduced by more than an order of magnitude.

***194. Audio frequencies*** can be measured with a frequency-sensitive bridge, such as the Wien bridge with parallel- and series-connected capacitance-resistance arms; or they can be conveniently observed with a cathode-ray oscilloscope, if a known reference frequency is available. One set of plates of the oscilloscope is excited by the known- and the other set by the unknown-frequency signals. If the two frequencies have an exact, simple fractional relation, the Lissajous figure formed on the screen is stationary. For a 1/1 relationship, the pattern is an ellipse; for other fractional relationships, the pattern is more complicated, the relationship being determined from the number of loops. If the relationship cannot be represented by a simple fraction, the pattern will change continuously and a count of the beat frequency is made over a measured time interval.

Electronic counters are widely used for frequency measurements. They work by counting the number of cycles of an input signal, or events, which occur in a very accurately known time interval (gate time). The gate time is based on the output of an internal standard oscillator (clock) or, optionally, on a reference frequency signal input to the counter. Most counters of laboratory quality can also be used to measure the period of low-frequency signals, time intervals, the ratio of the frequencies of two input signals, and a total number of events. They also afford control of triggering, thus enabling the user to set trigger levels and slopes, noise rejection levels, and input attenuation levels. Output is via digital display, ranging from six to nine digits, and (usually) high-speed digital computer interface. Accuracies of frequency measurements are usually stated by the manufacturer to be ± clock accuracy ± 1 count. Most laboratory-grade counters can be equipped with high-stability crystal-based clocks, mounted in temperature-controlled ovens, and are stated to have drift rates as low as $2 \times 10^{-8}$ per month. The frequency ranges covered are from nearly dc, directly or via period measurements, to as high as 500 MHz directly and to over 30 GHz using heterodyning techniques.

## Slip Measurements

***195. The slip*** of a rotating ac machine is the difference between its speed and the synchronous speed, divided by the synchronous speed; slip is usually expressed in percent. It may be computed from the measured speed of the machine and the synchronous speed, but direct methods are more accurate.

***196. Millivoltmeter Method.*** If sufficient stray field is produced by the current in the secondary of an induction motor, a dc millivoltmeter connected to an adjacent coil of wire or across the motor shaft or frame will oscillate at slip frequency, each swing being one pole slip. In motors with wire-wound rotors the millivoltmeter may be connected across the rotor slip rings.

***197. Dooley's Method of Measuring Slip.*** One form of device for indicating the slip of an induction motor is shown diagrammatically in Fig. 3-81. A small cylinder made of conducting material and in two parts, each insulated from the other, is mounted in a frame. Four small brushes, 1, 2, 3, and 4, bear upon the cylinder as shown. Brushes 3, 4 are connected through resistance $r$ across one phase of the supply circuit, and brushes 1, 2 are connected to a low-reading dc ammeter $I$. Each time the brushes 1, 2 bridge the insulating strip as the cylinder rotates, the circuit is completed in alternate directions through the ammeter. The cylinder should have as many segments as the motor has poles. The ammeter will indicate a constant current at synchronous speed and an oscillating current for any speed above or below synchronism, because the impulses of current through brushes 1, 2 will occur at the same point on the wave at synchronous speed and at constantly advancing or retarding points for other speeds. Thus, the ammeter will be reversed each time the motor loses one-half cycle and will reach a maximum positive value each time the motor loses one complete cycle. If the motor loses $n$ c/min, then the slip in percent $= 100n/60f$, where $f =$ frequency of the system in cycles per second (hertz).

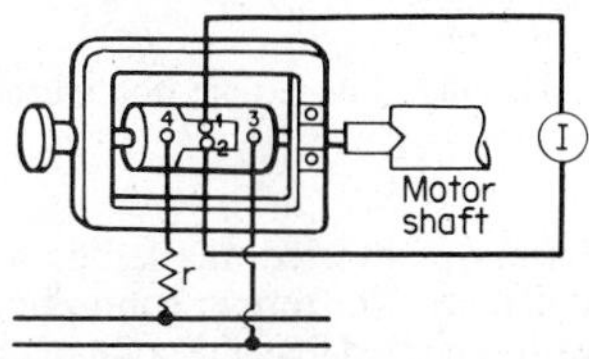


**FIG. 3-81** Slip-measuring device.

***198. Stroboscopic Method.*** In Fig. 3-82 a black disk with white sectors, equal in number to the number of poles of the induction motor, is attached to the induction-motor shaft. It is observed through another disk having an equal number of sector-shaped slits and carried on the shaft of a small self-starting synchronous motor, in turn fitted with a revolution counter which can be thrown in and out of gear at will. If $n$ is the number of passages of the sectors, then $100n/n_s n_r =$ slip in percent, where $n_s =$ number of sectors and $n_r =$ number of revolutions recorded by the counter during the interval of observation. For large values of slip the observations can be simplified by using only one sector ($n_s = 1$); then $n =$ slip in revolutions.

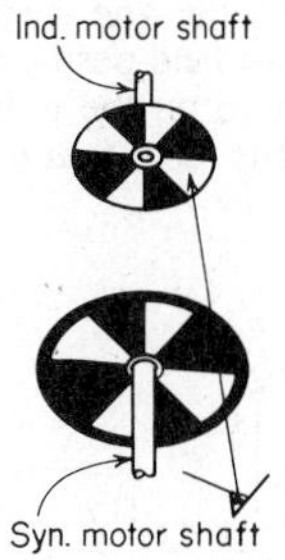


**FIG. 3-82** Slip measurement by stroboscopic method.

With a *synchronous light source* to illuminate the target on the induction-motor shaft, the synchronous motor is no longer necessary. An arc lamp connected across the ac supply may be used, but the carbons must be readjusted from time to time. A neon lamp makes a satisfactory source of light when the general illumination is not too bright. A portable stroboscope may consist of a gaseous discharge tube backed by a parabolic reflector to concentrate the light beam, and an adjustable-frequency voltage source to trigger the flashlamp synchronously. The flash can also be triggered externally. Light output measured 1 m from the lamp may exceed $10^6$ candela and flash duration may be as low as 0.5 $\mu$s.

***199. Synchronizing.*** In order to connect any synchronous machine in parallel with another machine or system, the two voltages must be made equal and the machines must be synchronized, that is, the speed so adjusted that corresponding instantaneous values on the two waves are reached at the same instant, when they will be in exact phase. Furthermore, with polyphase machines, the direction of phase rotation must assuredly be the same. This, however, is usually made right once and for all when the machines are installed, the phases being so connected to the switches that the phase rotation will always be correct.

***200. The lamp method of synchronizing*** is the simplest. The principle of lamp synchronizers is shown in Fig. 3-83, where $a$, $a_1$ are the sources being connected in parallel and $t$, $t_1$ are transformers, the secondaries of which are connected in opposition through incandescent lamps $l$, $l'$. When the two sources are in synchronism, the secondary emfs neutralize each other, and the lamps will be "dark." As the phase difference increases, the current through the lamps will increase, reaching a maximum at 180° of phase difference. If the machines run at different speeds, the lamps will "flicker." If the secondary of one transformer is reversed, the lamps will be brightest at synchronism and dark at 180° of phase difference. The former connection is preferable, because the point of total "darkness" is more easily detected than the point of maximum "brightness." A voltmeter may be substituted for the lamps by connecting it so that synchronism is indicated when the reading is a maximum. The disadvantage of this method is that it does not indicate which frequency is the higher. *Synchronism indicators* are instruments which not only overcome this objection but indicate the point of synchronism more accurately.

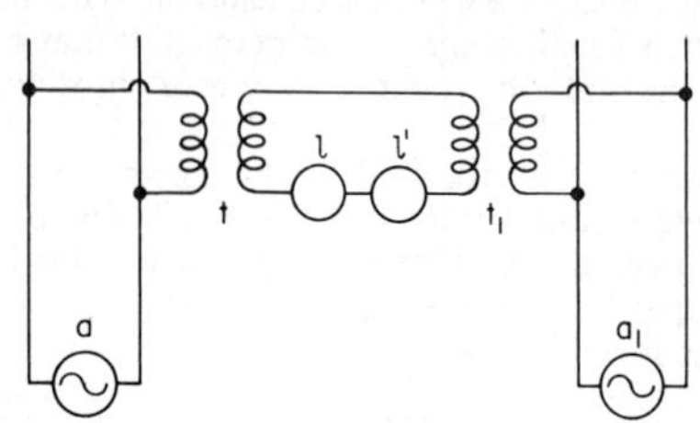


**FIG. 3-83** Connections for synchronizing with lamps.

***201. The principle of the Westinghouse synchroscope*** is shown in Fig. 3-84, where a rotating field is produced by the coils $M$ and $N$ connected to the buses through the reactance $P$ and the resistance $Q$, respectively. An iron vane $A$, free to rotate, is mounted in this rotating field and magnetized by the coil $C$, which in turn is connected across the incoming machine. As the vane is attracted or repelled by the rotating field from $M$ and $N$, it will take up a position where this field is zero at the same instant that the field from $C$ is zero. Hence the position at any instant indicates the difference in phase. When the two frequencies are different, this position is constantly changing, and the pointer will rotate in the "fast" or "slow" direction, coming to rest again at the zero-field position when the frequencies are equal. In a larger type, the split-phase winding is placed on the movable member, similar to the arrangement shown in Fig. 3-85, which represents the General Electric synchroscope.

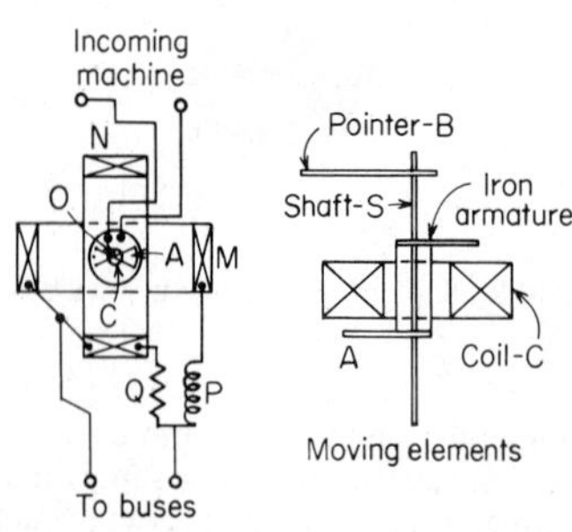


**FIG. 3-84** Circuits of Westinghouse synchroscope.

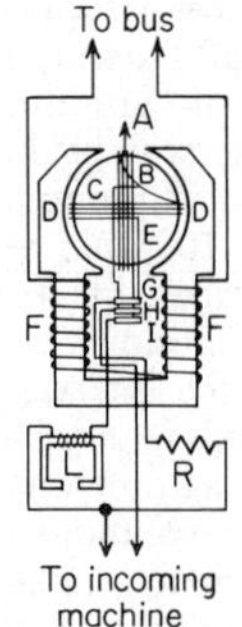


**FIG. 3-85** Circuit of General Electric synchronoscope.

***202. In the Weston synchroscope*** there is no iron in the instrument, and the moving element is not allowed to rotate. The elements are practically the same as in an electrodynamometer wattmeter. The fixed coils are connected in series with a resistance and to the buses. The moving coil is connected in series with a capacitor and the incoming machine. The two circuits are adjusted to exactly 90° difference in phase. At synchronism there is no torque, and the moving coil is held at the zero position by the control spring. If the frequencies are the same but there is a phase difference, a torque will be exerted and the coil will move to a position of balance at the right or left ("fast" or "slow"). If the frequencies are different, the torque will continually vary and the pointer will oscillate over the dial. A synchronizing lamp illuminates the scale simultaneously.

***203. The phase sequence of a 3-phase system*** is often desired. Figure 3-86 shows two lamp

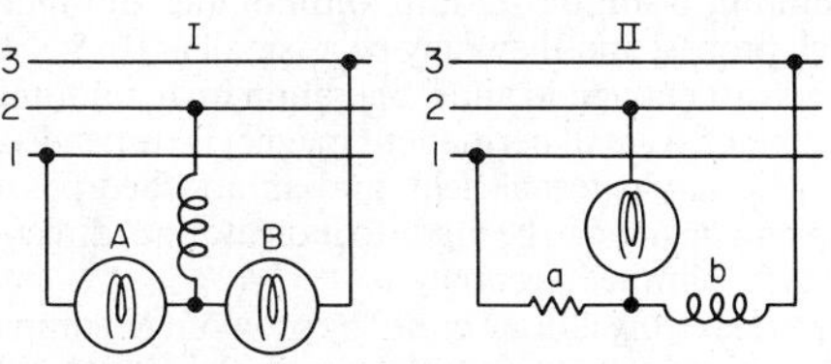


**FIG. 3-86** Phase-sequence indicators.

methods. In I, two lamps and a highly reactive coil, such as the potential coil of a watthour meter, are used. The bright lamp indicates the particular phase sequence. In arrangement II, a noninductive resistance and a reactive coil of equal impedance are used in conjunction with a lamp, the brightness of which indicates the sequence.

## Magnetic Measurements

***204. The two classes of magnetic measurements*** are: *field measurements,* such as the earth's field or the field in the air gap of a magnet, and *measurements to determine the characteristics* of magnetic materials.

***205. Magnetic field measurements*** are commonly made by induction methods in which a coil is placed with its plane perpendicular to the field. Removing the coil to a point of zero flux or reversing the coil will induce in it an emf that can be measured by the ballistic deflection of a galvanometer (see Par. **11**) in terms of its flux-linkage sensitivity (when operating in a circuit having the resistance of the search-coil circuit). In this measurement $\int e\,dt = N\,\Delta\phi/10^8$, where $\phi$ is the flux in maxwells enclosed by the coil, and $N$ is search-coil turns. The flux density $B$, in gauss, is $\phi/a$, where $a$ is the coil area in square centimeters.

The flux-linkage sensitivity of the galvanometer under the operating condition can be determined with the aid of a mutual inductor, with the galvanometer in the secondary circuit and a known current reversed in the inductor primary. Here $\int e\,dt = 2MI$ volt-seconds, where $M$ is mutual inductance in henrys and $I$ is primary current in amperes. A Grassot-type fluxmeter can be used in place of a ballistic galvanometer. This is essentially a ballistic instrument in which restoring torque is reduced substantially to zero so that the deflection remains steady after the change in flux linkages.

Low field measurements are also made with magnetometers. This instrument uses a strip of high-permeability, low-coercive-force material (usually supermalloy) with an ac excitation coil that drives the material into saturation each half cycle at a frequency of a few kilohertz. A second-harmonic detector coil on the same strip will sense a bias field to which the assembly is exposed. A third coil on the strip supplies a measured offset ampere-turns to return the detector to zero, providing a very sensitive field measurement device. This is widely used in earth's field and other low-level field measurements. A portable flux-gate magnetometer, in which the vector-magnetic-field component at the sensor is neutralized by a current in a solenoid surrounding the sensor, has a resolution of 1 gamma at the neutralizing control. The magnitude (in gamma) of the neutralizing field is indicated on manually operated digital dials, and any difference between ambient field-vector component and neutralizing field is indicated on a meter whose range may be selected between 25 and $10^4$ gammas. A nondirectional magnetometer system is based on proton gyromagnetic ratio and the functional relation between ambient field and resonance frequency in the sensor. This type of magnetometer is also used to sense small variations in the local earth's field.

Measurement of higher fields (20 to 20,000 G) and fields in spaces too confined for search coils are frequently made with Hall-effect gaussmeters. In a thin strip or film of a metal having a large Hall-effect coefficient and carrying a current, two points on opposite sides of the strip can be found between which there is no potential difference. If a magnetic field is then applied at a right angle to the plane of the strip, a potential will exist between these points which is proportional to the field. Germanium, bismuth, indium antimonide, and indium arsenide are the common materials for such probes; and they may be as small as $0.15 \times 1.2$ mm. Response of many of these instruments is fast enough to allow operation up to midrange audio frequencies.

In another type of gaussmeter, a small permanent magnet is suspended between taut bands. It will attempt to line up with any external field, and an attached pointer and scale can be calibrated in kilogausses. Such a device can be made to indicate both direction and magnitude of the external field to a somewhat limited accuracy.

***206. DC magnetic material testing*** is done either by providing a complete closed path of the sample material on which exciting and sensing windings can be placed, or by utilizing a "yoke" type of apparatus to furnish excitation to a small sample with its own sensing winding. Closed-loop samples may be a toroid composed of a stack of punched rings, a toroid made by wrapping tape into a spiral, or a closed loop made by stacks of strip samples assembled with overlapped ends in an Epstein frame.[1]

This arrangement, in the form of a square, has an excitation winding and a sensing winding distributed along the four sides of the square to enclose the sample. The geometry and construction of these coils is detailed in ASTM Standard A343, part 44 of the Annual Book of ASTM Standards. Punched-ring samples are not usually considered satisfactory for oriented materials, while either spiral-wrapped tape toroids or Epstein strip samples can be used in either oriented or nonoriented materials. In any of these closed-loop samples, the excitation can be determined in terms of the ampere-turns that supply it. If the mean diameter of the sample is large compared with its radial width, the excitation is calculated as $H = 0.4\pi NI/l$ oersteds, where $N$ is the number of turns in the magnetizing winding, $I$ is the current in amperes, and $l$ is the mean path length of the ring in centimeters. In using Epstein samples, it is necessary to make an assumption as to the actual magnetic-path length. This is normally taken as 94 cm in the 25-cm Epstein frame. Figure 3-87 shows the circuit for dc tests using either ring or Epstein samples. A mutual inductor is included for calibrating the ballistic galvanometer; the series and parallel resistors in the galvanometer circuit permit adjustment to make the system direct-reading in appropriate units while preserving a desired galvanometer damping; resistors in the excitation circuit permit reversal or step changes at a desired ampere-turn level. Both excitation and test windings on the sample should be uniformly distributed.

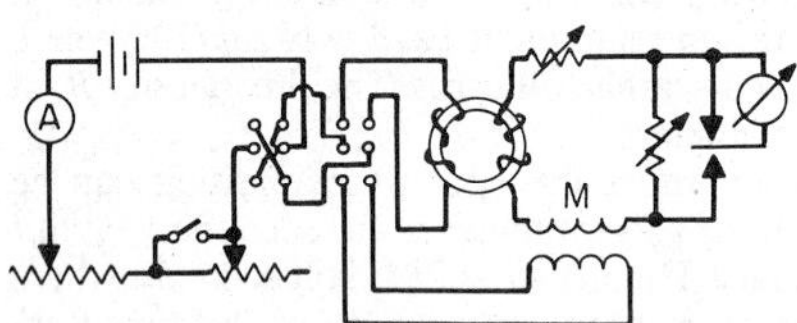

**FIG. 3-87** Circuit for testing ring samples.

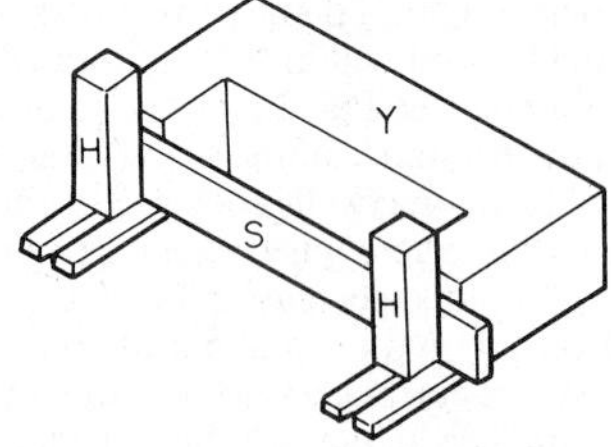


**FIG. 3-88** Magnetic circuit of Fahy permeameter.

***207. Permeameters*** are used for small samples and for "hard" magnetic materials which cannot be driven to a sufficiently high excitation by readily applied turns on closed-loop

[1] ASTM Standard A343.

specimens. Basically, all types of permeameters utilize heavy coils and large-cross-section yokes to provide a high excitation level in small samples. Figure 3-88 shows the magnetic circuit of the Fahy Simplex permeameter. The yoke $Y$ is composed of silicon-steel laminations and carries the excitation winding. Two soft-iron blocks $H$ make contact with the sample $S$. Clamps hold these blocks and the sample in place against the pole faces. The $B$ coil for measuring induction is wound on a brass form which surrounds the sample and extends over the entire length between the pole pieces. A uniformly wound $H$ coil on a nonmagnetic form extends horizontally between the upper ends of the $H$ blocks. Excitation furnished by a coil on the yoke is measured by the $H$ coil. Induction is measured by the $B$ coil surrounding the sample. Either a ballistic galvanometer or a fluxmeter can be used for these measurements. Several other forms of permeameter have been used, the medium-$H$ and the high-$H$ (Sanford-Bennet) and modifications of the Fahy. All use a large heavy yoke for excitation and some form of $B$ and $H$ coils for measurement. These devices all give individual point-by-point data from which hysteresis loops are plotted.

There is increasing use of complete plotting systems for drawing magnetization curves and dc hysteresis loops. Such systems use a magnet assembly with tapered pole pieces adjustable with a screw drive for excitation of the sample. Both $B$ and $H$ are obtained by means of electronic integrator circuits, as shown in Fig. 3-89. $H$ is obtained from a coil on a nonmagnetic form and $B$ from a coil directly on the sample. Excitation to 20,000 Oe is adequate for most "hard" magnetic materials as well as all "soft" materials. This type of equipment can also operate for soft materials in ring or Epstein samples without the use of the $H$ integrator, using the current in a primary as the $H$-determining quantity. A complete dc hysteresis loop and magnetization curve can be taken in approximately a minute. Samples with cross section as small as two layers of ¼-in audio recording tape can be satisfactorily plotted.

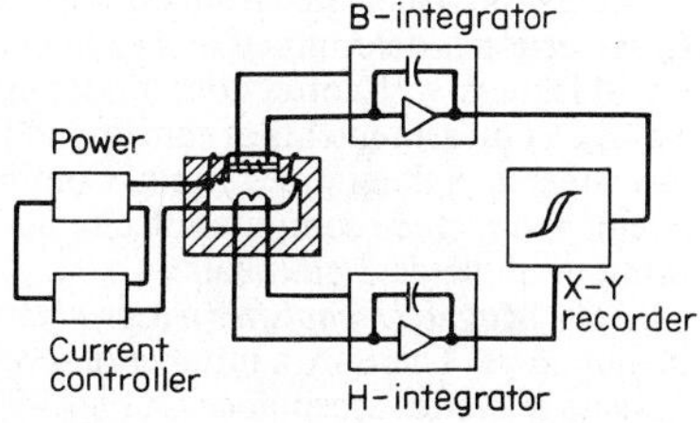

**FIG. 3-89** Loop tracer for hard materials.

***208. Magnetic susceptibility*** testing designates those measurements which require much more sensitive apparatus than the methods described above. Such tests are made by measuring the minute mechanical forces experienced when the sample is in a nonuniform field. All these systems—the Gouy, the Faraday, and the Thorpe-Senftle method—consist of a strong field in which the sample is placed and weighed. They differ in the method of obtaining a calculable nonuniform field.

***209. AC magnetic materials testing*** consists commercially in the determination of ac permeability and core loss in sheet materials. Substantially all such testing is done either in Epstein-frame samples or in EI-type laminations. Up to an induction of 6000 G, measurements are made with the modified Hay bridge[1] of Fig. 3-90. Above this level measurements are made by the voltmeter-wattmeter method; Fig. 3-91 shows the circuit of such a test system. $A$ is an ammeter of low impedance; $W$ is a wattmeter with low-current circuit impedance and designed for low-power-factor use; rms $V_m$ and av $V_m$ are, respectively, rms responding and average responding voltmeters of very high impedance; $L_m$ is a mutual inductance used with av $V_m$ to read $I_{peak}$ currents; $L_{mc}$ is a mutual inductance to compensate for the empty-frame mutual inductance of the Epstein frame. In operation, the flux density $B$ is set using the average-responding voltmeter and calculating from the equation $4.444ANfB_{max}/10^8 = 1.11E_{av}$, where $B_{max}$ is the maximum induction in gauss, $A$ is the cross section of the sample, $N$ is the number of turns in the secondary (700 for the standard Epstein frame),

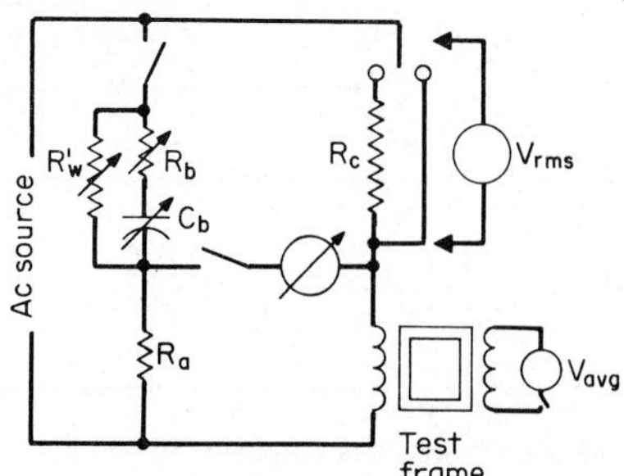

**FIG. 3-90** Modified Hay bridge.

[1] For specification of bridge components, details of operating procedure, and calculations of sample characteristics, see ASTM Standard A347, Part 44 of ASTM Annual Book of Standards.

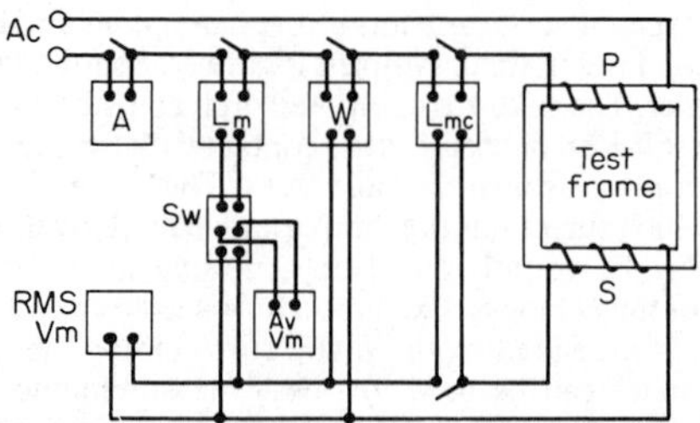

**FIG. 3-91** Voltmeter-wattmeter core-loss test system.

and $f$ is frequency in hertz. The value of $H$ is determined by the formula $0.4NI_{\text{peak}}/L = H$ oersteds, where $N$ is the number of turns in the magnetizing winding (700 for Epstein frame), $I_{\text{peak}}$ is the peak current in amperes (derived from the reading of the voltmeter on the secondary of $L_m$), and $L$ is the magnetic path length (94 cm for the 25-cm Epstein frame).

Core loss is calculated from the wattmeter reading divided by the active weight of the sample. Cross section is determined by weight of sample rather than an actual measurement of lamination thickness, with corrections for density of the material and assumed path length. Voltmeter-wattmeter measurements of core loss and ac permeability ($B_{\text{max}}/H_{\text{peak}}$) are made with the actual instruments in the simple system. Commercial units for high-level production follow the basic circuit and include computation circuits to provide readings directly in the desired units, with printout of the data optional.

***210. Magnetic amplifier material testing*** is a specialized procedure for materials to be used in amplifiers. There are a number of special tests in use on a supplier-user agreement basis that have no universal acceptance. In Fig. 3-92 is shown a circuit which will allow CCFR (constant-

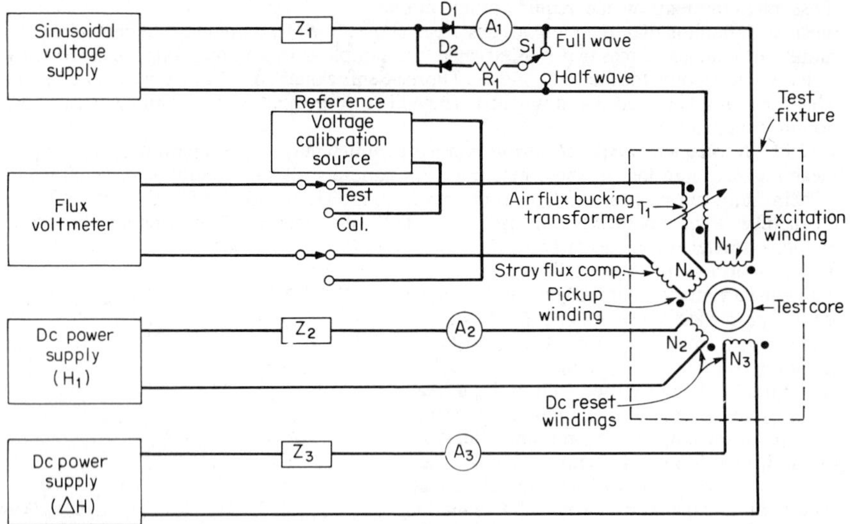

**FIG. 3-92** Magnetic amplifier test circuit.

current flux reset) testing of such cores. ASTM Bulletin A598-69 specifies a number of recommended test points for various materials, using this basic circuit. These tests have the largest acceptance of any presently in use, and most suppliers are equipped to furnish material based on this type of testing. Test frequencies most commonly used are 60, 400, and 1600 Hz.

## Bibliography

***211. Selected List of Reference Literature on Electric and Magnetic Measurements***

Karapetoff, V. (rev. by Boyd C. Dennison) "Experimental Electrical Engineering," 4th ed.; New York, John Wiley & Sons, Inc., Vol. 1, 1923; Vol. 2, 1941.

Radio Instruments and Measurements, 2d ed., *NBS Circ.* 74, 1924.

Harris, Forest K. "Electrical Measurements"; New York, John Wiley & Sons, Inc., 1952.

Drysdale, C. V., Jolley, A. C., and Tagg, G. F. "Electrical Measuring Instruments"; New York, John Wiley & Sons, Inc., 1952.

Banner, E. H. W. "Electronic Measuring Instruments"; London, Chapman & Hall, Ltd., 1954.

Moullin, E. B. "Radio-frequency Measurement," 2d ed.; London, Charles Griffin & Co., Ltd., 1931.

Hartshorn, L. "Radio-frequency Measurements by Bridge and Resonance Methods"; New York, John Wiley & Sons, Inc., 1940.

Batcher, Ralph R., and Moulic, William "Electronic Engineering Handbook"; New York, Electronic Development Associates, 1944.

Code for Electricity Metering, 6th ed.; New York, American National Standards Institute, 1975.

Keinath, G. "Die Technik elektrischer Messgeraete," 3d ed.; Munich, R. Oldenbourg KG, 1929.

Hund, August "High-frequency Measurements," 2d ed.; New York, McGraw-Hill Book Company, 1951.

Knowlton, A. E. "Electric Power Metering"; New York, McGraw-Hill Book Company, 1934.

Terman, F. E. "Measurements in Radio Engineering"; New York, McGraw-Hill Book Company, 1935.

Hague, B. "Instrument Transformers"; London, Sir Isaac Pitman & Sons, Ltd., 1936.

Hague, B. "Alternating-current Bridge Methods," 6th ed.; New York, Pitman Publishing Corporation, 1971.

Canfield, Donald T. "Measurement of Alternating-current Energy"; New York, McGraw-Hill Book Company, 1939.

NBS Monograph 47, Basic Magnetic Quantities and the Measurement of the Magnetic Properties of Materials, 1962.

NBS Special Publication 300, Precision Measurements and Calibration, Vol. 3, Low Frequency; Vol. 4, Radio Frequency, 1968.

Descriptions of new methods of measurement and new measuring apparatus usually appear first in the following technical periodicals [generally, articles sought can be located most quickly with the aid of *Science Abstracts,* which appears annually in two volumes: (A) Physics and (B) Electrical Engineering]:

*Scientific Papers* and *Journal of Research,* National Bureau of Standards, Washington; *Transactions,* American Institute of Electrical Engineers, New York; *Electrical Engineering* (formerly *Journal AIEE*) monthly; *Proceedings,* Institute of Radio Engineers, New York; *Journal,* Optical Society of America, Menasha, Wis.; *Journal,* Franklin Institute, Philadelphia; *Physical Review,* Lancaster, Pa.; *Science Abstracts,* Spon & Chamberlain, New York; *Electrical World,* New York; *General Electric Review,* Schenectady, N.Y.; *Bell System Technical Journal,* New York; *Instruments,* Pittsburgh; *Review of Scientific Instruments,* Lancaster, Pa.; *RCA Review,* New York.

*Journal,* Institution of Electrical Engineers, London; *Proceedings,* Physical Society, London; *Proceedings,* Royal Society, London; *Philosophical Transactions,* Royal Society, London; *Philosophical Magazine and Journal of Science,* London; *Journal of Scientific Instruments,* London; *Archiv für Elektrotechnik,* Berlin; *Annalen der Physik,* Leipzig; *Elektrotechnische Zeitschrift für Instrumentenkunde,* Berlin; *Archiv für technisches Messen,* Munich; *Comptes rendus,* Paris; *Annales de physique,* Paris; *Revue générale de l'électricité,* Paris; *L'Elettrotecnica,* Milan.

# MECHANICAL POWER MEASUREMENTS

## Torque Measurements

***212. Torque is best measured with dynamometers,*** of which there are two classes: absorption and transmission. Absorption dynamometers absorb the total power delivered by the machine being tested, whereas transmission dynamometers absorb only that part represented by friction in the dynamometer itself. Made in a wide variety of forms, typical forms are described in the following paragraphs.

***213. The Prony brake*** is the most common type of *absorption dynamometer.* The torque developed by the machine to overcome the friction is determined from the product of force required to prevent rotation of the brake and the lever arm. The load is applied by tightening the brake band or adding weights.

***214. Dissipation of Heat in Friction Brakes.*** The energy dissipated in the brake appears in the form of heat. In small brakes, natural cooling is sufficient, but in large brakes, special provisions have to be made to dissipate the heat. Water cooling is the usual method, one common scheme employing a pulley with flanges at the edges of the rim which project inward. Water from a hose is played on the inside surface of the pulley and collected again by means of a suitable scooping arrangement. About 100 $in^2$ of rubbing surface of brake should be allowed with air cooling or about 25 to 50 $in^2$ with water cooling per horsepower.

***215. The Westinghouse turbine brake*** employs the principle of the water turbine and is capable of absorbing several thousand horsepower at very high speeds.

In the *magnetic brake,* a metallic disk on the shaft of the machine being tested is rotated between the poles of magnets mounted on a yoke which is free to move. The pull due to the eddy currents induced in the disk is measured in the usual manner by counteracting the tendency of the yoke to revolve. This form of brake can be made in very small sizes and is therefore convenient for very small motors.

***216. The principal forms of transmission dynamometers*** are the torsion and the cradle types.

*In torsion dynamometers,* the deflection of a shaft or spiral spring, which mechanically connects the driving and driven machines, is used to measure the torque. The spring or shaft can be calibrated statically by noting the angular twist corresponding to a known weight at the end of a known lever arm perpedicular to the axis. When in use, the angle can be measured by various electrical and optical methods..

The *cradle dynamometer* is a convenient and accurate device which is extensively used for routine measurements of the order of 100 hp or less. An electric generator is mounted on a "cradle" supported on trunnions and mechanically connected to the machine being tested. The pull exerted between the armature and field tends to rotate the field. This torque is counterbalanced and measured with weights moved along an arm in the usual manner.

## Speed Measurements[1]

***217. Tachometers, or speed indicators,*** indicate the speed directly and thus include the time element. The principal types are centrifugal, liquid, reed, and electrical. In the *centrifugal type,* a revolving weight on the end of a lever moves under the action of centrifugal force in proportion to the speed, as in a flyball governor. This movement is indicated by a pointer which moves over a graduated scale. In the portable or hand type, the tachometer shaft is held in contact with the end of the shaft being measured, and in the stationary type, the instrument is either geared or belted. In the *liquid tachometer* of the Veeder type, a small centrifugal pump is driven by a belt consisting of a light cord or string. This pump discharges a colored liquid into a vertical tube, the height of the column being a measure of the speed.

*Reed tachometers* are similar to reed-type frequency indicators (Par. **188**), the reeds being set in resonant vibration corresponding to the speed of the machine. The instrument may be set on the bed frame of the machine, where any slight vibration due to the unbalancing of the reciprocating or evolving member will set the corresponding reed in vibration. Some forms are belted to the revolving shaft and the vibrations imparted by a mechanical device. *Electrical tachometers* may be either reed instruments operated electrically from small alternators geared or belted to the machine being measured or ordinary voltmeters connected to small permanent-magnet dc generators driven by the machine being tested.

***218. Chronographs*** are speed-recording instruments in which a graphic record of speed is made. In the usual forms, the record paper is placed on the surface of a drum which is driven at a certain definite and exact speed by clockwork or weights, combined with a speed-control device so that 1 in on the paper represents a definite time. The pens which make the record are attached to the armature of electromagnets. With the pens in contact with the paper and making a straight line, an impulse of current causes the pen to make a slight lateral motion and, therefore,

[1] A comprehensive discussion of speed measurements will be found in ASME Power Test Code, Instruments and Apparatus, Pt. 13, entitled Speed Measurements, issued September 1930.

a sharp indication in the record. This impulse can be sent automatically by a suitable contact mechanism on the shaft of the machine or by a key operated by hand. The time per revolution is then determined directly from the distance between marks.

***219. Stroboscopic methods*** are especially suitable for measuring the speed of small-power rotating machines where even the small power required to drive an ordinary speed counter or tachometer would change the speed, also for determining the speed of machine parts which are not readily accessible or where it is not practicable to use mechanical methods or where the speed is variable (see Par. **198** for description of stroboscopic principle).

One convenient form of stroboscopic tachometer employs a neon lamp connected to an oscillating circuit supplied from a 60-Hz circuit, which is adjusted to "flash" the neon lamp at the frequency necessary to make the moving part which the lamp illuminates appear to stand still. Speeds from a few hundred to many thousands of revolutions per minute can be very conveniently measured.

## TEMPERATURE MEASUREMENT

***220. Temperature Scale.*** There is an international temperature scale, IPTS-68 (International Practical Temperature Scale—1968) to which all temperature measurements may be referred. The scale is defined by a set of fixed points to which temperature values have been assigned, and a procedure for calibration and interpolation based on those fixed points. The scale is maintained in the United States by the National Bureau of Standards, and any laboratory may obtain calibrations from NBS based on this scale. In the region from 13 to 904 K, platinum resistance thermometers are calibrated to provide a very precise rendition (0.003 K or better) of IPTS-68. From 904 to 1337 K, a standard thermocouple is calibrated versus the scale with an imprecision of approximately 0.1 K. Above 1337 K, IPTS-68 is realized by calibrating an optical pyrometer at the melting point of gold (1337 K) and extrapolating to high temperatures.

The accuracy of IPTS-68 and the procedures used to calibrate the thermometers described above may be found in the references in the bibliography. The remainder of this section on thermometry will be devoted to thermometry at a less accurate, but more practical level.

***221. Thermoelectric Thermometers (Thermocouples).*** By far the most commonly used thermometer in practical situations is the thermocouple. It consists of a pair of dissimilar electrical conductors (usually wires) joined at two junctions. One junction is maintained at a reference temperature $T_0$ (usually the melting point of ice), while the other is maintained at the unknown temperature $T$. The temperature difference produces a thermal emf which is measured by a potentiometer (see Par. **31**) or a precise digital voltmeter. The latter is especially appealing since it is automatic (i.e., self-balancing), of sufficient resolution, and may easily be interfaced to an automatic data acquisition system.

***222. Metals Used for Thermocouples.*** There are seven combination of metals and alloys most extensively used, and they are designated as type B, E, J, K, R, S, and T. Table 3-4 gives

**TABLE 3-4** Standardized Thermocouples

| Type designation | Nominal composition | Range, °C | Highest $T$ for short-term service, °C |
|---|---|---|---|
| Type B | Pt–30% Rh vs. Pt–6% Rh | 0–1820 | 1700 |
| Type E | Ni–10% Cr vs. Cu–Ni(Constantan)* | −270–400 | 430–870† |
| Type J | Fe vs. Cu–Ni (Constantan)* | −210–760 | 370–760† |
| Type K | Ni–Cr (Chromel)* vs. Ni–Al (Alumel)* | −270–1400 | 870–1260† |
| Type R | Pt–13% Rh vs. Pt | −50–1600 | 1480 |
| Type S | Pt–10% Rh vs. Pt | −50–1700 | 1480 |
| Type T | Cu vs. Cu–Ni (Constantan)* | −270–1400 | 200–370† |

* Proprietary alloys; commercial designation.

† The highest temperature depends on the diameter of the wire. See ANSI-MC96.1-1982, table 7, for further explanation.

their nominal composition, temperature range, and highest suitable temperature for short-term use without significant degradation in performance.

Types R and S may be used for temperatures up to 1480°C, and Type B to 1700°C. It is recommended that the wire diameters exceed 0.5 mm if the thermocouple is to be used for long times at the upper temperature. These thermocouples are recommended for use in air since they are made from noble metals which are resistant to oxidation. They are easily degraded by other conditions, however, so they should be enclosed in a protective sheath. The IPTS-68 is defined in terms of the Type S thermocouple.

Type J may be used in a vacuum, inert, oxidizing, or reducing atmosphere. Again, a large diameter wire (at least 3 mm) is necessary for use at long times in an oxidizing atmosphere.

Type K is used up to 1200°C in inert or oxidizing atmospheres. Type E thermocouples are especially suitable for cryogenic use and may be used in vacuums, inert, oxidizing, or reducing atmospheres. The Type T thermocouple is similar to Type E.

***223. Temperature-EMF Relations for Various Thermocouples.*** Standard emf vs. $T$ tables have been generated and published (ANSI-MC96.1-1982) for the standardized thermocouples discussed above. An abbreviated version is given in Table 3-5. Most manufacturers produce wires of sufficient quality that a thermocouple may be fabricated from the materials given in Table 3-4 and their emf-$T$ relation will deviate only slightly from that given in Table 3-5. The magnitude of the deviation is given in Table 3-6. If greater accuracy than given in Table 3-6 is needed, the thermocouple must be calibrated. A calibration table, accurate to 0.1°C, is provided up to 300°C. The calibration accuracy is typically ±0.5°C for types S, R, and B to 1100°C increasing to ±2°C at 1450°C. Types E, J, K calibrations are roughly ±1°C at 1000°, 760°C, and 1100°C, respectively.

It must be understood that this performance will degrade with use. There are a number of factors which cause decalibration, such as the atmosphere to which they are held at temperature and the highest temperature used. These effects are discussed in detail in the bibliography references.

***224. Cold-Junction Corrections.*** The tables given above are appropriate for the situation in which the cold junction is maintained at the ice point ($T_0 = 0$°C). If the cold junction is not maintained at that temperature, an emf correction must be applied to the measured emf to account for this. Refer to ANSI-MC96.1-1982 for a discussion of this correction.

**TABLE 3-5** Abbreviated Tables for Standard Thermocouples ($T_0 = 0$°C)

| $T$, °C | Type B, mV | Type J, mV | Type K, mV | Type R, mV | Type S, mV | Type T, mV | Type E, mV |
|---|---|---|---|---|---|---|---|
| −100 | — | −4.632 | −3.553 | — | — | −3.378 | −5.237 |
| 0 | 0.000 | 0.000 | 0.000 | 0.000 | 0.000 | 0.000 | 0.000 |
| 100 | 0.033 | 5.268 | 4.095 | 0.647 | 0.645 | 4.277 | 6.317 |
| 200 | 0.178 | 10.777 | 9.137 | 1.468 | 1.440 | 9.286 | 13.419 |
| 300 | 0.431 | 16.325 | 12.207 | 2.400 | 2.323 | 14.860 | 21.033 |
| 400 | 0.786 | 21.846 | 16.395 | 3.407 | 3.260 | 20.869 | 28.943 |
| 500 | 1.241 | 27.388 | 20.640 | 4.471 | 4.234 | — | 36.999 |
| 600 | 1.791 | 33.096 | 24.902 | 5.582 | 5.237 | — | 45.085 |
| 700 | 2.430 | 39.130 | 29.128 | 6.741 | 6.274 | — | 53.110 |
| 800 | 3.154 | — | 33.277 | 7.949 | 7.345 | — | 61.022 |
| 900 | 3.957 | — | 37.325 | 9.203 | 8.448 | — | 68.783 |
| 1000 | 4.833 | — | 41.269 | 10.503 | 9.585 | — | 76.358 |
| 1100 | 5.777 | — | 45.108 | 11.846 | 10.754 | — | — |
| 1200 | 6.783 | — | 48.828 | 13.224 | 11.947 | — | — |
| 1300 | 7.845 | — | 52.398 | 14.624 | 13.155 | — | — |
| 1400 | 8.952 | — | — | 16.035 | 14.368 | — | — |
| 1500 | 10.094 | — | — | 17.445 | 15.576 | — | — |
| 1600 | 11.257 | — | — | 18.842 | 16.771 | — | — |
| 1700 | 12.426 | — | — | 20.215 | 17.942 | — | — |
| 1800 | 13.585 | — | — | — | — | — | — |

**TABLE 3-6** Initial Calibration Tolerances for Standard Type Thermocouples ($T_0 = 0°C$)

| Thermocouple type | Temperature range, °C | Standard tolerance (whichever is larger) | Special tolerance (whichever is larger) |
|---|---|---|---|
| B | 870–1700 | 0.5% | — |
| E | 0–900 | ±1.7°C or ±0.5% | ±1.0°C or ±0.4% |
| J | 0–750 | ±2.2°C or ±0.75% | ±1.1°C or ±0.4% |
| K | 0–1250 | ±2.2°C or ±0.75% | ±1.1°C or ±0.4% |
| R | 0–1450 | ±1.5°C or ±0.25% | ±0.6°C or ±0.1% |
| S | 0–1450 | ±1.5°C or ±0.25% | ±0.6°C or ±0.1% |
| T | 0–350 | ±1°C or ±0.75% | ±0.5°C or ±0.4% |

***225. Extension (or Compensating) Wires.*** In many situations it may be necessary for the reference junctions to be very distant (as much as several hundred feet) from the junction measuring $T$. Since most of the total emf in the thermocouple is generated by the short section at elevated temperatures, very little measurement error will occur if the remaining length at room temperature is replaced by thermocouple "extension" wires. Extension wires are made from materials having nearly the same thermal emf properties as the original thermocouple but which can be made at considerably less cost. For Types E, J, K, and T thermocouples the extension wires are made from the same alloy as the thermocouple wire, but with less stringent requirements for the composition. For Types R and S thermocouples, copper wire is used for one arm of the thermocouple, while a Cu–Ni alloy wire is used for the other.

***226. Thermocouples: Summary.*** Thermocouples are relatively inexpensive, small, have rapid thermal response, and produce a signal (i.e., the emf) which is easily measured by a digital voltmeter. Spools of the thermocouple wire may be purchased and many thermocouples may be made from them. Furthermore, each thermocouple will have an emf vs. temperature given to within specified tolerance (see Table 3-6) of the standard table, so that calibration is not necessary. There are disadvantages to thermocouples, however: the emf is sensitive to the temperature distribution along the wire, strains, thermal history, and degradation at elevated temperatures. If these latter problems outweigh the advantages, other thermometers described below may be more appropriate.

***227. Liquid-in-Glass Thermometry.*** A liquid-in-glass (LIG) thermometer consists of a reservoir of liquid and a stem with a temperature scale marked on it. The liquid has a very large thermal expansion compared to the reservoir and stem, and thus small temperature changes cause the liquid to expand into the stem where the length indicates $T$. A wide variety of liquids and glasses are used, but the most common liquid is mercury enclosed in borosilicate glass. If properly treated, LIG thermometers are capable of repeatedly measuring $T$ to within 0.01°C. The major cause of catastrophic failure is breakage; of noncatastrophic failure, heating the thermometer beyond its specified range. LIG thermometers are widely used throughout industry because they are inexpensive and easy to read with the human eye. They are not amenable to automation or continuous monitoring, however. In many applications they are being replaced by thermocouples or resistance temperature detectors.

***228. Resistance Temperature Detectors (RTDs).*** Since resistance is a physical property that is easy to measure and automate with modern instrumentation, RTDs are finding more general acceptance in temperature measurement. The two major classes of resistors with strong temperature dependence are thermistors and platinum resistors.

***229. Thermistors.*** Thermistors are made by sintering mixtures of oxides of Mn, Fe, Co, Cu, Mg, or Ti, bonding two electrical leads to the sintered material, and enclosing the unit in a protective coating. The devices are made in a wide variety of shapes (beads, disks, rods, and flakes), are very inexpensive, and are very compact (one bead type commercially available is only 0.07 mm in diameter). The resistance of the device is generally high (1 to 100 kΩ), so that lead resistance is not a significant source of measurement error. If thermistors are used below 200°C, they are quite stable. (Commercial units are available which drift by no more than 0.01°C per year.)

Thermistors are semiconducting devices whose resistance depends exponentially on temperature (see Fig. 3-93*a*). This means that the thermistor is very sensitive to temperature, but it

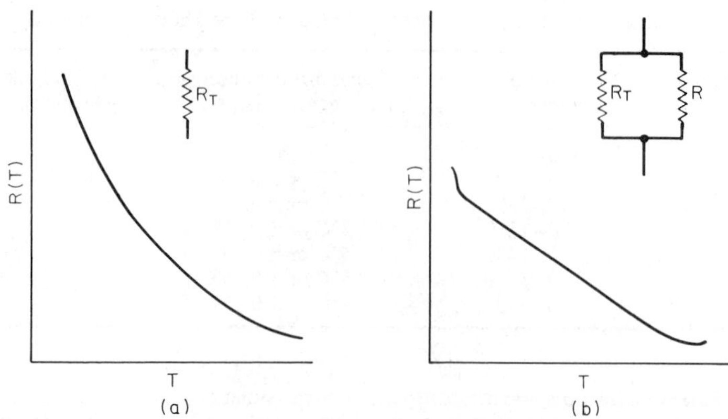


**FIG. 3-93** *(a)* Temperature dependence of a thermistor, $R_T(T)$. *(b)* Temperature dependence of a "linearized" thermistor in which a temperature-independent resistor is connected in parallel to $R_T$.

also means that its temperature range is limited (i.e., if $T$ becomes too low, the resistance becomes too high to measure. If $T$ is too high, the resistance becomes too low to measure). Thermistors may be chosen for use as low as 4 K, while others may be used in the region near 900 K. A technique, referred to as "linearization," may be used to extend the operating range of a thermistor. This consists of connecting a temperature-independent resistor $R$ in parallel to the thermistor (see Fig. 3-93*b*). If the value of the resistor is equal to that of the thermistor in the center of its operating range, the resistance of the circuit will be roughly linear in temperature rather than exponential.

Thermistor-based measurement systems with digital readouts which read directly in temperature are widely available. These consist of a sensor, a digital ohmmeter, and a logic unit, generally a microprocessor. The microprocessor is used to perform resistance-to-temperature conversion and perhaps integration, control timing, run a display, and provide digital output for computer analysis (see Pars. **46** to **55**).

***230. Platinum Resistance Thermometers (PRTs).*** The resistance of platinum is roughly linear in temperature over a very wide temperature range and thus PRTs may be used over a greater temperature range than thermistors. PRTs are more reproducible and are capable of much greater temperature accuracy. They are, however, more sensitive to mechanical shock and less sensitive to temperature change than are thermistors.

For the highest-accuracy temperature measurements, two types of "standard" PRTs are used. From 90 to 904 K a well-characterized, fine Pt wire is supported by insulators and enclosed in a glass casing. The assembled unit is 600 mm long and 7 mm in diameter. From 13 to 90 K a "capsule" version 60 mm long, 6 mm in diameter, with similar internal construction, is used. When properly used (see NBS Monograph 126), PRTs may be used to measure temperature with an imprecision not exceeding $\pm 0.001$ K over the range 13 to 904 K. Such standard PRTs are used to realize the IPTS-68 in this range. As great care must be exercised in measuring and handling these devices in order to achieve this performance, standard PRTs are generally restricted to the primary standards laboratory of any organization.

***231. Industrial PRTs (IPRTs).*** For situations requiring more robust resistance thermometers and for which less accuracy in temperature is acceptable, the IPRT is used. This class of devices comes in a wide variety of configurations which makes them comparatively immune to vibration and corrosive environments and more responsive to rapid temperature changes. Some IPRT sensors are made with thin Pt films and are quite compact (approximately $1 \times 1 \times 0.1$ cm). The resistance elements are embedded in an epoxy or compacted powder and enclosed in a stainless steel sheath. Such a wide variety of IPRTs precludes a general discussion of their temperature accuracy, long-term stability, and measurement techniques. The reader is referred to the manufacturer for this information.

There is now an accepted resistance vs. temperature table, adopted by the International

**TABLE 3-7** Standard $R$ vs. $T$ Table for IPRTs

| | | Tolerance | | | |
|---|---|---|---|---|---|
| | | Class A | | Class B | |
| $T$, °C | Resistance, Ω | Ω | °C | Ω | °C |
| −200 | 18.49 | ±0.24 | ±0.55 | ±0.56 | ±1.3 |
| −100 | 60.25 | ±0.14 | ±0.35 | ±0.32 | ±0.8 |
| 0 | 100.00 | ±0.06 | ±0.15 | ±0.12 | ±0.3 |
| 100 | 138.50 | ±0.13 | ±0.35 | ±0.30 | ±0.8 |
| 200 | 175.84 | ±0.20 | ±0.55 | ±0.48 | ±1.3 |
| 300 | 212.02 | ±0.27 | ±0.75 | ±0.64 | ±1.8 |
| 400 | 247.04 | ±0.33 | ±0.95 | ±0.79 | ±2.3 |
| 500 | 280.90 | ±0.38 | ±1.15 | ±0.93 | ±2.8 |
| 600 | 313.59 | ±0.43 | ±1.35 | ±1.06 | ±3.3 |
| 700 | 345.13 | — | — | ±1.17 | ±3.8 |
| 800 | 375.51 | — | — | ±1.28 | ±4.3 |
| 850 | 390.26 | — | — | ±1.34 | ±4.6 |

Electrotechnical Commission (IEC) for IPRTs. A resistor manufactured in conformity with this standard will fall within two tolerances (Class A or Class B) and will not need calibration to within this tolerance. This standard is being adopted by the ASTM and OIML.

An abbreviated version of the table from the IEC standard is given in Table 3-7. Note that, for industrial use, resistors made from Cu and Ni are also available. Copper resistors are the most linear of the three, while Ni resistors are the most sensitive.

***232. Radiation.*** The temperature of bodies may be estimated from the radiant energy, which they send out in the form of visible light, or the longer, infrared rays, which may be detected by their thermal effects. Since the intensity of radiation increases very rapidly with a rise in temperature, it would appear that a system of pyrometry based on the intensity of the light or total radiation from a hot body would be an ideal and simple one. However, different substances at the same temperature show vastly different intensities at a given wavelength; in other words, the *absorbing* or *emissive powers* may vary with the substance, with the wavelength, and also with the temperature.

***233. Blackbody Radiation.*** A substance which absorbs all the radiation of any wavelength falling upon it is known as a *blackbody.* Such a body will emit the maximum intensity of radiation for any given temperature and wavelength. No such material exists, but a very close approximation is obtained by heating the walls of a hollow opaque enclosure as uniformly as possible and observing the radiation coming from the inside through a very small opening in the wall.

***234. Stefan-Boltzmann Law.*** The relation between the total energy radiated by a blackbody and its temperature is expressed by the equation

$$J = \sigma(T^4 - T_0^4)$$

where $J$ = energy of all wavelength emitted per second per square centimeter of surface, $T$ and $T_0$ = absolute temperatures of radiator and surroundings, respectively, $\sigma$ = a constant of the value $5.7 \times 10^{-12}$ W/cm². In general $T_0^4$ is negligible in comparison with $T^4$, so that the above relation becomes $J = \sigma T^4$. Although the total energy emitted by any substance is not that emitted by a blackbody at the same temperature, it may be considered as some fractional part of that from the ideal radiator, this fraction $E$ being known as the *total emissivity.* If $S$ denotes the apparent absolute temperature, that is, the temperature on the blackbody scale corresponding to an amount of energy equivalent to that emitted by the nonblack substances at a true temperature $T°$ abs, the relation between its total emissivity $E$ and the quantities $S$ and $T$ is

$$\log E = 4(\log S - \log T)$$

***235. Radiation Pyrometry.*** In the radiation pyrometer, the radiant energy of all wavelength is brought to a focus by means of a quartz lens or a concave mirror upon the hot junctions of a minute thermopile. The cold junctions are suitably screened from the radiation of the hot body. The radiant energy focused upon the hot junctions develops an emf which may be measured by a potentiometer or automatically recorded. In practice the indicating or recording instrument is calibrated to read temperature directly. The relation between the emf $e$ and the absolute temperature $T$ may be expressed by the equation

$$e = aT^b \qquad \text{or} \qquad \log e = k + b \log T$$

where $a$ (or $k$) and $b$ = empirical const. The constant $b$ is usually between 4 and 5.

*Leeds & Northrup Rayotube* is a radiation pyrometer made either with a quartz lens or with a mirror. The mirror type is suitable for the lower temperatures at which the energy radiated is largely in the long wavelengths which would be too greatly absorbed by a lens.

***236. Optical Pyrometry.*** Optical pyrometers are based upon the photometric principle of comparing the intensity of visible monochromatic radiation emitted by a body with that of the same wavelength or color from a constant and reproducible comparison source, such as an electric lamp. The light from the comparison source must be calibrated in terms of the formula

$$\ln \frac{J_2}{J_1} = \frac{1.432}{\lambda}\left(\frac{1}{1336} - \frac{1}{t + 273}\right)$$

specified in the International Temperature Scale either by direct observations on a crucible of freezing gold and by use of sector disks to determine ratios of brightness or by checking against a standard pyrometer which has been previously so calibrated.

The above formula comes directly from applying Wien's law

$$J_\lambda = C_1\lambda^{-5} \exp\left(\frac{1.432}{\lambda(t + 273)}\right)$$

to the ratio of monochromatic radiation of wavelength $\lambda$ from a blackbody at two different temperatures. Practically all optical pyrometers now in use are of the disappearing-filament type.

***237. Disappearing-Filament Pyrometers.*** The filament of a small electric lamp $F$ (Fig. 3-94) is placed at the focal point of an objective $L$ and ocular, forming an ordinary telescope

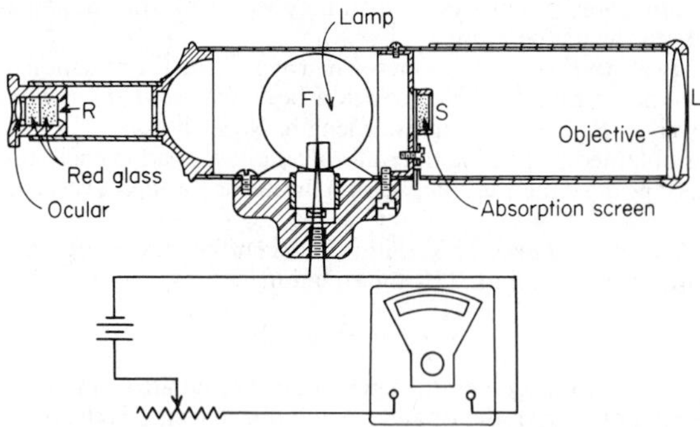


**FIG. 3-94** Disappearing-filament optical pyrometer.

which superposes upon the lamp the image of the source viewed. Red glass, such as Corning "high-transmission red," is mounted at the ocular to produce approximately monochromatic light. In making a setting, the current through the lamp is adjusted by rheostat until the tip or some definite part of the filament is of the same brightness as the source viewed.

The lamps should be operated at temperatures no higher than 1500°C, because of deterioration of the tungsten filament. If this temperature is not exceeded, the calibration of the

**TABLE 3-8** Monochromatic Emissivity for Red Light ($\lambda = 0.65\ \mu m$)

| Material | $E_\lambda$ | Material | $E_\lambda$ |
|---|---|---|---|
| Silver | 0.07 | Nichrome (oxidized) | 0.90 |
| Gold, solid | 0.14 | Cuprous oxide | 0.70 |
| Gold, liquid | 0.22 | Iron oxide | 0.80 |
| Platinum, solid | 0.30 | Nickel oxide, 800°C | 0.96 |
| Platinum, liquid | 0.38 | Nickel oxide, 1300°C | 0.85 |
| Palladium, solid | 0.33 | Nickel, solid and liquid | 0.36 |
| Palladium, liquid | 0.37 | Iridium | 0.30 |
| Copper, solid | 0.10 | Rhodium | 0.24 |
| Copper, liquid | 0.15 | Graphite powder (estimated) | 0.95 |
| Tantalum, 1100°C | 0.60 | Carbon | 0.85 |
| Tantalum, 2600°C | 0.48 | Porcelain | 0.25–0.50 |
| Tungsten, 1000°C | 0.46 | | |
| Tungsten, 2000°C | 0.43 | | |
| Tungsten, 3000°C | 0.41 | | |

lamp is good for hundreds of hours of ordinary use. For higher temperatures, absorption glasses *S* (Fig. 3-94) are placed between the lamp and the objective, or in front of the objective, to diminish the observed intensity of the source. The relation between the temperature of the source, $T°$ abs, and the observed temperature $T_0°$ abs, measured with the absorption glass interposed, is as follows: $1/T - 1/T_0 = A$, where $A$ = for most practical purposes a constant. Sometimes an instrument is calibrated to read the lamp current, and a table is furnished showing the relation between the current and the temperature both with and without the absorption glass. Often, however, the instrument reads directly in temperature and is provided with a double scale to take care of the high and low ranges.

Instruments may be obtained with an additional special glass for use when sighting upon iron or steel. This special glass differs from the regular absorption glass in such a way that its substitution for the regular screen automatically compensates for the emissivity of the metal. The instrument will thus read on the same scale either the true temperature of the metal (with the special glass in place) or the temperature of a blackbody (with the regular glass in place) (see Table 3-8).

***238. Emissivity Corrections for Optical Pyrometers.*** Optical pyrometers will indicate true temperatures when sighted upon a blackbody. Blackbody conditions are approximated in practice by a peephole in the side of a furnace or kiln or a closed porcelain tube thrust into molten metals or salts. When sighting upon objects in the open, certain corrections must be applied. Table 3-8 gives the emissivity of various substances for red light ($\lambda = 0.65\ \mu m$).

## Bibliography

***Temperature Scale and Discussion of Thermometers***

1. Quinn, T. J. *Temperature,* New York, Academic Press, 1983.
2. Benedict, Robert P. *Fundamentals of Temperature and Pressure and Flow Measurements,* New York, John Wiley and Sons, 1977.

***Thermocouples***

1. Quinn, T. J. *Temperature,* New York, Academic Press, 1983, Chap. 6.
2. Benedict, Robert P. *Fundamentals of Temperature and Pressure and Flow Measurements,* New York, John Wiley and Sons, 1977, Chap. 7.
3. Manual on the Use of Thermocouples in Temperature Measurement, SRP-470B, American Society for Testing and Materials (ASTM), 1974.
4. Powell, R. L. et al. Reference Tables for Thermocouples; NBS Monograph 125, 1974..
5. Temperature Measurement Thermocouples; ANSI-MC.96.1-1982, Research Triangle Park, N.C., Instrument Society of America, 1982.

***Liquid-in-Glass Thermometers***

1. Quinn, T. J. *Temperature,* New York, Academic Press, 1983, Chap. 8.
2. Wise, J. Liquid-in-Glass Thermometry; NBS Monograph 150, 1976.

**_Resistance Temperature Detectors_**

*Thermistors*

1. Quinn, T. J. *Temperature,* New York, Academic Press, 1983, Chap. 5.
2. Sachse, H. B. *Semiconducting Temperature Sensors and Their Applications,* London, Wiley, 1975.

*Platinum Resistance Thermometers*

1. Quinn, T. J. *Temperature,* New York, Academic Press, 1983, Chap. 5.
2. Riddle, J. L., Furukawa, G. T., Plumb, H. H. NBS Monograph 126, 1973.
3. International Electrotechnical Commission (IEC) Publication 751. Industrial Platinum Resistance Thermometer Sensors, Bureau Central de la Commission Electrotechnique Internationale, Geneva, Switzerland, 1983.

**_Radiometry_**

1. Quinn, T. J., *Temperature,* New York, Academic Press, 1983, Chap. 7.

## ELECTRICAL MEASUREMENT OF NONELECTRICAL QUANTITIES[1]

***239. A transducer*** is a device in which variations in energy of one form produce corresponding variations in energy of another form. In common usage, either the input or output of a transducer is electrical. Thermocouples and thermistors fall into that category, as does the *thermal converter,* whose electrical output (dc millivolts) is derived from a thermal effect that represents an electrical quantity (ac volts, current, watts, vars) that differs in nature from the output. A variety of methods is often available for the measurement of a specific variable. "Frequently, operational considerations will indicate the choice of transducer; for instance, piezoelectric transducers may not perform well if long cables are required; capacitive devices, although quite sensitive, may require intermediate electronic circuitry; and magnetic transducers should not be used in the presence of strong magnetic fields."[2]

***240. Mechanical displacement*** may be converted into an electrical variable by the simple expedient of adjusting resistance in an electrical circuit. A slide-wire resistor, having a movable contact attached to the part whose displacement is to be measured, may be connected through a 2-conductor circuit to a steady-voltage source in series with an ammeter (or milliammeter) calibrated in terms of the displacement. If the resistor is connected as a voltage divider, the need for a regulated supply is eliminated, and with a 3-conductor circuit the display instrument may be a ratio meter or a potentiometer. Such combinations are common and are available for both dc and ac operation. Where deflections are small—less than 0.1 in—measurement may be made by use of a differential transformer.

***241. Strain Gage.*** In the *strain gage,* microscopic relative displacements are electrically magnified and are displayed on an indicating or a recording meter or on an oscillograph. Modern resistance-type strain gages comprise fine wire windings arranged to be more or less elongated when subjected to deformation. The units may be used singly, in pairs, or in sets of four constituting a complete Wheatstone bridge. There are two main classes of wire-wound strain gages, *(a)* "bonded" and *(b)* "unbonded."

**a.** The *bonded strain gage* is composed of fine wire, wound and cemented on a resilient insulating support, usually a wafer unit. Such units may be mounted upon or incorporated in mechanical elements or structures whose deformations under stress are to be determined. While there are no limits to the basic values which may be selected for strain-gage resistances, a typical example may be taken as of the order of 100 to 500 Ω.

**b.** In the *unbonded strain gage,*[3] the resistance structure comprises a fine wire winding

[1] *InTech* (Instrumentation Technology), a monthly publication of the Instrument Society of America, has in-depth articles on various instrumentation and measurement problems.

[2] Paraphrased from "Transducers," by David B. Kret. Published 1953 by Allen B. DuMont Laboratories, Inc., Clifton, N.J. This work is a comprehensive treatment of transducers, especially, but not wholly, for use with oscillographic display devices. It includes a list of more than 800 transducers and their applications, as well as a tabulation of over 200 accessory devices and a bibliography of 200 references.

[3] Helfand, B. B. Developments in Unbonded Strain Gage Transducers; *Rept. AIEE-IRE Conf. Telemetering,* April 1953, p. 27.

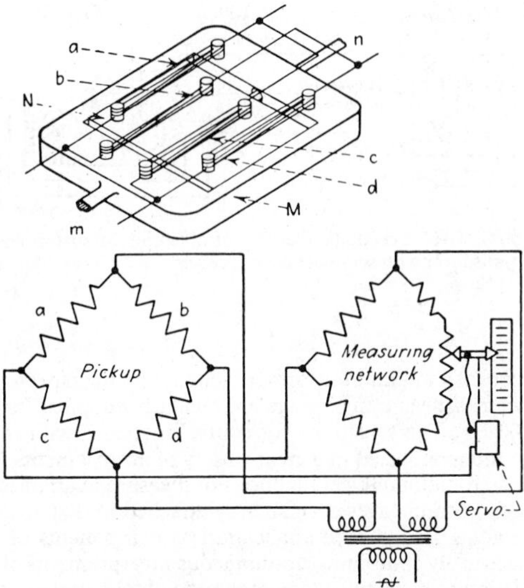


**FIG. 3-95** Diagram of unbonded wire strain gage. Supports *M* and *N* are attached by rods *m* and *n*, respectively, to points between which displacement is to be measured. Pickup and measurement networks are energized from similar but isolated sources. Unbalance originating in the pickup is detected and balanced by a servo-actuated measuring network, providing a reading of strain on a graduated scale.

stretched between insulating supports mounted alternately on the two members between which displacement is to be measured (see Fig. 3-95). These wires comprise the four arms of a Wheatstone-bridge network of which two opposite arms are tightened and the other two slackened by the displacement. While a bonded gage tends to respond to the average strain in the surface to which it is cemented, the unbonded form measures displacement between the two points to which the respective supports are attached. Unbonded wire strain gages are usually operated on input potentials ranging up to 35 V direct or alternating current. Under conditions of extreme unbalance, corresponding to full operating range, the open-circuit emf may be of the order of 8 to 10 mV and the closed-circuit current up to 100 $\mu$A.

Recently developed types of conductive rubber are used in resistive transducers capable of wider ranges of deformation than are those using wire or foil. Where the strain gage must operate over a temperature range, dummy gages exposed to the temperature but not the strain may be employed for temperature compensation, or alloys having a low temperature coefficient of resistance may be used. Piezoelectric strain gages are also available for applications in pressure, force, torque, and displacement measurement. Strain gages for use on ac circuits are supplied in both capacitive and inductive forms, wherein the corresponding characteristics of ac circuit components are varied by the displacements to be measured.

***242. Small Displacements.*** A popular means for measuring *small displacements* in the range from a millimeter to a micron is the *linear,* or *differential, transformer.*[1] This device (Fig. 3-96) is generally produced with a single primary winding and two secondaries, all disposed along a common axis and having in the common magnetic circuit a movable iron core longitudinally displaceable with the motion to be measured. The secondaries may be connected additively or differentially and may be included in the circuit of a null-type instrument balanced

[1] Differential transformers may be obtained for direct displacements as great as 10 in.

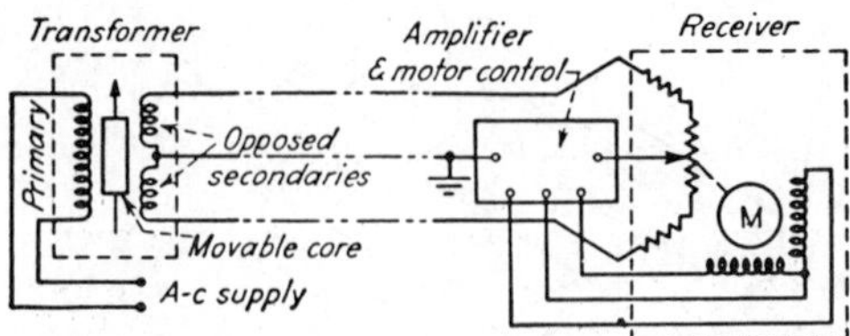


**FIG. 3-96** Schematic diagram of differential-transformer transducer with servo-actuated receiver.

either by shifting the core of a similar transformer excited from the same source or by the use of a slide-wire potentiometer. Linear transformers are regularly supplied for operation at all frequencies up to 30,000 Hz. The sensitivity, of course, increases with the frequency. Linear transformers[1] may be interconnected in a great variety of arrangements to perform computations or to express desired mathematical functions of measured variables.

***243. Pressure Gage.*** Strain gages permanently attached to diaphragms, tubes, and other pressure-sensitive elements find a wide application as components of *electrically actuated pressure gages.* By electrically combining simultaneous measurements of torque and velocity, continuous determination of mechanical power may be obtained, the combination becoming an electrical-transmission dynamometer.

***244. Vibration*** may be determined by a strain gage, but the fact that this magnitude involves motion renders it generally preferable to utilize alternating potentials developed by periodic change in the geometry of the measuring circuit. This may be embodied in either a *capacitor* or an *inductive device.* In a recently developed apparatus there are no moving parts except the object being shaken, and the vibration displacement is sensed by its effect on an electrostatic field between the pickup and the moving part. *Piezoelectric crystals* are particularly adapted to the measurement of vibration. The emf so obtained is proportional to the amplitude of deflection multiplied by the frequency squared.

***245. Air velocities*** and the *flow of gases* in general may be measured by the *hot-wire anemometer.* In its simplest form, this device utilizes the cooling effect of the gas stream to establish a temperature difference between exposed and protected bridge arms. Where the flow is in an enclosed conduit, a heating element may be introduced and the volume of flow determined by the amount of heat transferred between the heater and the temperature-sensitive bridge wires.

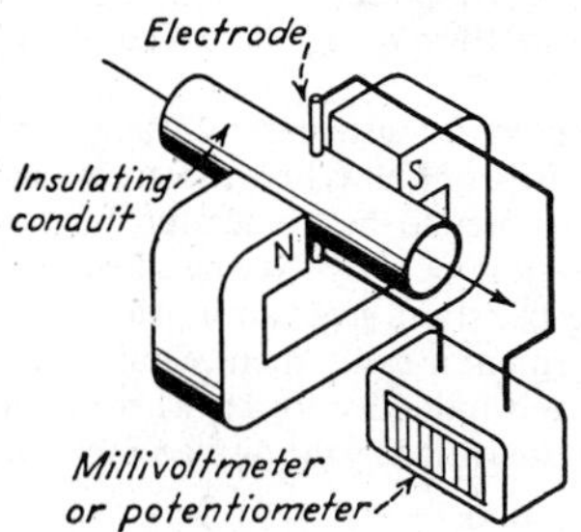


**FIG. 3-97** Electromagnetic flowmeter.

***246. Flow of an electrically conducting liquid*** may be determined by measuring the emf developed between a pair of electrodes set in opposite sides of an insulating conduit due to the movement of the liquid through a magnetic field established transversely of the conduit and perpendicular to both the flow and the line joining the electrodes (Fig. 3-97). By using an alternating field the effects of electrode polarization may be eliminated. Null measurement of the generated voltage renders the apparatus independent of the resistance of the liquid.

***247. Liquid level*** may be expressed electrically by the use of a transducer responsive to the vertical position of a float or by a pressure-sensitive strain gage immersed in the liquid below its lowest level. Variation in resistance of an immersed conductor is a widely accepted principle, especially in fuel tanks. If the liquid is an electrical insulator and of constant characteristics, its depth may be determined by its dielectric effect between a pair of vertically disposed

[1] Hornfeck, A. J. Computing Circuits and Devices for Industrial Process Functions; *Tech. Paper* 52-191, *Trans. AIEE,* Communication and Electronics, July 1952, p. 183.

capacitor plates. On the other hand, if the liquid is a conductor and very small changes in level are to be detected or regulated, the liquid may be made one electrode of a capacitor whose other electrode is a horizontal plate positioned above the surface.

***248. Levels of corrosive liquids*** or those operating under extreme pressures, temperatures, or other conditions rendering them inaccessible for measurement by conventional means may be determined by the use of *gamma radiation.* Several gamma-ray sources are spaced at equal vertical intervals in the tank or reactor containing the liquid to be measured but are positioned so that none of them obstructs the line of sight of a Geiger counter tube placed at the top of the container. The response of the Geiger tube depends upon the depth of the process material, and the output is measured on a null-type recording instrument.

***249. Vacuum*** may be measured by determining either *energy dissipation* or *electron emission* in the space under test. The former principle provides the basis of the *Pirani gage,* wherein two similar heated filaments forming arms of a bridge are located, respectively, in a reference bulb and a bulb connected to the evacuated space. Heat dissipation will vary with the degree of evacuation, while conditions in the reference bulb remain constant. The electrical condition of the bridge then provides a continuous measure of the vacuum. The normal range of operation of the Pirani gage is from $10^{-7}$ to 5 mm Hg. As the performance of a thermionic tube is highly responsive to the degree of vacuum, its action under controlled electrical conditions is a criterion of internal atmosphere. This principle forms the basis of a number of *electronic vacuum gages.* The normal range of operation lies between $10^{-7}$ and $10^{-3}$ mm Hg.

***250. Gas Analysis.*** Electrical methods for *analyzing gases,* while essentially thermal in their nature, are made practicable only by the application of electrical principles in determining thermal relationships. In the *thermal-conductivity method,* as best exemplified in the $CO_2$ recorder, two cells or sections of conduit containing, respectively, a standard sample and the gas under test have in them adjacent arms of a bridge network composed of wires having known resistance variation with temperature and carrying sufficient current to raise their temperatures appreciably above their surroundings. As more or less heat is dissipated in the test cell as compared with the reference cell, the relative resistance of the bridge arms varies, providing an electrical basis for measurement of the gas composition.

***251. Detection of Flammable Gases.*** The *catalytic-combustion method* is especially adapted to detection of flammable gases or determination of explosibility. The arrangement of cells and bridge wires may be similar to that of the thermal-conductivity type, but the filament is composed usually of activated platinum and is operated at a temperature sufficient to ignite the gas when a critical proportion is attained (Fig. 3-98). The increased heating of the bridge wire due to combustion abruptly disturbs the balance and provides a positive indication of explosibility. In some forms of this instrument the temperature rise is determined by thermocouples. The catalytic-combustion method is useful in determining mixtures containing such gases as propane, acetone vapor, carbon disulfide, and carbon monoxide. The equipment finds use in (1) solvent-recovery processes, (2) solvent-evaporating ovens, (3) combustible-gas storage rooms, (4) storage vaults, (5) gas-generating plants, (6) refineries, and (7) mines.

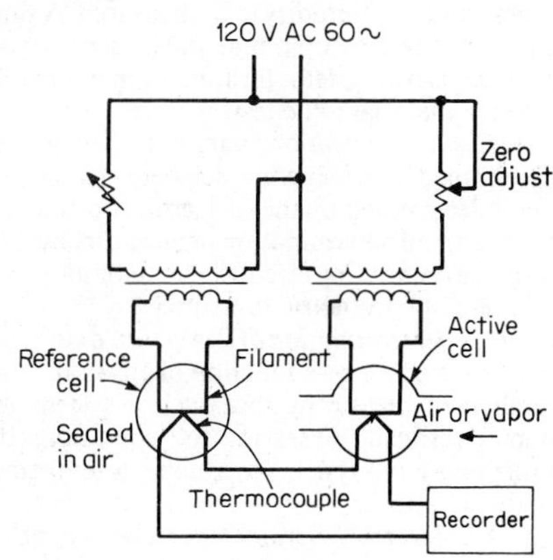


**FIG. 3-98** Basic thermocouple-type gas analyzer.

***252. Oxygen Content of Gases.*** Both the conventional thermal-conductivity method and the catalytic combustion method are applicable. In addition to these, use is made of the *magnetic susceptibility* of oxygen as a basis of operation. In one such instrument, a hot-wire bridge similar to that of a $CO_2$ recorder is employed, one of the gas chambers being placed in a strong magnetic field. This stimulates the flow of oxygen-containing gas through that chamber, thereby unbalancing the bridge by a measurable amount. In the other magnetic analyzer, a test chamber contains a small magnetic member rotatable in a distorted field whose conformation depends upon the amount of oxygen present. The resultant angular displacement of the test member may be used similarly to that of a galvanometer in either a direct-deflecting or a null-type instrument.[1]

***253. Toxic Gases.*** An analyzer especially suited to measurement of *toxic ionizable gases* or vapors to and beyond the toxic limits utilizes the electrical conductivity of an aqueous solution of the gas. The vapor under test is bubbled through distilled water at a fixed rate, and the conductivity of the solution becomes a measure of gas concentration. A typical use is the continuous recording of small quantities of substances like sulfur dioxide, hydrogen sulfide, chlorine, and carbon disulfide in the air.

***254. Atmospheric contamination*** may be determined by an *electronic leak detector,* utilizing emission of positive ions from an incandescent filament exposed to the air. The filament is enclosed in an open inner cyclinder and heated by alternating current. The atmosphere under test is forced through the annular space between the inner and an outer cylinder at a predetermined rate, and the electron flow due to a dc potential maintained between the cylinders is measured as an index of the amount of contaminant. Presence of extremely small proportions of halogen vapor compounds, of which Freon, chloroform, and carbon tetrachloride are good examples, greatly increases the emission. At room temperatures the device does not respond to Pyranol, but if this material is heated sufficiently to give off vapor, a response is obtained. It also responds to solid particles of the halogens and therefore will detect *smoke* from burning materials containing these elements. The instrument is also available as a recorder and/or a controller.[2]

***255. Relative humidity*** is determinable electrically by methods involving either of two basic principles: (1) variation of electrical conductivity or of dielectric constant of a hydrophilic element and (2) computation based on "dry-bulb" and "wet-bulb" temperatures of the atmosphere whose moisture content is to be determined. The most common embodiment of the former method consists in an insulating card, plate, or cylinder carrying a bifilar winding of conductive wire and having a relatively large surface exposed to the atmosphere. The two strands of wire are bridged by a coating of material such as lithium chloride or colloidal graphite, having a high affinity for moisture. This material quickly assumes a water content corresponding to that of the atmosphere, and the electrical resistance between the conductors becomes a function of the humidity to be measured. A similar principle is used in determining the moisture content of hygroscopic materials, such as wood, grain, or pulp. In such applications a resistance-measuring circuit terminates in electrodes or probes which are pressed against or inserted into the material to be tested.

*Moisture content* of material in a web or sheet form, such as paper, may be continuously determined by passing the web between the plates of a capacitor and thus obtaining a measurement determined by the dielectric constant of the material as affected by its water content.

Electrical determination of humidity by the "wet-and-dry-bulb" method requires somewhat intricate computing circuits which for accurate results must take account of absolute temperature and of barometric pressure.[3]

***256. Determination of dew point,*** or the temperature at which condensation takes place on a polished surface, as a function of absolute humidity, employs essentially a thermal and optical method of measuring, but such a system may be rendered continuous and automatic by photoelectrically observing the conditions of a polished surface in the tested atmosphere, utilizing a servo system to regulate its temperature, and thus obtaining an indication or a record of the dew point.

***257. Electric Micrometers.*** The two most popular types are (1) that utilizing the magnified output of a strain gage and (2) that based upon precise determination of capacitance between two electrodes whose spacing corresponds to the measured dimension.

***258. Ultrasonic thickness gages*** may be used to measure steel walls ranging in thickness from ⅛ in to 1 ft, utilizing the fact that sound vibrations tend to establish standing waves within the mass of the material upon which they are impressed. This device combines a variable-frequency oscillator with a piezoelectric crystal which is pressed against the wall to be tested. The

[1] Rigg, O. W. Oxygen Recorders; *Instruments,* February 1953, p. 248. Seffern, O. K. Oxygen Measurement; *Instruments,* September 1953, p. 1210.

[2] White, W. C., and Hickey, J. J. Electronics Simulates Sense of Smell, *Electronics,* March 1948, vol. 21, p. 100.

[3] Behr, Leo. A New Relative Humidity Recorder, *J. Opt. Soc. Am.,* 1925, vol. 12, p. 623.

circuit is tuned until the metal oscillates, causing a sharp increase in the loading. The frequency of this resonance indicates the thickness of the material.[1]

Selection of a method for determining the *thickness of sheet material* in process will depend primarily upon the inherent electrical conductivity of that material. If it is essentially a nonconductor, as rubber, plastic, or paper, measurement may be continuously performed by passing the sheet or web between the plates of a capacitor. (In such measurements on hygroscopic materials, moisture content may become a dominating factor.)

***259. Sheet Thickness.*** Sheet materials, whether conducting or insulating, may be measured by the *beta-ray gage* (see Fig. 3-99). In this device a stream of beta rays passes through the sheet to a pickup head whose response is amplified and continuously recorded and, if desired, made the controlling influence in automatic regulation. Provision is made for the combined radiation source and pickup to traverse the strip of material and scan its whole width.

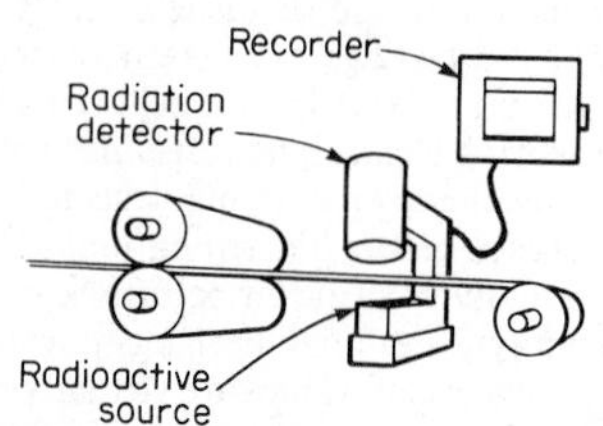


FIG. 3-99 Simplified diagram of beta-ray gage.

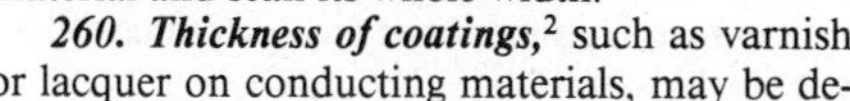

***260. Thickness of coatings,***[2] such as varnish or lacquer on conducting materials, may be determined by a continuous measurement of capacitance between the base and a reference electrode, the coating being included as a dielectric. With a magnetic base, such measurement may be performed effectively by determining the effect of the coating upon the gap in a magnetic circuit.

***261. Surface roughness*** may be determined either on an absolute basis or by comparison with a "standard" surface. A common method involves passing a small stylus systematically over the surface, similarly to a phonograph needle, and measuring the resulting vibration. The stylus may be attached to a strain gage, piezoelectric crystal, or a magnetic pickup. The resulting alternating emf may be amplified and displayed on an oscillograph, or it may be rectified and measured with a millivoltmeter. A basis for quantitative determination of surface roughness is found in USAS B46.2.

An absolute method of determining roughness utilizes the electrical capacitance of the tested surface in contact with an electrolyte as compared with that of an ideal (mercury) surface. On the assumption that the capacitance varies as the surface area, the comparison provides a figure representing the ratio of the tested surface to one of perfect smoothness.

***262. Radiant energy*** may be measured either by a *thermocouple* or by a *photocell.* Response of the former, restricted almost wholly to the longer waves, is manifested in a thermoelectric effect. Photocells may be either photovoltaic, relying upon a barrier layer to develop an emf depending upon light intensity, or photoresistive, wherein the resistance of the cell varies with the light. Since photocells of both classes are selective with respect to wavelength and are not essentially linear in response, quantitative measurement of radiation by such means becomes complicated and subject to a number of variables. Thus, in precision photometry, the photocell is usually employed essentially as a detector in a null-balance system.

***263. Transparency*** (or opacity) determination of materials and continuous monitoring of smoke density involve passing the substance to be examined through the path of a light beam directed upon a photocell. Uninterrupted measurement is made by means of a potentiometer or a bridge, according to the class of cell employed.

***264. Viscosity measurement*** is essentially mechanical in its nature, and the application of electrical methods consists in determination of stress or displacement set up in the measuring apparatus owing to the characteristic of the fluid. One method involves measuring the electrical input to a small motor driving an impeller or stirrer in the fluid. Another method is based on electrical determination of the angle of lag (torque measurement) in a resilient mechanism through which an impeller is driven. A further method utilizes *magnetostriction* to produce

[1] Branson, N. G. Portable Ultrasonic Thickness Gage; *Electronics,* January 1948, p. 88.

[2] Three Electronic Thickness Gages for Metallic Coatings; *NBS Tech. News Bull.* 38, September 1941, No. 9. Clarke, E., Carlin, J. R., and Barbour, W. E. Measuring the Thickness of Thin Coatings with Radiation Backscattering; *Elec. Eng.,* January 1951, p. 35. Thickness Meters, "The Instrument Manual"; London, United Trade Press, 1953, p. 12.

longitudinal oscillations in a steel rod carrying a diaphragm immersed in the liquid. Determination of the electrical loading on the exciting circuit provides a measure of viscosity.

***265. Chemical Magnitudes.*** Electrical measurement has superseded many of the older methods of quantitative determination of chemical magnitudes. The two best-known methods are based, respectively, on the electrical *conductivity* of solutions and on the *voltaic effect* in specific cells. The basic principles of these measurements are wholly different, as are their applications. In the *conductivity* cell every precaution must be observed to avoid electrolytic effects, the prime requisite being that the respective electrodes be of identical material. Even then, the passage of current or the application of the potential tends to produce internal polarization emf in the cell. This undesirable effect may be almost wholly eliminated by measuring electrolytic resistance with alternating current, and the highly sensitive ac detectors now available enable such tests to be made with precision. Outstanding among the uses of the resistance cell is determination of the *purity of water* for domestic and industrial purposes. Conductivity of water solutions usually increases in proportion to the amount of dissolved electrolytic material. Perfectly pure water has a specific resistance of 18 to 20 million $\Omega/cm^3$, but in practice such values are virtually unobtainable. Only by careful distillation or deionization is it possible to obtain water of 400,000 to 800,000 specific Ω at a reference temperature of 70°F. Continuously operating water-conductivity recorders are supplied for use with commercial ac power supply, and a typical range is 100,000 specific Ω to infinity.

***266. Electrolytic cells*** utilize measurement of emf developed between a standard combination of electrodes by the solution under test. Development of the principle has reached its highest refinement in the measurement of pH, or hydrogen-ion concentration, which is a criterion of the activity with which the solution will enter as an acid into a chemical reaction. The pH value is a logarithmic function of the emf developed with a given strength of the solution in a specified cell. For pure water, which is "neutral" in its reaction lying midway between the acids and the bases, the pH value is 7. pH measurement is essential in practically every industry involving any chemical process, as well as in waterworks, sewage systems, biological laboratories, and agricultural experiment stations.

## Bibliography

***267. Electrical Measurement of Nonelectrical Quantities***

Borden, P. A. Electrical Measurement of Physical Values; *Trans. AIEE,* February 1925, Vol. 44, pp. 238–263. (330 refs.)

Supplementary bibliographies to above:

*Trans. AIEE,* 1927, Vol. 46, pp. 709–712.

*Trans. AIEE,* 1928, Vol. 47, pp. 1168–1171.

Pflier, P. M. "Elektrische Messung mechanischer Grossen"; Berlin, Springer-Verlag OHG, 1943. (Extensive bibliography.)

Roberts, H. C. "Mechanical Measurements by Electrical Methods"; Pittsburgh, Instruments Publishing Co., 1946. (470 refs.)

Beckwith, T. G., and Buck, N. L. Mechanical Measurements"; 2d ed., Addison-Wesley, 1969.

Norton, H. N. "Handbook of Transducers for Electrical Measurements"; Prentice-Hall, 1969.

Keast, D. H. "Measurements in Mechanical Dynamics"; McGraw-Hill, 1967.

## *TELEMETERING*

***268. Telemetering*** is measurement with the aid of intermediate means which permit the measurement to be interpreted at a distance from the primary detector.

NOTE: The distinctive feature of telemetering is the nature of the translating means, which includes provision for converting the measurand into a representative quantity of another kind that can be transmitted conveniently for measurement at a distance. The actual distance is irrelevant.

***269. Electric telemetering*** is telemetering performed by deriving from the measurand or from an end device a quantitatively related separate electrical quantity or quantities as a translating means.

A *measurand* is a physical quantity, property, or condition which is to be measured.

Telemetering has been practiced many years in the central station industry and in the transmission and distribution of electric power, but until lately only to a limited extent in the nonelectrical fields. With the phenomenal expansion of pipelines for gas and for oil, the need has vastly increased, and electric telemetering installations have become indispensable in the remote measurement, totalization, regulations, and dispatching of these utilities. Telemetering has also found wide application in extensive industrial plants, such as refineries, steel mills, and large chemical plants, and in these installations it often forms an essential part of remote regulating apparatus.

***270. Field of Application.*** The past decade has seen a rapidly increasing use of telemetering in aircraft, meteorology, ordnance, and guided missiles. This has led to a sharp demarcation of telemetering philosophies and techniques into two classes, "mobile" and "stationary." In the former, the apparatus is expected to operate for a very short period of time—often only a matter of seconds. The transmitting unit at least must be considered as expendable, and the combination is generally subject to an overall calibration for each isolated test in which it is used. Obviously, there can be no interconnecting physical circuit, and a radio link is an essential part of the system.[1]

***271. Stationary systems*** in general involve transmitting and receiving units at fixed locations. These are usually of a permanent nature and are intended for operation over extended periods of time. Signal transmission between the stations usually involves a physical circuit, and even where radio principles are utilized, the most common practices require guiding of the signal by means of a more or less continuous conducting path.

***272. A telemetering system*** incorporates the same three essential elements as are required in a system for measurement of nonelectrical quantities by electrical means, namely, a *transmitting unit* (transducer or pickup), a *receiving unit* (an instrument for measuring an electrical variable), and an *interconnecting circuit* or channel by which the electrical variable (signal) originating at the transmitter is carried to and impressed upon the receiver.

***273. Channel.*** In transmission of measurement over considerable distances the *circuit or channel* may become the predominating factor in the system. In the ideal telemetering system the terminal apparatus would be inherently self-compensating, so that variations in circuit conditions would not adversely modify the signal. Merit of a telemetering system is directly related to the degree to which it approaches this ideal. Distance criterion of a telemetering system is not so much the number of feet or miles over which it will operate as it is the nature and magnitude of circuit impedance through which its signals will maintain their identity and proportionality. Since the data have been determined for specific types of circuits and channels, such magnitudes may generally be expressed in units of distance. A continually increasing proportion of telemetering is being carried out over circuits and channels leased from communication companies. With information available respecting the type of signal to be transmitted, the telephone or telegraph company provides a suitable circuit and assumes responsibility for its operation. Where privately owned circuits are used for telemetering, their maintenance and protection correspond to those for comparable communication circuits.

***274. Classification of Telemeters.*** In classifying telemetering systems the ANSI has adopted a grouping recommended by the AIEE and based on the nature of the electrical variable transmitted through the interconnecting circuit or channel. The names of the five classes are more or less self-explanatory and are as follows: *current, voltage, frequency, position,* and *impulse* types. In each of the first three of these classes the corresponding characteristic of the electrical output of transducer comprising the transmitting unit is varied with variations in the measurand. In the *position* system, the quantitative ratio, or the phase relationship, between two electrical voltages or currents determines the nature of the transmitted signal, usually requiring a circuit of three or more conductors. There are several *impulse* systems, in all of which the transmitting instrument acts to "key" a signal impressed upon the circuit, producing

[1] Treatment of mobile telemetering systems lies beyond the scope of this handbook, and until the appearance of an authoritative textbook on the subject, reference may be made to the *Reports* of the Annual Conferences on Telemetering sponsored by the IEEE in cooperation with other technical organizations.

**TABLE 3-9.** Typical Telemetering Systems
(Data selected from AIEE *Report* on telemetering, 1948)

| Designation | ASA class | Nature of transmitted current | | | No. of wires | Variables for which especially suited |
|---|---|---|---|---|---|---|
| | | Kind | Volts (max.) | Current, mA | | |
| Torque balance..... | Current | D-c | 125 | 8 | 2 | Electric power, or other quantities with specially adapted measuring elements |
| Photoelectric....... | Current | D-c | 250 | 5 | 2 | |
| Current balance.... | Current | D-c | 250 | 25 | 2 | |
| Thermal converter.. | Voltage | D-c | 1 | 50 | 2 | Electric power |
| Position motor..... | Position | A-c | 22 | 10 | 3 or 5 | Variables as measured |
| Bridge............. | Position | D-c or a-c | 6 | 5 | 3 | Variables as measured |
| Electronic......... | Position | A-c | 48 | 65 | 3 | Variables as measured |
| Impulse duration ... | Impulse | D-c or a-c | 115 | 50 | 2* | Variables as measured |
| | | | Frequency range, Hz | | | |
| "Frequency"...... | Frequency | A-c | 6– 27<br>20– 25<br>80–100 | | 2† | Electric power |

* Values given represent unrelayed signal. May be amplified or converted to any type of signal suited to channel.

† Signal voltage and current value for frequency systems not given. These are normally adapted to the specific installation.

a series of successive pulses which, according to their nature, are interpreted by the receiving instrument and expressed in terms of the measurand. See Table 3-9.

Telemetering systems are not always mutually exclusive. A single installation may represent a combination of several of the named systems. In some instances it becomes difficult to decide into which of the specified classes a particular method of telemetering may fall.

***275. Telemetering of electrical quantities,*** such as volts, watts, and vars, presents a problem owing to the inherently low torque of direct-deflecting instruments, whereas devices for measuring such magnitudes as position, flow, and liquid level are not subject to such restrictions. Accordingly, where measurements of electric units are to be transmitted, practice favors those systems which place a minimum of burden upon the primary measuring instruments and preferably those adapted to transmitters having no moving parts. Thus, photoelectric, thermoelectric, and capacitive transmitters have found considerable favor in the electric industry.

***276. Integrated Quantities.*** In transmitting measurements originating in integrating meters, such as watthour or varhour meters, the mechanism of the meter, either by photoelectric or electronic means or by a contact arrangement, is caused to develop a series of electrical pulses whose frequency of occurrence is proportional to the instantaneous value of the measured load. By a simple electronic network including capacitors charged and discharged at the frequency of the pulses, there is produced a direct current whose value is proportional to that frequency, the telemetering system being thus placed in the *current* class. On the other hand, the pulses may be directly impressed on the communication channel, whereupon the system falls into the *frequency* group.

***277. Torque Balance.*** Where the basic measurement is performed by a low-torque instrument of the direct-deflecting class, such as a wattmeter, common telemetering practice involves either *balancing the torque* or matching the deflection of the instrument by the effect of an automatically regulated direct current in the winding of a permanent-magnet moving-coil mechanism (Fig. 3-100). This current, remaining proportional to the instrument torque, is transmitted through a metallic circuit for measurement at the receiving station and, if desired, may be included with other and similar currents in a load totalization.

***278. Thermal Converter.*** A most flexible method for the transmission and totalization of electric power measurements involves the use of a *thermal converter* (Fig. 3-101). The several commercial forms of this device operate on a long-known but only recently applied principle combining the circuit of the thermal wattmeter with that of the thermocouple. In the former, the temperatures of two resistors are caused to assume values differing by an amount proportional to the power in the measured circuit. In the latter, there is developed an emf proportional to the temperature difference or to the *power* in the measured circuit, irrespective of power factor, frequency, or waveform. Thermal converters are supplied in single-element, two-element, and three-element forms, and the ac input circuits may be wired into the instrument-transformer secondaries on any conventional polyphase power system. The output from the dc terminals is either directly measured or interconnected with that of other converters to provide totals of measured loads. The full-load potentials are usually rated at 50 or 100 mV, according to make and type, and measurement is preferably made with a self-balancing potentiometer. For best results, thermal-converter output circuits, which, of course, must be wholly metallic, should be well shielded from parasitic electrical effects and should preferably be in a sheathed cable. An advantage of thermal-converter installations, even for relatively short distances within the plant, is that the seven or eight conductors necessary for connecting instrument-transformer secondaries to wattmeters or varmeters are replaced by two small wires operating at a negligible power level. Furthermore, physical damage to the output wiring, whether in the nature of an open circuit or a short circuit, is not hazardous to equipment or personnel, and upon restoration of the circuit, normal operation will be resumed without loss of accuracy.

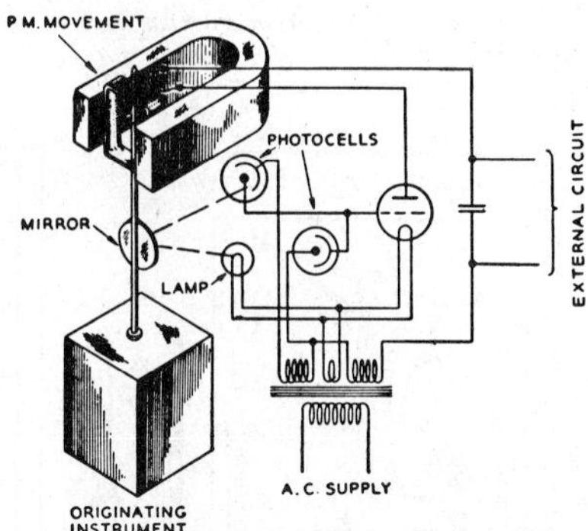


**FIG. 3-100** Photoelectric torque balance.

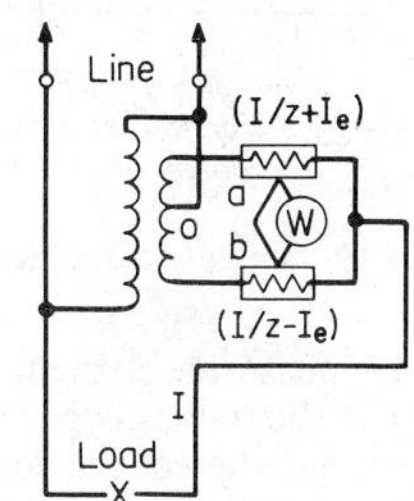


**FIG. 3-101** Thermal watt converter.

***279. Electrical impulses*** may be used as signals for telemetering in a number of ways, the most important in stationary installations being that based on *frequency* and that based on *duration* of successive impulses. Impulse systems are to telemetering what telegraphy is to other forms of communication. The function of the transmitting instrument is essentially one of "keying" a circuit. Since the significance of the transmitted signal is based on time only, it follows that the method is most nearly immune to circuit conditions, such as voltage variation, impedance changes, attenuation, poor connections, and pickup from adjacent disturbing influences. Impulses whose frequency represents the measured variable may be transmitted as such, then falling into the category of the *frequency system of telemetering,* or they may be converted into a proportional direct current and be classified with the *current* systems.

***280. Impulse-Duration Telemetering.*** Signals recur at uniform intervals, and each has a duration corresponding to the then existing value of the measured magnitude. The transmitting instrument includes a constantly running cam or scroll plate having a spiral trailing edge and operating in the plane of the pointer but perpendicular to the line of excursion (Fig. 3-102). At a fixed point in each revolution of the cam, the pointer is engaged and brought against the cam face until subsequently released by the trailing edge. With engagement and disengagement, the pointer is slightly deflected perpendicular to its line of travel and actuates a contact in a signal circuit. Because of the spiral form of the trailing edge, the length of the signal depends upon the position of the pointer and thus represents the measured variable.

The receiving unit includes the equivalent of a pair of electromagnetic clutches continuously driven by a constant-speed motor. These clutches are actuated by the incoming signals, one in an "upscale" and the other in a "downscale" sense, according to whether the transmitter pointer is on or off the cam. The receiver pointer or pen is frictionally retained in position and is "nudged" alternately toward one end or the other of its range by impellers or dogs carried by the clutches and respectively reset to zero as the corresponding clutch is released. Thus, with each signal, the receiver pointer finds or maintains a position corresponding to that of the transmitter pointer.

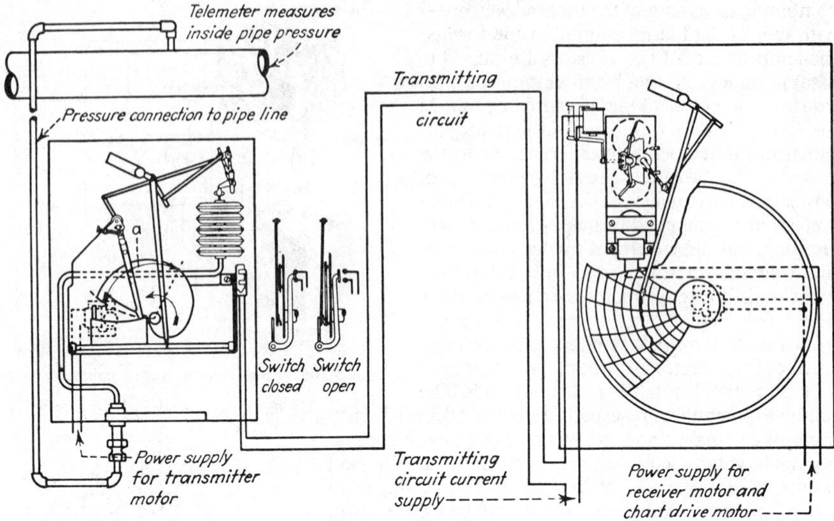


**FIG. 3-102** Impulse-duration telemetering.

If the measuring element follows a linear law, the cam is shaped to an arithmetical spiral. If nonlinear, the trailing edge may be made to provide the needed compensation. For example, in a flowmeter, wherein the deflection is basically proportional to the square of the rate of flow, correction for the quadratic law may be incorporated in the cam contour, with resulting linear signals. Not only does this produce a uniform scale in the receiving instrument, but it enables integration to be accomplished by driving a timing train in response to the upscale signals. The system being inherently "sampling," the transmitter pointer is free to deflect during those intervals when it is not in contact with the cam. Thus, the impulse-duration system may be applied to even a delicate electrical instrument mechanism without loss of accuracy. Since each operating cycle of the receiver is established by the incoming impulses from the transmitter, such synchronism as is necessary is inherent in the principle, it being required only that both units run at constant and properly related speeds. Furthermore, upon circuit trouble or power interruption the system is self-restoring as soon as normal service is resumed. The signals, being in the nature of discrete impulses, may, of course, be transmitted by carrier, by radio, or by microwave.

***281. Position Telemetering.*** In the *position system* of telemetering, the characteristic signal involves the relationship between two electrical quantities of a similar nature, that is, two voltages or two currents. Unless carrier is used, position systems (with one exception[1]) require an interconnecting circuit of three or more conductors. The simplest position-telemetering arrangements are those of the rheostatic, or bridge, type, either direct or alternating current. Mechanical attachment of the measuring element to a voltage-dividing resistor provides a transmitting unit wherein the relative value of two voltages may be made proportional to the measured quantity. The receiving instrument may take the form of a ratio meter or may be a self-balancing bridge. The accuracy of such systems is affected by the impedance of the interconnecting circuit, but by maintaining this value small in comparison with that of the terminal instruments, the error may be made negligible for considerable lengths of line.

***282. Selsyn.*** The inductive type of position system is best exemplified in the "selsyn" position motor or any one of its several equivalents. The transmitting and receiving units (Fig. 3-103) may be identical in structure. Each involves a stator and a rotor, one being provided with a single-phase and the other with a polyphase winding. The single-phase windings are excited from a common ac source, and the polyphase windings are interconnected. The rotors of the two units will tend to assume duplicate angular positions, so that if one is attached to a

---

[1] The *rectifier* system: see Borden and Thynell (p. 70) in Bibliography (Par. **284**).

measuring element, the other will provide a remote indication of its position. This system requires three line conductors in addition to the pair comprising the common power supply. The versatility and flexibility of the *differential transformer* (Par. 242) render it particularly adaptable to telemetering of mechanical displacements.

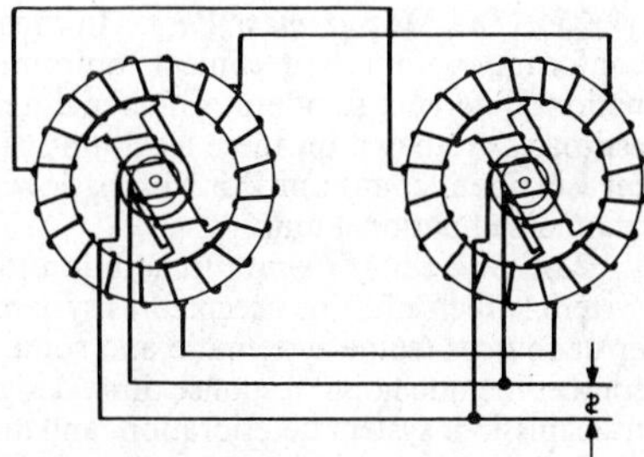

**FIG. 3-103** Position motor.

***283. Totalization*** of power loads and of other measured quantities may readily be effected in the current or voltage systems by connecting the outputs of the respective transmitters in parallel or in series, as the case may be. Subtotals and other mathematical functions also may be obtained. Telemetering, especially totalization and retransmission, is greatly facilitated by the power and flexibility of servo-actuated potentiometers and bridges. With these instruments available, there is practically no limit to the possibilities of telemetering, not only in the electrical-utility field but in association with pipelines and large industrial plants. By *multiplexing*[1] the circuits, it is possible for several telemetering transmitters and receivers to share a common communication channel. The most common systems of multiplexing are those based on *frequency* and those based on *time.* The frequency method transmits the signals on carriers having a specific frequency allotted to each transmitter and receiver combination. Time multiplexing involves the use of a multiple-point switch at each end of the circuit. These switches are progressively advanced at definite intervals, providing connection successively between each receiver and its corresponding transmitter. After a predetermined number of operations a distinct synchronizing signal checks and, if necessary, adjusts the relative position of the switches at the transmitting and the receiving stations.

## Bibliography

***284. Publications Containing Telemetering Bibliographies***


Stabelin, W. "Die Technik der Fernwirkanlagen"; Munich, R. Oldenbourg KG, 1934. (530 refs.)

Telemetering, Supervisory Control and Associated Circuits: Joint Subcommittee Report, AIEE, 1948. (43 refs.) (Under revision, 1956.)

Borden, P. A., and Thynell, G. M. "Principles and Methods of Telemetering"; New York, Reinhold Publishing Corporation, 1948. (80 refs. and extensive list of patents.)

Mabey, C. A. Bibliography on Telemetering; *AIEE Publ.* S68, December 1954. (850 refs.)

Perry, C. C., and Lissner, H. R. "The Strain Gage Primer"; New York, McGraw-Hill Book Company, 1955.


## *MEASUREMENT ERRORS*

***285. General.*** The complete statement of any measurement result has three elements: the *unit* in terms of which the result is stated; a *numeric* which states the magnitude of the result in terms of the chosen unit; its *uncertainty,* the experimenter's estimate of the range within which the result may differ from the actual value of the quantity. Any physical measurement is uncertain to some degree and errors are present in all phases of the measurement process, including the standards used to calibrate the system. Values assigned to local reference standards have uncertainties accumulated from the entire measurement chain extending back to the national reference standards that maintain a common measurement base. These national reference standards are themselves the experimental realization of the units defined in terms of the base mechanical units of length, mass, and time interval; and their assignments include an uncertainty estimate.

[1] Jacobson, A. W. A Time-Multiplex System for Impulse-Duration Telemetering; *Proc. Natl. Telemetering Conf.,* Chicago, 1953, p. 86.

***286. The Measurement Base.*** In most measurements, we are concerned only with their conformity within the technical community in which we work; our error chain stops at the national reference standards which maintain the *legal* units of the country, and our uncertainty estimates are based on these legal units. Rarely, when our concern is with the international measurement community or with basic science, must our uncertainty also include that of the maintained national unit.

***287. Sources of Error.*** In addition to uncertainties in the calibration of a measurement system (which must be accepted as systematic errors in its operation), there are a number of error sources (some systematic and some random) in its operation. These operational error sources include noise, response time, design limitations, energy required by the system, signal transmission, system deterioration, and ambient influences.

*Noise* is any signal that does not convey measurement information. Disturbances generated within the system or coming from outside make up the background against which the desired signal must be read. Noise signals may be picked up by electrical or mechanical coupling between an external source and an element of the system and may be amplified within the system. Under the most favorable circumstances, where noise has been minimized by filtering, by component selection, and by shielding and isolation of the system, there are still certain sources of noise present, resulting from the granular nature of matter and energy; the structure of phenomena is not infinitely fine-grained.

These fluctuations may be small compared with the total energy transfer involved in most measurements; yet they do give rise to a noise background that limits the ultimate sensitivity to which a measurement can be carried. Such sensitivity-limiting mechanisms include the brownian motion of a mechanical system, the Johnson noise in a resistance element, the Barkhausen effect in a magnetic element, and others.

The *response time* of a measuring system may contribute to measurement error. If the measured signal is not constant, lag in response results in an indication that depends on a sequence of values over a previous time interval.

*Design limitations* which contribute to measurement uncertainty include friction and resolution. Because a certain minimum force is needed to overcome friction and initiate motion, there results uncertainty in the rest position of an indicator. Resolution is the ability of the observer to distinguish between nearly equal quantities.

In an optical system, resolution is stated as the smallest angle at which points can be iguished as separate. If the components of the optical train were perfect, resolution would mited by the effective aperture of the system and the wavelength of the light used. If a scale is e read to determine magnitudes, resolution is limited to the smallest fraction of a scale ision that can be read with certainty. Most observers will attempt to estimate tenths of a vision, but they generally have individual bias patterns that make a reading uncertain by 0.1 to .2 division.

*Energy* extracted from the measurand to operate the system alters the measurand to some extent, and if the available energy is small, this contributes to error in the result. Where energy is supplied from an auxiliary source, coupling or feedback may alter the measurement result.

In the *transmission* of information from sensor to indicator, the signal may be distorted by selective attenuation or resonance in a communication channel, or it may suffer loss by leakage.

Physical or chemical *deterioration* or other alterations of elements in the system can contribute to measurement error.

Of the *ambient influences* affecting a measurement, temperature is the most pervasive. Other influences, not so universally important, include humidity, smoke and other air contaminants, barometric pressure, and the effect of gravity on an unbalanced system.

***288. Classes of Errors.*** In estimating the uncertainty of a measurement result, two classes of error must be considered: systematic (which bias the result) and random (which produce scatter).

*Systematic errors* are those that are repeated consistently with repetition of the measurement. Errors in the calibration of the system are systematic; uncertainty in the assigned value of a standard used in calibration must be accepted by the user as systematic. Changes of components through aging or deterioration produce systematic errors, as does failure to take into account energy extracted from a low-level source by the system. In attempting to search out and evaluate systematic errors, repetition of the measurement with definite, known changes in those parameters that are under the operator's control can be helpful, as is the use of different instrumentation or a different method. In some instances it is possible to measure something similar to the measurand, which is independently and accurately known.

*Random errors* are accidental, fluctuating in an unpredictable manner. In any repetitive measurement, observations are influenced by many factors, the parameters that the observer cannot control and the residue of those he or she attempts to control. It is reasonable that combinations of these influences that add to produce large excursions are less frequent than those which partly compensate to produce small excursions, since each is equally likely to produce a positive or a negative departure. In effect, the results of a repetitive measurement process approximate a probability distribution, and it is convenient to treat the scatter of such a process as though it followed the laws of probability.

***289. Evaluation of Data.*** Assuming that the data from a repetitive measurement approximate a normal statistical distribution, we say that (excluding systematic errors) the mean of a group of observations of a measurand is the best approximation of its actual value we can make from those data. Further, we estimate the imprecision of the result by certain statistical procedures. These procedures have validity only for random errors; systematic errors are not amenable to statistical treatment.

***290. Standard Deviation.*** A measure of the dispersion of a set of observations is the root mean square of the deviations of individual observations from the mean of the set. $s = \sqrt{\Sigma d_m^2/(n-1)}$, where $n$ is the number of observations and $d_m$ is the departure of an individual from the group mean. If the number is large, the standard deviation is $\sigma = \sqrt{\Sigma d_m^2/n}$. If the number of observations is small, a reasonable approximation of $s$ can be calculated easily and quickly from the range $r$, the difference between the largest and smallest observation of the set $s \cong r/\sqrt{n} (3 \leq n \leq 12)$.

***291. Probable Error.*** The probable error (pe) of an observation is that deviation from the mean for which the chances are equal that it will or will not be exceeded. If the number of observations is large, pe $= \pm 0.6745\sigma$. While this figure correctly expresses the range in which the chances are equally good that the actual value of the measurand will or will not be found (excluding systematics), it has actually no more significance as a precision index than the standard deviation from which it is derived. Thus, pe has fallen into disuse in current practice, although it was much used in the earlier literature as an index of precision. The pe of the mean of a set of observations is the amount by which the group mean can be expected to differ from the actual value of the measurand (excluding systematics) with a 50% probability. It may be calculated as $\pm 0.6745\sigma/\sqrt{n}$.

***292. Confidence Intervals.*** Probable error is a special case of a broader concept. A confidence interval is the range of deviation from the mean within which a certain fraction of the observed values may be expected to lie, and the probability that the value of a randomly selected observation will lie within this range is called the confidence level. If the number of observations is large and follows a normal distribution, various confidence levels and their confidence intervals about the mean ($\mu$) are given in Table 3-10. If the number of observations is small, these intervals must be broadened. On the same assumptions of randomness and normal distribution as before, these intervals are obtained by multiplying $s$ (the small set estimate of standard deviation) by the factors of Table 3-11.

***293. Comparison of Averages.*** If two sets of measurements taken under different conditions yield different average values ($A$ and $B$) of a quantity with standard deviations ($\sigma_A$ and $\sigma_B$),

**TABLE 3-10.** Confidence Level and Confidence Interval Values Where Number of Observations is Large

| Confidence level | Confidence interval | Values lying outside confidence interval |
|---|---|---|
| 0.50 | $\mu \pm 0.674\sigma$ | 1 in 2 |
| 0.80 | $\mu \pm 1.282\sigma$ | 1 in 5 |
| 0.90 | $\mu \pm 1.645\sigma$ | 1 in 10 |
| 0.95 | $\mu \pm 1.960\sigma$ | 1 in 20 |
| 0.99 | $\mu \pm 2.576\sigma$ | 1 in 100 |
| 0.999 | $\mu \pm 3.291\sigma$ | 1 in 1,000 |

**TABLE 3-11.** Factors for Establishing Confidence Interval Where Number of Observations is Small

| Number of degrees of freedom | Number of observations | Confidence level | | | |
|---|---|---|---|---|---|
| | | 0.5 | 0.9 | 0.95 | 0.99 |
| | | Confidence interval | | | |
| 1 | 2 | $\mu \pm 1.00s$ | $\mu \pm 6.31s$ | $\mu \pm 12.71s$ | $\mu \pm 63.66s$ |
| 2 | 3 | $\mu \pm 0.82s$ | $\mu \pm 2.92s$ | $\mu \pm 4.30s$ | $\mu \pm 9.92s$ |
| 3 | 4 | $\mu \pm 0.77s$ | $\mu \pm 2.35s$ | $\mu \pm 3.18s$ | $\mu \pm 5.84s$ |
| 4 | 5 | $\mu \pm 0.74s$ | $\mu \pm 2.13s$ | $\mu \pm 2.78s$ | $\mu \pm 4.60s$ |
| 5 | 6 | $\mu \pm 0.73s$ | $\mu \pm 2.02s$ | $\mu \pm 2.57s$ | $\mu \pm 4.03s$ |
| 6 | 7 | $\mu \pm 0.72s$ | $\mu \pm 1.94s$ | $\mu \pm 2.45s$ | $\mu \pm 3.71s$ |
| 7 | 8 | $\mu \pm 0.71s$ | $\mu \pm 1.90s$ | $\mu \pm 2.37s$ | $\mu \pm 3.50s$ |
| 8 | 9 | $\mu \pm 0.71s$ | $\mu \pm 1.86s$ | $\mu \pm 2.31s$ | $\mu \pm 3.36s$ |
| 9 | 10 | $\mu \pm 0.70s$ | $\mu \pm 1.83s$ | $\mu \pm 2.26s$ | $\mu \pm 3.25s$ |
| 10 | 11 | $\mu \pm 0.70s$ | $\mu \pm 1.81s$ | $\mu \pm 2.23s$ | $\mu \pm 3.17s$ |
| 15 | 16 | $\mu \pm 0.69s$ | $\mu \pm 1.75s$ | $\mu \pm 2.13s$ | $\mu \pm 2.95s$ |
| ∞ | ∞ | $\mu \pm 0.67s$ | $\mu \pm 1.64s$ | $\mu \pm 1.96s$ | $\mu \pm 2.58s$ |

NOTE: This table is a modification and abridgment of Table IV in Fisher and Yates, "Statistical Tables for Biological, Agricultural, and Medical Research," Edinburgh and London, Oliver & Boyd, Ltd.

their consistency may be questioned. Is the difference $A - B$ consistent with the assumption that random errors alone are operative; or does the difference indicate the presence of systematic errors that operate differently in sets $A$ and $B$? A rough but simple test is to take the sum of their standard deviations. Thus $A$ and $B$ are not consistent (but show a different operation of systematic errors) if $A - B > 2(\sigma_B + \sigma_B)$. There are other tests of consistency that are more sensitive and that may be worth using in some instances. For these the reader should refer to a text on the statistical processing of data.

***294. Bibliography.***

ASTM Manual on Presentation of Data.

Wilson, E. B., Jr. "An Introduction to Scientific Research"; McGraw-Hill, 1952.

Youden, W. J. "Statistical Methods for Chemists"; Wiley, 1951.

National Bureau of Standards Handbook 91, "Experimental Statistics," 1963.

Crochiere, R. E. and Rabiner, L. R. Interpolation and Decimation of Digital Signals—A Tutorial Review; *Proc. IEEE,* March, 1981, vol. 69, no. 3, p. 300.

# SECTION 4
# PROPERTIES OF MATERIALS

**Mark A. Baker**
*Vice President, Technology, Alcoa Conductor Products Company*

**Anthony L. Von Holle**
*Principal Research Engineer, Research and Technology, Armco, Inc.*

**K. L. Latimer**
*Senior Engineer, The Arnold Engineering Company*

**T. W. Dakin**
*Consultant; formerly Manager, Insulation and Chemical Technology Department, Westinghouse Research Laboratories; Fellow, IEEE; National Academy of Engineering*

**E. J. Croop**
*Consulting Scientist, Polymer and Composite Research Department, Westinghouse Research Laboratories*

**Charles A. Harper**
*Manager, Materials Engineering and Technologies, Westinghouse Electric Corporation*

**Robert W. Bohl**
*Professor, Metallurgy and Nuclear Engineering, University of Illinois; Member, American Society for Metals, American Institute of Mining, Metallurgical and Petroleum Engineers, American Nuclear Society, and American Society for Engineering Education*

**Duane E. Lyon**
*Professor of Wood Science, Forest Products Laboratory, Mississippi State University*

## CONTENTS

*Numbers refer to paragraphs*

## CONDUCTOR MATERIALS

*By MARK A. BAKER*

### General Properties

***1. Conducting Materials.*** A conductor of electricity is any substance or material which will afford continuous passage to an electric current when subjected to a difference of electric potential. The greater the density of current for a given potential difference, the more efficient the conductor is said to be. Virtually all substances in solid or liquid state possess the property of electric conductivity in some degree, but certain substances are relatively efficient conductors, while others are almost totally devoid of this property. The metals, for example, are the best conductors, while many other substances, such as metal oxides and salts, minerals, and fibrous materials, are relatively poor conductors, but their conductivity is beneficially affected by the absorption of moisture. Some of the less efficient conducting materials, such as carbon and certain metal alloys, as well as the efficient conductors such as copper and aluminum, have very useful applications in the electrical arts.

Certain other substances possess so little conductivity that they are classed as nonconductors, a better term being insulators or dielectrics. In general, all materials which are used commercially for conducting electricity for any purpose are classed as conductors.

***2. Definition of Conductor.*** A conductor is a body so constructed from conducting material that it may be used as a carrier of electric current. In ordinary engineering usage, a conductor is a material of relatively high conductivity.

***3. Definition of Circuit.*** An electric circuit is the path of an electric current, or, more specifically, it is a conducting part or a system of parts through which an electric current is intended to flow.

***4. General Properties of Conductors.*** Electric circuits in general possess four fundamental electrical properties, consisting of resistance, inductance, capacitance, and leakance. That portion of a circuit which is represented by its conductors will also possess these four properties, but only two of them are related to the properties of the conductor considered by itself. Capacitance and leakance depend in part upon the external dimensions of the conductors and their distances from one another and from other conducting bodies and in part upon the dielectric properties of the materials employed for insulating purposes. The inductance is a function of the magnetic field established by the current in a conductor, but this field as a whole is divisible into two parts, one being wholly external to the conductor and the other being wholly within the conductor; only the latter portion can be regarded as corresponding to the magnetic properties of the conductor material. The resistance is strictly a property of the conductor itself. Both the resistance and the internal inductance of conductors change in effective values when the current changes with great rapidity, as in the case of high-frequency alternating currents; this is termed the "skin effect."

In certain cases, conductors are subjected to various mechanical stresses. Consequently their weight, tensile strength, and elastic properties require consideration in all applications of this character. Conductor materials as a class are affected by changes in temperature and by the conditions of mechanical stress to which they are subjected in service. They are also affected by the nature of the mechanical working and the heat-treatment which they receive in the course of manufacture or fabrication into finished products.

***5. Types of Conductor.*** In general, a conductor consists of a solid wire or a multiplicity of wires stranded together, made of a conducting material and used either bare or insulated. Only bare conductors are considered in this subsection. Usually the conductor is made of copper or aluminum, but for applications requiring higher strength, such as overhead transmission lines, bronze, steel, and various composite constructions are used. For conductors having very low conductivity and used as resistor materials, a group of special alloys is available.

## Metal Properties

***6. Specific Gravity and Density.*** Specific gravity is the ratio of mass of any material to that of the same volume of water at 4°C. Density is the unit weight of material expressed as pounds per cubic inch, grams per cubic centimeter, etc., at some reference temperature, usually 20°C. For all practical purposes, the numerical values of specific gravity and density are the same, expressed in g/cm$^3$.

***7. Density and Weight of Copper.*** Pure copper, rolled, forged, or drawn and then annealed, has a density of 8.89 g/cm$^3$ at 20°C or 8.90 g/cm$^3$ at 0°C. Samples of high-conductivity copper will vary usually from 8.87 to 8.91 and occasionally from 8.83 to 8.94. Variations in density may be caused by microscopic flaws or seams or the presence of scale or some other defect; the presence of 0.03% oxygen will cause a reduction of about 0.01 in density. Hard-drawn copper has about 0.02% less density than annealed copper, on the average, but for practical purposes the difference is negligible.

The international standard of density, 8.89 at 20°C, corresponds to a weight of 0.32117 lb/in$^3$ or $3.0270 \times 10^{-6}$ lb/(cmil)(ft) or $15.982 \times 10^{-3}$ lb/(cmil)(mile). Multiplying either of the last two figures by the square of the diameter of the wire in mils will produce the total weight of wire in pounds per foot or per mile, respectively.

***8. Density and weight of copper alloys*** varies with the composition. For hard-drawn wire covered by ASTM Specification B105, the density of alloys 85 to 20 is 8.89 g/cm$^3$ (0.32117 lb/in$^3$) at 20°C; alloy 15 is 8.54 (0.30853); alloys 13 and 8.5 are 8.78 (0.31720).

***9. Density and weight of copper-clad steel*** wire is a mean between the density of copper and the density of steel, which can be calculated readily when the relative volumes or cross sections of copper and steel are known. For practical purposes a value of 8.15 g/cm$^3$ (0.29444 lb/in$^3$) at 20°C is used.

***10. Density and weight of aluminum wire*** (commercially hard-drawn) is 2.705 g/cm$^3$ (0.0975 lb/in$^3$) at 20°C. The density of electrolytically refined aluminum (99.97% Al) and for hard-drawn wire of the same purity is 2.698 at 20°C. With less pure material there is an appreciable decrease in density on cold working. Annealed metal having a density of 2.702 will have a density of about 2.700 when in the hard-drawn or fully cold-worked conditions (see *NBS Circ.* 346, pp. 68 and 69).

***11. Density and weight of aluminum-clad wire*** is a mean between the density of aluminum and the density of steel, which can be calculated readily when the relative volumes or cross sections of aluminum and steel are known. For practical purposes a value of 6.59 g/cm$^3$ (0.23808 lb/in$^3$) at 20°C is used.

***12. Density and weight of aluminum alloys*** varies with type and composition. For hard-drawn aluminum alloy wire 5005-H19 and 6201-T81, a value of 2.703 g/cm$^3$ (0.09765 lb/in$^3$) at 20°C is used.

***13. Density and weight of pure iron*** is 7.90 g/cm$^3$ [2.690 × $10^{-6}$ lb/(cmil)(ft)] at 20°C.

***14. Density and weight of galvanized steel wire*** (EBB, BB, HTL-85, HTL-135, and HTL-195) with Class A weight of zinc coating is 7.83 g/cm$^3$ (0.283 lb/in$^3$) at 20°C, with Class B is 7.80 g/cm$^3$ (0.282 lb/in$^3$), and with Class C is 7.78 g/cm$^3$ (0.281 lb/in$^3$).

***15. Percent Conductivity.*** It is very common to rate the conductivity of a conductor in terms of its percentage ratio to the conductivity of chemically pure metal of the same kind as the conductor is primarily constituted or in ratio to the conductivity of the international copper standard. Both forms of the conductivity ratio are useful for various purposes.

This ratio can also be expressed in two different terms, one where the conductor cross sections are equal and therefore termed the *volume-conductivity ratio* and the other where the conductor masses are equal and therefore termed the *mass-conductivity ratio.*

***16. The International Annealed Copper Standard (IACS)*** is the internationally accepted value for the resistivity of annealed copper of 100% conductivity. This standard is expressed in terms of mass resistivity as 0.15328 Ω·g/m$^2$, or the resistance of a uniform round wire 1 m long weight 1 g at the standard temperature of 20°C. Equivalent expressions of the annealed copper standard, in various units of mass resistivity and volume resistivity, are as follows:

| | | |
|---|---|---|
| 0.15328 | Ω·g/m$^2$ | |
| 875.20 | Ω·lb/mi$^2$ | |
| 1.7241 | μΩ·cm | |
| 0.67879 | μΩ·in | at 20°C |
| 10.371 | Ω·cmil/ft | |
| 0.017241 | Ω·mm$^2$/m | |

The above values are the equivalent of $\frac{1}{58}$ Ω·mm$^2$/m, so that the volume conductivity can be expressed as 58 S·mm$^2$/m at 20°C.

***17. Conductivity of conductor materials*** varies with chemical composition and processing. For industry specification values, see Table 4-1.

***18. Electrical resistivity*** is a measure of the resistance of a unit quantity of a given material. It may be expressed in terms of either mass or volume; mathematically,

Mass resistivity:
$$\delta = \frac{Rm}{l^2} \tag{4-1}$$

Volume resistivity:
$$\rho = \frac{RA}{l} \tag{4-2}$$

where $R$ = resistance, $m$ = mass, $A$ = cross-sectional area, $l$ = length.

***19. Electrical resistivity of conductor materials*** varies with chemical composition and processing. For industry specification values, see Tables 4-1 and 4-2.

***20. Electrical Conductivity and Resistivity: Nonferrous Conductors.*** See Table 4-1.

***21. Electrical Resistivity: Ferrous Conductors.*** See Table 4-2.

***22. Conversion Factors for Electrical Resistivity and Conductivity.*** See Table 1-14.

***23. Effects of Temperature Changes.*** Within the temperature ranges of ordinary service there is no appreciable change in the properties of conductor materials, except in electrical resistance and physical dimensions. The change in resistance with change in temperature is sufficient to require consideration in many engineering calculations. The change in physical dimensions with change in temperature is also important in certain cases, such as in overhead spans and in large units of apparatus or equipment.

***24. Temperature Coefficient of Resistance.*** Over moderate ranges of temperature, such as 100°C, the change of resistance is usually proportional to the change of temperature. Resistivity is always expressed at a standard temperature, usually 20°C (68°F). In general if $R_{t1}$ is the resistance at a temperature $t_1$, and $\alpha_{t1}$ is the temperature coefficient at that tem-

**TABLE 4-1** Electrical Conductivity and Resistivity*—Nonferrous Conductors

| Specification† | Temper and shape | Size limits, in | Conductivity at 20°C (68°F), IACS, % | Weight resistivity at 20°C (68°F) | | Volume resistivity at 20°C (68°F) | | | |
|---|---|---|---|---|---|---|---|---|---|
| | | | | Ω·g/m² | Ω·lb/mi² | Ω·cmil/ft | Ω·mm²/m | μΩ·cm | μΩ·in |
| **Copper and copper alloy—specific gravity 8.89:** | | | | | | | | | |
| ........ | ........ | ........ | 100.17 | 0.15302‡ | 873.75 | 10.354 | 0.017213 | 1.7213 | 0.67767 |
| B4<br>B5<br>B170<br>B187<br>B188<br>B298<br>B75 | Low-resistance Lake wirebar<br>Electrolytic wirebar<br>Oxygen-free wirebar<br>Soft bus bar, rod and shape<br>Soft bus tube<br>Silver-coated soft, round<br>Tube, soft, OF copper | All | 100.00 | **0.15328**§ | 875.20 | 10.371 | 0.017241 | 1.7241 | 0.67879 |
| B49<br>B3<br>B48 | Hot-rolled rod<br>Soft, round<br>Soft, rectangular | 1.375–0.250<br>All<br>All | 100.00 | 0.15328 | **875.20** | 10.371 | 0.017241 | 1.7241 | 0.67879 |
| ........<br>........<br>........ | ........<br>........<br>........ | ........<br>........<br>........ | 99.50<br>99.00<br>98.50 | 0.15405<br>0.15482<br>0.15561 | 879.60<br>884.04<br>888.53 | 10.423<br>10.476<br>10.529 | 0.017328<br>0.017416<br>0.017504 | 1.7328<br>1.7416<br>1.7504 | 0.68220<br>0.68565<br>0.68913 |
| B187 | Hard bus bar, rod and shape | Over 1 in OD<br>Over 0.375 × 4 | 98.40 | **0.15577** | 889.42 | 10.539 | 0.017521 | 1.7521 | 0.68981 |
| B188 | Hard bus tube, rectangular or square | Over 6 in | | | | | | | |
| B75 | Tube, soft, DLP copper | All | 98.16 | **0.15614** | | | | | |
| ........<br>B188 | ........<br>Hard bus tube, rectangular or square | ........<br>Up to 6 × 3/16 in wall | 98.00<br>97.80 | 0.15640<br>**0.15673** | 893.06<br>894.90 | 10.583<br>10.604 | 0.017593<br>0.017629 | 1.7593<br>1.7629 | 0.69265<br>0.69406 |
| B2<br>B33<br>B189 | Medium hard, round<br>Tinned soft, round<br>Lead-coated soft, round | 0.460–0.325<br>0.460–0.290<br>0.460–0.290 | 97.66 | 0.15694 | **896.15** | 10.619 | 0.017654 | 1.7654 | 0.69504 |
| ........ | ........ | ........ | 97.50 | 0.15721 | 897.64 | 10.637 | 0.017683 | 1.7683 | 0.69620 |
| B187 | Hard bus bar and rod | Up to 1 in OD<br>Up to 0.375 × 4 | | | | | | | |
| B188 | Hard bus tube, round<br>Hard bus pipe, IPS and extra strong<br>Hard bus tube, rectangular or square | Over 1 in OD<br>Over 4 in OD<br>Up to 6 in, over 3/16 in wall | 97.40 | **0.15737** | 898.55 | 10.648 | 0.017701 | 1.7701 | 0.69690 |
| B75<br>B372 | Tube, hard, OF copper<br>Waveguide tube, OF copper | All | | | | | | | |

**TABLE 4-1** Electrical Conductivity and Resistivity*—Nonferrous Conductors (*Continued*)

| Specification† | Temper and shape | Size limits, in | Conductivity at 20°C (68°F), IACS, % | Weight resistivity at 20°C (68°F) | | Volume resistivity at 20°C (68°F) | | | |
|---|---|---|---|---|---|---|---|---|---|
| | | | | Ω·g/m² | Ω·lb/mi² | Ω·cmil/ft | Ω·mm²/m | μΩ·cm | μΩ·in |
| ..... | ..... | ..... | 97.30 | 0.15753 | 899.49 | 10.659 | 0.017720 | 1.7720 | 0.69763 |
| B1 | Hard, round | 0.460–0.325 | | | | | | | |
| B47 and B116 | Hard trolley wire | All | 97.16 | 0.15775 | **900.77** | 10.674 | 0.017745 | 1.7745 | 0.69863 |
| B33<br>B189 | Tinned soft, round<br>Lead-coated soft, round | 0.289–0.103 | | | | | | | |
| ..... | ..... | ..... | 97.00 | 0.15802 | 902.27 | 10.692 | 0.017775 | 1.7775 | 0.69979 |
| B2 | Medium hard, round | 0.324–0.0403 | 96.66 | 0.15857 | **905.44** | 10.729 | 0.017837 | 1.7837 | 0.70224 |
| B188 | Hard bus pipe, SPS and extra strong | Up to 4 in OD | 96.60 | **0.15865** | 905.86 | 10.734 | 0.017845 | 1.7845 | 0.70257 |
| ..... | ..... | ..... | 96.50 | 0.15884 | 906.94 | 10.747 | 0.017867 | 1.7867 | 0.70341 |
| B1<br>B33<br>B189 | Hard, round<br>Tinned soft, round<br>Lead-coated soft, round | 0.324–0.0403<br>0.102–0.0201<br>0.102–0.0201 | 96.16 | 0.15940 | **910.15** | 10.785 | 0.017930 | 1.7930 | 0.70590 |
| B75<br>B372 | Tube, hard, DLP copper<br>Waveguide tube, DLP copper | All | 96.16 | **0.15940** | 910.15 | 10.785 | 0.017930 | 1.7930 | 0.70590 |
| B355 | Nickel-coated soft, round, Class 2 | All | 96.00 | 0.15966 | **911.67** | 10.803 | 0.017960 | 1.7960 | 0.70708 |
| B33<br>B189 | Tinned soft, round<br>Lead-coated soft, round | 0.0200–0.0111 | 94.16 | 0.16279 | **929.52** | 11.015 | 0.018312 | 1.8312 | 0.72092 |
| B355 | Nickel-coated soft, round, Class 4 | All | 94.0 | 0.16306 | **931.06** | 11.033 | 0.018342 | 1.8342 | 0.72212 |
| B246 | Tinned medium hard, round | 0.2043–0.103 | 93.22 | 0.16443 | **938.85** | 11.125 | 0.018495 | 1.84949 | 0.72816 |
| B33<br>B189 | Tinned soft, round<br>Lead-coated soft, round | 0.0110–0.0030 | 93.15 | 0.16454 | **939.51** | 11.133 | 0.018508 | 1.8508 | 0.72867 |
| B246 | Tinned hard, round | 0.2043–0.103 | 92.72 | 0.16532 | **943.92** | 11.185 | 0.018595 | 1.85947 | 0.73209 |
| B246 | Tinned medium hard, round | 0.103–0.0508 | 92.51 | 0.16569 | **946.06** | 11.211 | 0.018637 | 1.86369 | 0.73375 |
| B246 | Tinned hard, round | 0.103–0.0508 | 91.96 | 0.16668 | **951.72** | 11.278 | 0.018748 | 1.87484 | 0.73814 |
| B355 | Nickel-coated soft, round, Class 7 | All | 91.0 | 0.16844 | **961.76** | 11.397 | 0.018947 | 1.8947 | 0.74593 |
| ..... | ..... | ..... | 90.00 | 0.17031 | 972.45 | 11.524 | 0.019157 | 1.9157 | 0.75421 |
| B355 | Nickel-coated soft, round, Class 10 | All | 88.0 | 0.17418 | **994.55** | 11.785 | 0.019592 | 1.9592 | 0.77136 |
| B105 | Hard round, alloy 85 | All | 85.00‖ | 0.18039 | **1,030** | 12.206 | 0.020291 | 2.0291 | 0.79885 |
| B105<br>B9 | Hard round, alloy 80<br>Trolley wire, alloy 80 | All | 80.00‖ | 0.19160 | **1,094** | 12.964 | 0.021551 | 2.1551 | 0.84849 |
| B355 | Nickel-coated soft, round, Class 27 | All | 71.0 | 0.21588 | **1,232.7** | 14.607 | 0.024284 | 2.4284 | 0.95605 |
| B105<br>B9 | Hard round, alloy 65<br>Trolley wire, alloy 65 | All | 65.00‖ | 0.23573 | **1,346** | 15.950 | 0.026516 | 2.6516 | 1.0439 |

| | | | | | | | | | |
|---|---|---|---|---|---|---|---|---|---|
| B105<br>B9 | Hard round, alloy 55<br>Trolley wire, alloy 55 | All | 55.00‖ | 0.27864 | 1,591 | 18.854 | 0.031343 | 3.1343 | 1.2340 |
| B105<br>B9 | Hard round, alloy 40<br>Trolley wire, alloy 40 | All | 40.00‖ | 0.38320 | 2,188 | 25.928 | 0.043103 | 4.3103 | 1.6970 |
| B105 | Hard round, alloy 30 | All | 30.00‖ | 0.51086 | 2,917 | 34.567 | 0.057465 | 5.7465 | 2.2624 |
| B105 | Hard round, alloy 20 | All | 20.00‖ | 0.76638 | 4,376 | 51.856 | 0.086207 | 8.6207 | 3.3940 |
| *Resistivity temperature constant* | | | | 0.000597 | 3.41 | 0.0409 | 0.0000681 | 0.00681 | 0.00268 |
| **Copper alloy—specific gravity 8.54:** | | | | | | | | | |
| B105 | Hard, round, alloy 15 | All | 15.00‖ | 0.98162 | **5,605** | 69.142 | 0.11494 | 11.494 | 4.5253 |
| *Resistivity temperature constant* | | | | 0.000597 | 3.41 | 0.0409 | 0.0000681 | 0.00681 | 0.00268 |
| **Copper alloy—specific gravity 8.78:** | | | | | | | | | |
| B105 | Hard, round, alloy 13 | All | 13.00‖ | 1.1645 | **6,649** | 79.778 | 0.13263 | 13.263 | 5.2215 |
| B105 | Hard, round, alloy 8.5 | All | 8.50‖ | 1.7809 | **10,169** | 122.01 | 0.20284 | 20.284 | 7.9857 |
| *Resistivity temperature constant* | | | | 0.000597 | 3.41 | 0.0409 | 0.0000681 | 0.00681 | 0.00268 |
| **Aluminum and aluminum alloy—specific gravity 2.703:** | | | | | | | | | |
| B233 | Redraw rod 1350-0 | 1.000 | 61.8 | 0.075410 | **430.59** | 16.782 | 0.027899 | 2.7899 | 1.0983 |
| | Redraw rod 1350-H12 and -H22 | | 61.5 | 0.075778 | **432.69** | 16.864 | 0.028035 | 2.8035 | 1.1037 |
| | Redraw rod 1350-H14 and EC-H24 | | 61.4 | 0.075901 | **433.39** | 16.891 | 0.028080 | 2.8080 | 1.1055 |
| | Redraw rod 1350-H16 and EC-H26 | | 61.3 | 0.076025 | **434.10** | 16.919 | 0.028126 | 2.8126 | 1.1073 |
| B236 | Bus bar | All | 61 | **0.07640** | 436.24 | 17.002 | 0.028264 | 2.8624 | 1.1128 |
| B230<br>B262<br>B323<br>B314 | Hard, round<br>Three-quarter hard, round<br>Half hard, round<br>Half hard, round | All | 61.0 | 0.076399 | 436.23 | **17.002** | 0.028265 | 2.8625 | 1.1128 |
| B324 | All tempers, rectangular | All | 61.0 | 0.076397 | **436.24** | 17.002 | 0.028264 | 2.8624 | 1.1128 |
| B317 | T64 temper, extruded alloy 6101 | All | 59.5 | **0.0782** | 446.74 | 17.430 | 0.028976 | 2.8976 | 1.1408 |
| B317 | H111 temper, extruded alloy 6101 | All | 59.0 | **0.0789** | 450.52 | 17.578 | 0.029222 | 2.9222 | 1.1505 |
| B317 | T61 temper, extruded alloy 6101 | All | 57.0 | **0.0817** | 466.33 | 18.195 | 0.030247 | 3.0247 | 1.1909 |
| B317 | T65 temper, extruded alloy 6201 | | 56.5 | 0.082401 | 470.50 | 18.356 | 0.030518 | 1.2015 | 3.0518 |
| B317 | T63 temper, extruded allov 6101 | All | 56.0 | **0.0831** | 474.66 | 18.520 | 0.030788 | 3.0788 | 1.2121 |
| B317 | T6 temper, extruded alloy 6101 | All | 55.0 | **0.0846** | 483.29 | 18.856 | 0.031347 | 3.1347 | 1.2342 |
| B396 | Hard round, alloy 5005 | All | 53.5 | 0.087106 | 497.38 | **19.385** | 0.032226 | 3.2226 | 1.2687 |
| B398 | Hard round, alloy 6201 | All | 52.5 | 0.088764 | 506.85 | **19.754** | 0.032839 | 3.2839 | 1.2929 |
| *Resistivity temperature constant* | | | | 0.000305 | 1.74 | 0.0689 | 0.000115 | 0.0115 | 0.00451 |
| **Copper-clad steel—specific gravity 8.15:** | | | | | | | | | |
| B227 | Hard, round, grades 40 HS and EHS | All | 40‖ | 0.035837 | 2,046.3 | **26.45** | 0.043971 | 4.3971 | 1.7311 |
| B227 | Hard, round, grades 30 HS and EHS | All | 30‖ | 0.047773 | 2,727.8 | **35.26** | 0.058617 | 5.8617 | 2.3078 |
| *Resistivity temperature constant* (40%) | | | | 0.00135 | 7.68 | 0.100 | 0.000167 | 0.0167 | 0.00657 |
| *Resistivity temperature constant* (30%) | | | | 0.00179 | 10.24 | 0.134 | 0.000222 | 0.0222 | 0.00875 |
| **Aluminum-clad steel—specific gravity 6.59:** | | | | | | | | | |
| B415 | Hard, round | 0.204–0.080 | 20‖ | 0.55886 | 3,191.0 | **51.01** | 0.084805 | 8.4805 | 3.3384 |
| *Resistivity temperature constant* | | | | 0.00200 | 11.40 | 0.184 | 0.000306 | 0.0306 | 0.0121 |

* The value established as standard is indicated in boldface type; other values given are calculated.
† ASTM unless otherwise noted.
‡ Matthiessen's standard.
§ International Annealed Copper Standard (IACS).
‖ Nominal value.
NOTE: 1 in = 2.54 cm.

**TABLE 4-2** Electrical Resistivity—Ferrous Conductors

| Material | ASTM specification | Weight resistivity at 20°C (68°F) | | Volume resistivity at 20°C (68°F) | | | |
|---|---|---|---|---|---|---|---|
| | | Maximum | Average | Range | Average | Range | Average |
| | | Ω·lb/mi² | | Ω·cmil/ft | | μΩ·cm | |
| Pure iron | ........ | 4,410 | ..... | ....... | 58.83 | ......... | 9.78 |
| Contact rails: | | | | | | | |
| Open hearth | ........ | ..... | ..... | 100–135 | ...... | 16.6–22.4 | |
| Mild to soft steel | ........ | ..... | ..... | 77–95 | ...... | 12.8–15.8 | |
| Telephone and telegraph (galvanized): | | | | | | | |
| Extra Best Best (EBB) | A111 | 5,000 | | | | | |
| Best Best (BB) | A111 | 5,600 | | | | | |
| Grade 85 | A326 | 5,800 | | | | | |
| Grade 135 and 195 | A326 | 6,500 | | | | | |
| Commercial galvanized: | | | | | | | |
| Siemens-Martin | ........ | ..... | 7,280 | ....... | 97.9* | ......... | 16.3* |
| High strength | ........ | ..... | 9,000 | ....... | 121* | ......... | 20.1* |
| Extra-high strength | ........ | ..... | 9,360 | ....... | 126* | ......... | 20.9* |

*Calculated from average weight resistivity, with average specific gravity 7.83.
NOTE: 1 lb = 0.4536 kg; 1 mi² = 2.589 × 10⁶ m².

perature, the resistance at some other temperature $t_2$ is expressed by the formula

$$R_{t2} = R_{t1}[1 + \alpha_{t1}(t_2 - t_1)] \tag{4-3}$$

Over wide ranges of temperature the linear relationship of this formula is not usually applicable, and the formula then becomes a series involving higher powers of $t$, which is unwieldy for ordinary use.

When the temperature of reference $t_1$ is changed to some other value, the coefficient changes also. Upon assuming the general linear relationship between resistance and temperature previously mentioned, the new coefficient at any temperature $t$ within the linear range is expressed

$$\alpha_t = \frac{1}{(1/\alpha_{t1}) + (t - t_1)} \tag{4-4}$$

The reciprocal of $\alpha$ is termed the inferred absolute zero of temperature. Equation (4-3) takes no account of the change in dimensions with change in temperature and therefore applies to the case of conductors of constant mass, usually met in engineering work. For a more extended discussion of this subject see J. H. Dellinger, The Temperature Coefficient of Resistance of Copper, *NBS Bull.*, 1911, Vol. 8, pp. 71–101; also see *NBS Handbook* 100, Copper Wire Tables.

*The coefficient for copper of less than standard (or 100%) conductivity* is proportional to the actual conductivity, expressed as a decimal percentage. Thus, if $n$ is the percentage conductivity (95% = 0.95), the temperature coefficient will be $\alpha'_t = n\alpha_t$, where $\alpha_t$ is the coefficient of the annealed copper standard.

The coefficients given in Table 4-3 were computed from the formula

$$\alpha_1 = \frac{1}{[1/n(0.00393)] + (t_1 - 20)} \tag{4-5}$$

*The inferred absolute zero of temperature,* upon assuming a linear relationship between resistance and temperature, is given as quantity $-T$ in the last column of Table 4-3. At the absolute zero of temperature, the resistance would be zero (see Fig. 4-1).

The coefficient changes with the temperature of reference as shown in Table 4-3.

***25. Temperature-Resistance Coefficients for Copper.*** See Table 4-3.

***26. Temperature-resistance coefficients for copper alloys*** usually can be approximated by multiplying the corresponding coefficient for copper (100% IACS) by the alloy conductivity

**TABLE 4-3** Temperature-Resistance Coefficients for Aluminum and Copper

| Conductivity IACS, % | Temperature, deg C | | | | | | Temperature − *T* for inferred-zero resistance,* deg C |
|---|---|---|---|---|---|---|---|
| | 0 | 15 | 20 | 25 | 30 | 50 | |
| | Temperature coefficient of resistance, $\alpha_1$ per deg C | | | | | | |
| **Aluminum** | | | | | | | |
| 55 | 0.00392 | 0.00370 | 0.00363 | 0.00357 | 0.00351 | 0.00328 | 255.2 |
| 56 | 0.00400 | 0.00377 | 0.00370 | 0.00363 | 0.00357 | 0.00333 | 250.3 |
| 57 | 0.00407 | 0.00384 | 0.00377 | 0.00370 | 0.00363 | 0.00338 | 245.6 |
| 58 | 0.00415 | 0.00391 | 0.00383 | 0.00376 | 0.00369 | 0.00344 | 241.0 |
| 59 | 0.00423 | 0.00398 | 0.00390 | 0.00382 | 0.00375 | 0.00349 | 236.6 |
| 60 | 0.00431 | 0.00404 | 0.00396 | 0.00389 | 0.00381 | 0.00354 | 232.3 |
| 60.6 | 0.00435 | 0.00409 | 0.00400 | 0.00393 | 0.00385 | 0.00357 | 229.8 |
| 60.97 | 0.00438 | 0.00411 | 0.00403 | 0.00395 | 0.00387 | 0.00359 | 228.3 |
| 61.0 | 0.00438 | 0.00411 | **0.00403** | 0.00395 | 0.00387 | 0.00360 | 228.1 |
| 61.2 | 0.00440 | 0.00412 | 0.00404 | 0.00396 | 0.00388 | 0.00360 | 227.3 |
| 61.3 | 0.00441 | 0.00413 | 0.00405 | 0.00397 | 0.00389 | 0.00361 | 226.9 |
| 61.4 | 0.00441 | 0.00414 | 0.00406 | 0.00398 | 0.00390 | 0.00362 | 226.5 |
| 61.5 | 0.00442 | 0.00415 | 0.00406 | 0.00398 | 0.00390 | 0.00362 | 226.1 |
| 61.8 | 0.00445 | 0.00417 | 0.00408 | 0.00400 | 0.00392 | 0.00364 | 224.9 |
| 62.0 | 0.00446 | 0.00418 | 0.00410 | 0.00401 | 0.00393 | 0.00365 | 224.1 |
| 63 | 0.00454 | 0.00425 | 0.00416 | 0.00408 | 0.00400 | 0.00370 | 220.3 |
| 64 | 0.00462 | 0.00432 | 0.00423 | 0.00414 | 0.00406 | 0.00375 | 216.5 |
| 65 | 0.00470 | 0.00439 | 0.00429 | 0.00420 | 0.00412 | 0.00380 | 212.9 |
| **Copper** | | | | | | | |
| 95 | 0.00403 | 0.00380 | 0.00373 | 0.00367 | 0.00360 | 0.00336 | 247.8 |
| 96 | 0.00408 | 0.00385 | 0.00377 | 0.00370 | 0.00364 | 0.00339 | 245.1 |
| 97 | 0.00413 | 0.00389 | 0.00381 | 0.00374 | 0.00367 | 0.00342 | 242.3 |
| 97.5 | 0.00415 | 0.00391 | 0.00383 | 0.00376 | 0.00369 | 0.00344 | 241.0 |
| 98 | 0.00417 | 0.00393 | 0.00385 | 0.00378 | 0.00371 | 0.00345 | 239.6 |
| 99 | 0.00422 | 0.00397 | 0.00389 | 0.00382 | 0.00374 | 0.00348 | 237.0 |
| 100 | 0.00427 | 0.00401 | **0.00393** | 0.00385 | 0.00378 | 0.00352 | 234.5 |
| 101 | 0.00431 | 0.00405 | 0.00397 | 0.00389 | 0.00382 | 0.00355 | 231.9 |
| 102 | 0.00436 | 0.00409 | 0.00401 | 0.00393 | 0.00385 | 0.00358 | 229.5 |

* See Par. 24.

Conductivities 95 to 102 from *NBS Handbook* 100.

Coefficient 0.00403 at 20°C for 61.0% conductivity from ASTM Designation B193; others calculated on same basis.

**Boldface** type indicates standard values.

expressed as a decimal. For some complex alloys, however, this relation does not hold even approximately, and suitable values should be obtained from the supplier.

***27. Temperature-resistance coefficient for copper-clad steel wire*** is 0.00378/°C at 20°C.

***28. Temperature-Resistance Coefficients for Aluminum.*** See Table 4-3.

***29. Temperature-resistance coefficients for aluminum-alloy wires*** are: for 5005-H19, 0.00353/°C; for 6201-T81, 0.00347/°C at 20°C.

***30. Temperature-resistance coefficient for aluminum-clad wire*** is 0.0036/°C at 20°C.

***31. Temperature-resistance coefficient for pure iron*** is 0.0064/°C at 20°C. The coefficient, determined by extrapolation from tests on galvanized telephone and telegraph wire, for wire of 100% conductivity at 20°C, is 0.0061/°C.

***32. Temperature-resistance coefficients for galvanized-steel conductors*** are as follows:

| Commercial grade | Approximate temperature coefficient per °C |
|---|---|
| Extra Best Best (EBB) | 0.0056 |
| Best Best (BB), HTL-85 | 0.0046 |
| HTL-135, HTL-195 | 0.0043 |
| High strength | 0.0032 |
| Extra-high strength | 0.0031 |

***33. Temperature-resistance coefficients for typical composite conductors*** are as follows:

| Type | Approximate temperature coefficient per °C at 20°C |
|---|---|
| Copper—copper-clad steel | 0.00381 |
| ACSR (aluminum-steel) | 0.00403 |
| Aluminum-aluminum alloy | 0.00394 |
| Aluminum—aluminum-clad steel | 0.00396 |

***34. Reduction of Observations to Standard Temperature.*** A table of convenient corrections and factors for reducing resistivity and resistance to standard temperature, 20°C, will be found in Copper Wire Tables, *NBS Handbook* 100.

***35. Resistivity-Temperature Constant.*** The *change of resistivity per degree* may be readily calculated, taking account of the expansion of the metal with rise of temperature. The proportional relation between temperature coefficient and conductivity may be put in the following convenient form for reducing *resistivity* from one temperature to another: *The change of resistivity of copper per degree Celsius is a constant, independent of the temperature of reference and of the sample of copper. This "resistivity-temperature constant" may be taken, for general purposes, as* 0.00060 Ω *(meter, gram), or* 0.0068 *μΩ·cm.* More exact values for this constant are given in Table 4-1.

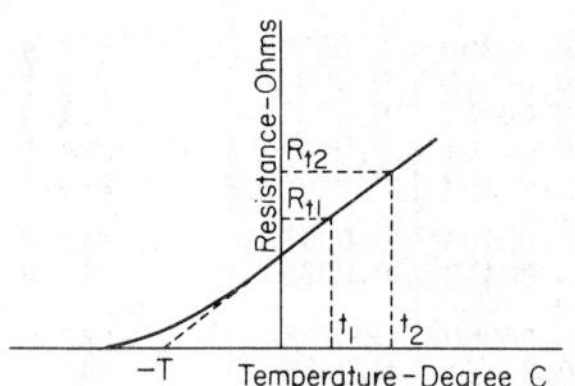


**FIG. 4-1** Resistance-temperature relationship.

Details of the calculation of the resistivity-temperature constant will be found in Copper Wire Tables, *NBS Handbook* 100; also see this reference for expressions for the temperature coefficients of resistivity and their derivation.

***36. Calculation of Percent Conductivity.*** The percent conductivity of a sample of copper is calculated by dividing the resistivity of the International Annealed Copper Standard at 20°C by the resistivity of the sample at 20°C. Either the mass resistivity or the volume resistivity may be used. Inasmuch as the temperature coefficient of copper varies with the conductivity, it is to be noted that a different value will be found if the resistivity at some other temperature is used. This difference is of practical moment in some cases. In order that such differences may not arise, it is best always to use the 20°C value of resistivity in computing the percent conductivity of copper. When the resistivity of the sample is known at some other temperature $t$, it is very simply reduced to 20°C by adding the quantity $20 - t$ multiplied by the resistivity-temperature constant, given in Table 4-1.

***37. Temperature coefficient of expansion*** (linear) of pure metals over a range of several hundred degrees is not a linear function of the temperature but is well expressed by a quadratic equation

$$\frac{L_{t2}}{L_{t1}} = 1 + [\alpha(t_2 - t_1) + \beta(t_2 - t_1)^2] \tag{4-6}$$

Over the temperature ranges for ordinary engineering work (usually 0 to 100°C), the coefficient can be taken as a constant (assumed linear relationship) and a simplified formula employed:

$$L_{t2} = L_{t1}[1 + \alpha_{t1}(t_2 - t_1)] \tag{4-7}$$

Changes in linear dimensions, superficial area, and volume take place in most materials with changes in temperature. In the case of linear conductors, only the change in length is ordinarily important.

The coefficient for changes in superficial area is approximately twice the coefficient of linear expansion for relatively small changes in temperature. Similarly the volume coefficient is three times the linear coefficient, with similar limitations.

***38. Temperature-Expansion Coefficients for Conductors.*** See Table 4-4.

***39. Specific heat of electrolytic tough pitch copper*** is 0.092 cal/(g)(°C) at 20°C (see *NBS Circ.* 73).

***40. Specific heat of aluminum*** is 0.226 cal/(g)(°C) at room temperature (see *NBS Circ.* C447, Mechanical Properties of Metals and Alloys).

***41. Specific heat of iron*** (wrought) or very soft steel from 0 to 100°C is 0.114 cal/(g)(°C); the true specific heat of iron at 0°C is 0.1075 cal/(g)(°C) (see *International Critical Tables,* vol. II, p. 518; also ASM, *Metals Handbook*).

***42. Thermal conductivity of electrolytic tough pitch copper*** at 20°C is 0.934 cal/($cm^2$)(cm)(s)(°C), adjusted to correspond to an electrical conductivity of 101% (see *NBS Circ.* 73).

***43. Thermal-Electrical Conductivity Relation of Copper.*** The Wiedemann-Franz-Lorenz law, which states that the ratio of the thermal and electrical conductivities at a given temperature is independent of the nature of the conductor, holds closely for copper. The ratio $K/\lambda T$ (where $K$ = thermal conductivity, $\lambda$ = electrical conductivity, $T$ = absolute temperature) for copper is 5.45 at 20°C.

***44. Thermal Conductivity of Copper Alloys.***

| ASTM alloy (Spec. B105) | Thermal conductivity (volumetric) at 20 C | |
|---|---|---|
| | Btu per sq ft per ft per hr per deg F | Cal per sq cm per cm per sec per deg C |
| 8.5 | 31 | 0.13 |
| 15 | 50 | 0.21 |
| 30 | 84 | 0.35 |
| 55 | 135 | 0.56 |
| 80 | 199 | 0.82 |
| 85 | 208 | 0.86 |

***45. Thermal Conductivity of Aluminum.*** The determination made by the Bureau of Standards at 50°C for aluminum of 99.66% purity is 0.52 cal/($cm^2$)(cm)(s)(°C) (*Circ.* 346; also see *Smithsonian Physical Tables* and *International Critical Tables*).

***46. Thermal conductivity of iron*** (mean) from 0 to 100°C is 0.143 cal/($cm^2$)(cm)(s)(°C); with increase of carbon and manganese content, it tends to decrease and may reach a figure

**TABLE 4-4** Temperature-Expansion Coefficients for Conductors

| Conductor type | Temperature coefficient | |
|---|---|---|
| | Per °F | Per °C |
| Copper | $9.4 \times 10^{-6}$ | $16.92 \times 10^{-6}$ |
| Copper alloy | $9.4 \times 10^{-6}$ | $16.92 \times 10^{-6}$ |
| Copper-clad steel | $7.2 \times 10^{-6}$ | $12.96 \times 10^{-6}$ |
| Aluminum | $12.8 \times 10^{-6}$ | $23.0 \times 10^{-6}$ |
| Aluminum alloy | $12.8 \times 10^{-6}$ | $23.0 \times 10^{-6}$ |
| Aluminum-clad steel | $7.2 \times 10^{-6}$ | $13.0 \times 10^{-6}$ |
| Pure iron | $6.72 \times 10^{-6}$ | $12.1 \times 10^{-6}$ |
| Galvanized steel: | | |
| Mild steel | $6.22 \times 10^{-6}$ | $11.2 \times 10^{-6}$ |
| ACSR core wire | $6.4 \times 10^{-6}$ | $11.52 \times 10^{-6}$ |
| HTL-85, HTL-135, HTL-195 | $5.7 \times 10^{-6}$ | $10.26 \times 10^{-6}$ |
| Composite conductors: | | |
| Copper–copper-clad: | | |
| Type 2A to 6A | $8.5 \times 10^{-6}$ | $15.30 \times 10^{-6}$ |
| Type F | $9.0 \times 10^{-6}$ | $16.20 \times 10^{-6}$ |
| Type E | $8.4 \times 10^{-6}$ | $15.12 \times 10^{-6}$ |
| Type EK | $8.8 \times 10^{-6}$ | $15.84 \times 10^{-6}$ |
| Copper–copper alloy | $9.4 \times 10^{-6}$ | $16.92 \times 10^{-6}$ |
| Aluminum–aluminum clad: | | |
| AWAC 5/2 | $10.0 \times 10^{-6}$ | $18.00 \times 10^{-6}$ |
| AWAC 4/3 | $9.3 \times 10^{-6}$ | $16.74 \times 10^{-6}$ |
| AWAC 3/4 | $8.6 \times 10^{-6}$ | $15.48 \times 10^{-6}$ |
| AWAC 2/5 | $8.0 \times 10^{-6}$ | $14.40 \times 10^{-6}$ |

of approximately 0.095 with about 1% carbon, or only about half that figure if the steel is hardened by water quenching (see *International Critical Tables,* vol. II, p. 518).

***47. Influence of Chemical Composition.*** The resistivity of most metals is very sensitive to changes in chemical composition or constituents. This applies particularly to the case of relatively small amounts of impurities present in metals which approach closely the chemically pure state. Such impurities may consist of oxides, slag, or traces of various foreign ingredients which escape elimination in the process of reducing the original ores to the metallic state, refining these intermediate products, and fabricating the ingots into the form of conductors by various processes of hot and cold working. The combined effects of such impurities are usually summed up in the percentage of conductivity of the conductor expressed as the ratio to the conductivity of chemically pure metal.

***48. Influence of Mechanical Treatment.*** The fabrication of conductors, from the ingot to the finished state, normally starts with hot rolling and finishes with cold drawing. Annealing operations may or may not take place at intermediate stages or at the finished state. The cold-working operations tend in general to harden the material, reduce its ductility, increase its tensile strength, and very slightly increase its resistivity. The increase in tensile strength is frequently very useful, and consequently many types of conductor are finished by cold working, in which condition they are usually described as hard-drawn.

***49. Copper*** is a highly malleable and ductile metal, of reddish color. It can be cast, forged, rolled, drawn, and machined. Mechanical working hardens it, but annealing will restore it to the soft state. The density varies slightly with the physical state, 8.9 being an average value. It melts at 1083°C (1981°F) and in the molten state has a sea-green color. When heated to a very high temperature, it vaporizes and burns with a characteristic green flame. Copper readily alloys with many other metals. In ordinary atmospheres it is not subject to appreciable corrosion. Its electrical conductivity is very sensitive to the presence of slight impurities in the metal.

Copper when exposed to ordinary atmospheres becomes oxidized, turning to a black color, but the oxide coating is protective, and the oxidizing process is not progressive. When exposed to moist air containing carbon dioxide, it becomes coated with green basic carbonate, which is also protective. At temperatures above 180°C it oxidizes in dry air. In the presence of ammonia it is readily oxidized in air, and it is also affected by sulfur dioxide. Copper is not readily attacked at high temperatures below the melting point by hydrogen, nitrogen, carbon monoxide, carbon dioxide, or steam. Molten copper readily absorbs oxygen, hydrogen, carbon monoxide, and sulfur dioxide, but on cooling, the occluded gases are liberated to a great extent, tending to produce blowholes or porous castings. Copper in the presence of air does not dissolve in dilute hydrochloric or sulfuric acid but is readily attacked by dilute nitric acid. It is also corroded slowly by saline solutions and sea water.

***50. Commercial grades of copper*** in the United States are electrolytic, oxygen-free, Lake, fire-refined, and casting. *Electrolytic copper* is that which has been electrolytically refined from blister, converter, black, or Lake copper. *Oxygen-free copper* is produced by special manufacturing processes which prevent the absorption of oxygen during the melting and casting operations or by removing the oxygen by reducing agents. It is used for conductors subjected to reducing gases at elevated temperature where reaction with the included oxygen would lead to the development of cracks in the metal. *Lake copper* is electrolytically or fire-refined from Lake Superior native copper ores and is of two grades, low resistance and high resistance. *Fire-refined copper* is a lower purity grade intended for alloying or for fabrication into products for mechanical purposes; it is not intended for electrical purposes. *Casting copper* is the grade of lowest purity and may consist of furnace-refined copper, rejected metal not up to grade, or melted scrap; it is exclusively a foundry copper.

***51. Copper content of commercial grades*** is given in the following table:

| Commercial grade | ASTM Designation | Copper content, minimum % |
|---|---|---|
| Electrolytic | B5 | 99.900 |
| Oxygen-free electrolytic | B170 | 99.95 |
| Lake, low resistance | B4 | 99.900 |
| Lake, high resistance | B4 | 99.900 |
| Fire-refined | B216 | 99.88 |
| Casting | B119 | 98 |

***52. Hardening and Heat-treatment of Copper.*** There are but two well-recognized methods for hardening copper; one is by mechanically working it, and the other is by the addition of an alloying element. The properties of copper are not affected by a rapid cooling after annealing or rolling, as are those of steel and certain copper alloys.

***53. Annealing of Copper.*** Cold-worked copper is softened by annealing, with decrease of tensile strength and increase of ductility. In the case of pure copper hardened by cold reduction of area to one-third of its initial area, this softening takes place with maximum rapidity between 200 and 325°C. However, this temperature range is affected in general by the extent of previous cold reduction and the presence of impurities. The greater the previous cold reduction, the lower is the range of softening temperatures. The effect of iron, nickel, cobalt, silver, cadmium, tin, antimony, and tellurium is to lower the conductivity and raise the annealing range of pure copper in varying degrees (see Effect of Iron, Cobalt and Nickel on Some Properties of High-purity Copper and Effect of Certain Fifth-period Elements on Some Properties of High-Purity Copper, by J. S. Smart, Jr., and A. A. Smith, Jr., *Trans. AIME,* 1942, vol. 147, p. 48, and 1943, vol. 152, p. 103). Oxygen content lowers the annealing range.

Trade-named coppers, such as *Hy-Therm, Tensilok,* and *High Thermo,* are produced by adding minute amounts of hardening agents to pure electrolytic copper. These coppers meet industry specifications for purity and conductivity but have higher annealing characteristics than commercial brands, as illustrated in Fig. 4-2. Their higher resistance to annealing permits higher continuous and short-time emergency overload current-carrying capacity of conductors used in overhead lines (see Hy-Therm Copper—An Improved Overhead-Line Conductor, by L. F. Hickernell, A. A. Jones, and C. J. Snyder, *Trans. AIEE,* 1949, vol. 68, pt. 1, p. 22).

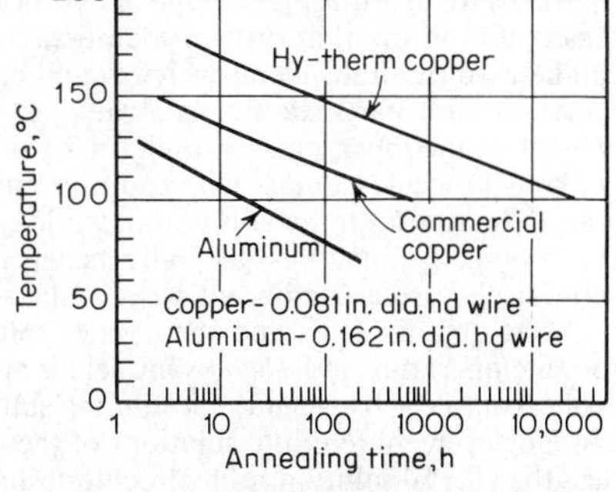


**FIG. 4-2** Annealing characteristics of copper and aluminum for 5% loss in strength, no tension.

***54. Alloying of Copper.*** Elements that are soluble in moderate amounts in a solid solution of copper, such as manganese, nickel, zinc, tin, and aluminum, generally harden it and diminish its ductility but improve its rolling and working properties. Elements that are but slightly soluble, such as bismuth and lead, do not harden it but diminish both the ductility and the toughness and impair its hot-working properties. See *NBS Circ.* 73, 2d ed., 1922, pp. 53–68, for extended discussion. Small additions (up to 1.5%) of manganese, phosphorus, or tin increase the tensile strength and hardness of cold-rolled copper.

*Brass* is usually a binary alloy of copper and zinc, but brasses are seldom employed as electrical conductors, as they have relatively low conductivity though comparatively high tensile strength. In general, brass is not suitable for use where exposed to the weather, owing to the difficulty from stress-corrosion cracking; the higher the zinc content, the more pronounced does this become.

*Bronze* in its simplest form is a binary alloy of copper and tin, in which the latter element is the hardening and strengthening agent. This material is rather old in the arts and has been used to some extent for electrical conductors for many years past, especially abroad. Modern bronzes are frequently ternary alloys, containing as the third constituent such elements as phosphorus, silicon, manganese, zinc, aluminum, or cadmium; in such cases the third element is usually given in the name of the alloy, as phosphor bronze, silicon bronze. Certain bronzes are quaternary alloys or contain two other elements in addition to copper and tin.

In bronzes for use as electrical conductors the content of tin and other metals is usually less than in bronzes for structural or mechanical applications where physical properties and resistance to corrosion are the governing considerations. High resistance to atmospheric corrosion is always an important consideration in selecting bronze conductors for overhead service.

***55. Commercial Grades of Bronze.*** Various bronzes have been developed for use as conductors, and these are now covered by ASTM Specification B105. They all have been designed to provide conductors having high resistance to corrosion and tensile strengths greater than hard-drawn copper conductors. The standard specification covers 10 grades of bronze, designated by numbers according to their conductivities.

***56. Copper-Chromium Alloy.*** A patented alloy of this type contains from 0.5 to 3.0% of chromium. When cast, rolled, heat-treated, and drawn it may have an electrical conductivity of 80% and a tensile strength of 72,000 to 80,000 lb/in$^2$.

***57. Copper-beryllium alloy*** containing 0.4% of beryllium may have an electrical conductivity of 48% and a tensile strength (in 0.128-in wire) of 86,000 lb/in$^2$. A content of 0.9% of beryllium may give a conductivity of 28% and a tensile strength of 122,000 lb/in$^2$. The effect of this element in strengthening copper is about ten times as great as that of tin.

***58. Copper-clad steel wires*** have been manufactured by a number of different methods. The general object sought in the manufacture of such wires is the combination of the high conductivity of copper with the high strength and toughness of iron or steel.

The principal manufacturing processes now in commercial use are: (*a*) coating a steel billet with a special flux, placing it in a vertical mold closed at the bottom, heating the billet and mold to yellow heat, and then casting molten copper around the billet, after which it is hot-rolled to rods and cold-drawn to wire, and (*b*) electroplating a dense coating of copper on a steel rod and then cold drawing to wire.

***59. Aluminum*** is a ductile metal, silver-white in color, which can be readily worked by rolling, drawing, spinning, extruding, and forging. Its specific gravity is 2.703. Pure aluminum melts at 660°C (1220°F). Aluminum has relatively high thermal and electrical conductivities. The metal is always covered with a thin, invisible film of oxide which is impermeable and protective in character. Aluminum, therefore, shows stability and long life under ordinary atmospheric exposure.

Exposure to atmospheres high in hydrogen sulfide or sulfur dioxide does not cause severe attack of aluminum at ordinary temperatures, and for this reason aluminum or its alloys can be used in atmospheres which would be rapidly corrosive to many other metals.

Aluminum parts should, as a rule, not be exposed to salt solutions while in electrical contact with copper, brass, nickel, tin, or steel parts, since galvanic attack of the aluminum is likely to occur. Contact with cadmium in such solutions results in no appreciable acceleration in attack on the aluminum, while contact with zinc (or zinc-coated steel as long as the coating is intact) is generally beneficial, since the zinc is attacked selectively and cathodically protects adjacent areas of the aluminum.

Most organic acids and their water solutions have little or no effect on aluminum at room temperature, although oxalic acid is an exception and is corrosive. Concentrated nitric acid (about 80% by weight) and fuming sulfuric acid can be handled in aluminum containers. However, more dilute solutions of these acids are more active. All but the most dilute (less than 0.1%) solutions of hydrochloric and hydrofluoric acids have a rapid etching action on aluminum.

Solutions of the strong alkalies, potassium, or sodium hydroxides dissolve aluminum rapidly. However, ammonium hydroxide and many of the strong organic bases have little action on aluminum and are successfully used in contact with it (see *NBS Circ.* 346).

Aluminum in the presence of water and limited air or oxygen rapidly converts into aluminum hydroxide, a whitish powder.

***60. Commercial grades of aluminum*** in the United States are designated by their purity, such as 99.99, 99.95, 99.90%.

***61. Electrical conductor alloy aluminum 1350,*** having a purity of approximately 99.5% and a minimum conductivity of 61.0% IACS, is used for conductor purposes. Specified physical properties are obtained by closely controlling the kind and amount of certain impurities.

***62. Annealing of Aluminum.*** Cold-worked aluminum is softened by annealing, with decrease of tensile strength and increase of ductility. The annealing temperature range is affected in general by the extent of previous cold reduction and the presence of impurities. The greater the previous cold reduction, the lower is the range of softening temperatures. Typical annealing characteristics of 1350 aluminum alloy are shown in Fig. 4-2.

***63. Alloying of Aluminum.*** Aluminum can be alloyed with a variety of other elements, with a consequent increase in strength and hardness. With certain alloys, the strength can be further increased by suitable heat-treatment. The alloying elements most generally used are copper, silicon, manganese, magnesium, chromium, and zinc. Some of the aluminum alloys, particularly those containing one or more of the following elements—copper, magnesium, silicon, and zinc—in various combinations, are susceptible to heat-treatment.

Pure aluminum, even in the hard-worked condition, is a relatively weak metal for construction purposes. Strengthening for castings is obtained by alloying elements. The alloys most suitable for cold-rolling seldom contain less than 90 to 95% aluminum. By alloying, working, and heat-treatment it is possible to produce tensile strengths ranging from 8500 $lb/in^2$ for pure annealed aluminum up to 82,000 $lb/in^2$ for special wrought heat-treated alloy, with densities ranging from 2.65 to 3.00.

***64. Electrical conductor alloys of aluminum*** are principally alloys 5005 and 6201 covered by ASTM Specifications B396 and B398.

***65. Aluminum-clad steel wires*** have a relatively heavy layer of aluminum surrounding and bonded to the high-strength steel core. The aluminum layer can be formed by compacting and sintering a layer of aluminum powder over a steel rod, by electroplating a dense coating of aluminum on a steel rod, or extruding a coating of aluminum on a steel rod, and then cold drawing to wire.

***66. Magnesium*** is a ductile metal, silver-white in color, which is distinguished by its light weight (sp. gr. 1.74) and ease of machining.

Pure magnesium has relatively low strength; so its uses are limited to applications where strength is of little importance. However, by alloying magnesium with small amounts of other metals, particularly aluminum, zinc, or manganese or combinations of these, alloys have been developed which show excellent mechanical properties and lead to the high strength-weight ratios of the magnesium products. In a general way the mechanical properties compare favorably with those of the commercial aluminum alloys. The various alloys are divided into those suitable for castings and for wrought products. A number of the standard alloys are susceptible to heat-treatment to improve the properties in general or for the improvement of some particular property.

Magnesium alloys are resistant to attack by alkalies and many organic chemicals. They are attacked by acids of any strength with the exception of pure hydrofluoric and chromic acids. Salt solutions in general corrode the metal, and applications involving continuous contact with saline solutions are not recommended. The alloys are commercially stable against ordinary atmospheric conditions, and for many uses the surface needs no protection.

A number of coatings applied by chemical treatment have been developed to provide a surface stability suitable for more difficult situations.

Pure magnesium is used for ingot, powder, shavings, extruded wire and strip, and rolled ribbon. Sand, permanent-mold, and die castings are available in magnesium alloys. Wrought magnesium alloys are available in extruded round, square, hexagonal, and rectangular bar; as special shapes, moldings, and structural sections; as rolled plate and sheet; and as hammer and hot-pressed forgings.

***67. Silicon*** is a light metal having a specific gravity of approximately 2.34. There is lack of accurate data on the pure metal, because its mechanical brittleness bars it from most industrial uses. However, it is very resistant to atmospheric corrosion and to attack by many chemical reagents. Silicon is of fundamental importance in the steel industry, but for this purpose it is obtained in the form of ferrosilicon, which is a coarse granulated or broken product. It is very useful as an alloying element in steel for electrical sheets and substantially increases the electrical resistivity and thereby reduces the core losses. Silicon is peculiar among metals in the respect that its temperature coefficient of resistance may change sign in some temperature range, the exact behavior varying with the impurities.

***68. Beryllium*** is a light metal having a specific gravity of approximately 1.84 or nearly the same as magnesium. It is normally hard and brittle and difficult to fabricate. Copper is materially strengthened by the addition of small amounts of beryllium, without very serious loss of electrical conductivity. The principal uses for this metal appear to be as an alloying element with other metals, such as aluminum and copper.

***69. Sodium*** is a soft, bright, silvery metal obtained commercially by the electrolysis of absolutely dry fused sodium chloride. It is the most abundant of the alkali group of metals, is extremely reactive, and is never found free in nature. It oxidizes readily and rapidly in air. In the presence of water (it is so light that it floats) it may ignite spontaneously, decomposing the water with evolution of hydrogen and formation of sodium hydroxide. This can be explosive. Sodium should be handled with respect, as it can be dangerous when improperly handled. It melts at 97.8°C, below the boiling point of water, and in the same range as many fuse metal alloys. Sodium is approximately one-tenth as heavy as copper and has roughly three-eighths the conductivity; hence 1 lb of sodium is about equal electrically to 3½ lb of copper. Interest in sodium as an electrical conductor recently was renewed with the development of a means of extruding a polyethylene insulating tube, simultaneously filling it with liquid sodium fed through a tube from a closed container, and cooling them together.

***70. Iron and Steel.*** Iron is a hard tenacious metal which has a silvery-white luster and takes a high polish. It is strongly attracted by a magnet but retains practically no magnetism. It softens at a red heat and may be readily welded at a white heat. Its melting point is higher than that of steel or wrought iron. Iron is very reactive chemically and dissolves in most dilute acids with liberation of hydrogen. In dry air it undergoes no change, but when exposed to atmospheres containing moisture it corrodes more or less rapidly and forms rust (hydrated ferric oxide); such corrosion is accelerated by the presence of carbon dioxide and sulfur dioxide. There are three oxides of iron: FeO, or ferrous oxide, is a black powder; $Fe_2O_3$, or ferric oxide, is known also as hematite and is a steel-gray crystalline substance with considerable luster; $Fe_3O_4$, or ferroso-ferric oxide, is also known as magnetite and is attracted by a magnet but not always magnetic of itself. Iron forms numerous compounds with carbon, sulfur, phosphorus, oxygen, hydrogen, nitrogen, and other elements; it alloys readily with numerous metallic elements such as manganese, silicon, nickel, cobalt, chromium, and tungsten.

The element iron is the base of all commercial iron and steel products, but in order to define these products in reasonably complete terms it is desirable to state both their constituents and the processes by which they were made. Iron and steel are always different substances in the commercial sense and moreover are manufactured in many different commercial varieties. Even the common varieties of steel are somewhat complicated substances; it is insufficient to describe them as alloys of iron and certain other elements, because the alloys take many different forms and impart certain distinctive properties, depending on both the alloy proportions and the processes of manufacture.

***71. Protective Coatings for Iron and Steel.*** Iron and steel wires for outdoor service as conductors or guys are protected from corrosion by the application of zinc or aluminum coatings. Such coatings may be applied by the hot-dip process or by electroplating. These coatings themselves are not impregnable against atmospheric attack but afford the best protection which has yet been found practicable.

***72. Galvanized-iron telephone- and telegraph-line wire*** is available commercially in two grades, EBB and BB, and with three weights of zinc coating. Characteristics are given in ASTM Specification A111. See Tables 4-2 and 4-25.

***73. Galvanized high-tensile steel telephone- and telegraph-line wire*** is available commercially in three grades, 85, 135, and 195, and with three weights of zinc coating. Characteristics are given in ASTM Specification A326. See Tables 4-2 and 4-25.

***74. Steel Rails.*** The resistivities of steel rails are given in Table 4-2.

The effective resistance of steel rails to alternating currents is increased on account of skin effect, which in turn is a function of magnetic permeability (see data in "Report of the Electric Railway Test Commission," New York, McGraw-Hill Book Company, 1906; also Experimental Researches on the Skin Effect in Steel Rails, by A. E. Kennelly, F. H. Achard, and A. S. Dana, *J. Franklin Inst.,* August 1916).

## Conductor Properties

***75. Electrical conductors*** are manufactured in various forms and shapes for various purposes. These may be wires, cables, flat straps, square or rectangular bars, angles, channels, or special designs for particular requirements. The most extensive use of conductors, however, is in the form of round solid wires and stranded conductors. The following terminology describes properly the various terms relating to conductors.

***76. Definitions of Electrical Conductors***

*Wire.* A rod or filament of drawn or rolled metal whose length is great in comparison with the major axis of its cross section.

The definition restricts the term to what would ordinarily be understood by the term "solid wire." In the definition, the word "slender" is used in the sense that the length is great in comparison with the diameter. If a wire is covered with insulation, it is properly called an "insulated wire"; while primarily the term "wire" refers to the metal, nevertheless when the context shows that the wire is insulated, the term "wire" will be understood to include the insulation.

*Conductor.* A wire or combination of wires not insulated from one another, suitable for carrying an electric current.

The term "conductor" is not to include a combination of conductors insulated from one another, which would be suitable for carrying several different electric currents.

Rolled conductors (such as busbars) are, of course, conductors but are not considered under the terminology here given.

*Stranded Conductor.* A conductor composed of a group of wires, usually twisted, or any combination of groups of wires.

The wires in a stranded conductor are usually twisted or braided together.

*Cable.* A stranded conductor (single-conductor cable) or a combination of conductors insulated from one another (multiple-conductor cable).

The component conductors of the second kind of cable may be either solid or stranded, and this kind of cable may or may not have a common insulating covering. The first kind of cable is a single conductor, while the second kind is a group of several conductors. The term "cable" is applied by some manufacturers to a solid wire heavily insulated and lead covered; this usage arises from the manner of the insulation, but such a conductor is not included under this definition of "cable." The term "cable" is a general one and in practice it is usually applied only to the larger sizes. A small cable is called a "stranded wire" or a "cord," both of which are defined below. Cables may be bare or insulated, and the latter may be armored with lead or with steel wires or bands.

*Strand.* One of the wires of any stranded conductor.

*Stranded Wire.* A group of small wires used as a single wire.

A wire has been defined as a slender rod or filament of drawn metal. If such a filament is subdivided into several smaller filaments or strands and is used as a single wire, it is called "stranded wire." There is no sharp dividing line of size between a "stranded wire" and a "cable." If used as a wire, for example, in winding inductance coils or magnets, it is called a stranded wire and not a cable. If it is substantially insulated, it is called a "cord," defined below.

*Cord.* A small cable, very flexible and substantially insulated to withstand wear.

There is no sharp dividing line in respect to size between a cord and a cable, and likewise no sharp dividing line in respect to the character of insulation between a cord and a stranded wire. Usually the insulation of a cord contains rubber.

*Concentric Strand.* A strand composed of a central core surrounded by one or more layers of helically laid wires or groups of wires.

*Concentric-Lay Conductor.* Conductor constructed with a central core surrounded by one or more layers of helically laid wires.

*Rope-Lay Conductor.* Conductor constructed of a bunch-stranded or a concentric-stranded member or members, as a central core, around which are laid one or more helical layers of such members.

*N-Conductor Cable.* A combination of *N* conductors insulated from one another.

It is not intended that the name as here given be actually used. One would instead speak of a "3-conductor cable," a "12-conductor cable," etc. In referring to the general case, one may speak of a "multiple-conductor cable."

*N-Conductor Concentric Cable.* A cable composed of an insulated central conducting core with *N*-1 tubular-stranded conductors laid over it concentrically and separated by layers of insulation.

This kind of cable usually has only two or three conductors. Such cables are used in carrying alternating currents.

The remark on the expression "*N* conductor" given for the preceding definition applies here also. (Additional definitions can be found in ASTM B354.)

***77. Wire sizes*** have been for many years indicated in commercial practice almost entirely by gage numbers, especially in America and England. This practice is accompanied by some confusion because numerous gages are in common use. The most commonly used gage for electrical wires, in America, is the *American wire gage.* The most commonly used gage for steel wires is the *Birmingham wire gage.*

There is no legal standard wire gage in this country, although a gage for sheets was adopted by Congress in 1893. In England there is a legal standard known as the *Standard wire gage.* In Germany, France, Austria, Italy, and other Continental countries practically no wire gage is used, but wire sizes are specified directly in millimeters. This system is sometimes called the *Millimeter wire gage.* The wire sizes used in France, however, are based to some extent on the old Paris gage (*jauge de Paris de* 1857) (for a history of wire gages see *NBS Handbook* 100, Copper Wire Tables; also see *Circ.* 67, Wire Gages, 1918).

There is a tendency to *abandon gage numbers* entirely and specify wire sizes by the *diameter in mils* (thousandths of an inch). This practice holds particularly in writing specifications and has the great advantages of being both simple and explicit. A number of the wire manufacturers also encourage this practice, and it was definitely adopted by the U.S. Navy Department in 1911.

***78. Mil*** is a term universally employed in this country to measure wire diameters and is a unit of length equal to one-thousandth of an inch.

***79. Circular mil*** is a term universally used to define cross-sectional areas, being a unit of area equal to the area of a circle 1 mil in diameter. Such a circle, however, has an area of 0.7854 (or $\pi/4$) *mil*$^2$. Thus a wire 10 mils in diameter has a cross-sectional area of 100 cmils or 78.54 mils$^2$. Hence, a cmil equals 0.7854 mil$^2$.

***80. American wire gage,*** also known as the *Brown & Sharpe gage,* was devised in 1857 by J. R. Brown. It is usually abbreviated AWG. This gage has the property, in common with a number of other gages, that its sizes represent approximately the successive steps in the process of wire drawing. Also, like many other gages, its numbers are retrogressive, a larger number denoting a smaller wire, corresponding to the operations of drawing. These gage numbers are not arbitrarily chosen, as in many gages, but follow the mathematical law upon which the gage is founded.

*Basis of the AWG* is a simple mathematical law. The gage is formed by the specification of two diameters and the law that a given number of intermediate diameters are formed by geometrical progression. Thus, the diameter of No. 0000 is defined as 0.4600 in and of No. 36 as 0.0050 in. There are 38 sizes between these two; hence the ratio of any diameter to the diameter of the next greater number is given by this expression:

$$\sqrt[39]{\frac{0.4600}{0.0050}} = \sqrt[39]{92} = 1.122\,932\,2 \qquad (4\text{-}8)$$

The square of this ratio = 1.2610. The sixth power of the ratio, that is, the ratio of any diameter to the diameter of the sixth greater number, = 2.0050. The fact that this ratio is so nearly 2 is the basis of numerous useful relations or short cuts in wire computations.

There are a number of approximate rules applicable to the AWG which are useful to remember:

1. An increase of three gage numbers (for example, from No. 10 to 7) doubles the area and weight and consequently halves the dc resistance.
2. An increase of six gage numbers (for example, from No. 10 to 4) doubles the diameter.
3. An increase of 10 gage numbers (for example, from No. 10 to 1/0) multiplies the area and weight by 10 and divides the resistance by 10.
4. A No. 10 wire has a diameter of about 0.10 in, an area of about 10,000 cmils, and (for standard annealed copper at 20°C) a resistance of approximately 1.0 Ω/1000 ft.
5. The weight of No. 2 copper wire is very close to 200 lb/1000 ft.

***81. Steel wire gage,*** also known originally as the *Washburn & Moen gage* and later as the *American Steel & Wire Co.'s gage,* was established by Ichabod Washburn about 1830. This gage also, with a number of its sizes rounded off to thousandths of an inch, is known as the *Roebling gage.* It is used exclusively for steel wire and is frequently employed in wire mills.

***82. Birmingham wire gage,*** also known as *Stubs' wire gage* and *Stubs' iron wire gage,* is said to have been established early in the eighteenth century in England, where it was long in use. This gage was used to designate the Stubs soft-wire sizes and should not be confused with Stubs' steel-wire gage. The numbers of the Birmingham gage were based upon the reductions of size made in practice by drawing wire from rolled rod. Thus, a wire rod was called "No. 0," "first drawing No. 1," and so on. The gradations of size in this gage are not regular, as will appear from its graph. This gage is generally in commercial use in the United States for iron and steel wires.

***83. Standard wire gage,*** which more properly should be designated *(British) Standard wire gage,* is the legal standard of Great Britain for all wires, adopted in 1883. It is also known as the *New British Standard gage,* the *English legal standard gage,* and the *Imperial wire gage.* It was constructed by so modifying the Birmingham gage that the differences between consecutive sizes become more regular. This gage is largely used in England but never has been used extensively in America.

***84. Old English wire gage,*** also known as the *London wire gage,* differs very little from the Birmingham gage. It formerly was used to some extent for brass and copper wires but is now nearly obsolete.

***85. Millimeter wire gage,*** also known as the *metric wire gage,* is based on giving progressive numbers to the progressive sizes, calling 0.1 mm diameter "No. 1," 0.2 mm "No. 2," etc.

***86. German wire gage,*** in which the diameter or thickness is expressed in millimeters, is retrogressive and contains 25 sizes.

**German Wire Gage Table**

(Diameters in millimeters)

| No. | Diam. | No. | Diam. | No. | Diam. | No. | Diam. | No. | Diam. |
|---|---|---|---|---|---|---|---|---|---|
| 1 | 5.50 | 6 | 3.75 | 11 | 2.50 | 16 | 1.375 | 21 | 0.750 |
| 2 | 5.00 | 7 | 3.50 | 12 | 2.25 | 17 | 1.250 | 22 | 0.625 |
| 3 | 4.50 | 8 | 3.25 | 13 | 2.00 | 18 | 1.125 | 23 | 0.562 |
| 4 | 4.25 | 9 | 3.00 | 14 | 1.75 | 19 | 1.000 | 24 | 0.500 |
| 5 | 4.00 | 10 | 2.75 | 15 | 1.50 | 20 | 0.875 | 25 | 0.438 |

***87. Conductor-Size Designation.*** *America* uses, for sizes up to 4/0, mil, decimals of an inch, or AWG numbers for solid conductors and AWG numbers or circular mils for stranded conductors; for sizes larger than 4/0, circular mils is used throughout. Other countries ordinarily use square millimeter area.

***88. Conductor-size conversion*** can be accomplished from the following relation:

$$\text{cmils} = \text{in}^2 \times 1{,}273{,}200 = \text{mm}^2 \times 1973.5 \tag{4-9}$$

***89. Measurement of wire diameters*** may be accomplished in many ways, but most commonly by means of a micrometer caliper. Stranded cables usually are measured by means of a circumference tape calibrated directly in diameter readings.

***90. Comparison of Wire Gages.*** See Table 4-11.

***91. Stranded conductors*** are used generally because of their increased flexibility and consequent ease in handling. The greater the number of wires in any given cross section, the greater will be the flexibility of the finished conductor. Most conductors above 4/0 AWG in size are stranded. Generally, in a given concentric-lay stranded conductor, all wires are of the same size and the same material, although special conductors are available embodying wires of different sizes and of different materials. The former will be found in some insulated cables, and the latter in overhead stranded conductors combining high-conductivity and high-strength wires.

The flexibility of any given size of strand obviously increases as the total number of wires increases. It is common practice to increase the total number of wires as the strand diameter increases, in order to provide reasonable flexibility in handling. So-called *flexible concentric strands* for use in insulated cables have about one to two more layers of wires than the standard type of strand for ordinary use.

***92. Number of Wires in Stranded Conductors.*** Each successive layer in a concentrically stranded conductor contains six more wires than the preceding one. The total number of wires in a conductor is

For 1-wire core constructions (1, 7, 19, etc.),

$$N = 3n(n + 1) + 1 \qquad (4\text{-}10)$$

For 3-wire core constructions (3, 12, etc.),

$$N = 3n(n + 2) + 3 \qquad (4\text{-}11)$$

when $n$ is number of layers over core, which is not counted as a layer.

***93. Wire size in stranded conductors*** is

$$d = \sqrt{\frac{A}{N}} \qquad (4\text{-}12)$$

where $A$ = total conductor area in circular mils and $N$ = total number of wires.

Copper cables are manufactured usually to certain cross-sectional sizes specified in total circular mils or by gage numbers in AWG. This necessarily requires individual wires drawn to certain prescribed diameters, which are different as a rule from normal sizes in AWG (see Table 4-17).

***94. Diameter of stranded conductors*** (circumscribing circle) is

$$D = d(2n + k) \qquad (4\text{-}13)$$

where $d$ = diameter of individual wire, $n$ = number of layers over core which is not counted as a layer, $k$ = 1 for constructions having 1-wire core (1, 7, 19, etc.), and $k$ = 2.155 for constructions having 3-wire core (3, 12, etc.).

For standard concentric-lay stranded conductors, the following rule gives a simple method of determining the outside diameter of a stranded conductor from the known diameter of a solid wire of the same cross-sectional area.

*To obtain the diameter of concentric-lay stranded conductor, multiply the diameter of the solid wire of the same cross-sectional area by the appropriate factor as follows:*

| Number of wires | Factor | Number of wires | Factor |
|---|---|---|---|
| 3 | 1.244 | 91 | 1.153 |
| 7 | 1.134 | 127 | 1.154 |
| 12 | 1.199 | 169 | 1.154 |
| 19 | 1.147 | 217 | 1.154 |
| 37 | 1.151 | 271 | 1.154 |
| 61 | 1.152 | | |

***95. Area of stranded conductors*** is

$$A = Nd^2 \text{ cmils} = \tfrac{1}{4}\pi Nd^2 \times 10^{-6} \text{ in}^2 \qquad (4\text{-}14)$$

where $N$ = total number of wires and $d$ = individual wire diameter in mils.

***96. Effects of Stranding.*** All wires in a stranded conductor except the core wire form continuous helices, of slightly greater length than the axis or core. This causes slight increase in weight and electrical resistance and slight decrease in tensile strength and sometimes affects the internal inductance, as compared theoretically with a conductor of equal dimensions but composed of straight wires parallel with the axis.

***97. Lay, or Pitch.*** The axial length of one complete turn, or helix, of a wire in a stranded conductor is sometimes termed the lay, or pitch. This is often expressed as the *pitch ratio,* which is the ratio of the length of the helix to its *pitch diameter* (diameter of the helix at the centerline of any individual wire or strand equals the outside diameter of the helix minus the thickness of one wire or strand). If there are several layers, the pitch expressed as an axial length may increase with each additional layer, but when expressed as the ratio of axial length to pitch diameter of helix, it is usually the same for all layers, or nearly so. In commercial practice, the pitch is commonly expressed as the ratio of axial length to outside diameter of helix, but this is an arbitrary designation made for convenience of usage. The *pitch angle* is shown in Fig. 4-3, where *ac* represents the axis of the stranded conductor and $l$ is the axial length of one complete turn or helix, *ab* is the length of any individual wire $l + \Delta l$ in one complete turn, and *bc* is equal to the circumference of a circle corresponding to the pitch diameter $d$ of the helix. The angle *bac*, or $\theta$, is the pitch angle, and the pitch ratio is expressed by $p = l/d$. There is no standard pitch ratio used by manufacturers generally, since it has been found desirable to vary this depending on the type of service for which the conductor is intended. Applicable lay lengths generally are included in industry specifications covering the various stranded conductors. For bare overhead conductors, a representative commercial value for pitch length is 13.5 times the outside diameter of each layer of strands.

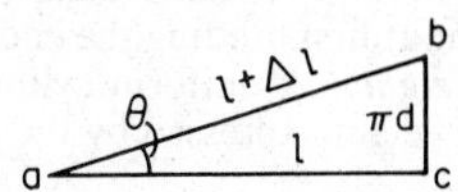


**FIG. 4-3** Pitch angle in concentric-lay cable.

***98. Direction of Lay.*** The direction of lay is the lateral direction in which the individual wires of a cable run over the top of the cable as they recede from an observer looking along the axis. *Right-hand lay* recedes from the observer in clockwise rotation or like a right-hand screw thread; *left-hand lay* is the opposite. The outer layer of a cable is ordinarily applied with a right-hand lay for bare overhead conductors and left-hand lay for insulated conductors, although the opposite lay can be used if desired.

***99. Increase in Weight Due to Stranding.*** Referring to Fig. 4-3, the increase in weight of the spiral members in a cable is proportional to the increase in length:

$$\frac{l + \Delta l}{l} = \sec\theta = \sqrt{1 + \tan^2\theta} = \sqrt{1 + \frac{\pi^2}{p^2}} = 1 + \frac{1}{2}\frac{\pi^2}{p^2} - \frac{1}{8}\left(\frac{\pi^2}{p^2}\right)^2 + \cdots \tag{4-15}$$

As a first approximation this ratio equals $1 + 0.5(\pi^2/p^2)$, and a pitch of 15.7 produces a ratio of 1.02. This correction factor should be computed separately for each layer if the pitch $p$ varies from layer to layer. Practical correction factors for stranded copper conductors are given in Table 4-17.

***100. Increase in Resistance Due to Stranding.*** If it were true that no current flows from wire to wire through their lineal contacts, the proportional increase in the total resistance would be the same as the proportional increase in total weight. If all the wires were in perfect and complete contact with each other, the total resistance would decrease in the same proportion that the total weight increases, owing to the slightly increased normal cross section of the cable as a whole. The contact resistances are normally sufficient to make the actual increase in total resistance nearly as much, proportionately, as the increase in total weight, and for practical purposes they are usually assumed to be the same. See Table 4-17.

***101. Decrease in Strength Due to Stranding.*** When a concentric-lay cable is subjected to mechanical tension, the spiral members tend to tighten around those layers under them and thus produce internal compression, gripping the inner layers and the core. Consequently the individual wires, taken as a whole, do not behave as they would if they were true linear conductors acting independently. Furthermore, the individual wires are never exactly alike in diameter or in strength or in elastic properties. For these reasons there is ordinarily a loss of about 4 to 11% in total tensile efficiency, depending on the number of layers. This reduction tends to increase as the pitch ratio decreases. Actual tensile tests on cables furnish the most dependable data on their ultimate strength.

***102. Tensile efficiency of a stranded conductor*** is the ratio of its breaking strength to the sum of the tensile strengths of all its individual wires. Concentric-lay cables of 12 to 16 pitch ratio have a normal tensile efficiency of approximately 90%; rope-lay cables, approximately 80%.

***103. Preformed Cable.*** This type of cable is made by preforming each individual wire (except the core) into a spiral of such length and curvature that the wire will fit naturally into its normal position in the cable instead of being forced into that shape under the usual tension in the stranding machine. This method has the advantage in cable made of the stiffer grades of wire that the individual wires do not tend to spread or untwist if the strand is cut in two without first binding the ends on each side of the cut.

***104. Weight.*** A uniform cylindrical conductor of diameter $d$, length $l$, and density $\delta$ has a total weight expressed by the formula

$$W = \delta l \frac{\pi d^2}{4} \tag{4-16}$$

The weight of any conductor is commonly expressed in pounds per unit of length, such as 1 ft, 1000 ft, or 1 mi. The weight of stranded conductors can be calculated using Eq. (4-16), but allowance must be made for increase in weight due to stranding (see Par. **99**). Rope-lay stranding has greater increase in weight because of the multiple stranding operations. As an example, industry standards for increase in weight of *stranded copper conductors* are as follows:

| ASTM stranding classes: AA, A, B, C, D | Increase, % | ASTM stranding classes: G, H | Increase, % | ASTM stranding classes: I, J, K, L, M, O, P, Q | Increase, % |
|---|---|---|---|---|---|
| 1 to 4 AWG (AA only) | 1 | 49 wires or less.... | 3 | Single bunched strand. | 2 |
| Up to 2,000 Mcm..... | 2 | 133 wires.......... | 4 | 7 ropes of bunched strand............ | 4 |
| Over 2,000–3,000 Mcm. | 3 | 259 wires.......... | 4.5 | 19 ropes of bunched strand............ | 5 |
| Over 3,000–4,000 Mcm. | 4 | 427 wires.......... | 5 | 7 × 7 ropes of bunched strand............ | 6 |
| Over 4,000–5,000 Mcm. | 5 | Over 427 wires.... | 6 | 19, 37 or 61 × 7 ropes of bunched strand... | 7 |

***105. Breaking Strength.*** The maximum load that a conductor attains when tested in tension to rupture.

***106. Total Elongation at Rupture.*** When a sample of any material is tested under tension until it ruptures, measurement is usually made of the total elongation in a certain initial test length. In certain kinds of testing, the initial test length has been standardized, but in every case, the total elongation at rupture should be referred to the initial test length of the sample on which it was measured. Such elongation is usually expressed in percentage of original unstressed length and is a general index of the ductility of the material. Elongation is determined on solid conductors or on individual wires before stranding; it is rarely determined on stranded conductors.

***107. Elasticity.*** All materials are deformed in greater or lesser degree under application of mechanical stress. Such deformation may be either of two kinds, known, respectively, as "elastic deformation" and "permanent deformation." When a material is subjected to stress and undergoes deformation but resumes its original shape and dimensions when the stress is removed, the deformation is said to be elastic. If the stress is so great that the material fails to resume its original dimensions when the stress is removed, the permanent change in dimensions is termed permanent deformation or "set." In general the stress at which appreciable permanent deformation begins is termed the "working elastic limit." Below this limit of stress the behavior of the material is said to be elastic, and in general the deformation is proportional to the stress.

***108. Stress and Strain.*** The stress in a material under load, as in simple tension or compression, is defined as the total load divided by the area of cross section normal to the direction of the load, assuming the load to be uniformly distributed over this cross section. It is commonly expressed in pounds per square inch. The strain in a material under load is defined as the total deformation measured in the direction of the stress, divided by the total unstressed length in which the measured deformation occurs, or the deformation per unit length. It is expressed as a decimal ratio or numeric.

In order to show the complete behavior of any given conductor under tension, it is customary to make a graph in terms of loading or stress as the ordinates and elongation or strain as the abscissas. Such graphs or curves are useful in determining the elastic limit and the yield point if the loading is carried to the point of rupture. Graphs showing the relationship between stress and strain in a material tested to failure are termed load-deformation or stress-strain curves.

***109. Hooke's law*** consists of the simple statement that the stress is proportional to the strain. It obviously implies a condition of perfect elasticity, which is true only for stresses less than the elastic limit.

***110. Stress-Strain Curves.*** A typical stress-strain diagram of hard-drawn copper wire is shown in Fig. 4-4, which represents No. 9 AWG. The curve *ae* is the actual stress-strain curve; *ab* represents the portion which corresponds to true elasticity, or for which Hooke's law holds rigorously; *cd* is the tangent *ae* which fixes the Johnson elastic limit (see Par. **114**); and the curve *af* represents the set, or permanent elongation due to flow of the metal under stress, being the difference between *ab* and *ae.*

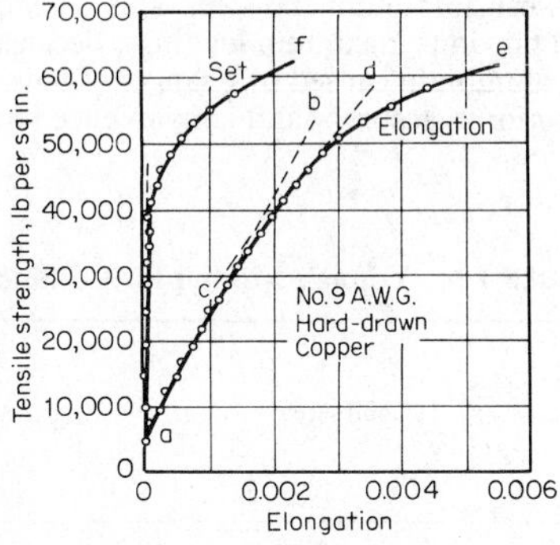


**FIG. 4-4** Stress-strain curves of No. 9 AWG hard-drawn copper wire (Watertown Arsenal Test).

A typical stress-strain diagram of hard-drawn aluminum wire, based on data furnished by the Aluminum Company of America is shown in Fig. 4-5.

***111. Modulus (or coefficient) of elasticity*** is the ratio of internal stress to the corresponding strain or deformation. It is a characteristic of each material, form (shape or structure), and type of stressing. For deformations involving changes both in volume and in shape, special coefficients are used. For conductors under axial tension, the ratio of stress to strain is called *Young's modulus.*

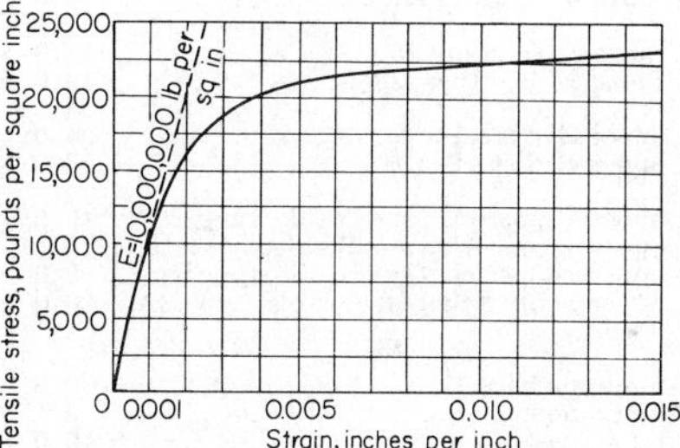


**FIG. 4-5** Typical stress-strain curve of hard-drawn aluminum wire.

If $F$ is the total force or load acting uniformly on the cross section $A$, the stress is $F/A$. If this magnitude of stress causes an elongation $e$ in an original length $l$, the strain is $e/l$. Young's modulus is then expressed

$$M = \frac{Fl}{Ae} \qquad (4\text{-}17)$$

If a material were capable of sustaining an elastic elongation sufficient to make $e$ equal to $l$, or such that the elongated length is double the original length, the stress required to produce this result would equal the modulus. This modulus is very useful in computing the sags of overhead conductor spans under loads of various kinds. It is usually expressed in pounds per square inch.

Stranding usually lowers the Young's modulus somewhat, rope-lay stranding to a greater extent than concentric-lay stranding.

When a new cable is subjected initially to tension and the loading is carried up to the maximum working stress, there is an apparent elongation which is greater than the subsequent elongation under the same loading. This is apparently due to the removal of a very slight slackness in the individual wires, causing them to fit closely together and adjust themselves to the conditions of tension in the strand. When a new cable is loaded to the working limit, unloaded, and then reloaded, the value of Young's modulus determined on initial loading may be on the order of one-half to two-thirds of its true value on reloading. The latter figure should approach within a few percent of the modulus determined by test on individual straight wires of the same material.

For those applications where elastic stretching under tension needs consideration, the stress-strain curve should be determined by test, with the precaution not to prestress the cable before test unless it will be prestressed when installed in service.

Commercially used values of Young's modulus for conductors are given in Table 4-5.

***112. Young's Moduli for Conductors.*** See Table 4-5.

***113. Young's Modulus for ACSR.*** The permanent modulus of ACSR is dependent upon the proportions of steel and aluminum in the cable and upon the distribution of stress between aluminum and steel. This latter condition is dependent upon temperature, tension, and previous maximum loadings. Because of the interchange of stress between the steel and the aluminum caused by changes of tension and temperature, graphical methods are much superior in accuracy and convenience to analytical methods for sag-tension calculations.

**TABLE 4-5** Young's Moduli for Conductors

| Conductor | Young's modulus,* lb/in² | | Reference |
|---|---|---|---|
| | Final† | Virtual initial‡ | |
| Copper wire, hard-drawn | $17.0 \times 10^6$ | $14.5 \times 10^6$ | Copper Wire Engineering Assoc. |
| Copper wire, medium hard-drawn | $16.0 \times 10^6$ | $14.0 \times 10^6$ | Anaconda Wire and Cable Co. |
| Copper cable, hard-drawn, 3 and 12 wire | $17.0 \times 10^6$ | $14.0 \times 10^6$ | Copper Wire Engineering Assoc. |
| Copper cable, hard-drawn, 7 and 19 wire | $17.0 \times 10^6$ | $14.5 \times 10^6$ | Copper Wire Engineering Assoc. |
| Copper cable, medium hard-drawn | $15.5 \times 10^6$ | $14.0 \times 10^6$ | Anaconda Wire and Cable Co. |
| Bronze wire, alloy 15 | $14.0 \times 10^6$ | $13.0 \times 10^6$ | Anaconda Wire and Cable Co. |
| Bronze wire, other alloys | $16.0 \times 10^6$ | $14.0 \times 10^6$ | Anaconda Wire and Cable Co. |
| Bronze cable, alloy 15 | $13.0 \times 10^6$ | $12.0 \times 10^6$ | Anaconda Wire and Cable Co. |
| Bronze cable, other alloys | $16.0 \times 10^6$ | $14.0 \times 10^6$ | Anaconda Wire and Cable Co. |
| Copper-clad steel wire | $24.0 \times 10^6$ | $22.0 \times 10^6$ | Copperweld Steel Co. |
| Copper-clad steel cable | $23.0 \times 10^6$ | $20.5 \times 10^6$ | Copperweld Steel Co. |
| Copper–copper-clad steel cable, type E | $19.5 \times 10^6$ | $17.0 \times 10^6$ | Copperweld Steel Co. |
| Copper–copper-clad steel cable, type EK | $18.5 \times 10^6$ | $16.0 \times 10^6$ | Copperweld Steel Co. |
| Copper–copper-clad steel cable, type F | $18.0 \times 10^6$ | $15.5 \times 10^6$ | Copperweld Steel Co. |
| Copper–copper-clad steel cable, type 2A to 6A. | $19.0 \times 10^6$ | $16.5 \times 10^6$ | Copper Wire Engineering Assoc. |
| Aluminum wire | $10.0 \times 10^6$ | .......... | Reynolds Metals Co. |
| Aluminum cable | $9.1 \times 10^6$ | $7.3 \times 10^6$ | Reynolds Metals Co. |
| Aluminum-alloy wire | $10.0 \times 10^6$ | .......... | Reynolds Metals Co. |
| Aluminum-alloy cable | $9.1 \times 10^6$ | $7.3 \times 10^6$ | Reynolds Metals Co. |
| Aluminum-steel cable, aluminum wire | $10.0 \times 10^6$ | .......... | Aluminum Co. of America |
| Aluminum-steel cable, steel wire | $29.0 \times 10^6$ | .......... | Aluminum Co. of America |
| Aluminum-clad steel wire | $23.5 \times 10^6$ | $22.0 \times 10^6$ | Copperweld Steel Co. |
| Aluminum-clad steel cable | $23.0 \times 10^6$ | $21.5 \times 10^6$ | Copperweld Steel Co. |
| Aluminum-clad steel–aluminum cable: | | | |
| AWAC 5/2 | $13.5 \times 10^6$ | $12.0 \times 10^6$ | Copperweld Steel Co. |
| AWAC 4/3 | $15.5 \times 10^6$ | $14.0 \times 10^6$ | Copperweld Steel Co. |
| AWAC 3/4 | $17.5 \times 10^6$ | $16.0 \times 10^6$ | Copperweld Steel Co. |
| AWAC 2/5 | $19.0 \times 10^6$ | $18.0 \times 10^6$ | Copperweld Steel Co. |
| Galvanized-steel wire, Class A coating | $28.5 \times 10^6$ | .......... | Indiana Steel & Wire Co. |
| Galvanized-steel cable, Class A coating | $27.0 \times 10^6$ | .......... | Indiana Steel & Wire Co. |

*For stranded cables the moduli are usually less than for solid wire and vary with number and arrangement of strands, tightness of stranding, and length of lay. Also, during initial application of stress, the stress-strain relation follows a curve throughout the upper part of the range of stress commonly used in transmission-line design.

†Final modulus is the ratio of stress to strain (slope of the curve) obtained after fully prestressing the conductor. It is used in calculating design or final sags and tensions.

‡Virtual initial modulus is the ratio of stress to strain (slope of the curve) obtained during initial sustained loading of new conductor. It is used in calculating initial or stringing sags and tensions.

NOTE: 1 lb/in² = 6.895 kPa.

Because ACSR is a composite cable made of aluminum and steel wires, additional phenomena occur which are not found in tests of cable composed of a single material. As shown in Fig. 4-6, the part of the curve obtained in the second stress cycle contains a comparatively large "foot" at its base, which is caused by the difference in extension at the elastic limits of the aluminum and steel.

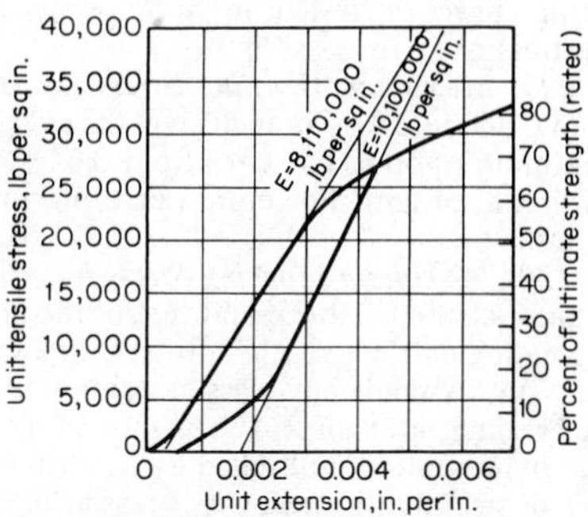


**FIG. 4-6** Repeated stress-strain curve, 795,000 cmils ACSR; 54 × 0.1212 aluminum strands; 7 × 0.1212 steel strands.

***114. Elastic Limit.*** This is variously defined as the limit of stress beyond which permanent deformation occurs or the stress limit beyond which Hooke's law ceases to apply or the limit beyond which the stresses are not proportional to the strains or the *proportional limit.* In some materials the elastic limit occurs at a point which is readily determined, but in others it is quite difficult to determine because the stress-strain curve deviates from a straight line but very slightly at first, and the point of departure from true linear relationship between stress and strain is somewhat indeterminate.

The late Dean J. B. Johnson of the University of Wisconsin, well-known authority on materials of construction, proposed the use of an arbitrary determination referred to frequently as the "Johnson definition of elastic limit." This proposal, which has been quite largely used, was that an "apparent elastic limit" be employed, defined as that point on the stress-strain curve at which the rate of deformation is 50% greater than at the origin. The apparent elastic limit thus defined is a practical value, which is suitable for engineering purposes, as it involves negligible permanent elongation.

The *Johnson elastic limit* is that point on the stress-strain curve at which the natural tangent is equal to 1.5 times the tangent of the angle of the straight or linear portion of the curve, with respect to the axis of ordinates, or *Y* axis.

***115. Yield Point.*** In many materials a point is reached on the stress-strain diagram at which there is a marked increase in strain or elongation without an increase in stress or load. The point at which this occurs is termed the yield point. It is usually quite noticeable in ductile materials but may be scarcely perceptible or possibly not present at all in certain hard-drawn materials such as hard-drawn copper.

***116. Maximum working stresses*** of conductor materials must be determined by tests on samples or specimens comparable in size and condition with the shapes or members intended for use under service conditions. The ratio of the safe maximum working stress to the ultimate breaking strength is from 50 to 60% in many classes of materials, but in others it may range as high as 75 to 80%. The maximum working stresses of any material should be determined only from complete knowledge of its properties and the conditions under which it will be used in service.

The working strength of strands is also affected by the mode of attachment to their supports. When gripped in clamping devices, the edges of the clamps should be rounded to prevent injury to the wires, and the grooves in the clamps should be of suitable diameter to fit the strand very closely and of sufficient length to grip all the wire surfaces firmly. If these precautions are not followed, the strength of any strand as a whole may be appreciably impaired at the clamps. In some forms of construction, strands are supported by means of tapered sockets, to which their ends are made fast by having a matrix, such as zinc, cast in the open end of the socket so as to fill it and grip the individual wires.

***117. Maximum Working Stress for Annealed-Copper Conductors.*** The stress-strain curves for annealed copper show that it has no definite elastic limit and starts to take a permanent deformation, or set, at comparatively small stresses. It is characteristic of such wire to stretch slowly but permanently under relatively moderate stresses, but in so doing it hardens and tends to increase its own elastic limit. The actual condition of any given wire depends on its previous loading. In overhead spans, where the slack has been pulled up repeatedly, the maximum working stress may approach a high percentage of the original ultimate strength; according to one authority this may become as much as 85%.

***118. Maximum Working Stress for Medium-hard-drawn Copper Conductors.*** The average is 50% of the ultimate tensile strength, but where more exact information is desired, stress-strain curves should be determined by test. If the wire will be prestressed at the time it is installed for service, it should be similarly prestressed before determining the stress-strain curve or tested in appropriate manner to show the stress-strain curve after prestressing.

***119. Maximum Working Stress for Hard-drawn Copper Conductors.*** For wire sizes from 0.460 to 0.325 in, inclusive, the average is 55% of the ultimate tensile strength, with a minimum of 50%. For sizes from 0.324 to 0.040 in, inclusive, the average is 60%, with a minimum of 55%. For more exact information, stress-strain curves should be determined by test.

***120. Maximum Working Stress for ACSR.*** Although the ultimate strength of the steel strands is 200,000 lb/in$^2$ and that of the aluminum strands is from 23,000 to 30,000 lb/in$^2$, depending on the size, the ultimate stresses do not occur at the same elongation. Since the aluminum strands have less elongation at rupture than the steel strands, it is evident that the breaking strength of ACSR is the sum of two quantities, the first being the cross-sectional area of aluminum multiplied by its ultimate stress and the second being the cross-sectional area of steel multiplied by its stress at the ultimate elongation of the aluminum wires. The ultimate strength of ACSR thus obtained is the tension at which the aluminum strands fail; the steel core will then stretch, thus reducing the tension. The safe maximum working tension may be defined as that value of tension which can be experienced repeatedly without altering the tensile properties which obtain after the first application of that tension.

***121. Maximum Working Stresses for Copper-clad, Aluminum-clad, and Galvanized-Steel Conductors.*** For sizes customarily employed, 60% of the breaking strength is used. Under some special conditions, this value may be exceeded.

***122. Prestressed Conductors.*** In the case of some materials, especially those of considerable ductility, which tend to show permanent elongation or "drawing" under loads just above the initial elastic limit, it is possible to raise the working elastic limit by loading them to stresses somewhat above the elastic limit as found on initial loading. After such loading, or prestressing, the material will behave according to Hooke's law at all loads less than the new elastic limit. This applies not only to many ductile materials, such as soft or annealed copper wire, but also to cables or stranded conductors, in which there is a slight inherent slack or looseness of the individual wires that can be removed only under actual loading. It is sometimes the practice, when erecting such conductors for service, to prestress them to the working elastic limit or safe maximum working stress and then reduce the stress to the proper value for installation at the stringing temperature without wind or ice.

***123. Resistance*** is the property of an electric circuit or of any body that may be used as part of an electric circuit which determines for a given current the average rate at which electrical energy is converted into heat. The term is properly applied only when the rate of conversion is proportional to the square of the current and is then equal to the power conversion divided by the square of the current. A uniform cylindrical conductor of diameter $d$, length $l$, and *volume resistivity* $\rho$ has a total resistance to continuous currents expressed by the formula

$$R = \frac{\rho l}{\pi d^2/4} \tag{4-18}$$

The resistance of any conductor is commonly expressed in ohms per unit of length, such as 1 ft, 1000 ft, or 1 mi. When used for conducting alternating currents, the effective resistance may be higher than the dc resistance defined above. In the latter case it is common practice to apply the proper factor, or ratio of effective ac resistance to dc resistance, sometimes termed the "skin-effect resistance ratio" (see Par. **127**). This ratio may be determined by test, or it may be calculated if the necessary data are available.

***124. Magnetic permeability*** applies to a field in which the flux is uniformly distributed over a cross section normal to its direction or to a sufficiently small cross section of a nonuniform field so that the distribution can be assumed as substantially uniform. In the case of a cylindrical conductor, the magnetomotive force (mmf) due to the current flowing in the conductor varies from zero at the center or axis to a maximum at the periphery or surface of the conductor and sets up a flux in circular paths concentric with the axis and perpendicular to it but of nonuniform distribution between the axis and the periphery. If the permeability is nonlinear with respect to the mmf, as is usually true with magnetic materials, there is no correct single value of permeability which fits the conditions, although an apparent or equivalent average value can be determined. In the case of other forms of cross section, the distribution is still more complex, and the equivalent permeability may be difficult or impossible to determine except by test.

***125. Internal Inductance.*** A uniform cylindrical conductor of nonmagnetic material, or of unit permeability, has a constant magnitude of internal inductance per unit length, independent of the conductor diameter. This is commonly expressed in microhenrys or millihenrys per unit of length, such as 1 ft, 1000 ft, or 1 mi. When the conductor material possesses magnetic susceptibility, and when the magnetic permeability $\mu$ is constant and therefore independent of the current strength, the internal inductance is expressed in absolute units by the formula

$$L = \frac{\mu l}{2} \tag{4-19}$$

In most cases $\mu$ is not constant but is a function of the current strength. When this is true, there is an effective permeability, one-half of which ($\mu/2$) expresses the inductance per centimeter of length, but this figure of permeability is virtually the ratio of the effective inductance of the conductor of susceptible material to the inductance of a conductor of material which has a permeability of unity. When used for conducting alternating currents, the effective inductance may be less than the inductance with direct current; this is also a direct consequence of the same skin effect which results in an increase of effective resistance with alternating currents, but the overall effect is usually included in the figure of effective permeability. It is usually the practice to determine the effective internal inductance by test, but it may be calculated if the necessary data are available.

***126. Skin effect*** is a phenomenon which occurs in conductors carrying currents whose intensity varies rapidly from instant to instant but does not occur with continuous currents. It arises from the fact that elements or filaments of variable current at different points in the cross section of a conductor do not encounter equal components of inductance, but the central or axial filament meets the maximum inductance, and in general the inductance offered to other filaments of current decreases as the distance of the filament from the axis increases, becoming a minimum at the surface or periphery of the conductor. This, in turn, tends to produce unequal current density over the cross section as a whole; the density is a minimum at the axis and a maximum at the periphery. Such distribution of the current density produces an increase in effective resistance and a decrease in effective internal inductance; the former is of more practical importance than the latter. In the case of large copper conductors at commercial power frequencies, and in the case of most conductors at carrier and radio frequencies, the increase in resistance should be considered (see Pars. **131** and **132**).

***127. Skin-Effect Ratios.*** If $R'$ is the effective resistance of a linear cylindrical conductor to sinusoidal alternating current of given frequency and $R$ is the true resistance with continuous current, then

$$R' = KR \qquad \text{ohms} \tag{4-20}$$

where $K$ is determined from Table 4-6 in terms of $x$. The value of $x$ is given by

$$x = 2\pi a \sqrt{\frac{2f\mu}{\rho}} \tag{4-21}$$

where $a$ = radius of conductor in centimeters, $f$ = frequency in cycles per second, $\mu$ = magnetic permeability of conductor (here assumed to be constant), $\rho$ = resistivity in abohm-centimeters (abohm = $10^{-9}\ \Omega$).

For practical calculation, Eq. (4-21) can be written

$$x = 0.063598 \sqrt{\frac{f\mu}{R}} \tag{4-22}$$

where $R$ = dc resistance at operating temperature in ohms per mile.

**TABLE 4-6** Skin-Effect Ratios
(Bur. Std. *Bull.* 169, pp. 226–228)

| $x$ | $K$ | $K'$ | $x$ | $K$ | $K'$ | $x$ | $K$ | $K'$ | $x$ | $K$ | $K'$ |
|---|---|---|---|---|---|---|---|---|---|---|---|
| 0.0 | 1.00000 | 1.00000 | 2.9 | 1.28644 | 0.86012 | 6.6 | 2.60313 | 0.42389 | 17.0 | 6.26817 | 0.16614 |
| 0.1 | 1.00000 | 1.00000 | 3.0 | 1.31809 | 0.84517 | 6.8 | 2.67312 | 0.41171 | 18.0 | 6.62129 | 0.15694 |
| 0.2 | 1.00001 | 1.00000 | 3.1 | 1.35102 | 0.82975 | 7.0 | 2.74319 | 0.40021 | 19.0 | 6.97446 | 0.14870 |
| 0.3 | 1.00004 | 0.99998 | 3.2 | 1.38504 | 0.81397 | 7.2 | 2.81334 | 0.38933 | 20.0 | 7.32767 | 0.14128 |
| 0.4 | 1.00013 | 0.99993 | 3.3 | 1.41999 | 0.79794 | 7.4 | 2.88355 | 0.37902 | 21.0 | 7.68091 | 0.13456 |
| 0.5 | 1.00032 | 0.99984 | 3.4 | 1.45570 | 0.78175 | 7.6 | 2.95380 | 0.36923 | 22.0 | 8.03418 | 0.12846 |
| 0.6 | 1.00067 | 0.99966 | 3.5 | 1.49202 | 0.76550 | 7.8 | 3.02411 | 0.35992 | 23.0 | 8.38748 | 0.12288 |
| 0.7 | 1.00124 | 0.99937 | 3.6 | 1.52879 | 0.74929 | 8.0 | 3.09445 | 0.35107 | 24.0 | 8.74079 | 0.11777 |
| 0.8 | 1.00212 | 0.99894 | 3.7 | 1.56587 | 0.73320 | 8.2 | 3.16480 | 0.34263 | 25.0 | 9.09412 | 0.11307 |
| 0.9 | 1.00340 | 0.99830 | 3.8 | 1.60314 | 0.71729 | 8.4 | 3.23518 | 0.33460 | 26.0 | 9.44748 | 0.10872 |
| 1.0 | 1.00519 | 0.99741 | 3.9 | 1.64051 | 0.70165 | 8.6 | 3.30557 | 0.32692 | 28.0 | 10.15422 | 0.10096 |
| 1.1 | 1.00758 | 0.99621 | 4.0 | 1.67787 | 0.68632 | 8.8 | 3.37597 | 0.31958 | 30.0 | 10.86101 | 0.09424 |
| 1.2 | 1.01071 | 0.99465 | 4.1 | 1.71516 | 0.67135 | 9.0 | 3.44638 | 0.31257 | 32.0 | 11.56785 | 0.08835 |
| 1.3 | 1.01470 | 0.99266 | 4.2 | 1.75233 | 0.65677 | 9.2 | 3.51680 | 0.30585 | 34.0 | 12.27471 | 0.08316 |
| 1.4 | 1.01969 | 0.99017 | 4.3 | 1.78933 | 0.64262 | 9.4 | 3.58723 | 0.29941 | 36.0 | 12.98160 | 0.07854 |
| 1.5 | 1.02582 | 0.98711 | 4.4 | 1.82614 | 0.62890 | 9.6 | 3.65766 | 0.29324 | 38.0 | 13.68852 | 0.07441 |
| 1.6 | 1.03323 | 0.98342 | 4.5 | 1.86275 | 0.61563 | 9.8 | 3.72812 | 0.28731 | 40.0 | 14.39545 | 0.07069 |
| 1.7 | 1.04205 | 0.97904 | 4.6 | 1.89914 | 0.60281 | 10.0 | 3.79857 | 0.28162 | 42.0 | 15.10240 | 0.06733 |
| 1.8 | 1.05240 | 0.97390 | 4.7 | 1.93533 | 0.59044 | 10.5 | 3.97477 | 0.26832 | 44.0 | 15.80936 | 0.06427 |
| 1.9 | 1.06440 | 0.96795 | 4.8 | 1.97131 | 0.57852 | 11.0 | 4.15100 | 0.25622 | 46.0 | 16.51634 | 0.06148 |
| 2.0 | 1.07816 | 0.96113 | 4.9 | 2.00710 | 0.56703 | 11.5 | 4.32727 | 0.24516 | 48.0 | 17.22333 | 0.05892 |
| 2.1 | 1.09375 | 0.95343 | 5.0 | 2.04272 | 0.55597 | 12.0 | 4.50358 | 0.23501 | 50.0 | 17.93032 | 0.05656 |
| 2.2 | 1.11126 | 0.94482 | 5.2 | 2.11353 | 0.53506 | 12.5 | 4.67993 | 0.22567 | 60.0 | 21.46541 | 0.04713 |
| 2.3 | 1.13069 | 0.93527 | 5.4 | 2.18389 | 0.51566 | 13.0 | 4.85631 | 0.21703 | 70.0 | 25.00063 | 0.04040 |
| 2.4 | 1.15207 | 0.92482 | 5.6 | 2.25393 | 0.49764 | 13.5 | 5.03272 | 0.20903 | 80.0 | 28.53593 | 0.03535 |
| 2.5 | 1.17538 | 0.91347 | 5.8 | 2.32380 | 0.48086 | 14.0 | 5.20915 | 0.20160 | 90.0 | 32.07127 | 0.03142 |
| 2.6 | 1.20056 | 0.90126 | 6.0 | 2.39359 | 0.46521 | 14.5 | 5.38560 | 0.19468 | 100.0 | 35.60666 | 0.02828 |
| 2.7 | 1.22753 | 0.88825 | 6.2 | 2.46338 | 0.45056 | 15.0 | 5.56208 | 0.18822 | ∞ | ∞ | 0 |
| 2.8 | 1.25620 | 0.87451 | 6.4 | 2.53321 | 0.43682 | 16.0 | 5.91509 | 0.17649 | | | |

If $L'$ is the effective inductance of a linear conductor to sinusoidal alternating current of a given frequency,

$$L' = L_1 + K'L_2 \tag{4-23}$$

where $L_1$ = external portion of inductance, $L_2$ = internal portion (due to the magnetic field within the conductor), and $K'$ is determined from Table 4-6 in terms of $x$. Thus the total effective inductance per unit length of conductor is

$$L' = 2 \ln \frac{d}{a} + K' \frac{\mu}{2} \tag{4-24}$$

The inductance is here expressed in abhenrys per centimeter of conductor, in a linear circuit; $a$ is the radius of the conductor and $d$ is the separation between the conductor and its return conductor, expressed in the same units.

Values of $K$ and $K'$ in terms of $x$ are shown in Table 4-6 and Figs. 4-7 and 4-8 (see *NBS Circ.* 74, pp. 309–311, for additional tables, and *Sci. Paper* 374).

Value of $\mu$ for nonmagnetic materials (copper, aluminum, etc.) is 1; for magnetic materials, it varies widely with composition, processing, current density, etc., and should be determined by test in each case.

***128. Skin-Effect Ratios.*** See Table 4-6.

***129. Skin-Effect Ratios—Copper Conductors NOT in Close Proximity.*** See Table 4-7.

***130. Alternating-Current Resistance.*** For small conductors at power frequencies, the frequency has a negligible effect, and dc resistance values can be used. For large conductors, frequency must be taken into account in addition to temperature effects. To do this, first calculate the dc resistance at the operating temperature, then determine the skin-effect ratio $K$, and finally determine the ac resistance at operating temperature (see Par. **127**).

***131. AC resistance for copper conductors NOT in close proximity*** can be obtained from

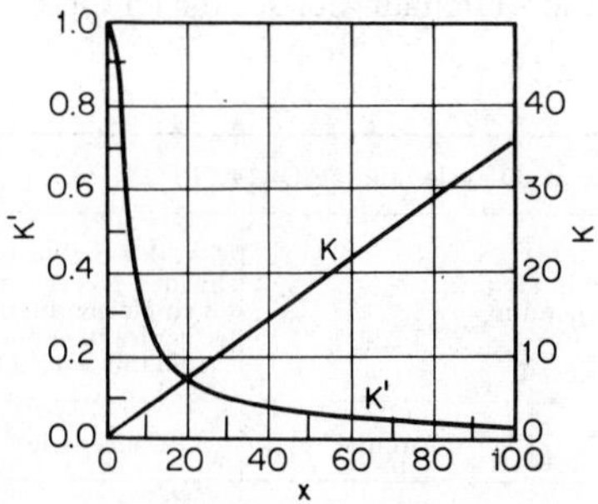


FIG. 4-7 *K* and *K'* for values of *x* from 0 to 100.

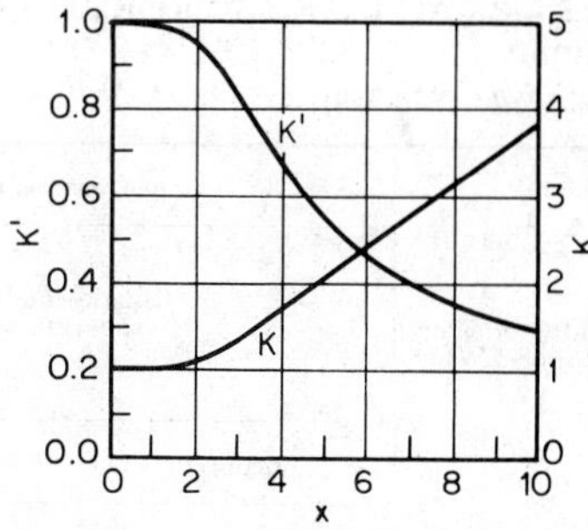


FIG. 4-8 *K* and *K'* for values of *x* from 0 to 10.

the skin-effect ratios given in Tables 4-6 and 4-7. Practical values for usual conductor sizes are given in Table 4-27.

***132. AC resistance for copper conductors in close proximity*** or for insulated copper conductors installed in conduit may be calculated from Table 4-8.

***133. AC/DC Resistance Ratios: Copper Conductors in Close Proximity.*** See Table 4-8.

***134. AC resistance for copper-clad and aluminum-clad steel conductors*** is dependent on several variables and should be determined by test. Practical values for usual conductor sizes are given in Tables 4-28, 4-29, and 4-31.

***135. AC Resistance for Aluminum Conductors.*** The increase in resistance and decrease in internal inductance of cylindrical aluminum conductors can be determined from the data in Par. **127.** It is not the same as for copper conductors of equal diameter but is slightly less because of the higher volume resistivity of aluminum.

***136. AC/DC Resistance Ratios: Aluminum Conductors in Close Proximity.*** See Table 4-8.

**TABLE 4-7** Skin-Effect Ratios—Copper Conductors NOT in Close Proximity
(See also Table 4-8; AC/DC Resistance Ratios—Copper and Aluminum Conductors IN Close Proximity)

| Conductor size, Mcm | Skin-effect ratio **K** at 60 cycles and 65 C (149 F) | | | | | | | | | | | | | | | |
|---|---|---|---|---|---|---|---|---|---|---|---|---|---|---|---|---|
| | Inside conductor diameter, in. | | | | | | | | | | | | | | | |
| | 0* | | 0.25 | | 0.50 | | 0.75 | | 1.00 | | 1.25 | | 1.50 | | 2.00 | |
| | Outside diameter, in | **K** | Outside diameter, in | **K** | Outside diameter, in | **K** | Outside diameter, in | **K** | Outside diameter, in | **K** | Outside diameter, in | **K** | Outside diameter, in | **K** | Outside diameter, in | **K** |
| 3000 | 1.998 | 1.439 | 2.02 | 1.39 | 2.08 | 1.36 | 2.15 | 1.29 | 2.27 | 1.23 | 2.39 | 1.19 | 2.54 | 1.15 | 2.87 | 1.08 |
| 2500 | 1.825 | 1.336 | 1.87 | 1.28 | 1.91 | 1.24 | 2.00 | 1.20 | 2.12 | 1.16 | 2.25 | 1.12 | 2.40 | 1.09 | 2.75 | 1.05 |
| 2000 | 1.631 | 1.239 | 1.67 | 1.20 | 1.72 | 1.17 | 1.80 | 1.12 | 1.94 | 1.09 | 2.09 | 1.06 | 2.25 | 1.05 | 2.61 | 1.02 |
| 1500 | 1.412 | 1.145 | 1.45 | 1.12 | 1.52 | 1.09 | 1.63 | 1.06 | 1.75 | 1.04 | 1.91 | 1.03 | 2.07 | 1.02 | 2.47 | 1.01 |
| 1000 | 1.152 | 1.068 | 1.19 | 1.05 | 1.25 | 1.03 | 1.39 | 1.02 | 1.53 | 1.01 | 1.72 | 1.01 | | | | |
| 800 | 1.031 | 1.046 | 1.07 | 1.04 | 1.16 | 1.02 | 1.28 | 1.01 | 1.45 | 1.01 | | | | | | |
| 600 | 0.893 | 1.026 | 0.94 | 1.02 | 1.04 | 1.01 | | | | | | | | | | |
| 500 | 0.814 | 1.018 | 0.86 | 1.01 | 0.97 | 1.01 | | | | | | | | | | |
| 400 | 0.728 | 1.012 | 0.78 | 1.01 | | | | | | | | | | | | |
| 300 | 0.630 | 1.006 | | | | | | | | | | | | | | |

*For standard concentric-stranded conductors (i.e., inside diameter = 0).
NOTE: 1 in = 2.54 cm.

**TABLE 4-8** AC/DC Resistance Ratios—Copper and Aluminum Conductors IN Close Proximity

(*IPCEA Publ.* P-34-359)

| Conductor size, Mcm or AWG | A-c/d-c resistance ratio at 60 cycles and 65°C (149°F) | | | | | | | |
|---|---|---|---|---|---|---|---|---|
| | Single-conductor cable in air or separate **nonmetallic** conduit | | | | | | 5–15-kV nonleaded shielded power cable, 3 single conductor cables in same **metallic** conduit | |
| | Concentric | | Segmental | | Annular | | Concentric | |
| | Copper | Aluminum | Copper | Aluminum | Copper | Aluminum | Copper | Aluminum |
| 5,000 | 1.77 | 1.42 | .... | .... | 1.12 | 1.05 | | |
| 4,500 | 1.69 | 1.36 | .... | .... | 1.12 | 1.05 | | |
| 4,000 | 1.61 | 1.31 | 1.18 | 1.08 | 1.12 | 1.05 | | |
| 3,500 | 1.52 | 1.25 | 1.15 | 1.06 | 1.11 | 1.04 | | |
| 3,000 | 1.43 | 1.20 | 1.11 | 1.04 | 1.11 | 1.04 | | |
| 2,500 | 1.33 | 1.14 | 1.08 | 1.03 | 1.07 | 1.03 | | |
| 2,250 | 1.28 | 1.12 | 1.06 | 1.03 | .... | .... | | |
| 2,000 | 1.24 | 1.10 | 1.05 | 1.02 | 1.05 | 1.02 | | |
| 1,750 | 1.19 | 1.08 | 1.04 | 1.02 | 1.04 | 1.02 | | |
| 1,500 | 1.14 | 1.06 | 1.03 | 1.01 | 1.04 | 1.01 | | |
| 1,250 | 1.10 | 1.04 | 1.02 | 1.01 | 1.04 | 1.01 | | |
| 1,000 | 1.07 | 1.03 | 1.01 | 1.01 | 1.03 | 1.01 | 1.36 | 1.17 |
| 900 | 1.06 | 1.02 | .... | .... | 1.03 | 1.01 | 1.30 | 1.14 |
| 800 | 1.04 | 1.02 | .... | .... | 1.02 | 1.01 | 1.24 | 1.11 |
| 750 | 1.04 | 1.01 | .... | .... | 1.02 | 1.01 | 1.22 | 1.10 |
| 700 | 1.03 | 1.01 | .... | .... | .... | .... | 1.19 | 1.09 |
| 600 | 1.03 | 1.01 | .... | .... | .... | .... | 1.14 | 1.07 |
| 500 | 1.02 | 1.01 | .... | .... | .... | .... | 1.10 | 1.05 |
| 400 | 1.01 | 1.01* | .... | .... | .... | .... | 1.07 | 1.03 |
| 350 | 1.01 | .... | .... | .... | .... | .... | 1.05 | 1.03 |
| 300 | 1.01 | .... | .... | .... | .... | .... | 1.04 | 1.02 |
| 250 | 1.01* | .... | .... | .... | .... | .... | 1.03 | 1.01 |
| 4/0 | 1.01* | .... | .... | .... | .... | .... | 1.02 | 1.01 |
| 3/0 | .... | .... | .... | .... | .... | .... | 1.01 | |
| 2/0 | .... | .... | .... | .... | .... | .... | 1.01 | |

*Conductor skin effect less than 1%.

***137. AC Resistance for ACSR.*** In the case of ACSR conductors, the steel core is of relatively high resistivity, and therefore its conductance is usually neglected in computing the total resistance of such strands. The effective permeability of the grade of steel employed in the core is also relatively small. It is approximately correct to assume that such a strand is hollow and consists exclusively of its aluminum wires; in this case the laws of skin effect in tubular conductors will be applicable. Conductors having a single layer of aluminum wires over the steel core have higher ac/dc ratios than those having multiple layers of aluminum wires. Practical values for usual conductor sizes are given in Table 4-30.

***138. AC Resistance for Steel Conductors.*** The increase in effective resistance with alternating currents depends fundamentally upon the circular permeability; there is also an increase, of much smaller proportions, caused by hysteresis. The permeability is very sensitive to variations in composition, heat-treatment, and working of the metal; for this reason it is not usually feasible to compute the skin effect except as an approximation. The results of tests show great variations depending upon the factors just mentioned. Typical test results are shown in Figs. 4-9 and 4-10. Practical values for small conductors are given in Table 4-26.

***139. Inductive Reactance.*** Present practice is to consider inductive reactance as split into two components: (1) that due to flux within a radius of 1 ft including the internal reactance within the conductor of radius $r$ and (2) that due to flux between 1 ft radius and the equivalent conductor spacing $D_s$ or geometric mean distance (GMD).

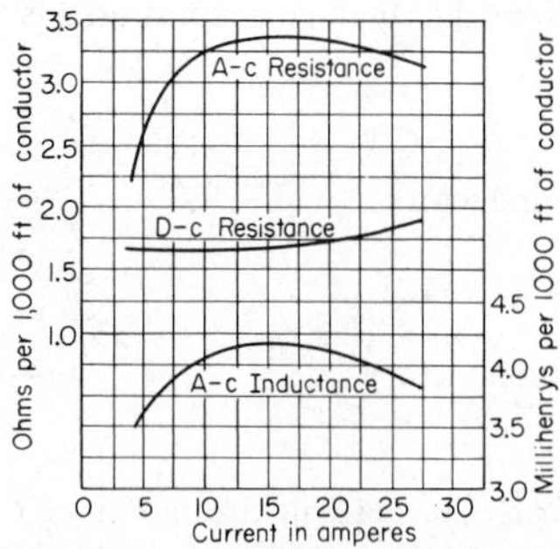


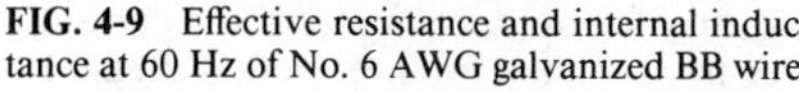
**FIG. 4-9** Effective resistance and internal inductance at 60 Hz of No. 6 AWG galvanized BB wire.

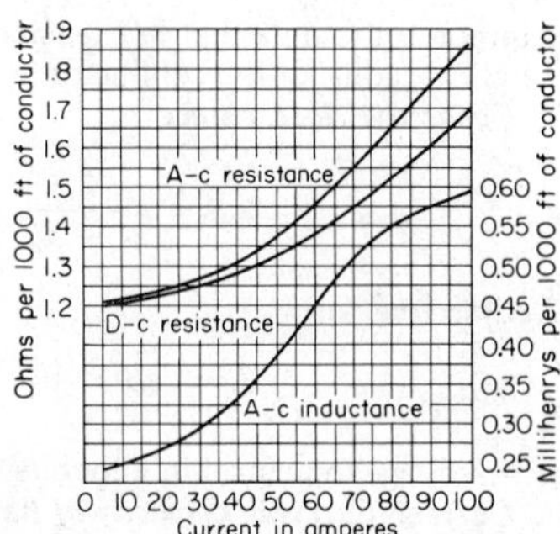


**FIG. 4-10** Effective resistance and internal inductance at 60 Hz of ⅜-in 7-wire Class A galvanized high-strength steel strand.

The fundamental inductance formula is

$$L = 2 \ln \frac{D_s}{r} + \frac{\mu}{2} \qquad \text{abH/(cm)(conductor)} \tag{4-25}$$

This can be rewritten

$$L = 2 \ln \frac{D_s}{1} + 2 \ln \frac{1}{r} + \frac{\mu}{2} \tag{4-26}$$

where the term 2 ln $(D_s/1)$ represents inductance due to flux between 1 ft radius and the equivalent conductor spacing and 2 ln $(1/r) + (\mu/2)$ represents the inductance due to flux within 1 ft radius [2 ln $(1/r)$ represents inductance due to flux between conductor surface and 1 ft radius, and $\mu/2$ represents internal inductance due to flux within the conductor].

By definition, geometric mean radius (GMR) of a conductor is the radius of an infinitely thin tube having the same internal inductance as the conductor. Therefore,

$$L = 2 \ln \frac{D_s}{1} + 2 \ln \frac{1}{\text{GMR}} \tag{4-27}$$

Since inductance reactance = $2\pi fL$, for practical calculation Eq. (4-27) can be written

$$X = 0.004657f \log \frac{D_s}{1} + 0.004657f \log \frac{1}{\text{GMR}} \qquad \Omega\text{/(mi)(conductor)} \tag{4-28}$$

In the conductor tables in this section, inductive reactance is calculated from Eq. (4-28), considering that

$$X = x_a + x_d \tag{4-29}$$

***140. Inductive reactance for conductors using steel*** varies in a manner similar to ac resistance (see Pars. **134** to **138**). Practical values for usual conductor sizes are given in Tables 4-30, 4-31, and 4-32.

***141. Capacitive Reactance.*** By the same reasoning used in Par. **139**, the capacitive reactance can be considered in two parts also, giving

$$X = \frac{4.099}{f} \log \frac{D_s}{1} + \frac{4.099}{f} \log \frac{1}{r} \qquad \text{M}\Omega\text{/(mi)(conductor)} \tag{4-30}$$

In the conductor tables in this section, capacitive reactance is calculated from Eq. (4-30), it being considered that

$$X' = x'_a + x'_d \tag{4-31}$$

It is important to note that in capacitance calculations the conductor radius used is the actual physical radius of the conductor.

***142. Capacitive Susceptance***

$$B = \frac{1}{x'_a + x'_d} \qquad \mu\text{S/(mi)(conductor)} \tag{4-32}$$

***143. Charging Current***

$$I_C = eB \times 10^{-3} \qquad \text{A/(mi)(conductor)} \tag{4-33}$$

where $e$ = voltage to neutral in kilovolts.

***144. Current-Carrying Capacity of Bare Conductors.*** The IEEE publication titled *Calculation of Bare Overhead Conductor Temperature and Ampacity for Steady-State Conditions,* 1984, has been accepted generally by the industry for the calculation of the current-carrying capacity of conductors for overhead power-transmission lines. For other methods in use, see the following:

Schurig, O. R., and Frick, C. W.: Heating and Current-Carrying Capacity of Bare Conductors for Outdoor Service; *Gen. Electr. Rev.,* March 1930, vol. 33, no. 3, p. 141.

Frick, C. W.: Current-Carrying Capacity of Bare Cylindrical Conductors for Indoor and Outdoor Service; *Gen. Electr. Rev.,* August 1934, vol. 34, no. 8, p. 464.

Kidder, A. H., and Woodward, C. B.: Ampere Load Limits for Copper in Overhead Lines; *Trans. AIEE,* 1943, March section, vol. 62, p. 149.

House, H. E., and Tuttle, P. D.: Current Carrying Capacity of ACSR; *Trans. AIEE,* 1958, vol. 77, pt. III, p. 1169.

***145. Current-Carrying Capacity of Insulated Conductors.*** For *power transmission* and distribution cables in air, in enclosed and exposed conduit and in underground ducts, the current-carrying capacities sponsored by the Insulated Power Cable Engineers Association (IPCEA) are in general use. See Power Cable Ampacities, *AIEE Publ.* S-135-1, 1962, vol. 1, Copper Conductors, vol. 2, Aluminum Conductors (available from IEEE, 345 E. 47th St., N.Y. 10017).

For *interior wiring* under jurisdiction of the National Electrical Code, the latest issue of the Standard of the National Board of Fire Underwriters for Electric Wiring and Apparatus, *NBFU Pamphlet* 70, should be consulted.

***146. Contact (Trolley) Wires.*** The special requirements for trolley contact wires have resulted in the development of materials having special characteristics. For general use, hard-drawn copper wire manufactured under particular specifications gives satisfactory service. Under unusual conditions of service, however, harder materials are needed to avoid constant wire replacements.

**FIG. 4-11** Grooved section.

**FIG. 4-12** Figure 8 section.

**FIG. 4-13** Figure 9 deep section.

Trolley contact wires are designed and manufactured for the dual purpose of providing current-carrying capacity and resisting the constant abrasive effect of the trolley wheel, shoe, or pantograph. They must be of good-quality material with a hard-wearing surface, and special care is necessary in manufacture to obtain a surface smooth and free from imperfections.

Because of the need for special methods of supporting trolley contact wires, the use of grooved wires is very common. These as well as other special designs are illustrated in Figs. 4-11 to 4-13.

***147. Bus conductors*** require that greater attention be given to certain physical and electrical characteristics of the metals than is usually necessary in designing line conductors. These characteristics are current-carrying capacity, emissivity, skin effect, expansion, and mechanical deflection. To obtain the most satisfactory and economical designs for busbars in power stations and substations, where they are used extensively, consideration must be given to choice not only of material but also of shape. Both copper and aluminum are used for busbars, and in certain outdoor substations, steel has proved satisfactory. The most common busbar form for carrying heavy current, especially indoors, is flat copper bar. Busbars in the form of angles, channels, and tubing have been developed for heavy currents and, because of better distribution of the conducting material, make more efficient use of the metal both electrically and mechanically. All such designs are based upon the need for proper current-carrying capacity without excess busbar temperatures and upon the necessity for adequate mechanical strength.

***148. Solid conductors (wire)*** for bare power conductors are usually made of copper or copper alloy, although for particular applications aluminum, aluminum alloy, copper-clad steel, aluminum-clad steel, or steel is used.

***149. Copper solid conductors*** for overhead line applications are supplied normally in hard-drawn or medium hard-drawn tempers in sizes No. 10 to 4 AWG. Number 2 AWG is occasionally used, but for this size and larger, stranded conductors are preferable. Soft or annealed conductors are available in all sizes for use in insulated conductors, grounding or bonding wires, weather-resisting (weatherproof) wire, etc.

***150. Hard-drawn copper wire*** is drawn through dies from rod to finished product without intermediate annealing. This results in a wire of high strength and low elongation. Both these characteristics can be controlled to some extent by choice of size of rod and by modification of certain details of the drawing process.

***151. Medium-hard-drawn copper wire*** is essentially and necessarily a special product, because when wire has once started on its course through the drawing operations, it can finish only as a hard-drawn wire to be used as such or to be annealed and become soft or annealed wire. Medium-hard-drawn wire is annealed wire drawn to a slightly smaller diameter.

Medium-hard-drawn wire approaches hard-drawn wire in its characteristics, but from the very nature of the product, exact uniformity in tensile strength cannot be obtained; hence the necessity for establishing a range of tensile strength within which standard medium-hard-drawn wire must be expected to be found.

***152. Soft or annealed copper wire*** is drawn by customary operations and annealed, finished by cleaning when necessary to remove scale or oxide. The wire is so soft and ductile that it is easily marred and even stretched by careless handling in the operations of winding or cabling; hence the necessity for confining specifications and inspection to wire in packages as it leaves the manufacturer and before being put through processes incident to its use by the purchaser.

***153. Copper-alloy solid conductors*** for overhead line applications normally are supplied only in hard-drawn temper. The usual range of sizes is No. 10 to 4 AWG. Wires are available in any of the several alloys designated in ASTM B105 as alloys 8.5 to 85 in accordance with their conductivities.

***154. Copper-clad steel solid conductors*** are rarely used for purposes other than signal or communication wire (weatherproofed or insulated) or telephone drop wire. Usual sizes are No. 17 to 10 AWG.

*Aluminum-clad steel solid conductors* are used primarily for signal and communication wire (either bare or insulated). The usual sizes employed are No. 8 through No. 12 AWG.

***155. Aluminum solid conductors*** are practically never used for overhead-line applications because of their relatively poor physical characteristics. Intermediate temper or hard-drawn conductors are used in all permitted sizes of insulated solid conductors.

***156. Iron and steel solid conductors*** are used occasionally for telephone and telegraph lines in Nos. 16, 10, and 8 BWG and Nos. 14, 13, and 12 SWG (see Table 4-11).

***157. Concentric-lay stranded conductors*** are made up of successive layers of helically laid solid members (wires). Usual strandings have 6 wires more in each added layer and start with either a 1- or a 3-wire core.

***158. Bunch-stranded conductors*** are made up of any number of wires grouped together without regard to their accurate geometric arrangement. Usually they are twisted together; if not, they are generally referred to as "parallel-strand."

**TABLE 4-9** Industry Specifications for Conductors

| Sponsoring organization | Specification designation | Specification title | Corresponding ANSI specification |
|---|---|---|---|
| | | General definitions | |
| AIEE | Book | Definitions of electrical terms | C42 |
| AIEE/IEEE | No. 30 | Definitions and general standards for wire and cable | C8.1 |
| ASTM | B354 | Definitions of terms relating to uninsulated metallic electrical conductors | |
| | | Conductor materials | |
| ASTM | B258 | Standard nominal diameters and cross-sectional areas of AWG sizes of solid round wires used as electrical conductors | C7.36 |
| ASTM | B5 | Electrolytic copper wire, bars, cakes, slabs, billets, ingots, and ingot bars | H17.2 |
| ASTM | B4 | Lake copper wire, bars, cakes, slabs, billets, ingots, and ingot bars | H17.1 |
| ASTM | B49 | Hot-rolled copper rods for electrical purposes | C7.7 |
| ASTM | B170 | Oxygen-free electrolytic copper wire bars, billets, and cakes | H23.11 |
| ASTM | B224 | Classification of coppers | |
| ASTM | B263 | Method for determination of cross-sectional area of stranded conductors | C7.29 |
| ASTM | B233 | Aluminum-alloy 1350 redraw rod for electrical purposes | C7.23 |
| ASTM | B296 | Recommended practice for temper designation of aluminum and magnesium alloys, cast and wrought | |
| | | Bare solid copper wire | |
| ASTM | B1 | Hard-drawn copper wire | C7.2 |
| ASTM | B2 | Medium hard-drawn copper wire | C7.3 |
| ASTM | B3 | Soft or annealed copper wire | C7.1 |
| ASTM | B48 | Soft rectangular and square bare copper wire for electrical conductors | C7.9 |
| ASTM | B47 | Copper trolley wire | C7.6 |
| ASTM | B116 | ASTM figure 9 deep-section grooved and figure 8 copper trolley wire for industrial haulage | C711 |
| | | Coated-copper wire | |
| ASTM | B33 | Tinned soft or annealed copper wire for electrical purposes | C7.4 |
| ASTM | B246 | Tinned hard-drawn and medium-hard-drawn copper wire for electrical purposes | C7.37 |
| ASTM | B189 | Lead-coated and lead-alloy-coated soft copper wire for electrical purposes | C7.15 |
| ASTM | B298 | Silver-coated soft or annealed copper wire | C7.38 |
| ASTM | B355 | Nickel-coated soft or annealed copper wire | C7.48 |
| | | Copper-clad-steel wire | |
| ASTM | B227 | Hard-drawn copper-clad steel wire | C7.17 |
| | | Copper-alloy wires | |
| ASTM | B9 | Bronze trolley wire | C7.5 |
| ASTM | B105 | Hard-drawn copper-alloy wires for electrical conductors | C7.10 |
| | | Bare solid aluminum wire | |
| ASTM | B230 | Aluminum wire, 1350-H19, for electrical purposes | C7.20 |

**TABLE 4-9** Industry Specifications for Conductors (*Continued*)

| Sponsoring organization | Specification designation | Specification title | Corresponding ANSI specification |
|---|---|---|---|
| | | General definitions | |
| ASTM | B262 | Aluminum wire, 1350-H16 or -H26 for electrical purposes | C7.35 |
| ASTM | B314 | Aluminum wire for communication cable | C7.40 |
| ASTM | B323 | Aluminum wire, 1350-H14 or -H24, for electrical purposes | C7.42 |
| ASTM | B324 | Aluminum rectangular and square wire for electrical purposes | C7.43 |
| | | Aluminum-alloy wire | |
| ASTM | B396 | Aluminum alloy 5005-H19 wire for electrical purposes | C7.49 |
| ASTM | B398 | Aluminum alloy 6201-T81 wire for electrical purposes | C7.51 |
| | | Aluminum-clad steel wire | |
| ASTM | B415 | Hard-drawn aluminum-clad steel wire | C7.55 |
| | | Copper cable | |
| ASTM | B8 | Concentric-lay-stranded copper conductors, hard, medium hard, or soft | C7.8 |
| ASTM | B172 | Rope-lay-stranded copper conductors having bunch-stranded members for electrical conductors | C7.12 |
| ASTM | B173 | Rope-lay-stranded copper conductors having concentric-stranded members for eletrical conductors | C7.13 |
| ASTM | B174 | Bunch-stranded copper conductors for electrical conductors | C7.14 |
| ASTM | B226 | Cored, annular, concentric-lay-stranded copper conductors | C7.16 |
| ASTM | B286 | Copper conductors for use in hookup wire for electronic equipment | C7.39 |
| | | Copper-clad steel and composite cables | |
| ASTM | B228 | Concentric-lay-stranded copper-clad steel conductors | C7.18 |
| ASTM | B229 | Concentric-lay-stranded copper and copper-clad steel composite conductors | C7.19 |
| | | Aluminum and ACSR cables | |
| ASTM | B231 | Alúminum conductors, concentric-lay-stranded 1350 | C7.21 |
| ASTM | B232 | Aluminum conductors, concentric-lay-stranded conductors, coated steel-reinforced (ACSR) | C7.22 |
| ASTM | B341 | Aluminum-coated (aluminized) steel-core wire for aluminum conductors, steel-reinforced (ACSR/AZ) | C7.47 |
| ASTM | B400 | Compact round concentric-lay-stranded aluminum 1350 conductors | C7.53 |
| ASTM | B401 | Compact round concentric-lay-stranded aluminum conductors, steel-reinforced (ACSR/COMP) | C7.54 |
| ASTM | B498 | Zinc-coated (galvanized) steel-core wire for aluminum conductors, steel-reinforced (ACSR) | |
| ASTM | B502 | Aluminum-clad steel-core wire for aluminum conductors, aluminum clad steel reinforced. | |
| ASTM | B524 | Concentric-lay stranded aluminum conductors, aluminum alloy reinforced (ACAR) | |

**TABLE 4-9** Industry Specifications for Conductors (*Continued*)

| Sponsoring organization | Specification designation | Specification title | Corresponding ANSI specification |
|---|---|---|---|
| ASTM | B549 | Aluminum conductors, concentric-lay stranded aluminum-clad, steel-reinforced (ACSR/AW) | |
| ASTM | B701 | Concentric-lay stranded self-damping aluminum conductors, steel-reinforced (ACSR/SD) | |
| | | Aluminum-alloy cables | |
| ASTM | B397 | Concentric-lay-stranded 5005-H19 aluminum alloy conductors | C7.50 |
| ASTM | B399 | Concentric-lay-stranded 6201-T81 aluminum alloy conductors | C7.52 |
| | | Aluminum-clad steel cables | |
| ASTM | B416 | Concentric-lay-stranded aluminum-clad steel conductors | |
| | | Bus conductors | |
| ASTM | B187 | Copper busbar, rod, and shapes | C7.25 |
| ASTM | B188 | Seamless copper bus pipe and tube | C7.26 |
| ASTM | B236 | Aluminum bars for electrical purposes (bus bars) | C7.27 |
| ASTM | B75 | Seamless copper tube | H23.3 |
| ASTM | B317 | Aluminum-alloy extruded bar, rod, pipe, and structural shapes for electrical purposes (bus conductos) | C7.45 |
| | | Waveguide tube | |
| ASTM | B372 | Seamless copper and copper-alloy rectangular waveguide tube | H37.1 |
| | | Stainless-steel strand | |
| ASTM | A368 | Stainless-steel wire strand | |
| | | Galvanized steel wire and strand | |
| ASTM | A111 | Zinc-coated (galvanized) "iron" telephone and telegraph line wire | C7.31 |
| ASTM | A326 | Zinc-coated (galvanized) high tensile steel telephone- and telegraph-line wire | C7.30 |
| ASTM | A363 | Zinc-coated (galvanized) steel overhead ground wire strand | |
| ASTM | A411 | Zinc-coated (galvanized) low-carbon steel armor wire | |
| ASTM | A475 | Zinc-coated steel wire strand | C7.46 |
| | | Methods of test | |
| ASTM | B193 | Resistivity of electrical conductor materials | C7.24 |
| ASTM | E8 | Tension testing of metallic materials | |
| ASTM | B279 | Stiffness of bare soft square and rectangular copper wire for magnet wire fabrication | C7.41 |
| ASTM | A90 | Weight of coating or zinc-coated (galvanized) iron or steel articles | G8.12 |
| ASTM | A239 | Uniformity of coating by the Preece test (copper sulfate dip) on zinc-coated (galvanized) iron or steel articles | |
| ASTM | B342 | Electrical conductivity by use of eddy currents | C7.44 |

## TABLE 4-10 Conductors—Physical and Electrical Properties

| | Conductivity % IACS, min. % | Resistivity, Ω | | | | Temp. coefficient of resistance per °C | | Weight at 20°C | | | Coefficient of linear expansion per | | Modulus of elasticity | |
|---|---|---|---|---|---|---|---|---|---|---|---|---|---|---|
| | | cmil/ft | | mm²/m | | | | | | lb/million cmil/1000 ft | | | | |
| | 20°C | 20°C | 25°C | 20°C | 25°C | 20°C | 25°C | g/cm³ | lb/in³ | | °F | °C | lb/in² | kg/mm² |
| Commercial 1350 aluminum wire | 61.0 | 17.002 | 17.345 | 0.028265 | 0.028834 | 0.00403 | 0.00395 | 2.705 | 0.09750 | 920.3 | 0.0000128 | 0.000023 | $10 \times 10^6$ | 7,030 |
| Aluminum alloy wire 6201 | 52.5 | 19.754 | 20.097 | 0.032840 | 0.033373 | 0.00347 | 0.00340 | 2.703 | 0.09765 | 920.3 | 0.0000128 | 0.000023 | $10 \times 10^6$ | 7,030 |
| Commercial hard-drawn copper wire | 97.0 | 10.692 | 10.895 | 0.017774 | 0.018113 | 0.00381 | 0.00374 | 8.89 | 0.32 | 3027 | 0.0000094 | 0.0000169 | $17 \times 10^6$ | 11,950 |
| Standard annealed copper wire | 100.0 | 10.371 | 10.575 | 0.017241 | 0.017579 | 0.00393 | 0.00385 | 8.89 | 0.321 | 3027 | 0.0000094 | 0.0000169 | $17 \times 10^6$ | 11,950 |
| Aluminum-coated steel-core wire | 9.0* | 115.23* | | 0.19157* | | | | 7.78 | 0.281 | 2649 | 0.0000064 | 0.0000115 | $29 \times 10^6$ | 20,400 |
| Zinc-coated steel-core wire | 9.0* | 115.23* | | 0.19157* | | | | 7.78 | 0.281 | 2649 | 0.0000064 | 0.0000115 | $29 \times 10^6$ | 20,400 |
| Aluminum-clad steel-core wire | 20.33 | 51.01 | 51.52 | 0.0848 | 0.08563 | 0.0036 | 0.00356 | 6.59 | 0.2380 | 2243 | 0.0000072 | 0.0000130 | $23.5 \times 10^6$ | 16,500 |

*Typical.

**TABLE 4-11** Copper Wire—Wire Gages, Diameter, Area, Weight

| Gage name | | | | | | Diameter at 20 C (68 F) | | Area at 20 C (68 F) | | | Weight at 20 C (68 F) bare copper wire | | |
|---|---|---|---|---|---|---|---|---|---|---|---|---|---|
| American (B & S) wire gage, AWG | Steel wire gage, Stl. wg | Birmingham (Stubs' iron) wire gage, BWG | Old English (London) wire gage | (British) standard wire gage, SWG | Metric wire gage | Mils | Mm | Sq mils | Cir mils | Sq mm | Lb per 1,000 ft | Lb per mile | Kg per km |
| Gage No. | | | | | | | | | | | | | |
| ... | ... | ... | ... | 7/0 | ... | 500 | 12.70 | 196,300 | 250,000 | 126.7 | 756.7 | 3996 | 1126 |
| ... | 7/0 | ... | ... | ... | ... | 490.0 | 12.45 | 188,600 | 240,100 | 121.7 | 726.8 | 3837 | 1082 |
| ... | ... | ... | ... | 6/0 | ... | 464 | 11.79 | 169,100 | 215,000 | 109.1 | 651.7 | 3441 | 969.8 |
| ... | 6/0 | ... | ... | ... | ... | 461.5 | 11.72 | 167,300 | 213,000 | 107.9 | 644.7 | 3404 | 959.4 |
| 4/0 | ... | ... | ... | ... | ... | 460.0 | 11.68 | 166,200 | 211,600 | 107.2 | 640.5 | 3382 | 953.2 |
| ... | ... | 4/0 | 4/0 | ... | ... | 454 | 11.53 | 161,900 | 206,100 | 104.4 | 623.9 | 3294 | 928.5 |
| ... | ... | ... | ... | 5/0 | ... | 432 | 10.97 | 146,600 | 186,600 | 94.56 | 564.9 | 2983 | 840.7 |
| ... | 5/0 | ... | ... | ... | ... | 430.5 | 10.93 | 145,600 | 185,300 | 93.91 | 561.0 | 2962 | 834.8 |
| ... | ... | 3/0 | 3/0 | ... | ... | 425 | 10.80 | 141,900 | 180,600 | 91.52 | 546.8 | 2887 | 813.7 |
| 3/0 | ... | ... | ... | ... | ... | 409.6 | 10.40 | 131,800 | 167,800 | 85.01 | 507.8 | 2681 | 755.7 |
| ... | ... | ... | ... | 4/0 | ... | 400 | 10.16 | 125,700 | 160,000 | 81.07 | 484.3 | 2557 | 720.8 |
| ... | 4/0 | ... | ... | ... | ... | 393.8 | 10.00 | 121,800 | 155,100 | 78.58 | 469.4 | 2479 | 698.6 |
| ... | ... | ... | ... | ... | 100 | 393.7 | 10.00 | 121,700 | 155,000 | 78.54 | 469.2 | 2477 | 698.2 |
| ... | ... | 2/0 | 2/0 | ... | ... | 380 | 9.652 | 113,400 | 144,400 | 73.17 | 437.1 | 2308 | 650.5 |
| ... | ... | ... | ... | 3/0 | ... | 372 | 9.449 | 108,700 | 138,400 | 70.12 | 418.9 | 2212 | 623.4 |
| 2/0 | ... | ... | ... | ... | ... | 364.8 | 9.266 | 104,500 | 133,100 | 67.43 | 402.8 | 2127 | 599.5 |
| ... | 3/0 | ... | ... | ... | ... | 362.5 | 9.208 | 103,200 | 131,400 | 66.58 | 397.8 | 2100 | 591.9 |
| ... | ... | ... | ... | ... | 90 | 354.3 | 9.0 | 98,589 | 125,500 | 63.62 | 380.0 | 2007 | 565.6 |
| ... | ... | ... | ... | 2/0 | ... | 348 | 8.839 | 95,115 | 121,100 | 61.36 | 366.6 | 1936 | 545.5 |
| ... | ... | 1/0 | 1/0 | ... | ... | 340 | 8.636 | 90,790 | 115,600 | 58.58 | 349.9 | 1848 | 520.7 |
| ... | 2/0 | ... | ... | ... | ... | 331.0 | 8.407 | 86,050 | 109,600 | 55.52 | 331.6 | 1751 | 493.5 |
| 1/0 | ... | ... | ... | ... | ... | 324.9 | 8.252 | 82,891 | 105,600 | 53.49 | 319.5 | 1687 | 477.6 |
| ... | ... | ... | ... | 1/0 | ... | 324 | 8.230 | 82,450 | 105,000 | 53.19 | 317.8 | 1678 | 472.9 |
| ... | ... | ... | ... | ... | 80 | 314.96 | 8.0 | 77,931 | 99,200 | 50.27 | 300.3 | 1585 | 446.9 |
| ... | 1/0 | ... | ... | ... | ... | 306.5 | 7.785 | 73,780 | 93,940 | 47.60 | 284.4 | 1501 | 423.2 |
| ... | ... | 1 | 1 | 1 | ... | 300 | 7.620 | 70,690 | 90,000 | 45.60 | 272.4 | 1438 | 405.4 |
| 1 | ... | ... | ... | ... | ... | 289.3 | 7.348 | 65,730 | 83,690 | 42.41 | 253.3 | 1338 | 377.0 |
| ... | ... | 2 | 2 | ... | ... | 284 | 7.214 | 63,350 | 80,660 | 40.87 | 244.4 | 1289 | 363.3 |
| ... | 1 | ... | ... | ... | ... | 283.0 | 7.188 | 62,900 | 80,090 | 40.58 | 242.4 | 1280 | 360.8 |
| ... | ... | ... | ... | 2 | ... | 276 | 7.010 | 59,830 | 76,180 | 38.60 | 230.6 | 1217 | 343.1 |
| ... | ... | ... | ... | ... | 70 | 275.59 | 7.0 | 59,650 | 75,950 | 38.49 | 229.9 | 1214 | 342.1 |
| ... | 2 | ... | ... | ... | ... | 262.5 | 6.668 | 54,120 | 68,910 | 34.92 | 208.6 | 1101 | 310.4 |
| ... | ... | 3 | 3 | ... | ... | 259 | 6.579 | 52,690 | 67,080 | 33.99 | 203.1 | 1072 | 302.2 |
| 2 | ... | ... | ... | ... | ... | 257.6 | 6.543 | 52,120 | 66,360 | 33.62 | 200.9 | 1061 | 298.9 |
| ... | ... | ... | ... | 3 | ... | 252 | 6.401 | 49,880 | 63,500 | 32.18 | 192.2 | 1015 | 286.1 |

| | | | | | | | | | | | | | |
|---|---|---|---|---|---|---|---|---|---|---|---|---|---|
| ... | 3 | ... | ... | ... | ... | 243.7 | 6.190 | 46,640 | 59,390 | 30.09 | 179.8 | 949.2 | 267.5 |
| ... | ... | 4 | 4 | ... | ... | 238 | 6.045 | 44,490 | 56,640 | 28.70 | 171.5 | 905.3 | 255.2 |
| ... | ... | ... | ... | ... | 60 | 236.2 | 6.000 | 43,830 | 55,800 | 28.27 | 168.9 | 891.8 | 251.4 |
| ... | ... | ... | ... | 4 | ... | 232 | 5.893 | 42,270 | 53,820 | 27.27 | 162.9 | 860.3 | 242.5 |
| 3 | ... | ... | ... | ... | ... | 229.4 | 5.827 | 41,330 | 52,620 | 26.67 | 159.3 | 841.1 | 237.1 |
| ... | 4 | ... | ... | ... | ... | 225.3 | 5.723 | 39,870 | 50,760 | 25.72 | 153.7 | 811.3 | 228.7 |
| ... | ... | 5 | 5 | ... | ... | 220 | 5.588 | 38,010 | 48,400 | 24.52 | 146.5 | 773.6 | 218.0 |
| ... | ... | ... | ... | 5 | ... | 212 | 5.385 | 35,300 | 44,940 | 22.77 | 136.0 | 718.3 | 202.5 |
| ... | 5 | ... | ... | ... | ... | 207.0 | 5.258 | 33,605 | 42,850 | 21.71 | 129.7 | 684.9 | 193.0 |
| 4 | ... | ... | ... | ... | ... | 204.3 | 5.189 | 32,780 | 41,740 | 21.15 | 126.3 | 667.1 | 188.0 |
| ... | ... | 6 | 6 | ... | ... | 203 | 5.156 | 32,370 | 41,210 | 20.88 | 124.7 | 658.6 | 185.6 |
| ... | ... | ... | ... | ... | 50 | 196.8 | 5.0 | 30,430 | 38,750 | 19.63 | 117.3 | 619.3 | 174.6 |
| ... | 6 | ... | ... | 6 | ... | 192.0 | 4.877 | 28,950 | 36,860 | 18.68 | 111.6 | 589.2 | 166.1 |
| 5 | ... | ... | ... | ... | ... | 181.9 | 4.620 | 25,990 | 33,090 | 16.77 | 100.2 | 528.8 | 149.0 |
| ... | ... | 7 | 7 | ... | ... | 180 | 4.572 | 25,450 | 32,400 | 16.42 | 98.07 | 517.8 | 146.0 |
| ... | ... | ... | ... | ... | 45 | 177.2 | 4.500 | 24,650 | 31,390 | 15.90 | 95.01 | 501.7 | 141.4 |
| ... | 7 | ... | ... | ... | ... | 177.0 | 4.496 | 24,610 | 31,330 | 15.87 | 94.83 | 500.7 | 141.1 |
| ... | ... | ... | ... | 7 | ... | 176 | 4.470 | 24,330 | 30,980 | 15.70 | 93.76 | 495.1 | 139.5 |
| ... | ... | 8 | 8 | ... | ... | 165 | 4.191 | 21,380 | 27,220 | 13.80 | 82.41 | 435.1 | 122.6 |
| 6 | 8 | ... | ... | ... | ... | 162.0 | 4.115 | 20,610 | 26,240 | 13.30 | 79.44 | 419.4 | 118.2 |
| ... | ... | ... | ... | 8 | ... | 160 | 4.064 | 20,110 | 25,600 | 12.97 | 77.49 | 409.2 | 115.3 |
| ... | ... | ... | ... | ... | 40 | 157.5 | 4.000 | 19,480 | 24,800 | 12.57 | 75.09 | 396.4 | 111.7 |
| ... | 9 | ... | ... | ... | ... | 148.3 | 3.767 | 17,270 | 21,990 | 11.14 | 66.57 | 351.5 | 99.07 |
| ... | ... | 9 | 9 | ... | ... | 148 | 3.759 | 17,200 | 21,900 | 11.10 | 66.30 | 350.1 | 98.67 |
| 7 | ... | ... | ... | ... | ... | 144.3 | 3.665 | 16,350 | 20,820 | 10.55 | 63.03 | 332.8 | 93.80 |
| ... | ... | ... | ... | 9 | ... | 144 | 3.658 | 16,290 | 20,740 | 10.51 | 62.77 | 331.4 | 93.41 |
| ... | ... | ... | ... | ... | 35 | 137.8 | 3.5 | 14,910 | 18,990 | 9.622 | 57.47 | 303.5 | 85.53 |
| ... | 10 | ... | ... | ... | ... | 135.0 | 3.429 | 14,310 | 18,220 | 9.235 | 55.17 | 291.3 | 82.10 |
| ... | ... | 10 | 10 | ... | ... | 134 | 3.404 | 14,100 | 17,960 | 9.098 | 54.35 | 287.0 | 80.89 |
| 8 | ... | ... | ... | ... | ... | 128.5 | 3.264 | 12,970 | 16,510 | 8.367 | 49.98 | 263.9 | 74.38 |
| ... | ... | ... | ... | 10 | ... | 128 | 3.251 | 12,870 | 16,380 | 8.302 | 49.59 | 261.9 | 73.80 |
| ... | 11 | ... | ... | ... | ... | 120.5 | 3.061 | 11,400 | 14,520 | 7.358 | 43.95 | 232.1 | 65.41 |
| ... | ... | 11 | 11 | ... | ... | 120 | 3.048 | 11,310 | 14,400 | 7.297 | 43.59 | 230.2 | 64.87 |
| ... | ... | ... | ... | ... | 30 | 118.1 | 3.0 | 10,960 | 13,950 | 7.068 | 42.22 | 222.9 | 62.83 |
| ... | ... | ... | ... | 11 | ... | 116 | 2.946 | 10,570 | 13,460 | 6.818 | 40.73 | 215.1 | 60.61 |
| 9 | ... | ... | ... | ... | ... | 114.4 | 2.906 | 10,280 | 13,090 | 6.631 | 39.61 | 209.2 | 58.95 |
| ... | ... | 12 | 12 | ... | ... | 109 | 2.769 | 9,331 | 11,880 | 6.020 | 35.96 | 189.9 | 53.52 |
| ... | 12 | ... | ... | ... | ... | 105.5 | 2.680 | 8,742 | 11,130 | 5.640 | 33.69 | 177.9 | 50.14 |
| ... | ... | ... | ... | 12 | ... | 104 | 2.642 | 8,495 | 10,820 | 5.481 | 32.74 | 172.9 | 48.72 |
| 10 | ... | ... | ... | ... | ... | 101.9 | 2.588 | 8,155 | 10,380 | 5.261 | 31.43 | 166.0 | 46.77 |
| ... | ... | ... | ... | ... | 25 | 98.42 | 2.5 | 7,609 | 9,687 | 4.908 | 29.32 | 154.8 | 43.63 |
| ... | ... | 13 | 13 | ... | ... | 95 | 2.413 | 7,088 | 9,025 | 4.573 | 27.32 | 144.2 | 40.65 |
| ... | ... | ... | ... | 13 | ... | 92 | 2.337 | 6,648 | 8,464 | 4.289 | 25.62 | 135.3 | 38.13 |
| ... | 13 | ... | ... | ... | ... | 91.5 | 2.324 | 6,576 | 8,372 | 4.242 | 25.34 | 133.8 | 37.71 |
| 11 | ... | ... | ... | ... | ... | 90.7 | 2.304 | 6,460 | 8,230 | 4.17 | 24.90 | 131 | 37.1 |

**TABLE 4-11** Copper Wire—Wire Gages, Diameter, Area, Weight (*Continued*)

| Gage name | | | | | | Diameter at 20 C (68 F) | | Area at 20 C (68 F) | | | Weight at 20 C (68 F) bare copper wire | | |
|---|---|---|---|---|---|---|---|---|---|---|---|---|---|
| American (B & S) wire gage, AWG | Steel wire gage, Stl. wg | Birmingham (Stubs' iron) wire gage, BWG | Old English (London) wire gage | (British) standard wire gage, SWG | Metric wire gage | Mils | Mm | Sq mils | Cir mils | Sq mm | Lb per 1,000 ft | Lb per mile | Kg per km |
| Gage No. | | | | | | | | | | | | | |
| ... | ... | 14 | 14 | ... | ..... | 83 | 2.108 | 5,411 | 6,889 | 3.491 | 20.85 | 110.1 | 31.03 |
| 12 | ... | ... | ... | ... | ..... | 80 | 2.05 | 5,130 | 6,530 | 3.310 | 19.8 | 104 | 29.4 |
| ... | 14 | ... | ... | 14 | ..... | 80.0 | 2.032 | 5,029 | 6,400 | 3.243 | 19.37 | 102.3 | 28.83 |
| ... | ... | ... | ... | ... | 20 | 78.74 | 2.0 | 4,869 | 6,200 | 3.142 | 18.77 | 99.09 | 27.93 |
| ... | 15 | 15 | 15 | 15 | ..... | 72.0 | 1.829 | 4,072 | 5,184 | 2.627 | 15.69 | 82.85 | 23.35 |
| 13 | ... | ... | ... | ... | ..... | 72.0 | 1.83 | 4,070 | 5,180 | 2.63 | 15.7 | 82.9 | 23.4 |
| ... | ... | ... | ... | ... | 18 | 70.87 | 1.8 | 3,944 | 5,022 | 2.545 | 15.20 | 80.27 | 22.62 |
| ... | ... | 16 | 16 | ... | ..... | 65 | 1.651 | 3,318 | 4,225 | 2.141 | 12.79 | 67.53 | 19.03 |
| 14 | ... | ... | ... | ... | ..... | 64.1 | 1.63 | 3,230 | 4,110 | 2.08 | 12.4 | 65.7 | 18.5 |
| ... | ... | ... | ... | 16 | ..... | 64 | 1.626 | 3,217 | 4,096 | 2.075 | 12.40 | 65.46 | 18.45 |
| ... | ... | ... | ... | ... | 16 | 62.99 | 1.6 | 3,116 | 3,968 | 2.011 | 12.01 | 63.41 | 17.87 |
| ... | 16 | ... | ... | ... | ..... | 62.5 | 1.588 | 3,068 | 3,906 | 1.979 | 11.82 | 62.43 | 17.60 |
| ... | ... | 17 | 17 | ... | ..... | 58 | 1.473 | 2,642 | 3,364 | 1.705 | 10.18 | 53.77 | 15.15 |
| 15 | ... | ... | ... | ... | ..... | 57.1 | 1.45 | 2,560 | 3,260 | 1.650 | 9.87 | 52.1 | 14.7 |
| ... | ... | ... | ... | 17 | ..... | 56 | 1.422 | 2,463 | 3,136 | 1.589 | 9.493 | 50.12 | 14.13 |
| ... | ... | ... | ... | ... | 14 | 55.12 | 1.4 | 2,386 | 3,038 | 1.539 | 9.196 | 48.56 | 13.69 |
| ... | 17 | ... | ... | ... | ..... | 54.0 | 1.372 | 2,290 | 2,916 | 1.478 | 8.827 | 46.60 | 13.14 |
| 16 | ... | ... | ... | ... | ..... | 50.8 | 1.29 | 2,030 | 2,580 | 1.31 | 7.81 | 41.2 | 11.6 |
| ... | ... | 18 | 18 | ... | ..... | 49 | 1.245 | 1,886 | 2,401 | 1.217 | 7.268 | 38.37 | 10.82 |
| ... | ... | ... | ... | 18 | ..... | 48 | 1.219 | 1,810 | 2,304 | 1.167 | 6.974 | 36.82 | 10.38 |
| ... | 18 | ... | ... | ... | ..... | 47.5 | 1.207 | 1,772 | 2,256 | 1.143 | 6.830 | 36.06 | 10.16 |
| ... | ... | ... | ... | ... | 12 | 47.24 | 1.200 | 1,753 | 2,232 | 1.131 | 6.756 | 35.67 | 10.05 |
| 17 | ... | ... | ... | ... | ..... | 45.3 | 1.150 | 1,610 | 2,050 | 1.040 | 6.21 | 32.8 | 9.24 |
| ... | ... | 19 | ... | ... | ..... | 42 | 1.067 | 1,385 | 1,764 | 0.8938 | 5.340 | 28.19 | 7.946 |
| ... | 19 | ... | ... | ... | ..... | 41.0 | 1.041 | 1,320 | 1,681 | 0.8518 | 5.088 | 26.87 | 7.572 |
| 18 | ... | ... | ... | ... | ..... | 40.3 | 1.02 | 1,280 | 1,620 | 0.823 | 4.92 | 26.0 | 7.32 |
| ... | ... | ... | 19 | 19 | ..... | 40 | 1.016 | 1,257 | 1,600 | 0.8107 | 4.843 | 25.57 | 7.207 |
| ... | ... | ... | ... | ... | 10 | 39.37 | 1 | 1,217 | 1,550 | 0.7854 | 4.692 | 24.77 | 6.982 |
| ... | ... | ... | ... | 20 | ..... | 36 | 0.9144 | 1,018 | 1,296 | 0.6567 | 3.923 | 20.71 | 5.838 |
| 19 | ... | ... | ... | ... | ..... | 35.9 | 0.912 | 1,010 | 1,290 | 0.653 | 3.90 | 20.6 | 5.81 |

| | | | | | | | | | | | | | |
|---|---|---|---|---|---|---|---|---|---|---|---|---|---|
| ... | ... | ... | ... | ... | 9 | 35.43 | 0.90 | 986.1 | 1,255 | 0.6362 | 3.800 | 20.07 | 5.656 |
| ... | ... | 20 | 20 | ... | ..... | 35 | 0.8890 | 962.1 | 1,225 | 0.6207 | 3.708 | 19.58 | 5.520 |
| ... | 20 | ... | ... | ... | ..... | 34.8 | 0.8839 | 951.1 | 1,211 | 0.6136 | 3.666 | 19.36 | 5.455 |
| ... | ... | 21 | ... | 21 | ..... | 32 | 0.8128 | 804.2 | 1,024 | 0.5189 | 3.100 | 16.37 | 4.613 |
| **20** | ... | ... | ... | ... | ..... | 32.0 | 0.813 | 804.0 | 1,020 | 0.519 | 3.10 | 16.4 | 4.61 |
| ... | 21 | ... | ... | ... | ..... | 31.7 | 0.8052 | 789.2 | 1,005 | 0.5092 | 3.042 | 16.06 | 4.527 |
| ... | ... | ... | 21 | ... | 8 | 31.5 | 0.80 | 779.1 | 992 | 0.5027 | 3.003 | 15.85 | 4.469 |
| ... | ... | ... | 22 | ... | ..... | 29.50 | 0.7493 | 683.3 | 870 | 0.4410 | 2.634 | 13.91 | 3.920 |
| ... | 22 | ... | ... | ... | ..... | 28.6 | 0.7264 | 642.4 | 818.0 | 0.4145 | 2.476 | 13.07 | 3.685 |
| **21** | ... | ... | ... | ... | ..... | 28.5 | 0.724 | 638.0 | 812.0 | 0.412 | 2.46 | 13.0 | 3.66 |
| ... | ... | 22 | ... | 22 | ..... | 28 | 0.7112 | 615.8 | 784.0 | 0.3973 | 2.373 | 12.53 | 3.532 |
| ... | ... | ... | ... | ... | 7 | 27.56 | 0.70 | 596.5 | 759.5 | 0.3848 | 2.299 | 12.14 | 3.421 |
| ... | ... | ... | 23 | ... | ..... | 27.00 | 0.6858 | 572.6 | 729.0 | 0.3694 | 2.207 | 11.65 | 3.284 |
| ... | 23 | ... | ... | ... | ..... | 25.8 | 0.6553 | 522.8 | 665.6 | 0.3373 | 2.015 | 10.64 | 2.998 |
| **22** | ... | ... | ... | ... | ..... | 25.3 | 0.643 | 503.0 | 640.0 | 0.324 | 1.94 | 10.2 | 2.88 |
| ... | ... | 23 | 24 | ... | ..... | 25 | 0.6350 | 490.9 | 625.0 | 0.3167 | 1.892 | 9.989 | 2.815 |
| ... | ... | ... | ... | 23 | ..... | 24 | 0.6096 | 452.4 | 576.0 | 0.2919 | 1.744 | 9.206 | 2.595 |
| ... | ... | ... | ... | ... | 6 | 23.62 | 0.60 | 438.3 | 558.0 | 0.2827 | 1.689 | 8.918 | 2.514 |
| ... | 24 | ... | 25 | ... | ..... | 23.0 | 0.5842 | 415.5 | 529.0 | 0.2680 | 1.601 | 8.455 | 2.383 |
| **23** | ... | ... | ... | ... | ..... | 22.6 | 0.574 | 401.0 | 511.0 | 0.259 | 1.55 | 8.16 | 2.30 |
| ... | ... | 24 | ... | 24 | ..... | 22 | 0.5588 | 380.1 | 484.0 | 0.2452 | 1.465 | 7.736 | 2.180 |
| ... | ... | ... | 26 | ... | ..... | 20.5 | 0.5207 | 330.1 | 420.3 | 0.2129 | 1.272 | 6.717 | 1.893 |
| ... | 25 | ... | ... | ... | ..... | 20.4 | 0.5182 | 326.9 | 416.2 | 0.2109 | 1.260 | 6.651 | 1.875 |
| **24** | ... | ... | ... | ... | ..... | 20.1 | 0.511 | 317.0 | 404.0 | 0.205 | 1.22 | 6.46 | 1.82 |
| ... | ... | 25 | ... | 25 | ..... | 20 | 0.5080 | 314.2 | 400.0 | 0.2027 | 1.211 | 6.393 | 1.802 |
| ... | ... | ... | ... | ... | 5 | 19.68 | 0.50 | 304.3 | 387.5 | 0.1963 | 1.173 | 6.193 | 1.746 |
| ... | ... | ... | 27 | ... | ..... | 18.75 | 0.4763 | 276.1 | 351.6 | 0.1781 | 1.064 | 5.619 | 1.584 |
| ... | 26 | ... | ... | ... | ..... | 18.1 | 0.4597 | 257.3 | 327.6 | 0.1660 | 0.9916 | 5.236 | 1.476 |
| ... | ... | 26 | ... | 26 | ..... | 18 | 0.4572 | 254.5 | 324.0 | 0.1642 | 0.9807 | 5.178 | 1.460 |
| **25** | ... | ... | ... | ... | ..... | 17.9 | 0.455 | 252.0 | 320.0 | 0.162 | 0.970 | 5.12 | 1.44 |
| ... | ... | ... | ... | ... | 4.5 | 17.72 | 0.45 | 246.5 | 313.9 | 0.1590 | 0.9501 | 5.016 | 1.414 |
| ... | 27 | ... | ... | ... | ..... | 17.3 | 0.4394 | 235.1 | 299.3 | 0.1517 | 0.9059 | 4.783 | 1.348 |
| ... | ... | ... | 28 | ... | ..... | 16.50 | 0.4191 | 213.8 | 272.3 | 0.1380 | 0.8241 | 4.351 | 1.226 |
| ... | ... | ... | ... | 27 | ..... | 16.4 | 0.4166 | 211.2 | 269.0 | 0.1363 | 0.8141 | 4.299 | 1.212 |
| ... | 28 | ... | ... | ... | ..... | 16.2 | 0.4115 | 206.1 | 262.4 | 0.1330 | 0.7944 | 4.194 | 1.182 |
| ... | ... | 27 | ... | ... | ..... | 16 | 0.4064 | 201.1 | 256.0 | 0.1297 | 0.7749 | 4.092 | 1.153 |
| **26** | ... | ... | ... | ... | ..... | 15.9 | 0.404 | 199.0 | 253.0 | 0.128 | 0.765 | 4.04 | 1.14 |
| ... | ... | ... | ... | ... | 4 | 15.75 | 0.40 | 194.8 | 248.0 | 0.1257 | 0.7507 | 3.964 | 1.117 |
| ... | ... | ... | 29 | ... | ..... | 15.50 | 0.3937 | 188.7 | 254.1 | 0.1217 | 0.7272 | 3.840 | 1.082 |
| ... | 29 | ... | ... | ... | ..... | 15.0 | 0.3810 | 176.7 | 225.0 | 0.1140 | 0.6811 | 3.596 | 1.014 |
| ... | ... | ... | ... | 28 | ..... | 14.8 | 0.3759 | 172.0 | 219.0 | 0.1110 | 0.6630 | 3.501 | 0.9867 |
| **27** | ... | ... | ... | ... | ..... | 14.2 | 0.361 | 158.0 | 202.0 | 0.102 | 0.610 | 3.22 | 0.908 |
| ... | 30 | 28 | ... | ... | ..... | 14.0 | 0.3556 | 153.9 | 196.0 | 0.09932 | 0.5933 | 3.133 | 0.8829 |
| ... | ... | ... | ... | ... | 3.5 | 13.78 | 0.35 | 149.1 | 189.9 | 0.09621 | 0.5747 | 3.035 | 0.8553 |
| ... | ... | ... | 30 | ... | ..... | 13.75 | 0.3493 | 148.5 | 189.1 | 0.09580 | 0.5723 | 3.022 | 0.8517 |

**TABLE 4-11** Copper Wire—Wire Gages, Diameter, Area, Weight (*Continued*)

| Gage name | | | | | | Diameter at 20 C (68 F) | | Area at 20 C (68 F) | | | Weight at 20 C (68 F) bare copper wire | | |
|---|---|---|---|---|---|---|---|---|---|---|---|---|---|
| American (B & S) wire gage, AWG | Steel wire gage, Stl. wg | Birmingham (Stubs' iron) wire gage, BWG | Old English (London) wire gage | (British) standard wire gage, SWG | Metric wire gage | Mils | Mm | Sq mils | Cir mils | Sq mm | Lb per 1,000 ft | Lb per mile | Kg per km |
| Gage No. | | | | | | | | | | | | | |
| ... | ... | ... | ... | 29 | ..... | 13.6 | 0.3454 | 145.3 | 185.0 | 0.09372 | 0.5599 | 2.956 | 0.8332 |
| ... | 31 | ... | ... | ... | ..... | 13.2 | 0.3353 | 136.8 | 174.2 | 0.08829 | 0.5274 | 2.785 | 0.7849 |
| ... | ... | 29 | ... | ... | ..... | 13 | 0.3302 | 132.7 | 169.0 | 0.08563 | 0.5116 | 2.701 | 0.7613 |
| ... | 32 | ... | ... | ... | ..... | 12.8 | 0.3251 | 128.7 | 163.8 | 0.08302 | 0.4959 | 2.619 | 0.7380 |
| 28 | ... | ... | ... | ... | ..... | 12.6 | 0.320 | 125 | 159.0 | 0.0804 | 0.481 | 2.54 | 0.715 |
| ... | ... | ... | ... | 30 | ..... | 12.4 | 0.3150 | 120.8 | 153.8 | 0.07791 | 0.4654 | 2.457 | 0.6926 |
| ... | ... | ... | 31 | ... | ..... | 12.25 | 0.3112 | 117.9 | 150.1 | 0.07604 | 0.4542 | 2.398 | 0.6760 |
| ... | ... | 30 | ... | ... | ..... | 12 | 0.3048 | 113.1 | 144.0 | 0.07297 | 0.4359 | 2.301 | 0.6487 |
| ... | ... | ... | ... | ... | 3 | 11.81 | 0.30 | 109.6 | 139.5 | 0.07069 | 0.4223 | 2.230 | 0.6284 |
| ... | 33 | ... | ... | ... | ..... | 11.8 | 0.2997 | 109.4 | 139.2 | 0.07055 | 0.4215 | 2.225 | 0.6272 |
| ... | ... | ... | ... | 31 | ..... | 11.6 | 0.2946 | 105.7 | 134.6 | 0.06818 | 0.4073 | 2.151 | 0.6061 |
| 29 | ... | ... | ... | ... | ..... | 11.3 | 0.287 | 100.0 | 128.0 | 0.0647 | 0.387 | 2.04 | 0.575 |
| ... | ... | ... | 32 | ... | ..... | 11.25 | 0.2858 | 99.40 | 126.6 | 0.06413 | 0.3831 | 2.023 | 0.5701 |
| ... | ... | ... | ... | 32 | ..... | 10.8 | 0.2743 | 91.61 | 116.6 | 0.05910 | 0.3531 | 1.864 | 0.5254 |
| ... | 34 | ... | ... | ... | ..... | 10.4 | 0.2642 | 84.95 | 108.2 | 0.05481 | 0.3274 | 1.729 | 0.4872 |
| ... | ... | ... | 33 | ... | ..... | 10.25 | 0.2604 | 82.52 | 105.1 | 0.05324 | 0.3180 | 1.679 | 0.4733 |
| 30 | ... | ... | ... | ... | ..... | 10.0 | 0.254 | 78.50 | 100.0 | 0.0507 | 0.303 | 1.60 | 0.450 |
| ... | ... | 31 | ... | 33 | ..... | 10.0 | 0.2540 | 78.54 | 100.0 | 0.05067 | 0.3027 | 1.598 | 0.4505 |
| ... | ... | ... | ... | ... | 2.5 | 9.842 | 0.25 | 76.09 | 96.87 | 0.04909 | 0.2932 | 1.548 | 0.4364 |
| ... | 35 | ... | 34 | ... | ..... | 9.5 | 0.2413 | 70.88 | 90.25 | 0.04573 | 0.2732 | 1.442 | 0.4065 |
| ... | ... | ... | ... | 34 | ..... | 9.2 | 0.2337 | 66.48 | 84.64 | 0.04289 | 0.2562 | 1.353 | 0.3813 |
| ... | 36 | 32 | 35 | ... | ..... | 9.0 | 0.2286 | 63.62 | 81.00 | 0.04104 | 0.2452 | 1.295 | 0.3649 |
| 31 | ... | ... | ... | ... | ..... | 8.9 | 0.226 | 62.20 | 79.2 | 0.0401 | 0.240 | 1.27 | 0.357 |
| ... | 37 | ... | ... | ... | ..... | 8.5 | 0.2159 | 56.75 | 72.25 | 0.03661 | 0.2187 | 1.155 | 0.3255 |
| ... | ... | ... | ... | 35 | ..... | 8.4 | 0.2134 | 55.42 | 70.56 | 0.03575 | 0.2139 | 1.128 | 0.3178 |
| ... | 38 | 33 | ... | ... | ..... | 8.0 | 0.2032 | 50.27 | 64.00 | 0.03243 | 0.1937 | 1.023 | 0.2883 |
| 32 | ... | ... | ... | ... | ..... | 8.0 | 0.203 | 50.30 | 64.0 | 0.0324 | 0.194 | 1.02 | 0.288 |
| ... | ... | ... | ... | ... | 2 | 7.874 | 0.20 | 48.69 | 62.00 | 0.03142 | 0.1877 | 0.9909 | 0.2793 |
| ... | ... | ... | ... | 36 | ..... | 7.6 | 0.1930 | 45.36 | 57.76 | 0.02927 | 0.1748 | 0.9231 | 0.2602 |
| ... | 39 | ... | 36 | ... | ..... | 7.5 | 0.1905 | 44.18 | 56.25 | 0.02850 | 0.1703 | 0.8990 | 0.2534 |
| ... | ... | ... | ... | ... | 1.8 | 7.087 | 0.1800 | 39.44 | 50.22 | 0.02545 | 0.1520 | 0.8026 | 0.2262 |
| 33 | ... | ... | ... | ... | ..... | 7.1 | 0.180 | 39.60 | 50.40 | 0.0255 | 0.153 | 0.806 | 0.227 |
| ... | 40 | 34 | ... | ... | ..... | 7.0 | 0.1778 | 38.48 | 49.00 | 0.02483 | 0.1483 | 0.7831 | 0.2207 |
| ... | ... | ... | ... | 37 | ..... | 6.8 | 0.1727 | 36.32 | 46.24 | 0.02343 | 0.1400 | 0.7390 | 0.2083 |
| ... | 41 | ... | ... | ... | ..... | 6.6 | 0.1676 | 34.21 | 43.56 | 0.02207 | 0.1319 | 0.6962 | 0.1962 |

| | | | | | | | | | | | | | |
|---|---|---|---|---|---|---|---|---|---|---|---|---|---|
| ... | ... | ... | 37 | ... | ..... | 6.50 | 0.1651 | 33.18 | 42.25 | 0.02141 | 0.1279 | 0.6753 | 0.1903 |
| 34 | ... | ... | ... | ... | ..... | 6.3 | 0.160 | 31.20 | 39.7 | 0.0201 | 0.120 | 0.634 | 0.179 |
| ... | ... | ... | ... | ... | 1.6 | 6.299 | 0.16 | 31.16 | 39.68 | 0.02011 | 0.1201 | 0.6342 | 0.1787 |
| ... | 42 | ... | ... | ... | ..... | 6.2 | 0.1575 | 30.19 | 38.44 | 0.01948 | 0.1164 | 0.6144 | 0.1732 |
| ... | 43 | ... | ... | 38 | ..... | 6.0 | 0.1524 | 28.27 | 36.00 | 0.01824 | 0.1090 | 0.5754 | 0.1622 |
| ... | ... | ... | ... | ... | 1.5 | 5.906 | 0.15 | 27.39 | 34.87 | 0.01767 | 0.1056 | 0.5575 | 0.1571 |
| ... | 44 | ... | ... | ... | ..... | 5.8 | 0.1473 | 26.42 | 33.64 | 0.01705 | 0.1018 | 0.5377 | 0.1515 |
| ... | ... | ... | 38 | ... | ..... | 5.75 | 0.1461 | 25.97 | 33.06 | 0.01675 | 0.1001 | 0.5284 | 0.1489 |
| 35 | ... | ... | ... | ... | ..... | 5.6 | 0.142 | 24.60 | 31.4 | 0.0159 | 0.0949 | 0.501 | 0.141 |
| ... | ... | ... | ... | ... | 1.4 | 5.512 | 0.14 | 23.86 | 30.38 | 0.01539 | 0.09196 | 0.4855 | 0.1369 |
| ... | 45 | ... | ... | ... | ..... | 5.5 | 0.1397 | 23.76 | 30.25 | 0.01533 | 0.09157 | 0.4835 | 0.1363 |
| ... | 46 | ... | ... | 39 | ..... | 5.2 | 0.1321 | 21.24 | 27.04 | 0.01370 | 0.08185 | 0.4322 | 0.1218 |
| 36 | 47 | 35 | 39 | ... | ..... | 5.0 | 0.127 | 19.60 | 25.00 | 0.0127 | 0.0757 | 0.400 | 0.113 |
| ... | 48 | ... | ... | 40 | ..... | 4.8 | 0.1219 | 18.10 | 23.04 | 0.01167 | 0.06974 | 0.3682 | 0.1038 |
| ... | ... | ... | ... | ... | 1.2 | 4.724 | 0.12 | 17.53 | 22.32 | 0.01131 | 0.06756 | 0.3567 | 0.1005 |
| ... | 49 | ... | ... | ... | ..... | 4.6 | 0.1168 | 16.62 | 21.16 | 0.01072 | 0.06405 | 0.3382 | 0.09532 |
| ... | ... | ... | 40 | ... | ..... | 4.50 | 0.1143 | 15.90 | 20.25 | 0.01026 | 0.06130 | 0.3236 | 0.09122 |
| 37 | ... | ... | ... | ... | ..... | 4.50 | 0.114 | 15.90 | 20.20 | 0.0103 | 0.0613 | 0.324 | 0.0912 |
| ... | 50 | ... | ... | 41 | ..... | 4.4 | 0.1118 | 15.21 | 19.36 | 0.009810 | 0.05860 | 0.3094 | 0.08721 |
| ... | ... | 36 | ... | 42 | ..... | 4.0 | 0.1016 | 12.57 | 16.00 | 0.008107 | 0.04843 | 0.2557 | 0.07207 |
| 38 | ... | ... | ... | ... | ..... | 4.0 | 0.102 | 12.60 | 16.0 | 0.00811 | 0.0484 | 0.256 | 0.0721 |
| ... | ... | ... | ... | ... | 1 | 3.937 | 0.10 | 12.17 | 15.50 | 0.007854 | 0.04692 | 0.2477 | 0.06982 |
| ... | ... | ... | ... | 43 | ..... | 3.6 | 0.09144 | 10.18 | 12.96 | 0.006567 | 0.03923 | 0.2071 | 0.05838 |
| 39 | ... | ... | ... | ... | ..... | 3.5 | 0.0889 | 9.62 | 12.2 | 0.00621 | 0.0371 | 0.196 | 0.0552 |
| ... | ... | ... | ... | 44 | ..... | 3.2 | 0.08128 | 8.042 | 10.24 | 0.005189 | 0.0310 | 0.1637 | 0.04613 |
| 40 | ... | ... | ... | ... | ..... | 3.1 | 0.0787 | 7.55 | 9.61 | 0.00487 | 0.0291 | 0.154 | 0.0433 |
| 41 | ... | ... | ... | ... | ..... | 2.8 | 0.0711 | 6.16 | 7.84 | 0.00397 | 0.0237 | 0.125 | 0.0353 |
| ... | ... | ... | ... | 45 | ..... | 2.8 | 0.07112 | 6.158 | 7.840 | 0.003973 | 0.02373 | 0.1253 | 0.03532 |
| 42 | ... | ... | ... | ... | ..... | 2.5 | 0.0635 | 4.91 | 6.25 | 0.00317 | 0.0189 | 0.0999 | 0.0282 |
| ... | ... | ... | ... | 46 | ..... | 2.4 | 0.06096 | 4.524 | 5.760 | 0.002919 | 0.01744 | 0.09206 | 0.02595 |
| 43 | ... | ... | ... | ... | ..... | 2.2 | 0.0559 | 3.80 | 4.84 | 0.00245 | 0.0147 | 0.0774 | 0.0218 |
| ... | ... | ... | ... | 47 | ..... | 2.0 | 0.05080 | 3.142 | 4.000 | 0.002027 | 0.01211 | 0.06393 | 0.01802 |
| 44 | ... | ... | ... | ... | ..... | 2.0 | 0.0508 | 3.14 | 4.00 | 0.00203 | 0.0121 | 0.0639 | 0.0180 |
| ... | ... | ... | ... | ... | 0.5 | 1.969 | 0.05 | 3.043 | 3.875 | 0.001963 | 0.01173 | 0.06193 | 0.01746 |
| 45 | ... | ... | ... | ... | ..... | 1.8 | 0.0457 | 2.54 | 3.24 | 0.00164 | 0.00981 | 0.0519 | 0.0146 |
| ... | ... | ... | ... | 48 | ..... | 1.6 | 0.04064 | 2.011 | 2.560 | 0.001297 | 0.007749 | 0.04092 | 0.01153 |
| 46 | ... | ... | ... | ... | ..... | 1.6 | 0.0406 | 2.01 | 2.56 | 0.00130 | 0.00775 | 0.0409 | 0.0115 |
| 47 | ... | ... | ... | ... | ..... | 1.4 | 0.0356 | 1.54 | 1.96 | 0.000993 | 0.00593 | 0.0313 | 0.00883 |
| 48 | ... | ... | ... | ... | ..... | 1.2 | 0.0305 | 1.13 | 1.44 | 0.000730 | 0.00436 | 0.0230 | 0.00649 |
| ... | ... | ... | ... | 49 | ..... | 1.2 | 0.03048 | 1.131 | 1.440 | 0.0007297 | 0.004359 | 0.02301 | 0.006487 |
| 49 | ... | ... | ... | ... | ..... | 1.1 | 0.0279 | 0.950 | 1.21 | 0.000613 | 0.00366 | 0.0193 | 0.00545 |
| ... | ... | ... | ... | 50 | ..... | 1.0 | 0.02540 | 0.7854 | 1.000 | 0.0005067 | 0.003027 | 0.01598 | 0.004505 |
| 50 | ... | ... | ... | ... | ..... | 1.0 | 0.0254 | 0.785 | 1.00 | 0.000507 | 0.00303 | 0.0160 | 0.00450 |

***159. Rope-lay stranded conductors*** are made up of successive layers of helically laid stranded members. The members may be concentric-lay or bunch-stranded.

***160. Special Stranded Conductor Shapes.*** For use in multiconductor insulated cables whose finished cross section must be round, special shapes such as D shape (hemispherical), sector shape (triangular), and semisector (oval) are commonly used.

*Annular* conductors are formed by stranding helically laid wires over a central core, which may be (1) rope or fibrous material, (2) copper helix, or (3) twisted copper I beam. This construction reduces the skin-effect ratio and is desirable in order to obtain economical use of the copper at high currents (see Table 4-8).

*Segmental* conductors are single conductors composed of either four or three segments which are combined to give a substantially circular cross section. The segments are electrically separated (usually by means of paper tape), and each strand of the individually stranded segments is alternately transposed between the inner and outer positions in the complete conductor due to its concentric lay in its segment. This construction reduces the skin-effect ratio and is desirable where high current-carrying capacity must be combined with small diameter (see Table 4-8).

***161. Copper stranded conductors*** for overhead-line applications are normally supplied in hard-drawn or medium-hard-drawn tempers in No. 4 AWG and larger. Soft or annealed conductors are used in all sizes for insulated conductors and to some extent for weather-resisting (weatherproof) conductors in overhead distribution systems.

***162. Copper-alloy stranded conductors*** are available in the same grades as copper-alloy solid conductors (see Par. **153**). Generally they are used for high strength together with conductance, where corrosion conditions do not permit the use of cheaper constructions; for applications such as overhead ground wires, messengers, railway catenaries; etc.

***163. Copper-clad steel stranded conductors*** are used in the same manner as copper-alloy stranded conductors, where their higher strength and limited range of conductance are suitable.

*Aluminum-clad steel stranded conductors* are used where high strength, limited conductance, and good corrosion resistance are required. They are widely used for overhead ground wires, neutral messenger and messenger strands, antenna conductors, and guy wires.

***164. Aluminum stranded conductors*** for bare and weatherproof overhead-line applications are normally supplied hard-drawn in No. 6 AWG and larger. Hard-drawn or three-quarter hard-drawn conductors are used in all sizes for insulated conductors.

***165. Iron and steel stranded conductors*** are used for overhead ground wires, messengers, and guy wires.

***166. Hollow (expanded) conductors*** are used on high-voltage transmission lines when, in order to reduce corona loss, it is desirable to increase the outside diameter without increasing the area beyond that needed for maximum line economy. Not only is the initial corona voltage considerably higher than for conventional conductors of equal cross section, but the current-carrying capacity for a given temperature rise is also greater because of the larger surface area available for cooling and the better disposition of the metal with respect to skin effect when carrying alternating currents.

*Air-expanded ACSR* is a conductor whose diameter has been increased by aluminum skeletal wires between the steel core and the outer layers of aluminum strands creating air spaces. A conductor having the necessary diameter to minimize corona effects on lines operating above 300 kV will, many times, have more metal than is economical if the conductor is made conventionally.

***167. Composite conductors*** are those made up of usually two different types of wire having differing characteristics. They are generally designed for a ratio of physical and electrical characteristics different from those found in homogeneous materials.

Cables of this type are particularly adaptable to long-span construction or other service conditions requiring more than average strength combined with liberal conductance. They lend themselves readily to economical, dependable use on transmission lines, rural distribution lines, railroad electrification, river crossings, and many kinds of special construction.

***168. Self-damping ACSR*** conductors are used to limit aeolian vibration to a safe level regardless of conductor tension or span length. They are concentrically stranded conductors composed of two layers of trapezoidal-shaped wires or two layers of trapezoidal-shaped wires and one layer of round wires of 1350 (EC) alloy with a high-strength, coated steel core (Fig. 4-14).

***169. Conductor Data Tables***

Table 4-9 Industry Specifications for Conductors

Table 4-10 Conductors—Physical and Electrical Properties

Table 4-11 Copper Wire—Wire Gages, Diameter, Area, Weight

Table 4-12 Copper Wire—Tensile Strength, Elongation

Table 4-13 Copper Wire—Weight, Breaking Strength, DC Resistance

Table 4-14 Copper Cable—Stranding Classes, Uses

Table 4-15 Copper Cable, Concentric Lay—Dimensions, Weight

**FIG. 4-14** Self-damping ACSR conductor.

Table 4-16 Copper Cable, Rope Lay—Dimensions, Weight

Table 4-17 Copper Cable, Classes AA, A, B—Weight, Breaking Strength, DC Resistance

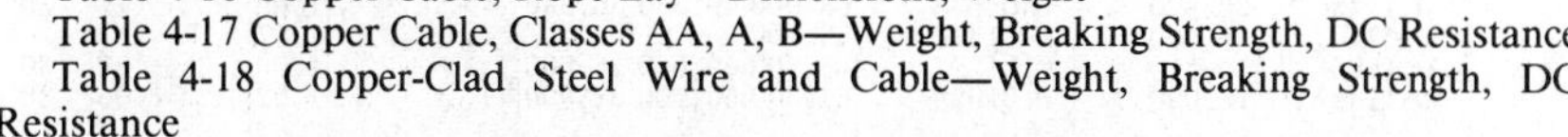

Table 4-18 Copper-Clad Steel Wire and Cable—Weight, Breaking Strength, DC Resistance

Table 4-19 Copper-Clad Steel-Copper Cable—Weight, Breaking Strength, DC Resistance

Table 4-20 Aluminum Wire—Dimensions, Weight, DC Resistance

Table 4-21 Aluminum Wire—Stranding Classes, Uses

Table 4-22 Aluminum Conductor—Physical Characteristics (EC-H19)

Table 4-23 ACSR Conductor—Physical Characteristics

Table 4-24 Aluminum-Clad Steel Wire and Cable—Weight, Breaking Strength, DC Resistance

Table 4-25 Galvanized-Steel Wire—Weight, Breaking Strength, DC Resistance

Table 4-26 Galvanized-Steel Strand—Dimensions, Weight, Breaking Strength

Table 4-27 Copper Wire and Cable—Electrical Characteristics

Table 4-28 Copper-Clad-Steel Cable—Electrical Characteristics

Table 4-29 Copper-Clad-Steel Copper Cable—Electrical Characteristics

Table 4-30 ACSR Conductor—Electrical Characteristics

Table 4-31 Aluminum-Clad Steel Cable—Electrical Characteristics

Table 4-32 Inductive-Reactance Spacing Factors

Table 4-33 Capacitive-Reactance Spacing Factors

Table 4-34 Properties of Resistance Metals and Alloys

## Fusible Metals and Alloys

***170. Fusible alloys*** having melting points in the range from about 60 to 200°C are made principally of bismuth, cadmium, lead, and tin in various proportions. Many of these alloys have been known under the names of their inventors (see index of alloys in "International Critical Tables," vol. 2), but typical compositions and melting points are shown in Table 4-35.

***171. Compositions and Melting Points of Fusible Alloys.*** See Table 4-35.

***172. Fuse metals*** for electric fuses of the open-link enclosed and expulsion types are ordinarily made of some low-fusible alloy; aluminum also is used to some extent. The resistance of the fuse causes dissipation of energy, liberation of heat, and rise of temperature. Sufficient current will obviously melt the fuse and thus open the circuit if the resultant arc is self-extinguishing. Metals which volatilize readily in the heat of the arc are to be preferred to those which leave a residue of globules of hot metal. The rating of any fuse depends critically upon its shape, dimensions, mounting, enclosure, and any other factors which affect its heat-dissipating capacity.

*(Numbered paragraphs resume on page 4-74.)*

**TABLE 4-12** Copper Wire—Tensile Strength, Elongation

(ASTM Specifications B1, B2, B3)

| Size,* AWG | Diameter at 20 C (68 F),† in | Area at 20 C (68 F) | | Tensile strength, psi | | | | Elongation, min % | | | | |
|---|---|---|---|---|---|---|---|---|---|---|---|---|
| | | | | Hard | Medium | | Soft§ | Hard | | Medium | | Soft |
| | | Cir mils | Sq in | Min | Min | Max | | In 10 in | In 60 in | In 10 in | In 60 in | In 10 in |
| 4/0 | 0.4600 | 211,600 | 0.1662 | 49,000 | 42,000 | 49,000 | ... | 3.75 | ... | 3.75 | ... | 35 |
| 3/0 | 0.4096 | 167,800 | 0.1318 | 51,000 | 43,000 | 50,000 | ... | 3.25 | ... | 3.60 | ... | 35 |
| 2/0 | 0.3648 | 133,100 | 0.1045 | 52,800 | 44,000 | 51,000 | ... | 2.80 | ... | 3.25 | ... | 35 |
| 1/0 | 0.3249 | 105,600 | 0.08291 | 54,500 | 45,000 | 52,000 | ... | 2.40 | ... | 3.00 | ... | 35 |
| 1 | 0.2893 | 83,690 | 0.06573 | 56,100 | 46,000 | 53,000 | ... | 2.17 | ... | 2.75 | ... | 30 |
| 2 | 0.2576 | 66,360 | 0.05212 | 57,600 | 47,000 | 54,000 | ... | 1.98 | ... | 2.50 | ... | 30 |
| 3 | 0.2294 | 52,620 | 0.04133 | 59,000 | 48,000 | 55,000 | ... | 1.79 | ... | 2.25 | ... | 30 |
| 4 | 0.2043 | 41,740 | 0.03278 | 60,100 | 48,330 | 55,330 | ... | ... | 1.24 | ... | 1.25 | 30 |
| 5 | 0.1819 | 33,090 | 0.02599 | 61,200 | 48,660 | 55,660 | ... | ... | 1.18 | ... | 1.20 | 30 |
| ... | 0.1650‡ | 27,220 | 0.02138 | 62,000 | ... | ... | ... | ... | 1.14 | | | |
| 6 | 0.1620 | 26,240 | 0.02061 | 62,100 | 49,000 | 56,000 | ... | ... | 1.14 | ... | 1.15 | 30 |
| 7 | 0.1443 | 20,820 | 0.01635 | 63,000 | 49,330 | 56,330 | ... | ... | 1.09 | ... | 1.11 | 30 |
| ... | 0.1340‡ | 17,960 | 0.01410 | 63,400 | ... | ... | ... | ... | 1.07 | | | |
| 8 | 0.1285 | 16,510 | 0.01297 | 63,700 | 49,660 | 56,660 | ... | ... | 1.06 | ... | 1.08 | 30 |
| 9 | 0.1144 | 13,090 | 0.01028 | 64,300 | 50,000 | 57,000 | ... | ... | 1.02 | ... | 1.06 | 30 |
| ... | 0.1040‡ | 10,820 | 0.008495 | 64,800 | ... | ... | ... | ... | 1.00 | | | |
| 10 | 0.1019 | 10,380 | 0.008155 | 64,900 | 50,330 | 57,300 | ... | ... | 1.00 | ... | 1.04 | 25 |
| ... | 0.0920‡ | 8,460 | 0.00665 | 65,400 | ... | ... | ... | ... | 0.97 | | | |
| 11 | 0.0907 | 8,230 | 0.00646 | 65,400 | 50,660 | 57,660 | ... | ... | 0.97 | ... | 1.02 | 25 |
| 12 | 0.0808 | 6,530 | 0.00513 | 65,700 | 51,000 | 58,000 | ... | ... | 0.95 | ... | 1.00 | 25 |
| ... | 0.0800‡ | 6,400 | 0.00503 | 65,700 | ... | ... | ... | ... | 0.94 | | | |
| 13 | 0.0720 | 5,180 | 0.00407 | 65,900 | 51,330 | 58,330 | ... | ... | 0.92 | ... | 0.98 | 25 |
| ... | 0.0650‡ | 4,220 | 0.00332 | 66,200 | ... | ... | ... | ... | 0.91 | | | |
| 14 | 0.0641 | 4,110 | 0.00323 | 66,200 | 51,660 | 58,660 | ... | ... | 0.90 | ... | 0.96 | 25 |
| 15 | 0.0571 | 3,260 | 0.00256 | 66,400 | 52,000 | 59,000 | ... | ... | 0.89 | ... | 0.94 | 25 |
| 16 | 0.0508 | 2,580 | 0.00203 | 66,600 | 52,330 | 59,330 | ... | ... | 0.87 | ... | 0.92 | 25 |
| 17 | 0.0453 | 2,050 | 0.00161 | 66,800 | 52,660 | 59,660 | ... | ... | 0.86 | ... | 0.90 | 25 |
| 18 | 0.0403 | 1,620 | 0.00128 | 67,000 | 53,000 | 60,000 | ... | ... | 0.85 | ... | 0.88 | 25 |
| 19 | 0.0359 | 1,290 | 0.00101 | ... | ... | ... | ... | ... | ... | ... | ... | 25 |
| 20 | 0.0320 | 1,020 | 0.000804 | ... | ... | ... | ... | ... | ... | ... | ... | 25 |
| 21 | 0.0285 | 812 | 0.000638 | ... | ... | ... | ... | ... | ... | ... | ... | 25 |
| 22 | 0.0253 | 640 | 0.000503 | ... | ... | ... | ... | ... | ... | ... | ... | 25 |
| 23 | 0.0226 | 511 | 0.000401 | ... | ... | ... | ... | ... | ... | ... | ... | 25 |
| 24 | 0.0201 | 404 | 0.000317 | ... | ... | ... | ... | ... | ... | ... | ... | 20 |
| 25 | 0.0179 | 320 | 0.000252 | ... | ... | ... | ... | ... | ... | ... | ... | 20 |
| 26 | 0.0159 | 253 | 0.000199 | ... | ... | ... | ... | ... | ... | ... | ... | 20 |
| 27 | 0.0142 | 202 | 0.000158 | ... | ... | ... | ... | ... | ... | ... | ... | 20 |
| 28 | 0.0126 | 159 | 0.000125 | ... | ... | ... | ... | ... | ... | ... | ... | 20 |
| 29 | 0.0113 | 128 | 0.000100 | ... | ... | ... | ... | ... | ... | ... | ... | 20 |
| 30 | 0.0100 | 100 | 0.0000785 | ... | ... | ... | ... | ... | ... | ... | ... | 15 |
| 31 | 0.0089 | 79.2 | 0.0000622 | ... | ... | ... | ... | ... | ... | ... | ... | 15 |
| 32 | 0.0080 | 64.0 | 0.0000503 | ... | ... | ... | ... | ... | ... | ... | ... | 15 |
| 33 | 0.0071 | 50.4 | 0.0000396 | ... | ... | ... | ... | ... | ... | ... | ... | 15 |
| 34 | 0.0063 | 39.7 | 0.0000312 | ... | ... | ... | ... | ... | ... | ... | ... | 15 |
| 35 | 0.0056 | 31.4 | 0.0000246 | ... | ... | ... | ... | ... | ... | ... | ... | 15 |
| 36 | 0.0050 | 25.0 | 0.0000196 | ... | ... | ... | ... | ... | ... | ... | ... | 15 |
| 37 | 0.0045 | 20.2 | 0.0000159 | ... | ... | ... | ... | ... | ... | ... | ... | 15 |
| 38 | 0.0040 | 16.0 | 0.0000126 | ... | ... | ... | ... | ... | ... | ... | ... | 15 |
| 39 | 0.0035 | 12.2 | 0.00000962 | ... | ... | ... | ... | ... | ... | ... | ... | 15 |
| 40 | 0.0031 | 9.61 | 0.00000755 | ... | ... | ... | ... | ... | ... | ... | ... | 15 |
| ASTM Specification Designation............ | | | | B1 | B2 | | B3 | B1 | | B2 | | B3 |

*The use of gage numbers to specify wire sizes is not recognized in these specifications, because of the possibility of confusion.

†The value of wire diameters in this table which correspond to gage numbers of the AWG are in agreement with the standard nominal diameters prescribed in ASTM Specification B258. For wire whose nominal diameter is more than 0.001 in (1 mil) greater than a size listed in the table and less than that of the next larger size, the requirements of the next larger size shall apply.

‡Diameters often employed by purchasers for communication lines, but not in the AWG (B. & S. wire gage) series. They correspond to certain of the numbers of the Birmingham wire gage or of the (British) standard wire gage.

§No requirements for tensile strength are specified.

NOTE: 1 in = 2.54 cm; 1 in$^2$ = 64.5 cm$^2$; 1 lb/in$^2$ = 6.895 kPa.

**TABLE 4-13** Copper Wire—Weight, Breaking Strength, DC Resistance
(Based on ASTM Specifications B1, B2, B3)

| Size, AWG | Diameter, in. | Area | | Weight | | Hard | | Medium | | Soft | |
|---|---|---|---|---|---|---|---|---|---|---|---|
| | | Cir mils | Sq in. | Lb per 1,000 ft | Lb per mile | Breaking strength, minimum,* lb | D-c resistance at 20 C (68 F) maximum,† ohms per 1,000 ft | Breaking strength, minimum,* lb | D-c resistance at 20 C (68 F) maximum,† ohms per 1,000 ft | Breaking strength, maximum,‡ lb | D-c resistance at 20 C (68 F) maximum,† ohms per 1,000 ft |
| 4/0 | 0.4600 | 211,600 | 0.1662 | 640.5 | 3382 | 8143 | 0.05045 | 6980 | 0.05019 | 5983 | 0.04901 |
| 3/0 | 0.4096 | 167,800 | 0.1318 | 507.8 | 2681 | 6720 | 0.06362 | 5666 | 0.06330 | 4744 | 0.06182 |
| 2/0 | 0.3648 | 133,100 | 0.1045 | 402.8 | 2127 | 5519 | 0.08021 | 4599 | 0.07980 | 3763 | 0.07793 |
| 1/0 | 0.3249 | 105,600 | 0.08291 | 319.5 | 1687 | 4518 | 0.1022 | 3731 | 0.1016 | 2985 | 0.09825 |
| 1 | 0.2893 | 83,690 | 0.06573 | 253.3 | 1338 | 3688 | 0.1289 | 3024 | 0.1282 | 2432 | 0.1239 |
| 2 | 0.2576 | 66,360 | 0.05212 | 200.9 | 1061 | 3002 | 0.1625 | 2450 | 0.1617 | 1928 | 0.1563 |
| 3 | 0.2294 | 52,620 | 0.04133 | 159.3 | 841.1 | 2439 | 0.2050 | 1984 | 0.2039 | 1529 | 0.1971 |
| 4 | 0.2043 | 41,740 | 0.03278 | 126.3 | 667.1 | 1970 | 0.2584 | 1584 | 0.2571 | 1213 | 0.2485 |
| 5 | 0.1819 | 33,090 | 0.02599 | 100.2 | 528.8 | 1590 | 0.3260 | 1265 | 0.3243 | 961.5 | 0.3135 |
| 6 | 0.1620 | 26,240 | 0.02061 | 79.44 | 419.4 | 1280 | 0.4110 | 1010 | 0.4088 | 762.6 | 0.3952 |
| 7 | 0.1443 | 20,820 | 0.01635 | 63.03 | 332.8 | 1030 | 0.5180 | 806.7 | 0.5153 | 605.1 | 0.4981 |
| 8 | 0.1285 | 16,510 | 0.01297 | 49.98 | 263.9 | 826.1 | 0.6532 | 644.0 | 0.6498 | 479.8 | 0.6281 |
| 9 | 0.1144 | 13,090 | 0.01028 | 39.61 | 209.2 | 660.9 | 0.8241 | 513.9 | 0.8199 | 380.3 | 0.7925 |
| 10 | 0.1019 | 10,380 | 0.008155 | 31.43 | 166.0 | 529.3 | 1.039 | 410.5 | 1.033 | 314.0 | 0.9988 |
| 11 | 0.0907 | 8,230 | 0.00646 | 24.9 | 131 | 423 | 1.31 | 327 | 1.30 | 249 | 1.26 |
| 12 | 0.0808 | 6,530 | 0.00513 | 19.8 | 104 | 337 | 1.65 | 262 | 1.64 | 197 | 1.59 |
| 13 | 0.0720 | 5,180 | 0.00407 | 15.7 | 82.9 | 268 | 2.08 | 209 | 2.07 | 157 | 2.00 |
| 14 | 0.0641 | 4,110 | 0.00323 | 12.4 | 65.7 | 214 | 2.63 | 167 | 2.61 | 124 | 2.52 |
| 15 | 0.0571 | 3,260 | 0.00256 | 9.87 | 52.1 | 170 | 3.31 | 133 | 3.29 | 98.6 | 3.18 |
| 16 | 0.0508 | 2,580 | 0.00203 | 7.81 | 41.2 | 135 | 4.18 | 106 | 4.16 | 78.0 | 4.02 |
| 17 | 0.0453 | 2,050 | 0.00161 | 6.21 | 32.8 | 108 | 5.26 | 84.9 | 5.23 | 62.1 | 5.05 |
| 18 | 0.0403 | 1,620 | 0.00128 | 4.92 | 26.0 | 85.5 | 6.64 | 67.6 | 6.61 | 49.1 | 6.39 |
| 19 | 0.0359 | 1,290 | 0.00101 | 3.90 | 20.6 | 68.0 | 8.37 | 54.0 | 8.33 | 39.0 | 8.05 |
| 20 | 0.0320 | 1,020 | 0.000804 | 3.10 | 16.4 | 54.2 | 10.5 | 43.2 | 10.5 | 31.0 | 10.1 |
| 21 | 0.0285 | 812 | 0.000638 | 2.46 | 13.0 | 43.2 | 13.3 | 34.4 | 13.2 | 24.6 | 12.8 |
| 22 | 0.0253 | 640 | 0.000503 | 1.94 | 10.2 | 34.1 | 16.9 | 27.3 | 16.8 | 19.4 | 16.2 |
| 23 | 0.0226 | 511 | 0.000401 | 1.55 | 8.16 | 27.3 | 21.1 | 21.9 | 21.0 | 15.4 | 20.3 |
| 24 | 0.0201 | 404 | 0.000317 | 1.22 | 6.46 | 21.7 | 26.7 | 17.5 | 26.6 | 12.7 | 25.7 |
| 25 | 0.0179 | 320 | 0.000252 | 0.970 | 5.12 | 17.3 | 33.7 | 13.9 | 33.5 | 10.1 | 32.4 |
| 26 | 0.0159 | 253 | 0.000199 | 0.765 | 4.04 | 13.7 | 42.7 | 11.1 | 42.4 | 7.94 | 41.0 |

**TABLE 4-13** Copper Wire—Weight, Breaking Strength, DC Resistance (*Continued*)

(Based on ASTM Specifications B1, B2, B3)

| Size, AWG | Diameter, in. | Area | | Weight | | Hard | | Medium | | Soft | |
|---|---|---|---|---|---|---|---|---|---|---|---|
| | | Cir mils | Sq in. | Lb per 1,000 ft | Lb per mile | Breaking strength, minimum,* lb | D-c resistance at 20 C (68 F) maximum,† ohms per 1,000 ft | Breaking strength, minimum,* lb | D-c resistance at 20 C (68 F) maximum,† ohms per 1,000 ft | Breaking strength, maximum,‡ lb | D-c resistance at 20 C (68 F) maximum,† ohms per 1,000 ft |
| **27** | 0.0142 | 202 | 0.000158 | 0.610 | 3.22 | 10.9 | 53.5 | 8.87 | 53.2 | 6.33 | 51.4 |
| **28** | 0.0126 | 159 | 0.000125 | 0.481 | 2.54 | 8.64 | 67.9 | 7.02 | 67.6 | 4.99 | 65.3 |
| **29** | 0.0113 | 128 | 0.000100 | 0.387 | 2.04 | 6.97 | 84.5 | 5.68 | 84.0 | 4.01 | 81.2 |
| **30** | 0.0100 | 100 | 0.0000785 | 0.303 | 1.60 | 5.47 | 108 | 4.48 | 107 | 3.14 | 104 |
| **31** | 0.0089 | 79.2 | 0.0000622 | 0.240 | 1.27 | 4.35 | 136 | 3.6 | 135 | 2.49 | 131 |
| **32** | 0.0080 | 64.0 | 0.0000503 | 0.194 | 1.02 | 3.53 | 169 | 2.90 | 168 | 2.01 | 162 |
| **33** | 0.0071 | 50.4 | 0.0000396 | 0.153 | 0.806 | 2.79 | 214 | 2.30 | 213 | 1.58 | 206 |
| **34** | 0.0063 | 39.7 | 0.0000312 | 0.120 | 0.634 | 2.20 | 272 | 1.82 | 270 | 1.25 | 261 |
| **35** | 0.0056 | 31.4 | 0.0000246 | 0.0949 | 0.501 | 1.75 | 344 | 1.44 | 342 | 0.985 | 331 |
| **36** | 0.0050 | 25.0 | 0.0000196 | 0.0757 | 0.400 | 1.40 | 431 | 1.16 | 429 | 0.785 | 415 |
| **37** | 0.0045 | 20.2 | 0.0000159 | 0.0613 | 0.324 | 1.13 | 533 | 0.944 | 530 | 0.636 | 512 |
| **38** | 0.0040 | 16.0 | 0.0000126 | 0.0484 | 0.256 | 0.898 | 674 | 0.750 | 671 | 0.503 | 648 |
| **39** | 0.0035 | 12.2 | 0.00000962 | 0.0371 | 0.196 | 0.691 | 880 | 0.577 | 876 | 0.385 | 847 |
| **40** | 0.0031 | 9.61 | 0.00000755 | 0.0291 | 0.154 | 0.543 | 1120 | 0.455 | 1120 | 0.302 | 1080 |
| **41** | 0.0028 | 7.84 | 0.00000616 | 0.0237 | 0.125 | ........ | 1380 | ........ | 1370 | 0.246 | 1320 |
| **42** | 0.0025 | 6.25 | 0.00000491 | 0.0189 | 0.0999 | ........ | 1730 | ........ | 1720 | 0.196 | 1660 |
| **43** | 0.0022 | 4.84 | 0.00000380 | 0.0147 | 0.0774 | ........ | 2230 | ........ | 2220 | 0.152 | 2140 |
| **44** | 0.0020 | 4.00 | 0.00000314 | 0.0121 | 0.0639 | ........ | 2700 | ........ | 2680 | 0.126 | 2590 |
| ASTM Specification Designation........ | | | | | | B1 | | B2 | | B3 | |

*No. 19 AWG and smaller, based on Anaconda data.

†Based on nominal diameter and ASTM resistivities.

‡No requirements for tensile strength are specified in ASTM B3. Values given here based on Anaconda data.

NOTE: 1 in = 2.54 cm; 1 in$^2$ = 64.5 cm$^2$; 1 lb = 0.4536 kg; 1 ft = 0.3048 m; 1 mi = 1.61 km.

**TABLE 4-14** Copper Cable—Stranding Classes, Uses
(ASTM Specifications)

| ASTM Designation | Construction | Class | Application | | | |
|---|---|---|---|---|---|---|
| B8 | Concentric lay | AA | For bare conductors usually used in overhead lines | | | |
| | | A | For weather-resistant (weatherproof), slow-burning conductors<br>For bare conductors where greater flexibility than is afforded by Class AA is required | | | |
| | | B | For conductors insulated with various materials such as rubber, paper, varnished-cambric, etc.<br>For the conductors indicated under Class A where greater flexibility is required | | | |
| | | C<br>D | For conductors where greater flexibility is required than is provided by Class B | | | |
| B173 | Rope lay with concentric-stranded members | G | Conductor constructions having a range of areas from 5,000,000 cir mils and employing 61 stranded members of 19 wires each down to No. 14 AWG containing 7 stranded members of 7 wires each<br>(Typical uses are for rubber-sheathed conductors, apparatus conductors, portable conductors, and similar applications) | | | |
| | | H | Conductor constructions having a range of areas from 5,000,000 cir mils and employing 91 stranded members of 19 wires each down to No. 9 AWG containing 19 stranded members of 7 wires each<br>(Typical uses are for rubber-sheathed cords and conductors where greater flexibility is required, such as for use on take-up reels over sheaves and extra-flexible apparatus conductors) | | | |
| B226 | Annular stranded | | For bare conductors, or covered with weather-resistant (weatherproof) materials or insulated with rubber, varnished-cambric or solid-type impregnated paper | | | |
| | | | Conductor size, AWG | Individual wire size | | |
| | | | | In. | AWG | |
| B174 | Bunch stranded | I | 7, 8, 9, 10 | 0.0201 | 24 | Rubber-covered, varnished-cambric, and paper-insulated conductors |
| | | J | 10, 12, 14, 16, 18, 20 | 0.0126 | 28 | Fixture wire |
| | | K | 10, 12, 14, 16, 18, 20, 22 | 0.0100 | 30 | Fixture wire, flexible cord, and portable cord |
| | | L | 10, 12, 14, 16, 18, 20, 24 | 0.0080 | 32 | Fixture wire and portable cord with greater flexibility than Class K |
| | | M | 14, 16, 18, 20, 22, 26 | 0.0063 | 34 | Heater cord and light portable cord |
| | | O | 16, 18, 20 24. 28 | 0.0050 | 36 | Heater cord with greater flexibility than Class M |
| | | P | 16, 18, 20 | 0.0040 | 38 | More flexible conductors than provided in preceding classes |
| | | Q | 18, 20 | 0.0031 | 40 | Oscillating fan cord. Very great flexibility |
| | | | Conductor size, cir mils | Individual wire size | | |
| | | | | In. | AWG | |
| B172 | Rope lay with bunched-stranded members | I | Up to 2,000,000 | 0.0201 | 24 | Typical use is for special apparatus cable |
| | | K | Up to 1,000,000 | 0.0100 | 30 | Typical use, special portable cord and conductors |
| | | M | Up to 1,000,000 | 0.0063 | 34 | Typical use is for welding conductor |

NOTE: 1 in = 2.54 cm.

**TABLE 4-15** Copper Cable, Concentric Lay—Dimensions, Weight
(ASTM Specification B8)

| Conductor size, Mcm or Awg | Class AA | | | Class A | | | Class B | | | Class C† | | Class D† | | Approximate weight, lb per 1000 ft |
|---|---|---|---|---|---|---|---|---|---|---|---|---|---|---|
| | No. of wires | Diameter each wire, mils | Nominal conductor diam, in. | No. of wires | Diameter each wire, mils | Nominal conductor diam, in. | No. of wires | Diameter each wire, mils | Nominal conductor diam, in. | No. of wires | Diameter each wire, mils | No. of wires | Diameter each wire, mils | |
| 5,000* | .. | ..... | ..... | 169 | 172.0 | 2.580 | 217 | 151.8 | 2.581 | 271 | 135.8 | 271 | 135.8 | 15,890 |
| 4,500 | .. | ..... | ..... | 169 | 163.2 | 2.448 | 217 | 144.0 | 2.448 | 271 | 128.9 | 271 | 128.9 | 14,300 |
| 4,000 | .. | ..... | ..... | 169 | 153.8 | 2.307 | 217 | 135.8 | 2.309 | 271 | 121.5 | 271 | 121.5 | 12,590 |
| 3,500 | .. | ..... | ..... | 127 | 166.0 | 2.158 | 169 | 143.9 | 2.159 | 217 | 127.0 | 271 | 113.6 | 11,020 |
| 3,000* | .. | ..... | ..... | 127 | 153.7 | 1.998 | 169 | 133.2 | 1.998 | 217 | 117.6 | 271 | 105.2 | 9,353 |
| 2,500* | .. | ..... | ..... | 91 | 165.7 | 1.823 | 127 | 140.3 | 1.824 | 169 | 121.6 | 217 | 107.3 | 7,794 |
| 2,000* | .. | ..... | ..... | 91 | 148.2 | 1.630 | 127 | 125.5 | 1.632 | 169 | 108.8 | 217 | 96.0 | 6,175 |
| 1,900 | .. | ..... | ..... | 91 | 144.5 | 1.590 | 127 | 122.3 | 1.590 | 169 | 106.0 | 217 | 93.6 | 5,866 |
| 1,800 | .. | ..... | ..... | 91 | 140.6 | 1.547 | 127 | 119.1 | 1.548 | 169 | 103.2 | 217 | 91.1 | 5,558 |
| 1,750* | .. | ..... | ..... | 91 | 138.7 | 1.526 | 127 | 117.4 | 1.526 | 169 | 101.8 | 217 | 89.8 | 5,403 |
| 1,700 | .. | ..... | ..... | 91 | 136.7 | 1.504 | 127 | 115.7 | 1.504 | 169 | 100.3 | 217 | 88.5 | 5,249 |
| 1,600 | .. | ..... | ..... | 91 | 132.6 | 1.459 | 127 | 112.2 | 1.459 | 169 | 97.3 | 217 | 85.9 | 4,940 |
| 1,500* | .. | ..... | ..... | 61 | 156.8 | 1.411 | 91 | 128.4 | 1.412 | 127 | 108.7 | 169 | 94.2 | 4,631 |
| 1,400 | .. | ..... | ..... | 61 | 151.5 | 1.364 | 91 | 124.0 | 1.364 | 127 | 105.0 | 169 | 91.0 | 4,323 |
| 1,300 | .. | ..... | ..... | 61 | 146.0 | 1.314 | 91 | 119.5 | 1.315 | 127 | 101.2 | 169 | 87.7 | 4,014 |
| 1,250* | .. | ..... | ..... | 61 | 143.1 | 1.288 | 91 | 117.2 | 1.289 | 127 | 99.2 | 169 | 86.0 | 3,859 |
| 1,200 | .. | ..... | ..... | 61 | 140.3 | 1.263 | 91 | 114.8 | 1.263 | 127 | 97.2 | 169 | 84.3 | 3,705 |
| 1,100 | .. | ..... | ..... | 61 | 134.3 | 1.209 | 91 | 109.9 | 1.209 | 127 | 93.1 | 169 | 80.7 | 3,396 |
| 1,000* | 37 | 164.4 | 1.151 | 61 | 128.0 | 1.152 | 61 | 128.0 | 1.152 | 91 | 104.8 | 127 | 88.7 | 3,088 |
| 900 | 37 | 156.0 | 1.092 | 61 | 121.5 | 1.094 | 61 | 121.5 | 1.094 | 91 | 99.4 | 127 | 84.2 | 2,779 |
| 800* | 37 | 147.0 | 1.029 | 61 | 114.5 | 1.031 | 61 | 114.5 | 1.031 | 91 | 93.8 | 127 | 79.4 | 2,470 |
| 750* | 37 | 142.4 | 0.997 | 61 | 110.9 | 0.998 | 61 | 110.9 | 0.998 | 91 | 90.8 | 127 | 76.8 | 2,316 |
| 700* | 37 | 137.5 | 0.963 | 61 | 107.1 | 0.964 | 61 | 107.1 | 0.964 | 91 | 87.7 | 127 | 74.2 | 2,161 |
| 650 | 37 | 132.5 | 0.928 | 61 | 103.2 | 0.929 | 61 | 103.2 | 0.929 | 91 | 84.5 | 127 | 71.5 | 2,007 |
| 600* | 37 | 127.3 | 0.891 | 37 | 127.3 | 0.891 | 61 | 99.2 | 0.893 | 91 | 81.2 | 127 | 68.7 | 1,853 |
| 550 | 37 | 121.9 | 0.853 | 37 | 121.9 | 0.853 | 61 | 95.0 | 0.855 | 91 | 77.7 | 127 | 65.8 | 1,698 |
| 500* | 19 | 162.2 | 0.811 | 37 | 116.2 | 0.813 | 37 | 116.2 | 0.813 | 61 | 90.5 | 91 | 74.1 | 1,544 |
| 450 | 19 | 153.9 | 0.770 | 37 | 110.3 | 0.772 | 37 | 110.3 | 0.772 | 61 | 85.9 | 91 | 70.3 | 1,389 |
| 400* | 19 | 145.1 | 0.726 | 19 | 145.1 | 0.726 | 37 | 104.0 | 0.728 | 61 | 81.0 | 91 | 66.3 | 1,235 |
| 350* | 12 | 170.8 | 0.710 | 19 | 135.7 | 0.679 | 37 | 97.3 | 0.681 | 61 | 75.7 | 91 | 62.0 | 1,081 |

| Conductor size, Mcm or Awg | Class AA | | | Class A | | | Class B | | | Class C† | | Class D† | | Approximate weight, lb per 1000 ft |
|---|---|---|---|---|---|---|---|---|---|---|---|---|---|---|
| | No. of wires | Diameter each wire, mils | Nominal conductor diam, in. | No. of wires | Diameter each wire, mils | Nominal conductor diam, in. | No. of wires | Diameter each wire, mils | Nominal conductor diam, in. | No. of wires | Diameter each wire, mils | No. of wires | Diameter each wire, mils | |
| 300* | 12 | 158.1 | 0.657 | 19 | 125.7 | 0.629 | 37 | 90.0 | 0.630 | 61 | 70.1 | 91 | 57.4 | 926.3 |
| 250* | 12 | 144.3 | 0.600 | 19 | 114.7 | 0.574 | 37 | 82.2 | 0.575 | 61 | 64.0 | 91 | 52.4 | 771.9 |
| 4/0* | 7 | 173.9 | 0.522 | 7 | 173.9 | 0.552 | 19 | 105.5 | 0.528 | 37 | 75.6 | 61 | 58.9 | 653.3 |
| 3/0* | 7 | 154.8 | 0.464 | 7 | 154.8 | 0.464 | 19 | 94.0 | 0.470 | 37 | 67.3 | 61 | 52.4 | 518.1 |
| 2/0* | 7 | 137.9 | 0.414 | 7 | 137.9 | 0.414 | 19 | 83.7 | 0.419 | 37 | 60.0 | 61 | 46.7 | 410.9 |
| 1/0* | 7 | 122.8 | 0.368 | 7 | 122.8 | 0.368 | 19 | 74.5 | 0.373 | 37 | 53.4 | 61 | 41.6 | 325.8 |
| 1 | 3 | 167.0 | 0.360 | ... | ..... | ..... | ... | ..... | ..... | ... | ..... | ... | ..... | 255.9 |
| 1* | .. | ..... | ..... | 7 | 109.3 | 0.328 | 19 | 66.4 | 0.332 | 37 | 47.6 | 61 | 37.0 | 258.4 |
| 2 | 3 | 148.7 | 0.320 | ... | ..... | ..... | ... | ..... | ..... | ... | ..... | ... | ..... | 202.9 |
| 2* | .. | ..... | ..... | 7 | 97.4 | 0.292 | 7 | 97.4 | 0.292 | 19 | 59.1 | 37 | 42.4 | 204.9 |
| 3 | 3 | 132.5 | 0.285 | ... | ..... | ..... | ... | ..... | ..... | ... | ..... | ... | ..... | 160.9 |
| 3* | .. | ..... | ..... | 7 | 86.7 | 0.260 | 7 | 86.7 | 0.260 | 19 | 52.6 | 37 | 37.7 | 162.5 |
| 4 | 3 | 118.0 | 0.254 | ... | ..... | ..... | ... | ..... | ..... | ... | ..... | ... | ..... | 127.6 |
| 4* | .. | ..... | ..... | 7 | 77.2 | 0.232 | 7 | 77.2 | 0.232 | 19 | 46.9 | 37 | 33.6 | 128.9 |
| 5* | .. | ..... | ..... | ... | ..... | ..... | 7 | 68.8 | 0.206 | 19 | 41.7 | 37 | 29.9 | 102.2 |
| 6* | .. | ..... | ..... | ... | ..... | ..... | 7 | 61.2 | 0.184 | 19 | 37.2 | 37 | 26.6 | 81.05 |
| 7* | .. | ..... | ..... | ... | ..... | ..... | 7 | 54.5 | 0.164 | 19 | 33.1 | 37 | 23.7 | 64.28 |
| 8* | .. | ..... | ..... | ... | ..... | ..... | 7 | 48.6 | 0.146 | 19 | 29.5 | 37 | 21.1 | 50.97 |
| 9* | .. | ..... | ..... | ... | ..... | ..... | 7 | 43.2 | 0.130 | 19 | 26.2 | 37 | 18.8 | 40.42 |
| 10* | .. | ..... | ..... | ... | ..... | ..... | 7 | 38.5 | 0.116 | 19 | 23.4 | 37 | 16.7 | 32.06 |
| 12* | .. | ..... | ..... | ... | ..... | ..... | 7 | 30.5 | 0.0915 | 19 | 18.5 | 37 | 13.3 | 20.16 |
| 14* | .. | ..... | ..... | ... | ..... | ..... | 7 | 24.2 | 0.0726 | 19 | 14.7 | 37 | 10.5 | 12.68 |
| 16* | .. | ..... | ..... | ... | ..... | ..... | 7 | 19.2 | 0.0576 | 19 | 11.7 | ... | ..... | 7.974 |
| 18* | .. | ..... | ..... | ... | ..... | ..... | 7 | 15.2 | 0.0456 | 19 | 9.2 | ... | ..... | 5.015 |
| 20* | .. | ..... | ..... | ... | ..... | ..... | 7 | 12.1 | 0.0363 | 19 | 7.3 | ... | ..... | 3.154 |

* The sizes of conductors which have been marked with an asterisk provide for one or more schedules of preferred series and are commonly used in the industry. Those not marked are given simply as a matter of reference, and it is suggested that their use be discouraged.

† To calculate the nominal diameters of Class C or Class D conductors or of any concentric-lay-stranded conductors made from round wires of uniform diameters, multiply the diameter of an individual wire by that one of the following factors which applies:

| Number of wires in conductor | 3 | 7 | 12 | 19 | 37 | 61 | 91 | 127 | 169 | 217 | 271 |
|---|---|---|---|---|---|---|---|---|---|---|---|
| Diameter calculation factor | 2.155 | 3 | 4.155 | 5 | 7 | 9 | 11 | 13 | 15 | 17 | 19 |

NOTE: 1 in = 2.54 cm; 1 lb = 0.4536 kg; 1 ft = 0.3048 m.

**TABLE 4-16** Copper Cable, Rope Lay—Dimensions, Weight
(ASTM Specification B173)

| Conductor size, Mcm or Awg | Class G | | | | | | Class H | | | | | |
|---|---|---|---|---|---|---|---|---|---|---|---|---|
| | No. of wires | No. of members | No. of wires in each member | Diameter each wire, mils | Nominal conductor diameter, in. | Approx weight, lb per 1000 ft | No. of wires | No. of members | No. of wires in each member | Diameter each wire, mils | Nominal conductor diameter, in. | Approx weight, lb per 1000 ft |
| 5000 | 1159 | 61 | 19 | 65.7 | 2.957 | 16,050 | 1729 | 91 | 19 | 53.8 | 2.959 | 16,060 |
| 4500 | 1159 | 61 | 19 | 62.3 | 2.804 | 14,435 | 1729 | 91 | 19 | 51.0 | 2.805 | 14,430 |
| 4000 | 1159 | 61 | 19 | 58.7 | 2.642 | 12,820 | 1729 | 91 | 19 | 48.1 | 2.646 | 12,840 |
| 3500 | 1159 | 61 | 19 | 55.0 | 2.475 | 11,255 | 1729 | 91 | 19 | 45.0 | 2.475 | 11,235 |
| 3000 | 1159 | 61 | 19 | 50.9 | 2.291 | 9,635 | 1729 | 91 | 19 | 41.7 | 2.294 | 9,650 |
| 2500 | 703 | 37 | 19 | 59.6 | 2.086 | 8,015 | 1159 | 61 | 19 | 46.4 | 2.088 | 8,010 |
| 2000 | 703 | 37 | 19 | 53.3 | 1.866 | 6,415 | 1159 | 61 | 19 | 41.5 | 1.868 | 6,400 |
| 1900 | 703 | 37 | 19 | 52.0 | 1.820 | 6,100 | 1159 | 61 | 19 | 40.5 | 1.823 | 6,100 |
| 1800 | 703 | 37 | 19 | 50.6 | 1.771 | 5,775 | 1159 | 61 | 19 | 39.4 | 1.773 | 5,770 |
| 1750 | 703 | 37 | 19 | 49.9 | 1.747 | 5,620 | 1159 | 61 | 19 | 38.9 | 1.751 | 5,625 |
| 1700 | 703 | 37 | 19 | 49.2 | 1.722 | 5,460 | 1159 | 61 | 19 | 38.3 | 1.724 | 5,455 |
| 1600 | 703 | 37 | 19 | 47.7 | 1.670 | 5,130 | 1159 | 61 | 19 | 37.2 | 1.674 | 5,145 |
| 1500 | 427 | 61 | 7 | 59.3 | 1.601 | 4,775 | 703 | 37 | 19 | 46.2 | 1.617 | 4,815 |
| 1400 | 427 | 61 | 7 | 57.3 | 1.547 | 4,460 | 703 | 37 | 19 | 44.6 | 1.561 | 4,485 |
| 1300 | 427 | 61 | 7 | 55.2 | 1.490 | 4,135 | 703 | 37 | 19 | 43.0 | 1.505 | 4,170 |
| 1250 | 427 | 61 | 7 | 54.1 | 1.461 | 3,975 | 703 | 37 | 19 | 42.2 | 1.477 | 4,015 |
| 1200 | 427 | 61 | 7 | 53.0 | 1.431 | 3,810 | 703 | 37 | 19 | 41.3 | 1.446 | 3,845 |
| 1100 | 427 | 61 | 7 | 50.8 | 1.372 | 3,500 | 703 | 37 | 19 | 39.6 | 1.386 | 3,535 |
| 1000 | 427 | 61 | 7 | 48.4 | 1.307 | 3,180 | 703 | 37 | 19 | 37.7 | 1.320 | 3,205 |
| 900 | 427 | 61 | 7 | 45.9 | 1.239 | 2,860 | 703 | 37 | 19 | 35.8 | 1.253 | 2,895 |
| 800 | 427 | 61 | 7 | 43.3 | 1.169 | 2,545 | 703 | 37 | 19 | 33.7 | 1.180 | 2,560 |
| 750 | 427 | 61 | 7 | 41.9 | 1.131 | 2,385 | 703 | 37 | 19 | 32.7 | 1.145 | 2,410 |
| 700 | 427 | 61 | 7 | 40.5 | 1.094 | 2,230 | 703 | 37 | 19 | 31.6 | 1.106 | 2,255 |
| 650 | 427 | 61 | 7 | 39.0 | 1.053 | 2,070 | 703 | 37 | 19 | 30.4 | 1.064 | 2,085 |
| 600 | 427 | 61 | 7 | 37.5 | 1.013 | 1,910 | 703 | 37 | 19 | 29.2 | 1.022 | 1,920 |
| 550 | 427 | 61 | 7 | 35.9 | 0.969 | 1,750 | 703 | 37 | 19 | 28.0 | 0.980 | 1,770 |
| 500 | 259 | 37 | 7 | 43.9 | 0.922 | 1,585 | 427 | 61 | 7 | 34.2 | 0.923 | 1,590 |
| 450 | 259 | 37 | 7 | 41.7 | 0.876 | 1,425 | 427 | 61 | 7 | 32.5 | 0.878 | 1,435 |
| 400 | 259 | 37 | 7 | 39.3 | 0.825 | 1,265 | 427 | 61 | 7 | 30.6 | 0.826 | 1,270 |
| 350 | 259 | 37 | 7 | 36.8 | 0.773 | 1,110 | 427 | 61 | 7 | 28.6 | 0.772 | 1,110 |
| 300 | 259 | 37 | 7 | 34.0 | 0.714 | 945 | 427 | 61 | 7 | 26.5 | 0.716 | 953 |
| 250 | 259 | 37 | 7 | 31.1 | 0.653 | 795 | 427 | 61 | 7 | 24.2 | 0.653 | 795 |
| 4/0 | 133 | 19 | 7 | 39.9 | 0.599 | 668 | 259 | 37 | 7 | 28.6 | 0.601 | 670 |
| 3/0 | 133 | 19 | 7 | 35.5 | 0.533 | 529 | 259 | 37 | 7 | 25.5 | 0.536 | 533 |
| 2/0 | 133 | 19 | 7 | 31.6 | 0.474 | 419 | 259 | 37 | 7 | 22.7 | 0.477 | 422 |
| 1/0 | 133 | 19 | 7 | 28.2 | 0.423 | 334 | 259 | 37 | 7 | 20.2 | 0.424 | 334 |
| 1 | 133 | 19 | 7 | 25.1 | 0.377 | 264 | 259 | 37 | 7 | 18.0 | 0.378 | 266 |
| 2 | 49 | 7 | 7 | 36.8 | 0.331 | 207 | 133 | 19 | 7 | 22.3 | 0.335 | 208 |
| 3 | 49 | 7 | 7 | 32.8 | 0.295 | 164 | 133 | 19 | 7 | 19.9 | 0.299 | 167 |
| 4 | 49 | 7 | 7 | 29.2 | 0.263 | 130 | 133 | 19 | 7 | 17.7 | 0.266 | 132 |
| 5 | 49 | 7 | 7 | 26.0 | 0.234 | 103 | 133 | 19 | 7 | 15.8 | 0.237 | 105 |
| 6 | 49 | 7 | 7 | 23.1 | 0.208 | 82 | 133 | 19 | 7 | 14.0 | 0.210 | 82 |
| 7 | 49 | 7 | 7 | 20.6 | 0.185 | 65 | 133 | 19 | 7 | 12.5 | 0.188 | 65 |
| 8 | 49 | 7 | 7 | 18.4 | 0.166 | 51 | 133 | 19 | 7 | 11.1 | 0.167 | 52 |
| 9 | 49 | 7 | 7 | 16.4 | 0.148 | 40.8 | 133 | 19 | 7 | 9.9 | 0.149 | 41 |
| 10 | 49 | 7 | 7 | 14.6 | 0.131 | 32.3 | | | | | | |
| 12 | 49 | 7 | 7 | 11.6 | 0.104 | 20.3 | | | | | | |
| 14 | 49 | 7 | 7 | 9.2 | 0.083 | 12.8 | | | | | | |

**TABLE 4-17** Copper Cable, Classes AA, A, B—Weight, Breaking Strength, DC Resistance

(ASTM Specifications B1, B2, B3, B8)

| Conductor size, Mcm or Awg | No. of wires (ASTM stranding class) | Wire diameter, in. | Conductor diameter, in. | Conductor area, sq in. | Conductor weight, lb | | Hard | | Medium | | Soft | |
|---|---|---|---|---|---|---|---|---|---|---|---|---|
| | | | | | Per 1000 ft | Per mile | Breaking strength, minimum,* lb | D-c resistance at 20 C (68 F), ohms per 1000 ft | Breaking strength, minimum,* lb | D-c resistance at 20 C (68 F), ohms per 1000 ft | Breaking strength, maximum,† lb | D-c resistance at 20 C (68 F), ohms per 1000 ft |
| 5000 | 169 (A) | 0.1720 | 2.580 | 3.927 | 15,890 | 83,910 | 216,300 | 0.002265 | 172,000 | 0.002253 | 145,300 | 0.002178 |
| 5000 | 217 (B) | 0.1518 | 2.581 | 3.927 | 15,890 | 83,910 | 219,500 | 0.002265 | 173,200 | 0.002253 | 145,300 | 0.002178 |
| 4500 | 169 (A) | 0.1632 | 2.448 | 3.534 | 14,300 | 75,520 | 197,200 | 0.002517 | 154,800 | 0.002504 | 130,800 | 0.002420 |
| 4500 | 217 (B) | 0.1440 | 2.448 | 3.534 | 14,300 | 75,520 | 200,400 | 0.002517 | 156,900 | 0.002504 | 130,800 | 0.002420 |
| 4000 | 169 (A) | 0.1538 | 2.307 | 3.142 | 12,590 | 66,490 | 175,600 | 0.002804 | 138,500 | 0.002790 | 116,200 | 0.002697 |
| 4000 | 217 (B) | 0.1358 | 2.309 | 3.142 | 12,590 | 66,490 | 178,100 | 0.002804 | 139,500 | 0.002790 | 116,200 | 0.002697 |
| 3500 | 127 (A) | 0.1660 | 2.158 | 2.749 | 11,020 | 58,180 | 153,400 | 0.003205 | 120,400 | 0.003188 | 101,700 | 0.003082 |
| 3500 | 169 (B) | 0.1439 | 2.159 | 2.749 | 11,020 | 58,180 | 155,900 | 0.003205 | 122,000 | 0.003188 | 101,700 | 0.003082 |
| 3000 | 127 (A) | 0.1537 | 1.998 | 2.356 | 9,353 | 49,390 | 131,700 | 0.003703 | 103,900 | 0.003684 | 87,180 | 0.003561 |
| 3000 | 169 (B) | 0.1332 | 1.998 | 2.356 | 9,353 | 49,390 | 134,400 | 0.003703 | 104,600 | 0.003684 | 87,180 | 0.003561 |
| 2500 | 91 (A) | 0.1657 | 1.823 | 1.963 | 7,794 | 41,150 | 109,600 | 0.004444 | 85,990 | 0.004421 | 72,650 | 0.004273 |
| 2500 | 127 (B) | 0.1403 | 1.824 | 1.963 | 7,794 | 41,150 | 111,300 | 0.004444 | 87,170 | 0.004421 | 72,650 | 0.004273 |
| 2000 | 91 (A) | 0.1482 | 1.630 | 1.571 | 6,175 | 32,600 | 87,790 | 0.005501 | 69,270 | 0.005472 | 58,120 | 0.005289 |
| 2000 | 127 (B) | 0.1255 | 1.632 | 1.571 | 6,175 | 32,600 | 90,050 | 0.005501 | 70,210 | 0.005472 | 58,120 | 0.005289 |
| 1750 | 91 (A) | 0.1387 | 1.526 | 1.374 | 5,403 | 28,530 | 77,930 | 0.006286 | 61,020 | 0.006254 | 50,850 | 0.006045 |
| 1750 | 127 (B) | 0.1174 | 1.526 | 1.374 | 5,403 | 28,530 | 78,800 | 0.006286 | 61,430 | 0.006254 | 50,850 | 0.006045 |
| 1500 | 61 (A) | 0.1568 | 1.411 | 1.178 | 4,631 | 24,450 | 65,840 | 0.007334 | 51,950 | 0.007296 | 43,590 | 0.007052 |
| 1500 | 91 (B) | 0.1284 | 1.412 | 1.178 | 4,631 | 24,450 | 67,540 | 0.007334 | 52,650 | 0.007296 | 43,590 | 0.007052 |
| 1250 | 61 (A) | 0.1431 | 1.288 | 0.9817 | 3,859 | 20,380 | 55,670 | 0.008801 | 43,590 | 0.008755 | 36,320 | 0.008463 |
| 1250 | 91 (B) | 0.1172 | 1.289 | 0.9817 | 3,859 | 20,380 | 56,280 | 0.008801 | 43,880 | 0.008755 | 36,320 | 0.008463 |
| 1000 | 37 (AA) | 0.1644 | 1.151 | 0.7854 | 3,088 | 16,300 | 43,830 | 0.01100 | 34,400 | 0.01094 | 29,060 | 0.01058 |
| 1000 | 61 (A-B) | 0.1280 | 1.152 | 0.7854 | 3,088 | 16,300 | 45,030 | 0.01100 | 35,100 | 0.01094 | 29,060 | 0.01058 |
| 900 | 37 (AA) | 0.1560 | 1.092 | 0.7069 | 2,779 | 14,670 | 39,510 | 0.01222 | 31,170 | 0.01216 | 26,150 | 0.01175 |
| 900 | 61 (A-B) | 0.1215 | 1.094 | 0.7069 | 2,779 | 14,670 | 40,520 | 0.01222 | 31,590 | 0.01216 | 26,150 | 0.01175 |
| 850 | 37 (AA) | 0.1516 | 1.061 | 0.6676 | 2,624 | 13,860 | 37,310 | 0.01294 | 29,440 | 0.01288 | 24,700 | 0.01245 |
| 850 | 61 (A-B) | 0.1180 | 1.062 | 0.6676 | 2,624 | 13,860 | 38,270 | 0.01294 | 29,840 | 0.01288 | 24,700 | 0.01245 |
| 800 | 37 (AA) | 0.1470 | 1.029 | 0.6283 | 2,470 | 13,040 | 35,120 | 0.01375 | 27,710 | 0.01368 | 23,250 | 0.01322 |
| 800 | 61 (A-B) | 0.1145 | 1.031 | 0.6283 | 2,470 | 13,040 | 36,360 | 0.01375 | 28,270 | 0.01368 | 23,250 | 0.01322 |
| 750 | 37 (AA) | 0.1424 | 0.997 | 0.5890 | 2,316 | 12,230 | 33,400 | 0.01467 | 26,150 | 0.01459 | 21,790 | 0.01410 |
| 750 | 61 (A-B) | 0.1109 | 0.998 | 0.5890 | 2,316 | 12,230 | 34,090 | 0.01467 | 26,510 | 0.01459 | 21,790 | 0.01410 |

**TABLE 4-17** Copper Cable, Classes AA, A, B—Weight, Breaking Strength, DC Resistance (*Continued*)

| Conductor size, Mcm or Awg | No. of wires (ASTM stranding class) | Wire diameter, in. | Conductor diameter, in. | Conductor area, sq in. | Conductor weight, lb | | Hard | | Medium | | Soft | |
|---|---|---|---|---|---|---|---|---|---|---|---|---|
| | | | | | Per 1000 ft | Per mile | Breaking strength, minimum,* lb | D-c resistance at 20 C (68 F), ohms per 1000 ft | Breaking strength, minimum,* lb | D-c resistance at 20 C (68 F), ohms per 1000 ft | Breaking strength, maximum,† lb | D-c resistance at 20 C (68 F), ohms per 1000 ft |
| 700 | 37 (AA) | 0.1375 | 0.963 | 0.5498 | 2,161 | 11,410 | 31,170 | 0.01572 | 24,410 | 0.01563 | 20,340 | 0.01511 |
| 700 | 61 (A-B) | 0.1071 | 0.964 | 0.5498 | 2,161 | 11,410 | 31,820 | 0.01572 | 24,740 | 0.01563 | 20,340 | 0.01511 |
| 650 | 37 (AA) | 0.1325 | 0.928 | 0.5105 | 2,007 | 10,600 | 29,130 | 0.01692 | 22,670 | 0.01684 | 18,890 | 0.01627 |
| 650 | 61 (A-B) | 0.1032 | 0.929 | 0.5105 | 2,007 | 10,600 | 29,770 | 0.01692 | 22,970 | 0.01684 | 18,890 | 0.01627 |
| 600 | 37 (AA-A) | 0.1273 | 0.891 | 0.4712 | 1,853 | 9,781 | 27,020 | 0.01834 | 21,060 | 0.01824 | 17,440 | 0.01763 |
| 600 | 61 (B) | 0.0992 | 0.893 | 0.4712 | 1,853 | 9,781 | 27,530 | 0.01834 | 21,350 | 0.01824 | 18,140 | 0.01763 |
| 550 | 37 (AA-A) | 0.1219 | 0.853 | 0.4320 | 1,698 | 8,966 | 24,760 | 0.02000 | 19,310 | 0.01990 | 15,980 | 0.01923 |
| 550 | 61 (B) | 0.0950 | 0.855 | 0.4320 | 1,698 | 8,966 | 25,230 | 0.02000 | 19,570 | 0.01990 | 16,630 | 0.01923 |
| 500 | 19 (AA) | 0.1622 | 0.811 | 0.3927 | 1,544 | 8,151 | 21,950 | 0.02200 | 17,320 | 0.02189 | 14,530 | 0.02116 |
| 500 | 37 (A-B) | 0.1162 | 0.813 | 0.3927 | 1,544 | 8,151 | 22,510 | 0.02200 | 17,550 | 0.02189 | 14,530 | 0.02116 |
| 450 | 19 (AA) | 0.1539 | 0.770 | 0.3534 | 1,389 | 7,336 | 19,750 | 0.02445 | 15,590 | 0.02432 | 13,080 | 0.02351 |
| 450 | 37 (A-B) | 0.1103 | 0.772 | 0.3534 | 1,389 | 7,336 | 20,450 | 0.02445 | 15,900 | 0.02432 | 13,080 | 0.02351 |
| 400 | 19 (AA-A) | 0.1451 | 0.726 | 0.3142 | 1,235 | 6,521 | 17,810 | 0.02750 | 13,950 | 0.02736 | 11,620 | 0.02645 |
| 400 | 37 (B) | 0.1040 | 0.728 | 0.3142 | 1,235 | 6,521 | 18,320 | 0.02750 | 14,140 | 0.02736 | 11,620 | 0.02645 |
| 350 | 12 (AA) | 0.1708 | 0.710 | 0.2749 | 1,081 | 5,706 | 15,140 | 0.03143 | 12,040 | 0.03127 | 10,170 | 0.03022 |
| 350 | 19 (A) | 0.1357 | 0.679 | 0.2749 | 1,081 | 5,706 | 15,590 | 0.03143 | 12,200 | 0.03127 | 10,170 | 0.03022 |
| 350 | 37 (B) | 0.0973 | 0.681 | 0.2749 | 1,081 | 5,706 | 16,060 | 0.03143 | 12,450 | 0.03127 | 10,580 | 0.03022 |
| 300 | 12 (AA) | 0.1581 | 0.657 | 0.2356 | 926.3 | 4,891 | 13,170 | 0.03667 | 10,390 | 0.03648 | 8,718 | 0.03526 |
| 300 | 19 (A) | 0.1257 | 0.629 | 0.2356 | 926.3 | 4,891 | 13,510 | 0.03667 | 10,530 | 0.03648 | 8,718 | 0.03526 |
| 300 | 37 (B) | 0.0900 | 0.630 | 0.2356 | 926.3 | 4,891 | 13,870 | 0.03667 | 10,740 | 0.03648 | 9,071 | 0.03526 |
| 250 | 12 (AA) | 0.1443 | 0.600 | 0.1963 | 771.9 | 4,076 | 11,130 | 0.04400 | 8,717 | 0.04378 | 7,265 | 0.04231 |
| 250 | 19 (A) | 0.1147 | 0.574 | 0.1963 | 771.9 | 4,076 | 11,360 | 0.04400 | 8,836 | 0.04378 | 7,265 | 0.04231 |
| 250 | 37 (B) | 0.0822 | 0.575 | 0.1963 | 771.9 | 4,076 | 11,560 | 0.04400 | 8,952 | 0.04378 | 7,559 | 0.04231 |
| 4/0 | 7 (AA-A) | 0.1739 | 0.522 | 0.1662 | 653.3 | 3,450 | 9,154 | 0.05199 | 7,278 | 0.05172 | 6,149 | 0.04999 |
| 4/0 | 12 — | 0.1328 | 0.552 | 0.1662 | 653.3 | 3,450 | 9,483 | 0.05199 | 7,378 | 0.05172 | 6,149 | 0.04999 |
| 4/0 | 19 (B) | 0.1055 | 0.528 | 0.1662 | 653.3 | 3,450 | 9,617 | 0.05199 | 7,479 | 0.05172 | 6,149 | 0.04999 |
| 3/0 | 7 (AA-A) | 0.1548 | 0.464 | 0.1318 | 518.1 | 2,736 | 7,366 | 0.06556 | 5,812 | 0.06522 | 4,876 | 0.06304 |
| 3/0 | 12 — | 0.1183 | 0.492 | 0.1318 | 518.1 | 2,736 | 7,556 | 0.06556 | 5,890 | 0.06522 | 4,876 | 0.06304 |
| 3/0 | 19 (B) | 0.0940 | 0.470 | 0.1318 | 518.1 | 2,736 | 7,698 | 0.06556 | 5,970 | 0.06522 | 5,074 | 0.06304 |
| 2/0 | 7 (AA-A) | 0.1379 | 0.414 | 0.1045 | 410.9 | 2,169 | 5,926 | 0.08267 | 4,640 | 0.08224 | 3,867 | 0.07949 |
| 2/0 | 12 — | 0.1053 | 0.438 | 0.1045 | 410.9 | 2,169 | 6,048 | 0.08267 | 4,703 | 0.08224 | 3,867 | 0.07949 |
| 2/0 | 19 (B) | 0.0837 | 0.419 | 0.1045 | 410.9 | 2,169 | 6,152 | 0.08267 | 4,765 | 0.08224 | 4,024 | 0.07949 |
| 1/0 | 7 (AA-A) | 0.1228 | 0.368 | 0.08289 | 325.8 | 1,720 | 4,752 | 0.1042 | 3,705 | 0.1037 | 3,067 | 0.1002 |
| 1/0 | 12 — | 0.0938 | 0.390 | 0.08289 | 325.8 | 1,720 | 4,841 | 0.1042 | 3,755 | 0.1037 | 3,191 | 0.1002 |
| 1/0 | 19 (B) | 0.0745 | 0.373 | 0.08289 | 325.8 | 1,720 | 4,901 | 0.1042 | 3,805 | 0.1037 | 3,191 | 0.1002 |

| | | | | | | | | | | | | |
|---|---|---|---|---|---|---|---|---|---|---|---|---|
| **1** | 3 (AA) | 0.1670 | 0.360 | 0.06573 | 255.9 | 1,351 | 3,621 | 0.1302 | 2,879 | 0.1295 | 2,432 | 0.1252 |
| **1** | 7 (A) | 0.1093 | 0.328 | 0.06573 | 258.4 | 1,364 | 3,804 | 0.1314 | 2,958 | 0.1308 | 2,432 | 0.1264 |
| **1** | 19 (B) | 0.0664 | 0.332 | 0.06573 | 258.4 | 1,364 | 3,899 | 0.1314 | 3,037 | 0.1308 | 2,531 | 0.1264 |
| **2** | 3 (AA) | 0.1487 | 0.320 | 0.05213 | 202.9 | 1,071 | 2,913 | 0.1641 | 2,299 | 0.1633 | 1,929 | 0.1578 |
| **2** | 7 (A-B) | 0.0974 | 0.292 | 0.05213 | 204.9 | 1,082 | 3,045 | 0.1657 | 2,361 | 0.1649 | 2,007 | 0.1594 |
| **3** | 3 (AA) | 0.1325 | 0.285 | 0.04134 | 160.9 | 849.6 | 2,359 | 0.2070 | 1,835 | 0.2059 | 1,530 | 0.1990 |
| **3** | 7 (A-B) | 0.0867 | 0.260 | 0.04134 | 162.5 | 858.0 | 2,433 | 0.2090 | 1,885 | 0.2079 | 1,592 | 0.2010 |
| **4** | 3 (AA) | 0.1180 | 0.254 | 0.03278 | 127.6 | 673.8 | 1,879 | 0.2610 | 1,465 | 0.2596 | 1,213 | 0.2509 |
| **4** | 7 (A-B) | 0.0772 | 0.232 | 0.03278 | 128.9 | 680.5 | 1,938 | 0.2636 | 1,505 | 0.2622 | 1,262 | 0.2534 |
| **5** | 7 (B) | 0.0688 | 0.206 | 0.02600 | 102.2 | 539.6 | 1,542 | 0.3323 | 1,201 | 0.3306 | 1,001 | 0.3196 |
| **6** | 7 (B) | 0.0612 | 0.184 | 0.02062 | 81.05 | 427.9 | 1,288 | 0.4191 | 958.6 | 0.4169 | 793.8 | 0.4030 |
| **7** | 7 (B) | 0.0545 | 0.164 | 0.01635 | 64.28 | 339.4 | 977.1 | 0.5284 | 765.2 | 0.5257 | 629.5 | 0.5081 |
| **8** | 7 (B) | 0.0486 | 0.146 | 0.01297 | 50.97 | 269.1 | 777.2 | 0.6663 | 610.7 | 0.6629 | 499.2 | 0.6408 |
| **9** | 7 (B) | 0.0432 | 0.130 | 0.01028 | 40.42 | 213.4 | 618.2 | 0.8402 | 487.4 | 0.8359 | 395.9 | 0.8080 |
| **10** | 7 (B) | 0.0385 | 0.116 | 0.008155 | 32.06 | 169.3 | 491.7 | 1.060 | 388.9 | 1.054 | 314.0 | 1.019 |
| **12** | 7 (B) | 0.0305 | 0.0915 | 0.005129 | 20.16 | 106.5 | 311.1 | 1.685 | 247.7 | 1.676 | 197.5 | 1.620 |
| **14** | 7 (B) | 0.0242 | 0.0726 | 0.003225 | 12.68 | 66.95 | 197.1 | 2.679 | 157.7 | 2.665 | 124.2 | 2.576 |
| **16** | 7 (B) | 0.0192 | 0.0576 | 0.002028 | 7.974 | 42.10 | 124.7 | 4.259 | 100.4 | 4.237 | 81.14 | 4.096 |
| **18** | 7 (B) | 0.0152 | 0.0456 | 0.001276 | 5.015 | 26.48 | 78.99 | 6.773 | 63.91 | 6.738 | 51.03 | 6.513 |
| **20** | 7 (B) | 0.0121 | 0.0363 | 0.0008023 | 3.154 | 16.65 | 50.04 | 10.77 | 40.67 | 10.71 | 32.09 | 10.36 |
| ASTM Designation | B8 | | | | | | B1 & B8 | | B2 & B8 | | B3 & B8 | |

*No. 10 AWG and smaller, based on Anaconda data.

†No requirements for tensile strength are specified in ASTM B3. Values given here based on Anaconda data.

## Weight and Resistance

| Stranding class | Conductor size, Mcm or Awg | Increment of resistance and weight, % |
|---|---|---|
| AA | 4–1 | 1 |
| | 1/0–1000 | 2 |
| A, B, C, D | 2000 and under | 2 |
| | Over 2000–3000 | 3 |
| | Over 3000–4000 | 4 |
| | Over 4000–5000 | 5 |

## Resistance

(ASTM requirements)

| Temper | Conductivity at 20 C (68 F), IACS, % | Resistivity at 20 C (68 F), ohms (mile, lb) |
|---|---|---|
| Hard | 96.16 | 910.15 |
| Medium | 96.66 | 905.44 |
| Soft | 100 | 875.20 |

The resistance values in this table are trade maximums and are higher than the average values for commercial cable.

NOTE: 1 in = 2.54 cm; 1 in² = 64.5 cm²; 1 ft = 0.3048 m; 1 lb = 0.4536 kg; 1 mi = 1.61 km.

**TABLE 4-18** Copper-Clad Steel Wire and Cable—Weight, Breaking Strength, DC Resistance

(Based on ASTM Specifications B227 and B228)

| Conductor size,* AWG or in. | Conductor stranding | | Conductor diam., in. | Conductor area | | Conductor weight, lb | | Breaking strength, min., lb | | | D-c resistance at 20 C (68 F), ohms per 1,000 ft | |
|---|---|---|---|---|---|---|---|---|---|---|---|---|
| | | | | | | | | High strength | | Extra-high strength | Conductivity, IACS | |
| | | | | | | | | Conductivity, IACS | | | | |
| | No. of wires | Wire size, AWG | | Cir mils | Sq in. | Per 1,000 ft | Per mile | 40 % | 30 % | 30 % | 40 % | 30 % |
| **Solid (B227)** | | | | | | | | | | | | |
| 4 | .. | .. | 0.2043 | 41,740 | 0.03278 | 115.8 | 611.6 | 3,541 | 3,934 | 4,672 | 0.6337 | 0.8447 |
| 5 | .. | .. | 0.1819 | 33,090 | 0.02599 | 91.86 | 485.0 | 2,938 | 3,250 | 3,913 | 0.7990 | 1.065 |
| 0.165 | .. | .. | 0.1650 | 27,230 | 0.02138 | 75.55 | 398.9 | 2,523 | 2,780 | 3,368 | 0.9715 | 1.295 |
| 6 | .. | .. | 0.1620 | 26,240 | 0.02061 | 72.85 | 384.6 | 2,433 | 2,680 | 3,247 | 1.008 | 1.343 |
| 7 | .. | .. | 0.1443 | 20,820 | 0.01635 | 57.77 | 305.0 | 2,011 | 2,207 | 2,681 | 1.270 | 1.694 |
| 8 | .. | .. | 0.1285 | 16,510 | 0.01297 | 45.81 | 241.9 | 1,660 | 1,815 | 2,204 | 1.602 | 2.136 |
| 0.128 | .. | .. | 0.1280 | 16,380 | 0.01287 | 45.47 | 240.1 | 1,647 | 1,802 | 2,188 | 1.614 | 2.152 |
| 9 | .. | .. | 0.1144 | 13,090 | 0.01028 | 36.33 | 191.8 | 1,368 | 1,491 | 1,790 | 2.020 | 2.693 |
| 0.104 | .. | .. | 0.1040 | 10,820 | 0.008495 | 30.01 | 158.5 | 1,177 | 1,283 | 1,487 | 2.445 | 3.260 |
| 10 | .. | .. | 0.1019 | 10,380 | 0.008155 | 28.81 | 152.1 | 1,130 | 1,231 | 1,460 | 2.547 | 3.396 |
| 12 | .. | .. | 0.0808 | 6,530 | 0.005129 | 18.12 | 95.68 | 785 | ...... | ...... | 4.051 | |
| 0.080 | .. | .. | 0.0800 | 6,400 | 0.005027 | 17.76 | 93.77 | 770 | ...... | 900 | 4.133 | 5.509 |
| **Stranded (B228)** | | | | | | | | | | | | |
| 7/8 | 19 | 5 | 0.910 | 628,900 | 0.4940 | 1770 | 9344 | 50,240 | 55,570 | 66,910 | 0.04264 | 0.05685 |
| 13/16 | 19 | 6 | 0.810 | 498,800 | 0.3917 | 1403 | 7410 | 41,600 | 45,830 | 55,530 | 0.05377 | 0.07168 |
| 23/32 | 19 | 7 | 0.721 | 395,500 | 0.3107 | 1113 | 5877 | 34,390 | 37,740 | 45,850 | 0.06780 | 0.09039 |
| 21/32 | 19 | 8 | 0.642 | 313,700 | 0.2464 | 882.7 | 4660 | 28,380 | 31,040 | 37,690 | 0.08550 | 0.1140 |
| 9/16 | 19 | 9 | 0.572 | 248,800 | 0.1954 | 700.0 | 3696 | 23,390 | 25,500 | 30,610 | 0.1078 | 0.1437 |
| 5/8 | 7 | 4 | 0.613 | 292,200 | 0.2295 | 818.9 | 4324 | 22,310 | 24,780 | 29,430 | 0.09143 | 0.1219 |
| 9/16 | 7 | 5 | 0.546 | 231,700 | 0.1820 | 649.4 | 3429 | 18,510 | 20,470 | 24,650 | 0.1153 | 0.1537 |
| 1/2 | 7 | 6 | 0.486 | 183,800 | 0.1443 | 515.0 | 2719 | 15,330 | 16,890 | 20,460 | 0.1454 | 0.1938 |
| 7/16 | 7 | 7 | 0.433 | 145,700 | 0.1145 | 408.4 | 2157 | 12,670 | 13,910 | 16,890 | 0.1833 | 0.2444 |
| 3/8 | 7 | 8 | 0.385 | 115,600 | 0.09077 | 323.9 | 1710 | 10,460 | 11,440 | 13,890 | 0.2312 | 0.3081 |
| 11/32 | 7 | 9 | 0.343 | 91,650 | 0.07198 | 256.9 | 1356 | 8,616 | 9,393 | 11,280 | 0.2915 | 0.3886 |
| 5/16 | 7 | 10 | 0.306 | 72,680 | 0.05708 | 203.7 | 1076 | 7,121 | 7,758 | 9,196 | 0.3676 | 0.4900 |
| ..... | 3 | 5 | 0.392 | 99,310 | 0.07800 | 277.8 | 1467 | 8,373 | 9,262 | 11,860 | 0.2685 | 0.3579 |
| ..... | 3 | 6 | 0.349 | 78,750 | 0.06185 | 220.3 | 1163 | 6,934 | 7,639 | 9,754 | 0.3385 | 0.4513 |
| ..... | 3 | 7 | 0.311 | 62,450 | 0.04905 | 174.7 | 922.4 | 5,732 | 6,291 | 7,922 | 0.4269 | 0.5691 |
| ..... | 3 | 8 | 0.277 | 49,530 | 0.03890 | 138.5 | 731.5 | 4,730 | 5,174 | 6,282 | 0.5383 | 0.7176 |
| ..... | 3 | 9 | 0.247 | 39,280 | 0.03085 | 109.9 | 580.1 | 3,898 | 4,250 | 5,129 | 0.6788 | 0.9049 |
| ..... | 3 | 10 | 0.220 | 31,150 | 0.02446 | 87.13 | 460.0 | 3,221 | 3,509 | 4,160 | 0.8559 | 1.141 |
| ..... | 3 | 12 | 0.174 | 19,590 | 0.01539 | 54.80 | 289.3 | 2,236 | .... | ...... | 1.361 | |

*To determine copper equivalent of copper-clad steel conductor, multiply circular-mil area by percent conductivity expressed as a decimal.

NOTE: 1 in = 2.54 cm; 1 in$^2$ = 64.5 cm$^2$; 1 ft = 0.3048 m; 1 mi = 1.61 km; 1 lb = 0.4536 kg.

**TABLE 4-19** Copper-Clad Steel-Copper Cable—Weight, Breaking Strength, DC Resistance

(ASTM Specification B229)

| Hard-drawn copper equivalent,* Mcm or AWG | Conductor type | Conductor stranding | | | | Conductor diam., in | Conductor area, in² | Conductor weight, lb | | Breaking strength, min., lb | D-c resistance at 20°C (68°F), Ω/1,000 ft |
|---|---|---|---|---|---|---|---|---|---|---|---|
| | | EHS 30% copper-clad wires | | Hard-drawn copper wires | | | | Per 1,000 ft | Per mi | | |
| | | No. | Diam., in | No. | Diam., in | | | | | | |
| 350 | E | 7 | 0.1576 | 12 | 0.1576 | 0.788 | 0.3706 | 1403 | 7409 | 32,420 | 0.03143 |
| 350 | EK | 4 | 0.1470 | 15 | 0.1470 | 0.735 | 0.3225 | 1238 | 6536 | 23,850 | 0.03143 |
| 300 | E | 7 | 0.1459 | 12 | 0.1459 | 0.729 | 0.3177 | 1203 | 6351 | 27,770 | 0.03667 |
| 300 | EK | 4 | 0.1361 | 15 | 0.1361 | 0.680 | 0.2764 | 1061 | 5602 | 20,960 | 0.03667 |
| 250 | E | 7 | 0.1332 | 12 | 0.1332 | 0.666 | 0.2648 | 1002 | 5292 | 23,920 | 0.04400 |
| 250 | EK | 4 | 0.1242 | 15 | 0.1242 | 0.621 | 0.2302 | 884.2 | 4669 | 17,840 | 0.04400 |
| 4/0 | E | 7 | 0.1225 | 12 | 0.1225 | 0.613 | 0.2239 | 848.3 | 4479 | 20,730 | 0.05199 |
| 4/0 | EK | 4 | 0.1143 | 15 | 0.1143 | 0.571 | 0.1950 | 748.4 | 3951 | 15,370 | 0.05199 |
| 4/0 | F | 1 | 0.1833 | 6 | 0.1833 | 0.550 | 0.1847 | 710.2 | 3750 | 12,290 | 0.05199 |
| 3/0 | E | 7 | 0.1091 | 12 | 0.1091 | 0.545 | 0.1776 | 672.7 | 3552 | 16,800 | 0.06556 |
| 3/0 | EK | 4 | 0.1018 | 15 | 0.1018 | 0.509 | 0.1546 | 593.5 | 3134 | 12,370 | 0.06556 |
| 3/0 | F | 1 | 0.1632 | 6 | 0.1632 | 0.490 | 0.1464 | 563.2 | 2974 | 9,980 | 0.06556 |
| 2/0 | F | 1 | 0.1454 | 6 | 0.1454 | 0.436 | 0.1162 | 446.8 | 2359 | 8,094 | 0.08265 |
| 1/0 | F | 1 | 0.1294 | 6 | 0.1294 | 0.388 | 0.09206 | 354.1 | 1870 | 6,536 | 0.1043 |
| 1 | F | 1 | 0.1153 | 6 | 0.1153 | 0.346 | 0.07309 | 280.9 | 1483 | 5,266 | 0.1315 |
| 2† | A | 1 | 0.1699 | 2 | 0.1699 | 0.366 | 0.06801 | 256.8 | 1356 | 5,876 | 0.1658 |
| 2 | F | 1 | 0.1026 | 6 | 0.1026 | 0.308 | 0.05787 | 222.8 | 1176 | 4,233 | 0.1658 |
| 4† | A | 1 | 0.1347 | 2 | 0.1347 | 0.290 | 0.04275 | 161.5 | 852 | 3,938 | 0.2636 |
| 6† | A | 1 | 0.1068 | 2 | 0.1068 | 0.230 | 0.02688 | 101.6 | 536.3 | 2,585 | 0.4150 |
| 8† | A | 1 | 0.1127 | 2 | 0.07969 | 0.199 | 0.01995 | 74.27 | 392.2 | 2,233 | 0.6598 |

*Area of hard-drawn copper cable having the same dc resistance as that of the composite cable.
†Sizes commonly used for rural distribution.
NOTE: 1 in = 2.54 cm; 1 in$^2$ = 64.5 cm$^2$; 1 ft = 0.3048 m; 1 lb = 0.4536 kg; 1 mi = 1.61 km.

**TABLE 4-20** Aluminum Wire—Dimensions, Weight, DC Resistance
(Based on ASTM Specifications B230, B262, and B323)

| Conductor size, AWG | Diam. at 20 C (68 F), mils | Area at 20 C (68 F) | | D-c resistance at 20 C (68 F),* ohms per 1,000 ft | Weight at 20 C (68 F),† lb | | Length at 20 C (68 F), ft per ohm |
|---|---|---|---|---|---|---|---|
| | | Cir mils | Sq in. | | Per 1,000 ft | Per ohm | |
| 2 | 257.6 | 66,360 | 0.05212 | 0.2562 | 61.07 | 238.4 | 3903 |
| 3 | 229.4 | 52,620 | 0.04133 | 0.3231 | 48.43 | 149.9 | 3095 |
| 4 | 204.3 | 41,740 | 0.03278 | 0.4074 | 38.41 | 94.30 | 2455 |
| 5 | 181.9 | 33,090 | 0.02599 | 0.5139 | 30.45 | 59.26 | 1946 |
| 6 | 162.0 | 26,240 | 0.02061 | 0.6479 | 24.15 | 37.28 | 1544 |
| 7 | 144.3 | 20,820 | 0.01635 | 0.8165 | 19.16 | 23.47 | 1225 |
| 8 | 128.5 | 16,510 | 0.01297 | 1.030 | 15.20 | 14.76 | 971.2 |
| 9 | 114.4 | 13,090 | 0.01028 | 1.299 | 12.04 | 9.272 | 769.7 |
| 10 | 101.9 | 10,380 | 0.008155 | 1.637 | 9.556 | 5.836 | 610.7 |
| 11 | 90.7 | 8,230 | 0.00646 | 2.07 | 7.57 | 3.66 | 484 |
| 12 | 80.8 | 6,530 | 0.00513 | 2.60 | 6.01 | 2.31 | 384 |
| 13 | 72.0 | 5,180 | 0.00407 | 3.28 | 4.77 | 1.45 | 305 |
| 14 | 64.1 | 4,110 | 0.00323 | 4.14 | 3.78 | 0.914 | 242 |
| 15 | 57.1 | 3,260 | 0.00256 | 5.21 | 3.00 | 0.575 | 192 |
| 16 | 50.8 | 2,580 | 0.00203 | 6.59 | 2.38 | 0.361 | 152 |
| 17 | 45.3 | 2,050 | 0.00161 | 8.29 | 1.89 | 0.228 | 121 |
| 18 | 40.3 | 1,620 | 0.00128 | 10.5 | 1.49 | 0.143 | 95.5 |
| 19 | 35.9 | 1,290 | 0.00101 | 13.2 | 1.19 | 0.0899 | 75.8 |
| 20 | 32.0 | 1,020 | 0.000804 | 16.6 | 0.942 | 0.0568 | 60.2 |
| 21 | 28.5 | 812 | 0.000638 | 20.9 | 0.748 | 0.0357 | 47.8 |
| 22 | 25.3 | 640 | 0.000503 | 26.6 | 0.589 | 0.0222 | 37.6 |
| 23 | 22.6 | 511 | 0.000401 | 33.3 | 0.470 | 0.0141 | 30.0 |
| 24 | 20.1 | 404 | 0.000317 | 42.1 | 0.372 | 0.00884 | 23.8 |
| 25 | 17.9 | 320 | 0.000252 | 53.1 | 0.295 | 0.00556 | 18.8 |
| 26 | 15.9 | 253 | 0.000199 | 67.3 | 0.233 | 0.00346 | 14.9 |
| 27 | 14.2 | 202 | 0.000158 | 84.3 | 0.186 | 0.00220 | 11.9 |
| 28 | 12.6 | 159 | 0.000125 | 107 | 0.146 | 0.00136 | 9.34 |
| 29 | 11.3 | 128 | 0.000100 | 133 | 0.118 | 0.000883 | 7.51 |
| 30 | 10.0 | 100 | 0.0000785 | 170 | 0.0920 | 0.000541 | 5.88 |

*Conductivity = 61.0% IACS.
†Density = 2.703 g per cu cm (0.09765 lb per cu in).
NOTE: 1 $in^2$ = 64.5 $cm^2$; 1 ft = 0.3048 m; 1 lb = 0.4536 kg.

**TABLE 4-21** Aluminum Cable—Stranding Classes, Uses
(ASTM Specification B231)

| Construction | Class | Application |
|---|---|---|
| Concentric lay | AA | For bare conductors usually used in **overhead** lines |
| | A | For conductors to be covered with **weather-resistant** (weatherproof), slow-burning materials and for bare conductors where greater flexibility than is afforded by Class AA is required. Conductors intended for further fabrication into tree wire or to be insulated and laid helically with or around aluminum or ACSR messengers shall be regarded as Class A conductors with respect to direction of lay only |
| | B | For conductors to be **insulated** with various materials such as rubber, paper, varnished cloth, etc., and for the conductors indicated under Class A where greater flexibility is required |
| | C, D | For conductors where greater flexibility is required than is provided by Class B conductors |

**TABLE 4-22** Aluminum Conductor—Physical Characteristics 1350-H19 Classes AA and A

| Cable code word | Conductor size | | Current-carrying capacity* | Stranding | | Conductor diam, in | Rated strength, lb | Nominal weight, lb† | |
|---|---|---|---|---|---|---|---|---|---|
| | cmils or AWG | in² | A | Class | No. and diam of wires, in | | | Per 1000 ft | Per mile |
| Peachbell | 6 | 0.0206 | 95 | A | 7 × 0.0612 | 0.184 | 563 | 24.6 | 130 |
| Rose | 4 | 0.0328 | 130 | A | 7 × 0.0772 | 0.232 | 881 | 39.2 | 207 |
| Iris | 2 | 0.0522 | 175 | AA, A | 7 × 0.0974 | 0.292 | 1,350 | 62.3 | 329 |
| Pansy | 1 | 0.0657 | 200 | AA, A | 7 × 0.1093 | 0.328 | 1,640 | 78.5 | 414 |
| Poppy | 1/0 | 0.0829 | 235 | AA, A | 7 × 0.1228 | 0.368 | 1,990 | 99.1 | 523 |
| Aster | 2/0 | 0.1045 | 270 | AA, A | 7 × 0.1379 | 0.414 | 2,510 | 124.9 | 659 |
| Phlox | 3/0 | 0.1317 | 315 | AA, A | 7 × 0.1548 | 0.464 | 3,040 | 157.5 | 832 |
| Oxlip | 4/0 | 0.1663 | 365 | AA, A | 7 × 0.1739 | 0.522 | 3,830 | 198.7 | 1,049 |
| Sneezewort | 250,000 | 0.1964 | 405 | AA | 7 × 0.1890 | 0.567 | 4,520 | 234.7 | 1,239 |
| Valerian | 250,000 | 0.1963 | 405 | A | 19 × 0.1147 | 0.574 | 4,660 | 234.6 | 1,239 |
| Daisy | 266,800 | 0.2097 | 420 | AA | 7 × 0.1953 | 0.586 | 4,830 | 250.6 | 1,323 |
| Laurel | 266,800 | 0.2095 | 425 | A | 19 × 0.1185 | 0.593 | 4,970 | 250.4 | 1,322 |
| Peony | 300,000 | 0.2358 | 455 | A | 19 × 0.1257 | 0.629 | 5,480 | 281.8 | 1,488 |
| Tulip | 336,400 | 0.2644 | 495 | A | 19 × 0.1331 | 0.666 | 6,150 | 316.0 | 1,668 |
| Daffodil | 350,000 | 0.2748 | 506 | A | 19 × 0.1357 | 0.679 | 6,390 | 328.4 | 1,734 |
| Canna | 397,500 | 0.3124 | 550 | AA, A | 19 × 0.1447 | 0.724 | 7,110 | 373.4 | 1,972 |
| Goldentuft | 450,000 | 0.3534 | 545 | AA | 19 × 0.1539 | 0.770 | 7,890 | 422.4 | 2,230 |
| Cosmos | 477,000 | 0.3744 | 615 | AA | 19 × 0.1584 | 0.793 | 8,360 | 447.5 | 2,363 |
| Syringa | 477,000 | 0.3743 | 615 | A | 37 × 0.1135 | 0.795 | 8,690 | 447.4 | 2,362 |
| Zinnia | 500,000 | 0.3926 | 635 | AA | 19 × 0.1622 | 0.811 | 8,760 | 469.2 | 2,477 |
| Hyacinth | 500,000 | 0.3924 | 635 | A | 37 × 0.1162 | 0.813 | 9,110 | 469.0 | 2,476 |
| Dahlia | 556,500 | 0.4369 | 680 | AA | 19 × 0.1711 | 0.856 | 9,750 | 522.1 | 2,757 |
| Mistletoe | 556,500 | 0.4368 | 680 | AA, A | 37 × 0.1226 | 0.858 | 9,940 | 522.0 | 2,756 |
| Meadowsweet | 600,000 | 0.4709 | 715 | AA, A | 37 × 0.1273 | 0.891 | 10,700 | 562.8 | 2,972 |
| Orchid | 636,000 | 0.4995 | 745 | AA, A | 37 × 0.1311 | 0.918 | 11,400 | 596.9 | 3,152 |
| Heuchera | 650,000 | 0.5102 | 755 | AA | 37 × 0.1325 | 0.928 | 11,600 | 609.8 | 3,220 |
| Verbena | 700,000 | 0.5494 | 790 | AA | 37 × 0.1375 | 0.963 | 12,500 | 656.6 | 3,467 |
| Flag | 700,000 | 0.5495 | 790 | A | 61 × 0.1071 | 0.964 | 12,900 | 656.8 | 3,468 |
| Violet | 715,500 | 0.5622 | 800 | AA | 37 × 0.1391 | 0.974 | 12,800 | 672.0 | 3,548 |
| Nasturtium | 715,500 | 0.5619 | 800 | A | 61 × 0.1083 | 0.975 | 13,100 | 671.6 | 3,546 |
| Petunia | 750,000 | 0.5892 | 825 | AA | 37 × 0.1424 | 0.997 | 13,100 | 704.3 | 3,719 |
| Cattail | 750,000 | 0.5892 | 825 | A | 61 × 0.1109 | 0.998 | 13,500 | 704.2 | 3,718 |
| Arbutus | 795,000 | 0.6245 | 855 | AA | 37 × 0.1466 | 1.026 | 13,900 | 746.4 | 3,941 |
| Lilac | 795,000 | 0.6248 | 855 | A | 61 × 0.1142 | 1.028 | 14,300 | 746.7 | 3,943 |

**TABLE 4-22** Aluminum Conductor—Physical Characteristics 1350-H19 Classes AA and A (*Continued*)

| Cable code word | Conductor size: cmils or AWG | Conductor size: in² | Current-carrying capacity* A | Stranding: Class | Stranding: No. and diam of wires, in | Conductor diam, in | Rated strength, lb | Nominal weight, lb†: Per 1000 ft | Nominal weight, lb†: Per mile |
|---|---|---|---|---|---|---|---|---|---|
| Cockscomb | 900,000 | 0.7072 | 925 | AA | 37 × 0.1560 | 1.092 | 15,400 | 845.2 | 4,463 |
| Snapdragon | 900,000 | 0.7072 | 925 | A | 61 × 0.1215 | 1.094 | 15,900 | 845.3 | 4,463 |
| Magnolia | 954,000 | 0.7495 | 960 | AA | 37 × 0.1606 | 1.124 | 16,400 | 895.8 | 4,730 |
| Goldenrod | 954,000 | 0.7498 | 960 | A | 61 × 0.1251 | 1.126 | 16,900 | 896.1 | 4,731 |
| Hawkweed | 1,000,000 | 0.7854 | 990 | AA | 37 × 0.1644 | 1.151 | 17,200 | 938.7 | 4,956 |
| Camellia | 1,000,000 | 0.7849 | 990 | A | 61 × 0.1280 | 1.152 | 17,700 | 938.2 | 4,954 |
| Bluebell | 1,033,500 | 0.8124 | 1015 | AA | 37 × 0.1672 | 1.170 | 17,700 | 970.9 | 5,126 |
| Larkspur | 1,033,500 | 0.8122 | 1015 | A | 61 × 0.1302 | 1.172 | 18,300 | 970.6 | 5,125 |
| Marigold | 1,113,000 | 0.8744 | 1040 | AA, A | 61 × 0.1351 | 1.216 | 19,700 | 1,045 | 5,518 |
| Hawthorn | 1,192,500 | 0.9363 | 1085 | AA, A | 61 × 0.1398 | 1.258 | 21,100 | 1,119 | 5,908 |
| Narcissus | 1,272,000 | 0.999 | 1130 | AA, A | 61 × 0.1444 | 1.300 | 22,000 | 1,194 | 6,304 |
| Columbine | 1,351,500 | 1.062 | 1175 | AA, A | 61 × 0.1489 | 1.340 | 23,400 | 1,269 | 6,700 |
| Carnation | 1,431,000 | 1.124 | 1220 | AA, A | 61 × 0.1532 | 1.379 | 24,300 | 1,344 | 7,096 |
| Galdiolus | 1,510,500 | 1.187 | 1265 | AA, A | 61 × 0.1574 | 1.417 | 25,600 | 1,419 | 7,492 |
| Coreopsis | 1,590,000 | 1.250 | 1305 | AA | 61 × 0.1615 | 1.454 | 27,000 | 1,493 | 7,883 |
| Jessamine | 1,750,000 | 1.375 | 1385 | AA | 61 × 0.1694 | 1.525 | 29,700 | 1,643 | 8,675 |
| Cowslip | 2,000,000 | 1.570 | 1500 | A | 91 × 0.1482 | 1.630 | 34,200 | 1,876 | 9,911 |
| Sagebrush | 2,250,000 | 1.766 | 1600 | A | 91 × 0.1572 | 1.729 | 37,700 | 2,132 | 11,257 |
| Lupine | 2,500,000 | 1.962 | 1700 | A | 91 × 0.1657 | 1.823 | 41,800 | 2,368 | 12,503 |
| Bitterroot | 2,750,000 | 2.159 | 1795 | A | 91 × 0.1738 | 1.912 | 46,100 | 2,606 | 13,760 |
| Trillium | 3,000,000 | 2.356 | 1885 | A | 127 × 0.1537 | 1.996 | 50,300 | 2,844 | 15,016 |
| Bluebonnet | 3,500,000 | 2.749 | 2035 | A | 127 × 0.1660 | 2.158 | 58,700 | 3,350 | 17,688 |

*Class of stranding.* The class of stranding must be specified on all orders. Class AA stranding is usually specified for bare conductors used on overhead lines. Class A stranding is usually specified for conductors to be covered with weather-resistant (weatherproof) materials and for bare conductors where greater flexibility than afforded by Class AA is required.

*Lay.* The direction of lay of the outside layer of wires with Class AA and Class A stranding will be right hand unless otherwise specified.

*Ampacity for conductor temperature rise of 40°C over 40°C ambient with a 2 ft/s crosswind and an emissivity factor of 0.5 without sun.

†Nominal conductor weights are based on ASTM standard stranding increments. Actual weights will vary with lay lengths. Invoicing will be based on actual weights.

NOTE: 1 in$^2$ = 64.5 cm$^2$; 1 in = 2.54 cm; 1 lb = 0.4536 kg; 1 ft = 0.3048 m; 1 mi = 1.61 km.

**TABLE 4-23** ACSR Conductor—Physical Characteristics

| | ACSR | | | | | | | | | | | | Rated strength, lb | | | |
|---|---|---|---|---|---|---|---|---|---|---|---|---|---|---|---|---|
| | Cross section | | | | Stranding No. and diam of strand, in | | Diameter, in | | Nominal weight, lb† | | | Zinc-coated core | | | | |
| | Aluminum | | Total | Current-carrying capacity,* | | | | | Per 1000 ft | | | Standard | | | | |
| Code word | cmils or AWG | in² | in² | A | Aluminum | Steel | Complete cond. | Steel core | Total | Al | Steel | weight coating | Class B coating | Class C coating | Aluminum-coated core |
| Turkey | 6 | 0.0206 | 0.0240 | 95 | 6 × 0.0661 | 1 × 0.0661 | 0.198 | 0.0661 | 36.1 | 24.5 | 11.6 | 1,190 | 1,160 | 1,120 | 1,120 |
| Swan | 4 | 0.0328 | 0.0382 | 130 | 6 × 0.0834 | 1 × 0.0834 | 0.250 | 0.0834 | 57.4 | 39.0 | 18.4 | 1,860 | 1,810 | 1,760 | 1,760 |
| Swanate | 4 | 0.0328 | 0.0411 | 130 | 7 × 0.0772 | 1 × 0.1029 | 0.257 | 0.1029 | 67.0 | 39.0 | 28.0 | 2,360 | 2,280 | 2,200 | 2,160 |
| Sparrow | 2 | 0.0522 | 0.0608 | 175 | 6 × 0.1052 | 1 × 0.1052 | 0.316 | 0.1052 | 91.3 | 62.0 | 29.3 | 2,850 | 2,760 | 2,680 | 2,640 |
| Sparate | 2 | 0.0522 | 0.0654 | 175 | 7 × 0.0974 | 1 × 0.1299 | 0.325 | 0.1299 | 106.7 | 62.0 | 44.7 | 3,640 | 3,510 | 3,390 | 3,260 |
| Robin | 1 | 0.0657 | 0.0767 | 200 | 6 × 0.1181 | 1 × 0.1181 | 0.355 | 0.1182 | 115.1 | 78.2 | 36.9 | 3,550 | 3,450 | 3,340 | 3,290 |
| Raven | 1/0 | 0.0830 | 0.0968 | 230 | 6 × 0.1327 | 1 × 0.1327 | 0.398 | 0.1327 | 145.3 | 98.7 | 46.6 | 4,380 | 4,250 | 4,120 | 3,980 |
| Quail | 2/0 | 0.1046 | 0.1221 | 265 | 6 × 0.1490 | 1 × 0.1490 | 0.447 | 0.1490 | 183.2 | 124.4 | 58.8 | 5,310 | 5,130 | 5,050 | 4,720 |
| Pigeon | 3/0 | 0.1317 | 0.1537 | 310 | 6 × 0.1672 | 1 × 0.1672 | 0.502 | 0.1672 | 230.8 | 156.7 | 74.1 | 6,620 | 6,410 | 6,300 | 5,880 |
| Penguin | 4/0 | 0.1662 | 0.1939 | 350 | 6 × 0.1878 | 1 × 0.1878 | 0.563 | 0.1878 | 291.1 | 197.7 | 93.4 | 8,350 | 8,080 | 7,950 | 7,420 |
| Waxwing | 266,800 | 0.2094 | 0.2210 | 430 | 18 × 0.1217 | 1 × 0.1217 | 0.609 | 0.1217 | 289.5 | 250.3 | 39.2 | 6,880 | 6,770 | 6,650 | 6,540 |
| Owl | 266,800 | 0.2096 | 0.2368 | 410 | 6 × 0.2109 | 7 × 0.0703 | 0.633 | 0.2109 | 342.4 | 250.5 | 91.9 | 9,680 | 9,420 | 9,160 | 9,160 |
| Partridge | 266,800 | 0.2095 | 0.2436 | 440 | 26 × 0.1013 | 7 × 0.0788 | 0.642 | 0.2364 | 367.3 | 251.7 | 115.6 | 11,300 | 11,000 | 10,600 | 10,640 |
| Merlin | 336,400 | 0.2642 | 0.2789 | 500 | 18 × 0.1367 | 1 × 0.1367 | 0.684 | 0.1367 | 365.2 | 315.7 | 49.5 | 8,680 | 8,540 | 8,400 | 8,260 |
| Linnet | 336,400 | 0.2640 | 0.3070 | 510 | 26 × 0.1137 | 7 × 0.0884 | 0.720 | 0.2652 | 462.5 | 317.0 | 145.5 | 14,100 | 13,700 | 13,300 | 13,300 |
| Oriole | 336,400 | 0.2642 | 0.3259 | 515 | 30 × 0.1059 | 7 × 0.1059 | 0.741 | 0.3177 | 527.1 | 318.1 | 209.0 | 17,300 | 16,700 | 16,200 | 15,900 |
| Chickadee | 397,500 | 0.3121 | 0.3295 | 555 | 18 × 0.1486 | 1 × 0.1486 | 0.743 | 0.1486 | 431.6 | 373.1 | 58.5 | 9,940 | 9,780 | 9,690 | 9,530 |
| Brant | 397,500 | 0.3122 | 0.3527 | 565 | 24 × 0.1287 | 7 × 0.0858 | 0.772 | 0.2574 | 512.1 | 375.0 | 137.1 | 14,600 | 14,300 | 13,900 | 13,900 |
| Ibis | 397,500 | 0.3119 | 0.3627 | 570 | 26 × 0.1236 | 7 × 0.0961 | 0.783 | 0.2883 | 546.6 | 374.7 | 171.9 | 16,300 | 15,800 | 15,300 | 15,100 |
| Lark | 397,500 | 0.3121 | 0.3849 | 575 | 30 × 0.1151 | 7 × 0.1151 | 0.806 | 0.3453 | 622.7 | 375.8 | 246.9 | 20,300 | 19,600 | 18,900 | 18,600 |
| Pelican | 477,000 | 0.3747 | 0.3955 | 625 | 18 × 0.1628 | 1 × 0.1628 | 0.814 | 0.1628 | 518.0 | 447.8 | 70.2 | 11,800 | 11,600 | 11,500 | 11,100 |
| Flicker | 477,000 | 0.3747 | 0.4233 | 635 | 24 × 0.1410 | 1 × 0.1628 | 0.846 | 0.2820 | 614.6 | 450.1 | 164.5 | 17,200 | 16,700 | 16,200 | 16,000 |
| Hawk | 477,000 | 0.3744 | 0.4354 | 640 | 26 × 0.1354 | 7 × 0.1053 | 0.858 | 0.3159 | 656.0 | 449.6 | 206.4 | 19,500 | 18,900 | 18,400 | 18,100 |
| Hen | 477,000 | 0.3747 | 0.4621 | 645 | 30 × 0.1261 | 7 × 0.1261 | 0.883 | 0.3783 | 747.4 | 451.1 | 296.3 | 23,800 | 23,000 | 22,100 | 21,300 |
| Osprey | 556,500 | 0.4369 | 0.4612 | 690 | 18 × 0.1758 | 1 × 0.1758 | 0.879 | 0.1758 | 604.1 | 522.2 | 81.9 | 13,700 | 13,500 | 13,400 | 12,900 |
| Parakeet | 556,500 | 0.4372 | 0.4938 | 700 | 24 × 0.1523 | 7 × 0.1015 | 0.914 | 0.3045 | 716.9 | 525.1 | 191.8 | 19,800 | 19,300 | 18,700 | 18,500 |
| Dove | 556,500 | 0.4371 | 0.5083 | 710 | 26 × 0.1463 | 7 × 0.1138 | 0.927 | 0.3414 | 766.0 | 524.9 | 241.1 | 22,600 | 21,900 | 21,200 | 20,900 |
| Eagle | 556,500 | 0.4371 | 0.5391 | 710 | 30 × 0.1362 | 7 × 0.1362 | 0.953 | 0.4086 | 871.9 | 526.2 | 345.7 | 27,800 | 26,800 | 25,800 | 24,800 |
| Peacock | 605,000 | 0.4753 | 0.5370 | 740 | 24 × 0.1588 | 7 × 0.1059 | 0.953 | 0.318 | 779.7 | 570.9 | 208.8 | 21,600 | 21,000 | 20,400 | 20,100 |
| Squab | 605,000 | 0.4749 | 0.5522 | 745 | 26 × 0.1525 | 7 × 0.1186 | 0.966 | 0.356 | 832.3 | 570.4 | 261.9 | 24,300 | 23,600 | 22,800 | 22,500 |
| Teal | 605,000 | 0.4751 | 0.5834 | 750 | 30 × 0.1420 | 19 × 0.0852 | 0.994 | 0.426 | 939.5 | 572.0 | 367.5 | 30,000 | 29,000 | 28,000 | 28,000 |
| Swift | 636,000 | 0.4994 | 0.5133 | 745 | 36 × 0.1329 | 1 × 0.1329 | 0.930 | 0.1329 | 643.7 | 596.9 | 46.8 | 13,800 | 13,600 | 13,500 | 13,400 |
| Kingbird | 636,000 | 0.4997 | 0.5275 | 750 | 18 × 0.1880 | 1 × 0.1880 | 0.940 | 0.1880 | 690.8 | 597.2 | 93.6 | 15,700 | 15,400 | 15,300 | 14,800 |
| Rook | 636,000 | 0.4996 | 0.5643 | 765 | 24 × 0.1628 | 7 × 0.1085 | 0.977 | 0.326 | 819.2 | 600.0 | 219.2 | 22,600 | 22,000 | 21,400 | 21,100 |
| Grosbeak | 636,000 | 0.4995 | 0.5808 | 775 | 26 × 0.1564 | 7 × 0.1216 | 0.990 | 0.365 | 875.2 | 599.9 | 275.3 | 25,200 | 24,400 | 23,600 | 22,900 |

**TABLE 4-23** ACSR Conductor—Physical Characteristics (*Continued*)

| | ACSR | | | | | | | | | | | | Rated strength, lb | | | |
|---|---|---|---|---|---|---|---|---|---|---|---|---|---|---|---|---|
| | Cross section | | | | Stranding No. and diam of strand, in | | Diameter, in | | Nominal weight, lb† | | | Zinc-coated core | | | |
| | Aluminum | | Total | Current-carrying capacity,* | | | | | Per 1000 ft | | | | | | |
| Code word | cmils or AWG | $in^2$ | $in^2$ | A | Aluminum | Steel | Complete cond. | Steel core | Total | Al | Steel | Standard weight coating | Class B coating | Class C coating | Aluminum-coated core |
| Egret | 636,000 | 0.4995 | 0.6135 | 775 | 30 × 0.1456 | 19 × 0.0874 | 1.019 | 0.437 | 988.2 | 601.4 | 386.8 | 31,500 | 30,500 | 29,400 | 29,400 |
| | 653,900 | 0.5136 | 0.5321 | 760 | 18 × 0.1906 | 3 × 0.0885 | 0.953 | 0.1906 | 676.2 | 613.8 | 62.4 | 14,800 | 14,700 | 14,500 | 14,500 |
| Flamingo | 666,600 | 0.5238 | 0.5917 | 790 | 24 × 0.1667 | 7 × 0.1111 | 1.000 | 0.333 | 858.9 | 629.1 | 229.8 | 23,700 | 23,100 | 22,400 | 22,100 |
| Gannet | 666,600 | 0.5234 | 0.6086 | 795 | 26 × 0.1601 | 7 × 0.1245 | 1.014 | 0.373 | 917.3 | 628.7 | 288.6 | 26,400 | 25,600 | 24,800 | 24,000 |
| Starling | 715,500 | 0.5620 | 0.6535 | 835 | 26 × 0.1659 | 7 × 0.1290 | 1.051 | 0.387 | 984.8 | 675.0 | 309.8 | 28,400 | 27,500 | 26,600 | 25,700 |
| Redwing | 715,500 | 0.5617 | 0.6896 | 840 | 30 × 0.1544 | 19 × 0.0926 | 1.081 | 0.463 | 1110 | 676 | 434 | 34,600 | 33,400 | 32,700 | 31,600 |
| Coot | 795,000 | 0.6243 | 0.6416 | 860 | 36 × 0.1486 | 1 × 0.1486 | 1.040 | 0.1486 | 804.7 | 746.2 | 58.5 | 16,800 | 16,600 | 16,500 | 16,300 |
| Tern | 795,000 | 0.6242 | 0.6674 | 875 | 45 × 0.1329 | 7 × 0.0886 | 1.063 | 0.266 | 895.8 | 749.7 | 146.1 | 22,100 | 21,700 | 21,200 | 21,200 |
| Cuckoo | 795,000 | 0.6244 | 0.7053 | 885 | 24 × 0.1820 | 7 × 0.1213 | 1.092 | 0.364 | 1024 | 750 | 274 | 27,900 | 27,100 | 26,400 | 25,600 |
| Condor | 795,000 | 0.6240 | 0.7049 | 885 | 54 × 0.1213 | 7 × 0.1213 | 1.093 | 0.364 | 1024 | 750 | 274 | 28,200 | 27,400 | 26,600 | 25,800 |
| Drake | 795,000 | 0.6247 | 0.7264 | 890 | 26 × 0.1749 | 7 × 0.1360 | 1.108 | 0.408 | 1094 | 750 | 344 | 31,500 | 30,500 | 29,600 | 28,600 |
| Mallard | 795,000 | 0.6245 | 0.7669 | 900 | 30 × 0.1628 | 19 × 0.0977 | 1.140 | 0.489 | 1235 | 752 | 483 | 38,400 | 37,100 | 35,800 | 35,100 |
| Ruddy | 900,000 | 0.7066 | 0.7555 | 945 | 45 × 0.1414 | 7 × 0.0943 | 1.131 | 0.283 | 1015 | 849 | 166 | 24,400 | 24,000 | 23,500 | 23,300 |
| Canary | 900,000 | 0.7068 | 0.7984 | 955 | 54 × 0.1291 | 7 × 0.1291 | 1.162 | 0.387 | 1159 | 849 | 310 | 31,900 | 31,000 | 30,200 | 29,300 |

*Ampacity for conductor temperature rise of 40°C over 40°C ambient with a 2 ft/s crosswind and an emissivity factor of 0.5 without sun.

†Nominal conductor weights are based on ASTM standard stranding increments. Actual weights will vary within standard tolerances for wire diameters and lay lengths. Invoicing will be based on actual weights.

NOTE: 1 $in^2$ = 64.5 $cm^2$; 1 in = 2.54 cm; 1 lb = 0.4536 kg; 1 ft = 0.3048 m.

**TABLE 4-24** Aluminum-Clad Steel Wire and Cable—Weight, Breaking Strength, DC Resistance

(Based on ASTM Specifications B415 and B416)

| Conductor stranding | | Conductor diam., in | Conductor area | | Conductor weight, lb | | Breaking strength, min., lb | D-c resistance at 20°C (68°F) Ω/1,000 ft |
|---|---|---|---|---|---|---|---|---|
| No. of wires | Wire size, AWG | | Cmils | In² | Per 1,000 ft | Per mi | | |
| **Solid (B415)** | | | | | | | | |
| 1 | 4 | 0.2043 | 41,740 | 0.03278 | 93.63 | 494.3 | 5,081 | 1.222 |
| 1 | 5 | 0.1819 | 33,100 | 0.02600 | 74.25 | 392.0 | 4,290 | 1.541 |
| 1 | 6 | 0.1620 | 26,250 | 0.02062 | 58.88 | 310.9 | 3,608 | 1.943 |
| 1 | 7 | 0.1443 | 20,820 | 0.01635 | 46.69 | 246.6 | 3,025 | 2.450 |
| 1 | 8 | 0.1285 | 16,510 | 0.01297 | 37.03 | 195.6 | 2,529 | 3.089 |
| 1 | 9 | 0.1144 | 13,090 | 0.01028 | 29.37 | 155.1 | 2,005 | 3.896 |
| 1 | 10 | 0.1019 | 10,380 | 0.008155 | 23.29 | 123.0 | 1,590 | 4.912 |
| 1 | 11 | 0.09074 | 8,234 | 0.006467 | 18.47 | 97.52 | 1,261 | 6.194 |
| 1 | 12 | 0.08081 | 6,530 | 0.005129 | 14.65 | 77.33 | 1,000 | 7.811 |
| **Stranded (B416)** | | | | | | | | |
| 19 | 5 | 0.910 | 628,900 | 0.4940 | 1,430 | 7,552 | 73,350 | 0.08224 |
| 19 | 6 | 0.810 | 498,800 | 0.3917 | 1,134 | 5,990 | 61,700 | 0.1037 |
| 19 | 7 | 0.721 | 395,500 | 0.3107 | 899.5 | 4,750 | 51,730 | 0.1308 |
| 19 | 8 | 0.642 | 313,700 | 0.2464 | 713.5 | 3,767 | 43,240 | 0.1649 |
| 19 | 9 | 0.572 | 248,800 | 0.1954 | 565.8 | 2,987 | 34,290 | 0.2079 |
| 19 | 10 | 0.509 | 197,300 | 0.1549 | 448.7 | 2,369 | 27,190 | 0.2622 |
| 7 | 5 | 0.546 | 231,700 | 0.1820 | 524.9 | 2,772 | 27,030 | 0.2264 |
| 7 | 6 | 0.486 | 183,800 | 0.1443 | 416.3 | 2,198 | 22,730 | 0.2803 |
| 7 | 7 | 0.433 | 145,700 | 0.1145 | 330.0 | 1,743 | 19,060 | 0.3535 |
| 7 | 8 | 0.385 | 115,600 | 0.09077 | 261.8 | 1,382 | 15,930 | 0.4458 |
| 7 | 9 | 0.343 | 91,650 | 0.07198 | 207.6 | 1,096 | 12,630 | 0.5621 |
| 7 | 10 | 0.306 | 72,680 | 0.05708 | 164.7 | 869.4 | 10,020 | 0.7088 |
| 7 | 11 | 0.272 | 57,640 | 0.04527 | 130.6 | 689.4 | 7,945 | 0.8938 |
| 7 | 12 | 0.242 | 45,710 | 0.03590 | 103.6 | 546.8 | 6,301 | 1.127 |
| 3 | 5 | 0.392 | 99,310 | 0.07800 | 224.5 | 1,186.0 | 12,230 | 0.5177 |
| 3 | 6 | 0.349 | 78,750 | 0.06185 | 178.1 | 940.2 | 10,280 | 0.6528 |
| 3 | 7 | 0.311 | 62,450 | 0.04905 | 141.2 | 745.6 | 8,621 | 0.8232 |
| 3 | 8 | 0.277 | 49,530 | 0.03890 | 112.0 | 591.3 | 7,206 | 1.038 |
| 3 | 9 | 0.247 | 39,280 | 0.03085 | 88.81 | 468.9 | 5,715 | 1.309 |
| 3 | 10 | 0.220 | 31,150 | 0.02446 | 70.43 | 371.8 | 4,532 | 1.651 |

NOTE: 1 in = 2.54 cm; 1 in² = 64.5 cm²; 1 lb = 0.4536 kg; 1 ft = 0.3048 m; 1 mi = 1.61 km.

**TABLE 4-25** Galvanized-Steel Wire—Weight, Breaking Strength, DC Resistance

(ASTM Specifications A111 and A326)

| Conductor size, BWG | Conductor diam., in | Conductor area, in² | Weight at 20°C* (68°F), lb/mi | Breaking strength, min., lb | | | | | D-c resistance at 20°C (68°F), max., Ω/mi | | | | |
|---|---|---|---|---|---|---|---|---|---|---|---|---|---|
| | | | | Grade EBB† | Grade BB† | Grade 85 | Grade 135 | Grade 195 | Grade EBB | Grade BB | Grade 85 | Grade 135 | Grade 195 |
| 4 | 0.238 | 0.04449 | 797 | 2,028 | 2,270 | ..... | ..... | ..... | 6.27 | 7.02 | | | |
| 6 | 0.203 | 0.03237 | 580 | 1,475 | 1,650 | ..... | ..... | ..... | 8.62 | 9.65 | | | |
| 8 | 0.165 | 0.02138 | 383 | 975 | 1,090 | ..... | ..... | ..... | 13.0 | 14.6 | | | |
| 9 | 0.148 | 0.01720 | 308 | 785 | 880 | 1,462 | ..... | ..... | 16.2 | 18.2 | 18.8 | | |
| 10 | 0.134 | 0.01410 | 253 | 645 | 720 | 1,199 | ..... | ..... | 19.8 | 22.2 | 22.9 | | |
| 11 | 0.120 | 0.01131 | 203 | 515 | 575 | ..... | ..... | ..... | 24.7 | 27.6 | | | |
| 12 | 0.109 | 0.009331 | 167 | 425 | 475 | 793 | 1,213 | 1,800 | 29.9 | 33.5 | 34.7 | 38.9 | 38.9 |
| 14 | 0.083 | 0.005411 | 97.0 | 247 | 275 | 460 | ..... | ..... | 51.6 | 57.7 | 59.8 | | |

*Density = 7.83 g per cu cm at 20°C.

†ASTM designation: Extra Best Best (EBB), Best Best (BB).

NOTE: 1 in = 2.54 cm; 1 in² = 64.5 cm²; 1 lb = 0.4536 kg; 1 mi = 1.61 km.

**TABLE 4-26** Galvanized-Steel Strand—Dimensions, Weight, Breaking Strength (ASTM Specifications A363, A475)

| Strand diameter, in. | | Stranding | | Strand area, sq in. | Strand weight, lb per 1000 ft | Breaking strength, minimum, lb | | | | | | | |
|---|---|---|---|---|---|---|---|---|---|---|---|---|---|
| | | | | | | Utilities grade* | | | | | | | |
| Nominal | Actual | No. of wires | Diameter of coated wires, in. | | | 1 | 2 | 3 | 4 | Common | Siemens-Martin | High strength | Extra-high strength |
| 1¼ | 1.253 | 37 | 0.179 | 0.9311 | 3248 | .... | .... | .... | ...... | 44,600 | 73,000 | 113,600 | 162,200 |
| 1⅛ | 1.127 | 37 | 0.161 | 0.7533 | 2691 | .... | .... | .... | ...... | 36,000 | 58,900 | 91,600 | 130,800 |
| 1 | 1.001 | 37 | 0.143 | 0.5942 | 2057 | .... | .... | .... | ...... | 28,300 | 46,200 | 71,900 | 102,700 |
| 1 | 1.000 | 19 | 0.200 | 0.5969 | 2073 | .... | .... | .... | ...... | 28,700 | 47,000 | 73,200 | 104,500 |
| ⅞ | 0.885 | 19 | 0.177 | 0.4675 | 1581 | .... | .... | .... | ...... | 21,900 | 35,900 | 55,800 | 79,700 |
| ¾ | 0.750 | 19 | 0.150 | 0.3358 | 1155 | .... | .... | .... | ...... | 16,000 | 26,200 | 40,800 | 58,300 |
| ⅝ | 0.625 | 19 | 0.125 | 0.2332 | 796 | .... | .... | .... | ...... | 11,000 | 18,100 | 28,100 | 40,200 |
| ⅝ | 0.621 | 7 | 0.207 | 0.2356 | 813 | .... | .... | .... | ...... | 11,600 | 19,100 | 29,600 | 42,400 |
| 9/16 | 0.565 | 19 | 0.113 | 0.1905 | 637 | .... | .... | .... | ...... | 9,640 | 16,100 | 24,100 | 33,700 |
| 9/16 | 0.564 | 7 | 0.188 | 0.1943 | 671 | .... | .... | .... | ...... | 9,600 | 15,700 | 24,500 | 35,000 |
| ½ | 0.500 | 19 | 0.100 | 0.1492 | 504 | .... | .... | .... | ...... | 7,620 | 12,700 | 19,100 | 26,700 |
| ½ | 0.495 | 7 | 0.165 | 0.1497 | 517 | .... | .... | .... | **25,000** | **7,400** | **12,100** | **18,800** | 26,900 |
| 7/16 | 0.435 | 7 | 0.145 | 0.1156 | 399 | .... | .... | .... | **18,000** | **5,700** | **9,350** | **14,500** | 20,800 |
| ⅜ | 0.360 | 7 | 0.120 | 0.07917 | 273 | .... | .... | .... | **11,500** | **4,250** | **6,950** | 10,800 | 15,400 |
| ⅜ | 0.356 | 3 | 0.165 | 0.06415 | 220.3 | .... | .... | **8500** | | | | | |
| 5/16 | 0.327 | 7 | 0.109 | 0.06532 | 225 | **6000** | | | | | | | |
| 5/16 | 0.312 | 7 | 0.104 | 0.05946 | 205 | .... | .... | .... | ...... | **3,200** | 5,350 | **8,000** | 11,200 |
| 5/16 | 0.312 | 3 | 0.145 | 0.04954 | 170.6 | .... | .... | **6500** | | | | | |
| 9/32 | 0.279 | 7 | 0.093 | 0.04755 | 164 | **4600** | .... | .... | ...... | 2,570 | 4,250 | 6,400 | 8,950 |
| ¼ | 0.240 | 7 | 0.080 | 0.03519 | 121 | .... | .... | .... | ...... | **1,900** | **3,150** | 4,750 | 6,650 |
| ¼ | 0.259 | 3 | 0.120 | 0.03393 | 116.7 | .... | **3150** | **4500** | | | | | |
| 7/32 | 0.216 | 7 | 0.072 | 0.02850 | 98.3 | .... | .... | .... | ...... | 1,540 | 2,560 | 3,850 | 5,400 |
| 3/16 | 0.195 | 7 | 0.065 | 0.02323 | 80.3 | **2400** | | | | | | | |
| 3/16 | 0.186 | 7 | 0.062 | 0.02113 | 72.9 | .... | .... | .... | ...... | **1,150** | 1,900 | 2,850 | 3,990 |
| 5/32 | 0.156 | 7 | 0.052 | 0.01487 | 51.3 | .... | .... | .... | ...... | **870** | 1,470 | 2,140 | 2,940 |
| ⅛ | 0.123 | 7 | 0.041 | 0.00924 | 31.8 | .... | .... | .... | ...... | **540** | 910 | 1,330 | 1,830 |
| **Elongation** in 24 in.: | | | | | | % | | | | | | | |
| .... | ..... | .. | .... | ...... | ...... | 10 | 8 | 5 | 4 | 10 | 8 | 5 | 4 |

*Used principally by communication and power and light industries.

NOTE: Sizes and grades in bold-faced type are those most commonly used and readily available. 1 in = 2.54 cm; 1 $in^2$ = 64.5 $cm^2$; 1 lb = 0.4536 kg; 1 ft = 0.3048 m.

**TABLE 4-27** Copper Wire and Cable—Electrical Characteristics

(Compiled from tables published by Westinghouse Electric Corp., "Electrical Transmission and Distribution Reference Book")

| Conductor size, Awg or Mcm | Stranding: No. of wires | Stranding: Wire diameter, in | Conductor diameter, in | Breaking strength, lb | Conductor weight, lb per mile | Geometric mean radius at 60 cycles, ft | Resistance at 25°C (77°F)* D-c | Resistance at 25°C (77°F)* 25 cycles | Resistance at 25°C (77°F)* 50 cycles | Resistance at 25°C (77°F)* 60 cycles | Resistance at 50°C (122°F)* D-c | Resistance at 50°C (122°F)* 25 cycles | Resistance at 50°C (122°F)* 50 cycles | Resistance at 50°C (122°F)* 60 cycles | Inductive reactance (series) at 1 ft spacing ($x_a$)† 25 cycles | Inductive reactance 50 cycles | Inductive reactance 60 cycles | Capacitive reactance (shunt) at 1 ft spacing ($x_a'$)‡ 25 cycles | Capacitive reactance 50 cycles | Capacitive reactance 60 cycles | Current-carrying capacity at 60 cycles§ (approx), amp |
|---|---|---|---|---|---|---|---|---|---|---|---|---|---|---|---|---|---|---|---|---|---|
| | | | | | | | Ohms per conductor per mile | | | | | | | | Ohms per conductor per mile | | | Megohms per conductor per mile | | | |
| **Solid conductors:** | | | | | | | | | | | | | | | | | | | | | |
| 2 | 1 | ..... | 0.258 | 3,003 | 1,061 | 0.00836 | 0.864 | 0.864 | 0.864 | 0.864 | 0.945 | 0.945 | 0.945 | 0.945 | 0.242 | 0.484 | 0.581 | 0.323 | 0.1614 | 0.1345 | 220 |
| 3 | 1 | ..... | 0.229 | 2,439 | 841 | 0.00745 | 1.090 | 1.090 | 1.090 | 1.090 | 1.192 | 1.192 | 1.192 | 1.192 | 0.248 | 0.496 | 0.595 | 0.331 | 0.1656 | 0.1380 | 190 |
| 4 | 1 | ..... | 0.204 | 1,970 | 667 | 0.00663 | 1.374 | 1.374 | 1.374 | 1.374 | 1.503 | 1.503 | 1.503 | 1.503 | 0.254 | 0.507 | 0.609 | 0.339 | 0.1697 | 0.1415 | 170 |
| 5 | 1 | ..... | 0.1819 | 1,591 | 529 | 0.00590 | 1.733 | 1.733 | 1.733 | 1.733 | 1.895 | 1.895 | 1.895 | 1.895 | 0.260 | 0.519 | 0.623 | 0.348 | 0.1738 | 0.1449 | 140 |
| 6 | 1 | ..... | 0.1620 | 1,280 | 420 | 0.00526 | 2.18 | 2.18 | 2.18 | 2.18 | 2.39 | 2.39 | 2.39 | 2.39 | 0.265 | 0.531 | 0.637 | 0.356 | 0.1779 | 0.1483 | 120 |
| 7 | 1 | ..... | 0.1443 | 1,030 | 333 | 0.00468 | 2.75 | 2.75 | 2.75 | 2.75 | 3.01 | 3.01 | 3.01 | 3.01 | 0.271 | 0.542 | 0.651 | 0.364 | 0.1821 | 0.1517 | 110 |
| 8 | 1 | ..... | 0.1285 | 826 | 264 | 0.00417 | 3.47 | 3.47 | 3.47 | 3.47 | 3.80 | 3.80 | 3.80 | 3.80 | 0.277 | 0.554 | 0.665 | 0.372 | 0.1862 | 0.1552 | 90 |
| **Stranded conductors:** | | | | | | | | | | | | | | | | | | | | | |
| 1000 | 37 | 0.1644 | 1.151 | 43,830 | 16,300 | 0.0368 | 0.0585 | 0.0594 | 0.0620 | 0.0634 | 0.0640 | 0.0648 | 0.0672 | 0.0685 | 0.1666 | 0.333 | 0.400 | 0.216 | 0.1081 | 0.0901 | 1300 |
| 900 | 37 | 0.1560 | 1.092 | 39,510 | 14,670 | 0.0349 | 0.0650 | 0.0658 | 0.0682 | 0.0695 | 0.0711 | 0.0718 | 0.0740 | 0.0752 | 0.1693 | 0.339 | 0.406 | 0.220 | 0.1100 | 0.0916 | 1220 |
| 800 | 37 | 0.1470 | 1.029 | 35,120 | 13,040 | 0.0329 | 0.0731 | 0.0739 | 0.0760 | 0.0772 | 0.0800 | 0.0806 | 0.0826 | 0.0837 | 0.1722 | 0.344 | 0.413 | 0.224 | 0.1121 | 0.0934 | 1130 |
| 750 | 37 | 0.1424 | 0.997 | 33,400 | 12,230 | 0.0319 | 0.0780 | 0.0787 | 0.0807 | 0.0818 | 0.0853 | 0.0859 | 0.0878 | 0.0888 | 0.1739 | 0.348 | 0.417 | 0.226 | 0.1132 | 0.0943 | 1090 |
| 700 | 37 | 0.1375 | 0.963 | 31,170 | 11,410 | 0.0308 | 0.0836 | 0.0842 | 0.0861 | 0.0871 | 0.0914 | 0.0920 | 0.0937 | 0.0947 | 0.1759 | 0.352 | 0.422 | 0.229 | 0.1145 | 0.0954 | 1040 |
| 600 | 37 | 0.1273 | 0.891 | 27,020 | 9,781 | 0.0285 | 0.0975 | 0.0981 | 0.0997 | 0.1006 | 0.1066 | 0.1071 | 0.1086 | 0.1095 | 0.1799 | 0.360 | 0.432 | 0.235 | 0.1173 | 0.0977 | 940 |
| 500 | 37 | 0.1162 | 0.814 | 22,510 | 8,151 | 0.0260 | 0.1170 | 0.1175 | 0.1188 | 0.1196 | 0.1280 | 0.1283 | 0.1296 | 0.1303 | 0.1845 | 0.369 | 0.443 | 0.241 | 0.1205 | 0.1004 | 840 |
| 500 | 19 | 0.1622 | 0.811 | 21,590 | 8,151 | 0.0256 | 0.1170 | 0.1175 | 0.1188 | 0.1196 | 0.1280 | 0.1283 | 0.1296 | 0.1303 | 0.1853 | 0.371 | 0.445 | 0.241 | 0.1206 | 0.1005 | 840 |
| 450 | 19 | 0.1539 | 0.770 | 19,750 | 7,336 | 0.0243 | 0.1300 | 0.1304 | 0.1316 | 0.1323 | 0.1422 | 0.1426 | 0.1437 | 0.1443 | 0.1879 | 0.376 | 0.451 | 0.245 | 0.1224 | 0.1020 | 780 |
| 400 | 19 | 0.1451 | 0.726 | 17,560 | 6,521 | 0.0229 | 0.1462 | 0.1466 | 0.1477 | 0.1484 | 0.1600 | 0.1603 | 0.1613 | 0.1619 | 0.1909 | 0.382 | 0.458 | 0.249 | 0.1245 | 0.1038 | 730 |
| 350 | 19 | 0.1357 | 0.679 | 15,590 | 5,706 | 0.0214 | 0.1671 | 0.1675 | 0.1684 | 0.1690 | 0.1828 | 0.1831 | 0.1840 | 0.1845 | 0.1943 | 0.389 | 0.466 | 0.254 | 0.1269 | 0.1058 | 670 |
| 350 | 12 | 0.1708 | 0.710 | 15,140 | 5,706 | 0.0225 | 0.1671 | 0.1675 | 0.1684 | 0.1690 | 0.1828 | 0.1831 | 0.1840 | 0.1845 | 0.1918 | 0.384 | 0.460 | 0.251 | 0.1253 | 0.1044 | 670 |
| 300 | 19 | 0.1257 | 0.629 | 13,510 | 4,891 | 0.01987 | 0.1950 | 0.1953 | 0.1961 | 0.1966 | 0.213 | 0.214 | 0.214 | 0.215 | 0.1982 | 0.396 | 0.476 | 0.259 | 0.1296 | 0.1080 | 610 |
| 300 | 12 | 0.1581 | 0.657 | 13,170 | 4,891 | 0.0208 | 0.1950 | 0.1953 | 0.1961 | 0.1966 | 0.213 | 0.214 | 0.214 | 0.215 | 0.1957 | 0.392 | 0.470 | 0.256 | 0.1281 | 0.1068 | 610 |
| 250 | 19 | 0.1147 | 0.574 | 11,360 | 4,076 | 0.01813 | 0.234 | 0.234 | 0.235 | 0.235 | 0.256 | 0.256 | 0.257 | 0.257 | 0.203 | 0.406 | 0.487 | 0.266 | 0.1329 | 0.1108 | 540 |

**TABLE 4-27** Copper Wire and Cable—Electrical Characteristics (*Continued*)

(Compiled from tables published by Westinghouse Electric Corp., "Electrical Transmission and Distribution Reference Book")

| Conductor size, Awg or Mcm | Stranding: No. of wires | Stranding: Wire diameter, in | Conductor diameter, in | Breaking strength, lb | Conductor weight, lb per mile | Geometric mean radius at 60 cycles, ft | Resistance at 25°C (77°F)*: D-c | 25 cycles | 50 cycles | 60 cycles | Resistance at 50°C (122°F)*: D-c | 25 cycles | 50 cycles | 60 cycles | Inductive reactance (series) at 1 ft spacing ($x_a$)†: 25 cycles | 50 cycles | 60 cycles | Capacitive reactance (shunt) at 1 ft spacing ($x_a'$)‡: 25 cycles | 50 cycles | 60 cycles | Current-carrying capacity at 60 cycles§ (approx), amp |
|---|---|---|---|---|---|---|---|---|---|---|---|---|---|---|---|---|---|---|---|---|---|
| | | | | | | | Ohms per conductor per mile | | | | | | | | Ohms per conductor per mile | | | Megohms per conductor per mile | | | |
| **Stranded conductors:—(Concluded)** | | | | | | | | | | | | | | | | | | | | | |
| 250 | 12 | 0.1443 | 0.600 | 11,130 | 4,076 | 0.01902 | 0.234 | 0.234 | 0.235 | 0.235 | 0.256 | 0.256 | 0.257 | 0.257 | 0.200 | 0.401 | 0.481 | 0.263 | 0.1313 | 0.1094 | 540 |
| 4/0 | 19 | 0.1055 | 0.528 | 9,617 | 3,450 | 0.01668 | 0.276 | 0.277 | 0.277 | 0.278 | 0.302 | 0.303 | 0.303 | 0.303 | 0.207 | 0.414 | 0.497 | 0.272 | 0.1359 | 0.1132 | 480 |
| 4/0 | 12 | 0.1328 | 0.552 | 9,483 | 3,450 | 0.01750 | 0.276 | 0.277 | 0.277 | 0.278 | 0.302 | 0.303 | 0.303 | 0.303 | 0.205 | 0.409 | 0.491 | 0.269 | 0.1343 | 0.1119 | 490 |
| 4/0 | 7 | 0.1739 | 0.522 | 9,154 | 3,450 | 0.01579 | 0.276 | 0.277 | 0.277 | 0.278 | 0.302 | 0.303 | 0.303 | 0.303 | 0.210 | 0.420 | 0.503 | 0.273 | 0.1363 | 0.1136 | 480 |
| 3/0 | 12 | 0.1183 | 0.492 | 7,556 | 2,736 | 0.01559 | 0.349 | 0.349 | 0.349 | 0.350 | 0.381 | 0.381 | 0.382 | 0.382 | 0.210 | 0.421 | 0.505 | 0.277 | 0.1384 | 0.1153 | 420 |
| 3/0 | 7 | 0.1548 | 0.464 | 7,366 | 2,736 | 0.01404 | 0.349 | 0.349 | 0.349 | 0.350 | 0.381 | 0.381 | 0.382 | 0.382 | 0.216 | 0.431 | 0.518 | 0.281 | 0.1405 | 0.1171 | 420 |
| 2/0 | 7 | 0.1379 | 0.414 | 5,926 | 2,170 | 0.01252 | 0.440 | 0.440 | 0.440 | 0.440 | 0.481 | 0.481 | 0.481 | 0.481 | 0.222 | 0.443 | 0.532 | 0.289 | 0.1445 | 0.1205 | 360 |
| 1/0 | 7 | 0.1228 | 0.368 | 4,752 | 1,720 | 0.01113 | 0.555 | 0.555 | 0.555 | 0.555 | 0.606 | 0.607 | 0.607 | 0.607 | 0.227 | 0.455 | 0.546 | 0.298 | 0.1488 | 0.1240 | 310 |
| 1 | 7 | 0.1093 | 0.328 | 3,804 | 1,364 | 0.00992 | 0.699 | 0.699 | 0.699 | 0.699 | 0.765 | 0.765 | 0.765 | 0.765 | 0.233 | 0.467 | 0.560 | 0.306 | 0.1528 | 0.1274 | 270 |
| 1 | 3 | 0.1670 | 0.360 | 3,620 | 1,351 | 0.01016 | 0.692 | 0.692 | 0.692 | 0.692 | 0.757 | 0.757 | 0.757 | 0.757 | 0.232 | 0.464 | 0.557 | 0.299 | 0.1495 | 0.1246 | 270 |
| 2 | 7 | 0.0974 | 0.292 | 3,045 | 1,082 | 0.00883 | 0.881 | 0.882 | 0.882 | 0.882 | 0.964 | 0.964 | 0.964 | 0.964 | 0.239 | 0.478 | 0.574 | 0.314 | 0.1570 | 0.1308 | 230 |
| 2 | 3 | 0.1487 | 0.320 | 2,913 | 1,071 | 0.00903 | 0.873 | 0.873 | 0.873 | 0.873 | 0.955 | 0.955 | 0.955 | 0.955 | 0.238 | 0.476 | 0.571 | 0.307 | 0.1537 | 0.1281 | 240 |
| 3 | 7 | 0.0867 | 0.260 | 2,433 | 858 | 0.00787 | 1.112 | 1.112 | 1.112 | 1.112 | 1.216 | 1.216 | 1.216 | 1.216 | 0.245 | 0.490 | 0.588 | 0.322 | 0.1611 | 0.1343 | 200 |
| 3 | 3 | 0.1325 | 0.285 | 2,359 | 850 | 0.00805 | 1.101 | 1.101 | 1.101 | 1.101 | 1.204 | 1.204 | 1.204 | 1.204 | 0.244 | 0.488 | 0.585 | 0.316 | 0.1578 | 0.1315 | 200 |
| 4 | 3 | 0.1180 | 0.254 | 1,879 | 674 | 0.00717 | 1.388 | 1.388 | 1.388 | 1.388 | 1.518 | 1.518 | 1.518 | 1.518 | 0.250 | 0.499 | 0.599 | 0.324 | 0.1619 | 0.1349 | 180 |
| 5 | 3 | 0.1050 | 0.226 | 1,505 | 534 | 0.00638 | 1.750 | 1.750 | 1.750 | 1.750 | 1.914 | 1.914 | 1.914 | 1.914 | 0.256 | 0.511 | 0.613 | 0.332 | 0.1661 | 0.1384 | 150 |
| 6 | 3 | 0.0935 | 0.201 | 1,205 | 424 | 0.00568 | 2.21 | 2.21 | 2.21 | 2.21 | 2.41 | 2.41 | 2.41 | 2.41 | 0.262 | 0.523 | 0.628 | 0.341 | 0.1703 | 0.1419 | 130 |

*Resistance is based on conductivity = 97.3% IACS.

Resistance is increased to allow for stranding: 3-wire conductors = 1%; all others = 2%. For resistance temperature conversion see Par. **24**.

†See Table 4-32.

‡See Table 4-33.

§For conductor at 75°C, air at 25°C, wind 2 ft/s (1.4 mi/h), average tarnished surface.

NOTE: 1 in = 2.54 cm; 1 lb = 0.4536 kg; 1 mi = 1.61 km; 1 ft = 0.3048 m.

**TABLE 4-28** Copper-Clad-Steel Cable—Electrical Characteristics

(Compiled from tables published by Westinghouse Electric Corp., "Electrical Transmission and Distribution Reference Book"; and Copperweld Steel Co.)

| Nominal conductor size, in | Conductor stranding | | Conductor diam., in | Conductor area, cir mils | Breaking strength, rated, lb | | Conductor weight, lb per mile | Geometric mean radius at 60 cycles, average currents, ft | Resistance at 25°C (77°F),* small currents | | | | Inductive reactance (series) at 1 ft spacing, average currents ($x_a$)† | | | Capacitive reactance (shunt) at 1 ft spacing ($x_a'$)‡ | | | Current-carrying capacity at 60 cycles§ (approx), amp |
|---|---|---|---|---|---|---|---|---|---|---|---|---|---|---|---|---|---|---|---|
| | No. of wires | Wire size, AWG | | | High strength | Extra high strength | | | D-c | 25 cycles | 50 cycles | 60 cycles | 25 cycles | 50 cycles | 60 cycles | 25 cycles | 50 cycles | 60 cycles | |
| | | | | | | | | | Ohms per conductor per mile | | | | Ohms per conductor per mile | | | Megohms per conductor per mile | | | |
| **30% conductivity** | | | | | | | | | | | | | | | | | | | |
| 7/8 | 19 | 5 | 0.910 | 628,900 | 56,570 | 66,910 | 9344 | 0.00758 | 0.306 | 0.316 | 0.326 | 0.331 | 0.261 | 0.493 | 0.592 | 0.233 | 0.1165 | 0.0971 | 620 |
| 13/16 | 19 | 6 | 0.810 | 498,800 | 45,830 | 55,530 | 7410 | 0.00675 | 0.386 | 0.396 | 0.406 | 0.411 | 0.267 | 0.505 | 0.606 | 0.241 | 0.1206 | 0.1005 | 540 |
| 23/32 | 19 | 7 | 0.721 | 395,500 | 37,740 | 45,850 | 5877 | 0.00601 | 0.486 | 0.496 | 0.506 | 0.511 | 0.273 | 0.517 | 0.621 | 0.250 | 0.1248 | 0.1040 | 470 |
| 21/32 | 19 | 8 | 0.642 | 313,700 | 31,040 | 37,690 | 4660 | 0.00535 | 0.613 | 0.623 | 0.633 | 0.638 | 0.279 | 0.529 | 0.635 | 0.258 | 0.1289 | 0.1074 | 410 |
| 9/16 | 19 | 9 | 0.572 | 248,800 | 25,500 | 30,610 | 3696 | 0.00477 | 0.773 | 0.783 | 0.793 | 0.798 | 0.285 | 0.541 | 0.649 | 0.266 | 0.1330 | 0.1109 | 360 |
| 5/8 | 7 | 4 | 0.613 | 292,200 | 24,780 | 29,430 | 4324 | 0.00511 | 0.656 | 0.664 | 0.672 | 0.676 | 0.281 | 0.533 | 0.640 | 0.261 | 0.1306 | 0.1088 | 410 |
| 9/16 | 7 | 5 | 0.546 | 231,700 | 20,470 | 24,650 | 3429 | 0.00455 | 0.827 | 0.835 | 0.843 | 0.847 | 0.287 | 0.548 | 0.654 | 0.269 | 0.1347 | 0.1122 | 360 |
| 1/2 | 7 | 6 | 0.486 | 183,800 | 16,890 | 20,460 | 2719 | 0.00405 | 1.043 | 1.050 | 1.058 | 1.062 | 0.293 | 0.557 | 0.668 | 0.278 | 0.1388 | 0.1157 | 310 |
| 7/16 | 7 | 7 | 0.433 | 145,700 | 13,910 | 16,890 | 2157 | 0.00361 | 1.315 | 1.323 | 1.331 | 1.335 | 0.299 | 0.569 | 0.683 | 0.286 | 0.1429 | 0.1191 | 270 |
| 3/8 | 7 | 8 | 0.385 | 115,600 | 11,440 | 13,890 | 1710 | 0.00321 | 1.658 | 1.666 | 1.674 | 1.678 | 0.305 | 0.581 | 0.697 | 0.294 | 0.1471 | 0.1226 | 230 |
| 11/32 | 7 | 9 | 0.343 | 91,650 | 9,393 | 11,280 | 1356 | 0.00286 | 2.09 | 2.10 | 2.11 | 2.11 | 0.311 | 0.592 | 0.711 | 0.303 | 0.1512 | 0.1260 | 200 |
| 5/16 | 7 | 10 | 0.306 | 72,680 | 7,758 | 9,196 | 1076 | 0.00255 | 2.64 | 2.64 | 2.65 | 2.66 | 0.316 | 0.604 | 0.725 | 0.311 | 0.1553 | 0.1294 | 170 |
| .. | 3 | 5 | 0.392 | 99,310 | 9,262 | 11,860 | 1467 | 0.00457 | 1.926 | 1.931 | 1.936 | 1.938 | 0.289 | 0.545 | 0.654 | 0.293 | 0.1465 | 0.1221 | 220 |
| .. | 3 | 6 | 0.349 | 78,750 | 7,639 | 9,754 | 1163 | 0.00407 | 2.43 | 2.43 | 2.44 | 2.44 | 0.295 | 0.556 | 0.668 | 0.301 | 0.1506 | 0.1255 | 190 |
| .. | 3 | 7 | 0.311 | 62,450 | 6,291 | 7,922 | 922.4 | 0.00363 | 3.06 | 3.07 | 3.07 | 3.07 | 0.301 | 0.568 | 0.682 | 0.310 | 0.1547 | 0.1289 | 160 |
| .. | 3 | 8 | 0.277 | 49,530 | 5,174 | 6,282 | 731.5 | 0.00323 | 3.86 | 3.87 | 3.87 | 3.87 | 0.307 | 0.580 | 0.696 | 0.318 | 0.1589 | 0.1324 | 140 |
| .. | 3 | 9 | 0.247 | 39,280 | 4,250 | 5,129 | 580.1 | 0.00288 | 4.87 | 4.87 | 4.88 | 4.88 | 0.313 | 0.591 | 0.710 | 0.326 | 0.1629 | 0.1358 | 120 |
| .. | 3 | 10 | 0.220 | 31,150 | 3,509 | 4,160 | 460.0 | 0.00257 | 6.14 | 6.14 | 6.15 | 6.15 | 0.319 | 0.603 | 0.724 | 0.334 | 0.1671 | 0.1392 | 100 |
| **40% conductivity** | | | | | | | | | | | | | | | | | | | |
| 7/8 | 19 | 5 | 0.910 | 628,900 | 50,240 | ...... | 9344 | 0.01175 | 0.229 | 0.239 | 0.249 | 0.254 | 0.236 | 0.449 | 0.539 | 0.233 | 0.1165 | 0.0971 | 690 |
| 13/16 | 19 | 6 | 0.810 | 498,800 | 41,600 | ...... | 7410 | 0.01046 | 0.289 | 0.299 | 0.309 | 0.314 | 0.241 | 0.461 | 0.553 | 0.241 | 0.1206 | 0.1005 | 610 |
| 23/32 | 19 | 7 | 0.721 | 395,500 | 34,390 | ...... | 5877 | 0.00931 | 0.365 | 0.375 | 0.385 | 0.390 | 0.247 | 0.473 | 0.567 | 0.250 | 0.1248 | 0.1040 | 530 |
| 21/32 | 19 | 8 | 0.642 | 313,700 | 28,380 | ...... | 4660 | 0.00829 | 0.460 | 0.470 | 0.480 | 0.485 | 0.253 | 0.485 | 0.582 | 0.258 | 0.1289 | 0.1074 | 470 |
| 9/16 | 19 | 9 | 0.572 | 248,800 | 23,390 | ...... | 3696 | 0.00739 | 0.580 | 0.590 | 0.600 | 0.605 | 0.259 | 0.496 | 0.595 | 0.266 | 0.1330 | 0.1109 | 410 |

**TABLE 4-28** Copper-Clad-Steel Cable—Electrical Characteristics (*Continued*)

(Compiled from tables published by Westinghouse Electric Corp., "Electrical Transmission and Distribution Reference Book"; and Copperweld Steel Co.)

| Nominal conductor size, in | Conductor stranding | | Con-ductor diam., in | Con-ductor area, cir mils | Breaking strength, rated, lb | | Con-ductor weight, lb per mile | Geometric mean radius at 60 cycles, average currents, ft | Resistance at 25°C (77°F),* small currents — Ohms per conductor per mile | | | | Inductive reactance (series) at 1 ft spacing, average currents ($x_a$)† — Ohms per conductor per mile | | | Capacitive reactance (shunt) at 1 ft spacing ($x_a'$)‡ — Megohms per conductor per mile | | | Current-carrying capacity at 60 cycles§ (approx), amp |
|---|---|---|---|---|---|---|---|---|---|---|---|---|---|---|---|---|---|---|---|
| | No. of wires | Wire size, AWG | | | High strength | Extra high strength | | | D-c | 25 cycles | 50 cycles | 60 cycles | 25 cycles | 50 cycles | 60 cycles | 25 cycles | 50 cycles | 60 cycles | |
| **40% conductivity—(*Concluded*)** | | | | | | | | | | | | | | | | | | | |
| 5/8 | 7 | 4 | 0.613 | 292,200 | 22,310 | ...... | 4324 | 0.00792 | 0.492 | 0.500 | 0.508 | 0.512 | 0.255 | 0.489 | 0.587 | 0.261 | 0.1306 | 0.1088 | 470 |
| 9/16 | 7 | 5 | 0.546 | 231,700 | 18,510 | ...... | 3429 | 0.00705 | 0.620 | 0.628 | 0.636 | 0.640 | 0.261 | 0.501 | 0.601 | 0.269 | 0.1347 | 0.1122 | 410 |
| 1/2 | 7 | 6 | 0.486 | 183,800 | 15,330 | ...... | 2719 | 0.00628 | 0.782 | 0.790 | 0.798 | 0.802 | 0.267 | 0.513 | 0.615 | 0.278 | 0.1388 | 0.1157 | 350 |
| 7/16 | 7 | 7 | 0.433 | 145,700 | 12,670 | ...... | 2157 | 0.00559 | 0.986 | 0.994 | 1.002 | 1.006 | 0.273 | 0.524 | 0.629 | 0.286 | 0.1429 | 0.1191 | 310 |
| 3/8 | 7 | 8 | 0.385 | 115,600 | 10,460 | ...... | 1710 | 0.00497 | 1.244 | 1.252 | 1.260 | 1.264 | 0.279 | 0.536 | 0.644 | 0.294 | 0.1471 | 0.1226 | 270 |
| 11/32 | 7 | 9 | 0.343 | 91,650 | 8,616 | ...... | 1356 | 0.00443 | 1.568 | 1.576 | 1.584 | 1.588 | 0.285 | 0.548 | 0.658 | 0.303 | 0.1512 | 0.1260 | 230 |
| 5/16 | 7 | 10 | 0.306 | 72,680 | 7,121 | ...... | 1076 | 0.00395 | 1.978 | 1.986 | 1.994 | 1.998 | 0.291 | 0.559 | 0.671 | 0.311 | 0.1553 | 0.1294 | 200 |
| .. | 3 | 5 | 0.392 | 99,310 | 8,373 | ...... | 1467 | 0.00621 | 1.445 | 1.450 | 1.455 | 1.457 | 0.269 | 0.514 | 0.617 | 0.293 | 0.1465 | 0.1221 | 250 |
| .. | 3 | 6 | 0.349 | 78,750 | 6,934 | ...... | 1163 | 0.00553 | 1.821 | 1.826 | 1.831 | 1.833 | 0.275 | 0.526 | 0.631 | 0.301 | 0.1506 | 0.1255 | 220 |
| .. | 3 | 7 | 0.311 | 62,450 | 5,732 | ...... | 922.4 | 0.00492 | 2.30 | 2.30 | 2.31 | 2.31 | 0.281 | 0.537 | 0.645 | 0.310 | 0.1547 | 0.1289 | 190 |
| .. | 3 | 8 | 0.277 | 49,530 | 4,730 | ...... | 731.5 | 0.00439 | 2.90 | 2.90 | 2.91 | 2.91 | 0.286 | 0.549 | 0.659 | 0.318 | 0.1589 | 0.1324 | 160 |
| .. | 3 | 9 | 0.247 | 39,280 | 3,898 | ...... | 580.1 | 0.00391 | 3.65 | 3.66 | 3.66 | 3.66 | 0.292 | 0.561 | 0.673 | 0.326 | 0.1629 | 0.1358 | 140 |
| .. | 3 | 10 | 0.220 | 31,150 | 3,221 | ...... | 460.0 | 0.00348 | 4.61 | 4.61 | 4.62 | 4.62 | 0.297 | 0.572 | 0.687 | 0.334 | 0.1671 | 0.1392 | 120 |
| .. | 3 | 12 | 0.174 | 19,590 | 2,236 | ...... | 289.5 | 0.00276 | 7.32 | 7.33 | 7.33 | 7.34 | 0.310 | 0.596 | 0.715 | 0.351 | 0.1754 | 0.1462 | 90 |

*For resistance temperature conversion see Pars. **24** and **27**.

†See Table 4-32.

‡See Table 4-33.

§For conductor at 125°C, air at 25°C, wind 2 ft/s (1.4 mi/h), average tarnished surface.

NOTE: 1 in = 2.54 cm; 1 lb = 0.4536 kg; 1 mi = 1.61 km; 1 ft = 0.3048 m.

**TABLE 4-29** Copper-Clad-Steel Copper Cable—Electrical Characteristics

(Compiled from tables published by Westinghouse Electric Corp., "Electrical Transmission and Distribution Reference Book"; and Copperweld Steel Co.)

| Conductor size | Hard-drawn copper equivalent | Conductor stranding | | | | Conductor diam., in | Breaking strength, rated, lb | Conductor weight, lb/mi | Geometric mean radius at 60 cycles, ft | Resistance at 25°C (77°F),* small currents | | | | Resistance at 50°C (122°F),* current approximately 75% capacity | | | | Inductive reactance (series) at 1-ft spacing ($x_a$)† | | | Capacitive reactance (shunt) at 1-ft spacing ($x_a'$)‡ | | | Current-carrying capacity at 60 cycles§ (approx), amp |
|---|---|---|---|---|---|---|---|---|---|---|---|---|---|---|---|---|---|---|---|---|---|---|---|---|
| Type | | EHS 30% Copper-clad steel wires | | Hard-drawn copper wires | | | | | | D-c | 25 cycles | 50 cycles | 60 cycles | D-c | 25 cycles | 50 cycles | 60 cycles | 25 cycles | 50 cycles | 60 cycles | 25 cycles | 50 cycles | 60 cycles | |
| Mcm or AWG | | No. | Diam., in | No. | Diam., in | | | | | Ω/(conductor) (mi) | | | | | | | | Ω/(conductor) (mi) | | | MΩ/(conductor) (mi) | | | |
| 350 E | 350 | 7 | 0.1576 | 12 | 0.1576 | 0.788 | 32,420 | 7,409 | 0.0220 | 0.1658 | 0.1728 | 0.1789 | 0.1812 | 0.1812 | 0.1915 | 0.201 | 0.204 | 0.1929 | 0.386 | 0.463 | 0.243 | 0.1216 | 0.1014 | 660 |
| 350 EK | 350 | 4 | 0.1470 | 15 | 0.1470 | 0.735 | 23,850 | 6,536 | 0.0245 | 0.1658 | 0.1682 | 0.1700 | 0.1705 | 0.1812 | 0.1845 | 0.1873 | 0.1882 | 0.1875 | 0.375 | 0.450 | 0.248 | 0.1241 | 0.1034 | 680 |
| 300 E | 300 | 7 | 0.1459 | 12 | 0.1459 | 0.729 | 27,770 | 6,351 | 0.0204 | 0.1934 | 0.200 | 0.207 | 0.209 | 0.211 | 0.222 | 0.232 | 0.235 | 0.1969 | 0.394 | 0.473 | 0.249 | 0.1244 | 0.1037 | 600 |
| 300 EK | 300 | 4 | 0.1361 | 15 | 0.1361 | 0.680 | 20,960 | 5,602 | 0.0227 | 0.1934 | 0.1958 | 0.1976 | 0.1981 | 0.211 | 0.215 | 0.218 | 0.219 | 0.1914 | 0.383 | 0.460 | 0.254 | 0.1269 | 0.1057 | 610 |
| 250 E | 250 | 7 | 0.1332 | 12 | 0.1332 | 0.666 | 23,920 | 5,292 | 0.01859 | 0.232 | 0.239 | 0.245 | 0.248 | 0.254 | 0.265 | 0.275 | 0.279 | 0.202 | 0.403 | 0.484 | 0.255 | 0.1276 | 0.1064 | 540 |
| 250 EK | 250 | 4 | 0.1242 | 15 | C.1242 | 0.621 | 17,840 | 4,669 | 0.0207 | 0.232 | 0.235 | 0.236 | 0.237 | 0.254 | 0.258 | 0.261 | 0.261 | 0.1960 | 0.392 | 0.471 | 0.260 | 0.1301 | 0.1084 | 540 |
| 4/0 E | 4/0 | 7 | 0.1225 | 12 | 0.1225 | 0.613 | 20,730 | 4,479 | 0.01711 | 0.274 | 0.281 | 0.287 | 0.290 | 0.300 | 0.312 | 0.323 | 0.326 | 0.206 | 0.411 | 0.493 | 0.261 | 0.1306 | 0.1088 | 480 |
| 4/0 EK | 4/0 | 4 | 0.1143 | 15 | 0.1143 | 0.571 | 15,370 | 3,951 | 0.01903 | 0.274 | 0.277 | 0.278 | 0.279 | 0.300 | 0.304 | 0.307 | 0.308 | 0.200 | 0.401 | 0.481 | 0.266 | 0.1331 | 0.1109 | 490 |
| 4/0 F | 4/0 | 1 | 0.1833 | 6 | 0.1833 | 0.550 | 12,290 | 3,750 | 0.01558 | 0.273 | 0.280 | 0.285 | 0.287 | 0.299 | 0.309 | 0.318 | 0.322 | 0.210 | 0.421 | 0.505 | 0.269 | 0.1344 | 0.1120 | 470 |
| 3/0 E | 3/0 | 7 | 0.1091 | 12 | 0.1091 | 0.545 | 16,800 | 3,552 | 0.01521 | 0.346 | 0.353 | 0.359 | 0.361 | 0.378 | 0.391 | 0.402 | 0.407 | 0.212 | 0.423 | 0.508 | 0.270 | 0.1348 | 0.1123 | 420 |
| 3/0 EK | 3/0 | 4 | 0.1018 | 4 | 0.1018 | 0.509 | 12,370 | 3,134 | 0.01697 | 0.346 | 0.348 | 0.350 | 0.351 | 0.378 | 0.382 | 0.386 | 0.386 | 0.206 | 0.412 | 0.495 | 0.274 | 0.1372 | 0.1143 | 420 |
| 3/0 F | 3/0 | 1 | 0.1632 | 6 | 0.1632 | 0.490 | 9,980 | 2,974 | 0.01388 | 0.344 | 0.351 | 0.356 | 0.358 | 0.377 | 0.388 | 0.397 | 0.401 | 0.216 | 0.432 | 0.519 | 0.277 | 0.1385 | 0.1155 | 410 |
| 2/0 F | 2/0 | 1 | 0.1454 | 6 | 0.1454 | 0.436 | 8,094 | 2,359 | 0.01235 | 0.434 | 0.441 | 0.446 | 0.448 | 0.475 | 0.487 | 0.497 | 0.501 | 0.222 | 0.444 | 0.533 | 0.285 | 0.1427 | 0.1189 | 350 |
| 1/0 F | 1/0 | 1 | 0.1294 | 6 | 0.1294 | 0.388 | 6,536 | 1,870 | 0.01099 | 0.548 | 0.554 | 0.559 | 0.562 | 0.599 | 0.612 | 0.622 | 0.627 | 0.228 | 0.456 | 0.547 | 0.294 | 0.1469 | 0.1224 | 310 |
| 2 A | 2 | 1 | 0.1699 | 2 | 0.1699 | 0.366 | 5,876 | 1,356 | 0.00763 | 0.869 | 0.875 | 0.880 | 0.882 | 0.950 | 0.962 | 0.973 | 0.979 | 0.247 | 0.493 | 0.592 | 0.298 | 0.1489 | 0.1241 | 240 |
| 2 F | 2 | 1 | 0.1026 | 6 | 0.1026 | 0.308 | 4,233 | 1,176 | 0.00873 | 0.871 | 0.878 | 0.884 | 0.885 | 0.952 | 0.967 | 0.979 | 0.985 | 0.230 | 0.479 | 0.575 | 0.310 | 0.1551 | 0.1292 | 230 |
| 4 A | 4 | 1 | 0.1347 | 2 | 0.1347 | 0.290 | 3,938 | 853 | 0.00604 | 1.382 | 1.388 | 1.393 | 1.395 | 1.511 | 1.525 | 1.540 | 1.545 | 0.258 | 0.517 | 0.620 | 0.314 | 0.1572 | 0.1310 | 180 |
| 6 A | 6 | 1 | 0.1068 | 2 | 0.1068 | 0.230 | 2,585 | 536 | 0.00479 | 2.20 | 2.20 | 2.21 | 2.21 | 2.40 | 2.42 | 2.44 | 2.44 | 0.270 | 0.540 | 0.648 | 0.331 | 0.1655 | 0.1379 | 140 |
| 8 A | 8 | 1 | 0.1127 | 2 | 0.0797 | 0.199 | 2,233 | 392 | 0.00394 | 3.49 | 3.50 | 3.51 | 3.51 | 3.82 | 3.84 | 3.86 | 3.87 | 0.280 | 0.560 | 0.672 | 0.341 | 0.1706 | 0.1422 | 100 |

*For resistance temperature conversion see Pars. **24** and **27**.

Resistance at 50°C total temperature, based on ambient of 25°C plus 25°C rise due to heating effect of current. The approximate magnitude of current necessary to produce the 25°C rise is 75% of the "approximate current-carrying capacity at 60 cycles."

†See Table 4-32.

‡See Table 4-33.

§For conductor at 75°C, air at 25°C, wind 2 ft/s (1.4 mi/h), average tarnished surface.

NOTE: 1 in = 2.54 cm; 1 lb = 0.4536 kg; 1 mi = 1.61 km; 1 ft = 0.3048 m.

**TABLE 4-30** ACSR Conductor—Electrical Characteristics

(Compiled from tables published by Westinghouse Electric Corp., "Electrical Transmission and Distribution Reference Book"; and Aluminum Co. of America)

| Conductor size, kcmil or Awg | Hard-drawn copper equivalent,* kcmil or Awg | Geometric mean radius at 60 Hz, ft | Resistance at 25°C (77°F),† small currents | | | | Resistance at 50°C (122°F)† current approx 75% capacity | | | | Inductive reactance (series) at 1 ft spacing ($x_a$)‡ | | | Capacitive reactance (shunt) at 1 ft spacing ($x_a'$)§ | | | Current-carrying capacity at 60 Hz¶ (approx), A |
|---|---|---|---|---|---|---|---|---|---|---|---|---|---|---|---|---|---|
| | | | DC | 25 Hz | 50 Hz | 60 Hz | DC | 25 Hz | 50 Hz | 60 Hz | 25 Hz | 50 Hz | 60 Hz | 25 Hz | 50 Hz | 60 Hz | |
| | | | Ω/conductor/mile | | | | | | | | Ω/conductor/mile | | | MΩ/conductor/mile | | | |
| Multilayer conductors | | | | | | | | | | | | | | | | | |
| 1590 | 1000 | 0.0520 | .0587 | .0588 | .0590 | .0591 | .0646 | .0656 | .0675 | .0684 | .1495 | .299 | .359 | .1953 | .0977 | .0814 | 1380 |
| 1510.5 | 950 | 0.0507 | .0618 | .0619 | .0621 | .0622 | .0680 | .0690 | .0710 | .0720 | .1508 | .302 | .362 | .1971 | .0986 | .0821 | 1340 |
| 1431 | 900 | 0.0493 | .0652 | .0653 | .0655 | .0656 | .0718 | .0729 | .0749 | .0760 | .1522 | .304 | .365 | .1991 | .0996 | .0830 | 1300 |
| 1351 | 850 | 0.0479 | .0691 | .0692 | .0694 | .0695 | .0761 | .0771 | .0792 | .0803 | .1536 | .307 | .369 | .201 | .1006 | .0838 | 1250 |
| 1272 | 800 | 0.0465 | .0734 | .0735 | .0737 | .0738 | .0808 | .0819 | .0840 | .0851 | .1551 | .310 | .372 | .203 | .1016 | .0847 | 1200 |
| 1192.5 | 750 | 0.0450 | .0783 | .0784 | .0786 | .0788 | .0862 | .0872 | .0894 | .0906 | .1568 | .314 | 376 | .206 | .1028 | .0857 | 1160 |
| 1113 | 700 | 0.0435 | .0839 | .0840 | .0842 | .0844 | .0924 | .0935 | .0957 | .0969 | .1585 | .317 | .380 | .208 | .1040 | .0867 | 1110 |
| 1033.5 | 650 | 0.0420 | .0903 | .0905 | .0907 | .0909 | .0994 | .1005 | .1025 | .1035 | .1603 | .321 | .385 | .211 | .1053 | .0878 | 1060 |
| 954 | 600 | 0.0403 | .0979 | .0980 | .0981 | .0982 | .1078 | .1088 | .1118 | .1128 | .1624 | .325 | .390 | .214 | .1068 | .0890 | 1010 |
| 900 | 566 | 0.0391 | .104 | .104 | .104 | .104 | .1145 | .1155 | .1175 | .1185 | .1639 | .328 | .393 | .216 | .1078 | .0898 | 970 |
| 874.5 | 550 | 0.0386 | .107 | .107 | .107 | .108 | .1178 | .1188 | .1218 | .1228 | .1646 | .329 | .395 | .217 | .1083 | .0903 | 950 |
| 795 | 500 | 0.0368 | .117 | .118 | .118 | .119 | .1288 | .1308 | .1358 | .1378 | .1670 | .334 | .401 | .220 | .1100 | .0917 | 900 |
| 795 | 500 | 0.0375 | .117 | .117 | .117 | .117 | .1288 | .1288 | .1288 | .1288 | .1660 | .332 | .399 | .219 | .1095 | .0912 | 900 |
| 795 | 500 | 0.0393 | .117 | .117 | .117 | .117 | .1288 | .1288 | .1288 | .1288 | .1637 | .327 | .393 | .217 | .1085 | .0904 | 910 |
| 715.5 | 450 | 0.0349 | .131 | .131 | .131 | .132 | .1442 | .1452 | .1472 | .1482 | .1697 | .339 | .407 | .224 | .1119 | .0932 | 830 |
| 715.5 | 450 | 0.0355 | .131 | .131 | .131 | .131 | .1442 | .1442 | .1442 | .1442 | .1687 | .337 | .405 | .223 | .1114 | .0928 | 840 |
| 715.5 | 450 | 0.0372 | .131 | .131 | .131 | .131 | .1442 | .1442 | .1442 | .1442 | .1664 | .333 | .399 | .221 | .1104 | .0920 | 840 |
| 666.6 | 419 | 0.0337 | .140 | .140 | .141 | .141 | .1541 | .1571 | 1591 | .1601 | .1715 | .343 | .412 | .226 | .1132 | .0943 | 800 |
| 636 | 400 | 0.0329 | .147 | .147 | .148 | .148 | .1618 | .1638 | .1678 | .1688 | .1726 | .345 | .414 | .228 | .1140 | .0950 | 770 |
| 636 | 400 | 0.0335 | .147 | .147 | .147 | .147 | .1618 | .1618 | .1618 | .1618 | .1718 | .344 | 412 | .227 | .1135 | .0946 | 780 |
| 636 | 400 | 0.0351 | .147 | .147 | .147 | .147 | .1618 | .1618 | .1618 | .1618 | .1693 | .339 | .406 | .225 | .1125 | .0937 | 780 |
| 605 | 380.5 | 0.0321 | .154 | .155 | .155 | .155 | .1695 | .1715 | .1755 | .1775 | .1739 | .348 | .417 | .230 | .1149 | .0957 | 750 |
| 605 | 380.5 | 0.0327 | .154 | .154 | .154 | .154 | .1700 | .1720 | .1720 | .1720 | .1730 | .346 | .415 | .229 | .1144 | .0953 | 760 |
| 556.5 | 350 | 0.0313 | .168 | .168 | .168 | .168 | .1849 | .1859 | .1859 | .1859 | .1751 | .350 | .420 | .232 | .1159 | .0965 | 730 |

| Conductor size, kcmil or Awg | Hard-drawn copper equivalent,* kcmil or Awg | Geometric mean radius at 60 Hz, ft | Resistance at 25°C (77°F),† small currents | | | | Resistance at 50°C (122°F)† current approx 75% capacity | | | | Inductive reactance (series) at 1 ft spacing ($x_a$)‡ | | | Capacitive reactance (shunt) at 1 ft spacing ($x_a'$)§ | | | Current-carrying capacity at 60 Hz¶ (approx), A |
|---|---|---|---|---|---|---|---|---|---|---|---|---|---|---|---|---|---|
| | | | DC | 25 Hz | 50 Hz | 60 Hz | DC | 25 Hz | 50 Hz | 60 Hz | 25 Hz | 50 Hz | 60 Hz | 25 Hz | 50 Hz | 60 Hz | |
| | | | Ω/conductor/mile | | | | | | | | Ω/conductor/mile | | | MΩ/conductor/mile | | | |
| | | | | | | Multilayer conductors | | | | | | | | | | | |
| 556.5 | 350 | 0.0328 | .168 | .168 | .168 | .168 | .1849 | .1859 | .1859 | .1859 | .1728 | .346 | .415 | 230 | .1149 | .0957 | 730 |
| 500 | 314.5 | 0.0311 | .187 | .187 | .187 | .187 | .206 | .206 | .206 | .206 | .1754 | .351 | .421 | .234 | .1167 | .0973 | 690 |
| 477 | 300 | 0.0290 | .196 | .196 | .196 | .196 | .216 | .216 | .216 | .216 | .1790 | .358 | .430 | .237 | .1186 | .0988 | 670 |
| 477 | 300 | 0.0304 | .196 | .196 | .196 | .196 | .216 | .216 | .216 | .216 | .1766 | .353 | .424 | .235 | .1176 | .0980 | 670 |
| 397.5 | 250 | 0.0265 | .235 | .235 | .235 | .235 | .259 | .259 | .259 | .259 | .1836 | .367 | .441 | .244 | .1219 | .1015 | 590 |
| 397.5 | 250 | 0.0278 | .235 | .235 | .235 | .235 | .259 | .259 | .259 | .259 | .1812 | .362 | .435 | .242 | .1208 | .1006 | 600 |
| 336.4 | 4/0 | 0.0244 | .278 | .278 | .278 | .278 | .306 | .306 | .306 | .306 | .1872 | .376 | .451 | .250 | .1248 | .1039 | 530 |
| 336.4 | 4/0 | 0.0255 | .278 | .278 | .278 | .278 | .306 | .306 | .306 | .306 | .1855 | .371 | .445 | .248 | .1238 | .1032 | 530 |
| 300 | 188.7 | 0.0230 | .311 | .311 | .311 | .311 | .342 | .342 | .342 | .342 | .1908 | .382 | .458 | .254 | .1269 | .1057 | 490 |
| 300 | 188.7 | 0.0241 | .311 | .311 | .311 | .311 | .342 | .342 | .342 | .342 | .1883 | .377 | .452 | .252 | .1258 | .1049 | 500 |
| 266.8 | 3/0 | 0.0217 | .350 | .350 | .350 | .350 | .385 | .385 | .385 | .385 | .1936 | .387 | .465 | .258 | .1289 | .1074 | 460 |

**TABLE 4-30** ACSR Conductor—Electrical Characteristics (*Continued*)

| Conductor size, kcmil or Awg | Hard-drawn copper equivalent,* kcmil or Awg | Current approx 75% capacity | Resistance at 25°C (77°F),† small currents | | | | Resistance at 50°C (122°F)† current approx 75% capacity | | | | Small currents, Hz | | | Current approx 75% capacity, Hz | | | Capacitive reactance (shunt) at 1 ft spacing ($x_a'$)§ | | | Current-carrying capacity at 60 Hz¶ (approx), A |
|---|---|---|---|---|---|---|---|---|---|---|---|---|---|---|---|---|---|---|---|---|
| | | | DC | 25 Hz | 50 Hz | 60 Hz | DC | 25 Hz | 50 Hz | 60 Hz | 25 | 50 | 60 | 25 | 50 | 60 | 25 Hz | 50 Hz | 60 Hz | |
| | | | Ω/conductor/mile | | | | | | | | | | | | | | MΩ/conductor/mile | | | |
| Single-layer conductors | | | | | | | | | | | | | | | | | | | | |
| 266.8 | 3/0 | 0.00684 | 0.351 | 0.351 | 0.351 | 0.352 | 0.386 | 0.430 | 0.510 | 0.552 | .194 | .388 | .466 | .252 | .504 | .605 | .259 | .1294 | .1079 | 460 |
| 4/0 | 2/0 | 0.00814 | 0.441 | 0.442 | 0.444 | 0.445 | 0.485 | 0.514 | 0.567 | 0.592 | .218 | .437 | .524 | .242 | .484 | .581 | .267 | .1336 | .1113 | 340 |
| 3/0 | 1/0 | 0.00600 | 0.556 | 0.557 | 0.559 | 0.560 | 0.612 | 0.642 | 0.697 | 0.723 | .225 | .450 | .540 | .259 | .517 | .621 | .275 | .1377 | .1147 | 300 |
| 2/0 | 1 | 0.00510 | 0.702 | 0.702 | 0.704 | 0.706 | 0.773 | 0.806 | 0.866 | 0.895 | .231 | .462 | .554 | .267 | .534 | .641 | .284 | .1418 | .1182 | 270 |
| 1/0 | 2 | 0.00446 | 0.885 | 0.885 | 0.887 | 0.888 | 0.974 | 1.01 | 1.08 | 1.12 | .237 | .473 | .568 | .273 | .547 | .656 | .292 | .1460 | .1216 | 230 |
| 1 | 3 | 0.00418 | 1.12 | 1.12 | 1.12 | 1.12 | 1.23 | 1.27 | 1.34 | 1.38 | .242 | .483 | .580 | .277 | .554 | .665 | .300 | .1500 | .1250 | 200 |
| 2 | 4 | 0.00418 | 1.41 | 1.41 | 1.41 | 1.41 | 1.55 | 1.59 | 1.66 | 1.69 | .247 | .493 | .592 | .277 | .554 | .665 | .308 | .1542 | .1285 | 180 |
| 2 | 4 | 0.00504 | 1.41 | 1.41 | 1.41 | 1.41 | 1.55 | 1.59 | 1.62 | 1.65 | .247 | .493 | .592 | .267 | .535 | .642 | .306 | .1532 | .1276 | 180 |
| 3 | 5 | 0.00430 | 1.78 | 1.78 | 1.78 | 1.78 | 1.95 | 1.98 | 2.04 | 2.07 | .252 | .503 | .604 | .275 | .551 | .661 | .317 | .1583 | .1320 | 160 |
| 4 | 6 | 0.00437 | 2.24 | 2.24 | 2.24 | 2.24 | 2.47 | 2.50 | 2.54 | 2.57 | .257 | .514 | .611 | .274 | .549 | .659 | .325 | .1627 | .1355 | 140 |
| 4 | 6 | 0.00452 | 2.24 | 2.24 | 2.24 | 2.24 | 2.47 | 2.50 | 2.53 | 2.55 | .257 | .515 | .618 | .273 | .545 | .655 | .323 | .1615 | .1346 | 140 |
| 5 | 7 | 0.00416 | 2.82 | 2.82 | 2.82 | 2.82 | 3.10 | 3.12 | 3.16 | 3.18 | .262 | .525 | .630 | .279 | .557 | .665 | .333 | .1666 | .1388 | 120 |
| 6 | 8 | 0.00394 | 3.56 | 3.56 | 3.56 | 3.56 | 3.92 | 3.94 | 3.97 | 3.98 | .268 | .536 | .643 | .281 | .561 | .673 | .342 | .1708 | .1423 | 100 |

*Area of hard-drawn copper cable (conductivity = 97% IACS) having the same dc resistance as that of the ACSR aluminum (conductivity = 61% IACS).

†For resistance temperature conversion see Pars. **24** and **28.**

Resistances at 50°C total temperature, based on ambient of 25°C plus 25°C rise due to heating effect of current. The approximate magnitude of current necessary to produce the 25°C rise is 75% of the "approximate current-carrying capacity at 60 Hz."

‡See Table 4-32.

§See Table 4-33.

¶For conductor at 75°C, air at 25°C, wind 2 ft/s (1.4 mi/h), average tarnished surface.

NOTE: 1 ft = 0.3048 m; 1 mi = 1.61 km.

**TABLE 4-31** Aluminum-Clad Steel Cable—Electrical Characteristics
(Compiled from tables published by Copperweld Steel Company)

| Conductor stranding | | Geometric mean radius at 60 cycles, average currents, ft | Resistance at 25°C (77°F),* small currents | | | Resistance at 75°C (167°F), current approx 75% of capacity | | | Reactance at 1-ft spacing: Inductive (series) ($x_a$)† | | Reactance at 1-ft spacing: Capacitive (shunt) ($x_a'$)‡ | | Current-carrying capacity at 60 cycles§ (approx), A |
|---|---|---|---|---|---|---|---|---|---|---|---|---|---|
| No. of wires | Wire size, AWG | | D-c | 50 cycles | 60 cycles | D-c | 50 cycles | 60 cycles | 50 cycles | 60 cycles | 50 cycles | 60 cycles | |
| | | | Ω/(conductor) (mi) | | | Ω/(conductor) (mi) | | | Ω/(conductor) (mi) | | MΩ/(conductor) (mi) | | |
| 19 | 5 | 0.004929 | 0.4420 | 0.4507 | 0.4507 | 0.5202 | 0.7203 | 0.7585 | 0.537 | 0.645 | 0.1165 | 0.0971 | 485 |
| 19 | 6 | 0.004387 | 0.5574 | 0.5683 | 0.5683 | 0.6559 | 0.8517 | 0.8886 | 0.548 | 0.658 | 0.1206 | 0.1005 | 425 |
| 19 | 7 | 0.003905 | 0.7030 | 0.7171 | 0.7171 | 0.8273 | 1.027 | 1.064 | 0.561 | 0.673 | 0.1248 | 0.1040 | 380 |
| 19 | 8 | 0.003478 | 0.8864 | 0.9038 | 0.9038 | 1.043 | 1.243 | 1.280 | 0.572 | 0.687 | 0.1289 | 0.1074 | 335 |
| 19 | 9 | 0.003098 | 1.118 | 1.140 | 1.140 | 1.315 | 1.518 | 1.554 | 0.584 | 0.701 | 0.1331 | 0.1109 | 295 |
| 10 | 10 | 0.002757 | 1.409 | 1.437 | 1.437 | 1.658 | 1.861 | 1.896 | 0.596 | 0.715 | 0.1360 | 0.1133 | 260 |
| 7 | 5 | 0.002958 | 1.217 | 1.240 | 1.240 | 1.432 | 1.634 | 1.669 | 0.589 | 0.707 | 0.1346 | 0.1122 | 280 |
| 7 | 6 | 0.002633 | 1.507 | 1.536 | 1.536 | 1.773 | 1.977 | 2.01 | 0.601 | 0.721 | 0.1388 | 0.1157 | 250 |
| 7 | 7 | 0.002345 | 1.900 | 1.937 | 1.937 | 2.24 | 2.44 | 2.47 | 0.612 | 0.735 | 0.1429 | 0.1191 | 220 |
| 7 | 8 | 0.002085 | 2.40 | 2.44 | 2.44 | 2.82 | 3.03 | 3.06 | 0.624 | 0.749 | 0.1471 | 0.1226 | 190 |
| 7 | 9 | 0.001858 | 3.02 | 3.08 | 3.08 | 3.56 | 3.77 | 3.80 | 0.636 | 0.763 | 0.1512 | 0.1260 | 160 |
| 7 | 10 | 0.001658 | 3.81 | 3.88 | 3.88 | 4.48 | 4.70 | 4.73 | 0.647 | 0.777 | 0.1552 | 0.1294 | 140 |
| 3 | 5 | 0.002940 | 2.78 | 2.78 | 2.78 | 3.27 | 3.52 | 3.56 | 0.589 | 0.707 | 0.1465 | 0.1221 | 170 |
| 3 | 6 | 0.002618 | 3.51 | 3.51 | 3.51 | 4.13 | 4.36 | 4.41 | 0.601 | 0.721 | 0.1506 | 0.1255 | 150 |
| 3 | 7 | 0.002333 | 4.42 | 4.42 | 4.42 | 5.21 | 5.43 | 5.47 | 0.612 | 0.735 | 0.1547 | 0.1289 | 130 |
| 3 | 8 | 0.002078 | 5.58 | 5.58 | 5.58 | 6.57 | 6.78 | 6.82 | 0.624 | 0.749 | 0.1589 | 0.1324 | 110 |

*For resistance temperature conversion see Pars. **24** and **30.**
†See Table 4-32.
‡See Table 4-33.
§For conductor at 125°C, air at 25°C, wind 2 ft/s (1.4 mi/h), average tarnished surface.
NOTE: 1 ft = 0.3048 m; 1 mi = 1.61 km.

**TABLE 4-32** Inductive-Reactance Spacing Factors ($x_d$)

| Frequency, cycles | Tens \ Units | 0 | 1 | 2 | 3 | 4 | 5 | 6 | 7 | 8 | 9 |
|---|---|---|---|---|---|---|---|---|---|---|---|
| | | Equivalent conductor spacing, ft — Ohms per conductor per mile | | | | | | | | | |
| 25 | 0 | ...... | 0 | 0.0350 | 0.0555 | 0.0701 | 0.0814 | 0.0906 | 0.0984 | 0.1051 | 0.1111 |
| | 1 | 0.1164 | 0.1212 | 0.1256 | 0.1297 | 0.1334 | 0.1369 | 0.1402 | 0.1432 | 0.1461 | 0.1489 |
| | 2 | 0.1515 | 0.1539 | 0.1563 | 0.1585 | 0.1607 | 0.1627 | 0.1647 | 0.1666 | 0.1685 | 0.1702 |
| | 3 | 0.1720 | 0.1736 | 0.1752 | 0.1768 | 0.1783 | 0.1798 | 0.1812 | 0.1826 | 0.1839 | 0.1852 |
| | 4 | 0.1865 | 0.1878 | 0.1890 | 0.1902 | 0.1913 | 0.1925 | 0.1936 | 0.1947 | 0.1957 | 0.1968 |
| 50 | 0 | ...... | 0 | 0.0701 | 0.1111 | 0.1402 | 0.1627 | 0.1812 | 0.1968 | 0.2103 | 0.2222 |
| | 1 | 0.2328 | 0.2425 | 0.2513 | 0.2594 | 0.2669 | 0.2738 | 0.2804 | 0.2865 | 0.2923 | 0.2977 |
| | 2 | 0.3029 | 0.3079 | 0.3126 | 0.3170 | 0.3214 | 0.3255 | 0.3294 | 0.3333 | 0.3369 | 0.3405 |
| | 3 | 0.3439 | 0.3472 | 0.3504 | 0.3536 | 0.3566 | 0.3595 | 0.3624 | 0.3651 | 0.3678 | 0.3704 |
| | 4 | 0.3730 | 0.3755 | 0.3779 | 0.3803 | 0.3826 | 0.3849 | 0.3871 | 0.3893 | 0.3914 | 0.3935 |
| 60 | 0 | ...... | 0 | 0.0841 | 0.1333 | 0.1682 | 0.1953 | 0.2174 | 0.2361 | 0.2523 | 0.2666 |
| | 1 | 0.2794 | 0.2910 | 0.3015 | 0.3112 | 0.3202 | 0.3286 | 0.3364 | 0.3438 | 0.3507 | 0.3573 |
| | 2 | 0.3635 | 0.3694 | 0.3751 | 0.3805 | 0.3856 | 0.3906 | 0.3953 | 0.3999 | 0.4043 | 0.4086 |
| | 3 | 0.4127 | 0.4167 | 0.4205 | 0.4243 | 0.4279 | 0.4314 | 0.4348 | 0.4382 | 0.4414 | 0.4445 |
| | 4 | 0.4476 | 0.4506 | 0.4535 | 0.4564 | 0.4592 | 0.4619 | 0.4646 | 0.4672 | 0.4697 | 0.4722 |

Total inductive reactance = $x_a + x_d$.
See Par. **139.**
NOTE: 1 ft = 0.3048 m; 1 mi = 1.61 km.

**TABLE 4-33** Capacitive-Reactance Spacing Factors ($x'_d$)

| Frequency, cycles | Tens \ Units | Equivalent conductor spacing, ft | | | | | | | | | |
|---|---|---|---|---|---|---|---|---|---|---|---|
| | | 0 | 1 | 2 | 3 | 4 | 5 | 6 | 7 | 8 | 9 |
| | | Megohms per conductor per mile | | | | | | | | | |
| 25 | 0 | ...... | 0 | 0.0494 | 0.0782 | 0.0987 | 0.1146 | 0.1276 | 0.1386 | 0.1481 | 0.1565 |
| | 1 | 0.1640 | 0.1707 | 0.1769 | 0.1826 | 0.1879 | 0.1928 | 0.1974 | 0.2017 | 0.2058 | 0.2097 |
| | 2 | 0.2133 | 0.2168 | 0.2201 | 0.2233 | 0.2263 | 0.2292 | 0.2320 | 0.2347 | 0.2373 | 0.2398 |
| | 3 | 0.2422 | 0.2445 | 0.2468 | 0.2490 | 0.2511 | 0.2532 | 0.2552 | 0.2571 | 0.2590 | 0.2609 |
| | 4 | 0.2627 | 0.2644 | 0.2661 | 0.2678 | 0.2695 | 0.2711 | 0.2726 | 0.2742 | 0.2756 | 0.2771 |
| 50 | 0 | ...... | 0 | 0.0247 | 0.0391 | 0.0494 | 0.0573 | 0.0638 | 0.0693 | 0.0740 | 0.0782 |
| | 1 | 0.0820 | 0.0854 | 0.0885 | 0.0913 | 0.0940 | 0.0964 | 0.0987 | 0.1009 | 0.1029 | 0.1048 |
| | 2 | 0.1067 | 0.1084 | 0.1100 | 0.1116 | 0.1131 | 0.1146 | 0.1160 | 0.1173 | 0.1186 | 0.1199 |
| | 3 | 0.1211 | 0.1223 | 0.1234 | 0.1245 | 0.1255 | 0.1266 | 0.1276 | 0.1286 | 0.1295 | 0.1304 |
| | 4 | 0.1313 | 0.1322 | 0.1331 | 0.1339 | 0.1347 | 0.1355 | 0.1363 | 0.1371 | 0.1378 | 0.1386 |
| 60 | 0 | ...... | 0 | 0.0206 | 0.0326 | 0.0411 | 0.0478 | 0.0532 | 0.0577 | 0.0617 | 0.0652 |
| | 1 | 0.0683 | 0.0711 | 0.0737 | 0.0761 | 0.0783 | 0.0803 | 0.0823 | 0.0841 | 0.0858 | 0.0874 |
| | 2 | 0.0889 | 0.0903 | 0.0917 | 0.0930 | 0.0943 | 0.0955 | 0.0967 | 0.0978 | 0.0989 | 0.0999 |
| | 3 | 0.1009 | 0.1019 | 0.1028 | 0.1037 | 0.1046 | 0.1055 | 0.1063 | 0.1071 | 0.1079 | 0.1087 |
| | 4 | 0.1094 | 0.1102 | 0.1109 | 0.1116 | 0.1123 | 0.1129 | 0.1136 | 0.1142 | 0.1149 | 0.1155 |

Total capacitive reactance = $x'_a + x'_d$.
See Pars. **141** and **142.**
NOTE: 1 ft = 0.3048 m; 1 mi = 1.61 km.

***173. Fusing currents of different kinds of wire*** were investigated by W. H. Preece, who developed the formula

$$I = ad^{3/2} \tag{4-34}$$

where $I$ = fusing current in amperes, $d$ = diameter of the wire in inches, $a$ = a constant depending upon the material. He found the following values for $a$:

| | | | |
|---|---|---|---|
| Copper | 10,244 | Iron | 3,148 |
| Aluminum | 7,585 | Tin | 1,642 |
| Platinum | 5,172 | Alloy (2Pb-1Sn) | 1,318 |
| German silver | 5,230 | Lead | 1,379 |
| Platinoid | 4,750 | | |

Although this formula has been used to a considerable extent in the past, it gives values that usually are erroneous in practice, because it is based on the assumption that all heat loss is due to radiation. A formula of the general type

$$I = kd^n \tag{4-35}$$

can be used with accuracy if $k$ and $n$ are known for the particular case (material, wire size, installation conditions, etc.).

***174. Fusing current-time for copper conductors and connections*** may be determined by an equation developed by I. M. Onderdonk:

$$33\left(\frac{I}{A}\right)^2 S = \log\left(\frac{T_m - T_a}{234 + T_a} + 1\right) \tag{4-36}$$

$$I = A\sqrt{\frac{\log\left(\dfrac{T_m - T_a}{234 + T_a} + 1\right)}{33S}} \tag{4-37}$$

**TABLE 4-34** Properties of Resistance Metals and Alloys

| Material | Chemical composition | Resistivity at 20°C (68°F) | Resistance | | Linear expansion | | Melting point, approx, °C | Tensile strength at 20°C (68°F), min, psi | Specific gravity | Weight, lb/in³ |
|---|---|---|---|---|---|---|---|---|---|---|
| | | Ω·cmils/ft | Temperature coefficient, per °C | Temperature range, °C | Temperature coefficient, per °C | Temperature range, °C | | | | |
| **Driver-Harris Co., Harrison, N. J.** | | | | | | | | | | |
| Karma* | Ni 73%-Cr 20% + Al + Fe | 800 | .......... | −50–105 | 0.00001 | 20–100 | 1400 | 130,000 | 8.105 | 0.292 |
| Nichrome* | Ni 60%-Cr 16%-balance Fe | 675 | 0.00015 | 20–500 | 0.000017 | 20–1000 | 1350 | 95,000 | 8.247 | 0.2979 |
| Nichrome V* | Ni 80%-Cr 20% | 650 | 0.00011 | 20–500 | 0.000017 | 10–1000 | 1400 | 100,000 | 8.412 | 0.3039 |
| Chromax* | Ni 35%-Cr 20%-balance Fe | 600 | 0.00036 | 20–500 | 0.0000158 | 20–500 | 1380 | 70,000 | 7.950 | 0.2872 |
| Nilvar* | Ni 36%-balance Fe | 484 | 0.00135 | 20–100 | 0.000001 | 20–100 | 1425 | 70,000 | 8.08 | 0.292 |
| Stainless type 304 | Cr 18%-Ni 8%-balance Fe | 438 | 0.00094 | 20–500 | 0.000020 | 0–1000 | 1399 | 100,000 | 7.93 | 0.286 |
| 142 alloy | Ni 42%-balance Fe | 400 | 0.0012 | 20–500 | 0.0000053 | 20–400 | 1425 | 70,000 | 8.12 | 0.293 |
| Advance* | Ni 43%-balance Cu | 294 | ±0.00002 | 20–100 | 0.0000149 | 20–100 | 1210 | 60,000 | 8.9 | 0.321 |
| Therlo* | Ni 29%-Co 17%-balance Fe | 294 | 0.0038 | 0–100 | 0.000006 | 30–500 | 1450 | 75,000 | 8.36 | 0.302 |
| Manganin | Mn 13%-balance Cu | 290 | ±0.000015 | 15–35 | 0.0000187 | 15–35 | 1020 | 40,000 | 8.192 | 0.296 |
| 146 alloy | Ni 46%-balance Fe | 275 | 0.0027 | 20–500 | 0.000008 | 25–425 | 1425 | 70,000 | 8.17 | 0.295 |
| 152 alloy (52) | Ni 51%-balance Fe | 260 | 0.0029 | 20–500 | 0.0000095 | 20–500 | 1425 | 70,000 | 8.247 | 0.2979 |
| Duranickel | Nickel plus additions | 260 | 0.001 | 20–100 | 0.000014 | 20–500 | 1435 | 90,000 | 8.75 | 0.316 |
| Midohm* | Ni 23%-balance Cu | 180 | 0.00018 | −50–150 | 0.0000175 | 20–500 | 1100 | 50,000 | 8.9 | 0.321 |
| R-63 alloy | Mn 4%-Si 1%-balance Ni | 130 | 0.003 | 20–250 | 0.0000152 | 20–500 | 1425 | 70,000 | 8.72 | 0.315 |
| Hytemco* | Ni 72%-balance Fe | 120 | 0.0042 | 20–100 | 0.000015 | 20–1000 | 1425 | 70,000 | 8.46 | 0.305 |
| Permanickel | Nickel plus additions | 100 | 0.0036 | 30–500 | 0.000014 | 30–1000 | 1450 | 90,000 | 8.75 | 0.316 |
| 90 alloy | Ni 11%-balance Cu | 90 | 0.00049 | −50–150 | 0.0000175 | 20–500 | 1100 | 35,000 | 8.9 | 0.321 |
| Gr. A nickel | Ni 99% | 60 | 0.0050 | 0–100 | 0.000015 | 20–500 | 1450 | 60,000 | 8.9 | 0.321 |
| Lohm* | Ni 6%-balance Cu | 60 | 0.0008 | −50–150 | 0.000018 | 20–500 | 1100 | 50,000 | 8.9 | 0.321 |
| 99 alloy | Ni 99.8% | 48 | 0.0060 | 0–100 | | | | | | |
| 30 Alloy | Ni 2.25%-balance Cu | 30 | 0.0015 | −50–150 | 0.0000175 | 20–500 | 1100 | 30,000 | 8.9 | 0.321 |
| **Hoskins Manufacturing Co., Detroit, Mich.** | | | | | | | | | | |
| Chromel AA* | Ni 68%-Cr 20%-Fe 8% | 700 | 0.00011 | 20–500 | 0.0000135 | 20–1000 | 1390 | 120,000 | 8.33 | 0.301 |
| Chromel A* | Ni 80%-Cr 20% | 650 | 0.00011 | 20–500 | 0.000017 | 10–1000 | 1400 | 100,000 | 8.412 | 0.3039 |
| Chromel C* | Ni 60%-Cr 16%-balance Fe | 675 | 0.00015 | 20–500 | 0.000017 | 20–1000 | 1350 | 95,000 | 8.247 | 0.2979 |
| Chromel D* | Ni 35%-Cr 20%-balance Fe | 600 | 0.00036 | 20–500 | 0.0000158 | 20–500 | 1380 | 70,000 | 7.950 | 0.2872 |
| Copel* | Ni 43%-balance Cu | 294 | ±0.00002 | 20–100 | 0.0000149 | 20–100 | 1210 | 60,000 | 8.9 | 0.321 |
| Alloy 875 | Cr 22.5%-Al 5.5%-balance Fe | 875 | 0.00002 | 20–500 | 0.0000174 | 20–1000 | 1520 | 110,000 | 7.10 | 0.256 |
| Alloy 815 | Cr 22.5%-Al 4.6%-balance Fe | 815 | 0.00008 | 20–500 | 0.0000159 | 20–1000 | 1520 | 110,000 | 7.25 | 0.262 |
| Alloy 750 | Cr 15%-Al 4%-balance Fe | 750 | 0.00015 | 20–500 | 0.0000150 | 20–1000 | 1520 | 110,000 | 7.43 | 0.268 |

**TABLE 4-34** Properties of Resistance Metals and Alloys (*Continued*)

| Material | Chemical composition | Resistivity at 20°C (68°F) | Resistance | | Linear expansion | | Melting point, approx, °C | Tensile strength at 20°C (68°F), min, psi | Specific gravity | Weight, lb/in³ |
|---|---|---|---|---|---|---|---|---|---|---|
| | | Ω·cmils/ft | Temperature coefficient, per °C | Temperature range, °C | Temperature coefficient, per °C | Temperature range, °C | | | | |
| **The Kanthal Corp., Bethel, Conn.** | | | | | | | | | | |
| Kanthal DR* | Fe 75%-Cr 20%-Al 4.5%-Co 0.5% | 812 | 0.00007 | 20–150 | 0.0000119 | 20–100 | 1505 | 100,000 | 7.2 | 0.262 |
| Nikrothal L* | Ni 75%-Cr 17%-balance Si + Mn | 800 | 0.000003 | 20–150 | 0.0000126 | 20–100 | 1410 | 150,000 | 8.1 | 0.292 |
| Nikrothal 6* | Ni 60%-Cr 16%-balance Fe | 675 | 0.000140 | 20–100 | 0.000013 | 20–100 | 1350 | 90,000 | 8.25 | 0.298 |
| Nikrothal 8* | Ni 80%-Cr 20% | 650 | 0.000080 | 20–100 | 0.000014 | 20–100 | 1400 | 95,000 | 8.41 | 0.304 |
| Cuprothal 294* | Ni 45%-balance Cu | 294 | 0.00002 | 20–100 | ......... | ....... | .......... | 60,000 | 8.9 | 0.321 |
| Cuprothal 180* | Ni 22%-balance Cu | 180 | 0.00018 | 20–100 | ......... | ....... | .......... | 50,000 | 8.9 | 0.321 |
| Cuprothal 90* | Ni 11%-balance Cu | 90 | 0.00045 | 20–100 | ......... | ....... | .......... | 35,000 | 8.9 | 0.321 |
| Cuprothal 60* | Ni 6%-balance Cu | 60 | 0.0008 | 20–100 | ......... | ....... | .......... | 35,000 | 8.9 | 0.321 |
| Cuprothal 30* | Ni 2%-balance Cu | 30 | 0.0014 | 20–100 | ......... | ....... | .......... | 30,000 | 8.9 | 0.321 |
| **Pure Metals** | | | | | | | | | | |
| Platinum | ........ | 63.80 | 0.00300 | 20 | 0.0000089 | 20 | 1773 | ....... | 21.45 | 0.7750 |
| Iron | ........ | 60.14 | 0.0050 | 20 | 0.0000117 | 20 | 1535 | 50,000 | 7.86 | 0.2840 |
| Molybdenum | ........ | 34.27 | 0.0033 | 20 | 0.000005 | 20 | 2625 | 100,000 | 10.2 | 0.3685 |
| Tungsten | ........ | 33.22 | 0.0045 | 18 | 0.000004 | 20 | 3410 ± 20 | 490,000 | 19.3 | 0.6973 |
| Aluminum | ........ | 16.06 | 0.00446 | ........ | 0.000024 | 20 | 660 | 35,000 | 2.7 | 0.0975 |
| Gold | ........ | 14.55 | 0.0034 | 20 | 0.0000142 | 20 | 1063 | ....... | 19.3 | 0.6973 |
| Copper | ........ | 10.37 | 0.00393 | 20 | 0.0000166 | 20 | 1083 | 35,000 | 8.92 | 0.3223 |
| Silver | ........ | 9.796 | 0.0038 | 20 | 0.0000189 | 20 | 960 | ....... | 10.5 | 0.3793 |

*Trademark.

NOTE: 1 ft = 0.3048 m; 1 lb/in$^2$ = 6.895 kPa; 1 lb/in$^3$ = 27,680 kg/m$^3$.

**TABLE 4-35** Compositions and Melting Points of Fusible Alloys
("International Critical Tables," vol. 2. p. 391)

| Chemical composition, % | | | | | Melting point, deg C | Chemical composition, % | | | | | Melting point, deg C |
|---|---|---|---|---|---|---|---|---|---|---|---|
| Bi | Pb | Sn | Cd | Hg | | Bi | Pb | Sn | Cd | Hg | |
| 20 | 20 | .. | .. | 60 | 20 | .. | 32 | 50 | 18 | .. | 145 |
| 50 | 27 | 13 | 10 | .. | 72 | 50 | 50 | .. | .. | .. | 160 |
| 52 | 40 | .. | 8 | .. | 92 | 15 | 41 | 44 | .. | .. | 164 |
| 53 | 32 | 15 | .. | .. | 96 | 33 | .. | 67 | .. | .. | 166 |
| 54 | 26 | .. | 20 | .. | 103 | 20 | .. | 80 | .. | .. | 200 |
| 29 | 43 | 28 | .. | .. | 132 | .. | .. | .. | .. | .. | |

Compositions and melting points are approximate.

where $I$ = current in amperes, $A$ = conductor area in circular mils, $S$ = time current applied in seconds, $T_m$ = melting point of copper in degrees Celsius, $T_a$ = ambient temperature in degrees Celsius.

***175. Copper Conductors.*** E. R. Stauffacher has prepared a chart of the fusing current for sizes from 30 AWG to 500,000 cmils from 0.1 to 10 s (see Fig. 4-15). This chart is based on the assumptions that (1) radiation may be neglected owing to the short time involved, that is, 10 s; (2) resistance of 1 cm cube of copper at 0°C is 1.589 $\mu\Omega$; (3) temperature-resistance coefficient of copper at 0°C is $1/234$; (4) melting point of copper is 1083°C; and (5) ambient temperature is 40°C.

For most practical purposes, Eq. (4-37) also may be applied where the melting temperature $T_m$ of solder or other materials used in making connections is the determining factor; for example:

*Soldered Connections.* Select a value of $T_m$ corresponding to the melting temperature for the composition of tin-lead alloy used. This may be determined by test or approximated from Fig. 4-16 prepared from the "Smithsonian Physical Tables" (for example, $T_m$ = 183°C for 70:30 solder).

*Brazed Connections.* A reasonable value of $T_m$ is 450°C.

*Bolted Connections.* Generally accepted value of $T_m$ is 250°C.

## Miscellaneous Metals and Alloys

***176. Contact metals*** may be grouped into three general classifications:

*Hard metals,* which have melting points, for example, tungsten and molybdenum. Contacts of these metals are employed usually where operations are continuous or very frequent and current has nominal value of 5 to 10 A. Hardness to withstand mechanical wear and high melting point to resist arc erosion and welding are their outstanding advantages. Tendency to form high-resistance oxides is a disadvantage, but this can be overcome by several methods, such as using high-contact force, a hammering or wiping action, and a properly balanced electric circuit.

*Highly conductive* metals, of which silver is the best for both electric current and heat. Its disadvantages are softness and a tendency to pit and transfer. In sulfurous atmosphere, a resistant sulfide surface will form on silver, which results in high contact-surface resistance. These disadvantages are overcome usually by alloying.

*Noncorroding* metals, which for the most part consist of the noble metals, such as gold and the platinum group. Contacts of these metals are used on sensitive devices, employing extremely light pressures or low currents in which clean contact surfaces are essential. Because most of these metals are soft, they are usually alloyed.

The metals commonly used are tungsten, molybdenum, platinum, palladium, gold, silver, and their alloys. Alloying materials are copper, nickel, cadmium, iron, and the rarer metals such as iridium and ruthenium. Some are prepared by powder metallurgy.

*Commercial grades* are available under trade names or alloy numbers from Baker & Co.,

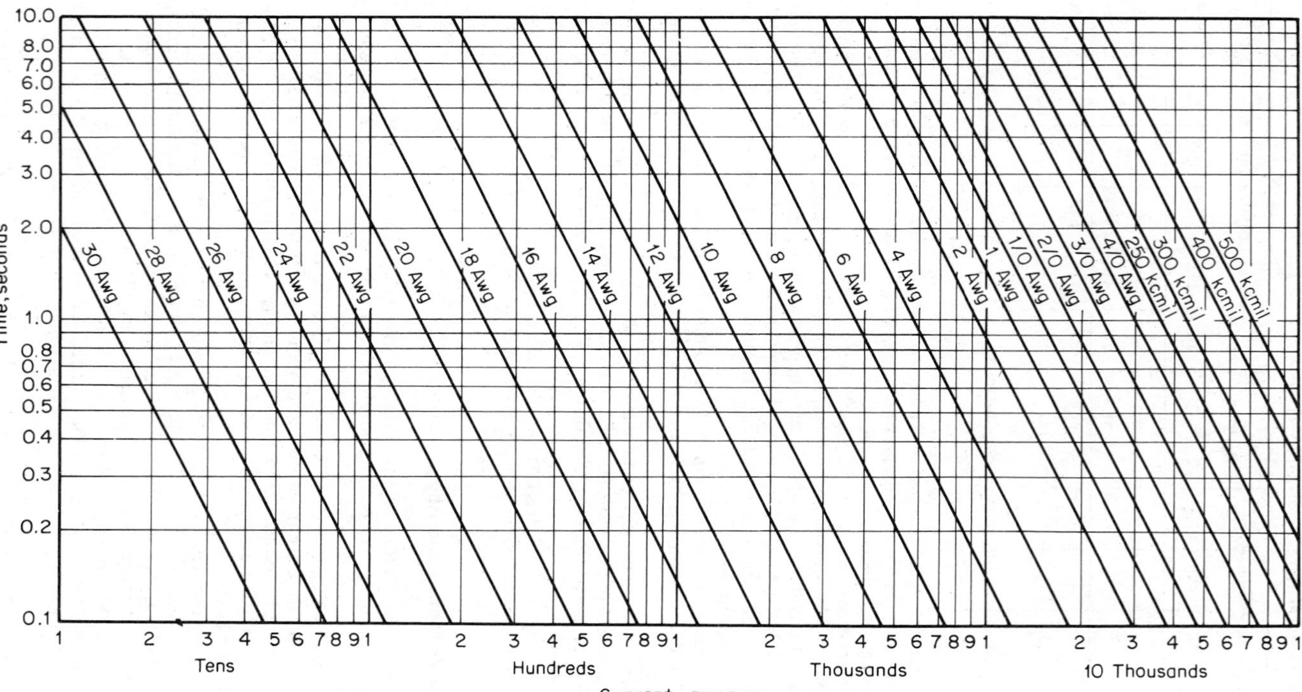


FIG. 4-15 Fusing current time for copper conductors.

Newark, N.J.; Fansteel Metallurgical Corp., North Chicago (Fasaloy, Fastell); North American Phillips Co., New York (Elmet); P. R. Mallory & Co., Indianapolis; and others.

***177. Tungsten*** (W) is a hard, dense, slow-wearing metal, a good thermal and electrical conductor, characterized by its high melting point and freedom from sticking or welding. It is manufactured in several grades having various grain sizes.

***178. Molybdenum*** (Mo) has contact characteristics about midway between tungsten and fine silver. It often replaces either metal where greater wear resistance than that of silver or lower contact-surface resistance than that of tungsten is desired.

***179. Platinum*** (Pt) is one of the most stable of all metals under the combined action of corrosion and electrical erosion. It has a high melting point and does not corrode, and surfaces remain clean and low in resistance under most adverse atmospheric and electrical conditions.

*Platinum alloys* of iridium (Ir), ruthenium (Ru), silver (Ag), or other metals are used to increase hardness and resistance to wear.

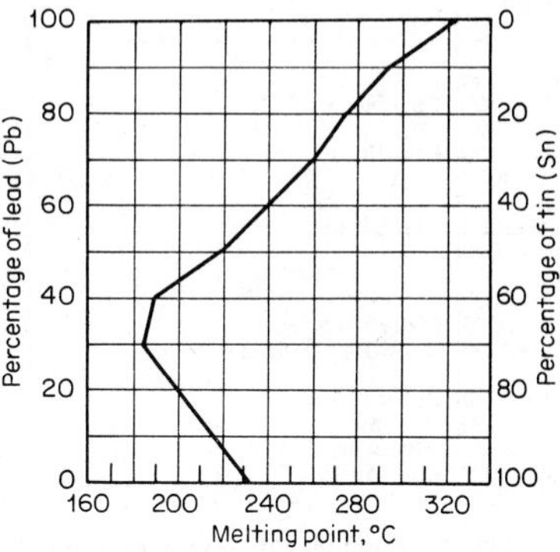


**FIG. 4-16** Melting temperatures ($T_m$) for sodder.

***180. Palladium*** (Pd) has many of the properties of platinum and frequently is used as an alternate for platinum and its alloys.

*Palladium alloys* of silver (Ag), ruthenium (Ru), nickel (Ni), and other metals are used to increase hardness and resistance to wear.

***181. Gold*** (Au) is similar to platinum in corrosion resistance but has a much lower melting point. Gold and its alloys are ductile and easily formed into a variety of shapes. Because of its softness it is usually alloyed.

*Gold alloys* of silver (Ag) and other metals are used to impart hardness and improve resistance to mechanical wear and electrical erosion.

***182. Silver*** (Ag) has the highest thermal and electrical conductivity (110%, IACS) of any metal. It has low contact-surface resistance, since its oxide decomposes at approximately 300°F. It is available commercially in three grades:

| Grade | Typical composition, % | |
|---|---|---|
| | Silver | Copper |
| Fine silver | 99.95+ | |
| Sterling silver | 92.5 | 7.5 |
| Coin silver | 90 | 10 |

*Fine* silver is used extensively under low contact pressure where sensitivity and low contact-surface resistance are essential or where the circuit is operated infrequently.

*Sterling and coin* silvers are harder than fine silver and resist transfer at low voltage (6 to 8 V) better than fine silver. Since their contact-surface resistance is greater than that of fine silver, higher contact-closing forces should be used.

*Silver alloys* of copper (Cu), nickel (Ni), cadmium (Cd), iron (Fe), carbon (C), tungsten (W), molybdenum (Mo), and other metals are used to improve hardness, resistance to wear and arc erosion, and for special applications.

***183. High-Melting-Point Metals.*** Table 4-36 shows the melting-point range of all the metals.

Table 4-37 contains the physical properties of four of the metals in the highest melting-point range.

***184. Selenium*** is a nonmetallic element chemically resembling sulfur and tellurium and

**TABLE 4-36** Melting Points of Metals
(Compiled from tables published by Fansteel Metallurgical Corp.)

| Melting point, deg C | Metal | Melting point, deg C | Metal |
|---|---|---|---|
| 3500–3400 | | 1500–1400 | Silicon, nickel, cobalt, yttrium |
| 3400–3300 | **Tungsten** | 1400–1300 | Beryllium |
| 3300–3200 | | 1300–1200 | Manganese |
| 3200–3100 | Rhenium | 1200–1100 | |
| 3100–3000 | | 1100–1000 | Gold, copper |
| 3000–2900 | **Tantalum** | 1000– 900 | Praseodymium, germanium, radium, silver |
| 2900–2800 | | 900– 800 | Cerium, arsenic, neodymium, calcium, lanthanum |
| 2800–2700 | | 800– 700 | Barium, strontium |
| 2700–2600 | **Molybdenum** | 700– 600 | Antimony, magnesium, aluminum |
| 2600–2500 | Osmium, iridium | 600– 500 | |
| 2500–2400 | **Columbium** | 500– 400 | Zinc, tellurium |
| 2400–2300 | Boron | 400– 300 | Cadmium, thallium, lead |
| 2300–2200 | Hafnium | 300– 200 | Selenium, tin, bismuth |
| 2200–2100 | Zirconium | 200– 100 | Indium, lithium |
| 2100–2000 | | 100– 0 | Cesium, gallium, rubidium, potassium, sodium |
| 2000–1900 | Chromium, ruthenium, rhodium | 0 to −100 | Mercury |
| 1900–1800 | Thorium | | |
| 1800–1700 | Vanadium, titanium, platinum | | |
| 1700–1600 | Uranium | | |
| 1600–1500 | Iron, palladium | | |

**TABLE 4-37** Properties of High-Melting-Point Metals
(Fansteel Metallurgical Corp.)

| Property | Tungsten | Tantalum | Molybdenum | Columbium |
|---|---|---|---|---|
| Specific gravity at 20 C | 19.3 | 16.6 | 10.2 | 8.57 |
| Electrical resistivity at 20 C, microhm-cm | 5.5 | 12.4 | 5.17 | 13.1 |
| Electrical resistance at 20 C, ohm (mil, ft) | 33.1 | 74.6 | 31.1 | 79.0 |
| Temperature coefficient of resistance at 20 C | 0.0051 | 0.0036 | 0.0047 | 0.00395 |
| Tensile strength, unannealed wire, lb per sq in. | 200,000 | 130,000 | 105,000 | 96,000 |
| Coefficient of linear expansion, per deg C | $4.3 \times 10^{-6}$ | $6.5 \times 10^{-6}$ | $4.9 \times 10^{-6}$ | $7.1 \times 10^{-6}$ |
| Specific heat at 20 C, cal per g atom per deg C | 6.18 | 6.512 | 6.24 | 6.012 |
| Melting point, deg C | 3410 | 2996 | 2625 | 2415 |

NOTE: 1 ft = 0.3048 in; 1 lb/in$^2$ = 6.895 kPa.

occurs in several allotropic forms varying in specific gravity from 4.3 to 4.8. It melts at 217°C and boils at 690°C. At 0°C it has a resistivity of approximately 60,000 Ω·cm. The dielectric constant ranges from 6.1 to 7.4. It has the peculiar property that its resistivity decreases upon exposure to light; the resistivity in darkness may be anywhere from 5 to 200 times the resistivity under exposure to light (see paper by W. J. Hammer, *Trans. AIEE,* 1903, vol. 21, pp. 372–393).

***185. Bibliography***

*Electric Utility Engineering Reference Book,* vol. 3; Westinghouse Electric Corporation, East Pittsburgh, Pa., 1959.

*Underground Systems Reference Book;* Edison Electric Institute, Transmission and Distribution Committee, 1957.

*Aluminum Electrical Conductor Handbook,* 2nd ed; The Aluminum Association, 818 Connecticut Avenue, N.W., Washington, D.C. 20006, 1982.

*Metals Handbook,* 9th ed.; American Society for Metals, 1979, vol. 1, Properties and Selection of Nonferrous Alloys, p. 1007.

Smith, C. S.: Thermal Conductivity of Copper Alloys, I. Copper-Zinc Alloys; *Trans. AIMME, Inst. Metals Div.,* 1930, vol. 89, p. 84.

Johnson, J. B.: *Materials of Construction;* New York, John Wiley & Sons, Inc., 1925.

Preece, W. H.: On the Heating Effects of Electric Currents; *Proc. Roy. Soc. (London),* April 1884; December 1887; April 1888.

Stauffacher, E. R.: Short-Time Current-Carrying Capacity of Copper Wires; *Gen. Elec. Rev.,* June 1928.

*Resistance and Reactance of Aluminum Conductors,* Alcoa Conductor Products Company, 510 One Allegheny Square, Pittsburgh, Pa. 15212.

*Current Temperature Characteristics of Aluminum Conductors,* Alcoa Conductor Products Company, Pittsburgh, Pa.

*Overload and Fault Current Limitations of Bare Aluminum Conductors,* Alcoa Conductor Products Company, Pittsburgh, Pa.

## CARBON AND GRAPHITE

*Prepared by THE CARBON AND MANUFACTURED GRAPHITE SECTIONS, NATIONAL ELECTRICAL MANUFACTURERS ASSOCIATION*

***186. Forms of Carbon.*** Carbon occurs in two forms, amorphous and crystalline. The crystalline forms include diamond and graphite. The amorphous forms include charcoal, coke, and carbon black; coal is an impure variety of amorphous carbon. Some of the typical properties of carbon and electrographites manufactured from these carbons are listed in Tables 4-38 and 4-39.

Most carbon used for electrical purposes is made from a mixture of powdered carbon and/or graphite (such as lampblack and petroleum coke) and binders (such as pitch and resins), which are mixed into a homogeneous mass, extruded or molded, and then baked. When the mixture is baked to approximately 900°C with the air excluded, the volatile part of the binding material is driven off and the remaining binder is carbonized. The resulting product can be converted into electrographite by furnacing it in the absence of oxygen to a temperature of not less than 2200°C, usually higher.

***187. Temperature Coefficient of Resistance.*** Carbon exhibits an increasing electrical and thermal conductivity with rising temperature. Graphite can exhibit a complicated change in electrical conductivity with rising temperature (see Fig. 4-17), and its thermal conductivity decreases markedly with rising temperature.

***188. Carbon-Brush Applications.*** The term *carbon brush* is used to designate all types of sliding electrical contacts that contain any appreciable percentage of carbon or graphite in their composition. Other ingredients may be metals and suitable binders.

Carbon and graphite brushes for commutator-type machines and collector rings of ac machines are made in various grades with appropriate characteristics for the types of service, including atmospheric conditions and load cycles.

The functions to be performed by carbon brushes on electrical machines are these:

Carbon brushes on a slip-ring machine have only to provide a suitable sliding electrical connection between the line and the rotor with reasonable life.

Carbon brushes on a dc machine have three entirely distinct functions to perform: (1) They must carry the current into and out of the commutator. (2) They participate in the reversal of all or part of the current in the armature coils during the time they are short-circuited. Performance of both functions simultaneously complicates the problem of design. In order that the current in the short-circuited armature coils may be reversed without arcing or sparking, there must be an appreciable resistance in the contact between the brush and commutator. The amount of resistance depends upon the magnitude of the reactance voltage. (3) And they must have reasonable life.

The carbon brush in its various types and compositions provides the characteristics needed for an ideal sliding electrical connection, namely, good conductivity, low coefficient of friction, and high durability.

The resistances of carbon contacts vary with pressure, current, and time (see *Bibliography and Abstracts on Electrical Contacts,* ASTM, Holm Conference, IEEE, 1934 to date, and R. Holm, *Electric Contacts,* Springer, 1967). The property of variable contact resistance with varying contact pressure is also very useful in carbon-pile resistors, which can be varied over

**TABLE 4-38** Typical Properties of Carbon

| Material | True specific gravity* | Specific heat at temperatures, g·cal/g·°C | | | | | | Threshold temp. of oxidation in air, °C |
|---|---|---|---|---|---|---|---|---|
| | | 26–76 °C | 26–282 °C | 26–538 °C | 36–902 °C | 47–1193 °C | 56–1450 °C | |
| Carbon, coke base, gas calcined | 1.98–2.10† | 0.168 | 0.200 | 0.199 | 0.315 | 0.352 | 0.387 | 350 |
| Carbon, coke base, graphitized | 2.20–2.24 | 0.168 | 0.200 | 0.199 | 0.315 | 0.352 | 0.387 | 400 |
| Carbon, lampblack base, gas calcined | 1.80–1.85† | | | | | | | 350 |
| Carbon, lampblack base, graphitized | 1.98–2.08 | | | | | | | 400 |
| Anthracite, gas calcined | 1.79† | | | | | | | — |
| Anthracite, electric calcined | 1.90–1.97† | | | | | | | — |
| Graphite, pure | 2.25 | 0.165 | 0.195 | 0.234 | 0.324 | 0.350 | 0.390 | — |
| Diamond | 3.51 | 0.160 | 0.315 | 0.415 | | | | — |

*As measured by pycnometer.
†Dependent on source and degree of calcination.

**TABLE 4-39** Properties of Manufactured Carbon and Graphite*

| Material | Form | Apparent density, g/cm$^3$ | Strength, lb/in$^2$ | | | Elastic modulus, lb/in$^2$ × 10$^6$ | Electrical resistivity at 20°C (68°F), Ω·in | Thermal expansion per °C × 10$^{-7}$ (RT-100°C)† |
|---|---|---|---|---|---|---|---|---|
| | | | Tensile | Compressive | Flexural | | | |
| Carbon rounds (coal based) | Up to 55 in diam. | 1.6 | 200 | 1800/2000 | 400/700 | 0.6/0.8 | 0.0012/0.0020 | 30 |
| Carbon blocks (coal based) | Up to 30 × 36 in | 1.6 | 200 | 2500/2700 | 600/750 | 0.8/0.9 | 0.0014/0.0020 | 35 |
| Graphite rounds (mold stock & electrodes) | Up to 3 in diam. | 1.6/1.8 | 800/2500 | 4000/8000 | 2000/5500 | 1.5/2.5 | 0.00025/0.00040 | 10/15 |
| | 4 to 10 in diam. | 1.6/1.8 | 500/2500 | 3000/6500 | 1500/3500 | 1.0/2.0 | 0.00025/0.00040 | 10/20 |
| | 12 to 28 in diam. | 1.6/1.8 | 450/2500 | 2000/6500 | 1000/3500 | 1.0/2.0 | 0.00025/0.00040 | 5/20 |
| | 30 in diam. & larger | 1.6/1.8 | 450/1500 | 3000/5000 | 1400/2000 | 1.0/2.0 | 0.00030/0.00040 | 20/30 |
| Graphite blocks | Up to 6 in thick | 1.6/1.8 | 700/2500 | 3000/8000 | 1700/4500 | 1.5/2.0 | 0.00025/0.00040 | 10/15 |
| | 6 to 20 in thick | 1.6/1.8 | 700/2500 | 3000/6000 | 1500/3500 | 1.0/2.0 | 0.00025/0.00040 | 10/20 |
| | Over 20 in thick | 1.6/1.8 | 550/2000 | 3000/5000 | 1500/2500 | 1.0/2.0 | 0.00030/0.00040 | 10/20 |

*Characteristics shown are typical values and will differ with individual grades and manufacturers.

†RT = room temperature.

NOTE: With grain, properties are listed as minimum/maximum values. 1 in = 2.54 cm; 1 lb/in$^2$ = 6.895 kPa.

**TABLE 4-40** Range of Properties and Characteristics of Typical Brush Materials*

| Physical properties and characteristics | Types of brush materials | | | | |
|---|---|---|---|---|---|
| | Carbon | Carbon graphite | Graphite (resin bonded) | Electrographitic | Metal graphite |
| Resistivity, Ω·in | 0.0010–0.0035 | 0.0005–0.0025 | 0.0003–0.050 | 0.0004–0.0035 | 0.000002–0.0001 |
| Scleroscope hardness | 45–85 | 40–75 | 10–35 | 12–75 | 7–35 |
| Flexural strength, lb/in² | 3500–10,000 | 3000–7500 | 1000–5000 | 1500–10,000 | 2500–10,000 |
| Current-carrying capacity, A/in² | 35–45 | 45–50 | 10–60 | 35–70 | 75–150 |
| Contact drop† | Medium–high | Low–medium | Low–very high | Medium–high | Very low–low |
| Coefficient of friction‡ | Medium–high | Low–medium | Low–medium | Low–medium | Very low–medium |
| Abrasiveness (cleaning action) | Pronounced | Medium–pronounced | Slight | Slight | Slight–medium |
| Peripheral speed, ft/min | 2000–4000 | 3000–5000 | 4000–12,000 | 3000–9000 | 3000–6000 |

*See NEMA Standard CB1—1984 for test procedures.

†These terms have the following meanings: high, over 2.5 V; medium, 1.8 to 2.5 V; low, 1.0 to 1.8 V; very low, below 1.0 V.

‡These terms have the following meanings: high, over 0.26; medium, 0.20 to 0.26; low, 0.15 to 0.20; very low, 0.15 and below.

NOTE: 1 in = 2.54 cm; 1 lb/in$^2$ = 6.895 kPa; 1 in$^2$ = 64.5 cm$^2$; 1 ft/min = 0.00508 m/s.

a wide range, from practically open circuit to very low resistance with manipulation of the pressure on the pile.

Carbon contacts for relays have the same characteristics as low-resistance carbon brushes (Table 4-40).

For references on electrical brushes and their design, see NEMA Standard CB1.

***189. Arc-Lamp Carbons.*** Carbons are produced in forms especially adapted for arc-lamp projectors of various types for the motion picture industry, for searchlights, and for irradiation applications. They contain various salts or compounds of metals such as cerium, calcium, cobalt, and strontium to control the wavelength of the arc output from ultraviolet through infrared. They can also be made to simulate the properties of sunlight.

The resistance of a ½-in-diameter by 12-in-long enclosed-arc carbon varies from 0.012 to 0.015 Ω/lin in. The resistances of other diameters vary according to their cross-sectional areas. For many applications, these carbons are copper-coated.

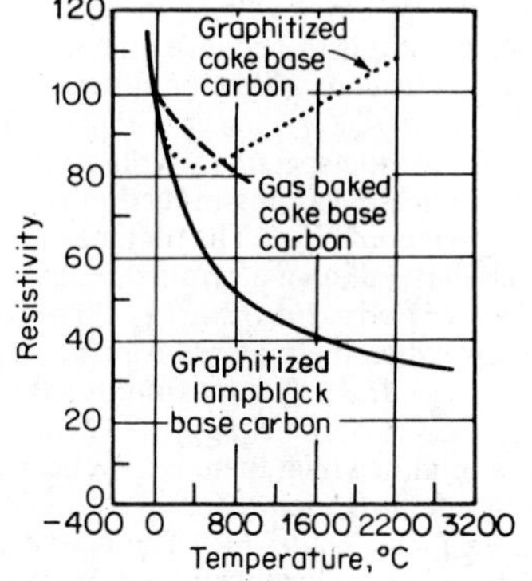


**FIG. 4-17** Temperature-resistance relationship of carbon. 100 = resistivity at 20°C.

## *MAGNETIC MATERIALS*

*By ANTHONY L. VON HOLLE and KENNETH L. LATIMER*

***190. Definitions.*** The following definitions of terms relating to magnetic materials and to the properties and testing of these materials have been selected from ASTM Standard A340.[1] Terms primarily related to magnetostatics are indicated by the symbol * and those related to magnetodynamics are indicated by the symbol **. General (nonrestricted) terms are not marked.

***AC Excitation* $N_1 I/l_1$. The ratio of the rms ampere-turns of exciting current in the primary winding of an inductor to the effective length of the magnetic path.

***Active (Real) Power* $P$. The product of the rms current $I$ in an electric circuit, the rms voltage $E$ across the circuit, and the cosine of the angular phase difference $\theta$ between the current and the voltage.

$$P = EI \cos \theta \tag{4-38}$$

NOTE: The portion of the active power that is expended in a magnetic core is the total core loss $P_c$.

*Aging, Magnetic.* The change in the magnetic properties of a material resulting from metallurgical change. This term applies whether the change results from a continued normal or a specified accelerated aging condition.

NOTE: This term implies a deterioration of the magnetic properties of magnetic materials for electronic and electrical applications, unless otherwise specified.

*Ampere-turn.* Unit of magnetomotive force in the rationalized mksa system. One ampere-turn equals $4\pi/10$ or 1.257 gilberts.

*Ampere-turn per Meter.* Unit of magnetizing force (magnetic field strength) in the rationalized mksa system. One ampere-turn per meter is $4\pi \times 10^{-3}$ or 0.01257 oersted.

*Anisotropic Material.* A material in which the magnetic properties differ in various directions.

[1] *1983 Annual Book of ASTM Standards,* sec. 3, vol. 03.04; ASTM, Philadelphia, Pa., 1983.

*Antiferromagnetic Material.* A feebly magnetic material in which almost equal magnetic moments are lined up antiparallel to each other. Its susceptibility increases as the temperature is raised until a critical (Neél) temperature is reached; above this temperature the material becomes paramagnetic.

*******Apparent Power $P_a$.* The product (volt-amperes) of the rms exciting current and the applied rms *terminal* voltage in an *electric* circuit containing inductive impedance. The components of this impedance due to the winding will be linear, while the components due to the magnetic core will be nonlinear.

*******Apparent Power; Specific, $P_{a(B,f)}$.* The value of the apparent power divided by the active mass of the specimen (volt-amperes per unit mass) taken at a specified maximum value of cyclically varying induction $B$ and at a specified frequency $f$.

**Coercive Force $H_c$.* The (dc) magnetizing force at which the magnetic induction is zero when the material is in a symmetrically cyclically magnetized condition.

**Coercive Force, Intrinsic, $H_{ci}$.* The (dc) magnetizing force at which the intrinsic induction is zero when the material is in a symmetrically cyclically magnetized condition.

**Coercivity $H_{cs}$.* The maximum value of coercive force.

***Core Loss, Specific, $P_{c(B,f)}$.* The active power (watts) expended per unit mass of magnetic material in which there is a cyclically varying induction of a specified maximum value $B$ at a specified frequency $f$.

***Core Loss (Total) $P_c$.* The active power (watts) expended in a magnetic circuit in which there is a cyclically alternating induction.

NOTE: Measurements of core loss are normally made with sinusoidally alternating induction, or the results are corrected for deviations from the sinusoidal condition.

*Curie Temperature $T_c$.* The temperature above which a ferromagnetic material becomes paramagnetic.

**Demagnetization Curve.* That portion of a normal (dc) hysteresis loop which lies in the second or fourth quadrant, that is, between the residual induction point $B_r$ and the coercive force point $H_c$. Points on this curve are designated by the coordinates $B_d$ and $H_d$.

*Diamagnetic Material.* A material whose relative permeability is less than unity.

NOTE: The intrinsic induction $B_i$, is oppositely directed to the applied magnetizing force $H$.

*Domains, Ferromagnetic.* Magnetized regions, either macroscopic or microscopic in size, within ferromagnetic materials. Each domain, per se, is magnetized to intrinsic saturation at all times, and this saturation induction is unidirectional within the domain.

***Eddy-Current Loss, Normal, $P_e$.* That portion of the core loss which is due to induced currents circulating in the magnetic material subject to an *SCM* excitation.

**Energy Product $B_dH_d$.* The product of the coordinate values of any point on a demagnetization curve.

**Energy-Product Curve, Magnetic.* The curve obtained by plotting the product of the corresponding coordinates $B_d$ and $H_d$ of points on the demagnetization curve as abscissa against the induction $B_d$ as ordinates.

NOTE 1: The maximum value of the energy product $(B_dH_d)_m$ corresponds to the maximum value of the external energy.

NOTE 2: The demagnetization curve is plotted to the left of the vertical axis and usually the energy-product curve to the right.

***Exciting Power, rms, $P_z$.* The product of the rms exciting current and the rms voltage induced in the exciting (primary) winding on a magnetic core.

NOTE: This is the apparent volt-amperes required for the excitation of the magnetic core only. When the core has a secondary winding, the induced primary voltage is obtained from the measured open-circuit secondary voltage multiplied by the appropriate turns ratio.

***Exciting Power, Specific $P_{z(B,f)}$.* The value of the rms exciting power divided by the active mass of the specimen (volt-amperes/unit mass) taken at a specified maximum value of cyclically varying induction $B$ and at a specified frequency $f$.

*Ferrimagnetic Material.* A material in which unequal magnetic moments are lined up antiparallel to each other. Permeabilities are of the same order of magnitude as those of ferromagnetic materials, but are lower than they would be if all atomic moments were parallel and in the same direction. Under ordinary conditions the magnetic characteristics of ferrimagnetic materials are quite similar to those of ferromagnetic materials.

*Ferromagnetic Material.* A material that, in general, exhibits the phenomena of hysteresis and saturation, and whose permeability is dependent on the magnetizing force.

*Gauss (Plural Gausses).* The unit of magnetic induction in the cgs electromagnetic system. The gauss is equal to 1 maxwell per square centimeter or $10^{-4}$ tesla. See *magnetic induction (flux density).*

*Gilbert.* The unit of magnetomotive force in the cgs electromagnetic system. The gilbert is a magnetomotive force of $10/4\pi$ ampere-turns. See *magnetomotive force.*

**Hysteresis Loop, Intrinsic.* A hysteresis loop obtained with a ferromagnetic material by plotting (usually to rectangular coordinates) corresponding dc values of intrinsic induction $B_i$ for ordinates and magnetizing force $H$ for abscissas.

**Hysteresis Loop, Normal.* A closed curve obtained with a ferromagnetic material by plotting (usually to rectangular coordinates) corresponding dc values of magnetic induction $B$ for ordinates and magnetizing force $H$ for abscissas when the material is passing through a complete cycle between equal definite limits of either magnetizing force $\pm H_m$ or magnetic induction $\pm B_m$. In general the normal hysteresis loop has mirror symmetry with respect to the origin of the $B$ and $H$ axes, but this may not be true for special materials.

**Hysteresis-Loop Loss $W_h$.* The energy expended in a single slow excursion around a normal hysteresis loop is given by the following equation:

$$W_h = \int H dB/4\pi \qquad \text{ergs} \tag{4-39}$$

where the integrated area enclosed by the loop is measured in gauss-oersteds.

***Hysteresis Loss, Normal, $P_h$.* *1.* The power expended in a ferromagnetic material, as a result of hysteresis, when the material is subjected to an *SCM* excitation.

*2.* The energy loss/cycle in a magnetic material as a result of magnetic hysteresis when the induction is cyclic (but not necessarily periodic).

*Hysteresis, Magnetic.* The property of a ferromagnetic material exhibited by the lack of correspondence between the changes in induction resulting from increasing magnetizing force and from decreasing magnetizing force.

*Induction B.* See *magnetic induction (flux density).*

**Induction, Intrinsic, $B_i$.* The vector difference between the magnetic induction in a magnetic material and the magnetic induction that would exist in a vacuum under the influence of the same magnetizing force. This is expressed by the equation

$$B_i = B - \Gamma_m H \tag{4-40}$$

NOTE: In the cgs-em system $B_i/4\pi$ is often called magnetic polarization.

*Induction, Maximum:*

**1.* $B_m$—The maximum value of $B$ in a hysteresis loop. The tip of this loop has the magnestostatic coordinates $H_m$, $B_m$, which exist simultaneously.

***2.* $B_{max}$—the maximum value of induction in a flux-current loop.

NOTE: In a flux-current loop, the magnetodynamic values $B_{max}$ and $H_{max}$ do not exist simultaneously; $B_{max}$ occurs later than $H_{max}$.

**Induction, Normal, B.* The maximum induction, in a magnetic material that is in a symmetrically cyclically magnetized condition.

NOTE: Normal induction is a magnetostatic parameter usually measured by ballistic methods.

**Induction, Remanent, $B_d$.* The magnetic induction that remains in a magnetic circuit after the removal of an applied magnetomotive force.

NOTE: If there are no air gaps or other inhomogeneities in the magnetic circuit the remanent induction $B_r$ will equal the residual induction $B_r$; if air gaps or other inhomogeneities are present, $B_d$ will be less than $B_r$.

**Induction, Residual, $B_r$.* The magnetic induction corresponding to zero magnetizing force in a magnetic material that is in a symmetrically cyclically magnetized condition.

**Induction, Saturation, $B_s$.* The maximum intrinsic induction possible in a material.

**Induction Curve, Intrinsic (Ferric).* A curve of a previously demagnetized specimen depicting the relation between intrinsic induction and corresponding ascending values of magnetizing force. This curve starts at the origin of the $B_i$ and $H$ axes.

**Induction Curve, Normal.* A curve of a previously demagnetized specimen depicting the relation between normal induction and corresponding ascending values of magnetizing force. This curve starts at the origin of the $B$ and $H$ axes.

*Isotropic Material.* Material in which the magnetic properties are the same for all directions.

*Magnetic Circuit.* A region at whose surface the magnetic induction is tangential.

NOTE: A practical magnetic circuit is the region containing the flux of practical interest, such as the core of a transformer. It may consist of ferromagnetic material with or without air gaps or other feebly magnetic materials such as procelain and brass.

*Magnetic Constant (Permeability of Space)* $\Gamma_m$. The dimensional scalar factor that relates the mechanical force between two currents to their intensities and geometrical configurations. That is,

$$dF = \Gamma_m I_1 I_2 \, dl_1 \times (dl_2 \times r_1)/nr^2 \tag{4-41}$$

where $\Gamma_m$ = magnetic constant when the element of force $dF$ of a current element $I_1 \, dl_1$ on another current element $I_2 \, dl_2$ is at a distance $r$

$r_1$ = unit vector in the direction from $dl_1$ to $dl_2$

$n$ = dimensionless factor; the symbol $n$ is unity in unrationalized systems and $4\pi$ in rationalized systems

NOTE 1: The numerical values of $\Gamma_m$ depend upon the system of units employed. In the cgs-em system $\Gamma_m = 1$, in the rationalized mksa system $\Gamma_m = 4\pi \times 10^{-7}\ h/m$.

NOTE 2: The magnetic constant expresses the ratio of magnetic induction to the corresponding magnetizing force at any point in a vacuum and therefore is sometimes called the permeability of space $\mu_r$.

NOTE 3: The magnetic constant times the relative permeability is equal to the absolute permeability.

$$\mu_{abs} = \Gamma_m \mu_r \tag{4-42}$$

*Magnetic Field Strength H.* See *magnetizing force.*

*Magnetic Flux φ.* The product of the magnetic induction $B$ and the area of a surface (or cross section) $A$ when the magnetic induction $B$ is uniformly distributed and normal to the plane of the surface.

$$\phi = BA \tag{4-43}$$

where $\phi$ = magnetic flux

$B$ = magnetic induction

$A$ = area of the surface

NOTE 1: If the magnetic induction is not uniformly distributed over the surface, the flux $\phi$ is the surface integral of the normal component of $B$ over the area.

$$\phi = \int\int_s B \, dA \tag{4-44}$$

NOTE 2: Magnetic flux is scalar and has no direction.

*Magnetic Flux Density B.* See *magnetic induction (flux density).*

*Magnetic Induction (Flux Density) B.* That magnetic vector quantity which at any point in a magnetic field is measured either by the mechanical force experienced by an element of electric current at the point, or by the electromotive force induced in an elementary loop during any change in flux linkages with the loop at the point.

NOTE 1: If the magnetic induction $B$ is uniformly distributed and normal to a surface or cross section, then the magnetic induction is

$$B = \phi/A \tag{4-45}$$

where $B$ = magnetic induction

$\phi$ = total flux

$A$ = area

NOTE 2: $B_{in}$ is the instantaneous value of the magnetic induction and $B_m$ is the maximum value of the magnetic induction.

*Magnetizing Force (Magnetic Field Strength) H.* That magnetic vector quantity at a point in a magnetic field which measures the ability of electric currents or magnetized bodies to produce magnetic induction at the given point.

NOTE 1: The magnetizing force $H$ may be calculated from the current and the geometry of certain magnetizing circuits. For example, in the center of a uniformly wound long solenoid

$$H = C(NI/l) \tag{4-46}$$

where $H$ = magnetizing force
$C$ = constant whose value depends on the system of units
$N$ = number of turns
$I$ = current
$l$ = axial length of the coil

If $I$ is expressed in amperes and $l$ is expressed in centimeters, then $C = 4\pi/10$ in order to obtain $H$ in the cgs = em unit, the oersted.

If $I$ is expressed in amperes and $l$ is expressed in meters, then $C = 1$ in order to obtain $H$ in the mksa unit, ampere-turn per meter.

NOTE 2: The magnetizing force $H$ at a point in air may be calculated from the measured value of induction at the point by dividing this value by the magnetic constant $\Gamma_m$.

***Magnetizing Force, AC* Three different values of dynamic magnetizing force parameters are in common use:

**a.** $H_L$—an assumed peak value computed in terms of peak magnetizing current (considered to be sinusoidal).

**b.** $H_x$—an assumed peak value computed in terms of measured rms exciting current (considered to be sinusoidal).

**c.** $H_p$—computed in terms of a measured peak value of exciting current, and thus equal to the value $H'_{max}$.

***Magnetodynamic.* The magnetic condition when the values of magnetizing force and induction vary, usually periodically and repetitively, between two extreme limits.

*Magnetomotive Force $\mathcal{F}$.* The line integral of the magnetizing force around any flux loop in space.

$$\mathcal{F} = \oint H\,dl \tag{4-47}$$

where $\mathcal{F}$ = magnetomotive force
$H$ = magnetizing force
$dl$ = unit length along the loop

NOTE: The magnetomotive force is proportional to the net current linked with any closed loop of flux or closed path.

$$\mathcal{F} = CNI \tag{4-48}$$

where $\mathcal{F}$ = magnetomotive force
$N$ = number of turns linked with the loop
$I$ = current in amperes
$C$ = constant whose value depends on the system of units. In the cgs system $C = 4\pi/10$. In the mksa system $C = 1$

**Magnetostatic.* The magnetic condition when the values of magnetizing force and induction are considered to remain invariant with time during the period of measurement. This is often referred to as a dc (direct-current) condition.

*Magnetostriction.* Changes in dimensions of a body resulting from magnetization.

*Maxwell.* The unit of magnetic flux in the cgs electromagnetic system. One maxwell equals $10^{-8}$ weber. See *magnetic flux*.

NOTE:

$$e = -N\, d\phi/dt \times 10^{-8} \tag{4-49}$$

where $e$ = induced instantaneous emf volts
$d\phi/dt$ = time rate of change of flux, maxwells per second
$N$ = number of turns surrounding the flux, assuming each turn is linked with all the flux

*Oersted.* The unit of magnetizing force (magnetic field strength) in the cgs electromagnetic system. One oersted equals a magnetomotive force of 1 gilbert/cm of flux path. One oersted equals $100/4\pi$ or 79.58 ampere-turns per meter. See *magnetizing force (magnetic field strength).*

*Paramagnetic Material.* A material having a relative permeability which is slightly greater than unity, and which is practically independent of the magnetizing force.

***Permeability, AC.* A generic term used to express various dynamic relationships between magnetic induction $B$ and magnetizing force $H$ for magnetic material subjected to a cyclic excitation by alternating or pulsating current. The values of ac permeability obtained for a given material depend fundamentally upon the excursion limits of dynamic excitation and induction, the method and conditions of measurement, and also upon such factors as resistivity, thickness of laminations, frequency of excitation, etc.

NOTE: The numerical value for any permeability is meaningless unless the corresponding $B$ or $H$ excitation level is specified. For incremental permeabilities not only the corresponding dc $B$ or $H$ excitation level must be specified, but also the dynamic excursion limits of dynamic excitation range ($\Delta B$ or $\Delta H$).

AC permeabilities in common use for magnetic testing are

*a.* ***Impedance (rms) Permeability* $\mu_z$. The ratio of the measured peak value of magnetic induction to the value of the apparent magnetizing force $H_z$ calculated from the measured rms value of the exciting current, for a material in the *SCM* condition.

NOTE: The value of the current used to compute $H_z$ is obtained by multiplying the measured value of rms exciting current by 1.414. This assumes that the total exciting current is magnetizing current and is sinusoidal.

*b.* ***Inductance Permeability* $\mu_L$. For a material in an *SCM* condition, the permeability is evaluated from the measured inductive component of the electric circuit representing the magnetic specimen. This circuit is assumed to be composed of paralleled linear inductive and resistive elements $\omega L_1$ and $R_1$.

*c.* ***Peak Permeability* $\mu_p$. The ratio of the measured peak value of magnetic induction to the peak value of the magnetizing force $H_p$, calculated from the measured peak value of the exciting current, for a material in the *SCM* condition.

Other ac permeabilities are

*d.* ***Ideal Permeability* $\mu_a$. The ratio of the magnetic induction to the corresponding magnetizing force after the material has been simultaneously subjected to a value of ac magnetizing force approaching saturation (of approximate sine waveform) superimposed on a given dc magnetizing force, and the ac magnetizing force has thereafter been gradually reduced to zero. The resulting ideal permeability is thus a function of the dc magnetizing force used.

NOTE: Ideal permeability, sometimes called anhysteretic permeability, is principally significant to feebly magnetic material and to the Rayleigh range of soft magnetic material.

*e.* ***Impedance, Permeability, Incremental,* $\mu_{\Delta z}$. Impedance permeability $\mu_z$ obtained when an ac excitation is superimposed on a dc excitation, *CM* condition.

*f.* ***Inductance Permeability, Incremental,* $\mu_{\Delta L}$. Inductance permeability $\mu_L$ obtained when an ac excitation is superimposed on a dc excitation, *CM* condition.

*g.* ***Initial Dynamic Permeability* $\mu_{0d}$. The limiting value of inductance permeability $\mu_L$ reached in a ferromagnetic core when, under *SCM* excitation, the magnetizing current has been progressively and gradually reduced from a comparatively high value to zero value.

NOTE: This same value, $\mu_{0d}$, is also equal to the initial values of both impedance permeability $\mu_x$ and peak permeability $\mu_p$.

*h.* ***Instantaneous Permeability (Coincident with $B_{max}$)* $\mu_t$. With *SCM* excitation, the ratio of the maximum induction $B_{max}$ to the instantaneous magnetizing force $H_t$, which is the value of apparent magnetizing force $H'$ determined at the instant when $B$ reaches a maximum.

*i.* ***Peak Permeability, Incremental,* $\mu_{\Delta p}$. Peak permeability $\mu_p$ obtained when an ac excitation is superimposed on dc excitation, *CM* condition.

**Permeability, DC.* Permeability is a general term used to express relationships between magnetic induction $B$ and magnetizing force $H$ under various conditions of magnetic excitation. These relationships are either (*1*) absolute permeability, which in general is the quotient of a change in magnetic induction divided by the corresponding change in magnetizing force, or (*2*) relative permeability, which is the ratio of the absolute permeability to the magnetic constant $\Gamma_m$.

NOTE 1: The magnetic constant $\Gamma_m$ is a scalar quantity differing in value and uniquely determined by each electromagnetic system of units. In the unrationalized cgs system $\Gamma_m$ is 1 gauss/oersted and in the mksa rationalized system $\Gamma_m = 4\pi \times 10^{-7}$ H/m.

NOTE 2: Relative permeability is a pure number which is the same in all unit systems. The value and dimension of absolute permeability depend on the system of units employed.

NOTE 3: For any ferromagnetic material permeability is a function of the degree of magnetization. However, initial permeability $\mu_0$ and maximum permeability $\mu_m$ are unique values for a given specimen under specified conditions.

NOTE 4: Except for initial permeability $\mu_0$, a numerical value for any of the dc permeabilities is meaningless unless the corresponding $B$ or $H$ excitation level is specified.

NOTE 5: For the incremental permeabilities $\mu_\Delta$ and $\mu_{\Delta i}$, a numerical value is meaningless unless both the corresponding values of mean excitation level ($B$ or $H$) and the excursion range ($\Delta B$ or $\Delta H$) are specified.

The following dc permeabilities are frequently used in magnetostatic measurements primarily concerned with the testing of materials destined for use with permanent or dc excited magnets.

*a.* **Absolute Permeability* $\mu_{abs}$. The sum of the magnetic constant and the intrinsic permeability. It is also equal to the product of the magnetic constant and the relative permeability:

$$\mu_{abs} = \Gamma_m + \mu_i = \Gamma_m \mu_r \tag{4-50}$$

*b.* **Differential Permeability* $\mu_d$. The absolute value of the slope of the hysteresis loop at any point, or the slope of the normal magnetizing curve at any point.

*c.* **Effective Circuit Permeability* $\mu_{eff}$. When a magnetic circuit consists of two or more components, each individually homogeneous throughout but having different permeability values, the effective (overall) permeability of the circuit is that value computed in terms of the total magnetomotive force, the total resulting flux, and the geometry of the circuit.

NOTE: For a symmetrical series circuit in which each component has the same cross-sectional area, reluctance values add directly, giving

$$\mu_{eff} = \frac{l_1 + l_2 + l_3 + \cdots}{l_1/\mu_1 + l_2/\mu_2 + l_3/\mu_3 + \cdots} \tag{4-51}$$

For a symmetrical parallel circuit in which each component has the same flux path length, permeance values add directly, giving

$$\mu_{eff} = \frac{\mu_1 A_1 + \mu_2 A_2 + \mu_3 A_3 + \cdots}{A_1 + A_2 + A_3 + \cdots} \tag{4-52}$$

*d.* **Incremental Intrinsic Permeability* $\mu_{\Delta i}$. The ratio of the change in intrinsic induction to the corresponding change in magnetizing force when the mean induction differs from zero.

*e.* **Incremental Permeability* $\mu_\Delta$. The ratio of a change in magnetic induction to the corresponding change in magnetizing force when the mean induction differs from zero. It equals the slope of a straight line joining the excursion limits of an incremental hysteresis loop.

NOTE: When the change in $H$ is reduced to zero, the incremental permeability $\mu_\Delta$ becomes the reversible permeability $\mu_{rev}$.

*f. *Initial Permeability* $\mu_0$. The limiting value approached by the normal permeability as the applied magnetizing force $H$ is reduced to zero. The permeability is equal to the slope of the normal induction curve at the origin of linear $B$ and $H$ axes.

*g. *Intrinsic Permeability* $\mu_i$. The ratio of intrinsic induction to the corresponding magnetizing force.

*h. *Maximum Permeability* $\mu_m$. The value of normal permeability for a given material where a straight line from the origin of linear $B$ and $H$ axes becomes tangent to the normal induction curve.

*i. *Normal Permeability* $\mu$ *(without subscript).* The ratio of the normal induction to the corresponding magnetizing force. It is equal to the slope of a straight line joining the extrusion limits of a normal hysteresis loop, or the slope of a straight line joining any point ($H_m$, $B_m$) on the normal induction curve to the origin of the linear $B$ and $H$ axes.

*j. *Relative Permeability* $\mu_r$. The ratio of the absolute permeability of a material to the magnetic constant $\Gamma_m$, giving a pure numeric parameter.

NOTE: In the cgs-em system of units the relative permeability is numerically the same as the absolute permeability.

*k. Reversible Permeability* $\mu_{rev}$. The limit of the incremental permeability as the change in magnetizing force approaches zero.

*l. Space Permeability* $\mu_0$. The permeability of space (vacuum), identical with the magnetic constant $\Gamma_m$.

***Reactive Power (Quadrature Power)* $P_q$. The product of the rms current in an electric circuit, the rms voltage across the circuit, and the sine of the angular phase difference between the current and the voltage.

$$P_q = EI \sin \theta \tag{4-53}$$

where $P_q$ = reactive power, vars
$E$ = voltage, volts
$I$ = current, amperes
$\theta$ = angular phase by which $E$ leads $I$

NOTE: The reactive power supplied to a magnetic core having an *SCM* excitation is the product of the magnetizing current and the voltage induced in the exciting winding.

**Remanence* $B_{dm}$. The maximum value of the remanent induction for a given geometry of the magnetic circuit.

NOTE: If there are no air gaps or other inhomogeneities in the magnetic circuit, the remanence $B_{dm}$ is equal to the retentivity $B_{rs}$; if air gaps or other inhomogeneities are present, $B_{dm}$ will be less than $B_{rs}$.

**Retentivity* $B_{rs}$. That property of a magnetic material which is measured by its maximum value of the residual induction.

NOTE: Retentivity is usually associated with saturation induction.

*Symmetrically Cyclically Magnetized Condition, SCM.* A magnetic material is in an *SCM* condition when, under the influence of a magnetizing force that varies cyclically between two equal positive and negative limits, its successive hysteresis loops or flux-current loops are both identical and symmetrical with respect to the origin of the axes.

*Tesla.* The unit of magnetic induction in the mksa (Giorgi) system. The tesla is equal to 1 $Wb/m^2$ or $10^4$ gausses.

*Var.* The unit of reactive (quadrature) power in the mksa (Giorgi) and the practical systems.

*Volt-Ampere.* The unit of apparent power in the mksa (Giorgi) and the practical systems.

*Watt.* The unit of active power in the mksa (Giorgi) and the practical systems. One watt is a power of one joule/second.

*Weber.* The unit of magnetic flux in the mksa and in the practical system. The weber is the magnetic flux whose decrease to zero when linked with a single turn induces in the turn a voltage whose time integral is one volt-second. One weber equals $10^8$ maxwells. See *magnetic flux.*

***191. Magnetic Properties and Their Application.*** The relative importance of the various magnetic properties of a magnetic material varies from one application to another. In general, properties of interest may include normal induction, hysteresis, dc permeability, ac permeability, core loss, and exciting power. It should be noted that there are various means of expressing ac permeability. The choice depends primarily on the ultimate use.

Techniques for the magnetic testing of many magnetic materials are described in the ASTM standards.[1] The magnetic and electric circuits employed in magnetic testing of a specimen are as free as possible from any unfavorable design factors which would prevent the measured magnetic data from being representative of the inherent magnetic properties of the specimen. The flux "direction" in the specimen is normally specified, since most magnetic materials are magnetically anisotropic. In most ac magnetic tests, the waveform of the flux is required to be sinusoidal.

As a result of the existence of unfavorable conditions, such as those listed and described below, the performance of a magnetic material in a magnetic device can be greatly deteriorated from that which would be expected from magnetic testing of the material. Allowances for these conditions, if present, must be made during the design of the device if the performance of the device is to be correctly predicted.

*Leakage.* A principal difficulty in the design of many magnetic circuits is due to the lack of a practicable material which will act as an insulator with respect to magnetic flux. This results in magnetic flux seldom being completely confined to the desired magnetic circuit. Estimates of leakage flux for a particular design may be made based on experience and/or experimentation.

*Flux Direction.* Some magnetic materials have a very pronounced directionality in their magnetic properties. Failure to utilize these materials in their preferred directions results in impaired magnetic properties.

*Fabrication.* Stresses introduced into magnetic materials by the various fabricating techniques often adversely affect the magnetic properties of the materials. This occurs particularly in materials having high permeability. Stresses may be eliminated by a suitable stress-relief anneal after fabrication of the material to final shape.

*Joints.* Joints in an electromagnetic core may cause a large increase in total excitation requirements. In some cores operated on ac, core loss may also be increased.

*Waveform.* When a sinusoidal voltage is applied to an electromagnetic core, the resulting magnetic flux is not necessarily sinusoidal in waveform, especially at high inductions. Any harmonics in the flux waveform cause increases in core loss and required excitation power.

*Flux Distribution.* If the maximum and minimum lengths of the magnetic path in an electromagnetic core differ too much, the flux density may be appreciably greater at the inside of the core structure than at the outside. For cores operated on ac, this can cause the waveform of the flux at the extremes of the core structure to be distorted even when the total flux waveform is sinusoidal.

***192. Types of Magnetism.*** Any substance may be classified into one of the following categories according to the type of magnetic behavior it exhibits:

1. Diamagnetic.
2. Paramagnetic.
3. Antiferromagnetic.
4. Ferromagnetic.
5. Ferrimagnetic.

Substances which fall into the first three categories are so weakly magnetic that they are commonly thought of as "nonmagnetic." In contrast, ferromagnetic and ferrimagnetic substances are strongly magnetic and are thereby of interest as "magnetic materials." The magnetic behavior of any ferromagnetic or ferrimagnetic material is a result of its spontaneously magnetized magnetic domain structure and is characterized by a nonlinear normal induction curve, hysteresis, and saturation.

The pure elements which are ferromagnetic are iron, nickel, cobalt, and some of the rare

[1] *1983 Annual Book of ASTM Standards,* sec. 3, vol. 03.04; ASTM, Philadelphia, Pa., 1983.

earths. Typical normal induction curves of annealed samples of iron, nickel, and cobalt of comparatively high purity are shown in Fig. 4-18 for the purpose of general comparison. Ferromagnetic materials of value to industry for their magnetic properties are almost invariably alloys of the metallic ferromagnetic elements with one another and/or with other elements.

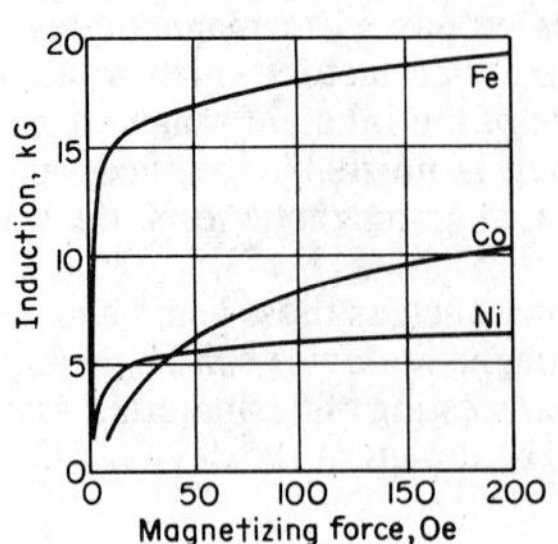


**FIG. 4-18** Typical normal-induction curves of annealed samples of iron, nickel, and cobalt.

Ferrimagnetism occurs mainly in the ferrites, which are chemical compounds having ferric oxide ($Fe_2O_3$) as a component. In recent years, some of the magnetic ferrites have become very important in certain magnetic applications. The magnetic ferrites saturate magnetically at lower inductions than do the great majority of metallic ferromagnetic materials. However, the electrical resistivities of ferrites are at least several orders of magnitude greater than those of metals.

***193. Commercial magnetic materials*** are generally divided into two main groups, each composed of ferromagnetic and ferrimagnetic substances:

1. Magnetically "soft" materials.
2. Magnetically "hard" materials.

The distinguishing characteristic of "soft" magnetic materials is high permeability. These materials are employed as core materials in the magnetic circuits of electromagnetic equipment.

"Hard" magnetic materials are characterized by a high maximum magnetic energy product $(BH)_{max}$. These materials are employed as permanent magnets to provide a constant magnetic field when it is inconvenient or uneconomical to produce the field by electromagnetic means.

Typical normal-induction curves for a wide range of commercial magnetic materials are shown in Fig. 4-19.

***194. "Soft" Magnetic Materials.*** A wide variety of "soft" magnetic materials have been developed to meet the many different requirements imposed on magnetic cores for modern electrical apparatus and electronic devices. The various soft magnetic materials will be considered under three classifications:

1. Materials for solid cores.
2. Materials for laminated cores.
3. Materials for special purposes.

***195. Materials for Solid Cores.*** These materials are used in dc applications such as yokes of dc dynamos, rotors of synchronous dynamos, and cores of dc electromagnets and relays. Proper annealing of these materials improves their magnetic properties. The principal magnetic requirements for the solid-core materials are high saturation, high permeability at relatively high inductions, and at times, low coercive force.

***196. Wrought iron*** is a ferrous material, aggregated from a solidifying mass of pasty particles of highly refined metallic iron, into which is incorporated, without subsequent fusion, a minutely and uniformly distributed quantity of slag. The better types of wrought iron are known as Norway iron and Swedish iron and are widely used in relays after being annealed to reduce coercive force and to minimize magnetic aging. A normal-induction curve for wrought iron is shown in Fig. 4-20.

***197. Cast irons*** are irons which contain carbon in excess of the amount which can be retained in solid solution in austenite at the eutectic temperature. The minimum carbon content is about 2%, while the practical maximum carbon content is about 4.5%. Cast iron was used in the yokes of dc dynamos in the early days of such machines.

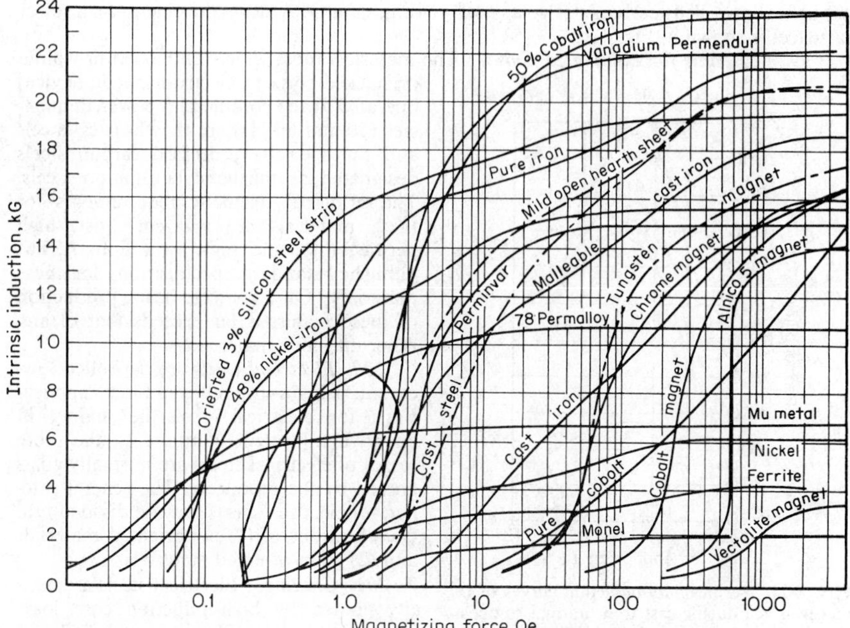


**FIG. 4-19** Typical normal-induction curves of a wide range of commercial magnetic materials.

*Gray cast iron* is a cast iron in which graphite is present in the form of flakes. It has very poor magnetic properties, inferior mechanical properties, and practically no ductility. It does lend itself well to the casting of complex shapes and is readily machinable. A normal-induction curve and a permeability curve for gray cast iron are shown in Figs. 4-20 and 4-21, respectively.

*Malleable cast iron* is a cast iron in which the graphite is present as temper carbon nodules. It is magnetically better than gray cast iron. A permeability curve for malleable cast iron is shown in Fig. 4-21.

*Ductile (nodular) cast iron* is a cast iron with the graphite essentially spheroidal in shape. It is magnetically better than gray cast iron. Ductile cast iron has the good castability and machinability of gray cast iron together with much greater strength, ductility, and shock resistance. Typical normal-induction curves and permeability curves for ductile cast iron are shown in Figs. 4-20 and 4-21, respectively.

***198. Carbon Steels.*** Carbon steels may contain from less than 0.1% carbon to more than 1% carbon. The magnetic properties of a carbon steel are greatly influenced by the carbon content and the disposition of the carbon. Low-carbon steels (less than 0.2% carbon) have magnetic properties which are similar to those of wrought iron and far superior to those of any of the cast irons.

*Wrought carbon steels* are widely used as solid-core materials. The low-carbon types are preferred in most applications.

*Cast carbon steels* replaced cast iron many years ago as the material used in the yokes of dc machines but have since been largely supplanted in this application by

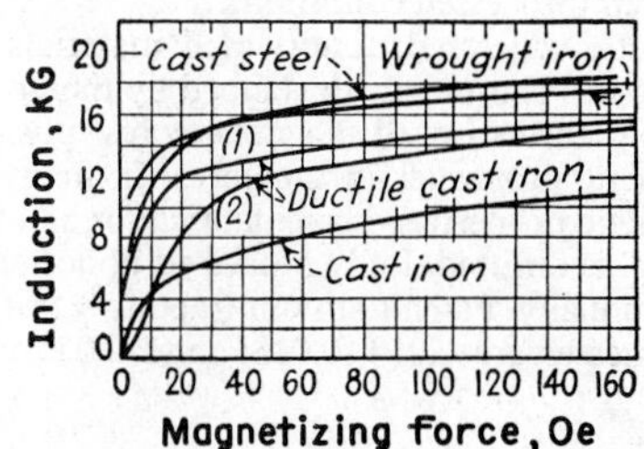


**FIG. 4-20** Normal-induction curves of typical wrought iron, cast iron, ductile cast iron (3% Si) as (1) cast and (2) annealed, and cast steel.

wrought (hot-rolled) carbon-steel plates of welding quality. A normal-induction curve for cast steel is shown in Fig. 4-20.

***199. Materials for Laminated Cores.*** The materials most widely employed in wound or stacked cores in electromagnetic devices operated at the commercial power frequencies (50 and 60 Hz) are the electrical steels and the specially processed carbon steels designated as magnetic lamination steels. The principal magnetic requirements for these materials are low core loss, high permeability, and high saturation. ASTM publishes standard specifications for these materials.[1] On a tonnage basis, production of these materials far exceeds that of any other magnetic material.

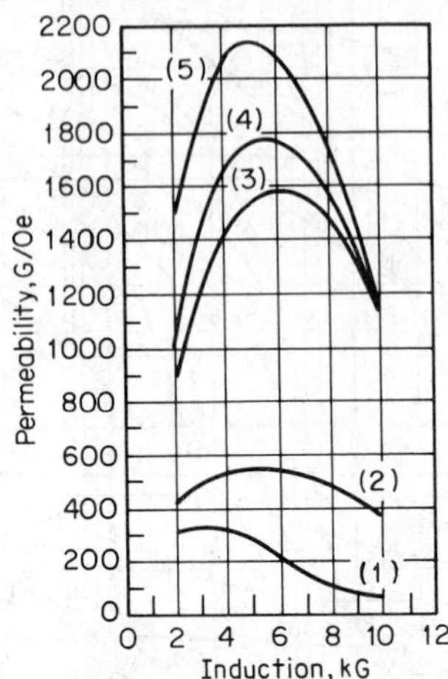


**FIG. 4-21** Permeability-induction curves of (1) cast iron, (2) ductile cast iron (normal composition) as cast, (3) annealed, (4) ductile iron (3% Si) annealed, and (5) malleable cast iron.

***200. Electrical steels*** are flat-rolled low-carbon silicon-iron alloys. Since applications for electrical steels lie mainly in energy-loss-limited equipment, the core losses of electrical steels are normally guaranteed by the producers. The general category of electrical steels may be divided into classifications of (1) nonoriented materials and (2) grain-oriented materials.

***201. Grading.*** Electrical steels are usually graded by high-induction core loss. Both ASTM and AISI have established and published designation systems for electrical steels based on core loss.[1,2]

The ASTM core loss type designation (ASTM Standard A664) consists of six or seven characters. The first two characters are 100 times the nominal thickness of the material in millimeters. The third character is a code letter which designates the class of the material and specifies the sampling and testing practices. The last three or four characters are 100 times the maximum permissible core loss in watts per pound at a specified test frequency and induction.

The AISI designation system has been discontinued but is still widely used. The AISI type designation for a grade consisted of the letter M followed by a number. The letter M stood for magnetic material, and the number was approximately equal to 10 times the maximum permissible core loss in watts per pound for 0.014-in material at 15 kG, 60 Hz in 1947.

*Nonoriented* electrical steels have approximately the same magnetic properties in all directions in the plane of the material (see Figs. 4-22 and 4-23). The common application is in punched laminations for large and small rotating machines and for small transformers. Today, nonoriented materials are always cold-rolled to final thickness. Hot rolling to final thickness is no longer practiced. Nonoriented materials are available in both fully processed and semiprocessed conditions.

Fully processed nonoriented materials (ASTM Standard A677) have their magnetic properties completely developed by the producer. Stresses introduced into these materials during fabrication of magnetic cores must be relieved by annealing to achieve optimum magnetic properties in the cores. In many applications, however, the degradation of the magnetic properties during fabrication is slight and/or can be tolerated and the stress-relief anneal is omitted. Fully processed nonoriented materials contain up to about 3.5% silicon. Additionally a small amount (about 0.5%) of aluminum is usually present. The common thicknesses are 0.014, 0.0185, and 0.025 in. Grade designations and maximum core-loss

---

[1] *1983 Annual Book of ASTM Standards,* sec. 3, vol. 03.04; ASTM, Philadelphia, Pa., 1983.
[2] *Steel Products Manual, Electrical Steel;* AISI, Washington, D.C., January 1983.

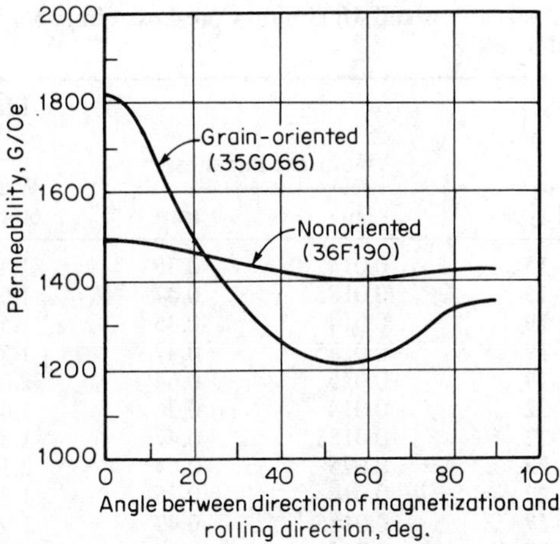


**FIG. 4-22** Effect of direction of magnetization on normal permeability at 10 Oe of fully processed electrical steels.

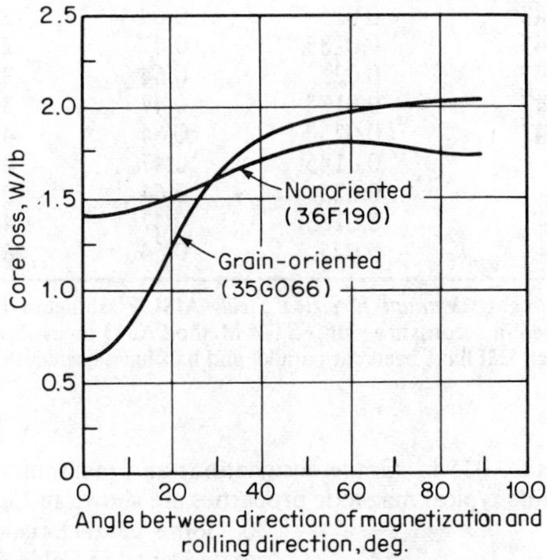


**FIG. 4-23** Effect of direction of magnetization on core loss at 15 kG, 60 Hz of fully processed electrical steel.

limits are given in Table 4-41. Some typical magnetic properties are shown in Figs. 4-24, 4-25, 4-26, and 4-27. Some general properties are presented in Table 4-44. Some characteristics and applications are described in Table 4-45.

Semiprocessed nonoriented materials (ASTM Standard A683) do not have their inherent magnetic properties completely developed by the producer and must be annealed properly to achieve both decarburization and grain growth. These materials are used primarily in high-volume production of small laminations and cores which would require stress-relief annealing if made from fully processed material. Semiprocessed nonoriented materials contain up to about 3% silicon. Additionally a small amount (about 0.5%) of aluminum is usually present. The carbon content may be as high as 0.05% but should be reduced to 0.005% or less by the required anneal. The common thicknesses of semiprocessed nonoriented

**TABLE 4-41** Grade Designations and Maximum Core-Loss Limits for Fully Processed Nonoriented Electrical Steels*

| ASTM type | Former AISI type | Nominal thickness | | Max. core loss at 15 kG† | |
|---|---|---|---|---|---|
| | | in | mm | W/lb, 60 Hz | W/kg, 60 Hz |
| 36F145 | M-15 | 0.014 | 0.36 | 1.45 | 2.53 |
| 47F168 | M-15 | 0.0185 | 0.47 | 1.68 | 2.93 |
| 36F158 | M-19 | 0.014 | 0.36 | 1.58 | 2.75 |
| 47F174 | M-19 | 0.0185 | 0.47 | 1.74 | 3.03 |
| 64F208 | M-19 | 0.025 | 0.64 | 2.08 | 3.62 |
| 36F168 | M-22 | 0.014 | 0.36 | 1.68 | 2.93 |
| 47F185 | M-22 | 0.0185 | 0.47 | 1.85 | 3.22 |
| 64F218 | M-22 | 0.025 | 0.64 | 2.18 | 3.80 |
| 36F180 | M-27 | 0.014 | 0.36 | 1.80 | 3.13 |
| 47F190 | M-27 | 0.0185 | 0.47 | 1.90 | 3.31 |
| 64F225 | M-27 | 0.025 | 0.64 | 2.25 | 3.92 |
| 36F190 | M-36 | 0.014 | 0.36 | 1.90 | 3.31 |
| 47F205 | M-36 | 0.0185 | 0.47 | 2.05 | 3.57 |
| 64F240 | M-36 | 0.025 | 0.64 | 2.40 | 4.18 |
| 47F230 | M-43 | 0.0185 | 0.47 | 2.30 | 4.01 |
| 64F270 | M-43 | 0.025 | 0.64 | 2.70 | 4.70 |
| 47F290 | M-45 | 0.0185 | 0.47 | 2.90 | 5.05 |
| 64F340 | M-45 | 0.025 | 0.64 | 3.40 | 5.92 |
| 47F380 | M-47 | 0.0185 | 0.47 | 3.80 | 6.62 |
| 64F470 | M-47 | 0.025 | 0.64 | 4.70 | 8.19 |
| 47F450 | — | 0.0185 | 0.47 | 4.50 | 7.84 |
| 64F550 | — | 0.025 | 0.64 | 5.50 | 9.58 |
| 47F490 | — | 0.0185 | 0.47 | 4.90 | 8.53 |
| 64F600 | — | 0.025 | 0.64 | 6.00 | 10.45 |

*Adapted from *Steel Products Manual, Electrical Steels;* AISI, Washington, D.C., January 1983.

†Core loss is determined in accordance with ASTM Method A343 on as-sheared Epstein specimens consisting of strips of which half have been cut parallel and half have been cut transverse to the rolling direction.

materials are 0.0185 and 0.025 in. Grade designations and maximum core-loss limits are given in Table 4-42. Some typical magnetic properties are shown in Figs. 4-28, 4-29, and 4-30. Some characteristics and applications are described in Table 4-45.

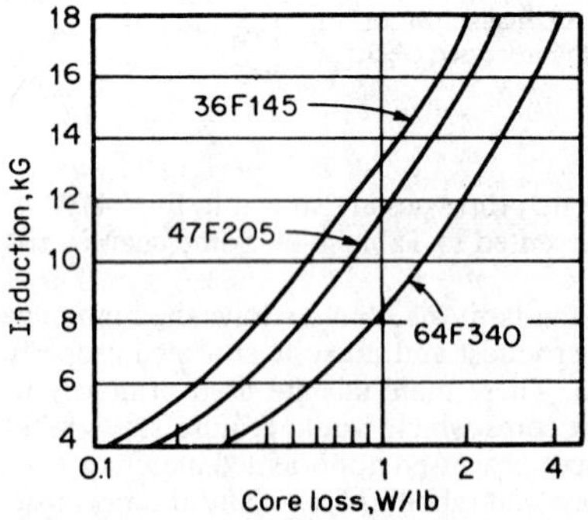

*Grain-oriented electrical steels* (ASTM Standards A665 and A725) have a pronounced directionality in their magnetic properties (Figs. 4-22 and 4-23). This directionality is a result of the "cube-on-edge" crystal structure achieved by proper composition and processing. Grain-oriented materials are employed most effectively in magnetic cores in which the flux path lies entirely or predominately in the rolling direction of the material. The common application is in cores of power and distribution transformers for electric utilities.

Grain-oriented materials are produced in a fully processed condition, either unflattened or thermally flattened, in thicknesses of 0.0090, 0.0106, 0.0118, and 0.0138 in.



**FIG. 4-24** Typical 60-Hz core loss of as-sheared 50/50 Epstein specimens of fully processed nonoriented electrical steels.

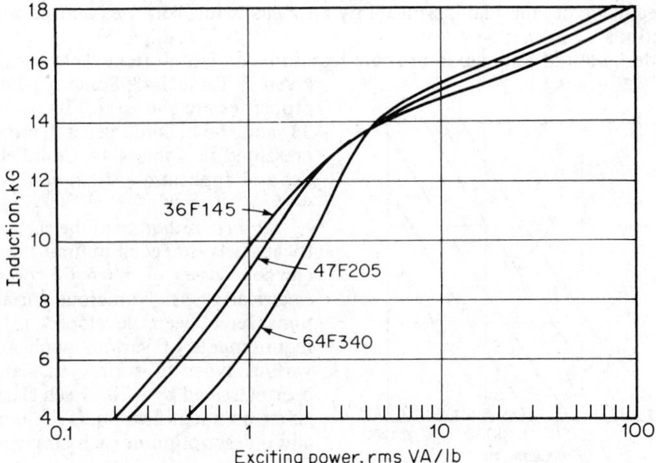


**FIG. 4-25** Typical 60-Hz exciting power of as-sheared 50/50 Epstein specimens of fully processed nonoriented electrical steels.

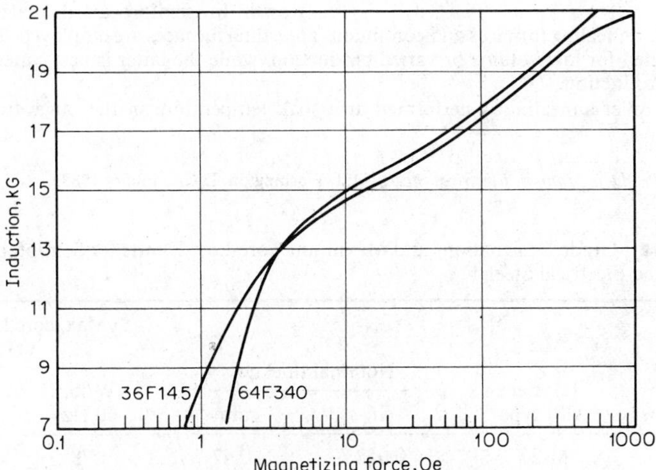


**FIG. 4-26** Typical normal-induction curves of as-sheared 50/50 Epstein specimens of fully processed nonoriented electrical steels.

Unflattened material has appreciable coil set or curvature. It is used principally in making spirally wound or formed cores. These cores must be stress-relief annealed to relieve fabrication stresses. Thermally flattened material is employed principally in making sheared or stamped laminations. Annealing of the laminations to remove both residual stresses from the thermal-flattening and fabrication stresses is usually recommended. However, special thermally flattened materials are available which do not require annealing when used in the form of wide flat laminations.

Two types of grain-oriented electrical steels are currently being produced commercially. The regular type, which was introduced many years ago, contains about 3.15% silicon and has grains about 3 mm in diameter. The high-permeability type, which was introduced more recently, contains about 2.9% silicon and has grains about 8 mm in diameter. In comparison with the regular type, the high-permeability type has better core loss and permeability at high inductions.

Grade designations and maximum core-loss limits for grain-oriented electrical steels are given in Table 4-43. Some typical magnetic properties are shown in Figs. 4-31, 4-32, 4-33, and 4-34. Some general properties are presented in Table 4-44. Some characteristics and applications are described in Table 4-45.

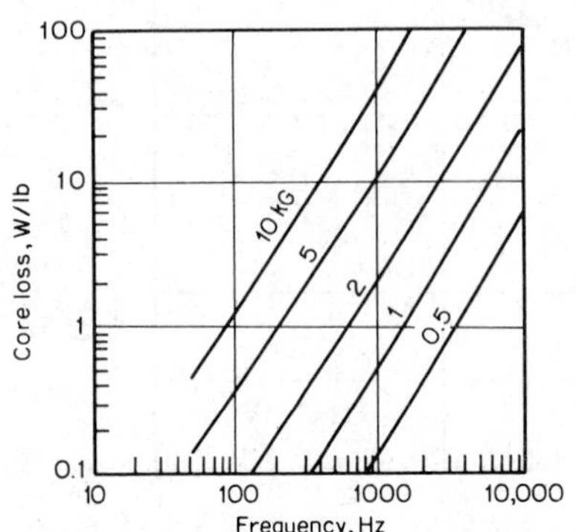


**FIG. 4-27** Effect of frequency on core loss of as-sheared 50/50 Epstein specimens of 36F145 fully processed nonoriented electrical steels.

*Surface insulation* of the surfaces of electrical steels is needed to limit the interlaminar core losses of magnetic cores made of electrical steels. Numerous surface insulations have been developed to meet the requirements of various applications. The various types of surface insulations have been classified by AISI.[1] Each class has been given an identification. The identification and a description of each class are given in Table 4-46.

*Annealing* of laminations or cores made from electrical steels is performed to accomplish either stress relief in fully processed material or decarburization and grain growth in semiprocessed material. Both batch-type annealing furnaces and continuous annealing furnaces are employed. The former is best suited for low-volume or varied production, while the latter is best suited for high-volume production.

Stress-relief annealing is performed at a soak temperature in the range from 730 to

[1]*Steel Products Manual, Electrical Steels;* AISI, Washington, D.C., January 1983.

**TABLE 4-42** Grade Designations and Maximum Core-Loss Limits for Semiprocessed Nonoriented Electrical Steels*

| ASTM type | Former AISI type | Nominal thickness | | Max. core loss at 15 kG† | |
|---|---|---|---|---|---|
| | | in | mm | W/lb, 60 Hz | W/kg, 60 Hz |
| 47S178 | M-27 | 0.0185 | 0.47 | 1.78 | 3.10 |
| 64S194 | M-27 | 0.025 | 0.64 | 1.94 | 3.38 |
| 47S188 | M-36 | 0.0185 | 0.47 | 1.88 | 3.27 |
| 64S213 | M-36 | 0.025 | 0.64 | 2.13 | 3.71 |
| 47S200 | M-43 | 0.0185 | 0.47 | 2.00 | 3.48 |
| 64S230 | M-43 | 0.025 | 0.64 | 2.30 | 4.01 |
| 47S230 | M-45 | 0.0185 | 0.47 | 2.30 | 4.01 |
| 64S260 | M-45 | 0.025 | 0.64 | 2.60 | 4.53 |
| 47S250 | — | 0.0185 | 0.47 | 2.50 | 4.35 |
| 64S280 | — | 0.025 | 0.64 | 2.80 | 4.88 |
| 47S300 | — | 0.0185 | 0.47 | 3.00 | 5.22 |
| 64S350 | — | 0.025 | 0.64 | 3.50 | 6.10 |

*Adapted from *Steel Products Manual, Electrical Steels;* AISI, Washington, D.C., January 1983.

†Core loss is determined in accordance with ASTM Method A343 on quality-evaluation-annealed Epstein specimens consisting of strips of which half have been cut parallel and half have been cut transverse to the rolling direction. The quality-evaluation-annealing is customarily done at a soak temperature of 845°C for approximately 1 h in a suitable decarburizing atmosphere.

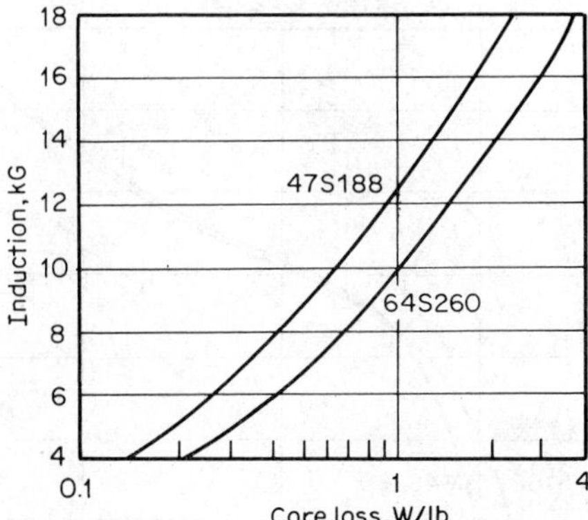


**FIG. 4-28** Typical 60-Hz core loss of annealed 50/50 Epstein specimens of semiprocessed nonoriented electrical steels.

845°C. The soak time need be no longer than that required for the charge to reach soak temperature. The heating and cooling rates must be slow enough that excessive thermal gradients in the material are avoided. The annealing atmosphere and other annealing conditions must be such that chemical contamination of the material is avoided.

Annealing for decarburization and grain growth is performed at a soak temperature in the range from 760 to 870°C. Atmospheres of hydrogen or partially combusted natural gas and containing water vapor are often used. The soak time required for decarburization depends not only on the temperature and atmosphere but also on the dimensions of the laminations or cores being annealed. If the dimensions are large, long soak times may be required.

***202. Magnetic lamination steels (ASTM Standard A726)*** are cold-rolled low-carbon steels intended for magnetic applications, primarily at power frequencies. The magnetic properties of magnetic lamination steels are not normally guaranteed and are generally inferior to those of electric steels. However, magnetic lamination steels are frequently used as

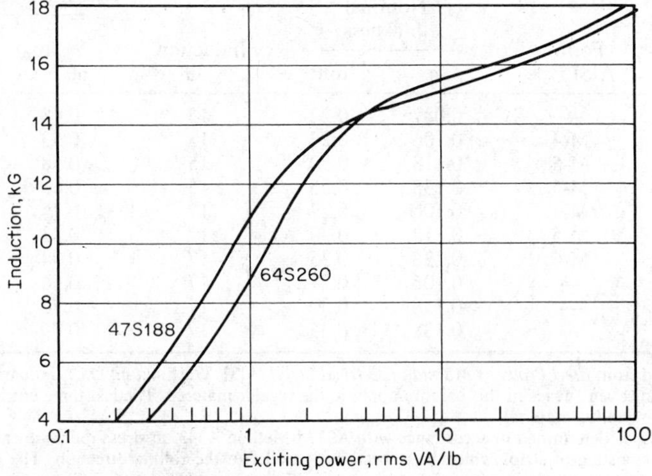


**FIG. 4-29** Typical 60-Hz exciting power of annealed 50/50 Epstein specimens of semiprocessed nonoriented electrical steels.

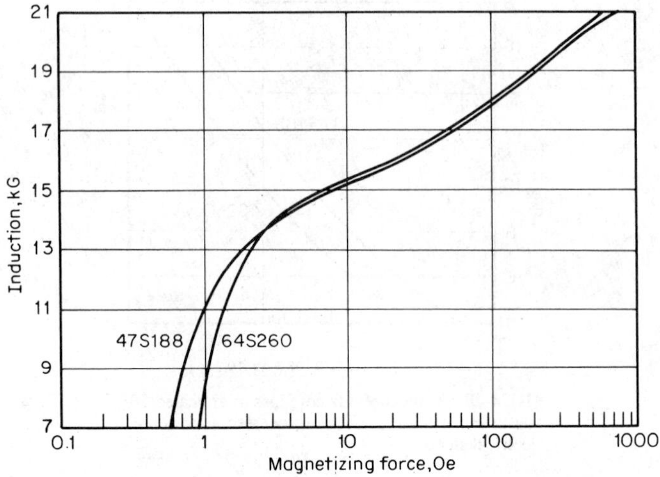


**FIG. 4-30** Typical normal-induction curves of annealed 50/50 Epstein specimens of semiprocessed nonoriented electrical steels.

core materials in small electrical devices especially when the cost of the core material is a more important consideration than the magnetic performance.

Usually, but not always, stamped laminations or assembled core structures made from magnetic lamination steels are given a decarburizing anneal to enhance the magnetic properties. Optimum magnetic properties are obtained when the carbon content is reduced to 0.005% or less from its initial value, which may approach 0.1%. The soak temperature of the anneal is in the range from 730 to 790°C. The atmosphere most often used at the present

**TABLE 4-43** Grade Designations and Maximum Core-Loss Limits for Fully Processed Grain-Oriented Electrical Steels*

| ASTM type† | Former AISI type | Nominal thickness, in | Nominal thickness, mm | Induction, kG | Max. core loss‡ W/lb, 60 Hz | Max. core loss‡ W/kg, 60 Hz |
|---|---|---|---|---|---|---|
| 23G048 | — | 0.087 | 0.22 | 15 | 0.48 | 0.80 |
| 27G053 | M-4 | 0.106 | 0.27 | 15 | 0.53 | 0.89 |
| 30G058 | M-5 | 0.118 | 0.30 | 15 | 0.58 | 0.97 |
| 35G066 | M-6 | 0.138 | 0.35 | 15 | 0.66 | 1.11 |
| 27H076 | M-4 | 0.106 | 0.27 | 17 | 0.76 | 1.27 |
| 30H083 | M-5 | 0.118 | 0.30 | 17 | 0.83 | 1.39 |
| 35H094 | M-6 | 0.138 | 0.35 | 17 | 0.94 | 1.57 |
| 27P066 | — | 0.106 | 0.27 | 17 | 0.66 | 1.11 |
| 30P070 | — | 0.118 | 0.30 | 17 | 0.70 | 1.17 |
| 35P076 | — | 0.138 | 0.35 | 17 | 0.76 | 1.27 |

*Adapted from *Steel Products Manual, Electrical Steels;* AISI, Washington, D.C., January 1983.

†The first seven entries in this column apply to the regular material. The last three entries apply to the high-permeability material.

‡Core loss is determined in accordance with ASTM Method A343 on stress-relief-annealed Epstein specimens consisting of strips which have been cut parallel to the rolling direction. The stress-relief annealing is customarily done at a soak temperature in the range of 790 to 845°C for approximately 1 h in an atmosphere comprised of a mixture of pure nitrogen and pure hydrogen with a dew point not greater than −20°C.

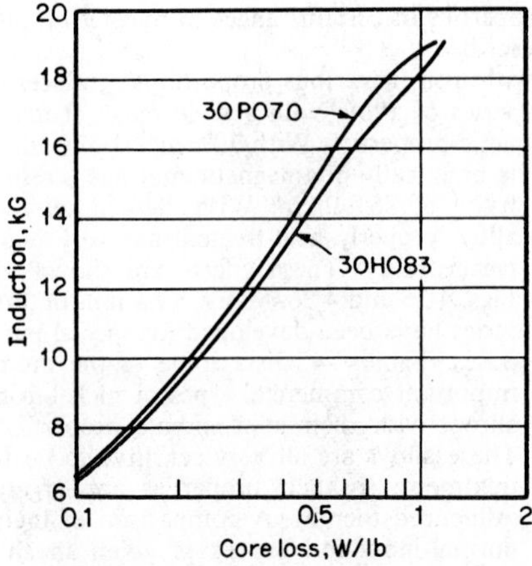


FIG. 4-31 Typical 60-Hz core loss of annealed parallel-grain Epstein specimens of fully processed grain-oriented electrical steels.

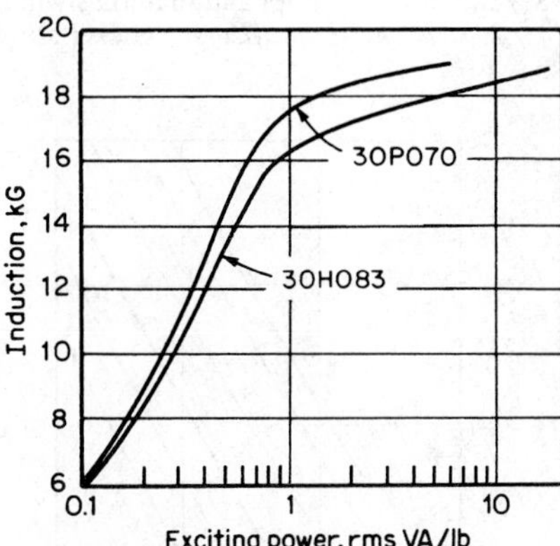


FIG. 4-32 Typical 60-Hz exciting power of annealed parallel-grain Epstein specimens of fully processed grain-oriented electrical steels.

time is partially combusted natural gas with a suitable dew point. Soak time is dependent to a considerable degree upon the dimensions of the laminations or core structures being annealed.

Three types of magnetic lamination steels are produced. Type 1 is usually made to a controlled chemical composition and is furnished in the full-hard or annealed condition without guaranteed magnetic properties. Type 2 is made to a controlled chemical composition, given special processing, and furnished in the annealed condition without guaranteed magnetic properties. After a suitable anneal, the magnetic properties of Type 2 are superior to those of Type 1. Type 2S is similar to Type 2, but the core loss is guaranteed. The maximum core-loss limits for Type 2S are shown in Table 4-47.

***203. Materials for Special Purposes.*** For certain applications of soft, or nonretentive, materials, special alloys and other materials have been developed which, after proper fab-

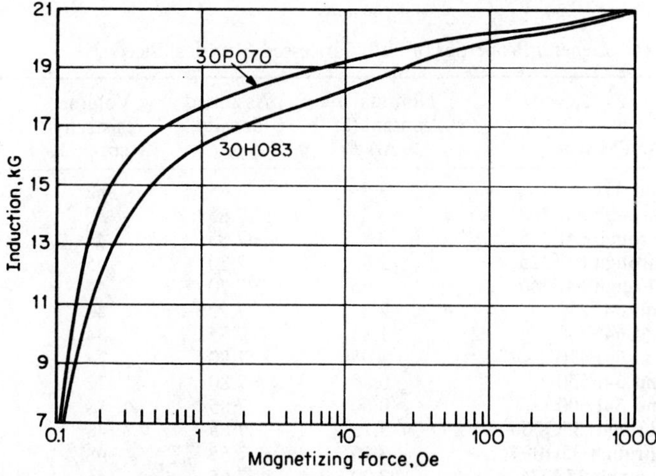


FIG. 4-33 Typical normal-induction curves of annealed parallel-grain Epstein specimens of fully processed grain-oriented electrical steels.

rication and heat-treatment, have superior properties in certain ranges of magnetization. Several of these alloys and materials will be described.

***204. Nickel-Iron Alloys.*** Nickel alloyed with iron in various proportions produces a series of alloys with a wide range of magnetic properties. With 30% nickel, the alloy is practically nonmagnetic and has a resistivity of 86 μΩ/cm. With 78% nickel the alloy, properly heat-treated, has very high permeability. These effects are shown in Figs. 4-35 and 4-36. Many variations of this series have been developed for special purposes. Table 4-48 lists some of the more important commercial types of nickel-iron alloys, with their approximate properties. These alloys are all very sensitive to heat-treatment; so their properties are largely influenced thereby. A comparison of their normal-induction curves is given in the curves of Fig. 4-37 and Fig. 4-19.

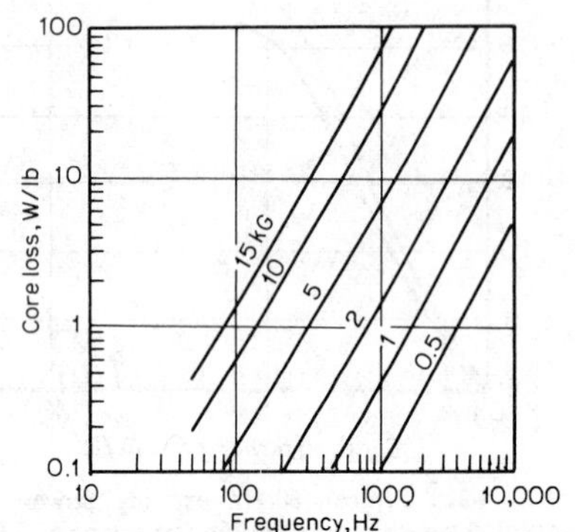


**FIG. 4-34** Effect of frequency on core loss of annealed parallel-grain Epstein specimens of 30H083 fully processed grain-oriented electrical steels.

***205. Permalloy***[1] is a term applied to a number of nickel-iron alloys developed by the Bell Laboratories, each specified by a prefix number indicating the nickel content. The term is usually associated with the 78.5% nickel-iron alloys, the important properties of which are high permeability and low hysteresis loss in relatively low magnetizing fields. These properties are obtained by a unique heat-treatment consisting of a high-temperature anneal, preferably in hydrogen, with slow cooling followed by rapid cooling from about 625°C. The alloy is very sensitive to mechanical strain; so it is desirable to heat-treat the alloy in its final form. The addition of 3.8% chromium or molybdenum increases the resistivity from 16 to 65 and 55 μΩ·cm, respectively, without seriously impairing the magnetic quality. In fact, low-density permeabilities are better with these additions. These alloys have found their principal application as a material for the continuous loading of submarine cables and in loading coils for land lines (see Figs. 4-37 and 4-38 and Table 4-48).

[1] *Elem. Bell Syst. Tech. J.*, 1936, vol. 15, p. 113.

**TABLE 4-44** General Properties of Fully Processed Electrical Steels

| ASTM type | Nominal alloy content (Si + Al), % | Assumed density, g/cm³ | Volume resistivity, Ω·cm × 10⁶ | Saturation induction, kG |
|---|---|---|---|---|
| 36F145 and 47F168 | 3.5 | 7.65 | 52 | 19.8 |
| 36F158 through 64F208 | 3.3 | 7.65 | 50 | 19.9 |
| 36F168 through 64F218 | 3.2 | 7.65 | 48 | 20.0 |
| 36F180 through 64F225 | 2.8 | 7.70 | 45 | 20.2 |
| 36F190 through 64F240 | 2.65 | 7.70 | 44 | 20.2 |
| 47F230 and 64F270 | 2.35 | 7.70 | 40 | 20.4 |
| 47F290 and 64F340 | 1.85 | 7.75 | 34 | 20.7 |
| 47F380 and 64F470 | 1.05 | 7.80 | 24 | 21.1 |
| 47F450 and 64F550 | 0.80 | 7.80 | 21 | 21.2 |
| 47F490 and 64F600 | 0.50 | 7.85 | 18 | 21.3 |
| 23G048 through 35G066 | 3.15 | 7.65 | 48 | 20.0 |
| 27H076 through 35H094 | 3.15 | 7.65 | 48 | 20.0 |
| 27P066 through 35P076 | 2.90 | 7.65 | 45 | 20.1 |

**TABLE 4-45** Some Characteristics and Typical Applications for Specific Types of Electrical Steels*

| ASTM type | Some characteristics | Typical applications |
|---|---|---|
| **Oriented types** | | |
| 23G048 through 35G066 or 27H076 through 35H094 or 27P066 through 35P076 | Highly directional magnetic properties due to grain orientation. Very low core loss and high permeability in rolling direction. | Highest-efficiency power and distribution transformers with lower weight per kVA. Large generators and power transformers. |
| **Nonoriented types** | | |
| 36F145 and 47F168 | Lowest core loss, conventional grades. Excellent permeability at low inductions. | Small power transformers and rotating machines of high efficiency. |
| 36F158 through 64F225 or 47S178 and 64S194 | Low core loss, good permeability at low and intermediate inductions. | High-reactance cores, generators, stators of high-efficiency rotating equipment. |
| 36F190 through 64F270 or 47S188 through 64S260 | Good core loss, good permeability at all inductions, and low exciting current. Good stamping properties. | Small generators, high-efficiency, continuous duty rotating ac and dc machines. |
| 47F290 through 64F600 or 47S250 through 64S350 | Ductile, good stamping properties, good permeability at high inductions. | Small motors, ballasts, and relays. |

*Adapted from *Steel Products Manual, Electrical Steels;* AISI, Washington, D.C., January 1983.

By special long-time high-temperature treatments, maximum permeability values greater than 1 million have been obtained. The double treatment required by the 78% Permalloy is most effective when the strip is thin, say under 10 mils. For greater thicknesses, the quick cooling from 625°C is not uniform throughout the section, and loss of quality results.

A 48% nickel-iron was developed for applications requiring a moderately high-permeability alloy with higher saturation density than 78 Permalloy. The same general composition is marketed under many names, such as Hyperm 50, Hipernik, Audiolloy, Allegheny Electric Metal, 4750, Carpenter 49 alloy. Annealing after all mechanical operations are completed is recommended. These alloys have found extensive use in radio, radar, instrument, and magnetic-amplifier components (see Fig. 4-37).

*Deltamax.* By the use of special techniques of cold reduction and annealing, the 48% nickel-iron alloy develops directional properties resulting in high permeability and a square hysteresis loop in the rolling direction (see Fig. 4-38). A similar product is sold under the name of Orthonic. For optimum properties, these materials are rapidly cooled after a 2-h anneal in pure hydrogen at 1100°C. They are generally used in wound cores of thin tape for applications such as pulse transformers and magnetic amplifiers.

***206. Iron-Nickel-Copper-Chromium.*** The addition of copper and chromium to high nickel-iron alloys has the effect of raising the permeability at low flux density. Alloys of this type are marketed under the names of Mumetal, 1040 alloy, Hymu 80. A typical induction characteristic is curve *A* in Fig. 4-37; for optimum properties they are annealed after cutting

**TABLE 4-46** Descriptions of Flat-Rolled Electrical Steel Insulations or Core Plates*

| Identification | Description |
|---|---|
| C-0 | This identification is merely for the purpose of describing the natural oxide surface which occurs on flat-rolled silicon steel which gives a slight but effective insulating layer sufficient for most small cores and will withstand normal stress-relief annealing temperatures. This oxidized surface condition may be enhanced in the stress-relief anneal of finished cores by controlling the atmosphere to be more or less oxidizing to the surface. |
| C-2† | This identification is for the purpose of describing an inorganic insulation which consists of a glasslike film which forms during high-temperature hydrogen anneal of grain-oriented silicon steel as the result of the reaction of an applied coating of MgO and silicates in the surface of the steel. This insulation is intended for air-cooled or oil-immersed cores. It will withstand stress-relief annealing temperatures and has sufficient interlamination resistance for wound cores of narrow width strip such as used in distribution transformer cores. It is not intended for stamped laminations because of the abrasive nature of the coating. |
| C-3 | This insulation consists of an enamel or varnish coating intended for air-cooled or oil-immersed cores. The interlamination resistance provided by this coating is superior to the C-1 type coating which is primarily utilized as a die lubricant. The C-3 coating will also enhance punchability, is resistant to normal operating temperatures, but will not withstand stress-relief annealing (see Note). |
| C-4 | This insulation consists of a chemically treated or phosphated surface intended for air-cooled or oil-immersed cores requiring moderate levels of insulation resistance. It will withstand stress-relief annealing and serves to promote punchability. |
| C-5 | This is an inorganic insulation similar to C-4 but with ceramic fillers added to enhance the interlamination resistance. It is typically applied over the C-2 coating on grain-oriented silicon steel. It is principally intended for air-cooled or oil-immersed cores which utilize sheared laminations and operate at high volts per turn, but finds application in all apparatus requiring high levels of interlaminar resistance. Like C-2, it will withstand stress-relief annealing in a neutral or slightly reducing atmosphere. |

*From *Steel Products Manual, Electrical Steels;* AISI, Washington, D.C., January 1983.

†C-1 has been deleted from this table and is generally superseded by C-3.

NOTE: In fabricating operations involving the application of heat, such as welding and die casting, it may be desirable that a thinner than normal coating be used to leave as little residue as possible. These coatings can enhance punchability, and the producers should be consulted to obtain a correct weight of coating. To identify these coatings, various letter suffixes have been adopted, and the producer should be consulted for the proper suffix.

and forming for 4 h at 1100°C in pure hydrogen and cooled slowly. Important applications are as magnetic shielding for instruments and electronic equipment and as cores in magnetic amplifiers.

***207. Constant-permeability alloys*** having a moderate permeability which is quite constant over a considerable range of flux densities are desirable for use in circuits in which waveform distortion must be kept at a minimum. Isoperm and Conpernik are two alloys of this type. They are nickel-iron alloys containing 40 to 55% nickel which have been severely cold-worked. Perminvar is the name given to a series of cobalt-nickel-iron alloys (for exam-

**TABLE 4-47** Maximum Core-Loss Limits for Type 2S Magnetic Lamination Steel*

| Actual thickness by micrometer† | | Max. core loss at 15 kG, 60 Hz‡ | |
|---|---|---|---|
| in | mm | W/lb | W/kg |
| 0.016 | 0.41 | 3.3 | 7.3 |
| 0.018 | 0.46 | 3.6 | 7.9 |
| 0.021 | 0.53 | 4.1 | 9.0 |
| 0.025 | 0.64 | 4.9 | 10.8 |
| 0.028 | 0.71 | 5.6 | 12.3 |

*Adapted from *Steel Products Manual, Electrical Steels;* AISI, Washington, D.C., January 1983.

†Other thicknesses can be specified, and the assigned core loss values will be commensurate with these stated values.

‡Core loss is determined in accordance with ASTM Method A343 on quality-evaluation-annealed Epstein specimens consisting of strips of which half have been cut parallel and half have been cut transverse to the rolling direction. The quality-evaluation annealing is customarily done at a soak temperature of 790°C for approximately 1 h in a suitable decarburizing atmosphere.

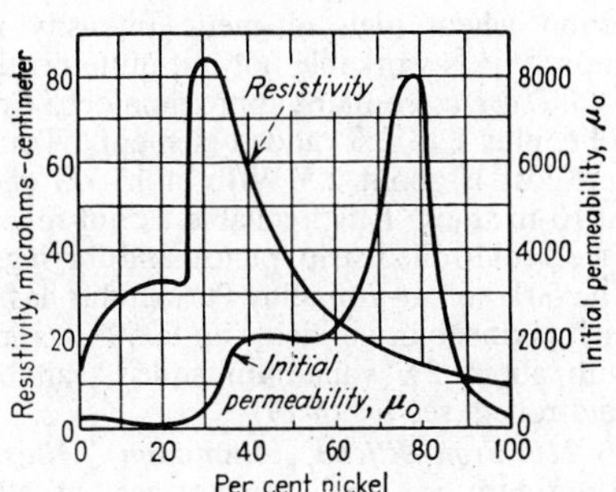


**FIG. 4-35** Electrical resistivity and initial permeability of iron-nickel alloys with various nickel contents.

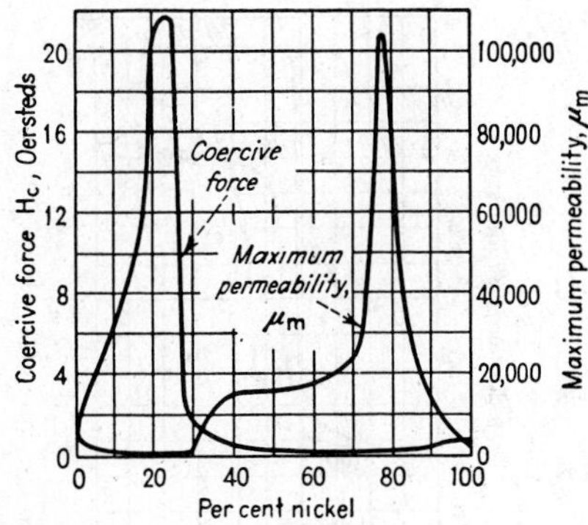


**FIG. 4-36** Maximum permeability and coercive force of iron-nickel alloys with various nickel contents.

**TABLE 4-48** Special-Purpose Materials

| Name | Approximate composition, % | Saturation, G | Maximum permeability | Coercivity (from saturation), Oe | Initial permeability | Resistivity, microhm-cm |
|---|---|---|---|---|---|---|
| 78 Permalloy | 78.5 Ni | 10,500 | 70,000 | −0.05 | 8,000 | 16 |
| MoPermalloy | 79 Ni, 4.0 Mo | 8,000 | 90,000 | −0.05 | 20,000 | 55 |
| Supermalloy | 79 Ni, 5 Mo | 7,900 | 900,000 | −0.002 | 100,000 | 60 |
| 48% nickel-iron | 48 Ni | 16,000 | 60,000 | −0.06 | 5,000 | 45 |
| Monimax | 47 Ni, 3 Mo | 14,500 | 35,000 | −0.10 | 2,000 | 80 |
| Sinimax | 43 Ni, 3 Si | 11,000 | 35,000 | −0.10 | 3,000 | 85 |
| Mumetal | 77 Ni, 5 Cu, 2 Cr | 6,500 | 85,000 | −0.05 | 20,000 | 60 |
| Deltamax | 50 Ni | 15,500 | 85,000 | −0.10 | ...... | 45 |

ple, 50% nickel, 25% cobalt, 25% iron) which also exhibit this characteristic of constant permeability over a low (≅800 G) density range. When magnetized to higher flux densities, they give a double loop constricted at the origin so as to give no measurable remanence or coercive force. The characteristics of the alloys in this group vary greatly with the chemical content and the heat-treatment. A sample containing approximately 45 Ni, 25 Co, and 30 Fe, baked for 24 h at 425°C and slowly cooled, had hysteresis losses as follows: At 100 G, 214 × $10^{-4}$ erg/($cm^3$)(cycle); at 1003 G, 15.27 ergs; at 1604 G, 163 ergs; at 4950 G, 1736 ergs; and at 13,810 G, 4430 ergs. Over the range of flux densities in which the permeability is constant (from 0 to 600 G), the hysteresis loss is very small, or on the order of the foregoing figure for 100 G. The resistivity of the sample was 19.63 μΩ·cm.

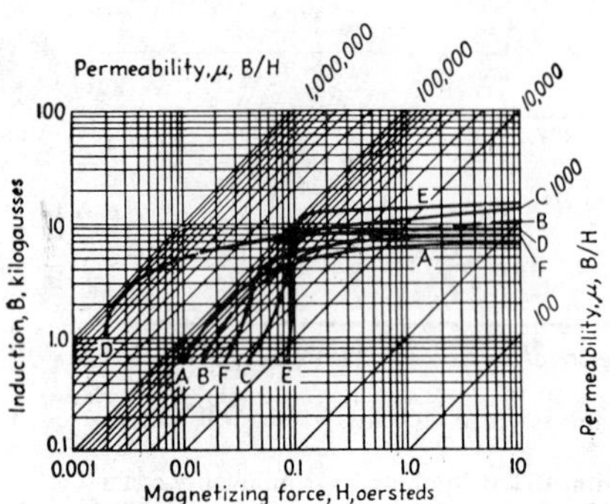


**FIG. 4-37** Induction-permeability curves of some high-nickel alloy strip (0.014 in). *A.* Mumetal. *B.* Permalloy. *C.* 48% nickel-iron. *D.* Supermalloy. *E.* Deltamax. *F.* MoPermalloy.

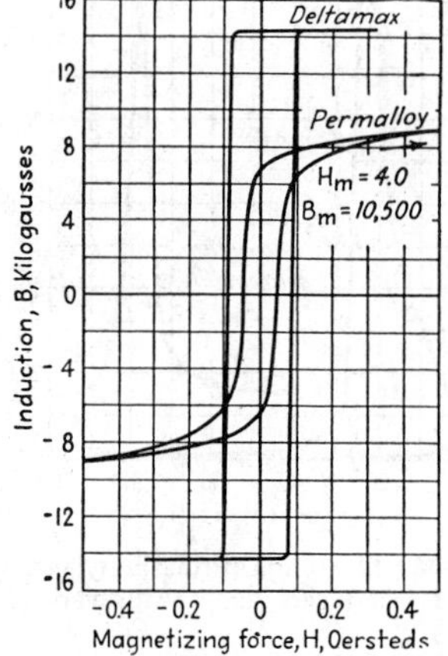


**Fig. 4-38** Hysteresis curves of 78 Permalloy and Deltamax from saturation.

***208. Monel metal*** is an alloy of 67% nickel, 28% copper, and 5% other metals. It is slightly magnetic below 95°C (see Fig. 4-19).

***209. Iron-Cobalt Alloys.*** The addition of cobalt to iron has the effect of raising the saturation intensity of iron up to about 36% cobalt ($Fe_2Co$). This alloy is useful for pole pieces of electromagnets and for any application where high magnetic intensity is desired. It is workable hot but quite brittle cold. *Hyperco* contains approximately ⅓ Co, ⅔ Fe, plus 1 to 2% "added element." Total core loss is about 2.5 W/lb at 15 kG and 0.010 in thick. It is available as hot-rolled sheet, cold-rolled strip, plates, and forgings. The 50% cobalt-iron alloy Permendur has a high permeability in fields up to 50 Oe and, with about 2% vanadium added, can be cold-rolled (see Fig. 4-19).

***210. Iron-Silicon Aluminum Alloys.*** Aluminum in small percentages, usually under 0.5%, is a valuable addition to the iron-silicon alloy. Its principal function appears to be as a deoxidizer. Masumoto[1] has investigated soft magnetic alloys containing much higher percentages of aluminum and found several that have high permeabilities and low hysteresis losses. Certain compositions have very low magnetostriction and anisotropy, high initial permeability, and high electrical resistivity. An alloy of 9.6% silicon and 6% aluminum with iron has better low-flux-density properties than the Permalloys. However, poor ductility has limited these alloys to dc applications in cast configurations or in insulated pressed-powder cores for high-frequency uses. These alloys are commonly known as Sendust. The material has been prepared in sheet form[2] by special processes.

[1]Masumoto: On a New Alloy "Sendust" and Its Magnetic and Electrical Properties; *Tohoku Imp. Univ. Rept.*, 1936, Anniversary Vol., Ser. I, p. 388.

[2]Helms and Adams: Sendust Sheet-processing Techniques and Magnetic Properties; *J. Appl. Phys.*, March 1964, vol. 35.

***211. Temperature-Sensitive Alloys.***[3] Inasmuch as the Curie point of metal may be moved up or down the temperature scale by the addition of other elements, it is possible to select alloys which lose their ferromagnetism at almost any desired temperature up to 1115°C, the change point in cobalt. Iron-base alloys are ordinarily used to obtain the highest possible permeability at points below the Curie temperature. Nickel, manganese, chromium, and silicon are the most effective alloy elements for this purpose;[2,4] and most alloys made for temperature-control application, such as instruments, reactors, and transformers, use one or more of these. Figure 4-39 shows the magnetization-temperature characteristics of a group of these alloys. The Carpenter Temperature Compensator 30 is a nickel-copper-iron alloy which loses its magnetism at 55°C and is used for temperature compensation in meters.[5]

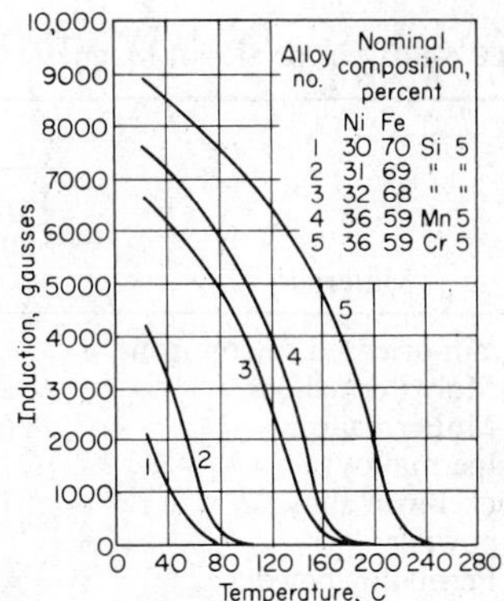


**FIG. 4-39** Thermosensitive alloys. Temperature-induction for $H = 20$.

***212. Heusler's alloys*** are ferromagnetic alloys composed of "nonmagnetic" elements. Copper, manganese, and aluminum are frequently used as the alloying elements. The saturation induction is about one-third that of pure iron.

***213. High-Frequency Materials Applications.*** Magnetic materials used in reactors, transformers, inductors, and switch-mode devices are selected upon the basis of magnetic induction, permeability, and associated material power losses at the design frequency. Control of eddy currents becomes of primary importance to reduce losses and minimize skin effect produced by eddy-current shielding. This is accomplished by the use of high-permeability alloys in the form of wound cores of thin tape, or compressed, insulated powder iron-alloy cores, or sintered ferrite cores.

Typically, the thin magnetic strip material is used in applications where operating frequencies range from 400 Hz to 20 kHz. Power conditioning equipment frequently operates at 10 kHz and up, and the magnetic materials used are compressed, powdered iron-alloy cores or sintered ferrite cores. Power losses in magnetic materials are of great concern, especially so when operated at high frequencies. Table 4-49 lists representative values of losses that may be developed by magnetic core materials operating in the range of 400 Hz to 100 kHz.

[3]Jackson and Russell: Temperature Sensitive Magnetic Alloys and Their Uses; *Instruments,* November 1938, p. 280.

[4]Shaw, J. L.: Curie Temperature Alloys; *Prod. Eng.,* June 1948.

[5]Eberly, W. S.: Temperature-Compensator Alloys; *Mach. Des.,* May 1954.

**TABLE 4-49** Comparative Power Losses of Magnetic Materials
(Core Losses in W/cm³ at 1.0 kG)

| Magnetic material | Test frequency 400 Hz | 10.0 kHz | 100.0 kHz |
|---|---|---|---|
| 3% oriented silicon-iron, 0.004 in thick | 0.00065 | 0.100 | — |
| 4-79 MoPermalloy, 0.004 in thick | 0.0001 | 0.020 | 1.20 |
| Iron powder (75 permeability) | 0.015 | 0.30 | — |
| 2-81 Permalloy powder (60 permeability) | 0.002 | 0.030 | 0.50 |
| Ferrite (2000 permeability) | 0.004 | 0.008 | 0.18 |

Other considerations are the ambient operating temperature of the magnetic component and the physical configurations in which the magnetic component can be supplied. A comparison of the dc properties of the more common commercially available cast and powdered alloys is shown in Table 4-50.

**TABLE 4-50** Properties of Magnetic Core Materials

| | dc properties | | | |
|---|---|---|---|---|
| Material | Permeability maximum, G/Oe | Magnetic induction, kG | Curie temp., C° | Electrical resistivity, $\mu\Omega \cdot$cm |
| 3% grain-oriented silicon iron | 30,000 | 20.0 | 740 | 47 |
| 4-79 Moly Permalloy | 250,000 | 7.8 | 454 | 58 |
| 5-79 MoPermalloy (Supermalloy) | 600,000 | 8.2 | 400 | 60 |
| 48 nickel-iron alloy | 100,000 | 15.5 | 482 | 47 |
| Iron powder | 75* | 12.0 | 770 | $10^7$ |
| 2-81 Permalloy powder | 125* | 7.0 | 400 | $10^6$ |
| Ferrite (MgZn) | 2,500* | 4.0 | 180 | $10^2$–$10^8$* |

*Varies by composition.

***214. 3% Silicon-iron alloys*** for high-frequency use are available in an insulated 0.001- to 0.006-in-thick strip that exhibits high effective permeability and low losses at relatively high flux densities. This alloy, as well as other rolled to strip soft magnetic alloys, is used to make laminated magnetic cores by various methods, including (1) the wound-core approach for winding toroids and C and E cores; (2) stamped or sheared-to-length laminations for laid-up transformers; and (3) stamped laminations of various configurations (rings E, I, F, L, DU, etc.) for assembly into transformer cores. Laminated core materials usually are annealed after all fabricating and stamping operations have been completed in order to develop the desired magnetic properties of the material. Subsequent forming, bending, or machining may impair the magnetic characteristics developed by the anneal.

***215. Nickel-alloy tape*** of high permeability is used in thickness of 0.000125 to 0.006 in for tape-wound cores designed for the frequency range of 0.1 to 100 kHz. Tapes less than 0.001 in are usually wound on nonmagnetic stainless steel or ceramic bobbins for support of the tape and to ensure stability of the magnetic properties. Commonly used nickel-iron alloys are 4-79 MoPermalloy, Mumetal, 48 nickel-iron, and Supermalloy with thickness and construction chosen to provide the desired permeability at the application frequency. Figure 4-40 shows the effect of tape thickness on the initial permeability of some of these materials. The decrease in permeability with increasing frequency is a characteristic of all magnetic materials. Applications include uses as current transformers, transformers for inverter-converters, high-power pulse transformers, and magnetic pick-up heads. Magnetic core materials less than 0.001 in thick are used in timer circuits, high-frequency inverters, digital memory devices, pulse transformers, and magnetometer sensing.

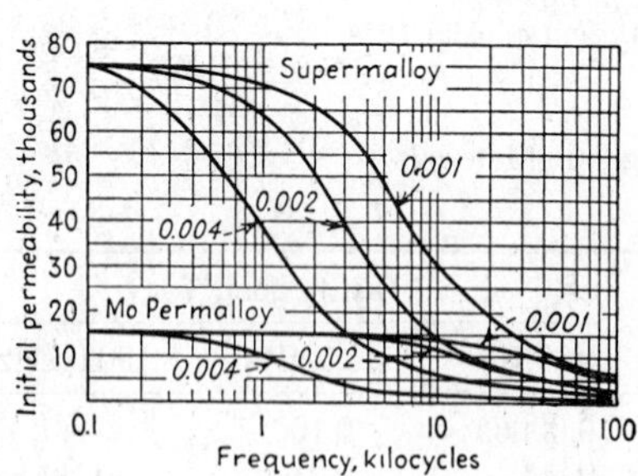


**FIG. 4-40** Effect of tape thickness on the initial permeability of Supermalloy and MoPermalloy at various frequencies.

*Other tape materials* of commercial significance are permendur (35 to 50% Co), vanadium permendur (49% Co, 2% V), and the amorphous metal alloys. Permendurs are high-induction (23.0 kG) materials which can be rolled to a 0.001-in-thick strip for use as recording head laminations.

Vanadium permendur is used in tape-wound transformer cores where size and weight reduction is important and operating frequencies are below 3.0 kHz. A high degree of magnetostriction may be developed by annealing, thus enabling these materials to be used in transducer applications.

***216. Amorphous metal alloys*** are made using a new technology which produces a thin (0.001 to 0.003 in) ribbon from rapidly quenched molten metal. The alloy solidifies before the atoms have a chance to segregate or crystallize, resulting in a glasslike atomic structure material of high electrical resistivity, 125 to 130 $\mu\Omega\cdot$cm. A range of magnetic properties may be developed in these materials by using different alloying elements. Amorphous metal alloys may be used in the same high-frequency applications as the cast, rolled to strip, silicon-iron and nickel-iron alloys.

***217. Nickel-iron powder cores*** are made of insulated alloy powder which is compressed to shape and heat-treated. The alloy composition most widely used is 2-81 Permalloy powder composed of 2% molybdenum, 81% nickel, and balance iron. Another less widely used powder, Sendust, is made of 7 to 13% silicon, 4 to 7% aluminum, and balance iron. Prior to pressing, the powder particles are thinly coated with an inorganic, high-temperature insulation which can withstand the high compacting pressures and the high-temperature (650°C) hydrogen atmosphere anneal. The insulation of the particles lowers eddy-current loss and provides a distributed air gap which can be controlled to provide cores in a range of permeabilities. The 2-81 Permalloy cores are commercially available in permeability ranges of 14 to 300, and Sendust cores have permeabilities ranging from 10 to 140.

These types of nickel-iron powder cores find use in applications where inductance must remain relatively constant when the magnetic component experiences changes in dc current or temperature. Additional stability over temperature can also be achieved by the addition of low Curie temperature powder materials to neutralize the naturally positive permeability-temperature coefficient of the alloy powder. Some applications are in telephone loading coils or filter chokes for power conditioning equipment where output voltage ripple must be minimized. Other uses are for pulse transformers and switch-mode power supplies where low power losses are desired. Operating frequencies can range from 1.0 kHz for 300 permeability materials to 500 kHz for the 14 permeability materials.

***218. Powdered-iron cores*** are manufactured from various types of iron powders whose particle size range from 2 to 100 $\mu$m. The particles are electrically insulated from one another using special insulating materials. The insulated powder is blended with phenolic or epoxy binders and a mold release agent. The powder is then dry-pressed in a variety of shapes including toroids, E cores, threaded tuning cores, cups, sleeves, slugs, bobbins, and other special shapes. A low-temperature bake of the pressed product produces a solid component in which the insulated particles provide a built-in air gap, reducing eddy-current losses, increasing electrical $Q$, and thus allowing higher operating frequencies. The use of different iron powder blends and insulation systems provides a range of permeability, from 4 to 90, for use over the frequency spectrum of 50 Hz to 250 MHz. Applications include high-frequency transformers, tuning coils, variable inductors, rf chokes, and noise suppressors for power supply and power control circuits.

***219. Ferrite cores*** are molded from a mixture of metallic oxide powders such that certain iron atoms in the cubic crystal of magnetite (ferrous ferrite) are replaced by other metal atoms, such as Mn and Zn, to form manganese zinc ferrite, or by Ni and Zn to form nickel zinc ferrite. Manganese zinc ferrite is the material most commercially available[1] and is used in devices operating below 1.5 MHz. Nickel zinc ferrites are used mainly for filter applications above that frequency. They resemble ceramic materials in production processes and physical properties. The electrical resistivities correspond to those of semiconductors, being at least 1 million times those of metals. Magnetic permeability $\mu_0$ may be as high as 10,000. The Curie point is quite low, however, in the range 100 to 300°C. Saturation flux density is generally below 5000 G (see Fig. 4-41.) Ferrite materials are available in several compositions which, through processing, can improve one or two magnetic parameters (magnetic induction, permeability, low hysteresis loss, Curie temperature) at the expense of the other

[1]Roess, E.: Soft Magnetic Ferrites and Applications in Telecommunications and Power Converters; IEEE Trans. Magnetics, vol. MAG-18, no. 6, November 1982.

parameters. The materials are fabricated into shapes such as toroids; E, U, and I cores; beads; and self-shielding pot cores. Ferrite cores find use in filter applications up to 1.0 MHz, high-frequency power transformers operating at 10 to 100 kHz, pulse transformer delay lines, adjustable air-gap inductors, recording heads, and filters used in high-frequency electronic circuits.

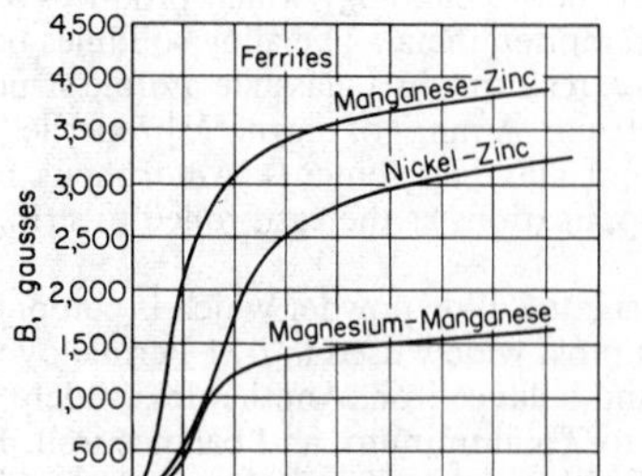

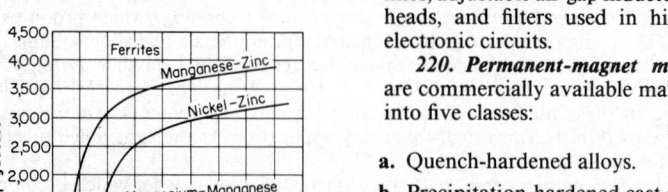

**FIG. 4-41** Typical normal-induction curves for "soft" ferrites.

***220. Permanent-magnet materials*** that are commercially available may be grouped into five classes:

**a.** Quench-hardened alloys.

**b.** Precipitation-hardened cast alloys.

**c.** Ceramic materials.

**d.** Powder compacts and elongated single-domain materials.

**e.** Ductile alloys.

***221. Quench-Hardened Alloys.*** Early permanent magnets were made of low carbon steel (1% C) that was hardened by heat-treatment. Later developments saw improvements in the magnetic properties through the use of alloying elements of tungsten, chromium, and cobalt. The magnetic properties of these alloys and others discussed in this section are listed in Table 4-51. The chrome steels are less expensive than the cobalt steels, and both find use in hysteresis clutch and motor applications.

***222. Precipitation-hardened cast alloys*** for magnetic applications are available in a wide range of magnetic properties. See the Alnicos listed in Table 4-51. Formed by conventional casting techniques, Alnico magnets are structurally weak and brittle and are not readily machined except by grinding. Alnico 5 DG and 5 Columnar grades are directionally solidified during casting. The Alnico 5 DG utilizes a chill plate in the mold to obtain a partially columnar crystallization in the casting, and the Alnico 5 Columnar uses a hot mold and chill plate procedure to develop a nearly 100% columnar structure throughout the magnet. All the Alnico grades listed, except grades 1, 2, 3, and 4, are heat-treated in a shape-oriented magnetic field which attains magnetic precipitate alignment and develops the excellent magnet anisotropic properties. Due to process limitations, Alnico 5 DG, 5 Columnar, 8, and 9 should be used only in straight flux line magnet designs. Alnico magnets find use in applications in loud speakers, microwave devices, motors, generators, meters, magnetos, separators, communication devices, and vending machines.

***223. Ceramic magnet material*** usage is increasing yearly because of improved magnetic properties and the high cost of cobalt used in metallic alloy magnets. The basic raw material used in these magnets is iron oxide in combination with either strontium carbonate or barium carbonate. The iron oxide and carbonate mixture is calcined, and then the aggregate is ball-milled to a particle size of about 1.0 $\mu$m. The material is compacted in dies using the dry powder or a water-based slurry of the powder. High pressures are needed to press the parts to shape. In some ceramic grades, a magnetic field is applied during pressing to orient the material in order to obtain a preferred magnetic orientation. Parts are sintered at high temperatures and ground to finished size using diamond grinding wheels with suitable coolants. Ceramic magnets are hard and brittle, exhibit high electrical resistivities, and have lower densities than cast magnet alloys.

Made in the form of rings, blocks, and arcs, ceramic magnets find use in applications for loudspeakers, dc motors, microwave oven magnetron tubes, traveling wave tubes, holding magnets, chip collectors, and magnetic separator units. Ceramic magnet arcs find wide use in the auto industry in engine coolant pumps, heating-cooling fan motors, and window lift motors. As with other magnets, they are normally supplied nonmagnetized and are magnetized in the end-use structure using magnetizing fields of the order of 10,000 Oe to saturate the magnet. The brittleness of the material necessitates proper design of the magnet support structure so as not to impart mechanical stress to the magnet.

**TABLE 4-51** Magnetic Properties and Chemical Compositions

| Magnet material | Nominal chemical composition | Residual flux density $B_r$, G, nominal* | Coercive force $H_c$, Oe, nominal* | Max. energy prod $(BH)$ max, MG·Oe, nominal* |
|---|---|---|---|---|
| 3½% Cr steel | 3.5 Cr, 1 C, bal. Fe | 10,300 | 60 | 0.3 |
| 3% Co steel | 3.25 Co, 4 Cr, 1 C, bal. Fe | 9700 | 80 | 0.38 |
| 17% Co steel | 18.5 Co, 3.75 Cr, 5 W, .75 C, bal. Fe | 10,700 | 160 | 0.69 |
| 36% Co steel | 38 Co, 3.8 Cr, 5 W, .75 C, bal. Fe | 10,400 | 230 | 0.98 |
| Alnico 1 | 12 Al, 21 Ni, 5 Co, 3 Cu, bal. Fe | 7200 | 470 | 1.4 |
| Alnico 2 | 10 Al, 19 Ni, 13 Co, 3 Cu, bal. Fe | 7500 | 560 | 1.7 |
| Alnico 3 | 12 Al, 25 Ni, 3 Cu, bal. Fe | 7000 | 480 | 1.35 |
| Alnico 4 | 12 Al, 27 Ni, 5 Co, bal. Fe | 5600 | 720 | 1.35 |
| †Alnico 5 | 8 Al, 14 Ni, 24 Co, 3 Cu, bal. Fe | 12,800 | 640 | 5.5 |
| †Alnico 5 DG | 8 Al, 14 Ni, 24 Co, 3 Cu, bal. Fe | 13,300 | 670 | 6.5 |
| †Alnico 5 Col. | 8 Al, 14 Ni, 24 Co, 3 Cu, bal. Fe | 13,500 | 740 | 7.55 |
| †Alnico 6 | 8 Al, 16 Ni, 24 Co, 3 Cu, 1 Ti, bal. Fe | 10,500 | 780 | 3.9 |
| †Alnico 8 | 7 Al, 15 Ni, 35 Co, 4 Cu, 5 Ti, bal. Fe | 8200 | 1650 | 5.3 |
| †Alnico 8 HC | 8 Al, 14 Ni, 38 Co, 3 Cu, 8 Ti, bal. Fe | 7200 | 1900 | 5.0 |
| †Alnico 9 | 7 Al, 15 Ni, 35 Co, 4 Cu, 5 Ti, bal. Fe | 10,500 | 1500 | 9.0 |
| Ceramic 1 | $MO \cdot 6\,Fe_2O_3$ | 2300 | 1860/3250‡ | 1.05 |
| †Ceramic 2 | $MO \cdot 6\,Fe_2O_3$ M represents one or more of | 2900 | 2400/3000‡ | 1.8 |
| †Ceramic 3 | $MO \cdot 6\,Fe_2O_3$ the metals chosen from the | 3300 | 2200/2400‡ | 2.6 |
| †Ceramic 4 | $MO \cdot 6\,Fe_2O_3$ group barium, strontium, | 2500 | 2300/3800‡ | 1.45 |
| †Ceramic 5 | $MO \cdot 6\,Fe_2O_3$ lead. | 3800 | 2400 | 3.4 |
| †Ceramic 6 | $MO \cdot 6\,Fe_2O_3$ | 3200 | 2820/3300‡ | 2.45 |
| †Ceramic 7 | $MO \cdot 6\,Fe_2O_3$ | 3400 | 3250/4000‡ | 2.75 |
| †Ceramic 8 | $MO \cdot 6\,Fe_2O_3$ | 3850 | 2950/3050‡ | 3.5 |

**TABLE 4-51** Magnetic Properties and Chemical Compositions (*Continued*)

| Magnet material | Nominal chemical composition | Residual flux density $B_r$, G, nominal* | Coercive force $H_c$, Oe, nominal* | Max. energy prod $(BH)$ max, MG·Oe, nominal* |
|---|---|---|---|---|
| Sint. Alnico 2 | 10 Al, 19 Ni, 13 Co, 3 Cu, bal. Fe | 7100 | 550 | 1.5 |
| †Sint. Alnico 5 | 8 Al, 14 Ni, 24 Co, 3 Cu, bal. Fe | 10,900 | 620 | 3.95 |
| †Sint. Alnico 6 | 8 Al, 16 Ni, 24 Co, 3 Cu, 1 Ti, bal. Fe | 9400 | 790 | 2.95 |
| †Sint. Alnico 8 | 7 Al, 15 Ni, 35 Co, 4 Cu, 5 Ti, bal. Fe | 7400 | 1500 | 4.0 |
| †Sint. Alnico 8 HC | 7 Al, 14 Ni, 38 Co, 3 Cu, 8 Ti, bal. Fe | 6700 | 1800 | 4.5 |
| †Rare earth cobalt 12 | RE·Co RE represents one or more of the | 7200 | 6500/10,000‡ | 12.0 |
| †Rare earth cobalt 15 | RE·Co metals chosen from the rare earth | 8000 | 7000/14,000‡ | 15.0 |
| †Rare earth cobalt 16 | RE·Co group: samarium, praseodymium, | 8300 | 7500/18,000‡ | 16.0 |
| †Rare earth cobalt 18 | RE·Co cerium, yttrium, misch metal, etc. | 8700 | 8000/20,000‡ | 18.0 |
| †ESD 31 | 20.7 Fe, 11.6 Co, 67.7 Pb | 5000 | 1000 | 2.3 |
| †ESD 32 | 18.3 Fe, 10.3 Co, 72.4 Pb | 6800 | 960 | 3.0 |
| ESD 41 | 20.7 Fe, 11.6 Co, 67.7 Pb | 3600 | 970 | 1.1 |
| ESD 42 | 18.3 Fe, 10.3 Co, 72.4 Pb | 4800 | 830 | 1.25 |
| †Cunife 1 | 60 Cu, 20 Ni, 20 Fe | 5500 | 530 | 1.4 |
| Vicalloy 1 | 10 V, 52 Co, bal. Fe | 7500 | 250 | 0.8 |
| Remalloy | 12 Co, 15 Mo, bal. Fe | 9700 | 250 | 1.0 |

*Values derived from major hysteresis loop.
†Anisotropic.
‡Intrinsic coercive force, $H_{ci}$.
NOTE: MG·Oe = megagauss · oersteds; ESD = elongated single domain.
SOURCE: Adapted from *Standard Specification for Permanent Magnet Materials;* Magnetic Materials Producers Association, Evanston, Ill.

***224. Powder compacts*** for magnets are represented by the sintered Alnico and the rare earth cobalt magnets. These are listed in Table 4-51.

***225. Sintered Alnico magnets*** are made of fine powders of the alloy which are compressed to the desired shape and size, then sintered at about 1200°C in a hydrogen atmosphere. During sintering, there is 2 to 10% shrinkage in part dimensions which must be considered in the die design. All of the sintered Alnico magnets listed, except sintered Alnico 2, receive a postsintering heat-treatment in a magnetic field. This treatment produces a preferred magnetic orientation in the part which should be considered in the magnet design.

Sintered magnets are structurally stronger than their cast equivalent, but they are limited to small sizes of 15 g or less. Larger sizes are generally more economically made by casting. Also, the sintered magnets usually exhibit slightly lower magnetic properties than equivalent grades of cast Alnico alloys.

***226. Rare earth cobalt magnets*** have the highest energy product and coercivity of any commercially available magnetic material. Magnets are produced by powder metallurgy techniques from alloys of cobalt (65 to 77%), rare earth metals (23 to 35%), and sometimes copper and iron. The rare earth metal used is usually samarium, but other metals used are praseodymium, cerium, yttrium, neodymium, lathanum, and a rare earth metal mixture called misch metal. The rare earth alloy is ground to a fine particle size (1 to 10 $\mu$m), and the powder is then die-compacted in a strong magnetic field. The part is then sintered and abrasive-ground to finish tolerances.

Although this material uses comparatively expensive raw materials, the high value of coercive force (5500 to 9500 Oe) leads to small magnet size and good temperature stability. These magnets find use in miniature electronic devices such as motors, printers, electron beam focusing assemblies, magnetic bearings, and traveling wave tubes. Plastic-bonded rare earth magnets are also being made, but the magnetic value of the energy product is only a fraction of the sintered product.

***227. Elongated single-domain (ESD) materials*** are available as alloys of iron and cobalt in the form of anisotropic particles (needles, or plates). A special process develops the magnetic particles in a lead (Pb) matrix which is able to flow and fill a compacting die in the form of the magnet. The particle magnetization is field-aligned in the long axis of the particles. When frozen in the lead matrix, the particles resist demagnetization. This magnetic material finds limited use due to its low operating temperature (300°F) and the high mass density due to the lead content.

***228. Ductile alloys*** include the materials Cunife, Vicalloy, Remalloy, chromium-cobalt-iron (Cr-Co-Fe), and, in a limited sense, manganese-aluminum-carbon (Mn-Al-C). They are sufficiently ductile and malleable to be drawn, forged, or rolled into wire or strip forms. A final heat-treatment after forming develops the magnetic properties. Cunife has a directional magnetism developed as a result of cold working and finds wide use in meters and automotive speedometers. Vicalloy has been used as a high-quality and high-performance magnetic recording tape and in hysteresis clutch applications. Remalloy has been used extensively in telephone receivers but is now being replaced by a newer, less costly magnetic material.

New permanent-magnet materials that are now being produced are the Cr-Co-Fe alloy and the Mn-Al-C alloy. The Cr-Co-Fe alloy family contains 20 to 35% chromium and from 5 to 25% cobalt. This alloy is unique among permanent-magnet alloys due to its good hot and cold ductility, machinability, and excellent magnetic properties. The heat-treatment of the alloy involves a rapid cooling from approximately 1200°C to a spinoidal decomposition phase occurring at about 600°C. The magnetic phase developed in the spinoidal decomposition process may be oriented by a heat-treatment in a magnetic field, or the material may be magnetically oriented by "deformation aging" as would be accomplished in a wire drawing operation. The magnetic properties that can be developed are comparable to those of Alnico 5 and are superior to those of the other ductile alloys, Cunife, Vicalloy, and Remalloy. Western Electric has introduced a Cr-Co-Fe alloy which replaces Remalloy in the production of telephone receiver magnets and at a lower cost due to reduced cobalt.

***229. The Mn-Al-C alloy*** achieves permanent-magnetic properties ($B_r$, 5500 G; $H_c$, 2300 Oe; Mg·Oe energy product, 5 Mg·Oe) when mechanical deformation of the alloy takes place at a temperature of about 720°C. Mechanical deformation may be performed by warm extrusion. Magnet size is limited by the amount of deformation needed to develop and orient the magnetic phase in the alloy. The alloying elements are inexpensive, but the tooling and equipment needed in the deformation process is expensive and may be a factor in the economical production of this magnet alloy. Magnets of this alloy would find use in loudspeakers, motor applications, and microwave oven magnetron tubes. The low density, 5.1 g/cm$^3$, is desirable for motors where reduced inertia and weight savings are important. The low Curie temperature, 320°C, limits the use of this alloy to applications where the ambient temperature is less than 125°C.

***230. Permanent-magnet design*** involves the calculation of magnet area and magnet length to produce a specific magnetic flux density across a known gap, usually with the magnet having the smallest possible volume. Designs are developed from magnet material hysteresis loop data of the second quadrant, commonly called *demagnetization curves.* An example of demagnetization curve data is shown in Fig. 4-42.

Other considerations are the operating temperature of the magnetic assembly, magnet weight, and cost. Also, care should be exercised in the calculation of any steel return path cross section to ensure that it is adequate to carry the flux output of the magnet. Table 4-52 illustrates the range of magnetic characteristics that may be considered in the design. Detailed magnetic and material specifications may be obtained from the magnet manufacturer.

***231. Bibliography.*** Information on the properties of commercial magnetic materials may be obtained from technical bulletins and catalogs issued by the various producers of magnetic materials. Other sources of information include the following:

Bozorth, R. M.: *Ferromagnetism;* Princeton, N.J., D. Van Nostrand Company, Inc., 1951.

Cullity, B. D.: *Introduction to Magnetic Materials;* Reading, Mass., Addison-Wesley Publishing Company, 1972.

Heck, C.: *Magnetic Materials and Their Applications;* New York, Crane, Russack, and Company, Inc., 1974.

Parker, R. J. and Studders, R. J.: *Permanent Magnets and Their Applications;* New York, John Wiley and Sons, Inc., 1962.

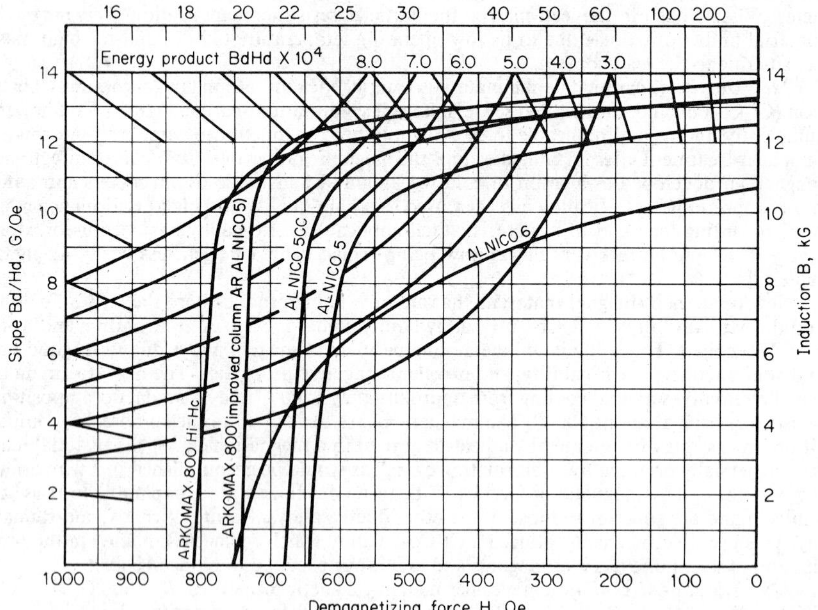


**FIG. 4-42** Typical demagnetization curves for Alnico permanent-magnetic materials. (Reprinted with permission from *PM-121C Alnico Permanent Magnets;* The Arnold Engineering Company, Marengo, Illinois.)

**TABLE 4-52** Comparison of Magnetic and Physical Properties of Selected Commercial Materials

| | Alnico 5 | Alnico 9 | Ferrite | Co5R |
|---|---|---|---|---|
| $B_r$, G | 12800 | 10500 | 4100 | 9500 |
| $H_c$, Oe | 640 | 1500 | 2900 | 6500 |
| $B_dH_d$, Mg·Oe | 5.5 | 9.0 | 4.0 | 22.0 |
| Curie point, °C | 850 | 815 | 470 | 740 |
| Temperature coefficient, %/°C | 0.02 | 0.02 | 0.19 | 0.03 |
| Density, g/cm$^3$ | 7.3 | 7.3 | 4.9 | 8.6 |
| Energy/unit weight | 0.8 | 1.2 | 0.8 | 2.6 |

*Steel Products Manual, Electrical Steel;* AISI, Washington, D.C., January 1983.

*1983 Annual Book of ASTM Standards,* sec. 3, vol. 03.04; ASTM, Philadelphia, Pa., 1983.

*Standard Specifications for Permanent Magnet Materials;* Magnetic Materials Producers Association, Evanston, Ill., 1983.

*Permanent Magnet Guidelines;* Magnetic Materials Producers Association, Evanston, Ill., 1983.

## *INSULATING MATERIALS*

### General Properties

*By T. W. DAKIN*

**232. *Electrical Insulation and Dielectric Defined.*** Electrical insulation is a medium or a material which, when placed between conductors at different potentials, permits only a small or negligible current in phase with the applied voltage to flow through it. The term dielectric is almost synonymous with electrical insulation, which can be considered the applied dielectric. A perfect dielectric passes no conduction current and only capacitive charging current between conductors. Only a vacuum at low stresses between uncontaminated metal surfaces satisfies this condition.

The range of resistivities of substances which can be considered insulators is from greater than $10^{20}$ Ω·cm downward to the vicinity of $10^6$ Ω·cm, depending on the application and voltage stress. There is no sharp boundary defined between low-resistance insulators and semiconductors. If the voltage stress is low and there is little concern about the level of current flow (other than that which would heat and destroy the insulation), relatively low-resistance insulation can be tolerated.

**233. *Circuit Analogy of a Dielectric or Insulation.*** Any dielectric or electrical insulation can be considered as equivalent to a combination of capacitors and resistors which will duplicate the current-voltage behavior at a particular frequency or time of voltage application. In the case of some dielectrics, simple linear capacitors and resistors do not adequately represent the behavior. Rather, resistors and capacitors with particular nonlinear voltage-current or voltage-charge relations must be postulated to duplicate the dielectric current-voltage characteristic.

The simplest circuit representation of a dielectric is a parallel capacitor and resistor, as shown in Fig 4-43 for $R_S = 0$. The perfect dielectric would be simply a capacitor. Another representation of a dielectric is a series-connected capacitor and resistor as in Fig 4-43 for $R_p = \infty$, while still another involves both $R_S$ and $R_p$.

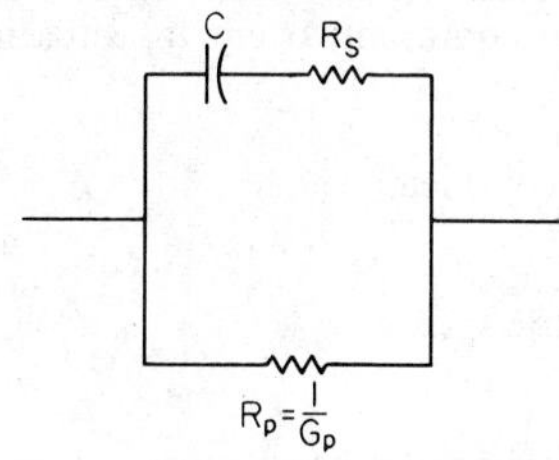


**FIG. 4-43** Equivalent circuit of a dielectric.

The ac dielectric behavior is indicated by the phase diagram (Fig. 4-44). The perfect dielectric capacitor has a current which leads the voltage by 90°, but the imperfect dielectric has a current which leads the voltage by less than 90°. The dielectric phase angle is $\theta$, and the difference, $90° - \theta = \delta$, is the loss angle. Most measurements of dielectrics give directly the tangent of the loss angle $\tan \delta$ (known as the *dissipation factor*) and the capacitance $C$. In Fig. 4-43, if $R_p = \infty$, the series $R_s - C$ has a $\tan \delta = 2\pi f C_S R_s$, and if $R_s = 0$, the parallel $R_p - C$ has a $\tan \delta = 1/2\pi f C_p R_p$.

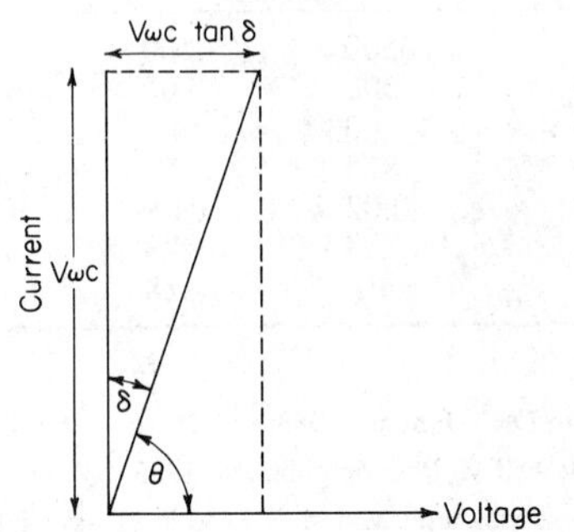


**FIG. 4-44** Current-voltage phase relation in a dielectric.

The ac power or heat loss in the dielectric is $V^2 2\pi f C \tan \delta$ watts, or $VI \sin \delta$ watts, where $\sin \delta$ is known as the power factor, $V$ is the applied voltage, $I$ is the total current through the dielectric, and $f$ is the frequency. From this it can be seen that the equivalent parallel conductance of the dielectric $\sigma$ (the inverse of the equivalent parallel resistance $\rho$) is $2\pi f C \tan \delta$. The ac conductivity is

$$\sigma = (5/9) f \varepsilon' \tan \delta \times 10^{-12}\ \Omega^{-1}\ \text{cm}^{-1} = 1/\rho \qquad (4\text{-}54)$$

where $\varepsilon'$ is the permittivity (or relative dielectric constant) and $f$ is the frequency. [The IEEE now recommends the symbol $\varepsilon'$ for the dielectric constant relative to a vacuum. The literature on dielectrics and insulation has also used $\kappa$ (kappa) for this dimensionless quantity or $\varepsilon'_r$. In some places, $\varepsilon'$ has been used to indicate the absolute dielectric constant, which is the product of the relative dielectric constant and the dielectric constant of a vacuum $\varepsilon_0$, which is equal to $8.85 \times 10^{-12}$ F/m.] $\kappa_0$ has also been used to represent the dielectric constant of a vacuum. While the ac conductivity theoretically increases in proportion to the frequency, in practice, it will depart from this proportionality insofar as $\varepsilon'$ and $\tan \delta$ change with frequency.

***234. Capacitance and Permittivity or Dielectric Constant.*** The capacitance between plane electrodes in a vacuum (with fringing neglected) is

$$C = \varepsilon' \varepsilon_0 A/t = 0.0884 \times 10^{-12} A/t \quad \text{farads} \qquad (4\text{-}55)$$

where $\varepsilon_0$ is the dielectric constant of a vacuum, $A$ the area in square centimeters, and $t$ the spacing of the plates in centimeters. $\varepsilon_0$ is $0.225 \times 10^{-12}$ F/in when $A$ and $t$ are expressed in inch units.

When a dielectric material fills the volume between the electrodes, the capacitance is higher by virtue of the charges within the molecules and atoms of the material, which attract more charge to the capacitor plates for the same applied voltage. The capacitance with the dielectric between the electrodes is

$$C = \varepsilon' \varepsilon_0 A/t \qquad (4\text{-}56)$$

where $\varepsilon'$ is the relative dielectric constant of the material. The capacitance relations for several other commonly occurring situations are

Coaxial conductors: $$C = \frac{2\pi\varepsilon'\varepsilon_0 L}{\ln(r_2/r_1)} \quad \text{farads} \qquad (4\text{-}57)$$

Concentric spheres: $$C = \frac{4\pi\varepsilon'\varepsilon_0 r_1 r_2}{r_2 - r_1} \quad \text{farads} \qquad (4\text{-}58)$$

Parallel cylindrical conductors: $$C = \frac{\pi\varepsilon'\varepsilon_0 L}{\cosh^{-1}(D/2r)} \quad \text{farads} \qquad (4\text{-}59)$$

In these equations, $L$ is the length of the conductors, $r_2$ and $r_1$ are the outer and inner radii and $D$ is the separation between centers of the parallel conductors with radii $r$. For dimensions in centimeters $\varepsilon_0$ is 0.0884 F/cm.

The value of $\varepsilon'$ depends on the number of atoms or molecules per unit volume and the ability of each to be polarized (that is, to have a net displacement of their charge in the direction of the applied voltage stress). Values of $\varepsilon'$ range from unity for vacuum to slightly greater than unity for gases at atmospheric pressure, 2 to 8 for common insulating solids and liquids, 35 for ethyl alcohol and 91 for pure water, and 1000 to 10,000 for titanate ceramics (see Table 4-53 for typical values).

The relative dielectric constant of materials is not constant with temperature, frequency, and many other conditions and is more appropriately called the dielectric permittivity. Refer to the volume by Smyth[1]* for a discussion of the relation of $\varepsilon'$ to molecular structure and to von Hippel[2,3] and other tables of dielectric materials[4] from the MIT Laboratory for Insulation Research. The permittivity of many liquids has been tabulated in *NBS Circ.* 514. The "Handbook of Chemistry and Physics" (Chemical Rubber Publishing Co.) also lists values for a number of plastics and other materials.

The permittivity of many plastics, ceramics, and glasses varies with the composition, which is frequently variable in nominally identical materials. In the case of some plastics, it varies with degree of cure and in the case of ceramics with the firing conditions. Plasticizers often have a profound effect in raising the permittivity of plastic compositions.

There is a force of attraction between the plates of a capacitor having an applied voltage. The stored energy is $1/2CV^2$J. The force equals the derivative of this energy with respect to the plate separation: $(1/2)\varepsilon'\varepsilon_0 E^2 \times 10^2 N/\text{cm}^2$ or $(1/2)\varepsilon'\varepsilon_0 E^2 \times 10$ bar, where $E$ is the electric field in volts per centimeter. The force increases proportionally to the capacitance or permittivity. This leads to a force of attraction of dielectrics into an electric field, that is, a net force which tends to move them toward a region of high field. If two dielectrics are present, the one with higher permittivity will displace the one with lower permittivity in the higher-field region. For example, air bubbles in a liquid are repelled from high-field regions. Correspondingly, elongated dielectric bodies are rotated into the direction of the electric field. In general, if the voltage on a dielectric system is maintained constant, the dielectrics move (if they are able) to create a higher capacitance.

***235. Resistance and Resistivity of Dielectrics and Insulation.*** The measured resistance

*Superscripts refer to bibliography for this subsection, Par. **246**.

**TABLE 4-53** Dielectric Permittivity (Relative Dielectric Constant), $\varepsilon'$

| | *k* | | *k* |
|---|---|---|---|
| Inorganic crystalline: | | Polymer resins: | |
| NaCl, dry crystal | 5.5 | Nonpolar resins: | |
| $CaCO_3$ (av) | 9.15 | Polyethylene | 2.3 |
| $Al_2O_3$ | 10.0 | Polystyrene | 2.5–2.6 |
| MgO | 8.2 | Polypropylene | 2.2 |
| BN | 4.15 | Polytetrafluoroethylene | 2.0 |
| $TiO_2$ (av) | 100 | Polar resins: | |
| $BaTiO_3$ crystal | 4,100 | Polyvinyl chloride (rigid) | 3.2–3.6 |
| Muscovite mica | 7.0–7.3 | Polyvinyl acetate | 3.2 |
| Fluorophlogopite (synthetic mica) | 6.3 | Polyvinyl fluoride | 8.5 |
| | | Nylon | 4.0–4.6 |
| | | Polyethylene terephthalate | 3.25 |
| Ceramics: | | Cellulose cotton fiber (dry) | 5.4 |
| Alumina | 8.1–9.5 | Cellulose Kraft fiber (dry) | 5.9 |
| Steatite | 5.5–7.0 | Cellulose cellophane (dry) | 6.6 |
| Forsterite | 6.2–6.3 | Cellulose triacetate | 4.7 |
| Aluminum silicate | 4.8 | Tricyanoethyl cellulose | 15.2 |
| Typical high-tension porcelain | 6.0–8.0 | Epoxy resins unfilled | 3.0–4.5 |
| Titanates | 50–10,000 | Methylmethacrylate | 3.6 |
| Beryl | 4.5 | Polyvinyl acetate | 3.7–3.8 |
| Zirconia | 8.0–10.5 | Polycarbonate | 2.9–3.0 |
| Magnesia | 8.2 | Phenolics (cellulose-filled) | 4–15 |
| Glass-bonded mica | 6.4–9.2 | Phenolica (glass-filled) | 5–7 |
| | | Phenolics (mica-filled) | 4.7–7.5 |
| Glasses: | | Silicones (glass-filled) | 3.1–4.5 |
| Fused silica | 3.8 | | |
| Corning 7740 (common laboratory Pyrex) | 5.1 | | |

$R$ of insulation depends upon the geometry of the specimen or system measured, which for a parallel-plate arrangement is

$$R = \rho t/A \qquad \text{ohms} \tag{4-60}$$

where $t$ is the insulation thickness in centimeters, $A$ is the area in square centimeters, and $\rho$ is the dielectric resistivity in ohm-centimeters. If $t$ and $A$ vary from place to place, the effective "insulation resistance" will be determined by the effective integral of the $t/A$ ratio over all the area under stress, on the assumption that the material resistivity $\rho$ does not change. If the material is not homogeneous and materials of different resistivities appear in parallel, the system can be treated as parallel resistors: $R = R_a R_b/(R_a + R_b)$. In this case, the lower-resistivity material usually controls the overall behavior. But if materials of different resistivities appear in series in the electric field, the higher-resistivity material will generally control the current and a majority of the voltage will appear across it, as in the case of series resistors.

The resistance of dielectrics and insulation is usually time-dependent and (for the same reason) frequency-dependent. The dc behavior of dielectrics under stress is an extension of the low-frequency behavior. The ac and dc resistance and permittivity can, in principle, be related for comparable times and frequencies.

Current flow in dielectrics can be divided into parts: (*a*) the true dc current, which is constant with time and would flow indefinitely, is associated with a transport of charge from one electrode into the dielectric, through the dielectric, and out into the other electrode; and (*b*) the polarization or absorption current, which involves, not charge flow through the interface between the dielectric and the electrode, but rather the displacement of charge within the dielectric. This is illustrated in Fig. 4-45, where it is shown that the displaced or absorbed charge is responsible for a reverse current when the voltage is removed.

Polarization current results from any of the various forms of limited charge displacement which can occur in the dielectric. The displacement occurring first (within less than nanoseconds) is the electronic and intramolecular charged atom displacement responsible for the very-high-frequency permittivity. The next slower displacement is the rotation of dipolar molecules and groups which are relatively free to move. The displacement most commonly observed in dc measurements, that is, currents changing in times of the order of seconds and minutes, is due to the very slow rotation of dipolar molecules and ions moving up to internal barriers in the material or at the conductor surfaces. When those slower displacement polarizations occur, the dielectric constant declines with increasing frequency and approaches the square of the optical refractive index $\eta^2$ at optical frequencies.

In composite dielectrics (material with relatively lower resistance intermingled with a material of relatively higher resistance) a large interfacial or "Maxwell-Wagner" type of polarization can occur.[5] A circuit model of such a situation can be represented by placing two of the circuits of Fig. 4-43 in series and making the parallel resistance of one much lower than the other. To get the effect, it is necessary that the time constant $R_p C$ be different for each material.

A simple model of the polarization current predicts an exponential decline of the current with time: $I_p = Ae^{-\alpha t}$, similar to the charging of a capacitor through a resistor. Composite materials are likely to have many different time constants, $\alpha = 1/RC$, superimposed. It is found empirically that the polarization or absorption current decreases inversely as a simple negative exponent of the time

$$I = At^{-n} \tag{4-61}$$

The ratio of the current at 1 min to that at 10 min has been called the polarization

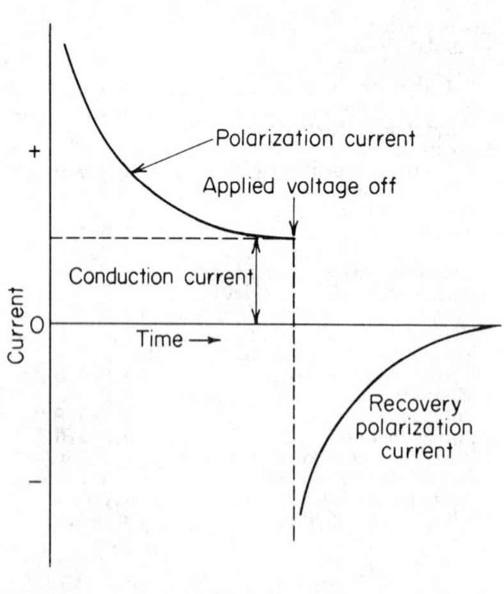

**FIG. 4-45** Typical dc dielectric current behavior.

index and is used to indicate the quality of composite machine insulation. A low polarization index associated with a low resistance sometimes indicates parallel current leakage paths through or over the surface of insulation (for example, in adsorbed water films).

The level of the conduction current which flows essentially continuously through insulation is an indication of the level of the ionic concentration and mobility in the material. Frequently, as with salt in water, the ions are provided by dissolved, adsorbed, or included impurity electrolytes in the material, rather than by the material itself. Purifying the material will therefore often raise the resistivity. If it is a liquid, purification can be done with adsorbent clays or ion-exchange resins.

The conductivity of ions in an insulation is given by the equation[6]

$$\sigma = \mu e c \qquad \Omega^{-1} \cdot \text{cm}^{-1} \tag{4-62}$$

where $\mu$ is the ion mobility, $e$ is the ionic charge in coulombs, and $c$ is the ionic concentration per cubic centimeter. The mobility, expressed in centimeters per second-volt per centimeter, decreases inversely with the effective internal viscosity and is very low for hard resins, but it increases with temperature and with softness of the resin, or fluidity of liquids. The ionic conductivity also varies widely with material purity. Among the polymers and resins, nonpolar resins such as polyethylene are likely to have high resistivities, of the order of $10^{16}$ or greater, since they do not readily dissolve or dissociate ionic impurities. Harder or crystalline polar resins have higher resistivity than do similar softer resins of similar dielectric constant and purity. Resins and liquids of higher dielectric constant usually have higher conductivities because they dissolve ionic impurities better, and the impurities dissociate to ions much more readily in a higher dielectric constant medium. Ceramics and glasses have lower resistivity if they contain alkali ions (sodium and potassium), since these ions are highly mobile.

Water is particularly effective in decreasing the resistivity by increasing the ionic concentration and mobility of materials, on the surface as well as internally. Water associates with impurity ions or ionizable constituents within or on the surface or interfaces. It helps to dissociate the ions by virtue of its high dielectric constant and provides a local environment of greater mobility, particularly as surface water films. Electrolyte conduction is discussed in Ref. 6. Table 4-54 indicates the effect of water on the surface resistivity of some insulating materials.

The ionic conductivity $\sigma$, exclusive of polarization effects, can be expected to increase exponentially with temperature according to the relation

$$\sigma = \sigma_0 e^{-B/T} \tag{4-63}$$

where $T$ is the Kelvin temperature and $\sigma_0$ and $B$ are constants. This relation, log $\sigma$ versus $1/T$, is shown in Fig. 4-46. It is often observed that at lower temperatures, where the resistivity is higher, the resistivity tends lower than the extrapolated higher temperature line

**TABLE 4-54** Surface Resistivities of Insulation and Effect of Humidity

| Material | Ω/square at 100% RH | % RH/decade decrease of Ω/square |
|---|---|---|
| Hydrocarbon wax | >20 × $10^{12}$ | |
| Silicone rubber | 10 × $10^{12}$ | |
| Polytetrafluoroethylene | 3.6 × $10^{12}$ | |
| Polystyrene | 8.4 × $10^{11}$ | |
| Cellulose acetate | 7 × $10^{9}$ | 6 |
| Polyvinyl chloride acetate | 5.7 × $10^{9}$ | 12 |
| Mica-filled phenolic | 5 × $10^{9}$ | 9 |
| Glazed porcelain | 3.7 × $10^{9}$ | 15 |
| Mica | 3 × $10^{9}$ | 12 |
| Steatite (L-4) | 2-6 × $10^{8}$ | |
| Cellulose-filled phenolic | 2.4 × $10^{6}$ | 10 |
| Quartz | 1.9 × $10^{6}$ | |

would predict. There are at least two possible reasons for this: the effect of adsorbed moisture and the contribution of a very slowly decaying polarization current.

***236. Variation of Dielectric Properties with Frequency.*** The permittivity of dielectrics invariably tends downward with increasing frequency, owing to the inability of the polarizing charges to move with sufficient speed to follow the increasing rate of alternations of the electric field. This is indicated in Fig. 4-47. The sharper decline in permittivity is known as a dispersion region. At the lower frequencies the ionic-interface polarization declines first; next the molecular dipolar polarizations decline. With some polar polymers two or more dipolar dispersion regions may occur owing to different parts of the molecular rotation.

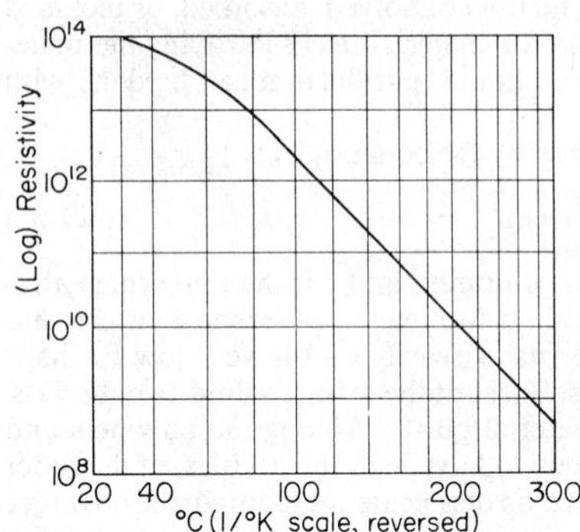


**FIG. 4-46** Typical dielectric resistivity-temperature dependence (Corning Glass 7740).

Figure 4-47 is typical of polymers and liquids, but not of glasses and ceramics. Glasses, ceramics, and inorganic crystals usually have much flatter permittivity-frequency curves, similar to that shown for the nonpolar polymer, but at a higher level, owing to their atom-ion displacement polarization, which can follow the electric field usually up to infrared frequencies.

The dissipation factor–frequency curve in Fig. 4-47 indicates the effect of ionic migration conduction at low frequency. It shows a maximum at a frequency corresponding to the permittivity dispersion region. This maximum is usually associated with a molecular dipolar rotation and occurs when the rotational mobility is such that the molecule rotation can just keep up with frequency of the applied field. Here it has its maximum movement in phase with the voltage, thus contributing to conduction current. At lower frequencies the molecule dipole can rotate faster than the field and contributes more to permittivity. At higher frequencies it cannot move fast enough. Such a dispersion region can also occur because of ionic migration and interface polarization if the interfaces are closely spaced, and if the frequency and mobility have the required values.

The frequency region where the dipolar dispersion occurs depends on the rotational mobility. In mobile, low-viscosity liquids it is in the 100- to 10,000-MHz range. In viscous liquids it occurs in the region of 1 to 100 MHz. In soft polymers it may occur in the audio-frequency range, and with hard polymers it is likely to be at very low frequency (indistin-

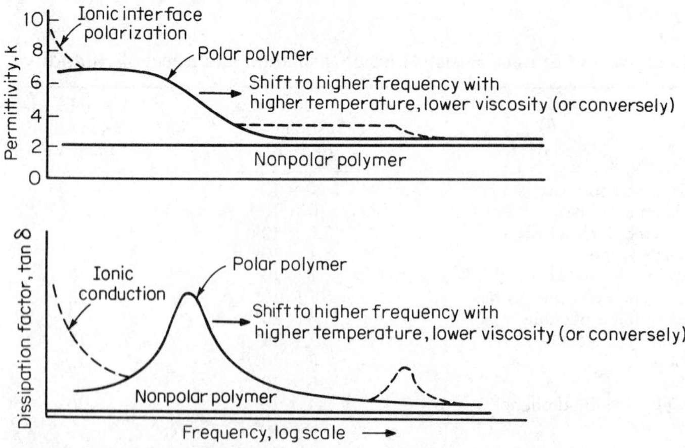


**FIG. 4-47** Typical variation in dielectric properties with frequency.

guishable from dc properties). Since the viscosity is affected by the temperature, increased temperature shifts the dispersion to higher frequencies.

***237. Variation of Dielectric Properties with Temperature.*** The trend in ac permittivity and conductivity, as measured by the dissipation factor, is controlled by the increasing ionic migrational and dipolar molecular rotational mobility with increasing temperature. This curve, which is indicated in Fig. 4-48, is in most respects a mirror image of the frequency trend shown in Fig. 4-47, since the two effects are interrelated.

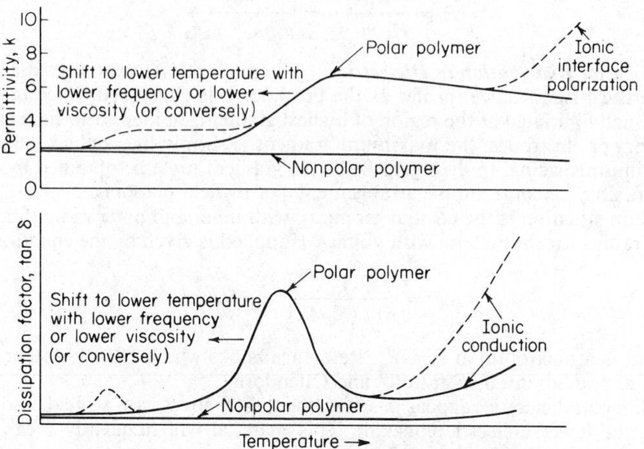


**FIG. 4-48** Typical variation in dielectric properties with temperature.

The permittivity-dispersion and dissipation-factor maximum region occurs below room temperature for viscous liquids, and still lower for mobile liquids. In fact mobile liquids may crystallize before they would show dispersion, except at high frequencies. With polymers the dissipation-factor maximum is likely to occur, at power frequencies, at a temperature close to a softening-point or internal second-order transition-point temperature. Dielectric dispersion and mechanical modulus dispersion can usually be correlated at the same temperature for comparable frequencies.

***238. Composite Dielectrics.*** The dielectric properties of composite dielectrics are generally a weighted average of the individual component properties, unless there is interaction, such as dissolving (as opposed to intermixing) of one material in another, or chemical reaction of one with another. Interfaces created by the mixing present a special factor, which can often lead to a higher dissipation factor and lower resistivity as a result of moisture and/or impurity concentration at the interface.

The ac properties of sheets of two dielectrics of dielectric constant $k_1$ and $k_2$ and of thickness $t_1$ and $t_2$ placed in series are related to the properties of the individual materials by the series of capacitance and impedance relations:

$$C = \frac{k_0 k_1 k_2 A}{k_1 t_2 + k_2 t_1} \tag{4-64}$$

$$\tan \delta = \frac{(t_1/t_2)\varepsilon_2' \tan \delta_1 + \varepsilon_1' \tan \delta_2}{\varepsilon_1' + \varepsilon_2'(t_1/t_2)} \tag{4-65}$$

Similarly the properties of two dielectrics in parallel are

$$C = \varepsilon_0 \left( \frac{\varepsilon_1' A_1}{t_1} + \frac{\varepsilon_2' A_2}{t_2} \right) \tag{4-66}$$

$$\tan \delta = \frac{t_2 \varepsilon_1' A_1 \tan \delta_1 + t_1 \varepsilon_2' A_2 \tan \delta_2}{t_2 \varepsilon_1' A_1 + t_1 \varepsilon_2' A_2} \tag{4-67}$$

With steady dc voltages, the resistivities control the current. With equal-area layer dielectrics in series,

$$R = R_1 + R_2 = \frac{1}{A}(\rho_1 t_1 + \rho_2 t_2) \tag{4-68}$$

When the dielectrics are in parallel and of equal thickness $t$,

$$R = \frac{R_1 R_2}{R_1 + R_2} = \frac{\rho_1 \rho_2 t}{\rho_1 A_2 + \rho_2 A_1} \tag{4-69}$$

***239. Potential Distribution in Dielectrics.*** The maximum potential gradient in dielectrics is of critical significance insofar as the breakdown is concerned, since breakdown or corona is usually initiated at the region of highest gradient. In a uniform-field arrangement of conductors or electrodes, the maximum gradient is simply the applied voltage divided by the minimum spacing. In divergent fields the gradient must be obtained by calculation (which is possible for some simple arrangements) or by field mapping.

A common situation is the coaxial geometry with inner and outer radii $R_1$ and $R_2$. The gradient at radius $r$ (centimeters) with voltage $V$ applied is given by the equation:

$$E = \frac{V}{r \ln (R_2/R_1)} \quad \text{V/cm} \tag{4-70}$$

The gradient is a maximum at $r = R_1$. Reference books which consider other geometries are Schwaiger and Sorensen,[7] Stratton,[8] and Ollendorff.[9]

When different dielectrics appear in series, the greater stress with ac fields is on the material having the lower dielectric constant. This material will frequently break down first unless its dielectric strength is much higher:

$$\frac{E_1}{E_2} = \frac{\varepsilon_2'}{\varepsilon_1'} \quad \text{and} \quad E_1 = \frac{V}{t_1 + t_2\varepsilon_1'/\varepsilon_2'} \tag{4-71}$$

The effect of the insulation thickness and dielectric constant (as well as the sharpness of the conductor edge) to create sufficient electric stress for local air breakdown (partial discharges) is shown in Fig 4-49.[17] With dc fields the stress distributes according to the resistivities of the materials, the higher stress being on the higher-resistivity material.

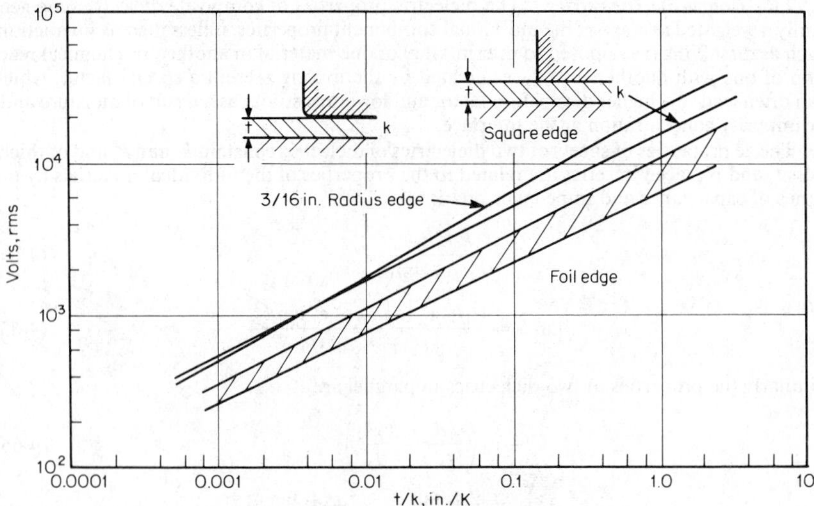


**FIG. 4-49** Corona threshold voltage at conductor edges in air as a function of insulation thickness.

***240. Dielectric Strength.*** This is defined by the ASA as the maximum potential gradient that the material can withstand without rupture. Practically, the strength is often reported as the breakdown voltage divided by the thickness between electrodes, regardless of electrode stress concentration.

Breakdown appears to require not only sufficient electric stress but also a certain minimum amount of energy. It is a property which varies with many factors such as thickness of the specimen, size and shape of electrodes used in applying stress, form or distribution of the field of electric stress in the material, frequency of the applied voltage, rate and duration of voltage application, fatigue with repeated voltage applications, temperature, moisture content, and possible chemical changes under stress.

The practical dielectric strength is decreased by defects in the material, such as cracks and included conducting particles and gas cavities. As will be shown in more detail in later sections on gases and liquids, the dielectric strength is quite adversely affected by conducting particles.

To state the dielectric strength correctly, the size and shape of specimen, method of test, temperature, manner of applying voltage, and other attendant conditions should be particularized as definitely as possible.

ASTM standard methods of dielectric strength testing should be used for making comparison tests of materials, but the levels of dielectric strength measured in such tests should not be expected to apply in service for long times. It is best to test an insulation in the same configuration in which it would be used. Also the possible decline in dielectric strength during long-time exposure to the service environment, thermal aging, and partial discharges (corona), if they exist at the applied service voltage, should be considered. ASTM has thermal life test methods for assessing the long-time endurance of some forms of insulation, such as sheet insulation (D2304), wire enamel (D2307) and others. There are IEEE thermal life tests for some systems such as random wound motor coils (IEEE-117).

The dielectric strength varies as the time and manner of voltage application as indicated in Fig. 4-50. With unidirectional pulses of voltage, having rise times of less than a few microseconds, there is a time lag of breakdown, which results in an apparent higher strength for

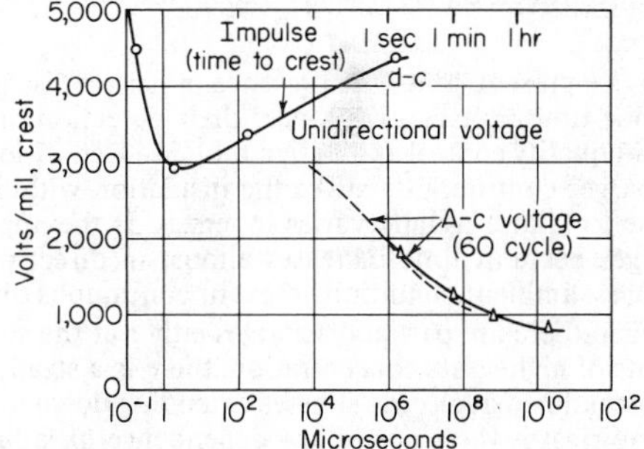


**FIG. 4-50** Dielectric strength of 0.032-in pressboard in oil as a function of time of voltage application.

very short pulses. In testing sheet insulation in mineral oil there is usually observed a higher strength for pulses of slow rise time and a still somewhat higher strength for dc voltages.

The trend in breakdown voltage with time, which is illustrated in Fig. 4-50, is typical of many solid insulation systems.

With ac voltages, the apparent strength declines steadily with time as a result of partial discharges[17] (in the ambient medium at the conductor or electrode edge). These penetrate the solid insulation. The discharges result from breakdown of the gas or liquid prior to the breakdown of the solid. See Fig. 4-50. The long-time strength with ac voltage declines, as shown in Fig. 4-51, and levels off at the partial discharge threshold (usually offset) voltage. Mica in particular, as well as other inorganic materials, is more resistant to such discharges. Organic resins should be used with caution where the ac voltage gradient is high and partial

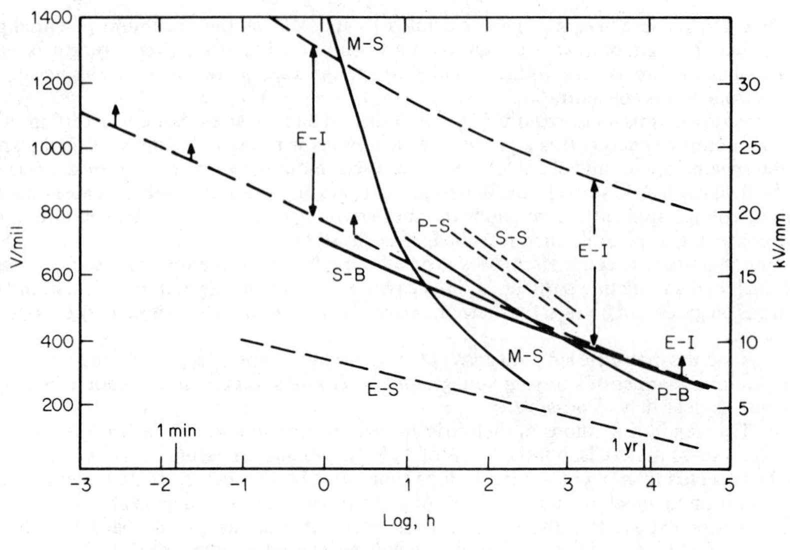


**FIG. 4-51** Alternating-current dielectric strength of various insulations vs. time under stress.

discharges (corona) may be present. Since the presence of partial discharges on insulation is so important to the long-time voltage endurance, their detection and measurement has become a very important quality control and design tool. See Ref. 27 for an extensive review of this subject. If discharges continuously strike the insulation within internal cavities, or on the surface, the time to failure usually varies inversely as the applied frequency, since the number of discharges per unit time increases almost in direct proportion to the frequency. But in some cases, ambient conditions prevent continuous discharges.

With ac voltages, when there are partial discharges either at the surface or internally in cavities or at local points of high stress concentration, there is a steady decline in dielectric strength and eventual breakdown at electric stresses extending down to the partial discharge threshold stress. An empirical inverse power law dependence of failure time $t$ on voltage gradient $E$, $t = AE^{-n}$, has been used to graph such data. A semilog relation between stress and log time has also been used. In the latter case, the data show an asymptotic approach to the partial discharge threshold stress as indicated in Fig. 4-51. The failure time appears to approach infinity at the partial discharge threshold. Below this level, discharges should have no effect on the failure time, but the material could of course fail by thermal or mechanical deterioration.

Figure 4-51 graphs voltage endurance data of several different types of insulation, including impregnated mica paper and mica splitting insulation, such as is used on high-voltage generator coils. Tests by Wichmann on this mica-filled insulation compare the insulation for two conditions: as sheets (S-S and P-S) with cylinder-to-plate electrodes on the surface (ASTM D2275), and wrapped on a high-voltage conductor bar, as the material would be used in service (S-B and P-B).

Figure 4-51 also compares tests on polyester organic film (M-S), which has a very high short time strength, but a steep decline in strength with time[17] when tested with surface electrodes (ASTM D2275).

When organic resin insulation is fabricated to avoid partial discharges using conductors or electrodes intimately bonded to the insulation, as in extruded polyethylene cables with a plastic semiconducting interface between the resin and the coaxial inner and outer metal conductors, respectively, the voltage endurance is greatly extended. Imperfections, however, in this "semicon"-resin interface, or at conducting particle inclusions in the resin, can lead to local discharges and the development of "electrical tree" growth.[28] Vacuum impregnating and casting electrodes or conductors into resin also tends to avoid cavities and surface discharges and greatly improves the voltage endurance at high stresses. This is illustrated by curves in Fig. 4-51 for silica-filled epoxy tested with vacuum cast-in electrodes (E-I) and with electrodes on the surface of a sheet of the epoxy (E-S).[29] The wide band of test values for the cast-in electrode system probably indicates varied degrees of success in avoiding partial discharge sites and imperfections. In this latter system, differences in the expansion coefficient of the resin and the metal conductor (or electrode) are important. In this case the percentage of silica filler in the resin was adjusted so that it matched the metal. See Ref. 28 for a more extensive review of "electrical tree" development.

In practical tests of the dielectric strength, the measured strength usually declines with increasing thickness of material. The breakdown gradient decreases approximately as the inverse half power of the thickness, or conversely, the breakdown voltage increases as the half power of the thickness. This is illustrated in Fig. 4-52. The value of the exponent may

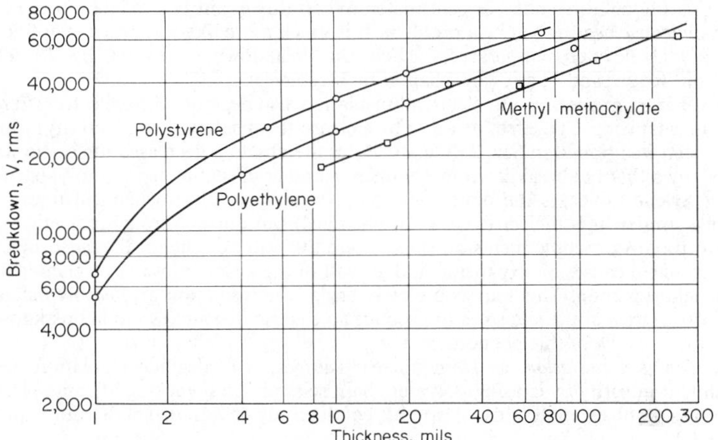


**FIG. 4-52** Variation in breakdown strength with thickness, 60-Hz ac voltage, 2 kV/s rate of rise.

vary somewhat with conditions. An exception to this behavior is noted with sheet or wrapper insulation having defects, which are covered by increased thickness. Figure 4-52 illustrates this effect for 1-mil film, which has small defects, giving it a somewhat reduced strength.

The dc strength of solid insulation is usually higher and declines much less with time than the ac strength, since corona discharges are infrequent.

The dielectric strength is much higher where surface discharges are avoided and when the electric field is uniform. This can be achieved with solid materials by recessing spherical cavities into the material and using conducting paint electrodes.

The "intrinsic" electric strength of solid materials measured in uniform fields, avoiding surface discharges, ranges from levels of the order of 0.5 to 1 MV/cm for alkali halide crystals, which are about the lowest, upward to somewhat more than 10 MV/cm. Polymers and some inorganic materials such as mica, aluminum oxide, etc., have strengths of 2 to 20 MV/cm for thin films. The strength decreases with increasing thickness and with temperature above a critical temperature (which is usually from 1 to 100°C), below which the strength has a level value or a moderate increase with increasing temperature. Below the critical temperature, the breakdown is believed to be strictly electronic in nature and is constant or increases slightly with temperature. Above this temperature, it declines owing to dielectric thermal heating.

The breakdown voltage of thin insulating materials containing defects, which give the minimum breakdown voltage, declines as the area under stress increases. The effect of area on the strength can be estimated from the standard deviation $S$ of tests on smaller areas by applying minimum value statistics:[10] $V_1 - V_2 = 1.497S \log (A_1/A_2)$, where $V_1$ and $V_2$ are the breakdown voltages of areas $A_1$ and $A_2$.

If the ac or dc conductivity of a dielectric is high, or the frequency is high, breakdown can occur as a result of dielectric heating, which raises the temperature of the material sufficiently to cause melting or decomposition, formation of gas, etc. This effect can be detected by measuring the conductivity as a function of applied electric stress. If the conductivity rises with time, with constant voltage, and at constant ambient temperature, this is evidence of an internal dielectric heating. If the heat transfer to the electrodes and ambient surroundings is adequate, the internal temperature may eventually stabilize, but if this heat transfer is inadequate, the temperature will rise until breakdown occurs. The criterion of this sort of breakdown is the heat balance between dielectric heat input and loss to the surroundings.

The dielectric heat input is given by the equation

$$\sigma E^2 = (5/9\varepsilon' f \tan \delta \times 10^{-12})E^2 \qquad \text{W/cm}^3 \tag{4-72}$$

where $E$ is the field in volts per centimeter. When this quantity is of the order of 0.1 or greater, dielectric heating can be a problem. It is much more likely to occur with thick insulation and at elevated temperatures. Dielectric breakdown is covered by Whitehead,[11] Peek,[12] and Roth[13] and in a review chapter by Mason.[14]

***241. Water penetration*** into electrical insulation also degrades the dielectric strength by several mechanisms. The effect of water to increase the insulation conductivity, which has already been mentioned in Sec. 235, contributes thereby to a decreased dielectric strength, probably by a thermal breakdown mechanism. Another effect noticed recently, particularly in polyethylene cables, is the development of "water" or "electrochemical trees." Water (and/or a similar high dielectric constant chemical) can diffuse through polyethylene and collect at tiny hygroscopic inclusion sites, where the water or chemical is adsorbed. Then, the electric field causes an expansion and growth of the adsorbed water or chemical in the electric field direction. This may completely bridge the insulation or possibly increase the local electric stress at the site so as to produce an electric tree and eventual breakdown. See Ref. 28 for a review of this phenomenon.

***242. Ionizing radiation,*** as from nuclear sources, may degrade insulation dielectric strength and integrity by causing polymer chain scission, and cracking of some plastics, as well as gas bubbles in liquids. Also the conductivity levels in solids and liquids are increased.

***243. Arc Tracking of Insulation.*** High current arc discharges between conductors across the surface of organic resin insulation may carbonize the material and produce a conducting track. In the presence of surface water films, formed from rain or condensation, etc., small arc discharges form between interrupted parts of the water film, which is fairly conducting, and conducting tracks grow progressively across the surface, eventually bridging between conductors and causing complete breakdown. Materials vary widely in their resistance to tracking, and there are a variety of dry and wet tests for this property. Table 4-55, from a survey paper by Mandelcorn and Sommerman,[15] indicates the difference between materials and the correlation or lack of correlation between the tests. With proper fillers, some organic resins can be made essentially nontracking. Some resins such as polymethyl methacrylate and polymethylene oxide burst into flame under arcing conditions.

Review references are by Mandelcorn and Sommerman[15] and Olyphant.[16]

***244. Thermal Aging.*** Organic resinous insulating materials in particular are subject in varying degrees to deterioration due to thermal aging, which is a chemical process involving decomposition or modification of the material to such an extent that it may no longer function adequately as the intended insulation.[25] The aging effects are usually accelerated by increased temperature, and this characteristic is utilized to make accelerated tests to failure or to an extent of deterioration considered dangerous. Such tests are made at appreciably higher than normal operating temperatures, if the expected life is to be several years or more, since useful accelerated tests should reasonably be completed in less than a year.

**TABLE 4-55** Comparison of Tracking Resistance of Various Materials Measured with Seven Test Procedures

| Column heading (reference) | A(15) | B(25, 26) | C(9, 28) | D(9, 27) | E(2) | F(11) | G(30) |
|---|---|---|---|---|---|---|---|
| Test method designation | D495-61 | IEC 113; VDE | D2132-62T | Lin.-Accel. | Differential | Inclined | Inclined |
| Units | Equiv s/10 | Drops, 0.9 kV | Std Dust-Fog | Dust-Fog | Wet Track | plane I | plane II |
| | | Nekal | h, 1.5 kV | H, 1.5 kV | W·min | V, kV | H, 2.5 kV |
| Symbol (data reference) | Liii(9) | L . . (27) | L. . . (9) | L. . . (27) | W. . . . (2, 32) | V . (11) | L . (32, 33) |
| Polyvinyl chloride | 0.5 Tr | | 0.5 Tr | 0.5 Tr | | | |
| Phenolic laminate, paper base | 0.5 Tr | 1 | 0.5 Tr | | 0.2* Tr | 1.5 Tr | |
| Epoxy resin, unfilled | 1.7 Tr | 60 + No Tr | 0.5 Tr | | 1.6 Tr | | |
| Polyamide resin | 58 + Er | 5 Tr | 0.5 Tr | | 1.3 Int | | |
| Silicone resin, glass cloth | 54 Tr | 10 Tr | 1.0 Tr | 1.0 Tr | 1.8 Tr | 1.5 Tr | |
| Melamine resin, glass cloth | 47 Tr | 6 Tr | 3.5 Tr | 2.5 Tr | 2.3 Tr | 2.3 Tr | 0.2 Tr |
| Polyethylene | 13 Tr | 60 + No Tr | 27 Tr | 10 Er + Tr | | | |
| Polyester, glass mat, h-m-f, 1 | 25 Tr | No Tr | 50 Tr | 12 Tr | 3.7† Tr | 2 Tr | 1.1 Tr |
| Polymethylmethacrylate | 100 + Er | No Tr | 90 Er | 33 Tr | 8.1 + Er | 6 F | |
| Polypropylene | 310 Er | No Tr | 180 Er | 40 Tr | 8.1 + Er | 3.8 Tr | |
| Epoxy resin, h-m-f | 100 + Er | No Tr | 200 Er | | | | |
| Polyester, glass mat, h-m-f, 2 | 51 Tr | No Tr | 350 Tr | 90 Tr | 6.4 Tr | 3 Tr | 11 Tr |
| Butyl rubber, h-m-f | 100 + Er | No Tr | 450 Er | 100 Er + Tr | 8.1 + Er | 6 F | |
| Silicone rubber, n-m-f | 5 Tr | | 750 Er | 120 Er + Tr | | 3.7 Tr | |
| Polytetrafluoroethylene | 310 + Er | No Tr | 2,700 Er | 330 Tr | 8.1 + Er | 7 F | |

h-m-f = hydrated-mineral filled; n-m-f = nonhydrated-mineral-filled; Tr = tracked; No Tr = no tracking; Er = eroded; Int = internal; C = carbonized; F = flame.

*Failed 1.3 W, 1 s

†Failed 5.5 W, 18 s.

Frequently other environmental factors influence the life, in addition to the temperature. These include presence or absence of oxygen, moisture, and electrolysis. Mechanical and electrical stress may reduce the life by setting a required level of performance at which the insulation must perform. If this level is high, less deterioration of the insulation is required to reach this level.

Sometimes a complete apparatus is life-tested, as well as smaller specimens involving only one insulation material or a simple combination of these in a simple model. New tests are being devised continually,[18] but there has been some standardization of tests by the IEEE and ASTM and internationally by the IEC.

It is important to note that frequently materials are assigned temperature ratings based on tests of the material alone. Often that material, combined with others in an apparatus or system, will perform satisfactorily at appreciably higher temperatures. Conversely, because of incompatibility with other materials, it may not perform at as high a temperature as it would alone. For this reason it is considered desirable to make functional operating tests on complete systems. These can also be accelerated at elevated temperatures and environmental exposure conditions such as humidification, vibration, cold-temperature cycling, etc., introduced intermittently.

The basis for temperature rating of apparatus and materials is discussed thoroughly in *IEEE Standards Publ.* 1. Tests for determining ratings are described in *IEEE Publs.* 98, 99, and 101.

***245. Application of Electrical Insulation.*** In applying an insulating material it is necessary to consider not only the electrical requirements but also the mechanical and environmental conditions of the application. Mechanical failure often leads to electrical failure, and mechanical failure is frequently the primary cause for failure of an aged insulation.

The initial properties of an insulation are frequently more than adequate for the application, but the effects of aging and environment may degrade the insulation rapidly to the point of failure. Thus, the thermal and environmental stability should be considered of equal importance. The effects of moisture and surface dirt contamination should be particularly considered, if these are likely to occur.

The application of insulation to shipboard insulation and rotating machinery generally is reviewed by Moses.[19] Application to a variety of apparatus is covered by Jackson.[20] A reference book[21] by Clark surveys the properties of a wide variety of materials and their application characteristics.

References which should be consulted for further details include the following: The ASTM Standards on Insulating Materials[22] are continually revised, and a single-volume collection of the electrical tests and specifications is published every two years. "Progress in Dielectrics" reviews the literature of the field.[23] The annual "Digest of Literature on Dielectrics"[24] is a reference source for each year's published papers in classified form. The "Reports of the Annual National Research Council Conference on Electrical Insulation" and the reports of the approximately biennial NEMA-IEEE "Electrical Insulation Conference" offer collections of papers covering the recent developments.

***246. General References on Insulating Materials***

1. Smyth, C. P.: "Dielectric Behavior and Structure"; New York, McGraw-Hill Book Company, 1955.
2. von Hippel, A.: "Dielectric Materials and Applications"; Cambridge, Mass., and New York, The Technology Press of the Massachusetts Institute of Technology and John Wiley & Sons, Inc., 1954.
3. von Hippel, A.: "Dielectrics and Waves"; New York, John Wiley & Sons, Inc., 1954.
4. Tables of Dielectric Materials, Vols. I–VI, *MIT Lab. Insulation Research Tech. Repts.*, 1944–1958.
5. Miner, D. F.: "Insulation of Electrical Apparatus"; New York, McGraw-Hill Book Company, 1941.
6. MacInnes, D. A.: "The Principles of Electrochemistry"; New York, Reinhold Publishing Corporation, 1939.
7. Schwiger, A., and Sorensen, R. W.: "The Theory of Dielectrics"; New York, John Wiley & Sons, Inc., 1932.
8. Stratton, J. A.: "Electromagnetic Theory"; New York, McGraw-Hill Book Company, 1941.
9. Ollendorff, F.: "Potential Felder der Elektrotechnik"; Berlin, Springer-Verlag, OHG, 1932.
10. Weber, K. H., and Endicott, H. S.: Area Effect and Its Extremal Basis for the Electric Breakdown of Transformer Oil; *Trans. AIEE*, 1956, Vol. 5–III, p. 371.

11. Whitehead, S.: "Dielectric Breakdown of Solids"; New York, Oxford University Press, 1951.
12. Peek, F. W.: "Dielectric Phenomena in High Voltage Engineering"; New York, McGraw-Hill Book Company, 1929.
13. Roth, A.: "Hochspannungstechnik"; Vienna, Springer-Verlag, OHG, 1959.
14. Mason, J. H.: "Progress in Dielectrics"; London, Heywood & Co., 1959, Vol. 1, Chap. 1, Breakdown of Solid Dielectrics.
15. Sommerman, G. M. L.: Electrical Tracking Resistance of Polymers; *Trans. AIEE,* 1960, Vol. 70–III, pp. 69–74. Mandelcorn, L., and Sommerman, G. M. L. "Collected Papers of the 1963 NEMA-IEEE Electrical Insulation Conference"; Chicago, Ill.
16. Olyphant, M.: "Arc Resistance I & II"; *ASTM Bull.,* 1952, Vol. 181, p. 60, Vol. 185, p. 31.
17. Dakin, T. W., Philofsky, H. M., and Divens, W. C.: Effect of Electrical Discharges on the Breakdown of Solid Insulation; *Trans. AIEE,* 1954, Vol. 73–I, pp. 155–162.
18. A Bibliography on Testing of Insulating Materials and Systems for Thermal Degradation, *AIEE Spec. Publ.* S–87.
19. Moses, G. L.: "Electrical Insulation, Its Application to Shipboard Electrical Equipment"; New York, McGraw-Hill Book Company, 1951.
20. Jackson, W.: "The Insulation of Electrical Equipment"; London, Chapman & Hall, Ltd., 1954.
21. Clark, Frank M.: "Insulating Materials for Design and Engineering Practice"; New York, John Wiley & Sons, Inc., 1962.
22. ASTM Standards, Pt. 29, Electrical Insulating Materials, 1964 and succeeding even years.
23. "Progress in Dielectrics"; 1959–1962, London and New York, Heywood & Co. and Academic Press, Inc., Vols. 1 to 4 issued.
24. "Digest of Literature on Dielectrics"; National Academy of Sciences—National Research Council, Conference on Electrical Insulation, Washington, D.C.
25. Dakin, T. W.: Electrical Insulation Deterioration Treated as a Chemical Rate Phenomenon; *Trans. AIEE,* 1948, Vol. 67, p. 113.
26. Field, R. F.: The Formation of Ionized Water Films on Dielectrics under Conditions of High Humidity; *J. Appl. Phys.,* 1946, Vol. 17, p. 318.
27. "Engineering Dielectrics," Vol. 1, "Corona Measurement and Interpretation," R. Bartnikas and E. J. McMahon, (eds); ASTM, Philadelphia, Pa., 1979.
28. "Engineering Dielectrics," Vol. 2, "Electrical Properties of Solid Insulating Materials, Molecular Structure and Electrical Behavior," R. Bartnikas and R. M. Eichhorn, (eds); ASTM, Philadelphia, Pa., 1983.
29. Studniarz, S. A., and Dakin, T. W.: "The Voltage Endurance of Cast Epoxy Resin—II"; Conference Record of 1982 IEEE International Symposium on Electrical Insulation, 82CH1780-6-EI, June 7–9, Philadelphia, pp. 19–25.
30. Wichmann, A., and Gruenwald, P.: "Proc. IEEE International Symposium on Electrical Insulation"; IEEE Conference Record, 76CH1088-4-EI, Montreal, 1976, pp. 88–92.


## Insulating Gases

*By T. W. DAKIN*

***247. General Properties of Gases.*** A gas is a highly compressible dielectric medium, usually of low conductivity and with a dielectric constant only a little greater than unity, except at high pressures. In high electric fields the gas may become conducting as a result of impact ionization of the gas molecules by electrons accelerated by the field, and by secondary processes, which produce partial breakdown (corona) or complete breakdown. Conditions which ionize the gas molecules, such as very high temperatures and ionizing radiation (ultraviolet rays, x-rays, gamma rays, high-velocity electrons, and ions, such as alpha particles), will also produce some conduction in a gas.

The gas density $d$ (grams per liter) increases with pressure $p$ (torrs or millimeters of mercury) and gram-molecular weight $M$ and decreases inversely with the absolute temperature

$T$ (degrees Celsius + 273), according to the relation

$$d = \frac{M}{22.4} \frac{p}{760} \frac{273}{T} \quad \text{g/L} \tag{4-73}$$

The above relation is exact for ideal gases but is only approximately correct for most common gases. For exact values, tables should be consulted as well as more exact equations such as the Van der Waals equation.[1]*

If the gas is a vapor in equilibrium with a liquid or solid, the pressure will be the vapor pressure of the liquid or solid. The logarithm of the pressure varies as $-\Delta H/RT$, where $\Delta H$ is the heat of vaporization in calories per mole and $R$ is the molar gas constant, 1.98 cal/(mol)(°C).[1] This relation also applies to all common atmospheric gases at low temperatures, below the points where they liquify.

***248. Dielectric Properties at Low Electric Fields.*** *Dielectric Constant.* The dielectric constant $k$ of gases is a function of the molecular electrical polarizability and the gas density. It is independent of magnetic and electric fields except when a significant number of ions is present. Values of the dielectric constant of some common gases are given in Table 4-56 at atmospheric pressure and 20°C. The increment above unity ($k - 1$) may be estimated[2] approximately by assuming that it varies proportionally to the pressure and inversely to the Kelvin temperature.

**TABLE 4-56** Dielectric Constant of Gases, 20°C, 1 atm gas

| | | | |
|---|---|---|---|
| Air* | 1.000536 | He | 1.000065 |
| $N_2$ | 1.000547 | A | 1.000517 |
| $O_2$ | 1.000495 | $SF_6$ | 1.002084 |
| $CO_2$ | 1.000921 | $H_2$ | 1.000254 |

*Dry, $CO_2$ free.

*Conduction.* The conductivity of a pure molecular gas at moderate electric stress and moderate temperature can be assumed, in the absence of any ionizing effect such as ionizing radiation, to be practically zero. Ionizing radiation induces conduction in the gas to a significant extent, depending on the amount absorbed and the volume of gas under stress.[3] The energy of the radiation must exceed, directly or indirectly, the ionization energy of the gas molecules and thus to produce an ion pair (usually an electron and positive ion). The threshold ionization energy is of the order of 10 to 25 electronvolts (eV)/molecule for common gases (10.86 eV for methyl alcohol, 12.2 for oxygen, 15.5 for nitrogen, and 24.5 for helium). Only very short wavelength ultraviolet light is effective directly in photoionization, since 10 eV corresponds to a photon of ultraviolet with a wavelength of 1240 Å. Since the photoelectric work function of metal surfaces is much lower (2 to 6 eV, for example, copper about 4 eV) the longer-wavelength ultraviolet commonly present is effective in ejecting electrons from a negative conductor surface. Such cathode-ejected electrons give the gas apparent conductivity.

High-energy radiation from nuclear disintegration is a common source of ionization in gases. Nuclear sources usually produce gamma rays of the order of $10^6$ eV energy. Only a small amount is absorbed in passing through a low-density gas. A flux of 1 R/h produces ion pairs corresponding to a saturation current (segment *ab* of Fig. 4-53) of $0.925 \times 10^{-13}$ A/cm$^3$ of air at 1 atm pressure, if all the ions formed are collected at the electrodes. The effect is proportional to the flux and the gas density.

At a voltage stress below about 100 V/cm, some of the ions formed will recombine before being collected and the current will be correspondingly less (segment *oa* of Fig. 4-53). Higher stresses do not increase the current if all the ions formed are collected. A very small current, of the order of $10^{-21}$ A/cm$^3$ of air, is attributable to cosmic rays and residual natural radioactivity.

*Superscripts refer to bibliography, Par. **257**.

Electrons (beta rays) produce much more ionization per path length than gamma rays, because they are slowed down by collisions and lose their energy more quickly. Correspondingly, the slower alpha particles (positive helium nuclei) produce a very dense ionization in air over a short range. For example, a 3-million-eV (MeV) alpha particle has a range in air of 1.7 cm and creates a total of $6.8 \times 10^5$ ion pairs. A beta particle (an electron) of the same energy creates only 40 ion pairs per centimeter and has a range of 13 m in air.

It should be noted that ionizing radiation of significant levels has only a small effect on gas dielectric strength. For example, the ionization current produced by a corona discharge from a needle point is typically much higher than that produced by a radiation flux of significant level, $10^{11}$ gamma photons per square centimeter.

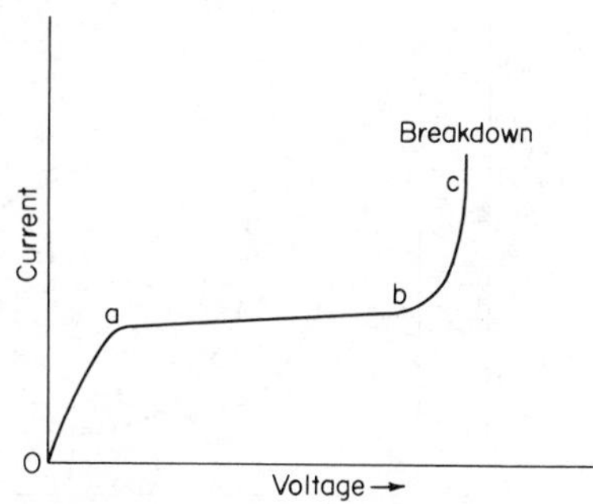


**FIG. 4-53** Current-voltage behavior of a slightly ionized gas.

Thermally induced conductivity occurs in gases, at very high temperatures, as a result of impact ionization by the very high velocity molecules in the gas. This ionization can be calculated from the Saha equation[4] if the ionization energy is known. Such conductivity, in air, becomes significant only above 2000°C. Introduction of quantities of "seed" atoms such as sodium and potassium, which have low ionization energies, has been used in MHD generators to increase the gas conductivity substantially at high temperatures. The chemical reactions in flames also produce significant quantities of ions, and these can carry currents.

At temperatures increasing above 600°C, it has been shown that thermionic electron emission from negative conductor surfaces produces significant currents compared with levels typical of electrical insulation. The order of magnitude of this effect can be estimated from the Richardson thermionic-emission equation.[4]

Since the rate of production of ions by the various sources mentioned above is limited, the current in the gas does not follow Ohm's law, unless the rate of collection of the ions at the electrodes is small compared with the rate of production of these ions, as in the initial part of segment *oa* in Fig. 4-53.

***249. Dielectric Breakdown.*** *Uniform Fields.* The dielectric breakdown of gases is a result of an exponential multiplication of free electrons induced by the field. It is indicated by segment *bc* of Fig. 4-53. It is generally assumed that the initiation of breakdown requires only one electron. However, if only a few electrons are present prior to breakdown, it is not easily possible to measure the trend of current shown in Fig. 4-53. If the breakdown is completed between metal electrodes, the spark develops extremely rapidly into an arc, involving copious emission of electrons from the cathode metal and, if the necessary current flow is permitted, vaporization of metal from the electrodes. Table 4-57 gives the dielectric strength of typical gases. References 17–22 are general references on gas breakdown.

In uniform electric fields breakdown occurs at a critical voltage which is a function of the product of the pressure *p* and spacing *d* (Paschen's law), as illustrated in Fig. 4-54 for

**TABLE 4-57** Relative Dielectric Strengths of Gases
(0.1 in gap)

| | | | |
|---|---|---|---|
| Air | 0.95 | $CF_4$ | 1.1 |
| $N_2$ | 1.0 | $C_2F_6$ | 1.9 |
| $CO_2$ | 0.90 | $C_3F_8$ | 2.3 |
| $H_2$ | 0.57 | $C_4F_8$ cyclic | 2.8 |
| A | 0.28 | $CF_2Cl_2$ | 2.4 |
| Ne | 0.13 | $C_2F_5Cl$ | 2.6 |
| He | 0.14 | $C_2F_4Cl_2$ | 3.3 |
| $SF_6$ | 2.3–2.5 | | |

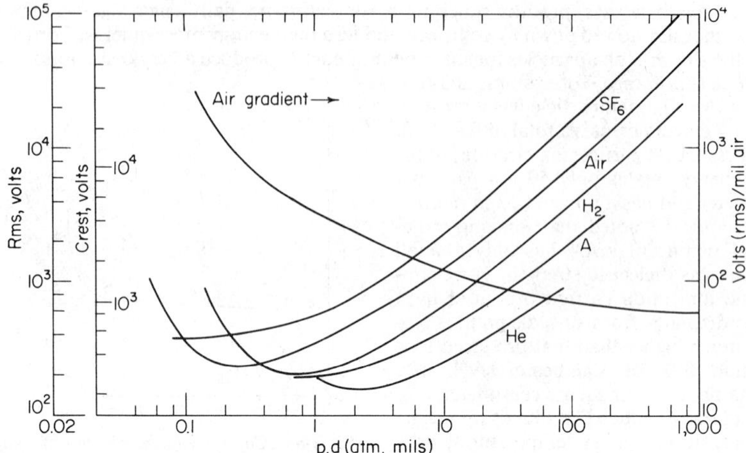


**FIG. 4-54** Pressure-spacing dependence of the dielectric strength of gases (Paschen's curves).

several gases at 20°C. New international summary curves have been published[23] by CIGRE for air, $N_2$, and $SF_6$, and curves for $H_2$, $CO_2$, and He are given in Ref. 24.

It would be more accurate to consider the gas density-spacing product, since the dielectric strength varies with the temperature only as the latter affects the gas density. It will be noted that the electric field at breakdown decreases as the spacing increases. This is typical of all gases and is due to the fact that a minimum amount of multiplication of electrons must occur before breakdown occurs. A single electron accelerated by the field creates an avalanche which grows exponentially as $e^{\alpha x}$, where $x$ is the distance and $\alpha$ is the Townsend ionization coefficient (electrons formed by collision per centimeter), which increases rapidly with electric field. At small spacings $\alpha$ and the field must be higher for sufficient multiplication. In divergent electric fields or large spacings, it has been found that when the integral $\int \alpha_E \, dx$ increases to about 18.4 ($10^8$ electrons), sufficient space charge develops to produce a streamer type of breakdown. It seems to be apparent that the final step in gas breakdown before arc development is the development of a branched filamentary streamer which proceeds more easily from the positive electrode toward the negative electrode.

The dielectric strength of air for larger sphere gaps is given in Table 4-58, selected from IEEE Standard 4 (revision of AIEE Standard 4). These values are used as voltage standards, but they indicate also the trend of the breakdown stress downward with increasing spacing. They also indicate that the impulse strength is almost indentical with the crest 60-Hz strength for smaller spacings but is a little greater for larger spacings. The positive (high-terminal) dc strength is the same as the positive-impulse strength. The higher values than the crest 60-Hz on the larger spacings are because of a slight asymmetry of the field due to the ground plane.

Pressure and (moderate) temperature affect the dielectric strength only as they affect the gas density according to Paschen's law (Fig. 4-54). IEEE Standard 4 gives correction factors for the relative pressure effect on sphere gap breakdown (see Table 4-59).

***250. Relative Dielectric Strengths of Gases.*** The relative dielectric strength, with few exceptions, tends upward with increasing molecular weight. There are a number of factors other than molecular or atomic size which influence the retarding effect on electrons. These include ability to absorb electron energy on collision and trap electrons to form negative ions. The noble atomic gases (helium, argon, neon, etc.) are poorest in these respects and have the lowest dielectric strengths. Table 4-57 gives the relative dielectric strengths of a variety of gases at 1 atm pressure at a $p \cdot d$ value of 1 atm × 0.25 cm. The relative strengths vary with the $p \cdot d$ value, as well as gap geometry, and particularly in divergent fields where corona begins before breakdown. It is best to consult specific references with regard to divergent field breakdown values.

## TABLE 4-58 Sphere Gap with One Sphere Grounded*

Peak values of disruptive-discharge voltages in kilovolts (50 % values for impulse tests)
Valid for:
alternating voltages
full negative standard impulses and impulses with longer tails
direct voltages of either polarity
Atmospheric reference conditions: 25°C and 101.3 kN/m² (760 mm Hg)

| Sphere-gap spacing, cm | Sphere diameter, cm | | | | | | | |
|---|---|---|---|---|---|---|---|---|
| | 6.25 | 12.5 | 25 | 50 | 75 | 100 | 150 | 200 |
| 1 | 31.4 | 31.2 | | | | | | |
| 2 | 57.5 | 58.0 | | | | | | |
| 3 | 78.0 | 84.0 | 84.5 | | | | | |
| 4 | (93.5) | 106 | 110 | | | | | |
| 5 | (105) | 127 | 135 | 136 | 136 | | | |
| 6 | (114) | 144 | 158 | | | | | |
| 8 | | (171) | 203 | | | | | |
| 10 | | (192) | 240 | 259 | 260 | 262 | 262 | 262 |
| 12 | | (208) | 271 | | | | | |
| 15 | | | (309) | 367 | 380 | 383 | | |
| 20 | | | (362) | 452 | 483 | 500 | 500 | 500 |
| 25 | | | (393) | 520 | 575 | 605 | | |
| 30 | | | | (575) | 655 | 700 | 730 | 735 |
| 40 | | | | (660) | (785) | 862 | 940 | 960 |
| 50 | | | | (720) | (880) | 1000 | 1110 | 1160 |
| 75 | | | | | (1025) | (1210) | 1420 | 1510 |
| 100 | | | | | | (1340) | (1630) | 1810 |
| 130 | | | | | | | (1840) | (2070) |
| 150 | | | | | | | (1930) | (2210) |
| 180 | | | | | | | | (2370) |
| 200 | | | | | | | | (2450) |

*Condensed from Table 1 in IEEE No. 4, 1969.

NOTE: The figures in parentheses, which are for spacings of more than $0.5D$, will be within ±5% if maximum clearances are met. On errors for direct current see paragraph 2.5.2.2 in IEEE No. 4.

For full positive standard impulses and impulses with longer tails, the values are zero to 7% higher, depending on the gap. For those values and for intermediate gaps, consult IEEE No. 4, 1969.

## TABLE 4-59 Air-Density Correction Factors for Sphere Gaps

| Relative air density | Sphere diameter, cm | | |
|---|---|---|---|
| | 6.25 | 12.5 | 25 |
| 0.50 | 0.547 | 0.535 | 0.527 |
| 0.55 | 0.595 | 0.583 | 0.575 |
| 0.60 | 0.640 | 0.630 | 0.623 |
| 0.65 | 0.686 | 0.677 | 0.670 |
| 0.70 | 0.732 | 0.724 | 0.718 |
| 0.75 | 0.777 | 0.771 | 0.766 |
| 0.80 | 0.821 | 0.816 | 0.812 |
| 0.85 | 0.866 | 0.862 | 0.859 |
| 0.90 | 0.910 | 0.908 | 0.906 |
| 0.95 | 0.956 | 0.955 | 0.954 |
| 1.00 | 1.000 | 1.000 | 1.000 |
| 1.05 | 1.044 | 1.045 | 1.046 |
| 1.10 | 1.090 | 1.092 | 1.094 |

***251. Corona and Breakdown in Nonuniform Fields between Conductors.*** In nonuniform fields, when the ratio of spacing to conductor radius of curvature is about 3 or less, breakdown occurs without prior corona. The breakdown voltage is controlled by the integral of the Townsend ionization coefficient $\alpha$ across the gap.[6] At larger ratios of spacing to radius of curvature, corona discharge occurs at voltage levels below complete gap breakdown, as shown in Fig. 4-55, Ref. 28.

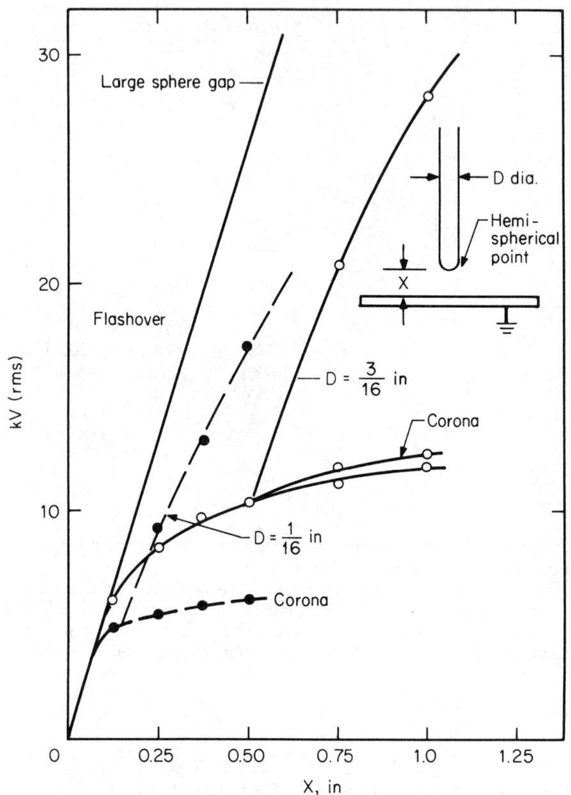


**FIG. 4-55** Alternating current breakdown and corona starting voltages for a hemispherical point to plane in air as a function of gap length (in air). See Ref. 28.

According to Peek,[7] corona in air at atmospheric pressure occurs before breakdown when the ratio of outer to inner radius of coaxial electrodes exceeds 2.72, or where the ratio of gap to sphere radius beween spheres exceeds 2.04. These discharges project some distance from the small-radii conductor but do not continue out into the weaker electric field region, until a higher voltage level is reached.

Such partial breakdowns are often characterized by rapid pulses of current and radio noise. With some conductors at intermediate voltages between onset and complete breakdown, they blend into a pulseless glow discharge around the conductor. When corona occurs before breakdown, it creates an ion space charge around the conductor, which modifies the electric field, reducing the stress at sharp conductor points in the intermediate voltage range. At higher voltages, streamers break out of the space-charge region and cross the gap.

The surface voltage stress at which corona begins increases above that for uniform field breakdown stress, since the field to initiate breakdown must extend over a finite distance. An empirical relation developed by Peek[7] is useful for expressing the maximum surface stress for corona onset in air for several geometries of radius $r$ cm:

For concentric cylinders: $E = 31\delta(1 + 0.308/\sqrt{\delta r})$ kV/cm (4-74)

For parallel wires: $E = 29.8\delta(1 + 0.301/\sqrt{\delta r})$ kV/cm (4-75)

For spheres: $E = 27.2\delta(1 + 0.54\sqrt{\delta r})$ kV/cm (4-76)

where $\delta$ is the density of air relative to that at 25°C and 1 atm pressure.

***252. Corona Discharges on Insulator Surfaces.*** It has been shown by a number of investigators that the discharge-threshold voltage stress on or between insulator surfaces is the same as between metal electrodes.[8] Thus, the threshold voltage for such discharges can be calculated from the series dielectric-capacitance relation for internal gaps of simple shapes, such as plane and coaxial gaps, insulated conductor surfaces, hollow spherical cavities, etc.

The corona-initiating voltage at a conductor edge on a solid barrier depends on the electric stress concentration and generally on the ratio of the barrier thickness to its dielectric constant[9] (see Fig. 4-49), except with low surface resistance. Any absorbed water or conducting film raises the corona threshold voltage by reducing stress concentration at the conductor edge on the surface.

It is sometimes possible to overvolt such gaps considerably prior to the first discharge, and the offset voltage may be below the proper voltage due to surface-charge concentration. With ac voltages, pulse discharges occur regularly back and forth each half cycle, but with dc voltage the first discharge deposits a surface charge on the insulator surface which must leak away, before another discharge can occur. Thus, corona on or between insulator surfaces is very intermittent with steady dc voltages, but discharges occur when the voltage is raised or lowered.

***253. Flashover on Solid Surfaces in Gases.*** As has been mentioned in the previous section on partial discharges, the breakdown in gases is influenced by the presence of solid insulation between conductors. This insulation increases the electric stress in the gas. A particular case of this is the complete breakdown between conductors across or around solid insulator surfaces. This can occur when the conductors are on the same side of the insulation or on opposite sides. A significant reduction in flashover voltage can occur whenever a significant part of the electric field passes through the insulation. The reduction is influenced by the percentage of electric flux which passes through the solid insulation and the dielectric constant of the insulation.

In the application of insulation in the outdoor environment, such as on transmission and distribution lines, or inside in severely polluted or wet locations, it is also important to recognize that conducting layers on the surface (such as from rain and fog with dissolved salts) can greatly reduce the surface flashover voltage in air (or any gas). This is illustrated by Fig. 4-56 for an epoxy strip.[29] The flashover voltage reduction increases with increasing

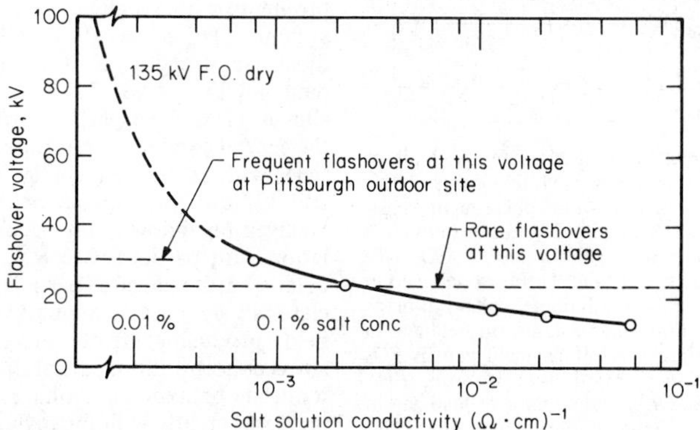


**FIG. 4-56** Flashover voltage of a 15-in long × 2-in wide insulator strip in a salt-solution fog.

surface conductivity. Suspension or post insulators and outdoor bushings are similarly affected as is the flat strip of Fig. 4-56. Reference 25 outlines standard methods of polluting insulator surfaces with controlled conducting layers to assess the flashover voltage in service. Several methods are used: salt fog exposure, clean fog with preapplied conducting coatings, and others. Nonwettable coatings such as silicone grease (which must be periodically replaced) help to maintain a higher flashover voltage in seacoast and polluted areas. More permanent silicone coatings are being tested. Cast epoxy and silicone insulators and bushings made with resins which are nontracking and durable (which are essential properties in outdoor and polluted applications) have performed as well as, or better than, ceramic or glass insulators in seacoast applications.

***254. High-Pressure Gas Breakdown.*** The dielectric strength of gases can be increased very considerably by increasing the pressure (and hence the density). At moderate pressures the increase in strength is slightly less than proportional to the pressure. At higher pressures the increase becomes appreciably less than proportional to the pressure, as indicated[5] in Fig. 4-57. In several cases such as $CO_2$, $N_2$, $SF_6$, and hexane vapor, the compressed gas strength[6] has been shown to approach that of the pure liquid. At very high pressures and the corresponding high stresses for breakdown, the breakdown voltage declines below that predicted by Paschen's law, Fig. 4-54. The departure from Paschen's law seems to begin for gases at stresses on the order of $2 \times 10^5$ volts/per centimeter, and the breakdown level becomes much more sensitive to surface roughness and conducting particles.

With nonuniform fields when corona occurs before breakdown, maxima are observed in the breakdown voltage-pressure curve for electronegative gases like $SF_6$.[26] With increasing pressure and an electric stress concentration point, the lower pressure part of the curve increases somewhat like that for the fixed wire particle in Fig. 4-58, but as the pressure is increased still further, the breakdown curve goes over a maximum and decreases toward a level nearly the same as the partial discharge (corona) threshold.

It should be noted that the high maximum in breakdown voltage shown in Fig. 4-58 for the fixed particle projection from the central conductor is typical for points in $SF_6$ and other electron-attaching gases. In this pressure region, corona discharges from a point create a cloud of ions about the point which electrostatically shield the point, raising the breakdown voltage with ac and 60 Hz. Because the ion space charge takes some time to develop, the impulse voltage breakdown is lower than the 60-Hz voltage breakdown in the pressure range. This is an important consideration in practice. This phenomenon also influences the breakdown with moving particles, which have a lower breakdown in this region, as illustrated in the Fig. 4-58.

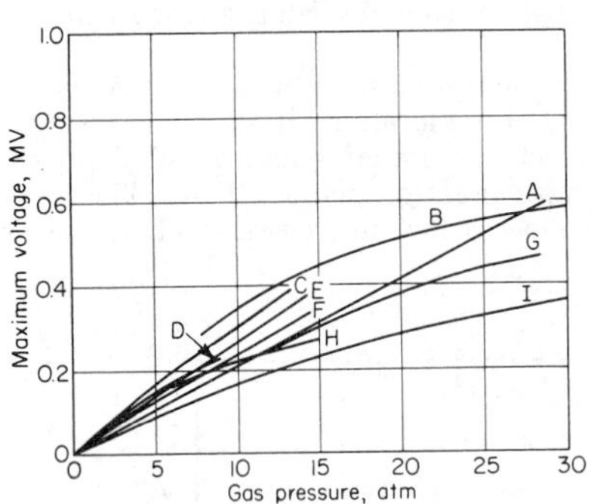


**FIG. 4-57** Breakdown of $N_2$, $CO_2$, and air at high-pressure, 12.7-mm gap. Philip (*Trans. AIEE*, 1963, vol. 82, p. 356): 64-mm sphere facing negative high-voltage terminal in $N_2 + CO_2$ (50%) (*A*); Kusko: Uniform field gap in $N_2 + CO_2$ (*B*); Trump, Stafford, and Cloud: Uniform field gap in air (*C*); Ganger: 50-mm-diameter sphere-to-plane gap in $N_2$ (*D*); Finkelmann: Uniform field gap in $N_2$ (*E*); Finkelmann: Uniform field gap in $CO_2$ (*F*); Trump, Cloud, Mann, and Hanson (*Elec. Eng.*, 1950, vol. 69, p. 961): Uniform field gap in $N_2 + CO_2$ (*G*); Palm: Uniform field gap in $N_2$ (*H*); Howell: Uniform field gap in air (*I*).

The increased use of high-pressure gas (particularly $SF_6$) insulation for enclosed coaxial transmission lines, substations, and power circuit breakers has led to increased problems with conducting particles in the system.[27] These may drastically reduce the dielectric strength, as is indicated by the data in Fig. 4-58. Three situations are shown: (1) a fixed particle projecting from the central conductor, multiple (16) free particles with (2) the test voltage raised slowly and (3) the test voltage held for 3 min. Delayed breakdown may occur in gas insulation with particles. See Ref. 27 for more details on the effect of particles. Free particles will be moved about by the electric field, producing greatly increased electric stress concentration, partial discharges, and reduced breakdown voltage. The effect increases greatly with the high applied average electric stresses used in high gas pressure

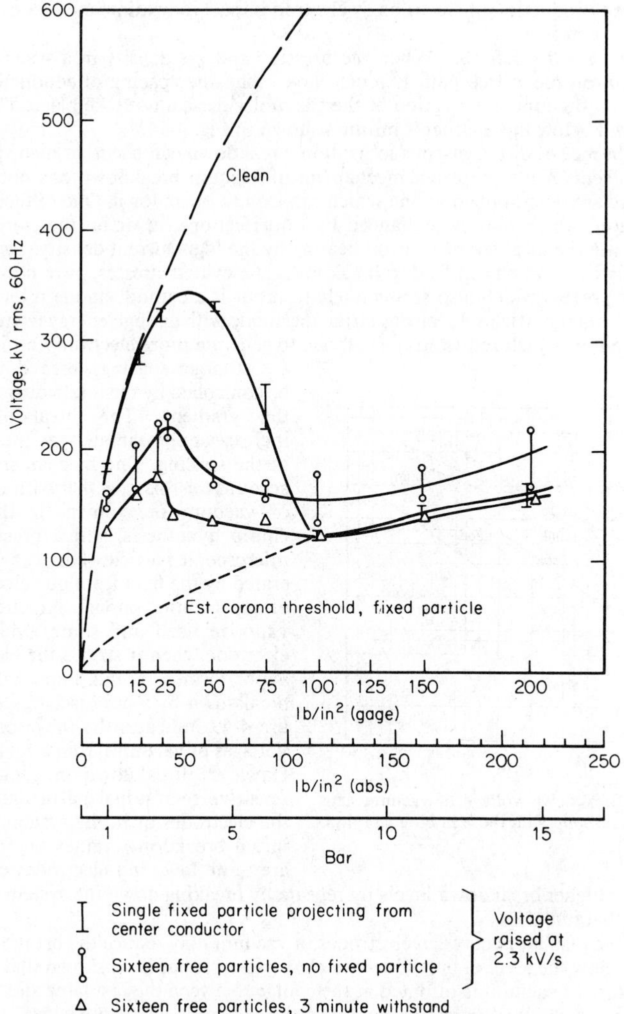


**FIG. 4-58** Effect of wire particles (0.64 × 0.45 cm diameter) on the breakdown voltage of compressed $SF_6$ in a coaxial system (7.6 cm inside diameter × 25 cm outside diameter) over a range of pressures.

systems. One corrective measure, which was not used in Ref. 27 tests, is the introduction of "particle traps," which are regions of low electric stress arranged so that moving particles will eventually fall into them and remain "trapped." Conductor particles may also cling to solid insulator surfaces which support high-voltage conductors and thereby produce stress concentration and reduced breakdown voltage.

***255. Breakdown at High Frequency.*** The ac dielectric strength of gases declines only slightly (6 to 10% at 600 kHz) as the frequency is increased, until the time of a half cycle is about the same as the transit time, first of positive ions, and then of electrons across the gap.[10] At these critical frequencies ($10^5$ to $10^7$ Hz) small maxima have been observed. Above the critical frequency, cumulative ionization occurs in the gap, and there is a sharp drop in breakdown voltage. At these high frequencies, the breakdown voltage is set by the equilibrium between production of electrons by electron impact ionization and loss by diffusion to the walls or electrodes.

***256. Vacuum Breakdown.*** When the pressure and gas density in a system are so low that the electron mean free path is much larger than the spacing of conductors, electron multiplication by impact ionization of the gas molecules cannot take place. This occurs at pressures well below the Paschen's minima shown in Fig. 4-54.

In the absence of direct gaseous ionization, breakdown can occur, at high stresses, from electrode effects. While the exact mechanism of vacuum breakdown has not been determined, there are several phenomena which can lead to breakdown. One of these is cathode field emission, which may be enhanced by imperfections, in or on the cathode surface, which increase the local stress, or even heating by the high current density. Steady cathode emission currents, which can lead to breakdown at elevated stresses, have been observed.

Another process which also seems likely to occur is a cathode-anode regeneration process of elementary particles. Electrons strike the anode with enough energy to create photons and positive ions which return to the cathode to generate more electrons and ions, etc.

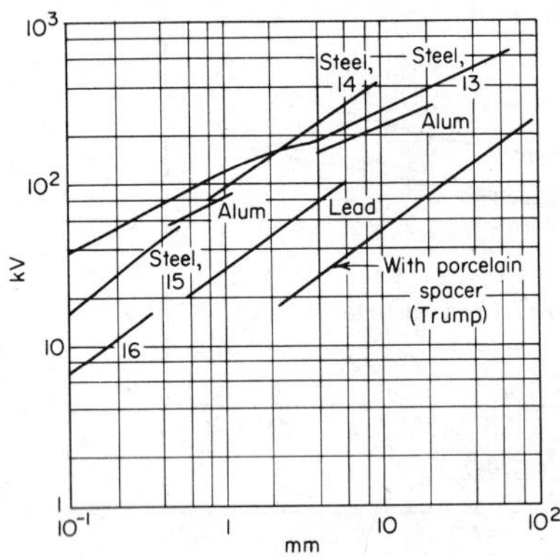


**FIG. 4-59** Breakdown voltage of vacuum gaps (numbers correspond with those in bibliography).

At larger spacings, breakdown seems to be controlled by the total voltage rather than the gradient. The breakdown voltage increases approximately as the square root of the spacing. One mechanism which can account for this, together with other aspects of vacuum breakdown, is the Cranberg clump hypothesis, which presumes that a microscopic particle of many atoms is accelerated by the field from one electrode to the other, gaining enough kinetic energy to vaporize itself and some atoms from the electrode when it strikes the electrode. The vapor formed by this impact then leads to breakdown by a gas discharge process. Figure 4-59 indicates the range of breakdown voltages of vacuum, from a review paper by Hawley.[11] Breakdown in vacuum is very sensitive to residual particulate matter on the electrodes or in the system. Frequently, initial breakdown values on fresh systems are quite low, and electrodes can be "conditioned" to higher breakdown levels by repeatedly breaking down the system with limited current discharges.[13–16]

Supporting insulators between electrodes in vacuum may reduce the breakdown voltage drastically below the level of breakdown in clear gaps.[12] It has been shown that flashover of such insulators in vacuum is initiated at the contact between the insulator and the cathode. If this region is shielded from the field and the insulator properly shaped, much higher breakdown voltage values can be obtained.

***257. References on Insulating Gases***


1. MacDougall, F. H.: "Physical Chemistry"; New York, The Macmillan Company, 1952.
2. Smyth, C.: "Dielectric Behavior and Structure"; New York, McGraw-Hill Book Company, 1955.
3. Curran, S. C., and Craggs, J. D.: "Counting Tubes"; London, Butterworth Scientific Publications, 1949.
4. Cobine, J. D.: "Gaseous Conductors"; New York, Dover Publications, Inc., 1958.
5. Trump, J. G., Cloud, R. W., Mann, J. G., and Hanson, E. P.: Influence of Electrodes on D-C Breakdown in Gases at High Pressures; *Electr. Eng.*, 1950, vol. 69, p. 961. Philp, S. F. Compressed Gas Insulation in the Million Volt Range, $SF_6$, $N_2$, $CO_2$; *Trans. IEEE*, 1963, vol. 82, p. 356.
6. Loeb, L. B.: Electrical Breakdown of Gases, "Encyclopedia of Physics"; Berlin, Springer-Verlag, OHG, 1956, vol. XXII.
7. Peek, F. W.: "Dielectric Phenomena in High Voltage Engineering"; New York, McGraw-Hill Book Company, 1929.
8. Hall, H. C., and Russek, R. M.: Discharge Inception and Extinction in Dielectric Voids; 1954, *Proc. IEE*, vol. 101–2, p. 47.

9. Dakin, T. W., Philofsky, H., and Divens, W.: Effect of Electrical Discharges on the Breakdown of Solid Insulation; *Trans. AIEE,* 1954, vol. 73–I, pp. 155–162.
10. Brown, S. C.: Breakdown in Gases, Alternating and High Frequency Fields, "Encyclopedia of Physics"; Berlin, Springer-Verlag, OHG, 1956, vol. XXII.
11. Hawley, R.: Vaccum as an Insulator; *Vacuum,* 1960, vol. 10, p. 310.
12. Kofoid, M. J.: Effect of Metal-Dielectric Junction Phenomena on High Voltage Breakdown over Insulators in Vacuum; 1960, *Trans. AIEE,* vol. 79–III, p. 999.
13. Trump, J. G., and Van de Graaf, R.: The Insulation of High Voltages in Vacuum; *J. Appl. Phys.,* 1947, vol. 18, p. 327.
14. Slivkov, I. N.: Mechanism for Electrical Discharge in Vacuum; *Sov. Phys. Tech. Phys.,* 1957, vol. 2, p. 1919.
15. Denholm, A. S.: The Electrical Breakdown of Small Gaps in Vacuum; *Can. J. Phys.,* 1958, vol. 36, p. 476.
16. Leader, D.: Electrical Breakdown in Vacuum; *Proc. IEE,* 1953, vol. 100–2A, p. 138.
17. Meek, J. M., and Craggs, J. D.: "Electrical Breakdown of Gases"; New York, Oxford University Press, 1953.
18. Llewellyn Jones, F.: "Ionization and Breakdown in Gases"; London, Methuen & Co., Ltd.
19. Ganger, B.: "Der Elektrische Durchschlag von Gasen"; Berlin, Springer-Verlag, OHG, 1953.
20. Gas Discharges I, "Encyclopedia of Physics"; Berlin, Springer-Verlag, OHG, 1956, vols. XXI and XXII.
21. Dakin, T. W., and Berg, D. Theory of Gas Breakdown, chapter in "Progress in Dielectrics"; London and New York, Heywood & Co., Ltd., and Academic Press, Inc., 1962, vol. 4.
22. Roth, A.: "Hochspannungstechnik," 4th ed.; Berlin, Springer-Verlag, OHG, 1959.
23. Dakin, T. W., with German and French members of CIGRE Group 15-03: Breakdown of Gases in Uniform Fields—Paschen Curves for Nitrogen, Air, and $SF_6$; *Electra* (published by CIGRE, Paris), No. 32, p. 61, January 1974.
24. Winkelnkemper, H., Krasucki, Z., Gerhold, J., and Dakin, T. W.: Breakdown of Gases in Uniform Fields, Paschen Curves for Hydrogen, Carbon Dioxide and Helium; *Electra,* No. 92, p. 67, May 1977 (published together with Ref. 23 as a booklet by CIGRE, Paris, France.)
25. "Artificial Pollution Tests on High Voltage Insulators to Be Used on A-C Systems"; IEC Standard 507, 1975.
26. Works, C. N. and Dakin, T. W.: Dielectric Breakdown of $SF_6$ in Non-uniform Fields; 1953, *Trans. AIEE,* vol. 72, pt. 1, pp. 682–687.
27. Wootton, R. E., Cookson, A. H., Emery, F. T., and Farish, O.: "Investigation of High Voltage Particle Initiated Breakdown in Gas Insulated Systems"; Electric Power Research Institute Report EL-1007.
28. Narbut, P., Berg, D., Works, C. N., and Dakin, T. W.: Factors Controlling Electric Strength of Gaseous Insulation; *Trans. AIEE,* 1959, vol. 78–III, pp. 59–74.
29. Dakin, T. W., and Mullen, G. A.: Continuous Recording of Outdoor Insulator Surface Conductance; *IEEE Trans.* on Electrical Insulation, EI-7, December 1972, pp. 169–175.

## Insulating Oils and Liquids

*By T. W. DAKIN*

***258. General Considerations.*** Typical insulating liquids are natural or synthetic organic compounds and frequently consist of mixtures of essentially isomeric compounds with some range of molecular weight. The mixture of very similar but not exactly the same molecules, with a range of molecular size and with chain and branched hydrocarbons, prevents crystallization and results in a low freezing point, together with a relatively high boiling point. Typical insulating liquids have permittivities (dielectric constants) of 2 to 7 and a wide range of conductivities, depending upon their purity. The dc conductivity in these liquids is usually due to dissolved impurities, which are ionized by dissociation. Higher ionized impurity and conductivity levels occur in liquids having higher permittivities and lower viscosities.

The function of insulating liquids is to provide electrical insulation and heat transfer. As insulation, the liquid is used to displace air in the system and provide a medium of high electric strength to fill pores, cracks, and gaps in insulation systems. It is usually necessary to fill and impregnate systems with liquid under vacuum, so that all air bubbles are eliminated. If air is completely displaced in all high electric field regions, the corona threshold voltage and breakdown voltage for the system are greatly increased. The viscosity selected for a liquid insulation is often a compromise to provide the best balance between electrical insulation and heat transfer and other limitations such as flammability, solidification at low temperatures, and pressure development at high temperatures in sealed systems.

The most commonly used insulating liquids are natural hydrocarbon mineral oils refined to give low conductivity and selected viscosity and vapor-pressure levels for transformer, circuit-breaker, and cable applications.

A variety of synthetic fluids are also used for particular applications where the higher cost above that of mineral oil is warranted by the requirements of the application or by the improved performance in relation to the apparatus design.

***259. Mineral Insulating Oils.*** Mineral insulating oils are hydrocarbons (compounds of hydrogen and carbon) refined from crude petroleum deposits from the ground.[1]* They consist partly of aliphatic compounds with the general formula $C_nH_{2n+2}$ and $C_nH_{2n}$, comprising a mixture of straight and branched chain and cyclic or partially cyclic compounds. Many oils also contain a sizable fraction of aromatic compounds related to benzene, naphthalene, and derivatives of these with aliphatic side chains. The ratio of aromatic to aliphatic components depends on the source of the oil and its refining treatment. The percent aromatics is of importance to the gas-absorption or evaluation characteristics under electrical discharges[2] and to the oxidation characteristics.[3]

The important physical properties of a mineral oil (as for other insulating liquids as well) are listed in Table 4-60 for three types of mineral oils. In addition to these properties, mineral oils which are exposed to air in their application have distinctive oxidation characteristics which vary with type of oil and additives and associated materials.[4]

**TABLE 4-60** Characteristic Properties of Insulating Liquids

| Type of liquid | Mineral oil: Transformer | Mineral oil: Cable and capacitor | Mineral oil: Solid cable |
|---|---|---|---|
| Specific gravity | 0.88 | 0.885 | 0.93 |
| Viscosity, Saybolt sec at 37.8°C | 57–59 | 0.100 | 100 |
| Flash point, °C | 135 | 165 | 235 |
| Fire point, °C | 148 | 185 | 280 |
| Pour point, °C | −45 | −45 | −5 |
| Specific heat | 0.425 | 0.412 | ........ |
| Coefficient of expansion | 0.00070 | ........ | 0.00075 |
| Thermal conductivity, cal/(cm)(s)(°C) | 0.39 | ........ | ........ |
| Dielectric strength,* kV~ | 30 | ........ | ........ |
| Permittivity at 25°C | 2.2 | ........ | ........ |
| Resistivity, Ω·cm × $10^{12}$ | 1–10 | 50–100 | 1–10 |

*ASTM D877.

Many manufacturers now approve the use of any of several brands of mineral insulating oil in their apparatus provided that they meet their specifications which are similar to ASTM D1040, values from which are tabulated in Table 4-60. Low values of dielectric strength may indicate water or dirt contamination. A high neutralization number will indicate acidity, developed very possibly from oxidation, particularly if the oil has already been used. Presence of sulfur is likely to lead to corrosion of metals in the oil.

The solubility of gases and water in mineral oil is of importance in regard to its function

*Superscripts refer to bibliography, Par. **268**.

in apparatus. Solubility is proportional to the partial pressure of the gas above the oil:

$$S = S_0(p/p_0) \tag{4-77}$$

where $S$ is the amount dissolved at pressure $p$ if the solubility is expressed as the amount $S_0$ dissolved at pressure $p_0$.

The solubility is frequently expressed in volume percent of the oil. Values for solubility of some common gases in transformer oil[5] at atmospheric pressure (760 torr) and 25°C are air 10.8%; nitrogen 9.0%; oxygen 14.5%; carbon dioxide 99.0%; hydrogen 7%; methane 30% by volume. The solubilities of all the gases, except $CO_2$, increase slightly with increasing temperature. Water is dissolved in new transformer oil to the extent of about 60 to 80 ppm at 100% relative humidity and 25°C. The amount dissolved is proportional to the relative humidity. Solubility of water increases with oxidation of the oil and the addition of polar impurities, with which the water becomes associated. Larger quantities of water can be suspended in the oil as fine droplets.

***260. Dielectric Properties of Mineral Oils.*** The permittivity of mineral insulating oils is low, since they are essentially nonpolar, containing only a few molecules with electric dipole moments. Some oils possess a minor fraction of polar constituents, which have not been identified. These contribute a dipolar character to the dielectric properties at low temperature and/or high frequency, similar to the trends shown in Figs. 4-47 and 4-48. A typical permittivity for American transformer oil at 60 Hz is 2.19 at 25°C, declining almost linearly to 2.11 at 100°C. At low temperatures and high frequencies, values of permittivity as high as 2.85 have been noted in oils with a relatively high level of polar constituents.

The dc conductivity levels of mineral oils range from about $10^{-15}\ \Omega^{-1} \cdot \text{cm}^{-1}$ for pure new oils up to $10^{-12}\ \Omega^{-1} \cdot \text{cm}^{-1}$ for contaminated used oils.[6] This conductivity is due to dissociated impurity ions or ions developed by oil oxidation.[7] It increases approximately exponentially with temperature about 1 decade in 80°C.

Alternating-current dissipation-factor values are nearly proportional to the dc conductivity, $10^{-13}\ \Omega^{-1} \cdot \text{cm}^{-1}$, corresponding to a $\tan \delta$ of 0.008. If no electrode polarization or interfacial polarization at solid barrier surface effects are present, the dc conductivity $\sigma$ should be related to the ac conductivity ($\tan \delta$) by

$$\sigma = \tfrac{5}{9}\varepsilon' f \tan \delta \times 10^{-12}$$

where $\varepsilon'$ is the dielectric permittivity (Table 4-53) and $f$ is the frequency.

***261. Dielectric Strength of Mineral Oils.*** The dielectric strength of mineral oils, as with all liquids, varies considerably with the state of purity, particularly with respect to particulate matter and moisture. Typical values (ASTM D877, D1816 standard test gaps) are shown in Fig. 4-60 as a function of moisture content.[8]

The dielectric strength of mineral oil has been shown by Weber and Endicott[10] to decrease with increasing area under stress according to the relation

$$V_1 - V_2 = 1.497S \log (A_2/A_1) \tag{4-78}$$

where $S$ is the standard deviation of tests with the smaller area. This relation is derived by application of minimum value statistics, assuming that the largest defect controls the breakdown. $A_1$ is the smaller area.

Typical ac values of the dielectric strength versus spacing and electrode geometry, which affects the maximum stress, are shown[9] in Fig. 4-61. It must not be assumed that the levels shown can be maintained indefinitely, since particulate matter may move into the field and reduce the strength. The dielectric strength is thus dependent upon the time of voltage application. This is illustrated in Fig. 4-62, where the dielectric strength is plotted as a function of the number of 60 Hz cycles using electrodes like those in ASTM D877. The single-cycle breakdown voltage is usually close to the impulse breakdown voltage; most of the decline in dielectric strength occurs in the time range from one cycle to a few thousand cycles, or about 1 min. Typical impulse breakdown voltages of transformer oil are shown in Fig. 4-63. Usually the impulse strength is about two to three times the crest 60-Hz 1-min strength. The difference decreases as the oil purity from particles increases.

The covering of metal conductor surfaces has been known to increase the ac strength of oil gaps.[11]

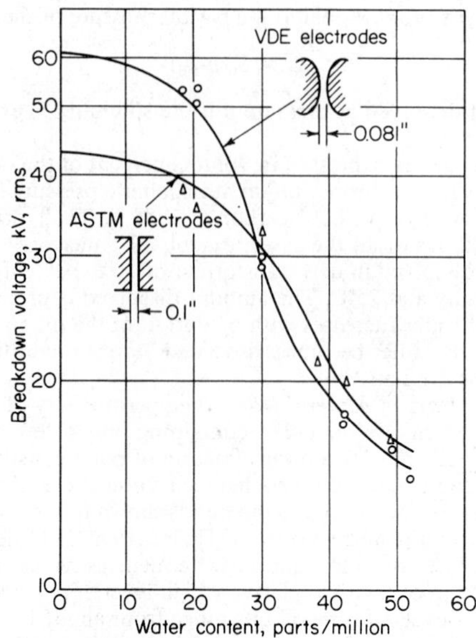


**FIG. 4-60** Electric strength of transformer oil versus water content with ASTM and VDE electrodes (rate of voltage rise, 2 kV rms/s).

Corona or partial breakdown can occur in mineral oil, as with any liquid or gas, when the electric stress is locally very high and complete breakdown is limited by a solid barrier or large oil gap (as with a needle point in a large gap). Such discharges produce hydrogen and methane gas, and sometimes carbon with larger discharges. Dissolved air is also sometimes released by the discharge. If the gas bubbles formed are not ejected away from the high field, they will reduce the subsequent discharge threshold voltage as much as 80%. The resistance of insulating oils to partial discharges is measured by two ASTM gassing tests: D2298 (Merrill test) and D2300 (modified Pirelli test). These tests measure the amount of decomposition gas evolved under specified conditions of exposure to partial discharges. A minimum amount of gas is, of course, preferred, particularly in applications for cables or capacitors. In fact, conventional mineral oils are inadequate in this respect for application in modern 60-Hz power capacitor designs.

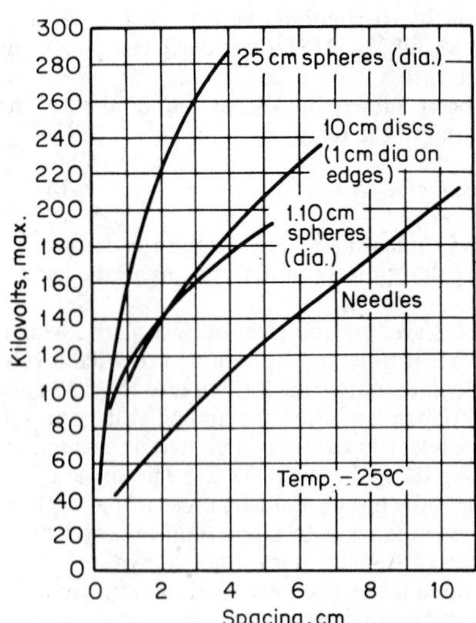


**FIG. 4-61** Sparkover of various shaped electrodes in oil at 60 Hz.

***262. Deterioration of Oil.*** Deterioration of oil in apparatus partially open or "breathing" is subject to air oxidation. This leads to acidity and sludge. There is no correlation between the amount of acid and the likelihood of sludging or the amount of sludge. Sludge clogs the ducts, reduces the heat transfer, and accelerates the rate of deterioration. ASTM tests for oxidation of oils are D1904, D1934, D1313, and D1314.

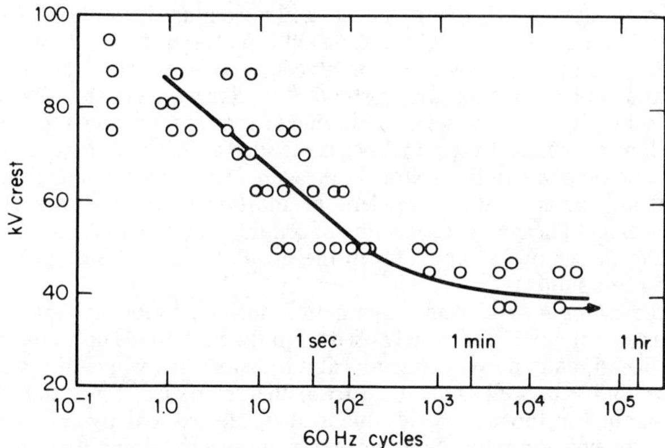


**FIG. 4-62** Time to breakdown of a 0.1-in oil gap with 1-in square-edge electrodes (ASTM D877). (See Ref. 15.)

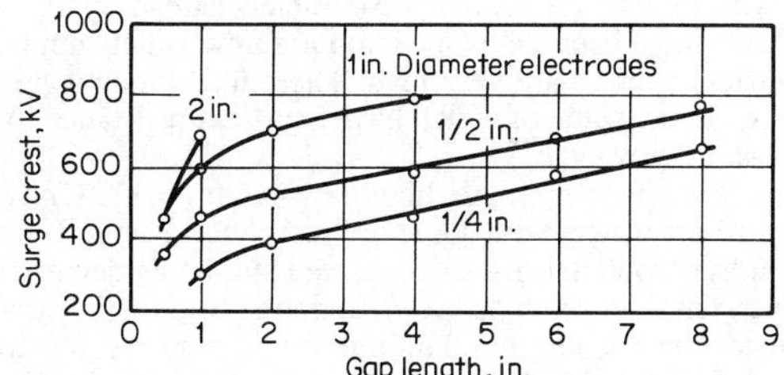


**FIG. 4-63** Curve showing relation of gap length to minimum surge crest voltage required for breakdown between cylindrical electrodes with hemispherical ends immersed in oil, 1½- × 40-μs wave.

Copper and lead and certain other metals accelerate the oxidation of mineral oils. Oils are considerably more stable in nitrogen atmospheres.

Inhibitors are now commonly added both to new and to used oils to delay the oxidation. Ditertiary butyl paracresol (DBPC) is the inhibitor most commonly used at present.

***263. Servicing, Filtering, and Treating.*** Oil in service is usually maintained by testing for acidity, dielectric strength, inhibitor content, interfacial tension, neutralization number, peroxide number, pour point, power factor, refractive index and specific optical dispersion, resistivity, saponification, sludge, corrosive sulfur, viscosity, and water content, as outlined in ASTM D117. These properties indicate various types of contamination or deterioration which might affect the operation of the insulating oil. It has been suggested that interfacial tension below a certain value indicates that sludging is imminent or has started (see *ASTM Spec. Tech. Publ.* 135, 1952).

Depending on the voltage rating of the apparatus, the oil is maintained above 16 to 22 kV (ASTM Test D877). The usual contaminants are water, sludge, acids, and, in circuit-breaker oils, carbon. The centrifuge is best suited for removing large quantities of water, heavier solid particles, etc. The blotter filter press is used for the removal of minute quantities of water, fine carbon, etc. In another method, after removing the larger particles the oil is heated and sprayed into a vacuum chamber, where the water and volatile acids are removed. Sludge and very fine solids are then taken out by a blotter filter press. All units are assembled together so that the process is completed in a single pass. Some work has been done in reclaiming oil by treating it to reduce acidity. One process is similar to the later stages in refining. Another treatment uses activated alumina, fuller's earth, or silica gel. The IEEE guide for Maintenance of Insulating Oil is published as *IEEE Standards Publ.* 64.

It has been found that analysis of the dissolved gas in oil or above the oil in oil-insulated transformers and cables is a good diagnostic tool to detect electrical faults, particularly, or deterioration, generally. For example, continuing or intermittent partial discharges produce hydrogen and low-molecular-weight hydrocarbons such as methane, ethane, ethylene, which accumulate in the oil and can be accurately measured to assess the magnitude of the fault. Higher-current arc faults produce acetylene in addition to $H_2$ and other low-molecular-weight hydrocarbons. Thermal deterioration of cellulosic or paper insulation is indicated by elevated concentrations of CO and $CO_2$ in the oil. Reference 16 should be consulted for more details on this subject.

***264. Synthetic Liquid Insulation.*** Synthetic chlorinated diphenyl and chlorinated benzene liquids (askarels) have been widely used from the mid-1930s up to the mid-1970s and are still in service in many power capacitors and transformers, where they were adopted for their nonflammability, as well as good electrical characteristics.[6] Since the mid-1970s their use has been banned in most countries due to their alleged toxicity and resistance to biodegradation in the environment. Now, when apparatus containing these fluids, which are commonly referred to as "PCBs," are taken out of service, environmental regulations in the United States require that the fluid not be released into the environment. Waste fluid should be incinerated at high temperature with HCl reactive absorbent scrubbers in the stack, since this acid gas is a product of the combustion. Methods of disposing of solid material (such as paper) saturated with PCBs have not been completely agreed upon.

New synthetic fluids have been developed and are now widely applied in power capacitors, where the electrical stresses are very high. These fluids include aromatic (containing benzene rings) hydrocarbons, some of which have excellent resistance to partial discharges. They are not fire-resistant, however.[17]

Very high boiling, low-vapor-pressure, high-flash-point (>300°C) hydrocarbon oils are being tried for power transformers with some fire resistance. Methods for assessing the risk of fire with such liquids, as well as with silicones, are still being debated.

Perchloroethylene (tetrachloroethylene), a nonpolar liquid, is now in use in sealed medium-power transformers, where nonflammability is required. With a boiling point at atmospheric pressure of 121°C, this fluid is completely nonflammable. It is also widely used in dry cleaning. Other important classes of synthetic insulating fluids are discussed in the following sections.

***265. Fluorocarbon Liquids.*** A variety of nonpolar nonflammable perfluorinated aliphatic compounds, in which the hydrogen has been completely replaced by fluorine, are available with different ranges of viscosity and boiling point from below room temperature to more than 200°C. These compounds have low permittivities (near 2.0) and very low conductivity. They are inert chemically and have low solubilities for most other materials. The chemical formula for these compounds is one of the following: $C_nF_{2n}$, $C_nF_{2n+2}$, and $C_nF_{2n}O$. The presence of the oxygen in the latter formula does not seem to reduce the stability. These compounds have been used for filling electronic apparatus[13] and large transformers to give high heat-transfer rates together with high dielectric strength. The vapors of these liquids also have high dielectric strengths.[14]

***266. Silicone Fluids.*** These fluids, chemically formed from Si—O chains with organic (usually methyl) side groups, have a high thermal stability, low temperature coefficient of viscosity, low dielectric losses, and high dielectric strength. They can be obtained with various levels of viscosity and correlated vapor pressures. Rated service temperatures extend from −65 to 200°C, some having short-time capability up to 300°C. Their permittivity is about 2.6 to 2.7, declining with increasing temperature. These fluids have a tendency to form heavier carbon tracks than other insulating liquids when breakdown occurs. They cannot be considered fireproof but will reduce the risk of fire due to their low vapor pressure.

***267. Ester Fluids.*** There are a few applications, mostly for capacitors, where organic ester compounds are used. These liquids have a somewhat higher permittivity, in the range of about 4 to 7, depending on the ratio of ester groups to hydrocarbon chain lengths. Their conductivities are generally somewhat higher than those of the other insulating liquids discussed here. The compounds are easily subject to hydrolysis with water to form acids and alcohols and should be kept dry, particularly if the temperature is raised. Their thermal stability is poor. Specifically dibutyl sebacate has been used in high-frequency capacitors and castor oil in energy-storage capacitors.

### *268. References on Insulating Oils and Liquids*


1. Gruse, W. A., and Stevens, D. R.: "Chemical Technology of Petroleum," 3d ed.; New York, McGraw-Hill Book Company, 1960.
2. Berberich, L. J.: Influence of Gaseous Electric Discharge on Hydrocarbon Oils; *Ind. Eng. Chem.*, 1938, vol. 30, p. 280. Blodgett, R. B, and Bartlett, S. C.: *Trans. AIEE,* 1961, vol. 80, p. 528. Olds, W. F., Feich, G., and Eich, E.: *Ann Rept. NRC* 1960 *Conf. on Electrical Insulation,* p. 93.
3. Berberich, L. J.: Oxidation Inhibitors in Electrical Insulating Oils; *ASTM Bull.* 149, pp. 65–73, 1947. Ford, J. G., and Sloat, T. K.: Inhibitors Lengthen Life of Transformer Oil; *Westinghouse Eng.*, 1950, p. 250.
4. Symposium on Insulating Oils: *ASTM Bulls.* 146 and 149, 1947. (Several authors.)
5. Kaufman, R. B., Shimanski, E. J., and MacFadyen, K. W.: Gas and Moisture Equilibriums in Transformer Oil; *Trans. AIEE,* 1955, vol. 74–III, p. 312.
6. Clark, F. M.: "Insulating Materials for Design and Engineering Practice"; New York, John Wiley & Sons, Inc., 1962.
7. Piper, J. D.: Chapter in "Dielectric Materials and Applications," A. von Hippel (ed.); New York, John Wiley & Sons, Inc., 1954.
8. Rohlfs, A. F., and Turner, F. J.: Correlation between the Breakdown Strength of Large Oil Gaps and Oil Quality Gauges; *Trans. AIEE,* 1956, vol. 75–III.
9. Peek, F. W.: "Dielectric Phenomena in High Voltage Engineering"; New York, McGraw-Hill Book Company, 1929.
10. Weber, K. H., and Endicott, H. S.: Area Effect and Its Extremal Basis for the Electric Breakdown of Transformer Oil; *Trans. AIEE,* 1956, vol. 75–III, p. 371.
11. Roth, A.: "Hochspannungstechnik," 4th ed.; Vienna, Springer-Verlag, OHG, 1959.
12. White, A. H., and Morgan, S. O.: The Dielectric Properties of Chlorinated Diphenyl; *J. Franklin Inst.*, 1933, vol. 216, p. 635.
13. Kilham, L. F., and Ursch, P. R.: *Proc. Natl. Electronics Computer Conf.*, Los Angeles, May 1955.
14. Berberich, L. J., Works, C. N., and Lindsay, E. W.: *Trans. AIEE,* 1955, vol. 74–I, p. 660.
15. Dakin, T. W., Studniarz, S. A., and Hummert, G. T.: *Annual Report, NRC-NAS Conference on Electrical Insulation and Dielectric Phenomena,* 1972.
16. *IEEE Guide for the Determination of Generated Gases in Oil Immersed Transformers and Their Relation to the Serviceability of the Equipment;* ANSI-IEEE C57, 104, 1978.
17. Mandelcorn, L., Dakin, T. W., Miller, R. L., and Mercier, G.: High-Voltage Power Capacitor Dielectrics, Recent Developments; *Proc. 14th IEEE Electrical/Electronics Insulation Conference,* Boston, Mass., October 1979, IEEE Publication No. 79CH1510-7-EI, p. 250


## Insulated Conductors

*By E. J. CROOP*

***269. Magnet-Wire Insulation.*** The term magnet wire includes an extremely broad range of sizes of both round and rectangular conductors used in electrical apparatus. Common round sizes for copper are AWG No. 42 (0.0025 in) to AWG No. 8 (0.1285 in). Ultrafine sizes of round wire, used in very small devices, range as low as AWG No. 60 for copper and AWG No. 52 for aluminum.

Approximately 30 different "enamels" are used commercially at present in insulating magnet wire. Enamel insulations generally are lowest in cost and best in space factor. The most widely used materials are based on polyvinyl acetals, polyesters, polyester amide-imides, and epoxy resins. The polyvinyl acetal and polyester types of materials possess good mechanical properties and good flexibility and perform well in automatic winding machines. Where low cost is important and winding conditions are not too severe, oleoresinous types and modified oleoresinous types are still used. Polyurethanes are employed where ease of solderability, without solvent or mechanical stripping, is required. These do not have high cut-through resistance, however. Epoxy enamels are used where resistance to chemicals and to moisture is important. Polyimide and other aromatic polymer types are employed for operation in the 200 to 220°C range. However, the aromatic types are among the most expensive of the enamels.

Table 4-61 lists some of the commonly used enameled wires by temperature class. It should be understood that this temperature rating is based on a thermal test[1]* and does not include other environmental factors such as exposure to high humidity or use with a varnish which may impair its thermal stability. Cycling tests which include humidification greatly reduce the lives at temperature of many insulation systems capable of undergoing hydrolytic as well as thermal and oxidative degradation.

Table 4-62 lists some fibrous insulations commonly used for insulating magnet wire conductors.[2] Fibrous insulations are employed where positive separation and high reliability are required. These are generally higher in cost and poorer in space factor than enamel insulations.

***270. Magnet-Strip and Foil Conductors.*** Magnet strip is a term generally employed to describe conductors, both copper and aluminum, with a width-to-thickness ratio greater than 50:1, while smaller ratios place the conductors in the category of rectangular magnet wire. If the thickness of the strip is less than 0.008 in, it is often referred to as "foil." Strip conductors are used in many electromagnetic devices including transformers, choke coils, welders, motor and generator fields, lift magnet coils, and electric clutches and brakes. Some of the advantages[3] of strip conductors are more uniform voltage distribution under surge or impulse conditions, better heat transfer, improved space factor, and stronger coil structure. For insulation, paper and polyester film (0.0005 in or less in thickness) have been used as interleaving materials. The width of the interleave is usually about 0.125 in wider than the strip. Other available interleaving materials are asbestos, polytetrafluoroethylene film, mica, and glass cloth. The most widely used insulation is enamel, which provides the best space factor and lowest cost. Many of the enamels used for insulating wire can be used also for strip, but the most widely used are epoxy and a modified polyester type. The enamel thickness generally ranges from 0.00025 to 0.0005 in on each side, or a build of 0.0005 to 0.0010 in.

***271. Wire and Cable Insulation.*** Many materials are used in wire and cable insulations. Some general references (6,7,8,11) which review the recent state of the art are included at the end of this section.

Polyvinyl chloride (PVC) is widely used for primary insulation or jacketing on communication wires, control cable, bell wire, building wire, hookup wire, fixture wire, appliance cords, power cables, motor leads, etc. Many formulations are available, including those with flame resistance. Dielectric strength is excellent, and flexibility is very good. PVC is one of the most versatile of the lower-cost conventional insulations. A conductive PVC can be used for both shielding and jacketing.

Butyl rubber, when properly compounded, is characterized by excellent resistance to oxidation and aging, exceptional ozone resistance, and very good electrical properties. Resistance to moisture and chemicals is also very good. Applications include low- and high-power cables, apparatus leads, and control cables. Ethylene-propyelene terpolymer rubbers (EPT) are replacing butyl in some applications.

Neoprene has been used as a cable-jacketing material for more than 30 years. Its application over lead-sheathed and rubber-insulated cables has grown rapidly during this time. Although the electrical properties of neoprene are inferior to many other insulations, they are adequate for low-voltage work.

Nitrile-butadiene rubber (NBR) offers excellent resistance to oils and solvents but has low electrical resistivity.

Polyethylene (PE) is used in wires and cables in very large amounts. Polyethylene has excellent electrical properties plus good abrasion resistance and solvent resistance (at temperatures below 50°C). It is employed for hookup wire, coaxial cable, communication cable, line wire, lead wire, high-voltage cable, etc. The high-molecular-weight polyethylene has been the predominant cable insulation for many years. Also available for primary insulation are propylene/ethylene copolymers which are highly compressive and abrasion-resistant. Their electrical characteristics are very similar to polyethylene. These materials are covered in REA Specifications PE-22 and PE-23. Chemically cross-linked filled polyethylene is growing in usage for hookup and lead wire. Properties are similar to those of conventional PE except for a marked improvement in heat resistance, mechanical properties, aging charac-

*Superscripts refer to bibliography, Par. **272**.

**TABLE 4-61** Some Typical Enamel-Insulated Wires

| Temperature class, °C | Type | NEMA Std. | Federal Spec.* J-W-1177/sub: | Advantages |
|---|---|---|---|---|
| 105 | Acrylic | MW-37 | /7, types SA | Resists refrigerants, low cost |
| 105 | Nylon | MW-6 | /3, types N | Excellent windability, solderable |
| 105 | Oleoresinous (plain enamel) | MW-1 | /1, types E | Low cost |
| 105 | Polyvinyl formal (Formvar) | MW-15, MW-18 | /4, /16, types T | Excellent windability |
| 105 | Polyvinyl formal, isocyanate modified, for hermetic use (Formetic) | None | None | Excellent resistance to R-22 |
| 105 | Polyvinyl formal with nylon overcoat | MW-17 | /5, types TN | Improved windability over Formvar and resistance to hot solvents |
| 105 | Polyvinyl formal with polyvinyl butyral overcoat | MW-19 | None | Can be self-bonded by heat or solvent |
| 105 | Cellulose lacquer | None | None | Can be applied in thin coatings to very fine wires. Bonds by solvent activation |
| 105 | Polyurethane | MW-2 | /2, types U | Solderable. Can be coated at high speeds |
| 130 | Epoxy | MW-14, MW-9 | None | Resistance to solvents, chemicals, hydrolysis |
| 130 | Epoxy with self-bonding overcoat | None | None | Used in making coils self-supporting. Eliminates varnish dip |
| 155–180 | Polyester | MW-5 | /10, types L | Good windability, heat shock, thermal stability |
| 200 | Modified polyester | MW-74 | None | Good windability, heat shock, thermal stability |

**TABLE 4-61** Some Typical Enamel-Insulated Wires (*Continued*)

| Temperature class, °C | Type | NEMA Std. | Federal Spec.* J-W-1177/sub: | Advantages |
|---|---|---|---|---|
| 180 | Polyester amide-imide | MW-72 | /12, types H | Good windability, heat shock, thermal stability, hermetic refrigerant resistance |
| >180 | Polytetrafluoroethylene | | None | Good thermal stability to 250°C. Solvent resistance |
| 220 | Polyimide, aromatic | MW-16, MW-20, | /15, /18, types M | Excellent thermal stability, solvent resistance, flexibility, scrape resistance; cut-through |
| 220 | Polamide-imide, aromatic (overcoat) | Used in MW-35 | None, used in /13, /14, types K | Somewhat lower cost than polyimide at some sacrifice in properties. Currently used chiefly as an overcoating to improve windability and hermetic refrigerant resistance |
| 220 | Ceramic with polytetrafluoroethylene overcoat | None | None | High cut-through resistance |
| 180 | Ceramic with silicone overcoat | None | None | High thermal stability. Radiation resistant |
| >220 | Ceramic with polyimide overcoat | None | None | High thermal stability |
| 650 | Ceramic | None | None | High thermal stability. Radiation resistant |

*See complete Federal Spec. J-W-1177/GEN. It supersedes the former MIL-W-583C, March 6, 1963, and Int. Fed. Spec. J-W-001177 (Navy-Ships) September 21, 1973.

**TABLE 4-62** Some Typical Fibrous-Covered Wires

| Temperature class, °C | Type | NEMA Std. | Federal Spec. J-W-1177/sub: | Advantages |
|---|---|---|---|---|
| 90–105 | Paper | MW-31, MW-33 | None | High electric strength when impregnated with oil (oil-filled transformers) |
| 90–105 | Cotton yarn | MW-11, MW-12 | None | Positive separation of conductors, good varnish absorption and bonding, good abrasion resistance |
| — | Cellulose-acetate fiber | None | None | Can be self-bonded by solvent activation |
| 130 | Asbestos with organic bond | None | None | High compressive strength. Being replaced* |
| 155 | Glass fibers, organic bond | MW-41, MW-42 | /19 types GV | Positive separation of conductors |
| 155 | Glass and polyester fibers, organic bond | MW-45, MW-46 | /20, types DgV | Positive separation of conductors, greatly improved abrasion resistance over glass alone |
| 200 | Glass and polyester fibers, silicone bond | MW-47 | /24, /25, types DgH | Positive separation. High-temperature capability |
| 200 | Polyamide fiber paper (Du Pont NOMEX) | MW-60 | None | Tough, high-temperature and moisture-resistant paper |
| 180 | Asbestos with silicone bond | None | None | High compressive strength. Being replaced* |
| 650 | Glass fibers, organic bond with dispersed ceramic filler | None | None | Windability and high thermal stability |

*Check with latest OSHA and EPA regulations.

teristics, and freedom from environmental stress cracking. Flame resistance can be provided by proper compounding. Uses include building wire, control cable, automotive wiring, and lead wire for motors and appliances. Polyethylene can also be cross-linked by irradiating the insulation on the wire. Advantages are similar to those of the chemically cross-linked material, but the process is generally limited to thin wall insulations, such as hookup wire (5 to 12 mils wall thickness). Foamed or cellular polyethylene represents a small but important part of the wire and cable insulation field. Dielectric constants of the order of 1.5 can be attained in this manner. In coaxial cables for community antenna television and closed-circuit television, the trend has been away from solid polyethylene to foamed polyethylene cable. Coaxial cables for military applications have either a solid low-density polyethylene insulation or polytetrafluoroethylene (TFE) in solid, semisolid, or tape-wrap form. PE and TFE have dielectric constants and dissipation factors which vary little over wide frequency and temperature ranges.

Polypropylene is the lightest of all plastics. It is similar to polyethylene in electrical properties but offers better heat resistance, tensile strength, and abrasion resistance. The material may be extruded, foamed, and made into cast and biaxially oriented films. Polypropylene film is being used as a cable wrap.

Fluorinated ethylene propylene (FEP) and TFE are used in critical applications where heat resistance, solvent resistance, and reliability are important, for example, wiring in jet aircraft, military electronic equipment, and supervisory wiring for steam-turbine generators.

Polyimide film laminated to FEP film (HF film) is a heat-sealable material which offers possibilities of savings in weight and space for wire insulation. It is rated for continuous use at 200°C.

Some properties of typical wire insulation materials are shown in Table 4-63 (from Ref. 5).

For a review of insulation for integrated microelectronic circuits see Ref. 6.

***272. References on Insulated Conductors***


1. Tentative Method of Test for Relative Thermal Endurance of Film Insulated Round Magnet Wire, ASTM D2307 (based on former IEEE No. 57).
2. Saums, H. L.: Magnet Wire, Strip, Hollow Conductors and Superconductors; *Insulation Directory/Encyclopedia Issue,* May 1965, pp. 332–352, Lake Publishing Corp., Libertyville, Ill.
3. Edge Conditioned Aluminum Strip Conductor, *Publ.* 731-1-8(5–665), 1965, Reynolds Metals Co., Richmond, Va.
4. Noble, M. G., and Savage, R. M.: A Status Report on Silicone Rubber for Wire and Cable Insulation; *Insulation,* November 1965, vol. 11, no. 12, p. 51.
5. Adams, H. S.: Problems in Insulated Wire and Cable in Space-Vehicle Systems; *Electrotechnol.,* 1963, vol. 72, no. 3, pp. 133–135.
6. Staff Report, "Where Does Insulation Technology Stand Today for Integrated Microelectronic Circuits?"; *Insulation,* September 1965, vol. 11, no. 10, p. 108.
7. Harper, C. A.: *Handbook of Wiring, Cabling, and Interconnecting for Electronics;* McGraw-Hill, 1972, Chaps. 5 and 7.


## THERMAL CONDUCTIVITY OF ELECTRICAL INSULATING MATERIALS

*By CHARLES A. HARPER*

***273. General.*** One of the general characteristics of electrical insulating materials is that they are also good thermal insulating materials. This is true, in varying degrees, for the entire spectrum of insulating materials, including air, fluids, plastics, glasses, and ceramics. While the thermal insulating properties of electrical insulating materials are not especally important for electrical and electronic designs which are not heat-sensitive, modern designs are increasingly heat-sensitive. This is often because higher power levels are being dissipated from smaller part volumes, thus tending to raise the temperature of critical elements of the product design. This results in several adverse effects, including degradation of electrical

**TABLE 4-63** Some Properties of Common Wire-Insulation Materials

| Physical properties | Unplasticized PVC | Plasticized PVC | Silicone rubber | Nylon | TFE fluorocarbon | FEP fluorocarbon | Polyethylene | Irradiated polyolefin |
|---|---|---|---|---|---|---|---|---|
| Specific gravity* | 1.40 | 1.2–1.5 | 1.9 | 1.13 | 2.15 | 2.15 | 0.930 | 1.2 |
| Tensile strength, lb/in$^2$ | 6,000–9,000 | 1,000–3,000 | 4,200 | 4,000–7,000 | 2,500 | 2,500 | 1,900–2,600 | 2,500 |
| Elongation, % | 2–40 | 200–400 | . . . | 300–600 | 200–300 | 250–330 | 200 | 250 |
| Abrasion resistance* | Good | Good | Poor | Excellent | Good | Fair | Good | Good |
| Maximum continuous operating temperature, °C* | 105 | 105 | 200 | 150 | 260 | 260 | 80 | 135 |
| Melting point, °C | 200 | 200 | >375 | 300 | 327 | 275 | 120 | Not thermoplastic |
| Flexibility at −180°C* | Cracks | Cracks | Cracks | . . . | Good | Good | Cracks | Fair |
| Cut-through resistance* | Good | Fair | Fair | Excellent | Fair | Fair | Good | Good |
| Flammability, in/min | Self-extinguishing | Self-extinguishing | 10–78 | Self-extinguishing | Nonflammable | Nonflammable | 1.0 | Self-extinguishing |
| Dielectric strength, V/mil (short time) | 425–1,300 | 1,000 | 375 | 385 | 480 | 550 | 480 | 1,000 |
| Dielectric constant, 1,000 c/s | 5–10 | 2–4 | 4.2 | 4–10 | 2.0 | 2.1 | 2.3 | 2.6 |
| Volume resistivity, Ω·cm | $2 \times 10^{12}$ | $2 \times 10^{14}$ | $>3 \times 10^{13}$ | $4.56 \times 10^{16}$ | Approx. $10^{19}$ | $>2 \times 10^{13}$ | $10^{16}$ | $>10^{16}$ |

*These properties are of particular importance in aerospace applications.
NOTE: Data compiled by Hughes Aircraft Company.
1 lb/in$^2$ = 6.895 kPa.
SOURCE: From Adams, H. S.: *Electrotechnol.*, 1963, vol. 72, no. 3, p. 133.

performance and degradation of many insulating materials, especially insulating papers and plastics. The net result is reduced life and/or reduced reliability of the electrical or electronic part. To maximize life and reliability, much effort has been devoted to data and guidelines for gaining the highest possible thermal conductivity, consistent with optimization of product design limitations, such as fabrication, cost, and environmental stresses.[1,2,3,4,*] This section will present data and guidelines which will be useful to electrical and electronic designers in selection of electrical insulating materials for best meeting thermal design requirements. Also, methods of determining thermal conductivity $K$ will be described.

***274. Basic Thermal-Conductivity Data.*** The thermal-conductivity values for a range of materials commonly used in electrical design are shown in Table 4-64. These data show the ranking of the range of materials, both conductors and insulating materials, from high to low. The magnitude of the differences in conductor and plastic thermal-conductivity values can be seen. Note that one ceramic, 95% beryllia, has a higher thermal-conductivity value than some metals—thus making beryllia highly considered for high-heat-dissipating designs which allow its use. Thermal-conductivity values for a range of plastic materials are shown in Table 4-65. Thermal conductivity is variously reported in many different units, and convenient conversions are shown in Table 4-66.

Values of thermal conductivity do not change drastically up to 100°C or higher, and

**TABLE 4-64** Thermal Conductivity of Materials Commonly Used for Electrical Design

| Material | Thermal conductivity | |
|---|---|---|
| | W/(in)(°C) | Btu/(h)(ft)(°F) |
| Silver | 10.6 | 241 |
| Copper | 9.6 | 220 |
| Eutectic bond | 7.50 | 171.23 |
| Gold | 7.5 | 171 |
| Aluminum | 5.5 | 125 |
| Beryllia 95% | 3.9 | 90.0 |
| Molybdenum | 3.7 | 84 |
| Cadmium | 2.3 | 53 |
| Nickel | 2.29 | 52.02 |
| Silicon | 2.13 | 48.55 |
| Palladium | 1.79 | 40.46 |
| Platinum | 1.75 | 39.88 |
| Chromium | 1.75 | 39.88 |
| Tin | 1.63 | 36.99 |
| Steel | 1.22 | 27.85 |
| Solder (60-40) | 0.91 | 20.78 |
| Lead | 0.83 | 18.9 |
| Alumina 95% | 0.66 | 15.0 |
| Kovar | 0.49 | 11.1 |
| Epoxy resin, BeO-filled | 0.088 | 2.00 |
| Silicone RTV, BeO-filled | 0.066 | 1.5 |
| Quartz | 0.05 | 1.41 |
| Silicon dioxide | 0.035 | 0.799 |
| Borosilicate glass | 0.026 | 0.59 |
| Glass frit | 0.024 | 0.569 |
| Conductive epoxy | 0.020 | 0.457 |
| Sylgard resin | 0.009 | 0.21 |
| Epoxy glass laminate | 0.007 | 0.17 |
| Doryl cement | 0.007 | 0.17 |
| Epoxy resin, unfilled | 0.004 | 0.10 |
| Silicone RTV, BeO-filled | 0.004 | 0.10 |
| Air | | 0.016 |

*Superscripts refer to bibliography, Par. 277.

**TABLE 4-65** Thermal Conductivity for Various Types of Plastics

| Material | Specific gravity | Btu/(h)(ft²)(°F)(ft)* |
|---|---|---|
| Thermoplastics: | | |
| Polyethylene | 0.92–0.96 | 0.1–0.4 |
| Polytetrafluoroethylene | 2.15–2.25 | 0.1–0.2 |
| Molded thermosets: | | |
| Phenolic, wood-flour-filled | 1.32–1.45 | 0.10–0.19 |
| Phenolic, mineral-filled | 1.65–1.92 | 0.24–0.34 |
| Diallyl phthalate, acrylic fiber | 1.31–1.45 | 0.18–0.19 |
| Laminates, perpendicular-to-face: | | |
| XXXP, paper-phenolic | 1.3–1.4 | 0.04–0.12 |
| G-7, silicone-glass | 1.6–1.8 | 0.07–0.17 |
| G-10, epoxy-glass | 1.7–1.8 | 0.10–0.17 |
| PTFE glass-cloth | 2.1–2.2 | 0.02–0.05 |
| Casting resins and foams: | | |
| Epoxy, unfilled | 1.16 | 0.13–0.20 |
| Epoxy, 73% alumina by weight | | 0.82 |
| Epoxy, 50–55% silica by weight | 1.6–1.7 | 0.29–0.53 |
| Epoxy, hollow phenolic spheres | 0.86 | 0.16 |
| Epoxy, hollow glass spheres | 0.95 | 0.38 |
| Polyester, unfilled | 1.23 | 0.10–0.15 |
| Polyester, 50% silica by weight | 1.6 | 0.19 |
| Polyurethane foam (10 lb/ft³) | 0.16 | 0.02–0.03 |

*To obtain: Btu/(h)(ft²)(°F)(in), multiply by 12.

**TABLE 4-66** Thermal-Conductivity Conversion Factors

| From \ To | (cal)(cm)/(s)(cm²)(°C) | (W)(cm)/(cm²)(°C) | (W)(in)/(in²)(°C) | (Btu)(ft)/(h)(ft²)(°F) |
|---|---|---|---|---|
| (cal)(cm)/(s)(cm²)(°C) | 1 | 4.18 | 10.62 | 241.9 |
| (W)(cm)/(cm²)(°C) | $2.39 \times 10^{-1}$ | 1 | 2.54 | 57.8 |
| (W)(in)/(in²)(°C) | $9.43 \times 10^{-2}$ | $3.93 \times 10^{-1}$ | 1 | 22.83 |
| (Btu)(ft)/(h)(ft²)(°F) | $4.13 \times 10^{-3}$ | $1.73 \times 10^{-2}$ | $4.38 \times 10^{-2}$ | 1 |

hence only a single value is usually given for plastics. For higher-temperature applications, such as with ceramics, the temperature effect should be considered. This is shown for ceramics in Fig. 4-64. In the case of ceramics, the composition of the ceramic also strongly influences thermal conductivity, as shown in Table 4-67 and Fig. 4-65.

In addition to bulk insulating materials, insulating coatings are frequently used. Thermal-conductivity values for coatings are given in Table 4-68.

***275. Use of Fillers to Increase Thermal Conductivity.*** Many of the plastics used for electrical insulation are liquid casting resins, such as epoxies, polyesters, silicones, and urethanes. Being liquid, and easy to formulate, much work has been done with fillers to increase the thermal conductivity of these materials. As the many available unfilled casting resins are not broadly different in their thermal conductivities, it is evident that a study of fillers and filled compounds will be the key to obtaining the highest possible thermal conductivity.

A survey of fillers with high thermal conductivities shows the following thermal-conductivity data, expressed in calories per centimeter per second per square centimeter per degree Celsius: mica, 0.0012; sand, 0.0028; aluminum, 0.497; copper, 0.918.

It might be deduced from these data that the use of powdered metallic fillers such as copper or aluminum would, when compounded into a casting-resin system, yield a compound that would give the highest possible thermal conductivity. Such is not entirely the case, as can be seen from Table 4-69, which shows the thermal conductivity of a group of filled compounds. For instance, the compounds prepared with coarse-grain sand and tabular alumina have a slightly higher thermal conductivity than does the compound that uses fine-mesh aluminum as filler, despite the much better thermal conductivity of aluminum. On the other hand, the compound filled with 30-mesh aluminum has an appreciably improved thermal conductivity over the sand- and fine-mesh aluminum-filled compounds. This might raise the question of how one aluminum filler can be better than another, when they are both basically the same material. Also note that the copper-filled compound has a lower thermal conductivity than the 30-mesh aluminum compound, even though copper has a much higher thermal conductivity than does aluminum.

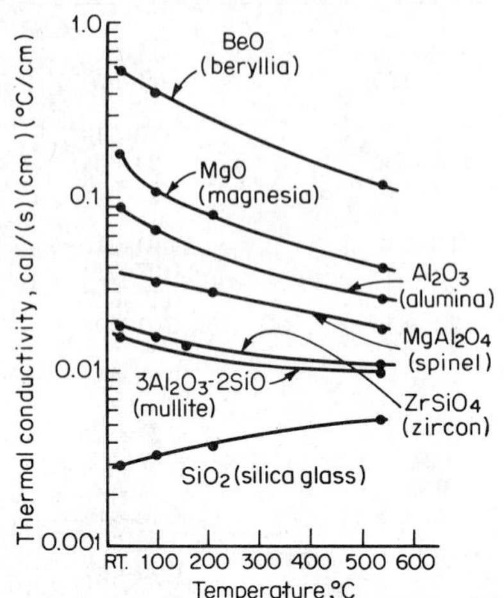


**FIG. 4-64** Effect of temperature on thermal conductivity of ceramics (Ref. 5).

The reason for these variations, from what might be expected from a comparative study of filler thermal conductivities alone, lies in the particle type and size of the filler. This is because the particle type and size determine the amount of filler that can be mixed with a given resin and still give a compound of sufficiently low viscosity to be pourable and fluid for embedded-packaging applications. Also involved in the considerations is the specific gravity of the filler; it will be explained below that volume concentration of filler is more meaningful than weight concentration in determining how much the thermal conductivity can be increased. The end result is that maximum thermal conductivity is achieved through the use of the highest-thermal-conductivity filler that will allow the highest volume concentration of filler in the filled compound. This leads to what might be considered the bulk effect.

*The Bulk Effect.* Perhaps the primary reason why the sand-filled compound has a better thermal conductivity than the aluminum-filled compound in Table 4-69 is the bulk effect. This is best described by the statement that, in general, increase in thermal conductivity of a filled epoxy compound depends more on the quantity of filler added to the compound than on the type of filler used, so long as the filler is of the same particle type. It is observed from Table 4-69 that the mica-filled compound contains 40% of mica by weight, as does the aluminum-filled compound. As can be seen, the weight concentration being the same, the aluminum-filled material has a much higher thermal conductivity than does the mica-filled material. Note, however, that the sand-filled compound contains 70% of sand by weight; thus, there is a much higher concentration of filler in the sand-filled compound on the basis of both weight and volume.

**TABLE 4-67** Effect of Alumina Content on the Thermal Conductivity of Alumina Substrates

| Alumina, % | Thermal conductivity, cal/(s)(cm²)(°C/cm) | Change in thermal conductivity, % |
|---|---|---|
| 99 | 0.070 | |
| 98 | 0.061 | −13 |
| 96 | 0.043 | −39 |
| 85 | 0.035 | −50 |

Practical considerations enter at this point. Fine-mesh fillers, such as fine mica powder and fine-mesh aluminum powder, give a high-viscosity compound with poor flow properties of 40% weight concentration. (Viscosity considerations are at 70°C for all cases given here, as the base epoxy viscosity is too high for use at room temperature.) However, a sand-filled resin is still very workable at a 70% weight concentration. Actually, a 40% concentration of fine-mesh aluminum filler makes the compound almost unworkable for embedment use, as the compound is approaching a paste at this point. On the other hand, sand can be used in concentrations up to 80% or so. The reason for this is that sand has a much larger particle size than mica or fine-mesh aluminum. Thus, results similar to those obtained with a high concentration of sand filler can also be obtained with a high concentration of such fillers as tabular alumina and other high-conductivity, large-grain-size fillers. An electrical grade of sand is usually much more economical, however. Although Table 4-69 shows that the copper-filled compound has a relatively high thermal conductivity, copper powder is not so conveniently used because of the higher density of the copper filler. The reason again is a practical one. It is often quite difficult to keep powdered copper in suspension in normal embedding compounds long enough for the compound to cure. The powdered copper will rapidly settle to the bottom. The lower the compound viscosity and the longer the pot life, the greater this settling problem becomes.

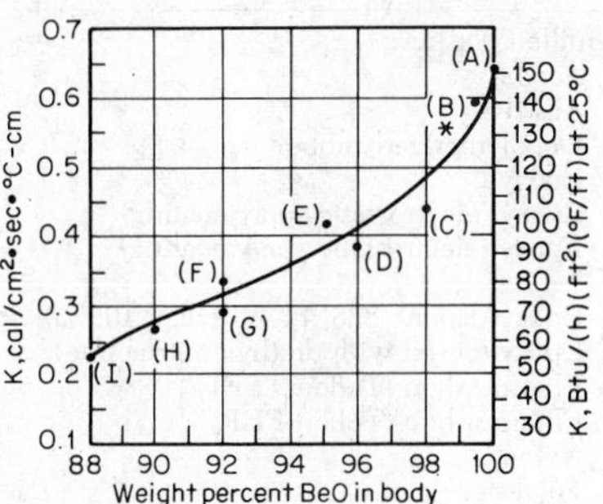


A. Theoretical density 100% BeO
B. Berlox® 99.5% BeO, 95% density
C. Commercial 98% BeO body
D. Commercial 96% BeO body
E. BeO body N*
F. BeO body A4
G. BeO body N4
H. BeO body O
I. BeO body N8

*Reference: J. Amer. Ceramic Soc., vol. 33, No. 4, 1950

® Berlox is the registered trademark of National Beryllia Corp. (beryllium oxide)

**FIG. 4-65** Thermal conductivity of beryllia as a function of BeO content at 25°C (Ref. 6).

Probably the most likely explanation for the above-described bulk effect is that the casting resins, being good insulators, form an insulating barrier between each particle and the thermal conductivity through the filler particles contributes less than the lower conductivity through the resin layers. Thus, overall thermal conductivity is best increased by adding more filler particles or, more specifically, by increasing the volume concentration of the filler in the compound. The result of the increased concentration is a shorter conductivity path through the resin, a longer conductivity path through the filler, and an overall reduction in the resistance to conductive heat flow through the material. Therefore, in general, improved thermal conductivity is obtained by adding an additional amount of filler material. The upper limitation on filler concentration is controlled by the practical working viscosity of the material. With any filler, at some concentration, the compound will be too thick for pouring and flowing around components in electronic packages.

Typical curves showing the effect of filler content on the thermal conductivity of liquid resin systems are shown in Fig. 4-66.

There is one case, however, where addition of more filler decreases the thermal conductivity. This is where the filler is a hollow, spherical type rather than a solid-particle type. Note from Table 4-69 the extremely low thermal conductivity of the spheroid-filled epoxy. The hollow, spherical fillers are usually filled with air or gas. Here, added spheroid filler essentially has the effect of reducing the conductivity through the insulator, when we consider air and gas to be optimum insulators. Thus, in this case, increasing the quantity of the spherical filler actually increases the resistance to conductive heat flow. Spherical fillers are generally used for reducing the specific gravity of an embedment compound, rather than increasing the thermal conductivity. Also shown in Table 4-69 for comparison are data for the thermal conductivity of an unfilled epoxy resin and for a low-density urethane foam. As would be expected, a very low-density foam material has an extremely low thermal conductivity.

**TABLE 4-68** Thermal Conductivity of Various Plastic Coatings

| Material | $k$ value,* cal/(s)(cm²)(°C/cm) × 10⁴ | Source of information |
|---|---|---|
| Unfilled plastics: | | |
| Acrylic | 4–5 | [a] |
| Alkyd | 8.3 | [a] |
| Depolymerized rubber | 3.2 | H. V. Hardman, DPR Subsidiary |
| Epoxy | 3–6 | [b] |
| Epoxy (electrostatic spray coating) | 6.6 | Hysol Corp., DK-4 |
| Epoxy (electrostatic spray coating) | 2.9 | Minnesota Mining & Mfg., No. 5133 |
| Epoxy (Epon† 828, 71.4% DEA, 10.7%) | 5.2 | |
| Epoxy (cured with diethylenetriamine) | 4.8 | |
| Fluorocarbon (Teflon TFE) | 7.0 | Du Pont |
| Fluorocarbon (Teflon FEP) | 5.8 | Du Pont |
| Nylon | 10 | [d] |
| Polyester | 4–5 | [a] |
| Polyethylenes | 8 | [a] |
| Polyimide (Pyre-M.L. enamel) | 3.5 | [e] |
| Polyimide (Pyre-M.L. varnish) | 7.2 | [f] |
| Polystyrene | 1.73–2.76 | [g] |
| Polystyrene | 2.5–3.3 | [a] |
| Polyurethane | 4–5 | [n] |
| Polyvinyl chloride | 3–4 | [a] |
| Polyvinyl formal | 3.7 | [a] |
| Polyvinylidene chloride | 2.0 | [a] |
| Polyvinylidene fluoride | 3.6 | [h] |
| Polyxylylene (Parylene N) | 3 | Union Carbide |
| Silicones (RTV types) | 5–7.5 | Dow Corning Corp. |
| Silicones (Sylgard types) | 3.5–7.5 | Dow Corning Corp. |
| Silicones (Sylgard varnishes and coatings) | 3.5–3.6 | Dow Corning Corp. |
| Silicone (gel coating) | 3.7 | Dow Corning Corp. |
| Silicone (gel coating) | 7 (150°C) | Dow Corning Corp. |
| Filled plastics: | | |
| Epon 828/diethylenetriamine = A | 4 | [b] |
| A + 50% silica | 10 | [b] |
| A + 50% alumina | 11 | [b] |
| A + 50% beryllium oxide | 12.5 | [b] |
| A + 70% silica | 12 | [b] |
| A + 70% alumina | 13 | [b] |
| A + 70% beryllium oxide | 17.8 | [b] |
| Epoxy, flexibilized = B | 5.4 | [i] |
| B + 66% by weight tabular alumina | 18.0 | [i] |
| B + 64% by volume tabular alumina | 50.0 | [i] |
| Epoxy, filled | 20.2 | Emerson & Cunning, 2651 ft |
| Epoxy (highly filled) | 15–20 | Wakefield Engineering Co. |
| Polyurethane (highly filled) | 8–11 | International Electronic Research Co. |

*All values are at room temperature unless otherwise specified.

†Trademark of Shell Chemical Co., New York, N.Y.

[a] *Mater. Eng.*, Materials Selector Issue, vol. 66, no. 5, Chapman-Reinhold Publication, mid-October 1967.

[b] Wolf, D. C.: *Proc. Nat. Electron. and Packag. Symp.*, New York, June 1964.

[c] Lee, H., and Neville, K.: *Handbook of Epoxy Resins;* New York, McGraw-Hill Book Company, 1966.

[d] Davis, R.: *Reinf. Plast.*, October 1962.

[e] *Du Pont Tech. Bull.* 19, Pyre-M.L. Wire Enamel, August 1967.

[f] *Du Pont Tech. Bull.* 1, Pyre-M.L. Varnish RK-692, April 1966.

[g] Teach, W. C., and Kiessling, G. C.: *Polystyrene;* New York, Reinhold Publishing Corporation, 1960.

[h] Barnhart, W. S., Ferren, R. A., and Iserson, H.: *17th ANTEC of SPE*, January 1961.

[i] Gershman, A. J., and Andreotti, J. R.: *Insulation*, September 1967.

**TABLE 4-69** Thermal Conductivity of Embedding Compounds Using Various Fillers

| Filler | Filler in compound by weight, % | Thermal conductivity of filled compound, W/(in²)(°C)(in) |
|---|---|---|
| Copper powder (Venus A, U.S. Bronze) | 90 | 0.040 |
| 30-mesh aluminum | 80 | 0.064 |
| Fine-mesh aluminum | 40 | 0.022 |
| Coarse-grain sand | 70 | 0.025 |
| Tabular alumina | 80 | 0.026 |
| 325-mesh mica | 45 | 0.013 |
| 325-mesh silica | 55 | 0.019 |
| None (epoxy resin)* | 0 | 0.005 |
| Hollow phenolic spheres | 15 | 0.003 |
| Unfilled urethane foam, 5 lb/ft² density* | | 0.001 |

*Presented for comparison.
NOTE: 1 lb/ft² = 4.882 kg/m².

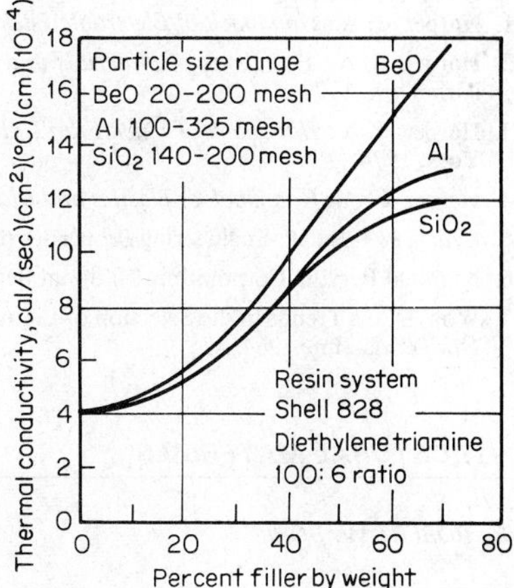


**FIG. 4-66** Effect of filler content on thermal conductivity for three filled liquid-resin systems (Ref. 7).

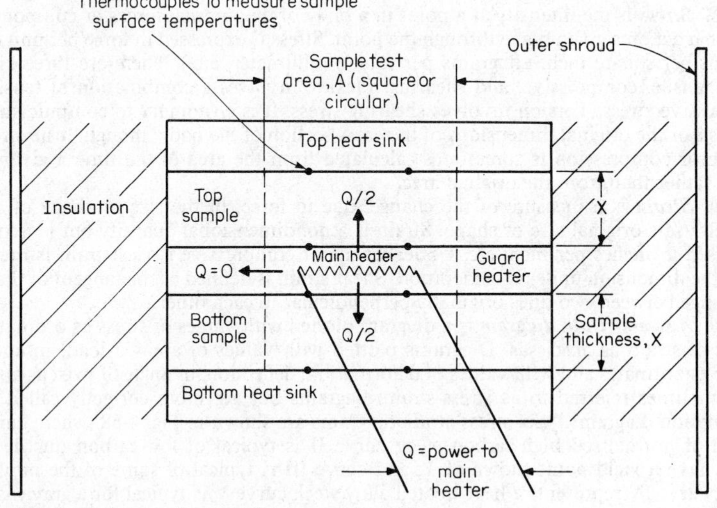


**FIG. 4-67** Schematic assembly of guarded hot plate.

***276. Thermal-Conductivity Measurements.*** The recognized primary technique for measuring thermal conductivity of insulating materials is the guarded-hot-plate method (ASTM C177). A schematic of the apparatus is shown in Fig. 4-67. The purpose of the guard heater is to prevent heat flow in all but the axial (up and down in the schematic) direction by establishing isothermal surfaces on the specimen's hot side. With this condition established and by measuring the temperature difference across the sample, the electrical power to the main heater area, and the sample thickness, the $K$ factor can be calculated as

$$K = \frac{QX}{2A\,\Delta T} \tag{4-79}$$

Instruments are available for this test which use automatic means to control the guard temperature and record the sample $\Delta T$. Unfortunately this test is fairly expensive.

Another technique uses a heat-flow sensor, which is a calibrated thermopile, in series with the heater, specimen, and cold sink. This method avoids the guard heater and requires only one specimen. This secondary technique is described in ASTM C518.

***277. References***


1. Harper, C. A.: *Handbook of Electronic Packaging,* McGraw-Hill Book Company, New York, 1969.
2. Harper, C. A.: *Handbook of Materials and Processes for Electronics,* McGraw-Hill Book Company, New York, 1970.
3. Harper, C. A.: *Handbook of Thick Film Hybrid Microelectronics,* McGraw-Hill Book Company, New York, 1974.
4. Harper, C. A.: *Handbook of Plastics and Elastomers,* McGraw-Hill Book Company, New York, 1975.
5. Lynch, J. F., et al.: Engineering Properties of Ceramics, *U.S. Dept. Comm. Bull.* AD803-765.
6. National Beryllia Corporation: National Beryllia Technical Bulletin on Berlox BeO.
7. Wolf, D. C.: Trends in the Selection of Liquid Resins for Electonic Packaging, *Proceedings, NEPCON,* New York, June 1964.


## STRUCTURAL MATERIALS

*By ROBERT W. BOHL*

### Definitions of Properties of Structural Materials

***278. Stress*** is the intensity at a point in a body of the internal forces or components of force that act on a given plane through the point. Stress is expressed in force per unit of area (pounds per square inch, kilograms per square millimeter, etc.). There are three kinds of stress: tensile, compressive, and shearing. Flexure involves a combination of tensile and compressive stress. Torsion involves shearing stress. It is customary to compute stress on the basis of the original dimensions of the cross section of the body, though "true stress" in tension or compression is sometimes calculated from the area of the time a given stress exists, rather than from the original area.

***279. Strain*** is a measure of the change, due to force, in the size or shape of a body referred to its original size or shape. Strain is a nondimensional quantity but is frequently expressed in inches per inch, etc. Under tensile or compressive stress, strain is measured along the dimension under consideration. Shear strain is defined as the tangent of the angular change between two lines originally perpendicular to each other.

***280. A stress-strain diagram*** is a diagram plotted with values of stress as ordinates and values of strain as abscissas. Diagrams plotted with values of applied load, moment, or torque as ordinates and with values of deformation, deflection, or angle of twist as abscissas are sometimes referred to as stress-strain diagrams but are more correctly called "load-deformation diagrams." Six stress-strain diagrams are shown in Fig. 4-68, where curve I is typical of normalized high-carbon steel; curve II is typical of low-carbon ductile steels, which have a yield point shown at $Y$; and curve III is typical of some of the nonferrous alloys; curve IV represents a heat-treated alloy steel; curve V is typical for a gray cast iron;

and curve VI shows approximately the type of curve obtained for timber. The stress-strain diagram for some materials is affected by the rate of application of the load, by cycles of previous loading, and again by the time during which the load is held constant at specified values; for precise testing, these conditions should be stated definitely in order that the complete significance of any particular diagram may be clearly understood.

***281. The modulus of elasticity*** is the ratio of stress to corresponding strain below the proportional limit. For many materials the stress-strain diagram is approximately a straight line below a more or less well-defined stress known as the "proportional limit." As there are three kinds of stress, there are three moduli of elasticity for a material, that is, the modulus in tension, the modulus in compression, and the modulus in shear. The value in tension is practically the same, for most ductile metals, as the modulus in compression; the modulus in shear is only about 0.36 to 0.42 of the modulus in tension. The modulus is expressed in pounds per square inch (or kilograms per square millimeter) and measures the elastic *stiffness* (the ability to resist elastic deformation under stress) of the material.

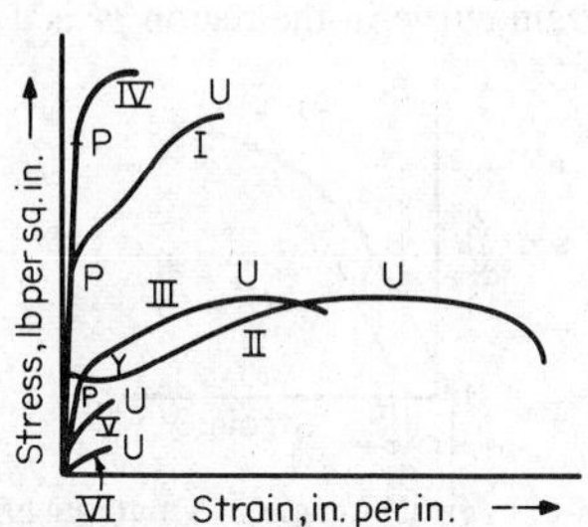


**FIG. 4-68** Typical stress-strain diagrams for tensile stress.

***282. Elastic Strength.*** To the user and the designer of machines or structures one significant value to be determined is a *limiting stress below which the permanent distortion of the material is so small that the structural damage is negligible and above which it is not negligible.* The amount of plastic distortion which may be regarded as negligible varies widely for different materials and for different structural or machine parts. In connection with this limiting stress for elastic action a number of technical terms are in use; some of them are

**a.** *Elastic Limit.* The greatest stress which a material is capable of withstanding without a permanent deformation remaining upon release of stress. Determination of the elastic limit involves repeated application and release of a series of increasing loads until a set is observed upon release of load. Since the elastic limit of many materials is fairly close to the proportional limit, the latter is sometimes accepted as equivalent to the elastic limit for certain materials. There is, however, no fundamental relation between elastic limit and proportional limit. Obviously the value of the elastic limit determined will be affected by the sensitivity of apparatus used.

**b.** *Proportional Limit.* The greatest stress which a material is capable of withstanding without a deviation from proportionality of stress to strain. The statement that the stresses are proportional to strains below the proportional limit is known as *Hooke's law.* Proportional limits for the metals in Fig. 4-68 are located at the points *P*; however, the numerical values of the proportional limit are influenced by methods and instruments used in testing and the scales used for plotting diagrams.

**c.** *Yield point* is the lowest stress at which marked increase in strain of the material occurs *without increase in load.* It is indicated at point *Y* on the stress-strain curve II in Fig. 4-68. If the stress-strain curve shows no abrupt or sudden yielding of this nature, then there is no yield point; for example, curve I in Fig. 4-68 exhibits no yield point. Iron and low-carbon steels have yield points, but most metals do not, including iron and low-carbon steels immediately after they have been plastically deformed at ordinary temperatures.

**d.** *Yield strength* is the stress at which a material exhibits a specified limiting permanent set. Its determination involves the selection of an amount of permanent set that is considered the maximum amount of plastic yielding which the material can exhibit, in the particular service condition for which the material is intended, without appreciable structural damage. A set of 0.2% has been used for several ductile metals, and values of yield strength for various metals are for 0.2% set unless otherwise stated. On the stress-strain diagram for the material (see Fig. 4-69) this arbitrary set is laid off as $q$ along the strain axis, and the line $mn$ drawn parallel to $OA$, the straight portion of the diagram. Since the stress-strain diagram for release of load is approximately parallel to $OA$, the intersection $r$ may be regarded as determining the stress at the yield strength. The yield strength is generally used to determine the elastic strength for materials whose stress-strain curve in the region $pr$ is a smooth curve of gradual curvature. See discussion in ASTM Designation E6.

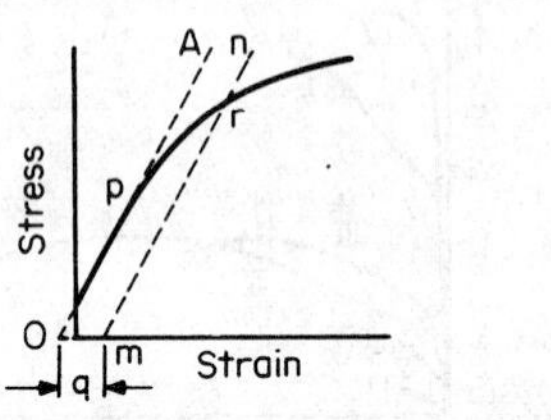


**FIG. 4-69** Yield strength of a material having no well-defined yield point.

***283. Ultimate strength (tensile strength or compressive strength)*** is the maximum stress which a material will sustain when slowly loaded to rupture. Ultimate strength is computed from the maximum load carried during a test and the original cross-sectional area of the specimen. For materials that fail in compression with a shattering fracture, the compressive strength has a definite value, but for materials that do not fracture, the compressive strength is an arbitrary value depending on the degree of distortion which is regarded as indicating complete failure of the material. In tensile tests of many materials, especially those having appreciable ductility, failure does not occur at the stress corresponding to the ultimate strength. For such materials, localized deformation, or necking, occurs and the nominal stress decreases because of the rapidly decreasing cross-sectional area until failure occurs.

***284. Shearing strength*** is the maximum shearing stress which a material is capable of developing. The general remarks in Par. **283** regarding methods of failure are also applicable to failures in shear. Owing to experimental difficulties of obtaining true shearing strength, the values of modulus of rupture in torsion are usually reported as indicative of the shearing strength.

***285. Modulus of rupture*** in flexure (or torsion) is the term applied to the computed stress, in the extreme fiber of a specimen tested to failure under flexure (or torsion), when computed by the arbitrary application of the formula for stress with disregard of the fact that the stresses exceed the proportional limit. Hence the modulus of rupture does not give the true stress in the member but is useful only as a basis of comparison of relative strengths of materials.

***286. Ductility*** is that property of a material which enables it to acquire large permanent deformation and at the same time develop relatively large stresses (as drawing into a wire). Though ductility is a highly desirable property required by almost all specifications for metals, the quantitative amount needed for structural applications is not entirely clear but probably does not exceed about 3% elongation after the structure is fabricated (see *Proc. ASTM*, vol. 40, p. 551). The commonly used measures of ductility are

**a.** *Elongation* is the ratio of the increase of length of a specimen, after rupture under tensile stress, to the original gage length; it is usually expressed in percent. The percentage of elongation for any given material depends upon the gage length, which should always be specified.

**b.** *Reduction of area or contraction of area* is the ratio of the difference between the original and the fractured cross section to the original cross-sectional area; it is usually expressed in percent.

**c.** *Bend test* measures the angle through which a given specimen of material can be bent, at a specified temperature, without cracking. In some cases the maximum angle through which the specimen can be bent around a certain diameter or the number of bendings back and forth through a stated angle are measured. In other cases the elongation in a given gage length across the crack on the tension side of the bend specimen is measured. See ASTM Standard E16.

***287. Plasticity*** permits a material to assume permanent deformations under loads without recovery of the strain when the loads are removed. Plasticity permits shaping of metal parts by plastic deformation; plastic materials deform instead of fracturing under load.

***288. Brittleness*** is defined as the ability of a material to fracture under stress with little or no plastic deformation. Brittleness implies a lack of plasticity.

***289. Resilience*** is the amount of strain energy (or work) which may be recovered from a stressed body when the loads causing the stresses are removed. Within the elastic limit the work done in deforming the bar is completely recovered upon removal of the loads; the total amount of work done in stressing a unit volume of the material to the elastic limit is called the *modulus of resilience.*

***290. Toughness*** is the ability to withstand large stresses accompanied by large strains before fracture. The toughness is usually measured by the total work done in stressing a unit volume of the material to complete fracture and may be interpreted as the total area under the stress-strain curve (Fig. 4-68). Ductility differs from toughness in that it deals only with the ability of the material to deform, whereas toughness is measured by the energy-absorbing capacity of the material.

***291. Impact Resistance.*** The ability of a material to resist impact or energy loads without permanent distortion is measured by the modulus of resilience. The ultimate resistance to impact before fracture is measured by the toughness of the material. For members with abrupt changes of section (holes, keyways, fillets, etc.), the resistance to a rapidly applied load depends greatly on the "notch sensitivity" (the resistance to the formation and spread of a crack); above certain critical velocities of loading and below certain critical temperatures, the impact strength is greatly reduced. Relative notch sensitivity under repeated loads is not the same as that in a single-blow notched-bar test. Impact values are influenced by speed of straining, shape and size of specimen, and type of testing machine.

*Charpy* or *Izod* impact bend tests measure the energy required to fracture small notched specimens (1 cm square) under a single blow. These tests are used as an indication of toughness, a property that is very sensitive to the composition and thermal-mechanical history of the material. Tests should be carried out over a range of temperatures to determine the temperature at which the alloy fails by brittle rather than ductile failure. (See ASTM Standard E23.)

***292. Fracture Mechanics.*** Three primary factors have been identified that control the susceptibility of a structure to brittle failure: material toughness (affected by composition and metallurgical structure as well as temperature, strain rate, and constraints to plastic yielding); flaw size (internal discontinuities such as porosity or small cracks from welding, fatigue, fabrication, etc.); and stress level (applied or residual). Fracture mechanics attempts to interrelate these variables in order to predict the occurrence of brittle fracture on a quantitative design basis rather than depend upon qualitative relationships between experience and results of impact tests such as Izod and Charpy. Fracture mechanics has had excellent success when applied to high-strength materials. The material parameter defined, called the "fracture toughness $K_C$," can be measured experimentally and utilized to specify safe loading conditions in the presence of a given size and geometry of flaw. See "Fracture Toughness," ASTM Spec. Tech. Publ. 514, 1971.

***293. Hardness*** is the resistance which a material offers to small, localized plastic deformations developed by specific operations such as scratching, abrasion, cutting, or penetration of the surface. Hardness does not imply brittleness, as a hard steel may be tough and ductile. The standard Brinell hardness test is made by pressing a hardened steel ball against a smooth, flat surface under certain standard conditions; the Brinell hardness number is the quotient of the applied load divided by the area of the surface of the impression. A different method of test is employed in the Shore scleroscope, in which a small, pointed hammer is allowed to fall from a definite height onto the material, and the hardness is measured by the height of the rebound, which is automatically indicated on a scale. The Rockwell hardness machine measures the depth of penetration in the metal produced by a definite load on a small indenter of spherical or conical shape. Vickers or Tukon hardness machines measure hardness on a microscopic scale. The dimensions of the impression of a lightly loaded diamond pyramid indenter on a polished surface are related to hardness number. (For further data see ASTM Standards E10, E18, and E92; also Metals Handbook, ASM, 8th ed., vol. 11, 1976, pp. 1–20.)

***294. Fatigue strength (fatigue limit)*** is a limiting stress below which no evidence of failure by progressive fracture can be detected after the completion of a very large number of repetitions of a definite cycle of stress. The fatigue limits usually reported are those for completely reversed cycles of flexural stress in polished specimens. For stress cycles in which an alternating stress is superimposed on a steady stress the endurance limit (based on the maximum stress in the cycle) is somewhat higher. Most ferrous metals have well-defined limits, whereas the fatigue strength of many nonferrous metals is arbitrarily listed as the maximum stress that is just insufficient to cause fracture after some definite number of cycles of stress, which should always be stated. The fatigue strength of actual members containing notches (holes, fillets, surface scratches, etc.) is greatly reduced and depends entirely on the "stress-raising" effect of these discontinuities and the sensitivity of the material to the localized stresses at the notch.

***295. Composition and Structure.*** Chemical analysis is employed to determine whether component elements are present within specified amounts and impurity elements are held below specified limits. Mechanical and physical properties, however, depend on the size, shape, composition, and distribution of the crystalline constituents that make up the structure of the alloy. Chemical analysis does not reveal these features of the structure. Metallographic techniques, which involve examination of carefully polished and etched surfaces by optical and electron microscopy or x-ray methods, are required to provide this vital information. Nondestructive testing (NDT) methods are useful in detecting the presence of flaws of various kinds in finished parts and structures. These techniques depend on the interference of the defect with some easily measured physical property, such as x-ray absorption, magnetic susceptibility, propagation of acoustical waves, or electrical conductivity. NDT techniques have particular application where defects are difficult to detect and quite likely to occur (as in welded structures) and where high-integrity performance requires 100% inspection.

***296. Aging*** is a spontaneous change in properties of a metal with time after a heat-treatment or a cold-working operation. Aging tends to restore the material to an equilibrium condition and to remove the unstable condition induced by the prior operation and usually results in increased strength of the metal with corresponding loss of ductility. The fundamental action involved is generally one of precipitation of hardening elements from the solid solution, and the process can usually be hastened by slight increase in temperature. This is a very important strengthening mechanism in a variety of ferrous and nonferrous alloys, e.g., high-strength aluminum alloys.

***297. Corrosion Resistance.*** There is no universal method of determining corrosion resistance, because different types of exposure ordinarily produce entirely dissimilar results on the same material. In general the subject of corrosion is rather complicated; in some cases corrosive attack appears to be chiefly chemical in its nature, while in others the attack is by electrolysis. Owing to the great diversity of materials exposed to corrosive influences in service and the wide range of service conditions, it is impracticable to formulate any universal measure of corrosion resistance. If the service life is likely to be determined by corrosion resistance, the degree of impairment which marks the end of usefulness will ordinarily be established by considerations of safety and reliability or perhaps of appearance. Corrosion testing is conducted in general by two methods: (a) normal exposure in service with periodic observations of corrosive action as it progresses under such conditions; (b) some type of artificially accelerated test, which may serve merely to obtain comparative results or, again, may simulate the conditions of service exposure. For specific information see H. H. Uhlig, "Corrosion and Corrosion Control," New York, John Wiley & Sons, Inc., 1963; ASTM STP290, Twenty Year Atmospheric Corrosion Investigation of Zinc-Coated and Uncoated Wire and Wire Products, 1961; and M. Fontana and N. Greene, *Corrosion Engineering*, McGraw-Hill, 1967.

***298. Powder Metallurgy.*** Many alloys and metallic aggregates having unusual and very valuable properties are being produced commercially by mixing metal powders, pressing in dies to desired shapes, and sintering at high temperatures. Parts may be produced to close dimensional tolerance, and the process enables the mixing of dissimilar materials which will not normally alloy or which cannot be cast because of insolubility of the constituents. Wide use of powder metallurgy is made in producing copper-molybdenum alloys for contact electrodes for spot welding, extremely hard cemented tungsten carbide tips for use in metal cutting tools, and copper-base alloys containing either graphite particles or a controlled dispersion of porosity for bearings of the "oilless" or oil-retaining types. Silver-nickel and silver-molybdenum alloys (tungsten or graphite may be added) for contact materials having high conductivity but good resistance to fusing can be produced by the method. Powdered iron is being used to manufacture gears and small complex parts where the savings in weight of metal and machining costs are able to offset the additional cost of metal and processing in the powdered form. Small Alnico magnets of involved shape which are exceedingly difficult to cast or machine can be produced efficiently from metallic powders and require little or no finishing. Solid mixtures of metals and nonmetals, such as asbestos, can be produced to meet special requirements. The size and shape of powder particles, pressing temperature and pressure, sintering temperature and time, all affect the final density, structure, and physical properties. For further details see W. D. Jones, *Fundamental Principles of Powder Metallurgy,* London, E. Arnold, 1960.

## Structural Iron and Steel

***299. Classification of Ferrous Materials.*** Iron and steel may be classified on the basis of composition, use, shape, method of manufacture, etc. Some of the more important ferrous alloys are described in the sections below.

***300. Ingot iron*** is commercially pure iron, and contains a maximum of 0.15% total impurities. It is very soft and ductile and can undergo severe cold-forming operations. It has a wide variety of applications based on its formability. Its purity results in good corrosion resistance and electrical properties, and many applications are based on these features. The average tensile properties of Armco ingot iron plates are: tensile strength 320 MPa (46,000 lb/in$^2$); yield point 220 MPa (32,000 lb/in$^2$); elongation in 8 in, 30%; Young's modulus 200 GPa ($29 \times 10^6$ lb/in$^2$) (see ASTM Designation A345).

***301. Plain carbon steels*** are alloys of iron and carbon containing small amounts of manganese (up to 1.65%) and silicon (up to 0.50%) in addition to impurities of phosphorus and sulfur. Additions up to 0.30% copper may be made in order to improve corrosion resistance. The carbon content may range from 0.05 to 2%, although few alloys contain more than 1.0%, and the great bulk of steel tonnage contains from 0.08 to 0.20% and is used for structural applications. Medium-carbon steels contain around 0.40% carbon and are used for constructional purposes—tools, machine parts, etc. High-carbon steels have 0.75% carbon or more, and may be used for wear and abrasion-resistance applications such as tools, dies, and rails. Strength and hardness increase in proportion to the carbon content while ductility decreases. Phosphorus has a significant hardening effect in low-carbon steels, while the other components have relatively minor effects within the limits they are found. It is difficult to generalize the properties of steels, however, since they can be greatly modified by cold working or heat-treatment.

***302. High-strength low-alloy steels*** are low-carbon steels (0.10 to 0.15%) to which alloying elements such as phosphorus, nickel, chromium, vanadium, and niobium have been added to obtain higher strength. This class of steel was developed primarily by the transportation industry to decrease vehicle weight, but the steels are widely used. Since thinner sections are used, corrosion resistance is more important, and copper is added for this purpose. See ASTM Designation A242.

***303. Free-Machining Steels.*** Additions of manganese, phosphorus, and sulfur greatly improve the ease with which low-carbon steels are machined. The phosphorus hardens the ferrite, and the manganese and sulfur combine to form nonmetallic inclusions that help form and break up machining chips. The improvement in machinability is gained at some loss of mechanical properties, and these steels should be used for noncritical applications. Small amounts of lead also improve machining characteristics of steel by helping break up chips as well as providing a self-lubricating effect. Lead is more often added to higher-carbon steels where the effect on mechanical properties is less detrimental than that caused by sulfide inclusions. (See ASM *Metals Handbook,* 8th ed., vol. 3.)

***304. Alloy Steels.*** When alloying ingredients (in addition to carbon) are added to iron to improve its mechanical properties, the product is known as an alloy steel. Heat-treatment is a necessary part of the manufacture and use of alloy steels; only through proper quenching and tempering can the full beneficial effects of the alloys be obtained. The chief advantages obtained from the addition of alloys to steel are (a) to increase the depth of hardening on quenching, thus making it possible to produce more uniform properties throughout thick sections with a minimum of distortion; (b) to form chemical compounds which when properly distributed develop desirable properties in the steel, that is, extreme hardness, corrosion or heat resistance, high strength without excessive brittleness.

The most commonly used alloy steels have been classified by the American Iron and Steel Institute and the Society of Automotive Engineers and are identified by a nomenclature system that is partially descriptive of the composition. The system of steel designations is shown in Table 4-70. Approximate strengths of several alloy steels after specific heat-treatments are given in Table 4-73.

Mechanical properties of the alloy steels vary over a wide range depending upon size, composition, and thermomechanical treatment. (See also E. C. Bain and H. Paxton, *Alloying Elements in Steel,* 2d ed.; American Society for Metals, 1961.)

***305. Cast Iron.*** Iron ore is reduced to the metallic form in a blast furnace yielding a product of molten iron saturated in carbon (about 4%). Most commonly, this "hot metal" is immediately processed to steel by a refining process without allowing it to solidify. Occasionally it is cast into bars; this product is called pig iron. Cast iron is made by remelting pig iron and/or scrap steel in a cupola or electric furnace, and casting it into molds to the desired shape of the finished part. Cast iron has a much higher carbon content than steel, usually between 2.5 and 3.75%.

***306. Gray Cast Iron.*** In gray cast iron, the excess carbon beyond that soluble in iron is present as small flake-shaped particles of graphite. The flakes of graphite account for some of the unique properties of gray iron; in particular, its low tensile strength and ductility, its ability to absorb vibrational energy (damping capacity), and its excellent machinability. Cast iron is easy to cast because it has a lower melting point than steel, and the formation of the low-density graphite offsets solidification shrinkage so that minimal dimensional changes occur on freezing. Other elements in the composition of ordinary gray cast iron are important chiefly insofar as they affect the tendency of carbon to form as graphite rather than in chemical combination with the iron as iron carbide ($Fe_3C$). Silicon is most effective in promoting the formation of graphite. Slower cooling rates during freezing also favor the formation of graphite as well as increase the size of the flakes. Cooling rate also affects the mode of decomposition of the carbon retained in solution during freezing. Slow cooling favors complete precipitation as graphite, leaving a soft ferrite matrix, while fast cooling produces a stronger matrix containing $Fe_3C$ (as pearlite). The tensile strength of gray cast iron typically ranges from 140 to 410 MPa (20,000 to 60,000 lb/in$^2$). Corresponding compressive strengths are 575 to 1300 MPa (85,000 to 190,000 lb/in$^2$). Young's modulus may range from 70 to 150 GPa ($10 \times 10^6$ to $20 \times 10^6$ lb/in$^2$), depending on the microstructure. See ASTM Specifications A48 and A126, ASM *Metals Handbook,* 9th ed., vol. 1, 1978.

***307. White Cast Iron.*** Careful adjustment of composition and cooling rate can cause all the carbon in a cast iron to appear in the combined form as pearlite or free carbide. This structure is very hard and brittle and has few engineering applications beyond resistance to abrasion. This product does serve, however, as an intermediate product in the production of malleable cast iron described in the following paragraph.

***308. Malleable Cast Iron.*** By annealing white cast iron at about 950°C, the combined carbon will decompose to graphite. This graphite grows in a spheroidal shape rather than the flakelike shape that forms during the freezing of gray cast iron. Because of this difference in graphite shape, malleable iron is much tougher and stronger. If the castings are slowly cooled from the malleabilizing temperature, the matrix can be converted to ferrite, with all the carbon appearing as graphite; this is a very tough product. Faster cooling will yield a pearlitic matrix with greater strength and hardness. It is also possible to quench and temper malleable iron for optimum combinations of strength and toughness. By careful control of composition, the malleabilizing cycle can be carried out in 8 to 20 h. See ASM *Metals Handbook,* 9th ed., vol. 1; ASTM Specification A48; "Malleable Iron Castings," Malleable Founders Society, Cleveland, 1960 (see Table 4–71).

***309. Nodular cast iron*** has the same carbon content as gray iron; however, the addition of a few hundreds of 1% of either magnesium or cerium causes the uncombined carbon to form spheroidal particles during solidification instead of graphite flakes. Strength properties comparable with those of steel may be achieved in the pearlitic iron. The softer ferritic and pearlitic as-cast irons exhibit considerable ductility, 10% elongation or more. As the hardness and strength are increased by appropriate heat-treatment or the thickness of the casting decreased below approximately ¼ in, the ductility decreases. An austenitic form of nodular iron may be obtained by adding various amounts of silicon, nickel, manganese, and chromium. For many purposes nodular iron exhibits properties superior to those of either gray or malleable cast iron. For more complete information see ASM *Metals Handbook,* 9th ed., vol. 1.

**TABLE 4-70** AISI-SAE System of Designations

| Numerals and digits | Type of steel and nominal alloy content |
|---|---|
| **Carbon steels:** | |
| 10XX(a) | Plain carbon (Mn 1.00% max) |
| 11XX | Resulfurized |
| 12XX | Resulfurized and rephosphorized |
| 15XX | Plain carbon (max Mn range—1.00 to 1.65%) |
| **Manganese steels:** | |
| 13XX | Mn 1.75 |
| **Nickel Steels:** | |
| 23XX | Ni 3.50 |
| 25XX | Ni 5.00 |
| **Nickel-chromium steels:** | |
| 31XX | Ni 1.25; Cr 0.65 and 0.80 |
| 32XX | Ni 1.75; Cr 1.07 |
| 33XX | Ni 3.50; Cr 1.50 and 1.57 |
| 34XX | Ni 3.00; Cr 0.77 |
| **Molybdenum steels:** | |
| 40XX | Mo 0.20 and 0.25 |
| 44XX | Mo 0.40 and 0.52 |
| **Chromium-molybdenum steels:** | |
| 41XX | Cr 0.50, 0.80 and 0.95; Mo 0.12, 0.20, 0.25 and 0.30 |
| **Nickel-chromium-molybdenum steels:** | |
| 43XX | Ni 1.82; Cr 0.50 and 0.80; Mo 0.25 |
| 43BVXX | Ni 1.82; Cr 0.50; Mo 0.12 and 0.25; V 0.03 min |
| 47XX | Ni 1.05; Cr 0.45; Mo 0.20 and 0.35 |
| 81XX | Ni 0.30; Cr 0.40; Mo 0.12 |
| 86XX | Ni 0.55; Cr 0.50; Mo 0.20 |
| 87XX | Ni 0.55; Cr 0.50; Mo 0.25 |
| 88XX | Ni 0.55; Cr 0.50; Mo 0.35 |
| 93XX | Ni 3.25; Cr 1.20; Mo 0.12 |
| 94XX | Ni 0.45; Cr 0.40; Mo 0.12 |
| 97XX | Ni 0.55; Cr 0.20; Mo 0.20 |
| 98XX | Ni 1.00; Cr 0.80; Mo 0.25 |
| **Nickel-molybdenum steels:** | |
| 46XX | Ni 0.85 and 1.82; Mo 0.20 and 0.25 |
| 48XX | Ni 3.50; Mo 0.25 |
| **Chromium steels:** | |
| 50XX | Cr 0.27, 0.40, 0.50 and 0.65 |
| 51XX | Cr 0.80, 0.87, 0.92, 0.95, 1.00 and 1.05 |
| **Chromium steels:** | |
| 50XXX | Cr 0.50 } C 1.00 min |
| 51XXX | Cr 1.02 } C 1.00 min |
| 52XXX | Cr 1.45 } C 1.00 min |
| **Chromium-vanadium steels:** | |
| 61XX | Cr 0.60, 0.80 and 0.95; V 0.10 and 0.15 min |
| **Tungsten-chromium steel:** | |
| 72XX | W 1.75; Cr 0.75 |
| **Silicon-manganese steels:** | |
| 92XX | Si 1.40 and 2.00; Mn 0.65, 0.82 and 0.85; Cr 0.00 and 0.65 |
| **High-strength low-alloy steels:** | |
| 9XX | Various SAE grades |
| **Boron steels:** | |
| XXBXX | B denotes boron steel |
| **Leaded steels:** | |
| XXLXX | L denotes leaded steel |

(a) XX in the last two digits of these designations indicates that the carbon content (in hundredths of a percent) is to be inserted.

**TABLE 4-71** Average Properties of Three Grades of Cupola Malleable Iron

| Properties | No. 1 | No. 2 | No. 3 |
|---|---|---|---|
| Tensile strength, MPa (lb/in$^2$) | 330 (49,700) | 285 (43,000) | 285 (43,000) |
| Yield strength, MPa (lb/in$^2$) | 375 (41,000) | 220 (33,000) | 205 (31,000) |
| Elongation in 2 in | 8.1 | 7.0 | 6.5 |

***310. Chilled cast iron*** is made by pouring cast iron into a metallic mold which cools it rapidly near the surfaces of the casting, thus forming a wear-resisting skin of harder material than the body of the metal. The rapid cooling decreases the proportion of graphite and increases the combined carbon, resulting in the formation of white cast iron.

***311. Alloy cast iron*** contains specially added elements in sufficient amount to produce measurable modification of the physical properties. Silicon, manganese, sulfur, and phosphorus, in quantities normally obtained from raw materials, are not considered alloy additions. Up to about 4% silicon increases the strength of pure iron; greater content produces a matrix of dissolved silicon that is weak, hard, and brittle. Cast irons with 7 to 8% silicon are used for heat-resisting purposes and with 13 to 17% silicon form acid- and corrosion-resistant alloys, which, however, are extremely brittle. Manganese up to 1% has little effect on mechanical properties but tends to inhibit the harmful effects of sulfur. Nickel, chromium, molybdenum, vanadium, copper, and titanium are commonly used alloying elements. The methods of processing or of making the alloy additions to the iron influence the final properties of the metal; hence a specified chemical analysis is not sufficient to obtain required qualities. Heat-treatment is also employed on alloy irons to enhance the physical properties. (See *Cast Metals Handbook,* American Foundrymen's Association, 1957.)

***312. Density of cast iron*** varies considerably depending upon the carbon content and the proportion of the carbon that is present as graphite. Using the density of pure iron, 7.86, as a reference, the density of cast iron may range from 7.60 for white cast iron to as low as 6.80 for gray cast iron.

***313. Thermal Properties of Cast Iron.*** Thermal properties vary somewhat with the composition and the proportions of graphitic carbon. The average specific heat from 20 to 110°C is 0.119; thermal conductivity, 0.40 W/(cm$^3$)(°C); coefficient of linear expansion, 0.0000106/(°C) at 40°C.

***314. Values of modulus of elasticity for ferrous metals*** may be assumed approximately as shown in Table 4-72. The values for all steels are fairly constant, whereas for cast irons the modulus increases somewhat with increased strength of material. Alloy steels have practically the same modulus as plain carbon steels unless large amounts, say 10%, of alloying material are added; for large percentages of alloying elements the modulus decreases slightly. The modulus of steels is not affected by heat-treatment.

***315. Heat-Treatment of Steel.*** The properties of steels can be greatly modified by thermal treatments which change the internal crystalline structure of the alloy. *Hardening* of steel is based on the fact that iron undergoes a change in crystal structure when heated above its "critical" temperature. Above this critical transformation temperature, the structure is called austenite, a phase capable of dissolving carbon up to 2%. Below the critical temperature, the steel transforms to ferrite, in which carbon is insoluble and precipitates as an iron carbide compound, $Fe_3C$ (sometimes called cementite). If a steel is cooled rapidly from

**TABLE 4-72** Approximate Modulus of Elasticity for Ferrous Metals

| Metal | Modulus in tension-compression, GPa (lb/in$^2$ × $10^{-6}$) | Modulus in shear, GPa (lb/in$^2$ × $10^{-6}$) |
|---|---|---|
| All steels | 206 (30.0) | 83 (12.0) |
| Wrought iron | 186 (27.0) | 75 (10.8) |
| Malleable cast iron | 158 (23.0) | 63 ( 9.2) |
| Gray cast iron, ASTM No. 20 | 103 (15.0) | 41 ( 6.0) |
| Gray cast iron, ASTM No. 60 | 138 (20.0) | 55 ( 8.0) |

above the critical temperature, the carbon is unable to diffuse to form cementite and the austenite transforms instead to an extremely hard metastable constituent called martensite in which the carbon is held in supersaturation. The hardness of the martensite depends sensitively on the carbon content. Low-carbon steels (below about 0.20%) are seldom quenched, while steels above about 0.80% carbon are brittle and liable to crack on quenching. Plain carbon steels must be quenched at very fast rates in order to be hardened. Alloying elements can be added to decrease the necessary cooling rates to cause hardening; some alloy steels will harden when cooled in air from above the critical temperature. It should be noted, however, that it is the amount of carbon that primarily determines the properties of the alloy; the alloying elements serve to make the response to heat-treatment possible.

*Normalizing* is a treatment in which the steel is heated over the critical temperature and allowed to cool in still air. The purpose of normalizing is to homogenize the steel. The carbon in the steel will appear as a fine lamellar product of cementite and ferrite called pearlite.

*Annealing* is similar to normalizing, except the steel is very slowly cooled from above the critical. The carbides are now coarsely divided and the steel is in its softest state, as may be desired for cold-forming or machining operations.

*Process annealing* is a treatment carried out below the critical temperature designed to recrystallize the ferrite following a cold-working operation. Metals become hardened and embrittled by plastic deformation, but the original state can be restored if the alloy is heated high enough to cause new strain-free grains to nucleate and replace the prior strained structure. This treatment is commonly applied as a final processing for low-carbon steels where ductility and toughness are important, or as an intermediate treatment for such products as wire that are formed by cold working.

*Stress-relief annealing* is a thermal treatment carried out at a still lower temperature. No structural changes take place, but its purpose is to reduce residual stresses that may have been introduced by previous nonuniform deformation or heating.

*Tempering* is a treatment that always follows a hardening (quenching) treatment. After hardening, steels are extremely hard, but relatively weak owing to their brittleness. When reheated to temperatures below the critical, the martensitic structure is gradually converted to a ferrite-carbide aggregate that optimizes strength and toughness. Figure 4-70 illustrates the effect of tempering on the properties of a typical medium-carbon alloy steel. When steels are tempered at about 260°C, a particularly brittle configuration of precipitated carbides forms; steels should be tempered above or below this range. Another phenomenon causing embrittlement occurs in steels particularly containing chromium and manganese that are given a tempering cycle that includes holding at, or cooling through, temperatures around

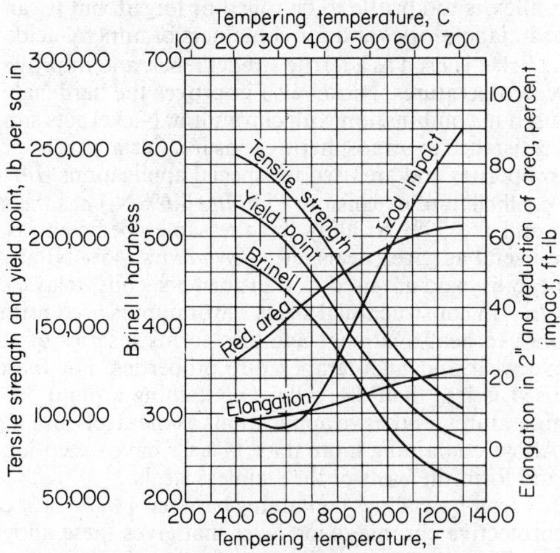


**FIG. 4-70** Mechanical properties of SAE 3140 (in small sizes, ½ to 1½ in diameter or thickness). Quenched from 1475 to 1525°F, in oil. Tempered as indicated.

567 to 621°C. Small molybdenum additions retard this effect, called "temper brittleness." It is believed to be caused by a segregation of trace impurity elements to the grain boundaries.

***316. Mechanical Properties of Iron and Steel.*** Representative properties of selected ferrous alloys given various heat-treatments are given in Table 4-73.

***317. Manganese Steels.*** Manganese is present in all steels as a scavenger for sulfur, an unavoidable impurity; otherwise the sulfur would form a low-melting constituent containing FeS and it would be impossible to hot-work the steel. The manganese content should be about five times the sulfur to provide protection against this "hot shortness." Beyond this amount, manganese increases the hardness of the steel and also has a strong effect on improving response to hardening treatment, but increases susceptibility to temper brittleness. Manganese can be specified up to 1.65% without the steel being classified as an alloy steel. The alloy containing 12 to 14% Mn and around 1% carbon is called Hadfield's manganese steel. This alloy can be quenched to retain the austenite phase and is quite tough in this condition. When deformed, this austenite transforms to martensite which confers exceptional wear and abrasion resistance. Applications for this unique steel include railroad switches, crushing and grinding equipment, dipper bucket teeth, etc. See ASTM Specification A128, *ASM Trans.,* vol. 49, 1957.

***318. Vanadium Steels.*** In amounts up to 0.01%, vanadium has a powerful strengthening effect in microalloyed high-strength, low-alloy steels. In alloy steels, 0.1 to 0.2% vanadium is used as a deoxidizer and carbide-forming addition to promote fine-grained tough steels with deep hardening characteristics. Vanadium accentuates the benefits derived from other alloying elements such as manganese, chromium, or nickel, and it is used in a variety of quaternary alloys containing these elements. (See SAE 6120, Table 4-117.) Vanadium in amounts of 0.15 to 2.50% is an important element in a large number of tool steels. (For detailed data see "Vanadium Steels and Irons," New York, Vanadium Corporation of America, 1950; *Metallurgist Metals Tech.,* 1977, vol. 9, p. 375.)

***319. Silicon Steels.*** Silicon is present in most constructional steels in amounts up to 0.35% as a deoxidizer to enhance production of sound ingot structures. Silicon increases the hardenability of steel slightly and also acts as a solid solution hardener with little loss of ductility in amounts up to 2.5%. Silicon in amounts of about 4.5% is a major ingredient in electrical steel sheets. Silicon improves the magnetic properties of iron, but even more importantly, these steels can be fabricated to produce controlled grain size and orientation. Since permeability is dependent upon crystal orientation, exceptionally small core losses are obtained by using grain-oriented silicon steel in motors, transformers, etc. (See ASM *Metals Handbook,* 9th ed., vol. 3.) Alloys containing 12 to 14% Si are exceptionally resistant to corrosion by acids. This alloy is too brittle to be rolled or forged, but it can be cast, and is widely used as drainpipe in laboratories and for containers of mineral acids.

***320. Nickel Steels.*** Nickel is used as a ferrite strengthener, and improves the toughness of steel, especially at low temperatures. Nickel also improves the hardenability and is particularly effective when used in combination with chromium. Nickel acts similarly to copper in improving corrosion resistance to atmospheric exposure. Certain iron-nickel alloys have particularly interesting properties and are used for special applications: *Invar* (36% Ni) has a very low temperature coefficient of expansion; *Platinite* (46% Ni) has the same expansion coefficient as platinum, and the 39% Ni alloy has the same coefficient as low-expansion glasses. These alloys are useful as gages, seals, etc. *Permalloys* (45% Ni and 76% Ni) have exceptionally high permeability and are used in transformers, coils, relays, etc.

***321. Chromium Steels.*** In constructional steels, chromium is used primarily as a hardener. It improves response to heat-treatment and also forms a series of complex carbide compounds that improve wear and high-temperature properties. For these purposes, the amount of chromium used is less than 2%. Alloys containing around 5% Cr retain high hardness at elevated temperatures, and have applications as die steels and high-temperature processing equipment. Alloys containing more than 11% Cr have exceptional resistance to atmospheric corrosion and form the basis of the stainless steels.

***322. Stainless Steels.*** Iron-base alloys containing between 11 and 30% chromium form a tenacious and highly protective chrome oxide layer that gives these alloys excellent corrosion-resistant properties. There are a great number of alloys that are generally referred to as "stainless steels," and they fall into three general classifications.

*Austenitic* stainless steels contain usually 8 to 12% nickel, which stabilizes the austenitic

**TABLE 4-73** Approximate Mechanical Properties of Iron and Steel

(Based on test data from various materials testing laboratories)

| Metal | Strength in tension, lb/in$^{2i}$ | | Strength in compression, lb/in$^{2i}$ | | Yield[a] strength in shear, lb/in$^{2b,i}$ | Endurance limit for reversed bending stress, lb/in$^{2i}$ | Brinell hardness No.[c] | Elongation in 2 in., % |
|---|---|---|---|---|---|---|---|---|
| | Yield[a] | Ultimate | Yield[a] | Ultimate | | | | |
| Gray cast iron: | | | | | | | | |
| ASTM 20 | *d* | 20,000 | *d* | 80,000 | *e* | 9,000 | *h* | Less than 1 |
| ASTM 35 | *d* | 35,000 | *d* | 125,000 | *e* | 15,000 | *h* | Less than 1 |
| ASTM 60 | *d* | 60,000 | *d* | 145,000 | *e* | 24,000 | *h* | Less than 1 |
| Gray cast iron with 1.15% nickel | *d* | 50,000 | *d* | 156,000 | *e* | 20,000 | *h* | Less than 1 |
| Malleable cast iron | 30,000 | 50,000 | 30,000 | *f* | 16,000 | 25,000 | 110 | 10 |
| Commercial pure iron, annealed | 19,000 | 42,000 | 19,000 | *f* | 12,000 | 26,000 | 69 | 48 |
| Commercial wrought iron, as rolled | 30,000 | 50,000 | 30,000 | *f* | 18,000 | 25,000 | 100 | 35 |
| Structural steel, as rolled, and SAE 1020 steel, as rolled | 35,000 | 60,000 | 35,000 | *f* | 21,000 | 30,000 | 120 | 35 |
| SAE 1040 steel, water quenched, 1050°F temper | 87,000 | 102,000 | 87,000 | *f* | 52,000 | 57,000 | 210 | 23 |
| SAE 1095 steel, oil quenched, 850°F temper | 97,000 | 188,000 | 97,000 | *f* | 58,000 | 98,000 | 380 | 10 |
| SAE 2340 steel, oil quenched, 1200°F temper | 91,000 | 112,000 | 91,000 | *f* | 54,000 | 67,000 | 248 | 24 |
| Oil quenched, 400°F temper | 174,000 | 282,000 | 174,000 | *f* | 96,000 | 112,000 | 488 | 8 |
| SAE 3325 steel, oil quenched, 700°F temper | 128,000 | 139,000 | 128,000 | *f* | 70,000 | 68,000 | 291 | 18 |
| SAE 4140 steel, water quenched, 1100°F temper | 116,000 | 140,000 | 116,000 | *f* | 63,000 | 64,000 | 250 | 16 |
| SAE 5150 steel, oil quenched, 800°F temper | 210,000 | 235,000 | 210,000 | *f* | 115,000 | 90,000 | 455 | 13 |
| SAE 6120 steel, water quenched, 1100°F temper | 130,000 | 164,000 | 130,000 | *f* | 72,000 | 92,000 | 350 | 16 |
| SAE 9260 steel, oil quenched | 100,000 | 158,000 | 100,000 | *f* | 60,000 | 62,000 | 240 | 16 |
| AISI 8650, 1 in. diam., oil quenched 1000°F temper | 158,000 | 170,000 | 158,000 | *f* | *h* | *h* | 350 | 14 |

**TABLE 4-73** Approximate Mechanical Properties of Iron and Steel (*Continued*)
(Based on test data from various materials testing laboratories)

| Metal | Strength in tension, lb/in²[i] | | Strength in compression, lb/in²[i] | | Yield[a] strength in shear, lb/in²[b,i] | Endurance limit for reversed bending stress, lb/in²[i] | Brinell hardness No.[c] | Elongation in 2 in., % |
|---|---|---|---|---|---|---|---|---|
| | Yield[a] | Ultimate | Yield[a] | Ultimate | | | | |
| AISI E 8740, 1 in. diam., quenched from 1525 F in oil tower, 1100°F temper | 134,000 | 149,000 | 134,000 | *f* | *h* | *h* | 302 | 18.3 |
| AISI E 9310, 1 in. diam., oil quenched from 1425 F, 300°F temper | 118,000 | 145,000 | 118,000 | *f* | *h* | *h* | 302 | 15.5 |
| AISI 9840, 1¼ in. diam., oil quenched from 1525 F, 800°F temper | 205,000 | 220,000 | 205,000 | *f* | *h* | *h* | 430 | 12 |
| 18-8 stainless steel, 18% chromium, 8% nickel, water quenched | 33,000 | 75,000 | 33,000 | *f* | 18,000 | 35,000 | 140 | 55 |
| Steel casting, 0.35% C, 1.71% Mn, annealed | 60,000 | 104,000 | 60,000 | *f* | 35,000 | 45,000 | 188 | 22 |
| Steel casting, 0.25% C, 0.68% Mn, annealed | 43,000 | 77,000 | 43,000 | *f* | 24,000 | 35,000 | 136 | 30 |
| Cold-drawn steel rod, 0.20% C | 60,000 | 80,000 | 60,000 | *f* | 36,000 | 38,000 | 150 | 18 |
| Drawn wire, iron or soft steel | 70,000 | 85,000 | *g* | *g* | 40,000 | *h* | *h* | *h* |
| High-carbon steel wire | 150,000 | 275,000 | *g* | *g* | 80,000 | *h* | *h* | *h* |

[a]Yield strength taken as yield point, or at 0.2% nominal set.
[b]Accurate data on ultimate strength in shear not available.
[c]See Par. **293** for description of Brinell hardness test.
[d]No well-defined yield strength.
[e]Shearing yield strength greater than tensile yield strength.
[f]For ductile metal the ultimate in compression is only slightly greater than yield strength.
[g]Wire can offer resistance only in tension and in shear.
[h]Data lacking.
[i]1 lb/in² = 6.895 kPa.
NOTE: $t_{°C} = (t_{°F} - 32)/1.8$.

phase. These are the most popular of the stainless steels. With 18 to 20% chromium, they have the best corrosion resistance, and are very tough and can undergo severe forming operations. These alloys are susceptible to embrittlement when heated in the range of 593 to 816°C. At these temperatures, carbides precipitate at the austenite grain boundaries, causing a local depletion of the chromium content in the adjacent region, so that this region loses its corrosion resistance. Use of "extra low carbon" grades and grades containing stabilizing additions of strong carbide-forming elements such as niobium minimizes this problem. These alloys are also susceptible to stress corrosion in the presence of chloride environments.

*Ferritic* stainless steels are basically straight Fe-Cr alloys. Chromium in excess of 14% stabilizes the low-temperature ferrite phase all the way to the melting point. Since these alloys do not undergo a phase change, they cannot be hardened by heat-treatment. They are the least expensive of the stainless alloys.

*Martensitic* stainless steels contain around 12% Cr. They are austenitic at elevated temperatures, but ferritic at low; hence they can be hardened by heat-treatment. To obtain a significant response to heat-treatment, they have higher carbon contents than the other stainless alloys. Martensitic alloys are used for tools, machine parts, cutting instruments, and other applications requiring high strength. The austenitic alloys are nonmagnetic, but the ferritic and martensitic grades are ferromagnetic. Properties of some representative stainless steels are given in Table 4-74. See ASM *Metals Handbook,* 9th ed., vol. 3, 1980; C. Zapfee, *Stainless Steels,* ASM, 1949; *Enduro Stainless Steels,* Republic Steel, Cleveland.

***323. Heat-resistant alloys*** are capable of continuous or intermittent service at temperatures in excess of 649°C. There are a great number of these alloys; they are best considered by class. *Iron-chromium alloys* contain between 10 and 30% chromium. The higher the chromium, the higher the service temperature at which they can operate. They are relatively low-strength alloys, and are used primarily for oxidation resistance. *Iron-chromium-nickel alloys* have chromium in excess of 18%, nickel in excess of 7%, and always more chromium

**TABLE 4-74** Properties of Typical Stainless Steels
(From *Metals Handbook*)

| Property | Austenitic | | Martensitic | Ferritic |
|---|---|---|---|---|
| | AISI 309 Cr. 22–24% Ni. 12–15% annealed | AISI 321 Cr. 17–19% Ni. 8–11% annealed | AISI 410 Cr. 11.5–13.5% quenched and tempered at 1000°F | AISI 430 Cr. 14–18% annealed |
| Ultimate strength, lb/in$^2$ | 95,000 | 85,000 | 145,000 | 75,000 |
| Yield strength, lb/in$^2$, 0.2% offset | 40,000 | 30,000 | 115,000 | 40,000 |
| Elongation in 2 in, % | 45 | 55 | 20 | 30 |
| Reduction of area, % | 50–65 | 55–65 | 65 | 40–55 |
| Hardness: | | | | |
| Rockwell | B78–90 | B75–90 | C31 | B79–90 |
| Brinell | 140–185 | 135–185 | 300 | 145–185 |
| Density | 7.9 | 7.9 | 7.7 | 7.7 |
| Weight, kg/m$^3$ | 0.80 | 0.80 | 0.775 | 0.775 |
| Thermal conductivity at 100°C, W/m$^2 \cdot$K | 45.4 | 52.7 | 81.5 | 85.6 |
| Coefficient of expansion per deg F (mean value from 32 to 1000°F) | $9.6 \times 10^{-6}$ | $10.3 \times 10^{-6}$ | $7.2 \times 10^{-6}$ | $6.3 \times 10^{-6}$ |
| Elastic modulus, lb/in$^2$ | $29 \times 10^6$ | $28 \times 10^6$ | $29 \times 10^6$ | $29 \times 10^6$ |
| Scaling temp, deg F | 2000 | 1650 | 1250 | 1500 |

NOTE: 1 lb/in$^2$ = 6.895 kPa; $t_{°C} = (t_{°F} - 32)/1.8$.

than nickel. They are austenitic alloys and have better strength and ductility than the straight Fe-Cr alloys. They can be used in both oxidizing and reducing environments and in sulfur-bearing atmospheres. Iron-nickel-chromium alloys have more than 10% Cr and more than 25% Ni. These are also austenitic alloys, and are capable of withstanding fluctuating temperatures in both oxidizing and reducing atmospheres. They are used extensively for furnace fixtures and components and parts subjected to nonuniform heating. They are also satisfactory for electric resistance-heating elements.

*Nickel-base alloys* contain about 50% Ni, and also contain some molybdenum. They are more expensive than iron-base alloys, but have better high-temperature mechanical properties.

*Cobalt-base alloys* contain about 50% cobalt and have especially good creep and stress-rupture properties. They are widely used for gas-turbine blades. Most of these alloys are available in both cast and wrought form; the castings usually have higher carbon contents and often small additions of silicon and/or manganese to improve casting properties. See ASM *Metals Handbook,* 9th ed., vol. 3; and publications of the International Nickel Company.

***324. Creep Strength.*** Metals subjected to static loading at elevated temperatures continue to elongate (creep) with time. After an initial period of adjustment to a fairly constant velocity of flow the time rate of deformation under constant stress and temperature (expressed as percentage elongation per hour) is called the *creep rate.* Short-time tensile-test values are not reliable design criteria for metals used at elevated temperatures. The useful strength is limited to the stress that will not produce a damaging amount of deformation during the normal life of the structure. The *creep limit* for a material is the stress that will not produce more than a specified elongation (usually 1%) in a definite time interval (often taken as 10,000 or 100,000 h) at the given temperature. Determination of the creep limit involves long-time testing of a series of specimens to determine initial deformations and creep rates for various stresses. The data are plotted and arbitrarily extrapolated to obtain approximate total creep at future times. Some nonferrous materials such as lead and zinc creep at room temperatures (see Par. **361**), whereas, except for stresses nearly up to the ultimate, no appreciable creep has been observed for steels until temperatures above about 260°C (500°F) are exceeded. Variations in data reported on steels have led to the conclusion that creep characteristics are too sensitive an index of strength to permit exact duplication either in different laboratories or in duplicate tests in the same laboratory. Figure 4-71 shows the approximate variation of creep stress with temperature for several steels. For boilers, piping, etc., operating at temperatures above 538°C (1000°F), the maximum working stresses that can be used (without excessive creep) are so small as to make it difficult to produce economical and safe designs until better creep-resistant materials are available. Eleven grades of alloy steels for service at temperatures from 399 to 593°C (750 to 1000°F) are covered by ASTM Specifications A351, A335, and A193 for specific applications to castings, bolting materials, and seamless pipe. These range from ordinary carbon-molybdenum steels to 18% Cr, 8% Ni austenitic steels. In general, 0.4 to 1.5% molybdenum is used in steels for high-temperature service, since it is the only element thus far proved to be effective in increasing creep resistance when present in only small amounts. (For a comprehensive tabulation of creep data see "High Temperature Strength Data of Metals and Alloys," Punched Cards, ASTM, 1965; and annual reports of the Joint Research Committee on Effect of Temperature on the Properties of Metals in *Proc. ASTM.*)

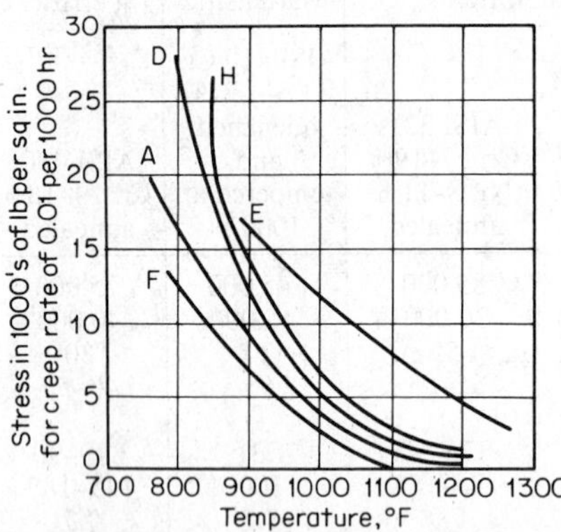


**FIG. 4-71** Creep stress for a creep rate of 0.01% per 1000 h for several steels. *A.* Wrought 0.01 to 0.20% carbon steels. *D.* Wrought chromium-molybdenum bolt steel (0.40% carbon). *E.* Wrought 18% chromium, 8% nickel steel. *F.* Wrought carbon steels (carbon above 0.20%). *H.* Wrought 1.0 to 2.5% chromium. 0.50% molybdenum steel (0.20% carbon maximum).

***325. General References on Iron and Steel***

Bain, E., and Paxton, H.: *Alloying Elements in Steel;* Cleveland, ASM, 1961.

Campbell, J.E., et al.: *Application of Fracture Mechanics for Selection of Metallic Structural Materials,* Metals Park, Ohio, ASM, 1982.

Horger, O.J.: *Metals Engineering—Design,* ASME Handbook, 2d ed.; New York, McGraw-Hill, 1965.

Roberts, G. A., et al.: *Tool Steels,* 4th ed., Cleveland, Ohio, ASM, 1980.

American Society for Metals: *Sourcebook on Industrial Alloys and Engineering Data,* ASM, 1978.

*Advances in the Technology of Stainless Steel and Related Alloys,* ASTM STP 369, 1965.

*Fracture Toughness,* ASTM STP 514, 1971.

*Making, Shaping, and Treating of Steel,* 9th ed.; Pittsburgh, United States Steel Co., 1971.

## Steel Strand and Rope

***326. Iron and Steel Wire.*** Annealed wire of iron or very mild steel has a tensile strength in the range of 310 to 415 MPa (45,000 to 60,000 lb/in$^2$); with increased carbon content, varying amounts of cold drawing, and various heat-treatments the tensile strength ranges all the way from the latter figures up to about 3450 MPa (500,000 lb/in$^2$), but a figure of about 1725 MPa (250,000 lb/in$^2$) represents the ordinary limit for wire for important structural purposes. For example, see the following paragraph on bridge wire. Wires of high carbon content can be tempered for special applications such as spring wire. The yield strength of cold-drawn steel wire is 65 to 80% of its ultimate strength. For examples showing the effects of drawing and carbon content on wire, see *Making, Shaping, and Treating of Steel,* U.S. Steel (Par. **325**).

***327. Galvanized-Steel Bridge Wire.*** The manufacture of high-strength bridge wire like that used for the cables and hangers of suspension bridges such as the San Francisco–Oakland Bay Bridge, the Mackinac Bridge in Michigan, and the Narrows Bridge in New York is an excellent example of careful control of processing to produce a quality material. The wire is a high-carbon product containing 0.75 to 0.85% carbon with maximum limits placed on potentially harmful impurities. Rolling temperatures are carefully specified, and the wire is subjected to a special heat-treatment called "patenting." The steel is transformed in a controlled-temperature molten lead bath to ensure an optimum microstructure. This is followed by cold drawing to a minimum tensile strength of 1550 MPa (225,000 lb/in$^2$) and a 4% elongation. The wire is given a heavy zinc coating to protect against corrosion. Joints or splices are made with cold-pressed sleeves which develop practically the full strength of the wire. Fatigue tests of galvanized bridge wire in reversed bending indicate that the endurance limit of the coated wire is only about 345 to 415 MPa (50,000 to 60,000 lb/in$^2$).

***328. Wire rope*** is made of wires twisted together in certain typical constructions and may be either flat or round. Flat ropes consist of a number of strands of alternately right and left lay, sewed together with soft iron to form a band or belt; they are sometimes of advantage in mine hoists. Round ropes are composed of a number of wire strands twisted around a hemp core or around a wire strand or wire rope. The standard wire rope is made of six strands twisted around a hemp core, but for special purposes, four, five, seven, eight, nine, or any reasonable number of strands may be used. The hemp is usually saturated with a lubricant, which should be free from acids or corrosive substances; this provides little additional strength, but acts as a cushion to preserve the shape of the rope and helps to lubricate the wires. The number of wires commonly used in the strands are 4, 7, 12, 19, 24, and 37, depending upon the service for which the ropes are intended. When extra flexibility is required, the strands of a rope sometimes consist of ropes, which in turn are made of strands around a hemp core. Ordinarily the wires are twisted into strands in the opposite direction to the twist of the strands in the rope. The make-up of standard hoisting rope is 6 $\times$ 19; extra-pliable hoisting rope is 8 $\times$ 19 or 6 $\times$ 37; transmission or haulage rope is 6 $\times$ 7; hawsers and mooring lines are 6 $\times$ 12 or 6 $\times$ 19 or 6 $\times$ 24 or 6 $\times$ 37, etc.; tiller or hand rope is 6 $\times$ 7; highway guard-rail strand is 3 $\times$ 7; galvanized mast-arm rope is 9 $\times$ 4 with a cotton center. The tensile strength of the wire ranges, in different grades, from 415 to 2415 MPa (60,000 to 350,000 lb/in$^2$), depending on the material, diameter, and treatment. The maximum tensile efficiency of wire rope is 90%; the average is about 82.5%, being higher for 6 $\times$ 7 rope and lower for 6 $\times$ 37 construction. The apparent modulus of elasticity for steel cables in service may be assumed to be 62 to 83 $\times$ $10^6$ kPa (9 to 12 $\times$ $10^6$ lb/in$^2$) of

**TABLE 4-75** Diameters of Sheaves and Drums for Wire Rope

| Rope | Sheave diameter ÷ rope diameter | | |
|---|---|---|---|
| | Average | Minimum | Larger installations |
| 6 × 7 | 72 | 42 | 96 |
| 6 × 19 | 45 | 30 | 90 |
| 6 × 37 | 27 | 18 | |
| 8 × 19 | 31 | 21 | |

**TABLE 4-76** Nominal Strengths of Wire Rope

[6 × 19 Classification/Bright (Uncoated), Independent Wire Rope Core]

| Nominal diameter | | Approximate mass | | Nominal strength* | | | |
|---|---|---|---|---|---|---|---|
| | | | | Improved plow steel† | | Extra improved plow steel† | |
| in | mm | lb/ft | kg/m | tons | metric tonnes | tons | metric tonnes |
| ¼ | 6.4 | 0.12 | 0.17 | 2.94 | 2.67 | 3.40 | 3.08 |
| 5/16 | 8 | 0.18 | 0.27 | 4.58 | 4.16 | 5.27 | 4.78 |
| ⅜ | 9.5 | 0.26 | 0.39 | 6.56 | 5.95 | 7.55 | 6.85 |
| 7/16 | 11.5 | 0.35 | 0.52 | 8.89 | 8.07 | 10.2 | 9.25 |
| ½ | 13 | 0.46 | 0.68 | 11.5 | 10.4 | 13.3 | 12.1 |
| 9/16 | 14.5 | 0.59 | 0.88 | 14.5 | 13.2 | 16.8 | 15.2 |
| ⅝ | 16 | 0.72 | 1.07 | 17.7 | 16.2 | 20.6 | 18.7 |
| ¾ | 19 | 1.04 | 1.55 | 25.6 | 23.2 | 29.4 | 26.7 |
| ⅞ | 22 | 1.42 | 2.11 | 34.6 | 31.4 | 39.8 | 36.1 |
| 1 | 26 | 1.85 | 2.75 | 44.9 | 40.7 | 51.7 | 46.9 |
| 1⅛ | 29 | 2.34 | 3.48 | 56.5 | 51.3 | 65.0 | 59.0 |
| 1¼ | 32 | 2.89 | 4.30 | 69.4 | 63.0 | 79.9 | 72.5 |
| 1⅜ | 35 | 3.5 | 5.21 | 83.5 | 75.7 | 96.0 | 87.1 |
| 1½ | 38 | 4.16 | 6.19 | 98.9 | 89.7 | 114 | 103 |
| 1⅝ | 42 | 4.88 | 7.26 | 115 | 104 | 132 | 120 |
| 1¾ | 45 | 5.67 | 8.44 | 133 | 121 | 153 | 139 |
| 1⅞ | 48 | 6.5 | 9.67 | 152 | 138 | 174 | 158 |
| 2 | 52 | 7.39 | 11.0 | 172 | 156 | 198 | 180 |
| 2⅛ | 54 | 8.35 | 12.4 | 192 | 174 | 221 | 200 |
| 2¼ | 57 | 9.36 | 13.9 | 215 | 195 | 247 | 224 |
| 2⅜ | 60 | 10.4 | 15.5 | 239 | 217 | 274 | 249 |
| 2½ | 64 | 11.6 | 17.3 | 262 | 238 | 302 | 274 |
| 2⅝ | 67 | 12.8 | 19.0 | 288 | 261 | 331 | 300 |
| 2¾ | 70 | 14.0 | 20.8 | 314 | 285 | 361 | 327 |

*To convert to kilonewtons (kN), multiply tons (nominal strength) by 8.896; 1 lb = 4.448 newtons (N).

†Available with galvanized wires at strengths 10% lower than listed, or at equivalent strengths on special request.

cable section. Grades of wire rope are (from historic origins) referred to as traction, mild plow, plow, improved plow, and extra improved plow steel. The most common finish for steel wire is "bright" or uncoated, but various coatings, particularly zinc (galvanized), are used. See Wire Rope Users Manual, AISI, Washington, D.C., 2nd ed., 1981.

***329. Diameter of Sheaves and Drums for Wire Rope.*** The average and minimum tread diameters, in accordance with the practice recommended by the American Steel & Wire Co., are shown in Table 4-75; also higher values should be used for larger hoisting installations. Diameters larger than those listed as minimum will give increased rope life.

***330. Nominal Properties of Wire Rope.*** See Table 4-76.

## Corrosion of Iron and Steel

***331. Principles of Corrosion.*** Corrosion may take place by direct chemical attack or by electrochemical (galvanic) attack; the latter is by far the most common mechanism. When two dissimilar metals that are in electrical contact are connected by an electrolyte, an electromotive potential is developed, and a current flows. The magnitude of the current depends on the conductivity of the electrolyte, the presence of high-resistance "passivating" films on the electrode surfaces, the relative areas of electrodes, and the strength of the potential difference. The metal that serves as the anode undergoes oxidation and goes into solution (corrodes).

When different metals are ranked according to their tendency to go into solution, the galvanic series, or electromotive series, is obtained. Metals at the bottom will corrode when in contact with those at the top; the greater the separation, the greater the attack is likely to be. Table 4-77 is such a ranking, based on tests by the International Nickel Company, in which the electrolyte was seawater. The nature of the electrolyte may affect the order to some extent. It should also be recognized that very subtle differences in the nature of the metal may result in the formation of anode-cathode galvanic cells: slight differences in composition of the electrolyte at different locations on the metal surface; minor segregation of impurities in the metal; variations in the degree of cold deformation undergone by the metal; etc. It is possible for anode-cathode couples to exist very close to each other on a metal surface. The electrolyte is a solution of ions; a film of condensed moisture will serve.

***332. Corrosion Prevention.*** An understanding of the mechanism of corrosion suggests possible ways of minimizing corrosion effects. Some of these include: (1) avoidance of metal combinations that are not compatible; (2) electrical insulation between dissimilar metals that have to be used together; (3) use of a sacrificial anode placed in contact with a structure to be protected (this is an expensive technique but can be justified in order to protect such structures as buried pipelines and ship hulls); (4) use of an impressed *emf* from an external power source to buck out the corrosion current (called "cathodic protection"); (5) avoiding the presence of an electrolyte—especially those with high conductivities; and (6) application of a protective coating to either the anode or the cathode. The problems of corrosion control are complex beyond these simple concepts, but since the use of protective coatings on iron and steel is extensive, this subject is treated in the following sections.

***333. Protective coatings*** may be selected to be inert to the corrosive environment and isolate the base metal from exposure, or the coating may be selected to have reasonable resistance to attack but act sacrificially to protect the base metal. Protective coatings may be considered in four broad classes: paints, metal coatings, chemical coatings, and greases. Painting is commonly used for the protection of structural iron and steel but must be maintained by periodic renewal. Metal coatings take various ranks in protective effectiveness, depending on the metal used and its characteristics as a coating material. A wide variety of metals are used to coat steels: zinc, tin, copper, nickel, chromium, cobalt, lead, cadmium, and aluminum; coatings of gold and silver are also used for decorative purposes. Coatings may be applied by these principal methods: hot dipping, cementation, spraying, electroplating, and vapor deposition. The latter may involve simply evaporation and condensation of the deposited metal, or may include a chemical reaction between the vapor and the metal to be coated.

***334. Zinc coatings*** are more widely used for the protection of structural iron and steel than coatings of any other type. The hot-dip process is the earliest type known and is very extensively used at the present time; two improvements, the Crapo process and the Herman, or "galvannealed," process, are used in galvanizing wire. The cementation, or sherardizing, process consists in heating the articles for several hours in a packing of zinc dust in a slowly

**TABLE 4-77** Galvanic Series of Alloys in Seawater

| | Alloy |
|---|---|
| ↑ Noble or cathodic | Platinum |
| | Gold |
| | Graphite |
| | Titanium |
| | Silver |
| | ⎡ Chlorimet 3 (62 Ni, 18 Cr, 18 Mo) |
| | ⎣ Hastelloy C (62 Ni, 17 Cr, 15 Mo) |
| | ⎡ 18-8 Mo stainless steel (passive) |
| | ⎢ 18-8 stainless steel (passive) |
| | ⎣ Chromium stainless steel 11–30% Cr (passive) |
| | ⎡ Inconel (passive) (80 Ni, 13 Cr, 7 Fe) |
| | ⎣ Nickel (passive) |
| | Silver solder |
| | ⎡ Monel (70 Ni, 30 Cu) |
| | ⎢ Cupronickels (60–90 Cu, 40–10 Ni) |
| | ⎢ Bronzes (Cu-Sn) |
| | ⎢ Copper |
| | ⎣ Brasses (Cu-Zn) |
| | ⎡ Chlorimet 2 (66 Ni, 32 Mo, 1 Fe) |
| | ⎣ Hastelloy B (60 Ni, 30 Mo. 6 Fe, 1 Mn) |
| | ⎡ Inconel (active) |
| | ⎣ Nickel (active) |
| | Tin |
| | Lead |
| | Lead-tin solders |
| | ⎡ 18-8 Mo stainless steel (active) |
| | ⎣ 18-8 stainless steel (active) |
| | Ni-Resist (high Ni cast iron) |
| | Chromium stainless steel, 13% Cr (active) |
| | ⎡ Cast iron |
| | ⎣ Steel or iron |
| | 2024 aluminum (4.5 Cu, 1.5 Mg, 0.6 Mn) |
| Active or anodic ↓ | Cadmium |
| | Commercially pure aluminum (1100) |
| | Zinc |
| | Magnesium and magnesium alloys |

NOTE: Alloys will corrode in contact with those higher in the series. Brackets enclose alloys so similar that they can be used together safely.

SOURCE: Fontana and Green, *Corrosion Engineering;* New York, McGraw-Hill Book Company.

rotating container. Electroplating is also employed, and heavier coatings can be obtained than are usual with the hot-dip process, but adherence is difficult to obtain, and this process is not often used. See ASTM specifications for zinc-coated iron and steel products.

***335. Aluminum coatings*** are applied by a cementation process which is commercially known as "calorizing." The articles to be coated are packed in a drum in a mixture of powdered aluminum, aluminum oxide, and a small amount of ammonium chloride. The articles are then slowly rotated and heated in an inert atmosphere, usually of hydrogen. Such coatings are very resistant to oxidation and sulfur attack at high temperatures. Aluminum coatings can also be applied by the hot-dipping method and then are heat-treated to improve the alloy bond. Aluminum can also be applied by spraying. Aluminum-coated steel is extensively used for oxidation protection, for example, for heat ducts and automobile mufflers. Aluminum-zinc coatings, applied by hot dipping, have been developed that combine the high temperature protection of aluminum with the sacrificial protection of zinc.

***336. Tin Coatings.*** Almost all tin coatings are now applied by electrolytic deposition methods. The accurate control obtained by electrolytic deposition is important because of the high cost of tin. Unlike zinc, tin is electropositive to iron. The coating must remain intact; once penetrated, corrosion of the iron will be accelerated. If a zinc coating is penetrated, the zinc will still sacrificially protect the adjacent exposed iron. Tin has good corrosion resistance, is nontoxic, readily bonds to steel, is easily soldered, and is extensively utilized by the container industry for food and other substances.

***337. Lead and Lead-Tin Coatings.*** The objective of lead coating of steel is to obtain an inexpensive corrosion-resistant coating. Lead alone will not alloy with iron; it is necessary to add some tin to the lead to obtain a smooth, continuous, adherent coating. Originally, about 25% tin (called "terne metal") was used, but the tin content has been reduced as the price of tin has increased. Since corrosion protection is less effective than with tin or zinc, and the surface is soft and easily scratched, terne-coated steel is not extensively used. Applications include uses where corrosion is not too critical or likely, such as gasoline tanks and roofing sheets, or where the lubricating properties of the soft lead surface helps forming operations.

***338. Metal-spray coatings*** are applied by passing metal wire through a specially constructed spray gun which melts and atomizes the metal to be used as coating. The surface to be sprayed must be roughened to afford good adhesion of the deposited metal. Nearly all the commonly used protective metals can be applied by spraying, and the process is especially useful for coating large members or repairing coatings on articles already in place. Sprayed coatings can also be applied that will resist wear and can be used to build up worn parts such as armature shafts and bearing surfaces or to apply copper coatings to carbon brushes and resistors.

***339. Chromium coatings*** can be applied by cementation or electroplating. In electroplating, the best results are secured by first plating on a base coating of nickel or nickel copper to receive the chromium. The great hardness of chromium gives it important applications for protection against wear or abrasion; it will also take and retain a high polish. Very thin coatings have a tendency to be inefficient as a result of the presence of minute pinholes.

***340. Electroplated Coatings.*** Electroplating is employed in the application of coatings of nickel, brass, copper, chromium, cadmium, cobalt, lead, and zinc. Only cadmium, chromium, and zinc are electronegative to iron. The other metals mentioned are employed because of their own corrosion-resistant properties and because they afford surface finishes having certain desirable characteristics.

***341. Protective paints*** are extensively employed to protect heavily exposed structures of iron and steel, such as bridges, tanks, and towers. The protection is not permanent but gradually wears away under weather exposure and must be periodically renewed. Various specially prepared paints are used for protecting the surface from dampness, oxidizing gases, and smoke. No one paint is suitable for all purposes but the choice depends on the nature of the corrosive influence present. Asphaltum and tar protect the surface by formation of an impervious film. A chemical protective action is exerted by paints containing linseed oil as the vehicle and red lead as the pigment; linseed oil absorbs oxygen from the atmosphere and forms a thick elastic covering, a formation hastened by adding salts of manganese or lead to the oil. All dryers, vehicles, and pigments used in paint must be inert to the steel; otherwise corrosion will be hastened instead of prevented. Graphite- and aluminum-flake pigments give very impermeable films but do not show the inhibitive action of red lead or zinc when the films are scratched. Aluminum has the advantage of reflecting both infrared and ultraviolet rays of the sun; hence it protects the vehicle from a source of deterioration and is used to paint gasoline-storage tanks to prevent excessive heating due to the sun's rays. A large number of new protective coatings have recently been developed from synthetic materials, such as silicones, artificial rubbers, and phenolic plastics. Many of these are tightly adhering compounds in the form of paints or varnishes which offer rather good protection against a wide variety of chemical attacks. The majority of these new coatings, however, are sensitive to abrasion, and many of them must be baked on to secure full effectiveness. (See Reports of Committee D1 in *ASTM Proc.;* also bibliography, Par. **347**.)

***342. Chemical coatings*** which are corrosion-resistant include Parkerizing and Bower-Barff finish. *Parkerizing* consists in immersing the steel in a solution of manganese dihydrogen phosphate, then heating to about the boiling point. The pieces are allowed to remain in the bath until effervescence ceases. After oiling, the surface produced has the appearance of gunmetal. These phosphate coatings are used mainly as bases for finishing enamels and paints. *Bower-Barff finish* consists in heating the pieces in a closed retort to a temperature of 871°C. Alternate injections of superheated steam and carbon monoxide then reduce the oxides formed on the surface to $Fe_3O_4$. The operations may be repeated several times until a sufficient depth of oxide is obtained. Other types of chemical surface which develop increased surface hardness and resistance to corrosion are produced commercially by treating a steel with nitrogen or silicon for prolonged periods of time at high temperature.

***343. Greases and oils of various grades*** are used to protect the surface by applying a thin film to parts of machines, tools, bearings, and steels which are to be put in storage or are to be shipped. These slushing compounds are usually mineral oils, fats and waxes, lanolin, or greases; oil-soluble chromates are sometimes added to aid in preventing corrosion. Silicone oils and greases are particularly effective because of their imperviousness and repellency of moisture.

***344. Corrosion-resistant ferrous alloys*** such as rustless or stainless iron and steel have come into use for both structural and ornamental purposes but on account of their chromium and nickel contents are relatively expensive in comparison with the ordinary structural steels (see Par. 322). Copper-bearing iron and steel, containing about 0.15 to 0.25% copper, are used extensively; the copper content tends to retard corrosion slightly but does not prevent it, and some protective coating is usually necessary. Some structural uses have been made of these steels without applying special protective coatings. A tightly adherent brown oxide surface film forms from weathering to serve as the future "protective coating."

***345. Stainless-Clad Sheets.*** Plates and sheets of steel faced with stainless steels are widely used in chemical processing, paper mills, nuclear reactors, and transportation or storage of corrosive liquids, food products, etc. The stainless-steel layer usually constitutes 10 to 20% of the thickness of the plate, and either one or both sides may be coated, but the surfacing must be strongly bonded to the base metal. One method of production consists in casting two stainless-steel sheets in the center of a steel ingot with a separating material between. The composite "sandwich" ingot is hot-rolled to the desired size and then separated into two sheets each clad on one face with stainless steel. Another method of production is to bond stainless sheet to a mild-steel plate by means of a great number of resistance welds closely spaced all over the sheet. These methods provide a stainless-clad surface on either thick or thin plates without the added expense of a solid stainless-steel sheet.

***346. Copperweld.*** A series of steel products, including wire, wire rope, bars, clamps, ground rods, and nails, that contain a copper-clad surface are made by the Copperweld process. The copper coating is intimately bonded to the steel by pouring a ring of molten copper about a heated steel billet fastened in the center of a refractory mold. The solidified composite ingot is then hot-rolled to bar stock and subsequently cold-drawn to the various wire sizes. The thickness of the copper coating on wire is 10 to 12½% of the wire radius and produces a high-strength steel wire with a resistance to corrosion similar to that of a solid copper wire. Their increased electrical conductivity over that of a solid steel wire or rod makes the Copperweld products suitable for high-strength conductors, ground rods, aerial cable messengers, etc.

***347. References to Technical Literature on Corrosion and Protective Coatings***

Uhlig, H. H. (ed.): *The Corrosion Handbook;* New York, John Wiley & Sons, Inc., 1948.

Evans, U. R.: *An Introduction to Metallic Corrosion,* 3d ed.; London, Arnold, 1981.

*Metal Finishing Guidebook and Directory;* Hackensack, N.J., Metals and Plastics Publications, Inc. Published annually.

Reports of Committee A5, Corrosion of Iron and Steek, *Ann. Proc. ASTM.*

Payne, H. F.: *Organic Coating Technology;* New York, John Wiley & Sons, Inc., 1961.

Fontana, M. G., and Greene, N. D.: *Corrosion Engineering;* New York, McGraw-Hill, 1967.

ASM *Metals Handbook,* 9th ed., vol. 5, 1982.

## Nonferrous Metals and Alloys

***348. Copper.*** Numerous commercial "coppers" are available. The standard product is "tough-pitch" copper, which contains about 0.04% oxygen. If it has been electrolytically refined, it is called "electrolytic tough pitch." This copper cannot be heated in reducing atmospheres because the oxygen will react with hydrogen and severely embrittle the alloy. Various deoxidized varieties are made. When deoxidized with phosphorus, there is some loss of electrical conductivity depending on the amount of residual phosphorus and the extent to which other impurities are reduced and redissolve in the metal. As a general principle, alloying elements that dissolve in copper reduce conductivity sharply; those that are insoluble have little effect.

*Copper castings* are improved by using special deoxidizers such as Boroflux and silicocalcium copper alloy. By the use of these deoxidizers the castings are improved structurally, and the electrical conductivity can be increased to about 80 to 90% of standard annealed copper. Boroflux is a mixture of boron suboxide, boric anhydride, magnesia, and magnesium; for data on its use, see publications of the General Electric Company.

*Oxygen-free high-conductivity copper* is deoxidized with carbon, and thus is free of residual oxide or deoxidizer. It is a more expensive product but does not suffer the potential embrittlement of tough pitch, and is capable of more severe cold-forming operations at the cost of a slight loss of electrical conductivity. Free-machining copper contains lead or tellurium that drops conductivity 3 to 5%. Since copper is a very difficult material to machine, this may be a small sacrifice for certain applications. Small amounts of silver improve resistance to elevated-temperature softening with no loss of physical or mechanical properties.

***349. Brass.*** Brasses are alloys of copper and zinc; commercial brasses contain from 5 to 45% zinc. A wide variety of properties are obtainable in the brasses. In general, the alloys have excellent corrosion resistance, good mechanical properties, colors ranging from red to gold to yellow to white, and are available in a wide variety of cast and wrought shapes. The alloy of 30% zinc has an optimum combination of strength and ductility. It is called "cartridge brass," since an early application was drawing of cartridge shells. It is the most commonly used brass alloy.

*Muntz metal* contains nominally 40% zinc, and is a two-phase alloy that is readily hot-worked in the high-temperature form and develops good strength when cooled. It is used for extruded shapes and for bolts, fasteners, and other high-strength applications. The properties of brass can be modified by small additions of numerous alloying elements; those commonly used include silicon, aluminum, manganese, iron, lead, tin, and nickel. The addition of 1% tin to cartridge brass results in an alloy called Admiralty brass which has very good corrosion resistance and is extensively used in heat exchangers.

Brasses, especially the high zinc-bearing alloys, are subject to a corrosion phenomenon called *dezincification.* It involves a selective loss of zinc from the surface and the formation of a spongy copper layer accompanied by deterioration of mechanical properties. It is more likely to occur with the high-zinc brasses in contact with water containing dissolved $CO_2$ at elevated temperatures. Like many other metals, the brasses are susceptible to stress-corrosion cracking—an embrittlement due to the combined action of stress and a selective corrosive agent. In the case of brass, the particular agent responsible for stress-corrosion cracking is ammonia and its compounds. Brass products that might be exposed to such environments should be stress-relief annealed before being placed in service. For details on compositions and properties of brasses, see the appropriate ASTM specifications and the publications of brass producers.

***350. Bronze, or Copper-Tin, Alloys.*** Bronze is an alloy consisting principally of copper and tin and sometimes small proportions of zinc, phosphorus, lead, manganese, silicon, aluminum, magnesium, etc. The useful range of composition is from 3 to 25% tin and 95 to 75% copper. Bronze castings have a tensile strength of 195 to 345 MPa (28,000 to 50,000 lb/in$^2$), with a maximum at about 18% of tin content. The crushing strength ranges from about 290 MPa (42,000 lb/in$^2$) for pure copper to 1035 MPa (150,000 lb/in$^2$) with 25% tin content. *Cast bronzes* containing about 4 to 5% tin are the most ductile, elongating about 14% in 5 in. Gunmetal contains about 10% tin and is one of the strongest bronzes. Bell metal contains about 20% tin. *Copper-tin-zinc alloy* castings containing 75 to 85% copper, 17 to 5% zinc, and 8 to 10% tin have a tensile strength of 240 to 275 MPa (35,000 to 40,000 lb/in$^2$), with 20 to 30% elongation. *Government bronze* contains 88% copper, 10% tin, and 2% zinc; it has a tensile strength of 205 to 240 MPa (30,000 to 35,000 lb/in$^2$), yield strength of about 50% of the ultimate, and about 14 to 16% elongation in 2 in; the ductility is much increased by annealing for ½ h at 700 to 800°C, but the tensile strength is not materially affected. *Phosphor bronze* is made with phosphorus as a deoxidizer; for malleable products, such as wire, the tin should not exceed 4 or 5%, and the phosphorus should not exceed 0.1%. *United States Navy bronze* contains 85 to 90% copper, 6 to 11% tin, and less than 4% zinc, 0.06% iron, 0.2% lead, and 0.5% phosphorus; the minimum tensile strength is 310 MPa (45,000 lb/in$^2$), and elongation at least 20% in 2 in. *Lead bronzes* are used for bearing metals for heavy duty; an ordinary composition is 80% copper, 10% tin, and 10% lead, with less than 1% phosphorus. *Steam or valve bronze* contains approximately 85% copper, 6.5% tin, 1.5% lead, and 4% zinc; the tensile strength is 235 MPa (34,000 lb/in$^2$), minimum, and elongation 22% minimum in 2 in (ASTM Specification B61). The bronzes have a great many industrial applications where their combination of tensile properties and corrosion resistance is especially useful.

***351. Beryllium-copper alloys*** containing up to 2.75% beryllium can be produced in the form of sheet, rod, wire, and tube. The alloys are hardenable by a heat-treatment consisting of quenching from a dull red heat, followed by reheating to a low temperature to hasten the precipitation of the hardening constituents. Depending somewhat on the heat-treatment, the alloy of 2.0 to 2.25% beryllium has a tensile strength of 415 to 650 MPa (60,000 to 193,000 lb/in$^2$), elongation 2.0 to 10.0% in 2 in, modulus of elasticity 125 × GPa (18 × $10^6$ lb/in$^2$), and endurance limit of about 240 to 300 MPa (35,000 to 44,000 lb/in$^2$). An outstanding quality of this alloy is its high endurance limit and corrosion resistance; it can be hardened by heat-treatment to give great wear resistance and has high electrical conductivity. Typical applications include nonsparking tools for use where serious fire or explosion hazards exist and many electrical accessories such as contact clips and springs or instrument and relay parts. Beryllium is a toxic substance, and care should be taken to avoid ingesting airborne particles during such operations as machining and grinding. For details on properties and uses of beryllium bronzes, see publications of Kawecki Berylco Industries, New York.

***352. General References on Copper and Copper Alloys***

Mendenhall, J. H.: *Understanding Copper Alloys,* Olin Corp, New York, 1977.

*Copper and Its Alloys;* London, Institute of Metals, 1970.

*Bridgeport Copper Metals Handbook;* Bridgeport, Conn., Bridgeport Brass Company, 1964.

*Source Book on Copper and Copper Alloys,* Am. Soc. for Metals, 1979.

Flinn, R. A.: *Copper, Brass, and Bronze Castings,* Cleveland, NonFerrous Founders Society, 1963.

***353. Nickel*** is a brilliant metal which approaches silver in color. It is more malleable than soft steel and when rolled and annealed is somewhat stronger and almost as ductile. The tensile strength ranges from 415 MPa (60,000 lb/in$^2$) for cast nickel to 795 MPa (115,000 lb/in$^2$) for cold-rolled full-hard strip; yield strength 135 to 725 MPa (20,000 to 105,000 lb/in$^2$); elongation in 2 in, 2% when full hard to about 50% when annealed; modulus in tension about 205 GPa (30 × $10^6$ lb/in$^2$). Nickel takes a good polish and does not tarnish or corrode in dry air at ordinary temperatures. It has various industrial uses in sheets, pipes, tubes, rods, containers, and the like, where its corrosion resistance makes it especially suitable.

The greatest tonnage use of nickel is as an alloying element in steels, principally stainless and heat-resisting steels. There are also a variety of copper-nickel alloys whose main applications are based on their excellent corrosion resistance, for example, condenser tubes. Additions of aluminum and titanium to nickel-base alloys result in age-hardening characteristics, and they can be heat-treated to exceptionally high strengths that are retained to high temperatures. The International Nickel Company publishes an extensive list of bulletins describing the characteristics of nickel and nickel alloys.

***354. Monel metal*** is a silvery-white alloy containing approximately 66 to 68% nickel, 2 to 4% iron, 2% manganese, and the remainder copper. It can be cast, forged, rolled, drawn, welded, and brazed and is easily machined. It melts at 1360°C and has a density of 8.80, coefficient of expansion of 14 × $10^{-6}$ per degree Celsius, thermal conductivity of 0.06 cgs unit, specific heat of 0.127 cal/(g)(°C), and modulus of 175 GPa (25 × $10^6$ lb/in$^2$). The tensile strength ranges from 450 MPa (65,000 lb/in$^2$) for cast monel metal to 860 MPa (125,000 lb/in$^2$) in cold-rolled full-hard strip; yield strength 175 to 725 MPa (25,000 to 115,000 lb/in$^2$). It is highly resistant to corrosion and the action of seawater or mine waters. The industrial uses for it include many applications where its combination of physical properties and corrosion resistance gives it special advantages. (See technological data published by the International Nickel Co.)

***355. Magnesium Alloys.*** The outstanding feature of magnesium alloys is their light weight (specific gravity of about 1.8). Alloys containing thorium and rare-earth additions have been developed that retain good strength at temperatures between 260 and 371°C. The correspondingly high strength/weight ratio makes them particularly useful to the aircraft industry. Less exotic alloys, based mainly on alloying with aluminum (up to 10%) and zinc (up to 6%) still have excellent strengths, and are heat-treatable. These alloys have many uses where low density is desired: portable tools, ladders, structural members for trucks and buses, housings, etc. Magnesium alloys are available as castings, forgings, extrusions, and rolled mill products in a variety of shapes. Their thermal coefficient of expansion is about 0.000029/°C, and melting point about 620°C. Tensile strengths of castings range from 145 to 235 MPa (21,000 to 34,000 lb/in$^2$), yield strengths from 62 to 150 MPa (9,000 to 22,000 lb/in$^2$), and elongation from 1 to 10% in 2 in. Forged or extruded alloys have tensile strengths of 225 to 300 MPa (33,000 to 43,000 lb/in$^2$), yield strengths 125 to 205 MPa (18,000 to 30,000 lb/in$^2$), and elongations of 5 to 17% in 2 in. The Brinell hardness ranges from 35 to 78, and the endurance limit from 40 to 115 Mpa (6,000 to 17,000 lb/in$^2$) depending on the alloy and heat-treatment. Since magnesium is highly anodic to other common metals, care must be taken in designing with this metal. Protective coatings are used, and care must be taken to avoid forming galvanic couples. Finely divided magnesium will burn but massive sections are safely melted and welded. See *Magnesium Alloys and Products,* Dow Chemical Company, Midland, Mich.

***356. Lead*** is a heavy, soft, malleable metal with a blue-gray color; it shows a metallic luster when freshly cut, but the surface is rapidly oxidized in moist air. It can be easily rolled into thin sheets and foil or extruded into pipes and cable sheaths but cannot be drawn into fine wire. Although in an ordinary tensile test lead may develop a tensile strength of 17 MPa (2400 lb/in$^2$), it may creep at ordinary room temperatures at stresses as low as 0.34 MPa (50 lb/in$^2$). Owing to this tendency to creep, it may fracture under long-continued load at stresses as low as 5.5 MPa (800 lb/in$^2$), and the ordinary static tensile properties do not have much significance. The resistance of pure lead to corrosion makes it useful in the form of sheets, pipes, and cable coverings, and large quantities of lead are used in the manufacture of various alloys, particularly in alloys for bearings. Common alloys of lead for cable sheathings contain (approximately) 0.04% Cu, 0.75% Sb, or 0.03% Ca. The greatest use of metallic lead is in the manufacture of storage batteries. See ASM *Metals Handbook,* 9th ed., vol. 2, 1979.

***357. Tin*** is a silvery-white, lustrous metal, very soft and malleable and of very low tensile strength. It has a density of about 7.3 and melts at 232°C. In ductility it equals soft steel. The tensile strength varies with the speed of testing. As a metal it has few uses except in sheets, but large quantities of it are used in various industrial alloys. Its chief uses are in tin- and terneplate, solder, babbitt and other bearing metals, brass, and bronze. Tin is very resistant to atmospheric corrosion, and water hardly affects it at all; however, it is electronegative to iron and therefore is not an efficient protective coating under atmospheric exposures. See Symposium on Tin, *ASTM Spec. Tech. Publ.* 141, 1953; ASM *Metals Handbook,* 9th ed., vol. 2, 1979.

***358. Bearing Metals.*** Bearing metals are designed to serve as sleeve bearings—to carry loads between surfaces undergoing relative motion. Traditionally, the metallic structure best suited for such service is one in which hard particles are embedded in a soft matrix. The hard particles provide wear resistance and carry the load, while the soft matrix gives conformability to variations in dimensions due to manufacturing tolerances and deflections due to the load, and also embed abrasive particles that otherwise might score the shaft. *White metal or Babbitt bearings* have served this function successfully for years. These are tin- or lead-base alloys containing additions of antimony and copper to form hard precipitate particles in the soft matrix.

Tin babbitts are more expensive than lead, but have better corrosion resistance. In more recent years, other materials have been developed that improve on one or another of the deficiencies of the white metals: low elevated-temperature strength, low thermal conductivity, corrosion resistance, and fatigue strength. Copper-, cadmium-, and aluminum-base alloys in particular are now being used for many applications.

Lead is a common addition to these alloys, and provides softness, conformability, embeddability, and self-lubricating properties. Sintered copper-tin powder-metallurgy bearings are also extensively used as "oilless" bearings for sealed motors. They are made with a high proportion of porosity, then impregnated with oil. In use, they heat up and the oil flows out to the shaft. When they cool, the oil is drawn back into the pores ready for the next period of service. See ASTM Specifications B23 for properties and compositions of babbitt bearings. For general study of bearing materials, see *Bearings;* Cleveland, ASM, 1969.

***359. Zinc*** is a bluish-white metal, which has a metallic luster on a new fracture. The density of cast zinc ranges from 7.04 to 7.16. At ordinary temperature it is brittle, but in the range of about 100 to 150°C it becomes malleable and can be rolled into sheets and drawn into wire. At 200°C it becomes so brittle that it can be pulverized. The tensile strength of cast zinc ranges from about 55 to 95 MPa (8000 to 14,000 lb/in$^2$) in an ordinary testing-machine test, and that of drawn zinc from about 150 to 200 MPa (22,000 to 30,000 lb/in$^2$); it has a poorly defined proportional limit of about 35 MPa (5000 lb/in$^2$) and exhibits a certain amount of creep at room temperatures; hence it may fracture in service under constant stresses below its testing-machine strength. It strongly resists atmospheric corrosion but is readily attacked by acids. The principal industrial uses for it are for galvanizing iron and steel, for plates and sheets for roofing and other applications, and for alloying with copper, tin, and other metals; very large quantities are used in the various types of brass. Next to galvanizing, the greatest use of zinc is in the production of die castings. Because of its moderate melting point, good mechanical properties, and especially because it does not attack steel melting pots and dies, it is the most popular die-casting material (although closely rivaled by aluminum). Zinc alloys for die casting contain some aluminum, copper, and magnesium; all ingredients must be very pure or the casting will have poor corrosion resistance and dimensional stability. See Morgan, S.: *Zinc and Its Alloys;* Plymouth, Macdonald & Evans, 1977; and publications of the New Jersey Zinc Company, New York.

***360. Titanium and Titanium Alloys.*** Titanium alloys are important industrially because of their high strength-weight ratio, particularly at temperatures up to 427°C. The density of the commercial titanium alloys ranges from 4.50 to 4.85 g/cm$^3$, or approximately 70% greater than aluminum alloy and 40% less than steel. The purest titanium currently produced (99.9% Ti) is a soft, white metal. The mechanical strength increases rapidly, however, with an increase of the impurities present, particularly carbon, nitrogen, and oxygen. The commercially important titanium alloys, in addition to these impurities, contain small percentages (1 to 7%) of (1) chromium and iron, (2) manganese, and (3) combinations of aluminum, chromium, iron, manganese, molybdenum, tin, or vanadium. The thermal conductivity of the titanium alloys is low, about 15 W/m·K at 25°C, and the electrical resistivity is high, ranging from 54 $\mu\Omega \cdot$ cm for the purest titanium to approximately 150 $\mu\Omega \cdot$ cm for some of the alloys. The coefficient of thermal expansion of the titanium alloys varies from 2.8 to 3.6 $\times 10^{-6}$ per degree Celsius, and the melting-point range is from 1371 to 1704°C for the purest titanium. The tensile modulus of elasticity varies between 100 to 120 GPa (15 to 17 $\times 10^6$ lb/in$^2$). The mechanical properties, at room temperature, for annealed commercial alloys range approximately as follows: yield strength 760 to 965 MPa (110,000 to 140,000 lb/in$^2$); ultimate strength 800 to 1100 MPa (116,000 to 160,000 lb/in$^2$); elongation 5 to 18%; hardness 300 to 370 Brinell. On the basis of the strength-weight ratio many of the titanium alloys exhibit superior short-time tensile properties as compared with many of the stainless and heat-resistant alloys up to approximately 427°C. However, at the same stress and elevated temperature, the creep rate of the titanium alloys is generally higher than that of the heat-resistant alloys. Above about 482°C the strength properties of titanium alloys decrease rapidly. The corrosion resistance of the titanium alloys in many media is excellent; for most purposes, it is the equivalent or superior to stainless steel. See *Titanium and Titanium Alloys;* ASM Source Book, American Society for Metals, 1982.

***361. Mechanical Properties of Nonferrous Metals and Alloys.*** See Table 4-78.

***362. Creep Stress for Nonferrous Metals.*** Figures 4-72 and 4-73 show for several metals the approximate creep strengths to produce a given creep rate per 1000 h. Creep characteristics are influenced by many factors, such as melting practice and grain size as well as chemical composition. Materials of the same composition will not necessarily have the same creep rate. Test data should therefore be regarded only as qualitative.

***363. Aluminum*** is an important commercial metal possessing some very unique properties. It is very light (density about 2.7) and some of its alloys are very strong, so its strength-weight ratio makes it very attractive for aeronautical uses and other applications in which weight saving is important. Aluminum, especially in the pure form, has very high electrical and thermal conductivities, and is used as an electrical conductor in heat exchangers, etc. Aluminum has good corrosion resistance, is nontoxic, and has a pleasing silvery white color; these properties make it attractive for applications in the food and container industry, architectural, and general structural fields.

Aluminum is very ductile and easily formed by casting and mechanical forming methods. Aluminum owes its good resistance to atmospheric corrosion to the formation of a tough, tenacious, highly insulating, thin oxide film, in spite of the fact that the metal itself is very anodic to other metals. In moist atmospheres, this protective oxide may not form, and some caution must be taken to maintain this film protection. Although aluminum can be joined by all welding processes, this same oxide film can interfere with the formation of good bonds during both fusion and resistance welding, and special fluxing and cleaning must accompany welding operations.

**TABLE 4-78** Approximate Mechanical Properties of Miscellaneous Nonferrous Metals and Alloys
(Compiled from Various Authorities)

| Metal | Condition | Approximate composition, % | Strength in tension,[a] 1000 lb/in²* Yield[b] | Strength in tension,[a] 1000 lb/in²* Ultimate | Endurance limit,[c] 1000 lb/in²* | Brinell hardness no. | Elongation, 2 in, % | Weight, lb/in³† |
|---|---|---|---|---|---|---|---|---|
| Aluminum: | | | | | | | | |
| EC-O | Wrought, annealed | 99.60% min. Al | 4 | 10 | [d] | [d] | 23[k] | 0.098 |
| EC-H19 | Wrought, extra hard | 99.60% min. Al | 24 | 27 | 7 | [d] | 1.5[k] | 0.098 |
| 1199-O | Wrought, annealed | 99.99 Al | 1.5 | 6.5 | [d] | [d] | 50 | 0.098 |
| 1100-H14 | Wrought, half hard | 99.0% min. Al | 17 | 18 | 7 | 32 | 20 | 0.098 |
| 2024-T4 | Quenched | Al 93, Cu 4.5, Mn 0.6, Mg 1.5 | 47[i] | 68[i] | 20 | 120 | 19 | 0.100 |
| 5052-H34 | Wrought, half hard | Al 97, Mg 2.5, Cr 0.25 | 31 | 38 | 18 | 68 | 16 | 0.097 |
| 6061-T6 | Quenched and aged | Al 98, Mg 1, Si 0.6, Cu 0.25, Cr 0.25 | 40 | 45 | 14 | 95 | 17 | 0.098 |
| 6063-T5 | Quenched and aged | Al 99, Mg 0.7, Si 0.4 | 21 | 27 | 17 | 60 | 22 | 0.097 |
| 6101-T6 | Quenched and aged | Al 99, Mg 0.6, Si 0.5 | 28 | 32 | 9 | 71 | 20 | 0.097 |
| 7075-T6[j] | Quenched and aged | Al 90, Zn 5.6, Mg 2.5, Cu 1.6, Cr 0.3 | 73 | 83 | 22 | 150 | 11 | 0.101 |
| A356-T61 | Permanent mold, quenched and aged | Al 9, Si 7, Mg 0.3 | 30 | 41 | 13 | 90 | 10 | 0.097 |
| Brass | Annealed, cold-drawn | Cu 60; Zn 40 | 18 | 54 | 22 | 72 | 56 | 0.30 |
| | | Cu 60; Zn 40 | 49 | 97 | 26 | 179 | 13 | 0.30 |
| Bronze | Annealed, cold-drawn | Cu 95; Sn 5 | 13 | 46 | 23 | 74 | 67 | 0.32 |
| | | Cu95; Sn 5 | 59 | 85 | 27 | 166 | 12 | 0.32 |
| Bronze, phosphor | Rolled sheet | Sn 8; Zn 0.2; P 0.1, Cu 91 | [d] | 50–100 | [d] | 35–90[g] | 5–50 | [d] |
| Bronze, aluminum | As rolled | Cu 88; Al 9; Fe 3 | 28 | 70 | [d] | 125 | 30 | 0.29 |

**TABLE 4-78** Approximate Mechanical Properties of Miscellaneous Nonferrous Metals and Alloys (*Continued*)
(Compiled from Various Authorities)

| Metal | Condition | Approximate composition, % | Strength in tension,[a] 1000 lb/in²* Yield[b] | Strength in tension,[a] 1000 lb/in²* Ultimate | Endurance limit,[c] 1000 lb/in²* | Brinell hardness no. | Elongation, 2 in, % | Weight, lb/in³† |
|---|---|---|---|---|---|---|---|---|
| Bronze, manganese | Cast | Cu 57, Zn 41, +Mn, Al, Fe & Sn | 30 | 70 | 17 | 115 | 33 | 0.30 |
| Copper | Annealed | Commercially pure | 5 | 32 | 10 | 47 | 56 | 0.32 |
| Copper alloy | Wire | Cu 95; +Si, Sn, Al, Fe | [d] | 68–120 | [d] | [d] | 0.8–3.8[e] | 0.32 |
| Copper-silicon alloy | Half hard | Cu 94, Si 3, +Mn, Zn, & Fe | 40–50 | 71–81 | [d] | 80–90[g] | 10–20 | 0.32 |
| Copper, beryllium | Heat-treat, half hard | Be 2, Ni 0.5, Fe 0.2, Cu 97 | 132 | 173 | 35 | 340 | 4.8 | 0.297 |
| Gunmetal | Cast | Cu 88, Sn 10, Zn 2–4 | 14–20 | 30–45 | [d] | 50–80 | 15–40 | 0.314 |
| Magnesium alloy: | | | | | | | | |
| AZ31C | Sheet, hard | Mg 96, Al 3, Zn 1, Mn 0.2 | 32 | 42 | 14 | 73 | 15 | 0.064 |
| AZ80A-T5 | Extrusions, aged | Mg 91, Al 8.5, Zn 0.5, Mn 0.2 | 39 | 54 | 19 | 82 | 7 | 0.065 |
| ZK60A | Extrusions, aged | Mg 94, Zn 5.5, Zr 0.6 | 43 | 52 | 16 | 82 | 12 | 0.066 |

| | | | | | | | | |
|---|---|---|---|---|---|---|---|---|
| AZ63A-T6 | Sand cast, quenched, aged | Mg 91, Al 6, Zn 3, Mn 0.2 | 19 | 40 | 17 | 73 | 5 | 0.067 |
| AZ92A-T6 | Sand cast, quenched and aged | Mg 89, Al 9, Zn 2, Mn 0.1 | 23 | 40 | 13 | 84 | 2 | 0.066 |
| Monel metal (B) | Cold-drawn | Ni 67, Cu 30, +Fe & Mn | 60–&95 | 85–125 | [d] | 160–220 | 35–15 | 0.318 |
| | Annealed | | 30–40 | 70–85 | | 120–160 | 50–35 | |
| Nickel | Cast | Ni 97, Si 1.2, +Mn & Fe | 21 | 55 | [d] | 120 | 22 | 0.319 |
| | Hot-rolled | Ni 99, +Fe, Cu, Mn, C, & Co | 27 | 73 | 33 | 109 | 46 | 0.319 |
| Zinc | Die-cast | Zn 95, Al 4, Cu 1 | [f] | 45[h] | [d] | 70–85 | 3 | 0.242 |
| | Rolled sheet | Zn 99.9 | [f] | 16–19[h] | [d] | [d] | 40-70 | 0.242 |

*1 lb/in$^2$ = 6.895 kPa.
†g/c$^3$ = 27.68 × lb/in$^3$.
[a]The yield strength in compression may be usually assumed approximately equal to the yield strength in tension; the ultimate in compression may be considered slightly above the yield strength; the strength in shear may be considered about 0.6 the strength in tension.
[b]Yield strength determined as stress corresponding to approximately 0.2% permanent set.
[c]For completely reversed cycles of flexural stress; polished specimens.
[d]Test data lacking.
[e]In 152 cm (60 in).
[f]Not clearly defined owing to creep at low loads.
[g]Rockwell B scale hardness.
[h]Values for ordinary testing machine tests; much lower values probable for long-time tests under steady load.
[i]The strengths of extrusions more than about 20 cm (¾ in) thick will be 15 to 20% higher.
[j]These values are for other products than extrusions, which will have strengths 8 to 10% higher.
[k]Value for wire in 254-cm (10-in) gage length.

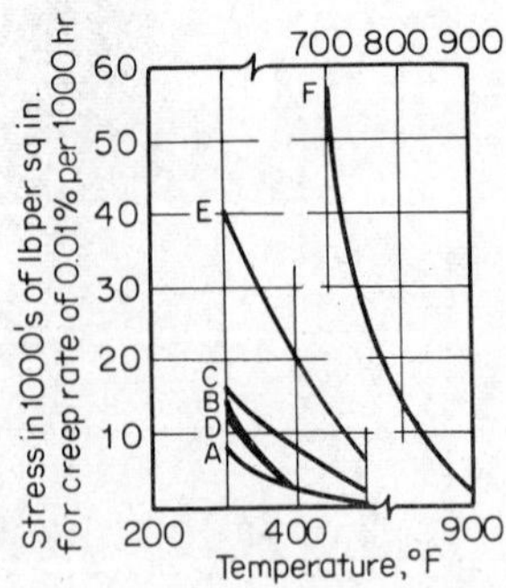


**FIG. 4-72** Creep stress for a creep rate of 0.01% per 1000 h for several nonferrous metals. *A.* Copper, deoxidized and annealed, grain size 0.013 mm. *B.* Aluminum brass (76% Cu, 22% Zn, and 2% A.), annealed grain size 0.015 mm. *C.* 70-30 brass, annealed, grain size 0.016 mm. *E.* Copper-nickel-phosphorus-tellurium alloy, quenched and aged (98.1% Cu, 1.11% Ni, 0.28% P, 0.51% Te). *F.* Monel metal, cold-drawn 40%.

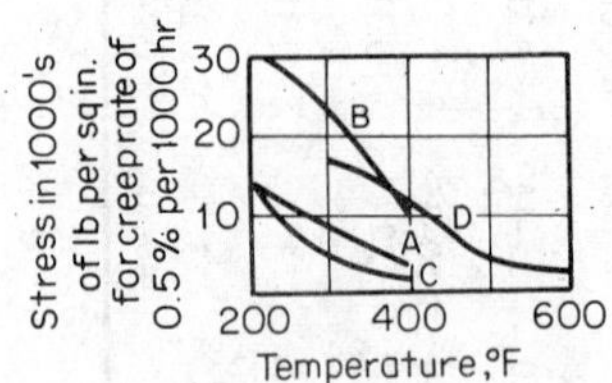


**FIG. 4-73** Creep stress for a creep rate of 0.5% per 1000 h for several aluminum and magnesium alloys. *A.* 3003-H14 aluminum alloy. *B.* 6061-T6 aluminum alloy (see Par. **365**). *C.* AZ-63 HTS magnesium alloy (5.3% to 6.7% Al, 2.5% to 3.5% Zn, 0.15% Mn). *D.* EM 51 HTA magnesium alloy (1.2% Mn, 3.8 to 6.2% Ce).

Commercially pure aluminum (99+%) is very weak and ductile: tensile strength of 90 MPa (13,000 lb/in$^2$), yield strength of 34.5 MPa (5000 lb/in$^2$), and shearing strength of 62 MPa (9500 lb/in$^2$). Extra-pure grades (electrical conductor grade) are 99.7+% pure, and are even weaker, but have better conductivity. A very comprehensive review of production, properties, processing, design, and applications of aluminum is given in a three-volume series, *Aluminum,* edited by K. R. Van Horn, Cleveland, American Society for Metals, 1966.

***364. Structural aluminum alloys*** are alloys of aluminum with relatively small additions of other elements, including copper, manganese, silicon, chromium, and zinc. Very small amounts of boron, titanium, and zirconium are sometimes added for grain-size control. The approximate compositions of alloys frequently used for structural applications are listed in Table 4-79 (see ASTM Specifications B209 and B211). The aluminum industry uses a four-part designation system to identify wrought alloys. The first number indicates the alloy type (pure aluminum, copper-bearing, etc.), the second number identifies alloy modifications, and the last two digits indicate the purity of the specific alloy. Following these four digits is a letter to show the temper of the alloy: O means annealed, F mean as fabricated, H means the alloy has been cold-worked, and T means the alloy has been heat-treated. Numbers following these letters specify the details of the treatment—the extent of cold working, or the type of heat-treatment.

***365. Heat-Treatment of Aluminum Alloys.*** Alloys of the 1000, 3000, and 5000 series cannot be hardened by heat-treatment. They can be hardened by cold working and are available in annealed (recrystallized) and cold-worked tempers. The 5000-series alloys are the strongest non-heat-treatable alloys, and are frequently used where welding is to be employed, since welding will generally destroy the effects of hardening heat-treatment. The remaining wrought alloys can be hardened by controlled precipitation of alloy phases. The precipitation is accomplished by first heating the alloy to dissolve the alloying elements, followed by quenching to retain the alloy in supersaturation. The alloys are then "aged" to develop a controlled size and distribution of precipitate that produces the desired level of hardening. Some alloys naturally age at room temperature; others must be artificially aged at elevated temperatures. Table 4-80 specifies recommended heat-treatments for hardening selected representative alloys. Additional hardening can be obtained in these alloys by cold working the quenched alloy before the aging begins in order to realize the combined strengthening of the two hardening mechanisms. The alloy additions that produce hardening have a deleterious effect on corrosion resistance; the phases introduced by alloying are generally cathodic to the metal, and set up galvanic corrosion cells. To protect high-strength alloys, they are commonly clad with a higher-purity aluminum alloy. The clad composition

**TABLE 4-79** Nominal Compositions of Structural Aluminum Alloys
(Aluminum Company of America)

| Alloy designation | Other elements added to aluminum, % | | | | | |
|---|---|---|---|---|---|---|
| | Copper | Manganese | Magnesium | Silicon | Chromium | Zinc |
| Wrought: | | | | | | |
| 1100 | (Minimum 99.0 aluminum) | | | | | |
| 3003 | .... | 1.2 | | | | |
| 3004 | .... | 1.2 | 1.0 | | | |
| 2014 | 4.4 | 0.8 | 0.5 | 0.8 | | |
| 2024 | 4.5 | 0.6 | 1.5 | | | |
| 5052 | .... | .... | 2.5 | ... | 0.25 | |
| 6061 | 0.28 | .... | 1.0 | 0.6 | 0.2 | |
| 6063 | .... | .... | 0.7 | 0.4 | | |
| 6070 | 0.28 | 0.7 | 0.8 | 1.4 | | |
| 7075 | 1.6 | .... | 2.5 | ... | 0.23 | 5.6 |
| Cast: | | | | | | |
| A443 | 0.3 max | .... | .... | 5.2 | | |
| A356 | 0.25 max | 0.35 max | 0.32 | 7.0 | | 0.35 max |
| 380 | 3.5 | .... | .... | 8.5 | | |

**TABLE 4-80** Age-Hardening Heat-Treatment of Aluminum Alloys

| Alloy designation | Heat-treating temperature, deg F | Approx duration of heating,* min | Quench† | Aging temperature, deg F | Time of aging |
|---|---|---|---|---|---|
| 2014 | 930–950 | 15–60 | Water | 340 | 10 hr |
| 2024 | 910–930 | 15–60 | Cold water | Room | 4 days‡ |
| 6061 | 980-990 | 15–60 | Cold water | 315–325 | 18 hr |
| 7075 | 870-900 | 15–60 | Cold water | 250 | 24 hr |

*This depends on the size and amount of material. In some cases, even longer times may be needed.
†The quench should be made with minimum time loss in transfer from furnace.
‡More than 90% of the maximum properties are obtained during the first day of aging.
NOTE: $t_{°C} = t_{°F} - 32)/1.8$.

should be designed to be anodic to the base metal in order to provide sacrificial protection in case the clad is penetrated and the core metal exposed.

## Stone, Brick, Concrete, and Glass Brick

***366. Building Stone.*** Stone is any natural rock deposit or formation of igneous, sedimentary, and/or metamorphic origin, in either its original or its altered form. Building stone is the quarried product of such deposit or formation which is suitable for structural and ornamental purposes. Igneous or volcanic rock, such as granite or basalt, is rock of plutonic or volcanic origin, formed from a fused condition and crystalline in structure. Sedimentary rock, such as limestone, dolomite, and sandstone, is formed by the deposition of particles from water and laminated in structure. Metamorphic rock, such as gneiss, marble, and slate, is rock formation which, in the natural ledge, has undergone marked change in microstructure or character due to heat, pressure, or moisture and therefore exists in form different from the original. (See ASTM Designation C119 for additional definitions.)

***367. Weight and Strength of Stone.*** The properties of stone from different quarries vary over a considerable range. Average values, from Everett: *Materials;* New York, John Wiley & Sons, Inc., 1978, are listed in Table 4-81. These values should be used only with great caution; those of limestones and sandstones particularly depend on the composition and location of the deposit. (See Mantell, C. L., *Engineering Materials Handbook,* New York, McGraw-Hill, 1958.)

***368. Building brick*** made from clay or shale is required to have certain physical properties (Table 4-82) under ASTM Specification C62-58: for sand-lime building brick see ASTM Specification C73-51. (Also see Miner, D. F., and Seastone, J. B.: *Handbook of Engineering Materials;* New York, John Wiley & Sons, Inc., 1955.)

***369. Clay firebrick*** for stationary and marine boiler service is covered by ASTM Standard Specification C64-61. *Refractory brick* for resisting high temperatures and the effects of very hot gases and molten slag and clinker is made of various special compositions known as "firebrick," "silica brick," "magnesia brick," "bauxite brick," "chromite brick," etc. (See ASTM specifications for brick and clay products; also see Dagostino, F. R.: *Materials of Construction;* Englewood Cliffs, N.J., Prentice-Hall, 1982.)

**TABLE 4-81** Summary of Properties of Building Stone

| | Density, kg/m³ | Failing stress in compression, N/mm³ | Thermal movement, mm/m per 90°C, % approx. | Moisture movement, mm/m for dry-wet change |
|---|---|---|---|---|
| Granites | 2560–3200 | 105<br>335 | 0·93 | None |
| Sandstones | 2130<br>2750 | 27·5<br>195 | 1·0 | Approx. 0·7 |
| Limestones | 1950–2400 | 16·5<br>42·5 | 0·25<br>(porous limestone)<br>0·34<br>(dense limestone) | 0·8<br><br>Negligible |
| Slates | 2800–3040 | 42·5<br>216 | 0·93 | Negligible |
| Marbles | 2880 | | 0·34 | Negligible |
| Quartzites | 2630 | | 0·90 | None |

**TABLE 4-82** Physical Properties of Building Bricks

| Designation | Minimum compressive strength (brick flatwise), lb per sq in., average, gross area | | Maximum water absorption by 5-hr boiling, % | | Maximum saturation coefficient* | |
|---|---|---|---|---|---|---|
| | Average of 5 bricks | Individual | Average of 5 bricks | Individual | Average of 5 bricks | Individual |
| Grade SW | 3000 | 2500 | 17 | 20 | 0.78 | 0.80 |
| Grade MW | 2500 | 2200 | 22 | 25 | 0.88 | 0.90 |
| Grade NW | 1400 | 1250 | No limit | No limit | No limit | No limit |

*The saturation coefficient is the ratio of absorption by 24-hr submersion in cold water to that after 5-hr submersion in boiling water.

***370. Structural gypsum products*** include gypsum tile, plaster board, wallboard, and plain or reinforced members cast in place. (For data on physical properties and specifications see ASTM specifications for gypsum and various gypsum products, and *Testing and Inspection of Engineering Materials,* McGraw-Hill, 1964.)

***371. Portland cement*** is produced by sintering a proportional mixture of lime and clay, which is subsequently ground with the addition of gypsum (to retard the rate of setting). The properties of the clay and limestone determine the principal characteristics: fineness, soundness, time of set, and strength. Strength is measured from briquettes made of a mortar of 1 part cement to 3 parts sand, and measured under specified conditions. Minimum tensile strengths are 2 MPa (275 lb/in$^2$) at 7 days, and 2.6 MPa (350 lb/in$^2$) at 28 days. See ASTM Specs. C109, C115, C151, C170, C191, and C126.

***372. Mortar*** is a mixture of sand, screenings, or similar inert particles, with cement and water, which has the capacity of hardening into a rocklike mass. The inert particles are usually less than ¼ in in size. The proportions of cement to sand range all the way from 1:0 to 1:4 for various purposes.

***373. Concrete*** is a mixture of crushed stone, gravel, or similar inert material with a mortar. The maximum size of inert particles is variable but usually less than 2 in. These inert constituents of mortar and concrete are known as the "aggregate." In making good concrete, the properties of the aggregate are as important as those of the cement. The fine aggregates consist of sand, screenings, mine tailings, pulverized slag, etc., with particle sizes less than ¼ in; the coarse aggregates consist of crushed stone, gravel, cinders, slag, etc. Rubble concrete is made by embedding a considerable proportion of boulders or stone blocks in concrete. The proportions by volume of cement, sand, and coarse aggregate range all the way from 1:1:2 for high compressive strength to 1:4:8 for structures requiring mass more than strength. In general, the strength of concrete increases with the density and richness of mix (proportion of cement) but is decreased in proportion to the amount of mixing water that is added beyond that required to produce a plastic workable mixture. In controlling the quality of a concrete of given mix, the ratio of the volume of mixing water to the volume of cement (water-cement ratio) is often used as a criterion of the strength. For proper curing the concrete should be kept moist for at least a week after placing, and care should be taken to prevent its freezing in cold weather during the early stages of curing. Freshly poured concrete gains strength very slowly in cold weather. Various admixtures are often added in small amounts to modify pouring characteristics and setting time as well as physical characteristics such as resistance to freezing and thawing cycles, wear, abrasion, and permeability. (For systematic information see publications of the Portland Cement Association and ASTM Standards.)

***374. Compressive Strength of Concrete.*** For concrete of common proportions cured under good conditions, 25 to 40% of the 2-year strength is developed in 7 days, 50 to 65% in 1 month, and 70 to 90% in 6 months. The tensile strength of concrete is very low (about one-tenth its compressive strength), and hence in structural members the concrete is usually designed to resist the compressive stresses only, the tensile strength of the concrete being considered negligible. For flexural members, steel reinforcing bars are usually inserted on the tensile side of the beam to resist the tensile stresses. The curves in Fig. 4-74 show typical variations of the compressive strength of concrete with the mix, density, and amount of mixing water used. The mix in each case is given as the proportion of cement to the total volume of aggregate used, and the water-cement ratio is plotted as the number of gallons of water per sack of cement. (One sack is assumed to be 1 ft$^3$ of cement.)

***375. Flexural Strength of Concrete.*** Slabs and pavements are often designed on the basis of the flexural strength of the concrete. Typical variation of the modulus of rupture with the compressive strength of concrete for a number of different mixes and water-cement ratios is shown in Fig. 4-75. This curve is adapted from data compiled by the Portland Cement Association on beams 7 in deep and 10 in wide, loaded with equal loads at the ⅓ points of a 36-in span.

***376. Glass bricks*** are not primarily intended as load-bearing structural units but are used mainly for partition walls and outside walls of buildings in which the main loads are carried by other structural units. Safe working stresses in compression (obtained from data published by the manufacturers) would probably be less than about 3 MPa (400 lb/in$^2$). The blocks are ribbed in various ways to transmit a large proportion of the incident light with almost complete diffusion to eliminate glare. The interior of these blocks contain a thoroughly dry air under partial vacuum, and they are therefore excellent thermal insulating materials for wall construction.

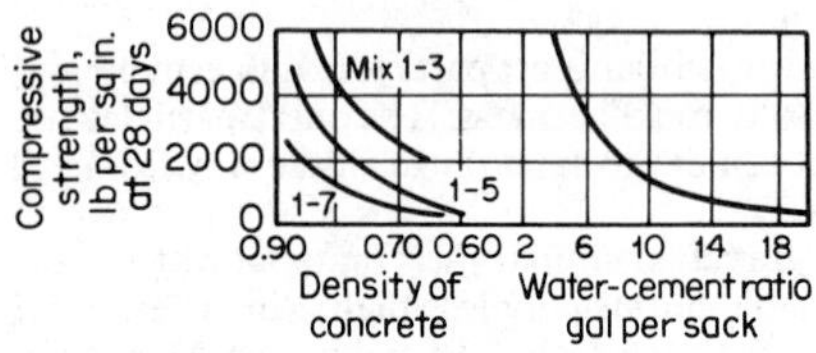


**FIG. 4-74** Typical variations of strength of concrete with density and water-cement ratio.

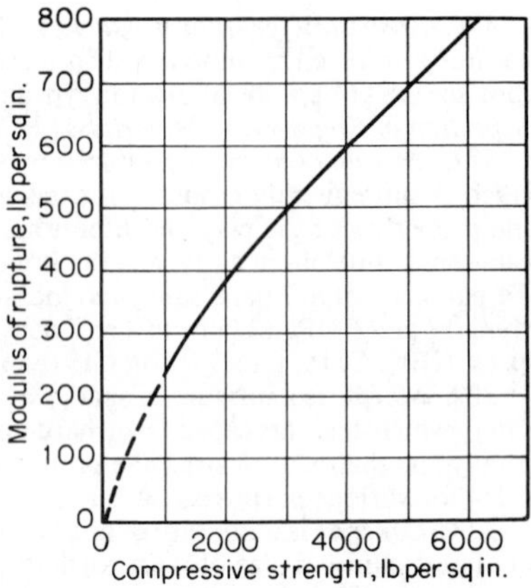


**FIG. 4-75** Average relation of modulus to rupture to compressive strength of concrete. Mix 1:4 by volume. Age of test, grading, and source of aggregate variable.

## *WOOD PRODUCTS*

*By DUANE E. LYON*

***377. Wood*** is the vascular tissue of trees. Woods presently considered to have major commercial value in the United States are obtained from approximately 60 native tree species and 30 imported species. The varied structure and physical characteristics of these species make wood suitable for a wide variety of products ranging from decorative veneers to utility poles.

***378. Hardwoods and Softwoods.*** Wood is obtained from two classes of trees, commonly referred to as hardwoods and softwoods. The hardwoods are angiosperms and are characterized by broad leaves and seeds enclosed in a fruit. Most hardwoods lose their leaves in the fall or winter. Oaks, maple, ash, and walnut are examples of common hardwood trees. The softwoods are members of the gymnosperms classified as conifers. The conifers are characterized by exposed seeds, usually in cones. Typical softwoods include the pines, spruces, and cedars. The terms hardwood and softwood do not indicate the relative hardness of the two types of wood, since many conifers are actually harder than trees of the angiosperm type.

***379. Gross wood structure*** may be used to differentiate between the hardwood and softwood classes, and to identify species within each group. The cross section of a log shows several well-defined features from the outer bark, through the wood, to the central pith. The purpose of the bark is to protect the inner living tissues from injury. The inner portion of the bark is a conductive tissue that transports foodstuff from the leaves to the living cells. The wood portion of the tree has outer sapwood and inner heartwood regions, which often are distinguishable by a difference in color. The lighter-colored sapwood is a living tissue. The darker-colored heartwood consists of entirely dead cells, and serves primarily for mechanical support in the tree. The heartwood contains deposits of gums, oils, and other organic infiltrations in the cells. These materials are called extractives, and import the darker color (ranging from light brown to black) to heartwood. Other differences include durability and permeability. Sapwood of all species is readily destroyed by insects and the decay fungi. The heartwoods of many species are also nondurable, but others are very durable (for example cypress, the cedars, and redwood). Sapwood is usually more permeable to liquids, owing largely to the absence of extractives. For this reason, sapwood is more easily treated with preservatives. Wood near the center of the tree is often characterized by fast growth, and is termed juvenile wood. The properties of juvenile wood are inferior to those of normal heartwood. The strength properties of heartwood and sapwood are the same.

***380. Annual Rings.*** In temperate zones, the tree increases in diameter and height by division of cells in a unicell layer called the cambium. The growth increment for each year is called an annual ring. Many North American woods have well-defined annual rings consisting of an inner light-colored zone called springwood and an outer darker-colored zone called summerwood. The summerwood, formed in the later part of the growing season, is denser than the springwood. The relative portion of summerwood present in an annual ring is an important factor in determining the mechanical behavior of wood. Many commercial species of wood imported from tropical zones do not have a well-defined annual ring structure.

***381. Minute wood structure*** differs for hardwood and softwood species. The conifers are composed primarily of long hollow fibers oriented with their long axis parallel to the length of the tree. The ends of these fibers are tapered and the cells overlap at the ends. The size, shape, and structure of these fibers vary considerably from species to species, which accounts for much of the variation in properties of wood. They also give rise to wood's anisotropic nature, having different properties along its three principal directions, longitudinal, tangential, and radial. The minute structure of the hardwood is more varied and consists of many types of fiberlike cells. Individual fibers are composed of a multilayer cell wall enclosing a central cavity. For more information on wood, refer to Ref. 13.

***382. Chemically*** wood consists of approximately 70% cellulose, 25% lignin, and about 5% extractives. The strength of wood may be attributed almost entirely to the cellulose and lignin present in the cell wall. The extractives do not contribute to the cell-wall structure, but do contribute to such properties as color, odor, taste, and resistance to decay. Additional information of the chemistry of wood may be found in Ref. 12.

***383. Specific gravity*** is a measure of the amount of material contained in a piece of wood. It is calculated by dividing the weight of a given volume of wood by the weight of an equal volume of water. Specific gravity varies according to the amount of water present in wood. For this reason the weight of bone-dry wood is usually used for reported values of specific gravity. The specific gravity of some commercially imported woods is listed in Table 4-83, which was abstracted from Ref. 12. For clear, straight-grained wood at a known moisture content, specific gravity is positively correlated with several important properties, including strength and stiffness in bending, tension, and compression. Approximate functions for predicting the mechanical properties of wood for a known specific gravity are given in the Wood Handbook (Ref. 12). When wood contains defects or natural growth imperfections, the relationship between properties and specific gravity may be less pronounced.

***384. Moisture in Wood.*** Wood is a hygroscopic material. Moisture in wood occurs in three forms: water vapor in air spaces in the cell cavities, capillary water in the cell cavities,

**TABLE 4-83** Specific Gravity and Shrinkage of Common Woods

| Species | Average specific gravity* | Shrinkage, % | |
|---|---|---|---|
| | | Radial | Tangential |
| Douglas fir, various regions | 0.43–0.48 | 3.6–5.0 | 6.2–7.8 |
| White ash | 0.60 | 4.8 | 7.8 |
| Aspen | 0.38 | 3.3 | 7.9 |
| Yellow birch | 0.55 | 7.2 | 9.2 |
| White fir | 0.37 | 3.2 | 7.1 |
| Western red cedar | 0.33 | 2.4 | 5.0 |
| Northern white cedar | 0.31 | 2.2 | 4.9 |
| Western hemlock | 0.42 | 4.3 | 7.9 |
| Southern yellow pines | 0.51–0.61 | 4.4–5.5 | 7.4–7.8 |
| Tamarack | 0.53 | 3.7 | 7.4 |
| White oak species | 0.63–0.68 | 4.1–5.5 | 7.2–10.8 |
| Red oak species | 0.59–0.69 | 4.0–5.5 | 8.2–10.6 |
| Hickory species | 0.69–0.75 | 7.0–7.8 | 10.0–12.6 |

*Weight ovendry and volume at 12% moisture content.
†From green to ovendry condition, based on green dimension.

and water molecules bound to the hydroxyl groups of the cellulose in the cell wall. In most end-use conditions, when wood is not in contact with water, nearly all the moisture present is bound water and is usually between 3 and 30% of the dry weight of the wood. Since this bound water tends to be at equilibrium with the vapor pressure of the surrounding atmosphere, the maximum amount of bound water in wood occurs in a saturated atmosphere. Any increase in moisture content above this maximum is due to capillary water, acquired from contact with liquid water.

The moisture content of wood is expressed as a percent of the ovendry weight of wood. It can be measured by weighing a wood sample before and after drying to constant weight at 210°F, using the relationship:

$$\text{Moisture content, \%} = \frac{\text{moist weight} - \text{dry weight}}{\text{dry weight}} \times 100$$

When moist wood dries, the liquid water present in the cell capillaries evaporates before the bound water leaves the cell wall. The fiber-saturation point is defined as the moisture content of wood that is in equilibrium with a saturated atmosphere, with no free water in the cell cavities. This moisture content is around 30% but varies with species.

***385. Volumetric Changes.*** Below the fiber-saturation point, water that evaporates from wood results in a reduction in wood volume. The amount of volumetric change is positively related to the changes in moisture content and density. The anisotropic nature of wood results in unequal shrinkage in the three principal grain directions. Shrinkage is greatest in the transverse grain direction parallel to the growth rings (tangential), and the total shrinkage from green to the ovendry condition ranges from 4 to 13%, depending upon the species and density. In general, shrinkage increases with density. The transverse shrinkage perpendicular to the growth rings (radial) is usually about one-half that parallel to the growth rings and ranges from 2 to 8%. Total shrinkage along the grain ranges from 0.1 to 0.3%. Some representative shrinkage values are given in Table 4-83. In poles, the tangential shrinkage results in seasoning checks, while radial shrinkage is the dominant factor in reducing pole circumference. Circumferential shrinkage is about 1% for poles dried to approximately 20% moisture content.

***386. Wood Seasoning.*** Most wood products are dried prior to use to remove the large amount of moisture present in freshly cut wood. Wood that has been dried offers a number of advantages, including reduced weight and shrinkage, and increased strength and durability. Drying may be accomplished by one of several procedures. Two of the most common are air drying and kiln drying. A number of defects may develop during drying if the process is not carefully controlled. These defects are the result of drying stresses due to unequal shrinkage. Most kiln-drying procedures include moisture-equalizing and conditioning treatments to improve moisture uniformity throughout the thickness of the wood product, and to relieve residual stresses. Improper drying may result in warping, checking, or more severe defects. For a more complete discussion of wood drying, see Ref. 10.

***387. Thermal Properties of Wood.*** Temperature affects several properties of wood. As wood is heated, it expands. The coefficient of thermoexpansion for wood averages near $1.1 \times 10^{-6}$ per degree Celsius for most native species. Wood is a good insulator, and does not respond very fast to a change in environmental temperature. The coefficient of thermoconductivity for wood ranges from 0.4 to 0.7 Btu/(h)(°C) for a 1 $\text{ft}^2$ area 1 in in thickness at a moisture content of 12%. The thermoconductivity of wood increases with increasing specific gravity and moisture content.

***388. Electrical Properties of Wood.*** Three important electrical properties of wood are resistivity, dielectric constant, and power factor. Wood is an excellent insulator. The resistivity of dry wood to the flow of direct current is high, approximately $3 \times 10^{17}\ \Omega\cdot\text{cm/cm}^3$ parallel to the grain. The presence of moisture lowers resistivity. The dielectric constant for wood determines the amount of stored electric potential energy when it is placed in a high-frequency alternating current. The dielectric constant for wood varies over a range from 2.0 for dry wood to 8 for wood above the fiber-saturation point. The dielectric constant is affected by density and grain direction. The power factor determines the amount of energy that is dissipated as heat when wood absorbs power in a high-frequency dielectric field. The power factor for wood is about 2 to 6% at low moisture contents for frequencies between 2 and 15 Hz. Additional information on the electrical properties may be found in Refs. 12 and 13.

***389. Specific Gravity and Fiber-Saturation Point.*** The specific gravity of wood is the weight of wood divided by the weight of an equal volume of water. As both the weight and volume of wood vary with the moisture content, specific gravity of wood is an indefinite quantity unless the conditions under which it is obtained are clearly specified. Most commonly, specific gravity of wood is based on weight ovendry and volume green, ovendry, or some intermediate moisture content. As wood dries, most of the liquid water held in the capillaries evaporates before the bound-water molecules begin to leave the cell wall. The fiber-saturation point is defined as the moisture content of the wood at this transition point and is the moisture content at which shrinkage begins. The radial and tangential shrinkage and specific gravity of a number of woods are listed in Table 4-83, which was abstracted from *U.S. Dept. Agr.*, 72, Wood Handbook, Forest Products Laboratory, Forest Service.

***390. Effect of Moisture on Strength.*** Clear wood increases in strength as it dries below the fiber-saturation point. The change in strength resulting from a 1% change in moisture content, expressed in percent, is approximately 4 for modulus of rupture, 2 for modulus of elasticity, and 5 for compression parallel to the grain. More exact relationships may be found in Ref. 12.

For structural lumber, the increase in strength of the clear wood is partly offset by the development of seasoning defects such as checks and splits. For this reason, the properties in the green condition are generally used as the base for the development of design stresses for wood. For lumber that is nominally 2 in in thickness, the design stress is increased by up to 25% in bending, 20% in modulus of elasticity, and 37.5% in compression parallel to grain when the moisture content is at or less than 15%.

***391. Effect of Temperature on Strength.*** In general, heating reduces and cooling increases the mechanical properties of wood. The change is immediate, and irreversible for temperatures remaining above 93°C for any appreciable period of time. The adverse effect of high temperature is more pronounced at high moisture contents. For elevated temperatures below 93°C, the immediate loss of strength is recovered when the wood is cooled to ambient conditions. When wood is repeatedly exposed to high temperature, the adverse effect on properties is cumulative.

***392. The mechanical properties of the commercially important woods of the United States*** have been evaluated in accordance with ASTM Standard D143, which specifies small, clear specimens to eliminate the influence of naturally occurring physical defects in the wood.

Tables 4-84 and 4-85 show some mechanical properties for wood in the green condition and at 12% moisture content, respectively. The green properties are obtained from specimens at essentially the same moisture content as in the living tree, well above the fiber-saturation point.

Tables 4-84 and 4-85 have been abstracted to include several of the more important hardwood and softwood species and the mechanical properties of each which are likely to be uniquely important for specific uses encountered in electrical engineering applications. For additional data on other strength properties and other species, see the references.

***393. Decay and Its Prevention.*** At ordinary temperatures, wood is very stable and unless attacked by living organisms remains the same for centuries, either in air or under water. Fungi are the chief enemies of wood, and they thrive best with warmth and abundance of moisture and air, for example, in contact with the ground. Higher temperatures near the surface of the ground, together with adequate air and a greater prevalence of fungi, cause decay to progress faster near the ground line than at several feet below. Proper seasoning, together with protection against the entrance of moisture and impregnating with fungus-inhibiting compounds (see Par. **394**), which prevent fungi from feeding on the wood, is the best means of preservation. Only the heartwood is resistant, however; consequently, the sapwood should be preservative-treated, irrespective of the species of wood, if decay resistance is needed. Various species differ materially in their natural resistance to decay. For systematic presentation of the subject of wood preservation, see the references.

***394. Wood Preservatives.*** Wood preservatives fall into two main classes: (1) oil-borne preservatives and (2) water-borne metallic salts. The former may be further subdivided into (a) coal-tar creosote with and without the mixture of cheaper materials such as petroleum or coal tar and (b) solutions of toxic organic chemicals such as pentachlorophenol dissolved in petroleum oils. Oil-type preservatives are used extensively for products that are exposed to ground contact whereby resistance to leaching is an important requirement of the preservative. These products include poles, crossties, piling, bridge timbers, and fence posts.

**TABLE 4-84** Mechanical Properties of Various Woods in the Green Condition Grown in the United States

| Species | Moisture content, % | Specific gravity* | Static bending: Modulus of rupture, lb/in² | Static bending: Modulus of elasticity, 1,000 lb/in² | Compression parallel to grain maximum crushing strength, lb/in² | Compression perpendicular to grain stress at proportional limit, lb/in² | Tension perpendicular to grain maximum tensile strength, lb/in² | Hardness†: End, lb | Hardness†: Side, lb | Maximum shearing strength parallel to grain, lb/in² |
|---|---|---|---|---|---|---|---|---|---|---|
| Ash, black | 85 | 0.45 | 6,000 | 1,040 | 2,300 | 350 | 490 | 590 | 520 | 860 |
| Ash, white | 42 | 0.55 | 9,600 | 1,460 | 3,990 | 670 | 590 | 1,010 | 960 | 670 |
| Aspen | 94 | 0.35 | 5,100 | 860 | 2,140 | 180 | 230 | 280 | 300 | 660 |
| Basswood | 105 | 0.32 | 5,000 | 1,040 | 2,220 | 170 | 280 | 290 | 250 | 600 |
| Beech | 54 | 0.56 | 8,600 | 1,380 | 3,550 | 540 | 720 | 970 | 850 | 1,290 |
| Birch, yellow | 67 | 0.55 | 8,300 | 1,500 | 3,380 | 430 | 430 | 810 | 780 | 1,110 |
| Cottonwood, eastern | 111 | 0.37 | 5,300 | 1,010 | 2,280 | 200 | 410 | 380 | 340 | 680 |
| Elm, American | 89 | 0.46 | 7,200 | 1,110 | 2,910 | 360 | 590 | 680 | 620 | 1,000 |
| Elm, slippery | 85 | 0.48 | 8,000 | 1,230 | 3,320 | 420 | 640 | 750 | 660 | 1,110 |
| Hickory, shagbark | 60 | 0.64 | 11,000 | 1,570 | 4,580 | 840 | ..... | ..... | ..... | 1,520 |
| Locust, black | 40 | 0.66 | 13,800 | 1,850 | 6,800 | 1,160 | 770 | 1,640 | 1,570 | 1,760 |
| Maple, silver | 66 | 0.44 | 5,800 | 940 | 2,490 | 370 | 560 | 670 | 590 | 1,050 |
| Maple, sugar | 58 | 0.56 | 9,400 | 1,550 | 4,020 | 640 | ..... | 1,070 | 970 | 1,460 |
| Oak, red | 80 | 0.56 | 8,300 | 1,350 | 3,440 | 610 | 750 | 1,060 | 1,000 | 1,210 |
| Oak, white | 68 | 0.60 | 8,300 | 1,250 | 3,560 | 670 | 770 | 1,120 | 1,060 | 1,250 |
| Sweetgum | 115 | 0.46 | 7,100 | 1,200 | 3,040 | 370 | 540 | 670 | 600 | 990 |
| Sycamore | 83 | 0.46 | 6,500 | 1,060 | 2,920 | 360 | 630 | 700 | 610 | 1,000 |
| Yellow poplar | 83 | 0.40 | 6,000 | 1,220 | 2,660 | 270 | 510 | 480 | 440 | 790 |
| Baldcypress | 91 | 0.42 | 6,600 | 1,180 | 3,580 | 400 | 300 | 440 | 390 | 810 |
| Cedar, northern white | 55 | 0.29 | 4,200 | 640 | 1,990 | 230 | 240 | 320 | 230 | 620 |
| Cedar, Port Orford | 43 | 0.40 | 6,200 | 1,420 | 3,130 | 280 | 180 | 460 | 400 | 830 |
| Cedar, western red | 37 | 0.31 | 5,100 | 920 | 2,750 | 270 | 230 | 430 | 270 | 710 |
| Douglas fir, coast‡ | 38 | 0.45 | 7,700 | 1,560 | 3,780 | 380 | 300 | 570 | 500 | 900 |
| Fir, white | 110 | 0.37 | 5,900 | 1,160 | 2,900 | 280 | 300 | 410 | 340 | 760 |
| Hemlock, western | 77 | 0.42 | 6,600 | 1,310 | 3,360 | 280 | 290 | 500 | 410 | 860 |
| Larch, western | 58 | 0.48 | 7,700 | 1,460 | 3,760 | 400 | 330 | 580 | 510 | 870 |
| Pine, lodgepole | 65 | 0.38 | 5,500 | 1,080 | 2,610 | 250 | 220 | 320 | 330 | 680 |
| Pine, ponderosa | 91 | 0.38 | 5,100 | 1,000 | 1,940 | 280 | 310 | 310 | 320 | 700 |
| Pine, loblolly | 81 | 0.47 | 7,300 | 1,410 | 3,490 | 390 | 260 | 420 | 450 | 850 |
| Pine, longleaf | 62 | 0.54 | 8,700 | 1,600 | 4,300 | 480 | 330 | 550 | 590 | 1,040 |
| Pine, shortleaf | 81 | 0.46 | 7,300 | 1,390 | 3,430 | 350 | 320 | 410 | 440 | 850 |
| Pine, western white | 54 | 0.36 | 5,200 | 1,170 | 2,650 | 240 | 260 | 310 | 310 | 640 |
| Spruce, Engelmann | 80 | 0.32 | 4,500 | 960 | 2,190 | 220 | 240 | 310 | 260 | 590 |
| Spruce, Sitka | 42 | 0.37 | 5,700 | 1,230 | 2,670 | 280 | 250 | 430 | 350 | 760 |

*Specific gravity based on green volume and ovendry weight.

†Load required to embed a 0.444-in ball to half its diameter.

‡Coast Douglas fir is defined as that coming from counties in Oregon and Washington west of the summit of the Cascade Mountains. For Douglas fir from other sources, see Western Wood Density Survey, *U.S. Forest Service Res. Paper* FPL 27.

NOTE: 1 lb/in² = 6.895 kPa; 1 lb = 0.4536 kg.

**TABLE 4-85** Mechanical Properties of Various Woods in the Air-Dry Condition Grown in the United States

| Species | Moisture content, % | Specific gravity* | Static bending: Modulus of rupture, lb/in² | Static bending: Modulus of elasticity, 1,000 lb/in² | Compression parallel to grain maximum crushing strength, lb/in² | Compression perpendicular to grain stress at proportional limit, lb/in² | Tension perpendicular to grain maximum tensile strength, lb/in² | Hardness†: End, lb | Hardness†: Side, lb | Maximum shearing strength parallel to grain, lb/in² |
|---|---|---|---|---|---|---|---|---|---|---|
| Ash, black | 12 | 0.49 | 12,600 | 1,600 | 5,970 | 760 | 700 | 1,150 | 850 | 1,570 |
| Ash, white | 12 | 0.60 | 15,400 | 1,770 | 7,410 | 1,160 | 940 | 1,720 | 1,320 | 1,160 |
| Aspen | 12 | 0.38 | 8,400 | 1,180 | 4,250 | 370 | 260 | 510 | 350 | 850 |
| Basswood | 12 | 0.37 | 8,700 | 1,460 | 4,730 | 370 | 350 | 520 | 410 | 990 |
| Beech | 12 | 0.64 | 14,900 | 1,720 | 7,300 | 1,010 | 1,010 | 1,590 | 1,300 | 2,010 |
| Birch, yellow | 12 | 0.62 | 16,600 | 2,010 | 8,170 | 970 | 920 | 1,480 | 1,260 | 1,880 |
| Cottonwood, eastern | 12 | 0.40 | 8,500 | 1,370 | 4,910 | 380 | 580 | 580 | 430 | 930 |
| Elm, American | 12 | 0.50 | 11,800 | 1,340 | 5,520 | 690 | 660 | 1,110 | 830 | 1,510 |
| Elm, slippery | 12 | 0.53 | 13,000 | 1,490 | 6,360 | 820 | 530 | 1,120 | 860 | 1,630 |
| Hickory, shagbark | 12 | 0.72 | 20,200 | 2,160 | 9,210 | 1,760 | ..... | ..... | ..... | 2,430 |
| Locust, black | 12 | 0.69 | 19,400 | 2,050 | 10,180 | 1,830 | 640 | 1,580 | 1,700 | 2,480 |
| Maple, silver | 12 | 0.47 | 8,900 | 1,140 | 5,220 | 740 | 500 | 1,140 | 700 | 1,480 |
| Maple, sugar | 12 | 0.63 | 15,800 | 1,830 | 7,830 | 1,470 | ..... | 1,840 | 1,450 | 2,330 |
| Oak, red | 12 | 0.63 | 14,300 | 1,820 | 6,760 | 1,010 | 800 | 1,580 | 1,290 | 1,780 |
| Oak, white | 12 | 0.68 | 15,200 | 1,780 | 7,440 | 1,070 | 800 | 1,520 | 1,360 | 2,000 |
| Sweetgum | 12 | 0.52 | 12,500 | 1,640 | 6,320 | 620 | 760 | 1,080 | 850 | 1,600 |
| Sycamore | 12 | 0.49 | 10,000 | 1,420 | 5,380 | 700 | 720 | 920 | 770 | 1,470 |
| Yellow poplar | 12 | 0.42 | 10,100 | 1,580 | 5,540 | 500 | 540 | 670 | 540 | 1,190 |
| Baldcypress | 12 | 0.46 | 10,600 | 1,440 | 6,360 | 730 | 270 | 660 | 510 | 1,000 |
| Cedar, northern white | 12 | 0.31 | 6,500 | 800 | 3,960 | 310 | 240 | 450 | 320 | 850 |
| Cedar, Port Orford | 12 | 0.42 | 11,300 | 1,730 | 6,470 | 620 | 400 | 730 | 560 | 1,080 |
| Cedar, western red | 12 | 0.33 | 7,700 | 1,120 | 5,020 | 490 | 220 | 660 | 350 | 860 |
| Douglas fir, coast‡ | 12 | 0.48 | 12,400 | 1,950 | 7,240 | 800 | 340 | 900 | 710 | 1,130 |
| Fir, white | 12 | 0.39 | 9,800 | 1,490 | 5,810 | 530 | 300 | 780 | 480 | 1,100 |
| Hemlock, western | 12 | 0.45 | 11,300 | 1,640 | 7,110 | 550 | 340 | 900 | 540 | 1,250 |
| Larch, western | 12 | 0.52 | 13,100 | 1,870 | 7,640 | 930 | 430 | 1,120 | 830 | 1,360 |
| Pine, lodgepole | 12 | 0.41 | 9,400 | 1,340 | 5,370 | 610 | 290 | 530 | 480 | 880 |
| Pine, ponderosa | 12 | 0.40 | 9,400 | 1,290 | 5,320 | 580 | 420 | 570 | 460 | 1,130 |
| Pine, loblolly | 12 | 0.51 | 12,800 | 1,800 | 7,080 | 800 | 470 | 750 | 690 | 1,370 |
| Pine, longleaf | 12 | 0.58 | 14,700 | 1,990 | 8,440 | 960 | 470 | 920 | 870 | 1,500 |
| Pine, shortleaf | 12 | 0.51 | 12,800 | 1,760 | 7,070 | 810 | 470 | 750 | 690 | 1,310 |
| Pine, western white | 12 | 0.38 | 9,500 | 1,510 | 5,620 | 440 | ..... | 440 | 370 | 850 |
| Spruce, Engelmann | 12 | 0.34 | 8,700 | 1,280 | 4,770 | 470 | 350 | 560 | 350 | 1,030 |
| Spruce, Sitka | 12 | 0.40 | 10,200 | 1,570 | 5,610 | 580 | 370 | 760 | 510 | 1,150 |

*Specific gravity based on green volume and ovendry weight.

†Load required to embed a 0.444-in ball to half its diameter.

‡Coast Douglas fir is defined as that coming from counties in Oregon and Washington west of the summit of the Cascade Mountains. For Douglas fir from other sources, see Western Wood Density Survey, *U.S. Forest Service Res. Paper* FPL 27.

NOTE: 1 lb/in$^2$ = 6.895 kPa; 1 lb = 0.4536 kg.

Water-borne preservatives are used mainly for the treatment of lumber. Wood treated with a water-borne preservative is clean, paintable, and odorless.

*Creosote* is a distillate of coal tar formed during the coking of coal. On the basis of the quantity of wood treated, it is the most important preservative. Much of the treated wood is used in ground contact; appreciable amounts are also used in coastal waters infested with marine organisms that bore into and destroy untreated wood.

*Pentachlorophenol* dissolved in petroleum oils of varied nature has come into wide use. As a general rule, the effectiveness is highest when high-boiling oils are used as solvents, but relatively low-boiling oils are sometimes used in the treatment of products, such as millwork, having high cleanliness requirements. Water-repellent materials are generally added to the preservative in millwork treatments to minimize the dimensional changes that accompany fluctuations in the moisture content of the wood. A relatively new type of treatment comprises the solution of pentachlorophenol in a liquefied petroleum gas which is subject to practically complete removal by evaporation, leaving the treated wood very clean and readily paintable. Other standard oil-borne preservatives are copper-8-quinolinolate, and tributyltin oxide.

*Water-borne preservatives* are generally mixtures of several inorganic salts, the most important of which are salts of copper, chromium, arsenic, and zinc. Sodium fluroide is an ingredient of two widely used commercial preservatives. Traditionally, the use of water-borne preservatives has been restricted to situations where resistance to leaching is not required; however, several formulations now available comprise mixtures of salts that undergo chemical reaction within the wood with the formation of relatively insoluble toxic compounds. Such preservatives give good protection to wood exposed to wet conditions. Wood treated with water-borne preservatives is clean and paintable after drying to below 25% moisture content.

*Paints, varnishes, and stains* are used for decorative effects, but they also afford surface protection by retarding moisture changes and thus decreasing checking, warping, and weathering. Such protection is only superficial, however, and internal decay may be expected unless the wood is kept dry.

*Fire-retardant chemicals* such as ammonium phosphate and sulfate and salts of zinc and boron are used to decrease the flammability of wood. Some fire-retardant formulations also give protection against decay.

***395. Methods of Treating Wood.*** The methods of preservative treatment may be divided into two classes, pressure and nonpressure. Pressure methods are by far the most effective for protecting wood. In pressure methods the wood is enclosed in a vessel, and the liquid preservative is forced into the wood under considerable hydrostatic pressure. Nonpressure methods do not utilize artificial pressure, the preservative being applied by dipping, soaking, brushing, or spraying. A third method, somewhat distinct from the others and called the thermal method, may be mentioned. It consists in heating the wood to expel air and then allowing the wood to cool in the liquid, whereby a partial vacuum forms in the internal spaces. Although movement of the liquid into the wood is due to atmospheric pressure, the process is not classed among pressure processes.

There are several modifications of the pressure process. In the full-cell process, the wood is subjected to a vacuum in order to evacuate the internal cavities used for the treatment of marine piling, which requires high retention of creosote for protection against wood-boring animals. The process is also used commonly in treatments of lumber with water-borne preservatives. Much wood for land use is treated with oil-type preservatives by one of the so-called empty-cell methods, whereby it is possible to increase the depth of penetration obtained with a limited retention of preservative. In the Rueping process, air is first injected to create within the wood a pressure greater than atmospheric. The cylinder is filled with preservative in such a way that the injected air is trapped in the wood. The pressure is then increased to force preservative into the wood. After the pressure is released and the cylinder drained, the compressed air in the wood expands to expel some of the preservative. The recovered preservative is called kickback. The Lowry process differs from the Rueping process in that no initial air pressure is applied. The air normally present is compressed during the pressure cycle and produces some kickback when pressure is released.

The conditioning of the wood prior to treatment is an important step. Air seasoning, kiln drying, and various processes of cylinder conditioning are employed. The latter include steaming plus vacuum, boiling in oil under vacuum, and vapor drying, in which green wood is surrounded by hot vapors of distillates of coal tar or petroleum.

When oil preservatives are applied by simple soaking methods, the wood should be well seasoned in order to provide air spaces into which the oil may move. Oil preservatives of low viscosity are preferable. The results attainable vary greatly with the species of wood.

Diffusion methods depend upon the diffusion of water-soluble chemicals into the moisture present in green wood. Here again, the species of wood is an important factor, but the results are affected by other factors such as the nature of the chemical, the concentration of the solution, and the duration of the soaking period.

***396. Applications of Preservative Treatment.*** Preservative treatments are applied to many wood products, the most important being poles, crossties, lumber and structural timbers, fence posts, piling, and crossarms. Approximately 85% of all crossties treated in 1970 were of hardwood species, with oak accounting for 53%. The coniferous species dominated the treatment of other wood items, with southern pine being the most important, followed by Douglas fir.

***397. Advantages of Preservative Treatment.*** In addition to the conservation of a natural resource, preservative treatment results in economic savings due to increased service life and reduced maintenance costs. This has been recognized for many years by railroad companies, utility companies, and other large users of wood products. Because of demonstrated savings, practically all crossties and poles are now given a preservative treatment before installation. There has been a gradual increase in the volume of lumber treated annually, due to more widespread knowledge of the need for such treatment when the wood is to be used under conditions favorable to attack by decay or insects. For best performance it is desirable that all machining operations be completed before treatment.

***398. Strength of Treated Lumber.*** The effect of a preservative such as creosote or pentachlorophenol, in and of itself, on the strength of treated lumber appears to be negligible. It may be necessary, however, in establishing design stress values, to take into account possible reductions in strength that may result from temperatures or pressures used in the conditioning or treating processes. Results of tests of treated wood show reductions of stress in extreme fiber in bending and in compression perpendicular to grain, ranging from a few percent up to 25%, depending on the processes used. Compression parallel to grain is affected less and modulus of elasticity very little. The effect on resistance to horizontal shear can be estimated by inspection for shakes and checks after treatment. Strength reductions for wood poles agreed upon in formulating fiber-stress recommendations in American Standard Specifications and Dimensions for Wood Poles, ANSI 05.1-1979, range from 0 to 15% in various species, depending upon the conditioning and treating processes. Treating conditions specified by the American Wood Preservers' Association should never be exceeded. Reductions of strength can be minimized by restricting temperatures, heating periods, and pressures as much as is consistent with obtaining the absorption and penetration required for proper treatment.

***399. Effect of Preservative Treatment on Electrical Resistivity.*** The electrical resistivity of wood depends on its moisture content to a much greater degree than any other single variable. Ovendry wood is an excellent insulator, but as the wood absorbs moisture, its resistivity decreases rapidly. Wood in normal use, however, where its moisture content may range from about 6 to 14%, is still a good enough insulator for many electrical applications.

When wood has been treated with salts for preservative or fire-retardant purposes, its electrical resistivity may be markedly reduced. The effect of such salt treatment is small when the wood moisture is below about 8% but increases rapidly as the moisture content exceeds about 10%. Treatment with creosote or pentachlorophenol has practically no effect on the resistivity of wood.

The resistivity of wood decreases by about a factor of 2 for each increase of 10°C in the temperature and is about half as great for current flow along the grain as across the grain.

***400. American Lumber Standards.*** Simplified Practice Recommendation 16, American Lumber Standards for Softwood Lumber, is a voluntary standard of manufacturers, distributors, and users, promulgated in cooperation with the U.S. Department of Commerce. It provides for use classifications of (1) yard lumber, (2) structural lumber, and (3) factory and shop lumber. Different grading rules apply to each class. Size standards and generalized grade descriptions are part of SPR 16, but details of grading rules are left to the organized agencies of the lumber manufacturing industry. The grades and working stresses for structural lumber are referred by SPR 16 to the authority of ASTM D245, Methods for Establishing Structural Grades of Lumber, or D2018, Recommended for Determining Design Stresses for Load-Sharing Lumber Members.

***401. Standard Commercial Names.*** Standard commercial names of the most commonly used structural softwood from ASTM D1165, Standard Nomenclature of Domestic Hardwood and Softwoods, are as follows:

Cedar:
- Alaska cedar
- Port Orford cedar
- Western red cedar

Fir:
- Douglas fir
- White fir

Hemlock:
- Eastern hemlock
- West Coast hemlock

Larch, Western

Pine:
- Jack pine
- Lodgepole pine
- Norway pine
- Ponderosa pine
- Southern yellow pine

Redwood

Spruce:
- Eastern spruce
- Engelmann spruce
- Sitka spruce

***402. Standard Structural Grades.*** Detailed descriptions of the standard structural grades are published in the grading rule books of the organized regional agencies of the lumber manufacturing industry. These are subject to review for compliance with the general requirements of SPR 16, American Lumber Standards for Softwood Lumber. The principal-use classes of structural lumber are: (1) *joists and planks,* pieces of rectangular cross section 2 to 4 in thick and 4 or more in wide (nominal dimensions), graded primarily for bending strength edgewise or flatwise; (2) *beams and stringers,* pieces of rectangular cross section 5 by 8 in (nominal dimensions) and up, graded for strength in bending when loaded on the narrow face; and (3) *posts and timbers,* pieces of square or nearly square cross section, 5 by 5 in (nominal dimensions) and larger, graded primarily for use as posts and columns.

***403. Working Stresses.*** Working stresses recommended by the lumber industry for their structural grades are found with the detailed grade descriptions in the grading rule books of the organized regional agencies of the industry. A complete listing of all structural grades and their working stresses is found in the "National Design Specification for Stress-Grade Lumber and Its Fastenings," published by the National Forest Products Association. Values for a few typical grades are shown in Table 4-86. Working stresses vary according to the grades and sizes of lumber and their condition with respect to moisture content. Stresses are adjustable also for duration of load and for special conditions such as extreme temperature. Stress increases are provided for "load-sharing members" in which the safety of the structure depends upon the strength of the assemblage of members rather than upon the lowest strength value for any single member. These stress modifications are described in ASTM standards.

Allowable working stresses for the structural grades of lumber are also a part of certain use specifications, such as the Minimum Property Standards of the Federal Housing Administration, the American Railway Engineering Association Manual, and various local or regional building codes. These allowable values may or may not coincide with the lumber industry stress recommendations for the same species and grade.

***404. Wood-Base Panel Materials.*** Included in this category are plywood, insulating board, hardboard, particle board, waferboard, and the medium-density building fiberboards. Plywood, normally fabricated by bonding an odd number of layers of veneers together with the grain direction in adjacent plies at right angles to each other, is more dimensionally stable and more uniform in strength in the plane of the sheet than wood. Qualities of glue line and veneer permitted are set by the various commercial standards for plywood and determine the grades under which plywood is sold. In general, glue-line quality determines whether plywood is classed as being suitable for interior or exterior use.

U.S. Product Standard PS1-83 covers the basic specifications for the manufacture of construction plywood. Decorative hardwood plywood is described by U.S. Product Standard PS51(5.2). Plywood manufactured according to this standard will carry a grade trademark of a qualified testing agency.

Insulation board, hardboard, and medium-density building fiberboard are panel products made by reducing wood substance to particles or fiber and reconstituting the fiber into stiff panels 4 by 8 ft in area or larger. Insulation board is of either interior or water-resistant quality and is usually manufactured for use where combinations of thermal and sound-insulating properties and stiffness and strength are desired. Hardboard with a density of 50 lb/

**TABLE 4-86** Typical Stress Grades and Working Stresses for Structural Lumber*
(Normal duration of load and dry conditions of use)

| Species | Grade | Allowable working stress | | | | |
|---|---|---|---|---|---|---|
| | | Bending or tension parallel, lb/in² | Horizontal shear, lb/in² | Compression perpendicular to grain, lb/in² | Compression parallel to grain, lb/in² | Modulus of elasticity, lb/in² |
| Douglas fir | Select structural beams and stringers | 1,900 | 120 | 415 | 1,400 | 1,760,000 |
| | Construction joists and planks | 1,500 | 120 | 390 | 1,200 | 1,760,000 |
| | Standard joists and planks | 1,200 | 95 | 390 | 1,000 | 1,760,000 |
| | Construction posts and timbers | 1,200 | 120 | 390 | 1,200 | 1,760,000 |
| West Coast hemlock | Construction joists and planks | 1,500 | 100 | 365 | 1,100 | 1,540,000 |
| | Construction, MC joists and planks | 1,650 | 105 | 365 | 1,250 | 1,540,000 |
| | Standard joists and planks | 1,200 | 80 | 365 | 1,000 | 1,540,000 |
| Western larch | Standard joists and planks | 1,200 | 95 | 390 | 1,000 | 1,760,000 |
| Southern pine | Dense structural 72 beams and stringers | 2,000 | 135 | 455 | 1,550 | 1,760,000 |
| | Dense structural 58 beams and stringers | 1,600 | 105 | 455 | 1,300 | 1,760,000 |
| | No. 1 dimension, 2-in | 1,500 | 120 | 390 | 1,350 | 1,760,000 |
| | No. 2 dimension, 2-in | 1,200 | 105 | 390 | 900 | 1,760,000 |
| Norway pine | Common structural joists and planks | 1,100 | 75 | 360 | 775 | 1,320,000 |
| Redwood | Heart structural joists and planks | 1,300 | 95 | 320 | 1,100 | 1,320,000 |
| Eastern spruce | 1300f structural joists and planks | 1,300 | 95 | 300 | 975 | 1,320,000 |

*Compiled from "National Design Specification for Stress-Grade Lumber and Its Fastenings"; Washington, D.C., National Forest Products Association, 1962.
NOTE: 1 lb /in$^2$ = 6.895 kPa.

ft$^3$ or more is used in many applications where a relatively thin, hard, uniform panel material is required. Of great importance in the electrical field are special high-density hardboard products expressly manufactured with high dielectric properties. Medium-density fiberboards with a density between that of insulation board and hardboard are new products.

Particle boards are panel products made by gluing small pieces of wood in a form such as flakes (this product is called waferboard) and shavings into relatively thick, rigid panels. Thermosetting resins, usually urea or phenolformaldehyde, are used to provide bonds of either interior or water-resistant quality. Standards ASTM C208-72 and ANSF A208.1 and PS58-73 govern minimum qualities of regular insulation board, particle board, and hardboard. The important physical and strength properties of various board products are indicated by Table 4-87.

***405. Wood Poles and Crossarms.*** Western red cedar and southern yellow pine are two species most commonly used in the United States for poles to support electric supply and communication equipment. In the northeast part of the country there are still a number of chestnut and northern white cedar poles in service, but these species are no longer available for purchase. Douglas fir, lodgepole pine, and western larch are used in considerable numbers, particularly in the western states. Other species that can be used for poles but not considered so desirable as western cedar or southern yellow pine are eastern hemlock, eastern larch, jack pine (large usage in Canada), northern white pine, ponderosa pine, red (Norway) pine, southern white cedar, spruce, sugar pine, western helmock, western white pine, and white fir.

***406. Standards for Wood Poles.*** The ANSI specifications for wood poles serve as a basis for purchasing and use. The ANSI specifications cover fiber stresses, dimensions, defect limitations, and manufacturing requirements. These specifications are also the basis for standards and specifications of using organizations such as Edison Electric Institute and American Telephone and Telegraph Co. EEI Specification TD-100 for non-pressure-treated cedar poles and AT&T Specification AT-7312 are good examples of users' pole-purchasing specifications.

**TABLE 4-87** Strength and Mechanical Properties of Wood-Base Fiber and Particle Panel Materials*

| Material | Density, lb/ft | Specific gravity | Modulus of rupture, lb/in² | Modulus of elasticity (bending), 1,000 lb/in² | Tensile strength parallel to surface, lb/in² | Tensile strength perpendicular to surface, lb/in² | Compression strength parallel to surface, lb/in² | 24-hr water absorption |  | Thickness swelling, 24-h soak, % | Maximum linear expansion,† % | Thermal conductivity, Btu/(ft²)(h)(°F)(in thickness) |
|---|---|---|---|---|---|---|---|---|---|---|---|---|
|  |  |  |  |  |  |  |  | % by volume | % by weight |  |  |  |
| Fibrous-felted boards: |  |  |  |  |  |  |  |  |  |  |  |  |
| 1. Structural insulating board | 10–26 | 0.16–0.42 | 200–800 | 25–125 | 200–500 | 10–25 | ........... | 1–10 | ....... | ..... | 0.5 | 0.27–0.45 |
| 2. Medium-density building fiberboard | 26–50 | 0.42–0.80 | 400–4,000 | 90–700 | 800–2,000 | ....... | 500–3,400 | .... | 6–150 | ..... | 0.2–1.30‡ | 0.50–0.60 |
| 3. Hardboard: |  |  |  |  |  |  |  |  |  |  |  |  |
| *a.* Untempered | 50–80 | 0.80–1.28 | 3,000–7,000 | 400–800 | 3,000–6,000 | ....... | 1,800–6,000 | .... | 3–30 | 10–25 | 0.6 | 0.80–1.40 |
| *b.* Tempered | 60–80 | 0.96–1.28 | 6,500–10,000 | 800–1,000 | 4,000–7,800 | ....... | 4,200–6,000 | .... | 3–20 | 8–15 | 0.4 | 1.10–1.50 |
| 4. Super hardboard | 85–90 | 1.36–1.44 | 10,000–12,500 | 1,250 | 7,800 | 500 | 26,500 | .... | 0.3–1.2 | ..... | ......... | 1.85 |
| Particle boards: |  |  |  |  |  |  |  |  |  |  |  |  |
| 1. Insulating type | 10–26 | 0.16–0.42 | 700 | ......... | ............ | ....... | ............ | .... | ..... | ..... | .... | 0.36 |
| 2. Medium-density type | 26–50 | 0.42–0.80 |  |  |  |  |  |  |  |  |  |  |
| *a.* Extrusion | ..... | ......... |  | Values not presented because extruded boards are always used and tested with facings applied |  |  |  |  |  |  |  |  |
| *b.* Flat-platen pressed | ..... | ......... | 1,500–8,000 | 150–700 | 500–4,000 | 40–400 | 1,400–2,800 | .... | 20–75 | 20–75 | 0.6 | 0.40–1.00 |
| 3. Hard-pressed type | 50–80 | 0.80–1.28 | 3,000–7,500 | 400–1,000 | 1,000–5,000 | 275–400 | 3,500–4,000 | .... | 15–40 | 15–40 | 0.85 | 1.10–1.50 |
| 4. Waferboard | 26–50 | 0.42–0.80 | 2,500–3,200 | 450–650 | 500–5,000 | 50 | 2,000–5,000 | .... | ..... | 20–25 | 0.20 | ........ |

*The data presented are general round-figure values, accumulated from numerous sources; for more exact figures on a specific product, individual manufacturers should be consulted or actual tests made. Values are for general laboratory conditions of temperature and relative humidity.

†Expansion resulting from a change in moisture content from equilibrium at 50% relative humidity to equilibrium at 90% relative humidity.

‡For homogeneous and laminated boards, respectively.

NOTE: 1 lb/ft = 1.488 kg/m; 1 lb/in² = 6.895 kPa.

***407. Ultimate Fiber Stresses.*** The ultimate fiber stresses approved by the ANSI and contained in its Standard 05.1 are as shown in Tables 4-88, 4-89, and 4-90.

These tables cover all species of poles normally used in communication and electrical power construction.

***408. Pole Dimensions.*** The circumference at "6 ft from butt" in Standard 05.1 is based on the following principles:

**a.** The classes from the lowest to the highest were arranged in approximate geometric progression, the increments in breaking load between classes being about 25%.

**b.** The dimensions were specified in terms of circumference in inches at the top and circumference in inches at 6 ft from the butt for poles of the respective classes and lengths, except for three classes having no requirement for butt circumference.

**c.** All poles of the same class and length were to have, when new, approximately equal strength or, in more precise terms, equal moments of resistance at the ground line.

**d.** All poles of different lengths within the same class were of sizes suitable to withstand approximately the same breaking load, on the assumption that the load is applied 2 ft from the top and that the break (failure) would occur at the ground line.

The breaking loads referred to in (d) above for the classes for which "6 ft from butt" circumferences are given are as follows: Class 1, 4,500 lb; Class 2, 3,700 lb; Class 3, 3,000 lb; Class 4, 2,400 lb; Class 5, 1,900 lb; Class 6, 1,500 lb; Class 7, 1,200 lb; Class 9, 740 lb; Class 10, 370 lb.

**TABLE 4-88** Dimensions of Northern White Cedar and Engelmann Spruce Poles

| | Class | | 1 | 2 | 3 | 4 | 5 | 6 | 7 | 9 | 10 |
|---|---|---|---|---|---|---|---|---|---|---|---|
| | Minimum circumference at top, in | | 27 | 25 | 23 | 21 | 19 | 17 | 15 | 15 | 12 |
| | Length of pole, ft | Ground-line distance from butt,* ft | Minimum circumference at 6 ft from butt, in | | | | | | | | |
| Northern white cedar poles (based on a fiber stress of 4,000 lb/in²) | 20 | 4 | 38.0 | 35.5 | 33.0 | 30.5 | 28.0 | **26.0** | **24.0** | **22.0** | **17.5** |
| | 25 | 5 | 42.0 | 39.5 | 36.5 | 34.0 | 31.5 | **29.0** | **27.0** | **24.0** | **19.5** |
| | 30 | 5½ | 45.5 | **43.0** | **40.0** | **37.0** | **34.5** | **32.0** | **29.5** | **26.0** | |
| | 35 | 6 | 49.0 | **46.0** | **42.5** | **39.5** | **37.0** | **34.0** | **31.5** | | |
| | 40 | 6 | **51.5** | **48.5** | **45.0** | **42.0** | **39.0** | **36.0** | | | |
| | 45 | 6½ | **54.5** | **51.0** | **47.5** | **44.0** | **41.0** | | | | |
| | 50 | 7 | **57.0** | **53.5** | **49.5** | 46.0 | 43.0 | | | | |
| | 55 | 7½ | **59.0** | **55.5** | **51.5** | 48.0 | 44.5 | | | | |
| | 60 | 8 | **61.0** | **57.5** | **53.5** | 50.0 | | | | | |
| Engelmann spruce poles (based on a fiber stress of 5,600 lb/in²) | 20 | 4 | 34.5 | 32.0 | 30.0 | 28.0 | 25.5 | **23.5** | **22.0** | **19.0** | **15.0** |
| | 25 | 5 | 38.0 | 35.5 | 33.0 | 30.5 | 28.5 | **26.0** | **24.5** | **21.0** | **16.5** |
| | 30 | 5½ | 41.0 | **38.5** | **35.0** | **33.0** | **30.5** | **28.5** | **26.5** | **22.5** | |
| | 35 | 6 | 43.5 | **41.0** | **38.0** | **35.5** | **32.5** | **30.5** | **28.0** | | |
| | 40 | 6 | **46.0** | **43.5** | **40.5** | **37.5** | **34.5** | **32.0** | | | |
| | 45 | 6½ | **48.5** | **45.5** | **42.5** | **39.5** | **36.5** | | | | |
| | 50 | 7 | **50.5** | **47.5** | **44.5** | 41.0 | 38.0 | | | | |
| | 55 | 7½ | **52.5** | **49.5** | **46.0** | 42.5 | 39.5 | | | | |
| | 60 | 8 | **54.5** | **51.0** | **47.5** | 44.0 | | | | | |
| | 65 | 8½ | **56.0** | **52.5** | **49.0** | 45.5 | | | | | |
| | 70 | 9 | **57.5** | **54.0** | **50.5** | 47.0 | | | | | |
| | 75 | 9½ | **59.5** | **55.5** | **52.0** | 48.5 | | | | | |
| | 80 | 10 | **61.0** | **57.0** | 53.5 | 49.5 | | | | | |
| | 85 | 10½ | **62.5** | **58.5** | 54.5 | | | | | | |
| | 90 | 11 | **63.5** | **60.0** | 56.0 | | | | | | |
| | 95 | 11 | **65.0** | **61.0** | 57.0 | | | | | | |
| | 100 | 11 | **66.0** | **62.0** | 58.0 | | | | | | |

*The figures in this column are intended for use only when a definition of ground line is necessary in order to apply requirements relating to scars, straightness, etc.

NOTES: Classes and lengths for which circumferences at 6 ft from the butt are listed in boldface type are the preferred standard sizes. Those shown in light type are included for engineering purposes only.

1 in = 2.54 cm; 1 ft = 0.3048 m; 1 lb/in² = 6.895 kPa.

**TABLE 4-89** Western Red Cedar, Ponderosa Pine, Douglas Fir, and Southern Pine

| | Class | | 1 | 2 | 3 | 4 | 5 | 6 | 7 | 9 | 10 |
|---|---|---|---|---|---|---|---|---|---|---|---|
| | Minimum circumference at top, in | | 27 | 25 | 23 | 21 | 19 | 17 | 15 | 15 | 12 |
| | Length of pole, ft | Ground-line distance from butt,* ft | Minimum circumference at 6 ft from butt, in | | | | | | | | |
| Western red cedar and ponderosa pine poles (based on a fiber stress of 6,000 lb/in²) | 20 | 4 | 33.5 | 31.5 | 29.5 | 27.0 | 25.0 | **23.0** | **21.5** | **18.5** | **15.0** |
| | 25 | 5 | 37.0 | 34.5 | 32.5 | 30.0 | 28.0 | **25.5** | **24.0** | **20.5** | **16.5** |
| | 30 | 5½ | 40.0 | **37.5** | **35.0** | **32.5** | **30.0** | **28.0** | **26.0** | **22.0** | |
| | 35 | 6 | 42.5 | **40.0** | **37.5** | **34.5** | **32.0** | **30.0** | **27.5** | | |
| | 40 | 6 | **45.0** | **42.5** | **39.5** | **36.5** | **34.0** | **31.5** | 29.5 | | |
| | 45 | 6½ | **47.5** | **44.5** | **41.5** | **38.5** | **36.0** | 33.0 | 31.0 | | |
| | 50 | 7 | **49.5** | **46.5** | **43.5** | 40.0 | 37.5 | 34.5 | 32.0 | | |
| | 55 | 7½ | **51.5** | **48.5** | **45.0** | 42.0 | 39.0 | 36.0 | | | |
| | 60 | 8 | **53.5** | **50.0** | **46.5** | 43.5 | 40.0 | 37.0 | | | |
| | 65 | 8½ | **55.0** | **51.5** | **48.0** | 45.0 | 41.5 | | | | |
| | 70 | 9 | **56.5** | **53.0** | **49.5** | 46.0 | 42.5 | | | | |
| | 75 | 9½ | **58.0** | **54.5** | **51.0** | 47.5 | | | | | |
| | 80 | 10 | **59.5** | **56.0** | 52.0 | 48.5 | | | | | |
| | 85 | 10½ | **61.0** | **57.0** | 53.5 | | | | | | |
| | 90 | 11 | **62.5** | **58.5** | 54.5 | | | | | | |
| | 95 | 11 | **63.5** | **59.5** | 56.0 | | | | | | |
| | 100 | 11 | **65.0** | **61.0** | 57.0 | | | | | | |
| | 105 | 12 | **66.0** | **62.0** | 58.0 | | | | | | |
| | 110 | 12 | **67.5** | **63.0** | 59.0 | | | | | | |
| | 115 | 12 | **68.5** | **64.0** | | | | | | | |
| | 120 | 12 | **69.5** | **65.0** | | | | | | | |
| | 125 | 12 | **70.5** | **66.0** | | | | | | | |
| Douglas fir and southern pine poles (based on a fiber stress of 8,000 lb/in)² | 20 | 4 | 31.0 | 29.0 | 27.0 | 25.0 | 23.0 | **21.0** | **19.5** | **17.5** | **14.0** |
| | 25 | 5 | 33.5 | 31.5 | 29.5 | 27.5 | 25.5 | **23.0** | **21.5** | **19.5** | **15.0** |
| | 30 | 5½ | 36.5 | **34.0** | **32.0** | **29.5** | **27.5** | **25.0** | **23.5** | **20.5** | |
| | 35 | 6 | 39.0 | **36.5** | **34.0** | **31.5** | **29.0** | **27.0** | **25.0** | | |
| | 40 | 6 | **41.0** | **38.5** | **36.0** | **33.5** | **31.0** | **28.5** | 26.5 | | |
| | 45 | 6½ | **43.0** | **40.5** | **37.5** | **35.0** | **32.5** | 30.0 | 28.0 | | |
| | 50 | 7 | **45.0** | **42.0** | **39.0** | 36.5 | 34.0 | 31.5 | 29.0 | | |
| | 55 | 7½ | **46.5** | **43.5** | **40.5** | 38.0 | 35.0 | 32.5 | | | |
| | 60 | 8 | **48.0** | **45.0** | **42.0** | 39.0 | 36.0 | 33.5 | | | |
| | 65 | 8½ | **49.5** | **46.5** | **43.5** | 40.5 | 37.5 | | | | |
| | 70 | 9 | **51.0** | **48.0** | **45.0** | 41.5 | 38.5 | | | | |
| | 75 | 9½ | **52.5** | **49.0** | **46.0** | 43.0 | | | | | |
| | 80 | 10 | **54.0** | **50.5** | 47.0 | 44.0 | | | | | |
| | 85 | 10½ | **55.0** | **51.5** | 48.0 | | | | | | |
| | 90 | 11 | **56.0** | **53.0** | 49.0 | | | | | | |
| | 95 | 11 | **57.0** | **54.0** | 50.0 | | | | | | |
| | 100 | 11 | **58.5** | **55.0** | 51.0 | | | | | | |
| | 105 | 12 | **59.5** | **56.0** | 52.0 | | | | | | |
| | 110 | 12 | **60.5** | **57.0** | 53.0 | | | | | | |
| | 115 | 12 | **61.5** | **58.0** | | | | | | | |
| | 120 | 12 | **62.5** | **59.0** | | | | | | | |
| | 125 | 12 | **63.5** | **59.5** | | | | | | | |

*The figures in this column are intended for use only when a definition of ground line is necessary in order to apply requirements relating to scars, straightness, etc.

NOTES: Classes and lengths for which circumferences at 6 ft from the butt are listed in boldface type are the preferred standard sizes. Those shown in light type are included for engineering purposes only.

1 in = 2.54 cm; 1 ft = 0.3048 m; 1 lb/in² = 6.895 kPa.

Minimum top circumferences and minimum circumferences at 6 ft from butt are given in Tables 4-88, 4-89, and 4-90.

*Length.* Poles under 50 ft in length should not be more than 3 in shorter or 6 in longer than nominal length. Poles 50 ft or over in length should not be more than 6 in shorter or 12 in longer than nominal length.

Length should be measured between the extreme ends of the pole.

*Circumference.* The minimum circumference at 6 ft from the butt and at the top, for each length and class of pole, is listed in the tables of dimensions. The circumference at 6 ft from the butt of poles should be not more than 7 in or 20% larger than the specified minimum, whichever is greater.

The top dimensional requirement should apply at a point corresponding to the minimum length permitted for the pole.

## TABLE 4-90 Additional Wood Species

| | Class | | 1 | 2 | 3 | 4 | 5 | 6 | 7 | 9 | 10 |
|---|---|---|---|---|---|---|---|---|---|---|---|
| | Minimum circumference at top, in | | 27 | 25 | 23 | 21 | 19 | 17 | 15 | 15 | 12 |
| | Length of pole, ft | Ground-line distance from butt,* ft | Minimum circumference at 6 ft from butt, in | | | | | | | | |
| Jack pine, lodgepole pine, red pine, redwood, Sitka spruce, western fir, and white spruce poles (based on a fiber stress of 6,600 lb/in²) | 20 | 4 | 32.5 | 30.5 | 28.5 | 26.5 | 24.5 | **22.5** | **21.0** | **18.0** | **14.5** |
| | 25 | 5 | 36.0 | 33.5 | 31.0 | 29.0 | 27.0 | **25.0** | **23.0** | **20.0** | **15.5** |
| | 30 | 5½ | 39.0 | **36.5** | **34.0** | **31.5** | **29.0** | **27.0** | **25.0** | **21.0** | |
| | 35 | 6 | 41.5 | **38.5** | **36.0** | **33.5** | **31.0** | **28.5** | **26.5** | | |
| | 40 | 6 | **44.0** | **41.0** | **38.0** | **35.5** | **33.0** | **30.5** | 28.0 | | |
| | 45 | 6½ | **46.0** | **43.0** | **40.0** | **37.0** | **34.5** | 32.0 | 29.5 | | |
| | 50 | 7 | **48.0** | **45.0** | **42.0** | 39.0 | 36.0 | 33.5 | 31.0 | | |
| | 55 | 7½ | **49.5** | **46.5** | **43.5** | 40.5 | 37.5 | 34.5 | | | |
| | 60 | 8 | **51.5** | **48.0** | **45.0** | 42.0 | 38.5 | 36.0 | | | |
| | 65 | 8½ | **53.0** | **49.5** | **46.0** | 43.0 | 40.0 | | | | |
| | 70 | 9 | **54.5** | **51.0** | **47.5** | 44.5 | 41.0 | | | | |
| | 75 | 9½ | **56.0** | **52.5** | **49.0** | 45.5 | | | | | |
| | 80 | 10 | **57.5** | **54.0** | 50.5 | 47.0 | | | | | |
| | 85 | 10½ | **58.5** | **55.0** | 51.5 | | | | | | |
| | 90 | 11 | **60.0** | **56.5** | 52.5 | | | | | | |
| | 95 | 11 | **61.5** | **57.5** | 54.0 | | | | | | |
| | 100 | 11 | **62.5** | **58.5** | 55.0 | | | | | | |
| | 105 | 12 | **63.5** | **60.0** | 56.0 | | | | | | |
| | 110 | 12 | **65.0** | **61.0** | 57.0 | | | | | | |
| | 115 | 12 | **66.0** | **62.0** | | | | | | | |
| | 120 | 12 | **67.0** | **63.0** | | | | | | | |
| | 125 | 12 | **68.0** | **64.0** | | | | | | | |
| Alaska yellow cedar and western hemlock poles (based on a fiber stress of 7,400 lb/in²) | 20 | 4 | 31.5 | 29.5 | 27.5 | 25.5 | 23.5 | **22.0** | **20.0** | **17.5** | **14.0** |
| | 25 | 5 | 34.5 | 32.5 | 30.0 | 28.0 | 26.0 | **24.0** | **22.0** | **19.5** | **15.0** |
| | 30 | 5½ | 37.5 | **35.0** | **32.5** | **30.0** | **28.0** | **26.0** | **24.0** | **20.5** | |
| | 35 | 6 | 40.0 | **37.5** | **35.0** | **32.0** | **30.0** | **27.5** | **25.5** | | |
| | 40 | 6 | **42.0** | **39.5** | **37.0** | **34.0** | **31.5** | **29.0** | 27.0 | | |
| | 45 | 6½ | **44.0** | **41.5** | **38.5** | **36.0** | **33.0** | 30.5 | 28.5 | | |
| | 50 | 7 | **46.0** | **43.0** | **40.0** | 37.5 | 34.5 | 32.0 | 29.5 | | |
| | 55 | 7½ | **47.5** | **44.5** | **41.5** | 39.0 | 36.0 | 33.5 | | | |
| | 60 | 8 | **49.5** | **46.0** | **43.0** | 40.0 | 37.0 | 34.5 | | | |
| | 65 | 8½ | **51.0** | **47.5** | **44.5** | 41.5 | 38.5 | | | | |
| | 70 | 9 | **52.5** | **49.0** | **46.0** | 42.5 | 39.5 | | | | |
| | 75 | 9½ | **54.0** | **50.5** | **47.0** | 44.0 | | | | | |
| | 80 | 10 | **55.0** | **51.5** | 48.5 | 45.0 | | | | | |
| | 85 | 10½ | **56.5** | **53.0** | 49.5 | | | | | | |
| | 90 | 11 | **57.5** | **54.0** | 50.5 | | | | | | |
| | 95 | 11 | **58.5** | **55.0** | 51.5 | | | | | | |
| | 100 | 11 | **60.0** | **56.0** | 52.5 | | | | | | |
| | 105 | 12 | **61.0** | **57.0** | 53.5 | | | | | | |
| | 110 | 12 | **62.0** | **58.0** | 54.5 | | | | | | |
| | 115 | 12 | **63.0** | **59.0** | | | | | | | |
| | 120 | 12 | **64.0** | **60.0** | | | | | | | |
| | 125 | 12 | **65.0** | **61.0** | | | | | | | |
| Western larch poles (based on a fiber stress of 8,400 lb/in²) | 20 | 4 | 30.0 | 28.5 | 26.5 | 24.5 | 22.5 | **21.0** | **19.0** | **17.0** | **13.5** |
| | 25 | 5 | 33.0 | 31.0 | 29.0 | 26.5 | 24.5 | **23.0** | **21.0** | **18.5** | **14.5** |
| | 30 | 5½ | 35.5 | **33.5** | **31.0** | **29.0** | **26.5** | **24.5** | **23.0** | **19.5** | |
| | 35 | 6 | 38.0 | **35.5** | **33.0** | **31.0** | **28.5** | **26.5** | **24.5** | | |
| | 40 | 6 | **40.0** | **37.5** | **35.0** | **32.5** | **30.0** | **28.0** | 26.0 | | |
| | 45 | 6½ | **42.0** | **39.5** | **37.0** | **34.0** | **31.5** | 29.0 | 27.0 | | |
| | 50 | 7 | **44.0** | **41.0** | **38.5** | 35.5 | 33.0 | 30.5 | 28.5 | | |
| | 55 | 7½ | **45.5** | **42.5** | **40.0** | 37.0 | 34.5 | 31.5 | | | |
| | 60 | 8 | **47.0** | **44.0** | **41.0** | 38.5 | 35.5 | 33.0 | | | |
| | 65 | 8½ | **48.5** | **46.0** | **42.5** | 39.5 | 36.5 | | | | |
| | 70 | 9 | **50.0** | **47.0** | **44.0** | 41.0 | 38.0 | | | | |
| | 75 | 9½ | **51.5** | **48.0** | **45.0** | 42.0 | | | | | |
| | 80 | 10 | **52.5** | **49.5** | 46.0 | 43.0 | | | | | |
| | 85 | 10½ | **54.0** | **50.5** | 47.0 | | | | | | |
| | 90 | 11 | **55.0** | **51.5** | 48.5 | | | | | | |
| | 95 | 11 | **56.5** | **53.0** | 49.5 | | | | | | |
| | 100 | 11 | **57.5** | **54.0** | 50.5 | | | | | | |
| | 105 | 12 | **58.5** | **55.0** | 51.5 | | | | | | |
| | 110 | 12 | **59.5** | **56.0** | 52.5 | | | | | | |
| | 115 | 12 | **60.5** | **57.0** | | | | | | | |
| | 120 | 12 | **61.5** | **58.0** | | | | | | | |
| | 125 | 12 | **62.5** | **58.5** | | | | | | | |

*The figures in this column are intended for use only when a definition of ground line is necessary in order to apply requirements relating to scars, straightness, etc.

NOTES: Classes and lengths for which circumferences at 6 ft from the butt are listed in boldface type are the preferred standard sizes. Those shown in light type are included for engineering purposes only.

1 in = 2.54 cm; 1 ft = 0.3048 m; 1 lb/in$^2$ = 6.895 kPa.

*Classification.* The true circumference class should be determined as follows: Measure the circumference at 6 ft from the butt. This dimension will determine the true class of the pole, provided that its top (measured at the minimum length point) is large enough. Otherwise the circumference at the top will determine the true class, provided that the circumference at 6 ft from the butt does not exceed the specified minimum by more than 7 in or 20%, whichever is greater.

The above information relating to the pole standards approved by the ANSI does not constitute the complete standards. For further information, consult the standards, which may be obtained at a nominal charge.

***409. Machine shaving*** of poles has increased as a practice of producers. Approximately 85% of present production is so shaved. Some producers also turn the pole down in the process, thereby obtaining a straighter pole with a specific taper. The machine-processed poles season more rapidly, which is particularly important with species like southern yellow pine which are susceptible to fungus attack before treatment. Machine shaving makes for easier detection of defects and provides a pole of improved appearance. If poles having normally thin sapwood (such as western red cedar and larch) are to be full-length when treated with preservative, it is undesirable to reduce the thickness of the sapwood more than necessary to obtain a dressed pole.

***410. Preservative Treatment.*** For a nominal cost the service life of wood can be greatly increased by the use of preservative treatment. Creosote and pentachlorophenol are extensively used for the protection of poles and crossarms. Southern yellow pine because of its thick sapwood requires a pressure treatment. Species with intermediate sapwood thickness such as Douglas fir, lodgepole pine, and jack pine are treated by either pressure or nonpressure processes. Thin sapwood species such as western red cedar and larch are generally treated by nonpressure processes.

The American Wood-Preservers' Association Standards are used to specify preservative chemicals and treatment methods for wood products.

***411. Inspection.*** Poles are inspected prior to treatment for physical defects and decay and after treatment for penetration and retention of preservative and for cleanliness. Inspection is most effective when made at vendors' plants, because defects that may be hidden by preservative are detected and freight is saved on rejects. Commercial inspection agencies are available at most producing locations, and it is normally economical to utilize their services. Quantity users may have their own trained inspectors. *Crossarm inspection* is important because safety of linemen is a consideration in addition to quality of timber. As with poles, inspection should be made before treatment for defects and after treatment for penetration, retention, and cleanliness. Inspection should be done by qualified timber specialists.

***412. Conductivity*** is of concern to many electric-utility companies. Pole resistance varies greatly with moisture content. Dry wood of all species exhibits high resistance. Surface absorption of rainwater by untreated wood may vary the resistance over a wide range. Full-length-treated poles thoroughly dried before treatment generally show only moderate reduction in resistance following a rain. A rough correlation between resistance and moisture will show that 500,000 Ω over a 20-ft length of pole between contacts driven 3½ in deep corresponds to a moisture content of about 25%. Other average points on the curve band are 50,000,000 Ω 15% moisture and 20,000 Ω 40% moisture.

***413. Depth of Pole Setting.*** The values in the column headed "Ground-line distance from butt" in Tables 4-88 to 4-90 may be accepted as a guide for a satisfactory depth of pole settings in ordinary firm soil. In marshy soil and at unguyed angles in lines, setting depths should be increased 1 to 2 ft. In rock, the indicated settings may be reduced one-half for that part of the pole set in rock. Rock backfill in ordinary earth locations is not considered as set in rock.

***414. Pole stubbing*** can frequently be employed to effect substantial money savings. An otherwise good pole that is decayed at or below the ground line is fastened securely to a new preservative-treated stub set in the ground alongside it. The major part of the savings resulted from avoidance of transferring wires and equipment.

***415. Salvaging.*** Poles removed for any reason can frequently be salvaged for future use. Users of large quantities can economically do this. One or more of the following operations may be employed: cut off top, cut off butt, remove old hardware, shave, reframe, retreat.

**TABLE 4-91** Bending Load and Crushing Strength of Crossarms

| Species | Rings per in. | Per cent | | | Density (dry) | Max. bending load, lb | Maximum crushing strength lb per sq in. |
|---|---|---|---|---|---|---|---|
| | | Summer wood | Sap wood | Moisture | | | |
| Douglas fir | 20 | 40 | 0 | 11.5 | 0.48 | 7,590 | 7,080 |
| Longleaf pine (50% heart) | 18 | 44 | 55 | 13.4 | 0.54 | 8,984 | 5,425 |
| Longleaf pine (75% heart) | 19 | 53 | 32 | 13.5 | 0.63 | 10,180 | 8,950 |
| Longleaf pine (100% heart) | 16 | 44 | 1 | 12.8 | 0.63 | 9,782 | 8,940 |
| Shortleaf pine | 11 | 46 | 79 | 13.3 | 0.52 | 9,260 | 7,300 |
| Shortleaf pine, creosoted | 11 | 49 | .. | .... | .... | 7,649 | 5,770 |
| White cedar | 12 | 45 | 2 | 14.3 | 0.36 | 5,200 | 4,700 |

NOTE: 1 in = 2.54 cm; 1 lb = 0.4536 kg; 1 lb/in$^2$ = 6.895 kPa.

***416. Kinds of Timber for Crossarms.*** Two kinds of timber are in general use for crossarms, Douglas fir and southern yellow pine. All pine crossarms are treated with creosote or pentachlorophenol. The practice of treating fir crossarms is increasing rapidly as users recognize the need for arm life to match pole life.

Most Douglas fir arms used for communication and power-distribution lines are manufactured from timber selected for the purpose. Dense and close-grain lumber is used. *Publication* 14 of the West Coast Lumberman's Association sets forth grading and dressing requirements.

There is no grade of southern pine timber designated as crossarm stock, and crossarm users depend on the limitations set forth in their specifications to obtain a satisfactory quality of product. Pine arms are usually small boxed heart timbers.

Laminated arms are coming into use, but a buying specification is not available. Large transmission-line arms may be laminated structures or framed, treated round poles.

***417. Crossarm Specifications.*** The most widely used specifications for power-distribution crossarms have been prepared by the Transmission and Distribution Committee of the Edison Electric Institute. For fir crossarms: Specification TD-90, which combines both dense and close-grain grades; Specification TD-92, Heavy-Duty Douglas Fir Crossarms; Specification TD-93, Heavy-Duty Douglas Fir Braces. For pine crossarms: Specification TD-91, Dense Southern Pine Crossarms Preservative Treated. Widely used specifications for communication crossarms are American Telephone and Telegraph Co. Specification AT-7298, Crossarms.

***418. Strength of Crossarms.*** The most reliable source of information on the strength comes from tests made under conditions to simulate crossarms in service. Some tests have been made, and others are under consideration. Theoretical considerations, treating a crossarm as a beam, are valuable if those factors which control the actual strength are taken into account. Tests made several years ago on 84 six-pin, 3¼ by 4¼-in by 6-ft crossarms, with a uniformly distributed vertical load, gave average results shown in Table 4-91 (*U.S. Forest Service Circ.* 204, by T. R. C. Wilson). The maximum bending load shown in Table 4-91 is the total distributed vertical load. The maximum crushing strength is under compression parallel to the grain. Methods of tests are covered by ASTM specifications.

***419. References on Wood Products***

1. American Wood-Preservers' Association, *Manual of Recommended Practice.*
2. Appropriate standards of the ASTM.
3. Eggleston, R. C.: Evaluating the Relative Bending Strength of Crossarms; *Bell Systems Tech. J.,* January 1945.
4. Gurfinkel, G.: *Wood Engineering:* New Orleans, Southern Forest Products Association, 1973.
5. Nicholas, D. D. (ed.): *Wood Deterioration and its Prevention by Preservative Treatments;* Syracuse, New York, University Press, 1973.
6. MacLean, V. P.: Preservative Treatment of Wood by Pressure Methods; *U.S. Dept. Agr., Agr. Handbook* 40, 1960.
7. *Manual of Recommended Practice;* Washington, D.C., American Wood-Preservers' Association.
8. Ostman, H. F.: Crossarm Loading Studies; *Bull. EEI,* June 1945.

9. Publications of the U.S. Forest Products Laboratory, Madison, Wis.
10. Rasmussen, E. F.: Dry Kiln Operators Manual; *U.S. Dept Agr., Agr. Handbook* 188, Forest Products Laboratory, 1968.
11. *Timber Design and Construction Manual;* American Institute of Timber Construction, Current Edition.
12. Wood Handbook; *U.S. Dept. Agr., Agr. Handbook* 72, Forest Products Laboratory, 1974.
13. *Wood Structures;* New York, American Society of Civil Engineers, 1975.

# SECTION 5
# GENERATION


**Jack A. Makuch**
*Director, Product Design, Fossil Power Systems, Combustion Engineering, Inc.*

**Loering M. Johnson**
*President, Johnson Consulting Engineers*

**A. L. Gaines**
*President, A. Gaines Company*

**R. C. Miller**
*Section Manager, Water Resources and Hydraulic Engineering Group, Stone & Webster Engineering Corporation*

**Allen L. Clapp, P.E.**
*Managing Director, Clapp Research Associates; Member, IEEE-PES/IAS, AEE*


## CONTENTS

*Numbers refer to paragraphs*

## FOSSIL-FUELED PLANTS

By JOHN A. MAKUCH

***1. Introduction.*** Traditionally, fossil fuels have supplied most of the world's energy requirements. Current predictions are that the United States—as well as many other countries—will continue to rely heavily on these fuels well into the next century.

One of the largest users of fossil fuels is the electric utility industry. The combustion of fossil fuels produces high-pressure (typically 2400 to 3500 psig) and high-temperature (most commonly 1000°F) steam, which is used to drive a turbine at 3600 r/min. The turbine drives an electrical generator. There is more than 600,000 MW of electrical generating capacity installed in the United States. Well over half of this capacity is powered by fossil fuels.

***2. Thermodynamic Cycles.*** *Rankine Cycle.* The cornerstone of the modern steam power plant is a modification of the Carnot cycle proposed by W. J. M. Rankine, a distinguished Scottish engineering professor of thermodynamics and applied mechanics. The temperature-entropy and enthalpy-entropy diagrams of Fig. 5-1 illustrate the state changes for the Rankine

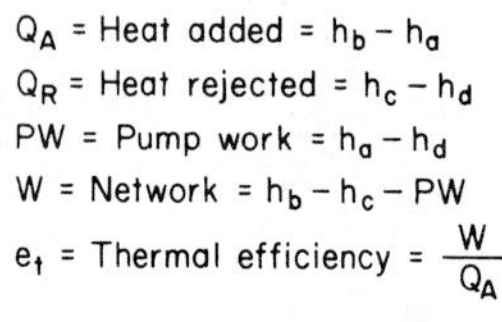


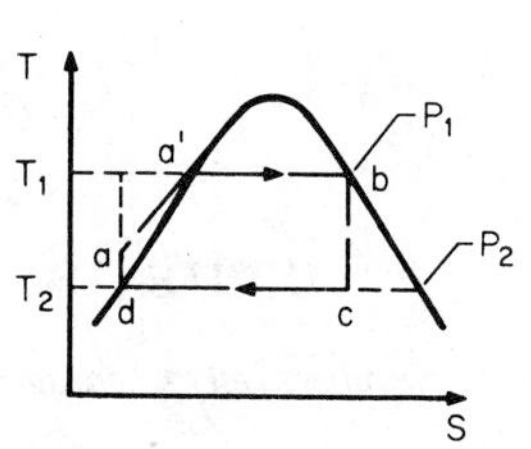


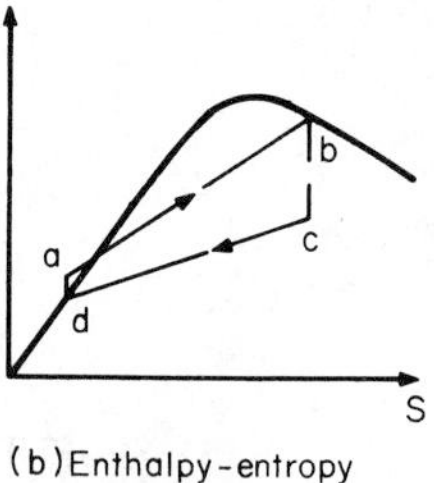


**FIG. 5-1** Simple Rankine cycle (without superheat).

cycle. With the exception that compression terminates (state *a*) at boiling pressure rather than the boiling temperature (state *a′*), the cycle resembles a Carnot cycle. The triangle bounded by *a-a′* and the line connecting to the temperature-entropy curve in Fig. 5-1*a* signify the loss of cycle work because of the irreversible heating of the liquid from state *a* to saturated liquid. The lower pressure at state *a*, compared to *a′*, makes possible a much smaller work of compression between *d-a*. For operating plants it amounts to 1% or less of the turbine output.

This modification eliminates the two-phase vapor compression process, reduces compression work to a negligible amount, and makes the Rankine cycle less sensitive than the Carnot cycle to the irreversibilities bound to occur in an actual plant. As a result, when compared with a Carnot cycle operating between the same temperature limits and with realistic component efficiencies, the Rankine cycle has a larger network output per unit mass of fluid circulated, smaller size, and lower cost of equipment. In addition, because of its relative insensitivity to irreversibilities, its operating plant thermal efficiencies will exceed those of the Carnot cycle.

*Regenerative Rankine Cycle.* Refinements in component design soon brought power plants based on the Rankine cycle to their peak thermal efficiencies, with further increases realized by modifying the basic cycle. This occurred through increasing the temperature of saturated steam supplied to the turbine, by increasing the turbine inlet temperature through constant-pressure superheat, by reducing the sink temperature, and by reheating the working vapor after partial expansion followed by continued expansion to the final sink temperature. In practice, all of these are employed with yet another important modification. The irreversibility associated with the heating of the compressed liquid to saturation by a finite temperature difference is the primary thermodynamic cause of lower thermal efficiency for the Rankine cycle. The regenerative cycle attempts to eliminate this irreversibility by using as heat sources other parts of the cycle with temperatures slightly above that of the compressed liquid being heated.

This procedure of transferring heat from one part of a cycle to another in order to eliminate or reduce external irreversibilities is called "regenerative heating" which is basic to all regenerative cycles.

The scheme shown in Fig. 5-2 is a practical approach to regeneration. Extraction or "bleeding" of steam at state *c* for use in the "open" heater avoids excessive cooling of the vapor during turbine expansion; in the heater, liquid from the condenser increases in temperature by $\Delta T$. (Regenerative cycle heaters are called "open" or "closed" depending on whether hot and cold fluids are mixed directly to share energy or kept separate with energy exchange occurring by the use of metal coils.)

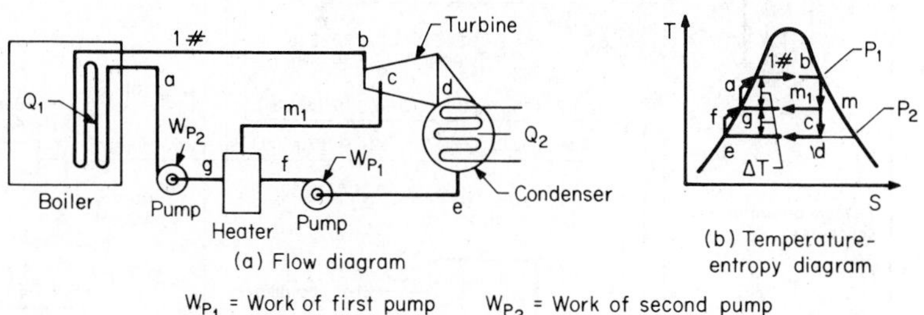


**FIG. 5-2** Single extraction regenerative cycle.

The extraction and heating substitute the finite temperature difference $\Delta T$ for the infinitesimal $dT$ used in the theoretical regeneration process. This substitution, while failing to realize the full potential of regeneration, halves the temperature difference through which the condensate must be heated in the basic Rankine cycle. Additional extractions and heaters permit a closer approximation to the maximum efficiency of the idealized regenerative cycle, with further improvement over the simple Rankine cycle shown in Fig. 5-1.

Reducing the temperature difference between the liquid entering the boiler and that of the saturated fluid increases the cycle thermal efficiency. The price paid is a decrease in net work produced per pound of vapor entering the turbine and an increase in the size, complexity, and initial cost of the plant. Additional improvements in cycle performance may be realized by continuing to accept the consequences of increasing the number of feedwater heating stages. Balancing cycle thermal efficiency against plant size, complexity, and cost for production of power at minimum cost determines the optimum number of heaters.

*Reheat Cycle.* The use of *superheat* offers a simple way to improve the thermal efficiency of the basic Rankine cycle and reduce vapor moisture content to acceptable levels in the low-pressure stages of the turbine. But with continued increase of higher temperatures and pressures to achieve better cycle efficiency, in some situations available superheat temperatures are insufficient to prevent excessive moisture from forming in the low-pressure turbine stages.

The solution to this problem is to interrupt the expansion process, remove the vapor for *reheat* at constant pressure, and return it to the turbine for continued expansion to condenser pressure. The thermodynamic cycle using this modification of the Rankine cycle is called the "reheat cycle." Reheating may be carried out in a section of the boiler supplying primary steam, in a separately fired heat exchanger, or in a steam-to-steam heat exchanger. Most present-day utility units combine superheater and reheater in the same boiler.

Usual central-station practice combines both regenerative and reheat modifications to the basic Rankine cycle. For large installations, reheat makes possible an improvement of approximately 5% in thermal efficiency and substantially reduces the heat rejected to the condenser cooling water. The operating characteristics and economics of modern plants justify the installation of only one stage of reheat except for units operating at supercritical pressure.

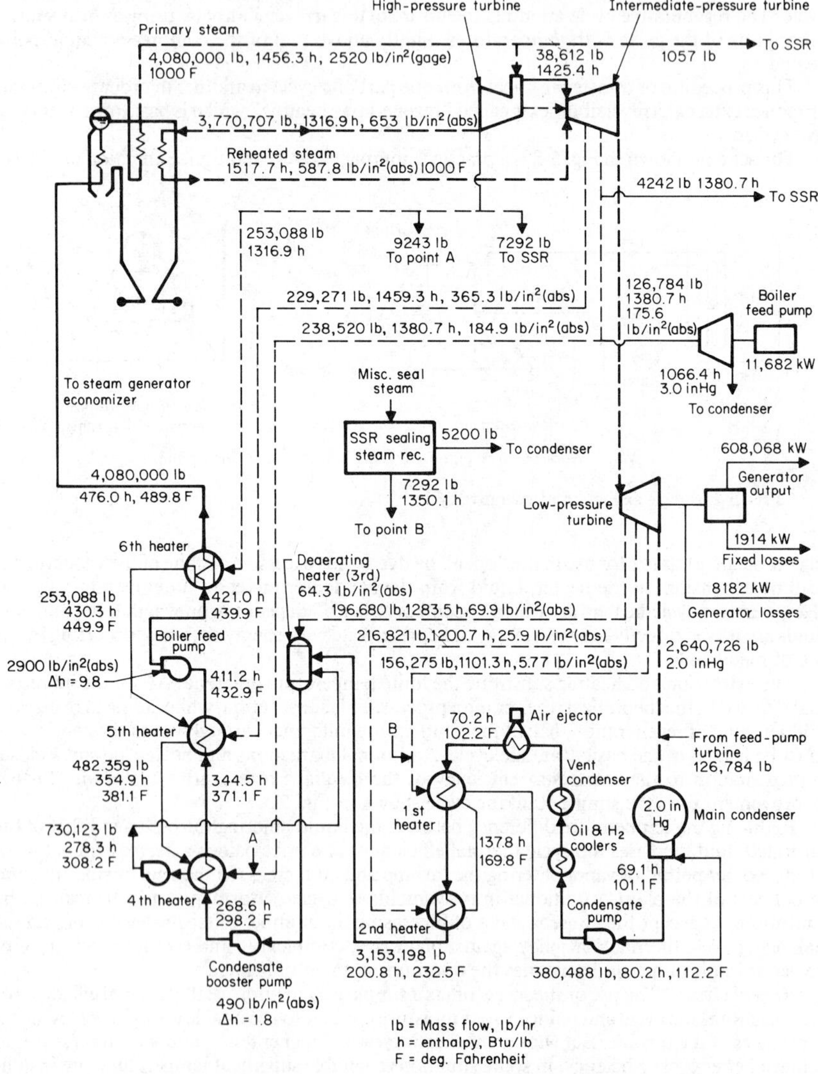


**FIG. 5-3** Reheat regenerative cycle, 600-MW subcritical-pressure fossil-fuel power plant.

Figure 5-3 shows a flow diagram for a 600-MW fossil-fueled reheat cycle designed for initial turbine conditions of 2520-psig and 1000°F steam. Six feedwater heaters are supplied by exhaust steam from the high-pressure turbine and extraction steam from the intermediate and low-pressure turbines. Except for the deaërating heater (third), all heaters shown are closed heaters. Three pumps are shown: (1) the condensate pump which pumps the condensate through oil and hydrogen gas coolers, vent condenser, air ejector, first and second heaters, and deaerating heater; (2) the condensate booster pump which pumps the condensate through fourth and fifth heaters; and (3) the boiler feed pump which pumps the condensate through the sixth heater to the economizer and boiler. The mass flows noted on the diagram are in pounds per hour at the prescribed conditions for full-load operation.

***3. Reheat Steam Generators.*** The boiler designer must proportion heat-absorbing and heat-recovery surfaces to make best use of the heat released by the fuel. Waterwalls, superheaters, and reheaters are exposed to convection and radiant heat, whereas convection heat transfer predominates in air heaters and economizers.

The relative amounts of such surfaces vary with the size and operating conditions of the boiler. A small low-pressure heating plant with no heat-recovery equipment has quite a different arrangement from a large high-pressure unit operating on a reheat regenerative cycle and incorporating heat-recovery equipment.

*Factors Influencing Boiler Design.* In addition to the basics of unit size, steam pressure, and steam temperature, the designer must consider other factors that influence the overall design of the steam generator.

*Fuels.* Coal, although the most common fuel, is also the most difficult to burn. The ash in coal consists of a number of objectionable chemical elements and compounds. Because of the high percentage of ash that can occur in coal, it has a serious effect on furnace performance.

At the high temperatures resulting from the burning of fuel in the furnace, fractions of ash can become partially fused and sticky. Depending on the quantity and fusion temperature, the partially fused ash may adhere to surfaces contacted by the ash-containing combustion gases, causing objectionable buildup of slag on or bridging between tubes. Chemicals in the ash may attack materials such as the alloy steel used in superheaters and reheaters.

In addition to the deposits in the high-temperature sections of the unit, the air heater (the coolest part) may be subject to corrosion and plugging of gas passages from sulfur compounds in the fuel acting in combination with moisture present in the flue gas.

*The Furnace.* Heat generated in the combustion process appears as furnace radiation and sensible heat in the products of combustion. Water circulating through tubes that form the furnace wall lining absorbs as much as 50% of this heat which, in turn, generates steam by the evaporation of part of the circulated water.

Furnace design must consider water heating and steam generation in the wall tubes as well as the processes of combustion. Practically all large modern boilers have walls comprised of water-cooled tubes to form complete metal coverage of the furnace enclosure. Similarly, areas outside of the furnace which form enclosures for sections of superheaters, reheaters, and economizers also use either water- or steam-cooled tube surfaces. Present practice is to use tube arrangements and configurations which permit practically complete elimination of refractories in all areas that are exposed to high-temperature gases.

Waterwalls usually consist of vertical tubes arranged in tangent or approximately so, connected at top and bottom to headers. These tubes receive their water supply from the boiler drum by means of downcomer tubes connected between the bottom of the drum and the lower headers. The steam, along with a substantial quantity of water, is discharged from the top of the waterwall tubes into the upper waterwall headers and then passes through riser tubes to the boiler drum. Here the steam is separated from the water, which together with the incoming feedwater is returned to the waterwalls through the downcomers.

Tube diameter and thickness are of concern from the standpoints of circulation and metal temperatures. Thermosyphonic (also called thermal or natural) circulation boilers generally use larger-diameter tubes than positive (pumped) circulation or once-through boilers. This practice is dictated largely by the need for more liberal flow area to provide the lower velocities necessary with the limited head available. The use of small-diameter tubes is an advantage in high-pressure boilers because the lesser tube thicknesses required result in lower outside tube-metal temperatures. Such small-diameter tubes are used in recirculation boilers in which pumps provide an adequate head for circulation and maintain the desired velocities.

*Superheaters and Reheaters.* The function of a superheater is to raise the boiler steam temperature above the saturated temperature level. As steam enters the superheater in an essentially dry condition, further absorption of heat sensibly increases the steam temperature.

The reheater receives superheated steam which has partly expanded through the turbine. As described earlier, the role of the reheater in the boiler is to re-superheat this steam to a desired temperature.

Superheater and reheater design depends on the specific duty to be performed. For relatively low final outlet temperatures, superheaters solely of the convection type are generally used. For higher final temperatures, surface requirements are larger and, of necessity, superheater elements are located in very high gas-temperature zones. Wide-spaced platens or panels, or wall-type superheaters or reheaters of the radiant type, can then be used. Figure 5-4 shows an

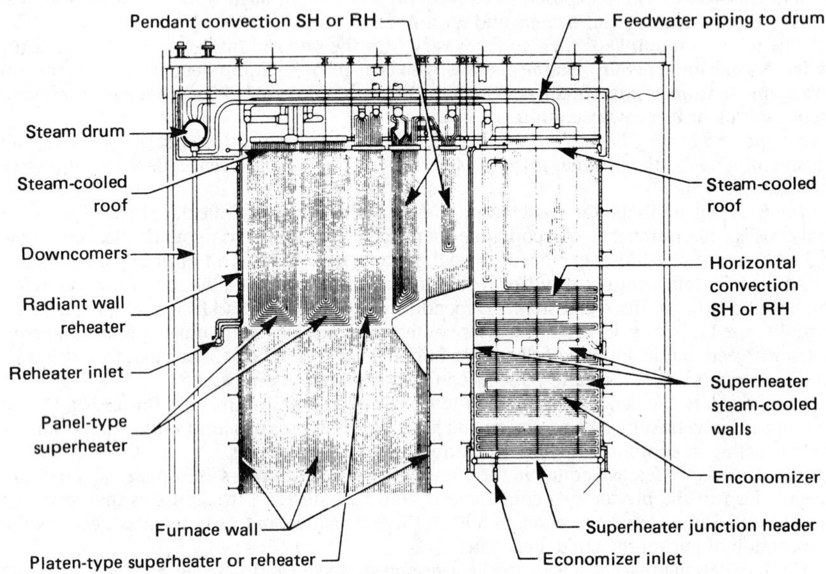


**FIG. 5-4** Arrangement of superheater, reheater, and economizer of a large coal-fired steam generator.

arrangement of such platen and panel surfaces. A relatively small number of panels are located on horizontal centers of 5 to 8 ft to permit substantial radiant heat absorption. Platen sections, on 14- to 28-in centers, are placed downstream of the panel elements; such spacing provides high heat absorption by both radiation and convection.

*Economizers.* Economizers help to improve boiler efficiency by extracting heat from flue gases discharged from the final superheater section of a radiant/reheat unit (or the evaporative bank of an industrial boiler). In the economizer, heat is transferred to the feedwater, which enters at a temperature appreciably lower than that of saturated steam. Generally economizers are arranged for downward flow of gas and upward flow of water.

Water enters from a lower header and flows through horizontal tubing comprising the heating surface. Return bends at the ends of the tubing provide continuous tube elements, whose upper ends connect to an outlet header that is in turn connected to the boiler drum by means of tubes or large pipes.

As shown in Fig. 5-4, economizers of a typical utility-type boiler are located in the same pass as the primary or horizontal sections of the superheater, or superheater and reheater, depending on the arrangement of the surface. Tubing forming the heating surface is generally low-carbon steel. Because steel is subject to corrosion in the presence of even extremely low concentrations of oxygen, it is necessary to provide water that is practically 100% oxygen-free. In central stations and other large plants, it is common practice to use deaerators for oxygen removal.

*Air Heaters.* Steam-generator air heaters have two important and concomitant functions: they cool the gases before they pass to the atmosphere, thereby increasing fuel-firing efficiency; at the same time, they raise the temperature of the incoming air of combustion. Depending on the pressure and temperature cycle, the type of fuel, and the type of boiler involved, one of the two functions will have prime importance.

For instance, in a low-pressure gas- or oil-fired industrial or marine boiler, combustion-gas temperature can be lowered in several ways—by a boiler bank, by an economizer, or by an air heater. Here, an air heater has principally a gas-cooling function, as no preheating is required to burn the oil or gas. If the boiler is a high-pressure reheat unit burning a high-moisture subbituminous or lignitic coal, high preheated-air temperatures are needed to evaporate the moisture in the coal before ignition can take place. Here, the air-heating function becomes primary. Without exception, then, large pulverized-coal boilers either for industry or electric power generation use air heaters to reduce the temperature of the combustion products from the 600 to 800°F level to final exit-gas temperatures of 275 to 350°F. In these units, the combination air is heated from about 80°F to between 500 to 750°F, depending on coal calorific value and moisture content.

In theory, only the primary air must be heated: that is, air used to actually dry the coal in the pulverizers. Ignited fuel can burn without preheating the secondary and tertiary air. However, there is considerable advantage to the furnace heat-transfer process from heating *all* the combustion air; it increases the rate of burning and helps raise adiabatic temperature.

***Fossil Fuels.*** Fossil fuels used for steam generation in utility and industrial power plants may be classified into solid, liquid, and gaseous fuels as in Table 5-1. Each fuel may be further classified as a natural, manufactured, or by-product fuel. Not mutually exclusive, these classifications necessarily overlap in some areas. Obvious examples of natural fuels are coal, crude oil, and natural gas.

Of all the fossil fuels used for steam generation in electric-utility and industrial power plants today, coal is the most important. It is widely available throughout much of the world, and the quantity and quality of coal reserves are better known than those of other fuels.

***5. Classification of Coal.*** Coals are grouped according to rank. For the purposes of the power-plant operator, there are several suitable ranks of coal:

Anthracite
Bituminous
Subbituminous
Lignite

The following description of coals by rank gives some of their physical characteristics.

*Anthracite.* Hard and very brittle, anthracite is dense, shiny black, and homogeneous with no marks or layers. Unlike the lower-rank coals, it has a high percentage of fixed carbon and a low percentage of volatile matter. Anthracites include a variety of slow-burning fuels merging into graphite at one end and into bituminous coal at the other. They are the hardest coals on the market, consisting almost entirely of fixed carbon, with the little volatile matter present in them chiefly as methane, $CH_4$. Anthracite is usually graded into small sizes before being burned on stokers. The "metaanthracites" burn so slowly as to require mixing with other coals, while the "semianthracites," which have more volatile matter, are burned with relative ease if properly fired. Most anthracites have a lower heating value than the highest-grade bituminous coals. Anthracite is used principally for heating homes and in gas production.

Some semianthracites are dense, but softer than anthracite, shiny gray, and somewhat granular in structure. The grains have a tendency to break off in handling the lump, and produce a coarse, sandlike slack. Other semianthracites are dark gray and distinctly granular. The grains break off easily in handling and produce a coarse slack. The granular structure has been produced by small vertical cracks in horizontal layers of comparatively pure coal separated by very thin partings. The cracks are the result of heavy downward pressure, and probably shrinkage of the pure coal because of a drop in temperature.

*Bituminous.* By far the largest group, bituminous coals derive their name from the fact that on being heated they are often reduced to a cohesive, binding, sticky mass. Their carbon content is less than that of anthracites, but they have more volatile matter. The character of their volatile matter is more complex than that of anthracites and they are higher in calorific value. They burn

**TABLE 5-1** Classification of Coals of Rank[a]

| Class and group | Fixed carbon limits, % (dry, mineral-matter-free basis) | | Volatile matter limits, % (dry, mineral-matter-free basis) | | Calorific value limits, Btu/lb (moist,[b] mineral-matter-free basis) | | Agglomerating character |
|---|---|---|---|---|---|---|---|
| | Equal or greater than | Less than | Equal or greater than | Less than | Equal or greater than | Less than | |
| Anthracitic | | | | | | | |
| Metaanthracite | 98 | ... | ... | 2 | ... | ... | Nonagglomerating |
| Anthracite | 92 | 98 | 2 | 8 | ... | ... | |
| Semianthracite[c] | 86 | 92 | 8 | 14 | ... | ... | |
| Bituminous | | | | | | | |
| Low-volatile bituminous coal | 78 | 86 | 14 | 22 | ... | ... | |
| Medium-volatile bituminous coal | 69 | 78 | 22 | 31 | ... | ... | |
| High-volatile A bituminous coal | ... | 69 | 31 | ... | 14,000[d] | ... | Commonly agglomerating[e] |
| High-volatile B bituminous coal | ... | ... | ... | ... | 13,000[d] | 14,000 | |

| | | | | | | | |
|---|---|---|---|---|---|---|---|
| High-volatile C bituminous coal | ... | ... | ... | ... | 11,500 | 13,000 | |
| | | | | | 10,500 | 11,500 | Agglomerating |
| Subbituminous | | | | | | | |
| Subbituminous A coal | ... | ... | ... | ... | 10,500 | 11,500 | |
| Subbituminous B coal | ... | ... | ... | ... | 9,500 | 10,500 | Nonagglomerating |
| Subbituminous C coal | ... | ... | ... | ... | 8,300 | 9,500 | |
| Lignitic | | | | | | | |
| Lignite A | ... | ... | ... | ... | 6,300 | 8,300 | |
| Lignite B | ... | ... | ... | ... | ... | 6,300 | |

[a] This classification does not include a few coals, principally nonbanded varieties, which have unusual physical and chemical properties and which come within the limits of fixed carbon or calorific value of the high-volatile bituminous and subbituminous ranks. All of these coals either contain less than 48% dry, mineral-matter-free fixed carbon or have more than 15,500 moist, mineral-matter-free Btu per pound.

[b] Moist refers to coal containing its natural inherent moisture but not including visible water on the surface of the coal.

[c] If agglomerating, classify in low-volatile group of the bituminous class.

[d] Coals having 69% or more fixed carbon on the dry, mineral-matter-free basis shall be classified by fixed carbon, regardless of calorific value.

[e] It is recognized that there may be nonagglomerating varieties in these groups of the bituminous class, and there are notable exceptions in the high-volatile C bituminous group.

NOTE: 1 Btu/lb = 2326 J/kg.

Source: ASTM Standards D 388, Classification of Coals by Rank.

easily, especially in pulverized form, and their high volatile content makes them good for producing gas. Their binding nature enables them to be used in the manufacture of coke, while the nitrogen in them is utilized in processing ammonia.

The low-volatile bituminous coals are grayish-black and distinctly granular in structure. The grain breaks off very easily, and handling reduces the coal to slack. Any lumps that remain are held together by thin partings. Because the grains consist of comparatively pure coal, the slack is usually lower in ash content than are the lumps.

Medium-volatile bituminous coals are the transition from high-volatile to low-volatile coal and, as such, have the characteristics of both. Many have a granular structure, are soft, and crumble easily. Some are homogeneous with very faint indications of grains or layers. Others are of more distinct laminar structure, are hard, and stand handling well.

High-volatile A bituminous coals are mostly homogeneous with no indication of grains, but some show distinct layers. They are hard and stand handling with little breakage. The moisture, ash, and sulfur contents are low, and the heating value high.

High-volatile B bituminous coals are of distinct laminar structure; the layers of black, shiny coal alternate with dull, charcoallike layers. They are hard and stand handling well. Breakage occurs generally at right angles and parallel to the layers, so that the lumps generally have a cubical shape.

High-volatile C bituminous coals are of distinct laminar structure, are hard, and stand handling well. They generally have high moisture, ash, and sulfur contents and they are considered to be free-burning coals.

*Subbituminous.* These coals are brownish black or black. Most are homogeneous with smooth surfaces, and with no indication of layers. They have high moisture content, as much as 15 to 30%, although appearing dry. When exposed to air they lose part of the moisture and crack with an audible noise. On long exposure to air, they disintegrate. They are free-burning, entirely noncoking, coals.

*Lignite.* Lignites are brown and of a laminar structure in which the remnants of woody fibers may be quite apparent. The word *lignite* comes from the Latin word *lignum* meaning wood. Their origin is mostly from plants rich in resin, so they are high in volatile matter. Freshly mined lignite is tough, although not hard, and it requires a heavy blow with a hammer to break the large lumps. But on exposure to air, it loses moisture rapidly and disintegrates. Even when it appears quite dry, the moisture content may be as high as 30%. Owing to the high moisture and low heating value, it is not economical to transport it long distances.

Unconsolidated lignite (B in Table 5-1) is also known as "brown coal." Brown coals are generally found close to the surface, contain more then 45% moisture, and are readily won by strip mining.

***6. Impact of Fuel on Boiler Design.*** The most important item to consider when designing a utility or large industrial steam generator is the fuel the unit will burn. The furnace size, the equipment to prepare and burn the fuel, the amount of heating surface and its placement, the type and size of heat-recovery equipment, and the flue-gas-treatment devices are all duel-dependent.

The major differences among those boilers that burn coal or oil or natural gas result from the ash in the products of combustion. Firing oil in a furnace results in relatively small amounts of ash; there is no ash from natural gas. For the same output, because of the ash, coal-burning boilers must have larger furnaces and the velocities of the combustion gases in the convection passes must be lower. In addition, coal-burning boilers need ash-handling and particulate-cleanup equipment that costs a great deal and requires considerable space.

Table 5-2 lists the variation in calorific values and moisture contents of several coals, and the mass of fuel that must be handled and fired to generate the same electrical-power output. These values are important because the quantity of fuel required helps determine the size of the coal-storage yard, as well as the handling, crushing, and pulverizing equipment for the various coals.

*Furnace Sizing.* The most important step in coal-fired unit design is to properly size the furnace. Furnace size has a first-order influence on the size of the structural-steel framing, the boiler building and its foundations, as well as on the sootblowers, platforms, stairways, steam piping, and duct work. The fuel-ash properties that are particularly important when designing and establishing the size of coal-fired furnaces include:

The ash fusibility temperatures (both in terms of their absolute values and the spread or difference between initial deformation temperature and fluid temperature)

**TABLE 5-2** Representative Coal Analyses

| | Medium-volume bituminous | High-volume bituminous | Subbituminous C | Low-sodium lignite | Medium-sodium lignite | High-sodium lignite |
|---|---|---|---|---|---|---|
| Total $H_2O$, % | 5.0 | 15.4 | 30.0 | 31.0 | 30.0 | 39.6 |
| Ash, % | 10.3 | 15.0 | 5.8 | 10.4 | 28.4 | 6.3 |
| VM, % | 31.6 | 33.1 | 32.6 | 31.7 | 23.2 | 27.5 |
| FC, % | 53.1 | 36.5 | 36.6 | 26.9 | 18.4 | 26.6 |
| Btu/lb, as fired | 13,240 | 10,500 | 8,125 | 7,590 | 5,000 | 6,523 |
| Btu/lb, MAF | 15,640 | 15,100 | 12,650 | 12,940 | 12,020 | 12,050 |
| Fusion (reducing), °F | | | | | | |
| Initial def. | 2,170 | 1,990 | 2,200 | 2,075 | 2,120 | 2,027 |
| Softening | 2,250 | 2,120 | 2,250 | 2,200 | 2,380 | 2,089 |
| Fluid | 2,440 | 2,290 | 2,290 | 2,310 | 2,700 | 2,203 |
| Ash analysis, % | | | | | | |
| $SiO_2$ | 40.0 | 46.4 | 29.5 | 46.1 | 62.9 | 23.1 |
| $Al_2O_3$ | 24.0 | 16.2 | 16.0 | 15.2 | 17.5 | 11.3 |
| $Fe_2O_3$ | 16.8 | 20.0 | 4.1 | 3.7 | 2.8 | 8.5 |
| CaO | 5.8 | 7.1 | 26.5 | 16.6 | 4.8 | 23.8 |
| MgO | 2.0 | 0.8 | 4.2 | 3.2 | 0.7 | 5.9 |
| $Na_2O$ | 0.8 | 0.7 | 1.4 | 0.4 | 3.1 | 7.4 |
| $K_2O$ | 2.4 | 1.5 | 0.5 | 0.6 | 2.0 | 0.7 |
| $TiO_2$ | 1.3 | 1.0 | 1.3 | 1.2 | 0.8 | 0.5 |
| $P_2O_5$ | 0.1 | 0.1 | 1.1 | 0.1 | 0.1 | 0.2 |
| $SO_3$ | 5.3 | 6.0 | 14.8 | 12.7 | 4.6 | 17.7 |
| Sulfur, % | 1.8 | 3.2 | 0.3 | 0.6 | 1.7 | 0.8 |
| Lb $H_2O$/million Btu | 3.8 | 14.7 | 36.9 | 40.8 | 60.0 | 60.7 |
| Lb ash/million Btu | 7.8 | 14.3 | 7.1 | 13.7 | 56.8 | 9.7 |
| Fuel fired,* 1000 lb/h | 405 | 520 | 705 | 750 | 1,175 | 900 |

* Constant heat output, nominal 600-MW unit, adjusted for efficiency.

NOTE: 1 Btu/lb = 2326 J/kg; $t_{°C} = (t_{°F} - 32)/1.8$; 1 lb = 0.4536 kg; 1 Btu = 1055 J.

The ratio of basic to acidic ash constituents

The iron/calcium ratio

The fuel-ash content in terms of pounds of ash per million Btu's

The ash friability

These characteristics and others translate into the furnace sizes in Fig. 5-5, which are based on the six coal ranks shown in Table 5-2. This size comparison illustrates the philosophy of increasing the furnace plan area, volume, and the fuel burnout zone (the distance from the top fuel nozzle to the furnace arch), as lower grade coals with poorer ash characteristics are fired.

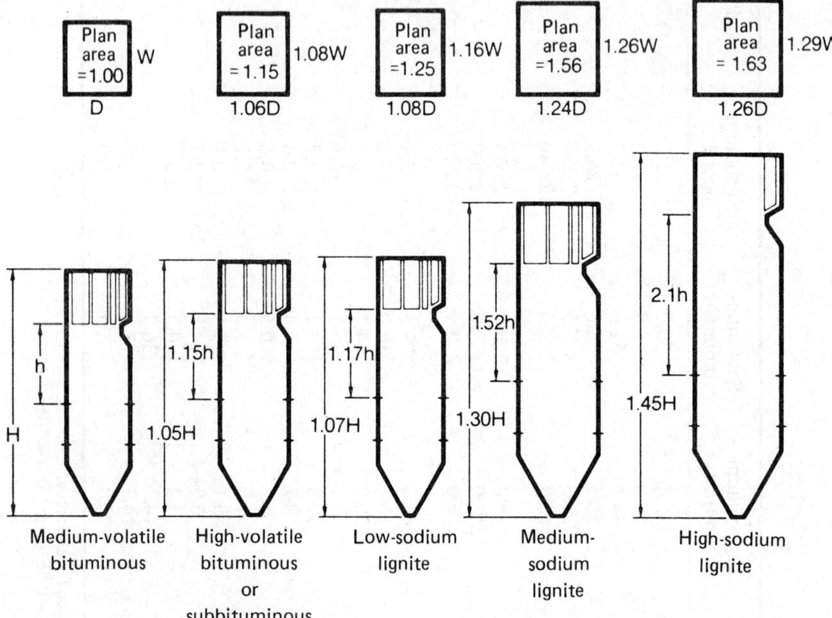


**FIG. 5-5** Effect of coal rank on furnace sizing (constant heat output).

Figure 5-5 is a simplified characterization of actual furnaces built to burn the fuels listed in Table 5-2. Wide variations exist in fuel properties within coal ranks, as well as within several subclassifications (e.g., subbituminous A, B, C), each of which may require a different size furnace.

Among the most important design criteria in large pulverized-fuel furnaces are net heat input in Btu's per hour per square foot of furnace plan area (NHI/PA) and the vertical distance from the top fuel nozzle to the furnace arch. Furnace dimensions must be adequate to establish the necessary furnace retention time to properly burn the fuel as well as to cool the gaseous combustion products. This is to ensure that the gas temperature at the entrance to the closely spaced convection surface is well below the ash-softening temperature of the lowest-quality coal burned. Heat-absorption characteristics of the walls are maintained using properly placed wall blowers to control the furnace outlet gas temperature by removing ash deposited on the furnace walls below the furnace outlet plane.

*7. **Environmental Considerations.*** Over the past 15 years, concerns for the control of air quality have probably had the largest single impact on power plant site selection, design, operation, and cost. The three classes of emissions which are of major concern are nitrogen oxides, sulfur oxides, and particulate matter.

*Nitrogen Oxides.* In the United States, nitrogen oxides can be controlled within regulatory limits by proper design and control of the firing system. Each steam generator manufacturer has developed a specific firing system design. The common characteristics of all of these designs, however, included a careful regulation of the fuel/air ratio in the firing zone where the major fraction of the fuel nitrogen compounds are liberated and control of the heat liberation pattern in the furnace.

*Particulate Control.* The traditional particulate control device in power plant applications has been the electrostatic precipitator. In recent years, fabric filters (also called baghouses) have become increasingly popular.

In electrostatic precipitation, suspended particles in the gas are electrically charged, then driven to collecting electrodes by an electrical field; the electrodes are rapped to cause the particles to drop into collecting hoppers. This process differs from mechanical or filtering processes in which forces are exerted directly on the particulates rather than the gas as a whole. Effective separation of particles can be achieved with lower power expenditure, with negligible draft loss, and with little or no effect on the composition of the gas.

As initially applied, precipitators were designed to provide a minimum plate area at low cost; designs used interlocking or opzel collection plates and hanging weighted-wire discharge electrodes (Fig. 5-6). Roof-mounted gang-rapping vibrators removed particulate from the collectors.

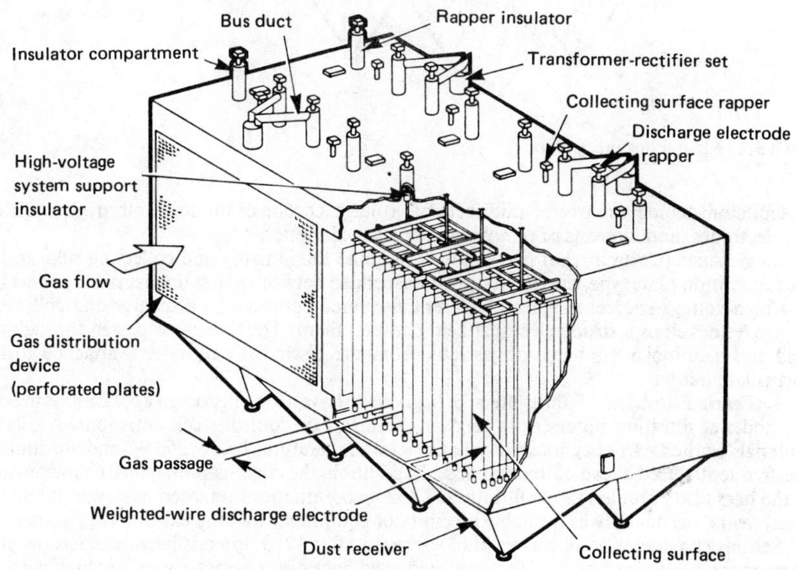


**FIG. 5-6** Schematic arrangement of weighted-wire precipitator.

To meet a demand for ultra-high-efficiency collectors of rugged construction and high reliability, European manufacturers in possession of the basic patents disseminated by Frederick Gardner Cottrell developed the rigid-frame precipitator (Fig. 5-7). Actually, this design more closely approximated Cottrell's original design than did the U.S.-style weighted-wire designs. The term *rigid-frame* refers to the rugged pipe-frame or mast-construction discharge electrode, which largely precludes wire breakage. The basic design incorporates: segmented collecting plate configurations and profiles for close fabrication tolerances over heights more than 40 ft, rigid discharge electrodes, and much greater division of rapping, often with individual rapping for each discharge frame.

The principle of electrostatic precipitation is relatively simple. The process applies an electrostatic charge to dust particles with a corona discharge and passes them through an electrical field where the particles are attracted to a collecting surface. The basic elements of a

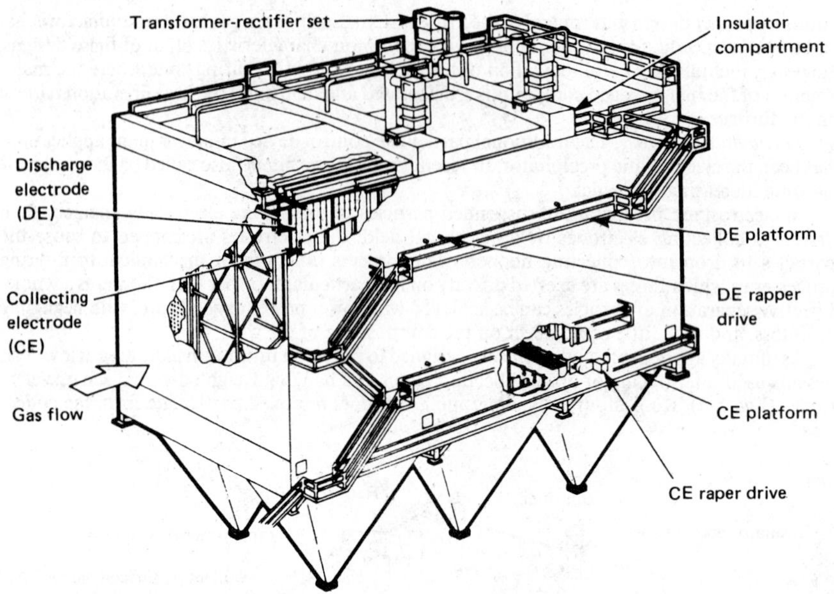


**FIG. 5-7** Rigid-frame precipitator.

precipitator include a source of unidirectional voltage, corona or discharge electrodes, collecting electrodes, and a means of removing the collected matter.

Single-stage (Cottrell-type) precipitators combine the ionizing and collecting step. In the more common plate type, the electrodes are suspended between plates on insulators connected to a high-voltage source. A voltage differential created between the discharge and collecting electrodes develops a strong electrical field between them. The flue gas is passed through the field and a unipolar discharge of gas ions, from the discharge electrode, is attached to the particulate matter.

***8. Fabric Filtration.*** Fabric filters, or baghouses, have a long history of applications in both dry and wet filtration processes to recover chemicals or control stack emissions. Available materials limited early baghouse installations to temperatures below 250°F, and air dilution was frequently used ahead of the baghouse. In addition, the chemical-resistance characteristics of the bags also curtailed fabric filtration. These two limitations retarded its development for many years, particularly as available precipitator equipment met the existing regulations.

Serious consideration of this technology began after 1970; interest heightened as installations on large coal-fired boilers demonstrated good operating characteristics and high particulate-removal efficiencies.

*Types of Fabric Filters.* Baghouses have a relatively constant collection efficiency but varying pressure drop, whereas precipitators have a relatively constant pressure drop but varying efficiency. When dirty gas flows through a fabric, the particulate matter in the gas forms a "cake" on the fabric. This deposit increases both the filtration efficiency of the cloth and its resistance to gas flow. Thus, for continuous operation, a fabric filter must have some provision to periodically clean the deposit from the cloth.

Various fabric filter designs are available. Selection depends on gas temperature, particulate characteristics, and the type of cleaning mechanism for removing the collected dust from the cloth.

This oldest of filter designs uses vigorous shaking to remove the filter cake (captured particulate). The shaking causes the cake to fracture and fall into the collection and disposal hopper. This method of cleaning has been applied to both inside collectors (those collecting particulate matter on the inside of the individual filter bags) and to outside collectors (those collecting the particulate matter on the outside of the individual filter bags).

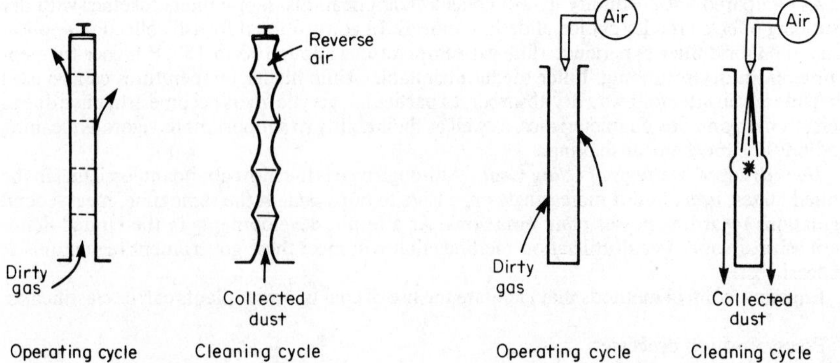


**FIG. 5-8** Reverse-air type of fabric collector.

**FIG. 5-9** Pulse-jet-cleaning type of fabric collector.

In the reverse-air system (Fig. 5-8), clean flue gas taken from the outlet duct work is blown in the direction counter to normal flow through the fabric of one or two compartments not in service. The cleaning frequency depends upon the porosity of the cake being formed, the inlet grain loading, and the predetermined allowable system pressure drop.

The pulse-jet (Fig. 5-9) outside collector filters the gases through the bag from the outside to inside. Wire or mesh frame cages support the bags to prevent their collapse during the filtering period. Pulse-jet cake removal uses a short-duration pulse of compressed air injected into the open end of the filter bag. This air pulse causes a momentary expansion of the bag as the pulse wave travels the bag length. The action fractures the cake, which falls into the collection and disposal hopper.

***9. Flue-Gas Desulfurization Systems.*** Flue-gas desulfurization (FGD) began in England in 1935. The technology remained dormant until the mid-1960s when it became active primarily in the United States and Japan. Since then, over 50 FGD processes have been developed, differing in the chemical reagents and the resultant end products.

The most common FGD system is a lime/limestone wet scrubber. After the flue gas has been treated in the precipitation (or baghouse), it passes through the induced fans and enters the $SO_2$ scrubber. If the required $SO_2$ removal efficiency is less than 85%, a fraction of the flue gas can be treated while bypassing the rest to mix with and reheat the saturated flue gas leaving the scrubber.

For higher-sulfur fuels requiring $SO_2$ removal efficiencies of 90% or greater, the entire flue-gas stream must be treated. Upon leaving the $SO_2$ absorption section, the flue gas is passed through entrainment separators to remove any slurry droplets mixed with the gas. The saturated flue gas is then reheated approximately 25 to 50°F above the water dewpoint before it is vented to the stack.

For low- to medium-sulfur fuels, an alternate scrubbing technology is dry scrubbing. This process minimizes water consumption and eliminates the requirement for flue-gas reheating but requires more expensive additives than the wet limestone systems.

The typical dry $SO_2$ absorber is a cocurrent classifying spray dryer. Flue gas enters the top of the absorber through inlet assemblies containing swirl vanes. The absorbent is injected pneumatically into the center of each swirler assembly by ultrasonic atomizing nozzles that require an air pressure of about 60 psig. Slurry feed pressures are 10 to 15 psig. The compressed air induces primary dispersion of the absorbent slurry by mechanical shear forces produced by the two fluid streams. Final dispersion is accomplished by shattering the droplets with ultrasonic energy produced by the compressed air used with a proprietary nozzle design. Then ultrasonic nozzles generate extremely fine droplets, which photographic studies show have diameters that range from 10 to 50 $\mu$m.

The flue-gas outlet design requires that effluent gases make a 180° turn before leaving the absorber. Besides eliminating product accumulation in the outlet duct, the abrupt directional change also allows the larger particles to drop out in the absorber product hopper. This design curtails the particulate loading to the fabric filter. Consequently, the number of cleaning cycles as well as abrasion of the filter medium are reduced.

As compared with ordinary fly-ash collection applications, fabric filters together with dry scrubbing offer a broader choice of design options. In conventional fly-ash collection applications, the fabric filter experiences flue-gas temperatures about 100 to 150°F higher than encountered in dry scrubbing. Filter media unsuitable at the higher temperatures can be used when the fabric filter follows a dry absorber. In particular, acrylic fibers become attractive due to their strength and flex characteristics, as well as their ability to support more vigorous cleaning methods like mechanical shaking.

***10. Advanced Methods of Using Coal.*** Although coal is the most abundant fossil fuel in the United States, users find it increasingly expensive to burn and, at the same time, meet federal regulations regarding power-plant emissions. As a result, developments in the United States emphasize advanced coal-utilization methods that will meet these government regulations at the least cost.

Emission-control methods that facilitate the use of coal in power plants can be classified as:

Precombustion processes

Concurrent-combustion processes

Postcombustion processes

Pollution control before combustion is accomplished either by partial oxidation of the coal to produce a clean gas or by producing a "clean" fuel through coal-liquefaction processes. Sulfur and ash are removed in these processes. The use of coal to produce a gas is not a new idea; it has been used in one form or another for over 200 years. But its use in the United States had almost disappeared by 1930, because natural gas was abundant and cost very little. Diminishing supplies and rising costs have now renewed the interest in coal gasification to produce substitute natural gas (SNG) and low- and medium-calorific-value (LCV and MCV) gas. Likewise, coal liquefaction is not a new technology, but it is only in limited commerical use in the United States. Large-scale production of synthetic liquid fuels from coal began about 1910 in Germany with the Fischer-Tropsch process, which is still used to produce a variety of fuels.

In fluidized-bed combustion, a concurrent-combustion emission-control process, much of the sulfur is removed during combustion by a sorbent material (limestone). In this process $NO_x$ production is low because of the low temperature at which the combustion reaction takes place. Fluidized-bed combustion has been under development since the 1950s, with the principal intent to demonstrate the economy of the coal and limestone system for sulfur oxide control as compared with postboiler wet scrubbing.

A coal-fired magnetohydrodynamic (MHD) system also is a control process that takes place concurrently with combustion of the coal. It is the direct conversion of kinetic and thermal energy in an ionized gas to electrical energy. Although the least developed of the new systems, it has the potential for maximum efficiency for the generation of electric power and thus will use less coal. A potassium seed used in the MHD power generation captures the sulfur, which then is separated chemically when the seed material is recovered. Oxides of nitrogen are controlled by staged combustion similar to that used in a pulverized-coal-fired boiler.

Postcombustion control processes are in extensive use today for $SO_2$ and particulate control. Lime-limestone scrubbers for $SO_2$ removal and equipment for particulate removal are described in Par. **9**.

Processes and equipment for removing the $NO_x$ from flue gases leaving a boiler are developmental. In the United States, most new steam generators do not require postcombustion $NO_x$ removal because it is possible to meet the current $NO_x$ standard with proper firing-system design.

To gasify coal economically for electric-power production, the coal must be transported to the electric generating plant for on-side gasification. The primary reason is that, in addition to producing a clean fuel, the gasification process generates large amounts of heat. To attain the high efficiencies necessary for economic base-load power generation, this heat must be recovered by functionally integrating the gasification plant with the electric generating plant (Fig. 5-10).

In an integrated electric power-plant application, the gasification system is part of a two-stage coal-combustion process. In the first, or gasification stage, the coal is partially reacted with a deficiency of oxygen to produce a low-heating-value fuel gas that can be readily cleaned. In the second stage, the cleaned fuel gas is burned in a boiler and/or gas turbine for the generation of electric power.

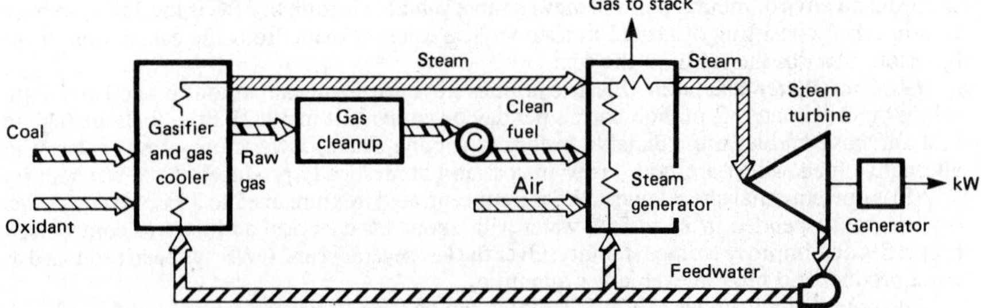


**FIG. 5-10** Integrated coal-gasification and steam-cycle power plant.

***11. Fluidized-Bed Combustion.*** For decades fluidized-bed reactors have been used in noncombustion reactions in which the thorough mixing and intimate contact of the reactants in a fluidized bed result in high product yield with improved economy of time and energy. Although conventional methods of burning coal also can generate energy with very high efficiency, fluidized-bed combustion can burn coal efficiently at a temperature low enough to avoid many of the problems of conventional combustion.

The outstanding advantage of fluidized-bed combustion (FBC) is its ability to burn high-sulfur coal in an environmentally acceptable manner without the use of flue-gas scrubbers. A secondary benefit is the formation of lower levels of nitrogen oxides compared to other combustion methods. In addition, low bed temperatures help eliminate the potential for slag formation on the water-cooled walls of the furnace (an operational advantage) and high heat transfer in the fluidized bed permits a more economical boiler design.

A fluidized bed is a layer of solid particles kept in turbulent motion by bubbles created by air being forced into the bed from below (Fig. 5-11). A fuel may be added and burned in this bed.

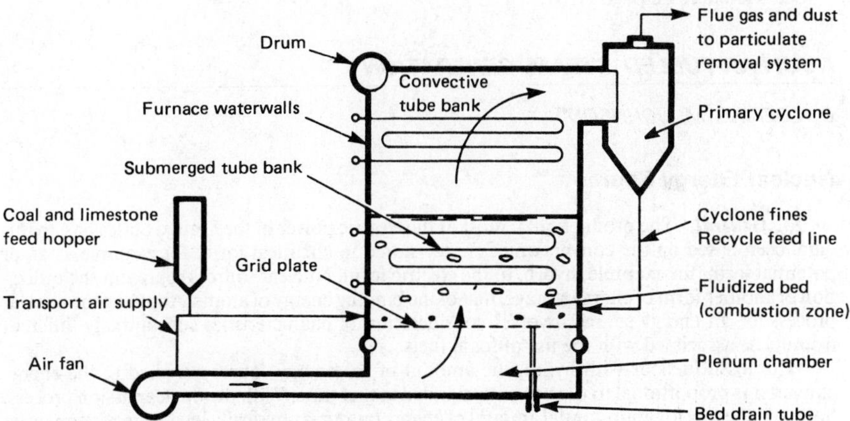


**FIG. 5-11** Schematic diagram of fluidized-bed combustion boiler.

Although the fuel will make up less than 1% of the bed, all of the solid particles introduced are quickly heated by the churning action of the bed. If heat-absorbing surface is submerged in the bed, the temperature of the solid-bed particles can be controlled at a predetermined level. The hot solids and gases surround the heat-absorbing tubes, and their intimate contact results in a high heat-transfer coefficient. This is the fundamental design principle of a fluidized-bed boiler. An FBC unit can be fired with many types of fuel, including the lower-grade coals, municipal solid wastes, coal-cleaning wastes, agricultural wastes, peat, and other fuels difficult to burn by conventional methods.

From an environmental point of view, a most valuable feature of FBC is the ability to burn coal in a bed consisting of heated limestone. The calcium oxide from the calcination of the limestone absorbs the sulfur in the coal.

*12. Coal-Water Mixtures.* Electric utilities still burn significant amounts of oil and natural gas (approximately 2 million barrels per day oil equivalent in 1982). Both fuels are high in cost and have had an unpredictable history of supply. There is, therefore, strong interest in alternative fuels which are both lower in cost and domestically produced. Coal-water slurry (CWS) is one potential alternative. CWS is a concentrated mixture of 65 to 75% pulverized coal (by weight) suspended in 25 to 35% water with about 1% chemical additives to control flow properties and improve storage stability. Over the last several years, CWS has been produced as a test product and fired in ever larger quantities.

The potential advantage of CWS is that is can be used in boilers designed to fire oil with relatively limited modifications to the existing plant. CWS is a liquid fuel which can be handled in conventional liquid-fuel systems with modifications such as pump changes and the addition of agitators to storage tanks. Although CWS contains 25 to 35% water, stable combustion has been demonstrated and the boiler efficiency loss is only 2 to 4% as compared to oil.

The major difficulty with using CWS is the presence of ash. Most oil and gas boilers were designed for a fuel which is virtually ash-free, but CWS produced from even a very clean coal has a significant quantity of ash. Ash deposits can accumulate in a boiler, reducing heat transfer and, in some cases, blocking gas passages. Ash, which passes through the boiler as fly ash, can cause erosion of boiler tubes. These problems generally require that boiler output when firing CWS be reduced (derated) from output levels when firing oil or gas. This can be significant, amounting to as much as 60% derating.

Derating can be minimized by reducing the amount of ash. This can be accomplished by cleaning (beneficiating) the coal. One important advantage of CWS is that physical coal-cleaning steps can be incorporated into the preparation process, and the energy-intensive drying of the clean coal can be eliminated.

Widespread commerical use of CWS will depend on its cost effectiveness. CWS must be sufficiently less expensive than oil to pay for the cost of boiler and plant modifications and the potential loss of capacity due to derating.

See references on page 5-109.

## NUCLEAR-FUELED STEAM GENERATION

*By LOERING M. JOHNSON*

### Nuclear Energy Source

*13. General.* The production of most of the electric power in the United States has, to this date, been based on the conversion of energy stored in chemical form, for example, coal, or potential form, for example, hydro, to the electric form. Nuclear sources represent the utilization of another form of energy storage, that of the binding energy of atoms. Applying the nuclear process for an energy source involves consideration of characteristics substantially different from those associated with the use of fossil fuels.

With fossil fuels or with hydro, the amount of the energy source supplied to the energy converter is proportional to the power demanded at that time. With the nuclear fission process, however, the fuel for a substantial amount of energy output is physically located in the converter at any time.

A second important characteristic of the nuclear process is the energy density. The thermal energy density in a typical fossil boiler (heated volume or core volume) is in the range of 0.20 kW/L; in a typical nuclear power generator it is in the range of 80 kW/L.

A third important difference is that of continued heat generation when the nuclear process is shut down following power operation.

A fourth important difference is that of emanations. The fossil process requires the intake of large volumes of air and fuel and the corresponding exhaust of large volumes of waste gas, some particulate matter, and in the case of coal-fired boilers, substantial quantities of ash. The nuclear process, however, requires only the input of the material placed in the core; its output is radioactive residue and corresponding radiation. The residue includes small quantities of gases which may be released or may be stored and solids which are contained within the fuel.

These differences, as well as many others, introduce many new considerations in the equipping and regulation of the nuclear process.

*14. Mass-Energy Relationships.* One of the first applications of the special theory of relativity proposed by Einstein in 1905 was the concept of the proportionality between mass and energy. This relationship, expressed by the equation $E = mc^2$, indicates that a change in nuclear mass can be converted to energy. If the mass $m$ is expressed in kilograms and the velocity of light $c$ in meters per second, the energy $E$ is in joules.

$$\begin{aligned} E(\text{J}) &= \text{mass (kg)} \times (2.998 \times 10^8 \text{ m/s})^2 \\ &= \text{mass (kg)} \times 8.99 \times 10^{16} \text{ m}^2\text{s}^2 \end{aligned} \tag{5-1}$$

In atomic and nuclear calculations, the joule is an inconvenient unit of energy because the amounts of energy involved in single nuclear events are usually very small. For convenience the electron volt, the energy acquired by any charged particle carrying a unit electronic charge falling through a potential of 1 V, is often used. One electron volt (eV) = $1.602 \times 10^{-19}$ J = 0.1602 attojoule and, correspondingly, 1 keV = $1.602 \times 10^{-16}$ J = 0.1602 femtojoule. One MeV = $1.602 \times 10^{-13}$ J = 0.1602 picojoule.

The mass-energy relationships become

$$\begin{aligned} E(\text{eV}) &= \text{mass(kg)} \times \frac{8.99 \times 10^{16} \text{ m}^2/\text{s}^2}{1.602 \times 10^{-19} \text{ J/eV}} \\ &= \text{mass(kg)} \times 5.61 \times 10^{35} \text{ keV/kg} \\ E\,(\text{keV}) &= \text{mass(kg)} \times 5.61 \times 10^{32} \text{ keV/kg} \\ E\,(\text{MeV}) &= \text{mass(kg)} \times 5.61 \times 10^{29} \text{ MeV/kg} \end{aligned}$$

where $1 \text{ J} = 1 \text{ m}^2 \cdot \text{kg/s}^2$

It is often convenient to use the energy corresponding to 1 atomic mass unit (amu). One amu = $1.657 \times 10^{-27}$ kg (1 amu = 1/12 of the mass of a neutral atom of $^{12}$C.

$$\begin{aligned} E_{\text{amu}} &= 1.66 \times 10^{-27} \text{ kg} \times 5.61 \times 10^{29} \text{ MeV/kg} \\ &= 931 \text{ MeV/amu} \end{aligned} \tag{5-2}$$

The atomic mass of a nuclide can be evaluated in terms of the masses of its constituent particles and a quantity called the *binding energy* (Fig. 5-12). A survey of the atomic masses shows that the mass of the nuclide is less than the sum of its constituent particles in the free state. To account for this difference in mass, the principle of equivalence of mass and energy is used. If $\Delta M$ is the decrease in mass when a number of protons, neutrons, and electrons combine to form an atom, then the mass-energy equivalence principle states that an amount of energy equal to $\Delta E = c^2 \Delta M$ is released in the process. The difference in mass $\Delta M$ is called the *mass defect;* it is the amount of mass which would be converted to energy if a particular atom or nuclide were to be assembled from the requisite number of protons, neutrons, and electrons. The same amount of energy would be needed to break the atom into its constituent particles, and the energy equivalent of the mass defect is therefore a measure of the binding energy of the nuclei. The mass of the constituent particles is the sum of $Z$ proton masses, $Z$ electrons, and $A - Z$ neutrons, where $A$ refers to the mass number of the element. Pairs of protons and electrons can be represented by hydrogen atoms; the loss in mass which accompanies the formation of the hydrogen atom from the proton and an electron is negligible. The mass defect can then be written $\Delta M = ZM_H + (A - Z)M_n - M_{ZA}$ where $M_H$ is the mass of the hydrogen atom, 1.008142 amu, and $M_n$ is the mass of the neutron, 1.008982 amu, and $M_{ZA}$ is the mass of nuclide of concern.

The curve of Fig. 5-12 does not give a complete picture of the nuclear binding energy. In the higher mass numbers the actual binding energy is not the same for each particle in the nucleus. After the maximum of the curve, almost every successive particle (proton or neutron) is bound less tightly than those already present, and the overall average decreases. The binding energy represented, however, is sufficiently accurate for engineering evaluation.

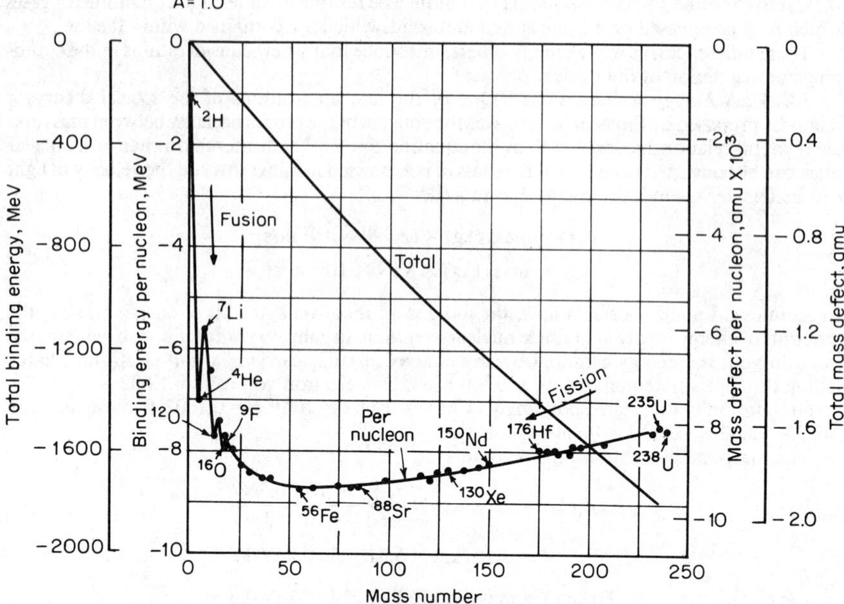


FIG. 5-12 Mass defects and binding energies of nuclei. *(From Ref. 9.)*

***15. The Fission Process.*** In the higher mass numbers, several of the naturally occurring elements are radioactive or have a characteristic which enables them to emit nuclear particles and be transmuted to different elements as a function of time. The various naturally occurring series are designated the thorium, uranium, and actinium series. These designations are related to the elements at or near the head of the series and can be expressed as multiples of a number $N$, where $N$ is an integer. The series are indicated by $4N$, $4N + 2$, and $4N + 3$, respectively. There is no naturally occurring $4N + 1$ element; such an element has been created in the process of artificial nuclear transmutation. This element is designated neptunium and has the mass characteristic of $4N + 1$. It, too, heads a radioactive series. The four radioactive series are shown in Fig. 5-13.

A number of elements with high mass numbers, both natural and artificially produced, undergo a process of nuclear fission. In the fission process, a nucleus absorbs a neutron and the resulting compound nucleus is so unstable that it immediately breaks up into parts. Many of the heavy nuclides can be induced to fission, but most only with neutrons of high energy. Naturally occurring heavy nuclides that fission with neutrons of energy in the range of the neutrons produced by the fission are uranium isotopes $^{235}U$ and $^{238}U$ and thorium 232. In addition, artificially produced nuclides $^{233}U$ and $^{239}Pu$, produced by $(n,\beta)$ reactions in $^{232}Th$ and $^{238}U$, respectively, are capable of fission. The fission process, in a nuclear reactor, is initiated by neutrons which are generated as part of the process. The general fission process may be expressed by

$$^{m}F + {}^{1}n \rightarrow {}^{x}A + {}^{m-(x+C)}B + C\,{}^{1}n \tag{5-3}$$

where $F$ = fuel nuclide, mass number $m$

$n$ = neutron

$A$, $B$ = fragment nuclides

$C$ = number of neutrons produced

$x$ = atomic number

The process of fission of heavy nuclides produces fission fragment nuclides which are predominantly in the center of the mass number range. Consequently there is a loss of mass in the process; this mass is converted to energy. The percentage of nuclide production as a function of mass number is shown in Fig. 5-14.

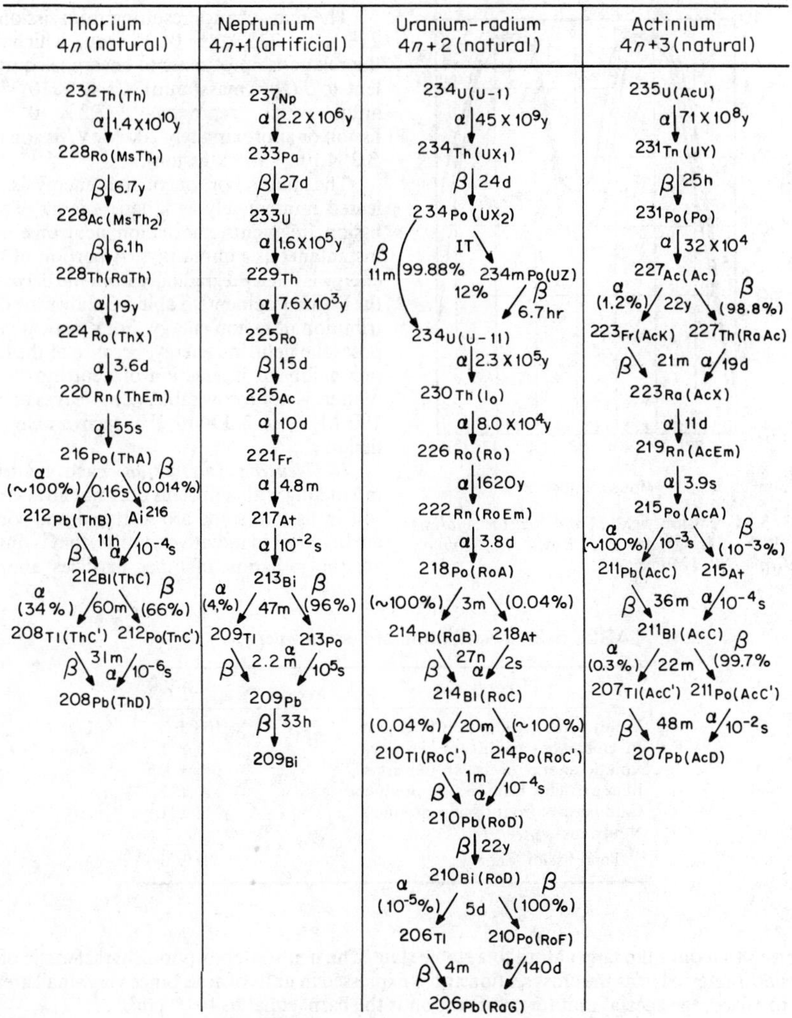

**Thorium 4n (natural)**

$^{232}$Th(Th)
α ↓ 1.4×10$^{10}$y
$^{228}$Ra(MsTh$_1$)
β ↓ 6.7y
$^{228}$Ac(MsTh$_2$)
β ↓ 6.1h
$^{228}$Th(RaTh)
α ↓ 1.9y
$^{224}$Ra(ThX)
α ↓ 3.6d
$^{220}$Rn(ThEm)
α ↓ 55s
$^{216}$Po(ThA)
α (~100%) ↙ 0.16s ↘ β (0.014%)
$^{212}$Pb(ThB) At$^{216}$
β ↘ 11h ↙ α 10$^{-4}$s
$^{212}$Bi(ThC)
α (34%) ↙ 60m ↘ β (66%)
$^{208}$Tl(ThC') $^{212}$Po(ThC')
β ↘ 31m ↙ α 10$^{-6}$s
$^{208}$Pb(ThD)

**Neptunium 4n+1 (artificial)**

$^{237}$Np
α ↓ 2.2×10$^{6}$y
$^{233}$Pa
β ↓ 27d
$^{233}$U
α ↓ 1.6×10$^{5}$y
$^{229}$Th
α ↓ 7.6×10$^{3}$y
$^{225}$Ra
β ↓ 15d
$^{225}$Ac
α ↓ 10d
$^{221}$Fr
α ↓ 4.8m
$^{217}$At
α ↓ 10$^{-2}$s
$^{213}$Bi
α (4%) ↙ 47m ↘ β (96%)
$^{209}$Tl $^{213}$Po
β ↘ 2.2m ↙ α 10$^{5}$s
$^{209}$Pb
β ↓ 3.3h
$^{209}$Bi

**Uranium-radium 4n+2 (natural)**

$^{234}$U(U-1)
α ↓ 4.5×10$^{9}$y
$^{234}$Th(UX$_1$)
β ↓ 24d
$^{234}$Pa(UX$_2$)
β 1.1m 99.88% ; IT 12% → $^{234m}$Pa(UZ) ; β ↙ 6.7hr
$^{234}$U(U-II)
α ↓ 2.3×10$^{5}$y
$^{230}$Th(Io)
α ↓ 8.0×10$^{4}$y
$^{226}$Ra(Ra)
α ↓ 1620y
$^{222}$Rn(RaEm)
α ↓ 3.8d
$^{218}$Po(RaA)
(~100%) ↙ 3m ↘ (0.04%)
$^{214}$Pb(RaB) $^{218}$At
β ↘ 27m ↙ α 2s
$^{214}$Bi(RaC)
(0.04%) ↙ 20m ↘ (~100%)
$^{210}$Tl(RaC') $^{214}$Po(RaC')
β ↘ 1m ↙ α 10$^{-4}$s
$^{210}$Pb(RaD)
β ↓ 22y
$^{210}$Bi(RaD)
α (10$^{-5}$%) ↙ 5d ↘ β (100%)
$^{206}$Tl $^{210}$Po(RaF)
β ↘ 4m ↙ α 140d
$^{206}$Pb(RaG)

**Actinium 4n+3 (natural)**

$^{235}$U(AcU)
α ↓ 7.1×10$^{8}$y
$^{231}$Th(UY)
β ↓ 25h
$^{231}$Pa(Pa)
α ↓ 3.2×10$^{4}$
$^{227}$Ac(Ac)
α (1.2%) ↙ 22y ↘ β (98.8%)
$^{223}$Fr(AcK) $^{227}$Th(RaAc)
β ↘ 21m ↙ α 19d
$^{223}$Ra(AcX)
α ↓ 11d
$^{219}$Rn(AcEm)
α ↓ 3.9s
$^{215}$Po(AcA)
α (~100%) ↙ 10$^{-3}$s ↘ β (10$^{-3}$%)
$^{211}$Pb(AcC) $^{215}$At
β ↘ 36m ↙ α 10$^{-4}$s
$^{211}$Bi(AcC)
α (0.3%) ↙ 22m ↘ β (99.7%)
$^{207}$Tl(AcC') $^{211}$Po(AcC')
β ↘ 48m ↙ α 10$^{-2}$s
$^{207}$Pb(AcD)

**FIG. 5-13** The four radioactive series. *(From Ref. 11.)*

A typical example is the fission of $^{235}$U with the production of two most likely fission fragments.

$$^{235}U + {}^{1}n \rightarrow {}^{95}A + {}^{139}B + 2\,{}^{1}n \tag{5-4}$$

The mass balances of this equation are

| Before fission | | After fission | |
|---|---|---|---|
| $^{235}$U – | 235.124 amu | $^{95}A$ | –94.945 amu |
| $^{1}n$ | 1.009 amu | $^{139}B$ | 138.955 amu |
| | | $2\,^{1}n$ | 2.018 amu |
| | 236.133 amu | | 235.918 amu |

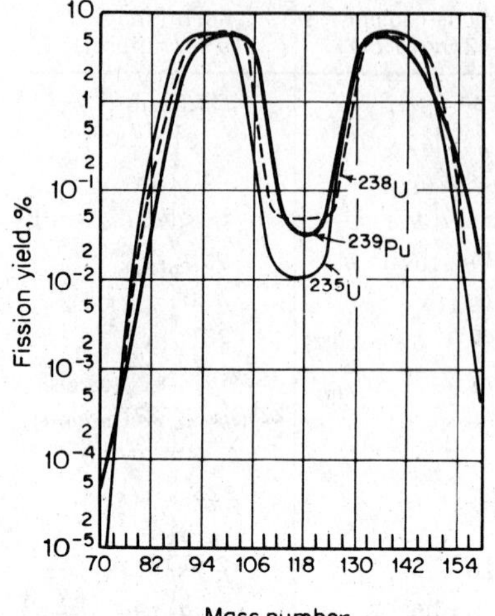


**FIG. 5-14** Fission yield. *(From Reactor Analysis by R. V. Megreblian and David K. Holmes, McGraw-Hill, 1960.)*

The mass change resulting from fission is 236.133 − 235.918 = 0.215 amu, which by the relationship of mass to energy is equivalent to $E(\mathrm{J}) = \mathrm{mass(amu)} \times 1.49 \times 10^{-10}$ J/amu, which represents $\sim 3.2 \times 10^{-11}$ J/fission or approximately 200 MeV/fission (or $3.2 \times 10^{-11}$ W · s/fission).

The major portion of this energy is released immediately as kinetic energy of the fission fragments, the fission neutrons, and instantaneous gamma rays. A portion of the energy is released gradually from the decay of the fission fragments. Table 5-3 shows the distribution of fission energy. For practical purposes the neutrino energy, because of the low probability of interaction of neutrinos with matter, is not recoverable. (This leaves about 190 MeV, or $3.0 \times 10^{-11}$ J, recoverable per fission.)

***16. Neutron Interaction.*** Each neutron interacting with a nucleus does not always result in fission; some are scattered and some are involved in radiative capture, that is, initiate the radiation of other particles and/or photons to reduce the target atom to a stable state. The neutron-absorption characteristic of a nuclide is referred to as the cross section and is expressed in units of area. Since very small areas are involved, the special unit for cross section is the barn, equal to $10^{-24}$ cm$^2$.

**TABLE 5-3** Distribution of Fission Energy

| Energy | MeV |
|---|---|
| Kinetic energy of fission fragments | 168 ± 5 |
| Instantaneous gamma-ray energy | 5 ± 1 |
| Kinetic energy of fission neutrons | 5 ± 0.5 |
| Beta particles from fission products | 7 ± 1 |
| Gamma rays from fission products | 6 ± 1 |
| Neutrinos | ~10 |
| Total fission energy | 201 ± 6 |

The cross section may be considered as the probability that a neutron passing through a material (an assembly of molecules, usually structured, for example, face-centered cubic lattice) will interact with an atom's nucleus. Table 5-4 shows the relative diameters of the entities involved.

These relative diameters indicate that there is much "open space" in the material through

**TABLE 5-4** Relative Diameters of Atom Entities

| Item | Order of magnitude of diameter |
|---|---|
| Neutron | $10^{-8}$ to $10^{-12}$ mm |
| Atom | $10^{-7}$ mm |
| Nucleus | $10^{-12}$ mm |

**TABLE 5-5** Thermal-Neutron Cross Sections of Fissionable Nuclei

| Nucleus | Fission barns | Radiative capture, barns | Total absorption, barns |
|---|---|---|---|
| Uranium 235 | 549 | 101 | 650 |
| Plutonium 239 | 664 | 361 | 1225 |
| Uranium 238 | 0 | 2.8 | 2.8 |
| Uranium (natural) | 3.92 | 3.5 | 7.42 |

which the neutron can pass. As a result, the probability of interaction or cross section is quite small.

The effective diameter of the neutron varies with energy. One explanation of this is that, in quantum mechanics, the wavelength $\lambda$ of the neutron is inversely proportional to its energy $E$ or velocity, and may be expressed by

$$\lambda = \frac{2.86 \times 10^{-8}}{\sqrt{E}(\text{ev})} \quad \text{mm} \tag{5-5}$$

For fast neutrons (about 1 MeV), $\lambda$ is of the order of $10^{-11}$ mm, and for thermal neutrons (about 0.03 MeV), $\lambda$ is about $1.7 \times 10^{-7}$ mm. The slower neutrons behave as though they had a diameter approaching that of the atom and thus have a larger probability of interaction.

The neutron cross sections of the principal fuel nuclides for low- and high-energy neutrons are shown in Table 5-5. Of greatest interest is $^{235}$U, whose cross section is largest for low-energy neutrons and which is a naturally occurring element.

The cross section of the fuel nuclide to an incident neutron varies and is a function of the energy of the neutron (see Fig. 5-15), generally decreasing with increasing energy (called the $1/v$

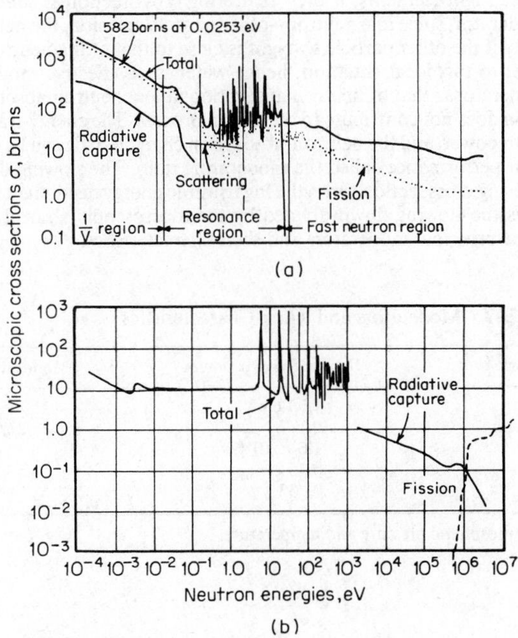


**FIG. 5-15** Neutron cross sections for *(a)* $^{235}$U and *(b)* $^{238}$U. *(From Ref. 9.)*

law, where $v$ = neutron velocity, m/s). As the energy of the incident neutrons increases through the intermediate energy range, the absorption cross section for the materials of interest exhibits numerous abrupt increases and decreases or resonances. As the temperature of the material increases, these resonance peaks widen, causing increased absorption of neutrons. This Doppler effect is important in the regulation of the fission process.

When neutrons are produced in the fission process, they are released with energies from 1 to 10 million electronvolts (MeV); with an average energy of $2.0 \pm 0.1$ MeV. These neutrons move with velocities of up to $10^7$ m/s and are referred to as fast neutrons. Elastic collisions with light nuclei cause the neutrons to lose energy until they reach the equilibrium energy of the temperature of the medium, the thermal energy. Average energies and speeds of neutrons versus temperature are shown in Table 5-6.

**TABLE 5-6** Average Speeds and Energies of Thermal Neutrons

| Temperature, °C | Energy, eV | Speed, m/s |
|---|---|---|
| 25 | 0.026 | $2.2 \times 10^3$ |
| 200 | 0.041 | 2.8 |
| 400 | 0.058 | 3.4 |
| 600 | 0.075 | 3.8 |
| 800 | 0.092 | 4.2 |

Since, for efficiency, it is desirable to provide a fuel cross section as large as possible, commercial reactors utilizing $^{235}U$ as an energy source operate with thermal neutrons. The fission neutrons are brought to thermal energy by elastic collisions with atoms in a material called a moderator. Light elements, if their scattering cross section is sufficient, serve most effectively as a moderator, since in a neutron–other particle collision, the neutron will transfer energy most rapidly if the other particle's weight is close to that of the neutron. The hydrogen atom is the nearest to this ideal situation; hence water is an effective moderator. A second criterion for a moderator is that of neutron absorption; a low neutron absorption is desirable since the absorption does not contribute to the fission process. The energy-transfer characteristic, or slowing-down power, and the neutron-absorption characteristic can be combined into an index of moderator performance called the moderating ratio. The slowing-down power is the macroscopic scattering cross section times the logarithmic energy decrement per collision. The moderating ratio is the slowing down divided by the macroscopic absorption cross section. Table 5-7 shows the principal moderators and their characteristics.

**TABLE 5-7** Moderators and Their Characteristics

| Moderator | Slowing-down power | Moderating ratio |
|---|---|---|
| Water | 1.53 $cm^{-1}$ | 70 |
| Heavy water | 0.177 | 21,000 |
| Helium* | $1.6 \times 10^{-5}$ | 83 |
| Beryllium | 0.16 | 150 |
| Carbon | 0.063 | 170 |

* At atmospheric pressure and temperature.

## Radiation

***17. Nuclide Composition.*** The elements of the periodic table, both naturally occurring and artificial, are composed (except for hydrogen) of protons, neutrons, and electrons. Many of the

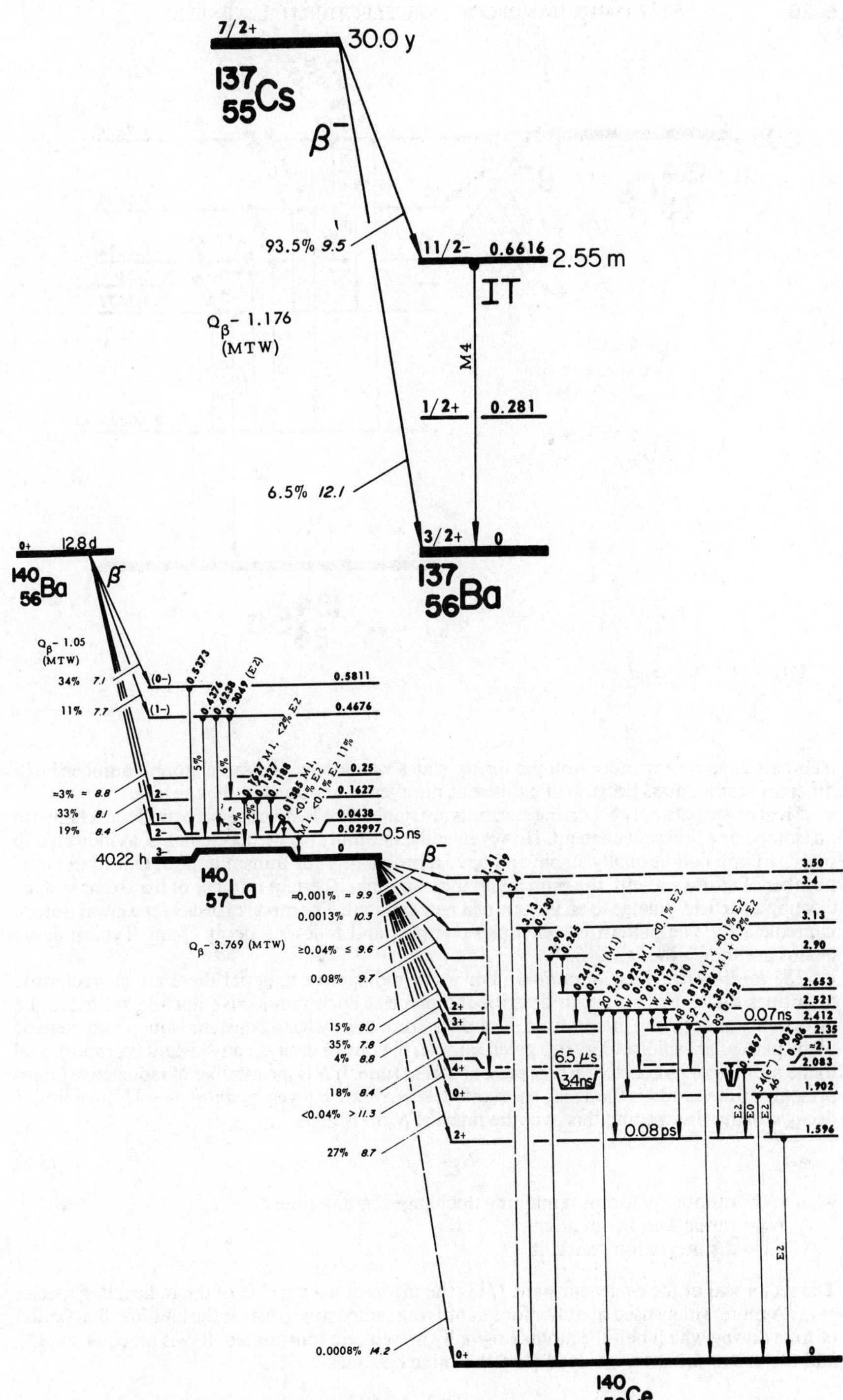


FIG. 5-16 Examples of radioactive decay modes. *(From Ref. 13.)*

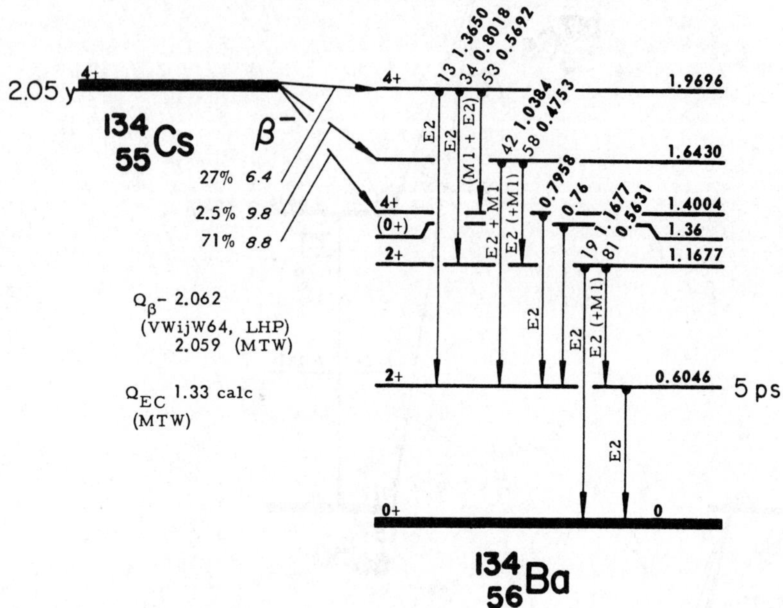


**FIG. 5-16** *Continued*

elements have two or more isotopic forms, states which have the same atomic number but a different atomic mass because of a different number of neutrons in the nucleus.

Most of the naturally occurring elements are stable, that is, do not eject particles to change to an isotope or a different element. However, some naturally occurring elements, as indicated in Fig. 5-13, are conditionally stable and have a probability for transmutation. Out of the total number of atoms present, the probability indicates that a certain number of the atoms will, by ejecting a particle, change to an isotope or a new element. The mode of decay for a given isotope is predictable. The pattern is sometimes complex and follows a decay chain. Typical decay chains are shown in Fig. 5-16.

***18. Radioactive Transmutation.*** For every radioactive material there are characteristic quantities that may be used to describe the process. Each radioactive nuclide has a definite probability of decaying in unit time. This decay probability has a constant value, characteristic of the particular radioisotope. In a given sample, the rate of decay at any instant is proportional to the number of radioactive atoms present at that time. If $N$ is the number of radioactive atoms present at time $t$ and $\lambda$ is the decay constant, the decay rate is given by $dn/dt = -\lambda N$ for a simple decay scheme. Integrating this over the interval $N_0$ to $N$ gives

$$N = N_0 e^{-\lambda/t} \tag{5-6}$$

where $N$ = number of atoms remaining unchanged at any time $t$
$N_0$ = initial number of atoms
$\lambda$ = disintegration constant

The reciprocal of the decay constant $1/\lambda$ is the mean or average life of the radioactive species $= t_m$. A more widely used quantity for quantifying radioactive decay is the half-life, that period of time during which half the atoms originally present are transmuted. If $N$ is set equal to $\frac{1}{2}N_0$ and the above equation is solved for $t$, the value becomes

$$t_{1/2} = \frac{\ln 2}{\lambda} = \frac{0.6931}{\lambda} \tag{5-7}$$

In a radioactive species, a nuclide may undergo successive decay before reaching the ground state. For a compound decay scheme involving two states *A* and *B*, the net rate of change of *B* with time is given by

$$\frac{dNb}{dt} = \lambda_A N_A - \lambda_B N_B \tag{5-8}$$

where the solution is $N = \frac{\lambda_A N_{A0}}{\lambda_B - \lambda_A}(e^{-\lambda_A t} - e^{-\lambda_B t})$. The first term on the right represents the production of *B* from the decay of *A*; the second term is the production of *B*. $N_{A0}$ is the number of parent atoms at time $t = 0$.

Sample decay curves in Fig. 5-17 show both a simple decay and a compound two-stage decay.

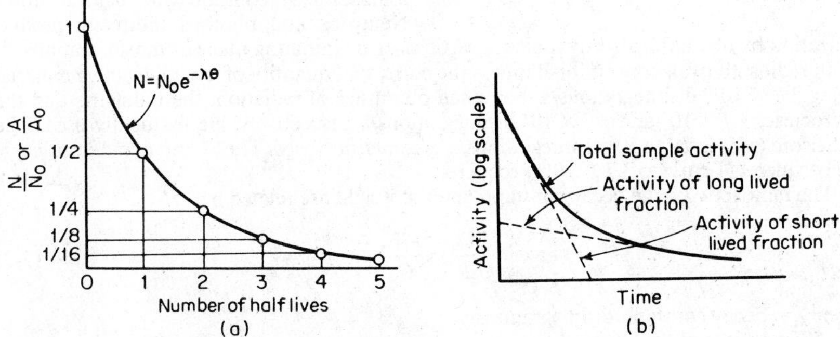


**FIG. 5-17** *(a)* Radio decay of a single radionuclide as a function of half-life. *(b)* Decay of mixture of independent radionuclides.

If the radiation occurs by the emission of a quantity of energy (photon), the nuclide retains its atomic weight and number. If the decay occurs by emission of a particle, the nuclide changes to an isotope (same atomic number), an isobar (same mass number), or a different element. The position of the nuclide in a plot of the nuclides (*A* versus *Z*) changes as shown in Fig. 5-18 for various particles emitted.

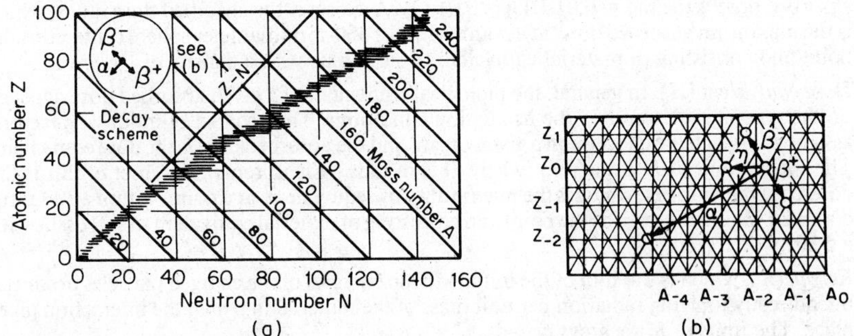


**FIG. 5-18** *(a) Z-N* chart of the known isotopes. *(b)* Change in mass number and atomic numbers for various decay modes. *(From Ref. 9.)*

Artificial elements, including those resulting from the fission process, are very likely to be radioactive. In some cases, this activity results in the emission of a photon of energy to allow the

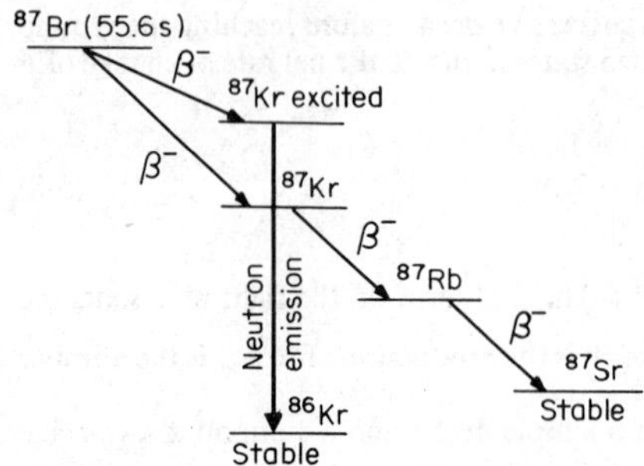


**FIG. 5-19** Radioactive decay showing delayed neutron emission. *(From Ref. 5.)*

atom to reach a lower energy state. In other cases, a particle is emitted; the particle emitted for some decaying nuclides is a neutron (see Fig. 5-19). These delayed neutrons are important to the regulation of the fission process.

***19. Types of Radiation.*** There are three categories of radiation emanations of biological concern in nuclear power. The first category is that of charged particles, principally alpha particles and beta particles. The second is that of uncharged particles, chiefly neutrons. The third is that of photons or gamma rays. The charged particles directly produce ionization by collision with neutral atoms. Neutrons and photons indirectly produce ionization by liberating directly ionizing particles or by initiating nuclear transformations.

In radioactivity, a conventional unit is the curie, that quantity of any radioactive material giving $3.7 \times 10^{10}$ disintegrations/s. For small quantities of radiation, the millicurie and the microcurie, $3.7 \times 10^{7}$ and $3.7 \times 10^{4}$ disintegrations/s, respectively, are frequently used. The rutherford (rd), equal to $10^{6}$ disintegrations/s, is sometimes used. The SI unit of radioactivity is the becquerel (1 curie = $3.7 \times 10^{10}$ becquerel).

The radioactivity, the decay constant, and the weight are related by

$$\frac{dN}{dt} = \frac{\lambda W_A}{G_w} \tag{5-9}$$

where $\lambda$ = decay constant, disintegrations/s
$W$ = weight of the material, g
$A$ = Avogadro's number
$G_w$ = gram atomic weight of the material
$N$ = number of atoms

This equation shows that a given amount of radioactivity may occur from a large mass with a small decay rate or a small mass which has a high decay rate.

Radiation dosage is expressed in four ways:

1. *Absorbed dose* ($D$), which is the energy absorbed per unit mass at a specific place in a material. The standard of absorbed dose is the gray. 1 Gy = 1 J/kg. The special unit of absorbed dose is the rad = 0.01 J/kg = 0.01 Gy. A subset is the absorbed-dose index, which is the maximum absorbed dose, at a point, within a 300-mm-diameter sphere centered at the point and consisting of material equivalent to soft tissue with a density of 1 g/cm$^3$.
2. *Dose equivalent* ($H$). In general, the biological equivalent of a given absorbed dose depends on the type of radiation and the irradiation conditions. The product of modifying factors, assigned to weigh the effect upon a given organ, and the absorbed dose is the dose equivalent. The special unit of $H$ is the rem (where $D$ is in rads, $H$ is in rems). A subset of this is the dose-equivalent index, which is the maximum dose equivalent, at a point, within a 300-mm-diameter sphere centered at the point and consisting of material equivalent to soft tissue with a density of 1 g/cm$^3$.
3. *Kerma* ($K$), which is the sum of the initial kinetic energies of the charged particles produced by indirectly ionizing radiation per unit mass of the material in which the interaction takes place. The units of $K$ are grays or rads.
4. *Exposure* ($X$) is the measure of a particular field of electromagnetic radiation (x- or gamma rays) to ionize air. The special unit of exposure is the roentgen (R) = $2.58 \times 10^{-4}$ coulombs/kg of air.

***20. Nuclear Plant Safety.*** The nuclear-powered steam supply system characteristics of substantial energy potential present in the reactor, radiation production during the fission process, and continued radiation production and heat generation after shutdown require that special safety precautions be taken in design and operation of a nuclear plant. The health and welfare of the public depends on both the continuation of the plant's power production and the avoidance of any incident which would endanger the environment. In order to achieve the latter goal and to aid the former, special regulations relating to nuclear plants have been formulated.

***21. Federal Regulations.*** Title 10 of the Code of Federal Regulations (10 CFR) has the following parts which are of primary importance to nuclear power facilities:

Part 20 Standards for Protection against Radiation
Part 50 Licensing of Production and Utilization Facilities
Part 55 Operators Licenses
Part 70 Special Nuclear Material
Part 100 Reactor Site Criteria

There are several other parts of 10 CFR which relate to the usage or handling of radioactive material. Most of the parts previously listed have appendixes which treat requirements for specific subjects. Authority for regulation of commercial, nuclear-powered plants is vested with the U.S. Nuclear Regulatory Commission. This authority includes the licensing of new facilities and the surveillance of operating facilities. An applicant for a nuclear-powered plant is required to apply for a license to construct and operate the facility. Such application includes the submission of Safety Analysis Reports which describe the design bases, the design, and the analyses performed to show that plant performance and conditions will be within established limits.

***22. Standards.*** Appendix A of 10 CFR Part 50 provides general design criteria for nuclear power plants. Criterion 1 requires that structures, systems, and components important to safety be designed, fabricated, erected, and tested to quality standards. The nuclear standards program of the American National Standards Institute has developed a sizable group of standards for this requirement. The principal design, systems, and operation standards are those developed by ASME, IEEE, ANS, and ISA. Examples of these standards are listed in Sec. 28 (Standards in Electrotechnology), Par. 140.

Many other documents providing criteria, standard practices, or guidance are available. In the nuclear area specific designs have not been repeated frequently enough to accumulate a significant backlog of experience. As a result, many of the "standards" are developed to provide leadership in addressing given areas.

The Nuclear Regulatory Commission provides guidance in many areas of design, construction, and operation through Regulatory Guides. Individual guides may cite a standard as an acceptable method of addressing the area concerned.

***23. Quality Assurance.*** The best defense against incidents which endanger the public is to prevent them. In a similar way, the best system performance is effected when malfunctions are eliminated. Reliability is the interface between quality assurance and safety. Reliability can neither be tested nor legislated into equipment; it must be built in. High quality in design, procurement, installation, and operation will lead to a system that has high availability, good reliability, and a low probability of incurring an accident. Quality assurance is a total systems approach to achieving these aims. Quality assurance does represent an increase in costs; this increase must be balanced against safer operation and savings resulting from less time lost, fewer repairs, and better control. The prime responsibility for an effective quality assurance program lies with the owner/operator of the plant, who may delegate portions of the program to major suppliers.

## Nuclear Energy System

***24. Reactor-System Assembly.*** In order to make it possible for the fission process to be self-sustaining, but regulated, and for the energy released to be transferred to a device which converts it to the electrical form, a reactor system is constructed.

The nuclear fuel, usually uranium, is fabricated into fuel elements. The typical design for the fuel of a light-water power reactor involves the fuel in oxide form. Where the fuel is uranium, the uranium dioxide is fabricated into pellets, right circular cylinders approximately 19 mm high and 8 mm in diameter.

In light-water reactors, the uranium dioxide material will have been enriched to a low value, approximately 3%, in the fissionable isotope $^{235}U$. This enrichment is necessary because light water has an appreciable neutron-absorption characteristic. The extra neutrons available from the added fissile material compensate for the absorption in the moderator. The fuel-pellet material is of a ceramic nature; the pellets are dished at both ends to allow for differential thermal expansion and fuel volumetric growth with burnup.

The pellets are inserted into fuel tubes, typically thin-walled tubes of stainless steel or Zircalloy. An open space (with the column of pellets spring-loaded) at the top of the tube is provided to accommodate generation of gases during the fission process. The tubes are sealed top and bottom and are assembled into a configuration involving fixed spacing in a fuel assembly. A representative fuel assembly is shown in Fig. 5-20. This assembly has an overall length of approximately 4.5 m with an active length of approximately 3.8 m.

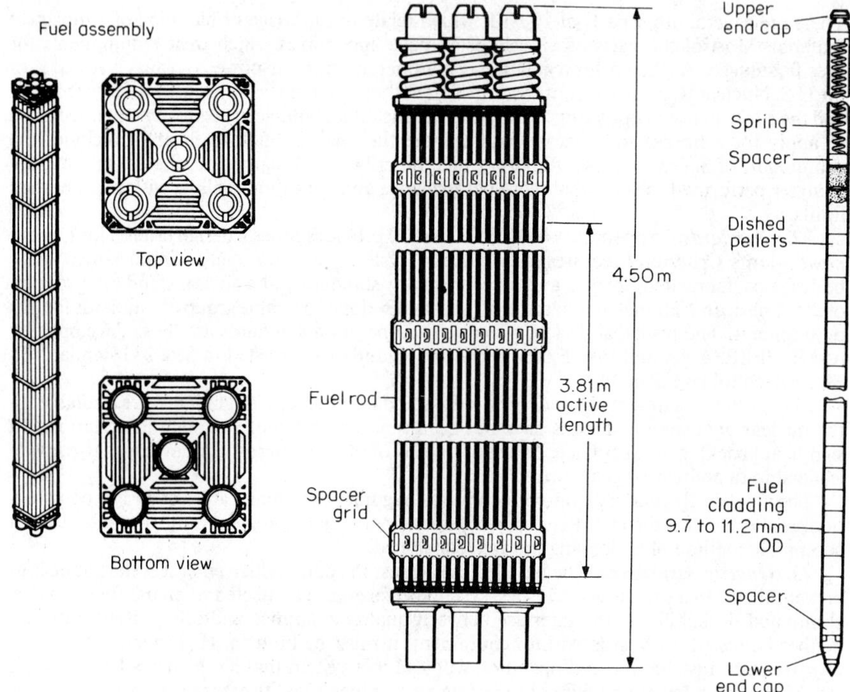


**FIG. 5-20** PWR fuel assembly.

Plutonium, $^{239}P$, is produced in the fuel elements during power operation by the absorption of neutrons in the $^{238}U$. This material is fissionable and may be recovered during fuel reprocessing and fabricated into fuel elements.

In gas-cooled reactors in the United States, the fuel-element design differs from that of light-water reactors. The recent gas-cooled reactor elements are hexagonal graphite blocks into which blind longitudinal holes are drilled to receive rods of fuel particles. The fissile material is enriched uranium carbide, $UC_2$. Kernels of this material are coated with a pyrolytic carbon–silicon carbide–pyrolytic carbon sandwich. Fertile material in the form of thorium oxide, $ThO_2$, kernels is also used. A fertile material is one which, by absorption of neutrons, is changed to a material which can be fissioned. In this case $^{232}Th$ is converted to $^{233}U$, which has superior characteristics for fission reactions. The kernels are coated with two layers of pyrolitic carbon. The two types of fuel particles are mixed in the proper proportions and are formed with a carbon matrix into fuel "rods" about 15.6 mm in diameter and about 60 mm long. These rods are inserted into the holes in the graphite blocks. Through holes are provided in the block for the helium coolant flow. These loaded graphite blocks or "fuel assemblies" form the basic module for the core of the gas reactor.

The required number of fuel assemblies to produce a power output desired for the reactor plant are assembled into a reactor-core configuration approximating a right circular cylinder. This configuration provides a high volume-to-surface ratio which minimizes the neutron leakage and conserves the neutrons produced for further fission action. A typical core cross section is shown in Fig. 5-21. For a 1300-MW (electrical) light-water nuclear plant, a representative core

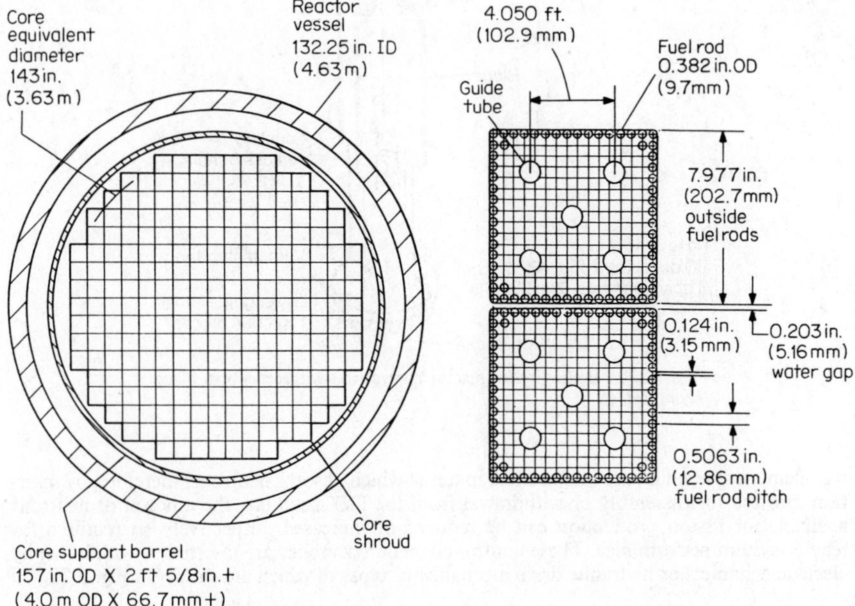


**FIG. 5-21** PWR reactor core cross section.

assembly might involve 241 fuel assemblies each weighing approximately 660 kg, for a core equivalent diameter of 3.6 m, a core height of approximately 5 m, and a total core weight of approximately 160 metric tons.

***25. Control-Element Assemblies.*** In each fuel assembly several holes are shown. These open holes are spaces into which control elements are inserted for regulation of the fission process. The individual control elements may be grouped typically into control-element assemblies as shown in Fig. 5-22. Control elements for current PWRs (Fig. 5-22*a*) are located in the fuel elements in this fashion. Control elements of BWRs are blade-type cruciform units as shown in Fig. 5-22*b*. These units are inserted into or withdrawn from the spaces between the fuel assemblies. The con-

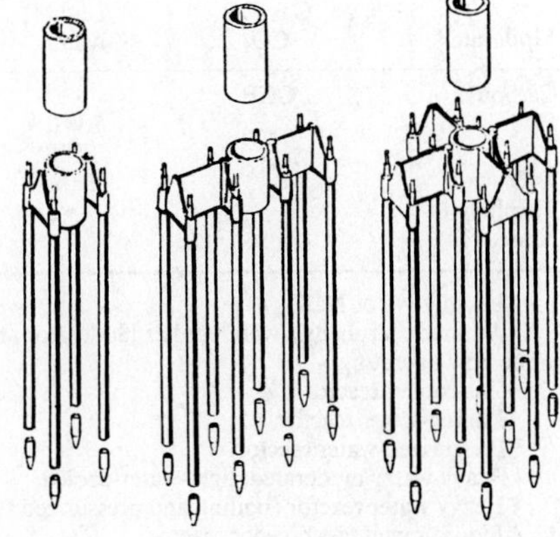

**FIG. 5-22a** Control element assemblies for PWR.

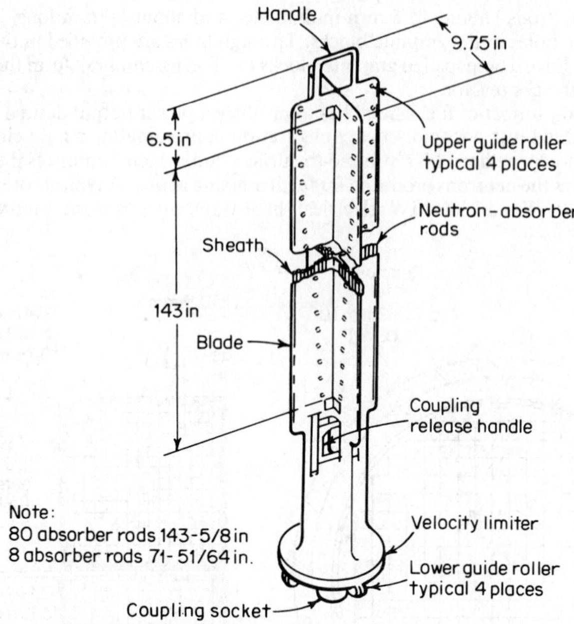


**FIG. 5-22b** Boiling water reactor control rod with cruciform blade. *(From Ref. 4.)*

trol-element assemblies are selected of a material which absorbs neutrons; therefore, by insertion into the fuel assembly or withdrawal from the fuel assembly, the amount of neutrons available for fission production can be reduced or increased, respectively, as required for reactor-system performance. These control-element assemblies are inserted or withdrawn by electromechanical or hydraulic drive mechanisms, types of which are shown in Fig. 5-23.

**TABLE 5-8** Moderators and Heat-Transfer Systems

| | Coolant | | | | |
|---|---|---|---|---|---|
| Moderator | Gas He, $CO_2$ | Light water | Heavy water | Light metal[a] | Organic[b] |
| Carbon | GCR[c] | | | | |
| $H_2O$ | | BWR[d] | | | |
| | | PWR[e] | | | |
| $D_2O$ | | Tube[f] | HWR[g] | | |
| Terphenyl | | | | | Piqua[i] |
| None | | | | LMFBR[h] | |

[a] Usually Na or NaK.
[b] Aromatic terphenyls were used as moderator and coolant in some of the early demonstration reactors (none now operating).
[c] Gas-cooled reactor.
[d] Boiling-water reactor.
[e] Pressurized-water reactor.
[f] Heavy-water-moderated, light-water-cooled.
[g] Heavy-water reactor (boiling and pressurized types used).
[h] Liquid-metal fast-breeder reactor.
[i] City of Piqua reactor (12.5 MWe).

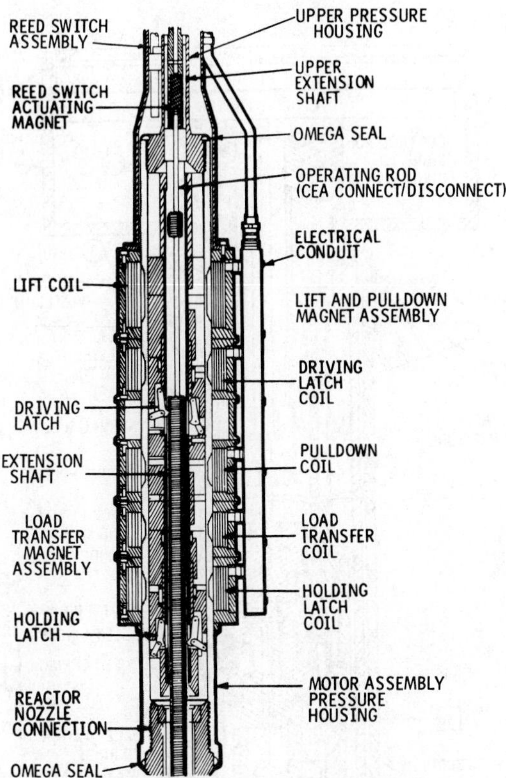


**FIG. 5-23a** PWR control element drive mechanism.

***26. Moderator and Heat-Transfer Medium.*** The thermal energy released from the core must be conveyed to the electric generator at a rate and in a fashion which meets the requirements. Some consideration has been given to the use of a reactor core to heat gas which is supplied to a magnetohydrodynamic generator of electricity. Commerical systems for the present and near future will continue to use steam turbines as the motive power for the generator.

If fuel of low enrichment (e.g., 3 to 4%) in $^{235}U$ or other fissile isotope is used, a moderator is needed to take advantage of the larger cross section at thermal neutron energy. If enrichments greater than 20% are used, sufficient fissile material is present to overcome the nonfission capture effects of $^{238}U$, and a moderator is not needed. In this case the reactor is said to be "fast" (referring to the neutron velocity), and liquid metals (selected for low neutron absorption) are used as a coolant.

The systems currently in use or scheduled for operation in the United States are the GCR, BWR, PWR, and LMFBR types (Table 5-8).

## Plant Arrangement

***27. Primary-System Configuration.*** The fuel elements, assembled into the core arrangement, are positioned within a reactor vessel by support structures also referred to as reactor internals. The reactor vessel also contains, guides, and directs the primary coolant. Elementary configurations for the various systems are shown in Fig. 5-24.

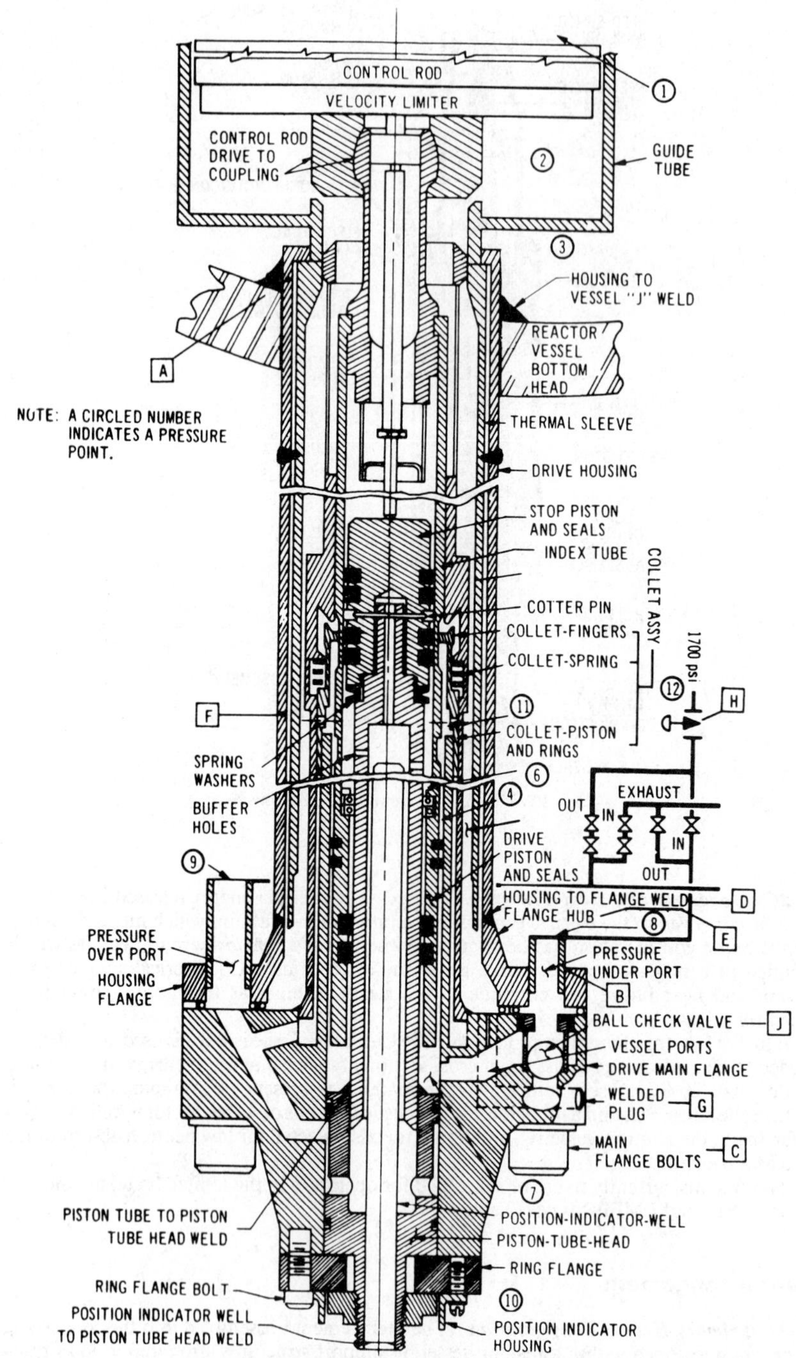


**FIG. 5-23b** BWR control rod drive unit (schematic). *(General Electric.)*

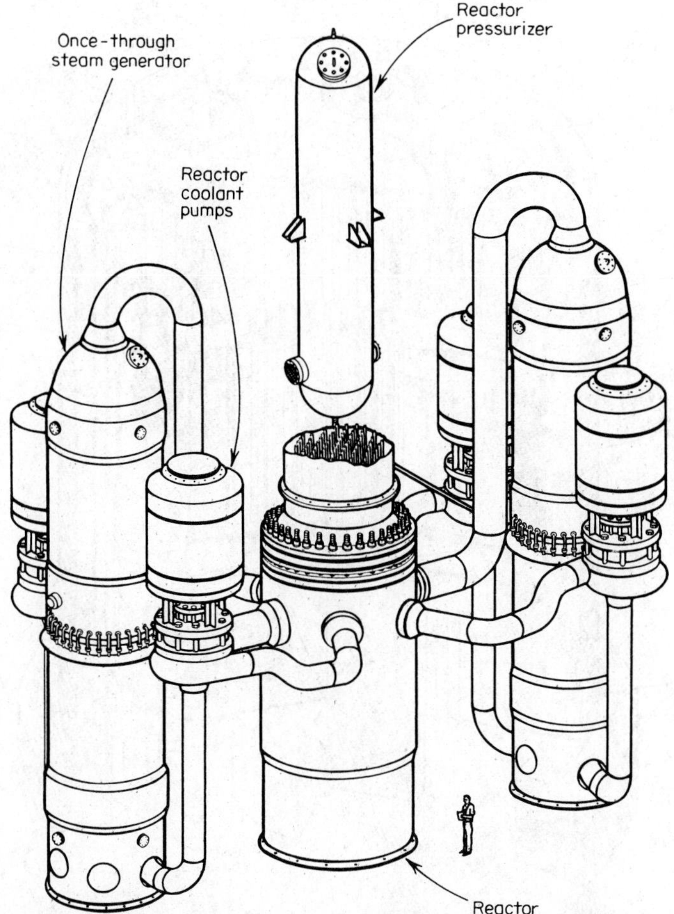


**FIG. 5-24a** PWR arrangement with once-through steam generators. *(Reprinted with permission from the 1974 Generation Planbook, by Power Magazine.)*

In the BWRs, the reactor vessel also contains the steam-separation apparatus, since the coolant is converted to steam in the core. Steam is piped from the reactor vessel to the turbine and condensate is returned from the hotwell through feedwater systems to the reactor vessel. Recirculation loops on the reactor vessel increase the recirculation ratio to effect better steam separation and provide better control of void fraction within the core (reactivity control). Access to the core for control rods and for in-core instrumentation is provided from the bottom of the reactor vessel to avoid the steam-separation area at the top.

In PWRs, the primary system is maintained at a subcooled condition by operating at a pressure greater than saturation. Conventionally this pressure is in the range of 12 to 16 MPa. This pressure is maintained by a pressurizer connected to the primary piping. A steam-water interface is maintained in the pressurizer by the action of electric resistance heaters which boil water to raise the pressure or spray flow to quench steam and lower the pressure. The reactor vessel is connected, by heavy piping, to one or more steam generators, and coolant is circulated through this primary system by large pumps. On the secondary side of the steam generator are located the customary steam piping complex, the turbine condenser and feedwater system.

In liquid-metal reactors, the concerns associated with coolant metal–water reaction in the

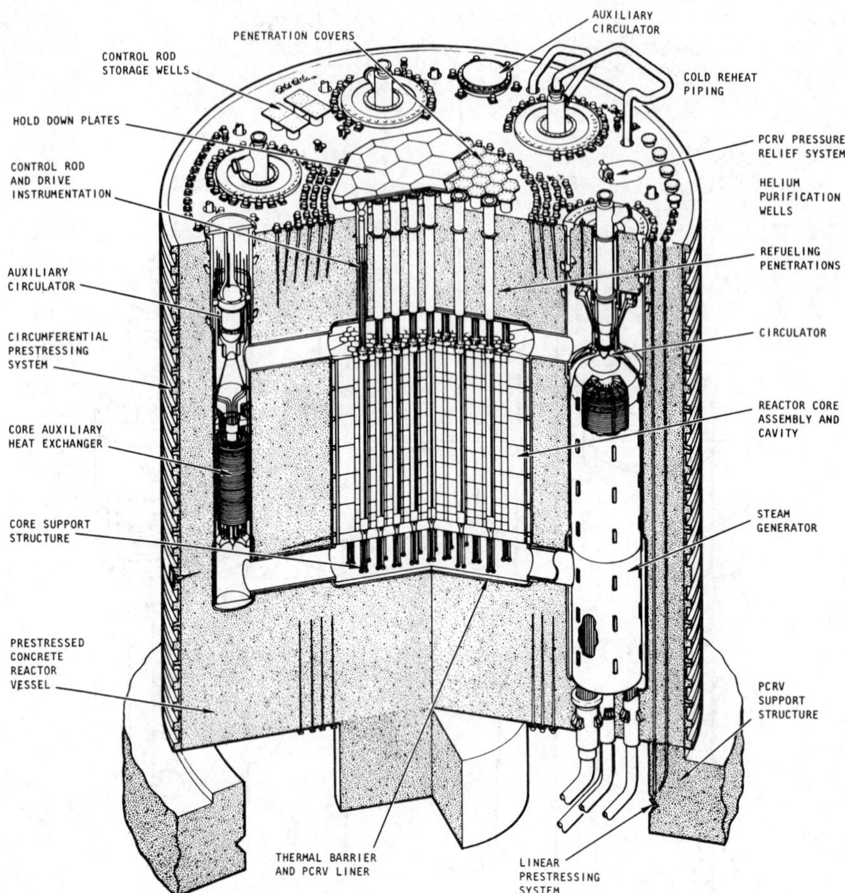


**FIG. 5-24b** GCR design arrangement. *(General Atomic Company.)*

event of leakage and the induced activity in the primary metal sometimes direct the design to an intermediate heat-exchange system between the primary-metal coolant and the steam system.

***28. Containment Structure.*** To provide a secondary barrier for radioactivity in the fluid systems and a fourth barrier for the activity in the fuel (the fuel matrix or ceramic, the fuel sheath, and the primary-system boundary are the other three), the primary-system components are located within a containment structure. This structure is designed to confine gases and materials that might occur in the event of an inadvertent release from the primary barrier(s). Examples of containments are shown in Fig. 5-25.

The principal types of containments that have been used are

1. The steel sphere made of sections of steel sheet welded together
2. The "inverted light bulb" of the BWR, which is a concrete cavity in the shape described with a pressure-retaining steel liner
3. The domed cylinder for the PWR; the cylinder of reinforced concrete with an impervious internal steel membrane
4. The cylindrical prestressed-concrete reactor vessel (PCRV) for the GCRs

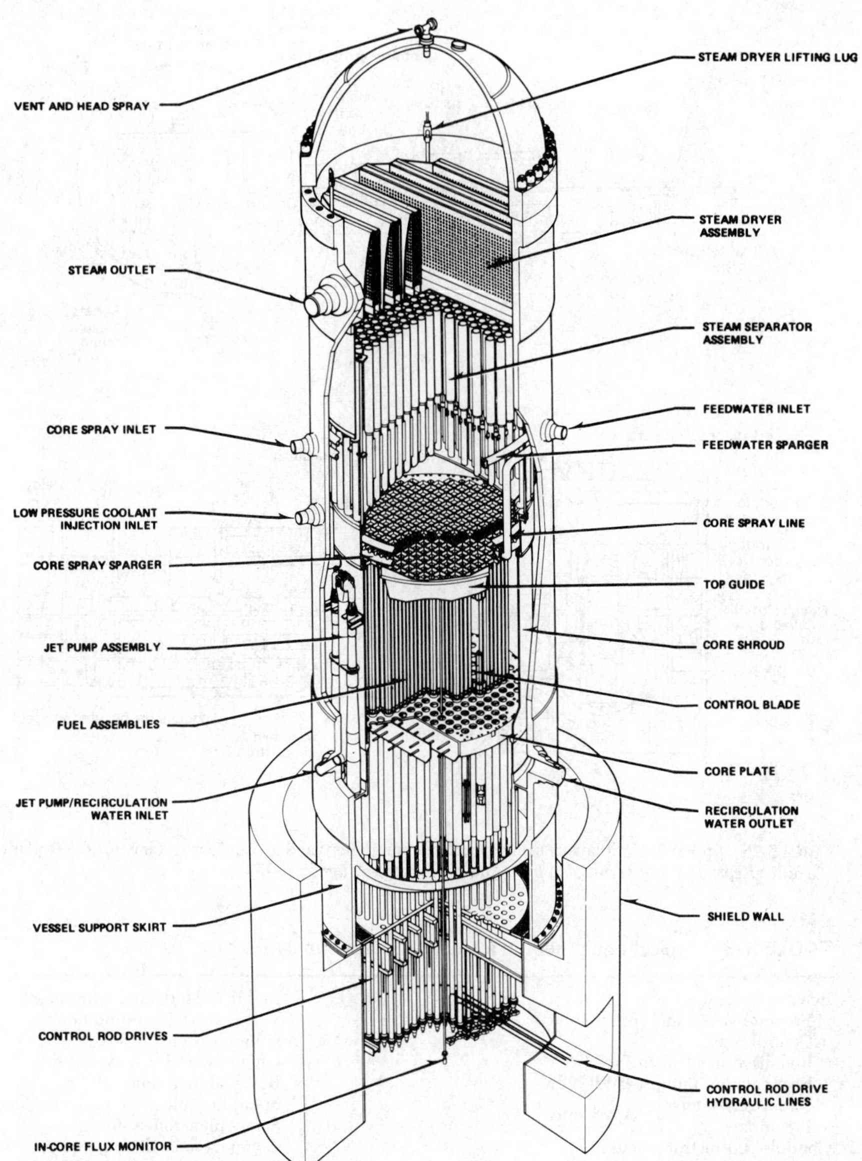


**FIG. 5-24c** BWR arrangement. *(General Electric.)*

These structures serve to provide isolation and shielding for the primary-system components. Instrumentation, control, and electric power conductors must pass through the pressure seal while maintaining its integrity. To provide this capability, banks of containment penetrations are provided. The requirements for these items are described by IEEE 317-1983, "Electric Penetration Assemblies in Containment Structures for Nuclear Power Generating Stations."

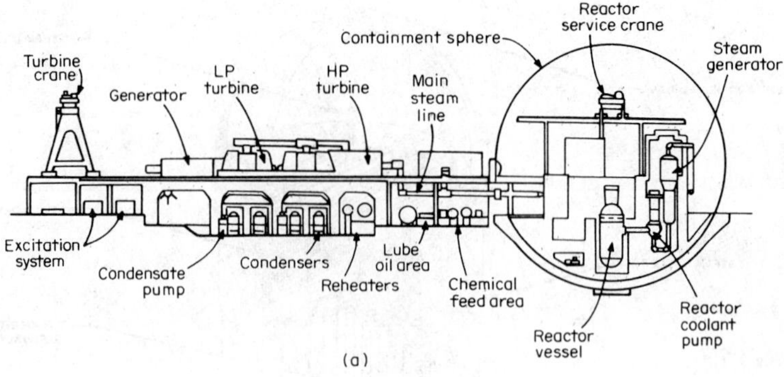


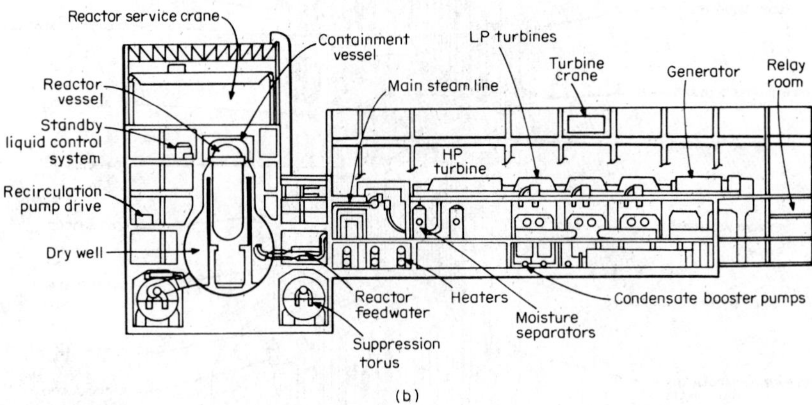


**FIG. 5-25** *(a)* **PWR plant layout using spherical containment and an outdoor turbine;** *(b)* **BWR plant layout.** *(Reprinted with permission from Power magazine, August, 1967.)*

**TABLE 5-9** Typical Equipment in a Nuclear Steam-Supply Package

Reactor system:
- Pressure vessel and internals
- Control rods
- Rod-drive mechanisms
- Recirculation pumps, BWR only
- Steam generators } PWR only
- Pressurizer }
- Rod-drive servicing equipment
- Vessel-servicing equipment

Controls and instrumentation:
- Reactor controls and instrumentation
- Rod-drive position indicators
- In-core instrumentation
- Ex-core nuclear instrumentation
- Plant protection system
- Plant monitoring and supervisory system
- Steam bypass control equipment
- Auxiliary system controls and instrumentation
- Control-room panels
- Feedwater regulating equipment
- Tools and servicing equipment
- Fuel-handling equipment

Auxiliary systems, PWR:
- Chemical and volume control
- Residual-heat removal
- Spent-fuel pit
- Safety-injection system
- Component-cooling system
- Radioactive-waste processing
- Sampling system

Auxiliary systems, BWR:
- Reactor-water cleanup
- Standby-liquid control
- Core-isolation cooling
- Residual-heat removal
- Core spray system
- High-pressure coolant injection
- Fuel-pool cooling and filtering
- Radioactive-waste processing

***29. Nuclear-Plant Costs.*** A large amount of equipment and capital investment is required for a nuclear plant. A nuclear steam-supply system (NSSS) is arranged and equipped to initiate, sustain, and regulate the fission process in the reactor fuel, transfer heat from the fuel to the steam generators, and produce steam to supply a turbine. Auxiliary fluid systems to provide chemical cleaning and conditioning of the primary-system fluid, control of chemical shim (if used), supply and purification of secondary water (if required), and processing of radioactive effluents (liquid and gas) are associated with the NSSS. Equipment for instrumentation, control, protection, and electric power distribution is included. Typical equipment supplied in a nuclear steam-supply package is shown in Table 5-9. Cost trends for nuclear-fueled plants are shown in Fig. 5-26.

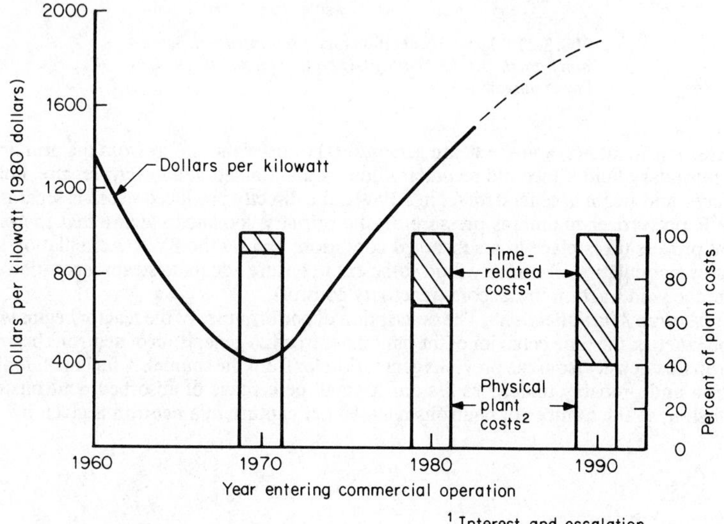


**FIG. 5-26** Cost trends for nuclear-fueled plants.

This large investment in equipment encourages the development of nuclear plants of large capacity to reduce the per megawatt cost. Plants in the current generation have been from 500 to 1300 MW in electrical capacity. Depending on the size of the system, a single nuclear plant can represent 10 to 20% or more of the system operating capacity. This large size, the high capital investment, plus a low fuel cost of the nuclear plant direct the base loading of the nuclear plant.

***30. Loading of the Nuclear Station.*** Figure 5-27 shows the position occupied by energy supplied by nuclear sources in a projected week of energy usage for one system. The decision on how to utilize a nuclear plant in a system is based on many considerations, including economics of production, operational capability, service and maintenance schedules, fuel availability, and load demand.

***31. Performance Evaluation.*** Evaluating the nuclear steam-supply system for performance as a source of steam energy often requires that the system be modeled, that is, the system equations be developed. The principal components whose characteristics are important for transient analysis are the reactor fission process and thermal process and the primary coolant piping and the steam generator(s) in a PWR along with the associated control and protection systems.

The functions of these components, in combination, are: *(a)* The reactor core through the fission process converts potential energy in the uranium atoms to thermal energy in the fuel elements; *(b)* in a PWR, GCR, LMFBR, the primary coolant piping conveys the thermal energy

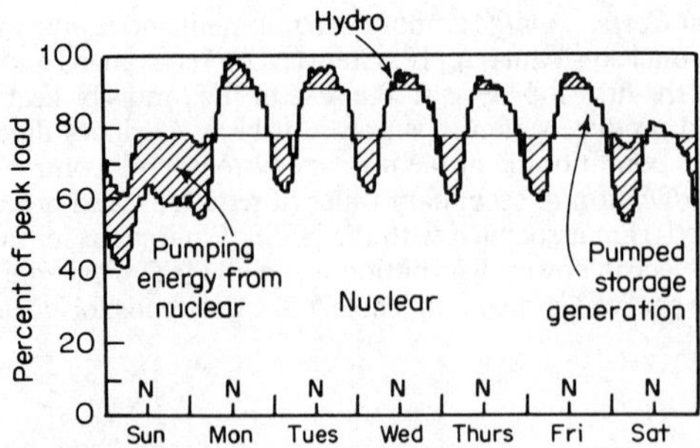


**FIG. 5-27** Typical generation mix. *(Reprinted with permission from the Electric Utility Generation Planbook, 1974, by Power magazine.)*

to the steam generator(s) and the steam generator(s) transfer the energy from the primary fluid to the secondary fluid. Since the secondary fluid is maintained at a lower pressure, boiling is introduced and steam is generated; *(c)* in a BWR, the directly produced steam is separated; *(d)* the PWR pressurizer maintains pressure on the primary coolant to assure that the primary thermal process takes place in a subcooled condition; and *(e)* the BWR recirculation system maintains a circulation of primary fluid sufficient to ensure adequate steam separation and to regulate the void fraction in the core (reactivity control).

***32. Neutron Multiplication.*** The description of performance of the reactor begins with the neutron kinetics, the time behavior of the fission neutrons. A generation of neutrons begins with a neutron flux density from the previous generation (or from the source). A fuel nuclide absorbs a neutron and probably undergoes fission. A small percentage of absorbed neutrons do not cause fission; so the number of neutrons released per capture of a neutron is given by

$$\eta = \nu \frac{\Sigma_f}{\Sigma_u} \tag{5-10}$$

where $\eta$ = number of neutrons released per capture
$\nu$ = number of neutrons released per fission
$\Sigma_f$ = cross section of the fuel for fission
$\Sigma_u$ = absorption cross section (fission and nonfission) in the fuel

In a reactor assembly where the fissions are initiated principally by thermal neutrons, some fissioning will be introduced by the fast neutrons before they have been thermalized. The ratio of the total number of fast neutrons produced by neutrons of all energies to the number produced by thermal neutrons is given by $\epsilon$.

During the slowing-down (thermalizing) process, some neutrons are captured in nonfission processes; the fraction escaping such capture is $\rho$. When the neutrons have been thermalized, they will diffuse in the core region until absorbed in the fuel or in some other material, structure, moderator, poison, etc. The fraction absorbed in the fuel is given by

$$f = \frac{\text{thermal neutrons absorbed in fuel}}{\text{total thermal neutrons absorbed}}$$

If $n$ neutrons are present in one generation (in an infinite system), the multiplication factor, $k_{\text{inf}}$, or the ratio of the neutrons in one generation to those in the next generation, is given by:

$$k_{\text{inf}} = \frac{n\eta\epsilon\rho f}{n} = \eta\epsilon\rho f$$

This is also known as the four-factor formula.

In a reactor system of finite dimensions, there is also leakage out of the system, so that the infinite multiplication factor must be adjusted to provide for leakage. Considering the diffusion process and the boundary conditions, it is evident that, for a reactor or finite physical dimensions there will be some leakage (loss) of neutrons from the boundary. In describing the process of slowing down the neutrons, two quantities are developed. The first of these is the diffusion length $L$, which is equal to one-sixth of the net vector distance that a monoenergetic neutron travels from its source to the point where it is absorbed by a nucleus. A second quantity is the buckling $B$, which represents the "bending" or appreciable reduction of the value of neutron flux at any point in the reactor.

These quantities may be used to develop two factors which take into account the finite size of the reactor and the leakage which occurs. For the first factor, the term $e^{-B^2L_s}$ represents the nonleakage probability of the neutrons as they slow down, where $L_s$ is a slowing-down length. The algebraic loss, by diffusion, of thermal neutrons in a volume element is $-D\nabla^2\phi = DB^2\phi$. The ratio of thermal leakage to thermal absorption is

$$\frac{\text{Thermal leakage}}{\text{Thermal absorption}} = \frac{DB^2\phi}{\Sigma_\alpha\phi} = L^2B^2 \quad \text{so} \frac{D}{\Sigma_\alpha} = L^2 \tag{5-11}$$

Adding the thermal absorption to the thermal leakage, effectively adding unity to both sides of the equation, and inverting, gives, for the second factor, the ratio

$$\frac{\text{Thermal absorption}}{\text{Thermal leakage} + \text{thermal absorption}} = \frac{1}{1 + L^2B^2}$$

which accounts for the nonleakage probability at thermal energy.

For a finite reactor then, the effective multiplication factor may be expressed by a combination of these two factors, namely,

$$k_{\text{eff}} = k_{\text{inf}} \frac{e^{-B^2L_s}}{1 + L^2B^2} \tag{5-12}$$

When fission occurs, more than 99% of the resulting neutrons are produced within $10^{-3}$ s. The remaining neutrons are produced during the decay of the fission fragments. The time required for their production varies; they may be separated into groups for convenience. A typical grouping is shown in Table 5-10. These delayed neutrons are essential to the regulation of the fission process.

***33. Reactor Kinetics.*** For development of control equations, a single delayed group model (Fig. 5-28) may be used for approximation of the neutron production.

**TABLE 5-10** Delayed-Neutron Half-Lives and Yields in Thermal-Neutron Fission

| Isotope | Delayed neutrons/fission | Group index $i$ | Half-life $T_{1/2}$, s | Decay constant* $\lambda$, $s^{-1}$ | Relative abundance $a$ | Absolute group yield, % |
|---|---|---|---|---|---|---|
| $^{233}U$ | 0.0066 ± 0.0003 | 1 | 55.00 ± 0.54 | 0.0126 ± 0.0002 | 0.086 ± 0.003 | 0.057 ± 0.003 |
| | | 2 | 20.57 ± 0.38 | 0.0337 ± 0.0006 | 0.299 ± 0.004 | 0.197 ± 0.009 |
| | | 3 | 5.00 ± 0.21 | 0.139 ± 0.006 | 0.252 ± 0.040 | 0.166 ± 0.027 |
| | | 4 | 2.13 ± 0.20 | 0.325 ± 0.030 | 0.278 ± 0.020 | 0.184 ± 0.016 |
| | | 5 | 0.615 ± 0.242 | 1.13 ± 0.40 | 0.051 ± 0.024 | 0.034 ± 0.016 |
| | | 6 | 0.277 ± 0.047 | 2.50 ± 0.42 | 0.034 ± 0.014 | 0.022 ± 0.009 |
| $^{235}U$ | 0.0158 ± 0.0005 | 1 | 55.72 ± 1.28 | 0.0124 ± 0.0003 | 0.033 ± 0.003 | 0.052 ± 0.005 |
| | | 2 | 22.72 ± 0.71 | 0.0305 ± 0.0010 | 0.219 ± 0.009 | 0.346 ± 0.018 |
| | | 3 | 6.22 ± 0.23 | 0.111 ± 0.004 | 0.196 ± 0.022 | 0.310 ± 0.036 |
| | | 4 | 2.30 ± 0.09 | 0.301 ± 0.012 | 0.395 ± 0.011 | 0.624 ± 0.026 |
| | | 5 | 0.61 ± 0.083 | 1.13 ± 0.15 | 0.115 ± 0.009 | 0.182 ± 0.015 |
| | | 6 | 0.23 ± 0.025 | 3.00 ± 0.33 | 0.042 ± 0.008 | 0.066 ± 0.008 |
| $^{239}Pu$ | 0.0061 ± 0.0003 | 1 | 54.28 ± 2.34 | 0.0128 ± 0.0005 | 0.035 ± 0.009 | 0.021 ± 0.006 |
| | | 2 | 23.04 ± 1.67 | 0.0301 ± 0.0022 | 0.298 ± 0.035 | 0.182 ± 0.023 |
| | | 3 | 5.60 ± 0.40 | 0.124 ± 0.009 | 0.211 ± 0.048 | 0.129 ± 0.030 |
| | | 4 | 2.13 ± 0.24 | 0.325 ± 0.036 | 0.326 ± 0.033 | 0.199 ± 0.022 |
| | | 5 | 0.618 ± 0.213 | 1.12 ± 0.39 | 0.086 ± 0.029 | 0.052 ± 0.018 |
| | | 6 | 0.257 ± 0.045 | 2.69 ± 0.47 | 0.044 ± 0.016 | 0.027 ± 0.010 |

* The decay constants are related to the half-lives by the equation $\lambda = (\ln 2)/T_{1/2} = 0.693/T_{1/2}$.

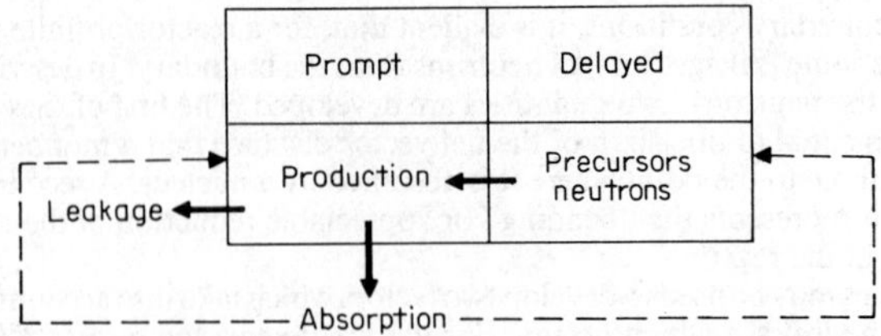


**FIG. 5-28** Single neutron delayed group model.

The production of neutrons for fission initiation including generation and thermalization is given by $k_{\text{inf}}\,\Sigma_\alpha \phi e^{-B^2L_s}$, where $\phi$ is the neutron population, $\Sigma_\alpha$ is the absorption cross section, and $e^{-B^2L_s}$ is the nonleakage probability during thermalization.

The leakage of neutrons is $D\nabla^2\phi = -DB^2\phi$. Assuming $\beta$ is the fraction of neutrons delayed, the neutron balance for the main group is

$$\dot{n} = (1-\beta)k_{\text{inf}}\Sigma_\alpha \phi e^{-B^2L_s} + \lambda C - DB^2\phi - \Sigma_\alpha\phi \tag{5-13}$$

where $\lambda$ = decay constant of the delayed neutron precursors
$C$ = population of delayed neutron precursors
$\phi$ = neutron fluence rate (flux)

Since $\phi = nv$, the equation becomes

$$\dot{n} = n[(1-\beta)k_{\text{inf}}v\,\Sigma_\alpha e^{-B^2L_s} - vDB^2 - v\Sigma_\alpha] + \lambda C$$

But $k_{\text{eff}} = \dfrac{k_{\text{inf}}\,e^{-B^2L_s}}{1+L^2B^2}\;\dfrac{1}{\bar{l}} = v\Sigma_\alpha(1+L^2B^2)$, and $D = \Sigma_\alpha L^2$ where $\bar{l}$ is the average lifetime of the neutrons.

Substituting for $k_{inf}$ and $D$ gives

$$\dot{n} = n[(1-\beta)k_{\text{eff}}v\Sigma_\alpha(1+L^2B^2) - v\Sigma_\alpha(1+L^2B^2)] + \lambda C$$

Substituting $1/\bar{l}$ for $v\Sigma_\alpha\,(1+L^2B^2)$ gives

$$\dot{n} = n\left[(1-\beta)\frac{k_{\text{eff}}}{\bar{l}} - \frac{1}{\bar{l}}\right] + \lambda C$$

$$\dot{n} = n\,\frac{(1-\beta)k_{\text{eff}} - 1}{\bar{l}} + \lambda C$$

The balance equation for the delayed group is

$$\dot{c} = n(\beta k_{\text{inf}}v\Sigma_\alpha e^{-B^2L_s}) - \lambda c \tag{5-14}$$

Substituting $\dfrac{k_{\text{inf}}e^{-B^2L_s}}{1+L^2B^2} = k_{\text{eff}}$ gives

$$\dot{c} = n[\beta k_{\text{eff}}v\Sigma_\alpha(1+L^2B^2)] - \lambda c$$

and substituting $1/\bar{l} = v\Sigma_\alpha(1+L^2B^2)$,

$$\dot{c} = n\,\frac{\beta k_{\text{eff}}}{\bar{l}} - \lambda c$$

Rearranging the equation for $\dot{n}$ gives

$$\dot{n} = n\,\frac{(k_{\text{eff}}-1) - \beta k_{\text{eff}}}{\bar{l}} - \lambda\dot{C} = n\,\frac{k_{\text{eff}}(1-\beta) - 1}{\bar{l}} - \lambda C$$

Since $k_{eff}$ is very close to 1, $\beta k_{eff} \approx \beta$ and reactivity $\rho$ is the ratio of the excess multiplication factor to the effective multiplication or

$$\rho = \frac{k_{eff} - 1}{k_{eff}} = \frac{k_{ex}}{k_{eff}} \quad \text{or} \quad \frac{\delta k}{k_{eff}}$$

Where the deviations from criticality are small, $k_{eff} \approx 1$ and

$$\rho \approx k_{ex} = k_{eff} - 1 = \delta k \qquad \text{so } \dot{n} = n[(\delta k - \beta)/\bar{l}] - \lambda C.$$

The power level $P$ is proportional to the neutron concentration. There are typically six delayed neutron groups. The balance equations then become

$$\dot{P} = \frac{P(\delta k - \beta)}{\bar{l}} + \sum_{j=1}^{\phi} \lambda_j C_j \tag{5-15}$$

$$\dot{C}_j = P\beta_j/\bar{l} - \lambda_j C_j \tag{5-16}$$

where $j$ represents the delayed neutron group and $\beta_j$, $\lambda_j$, $C_j$ are the fraction, decay constant, and concentration of delayed neutrons, respectively, of the $j$th group.

The balance equations show that, in the steady state, the effective multiplication factor is equal to 1. If the $k_{eff}$ increases above 1, the multiplication will increase with time; if $k_{eff}$ decreases below 1, the multiplication will decrease below 1. If $k_{eff}(1 - \beta) - 1$ increases to a value of 1 or greater, the reactor is said to be prompt critical and the rate of power increase depends on the ratio of $k_{eff}(1 - \beta) - 1$ to $l$. Since $l$ is so small (about $10^{-4}$ s or less for a thermal reactor; $10^{-7}$ s or less for a fast reactor), $P$ increases very rapidly with time for any appreciable value of $k_{eff}(1 - \beta) - 1$. Regulation of the process at these rates with conventional apparatus is very difficult. For this reason, $k_{eff}$ in power reactors is kept below the value $1/(1 - \beta)$ when the reactor is operating.

***34. Reactivity Control.*** The reactivity is affected by neutron absorption. The absorption occurs principally from control-element-type absorbers, dissolved-chemical control absorbers, resonance absorption in the fuel, absorption in the moderator, and absorption by fission products.

The absorption initiated by the control elements and the chemical shim are varied by the operators and are characterized by $\delta k_c$.

The reactivity effect from the fuel is caused by the widening of the resonance peaks (Fig. 5-15), with temperature which increases the nonfission capture of neutrons. With cores containing large amounts of $^{238}$U and $^{232}$Th, this Doppler effect is negative; that is, increasing the power level introduces a reactivity change which opposes the increase. The Doppler coefficient varies with coolant and fuel temperature and with moderator voids. Typical values of the Doppler coefficient are shown in Fig. 5-29. The curves may be approximated by the equation

$$\delta k_F = AT_F - BT_F^2 \tag{5-17}$$

The reactivity effect of the moderator depends on the type of moderator and the type of reactor. In water-cooled and -moderated reactors, the reactivity effects are initiated by changes in density which affect the slowing-down power and the absorption. Boiling-water reactors are operated at a relatively constant pressure and saturated conditions, which corresponds to a relatively constant temperature. Density changes due to temperature are small. The steam production varies with power level so that density variation by voids is appreciable. In a pressurized-water reactor, the coolant in the core is subcooled and voids are suppressed. Coolant temperature may be varied with power, which would cause density changes as a function of temperature.

A reactivity change from the moderator based on the density change also occurs if there is dissolved neutron absorber (chemical shim) in the moderator. A density decrease causes less of the absorber to be present in a given volume and a decrease in neutron absorption with increasing temperature.

The reactivity effect of voids and of temperature change of the water, therefore, is to oppose a change in power while the reactivity effect, due to temperature change, of dissolved chemical shim is to aid a change in power. Since it is desirable to have a negative moderator temperature coefficient of reactivity, that is, a coefficient that with decreasing moderator density acts to

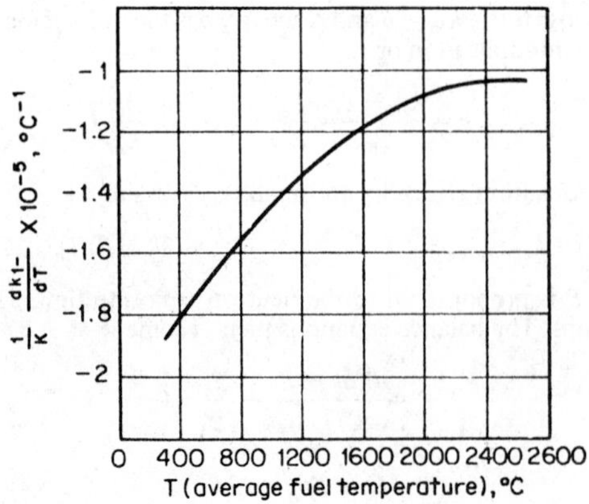


**FIG. 5-29** Values of the Doppler coefficient of reactivity. *(General Electric.)*

retard the fission process, the amount of dissolved chemical shim is usually limited to that which will do no more than reduce the temperature coefficient of reactivity to zero.

Another reactivity effect is produced by the buildup of fission products. Some of these are neutron absorbers and will act as a retardant of the fission process by removing active neutrons. Two of the strongest absorbers are xenon, $^{135}$Xe, and samarium, $^{149}$Sm, whose absorption cross sections are $3 \times 10^6$ barns and $5 \times 10^4$ barns, respectively. $^{135}$Xe is produced directly as a fission fragment (fission yield 0.3%), and in the decay chain of the fission fragment $^{135}$Te (fission yield 5.6%) is

$$^{135}\text{Te} \xrightarrow[\text{1 min}]{} {}^{135}\text{I} \xrightarrow[\text{6.7 h}]{} {}^{135}\text{Xe} \xrightarrow[\text{9.2 h}]{} {}^{135}\text{Cs} \xrightarrow[2.1 \times 10^6\text{yr}]{} {}^{135}\text{Ba}$$

On neutron absorption, $^{135}$Xe is converted to $^{136}$Xe, which is stable and has a low neutron cross section. $^{135}$Xe, therefore, has two modes of production and two modes of elimination. This can be represented by the equation

$$\frac{dX}{dt} = (\gamma_x \Sigma_f - \sigma_x X)\phi + \lambda_x I - \lambda_x X$$
$$\frac{dI}{dt} = \gamma_I \Sigma_f \phi - \lambda_1 I \qquad (5\text{-}18)$$

where $X$ = number of atoms of $^{135}$Xe present per cm$^3$ at any time $t$
$\gamma_x$ = fractional yield of xenon as a direct fission product
$\sigma_x$ = microscopic thermal-neutron absorption cross section of $^{135}$Xe
$\phi$ = thermal-neutron flux
$\lambda_1$ = decay constant of $^{135}$I
$I$ = number of atoms of $^{135}$I present per cm$^3$ at any time $t$
$\lambda_x$ = decay constant of $^{135}$Xe
$\Sigma_f$ = macroscopic fission cross section of fuel in reactor
$\gamma_1$ = fractional yield of $^{135}$I from direct fission process

The concentration reaches an equilibrium value during steady-state operation of the reactor but undergoes transients as the power changes. Of special concern to reactor regulation is the variation that occurs when the core is made subcritical following a power history. With the resulting large decrease in Xe removal by neutron absorption, the concentration of Xe increases because the difference in the $^{135}$I decay and the $^{135}$Xe decay allows a buildup. The peak concentration of $^{135}$Xe is proportional to the preshutdown power level, as shown in Fig. 5-30. This absorbing effect must be overridden by control rods or elements if start-up is to occur.

***35. Thermal Reaction.*** The heat from the core is generated in the fuel material as a result of the fissions initiated by the impinging neutrons. Although the center of a fuel element is subject to self-shielding, the fuel-element radial temperature distribution can be obtained by assuming a uniform volumetric heat source in a conduction problem. The gradient from the centerline of the fuel element through the gas gap and cladding to the coolant might be as shown in Fig. 5-31*a*. The power distributions (Fig. 5-31*b* and *c*) axially and radially in the core also vary, generally being highest in the center and decreasing toward the top and bottom and outside.

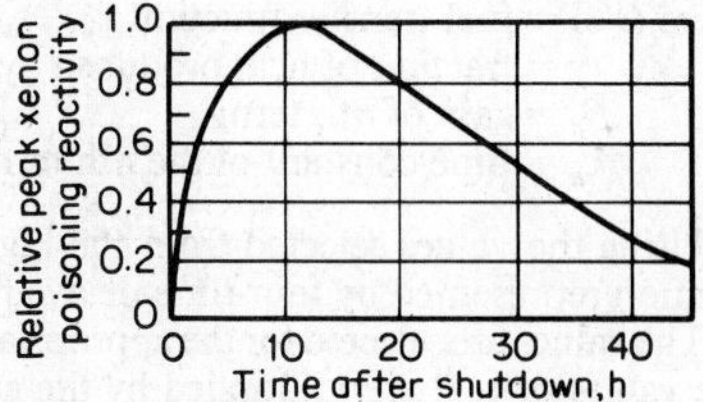


**FIG. 5-30** Relative peak xenon-poisoning reactivity. *(From Ref. 10.)*

Control elements are adjusted to provide some degree of flattening. This makes a higher average power possible from the reactor for a given peak power.

For instrumentation and control purposes, the fuel may be considered to be mathematically represented by a series of time lags. These lags are important because the protection and control are strongly dependent on the time involved in transferring the heat out of the fuel. Since the elements are very long compared with their radius, conduction of heat in the axial direction can be neglected. In cylindrical coordinates, the Laplace equation for heat flow (neglecting the $Z$ direction) is

$$\frac{\partial^2 T}{\partial \gamma^2} + \frac{1}{\gamma}\frac{\partial T}{\partial \gamma} + \frac{1}{\gamma^2}\frac{\partial^2 T}{\partial \phi^2} = \frac{\rho C}{k}\frac{\partial T}{\partial t} \tag{5-19}$$

where $T$ = temperature, °C
$t$ = time, s
$k$ = thermal conductivity, W/(m)(°C)
$C$ = specific heat, J/(kg)(°C)
$\rho$ = density, kg/m³
$\gamma$ and $\phi$ = cylindrical coordinates, m

Solving this equation, assuming that the heat capacity of the cladding can be neglected, the thermal resistance from the surface of the fuel pellet to the cooling channel gives the fuel-element transfer function

$$G(s) = (1-\gamma)\sum_{\eta=1}^{4}\frac{F_n}{1+\tau_n S} + \gamma \tag{5-20}$$

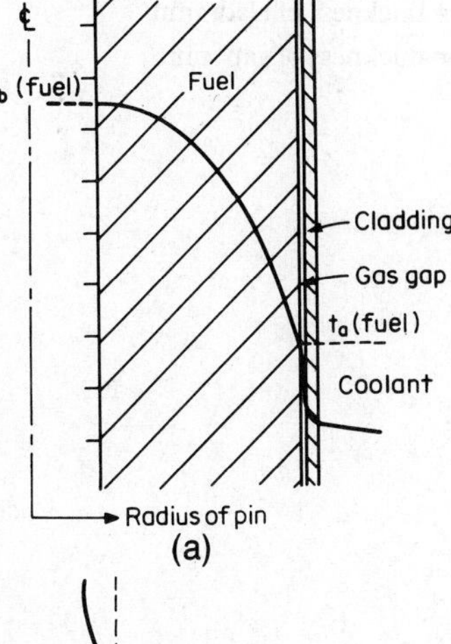


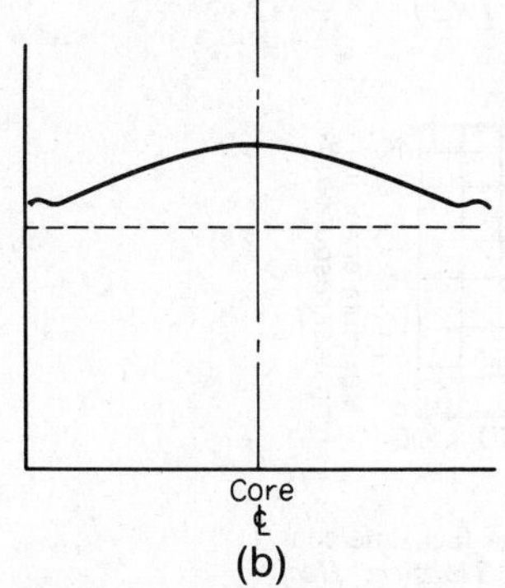


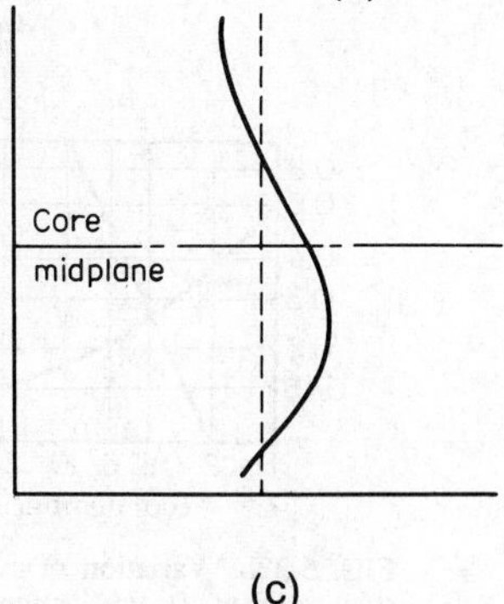


**FIG. 5-31** *(a)* Temperature distribution in fuel rod; *(b)* radial power profile; *(c)* axial power profile.

where $G(s)$ = fuel transfer function
$\gamma$ = fraction of heat produced by photons
$F_n$ = gain of $n$th term
$\tau_n$ = time constant of the $n$th term

Using the values selected from the Fig. 5-32 curves and from Table 5-11, a fuel transfer function represented by four time-delay groups can be generated.

The values are selected for the appropriate value of the Biot number $N_{B1}$, where $N_{B1} = rH/k$. The values of $\tau_n/t$ are multiplied by the appropriate value of $t_s$, where $t_s = r^2\rho C/k$, the time constant of an equivalent slab.

$C$ = specific heat of fuel, J/(kg)(°C)
$H$ = total heat-transfer coefficient, fuel to coolant
$= 1/(X_c/k_c + X_g/k_g + 1/h_f)$
$k$ = thermal conductivity of fuel, W/(m)(°C)
$k_c$ = thermal conductivity of clad, W/(m)(°C)
$k_g$ = thermal conductivity of gas, W/(m)(°C)
$r$ = radius of the fuel, mm
$\rho$ = density of fuel, kg/m³
$X_c$ = thickness of clad, mm
$X_g$ = thickness of gap, mm

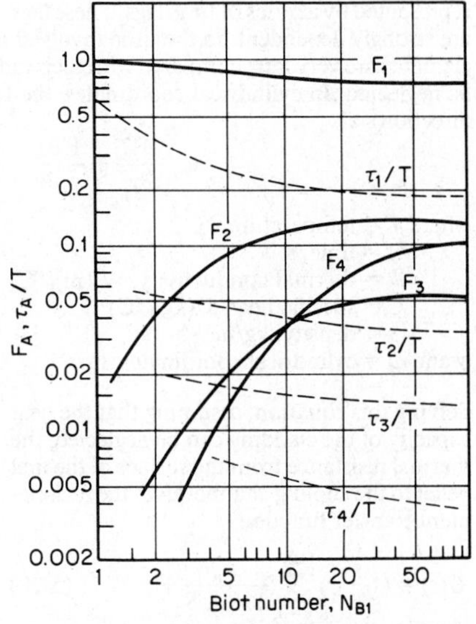


**FIG. 5-32a** Variation of ceramic-fuel transfer. *(From D. H. Crimmins, Transient Heat Transfer from Ceramic Fuel Pins; Nucleonics, vol. 20, no. 8, August 1962.)*

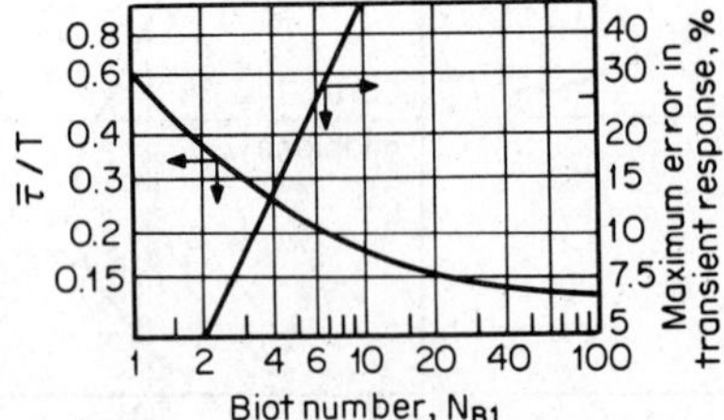


**FIG. 5-32b** Variation of average fuel time constant. *(From D. H. Crimmins, Transient Heat Transfer from Ceramic Fuel Pins, Nucleonics, vol. 20, no. 8, August 1962.)*

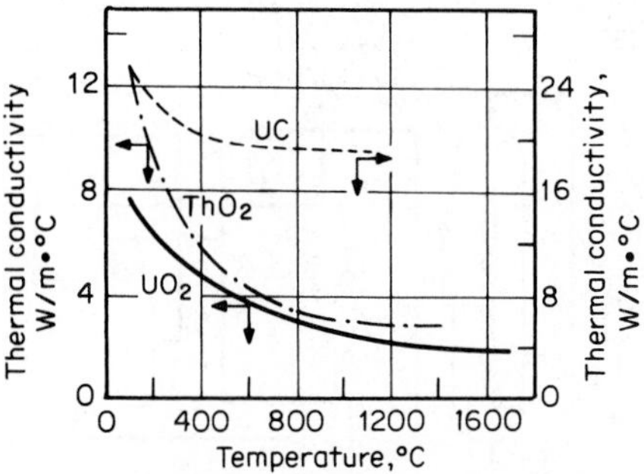


**FIG. 5-32c** Thermal conductivity of fuel materials.

**TABLE 5-11** Fuel-Material and Steam-Generator-Tube Data

| | Property | |
|---|---|---|
| Material | Density, g/cm³* | Specific heat, kJ/(kg)(°C) |
| $UO_2$ | 10.50 (10.97 theoretical) | 0.247 (0–1200°C) |
| $ThO_2$ | 10.01 | 0.243 (0–760°C) |
| UC | 12.97 (13.63 theoretical) | 0.147 (100°C) |
| Types 304 and 394 L SS | 7.9 | 0.502 (0–100°C) |
| Type 347 SS | 8.0 | 0.502 (0–100°C) |
| Inconel | 8.51 (at 20°C) | 0.456 (25–100°C) |
| Zircalloy 2 | 6.55 | 0.33 (0–800°C) |

* Multiply $g/cm^3 \times 1000$ to obtain $kg/m^3$.

***36. Application of Performance Equations.*** The equations of performance for the various portions of the NSSS may be used for mathematical analyses or for development of simulations. Simulation of an NSSS is frequently desired. Such simulation enables: (*a*) evaluation of system performance to assure that plant performance requirements are met, (*b*) performance evaluation of monitoring or control equipment, and (*c*) analyses of protection action to assure protective-system adequacy.

The simulation may be performed with an analog computer, a digital computer, or a hybrid computer; selection of the appropriate method should be based on the equipment to be simulated and the objectives of the tests to be performed.

The level of detail of the simulation also varies with the objectives of the test. For instance, modeling of the reactor core may be one node for gross thermal input to another component or be multinode to evaluate the performance of in-core instrumentation. One-, four-, seven-, and thirty-eight-node models are examples of core simulations that have been found to be useful.

Elementary thermal cycles of a PWR and a BWR are shown in Fig. 5-33. For purposes of illustration, a simplified model of a PWR is considered below. The system is diagramed as shown in Fig. 5-34. Reactor kinetics and core heat transfer are assumed to occur at a single point. Heat transfer in the steam generator also occurs at a single node. Flow between the reactor and steam generator is represented by appropriate time delays. The equations for this model are

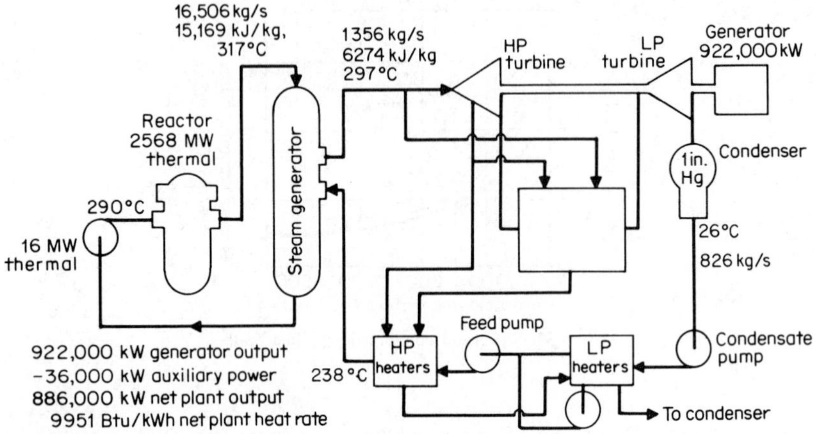


**FIG. 5-33a** Simplified thermal cycle of PWR system. *(Reprinted with permission from the 1974 Generation Planbook, by Power magazine.)*

Reactivity:

$$\dot{p} = \frac{\delta k - \beta}{\bar{l}} P + \sum_{1}^{n} \lambda_j C_j$$

$$\dot{C}_j = \frac{\beta i}{\bar{l}} P - \lambda_j C_j$$

$$\delta k = \delta k_c + \delta k_w + \delta k_D$$

$$= \delta k_c f(p) + \delta k_w (Tw_i - Tw_0) + \delta k_D (T_f - T_f^2)$$

Reactor thermal:

$$\rho_f C_f V_f \frac{dT_f}{dt} = P - kA_f (T_f - T_{av})$$

$$\rho_w C_w V_w \frac{dT_{av}}{dt} = k\,A_f (t_f - T_{av}) - F_w C_w (T_h - T_c)$$

$$T_c = 2T_{av} - T_H$$

Steam generation, thermal:

$$\rho_m C_m V_m \frac{dT_m}{dt} = FC_w (T_p - T_p) - U_\beta A_\beta (T_m - T_s)$$

$$\rho \psi V_s \frac{dT_s}{dt} = U_\beta A_\beta (T_m - T_s) - F_s (h_{se} - h_{si})$$

where $P$ = reactor power, J/s
$\beta$ = delayed neutron fraction
$C_j$ = delayed neutron precurser concentration
$\lambda_j$ = decay constant for neutron precurser, $\delta^{-1}$
$\bar{l}$ = mean neutron lifetime, s
$\delta k$ = control reactivity
$\delta k_c$ = coefficient of control-element worth
$\delta k_w$ = reactivity coefficient of water temperature

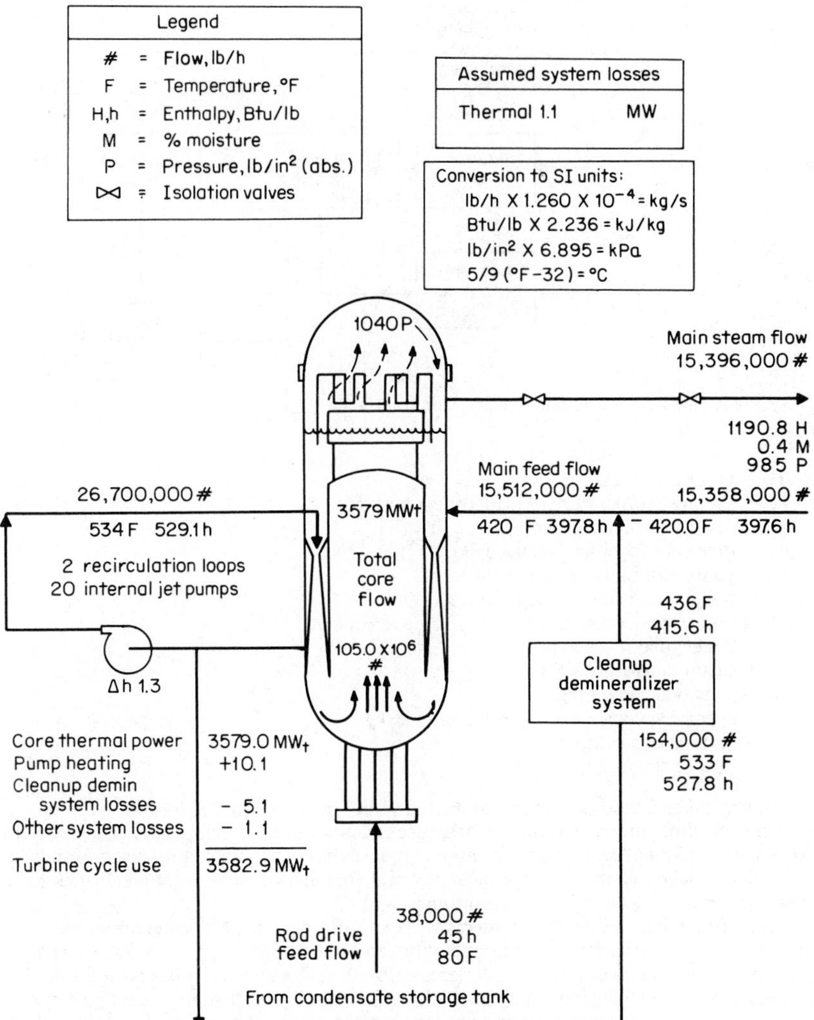


**FIG. 5-33b** BWR thermal diagram.

$\delta k_D$ = reactivity coefficient of fuel temperature (Doppler coefficient)
$\rho_f$ = density of fuel, kg/m³
$C_f$ = thermal capacity of fuel, J/(kg)(°C)
$V_f$ = volume of fuel, m³
$T_f$ = temperature of fuel, °C
$F_w$ = flow of water, kg/s
$h_{se}$ = inlet enthalpy
$h_{si}$ = exit enthalpy
$A_f$ = area of fuel heat transfer, m²
$K$ = thermal conductivity, J/(s)(m²)(°C)
$\rho_w$ = density of water, kg/m³
$C_w$ = thermal capacity of water, J/(kg)(°C)

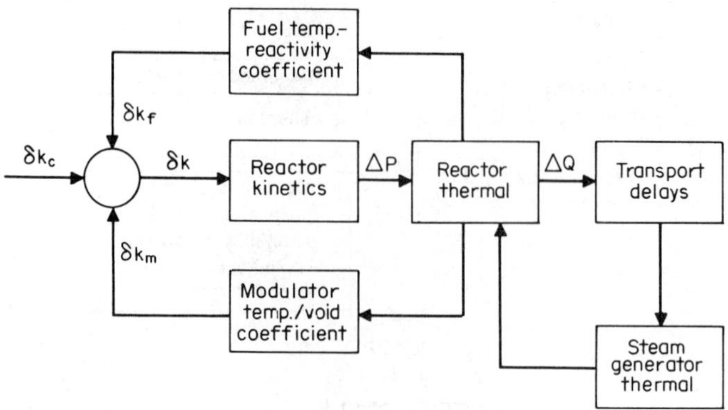


**FIG. 5-34** Block diagram for simulation.

$V_w$ = volume of water, m³
$T_{av}$ = average temperature of primary coolant, °C
$\rho_m$ = density of metal, kg/m³
$C_m$ = thermal capacity of metal, J/(kg)(°C)
$V_m$ = volume of metal in boiler tubes, m³
$U_\beta$ = boiler heat-transfer coefficient, J/(s)(m²)
$A_\beta$ = area of boiler tubes, m²
$T_m$ = boiler-tube temperature, °C
$T_s$ = steam temperature, °C
$\rho_s$ = density of steam, kg/m³
$\psi$ = average specific heat of steam water, J/kg
$V_s$ = volume of steam, m³
$F_s$ = flow of steam, kg/s

Values for metal and fluid properties can be found in Fig. 5-35 and Table 5-11.

From such simulations, various performance indexes can be developed for use in system synthesis or component evaluation. Traditional indexes have been frequently-response-related such as Bode diagrams and Nichols plots. Typical frequency-response characteristics for a nuclear system are shown in Figs. 5-36 through 5-39.

Since commercial power-reactor systems are large and are oriented to power production, it is inconvenient to encumber them with apparatus and operations directed to obtaining their time-response characteristics, which interferes with normal operation. It has been found that the discrete nature of neutrons and the statistical nature of the fission process give rise to random fluctuations in neutron population or "reactor noise." The production, absorption, and leakage of neutrons can be considered analogous to the random flow, from emitter to collector, of electrons in a diode. Using power-spectral-density techniques, the reactor transfer function can be developed from an analysis of reactor noise.

## Control Systems

***37. Fission Regulation.*** The fission process is regulated by the absorption, in a controlled manner, of some of the neutrons which cause fission. The controlled absorption may be provided by the control rods or control elements which are mechanically inserted into, or withdrawn from, the reactor core or by absorber material dissolved within the reactor coolant. Because steam is produced directly from the coolant in a BWR, dissolved poisons are usually not used in normal operation of the BWR. In a gas-cooled reactor where the coolant gas is either helium or carbon dioxide, it is not feasible to provide dissolved absorbers in the coolant. Dissolved poisons, therefore, are used principally in PWRs. Where dissolved poisons or fixed poisons within the core are used, they are generally used for the purpose of accommodating core

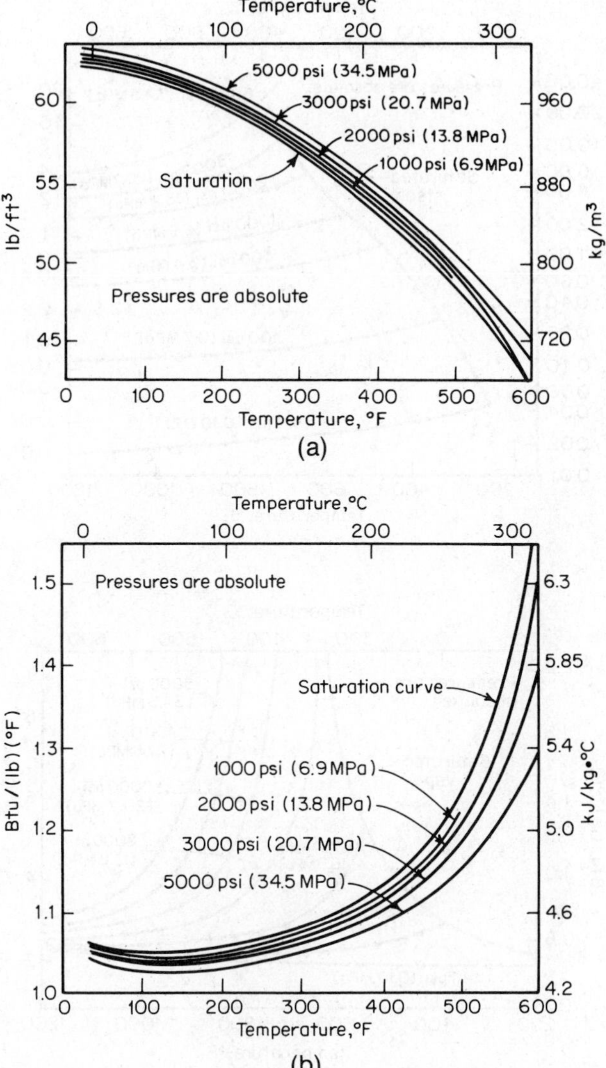


**FIG. 5-35** Fluid properties: *(a)* density and *(b)* specific heat of liquid light water.

burnup. Changes in the reactivity for normal operation, initial start-up, planned shutdown, and restart are normally accomplished by control rods or control elements. Since the rate of change available from dissolved poisons or fixed burnable poisons is normally quite slow, the mechanically inserted control rods or control elements are also used for emergency shutdown.

Considerations involved in determining the characteristics of the control rods or elements are the amount of reactivity that has to be controlled, the position accuracy (which corresponds to the minimum increment of the reactivity) to be provided, the rate of reactivity that must be provided for operation, and the reliability of components. The reactivity requirements of a nuclear system are based on the planned rate of fuel depletion, the fission-product buildup,

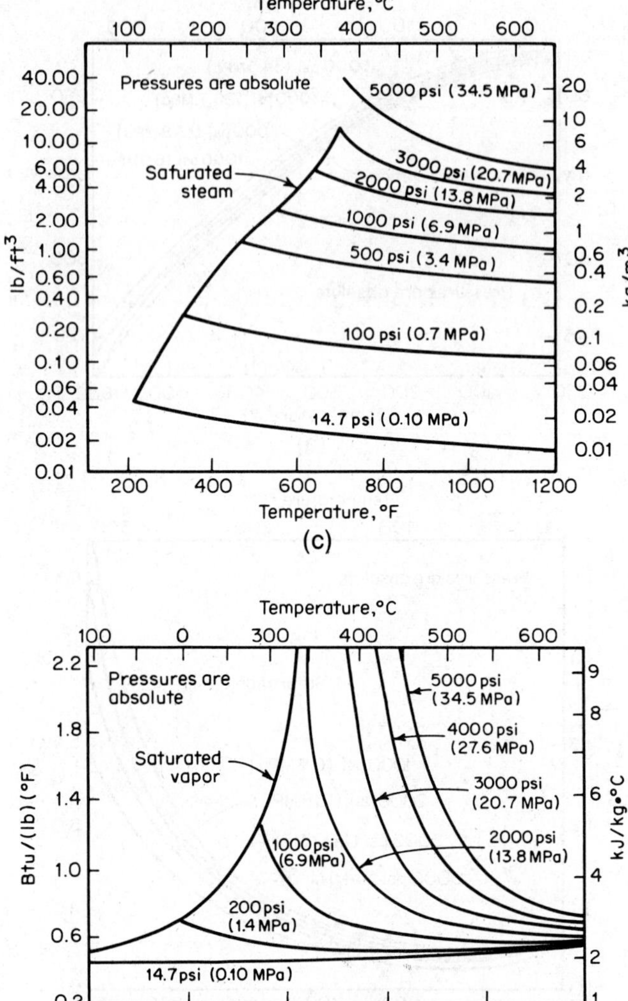


**FIG. 5-35** *(Continued)* Fluid properties: *(c)* density and *(d)* specific heat of water vapor.

including the principal poisons xenon and samarium, inherent reactivity effects such as temperature or void changes, and the control range necessary for maneuvering to which the plant may be subjected.

The control rods or elements are also used to flatten the power developed radially and axially within the reactor to raise the average power and increase the output from the system. To accomplish this purpose, the rods or elements are normally assigned to groups or banks which are operated together. The capability of a rod or element to perform its neutron absorption or the worth of the control unit is dependent upon the position of the element within the core. As the element progresses from the bottom, its worth increases from a low value to a peak and then

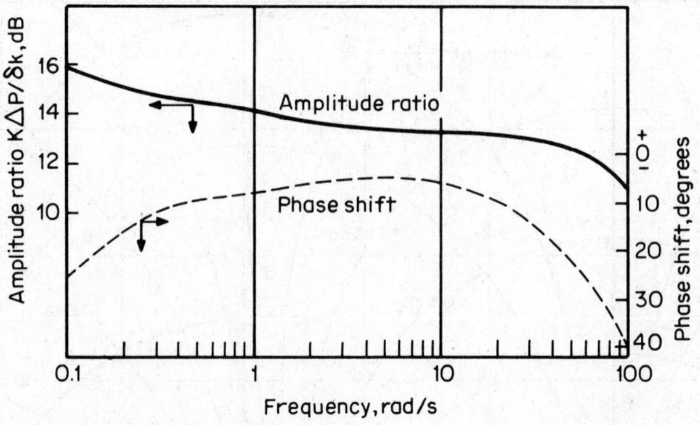


**FIG. 5-36** Amplitude ratio and phase shift versus frequency diagram for reactor kinetics transfer function.

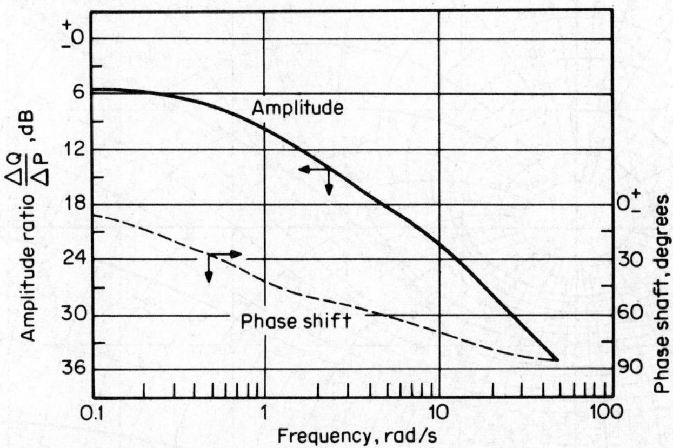


**FIG. 5-37** Amplitude ratio and phase shift versus frequency diagram for fuel transfer function.

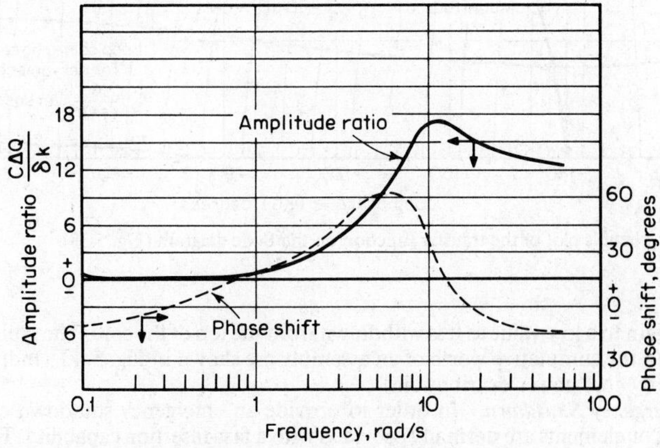


**FIG. 5-38** Amplitude ratio and phase shift versus frequency diagram for overall, closed-loop transfer function.

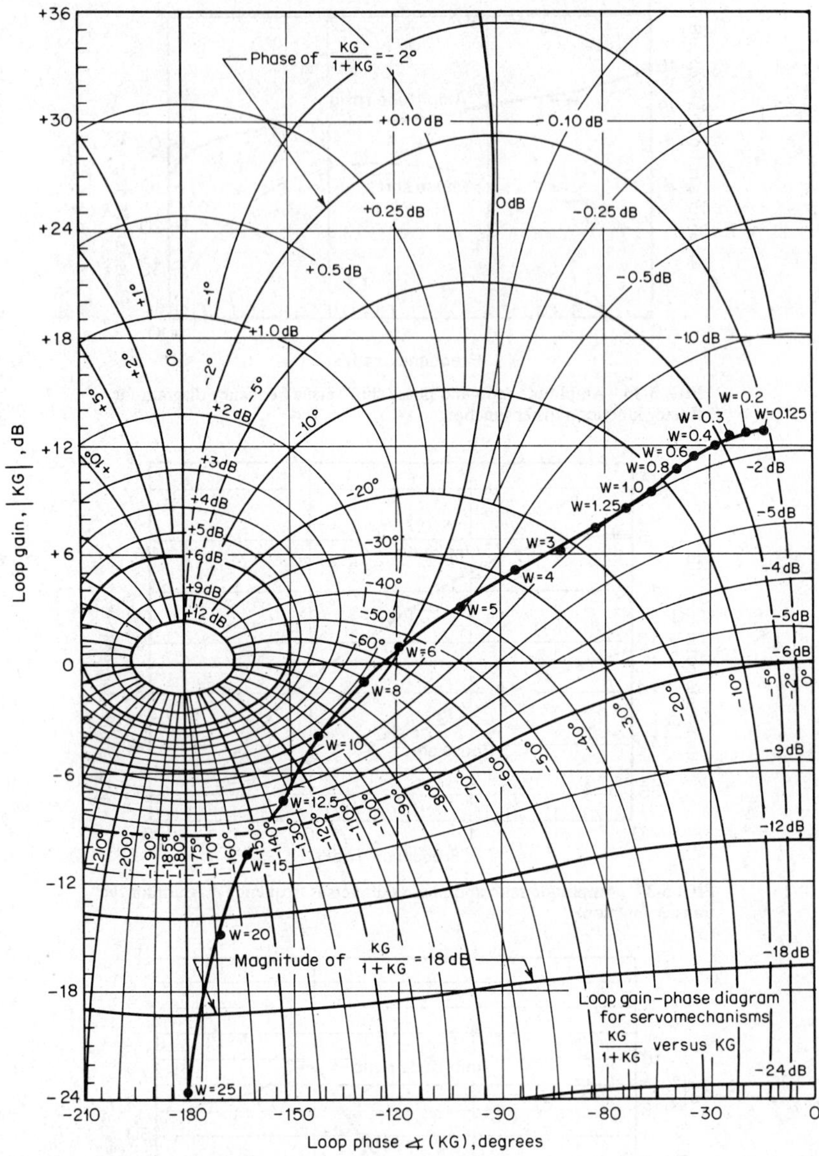


FIG. 5-39 Nichols plot of the transfer function on the Bode diagram (Fig. 5-38).

decreases again to a low value as it is withdrawn from the top of the core. The typical incremental worth and the cumulative worth of an assembly are shown in Fig. 5-40. (Individual worths are affected by core power distribution.)

***38. Emergency Shutdown.*** In order to provide an emergency shutdown capability, the control rods or elements are normally provided with a fast-insertion capability. This characteristic, also referred to as scram, is provided for control drives mounted on top of the reactor by a delatching capability with the rods or elements free-falling into the reactor core. With rods or elements that are mounted on the bottom of the reactor, a capability is provided to drive the rods rapidly up to their full insertion position. The latter mechanism is typically hydraulically actuated. Since the position of the control rods or elements is important information in determining core power distribution and represents a knowledge of the rate at which the fission process is proceeding, it is desirable to provide indication of the position of the control elements and to provide input from the rod-position sensors to core-power-distribution calculations. Position indication can be provided by analog meters, digital indicators, analog presentations on a cathode-ray tube, position logged by a digital computer with printout, or other types of display. Because of the importance of the position of the control element in reactor regulation and reactor shutdown, two systems of indication for the control elements are normally provided.

*39. Controlling Fluid Processes.* The primary system of most U.S. reactors involves either light water or helium as a coolant. Other fluid systems are provided to supply makeup, effect cleanup, process waste, etc. Conventional instrumentation is used to monitor these variables, and control signals in accordance with preselected control program are applied to appropriate actuator elements. An example of control programs for a PWR system is shown in Fig. 5-41.

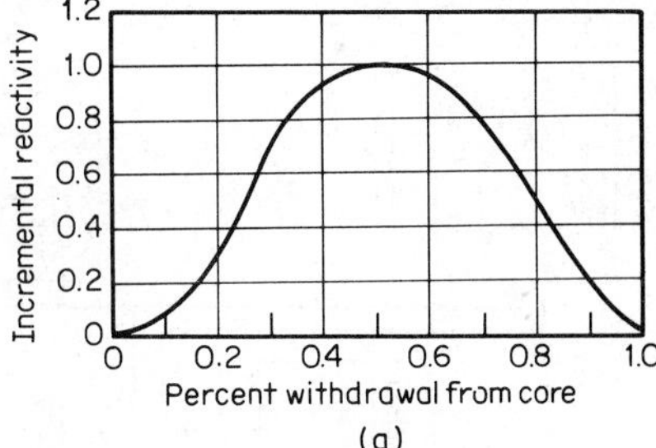


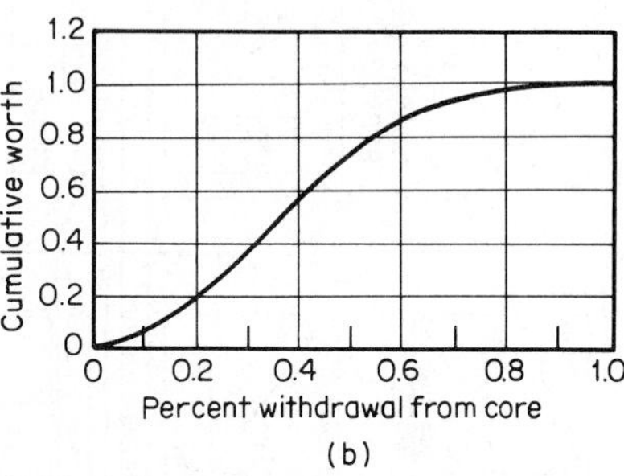


**FIG. 5-40** *(a)* Incremental reactivity versus core withdrawal; *(b)* cumulative worth versus core withdrawal.

The special requirements of nuclear systems for continuity of cooling may require that extra care be used in the application of control equipment to assure high reliability. Establishment of set points and alarm points should also be done with due care.

*40. Protection Systems.* The high specific power of nuclear reactor sources coupled with the potential for the release of radioactivity, which might be a hazard to human beings, requires that additional systems be provided for regulation. In addition to the instrumentation provided for the control of the fission process and the control of the fluid processes, systems are specifically supplied to initiate protective action in the event that preselected limits are exceeded. The relation of protection systems to other instrumentation and control systems is shown in Fig. 5-42.

Basically a protection system must provide a functional capability to initiate action in the event of a design-basis event which, if unchecked, could lead to unacceptable consequences. Because of this requirement, the protection system must operate when required and must operate correctly. The development of functional and reliability requirements is very important in this arrangement.

The consequence of greatest concern associated with a nuclear system is the release of radioactivity. The nuclear fuel itself has multiple barriers between the fuel material and the outside boundaries for the public. These barriers are the fuel matrix, the fuel jacket or cladding, the primary system, and the containment structure. For any given condition it is the duty of the protection systems to prevent a situation which would lead to an excessive release of radioactivity past a given boundary from occurring.

The protection systems therefore include the reactor-shutdown system and other systems that effect the containment of the radioactivity such as emergency core cooling, containment isolation, containment pressure reduction, emergency power sources, and air filtration. A protection system itself includes the instruments, logic systems, actuators, protective interlocks, and mechanisms which carry out the necessary functions. That system which effects reactor shutdown, normally referred to as the reactor protection system, includes all electrical and mechanical devices and circuits used to initiate a reduction of the fission process below criticality. The engineered safety features or engineered safeguards systems include everything else associated with protection except the reactor protection system.

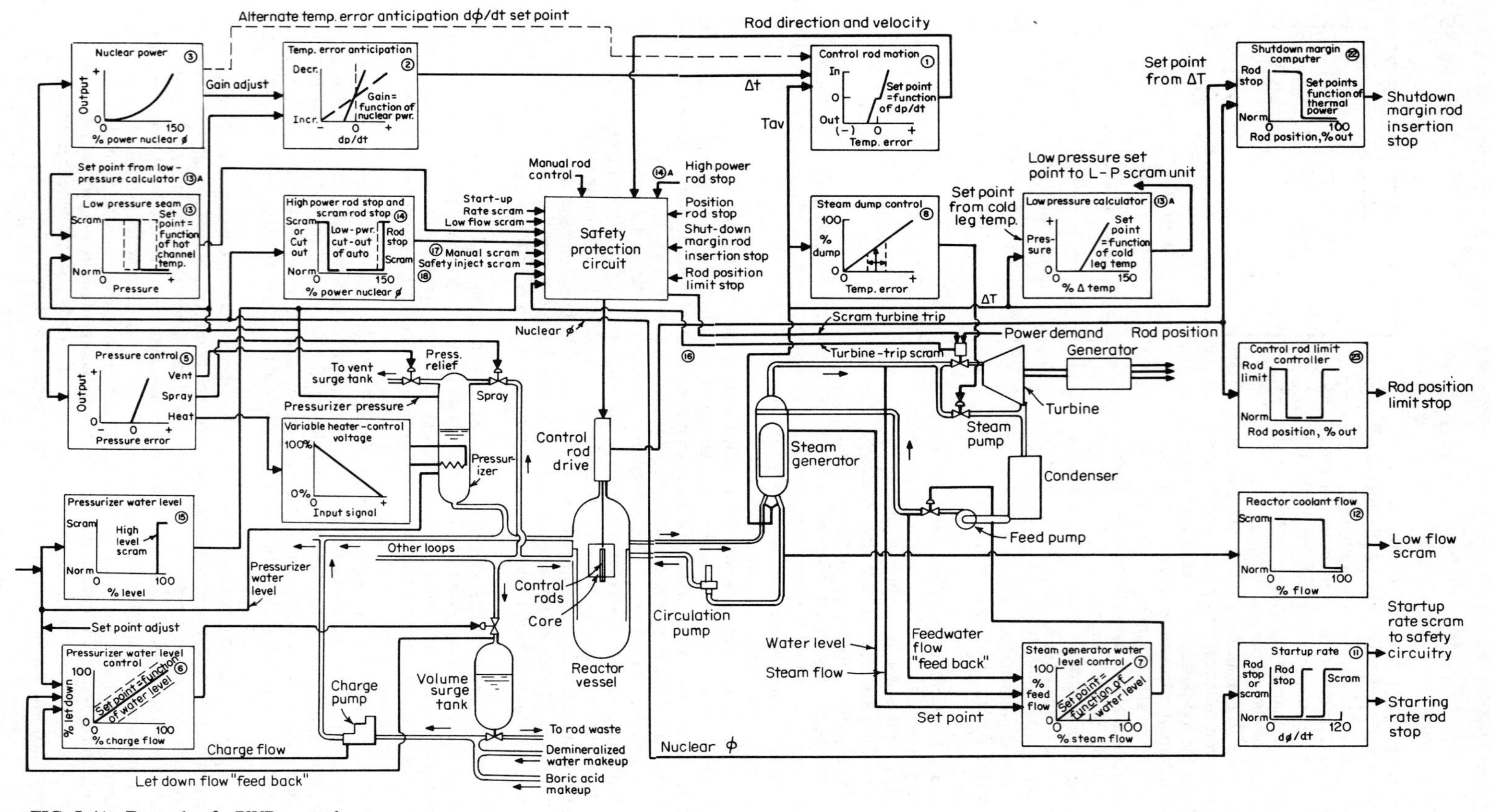


**FIG. 5-41** Example of a PWR control system.

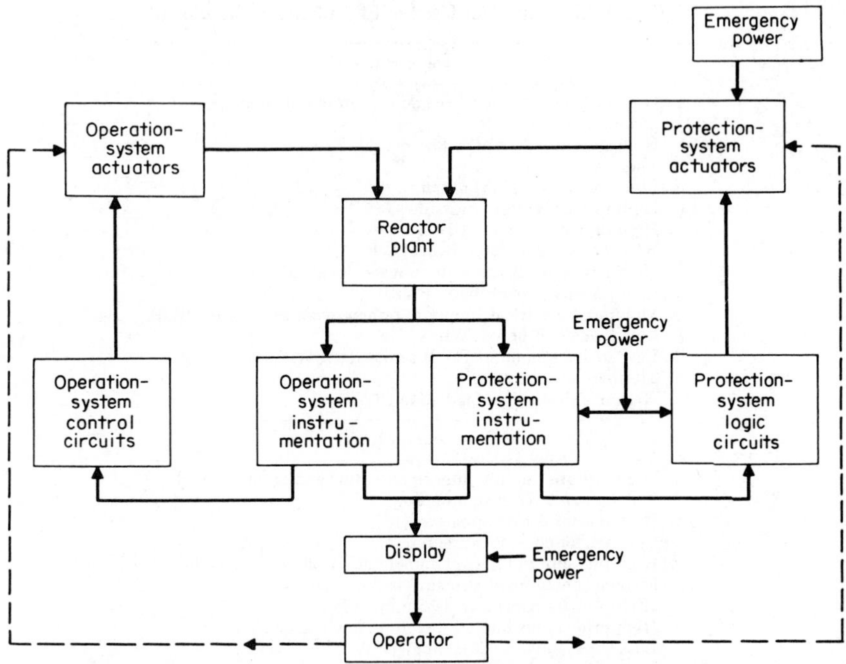


**FIG. 5-42** Relation of protection system to other instrumentation and control. *(From Ref. 8.)*

In order to accomplish a high availability, that is, the capability to act when needed, it is necessary to provide multiple channels to perform the same action. The probability of any one channel not being able to act at a given time, therefore, is offset by having other channels capable of performing the action. In order to avoid spurious action, that is, to initiate a protection when none is required, it is usually the practice to provide a logic arrangement and initiate action only when there is a coincidence of two or more channels. The consequences of spurious action, for example, loss of a power source, in a large system can also be rather severe in terms of impact not only at the plant site but also in the area of public usage of the plant output, and such false action should therefore be precluded. Conventional protection systems have used a logic arrangement involving three or four channels and requiring the coincidence of two of three or two of four in order to initiate action.

In the selection of the plant variables that are sensed by the protection system, it is usual to evaluate those plant conditions which could occur as a result of some event. In some cases, such as those involved with protection of the clad integrity, the variables cannot be measured directly and an inferred or computed variable must be used to initiate the protective action. Such plant variables used to evaluate the fuel design limits are the departure from nucleate boiling (DNB) ratio in the case of PWRs and minimum critical heat-flux ratio (MCHFR) in the case of BWRs. In order to maintain these critical values below safety limits, it is necessary to monitor the observable parameters which affect DNB or MCHFR such as thermal power, coolant flow, coolant temperature, coolant pressure, and core power distribution (from the nuclear instrumentation and control-element-position sensors). These values are translated into systems relating to desired protection, and reactor shutdown is initiated if the measured variables approach boundaries of regions established by these systems. The earlier nuclear plants utilized parameters directly and the shutdown values or limits were set on the basis of calculated values relating to a set of curves. Current nuclear plants involve, to a certain degree, the use of on-line digital calculators permitting the protection system to provide continual computations of the relation of the variables. Shutdown limits are initiated as a function of the instantaneous value

**TABLE 5-12** Variables Used to Effect Reactor Shutdown

| Light-water reactors |
|---|
| *a.* High neutron flux for source range, intermediate range, and power range |
| *b.* High start-up rate and/or low period |
| *c.* Thermal margin* |
| *d.* Low primary coolant pressure |
| *e.* High primary system pressure |
| *f.* High drywell pressure (BWR) |
| *g.* High pressurizer water level (PWR) |
| *h.* High scram discharge volume water level (BWR) |
| *i.* Low primary coolant flow (PWR) |
| *j.* Monitored electrical supply to primary coolant pumps (PWR) |
| *k.* Low feedwater flow (PWR) |
| *l.* Low steam generator (PWR) or reactor (BWR) water level |
| *m.* Turbine-generator trip† |
| *n.* High main-steam-line radiation (BWR) |
| **Gas-cooled reactors** |
| *a.* High average neutron fluence rate (flux) and/or ratio of high fluence rate to core cooling flow |
| *b.* High reactor coolant pressure |
| *c.* Low reactor coolant pressure |
| *d.* High pressure of the containment atmosphere, when a low leakage-containment structure is provided |
| *e.* High primary coolant moisture in a loop |
| *f.* High primary coolant temperature, low coolant flow |
| *g.* Loss of essential electrical power |
| *h.* Loss of feedwater flow |
| **Liquid-metal reactors** |
| *a.* High average neutron fluence rate (flux) |
| *b.* High and low coolant level |
| *c.* High core outlet temperature |
| *d.* Low core coolant flow |
| *e.* High subassembly outlet temperature |
| *f.* Low subassembly coolant flow |
| *g.* Low primary pump speed |
| *h.* Incipient or actual boiling of coolant |
| *i.* High rate of change of neutron fluence rate (flux) period at start-up |
| *j.* High noise of neutron fluence rate (flux) |
| *k.* High activity in ventilated containment system |
| *l.* High rate of change of temperature of coolant outlet |
| *m.* High seismic acceleration |
| *n.* High activity of coolant |

* Thermal-margin trip may be used to protect against excessive boiling in the core and against DNB (or MCHFR) and against excessive power density.

† Turbine-generator trip may be unnecessary in plants where sufficient steam-discharge capability or fast power cutback is provided.

of the measured variables. Table 5-12 is a list of the monitored variables used for reactor protection.

The locations where the monitoring of these variables occurs are critical. The sensors should be located where a timely and linear relationship, if possible, of the measured variable exist. Due consideration must be given to the condition of the variable, which might preclude accurate information on its condition such as flow distribution, transient effects, and time lags. Additional concerns occur when there is a large area to be measured and spatial effects can be present.

In this case, it may be that sensors must be distributed to monitor the spatial-dependent variables such that a true relationship of the average condition can be obtained. Such spatial measurements are now customarily provided in the monitoring of core power and in some cases the monitoring of coolant temperature conditions within the large pipes that are provided to conduct the coolant flow.

A set of postulated events, referred to as design-basis events (DBE), and corresponding criteria for acceptable consequences are defined to provide a basis for the plant safety systems. Based on the origin, progress, and consequences of the events, specific instrumentation types and locations must be chosen to provide signals which can be utilized to issue commands to the protective actuators. The system developed to initiate protective action must have a high availability, an independence from common-mode failures, and an ability to be tested and must accommodate other concerns. (Protection-system criteria are delineated in IEEE 279-1971 (R1978).) Special conditions under which the protection system must function include the requirements that it must not be paralyzed by the accident with which it was designed to cope. Changes in plant configuration or damage to given equipment areas might remove or distort information that was being supplied from a given area. Consideration must be given to providing the necessary information for the period required to initiate protection action. The protection system must also be equipped to withstand the effects of hazards that might serve as a disrupting influence. These include natural phenomena such as earthquakes, tornados, and floods, and man-made phenomena such as aircraft crashes into the plant.

High availability and serviceability characteristics are required of protection-systems equipment. Because it is very difficult to obtain equipment which in itself has very low probability of failure and at the same time a low probability of spurious output, it is conventional to provide redundant instrumentation with coincidence logic. The coincidence logic requires that at least two of the detectors provide a signal about the condition. Such signals must be in agreement before that particular channel initiates the shutdown signal. Two types of logic are generally

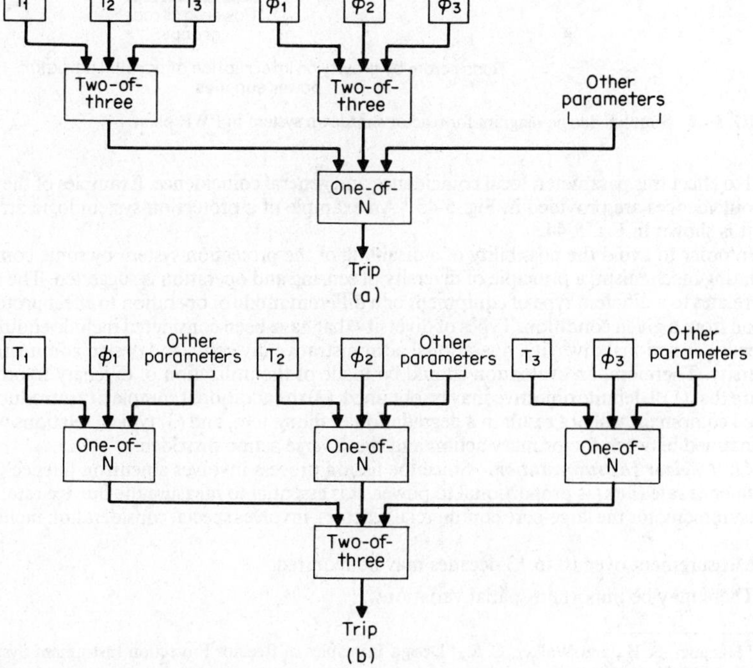


**FIG. 5-43** Types of coincidence: *(a)* local; *(b)* general.

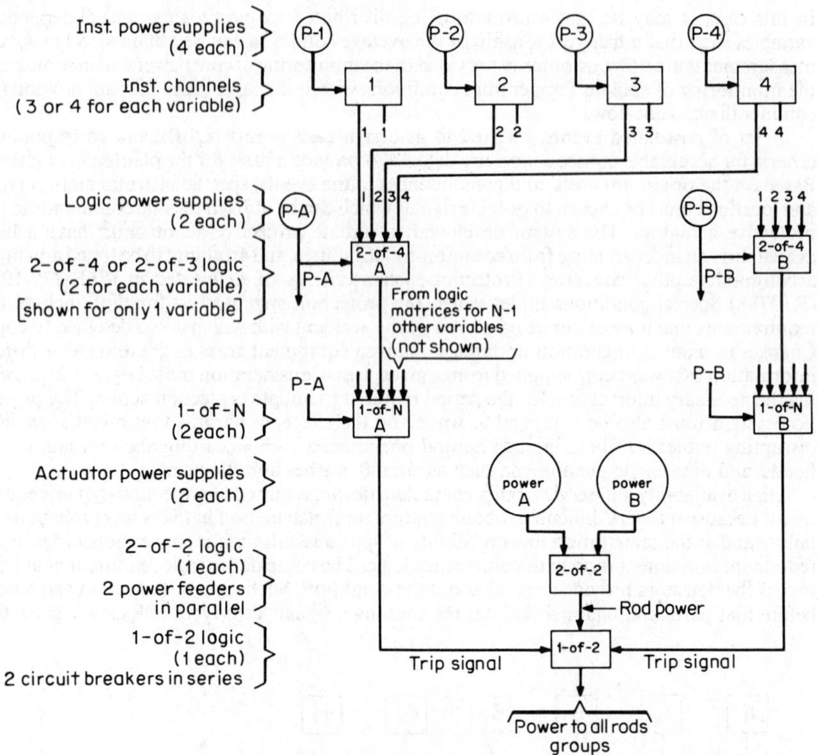


**FIG. 5-44** Simplified logic diagram for reactor shutdown system in PWR plant.

used to effect this parameter, local coincidence and general coincidence. Examples of the types of coincidences are provided in Fig. 5-43.[1] An example of a protection-system logic arrangement is shown in Fig. 5-44.[2]

In order to avoid the possibility of a disabling of the protection system by some common initiating mechanism, a principle of diversity of sensing and operation is suggested. The diversity relates to a different type of equipment or a different mode of operation to effect protection action from a given condition. Types of diversity that have been considered include equipment diversity, functional diversity, operational administrative diversity, and design administrative diversity. Therefore, an evaluation should be made of the utilization of diversity in order to assure that (1) a definite objective may be obtained, (2) the additional complexity introduced by added equipment will not result in a degradation of the system, and (3) typical relations will be maintained between the primary action and the diverse action provided.

***41. Nuclear Instrumentation.*** Since the fission process involves a neutron fluence where the fluence rate (flux) is proportional to power, it is essential to measure the fluence rate. Such measurement, for the large-core commercial reactors, involves special consideration including:

Measurement over 10 to 13 decades may be required.

There may be important spatial variations.

---

[1] Hanauer, S. H., and Walker, C. S., "Design Principles of Reactor Protection Instrument Systems," ORNL-NSIC-51, (1968), p 42.

[2] O'Brien, H. G., and Walker, C. S., "Protection Instrumentation Systems in Light-Water-Cooled Power Reactor Plants," ORNL-NSIC-29 (1969), p. 15.

The measurement is of uncharged nuclear particles (neutrons).

The measurements have to be made in a background of substantial gamma radiation.

In addition to the monitoring of neutrons, a nuclear steam system involves the monitoring of radiation, principally gamma, from process lines and fuel.

***42. Ex-Core Neutron Monitoring.*** Ex-core, or out-of-core, detectors are those which are located external to the reactor core and usually external to the primary pressure boundary. The fast neutron flux leaking from the core provides a neutron flux spectrum for some distance beyond the vessel wall. This flux, proportional to a spatially averaged core power, becomes thermalized in the shielding, usually hydrogenous material surrounding the reactor.

The environment in which the detectors are located involves, typically, neutron fluxes from up to $10^{11}$ neutrons/(cm$^2$)(s), gamma dose rate up to $10^7$ R/h, and temperatures to 200°C.

The detectors used are devices which produce a current pulse when subjected to the passage of a nuclear emission. The incident radiation drops some or all of its energy within the detector, causing ions to be produced. The ions produced are attracted to electrodes, within the detector, by the effect of voltage across the electrodes. The number of ions collected is a function of applied voltage. The chambers are filled with a gas selected to enhance the performance for a particular type of operation. The chambers are operated with voltages between the electrodes to effect the collection of charged particles as pulses or current (continuous pulses). The relative pulse-amplitude variation with voltage is shown in Fig. 5-45. Operation for ex-core ion-

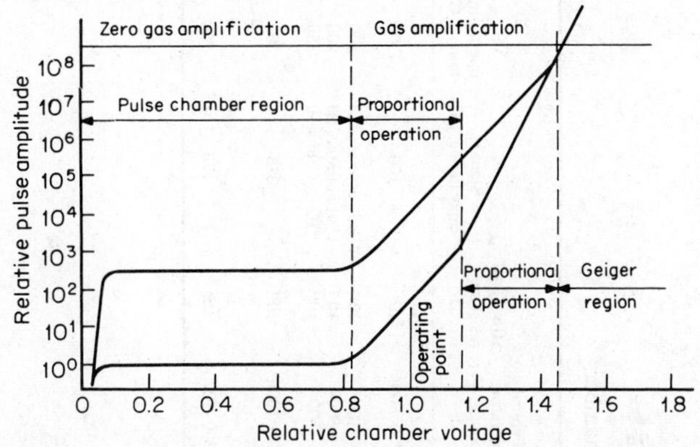


**FIG. 5-45** Nuclear detector operating regions versus voltage. *(From Ref. 10.)*

chamber detectors is normally in the flat or plateau region. The "knee" at the left of the curve moves to the right with increasing ambient radiation, so the selected voltage operating point must be sufficiently high so as to remain on the plateau for all conditions.

The chambers can be made sensitive to neutrons by coating the electrodes with a film of material containing an element with which neutrons interact, such as $^{235}$U or $^{10}$B. Enrichment of the isotope can be selected to effect desired performance in the range of neutron flux. In addition to pulses from incident neutrons, there are pulses from other incident ionizing radiation such as gamma rays. The contribution from gamma rays can be countered or compensated for by supplying two identical volumes, one with neutron-sensitive coating and one without, and subtracting their output. Since the neutron fluence rate and the gamma level do not increase together, that is, maintain the same ratio, this compensation can be completely canceled only at a given power level. Over the range of operation, about 97 to 98% compensation can be effected. The compensated ion chambers (CIC) can be used in a fluence rate about two decades below that of the uncompensated ion chambers (UIC).

Ion chambers are applied for both in-core and ex-core monitoring. Typical characteristics of ion chambers for reactor neutron monitoring are given in Table 5-13.

Low neutron fluence rates produce currents too low to be accurately measured with an ion chamber. Proportional counters operating in a pulse-counting mode are conventionally used in

## TABLE 5-13 Typical Characteristics for Ion Chambers

| Thermal-neutron sensitivity | Max operating thermal-neutron flux $nv$[a] | Typical operating voltage, V dc | Min signal resistance, Ω | Signal capacitance, pF | Max operating temp., °C | Gas fill pressure, cm Hg | Gamma sensitivity, A/(R/h) | Insulator type[b] detector | Nominal dimensions, mm: Length: Sensitive | Nominal dimensions, mm: Length: Overall | Nominal dimensions, mm: Detector OD |
|---|---|---|---|---|---|---|---|---|---|---|---|
| Fission counters:[c] | | | | | | | | | | | |
| 0.7 | $1.4 \times 10^5$ | 200–800 | $10^9$ | 160 | 300 | | | $Al_2O_3$ | 178 | 346 | 76 |
| Proportional counters:[c] | | | | | | | | | | | |
| 35 | $1.4 \times 10^4$ | 2100 | $10^{12}$ | $BF_3$ counters 10 | 150 | 70 | | $Al_2O_3$ | 305 | 391 | 50 |
| 10 | $5.0 \times 10^4$ | 800 | $10^{12}$ | $^{10}B$ counters 20 | 200 | 20 | | $Al_2O_3$ | 660 | 771 | 25 |
| 22 | $2.5 \times 10^4$ | 4500 | $10^{11}$ | $^{3}He$ counters 10 | 150 | 760 | | $Al_2O_3$ | 152 | 263 | 25 |
| Compensated ionization chambers:[d] | | | | | | | | | | | |
| $4.4 \times 10^{14}$ | $2.5 \times 10^{-10}$ | 300–1000 | $10^{12}$ | 315 | 300 | | $2.5 \times 10^{-11}$ (uncompen.) | $Al_2O_3$ | 355 | 603 | 79 |
| Uncompensated ionization chambers: | | | | | | | | | | | |
| $2.8 \times 10^{-14}$ | $5.0 \times 10^{10}$ | 300–1000 | $10^9$ | 170 | 260 | | $4.0 \times 10^{-11}$ | $Al_2O_3$ | 178 | 346 | 76 |

[a] $nv$ is expressed in neutrons/(cm$^2$)(s).
[b] $Al_2O_3$ is a high-alumina-content ceramic.
[c] (Counts/s)/$nv$.
[d] A/$nv$.

this application. These detectors are filled with a gas, such as $BF_3$, which interacts with incident neutrons to produce ionized particles. Gas amplification is used to increase the output for a given event. Voltage applied is typically in the center of the proportional range. A well-regulated voltage supply is necessary.

Ion chambers using a sensitive coating involving $^{235}U$ absorb neutrons and undergo fission which generates the ions. Because of the substantial energy imparted to the ions by the fission process, these detectors are used satisfactorily for both current and pulse generation. Operation over 10 decades of neutron fluence rate may be satisfactorily achieved.

Another type of detector which has been developed for reactor monitoring is the self-powered type (Fig. 5-46). These detectors operate by means of the characteristic of some materials such as aluminum, vanadium, manganese, rhodium, and silver to emit beta particles in the radioactive decay after being activated by neutrons. The beta particles drift from the central wire emitter to the collector on the outer shell and are passed through a readout device, where the resultant current is displayed. The self-powered detectors are usually of low sensitivity and are therefore used principally for in-core monitoring.

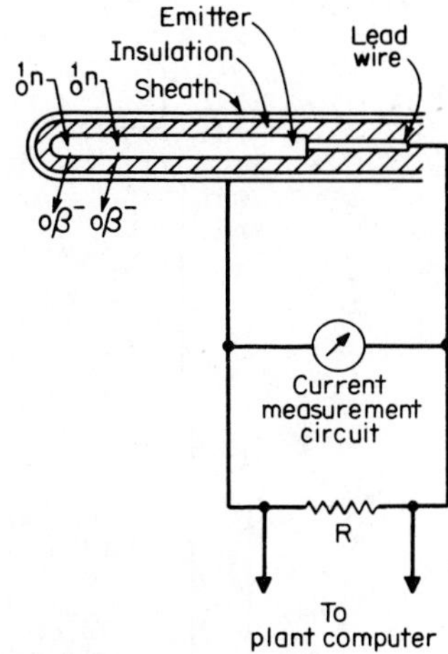


**FIG. 5-46** Self-powered detector schematic.

***43. Neutron Detector Placement.*** The fission process levels extend over 10 or more decades. The process is also subject to local variations, so that spatial monitoring must be achieved. In-core detectors may be located axially and radially in sufficient numbers to assure adequate monitoring. Ex-core detectors are usually located in tubes in the biological shielding adjacent to the reactor and are spaced around the circumference; several detectors may be also stacked axially in each location. Figure 5-47 shows typical ex-core detector locations. This arrangement provides spatial monitoring for the ex-core instruments. Current BWRs are using only in-core detectors; PWRs are using a combination of in-core and ex-core detectors.

A neutron start-up source is usually installed in the core, because: (1) The natural activity of the core produces a low fluence rate; (2) a count rate below 1 count/s is not desirable for process monitoring; and (3) it is preferable to "see" the fission process from the beginning of ascent to operating power. The detectors intended for monitoring the lower levels of start-up should be located so that they are reasonably close to the source but receive incident neutrons from reactions in the core.

The sources produce neutrons by a $\gamma$-$\eta$ reaction. For example, an antimony-beryllium source containing about 30 g of antimony, activated by neutron irradiation to saturation, will produce about $10^7$ neutrons/s. The neutron flux at the instrument can be approximated by

$$\text{Instrument } nv = \frac{SAvM\bar{l}}{V} \qquad \text{neutrons/(m}^2\text{)(s)}$$

where $S$ = source strength, neutrons/s
$V$ = core volume, $m^3$
$A$ = attenuation factor (about $10^{-4}$ for a typical power reactor)
$v$ = neutron velocity, m/s (about $2 \times 10^3$ for thermal reactor)
$M$ = multiplication factor
$\bar{l}$ = mean effective neutron lifetime, s

In selecting and arranging detectors, the temperature, humidity, and gamma-radiation ambients must be considered and the detector burnup provided for.

A typical neutron channel consists of a detector, associated cabling and connectors, power supplies if required, and readout. Coaxial cable is used for ion chambers and proportional counters; the cabling and connectors must be adequate for the environment and must maintain

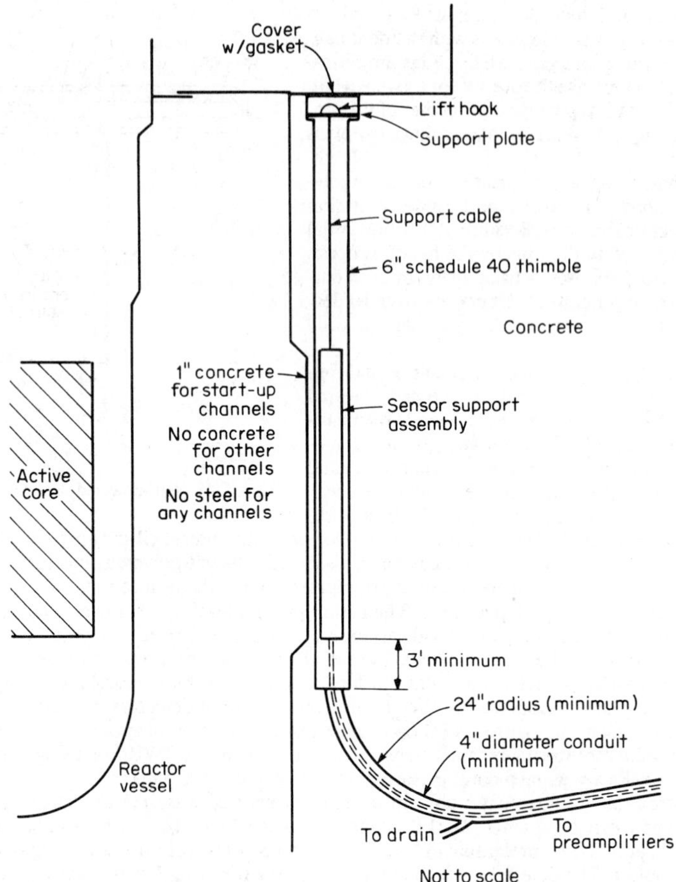


**FIG. 5-47a** Recommended thimble details.

high insulation resistance ($\approx 10^6\ \Omega$). In some cases, especially for pulse counters in low-count-rate channels, preamplifiers are needed to drive the transmission cable and supply sufficient signal for the readout equipment.

In pulse-counting channels, a discriminator is frequently used; this unit enables rejection of pulses from ionization events below a selected energy such as gamma background.

The readout from the neutron monitors is in terms of (1) rate of change of neutron fluence rate; (2) neutron fluence rate itself, presented on a logarithmic basis, from start-up to a value greater than the overpower trips; and (3) neutron fluence rate, presented on a linear basis, in the power range (i.e., from about 1 to 125% of design power).

Display of these readouts may be on individual analog meters, on individual digital meters, on CRT displays, by digital-computer printout, or by a combination of these methods. Readouts of those detectors involved in spatial monitoring are usually summed or combined in a manner appropriate to presentation of operating conditions in the section of the core monitored. Those channels used for reactor protection must have the necessary separation and independence described in protection systems.

***44. Process Instrumentation.*** Apart from the fission process which is monitored by the nuclear instrumentation, the nuclear steam-supply system is basically a heat-transfer process and requires the instrumentation associated with such a process. The primary fluid processes

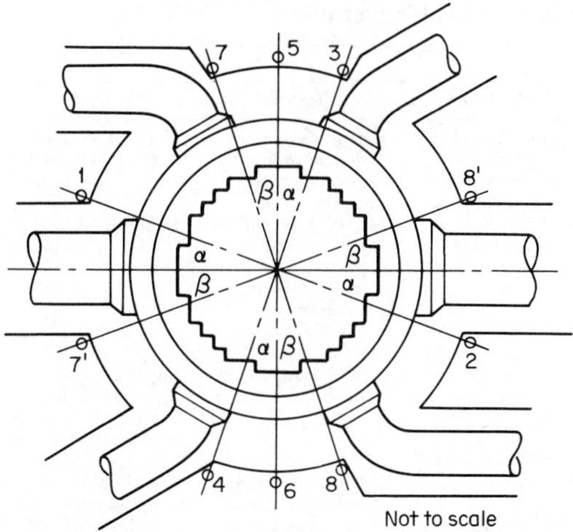


Safety channels – wells 1,2,3,4
Start-up channels – wells 5,6
Control channels – wells 7, 8 or 7', 8'

Angular requirements:

$15° \leqslant \alpha \leqslant 22°$ with 1, 2, 3, 4 each 90° apart

$15° \leqslant \beta \leqslant 22°$ with 7, 8 (or 7', 8') ~ 180° apart

Wells 5, 6 exactly on core flats 180° apart

Total of 8 wells required

**FIG. 5-47b** Detection-location criteria.

utilize channels for sensing temperature, pressure, flow of coolant, liquid level, steam properties, water properties, and gas properties. Characteristics for some of the basic variables are listed in Table 5-14.

Application of the process instrumentation in a nuclear system may impose, depending on the performance requirements and the ambient environment, unique or increased requirements. Selection, installation, calibration, and maintenance of the instrumentation must consider these variables:

Performance requirements, including speed of response, sensitivity, range, stability, reliability, and maintainability.

Ability to withstand ambient environmental conditions such as radiation, heat, humidity, vibration, and pressure for the extent of time the information is required.

Separation of the installed equipment to provide for accessibility, independence of redundant channels, and protection from missiles and fires and other dangers.

Readout from the process instrumentation may be displayed on analog meters, digital meters, or CRT displays, logged by the plant computer, or handled with a combination of these methods.

***45. Instrument Requirements.*** Because of the need for protective-active availability, selected instruments used for protective-system inputs have special requirements imposed upon them.

*Reliability.* Selection and procurement of the equipment must include consideration of minimum maintenance and low-failure-rate characteristics. The steps taken to achieve quality levels include good design practices, quality control, qualification testing, calibration, and system testing.

*Independence.* The equipment is to be installed so that independence of redundant chan-

**TABLE 5-14** Basic Variables Monitored

| Characteristic | Basic variable | | | |
|---|---|---|---|---|
| | Temperature | Pressure | Flow | Liquid level |
| Response time | Appreciable; system often has substantial thermal capacitance | Can be slow or rapid depending on process rate | Fast; little capacitance exhibited | Appreciable; system often has substantial fluid capacitance |
| Process rate | Generally slow because of system capacitance | May be appreciable for pumping or heating or be rapid as in void collapse | Rapid because of low capacitance | Generally slow because of capacitance. Rapid changes may occur in some cases, e.g., void collapse |
| Dead time | Often considerable because of separation of generation and measurement points | Small; changes have small transfer lags | Small for liquids; somewhat longer for gases because of compressibility | Small transfer lags so time depends on system capacitance |
| Inherent noise | Low because of system capacitance | May be subject to pulsations from pumps and compressors or to valve transients | Transients or pulsations from valves and pumps may occur | Low because of capacitance; voids in fluid may introduce noise |
| Other problem areas | Spatial variations may occur in large areas | May be subjected to shock, e.g., water hammer | Flow profiles and nonuniformities may require accommodation | Readings may be affected by fluid-density changes |

nels is preserved. This requires that: (1) components and circuits be electrically separated to prevent the propagation of electrical faults; (2) components and circuits be physically protected from destructive factors such as missiles and water or steam jets; and (3) steps be taken to avoid loss of protective action in the event of common-mode events such as fire or high temperature.

*Signal Validation.* The equipment design and arrangement is to be such that there are means of verifying that the signal represents the actual condition of the variable monitored. Such verification may include:

1. Calibration
2. Cross checking between channels
3. Introducing and measuring known perturbation in the variable

*Maintenance.* In order to assure high system availability, redundant parts of the systems must be able to be repaired and adjusted. Special consideration must be given to access, bypassing, removal of modules, and calibration.

*Information Readout.* The system readouts are to be designed to provide operators with accurate, complete, and timely information. Consideration must be given to sequence and trend indication and to indication of related conditions.

***46. Wiring.*** Some of the power, control, and signal wiring in a nuclear plant is also subject to special requirements because of the safety circuits, the need for pressure tightness of the containment, and the ambient radiation.

As the protection and safeguards systems are made up of redundant channels which are physically and electrically separate, so also the control and signal cabling for the channels are separated. The separation must be such that the independence of a sufficient number of circuits to effect protective action is maintained. The degree of separation in a given area varies with the potential hazards in that area. Clear marking or color coding of the wires and cables is necessary to assure proper location of all the wiring.

Because of the necessity for continuity and availability of service, prime consideration must be given to safeguarding the wiring from damage from events such as fire. Physical separation and shielding and use of insulation which will not sustain combustion are techniques which may be used.

Circuits connecting safety-class equipment with non-safety-class equipment are associated circuits and are treated in the same manner as the principal circuit unless and until they pass through an isolating device. The connection of safety-class equipment to non-safety-class equipment should, as a rule, be avoided. High-energy equipment or other potential sources of missiles should not be located near unprotected cable runs or in cable spreading areas.

***47. Control-Room Arrangements.*** The plant control room is the location of the principal operator-plant interface. The control room is continuously staffed during operation, and decisions essential to safe operation are being made on the basis of the information presented. The assignment of functions to an operator and the degree to which automatic control is utilized are important decisions which must be made. The type, layout, and accessibility of the necessary information and action devices must be considered.

The control room, the control switchboards, and the readouts and controls provided must be designed as a coordinated system which minimizes the potential for erroneous, inadequate, or untimely action.

Controls and displays should be grouped functionally and located for the operator in order of importance. Audible and visual alarms should be used to provide indication of approaches to design limits. Use of the audible alarm must be judicious to avoid distracting the operator unnecessarily. Such alarm indications must also be provided with manual reset so that momentary indications are not missed.

Control-room practice has traditionally been to provide displays of the principal variables monitored, such displays being primarily analog meters and recorders, status lights, and alarm indicators. A control board of this type is shown in Fig. 5-48. With increasing complexity of the

**FIG. 5-48** Example of conventional control board for NSSS. *(Northeast Utilities.)*

large power systems and the redundancy and independence requirement of the nuclear systems, control-board arrangement using traditional methods consistent with maintaining adequate operator perspective has become increasingly difficult.

The increased use of digital-computer systems in the control room has afforded the means to make control-board arrangement more compact and more efficient. Data-acquisition systems can monitor large numbers of conditions or status and present, on a preselected schedule or on demand, the desired information. Presentation can be on analog meters, digital indicators, or CRTs, or in the form of recorded output.

This concept permits information to be presented on a preselected basis and in a manner and location most suited to providing the operator with timely, relevant, and unambiguous data. Colors can be used on CRTs to illustrate different conditions effectively. Complex relations between variables can be computed before presentation to aid the operator in analysis of conditions. System diagrams, reference data, and limit conditions can be stored in the computer for presentation on demand. A control-room arrangement utilizing this type of presentation is shown in Figs. 5-49 and 5-50.

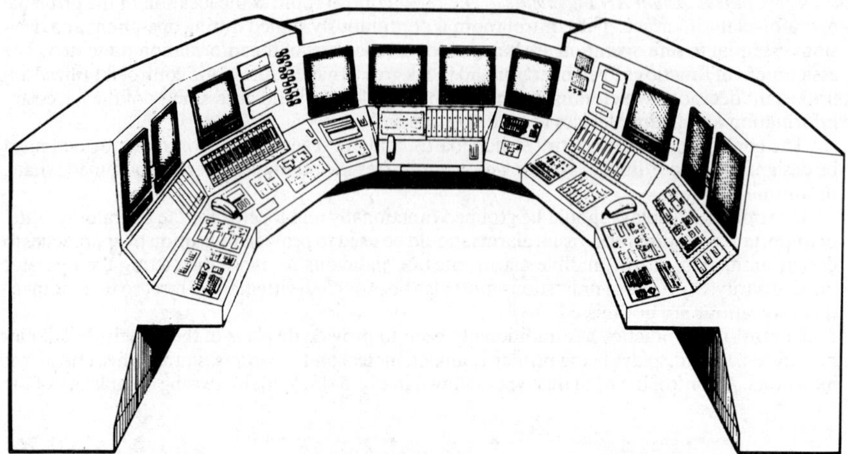

**FIG. 5-49** Example of master control board of NSSS using computer-supplied CRTs.

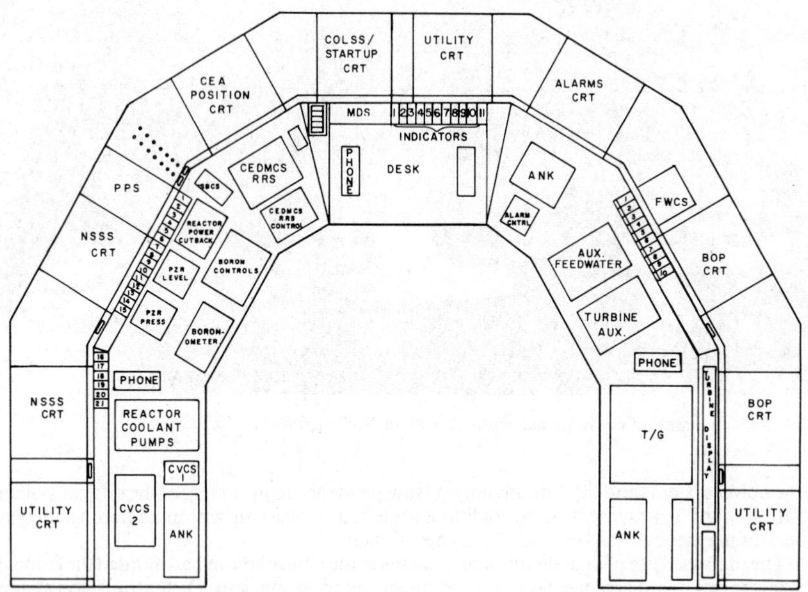


**FIG. 5-50** Schematic of Fig. 5-49.

***48. Emergency Power Systems.*** Because of the need for protective action to be available at all times when the reactor is operating and the need for continued cooling and monitoring when the reactor is shut down, systems must be provided to assure high availability of electric power.

*Primary Coolant Circulators.* The largest single plant load is the drives for primary coolant circulation. Since it is important to maintain coolant circulation and since these drives are generally too large to be supplied by engine-driven sources, provisions should be made to be able to supply the coolant circulator drives from two or more sources. Frequently arrangements are made for the main generator to supply two or more power lines. Provisions in the switchyard enable the plant distribution system to be supplied from the plant generator or from one or more of the outside lines. A typical plant electrical diagram is shown in Fig. 5-51.

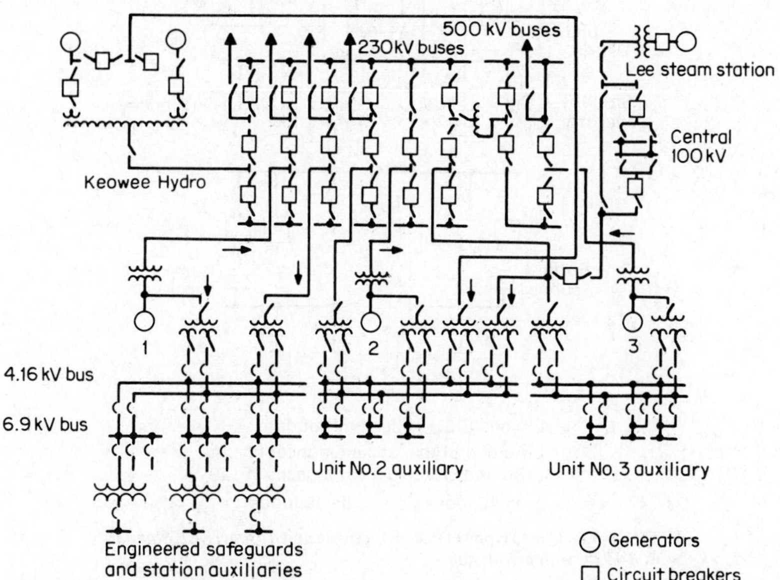


**FIG. 5-51** Example of a transmission and distribution arrangement for a nuclear plant. *(Reprinted with permission from the 1974 Generation Planbook, by Power magazine.)*

In spite of possible connection of plant loads to multiple external power sources, it is possible to lose all external lines, for instance, by a tornado. In this event a local source of power to supply critical ac loads is required. For these purposes, engine (diesel)-driven generators are usually used. Credit can sometimes be taken for local hydro generators or gas-turbine generators if these sources can meet the requirements. These power systems must be designed so that they provide power to the station following a design-basis event. An ac power system (generation and distribution), a dc power system, and a vital instrumentation and control power system are provided. An example of a safety-grade power system is shown in Fig. 5-52.

In the ac system, each of the redundant load groups must have access to both a preferred and a standby power supply. The units of the standby supply must have sufficient independence from the preferred supply and from one another to preclude a common failure mode. Load assignment must be such that the safety actions of each group are redundant and independent. Protective devices must be provided to limit the degradation of the system and maintain the power quality (voltage and frequency) within acceptable limits. Following a demand for the standby power supply, it must be available within a time consistent with the requirements of the engineered safeguards features and the shutdown systems.

In the dc system, batteries, distribution equipment, and load groups are arranged to supply critical dc loads and switching and control power. Redundant load groups, and corresponding battery sections, must be sufficiently independent to preclude common failure modes. Each of the redundant load groups must have access to one or more battery chargers; the batteries are to be kept charged. The battery supplies must be sized to be able to start and operate their assigned loads in the expected loading sequence for a length of time commensurate with the protection provided.

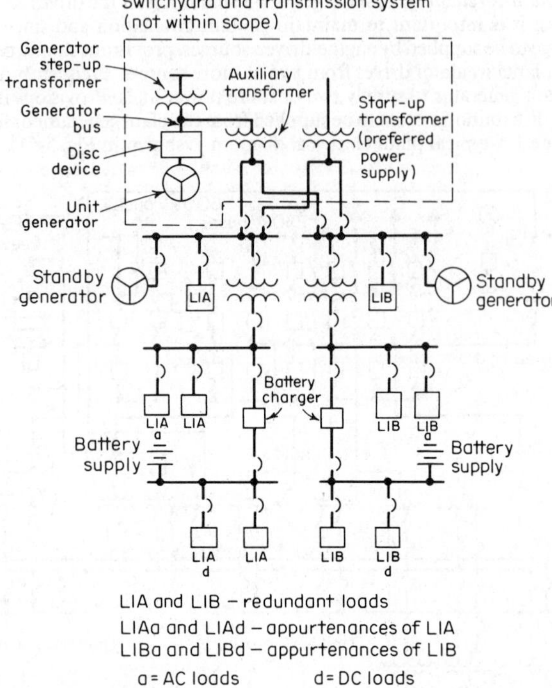


**FIG. 5-52** Class 1E power system for single unit. *(From IEEE Standard 308-1974, with permission.)*

Battery chargers supplying the redundant load groups must have sufficient capacity to restore the battery from its design minimum charge to its fully charged state while supplying normal and postaccident loads. Each charger supply must have a disconnecting device in its ac feeder and one in its dc output line.

The dc system must be equipped with surveillance equipment to monitor its status and to indicate actions.

The vital instrument system is provided to power the instrumentation needed for reactor protection and engineered safety features. Since there may be considerable variation in the instrumentation in various plants, the vital system may be required to supply ac or dc or both. To preserve freedom from common-mode failure, the vital supply must be divided into redundant and independent systems with adequate status indication. Provisions for testing, adjustment, and repair should be included in the parts of the emergency power systems to improve reliability and availability.

***49. Trends in Nuclear-Fueled Plant Development.*** The development of nuclear-fueled steam-electric plants underwent substantial change in the decade of the seventies. At the beginning of the decade, orders for nuclear-fueled plants were increasing to a peak of 38 per year. Following the oil crisis of 1973–1974, changes in the economy began to affect the cost of, and consequently the demand for, electric power. Opposition to the use of nuclear energy for electric power production increased; litigation was frequently employed. Near the end of the decade, sociopolitical aspects of nuclear-fueled plants became as involved and time consuming as the technical aspects. In order to participate effectively in the design, construction, and operation of nuclear-fueled plants, one must be familiar with the energy perspective; the concerns about the use of nuclear energy; and the functions of advocates, intervenors, and regulators. See Refs. 14, 15 and 16.

With the maturity of the nuclear-fueled plants, more emphasis was placed on project management (Ref. 17). Siting of the plants became a major task (Ref. 18). Because of the reduced demand for electric power, the increased cost of money, and the difficulty of resolving the objections raised, orders for nuclear-fueled plants began to decrease sharply after the middle of the decade. Some orders were canceled. Then in March, 1979, an incident occurred at the Three Mile Island plant, causing serious damage to the plant. This incident raised questions about the operation of nuclear-fueled plants and a review of the value of nuclear energy (Ref. 19). At the end of the decade, orders for new nuclear-fueled plants had been reduced to zero and a sizable number of plant orders had been canceled.

The beginning of the decade of the eighties saw reinforcement of the need for commercial use of nuclear energy (Ref. 20), but also heralded changes in the safety, control, and maintenance systems. In the electrical area, the most notable changes were the redesign of control rooms and stations and the increased use of computers in more sophisticated safety systems (Ref. 21). The study of incidents and malfunctions by means of computers has provided another means to inform and guide operators and to evaluate possible trouble spots (Ref. 22). The availability and capability of the microprocessor has provided new ways to improve the safety and performance of plant instrumentation, control, and safety systems.

***50. Bibliography***


1. Kaplan, I.: *Nuclear Physics;* Cambridge, Addison-Wesley, 1956.
2. Glasstone, S.: *Principles of Nuclear Reactor Engineering;* New York, D. Van Nostrand, 1955.
3. Loftnen, R. L.: *Nuclear Power Plants;* New York, D. Van Nostrand, 1964.
4. Sesonske, A.: *Nuclear Power Plant Design Analysis;* NTLS (TID-26241), Springfield, Va., 1973.
5. Glasstone, S., and Edlund, M. C.: *The Elements of Nuclear Reactor Theory;* New York, D. Van Nostrand, 1952.
6. Gloyna, E. F., and Ledbetter, J. O.: *Principles of Radiological Health;* New York, Marcel Dekker, 1969.
7. *Radiological Health Handbook* (Bureau of Radiological Health), rev. January 1970.
8. (*a*) Harrar, J. M., and Beckerly, J. G.: *Nuclear Power Reactor Instrumentation Handbook;* vol. I, AEC Office of Information Services. TID-25952-P1. (*b*) Harrer, J. M., and Beckerley, J. G.: *Nuclear Power Reactor Instrumentation Systems Handbook;* vol. 2, AEC Office of Information Services, TID-25952-P2, 1974.
9. El-Wakil, M. M.: *Nuclear Power Engineering;* New York, McGraw-Hill, 1962.
10. Etherington, H.: *Nuclear Engineering Handbook;* New York McGraw-Hill, 1958.
11. Peterson, S., et al.: *Fundamental Chemistry for Nuclear Reactor Engineers;* USAEC, QR TID-5620, 1955.
12. Meghreblian, R. V., and Holmes, David K.: *Reactor Analysis;* New York, McGraw-Hill, 1960.
13. Lederer, C. M., et al.: *Table of Isotopes;* New York, John Wiley & Sons, 1967.
14. Schmidt, F. H., and Bodansky, D.: *The Fight Over Nuclear Power;* San Francisco, Albion Publishing Co., 1976.
15. Warnock, D., and Bossong, K.: *Nuclear Power and Civil Liberties; Can We Have Both;* Washington, D.C., Citizens Energy Project, 1979.
16. Cook, C. E.: *Nuclear Power and Legal Advocacy;* Lexington, Mass., Lexington Books, 1980.
17. Pederson, E. S.: *Nuclear Power Project Management;* Ann Arbor, Ann Arbor Science, 1978.
18. Winter, J. V. and Connor, D. A.: *Power Plant Siting;* New York, Van Nostrand, 1978.
19. Rubenstein, E.: Three Mile Island and the Future of Nuclear Power; *Spectrum,* Nov. 1979, vol. 16, no. 11, pp. 30–111.
20. Greenhalgh, G.: *The Necessity for Nuclear Power,* London, Graham and Trotman, Ltd., 1980.
21. Hanes, L. F., O'Brien, J. F., and DiSalvo, R.: Control Room Designs; Lessons From TMI; *Spectrum;* June 1982, vol. 19, no. 6, pp. 46–52.
22. Kaplan, G.: Nuclear Power Plant Malfunction Analysis, *Spectrum,* June 1983, vol. 20, no. 6, pp. 53–58.


## *NUCLEAR POWER FOR THE FUTURE*

*By A. L. GAINES*

***51. Breeder Reactors.*** The breeder reactor is able to create more fissionable material than it consumes. A fast breeder is designed to have more nuclear reactions in the high-energy region than in the thermal-energy region used by the light-water reactors. This design is accomplished by not moderating, or slowing down, the energy of the neutrons created by one fission reaction prior to subsequent captures of, or fissions by, the neutrons created. In the thermal nuclear reactors previously described, the water coolant absorbs energy from the neutrons, creating a condition (called the thermal-energy region) in which more neutrons fission the $^{235}U_{92}$ isotopes than are captured by the $^{238}U_{92}$ isotopes. The capture of a neutron by the $^{238}U_{92}$ isotopes results in a decay chain that produces an atom of plutonium (see reaction chain following). The plutonium can be used to fuel nuclear reactors. Liquid-metal fast breeder reactors (LMFBRs) operate in the "fast," or high-energy region, where more of the neutrons are captured by fertile material than cause fission in fissionable material, and where the average number of neutrons created in fission reactions by fast neutrons is greater than the average number created by thermal neutrons. These additional neutrons permit the breeder to produce more fuel by the following reaction chain than is consumed by the fission reactions.

$$^{1}n_0 + {}^{238}U_{92} \rightarrow {}^{239}U_{92} \xrightarrow[23\text{ min}]{\beta^-} {}^{239}Np_{93} \xrightarrow[2.3\text{ days}]{\beta^-} {}^{239}Pu_{94}$$

This same reaction occurs in a light-water reactor, but, without the greater number of neutrons and higher ratio of capture cross section to fission cross section, less fuel is created than is consumed. Much of this bred fissionable material is fissioned in situ. For the excess fissionable material in a breeder to be of use in other reactors, or for refueling itself, the spent fuel and blanket assemblies require reprocessing to separate the fissionable material from the fission products created by the nuclear fissions. These fission products, and the damage to the material of the fuel tubes, also limit the life, or time, breeder fuel can remain in the core to produce power. The fission products compete for capture of the neutrons and eventually "poison" the fission reactions, shutting down the chain reaction.

Thermal breeder reactors are possible using a $^{233}U_{92}$ fuel. This isotope results from the decay chain started from the fertile material thorium, $^{232}Th_{90}$.

$$^{1}n_0 + {}^{232}Th_{90} \rightarrow {}^{233}Th_{90} \xrightarrow[23\text{ min}]{\beta^-} {}^{233}Pa_{91} \xrightarrow[27\text{ days}]{\beta^-} {}^{233}U_{92}$$

Experimental thermal breeder reactors have been operated using a seed-and-blanket core with water coolant or a molten-salt homogeneous core. The helium-cooled graphite-moderated reactor can theoretically perform as a breeder.

In light-water reactors, power is produced by expanding steam through a turbine generator. In the present boiling water reactors (BWRs), a direct cycle is used, with the condensate from the turbine serving as the feedwater that cools the reactor by producing the steam that drives the turbine. In the pressurized water reactors (PWRs), an indirect cycle is used, with a steam generator exchanging the heat between the pressurized reactor coolant and the condensate-feedwater-steam cycle just described. The present breeder reactor designs use a liquid metal (sodium) to cool the core without moderating the energy level of the neutrons. This liquid metal becomes radioactive (15.5 h half-life) and is isolated from the condensate-feedwater-steam by a secondary, or intermediate, liquid-metal coolant circuit. Thus, an additional heat exchanger and pump are necessary with the attendant impact on plant cost and operating complexity.

Because all fuel resources are finite, breeding of fuels will be necessary to sustain the present per capita use of power with an increasing population and diminishing reserve of fuel. Crude oil and natural gas fuels will be depleted for practical purposes in 100 years. Coal resources will be depleted in about 450 years. Without breeding, uranium will be depleted in less than 100 years, as only 0.7% of natural uranium is the fissionable isotope $^{235}U_{92}$. Breeders will permit most of the uranium and thorium resources to be used as fuel. Breeders can supply our energy needs for thousands of years. At a growth rate of 3% per year in power utilization, breeder technology needs to double the fuel every 23 years. Present experimental LMFBRs have demonstrated that even shorter doubling times are attainable.

In 1983, the Congress of the United States voted to withhold funds for completing the Clinch River breeder reactor, a power-demonstration fast breeder reactor. France, Russia, Germany, Great Britain, and Japan, however, are continuing their research and development support of breeder reactors.

With the resources of fossil fuels being depleted and the increasing demand for higher living standards by the "undeveloped" countries, there is an immediate need to understand and plan for the inevitable fuel shortages. The recent development of oil reserves, combined with conservation awareness by consumers, has resulted in an oversupply of petroleum for current needs, easing the urgency for assuring long-term fuel resources. Also, power from breeder reactors appears to be 10 to 15% more expensive than power from fossil fuels or the once-through cycle of enriched uranium that fuels the light-water reactors, as long as natural fuels are plentiful. Demand for and utilization of more power by the developing countries will cause the cost of existing fuels to increase, making breeders competitive in the next century.

***52. Fusion Reactors.*** Fusion liberates energy by combining two atoms of light elements into one atom of a heavier element. The resulting mass is less than that of the "fusing" atoms. Fission, as described in Par. **14** results in "splitting" one atom of a heavy element into two atoms of lighter elements. The resulting mass, again, is less than that of the "fissioning" atoms. Each fissioning reaction releases about 200 MeV of energy. Each fusion reaction releases much less energy than a fission reaction, but a given weight of material yields much more energy. The three most attainable fusion reactions are listed below. The optimum operating temperature in keV (1 keV = 11,600,000 K) for a self-sustaining thermonuclear reactor is indicated underneath the arrow. The energy of each resulting particle is shown in parentheses.

$$^2D_1 + {}^3T_1 \xrightarrow[\text{keV}]{@10} {}^4He_2(3.5\text{ MeV}) + {}^1n_0(14.1\text{ MeV})$$

$$^2D_1 + {}^2D_1 \xrightarrow{@\,50\text{ keV}} \begin{cases} \xrightarrow{50\%} {}^3He_2(0.82\text{ MeV}) + {}^1n_0(2.45\text{ MeV}) \\ \xrightarrow{50\%} {}^3T_1(1.01\text{ MeV}) + {}^1p_1(3.02\text{ MeV}) \end{cases}$$

$$^2D_1 + {}^3He_2 \xrightarrow[\text{keV}]{@100} {}^4He_2(3.6\text{ MeV}) + {}^1p_1\ (14.7\text{ MeV})$$

In any application of fusion each of these reactions will be occurring at rates dependent upon the energy distribution in the plasma and the cross sections of the ions at the energy level existing when collision occurs.

Fusion will enable use of an extremely large fuel resource—the deuterium isotope, $^2D_1$, which exists as 1 part to 6500 parts of $^1H_1$ in natural hydrogen. The deuterium available from the ocean could fuel the power requirements of the world for billions of years—even until the sun expands and envelops the earth. Unfortunately, the deuterium-deuterium fusion reaction requires a much more complex device to reach its ignition temperatures than the tritium-deuterium reactor. Therefore, tritium resources will be the limiting fuel for fusion. Tritium is almost nonexistent in nature (half-life of 12.4 years). It is created by nuclear reactions in lithium and a few other elements. The reserves of lithium in the ground compare to those of coal (will last for 450 years). Lithium, like deuterium, can be extracted from seawater. It is estimated that the lithium in seawater represents a fuel source 10,000 times larger than that in the ground.

The nuclear reactions that breed tritium from lithium are

$$^1n_0 + {}^6Li_3 \rightarrow {}^3T_1 + {}^4He_2\ (4.8\text{ MeV})$$

$$^1n_0 + {}^7Li_3 \rightarrow {}^3T_1 + {}^4He_2 + {}^1n_0\ (-2.5\text{ MeV})$$

## *CIRCULATING-WATER SYSTEMS AND COMPONENTS*

*By R. C. MILLER*

***53. Purpose.*** The primary purpose of the circulating-water system is to provide the cooling water to condense the steam exhausted from the turbine for reuse as feedwater by the steam generator. The system is also sometimes used as a source of cooling water for various auxiliary services in the turbine building.

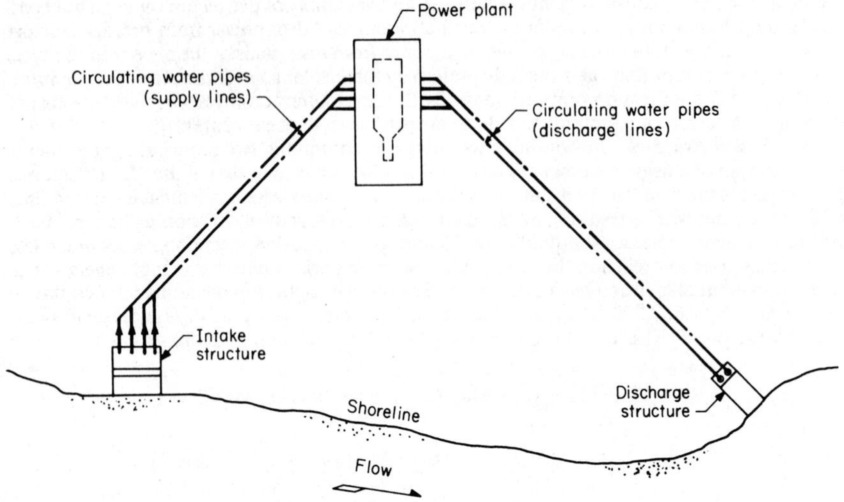


**FIG. 5-53** Once-through circulating-water system.

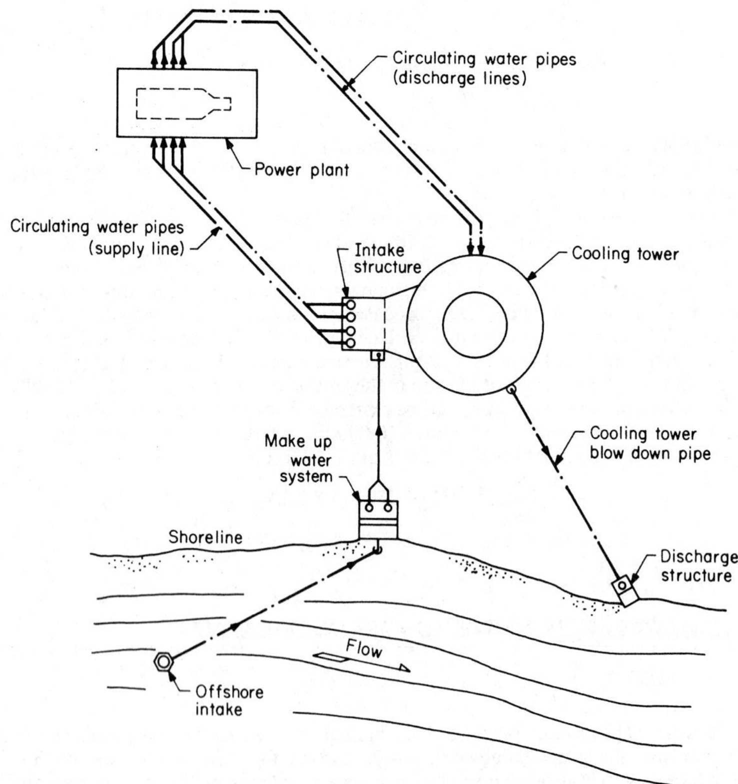


**FIG. 5-54** Closed-loop circulating-water system.

***54. Description.*** The circulating-water system may be either a once-through (open) or closed-loop system. The open system continuously draws the required flow from a source such as a river, lake, or ocean and returns the heated water to the same source. The source then becomes a natural heat sink (Fig. 5-53). The closed-loop system recirculates its cooling water in a continuous loop which contains a manmade heat sink. This sink may be a reservoir, cooling pond, spray pond, wet cooling tower, or dry cooling tower (Fig. 5-54). Hybrids (once-through systems containing a manmade heat sink on the discharge line to reduce the heat load on the system's water source) have been built to meet regulatory requirements which limit the temperature of thermal discharges (Fig. 5-55). However, the effectiveness of such arrangements is

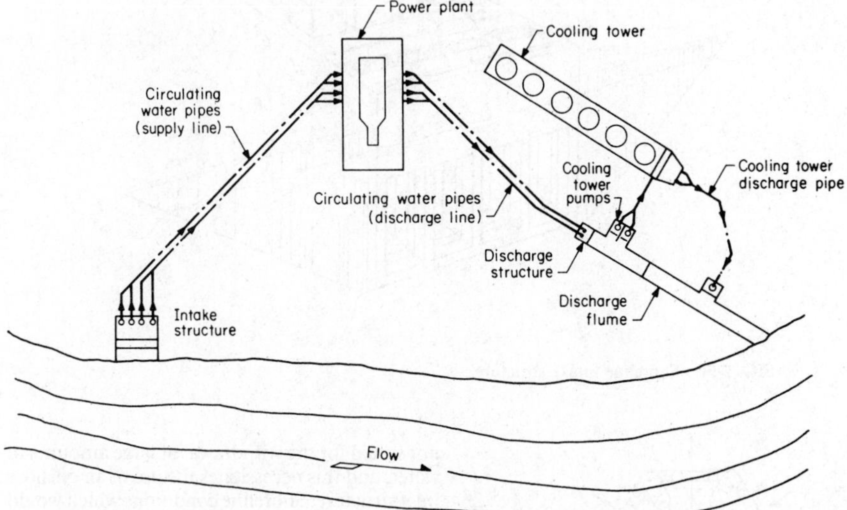


**FIG. 5-55** Once-through circulating-water system with supplementary cooling tower.

limited. During periods when the ambient air temperature is high relative to the temperature of the natural water body into which the system is discharging, the cooling capacity of the "supplementary" heat sink becomes negligible.

Flows of modern circulating-water systems range from approximately 200,000 to 1,000,000 gal/min. As a general rule, fossil and nuclear power stations require approximately 500 and 750 gal/min per megawatt of generating capacity, respectively. This is because of the higher steam flow requirements of the turbine in a nuclear power station.

A system typically consists of an intake structure which contains screening equipment and from two to six circulating-water pumps; several thousand feet of pipe, with diameters of up to 12 ft; a surface condenser; a discharge structure in the form of a flume or submerged diffuser; and associated valves, gates, expansion joints, etc. In addition, a closed-loop system will have a cooling tower or other manmade heat sink, a makeup-water supply system, and a blowdown system. Sources of makeup water normally include rivers and lakes, and, more recently, seawater. Some stations have used groundwater, sewage treatment plant effluent, and irrigation drainage for makeup water.

Cooling in a closed-loop system is accomplished primarily by evaporation (with the exception of dry cooling towers). In addition to the evaporation loss, a certain portion of the circulation water must be disposed of (blowdown) and replaced by fresh water to limit the concentration of dissolved solids in the water. Dissolved solids are generally maintained less than three to seven times that of the concentration in the makeup water. Makeup water requirements for evaporative losses, blowdown, and other minor losses typically amounts to less than 5% of the total circulating-water flow.

***55. Major Components.*** The following are the major components found in open- and closed-loop circulating-water systems.

***56. Intakes.*** The intake of a once-through circulating-water system, or makeup system for a closed-loop circulating-water system, generally consists of a shoreline or bankline structure which houses screening equipment and pumps (Fig. 5-56). Shoreline conditions sometimes are

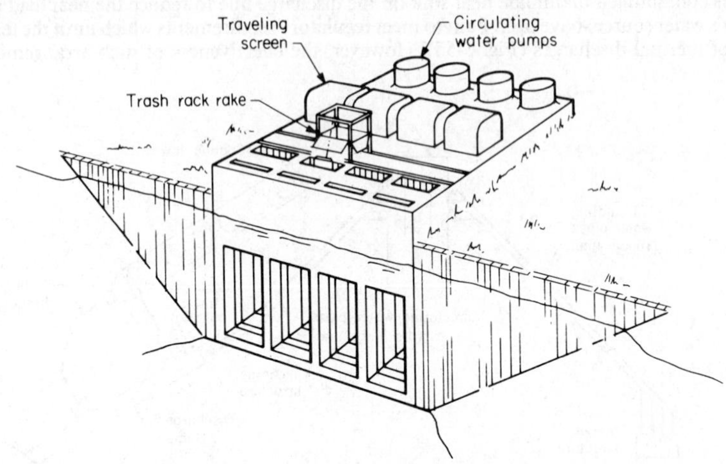


**FIG. 5-56** Shoreline intake structure.

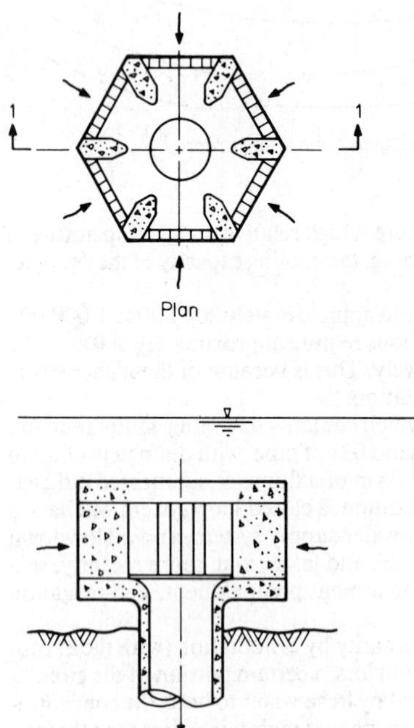


**FIG. 5-57** Offshore submerged inlet structure.

not suited for the withdrawal of large amounts of water, and this necessitates the use of an offshore inlet structure. Shoreline conditions which would preclude a shoreline installation include shallow coastlines, severe coastal icing, and abnormal quantities of shoreline debris or sediment.

When a power plant has an offshore inlet, equipment which requires maintenance, such as mechanical screens and pumps, are normally located in an easily accessible structure on shore. Water is conveyed to this onshore screen/pumpwell via a pipeline or tunnel leading from the offshore inlet. The offshore inlet is normally a submerged concrete structure with vertical openings protected by bars spaced so as to keep out large debris (Fig. 5-57). The bars may be electrically heated to prevent ice formation.

Offshore inlets to makeup-water systems drawing water from a river occasionally consist of a cylindrical or basket-shaped screen. The screen is piped directly to the makeup pumps and cleaning is accomplished by backwashing (Fig. 5-58).

The conventional onshore intake structure typically includes mechanically cleaned bar racks, traveling water screens, pumps, and hoisting equipment. A curtain wall is provided at the entrance to keep out cold air, floating debris, and ice (see Fig. 5-67). It is designed to operate under a range of water-level conditions based on histori-

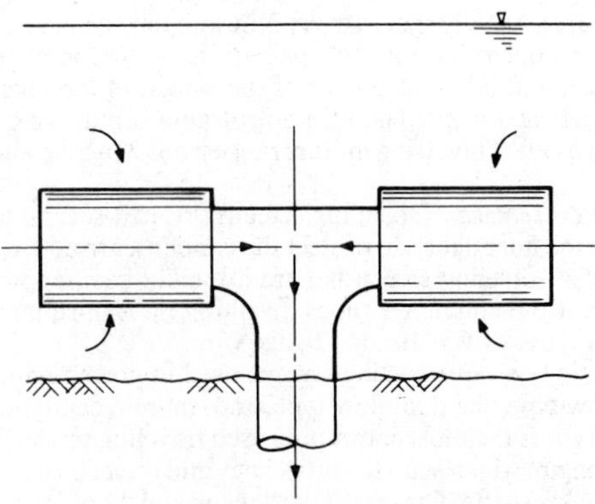
FIG. 5-58 Offshore cylindrical screen inlet.

cal low-water and flood records. Traveling screen and pump requirements normally limit entrance velocities to 1.0 to 1.5 ft/s. Environmental considerations regarding fish entrapment have at some sites limited entrance velocities to less than 0.5 ft/s.

***57. Trash Racks.*** The entrance to the conventional shoreline intake is protected by trash racks. The racks keep large, cumbersome debris from reaching and damaging the traveling

FIG. 5-59 Mechanical trash-rack rake. *(Envirex, Inc.)*

screens. The rack bars are generally spaced every 3 in. and may be cleaned by a mechanical rake. Racks can be vertical or sloped. Sloping the rack requires more room; however, this eliminates the need for rake guides and takes advantage of the weight of the rake. A single rake usually traverses several screens via a track. The rake hoist frame contains a cart into which the rake dumps the debris (Fig. 5-59). Traversing motors range from 2 to 5 hp and hoist motors from 10 to 15 hp.

***58. Traveling Water Screens.*** Traveling screens, located several feet downstream of the trash racks, screen out the finer debris to protect the condenser from clogging. The criterion for sizing the openings of a traveling screen has traditionally been approximately one-third to one-half the diameter of the condenser tubes, resulting in ⅜-in-square openings. The screen panels are made up of cross-woven 10- or 12-gage wire.

There are three basic types of traveling screens used in power plant intakes in the United States: the through-flow type, the dual-flow type, and, more recently, the center-flow type.

The through-flow type is the most commonly used traveling screen. It consists of a series of panels, up to 14 ft in length, which span the intake bay and travel on a continuous pair of chains around two sets of sprocket shafts. The axis of the "head" and "foot" sprocket shafts is normal to the direction of flow (Fig. 5-60). The direction of travel is upward on the upstream face and can vary from a few feet to 30 ft/min. Debris is flushed from the panels by a high-pressure backwash spray system consisting of a series of nozzles on a horizontal header. The debris is washed into a trough, which extends over the width of the intake structure, and is sluiced to a disposal area.

Screenwash water may be provided by separate screenwash pumps or by a connection to the discharge of the circulating-water, makeup-water, or service-water pumps. A disadvantage of this type of screen is that debris that is not washed off will be carried over into the pump bay.

The dual-flow, or so-called no-well screen, is similar in structure to the through-flow screen

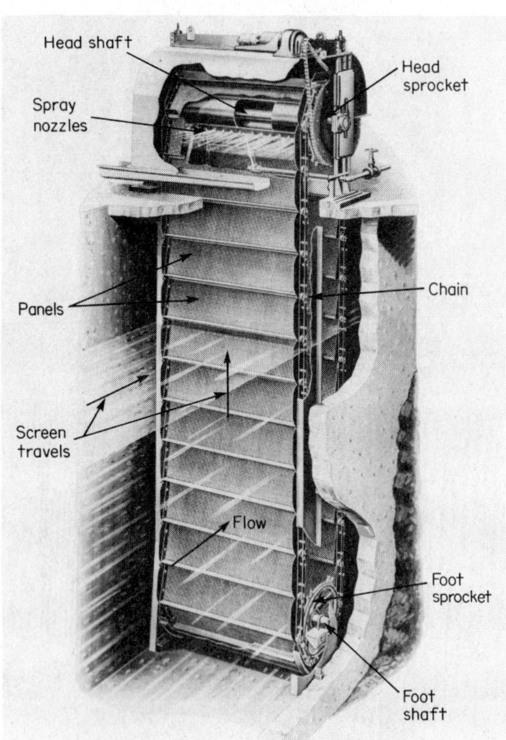


**FIG. 5-60** Through-flow traveling screen. *(FMC Corporation.)*

**FIG. 5-61** Dual-flow traveling screen. *(FMC Corporation.)*

(Fig. 5-61). However, the axis of the sprocket shafts is parallel to the flow, and the screen is normally directly connected to the pump via a transition piece and elbow leading into the pump suction. This arrangement lends itself to a "no well" scheme, in which the pump and screen combination are hung from a platform supported on piles. This can be an economical design; however, it is practical only in warm climates where ice is not a concern. Because the pump takes suction from the center of the screen, debris carryover is not possible. The screenwash system is similar to the through-flow screen.

In recent years, the center-flow screen, which originated in Europe, has gained favor at certain installations in the United States because of its ability to gently remove fish and other organisms. In principle, it is comparable to a dual-flow screen with its flow paths reversed. Flow enters the central area of the screen and exits outward through the screen panels. It is oriented with its axis parallel to the flow path, as with the dual-flow type.

The screen panels are semicircular in cross section, and the debris trough is located in the central area of the screen. As the screen rotates, much of the debris falls from the screen into the debris trough and there is no carryover. A backwash spray system similar to the through-flow and dual-flow screens is provided to remove the remainder of the debris (Fig. 5-62).

Traveling screens are normally designed to run periodically and automatically with the use of timers and to react to differential pressure. As debris collects on the panels and the water-level differential across the screens increases to a preset limit, usually greater than 4 in., the screens are automatically set into motion and a wash cycle is initiated. The screens generally have two or three separate speeds (e.g., 5, 10, and 20 ft/min), and the speed automatically increases if the differential continues to increase after the cleaning cycle is initiated. Speed variation is accomplished by single-speed motors with fluid couplings or geared transmissions, and by variable-speed motors. Power requirements of the motor are below 10 hp.

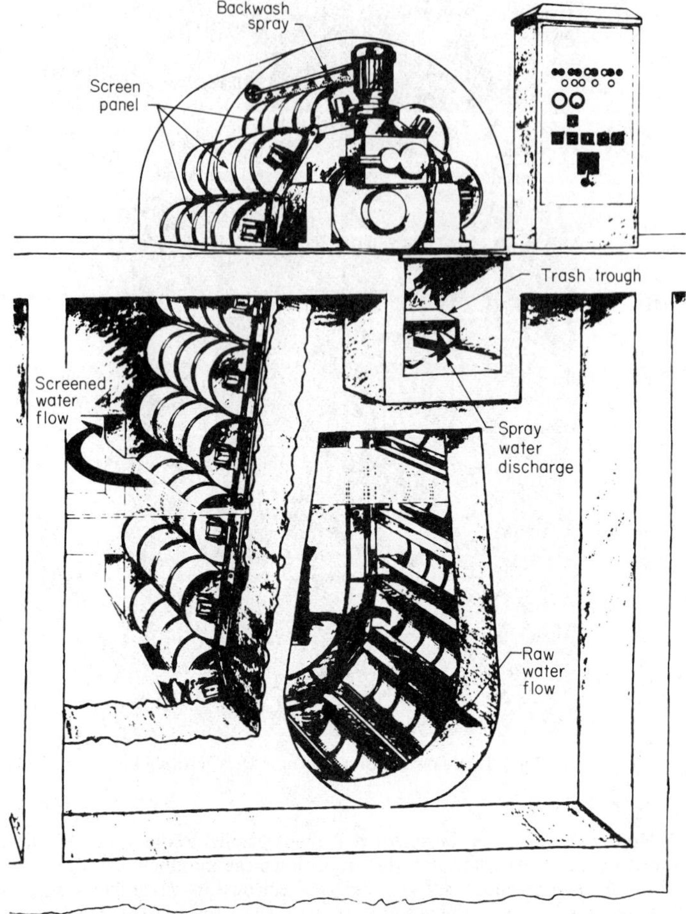


**FIG. 5-62** Center-flow traveling screen. *(Passavant Corporation.)*

Environmental considerations have resulted in the addition of many optional features on traveling screens. Among these are the following: lift buckets on through-flow and dual-flow screens to convey fish from the intake to a sluiceway for safe return of the water body; fine-mesh screens, with openings as small as 0.5 mm, to remove eggs and larvae from the circulating water; low-pressure screen wash sprays to reduce damage to fish washed from screens; and continuous operation of the screens to reduce residence time for fish and other organisms living in the intake area. Continuous operation has resulted in several changes in screen design. These have included tougher drive trains, better chains, improvements in the bearings at the head and foot shafts, and general strengthening of the framing. In addition, manufacturers have reduced their maximum recommended width from 14 to 10 ft if the screen is to run continuously.

***59. Pumps.*** Pumps provided to move the circulating water throughout the system are of the rotating type and may be axial-flow, mixed-flow, or centrifugal-flow in design. Axial-flow (propeller) pumps develop capacity by the lifting action of the impeller blades. They are typically used in low-head, high-flow situations. Pure centrifugal pumps develop capacity from centrifugal forces created by the rotation of the impeller. They are used in high-head, low-flow situations. Mixed-flow pumps develop capacity from a combination of lift and centrifugal

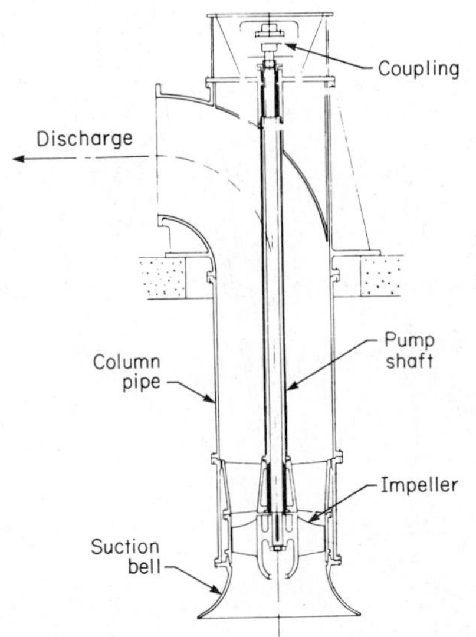


**FIG. 5-63** Axial-flow pump. *(Allis Chalmers Corporation.)*

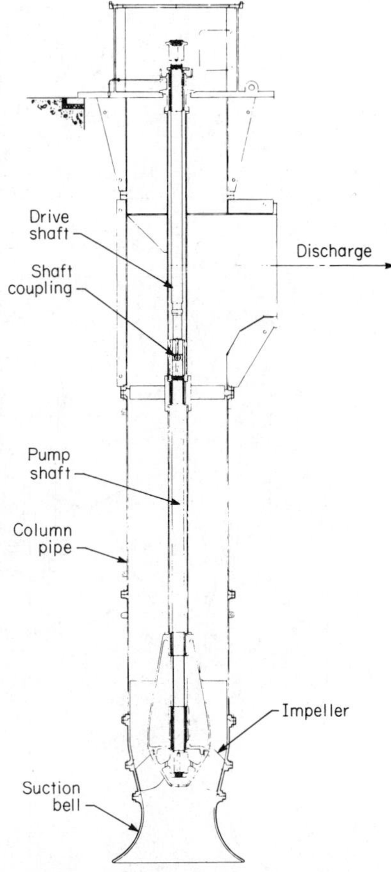


**FIG. 5-64** Mixed-flow vertical column pump. *(Allis Chalmers Corporation.)*

forces. Their application falls between axial- and centrifugal-flow pumps. The head-flow requirements of most circulating-water systems dictate the use of pumps of this type.

Axial-flow circulating-water pumps are generally of the vertical-column, water-lubricated type (Fig. 5-63). Mixed-flow pumps may be of the vertical-column water-lubricated type; the vertical-suction, oil-lubricated, volute type; or the horizontal, axially split case, double-suction, oil-lubricated type (Fig. 5-64 to 5-66). Centrifugal-flow pumps are generally of the horizontal, axially split case, double-suction, oil-lubricated type.

There are two basic types of pumpwell installations: wet pit and dry pit. Wet pit commonly refers to arrangements where the majority of the pump is located in the flooded area of the pumpwell. Dry pit commonly refers to arrangements where the majority of the pump is located in a dry area of the pumpwell. The axial-flow and mixed-flow vertical-column pumps may be installed in either wet- or dry-pit pumpwells (Figs. 5-67, 5-68). Vertical-suction, volute pumps and horizontal, double-suction pumps are installed in dry-pit pumpwells (Figs. 5-69, 5-70).

All of the above arrangements can be used in either a once-through or a closed-loop system. The trend in the industry seems to be toward using water-lubricated, axial-flow and mixed-flow vertical pump, wet- or dry-pit installations in both types of systems; oil-lubricated, vertical-suction, volute-pump dry-pit installations in once-through systems; and oil-lubricated, horizontal, double-suction, dry-pit installations in closed-loop systems.

Considering the types of pumps available and the installations possible, several arrangements are normally feasible for a particular system, and the final decision is usually based on economics, engineering judgment, and operator experience.

Pumping-head requirements vary from approximately 25 to 100 ft, and flows typically range from approximately 65,000 to 200,000 gal/min per pump. Most power plants use between two and four pumps, although six pumps have been used at large nuclear plants. Gener-

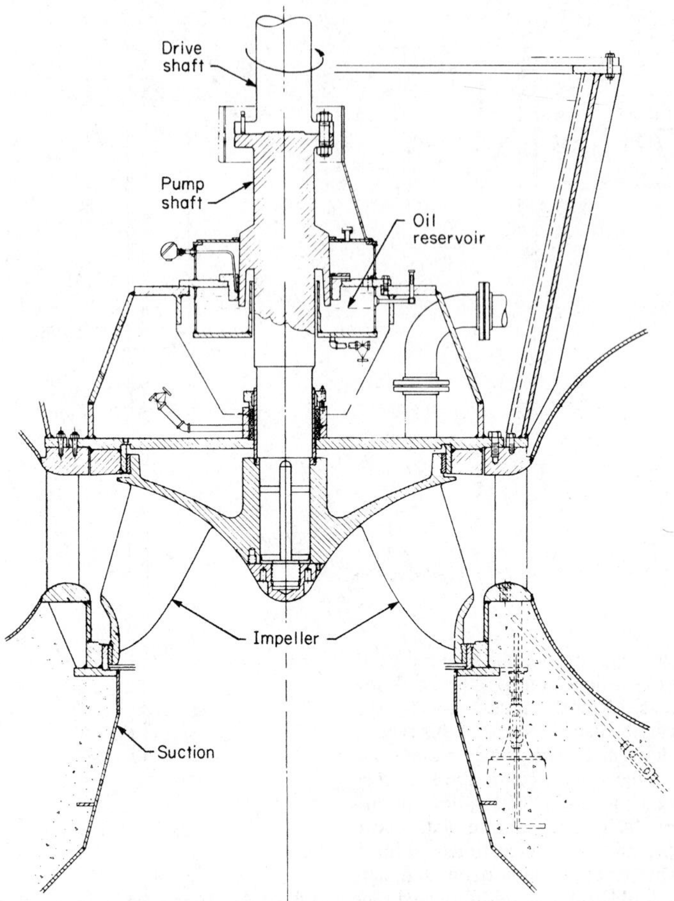


**FIG. 5-65** Mixed-flow vertical suction volute pump. *(Allis Chalmers Corporation.)*

ally, spare capacity is not provided. Horsepower requirements per pump are on the order of 750 to 5000 hp.

Pump motors are normally 3-phase, single-speed induction machines and are connected to the pump via a direct-drive shaft. Motors are typically close-coupled and mounted on the pump. Special circumstances, such as concern of flooding, may require motors to be mounted on a higher-level floor. Motors are generally sized to provide some margin in horsepower (±10%) over that required by the pumps at their normal operating point. During start-up, a pump motor must be capable of accelerating itself and the pump to full speed at reduced voltage.

***60. Condenser.*** The purpose of the circulating-water-system condenser, sometimes referred to as the main condenser in a power plant, is to remove waste heat from the turbine exhaust steam. The condensed steam collects in a "hot well" at the base of the condenser where it is pumped back to the steam generator as feedwater.

The condenser is located below the low-pressure turbine. The tubes may be arranged longitudinally (i.e., in line with the turbine axis) or transversely (i.e., at 90° to the turbine axis), depending on turbine configuration, tube length, and plant layout.

There are two basic types of condensers, the surface type and the direct-contact type. In the direct-contact type, the cooling water is sprayed directly into the exhaust steam. Because of condensate contamination, this design is not practical for cooling systems in which the cooling

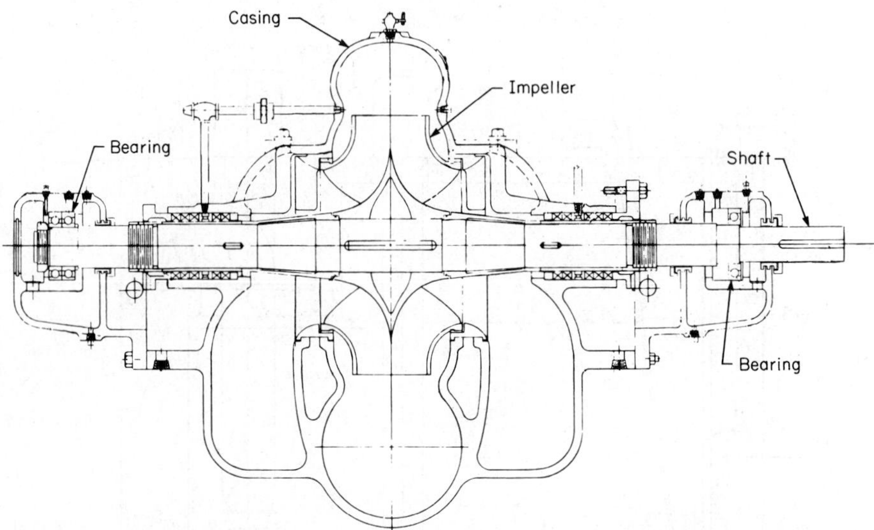


**FIG. 5-66** Horizontal double-suction pump. *(Allis Chalmers Corporation.)*

water comes in contact with the outside environment, such as with an evaporative cooling tower. This limits its use to sealed systems such as "dry" cooling systems in which the cooling water is cooled by sensible heat transfer via a finned-tube air-cooled heat exchanger. Such

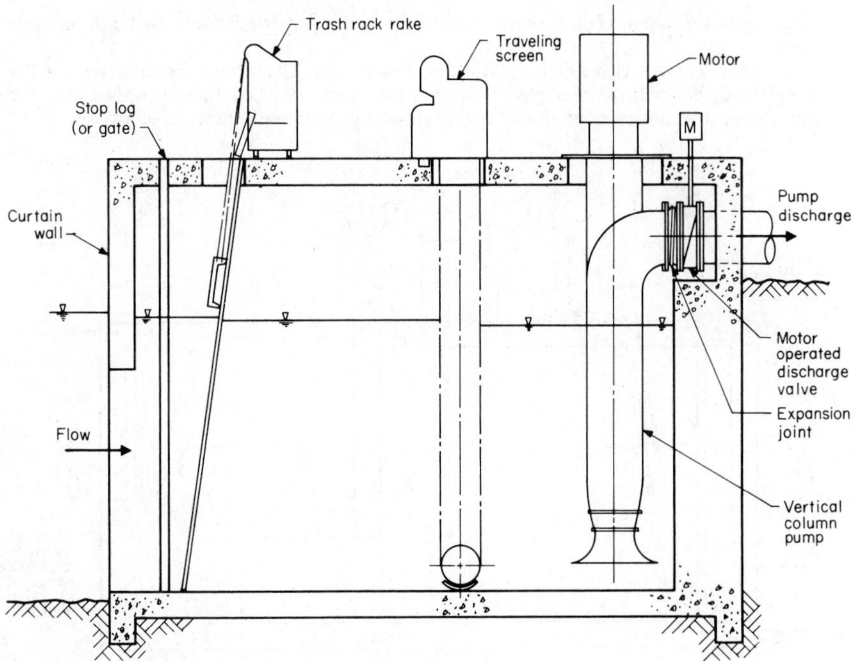


**FIG. 5-67** Vertical wet-pit pumpwell arrangement.

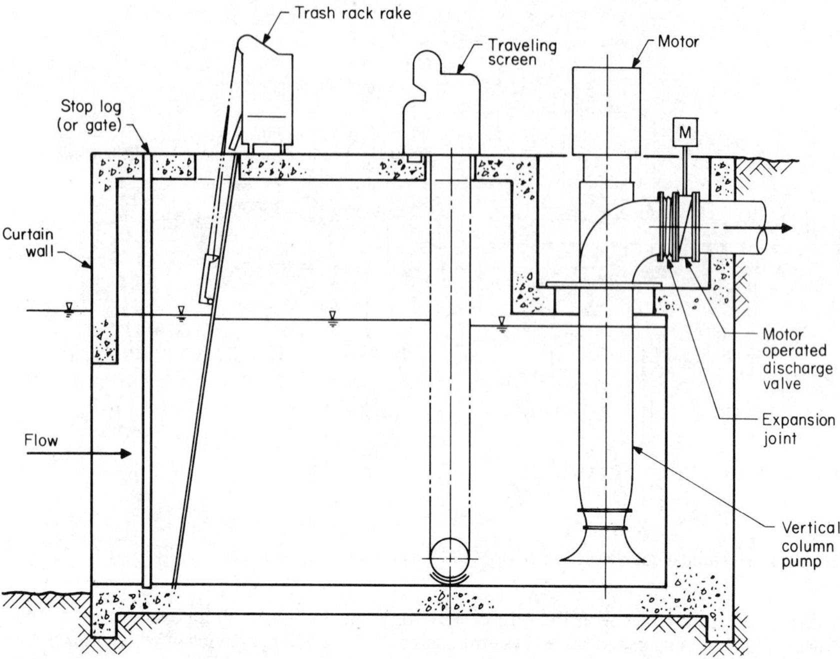


**FIG. 5-68** Vertical dry-pit pumpwell arrangement (column pump).

arrangements are rare at central power stations because of their relatively high cost and low efficiency.

The surface condenser is a shell-and-tube heat exchanger. The turbine exhausts steam to the shell side and the cooling water passes through the tubes, never coming in contact with the steam. The steam condenses on the tubes and collects in a hot well below the tubes (Fig. 5-71).

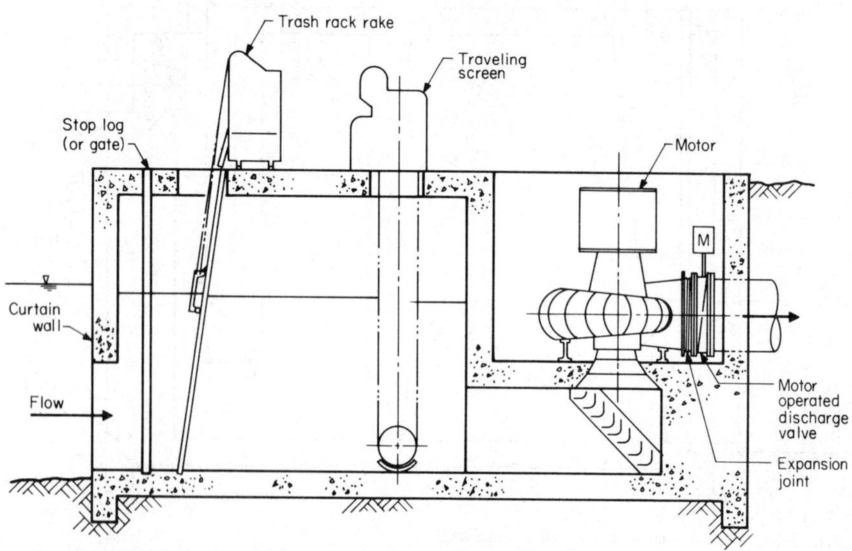


**FIG. 5-69** Vertical dry-pit pumpwell arrangement (volute pump).

Stop log
(or gate) slot
Stationary
screen slot
M
Flow
Inlet
isolation
valve
Expansion
joint
Motor operated
discharge valve

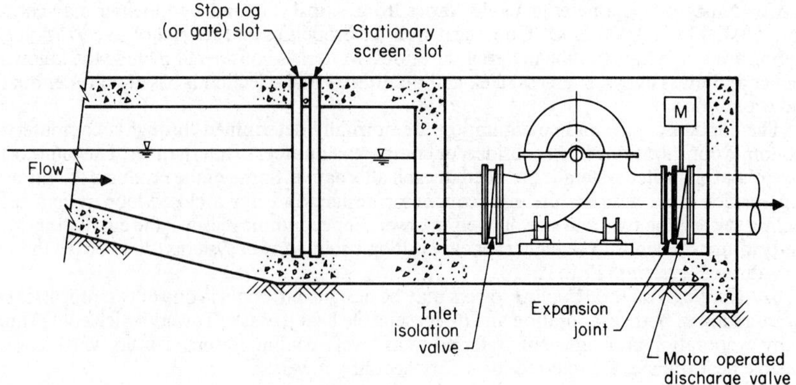

**FIG. 5-70** Horizontal dry-pit pumpwell arrangement.

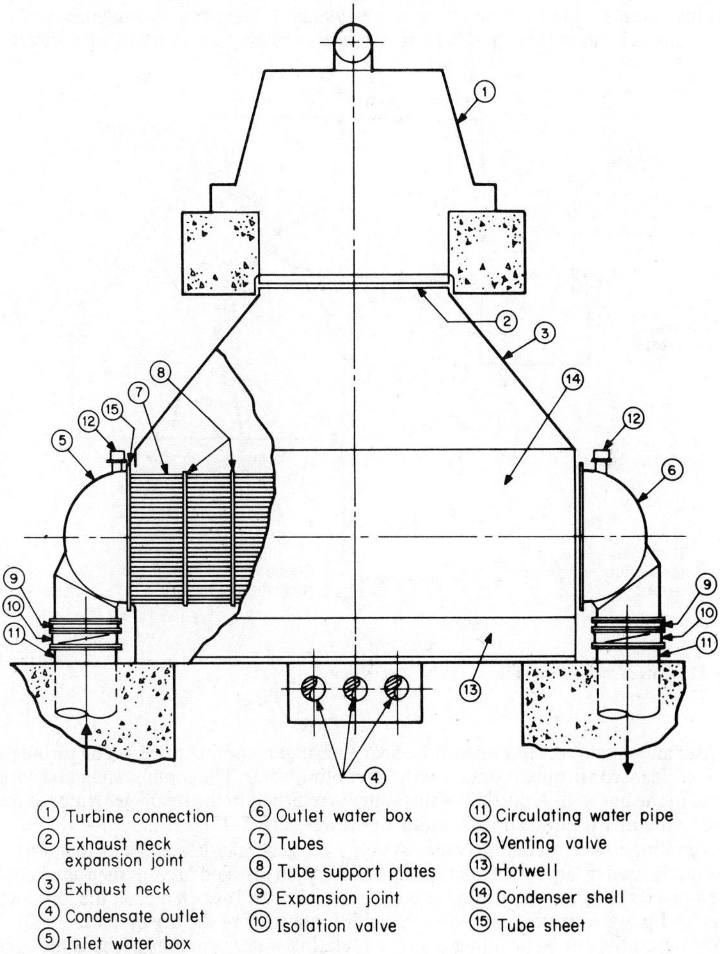

1
2
3
4
5
6
7
8
9
10
11
12
13
14
15
① Turbine connection
② Exhaust neck expansion joint
③ Exhaust neck
④ Condensate outlet
⑤ Inlet water box
⑥ Outlet water box
⑦ Tubes
⑧ Tube support plates
⑨ Expansion joint
⑩ Isolation valve
⑪ Circulating water pipe
⑫ Venting valve
⑬ Hotwell
⑭ Condenser shell
⑮ Tube sheet

**FIG. 5-71** Shell-and-tube surface condenser.

Condenser-tube diameter generally varies from ¾ to 1¼ in and a condenser may contain from 50,000 to 100,000 tubes. Good heat-transfer properties and resistance to corrosion and fouling are key factors in tube material selections. Materials considered include the following: copper and brass alloys, stainless steel, and titanium. Final selection is based on water quality and economics.

The condenser size and configuration are normally determined through a computerized economic optimization which considers or compares a number of alternatives. The total cost of the circulating-water system is included in each alternative. Some of the products of optimization include flow, temperature rise, condenser design, and, for a closed-loop system using cooling towers, the type and size of cooling tower. For economic reasons, the circulating-water temperature rise tends to be higher for closed-loop cooling-tower systems (20 to 30°F) than for once-through systems (15 to 25°F).

***61. Cooling Towers.*** Cooling towers may be designed to cool via one of two processes or a combination of both: evaporation and direct sensible heat transfer. Towers which cool primarily by evaporation are commonly referred to as "wet" cooling towers. Towers which cool by sensible heat transfer are referred to as "dry" cooling towers.

There are two basic types of *dry-cooling-tower systems:* the direct system, in which the cooling tower also functions as the condenser, and the indirect system, in which a portion of the condensate is cooled. In the direct system, the turbine exhaust steam is ducted directly to the cooling tower (an air-cooled finned-tube heat exchanger). The steam is condensed and returned to the steam generator (Fig. 5-72). In the indirect system, use is made of a direct-contact

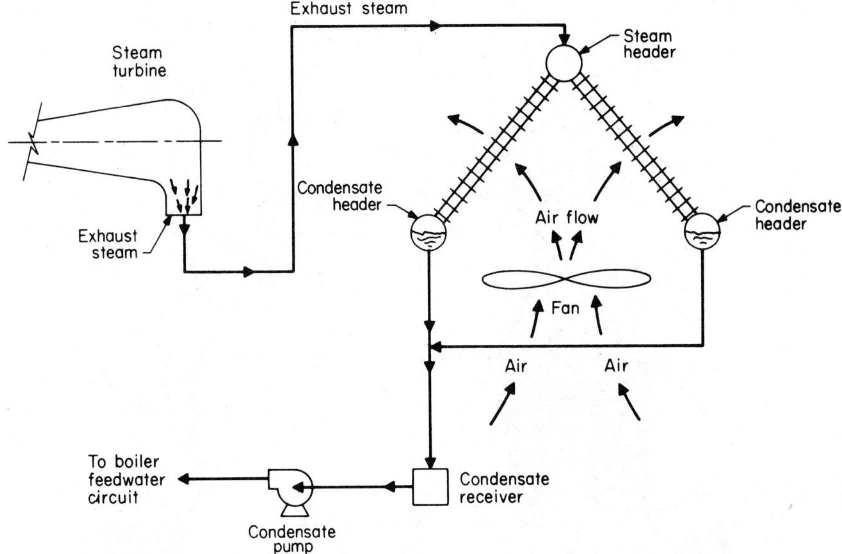


**FIG. 5-72** Direct dry-cooling-tower condensing system.

condenser and an air-cooled finned-tube heat exchanger (cooling tower). The turbine exhaust steam is condensed via direct contact with the cooling water. The cooling water and condensate combine in the hot well. A portion of this water is returned to the steam generator as feedwater and the remainder is pumped back to the dry tower (Fig. 5-73).

Dry-cooling-tower circulating-water systems are generally less efficient than wet systems and are rarely used in large power plants. They have been justified in cases such as where a heavy load center existed near a coal mine and water was scarce. In such a case, the low cost of fuel delivery and power transmission offset the additional costs of the dry system.

The evaporative, or wet-cooling-tower, circulating-water system is perhaps the most com-

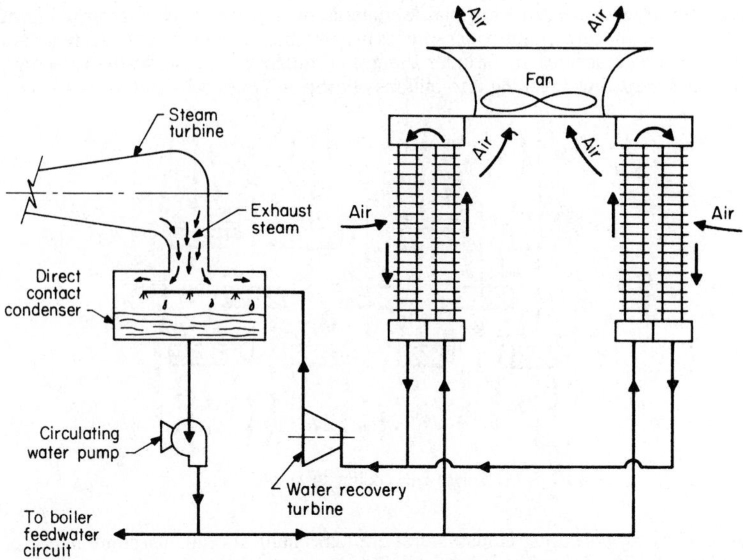


**FIG. 5-73** Indirect dry-cooling-tower condensing system.

mon type of closed-loop system. It is adaptable to most site conditions and requires considerably less space than cooling ponds and reservoirs. The water is introduced to a wet tower via pipes which, in the tower, are divided into thousands of distribution nozzles. From here the water falls into a cooling (fill) section. The fill or packing section typically consists of either

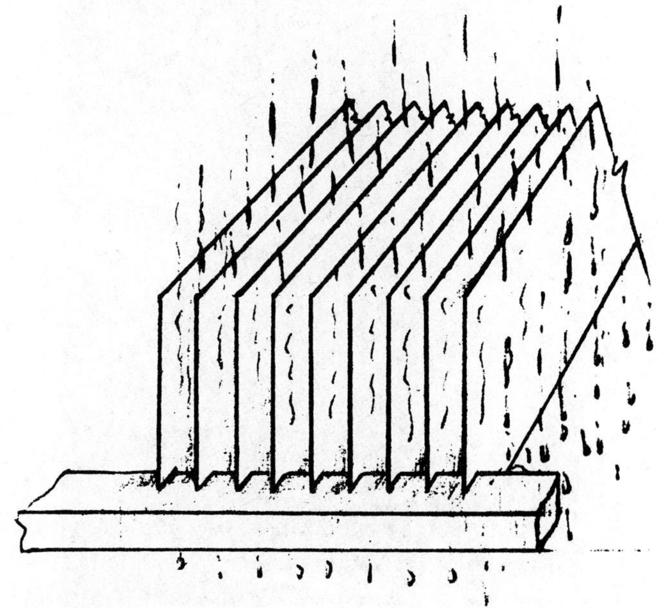

**FIG. 5-74** Film-type cooling-tower fill.

vertically positioned sheets of thin plastic material or of systematically arranged horizontal "splash bars." In the former, the water spreads in a thin film over the sheets, creating thousands of square feet of contact area. In the latter, the area of contact is increased by the water impinging on the splash bars, thus breaking into billions of droplets (Figs. 5-74 and 5-75).

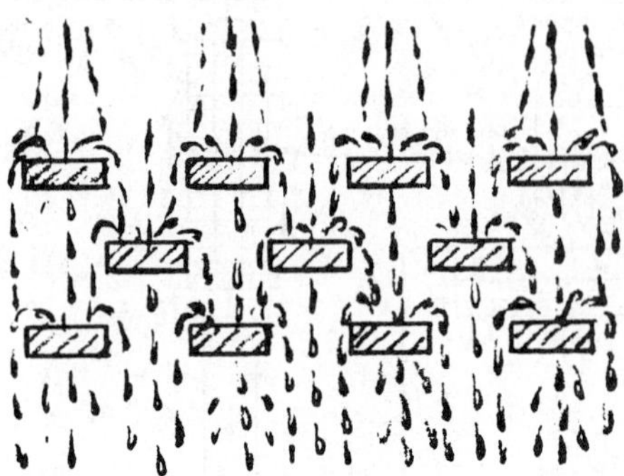

**FIG. 5-75** Splash-type cooling-tower fill.

Wet cooling towers may be either natural-draft, mechanical-draft, or a combination of both. They may be further defined as either cross-flow or counterflow.

Natural-draft towers are characterized by their tall hyperbolic concrete shells which, in some instances, have approached 600 ft in height (Fig. 5-76). Air flow through the tower is created by

**FIG. 5-76** Natural-draft cooling tower. *(Custodis-Hamon Constructor, Inc.)*

the chimney effect. Air in the tower receives heat from the cooling water and its density is reduced. The higher-density outside air flows into the tower to replace the heated air, forcing it up and out of the tower. The driving force is directly related to the height of the hyperbolic shell or stack. Natural-draft-tower performance is dependent on wet-bulb temperature as well as relative humidity. Towers in cool, humid climates require lower stack heights than those in relatively warm, dry climates. Natural-draft towers are not economically feasible in many parts of the world.

There are many types of mechanical-draft cooling towers. As the name implies, the required air flow is achieved by mechanical means. Power plant mechanical-draft towers in the United States are generally of the induced-draft-fan type. The fans are located downstream of the tower fill. Forced-draft-fan types exist but are not common in the United States because of the icing potential at the fans.

Traditionally, mechanical towers have been made up of cells, each cell containing a fan. The cells are arranged in line, making the tower rectangular in shape. More recently, manufacturers are offering round mechanical towers. In this design, the cooling section is arranged in a ring, with the fans grouped in a central area (Figs. 5-77 and 5-78).

**FIG. 5-77** Rectangular mechanical-draft cooling tower. *(The Marley Company.)*

**FIG. 5-78** Round mechanical-draft cooling tower. *(The Marley Company.)*

Because the mechanical tower relies on fans for its required air flow, it is not dependent on ambient relative humidity as is the natural-draft tower. As such, this type of tower is feasible at most locations. The power requirements of the fan motor average about 200 hp.

Motors are generally single-speed, 3-phase squirrel-cage machines and are connected to the fans via a drive shaft and gear reducer. When more control of air flow is required, two-speed motors with reversing capability have been supplied.

The cooling section of natural- and mechanical-draft towers may be either a cross-flow or counterflow arrangement. In the cross-flow type, the fill usually consists of splash bars as described above. The required air flow is drawn in through the sides of the tower and crosses the downward path of the water droplets. In the counterflow type, the fill usually consists of vertically positioned sheets as described above. The required air flow is drawn in at the bottom of the tower and moves up through the fill, counter to the downward movement of the water. In the recent past, the cross-flow design was usually chosen because of its lower capital cost. However, the counterflow design requires less circulating-water pumping head, and, with increasing fuel costs, it has become the more economical arrangement at many power plants (Figs. 5-79 and 5-80).

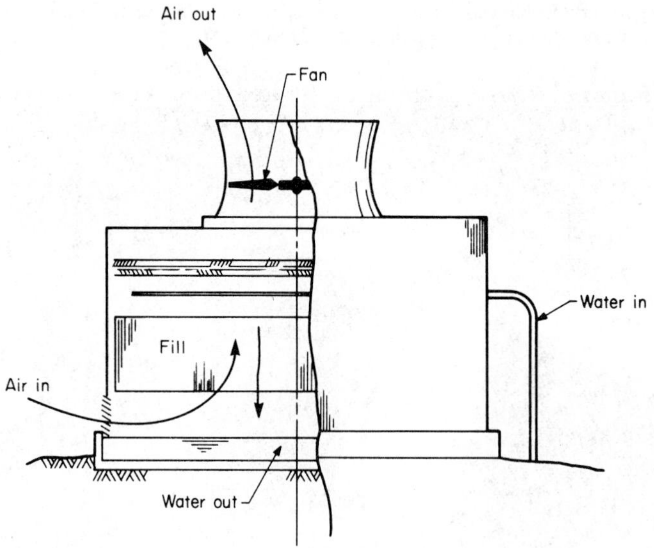


**FIG. 5-79** Section of a counterflow cooling tower.

***62. Conduit.*** Circulating-water-system conduit may be pipe, tunnels, or canals, depending on site conditions. Pipe is the most commonly used conduit. Circulating-water pipe is normally sized based on velocities of the order of 8 to 9 ft/s. Line sizes at recently built power stations range up to 12 ft in diameter. Pipe generally falls into three categories: concrete cyclinder, carbon steel, and fiberglass. Concrete-cylinder pipe typically consists of a steel cylinder coated on both sides with reinforced concrete. The reinforcing on the outside may be prestressed. It is suitable for both freshwater and brackish-water sites because of its excellent corrosion-resistant properties.

Carbon-steel pipe is extremely durable and easily fabricated into the special shapes often required in circulating-water systems. However, it cannot be used in brackish water without protective coatings.

Fiberglass is a relatively recent development in large-pipe materials. It is light, easy to handle, and appears to have good corrosion-resistant properties, but great care must be exercised during its installation. Differential loads near joints will lead to leaks; and because it is light, it must be anchored to resist bouyant forces.

***63. Valves.*** The predominant type of valve used in circulating-water systems is the butterfly valve (Fig. 5-81). Valves are generally found at pump discharge nozzles, condenser inlet and outlet water boxes, and at the entrance to cooling towers. Valves at the pumps are typically

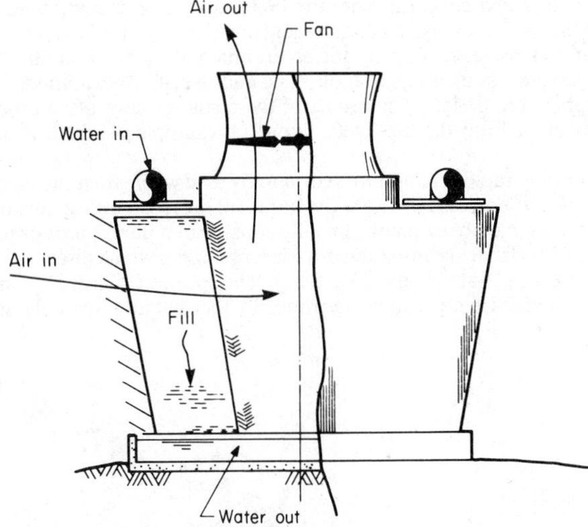


**FIG. 5-80** Section of a cross-flow cooling tower.

motor-operated with local and remote controls, and interlocked with the operation of the pump. Valves elsewhere tend to be locally operated, either manually or by motor operator.

More control is required for the valve at the pump since it may be used to smooth out system start-up and shutdown. Transient pressures can be severe, and use of this valve to control flow entering the system during start-up and control flow reduction during shutdown can reduce these pressures significantly. Valves can be interlocked with pumps such that a pump starts when its discharge valve is opened slightly and trips when its valve is closed. This minimizes reverse rotation of the pump. Valve opening and closing times are typically from 20 to 60 s.

When two or more pumps discharge into a common header, it is important to ensure that when a pump trips, its associated valve closes to prevent backflow into the failed pump. This may be accomplished by placing the pump and valve motors on separate electrical buses.

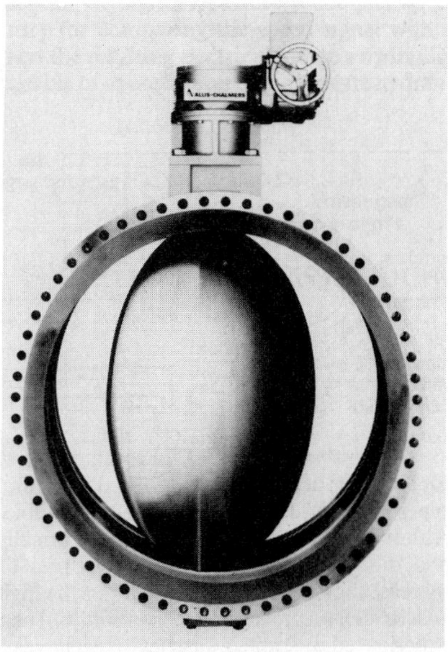

**FIG. 5-81** **Butterfly valve.** *(Allis Chalmers Corporation.)*

***64. Discharge Structures.*** The discharge structure of a once-through circulating-water system or of the blowdown line from a closed-loop circulating-water system is generally either an onshore surface type or an offshore submerged type.

Shoreline surface discharges consisting of a concrete headwall and sheetpile flume were quite common until the 1960s. Concerns were usually limited to such areas as providing sufficient distance from the intake to minimize thermal recirculation and restricting discharge velocity at the outlet to prevent interference with navigation. Thermal-discharge regulations existed in many areas but generally were not very restrictive.

In addition to the above concerns, since the 1960s, discharge structures must be designed to meet prescribed thermal-discharge regulations and must be shown to have a minimal impact on the environment. Extensive engineering and environmental studies, including the gathering of biological data, sometimes over a period of years, and use of physical models are required to develop and support the design. The results of these studies have often produced discharge systems quite different from the pre-1960s flume, an example of which is the offshore submerged diffuser.

A diffuser promotes rapid mixing and is commonly used where there are thermal restrictions that limit the allowable temperature rise at the surface of the receiving water body. A submerged diffuser system for a large power plant generally consists of a buried pipe or tunnel extending several thousand feet offshore terminating in a series of nozzles which discharge a few feet off the bottom of the receiving water body (Fig. 5-82). The location, number, configuration, and orientation of the nozzles is normally determined by thermal and hydraulic model studies.

**FIG. 5-82** Submersed diffuser discharge.

## *INDUSTRIAL COGENERATION*

*By ALLEN L. CLAPP*

### Impetus for Cogeneration

***65. Uncertain Future.*** In recent years, industries have been faced with rapidly rising costs for all forms of energy, including coal, natural gas, fuel oil, and electricity. The cost of the major types of fuels used by industry has doubled since 1975 on a dollar per million Btu basis. Although the cost of electricity has historically lagged behind the escalation in other energy costs and, in some areas, is today the same in real terms as it was in 1960, electricity cost is expected to increase rapidly in the next decade as the high costs of recent construction are placed into rates. Just as short-term supply disruptions of energy sources, such as oil and natural gas, have been experienced by some industries in recent years, the long-term availability of several fossil fuels is unclear at the present time.

These factors have had, and may be expected to continue to have, a significant impact upon the ability of many businesses to operate profitably. Their ability to invest in long-term capital projects is also affected. Although energy utilization measures have been recognized and implemented by companies over the past few years to control rising costs, the challenge facing business managers in the coming years will continue to be how to operate profitably in the face of escalating energy prices. In many cases, cogeneration of electricity and steam is not only a sound economic investment today but is also a hedge against future inflation.

***66. Cogeneration Defined.*** Cogeneration is the efficient production of two forms of useful energy from the same fuel resource, using the exhaust energy from one production system as the input for the other. Ordinarily the primary energy form is thermal (steam) and the secondary form is either electrical or mechanical. The electrical or mechanical energy can be used internally to run company equipment, or the electricity can be sold to a utility. The more energy-intensive the plant, or the more valuable electricity is to the area, the more likely it is that cogeneration will increase a firm's profits.

Before cogeneration is considered, the existing or planned process should first be examined carefully to reduce overall steam requirements. In many cases, high-efficiency industrial heat pumps can be economically used to increase the quality of low-enthalpy exhausts from one set of processes to make them suitable as inputs to the same or other processes, thus reducing overall steam production requirements.

While either electrical and/or mechanical energy can be produced by cogeneration systems along with the thermal energy, this discussion assumes that these systems will be used to generate electrical energy. Electrical energy has significant internal advantages over mechanical energy, such as flexibility, economy, security, and potential salability to utilities. It should be pointed out, however, that there are a number of applications, such as boiler feed pumps and other steady-state pumps, compressors, and absorption chillers, where a steam drive may be an economical substitute for electrical drives.

Since cogeneration is a sequential process, with the waste heat from one process being captured for use as a heat input in another process, cogeneration requires (1) some amount of common physical plant space for the two processes, and (2) a sharing of the energy content of the fuel. This joint system can reduce energy input to 10 to 30% below that required by separate systems to produce the same outputs. Total system efficiency can approach 90%, a significant improvement over the 50 to 90% efficiency of many industrial boilers and the 30 to 35% efficiency of electrical conversion when separate production is used. Typically, adding cogeneration to new construction allows the use of higher-pressure, more-efficient boilers than would be used to generate steam alone. The turbine characteristics are then matched to the plant steam requirements to assure that the turbine exhaust will be of a quality useful as the steam input to industrial processes. As a result, this simultaneous efficient production of two energy forms can significantly reduce total operating costs in many instances, even after paying for the increased capital costs.

***67. Application of Cogeneration.*** Many companies in energy-intensive industries have begun reevaluating the expected economic benefits from cogeneration. As a result, both larger and smaller companies are turning to cogeneration to reduce or offset their energy costs, reduce their reliance on oil and natural gas, improve their ability to withstand energy supply disruptions, and increase their overall system efficiency. In addition, some of these companies have found that their cogeneration systems can use less expensive, alternative sources of fuel, such as wood, wood wastes, and processed by-products.

In general terms, cogeneration is most likely appropriate for any industrial or commercial operation requiring the use of relatively significant amounts of steam or thermal energy. However, in order for cogeneration to be technically feasible, the firm's primary need for thermal energy must be compatible with the operational requirements of the secondary use. For example, the steam load consistency, steam quality, and hours of operation of a steam-using plant process must fit the input-output and operating demands of a turbine-generator set if the system is to work successfully. Successful cogeneration installations to date have had very high cogeneration equipment capacity factors, usually much greater than 50%. Industries likely to offer a good cogeneration potential include pulp, lumber, and paper mills, as well as many food processing, textile, chemical, furniture, and pharmaceutical plants.

The experiences of the present cogenerators show that, in the right technical and economic settings, a cogeneration system can improve a firm's short- and/or long-term profitability. This does not mean that cogeneration is right for every company. Adding a cogeneration system has costs and risks that should be carefully weighed against the projected benefits before committing to development. As a rough rule, the higher the quality of the waste heat from the existing process or the lower the quality of heat needed for the existing process, the more likely cogeneration is to succeed. As a result, firms with existing process applications having high overall system efficiencies—for example, 70% efficiency—are probably not good cogeneration candidates unless they have significant waste fuel resources or other problems for which the extra steam needed for cogeneration can provide a ready answer. Considering the magnitude of the potential benefits of cogeneration in the appropriate plant setting, it is valuable for firms to explore the merits of a cogeneration system wherever it seems technically feasible.

One of the major problems for a firm interested in cogeneration is the relative capital intensity of a cogeneration system. The capital cost of a cogeneration system may range from $500 to $1000 per thousand watts (kilowatt or kW) of installed capacity for oil and gas systems to $1000 to $1500 for coal, wood, or biomass systems. When in times of high interest rates and credit restrictions, such as those we have recently experienced, this capital intensity makes the financing of such systems a major consideration in their successful development. Where the firm's overall economic picture limits its ability to use the tax benefits associated with such projects, third-party ownership is becoming more widespread.

## Types of Cogeneration Systems

***68. Basic Types.*** There are two basic types of cogeneration systems, depending on whether thermal or electrical energy is produced first. In one, a topping-cycle cogeneration system, or *topping system,* the waste heat from the production of electricity is used to provide space heating, industrial process heating, absorption air-conditioning, or other thermal-related requirements. A bottoming-cycle cogeneration system, or *bottoming system,* on the other hand, uses the heat exhausted by an industrial process to produce electricity. Because of the high-quality steam (steam of sufficient temperature and pressure) desired for the generation of electricity, bottoming systems are not as likely to be as appropriate applications as topping systems. However, bottoming systems may be a perfect solution when the steam exhausted from a high-pressure operation, such as a pressboard press, is still of fairly high quality. This is especially true if there are also low-pressure applications, such as heating, available to take the turbine output, thus effectively turning the turbine into a highly efficient pressure reducer.

***69. Topping Systems.*** In topping systems, a steam turbine or internal combustion engine is used to drive a generator and produce electricity (Fig. 5-83 and 5-84). The waste heat resource is

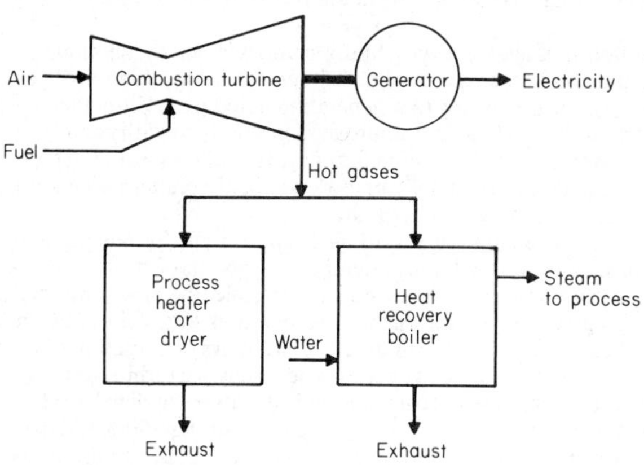


**FIG. 5-83** Gas turbine topping cycle.

either (1) steam from the exhaust of the steam turbine or (2) exhaust gases or jacket heat recovered from internal combustion engines. These heat resources are then used directly for processes such as the following:

Drying wood or other building materials

Reheating metal

Heating water, air, or other heat-transfer media such as commercially available oil

Producing steam in heat-recovery boilers

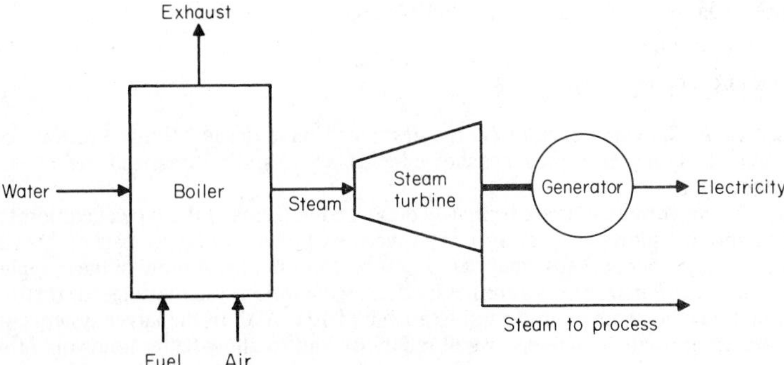


**FIG. 5-84** Steam topping cycle.

Gas turbines, diesels, spark-ignited engines, and various types of steam turbines can all be utilized in topping systems. Gas turbines, diesels, and spark-ignited engines may also be used, with some limitations, in combination with steam turbines. If the motive power is created by a gas turbine, and the exhaust gases have a high enough oxygen content, the exhaust can also be used to preheat the combustion air in boilers and decrease fuel required.

***70. Bottoming Systems.*** Waste heat for bottoming systems may be exhausted from furnaces, kilns, chemical reactions, and other processes, such as high-pressure steam presses. Probably the most common bottoming system utilizes the waste heat to generate steam in a recovery boiler in order to drive a turbine or generator (Fig. 5-85), but commercial experience

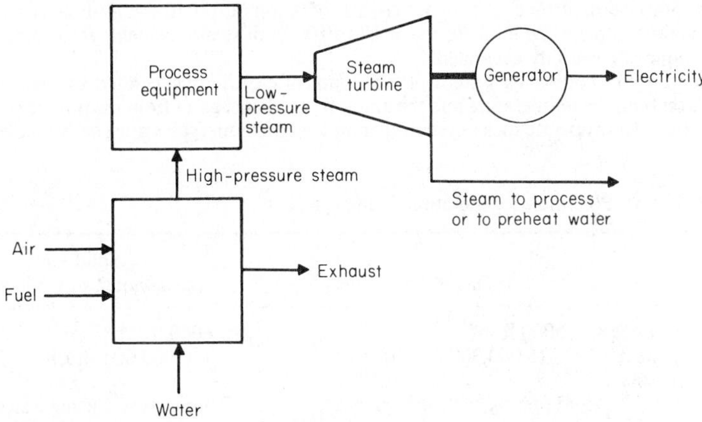


**FIG. 5-85** Steam turbine bottoming cycle.

with such systems is limited. Such a system has the advantage of requiring no additional fuel to generate the electricity, and should benefit from a high degree of reliability because it operates, other things being equal, at lower temperatures than do topping systems. It also has the advantage of being adaptable to a number of retrofit applications.

***71. Typical Configurations.*** For each type of cogeneration system, many different configurations can be designed using readily available equipment. The configuration of the system needed depends upon specific characteristics of the facility such as the following:

Plant size

Fuel type

Process heat or steam temperature and process requirements
Emission limitations
Economic factors

Based upon the experience of companies which have designed, installed, and operated cogeneration systems, the approaches shown in Table 5-15 and summarized thereafter are used most frequently.

The following discussions are primarily organized in terms of the type of equipment using the prime moving source (e.g., steam). Heat-recovery boilers, while not part of this classification, are discussed here because they can be a subsystem in one or more of these applications.

*72. **Steam Turbines.*** Steam turbines used in cogeneration systems range in size from those producing a few horsepower up to units capable of 105 MW. In the larger systems, multiple boilers are often used to increase overall reliability and to allow better following of seasonal steam demand changes. The amount of power they can produce depends on the temperature, pressure, and amount of steam the boiler produces. Steam turbines may be used in both topping and bottoming systems and are available in configurations able to utilize almost any fossil fuel, or a combination of fuels, such as natural gas and petroleum distillates, or wood and biomass.

Steam turbines, however, are most often used in topping systems where steam can be extracted for process applications, with the excess steam being used for steam-turbine generation (see Figure 5-84). The mix of process use and generation will vary as process demands dictate. In these applications, the incremental fuel cost of electricity is between 4500 to 6500 Btu. Depending upon steam quality, approximately 10,000 to 50,000 lb of steam per hour is required to operate this system economically.

As a rule, larger systems will obtain greater overall system efficiencies, mainly because of the turbine's efficiency. For small turbines of less than 100 kW, turbine efficiency is nearly always below 50%, whereas for the very large, multistage turbines, efficiencies can exceed 80%. Steam-turbine cogeneration systems produce considerably less electricity per pound of *process* steam than do combustion turbines. The expected ratio of power output to energy input is in the range of 10 to 60 kWh per 1000 lb of steam, or 17 to 100 lb of steam available for process use per kilowatthour of electricity generated.

When used in bottoming systems, the amount of electricity generated by steam turbines depends on both the temperature and the amount of exhaust gases from the primary industrial process. In order to operate such a system economically, exhaust gases must be available at a rate

**TABLE 5-15** Potential Cogeneration Configurations

| | Gas or liquid fuel | Solid fuel (coal, wood, or wood waste) |
|---|---|---|
| Minimum process steam or heat requirements | 5000 lb/h* (5,000,000 Btu/h) | 10,000 lb/h* (10,000,000 Btu/h) |
| Suggested configurations | Topping cycle using a gas combustion turbine, or engine, exhausting to a heat-recovery boiler | Topping cycle using a steam turbine, exhausting to a process application |
| | Topping cycle using a gas combustion turbine, or engine, exhausting to a process application | Bottoming cycle, exhausting to a steam turbine |
| Other configurations | Topping cycle exhausting to a recuperator | Topping cycle exhausting to a conventional boiler |
| | Topping cycle exhausting to a retrofit heater or boiler | Topping cycle exhausting to a recuperator |

* Depending upon steam pressure and temperature.
NOTE: 1 lb = 0.4536 kg; 1 Btu/h = 0.293 W.

of approximately 500,000 Btu/h. Above that level, such a system can be expected to produce 35 to 50 kW per 100,000 Btu/h, at an incremental fuel cost that is very low.

As a result of the high heat of vaporization of water, steam boilers can be made more efficient by generating steam at two or more pressures so that greater amounts of thermal energy are transferred. Dual-pressure boilers, which are particularly useful in a bottoming-cycle cogeneration system using steam turbines, may also be used to generate steam for direct process use.

*73. Gas Turbines.* Gas turbines come in sizes ranging from 6 kW to 100 MW and can be configured to operate on one or more fuels. Units that burn natural gas give the most reliable and continuous operation, while those which use low-grade oils, such as residual "No. 6" fuel oil, require more frequent maintenance and cleaning.

Gas turbines are most commonly found in topping systems which produce heat in the process of generating electricity. The waste heat in the exhaust gases is either used in some direct process application or is collected by a recovery boiler to produce steam that is then used in some industrial process (see Fig. 5-83). Because gas-turbine exhaust temperatures range between 800 and 1000°F, they are capable of producing relatively high quality steam when used with a recovery boiler. The result is a power-to-steam ratio which may be three to four times higher than that produced by a steam turbine. Ratios as high as 265 kWh per 1000 lb of process steam, or 3.8 lb of available process steam per kilowatthour of electricity generated, are possible. However, if supplemental firing of the boiler is necessary to produce higher-quality steam, the ratio will be lower. Generally, an additional 4000 to 6000 Btu of fuel above what is used to meet a system's process requirements will be needed to produce a kilowatthour of electricity.

There are two important performance characteristics of gas turbines. First, exhaust gas temperature and the amount of power generated depend highly on the ambient temperature. Second, the efficiency of the electrical generation decreases significantly as the output decreases from full load. If the decrease in electrical demand is accompanied by an increase in thermal demand, this decrease in electrical efficiency can be offset by the increased thermal recovery, and the overall system efficiency will thereby remain constant. However, if thermal demand decreases or remains constant with a decrease in electrical demand from the gas cogeneration system, then the overall fuel utilization will decrease, and there will be increased waste.

*74. Diesel Engines.* Industrial diesels offer a variety of cogeneration possibilities. Large, slow-speed diesel engines are available with the capability of producing up to 30 MW of electricity. Smaller high-speed diesels, which burn higher grades of fuel oil, typically have the capability of producing less than 5 MW. In any configuration, heat may be reclaimed from both the exhaust gases and from the engine's internal cooling system.

The temperature of exhaust gases released by diesel engines is below that of gas turbines. Consequently, diesels are likely to be used for drying raw materials, for space heating, or, in applications requiring low-quality steam, in conjunction with a heat-recovery boiler. If the oxygen content is high enough in the diesel exhaust, they may also be used to heat boiler combustion air. However, if supplementary firing of the recovery boiler is utilized in applications requiring higher steam pressures or temperatures, there will be a corresponding decrease in overall system efficiency.

The heat recovered from diesel cooling systems, though of a much lower temperature, should not be discounted; it can often be used for space heating or in low-temperature process applications, such as food processing or paper and textile manufacturing. Because of this second source of recoverable thermal energy, the comparative ratio of electricity produced per Btu of fuel burned by a diesel cogeneration system is a composite representing the overall potential system. Diesels are capable of 250 to 500 kWh/MBtu of fuel when used with an unfired recovery boiler (one in which no supplemental firing is necessary). When used to generate steam only, this ratio may become as high as 1500 kWh/MBtu.

Diesels are not as sensitive to ambient temperature as are gas turbines. Unlike gas turbines, however, the amount of heat produced is generally proportional to the load on the engine.

*75. Heat-Recovery Boilers.* Heat-recovery boilers transfer the thermal energy of the exhaust produced by gas turbines or other heat sources to water in order to produce steam. It is often advantageous to connect more than one heat source to a common boiler. Like all boilers, steam can be extracted at a number of places in the heating and pressurization process, depending upon the demands on that particular boiler application.

Unfired heat-recovery boilers transfer the thermal content of the exhaust gases directly to the water; however, if it is necessary to increase the thermal input to the boiler in order to produce steam of the desired quality, supplemental firing will be necessary. As already pointed out, the ability of dual-pressure boilers to provide an increase of 10 to 20% in overall system efficiency is valuable.

## Laws and Regulations Affecting the Purchase and Sale of Electricity

***76. Public Utility Regulatory Policies Act (PURPA) of 1978.*** PURPA provides significant incentives to develop *qualifying* cogeneration and small power production facilities (QFs). To be eligible for the PURPA incentives, cogeneration facilities must meet operating, efficiency, and ownership standards. Small power production facilities must meet size, fuel use, and ownership standards.

PURPA and the Federal Regulatory Commission (FERC) regulations implementing PURPA require electric utilities to purchase all energy and capacity made available to them by QFs and, in turn, the utilities are required to sell energy and capacity to QFs.

Utilities must pay the QF a rate for purchased power based on the utility's "avoided cost"—the incremental cost of power to the utility that the utility would either generate itself or purchase elsewhere if it were not purchasing from the QF. This power can be sold to the utility on an "as available" basis, or pursuant to contract or legally enforceable obligation. Power sales to the QF must be at nondiscriminatory rates. PURPA also exempts most QFs from regulation under almost all sections of the Federal Power Act, from certain sections of the Public Utility Holding Company Act, and from state public utility financial and organizational regulation.

Although PURPA is a federal statute, state public utility commissions are generally responsible for its implementation. While they must carry out PURPA in accordance with FERC regulations promulgated according to the act, the state public utility commissions have been given some discretion in the implementation process.

***77. Definitions and Qualifying-Facility Status.*** The following definitions are contained in FERC regulations.

**Cogeneration Facilities.** A "cogeneration facility" is defined by the FERC regulations as equipment used to produce electric energy and forms of useful thermal energy used for industrial or commercial heating or cooling purposes through the sequential use of energy. In order to be considered a QF, for the purposes of PURPA, a plant must meet the cogeneration facility criteria or the qualifying small power production facility criteria established by the FERC regulations. Qualifying cogeneration facilities must usually meet operating, efficiency and ownership standards; however, the FERC can waive any operating or efficiency standard that would produce significant energy savings. The standards normally required for a qualifying cogeneration facility are the following.

*Operating standard.* The cogeneration operating standard applies to new and existing topping-style facilities; there is no operating standard for bottoming facilities. This standard requires that at least 5% of a topping-cycle facility's total energy output be in the form of useful thermal energy output.

*Efficiency standard.* If the installation of a cogeneration facility began on or after March 13, 1980, and the facility is a topping-cycle facility using oil or natural gas to provide all or a part of its energy, or a bottoming-cycle facility using oil or natural gas for supplementary firing, it must meet the efficiency standard. If the facility is a topping-cycle facility during any calendar year periods, its useful power output plus one-half of the useful thermal energy output must be: (1) no less than 42.5% of the total energy input of natural gas or oil, or (2) no less than 45% of the total energy input of natural gas or oil if the useful thermal energy output is less than 15% of the total energy output of the facility. If the facility is a bottoming-cycle facility, the useful power output must be no less than 45% of the energy input of natural gas or oil during any calendar-year period.

*Ownership standard.* A qualifying cogeneration facility may not be owned by a person primarily engaged in the generation or sale of electric power, other than electric power solely from cogeneration or small power production facilities. This means that not more than 50% of the equity interest in a facility can be owned by an electric utility or utilities, an electric subsidiary, a public utility holding company, a public utility holding company subsidiary, or any combination thereof.

**Small Power Production Facilities.** Qualifying small power production facilities must meet size, fuel use, and ownership standards. These requirements are explained below.

*Size standard.* The total power production capacity of all facilities using the same energy source, owned by the same person, and located at the same site may not exceed 80 MW.

*Fuel use standard.* The primary energy source of the facility must be biomass, waste, renewable resources, or any combination thereof, and more than 75% of the total energy input must be from these sources. During any calendar year, use of oil, natural gas, and coal by a facility may not in the aggregate exceed 24% of the total energy input of the facility.

*Ownership standard.* The ownership standard is the same as for the qualifying cogeneration facilities.

***78. Sales of Electricity to Utilities.*** PURPA represents an attempt by the Congress to reorder the relationship between electric utilities and small power producers to encourage energy conservation and the development of renewable resources and cogeneration opportunities. The strategy adopted by Congress has two main components: first, the utility must purchase the power produced by those facilities that meet the definitional criteria of a QF; and second, the utility must pay a price for that power based on the utility's avoided costs and set by the state regulatory commission in accordance with the guidelines promulgated by the FERC.

***79. Avoided Cost and Power Purchase Contracts.*** In order to effectuate the PURPA mandate that utilities must buy the power produced by a QF, utilities are required to pay the QF a rate for that power based on the utility's avoided costs. Avoided cost is the incremental cost to the electric utility of power that the utility would either generate itself or purchase elsewhere if it did not purchase from the QF. The purpose of the avoided cost rate is to maintain the electric utility and its ratepayers in the same financial position it would have otherwise occupied had it not made a purchase from the QF. Thus, avoided costs are calculated with respect to the value of the power to the utility and are not related to the costs that the QF incurs in producing the power. This formulation of avoided costs is designed to avoid imposing a burden on the utility's ratepayers.

The regulations explaining avoided costs have broken the concept into two key elements: energy costs and capacity costs. Energy costs are the variable costs associated with the production of electric energy. They represent the costs of fuel, some operating and maintenance expenses, fuel inventory, and line losses. Capacity costs are the costs associated with the instantaneous capability to satisfy energy demands, and they consist primarily of the capital costs of facilities. The calculation of a utility's avoided energy and capacity costs may take into account the aggregate value of energy and capacity from all the QFs on the electric utility's system. To the extent that the purchase of power from a QF increases or supplements the electric utility's ability to meet demand, the QF is entitled to a purchase rate reflecting the avoided capacity costs associated with this increased ability.

A qualifying facility providing power on an "as available" basis will have the rates for such purchases based on the avoided cost calculated at the time of delivery. A QF that enters into a binding obligation with a utility often has the option of having a purchase rate based on avoided costs calculated at the time of delivery, or estimated for the term of the obligation at the time of the agreement. The fact that an estimated rate differs from actual avoided cost during the term of the obligation does not invalidate the rate or violate the regulations. The rules also provide the QF with the flexibility to enter into a long-term front-loaded contract with a utility, thereby receiving a greater percentage of the total avoided cost purchase price at the beginning of the obligation, as long as total payment does not exceed the estimated avoided cost.

Under PURPA, a QF can sell to a utility either with or without a contract. However, PURPA does not override the power of an electric utility and a QF to agree to terms, conditions, or rates related to purchases. PURPA merely expands the options available to QFs by creating the idea of a legally enforceable obligation. If a utility does not wish to negotiate a special contract with a QF, the utility is nevertheless required to purchase power from the QF, with the terms and conditions of the purchase supplied by the FERC regulations.

***80. Interconnection Costs.*** In addition to the requirement that the electric utility buy the power produced by a QF at the utility's avoided cost, the utility must make any interconnection with a QF necessary to facilitate purchases and sales. However, PURPA is designed so that the QF imposes no economic burden on either the utility or the consumers. Therefore, the QF is responsible for paying all interconnection costs above what the utility would normally pay for a nongenerating customer of a similar size. These costs can include such things as meters, safety devices, and administrative and reasonable insurance costs, but are limited to those costs actually incurred by the utility making the interconnection. Some state commissions have handled the issue of interconnection costs by providing for approval of plans for payment of such costs when the individual contract documents between the QF and the utility are filed by the utility.

***81. Simultaneous Transactions.*** Purchases by utilities, whether they be under color of contract or legally enforceable obligation, will take either of two forms: a surplus sale or a simultaneous purchase and sale. Under a surplus sale arrangement, the QF obtains the PURPA-mandated rate only for the surplus energy it can generate. Only the QF-generated power in excess of the QF's load, i.e., the net energy input to the utility grid, is purchased by the utility. At times when the QF cannot meet all of its own energy needs, the utility is required to sell power to the QF at the same price other nongenerating customers are charged.

In contrast to the surplus sale arrangement, the QF participating in a simultaneous purchase-and-sale arrangement sells all of the power it generates to the utility at the PURPA rate while purchasing all of the power for its own needs at the rate other nongenerating customers are charged. Therefore, if the avoided cost rate paid to a QF is greater than the rate the QF pays for power, a simultaneous purchase-and-sale arrangement can benefit a QF more than a surplus sale.

***82. State Statutes and Regulations.*** Several states have enacted statutes providing incentives to the development of certain small power facilities. Initially some states encouraged cogeneration by setting exceptionally high buyback rates and by allowing long-term, fixed-price contracts. At this writing, some of these states are decreasing the buyback rates and requiring annual contracts for at least one of three reasons. First, present expectations of fuel cost escalation rates are drastically reduced from those expected a few years ago; as a result, the expected avoided cost to utilities by purchasing from cogenerators has decreased. Second, PURPA regulations in effect essentially require the ratepayers to accept the full risk of construction of cogeneration plants. Under the regulations of the FERC the *full* avoided cost is required to be paid. Ratepayers do not participate in any fuel cost savings resulting from the use of cogeneration. At this writing, many utilities have excess generating capacity and will continue to do so for several years. Thus, for these utilities, no capacity cost can be avoided in the near term and the *amount* which might be avoided in the long term is uncertain. As a result, ratepayers may not save anything at all today if their utilities purchase power from cogenerators. If the payback rates are based on uncertain future avoided costs, depending upon future fuel and construction cost savings, ratepayers may never save and, in fact, may be subject to a substantial loss. As a result, regulators are tending to remove capacity credits from the buyback rates, prohibit long-term contracts, and only allow short-term fuel energy credits to be paid. Third, the initial cogenerators have, in many cases, signed long-term contracts for the initial block of high avoided costs; as more cogenerators come on line, the incremental avoided costs become less as they displace progressively more efficient, lower-cost generation. As a result, the avoided cost rates will be less for the next cogenerators. Even if long-term contracts are not allowed, the same phenomenon holds; as the amount of generation avoided becomes greater, the value of the incremental generation avoided will be less and the average price paid will, thus, fall until it reaches an equilibrium and no more cogenerators will come on line. Increasingly, then, cogeneration will only make economic sense to an industry when it more than offsets the cost of buying the same amount of electricity or when the industry is in a capacity-short state where buyback rates and normal industrial rates are higher.

If, however, the PURPA regulations change to allow a sharing of the benefits of cogeneration with ratepayers, industries can expect an enthusiastic partnership from many utilities and commissions. Both utilities and ratepayers can benefit in some cases from the decreased construction time and increased efficiency of industrial cogeneration systems. By decreasing the risk to utilities and their benefit ratepayers, a sharing arrangement would be expected to increase the value of cogeneration.

It should be pointed out that, while PURPA relieves the cogenerator from many aspects of state regulation of electricity sales, many states will still regulate the sale of steam by a cogenerator to another party.

## Laws and Regulations Affecting Construction of a Cogeneration Facility

***83. Requirement for Certificate of Public Convenience and Necessity or Report of Proposed Construction.*** Most states require any potential cogenerator planning a facility for the purpose of selling electrical power to obtain a certificate of public convenience and necessity or other permit prior to beginning construction. Those planning to use cogenerated electricity primarily within their own plant generally need not obtain a certificate, but often must file a report of proposed construction with the state utilities or energy commission.

***84. The Power Plant and Industrial Fuel Use Act of 1978.*** The Fuel Use Act is a congressional mandate intended to encourage industry to depend less on natural gas or oil products for its energy needs. The act covers four types of facilities: new and existing power plants and new and existing major fuel-burning installations (MFBIs). New facilities, with some exceptions, are prohibited from using natural gas or oil as a primary fuel. Restrictions are less severe for existing facilities until 1990. Existing power plants are prohibited from using natural gas as a primary fuel after January 1, 1990, and are limited to current level of oil consumption until that date. The act may affect cogeneration facilities if they fall within one of the above definitions. However, there are both permanent and temporary exemptions from the restrictions that may be applicable to certain cogenerators.

An important "exemption" is that cogeneration facilities meeting both of the following criteria will not be considered to be new or existing power plants for the purposes of the act. Thus, if a cogeneration facility meets these criteria and is not an MFBI, act restrictions will not apply.

If the facility produces electric power and another form of useful energy, where the electricity constitutes more than 10%, and less than 90%, of the total useful energy output of the facility *and*

If it consumes more than 50% of the net annual electric power that it generates

A variety of permanent and temporary exemptions are available; perhaps the most important are for plants demonstrating energy savings with oil or gas systems, those lacking an adequate supply of coal at a price less than that of imported petroleum, or those located in areas where the plant cannot burn coal without violating state or federal environmental requirements. However, the burden for obtaining an exemption is placed on the operator of the plant, and it is a heavy burden. Among other requirements, the operator must submit a report to the effect that alternative approaches have been "rigorously explored."

***85. Applicability of Fuel Use Act to Cogeneration Projects.*** The final regulations promulgated under the Fuel Use Act define a cogeneration facility as an electric power plant or MFBI that produces electric power and any other form of useful energy, but the electricity must constitute more than 10% and less than 90% of the useful energy output of the facility. In order to avoid regulation of certain cogeneration facilities as electric generating units under the act, an electric generating unit is defined so as not to include a cogeneration facility from which less than 50% of the net annual electric power generation is sold or exchanged for resale. Sales or exchanges among owners of the cogeneration facility and to or with an electric utility for resale to the cogenerating supplier are excluded.

A cogeneration facility may be permanently exempted from the provisions of the Fuel Use Act if it is shown that the economic and other benefits of cogeneration are unobtainable unless petroleum, natural gas, or both are used by demonstrating at least the following minimum criteria.

The oil or gas consumed by the facility will be less (calculated according to the rules set out in the regulations) than that which would be consumed in the absence of the cogeneration facility, *or*

It would be in the public interest to grant an exemption because of special circumstances, e.g., technical innovations or maintaining industry in urban areas, or other reasons approved by the Department of Energy, *and* the use of mixtures is not found to be feasible as required by the regulations

Detailed evidence is required to support an application for an exemption, and any exemption may be denied, even though it meets the eligibility criteria, if it is in the public interest not to grant it. It should be noted that, at this writing, the Congress is considering an amendment to PURPA to provide an automatic exemption from the Fuel Use Act for cogeneration facilities qualifying as QFs under the FERC regulations.

## Environmental Requirements

***86. Clean Air Act.*** The federal Clean Air Act regulates air quality by setting up and requiring the states to implement programs geared to attaining both national ambient air quality standards (NAAQS) and standards for emissions from particular air pollution sources. Within each state, the NAAQS designate areas as either "attainment" or "nonattainment," depending on whether or not the standards established for that area have been achieved.

The act requires that new sources of designated pollutants obtain permits prior to construction. The requirements for obtaining a permit vary depending on whether the source is located in an attainment or nonattainment area. In a nonattainment area, construction of a new source or a major modification to an existing source may be prohibited. Additionally, unless a waiver is obtained, both new sources of pollutants and existing sources with major modifications must comply with new-source performance standards (NSPS).

Air quality control requirements can be fairly complex; their applicability depends on many variables, including plant size, type, location, and major fuels. Most sources covered by state regulations must be permitted, and the permitting process may take from 3 months to more than a year. It is the source owner's responsibility to demonstrate compliance with all applicable permitting, monitoring, and reporting requirements. To minimize time constraints, contact should be made with the regulatory agency as early in the planning process as possible. See

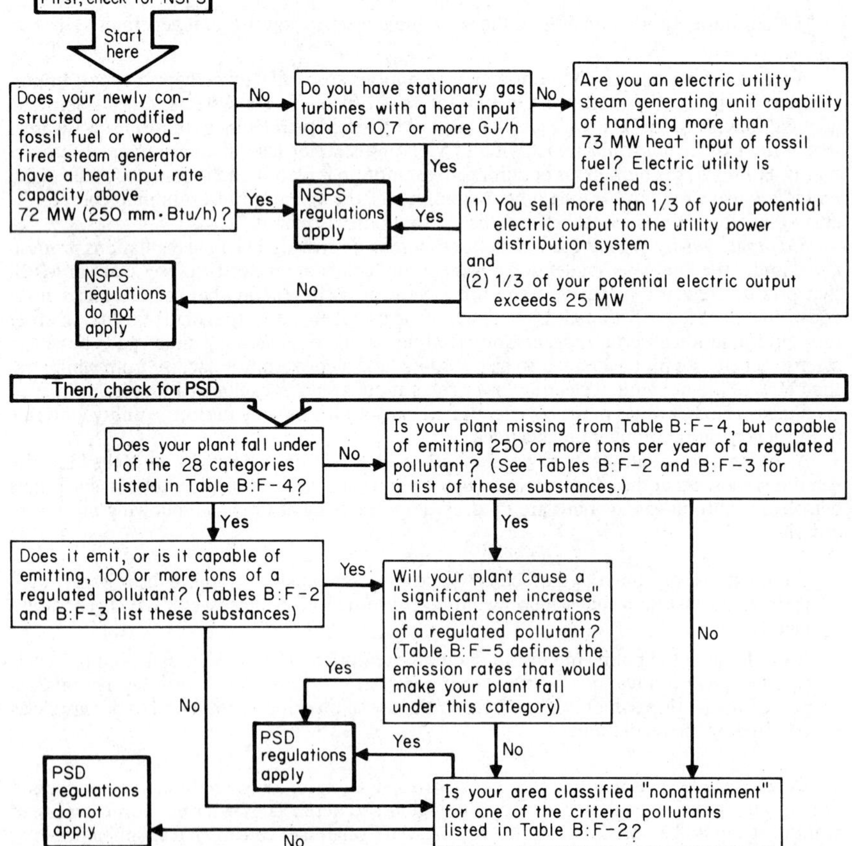


**FIG. 5-86** Applicability of air quality regulations.

Figure 5-86 for the relationships of New Source Performance Standards and the Prevention of Significant Deterioration program discussed below.

***87. NAAQS and Related Emissions Controls.*** The NAAQS mandate minimum air quality levels by establishing maximum ambient concentrations for six substances termed "criteria pollutants" (Table 5-16). State standards for meeting NAAQS requirements incorporate the

**TABLE 5-16** NAAQS Criteria Pollutants

| |
|---|
| Carbon monoxide |
| Nitrogen oxides |
| Particulate matter |
| Sulfur dioxide |
| Ozone (regulated through volatile organic compounds)* |
| Lead |

*Hydrocarbons.

New Source Performance Standards (NSPS) and the Prevention of Significant Deterioration (PSD) program outlined by the Clean Air Act and amendments. In addition, the states regulate open burning, visible emissions, odorous emissions, and noncriteria pollutants (Table 5-17).

**TABLE 5-17** NAAQS Noncriteria Pollutants

| |
|---|
| Asbestos |
| Beryllium |
| Mercury |
| Vinyl chloride |
| Fluorides |
| Sulfuric acid mist |
| Hydrogen sulfide ($H_2S$) |
| Total reduced sulfur (including $H_2S$) |
| Reduced sulfur compounds (including $H_2S$) |

***88. PSD.*** One of the major purposes of the Clean Air Act is to both prevent significant deterioration of the nation's air quality and allow simultaneous industrial growth. Accordingly, every airshed is classified based on prevailing levels of the six criteria pollutants listed in Table 5-16. Areas are termed attainment or nonattainment for each regulated pollutant, with attainment defined as compliance with the appropriate National Ambient Air Quality Standard. Emission standards are then developed to prevent further deterioration of each airshed. Allowable emissions vary with local conditions.

***89. Attainment.*** Allowable emissions in a state's attainment areas depend on the area's categorizations as Class I, Class II, or Class III. Emission limits are the most stringent in Class I areas, since these are considered pristine. Air quality in attainment areas is controlled by the provisions of the PSD section of the Clean Air Act and amendments. Major new sources and modifications in an attainment area are subject to PSD review if these sources "significantly increase" the emission of any of the regulated substances listed in Table 5-17. Under PSD, a source is considered *major* if:

It falls under one of the 28 categories listed in Table 5-18 *and* it emits, or is capable of emitting, 100 or more tons per year of a regulated pollutant; *or*

It is not listed in Table 5-18 but does, or could, emit 250 or more tons per year of a regulated pollutant

**TABLE 5-18** Specific PSD Source Categories

1. Fossil-fuel-fired steam electric plants of more than 250 MBtu/h heat input
2. Coal cleaning plants (with thermal dryers)
3. Kraft pulp mills
4. Portland cement plants
5. Primary zinc smelters
6. Iron and steel mill plants
7. Primary aluminum ore reduction plants
8. Primary copper smelters
9. Municipal incinerators capable of charging more than 250 tons of refuse per day
10. Hydrofluoric acid plants
11. Sulfuric acid plants
12. Nitric acid plants
13. Petroleum refineries
14. Lime plants
15. Phosphate rock processing plants
16. Coke oven batteries
17. Sulfur recovery plants
18. Carbon black plants (furnace process)
19. Primary lead smelters
20. Fuel conversion plants
21. Sintering plants
22. Secondary metal production plants
23. Chemical process plants
24. Fossil-fuel boilers (or combinations thereof) totaling more than 250 MBtu/h heat input
25. Petroleum storage and transfer units with a total storage capacity exceeding 300,000 barrels
26. Taconite ore processing plants
27. Glass fiber processing plants
28. Charcoal production plants

NOTE: 1 Btu/h = 0.293 W; 1 ton = 907.2 kg.

A modification is considered major if it results in a significant net emissions increase of a regulated pollutant. "Significant increase" is defined in Table 5-19.

If a facility falls under PSD review, a specific permitting process must be followed. The permit application will require analysis of best-available control technology (BACT), air quality impacts analysis (including air quality monitoring), and additional impacts analysis for each pollutant released in significant quantities. Statutory requirements for each part of the analysis are explicit and vary for each installation. The permit process itself may take from 3 months to more than a year, depending in part on whether or not sufficient data are available regarding local air quality. If these data are not available, a baseline air quality monitoring program may be required before issuance of any construction permits. A regular emission-monitoring and -reporting schedule may also be required once the facility is in operation.

Emissions of criteria pollutants already causing nonattainment are sharply constrained and fall under much stricter regulations than those found under PSD.

***90. NSPS.*** NSPS apply to new construction, reconstruction, and major modification of facilities meeting specific size, type, and fuel criteria. If a cogeneration plant falls under one of the following categories, NSPS applies.

Fossil-fuel-fired steam generators with a heat input rate above 72 MW (250 MBtu/h)

Fossil-fuel- or wood-fired steam generators capable of firing fuel at an input rate greater than 73 MW

Electric utility steam-generating units capable of handling more than 73 MW heat input of fossil fuel, whether that fuel is used alone or in combination with other sources (with electric utility defined as a steam-generating unit selling more than one-third of its potential electric output *and* more than 25 MW of electricity to a utility power distribution system)

Stationary gas turbines with a heat input load equaling or exceeding 10.7 GJ/h

**TABLE 5-19** Significant Emission Rates*

| Pollutant | Emissions rate, tons/yr |
|---|---|
| Carbon monoxide | 100 |
| Nitrogen oxides | 40 |
| Sulfur dioxide | 40 |
| Particulate matter | 25 |
| Ozone (VOC) | 40 (of VOCs) |
| Lead | 0.6 |
| Asbestos | 0.007 |
| Beryllium | 0.0004 |
| Mercury | 0.1 |
| Vinyl chloride | 1 |
| Fluorides | 3 |
| Sulfuric acid mist | 7 |
| Hydrogen sulfide ($H_2S$) | 10 |
| Total reduced sulfur (including $H_2S$) | 10 |
| Reduced sulfur compounds (including $H_2S$) | 10 |
| Any other pollutant regulated under the Clean Air Act | Any emission rate |
| Each regulated pollutant | Emission rate that causes an air quality impact of 1 pg/m³ or greater (24-h basis) in any Class I area located within 10 km of the source |

* Extracted from 40 CFR 52.21(b)(23).
NOTE: 1 ton = 907.2 kg.

Under NSPS, monitoring and emission requirements vary with the source type and must be determined through consultation with the proper permitting authorities. The technology required by NSPS usually satisfies the BACT provisions of PSD, and the allowable emissions are usually the same. However, if there are excessively high ambient concentrations of a regulated pollutant, PSD's tougher special provisions will supersede those of NSPS. Both programs may require the source owner to regularly monitor and report stack emission levels as well as ambient air quality.

***91. Regulations for Facilities Not Covered by PSD or NSPS.*** Many states regulate particulate and sulfur dioxide emissions from certain boilers too small to be covered by either PSD or NSPS; standards vary with the type of fuel used and the boiler's capacity. Permits are required for operating these sources, but extensive monitoring usually is not.

***92. Federal Water Pollution Control Act.*** Under the federal Water Pollution Control Act, every operator of a new point source must obtain a permit to discharge before beginning construction or operation. Direct dischargers must obtain a national pollutant discharge elimination system (NPDES) permit from the EPA or the state if the EPA has approved the state's permitting program. The act established compliance standards based on the technology of the point source, and regulations promulgated pursuant to the act require compliance with increasingly stringent standards by a series of deadlines. Discharge of oil and hazardous substances is prohibited. Again, the project developer depending upon water, even if for cooling purposes only, must contact the state's water pollution authorities early in the project's development period.

## Requirements for Utility Interconnection Protection

***93. Utility and Cogenerator Concerns.*** Electric utilities have traditionally been concerned with power flows from central station generation plants through the transmission and distribution system to the customer. The assumption of one-way power flow has been built into utility planning, operation, and protection schemes. Now, under PURPA, cogenerators are allowed to feed (and sell) power back to the utility grid. As a result, the traditional one-way power flow assumption may not necessarily be valid where cogenerators are present.

The parallel operation of a generating unit presents several concerns for a utility. These include safety, power quality, reliability, protection of facilities, and planning or operating problems. The utilities must maintain the integrity of their systems in order to ensure a reliable supply of electricity to their customers. Therefore, any interconnected cogeneration system may include some equipment dedicated to protecting the utility system from problems that may originate in the cogeneration system.

The cogenerator may have a slightly different set of concerns, such as safety, interconnection cost, protection of the cogeneration system, power factor cost, and impact on production. However, the cogenerator often depends on the utility to supply a major portion of the facility's electrical power. Therefore, it is in the cogenerator's best interest to help protect the integrity of the utility's system.

***94. Safety Issues.*** Safety concerns focus first on *isolating the generator from the utility when the utility line is opened.* Isolation can be both automatic and manual. When a utility circuit is damaged, such as when a power line falls to the ground in a storm, a hazard to the public exists until the utility and all other power sources are isolated from that circuit. Because speed is critical, this type of isolation must be automatic. The methods for accomplishing automatic isolation may vary, depending on size, electrical characteristics, and other factors of both the generator and the utility circuits. Other situations can require a manually operated switch or set of disconnects (in some cases this should be lockable). For example, when line maintenance is being performed, safety work rules require visible breaks between all energy sources and the circuits being maintained. The utility will need to keep location records or maps on each generator, place identification on utility breakers or switches that connect generation sources to the load side, and other administrative and operating procedures to effectively satisfy work safety rules.

Another problem with disconnecting cogeneration equipment is *self-excitation* of the generators. When an induction generator is isolated from the rest of the grid (because of a downed line or a breaker opening the line), the absence of grid-produced power usually will shut down the generators. However, if there is sufficient capacitance in the nearby circuits (e.g., power-factor-correcting capacitors), the induction generator may continue to operate independently of any power supplied to the grid. The power produced by this isolated self-excited induction generator will not be regulated by the grid, and the customers' electricity-using equipment may be damaged. More importantly, a self-excited, isolated induction generator that reenergizes a downed transmission or distribution line could endanger utility workers. Voltage and frequency relays and automatic disconnect circuit breakers can be used to protect both the utility workers and customer's equipment.

The third safety concern is *preventing the isolated generator from closing on a deenergized utility circuit.* Closure would endanger the public, rescue workers, utility workers, and others who have determined that the circuit was deenergized and were in contact with the circuit. In this situation, closure should be blocked for all automatic and manual isolation schemes.

***95. Reliability Issues.*** The reliability of utility service may be as important to the cogenerator as to the utility itself. The cogenerator is also an electricity customer and desires reliable service. Therefore, the cogenerator should ensure that the generation equipment does not impact the reliability of utility service by causing unnecessary interruption. This consideration may influence the equipment protection schemes (discussed later) and the selection of equipment operating points.

For example, the *appropriate setting of relay trip points* may not be immediately obvious. Many cogenerators would like to set wide limits of operation so that the generator can operate a greater portion of the time and trips (disconnections) would be infrequent. However, the utility may require a greater margin of protection for its facilities in order to maintain service to other customers. Thus, protective relay set points must be coordinated so that the safety and reliability of electrical service are not compromised.

Another consideration that impacts utility system reliability is the *location* of the cogeneration interconnection on the utility system. If the interconnection is at a transmission level, then interruptions of service have greater potential impact than those at a distribution level. The utility may require even more stringent protection requirements of cogenerators connected to the transmission system to protect other customers "downstream" of the cogenerator.

A final consideration is the *generator size* compared to the plant load and the load on the utility line. If the generator kilowatt output capacity is much smaller than the plant load, the impacts of disturbances created by the generator may not be electrically significant to the rest of the plant and to the utility. The more on-site load, the more damping there will be to minimize effects of generator disturbances. The result in this situation might be less stringent protective requirements. However, as generator kilowatt output capacity approaches or exceeds plant load, the impacts of generator disturbances may become significant to both the rest of the plant and the utility as well. This will usually result in more stringent protective requirements since the utility and cogenerator will again seek to protect utility and customer equipment from service disruptions.

***96. Quality of Service Issues.*** Utility customers expect electric power to meet certain tolerances (of voltage, frequency, etc.) so that computers, lights, motors, and other appliances will function efficiently and reliably and not be damaged under normal operating conditions. Power supplied to the grid by an interconnected cogenerator is also expected to be within certain tolerances so that the overall power quality of the utility system remains satisfactory.

One problem may be *low power factor.* Power factor can vary from 0 to 1.0. When it is less than 1.0, it will be either leading (when capacitance dominates) or lagging (when inductance dominates). If leading and lagging power factors of the same value occur on the same system, they cancel and the net power factor is 1.0. This fact allows the use of capacitors to correct a lagging power factor to near 1.0. If too much capacitance is used, the power factor will change from lagging to less than 1.0 in the leading direction. Careful placement of capacitors is necessary for proper power-factor correction.

Low power factor is likely with induction generators and some dc-ac inverters. The results of low power factor can be an increase in power losses, larger voltage drops in supply circuits, and equipment overheating.

The utility may include a power-factor-correction clause in its billing agreement and the cogenerator may ultimately incur additional expenses for low power factor. Capacitors installed nearby are used to improve power factor but may introduce other problems (see below and self-excitation mentioned above). In some cases the utility might require modification of the interconnection scheme before capacitors are installed.

*Harmonic currents* are another problem that may be present. Harmonic currents occur at frequencies other than the desired 60 Hz of the power signal. When transmitted along power transmission and distribution lines, these harmonics distort the power signal. Harmonic currents may cause improper operation of ground-fault and other protective relays, carrier communications systems, motors, sensitive electronic equipment, and computers; they also may cause capacitor overloading. Sources of these harmonics include solid-state switching devices (dc-to-ac inverters, variable-speed drives, ac-to-dc-to-ac converters, etc.), generator winding faults and generator windings pitched for stand-alone operations.

A third problem may be *voltage control* in the areas near cogeneration facilities. For example, voltage variations or "flicker" can occur when there are sudden and large changes of current flowing in a circuit. Some causes of voltage flicker may be generation coming on or off line suddenly or starting generators and motors. Voltage flicker can cause misoperation of sensitive equipment or visual irritation. As another example, resonant overvoltages can typically be two or three times normal voltage for long durations. This condition may occur when a utility circuit is isolated with light loads and there is sufficient generating capacity and power-factor-correcting capacitors (or other sources of capacitive reactance) to cause resonant currents to flow. Voltages of this level can damage utility and customer-owned equipment.

***97. Typical Conditions Requiring Protection.*** Both the cogenerator and the utility need to protect their respective personnel and facilities from injury and damage that could be caused by generator misoperation. This protection often utilizes automatic equipment to detect and isolate the faulty sections of a circuit so that the remainder of the network can continue to deliver power without interruption. The equipment also attempts to protect all electrical equipment from damage, and thereby improves and maintains the reliability and power quality of electric service to the cogenerator and other customers. Listed below are several electrical parameters that the cogenerator may consider monitoring and using in automatic protection schemes.

Both utility-owned and cogeneration-owned electrical equipment (including the generator) is designed to operate within specific *voltage* limits. If the voltage exceeds these limits—either under or over—then the operation of equipment will be affected and damage may occur. When a cogenerator causes or contributes to this problem, it must be isolated from the plant loads and utility service.

Large magnitude, temporary *overcurrents* can occur as a result of grounded wires or other electrical faults. Significant damage can occur before protective equipment responds. Small, steady-state overcurrents, if undetected by protective equipment, can result in overheating of electrical equipment connected to the generator as well as in the generator itself. This also leads to inefficient operation as well as potential damage.

*Ground faults* are typically high-impedance shorts to ground. When this happens, relatively small currents flow to the ground—usually through the metal enclosures covering the generator. This creates a safety hazard as well as potentially damaging the equipment. When a high-impedance fault to ground occurs and does not result in a large overcurrent condition, the ground faults may go undetected unless specific ground-fault protection schemes are used.

*Over-* or *underfrequency* conditions will usually indicate abnormal generator operation or an interruption of utility service. When this condition is sustained, non-60-Hz power will be transmitted on the power lines. This may affect operation of sensitive equipment. Again, automatic protection against this condition will prevent sustained abnormal operation.

*Reverse power flow* occurs when the generator acts like a motor and drives the turbine. In this condition, the generator is receiving current instead of generating current. This could lead to serious damage of the generator and should be protected against.

A *phase imbalance* usually occurs under single-phase-fault or open-circuit conditions. When a phase imbalance occurs, a generator may see unbalanced electromagnetic fields which could lead to transient overvoltages or overcurrents and possible damage.

*Other generator protection* schemes may include temperature and vibration detection. If the generator attempts to operate at levels above specified limits of either of these quantities, the generator might be tripped for its own protection.

The *transformer connections* between the cogenerator (at the generator itself or at plant entrance) and the utility service will play an important part in the overall protection schemes. These connections are crucial in determining appropriate grounding schemes as well.

Finally, for a synchronous generator, *synchronizing* with the utility before actually connecting to it each time prevents overvoltages and potential damage to both the generator and utility equipment. This synchronizing may be done manually, but automatic synchronization may be required on larger units as the margin for error becomes smaller.

***98. Coordination of Protective Devices.*** For each of the areas of concern mentioned above, out-of-limit conditions can be detected by various sensors and relays. The actual protection occurs when a circuit breaker is operated (tripped) to isolate the generator from the rest of the system. The next area of concern becomes the relationship between protective devices so that they work in a coordinated manner to minimize the extent and duration of the resulting outages.

On utility circuits, the duration of an outage is commonly reduced by automatic reclosing of protective devices. This allows short circuits which are temporary in nature, such as electrical arcs formed by lighting, to be extinguished by the protective device and the line reenergized. This may sometimes be accomplished in approximately 0.3 s. If a generator on the circuit continues to supply power to the arc during this process, the arc may not be extinguished and a longer outage will occur.

The extent of outages on a circuit is minimized by opening only the protective devices closest to the circuit segment needing isolation. This is accomplished by calculating the currents that would flow through a series of devices and placing successively more sensitive devices on the circuit as the distance from the power source increases. When the calculated value of short-circuit current flows through two overcurrent protective devices (fuses, relays, etc.), the planned operating point of the devices is such that only the one closest to the short circuit opens. When generators are added to the line, short-circuit currents may now flow from several different sources and the actual current values may vary widely from the expected current values as generation is turned off or on. The result may be the misoperation of protective devices, which needlessly put additional utility customers without electric service. Thus, the presence of cogenerators may require the utility to revise its coordination of protective devices.

***99. Planning and Operations.*** When cogenerators are present, several areas of utility planning may be affected. This impact stems from two reasons: the first is that a dispersed generator exists (as opposed to centralized generators), and the second, related, reason is that the assumption of unidirectional power flow from utility to customers may no longer be valid. These impacts may show up in the following areas.

*Voltage regulation.* A change in the direction of power flow in a circuit may result in excessive steady-state voltages on the circuit. Placement of a large generator at the end of a long supply line may create voltage-regulation problems.

*Load restoration after interruptions.* Systems with multiple sources connect only one source at a time to the circuit. As more dispersed generation is added, extra sectionalizing of loads may be necessary to restore service after an outage. Protective schemes become increasingly complex since each source must be protected from the other sources.

*Sufficient circuit capacity* must be provided to transfer generated power to loads as well as capacity to supply loads when generators are out of service for maintenance or repair.

Utilities will need to account for deviations from their expected *load surveys* and *load factors* caused by dispersed generation. Load surveys are done to determine the electrical demands the utility must meet at the present and in the future. These form the foundation of utility planning. Load factors reflect the utilization of utility equipment (ratio of average usage to peak usage). Both of these are affected by dispersed generation, which is usually not planned for and displaces some electrical demand as well.

The *operations* of both the utility and cogenerator will also be affected by the cogeneration system. First of all the cogenerator must decide how much operating information is needed to keep the system running and how much information is needed for long-term operations and maintenance. This will influence how much control, metering or monitoring, and data collection equipment is needed. These decisions will also influence how much attention and labor power the cogenerator is willing to give to this system.

The impact on utility operations may include the extra records and communications required to work safely on circuits with dispersed generation and the increased time to do repair work. The first of these was covered under safety concerns. The increased time to do repair work, particularly during forced outages, results from the need to isolate dispersed generating sources from the circuit before work can begin. Compliance with this safety rule could extend outage times as the number of sources increases on a given circuit.

In addition, if the cogenerator is very large, the utility may wish to include the system in its dispatch of generating units and retain some control over the cogenerator operation. The control system may use telemetering. Also, very large cogenerators may impact utility system stability (the ability of the utility system to stay synchronized after any disturbance), and the utility may require even more stringent protection requirements or special equipment to maintain stability margins.

***100. Ownership Issues.*** Changes to the utility system may be necessary and they may be expensive. Under PURPA, the cost of these changes must be borne by the cogenerator. Generally, the utility owns the metering equipment and other facilities required to protect its system and charges a monthly fee for operation and maintenance. In some cases, the utility may approve customer-owned protection equipment. Since the utility is ultimately responsible for the quality and safety of the electric grid (as required by the regulatory commissions), it sometimes retains inspection and maintenance rights and may charge for those services. Recent North Carolina Utilities Commission findings on interconnection issues are included under times retains inspection and maintenance rights and may charge for those services.

***References on Cogeneration***

C. F. R. Sections 292 and 500

U.S.C. Sections 1251 *et seq.*, 1311, 1321, 7401 *et seq.*, and 8301–8483

Making Interconnection Work; *Power,* June 1982.

Ringo, M. J., Wilson, W. H., Clapp, A. L., et al: *Cogeneration Project Evaluation Manual;* Research Triangle Park, N.C., North Carolina Alternative Energy Corporation, 1985.

***References on Fossil-fueled Plants***

Singer, Joseph G. (ed.): *Combustion: Fossil Power Systems,* Combustion Engineering, Inc., Windsor, Conn., 1981.

Kiefer, Kinney, and Stuart: *Principles of Engineering* Thermodynamics, John Wiley & Sons, Inc., New York.

# SECTION 6
# PRIME MOVERS

**Donald H. Hall**
*Senior Engineer, Thermal Design and Application, Medium Steam Turbine Department, General Electric Company; Member, American Society of Mechanical Engineers*

**Lawrence R. Mizen**
*Senior Engineer, Control Design and Application, Medium Steam Turbine Department, General Electric Company*

**Roy P. Allen**
*Manager, Advanced Development Engineering, Turbine Technology Department, General Electric Company*

## CONTENTS

*Numbers refer to paragraphs*

# STEAM PRIME MOVERS

By DONALD H. HALL AND LAWRENCE R. MIZEN

## Introduction

***1. Steam Engines and Steam Turbines.*** Steam prime movers are either reciprocating engines or turbines, the former being the older, dominant type until 1900. Reciprocating engines offer low speed (100 to 400 r/min), high efficiency in small sizes (less than 500 hp), and high starting torque, and are almost foolproof. In the Industrial Revolution they powered mills and steam locomotives. Steam turbines are a product of the twentieth century and have established a wide usefulness as prime movers. They completely dominate the field of power generation and are a major prime mover for variable-speed applications in ship propulsion (through gears), centrifugal pumps, compressors, and blowers. Single steam turbines can be built in greater capacities (over 1,000,000 kW) than any other prime mover. Turbines offer high speeds (1800 to 25,000 r/min) and high efficiencies (over 85% in larger units); require minimum floor space with relatively low weight; need no internal lubrication; and operate at high steam pressures [5000 lb/in² (gage)], high steam temperatures (1050°F), and low vacuums [0.5 inHg (abs.)]. Steam turbines have no reciprocating mass (with resulting vibrations) nor parts subject to friction wear (except bearings) and consequently provide very high reliability at low maintenance costs.

***2. Steam-Engine Types and Application.*** The former great diversity in engine types has been reduced so that (1) simple D-slide engines (less than 0.100 hp) are used for auxiliary drive and (2) single-cylinder counterflow and uniflow engines (less than 1000 hp), with Corliss or poppet-type valve gear, are used for generator or equipment drive in factories, office buildings, paper mills, hospitals, laundries, and process applications (where noncondensing by-product power operations prevail). Multiple-expansion, multicylinder constructions are largely obsolete except for some marine applications. Although engines as large as 7500 kW have been built and are still found in service, the field is generally limited to engines less than 500 kW in size. Engine governing is by flyball or flywheel types to (1) throttle steam supply or (2) vary cutoff.

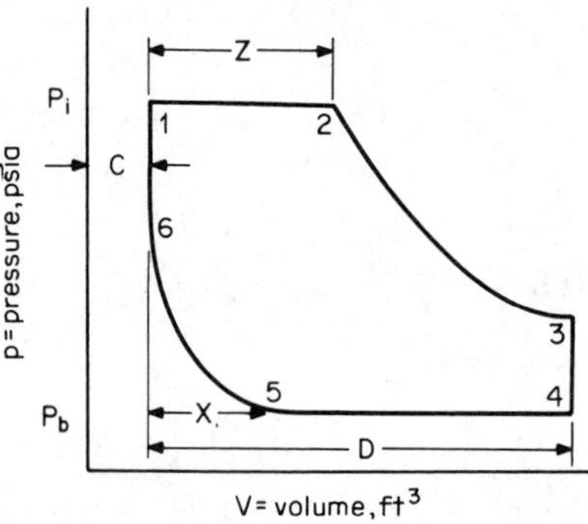


**FIG. 6-1** Pressure-volume diagram for a steam-engine cycle. Phase 1-2, constant-pressure admission at $p_i$; 2-3, expansion, $pv = C$; 3-4, release; 4-5, constant-pressure exhaust pipe at $p_b$; 5-6, compression, $pv = C$; 6-1, constant-volume admission.

***3. Steam-Engine Performance.*** The basic thermodynamic cycle is shown in Fig. 6-1. The net work of the cycle is represented by the area enclosed within the diagram and is represented by the mean effective pressure (mep), that is, the net work (area) divided by the length of the diagram. The power output is computed by the "plan" equation, viz.,

$$\text{hp} = p_m Lan/33{,}000$$

where hp = horsepower; $p_m$ = mep, pounds per square inch; $L$ = length of stroke, feet; $a$ = net piston area, square inches; and $n$ = number of cycles completed per minute.

The theoretical mep and horsepower are larger than the actual indicated values and are customarily related by a diagram factor ranging between 0.5 and 0.95. The shaft or brake mep and horsepower are lower still, with mechanical efficiency ranging between 0.8 and

| Where | Usual value |
|---|---|
| $D$ = displacement | 0.05–20 ft² |
| $C$ = clearance, | 0.03–0.2 |
| $Z$ = cutoff, fraction of $D$ | 0.1–0.6 |
| $X$ = compression, fraction of $D$ | 0.1–0.8 |
| $p_i$ = initial pressure | 100–300 lb/in² (abs.) |
| $p_b$ = back pressure | 2–30 lb/in² (abs.) |
| $p_m$ = mean effective pressure | 50–125 lb/in² (abs.) |

0.95. Representative actual water rates (pounds per kilowatthour) and Willans' lines (pounds steam per hour) are shown in Fig. 6-2.

**4. *Steam Turbines—General***

**a.** *Expansion of Steam Through Nozzles and Buckets.* Basically steam turbines are a series of calibrated nozzles through which heat energy is converted into kinetic energy which, in turn, is transferred to wheels or drums and delivered at the end of a rotating shaft as usable power.

**b.** *Impulse, Reaction, and Curtis Staging.* Turbines are built in two distinct types: (1) impulse and (2) reaction. *Impulse turbines* have stationary nozzles, and the total stage pressure drop is taken across them. The kinetic energy generated is absorbed by the rotating buckets at essentially constant static pressure (Fig. 6-3*A* and *B*). Increased pressure drop can be efficiently utilized in a single stage (at constant wheel speed) by adding a row of turning vanes or "intermediates" which are followed by a second row of buckets. This is commonly called a *Curtis* or 2-row stage (Fig. 6-3*C*).

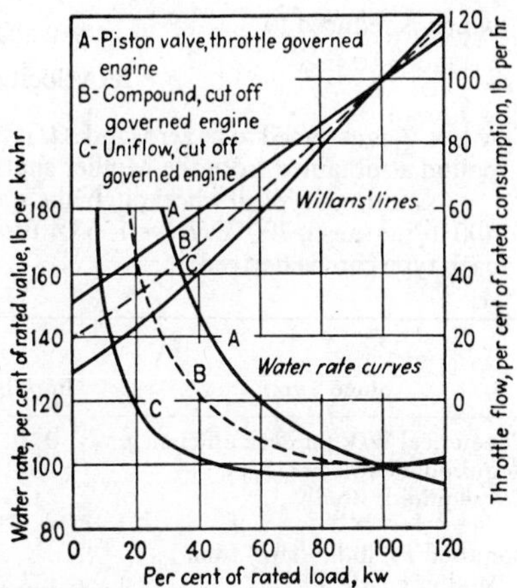


FIG. 6-2 Water-rate curves and Willans' lines for typical steam engines.

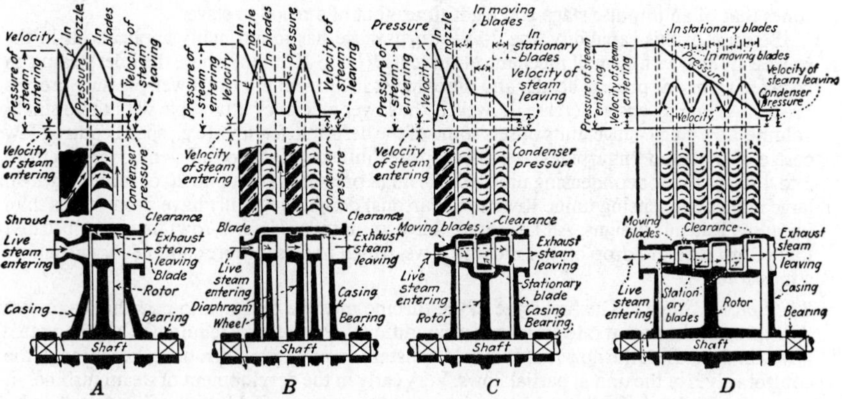


FIG. 6-3 Basic steam-turbine types. *(A)* Simple impulse turbine; *(B)* pressure-staged (Rateau) impulse turbine; *(C)* velocity-staged (Curtis) impulse turbine; *(D)* reaction (Parsons) turbine.

In the *reaction* design both the stationary and rotating parts contain nozzles, and an approximately equal pressure drop is taken across each (Fig. 6-3*D*). The pressure drop across the rotating parts of reaction-design turbines requires full circumferential admission and much closer leakage control.

To illustrate the variations in energy-absorbing capacities of an impulse stage, a 2-row impulse stage, and a reaction stage, one must start with the general energy equation as applied to a nozzle.

$$\frac{V_1^2}{2gJ} + H_1 = \frac{V_2^2}{2gJ} + H_2 \tag{6-2}$$

which is reduced to

$$\text{Jet velocity, ft/s} = 223.7\sqrt{\Delta H} \qquad (6\text{-}3)$$

where $V_1$ is assumed to be zero, and $\Delta H$ is the enthalpy drop (isentropic expansion) in Btu per pound as obtained from the Mollier chart for steam (Fig. 6-4).

Assuming a typical wheel pitch line speed ($W$) of 550 ft/s and initial steam conditions of 400 lb/in² (abs.), 700°F ($H_1$ = 1363.4 Btu/lb), the optimum energy-absorbing capacities of each type can be derived.

| Stage Type | Impulse | 2-row wheel | Reaction Sta. | Reaction Rot. | Combined |
|---|---|---|---|---|---|
| Theoretical $W/V$ for peak efficiency | 0.50 | 0.25 | — | — | 0.707 |
| Required jet velocity, ft/s | 1100 | 2200 | 550 | 550 | — |
| $\Delta H$ required, Btu/lb | 24.2 | 96.6 | 6.045 | 6.045 | 12.09 |
| Required $P_2$, lb/in² (abs.) from Mollier chart | 327 | 168 | 380.6 | 361.9 | 361.9 |
| Stage $P_1/P_2$ | 1.224 | 2.38 | 1.052 | 1.105 | 1.105 |

NOTE: 1 ft/s = 0.3048 m/s; 1 lb/in² = 0.06895 bar; 1 Btu/lb = 2.326 kJ/kg.

This example illustrates that the energy-absorbing capability of the Curtis stage is four times that of an impulse stage and eight times that of a reaction stage.

Because of this capability, the 2-row Curtis stage has found many applications in the process industries for small mechanical-drive use (up to 1000 hp) where the inlet steam can be taken from one process header and the exhaust steam sent out to a lower-pressure process header. As energy costs increase, however, the lower efficiency attainable with these small-volume-flow single-stage units offsets some of the desirable features (e.g., speed control, low cost, etc.). All modern turbines over 1000 hp are multistage for good efficiency, varying from 3 to 4 stages on noncondensing units with a small pressure ratio up to 20 or more stages on large reheat condensing units. Reaction (Parsons) designs generally have more stages than impulse (Rateau) designs. All large units have an impulse (1- or 2-row) first stage because there is no pressure drop on the moving rows, which makes it more suitable for partial-arc admission.

c. *The Control Stage.* The first stage of the turbine must be designed to pass the maximum flow through the unit at rated inlet steam conditions. The pressure required at less than rated flow will decrease if the nozzle area is held constant, resulting in a throttling loss through the control valves of the unit at partial flows. Very early in the development of steam turbines, it was recognized that if full throttle pressure could be made available to the first-stage nozzles across the load range, the maximum isentropic energy that would be available for work and overall efficiency would be increased at part load. Most first stages now use sectionalized first-stage nozzle plates (Fig. 6-5) with 4, 6, or 8 separate ports (depending on steam conditions, unit size, and manufacturer).

The flow to each port is controlled by its own valve, and the valves are opened sequentially. As each valve is opened to its governing point, the full throttle pressure (minus stop-valve and control-valve pressure loss) becomes available to the arc of nozzles fed by that valve. The overall result is a greater availability of energy to do work.

The efficiency of the control stage is poor at partial load because too much energy is released to the first stage in relation to the wheel speed available. But there is still a considerable increase in the total used energy of the turbine in contrast to the throttle-controlled unit in which full throttle pressure is not available to the stages at any flow below maximum. The choice of simple impulse or velocity-compounded (2-row) control stage depends upon the application and the manufacturer. The single-row impulse-control stage is used on large base-loaded reheat turbines for best efficiency. Small and medium-size nonreheat turbines generally use 2-row impulse-control stages to maintain efficiency over a wider range of operating loads. Noncondensing turbines are built with either 1-row or 2-row wheel control stages. (The 1-row wheel is more likely to be used where the total energy of the turbine is small, because of the higher efficiency level obtained.) Mechanical drive and marine propulsion-ahead elements which operate at high rpm generally use 1-row control stages but the reversing element of marine turbines usually consists of only a 2- or 3-row wheel to provide the required torque while limiting the windage loss when operating ahead.

**d.** *Steam-Path Design.* Condensing-turbine sizes increase with the development of longer last-stage buckets and, consequently, the last-stage dimensions (length and diameter) are the first to be determined; these dimensions fix the diameter of the L-1 stage and the optimum energy (pressure drop) which can be placed on that stage. This stage in turn defines the parameters of the L-2 stage and so on up to the first stage, and it can be said that steam paths of turbines are designed backwards except for the first stage. In the 1970s, the largest-capacity single-flow condensing turbine was approximately 120,000 kW. Larger ratings are obtained by multiplying the number of exhaust stages (usually the last 5 to 7 stages are involved) by 2, 4, 6, or 8 times to satisfy the rating requirements. This practice is limited to the larger blades to round out a product line to well over 1,000,000 kW.

P-V-T Values for Dry Saturated Steam

| Pressure, in.Hg abs | Temp., °F | Sp. vol., ft³/lb |
|---|---|---|
| 0.5 | 58.8 | 1250 |
| 1.0 | 79.0 | 652 |
| 1.5 | 91.7 | 445 |
| 2.0 | 101.1 | 339 |
| 3.0 | 115.1 | 232 |
| **Pressure, lb/in² (abs.)** | | |
| 14.7 | 212.0 | 26.8 |
| 50 | 281.0 | 8.52 |
| 100 | 327.8 | 4.43 |
| 200 | 381.8 | 2.29 |
| 300 | 417.3 | 1.543 |
| 400 | 444.6 | 1.161 |
| 600 | 486.2 | 0.770 |
| 900 | 532.0 | 0.501 |
| 1200 | 567.2 | 0.362 |
| 1800 | 621.0 | 0.218 |
| 2400 | 662.0 | 0.141 |
| 3206 | 705.4 | 0.050 |

NOTE: 1 in = 25.4 mm; $t_{°C} = (t_{°F} - 32)/1.8$; 1 ft³/lb = 0.0624 m³/kg.

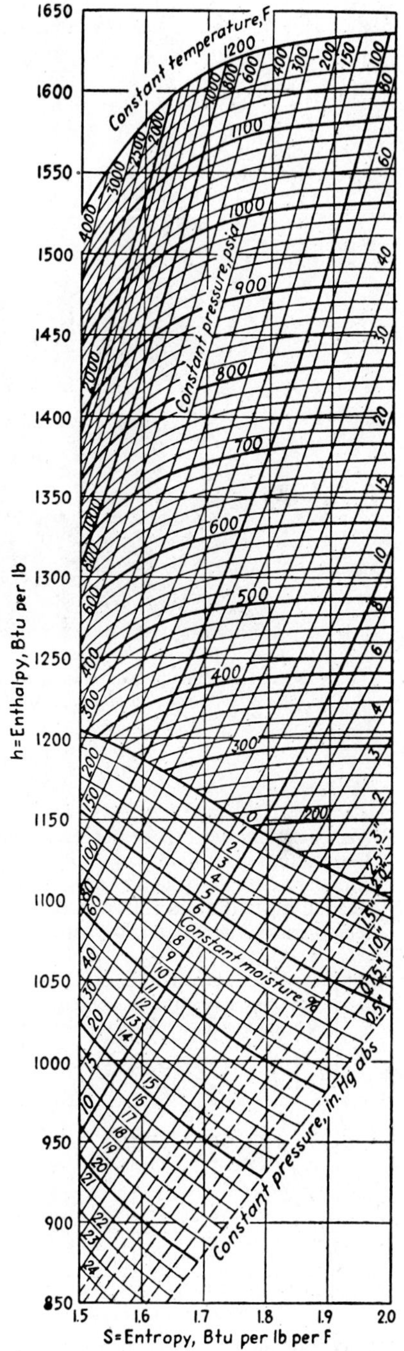


**FIG. 6-4** Mollier chart for steam. *(Adapted from Power Handbook, July, 1951, p. 93. Recommended publication for complete properties of steam, ASME Steam Tables, 1967.)*

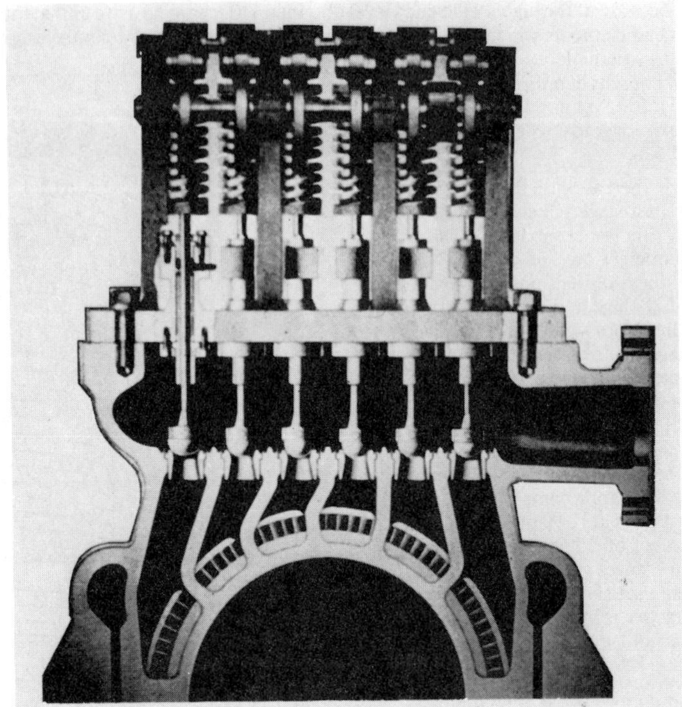

FIG. 6-5 High-pressure casing with first-stage nozzle control. (*From GE Publ. GEA-8510, Steam Turbine Generators for Industrial Applications, p.* E6.)

### 5. Turbine Efficiency

**a.** *Nozzle and Bucket.* The turbine stage efficiency is defined as the actual energy delivered to the rotating blades divided by the ideal energy released to the stage in an isentropic expansion from $P_1$ to $P_2$ of the stage. The most important factors determining the stage efficiency are the relationship of the mean blade speed to the theoretical steam velocity, the aspect ratio (blade length/passage width), and the aerodynamic shape of the passages. Figure 6-6 describes the typical variation in nozzle and bucket efficiencies with velocity ratio and nozzle height.

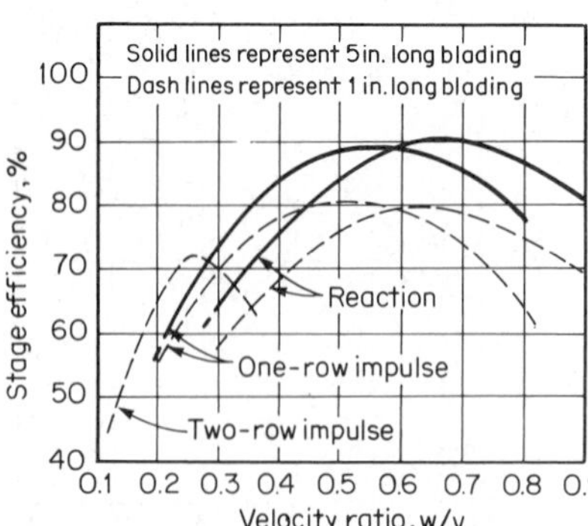


FIG. 6-6 Approximate relative efficiencies of turbine stage types.

**b.** *Losses. Clearance Leakage.* A 100% efficiency cannot be obtained because of friction in the blading and clearance between the stationary and rotating parts, and because the nozzle angle cannot be zero degrees. Axial clearance increases in the stages further from the thrust bearing to satisfy the need to maintain a minimum clearance at extreme operating conditions when the differential expansion between the light rotor and heavy casing is at its worst. To reduce this leakage, radial spill-bands are used. These thin, metal-strip seals may be attached to the diaphragm or casing and extend close to the shroud bands covering the rotating blades. This clearance can be kept quite close (0.020 to 0.060 in), and axial changes in the rotor position do not affect the clearance since the spillbands ride over the shrouds. The need to control the clearance leakage area is especially important on reaction stages with small blade heights because of the pressure drop across the moving blades.

*Nozzle Leakage.* Leakage around the nozzles between the bore of the blade ring or nozzle diaphragm and the drum or rotor must be kept to a minimum. This leakage is controlled through the use of a metallic labyrinth packing which consists of a single ring with multiple teeth arranged to change the direction of the steam as well as to minimize the leakage area. Labyrinth packings are also used at the shaft ends to step the pressure down at the high-pressure end and to seal the shaft at the vacuum end.

*Rotation Loss.* Rotation losses of the rotor consist of losses due to the rotation of the disks, the blades, and shrouds. Partial-arc impulse stages have a greater windage loss within the idle buckets. Rotation losses vary directly with the steam density, the fifth power of the pitch diameter, and the third power of the rpm. In general, the windage loss amounts to less than 1% of stage output at normal rated output. At no-load conditions windage loss for noncondensing turbines approximates 1.5% of the rating per 100 lb/in$^2$ exhaust pressure, and on condensing units approximates from 0.4 to 1.0% of the rating at 1.5 inHg (abs.) exhaust pressure.

*Carryover Loss.* A carryover loss (about 3%) occurs on certain stages when the kinetic energy of the steam leaving the rotating blades cannot be recovered by the following stage because of a difference in stage diameters or a large axial space between adjacent stages. Typically this happens in control stages and in the last stages of noncondensing sections. The last stages of condensing turbines have the largest carryover losses (normally referred to as exhaust loss) because of the large variations in exhaust volumetric flow with exhaust pressure and the large variation of stage pressure ratio with load. Stages preceding the last operate with essentially a constant pressure ratio down to very low loads and consequently can be designed for peak efficiency at a wide range of loads.

*Leaving Loss.* Condensing turbines are frequently "frame sized" by last-stage blade height. It is sometimes economical to size the unit with exhaust loss equal to 5% deterioration in overall turbine performance at the design point [valves wide-open throttle flow and 1.5 inHg (abs.) exhaust pressure] when the normal expected exhaust pressure will be higher or the unit will be operating at part load a large part of the time.

*Nozzle End Loss, Partial Arc.* Control stages and partial-arc impulse stages are subject to end losses at the interface of the active and inactive portions of the blading as the stagnant steam within the idle bucket passages enters the active arc of nozzles and must be accelerated. There is also a greater turbulence in the steam jet at both ends of the active arc. In partial-arc impulse stages the increase in efficiency due to larger blade heights (aspect ratio) is partially offset by increased rotation and end losses, and there is an optimum to this proportioning beyond which there is an overall loss.

*Supersaturation and Moisture Loss.* Moisture in the steam causes supersaturation and moisture losses in the stage. The acceleration of the moisture particles is less than that of the steam, causing a momentum loss as the steam strikes the particles. The moisture particles enter the moving blades (buckets) at a negative velocity relative to the blades, resulting in a braking force on the back of the blades. Supersaturation is a temporary state of supercooling as the steam is rapidly expanded from a superheated state to the wet region before any condensation has begun. The density is greater than when in equilibrium, resulting in a lower velocity as the steam leaves the nozzle. As soon as some condensation occurs at approximately 3.5% moisture, according to Yellot, a state of equilibrium is almost instantly achieved and supersaturation ceases.

c. *Turbine Efficiency.* The internal used energy of the stage is obtained by multiplying the isentropic energy available to the stage by the stage efficiency. The sum of the used energies of all stages in the turbine represents the total used energy of the turbine. The internal efficiency of the turbine can be obtained by dividing the total used energy by the overall isentropic available energy from throttle pressure and temperature conditions to the exhaust pressure. (NOTE: The sum of the available energies of the stages is greater than the overall available energy and represents the reheat factor or gain attributable to the unused energy of preceding stages becoming available to following stages.) The use of overall available energy will automatically account for pressure-drop losses occurring in stop valves, control valves, exhaust hood, and piping between HP and LP elements. Other losses which must be accounted for to arrive at the turbine overall efficiency include valve-stem and shaft-end packing leakages and bearing and oil pump losses. Determination of the overall efficiency of a turbine and its driven equipment must take into account the losses of gears or generators and their bearings as well.

***6. Turbine Construction.*** Prior to World War I, vertical-shaft steam turbines were favored and widely used. Since that time, however, horizontal-shaft units have been universally applied. Horizontal units may be single-shaft or double-shaft, with single, double, or triple steam cylinders on one shaft. These modern units may be throttle or multiple-nozzle governed, have one or more steam extraction points, and exhibit innumerable variations in construction. Figure 6-7 shows several of the more commonly used types of turbines in schematic cross sections.

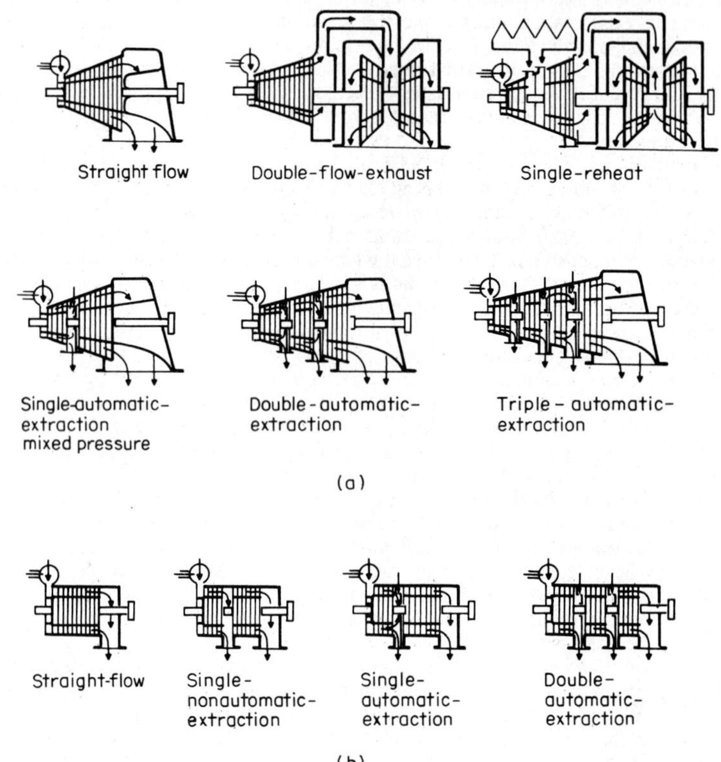


**FIG. 6-7** *(a)* Condensing turbines (exhaust at back pressures less than atmospheric); *(b)* noncondensing turbines (wide range of back pressures). *(Steam Turbines, Power Special Report, GER-430A, June 1962.)*

Steam turbines may be classified into several broad categories, according to the basic purpose and design of the steam path: (1) straight condensing, (2) straight noncondensing, (3) uncontrolled extraction, (4) single, double, or triple controlled extraction, and (5) reheat. Various combinations of these features may be present in a typical unit, and occasionally unusual variations on the above types may be seen.

Figure 6-8 is a cross section of a modern automatic-extraction turbine, showing the details of construction. A steam turbine consists of the following basic parts: (1 to 3) steam path made up of rotating and stationary blading (buckets and nozzles); (4) casing to contain the stationary parts and act as a steam pressure vessel; (5 to 8) controlling and protective valves, piping, and

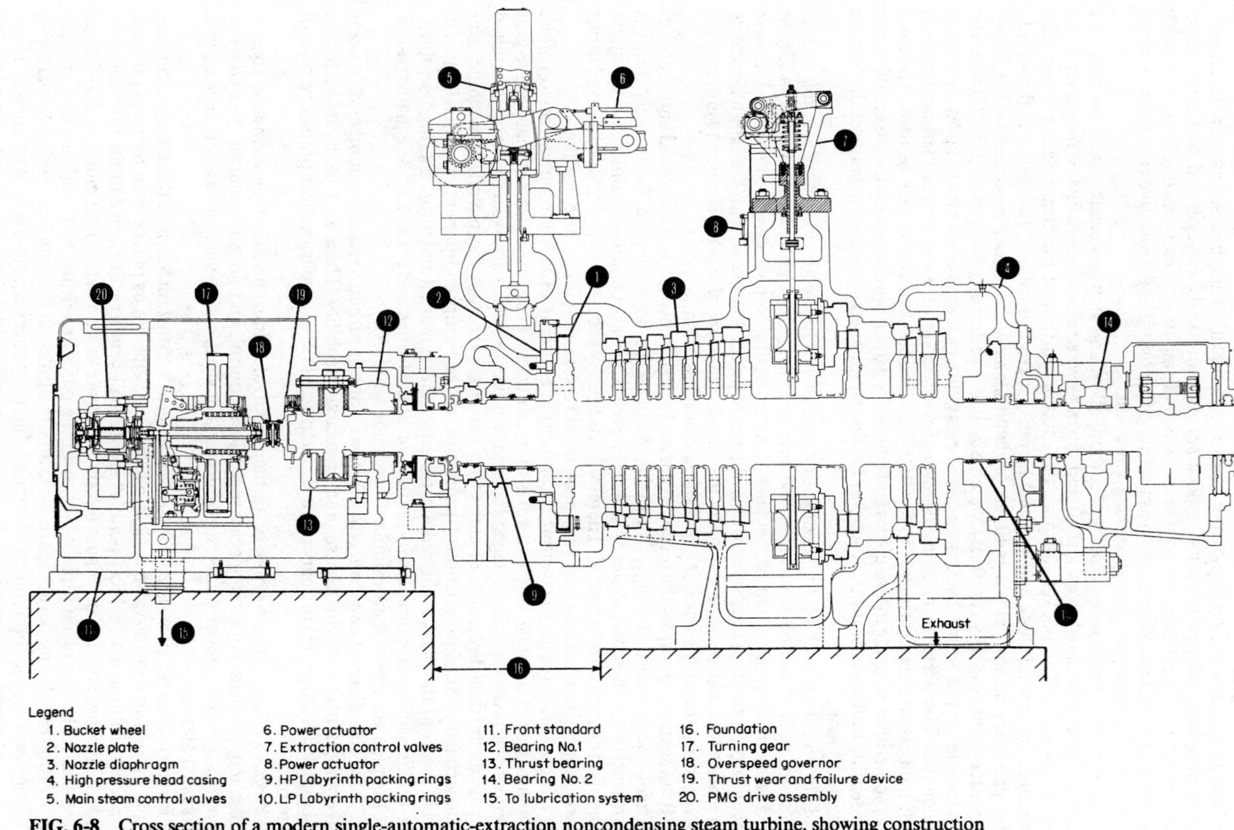


**FIG. 6-8** Cross section of a modern single-automatic-extraction noncondensing steam turbine, showing construction details. *(Medium Steam Turbine Department, General Electric Company.)*

associated components to accept and control the steam admitted to the steam path; (9 and 10) packing and sealing arrangement to prevent steam from escaping into the surrounding area; (11) front standard which houses lubrication, control, and protective equipment and supports part of the casing; (12 to 14) set of journal and thrust bearings to support the rotating elements and absorb all static and dynamic rotor loads; (15) lubrication and hydraulic system for supplying bearing lubrication and (when applicable) generator seals, control, and protective oil requirements; (16) supporting foundation on which the major stationary parts rest; and (17 to 20) various accessory components, such as turning gear, control and protective components, drain valves, etc., as required by the specific application.

Turbines are constructed chiefly of carbon, alloy, and stainless steels. The rotor may be a single forging, fabricated from a shaft and separate wheels, or constructed of forged elements welded together. The buckets forming the rotating portion of the steam path are generally machined from solid stock and attached by pins, or grooves called "dovetails," to the wheels. The stationary steam path is built up of diaphragms with nozzles mounted in the heavy, two-piece casing (usually cast steel), which is bolted together on a horizontal joint. If the unit is condensing or has a low back pressure, the exhaust casing may be made up as a separate assembly and bolted to the main casing through a vertical joint. To minimize thermal stresses in high-temperature applications (950 to 1050°F), a double shell casing may be used. The rotor may sometimes be made in two pieces and coupled together, particularly in the case of the larger condensing double-flow units. A solid or flexible coupling may be used to connect the turbine rotor to its load.

Labyrinth-type packing rings, consisting of high and low teeth, are arranged at the ends of the steam path to inhibit steam from escaping into the surrounding area. (Similar packing is used at each diaphragm, and particularly at stages having control valves, to prevent excessive leakage from one stage to another within the steam path.) Associated with the external packing is a seal system which draws a vacuum to exhaust a mixture of leaking steam and air and thus prevents any steam from leaking into the surrounding room.

Bearings for supporting the turbine rotor are located in pedestals at either end and consist of journals and a thrust assembly. Normally, when steam flows through the turbine, thrust is developed in the direction of steam flow. However, unusual operating conditions or configurations often cause a thrust reversal. Therefore, it is usually necessary to provide an "active" thrust bearing for normal loading and an "inactive" thrust bearing for reverse loading.

The control-valve gear-activating equipment in a turbine usually is mounted on top of the turbine casing at the stage where steam is to be admitted. There are at least as many valve-gear assemblies as there are control stages, sometimes more if a lower valve-gear assembly is required for passing the flow. Protective or emergency valves are generally located off the machine, near the associated steam piping. Control, protective, and accessory components are often located in part in the front standard, at the pedestal housing the first journal and thrust bearing assemblies. The unit is usually supported on its foundation at the front standard and at the exhaust casing. The coupled generator shares similar foundation supports.

***7. Turbine Control and Protective Systems.*** Steam turbines require a number of systems and components to provide control and protective capability. These may be divided into two functional categories: (1) primary control systems and (2) secondary and/or protective control systems.

*Primary control systems* may be further subdivided into the following elements: control valves and associated operating gear, speed/load control, and pressure control. *Secondary or protective systems* consist of overspeed limiting devices, emergency valves, trip devices, and associated alarm devices.

*Primary Control Systems. Control-valve gear.* Most modern turbine-generators use steam-admission control-valve designs which are as efficient as practical, in terms of pressure drop and throttling losses. Most popular are the ball-venturi valves used in the inlet stages of modern high-pressure units. (Figure 6-5 shows a cross section of a typical multiple ball-venturi valve gear.) The use of multiple valves, with the efficient venturi seat configuration and the tight-seating ball valves, permits partial-arc nozzle admission to the turbine with good part-load efficiency and a sequential opening action which produces nearly linear flow curves. These valves may be opened by one of two basic means: bar lift, with valves sequenced by stem lengths, or cam-operated by levers and rollers, to linearize the inherently nonlinear flow characteristics of ball-venturi valves. A common variation of the ball-venturi valve gear once widely used is the poppet-valve gear, with beveled valves and seats. This arrangement is not as efficient as a ball-venturi valve gear and is not presently very popular.

Another commonly employed valve gear, particularly on lower-pressure high-volume-flow applications, is the double-seated spool valve gear. This valve is not very efficient but passes high-volume flow and can be programmed to open in a manner similar to a ball-venturi valve.

A third scheme used somewhat in the past, but now less popular, is the grid valve, which consists of two plates with specially shaped holes arranged so that when one is rotated relative to the other, the flow area is developed as the holes coincide. High volume flow and short physical span are the grid valve's strong points, but its efficiency and accuracy are not good, and operating forces are high.

*Speed/load control systems.* After a choice has been made from the types of valves available for steam admission to the turbine, a means must be provided for positioning these valves to obtain basic speed/load control. The primary requirement is the maintenance of an accurate, predetermined rotating speed, since all turbines are designed to operate at a specific speed or over a specific range of speeds. Every turbine, therefore, has some type of speed governor or, more generally, "speed/load control system." Its purpose is to maintain a relationship between actual turbine speed and some reference value, over a wide range of load torques.

Land turbines used for power generation generally operate at a specific rated speed whereas marine and mechanical-drive turbines, because of the inherent coupling characteristics between rotating blades and fluids, operate over a range of speeds.

In most of the Western Hemisphere, the accepted operating frequency for turbine-generator machinery is 60 Hz. Such units having a 2-pole generator must, therefore, operate at 60 r/s, or 3600 r/min. A 4-pole unit operating at 60 Hz will rotate at half the speed of a 2-pole unit, or 1800 r/min. Most units in the United States operate at either 1800 or 3600 r/min. In most of the remainder of the world, the accepted electric frequency is 50 Hz, with common operating speeds of 1500 or 3000 r/min. In order to understand steam-turbine speed/load control, it is helpful to consider the example of constant-speed/load-based utility or industrial units.

Figure 6-9 shows the relationship designed into the speed/load control system of most such units built in the United States. The commonly accepted speed "droop," with load, for such a system is (−)5% for 100% load change, based on a given speed/load reference setting. Note that speed droops proportionally with increasing load, on a steady-state basis. The system should be so designed that a steady-state error in speed is required to provide the command signal to move the control valve gear to accept the required load.

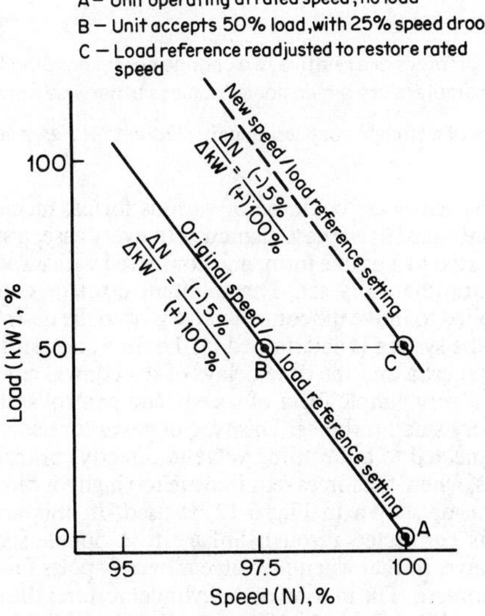


**FIG. 6-9** Steady-state speed/load regulation for a given reference setting in the speed/load system of a steam turbine. *(Medium Steam Turbine Department, General Electric Company.)*

If such a unit operates independently, the speed/load characteristic will be as shown by the solid line of Fig. 6-9 (point *A*). If the unit is tied to a system much larger than itself, and the same system load change occurs, obviously the effect on the unit will be much less, and speed will not vary as much. The system is said to be "stiff" compared with the unit. Since speed accuracy is very important if the operating unit is isolated, any speed droop experienced with a load change (point *B*) must be corrected by changing the speed/load reference setting. This is illustrated by the dashed line of Fig. 6-9, where a 50% load change was followed by a reference correction to restore rated speed (point *C*). If an operating unit is tied to a "stiff" system, and it must accept more of the system load, a similar adjustment will cause it to pick up load with no change in speed, as the dashed line shows.

Manual speed/load reset, therefore, permits a unit, whether isolated or tied to a system, to be set to hold speed, or carry load, as the operator desires. However, if such a unit is to operate for long periods of time, and under varying load conditions, manual load reset is an inadequate solution to the problem of maintaining speed accuracy. In such cases, the use of an automatic reset device or "speed corrector" to provide isochronous control is common.

Figure 6-10 is a generalized block diagram for a speed/load control system containing all the

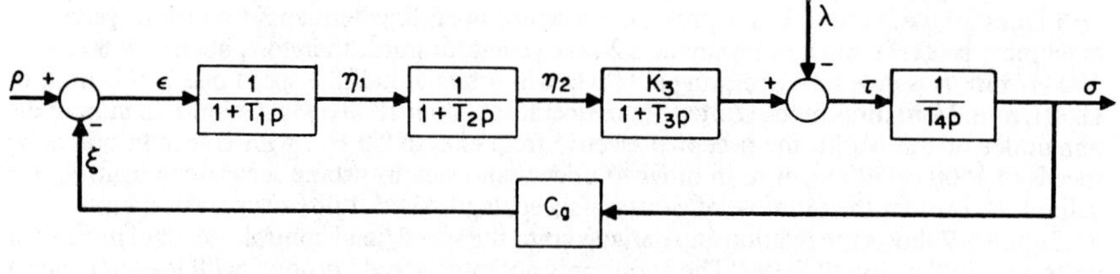


$C_g$ – normalized speed governor constant
$K_3$ – gain of values at no load point
$p$ – differential operator d( )/dt, $s^{-1}$
$T_1$ – speed relay time constant, s
$T_2$ – servomotor time constant, s
$T_3$ – steam bowl time constant, s
$T_4$ – characteristic time of turbine, s
$\epsilon$ – speed error signal
$\lambda$ – relative load change
$\eta_1$ – relative change of speed relay position
$\eta_2$ – relative change of servomotor position
$\rho$ – relative change of reference position
$\tau$ – relative torque change on turbine shaft
$\sigma$ – relative speed change
$\xi$ – relative change of speed governor stroke

Note: All parameters are relative to a change corresponding to a 5% speed change. All parameters are dimensionless unless otherwise noted.

**FIG. 6-10** Block diagram of a straight condensing unit. *(From ASM Paper 60-WA-34.)*

elements discussed. The hardware could be of various forms: mechanical, hydraulic, pneumatic, electrical, or combinations of these elements. In every case, a speed signal is taken from the turbine shaft, converted to a usable form, and compared with a speed/load reference which may be manually or automatically set. The resultant error or command signal is usually amplified and then applied to move the control-valve gear to the desired position. Note that the frequency response of the system is determined by the time constants of the various elements, most notably the rotor inertia and the time delays of the control components.

Figure 6-11 shows a very simple form of speed/load control system: a mechanical speed governor suitable for very small turbines. This type of governor uses a spring-load mechanical flyball mechanism connected to a throttling valve to directly control steam admission to the turbine. On larger units, where the forces required are too high for direct operation, a hydraulic relay governing system, as shown in Fig. 6-12, is used. In this arrangement a centrifugal flyball-type governor is connected through linkage to a double-spooled pilot valve. Oil is admitted to the pilot valve, so that when the valve moves, it ports fluid either into or out of an operating cylinder as required. The motion of the cylinder restores the pilot valve, through other linkage, to maintain a stable relationship between the pilot valve and its cylinder. Available force for operating the control valves is multiplied many times with this arrangement. On units larger than about 1000 kW, a mechanical hydraulic control system having two or more such

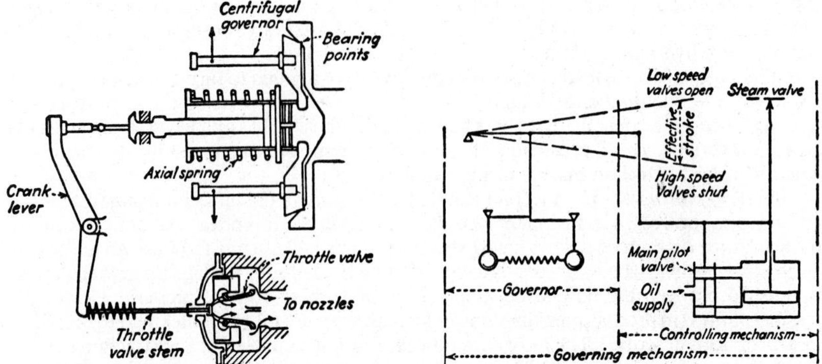


**FIG. 6-11** Mechanical governor for small turbines.

**FIG. 6-12** Governor with hydraulic power amplifier.

hydraulic relays or amplifiers is used to multiply available force and to operate multiple control-valve gear systems.

In the early 1960s, a new electrohydraulic control system made its appearance, offering greater accuracy, higher operating forces, remote and centralized control capability, and more options and flexibility than any previous system. Shown applied to speed/load control in Fig. 6-13, it consists of (1) a permanent magnet generator or digital-type reluctance pickup to provide a shaft-speed signal, (2) electronic circuitry for comparing the speed signal with a reference signal, (3) a high-gain servo valve to convert the resulting electric signal to a hydraulic signal, (4) a valve-gear power-actuator assembly capable of operating on high-pressure hydrau-

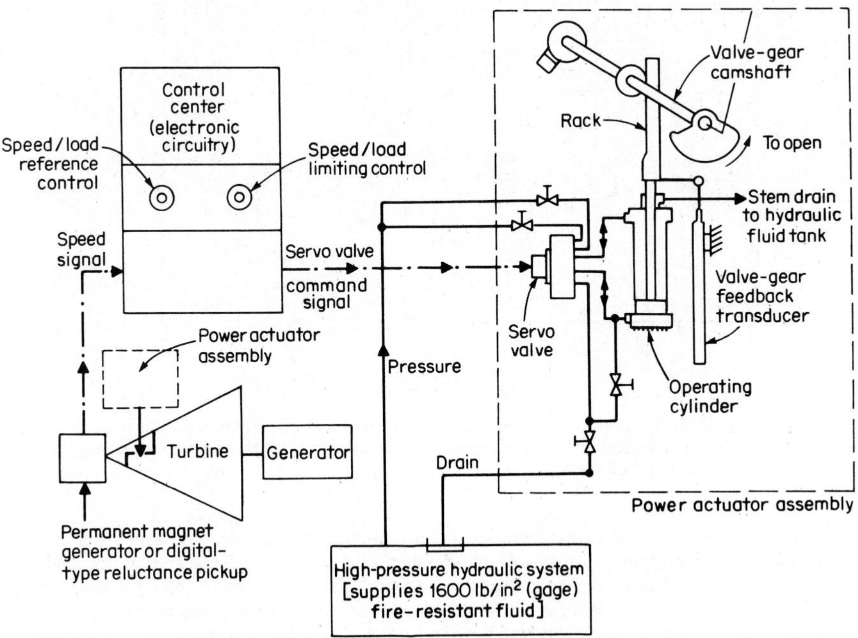


**FIG. 6-13** Schematic diagram of a basic electrohydraulic speed/load control system. *(Medium Steam Turbine Department, General Electric Company.)*

lics upon receipt of the servo-valve signal, (5) a feedback transducer on the power actuator to restore the servo valve to a stable condition when the desired valve position is reached, and (6) a high-pressure hydraulic system to provide the force required.

Other types of speed/load control systems have been applied to turbines from time to time. These include pneumatic, hydraulic, or electric devices. However, the two most common systems for turbine control are the mechanical hydraulic (MHC) and electrohydraulic (EHC) systems described. Another version of the EHC system was developed in the late 1960s to provide bridge control on marine turbine applications.

*Pressure control systems.* A second major area of control technology on steam turbines deals with process control. In industrial power plants particularly, it is often economical to generate and control several process flows, using steam from available steam turbines. As in the case of speed/load control, when a process is to be controlled, a definite relationship or "regulation" is established between the flow to be supplied by the turbine and the pressure. However, the possible options in process-pressure-control management are much greater than the speed/load control options described. The most common application control is for an extraction or exhaust flow from a turbine, which is to be controlled accurately in pressure and used in an industrial process. For this purpose, automatic-extraction pressure control systems and exhaust pressure control systems have been designed, using both the MHC and EHC technologies. Occasionally, particularly on waste-heat boiler applications, there is a need for initial pressure control as well.

Figure 6-14 is a greatly simplified schematic representation of a mechanical hydraulic

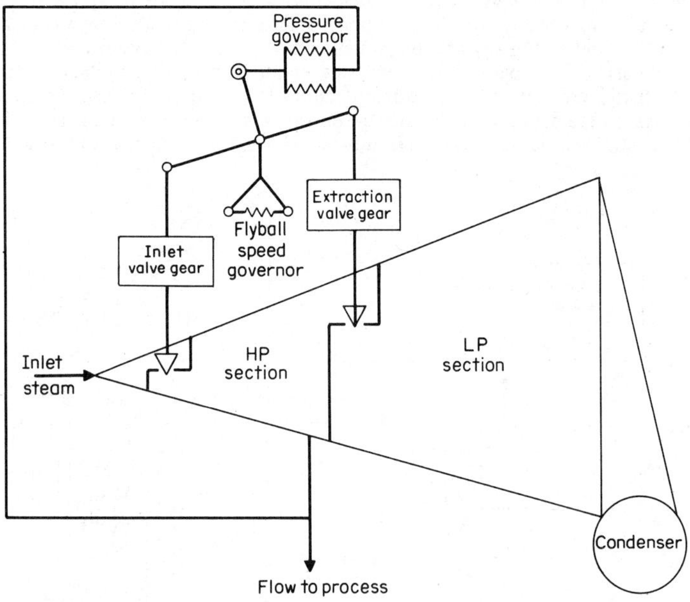


**FIG. 6-14** Simplified schematic diagram of a speed and pressure control system for a single-automatic-extraction condensing turbine. *(Medium Steam Turbine Department, General Electric Company.)*

control system on a single-automatic-extraction condensing turbine. The unit is really two turbines, an HP and an LP section (each supplied by a separate valve gear), on one shaft. A flyball speed governor is used to move the two sets of valves to control speed, or load, and a bellows-type pressure governor is employed to sense process pressures and move the valves in opposite directions to control process flow and pressure. (Actual hardware required for these actions would, of course, include either mechanical hydraulic relays and linkage or electrohydraulic components.) The system is usually so designed that load and process flow variations

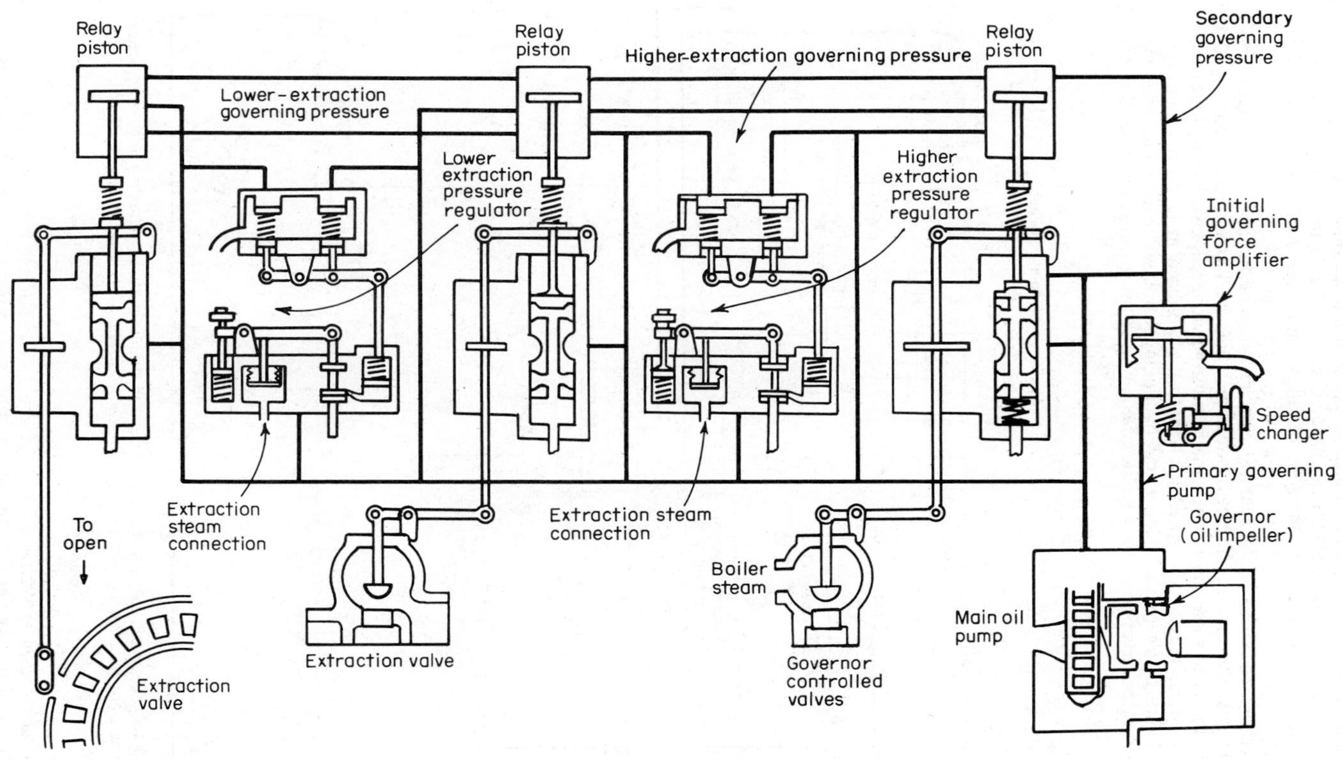


**FIG. 6-15** Double-automatic-extraction turbine using a combination of three signals on 3-relay pistons to hold constant pressure at each of its extraction openings and maintain a constant shaft speed. *(Steam Turbines, Power Special Report, GER-430A, June 1962.)*

can be satisfied at the same time with a minimum of interaction between the two variables. Figure 6-15 shows schematically the mechanical hydraulic hardware for a typical double-automatic-extraction turbine, where three sets of valves control speed and two sets control process pressures.

Often the need exists as well for control of exhaust pressure on a noncondensing turbine-generator. In this case, since the number of variables which can be controlled is only equal to the number of control-valve stages, one variable must be sacrificed. Usually, the unit is tied to a "stiff" electrical system, and speed/load control is sacrificed in favor of exhaust pressure control. Figure 6-16 shows a modern electrohydraulic control system for a single-automatic-extraction noncondensing turbine capable of controlling two process pressures for industrial needs.

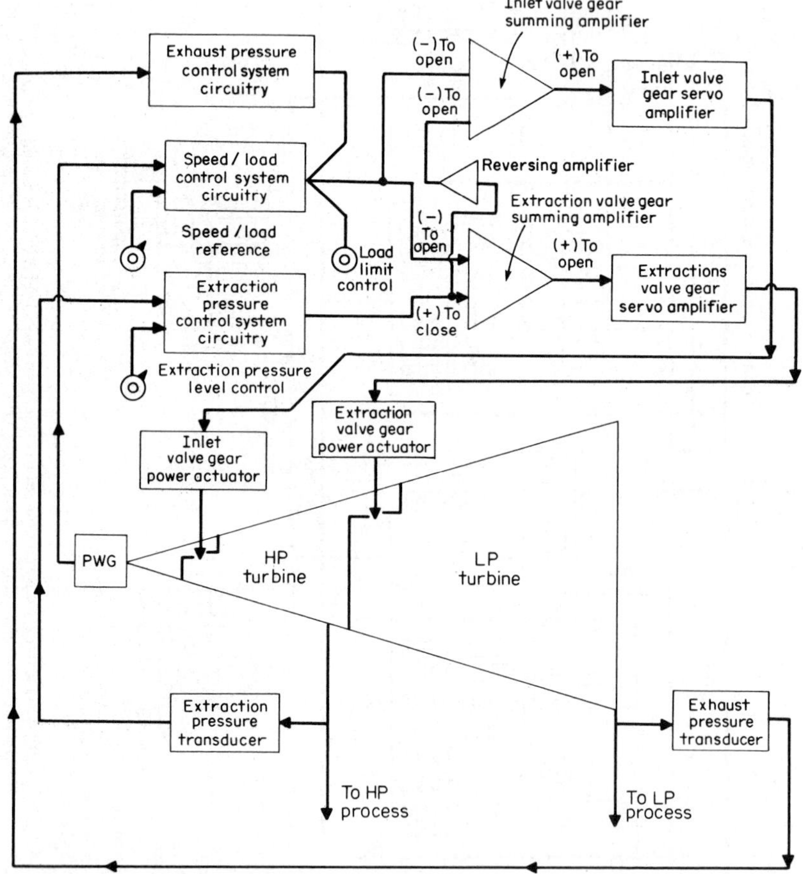


**FIG. 6-16** Schematic diagram of a modern electrohydraulic control system for a single-automatic-extraction noncondensing turbine. *(Medium Steam Turbine Department, General Electric Company.)*

*Secondary and/or Protective Control Systems.* In addition to the control valves and associated components described, there are a number of components and systems charged with protecting turbines against such problems as overspeed, electrical fault, inadequate lubrication, etc. Fundamentally, these devices provide a backup, or "second line of defense," against problems in situations where the basic control elements are assumed to be the primary line of defense. In most cases the protective action taken is to trip the unit out of service by closing all valves controlling admission of steam to the turbine control valves and steam path. In addition, there may be devices for dumping steam energy which is in the unit or in connected volumes at trip-out, and which could expand and cause overspeed and resultant damage.

Figure 6-17 shows an overspeed trip system applied to a modern EHC industrial unit that operates on fire-resistant high-pressure fluid. Note that the components include a main stop valve (which shuts off steam to the turbine inlet), an overspeed governor and trip device, an electrical trip device, and a low-bearing-pressure trip relay, all in series so that any one of them can trip the unit if a fault occurs. In parallel with the stop-valve trip scheme is an extraction nonreturn valve-trip relay, which trips that check valve closed to protect against possible admission at the extraction opening. Also shown is a thrust-failure device, one of a number of commonly applied devices for monitoring unit conditions and tripping electrically if a fault occurs.

Figure 6-18 shows a more complicated system for protecting a reheat turbine, having both a main stop valve and two reheat stop and intercept valves responsible for shutting off steam to the turbine. In addition to the trip devices described earlier, this unit has a packing blowdown valve to dump high-energy steam from the shaft packing located between the HP and reheat turbines, and to prevent its expansion into the unit and resultant overspeed.

***8. Lubrication and Hydraulic Systems.*** Forced-feed lubrication of turbines and generator bearings is normally used on units above approximately 200 hp in size. On such units, the lubrication system is sometimes used to supply low-pressure seal oil for a hydrogen-cooled generator as well. Also, it often is used to supply the higher-pressure oil for the turbine control and protective systems. This is normally the case on units having a mechanical hydraulic control system and operating on turbine oil at a pressure of 250 lb/in$^2$ (gage) or less. On units having electrohydraulic control systems, operating at higher hydraulic pressures up to 3000 lb/in$^2$ (gage), and using fire-resistant fluids, a separate hydraulic-fluid power unit supplies all fluid for the control systems and usually for the protective systems as well.

*Lubrication and Hydraulic System for a Mechanical Hydraulically Controlled Turbine.* Figure 6-19 shows lubrication and hydraulic systems typically applied to an MHC unit. The turbine bearing lubrication system is composed primarily of a main oil tank, piping to the respective bearings, seals, and hydraulic elements, and a turbine-shaft-driven main oil pump. The main oil tank contains the pumps, coolers, pressure regulator, and other items required for a completely integrated system. The main shaft pump may be of the centrifugal impeller type or worm-driven-gear type. The oil from the main pump usually pumps against a closed check valve in its discharge. A relief valve is also generally applied to the hydraulic header to prevent overpressure.

An ac-driven auxiliary oil pump is provided for start-up and also acts as an emergency pump if the oil pressure should fall below the normal level. An ac-motor-driven bearing- and seal-oil pump may be provided to supply oil for operating the machine when it is on turning gear. It then supplies oil to the bearings and hydrogen shaft seals as required. It is equipped for automatic starting in case of abnormally low bearing-oil pressure and, therefore, acts as a backup for the auxiliary pump.

A final (emergency) backup for the bearing and seals is often provided in the form of a dc-motor-driven oil pump, which will start automatically in case of abnormally low bearing pressure. Power for this pump is supplied by station batteries.

Two main coolers are provided on the main oil tank to cool the oil to 120°F before it is supplied to the turbine bearings. It is usually necessary to use only one cooler at a time, with the second in reserve. The oil coolers are usually the shell-and-tube type.

*Lubrication System for an Electrohydraulically Controlled Turbine.* Figure 6-20 shows a lubrication system typically applied to EHC units. The system is composed primarily of a main oil tank and the piping to the respective bearings. Generally there is no main shaft pump in this type of system. The main oil tank contains the pumps, coolers, pressure regulator, and other items required for a completely integrated lubrication system. Since the pumping system in this case is for bearings and hydrogen seals only, it operates at a single lower pressure level [about 60 lb/in$^2$ (gage)]. The two ac centrifugal-type oil pumps are arranged in parallel, with one operating and one on standby. Should the operating pump fail, a drop in oil pressure provides a signal to start the standby pump. Should both of the ac-motor-driven pumps fail to start, the oil pressure will drop to a lower level to signal the dc-motor starter to start that pump.

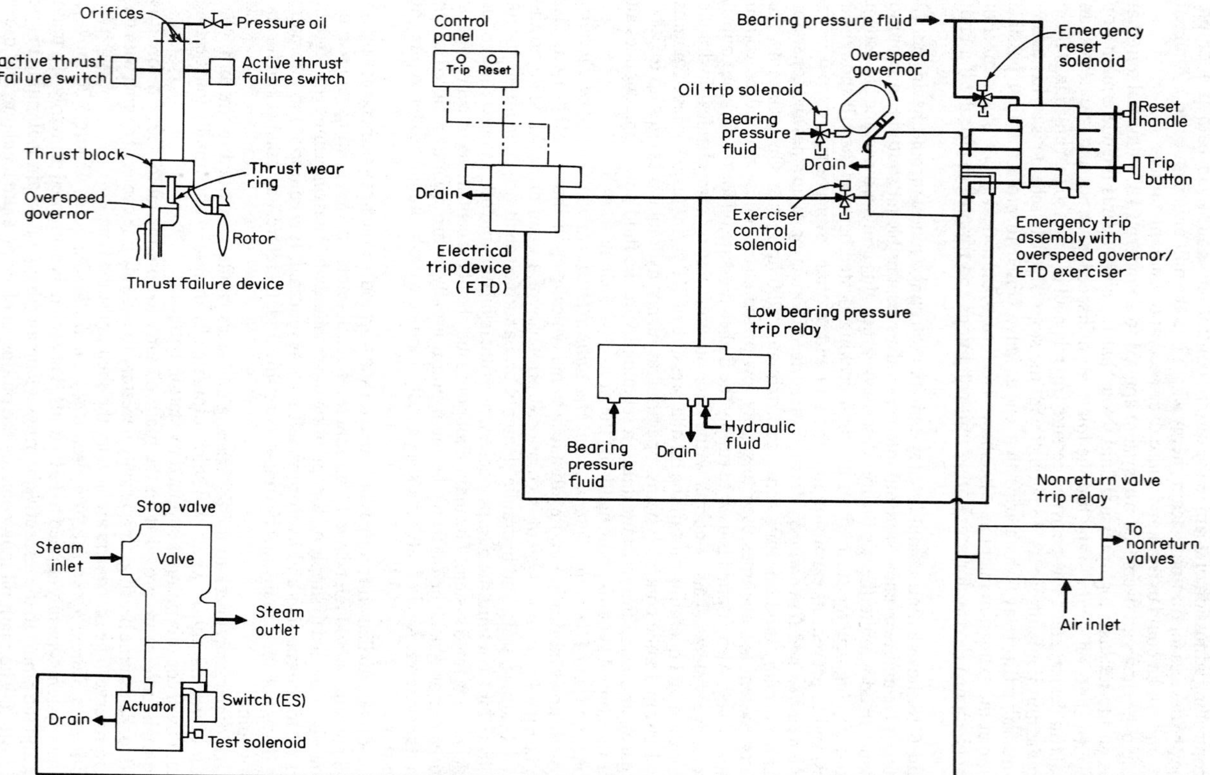


**FIG. 6-17** Emergency trip system for an EHC industrial unit. *(Medium Steam Turbine Department, General Electric Company.)*

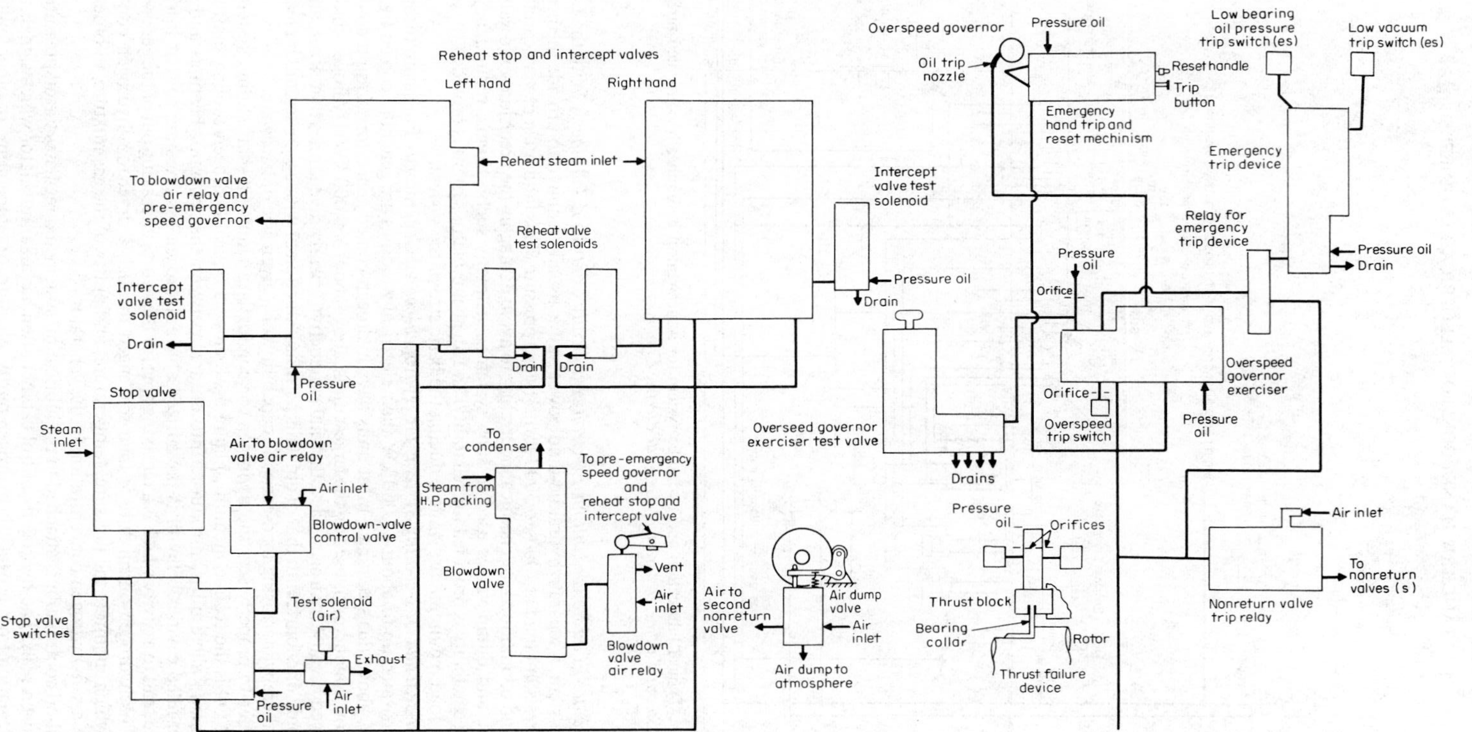


**FIG. 6-18** Emergency trip system for an MHC unit. *(Medium Steam Turbine Department, General Electric Company.)*

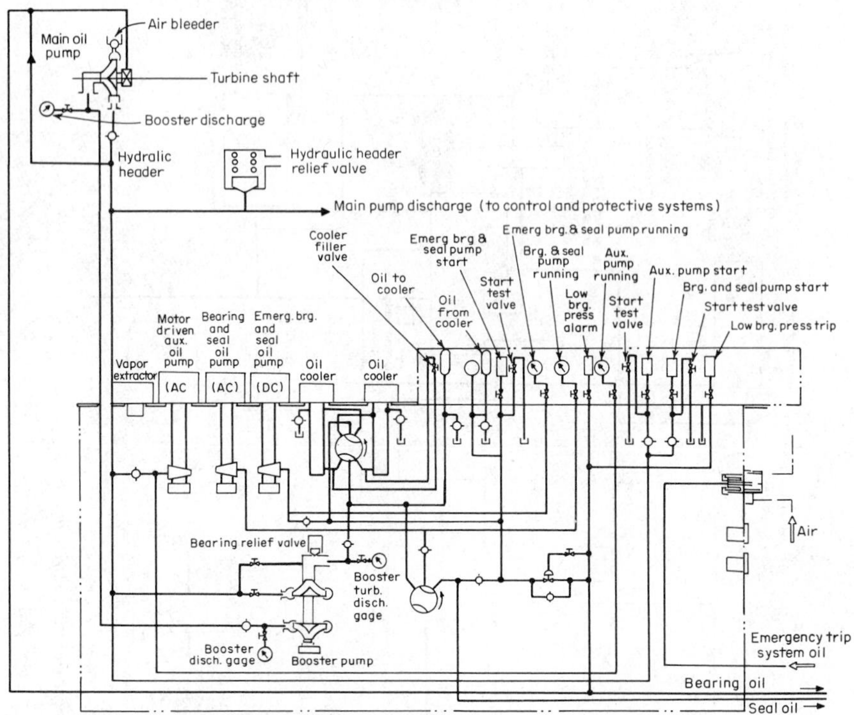


**FIG. 6-19** Typical lubrication and hydraulic system for a mechanical hydraulically controlled turbine. *(Medium Steam Turbine Department, General Electric Company.)*

*Hydraulic-Fluid Power Unit for Electrohydraulically Controlled Turbines.* With the EHC control system, a separate hydraulic-fluid power unit is usually required to provide the necessary fire-resistant fluid at 1000- to 3000-lb/in$^2$ (gage) pressure. This fluid must be very carefully filtered and conditioned for chemical neutrality, to avoid damaging the precision servo valves, actuators, and dump valves used in such systems. On many such systems, redundancy of pumping capability is provided for reliability, and an accumulator system is provided to carry transient load requirements, while reducing the basic size of the system.

***9. Oil-seal and Gas-cooling Systems for Hydrogen-cooled Generators.*** For steam turbine-generators rated up to about 40,000 kW their electrical windings are generally cooled by air. However, above this size range, most units have hydrogen-cooled generators. Liquid cooling with hollow conductors is used on the largest units, above about 300,000 kW. Hydrogen cooling is employed because hydrogen has a thermal conductivity nearly seven times that of air, and a density only one-fourteenth that of air. This permits reduction of windage losses and increased cooling, thereby increasing load-carrying capability for a given size of hardware.

In order to properly seal the hydrogen for cooling larger units, a shaft sealing system is required. Oil is the sealing medium. Figure 6-21 shows a typical hydrogen-cooling shaft seal arrangement, where oil is introduced into an area between the generator bearings and the hydrogen-filled cavities. The oil is pressurized above hydrogen gas pressure, so that it leaks across the seals to a cavity which is a receiving area for the hydrogen leaking out of the generator casing. The hydrogen-oil mixture is then scavenged through a dryer to a hydrogen control cabinet which monitors the pressure, temperature, and purity of the gas mixture, in order to maintain a safe hydrogen concentration in the generator.

A seal-oil control unit is attached to the lubrication system and regulates the oil pressure to the hydrogen seals. The hydrogen control cabinet often has an annunciator which signals any serious deviation from acceptable hydrogen purity, pressure, or temperature, or other faults in the system.

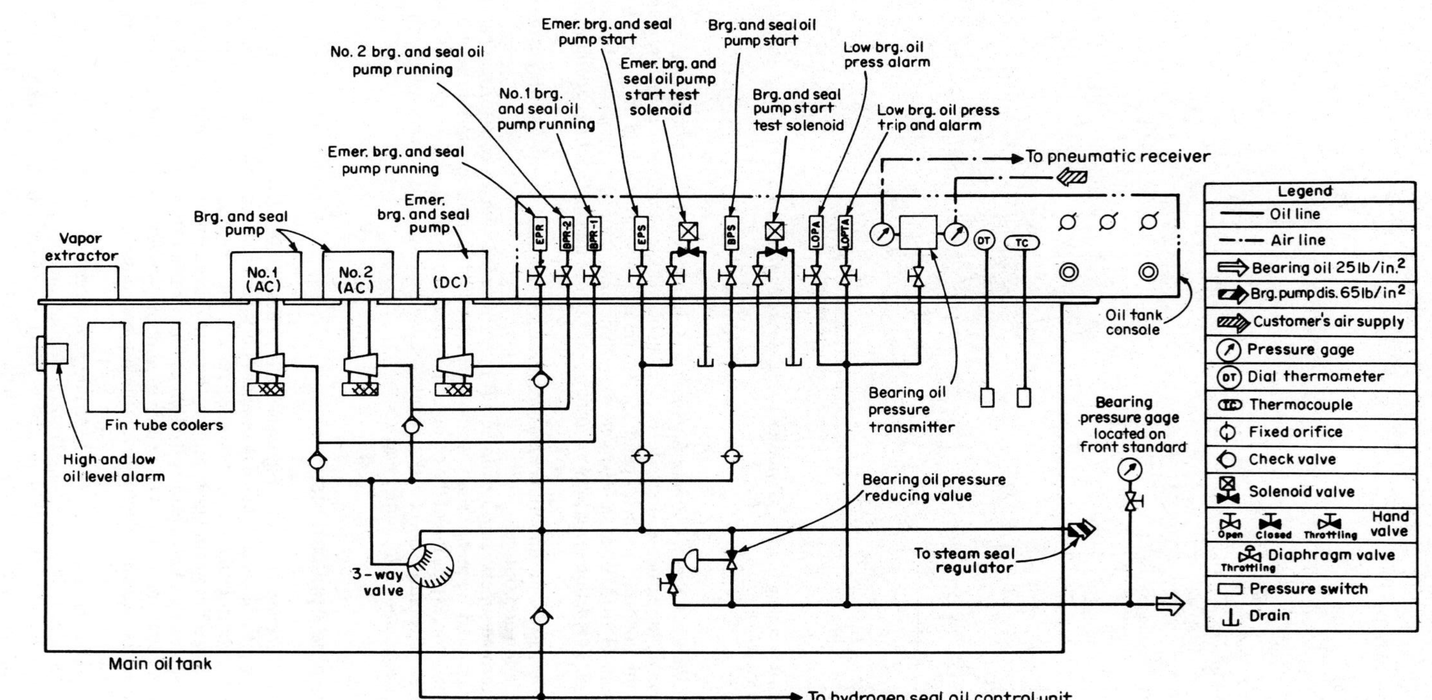


**FIG. 6-20** Schematic of a lubrication system used in an electrohydraulically controlled unit with a hydrogen-cooled generator. *(Medium Steam Turbine Department, General Electric Company.)*

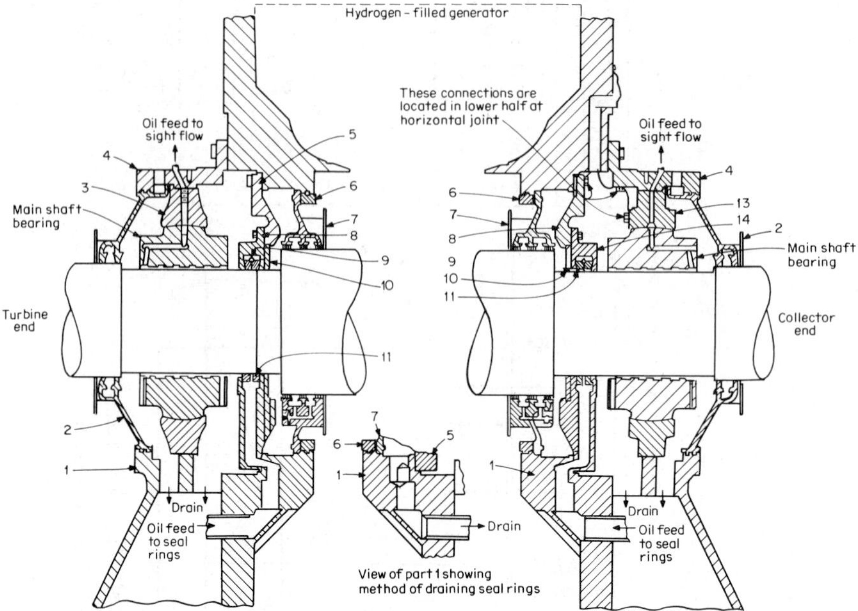


FIG. 6-21 Shaft at sealing rings for a hydrogen-cooled generator. *(Medium Steam Turbine Department, General Electric Company.)*

***10. Miscellaneous Steam-Turbine Components.*** In addition to the systems and components discussed in the earlier sections, steam turbines often have a number of accessory components which are important to their operation. A turning gear is provided on units rated larger than approximately 10 MW, to slowly rotate the turbine shaft before the unit is started and after it is shut down. This action helps prevent rotor bowing due to unequal heating or cooling of the rotor.

Another device of some importance is a lifting gear, for assembly and disassembly of the unit during installation and outages.

A set of turbine supervisory instruments is often included with a steam-turbine package. Typically monitored items are shaft vibration, differential thermal expansion between the casing and the rotor, expansion of the casing, eccentricity while on turning gear, thrust bearing position, speed of rotation, acceleration, control-valve position, and various other items, as required by design and/or customer needs.

## Steam-Turbine Applications

***11. Central-Station Turbines.*** Figure 6-22 illustrates a 60,000-kW, 3600-r/min *nonreheat steam turbine* typical of those installed in smaller utility plants. Steam flows into the steam chest and through the control valves to the first-stage nozzle. After expanding through a Curtis-type, 2-row control stage, the steam flows through 16 more Rateau (impulse) stages to the exhaust. During the expansion, some steam is bled off at four or five extraction points for feedwater heating. Larger-rated units, such as those used in combined-cycle plants, require double flowing of the last five or six stages in order to provide the last-stage annulus area necessary to maintain a low leaving loss.

Figure 6-23 illustrates a typical 600-M to 800-MW, *tandem-composed single-reheat steam turbine,* the type used in large fossil-fired central stations. Steam conditions are predominantly 2400 lb/in² (gage), 1000°/1000°F, but some applications are at 3500 lb/in² (gage) and a few utilize double reheat as well. Steam enters the high-pressure turbine element through four pipes

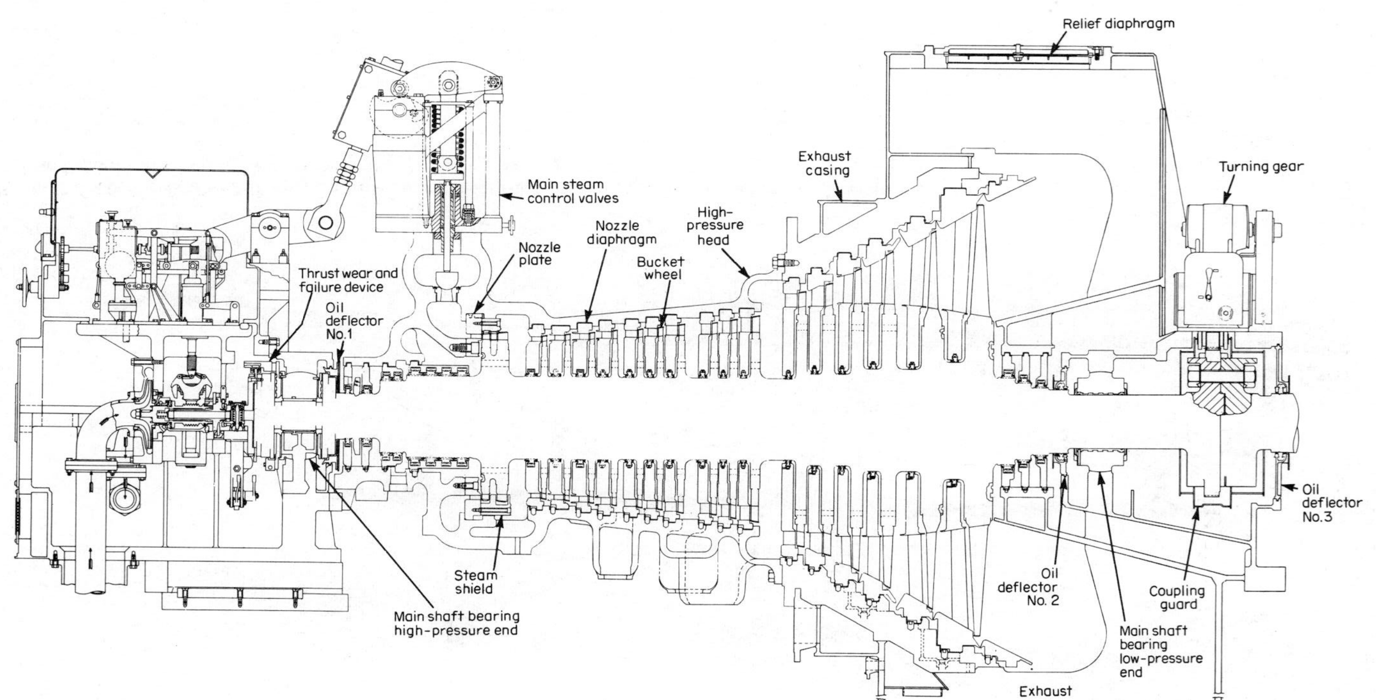


**FIG. 6-22** **60,000-kW nonreheat steam turbine.** *(Medium Steam Turbine Department, General Electric Company.)*

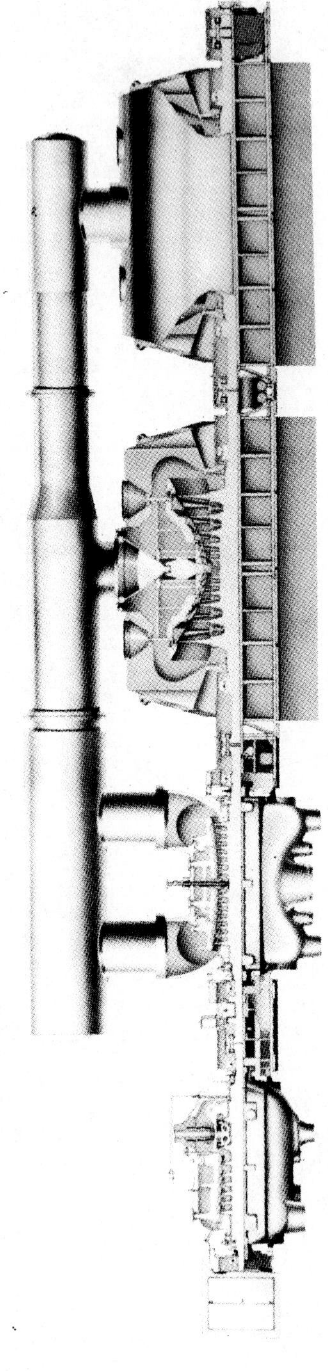

**FIG. 6-23** 600- to 800-MW, 3600-rpm tandem-compound four-flow four-casing reheat steam turbine. *(Medium Steam Turbine Department, General Electric Company.)*

leading from the off chest-control valves to the nozzle box and the double-flow first stage. The flow is expanded through 6 more impulse stages before exiting from the HP casing to the reheater. The reheated steam enters at the center of the intermediate turbine and expands through seven double-flow intermediate stages before exhausting to the crossover pipe. The crossover feeds the steam to the four-flow LP elements where it is expanded to completion through six more stages before exhausting to the condenser.

A typical *nuclear steam turbine* of 1000 to 1300 MW capacity is shown in Fig. 6-24. Steam

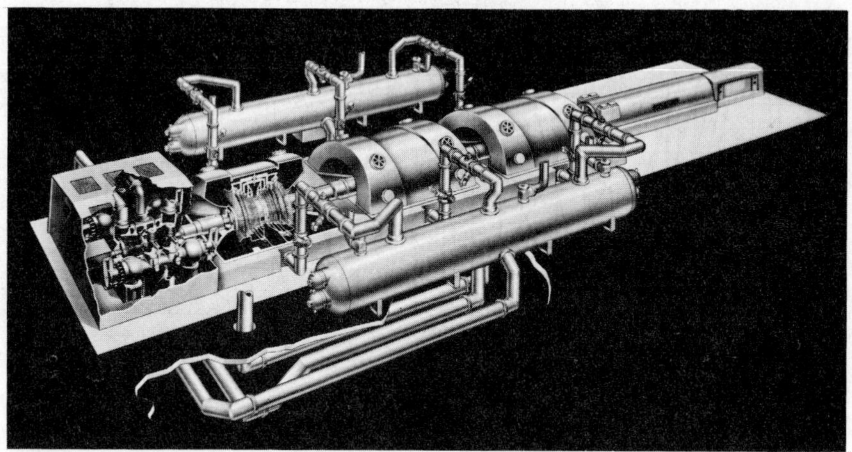

**FIG. 6-24** Westinghouse tandem-compound six-flow nuclear steam turbine-generator.

enters the double-flow high-pressure element at the left at 1000 lb/in$^2$ (gage), 546°F, and exhausts at about 200 lb/in$^2$ (abs.) to the two large combined moisture-separator reheaters which straddle the three double-flow low-pressure elements. After the moisture is removed and the steam slightly reheated, it is passed to the six-flow low-pressure element where it is expanded to completion. The low superheat available with the light water reactors and the large ratings encountered require the use of 1800-r/min machinery to keep blade speeds low, reducing the erosion from moisture, and to provide the large flow areas which are more easily obtained using larger wheel diameters. The moisture-separator reheater is provided to decrease the erosion in the low-pressure elements and improve their performance. The *condensing boiler-feedwater-pump-drive turbine* was introduced in the 1960s, and became accepted rapidly because the increase in condensing annulus area served to improve both the heat rate and capacity of the station. Figure 6-25 illustrates a 10,000-hp straight condensing boiler-feed-pump-drive turbine for a nuclear application. Steam is extracted from the main-unit cycle after it has gone through the moisture separator-reheater and is delivered to the inlet at approximately 150 lb/in$^2$ (abs.) and from 0 to 100°F superheat. It is expanded to completion in the six stages. For operation at light load, steam is taken from the main steam header and sent to the high-pressure inlet in the lower half of the first stage.

***12. Industrial Steam Turbines.*** In many industrial plants, particularly those in the pulp and paper, petrochemical, and related industries, the need exists for large amounts of electric power and process steam at various pressure levels. In this type of situation, industrial users can justify generating their own power and charging a large part of the cost to the process, because the steam is also needed. Plants have historically been built and expanded—and are still being built and expanded—with various condensing and noncondensing extraction turbine types, as the process requires, and in sizes ranging from 1 to 200 MW.

Figure 6-26 shows one example of a paper-mill installation having a mixture of both straight noncondensing, single-automatic-extraction and double-automatic-extraction turbines. In this plant, the user originally purchased the 5000-kVA straight-condensing unit, operating at 400 lb/in$^2$ (gage), for electric power needs. As the need for process steam developed, the user

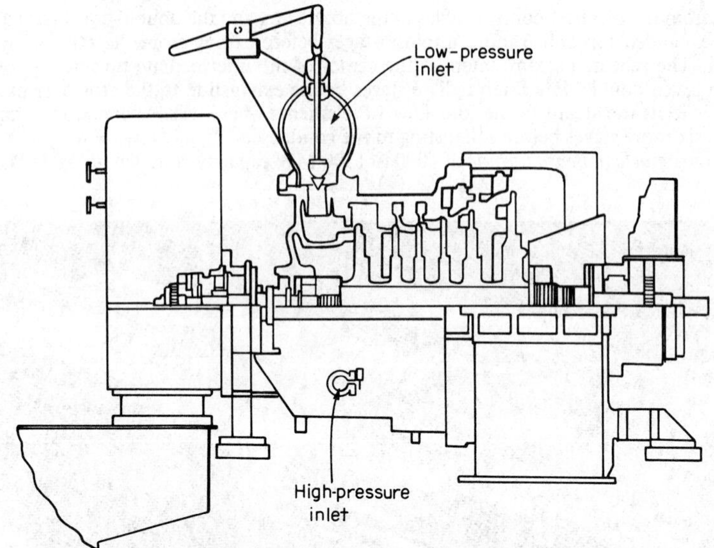


**FIG. 6-25** 10,000-hp, 5500-r/min steam-generator–steam feed-pump-drive turbine for nuclear application. *(Mechanical Drive Turbine Department, General Electric Company.)*

purchased the 5000-kVA single-automatic-extraction noncondensing unit, also operated on the 400-lb/in$^2$ (gage) header, extracting at 160 lb/in$^2$ (gage), and exhausting to the 60-lb/in$^2$ (gage) process.

At a later date, discovering that the electric-power and process needs were increasing and taking advantage of newer boiler technology, the user expanded by purchasing a 15,625-kVA single-automatic-extraction condensing unit, which operates at the new 600-lb/in$^2$ (gage) pressure and provides additional 160-lb/in$^2$ (gage) process steam. This relatively larger unit also provided needed flexibility for load swings, because of its condensing capability. Next, needing more low-pressure process steam and an additional block of electric power, a 15,625-kVA straight noncondensing unit operating on the new 600-lb/in$^2$ (gage) steam conditions and exhausting to 60 lb/in$^2$ (gage) was purchased. With this unit, the previous units, and the various pressure-reducing stations, the user had a flexible two-level process plant capable of generating power needs as well.

At the next state of expansion, needing more process steam at both levels, as well as more power, a 32,000-kVA double-automatic-extraction condensing unit, which is shown in cross section in Fig. 6-27, was purchased. This large block of process steam and power nearly doubled operating capabilities. Moreover, with the new unit and its more sophisticated control system, the user could now centrally control some of the existing units and pressure-reducing stations from the new unit control panel, thus centralizing the total plant control.

Finally, in the last expansion, because of energy-cost considerations, the user decided to take advantage of the latest steam-generator technology and purchased a 44,000-kVA double-automatic-extraction noncondensing unit. This provided a large increase in both important process capabilities, a large block of power, and the flexibility of admitting or extracting steam at 600 lb/in$^2$ (gage). It also added over 50% to the existing plant operating capability. Because of the size of the last unit, and the size and complexity of the total plant, the last expansion also dictated the addition of a sophisticated centralized pressure control system.

Such has been the growth of many industrial power plants. The steps between expansion have been growing even larger, and the processes and controls ever more complicated. Future energy considerations, utility power rates, and capital costs all bear on the construction and expansion of such plants.

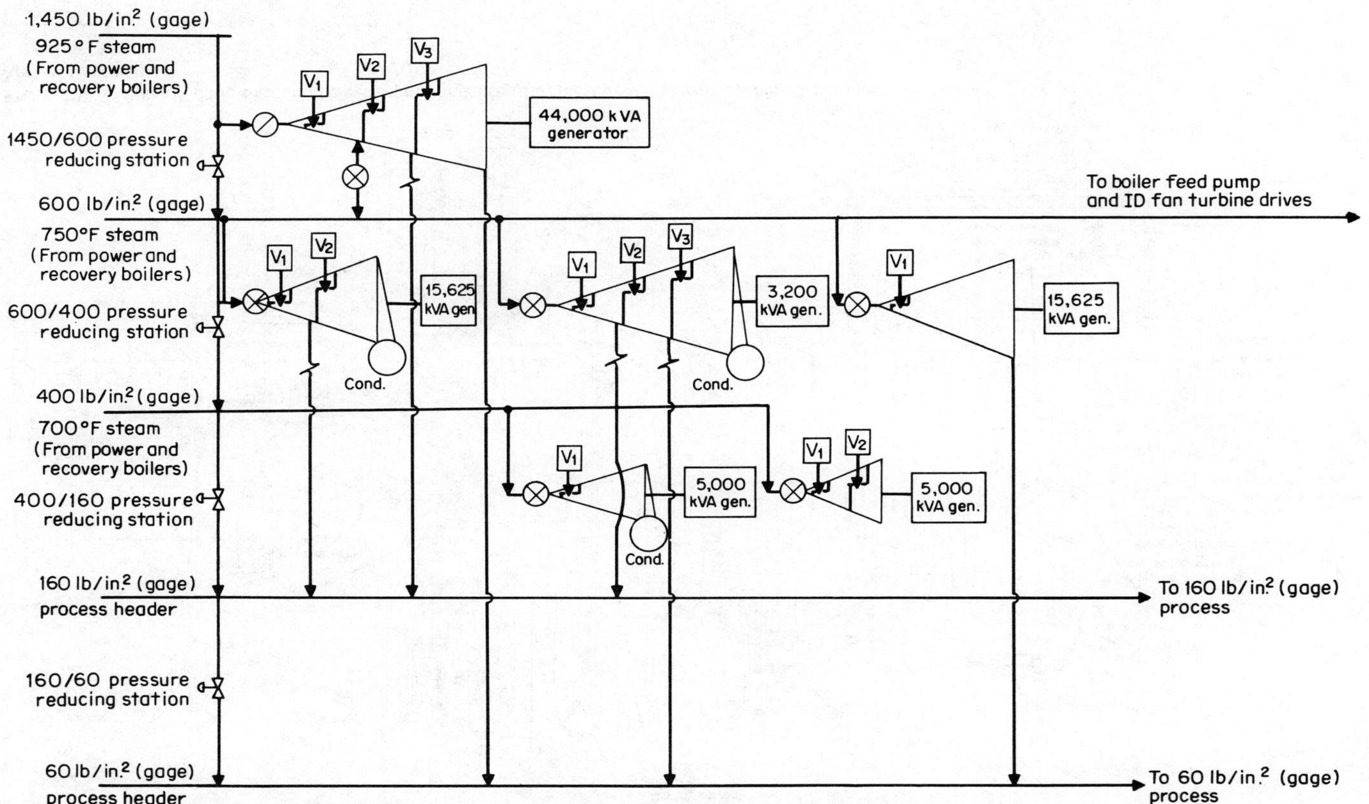


**FIG. 6-26** Schematic diagram showing a multiple-unit power plant and industrial-process unit plant. *(Medium Steam Turbine Department, General Electric Company.)*

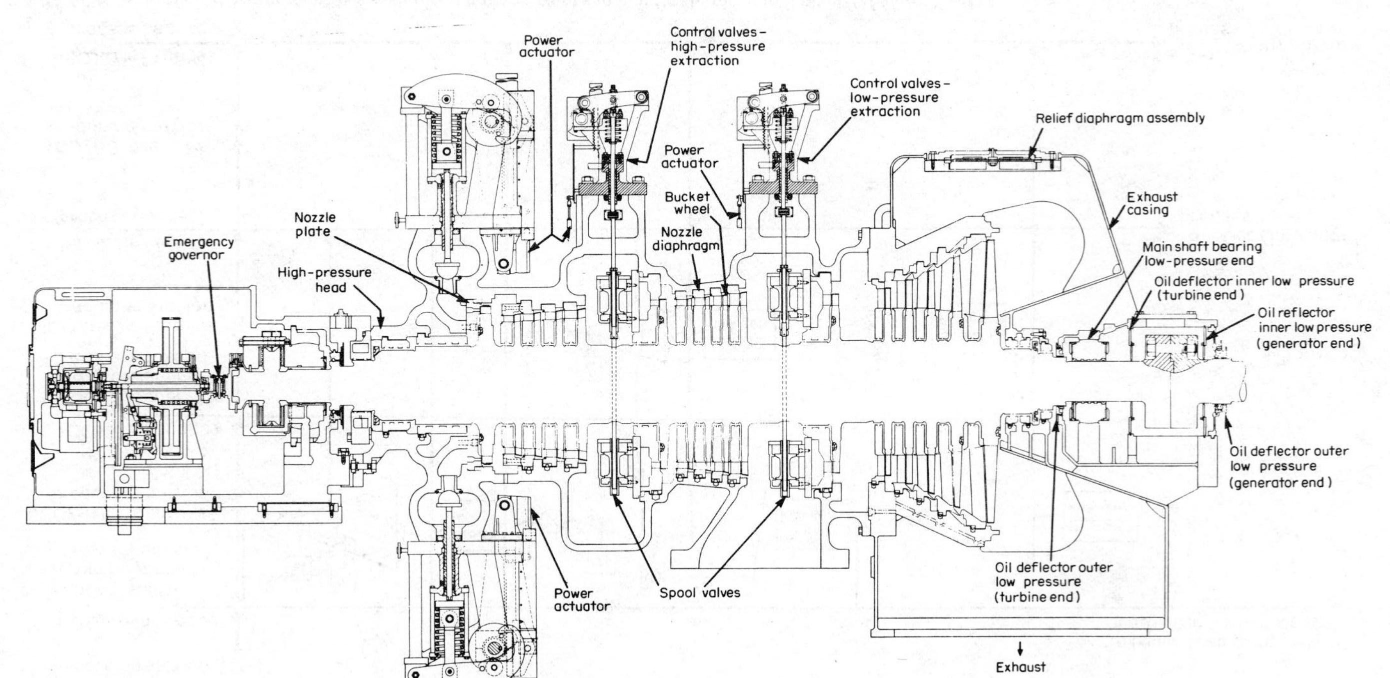


**FIG. 6-27** 25,000-kW double-automatic-extraction condensing turbine. *(Medium Steam Turbine Department, General Electric Company.)*

*13. **Variable-Speed Turbines.*** The steam turbine is used extensively as the prime mover for ship propulsion at ratings above 10,000 shp. The cross-compound design is almost universal as it provides emergency capability for getting back to port as well as providing two pinions which divide the load on the low-speed gear, reducing gear weight. The major applications are in high-powered, high-utilization ships, such as tankers and container ships. They are used almost exclusively in naval combat ships (aircraft carriers and nuclear submarines) as well as for large auxiliary supply ships.

Applications are predominantly nonreheat, but because of the steadily rising cost of fuel, reheat applications are gaining popularity. Figures 6-28 and 6-29 illustrate the HP and LP elements of a typical 20,000- to 50,000-shp nonreheat application at steam conditions of 850 lb/in$^2$ (gage), 950°F. The HP element (Fig. 6-28) supplies about 50% of the output at full power and is designed to operate at a speed up to 100% greater than the LP element for performance weight and space reasons. The LP element (Fig. 6-29) contains a reversing element consisting of two Curtis stages which supply 100% torque at zero speed and approximately 70% torque at 70% speed in reverse, providing excellent reversing capability. During normal ahead operation, the windage loss of the reversing element is held to approximately ½% by its location in the exhaust, where the vacuum is greatest.

Industry uses *mechanical-drive turbines* for a wide variety of purposes. These turbines drive paper machines; blast furnace blowers; and ethylene, ammonia, and liquefied natural-gas plant compressors because of their ability to follow the speed-output characteristics of this type of equipment without loss of efficiency from throttling, recirculation, or use of fluid couplings. Figure 6-30 illustrates a mechanical-drive turbine typical of that used in an ethylene plant. This unit is an automatic-extraction condensing turbine operating at an inlet pressure of 900 lb/in$^2$ (gage), with extraction at 225 lb/in$^2$ (gage). The unit responds to variations in extraction demands by maintaining extraction pressure. As extraction pressure decreases, the HP inlet

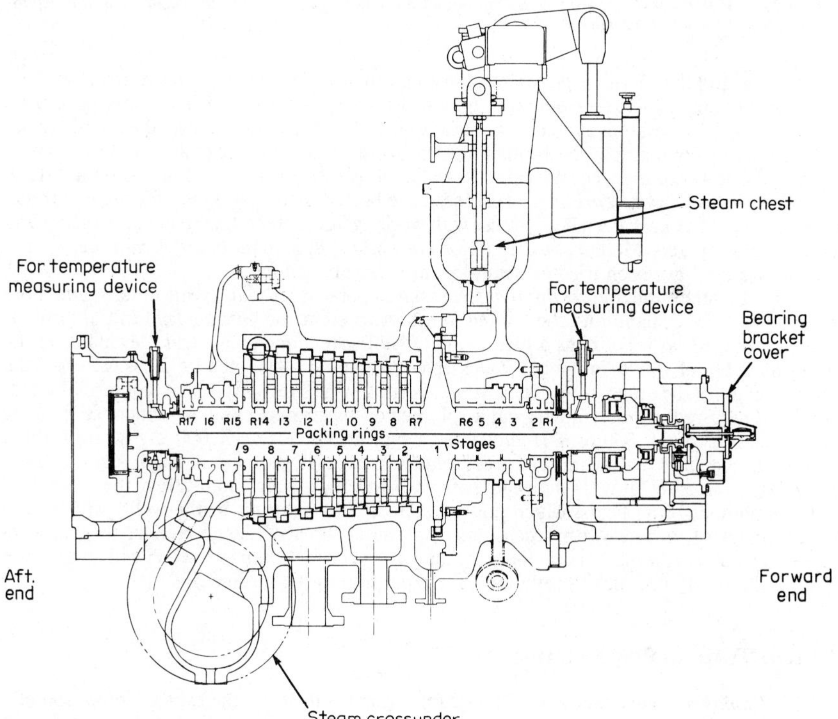


**FIG. 6-28** HP element of a marine cross-compound nonreheat turbine. *(Medium Steam Turbine Department, General Electric Company.)*

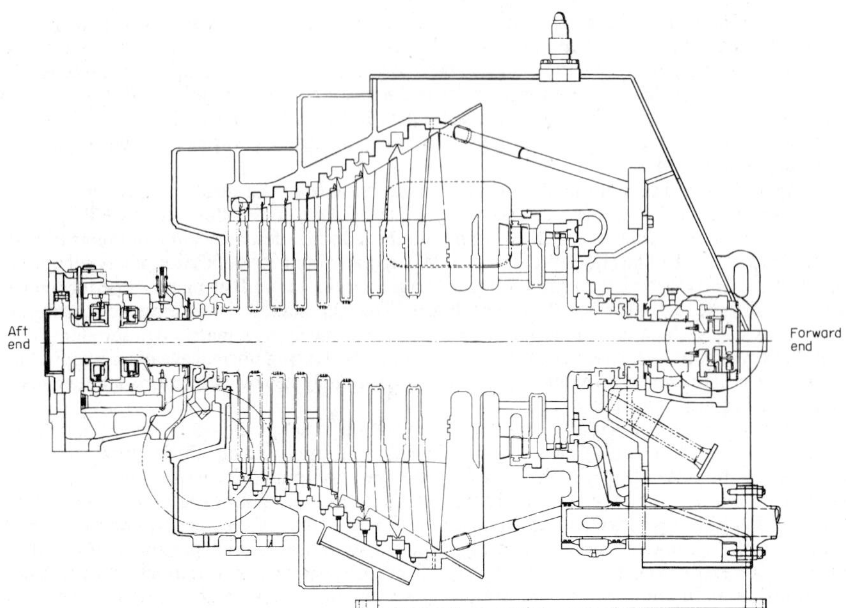


**FIG. 6-29** LP element of a marine cross-compound nonreheat turbine. *(Medium Steam Turbine Department, General Electric Company.)*

valves open and the LP valves close simultaneously to provide the increased extraction without changing the load. Increased output is obtained by incrementally opening both valve gears simultaneously. Operation at higher speeds as well as variable speed requires the use of heavier last stages than comparable generating units and consequently the use of double-flow exhausts is applied at lower ratings. The typical cross section illustrates the use of a double-flow last stage.

***14. Special-Purpose Turbines.*** Turbines have been designed and built for many unusual applications. They have been built for use with working fluids other than steam in refineries and petrochemical plants. Mercury has been used as a working fluid in the *binary steam power plant.* Steam turbines have been tried in steam-locomotive applications.

Steam turbines utilizing *geothermal steam* have operated for many years in Italy, and more recently in New Zealand and the United States. At present the Geysers field in California is being developed, as it contains a high grade of geothermal energy as steam, available at the wellhead at about 100 lb/in$^2$ (gage). The high cost of liquid fuels since 1973 has increased the attractiveness of the much more extensive geothermal brine sites as well. The geothermal-steam turbine is required to pass a much greater volume flow of steam per kilowatt generated than the central station plant. Figure 6-31 illustrates a typical 50,000- to 60,000-kW steam turbine operating on geothermal steam. The presence of impurities in geothermal steam requires much more extensive use of alloys in the steam path and protection against moisture in the steam. The development of brine fields, which contain 300 to 500°F liquid in the wells, will require development of turbines with much larger volume-flow capacities to recover the low-level energy available. Alternative development may utilize heat exchangers which will transfer the energy to other fluids, such as isobutane, for expansion in the turbine.

## Steam-Turbine Performance

***15. Rankine-Cycle Efficiency.*** The steam turbine constitutes the expansion portion of a vapor cycle, which requires separate devices, including a boiler, turbine, condenser, and feed-water pump, to complete the cycle. This vapor cycle for steam power plants is commonly called the Rankine cycle (Figs. 6-32 and 6-33) and is less efficient than the Carnot cycle because the exhaust vapor is completely liquefied to facilitate pumping, and because superheat is added at

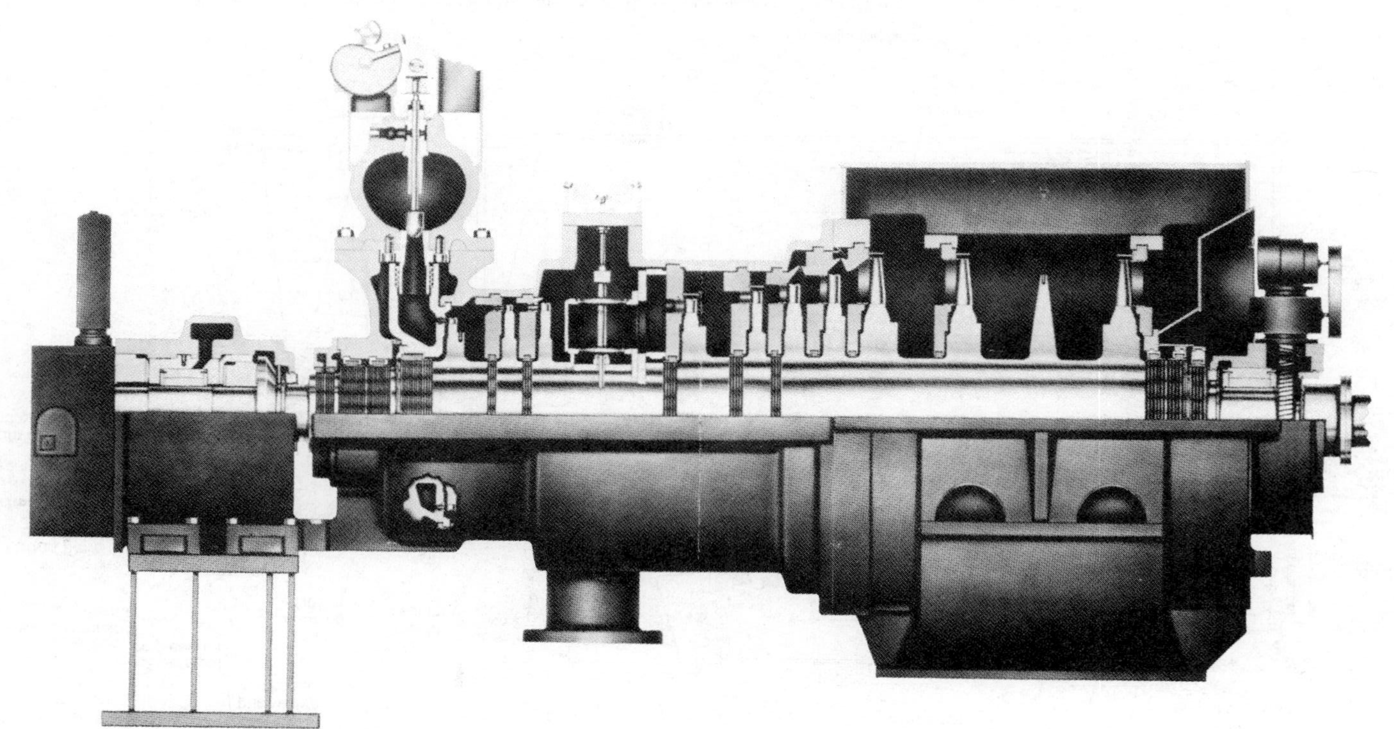

**FIG. 6-30** 20,000- to 30,000-hp, 4400-r/min single-automatic-extraction condensing mechanical-drive turbine with double-flow exhaust. *(Mechanical Drive Turbine Department, General Electric Company.)*

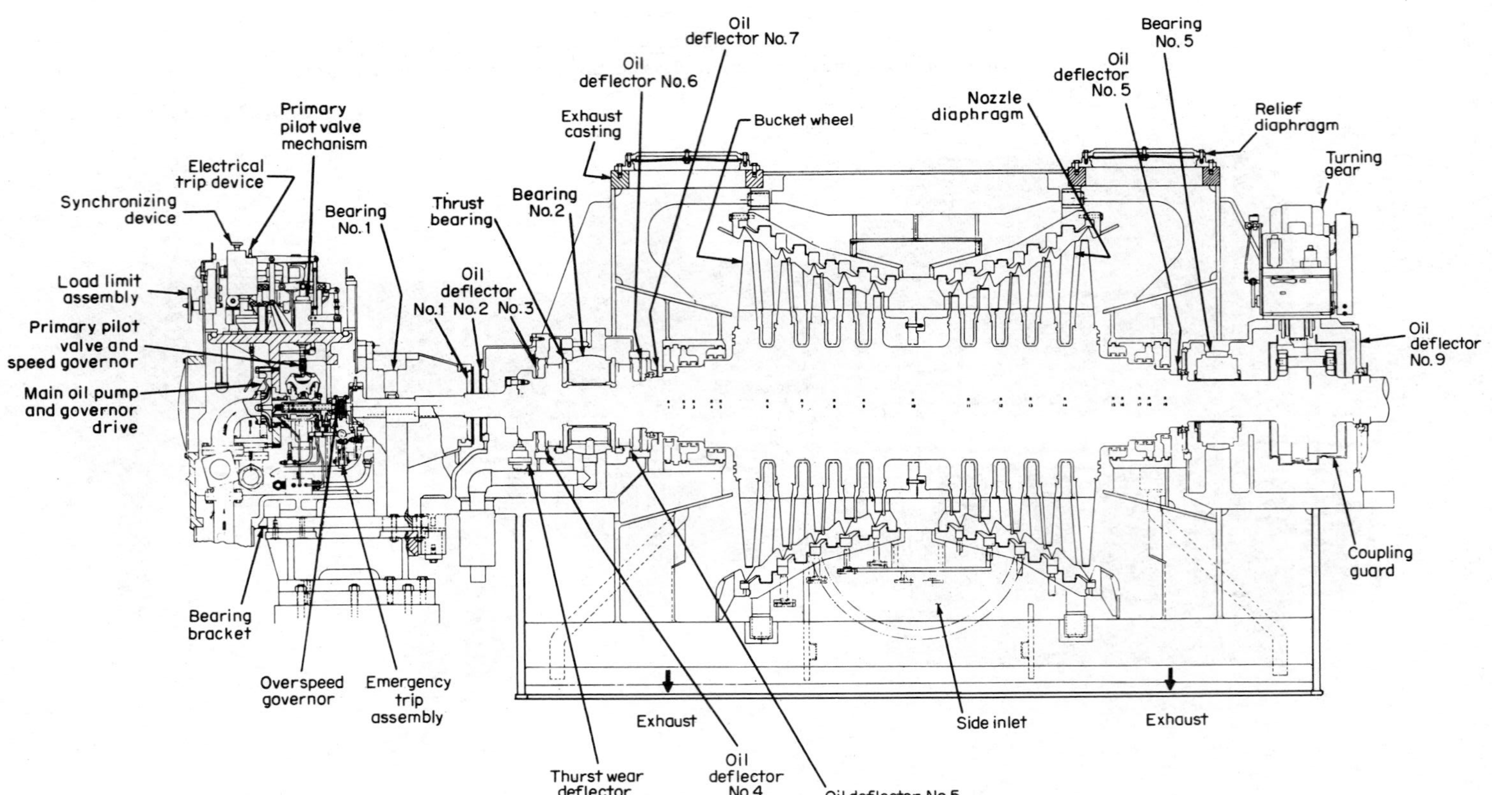


**FIG. 6-31** 60,000-kW, 3600-r/min geothermal-steam turbine. *(Medium Steam Turbine Department, General Electric Company.)*

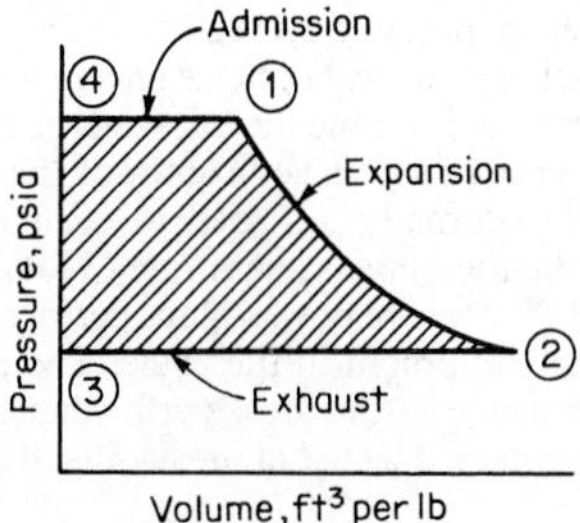


FIG. 6-32 Pressure-volume diagram for the Rankine cycle. Phase 4-1, constant-pressure admission; 1-2, complete isentropic expansion; 2-3, constant-pressure exhaust. Crosshatched area represents the work of the cycle.

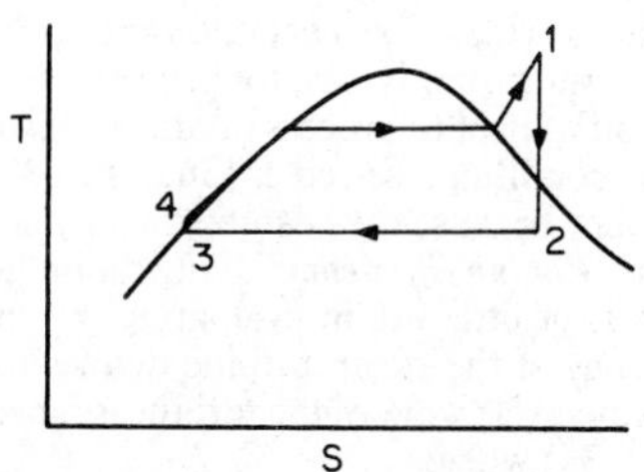


FIG. 6-33 *T-S* diagram; nonreheat, nonextracting turbine cycle.

increasing temperature. The work of the cycle is equal to $h_1 - h_2$ minus the small pump work $h_4 - h_3 = v_3(P_4 - P_3)/J$ required, and the heat added to the cycle is equal to $h_1 - h_4$. Therefore,

$$\text{Rankine-cycle efficiency} = \frac{(h_1 - h_2) - v_3(P_4 - P_3)/J}{h_1 - h_4} \tag{6-4}$$

Cycle efficiency is not commonly used when comparing plant efficiencies because it is only indirectly determined, compared with heat rate, which can be quickly measured. In central-station practice the station heat rate defines the heat required from fossil fuel, reactor energy, waste gases, etc., per kWh of station output (gross electric output minus all auxiliary power required within the plant). For example, a plant with a station heat rate of 10,000 Btu-fuel/kWH has a real cycle efficiency of

$$\text{Cycle efficiency} = \frac{\text{output}}{\text{input}} = \frac{3412.1}{10{,}000} \times 100 = 34.1\% \tag{6-5}$$

Power plants for marine propulsion drive commonly use the "ships all-purpose fuel rate" (lb fuel/shp-h) as a measure of the plant's efficiency. Besides accounting for all losses required to generate the propeller-shaft output work, this fuel rate includes the requirements for the ship's "hotel" electric load, fresh-water evaporators, and steam for heating and unloading cargo and cleaning tanks. A typical ship's fuel rate of 0.45 lb fuel/shp-h based on 18,500 Btu/lb fuel would indicate cycle efficiency for the ship as

$$\text{Cycle efficiency} = \frac{2544.1}{0.45 \times 18{,}500} \times 100 = 30.6\%$$

In the process industries, such as paper and petrochemical, large amounts of steam are used. Considerable by-product power can be generated by raising the boiler pressure above the process pressure and expanding the steam through a noncondensing turbine before exhausting it to the process. In this cycle no heat is rejected because the exhaust steam is required for process and the thermodynamic cycle efficiency of this power is affected only by the boiler efficiency, the auxiliary losses chargeable to the power generation (mostly extra boiler-feedpump work), and the mechanical and electrical losses of the turbine and generator.

The station heat rate of such by-product power generation ranges from 3900 to 4500

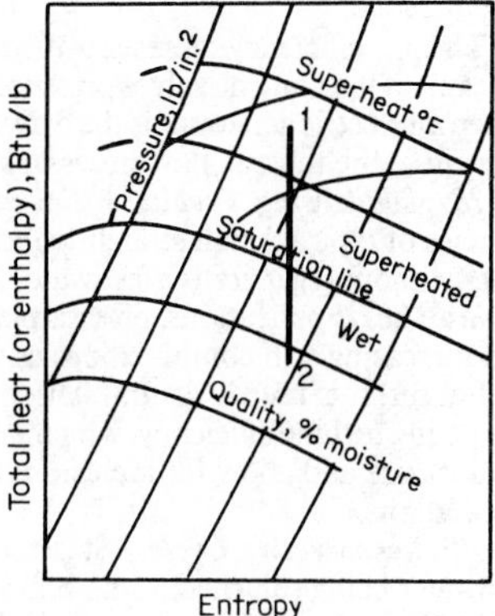


FIG. 6-34 Steam chart (Mollier diagram).

Btu/kWh, depending on the size of plant and the boiler efficiency, and cycle efficiencies of 80 to 85% are normal. This heat rate varies very little with turbine efficiency because energy not used to generate power is used for process. However, it is necessary to define the kilowatts generated per unit of heat to process in order to evaluate the influence of turbine efficiency or the initial steam conditions selected. Guaranteed steam rates are normally provided to evaluate the efficiency because they can be directly compared with the theoretical steam rate (TSR).

***16. Engine Efficiency.*** The station heat rate is used to measure power-plant performance, but it is of little use in evaluating the specific pieces of equipment in the cycle. The engine efficiency of the steam turbine defines its actual performance to the ideal performance. The Rankine-cycle work of the turbine is most conveniently obtained by use of the Mollier diagram (Fig. 6-34) where

$$\Delta W = h_1 - h_2 \tag{6-6}$$

and $\Delta W$ = Rankine-cycle work in Btu/lb, $h_1$ = steam enthalpy at throttle in Btu/lb, $h_2$ = steam enthalpy at exhaust in Btu/lb, and $h_1$ and $h_2$ are at the same entropy (vertical line). The actual work of a real turbine is less than the ideal Rankine-cycle work, with engine efficiency defined as

$$\text{Engine efficiency} = \frac{\text{actual work, Btu/lb}}{\text{Rankine-cycle work, Btu/lb}} \tag{6-7}$$

***17. Theoretical Steam Rates.*** The theoretical steam rate in lb/kWh for the Rankine-cycle work is expressed as

$$\text{TSR} = \frac{3412.14}{h_1 - h_2} \tag{6-8}$$

and the actual steam rate as

$$\text{ASR} = \frac{\text{TSR}}{\text{engine efficiency}} \tag{6-9}$$

Table 6-1 give theoretical steam rates for representative steam conditions.

The actual output of the turbine will be less than the isentropic work because of losses from nozzle and bucket friction, packing leakage, windage, carryover and exhaust losses, bearings, radiation, and throttling. The efficiency of the turbine, including its losses, is sometimes represented by the term $\eta_{evpm}$, to help identify the losses which are included in the efficiency. The subscript $e$ denotes exhaust loss (significant on condensing units only), $v$ denotes valve loss (control valves only), $p$ denotes external packing loss (valve stems and shaft end packings), and $m$ the mechanical losses. The efficiency of the generator or gear is represented by $\eta_g$.

This or similar terminology enables turbine designers to identify a particular efficiency for discussion. The stateline efficiency $\eta$ represents the actual Mollier-chart expansion line of a particular turbine for given inlet steam conditions and exhaust pressure without any exhaust loss included.

The $\eta_{evp}$ efficiency represents the internal efficiency corrected to include the exhaust loss, a "mean" of the control-valves pressure-drop loss, and the packing loss for the point in question. When an electric generator is the driven equipment, the overall engine efficiency $\eta_{evpmg}$ is used to determine the throttle flow necessary to produce a given electric output (or vice versa).

***18. Condensing-Turbine Efficiencies.*** Table 6-2 defines some approximate overall efficiencies of typical small straight condensing turbine-generators. The application of small turbines without regenerative feedwater heating is rare in central-station practice but is still found in waste-heat applications, process plants where feedwater heating is supplied by other sources, and increasingly in combined cycles where the stack gas is used to heat feedwater.

For turbines used in central stations there are more satisfactory methods (see References) for predicting turbine efficiency, which take into account the many variables of the steam path and exhaust size and allow for inclusion of the regenerative cycle and reheat cycle in heat-balance calculations.

***19. Regenerative Cycle.*** Steam can be extracted at several stages in the turbine to heat feedwater being returned to the boiler. In the Rankine cycle (Fig. 6-33) it was shown that the feedwater was heated from $h_4$ to $h_1$ in the boiler. By raising the temperature $h_4$ entering the boiler

**TABLE 6-1** Theoretical Steam Rates at Commonly Used Steam Conditions, lb/kWh
(From "Theoretical Steam Rate Tables—Compatible with the 1967 ASME Steam Tables," ASME, 1969)

| | Initial pressure, lb/in² (gage) | | | | | | | | | | | |
|---|---|---|---|---|---|---|---|---|---|---|---|---|
| | 150 | 400 | 600 | 600 | 850 | 850 | 1250 | 1250 | 1450 | 1450 | 1800 | 2400 |
| | Initial temperature, °F | | | | | | | | | | | |
| | 365.9 | 600 | 750 | 825 | 825 | 900 | 900 | 950 | 950 | 1000 | 1000 | 1000 |
| | Initial superheat, °FS | | | | | | | | | | | |
| | 0 | 151.9 | 261.2 | 336.2 | 297.8 | 372.8 | 326.1 | 376.1 | 357.0 | 407.0 | 377.9 | 337.0 |
| | Initial enthalpy, Btu/lb | | | | | | | | | | | |
| Exhaust pressure | 1195.5 | 1306.2 | 1379.6 | 1421.4 | 1410.6 | 1453.5 | 1438.4 | 1468.1 | 1461.2 | 1491.2 | 1480.1 | 1460.4 |
| inHg (abs.) | | | | | | | | | | | | |
| 1.5 | 10.10 | 7.83 | 6.88 | 6.58 | 6.41 | 6.12 | 5.99 | 5.80 | 5.76 | 5.59 | 5.54 | 5.51 |
| 2.5 | 10.88 | 8.30 | 7.25 | 6.92 | 6.72 | 6.42 | 6.26 | 6.06 | 6.01 | 5.83 | 5.77 | 5.73 |
| 3.5 | 11.49 | 8.66 | 7.53 | 7.17 | 6.96 | 6.63 | 6.46 | 6.25 | 6.20 | 6.00 | 5.94 | 5.89 |
| 5 | 12.25 | 9.09 | 7.85 | 7.48 | 7.23 | 6.89 | 6.69 | 6.48 | 6.42 | 6.21 | 6.14 | 6.08 |
| lb/in² (gage) | | | | | | | | | | | | |
| 0 | 19.38 | 12.59 | 10.40 | 9.82 | 9.32 | 8.81 | 8.41 | 8.10 | 7.98 | 7.69 | 7.54 | 7.41 |
| 5 | 21.69 | 13.55 | 11.05 | 10.42 | 9.84 | 9.29 | 8.82 | 8.49 | 8.35 | 8.04 | 7.87 | 7.71 |
| 10 | 23.97 | 14.42 | 11.64 | 10.95 | 10.30 | 9.71 | 9.18 | 8.83 | 8.67 | 8.35 | 8.16 | 7.98 |
| 20 | 28.63 | 16.02 | 12.68 | 11.90 | 11.10 | 10.43 | 9.80 | 9.42 | 9.23 | 8.87 | 8.64 | 8.42 |
| 30 | 33.69 | 17.52 | 13.63 | 12.75 | 11.80 | 11.08 | 10.34 | 9.92 | 9.70 | 9.32 | 9.06 | 8.80 |
| 40 | 39.39 | 18.98 | 14.51 | 13.54 | 12.46 | 11.66 | 10.83 | 10.38 | 10.13 | 9.72 | 9.43 | 9.14 |
| 50 | 46.00 | 20.42 | 15.36 | 14.30 | 13.07 | 12.22 | 11.28 | 10.80 | 10.53 | 10.09 | 9.77 | 9.44 |
| 60 | | 21.88 | 16.18 | 15.05 | 13.66 | 12.74 | 11.71 | 11.20 | 10.90 | 10.44 | 10.08 | 9.73 |
| 80 | | 24.87 | 17.80 | 16.54 | 14.78 | 13.77 | 12.52 | 11.95 | 11.60 | 11.09 | 10.67 | 10.25 |
| 100 | | 28.04 | 19.43 | 18.05 | 15.86 | 14.77 | 13.27 | 12.65 | 12.24 | 11.69 | 11.21 | 10.73 |
| 125 | | 32.43 | 21.56 | 20.03 | 17.22 | 16.04 | 14.17 | 13.51 | 13.01 | 12.42 | 11.84 | 11.28 |
| 150 | | 37.53 | 23.83 | 22.14 | 18.61 | 17.33 | 15.06 | 14.35 | 13.75 | 13.13 | 12.44 | 11.80 |
| 175 | | 43.64 | 26.29 | 24.43 | 20.04 | 18.66 | 15.94 | 15.20 | 14.49 | 13.83 | 13.03 | 12.29 |
| 200 | | 51.14 | 29.00 | 26.95 | 21.53 | 20.05 | 16.84 | 16.05 | 15.23 | 14.54 | 13.62 | 12.77 |
| 250 | | 73.29 | 35.40 | 32.89 | 24.78 | 23.08 | 18.68 | 17.81 | 16.73 | 15.97 | 14.78 | 13.69 |
| 300 | | | 43.72 | 40.62 | 28.50 | 26.53 | 20.62 | 19.66 | 18.28 | 17.44 | 15.95 | 14.59 |
| 400 | | | 72.2 | 67.0 | 38.05 | 35.43 | 24.99 | 23.82 | 21.64 | 20.65 | 18.39 | 16.41 |
| 500 | | | | | | | 30.29 | 28.87 | 25.52 | 24.35 | 21.06 | 18.29 |
| 600 | | | | | | | 37.03 | 35.30 | 30.16 | 28.78 | 24.06 | 20.29 |

NOTE: 1 lb/in² = 6.895 kPa; $t_{°C} = (t_{°F} - 32)/1.8$; 1 Btu/lb = 2326 J/kg; 1 in = 25.4 mm.

**TABLE 6-2** Typical Efficiencies of Straight Condensing Turbine-Generators

| | Base efficiency, $f_1$ | | | | |
|---|---|---|---|---|---|
| kW Rating | 5000 | 10,000 | 15,000 | 20,000 | 30,000 |
| 250 | 0.743 | 0.766 | | | |
| 400 | 0.733 | 0.757 | 0.769 | 0.777 | |
| 600 | 0.720 | 0.748 | 0.763 | 0.772 | 0.776 |
| 850 | | 0.742 | 0.758 | 0.768 | 0.773 |
| 1250 | | | 0.754 | 0.765 | 0.770 |
| Correction for initial superheat | | | | | |
| °FS | 0 | 100 | 200 | 300 | 400 |
| $f_2$ | 0.95 | 0.98 | 1.00 | 1.017 | 1.030 |
| Correction for exhaust pressure | | | | | |
| $P_f$, in Hg (abs.) | 1.0 | 1.5 | 2.0 | 3.0 | |
| $f_3$ | 0.98 | 1.00 | 1.01 | 1.02 | |

EXAMPLE: 15,000-kW turbine-generator
Steam conditions: 850 lb/in² (gage), 900°FTT, 2.5 inHg (abs.)
TSR = 6.42 lb/kWh Superheat = 372.8°F
100% load: $\eta_{evpmg} = f_1 \times f_2 \times f_3 = 0.758 \times 1.026 \times 1.015 = 0.789$

$$100\%\ \text{ASR} = \frac{6.42}{0.789} = 8.14\ \text{lb/kWh}$$

Throttle flow $F_r = 15{,}000 \times 8.14 = 122{,}100$ lb/h
SOURCE: Medium Steam Turbine Department, General Electric Company.

close to the saturation temperature in the drum, less fuel will be consumed in evaporating each pound of steam to $h_1$ conditions. The heat in the extracted steam is added to the feedwater without loss, and the heat rejected to the condenser decreases as extraction flow increases. The kilowatts do not decrease inversely with extraction flow, however, as partial expansion is made down to the extraction stages. The result is an improvement in heat rate which can be estimated by the use of Figs. 6-35 and 6-36.

*Example.* Using the 15,000-kW straight condensing turbine studied previously:

Steam conditions: 850 lb/in² (gage), 900°F, 2.5 inHg (abs.); ASR = 8.14 lb/kWh; $H_0 =$ 1453.0 Btu/lb. From the steam tables: At 850 lb/in² (gage), $h_{sat} = 521$, at 2.5 inHg (abs.), $h_{sat} = 77$ Btu/lb. With no FW heating the heat rate = 8.14(1453 − 77) = 11,200 Btu/kWh. If

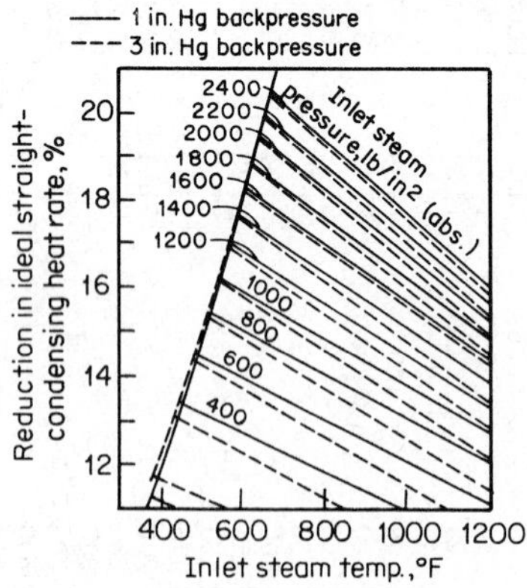


FIG. 6-35 Theoretical reduction in straight condensing heat rate.

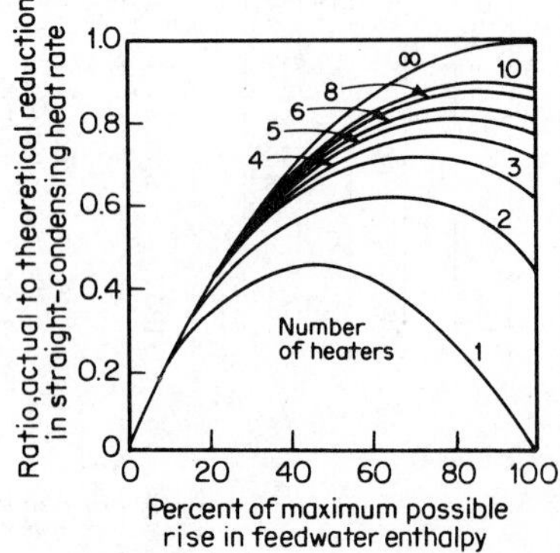


FIG. 6-36 Actual to theoretical reduction in straight condensing heat rate.

feedwater is heated to 400°F in four heaters, what would be the heat rate? At 400°F, $h_g = 375.0$, actual heat rise = 375 − 77 = 298 Btu/lb. Possible heat rate rise = 521 − 77 = 444 Btu/lb; actual/possible = 0.671. Theoretical reduction in straight condensing heat rate = 13.5% (see Fig. 6-35). Actual/theoretical reduction in nonextraction heat rate = 0.755 (see Fig. 6-36). Therefore, the reduction in the nonextraction heat rate = 13.5 × 0.755 = 10.2%, or 1140 Btu/kWh and approximate heat rate with four stages FWH to 400°F = 10,060 Btu/kWh.

The actual heat rate of a regenerative cycle must be determined from a heat balance prepared by using the extraction conditions available from the turbine and the heater characteristics as specified. Table 6-3 shows the influence of heater type, temperature difference, and piping pressure drop on the gross heat rate of a 50,000-kW unit.

***20. Reheat Cycle.*** Practically all large (over 100,000 kW) central-station plants built since 1950 have been of the reheat type. In this type of fossil-fuel plant the steam is expanded in the turbine down to about 25% of the initial pressure, then it is sent through a reheater where it is resuperheated back up to the original initial temperature (usually 1000°F) and returned to the turbine where it is expanded to completion. In general, reheating improves the heat rate by about 5% of which only 2% is due to a higher average cycle temperature and about 3% comes from improved turbine internal efficiency due to reduced moisture and increased reheat factor.

***21. Gross and Net Heat Rates.*** The heat-balance cycle for the turbine, condenser, feedwater pump, and heaters usually defines the gross heat rate of the cycle when a motor-driven

**TABLE 6-3** Heat-Rate Variation with Heater Cycle of 50,000-kW Nonreheat Turbine [1250 lb/in² (gage), 950°FTT, 1.5 inHg (abs.)]

Base cycle HR = 8852 Btu/kWh, FWT = 430°F

| | Heater no. | | | | | ΔHeat rate, Btu/kWh |
|---|---|---|---|---|---|---|
| | 5 | 4 | 3 | 2 | 1 | |
| TTD, °F | 5 | 5 | open | 5 | 5 | Base |
| DCTD, °F | 10 | 10 | Htr. | 10 | 10 | Base |
| %ΔP | 5 | 5 | 5 | 5 | 5 | Base |
| Variation in terminal temperature difference, TTD | | | | | | |
| | 0 | 5 | | 5 | 5 | −10 |
| | −5 | 5 | | 5 | 5 | −20 |
| | 5 | 5 | | 5 | 10 | +6 |
| | 10 | 10 | | 10 | 10 | +27 |
| Variation in drain cooler temperature difference, DCTD | | | | | | |
| | 5 | 5 | | 5 | 5 | −2 |
| | 20 | 20 | | 20 | 20 | +5 |
| | 10 | 10 | | 10 | 20 | +2 |
| Variation in pressure drop to heater, %ΔP | | | | | | |
| | 0 | 0 | 0 | 0 | 0 | −24 |
| | 10 | 10 | 10 | 10 | 10 | +26 |
| | 10 | 5 | 5 | 5 | 5 | +10 |
| Variation from drain cooled to pumped or cascaded drips | | | | | | |
| | 10 | 10 | | 10 | C | +12 |
| | C | C | | C | C | +30 |
| | 10 | PD | | 10 | PD | 0 |
| | 10 | C* | | 10 | 10 | +32 |

* Cascaded to No. 2 heater.

NOMENCLATURE: TTD = heater terminal temperature difference; DCTD = drain cooler approach difference; %ΔP = pressure drop from turbine flange to heater shell; C = cascaded heater drains; PD = drains pumped forward.

NOTE: $t_{°C} = (t_{°F} - 32)/1.8$.

SOURCE: Medium Steam Turbine Department, General Electric Company.

feed pump is used. The pump work on the feedwater is included, but the power consumed by the pump is not. When a turbine-driven boiler feed pump is used in the cycle (frequently in units above 300-MW rating) the steam for the feed-pump turbine is expanded in the main unit down to the crossover to the low-pressure elements [about 100 to 200 lb/in$^2$ (abs.)] where it is sent to the pump turbine. In these cases the pump power required is included in the heat balance, and the heat rate calculated is called the net heat rate. In this case only the losses from the boiler and auxiliaries (excluding the feedwater pump) must be accounted for to obtain the station heat rate.

Table 6-4 shows typical values of gross and net heat rates at rated load for nonreheat and reheat units of typical ratings and inlet steam conditions. Exhaust pressure is 2.0 inHg (abs.) in all cases, motor driven feed-pump drive efficiency is 90%, pump efficiency is 78%, and the exhaust annulus area is normal for the rating.

The overall station heat rates of these applications are 15 to 25% greater than the net heat rate, depending on the steam-generator (boiler) efficiency and other auxiliary losses.

***22. Nuclear Cycles.*** The turbines used in light-water nuclear cycles do not have the same freedom of steam conditions as fossil-fired cycles. The nuclear steam supply limits initial steam pressure to approximately 950 lb/in$^2$ (gage) at a saturated steam temperature of 540°F. The initial costs of these plants is so great that only the largest can be economically justified, and ratings are limited to about 800 MW minimum. Crossover pressures range from about 150 to 200 lb/in$^2$ (abs.), and regenerative feedwater heating improvement is optimized by using about six heaters. These constraints and optimizations of the nuclear power-plant cycles have resulted in a rather narrow band of heat-rate fluctuations. At rated load and 2 inHg exhaust pressure the turbine net heat rate varies from about 9800 Btu/kWh at an exhaust loading of 1500 kW per square feet of annulus area up to about 10,000 Btu/kWh at 2000 kW per square feet of exhaust loading.

Turbines used in the high-temperature, gas-cooled reactor cycle operate at steam conditions, including reheat similar to those used in fossil-fired plants. Because of the low moisture content and lower volume flow required, 3600-r/min turbine design speed can be utilized as well. At 2400 lb/in$^2$ (gage), 1000/1000°F, 2.0 inHg (abs.) design conditions, a heat rate of about 8300 Btu/kWh can be obtained, including the power required for gas recirculation. This represents a 15 to 17% improvement in cycle efficiency when the first plant becomes operational.

***23. Combined Cycles.*** Improvements in the steam cycle since the 1920s have been rapid. The advancement of steam conditions, regenerative feed heating, reheating, and size of unit have brought the overall station heat rate down from 16,000 Btu/kWh to less than 8800 Btu/kWh on the best stations. During the 1930s and 1940s attempts were made to use other fluids in combination or as a substitute for steam with little success. Considerable effort was made in the combined mercury-steam binary cycle, and several plants were built. At that time the low saturation pressure of mercury made it attractive for use in the high-temperature part of the cycle where it was expanded to completion and condensed at about 1.3 lb/in$^2$ (abs.), 475°F, in a condenser boiler which generated saturated steam at 350 lb/in$^2$ (abs.).

After superheating to 750°F in the mercury fossil-fuel boiler, the steam was expanded in the steam turbine. This binary cycle produced a station heat rate of about 9500 Btu/kWh and had a decided attraction where fuel costs were high in the Northeast, as conventional steam plants had progressed to only 12,000 to 13,000 Btu/kWh. Further advances in high-pressure and temperature steam turbines, and the drawbacks of the mercury cycle (mercury is poisonous, expensive, and has poor heat-transfer properties) led to its demise.

The gas turbine developed very rapidly as a prime mover after World War II. In the very beginning its potential for offering an improvement in the steam cycle was recognized. During the 1950s several exhaust-fired combined-cycle power plants were built, utilizing the exhaust gas from the gas turbine as the air supply for a fired main steam generator. After the Northeast blackout of 1965 a large number of gas turbines were installed in the United States to serve as black start and peaking capacity units. As a result they have become well established and accepted as a prime mover for peaking capacity.

Combined-cycle interest was renewed with the development of non-radiant-heat recovery steam generators, and the electric generation in combined-cycle plants changed from 80 to 90% steam-cycle power to 70% gas-cycle power. Since 1970 utilities have installed increasing numbers of this breed of combined-cycle plant, as they offer low initial cost, consume about one-third of the water used by straight steam plants, and provide a station heat rate 5 to 10% better than the most efficient steam plants. The major obstacle to universal acceptance of the steam-and-gas combined cycle is its fuel dependency on clean gaseous or liquid fuels.

**TABLE 6-4** Typical Heat Rates of Nonreheat and Reheat Turbine-Generators

| | Steam conditions | | | Last-stage buckets | | | | | Performance at 2.0 inHg (abs.) | | | |
|---|---|---|---|---|---|---|---|---|---|---|---|---|
| Rating, MW | $P_0$, lb/in² (gage) | $T_0$, °F | $T_{RH}$, °F | No. rows | Annulus area, ft² | Boiler feedpump drive | Number of feedwater heaters | FFWT, °F | Gross heat rate, Btu/kWh | Net heat rate, Btu/kWh | Throttle SR, lb/kWh | Condenser SR, lb/kWh |
| 15 | 600 | 825 | | 1 | 12 | Motor | 3 | 390 | 10610 | 10700 | 10.05 | 7.52 |
| 25 | 850 | 900 | | 1 | 14 | Motor | 4 | 365 | 9730 | 9850 | 8.73 | 6.69 |
| 40 | 1250 | 950 | | 1 | 19 | Motor | 5 | 444 | 9240 | 9410 | 8.86 | 6.23 |
| 50 | 1250 | 950 | | 1 | 26 | Motor | 5 | 429 | 9080 | 9240 | 8.56 | 6.10 |
| 75 | 1250 | 950 | | 1 | 41 | Motor | 5 | 437 | 8890 | 9040 | 8.44 | 5.96 |
| 75 | 1450 | 1000 | 1000 | 1 | 33 | Motor | 5 | 431 | 8320 | 8450 | 6.69 | 4.92 |
| 75 | 1800 | 1000 | 1000 | 1 | 33 | Motor | 5 | 448 | 8140 | 8290 | 6.70 | 4.80 |
| 100 | 1800 | 1000 | 1000 | 1 | 41 | Motor | 5 | 458 | 8110 | 8270 | 6.77 | 4.80 |
| 150 | 1800 | 1000 | 1000 | 2 | 66 | Motor | 6 | 448 | 7970 | 8130 | 6.51 | 4.66 |
| 200 | 2400 | 1000 | 1000 | 2 | 82 | Motor | 7 | 461 | 7750 | 7950 | 6.39 | 4.49 |
| 350 | 2400 | 1000 | 1000 | 2 | 132 | Motor | 7 | 473 | 7750 | 7950 | 6.49 | 4.45 |
| 350 | 2400 | 1000 | 1000 | 2 | 132 | Turbine | 7 | 473 | | 7890 | 6.62 | 4.27 |
| 500 | 2400 | 1000 | 1000 | 4 | 222 | Turbine | 7 | 473 | | 7830 | 6.57 | 4.26 |
| 700 | 2400 | 1000 | 1000 | 4 | 264 | Turbine | 7 | 473 | | 7870 | 6.58 | 4.27 |
| 1000 | 3500 | 1000 | 1000 | 6 | 334 | Turbine | 7 | 504 | | 7730 | 6.71 | 4.04 |

NOTE: 1 ft² = 0.0929 m²; 1 lb/in² = 6.895 kPa; $t_{°C} = (t_{°F} - 32)/1.8$; 1 in = 25.4 mm.
SOURCE: Medium Steam Turbine Department, General Electric Company.

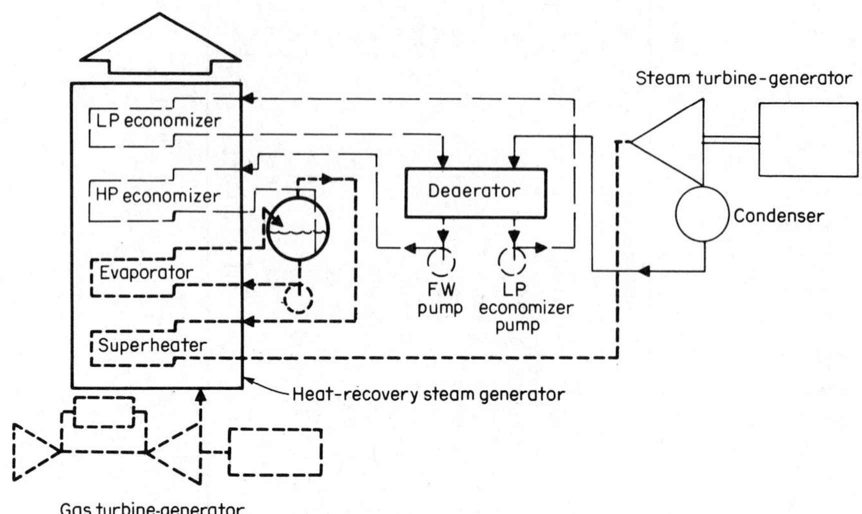


**FIG. 6-37** Combined-cycle diagram. *(GE Publ. GEA-100015.)*

Figure 6-37 describes a typical arrangement and flow path of a combined-cycle plant with an unfired steam generator. In this cycle the turbine condensate is mixed with sufficient LP economizer flow in the deaerator to raise the temperature sufficiently to avoid corrosion at the cold end of the heat-recovery steam generator. The use of regenerative feedwater heaters is avoided because of the availability of excess heat in the stack for that purpose. Typical performance of combined-cycle plants with unfired steam generators is shown in Table 6-5. Part-load heat rate can be maintained at greatly reduced load in the multigas turbine plants because the units can be put in service sequentially. A so-called hockey-stick curve of part-load performance is shown in Fig. 6-38 for a combined-cycle plant utilizing four gas turbines. Below about 75% load a gas turbine should be removed from service for best plant efficiency.

**TABLE 6-5** Combined-Cycle Performance

(ISO conditions, 59°F, sea level, base load on gas turbine using No. 2 distillate)

| Number GT | HRSG | Steam Turbine-Generator | Net plant output, MW | Net plant heat rate, Btu/kWh-HHV |
|---|---|---|---|---|
| 1 | 1 | 1 | 98 | 8150 |
| 4 | 4 | 1 | 396 | 8070 |
| 6 | 6 | 1 | 595 | 8060 |

NOTE: Transformer loss not included.
SOURCE: Medium Steam Turbine Department, General Electric Company.

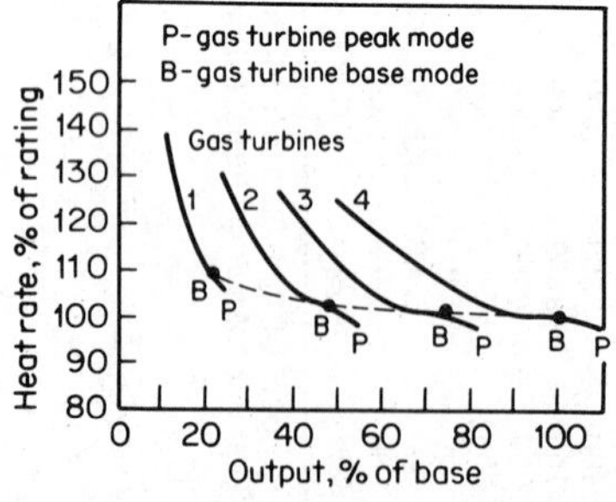


**FIG. 6-38** Output versus heat-rate chart. *(STAG Combined-Cycle Power System.)*

***24. Noncondensing-Turbine Efficiencies.*** The straight noncondensing turbine-generator is widely used in the process industries. Table 6-6 defines some approximate overall efficiencies at typical rating and steam

**TABLE 6-6** Estimating Straight Noncondensing Turbine-Generator Performance

Corrected efficiency at any load $\eta_{evpmg} = f_1 \times f_2 \times f_3 \times f_4$
Base efficiency, ($f_1$) at 100% load and $P_o/P_f$ = 6.0

| | Initial pressure, lb/in² (gage) | | | | | |
|---|---|---|---|---|---|---|
| | 400 | 600 | 850 | 1250 | 1450 | 1800 |
| kW Rating | | | | | | |
| 5,000 | 0.765 | 0.745 | 0.721 | | | |
| 10,000 | 0.789 | 0.779 | 0.766 | 0.750 | 0.744 | |
| 15,000 | 0.806 | 0.800 | 0.791 | 0.779 | 0.774 | |
| 20,000 | | 0.811 | 0.805 | 0.795 | 0.789 | |
| 25,000 air-cooled | | | 0.809 | 0.801 | 0.797 | |
| 30,000 $H_2$-cooled | | | 0.816 | 0.810 | 0.807 | 0.802 |
| 40,000 | | | | 0.818 | 0.816 | 0.812 |
| 50,000 | | | | 0.823 | 0.821 | 0.818 |

Correction for pressure ratio, $f_2$

| $P_o/P_f$ | 2.0 | 3.0 | 4.0 | 6.0 | 10.0 | 20.0 |
|---|---|---|---|---|---|---|
| $f_2$ | 0.94 | 0.97 | 0.99 | 1.00 | 1.00 | 1.00 |

Correction for superheat, $f_3$

| Superheat, °FS | 0 | 50 | 100 | 200 |
|---|---|---|---|---|
| $f_3$ | 0.95 | 0.975 | 0.99 | 1.00 |

Correction for part load, $f_4$

| TSR, % load | 100 | 80 | 60 | 40 |
|---|---|---|---|---|
| 10 | 1.000 | 0.985 | 0.952 | 0.881 |
| 20 | 1.000 | 0.981 | 0.932 | 0.833 |
| 30 | 1.000 | 0.977 | 0.923 | 0.808 |

NOTE: 1 lb/in² = 6.895 kPa.
SOURCE: Medium Steam Turbine Department, General Electric Company.

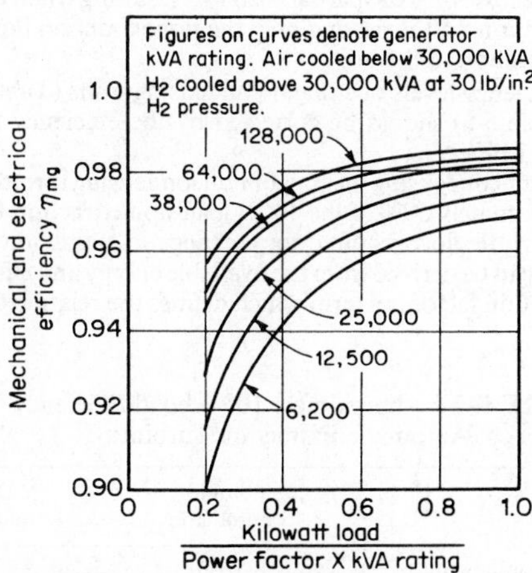


**FIG. 6-39** Typical mechanical and electrical efficiencies of turbine-generators. *(Medium Steam Turbine Department, General Electric Company.)*

conditions for these turbines. The flow through a noncondensing turbine is dependent upon the heat required in the process, and in order to determine the heat leaving the turbine it is also necessary to know the enthalpy of the exhaust steam. Figure 6-39 is a plot of the mechanical and electrical efficiency of several turbine generators from 20 to 100% load. This curve permits the derivation of the internal wheel efficiency ($\eta_{evp}$) of the turbine, and ultimately the exhaust enthalpy.

*Example.* 50,000-kW TG unit, 64,000-kVA generator, 0.9 pf, 30-lb $H_2$. Steam conditions: 1250 lb/in² (gage), 950°F, 150 lb/in² (gage). From Table 6-1: TSR = 14.35 lb/kWh, Superheat = 376.1°FS; $H_0$ = 1468.1 Btu/lb, $P_0/P_f$ = 1264.7/164.7 = 7.7; AE = 3412.1/14.35 = 237.8 Btu/lb.

| | | | | |
|---|---|---|---|---|
| kW Load (given) | 50,000 | 40,000 | 30,000 | 20,000 |
| % kW (kW load rating) | 100 | 80 | 60 | 40 |
| $f_1$ (Table 6-6) | 0.823 | → | → | → |
| $f_2$ (Table 6-6) | 1.00 | → | → | → |
| $f_3$ (Table 6-6) | 1.00 | → | → | → |
| $f_4$ (Table 6-6) | 1.00 | 0.983 | 0.942 | 0.857 |
| $\eta_{evpmg}$, ($f_1 \times f_2 \times f_3 \times f_4$) | 0.823 | 0.809 | 0.775 | 0.705 |
| ASR, lb/kWh [Eq. (6-9)] | 17.44 | 17.74 | 18.52 | 20.37 |
| Throttle flow, lb/h kW × ASR | 872,000 | 709,600 | 555,600 | 407,400 |
| kW/Pf × kVA Rating, kW / (0.9 × 64,000) | 0.868 | 0.694 | 0.521 | 0.347 |
| $\eta_{mg}$ (Fig. 6-39) | 0.983 | 0.981 | 0.977 | 0.968 |
| Wheel eff. $\eta_{evp}$, ($\eta_{evpmg}/\eta_{mg}$) | 0.837 | 0.825 | 0.793 | 0.728 |
| $h_1$, Btu/lb (given) | 1468.1 | → | → | → |
| UE, Btu/lb, (AE × $\eta_{evp}$) | 199.0 | 196.2 | 188.6 | 173.1 |
| $h_2$, Btu/lb ($h_1$-UE) | 1269.1 | 1271.9 | 1279.5 | 1295.0 |

The exhaust enthalpy ($h_2$) times the throttle flow as derived in the example represents the heat flow to process. Very little error is introduced by the neglect of external packing leakages.

***25. Automatic-Extraction-Turbine Efficiencies.*** The automatic-extraction turbine provides the capability of delivering extraction steam at more than one process pressure simultaneously. When a condensing element is used, the kilowatt output of the unit can be maintained if the process flow varies, and will permit generation in excess of by-product power capability. The base efficiency for an automatic-extraction turbine is less than that of a straight condensing or straight noncondensing turbine because of *(a)* the introduction of a second control stage and accompanying parasitic losses, *(b)* the partial-load loss resulting when the high-pressure section of the automatic-extraction unit is passing only the nonextraction flow, and *(c)* the decreased pressure ratio of each section of the unit.

The 100% load efficiency levels of straight condensing units (Table 6-2) and straight noncondensing units (Table 6-6) should be corrected by the efficiency factors in Table 6-7 for automatic-extraction turbines.

The 50% load straight condensing or straight noncondensing throttle flow for automatic-extraction units is approximately 59% of the 100% load nonextraction flow.

The variation in throttle flow required for a change in extraction flow while maintaining constant kilowatt load can be derived from the available energy and efficiency of each section of the turbine. The extraction factor is a term which defines the relationship and represents the Δ

**TABLE 6-7** Approximate 100% Load Efficiency Factors of Automatic-Extraction Turbines

| | Single automatic | Double automatic |
|---|---|---|
| Noncondensing | 0.93 | 0.89 |
| Condensing | 0.96 | 0.92 |

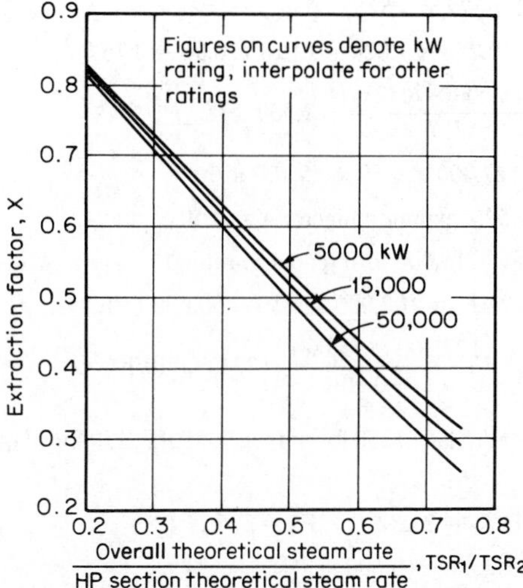


**FIG. 6-40** Extraction factors. *(Medium Steam Turbine Department, General Electric Company.)*

throttle flow ($\Delta F_T$) required for a $\Delta$ extraction flow ($\Delta F_x$) of 1 lb/h at constant kilowatt load. The term represents a comparison of the used energy in the turbine below the automatic-extraction point to the total used energy of the turbine. Figure 6-40 describes the approximate extraction factor variations with the ratio of the theoretical steam rates for the turbine.

As extraction flow increases, the HP section of the unit generates more of the kilowatts and the LP section generates less until the steam flow to the LP stages is at the minimum necessary for cooling purposes. At this point the maximum extraction for the load in question has been reached and further extraction flow must be accompanied at increasing output. The minimum cooling steam required varies with turbine size, extraction pressure, and the exhaust pressure. As an approximation, the minimum section flow in pounds per hour can be considered equal to the rating of the turbine in kilowatts.

Estimating performance of automatic-extraction turbines may be put in equation form:

$$100\%\text{ load } F_{T_{0x}} = \frac{\text{kW rating} \times \text{TSR}_1}{\text{S.C. or SNC eff.} \times \text{auto. extr. factor}} \tag{6-10}$$

$$50\%\text{ load } F_{T_{0x}} = 100\%\ F_{T_{0x}} \times 0.59 \tag{6-11}$$

$$\text{With extraction flow } F_T = F_{T_{0x}} \times F_x X \tag{6-12}$$

$$\text{Max. extraction at any load } F_{x_{\max}} = \frac{F_{t_{0x}} - F_{c_{\min}}}{1 - X} \tag{6-13}$$

Where $F_{T_{0x}}$ represents the zero extraction throttle flow, lb/h, at any load point, $F_x$ = extraction flow, $F_c$ = exhaust flow (LP section flow), and $F_{c_{\min}}$ = minimum exhaust flow. SC or SNC efficiencies are obtained for the total unit from Table 6-2 or 6-6, automatic-extractor factor from Table 6-7, TSRs from Table 6-1, and $f_x$ from Fig. 6-40.

Two illustrations of this approximate method of determining the performance of automatic-extraction turbines are described in the following cases.

*Case 1.* 15,000 kW SAXC; 850 lb/in$^2$ (gage), 900°F, 2.5 inHg (abs.); AE at 50 lb/in$^2$ (gage).

Overall TSR$_1$ = 6.42 °FS = 372.8

HP section TSR$_2$ = 12.22

$TSR_1/TSR_2 = 6.42/12.22 = 0.525$

100% kW OX eff. $= 0.758 \times 1.026 \times 1.015 \times 0.96 = 0.757$

At 15 MW $F_{T_{0x}} = \dfrac{15{,}000 \times 6.42}{.757} = 127{,}300$ lb/h

At 7.5 MW $F_{T_{0x}} = 127{,}300 \times 0.59 = 75{,}100$ lb/h

At $TSR_1/TSR_2 = 0.525$, extraction factor $X = 0.50$

Min. LP section flow = 15,000 lb/h (equal to rating)

per 100,000 lb/h $F_x$, $\Delta F_T = 100{,}000 \times 0.50 = 50{,}000$ lb/h

At 15 MW max. $F_x = \dfrac{127{,}300 - 15{,}000}{1 - 0.50} = 224{,}600$ lb/h

*Case 2.* 30,000 kW SAXNC; 1250 lb/in² (gage), 950°F, 150 lb/in² (gage); AE at 400 lb/in² (gage).

$TSR_1 = 14.35$   $TSR_2 = 23.82$   °FS = 376.1

$TSR_1/TSR_2 = 14.35/23.82 = 0.603$

Overall $P_0/P_f = 7.7$

100% kW OX eff. $= 0.810 \times 1.0 \times 1.0 \times 1.0 \times 0.93 = 0.753$

100% $F_{T_{0x}} = \dfrac{30{,}000 \times 14.35}{.753} = 572{,}000$ lb/h

50% $F_{T_{0x}} = 572{,}000 \times 0.59 = 337{,}000$ lb/h

At $TSR_1/TSR_2 = 0.603$, $X = 0.41$ (Fig. 6-40)

Min. LP section flow = 30,000 lb/h

per 100,000 lb/h $F_x$, $F_T = 100{,}000 \times 0.41 = 41{,}000$ lb/h

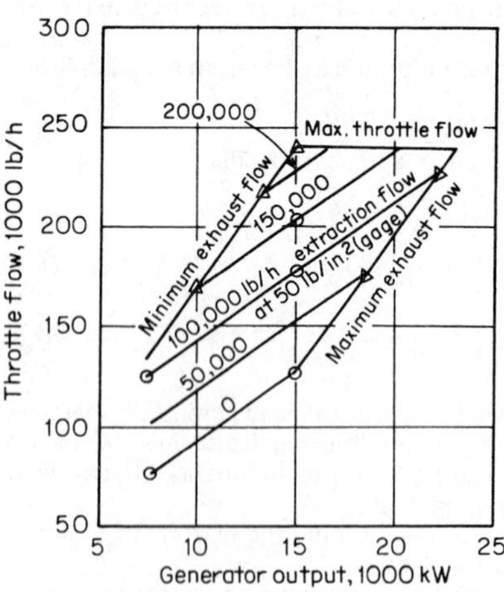


**FIG. 6-41** Turbine-generator performance (15 MW); 85 lb/in² (gage), 900 FTT, 215 inHg (abs.) AE at 50 lb/in² (gage). *(Medium Steam Turbine Department, General Electric Company.)*

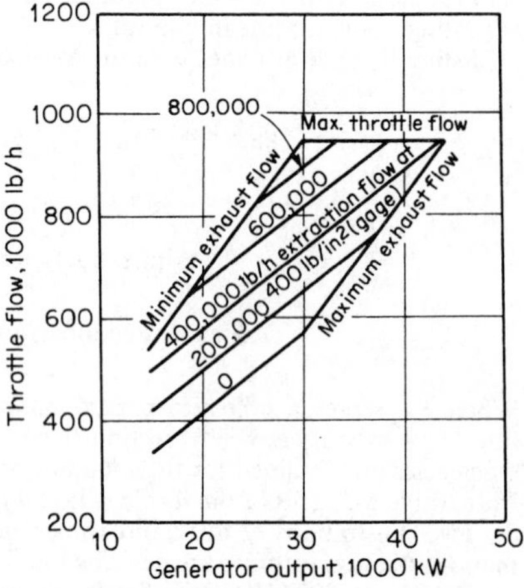


**FIG. 6-42** Turbine-generator performance (30 MW); 1250 lb/in² (gage), 950°F, 150 lb/in² (gage) AE at 400 lb/in² (gage). *(Medium Steam Turbine Department, General Electric Company.)*

$$\text{At 30 MW max. } F_x = \frac{572{,}000 - 30{,}000}{1 - 0.41} = 919{,}000 \text{ lb/h}$$

Performance curves for both cases are shown in Figs. 6-41 and 6-42 with extraction lines drawn parallel to the zero extraction performance at a vertical spacing equal to the extraction flow times extraction factor. Minimum and maximum section flow cutoffs are defined by $F_T$ $F_c + F_x$.

## References on Steam Turbines

***26. References***


Baumeister, T.: *Marks' Standard Handbook for Mechanical Engineers,* 7th ed., New York, McGraw-Hill Book Company, 1967.

ASME, Boiler and Pressure Vessel Code, Piping Code, Power Test Codes, Fluid Meters Committee Report, Bibliography on Gas Turbines, Annual Reports on Oil Engine Power Costs.

Ayres and Scarlott: *Energy Sources: The Wealth of the World,* New York, McGraw-Hill Book Company.

Babcock & Wilcox Company: Steam, Its Generation and Use.

Combustion Engineering, Inc.: Combustion Engineering.

Downs and Holley: Progress in the Design of Large Steam-turbine Generators, *Natl. Power Conf., IEEE* and *ASME,* 1965.

Edison Electric Institute: Prime Movers Committee Reports.

*Electrical World,* Annual Steam Station Design Survey.

Elston and Knowlton: Comparative Efficiencies of Central Station Reheat and Non-reheat Steam Turbine-generator Units, General Electric Company, *GER*-482.

Faires: *Applied Thermodynamics;* New York, The Macmillan Company.

Fiehn: Major Influences of Large Unit Size on Steam-electric Station Design; *Combustion,* July 1966.

Gaffert, G. A.: *Steam Power Stations,* 4th ed.; New York, McGraw-Hill Book Company, 1952.

General Electric Company: Turbine Generator Foundations, *GET*-1749.

Heat Exchange Institute: Standards.

Newman: *Modern Turbines;* New York, John Wiley & Sons, Inc.

*Power,* Annual Energy Systems Design Survey, Power Handbook, Pump Handbook, Steam Turbine Handbook.

Salisbury: *Steam Turbines and Their Cycles;* New York, John Wiley & Sons, Inc.

Skrotzki, B. G. A., and Vopat, W. A.: *Applied Energy Conversion;* New York, McGraw-Hill Book Company.

Spencer, Cotton, and Cannon: A Method for Predicting the Performance of Steam Turbine-generators 16500 kW and Larger, *Trans. ASME,* 1963.

Bailey, Cotton, and Spencer: Predicting the Performance of Large Steam Turbine-generators Operating with Saturated Steam and Low Superheat Conditions, *Am. Power Conf.,* 1967.

Heat Rates for Fossil Reheat Cycles using General Electric Steam Turbine Generators 150,000 kW and Larger, *Get* 2050C, 1974.

Theoretical Steam Rate Tables Compatible with the 1967 ASME Steam Tables, 1969.

Bartlett: *Steam Turbine Performance and Economics;* New York, McGraw-Hill Book Company, 1958.

Hall, D.: The Effect of Industrial Turbine Design on Plant Operating Costs, *TAPPI Eng. Conf.,* 1971.

Yellott: Supersaturated Steam, *Trans. ASME,* 1934.

Eggenberger, M. A.: Introduction to the Basic Elements of Control Systems for Large Steam Turbine Generators, *GET*-3096A, 1966.

Eggenberger, M. A.: A Simplified Analysis of the No-Load Stability of Mechanical-Hydraulic Speed Control Systems for Steam Turbines, *ASME Paper* 60-*WA*-34, 1960.

AIEE and ASME, Recommended Specification for Speed-Governing of Steam Turbines Intended to Drive Electric Generators Rated 500 kW and Larger, published by American Institute of Electrical Engineers, 1959.

Overspeed Trip Systems for Steam Turbine Generator Units, ASME PTC 20.2, 1965.

ASME Steam Tables—1967.

## GAS TURBINES

*By ROY P. ALLEN*

***27. Cycles.*** Mixture engines operate on the Otto cycle, injection engines operate on the diesel cycle, and the gas or combustion turbine operates on the Brayton cycle (Fig. 6-43). Most

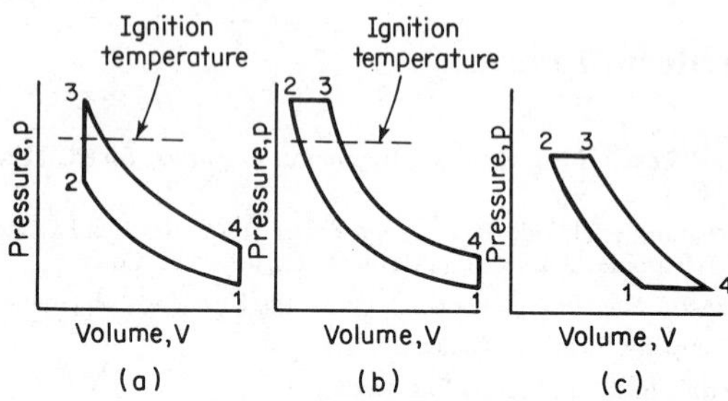


**FIG. 6-43** Ideal indicator cards (pressure-volume diagrams) for internal combustion cycles. *(a)* Otto cycle. *(b)* Diesel cycle. *(c)* Brayton cycle. In general, Phase 1-2 represents isentropic compression; 2-3 heat addition at constant pressure or volume; 3-4 isentropic expansion; 4-1 heat rejection at constant pressure or volume.

turbines are open cycle, where ambient air is the working fluid. An axial or centrifugal compressor (Fig. 6-44*a*) delivers the compressed air to the combustion system, and fuel is added to increase the fluid temperature. The gases expand through the turbine, producing sufficient power to drive the compressor and the load device. The gases are exhausted to atmosphere at a relatively high temperature. In a closed cycle (Fig. 6-44*b*) the fluid, usually air or helium, circulates continuously through the compressor, turbine, and heat exchangers. Heat is supplied from outside the cycle by a nuclear reactor, coal burner, or other source. An improvement in efficiency of both open and closed cycles can be obtained through use of a regenerator or recuperator (Fig. 6-44*c*), which exchanges heat from the exhaust to the combustor inlet to reduce the fuel required to heat the gas. The highest efficiencies are available from combined cycles, where the gas-turbine exhaust heat produces steam to power a steam generator (Fig. 6-44*d*). The steam-turbine output is obtained with no additional fuel input.

***28. Design.*** Nearly all gas turbines above 1000 kW have multistage axial-flow compressors and turbines. Lower-power units tend to have single-stage centrifugal compressors and radial turbines to minimize weight, size, and cost. Thermal efficiency improves with size as friction and tip leakages become a smaller percentage of the power produced, and multistage axial-flow components are more efficient than single radial stages. Figure 6-45 shows typical gas-turbine design features.

For applications where output shaft speed varies, as in a compressor drive, a multiple-shaft turbine is employed (Fig. 6-46). The high-pressure turbine drives the compressor; the remaining energy drives the load through a separate turbine stage (or stages). The single-shaft unit has an effective operating range of 85 to 105% of rated speed or less; the multiple-shaft arrangement allows the high-pressure set to operate in this band, while the low-pressure set speed can be controlled independently (50 to 105% speed). Much larger starting and part-load torques, which are needed to accelerate devices under load such as compressors and pumps, are available with this arrangement.

***29. Performance.*** Component efficiencies, air-flow, compression ratio, and turbine inlet (or firing) temperature ($T_f$) are the major factors affecting gas-turbine output and efficiency. Multistage axial component efficiencies are in the 88 to 92% range. Material developments and turbine cooling techniques permit $T_f$ to exceed 2000°F (1100°C). Table 6-8 shows typical thermal efficiencies and optimum compression ratios for various cycles.

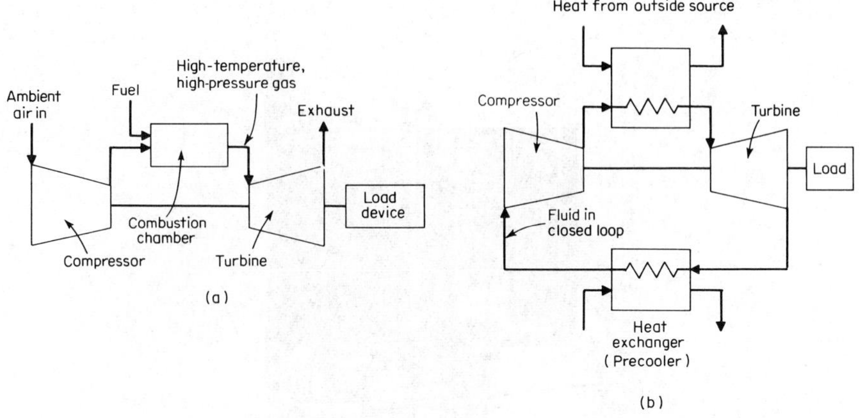


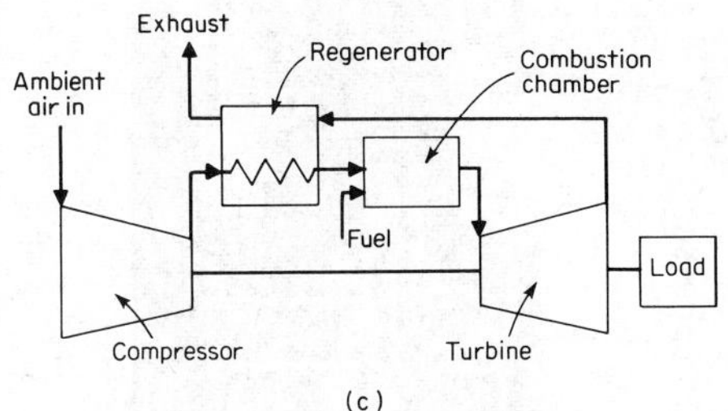


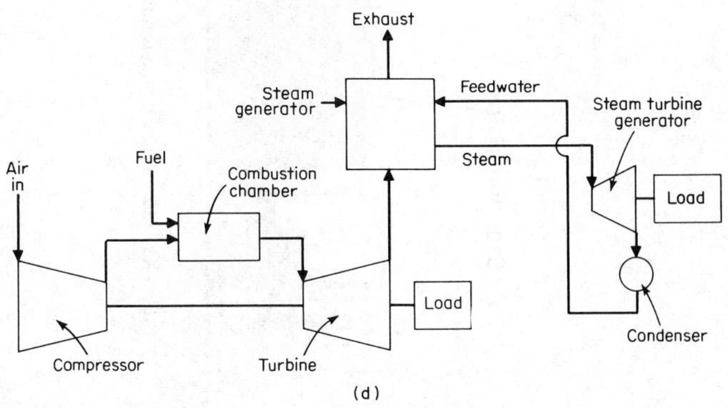


**FIG. 6-44** Typical gas turbine cycles. *(a)* Open; *(b)* closed; *(c)* regenerative; *(d)* combined.

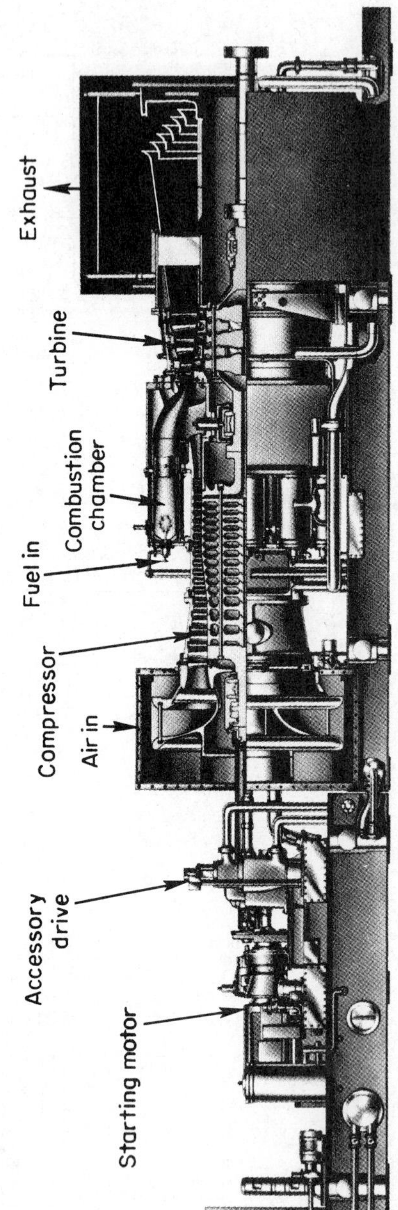


**FIG. 6-45** Gas turbine design features.

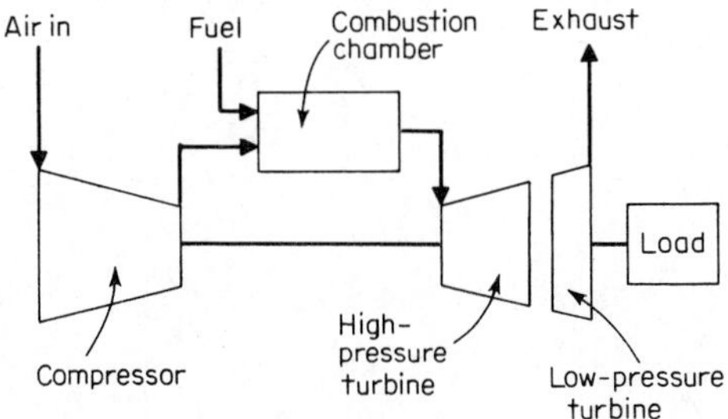


**FIG. 6-46** Multiple-shaft turbine.

**TABLE 6-8** Compression Ratio and Thermal Efficiencies

| Cycle | Optimum compression ratio | Cycle thermal efficiency, % |
|---|---|---|
| Simple | High 15 | 30–35 |
| Regenerative | Medium 6–8 | 35–38 |
| Closed | Low 3–4 | 30–33 |
| Combined | Moderate 10–15 | 40–45 |

***30. Applications.*** The most familiar application of gas turbines has been for aircraft propulsion, where the turbine drives only the compressor and the remaining energy is used for thrust. Generation of electric power and gas or oil pipeline compression are the major uses involving driven equipment. Extensive application of gas turbines to auto, marine, and locomotive propulsion have also been made, along with auxiliary power units for aircraft and aerospace use.

The ability to burn a variety of fuels has become an important attribute. Economics and supply of fuels are the leading factors that are causing users to change from the conventional natural gas and distillate oils to crude and residual oils, gas that is derived from oil and coal, methanol, and other materials. The prevention of turbine-section erosion and corrosion must be considered in using these alternate fuels.

# SECTION 7

# ALTERNATING-CURRENT GENERATORS

**R. E. Appleyard**
*Engineering Consultant, Formerly Manager Engineering and Development, Large Rotating Apparatus Division, Siemens–Allis, Inc.*

**L. T. Rosenberg, M.S., P.E.**
*Turbogenerator Consultant, Formerly Chief Generator Design Engineer, Steam Turbine Department, Allis-Chalmers Corp.*

## CONTENTS

*Numbers refer to paragraphs*

## GENERAL

***1. General Construction.*** An alternating-current generator consists principally of a magnetic circuit, dc field winding, ac armature winding, and mechanical structure, including cooling and lubricating systems. The magnetic circuit and field windings are arranged so that, as the machine rotates, the magnetic flux linking the armature winding changes cyclically, thereby inducing alternating voltage in the armature winding.

Many different geometrical arrangements of these elements are possible, and each has its own economical field of application. For high power generation, the most common types are (1) the salient-pole construction illustrated in Fig. 7-1 and characteristic of hydraulic turbine or

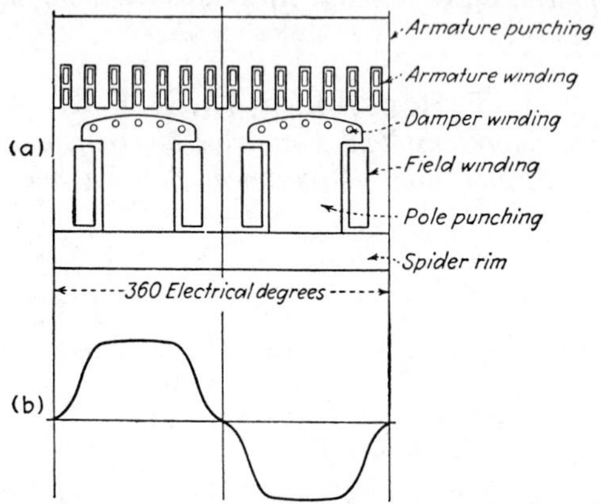


**FIG. 7-1** *(a)* Generalized sketch of one pair of poles for a salient-pole machine. *(b)* Flux form for typical pair of poles (on the basis of current in field winding only).

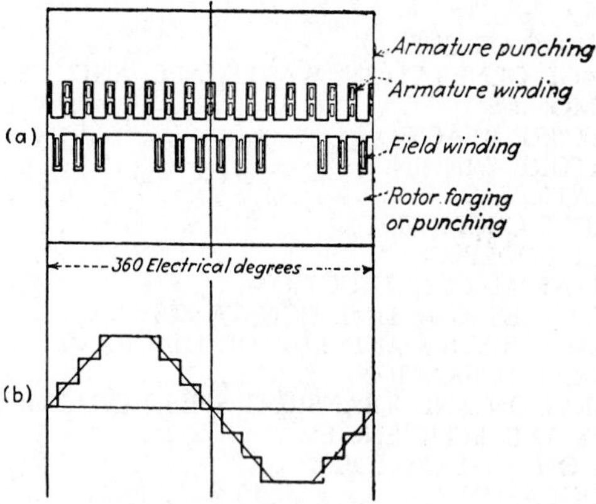


**FIG. 7-2** *(a)* Generalized sketch of one pair of poles for a cylindrical-rotor machine. *(b)* Flux form for a typical pair of poles (on the basis of current in field winding only).

large diesel-engine drives and (2) the cylindrical rotor machine illustrated in Fig. 7-2 and characteristic of steam-turbine drives. These machines normally have rotating fields and stationary armatures. Automotive alternators and high-frequency generators are frequently inductor-type machines which have a variety of forms.

Most of the principles to be dealt with in the following articles apply equally well to all these types, but the discussion will be related primarily to machines in the size range of 100 to 1,500,000 kVA unless otherwise indicated.

***2. Synchronous Speed.*** Two poles must pass a point on the stator to complete 1 cycle so that

$$\begin{aligned} \text{r/min} &= 60\,\frac{\text{hertz}}{\text{pairs of poles}} \\ &= \frac{7200}{\text{poles}} \qquad \text{for 60 Hz} \end{aligned} \tag{7-1}$$

***3. Electrical Degrees.*** Because 1 cycle of a sine wave is 360°, it is convenient to measure distance around the machine periphery in electrical degrees with two poles spanning 360 elec deg. Thus, elec deg = mech deg × pairs of poles.

Electrical degrees are also commonly used as a measure of time, with 360 elec deg corresponding to the time period of 1 cycle as illustrated in Fig. 7-3.

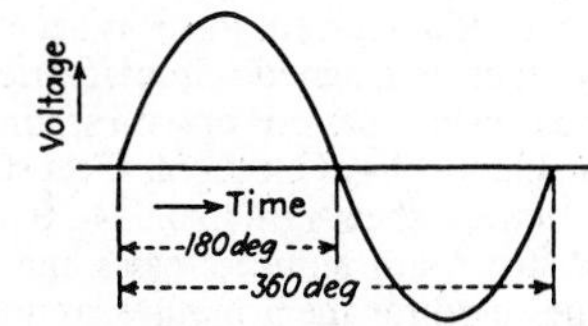

FIG. 7-3 One cycle of the voltage wave of a synchronous machine.

***4. Air Gap.*** The clearance between the stator and rotor is commonly called the air gap even though the machine may normally operate in an atmosphere of hydrogen or other gas, rather than air.

***5. Major standards*** pertaining to ac generators are now under the jurisdiction of the American National Standards Institute. New standards approved by ANSI will be designated as ANSI Standards. This organization replaces the American Standards Association (ASA) and the United States of America Standards Institute. The existing ASA and USASI Standards remain valid until they are replaced by ANSI Standards.

A similar situation exists with respect to American Institute of Electrical Engineers (AIEE) Standards, this organization having been replaced by the Institute of Electrical and Electronics Engineers (IEEE). See also Sec. **28**.

## MAGNETIC CIRCUIT

***6. Functions*** The magnetic circuit determines to a large extent the output ratings and performance characteristics that are possible for a particular machine. Because output results from the interaction of the current-carrying armature conductors and the air-gap flux and is proportional to their product, the magnetic circuit must provide space for the windings as well as a path for the magnetic flux. The objective of magnetic-circuit design is to provide an optimum division of machine volume between the flux-carrying and current-carrying parts.

***7. Flux Paths.*** The paths for the principal components of flux with load are indicated in Fig. 7-4. Additional paths, mainly for the leakage fluxes, exist at the ends of the machine.

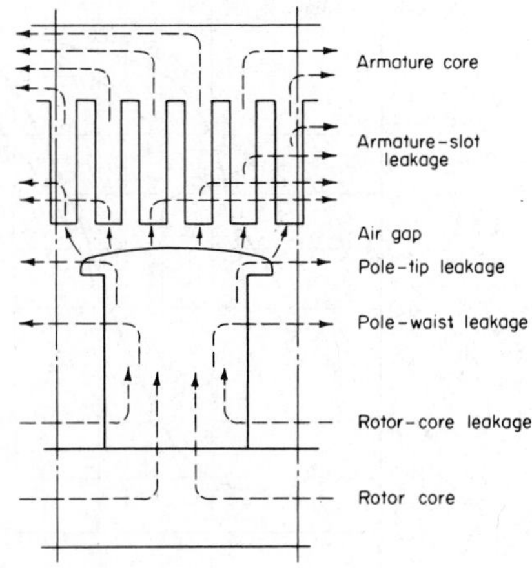

FIG. 7-4 Flux paths for a salient-pole machine with load.

***8. The proportions of the magnetic circuit*** are usually consistently related to the pole pitch for a particular number of poles. As the number of poles decreases, the restrictions in space available in the rotor result in most of the magnetic circuit dimensions being a smaller proportion of the pole pitch.

The armature slot width is determined principally by the insulation thickness required for the machine voltage and is commonly such that the resulting total copper width per slot is 40 to 60% of the slot width.

***9. Materials and Losses.*** The magnetic flux in the rotor is essentially unidirectional and varies only slightly with changes in load or terminal voltage during normal operation. This allows the rotor magnetic circuit to be made of solid steel, which is commonly done on turbine-generators and occasionally on some highly stressed salient-pole machines. However, the armature slots result in local variations of the air-gap flux density which cause eddy currents and losses in the rotor pole faces. Where solid pole faces are used, the air gap is usually relatively large, reducing these losses to acceptable values. Most salient-pole machines, on the other hand, have smaller air gaps relative to the armature slot width, and laminated poles are necessary to reduce the eddy-current losses at the pole faces. Pole laminations are commonly made of low-carbon steel, 1/16 in thick, with magnetization characteristics specified and controlled in production. Thinner steel, sometimes with some silicon content, may be used where further reduction of eddy-current losses is necessary.

The armature magnetic circuit carries alternating flux and is always laminated, either with complete ring laminations or with segmental laminations, depending on the machine size and the available widths of electrical sheet steel. The material most commonly used is about 3.5% silicon electrical sheet steel in 0.014-, 0.018-, or 0.025-in thickness for 60-Hz machines. Grain-oriented steel, with reduced losses and improved permeability in the direction of rolling, is sometimes used for the armature laminations of large turbine-generators. Orientation in the circumferential direction is advantageous in such machines because of the large proportion of steel and moderate flux densities in the ring portion of the armature. At high flux densities characteristic of the armature teeth, the advantages of grain orientation become less pronounced. Typical magnetization characterics are illustrated in Fig. 7-5.

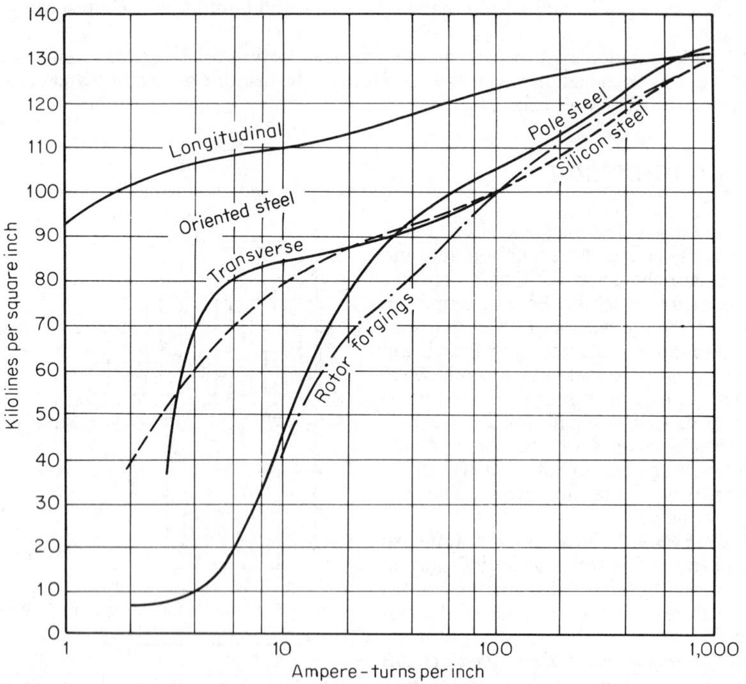


**FIG. 7-5** Magnetization curves for commonly used steels.

***10. Performance.*** The field current required for a particular load condition is determined by the magnetic circuit, in conjunction with its armature and field windings. This is calculated in design by evaluating the flux densities and the corresponding ampere-turns in all parts of the magnetic circuit. After the machine is built, the magnetic characteristics are represented by the test performance shown in the article "Characteristics and Phasor Diagrams."

Magnetic saturation significantly affects machine characteristics and performance under normal operating conditions. Unfortunately, much of the mathematical treatment of synchronous machine circuit theory is based on the assumption of negligible magnetic saturation because of the more manageable equations which then result. Equations or analogs intended to represent the machine generally must include the effects of saturation for acceptable accuracy. Computer programs have been developed that facilitate inclusion of saturation effects in generator behavior calculations.

***11. Size and Output.*** A commonly used relationship between the rated machine output and the armature inside diameter $D$ and gross core length $l$ is the output factor.

$$\text{Output factor} = \frac{\text{kVA} \times 10^5}{\text{r/min} \times D^2 l} \tag{7-2}$$

The quantities determining the output factor are more clearly indicated in the equation

$$\text{Output factor} = \frac{B_{gf}}{8600} \times kac/\text{in} \tag{7-3}$$

where $B_{gf}$ = peak of fundamental component of air-gap flux density, lines per square inch, and $kac/\text{in}$ = rms armature kiloampere conductors per inch of armature periphery at inside diameter $D$, modified to allow for pitch and distribution factors.

The output factor normally increases with pole pitch for a particular voltage and number of poles because the deeper armature slots and greater field coil space allow more kiloampere conductors per inch. However, the output factor normally decreases as the number of poles decreases because of the dimensional restrictions resulting from the mechanical angle between poles. Both $B_{gf}$ and the kiloampere conductors per inch are then reduced. Typical output factors for indirectly cooled salient-pole generators are shown in Fig. 7-6. Output factors for directly cooled machines are much higher, ranging up into the 30s for large turbine-driven 2- and 4-pole generators directly cooled with water or hydrogen at 75 psig. Further advances in output factor are anticipated with "super-conductor generators."*

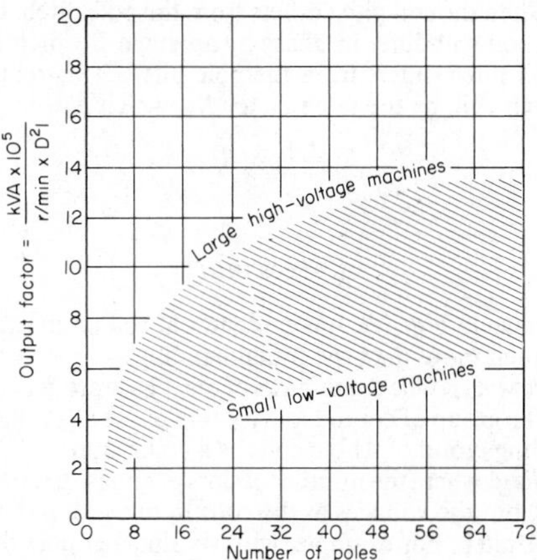


**FIG. 7-6** Typical output factors for salient-pole generators.

* See Par. **54** and Refs. 30 and 31.

## VOLTAGE GENERATION, WAVESHAPE, AND HARMONICS

*12. **Voltage Generation.*** Voltage is generated in the armature winding as a result of relative motion between the field and armature. The magnetic flux linking each armature coil changes as the machine rotates, causing induced voltages in accordance with the basic relationship

$$E/N = d\phi/dt \quad \text{instantaneous V/turn} \tag{7-4}$$

where $d\phi/dt$ = webers per second change in flux linking the turn.

The change in flux per turn occurs principally at the conductors in the armature slots, and it is convenient to consider each conductor separately as though it were cutting the air-gap flux. At a particular rotating speed, the instantaneous volts per conductor are proportional to the air-gap flux density at the conductor. The waveshape of the conductor voltage vs. time is therefore the same as that of the air gap flux density versus distance around the periphery.

*13. **Flux Forms.*** Typical flux forms resulting from field-winding magnetization and illustrated in Figs. 7-1 and 7-2 are generally not sinusoidal but rather are designed to provide a suitable compromise relationship between the fundamental component of flux and maximum flux density, average flux density, harmonic content, and iron losses in the pole faces.

The ratio of the peak value of the fundamental component of flux density to the actual maximum flux density is designated as $C_1$. Corresponding values for the $n$th harmonic are designated as $C_n$. Flat and wide flux forms may have a large value of $C_1$, approaching 1.27 as a maximum. Such waveforms result in lower air-gap and armature-teeth flux densities for a given fundamental component, but at the expense of greater total flux, higher harmonic content, and greater pole-face losses. Opposite effects with lower values of $C_1$ may result from narrow, peaked flux forms.

Reference 11 evaluates $C_1$ and other quantities for commonly used pole-face shapes and field-winding distributions. A more general approach is to apply Fourier analysis to the flux form determined from the machine dimensions. Even harmonics are not present in symmetric flux forms.

Although flux forms are commonly evaluated from the air-gap permeance, magnetic saturation in the pole tips, armature teeth, or elsewhere may substantially change the flux form at high flux densities.

Harmonics in the flux form generally do not result in harmonic voltages of corresponding magnitude at the armature-winding terminals because of the factors discussed in the following paragraphs.

*14. **Pitch Factor.*** When the coil pitch differs from the pole pitch, the voltage developed in the two sides of a single coil will differ in phase by an angle $B$ which is the angle in electrical degrees by which the coil pitch differs from the pole pitch. This reduces the coil voltage, as compared with a full-pitch coil, by the pitch factor $K_p$, which is

$$K_p = \cos (B/2) \tag{7-5}$$

For the $n$th harmonic,

$$K_{pn} = \cos (nB/2) \tag{7-6}$$

Values for $K_{pn}$ are shown in Fig. 7-7. It is evident that coil pitch can be chosen to reduce at least some harmonics much more than the fundamental.

*15. **Distribution Factor $K_d$.*** The conductors of one phase are generally distributed in more than one slot per pole. The group of conductors for one phase at a pole is referred to as a phase belt, and the corresponding group of whole coils as a coil group.

For *integral slot windings* where the number of slots per phase per pole is a whole number, all coil groups are identical, but the voltages of the coils in one group differ in phase by an angle corresponding to the slot pitch. For example, with six slots per pole the fundamental components of the two coil voltages of one group are 30° out of phase and the third harmonics are 90° out of phase, etc. The vector sum of the coil voltages is less than their arithmetic sum because of the phase differences. The factor by which these phase differences reduces the total voltage is the distribution factor $K_d$ and is given in Table 7-1 for integral numbers of slots per pole.

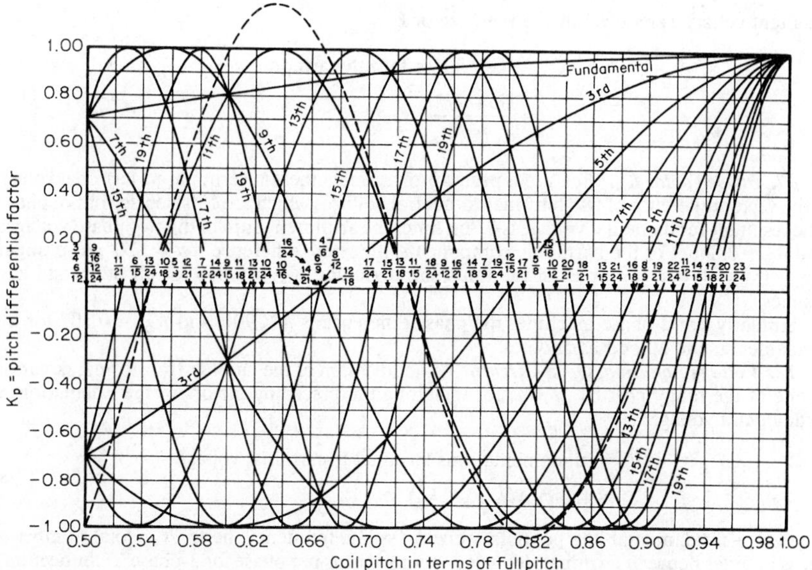


**FIG. 7-7** Curves of pitch factors, $K_{pn}$. NOTE: The dotted curve shows the ratio of $K_{pn}$ to $K_p$ for the eleventh harmonic. This is obtained by dividing the ordinates of the curve for the eleventh harmonic by the corresponding ordinates of the fundamental curve.

For *fractional slot windings,* where the number of slots per phase per pole is not a whole number, the number of coils per group is not the same at all poles. Differences in voltage phase angle exist among coil groups as well as among the coils of a group. Distribution factors for fractional slot windings are tabulated for 3-phase windings in Ref. 12. For most fractional slot windings with 60° phase belts, the coils of one phase occupy many different angular positions. The fundamental distribution factor is then very nearly the ratio of a 60° chord to a 60° arc, and $K_d = 0.955$.

***16. Skew Factor $K_s$.*** When a skewed relationship exists between the armature conductors and the axis of symmetry of the field flux form, the total voltage induced in an armature conductor is the integral of the incremental voltages over the total electrical angle of skew $\lambda$. The

**TABLE 7-1** Distribution Factors $K_{dn}$ for 3-phase Machines with 60° Phase Belts
$N_{sp}$ = number of slots per pole

| $N_{sp}$ | 3 | 6 | 9 | 12 | 15 | 18 | 21 | 24 | ∞ |
|---|---|---|---|---|---|---|---|---|---|
| $n = 1$ | 1.00 | +0.966 | +0.960 | +0.958 | +0.957 | +0.956 | +0.956 | +0.956 | +0.955 |
| $n = 3$ | 1.00 | +0.707 | +0.667 | +0.653 | +0.647 | +0.644 | +0.642 | +0.641 | +0.637 |
| $n = 5$ | 1.00 | +0.259 | +0.218 | +0.205 | +0.200 | +0.197 | +0.196 | +0.194 | +0.191 |
| $n = 7$ | 1.00 | −0.259 | −0.177 | −0.157 | −0.149 | −0.145 | −0.143 | −0.141 | −0.136 |
| $n = 9$ | 1.00 | −0.707 | −0.333 | −0.270 | −0.248 | −0.236 | −0.229 | −0.225 | −0.212 |
| $n = 11$ | 1.00 | −0.966 | −0.177 | −0.128 | −0.109 | −0.102 | −0.097 | −0.095 | −0.087 |
| $n = 13$ | 1.00 | −0.966 | +0.218 | +0.128 | +0.102 | +0.091 | +0.086 | +0.083 | +0.073 |
| $n = 15$ | 1.00 | −0.707 | +0.667 | +0.270 | +0.200 | +0.173 | +0.159 | +0.149 | +0.127 |
| $n = 17$ | 1.00 | −0.259 | +0.960 | +0.157 | +0.102 | +0.084 | +0.075 | +0.070 | +0.056 |
| $n = 19$ | 1.00 | +0.259 | +0.960 | −0.205 | −0.109 | −0.084 | −0.072 | −0.066 | −0.050 |
| $n = 21$ | 1.00 | +0.707 | +0.667 | −0.653 | −0.248 | −0.173 | −0.147 | −0.127 | −0.091 |
| $n = 23$ | 1.00 | +0.966 | +0.218 | −0.958 | −0.149 | −0.091 | −0.072 | −0.063 | −0.041 |
| $n = 25$ | 1.00 | +0.966 | −0.177 | −0.958 | +0.200 | +0.102 | +0.075 | +0.063 | +0.038 |
| $n = 27$ | 1.00 | +0.707 | −0.333 | −0.653 | +0.647 | +0.236 | +0.159 | +0.127 | +0.071 |

resultant voltage is reduced by the skew factor $K_s$,

$$K_{sn} = \text{skew factor for } n\text{th harmonic} = \frac{\sin (n\lambda/2)}{n\lambda/2} \tag{7-7}$$

***17. Phase Factor $K_\phi$.*** For Y-connected 3-phase machines the terminal-to-terminal voltage is the vector difference of two terminal-to-neutral voltages which are 120 elec deg out of phase. The resultant fundamental voltage is 0.866 times the arithmetic sum of the two phase voltages, and $K_\phi = 0.866$. For the harmonic components, the phase difference is $n \times 120°$, resulting in $K_{\phi n} = 0$ for all odd multiples of the third harmonic and $K_{\phi n} = 0.866$ for all other odd harmonics.

Similarly, for 2-phase machines the phase difference is $n \times 90°$, and $K_{\phi n} = 0.707$ for the fundamental and all odd harmonics.

***18. Voltage Equation, Open Circuit.*** A modification of the familiar transformer equation is one of the many possible means for including the preceding factors in the calculation of open-circuit voltage,

$$V_{tt} = \text{fundamental rms terminal-to-terminal voltage} = 4.44\phi f N(K_p K_d K_s K_\phi) \tag{7-8}$$

where $\phi$ = fundamental flux per pole, webers, $f$ = line frequency, hertz, $N$ = total number of turns in series between terminals (equals 2 × series turns per phase for 3-phase Y connection), and $K_p K_d K_s K_\phi$ = factors from preceding paragraphs.

For the $n$th harmonic, the fundamental voltage is multiplied by the reduction factor $C_n K_{pn} K_{dn} K_{sn} K_{\phi n} / C_1 K_p K_d K_s K_\phi$.

***19. Tooth Ripples.*** The flux forms illustrated in Figs. 7-1 and 7-2, and the preceding analysis, do not include the effects of the armature teeth and slots on the air-gap flux. The resulting harmonic fluxes do not directly generate voltages in the armature conductors but rather cause currents in the field and damper windings. These currents produce flux components which do generate voltages in the armature conductors at frequencies which are multiples of the number of slots per pair of poles plus or minus 1. For example, with 12 slots per pair of poles (6 slots per pole) the harmonics will be the eleventh, thirteenth, twenty-third, twenty-fifth, etc.

***20. Voltage Waveform Standards.*** The harmonic content of the voltage wave is generally specified in terms of (1) the maximum deviation of the voltage waveform from a pure sine wave and (2) the weighted average of all harmonics.

The deviation factor of the open-circuit terminal voltage wave is defined in ANSI C42.10, and limiting values are specified in ANSI C50.12, C50.13, and C50.14. This factor is the ratio of the maximum difference between corresponding ordinates of the wave and of the equivalent sine wave to the maximum ordinate of the equivalent sine wave when the waves are superimposed in such a way as to make this maximum difference as small as possible.

The *telephone influence factor* of the open-circuit terminal voltage waveform is defined in the ANSI C42.10 Standards, and limiting values are specified in the ANSI C50.12, C50.13, and C50.14 Standards. This factor is the weighted sum of all harmonics in the voltage wave. The weighting of each harmonic is intended to reflect the relative objectionable effect of inductive coupling at the harmonic frequency with telephone communication.

***21. Harmonics under Load Conditions.*** Load on the machine affects the harmonic content principally in the following ways: (1) Increased field current tends to increase all internal voltages, including harmonics. (2) Armature reaction generally reduces the fundamental more than the harmonics and may introduce additional harmonics. (3) Magnetic saturation changes the harmonic magnitudes. (4) The portion of the harmonic internal voltage appearing at the terminals depends on the relationship of the load impedance and the machine internal impedance at the harmonic frequency.

Generally, these effects tend to offset each other, but it is possible for series resonance to occur between machine internal inductance and load capacitance at some generated harmonic frequency, resulting in a very large harmonic voltage at the machine terminals. Also, the possible malfunction of some external device because of generator harmonics depends as much on the nature of the device and the associated electrical system as it does on the generator characteristics. For these reasons, generator voltage-waveform specifications are usually limited to open-circuit conditions where the generator is isolated from the influences of the load circuit.

## ARMATURE REACTION

***22. General Effect.*** Current in the armature conductors produces an mmf which has the same number of poles as the field structure and rotates at synchronous speed. This mmf may add to or subtract from the field mmf, depending on its angular displacement from the pole axis. For generators supplying reactive current to an inductive load, the net effect of the armature reaction is to oppose the field mmf, requiring additional field current to offset the armature reaction and sustain the flux voltage.

***23. Armature Ampere-Turns.*** The effective number of turns producing armature reaction is reduced by the same factors as apply for the generated voltage. The effects of the pitch and distribution factors are illustrated in Fig. 7-8 for 1 phase at one pole. The mmf's of the two

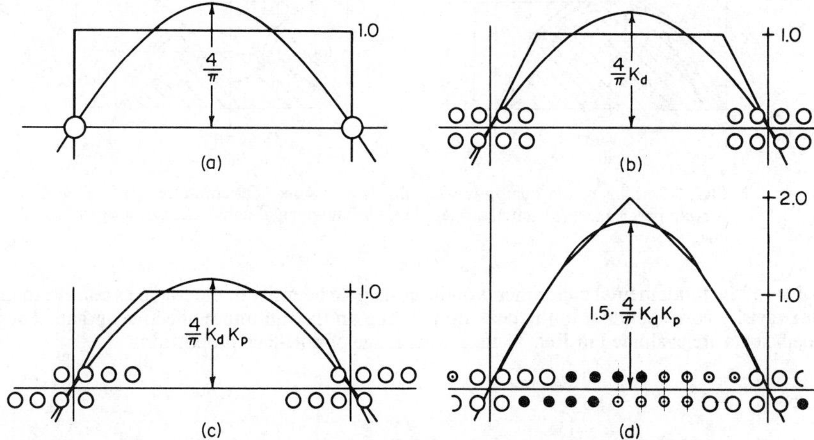


**FIG. 7-8** Effect of coil pitch and phase-belt distribution on fundamental armature-reaction mmf. *(a)* Single full-pitch coil; *(b)* full-pitch coils, 60° phase belts; *(c)* short-pitch coils, 60° phase belts; *(d)* three-phase mmf.

adjacent phase belts are displaced +60 and −60° in space and time, the effect being to increase the mmf in the axis of the phase illustrated by a factor of 1.5. Then

$$AT_d = \text{peak of fundamental component of armature At/pole}$$
$$= 1.5\sqrt{2}IN_{pp}(K_pK_dK_s)4/\pi \tag{7-9}$$

where $I$ = rms current, $N_{pp}$ = turns per phase per pole, and $K_pK_dK_s$ are from Eq. (7-8) and Pars. **14, 15,** and **16.**

***24. Harmonic MMF's and Losses.*** The distribution of the armature ampere-turns also produces space harmonic mmf's and, in the case of fractional slot windings, subharmonic mmf's,

$$AT_{dn} = \text{peak of } n\text{th harmonic of armature At/pole}$$
$$= (1.5/n)\sqrt{2}IN_{pp}(K_{pn}K_{dn}K_{sn})4/\pi \tag{7-10}$$

where $n$ = order of harmonic or subharmonic ($n$ is less than 1.0 for subharmonics).

All these mmf's result from fundamental current and rotate at the synchronous speed corresponding to their order $n$. The corresponding fluxes induce voltages of fundamental frequency in the armature winding. However, these fluxes rotate forward or backward with respect to the rotor and cause losses in the pole faces.

The fifth- and seventh-harmonic mmf's are often sources of substantial loss unless they are reduced by a suitable coil pitch such as 5/6. Another common source of loss is the steps in mmf from one slot to the next. The resulting harmonics are not reduced by $K_{pn}$ or $K_{dn}$, but more slots per pole or a larger air gap reduces the resulting flux and loss.

***25. Direct and Quadrature Axes.*** The fundamental armature mmf is commonly resolved into two components 90° displaced from each other for purposes of analysis. One of these is the direct-axis component acting in line with the poles, and the other is the quadrature-axis component acting in line with the axis of symmetry midway between poles. These axes are the direct and quadrature axes.

Because the permeance distributions are different in the two axes, the same mmf produces different fluxes in the two cases. Figure 7-9 illustrates these conditions. $C_{d1}$ and $C_{q1}$ are the peak

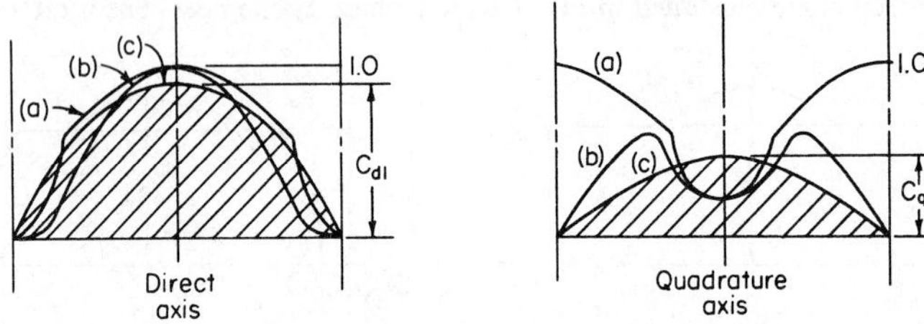


**FIG. 7-9** Direct- and quadrature-axis fluxes with sinusoidal armature mmf. *(a)* Air-gap permeance; *(b)* actual air-gap flux; *(c)* fundamental component of air-gap flux.

values of the fundamental components of the air-gap flux density in the two axes relative to the flux density corresponding to the peak mmf acting on the minimum effective air gap. These coefficients are evaluated in Ref. 11 for a wide range of pole-face proportions.

## ARMATURE WINDINGS

***26. Winding Types.*** A wide variety of winding types are possible to produce a desired voltage in the proper number of phases and with a suitable waveshape. Practical considerations, mainly economic, limit the usual alternator winding to a double-layer 3-phase winding, arranged in 60° phase belts in open slots. The number of coils, the number of turns per coil, the coil pitch, the number of circuits, and the connection of the phases are selected to give the desired voltage and waveform.

*Double-layer windings* in open slots permit the use of form-wound coils which are all alike in a given machine. These coils have the characteristic diamond shape in the end area. Three-phase windings may be either Δ- or Y-connected; Y-connected machines are much more common, particularly in the larger sizes. The winding may be arranged to be connected either Y or Δ, with leads brought out from both ends of each phase to make this possible. Lap windings, where both ends of the coil have the familiar diamond shape, are most commonly used. Wave windings may also be used, particularly on larger machines with single-turn coils, where they may offer economies in simplified connections among parallel circuits and in more efficient utilization of the winding. Lap- and wave-coil shapes are illustrated in Fig. 7-10.

***27. Number of Coils.*** For integral slot windings, where the number of slots per phase per pole is a whole number, all coil groups are identical, and the phase voltages are balanced with respect to magnitude and angle. Fractional slot windings, where the number of slots per phase per pole is not an integer, have unequal coil groups. These can be arranged to produce balanced voltages if the number of phases is a factor of the number of slots per phase at least as many times as it is a factor of the number of poles. For example, a balanced 3-phase winding for a 36-pole alternator will require a number of slots per phase which is a multiple of 3 times 3, since 3 appears twice as a factor of 36. Similarly, a balanced 2-phase winding for a 16-pole alternator will require a number of slots per phase which is a multiple of 16 (2 times 2 times 2 times 2).

Other numbers of slots than those which satisfy the above criterion may be used by making coils which do not all have the same number of turns. Special groupings of the coils may also permit a balanced winding when the number of slots does not conform to this criterion. These expedients may be used because of limitations in available punching dies.

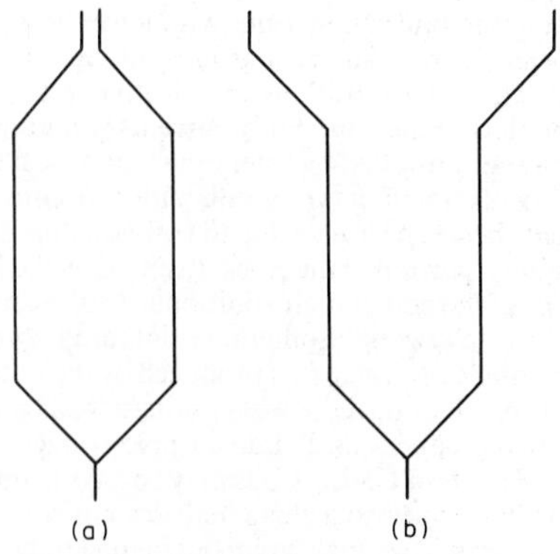


**FIG. 7-10** *(a)* Lap coil; *(b)* wave coil.

***28. Conductor Design.*** Conductors are stranded, except in the smallest ratings, to enable easier shaping of the coil and to limit eddy-current losses which result from the flux that crosses the slot. The effect of this flux is to produce a voltage within the strand which results in circulating currents. The thinner the strand, the less are the voltage and the resulting eddy currents. Since the strands ordinarily are all connected together at the joints between coils, there will also be eddy currents circulating between strands because of the difference in flux linked by the various strands. This source of loss can often be reduced sufficiently by the use of more and thinner conductors per coil, with a corresponding increase in the number of parallel circuits. As an alternative, some kind of transposition may be used to control eddy currents between strands. The purpose of the transposition is to arrange the strands in one part of the coil with respect to another so that the induced voltages cancel each other.

***29. Types of Transpositions.*** The Roebel type of transposition is often used where conductor strands are arranged two in width by several strands deep. This is illustrated in Fig. 7-11. The effect of the Roebel transposition is the same as twisting the bundle of strands 180 or 360°. This is accomplished without substantially increasing the width of the bundle of strands. For a multiturn coil, a 180° transposition usually suffices. For a single-turn coil, because of its greater depth, a 360° transposition (illustrated) may be necessary in which each strand occupies all possible positions in the slot.

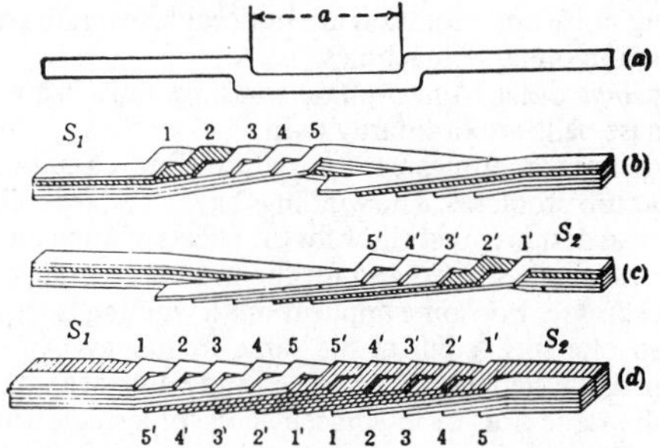


**FIG. 7-11** Roebel-type of transposition. *(a)* Typical offset conductor strand; *(b)* group of conductor strands composing half the conductor; *(c)* complementary group of strands; *(d)* completely assembled conductor.

Another type of transposition provides a 180° twist of the strand bundle in the end-turn portion of the winding, where the increased width dimension can be tolerated.

Sometimes it is satisfactory to make a transposition by groups of strands rather than by individual strands. This may be done in the end turns of the winding, by connecting upper groups in one coil in series with lower groups of adjacent coils. The groups of strands must be insulated from adjacent groups throughout the coils included in the transposition.

These transpositions are effective in limiting only eddy currents produced by flux crossing the slots in the core. Eddy currents resulting from voltages induced in the ends of the winding may be controlled by extending a transposition such as the Roebel type into the coil ends. This presents manufacturing difficulties. A convenient means of essentially eliminating eddy currents between strands due to coil-end flux is the 540° "Ringland" transposition entirely in the straight portion of the coils, invented by W. L. Ringland in 1956. A 180° transposition occupies half of the coil and an additional 180° occupies each remaining quarter.

***30. Skewing.*** Sometimes slots are skewed to minimize the effects of voltages resulting from the ripple in air-gap flux produced by the stator slots. Skewing a slot essentially a full slot pitch in the length of the core will eliminate voltages due to slot ripple. The effect of a number of slots which produces undesirable ripple voltages can be limited by some degree of skew.

***31. Dead Coils.*** Coils may be left unconnected in the armature circuit during manufacture to obtain operating characteristics not readily obtainable from adjustment of number of coils and turns. They may be cut out of machines in service in order to permit operation after one coil or more has been damaged. Where the winding has parallel circuits, the number of coils cut out must be the same in each circuit in order to avoid damaging circulating currents. If more than a few percent of the coils are cut out, it may also be necessary to cut out a like number of coils in the other phases of the machine. The controlling consideration is the voltage balance and the circulating currents. This situation would be most critical in a Δ-connected winding.

***32. Coil Pitch.*** Most economical use of the flux in the magnetic circuit results when the stator coil is full pitch, or as near to full pitch as a fractional slot winding will permit. In a fractional slot winding, the maximum pitch ordinarily used is full pitch for the integer; that is, for a 7⁵⁄₇ slots per pole winding, a coil pitch of 1 to 8 would be the maximum ordinarily used. Less than full-pitch coils are used to obtain adjustments in the voltage generated or to limit harmonics.

***33. Single-phase windings*** are usually 3-phase windings with 1 phase not used. Sometimes the coils for the third phase are omitted, but most often they are wound in the machine and might be considered as spares. In small sizes a special concentric winding may be used. A low-resistance damper winding is generally necessary on single-phase machines to reduce the flux pulsations that are set up by the single-phase armature reaction and to reduce the effective armature reactance. Single-phase machines have an inherent torque pulsation at twice rated frequency. The resulting noise and vibration are noticeable even on small machines and may require special construction on large machines.

***34. Two-phase windings*** differ from 3-phase windings only in the grouping of the stator coils. Ninety-degree phase belts are ordinarily used.

***35. Double windings*** are sometimes used to reduce short-circuit currents and to simplify switchgear and bus structure problems. The windings have standard coil design with special end connections. The electrical designs must allow for the effects of unbalanced armature reaction if the two windings are not equally loaded and for the mutual reactance between the windings.

***36. Multispeed Windings.*** For some applications a winding is required which will permit operation at more than one speed, but at the same frequency. An example is a hydraulic pump-turbine unit, which may generate at one speed and pump at another. Such machines may have two windings in the same slots or in adjacent slots, or a single armature winding may be reconnected by means of suitable switches to serve this purpose. Special rotor construction and field-winding reconnections are also necessary.

## INSULATION

***37. Classes of Insulation.*** The electrical insulation of the alternator windings is designed to operate satisfactorily at the specified voltages and temperature and to retain its dielectric and mechanical strength and dimensional stability over many years of operation. ANSI has defined various classes of insulating systems based on the maximum steady-state operating temperatures and has established voltage proof tests to demonstrate the dielectric capability of the insulation system. Other nondestructive test techniques have been developed to evaluate the dielectric capability and condition of the insulation systems.

Of the several insulation-system classes, four are most applicable to large rotating machines. These are Classes A, B, F, and H. These classes are sometimes designated as 105, 130, 155, and 180, respectively, where the numbers signify the design hot-spot temperatures in degrees Celsius. Class 105 and Class 130 are most frequently applied to alternators; most of the discussion in this article will refer to these two.

*Class A insulation systems* comprise organic materials such as cotton, silk, paper, and certain synthetic films. Varnishes and synthetic resins are used as binders. *Class B systems* comprise inorganic materials such as mica, glass fibers, asbestos, and synthetic films, with suitable binders. *Class F systems* comprise generally similar materials to those of Class 130, with binders selected for suitable life at the higher temperatures. *Class H systems* include the silicone elastomers as well as mica, glass fibers, asbestos, etc., and high-temperature binders. Any of these systems may include other materials or combinations of materials in limited quantities if by experience or accepted test they can be shown to have acceptable thermal life at the specified temperature.

The rapid growth of the field of synthetic chemistry has presented a continual flow of materials suitable for electrical insulation. The IEEE has prepared guides for test procedures for the thermal evaluation of electrical insulating materials and systems (see Ref. 5).

***38. Hot-spot Allowance.*** It is ordinarily not practicable to measure maximum hot-spot temperatures on alternator windings; consequently machine ratings are based on observable temperatures, and a suitable insulation class is used, based on the expected difference between the observed temperature and the maximum temperature. This temperature difference is called the hot-spot allowance.

Conventional hot-spot allowances have been established depending on the method of determining observed temperature. Although in a specific instance a manufacturer may deviate from these, they provide guidance for selecting suitable insulation systems and for establishing standard temperature rises.

***39. Measuring Winding Temperatures.*** Several methods of observing winding temperatures are defined in the ANSI Standards. The *thermometer method* consists in determining the temperature with a mercury thermometer or other suitable temperature-measuring device applied to the hottest parts ordinarily accessible to mercury thermometers. This method is commonly applied to armature windings of smaller alternators. Of the several methods listed here, the thermometer method gives readings furthest from the maximum winding temperature and has the largest hot-spot allowance associated with it. The *resistance method* consists in determining temperature from a comparison of the winding resistance at the operating temperatures with its resistance at a known temperature. For the *embedded-detector method,* thermocouples or resistance temperature detectors are built into the machine in locations inaccessible to mercury thermometers. In the commonest application they are placed between the coil sides in a two-layer armature winding near the middle or warmest region of the core. The resistance and embedded-detector methods give temperature readings closer to the maximum temperature and ordinarily have the same hot-spot allowance associated with them. The *applied-thermocouple method* uses a thermocouple in direct contact with the conductor or separated from it by only the integrally applied insulation of the conductor itself. Applied thermocouples may be located on or near the hottest parts of the winding. This method is not ordinarily associated with machine temperature ratings but is used for experimental testing to determine hot-spot allowances, etc.

***40. Temperature Ratings.*** Temperature ratings of alternator armature windings are usually based on the resistance method for smaller ratings and on the embedded-detector method for larger ratings. Standards have been established for rating machines, using these temperature-measuring methods and related to the class of insulation system provided. These are specified in ANSI C50.12, C50.13, and C50.14.

***41. Armature Winding Insulation.*** Standard armature voltages range from 220 to 18,000 V or more. Appropriate amounts of turn and ground insulation are provided to withstand normal operating voltages under both steady-state and transient conditions. In low-voltage systems the turn insulation may be applied directly to the conductor as a film or serving. In the higher voltage ranges (generally above 5000 V) special construction is used to control corona. This is an electrostatic discharge due to the voltage gradient within or at the surface of the coils exceeding the dielectric strength of the air. In the presence of moisture this discharge produces nitrous acid, which decomposes organic materials associated with the insulation system. Insulation for high-voltage coils is applied in such a way that internal voids are minimized. The outer surfaces of the slot portion of the highest-voltage windings are coated with a semiconducting medium to lower the voltage gradient between coil and core. Spacing between coils in the end windings is controlled to limit discharge which would be damaging to the cording and blocking of the coil-support structure.

***42. Field Winding Insulation.*** Field insulation systems, because of the lower voltages involved, present fewer design problems than armature systems. Operating voltages of field windings are in the range of 125 to 600 V, occasionally somewhat higher. Transient conditions, for example, the interruption of full-load field current, may impose voltages several times rated volts for a short time. The same insulation-system classes are applied to field insulation systems. Temperatures are usually measured by the resistance method.

Field windings are subjected to the centrifugal forces of rotation, and the assembly must be dimensionally stable so that the turns in the coil do not separate from each other and the coil does not become loose on the pole.

***43. Tests of Insulation Systems.*** The high-potential dielectric test and the insulation-resistance test are the principal methods of evaluating insulation capability and condition. The high-potential test prescribed by the ANSI is twice-rated terminal-to-terminal voltage, plus 1000 V for armature windings, and ten times rated voltage, but not less than 1500 V, for alternator field windings except for field windings rated over 500 V, which are to be tested at 4000 V plus 2 times the rated value. This is an ac test, of specified waveshape, applied under controlled conditions, and maintained at its top value for 1 min. It is a severe test and is applied only on new windings after ascertaining that they are dry and otherwise in good condition. The design breakdown level of the insulation may be several times the test voltage, to provide a suitable factor of safety over process control. Direct-voltage tests are sometimes substituted for ac tests, applying a multiplier of 1.7 to the rms ac value.

The *insulation-resistance test* is often used as a measure of the condition of the winding. Insulation resistance is the ratio of the applied voltage to the current at some specified time after the voltage is applied. Direct, rather than alternating, voltages are used for measuring insulation resistance.

The principal currents affecting insulation resistance after 1 min or 10 min of application of the test potential are (1) leakage current over the winding surface, (2) conduction through the insulation material, and (3) absorption currents in the insulation. The first two currents are essentially steady with time, but the last current decays approximately exponentially from an initial high value. Such insulation-resistance measurements are affected by surface condition (dirt or moisture on the winding surfaces), moisture within the insulation wall, and insulation temperature. The magnitude of the test potential may also affect the insulation value, especially if the insulation is not in good condition. Therefore, it is desirable in using insulation resistance as a measure of winding condition over a period of years to make readings under similar conditions each time.

The *dielectric-absorption characteristic,* which is displayed in the shape of the curve of insulation resistance against time when a constant test potential is applied for 10 or 15 min, is also frequently used as an indication of the winding condition. The resistance of a clean, dry winding will continue to rise as the test potential is maintained, becoming fairly steady after 10 or 15 min. A wet or dirty winding will reach its steady value much sooner.

The ratio of the 10-min reading to the 1-min reading is called the *polarization index.* Other things being equal, a high polarization index indicates good winding condition.

A common device for measuring insulation resistance is a direct-indicating ohmmeter with a self-contained hand- or power-driven generator. The megger is a typical device of this type. Measurements are made at voltages in the range of 500 to 2500 V, higher voltages sometimes being used in checking very high voltage windings.

Because of the many factors affecting insulation resistance it is impracticable to establish rigid standards for minimum values of either insulation resistance or polarization index. It has been recommended by the IEEE, however, that the minimum insulation-resistance value for alternator windings be 1 MΩ/1000 V rated voltage + 1 MΩ, measured with the winding at 40°C with a 500-V test potential. It is recommended also that the minimum value of polarization index be 1.5 for Class A insulation systems and 2.0 for Class B insulation systems.

Power-factor measurements of the armature coil insulation may be made to demonstrate relative freedom from internal voids in the insulation. Since change in power factor with voltage is more significant than power factor itself, measurements are made over a range of voltages up to at least the rated operating voltage to ground. Increases in power factor at higher voltages are indicative of the ionization of internal voids. The change in power factor from 25 to 100% of rated voltage to ground is referred to as "power-factor tip-up." Experience is being gathered which will become the basis for standards of maximum values of tip-up for various insulation systems.

Reference 6 provides more information for testing and interpreting insulation condition.

## INDIRECT COOLING

***44. Definition.*** Indirectly cooled machines dissipate their losses to a cooling medium which is entirely outside the coil insulation. All air-cooled machines, with rare exceptions, are cooled in this manner, as well as most hydrogen-cooled machines under 100 mVA. Turbine-generators rated above 100 mVA usually employ direct cooling, as described below.

***45. Cooling Media.*** The comparative characteristics of various gases that might be used for cooling are shown in Table 7-2. Air is most commonly used, for obvious reasons. Hydrogen provides better heat transfer with much less windage loss, which is nearly proportional to density. Frequently hydrogen is used at higher than atmospheric pressure, which further improves its heat-transfer capabilities, but with greater windage loss. Other advantages of hydrogen are the reduction of insulation oxidation and fire hazard. Hydrogen purity is normally maintained in the range of 95 to 99% where the mixture is nonexplosive and will not support combustion. The explosive range of hydrogen-air mixtures is 5 to 75% hydrogen.

**TABLE 7-2** Properties of Cooling Gases
(From Ref. 13)

| Characteristics | Air | $N_2$ | $CO_2$ | $NH_3$ | $H_2$ | He | Methane |
|---|---|---|---|---|---|---|---|
| Thermal conductivity | 1 | 1.08 | 0.638 | 0.868 | 6.69 | 6.40 | 1.29 |
| Density | 1 | 0.966 | 1.52 | 0.588 | 0.0696 | 0.1378 | 0.554 |
| Specific heat (const. press) | 1 | 1.046 | 0.848 | 2.185 | 14.35 | 5.25 | 2.495 |
| Heat capacity | 1 | 1.02 | 1.29 | 1.232 | 0.996 | 0.72 | 1.38 |
| Heat transfer | 1 | 1.03 | 1.132 | 1.228 | 1.51 | 1.18 | 1.43 |

Typical effects of different cooling gases and hydrogen pressure variation on temperature rise are illustrated in Figs. 7-12 and 7-13.

***46. Ventilating Paths.*** Stators are usually cooled by blowing air or hydrogen over the coil ends and through radial ducts in the armature core. The ducts are normally in the range of 0.25 to 0.375 in wide, with a spacing of about 2 in, but they may be omitted entirely on machines with short core lengths. In salient-pole machines, the gas normally enters at the ends of the rotor in the space between poles and flows radially outward through the stator ducts. In turbine-generators the cross section for axial flow into the machine is greatly restricted, and the cooling gas is usually ducted to the outer diameter of the stator core, where it is directed radially inward through some of the stator ducts to the air gap.

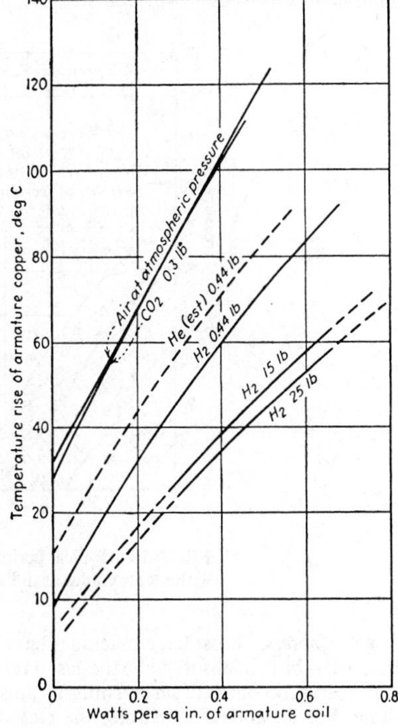


**FIG. 7-12** Armature heating of a 6250-kVA alternator with different cooling gases. *(From Ref. 14.)*

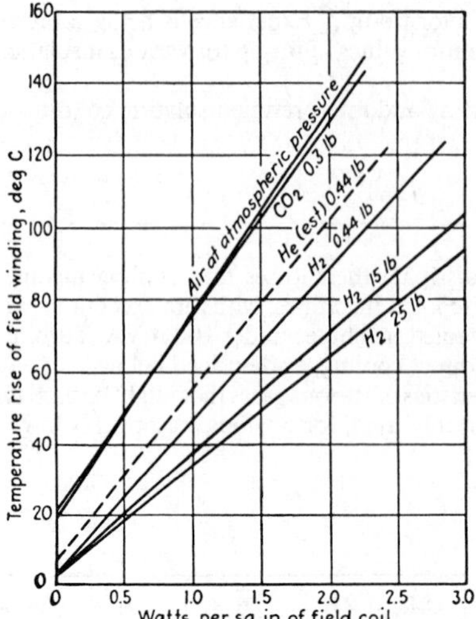


**FIG. 7-13** Field heating of a 6250-kVA alternator with different cooling gases. *(From Ref. 14.)*

Other radial ducts in the stator allow the gas to discharge radially back into the cooling system.

The field coils of salient-pole machines are normally cooled by the axial flow between poles and by the relative velocities at the ends resulting from rotation. For large salient-pole machines, the field coils may have strap copper conductors wound on edge with some turns extended, or the exposed edge of the strap may be beveled to increase the exposed surface and improve the cooling. Turbine-generator rotors normally have axial gas flow through ventilating slots or other passages to supplement the cooling at the exposed cylindrical surface.

***47. Fans and Enclosures.*** Machine enclosures have a significant effect on the fans and ventilating system. Small slow-speed salient-pole machines often have open end shields and are cooled largely by the unconfined air circulation at the ends produced by simple fan blades on the rotor. Air flow through the center of these machines may be produced largely by the centrifugal effects acting on the air columns between poles. For larger machines, solid end shields are generally used, providing static pressure chambers at the ends and more definite gas-flow paths. More effective fans are also characteristic of larger machines and may be either axial-flow propeller type or radial-flow centrifugal. Typical machine-fan characteristics are shown in Figs. 7-14 and 7-15.

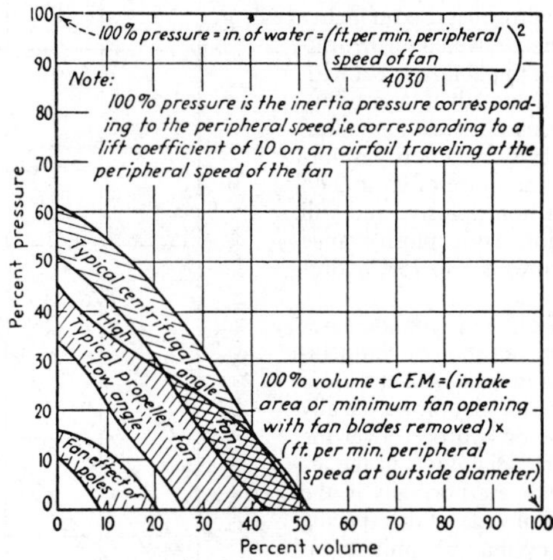


**FIG. 7-14** Typical performance of shrouded fans as used without air guides or diffusers on air-cooled machines.

***48. Coolers.*** Most large machines have closed recirculating systems with fin-tube coolers to transfer the heat absorbed by the gas to the cooling water. For hydrogen-cooled machines the coolers usually consist of several units arranged either vertically or horizontally within the stator frame. For air-cooled machines, the closed recirculating system provides clean, cooled air independent of the environment. Horizontal air-cooled machines commonly have a single cooler located in a horizontal position below the machine, and vertical waterwheel generators

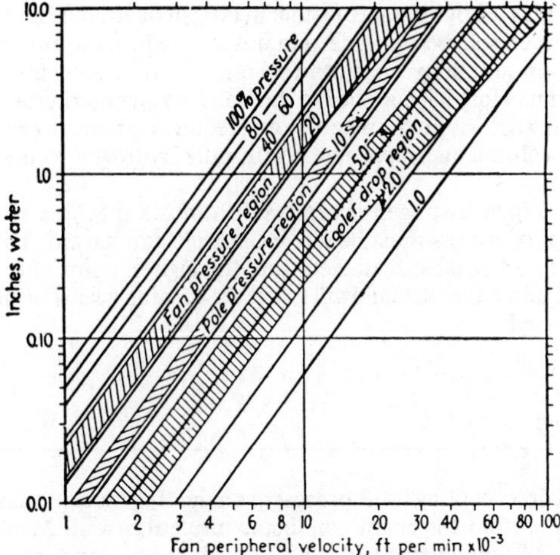


**FIG. 7-15** Curve for estimating permissible cooler drops and approximate pressure developed by fans and by poles for air-cooled machines.

usually have several coolers distributed around the outer periphery of the stator within an air housing.

Coolers are usually selected to have a gas pressure drop which is low enough so that the gas flow through the machine is not unduly restricted. Typical cooler drops for air-cooled machines are shown in Fig. 7-15.

***49. Gas Quantities.*** A major factor determining the required gas flow through the machine is the temperature rise of the gas resulting from the heat loss it absorbs. Air and hydrogen *at atmospheric pressure* are essentially the same in this respect, and

$$^\circ\text{C gas rise} = 1880\,\frac{\text{kW loss absorbed}}{\text{gas, ft}^3/\text{min}} \qquad (7\text{-}11)$$

A typical volume for operation at atmospheric pressure is 100 ft³/min per kW of loss absorbed, resulting in a gas temperature rise of 18.8°C in passing through the machine. Machines designed to operate at more than atmospheric pressure may require less volumetric flow because the degrees Celsius gas temperature rise is inversely proportional to gas density for a given cubic feet per minute per kilowatt of loss absorbed.

***50. Cooling Water.*** The temperature rise of the cooling water in absorbing the machine losses is usually quite small and is

$$^\circ\text{C water rise} = 3.78\,\frac{\text{kW loss absorbed}}{\text{water, gal/min}} \qquad (7\text{-}12)$$

The water quantity normally used is roughly 1.0 gal/min per kW loss, resulting in a water temperature rise of 3.78°C. Because the required cooler-gas outlet temperature is normally 40°C, the economical water temperature rise may vary considerably, depending on the available water temperature, which may be as low as 20°C (68°F) or as high as 35°C (95°F). Also, the required heat-transfer rates in the cooler along with the water-flow pattern and velocity may influence the gallons per minute required. Generally, water requirements are the result of an economic compromise with cooler cost.

***51. Hydrogen Seals.*** Shaft seals for hydrogen-cooled turbine-generators maintain an oil film under pressure in a small clearance between the rotating shaft and a stationary member. The construction may be similar to a journal bearing with a cylindrical oil film or similar to a spring-loaded thrust bearing with the oil film in a plane at right angles to the shaft axis. The oil film is maintained by a supply pressure higher than the hydrogen pressure.

Oil can absorb about 10% by volume of either hydrogen or air. It is important that the seal oil flow toward the hydrogen side be minimized in order to reduce the amount of air carried into the machine and the amount of hydrogen carried out. The oil is sometimes vacuum-treated to minimize its air content before it is supplied to the seal. Any air entering at the seal becomes part of the hydrogen-air mixture within the machine and requires exhausting a mixture volume of 20 to 100 times the air volume and replacing it with pure hydrogen to maintain the hydrogen purity.

***52. Scavenging.*** When hydrogen is removed from the machine to allow inspection or servicing, the scavenging process must be safe from explosion hazard. The usual method is to admit carbon dioxide and exhaust hydrogen from the highest point of the enclosure with the machine deenergized and either at standstill or on the turning gear. For filling with hydrogen, the gas flows are reversed.

## DIRECT COOLING

***53. Definition.*** Direct cooling is the process of dissipating the armature and field coil losses to a cooling medium within the main conductor insulation wall. Machines cooled in this manner are also called "supercharged," "inner-cooled," and "conductor-cooled" by various manufacturers. The cooling medium either is in direct contact with the conductor copper or is separated only by thin materials having little thermal resistance. Direct cooling eliminates the temperature differential resulting from heat flow through the coil insulation, providing greater current-carrying capability for the same hot-spot temperature rise.

***54. Cooling media*** normally used are hydrogen, oil, and water. Most directly cooled turbine-generators operate in a hydrogen atmosphere which provides cooling for all the machine except, in some instances, the armature coils. These coils may be cooled from the common hydrogen system or by a separate oil, water, or high-pressure hydrogen system. Because of the limited cross section available within the coils for cooling-medium flow, the temperature rise of the cooling medium in absorbing the heat loss is usually the most important factor determining the hot-spot temperature rise. Hydrogen-cooled machines are often operated at 60 lb/in$^2$ or more to increase the mass flow and reduce the hydrogen temperature rise. The surface heat-transfer rates are also improved, of course. Direct water cooling of the rotor windings of very large turbine-generators is also employed.

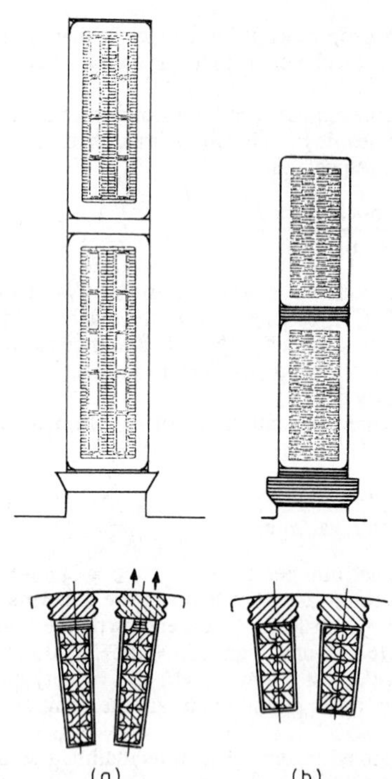


**FIG. 7-16** Cross section of stator and rotor slots of a directly cooled turbine-generator. *(a)* Hydrogen cooling; *(b)* water cooling.

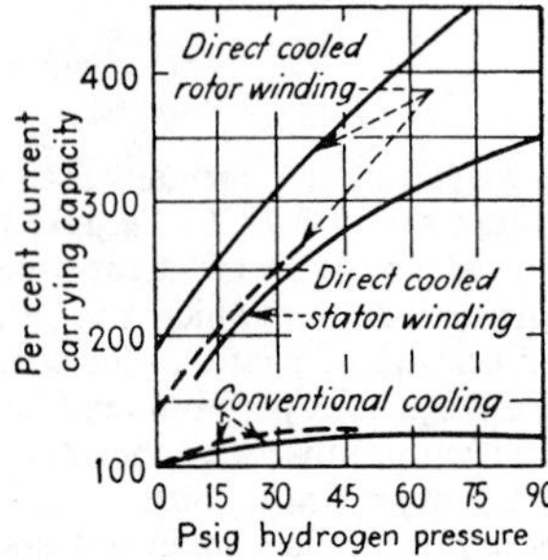


**FIG. 7-17** Current-carrying capacity of rotor and stator coils as a function of hydrogen pressure. *(Solid curves from Ref. 16; dotted curves from Ref. 17.)*

Advances in cryogenics have led to the construction of *super-conductor generators** in which liquid helium is circulated within the rotor conductors, reducing their temperature very nearly to 0 K. At this temperature the conductor resistance becomes almost negligible, thus permitting enormous field current in very small conductors.

Direct cooling is applied also to salient-pole synchronous capacitors and hydrogenerators of the largest ratings. The direct-cooling medium usually is water.

Electrical insulation associated with hydrogen direct cooling must allow for adequate creepage distances at the coil hydrogen inlets and outlets. Oil and water systems require, in addition, insulated piping. Water is deionized to maintain low electrical conductivity.

***55. Coil Construction and Flow Paths.*** Typical armature coil construction for hydrogen cooling is illustrated in Fig. 7-16*a*, and for water cooling in Fig. 7-16*b*. The hydrogen passages are usually thin-wall high-resistance material, such as stainless steel, to reduce eddy-current losses. German-silver or nickel-silver ducts also are used. Liquid-cooled machines may have similar internal tubes, or some or all of the conductor strands may be hollow to convey the liquid. The flow path may have the two sides of a coil in parallel or in series.

Rotor coil constructions and rotor cooling systems in use are more varied. Some salient-pole machines have hollow-conductor liquid-cooled field windings, but hydrogen is almost universally used for turbine-generators. Figure 7-16*b* shows comparative rotor cross sections of a turbine-generator with hydrogen *(a)* and water *(b)* cooling. With hydrogen cooling the flow path may be from end to end, from the ends to discharge openings near the center, or between alternate high- and low-pressure zones along the rotor length. These zones are separated by baffles in the air gap, with radial ducts in the stator arranged to allow the blower to circulate the hydrogen between the zones. The length of the machine is the major factor determining the optimum number of parallel paths required. Another means of providing these parallel-flow paths is the "air-gap pickup" construction, where passages inclined in the direction of rotation scoop hydrogen out of the air gap and convey it to the coil passages. Similar passages with openings inclined backward with respect to rotation provide for the hydrogen discharge. In directly water-cooled rotors of turbine-generators, the water enters a centered tube in the bore of the outboard end and is distributed in many parallel paths in the hollow rotor conductors. The water returns through the annular space between the central tube and the rotor bore.

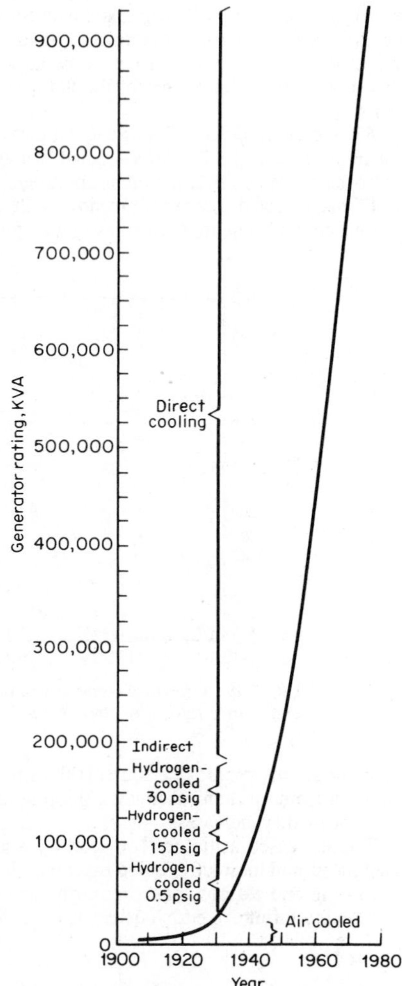


**FIG. 7-18** Maximum kVA ratings for 3600-r/min turbine-generators. *(From Refs. 18 and 23.)*

***56. Hydrogen Blowers.*** Blower requirements for circulating hydrogen in direct-cooled machines are more critical than for conventional cooling, for two reasons: The required pressure differential is greater, and the increased fan power output requires higher blower efficiency to avoid excessive fan power loss. The blowers are usually mounted on the rotor within the

* See Refs. 30 and 31.

hydrogen enclosure. Multistage axial blowers are most common. Multi- or single-stage centrifugal blowers are also used, sometimes with an impeller diameter considerably exceeding that of the rotor. In water-cooled turbine-generators a shaft-driven pump is usually provided which may supply the cooling water for the stator and rotor windings and other water-cooled components.

***57. Characteristics.*** The increased current-carrying capability of directly cooled coils is illustrated in Fig. 7-17. Liquid-cooled armature coils are comparable in performance with hydrogen-cooled coils at the higher hydrogen pressures.

Directly cooled machines are now built mainly for larger ratings than are possible with indirect cooling. The increase in available ratings in recent years shown in Figs. 7-18 and 7-19

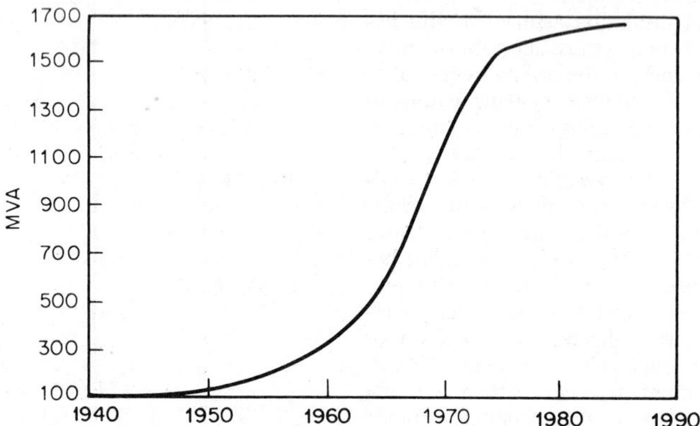


**FIG. 7-19** Growth of 4-pole generator ratings since 1940. *(From Refs. 23 and 24 and courtesy of Utility Power Corp.)*

has come about primarily because of this improved cooling method. Machine size has increased somewhat, but mechanical stresses, shipping clearances and available forgings are severe limitations for further increases.

The increased armature coil loading results in directly cooled machines having higher reactances than indirectly cooled machines. This increased loading also results in more leakage flux in the end regions, often requiring low-resistance nonmagnetic or laminated magnetic shielding to minimize eddy-current losses in the structural parts.

## *MECHANICAL CONSTRUCTION*

***58. Basic Types of Constructions.*** The magnetic circuit and the windings of the alternator are designed to function satisfactorily mechanically as well as electrically and are provided with a suitable supporting structure consisting of a stator yoke or frame, rotor body, shaft, bearings, and bearing supports. The design criteria are based not only on normal operating conditions but also on abnormal conditions such as short circuits and overspeeds.

Two fundamental variations in mechanical construction are distinguished by the field configuration: salient-pole and round-rotor. In all but the smallest modern alternators and some ac brushless exciters the field is the rotating element (rotor), and the armature is the stationary element (stator). Salient-pole construction, where the field windings are on pole pieces attached to a rotor body, is used on slower-speed machines, 1200 r/min and less, because of its relatively lower cost. Round-rotor construction, where the field windings are inserted in axial slots in a cylindrical rotor body, is used on essentially all 2-pole machines and on the larger 4-pole machines, because solutions to the problems of attaching salient poles to the rotor bodies of such high-speed machines become impracticable. The fundamentals of stator core and winding construction for these two types of designs are the same.

Two additional variations of construction are characterized by whether the shaft is horizontal or vertical. The prime-mover design ordinarily determines whether the machine will be horizontal or vertical, although special problems in generator design may affect the choice. Most steam-turbine-driven and engine-driven alternators have horizontal shafts; hydraulic-turbine-driven alternators may be either horizontal or vertical, the larger slower-speed units usually being vertical. The magnetic circuits and windings of horizontal and vertical machines are similar; the differences between these two types are in their structural members.

***59. Stator Construction.*** Armature cores are built up of thin laminations, produced as segments or rings, depending on size. Successive layers or groups of layers of the segmented laminations are staggered to minimize the effect of the joints in the magnetic circuit. The core is clamped between pressure plates and fingers to support it with sufficient pressure to prevent undue vibration of the laminations. Especially in long cores, the clamping arrangement may include some provision to compensate for compacting of the core after initial assembly.

The armature windings are fitted tightly in the slots and secured radially by slot sticks, or wedges, driven into suitable notches at the air-gap end of the slots. It is necessary that the stator coil ends be able to resist the abnormal forces associated with short circuits. A supporting structure may be employed for this purpose. There are many variations of support design; most of them provide filler blocks between the coil sides, strategically located to transmit the circumferential forces from coil to coil, and additional structure to counteract the radial forces.

Coil supports ordinarily are designed to suit the need of a particular machine. Large 2-pole machines require a quite elaborate structure; the combination of large short-circuit currents and coil ends inherently flexible because of their long length makes these machines particularly susceptible to coil-end movement. Low-speed machines with stiffer coil ends require less support; in the smallest ratings the coils may be capable of withstanding the short-circuit forces without any additional support.

Stator frames, or yokes, commonly are fabricated from structural steel, designed to support the core in proper alignment with the rotor and to suit the ventilating scheme used.

***60. Rotor Construction.*** The pole pieces of salient-pole alternators may be built up of steel laminations, both as a manufacturing convenience and as a means of limiting the loss in their air-gap surfaces due to pulsations in air-gap flux. The field coils, wound directly on the poles or preformed and then mounted on the poles, are suitably insulated from the poles for the voltages associated with normal and transient operation. The pole and coil assembly is bolted, dovetailed, or otherwise attached to the rotor body. It is the limitation of this attachment which usually dictates when round-rotor construction must be used rather than salient-pole construction.

The rotor body for a salient-pole machine may be a solid forging or assembly of heavy steel plates, for high-speed designs, or a spider-and-rim assembly, for low-speed designs. The shaft may be integral with the body, as in the case of a forging, or may be bolted to or inserted into the body.

When the spider-and-rim construction is used, the entire assembly may be an integral weldment or casting or the rim may be separate from the spider, as in the case of large waterwheel-driven generators. A common construction for this latter case is a rim built up of thin steel laminations, assembled around a cast or fabricated spider, bolted together between steel end plates, and keyed to the spider.

The rotor of a round-rotor machine is cylindrical in shape, with axial slots provided in its body for the field coils. The body is usually a steel forging with the shaft ends integral. In special applications, other constructions may be used, with this same general configuration. The field coils are wound in axial slots in the rotor body, held in place by heavy slot wedges and by retaining rings over the coil ends.

Rotors are designed for operation at overspeeds which depend on the characteristics of the prime mover. Overspeeds (the speed above rated at which the unit must be capable of safe operation) may be as low as 20% for a steam-turbine-driven unit or as high as 125% for some adjustable-blade axial-flow hydraulic-turbine-driven units.

***61. Critical Speeds.*** The shaft system of the entire unit (generator and prime mover) must be designed with regard to critical speeds. Both lateral and torsional critical speeds are considered. Lateral critical speed is the speed corresponding to the natural frequency of the shaft system in response to lateral or transverse forces such as residual unbalance forces. Torsional critical speed relates to the response of the shaft system to torsional forces. A lateral critical speed is associated with each mode of lateral vibration, the first critical speed corresponding to the lowest frequency mode. Critical speeds are affected by shaft support, including foundation (particularly lateral critical speeds), and by internal and external damping.

It is preferable that the operating speed be at least 20% away from the nearest critical speed. Low-speed rotors ordinarily operate below the first critical speed. High-speed rotors, especially 2- and 4-pole turbogenerator rotors, often operate above the first critical speed and sometimes, in the largest ratings, above the second critical speed. It is particularly important that such rotors be carefully balanced so that the forces, and resultant stresses, while passing through the critical speeds on start-up and shutdown are not excessive.

Torsional critical speeds are excited by external forces, such as a sudden load change or a short circuit or other system disturbances or cyclic variations in prime-mover torque, for example, as from an internal-combustion engine. Impulses from the buckets of a hydraulic turbine or from the gear teeth in a unit driven through a gear may also excite torsional vibrations.

***62. Bearings.*** Although antifriction bearings are occasionally used on alternators of the smaller ratings, the great majority are furnished with oil-lubricated babbitted bearings. For horizontal shafts these are self-contained ring-oiled bearings wherever design conditions permit. At higher shaft peripheral speeds and higher bearing loadings ring oiling is supplemented with recirculation of externally cooled oil. The rings may be eliminated, or they may be retained to afford some degree of emergency oil supply in the event of a failure of the external system. Lead-base babbitts are commonly used for journal bearings, although tin-base babbitt may be employed for some heavy-duty applications. Although bearing supports may be designed to afford some degree of self-alignment for the bearing bushing or shell, they must be sufficiently rigid so as to not affect unduly the lateral critical speeds of the shaft system.

Two principal types of *thrust bearings* are used on vertical alternators: the pivoted-shoe type and the spring type. The adjustable pivoted-shoe type, introduced in the United States by Albert Kingsbury, consists of a flat rotating collar or runner of steel or fine-grained cast iron resting on a stationary member consisting of several babbitted segmental shoes pivoted near their center on adjusting screws, which, by changing the elevation of the shoes, can provide equal loading on them. The screws also permit small adjustments in rotor elevation to correct generator and turbine clearances. The spring-type bearing manufactured by the General Electric Company consists of a flat rotating thrust collar, resting on a series of stationary babbitted segments supported on a number of precompressed springs.

The bearings are immersed in oil. In operation, a thin, wedge-shaped film of oil is formed between the runner and shoe. The oil is continuously circulated by the rotation of the runner and is cooled either by radiation or by water cooling, usually within the oil bath but occasionally by an external system. Some of the larger bearings are cooled by means of water circulated through tubes embedded below the babbit surface.

The spring-type bearing is inherently self-equalizing; that is, each shoe carries very nearly the same amount of load. A variation of the pivoted-shoe bearing, in which the shoes are supported on a system of interconnected levers, provides the same self-equalizing feature.

The *spherical bearing* is another variation of the pivoted-shoe thrust bearing, in which the runner is a part of a sphere and the shoes of corresponding shape. This type of bearing restrains lateral movement of the shaft, serving the dual function of thrust and guide bearing.

Horizontal-shaft alternators occasionally require thrust bearings, as, for example, a single-impeller reaction turbine having unbalanced hydraulic thrust which must be restrained by the bearing. Thrust-bearing designs for this application are generally of the pivoted-shoe type, either adjustable or equalizing.

Some thrust bearings, particularly of the adjustable pivoted-shoe type, may be provided with load cells for measuring and equalizing the thrust on the shoes. These may be of the hydraulic or the strain-gage type, the latter being more common in modern applications. In addition to providing a check on the adjustment of the shoe loadings, these devices provide information about the hydraulic-thrust characteristics of the turbine.

*Guide bearings* for vertical alternators (Fig. 7-20) are oil-lubricated babbitted rings. These are frequently segmental to facilitate assembly and may be composed of individual shoes which are radially adjustable. Guide bearings usually are partly immersed in an oil bath, with oil circulated by the pumping action of sloping grooves in the babbitt surface. Occasionally a separate lubrication system is provided which introduces oil at the top clearance of the bearing, collects it at the bottom, and recirculates it. It is common practice to place a guide bearing closely above the thrust bearing, in the same oil pot. In some instances the guide bearing is on the outside periphery of the thrust runner.

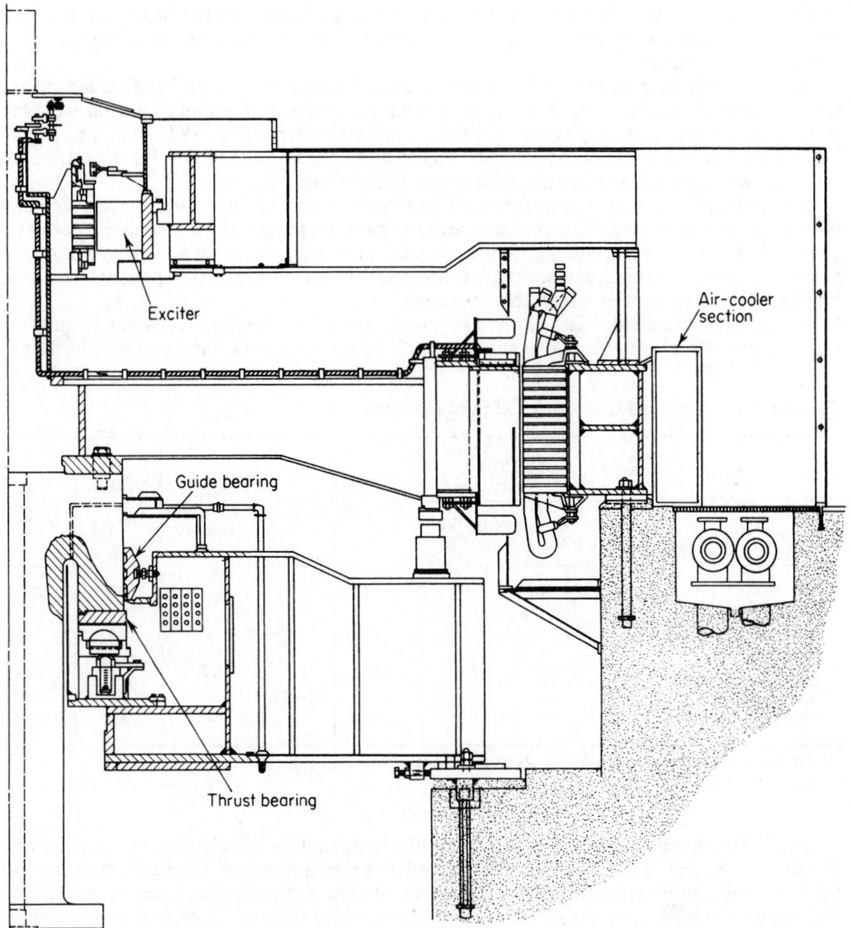


**FIG. 7-20** Cross section of a typical vertical alternator.

Guide-bearing clearances are on the order of 0.001 to 0.0005 in (diametral)/in of bearing-journal diameter, with this figure decreasing as diameter increases, and with a maximum on the order of 0.025 in.

***63. Bearing Arrangements.*** Vertical alternators are classified by their bearing arrangement. A suspended unit has the thrust bearing above the rotor and is provided with two guide bearings, one above the rotor and one below. The upper guide bearing is frequently placed just above and in the same oil pot as the thrust bearing. An umbrella unit has the thrust bearing below the rotor and one guide bearing also below the rotor, usually in the same oil pot and just above the thrust bearing. When umbrella arrangements are used, careful consideration must be given to the stability of the unit with respect to over-turning moments. Large-diameter slow-speed units, in which the ratio of rotor diameter to the height of rotor center of gravity above the thrust surface is relatively large, lend themselves to umbrella construction. A principal advantage of umbrella construction is a substantial reduction in required powerhouse headroom, since the relatively high bearing support structure is below the rotor in the turbine pit. Furthermore, this structure itself usually spans a shorter distance when placed below the rotor and is not so high because of this, and shaft length may be reduced. A modified umbrella arrangement, having a guide bearing above the rotor, may be used when mechanical stability precludes classic umbrella construction; this retains some of the advantage of reduction in headroom requirement. The thrust bearing may be supported from the turbine head cover to achieve a more compact arrangement.

***64. High-Pressure Systems.*** Horizontal journal bearings and thrust bearings are sometimes provided with high-pressure oil to reduce starting torque and minimize bearing wear on start-up. Oil at pressures on the order of 1500 $lbf/in^2$ is introduced at the bottom of the journal bearing—or at the center of each shoe or segment of a thrust bearing—to lift the rotor and introduce an oil film in the bearing clearance before the shaft rotates.

Steam turbine-generator units frequently are operated at very low speeds when they are disconnected from the system, to prevent the shaft system from sagging while cooling or while at standstill for extended periods. This "turning-gear" operation may be at 5 to 10 r/min, much below the speed at which the bearing would maintain an oil film. High-pressure or flood lubrication may be applied during this operation.

***65. Oil Specifications.*** The oil used in journal and thrust bearings is selected to suit the requirements of the particular application. Table 7-3 gives typical lubricating-oil specifications.

**TABLE 7-3** Typical Lubricating-Oil Specifications

| Application | Viscosity (SSU) | | Flash point (min.), °F | Pour point (max.), °F | Specific gravity at 60°F (max.) | Neutral No. (max.) | Vis. index (min.) |
|---|---|---|---|---|---|---|---|
| | At 100°F | Min. at 210°F | | | | | |
| Pivoted—shoe thrust bearings, and sleeve bearings on horizontal machines with speeds 1,800 r/min and below | 275–375 | 50 | 400 | +10 | 0.89 | 0.2 | 90 |
| Sleeve bearings on horizontal machines with speeds above 1,800 r/min to and including 3,600 r/min | 140–160 | 43 | 400 | +10 | 0.89 | 0.2 | 90 |

NOTES: Lubricants are high-quality petroleum oils having rust and oxidation inhibitors.
$t_{°C} = (t_{°F} - 32)/1.8$.

***66. Temperature Limitations.*** Safe operating temperatures of babbitted bearings are dependent on the ability of the babbitt to withstand plastic deformation. Babbitt temperatures of 95°C are safe, but most bearings are designed for operation at somewhat lower temperatures. Bearing temperature detectors are placed in the bearing shoe material as close as practicable to the babbitt surface, or in the oil discharging from the bearing. The trend in bearing temperature is more significant than the temperature itself; a bearing which suddenly rises from 50 to 85°C, for no apparent reason, is much more a matter for concern than one which operates consistently at any point within this range.

## REACTANCES AND TIME CONSTANTS

***67. General.*** To facilitate mathematical analysis of alternators operating alone or in conjunction with other machines and systems, the reactances and time constants of synchronous machines are defined under various operating conditions. Where variations in reactance with rotor position occur because of asymmetry, direct- and quadrature-axis values of reactance are required.

Reactances of synchronous machines generally are taken to be equal to their corresponding impedances. However, resistance has a major influence on time constants.

***68. Per Unit System.*** Voltages and currents are commonly expressed in percent or per unit of rated values. The corresponding base for expressing per unit reactances is the ohms which would produce a voltage drop of rated volts per phase when rated current flows through it. A principal advantage of this system of notation is the ease in comparing similar machines. As an example, if 1.0 per unit voltage is applied to 0.5 per unit reactance, the current will be 2.0 per unit regardless of the actual machine rating. Time constants are expressed in seconds.

***69. Principal Reactances.*** Each of the commonly used reactances, expressed in per unit terms, is equal to the fundamental voltage induced in the armature winding by the flux resulting from rated armature current acting on a particular combination of permeances.

*Stator leakage reactance* $x_l$ is a portion of all other machine reactances and is a result of flux produced across the armature slots and in the coil end region by armature current. Figure 7-4 shows the slot portion of this flux path. This component of flux is essentially independent of rotor position, but it is drastically affected by coil pitch in the case of zero-sequence reactance $x_0$, discussed later.

Synchronous reactances are applicable when the rotor is moving in synchronism with the mmf produced by steady-state armature current. The principal flux paths are then as shown in Fig. 7-21*a*. For the sake of simplification the stator slots and stator leakage fluxes are not

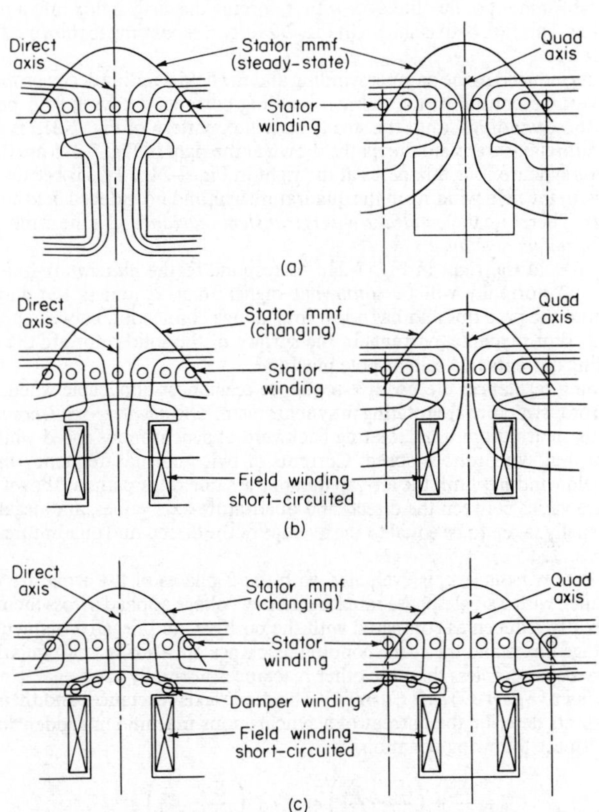


**FIG. 7-21** Principal flux paths for direct and quadrature axes. *(a)* Synchronous reactance; *(b)* transient reactance; *(c)* subtransient reactance.

included in this figure. The *direct-axis synchronous reactance* $x_d$ corresponds to the condition shown at the left in Fig. 7-21*a*, where the rotor-pole axis is in line with the maximum value of the armature mmf. The *quadrature-axis synchronous reactance* $x_q$ corresponds to the condition shown at the right in Fig. 7-21*a*, where the axis between the rotor poles is in line with the peak value of the armature mmf. The difference in air-gap flux between these two conditions is more clearly illustrated in Fig. 7-9. The considerably lower air-gap permeance in the interpole space results in much less actual and fundamental flux in the quadrature axis. The reactances corresponding to these two air-gap permeances are commonly called the magnetizing reactances, or reactances of armature reaction, and are designated as $x_{ad}$ and $x_{aq}$. Since the synchronous reactances include stator leakage,

$$x_d = x_l + x_{ad} \qquad \text{and} \qquad x_q = x_l + x_{aq}$$

Transient and subtransient reactances are applicable when the armature mmf is changing with respect to time. Currents will then be induced in the rotor windings which will affect the air-gap flux. If the direct-axis mmf is established suddenly and the field circuit is closed, current will be induced in the field winding which will oppose the sudden establishment of flux linkages with it and will force the air-gap flux to cross the space between the poles and not penetrate to the rotor spider. This flux path is shown in Fig. 7-21*b*. The total permeance of this flux path corresponds to the *direct-axis transient reactance* $x'_d$.

Furthermore, if there is a damper winding in the pole faces, current induced in it will oppose the sudden establishment of flux linkages with it, forcing the air-gap flux into a path above the damper winding. This flux path is shown in Fig. 7-21*c*, corresponding to the *direct-axis subtransient reactance* $x''_d$.

The currents induced in the damper winding and the field winding decay proportionately to their respective time constants, the damper-winding currents decaying first, permitting flux linkages with these windings, until the steady-state flux pattern of Fig. 7-21*a* is established.

The quadrature-axis transient flux paths shown at the right in Fig. 7-21*b* are the same as the quadrature-axis synchronous flux paths at the right in Fig. 7-21*a*. This is because there are no flux linkages with the field winding in the quadrature axis and no induced field current to affect the air-gap flux. Therefore the *quadrature-axis transient reactance* $x'_q$ is the same as the *quadrature-axis synchronous reactance* $x_q$.

The flux paths at the right in Fig. 7-21*c* correspond to the *quadrature-axis subtransient reactance* $x''_q$; $x''_q$ normally will be somewhat higher than $x''_d$ unless the damper winding extends well into the pole tips and has a low impedance connection between poles.

In a round-rotor machine, currents in the surface of the solid rotor are the equivalent of damper-winding currents in a salient-pole machine.

The foregoing reactances are positive-sequence reactances applicable when the armature mmf and the rotor structure are rotating in synchronism. The *negative-sequence reactance* $x_2$ is applicable with an armature mmf rotating backward at synchronous speed while the rotor is rotating forward at synchronous speed. Currents of twice normal frequency induced in the damper and field windings limit the air-gap flux to the same rotor paths as shown in Fig. 7-21*c*. The permeance varies between the direct- and quadrature-axis values, and negative-sequence reactance is usually taken to be equal to the average of the direct- and quadrature-axis subtransient reactances.

*Zero-sequence reactance* $x_0$ is evaluated with the 3 phases of the armature winding connected in parallel, with a single-phase rated frequency voltage applied across them. The air-gap flux is then greatly reduced as compared with the positive- and negative-sequence cases. The armature leakage fluxes and the corresponding reactance are also less, particularly with short-pitch armature coils. $x_0$ is less than the other reactances normally used.

***70. Equation of Short-Circuit Current.*** The direct-axis reactances and their corresponding time constants describe the current of a synchronous machine in sudden short circuit in accordance with the following equation,

$$I = \frac{E}{x_d} + \left(\frac{E}{x'_d} - \frac{E}{x_d}\right)\epsilon^{-t/T'_d} + \left(\frac{E}{x''_d} - \frac{E}{x'_d}\right)\epsilon^{-t/T''_d} \qquad (7\text{-}13)$$

where $I$ = the ac rms component of current following a 3-phase symmetrical short circuit from no load, per unit; $E$ = ac rms voltage prior to short circuit, per unit; $t$ = time after instant of short circuit, seconds; $T'_d$ = direct-axis transient time constant, seconds; and $T''_c$ = direct-axis subtransient time constant, seconds. Armature circuit resistances are neglected, and excitation is assumed to be constant.

Typical oscillograms of the currents and voltages of a sudden 3-phase short circuit are shown in Fig. 7-22. Detailed methods for analyzing the oscillograms of short-circuit currents to obtain values for reactances and time constants are given in Ref. 9.

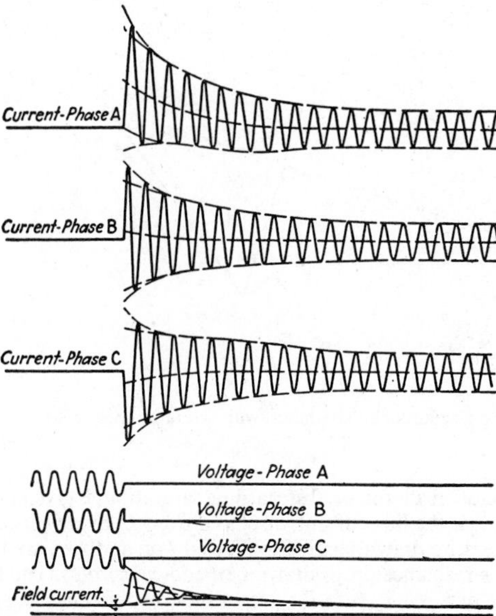


**FIG. 7-22** Typical oscillogram of a sudden 3-phase short circuit.

*71.* ***Offset Currents.*** At the instant of a short circuit from rated-voltage no load, each of the 3-phase windings will be linked by some portion of the air-gap flux. To maintain these flux linkages constant, each armature current will have an added dc component proportional to it. The dc components of current produce a stationary air-gap mmf which causes a fundamental-frequency current in the field winding, in order that the field flux linkages remain constant. The *armature time constant* $T_a$ is the time constant of the exponential rate of decay of the dc components.

The dc components of armature current, which cause the currents to be asymmetric with respect to the zero axis, are shown in Fig. 7-22. A phase could have zero initial flux linkages, in which case the short-circuit current would be symmetric. The maximum initial flux linkages would occur if the phase were directly in line with the rotor pole at the instant of short circuit. The dc component of armature current in that phase would then equal the peak value of the initial ac component. The total rms asymmetric current over the first cycle would then be the square root of the sum of the squares of the dc component and an equal rms ac component.

## CHARACTERISTICS AND PHASOR DIAGRAMS

*72.* ***Phasor Diagrams.*** The terminology and the convention of phase relationships used in this article are in common use in the power field, although others are also used in the literature on this subject.

Phasor diagrams of synchronous machines are based on the two-reaction theory, which treats the asymmetry of the rotor circuits by resolving voltage and current vectors into direct- and quadrature-axis components. The diagrams are valid only for unsaturated conditions, since saturation in the two axes is not the same and combining them by superposition is therefore invalid.

Saturation is here evaluated only in calculating load excitation, and then only in the direct axis. The method described in Par. **74** is generally accepted because of its good conformance with actual test data.

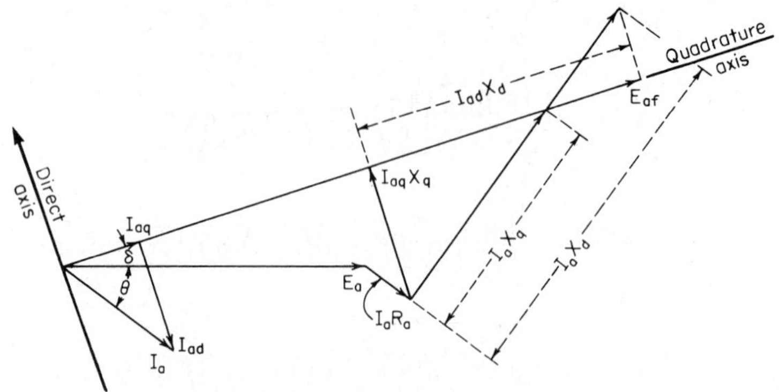


**FIG. 7-23** Phasor diagram of an alternator, with saturation neglected.

The phasor diagram of an alternator, disregarding saturation, is given in Fig. 7-23, where $E_a$ is the terminal voltage, $I_a$ is the line current, displaced from $E_a$ by power-factor angle $\theta$, $I_aR_a$ is the armature resistance drop drawn parallel to $I_a$, and $I_ax_q$ and $I_ax_d$ are the quadrature- and direct-axis synchronous reactance drops drawn perpendicular to $I_a$. The line from the origin through the vector sum of $E_a$, $I_aR_a$, and $I_ax_q$ establishes the reference axis for quadrature-axis voltages. The current may then be resolved into its quadrature- and direct-axis components and the corresponding voltage drops $I_{aq}x_q$ and $I_{ad}x_d$ added to $I_aR_a$ and $E_a$ to determine the magnitude of $E_{af}$, which is the internal voltage produced by the field current acting alone. The angle between the internal voltage $E_{af}$ and the terminal voltage $E_a$ is the displacement angle $\delta$.

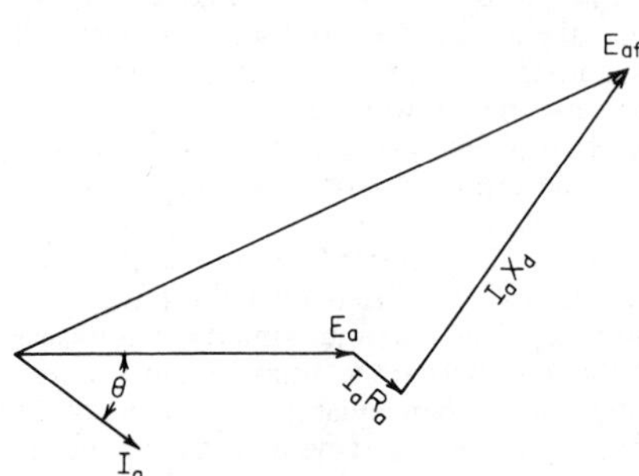


**FIG. 7-24** Phasor diagram for a round-rotor alternator, with saturation neglected.

If $x_q$ nearly equals $x_d$, as in a round-rotor machine, the diagram may be simplified to Fig. 7-24. It will be noted that, insofar as determining the magnitude of $E_{af}$ is concerned, the refinement of considering both direct and quadrature axis has a very minor effect. On this basis the ANSI has stipulated a method for calculating excitation requirements based on round-rotor theory and using data from the open- and short-circuit saturation curves and the zero-power-factor saturation curve at rated armature current.

***73. Characteristic Curves.*** Typical characteristic curves of an alternator are shown in Fig. 7-25. The ANSI method for calculating load excitation requires calculation of the Potier reactance $x_p$ from these characteristic curves in the following manner: The intersection of the rated-current zero-power-factor saturation curve with the rated-voltage line locates the point *d*. To the left of *d* on the rated-voltage line, the length *ad* is laid off equal to the field current ($I_{FSI}$) for zero voltage on the rated-current zero-power-factor saturation curve. This is the field current to produce rated-armature-current short circuit. Through *a* the line *ab* is drawn parallel to the air-gap line. The intersection of this line with the no-load saturation curve locates the point *b*. The vertical distance *bc* from the point *b* is the Potier reactance. If the zero-power-factor saturation curve is obtained for some current slightly different from rated current, the distance *bc* is divided by the per unit armature current at which the zero-power-factor saturation curve was actually obtained, to give the Potier reactance. Methods for calculating Potier reactance are given in Refs. 28 and 29.

***74. Calculating Excitation Requirement.*** The rated-load excitation requirement $I_{FL}$ is then determined as shown in Fig. 7-26. $I_{FG}$ is the excitation current corresponding to rated armature voltage on the air-gap line. $I_{FSI}$ is the excitation current to produce rated armature

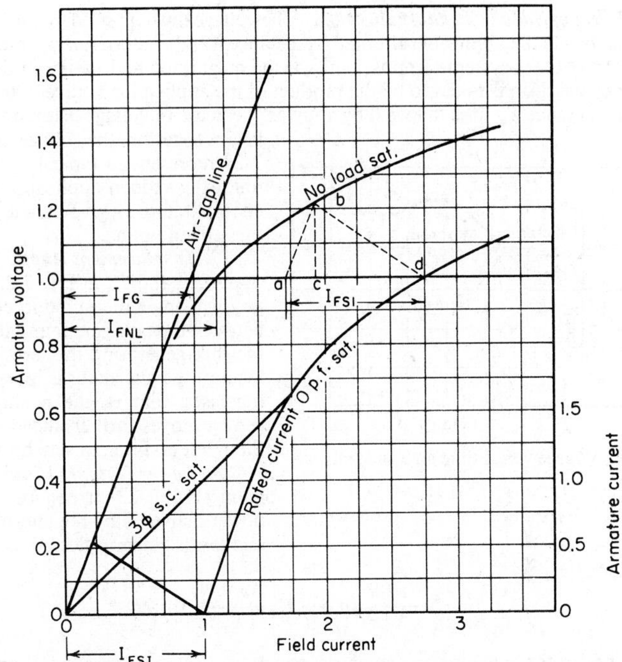


**FIG. 7-25** Typical characteristic curves of an alternator, showing graphic determination of Potier reactance. Quantities are in per unit.

current, short circuit. $\theta$ is the rated power-factor angle. $\theta$ is laid off clockwise for lagging power factor, that is, with the generator supplying an inductive current.

Armature resistance is normally neglected. $I_{FS}$ is the field current corresponding to saturation and is measured between the air-gap line and the actual saturation curve at the voltage, $E_{xp}$, behind Potier reactance. This voltage is the vector sum of rated armature voltage and the voltage drop corresponding to Potier reactance, $I_a x_p$, added at the power-factor angle $\theta$ (see Fig. 7-27). It will be noted that Fig. 7-26 corresponds to the phasor diagram for a round-rotor machine (Fig. 7-24), with resistance neglected and the effect of saturation added.

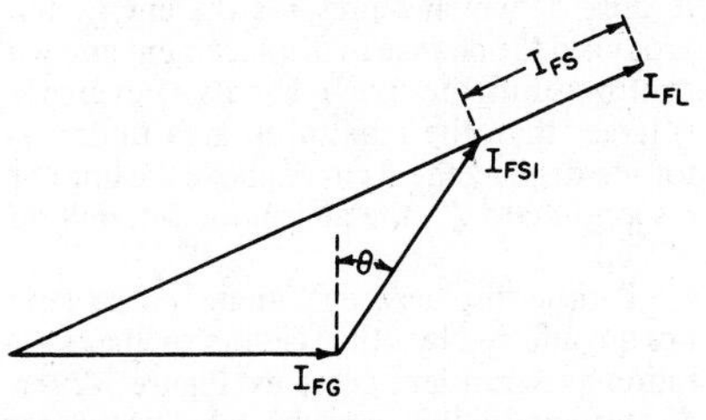

**FIG. 7-26** ANSI method for calculating load excitation.

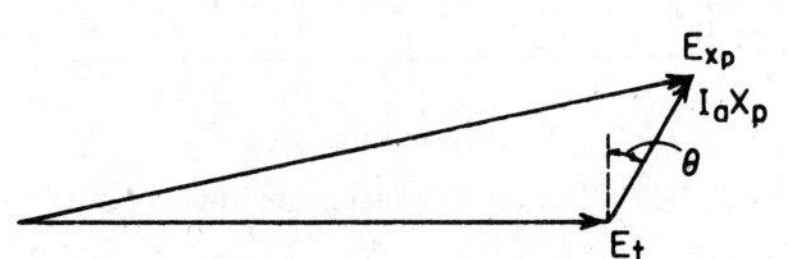


**FIG. 7-27** Calculation of voltage behind Potier reactance.

For loads other than rated load, and at voltages other than rated voltage, appropriate values of $I_{FG}$, $I_{FSI}$, $\theta$, and $I_{FS}$ are used.

***75. Regulation.*** The voltage regulation of an alternator is defined as the rise in voltage with constant rated-load field current when the load is reduced from rated load to zero, expressed as a percent of rated voltage. It is the difference between the voltage on the no-load saturation curve at rated-load field current and rated voltage, divided by rated voltage and multiplied by 100.

***76. Effect of Sudden Load Application.*** The voltage drop on sudden application of load is influenced by (1) the applicable alternator reactance, (2) the alternator open-circuit time constant, (3) the exciter-system response, and (4) the magnitude and nature of the load. In most cases the initial drop is taken to be the product of the applied load current and the alternator transient reactance $X'_d$. It is followed by a further decrease in voltage before the exciter system begins to restore the voltage to normal. Figure 7-28 represents a typical voltage-time curve following sudden application of load. References 20 and 21 give detailed discussions of these phenomena.

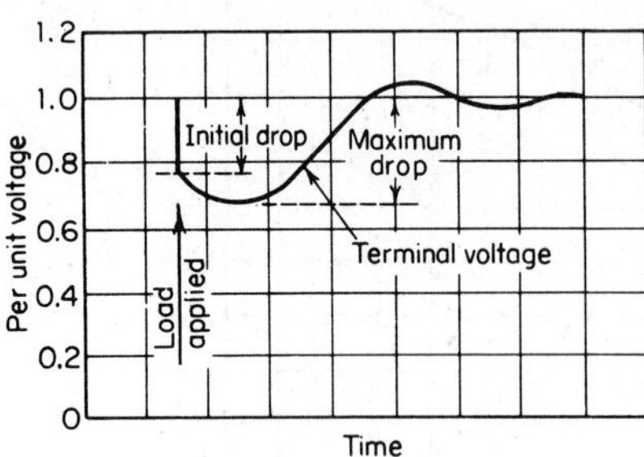


**FIG. 7-28** Voltage versus time for a suddenly applied load.

***77. Displacement Angle.*** The angle $\delta$ between $E_{af}$ and $E_a$ in Fig. 7-23 represents the physical angle measured in electrical degrees between the pole center and the peak value of the flux generating the terminal voltage. As load on the alternator is increased, this angle increases; that is, the position of the pole center moves farther ahead of this flux, although remaining in synchronism with it.

***78. Power Output.*** From the phasor diagram (Fig. 7-23) it can be shown that the power delivered by the alternator, saturation and losses being neglected, is

$$\frac{E_{af}E_a}{X_d}\sin\delta + E_a^2\frac{X_d - X_q}{2X_dX_q}\sin 2\delta$$

For a round-rotor machine in which $X_q$ equals $X_d$, the second term drops out and $P = (E_{af}E_a/X_d)\sin\delta$. The maximum power then is $E_{af}E_a/X_d$ when $\delta$ equals 90°. This is the stability limit under steady-state conditions.

***79. Stability.*** The relationship between power, or torques, and displacement angle represented by the preceding equation is shown in Fig. 7-29. If the machine is operating at a load corresponding to torque $T_0$ and displacement $\delta_0$ and the torque is suddenly changed to $T_1$, the rotor will start to increase its displacement toward a new value of $\delta_1$. But the rotor will overshoot with an energy equal to area $A_1$ which represents the energy that produced the change in displacement and will continue until the area $A_2$ equals $A_1$. If area $A_1$ is larger than the maximum area under the torque-displacement curve above the line corresponding to $T_1$, the alternator will pull out of step.

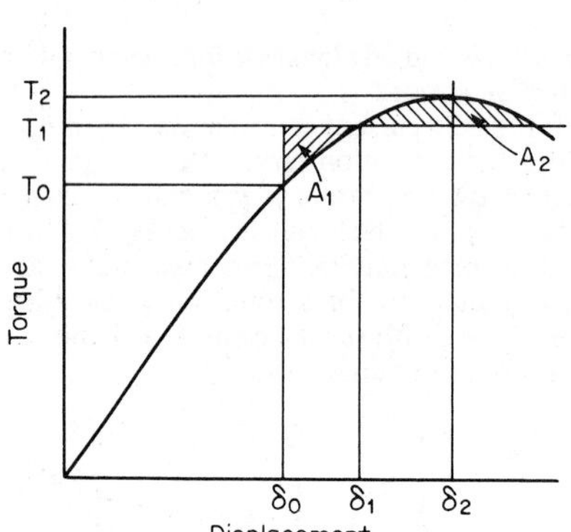


**FIG. 7-29** Torque-displacement angle curve.

Torque-displacement angle characteristics are affected by other factors such as saturation and transient currents. Figure 7-29 indicates only the general nature of the phenomenon. Consult the References for articles giving detailed treatment of the subject.

## PARALLEL OPERATION

***80. Conditions for Paralleling.*** In connecting two alternators in parallel it is desirable, in order to minimize current surges, that voltages and speeds be equal and corresponding voltages be in phase. Synchroscopes measure the difference between corresponding voltages of alternators to show when they are in synchronism and may be connected together. Various schemes have been devised to sense synchronism and initiate switching sequences automatically. If a synchroscope or automatic synchronizing device is not available, synchronism may be indicated with a system of lamps as shown in Fig. 7-30. Indicating lamps connected across the

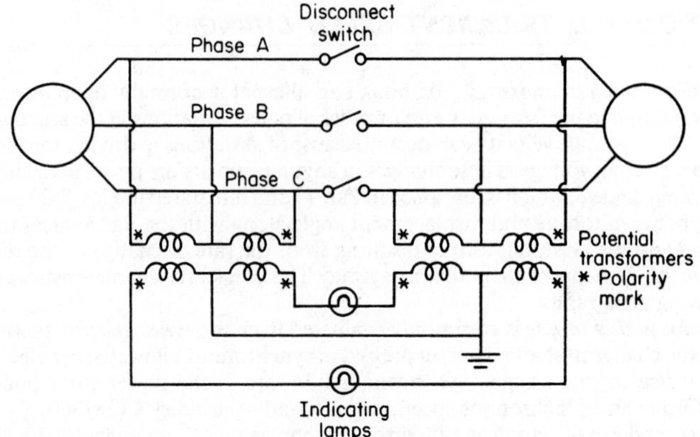


**FIG. 7-30** Connection diagram for indicating lamps to show synchronism.

disconnect switch between the two machines measure the difference between two of the terminal voltages. This difference ranges between zero (when the voltages are in phase) and twice rated voltage (when they are 180° out of phase). The lamps are alternately dark and bright as the phase position changes. The two sets of lamps go dark and bright together if the sequence of the terminal voltages is the same; if the sequence is opposite, they go dark and bright alternately.

***81. Division of Power.*** As load is increased on two alternators connected in parallel, there is a reduction in their speed which is sensed by the prime-mover speed-governing systems. The governors act to restore speed to normal. The division of load between the two alternators is determined by the characteristics of their prime-mover governing systems. If one system has speed characteristic *a* in Fig. 7-31 and another has characteristic *b*, they will divide the load in the proportion $P_a$ and $P_b$ when operated at speed *S*. Control of the load on a unit is obtained by adjusting the governor speed characteristic up or down.

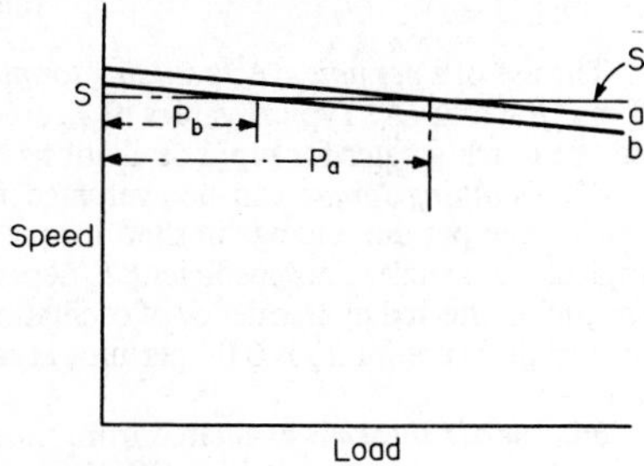


**FIG. 7-31** Diagram of division of load between two alternators in parallel.

***82. Division of Reactive Voltamperes.*** The voltage applied to a load connected to two alternators is determined by the total excitation on the alternators. Identical alternators with identical prime-mover speed-governing characteristics share the load equally and, with equal excitation, share the reactive voltamperes equally. Each alternator operates at the same power factor as the load power factor. An increase in excitation on one will increase the system voltage, and that alternator will supply a greater share of the reactive voltamperes. A decrease in excitation of the other alternator will restore the terminal voltage to the original value but will increase the difference in the division of the reactive voltamperes. Adjustment of alternator excitation, then, determines not only the voltage applied to the load but also the division of reactive voltamperes between the alternators.

Automatic voltage regulators used on alternators operating in parallel may be provided with cross-current compensation to adjust the excitation on machines having different saturation characteristics so that they will operate at the same power factor. A voltage regulator applied to an alternator connected to a relatively very large system controls the power factor of the alternator rather than its voltage.

## OSCILLATION AND TRANSIENT SPEED CHANGES

***83. Oscillatory Characteristics.*** Although an alternator normally operates at a constant speed corresponding to the frequency and number of poles, variations in the angular velocity of the rotor about its average velocity can occur because of variations in driving torque, load, field excitation, or terminal voltage. These changes in angular velocity are associated with changes in the displacement angle $\delta$, which is discussed in Par. 77 and illustrated in Figs. 7-23 and 7-29. The relationship between torque and displacement angle along with the mechanical inertia of the rotating parts and the damping torque resulting from the rate of change of the displacement angle provide the elements of an oscillatory system. These machine characteristics are discussed in the following paragraphs.

***84. Synchronizing torque*** is commonly evaluated from the *synchronizing power coefficient* $P_r$, which is the change in shaft power, expressed in synchronous kilowatts, per electrical radian of change in displacement angle. Synchronous kilowatts is the power corresponding to the product of torque and synchronous speed, and elec rad = elec deg × ($2\pi/360$).

With slow and small variations in displacement angle, $P_r$ approaches the slope of the steady-state torque-angle curve (Fig. 7-29) at the average operating point, with due regard for the units used. Since the torque-angle curve is determined in part by the field excitation, $P_r$ is influenced by this variable as well as by the average load. More rapid variations in the displacement angle increase $P_r$ because of currents induced in the rotor circuits.

As an approximation, $P_r$ at rated load is commonly taken as rated kilowatts divided by rated-load displacement angle. From Fig. 7-23, with $E_a = 1.0$, $I_z = 1.0$, and $X_q$ in per unit terms, neglecting $I_aR_a$,

$$\text{Rated-load } \delta = \tan^{-1} \frac{X_q \cos\theta}{X_q \sin\theta + 1.0} \quad \text{deg} \tag{7-14}$$

$$\text{Rated-load } P_r = \frac{\text{rated kW}}{\text{rated-load } \delta \times (2\pi/360)} \tag{7-15}$$

The use of a *per unit synchronizing torque coefficient* $T_s$ is frequently more convenient, and $T_s = P_r/\text{rated kVA}$. Typical values of $T_s$ are in the range of 1.2 to 2.0 for slow oscillations but may be much greater for rapid oscillations.

***85. Damping torque*** can be evaluated from the *per unit damping torque coefficient* $T_d$, which is the per unit change in shaft torque per electrical radian per second rate of change in displacement angle. This coefficient $T_d$ depends greatly on machine damper winding construction and is affected by frequency of oscillation, field excitation, and average load. An order-of-magnitude figure for $T_d$ is 0.02 per unit at rated load, but large deviations from this value are common.

***86. Inertia torque*** is evaluated from the mechanical rotational inertia or flywheel effect of the unit, commonly expressed as $WK^2$, where $W$ is the weight of the rotating parts in pounds and $K$ is the effective radius of gyration in feet. The rotational stored energy is also commonly expressed in per unit terms as the *inertia constant H,* which is the kilowattseconds of mechanical stored energy at synchronous speed divided by rated kVA. Then

$$H = \frac{0.231 \times 10^{-6}(WK^2)(\text{r/min})^2}{\text{rated kVA}} \tag{7-16}$$

Typical values of $H$ range from 0.5 to 4.0, the higher values being for larger, lower-speed machines.

***87. Basic Equations.*** The electromechanical oscillation and transient speed changes of an alternator can be expressed in the same mathematical forms common to other oscillatory systems, including electric circuits and mechanical spring-mass systems. In this case, the applicable differential equations of motion are based on the essential condition that the torque applied to the shaft must at all times equal the algebraic sum of the synchronizing, damping, and inertia torques.

As an example of the application of these principles, for the simple case of sustained oscillation of a single synchronous machine operating on an infinite system with a component of applied shaft torque varying sinusoidally with time, the equilibrium of torques is shown by the equation

$$T_0 = Y_0[T_s + j\omega_0 T_d - \omega_0^2(H/\pi f)] \qquad (7\text{-}17)$$

where $T_0$ = per unit applied shaft torque (phasor quantity), $Y_0$ = electrical radians displacement from the average displacement angle (phasor quantity), $\omega_0 = 2\pi \times$ frequency of oscillation in hertz and $f$ = line frequency in hertz.

Practical problems of sustained and transient oscillation often involve additional machines and other elements interconnected in a system of sufficient complexity to justify computer solutions. In such instances, the alternators may also be represented by analogs to allow for variation of $T_s$ and $T_d$, which are not always constant.

***88. The natural frequency of oscillation*** $f_{on}$, with damping neglected, occurs when $T_s = \omega_0^2(H/\pi f)$;

$$f_{on} = \omega_{on}/2\pi = \sqrt{\frac{fT_s}{4\pi H}} \quad \text{Hz} \qquad (7\text{-}18)$$

Note that $f$ is line frequency in hertz. Typical values of $f_{on}$ range from about 1.2 to 4.5 Hz, although added mechanical inertia in the alternator or the prime mover may result in a lower natural frequency.

The ANSI Standards express the natural frequency of oscillation in different, but equivalent, terms, as follows:

$$F = \frac{35{,}200}{\text{r/min}} \sqrt{\frac{fP_r}{WK^2}} \quad \text{c/min} \qquad (7\text{-}19)$$

***89. A reciprocating engine drive*** applies pulsating torque to the alternator shaft. The principal alternating component of the torque has a frequency equal to the product of r/min × (no. cylinders normally fired)/r. Torque components of lesser magnitude and higher frequency are also produced, and inequalities in the torque per cylinder may introduce lower-frequency components.

An alternator driven by such an engine and connected to a large system oscillates in accordance with the equation in Par. **87**. The per unit power variation is $\pm Y_0 T_s$, and the resulting current pulsation could cause objectionable voltage variation in the system, usually evident as light flicker. Since voltage variations caused by one unit can produce oscillation of other units, the analysis of possible oscillations in a system with several reciprocating engine-driven alternators may be quite complex.

One suitable means for limiting the oscillations is to provide flywheel effect in the new unit such that its natural frequency of oscillation for rated-load conditions is well below, perhaps by 20%, the lowest possible forced frequency produced by any of the engines, including that of the new unit. Where this is not practical, oscillation at resonance may be sufficiently limited under some conditions if suitable damping torque is provided, or system operating procedures may be altered to avoid load conditions on individual units resulting in excessive oscillation.

A single reciprocating-engine-driven alternator supplying an isolated load may have little or no synchronizing or damping torque because the line frequency and voltage may both vary with the oscillation in speed resulting from the engine torque pulsations. This speed variation, and the resulting effects on voltage and frequency, are then limited almost entirely by the flywheel effect provided.

***90. Governors and voltage regulators*** have oscillatory and transient characteristics of their own which sometimes influence the specification of alternator characteristics. Increased flywheel effect is usually beneficial from the standpoint of allowing these regulating elements more time to act in the event of load changes before the corresponding speed change becomes too great. Transient stability, as discussed in Par. **79**, is improved. Flywheel effect in a *hydraulic-tur-*

*bine-driven alternator* is often particularly important because the response of the speed-governing system to loss of load may be deliberately delayed to avoid excessive pressure rise in the penstock as the gates close. Added flywheel effect reduces the overspeed resulting from these conditions.

## LOSSES AND EFFICIENCIES

***91. Conventional efficiencies*** used to evaluate alternator performance are based on losses which can be readily measured by test procedures described in the following article. The efficiency is given by the following equation, where the losses are the sum of the individual losses described in the succeeding paragraphs.

$$\text{Efficiency in per unit} = \frac{\text{output}}{\text{input}} = 1 - \frac{\text{losses}}{\text{input}} \qquad (7\text{-}20)$$

***92. Losses.*** Five losses are usually evaluated for alternators. Two of them are considered to be fixed losses; that is, they are assumed to be fixed with respect to load. The other three are variable losses; they are variable with respect to load. The fixed losses are (1) windage and friction and (2) core loss. The variable losses are (3) field copper loss, (4) armature copper loss, and (5) stray loss or load loss. In addition to these, exciter and/or rheostat losses may be evaluated, but in determining operating efficiency these ordinarily are charged, not to the machine, but to the plant.

*Windage and friction loss* is affected by the size and shape of the rotating parts, fan design, bearing design, and enclosure arrangements. An approximate combined windage and friction loss is given by the equation

$$\text{Windage and friction} = K\left(\frac{V}{10{,}000}\right)^{2.5} D\sqrt{L} \quad \text{kW} \qquad (7\text{-}21)$$

where $V$ = rotor peripheral velocity, feet per minute, $D$ = rotor diameter, inches, $L$ = rotor effective length, inches, and $K$ = 0.08 to 0.11 for slow-speed salient-pole machines, 0.06 to 0.08 for higher-speed salient-pole machines, and 0.06 to 0.085 for air-cooled turbogenerators.

*Core loss* is caused by the main flux of the machine and occurs primarily in the stator teeth, in the portion of the stator core behind the teeth, and in the surface of the rotor poles, but it includes also the losses in structural parts of the machine which are exposed to stray alternating magnetic fields.

Stator cores ordinarily are constructed of thin laminations of a silicon steel, insulated from each other, to limit the hysteresis and eddy-current losses in the steel. The spacers in the ventilating ducts and the clamping structure at the ends of the cores may be of nonmagnetic materials to limit losses in these components. The end packets of the core may be stepped back from the air gap to decrease fringing flux which enters the core axially, or they may be provided with radial slits in the teeth to decrease eddy-current losses due to flux entering the ends of the core axially. Since the stator slots introduce a variation in the air-gap flux density which will produce eddy-current losses in the surfaces of the rotor poles, the poles on salient-pole machines are ordinarily built of relatively thin laminations, or the surface of a round-rotor machine may be grooved, to reduce these losses.

*Copper losses* are calculated with the winding resistance corrected to a temperature corresponding to the class of insulation on the winding, as follows:

| Class of insulation | Temperature, °C |
|---|---|
| A | 75 |
| B | 95 |
| F | 115 |
| H | 130 |

Where the rated temperature rise is that of a lower class of insulation, the temperature for resistance correction is that for the lower class.

*Field copper loss* is calculated from the field current and the dc resistance of the field winding at the appropriate temperature. Field current may be measured under actual operating conditions or calculated by the method described in Par. **74.** The voltage drop across the collector ring brushes ordinarily is neglected but may be included in the exciter loss.

*Armature copper loss* is calculated from the dc resistance of the armature winding at the appropriate temperature. For a 3-phase machine the loss is $1.5RI^2$ or $3R'I^2$, where $R$ is the armature winding resistance measured terminal to terminal, $I$ is the line current, and $R'$ is the armature resistance measured terminal to neutral.

*Stray loss or load loss* is caused by the flux produced by armature current and includes eddy-current loss in the armature conductors, core losses set up by the armature current, losses in the core supporting structure and end housings due to armature current, and field surface or damper-winding losses due to armature current. Stray loss ordinarily varies nearly as the square of the armature current and may be expressed as a percentage of the armature copper loss.

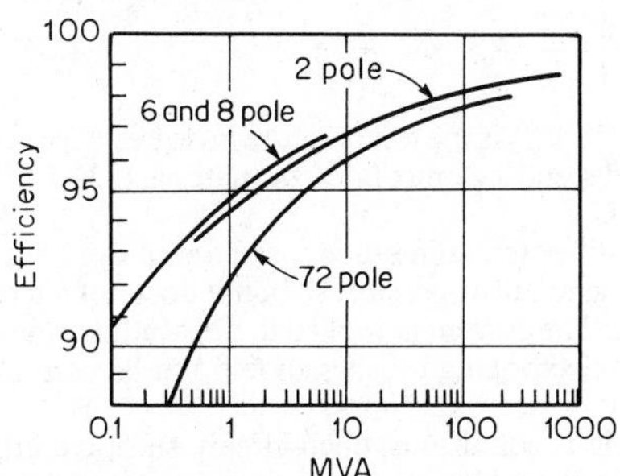


**FIG. 7-32** Trends in full-load efficiencies for 60-Hz 80% power-factor alternators.

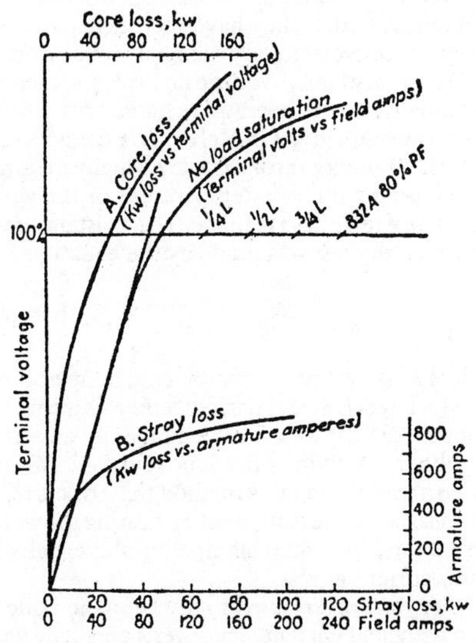


**FIG. 7-33** Typical 14-pole 80% power-factor salient-pole alternator losses.

**TABLE 7-4** Efficiency Calculation for the Typical Generator in Fig. 7-33

(Rating, 10,000 kVA, 80% pf, 6900 V, 3-phase, 60 Hz, 832 A, 514 r/min. Stator resistance, 0.060 Ω at 75°C. Field resistance, 1.82 Ω at 75°C.)

| Load, kW | 8,000 | 6,000 | 4,000 | 2,000 |
|---|---|---|---|---|
| Power factor, % | 80 | 80 | 80 | 80 |
| Field current, A | 250 | 206 | 164 | 125 |
| Core loss (see curve *A*, Fig. 7-33) | 48 | 48 | 48 | 48 |
| Field and winding loss | 45 | 45 | 45 | 45 |
| Field $I^2R$ | 52 | 35 | 22 | 13 |
| Stator $I^2R$ | 60 | 34 | 15 | 4 |
| Stray loss (see curve *B*, Fig. 7-33) | 59 | 33 | 15 | 4 |
| Exciter loss | 5 | 4 | 3 | 3 |
| Pilot exciter loss | 0.3 | 0.2 | 0.2 | 0.2 |
| Total loss | 269.3 | 199.2 | 148.2 | 117.2 |
| Output, kW | 8,000 | 6,000 | 4,000 | 2,000 |
| Output and loss, kW | 8,269.3 | 6,119.2 | 4,148.2 | 2,117.2 |
| Loss, % | 3.25 | 3.21 | 3.58 | 5.52 |
| Efficiency, % | 96.75 | 96.75 | 96.42 | 94.48 |

***93. Typical Efficiencies.*** Trends in full-load efficiencies of 80% power-factor synchronous machines are shown in Fig. 7-32. These curves are representative for normal machines. If voltage, short-circuit ratio, $WK^2$, or temperature ratings are higher than normal, or if reactances are lower than normal, the tendency will be to reduce these efficiencies. Special designs with higher-cost materials may be employed to increase them.

Losses may be furnished in the form of curves such as Fig. 7-33. A calculation of efficiency from these curves is given in Table 7-4.

## TESTS ON ALTERNATORS

***94. Requirements.*** Tests are performed on alternators, either at the factory or in the field, to demonstrate that they have met their required performance. The schedule of tests may vary from commercial tests, which may require only the measurement of winding resistance, excitation current at rated-voltage no load, and dielectric tests, to a complete schedule which evaluates essentially all the operating characteristics of the machine. Details of all the tests ordinarily performed are given in Ref. 9, *IEEE Publ.* 115, Test Procedures for Synchronous Machines.

***95. Winding resistances*** are usually measured with a Kelvin bridge or similar equipment. It is important that the temperature of the winding be accurately determined at the time the resistance readings are made. The resistance may be corrected to another temperature from that at which the test was made by the equation

$$R_1 = R_2 \frac{K + T_1}{K + T_2} \tag{7-22}$$

where $R_2$ is the known resistance at temperature $T_2$(°C), $T_1$ is the temperature in degrees Celsius at which resistance is desired, and $K$ is a constant depending on conductor material (234.5 for copper, 225 for electrical-conductivity aluminum).

When a Kelvin bridge is not available, the drop-of-potential method may be employed. It is important in using this method that the current be as small as possible so that it does not affect the resistance measurement by heating the winding. The current is applied to the winding being tested through cables clamped to the terminals; collector ring brushes should not be used to carry current at rest.

***96. Saturation curves*** may be made under various conditions: open-circuit, short-circuit, and with load current. These tests are made with the alternator operated at rated speed and the excitation varied to produce data for a curve of voltage or current versus excitation. If the open-circuit and short-circuit saturation curves are obtained by a method which enables measuring power input, core loss and stray loss may be obtained at the same time.

***97. The separate driving-motor method*** is commonly used to obtain saturation curves, particularly when losses are also desired. The alternator being tested is mechanically coupled to a driving motor and driven at rated speed. For an *open-circuit saturation curve* the alternator terminals are open-circuited, and its excitation is varied to produce terminal voltage over a range of about 30 to 130% voltage. Simultaneous readings of excitation current and terminal voltage are taken to provide data for plotting the open-circuit saturation curve. If the power input to the driving motor is measured and its losses are known, the input to the alternator can be determined. This input represents windage and friction and core loss. Data taken with zero excitation give the windage and friction, which may be subtracted from the alternator input at the various values of excitation to determine core loss.

A *short-circuit saturation curve* is obtained in much the same way, but with the terminals of the alternator short-circuited. Excitation is varied from a value which will give about 30% stator current to one which will give about 130% stator current. The input to the alternator may be determined from measurements of driving-motor input power and is the sum of windage and friction, armature copper loss, and stray loss. The windage and friction are determined from power-input data taken with zero excitation on the alternator, and the armature copper loss may be calculated from measured values of winding resistance and corrected to the temperature of the armature winding measured during the test. The stray loss is calculated as the alternator input minus the windage and friction and armature copper losses.

The *load saturation curve* most frequently determined by test is with rated armature current at zero power factor. It is obtained by operating the alternator as a motor without connected load, that is, as a synchronous capacitor. Its excitation and the excitation of the supply generator are adjusted to obtain a curve of terminal voltage versus excitation, with rated armature current. This saturation curve, in conjunction with the open-circuit and short-circuit saturation curves, provides data from which the excitation requirement for any other value of load may be estimated, as discussed in Par. **74.** Since the power requirement for this saturation curve is limited to the losses in the two machines, it can be obtained at the factory even for quite large machines, provided that a test machine is available capable of absorbing the reactive kVA from the machine being tested.

***98. The deceleration method*** of obtaining saturation curves and losses may also be used, particularly with large vertical generators. The machine is brought up to a speed above rated speed and allowed to decelerate with the terminal condition and excitation established at the desired values. Excitation is best supplied from a separate source. For an open-circuit saturation-curve and core-loss test, the armature terminals are open-circuited, and the excitation is adjusted to the value desired and held constant while the generator decelerates. The curve of voltage versus speed will be a straight line, intercepting rated speed at the point on the saturation curve corresponding to the excitation used. The rate of deceleration is proportional to the windage and friction and core loss corresponding to the value of excitation used. By measuring the rate of deceleration, the loss at rated speed may be calculated from the following equation. A deceleration run with zero excitation will give a value of windage and friction.

$$\text{Loss} = 0.462 WK^2 N \frac{dN}{dt} \times 10^{-6} \quad \text{kW} \tag{7-23}$$

where $WK^2$ is moment of inertia of rotating parts, pound-feet squared, $N$ is the speed at which loss is being evaluated, revolutions per minute, and $dN/dt$ is the slope of the speed-time curve at $N$, revolutions per minute per second.

A short-circuit saturation curve and stray-loss measurement is made by deceleration in a similar manner, except that the armature terminals are short-circuited during the deceleration. Points on the short-circuit saturation curve are determined from measurements of the short-circuit current at rated speed for the various values of constant excitation. It will be noted that there is very little variation in short-circuit current with speed. The total loss may also be calculated from the rate of deceleration and the $WK^2$ of the unit. This total loss is the sum of windage and friction, armature copper loss, and stray loss. A test with zero excitation will give the windage and friction loss; the armature copper loss may be calculated from the current, the measured winding resistance, and the measured winding temperature during the deceleration run.

***99. Tests as a Motor.*** An open-circuit saturation curve may also be obtained by operating the machine as a motor from a variable-voltage source. At a particular voltage, the excitation of the machine is adjusted so that armature current is minimum. The armature current of an unloaded motor may be as low as 2 or 3% of rated current; low-range current-measuring equipment will be required to determine this minimum point. At this point the power factor is unity, and the excitation is very nearly that for open-circuit conditions at the same voltage. A more sensitive indication of unity power factor can be obtained by connecting a single-phase wattmeter with its current coil in one line and voltage coil across the other two. The meter reading is zero at unity power factor. Similar tests at other values of voltage will produce data from which an open-circuit saturation curve is plotted. Measurements of input power may be made to determine open-circuit losses.

A very close approximation of the short-circuit saturation curve may be obtained in a similar manner. The machine is operated as a synchronous motor at a fixed voltage of the lowest value at which stable operation can be obtained. This will be on the order of 30% of rated voltage. Field current is varied from a value which will produce about 150% of rated armature current, down to a value which will produce near zero armature current, recording data at several points. Lower values of field current will produce increasing armature current. This curve of armature current versus field current is plotted. The short-circuit saturation curve is a curve parallel to it, but passing through the origin. Measurements of input power may be made to determine short-circuit losses, and stray loss can be separated from copper loss if winding temperatures are observed during the test.

These test methods based on operating the machine as a motor are often used on small- and medium-sized units to simplify the test equipment needed and minimize the cost of the testing.

***100. Heat runs*** are made to establish the capability of the machine to operate at its rated load without exceeding the guaranteed temperature rises. The types of heat runs most commonly made in the factory are open-circuit, short-circuit, and zero-power-factor. Heat runs under rated load conditions may be made in the field after the equipment has been installed but ordinarily are not conducted in the factory because of the large power requirements.

A heat run at zero power factor, with the machine operated at no load as a synchronous capacitor and with appropriate conditions of armature current, voltage, and frequency maintained until the machine reaches constant temperature, can give temperature data most nearly approximating that corresponding to a rated-load test. When the test is made at rated armature current, it may be desirable to run at less than rated armature voltage to compensate for the higher internal voltage associated with zero-power-factor overexcited operation or to avoid excessive field current and temperature. The field temperature rise may be corrected to the field current corresponding to rated load, by using the following equation from Ref. 9,

$$T_r = T_f + (I_f/I_t)^2(T_t - T_f)\frac{k + T_a + T_f}{k + T_t + T_{at} - (I_f/I_t)^2(T_t - T_f)} \tag{7-24}$$

where $T_r$ = temperature rise corrected to the desired field current, degrees Celsius, $I_f$ = desired field current, amperes, $I_t$ = test field current, amperes, $T_t$ = temperature rise at test field current, degrees Celsius, $T_f$ = temperature rise through the fan or blower, degrees Celsius (usually neglected for salient-pole machines), $T_a$ = ambient temperature at which corrected temperature rise is desired, degrees Celsius, $T_{at}$ = ambient temperature during test, degrees Celsius, and $k$ = constant of field winding material, 234.5 for copper and 225 for electrical-conductivity aluminum.

A heat run at rated voltage with the terminals open-circuited and one at rated armature current with the terminals short-circuited provide an alternative way to approximate rated-load temperatures. The armature temperature rise is nearly the sum of the temperature rises for the two tests. Since these tests are conducted with relatively low values of excitation, it may be desirable to run another open-circuit heat run with excitation corresponding to rated load to approximate the field temperature rise under rated-load conditions.

Heat runs are conducted until the temperatures become essentially constant. For ambient temperatures within the range of 10 to 40°C temperature rises above the ambient ordinarily are not corrected for ambient temperature but are used as observed to evaluate the performance of the machine compared with the guarantees.

***101. Reactances.*** Test values for most of the commonly used reactances can be determined. Synchronous reactances can be measured by applying a positive-sequence voltage to the armature windings with no field current, while the machine is rotated at slightly above or below rated speed, and measuring the maximum and minimum values of the varying current. Direct-axis synchronous reactance is the ratio of the voltage to the minimum value of the current; and quadrature-axis synchronous reactance is the ratio of the voltage to the maximum value of the current. Direct-axis synchronous reactance may also be calculated as the ratio of $I_{FSI}$ to $I_{FG}$ from Fig. 7-25. Direct-axis transient and subtransient reactances and their respective time constants can be calculated from oscillograms taken during symmetrical sudden short circuits, by using the equation given in Par. **70**. Negative-sequence reactance can be measured by applying a negative-sequence voltage to the armature windings while the machine is rotated at rated speed in the positive direction with its field winding short-circuited. Zero-sequence reactance can be measured by applying a single-phase voltage from the machine neutral to the three armature terminals in parallel, with the machine rotated at rated speed and with its field winding short-circuited. Reference 9 describes detailed procedures for making these and other reactance tests.

Procedures for obtaining synchronous machine parameters by frequency response testing with the machine at rest are described in IEEE Publication 115A, which is an addendum to IEEE Publication No. 115, 1981.

***102. Dielectric Tests.*** The principal dielectric tests are described in Par. **43**.

## Bibliography

*103.* NOTE: Because of the large amount of available literature on synchronous machinery, this bibliography is limited to general sources and references specifically cited in the text.

***104. General Bibliographies***


1. Bibliography of Rotating Electric Machinery 1886–1947; *AIEE Publ.* S-32, January, 1950.
2. Bibliography of Rotating Electric Machinery for 1948–1961; *Trans. IEEE*, Power Apparatus and Systems, June, 1964, vol. 83, no. 6, pp. 589–606, 650.

***105. Standards***

3. Definitions of Electrical Terms, Group 10 (Rotating Machinery); ANSI C42.10, 1951.
4. Synchronous Machines (General); ANSI C50.10. Synchronous Motors; ANSI C50.11. Salient Pole Synchronous Generators and Condensers; ANSI C50.12. Cylindrical Rotor Synchronous Generators; ANSI C50.13.
5. Guide for the Preparation of Test Procedures for the Thermal Evaluation and Establishment of Temperature Indices of Solid Electrical Insulation Materials; *IEEE Publ.* 98, 1972.
6. Recommended Guide for Testing Insulation Resistance of Rotating Machinery; ANSI C50.22, 1972.
7. Guide for Insulation Maintenance for Large Alternating-current Rotating Machinery; ANSI C50.25, 1972 and *IEEE Publ.* 56. Reaffirmed 1982.
8. Recommended Guide for Making Dielectric Measurements in the Field; *AIEE Publ.* 62, 1958.
9. Test Procedures for Synchronous Machines; *IEEE Publ.* 115, 1981.
10. Trial Procedures for Obtaining Synchronous Machine Parameters by Standstill Frequency Response Tests; *IEEE Publ.* 115A, 1983.

***106. Articles***

11. Wieseman, R. W.: Graphical Determination of Magnetic Fields—Practical Applications to Salient-pole Synchronous Machine Design; *Trans. AIEE,* 1927, vol. 46, pp. 141–154.
12. Calvert, J. F.: Amplitudes of MMF Harmonics for Fractional Slot Windings; *Trans. AIEE,* 1938, vol. 57, pp. 777–784. See also *Univ. Iowa Bull.* 142.
13. Fechheimer, C. J.: Hydrogen, a Successor to Air; *Electr. J.,* September, 1929, p. 405.
14. Knowlton, Rice, and Frieburghouse: Hydrogen as a Cooling Medium for Electrical Machinery; *Trans. AIEE,* 1925, vol. 44, pp. 922–932.
15. Beckwith and Rosenberg: A New Fully Supercharged Generator; *Trans. AIEE,* 1954, vol. 73, pt. 111-A, pp. 477–483.
16. Baudry and Heller: Ventilation of Inner-cooled Generators; *Trans. AIEE,* 1954, vol. 73, pt. 111-A, pp. 500–508.
17. Holley and Taylor: Direct Cooling of Turbine-Generator Field Windings; *Trans. AIEE,* 1954, vol. 73, pt. 111-A, pp. 542–550.
18. Baudry and King: Improved Cooling for Generators of Larger Rating; *Trans. IEEE,* Power Apparatus and Systems, February, 1965, vol. 84, no. 2, pp. 106–114.
19. Downs and Holley: Progress in the Design of Nuclear Turbine-Generators; *ASME-IEEE Joint Power Generation Conf.,* Denver, Sept. 19, 1966.
20. Harder, E. L., and Cheek, R. C.: Regulation of Alternating-current Generators, with Suddenly Applied Loads; *Trans. AIEE,* 1944, vol. 63; pp. 310–318.
21. Harder, E. L., Cheek, R. C., and Clayton, J. M.: Regulation of Alternating Current Generators with Suddenly Applied Loads, Part II; *Trans. AIEE,* 1950, vol. 69, pp. 395–405.
22. Joyce, Abolins, and Lambrecht: Factors Influencing Reliability of Large Generators for Nuclear Power Plants; *Trans. IEEE,* Jan./Feb., 1974, vol. PAS 93, pp. 210–219.
23. Joyce, Abolins, and Lambrecht: Maximum Capability of Two and Four-Pole Generators; *Proc. Am. Power Conf.,* 1973, vol. 35, pp. 378–393.
24. Armour and Gibney: Direct Conductor-Cooling of Large Steam Turbine Generator Four-Pole Rotors; *Trans. IEEE,* Mar./Apr., 1974, vol. PAS 93, pp. 477–486.
25. Ringland and Rosenberg: A New Stator Coil Transposition for Large Machines; *Trans. AIEE,* October, 1959, vol. PAS 78, pp. 743–747.
26. Bennington and Brenner: Transpositions in Turbogenerator Coil Sides Short Circuited at Each End; *Trans. IEEE,* Nov./Dec., 1970, vol. PAS 89, pp. 1915–1921.
27. Jones, Temoshok, and Winchester: Design of Conductor-cooled Steam Turbine-Generators and Application to Modern Power Systems; *Trans. IEEE,* February, 1965, vol. PAS 84, pp. 131–146.
28. Beckwith, S.: Approximating Potier Reactance; *Trans. AIEE,* 1937, vol. 56, pp. 813–819.
29. Mikhail, S. L.: Potier Reactance for Salient Pole Synchronous Machines; presented at the AIEE Winter General Meeting, New York, Jan. 30–Feb. 3, 1950, *AIEE Paper* 50-39.
30. Parker, J. H., Jr., and Towne, R. A.: Design of Large Super-Conducting Turbine Generators for Electric Utility Application, *Trans. IEEE,* Nov./Dec. 1979, vol. PAS 98, no. 6.
31. Laskaris, T. E., and Schoch, K. F.: Superconductor Rotor Development for a 20 MVA Generator. *Trans. IEEE,* Nov./Dec. 1980, vol. PAS 99, no. 6.
32. Fitzgerald, A. E., et al: *Electric Machinery.* New York, McGraw-Hill, 1983.

# SECTION 8

# DIRECT-CURRENT GENERATORS

**E. H. Myers**

*Consultant, Large Motor Engineering, Fellow, Institute of Electrical and Electronics Engineers*

## CONTENTS

*Numbers refer to paragraphs*

## THE DC MACHINE

***1. Applications.*** The most important role played by the dc generator is the power supply for the important dc motor. It supplies essentially ripple-free power and precisely held voltage at any desired value from zero to rated. This is truly dc power, and it permits the best possible commutation on the motor because it is free of the severe waveshapes of "dc" power from rectifiers. It has excellent response and is particularly suitable for precise output control by feedback control regulators. It is also well suited for supplying accurately controlled and responsive excitation power for both ac and dc machines.

The dc motor plays an ever-increasing vital part in modern industry, because it can operate at and maintain accurately any speed from zero to its top rating. For example, high-speed multistand steel mills for thin steel would not be possible without dc motors. Each stand must be held precisely at an exact speed which is higher than that of the preceding stand to suit the reduction in thickness of the steel in that stand and to maintain the proper tension in the steel between stands.

***2. General Construction.*** Figure 8-1 shows the parts of a medium or large dc generator. All sizes differ from ac machines in having a commutator and the armature on the rotor. They also have salient poles on the stator, and, except for a few small ones, they have commutating poles between the main poles.

***3. Construction and Size.*** Small dc machines have large surface-to-volume ratios and short paths for heat to reach dissipating surfaces. Cooling requires little more than means to blow air over the rotor and between the poles. Rotor punchings are mounted solidly on the shaft, with no air passages through them.

Larger units, with longer, deeper cores, use the same construction, but with longitudinal holes through the core punchings for cooling air.

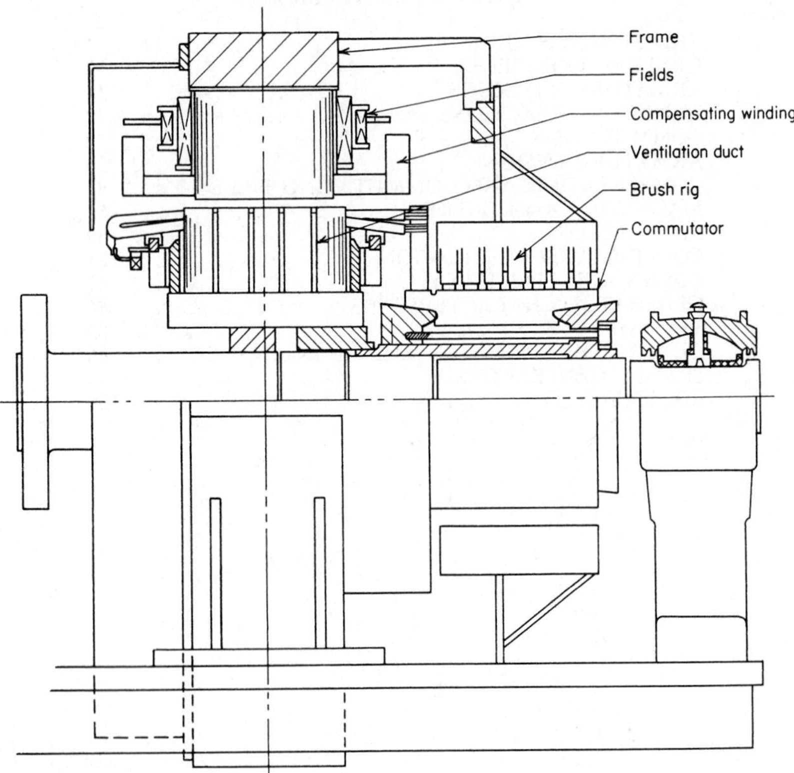


**FIG. 8-1** The dc machine.

Medium and large machines must have large heat-dissipation surfaces and effectively placed cooling air, or "hot spots" will develop. Their core punchings are mounted on arms to permit large volumes of cool air to reach the many core ventilation ducts and also the ventilation spaces between the coil end extensions.

***4. Armature-core punchings*** are usually of high-permeability electrical sheet steel, 0.017 to 0.025 in thick, and have an insulating film between them. Small and medium units use "doughnut" circular punchings, but large units, above about 45 in in diameter, use segmental punchings shaped as in Fig. 8-2, which also shows the fingers used to form the ventilating ducts.

***5. Main- and commutating-pole punchings*** are usually thicker than rotor punchings because only the pole faces are subjected to high-frequency flux changes. These range from 0.062 to 0.125 in thick, and they are normally riveted.

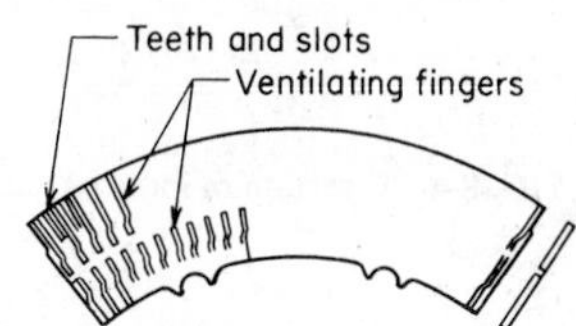


**FIG. 8-2** Armature segment for a dc generator, showing vent fingers applied.

***6. The frame yoke*** is usually made from rolled mild steel plate, but, on high-demand large generators for rapidly changing loads, laminations may be used. The solid frame has a magnetic time constant of ½ s or more, depending on the frame thickness. The laminated frame ranges from 0.05 to 0.005 s.

***7. The commutator*** is truly the heart of the dc machine. It must operate with temperature variations of at least 55°C and with peripheral speeds that may reach 7000 ft/min. Yet it must remain smooth concentrically within 0.002 to 0.003 in and true, bar to bar, within about 0.0001 in.

The commutator is made up of hard copper bars drawn accurately in a wedge shape. These are separated from each other by mica plate segments, whose thicknesses must be held accurately for nearly perfect indexing of the bars and for no skew. This thickness is 0.020 to 0.050 in, depending on the size of the generator and on the maximum voltage that can be expected between bars during operation. The mica segments and bars are clamped between two metal V-rings and insulated from them by cones of mica. On very high speed commutators of about 10,000 ft/min shrink rings of steel are used to hold the bars. Mica is used under the rings.

***8. Carbon brushes*** ride on the commutator bars and carry the load current from the rotor coils to the external circuit. The brush holders hold the brushes against the commutator surface by springs to maintain a fairly constant pressure and smooth riding.

## GENERAL PRINCIPLES

***9. Electromagnetic Induction.*** A magnetic field is represented by continuous lines of flux considered to emerge from a north pole and to enter a south pole. When the number of such lines linked by a coil is changed (Fig. 8-3), a voltage is induced in the coil equal to 1 V for a change of $10^8$ linkages/s (Mx/s) for each turn of the coil, or $E = (\Delta\phi T \times 10^{-8})/t$ V.

If the flux lines are deformed by the motion of the coil conductor before they are broken, the direction of the induced voltage is considered to be into the conductor if the arrows for the distorted flux are shown to be pointing clockwise and outward if counterclockwise. This is generator action (Fig. 8-4).

***10. Force on Current-carrying Conductors in a Magnetic Field.*** If a conductor carries current, loops of flux are produced around it (Fig. 8-5). The direction of the flux is clockwise if the current flows away from the viewer into the conductor and counterclockwise if the current in the conductor flows toward the viewer.

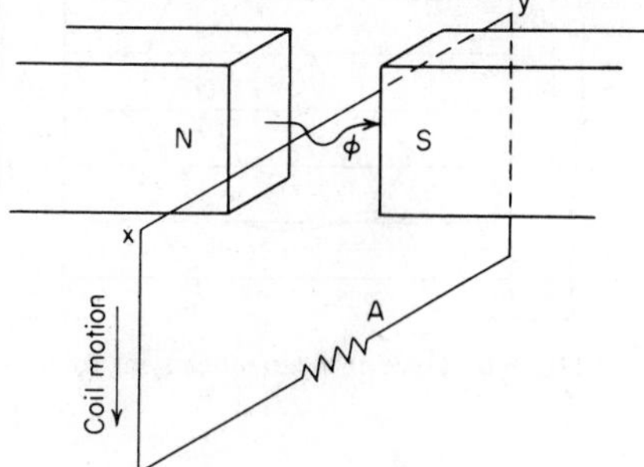


**FIG. 8-3** Generated emf by coil movement in a magnetic field.

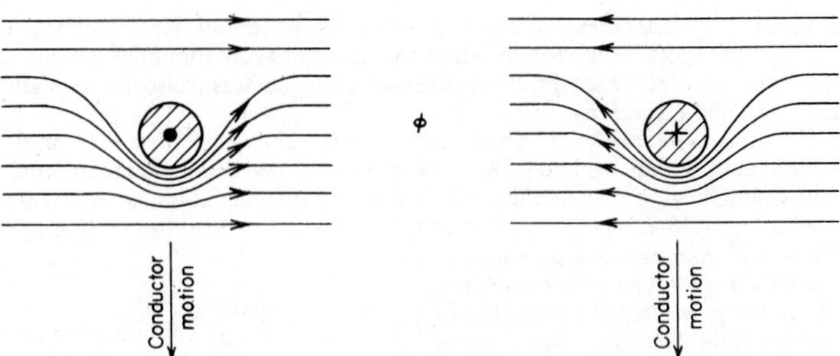


FIG. 8-4 Direction of induced emf by conductor movement in a magnetic field.

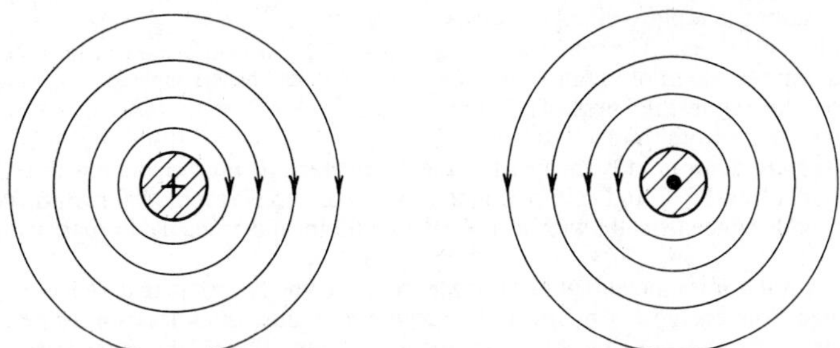

FIG. 8-5 Magnetic fields caused by current-carrying conductors.

If this conductor is in a magnetic field, the combination of the flux of the field and the flux produced by the conductor may be considered to cause a flux concentration on the side of the conductor where the two fluxes are additive and a diminution on the side where they oppose. A force on the conductor results that tends to move it toward the side with reduced flux (Fig. 8-6). This is motor action.

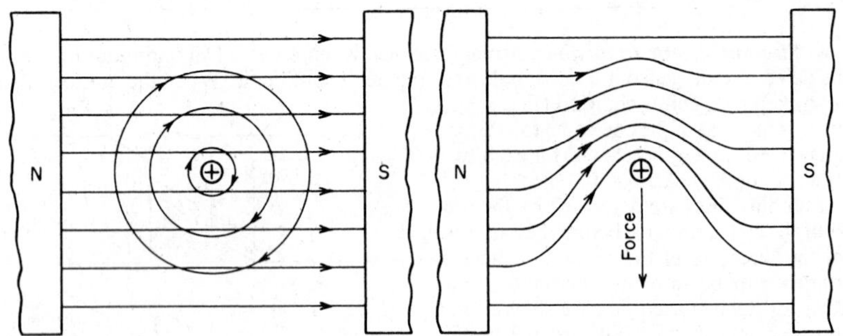


FIG. 8-6 Force on a current-carrying conductor in a magnetic field.

***11. Generator and Motor Reactions.*** It is evident that a dc generator will have its useful voltage induced by the reactions described above and an external driving means must be supplied to rotate the armature so that the conductor loops will move through the flux lines from the stationary poles. However, these conductors must carry current for the generator to be useful, and this will cause retarding forces on them, as described in Par. **10.** The prime mover must overcome these forces.

In the case of the dc motor, the conductor loops will move through the flux, and voltages will be induced in them as described in Par. **9.** These induced voltages are called the "counter emf," and they oppose the flow of currents which produce the forces that rotate the armature. Therefore, this emf must be overcome by an excess voltage applied to the coils by the external voltage source.

***12. Direct-Current Features.*** Direct-current machines require many conductors and two or more stationary flux-producing poles to provide the needed generated voltage or the necessary torque. The direction of current flow in the armature conductors under each particular pole must always be correct for the desired results (Fig. 8-7). Therefore, the current in the conductors must reverse at some time while the conductors pass through the space between adjacent north and south poles.

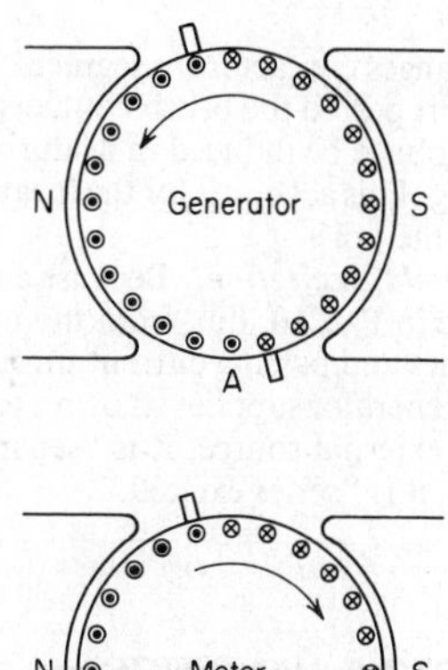


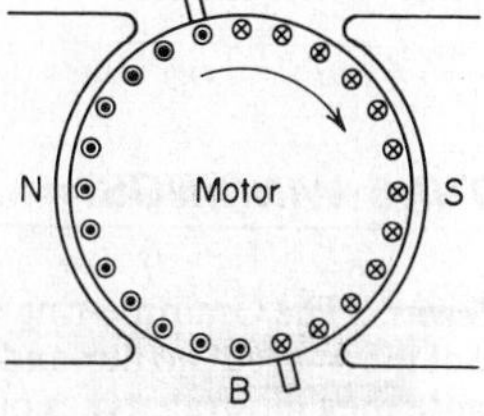


**FIG. 8-7** Direction of current in generator and motor.

This is accomplished by carbon brushes connected to the external circuit. The brushes make contact with the conductors by means of the commutator.

To describe commutation, the Gramme-ring armature winding (which is not used in actual machines) is shown in Fig. 8-8. All the conductors are connected in series and are wound around a steel ring. The ring provides a path for the flux from the north to the south pole. Note that only the outer portions of the conductors cut the flux as the ring rotates. Voltages are induced as shown. With no external circuit, no currents flow, because the voltages induced in the two halves are in opposition. However, if the coils are connected at a commutator *C* made up of copper blocks insulated from each other, brushes *B*− and *B*+ may be used to connect the two halves in parallel with respect to an external circuit and currents will flow in the proper direction in the conductors beneath the poles.

As the armature rotates, the coil *M* passes from one side of the neutral line to the other and the direction of the current in it is shown at three successive instants at *A, B,* and *C* in Fig. 8-9. As the armature moves from *A* to *C* and the

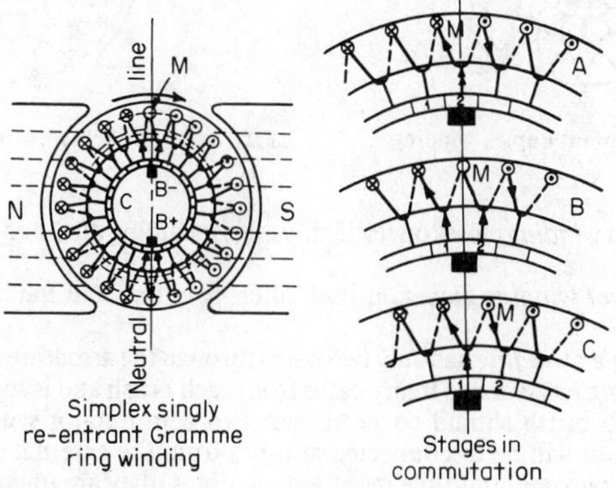


**FIG. 8-8** Principle of commutation.

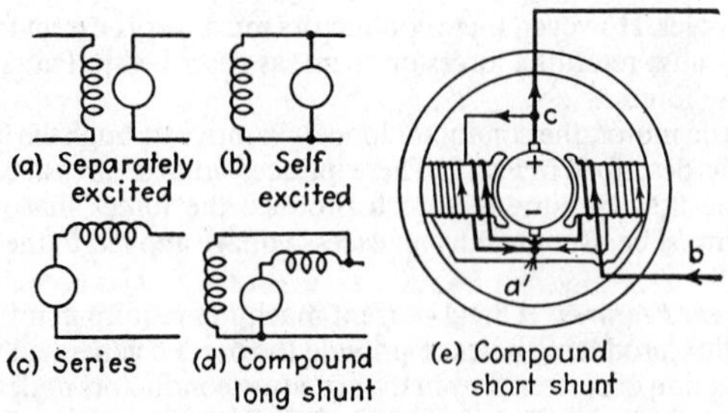


**FIG. 8-9** Methods of excitation.

brush changes contact from segment 2 to segment 1, the current in *M* is automatically reversed. For a short period the brush contacts both segments and short-circuits the coil. It is important that no voltage be induced in *M* during that time, or the resulting circulating currents could be damaging. This accounts for the location of the brushes so that *M* will be at the neutral flux point between the poles.

***13. Field Excitation.*** Because current-carrying conductors produce flux that links them as described in Par. **10**, flux from the main poles is obtained by winding conductors around the pole bodies and passing current through them. This current may be supplied in different ways. When a generator supplies its own exciting current, it is "self-excited." When current is supplied from an external source, it is "separately excited." When excited by the load current of the machine, it is "series excited."

## ARMATURE WINDINGS

***14. Terms.*** The Gramme-ring winding is not used, because half the conductors (those on the inside of the ring) cut no flux and are wasted. Figures 8-8, 8-10, and 8-11 show such windings only because they illustrate types of connections so well.

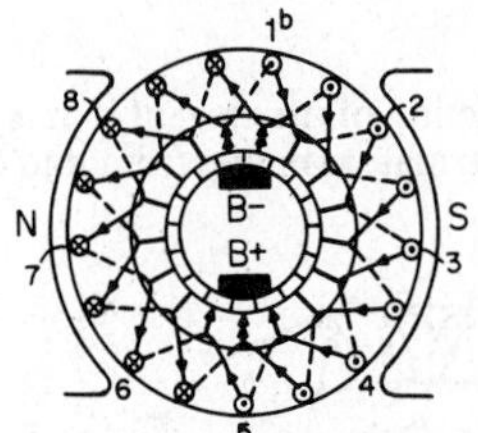


**FIG. 8-10** Singly reentrant duplex winding.

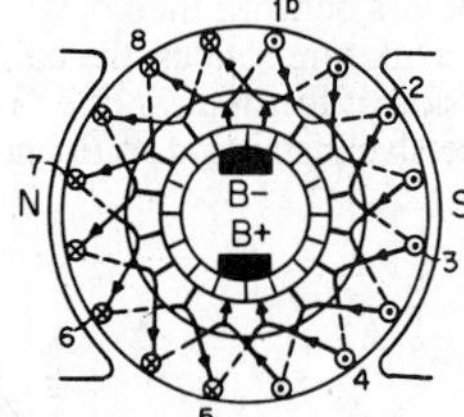


**FIG. 8-11** Doubly reentrant duplex winding.

A *singly reentrant winding* closes on itself only after including all the conductors, as shown in Figs. 8-8 and 8-10.

A *doubly reentrant winding* closes on itself after including half the conductors, as in Fig. 8-11.

As shown, a *simplex winding* has only two paths through the armature from each brush (Fig. 8-8). A *duplex winding* has twice as many paths from each brush and is shown in Figs. 8-10 and 8-11. Note that each brush should cover at least two commutator segments with a duplex winding, or one circuit will be disconnected at times from the external circuit. Although it is possible to use multiplex and multiple reentrant windings, they are uncommon in the United States. They are used in Europe in some large machines.

Modern dc machines have the armature coils in radial slots in the rotor. Nonmetallic wedges restrain the coils normally, but some wedgeless rotors use nonmetallic banding around the core, such as glass fibers in polyester resin. This permits shallower slots and helps to reduce commutation sparking. However, the top conductors are near the pole faces and may have high eddy losses. The coil ends outside the slots are held down on coil supports by glass polyester bands for both types.

***15. Multiple, or Lap, Windings.*** Figure 8-12 shows a *lap-winding coil.* The conductors shown on the left side lie in the top side of the rotor slot. Those on the right side lie in the bottom half of another slot approximately one pole pitch away. At any instant the sides are under adjacent poles, and voltages induced in the two sides are additive. Other coil sides fill the remaining portions of the slots. The coil leads are connected to the commutator segments, and this also connects the coils to form the armature winding. This is shown in Fig. 8-13. The pole faces are slightly shorter than the rotor core.

Almost all medium and large dc machines use simplex lap windings in which the number of parallel paths in the armature winding equals the number of main poles. This permits the current per path to be low enough to allow reasonable-sized conductors in the coils.

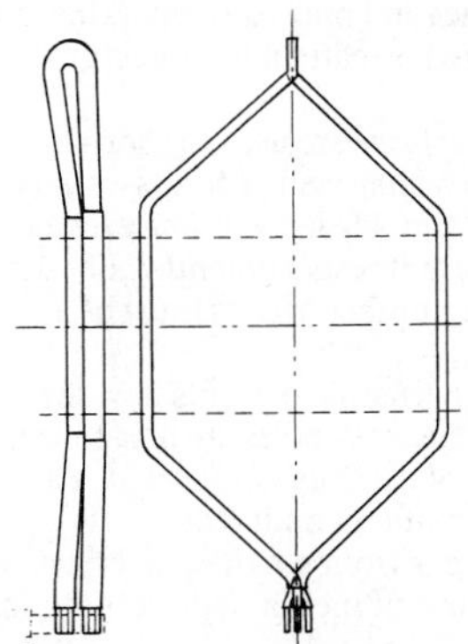

**FIG. 8-12** Coil for one-turn lap winding.

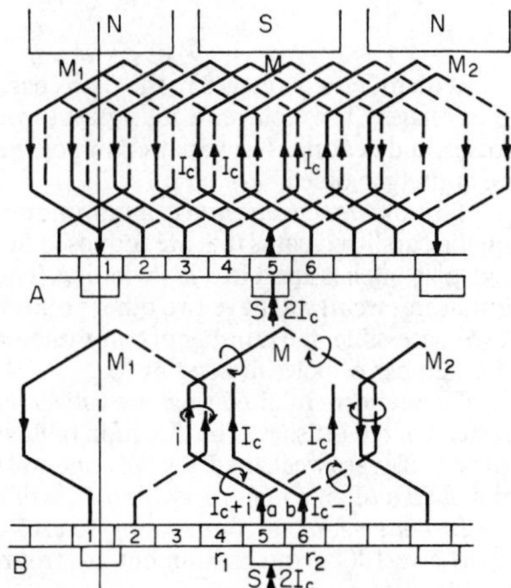


**FIG. 8-13** Multiple, or lap, winding.

*Windings.* Representations of dc windings are necessarily complicated. Figure 8-14 shows the lap winding corresponding to the Gramme-ring winding of Fig. 8-8. Unfortunately, the nonproductive end portions are emphasized in such diagrams, and the long, useful portions of the coils in the core slots are shown as radial lines. Conductors in the upper layers are shown as full lines, and those in the lower layers as dotted lines. The inside end connections are those connected to the commutator bars. For convenience, the brushes are shown inside the commutator.

Note that both windings have the same number of useful conductors but that the Gramme-ring winding requires twice the number of actual conductors and twice the number of commutator bars.

Figure 8-15 shows a 6-pole simplex lap winding. Study of this reveals the six parallel paths between the positive and negative terminals. The three positive brushes are connected outside the machine by a copper ring $T+$ and the negative brushes by $T-$.

The two sides of a lap coil may be full pitch (exactly a pole pitch apart), but most machines use a short pitch (less than a pole pitch apart), with the coil throw one-half slot pitch less than a pole pitch. This is done to improve commutation.

*Equalizers.* As shown in Fig. 8-15, the parallel paths of the armature circuit lie under different poles, and any differences in flux from the poles cause different voltages to be generated in the various paths. Flux differences can be caused by unequal air gaps, by a different number of turns on the main-pole field coils, or by different reluctances in the iron circuits.

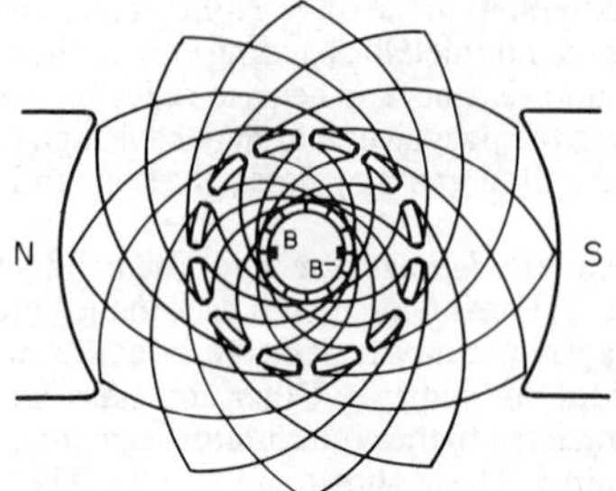


FIG. 8-14 Simplex lap winding.

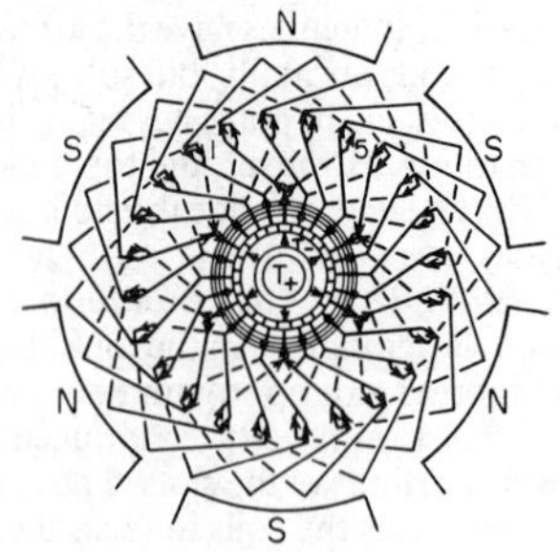


FIG. 8-15 Simplex singly reentrant full-pitch multiple winding with equalizers.

With different voltages in the paths paralleled by the brushes, currents will flow to equalize the voltages. These currents must pass through the brushes and may cause sparking, additional losses, and heating. The variation in pole flux is minimized by careful manufacture but cannot be entirely avoided.

To reduce such currents to a minimum, copper connections are used to short-circuit points on the paralleled paths that are supposed to be at the same voltage. Such points would be exactly two pole pitches apart in a lap winding. Thus in a 6-pole simplex lap winding each point in the armature circuit will have two other points that should be at its exact potential. For these points to be accessible, the number of commutator bars and the number of slots must be a multiple of the number of poles divided by 2.

These short-circuited rings are called "equalizers." Alternating currents flow through them instead of the brushes. The direction of flow is such that the weak poles are magnetized and the strong poles are weakened. Usually one coil in about 30% of the slots is equalized. The cross-sectional area of an equalizer is 20 to 40% of that of the armature conductor.

*Involute necks,* or connections, to each commutator bar from conductors two pole pitches apart give 100% equalization but are troublesome because of inertia and creepage insulation problems.

Figure 8-15 shows the equalizing connections behind the commutator connections. Normally they are located at the rear coil extensions, and so they are more accessible and less subject to carbon-brush dust problems.

***16. Two-Circuit, or Wave, Windings.*** Figure 8-16 shows a wave type of coil. Figure 8-17 gives a 6-pole wave winding. Study reveals that it has only two parallel paths between the positive and negative terminals. Thus, only two sets of brushes are needed. Each brush shorts $p/2$ coils in series. Because points *a, b,* and *c* are at the same potential (and, also, points *d, e* and *f*), brushes can be placed at each of these points to allow a commutator one-third as long.

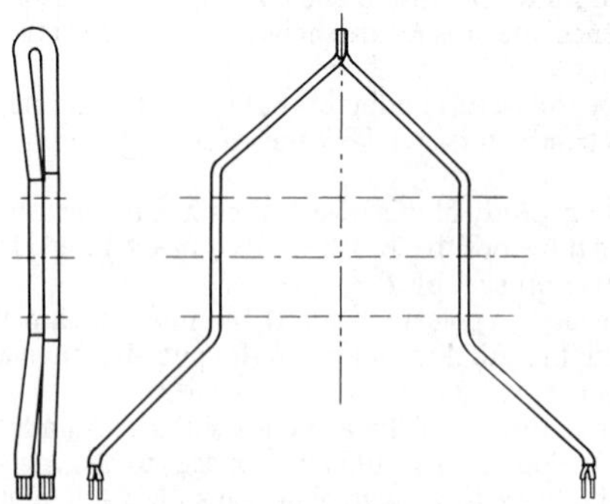

FIG. 8-16 One-turn wave winding.

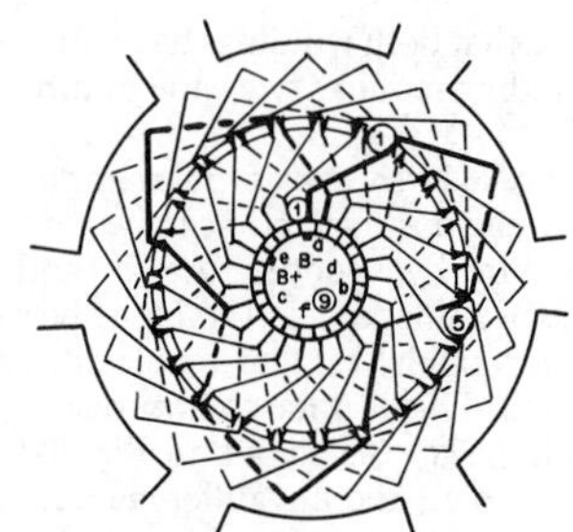

FIG. 8-17 Two-circuit progressive winding.

The winding must progress or retrogress by one commutator bar each time it passes around the armature for it to be singly reentrant. Thus, the number of bars must equal $(kp/2) \pm 1$, where $k$ is a whole number and $p$ is the number of poles. The winding needs no equalizers because all conductors pass under all poles.

Although most wave windings are 2-circuit, they can be multicircuit, as 4 or 16 circuits on a 4-pole machine or 6, 12, or 24 circuits on a 12-pole machine. Multicircuit wave windings with the same number of circuits as poles can be made by using the same slot and bar combinations as on a lap winding. For example, with an 8-pole machine with 100 slots and 200 commutator bars, the bar throw for a simplex lap winding would be from bar 1 to bar 2 and then from bar 2 to bar 3, etc. For an 8-circuit wave winding the winding must fail to close by circuits/2 bars, or 4. Thus, the throw would be bar 1 to 50, to bar 99, to bar 148, etc. The throw is (bars $\pm$ circuits/2)$(p/2)$, in this case, $(200 - 4)/4 = 49$. Theoretically such windings require no equalizers, but better results are obtained if they are used.

Since both lap and multiple wave windings can be wound in the same slot and bar combination simultaneously, this is done by making each winding of half-size conductors. This combination resembles a *frog's leg* and is called by that name. It needs no equalizers but requires more insulation space in the slots and is seldom used.

Some wave windings require *dead coils.* For instance, a large 10-pole machine may have a circle of rotor punchings made of 5 segments to avoid variation in reluctance as the rotor passes under the 5 pairs of poles. To avoid dissimilar slot arrangements in the segments, the total number of slots must be divisible by the number of segments, or 5 in this case. This requires the number of commutator bars to be also a multiple $k$ of 5. However, the bar throw for a simplex wave winding must be an integer and equal to (bars $\pm$ 1)$(P/2)$. Obviously $(5k \pm 1)/5$ cannot meet this requirement. Consequently one coil, called a *dead coil,* will not be connected into the winding, and its ends will be taped up to insulate it completely. No bar will be provided for it, and thus the bar throw will be an integer. Dead coils should be avoided because they impair commutation.

## ARMATURE REACTIONS

***17. Cross-Magnetizing Effect.*** Figure 8-18*a* represents the magnetic field produced in the air gap of a 2-pole machine by the mmf of the main exciting coils, and part *b* represents the magnetic field produced by the mmf of the armature winding alone when it carries a load

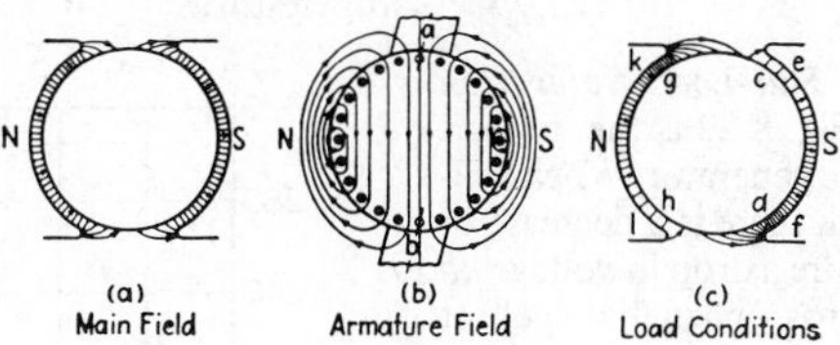


**FIG. 8-18** Flux distribution.

current. If each of the $Z$ armature conductors carries $I_c$ A, then the mmf between $a$ and $b$ is equal to $ZI_c/p$ At. That between $c$ and $d$ (across the pole tips) is $\psi ZI_c/p$ At, where $\psi$ = ratio of pole arc to pole pitch. On the assumption that all the reluctance is in the air gap, half the mmf acts at $ce$ and half at $fd$, and so the cross-magnetizing effect at each pole tip is

$$\psi ZI_c/2p \qquad \text{ampere-turns} \qquad (8\text{-}1)$$

for any number of poles.

***18. Field Distortion.*** Figure 8-18*c* shows the resultant magnetic field when both armature and main exciting mmf's exist together; the flux density is increased at pole tips $d$ and $g$ and is decreased at tips $c$ and $h$.

***19. Flux Reduction Due to Cross Magnetization.*** Figure 8-19 shows part of a large machine with $p$ poles. Curve $D$ shows the flux distribution in the air gap due to the main exciting mmf acting alone, with flux density plotted vertically. Curve $G$ shows the distribution of the armature mmf, and curve $F$ shows the resultant flux distribution with both acting. Since the armature teeth are saturated at normal flux densities, the increase in density at $f$ is less than the decrease at $e$, so that the total flux per pole is diminished by the cross-magnetizing effect of the armature.

***20. Demagnetizing Effect of Brush Shift.*** Figure 8-20 shows the magnetic field produced by the armature mmf with the brushes shifted through an angle $\theta$ to improve commutation. The armature field is no longer at right angles to the main field but may be considered the resultant of two components, one in the direction $OY$, called the "cross-magnetizing component," discussed in Par. **19** and the other in the direction $OX$, which is called the "demagnetizing component" because it directly opposes the main field. Figure 8-21 gives the armature divided to show the two components, and it is seen that the demagnetizing ampere-turns per pair of poles are

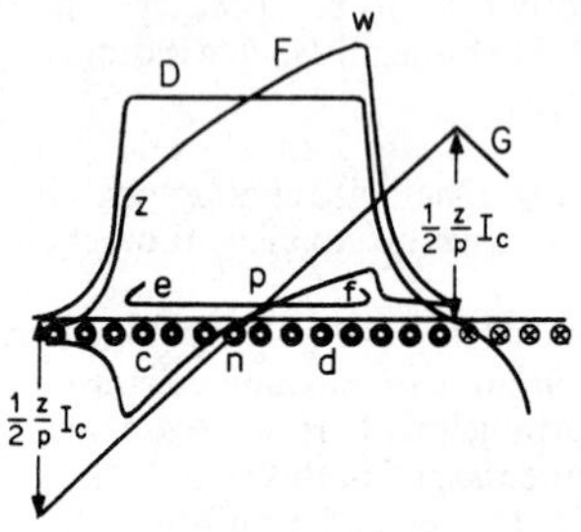


FIG. 8-19 Flux distribution.

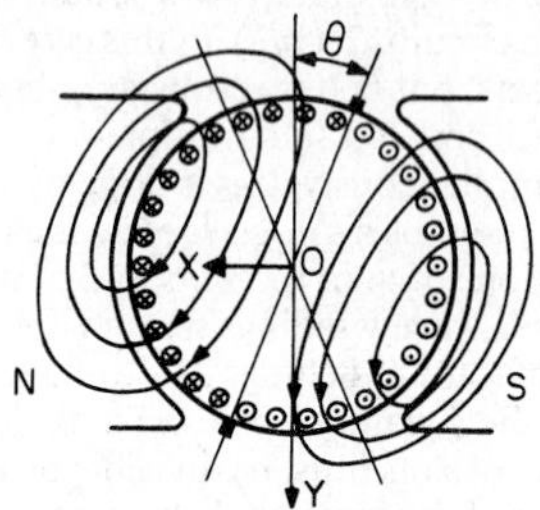


FIG. 8-20 Demagnetizing effect.

$$\frac{ZI_c}{p} \times \frac{2\theta}{180} \quad \text{ampere-turns} \tag{8-2}$$

where $2\theta/180$ is about 0.2 for small noncommutating pole machines where brush shift is used. The demagnetizing ampere-turns per pole would be

$$0.1ZI_c/p \quad \text{ampere-turns} \tag{8-3}$$

***21. No-Load and Full-Load Saturation Curves.*** Curve 1 of Fig. 8-22 is the no-load saturation curve of a dc generator. When full-load current is applied, there is a decrease in useful flux, and therefore a drop in voltage $ab$ due to the armature cross-magnetizing effect (see Par. **19**). A further voltage drop from

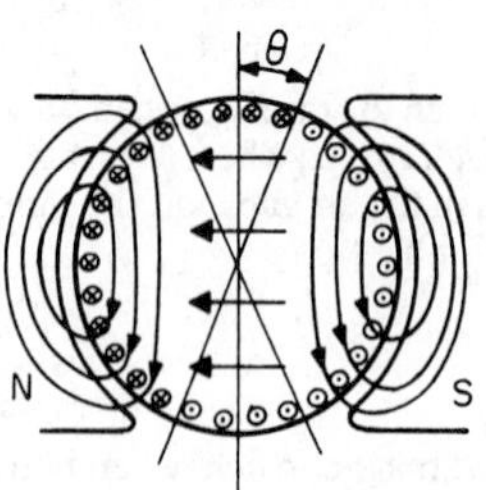


FIG. 8-21 Cross-magnetizing effect.

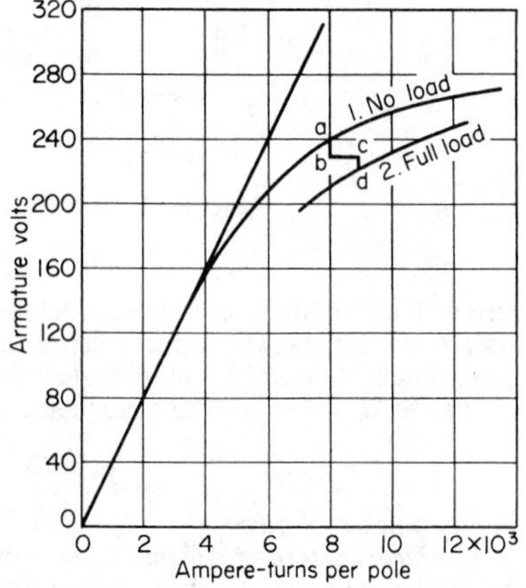


FIG. 8-22 Saturation curves—dc generator.

brush shift is counterbalanced by an increase in excitation $bc = 0.1ZI_c/p$; also a portion $cd$ of the generated emf is required in overcoming the voltage drop from the current in the internal resistance of the machine. The no-load voltage of 240 V requires 8000 At. At full load at that excitation the terminal voltage drops to 220 V. To have both no-load and full-load voltages equal to 240 V, a series field of 10,700 − 8000 = 2700 At would be required.

## COMMUTATION

***22. Commutation.*** The voltages generated in all conductors under a north pole of a dc generator are in the same direction, and those generated in the conductors under a south pole are all in the opposite direction (Fig. 8-23). Currents will flow in the same directions as induced

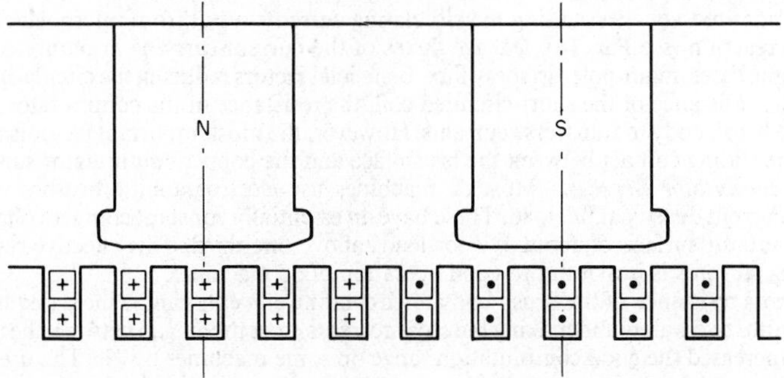


**FIG. 8-23** Conductor currents.

voltages in generators and in the opposite direction in motors. Thus, as a conductor of the armature passes under a brush, its current must reverse from a given value in one direction to the same value in the opposite direction. This is called "commutation."

***23. Conductor Current Reversal.*** If commutation is "perfect," the change of the current in a coil will be linear, as shown by the solid line in Fig. 8-24. Unfortunately, the conductors lie in steel slots, and self- and mutual inductances in Fig. 8-25 cause voltages in the coils short-circuited by the brushes. These result in circulating currents that tend to prevent the initial current

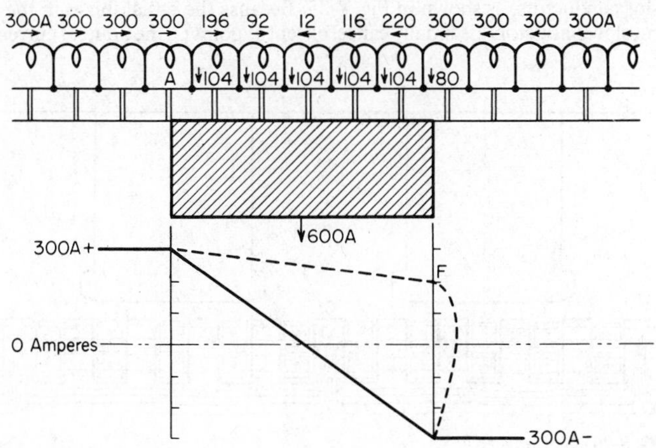


**FIG. 8-24** Commutation.

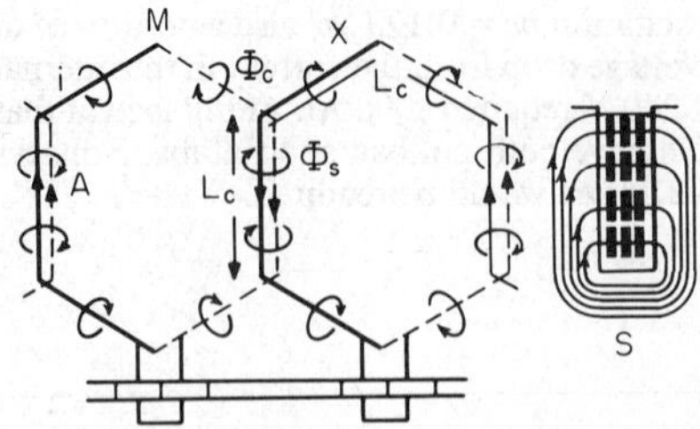


FIG. 8-25 Magnetic field surrounding short-circuited coils.

change, delaying the reversal. In extreme cases, the delay may be as severe as indicated by the dotted line of Fig. 8-24. Because the current must be reversed by the time the coil leaves the brush (when there is no longer any path for circulating currents), the current remaining to be reversed at $F$ must discharge its energy in an electric arc from the commutator bar to the heel of the brush. This is commutation sparking. It can burn the edges of the commutator bars and the brushes. However, most large and heavy-duty dc machines have some nondamaging sparking, and "sparkless" commutation is not required by accepted standards. However, commutation must not require undue maintenance.

The undesired voltages causing the circulating currents result from interpolar fluxes from armature reaction (see Par. **18**), leakage fluxes of the current-carrying armature conductors, and, in some cases, main-pole-tip spray flux. Beneficial factors reducing the circulating currents include the resistance of the short-circuited coil, the resistance of the commutator risers, and that of the brush body to transverse currents. However, the most important factor is the voltage drop at the sliding contact between the brush face and the copper commutator surface.

***24. Commutator Brushes.*** Most dc machines use electrographitic brushes with about 60 A/in$^2$ current density at full load. These have an essentially constant contact voltage drop at the commutator surface of about 1 V for loads above one-third. This effective resistance to circulating currents is important to good operation of dc machines.

The cross resistance of the brush body to circulating currents can be increased by splitting the brush into two wafers and making the cross currents cross the air gap between the two pieces. This has increased the good commutation range on some machines by 7%. The use of double brush holders which have metal dividers between two brushes in the holder is even more effective and has increased the good commutation range as much as 15% over single solid brushes.

Unless special brushes are used, machines should be operated for not more than a few hours at a time at brush densities below 30 A/in$^2$. If this is done, the commutator surface develops a hard glaze which makes the brushes chatter. This results in frayed shunts, chipped and broken brushes, and excessive brush-finger wear.

***25. Reactance Voltage of Commutation.*** The sum of the voltages induced in the armature coil while it is short-circuited by the brushes while undergoing commutation is called the *reactance voltage of commutation.* One of the most important of the fluxes causing this voltage is the *slot leakage flux* shown in Fig. 8-26. This is the resultant flux leakage from current in the individual slot conductors, as shown in Fig. 8-25. Because the radial fluxes in the rotor teeth from adjacent slot conductors essentially cancel except at point $C$ (the point of current reversal),

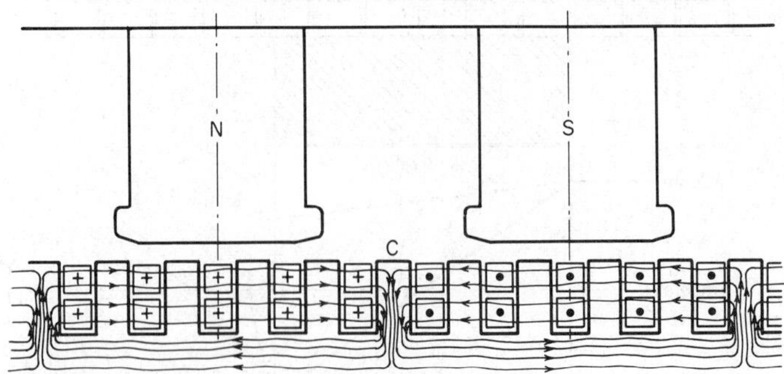


FIG. 8-26 Slot-leakage flux.

the resultant flux is as shown in Fig. 8-26. As the conductors commutate and pass through *C*, they cut the flux shown there and this generates the *reactance voltage of commutation.* Actually part of this voltage is also due to leakage-flux changes at the coil ends, to armature reaction flux, etc., but, for simplicity, only the important slot leakage flux is shown.

***26. Commutating Poles.*** The beneficial factors described in Pars. **24** and **25** that limit the circulating currents in coils being commutated are not adequate to prevent serious delays in current reversal. Other means must be taken to prevent sparking.

If the flux at *C* (Fig. 8-26) could be nullified by an equal flux in the opposite direction, the circulating currents due to the slot leakage flux would be prevented.

The location of *C* is fixed by the location of the brushes. If the brushes were shifted toward the south main pole, a position could be found where the main flux upward into the south pole would cancel the downward flux due to slot leakage at *C*.

This method was used early in the history of dc machines. Unfortunately, the slot leakage flux at *C* is proportional to conductor load current, whereas the flux into the south pole is not. Thus, a new brush position is needed for every change in load current.

A better solution is to provide stationary poles midway between the main poles, as shown in Fig. 8-27. Windings on these *commutating poles* carry the load current. Thus, the flux into the

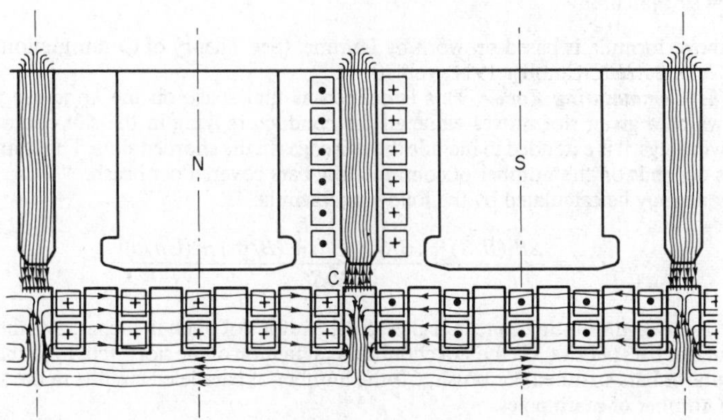


**FIG. 8-27** Slot-leakage flux and commutating-pole flux.

pole at *C* is proportional to the rotor conductor currents and, theoretically, can cancel the voltages induced in the coils being commutated by the slot leakage flux. In the case of the dc motor, the current reverses in both the armature and the commutating field, and proper canceling is maintained.

Note that the strength of the commutating-pole winding must be greater than the armature-winding ampere-turns per pole by the amount required to carry the needed flux across the commutating-pole air gap.

Almost all modern dc machines use commutating poles, although some small machines have only half as many as main poles.

The commutating-pole tip is usually shaped with tapered sides, to approximate the shape of the reactance voltage of commutation form (see Figs. 8-27 and 8-28).

***27. Reactance Voltage of Commutation Formula.*** To determine the useful flux needed across the commutating-pole air gap, it is useful to calculate the reactance voltage of commutation (the total of the voltages induced in the armature coil as it undergoes commutation). The approximate value of this volt-

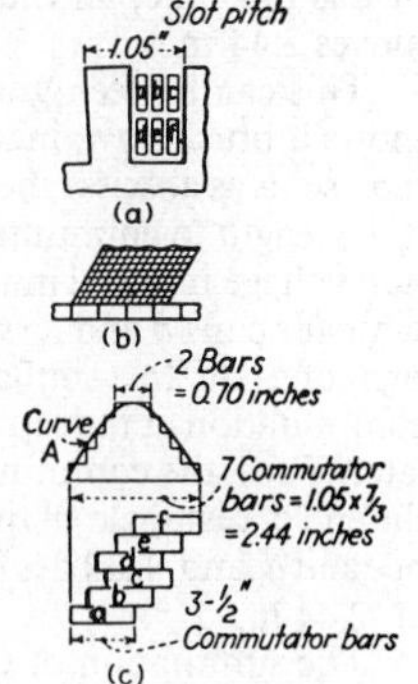


**FIG. 8-28** Commutating zone.

age may be calculated by the use of the following formula:

$$E_c = \frac{\text{poles}}{\text{paths}}(I_c ZT)(\text{r/min})(10^{-10})\left[(K_1 L_r) + K_2(PP)(4.5 + 0.2t_s) + \frac{L_r}{b_s}(3d_s + 2SP)\right] \text{ volts} \tag{8-4}$$

where $I_c$ = current per armature conductor, in A
$Z$ = total no. armature conductors
$T$ = no. of turns/coil between commutator bars
$L_r$ = gross armature-core length, in
$K_1$ = 18.5 for noncommutating-pole machines
= 0 for machines with commutating-pole length = $L_r$
$K_2$ = 1.0 for machines using nonmagnetic bands
= 1.7 for machines using magnetic bands
$PP$ = pole pitch, in
$t_s$ = coil throw, slots
$b_s$ = width of slot, in
$d_s$ = depth of slot, in
$SP$ = slot pitch, in

The above formula is based on work by Lamme. (See Theory of Commutation by B. G. Lamme, *Trans. AIEE,* October 1911, vol. 30.)

***28. The Commutating Zone.*** This is defined as that space on the armature periphery through which a given slot moves while all the conductors lying in the slot commutate. In chorded windings, it is extended to include the coil edges in the chorded slots. The commutating zone thus depends on the number of commutating bars covered per brush.

The zone may be calculated by the following formula:

$$CZ = \frac{SP[(B/S) + (B/S \times Ch) + (B/Br) - (Cir/p)]}{B/S} \tag{8-5}$$

where *CZ* is the commutating zone in inches, *SP* the rotor slot pitch in inches, *B/S* the number of commutator bars per slot, *Ch* the slot chording as a fraction of the slot pitch, *B/Br* the number of commutator bars spanned per brush, *Cir* the number of paralleled circuits in the armature, and *p* the number of main poles.

Consider an 8-pole simplex lap winding with three bars per slot, chording of ½ slot, 3½ bars per brush, and slot pitch of 1.05 in.

$$CZ = \frac{1.05 \times (3 + 1\tfrac{1}{2} + 3\tfrac{1}{2} - 8/8)}{3} = 2.44 \text{ in}$$

In this machine, all the conductors in a slot are commutated while the armature periphery moves 2.44 in.

This can be seen graphically in Fig. 8-28, where *(a)* shows a slot with six conductors, *(b)* shows a brush covering 3½ bars, and *(c)* shows the graphical solution. In *(c)* the rectangle *a* represents as abscissa the space of 3½ commutator bars if they were at the armature surface. This is the length to commutate coil *a.* The ordinate represents to a convenient scale the commutation voltage induced in this conductor while it is being commutated. Rectangles *b* and *c* are the same for coils *b* and *c.* Since *b* commutates 1 bar later than *a,* it is shown one bar space to the right of *a,* etc. In a similar manner *d, e,* and *f* are shown. Normally *d* would be expected to start commutation at the same time as *a,* but, because of chording, it starts later, in this case 1½ bars later. Thus, the commutating zone starts with the beginning of rectangle *a* and is completed at the end of rectangle *f.* Upon adding the spaces of the parts, this is 3½ bars for *f,* 2 bars for the steps of *e* and *d,* and 1½ bars for chording, or a total of 7 bars at the rotor surface, which is 1.05 × 7/3, or 2.44 in.

The summation of the individual rectangles as smoothed off by curve *A* of *(c)* is a rough representation of the reactance voltages induced in the coils during commutation.

***29. Single Clearance.*** The center line of the commutating zone and curve *A* of Fig. 8-28 lie midway between the adjacent main-pole tips if the brushes are not shifted off neutral. The arc on the rotor surface between the tips of adjacent main poles is called the *neutral zone.* If the commutating zone is centered in this arc, the spaces left at each end are called the *single clearance.* Thus, the single clearance is

$$SC = (\text{neutral zone} - \text{commutating zone})/2 \qquad (8\text{-}6)$$

The single clearance is an indication of the probability that spray flux from the mainpole tips might flow into the commutating zone. Such flux would not vary with load and would distort the form of the useful flux from the commutating pole. The commutating-pole useful flux form should closely resemble that of curve *A* in Fig. 8-28.

Noncompensated dc machines usually have main-pole tips with short radial dimensions and have limited spray flux into the neutral zone. The minimum single clearance for these should be not less than 0.6 in and not less than 0.9 with commutation voltages above 3 or 4 V.

Compensated-machine main poles usually have tips 2 to 3 in deep to accommodate the compensating slots and are more likely to spray flux into the commutating zone. These require single-clearance minimums of 1.2 to 1.4 in.

If there is any question about tip flux reaching the commutating zone, flux plots should be made.

***30. Commutating-pole Excitation.*** Figures 8-18*b* and 8-19 show that flux should normally be expected in the commutation area. It is caused by the armature-winding ampere-turns per pole. It could be reduced to zero if the commutating pole had ampere-turns equal and opposite to those of the armature winding. This is $ZI_c/2p$ At/pole, as explained in Par. **17.**

However, it is necessary that the commutating winding also produce useful flux across the commutating-pole gap to counteract the reactance voltage of commutation, as shown in Fig. 8-27. For this reason, the strength of the commutating field is usually 20 to 30% greater than the armature ampere-turns per pole. This difference is called the *excess ampere-turns.* These must be added to the dotted-line bar diagram of Fig. 8-29. The actual flux across the gap is set accurately during the factory test by adjusting the number of sheet-steel shims behind the commutating poles to set the reluctance of the gap for the exact flux needed.

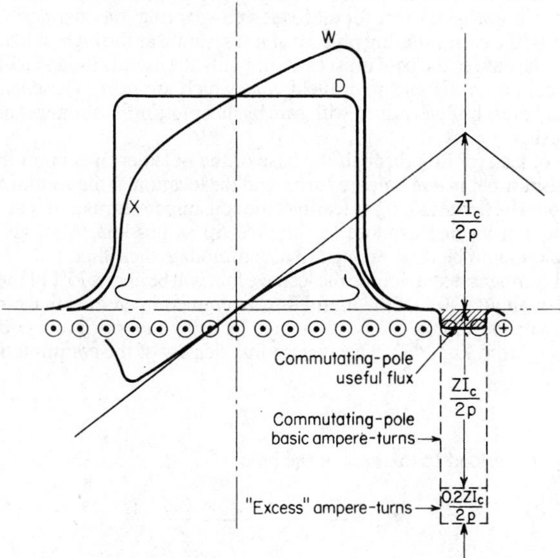


**FIG. 8-29** Commutating-pole ampere-turns.

***31. Calculation of Commutating-Pole Air Gaps.*** With fixed *excess ampere-turns* on the commutating-pole winding and a certain commutation voltage at rated current and speed, only one particular commutating-pole air gap will result in the most favorable compensation of the commutation voltage. The shape of the pole tip will determine the form of the flux density under it, but the length of the air gap will determine the magnitude of the density.

To counteract the reactance voltage of commutation $E_c$, the approximate maximum flux density needed in the commutating-pole air gap is

$$B_m = \frac{E_c \times 23 \times 10^8}{Z/\text{bars} \times DL_c \times \text{r/min}} \tag{8-7}$$

where $E_c$ is the full-load reactance voltage of commutation at speed r/min, $Z$ is the total number of armature conductors, bars is the total number of commutator bars, $D$ is the armature diameter in inches, $L_c$ is the axial length of the commutating poles in inches, and r/min is the revolutions per minute for which $E_c$ was calculated.

The approximate length of the needed commutating-pole single air gap may be calculated by the following formula:

$$\text{Gap} = \frac{3.19 \times \text{excess ampere-turns}}{B_m} \tag{8-8}$$

When the machine is on factory test, the excess ampere-turns can be adjusted to obtain the best commutation possible by placing another dc generator or a battery across the commutating winding to add to the load current flowing in it or to lower the excess by shunting out some of the load current. This is known as a "boost or buck" test. Afterward the commutating-pole air gap is changed to produce the "best" gap flux density with the actual excess ampere-turns. The new gap will be

$$\text{Gap}_2 = \frac{\text{excess At}_1}{\text{excess At}_2} \times \text{gap}_1 \tag{8-9}$$

***32. Dimensions of Commutating Poles.*** If the useful flux across a commutating-pole air gap is not proportional to the machine load current, the compensation of the reactance voltage of commutation will not be correct for all loads and sparking may damage the brushes and commutator. Thus, the commutating pole must not saturate at the highest load currents to be accommodated. The base of the pole must carry not only the useful air-gap flux but also leakage fluxes from the commutating and main field coils which are near. These leakage fluxes are relatively large and must be determined with care by flux plotting if the danger of commutating-pole saturation exists.

The amount of leakage flux through the base of the pole depends upon the length of the leakage paths, the number of coil ampere-turns, and the location of the commutating field. The leakage paths should be made as long as feasible, the coil ampere-turns as few as reasonable, and the commutating coil located as close to the pole tip as possible. Also, all sections of the commutating pole should be large enough to accommodate their flux.

For a normal compensated machine the leakage flux will be about 75% of the commutating-pole useful flux, or about 140% of the useful flux in a noncompensated machine (see Par. **33**).

The approximate useful flux can be calculated by using the maximum commutating-pole air-gap flux density from Eq. (8-7). The average flux density of the commutating zone will be approximately

$$B_a = 0.83B_m \tag{8-10}$$

The flux density at overload in the base of the pole is

$$B_{cp} = \frac{K_3 \times K_4 \times B_a \times CZ}{L_c \times W_c} \tag{8-11}$$

where $K_3$ is 1.75 for compensated machines and 2.40 for noncompensated machines, $K_4$ is the ratio of overload current to rated current, $B_a$ is the average flux density in the commutating zone, $CZ$ is the width of the commutating zone, $L_c$ is the axial length of the commutating pole, and $W_c$ is the circumferential width of the pole at its base. $B_{cp}$ should not exceed 80,000 to 90,000 lines/in$^2$ for good commutation.

***33. Compensating Windings.*** Although the commutating pole is a good solution for commutation, it does not prevent distortion of the main-pole flux by armature reaction, as explained in Par. **18.** The flux set up across the main-pole face by the armature mmf is shown in Fig. 8-30. If the pole face is provided with another winding, as shown in Fig. 8-31, and connected

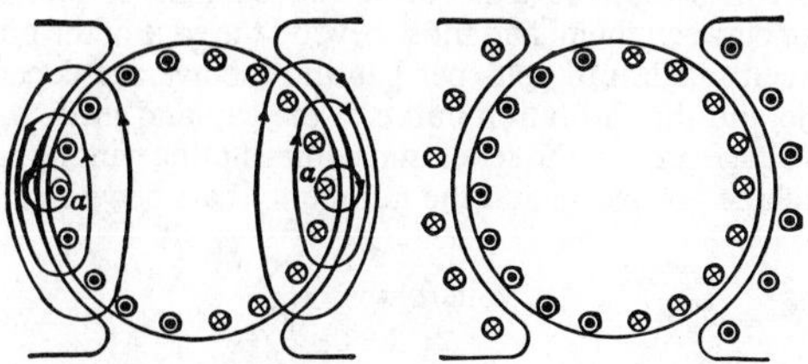


**FIGS. 8-30 and 8-31** Armature field without and with compensating windings.

in series with the load, it can set up an mmf equal and opposite to that of the armature. This would tend to prevent distortion of the air-gap field by armature reaction. Such windings are called *compensating windings* and are usually provided on medium-sized and large dc machines to obtain the best possible characteristics. They are also often needed to make machines less susceptible to flashovers (see Par. **34**).

The use of compensating windings reduces the number of turns required on the commutating-pole fields, and this materially reduces the leakage fluxes of the field and, in turn, the pole saturations at high currents. The ampere-turns on the commutating field are reduced by about 50% with the use of a compensating field. This new winding may be considered to be some of the turns taken off the commutating-pole winding and relocated in slots in the main-pole faces.

The number and location of the compensating slots must be carefully chosen to match, as closely as possible, the rotor ampere-turns per inch. However, the slot spacing must not correspond closely to that of the rotor. This would cause a major change in reluctance to the main-pole useful flux every time the rotor moved from a position where the rotor and stator slots all coincided to where the rotor slots coincided with the stator teeth. This would occur once for every slot-pitch movement. The resulting rapid changes in useful flux would cause ripples in the output voltage and also serious magnetic noise. If too few slots are used, local flux distortions occur and the compensating winding loses some of its effectiveness (see Fig. 8-33).

Compensation of armature reaction effectively reduces the armature circuit inductance. This makes the machine less susceptible to the bad effects of $L(di/dt)$ voltages caused by very fast load-current changes.

During manufacture it is possible to locate the compensating winding nonsymmetrically about the center line of the main pole. This causes a direct-axis flux, which will give a series field effect (Fig. 8-32). For generator cumulative compounding, the slots must be shifted in the direction of the machine rotation. This shift gives a motor differential compounding. The effect cannot be adjusted after manufacture. It seldom exceeds ½ in, and this does not materially reduce the effectiveness of the compensation.

***34. Volts per Bar.*** The mica thickness between the commutator segments depends on the machine design and varies from 0.020 in on small machines to 0.050 in on large units. Although several hundred volts would normally be required to jump these distances, the presence of ionized air from sparking and the presence of conducting carbon dust make it necessary that the voltage between segments be held to low values. If a low-resistance arc does jump between segments, it raises the voltages across the remaining bars. It also tends to ionize some air to form conducting paths across the rest of the bars. If this

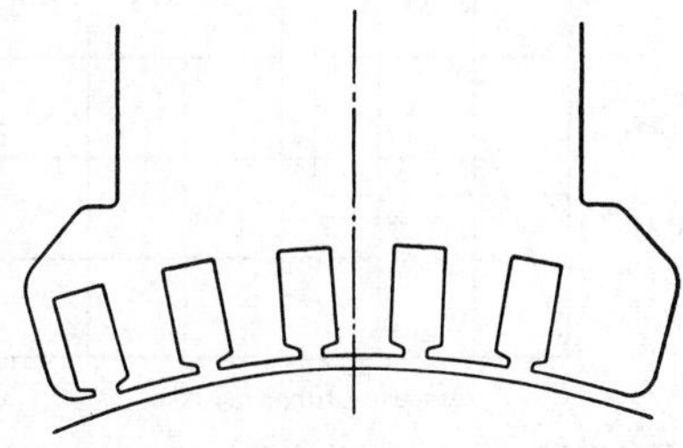

**FIG. 8-32** Offset compensating winding.

progresses until all the segments between brush arms of opposite polarity are bridged, then a *flashover* occurs and severe damage may result to the commutator, brushes, and brushholders.

Because the highest voltage between bars is the "trigger" that starts the flash, this is an important limit. The "average" volts per bar has little significance. Figure 8-29 shows that the maximum volts per bar depends on the field form. For the noncompensated machine shown, the maximum volts between segments exists at $w$. The segments connected to conductors at $x$ have much less voltage between them, and those beyond the edge of the pole have almost none.

The relation between maximum volts per bar and the average depends on the armature ampere-turns per pole and the saturation curve of the gap and teeth at the pole tips. Upon neglecting the small voltage drop in the series and commutating windings, the voltage between brush arms is the machine voltage $V$, and the number of bars between arms is $B/p$. Thus

$$\text{Av. volts/bar} = \frac{V \times p}{B} \tag{8-12}$$

where $B$ is the total number of commutator bars and $p$ the number of main poles.

Even if no distortion exists, only the conductors under the pole faces generate voltage, and so the corrected average volts per bar should be

$$\frac{V \times p}{B \times \Psi} \quad \text{volts}$$

where $\Psi$ is the ratio of pole arc to pole pitch, about 0.65. This is represented by $D$ in Fig. 8-29. However, the maximum volts per bar at $w$ is greater than this, as the height $w$ is greater than $D$, or

$$\text{Max. volts per bar} = \frac{V \times p}{B} \times \frac{w}{D \times \Psi} \tag{8-13}$$

In practice the value of $w/D$ for a noncompensated machine at full-strength main field varies from about 1.7 to 1.9. However, any reduction in saturation causes the effects of the armature ampere-turns (which cause the distortion) to be magnified. The designer must check the actual value of $w/D$, since it may be as high as 4.5 for a dc motor at a weak mainfield strength (high speed). This is evident in Fig. 8-33. The distorting effect for the high-speed (low-average-flux) condition $\phi_{02}$ raises the maximum flux to $\phi_{w2}$, which is over three times the change for the saturated (low-speed) condition $\phi_{01}$ to $\phi_{w1}$ with the same distorting ampere-turns $X$.

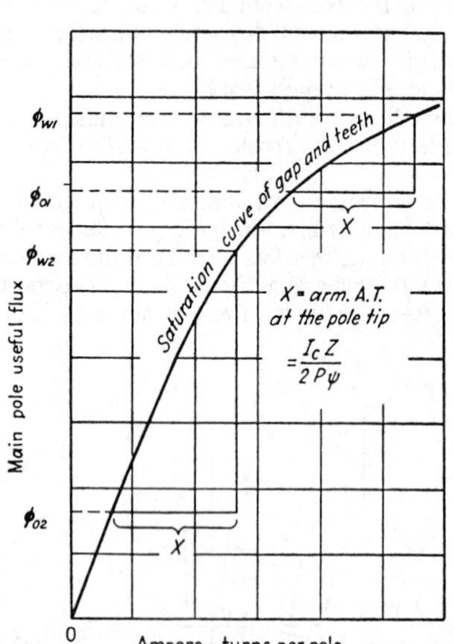


FIG. 8-33 Effect of flux-distortion armature ampere-turns at normal and low saturation.

The use of a compensating winding tends to eliminate the flux distortion, and for saturated conditions the flux curve coincides well with the no-load curve $D$ of Fig. 8-29. However, under low saturation conditions the stationary compensating windings permit localized flux distortions. These are shown in Fig. 8-34. Similar distortions occur at low main flux densities on dc generators, but the output voltage $V$ is reduced in the same proportion as the main flux, and the maximum voltage between bars is not affected seriously.

At full field on well-compensated motors or generators $w/D$ is about 1.4 to 1.5. Direct-current motors at weak field may have ratios of 2.0 or more. On any questionable machine the designer should check this value carefully.

Approximate safe limits of maximum volts per bar are 40 V for motors and 30 V for generators on machines having 0.040-in-thick mica between segments.

***35. Brush Potential Curves.*** When a dc machine develops some commutation sparking, the user may suspect that the commutating-pole air gap is not set correctly. "Brush potential curves" are often taken to prove or disprove such suspicions.

These are taken by measuring the voltage drops between the brush and commutator surface at four points while the machine is operating at constant speed and load current (see Fig. 8-35). The voltages at 1, 2, 3, and 4 are taken by touching the pointed lead of a wooden pencil to the commutator surface. The circuit is completed with leads and a low-reading voltmeter is shown.

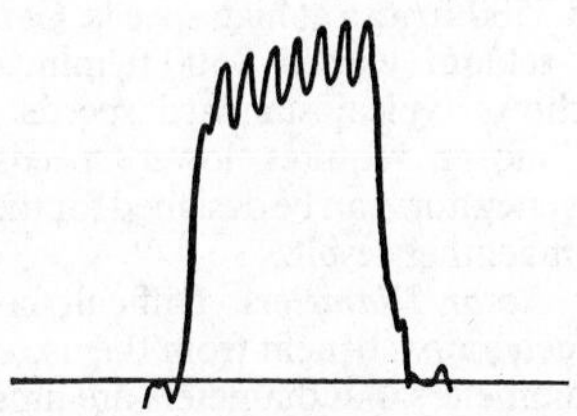

FIG. 8-34 Main-pole flux distortion on a compensated motor at full load and 2½ times base speed.

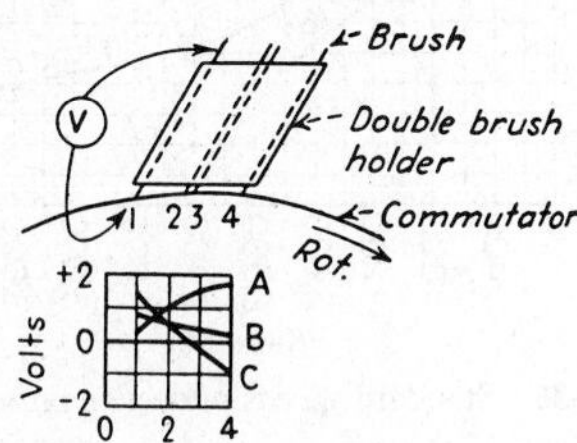


FIG. 8-35 Brush potential curves.

The voltages are then plotted. A curve such as *A* of Fig. 8-35 may indicate undercompensation due to a too large commutating-pole gap. Curve *C* may indicate overcompensation with too much flux density in the commutating-pole air gap. Curve *B* is typical of good compensation.

Justification for such conclusions is based on the theory that best commutation (coil current reversal) will be linear while the coil passes under the brush. This is possible only if there are no circulating currents. Undercompensation should cause circulating currents that would crowd the current to the leaving edge of the brush and cause a high voltage at point 4. Overcompensation would reverse the current too soon and would actually reverse the voltage drop at point 4.

Even to an expert, this test is only an indicator that more definitive tests, such as a buck-boost test, are needed [see Par. **31** and Eq. (8-8)]. Many other factors, including brush riding, commutator surface conditions, sparking, etc., influence the readings. Where machine changes may be required, the manufacturer should be consulted.

## ARMATURE DESIGN

***36. EMF Equation.*** If $10^8$ lines (Mx) of flux are cut by one conductor in 1 s, 1 V is induced in it (see Par. **9**). Therefore, the induced voltage of a dc machine is

$$E = \phi_t \times \frac{Z}{C} \times \frac{\text{r/min}}{60} \times 10^{-8} \tag{8-14}$$

where $\phi_t$ is the total flux in maxwells across the main air gaps and $Z/C$ is the number of conductors in series per circuit ($C$).

***37. Output Equation.*** Equation (8-14) is converted to watts output if both sides are multiplied by the load current $I_L$, $I_c \times C$. The formula can then be rearranged,

$$D^2L = \frac{\text{watts} \times 6.08 \times 10^8}{\text{r/min} \times B_g \times \psi \times q} \tag{8-15}$$

where $D$ is the armature diameter and $L$ is the armature gross core length, $B_g$ is the main-pole air-gap density in maxwells (lines), $\psi$ is the ratio of pole arc to pole pitch, $q$ is $ZI_c/\pi D$ (a useful loading factor), and $\phi_t$ is the total air-gap flux equal to

$$B_g \psi \pi D L \tag{8-16}$$

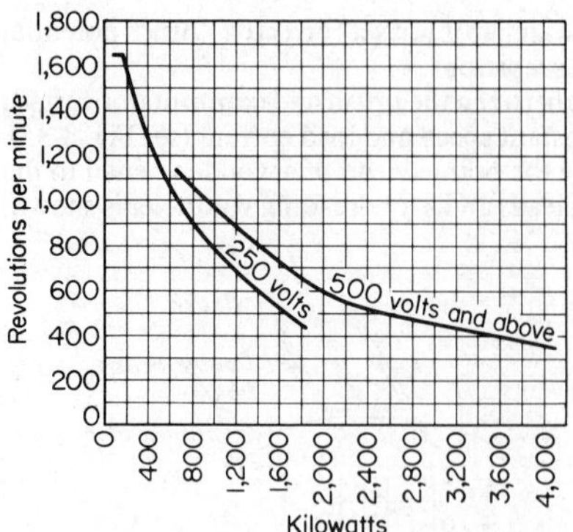


FIG. 8-36 Standard speeds of dc generators.

***38. Rotor Speeds.*** Standards list dc generator speeds as high as are reasonable to reduce their size and cost. This relation is seen from Eq. (8-15). The speeds may be limited by commutation, maximum volts per bar, or the peripheral speeds of the rotor or commutator. Generator commutators seldom exceed 5000 ft/min, although motor commutators may exceed 7500 ft/min at high speeds. Generator rotors seldom exceed 9500 ft/min. Figure 8-36 shows typical standard speeds. If the prime mover requires lower speeds than these, generators can be designed for them but larger machines result.

***39. Rotor Diameters.*** Difficult commutating generators benefit from the use of large rotor diameters, but diameters are limited by the same factors as rotor speeds listed above. The resultant armature length should be not less than 60% of the pole pitch, because such a small portion of the armature coil would be used to generate voltage. Typical generator diameters are shown in Fig. 8-37.

Direct-current motor speeds must suit the application, and often the rotor diameter is selected to meet the inertia requirements of the application. Core lengths may be as long as the diameter. Such motors are usually force-ventilated.

***40. Number of Poles.*** The rotor diameter usually fixes the number of main poles. Typical pole pitches range from 17.5 to 20.5 in on medium and large machines. When a choice is possible, high-voltage generators use fewer poles to allow more voltage space on the commutator between the brush arms. However, high-current generators need many poles to permit more current-carrying brush arms and shorter commutators. Commutators for 1000 to 1250 A/(brush arm)(polarity) are costly, and lower values should be used where existing dies will permit.

***41. The main-pole air-gap flux density $B_g$*** is limited by the density at the bottom of the rotor teeth. The reduced taper in the teeth of large rotors permits the higher gap densities, as shown in Fig. 8-38.

***42. Ampere conductors per inch of rotor circumference (q)*** is limited by rotor heating, commutation, and, at times, saturation of commutating poles. Approximate acceptable values of $q$ are shown in Fig. 8-39.

***43. The commutator diameter*** is usually about 55 to 85% of the rotor diameter, depending on the sizes available to the designer, the peripheral speed, and the resulting single clearances. Heating may also limit the choice (see Pars. **28** and **29**).

***44. Brushes and brush holders*** are chosen from designs available to limit the brush current density to 60 to 70 A/in$^2$ at full load, to obtain the needed single clearance, and to obtain acceptable commutator heating.

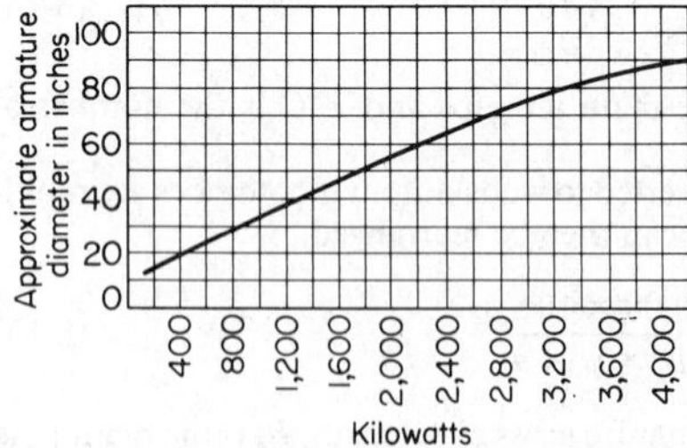


FIG. 8-37 Approximate rotor diameters for standard speeds of dc generators.

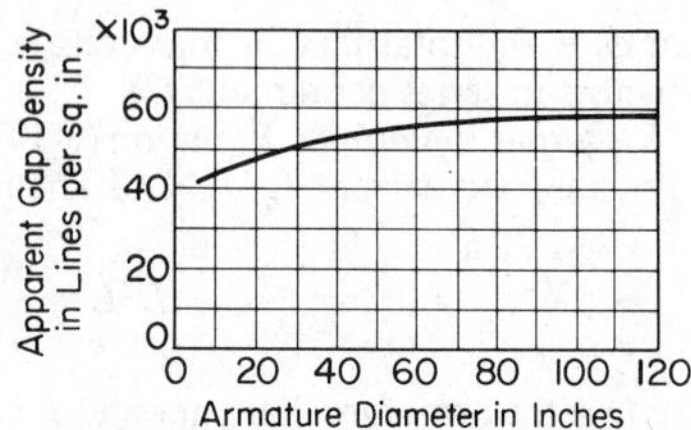


FIG. 8-38 Curve of apparent gap density vs. armature diameter.

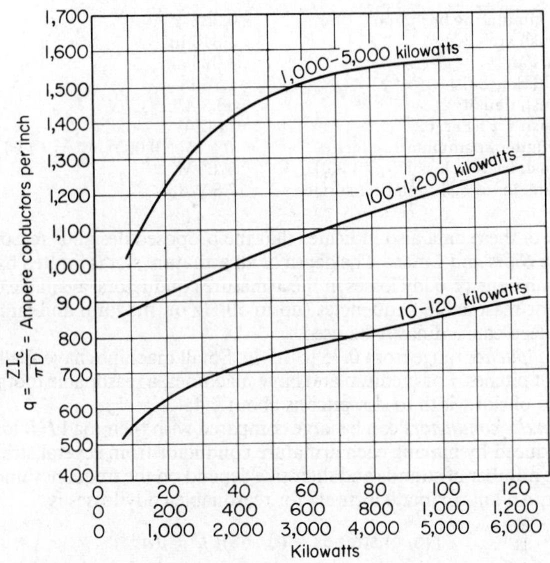


FIG. 8-39 Ampere conductors per inch of armature circumference.

***45. Selection of an Approximate Design.*** Consider a generator rated 2500 kW, 700 V, 3571 A, and 514 r/min. From Figs. 8-39 and 8-40

| | |
|---|---|
| Approx. dia. | $D$ = 62 in |
| Available dia. | $D$ = 56 in |
| No. of poles | 10 |
| Pole pitch | 17.59 in |
| Pole arc | 12.0 in |
| (Arc/pitch) | 0.687 |
| Neutral zone | 6.04 in |
| $B_g$ gap density @ 721 V | 58,500 lines/in² |
| Approx $q$ (Fig. 8-39) | 1480 A cond./in |
| $D^2L$ [Eq. (8-15)] | 50,200 in³ |
| $L$ (gross core) | 16 in |
| No. ⅜-in vents in core | 5 |
| Net core length | 14.125 in |
| $\phi$ [Eq. (8-14)] | $1.12 \times 10^8$ |
| Approx. total cond. Z | 752 (use 750) |
| Actual $q$ | 1520 A cond./in |
| No. commutating bars (1-turn lap) | 375 |
| No. slots | 125 |
| Slot pitch | 1.407 in |
| Slot throw | 12½ (use 12) |
| Chording | ½-slot pitches |

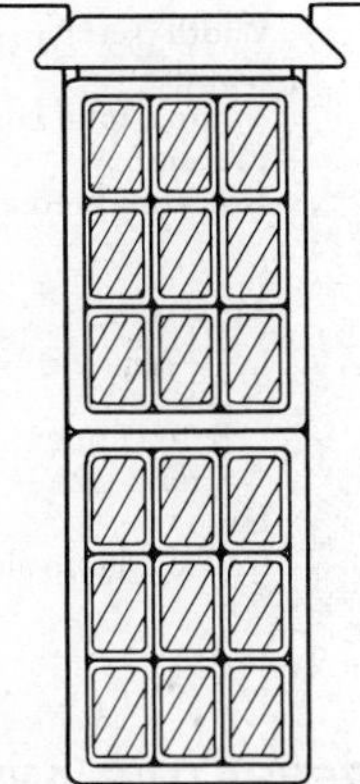

FIG. 8-40 Armature slot cross section.

Examination of the data indicates that the design appears feasible, and so we may continue.

| | |
|---|---|
| Commutating dia. (Par. 43) | 39 in |
| Brush size (35°) | 2(0.500 × 1.75) in |
| $CZ$ (commutating zone) [Eq. (8-5)] | 3.53 in |
| Brushes/arm (Par. 44) | $7b/a$ |
| Commutating speed | 5250 ft/min |

| | |
|---|---|
| Commutating bar pitch | 0.327 |
| Brush arc | 1.315 in |
| Bars/arc | 4.02 bars |
| $SC$ [Eq. (8-6)] | 1.26 in |
| Brush density | 58.3 A/in$^2$ |
| Brush $I^2R$ (Par. **69**) | 7142 W |
| Length of commutating face | 7(1.75 + 0.063) + 1 = 14.56 in |
| Brush friction loss [Eq. (8-33)] | 6760 W |
| Watts/in$^2$ of commutating surface | 7.8 W/in$^2$ |

Examination of these data also indicates that the proposed design is reasonable.

***46. Armature Slots and Coils.*** The depth of an armature slot is limited by several factors, including the tooth density, eddy losses in the armature conductors, available core depths, and commutation. For reasonable frequencies (up to 50 Hz on medium and large dc machines), slots about 2 in deep can ordinarily be used.

*Acceptable slot pitches* range from 0.75 to 1.5 in. Small machines have shallower slots and a lower range of slot pitches. For medium and large machines, a reasonable tooth density usually results if the ratio of slot width to slot pitch is about 0.4.

*Eddy losses in the conductors* can be large compared with their load $I^2R$ losses. Sometimes these must be reduced by making each armature conductor from several strands of insulated copper wire. The number of strands and their size depend on the frequency and the total depth of the conductor. An approximate formula for reasonable eddy losses is

$$\text{No. of strands} = (0.168)(f^{0.83})(d_c^{0.4}) \tag{8-17}$$

where $f$ is the frequency in hertz, (r/min × poles)/120, and $d_c$ is the total depth of a conductor.

The insulation space required depends on the type used. Typical conductor strands have about 0.018 in of glass strands and varnish total. Mica wrappers, binding tapes, and varnish and slot finish allowance (0.010 in) total about 0.085 in on the coil width. If the space for the wedge and its retainer is included, the two coils depthwise total about 0.315 in (see Fig. 8-40).

***47. Approximate Slot Design***

| | |
|---|---|
| Width (see Par. **46**) 0.4 × 1.407 | 0.563 in |
| Depth | 2.0 in |
| Approx. total cond. depth | 0.875 in |
| Frequency | 42.8 Hz |
| No. strands/conductor [Eq. (8-17)] | 3 |

| | Slot width | Depth |
|---|---|---|
| Approx. | | |
| Size | 0.563 | 2.000 in |
| Insulation | 0.139 (0.085 + 0.054) | 0.423 (0.315 + 0.108) |
| Bare copper | 0.424 | 1.577 in |
| Strand size | 0.141 | 0.263 in |
| Use | 3(0.144 | 0.289) in strands/conductor |
| Use available slot | 0.570 | 2.250 in |

## COMPENSATING AND COMMUTATING FIELDS

***48. Compensating Winding Data.*** See Par. **33**. The compensating winding should closely match the armature ampere-turns per inch, should avoid causing magnetic noise, and should result in an acceptable maximum volts per bar (see Par. **34**). Machines for 40°C temperature rise will have compensating bar densities of about 2500 to 3000 A/in$^2$. The pole tip section will limit the maximum depth of the compensating bar. Localized areas of high flux density must be avoided where flux must funnel between the pole "shoe" surface and the bottom of the compensating slot.

For single compensating bar-per-slot designs, the typical width required for insulation, varnish, and stacking factor is about 0.140 in. With the wedge space included, the insulation-depth requirement is about 0.400 in.

***49. Compensating Winding Calculations***

| | |
|---|---|
| $q$ (armature) | 1,520 A cond./in |
| Pole arc of 12.1 in covers | 18,400 A cond. |
| Approx. compensating At | 9200 At |
| Load current | 3,571 A |
| Approx. turns/pole 2.68 | use 2.5 turns/pole |
| Consider 5 slots/pole | 1 bar/slot = 2.5 turns/pole |
| Size of compensating bar | 0.688 × 2.0 in |
| Bar density | 2590 A/in² |
| Compensating slot width 0.828 | use 0.830 in |
| Compensating slot depth 2.400 | use 2.400 in |
| Compensating slot pitch (layout) | 2.25 in |
| Rotor slot pitch | 1.407 in |
| No magnetic noise | improbable |
| Maximum volts/bar | see Par. **58** |

***50. Commutating Winding Calculations.*** See Par. **30.** The total of the commutating and compensating ampere-turns per pole should be about 120 to 130% of those on the rotor.

| | |
|---|---|
| Armature At/pole = $ZI_c/2p$ = (750 × 357)(2 × 10) | 13,400 At/pole |
| Equiv. armature-turns/pole on line ampere basis | 3.75 turns/pole |
| Approx. commutating + compensating At/pole (1.2 × 13,400) | 16,100 At/pole |
| Commutating + compensating turns/pole 16,100/3571 | 4.5 turns/pole |
| Less compensating turns/pole | −2.5 turns/pole |
| Requires commutating winding of | 2.0 turns/pole |
| Excess At/pole, 16,100–13,400 | 2700 At/pole |

Well-ventilated commutating coils may have densities of 2000 to 2500 A/in² (see Fig. 8-49).

***51. Commutation Calculations***

| | |
|---|---|
| $E_c$ = reactance voltage of commutation | 5.42 V [see Eq. (8-4)] |
| Commutating-pole gap density $B_m$ | 13,550 $L$/in² [Eq. (8-7)] |
| Excess At | 2700 At/pole |
| Commutating-pole air gap | 0.609 in [Eq. (8-8)] |

## MAGNETIC CALCULATIONS

***52. Flux Paths.*** Figure 8-41 shows the paths of the main-pole flux for a typical medium-sized machine. The commutating poles and the compensating slots are not shown. Saturation calculations involve only half the length of a complete flux loop, because that is all that one field coil accommodates. Except for the main-pole air gap and the rotor teeth ampere-turns, the

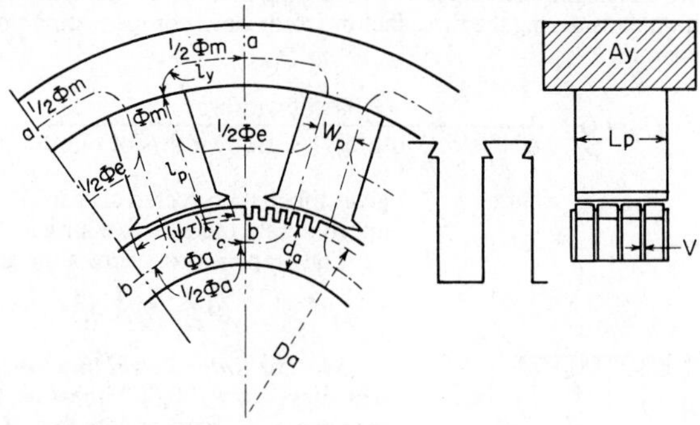


**FIG. 8-41** Paths of the main and of the leakage fluxes.

calculations are simple. They require (1) the determination of flux densities by dividing the flux in a section by its cross-sectional area, $\beta = \Phi/\text{area}$; (2) reading a magnetization curve for the material involved to find the ampere-turns per inch needed for the density; and (3) finding the total ampere-turns for the part by multiplying the length of the portion of the path by those ampere-turns per inch. Typical magnetization curves are shown in Fig. 8-42.

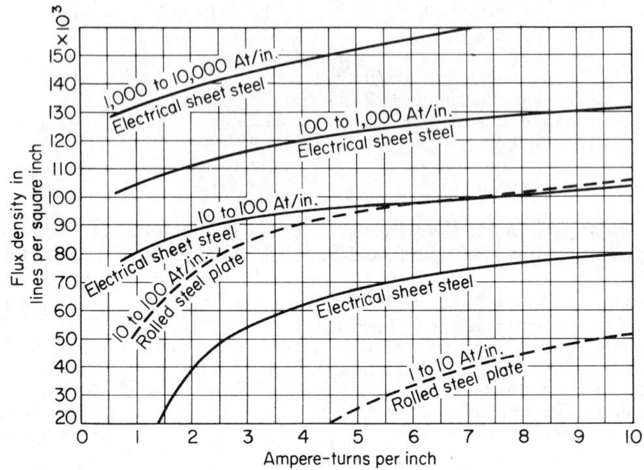


**FIG. 8-42** Magnetization curves.

The rotor core is usually built up of sheet steel laminations 0.017 to 0.025 in thick. Because of burrs and surface coatings, a stacking factor of 93% is common. The main poles use thicker laminations, and a factor of 95% is common. If the frame is also made up of laminations, a similar factor is necessary. Of course, a solid frame uses its full area.

The leakage flux ($\frac{1}{2}\phi_e$) in Fig. 8-41 from the main field coils must be included with the useful flux in the frame yoke and the pole body. Calculations depend on the actual machine dimensions and on the main field ampere-turns. However, the ampere-turns in these parts represent only a small part of the total required for the entire path, and it is usually accurate enough to estimate this leakage to be 12% of the useful flux normally and 20% at high saturations. For accurate calculations the actual leakage can be plotted. No leakage fluxes are considered in computing the gap, teeth, or core densities.

***53. The Carter Coefficient and Gap Ampere-Turns.*** The presence of rotor slots, compensating slots, and vent ducts in the generator causes the actual densities in the main-pole air gap to be greater than for a smooth, solid core. Also, the average lengths of the flux paths are longer (see Fig. 8-43). The two effects may be lumped by assuming that the air gap is larger than measured mechanically. Upon considering the three factors (rotor slots, compensating slots, and vents) in succession, the formula

$$G_1 = G \times \frac{G + (\text{slot width}/5)}{G + (\text{slot width}/5)(1 - \text{slot width}/\text{slot pitch})} \tag{8-18}$$

gives the first corrected air gap $G_1$; this will closely approximate the effective air gap.

The ampere-turns across the gap will be

$$At_g = \beta_g \times 0.313 \times G_1 \tag{8-19}$$

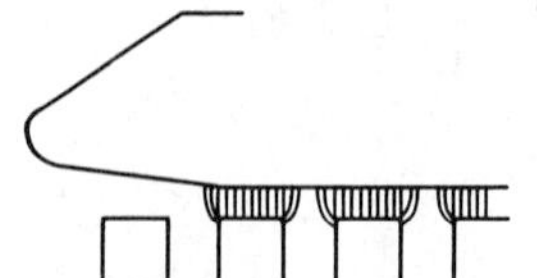

**FIG. 8-43** Distribution of flux in the air gap.

***54. The Rotor Teeth Ampere-Turns.*** For tooth densities below 100,000 lines/in², the ampere-turn drops in a tooth are so low that practically no flux will pass down the adjacent slot because the reluctance of air is so great. However, as tooth flux densities become larger, they produce very high ampere-turn drops from the top of the tooth to its bottom owing to saturation. Because these ampere-turns are also across the parallel flux path in the adjacent slot, when they are large enough, some useful flux will pass down the slot, relieve the tooth of some of its flux, and lower its actual density. If the tooth apparent density is calculated by assuming that all the flux across a slot pitch passes down the tooth, the actual density will be less than the apparent, depending on the amount of saturation.

The relation between the apparent tooth density $\beta_{ta}$ and the actual tooth density $\beta_t$ for different ratios of air area to iron area at any section of the tooth is shown in Fig. 8-44. The *K* of

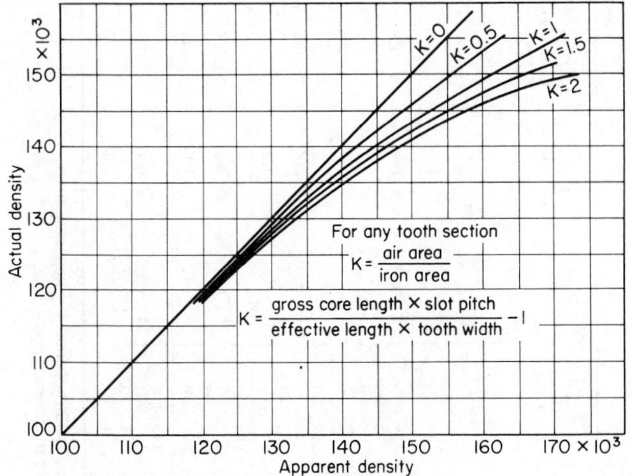


**FIG. 8-44** *K* curves.

these areas is

$$K = \frac{\text{air area}}{\text{iron area}} = \frac{(\text{gross core length}) \times (\text{slot pitch})}{(\text{eff. core length}) \times (\text{tooth width})} - 1 \qquad \text{(8-20)}$$

For accuracy in calculating tooth ampere-turns, it is desirable to divide the tooth into several parts, find the ampere-turns drop across each section, and total them. The flux density is found at the middle of each section, and the *K* ratio is calculated at the middle of each section.

***55. Calculation of No-Load Saturation Data.*** Considering the 2500-kW, 700-V, 3571-A, 514-r/min generator of Par. **45**, we have the values shown in Table 8-1. Using the magnetization

**TABLE 8-1** Magnetic Dimensions

| Section | *K* | Net area, in² | Eff. length, in |
|---|---|---|---|
| Frame yoke 6 × 17 | .... | 102 | 13.85 |
| Pole body 9½ × 15½ | .... | 140 | 10.35 |
| Comp. pole teeth (layout) | .... | 100 | 2.40 |
| Eff. air gap | .... | ..... | 0.268 |
| Tooth 1 (upper ⅓) | 0.92 | 10.8 | 0.75 |
| Tooth 2 (middle ⅓) | 0.96 | 10.3 | 0.75 |
| Tooth 3 (bottom ⅓) | 1.00 | 9.8 | 0.75 |
| Core | .... | 79.3 | 7.15 |

NOTE: 1 in = 25.4 mm; 1 in² = 645 mm².

curves of Fig. 8-42 and these data, the no-load saturation curve is calculated for several voltages. Note that 721 V is chosen in Table 8-2 on the assumption that the *IR* drop in the generator will

**TABLE 8-2** Calculated Ampere-turns per Pole

| Volts | $\phi_t$ | Gap, $L = 0.268$ | Tooth 1, $L = 0.75$, $K = 0.92$ | Tooth 2, $L = 0.75$, $K = 2.96$ | Tooth 3, $L = 0.75$, $K = 1.0$ | Core, $L = 7.15$ | Frame, $L = 13.85$ | Pole, $L = 10.35$ | C. tooth, $L = 2.40$ | Total ampere-turns |
|---|---|---|---|---|---|---|---|---|---|---|
| | | $\beta$<br>At | Apparent $\beta$<br>Actual $\beta$<br>At/in<br>At | Apparent $\beta$<br>Actual $\beta$<br>At/in<br>At | Apparent $\beta$<br>Actual $\beta$<br>At/in<br>At | $\beta$<br>At/in<br>At | $\beta$<br>At/in<br>At | $\beta$<br>At/in<br>At | $\beta$<br>At/in<br>At | |
| 420 | $65.5 \times 10^6$ | 33,800<br>2,850 | 70,500<br>70,500<br>5.4<br>5 | 74,600<br>74,600<br>7.2<br>5 | 78,000<br>78,000<br>9.0<br>5 | 41,300<br>2.1<br>15 | 36,000<br>6.5<br>90 | 52,400<br>2.8<br>30 | 65,500<br>4.2<br>10 | 3,018 |
| 630 | $98.3 \times 10^6$ | 50,800<br>4,280 | 106,000<br>106,000<br>130<br>100 | 112,000<br>111,300<br>210<br>160 | 117,000<br>116,200<br>320<br>240 | 62,000<br>3.7<br>25 | 54,000<br>11<br>150 | 78,600<br>9.2<br>95 | 98,300<br>6.2<br>15 | 5,065 |
| 721 | $112.5 \times 10^6$ | 58,200<br>4,900 | 121,200<br>120,500<br>440<br>330 | 128,100<br>126,500<br>660<br>495 | 134,000<br>132,300<br>1,000<br>750 | 71,000<br>5.7<br>40 | 61,800<br>14<br>195 | 90,000<br>26<br>270 | 112,500<br>225<br>540 | 7,520 |
| 770 | $120 \times 10^6$ | 62,100<br>5,230 | 129,500<br>128,000<br>710<br>535 | 137,000<br>134,600<br>1,200<br>900 | 143,000<br>139,000<br>1,850<br>1,390 | 75,800<br>7.9<br>55 | 66,000<br>16<br>220 | 96,000<br>46<br>475 | 120,000<br>410<br>985 | 9,790 |

NOTES: $L$ = length of flux path, in; $K$ = air area/iron area at particular position on tooth; apparent $\beta$ = apparent flux density, lines/in$^2$; actual $\beta$ = actual flux density, lines/in$^2$; At/in = ampere-turns per in; At = ampere-turns.

1 in = 25.4 mm; 1 in$^2$ = 645 mm$^2$.

not exceed 3%, or 21 V in this case. The generator (Fig. 8-45) must have this additional voltage induced in it for a 700-V terminal voltage. In the case of a motor, the induced voltage would be lower by the amount of the *IR* drop, or 679 V.

***56. Full-Load Saturation Curve for a Compensated Machine.*** Figure 8-46 shows the calculated no-load saturation curve. For a well-compensated machine, the brushes will have little or no shift, and essentially no useful flux will be lost because of armature reaction (see Par. **17**). Only the armature-circuit-resistance *IR* drop need be considered, and the full-load excitation ampere-turns required can be read directly from the no-load saturation curve at the induced voltage.

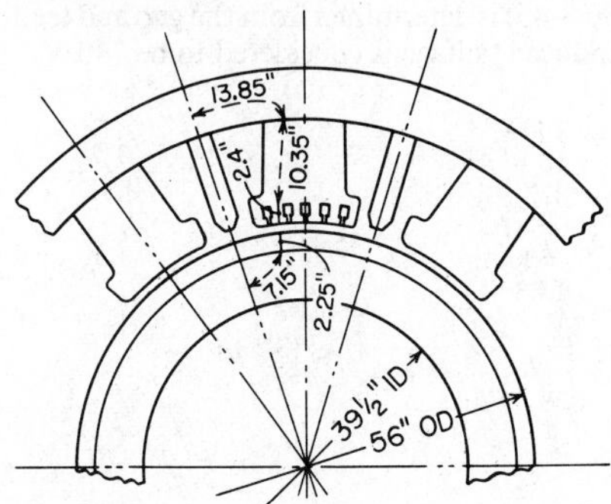


**FIG. 8-45** Cross section of a 2400-kW generator.

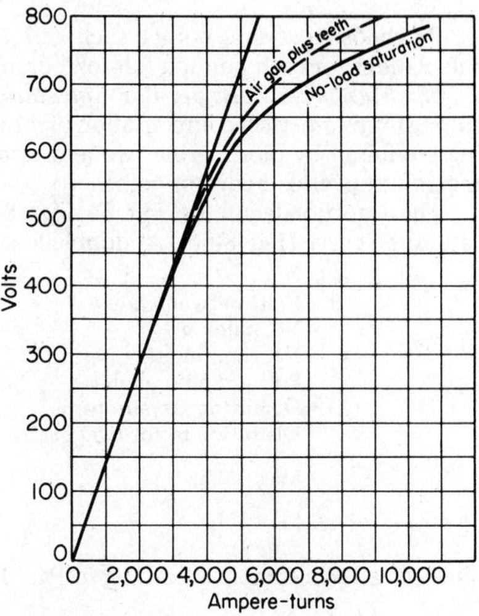


**FIG. 8-46** No-load saturation curves.

For the 2500-kW generator the excitation required at 721 V is 7520 At at full load.

***57. Full-Load Saturation Curve for a Noncompensated Machine.*** With commutating poles, there is no need for brush shift, but the uncompensated armature reaction (see Par. **17**) will result in loss of useful flux as the load is increased. Figure 8-47 shows a method of calculating the additional ampere-turns excitation to replace this lost flux.

*OBD* = saturation curve of air gap plus teeth and pole face

*BC* = *IR* drop in armature circuit plus the brush drop. *B* = any point chosen on curve *OBD*

*FB* = *BE* = full-load-armature At/pole arc, or At/$p \times \psi$, laid off on a horizontal line

Through *E* and *F* draw vertical lines of indefinite length. Move line *GI* vertically upward or downward parallel to *FBE* to a position *GHKI,* so that area *JGHOJ* = area HABDIKH.

Through *B* draw a vertical line *BCK*. Then *HK* = distortion ampere-turns for the load-current considered for point *B*.

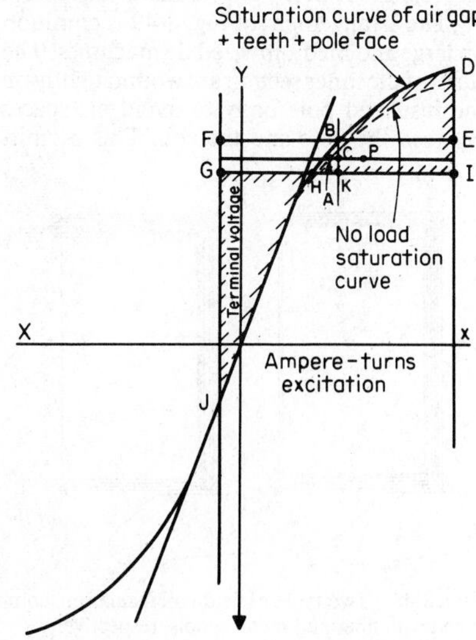


**FIG. 8-47** Calculation of load saturation curve.

Through $C$ draw a horizontal line of indefinite length cutting the no-load saturation curve at $A$.

$CP = HK$, to be extended from right at $C$

$AP$ = total ampere-turns required at load current considered to maintain load at same value as at no load

By choosing several points, such as $B$, along the saturation curve and making the same calculations for each point, a full-load or any load saturation curve can be produced.

***58. Maximum Volts per Bar Calculations.*** See Par. **34.** The distorting ampere-turns resulting from imperfect compensation of the armature ampere-turns by the compensating winding are found by plotting the two and noting the maximum difference. This is done at the maximum-overload-current point.

The distortion factor (see Par. **34,** Figs. 8-33 and 8-46) is determined from the gap and teeth saturation curve (Fig. 8-46). At double load the induced voltage is considered to be 740 V.

| | |
|---|---|
| Volts between arms | 700 V |
| No. poles $p$ | 10 |
| No. commutating bars | 375 |
| Pole arc/pole pitch $\psi$ | 0.687 |
| Distorting ampere-turns | 1600 At |
| Distortion factor $w/D$ of Fig. 8-33 | 1.06 |
| Max. V/bar | $\frac{V \times p}{B} \times \frac{w}{D \times \psi}$ [Eq. (8-13)] |
| Max. V/bar | 28.8 V/bar |

This value is acceptable according to Par. **34.**

## MAIN FIELDS

***59. Main Field and Main Field Heating.*** Figures 8-48 and 8-49 show three types of dc main fields. Small machines commonly use those of Fig. 8-48. They are wound on molds and then slipped on the poles. Type A is wound on an insulating spool, and type B uses an insulated steel spool for better heat transfer and mechanical protection.

The arrangement of Fig. 8-49 is common on large and medium-sized dc machines. The turns of the inner section are wound tightly on the insulated pole body to avoid air spaces between the pole and the coil. This permits

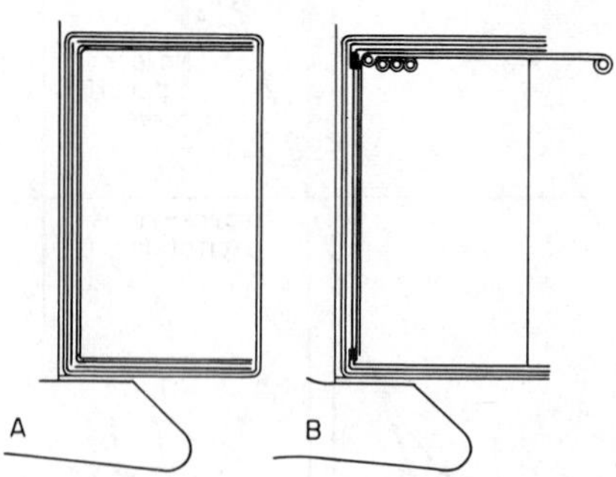


**FIG. 8-48** Two types of field-coil insulation, combined with fiber and metal spools, respectively.

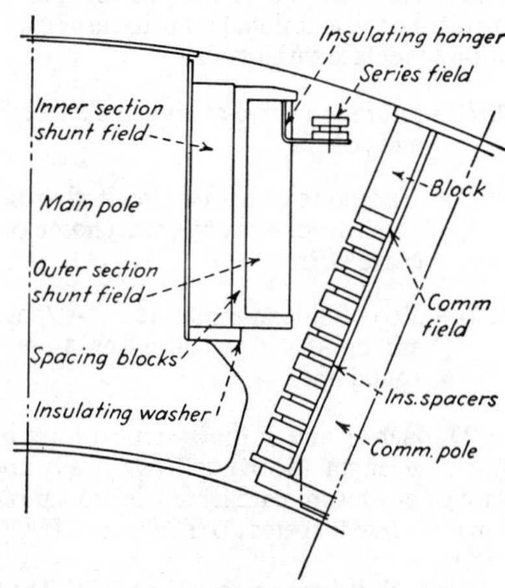


**FIG. 8-49** Ventilated field coils.

maximum heat transfer. The second section is spaced away from the inner coil to permit the cooling air to flow over the maximum surface area possible. The thickness of a coil section is limited to about 1¼ to 1¾ in for a small temperature gradient within the coil.

All three types may use wire insulated with varnish, double cotton covering, or glass slivers in varnish. Air pockets which act as barriers to transfer of heat must be avoided, and so rectangular wire is common. Also, varnish or resin is liberally applied during winding or applied by vacuum impregnation after the coil is wound.

Design criteria suitable for all dc machines cannot be established, because the field cooling depends on air pressures from the armature rotation, the air-passage areas through the fields, and the radiation of heat from adjacent parts. These factors vary with machine design. However, on medium and large self-ventilated dc generators (built as in Fig. 8-49) empirical data are useful.

The main fields receive heat, not only from their own $I^2R$ losses, but from heat radiated from the hot armature and the commutation coils. Also, the air cooling the coils is already heated by the rotor. This lowers the temperature gradient for cooling the coils. The temperature rise of the fields must be calculated, not on the basis of the actual air temperature, but on the basis of the cool ambient-air temperature outside the machine. Figure 8-50 shows empirical data for such typical self-ventilated medium and large machines, built as shown in Fig. 8-49.

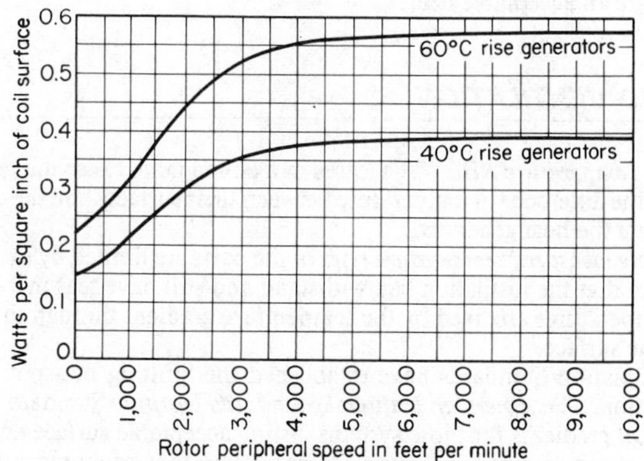


**FIG. 8-50** Main-field loss per surface area.

The "surface area" for these curves includes the entire periphery of the coil, because the heat transfer to the pole body is as effective as that to the air-cooled surfaces.

Little gain is made in cooling with increase in rotor velocities above 5000 ft/min, because most of the armature air must pass through the limited field structure area. At high rotor speeds the air is throttled owing to the high-velocity pressure drops.

***60. Main-Field Calculations.*** These are made by making a layout similar to that shown in Fig. 8-49. This permits the estimate of approximate mean length of turns ($MLT$) for the sections.

The means of excitation and the particular application usually determine the $IR$ drop of the main field. This is met in design by selection of the field wire cross-sectional area. This is calculated by Eq. (8-21).

$$\text{Conductor sectional area} = \frac{At/p \times MLT \times p \times 8.25 \times 10^{-7}}{IR} \tag{8-21}$$

where $At/p$ is the number of ampere-turns per pole needed, $MLT$ is the mean length of turns, $p$ is the number of coils in series, and $IR$ is the required voltage drop.

Typical field calculations are

| | |
|---|---|
| *At/p* | 7520 At |
| Approx. *MLT* | 55 in |
| *IR* drop needed | 90 V |
| Conductor area [Eq. (8-21)] | 0.038 |
| Insulation conductor | 0.018 in |
| Section of coil | 6.78 × 1.6 in |
| Actual *IR* | 86.5 V |
| Watts (*IR* × *I*) | 3,380 W |
| W/in² | 0.362 W/in² |
| W/in² allowed | 0.388 W/in² |
| Res. 75°C | 2.21 Ω |
| Coils in series | 10 |
| Copper | 0.162 × 0.258 = 0.04 in² |
| Coil | 24T high × 8lay. 192T/coil |
| Layout *MLT* | 55.65 in |
| *I* = At/t | 39.1 A |
| Surface $2(H + tk(MLT)p$ | 9350 in² |
| Rotor velocity | 7350 ft/min |
| Current density | 977 A/in² |

These data indicate an acceptable field.

## COOLING AND VENTILATION

***61. Cause of Temperature Rise.*** The losses in a dc machine cause the temperature of the parts to rise until the difference in temperature between their surfaces and the cooling air is great enough to dissipate the heat generated.

***62. Permissible measured temperature rises*** of the parts are limited by the maximum "hot-spot" temperature that the insulation can withstand and still have reasonable life. The maximum surface temperatures are fixed by the temperature gradient through the insulation from the hot spot to the surface.

The IEEE Insulation Standards have established the limiting hot-spot temperatures for systems of insulation. The *American National Standards Institute* Standard C50.4 for dc machines gives typical gradients for those systems, listing acceptable surface and average copper temperature rises above specified ambient-air temperatures for various machine enclosures and duty cycles. Typical values are 40°C for Class A systems, 60°C for Class B, and 80°C rise for Class F systems on armature coils. Class H systems usually contain silicones and are seldom used on medium and large dc machines. Silicone vapors can cause greatly accelerated brush wear at the commutator and severe sparking, particularly on enclosed machines.

***63. Temperature Gradients in Rotor Coils.*** Figure 8-51 represents a current-carrying conductor insulated from the core slot in which it is embedded. The hot spot is probably at the core center line and near the center of the conductor. Heat will probably travel along the conductor to the end turn and also through the insulation to the iron. The amount of heat flowing in each direction is difficult to calculate. Also, variations in the coils, such as resin fill and tightness in the slots, make heat conductivity factors difficult to predict.

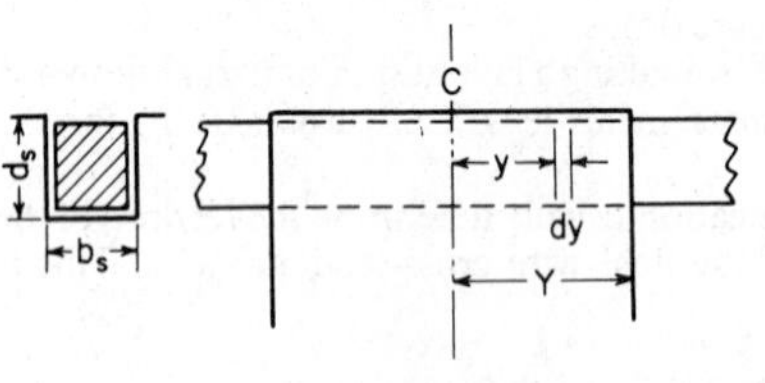


**FIG. 8-51** Heat paths in an armature conductor.

**a.** Assume that all the heat must travel down the conductor to the end turn. What will be the temperature difference in the conductor between the center of the core and its edge?

Resistivity of copper at 75°C = $8.25 \times 10^{-7}$ Ω/in³

$$\text{Thermal cond. copper} = 9.75\ \text{W/(in)(°C) for 1-in}^2\ \text{section}$$

Therefore, the energy crossing $dy$ of Fig. 8-51 is

$$\text{Watts} = (I_c)^2R_y = \frac{(I_c)^2(y)(8.25 \times 10^{-7})}{A} \tag{8-22}$$

where $I_c$ is the conductor amperes, $R_y$ the resistance of length $y$, and $A$ the conductor cross-section area.

The difference in temperature between two faces $dy$ apart is

$$°\text{C} = \frac{(I_c)^2(y)(8.25 \times 10^{-7})}{A} \times \frac{dy}{A} \times \frac{1}{9.75} \tag{8-23}$$

and the difference in temperature between the center $C$ and any point $y$ is

$$°\text{C} = \frac{(I_c)^2(8.25 \times 10^{-7})}{A} \int_0^y \frac{y\,dy}{A} \times \frac{1}{9.75}$$

or

$$°\text{C} = 4.22 \left(\frac{I_c}{A}\right)^2 (y)^2(10)^{-8} \tag{8-24}$$

Consider a current density of 2920 A/in² and a total core length of 16 in. Then the coil temperature gradient from the core center, with no ventilating ducts, to the edge is 28.8°C. This assumes that no heat passes through the insulation to the iron, and so medium and large machines normally use ventilating core ducts every few inches.

**b.** Assume that the end turns are so hot that no heat flows longitudinally down the coil. The $I^2R$ loss of each inch of conductor length is

$$\text{Watts} = \frac{(I_c)^2(8.25 \times 10^{7})}{A}$$

If the slot contains several conductors

$$\text{Watts} = (\text{ampere conductors})(\text{A/in}^2)(8.25 \times 10^{-7})$$

and the temperature difference between the bare conductor and the steel across the insulation is

$$°\text{C} = (\text{amp conductors})(\text{A/in}^2) \times \frac{\text{insulation thickness}}{2d_s + b_s} \times \frac{8.25 \times 10^{-7}}{0.003} \tag{8-25}$$

The factor 0.003 is the thermal conductivity of the insulation in watts per cubic inch per degree Celsius difference.

Thus, for 2142 ampere conductors per slot, 2920 A/in², a surface of two slot depths plus a slot width (times 1 in) = 5.07 in², and an insulation thickness of 0.051 in (data for the 2500-kW generator of Par. **45**), the temperature drop across the insulation is 17.65°C.

This figure cannot be considered precise because the thermal conductivity can vary widely with the insulation used and the presence of varying amounts of air in it. The conductivity figure for air is 0.0007, whereas that of mica is 0.007 W/(in³)(°C). Also, heat moves along the coil. Because of these difficulties, empirical data from actual machines are more reliable and easier to use.

***64. Heating of End Connections of Armature Windings.*** Small machines often have "solid" end windings banded down on insulated "shelf"-type coil supports. Larger machines are more heavily loaded per unit volume and usually have narrow coil supports, air spaces between the end turns, and ventilating air scouring both the top and bottom surfaces of the coil extensions.

With this construction the approximate allowable product of ampere conductors per inch of outer circumference times the amperes per square inch for various rotor velocities is shown in Fig. 8-52 for a 40°C rise on the end turns.

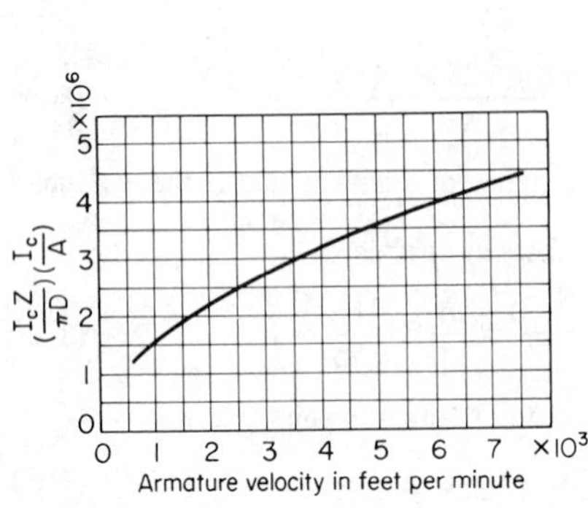


**FIG. 8-52** End-winding cooling.

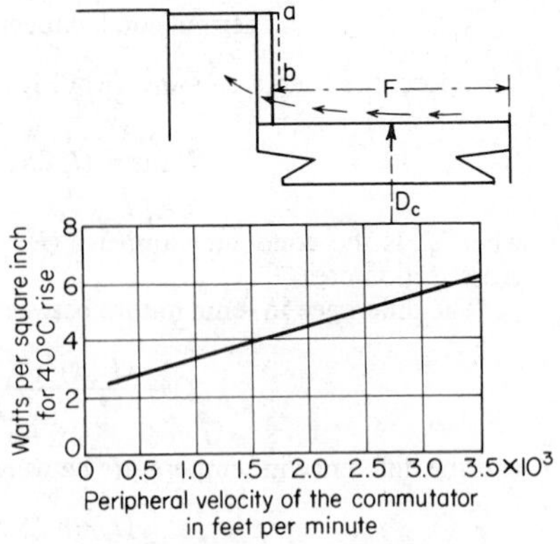

**FIG. 8-53** Temperature rise of a commutator.

***65. Commutator Heating.*** A modern dc armature is shown in Fig. 8-53. The commutator diameter ranges from 55 to 85% of the rotor core, and the commutator necks joining the bars with the rotor winding extensions are usually separated from one another by air spaces, so that, when the armature revolves, air circulation is set up as shown by the arrows.

A typical relation between permissible watts per square inch of commutator surface and its peripheral velocity is shown in Fig. 8-53. The radiating surface is the commutator circumference times its face length. Neck area is not included.

The heat to be dissipated is that due to brush friction and the brush contact $I^2R$ losses. There may be other losses due to poor commutation, brush chattering, and commutator surface, and, if so, the rise will be greater than indicated in Fig. 8-53. If commutation is very good and brush riding excellent, the temperature will be lower.

***66. Application of Heating Constants.*** The paragraphs covering the design of the armature, main fields, compensating windings, and commutating windings included typical loading data such as ampere conductors per inch, amperes per square inch, flux densities and watts per square inch of cooling surface. More accurate data depend on the exact arrangements used in a particular design. If possible, new design should be compared with similar machines which have already been tested. Any machine enclosure variation that restricts or increases the ventilation will affect the temperature rises.

## LOSSES AND EFFICIENCY

***67. Armature Copper* I²R *Loss.*** At 75°C the resistivity of copper is $8.25 \times 10^{-7}$ Ω/in³. Thus, for an armature winding of $Z$ conductors, each with a length of $MLT/2$ (half the mean length turn of the coil), each with a cross-sectional area of $A$ and arranged in several parallel circuits, the resistance is

$$R_a = Z \frac{MLT}{2A} \frac{8.25 \times 10^{-7}}{(\text{circuits})^2} \quad \text{ohms} \tag{8-26}$$

The $MLT$ is best found by layout, but an approximate value is

$$MLT = 2[(1.35)(\text{pole pitch}) + (\text{rotor length}) \times 3] \tag{8-27}$$

There are also eddy-current losses in the rotor coils, but these may be held to a minimum by conductor stranding in accordance with Eq. (8-17). Some allowance for these is included in the load loss.

***68. Compensating, Commutating, and Series Field* I²R *Losses.*** These fields also carry the line current, and the $I^2R$ losses are easily found when the resistance of the coils is known. Their

*MLT* is found from sketch layouts. At 75°C

$$R = T\frac{MLT}{A}\frac{8.25 \times 10^{-7}}{(\text{circuits})^2}p \qquad \text{ohms} \tag{8-28}$$

where $R$ is the field resistance in ohms, $T$ the number of turns per coil, $p$ the number of poles, $MLT$ the mean length of turn, and $A$ the area of the conductor. The total of these losses ranges from 60 to 100% of the armature $I^2R$ for compensated machines and is less than 50% for noncompensated machines.

***69. The brush* I²R *loss*** is caused by the load current passing through the contact voltage drop between the brushes and the commutator. The contact drop is assumed to be 1 V.

$$\text{Brush } I^2R \text{ loss} = 2(\text{line amperes}) \qquad \text{watts} \tag{8-29}$$

***70. Load Loss.*** The presence of load current in the armature conductors results in flux distortions around the slots, in the air gap, and at the pole faces. These cause losses in the conductors and iron that are difficult to calculate and measure. A standard value has been set at 1% of the machine output.

$$\text{Load loss} = 0.01(\text{machine output}) \tag{8-30}$$

***71. Shunt Field Loss.*** Heating calculations are concerned only with the field copper $I^2R$ loss. It is customary, however, to charge the machine with any rheostat losses in determining efficiency. Thus,

$$\text{Shunt field and rheostat loss} = I_f V_{ex} \qquad \text{watts} \tag{8-31}$$

where $I_f$ is the total field current and $V_{ex}$ is the excitation voltage.

***72. Core Loss.*** As seen from Fig. 8-54, the flux in any portion of the armature passes through $p/2$ c/r (cycles per revolution) or through $(p/2)[(\text{r/min})/60]$ Hz.

The *iron losses* consist of the *hysteresis loss,* which equals $K\beta^{1.6}fw$ watts, and the *eddy-current loss,* which equals $K_e(\beta ft)^2w$ watts. $K$ is the hysteresis constant of the iron used, $K_e$ is a constant inversely proportional to the electrical resistance of the iron, $\beta$ is the maximum flux density in lines per square inch, $f$ is the frequency in hertz, $w$ is the weight in pounds, and $t$ is the thickness of the core laminations in inches.

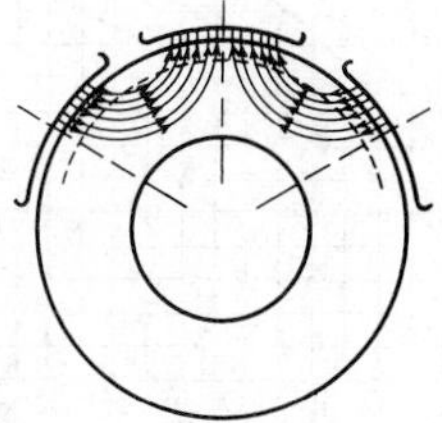

**FIG. 8-54** Distribution of flux in the armature.

The *eddy loss* is reduced by using iron with as high an electrical resistance as is feasible. Very high resistance iron has a tendency to have low flux permeability and to be mechanically brittle and expensive. It is seldom justified in dc machines. The loss is kept to an acceptable value by the use of thin core laminations, 0.017 to 0.025 in in thickness.

Another significant loss is the *pole-face loss.* Figure 8-43 shows the distribution of flux in the air gap of a dc machine. As the armature rotates and the teeth move past the pole face, emf's are induced which tend to cause currents to flow across the pole face. These losses are included in the core loss.

Unfortunately, there are other losses in the core that may differ widely even on duplicate machines and that do not lend themselves to calculation. These include:

**a.** Loss due to filing of slots. When the laminations have been assembled, it will be found in some cases that the slots are rough and must be filed to avoid cutting the coil insulation. This burrs the laminations and tends to short-circuit the interlaminar resistance.

**b.** Losses in the solid spider, core end plates, and coil supports from leakage fluxes may be appreciable.

c. Losses due to nonuniform distribution of flux in the rotor core are difficult to anticipate. In calculating core density, it is customary to assume uniform distribution over the core section. However, flux takes the path of least resistance and crowds behind the teeth until saturation forces it into the less used, longer paths below. As a result of the concentration, the core loss, which is about proportional to the square of the density, is greater than calculated.

Thus, it is not possible to predetermine the total core loss by the use of fundamental formulas. Consequently, core-loss calculations for new designs are usually based on the results from tests on similar machines built under the same conditions. Such test results are plotted in Fig. 8-55 for machines using ordinary laminations 0.017 in thick and a limited amount of filing. They do not include the pole-face losses, which would increase the values about 30%.

***73. Brush Friction Loss.*** This loss varies with the condition of the commutator surface and the grade of carbon brush used. A typical machine has about 8-W loss/(in$^2$ of brush contact surface)(1000 ft/min) of peripheral speed when normal brush pressure of 2½ lb/in$^2$ is used.

$$\text{Brush friction} = (8)(\text{contact area})(\text{peripheral velocity}/1000) \tag{8-32}$$

***74. Friction and Windage.*** Most large dc machines use babbitt bearings and many small machines use ball or roller bearings, although both types of bearing may be used in machines of any size. The bearing friction losses depend on the speed, the bearing load, and the lubrication. The windage losses depend on the construction of the rotor, its peripheral velocity, and the machine restrictions to air movement. The two losses are lumped in most estimates because it is not practical to separate them during machine testing.

Figure 8-56 shows typical values of friction and windage losses for various rotor diameters referred to rotor velocities.

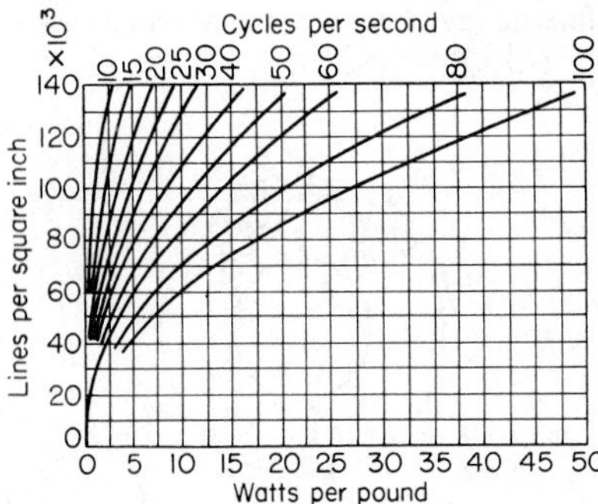


**FIG. 8-55** Iron-loss curves for a dc machine.

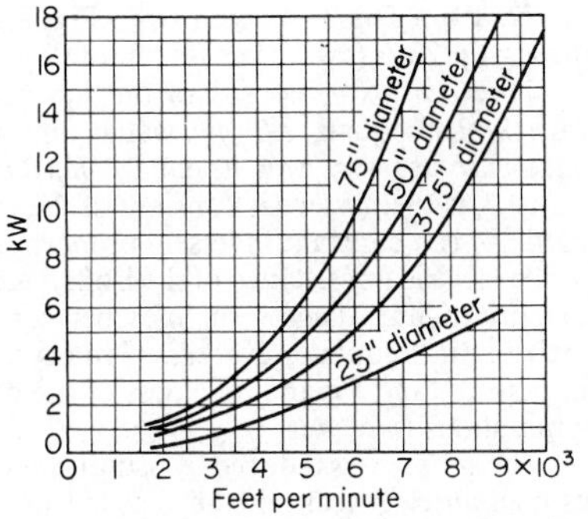

**FIG. 8-56** Friction and windage vs. rotor velocity.

***75. Efficiency.*** The efficiency of a generator is the ratio of the output to its input. The prime mover must supply the output and, in addition, the sum of the losses listed in Pars. **67** to **74**. This is the input.

$$\text{Efficiency} = \frac{\text{output}}{\text{input}} = \frac{\text{output}}{\text{output} + \text{losses}} \tag{8-33}$$

## *GENERATOR CHARACTERISTICS*

***76. The voltage regulation of a dc generator*** is the ratio of the difference between the voltage at no load and that at full load to the rated-load voltage. The characteristic is normally drooping as the load is increased, but it can rise because of series field effects or the action of circulating currents of communication at very low voltage operation.

For a dc generator, the terminal-voltage equation is

$$TV = E - IR = [K(\phi_t)(\text{r/min}) - IR] \tag{8-34}$$

where $E$ is the induced emf, $IR$ is the armature circuit drop, $K$ is a constant depending on the machine design, and $\phi_t$ is the total main-pole flux of the generator.

The regulation curves are easily calculated by using the no-load and full-load saturation curves determined in Pars. **56** and **57** and shown in Fig. 8-57. The effect of the excitation method is found by the use of the field and rheostat $IR$ line for self-excited machines and by the constant-ampere-turn line for separate excitation.

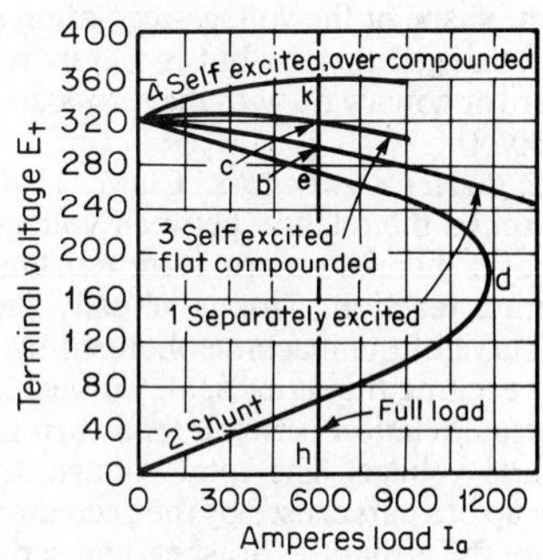


**FIG. 8-57** External characteristics vs. excitation methods.

***77. A separately excited compensated generator*** which is shunt-wound will have a voltage-load characteristic which will approach a straight line; it droops to full load an amount equal to the percent $IR$ drop. There is little or no flux loss due to armature reaction or brush shift.

At voltages 10% or less of rated, the main field strength is so weak that currents circulating in the coils short-circuited by the brushes at commutation may cause an increase in main-pole flux with load that causes a rising characteristic. These armature coils loop the main poles and their ampere-turns produce direct axis flux. A rising voltage characteristic can be undesirable, particularly if the generator supplies a dc motor whose speed is caused to rise with load, since this causes instability.

***78. A separately excited noncompensated dc generator*** which is shunt-wound has a nonlinear loss of flux due to armature reaction as the load current is increased (see Par. **19**). It can be seen from Eq. (8-34) that this causes a characteristic which droops at an ever-increasing rate with load increase, giving a curve which is concave downward.

***79. A self-excited noncompensated dc generator*** which is shunt-wound has its shunt field excitation decreased as the terminal voltage drops, as described in Par. **78.** This results in a reduction of main field ampere-turns and a loss of still more flux. This gives a severe droop which may be so great that, above a certain peak-load current, the terminal voltage will not be high enough to provide enough field current to maintain the voltage and load current and the voltage will collapse, as shown in $d$ of Fig. 8-58.

***80. Instability of Self-excited Generators.*** A self-excited dc generator is unstable if the rheostat line does not make a definite intersection with the load-saturation curve (see Fig. 8-57). The shunt field current is fixed by the terminal voltage, and the resistance is in the shunt field circuit. Instability will exist if the slope of the rheostat line is nearly equal to or greater than the slope of a line tangent to the operating point on the saturation curve. In Fig. 8-58, point $b$ is a stable operating condition, but point $c$ is not, because a decrease in voltage decreases the shunt field ampere-turns and this produces a further decrease in voltage.

If the field circuit resistance were set at $d$, the self-excited generator would never build up beyond residual voltage. Another cause of failure to build up may be the connection of the shunt field. If the current flow due to residual voltage is such that it tends to kill the flux producing the residual voltage, no build-up occurs.

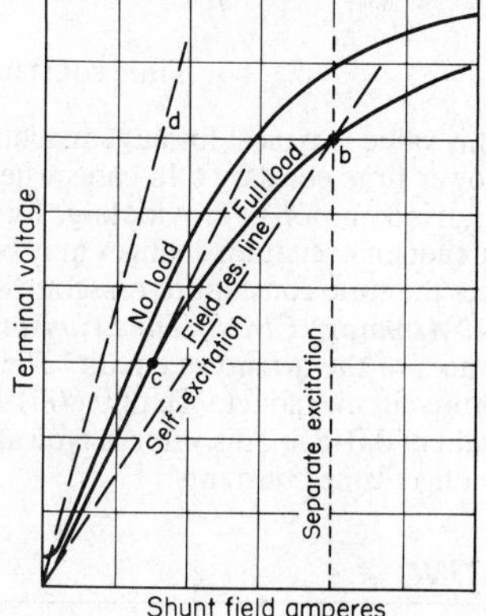


**FIG. 8-58** No-load and field-load saturation curves.

***81. Compound-wound DC Generators.*** The generators described above can be compounded by adding series fields excited by the load current. However, the resulting field strength of these fields is linear with load and the shape of the voltage-regulation curve is not changed thereby but is merely rotated upward or downward with the zero-load point as a pivot.

***82. Series Generators.*** Curve 1 of Fig. 8-59 shows the relation between voltage and current if there is no armature resistance or armature reaction. This is actually the no-load curve of the machine obtained by separately exciting the series field. Curve 2 shows the actual relation between load current and terminal voltage. The total voltage drop is made up of a part caused by the decrease in flux by armature reaction and a part caused by the $IR$ drop of the armature, brushes, and series fields.

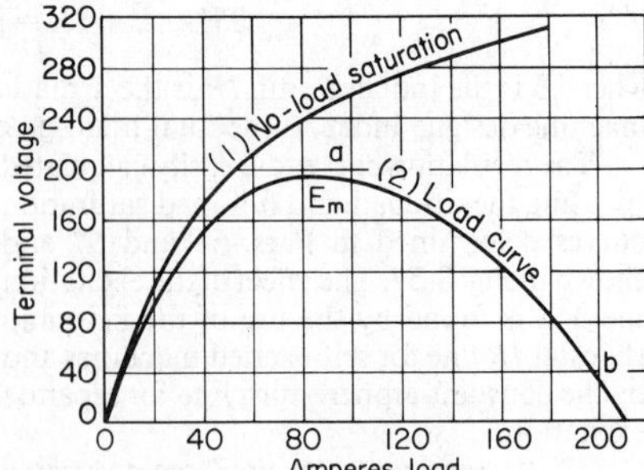


**FIG. 8-59** Characteristic curves of a series generator.

***83. Field Time Constants.*** The major delay in change of output voltage by an excitation change is caused by the inductance of the main fields. The time constant of the shunt field is the ratio of its inductance in henrys to its resistance in ohms, and this ratio represents the time in seconds required for 63% of a field current change to occur when the excitation voltage is suddenly changed. In the case of the 2500-kW generator whose armature was designed in Par. **45** and whose fields were developed in Par. **60,** a mean main-field inductance over the voltage range from zero to rated is 6.20 H. The main-field resistance is 2.21 Ω. The field time constant is therefore 2.8 s.

The inductance $L$ of a coil is the incremental change of flux linkages per incremental change in field current times $10^{-8}$. This is proportional to the slope of the saturation curve and is constant over the air-gap line. It is therefore a decreasing variable after the curve leaves the air-gap line (see Fig. 8-46). The overall inductance, as the voltage builds up from zero, is not so high as that of the air-gap portion or as low as at the rated-voltage point. A common compromise is the slope of a straight line drawn from zero voltage through the full-load point at rated voltage. For the 2500-kW generator the total flux at this point is $112.5 \times 10^6$ lines. With a leakage flux of 12%, each coil has a flux of $12.6 \times 10^6$ lines (see Table 8-2). As indicated in Par. **60** each coil has 192 turns and there are 10 coils in series. The field current is 39.1 A.

$$L = \frac{\phi T}{I_f} \times 10^{-8} = \frac{(12.6 \times 10^6)(192)(10)}{39.1} \times 10^{-8} = 6.2 \text{ H} \tag{8-35}$$

$$\text{Time constant} = L/R = 6.2/2.21 = 2.8 \text{ s} \tag{8-36}$$

This value is typical for large machines. Smaller generators have less copper in their fields and lower time constants. In cases where drive systems must have very rapid voltage adjustments, it is common to provide large forcing voltages on the field to overcome the inductive lag. These sudden excitation changes may be 4 to 10 times the $IR$ drop of the field. This effectively reduces the time constant to one-fourth or one-tenth its normal value.

***84. Armature-Circuit Time Constants.*** Compensating windings effectively lower the inductances of the armature circuit. The 2500-kW generator developed in this section has an armature-circuit inductance of 0.0001929 H and a circuit resistance of 0.00398 Ω for a time constant of 0.048 s. This value is typical for large dc machines. Smaller noncompensated units have longer time constants.

## TESTING

***85. Factory Tests.*** These depend on the size, application, and design of the dc generator. The American National Standards Institute (ANSI) C50.4 for dc machines includes lists of recommended tests for dc generators and motors. The IEEE Test Code for dc machines covers recommended methods to be used for these tests.

## *GENERATOR OPERATION AND MAINTENANCE*

***86. General.*** Despite its rugged construction, a dc machine is a delicate device. Factory tests on large units may cost thousands of dollars and must be performed carefully to adjust the generator to obtain the best possible characteristics and commutation. Owing to shipping requirements, the generator may then have to be disassembled and shipped in several pieces. If the final assembly is not correctly accomplished, not only have the factory tests been wasted but the machine may be damaged.

The manufacturer's instruction book should be studied carefully.

***87. Before Installation.*** Upon arrival, the generator should be inspected for damage and to be sure it is dry. If it is wet, consult the manufacturer. Drying out with heat should be done only by slowly raising the generator temperature to 100°C so that moisture can escape without forming gas pockets within the insulation.

If the generator is dry and clean, the windings should be checked with a megger for insulation resistance to ground measurements. If any readings less than 1 MΩ are found, check with the manufacturer.

***88. Alignment.*** After the machines are installed and grouted to the foundations, all couplings should be opened and alignments of all shafts finally checked. Regardless of whether solid or flexible couplings are used, the alignment should be as accurate as possible. The difference between the bottom and the top openings should not exceed 0.002 in for 12 in of flange diameter, and the large opening should be at the top. Regardless of the size of coupling, the difference should not exceed 0.004 in. Differences at the side should not exceed 0.001 in. Shafts should be rotated 180° and rechecked.

The frame should be set on the *magnetic center* of the core. This position can be located by setting the armature in rotation and forcing it to oscillate longitudinally the full end play of the bearing by pushing on the end of the shaft. While the rotor is coasting and oscillating freely, excite the main field. The stator can then be shifted so that the rotor position with excitation coincides with the center of bearing end play.

Air gaps between the rotor and poles should be uniform. A typical limit of variation is 0.010 in. The brushes should ride properly on the commutator surface at both extremes of bearing end play.

***89. Prerunning Checks.*** The circumferential position of the brushes on the commutator is important for commutation and also to provide the voltage characteristics set at the factory. Brushes should be on the factory test setting. The toes of the brushes should be aligned and should have no skew. The spacing between adjacent arms of brushes should be identical within 0.032 in. The brushes should move freely in their holders and should have a pressure against the commutator of 2 to 3 lb/in$^2$ on the basis of brush cross section. The faces of the brushes should accurately match the curvature of the commutator surface.

The polarity of the main fields may be checked by tracing the wiring around the frame or by lightly exciting the fields and using a compass around the frame behind the poles.

The oiling system for the bearings should be checked and the oil rings tested for freedom. The entire machine, particularly its air gaps, should be inspected for foreign material.

***90. Running Checks.*** Note any unusual noise as the unit is brought up to speed. Bearing temperatures should level out at acceptable values within a few hours.

The voltage should be slowly raised at no load and commutation observed. If satisfactory, the voltage should be raised to 110% of rated and then reduced.

The generator may then be loaded gradually while commutation is observed, until rated current is reached. If commutation remains satisfactory until stable temperatures are achieved, the generator is ready for work.

***91. Shunt-wound Generators in Parallel.*** *A* and *B* of Fig. 8-60 are two similar generators feeding the same bus bars *C* and *D*. If *A* tends to take more than its share of the total load, its voltage falls and more load is automatically thrown on *B*. Also, if the driver of one of the generators slows down to stop, the emf of the machine falls until the other generator starts to drive it as a motor. This continues until its driver takes over again.

The external characteristics of the two machines are shown in Fig. 8-61. At voltage *E*, the currents in the generators are $I_a$ and $I_b$, and the line current is $I_a + I_b$. To make machine *A* take more of the load, its excitation must be increased to raise its characteristic curve. If a 1000-kW generator and a 500-kW machine have the same regulation curves, the machines will divide the load according to their respective capacities, as shown in Fig. 8-62.

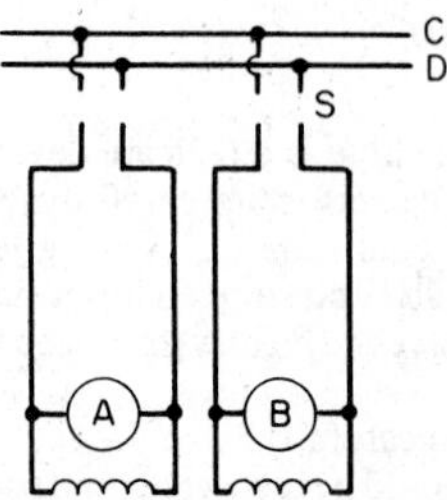


FIG. 8-60 Shunt generators in parallel.

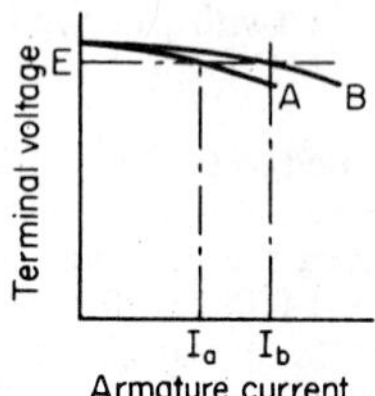


FIG. 8-61 Two shunt-wound generators in parallel: external characteristics.

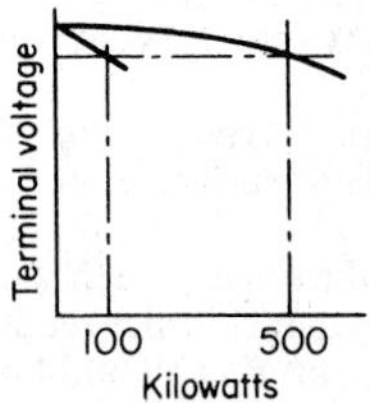


FIG. 8-62 Division of load between two shunt generators in parallel.

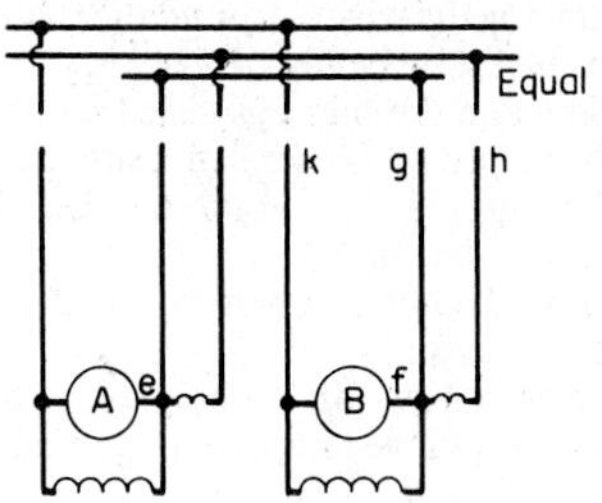


FIG. 8-63 Compound generators in parallel.

***92. Compound-Wound Generators in Parallel.*** *A* and *B* of Fig. 8-63 are two compound-wound machines. If *A* tends to take more than its share of the load, the series excitation of *A* increases, its voltage rises, and it takes still more of the load. Thus, the operation is unstable. If this continues until *A* takes all the load and the voltage of *B* drops to the point that *A* reverses the current in *B*, *B* will be driven as a motor. With the reversed current in the series field of *B* it becomes a differentially compounded motor, and the series weakens the flux to speed up the motor. This may progress to a point at which the unit may be damaged mechanically and electrically.

To prevent this, a bus bar of large section and of negligible resistance, called an *equalizer bus,* is connected from *e* to *f* (Fig. 8-63). Points *e* and *f* are then practically at the same potential. Therefore, the current in each series coil is independent of the current in its particular generator, is inversely proportional to the resistance of the coils, and is always in the same direction.

When a single compound generator has too much compounding, a shunt in parallel with the series field coils will reduce the current in these coils and so reduce the compounding. When compounded generators are operating in parallel using an equalizer bus, the current in the series field coils depends only on the resistance of the coils and a shunt connected across one of them is actually across all of them, reducing the compounding of all but not disturbing the relative compounding between the machines. To reduce the compounding of a single machine, it is necessary to place a resistance in series with the coils. This may require a large resistor to handle the large load current it must carry.

***93. Maintenance.*** Except for the commutator and its brushes, maintenance of dc machines differs little from that of other rotating electrical machines. Proper lubrication must be provided for the bearings, and the machine must be kept clean and dry. In addition, the brushes should be checked periodically for commutation, riding ability, freedom of motion in the holders, pressure, and length.

Because the commutator necks are not insulated and receive full voltage, conducting dust from brush wear or from ventilating air can cause creepage currents between the risers and ground over insulated surfaces. To avoid this, the dc generator must be cleaned and blown out with clean, dry air at regular intervals. Air pressures above 25 lb/in$^2$ should not be used because of the danger of lifting the edges of insulating tape. The effectiveness of the cleaning program should be verified occasionally by megger readings.

***94. Poor Commutation.*** Sparking and bar burning are usually due to one or more of the following causes:

**a.** *Brushes not in the proper position.*

**b.** *Incorrect spacing of brushes.* This may be checked by marking an adding-machine tape around the commutator.

**c.** *Projecting-bar-edge mica.* Mica between bars should be undercut about 0.063 in below the commutating surface, but occasionally slivers of mica are left inadvertently along the bar.

**d.** *Rough or burned commutator.* The commutator should be ground according to the manufacturer's instruction book.

**e.** *Grooved commutator.* This may be prevented by properly staggering the brush sets so that the spaces between the brushes of an arm are covered by brushes of the same polarity of other arms.

**f.** *Poor brush contact* due to improper fitting of the brushes to the commutator surface. To seat the brushes, sandpaper should be moved between the commutator and the brush face. Emery cloth should not be used, because its abrasive is conducting.

**g.** *Worn brushes replaced by others of wrong size or grade.*

**h.** *Sticking brushes,* which do not move freely in their holders so that they can follow the irregularities of the commutator.

**i.** *Chattering of the brushes.* This is usually due to operation at current densities below 35 A/in$^2$ and must be corrected by lifting brushes to raise the density or by using a special grade of brush.

**j.** *Vibration.* This may be due to poor line-up, inadequate foundations, or poor balance of the rotor.

**k.** *Short-circuited turns* on the commutating or compensating fields. These may be obvious on inspection but usually must be found by passing ac current through them for voltage-drop comparisons.

**l.** *Open or very high resistance joints* between the commutator neck and the coil leads. In this case the bar at the bad joint will usually be burned.

**m.** *An open armature coil.* A broken coil conductor produces an effect similar to that produced by the poor joints described in item *l*. For emergency operation, the open coil may be opened at both ends, insulated from the circuit, and a jumper placed across the two affected necks. Since some sparking will probably result, operation should be limited.

**n.** *Short-circuited main-field coils.* With the resulting unbalanced air-gap fluxes under the poles, large circulating currents must be expected even with good armature cross connections. The offending coil may be found by comparing voltage drops across the individual coils.

**o.** *Reversed main-field coil.* This is an extreme case of *n*.

**p.** *Overloading.*

## SPECIAL GENERATORS

***95. General.*** The adaptability of the dc generator for specific uses has led to the development of many special generators. These machines over the years made a significant contribution to industrial progress. However, most of these special applications have disappeared or are now being met with other devices such as silicon-controlled rectifiers or programmed control of field currents to the main dc generator.

***96. Synchronous Converters.*** Of all the special generators this was one of the earliest and most widely used. It was the principal dc power source for streetcars and interurban lines. It was a most ingenious device, combining in a single armature and winding an ac motor taking its current from the lines through slip rings at the rear and a dc generator providing dc power from a commutator on the front end. Because the flow of the currents was in opposition, the resulting rotor winding could be small in cross section. A single stator provided flux for both functions. With the decline of street railway systems the synchronous converter disappeared.

***97. Rotating Regulators.*** These dc machines had trade names like Rototrol, Regulex, and Amplidyne. They too have been replaced by solid-state devices. In addition to having fields for feedback intelligence, response was enhanced using self-excited shunt fields tuned to the air-gap line or by means of cross magnetization from armature reaction.

***98. Three-Wire Devices.*** Because three-wire dc circuits are no longer in use, balancer sets and three-wire generators are relics in school labs or museums.

***99. Homopolar or Acyclic DC Generators.*** The single-pole machine principle still fascinates electrical engineers and several research and development labs continue to study new arrangements of its basic parts. Fundamentally it consists of a single conductor moving through a uniform single-direction flux with a collector at each end of the conductor. The output is a steady ripple-free pure dc current and no commutation. Currents reaching 270,000 A at 8 V were provided by one commercial unit shown in Fig. 8-64. Recent efforts have been mainly to use liquid metals to take the large currents from the rotating collectors and to obtain higher voltages by connecting units in series. Some success has been possible, but restricting the sodium potassium to the collector area has proved difficult.

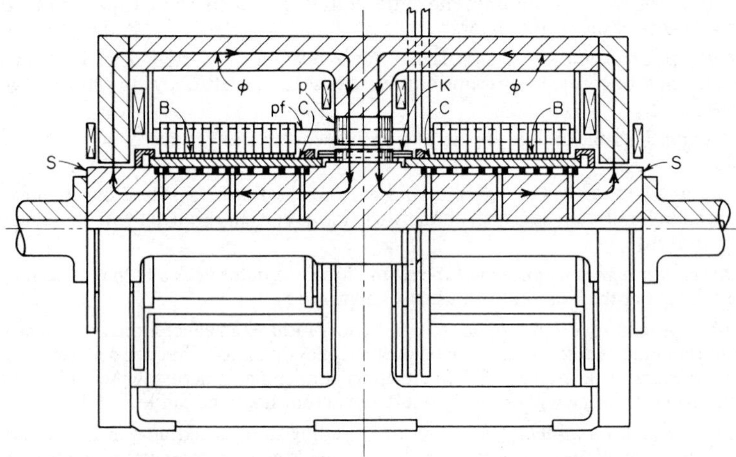


**FIG. 8-64** Brush-type homopolar generator.

***100. Details.*** More detailed information on these and other old dc generators popular in the past will be found in past editions of this handbook.

## Bibliography

***101. General***

Lamme B. G.: *Engineering Papers;* East Pittsburgh, Westinghouse Electric Corporation, 1919.

Kloeffler, R. G., Brenneman, J. L., and Kerchner, R. M.: *Direct Current Machinery;* New York, The Macmillan Company, 1950.

Woodson, H. H., and Melcher, J. R.: *Electromechanical Dynamics,* Part 1: *Discrete Systems;* New York, John Wiley & Sons, Inc., 1968, pp. 283–317.

Lewschitz and Ganick: *Fundamentals of DC Machines;* Princeton, N.J., D. Van Nostrand Company, Inc., 1951.

Pasculle, M. J.: Armature Tooth Pulsations Eddy Currents; *Trans. AIEE,* 1960, vol. 79, pt. III, pp. 612–618.

Erdelyi, E. A.: Calculation of Stray Load Losses in DC Machinery; *Trans. AIEE,* 1960, vol. 79, pt. III, pp. 129–138.

Darling, A. J., and Linville, T. M.: Rate of Rise of Short Circuit Current of DC Motors and Generators; *Trans. AIEE,* 1952, vol. 71, pt. III, pp. 314–325.

O'Connor, J. P., and Cybulski, J.: DC Machines—Short Circuit Calculations and Test Results; *Trans. AIEE,* 1955, vol. 74, pt. III, pp. 222–238.

Cybulski, J., Broncato, E. L., and O'Connor, J. P.: Transient Performance of DC Machinery; *Trans. AIEE,* 1953, vol. 72, pt. III, pp. 45–52.

Nitta, K., Okitsu, H., and Suzuki, T.: Measurement and Analysis of Dynamic Performances of DC Series Motor; *Electr. Eng. in Jpn.,* 1968, vol. 88, no. 3, pp. 29–38.

Bird, B. M., and Harlen, R. M.: Variable Characteristics DC Machines; *Proc. IEE,* 1966, vol. 113, no. 11, pp. 1813–1819.

Andrews, H. I.: Development of an Electronically Commutated Motor with Laminated Brushes; *Proc. IEE,* 1969, vol. 116, no. 5, pp. 763–768.

Dunaiski, R. M.: The Effect of Rectifier Power Supply on Large DC Motors; *Trans. AIEE,* 1960, vol. 79, pt. III, pp. 253–259.

Schmidt, J., and Smith, W. P.: Operation of Large DC Motors from Controlled Rectifiers; *Trans. AIEE,* 1948, vol. 67, pt. I, pp. 679–683.

Erdelyi, E. A., and Fuchs, E. A.: Nonlinear Magnetic Field Analysis of DC Machines, Parts I, II, and III: *IEEE Trans. Power Appar. Syst.,* 1970, vol. 89, pp. 1546–1583.

### *102. Commutation*

Alger, J. R. M., and Vewley, P. F.: An Analysis of DC Machine Commutation; *Trans. AIEE,* 1957, vol. 76, pt. III, pp. 399–416.

Linville, T. M., and Rosenberry, G. M., Jr.: Commutation of Large DC Motors and Generators; *Trans. AIEE,* 1925, vol. 71, pt. III, pp. 326–336.

McLean, H. J., and Coho, O. C.: DC Motor Flashover Torque; *Trans. AIEE,* 1961, vol. 80, pt. III, pp. 850–853.

Phillips, R., Thompson, J. E., and Luck, E. M.: Commutation Phenomena in Electrical Machines, Part 2, Further Experiments in Flashover; *Proc. IEEE,* 1968, vol. 115, no. 11, pp. 1649–1653.

Turner, M. J. B., and Swinnerton, B. R. G.: Sparking and Arcing in Electrical Machines; *Proc. IEE,* 1966, vol. 113, no. 8, pp. 1376–1386.

### *103. Homopolar Generators*

Myers, E. H.: The Unipolar Generator; March, 1956, *Westinghouse Eng.,* vol. 16, pp. 59–61.

Gigot, E. N.: Applying Unipolar Generators; 1962, *Allis-Chalmers Electr. Rev.,* 2d Quarter.

Jeisler, F. L.: A High Power Density Electric Machine Element; *IEEE Trans. Power Appar. Syst.,* 1967, vol. 86, no. 7, pp. 811–818.

Mole, C. J., Brenner, W. C., and Haller, H. E., III: Superconducting Electrical Machinery; *Proc. IEEE,* January 1973, vol. 61, no. 1, pp. 95–105.

# SECTION 9
# HYDROELECTRIC POWER GENERATION

**Grover F. Wachter**
*Manager, Repair and Rehabilitation Services, Allis-Chalmers Hydro Inc.; Member, Hydraulic Power Committee, Edison Electric Institute*

## CONTENTS

*Numbers refer to paragraphs*

## GENERAL

***1. Nomenclature.*** The nomenclature used throughout this section is based on National Electrical Manufacturers Association (NEMA) Publication HT1-1957, Hydraulic Turbines, Governors and Accessory Equipment, which contains illustrated terms and definitions.

A hydroelectric power plant converts the inherent energy of water under pressure into electric energy. Its main elements are

An *upper,* or *high-level, reservoir,* usually formed by building a dam across a river

An *intake,* consisting of a canal or concrete passageway to carry the water directly to low-head turbines or to the pressure conduit used for medium- and high-head turbines

A *pressure conduit,* consisting of a tunnel, pipeline, or penstock, or any combination thereof, to carry the water under pressure to medium- and high-head turbines

A *surge tank,* to prevent excessive pressure rises and drops during sudden load changes, installed somewhere along the pressure conduit when this conduit is quite long

*Trash racks* at the inlet to the intake or pressure conduit

*Intake* and *draft-tube gates*

A *penstock shutoff valve,* located near the downstream end of the penstock

A *hydraulic turbine* consisting primarily of a runner, connected to a shaft, for producing prime motive power from the inherent energy of the water under pressure, a mechanism for controlling the quantity of water flowing to the runner, and water passages leading to and away from the runner

A *governor* for operating the hydraulic-turbine control mechanism

An *electric generator* connected to the hydraulic-turbine shaft to convert the prime motive power of the turbine to electric power

A *pressure regulator,* sometimes used instead of a surge tank, to prevent excessive pressure rises and drops during sudden load changes in plants with long pressure conduits

A *powerhouse* to enclose and support the hydraulic turbine, generator, governor, pressure regulator (if used), and auxiliaries

A *draft tube,* usually a part of the powerhouse structure to carry the water away from the turbine runner

A *tailrace,* sometimes used to carry the water away from the draft tube to the tailrace reservoir

A *tail-water reservoir* which receives the water discharged from the draft tube or tailrace and is usually part of the original river at an elevation lower than the upper reservoir

The difference in elevation between the water level in the upper reservoir and the level of the water in the tailrace or tail-water reservoir is called the *gross head* on the plant.

The size, location, and type of power plant depend upon the topography, the geological conditions, and the amount of water and head available. Hydropower developments can be classified as low-head, medium-head, or high-head. Figure 9-1 shows in outline the most common arrangement and features of some of the elements listed above for the various developments.

## HYDRAULIC TURBINES

***2. Turbine Characteristics.*** Hydraulic turbines derive power from the pressure or force exerted by water falling through a given distance (the head).

The theoretical power usually expressed in horsepower, $P_t$, is determined by the equation

$$P_t = HQw/550 = HQ/8.82 \tag{9-1}$$

where $H$ = head in feet, $Q$ = flow of water in cubic feet per second, $w$ = weight of water in pounds per cubic foot. The head is established by the topography of the country and the location of the dam, intake works, powerhouse, and tailrace or tail-water reservoir. An analysis of the river-flow records, type of turbine, and type of load (whether base or peak) will fix the maximum and mean value of flow to be used for design.

The actual horsepower $P$ of a hydraulic turbine is the theoretical horsepower $P_t$ multiplied by the turbine efficiency $e$,

$$P = P_t e = HQe/8.82 \tag{9-2}$$

The efficiency varies depending upon type of turbine load and operating head. For general purposes it is usual to assume a mean efficiency of 90%. Maximums approaching 95% at the peak of the curve have been obtained, based on field tests.

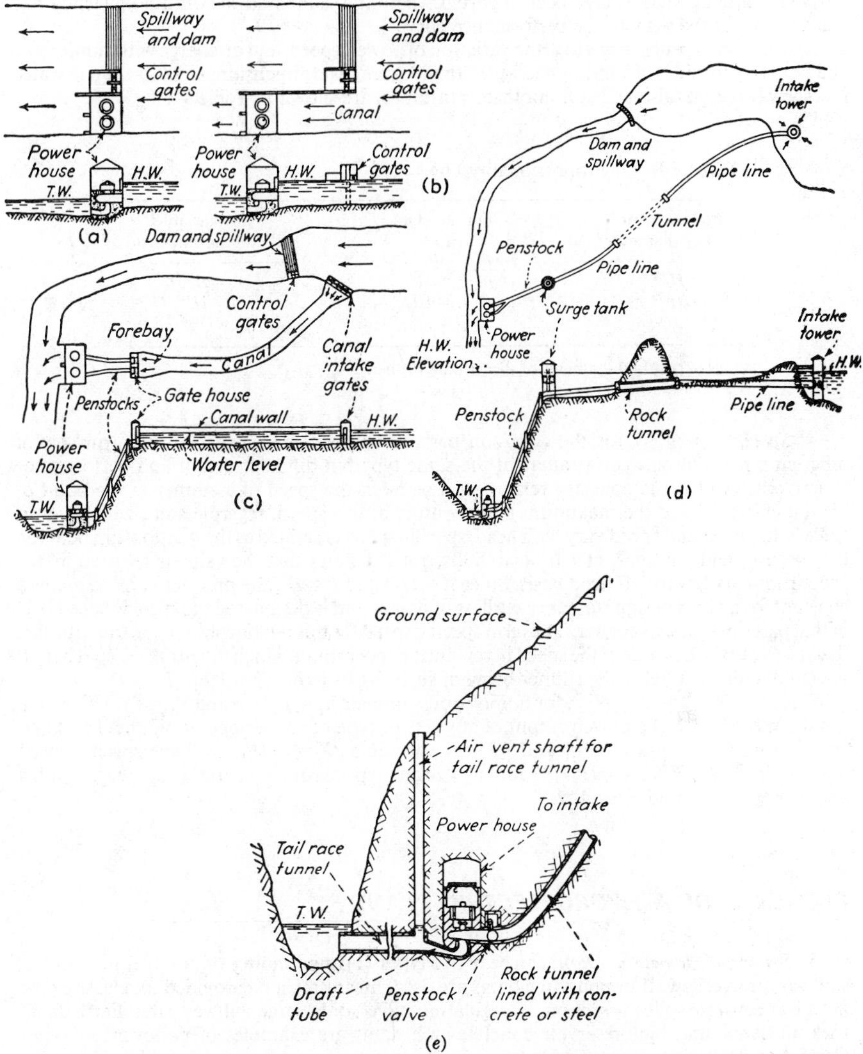


**FIG. 9-1** Outline sketches of several typical hydropower developments. *(a)* Low-head development with dam, spillway, and powerhouse as an integral unit. *(b)* Low-head development with a short intake canal and powerhouse separate from the dam. *(c)* Medium-head development with a long intake canal, gatehouse, and penstocks connecting the forebay with the powerhouse. *(d)* High-head development with a large storage reservoir, pipeline, and tunnel leading to a surge tank at the upper end of the penstocks. Powerhouse at the lower end of the penstocks is a considerable distance from the dam and spillway. *(e)* Outline sketch of underground power plant, showing penstock and tailrace tunnels.

The kilowatt capacity of a hydroelectric unit can be obtained by converting the turbine output in actual horsepower to generator output kilowatts by the following equation.

$$kW = 0.746Pe_g \tag{9-3}$$

where $e_g$ = generator efficiency, which will range from 94% in the smaller machines to over 98% in the larger machines.

A combined efficiency of both turbine and generator of 85% is conservative for general purposes, although cases have been reported where the combined maximum efficiency has reached 93% at the peak of the performance curve.

***3. The laws of proportionality*** (the variation of power, speed, and discharge with runner size and head) for turbines of varying size, but with the same basic dimensional relationship in water passageway design (also called homologous turbines) are shown in Table 9-1.

**TABLE 9-1** Proportionality Laws

| For constant runner diameter | For constant head | For variable diameter and head |
|---|---|---|
| $P \propto H^{3/2}$ | $P \propto D^2$ | $P \propto H^{3/2}D^2$ |
| $n \propto H^{1/2}$ | $n \propto 1/D$ | $n \propto H^{1/2}/D$ |
| $Q \propto H^{1/2}$ | $Q \propto D^2$ | $Q \propto H^{1/2}/D^2$ |

$D$ = nominal diameter of the turbine runner.

***4. Specific speed*** ($N_S$) is the common basis of comparison between turbine runners of different types and between runners of the same type but different design and performance characteristics. It is the constant relationship between the speed of a runner at the point of highest efficiency and the maximum power output at this speed, regardless of size. However, since both power and speed vary with head, specific speed is defined as the relationship between the speed $n_1$ and power $P_1$ at 1-ft head. Subscript 1 denotes that the value is reduced by the proportionality law to 1-ft head basis. Since $n \propto 1/D$ and $P \propto D^2$, the product $n_1\sqrt{P_1}$ remains a constant for a given design runner regardless of its size and is designated the specific speed ($N_S$) of that particular design runner. The term specific speed for this relationship stems from the fact that $n_1\sqrt{P_1}$ also is the value of the speed in revolutions per minute which the runner would have if operated under 1-ft head, the runner being of such size as to develop 1 hp ($P_1 = 1$).

Since $n \propto H^{1/2}$ and $P \propto H^{3/2}$, for homologous runners, $n_1 = n/H^{1/2}$ and $P_1 = P/H^{3/2}$, where $n$ is the speed and $P$ the power output of any size or type of runner operating under head $H$. Substituting these values for $n_1$ and $P_1$ in the formula $N_S = n_1\sqrt{P_1}$, we have specific speed $N_S = (n/H^{1/2})(\sqrt{P/H^{3/2}}) = n\sqrt{P}/H^{5/4}$ for any size and type of runner operating at speed $n$, with a power output of $P$ under head $H$.

## ELEMENTS OF A HYDROELECTRIC PLANT

***5. Principal Elements.*** *Dams* can be of two types: (1) impounding, or nonoverflow, and (2) spillway, or overflow. If impounding dams are used, means must be provided to release excess flow, by a separate spillway section, by regulating valves, or by large spillway gates. Earth dams, rock-fill dams, and high-reservoir concrete-arch dams are examples of nonoverflow types. Careful control of the reservoir elevation is needed to prevent overtopping these dams and causing damage or even failure. Spillway dams are always concrete, and for low-head installations the powerhouse usually forms part of the dam.

*Intakes* may consist of canals, flumes, or concrete passageways.

*Conduits* may consist of concrete or rock tunnels, steel pipelines, steel penstocks, or any combination thereof.

*Trash racks* are provided at the inlet to the intake or the conduit to protect the turbine against floating or other material. Cleaning devices such as rakes, either manual or motor-operated, are provided to remove debris from the racks.

*Head gates* or *stop logs* are provided at the inlet to the intake or conduit and at the outlet of the draft tube for shutting off the flow to the turbine for safety and for ease of maintenance. The head gates are usually of steel. The head gates or stop logs are lowered and raised by a motor-operated crane.

*Pipelines* and *tunnels* are the closed conduits connecting the upper reservoir to the surge tank or penstock.

*Penstocks* are the closed conduits connecting the upper reservoir, tunnel, or surge tank with the turbine casing. In medium-head installations, each turbine usually has its own penstock. In the case of high heads, a single penstock is frequently used and branch connections provided at the lower end to supply two or more turbines.

*Penstock valves* located at the intake to the turbine spiral case are usually provided when the conduit is of considerable length. This permits shutting off the flow to each turbine, for safety and maintenance and to reduce leakage losses during long turbine shutdowns, without having to drain and refill a long conduit. Penstock valves are also a necessity where more than one turbine is connected to a single conduit so that the flow can be shut off to each turbine individually.

*Butterfly valves* are most commonly used as turbine inlet valves for low- and medium-head turbines. Butterfly-valve disks (the part which moves in the water passageway to shut off flow) are constructed with either a lenticular shape or a lattice-type design, which permits flow to pass through the disk, thereby reducing the head loss of the butterfly valve in the water passageway.

In the past, gate valves have been used for high heads in connection with impulse turbines. However, these have been entirely superseded by spherical or plug valves, which are also sometimes used for medium heads where the loss through the butterfly valve is considered to be excessive owing to the obstruction of the valve disk to the flow of water.

***6. Powerhouse Structure.*** The powerhouse foundation and superstructure (see Fig. 9-12) contain the hydraulic turbine, the generator, the governor system, water passages, draft tube, basements, passageways for access to the turbine casing and draft tube, and sometimes the penstock valve. The electrical apparatus is usually housed in the superstructure.

Transformers and oil circuit breakers are located within the superstructure, on the roof, or, frequently, on a deck built over the draft-tube extension. The transformers and switchgear are usually located outdoors adjacent to the powerhouse and are not an integral part of it.

Cranes are provided in the powerhouse to handle the heaviest pieces of the turbine and generator and sometimes extend over the penstock valves. Alternative powerhouse designs have included separate cranes for the penstock valves.

***7. Outdoor Powerhouse.*** One common powerhouse design is the outdoor type where the operating floor is placed adjacent to the turbine pits with the generator located outdoors on the roof of a one-story structure. All superstructure is omitted and a watertight removable cover placed over the generator. A gantry crane can be used for erection and servicing of the equipment. This outdoor design reduces overall costs of the power plant.

***8. Powerhouse Auxiliaries.*** The hydroelectric powerhouse requires some basic *auxiliaries* such as controls, switchboards, exciters, cranes, circuit breakers, and transformers. In addition, some special apparatus may consist of:

1. *Service units.* A small hydraulic turbine and generator used for supplying power for internal plant use and as a source of independent power supply in case the power plant is electrically separated from the main system
2. *Casing drain valves* for draining the turbine
3. *Strainers,* or *filters,* for bearing or cooling-water supply
4. *Air compressors* for charging governor oil systems, generator brakes, tail-water depression systems, etc.
5. *Carbon dioxide systems* for fire protection
6. *DC service* for emergency power supply.

## TYPES OF HYDRAULIC TURBINES

***9. The hydraulic turbine,*** sometimes referred to as a waterwheel, is the most important element in a hydroelectric power plant. There are two general groups of hydraulic turbines: (1) reaction (Fig. 9-2), where the water enters the turbine with high potential energy (in the form of

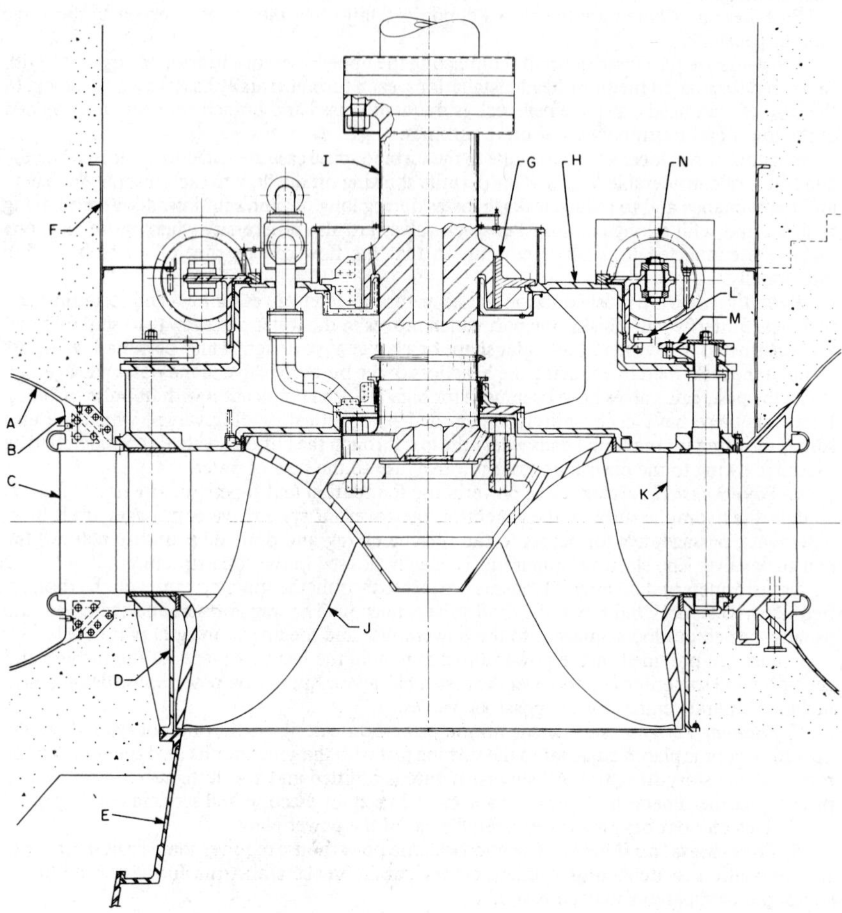


FIG. 9-2 Sectional elevation of a Francis reaction turbine. *(A)* Spiral case; *(B)* stay ring; *(C)* stay vane; *(D)* discharge ring; *(E)* draft-tube liner; *(F)* pit liner; *(G)* main-shaft bearing; *(H)* head cover; *(I)* main shaft; *(J)* runner; *(K)* wicket gates; *(L)* links; *(M)* gate levers; *(N)* servomotors.

pressure) and a lesser amount of kinetic energy (in the form of velocity); and (2) impulse, where the water enters the turbine with high kinetic energy and a relatively low value of potential energy (see Fig. 9-7). A further classification of reaction turbines is (1) Francis and (2) propeller, which can be further subdivided into:

1. Fixed-blade propeller turbines (Fig. 9-3)
2. Adjustable-blade propeller (Kaplan) turbines (Figs. 9-4 and 9-5)
3. Axial-flow propeller turbines (Tube, pit, or bulb)

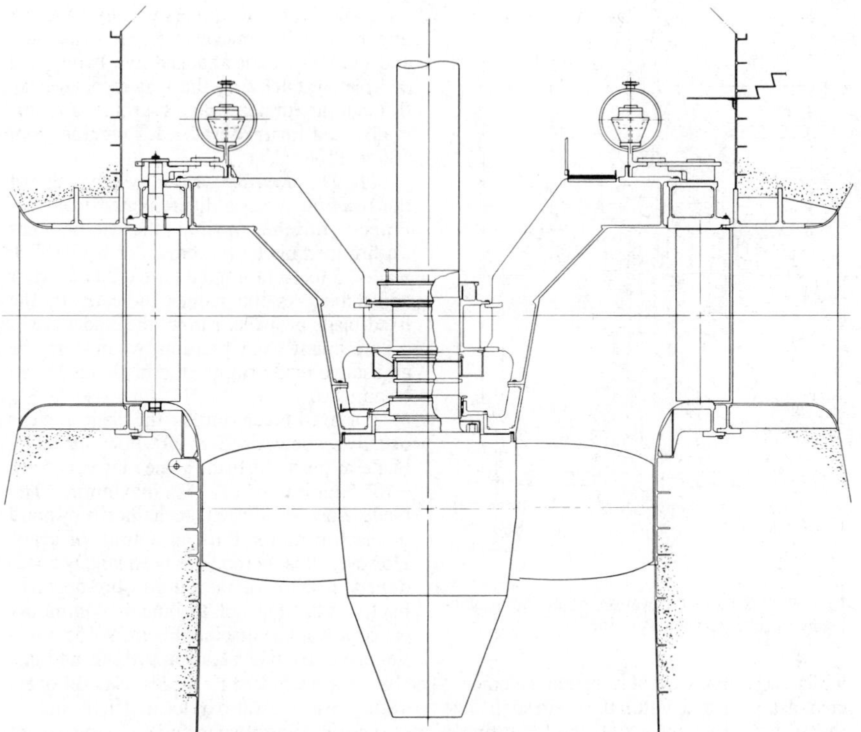

**FIG. 9-3** Vertical-shaft reaction turbine, fixed-blade-propeller type.

**4.** Diagonal-flow turbines (Fig. 9-6)

Reaction and impulse turbines have in common a stationary guide case and a revolving part, the runner. In the reaction turbine, the water enters the guide case of the turbine with high potential energy and relatively low kinetic energy. The potential energy, which is a function of the pressure difference between the runner inlet and exit, causes the fluid to flow through the runner buckets. As the fluid flows over the curved surface of the runner buckets, the fluid velocity on one side of the bucket is higher than on the opposite side. This difference in velocity on the surfaces of the buckets causes a pressure differential across the bucket which exerts a force on the bucket. This force at its respective radius in the runner then causes the runner to rotate and impart mechanical energy to the turbine shaft.

In the guide case (designated the nozzle pipe, needle nozzle, and nozzle tip) of the impulse turbine, the water enters the runner bucket with high kinetic energy (high velocity) and leaves with a relatively low kinetic energy (low velocity). The force on the bucket is a result of the impulse or momentum change of the water as its absolute velocity is reduced to near zero in the bucket. The impulse turbine thus utilizes the kinetic energy of the fluid entering the turbine to generate power.

***10. Reaction Turbines.*** Figure 9-2 shows a Francis-type inward-flow reaction turbine. Water enters the spiral case from intake passages or penstocks (see Figs. 9-9 and 9-12), passes through the stay ring, guided by the stationary stay-ring vanes, thence through the movable wicket gates through the runner and into the draft tube, through which it flows into the tailrace or tail-water reservoir. The movable wicket gates with axis parallel to the main shaft control the flow of water to the runner and thereby control the power output of the turbine.

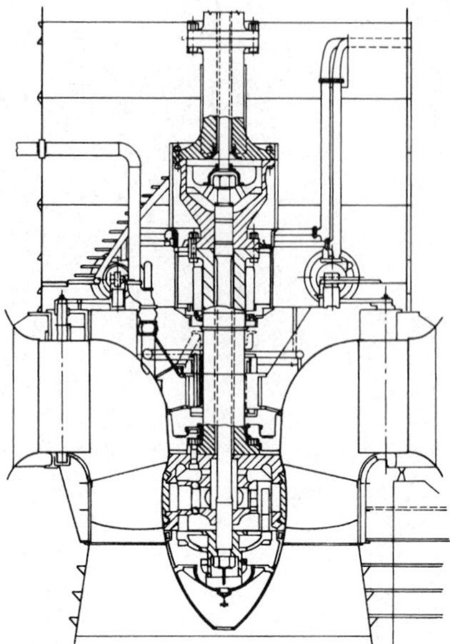

**FIG. 9-4** Sectional elevation of an adjustable-blade propeller (Kaplan) turbine.

Francis turbine runners usually have the upper ends of the buckets attached to a crown and the lower ends attached to a band, thus completely enclosing the water passageway through the runner. Francis turbines are normally used for medium heads ranging from 100 to 1500 ft.

***11. The propeller turbine,*** which is also of the reaction type, is differentiated from the Francis turbine in that the runner has unshrouded blades (no crown or band). The blades, 3 to 10 in number, are either fixed or adjustable. As the names indicate, in the fixed-blade propeller runner the blades are in a permanent fixed position, whereas in the adjustable-blade runner the blade angle can be adjusted.

For fixed-blade runners the blade angle is usually set between 20 to 28°. For adjustable-blade runners, the blade angle may vary from −10° minimum to 35° maximum. The blades may be adjusted mechanically by hand or electric motor through a train of gears. However, this method has been largely abandoned in favor of the oil-pressure-operated blades. This type of turbine is commonly called a Kaplan turbine. Figure 9-4 is a sectional elevation of a Kaplan turbine, and Fig. 9-5 is a sectional view of its rotating element. The blades are adjusted by means of an oil-operated piston located within the main shaft. The operating piston can also be located in the hub of the runner, either above or below the runner blades. The oil is admitted to and discharged from above and below the piston by means of an oil distributor located either on top of the generator shaft above the generator or surrounding the main shaft below the generator. The oil pressure is supplied from the governor oil-pressure system, and the flow of oil is controlled by the governor. The control has a cam so shaped and arranged that blade position will vary with the wicket-gate opening so as to produce a maximum-efficiency envelope curve (see Fig. 9-15). The greater the wicket-gate opening, the greater the angle of the blades and the greater the power output.

***12. Axial-flow turbines*** use the propeller-type runner with either fixed or adjustable blades. The axis of the blades is normally at right angles to the main shaft (Figs. 9-3 to 9-5). Another type of propeller runner, in which the axis of the blades is at approximately 45° with the main shaft, is known as the *diagonal-flow turbine.* The blades may be either fixed or adjustable. The axis of the wicket gates (either fixed or movable) are set at a 45° angle with the main shaft and the spiral case angled accordingly (Fig. 9-6). Some adjustable-blade diagonal-flow runners are so designed that the blades can be closed against one another to shut off the flow of water through the runner, thus eliminating the need for adjustable wicket gates for this purpose.

More recently, horizontal axial-flow turbines have been used in tidal and other hydro plants and are designed to operate either as pumps or turbines. Their principal characteristic feature is the straight-through or nearly straight-through water passageway from intake to draft-tube discharge. The shaft is therefore either slightly inclined (Fig. 9-10) or horizontal (Fig. 9-11).

The regular propeller turbines are used for low heads, ranging from the lowest head that is practical (one installation operates under a 7-ft head) to heads up to 200 ft, thus partly overlapping the range of heads for Francis turbines.

***13. Impulse Turbines.*** The impulse turbine in its modern form consists of one or more free jets of water discharging into an aerated space and impinging on a set of buckets attached around the periphery of a disk (see Figs. 9-7 and 9-13). The buckets vary in some details of their construction, but in general are bowl-shaped and have a central dividing wall, or splitter, extending radially outward from the shaft. This splitter divides the stream, and the bowl-shaped portions of the bucket turn the water back, imparting the full effect of the jet to the runner. The free jet is formed by the water passing through the nozzle pipe, the needle nozzle, and thence through the nozzle tip.

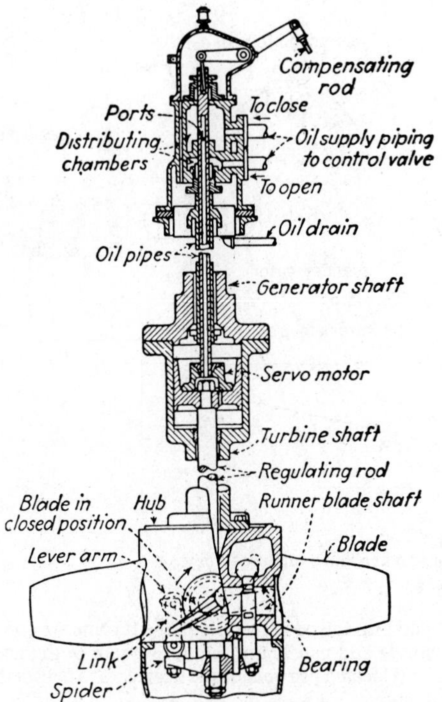


**FIG. 9-5** Sectional elevation of the runner and operating mechanism of an adjustable-blade propeller (Kaplan) turbine.

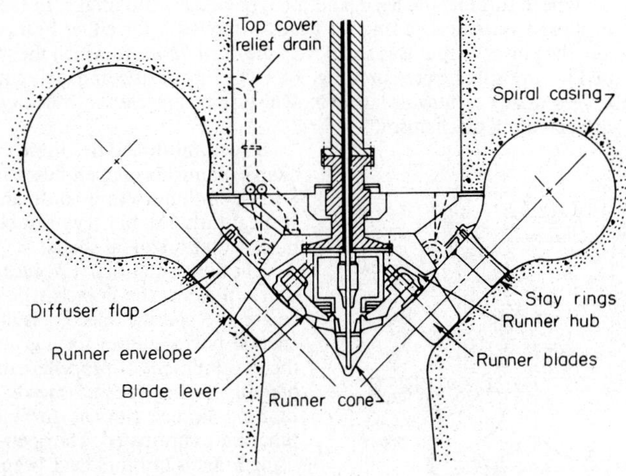


**FIG. 9-6** Sectional elevation of a diagonal-flow turbine.

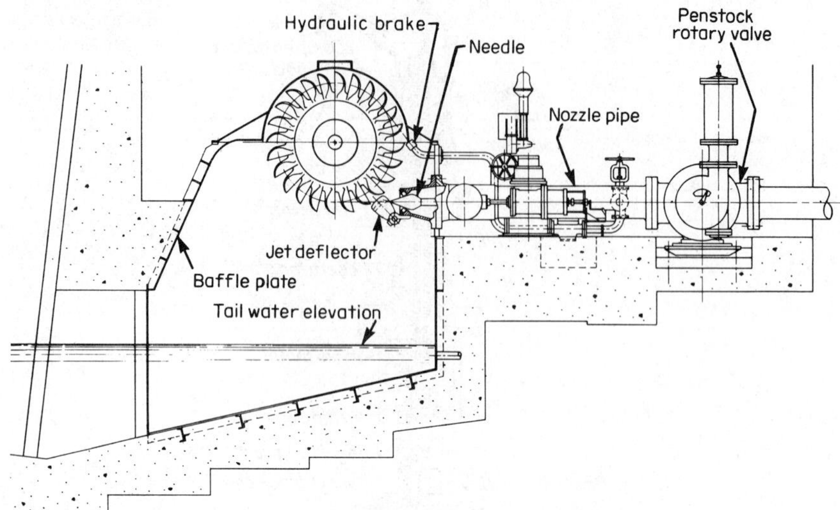


**FIG. 9-7** Section through a horizontal impulse turbine.

The size of the jet and thus the power output of the turbine are controlled by a needle in the center of the needle nozzle and needle tip. The movement of the needle is controlled by the governor. A jet deflector is located just outside the nozzle tip to deflect the jet from the buckets to effect sudden load reductions.

Impulse turbines are utilized when the head is too high for practical use of Francis turbines, which is normally any head exceeding 1000 to 1500 ft. They have been installed in projects having heads as high as 5800 ft.

## POWER-PLANT SETTINGS

***14. Plant Arrangement.*** The setting or arrangement of hydraulic turbines in a power plant varies with the type of turbine, the head, and the type of dam and intake. In the past the most common and most economical setting for heads below 40 ft for either Francis or propeller turbines, where the power output was small, was the *open flume,* in which the water has a free surface exposed to atmospheric pressure (Fig. 9-8). The turbine is completely submerged in an open chamber, essentially rectangular in form. One disadvantage is the difficulty of lubricating the wicket-gate-operating mechanism.

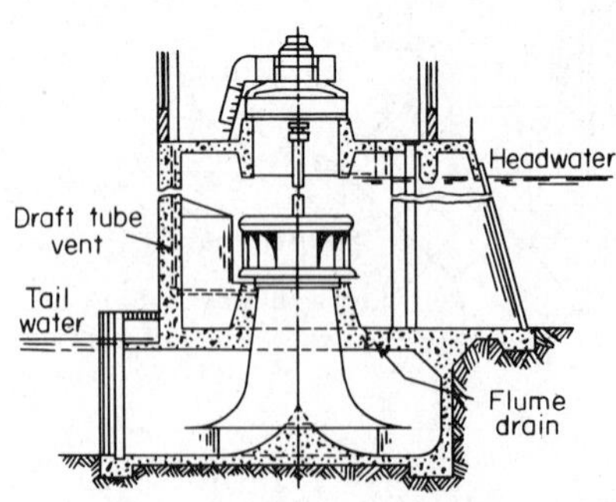


**FIG. 9-8** Vertical open-flume setting.

Structural difficulties limit the runner discharge diameters of open-flume turbines to 6 to 8 ft. Open-flume settings ordinarily are used for vertical turbines, but they can be used for turbines with horizontal shafts.

The horizontal-shaft turbines can use two or four runners of the Francis type in order to increase the specific speed, resulting in an increased power output for a given speed. With the development of propeller runners with inherently higher specific speeds, the need for multiple-Francis-runner turbines has been practically eliminated. The open-flume settings using Francis turbines have been largely superseded by axial-flow turbines having either fixed- or adjustable-blade runners.

The axial-flow turbines of large size with heads up to 150 ft usually have vertical settings (Fig. 9-9). Concrete semispiral cases are generally used with these units up to approximately 100 ft. For values greater than this, plate-steel spiral cases are used because of the higher velocities and the desire to minimize concrete erosion in the water passageways. In addition to the most common vertical arrangement of axial-flow turbines (Fig. 9-9), three other arrangements have been used: (1) tube, (2) pit, and (3) bulb.

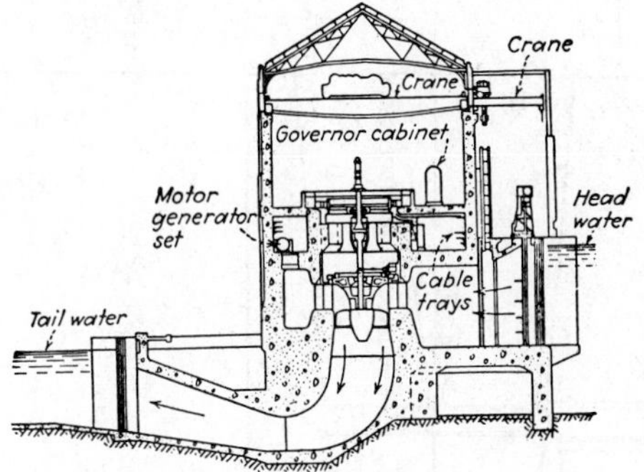


**FIG. 9-9** Cross section of a typical low-head concrete spiral-case setting with Kaplan turbine.

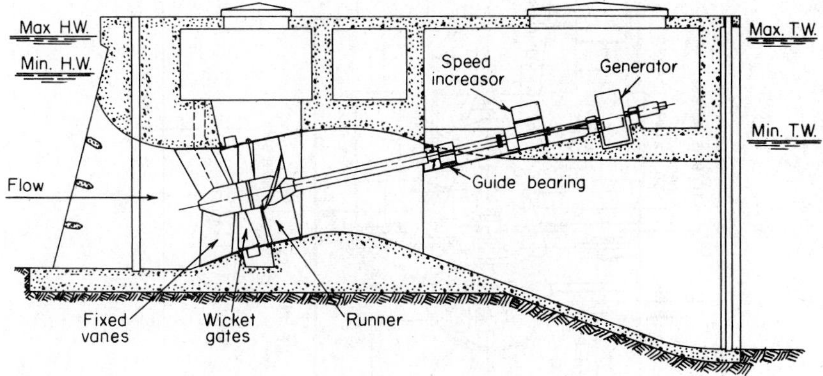


**FIG. 9-10** Sectional elevation of an axial-flow (tube) turbine.

The *tube turbine* (Fig. 9-10) has its generator located outside the water passages. With this type, a slight bend in the water passageway permits extending the turbine shaft externally. Although the unit can be arranged so that the generator is either upstream or downstream, the latter is more practical for large low-head units. To reduce excavation, the shaft may be inclined, thereby raising the generator higher with reference to tailwater elevations. Small-capacity tube turbines, because of their very low speed, are normally furnished with geared-type speed increasers to reduce the physical size and cost of the generator.

In the *pit-type turbine,* a watertight submerged enclosure is used to house the generator.

The third type is the *bulb turbine,* which has the generator mounted inside a steel bulb located within the powerhouse water passage (Fig. 9-11). The principal advantages of a horizontal shaft bulb unit, when compared to a vertical shaft propeller turbine with external generator mounted overhead, are as follows:

1. Flow through the turbine is more favorable, since the directional changes in the intake, semispiral casing, and draft tube are eliminated. As a result maximum turbine efficiencies are generally higher for the equivalent size unit.

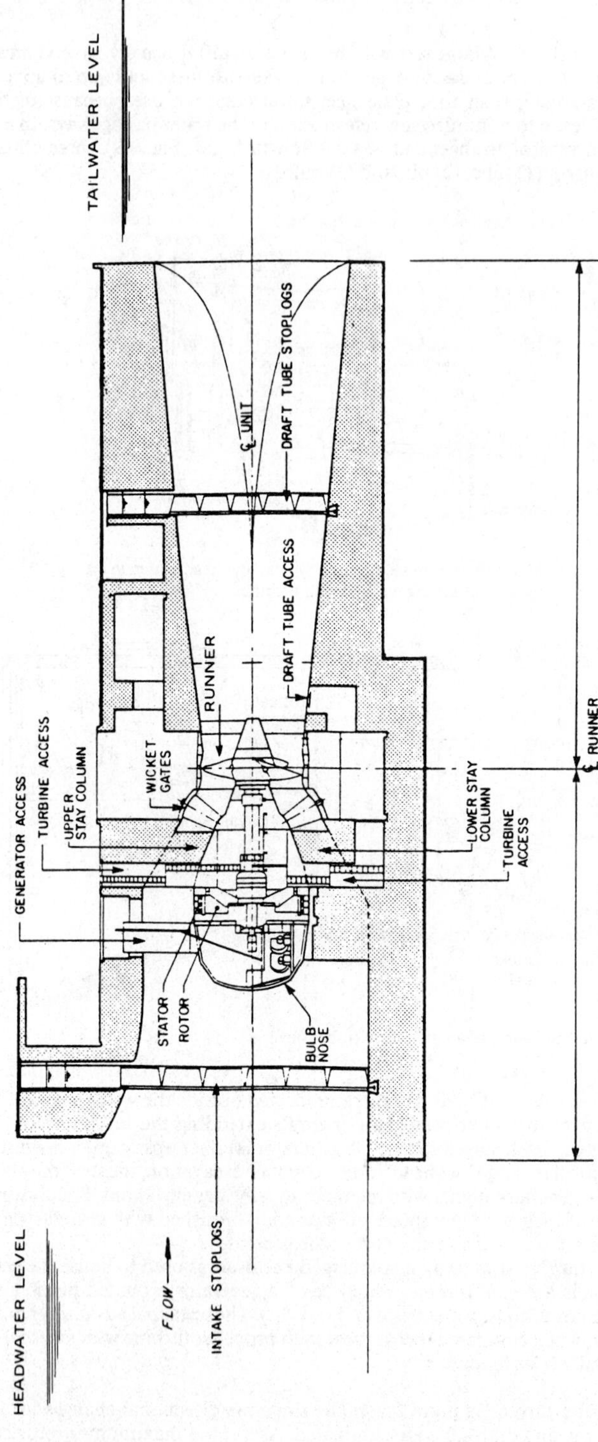


FIG. 9-11 Sectional elevation of an axial-flow (bulb) turbine.

2. The length, width, and height of the powerhouse structure can be reduced in most cases.
3. Normally, less excavation is required for the powerhouse.
4. Framework for the water passages is simpler and less costly.

Typically, bulb-turbine units have been built with high capacity and direct-driven generators to obtain additional flywheel effect, improved cooling, and higher efficiencies. This design (Fig. 9-11) generally requires two access shafts, one serving the upstream generator compartment, and the other serving the downstream turbine compartment.

Pit-, bulb-, and tube-type turbines are suitable for heads up to approximately 75 ft, with other limitations being basically the same as for the conventional vertical Kaplan or fixed-blade propeller turbines. Maximum unit capacity is limited, however, by maximum practical horizontal-generator capacities. Either fixed-position or movable radial wicket gates are used, depending on the range of load and head conditions, and the benefit-cost comparisons for the differences in equipment performance.

*15. Metal-spiral-case vertical settings* (Fig. 9-12) are most frequently used for reaction

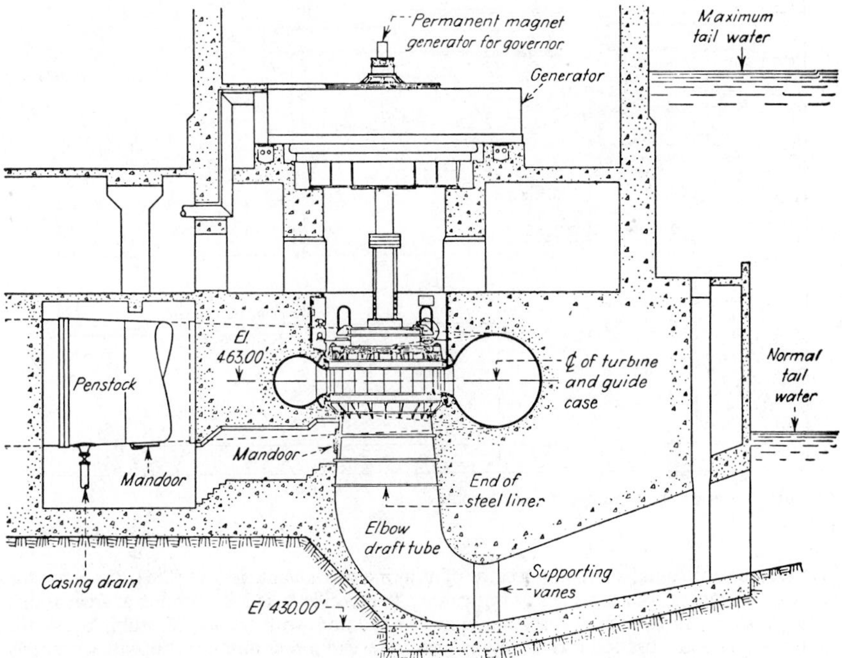


**FIG. 9-12** Plate steel spiral-case setting of vertical Francis turbine, welded casing.

turbines for heads above 100 ft up to an upper limit of about 2500 ft. Plate-steel welded construction is generally used for the cases, either partially shop-welded and completed in the field or entirely field-welded. Cast-steel cases are sometimes used for small turbines under high heads, where the forming of heavy plate steel would be difficult.

The metal spiral cases are usually directly connected to a steel penstock or penstock valve. Metal-spiral horizontal-shaft settings can be used for small turbines.

*16. The horizontal single-jet impulse-wheel* arrangement is shown in Fig. 9-7. To minimize the loss of available head, the lower edge of the buckets should be set as close to maximum tail water as possible, but not closer than 3 ft, to ensure that the runner revolves in air at all times.

Impulse turbines have relatively low specific speeds ($N_S = 3.5$ to 6.0), with resulting low unit speeds. This is overcome on horizontal units by the use of two runners (commonly known as the double-overhung unit) with the generator mounted between the two runners or by use of two jets per runner, thus doubling the output for a given size of turbine and increasing $N_S$ by $\sqrt{2}$ times, or about 41%. The double-overhung unit, or two jets per runner, requires a Y type of nozzle pipe to bring the water from the penstock to the needle nozzles.

***17. The vertical-shaft multijet high-capacity impulse turbine*** (Fig. 9-13) has generally superseded the horizontal-shaft unit because of its numerous advantages over the latter. Since the vertical unit can use four to six jets on one runner, the specific $N_S$ is 2 to 2½ times as great as that of the single-jet single-runner horizontal unit. In addition, with the use of multiple jets, the runner windage and friction loss are less as a percentage of power output. Also, with a properly designed housing for the vertical unit, there is less tendency for the discharged water to interfere with the runner. These features increase the efficiency over horizontal-unit performance 2 to 3%. Another advantage is that with multiple jets the unit can be operated at part load with a reduced number of jets, thus increasing part-load efficiency.

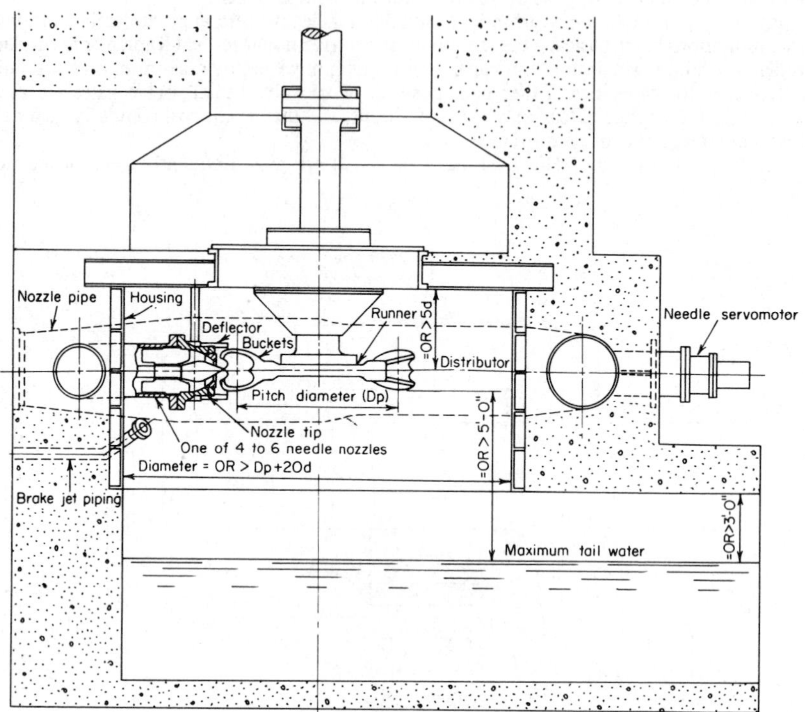


**FIG. 9-13** Vertical-shaft multijet impulse turbine.

Model tests have indicated that six jets are about the maximum number that can be used on one runner without their interfering with each other.

In vertical units the lower edge of the buckets should be at least 5 ft above maximum tail-water elevation. Any gain in head by setting the unit closer is more than offset by loss in performance. Impulse-turbine discharge entrains a large amount of air, and unless this is completely replaced, a vacuum will be produced if the outlet from the housing is sealed off by the tail water. The vacuum will then draw the tail water up until it drowns out the runner. Thus the roof of the discharge tunnel or passageway for both horizontal and vertical units should be at least 3 ft above the maximum operating tail-water elevation to permit the circulation of free air to the runner (Figs. 9-7 and 9-13). However, in both types of units where extremely high tail water is likely to occur for a short period of time, compressed air can be used to depress the water elevation, permitting the runner to be set closer to normal tail water.

## EFFICIENCY PERFORMANCE

***18. Efficiency Performance.*** In selecting the type of turbine for a given hydroelectric power plant, it is important to consider the efficiency performance of the various types available for the head contemplated. Not only is this true of the maximum efficiency obtainable, but the percent of full load where this maximum occurs and the efficiencies at part loads are also important.

Figure 9-14 shows the efficiency from 0 to 100% rated load of various specific-speed reaction

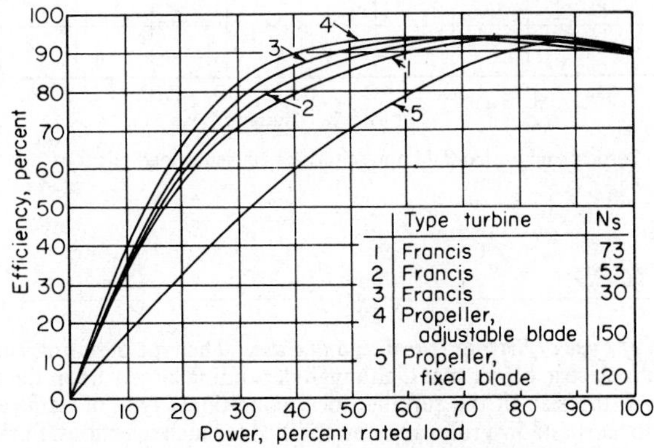


**FIG. 9-14** Efficiency-load relations for reaction turbines.

turbines, both Francis and propeller. As the specific speed increases, the percent of full load at which maximum efficiency occurs increases and part-load efficiencies drop. Although the maximum efficiency of the adjustable-blade propeller turbine is slightly less than the medium specific-speed Francis turbine, its part-load efficiencies are higher.

Figure 9-15 shows how the adjustable-blade turbine produces an envelope efficiency curve consisting of the maximum efficiencies of all the blade angles. Figure 9-15 also shows that the blade-tilt angle producing maximum efficiency on the envelope curve is at a lower tilt angle than the maximum for the normal adjustable-blade runner. Thus, since a fixed-blade runner usually has its blade-tilt angle set for maximum efficiency, its maximum output is about 15 to 25% less than that for an adjustable-blade runner of the same size.

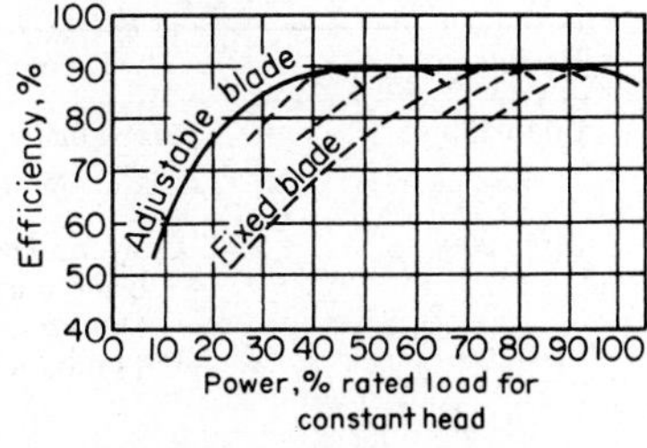


**FIG. 9-15** Efficiency-load relations for fixed- and adjustable-blade propeller turbines.

Figure 9-16 shows the efficiencies of a horizontal and a vertical impulse turbine. Although the maximum efficiency of both is somewhat lower than that for Francis-type turbines, the part-load efficiencies are higher, which is one of the advantages of impulse turbines. The efficiencies shown in Figs. 9-14 to 9-16 are all based on field tests of actual installations.

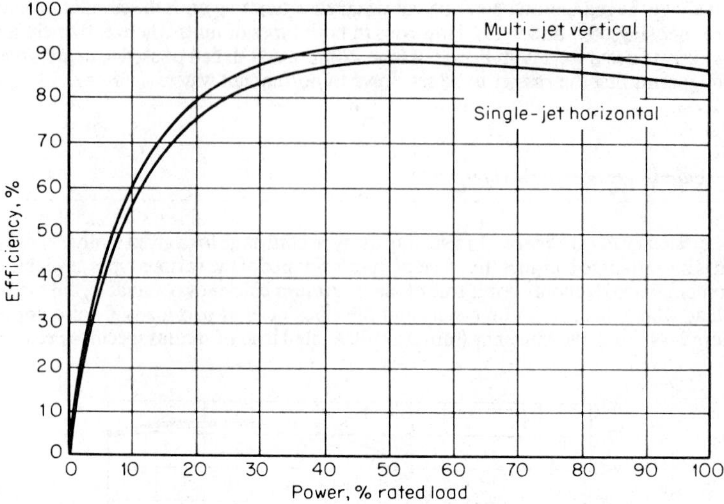


**FIG. 9-16** Vertical and horizontal impulse turbine efficiency load relations.

## DESIGN FACTORS

***19. Selection of Type of Setting, Turbine, and Case.*** The type of setting, turbine, and case selected for a hydroelectric power plant, although depending largely upon the available head, also depends upon the size of the turbine, local conditions, type of dam, and economics involved. Thus, no hard-and-fast rule can be established for such selections. Furthermore, there is considerable overlap in head where several types of turbines or cases can be used. Table 9-2 is therefore based upon general practice and should not be taken as limits in head above or below which the respective types cannot be used.

***20. Selection of Specific Speed and Unit Speed.*** The speed of the turbine should be as high as practical, as the higher the speed, the smaller the overall size of the turbine and the less costly.

**TABLE 9-2** General Arrangements of Turbine Installations and Usual Head Limits Employed

| Head range, ft | Type of turbine | General arrangement |
|---|---|---|
| Up to 150 | Fixed-blade propeller | Vertical with concrete semispiral or plate-steel spiral case |
| Up to 200 | Conventional adjustable-blade propeller and Deriaz | Vertical with concrete semispiral or plate-steel spiral case |
| Up to 75 | Pit, bulb, or tube | Horizontal or inclined, concrete and/or plate-steel intake |
| 100 to 1500 | Francis | Vertical or horizontal, plate-steel spiral case |
| 1000 to 5000 | Impulse | Vertical or horizontal plate-steel spiral case |

NOTE: 1 ft = 0.3048 m.

Also, since hydraulic turbines are usually connected to electric generators, the higher the speed, the less costly and more efficient are the generators.

The speed of the turbine is tied in with the specific speed $N_S$ which is a characteristic of each design of turbine runner (Par. **4**), which in turn varies with the head under which it will be used. In general, fixed-blade propeller-type runners are designed so that $N_S = 550/H^{1/3}$ and adjustable-blade propeller-type runners have an $N_S = 625/H^{1/3}$. For Francis-type runners, the value of $N_S$ may vary from $750/H^{1/2}$ to $950/H^{1/2}$. The adoption of these values of specific speed is predicated on the use of a reasonable setting of the unit with reference to the tail-water elevation to obtain the proper cavitation coefficient (Par. **42**). The head at which a runner of a given design and specific speed may be used may be increased by lowering the turbine in elevation with respect to the tail water, provided that the high-specific-speed runners can be made physically strong enough to operate under the higher heads.

For impulse turbines an $N_S$ per jet of 5.0 to 5.5 for heads around 1000 ft gives the best efficiency. For heads around 2000 to 3000 ft, $N_S$ for best efficiency should be 3.5 to 4.5 per jet. The specific speed of impulse units can be increased by increasing the number of jets used on a single runner or by increasing the number of runners per unit, since the specific speed of the unit is the specific speed per jet times the square root of the number of jets used. This indicates the advantage of the vertical impulse turbine, which can use up to six jets on one runner.

After having determined the specific speed for the type of turbine selected, the approximate speed of the turbine can be obtained from the equations $N_S = n\sqrt{P}/H^{5/4}$, $n = N_S H^{5/4}/\sqrt{P}$. Since hydraulic turbines are usually directly connected to alternating-current generators, the turbine speed must agree with one of the nearest synchronous speeds as determined from the system frequency. This speed is obtained from the equation $n = 120f/$(number poles in the generator), where $f$ is the system frequency (hertz). The number of poles must be an even number.

***21. Number of Units.*** From the standpoint of reducing the number of auxiliaries and the amount of associated equipment and also reducing initial and maintenance costs for the entire plant, the number of units should be a minimum. Also, the larger the unit, the higher is the efficiency. However, other considerations, such as flexibility of operation, higher-efficiency operation during low-load demands, and minimum loss of capacity during shutdown for repair or maintenance, might dictate the use of multiple units, where one unit would be feasible from the standpoint of physical size. For some projects using Francis or fixed-blade propeller runners, the physical size of the unit has been limited to the maximum size of runner that could be shipped in one piece, largely owing to the extra manufacturing costs involved in furnishing split runners. However, since split runners present no serious mechanical difficulties, the tendency in recent years has been to disregard this limitation. For the Bureau of Reclamation Grand Coulee Project, the large physical size units of 700-MW capacity required fabrication, stress relieving, and machining of the Francis runners at the job site instead of conventional shop manufacture.

***22. Runaway Speed.*** If a reaction-type turbine runner is allowed to revolve freely without load and with the wicket gates wide open, it will overspeed to a value called the runaway speed. The runaway speed of a turbine at normal head varies with the specific speed; for Francis turbines it ranges from about 155% (normal speed = 100%) at low specific speed ($N_S = 20$) to 195% at high specific speed ($N_S = 100$). For propeller turbines the runaway speed varies with blade angle; the steeper the blade angle, the lower is the runaway speed. For fixed-blade propellers with the blades set at 24 to 28°, the runaway speed will be about 220 to 190%, respectively. For adjustable-blade turbines, where the minimum blade angle is sometimes as low as 6° to −10° in order to obtain efficiency at low load, the maximum possible runaway speed is about 290%. However, with adjustable-blade propeller turbines there is, from the standpoint of efficiency, an optimum relationship between runner blade angle and wicket-gate opening, usually controlled by a cam in the operating mechanism; the higher the gate opening, the steeper the blade angle. Thus the combination of wide-open gate and minimum blade angle can occur only in the so-called off-cam position, which is an extremely rare possibility. In most adjustable-blade propeller units this maximum possible off-cam runaway speed is reduced by limiting the minimum blade angle to 14 to 16°.

The runaway speed for impulse turbines ranges from 180 to 190% of normal speed, depending upon the specific speed of the runner per jet. The higher the specific speed, the higher is the runaway speed.

For all turbines, the maximum head is normally higher than the rated head. Therefore the runaway speed will be increased in proportion to the square root of the head. For this reason, runaway speeds should be based on the maximum operating head rather than on the rated head.

Generators must be designed for the maximum runaway speed. The greater this speed, the greater is the cost of the generator.

*23. **Weight of Runner, Turbine Thrust, and WR².*** The approximate weight of any Francis runner is $0.030D^3$, where $D$ is the diameter of the runner in inches at the centerline of the distributor. For propeller-type runners with fixed blades the weight may be taken as $0.009D^3$ and for Kaplan-type runners as $0.014D^3$, where $D$ is the runner diameter in inches.

The hydraulic thrust on Francis runners varies considerably with type, design, specific speed, the pressure between the movable wicket gates and runner, seal design, seal clearance, and the method of venting. It is approximately between 25 and 40% of the weight of the full head of water acting on the discharge diameter $D_d$ of the runner. The higher the specific speed, the greater is the thrust. With propeller-type runners, the thrust is nearly equal to the weight of water on the full area.

Impulse turbines have no hydraulic thrust of any consequence.

The $WR^2$ of turbine runners, which is their weight times the square of the radius of gyration, varies widely with the type of runner and its design. Thus, to obtain reasonable values, general equations cannot be used, and the $WR^2$ must be calculated for each specific design, based upon its configuration and distribution of weight.

## REACTION-TURBINE ELEMENTS

*24. **Thrust Bearing.*** The weight of the rotating parts of the turbine and generator and the hydraulic thrust of a vertical unit are carried by a thrust bearing usually located just above or just below the generator rotor and furnished by the generator manufacturer. Occasionally this thrust bearing is located on the turbine head cover, in which case it is furnished by the turbine manufacturer. On horizontal units the weight of the rotating parts is carried by special heavy horizontal bearings.

*25. **Runner and Wearing Rings.*** The number of buckets for Francis runners varies from about 19 for low specific speed to 13 for high specific speed.

The number of blades for propeller-type turbines ranges from 10 for low specific speeds to 3 for high specific speeds.

Most runners are made of cast steel, which can readily be repaired by welding with either mild- or stainless-steel electrodes. Often built-up runners with cast-steel buckets welded to a cast-steel crown and band are used for Francis turbines. Bronze may be used for smaller Francis runners with low and medium operating heads. The use of cast stainless-steel runners is increasing for high heads and for conditions where pitting due to cavitation may be troublesome.

Propeller runners are practically always made of cast steel, and the surfaces over which pitting may be expected are overlaid with stainless-steel welding, before finishing to the final contour. In place of overlay welding, solid stainless-steel inserts are sometimes welded into the Francis buckets or propeller blades.

The functions of the runner seals for Francis turbines are to prevent excessive leakage loss and thus improve the efficiency, to reduce the hydraulic thrust, and to prevent seizure in operation.

Rolled, forged, and cast steels make excellent wearing rings for low- and medium-head units. To prevent seizure in case of contact, the rotating-ring material should differ from that of the stationary ring, with special care being taken to avoid having both the rotating and stationary rings contain nickel. Stainless steel, bronze, or steel with bronze inserts makes excellent wearing rings. For extremely high heads, stainless steel should be used to prevent undue wear and erosion. Wearing rings should be made renewable, or provision should be made for restoration of clearances by welding and remachining.

Seal clearances are made as small as practicable to reduce leakage, particularly on high-head units. Larger clearances are required with water-lubricated main bearings, subject to considerable wear before being readjusted.

*26. **Main Shaft and Bearing.*** The main shaft must be rigid and is made of a medium grade of forged steel, with torsional stress limited to 7000 lb/in² at maximum load.

For water-lubricated bearings a bronze or, preferably, stainless-steel sleeve is installed on the bearing surface. Such a sleeve is also put on the shaft adjacent to the packing box.

Turbines are usually provided with one main bearing located in the head cover as near the runner as practicable. The bearings are babbitted, either in halves or equipped with pivoted shoes, with an independent low-pressure oiling system. Self-lubricated babbitted bearings, containing an oil reservoir and pumping grooves in the babbitt, are also used.

Water-lubricated bearings of lignum vitae, rubber, or special composition materials are sometimes used, particularly for small and medium-sized propeller-type turbines where the bearing is located at the bottom of the head cover cone and where the packing box of an oil-lubricated bearing would be inaccessible and where it would be difficult to avoid water contamination. With the water-lubricated bearing the packing box is placed above the bearing.

***27. Spiral Case.*** The spiral case must be proportioned so as to cause relatively low friction losses, as well as to prevent eddying which would travel into the runner and affect its efficiency. The cases generally used are (1) metal case: cast steel or steel plate, and (2) concrete case. Metal cases are made as complete spirals, customarily with uniform or slightly increasing velocity from the throat to the small end. It is preferable that the water be accelerated as it approaches the case in order to suppress vortices.

Concrete cases are generally semispirals, rectangular or oval in cross section.

Rated-load velocities for general practice for cases of various types are shown in Fig. 9-17. Higher velocities are sometimes used with the larger units to reduce size and cost.

***28. Draft Tubes.*** The draft tube serves the double purpose of (1) allowing the turbine to be set above tail-water level, without loss of head, to facilitate inspection and maintenance, and (2) regaining, by diffuser action, the major portion of the kinetic energy delivered to it from the runner.

At rated load the velocity at the upstream end of the tube for modern units ranges from 24 to 30 ft/s. As the specific speed is increased and the head reduced, it becomes increasingly important to have an efficient draft tube. Good practice limits the velocity at the discharge end of the tube to 5 to 7 ft/s, representing less than 1-ft velocity head loss.

Two types of tubes are commonly used: (1) the straight conical or concentric tube and (2) the elbow type. Properly designed, the two types are about equally efficient, over 85%.

The *conical type* is used on low-powered units for all specific speeds. The side angle of flare ranges from 4 to 6°, the length from 3 to 4 $D_d$, and the discharge area from four to five times the throat area (see Fig. 9-8).

The *elbow type* of tube is used with most turbine installations. With this type, the vertical portion begins with a conical section which gradually flattens in the elbow section and then discharges horizontally through substantially rectangular sections to the tailrace. Most of the regain of energy takes place in the vertical portion, very little in the elbow section, which is shaped to deliver the water to the horizontal portion so that the regain may be efficiently completed. Figure 9-18 shows proportions of a good elbow tube, taken as the average proportions from a large number of recent installations. One or two vertical piers are placed in the horizontal portion of the tube, for structural and hydraulic reasons.

Small conical tubes are sometimes made entirely of steel plate. Most tubes are made of concrete with a steel-plate lining extending

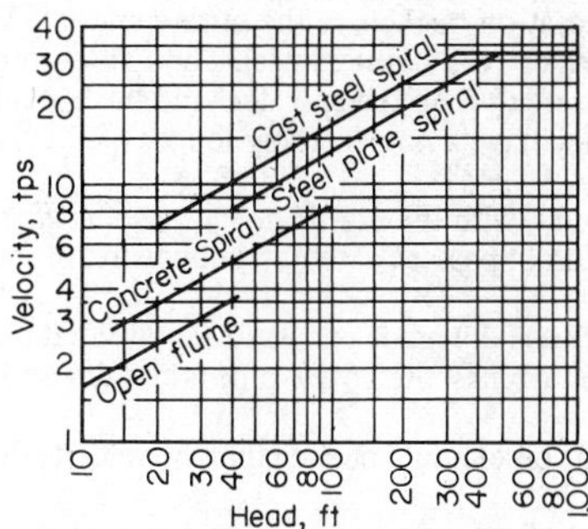


**FIG. 9-17** Case velocities, open flume and spiral.

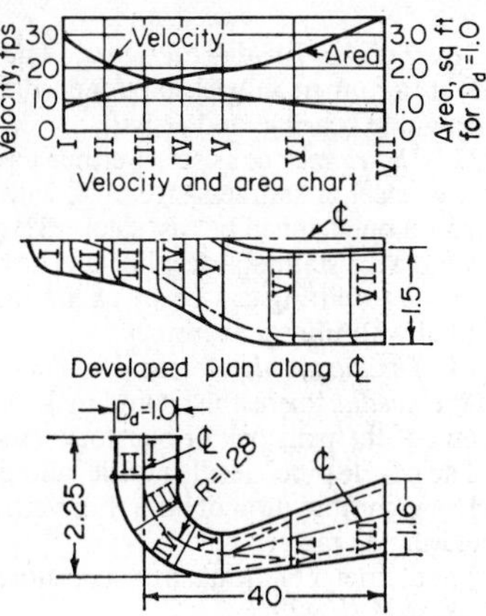


**FIG. 9-18** Elbow draft tube velocity, area, and layout.

from the upper end to a point where the velocity has been sufficiently reduced (say 20 ft/s) to prevent erosion of the concrete. The liner generally is carried around the elbow for low-specific-speed units. Pier noses are also lined where necessary to prevent erosion and for structural reasons.

***29. Stay Ring.*** That part of the guide apparatus between the spiral case and the wicket gates and containing stationary stay vanes is called the stay ring. The water is accelerated within this space as it approaches the gates. The number of stay vanes employed is usually equal to or one-half the number of gates. They are placed at the angle that will cause the least obstruction to the flow and in line with the wicket gates at the best gate position.

The stay ring is sometimes cast integral with cast-steel spiral cases and is almost always made separately of welded or cast steel. It should be very rigid since it serves as a foundation for the rest of the turbine.

***30. Wicket Gates and Operating Mechanism.*** Wicket gates control the power and speed of the turbine. The number of gates ranges from 16 to 28. The overall dimensions of the turbine decrease as the number of gates increases.

To prevent interference between the gates and the runner buckets, which may cause noise and vibration, the discharge tips of the fully open gates should be kept well away from the inlet edges of the runner buckets of Francis turbines, the radial clearance being large enough to prevent the gate tips from overhanging the curved part of the discharge ring.

The height and angular movement of the gates increase with the specific speed. The angular movement varies from 25° for low specific speed to 70° for high specific speed.

Wicket gates are generally made of cast steel for higher heads; for lower heads, the construction consists of weldments built from rolled materials and castings or forgings. To reduce wear, the gate tips and ends may be coated with stainless steel welded on before final finishing.

Each gate connection to the operating ring should be provided with a breaking element to protect the gate and other mechanism in case of an obstruction. Each gate should also be provided with stops to prevent it from striking the runner or reversing after the breaking element fails. In addition, gate restraining mechanisms are often furnished to prevent flailing of a gate after the breaking element fails.

One or more *vacuum breakers* or *air valves* are installed in the head cover to admit air to the runner or draft tube, to improve efficiency at low gate openings, or to alleviate draft-tube vortex cavitation. An air valve is also necessary on propeller-type units to break the vacuum under the head cover and to help prevent the backslap of a broken water column when the gates are suddenly closed. The air valve is piped to the outside of the powerhouse above the floodwater level.

## IMPULSE-TURBINE ELEMENTS

***31. Impulse-Turbine Elements.*** The essential elements of an impulse turbine (Figs. 9-7 and 9-13) are a runner, a nozzle pipe, a needle nozzle, a needle tip, a needle, a jet deflector, a housing with a guide bearing, and a shaft.

***32. The runner*** consists of either a plate-steel or forged-steel disk with bolted-on buckets or a cast-steel or stainless-steel disk with integrally cast-on buckets of the same material. The bolted-on buckets can be cast steel, 13% chrome steel, or 18-8 stainless steel. The 13% chrome-steel and 18-8 stainless-steel buckets have greater strength than the cast steel and have greater resistance to pitting caused by cavitation. Thus, they have a considerably longer life. This is essential with multijet turbines.

***33. The nozzle pipe*** forms the water passage which, with the *needle nozzle,* the *needle tip,* and the *needle,* forms the jet and includes that portion of passage extending downstream from the end of the penstock or penstock valve.

The nozzle pipe, needle nozzle, and needle tip are usually made of cast steel or welded plate steel, or a combination of both. For vertical impulse turbines, the nozzle pipe is circular and of decreasing diameter.

The diameter of the upstream portion of the nozzle pipe should be such that the velocity does not exceed $0.10\sqrt{2gH}$.

***34. The needle*** is a moving element inside the needle nozzle and needle tip, actuated by a governor, a servomotor, or a hand mechanism to control the size of jet impinging on the bucket, thus controlling the power output of the turbine.

The needle and its seat in the nozzle tip should be made of a material highly resistant to erosion.

***35. Jet Deflector.*** The inertia of the water flowing through the long penstocks usually employed with impulse turbines prohibits rapid reduction in velocity because of the pressure rise which would occur. Therefore, to minimize the speed rise following a sudden load rejection, it is necessary to reduce the hydraulic power delivered to the runner without changing the flow in the penstock too rapidly. Although pressure regulators (see Par. **52**) have been used for this purpose, they have been entirely superseded by placing a governor-controlled jet deflector between the needle tip and the runner. The governor moves this deflector rapidly into the jet, reducing the power output. It is not unusual for the deflector to cut off the entire jet in about 1½ s. Since the deflector acts on the jet after it leaves the nozzle, there is no change of flow in the penstock; hence, no pressure rise. The governor then moves the needle at a permissible rate (from the standpoint of pressure rise) with simultaneous automatic withdrawal of the deflector. The jet then is finally reduced the necessary amount to correspond to the reduced load. The needle must also move slowly in the opening direction for oncoming loads to avoid penstock collapse due to large pressure drops.

***36. The housing*** serves primarily to carry off the discharged water to the tail pit below and to support the nozzles. On horizontal-shaft units, at the place where the runner receives the impact from the jet, the housing width should be about 10 to 12 times the jet diameter. The more ample width results in higher efficiency. At the place where the runner has been cleared of the discharge, the housing should be as narrow as possible, to decrease windage. Suitable baffling should be used to carry the discharge away from the buckets. The housing should be adequately vented near the center of the runner, to permit the inflow of air to replace that entrained with the discharged water.

For vertical-shaft units, the housing should be of ample size, to prevent discharge water from interfering with the buckets. The distance from the head cover to the center line of the jet should be not less than five times the jet diameter (Fig. 9-13). The diameter of the housing should be not less than $D_p + 20d$, where $D_p$ is the pitch diameter and $d$ is the jet diameter. The housing should also be adequately vented near the center of the runner.

## REVERSIBLE PUMP/TURBINES

***37. Pumped Storage.*** Low-cost peaking capacity becomes increasingly important as peak loads grow, sizes of base-loaded thermal and nuclear units increase, and load demands become greater. The use of pumped-storage hydro facilities to meet peaking capacity has been growing at an increasing rate. In addition, pumped-storage facilities can contribute to the overall electrical system in performing one or more of the following functions:

**a.** Load regulation

**b.** Quick-response reserve capacity to offset short-term generation or transmission outages

**c.** Increase in overall system energy

Pumped-storage projects have also been constructed for multiple use with water-supply projects, flood control, and nuclear cooling capacity.

The concept of pumping to storage and later releasing the water to drive turbines is the true definition of pumped storage and applies regardless of the equipment used. Nevertheless, the term as commonly used implies the use of reversible, single-stage, vertical-shaft pump/turbine units of the Francis or adjustable-blade type (axial or diagonal flow). Although separate centrifugal pumps and hydraulic turbines of conventional design can and have been used for pumped-storage projects, single-stage pump/turbines are considerably less in initial cost.

Figure 9-19 is a cross section of a reversible pump/turbine. With its impeller-runner shaft connected to the generator motor, a spiral case to lead the water into the runner for generating power or directing it into the penstock when pumping, a reversible pump/turbine resembles a conventional Francis turbine. The draft tube for discharging water into the tailrace when generating power doubles as a suction tube for pumping the water from the tailrace and guiding it into the impeller runner.

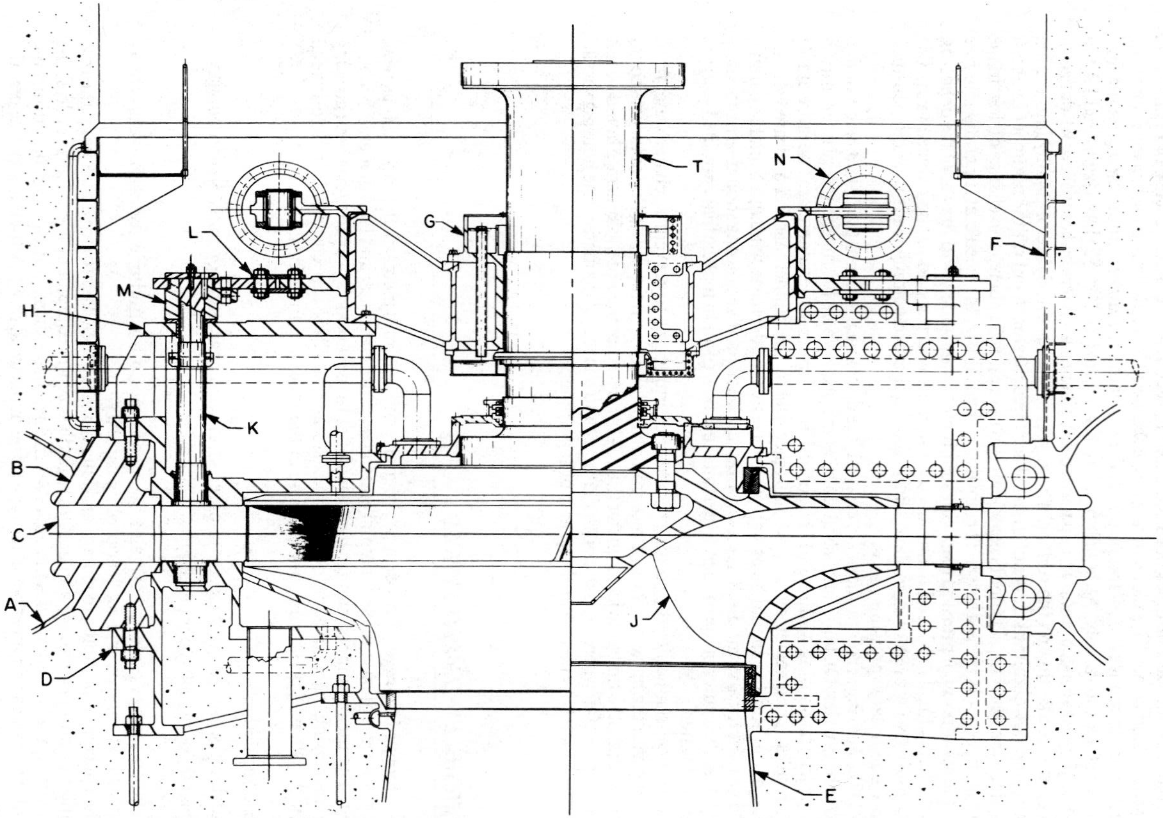


**FIG. 9-19** Sectional elevation of reversible pump turbine. *(A)* Spiral case; *(B)* stay ring; *(C)* stay vane; *(D)* combined bottom and discharge ring; *(E)* draft tube liner; *(F)* pit liner; *(G)* main-shaft bearing; *(H)* head cover; *(I)* main shaft; *(J)* impeller runner; *(K)* wicket gates; *(L)* links; *(M)* gate levers; *(N)* servomotors.

An impeller runner for a reversible pump/turbine is similar to a centrifugal-pump impeller but modified to produce optimum performance in both the pumping and generating direction. Conventional turbine runners, because of their short blades, are not well suited for pumping operating with reversible pump/turbines. An impeller runner for a reversible pump/turbine is usually larger in diameter than a Francis runner for the same generating capacity, and it has fewer blades at a wider spacing. A reversible pump/turbine, like a centrifugal pump, must have its impeller runner submerged in water to permit priming.

Most reversible pump/turbines have movable wicket gates, like conventional turbines, to control the flow of water for pumping or generating operation, and produce an optimum efficiency for both cycles of operation, as well as for use in starting and stopping the unit.

***38. Pump/Turbine Specific Speed.*** As in turbine runners, specific speed ($N_S$) is also a common basis of comparison of reversible pump/turbine impeller runners. References are used for either generating specific speed or pumping specific speed.

Generating specific speed for pump/turbines ($N_{ST}$) is derived in the same manner as that for conventional turbines. (See Par. **4**.)

Pumping specific speed for a pump/turbine ($N_{SP}$) is defined as the speed at which a geometrically similar impeller runner would run if it were sized to discharge one gallon per minute at one foot of head. For any pump/turbine, the computation of specific speed would be $N_{SP} = n\sqrt{Q}/H^{3/4}$ where $n$ is the shaft synchronous speed in revolutions per minute, $Q$ is the discharge in U.S. gallons per minute, and $H$ is the head in feet of water. The value of $N_{SP}$ for reversible pump/turbines may vary from $N_{SP} = 40{,}000/H$ to $N_{SP} = 70{,}000/H$, depending upon powerhouse excavation conditions, selection of synchronous speeds, and economic factors.

For generating specific speed ($N_{ST}$) the value varies from $N_{ST} = 800/H$ to $1200/H$.

A general trend with increased operating experience has been to gradually increase the specific speed from the low values listed above toward the maximum in order to take advantage of cost reductions associated with the reduced physical size of the equipment.

***39. Pump/Turbine Performance.*** The reversible pump/turbines have certain fundamental performance characteristics which are inherent in the design. The relationship between pumping and generating performance for a given specific speed is more or less fixed and can be modified only to a minor degree by alterations in the design. For example, if a certain generating capacity is desired, the pumping capacity will be fixed within certain limits. On the other hand, if a certain pumping capacity is desired, the maximum generating capacity is fixed. Figure 9-20 shows the expected generating performance, and Fig. 9-21 shows the expected pumping performance based on model tests of a reversible pump/turbine with a specific speed $N_S = 47.3$ generating and 2600 pumping.

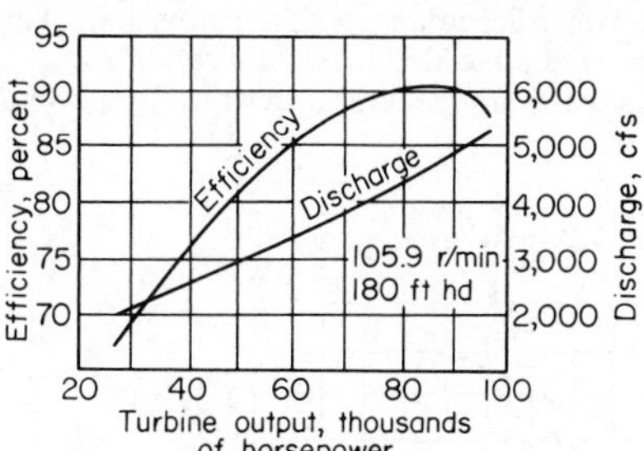


**FIG. 9-20** Reversible pump/turbine performance, generating.

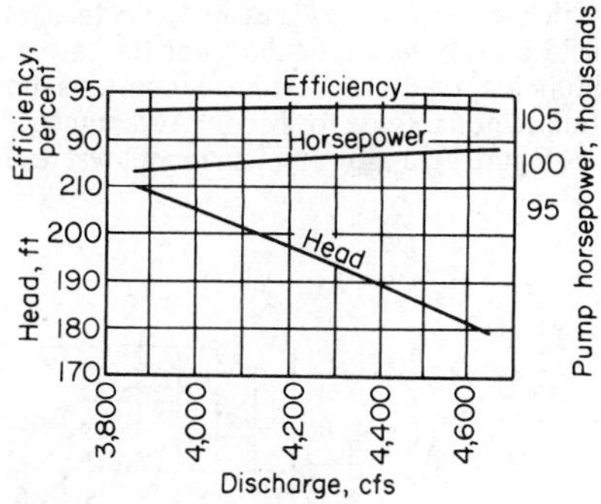


**FIG. 9-21** Reversible pump/turbine performance, pumping.

Because of the normal characteristics of reversible pump/turbines, the best-efficiency generating occurs at a lower speed than in pumping. The efficiency curve for the pumping cycle has a sharper peak than that for the turbine. It decreases sharply for higher pumping heads, and both efficiency and discharge become zero at the pump shutoff head. For this reason, a pumped-storage project having a wide head range makes it difficult to secure uniformly high performances over the full head range. There are instances where the head variation for a particular installa-

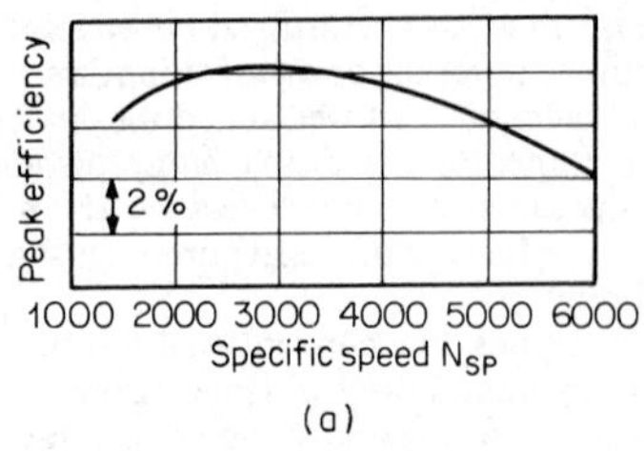


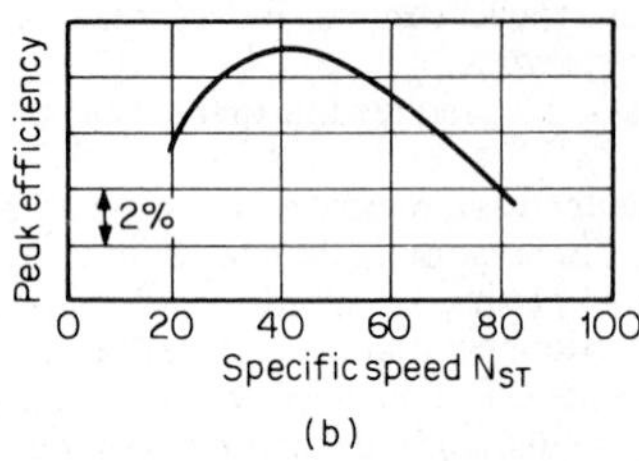


**FIG. 9-22** Specific speed of *(a)* pump and *(b)* turbine.

tion will make two-speed operation attractive. For example, in the case of large seasonal head variations, turbine efficiency and output will usually be considerably lower under low-head conditions than for normal-head operation. If, however, the revolutions per minute of the unit can be dropped to a speed giving best hydraulic performance for the lower head, overall efficiency will be appreciably improved. This improvement may be sufficient to justify the increased cost of a two-speed unit.

Figure 9-22 illustrates the envelope curve of peak efficiencies plotted on a relative basis versus specific speed for Francis-type pump/turbines.

Single-stage reversible pump/turbines can be built for heads up to 3000 ft. Beyond this, either multiple-stage reversible units or separate pumps and turbines should be used.

The runaway speed for pump/turbines is considerably lower than for conventional hydraulic turbines. It ranges from 150% of normal for low-specific-speed runners to 175% for high-specific-speed runners.

## MODEL TESTS

***40. Model tests*** serve several purposes. They are primarily used to check turbine runner, wicket gate, draft tube, casing, and sometimes inlet work designs for optimum performance. Correctly interpreted, they may also be used as a reliable indication of the performance of the units in the field. In many cases purchasers specify the performance of a homologous model test, which is then often used as an acceptance test of the unit instead of field tests. In such cases the field conditions, particularly for the casing and draft tube, must be reproduced in complete homology on the model unit. Model tests should be run in accordance with the provisions of the International Code for Model Acceptance Tests of Hydraulic Turbines, Publication 193.

Figure 9-23 shows typical model-test results of a reaction-type turbine in which the power *P*

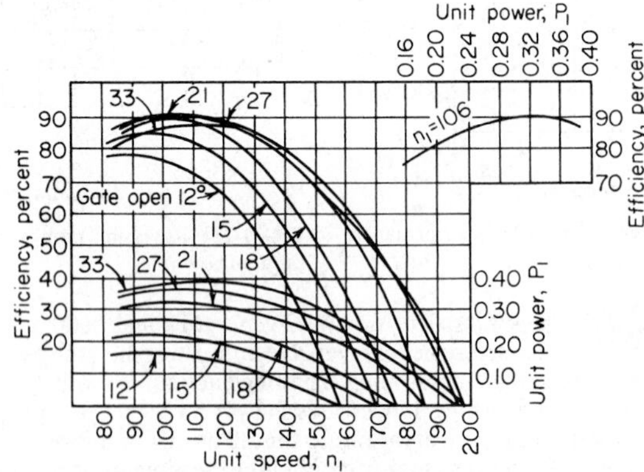


**FIG. 9-23** Model test curves for Francis runner; $N_S$ approximately 65.

at 1-ft head and the efficiency are each plotted against the speed $n$ at 1-ft head for each of several gate openings.

*41. **Laws of Proportionality for Homologous Turbines.*** The laws of proportionality shown in Table 9-1 are used to calculate, from model tests, the power, speed, and discharges of homologous turbines of various sizes for various heads. Actually the laws are employed only in computing the power and speed data for the field unit. The field unit will have a somewhat higher efficiency owing to proportionally smaller frictional and bearing losses. Expected field efficiency is customarily computed from the model efficiency by the Moody formula $e' = 1 - (1 - e)(D/D')^{1/5}$, in which the primed letters refer to the field installation and the other letters to model data. The step-up efficiency is computed for the point of best efficiency only, and the corresponding differential is applied as a constant value to all model efficiencies.

## CAVITATION

*42. **Cavitation*** occurs when the pressure at any point in the flowing water drops below the vapor pressure of the water.

The relationship among vapor pressure, barometric pressure, setting of the runner with respect to tail water, and net effective head on the turbine which produces cavitation is expressed by the *Thoma cavitation coefficient,*

$$\sigma = (H_b - H_v - H_s)/H \tag{9-4}$$

where $H_b$ is the barometric head, feet of water; $H_v$ is the vapor pressure of water, absolute; $H_s$ is the elevation, feet, of the runner above tail water, measured at the throat of a Francis runner and at the center line of the blades of a propeller runner (if the runner is submerged, $H_s$ becomes negative); and $H$ is the total or net effective head, in feet, on the turbine. From the above formula

$$H_s = H_b - H_v - \sigma H \tag{9-5}$$

Thus the setting of the runner depends upon the value of $\sigma$, which varies with specific speed $N_S$ of the runner and the individual characteristics of a particular runner design. In practice the model of the proposed runner is first tested with relatively high back pressure ($H_s$ small or negative). Then the back pressure is reduced in increments until the breaking points as indicated by a drop in power, efficiency, and discharge are reached. This breaking point is designated as the critical $\sigma$ and varies with gate opening and speed and for propeller turbines with blade angle. Consequently $\sigma$ must be determined for a range of limiting conditions.

In the absence of cavitation tests the value of $\sigma$ should not be lower than $\sigma = N_S^{2.6}/222{,}000$ for Francis runners, $\sigma = N_S^{1.4}/2180$ for propeller runners, and $\sigma = N_S^{1.85}/13{,}500$ for adjustable-blade propeller runners.

The value of $\sigma$ at which a plant operates, depending largely upon the setting of the runner with respect to tail water, is called the plant $\sigma$. To avoid excessive cavitation, the plant $\sigma$ should exceed the critical $\sigma$. The greater this margin is, the less the possibility of cavitation during operation. For a general discussion of cavitation phenomena see Knapp, Recent Investigations of the Mechanics of Cavitation and Cavitation Damage, *Trans. ASME,* October 1955. There is considerable difference in the resistance of various materials to cavitation erosion (pitting). Laboratory tests and experience have shown that materials having a high resistance and suitable for use in hydraulic turbines are the stainless steels and aluminum bronzes, especially when used as welding overlays (see Rheingans, Resistance of Various Materials to Cavitation Damage, *ASME Rep., Cavitation Symp.,* 1956).

## SPEED CONTROL

*43. **Speed control*** of a hydraulic turbine is provided by varying the flow of water through the turbine. This flow control is accomplished by changing the position of the wicket gates in the case of a reaction turbine and by the needle stroke and/or the deflector position in the case of an

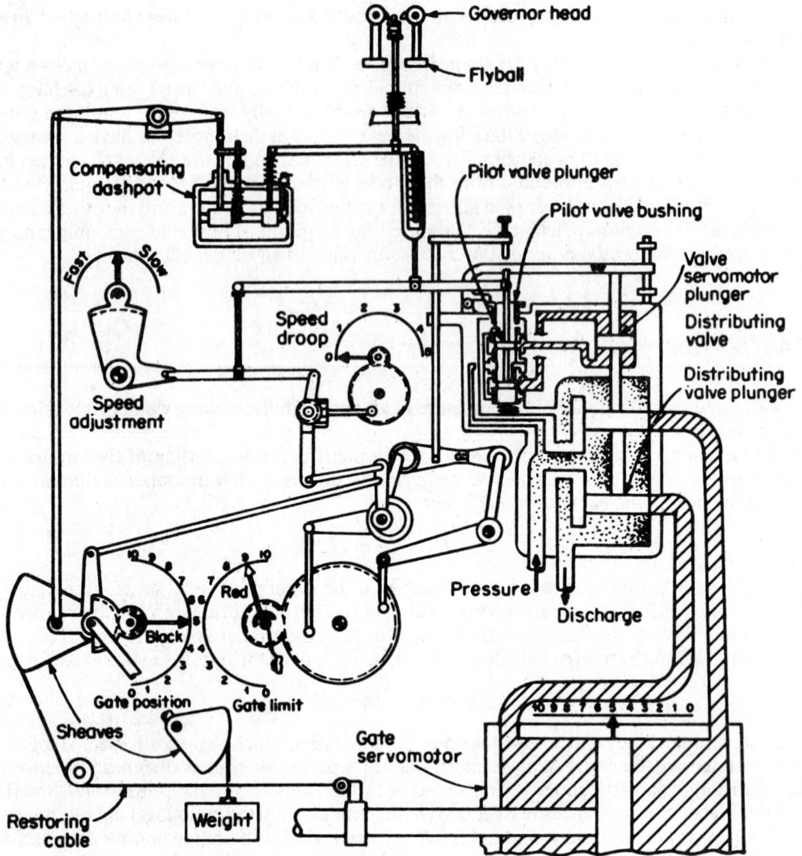


FIG. 9-24 Schematic diagram of mechanical hydraulic governor.

impulse turbine. The turbine wicket gates, needles, or deflectors are positioned by hydraulic servomotors which are controlled by the speed governor in response to changes in unit speed or governor speed setting.

***44. Mechanical Hydraulic Governor.*** A schematic of one type of mechanical hydraulic governor is shown in Fig. 9-24. The governor shown is made up of the following main components:

Flyball mechanism (ballhead)
Pilot-valve assembly
Distributing valve
Permanent feedback (droop)
Temporary feedback (dashpot)

Because of the stringent requirements of present-day utility systems, the ballhead is required to respond to system frequency changes of less than 0.01%. To achieve this extremely small dead band, the ballhead driving motor is supplied from a specially designed permanent magnet generator (PMG) which is directly driven from the generating-unit shaft.

The pilot valve system is used to amplify the output of the ballhead and to control the position of the distributing valve plunger. Oil to move the turbine servomotors is controlled by the distributing valve, which may vary in size over the range of 2- to 12-in pipe size.

In order to produce the permanent droop characteristic, which relates turbine wicket position to the corresponding speed error, an adjustable feedback mechanism is provided. The range of speed droop is generally taken to be 0 to 5%, although 0 to 10% is available.

For many types of prime movers, that is, diesel engines and steam turbines, the above described governor can provide a fast, stable speed control. The hydraulic turbine requires an additional feedback mechanism called a *compensating dashpot* to overcome the large delay caused by the inertia of the water column supplying the turbine. Values of droop from 30 to 100% are required to prevent the wicket gates from overtraveling, causing unstable speed control. Such high values of droop are undesirable because of the large speed variations caused by moderate changes in the generator loading. The dashpot mechanism shown provides the high range of droop required and is arranged so that the "temporary" droop will die out as unit speed returns to its normal steady-state value determined by the permanent droop for which the unit has been adjusted.

The transfer function of the governor shown in Fig. 9-24 is the following:

$$y = -\frac{1}{\sigma}\,\frac{(1 + T_rD)(x - c)}{\dfrac{T_rT_sD^2}{\sigma} + \dfrac{[T_s + T_r(\delta + \sigma)]D}{\sigma} + 1}$$

where $c$ = per unit speed reference
$D$ = derivative operator ($d/dt$)
$T_r$ = dashpot return time, s
$T_s$ = gate response time $= \beta t_g$
$\beta$ = speed error required to saturate distributing valve
$t_g$ = wicket-gate time for full stroke at maximum velocity, s
$\delta$ = temporary droop, per unit value
$\sigma$ = permanent droop, per unit value
$x$ = per unit speed
$y$ = per unit gate position

The normal range of values provided for the above parameters is

$$0.015 \leq \beta \leq 0.095$$
$$0.85 \leq c \leq 1.10$$
$$1 \leq T_r \leq 30$$
$$0 \leq \delta \leq 1.0$$
$$0 \leq \sigma \leq 0.10$$

To provide for faster governor response when used for load control on interconnected power systems, a two-level dashpot reset time ($T_r$) is available. The normal required value of $T_r$ may be reduced to some lower value by actuating a solenoid device which connects a second timing adjustment (needle valve) in parallel with the normal dashpot adjustment. This allows the dashpot to recenter faster and provide faster wicket-gate response. Care must be taken in using the faster-response capabilities of the governor as system stability may be affected.

***45. Electric Hydraulic Governors.*** Since the early 1960s, three-term, solid-state, electric hydraulic speed governors have been available for hydraulic turbines. In addition to the proportional and integral channels provided in the mechanical-hydraulic-governor design, the electric hydraulic provides a derivative channel. A schematic of the electric hydraulic governor is shown in Fig. 9-25.

A *speed-signal generator* is supplied with the governor and is mounted on, and directly driven from the shaft of, the unit whose speed is being controlled. The digital signal, proportional to unit speed, is converted to analog form in a hybrid digital-analog speed sensor and combined at this point with electrical signals representing desired speed setting and load feedback.

The proportional, integral, and derivative functions of the speed error are developed in parallel-connected channels as shown. Desired wicket-gate position is represented by the position of the electric-hydraulic transducer output servomotor.

The hydraulic amplifier comprises a pilot-valve and distributing-valve assembly with appro-

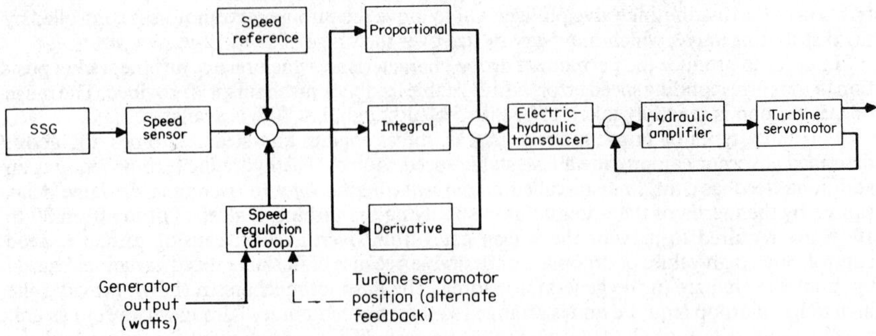


FIG. 9-25 Schematic diagram of electric-hydraulic governor.

priate feedback so that the turbine wicket-gate servomotor position matches that of the electric-hydraulic transducer output.

Unlike the mechanical hydraulic governor, which offers only wicket-gate servomotor feedback to produce a permanent speed droop characteristic, the electric hydraulic governor makes use of the generator output (watts) to produce a permanent speed-regulation characteristic.

The transfer function of the governor shown in Fig. 9-25 is

$$y = \frac{-(b_a + b_i/D + b_d D)(x - c + R_p)}{(0.15D + 1)^2(0.05D + 1)}$$

where $b_a$ = proportional gain
$b_d$ = derivative gain
$b_i$ = integral gain
$c$ = per unit speed reference
$D$ = derivative operator ($d/dt$)
$p$ = per unit power
$R$ = speed regulation
$x$ = per unit speed
$y$ = per unit gate position

Standard values of the governor gains are

| | Off line | On line |
|---|---|---|
| Proportional | $0.2 \leqslant b_a \leqslant 5.0$ | $0.5 \leqslant b_a \leqslant 18$ |
| Integral | $0.1 \leqslant b_i \leqslant 1.8$ | $0.3 \leqslant b_i \leqslant 7$ |
| Derivative | $0 \leqslant b_d \leqslant 1.8$ | $0 \leqslant b_d \leqslant 1.8$ |

The two ranges of gains shown are able to be switched electrically to provide suitable governor adjustments for synchronizing small-system operation or operation on a large interconnection.

The electric hydraulic governor has the following advantages over the conventional mechanical-hydraulic design.

1. Faster response to system speed changes and unit speed adjustment changes.
2. Wider range of stability adjustments.
3. Capability of accepting analog electric signals for load control.

4. Inherent provision for future update of electronics to match "state-of-the-art" advances in technology without replacing the long-life hydraulic power components.
5. The number of electric speed-sensitive relays is not restricted by available space in the speed-signal generator as for the mechanical-hydraulic type.

***46. Auxiliary Devices.*** A great many auxiliary devices have been made available for hydraulic turbine governors such as

1. Synchronizing motor for controlling the unit speed reference (speed adjustment from a control point remotely located).
2. Gate-limit device which will prevent the turbine wicket gates from opening beyond a predetermined maximum.
3. Gate-limit motor for control of gate-limit setting from a remote point.
4. Remote indication of gate-limit setting and wicket-gate position.
5. Automatic starting and shutdown. These functions are generally initiated by solenoid devices acting directly on the governor gate-limit mechanism.

In addition to the above devices which are generally applicable, a whole group of special governor auxiliary devices have been developed to meet the requirements of pumped-storage installations. Some examples are

1. Control of wicket-gate position as a function of plant head to *(a)* limit generator/motor input power or *(b)* achieve maximum pumping efficiency (pumping cycle)
2. Control of wicket-gate position to limit turbine torque to some desired value for control of generator/motor acceleration (used in back-to-back pump-starting sequence)

***47. Hydraulic Supply System.*** To meet the hydraulic power requirements of the governor, a supply system including an air/oil accumulator (pressure tank) and dual pumping units is normally furnished. The total pumping capacity $Q$ recommended is 0.25 times the volume rate use of oil by the governor; that is, $Q = 0.25\Sigma(V_s/t_g)$ where $V_s$ is servomotor volume and $t_g$ is servomotor travel time in seconds. For reasons of reliability, this capacity is split between two identical pumping units.

The pressure-tank volume, generally taken to be 20 servomotor volumes, will provide approximately 3 servomotor volumes of oil.

For further information on governors and auxiliary devices the reader is directed to the following publication: IEEE-125, Recommended Guide for Preparation of Equipment Specifications for Speed Governing of Hydraulic Turbines Intended to Drive Electric Generators.

***48. Speed-Control Requirements.*** Usually a sufficient measure of the regulation provided is the maximum speed rise resulting from sudden rejection of full load, as from the breaker tripping. A maximum speed rise of 35% of normal speed for this condition is a common limitation.

***49. Speed Rise Following Load Reduction.*** For sudden load reductions the approximate speed rise is

$$\frac{n_x}{n} = \left[1 + \frac{1{,}620{,}000 T_x P_x (1 + h/H)^{3/2}}{WR^2 n^2}\right]^{1/2} \tag{9-6}$$

where $n_x$ is the revolutions per minute at the end of time $T_x$; $n$ is the speed before the load decrease; $T_x$ is the time interval, seconds, for the governor to adjust the flow to the new load; $P_x$ is the reduction in load; $h$ is the head rise caused by the retardation of the flow; $H$ is the net effective head before the load change; $WR^2$ is the product of the revolving parts, pounds, and the square of their radius of gyration, feet. For values of $h$, see below. Very rapid gate closure produces a reduction of pressure in the draft tube and the possibility of breaking the water column, with subsequent violent resurge which may damage the turbine.

***50. Speed Drop Following Load Increase.*** For sudden load increases the approximate

$$\frac{n_x}{n} = \left[1 - \frac{1{,}620{,}000 T_x P_x}{WR^2 n^2 (1 - h/H)^{3/2}}\right]^{1/2} \tag{9-7}$$

where $P_x$ is the actual load increase and $h$ is the head drop caused by the increase of the flow. If the speed drop is to be determined for a given increase in gate opening, the governor time $T_x$ for making this increase and the normal change in load for the change in gate opening, under constant head $H$, can be used in the following formula:

$$\frac{n_x}{n} = \left[1 - \frac{1{,}620{,}000 T_x P_x (1 - h/H)^{3/2}}{WR^2 n^2}\right]^{1/2} \tag{9-8}$$

The actual change in load, however, will be $P_x(1 - h/H)^{3/2}$.

For derivation of the above speed-variation formulas and for a more accurate determination, see Strowger and Kerr, Speed Changes of Hydraulic Turbines for Sudden Changes of Load, *Trans. ASME,* 1926, and Rich, *Hydraulic Transients,* McGraw-Hill Book Company.

***51. Water Hammer in Penstocks.*** If a gate movement is considered as a series of instantaneous movements with a very small interval between each movement, the pressure variation in the penstock following the gate movement will be the effect of a series of pressure waves, each caused by one of the instantaneous small gate movements. For a steel penstock, the velocity of the pressure wave $\alpha = 4660/\sqrt{1 + (d/100t)}$, where $d$ is the penstock diameter, inches, and $t$ is the penstock wall thickness, inches. The pressure change at any point along the penstock at any time after the start of the gate movement may be calculated by summing up the effect of the individual pressure waves (see *ASME Symp. on Water Hammer,* 1933, and *Water Hammer Analysis* by John Parmakian, Prentice-Hall, Inc.).

Approximate formulas (De Sparre) for the increase in pressure $h$, feet, following gate closure are given below. They are quite accurate for pressure rises not exceeding 50% of the initial pressure, which includes most practical cases.

$$h = aV/g \qquad \text{for } K < 1 \text{ and } N < 1 \tag{9-9}$$

$$h = aV/g[N + K(N - 1)] \qquad \text{for } K < 1 \text{ and } N > 1 \tag{9-10}$$

$$h = aV/g(2N - K) \qquad \text{for } K > 1 \text{ and } N > 1 \tag{9-11}$$

where $K = aV/2gH$; $N = aT/2L$; $V$ and $H$ are the penstock velocity, feet per second, and head, feet, prior to closure; $L$ is the penstock length, feet; and $T$ is the time of gate closure. For full-load rejection, $T$ may be taken as 85% of the total gate traversing time to allow for nonuniform gate motion.

For pressure drop following a complete gate opening, the following formula (S. Logan Kerr) may be used with $T$ not less than $2L/a$:

$$h = \frac{aV}{g} \frac{-K + \sqrt{K^2 + N^2}}{N^2} = \text{pressure drop, ft} \tag{9-12}$$

Pressure variations exceeding 40% rise and about 25% drop should be avoided.

When control directly by the governor causes undesirable pressure variations, a surge tank, a pressure regulator, or a jet deflector may be used.

A surge tank is a standpipe with an atmospheric tank, attached to the penstock as close as possible to the casing inlet. The tank provides a reservoir and expansion chamber for the water demand or the water rejection following sudden gate movements, so that sudden accelerations or decelerations of the flow in the penstock are avoided.

***52. Pressure regulators*** may be either of the water-wasting or water-saving type. The *water-wasting type* is a synchronous bypass, generally attached to the turbine casing. It is operated directly from the governor, or the gate mechanism of the turbine, and wastes such an amount as to keep the total water discharge equal at all times to the full-load discharge of the turbine. The bypass is a needle nozzle, a cone valve, or a mushroom-shaped disk valve which opens and is partly balanced hydraulically by a piston under pipeline pressure.

The *water-saving type* permits the regulator to open upon rapid closure of the turbine gates, and then close slowly, so that the total water discharge is gradually reduced and finally limited to that through the turbine, adjusted for the new load.

## TURBINE TESTS

***53. Field testing of hydraulic turbines*** to determine the absolute efficiency and output involves careful and accurate measurement of the power available in the water supplied to the turbine (water horsepower) and the turbine output (developed horsepower). $e =$ (developed horsepower)/(water horsepower) $= 8.8P/QH$. The tests should be conducted in accordance with the provisions of the International Code for Field Acceptance Tests of Hydraulic Turbines and/or Storage Pumps, Publications 41 and 198.

Because of the difficulties and costs involved in making accurate measurements of horsepower, net head, and discharge in the field, there has been a trend in recent years to dispense with the field test, especially where a laboratory test on a homologous model turbine is available. Instead, an index test is made on the unit in the field, which measures the turbine output and relative discharge under various conditions. Index tests should be conducted in accordance with the ASME Test Code for Hydraulic Prime Movers.

## NOTATION

***54. Notation***

$D$ = diameter of runner, in
$D_d$ = diameter top of draft tube, in; discharge diameter of runner, in
$D_p$ = pitch diameter of impulse-turbine runner, in
$d$ = jet diameter of impulse turbine, in
$e$ = overall efficiency of turbine
$g$ = acceleration of gravity
$H$ = net effective head, ft
$h$ = head change due to load change, ft
$n$ = r/min
$n_1$ = r/min at 1-ft head $= n/\sqrt{H}$
$N_S$ = specific speed $= n\sqrt{P}/H^{5/4} = n_1\sqrt{P_1}$
$P$ = horsepower
$P_1$ = horsepower at 1-ft head $= P/H^{3/2}$
$Q$ = discharge, $ft^3/s$
$Q_1$ = discharge at 1-ft head $= Q/\sqrt{H}$, $ft^3/s$
$t$ = time, s
$V$ = absolute velocity of water, ft/s
$w$ = weight of water/$ft^3$
$\sigma$ = cavitation coefficient

## References

***55. References***

Daugherty, R. L.: *Hydraulic Turbines;* New York, McGraw-Hill Book Company, 1920.

Creager and Justin: *Hydroelectric Handbook;* New York, John Wiley & Sons, Inc.

Barrows, H. K.: *Water Power Engineering,* 3d ed.; New York, McGraw-Hill Book Company, 1943.

Daugherty, R. L., and Ingersoll: *Fluid Mechanics;* New York, McGraw-Hill Book Company.

Rheingans: Operating and Maintenance Experience with Pump-Turbines in the U.S., Italy, Japan and Brazil, *ASME J. Eng. Power,* July 1966.

Henry, L. F.: Selection of Reversible Pump/Turbine Specific Speeds; *Allis-Chalmers Corp. Bull.* 54R4586.

Detailed Hydrogovernor Representation for System Stability Studies, IEEE 69C2-PWR.

Recommended Guide for Preparation of Equipment Specifications for Speed-governing of Hydraulic Turbines Intended to Drive Electric Generators, IEEE 125.

Power Test Code for Speed-governing Systems for Hydraulic Turbine-Generating Units, ASME PTC29-1965.

International Code for Testing of Speed Governing Systems for Hydraulic Turbines, *Int. Electrotech. Comm. Publ.* 308.

Wylie, E. B. and Streeter, V. L.: *Fluid Transients;* New York, McGraw-Hill Book Company.

Chacour, S. A. and Deitz, R. E.: "Cora" Hydraulic Transient "Plus"; *Allis-Chalmers Corp. Bull.* 54P10462.

# SECTION 10
# POWER-SYSTEM COMPONENTS

**W.J. McNutt**
*Manager, Advance Development Engineering, Transformer Business Department, General Electric Co.*

**B. P. Boehle**
*Switchgear Division, Brown Boveri Company, Mannheim, Germany*

**W. J. Schmitt**
*Switchgear Division, Brown Boveri Company, Mannheim, Germany*

**Kelly A. Shaw**
*Manager of Marketing, Small Motor Division, Siemens-Allis Inc.*

**J. Lapp**
*Manager, Advanced Development, McGraw Edison Company; Senior Member, IEEE*

**J. R. Willy**
*Vice President, Engineering, Applied Systems, Inc.; Professional Engineer*

**J. H. Harlow**
*Manager, Development Engineering, Siemens-Allis, Inc.*

**Arthur D. Crino**
*Manager of Engineering, Power Switching Division, Siemens-Allis, Inc.*

**Donald L. Lott**
*Development Engineer, Power Switching Division, Siemens-Allis, Inc.*

**Ronald E. Mittelstaedt**
*Supervisor, Line Backer® Marketing, Power Switching Division, Siemens-Allis, Inc.*

**C. A. Popeck**
*Manager, Apparatus Engineering, A. B. Chance Company; Senior Member IEEE; Professional Engineer*

## CONTENTS

*Numbers refer to paragraphs*

# TRANSFORMERS

*By W. J. McNUTT*

## Transformer Theory

***1. Elementary theory*** given in Pars. **1** to **10** is developed from the viewpoint of a 3-phase three-leg concentric-cylindrical two-winding transformer, with the primary low-voltage winding next to the core and the secondary high-voltage winding outside the primary winding. This corresponds to a simple substation transformer or generator-step-up transformer of moderate kVA. Most of the information is also applicable to single-phase transformers with windings on two legs, 3-phase transformers with five-leg cores, transformers with the primary winding outside the secondary winding, three-winding transformers, etc.

***2. Sinusoidal voltage*** is induced in windings by sinusoidal variation of flux,

$$E = 4.44 \times 10^{-8} a_c B f N \tag{10-1}$$

where $a_c$ = square inches cross section of core, $B$ = lines per square inch peak flux density, $E$ = rms volts, $f$ = frequency in hertz, and $N$ = number of turns in winding.

The induced voltage in the primary (excited) winding approximately balances the applied voltage. The induced voltage in the secondary (loaded) winding approximately supplies the terminal voltage for the load.

*Voltage ratio* is the ratio of number of turns ("turn ratio") in the respective windings. The rated open-circuit (no-load) terminal voltages are proportional to the turns in the windings, but under load the primary voltage usually must be somewhat higher than the rated value if rated secondary voltage is to be maintained, because of regulation effects.

***3. Characteristics on Open Circuit.*** The core loss (no-load loss) of a power transformer may be obtained from an empirical design curve of watts per pound of core steel (Fig. 10-1). Such

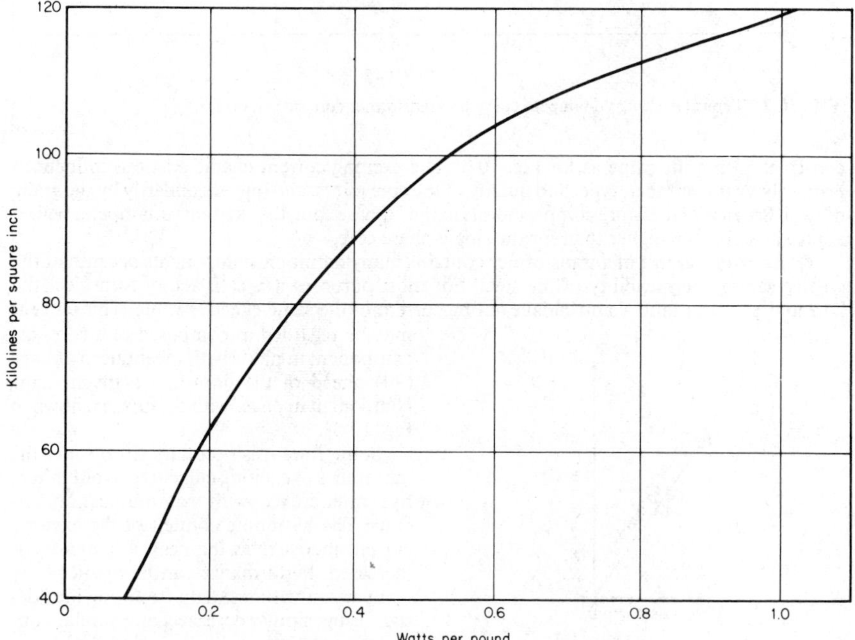


**FIG. 10-1** Typical core-loss curve for transformer core steel at 60 Hz.

curves are established by plotting data obtained from transformers of similar construction. The basic loss level is determined by the grade of core steel used and is further influenced by the number and type of joints employed in construction of the core. Figure 10-1 applies for 9-mil-thick, M-3 grade steel in a single-phase core with 45° mitered joints. Loss for the same grade of steel in a three-phase core would usually be 10 to 15% higher.

Exciting current for a power transformer may be established from a similar empirical curve of exciting voltamperes per pound of core steel as given in Fig. 10-2. The steel grade and core

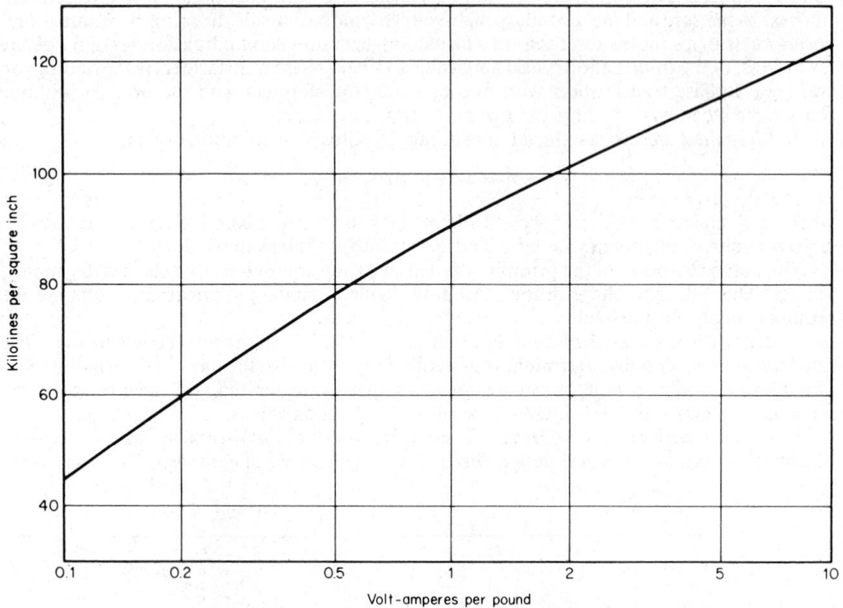


**FIG. 10-2** Typical exciting voltampere curve for transformer core steel at 60 Hz.

construction are the same as for Fig. 10-1. The exciting current characteristic is influenced primarily by the number, type, and quality of the core joints, and only secondarily by the grade of steel. Because of the more complex joints in the 3-phase core, the exciting voltamperes will be approximately 50% higher than for the single-phase core.

***4. Exciting current*** of a transformer contains many harmonic components because of the greatly varying permeability of the steel. For most purposes it is satisfactory to neglect the harmonics and assume a sinusoidal exciting current of the same effective value. This current may be regarded as composed of a core-loss component in phase with the induced voltage (90° ahead of the flux) and a magnetizing component in phase with the flux, as shown in Fig. 10-3.

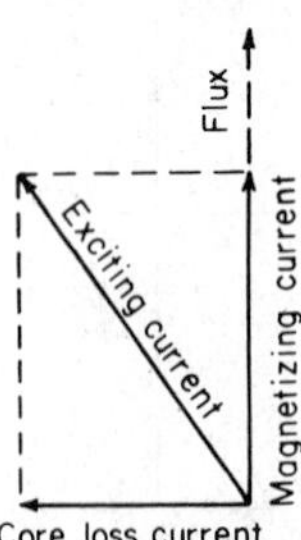


**FIG. 10-3** Phasor diagram of equivalent sinusoidal exciting current.

Sometimes it is necessary to consider the harmonics of exciting current to avoid inductive interference with communication circuits. The harmonic content of the exciting current increases as the peak flux density is increased. Performance can be predicted by comparison with test data from previous designs using similar core steel and similar construction.

The largest harmonic component of the

exciting current is the third. Higher-order harmonics are progressively smaller. For balanced 3-phase transformer banks the third-harmonic components (or multiples of the third) are displaced by 120 fundamental deg (or multiples of 120 fundamental deg) or 360 harmonic deg and therefore constitute a zero-phase-sequence system. Triple-harmonic currents may flow internally in delta-connected windings and externally in zero-phase-sequence paths in the connected system. The division of third-harmonic exciting current among available paths is not readily calculable.

***5. Magnetizing Inrush Current.*** If an idle transformer is energized at a time in the voltage cycle when the flux in the core would normally be other than the actual residual flux in the core, the sinusoidal flux curve will be initially offset, and the offset decreases gradually with time.[1] In extreme cases the peak flux may be more than doubled, exceeding saturation of the core, and causing peak magnetizing current several times rated load current. Magnetizing inrush current is important principally because of the possibility of false operation of transformer protective relays.

***6. Characteristics on Short Circuit.*** If the primary winding of a transformer with 1 : 1 turn ratio is excited with the secondary winding short-circuited, a small exciting current flows in the primary winding, producing mutual flux mostly in the core. In addition, a short-circuit current flows forward in the primary and reverses in the secondary, causing leakage flux which passes between the two windings and completes its path through the core. The mutual and leakage flux together make net flux linkages with the secondary to induce voltage to supply the resistance drop in the secondary and make net flux linkages with the primary to induce a countervoltage equal to the applied voltage less the resistance drop in the primary. Figure 10-4 shows the space

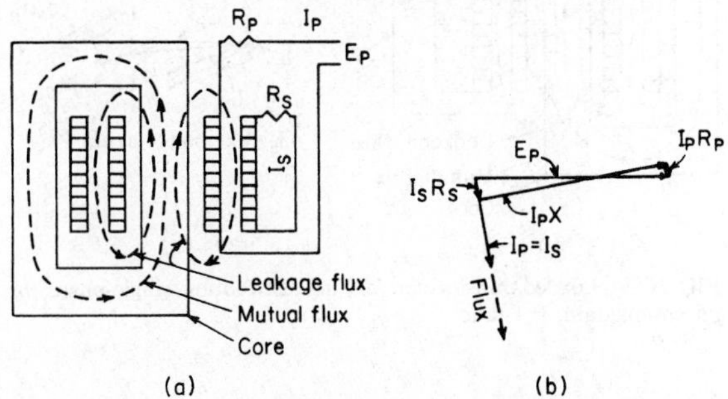


**FIG. 10-4** Short-circuited transformer. *(a)* Flux distribution, single-phase; *(b)* phasor diagram, 1 : 1 ratio.

relationships and the phase relationships neglecting the exciting current. It is apparent that

$$E_P = I_p(R_P + R_S + jX) = I_P Z \tag{10-2}$$

where $E_P$ = rms volts applied to primary (phasor), $I_P$ = rms amperes in primary (phasor), $R_P$ = ohms resistance of primary winding, $R_S$ = ohms resistance of secondary winding, $X$ = ohms reactance (corresponding to the voltage induced in the primary by the leakage flux), and $Z$ = ohms impedance $(R_P + R_S + jX)$.

***7. Resistance, Reactance, and Impedance.*** $R_P$ and $R_S$ are effective ac resistances. They are greater than the dc resistances as measured with direct current, because they include eddy loss in the conductor and stray loss in the core clamps, tank, etc. The reactance of the transformer is $X$, and the impedance is $Z = R_P + R_S + jX$.

***8. Load Loss.*** The loss on short-circuit test at rated current is the load loss at rated kVA,

$$L_L = I_R^2 R = I_R^2 Z_M \cos\theta \tag{10-3}$$

[1] T. R. Specht: Transformer Inrush Currents; *IEEE Trans. Power Appar. Syst., Paper* 68 TP 682-*PWR*, April 1969.

where $I_R$ = rms amperes rated current, $L_L$ = watts load loss at rated current, $R$ = ohms ac resistance $(R_P + R_S)$, $Z_M$ = ohms impedance magnitude $[(R^2 + X^2)^{1/2}]$, and $\theta$ = impedance angle of transformer.

The load loss at another current is

$$L = L_L I^2 / I_R^2 \tag{10-4}$$

where $I$ = rms amperes and $L$ = watts load loss.

***9. Characteristics under Load.*** Exciting current in the primary winding produces mutual flux mostly in the core. Opposing currents in the primary and secondary windings cause leakage flux which passes between the two windings and completes its path through the core. The magnitude and phase of the mutual flux depend on the voltage. The magnitude and phase of the leakage flux depend on the current. The mutual and leakage flux together generate in the primary a countervoltage equal to the applied voltage less the resistance drop in the primary and generate in the secondary a voltage equal to the terminal voltage plus the resistance drop in the secondary.

For most purposes the effect of the leakage flux can be represented by the effect of series reactance in the secondary-winding circuit. Figure 10-5 shows the space relationships and the

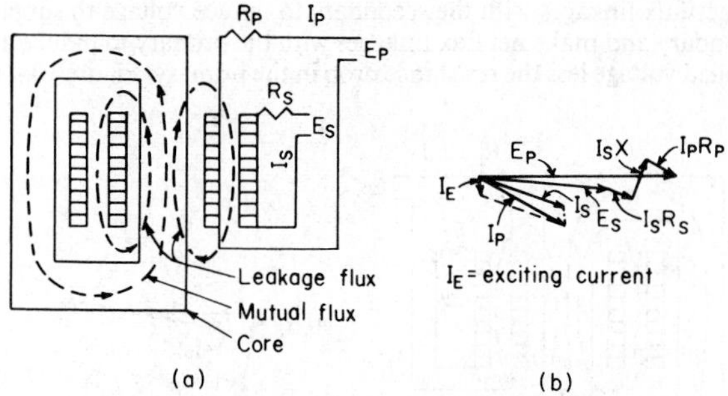


**FIG. 10-5** Loaded transformer. *(a)* Flux distribution, single-phase; *(b)* phasor diagram, 1 : 1 ratio.

phase relationships in a transformer of 1 : 1 ratio. It is apparent that

$$E_P = E_S + I_S(R_S + jX) + I_P R_P \tag{10-5}$$

where $E_P$ = rms volts at primary terminal (phasor), $E_S$ = rms volts at secondary terminal (phasor), $I_P$ = rms amperes in secondary (phasor), $I_S$ = rms amperes in secondary (phasor), $R_P$ = ohms resistance of primary winding, $R_S$ = ohms resistance of secondary winding, and $X$ = ohms reactance of transformer.

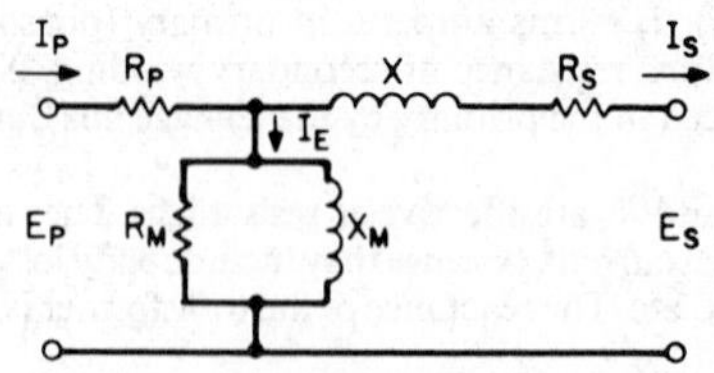


**FIG. 10-6** Equivalent circuit of a two-winding transformer considering exciting current.

***10. Equivalent Circuits.*** Figure 10-6 shows a circuit which for most practical purposes is equivalent to the transformer of Fig. 10-5. The exciting current, $I_E$, is made up of two components, a magnetizing component flowing through $X_M$ (the major component), and a loss component flowing through $R_M$. The values of $R_M$ and $X_M$ can be related to Figs. 10-1 and 10-2 if the core flux density at rated voltage is known. It will be found that these quantities vary with the voltage applied to the primary winding and they are usually determined for the rated voltage condition. For many purposes the exciting current can be neglected, and this leads to the simpler circuit of Fig. 10-7.

***11. Effect of Turn Ratio.*** Equation (10-5) and Fig. 10-7 represent a transformer of 1 : 1 turn

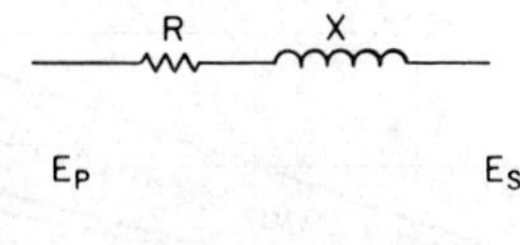


FIG. 10-7 Equivalent circuit of a two-winding transformer neglecting exciting current.

ratio. A transformer of turn ratio $T$ secondary to primary can be transformed into an equivalent 1 : 1 transformer by imagining the secondary winding replaced by a winding with the same number of turns as the primary winding, but using the same weight of conductor and occupying the same space as the secondary winding. $I_S$, $E_S$, and $R_S$ in the real secondary winding become $I_S T$, $E_S/T$, and $R_S/T^2$. The impedance of the load, $Z_L$, becomes $Z_L/T^2$. Thus, although Eqs. (10-2) to (10-5) and Fig. 10-4 to Fig. 10-7 were given for 1 : 1 turn ratio, they can be applied to any turn ratio. The fact that the simple series impedance of Fig. 10-7 may be used as equivalent to a transformer of any turn ratio is very helpful in the analysis of electric power systems. Secondary-winding characteristics corresponding to a fictitious secondary winding of 1 : 1 turn ratio are called secondary characteristics referred to the primary side. If more convenient, all characteristics can be referred to the secondary side by a reverse process.

*12. Percent and Per Unit.* Current, voltage, and kVA are frequently expressed as per unit or percent of rated value (25% = 0.25 per unit). The procedure is extended to resistance, reactance, and impedance by defining per unit impedance as (ohms impedance) × (rated current in amperes) ÷ (rated voltage in volts). Quantities expressed in percent or per unit are the same regardless of whether they are referred to the primary side or the secondary side.

*13. Regulation.* From Eq. (10-5) it is apparent that if the load current and the secondary voltage are at rated value, the primary voltage must exceed rated value. The excess is called regulation. Regulation in per unit is defined as the difference between primary and secondary voltage divided by secondary voltage. For rated load at lagging power factor and rated secondary voltage, regulation is given exactly by Eq. (10-6)[1] or approximately by Eq. (10-7).

$$G_r = [(R_r + P_r)^2 + (X_r + Q_r)^2]^{1/2} - 1 \qquad (10\text{-}6)$$

$$G_0 = 100\left[P_r R_r + Q_r X_r + \frac{(P_r X_r - Q_r R_r)^2}{2}\right] \qquad (10\text{-}7)$$

where $G_0$ = percent regulation, $G_r$ = per unit regulation, $P_r$ = per unit load power factor, $Q_r = (1 - P_r^2)^{1/2}$, $R_r$ = per unit resistance of transformer, and $X_r$ = per unit reactance of transformer.

The calculation of regulation of a three-winding transformer is considerably more complex, depending upon the load sharing between the two secondary windings. It will not be treated here.

*14. Impedance Data.* Resistance and reactance of transformers tend to follow normal patterns according to the ratings. Figure 10-8 shows resistance in percent (as determined by

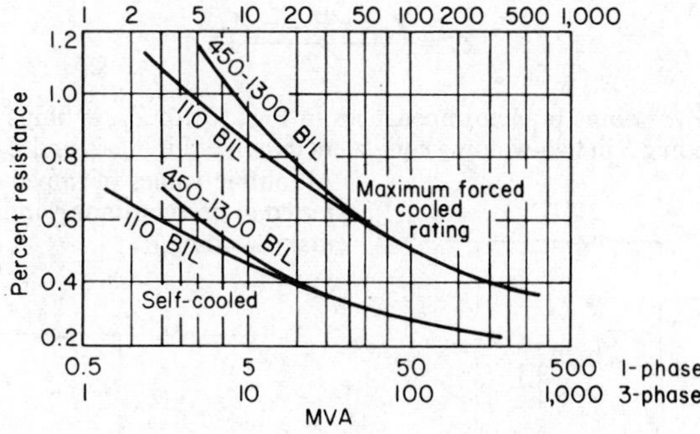


FIG. 10-8 Resistance of typical power transformers.

[1] ANSI/IEEE C57.12.90-1980.

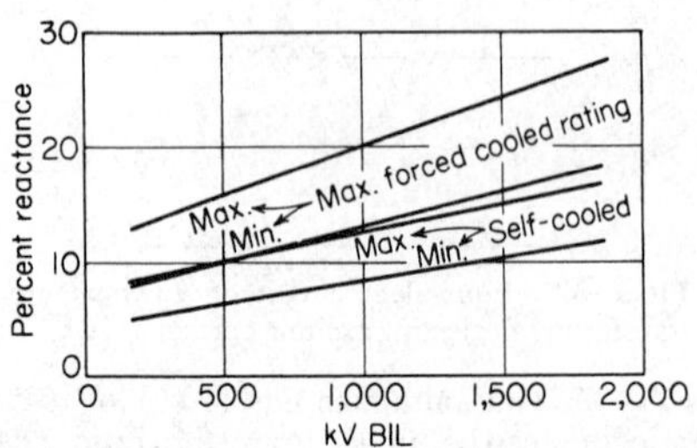


FIG. 10-9 Reactance of typical power transformers.

measurement of load loss on impedance test). Specific units may vary as much as ±30% depending largely on the evaluation of losses as compared with capital cost. Figure 10-9 shows ranges of reactance in percent. Special designs (transformers with all windings high-voltage, autotransformers, designs with overload ratings, etc.) may have reactances outside the limits shown.

***15. Efficiency*** is given by Eq. (10-8),

$$F_r = \frac{E_S I_S \cos\theta}{E_S I_S \cos\theta + L_{NS} + L_{LS}} \tag{10-8}$$

where $E_S$ = rms volts at secondary terminals, $F_r$ = per unit efficiency, $I_S$ = rms amperes in secondary, $L_{LS}$ = watts load loss at $I_S$, $L_{NS}$ = watts no-load loss at $E_S(I_S = 0)$, and $\theta$ = impedance angle of load.

***16. Three-Winding-Transformer Load Losses.*** The load losses of three-winding transformers, with all three windings carrying loads simultaneously, may be calculated from characteristics obtained by considering each pair of windings as a two-winding transformer.

$$L_T = \left(\frac{I_P}{I_A}\right)^2 \frac{L_{PS} + L_{PT} - L_{ST}}{2} + \left(\frac{I_{SP}}{I_A}\right)^2 \frac{L_{ST} + L_{PS} - L_{PT}}{2} + \left(\frac{I_{TP}}{I_A}\right)^2 \frac{L_{PT} + L_{ST} - L_{PS}}{2} \tag{10-9}$$

where $I_A$ = rms amperes reference current referred to winding $P$; $I_P$ = rms amperes in winding $P$; $I_{SP}$ = rms amperes in winding $S$ referred to winding $P$; $I_{TP}$ = rms amperes in winding $T$ referred to winding $P$; $L_{PS}$ = watts load loss in windings $P$ and $S$ as a two-winding transformer at $I_A$, $L_{PT}$, $L_{ST}$ = similar; and $L_T$ = watts total load loss.

The loss is usually computed at, or corrected to, a temperature of 75°C for 55°C average rise units and 85°C for 65°C average rise units.

***17. Three-Winding-Transformer Equivalent Circuit.*** The equivalent circuit of a three-winding transformer may be determined from the three impedances obtained by considering each pair of windings separately. One form is shown in Fig. 10-10, in which

$$Z_P = \frac{Z_{PS} + Z_{PT} - Z_{ST}}{2} \tag{10-10}$$

$$Z_S = \frac{Z_{PS} + Z_{ST} - Z_{PT}}{2} \tag{10-11}$$

$$Z_T = \frac{Z_{PT} + Z_{ST} - Z_{PS}}{2} \tag{10-12}$$

where $Z_P$, $Z_S$, $Z_T$ = ohms branch impedances in Fig. 10-10; $Z_{PS}$ = ohms impedance from winding $P$ to winding $S$ in two-winding equivalent circuit of Fig. 10-7; and $Z_{PT}$, $Z_{ST}$ = similar. All ohmic values of impedance must be referred to one common winding. (i.e., the primary winding.)

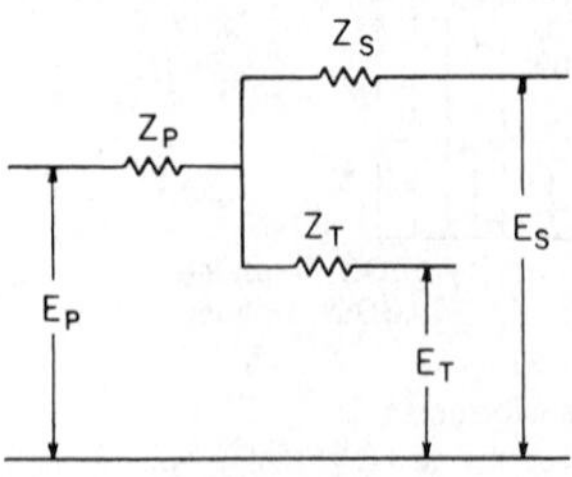


FIG. 10-10 Equivalent circuit of a three-winding transformer.

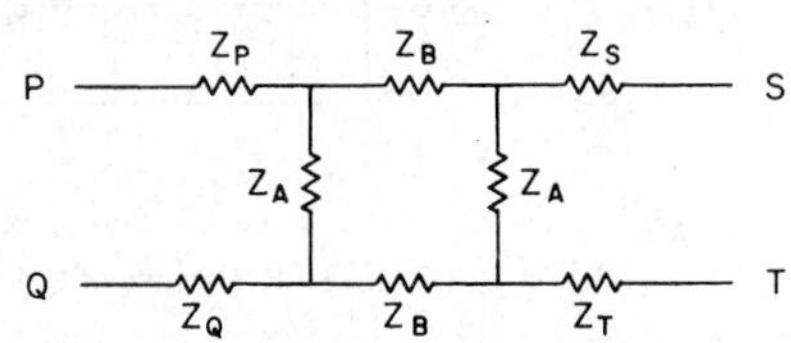


FIG. 10-11 Equivalent circuit of a four-winding transformer.

***18. Four-Winding-Transformer Equivalent Circuit.*** The equivalent circuit of a four-winding transformer may be determined from the six impedances obtained by considering each pair of windings separately. One form is shown in Fig. 10-11, in which

$$Z_P = \frac{Z_{PQ} + Z_{PS} - Z_{SQ}}{2} - \frac{Z_A Z_B}{2(Z_A + Z_B)} \tag{10-13}$$

$$Z_S = \frac{Z_{PS} + Z_{ST} - Z_{PT}}{2} - \frac{Z_A Z_B}{2(Z_A + Z_B)} \tag{10-14}$$

$$Z_T = \frac{Z_{ST} + Z_{TQ} - Z_{SQ}}{2} - \frac{Z_A Z_B}{2(Z_A + Z_B)} \tag{10-15}$$

$$Z_Q = \frac{Z_{TQ} + Z_{PQ} - Z_{PT}}{2} - \frac{Z_A Z_B}{2(Z_A + Z_B)} \tag{10-16}$$

$$Z_A = (K_1 K_2)^{1/2} + K_1 \tag{10-17}$$

$$Z_B = (K_1 K_2)^{1/2} + K_2$$

$$K_1 = Z_{PT} + Z_{SQ} - Z_{PS} - Z_{TQ} \tag{10-18}$$

$$K_2 = Z_{PT} + Z_{SQ} - Z_{PQ} - Z_{ST}$$

where $Z_A$ = ohms branch impedance in Fig. 10-11(complex); $Z_B$, $Z_p$, $Z_S$, $Z_T$, $Z_Q$ = similar; $Z_{PS}$ = ohms impedance (complex) from winding $P$ to winding $S$ in two-winding equivalent circuit of Fig. 10-7; and $Z_{PT}$, $Z_{PQ}$, $Z_{ST}$, $Z_{SQ}$, $Z_{TQ}$ = similar.

***19. Phase-interconnected transformers*** (i.e., with windings from more than one phase on a single core leg) can be represented by an equivalent circuit only if each winding on a leg is considered as if it were brought out to separate terminals.[1]

## Connections

***20. Parallel Operation.*** Two single-phase transformers will operate in parallel if they are connected with the same polarity. Two 3-phase transformers will operate in parallel if they have the same winding arrangement (for example, Y-delta), are connected with the same polarity, and have the same phase rotation. If two transformers (or two banks of transformers) have the same voltage ratings, the same turn ratios, the same impedances (in percent), and the same ratios of reactance to resistance, they will divide the load current in proportion to their kVa ratings, with no phase difference between the currents in the two transformers. If any of the above conditions are not met, the load current may not divide between the two transformers in proportion to their kVA ratings and there may be a phase difference between currents in the two transformers.

***21. Two unlike transformers connected in parallel*** will supply current to a load as follows:

$$I_L = \frac{E_P}{\dfrac{1}{(T_1/Z_1) + (T_2/Z_2)} + Z_L \dfrac{(1/Z_1) + (1/Z_2)}{(T_1/Z_1) + (T_2/Z_2)}} \tag{10-19}$$

where $E_P$ = rms volts on primary side (phasor), $I_L$ = rms amperes total load current (phasor), $T_1$ = turn ratio secondary to primary of unit 1, $T_2$ = turn ratio secondary to primary of unit 2, $Z_1$ = ohms impedance of unit 1 referred to secondary side (complex), $Z_2$ = ohms impedance of unit 2 referred to secondary side (complex), and $Z_L$ = ohms impedance of load (complex).

The magnitude of the current in unit 1 is

$$I_{r1} = \frac{\{[T_1 R_{r2} I_{rL} + (T_1 - T_2) E_{r1} \cos\theta]^2 + [T_1 X_{R2} I_{rL} + (T_1 - T_2) E_{r1} \sin\theta]^2\}^{1/2}}{[(T_1 R_{r2} + T_2 R_{r1})^2 + (T_1 X_{r2} + T_2 X_{r1})^2]^{1/2}} \tag{10-20}$$

[1] B. A. Cogbill, Sequence Impedance of Symmetrical Three Phase Transformer Connections; *Trans. AIEE Paper* 55-671, 1955.

where $E_{r1}$ = rms voltage of secondary terminals in per unit of unit 1, $I_{rL}$ = rms total load current in per unit of unit 1, $I_{r1}$ = rms current in secondary of unit 1 in per unit of unit 1, $T_1$ = ratio secondary turns to primary turns in unit 1, $T_2$ = ratio secondary turns to primary turns in unit 2, $R_{r1}$ = equivalent resistance of unit 1 in per unit of unit 1, $R_{r2}$ = equivalent resistance of unit 2 in per unit of unit 1, $X_{r1}$ = equivalent reactance of unit 1 in per unit of unit 1, $X_{r2}$ = equivalent reactance of unit 2 in per unit of unit 1, and $\theta$ = impedance angle of load (lagging current positive).

NOTE: Per unit means percent divided by 100, that is, 10% = 0.1 per unit.

The current in the second unit may be determined by using Eq. (10-20) with designation of first and second transformers reversed.

***22. Phase-interconnected transformers*** (i.e., with windings from more than one phase on a single core leg) offer special complication when unlike units are connected in parallel.[1]

***23. Three-Phase to Three-Phase Transformations.*** The delta-delta, the delta-Y, and the Y-Y connection are the most generally used; they are illustrated in Fig. 10-12. The Y-delta and

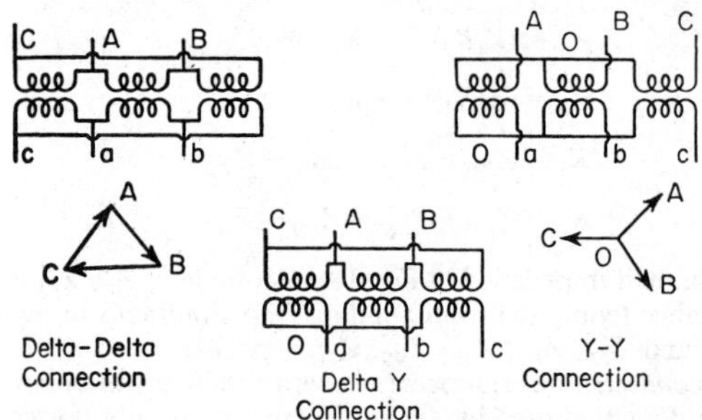


**FIG. 10-12** Standard 3-phase to 3-phase transformer systems.

delta-delta connections may be used as step-up transformers for moderate voltages. The Y-delta has the advantage of providing a good grounding point on the Y-connected side which does not shift with unbalanced load and has the further advantage of being free from third-harmonic voltages and currents; the delta-delta has the advantage of permitting operation in V in case of damage to one of the units. Delta connections are not the best for transmission at very high voltage; they may, however, be associated at some point with other connections which provide means for properly grounding the high-voltage system; but it is better, on the whole, to avoid mixed systems of connections. The delta-Y step-up and Y-delta step-down connections are without question the best for high-voltage transmission systems. They are economical in cost, and they provide a stable neutral whereby the high-voltage system may be directly grounded or grounded through resistance of such value as to damp the system critically and prevent the possibility of oscillation.

***24. The Y-Y connection*** (or Y-connected autotransformer) may be used to interconnect two delta systems and provide suitable neutrals for grounding both of them. A Y-connected autotransformer may be used to interconnect two Y systems which already have neutral grounds, for reasons of economy. In either case a delta-connected tertiary winding is frequently provided for one or more of the following purposes.

*Stabilization of the Neutral.* If a Y-connected transformer (or autotransformer) with a delta-connected tertiary is connected to an ungrounded delta system (or poorly grounded Y system), stability of the system neutral is increased. That is, a single-phase short circuit to ground on the transmission line will cause less drop in voltage on the short-circuited phase and less rise in voltage on the other two phases. A 3-phase three-leg Y-connected transformer without delta tertiary furnishes very little stabilization of the neutral, and the delta tertiary is generally needed. Other Y connections offer no stabilization of the neutral without a delta

[1] B. A. Cogbill, Sequence Impedance of Symmetrical Three Phase Transformer Connections; *Trans. AIEE Paper* 55-671, 1955.

tertiary. With increased neutral stabilization the fault current in the neutral on single-phase short circuit is increased, and this may be needed for improved relay protection of the system.

*Third-harmonic components of exciting* current find a relatively low impedance path in a delta tertiary on a Y-connected transformer, and less of the third-harmonic exciting current appears in the connected transmission lines, where it might cause interference with communication circuits. Failure to provide a path for third-harmonic current in Y-connected 3-phase shell-type transformers or banks of single-phase transformers will result in excessive third-harmonic voltage from line to neutral. The tank of a 3-phase, three-legged core-type Y-connected transformer acts as a delta winding with high impedance to the other windings. As a consequence, there is very little third-harmonic line-to-neutral voltage and a separate delta tertiary is not needed to reduce it.

*An external load can be supplied* from a delta tertiary. This may include synchronous or static capacitors to improve system operating conditions.

***25. Loading Y-Connected Transformers Line to Neutral.*** Load can be connected line to neutral only if (1) the source side of the transformer is delta-connected, or (2) the source side is Y-connected with the neutral connected back to the source neutral.

If one of these two conditions is not maintained, the neutral will shift, reducing the voltage of the loaded phase and increasing the voltage of the other phases.

***26. The open-delta connection, or V connection,*** is an unsymmetrical connection which is used if one transformer of a bank of three single-phase delta-connected units must be cut out because of failure. It is a connection that is sometimes resorted to as an emergency expedient or used as a temporary measure with the intention of completing the delta when conditions of load warrant the addition of a third unit. If one phase of a 3-phase delta-connected transformer of the shell type should fail, operation may be continued at reduced capacity by short-circuiting the damaged phase; if of the core type, operation may be continued by leaving the damaged phase open-circuited, provided that the windings are still capable of withstanding the voltage stresses. Since full-line currents flow in the windings out of phase with the transformer voltages, the normal capacity of the open-delta bank is reduced to 57.7% of its delta rating.

***27. The T connection*** uses two transformers, the first called the "main" transformer, connected from line to line; and the second, called the "teaser" transformer, connected from the midpoint of the first to the third line. It requires that the midpoint of both primary and secondary windings be available for connections. It has an advantage over the V connection in being more nearly symmetrical if the proper taps have been provided. As in the case of the V connection, two transformers of a bank of delta-connected transformers, one of which has failed, may be connected in T, and if 10% taps can be used for the teaser transformer, the transformation will be more nearly symmetrical than if the V connection were used. Where T-connected transformers are installed, they may later be changed to delta with the addition of one more transformer and an increase in rating of the bank of 73%. In the T connection (Fig. 10-13) the transformer *AD*, known as the teaser transformer, may be a duplicate of the main transformer so as to be interchangeable with it, and it may or may not be provided with an 86.6% tap. Its rated capacity will then be 15.5% more than actually necessary. The main transformer operates at a power factor of 0.866, and therefore, if the two transformers are duplicates, their total rated capacity will be 15.5% greater than the capacity of the load in kVA, or each transformer must have a rating of 0.577 of the kVA delivered. If the transformers are not interchangeable, the teaser may be reduced to a rating of one-half the kVA delivered.

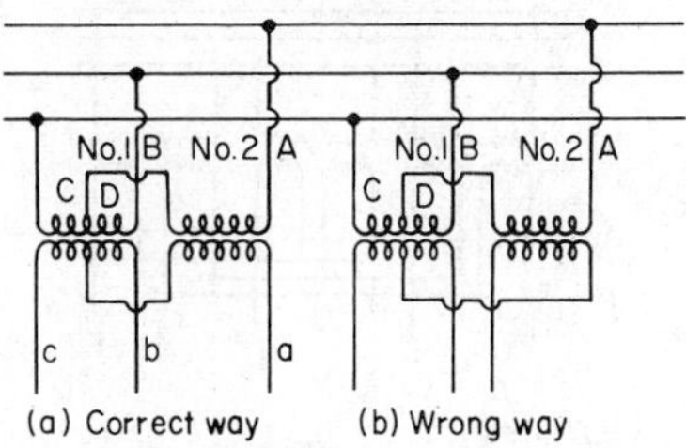


**FIG. 10-13** T-connected transformers.

In connecting transformers in T *care should be taken to keep the relative phase sequence of the windings the same;* otherwise the impedance of the main transformer may be excessively high and cause undue unbalance. Figure 10-13 illustrates the right and the wrong way.

***28. Three-Phase to Six-Phase.*** Six phases are commonly used for supply of rectifiers (also see Par. **206**). Six phases can be obtained as shown in Fig. 10-14 (double delta) or as in Fig. 10-15 (double Y). It is not necessary in this transformation, when the neutral connection is required, to have two secondary windings; instead a middle tap may be brought out, all the middle taps being connected together to form the neutral.

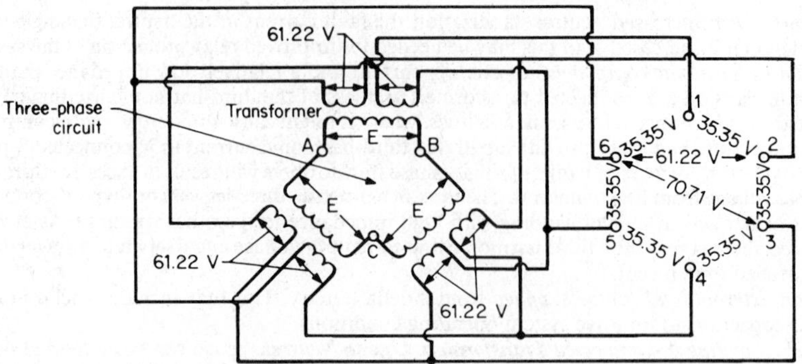


**FIG. 10-14** Three- to six-phase transformation, double delta.

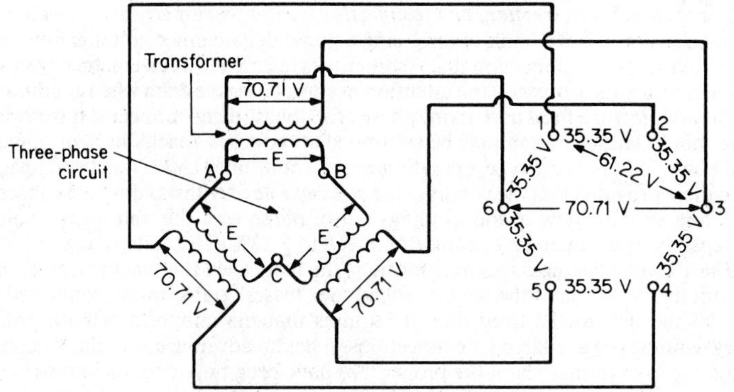


**FIG. 10-15** Three- to six-phase transformation, double Y.

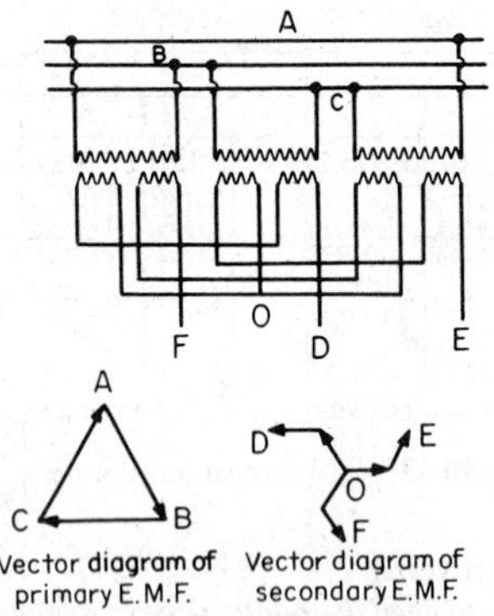


**FIG. 10-16** Interconnected Y connection.

***29. The interconnected Y connection*** shown in Fig. 10-16 is commonly referred to as the *zigzag* connection. It may be used with either a delta-connected winding as shown or a Y-connected winding for step-up or step-down operation. In either case, the zigzag winding produces the same angular displacement as a delta winding and, in addition, provides a neutral for grounding purposes. Owing to the angular relation of voltages of the zig and zag windings, the amount of conductor material required for such a connection is 15% greater than a corresponding Y or delta connection. If a transformer consists of zigzag and Y connections, a third winding, delta-connected, is usually necessary for reasons given under the Y-Y connection. If the delta-connected winding is included for purposes other than that of providing a third source of power, in some cases it is practical to design it for the same voltage as the zigzag winding and connect it in parallel with the zigzag winding to form the delta-grounded transformer connection.[1]

[1] E. T. B. Gross and K. J. Rao, Analysis of the Delta-grounded Transformer; *Trans. AIEE,* 1953, vol. 72, pp. 817–826.

The zigzag connection is used extensively for grounding transformers, the sole purpose of which is to establish a neutral point for grounding purposes; therefore no other windings are required.

## Power Transformers

***30. The information*** given in Pars. **30** to **159** deals particularly with power transformers, which may be defined as transformers used to transmit or distribute power in ratings larger than distribution transformers (usually over 500 kVA or over 67 kV). Some of the following information on power transformers is also applicable to some other transformers which are covered subsequent to Par. **159**.

***31. The Rated Constants of a Power Transformer.*** The kVA, terminal voltages and currents are defined in ANSI C57.12.80. They are all based on the rated winding voltages at no load, although it is recognized that the actual primary voltage in service must be higher than the rated value by the amount of the regulation, if the transformer is to deliver rated voltage to the load on the secondary.

### Design

***32. The design of successful commercial transformers*** requires the selection of a simple form of structure so that the coils may be easy to wind and the magnetic circuit easy to build. At the same time the mean length of the windings and of the magnetic circuit must be as short as possible for a given cross-sectional area, so that the amount of material required and the losses shall be as low as possible. The form of construction should permit the easy removal of heat by means of ventilating ducts, it should admit of being insulated in a simple and economical manner, and the windings should be of such forms as may be easily reinforced to withstand mechanical stresses.

***33. Two Types of Transformer in Common Use.*** When the magnetic circuit takes the form of a single ring encircled by two or more groups of primary and secondary windings distributed around the periphery of the ring, the transformer is termed a *core-type transformer.* When the primary and secondary windings take the form of a common ring which is encircled by two or more rings of magnetic material distributed around its periphery, the transformer is termed a *shell-type transformer* (Fig. 10-17). Actually, core-type (or "core-form") in U.S. power-transformer engineering usage means that the coils are cylindrical and concentric (the outer winding over the inner) whereas shell-type (or "form") denotes large pancake coils which are stacked or interleaved to make primary-secondary (P-S) groups. Except for certain extremes of current rating, the choice between the core- and shell-type construction is largely a matter of manufacturing facilities and of individual preference.

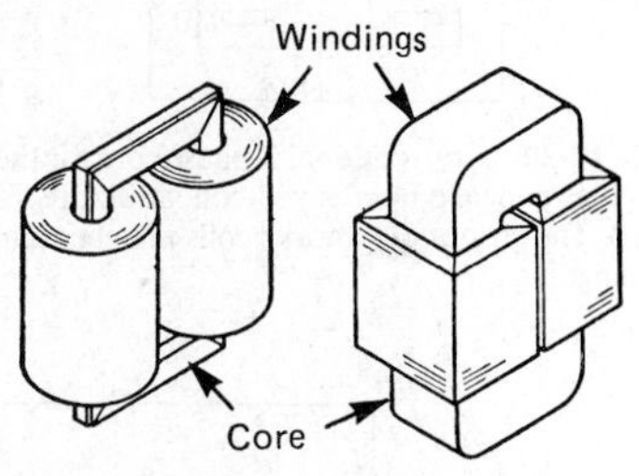


**FIG. 10-17** Forms of magnetic circuits for transformers.

***34. Core form transformer characteristic features*** are a long mean length of magnetic circuit and a short mean length of windings. Commonly used core constructions for single-phase and three-phase units are shown in Figs. 10-18 and 10-19, respectively. The three-leg (one active leg) and four-leg (two active) construction of single-phase cores and the five-leg (three active) construction of three-phase cores are used to reduce overall height. In these cases the core encloses the cylindrical windings in a similar fashion to the shell-form construction. The simple concentric primary (inside) and secondary (outside) winding arrangement is common for all small- and medium-power transformers. However, large MVA transformers frequently have some degree of interleaving of windings, such as secondary-primary-secondary (S-P-S). The core-form construction can be used throughout the full size range of power transformers.

Shell-form transformer characteristic features are short mean length of magnetic circuit and long mean length of windings. This results in the shell-form transformer having a larger area of

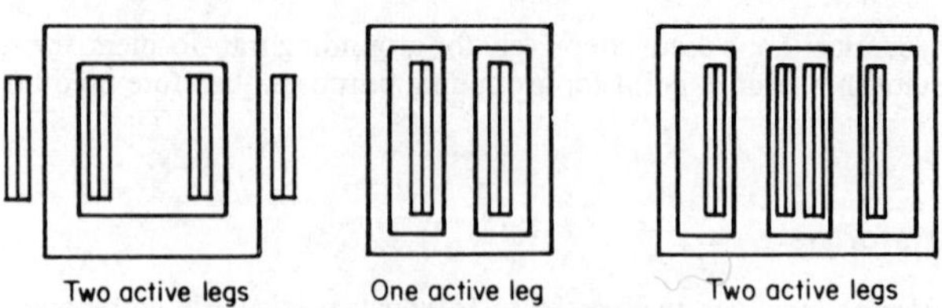


**FIG. 10-18** Single-phase core-form core constructions.

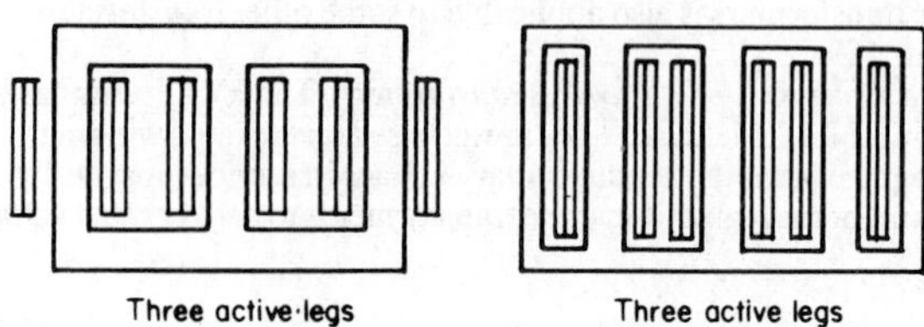


**FIG. 10-19** Three-phase core-form core constructions.

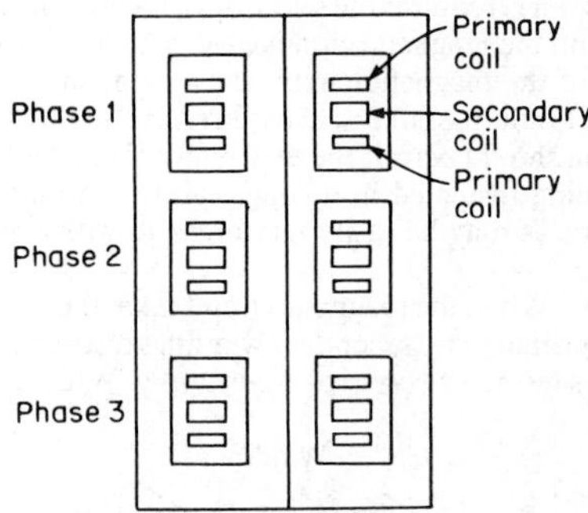


**FIG. 10-20** Conventional 3-phase core for the rectangular-pancake-interleaved-coil structure (shell-type). The groups of pancake coils may be round or rectangular.

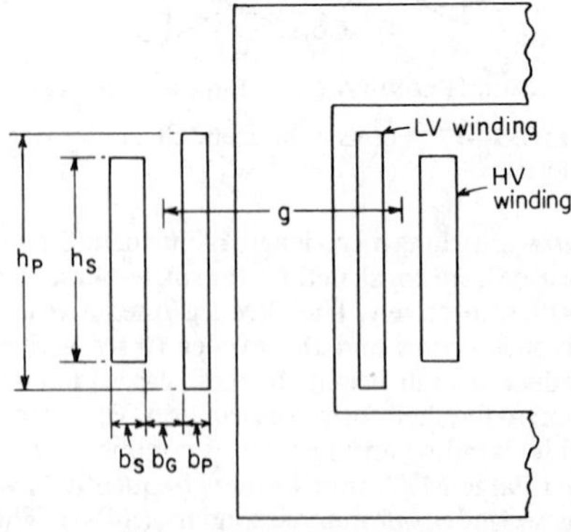


**FIG. 10-21** Dimensions of core-type concentric windings for reactance calculations.

core and smaller number of winding turns than the core form of same output and performance. Also, the shell form would typically have a greater ratio by weight of steel to copper. Figure 10-20 shows the conventional three-phase shell-form core with the coils in cross section. Primary-secondary-primary (P-S-P) coil grouping is most common, but P-S-P-S-P is often used.

***35. Design Process.*** Most power transformers are designed by assuming dimensions, calculating characteristics, comparing calculated characteristics with desired characteristics, and modifying the assumed dimensions better to meet the desired characteristics. Repeating the process leads to close agreement of calculated characteristics with desired characteristics. The repeated calculations, converging on the optimum design, are usually performed by computer. Closeness of agreement of calculated characteristics with tested characteristics depends upon the degree of refinement of the design process, the closeness of agreement of the physical properties of the materials used (particularly the dielectric properties of the insulating materials and the magnetic properties of the core steel) with the properties assumed in the design calculation, and the accuracy of the manufacturing procedures and processes.

Refinement of the design process results from comparison with test data obtained on similar transformers. This applies particularly to core loss, stray loss, noise level, reactance, and dielectric strength. The following calculation methods are mostly approximate.

***36. Number of Turns.*** With an assumed core cross section and flux density, the number of turns in each winding is established from Eq. (10-1). The flux density is adjusted to give an integral number of turns in the low-voltage winding, and then an acceptable ratio of open-circuit terminal voltage results from an integral number of turns in the high-voltage winding.

***37. Leakage flux density*** in the main gap (insulation space between windings) for a transformer with one core leg per phase, as shown in Fig. 10-21, is as follows:

$$B_L = 4.52 I_R N / h_E \tag{10-21}$$

where $B_L$ = lines per square inch peak leakage flux density, $h_E$ = inches effective length of leakage flux path, $I_R$ = rms amperes rated current of winding, and $N$ = number of turns in winding.

If there is more than one leg per phase, Eq. (10-21) applies to the portion of winding on one leg. The effective length of leakage path is difficult to evaluate accurately. For concentric cylindrical windings it is approximately

$$h_E = \frac{h_P + h_S}{2} + 0.8\left(\frac{b_P + b_S}{3} + b_G\right) \tag{10-22}$$

where $b_G$ = inches radial distance between windings, $b_P$ = inches radial width of winding $P$, $b_S$ = inches radial width of winding, $S$, $h_P$ = inches length of winding $P$, and $h_S$ = inches length of winding $S$.

***38. Leakage reactance*** may be calculated for a transformer with one set of coils per phase as follows:

$$X_r = \frac{2.01 \times 10^{-7} a_L f I_R N^2}{E_R h_E} \tag{10-23}$$

where $a_L$ = square inches effective cross section of leakage flux path, $E_R$ = rms volts rated voltage of winding, $f$ = frequency in hertz, $h_F$ = effective length of leakage flux path, in inches, $I_R$ = rms amperes rated current of winding, $N$ = number of turns in winding, and $X_r$ = per unit reactance.

The effective cross section of the leakage flux path is difficult to evaluate accurately. For concentric cylindrical windings it is approximately

$$a_L = \left(b_G + \frac{b_P + b_S}{3}\right)\pi g \tag{10-24}$$

where $g$ = mean diameter of main gap, in.

***39. Resistance loss in winding*** is

$$W_R = 2.57 M^2 \frac{234.5 + C}{309.5} \tag{10-25}$$

$$L_R = H_C W_R \tag{10-26}$$

where $L_R$ = watts resistance loss in winding, $M$ = rms kiloamperes per square inch current density, $H_C$ = pounds weight of copper in winding, $C$ = temperature in degrees Celsius, and $W_R$ = watts per pound resistance loss in winding.

***40. Eddy loss*** in the winding may be regarded as caused by circulating current induced in the strand by the magnetic flux passing through the strand. For a two-winding transformer

$$W_E = 2.06 \times 10^{-10} d^2 f^2 B_L^2 \frac{309.5}{234.5 + C} \tag{10-27}$$

$$L_E = H_C W_E \tag{10-28}$$

where $B_L$ = lines per square inch peak leakage flux density, from Eq. (10-21), $C$ = Celsius temperature, $d$ = inches thickness of strand perpendicular to flux, $f$ = hertz frequency, $L_E$ = watts eddy loss in winding, $H_C$ = pounds weight of copper in winding, and $W_E$ = watts per pound average eddy loss in winding.

***41. Load loss*** is the sum of resistance and eddy losses in all windings plus stray loss. The stray loss is in itself an eddy loss produced by the leakage flux penetrating the surface of other conducting components, such as the core, core clamps, and tank. Historically the stray loss has been predicted from test results on similar transformers, but finite-element computer solutions have recently been developed to define the leakage flux paths accurately and permit more exact calculation of both stray and eddy losses.

***42. No-load loss*** equals watts per pound determined from Fig. 10-1, multiplied by weight of the core multiplied by a correction factor depending on core configuration and processing and determined by experience.

***43. General Design Characteristics.*** The relationship of power-transformer characteristics to scale factor can be illuminated by considering the effect of increasing all dimensions in the ratio $S$, while retaining the same thickness of core lamination and thickness of conductor strand, but imagining the conductor turns to be reconnected for a terminal voltage proportionate to the insulation thickness. Similarly the effect of increasing the flux density in the ratio $B$ and the current density in the ratio $M$ can be examined. The results are shown in Table 10-1.

**TABLE 10-1** Scale Effects

| Characteristic | At scale factor $S$ | At flux density $B$* | At current density $M$ |
|---|---|---|---|
| Linear dimension | $S$ | | |
| Flux density | .... | $B$ | |
| Current density | .... | .... | $M$ |
| Rated current | $S^3$ | $B$ | $M$ |
| Rated voltage | $S$ | | |
| Rated kVA | $S^4$ | $B$ | $M$ |
| Weight | $S^3$ | | |
| KVA/lb | $S$ | $B$ | $M$ |
| Ω reactance | $S^{-1}$ | | |
| % reactance | $S$ | $B^{-1}$ | $M$ |
| W core loss | $S^3$ | $B^2$ | |
| % core loss | $S^{-1}$ | $B$ | $M^{-1}$ |
| W $I^2R$ loss | $S^3$ | .... | $M^2$ |
| % $I^2R$ loss | $S^{-1}$ | $B^{-1}$ | $M$ |
| W eddy loss | $S^5$ | .... | $M^2$ |
| % eddy loss† | $S$ | $B^{-1}$ | $M$ |
| W stray loss | $S^4$ | .... | $M^2$ |
| % stray loss† | .... | $B^{-1}$ | $M$ |

* Applies only to the range in which core loss varies with the square of $B$.
† As a percent of rated current times rated volts.

***44. Core dimensions*** are generally standardized in steps, with only a small number of dimensions varying to meet the requirement of the particular rating. Cold-rolled grain-oriented silicon steel strip in gages of 0.009 to 0.014 in is used with mitered corner joints to take advantage of the good characteristics of this material when carrying flux in the rolling direction.

## Insulation

***45. Insulation systems*** used in power transformers comprise liquid systems and gas systems. In both cases some solid insulation is used. Liquid systems include mineral oil, which is most frequently used, and various low-flammability fluids. Askarel, the fluid traditionally used to avoid combustability, can be found in transformers in field service, but it has been completely phased out of new transformer production because of environmental concerns. It has been replaced by any of a wide variety of high-flash-point fluids (silicones, high-flash-point hydrocarbons, chlorinated benzenes, or chlorofluorocarbons). The gas systems include nitrogen, air, and fluorogases. The fluorogases are used to avoid combustibility and limit secondary effects of internal failure.

Major insulation separates the high-voltage winding from the low-voltage winding and ground. This insulation carries the highest voltage and occupies the most limited space; hence it usually operates at the highest stress. Depending on the construction, layer insulation or coil insulation may be provided between parts of windings. Turn insulation is applied to each strand of conductor or to groups of strands forming a single turn.

***46. Oil-Insulated Transformers.*** Low cost, high dielectric strength, and ability to recover after dielectric overstress make mineral oil the most widely used transformer insulating material. The oil is reinforced with solid insulation in various ways. The major insulation usually includes barriers of wood-based paperboard (pressboard), the barriers usually alternating with oil spaces. Because the dielectric constant of the board is about 4.0, compared with 2.0 for oil, the dielectric stress on the oil is about double that on the pressboard, and the permissible stress on the oil usually limits the effective strength of the structure.

The insulation on the conductors in the winding may be enamel or wrapped paper which may be wood- or nylon-based. The use of high-dielectric-constant insulation directly on the conductor actually reduces the dielectric stress concentration at the conductor surface, and again the limit of dielectric strength is usually that of the oil.

Heavy paper wrapping is also usually used on the leads coming from the winding.

The relatively porous and hygroscopic paper-based insulation must be carefully dried and vacuum-impregnated with oil to remove moisture and gas to obtain the final required high dielectric strength, and to resist deterioration at operating temperature.[1]

Gas pockets or gas bubbles in the insulation are particularly destructive to the insulation strength because gas (usually air) has a low dielectric constant (about 1.0) which means that it will be stressed more highly than the other insulation while at the same time it has low dielectric strength.

***47. Askarel-Insulated Transformers.*** These transformers have constructions similar to oil-insulated transformers. The relatively high dielectric constant of the askarel aids in transferring dielectric stress to the solid elements. Askarel has limited ability to recover after dielectric overstress, and thus the strength is limited in nonuniform dielectric fields. Askarels are seldom used over 34.5 kV operating voltage. Askarels are powerful solvents; their products of decomposition are so harmful that they have been completely abandoned in new transformers.

***48. Fluorogas-Insulated Transformers.*** Fluorogases have better dielectric strength and heat-transfer capacity than nitrogen or air. Both dielectric strength and heat transfer capacity increase with density, and fluorogas transformers operate above atmospheric pressure, in some cases up to 3 atm gage pressure. The gas insulation is reinforced with solid insulation used in the form of barriers, layer insulation, turn insulation, and lead insulation.

It is usually economical to operate fluorogas-insulated transformers at higher temperatures than oil-insulated transformers. Suitable solid insulating materials include glass, asbestos, mica, high-temperature resins, ceramics, etc.

Dielectric stress on the gas is several times as high as the stress on the adjacent solid insulation in series in the dielectric structure. Care in designing is required to avoid overstressing the gas.

***49. Nitrogen- and air-insulated transformers*** are generally limited to 34.5 kV and lower operating voltages. Air-insulated transformers in clean locations are frequently ventilated to the atmosphere. In contaminated atmospheres a sealed construction is required, and nitrogen is generally used at approximately atmospheric pressure and somewhat elevated operating temperatures.

***50. Contamination*** in small traces, particularly in the presence of moisture, may seriously reduce the dielectric strength of insulation materials or structures. Extreme cleanliness is needed in transformer manufacture.

***51. Extreme dryness*** is necessary for the development of full dielectric strength in liquid-filled transformers. Appreciable quantities of water may decrease the dielectric strength and cause failure at operating voltage. Transformers are carefully vacuum-dried during manufacture before dielectric test. The initial level of dryness should be preserved through installation and subsequently in service.

***52. Effect of Transient Voltages.*** When a short-duration aperiodic impulse voltage is applied to the terminal of a transformer, the voltage does not divide uniformly throughout the turns of the winding. The initial voltage distribution is such that the line turns take much more than their share of the voltage. The steeper the applied wave, the greater is the concentration of

[1] F. M. Clark, Factors Affecting the Deterioration of Cellulose Insulation; *Trans. AIEE,* 1942, vol. 61, pp. 742–749.

voltage on the line end turns. This happens because initially the only current flow is through the capacitance between turns of the winding and from winding to ground. Since the total distributed shunt capacitance to ground is much greater than the series capacitance through the winding to ground, most of the impulse current flows through the shunt capacitance near the line end of the winding, thereby causing a large voltage drop across the line-end portions of the winding.

Following the initial period, electrical oscillations occur within the windings. These oscillations impose greater stresses from the middle part of the winding to ground for long waves than for short waves. On the other hand, steep chopped waves impose the greater stresses between turns and coil portions. Longer-duration switching-surge transient voltages are of two types, aperiodic and oscillatory. Aperiodic waves have front times in the order of tens to hundreds of microseconds and tails in excess of 1000 $\mu s$. Because of their long duration, the voltage tends to distribute linearly throughout the winding in proportion to turn ratio. Hence stresses in the major insulation are of more concern than internal winding stresses. In contrast, oscillatory switching voltages can excite winding natural frequencies and produce stresses of concern in the internal winding insulation.[1] Transformers having low natural frequencies are most vulnerable, because internal damping is more effective at high frequencies.

***53. Electrostatic shielding*** may be used to prevent excessive concentration of transient voltages on the line-end turns. The purpose of the electrostatic shielding is to increase the series capacitance of the winding, particularly near the line end, and thereby lower the voltage across the line-end coils and turns. The transient voltage behavior including effect of shielding can be predetermined with reasonable accuracy by calculation or by model tests so that the designer can then provide adequate insulation. Most modern high-voltage transformers utilize electrostatic shields in combination with desirable winding arrangements to limit the concentration of impulse voltages near the line end of the winding.

***54. Impulse insulation level*** may be demonstrated by factory impulse-voltage tests using $1.5 \times 50$ $\mu s$ full waves and chopped waves. The full wave demonstrates the basic-impulse insulation level (BIL) for traveling waves coming into the station over the transmission line. The chopped wave demonstrates strength against a wave traveling along the transmission line after flashing over an insulator some distance away from the transformer. These waves do not simulate direct lightning strokes on or near the transformer terminals, which would result in the application of a steep-front wave to the transformer winding. Such strokes are usually avoided by ground wires or protecting grounded structures.

***55. High-voltage dc stresses*** may be imposed on transformers used in terminal equipment for dc transmission lines. Direct-current voltage applied to a composite insulation structure divides between individual components in proportion to the resistivities of the materials. In general the resistivity of an insulating material is not a constant but varies over a range of 100 : 1 or more, depending on temperature, dryness, contamination, and stress.

## Cooling

***56. Removal of heat*** caused by losses is necessary to prevent excessive internal temperature which would shorten the life of the insulation. Paragraphs **57** to **67** cover the procedure for calculating the internal temperature of oil-insulated self-cooled power transformers of conventional core-type construction using radiators. Almost all modern power transformers have insulation systems designed for operation at 65°C average winding rise over ambient temperature and 80°C hottest spot winding rise over ambient in an average ambient of 30°C. Older power transformers were designed for 55°C average winding rise/65°C hottest spot winding rise over ambient.

***57. The average temperature of a winding*** is the temperature determined by measuring the dc resistance of the winding and comparing it with the measurement previously obtained at a

[1] Also see R. C. Degeneff, W. J. McNutt, W. Neugebauer, J. Panek, M. E. McCallum, and C. C. Honey, Transformer Response to System Switching Voltages; *IEEE Trans.* June 1982, vol. PAS-101, pp. 1457–1470.

known temperature. The rise of the average temperature of a winding above ambient temperature is

$$U = B + E + N + T \tag{10-29}$$

where $B$ = degrees Celsius rise of effective oil over ambient, $E$ = degrees Celsius rise of average oil over effective oil, $N$ = degrees Celsius rise of average coil surface over average oil, $T$ = degrees Celsius rise of conductor over coil surface, and $U$ = degrees Celsius rise of average conductor over ambient.

***58. Effective oil temperature*** is the equivalent uniform temperature with equal ability to dissipate heat to the air. The effective oil temperature is approximately the average of the oil entering the top of the radiator and the oil leaving the bottom of the radiator. The oil temperature is approximately the same as the temperature of the adjacent radiator surface exposed to air. A smooth, vertical transformer-tank surface will dissipate heat to the air as follows:

$$D_B = 1.40 \times 10^{-3} B^{1.25} + 1.75 \times 10^{-3}(1 + 0.011A)B^{1.19} \tag{10-30}$$

where $A$ = degrees Celsius ambient temperature, $B$, = degrees Celsius effective oil rise over ambient, and $D_B$ = watts per square inch dissipated to the air.

The first term of Eq. (10-30) covers heat transferred by convection. Usually the radiator consists of parallel flattened tubes with limited accessibility to cooling air, and it is therefore necessary to multiply the first term by an experimentally determined friction factor (less than 1). The second term of Eq. (10-30) covers heat transferred by radiation, on the assumption of low temperature emissivity of 0.95, which applies to most painted surfaces commonly encountered. For any other value of low-temperature emissivity this term should be multiplied by emissivity/0.95 (see Table 10-2). Usually the radiator consists of parallel flattened tubes which radiate heat to each other. The net radiation of heat can be determined by considering the transformer and radiators replaced by a nonreentrant enveloping surface. If the second term of Eq. (10-30) is multiplied by the ratio of the area of the enveloping surface to actual surface (less than 1), the effect of reabsorption of radiation is eliminated. When radiation is small compared with convection, it can be assumed that $A = 25°C$ and the $B^{1.19}$ can be replaced by $0.79B^{1.25}$, and Eq. (10-30) becomes

$$B = \frac{100D_B^{0.8}}{(0.44F + 0.56V)^{0.8}} \quad °C \tag{10-31}$$

where $V$ = ratio of envelope surface area to actual surface area and $F$ = friction factor determined by experiment.

***59. The temperature rise of average oil over effective oil,*** $E$, is usually negligible for normal transformer designs. It may become important if (1) the center of gravity of the radiators is not elevated sufficiently above the center of gravity of the core and coils, (2) there is unusual loss in the oil space over the core such as might result from high-current leads, (3) a winding has unusually restricted oil ducts, or (4) pumps are used to circulate oil through the radiator without channeling the pumped oil through the oil ducts of the coil. For such cases, $E$ is best evaluated by comparison with performance of previous designs.

***60. The temperature rise of average coil surface over average oil,*** $N$, carries the loss in the coil through a film of stationary oil into moving oil. For a horizontal pancake coil (vertical axis) most of the heat escapes through the thin oil film on the upper surface and very little heat escapes from the lower surface. On the assumption that all the heat escapes from the upper surface, the temperature rise is

$$N = 13.2D_N^{0.8} \quad °C \tag{10-32}$$

where $D_N$ = watts per square inch dissipated from the coil to the oil.

For a vertical pancake coil (axis horizontal) the heat leaves both sides equally, and

$$N = 14D_N \quad °C \tag{10-33}$$

***61. The temperature rise of conductor over coil surface,*** $T$, carries the heat from the copper through the solid insulation applied to the conductor and the coil,

$$T = R_T t D_N \qquad °\text{C} \tag{10-34}$$

where $D_N$ = watts per square inch dissipated from the coil to the oil, $R_T$ = degrees Celsius per watt per inch thermal resistivity, and $t$ = inch length of path.

***62. The components of the winding rise*** over ambient are determined from Eqs. (10-31), (10-32), or (10-33), and (10-34) by using values of watts per square inch determined from the calculated losses and the design geometry. Then the total rise is determined from Eq. (10-29).

***63. Oil Circulation.*** The oil moves generally upward through ducts in the core and coils, rising in temperature as it goes. It moves generally downward through the radiators, falling in temperature as it goes (see Fig. 10-22). The space above the core and coils is filled with hot oil, so

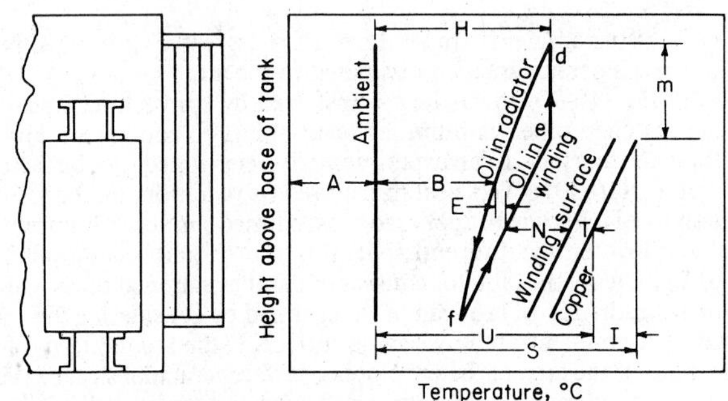


**FIG. 10-22** Oil-circulation diagram.

that the height-temperature curve of the circulating oil forms a triangle *def*. The difference in weight of the two columns of oil which furnishes the circulating force is proportional to the area of the triangle,

$$w = 2.50 \times 10^{-5} mI \tag{10-35}$$

where $m$ = inches headroom, $I$ = degrees Celsius top oil rise over average oil, and $w$ = pound-force per square inch circulating force.

$I$ is established by the following relations:

$$L = 222 I G_C \tag{10-36}$$

$$w = G_C R_H \tag{10-37}$$

$$I = 13.5 \left(\frac{L R_H}{m}\right)^{1/2} \tag{10-38}$$

where $G_C$ = gallons per minute rate of circulation of oil, $L$ = watts loss, and $R_H$ = friction opposing oil flow in pound-force per square inch per gallon per minute.

$R_H$ is not easily evaluated except by test, but Eq. (10-38) is useful in evaluating the effect of changing $L$ or $m$.

The limiting temperature rises are

$$H = B + E + I \tag{10-39}$$

$$S = B + E + N + T + I = U + I \tag{10-40}$$

where $H$ = degrees Celsius top oil rise over ambient temperature and $S$ = degrees Celsius hot-spot rise over ambient temperature.

Equation (10-40) gives the temperature of the top pancake coil. Values of $N$ and $T$ may need to be separately computed for this coil or for other coils if they have different loss density or different insulation than the main winding coils. If a coil other than the top coil is found to have higher conductor rise over adjacent oil, then appropriately reduced values of $I$ should be used to calculate $S$ for that coil.

***64. Variation of Temperature with Load.*** If temperature-rise conditions for rated load (or for any load) are known, temperature rises for any other load can be determined:

$$B_2 = B_1 \left(\frac{L_2}{L_1}\right)^{0.8} \qquad (10\text{-}41)$$

$$N_2 = N_1 \left(\frac{L_2}{L_1}\right)^{0.8} \qquad (10\text{-}42)$$

$$T_2 = T_1 \frac{L_2}{L_1} \qquad (10\text{-}43)$$

$$I_2 = I_1 \left(\frac{L_2}{L_{21}}\right)^{1/2} \qquad (10\text{-}44)$$

where $B_1$, $N_1$, $T_1$, $I_1$, and $L_1$ correspond to the known condition and $B_2$, $N_2$, $T_2$, $I_2$, and $L_2$ correspond to the new condition.

The total loss should be used for $L_1$ and $L_2$ in Eqs. (10-41) and (10-44). The resistance and eddy loss only should be used for $L_1$ and $L_2$ in Eqs. (10-42) and (10-43). The exponent in Eq. (10-42) should be 1.0 for vertical pancake coils. At constant voltage the resistance loss varies with the square of the load kVA and with the resistivity as affected by average temperature of copper according to Eq. (10-25). The eddy loss varies with the square of the load and inversely with the resistivity of the copper according to Eq. (10-27). The stray loss varies with the square of the load and may be assumed to vary inversely with the resistivity of the copper, like the eddy loss. The core loss may be assumed unaffected by load or temperature. For many purposes it is reasonable to assume that the entire load loss varies with the square of the load and with the resistivity.

***65. Example.*** What is the temperature rise in a 30°C ambient at 80% load of a transformer with the following characteristics?

Load loss is two times no-load loss at 85°C

Load loss is assumed all resistance loss 85°C

Full-load temperature rises with 85°C losses are

$$B_1 = 45$$

$$N_1 = 15 \qquad U_1 = 65$$

$$T_1 = 5 \qquad S_1 = 80$$

$$I_1 = 15 \qquad H_1 = 60$$

Upon assuming (for a trial) that the average copper temperature will be 80°C, the relative load loss is

$$(0.8)^2 \times \frac{234.5 + 80}{234.5 + 85} = 0.630$$

and the relative total loss is

$$\frac{0.630 \times 2 + 1}{3} = 0.753$$

Then

$$B_2 = 45(0.753)^{0.8} = 35.9°\text{C}$$

$$N_2 = 15(0.630)^{0.8} = 10.4°\text{C}$$

$$T_2 = 5 \times 0.630 = 3.2°\text{C}$$

$$I_2 = 15(0.753)^{0.5} = 12.0°\text{C}$$

$$U_2 = 35.9 + 10.4 + 3.2 = 49.5°\text{C}$$

$$S_2 = 35.9 + 10.4 + 3.2 + 12.0 = 61.5°\text{C}$$

$$H_2 = 35.9 + 12.0 = 47.9°\text{C}$$

The average copper temperature is 49.5 + 30 = 79.5°C, which is close enough to the assumed 80°C.

***66. Example.*** What is the temperature rise in a 30°C ambient at 140% load for the same transformer? Upon assuming (for a trial) that the average copper temperature will be 140°C, the relative load loss is

$$(1.4)^2 \frac{234.5 + 140}{234.5 + 85} = 2.30$$

and the relative total loss is

$$\frac{2.30 \times 2 + 1}{3} = 1.87$$

Then

$$B_2 = 45(1.87)^{0.8} = 74.2°\text{C}$$

$$N_2 = 15(2.30)^{0.8} = 29.2°\text{C}$$

$$T_2 = 3 \times 2.30 = 6.9°\text{C}$$

$$I_2 = 15(1.87)^{0.5} = 20.5°\text{C}$$

$$U_2 = 74.2 + 29.2 + 6.9 = 110.3°\text{C}$$

$$S_2 = 74.2 + 29.2 + 6.9 + 20.5 = 130.8°\text{C}$$

$$H_2 = 74.2 + 20.5 = 94.7°\text{C}$$

The average copper temperature is 110.3 + 30 = 140.3, which is close enough to the assumed 140°C. These temperatures would be considered excessive for continuous loading and should not be continued for any extended time period.

***67. Industry loading guides*** (ANSI/IEEE C57.92-1981) provide tables which give winding hottest-spot temperature and top oil temperature for representative transformers over a wide range of ambient temperature and per unit loading conditions. These guides also contain a cautionary note that operation at hottest-spot temperatures above 140°C may cause gassing in the solid insulation and oil which could reduce the dielectric integrity of the transformer.[1,2]

***68. Transient Thermal Conditions.*** Since transformer temperature takes many hours to stabilize after a change in loading, it is sometimes desirable to calculate the temperature during the transient period.

---

[1] F. W. Heinrichs, Bubble Formation in Power Transformer Windings at Overload Temperature; *IEEE* Trans., Sept./Oct., 1979, vol. PAS-98, pp. 1576–1582.

[2] W. J. McNutt, G. H. Kaufman, A. P. Vitols, J. D. MacDonald, Short Time Failure Mode Considerations Associated with Power Transformer Overloading; *IEEE* Trans., May/June 1980, vol. PAS-99, pp. 1186–1197.

$$B_H = B_U - (B_U - B_0)\epsilon^{-h/h_B} \quad (10\text{-}45)$$

$$h_B = \frac{K_B(B_U - B_0)}{L_D - L_0} \quad (10\text{-}46)$$

$$K_B = 0.06H_{CC} + 0.04H_T + 1.33G_T \quad \text{for nondirected flow cooling}$$

$$K_B = 0.06H_{CC} + 0.06H_T + 1.93G_T \quad \text{for directed flow cooling} \quad (10\text{-}47)$$

where $B_H$ = degrees Celsius effective oil rise after $h$ hours, $B_0$ = degrees Celsius initial oil rise, $B_U$ = degrees Celsius ultimate effective oil rise for the new load, $G_T$ = gallons of oil, $\epsilon = 2.718$, $h$ = hours after the change in load, $h_B$ = hours time constant of the transformer, $H_{CC}$ = pounds weight of core and coils, $H_T$ = pounds weight of tank and fittings, $K_B$ = watthours per degree Celsius thermal capacity of the transformer, $L_D$ = watts dissipated at the new load, $L_0$ = watts loss at the initial condition ($h = 0$).

Equation (10-45) applies whether $B_U$ is greater or less than $B_0$. However, if $B_0$ and $L_0$ are zero,

$$B_H = B_U(1 - \epsilon^{-h/h_B}) \quad (10\text{-}48)$$

$$h_B = \frac{K_B B_U}{L_D} \quad (10\text{-}49)$$

***69. The rise of average conductor over average oil*** ($N + T = Q$) is a single variable for transient calculations. The ultimate value is reached in 15 to 30 min.

$$Q_H = Q_U - (Q_U - Q_0)\epsilon^{-h/h_Q} \quad (10\text{-}50)$$

$$h_Q = \frac{K_Q(Q_U - Q_0)}{L_0 - L_D} \quad (10\text{-}51)$$

$$K_Q = 0.05H_C \frac{a/2 + b}{b} \quad (10\text{-}52)$$

where $a$ = square inches cross section of strand insulation, $b$ = square inches cross section of copper strand, $H_C$ = pounds weight of copper in winding, $K_Q$ = watthours per degree Celsius thermal capacity of winding, $L_D$ = watts dissipated at $h = 0$, $L_0$ = watts loss at $h = 0$, $h$ = hours, $h_Q$ = hours time constant of winding, $Q_H$ = degrees Celsius average conductor rise over average oil after $h$ hours, $Q_0$ = degrees Celsius initial average conductor rise over average oil, and $Q_U$ = degrees Celsius ultimate average conductor rise over average oil.

Equation (10-50) applies whether $Q_U$ is greater or less than $Q_0$. However, if $Q_U$ and $L_0$ are zero,

$$Q_H = Q_0\epsilon^{-hlh_Q} \quad (10\text{-}53)$$

$$h_Q = \frac{K_Q Q_0}{L_D} \quad (10\text{-}54)$$

***70. Example.*** If the transformer in Par. **65** operates in a 30°C ambient at 80% load until ultimate temperature conditions are established and then operates at 140% load, what will be the conditions after 4 h at 140% load? Assume nondirected flow cooling. It is now necessary to specify additional characteristics of the transformer as follows:

$$H_C = 20{,}000 \qquad a = 0.018$$

$$H_{CC} = 100{,}000 \qquad b = 0.090$$

$$H_T = 30{,}000$$

$$G_T = 10{,}000$$

Total loss = 150,000 W at 85°C.

Then

$$K_B = 0.06 \times 100{,}000 + 0.04 \times 30{,}000 + 1.33 \times 10{,}000 = 20{,}500$$

Assume that the averge winding temperature after 4 h is 125°C.

$$L_D = \frac{2 \times (1.4)^2(234.5 + 125)/319.5 + 1}{3} 150{,}000 = 270{,}540 \text{ W}$$

$$L_0 = 0.753 \times 150{,}000 = 112{,}950 \text{ W}$$

$$h_B = \frac{20{,}500\ (74.2 - 35.9)}{270{,}540 - 112{,}950} = 4.98 \text{ h}$$

$$B_H = 74.2 - (74.2 - 35.9)\epsilon^{-4/4.98} = 57.0\text{°C}$$

$$\left.\begin{aligned} N_H &= 29.2 \\ T_H &= 6.9 \\ I_H &= 20.5 \end{aligned}\right\} \text{(These reach ultimate value before 4 h)}$$

$$U_H = 57.0 + 29.2 + 6.9 = 93.1\text{°C}$$

$$S_H = 57.0 + 29.2 + 6.9 + 20.5 = 113.6\text{°C}$$

$$H_H = 57.0 + 20.5 = 76.5\text{°C}$$

***71. Example of Load Cycle.*** If the transformer examined in Par. **70** operates in a 30°C ambient on a daily cycle of 20 h at 80% load and 4 h at 140% load, what are the temperature rises at the end of the 140% load period? Since the transformer time constant in Par. **70** was approximately 5 h, it would be reasonable to assume that the oil temperature will have essentially stabilized after 20 h at 80% load. (The time constant for the transient from 140% load to 80% load will be essentially the same as the time constant for the transient from 80% load to 140% load.) As a result, the temperature rises at the end of the 140% load period during this cycle will be the same as those found in Par. **70.** Figure 10-23 shows the complete set of time-temperature curves for the 24 h cycle.

***72. The short-circuit temperature rise*** is calculated on the assumption that all heat generated in the copper is stored in the copper until the short circuit is over,

$$C_2 = 309.5 \times \left\{\left[\left(\frac{C_1 + 234.5}{309.5}\right)^2 + g\right] \epsilon^{M^2 s/10{,}800} - g\right\}^{1/2} - 234.5 \qquad (10\text{-}55)$$

where $C_1$ = degrees Celsius average temperature of winding at start of short circuit, $C_2$ = degrees Celsius average temperature of winding at end of short circuit, $M$ = rms kiloamperes per square inch current density, $s$ = seconds duration of short circuit, and $g$ = ratio of eddy loss to resistance loss in winding at 75°C.

The temperature resulting from any short circuit may be read directly from Fig. 10-24, which is plotted from Eq. (10-55).

***73. Example.*** If a short circuit resulting in 50 kA/in² current density is held for 3.5 s in a winding with 10% eddy loss ($g = 0.1$) at 75°C, and with a starting temperature of 75°C, what is the average temperature of the winding at the end of the short circuit?

On the curve for $g = 0.1$, at 75°C the value of $M^2 s$ is 5300.

$$5300 + 50^2 \times 3.5 = 14{,}050$$

On the curve for $g = 0.1$, at $M^2 s$ = 14,050, the temperature is 246°C.

***74. Fan-cooled transformers*** use external fans to improve heat dissipation from the radiators, and sometimes internal pumps to circulate the oil through the radiators (and sometimes also through cooling ducts in the core and coils). With fan cooling the effective oil-temperature rise $B$ is determined from test data on the particular arrangement, instead of Eq. (10-31). With oil pumps $I$ is determined from Eq. (10-38). It is usually possible to obtain 67% more capacity with fans and pumps running.

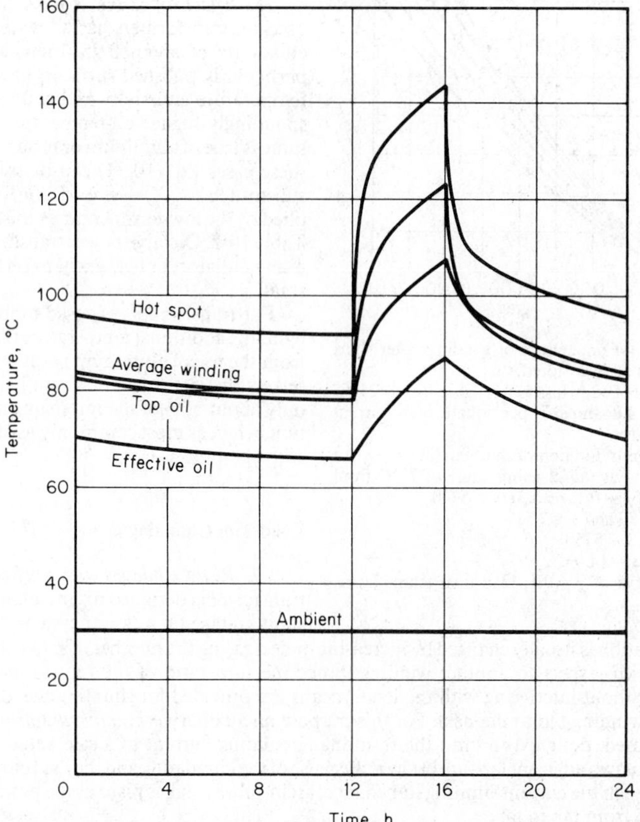


**FIG. 10-23** Transformer temperatures during the daily load cycle described in Par. 71.

*75. **Forced-cooled transformers*** use external oil-to-air heat exchangers requiring both air fans and oil pumps for all operating conditions. The effective oil rise $B$ is calculated from the characteristics of the oil-to-air heat exchanger, and $I$ is determined from Eq. (10-36). Forced-cooled transformers have no continuous load capacity without pump and fans.

*76. **Water-cooled transformers*** usually have the oil withdrawn from the transformers at the top of the tank, pumped through an external cooler, and returned to the bottom of the tank. The temperature drop of the oil in passing through the cooler is

$$Y = \frac{L}{G_c \times 111} \tag{10-56}$$

where $Y$ = degrees Celsius drop of oil in cooler, $L$ = watts loss, and $G_c$ = gallons-per-minute rate of circulation of oil.

The effective oil-temperature rise $B$ may be calculated from the characteristics of the oil-to-water heat exchanger. The top oil rise over average oil, $I$, is $Y/2$. The other components of temperature rise are calculated as for a self-cooled transformer.

*77. **Operation at high altitude*** increases the effective oil rise of air-cooled transformers. ANSI C57 provides for a compensating correction of 0.4% of rated kVA for self-cooled transformers or 0.5% of rated kVA for forced air-cooled or forced oil-cooled transformers for each 330 ft of additional altitude above 3300 ft altitude.

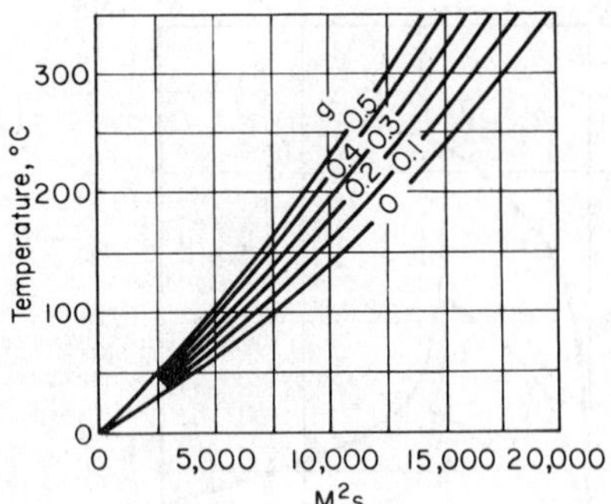


FIG. 10-24 Temperature of windings after short circuit, with all heat stored.

$g$ = ratio of eddy loss to resistance loss at 75°C
$M$ = rms kiloamperes per square inch current density
$s$ = seconds duration of short circuit

*Example:* For *initial* temperature of 75°C from the curve of $g = 0.1$, read $M^2s = 5300$

For $M = 50$ and $s = 3.5$:
$(50)^2 \times 3.5 = 8750$
Total $M^2s = 14{,}050$

From curve of $g = 0.1$, read final temperature: 246°C.

***78. Effect of Tank Color.*** Most paint used on transformers has a low-temperature emissivity of about 0.95. Metallic surfaces, particularly polished surfaces, have less low-temperature emissivity and will cause correspondingly higher oil-temperature rise. The same is true of aluminum or bronze paint. For these cases Eq. (10-31) can be used with $V'$ substituted for $V$, where $V'$ is $V/0.95$ multiplied by the low-termperature emissivity from Table 10-2. On large power transformers with many radiators or heat exchangers the effect is small.

For transformers exposed to intense sunlight, the additional temperature rise resulting from the use of aluminum paint is largely offset[1] by the fact that aluminum paint absorbs only about 55% of the impinging solar radiation, whereas most commonly used paints absorb about 95%.

## Load Tap Changing[2,3]

***79. Ratio Changes with Shifted Taps.*** In transformers designed for maintaining a constant voltage on a power system, the ratio of transformation is usually changed by increasing or decreasing the number of active turns in one winding with respect to another winding. Since the turn ratio of the transformer must be changed without interfering with the load, means are provided for shunting the load current from one winding tap to the next. For this purpose an auxiliary *preventive autotransformer* is generally used, designed to limit the resulting circulating current to a safe value during the interval that two adjacent taps are bridged. Because of the circulating and the load current which passes through the current-limiting impedance, arcing always takes place as the power circuit is transferred from tap to tap.

Although a variety of switching equipments and transformer connections have been used for the purpose of changing taps under load, the underlying principle remains unchanged and is shown by the transformer connection in Fig. 10-25.

**TABLE 10-2** Low-Temperature Total Emissivity

| | |
|---|---|
| Aluminum, highly polished | 0.08 |
| Copper | 0.15 |
| Cast iron | 0.25 |
| Aluminum paint | 0.55 |
| Oxidized copper | 0.60 |
| Oxidized steel | 0.70 |
| Bronze paint | 0.80 |
| Black gloss paint | 0.90 |
| White lacquer | 0.95 |
| White vitreous enamel | 0.95 |
| Green paint | 0.95 |
| Gray paint | 0.95 |
| Lampblack | 0.95 |

[1] V. M. Montsinger and L. Wetherill, Effect of Color of Tank on the Temperature of Self-cooled Transformers under Service Conditions; *Trans. AIEE,* 1930, vol. 49, p. 41.

[2] Some of the material under this title is reprinted with certain modifications from L. F. Blume, A. Boyajian, G. Camilli, T. C. Lennox, S. Minneci, and V. M. Montsinger, *Transformer Engineering,* 2d ed.; New York, John Wiley & Sons, Inc., 1951.

[3] *AIEE Paper* 57-48 O. P. McCarty and W. M. Johnson, Short-Circuit Capability Tests of Load Tap Changing Mechanisms.

*Example.* To move from transformer tap $A$ to $B$, it is first necessary to close the circuit to $B$, as shown in Fig. 10-25, before opening the circuit at $A$. During the interval when $A$ and $B$ are both closed on adjacent taps, a circulating current flows through and is limited by the impedance on the loop composed of the tap winding $AB$ and autotransformer $C$. With both ends of the autotransformer connected to $A$, the load current divides equally between the two halves of the autotransformer. Since the current flows in opposite directions, a negligible amount of reactance is introduced into the circuit and the only loss is the $I^2R$ due to the 50% load current in each half of the autotransformer winding. With $A$ closed and $B$ open, all the load current flows through one-half of the autotransformer, magnetizing the autotransformer and thereby introducing into the circuit the induced voltage. It is important, therefore, that the magnetizing reactance be kept as low as possible to avoid excessive arcing duty on the circuit-interrupting device.

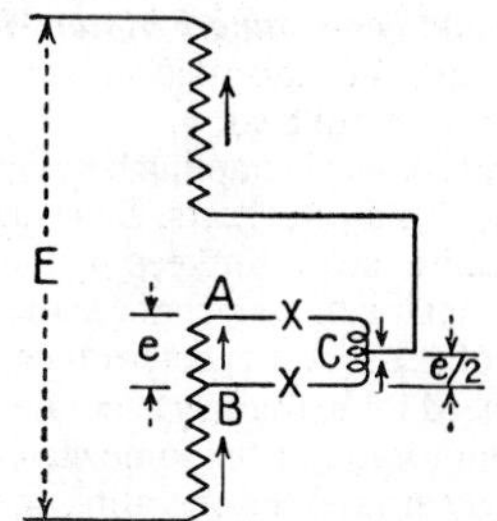


**FIG. 10-25** Bridging position for ratio change under load.

With $A$ and $B$ closed on adjacent taps, the tap voltage $e$ is impressed on the autotransformer $C$ and causes a circulating current to flow through the impedance loop. Because of the autotransformer action, a voltage midway between $A$ and $B$ is impressed in the circuit. The load current again divides equally through the autotransformer windings.

To avoid an excessive voltage drop through the autotransformer when one side is open and at the same time to keep the circulating current at a low level when in the bridging position, the autotransformer is usually designed with an air gap in the magnetic circuit to get a magnetizing current of approximately 60% of the normal full-load current.

***80. Voltage across Autotransformer.*** Figure 10-26 shows the voltage relations across an autotransformer and switching contacts during a tap-changing cycle using an autotransformer designed for 60% circulating current and with 100% load current at 80% power factor flowing through it. Perfect interlacing between the autotransformer halves is assumed, and the voltage drop due to resistance of the autotransformer winding is neglected.

A study of Fig. 10-26 will disclose the fact that increasing the magnetizing reactance of the autotransformer to reduce the circulating current will:

1. Increase the voltage across the full autotransformer winding
2. Increase the voltage to be ruptured
3. Introduce undue voltage fluctuations in the line

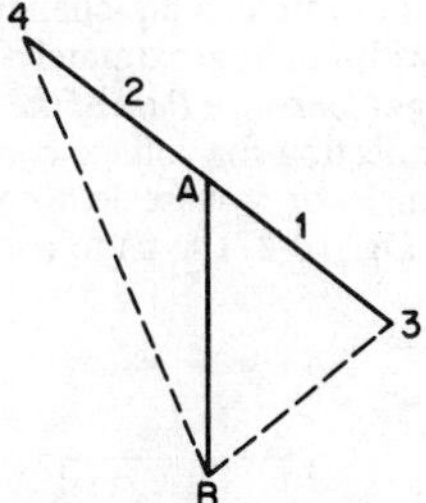


**FIG. 10-26** Vector relations for bridging position.

Since $B$-4 and $B$-3 represent the voltages appearing across the arcing contacts when the bridging position is opened at $A$ and $B$, the voltage-rupturing duty will increase with:

**a.** Increase in voltage between adjacent taps

**b.** Increase in load

**c.** Decrease in power factor of the load

**d.** Decrease in the magnetizing current for which the autotransformer is designed

***81. Load Tap-Changer Motor Mechanisms.*** The mechanism which drives the tap changer, and the control of this mechanism, must be designed so that a tap change, once initiated, is certain to be completed.

The mechanical coupling between the operating motor and the tap-changing switches may be through fixed-ratio gears, Geneva gears, cams, springs, or combinations of these. All mechanisms require means for keeping the motor energized until the change of tap is accomplished and for bringing the tap changer to rest on each operating position. The degree of permissible coasting of the motor is determined by the motor mechanism and the switch design.

The need for extremely accurate stopping of the motor is avoided by arranging the parts so that the motor may coast somewhat after the operating position is reached without moving the tap changer around the operating position. This may be accomplished by the inactive sectors of Geneva gears or cams or by the motor travel inherently involved in recharging a spring. Motor mechanisms are provided with limit switches and mechanical stops to prevent operation beyond the limit positions. Operation counters and position indicators are standard auxiliaries on most tap chargers. On large station-type units where the control devices are generally mounted on a remote-control panel, remote position indicators, either of the lamp type or of the self-synchronizing type, are generally provided.

***82. Automatic Control for Tap Changers.*** It is usual practice to use some sort of voltage-measuring device to control the operation of the motor which drives the tap changer. Such devices may be mechanical, balancing the force of a solenoid actuated by the voltage against weights or springs, or they may be an electrical network, usually a bridge circuit which balances against the voltage of a Zener diode. With either type of device, a voltage higher than a desired upper limit will start the tap-changer driving motor to change to the next lower tap voltage; similarly, a voltage lower than the desired lower limit will cause a change to the next higher tap.

The circuit usually includes a time delay to prevent tap changes which would occur unnecessarily during very short time variations in voltage.

The voltage-regulating relay (or contact-making voltmeter) should be adjusted so that the voltage bandwidth, or spread between voltages at which the raising and lowering contacts close, will be not less than the percentage transformer tap plus an allowance for irregular voltage variations. For example, a tap-changing transformer with 1¼% taps should have a minimum voltage bandwidth of approximately 1¼% + ½% = 1¾%.

***83. Voltage Control, a Part of the Power Transformer.*** The simplest and generally the least expensive connection for voltage control is to provide the necessary taps in the power transformer. For single- or 3-phase delta connection the taps are preferably located on the interior of the winding (Fig. 10-27) so as to avoid the abnormal voltage stresses to which end coils are

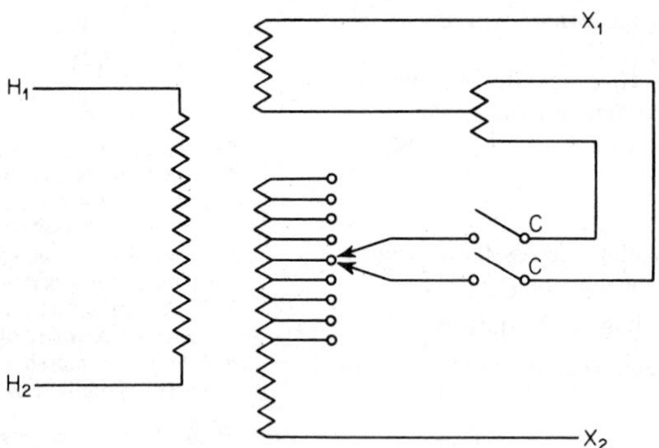


**FIG. 10-27** Tap-changing equipment in the middle of the winding.

usually subjected. In the Y connection the taps may be placed at the neutral end of the winding, and if the neutral is to be solidly grounded, it becomes possible, by locating the taps next to ground, to use load changers designed with greatly reduced insulation; thus, for example, 15-kV apparatus may be placed in the grounded neutral end of a much higher-voltage circuit.

If the rated current of the transformer exceeds that of the switching equipment, a series transformer may be used (Fig. 10-28). Excitation is derived from taps inserted in the secondary of the power transformers, and by means of the series transformer, the desired voltage is inserted into the circuit. Thus, if a ratio of 3 : 1 exists in the series transformer, the current handled by the switching equipment becomes one-third of the current in the line.

***84. Regulating Transformers (Single-Core).*** When a power transformer is not available or it is not desirable to equip the power unit with voltage control, regulating autotransformers are used. In the simplest of these, the necessary taps and switches are placed in the series winding of an autotransformer (Fig. 10-29). For 3-phase circuits, in order that the derived voltage may be in phase with circuit voltage, a Y connection is commonly used, and hence all the precautions necessary to safeguard the operation of Y-connected autotransformers should be observed. A tertiary winding may or may not be provided, depending upon circuit conditions. As the series winding is inserted in the line, adequate insulation must be provided for the tap-changing equipment and taps against the abnormal voltages to which the circuit is subjected.

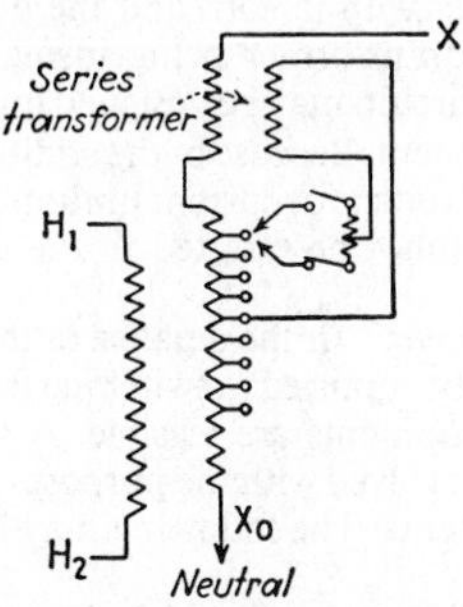


**FIG. 10-28** Tap-changing circuit with taps located in the interior of the transformer winding and an auxiliary series transformer to bring the current and voltage duty on equipment within rating limits.

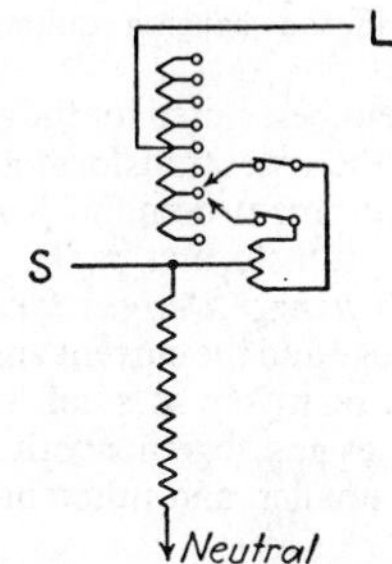


**FIG. 10-29** Regulating single-core autotransformer with taps located in the series winding and circuit connected for boost and buck.

For very high voltage autotransformers the contactor component of the load tap changer may be supported in a compartment on top of the high-voltage common line bushing. The insulation inside the compartment then becomes that required for one tap, and the major insulation to ground is provided by the high-voltage bushing.

Economy in transformer size may be obtained by means of a reversing switch which functions to reverse the connections to the series winding when the regulator is passing through the neutral position. The circuit is so designed and the mechanical sequence is such that the reversing switch operates without rupturing current. The connection diagram (Fig. 10-30) shows the load tap changer provided with nine taps, which gives 17 full-cycle or 33 half-cycle positions. The ratio adjuster is designed with contacts uniformly spaced on the circumference of the circle so as to permit motion through two revolutions.

***85. The Series Transformer.*** In many instances the voltage of the circuit is greater than that for which the switching equipment is designed, and in others the current to be handled exceeds the safe limits of operation. In either case voltage control can be obtained, without the design of special switching equipment, by using an insulating series transformer (Fig.

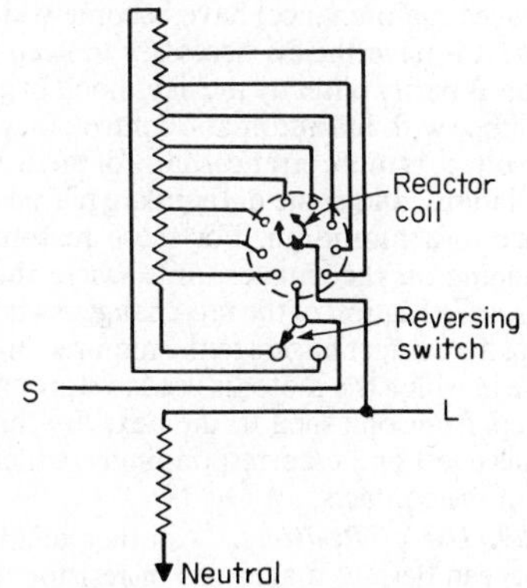


**FIG. 10-30** Regulating single-core autotransformer with a reversing switch to obtain buck and boost.

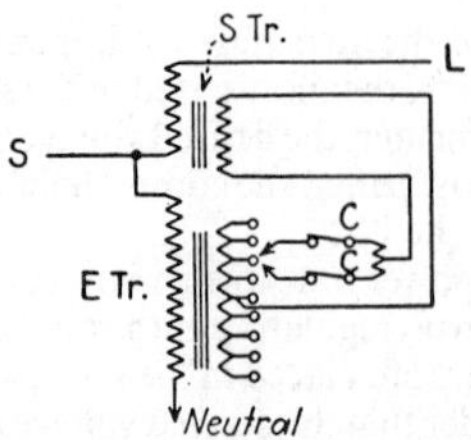


FIG. 10-31 Exciting transformer with taps in the secondary and a series transformer forming complete isolation for tap-changing equipment.

10-31) in addition to the exciting transformer, the combination functioning, as far as the circuit is concerned, like an autotransformer. The primary of the exciting transformer is generally connected in Y in order that the derived voltages may be in phase with circuit voltages. The secondary of the exciting transformer provided with the regulating taps is usually connected in delta. The local circuit, consisting of the secondary of the exciting transformer with its taps and the primary of the series transformer being insulated from the main circuit, may be designed for the voltage and current best suited for the available switching equipment. Because of the additional cost and losses of the series transformer, it is used only when the voltage or current limitations of the switching equipment demand it and when the control cannot be inserted in the grounded neutral of the transformer bank.

***86. Tap-Changer Designs for Moderate kVA and Current.*** In the smaller ratings where both the voltage and the current are moderate, the energy to be ruptured in switching from tap to tap becomes relatively so small that light and simple equipments are feasible. A variety of mechanical designs, together with special circuits, has been evolved with the purpose of providing simpler, smaller, and inherently less expensive equipments. The following may be noted:

1. Designing the tap changer so that it is capable of rupturing the current directly on the same switches which select the taps
2. Designing the circuit so that the tapped winding is reversed in going from maximum to minimum range, thereby securing a substantial reduction in the rating of core and coils for a given output
3. Using higher switching speed, by means of which the life of the arcing contacts is increased

***87. Tap Changers Designed to Interrupt Current.*** The contactors *C* (Fig. 10-27) operate to open the switching circuits so that there is no interrupting duty on the selector contacts which connect to the transformer taps. When the rated current is moderate, it becomes possible to rupture the current directly on the tap-selector switches and thus obtain a major economy in the cost of the mechanical equipment. This method is shown in Fig. 10-30.

***88. High-Speed Switching.*** Large units include contactors with high-speed contacts which serve as extinction devices that are specially designed to interrupt repeatedly the high currents and voltages encountered. They may be single- or multiple-break contactors operating in oil or in air with magnetic-arc chutes, or oil-blast contactors. Vacuum switches with their longer life (reduced maintenance) have become widely used. In small units, however, the arcing duty is mild. It is nevertheless necessary to keep in mind that mild arcing duty in the smaller equipments is partly offset by the likelihood of greater frequency of operation. Such units are usually equipped with full automatic control; they are likely to be located on distribution circuits where the voltage is more erratic. Many of them are located on the lines at considerable distances from substations, and some of them are placed on poles. It is desirable, therefore, to reduce maintenance to a minimum. For these reasons, it is necessary to provide means for high-speed switching on the smaller units where the tap-selector switches are used to rupture current. High-speed action of the tap-changer switches is obtained through Geneva gears or cams which bring the contact fingers to the required high speed at the moment of parting or through a spring drive in which the motor is used to store energy with the release of a spring snapping the contact fingers from one shelf to the next. By these means the duration of the switching arc may be reduced to 1 or 2 c, correspondingly reducing the amount of contact burning and increasing the life of the contacts.

***89. Use of Resistors.*** Another method, used more frequently in load tap changers of European design, makes use of resistors to bridge the tap instead of a preventive autotransformer. Figure 10-32 shows the tap selectors 1 and 2 connected to alternate taps in the winding. The contactor, or diverter switch, shown connected to 1 and $R_1$, progressively connects only to $R_1$, then to $R_1$ and $R_2$, then to $R_2$, and finally to $R_2$ and tap selector 2. For the next tap change, tap selector 1 moves to an adjacent tap and is then followed by the diverter switch operating in a reverse manner from $R_2$ and 2 back to $R_1$ and 1. Since the resistors are designed to carry current for only a very short time, the diverter switch is usually spring-actuated and moves through its sequence in a few cycles of 60-Hz current. This method has the advantage of relatively small-size resistors but requires a transformer tap for each operating voltage, while the autotransformer circuit uses the tap bridging position for an operating voltage and thus requires half the number of transformer taps.

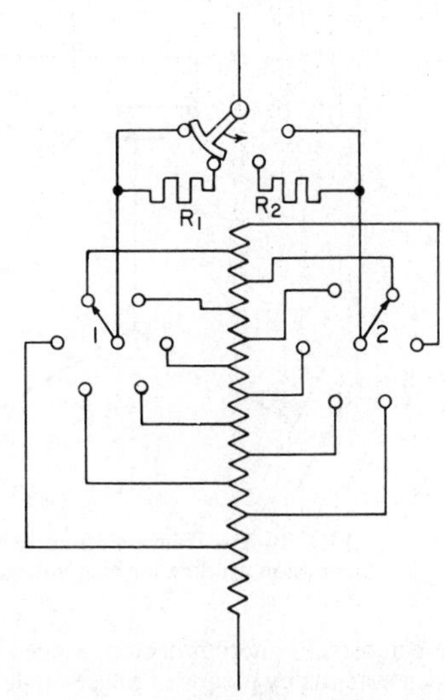


**FIG. 10-32** Tap-changing circuit employing tap selectors and contactor, or diverter switch, and resistors to bridge the taps during switchover.

***90. Applications for Voltage-Control and Equipments.*** The control of transformer ratio under load is a desirable means of regulating the voltage of high-voltage feeders and of primary networks. It may be used for the control of the bus voltage in large distributing substations. It finds a wide field of application in controlling the ratio on step-up transformers operating from power stations whose bus voltage must be varied to suit local distribution.

In industrial work, it is used for the control of current in a variety of furnace operations and electrolytic processes. It also furnishes a convenient means for voltage regulation of concentrated industrial loads.

Many load tap-changer equipments are installed at points of interconnection between systems or between power stations, in order to control the interchange of reactive current, or, in other words, to control the power factor in the tie line. This reactive current may be highly undesirable, especially as it may add to the burden on a fully loaded generating system. It can be increased, eliminated, or reversed by inserting a suitable small ratio of transformation between the systems. It can be varied in amount and in direction of flow to suit varying system conditions if this ratio is variable and under the control of a station operator. Inserting such a ratio of transformation in a tie line by means of a tap-changing equipment is equivalent in its effect on the flow of reactive current to raising or lowering the voltage on one of the systems. Current can be exchanged at any power factor from zero lag to zero lead, without interfering with the voltage maintained on either system.

***91. Transformers for Phase-Angle Control.*** Tap-changing equipment is sometimes used in a loop system for phase-angle control for the purpose of obtaining minimum losses in the loop due to unequal impedances in the various portions of the circuit.

Transformers used to derive phase-angle control do not differ materially, either mechanically or electrically, from those used for inphase control. In general, phase-angle control is obtained by interconnecting the phases, that is, by deriving a voltage from one phase and inserting it in another.

A simple arrangement given in Fig. 10-33*a* illustrates a single-core delta-connected autotransformer in which the series windings are so interconnected as to introduce into the line a quadrature voltage. One phase only is printed in solid lines so as to show more clearly how the quadrature voltage is obtained. The terminals of the common winding are connected to the midpoints of the series winding in order that the inphase voltage ratio between the primary lines *ABC* and secondary lines *XYZ* is unity for all values of phase angle introduced between them.

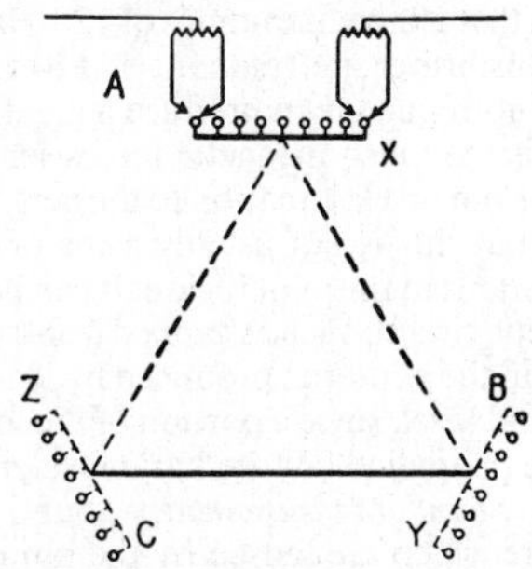


**FIG. 10-33*a*** Phase-shifting regulating transformers. Single-core delta-connected common winding for low-voltage systems.

As large high-voltage systems have be-

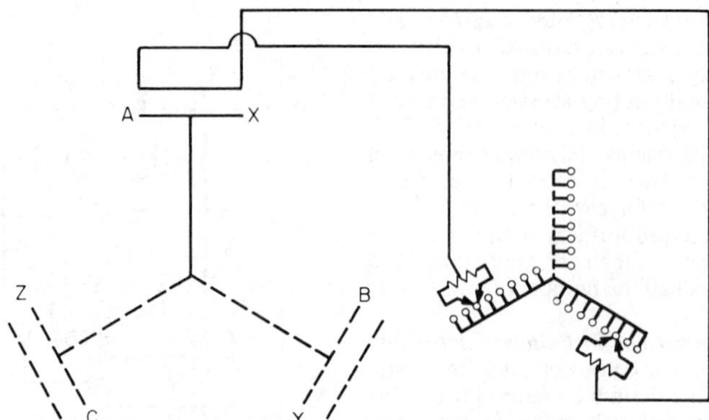


**FIG. 10-33*b*** Phase-shifting regulating transformer. Two-core Y-connected common winding for high-voltage systems.

come extensively interconnected, a need has developed to control the transfer of real power between systems by means of phase-angle-regulating transformers. The most commonly used circuit for this purpose is the two-core, four-winding arrangement shown in Fig. 10-33*b*. The high-voltage common winding is Y-connected, with reduced insulation at the neutral for economy of design, and a series transformer is employed so that low-voltage-switching equipment may be used.

## Audible Sound

***92. Source of Sound.*** Transformers, although they are classed as static apparatus, vibrate and radiate audible sound energy. There are two distinct and different sources. One source is the auxiliary cooling equipment, such as fans, blowers, coolers, and pumps, which are characterized by a broadband frequency spectrum of approximately equal amplitude, commonly called "white noise." The second and major source of transformer sound is the core. This source is characterized by a range of harmonic tones which are even multiples of the exciting frequency. Thus, for 60-Hz excitation, the core will radiate energy at 120, 240, 360, 480, 600 Hz, and so on.

When flux flows in a core lamination, the lamination changes in length by a very small amount, measured in parts per million, but nevertheless large enough to cause the core to vibrate and produce the very audible hum associated with transformers. The change in length is the same for either direction of the alternating flux, and, as the change in length then occurs at each half cycle, the resulting sound frequency is double the source frequency, that is, 120 Hz for a 60-Hz source. Furthermore, the extension in length is not proportional to the flux, and this means that all the harmonics of 120 Hz are generated. Finally, if any of the mechanical parts of the transformer are resonant at 120 Hz or any of the harmonics, they can be excited at the resonant frequency to produce a great deal of vibration and noise.

There are also magnetic forces set up between laminations at the joint region where flux passes from one lamination to the next. These forces may likewise produce motion and generate noise, but this is not usually a major source for core constructions which have overlapping laminations in the joint region. It can be a major source for devices which have butt joints in the magnetic circuit, such as gapped iron-core reactors. Of very minor importance are the magnetic forces in the windings produced by load current. In general, the load current has a minor effect on sound level, since a portion of the load-produced leakage flux passes through the core, but a change in applied voltage has considerable effect.

***93. Sound Measurement.*** Sound waves produce small fluctuations in the atmospheric pressure which are sensed by the human ear. The alternating portion of the total pressure is called "sound pressure." To handle the wide range of sound pressures which are encountered, a quantity called "sound-pressure level," which bears a logarithmic relation to sound pressure, has been introduced. Sound-pressure level is measured in decibels and is defined by[1]

$$P = 20 \log \frac{F}{0.0002} \tag{10-57}$$

where $F$ = rms dynes per square centimeter sound pressure and $P$ = decibels sound-pressure level.

Sound-level-measuring equipment as specified by ANSI Standard S1.4-1971 consists of a microphone, amplifier, frequency weighting network, and indicating meter. Three weighting networks A, B, and C have been standardized. The C network provides an essentially flat response over the frequency range of 20 to 10,000 Hz. The response of the human ear approaches this flat characteristic for very high sound-pressure levels (over 85 dB). At lower levels, the response of the human ear is not linear with frequency, and the A and B networks are weighted to simulate the response of the ear to single-frequency tones of 40 and 70 dB, respectively. Weighted values read on the A and B scales are referred to as "sound levels," rather than sound-pressure levels.

The A, or 40-dB, weighting network has been selected for all transformer sound measurements, since it represents the approximate sound level of a transformer at the nearest residence. With this weighting network, frequencies up to 1000 Hz and above 5000 Hz are given a negative decibel weighting and from 1000 to 5000 Hz slightly positive weighting.

***94. Standard Transformer Sound Level.*** ANSI/IEEE C57.12.90-1980 specifies the method for measuring the average sound level of a transformer. The measured sound level is the arithmetic average of a number of readings taken around the periphery of the unit. For transformers with a tank height of less than 8 ft, measurements are taken at one-half tank height. For taller transformers, measurements are taken at one-third and two-thirds tank height. Readings are taken at 3-ft intervals around the string periphery of the transformer, with the microphone located 1 ft from the string periphery and 6 ft from fan-cooled surfaces. The ambient must be at least 5 and preferably 10 dB below that of the unit being measured. There should be no acoustically reflecting surface, other than ground, within 10 ft of the transformer. The A weighting network is used for all standard transformer measurements regardless of sound level.

NEMA Publication TR 1 contains tables of standard sound levels. For oil-filled transformers from 1000 to 100,000 kVA self-cooled (400,000 kVA forced-oil-cooled) standard levels are given approximately by Eq. (10-58):

$$L = 10 \log E + K \tag{10-58}$$

where $E$ = equivalent two-winding, self-cooled kVA (for forced-oil-forced-air-cooled units, use $0.6 \times$ kVA), $K$ = constant, from Table 10-3, and $L$ = decibel sound level.

**TABLE 10-3** Values of $K$ for Eq. (10-58)

| High-voltage winding BIL, kV | Self-cooled and water-cooled ratings | Forced-air and forced-oil-forced-air-cooled 25 to 35% above self-cooled rating | Forced-air and forced-oil-forced-air-cooled 67% above self-cooled rating or without self-cooled rating |
|---|---|---|---|
| 350 and below | 28 | 30 | 31 |
| 450–650 | 30 | 32 | 33 |
| 750–825 | 31 | 33 | 34 |
| 900–1,050 | 32 | 34 | 35 |
| 1,175 | 33 | 35 | 36 |
| 1,300 and above | 34 | 36 | 37 |

*Example.* A transformer rated 50,000 kVA self-cooled, 66,667 kVA forced-air-cooled, 83,333 kVA forced-oil-forced-air-cooled, at 825 kV BIL, would have standard sound levels of 78, 80, and 81 dB on its respective ratings.

[1] C. M. Harris, *Handbook of Noise Control;* New York, McGraw-Hill Book Company, 1957, pp. 1–14.

***95. Public Response to Transformer Sound.*** The basic objective of a transformer noise specification is to avoid annoyance. In a particular application, the NEMA Standard level may or may not be suitable, but in order to determine whether it is, some criteria must be available. One such criterion is that of audibility in the presence of background noise.[1] A sound which is just barely audible should cause no complaint.

Studies of the human ear indicate that it behaves like a narrow-band analyzer, comparing the energy of a single frequency tone with the total energy of the ambient sound in a critical band of frequencies centered on that of the pure tone. If the energy in the single-frequency tone does not exceed the energy in the critical band of the ambient sound, it will not be significantly audible. This requirement should be considered separately for each of the frequencies generated by the transformer core.

The width of the ear-critical band is about 40 Hz for the principal transformer harmonics. The ambient sound energy in this band is 40 times the energy in a 1-Hz-wide band. The sound level for a 1-Hz bandwidth is known as the "spectrum level" and is used as a reference. The sound level of the 40-Hz band is 16 dB (10 log 40) greater than the sound level of the 1-Hz band. Thus, a pure tone must be raised 16 dB above the ambient spectrum level to be barely audible.

The transformer sound should be measured at the standard NEMA positions with a narrow-band analyzer. If only the 120- and 240-Hz components are significant, an octave-band analyzer can be used, since the 75- to 150-Hz and 150- to 300-Hz octave bands each contain only one transformer frequency. The attenuation to the position of the observer can be determined according to Par. **96.**

The ambient sound should be measured at the observer's position. For each transformer frequency component, the ambient spectrum level should be determined. An octave-band reading of ambient sound can be converted to spectrum level by the equation

$$S = B - 10 \log C \tag{10-59}$$

where $B$ = decibels octave-band reading, $C$ = hertz octave bandwidth, and $S$ = decibels spectrum level.

*Example.* Consider the following case:

Transformer sound at 120 Hz by NEMA method = 72 dB

Transformer-sound attenuation to observer (calc. according to Par. **96**) = 35 dB

Ambient sound at the 75 to 150-Hz octave band = 36 dB

$$72 - 35 = 37 \text{ dB at the observer's position}$$

$$36 - 10 \log (150 - 75) = 17.3\text{-dB ambient spectrum level}$$

The 120-Hz transformer sound at the observer's position exceeds the ambient spectrum level by 19.7 dB. This is 3.7 dB greater than the 16-dB differential which would result in bare audibility; thus the transformer sound will be audible to the observer.

When transformer sound exceeds the limits of bare audibility, public response is not necessarily strongly negative. Some attempts have been made to categorize public response on a quantitative basis when the sound is clearly audible.[2] For a case where specific knowledge of transformer- and ambient-sound-level frequency composition is not available, some more general guides are useful.[3] Typical average nighttime ambient-sound levels for certain types of communities have been established. These are 30 dB for a "quiet-suburban," 35 dB for a "residential-suburban," and 40 dB for a "residential-urban" community. All sound levels are based on the A scale of weighting. Calculations for typical transformer frequency distributions have been made to determine the nighttime transformer noise which will be audible 50% of the time in these communities. The results are 24 dB for quiet-suburban, 29 dB for residential-suburban, and 34 dB for residential-urban. The NEMA standard sound level can be corrected for

[1] *AIEE Committee Report,* Transformer Noise Measurement Methods; *Trans. AIEE,* 1954, vol. 73, p. 683

[2] M. W. Schultz, Jr., and R. J. Ringlee, Some Characteristics of Audible Noise of Power Transformers and Their Relationship to Audibility Criteria and Noise Ordinances; *Trans. AIEE,* 1960, vol. 79, p. 316.

[3] Power Transformer Noise—Prevention and Care; *Stanford Res. Inst. Rep.,* SRI Project S2410, 1960.

attenuation with distance to the nearest observer and checked against the above guides for audibility.

The broadband sound from fans, pumps, and coolers has the same character as ambient sound and tends to blend in with the ambient. While the noise from cooling equipment may be audible to a neighboring observer, it will seldom, if ever, cause a complaint.

***96. Sound Attenuation with Distance.*** A point source in a free field radiates sound in spherical waves. The resultant sound pressure varies inversely with the square of the distance from the source; thus the sound level is reduced by 6 dB for each doubling of distance. The sound of auxiliary cooling equipment follows this relation for decrement with distance, since it is the sum of point-source sound contributions.

The transformer tank, which radiates vibrational energy from the core, is a more complex sound source and does not appear as a point source except at substantial distance from the tank. The modes of tank vibration are complicated, and various parts of the tank may act as independent sources, with different amplitudes, phase relations, and frequencies. Studies of scale models[1] and full-size units have uncovered certain useful relationships as follows:

$$A = 20 \log \frac{2.83D}{Q} \tag{10-60}$$

$$Q = 1.7(WH)^{1/2} \tag{10-61}$$

where $A$ = decibels attenuation for distance exceeding $Q$, $D$ = distance from transformer to observer, $H$ = height of transformer tank, $Q$ = critical distance from transformer beyond which it appears as a point source, and $W$ = width of transformer tank perpendicular to a line from transformer to observer.

Equations (10-60) and (10-61) apply in the absence of wind, temperature gradients, and reflecting surfaces other than ground. Each of these factors may significantly influence the observed sound level at a distance from the source, but not always in predictable fashion.

***97. Selection of Site.*** There are a number of methods available for avoiding transformer-noise complaints. Some of the discussion in the previous paragraphs suggests that potential noise problems should be considered when the substation site is selected. It may be possible to take advantage of attenuation with distance to reduce the transformer sound at the nearest observer position to an inaudible level. It may also be possible to choose the site in a location where the normal ambient noise will mask the transformer sound. If these possibilities are kept in mind during the planning stages, more expensive solutions to noise problems may be avoided later.

***98. Design Measures.*** Manufacturers have at their disposal a variety of means of obtaining sound reduction. Most basic is the use of a reduced level of induction in the transformer-core steel. More steel is required, but the magnitude of the magnetostrictive motion is decreased, with a resultant decrease in radiated energy. Alternatively, better grades of steel having a reduced magnetostriction characteristic can be utilized, or improved corner joint configurations can be employed. These methods can be used individually or in combination to achieve reductions in the order of 10 to 12 dB. Moderate reductions can also be realized by the use of barriers within the tank. Some of these are "soft" barriers, which operate on the principle of absorbing vibrational energy from the core and reducing its transmission to the tank. Others are "mass" barriers, which operate on the principle of loading the tank to decrease its magnitude of vibration for given energy transmission from the core. To achieve large sound reductions (as much as 25 to 30 dB), some manufacturers employ complete external enclosures of steel. For smaller substation units, these enclosures can be preassembled and shipped in place over the transformer tank.[2]

When sound originating at the core is lowered by one of the above means, sound orginating at the cooling equipment may have to be reduced also. Fans on radiators or in coolers are the principal sources. Their contribution to total sound can be decreased by using a larger quantity of slower-speed fans.

---

[1] K. A. Johnson, R. J. Ringlee, and M. W. Schulz, Jr., Use of Scale Models for Studying Power Transformer Audible Noise; *Noise Control,* March, 1956, vol. 2, p. 54.

[2] M. W. Schulz, Jr., and W. J. McNutt, A Way to Get Low Sound Levels in Large Power Transformers—Preassembled Enclosures; *Trans. AIEE,* 1957, vol. 76, p. 1365.

***99. Improving Existing Installation.*** To reduce the sound level of an existing transformer, the most satisfactory method has been found to be the erection of barrier walls on one or more sides of the transformer. The attenuation which can be achieved depends on the transmission loss through the barrier, the diffraction over and around the barrier, and the pressure buildup between the tank and the barrier.

Transmission loss through a barrier wall is a function of the mass of the wall. Structural requirements of most practical masonry barriers ensure sufficient mass to produce 25 to 40 dB attenuation through the wall. The effectiveness is usually limited by diffraction around the edges of the barrier. A theoretical method for calculation of attenuation as limited by diffraction has been formulated[1] as follows:

$$N = \frac{2}{\lambda}[(M^2 - U^2)^{1/2} - M + (G^2 - U^2)^{1/2} - G] \tag{10-62}$$

where $M$, $U$, and $G$ are defined in Fig. 10-34, in any convenient unit (feet or inches), $\lambda$ = wavelength of harmonic under investigation, in units consistent with $M$, $U$, and $G$, and $N$ = dimensionless parameter given in Fig. 10-34.

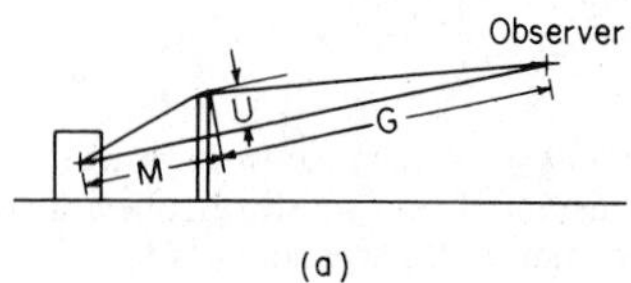


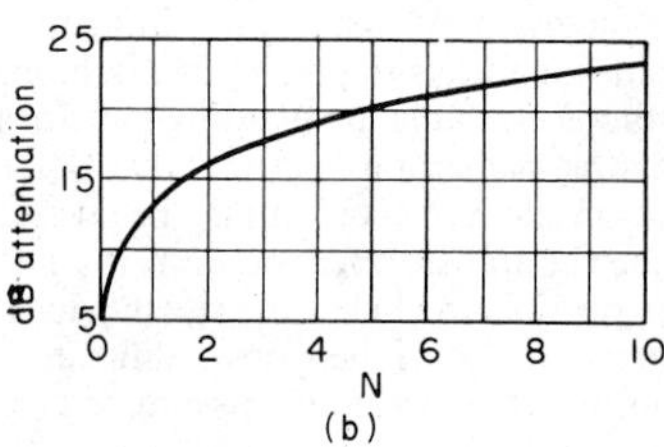


**FIG. 10-34** Effectiveness of a barrier in reducing noise level. *(a)* Identification of dimensions for calculation of the dimensionless parameter $N$ from Eq. (10-62); *(b)* determination of attenuation in decibels.

The calculation procedure is to determine $N$ from the equation and then find the corresponding attenuation from Fig. 10-34*b*.

Test results on models and full-size transformers with two- and three-wall barriers correlate reasonably with Eq. (10-62). Data on four-wall enclosures generally do not correlate. It has been found that approximately 10-dB attenuation can be achieved with a four-wall enclosure having walls 5 ft higher than the transformer.

Enclosures with fewer than four walls should extend at least a distance $M$ beyond the tank, so that attenuation will be limited by diffraction over the top rather than around the ends of the barrier. It should be noted that the sound level on the open side of this type of enclosure will be increased above what it was without the enclosure. Energy is redirected from the critical side of the transformer to the less critical side.

The effective attenuation of an enclosure can be reduced by pressure buildup between the tank and the barrier. The buildup is the result of reflection from hard wall surfaces and reinforcement of direct and reflected waves. Buildup will be most pronounced for spacings between tank and barrier walls which are multiples of the half wavelength of any of the principal sound frequencies. Such spacings should be avoided. Sound-absorbent lining on the interior surface of the barrier walls is helpful in reducing or eliminating buildup.

Masonry enclosures can also be used to hide substation transformers and associated equipment and in that way alleviate complaints which are based on appearance in addition to noise. Some utilities use a three-sided enclosure which resembles the houses in the neighborhood.[2,3] A casual observer on the street may not detect the presence of the substation.

## Partial Discharges

***100. Partial discharges*** may take place in liquid or gaseous dielectrics when the dielectric stress at the point of maximum stress concentration reaches the breakdown level but when

[1] AIEE Committee Report, Sound Barrier Walls for Transformers; *Trans. AIEE,* 1960, vol. 79, p. 932.
[2] *Ibid.*
[3] F. W. Buck, Residential Sub Has Novel Design; *Electr. World,* Dec. 14, 1959, p. 53.

complete breakdown of the dielectric is prevented because the dielectric stress decreases very rapidly away from the point of maximum stress concentration or because a solid dielectric intervenes. One form of such partial discharge, in air around a small conductor at high voltage, has been called "corona" because of its appearance as a visible glow around the conductor surface.

The local breakdown in the region of stress concentration ionizes a path (forms a streamer) in a very short time (microseconds), effectively short-circuiting a small region of the dielectric, and a pulse of current appears in the main dielectric circuit, reflecting the instantaneous short-circuiting of part of the circuit capacitance.

Partial discharges usually are accompanied by chemical decomposition of the liquid or gas, and sometimes they cause erosion of the adjacent solid insulation. A partial discharge in oil usually causes chemical breakdown, with the formation of carbon and gas, and unless the gas is immediately able to escape, more severe discharges in the gas itself may lead to complete breakdown of the insulation structure.

The presence of gas bubbles in the insulation of an oil-insulated transformer may result in partial discharges; this is the reason for particular attention to filling transformers with oil under vacuum.

Partial discharges may also be caused by wet fibers or any small conducting particles which distort the electric field and cause local points of stress concentration.

***101. Partial discharges can be detected*** when they occur within the insulation of a transformer by any of a number of schemes which detect or measure either the pulse of current or the momentary loss of voltage at the transformer terminal. The charge transfer at a terminal can be measured in picofarads but this generally does not give the actual transfer of charge which occurs somewhere within the transformer. *NEMA Publication* 107 describes a method for measuring the equivalent high-frequency voltage, usually at 1 MHz, which appears at the terminals of the transformer. For power transformers the coupling capacitor is replaced by the capacitance of the high-voltage bushing, using the potential tap, as the means for coupling to the high-voltage circuit, with the effect of the capacitive impedance of the bushing being reduced by an adjustable reactor connected to the bushing tap. The voltage-measuring instrument is described in ANSI C63.2-1979.

***102. BIL Reduction.*** The need to demonstrate absence of significant partial discharges in operation is increased for higher circuit voltages where improved surge-arrester characteristics have encouraged a continuing trend toward BIL reduction. Because of progressively decreasing margins between the conventional induced test voltage and operating voltage, new standards for transformers rated 345 kV and above require a 1-h induced voltage test with continuous monitoring of partial-discharge levels to demonstrate the soundness of the insulation. During this test all parts of the insulation system must be overstressed to a degree corresponding to 150% of maximum system voltage at the high-voltage terminals (see IEEE Std. 262B-1977). Similar standards are currently in preparation for lower-voltage transformers.

***103. Partial discharges in transformers*** may also be detected by acoustic transducers in the oil or on the tank wall. If a sensitive transducer shows no partial discharges, any partial discharges picked up on the bushing tap originate outside the transformer. If the transducer shows corona, it can be used to locate the source of partial discharges within the transformer tank by measuring the time interval after the partial discharges voltage appears at the bushing tap until the effect appears at the transducer. Then the distance from the transducer to the source of partial discharges is 1 in for each 15 $\mu$s of delay.

## Radio-Influence Voltage

***104. Excessive partial discharges*** may cause high-frequency voltages to appear at the terminals which can interfere with radio communication. Suitable maximum limits of voltage in compliance with Federal Communication Commission requirements have been established and are shown in *NEMA Publication* TR 1. For power transformers this limits the high-frequency voltage at 1 MHz, measured at about 110% of operating voltage, to 250 $\mu$V up to 14.4 kV operating, 650 $\mu$V up to 34.5 kV operating, 1250 $\mu$V up to 69 kV operating, and 5000 $\mu$V up to 345 kV operating.

## Testing

***105. Standard Tests.*** ANSI/IEEE C57.12.00-1980 defines routine, design, and other tests for liquid-immersed transformers. The following are listed as routine tests for transformers 501 kVA and larger.

1. Measurement of resistances of the windings
2. Measurement of turns ratio
3. Phase-relation tests: polarity, angular displacement, and phase sequence
4. No-load loss and exciting current
5. Load loss and impedance voltage
6. Low-frequency dielectric tests (applied voltage and induced voltage)
7. Leak test on the transformer tank

The following are listed as design tests for transformers 501 kVA and larger (required on only one unit of a given design):

1. Temperature rise tests (This test could be omitted if a unit which is essentially a thermal duplicate had been previously tested.)
2. Lightning-impulse tests (full wave and chopped wave)
3. Audible sound level
4. Mechanical tests of lifting and moving devices
5. Pressure test on the transformer tank

Other tests listed in ANSI/IEEE C57.12.00-1980 (including short-circuit tests and specialized dielectric tests) shall be made only when specified. Test procedures for all routine and design tests (and many other tests) are defined in the test code document ANSI/IEEE C57.12.90-1980.

***106. The regulation*** of a transformer may be determined by loading it according to the required conditions at rated secondary voltage and measuring the rise in secondary voltage when the load is disconnected. The rise in voltage when expressed as a percentage of the rated voltage is the percentage regulation of the transformer. This test is seldom made, because the regulation is easily calculated from the measured impedance characteristics (see Par. **13**).

***107. Efficiency*** of a transformer is seldom measured directly, because the procedure is inconvenient and the efficiency can be readily calculated (see Par. **15**).

***108. Accessories Tests.*** Appropriate tests are made on auxiliary equipment such as current transformers, winding-temperature indicators, load tap-changing equipment, fans, pumps, and the like, for the purpose of checking calibration, operation, and controls.

## Oil-Preservation Systems and Detection of Faults

***109. Oil-Preservation Systems.*** Although transformer oil is a highly refined product, it is not chemically pure. It is a mixture principally of hydrocarbons with other natural compounds which are not detrimental. There is some evidence that a few of these compounds are beneficial in retarding oxidation of the oil.

Although oil is not a "pure" substance, a few particular impurities are most destructive to its dielectric strength and dielectric properties. The most troublesome factors are water, oxygen, and the many combinations of compounds which are formed by the combined action of these at elevated temperatures. A great deal of study has been given to the formation of these compounds and their effects on the dielectric properties of oil, but there apparently is no clear relation between these compounds and the actual dielectric strength of the transformer insulation structure.

Oil will dissolve in true solution a very small quantity of water, about 70 ppm at 25°C and 360 ppm at 70°C. This water in true solution has relatively little effect on the dielectric strength of oil. If, however, acids are present in similar amounts, the capacity of oil to dissolve water is increased, and its dielectric strength is reduced by the dissolved water. Small amounts of water in suspension cause severe decreases in dielectric strength. The primary reason for concern over moisture in transformer oil, however, may not be for the oil itself but for the paper and pressboard which will quickly absorb it, increasing the dielectric loss and decreasing the dielectric strength as well as accelerating the aging of the paper.

It is generally recognized today that the best answer to the problem of air and water is to eliminate them and keep them out.

For this purpose, in American practice transformer tanks are completely sealed. About three basic schemes are used in sealed transformers to permit normal expansion and contraction of oil (0.00075 per unit volume expansion per degree Celsius) as follows:

**1.** A gas space above the oil large enough to absorb the expansion and contraction without excessive variation in pressure. Some air may unavoidably be present in the gas space at the time of installation but soon the oxygen mostly combines with the oil without causing significant deterioration, leaving an atmosphere which is mostly nitrogen.

**2.** A nitrogen atmosphere above the oil maintained in a range of moderate positive pressure by a storage tank of compressed nitrogen and automatic valving. This scheme has the advantage that entrance of air or moisture is prevented by the continuous positive internal pressure, and the disadvantage of somewhat higher cost.

**3.** A flexible synthetic-rubber diaphragm floating on top of the oil. This scheme has the advantage that the oil is never under pressure or vacuum or under variable pressure, and the disadvantage of higher cost. A number of mechanical variations and elaborations of this general idea have been devised.[1]

It is now generally recommended that the constant-pressure oil-preservation system of item 3 be employed on all high-voltage power transformers (345 kV and above). This is a consequence of unfavorable experience with transformers having gas-cushion systems, which inherently operate with large quantities of the cushion gas in solution in the hot oil under load. If the oil is suddenly cooled (reduction of ambient temperature or load), the oil volume contracts and the static pressure of gas over the oil drops rapidly, allowing free gas bubbles to come out of solution throughout the insulation system. The dielectric strength of the oil and cellulose insulation system is drastically weakened when it has free gas inclusions, and this has occasionally led to electrical failure of operating transformers.

***110. Detection of internal faults*** in transformers at an early stage of their development is most desirable to limit the extent of damage. Two levels of seriousness of faults are recognized. Incipient (or developing) faults have not yet progressed to the point where they affect the functional capability of the transformer, but it is likely that their seriousness will increase with time if not corrected. Examples would include partial discharge sites within the insulation, intermittent low-energy sparking, overheated conductor insulation, or hot metal parts in contact with oil only. More serious or permanent internal faults affect functionality immediately and must be removed quickly before their consequences can jeopardize the safety of personnel or other equipment.

Most commonly employed means of sensing incipient faults relate to detection of gases generated at the fault site. Automatically operating gas-detection devices which can be supplied on the transformer employ any of the following principles:

**a.** Free gas accumulation at the cover

**b.** Sensing of combustible gases within a gas cushion over the oil

**c.** Separation of certain gases dissolved in the oil

In addition, periodic manual sampling of the oil for laboratory analysis can be practiced. The composition of the gas dissolved in the oil is very useful for diagnosis of the nature of an incipient fault.

Permanent internal faults can be detected by fault-pressure relays or differential relays,

[1] Much of the above material has been adapted from Bean, Chackan, Moore, and Wentz, *Transformers for the Electric Power Industry;* presently published by the Transformer Division, Westinghouse Electric Corporation. (Formerly published by McGraw-Hill.)

either of which give a signal which can be used to trip circuit breakers and remove the transformer from the system. The fault-pressure relay senses the sudden buildup of pressure produced by arc-generated gases after a fault has occurred. Unfortunately such relays can also be operated by any other event which causes a rapid pressure change, so they cannot be set to be too sensitive. Differential relays sense that more current is flowing into the transformer than is flowing out, but relays which are insensitive to the initial inrush of exciting current should be used. After a transformer has been disconnected as a result or relay operation, it is always desirable to get it back into service as quickly as possible. Following differential-relay operation, circuit breakers may be reclosed to check whether the fault is self-healing. The penalty for reconnection of a damaged transformer is that if the fault recurs, the damage to the transformer and possibly associated equipment will be greater. Under no circumstances should a transformer be reconnected to the system following operation of a fault pressure relay without thorough investigation of the cause of the relay operation.

When a transformer has been taken out of service because of fault indications, the following procedure should be used:

1. Take samples of the gas from the gas space for analysis to determine whether products of decomposition are present.
2. If there is no gas space, take oil samples for extraction of dissolved gases for similar analysis.
3. Remove the manhole cover to see what can be observed visually. Sometimes the odor of burning is quite obvious.
4. Make insulation power factor, insulation resistance, and turns ratio tests to check whether their results conform to normal values.
5. Perform any other tests which seem to be indicated by the results of the first tests.
6. Check the operation and calibration of the protective relay.

### Overcurrent Protection

***111. Effects of Overcurrent.*** A transformer may be subjected to overcurrents ranging from just in excess of nameplate rating to as much as 10 or 20 times rating. Currents up to about twice rating normally result from overload conditions on the system, while higher currents are a consequence of system faults. When such overcurrents are of extended duration, they may produce either mechanical or thermal damage in a transformer, or possibly both. At current levels near the maximum design capability (worst-case through fault), mechanical effects from electromagnetically generated forces are of primary concern. The pulsating forces tend to loosen the coils, conductors may be deformed or displaced, and insulation may be damaged. Lower levels of current produce principally thermal heating, with consequences as described in Par. **144** on loading practices. For all current levels the extent of the damage is increased with time duration.

***112. Protective Devices.*** Whatever the cause, magnitude, or duration of the overcurrent, it is desirable that some component of the system recognize the abnormal condition and initiate action to protect the transformer. Fuses and protective relays are two forms of protective devices in common use. A fuse consists of a fusible conducting link which will be destroyed after it is subjected to an overcurrent for some period of time, thus opening the circuit. Typically fuses are employed to protect distribution transformers and small power transformers up to 5000 to 10,000 kVA. Traditional relays are electromagnetic devices which operate on a reduced current derived from a current transformer in the main transformer line to close or open control contacts which can initiate the operation of a circuit breaker in the transformer line circuit. Relays are used to protect all medium and large power transformers.

***113. Coordination.*** All protective devices, such as fuses and relays, have a defined operating characteristic in the current-time domain. This characteristic should be properly coordinated with the current-carrying capability of the transformer to avoid damage from prolonged overloads or through faults. Transformer capability has been defined in general terms in a new guide document, ANSI/IEEE C57.109-1984, "Transformer Through Fault Current Duration Guide." The format of the transformer capability curves is shown in Fig. 10-35. The solid curve, $A$, defines the thermal capability for all ratings, while the dashed curves, $B$ (appropriate to the

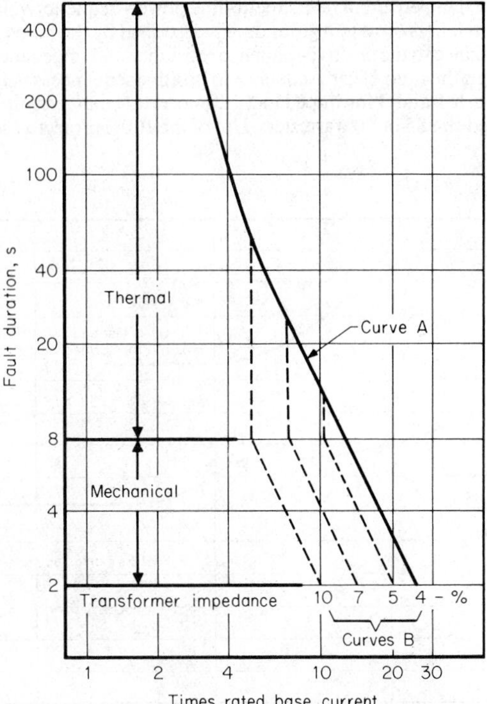


**FIG. 10-35** Transformer through-fault protection curves.

specific transformer impedance), define mechanical capability. For proper coordination on any power transformer, the protective-device characteristic should fall below both the mechanical and thermal portions of the transformer capability curve. (See ANSI/IEEE C57.109-1984 for details of application.)

## Protection against Lightning[1]

***114. A transformer may be subjected to severe lightning voltages*** as a result of a direct stroke to the transformer terminal, adjacent bus, or transmission line. Less severe voltages may result from strokes on a distant part of the system or from strokes to ground near the system. Since lightning voltage may exceed the insulation strength of the transformer, protection is necessary.

Volt-time curves are used in evaluating protection, because for short times the insulation strength changes significantly with duration of voltage. Protection is effective if the volt-time curve of the transformer is above the volt-time curve of the protective equipment, so that for any time duration the kilovolts insulation strength of the transformer exceeds the protective level at the same duration. The volt-time curves of transformer insulation have considerable "turn-up," that is, for durations under 10 $\mu$s the kilovolts insulation strength is much greater. Rod gaps in air are unsuitable for protecting transformers, because they have even more turn-up than transformers.

***115. Surge Arresters.*** The modern surge arrester has very little turn-up and is an essential adjunct to the transformer whenever there is lightning exposure.

The required surge arrester rating depends upon the effectiveness of the neutral grounding.

[1] ANSI C62.2-1981, Guide for the Application of Surge Arresters.

The rating is expressed in percent of rated line-to-line power-frequency voltage that the arrester will withstand. Effectiveness of system grounding is described by the ratios of the zero-sequence resistance and impedance to the positive-sequence resistance and impedance. An 80% arrester is commonly used when the ratio of zero sequence to positive sequence is between 0.5 and 1.5 for resistance and between 1 and 3 for impedance. Lower ratios may permit 75 or 70% arresters. Higher ratios may require 85 or 90% arresters. Use of the 100% protective level is not economical at high voltages.

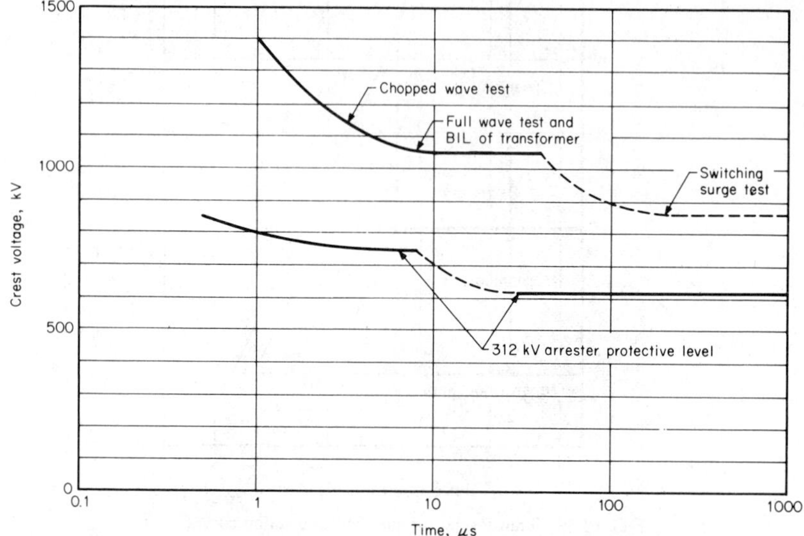


**FIG. 10-36** Surge-arrester protection margin for impulse and switching-surge conditions for a 345-kV transformer with BIL reduced two steps. The arrester curve shown is for an $8 \times 20$ μs current wave of 15-kA crest.

Figure 10-36 shows the volt-time curve of a 345-kV transformer, with 1050-kV BIL (reduced two steps) compared with the volt-time curve of a 90% metal-oxide surge arrester. Since this new type of arrester has no series gaps to spark over, the characteristic is described as a protective level rather than a sparkover level. The volt-time curve of an arrester depends on the amount of impulse current. The volt-time curve shown in Fig. 10-36 corresponds to 15,000-A crest, which is not likely to be exceeded except by direct strokes. For best protection, the arrester should be located as close as possible to the terminals of the transformer, and the arrester ground should be connected by a short, direct conductor to the transformer tank and substation grounding system.

***116. The steep rate of rise of lightning current*** such as may result from a nearby direct stroke will produce discharge voltages higher than shown in Fig. 10-36 and may damage the lightning arrester. Therefore the substation and the first half mile of connected transmission lines should be protected against direct strokes by a suitable combination of grounded masts and ground wires.[1]

## Installation and Maintenance

***117. Careful installation and maintenance*** of power transformers are needed for long life in service. General requirements from ANSI C57.93 and from manufacturer's instructions are

[1] H. Linck, Shielding of Modern Substations Against Direct Lightning Strokes; *IEEE Trans. Power Appar. Syst.*, September 1975, vol. PAS-94, no. 5, p. 1674.

summarized below. Many power transformers have in addition special instructions on installation and maintenance which should be carefully observed. The following information applies to the majority of power transformers without special features or unusual ratings.

***118. Shipment.*** Transformers are thoroughly dried, tested, and inspected before shipment. Power transformers are generally shipped as fully assembled as shipping limits permit. If the oil is shipped separately, the tank is filled with dry air or nitrogen. Windings are connected for maximum rated voltage, except that, if there are taps above rated voltage, the taps are connected for rated voltage.

***119. Inspection on Arrival.*** Before removal from the car, inspect for damage from rough handling. If damage is found, a claim should be filed with the carrier and the manufacturer should be notified.

***120. Transformers shipped oil-filled*** should be inspected for evidence of entrance of moisture during shipment. If the transformer is received in damaged condition, indicating that water or other foreign material has had an opportunity to enter the tank, tests should be made to check the transformer for dryness. Insulation power-factor measurement can be used for this purpose. Power factor of new transformers runs from 0.3 to 0.5% at 25°C. The transformer should be dried if the power factor is over 0.6% at 25°C. In any case a power-factor measurement may be helpful for comparison with subsequent readings.

***121. Transformers shipped gas-filled*** are fitted with a pipe connection on the cover to which a vacuum pressure gage with a range of $-10$ to $+10$ lb/in$^2$ may be connected. A positive or negative pressure indicates that the tank is tight; a continuous zero reading indicates that it leaks.

Inspect the core and coil assembly for signs of damage. CAUTION: If the transformer is filled with nitrogen, it must be purged before anyone enters the tank in order to avoid asphyxiation. The nitrogen gas can be purged with dry air or in dry weather blown out with a fan. Even in good weather with moderate humidity the exposure of the core and coil assembly to the atmosphere should be limited to a maximum of 24 h. Afterward the transformer should be vacuum-dried at an absolute pressure not to exceed 2 mmHg for 4 h longer than the period of exposure, to evaporate surface moisture which may have been deposited.

***122. Handling.*** Lifting cables should be held apart by a spreader to avoid damaging bushings or other parts in lifting a transformer. If a transformer cannot be handled by crane, it may be skidded or moved on rollers, provided that care is taken not to tip the transformer or damage the base. A transformer should never be lifted or moved by attaching jacks or tackle to the drain valve, cooler, or radiator connections or other accessories. If a tool or other foreign object is dropped into the transformer, it should be removed before the transformer is energized. Care should be exercised during the entire installation process to protect the transformer against the entrance of moisture.

***123. Storage.*** When a transformer is received from the manufacturer, it should be placed in its permanent location to minimize handling. A transformer shipped gas-filled may be stored gas-filled indefinitely provided that it is equipped with a pressure gage which continues to show a positive pressure in the tank. Otherwise the transformer should be filled with oil and the oil-preservation system made fully effective. Bushings and accessories that are shipped separately should be protected against absorbing moisture until they are installed in the transformer.

***124. Oil Sampling.*** Samples of oil should be taken from the bottom. An oil-sampling valve is provided at the bottom of the transformer tank for this purpose. A metal or glass thief tube can be conveniently used to obtain a bottom sample from an oil barrel. Test samples should be taken only after the oil has settled for some time, varying from 8 h for a barrel to several days for a large transformer. Cold oil is much slower in settling. In drawing samples of oil from a sampling valve, some oil should first be discarded so that the sample will come from the bottom of the container and not from the sampling pipe. Examine a sample in a clear glass container for free water, which in any quantity is readily observable. The sample container should be a large-mouthed glass bottle, 1 qt or larger, with cork or glass stopper. The bottle should be carefully cleaned and dried before being used. Bottles should be amber color if samples are to be stored to be tested later for color or sludge-forming characteristics. Refer to ASTM 923-81 for important details of oil-sampling technique.

***125. Testing for Dielectric Strength.*** The testing fixture should be cleaned thoroughly to remove any particles or fibers and rinsed out with a portion of the oil to be tested. The testing fixture should be filled with oil, both oil and fixture being at room temperature. Allow 3 min for air bubbles to escape before applying voltage. Tests are made by two methods. ASTM D877-82 uses 1-in-diameter square-edge electrodes spaced 0.10 in apart and a rate of voltage rise of 3000 V/s. ASTM D1816-82 uses special radiused-surface electrodes spaced 0.04 in apart, with continuous oil circulation, and a rate of voltage rise of 500 V/s. The latter test is more sensitive to slight moisture or particulate contamination. In either case, the average voltage for five breakdowns is taken as the dielectric strength of the oil. Strength of new oil should exceed the minimum value for good oil as shown in Table 10-4. (See also ANSI/IEEE C57.106-1977.)

**TABLE 10-4** Dielectric Strength of Oil

| KV average dielectric strength by ASTM D877-82 | KV average dielectric strength by ASTM D1816-82 | Condition of oil |
|---|---|---|
| 30 or over | 29 or over | Good |
| 26 to 29 | 23 to 28 | Usable |
| Under 26 | Under 23 | Poor |

***126. Filtering to Increase Dielectric Strength.*** If oil tests below "good," it should be filtered to remove impurities and moisture. It is best to discharge filtered oil into a clean, dry tank and avoid mixing with unfiltered oil. If the filtered oil must be discharged back into the transformer tank, the oil should be withdrawn from the bottom filter-press valve and, after filtering, returned through the top filter-press valve. Oil should not be filtered while the transformer is energized, because the dielectric strength may be temporarily reduced by aeration. If no facilities are available for making dielectric tests, send a sample to the manufacturer marked with the serial number of the transformer.

***127. Drying the core and coils*** should be necessary only if some accident occurs during shipping, storage, or service to break the seal and permit moisture or air to enter, or unless the transformer has been overloaded to the point where excessive temperature has broken down the paper or pressboard, with release of considerable quantities of water. However, if drying of the core and coils is found to be necessary, it may be accomplished by the following means.

1. Heat in oil
2. Heat in air
3. Vacuum
4. Heat and vacuum

(Refer also to ANSI/IEEE C57.12.11-1980.)

***128. Heat in Oil.*** The oil is heated to a temperature of 75 to 85°C by applying reduced voltage to one winding with another winding short-circuited. It may be desirable to close radiator valves, reduce flow of cooling water, or blanket the tank, to reduce the amount of heat required. If the current in the winding does not exceed three-quarters of the rating, the winding temperature will not be greatly in excess of the maximum oil temperature, and relatively high oil temperature can be obtained without overheating the winding. Ventilation should be obtained by slightly raising manhole covers and protecting the openings from the weather. The cover should be lagged with thermal insulation to prevent recondensation of moisture on the underside. The pumps of forced-oil-cooled transformers should be operated during drying to circulate the oil through the windings. The oil should not be filtered during the drying process. Table 10-5 shows the effect of oil temperature on the maximum short-circuit current that can be held without injury. These values should not be exceeded.

***129. When to Discontinue Drying.*** Drying should be continued until oil from both top and bottom of the tank meets the requirements for good oil of Table 10-4. The ventilating openings should then be closed and the oil kept at the same temperature for another 24 h while the oil is tested at 4-h intervals. If the dielectric strength of the oil decreases, moisture is still passing from the insulation into the oil, and the ventilators should be opened, the oil filtered, and the drying

**TABLE 10-5** Maximum Current for Drying in Oil

| Top Oil Temperature, °C | Maximum Percent of Rated Current |
|---|---|
| 75 | 100 |
| 80 | 85 |
| 85 | 50 |

process continued. If the dielectric strength of the oil remains constant or increases, the transformer should be operated at approximately two-thirds voltage, the same oil temperature being maintained, and the oil should be filtered every 24 h until 24 h elapses without a decrease in oil strength. After a satisfactory two-thirds voltage test, full voltage should be applied until 24 h elapses without a decrease in oil strength. Water-cooled and forced-oil-cooled transformers may require some operation of the cooling system in order to hold the top oil temperature within the 85°C limit during this process. This is a very slow method of drying and is not effective for transformers containing large amounts of insulation.

***130. Heat in Air.*** The transformer core and coil assembly should be placed in a box with holes near the top and bottom for air circulation. The clearance between the sides of the transformer and the box should be small so that most of the heated air will pass up through the ventilating ducts of the coils and not around the sides. The heat should be applied near the bottom of the box. With some types of transformers it is better to distribute the heat around the lower part of the coils. The best way to obtain heat is from resistance units. Air going into the enclosure should not exceed 90°C. The core and coils should be carefully protected against direct radiation from the heaters. There should be no flammable material near the heaters. The box should be lined with fire-proof insulation. When forced-air circulation is used, suitable baffles should be placed between the heater and the inlet to the transformer enclosure.

***131. Precautions to Be Observed.*** As the drying temperature approaches the point where fibrous materials deteriorate, great care must be taken to see that there are no points where the temperature exceeds 85°C. Several thermometers should be used. They should be placed in the coils near the top and screened from air currents. As the temperature rises rapidly at first, the thermometers must be read at intervals of about ½ h. In order to have a reference temperature for insulation-resistance measurements, one thermometer should be placed where it can be read without removing it or changing its position. The other thermometers should be shifted about until the hottest points are found and should then remain at these points throughout the drying period. When possible, the temperature should be checked by the increase-in-resistance method.

***132. Vacuum.*** A vacuum of approximately 50 μm is drawn on the core and coils in the tank without oil. A pump having a capacity of 50 ft³/min and 10-μm capability is recommended. The moisture is evaporated from the insulation and removed as vapor. A cold trap is placed in the vacuum line between the transformer and the vacuum pump, to freeze out the water vapor. The cold trap contains a thimble, which can be filled with liquid nitrogen or a mixture of acetone and dry ice, to lower the temperature and freeze the water vapor. The cold trap can be removed and its contents melted. The condensate will be a mixture of oil and water and should be allowed to separate so that the water can be decanted and measured. A curve showing the rate of water removal versus time can be plotted to show the progress in drying. A transformer of average size is dry when no more than 24 fluid ounces of water is removed in each of three successive 24-h periods. A typical drying curve is shown in Fig. 10-37.

***133. Heat and vacuum*** can be applied as in factory processing with a portable boiler to supply a vaporized petroleum fraction (like kerosene) under vacuum to the transformer tank. The vapor condenses on the core and coils, heating them to the desired temperature, and then drains out of the tank. Simulta-

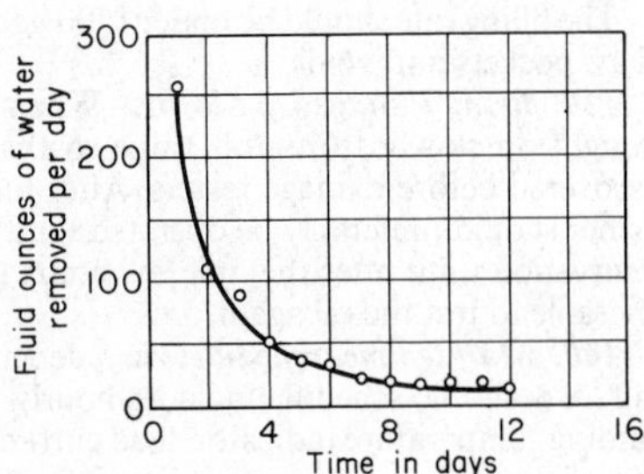


**FIG. 10-37** Typical drying curve showing water removed by a cold trap at a vacuum of 60 to 65 μm.

neously a vacuum connection to the top of the tank withdraws water vapor and evolved gases, which can be passed through a cold trap.

***134. Time required for drying*** is generally 1 to 3 weeks, depending on the condition and construction of the transformer and the method of drying. In general, the use of high vacuum and a cold trap is faster, safer, and more economical than the use of heat alone.

***135. Insulation resistance*** will indicate the degree of dryness only when the transformer is dried without oil. If the initial insulation resistance is measured at room temperature, it may be high, although the insulation is not dry, but as the transformer is heated up, it will drop rapidly.

As the drying proceeds at a constant temperature, the insulation resistance will generally increase gradually until toward the end of the drying period, when it increases quite rapidly and then levels off at a high value. The drying should continue until the resistance is constant for a period of 12 h.

***136. Insulation power-factor readings*** (at 60 Hz) will indicate the degree of dryness. The power factor will first increase as the temperature increases and then will gradually decrease as drying progresses. Drying should continue until the power factor is constant for a period of 12 h. If power factor is measured on transformers dried in oil by the short-circuit method, the power factor should be used to supplement oil tests as a measure of dryness.

***137. Filling without Vacuum.*** (Note: This method should only be practiced on low-voltage transformers. Check manufacturer's recommendations.) Use extreme care to keep moisture out of the core and coils. The tank should not be opened to the atmosphere until the core and coils are under oil, unless vacuum filling is available. The oil shipping tank or oil drums should not be opened until their temperature is the same as or higher than that of the surrounding air and the transformer is in place and ready to receive the oil. Metal or synthetic rubber hose should be used for filling, because transformer oil is contaminated by natural rubber. Oil should never be added to a transformer without passing through a filter press.

Static charges can be developed when transformer oil flows in pipes, hoses, or tanks. Oil leaving a filter press may be charged to over 50,000 V. To accelerate dissipation of the charge in the oil, ground the filter press, the tank, and all bushings or winding leads during oil flow into any tank. Conduction through oil is slow; therefore it is desirable to maintain these grounds for at least 1 hr after the oil flow has ended.

Avoid explosive gas mixtures in any container into which oil is flowing. Arcs can occur along the surface of the charged oil even though all metal is grounded.

***138. Filling with Vacuum.*** The vacuum line should be connected to a tapped opening on a cover-mounted shipping plate or to a valve near the top of the tank. An opening of 2 in minimum is recommended. The oil line can be connected to a suitable opening on a cover-mounted shipping plate or the top filter-press valve. The oil line should always be connected at the top of the tank so that the oil can be deaerated as it enters.

Transformers with operating voltage less than 161 kV and with core and coils not exposed to the atmosphere should be filled under vacuum better than 25 mm Hg absolute pressure. The vacuum should be held 4 h before filling and continued during filling until the core and coils are covered. The vacuum can then be removed for installation of bushings and the remaining oil added without vacuum. Transformers with an operating voltage of 161 kV and above or transformers with core and coils exposed to the atmosphere should be completely filled under a vacuum better than 2 mm Hg absolute pressure. A 2-mm vacuum should be held until the tank is filled to the 25°C level.

The filling rate should be under 1500 gal/h to facilitate evacuation and complete oil filling of all air pockets and voids.

***139. Bring Voltage Up Slowly.*** When the voltage is first applied, it should, if possible, be brought up slowly to its full value so that any wrong connection or other trouble will be discovered before damage results. After full voltage has been applied successfully, the transformer should preferably be operated for a short period without load. It should be kept under observation until after the first few hours that it delivers load. After 4 or 5 days' service it is advisable to test the oil again.

***140. Maintenance schedules vary,*** depending on the size, complexity, and importance of the unit. A desirable schedule includes hourly readings of ambient temperature, oil temperature, winding-temperature indicator, load current, voltage, and tank pressure and also daily observation of the tank pressure gage, liquid-level indicator, and automatic gas seal equipment and monthly check of control circuits, alarm circuits, and cooling fans. In addition fan-motor bearings should be lubricated every 2 years or after 6000 h of operation, and oil should be tested after a few days, after 6 months, after 12 months, and then once a year; oil-to-air heat exchangers should be cleaned as necessary; and load-tap-changing equipment should be serviced as called for in the manufacturer's instructions.

***141. Internal inspection*** of transformers in service should be made only after a specific indication of trouble, such as detection of combustible gas, either by a portable gas detector or by a permanent gas-detector relay. The greatest sensitivity is offered by a gas-detector relay on a transformer with the tank completely filled with oil and external provision for expansion of oil. Generation of combustible gas usually indicates internal trouble (not necessarily serious). Analysis of the gas sometimes helps to identify the source. If collection of combustible gas continues without discoverable cause, partial discharge voltage measurement may establish whether or not there is an internal fault and ultrasonic measurements may locate the fault (see Par. **103**).

***142. Idle Water Coolers.*** When a transformer is idle and exposed to freezing temperatures, water should be completely drained from the coolers.

***143. Operating without Cooling.*** A liquid-cooled transformer should not be run continuously, even at no load, without the cooling liquid. In an emergency, forced-oil air-cooled transformers may be operated without fans and pumps (1) at rated load for approximately 1 h, starting at full-load temperature rise, (2) at rated load for approximately 2 h, starting cold (at ambient temperature), (3) at rated voltage and no load for approximately 6 h, starting at full-load temperature rise, and (4) at rated voltage and no load for approximately 12 h, starting cold.

When only a portion of the cooling equipment is operating, the transformer may be operated at reduced load approximately as indicated in Table 10-6.

**TABLE 10-6** Operation with Limited Cooling Equipment

| Percent of Cooling Equipment in Operation | Percent of Rated Load That May Be Carried |
|---|---|
| 33 | 50 |
| 40 | 60 |
| 50 | 70 |
| 80 | 90 |

## Loading Practice

***144. Temperature Limitation of Loading.*** Ordinarily the kVA that a transformer should carry is limited by the effect of reactance on regulation or by the effect of load loss on system economy. At times it is desirable to ignore these factors and increase the kVA load until the effect of temperature on insulation life is the limiting factor. High temperature decreases the mechanical strength and increases the brittleness of fibrous insulation, making transformer failure increasingly likely, even though the dielectric strength of the insulation material may not be seriously decreased. Overloading of transformers should be limited by reasonable consideration of the effect on insulation life and the probable effect on transformer life.

***145. The insulation life of a transformer*** is defined as the time required for the mechanical strength of the insulation material to lose a specified fraction of its initial value. Loss of 50% of the tensile strength is the usual basis for evaluating conductor insulation for power transformers.

***146. The aging of insulation*** is a chemical process which occurs more rapidly at higher temperatures according to the Arrhenius reaction-rate theory, as expressed in Eq. (10-63),

$$\log h = \frac{K_1}{C + 273} + K_2 \tag{10-63}$$

where $C$ = temperature in degrees Celsius of insulation, $K_1, K_2$ = constants determined by test, and $h$ = hours of life.

Use of this equation permits results of relatively short duration tests at relatively high temperature to be extrapolated to indicate probable insulation life at moderate temperatures.

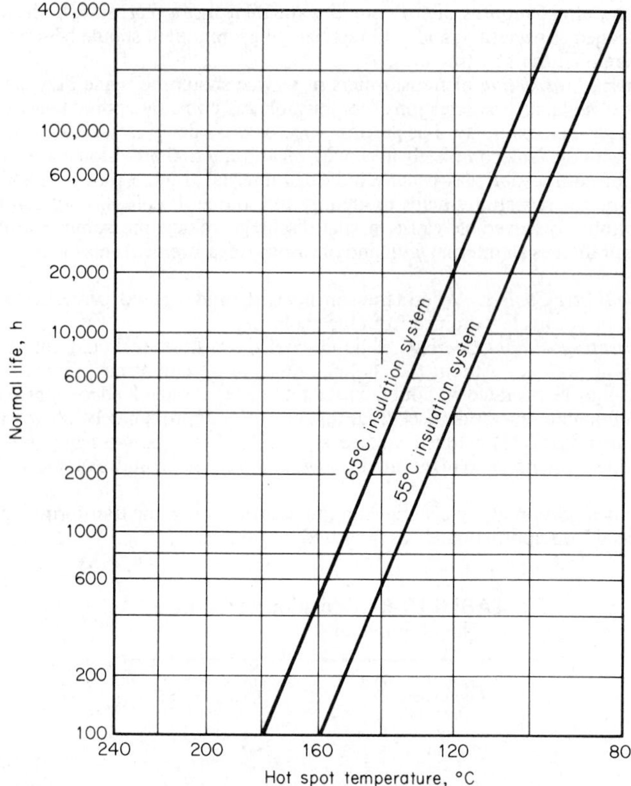


**FIG. 10-38** Loss of life as a function of temperature for power transformers up to 100 MVA under loading conditions in ANSI/IEEE C57.92-1981.

ANSI/IEEE C57.92-1981 contains loading recommendations for power transformers up to 100 MVA with 55°C and 65°C average winding-rise insulation systems based on extrapolated life tests. Figure 10-38 shows the corresponding curves of rate of loss of life as a function of temperature as defined in this document. The constants in Eq. (10-63) are

$$55°\text{C rise} \qquad K_1 = 6972.15 \qquad K_2 = -14.133$$

$$65°\text{C rise} \qquad K_1 = 6972.15 \qquad K_2 = -13.391$$

ANSI/IEEE C57.91-1981 provides similar loading recommendations for distribution transformers, including the following values for the constants in Eq. (10-63):

$$55°\text{C rise} \qquad K_1 = 6328.8 \qquad K_2 = -11.968$$

$$65°\text{C rise} \qquad K_1 = 6328.8 \qquad K_2 = -11.269$$

Accepted methods for functional life evaluation have been established for distribution transformers,[1] but they are just being explored for power transformers.[2,3]

---

[1] ANSI C57.100-1974, Test Procedure for Thermal Evaluation of Oil-Immersed Distribution Transformers (IEEE Standard 345).

[2] W. J. McNutt, A Proposed Functional Life Test Model for Power Transformers, *IEEE Trans.*, Sept./Oct. 1977, vol. PAS-96, pp. 1648–1656.

[3] W. J. McNutt, G. H. Kaufmann, Evaluation of a Functional Life Test Model for Power Transformers, *IEEE Trans.*, May 1983, vol. PAS-102, pp. 1151–1162.

***147. To determine the aging of the insulation resulting from a specific daily load cycle,*** (1) establish an approximately equivalent stepped load cycle, (2) calculate the resulting curve of hot-spot temperature by the methods of Par. **68**, (3) replace the hot-spot temperature curve by an approximately equivalent stepped curve, (4) calculate the percent aging for each step from the applicable curve of Fig. 10-38, and (5) add the aging for all the steps in the daily cycle. The result is the fraction of insulation life used up each day. The reciprocal is the number of days of total insulation life if the same load cycle repeats every day.

***148. Example.*** Consider the transformer used in the example of Par. **71**, with a daily load cycle of 4 h at 140% load and 20 h at 80% load in 30°C ambient. The hot-spot temperature curve shown in Fig. 10-23 is reproduced in Fig. 10-39, together with an equivalent stepped curve. The calculation of loss of life per day is shown in Table 10-7.

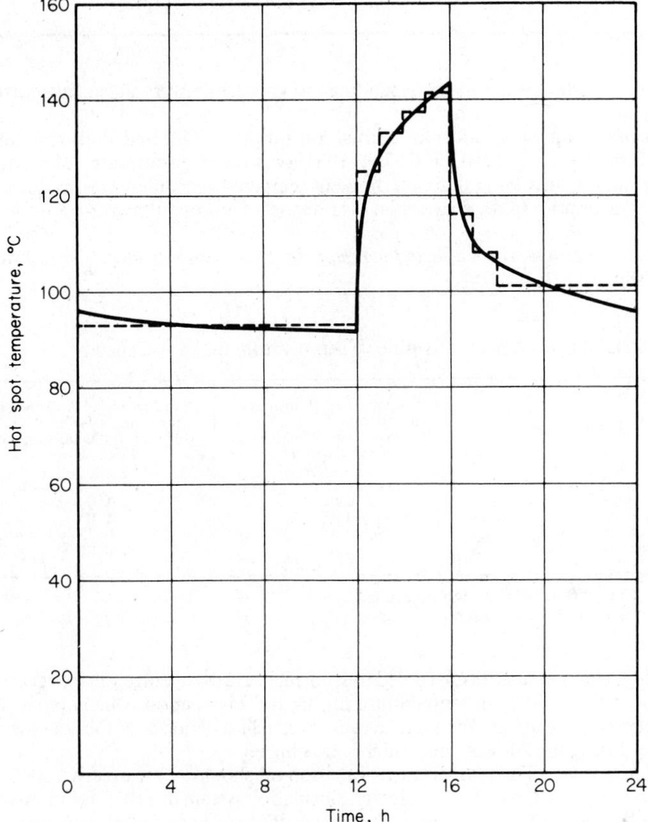


**FIG. 10-39** Equivalent stepped curve of hot-spot temperature for loss-of-life calculations of a daily load curve.

The normal life at every temperature can be determined from Fig. 10-38 or from Eq. (10-63) for the 65°C rise insulation system. The fraction of the life consumed during each time step can then be calculated and summed for all of the time steps in the 24-h period. In this case, 0.09% of the life is consumed in one day, so the total life would be 1,111 days or about 3 years if this load cycle were continued. For comparison, a transformer with a 65°C average winding-rise insulation system (80°C hot-spot rise) operating in a 30°C ambient would have a hot-spot temperature of 110°C and a normal life of 65,000 hours or 7.4 years. The shortening of the insulation life from 7.4 years to 3 years is a measure of the severity of the load cycle. The actual transformer life

**TABLE 10-7** Calculation of Loss of Life per Day on Daily Load Cycle

| Duration of step, *h* | Temp. of hot spot, °C | Life, h, at temp. | % Loss of life on step |
|---|---|---|---|
| 12 | 93 | 455,600 | 0.003 |
| 1 | 125 | 13,396 | 0.007 |
| 1 | 133 | 6,050 | 0.017 |
| 1 | 137 | 4,114 | 0.024 |
| 1 | 141 | 2,818 | 0.035 |
| 1 | 116 | 34,062 | 0.003 |
| 1 | 108 | 81,023 | 0.001 |
| 6 | 101 | 178,284 | 0.003 |
| | | | 0.093 |

may, of course, be shorter or longer, depending on exposure to overvoltage, overcurrent, shock, contamination, etc.

***149. Loading-capability tables*** for normal consumption of life and for moderate sacrifice of life are documented in ANSI/IEEE C57.92-1981 for power transformers. Information is provided for situations involving different ambient temperatures, different peak-load durations, and different loads prior to the peak for an assumed set of representative transformer characteristics.

***150. Ambient temperature affects load capacity*** by an amount dependent on the type of cooling, as shown in Table 10-8.

**TABLE 10-8** Effect of Ambient Temperature on kVA Capacity

| Type of cooling | % of rated kVA decrease in capacity for each °C increase over 30°C air or 25°C water | % of rated kVA increase in capacity for each °C decrease under 30°C air or 25°C water |
|---|---|---|
| Self-cooled* | 1.5 | 1.0 |
| Water-cooled† | 1.5 | 1.0 |
| Forced-air-cooled* | 1.0 | 0.75 |
| Forced-oil-cooled* | 1.0 | 0.75 |

* From 0 to 50°C air temperature.
† Up to 35°C water temperature.

For ambient temperature of air-cooled transformers use the average value over a 24-h period or 10°C under the maximum temperature during the 24-h period, whichever is higher. For ingoing water temperature use the average value over a 24-h period or 5°C under the maximum temperature during the 24-h period, whichever is higher.

***151. Limitations.*** The temperature of the top oil should never exceed 100°C for power transformers with a 55°C average winding-rise insulation system or 110°C for those with a 65°C average winding-rise insulation system. The consequence of exceeding these limits could be oil overflow or excessive pressure. The winding hot spot should not exceed 150°C for the 55°C average winding-rise insulation system or 180°C for the 65°C average winding-rise insulation system. These limitations are based principally on a concern for rates of insulation aging, but it should be noted that free gas bubbles may be evolved at temperatures above 140°C, with consequent weakening of dielectric strength. The peak short-duration loading should never exceed 200% of rating, except for transformers rated over 100 MVA a limit of 150% of rating is recommended.[1] This reflects a concern for stray flux heating in large units.

[1] IEEE Standard 756-1983, Guide for Loading Mineral-Oil-Immersed Power Transformers Rated in Excess of 100 MVA.

## Loss Evaluation[1]

*152.* ***Loss evaluation*** is a procedure by which the buyer and seller achieve an economic balance in adding material to the transformer design to get lower losses. It is achieved by establishing a value in dollars per kilowatt for load loss and a similar value for no-load loss.

An incremental investment in capacity is required to generate power to supply loss and bring it to the transformer. In addition there is a continuing expense for fuel to supply the lost power. The continuing expense is converted to present worth and added to the incremental investment to give the total present worth of the loss. This present worth of a kilowatt of loss is naturally higher for the no-load loss, which is continuous, than it is for the load loss, and the value is higher the farther the transformer is from the generator; the values will of course depend on the accounting rules and procedures in force at the particular location. The value of a no-load kilowatt can range from \$500 to \$10,000 and a load-loss kilowatt from \$300 to as much as \$8,000.

*153.* ***The following equations are commonly used*** to establish loss evaluations:

$$V_L = S + 8760EF_L/R \tag{10-64}$$

$$V_N = S + 8760EF_N/R \tag{10-65}$$

where $E$ = dollars per kilowatthour cost of energy (this can conceivably be very low for a hydro station but can range up to 0.02 or more for fuel-fired stations, depending on fuel cost, and, of course, the figure will be even higher at locations remote from the generating station), $F_L$ = ratio of average load loss to rated load loss, $F_N$ = ratio of average no-load loss to rated no-load loss (1.00 for continuous operation), $R$ = per unit (%/100) annual carrying charge on system investment (covers insurance, taxes, depreciation, and return on investment), $S$ = dollars per kilowatt system investment (200 and up, depending on the system investment out to the transformer location), $V_L$ = dollars per kilowatt evaluation of rated load loss, and $V_N$ = dollars per kilowatt evaluation of rated no-load loss.

*154.* ***Loss evaluation is an important factor*** in purchasing new transformers, as in many cases the evaluation of the total loss equals or exceeds the price of the transformer.

## Standard Power Transformers

*155.* ***Standard Transformer Ratings and Features.*** Many of the ratings, electrical characteristics, mechanical arrangements, and accessories for power transformers have been standardized by ANSI, and more progress can be expected.

## Autotransformers

*156.* ***Part of an autotransformer winding is common to both primary and secondary circuits.*** The common portion is called the common winding, and the remainder is called the series winding. The high-voltage terminal is called the series terminal, and the low-voltage terminal is called the common terminal. Part of the power passes from one winding to the other by transformation, and the rest passes directly through without transformation. Figure 10-40 shows an autotransformer compared with an equivalent two-winding transformer. Both have the same ratio of secondary voltage to primary voltage, $T$, and both have the same power output. The fraction $1 - T$ of the power is transformed, and the fraction $T$ passes through without transformation. The fraction $1 - T$, called the "co-ratio," is a mea-

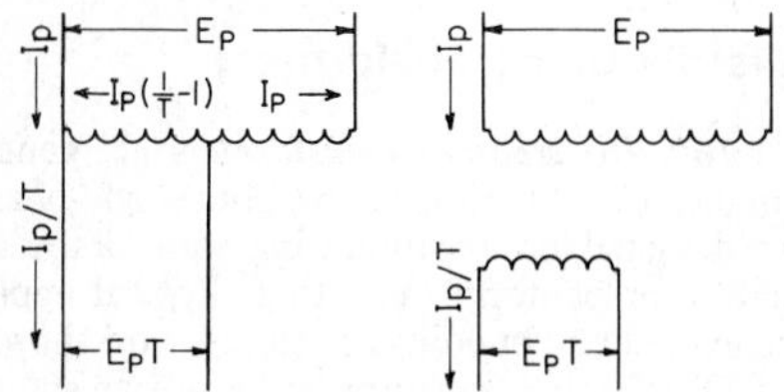


**FIG. 10-40** Comparison of an autotransformer with a two-winding transformer.

[1] H. J. Mason, *Distribution Mag.*, April, 1962.

sure of the required size of the core and coils as compared with a two-winding transformer. In addition the losses and reactance are reduced in approximately the same ratio. For a low value of $1 - T$ the economy of an autotransformer compared with a transformer is attractive.

***157. The following special characteristics*** of autotransformers may need consideration:

A metallic connection exists between the primary and secondary circuits; this is generally of little consequence with low-voltage circuits, but with high-voltage systems the neutral point must be grounded for safe operation.

*The impedance of an autotransformer is normally lower* than that of the equivalent two-winding transformer, and the short-circuit current is higher.

*Taps near the neutral* are relatively ineffective, because turns tapped out at the neutral come out of both circuits and increase core flux density without greatly changing the voltage ratio. The high-voltage rating is more effectively adjusted by taps in the series winding. The low-voltage rating is more effectively adjusted by tapping turns out of the common winding while turns are added to the series winding.

Inversion of the neutral may occur under abnormal conditions in Y-connected autotransformers with ungrounded neutrals. If the voltage on a series terminal is lower than the voltage on the corresponding common terminal, high voltage tends to appear on the neutral. Inversion of the neutral can occur on power-frequency voltage or on transient voltage. Grounding the autotransformer neutral, use of a delta tertiary, and use of three-leg 3-phase cores all help to prevent inversion of the neutral.

***158. Autotransformers*** are most commonly used to connect two transmission systems at different voltages, frequently with a delta tertiary winding. It is also possible to apply an autotransformer as a generator step-up transformer when it is desired to feed two different transmission systems. In this case the delta tertiary winding is a full-capacity winding connected to the generator, and the two transmission systems are connected to the autotransformer windings. Advantages of the autotransformer when compared to a normal transformer include lower impedance, lower losses, better regulation, smaller size, and lighter weight.

***159. References.***


Massachusetts Institute of Technology, Department of Electrical Engineering: *Magnetic Circuits and Transformers;* New York, John Wiley & Sons, Inc., 1943.

*Transformer Reference Book;* Milwaukee, Allis-Chalmers Manufacturing Company, 1951.

Blume, L. F., Boyajian, A., Camilli, G.,Lennox, T. C., Minneci, S., and Montsinger, V. M.: *Transformer Engineering,* 2d ed.; New York, John Wiley & Sons, Inc., 1951.

Bean, R. L., Chackan, N., Moore, H. R., and Wentz, E. C.: *Transformers for the Electric Power Industry;* Westinghouse Electric Corporation. (Formerly published by McGraw-Hill Book Company, 1959.)

Stigant, S. A., Lacey, H. M., and Franklin, A. C.: *J & P Transformer Book,* 9th ed.; London, Johnson & Phillips, Ltd., 1961.

Bibliography on Transformer Noise; *IEEE Comm. Rep. Power Appar. Syst.,* February 1968, vol. 87, pt. 2, p. 372.

*Standards*

NEMA Publ. TR-1.

NEMA Publ. 107.

ANSI C57.


## Distribution Transformers

***160. Distribution transformers*** are generally considered as transformers 500 kVA and smaller, 67,000 V and below, both single-phase and 3-phase. Although the majority of the units are designed for pole mounting, some of the larger kVA sizes above the 18-kV class are built for station or platform mounting. Typical applications are for supplying power to farms, residences, public buildings or stores, workshops, and shopping centers.

Distribution transformers have been standardized as to high- and low-voltage ratings, taps, type of bushings, size and type of terminals, mounting arrangements, nameplates, accessories, and a number of mechanical features, so that a good degree of interchangeability results for transformers in a certain kVA range of a given voltage rating. They are now normally designed for 65°C rise.

The most popular primary voltages are 12,470Y/7200, 13,200Y/7620, and 12,000 V delta.

Many of the 2400- and 4800-V primary systems have been converted to 7200 and 7620 V. There is also increasing use of higher-voltage distribution systems such as 24,900Y/14,400 and 34,500Y/19,900 V. Secondary voltage for pole-type units is usually 120/240 or 240/480.

***161. Magnetic cores,*** in general, are composed of cold-rolled silicon steel strip. They take various forms, all designed so that the magnetic flux will pass through the sheet in the direction of rolling in order to secure the maximum benefit of the superior magnetic quality of this material. For an appreciable portion of the 24-h day, the typical distribution transformer (particularly the pole-mounted 5- to 167-kVA range) is lightly loaded. Because of this, the loss in the core is a significant portion of the total daily loss. Cores for these units are therefore designed for low exciting current and for relatively low core loss to minimize the operating cost. Low-loss cold-rolled silicon strip has contributed materially to reduced losses, weights, and dimensions.

***162. Coils*** are usually wound in a concentric layer arrangement, with cooling ducts distributed periodically between the layers to maintain reasonable differentials between oil temperature and the average coil and hot-spot temperatures. As a matter of practical operating procedure, distribution transformers are subjected to considerable numbers of overloads for short time periods. Hot-spot temperatures must be limited on these overload excursions if the transformer is to have a long insulation life. It is now general practice to employ thermally upgraded materials in the insulation system to improve aging characteristics. Increased mechanical strength is achieved by using a special intermittent coating of a heat-reactive adhesive on the layer insulation to bond the coil into a rigid mass during the drying process. Heat and vacuum drying plus vacuum oil filling imparts good dielectric strength to the windings.

Aluminum conductor is now replacing copper in many windings. This is particularly so for the secondary windings, where full-width aluminum strip is frequently employed. Such coils are also mechanically stronger.

***163. To cool the unit,*** the radiating surface of the tank itself suffices in the smaller ratings. In the larger ratings, auxiliary cooling is provided by the addition of fins or tubes. By these means, the height, size, and weight are held to desirable minimums. Special attention is given to sealing the transformers from the atmosphere. Likewise, careful attention is given to the external finish and fittings to assure reliable service for many years of exposure to the elements. External connectors are good for either aluminum or copper conductors.

***164. The conventional pole type*** consists of core and coils securely mounted in an oil-filled tank, with the necessary terminals brought out through their appropriate bushings. The high-voltage bushings may be two in number but one bushing plus a ground terminal on the tank wall connected to the ground end of the high-voltage winding for use on multiple-grounded circuits is the most common usage. The conventional type includes just the basic transformer structure without any protective equipment. The desired overvoltage, overload, and short-circuit protection is obtained by using lightning arresters and primary fuse cutouts separately mounted on the pole or crossarm closely adjacent to the transformer. The primary fuse cutout provides a means of visually detecting blown fuses on the system primary and also serves to remove the transformer from the high-voltage line, either manually when desired or automatically in the event of an internal coil failure.

***165. The self-protected transformer*** (Figs. 10-41 and 10-42) has an internally mounted, thermally controlled secondary circuit breaker for overload and short-circuit protection; an internally mounted protective link in series with the high-voltage winding to disconnect the transformer from the line in the event of an internal coil failure; and a lightning arrester or arresters integrally mounted on the outside of the tank for overvoltage protection. On most of these transformers, except some 5-kVA ratings, the circuit breaker operates a signal light when a predetermined winding temperature has been reached, as a warning before tripping. If the signal is unheeded and the breaker trips, the breaker may be reset and the load restored by an external handle. Usually this can be accomplished with the normal breaker setting. If, however, the load has been a long-sustained one which has allowed the oil to reach a high temperature, the breaker may soon trip again; or it may be impossible to reset so that it will remain closed. In such cases, the trip temperature may be set up by an auxiliary external control handle to allow reclosing of the breaker for the emergency until a larger transformer can be installed.

***166. Three-phase self-protected transformers*** are similar to the single-phase units except that a 3-pole circuit breaker is used. The breaker is arranged to open all 3 poles in case of a serious overload or fault on one of the phases.

***167. The self-protected transformer for secondary banking*** is another variation. Such transformers are provided with the two secondary breakers to sectionalize the low-voltage circuits, confining the outage to just the faulted or overloaded section, leaving the entire transformer capacity available for supplying the remaining sections. These are also made for single- and 3-phase.

***168. "Station-type" distribution transformers*** are normally rated 250, 333, or 500 kVA. A "pole/station type" distribution transformer is shown in Fig. 10-43. For distribution to low-voltage ac networks in areas of high-load density *network transformers* are available in even higher ratings.

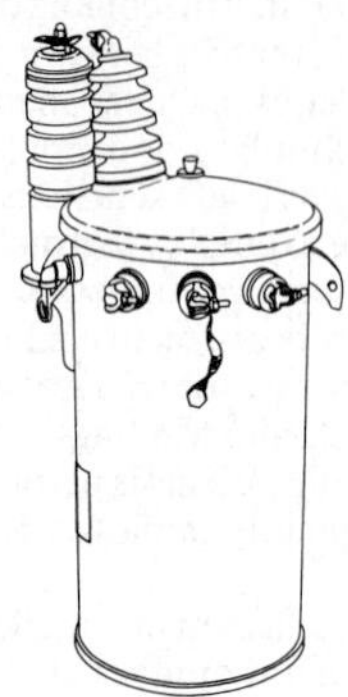

**FIG. 10-41** Completely self-protected 10-kVA distribution transformer. *(Westinghouse.)*

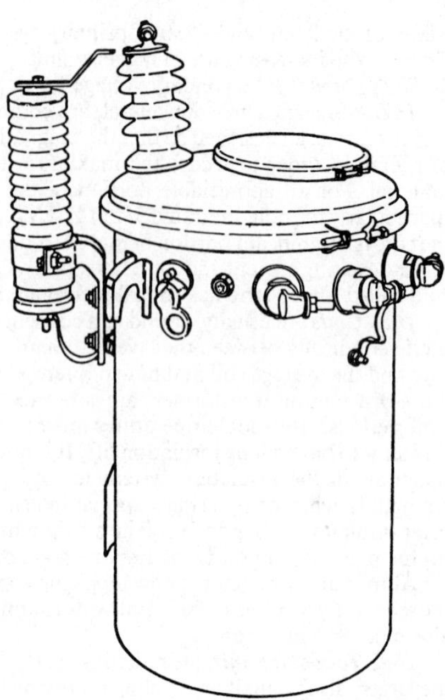

**FIG. 10-42** Self-protected pole-type distribution transformer rated 25 kVA, 12,740 Grd Y/7200-120/240 V, 65°C rise. *(General Electric Co.)*

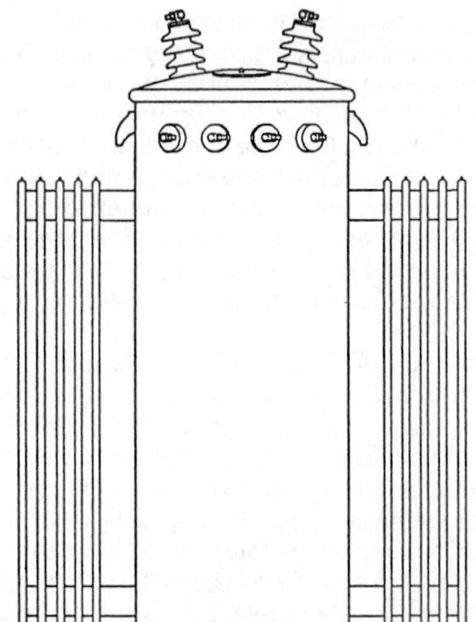

**FIG. 10-43** Pole/station-type distribution transformer rated 250 kVA, 7620/13200Y-240/480 V, 65°C rise. (*McGraw-Edison.*)

***169. Losses and Characteristics.***[1] For the pole-type ratings 100 kVA and smaller full-load efficiencies range from 97 to 99%, and impedance is generally less than 2%.

***170. Recent trends*** have been toward further reduction of no-load loss, exciting current, and sound level, and toward the replacement of copper with aluminum for the windings. Replacement of much of the round and rectangular conductors with full-width strip has resulted in more compact windings with greatly increased mechanical strength.

More effective utilization of distribution-transformer investment is being made possible through transformer-load-management programs.[2,3]

***171. Transformers for Underground Distribution Systems.*** Since more distribution circuits are being put underground, transformers have been especially developed to be used with such systems. The most widely used type is the pad-mounted transformer, so

[1] M. F. Beavers, Effect at Overload and Resulting Temperature on Load Losses in Distribution Transformers; *IEEE Trans.*, 1963, vol. 82, *Suppl.*, pp. 599–609.

[2] A. M. Lockie, Revolutionary Changes in Distribution Systems; *Trans. Distrib.*, March 1968, p. 96.

[3] A. P. Wurmlinger and D. T. Egly, Distribution Transformer Load Management; *IEEE Trans. Power Appar. Syst.*, July 1968, vol. 87, p. 67.

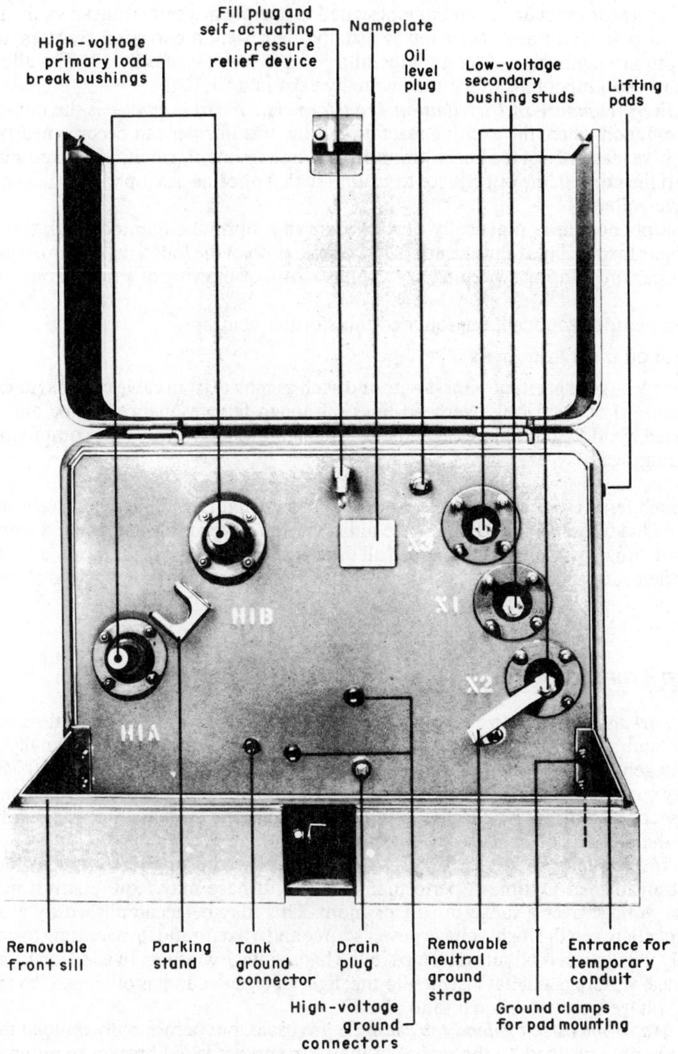


**FIG. 10-44** Typical pad-mounted distribution transformer.

called because it is designed to mount on a concrete-surface slab or *pad.* A typical transformer is shown in Fig. 10-44. The essential differences from the pole-type transformers of Figs. 10-41 and 10-42 are only in the mechanical arrangement:

1. A rectangular case in two compartments.
2. One compartment containing the conventional core-coil assembly.
3. A second compartment for cable termination and connection. The primary cable conductors are connected by plug-in connectors suitable for load make or break. The secondary conductors usually bolt to bushing terminals.
4. Fuses of various sorts provided for by a fuse holder placed in a well in the side of the tank so that the fuse holder can be drawn out.

Another transformer arrangement is designed to operate in a subterranean vault. This looks more like a pole-type transformer but is usually made with a corrosion-resistant steel tank, plug-in primary connectors, and a temperature rise in free air of only 55°C to allow for the higher ambient temperaure which may actually exist in a vault.

*172. **Ferroresonance in Distribution Transformers.*** *Ferroresonance* is the name given to the phenomenon where the exciting reactance of the transformer can become nearly equal to the capacitive reactance of the line to ground, forming a resonant circuit. Such a resonant circuit can distort the normal line impedance to ground so that one line of a 3-phase circuit can rise to a destructive voltage.[1-3]

Such a phenomenon practically never occurs in a normal circuit configuration with the transformers loaded, but it can exist under a combination of the following circumstances which usually occur only during switching of a 3-phase bank or blowing of a fuse in one line:

1. System neutral grounded, ungrounded transformer neutral
2. No load on the transformers
3. Relatively large capacitance line-to-ground such as may exist in cable circuits (underground distribution) or very long overhead lines (although ferroresonance can be and has been corrected by adding still more capacitance which presumably throws the combination out of resonance again)

Although ferroresonance has been studied at some length, it still does not seem possible to reliably predict its occurrence. Experience indicates that it is possible to prevent ferroresonance during switching on a transformer bank if all three transformers are resistance-loaded to 15% or more of their rating, or if special switches are used to assure that the three lines close simultaneously.

## Furnace Transformers

*173. **Furnace transformers supply power to electric furnaces*** of the induction, resistance, open-arc, and submerged-arc types. The secondary voltages are low, occasionally less than 100 V, but generally several hundred volts. Sizes range from a few kVA to over 50 MVA, with secondary currents over 60,000 A. High currents are obtained by parallel connection of many winding sections. Current is collected by internal bus bars and brought through the transformer cover by the bus bars or by high-current bushings.

*174. **The power input to the furnace is controlled*** by adjusting the output voltage of the furnace transformer. Optimum performance of the furnace may require adjustment of the secondary voltage over a range of 3 : 1 or more. This may be accomplished by a regulating transformer between the high-voltage power source and a fixed-ratio furnace transformer. More frequently, regulation is obtained by taps in the high-voltage winding. In addition to taps in the high-voltage winding, a delta-Y switch in the high-voltage winding is often used to extend the range of voltage by an additional ratio of 1.73.

*175. **Motor-operated off-load tap changers*** are usual, but occasionally on-load tap-changing equipment is justified by the saving in melt time and reduced breaker maintenance. The load-tap-changing duty is more severe than on the usual power transformer, not only with respect to frequency of operation, but also because of the extreme range, which results in large kVA increments per tap.

*176. **Circuit reactance furnishes current stability*** for ac arc furnaces. In the larger sizes the inherent impedance of the transformer and its associated secondary conductors is sufficient for adequate stability. This is not generally true for smaller arc furnaces. Consequently, it is customary in furnace transformers rated 7500 kVA and below to include a reactor in the tank with the

[1] F. S. Young, R. L. Schmid, and P. O. Fergestad, Laboratory Investigation of Ferro-resonance in Cable-connected Transformers; *IEEE Trans. Power Appar. Syst.*, May 1968, vol. PAS-87, p. 1240.

[2] R. H. Hopkinson, Ferro-resonance Control Based on Tests on Three-Phase Delta-Y Transformers; *IEEE Trans. Power Appar. Syst.*, October 1967, vol. PAS-86, p. 1258.

[3] D. R. Smith, S. R. Swanson, and J. D. Borst, Overvoltages with Remotely-switched Cable-fed Grounded Wye-Wye Transformers; *IEEE Trans. Power Appar. Syst.*, September 1975, vol. PAS-94, no. 5, p. 1843.

transformer. This reactor is connected in the high-voltage circuit and is furnished with taps to permit adjusting the total reactance to that required to maintain arc stability under the existing service conditions.

## Grounding Transformers

***177. A grounding transformer*** is intended primarily for the purpose of providing a neutral point for grounding purposes. It may be a two-winding unit with a delta-connected secondary winding and a Y-connected primary winding which provides the neutral for grounding purposes, or it may be a single-winding 3-phase autotransformer with windings in interconnected Y or zigzag. With the latter, the windings consist of six equal parts, each designed for one-third the line-to-line voltage; two of these parts are placed on each leg and connected as in Fig. 10-45. In

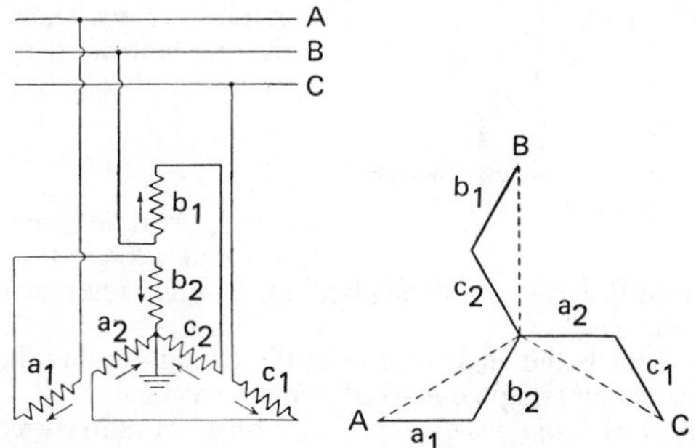


**FIG. 10-45** Grounding autotransformer with interconnected Y or "zigzag" windings.

the case of a ground fault on any line, the ground current flows equally in the three legs of the autotransformer, and the interconnection offers the minimum impedance to the flow of the single-phase fault current.

## Instrument Transformers

***178. Functions.*** Instrument transformers are used to insulate measuring and control devices connected in the secondary circuit from the primary-circuit operating voltages. To provide complete protection, the secondary circuit should be grounded at one point. Metal cases should also be grounded. They are likewise used to transform the primary current or voltage to values suitable for standard ratings for instruments, meters, relays, and other measuring or control devices. The normal secondary ratings are 5 A for current transformers and 115 or 120 V for voltage transformers.

*The primary winding of a current transformer* is connected in series with the load for which the current is to be measured or controlled; the primary winding of a voltage transformer is connected in parallel with the load for which the voltage is to be measured or controlled (see Fig. 10-46). The secondary windings provide a current or voltage that is substantially proportional to the primary values for the operation of measuring instruments and control devices.

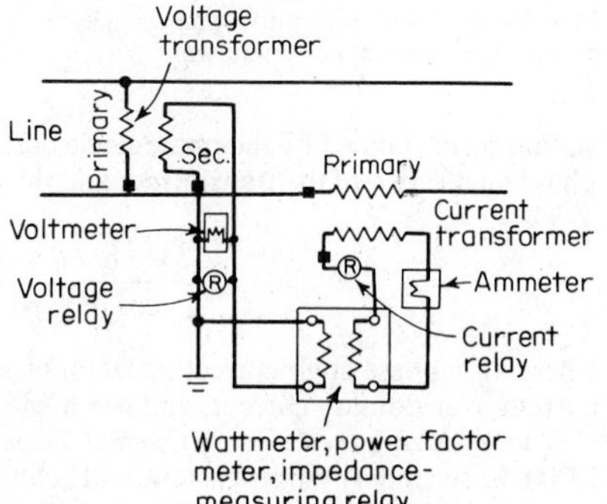


**FIG. 10-46** Voltage and current transformers as commonly connected to insulate meters and to transform current and voltage to convenient values.

***179. Polarity.*** When instrument transformers are used with measuring or control devices that respond only to the magnitude of the current or voltage, the direction of the current flow does not affect the response and the connections to the secondary terminals can be reversed without affecting the operation of the devices. When instrument transformers are used with measuring or control devices that respond to the interaction of two or more currents, the correct operation of the devices depends on the relative phase positions of the currents, in addition to the magnitudes. To show the relative instantaneous directions of current flow, one primary and one secondary terminal are identified with a distinctive polarity marker; these indicate that at the instant when the primary current is flowing into the marked primary terminal the secondary current is flowing out of the marked secondary terminal (see Fig. 10-47).

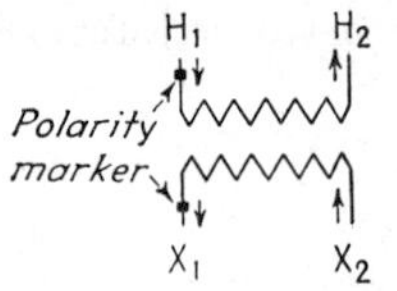

$H_1$ $H_2$ Polarity marker $X_1$ $X_2$

**FIG. 10-47** Polarity definition. Arrows indicate the instantaneous relative direction of currents in the windings.

***180. Errors in Current Transformers.*** There are two types of errors that affect the accuracy of the measurements made with current transformers. The ratio-correction factor is the true ratio of the primary to the secondary current, divided by the nameplate ratio,

$$F_{CR} = R_{CT}/R_{CN} \tag{10-66}$$

where $F_{CR}$ = ratio correction factor of current transformer, $R_{CR}$ = true ratio [(primary current)/(secondary current)], and $R_{CN}$ = nameplate ratio [(primary current)/(secondary current)] of current transformer.

The phase-angle error is the angle of lead of the current leaving the marked secondary terminal over the current entering the marked primary terminal.

*Relative Importance of Ratio and Phase-Angle Errors.* A ratio correction factor of 1.010 indicates that the secondary current is lower than the correct value by 1% and that all measuring or control devices connected in the secondary circuit will have 1% less current than the primary current divided by the marked ratio. The phase-angle error does not affect current-actuated devices, such as ammeters or overcurrent relays, but the operation of devices that respond to the products, the sums, or the differences of currents is affected by the phase-angle error.

A wattmeter is a device that responds to the product of the voltage applied to the potential terminals, the current through the current coils and the power factor, which is the cosine of the angle between the voltage and current. If the current is supplied from the secondary of a current transformer with unity nameplate ratio, unity ratio correction factor, a phase-angle error of $\beta$, and primary current lagging the voltage, then the wattmeter will not indicate the true watts, $EI \cos \theta$, but will indicate $EI \cos (\theta - \beta)$. If the sign of $\beta$ is plus, the cos $(\theta - \beta)$ will be larger than the cos $\theta$ and the wattmeter will read high (see Fig. 10-48). If the sign of $\beta$ is minus, the wattmeter will read low. To obtain the true watts, the apparent watts should be multiplied by the phase-angle correction factor $K_\beta$, which is dependent on both the phase-angle error of the transformer and the power factor of the load, as shown in Eq. (10-67):

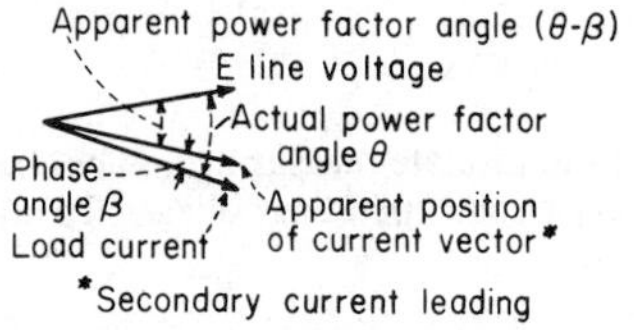

Apparent power factor angle $(\theta-\beta)$
E line voltage
Actual power factor angle $\theta$
Phase angle $\beta$
Apparent position of current vector*
Load current
*Secondary current leading

**FIG. 10-48** Effect of positive phase angle in increasing the apparent power factor.

$$K_\beta = \frac{\cos \theta}{\cos (\theta - \beta)} = \frac{1}{\cos \beta + \sin \beta \tan \theta} \tag{10-67}$$

where $K_\beta$ = phase-angle correction factor of current transformer, $\beta$ = angle of lead of secondary current over primary current, and $\theta$ = angle of lag of load current behind load voltage.

*Summarizing the Effect of Current-Transformer Errors.* Ratio correction factors: Above 1.000 the secondary current is low, and below 1.000 the secondary current is high. Phase-angle errors with lagging load current: Positive phase-angle errors cause wattmeter readings to be high if the load-current power-factor angle is greater than the phase-angle error, and negative phase-angle errors cause the wattmeter readings to be low.

In practical metering problems $\beta$ will be less than 30 min and $K_\beta$ can be written

$$K_\beta = 1 - \sin\beta \tan\theta = 1 - \tan\theta \qquad \text{where } \beta \text{ is in radians}$$
$$K_\beta = 1 - \beta \tan\theta/3438 \qquad \text{where } \beta \text{ is in minutes} \tag{10-68}$$

The uncertainty in knowledge of the exact values of $\beta$ and $\theta$ represents a greater error than will result from this simplification of Eq. (10-67).

***181. Transformer Correction Factor.*** The transformer correction factor to be applied to the reading of the wattmeter for both the ratio and phase-angle errors is given by Eq. (10-69):

$$F_T = F_{CR} K_\beta \tag{10-69}$$

where $F_T$ = transformer correction factor.

If $F_{CR}$ and $K_\beta$ are both between 0.985 and 1.015, Eq. (10-70) may be used with an error under ±0.0003:

$$F_T = F_{CR} + K_\beta - 1.000 \tag{10-70}$$

***182. Classification of Errors.*** ANSI C57.13 classifies current transformers as to accuracy by a method which limits the total error in a wattmeter or watthour-meter reading resulting from the combination of ratio and phase-angle errors, over a range of power factor of the metered load of 0.6 to 1.0. The most accurate classification (usually specified for use with watthour meters for billing metering) is 0.3, which means that the total error at the meter caused by the current transformer will not exceed 0.3% at rated current (or at maximum continuous current), 0.6% at 10% rated current. ANSI C57.13 also recognizes 0.6 and 1.2 accuracy classes. The accuracy class is specified in connection with one or more of the standard burdens (Par. **183**). For example, "0.3 B-0.2" describes a transformer of 0.3 accuracy class when loaded with a B-0.2 burden on the secondary terminals.

***183. Standard Burdens.*** ANSI C57.13 also recognizes a number of standard values of secondary impedance loading (called "burden" in instrument-transformer parlance) for use in describing current transformer performance. The burdens are designated as B0.1, B0.2, B0.5, B0.9, B1.8 to mean impedances of 0.1, 0.2, etc., ohms at 0.9 power factor, and also B1.0, B2.0, B4.0, and B8.0 for corresponding impedances at 0.5 power factor. (These standard burden impedances are only applicable if rated secondary current is 5 A.)

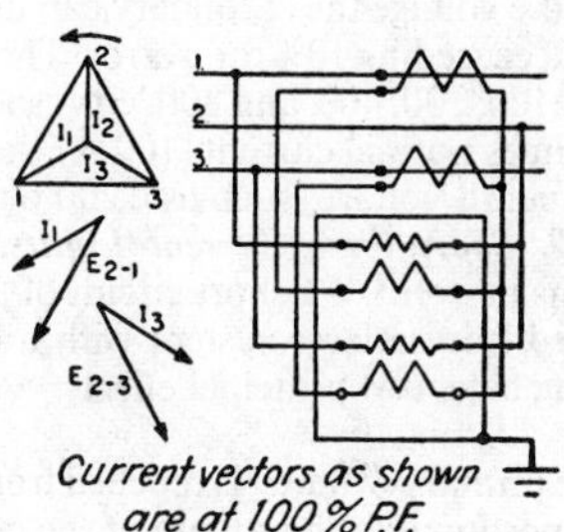


**FIG. 10-49** Metering connections for 3-phase 3-wire with two watthour-meter elements.

***184. Effect of Phase-Angle Errors on Metering 3-Phase 3-Wire Power with a Two-Stator Watthour Meter.*** With the meter connected as indicated in Fig. 10-49 with balanced load, if the marked ratio and ratio correction factor are both 1.000 and the phase-angle error $\beta$ is the same for both current transformers,

$$K_\beta = \frac{E_1 \cos(\theta + 30) + E_1 \cos(\theta - 30)}{[E_1 \cos(\theta + 30 - \beta) + E_1 \cos(\theta - 30 - \beta)]} = \frac{1}{\cos\beta + \sin\beta \tan\theta} \tag{10-71}$$

This is identical with Eq. (10-67) for a single-phase measurement.

***185. Causes of Errors.*** The errors in a current transformer are due to the energy required to produce the core flux which induces the secondary winding voltage that supplies the current through the secondary circuit. The total ampere-turns available to provide secondary current are vectorially equal to the primary ampere-turns minus the ampere-turns required to produce the core flux.

A change in secondary burden alters the flux required in the core and changes the core-exciting ampere-turns; leakage flux entering the core changes the magnetic characteristics of the core and affects the core-exciting ampere-turns.

***186. Calculation of ratio and phase angle*** in current transformers can, in theory, be done simply by first calculating the exciting ampere-turns required to magnetize the core and subtracting them from the primary ampere-turns to get the secondary ampere-turns (vectorial subtraction). The difficulty in calculation is that in many transformers the leakage flux which flows in only part of the core is as large as the working flux and its effect on exciting ampere-turns is most difficult to calculate.[1] For exact determination of ratio and phase angle, therefore, actual measurement of ratio and phase angle as described in ANSI C57.13 is necessary.

***187. Approximate calculation of ratio for relaying service*** is more practical and in fact often necessary because transformers supplying relays must often operate at currents up to 20 times normal or even higher, and exact measurements are extremely difficult and expensive. Accordingly ANSI C57.13 recognizes an adequate ratio-error calculation method for current transformers in which the leakage flux which enters the core has only a negligible effect on performance. This calculation method is described in the Test Methods Section of C57.13. The ratio error calculated by this method may in fact be greater than the actual ratio error, but it is certain that it will not be less, a conservative result.

Generally, the leakage flux in a current transformer can be said to be negligible only in a current transformer consisting of a toroidal core with the secondary winding fairly well distributed around the core, for example, an assembly intended to be placed over a bushing in a circuit breaker or transformer and then only if the return conductor is at a distance at least as great as in typical circuit breakers.[2] If this type of transformer is used with a primary coil or even a single primary turn placed around one side of the core, it is doubtful that the leakage flux will be negligible.

Transformers with negligible leakage flux (principally bushing type) are designated in C57.13 as Class C transformers, meaning that it is safe to calculate the ratio error.

***188. Classification of Current Transformers for Relaying Service.*** As described in Par. **187**, transformers for which calculation of ratio will give conservative results are designated as Class C; all other transformers are designated as Class T, meaning that their performance must be determined by test. The performance of both types of transformers is then classified according to the voltage the secondary can deliver to the burden at 20 times rated secondary current without exceeding 10% ratio error. The established standard secondary terminal voltages are 10, 20, 50, 100, 200, 400, and 800, corresponding to the voltage required by ANSI standard burdens at 20 times normal current (100 A). A current transformer is classified by the letter C or T plus the standard voltage, such as C200 or T400.

***189. Short-Time Current Limits.*** Current transformers may have to carry very large currents in event of a short circuit of the system, especially when a low-rated branch circuit is supplied from a large system with a high fault-current capability. The large currents in the winding have two principal effects:

1. The primary winding is repelled from the secondary by the electromagnetic forces, which are proportional to the square of the current.
2. The windings heat very rapidly at a rate nearly proportional to the square of the current.

In order to apply current transformers properly, it is necessary to give them ratings:

1. Mechanical short-time rating, the current, usually stated in terms of times normal, which the transformer can withstand mechanically even if the current is initially fully offset.
2. Thermal short-time rating, the current which will not heat the winding to more than 250°C for copper conductors or 200°C for EC aluminum conductors. This limit is not to be demonstrated by test but is calculated on the conservative assumption that all the heat generated by the current is stored in the conductor.

C57.13 may be consulted for additional detail and for the test method to be used to demonstrate the mechanical limit.

---

[1] F. C. Wentz, A Simple Method for Determination of Ratio Error and Phase Angle in Current Transformers; *Trans. AIEE,* 1941, vol. 60, p. 949.

[2] R. A. Pfuntner, The Accuracy of Current Transformers Adjacent to High Current Busses; *Trans. AIEE,* 1951, vol. 70, p. 1656.

***190. Types of Construction.*** The types of current transformers are the wound type which consists of primary and secondary windings completely installed and permanently assembled on the magnetic circuit; the bar type, which is similar to the wound type except that the primary is a single straight and fixed turn; the window type, which has a secondary winding completely insulated and permanently assembled on the magnetic circuit and a window through which a conductor can be passed to provide a primary winding; and the bushing type, which is a special window type designed to fit over apparatus bushings, with the conductor through the bushing as the primary winding.

Current transformers are classified in accordance with the major insulation used as dry-type, compound-filled, molded, or liquid-immersed.

***191. Safety Precautions.*** The secondary winding should always be short-circuited before disconnecting the burden. If the secondary circuit is open, with primary current flowing, all the primary ampere-turns are magnetizing ampere-turns and usually will produce an excessively high secondary voltage across the open circuit. All instrument-transformer secondary circuits should be connected to ground; when instrument-transformer secondaries are interconnected, only one point should be grounded. If the secondary circuit is not grounded, the secondary becomes, in effect, the middle plate of a capacitor, with the high-voltage winding and ground acting as the other two plates.

***192. Errors in Voltage Transformers.*** There are two types of errors that affect the accuracy of the measurements made with voltage transformers. The ratio error is the difference between the true ratio of the primary to secondary voltage and the ratio that is marked on the nameplate. The phase-angle error is the difference in the phase position of the voltage applied to the secondary burden and the voltage applied to the primary winding.

The ratio error is expressed as a ratio correction factor by which the secondary-voltage value should be multiplied to obtain a secondary voltage that is directly proportional to the primary voltage,

$$F_{PR} = R_{PT}/R_{PH} \tag{10-72}$$

where $F_{PR}$ = ratio correction factor of the voltage transformer, $R_{PT}$ = true ratio (primary/secondary) of the voltage transformer, and $R_{PN}$ = nameplate ratio (primary/secondary) of the voltage transformer.

The phase-angle error is designated by the symbol $\gamma$, is expressed in minutes, and is defined as positive when the voltage applied to the burden from the marked to the unmarked secondary terminal leads the voltage applied to the primary from the marked to the unmarked terminal.

***193. Relative Importance of Ratio and Phase-Angle Errors.*** The effect of the ratio and phase-angle errors of voltage transformers is the same as described for current transformers in Par. **180**, except that with a lagging power-factor load a positive voltage transformer phase-angle error will cause the wattmeter to read low. To obtain the true watts, the apparent watts should be multiplied by the phase-angle correction factor $K_\gamma$.

$$K_\gamma = \frac{\cos\theta}{\cos(\theta+\gamma)} = \frac{1}{\cos\gamma - \sin\gamma\tan\theta} \tag{10-73}$$

where $K_\gamma$ = phase-angle correction factor of the voltage transformer, $\theta$ = angle of lag of load current behind load voltage, and $\gamma$ = angle of lead of secondary voltage over primary voltage.

***194. Classification of Errors.*** The errors in a voltage transformer change with the current required by the burden connected across the secondary terminals, the frequency, and the magnitude of the secondary voltage. ANSI C57.13 classifies voltage transformers as to accuracy by a method essentially the same as that described in Par. **182** for current transformers, the principal difference being that the limits of error apply over the range of rated voltage from 90 to 110% and from zero burden up to the burden at which the rating is given. The standard burdens for rating purposes are given in Table 10-9. The complete ANSI accuracy classification of a voltage transformer must include the secondary burden, such as 0.3X or 0.6Z.

***195. Voltage transformers*** are made for all the standard rated circuit voltages. They are usually dry-type or molded for voltages below 23 kV and liquid-filled for the higher voltages.

***196. The measurement of power by using a wattmeter and both current and voltage transformers*** requires a correction for the ratio and phase-angle errors of both of the instrument transformers and a correction for the phase angle $\alpha$ in minutes of the potential circuit of the

**TABLE 10-9** Standard Burdens for Voltage Transformers
From Table 13-11.110, ANSI C57.13

| Burden designation | Secondary voltamperes* | Burden power factor |
|---|---|---|
| W | 12.5 | 0.10 |
| X | 25 | 0.70 |
| M | 35 | 0.20 |
| Y | 75 | 0.85 |
| Z | 200 | 0.85 |
| ZZ | 400 | 0.85 |

* At 120 or 69.3 secondary volts, if rated secondary voltage is from 90% to 110% of 120 or 69.3 V.

wattmeter. Figure 10-50 shows the relative phase positions of the primary and secondary voltages and currents of the instrument transformers if both $\beta$ and $\gamma$ are positive. If the potential circuit of the wattmeter is inductive by a small amount, then the current in this circuit will lag the voltage and the phase angle will have the same effect as a negative voltage-transformer phase angle. The total phase-angle correction factor is

$$K_S = \frac{\cos(\theta + \beta - \gamma + \alpha)}{\cos \theta} \tag{10-74}$$

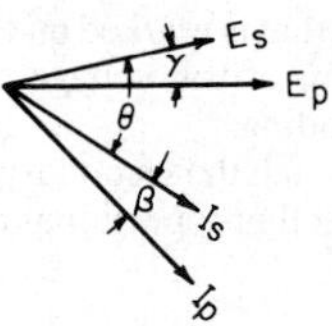


**FIG. 10-50** Vector relations in current and voltage transformers where $p$ = primary and $s$ = secondary.

where $K_S$ = total phase-angle correction factor, $\alpha$ = angle of lag of potential circuit of wattmeter, $\beta$ = angle of lead of secondary current over primary current, $\gamma$ = angle of lead of secondary voltage over primary voltage, and $\theta$ = angle of lag of secondary current behind secondary voltage.

The true watts, with all corrections included, is

$$P = WR_{CN}R_{PN}F_{CR}F_{PR}K_S \tag{10-75}$$

where $F_{CR}$ = ratio-correction factor of current transformer, $F_{PR}$ = ratio-correction factor of voltage transformer, $P$ = watts drawn by load, $R_{CN}$ = nameplate ratio of current transformer, $R_{PN}$ = nameplate ratio of voltage transformer, and $W$ = watts reading of wattmeter.

In watthour meters the phase angle corresponding to $\alpha$ is corrected for in the meter by the lag adjustment or by making an overall calibration and adjustment for all errors, including the ratio and phase-angle errors of the instrument transformers.

Figures 10-51 and 10-52 show typical ratio-correction-factor and phase-angle curves for a current and a voltage transformer.[1]

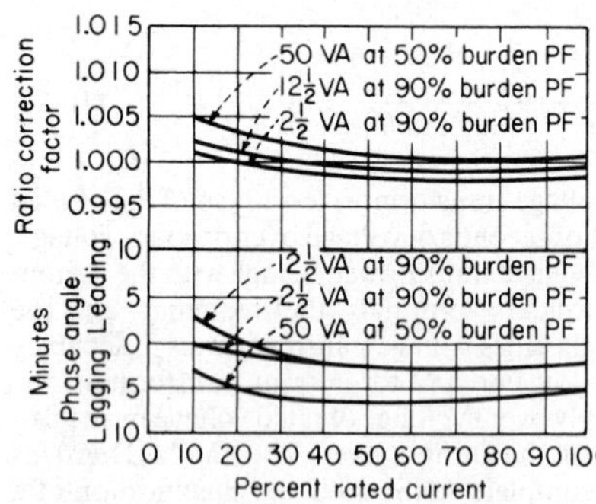


**FIG. 10-51** Typical curves of the ratio correction of phase angle for a current transformer.

***197. Example.*** Consider a load measured with 50/5-A current transformer, with a 12.5-VA 90% power-factor burden characteristic curve according to Fig. 10-51 with 2300:115 V voltage transformer rated 200 VA with a 75-VA 85% power-factor burden characteristic curve according to Fig.

[1] J. L. Settles, W. R. Farber, and E. E. Conner, the Analytical and Graphical Determination of Complete Potential Transformer Characteristics; *Trans. AIEE,* 1961, vol. 79, pt. 3, p. 1213.

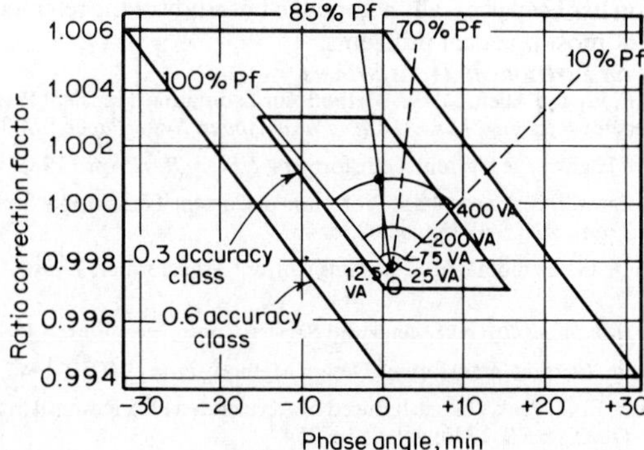


**FIG. 10-52** Typical curves of ratio correction factors and of phase angle for a voltage transformer.

10-52 and with a lag angle of 6 min in the wattmeter potential circuit. What is the load true watts when the instruments read 500 W, 115 V, and 5 A, calibration errors of instruments being neglected? From Fig. 10-51:

$$F_{CR} = 0.9997 \qquad \beta = -2'$$

From Fig. 10-52:

$$F_{PR} = 0.9984 \qquad \gamma = +1$$

Stated above,

$$\alpha = +6'$$

$$\theta = \cos^{-1}\frac{500}{5 \times 115} = \cos^{-1} 0.86957 = 29°35.47'$$

From Eq. (10-74),

$$K_S = \frac{\cos(29°35.47' - 2' - 1' + 6')}{\cos(29°35.47')} = \frac{0.86914}{0.86957} = 0.9995$$

From Eq. (10-75),

$$P = 500 \times 10 \times 20 \times 0.9997 \times 0.9984 \times 0.9995 = 99760$$

***198. Ferroresonance in voltage transformers*** can occur when a voltage transformer is connected line to ground on an ungrounded system; if the capacitive reactance, line to ground, can equal the exciting reactance of the voltage transformer, a condition which typically occurs at considerable overvoltage. A parallel resonant circuit of very high impedance is formed, raising the line-to-ground voltage considerably above normal; if the voltage is high enough, the voltage transformer will fail due to excessive exciting current or high voltage stresses.

Measures taken to reduce the incidence of ferroresonance include operation at low flux densities, resistance loading of the secondaries,[1] and insertion of resistance in the connection of the voltage transformer primary neutral to ground.

***199. Transient performance of current transformers*** with initially offset primary current has been given a great deal of study in recent years. It has been realized for many years that a transformer has difficulty in transforming the dc component of an offset current. No one

[1] R. F. Karlicek and E. R. Taylor, Ferroresonance of Grounded Potential Transformers on Ungrounded Systems; *Trans. AIEE,* 1959, vol. 78, pt. 3A, p. 607.

method of calculation has been generally acceptable but study of the references[1] will help one to arrive at a solution of most practical problems.

***200. References on Instrument Transformers***

Conner, E. E., Wentz, E. E., and Allen, D. W.: Methods for Estimating Transient Performance of Practical Current Transformers for Relaying; *IEEE Trans. Power Appar. Syst., Paper* T74 382-8, 1974, p. 116.

Park, J. H.: Accuracy of High-range Current Transformers; *NBS J. Res.*, April 1935, vol. 14, pp. 367–392.

Kaufmann, R. H., and Camilli, G.: Overvoltage Protection of Current Transformer Secondary Windings and Associated Circuits; *Trans. AIEE,* 1943, vol. 62, p. 467.

Boyajian, A., and Camilli, G.: Orthomagnetic Bushing Current Transformer; *Trans. AIEE,* 1945, vol. 64, p. 137.

*Electrical Metermen's Handbook,* 7th ed.; Meter and Service Committee, Edison Electric Institute, 1965.

Specht, T. R.: Biased-core Current Transformer Design Method; *Trans. AIEE,* 1945, vol. 64, p. 635.

Specht, T. R., and Wentz, E. C.: Peak Voltage Induced by Accelerated Flux Reversal in Cores Operated above Saturation Density; *Trans. AIEE,* 1946, vol. 65, p. 254.

Wiggins, A. M.: Parallel Operation of Current Transformers for Totalizing Two or More Circuits; *Electr. J.,* 1929, p. 379.

Agnew, P. G.: Accuracy of the Formulas for the Ratio, Regulation and Phase Angle of Transformers; *NBS Sci. Paper* 211, 1914.

Brooks, H. B.: Testing Potential Transformers; *NBS Sci. Paper* 217, 1914.

Price, L. D.: Potential Transformer Connections for Three-Phase, Four-Wire Metering; *Electr. J.,* August 1927, p. 377.

## *CIRCUIT BREAKERS*

*By B. P. BOEHLE and W. J. SCHMITT*

### Fundamentals

***201. Definitions.*** Circuit breakers are mechanical switching devices capable of making, carrying, and breaking currents under normal circuit conditions and also making, carrying for a specified time, and breaking currents under specified abnormal conditions such as those of short circuit. The medium in which circuit interruption is performed may be designated by a suitable prefix, for example, air-blast circuit breaker, gas circuit breaker, oil circuit breaker, or vacuum circuit breaker.

Circuit breakers are rated by voltage, insulation level, current, interrupting capabilities, transient recovery voltage, interrupting time, and trip delay.

For standard definitions, reference is made to ANSI C37.03 AC High-Voltage Circuit Breakers and ANSI C37.100 Definitions for Power Switchgear.

***202. History of Development.*** In the early days of electrification (1890) switches were of the hand-operated, knife-blade type.

*Air Switches.* With increasing currents and voltages, spring-action driving mechanisms were developed to reduce contact burning by faster-opening operation. Later, main contacts were fitted with arcing contacts of special material and shape, which opened after and closed before the main contacts. Further improvements of the air switch were the brush-type contact with a wiping and cleaning function, the insulating barriers leading to arc chutes, and blowout coils with excellent arc-extinguishing properties. These features, as well as the horn gap contact, are still in use in low-voltage ac and dc breakers.

*Oil Circuit Breaker.* Around 1900, in order to cope with the new requirement for "interrupting capacity," ac switches were immersed in a tank of oil. Oil is very effective in quenching the arc and establishing the open break after current zero. Deion grids, oil-blast features,

[1] E. E. Conner, E. C. Wentz, and D. W. Allen, Methods for Estimating Transient Performance of Practical Current Transformers for Relaying; *IEEE Trans. Power Appar. Syst., Paper* T74 382-8, 1974, p. 116 (and associated references).

pressure-tight joints and vents, new operating mechanisms, and multiple interrupters were introduced over several decades to make the oil circuit breaker a reliable apparatus for system voltages up to 362 kV.

*Minimum-Oil Circuit Breaker.* These breakers were developed after 1930, are used mainly in Europe, and make use of special low-oil-volume interrupting chambers of extralight weight. By means of current-dependent oil streams in different directions and supported by oil injection, the arc is cooled and extinguished effectively. The interrupters are mounted on porcelain or molded-resin supports, thus avoiding oil as an insulating medium to ground.

For both oil and minimum-oil circuit breakers, standard transformer oil can be used.

*Air-Blast Circuit Breaker.* Further increase of system voltages and generating capacities triggered the search for faster and stronger circuit breakers utilizing oilless arc interruption. After 1940, the air-blast circuit breaker was developed, making use of the good insulating and arc-quenching properties of dry and clean compressed air.

Figure 10-53 shows a typical air-blast circuit breaker of modular design, installed in 1950.

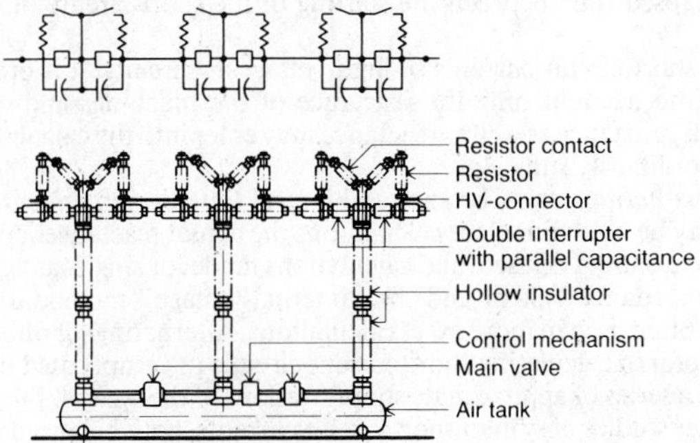


**FIG. 10-53** Outdoor air-blast circuit breaker 230 kV, 1000 A, 16 kA, 3-cycle interrupting time. *(Brown Boveri.)*

Further development of the air-blast breaker led to two-cycle interrupting time, extraheavy interrupters, and the constant-pressure control system.

*The magnetic air circuit breaker* uses a combination of a strong magnetic field (coil or soft-iron plates) with a special arc chute to lengthen the arc until the system voltage cannot maintain the arc circuit any longer. This interrupting principle is applied mainly in the distribution voltage range in metal-clad switchgear.

*$SF_6$ Circuit Breaker.* The excellent arc-quenching and insulating properties of $SF_6$ gas, sulfur hexafluoride (see Fig. 10-54), stimulated this breaker development around 1960. The $SF_6$ principle is used in conventional outdoor circuit breakers of 115 to 800 kV. The enclosed breaker design incorporated in gas-insulated substations (GIS) up to 800 kV received acceptance during the 1970s owing to its space-saving layout and improvements of complying with rising environmental, system, and service requirements. The $SF_6$ breaker further provides an oilless indoor solution for metal-clad and metal-enclosed switchgear up to 38 kV.

*Vacuum Circuit Breaker.* This has been the most recent advancement in new arc-interrupting and breaker development. Vacuum-bottle interrupters are designed for higher system voltage, current, and interrupting ratings. Increased application of vacuum circuit breakers spreads at distribution systems both in metal-clad and metal-enclosed switchgear.

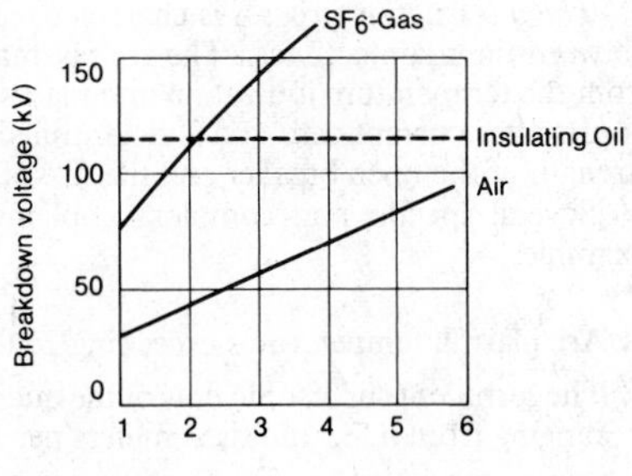


**FIG. 10-54** Breakdown voltage of oil, air, and $SF_6$ gas as a function of pressure at 38 mm (1½ in) electrode distance.

***203. Design Fundamentals.*** Regardless of the medium of arc quenching and insulation, each circuit breaker unit consists of the following *construction elements:* (1) main contact at system voltage, (2) insulation between main contact and ground potential (porcelain, oil, gas), (3) operating and supervisory devices as well as accessories out of reach of the system voltage-life zone, and (4) an insulated link between operating device and main contact.

*Tripping Facilities.* As part of the circuit-breaker control system, tripping facilities are vital to ensure proper function under all service conditions.

A circuit breaker is designed to cope not only with service or short-circuit currents but also to master the voltage stresses imposed on it when switching special equipment such as reactors, transformers, cables, etc.

The *short-circuit duty* is determined by the maximum short-circuit current that the rotating machinery connected to the system at the time of short circuit can pass through the breaker to a point just beyond the breaker, at the instant the breaker contacts open. The short-circuit current is determined by the characteristics of synchronous and induction machines connected to the system at the time of the short circuit, the impedance between them and the point of short circuit, and the elapsed time between the starting of the short circuit and the parting of the breaker contacts.

In *calculating* short-circuit currents of high-voltage ac circuits, it is ordinarily sufficiently accurate to take into account only the reactance of the machines and circuits, whereas in low-voltage circuits resistance as well as reactance may enter into the calculations. In dc circuits, resistance only is ordinarily sufficient.

For first *approximations,* the reactance and typical time-decrement curves of the synchronous machines may be used. For close calculations the actual reactances and time characteristics of the equipment should be used, and calculations made for single- as well as 3-phase faults. The "per unit" impedance system and the "internal-voltage" method, using "symmetrical components," are often used in more exact calculations. Alternating- or direct-current calculating boards are helpful in calculating short-current circuits in complicated networks. It is often necessary to make a series of approximate simplifications of the system. Programs are available for digital computer studies of system short-circuit currents, both balanced 3-phase and phase-to-ground.

The *interrupting capacity,* in kilovoltamperes, is the product of the phase-to-ground voltage, in kilovolts, of the circuit and the interrupting ability, in amperes, at stated intervals and for a specific number of operations. The current taken is the rms value existing during the first half cycle of arc between contacts during the opening stroke.

*Symmetrical Current Basis.* It has become a widely adopted practice to determine the interrupting capability of circuit breakers in kiloamperes symmetrical. The rated short-circuit current in rms kiloamperes is referred to the rated maximum voltage in kilovolts.

Reference is made to the following standards related to ac high-voltage circuit breakers rated on a symmetrical current basis.

ANSI C37.04-1979 Rating Structure

ANSI C37.06-1979 Preferred Ratings

*The interrupting process* is characterized by an arc appearing for a limited period of time between the main contacts. The arc plasma column contains a high conductivity originating from the temperature ionization and places a strain on the arc gap. Interruption of the circuit occurs after current zero and is determined by the race between the buildup of the dielectric strength of the open breaker and the rise of the initial transient recovery voltage.

Several specific and complex problems are involved with the interrupting process, for example;

1. Arc plasma temperatures exceeding 20,000 K
2. The turbulent supersonic flow of the quenching gas in a changing flow geometry with speeds ranging from a few hundred meters per second to several thousand meters per second
3. The interrupter-moving system and its drive accelerates the moving masses in a few thousandths of a second to 10 m/s while simultaneously compressing the quenching gas
4. The stress placed on the network system by the current interruption and the recovery voltage

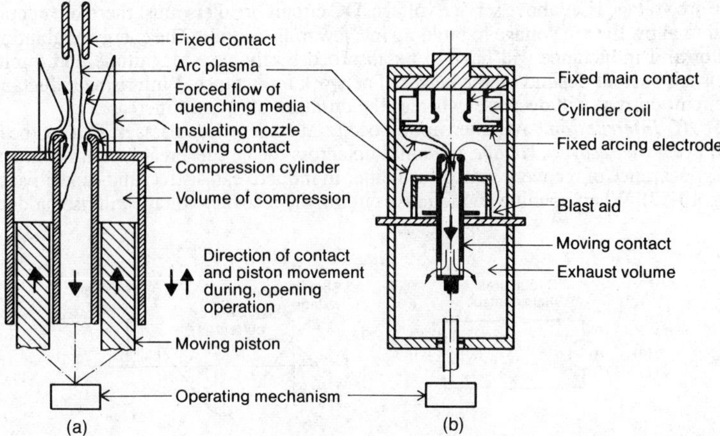


**FIG. 10-55** Principle of arc interruption. *(a)* Puffer-type interrupter; *(b)* self-extinction interrupter.

The *interrupting principle* of an $SF_6$ puffer-type interrupter is sketched in Fig. 10-55*a*. On opening, the fixed and moving contacts are pulled apart by the operating mechanism. Thus the fault current is forced to flow along the arc plasma. The contact movement combined with the compression cylinder movement in the opposite direction compresses the quenching gas inside the cylinder. The quenching gas is consequently forced to flow through the contact system, and the insulated nozzle toward the exhaust. This intensive flow of quenching medium along the arc rapidly removes the energy converted within the arc plasma and transforms the path between the open contacts into an insulating gap. The same effect is achieved by employing an interrupter of the self-extinction type (see Fig. 10-55*b*) for medium currents.

This principle makes use of a self-produced flow of quenching medium triggered by the magnetic field of a cylinder coil located at the arc-extinction zone.

***204. DC Interruption.*** DC interruption (see Fig. 10-56) is basically different from ac interruption. After contact parting, the arc is lengthened and cooled and consequently the arc voltage is rising. The current will extinguish after the change of current *di/dt* becomes negative

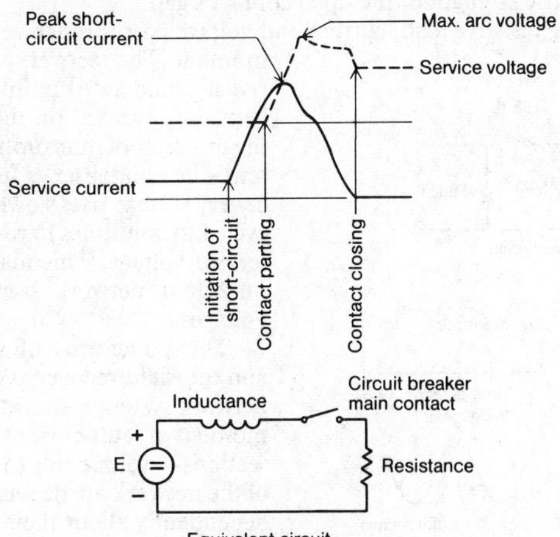


**FIG. 10-56** Direct-current interruption; typical shape of short-circuit current and recovery voltage.

and the arc voltage rises above service voltage. DC circuit breakers must therefore operate fast, in order to allow the arc voltage to build up in a few milliseconds. The energy originating from generator and inductance will have to be absorbed by the arc. Magnitude and duration of short-circuit current depend on the height of network inductance. With rising inductance, the short-circuit current will decrease, whereas the duration of arc will increase.

***205. AC Interruption.*** AC interruption occurs at current zero passage. During the following half cycle the *recovery voltage* will build up across the circuit breaker main contacts. The typical appearance of recovery voltage will differ in inductive, resistive, and capacitive circuits (see Fig. 10-57). When opening an inductive circuit, the recovery voltage will rise suddenly at a high rate because current interruption occurs at the moment of voltage peak. This case requires fast buildup of dielectric strength of the open contact's gap.

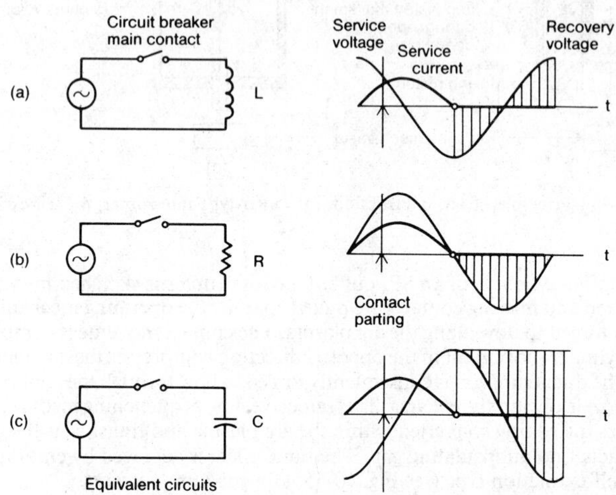


**FIG. 10-57** Typical shape of recovery voltage on interruption: *(a)* inductive current; *(b)* resistive current; *(c)* capacitive current.

When interrupting resistive load, current and voltage pass through zero at about the same moment. The recovery voltage will therefore rise at a moderate rate and no particular problems are imposed on the circuit breaker. At the moment of interruption of capacitive current, the condenser is fully charged. The recovery voltage rises slowly during the first half cycle but continues to rise to a value twice the service voltage. This may lead to reignition, undesired network oscillations, and overvoltages.

The waveforms of short-circuit current and transient recovery voltage in a simplified network system is shown in Fig. 10-58. At the moment of fault-current interruption the two sections—source side ($S$) and line side ($L$)—of the network are decoupled and oscillate independently about their driving voltage. The difference of these two transients appears across the breaker pole. The behavior of this transient recovery voltage is determined by the circuit parameters. The still-moving or al-

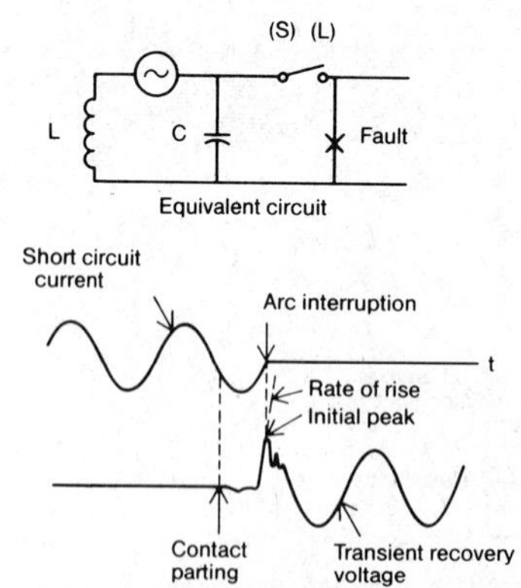


**FIG. 10-58** Alternating-current interruption; typical shape of short-circuit current and transient recovery voltage.

ready fully open breaker contacts must be able to withstand the recovery voltage. The most severe stress for the open contact gap is the *initial peak* and the *rate of rise* (kV/$\mu$s) of the recovery voltage.

If the recovery voltage exceeds the gap insulation, the restruck arc will continue until the next current zero, when interruption will again be attempted. The rate of rise of recovery voltage is a function of the constants of the circuits which supply power through the breaker. The larger the adjacent capacitance to ground before the major inductance limiting the fault current, the slower will be the rise of the recovery voltage. Some breakers modify the recovery-voltage characteristic by limiting the current, modifying its power factor, etc.

## Severe Interrupting Conditions

The following severe cases of circuit-breaker switching conditions have to be considered carefully: terminal fault, short-line fault, out-of-phase switching, switching of small inductive currents, switching of capacitive currents, closing on fault. The chosen examples are typical only; they are uniformly based on similar and simplified network configurations and are restricted to single-phase fault conditions. Other switching conditions may also be important.

***206. Terminal Fault.*** After interruption of short-circuit current, the recovery voltage oscillates toward the service-frequency driving voltage via an initial peak. The natural frequency is determined by the inductance and the capacitance of the driving system (Fig. 10-59).

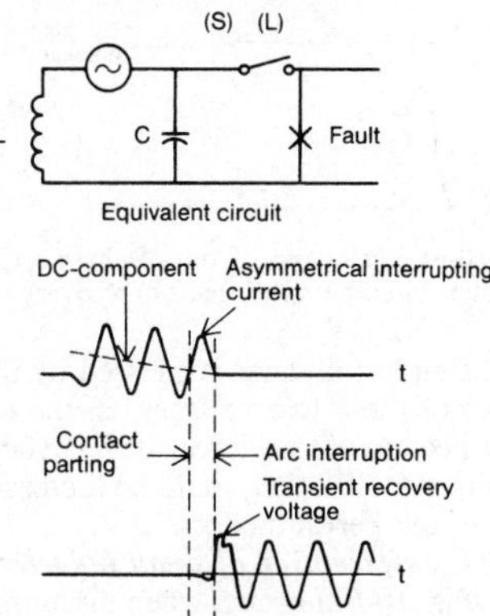


**FIG. 10-59** Principle of terminal fault interruption, equivalent circuit; typical shape of short-circuit current and transient recovery voltage.

The dc-component of the short-circuit current depends on the time constants of the network components like generators, transformers, cables, and high-voltage lines and their reactances of the zero-sequence and the positive-sequence systems. The recovery voltage will accordingly vary depending on the location of the circuit breaker within the network.

***207. Short-Line Fault.*** In the case of a short-line fault, a section of line lies between the breaker and the fault location (Fig. 10-60). After the short-circuit current has been interrupted, the oscillation at the line side (*L*) of the breaker assumes a superimposed "sawtooth" shape. The rate of rise of this line oscillation is directly proportional to the effective surge impedance and the change of current at current zero. The component on the supply side (*S*) basically exhibits the same waveform as a terminal fault. The circuit breaker is stressed by the difference between these two voltages. Because of the high frequency of the line oscillation, the transient recovery voltage has a very steep initial rate of rise. Since the initial rate of rise increases with increasing rate of current change, the interrupting capability of many breaker designs is determined by the short-line fault.

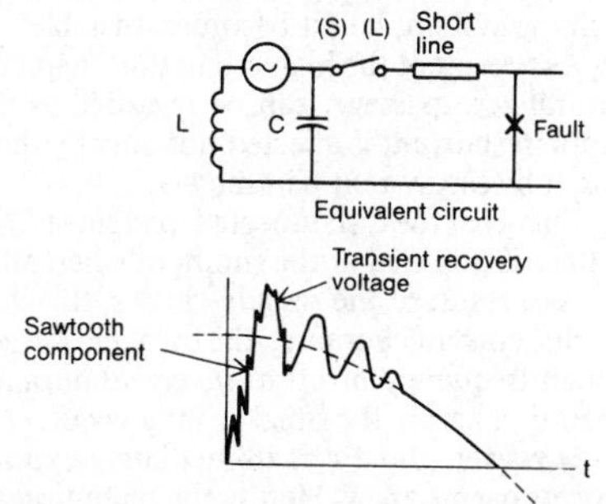


**FIG. 10-60** Principle of short-line fault interruption, equivalent circuit; typical shape of recovery voltage.

***208. Out-of-Phase Switching.*** Two network systems with driving voltages $E1$ and $E2$ are connected via a high-voltage transmission line (Fig. 10-61). Since the circuit is closed via the closed circuit breaker, the resulting driving voltage is equal to the sum of the two system voltages. Driving voltage $E2$ may, for example, exceed voltage $E1$ by the voltage drop along the transmission line. After opening the breaker, the transient recovery voltages of the disconnected networks oscillate independently.

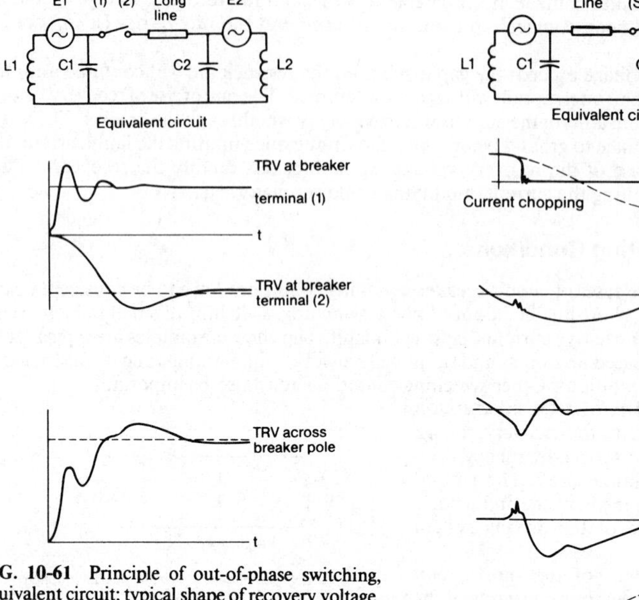


**FIG. 10-61** Principle of out-of-phase switching, equivalent circuit; typical shape of recovery voltage.

**FIG. 10-62** Principle of small-inductive-current interruption; equivalent circuit; typical shape of current and voltage.

The circuit breaker is stressed by the difference of these two voltages. In the case of disconnection of long lines, the recovery voltage across the breaker could be increased because of the Ferranti effect.

***209. Interruption of small inductive currents*** (Fig. 10-62) occurs when disconnecting unloaded transformers, reactors, or compensating coils. An arc is produced between the contacts when the circuit breaker is opened. The arc voltage is approximately constant at higher currents, since the arc energy is removed only by convection. With small currents, the arc voltage increases as a result of arc looping and a change in the cooling mechanism.

When approaching current zero, the arc current begins to oscillate as a result of interaction with the system; i.e., it becomes unstable.

As a result of the high oscillation frequency, the current interruption may occur prior to the natural zero passage, can be regarded as instantaneous, and is called *current chopping.* The chopping current is affected not only by the properties of the circuit breaker but also to a great extent by the system parameters.

Energy at the disconnected load side (*L*) oscillates with the natural frequency; the maximum voltage is attained at the moment when all energy is converted into capacitive energy.

As a result of the resistive losses, the voltage on the disconnected load side decays to zero. During current chopping, the breaker is stressed by the supply-side voltage which builds up with a high frequency of up to several thousand cycles per second. During this increasing stress, reignition across the breaker may occur.

However, the arc is immediately extinguished again because of the low current and the process begins anew. Hence, the reignition also helps to reduce the energy stored in the disconnected circuit.

***210. Interruption of Capacitive Currents.*** Capacitive currents occur during line dropping as well as during disconnecting open end cables or capacitor banks (Fig. 10-63). Although switching of capacitor banks is regarded as a special application, disconnecting of charged lines is a frequent switching operation.

Current chopping may occur at a low instantaneous value during interruption of capacitive currents, but this does not lead to overvoltages. After interruption of current, the voltage at the line capacitance (*L*) remains at the peak value of the power frequency voltage, whereas the voltage on the source side (*S*) oscillates about the driving voltage. The difference between the two voltages appears across the circuit breaker with an amplitude of more than double the rated voltage. If the circuit breaker cannot withstand this higher voltage, *restriking* may occur. Restriking is similar to closing transmission lines with trapped charge. After restriking, a transient current flows through the circuit breaker, which is of higher frequency than that of the system and which can again be interrupted during the reignition process. After *reextinction,* the line is charged to the potential of the peak value of the equalizing process, whereas the circuit-breaker terminal on the source side (*S*) recovers to the system voltage. A very high differential voltage appears across the breaker, which may lead to renewed restriking and even switching failures. Restrike-free interruption of capacitive currents is thus of utmost importance. Basically the same phenomenon occurs during disconnection of capacitor banks. To determine the voltage stresses of the circuit breaker, however, the grounding conditions of the supply system and capacitor bank and the arrangement of the bank have to be taken into account.

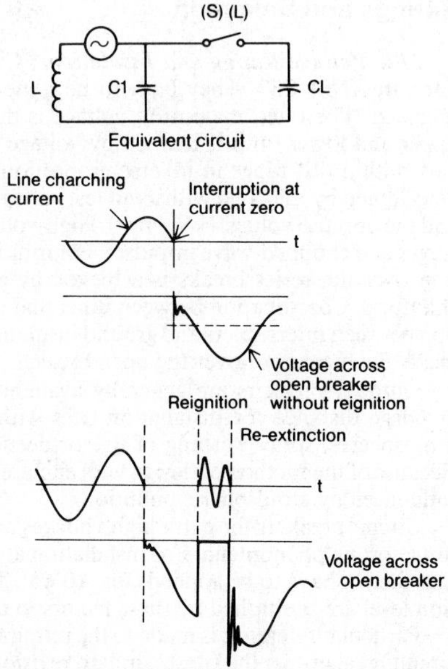


**FIG. 10-63** Principle of capacitive current interruption, equivalent circuit; typical shape of current and voltage.

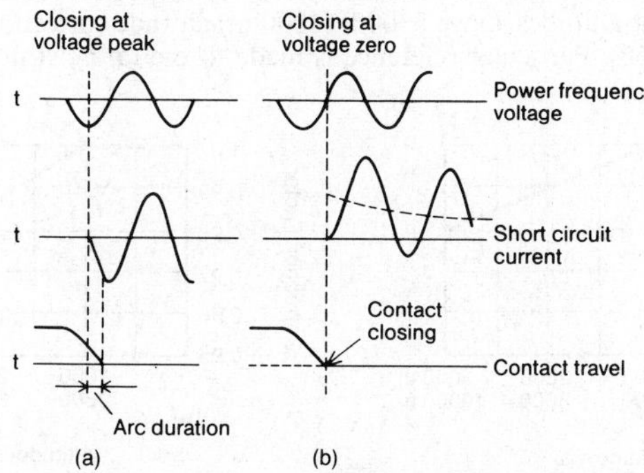


**FIG. 10-64** Stress on contact when closing on a fault, contact travel related to *(a)* symmetrical, *(b)* asymmetrical short-circuit current.

***211. Closing on a fault*** (Fig. 10-64) directs the stress onto the circuit-breaker contact system, particularly as regards the electrodynamic and thermal forces. The current and voltage stress is different during closing on *(a)* symmetrical or *(b)* asymmetrical short-circuit current. The deciding factor is the moment of contact touch relative to the phase angle of service voltage. In case contact touch and consequently ignition of the arc occurs at voltage maximum, the short-circuit current will appear symmetrical. The other extreme case takes place with the moment of closing at voltage zero. Here the asymmetrical short-circuit current contains the maximum dc component. A contact system designed for fast closing operation will be subjected to a shorter arcing time and consequently to reduced contact burning when closing on asymmetrical currents. Fast operation is therefore not only important for opening but also for circuit-breaker closing.

## Ratings and Selection

***212. Voltage Rating and Insulation.*** Circuit breakers are built for voltage ratings as defined in ANSI C37. They have to be dimensioned to withstand the maximum voltages as specified. The rated maximum voltage is the upper limit for operation. The range between upper and lower limit is defined by voltage range factor *K*. Current-interrupting capabilities vary within this range in inverse proportion to the operating voltage. The insulation level is determined by the rated withstand test voltages specifying the low-frequency voltage (kV, rms) and the impulse voltage (kV, crest). High-voltage breakers must essentially withstand switching surges and chopped-wave impulses. For multiple-break circuit breakers, equal voltage distribution over the series breaks is achieved by grading capacitors paralleled to the interrupting chambers. Coordination between inner and outside insulation, as well as insulation coordination between interrupters and ground insulation, has to be laid out properly to prevent flashover inside the breaker or over the open break.

Outdoor breakers are generally available with special porcelains that provide increased creepage distance for installation sites with highly contaminated air. For heavily polluted atmospheres, spray washing of live or deenergized breakers may be an additional measure. Because of the method of design with enclosed ground insulation, the GIS circuit breaker is not influenced by atmospheric pollution.

Circuit breakers for extra high voltages are fitted with protection armatures as standard to limit corona phenomena. For installation at altitudes above 3300 ft (1000 m), altitude correction factors have to be applied (Fig. 10-65). The values of rated maximum voltage and insulation level are multiplied by these factors to obtain the values for the application.

Particular reference is made to the rating structures and preferred ratings for ac high-voltage circuit breakers per the latest standard revisions of ANSI/IEEE C37.04-1979 and ANSI C37.06-1979.

***213. Current Rating.*** The rated continuous current is the current a circuit breaker has to be able to carry continuously without exceeding a specified temperature limitation in a specified ambient temperature. Attention must be paid to reduction factors arising from the kind and site of installation, class of insulating material, and electrical endurance requirements.

For installations at altitudes above 3300 ft (1000 m) altitude correction factors have to be applied (see Fig. 10-66). Particular reference is made to the rating structures and preferred ratings and capabilities for ac high-voltage circuit breakers rated on a symmetrical current basis per the latest standard revisions of ANSI/IEEE C37.04-1979 and ANSI C37.06-1979.

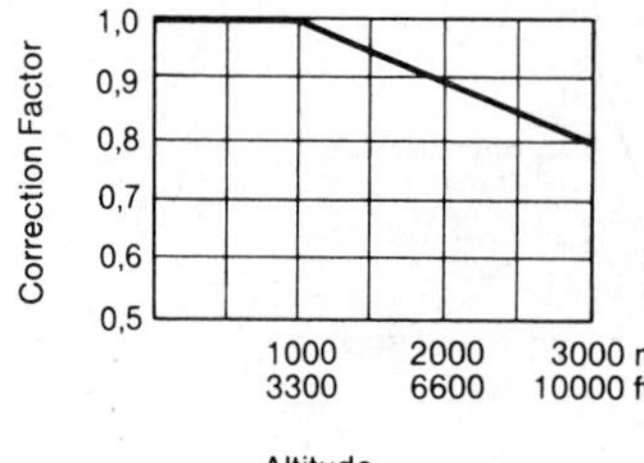

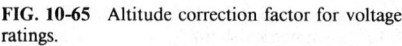

**FIG. 10-65** Altitude correction factor for voltage ratings.

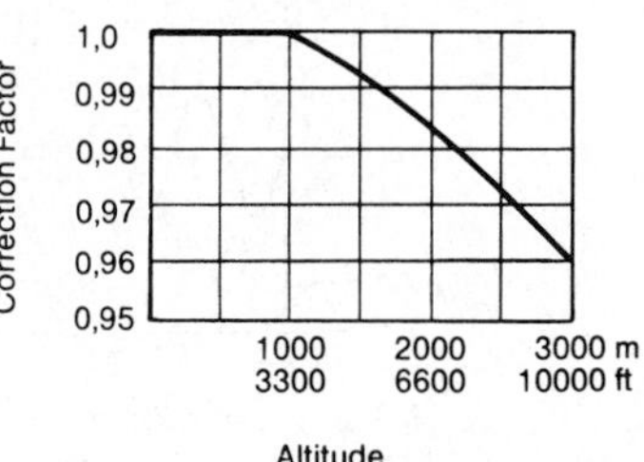


**FIG. 10-66** Altitude correction factor for current ratings.

AC high-voltage breakers must be able to interrupt the rated short-circuit current (kA, rms). They shall be capable of performing the required closing-latching-carrying-interrupting duties in immediate succession. The closing and latching capability shall be 1.6*K* times rated short-circuit current (kA, rms). The current-carrying capability is determined by the 3-s short time current (kA, rms).

***214. Selection and Application.*** The proper selection and application of circuit breakers is an extremely important element in the design of an electrical system. Breakers are relied upon to separate a defective portion of the system from the remainder to prevent the spread of damage and to permit the good portion to continue in service.

Application conditions and considerations for ac high-voltage circuit breakers are outlined in ANSI C37.010-1972.

Among others, the following criteria have to be considered when selecting a circuit breaker:

*System data,* such as maximum service voltage, insulation level, short-circuit requirements, line or cable parameters

*Switching conditions,* such as service currents; switching of unloaded transformers, unloaded lines and cables, choke coils, capacitors, generators, and motors; interrupting short-circuit currents and performing special duties like phase opposition, evolving fault, closing on fault, closing of long lines; duty cycle, reclosing, and operating times

*Service requirements,* such as special application for industrial plants, hazardous plants, furnace duty, railway duty, marine duty, maintenance, and operation

*Site of installation,* altitude above 3300 ft, climatic conditions, humidity, wind load, ice, air contamination, space requirements, environmental requirements, earthquake, connection to and function with other switchyard and network components, open installation, metal-clad or metal-enclosed

## Operating Functions

***215. Opening Operation and Duty Cycle.*** Speed of modern breakers has been increased to reach standard interrupting times of 2 or 3 cycles, measured from energizing of trip coil until extinguishing the arc; even higher speeds can be obtained where conditions warrant a special design. The generally accepted *duty cycle* is two openings, with a 15-s interval between them. The standards require that a power circuit breaker shall perform at or within its interrupting rating without emmitting flame; that, at the end of any performance within its interrupting rating, the circuit breakers shall be in substantially the same mechanical condition as at the beginning; that it shall then withstand system overvoltage, and its main current-carrying parts shall be in substantially the same condition as at the beginning. It is recognized, however, that after a breaker performs its duty cycle at or near its interrupting rating, the breaker may have its interrupting ability materially reduced and should be inspected and repaired if necessary.

For interrupting duties below 25% of the asymmetrical interrupting capability at rated maximum voltage, the interrupting time may be increased by 1 cycle for breakers having 3 cycles or less standard interrupting time. Operating times and contact travel of breakers have to be coordinated properly (see Fig. 10-67).

***216. Closing Operation.*** Circuit breakers are designed to perform the closing and reclosing operations as per standard requirements. When operated to close on long lines, extra-high-voltage circuit breakers require special measures to keep switching overvoltages within specified limits. Such measures may be either single- or multiple-step *closing resistors,* synchronously closing at the moment of voltage zero, or *polarity-controlled-closing,* which means closing during the period of equal polarity at the line and source side of the breaker.

The magnitude of overvoltages on energizing and reenergizing is influenced by the nature and variables of the power system. Parameters of supply side and line must be taken into account in order to compute the overvoltages or to determine them on transient network analyzers.

For a summary of the magnitude of overvoltages occurring when energizing high-voltage lines, based on numerous studies and measurements in high-voltage networks, see Table 10-10.

***217. Operating Mechanism.*** Opening and closing of power circuit breakers under service conditions is seldom performed manually, since most breakers are installed in systems designed for remote control providing specific redundancy. Various means of operation are used, such as *(a)* dc solenoids, *(b)* solenoids operated from an ac source through a dry-type rectifier, *(c)* compressed air, *(d)* high-pressure oil, *(e)* charged spring, and *(f)* electric motor. *Automatic reclosing* of breakers in overhead line feeders is frequently used to restore service quickly after a line trips out because of lightning or other transitory fault. Instantaneous or time-delay reclosing may be provided with a lockout to prevent more than one to several successive reclosures, as desired. If the fault is cleared before the lockout feature operates, the reclosing device resets itself, permitting a complete cycle of reclosing at a subsequent fault.

The circuit-breaker-operating device has to cope with the increasing requirements in inter-

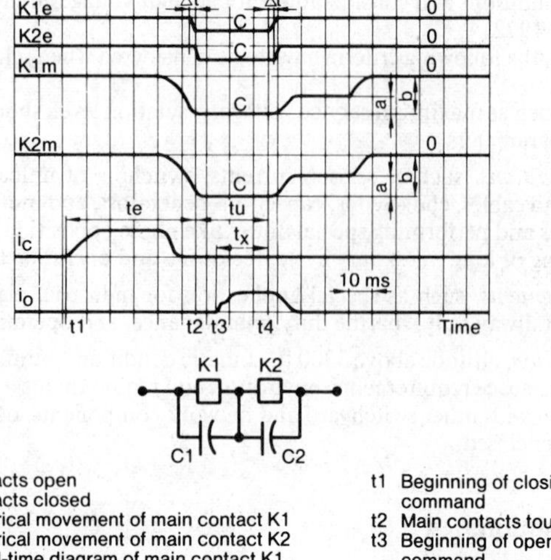


O Contacts open
C Contacts closed
K1e Electrical movement of main contact K1
K2e Electrical movement of main contact K2
K1m Travel-time diagram of main contact K1
K2m Travel-time diagram of main contact K2
a Extinction distance
b Isolating distance

t1 Beginning of closing command
t2 Main contacts touch
t3 Beginning of opening command
t4 Main contacts part
te Closing time
tu Make-break time
tx Opening time
Δt Time difference between main contacts K1–K2
$i_c$ Closing command
$i_o$ Opening command

**FIG. 10-67** Switching-time and contact-travel oscillograms on closing and opening of an air-blast double-interrupting chamber.

**TABLE 10-10** Overvoltages Occurring When Energizing High-Voltage Lines

| Prevailing conditions | Overvoltage factor (per unit) |
|---|---|
| 1. Line with trapped charge, no compensation, no means of reduction employed | >3 |
| 2. Line without trapped charge, no compensation, no closing resistors, or with trapped charge, no closing resistors, but polarity-dependent closing | 2,0 to 2,8 |
| 3. As 2, but with compensation | 2,0 to 2,5 |
| 4. Single-stage closing resistors, compensated line | ≤2,0 |
| 5. Two-stage closing resistors, optimum compensation | ≤1,7 |
| 6. Two-stage closing resistors, combined with polarity-dependent closing, or compensation with optimized multistage closing resistors | 1,5 |

rupting and current-carrying capability as well as with shorter operating times. Simplicity of design, robustness, and reliability have to ensure safe operation of this vital link between the electrical system controls and the interrupter. The principle of a pneumatic drive is sketched for an extra-high-voltage circuit breaker which functions according to the differential piston principle in Fig. 10-68. A pneumatic interlocking device in connection with the $SF_6$ gas system ensures that the breaker always remains in the defined open or closed position even on loss of air pressure. Besides opening and closing functions, effective damping of the highly accelerated moving parts is incorporated.

***218. Accessories.*** Circuit breakers may be equipped with a wide range of accessories, either required, like pressure controls, gas-density monitors, safety valves, and position indicator, or optional, like a choice of different release, alarms, or auxiliary contacts. To illustrate the

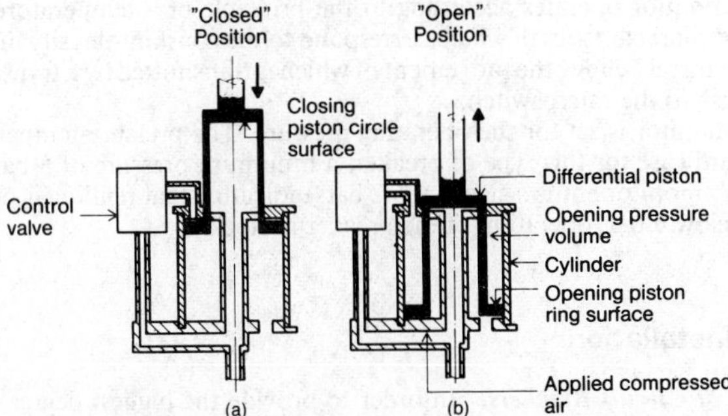


**FIG. 10-68** Principle of the drive system for an $SF_6$ outdoor breaker: *(a)* closed position; *(b)* open position. *(Brown Boveri.)*

importance of accessories for safe and reliable circuit-breaker operation, Fig. 10-69 shows the $SF_6$ gas monitoring system of a high-voltage $SF_6$ outdoor breaker.

The breaking capacity of an $SF_6$ breaker depends on the gas density. It is assumed that the volume remains constant during temperature variation, whereas the pressure of $SF_6$ is highly dependent on temperature change. Hence, to monitor the state of the gas, it is logical to supervise not the pressure but the density of the gas.

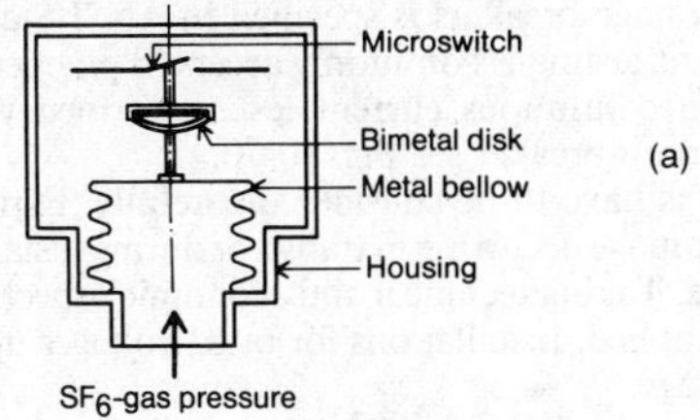


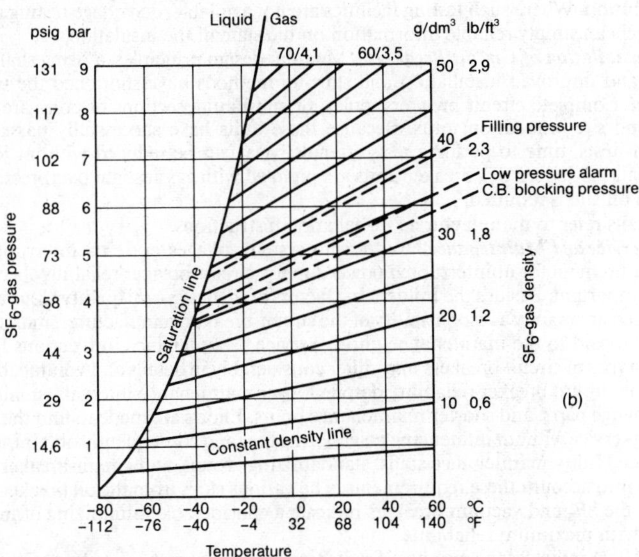


**FIG. 10-69** *(a)* Arrangement and *(b)* pressure-temperature diagram for $SF_6$ gas-density monitor system for an outdoor breaker. *(Brown Boveri.)*

The density monitor operates according to the principle of a temperature-compensated pressure gage, the characteristics of which correspond to the constant-density line. The $SF_6$ gas pressure acts on a metal bellow, the movement of which is transmitted by a transfer mechanism with a bimetal disk to the microswitch.

The density monitor is set for the operating pressure. The pressure-temperature diagram shows the standard case for this type of breaker, a minimum pressure of 5 bar, measured at 20°C. The density monitor emits a signal at 5.2 bar, indicating that refilling is necessary. If the pressure drops below 5 bar, operation of the breaker is blocked.

## Testing and Installation

***219. Testing of Circuit Breakers.*** In order to provide the highest degree of quality and reliability, the circuit breakers have to satisfy a wide range of tests, including

*Design tests* during development and early manufacturing stages with each design and rating of breaker for compliance with specified values and regulations

*Production tests* on each individual breaker or part of breaker

*Tests after delivers*

*Field tests*

*Conformance tests*

Test procedure for ac high-voltage breakers is specified in ANSI Standard C37.09-1979. Element testing or synthetic circuit testing is commonly accepted practice for extra-heavy-capacity breakers. For results of a continuous current test performed with an interrupting chamber of a high-voltage $SF_6$ circuit breaker see Fig. 10-70.

Field tests on GIS installations have to be considered carefully. Experience gained from commissioning tests shows that damage occurring in transit or during installation is small. Even so, a high-voltage test is advisable. Taking technical and economic aspects into consideration, the following guideline may be applied: Installations for rated voltages up to 242 kV are best tested with power-frequency voltage.

For installations with higher rated voltages, testing with oscillating switching surges proves the best solution. Where such testing facilities are not available, dc voltage testing at a suitable voltage level can supply reliable information on the state of the insulation.

***220. Installation of Circuit Breakers.*** Modular design principles, a rising share in factory assembly, and improved installation and start-up methods have shortened the time for site installation. Complete circuit breakers, poles, or substantial sections of poles are handled as shipping and site-installation units. Because these units have successfully passed extensive production tests, time to perform tests after delivery can be reduced without jeopardizing breaker quality. Modern $SF_6$ breaker units are prefilled with gas at slight overpressure, thus no evacuation on site is required.

For details refer to manufacturers' installation instructions.

***221. Service and Maintenance.*** With rising system voltages, currents, interrupting ratings, and the requirement for uninterrupted power supply, circuit-breaker reliability becomes more and more important. Besides the influencing factors of *(a)* design, *(b)* quality assurance, and *(c)* testing, which are mainly a responsibility of the circuit-breaker manufacturer, maximum attention must be paid to the maintenance during service. Maintenance instructions for different makes and types of circuit breakers may differ considerably in details and volume, but all strive to obtain maximum breaker reliability despite longer maintenance intervals, smaller inventories of exchange parts, and shorter maintenance hours. Efforts are made to find the easiest way of handling service without influencing neighboring gear and consequently obtaining the lowest service costs. Utility maintenance staffs, standardizing groups, and circuit-breaker developers have taken into account these requirements. The various steps from the oil breaker to air-blast and finally the $SF_6$ and vacuum breakers indicate a considerable minimizing of maintenance combined with maximum reliability.

Figure 10-71 indicates the number of switching operations with related interrupting current as one important maintenance criterion for a modern $SF_6$ circuit breaker.

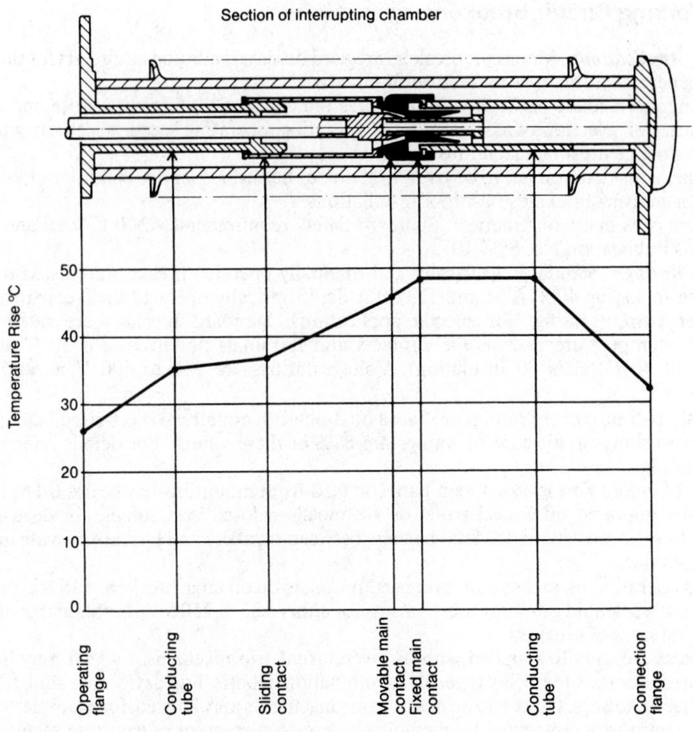


**FIG. 10-70** Temperature rise curve; current path of a 145-kV, 2000-A outdoor interrupter unit. *(Brown Boveri.)*

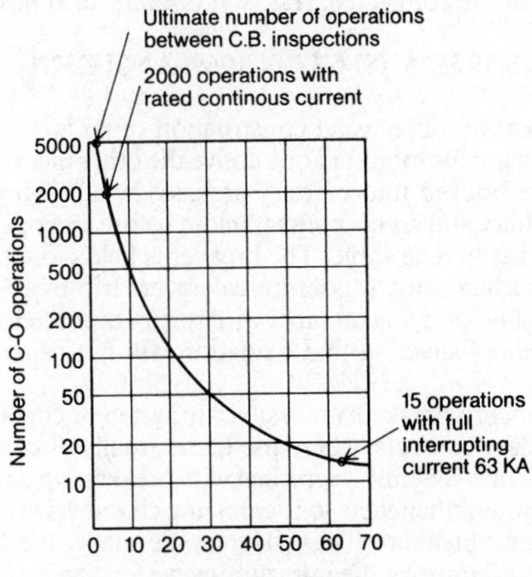


**FIG. 10-71** Maintenance as a function of switching operations and interrupting current; $SF_6$ outdoor circuit breaker. *(Brown Boveri.)*

## Low-Voltage Circuit Breakers

***222. Application.*** Air circuit breakers are used on dc circuits and ac circuits for the protection of general lighting, power, and motor circuits.

Distinction is made between various protection classes and different service and ambient conditions. For selection of a breaker, type and rating, operating speed, selectivity with fuses, and high voltage must be taken into account.

Further consideration has to be given to severe or hazardous service conditions like tropical climate or marine- or explosion-proof installations.

Reference is made to American Standard Safety requirements ANSI C37.19 and NEMA Standards Publication No. SG3-1971.

***223. Ratings.*** Standard electrically and manually operated breakers are listed in ratings up to and including 4000 A ac and 12,000 A dc. Electrically operated breakers are available in higher current ratings for special applications. Standard breakers are rated on the basis of a temperature rise on the contacts and terminals not to exceed 50°C above an ambient of 40°C (class 90 insulation). Voltage ratings are 250 to 600 V ac and 250 to 750 V dc.

The short-time current ratings are based on 3-phase symmetrical short-circuit currents; the single-phase short-circuit current ratings are 87% of these values. For details refer to ANSI C37.16-1973.

***224. Assembly Variations.*** Open-panel or dead-front mounting may be used. The breakers are usually mounted on a steel panel or are metal-enclosed in a cubicle for dead-front or drawout type of construction. Metal barriers between breakers and busbars provide increased safety in service.

Hand operation by means of a lever is common, even on large breakers. Electric operation by means of a solenoid or motor mechanisms for either 125 or 250 V dc is obtainable on all but the smallest sizes of breakers.

Breakers are usually supplied with an overcurrent trip mechanism which may be of the instantaneous or the time-delay type, or a combination of both. Trip devices are adjustable over a wide range of ratings. Other trip devices and arrangements may be used, for example, low-voltage trips, shunt trips connected to overvoltage, reverse current, or overcurrent relays.

Multiple-pole circuit breakers are commonly used in practically all capacities, one pole being used for each ungrounded line of a circuit, that is, a 2-pole breaker for a 3-wire grounded circuit or a single-pole breaker for a 2-wire grounded circuit.

Breakers can usually be equipped with auxiliary contacts, alarm contact, push-button control, position indicator, and key interlock. The widely used drawout type of breaker may be moved into and locked in the connected, test, and disconnected positions and/or completely withdrawn.

Refer to ANSI C37.13-1973, ANSI C37.14-1969, ANSI C37.17-1972, and ANSI C37.19-1963.

***225. Air Circuit Breaker.*** The usual construction of an air circuit breaker (Fig. 10-72) makes use of two fixed terminals mounted one above the other in a vertical plane, which, when the breaker is closed, are bridged under heavy pressure by a bridging member operated by a system of linkages. Auxiliary and arcing contacts close before and open after the main contacts. The arcing contacts are easily renewable. The breaker is held closed by a latch which may be tripped electrically or mechanically. Modern breakers are trip-free.

Many breakers use a solid bridging member with spring-mounted self-aligning contacts. The contact surfaces are made of silver so that oxidation will not cause excessive resistance and overheating.

Arcing contacts of modern breakers use a silver-tungsten or copper-tungsten alloy which is arc-resisting. The secondary contacts, where used, are usually of copper or silver alloy.

Barriers between poles are generally furnished with breakers on ac and dc circuits 250 V and above, and special arc chutes, quenchers, or deionizing chambers are also used throughout the available lines of air circuit breakers. These devices are made in different forms by different manufacturers and serve to improve the interrupting performance of the breaker and shortening the arcing time.

***226. Molded-Case Circuit Breaker.*** This circuit breaker is completely enclosed within a ruggedly constructed molded case of insulating material. It has received wide acceptance in the industry and is particularly adaptable in large buildings and industrial plants. The molded-case

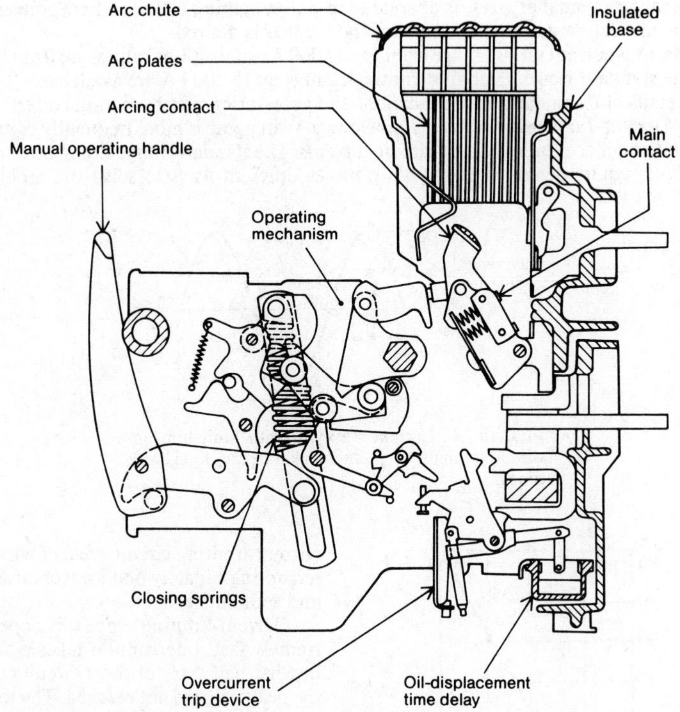


**FIG. 10-72** Typical low-voltage air circuit breaker with magnetic air chutes; breaker in the open position. *(I-T-E; GE; Westinghouse; Brown Boveri; etc.)*

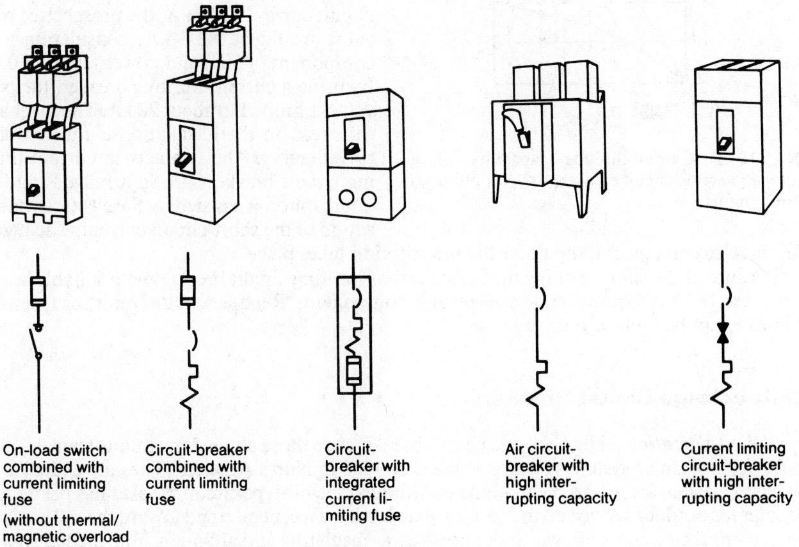


**FIG. 10-73** Methods of current-limiting in low-voltage circuits.

circuit breaker, in smaller sizes, is adaptable in home lighting circuits where convenience of automatic protection with manual reset of the breaker is desired.

Continuous current ratings range from 15 to 4000 A, interrupting ratings are from 5 to 45 kA within the standard range. High interrupting ratings up to 200 kA are available.

For details of technical data, application, and accessories refer to manufacturers' catalogs.

***227. Current-Limiting Breaker.*** Low-voltage switchgear is more frequently connected to systems with high or extrahigh short-circuit currents. The standard-range circuit breaker cannot satisfy these requirements. Figure 10-73 outlines different ways to solve the problem. The

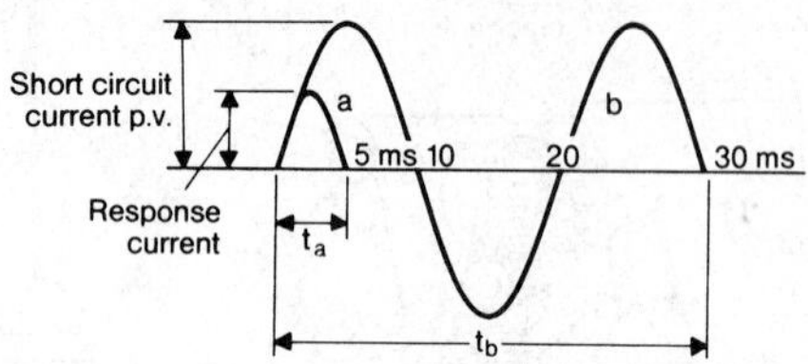


**FIG. 10-74** Current wave *(a)* with limitation, *(b)* without limitation; $t_a$-total break time *a*; $t_b$-total break time *b*.

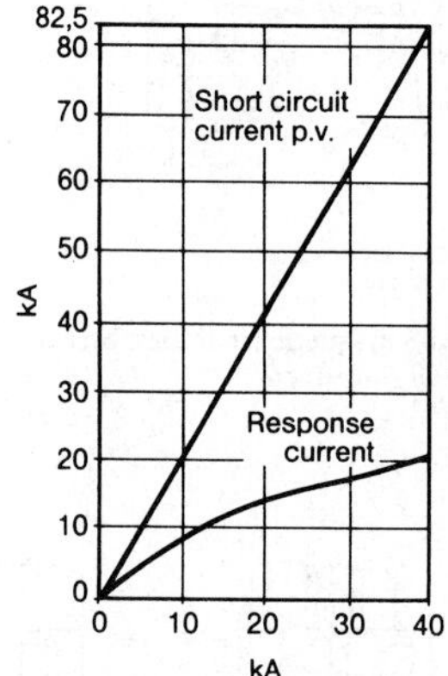


**FIG. 10-75** Current-limiting capability of a motor-protection circuit breaker; 100-A continuous current rating.

current-limiting circuit breaker with high interrupting capacity offers a technically sound and economical solution.

Current-limiting breakers operate extremely fast. Interruption takes place within the first half cycle of short-circuit current, so the peak value is not reached. The total break time is less than 5 ms. Figure 10-74 illustrates the current curve, and Fig. 10-75 shows the current-limiting characteristic of a 100-A breaker. With an initial symmetrical short-circuit current of 40 kA, the prospective peak value would be 82.5 kA, considering a dc component of 50% and power factor of 0.25. By using a current-limiting breaker, the peak value is limited to about 20 kA. The mechanical stress on the installation is thus reduced considerably. The contacts in current-limiting circuit breakers are so arranged that the interruption is assisted by the electrodynamic action of the short-circuit current. The higher the short-circuit current, the faster the interruption takes place.

Because of the short opening time, the current-limiting circuit breaker, with suitable accessories, can be used to protect power electronic components. Rectifier circuits omitting fuses, for example, can be built in this way.

## High-Voltage Circuit Breakers

***228. Application.*** High-voltage circuit breakers are those applied in circuits from 1000 V to the maximum ac system voltage, which is 800 kV in North America. They are available in two basic designs known as oil and oilless. Although the oil-type circuit breaker has been most popular for outdoor service up to 362 kV, there has been a general trend toward the oilless types using compressed air or $SF_6$ gas under pressure as insulating and arc-quenching media. At 550 and 800 kV only oilless breakers, and to some extent minimum-oil breakers, are used. For indoor service, mainly oilless circuit breakers are used in new installations.

Indoor minimum-oil breakers up to 34.5 kV are widely used in metal-clad switchgear overseas and to some extent in North America. Indoor oilless circuit breakers used in the United States are principally magnetic-air and air-blast types, and in recent years $SF_6$ type. The vacuum interrupting principle made its appearance in circuit breakers and circuit reclosers during the 1970s. For 2.5 to 34.5 kV, indoor-type breakers of the magnetic-air, air-blast, and minimum-oil designs have been applied outdoors by placing them in metal housing. This practice has been found to be both economical and practical.

The basic reasons for the swing to oilless circuit breakers are *(a)* elimination of oil fire hazard, *(b)* elimination of bulk-oil handling, *(c)* shorter contact-maintenance time and breaker-outage time, *(d)* cleanliness, *(e)* higher performance speeds, and *(f)* rapid growth of transmission system voltages up to 800 kV.

For circuit-breaker selection and application, refer to the latest revisions of the following standards: ANSI C37.010-1972, ANSI C37.011-1979, ANSI C37.012-1979.

***229. Ratings.*** Continuous current ratings range from 600 to 3000 A; higher values are available for special applications, in particular for generator circuit breakers or extra-high-voltage GIS breakers. Standard short-circuit currents range from 8.8 to 40 kA; special designs reach up to 80 kA. Ratings differentiate between indoor or outdoor service and between oil or oilless design. Preferred ratings are further established for capacitive current switching, dielectric test values, transient recovery voltage capabilities, switching surge factors for line closing, and operation endurance capabilities. For details refer to the following standards: ANSI C37.04-1979 (rating structure) and ANSI C37.06-1979 (preferred ratings).

***230. Oil Circuit Breakers.*** These breakers can be classified as two general types, dead-tank construction and live-tank construction. The former is available for many voltage and interrupting ratings for indoor and outdoor application, whereas the latter has generally been restricted to voltages of 14.4 kV and below, although applications up to 34.5 kV have been made.

A *dead-tank breaker* (see Fig. 10-76) consists of a steel tank partly filled with oil, through the

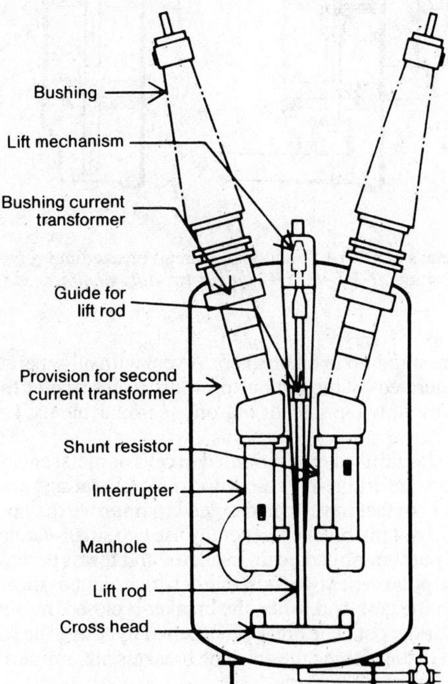


**FIG. 10-76** Outline of a 161-kV outdoor oil circuit breaker. *(I-T-E, GE, Westinghouse, Allis-Chalmers, etc.)*

cover of which are carried porcelain or composition bushings. Contacts at the bottom of the bushings are bridged by a conducting crosshead carried by a wood or composition lift rod, which, in common designs, drops by gravity following contact separation by spring action, thus opening the breaker. Accelerating springs are used to increase the speed of opening. In some designs, the crosshead is opened with a rotary motion by springs.

Breakers with the three poles in one tank are available for ratings up to 69 kV and 40 kA interrupting current. Insulating fibrous barriers are inserted between the phases in the tank if required.

Multitank breakers with each pole in a separate tank are used for higher voltages and interrupting ratings, and generally for the larger outdoor breakers. Figure 10-77 shows the principle of a typical oil breaker interrupter.

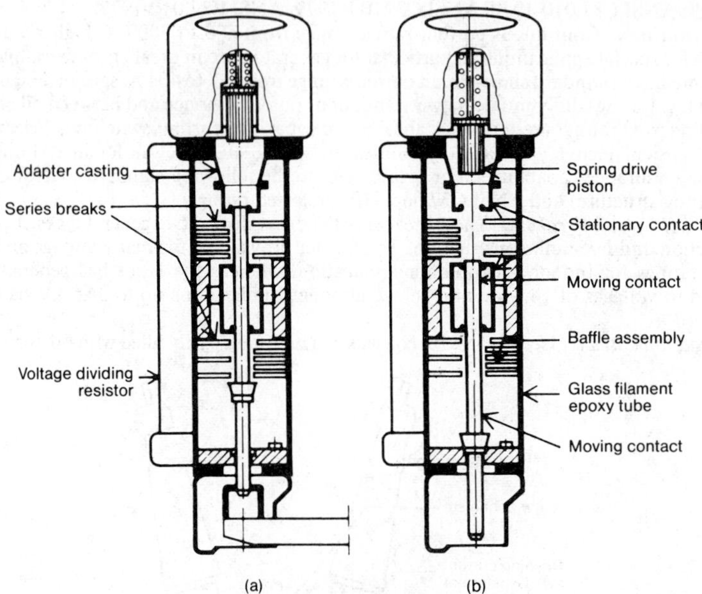


**FIG. 10-77** Details of a 161-kV outdoor oil circuit breaker interrupter. *(a)* Closed position; *(b)* open position. *(I-T-E, GE, Westinghouse, Allis-Chalmers, etc.)*

Oil-tight features are standard in oil breakers. A vent with oil-separating features permits the escape of the gases generated by the arc but prevents the escape of the entrained oil. Indoor breakers are generally mounted in steel or masonry compartments. Large dead-tank breakers are frame-mounted.

*Live-tank breakers* should always be mounted in cells or metal enclosures. Outdoor-station-type oil circuit breakers are frame-mounted up to 69 kV ratings and floor-mounted above 69 kV. Various types of contact have been designed to improve the operation and increase the interrupting capacity. Live-tank breakers generally use two small-diameter cylindrical tanks, or "pots," per pole. Each pot is mounted on an insulator and forms part of the circuit. A conducting crosshead above the pots carries two rods, one of which extends through a porcelain sleeve in the top of each pot into the tank and, when the breaker is closed, makes contact with a flexible contact in the bottom of the pot. The breaker is opened by lifting the rods. Insulating baffles or oil-blast features aid in extinguishing the arc. The breakers are ordinarily operated by a combination of spring and motor mechanisms.

Bushings on oil circuit breakers are usually plain porcelain, composition, or both. Special construction is used for bushings on the higher voltages, for example, condenser bushings or Herkolite bushings (General Electric Company). Bushing potential devices can be applied to circuit-breaker bushings of 115 kV and above. The bushing is equipped with a capacitance tap connected to a concentric cylindrical metal electrode or capacitance divider inside the bushing. This provides a voltage supply for operating instruments and relays from high-voltage circuits with reasonable accuracy and burden adjustment.

Oil-circuit breakers are generally equipped with bushing-type current transformers usually designed for 5-A secondary current.

***231. Minimum-Oil Circuit Breaker.*** These breakers were developed mainly in Europe, aiming to eliminate oil as an insulating medium and thus reduce the quantity of oil in switchgear installations to an amount that would not cause hazard. The excellent arc-quenching properties of oil, however, were used in specially developed oil- and pressure-tight arc-interrupting chambers. Minimum-oil circuit breakers, also known as low-oil-volume circuit breakers, are available for indoor applications for voltages up to 38 kV and for outdoor applications for voltages up to 800 kV. Continuous current ratings are 600 to 3000 A and higher. Interrupting capacities cover the lower and medium ranges; details may be taken from manufacturer lists.

Minimum-oil circuit breakers are used mainly in the medium-voltage range for indoor service. The breakers are either fixed or, for integration in metal-clad switchgear, mounted on trucks together with appropriate disconnecting and interlock facilities. Most breakers have a manually tensioned stored-energy operating mechanism. If required, the operating mechanism can also be driven by a motor, fitted with autoreclosing facilities and with various kinds of release. The layout of a medium-voltage minimum-oil circuit breaker is shown in Fig. 10-78.

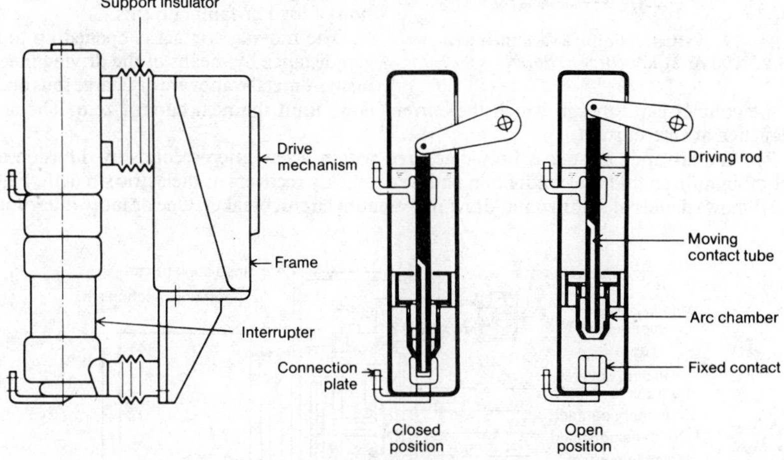


**FIG. 10-78** Outline and interrupter details of a 15-kV 3-pole minimum-oil circuit breaker. *(Brown Boveri.)*

***232. Circuit-breaker oil*** should be of proper quality and carefully maintained. In modern oil-tight breakers, transformer oil can generally be used. It should have the following characteristics:

| | |
|---|---|
| Flash point | 133°C |
| Burning point | 148°C |
| Freezing point | −40°C |
| Viscosity point | 57 s |
| Color | Pale amber, clear |

For extra-low-temperature conditions, oil with increased viscosity may become necessary. The oil level should be carefully maintained, and the oil should be filtered or otherwise reconditioned if its dielectric strength drops to 15,500 V ac as measured by breakdown between 1-in disks 0.1 in apart. New oil should have a dielectric strength of at least 30 kV. Oil should be changed or reconditioned if excessively carbonized. Some operators check oil and contact condition as soon as possible after each severe operation of large, important breakers. Periodic checks of oil and breaker operation and condition are recommended, the frequency depending upon the service conditions. Generally, annual periodic checks are adequate, but even longer periods have been used successfully. Where operating duties are frequent, a shorter period of time should be used.

***233. Vacuum Circuit Breaker.*** Progress in high-vacuum technology and breaker development, combined with improved manufacturing and testing methods, has opened a growing area for vacuum breaker application, concentrating, but not limited to voltages up to 38 kV, continuous current ratings up to 3000 A, and covering all standard interrupting ranges. The principal design of a vacuum interrupter is shown in Fig. 10-79.

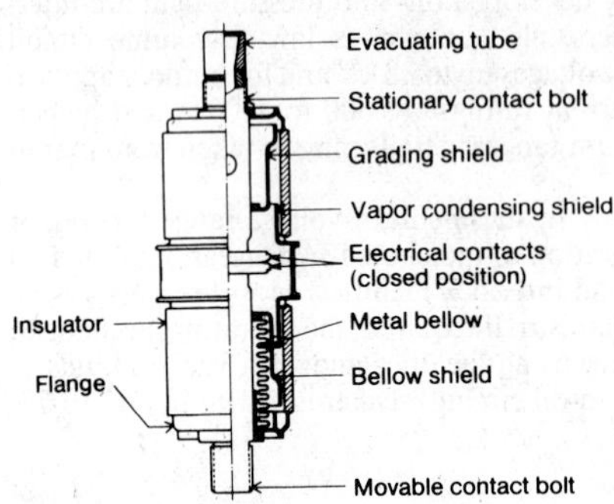


**FIG. 10-79** Partial section of a vacuum interrupter 23 kV, 2000 A, 21 kA. *(Brown, Boveri.)*

Two contacts are mounted on an insulating envelope from which virtually all air has been evacuated. One contact is stationary, the other movable. Vacuum interruption has the inherent advantage of moving a lightweight contact only a very small distance in an almost perfect dielectric medium. This results in safe, quiet, and fast switching or interruption of load or fault currents.

The moving contact is opened up to full gap distance by means of the driving mechanism. A metal-vapor arc discharge thus occurs in the contact gap through which the current flows until the next current zero. The arc is quenched at zero current.

The metal-vapor plasma is fully deionized within a few microseconds by diffusion and recombination so that the conduction path very quickly recovers its dielectric strength. Figure 10-80 shows details of a horizontal-draw-out vacuum circuit breaker. One or more interrupters may be utilized in series per pole. The vacuum interrupters are additionally protected against outside influences by an insulating casing. The vacuum circuit breakers are therefore completely insulated. They may be fitted with hand- or motor-charged stored-energy-operated mechanisms.

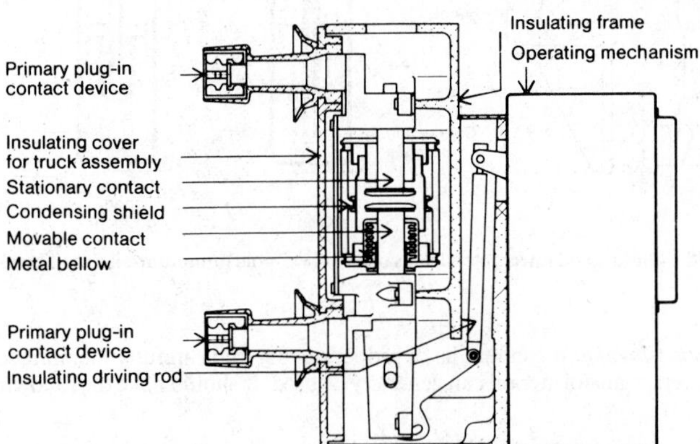


**FIG. 10-80** Outline and interrupter details of a 15-kV horizontal-drawout vacuum circuit breaker 2000 A, 28 kA. *(Brown, Boveri.)*

Because of their short closing and opening times, vacuum breakers are particularly suitable for autoreclosure and synchronizing duty. Breaking of short-circuit currents with very steep initial rise of transient recovery voltage is possible due to the restoration of the contact gap within a few microseconds. The steep rise of dielectric strength over the whole current range offers a high capacitive-current -switching capability. Switching of unloaded transmission lines and cables can therefore reliably be performed.

***234. Magnetic Air Circuit Breaker.*** This type of circuit breaker is usually stored-energy-mechanism-operated and interrupts the main circuit in the normal atmosphere under the influence of a strong magnetic field which acts to force the arc deep into a specially designed arc chute. Solenoid operating mechanisms are available. The chute cools and lengthens the arc to a point where the circuit cannot be maintained by the voltages of the system, and interruption is accomplished. The zone between the main contacts is clear of ionized air by the time interruption is obtained in the arc chute, and so restriking at this point is no problem. Since the magnetic effect is not great at low currents such as load, transformer magnetizing, and cable-charging current, all designs use an air-pump "puffer" actuated by the operating mechanism that blows a blast of air across the arc and thereby assures its entering the arc chute and giving rapid interruption at the low-current values also. When the circuit breaker is opened, the arc transfers from the main arcing contacts to fixed arcing horns which are within the arc chute. The magnetic field is produced by coils in the main-current circuit, in some cases wound around a magnetic core which magnetizes soft-iron plates in the sides of each arc chute. Some designs do not require the iron core.

Magnetic air breakers may be obtained in any of the ratings of Table 2 of ANSI Standard C37.06 through the 15-kV ratings. All are designed for use in metal-clad enclosures. Figure 10-81 shows the horizontal-drawout type of breaker in a metal-clad enclosure. Although the

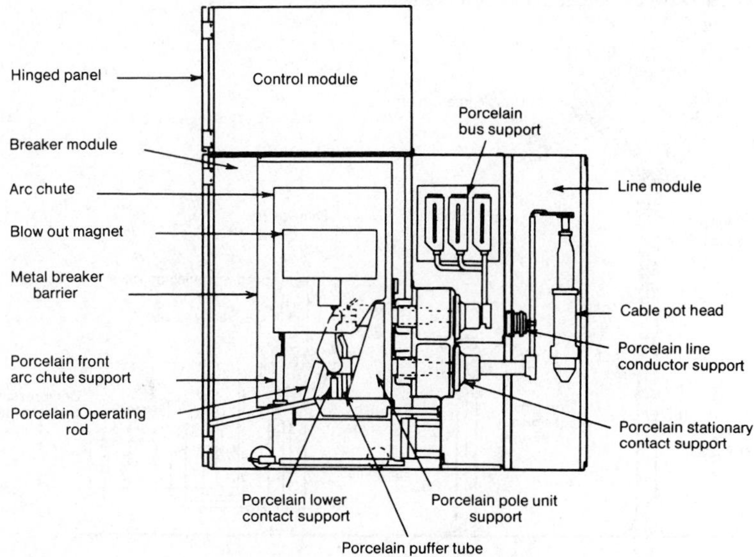


**FIG. 10-81** Horizontal-drawout metal-clad magnetic circuit breaker in service position.

design shown is for indoor use only, the same circuit breakers are placed in weatherproof housings for outdoor service. When they are so used, suitable heaters are put in the housings to avoid internal moisture condensation.

***235. Air-Blast Circuit Breakers.*** These breakers fulfill the heavy-duty requirements of circuit breakers in high-voltage systems. They are used to provide the indoor ratings in Table 2 of ANSI Standard C37.6 up to 38 kV. They are, however, mainly used in outdoor applications up to 800 kV for ratings given in Tables 4 to 6 of ANSI Standard C37.06.

Air-blast circuit breakers are further available for special applications as *(a)* generator breakers with continuous current ratings of up to 42 kA and higher, *(b)* furnace breakers with an extrahigh number of switching operations (20 to 50 duty cycles per hour), *(c)* as single- or 3-pole breakers for traction systems, and *(d)* extrahigh interrupting currents. Air-blast circuit breakers are usually fixed-mounted, but a variety of breaker types may also be truck-mounted for application in drawout metal-clad switchgear. All air-blast circuit breakers make use of dry and clean air compressed to a certain pressure, which may differ for the various makes and types of breakers; for details see manufacturer lists. The compressed air is used to operate the breaker as well as to serve as the medium for arc quenching and insulation.

Continuous current ratings of up to 4000 A are possible. Total breaking time of 2 cycles (from energizing of trip coil until arc extinguishing) is standard; special designs may allow even shorter breaking time. Some 69-kV breakers are equipped with sequential isolators, but the bulk of designs have not integrated the isolator to form part of the circuit breaker. Some older designs employed separate chambers for opening and closing operation, but modern air-blast breakers perform opening and closing with the same contact system. Closing resistors and/or, with some designs, opening resistors may be used. Equal voltage distribution over the multiple breaks of one pole is usually achieved by parallel capacitors.

The modular principle has been accepted widely for manufacture and assembly because of quality and economic reasons. The Brown Boveri type DLF circuit breaker, for example, is constructed on this modular principle for voltages up to 1100 kV and symmetrical interrupting ratings of up to 80 kA. With six interrupter modules of different size and creepage distance and also six pressurized support columns of different lengths and creepage distances, this breaker meets a wide range of applications. Air tanks of two different sizes cover the complete range of circuit-breaker arrangements. The tanks are either of the vertical-support design or horizontally flange-mounted onto the operating-mechanism housing (see Fig. 10-82). The operating pressure varies between 300 and 440 lb/in$^2$ (gage).

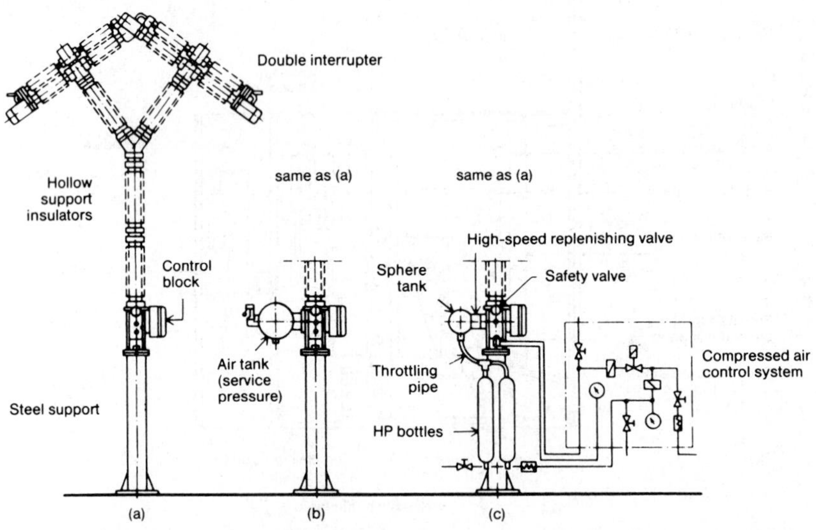


**FIG. 10-82** Modular setup of a 362-kV outdoor air-blast circuit breaker with two uprating steps: *(a)* without air tank—low breaking capability; *(b)* with air tank-standard breaking capability; *(c)* with constant-pressure air supply and high breaking capability. *(Brown Boveri.)*

For uprating, a constant-pressure system is added to the breaker. The high-pressure air of 2150 lb/in$^2$ (gage) is fed to the breaker and stored in a specially dimensioned spherical tank. Air for the second and third opening of a duty cycle is stored in two high-pressure bottles, mounted on the supporting frame. When the breaker opens, a dump valve allows the contents of the spherical tank to pass instantly into the extinction chambers. No specific time is necessary because the spherical tank empties until the pressures are fully equalized. Figure 10-83 illustrates the difference of pressures in the contact zone between a conventional air-blast breaker and a constant-pressure breaker during an O-C-O duty cycle. Thus, maintaining the rated pressure in the interrupter during the opening operation results in a very substantial increase in interrupting capacity without the need for additional interrupters or resistors.

Modern air-blast breakers can be equipped with noise mufflers, if so desired, to comply with environmental requirements. For circuit-breaker installation in earthquake-prone areas, high-grade support insulators and earthquake dampers have been developed and have proved reliable. Factory-assembled compressor plants for indoor or outdoor installations, either of the unit or central arrangement, have considerably increased breaker reliability. High-pressure storage at some 3000 lb/in$^2$ (gage) is standard. For special climatic conditions, air dryers are available. Compressor maintenance intervals have been extended to 2000 h running time.

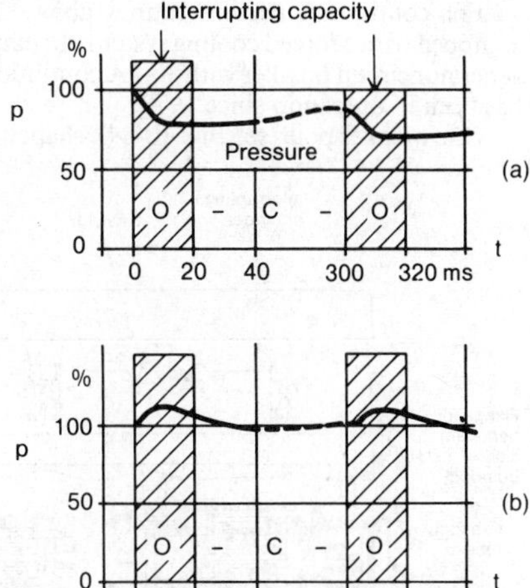


**FIG. 10-83** Interrupting capacity depending on pressure change during contact opening: *(a)* conventional air-blast breaker; *(b)* constant-pressure design. *(Brown Boveri.)*

***236. Generator Circuit Breaker.*** An increasingly important field of application of air-blast circuit breakers is their use as generator circuit breakers incorporated in generator bus ducts. Employing a generator breaker has economic, operational, and technical advantages (see Table 10-11).

Generator breakers of the Brown Boveri type DR are used for continuous current ratings from 8 up to 42 kA and higher. The breakers are enclosed, of tubular shape, and mounted in line with the generator bus duct. The range of breakers includes load switches as well as circuit breakers of different ratings, all assembled from the same module components. The DR breaker is proving particularly suitable for use in big power stations, where a reliable station-service supply is especially important. There is considerable advantage in the application of a generator breaker in pump storage plants with generator or motor operation.

**TABLE 10-11** Service Criteria of Generator Circuit Breaker

| Function | Advantages |
|---|---|
| Disconnect generator from station-services supply during start-up. | Station services fed via generator transformer. No need for starting transformers and associated switchgear and changeover facilities. |
| Synchronize on low-voltage side of generator transformer. | Eliminates voltage transformers on high-voltage side of generator transformer. Possibility of connecting two generators to one overhead line via two separate transformers or a three-winding transformer. |
| Isolate fault in generator transformer or in station services transformer. | Effects of fault less than with rapid deexcitation. |
| Isolate fault in generator. | Station services remain continuously connected to network. |
| Isolate fault on line from power plant to next transformer or switching station. | No need for high-voltage breaker in generating station. |

For continuous current ratings above 20 kA, the generator circuit breaker is usually equipped with a forced cooling system; for extrahigh ratings, water cooling is preferred. The first generator circuit breaker with 30-kA continuous current and 150-kA interrupting current (rms) has been in operation since 1973.

One breaker pole (see Fig. 10-84), shaped to match the isolated-phase bus duct, consists of

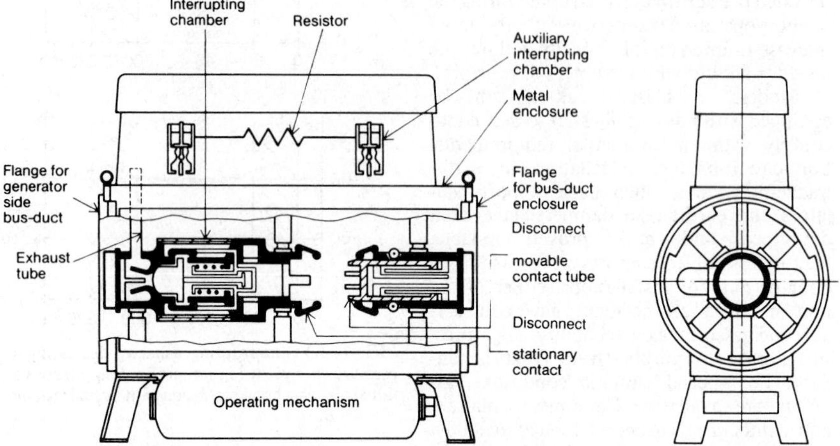


**FIG. 10-84** Outline and interrupter details of a generator air-blast circuit breaker type DR, 36 kV, up to 42,000 A with forced cooling, 200 kA. *(Brown Boveri.)*

one interrupting chamber and one isolator connected in series and supported by insulators centered within the enclosure, which is dimensioned to carry a current equivalent to the full-phase current. The operating mechanism and air storage are located at the lower breaker part. Inside the interrupting chamber different contact systems serve for current carrying and current switching. The sliding isolator is mechanically locked in both the open and closed positions and can be viewed from outside. For interrupting currents up to 150 kA, the breaker is equipped with one interrupting chamber, whereas for higher ratings two interrupters may be connected in series.

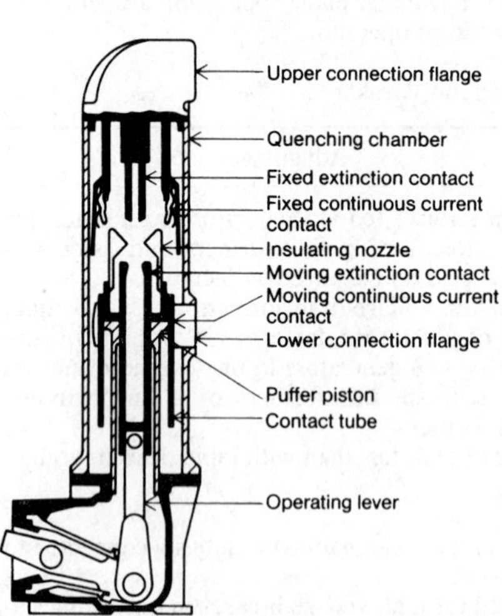


**FIG. 10-85** Section of a $SF_6$ puffer-piston indoor circuit breaker, 23 kV. *(Brown Boveri.)*

The interrupting capacity of the DR generator circuit breaker can be increased by application of a constant-pressure air system, similar to that for outdoor air-blast circuit breakers (see Fig. 10-82). For integration of the gènerator circuit breaker in the bus duct, refer to Par. **238**, "Switchgear Assemblies."

***237. $SF_6$ Circuit Breakers.*** Sulfur hexafluoride ($SF_6$) gas has proved to be an excellent arc-quenching and insulating medium for circuit breakers. $SF_6$ is a very stable compound, inert up to about 500°C, nonflammable, nontoxic, odorless, and colorless. At a temperature of about 2000 K, $SF_6$ has a very high specific heat and, accordingly, a high thermal conductivity which promotes cooling of the arc plasma just before and at current zero, and thus facilitates quenching of the arc.

The electronegative behavior of the $SF_6$, i.e., the property of capturing free electrons and forming negative ions, is primarily responsible for the high electric breakdown strength but also promotes rapid recovery of dielectric strength of the arc channel immediately after extinction of the arc.

$SF_6$ circuit breakers are available for all voltages up to 800 kV, continuous currents up to 4000 A, and symmetric interrupting ratings up to 63 kA. $SF_6$ circuit breakers are of either the *indoor design* (see Fig. 10-85) to fit into metal-clad switchgear, of the *tank or metal-enclosed design* (see Fig. 10-86) for GIS or conventional substation arrangement or of the *outdoor design* (see Fig. 10-87) for conventional substation layout. Most $SF_6$ circuit breakers operate according

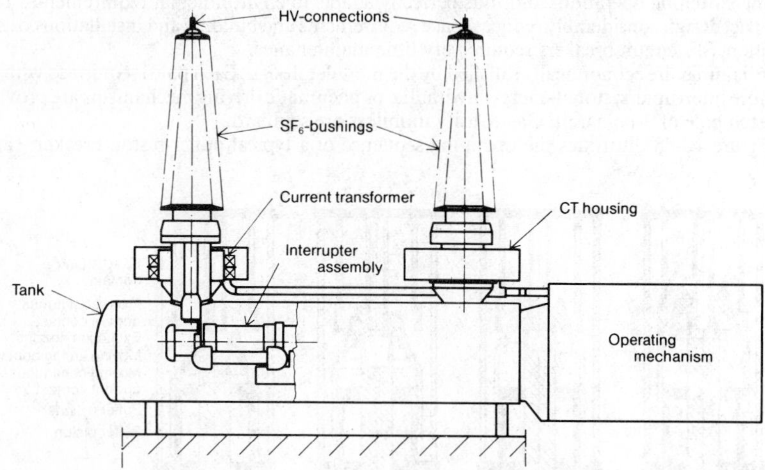


**FIG. 10-86** Partial section of a $SF_6$ outdoor breaker type 242 PA 63, 242 kV, 3000 A, 63 kA. *(Brown Boveri; I-T-E.)*

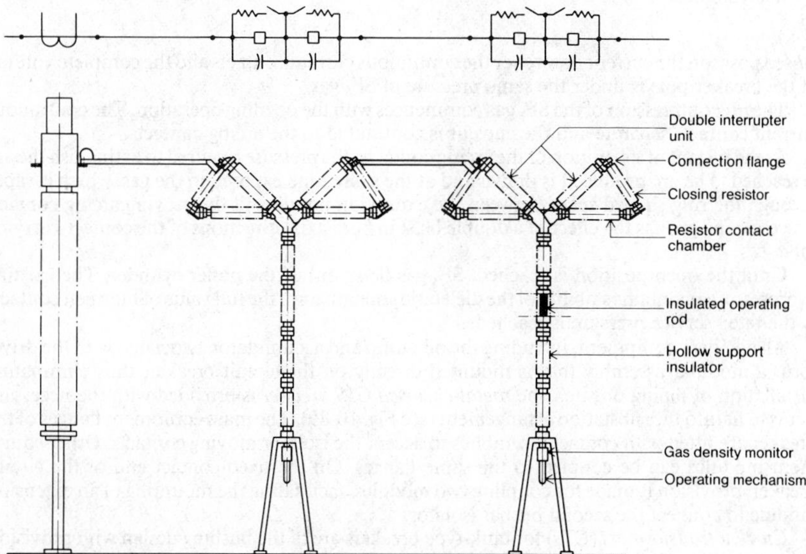


**FIG. 10-87** $SF_6$ outdoor circuit breaker and current transformer arrangement, 800 kV, 3000 A, 40 kA. *(Brown Boveri.)*

to the *piston (puffer) system*. Others operate on the *self-extinction* system. The *dual pressure* design is used to a lesser extent. All systems operate fully independently from the auxiliary gas supply. To prevent liquefaction of $SF_6$ gas at low temperatures, many breakers are equipped with electric heating systems. Depending on the make and type of breaker, the service pressure varies between about 20 lb/in² (gage) (insulation) and 250 lb/in² (gage); for details refer to manufacturer lists.

Over the years, $SF_6$ circuit breakers have reached a high degree of reliability; thus they can cope with all known switching phenomena. Their closed-gas system eliminates external exhaust during switching operations and thus perfectly adapts to environmental requirements. Their compact design considerably reduces space requirements and building and installation costs. In addition, $SF_6$ circuit breakers require very little maintenance.

All ratings are economically satisfied by the modular design. Each pole is equipped with one or more interrupters; stored-energy, hydraulic, or pneumatic driving mechanisms are provided for each pole or 3-pole unit. Gas-density monitors are standard.

Figure 10-88 illustrates the operating sequence of a typical puffer-piston breaker. In the

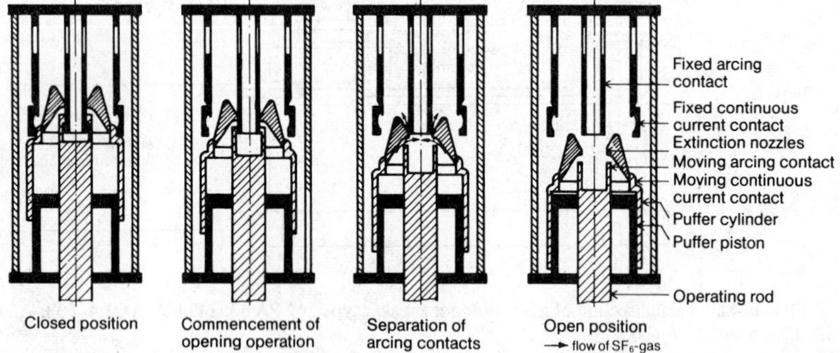


**FIG. 10-88** Principle of $SF_6$ puffer-type outdoor interrupter; four positions during opening operation. *(Brown Boveri.)*

closed position the current flows over the continuous current contacts and the complete volume of the breaker pole is under the same pressure of $SF_6$ gas.

The precompression of the $SF_6$ gas commences with the opening operation. The continuous current contacts separate and the current is commuted to the arcing contacts.

At the instant of separation of the arcing contacts, the pressure required to extinguish the arc is reached. The arc produced is drawn and at the same time exposed to the gas, which escapes through the ring-shaped space between the extinction nozzle and the moving arcing contact. The escaping gas has the effect of a double blast in both axial directions of the contact-carrying tube.

Until the open position is reached, $SF_6$ gas flows out of the puffer cylinder. The existing overpressure maintains stability of the dielectric strength until the full value of the open contacts at the rated service pressure is reached.

All ancillary equipment, including the oil pump and accumulator associated with the drive, form a modular assembly that is mounted directly on the circuit breaker, thus eliminating installation of piping on site. The *metal-enclosed GIS breaker* is provided with the necessary items to fit into the substation arrangement (see Fig. 10-89). The main equipment flanges of the breaker are fitted with contact assemblies to accept the isolator moving contacts. Other equipment modules can be coupled to the same flanges. On the fixed-contact end of the circuit breaker, provision is made for coupling two modules, facilitating the mounting of an extension module to connect the second busbar isolator.

*Current transformers* (CTs) for tank-type breakers are of the bushing design with provision made for potential taps. Outdoor breakers of the conventional layout are generally provided with CTs of the paper-oil-insulated post-type design (see Fig. 10-87). The CT hermetical seal of oil is either of the fixed design with gas cushion or of the pressure-free bellow type. Up to six

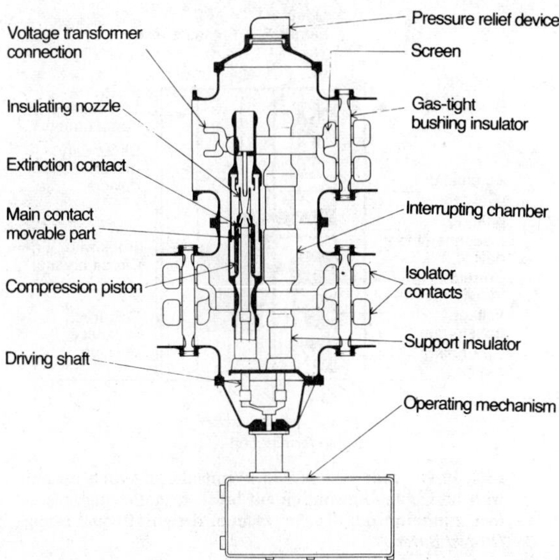


**FIG. 10-89** Section of a 145-kV $SF_6$ circuit breaker for gas-insulated-substation (GIS) type ELK. *(Brown Boveri.)*

magnetic cores can be provided per CT unit, generally in multiratios for 5- or 1-A secondary by means of secondary taps. Primary-current ratings up to 2000 A normally employ the wound-type design with two or more turns. Higher primary currents up to 6000 A require the inverted or head design, with a straight tube as single-turn primary winding and the core and secondary-winding assembly arranged at the CT top to limit temperature rise and to increase the mechanical withstand capability of the CT. The latter design has its full main insulation on the secondary winding. Post-type CTs are available for all output and accuracy requirements for modern system relaying and measuring for voltages up to 800 kV. For the upper voltage ranges, post-type CTs are normally provided with separate potential layers. The CTs are generally dimensioned for the same dielectric and mechanical characteristics chosen for the related circuit breakers.

## Switchgear Assemblies

Switchgear assemblies cover a wide range of low-voltage and high-voltage structures that are generally factory-assembled and are divided into three main groups (*a*) metal-enclosed power switchgear, (*b*) metal-enclosed bus, and (*c*) switchboards. Only three types of assemblies will be referred to in the following paragraphs. ANSI C37.20 and NEMA SG5-1967 apply.

***238. Metal-Clad Switchgear.*** The term "metal-clad switchgear" indicates a design in which all the equipment required to control an individual circuit, including bus, circuit breaker, disconnecting devices, current and voltage transformers, controls, instruments, and relays, is assembled in separate metal compartments and the circuit breaker is provided with a means for ready removal from the cubicle. Circuit breakers are preferably the oilless type, air or air-blast. Progress in development of $SF_6$ indoor breakers and vacuum breakers offers improved alternatives in metal-clad switchgear.

Circuit-breaker disconnection is accomplished by vertical-lift or horizontal-drawout designs, the latter illustrated in Figs. 10-90 and 10-91. Interlocks are provided in metal-clad assemblies to prevent disconnecting or connecting the circuit breaker while in the closed position and to prevent breaker operation while moved between disconnected and connected position or vice versa. The metal-clad assembly is equipped with metal shutters to protect

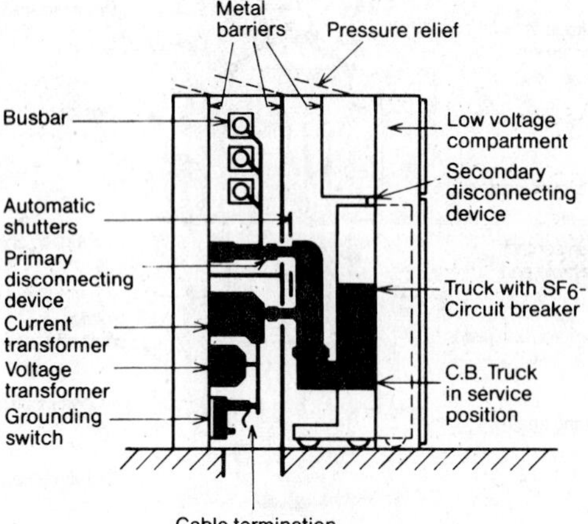


**FIG. 10-90** Side view of a 15-kV metal-clad switchgear unit with horizontal-drawout circuit breaker; interchangeable either minimum oil, $SF_6$, or vacuum design of equal rating. *(Brown Boveri.)*

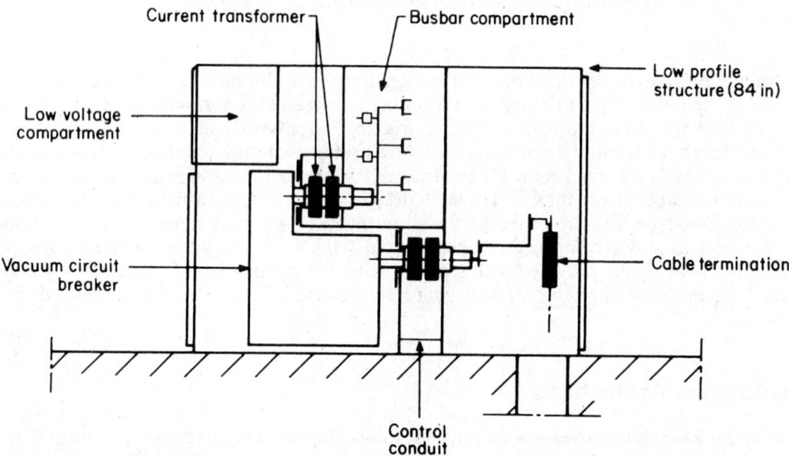


**FIG. 10-91** Side view of a 38-kV metal-clad switchgear unit with horizontal-drawout vacuum circuit breaker; type HKV, up to 3000 A (fan-cooled), 22 kA. *(Brown Boveri; I-T-E.)*

personnel from coming in contact with the high-voltage circuits when the circuit breaker is removed from the cubicle.

Metal-clad switchgear is used for low- and medium-capacity circuits, for indoor and outdoor installations with nominal voltages of 4.16 to 34.5 kV. A circuit-breaker test position is standard to allow breaker control while the main plug-in contacts are open but while auxiliary and ground contacts between cubicle and breaker truck are still closed.

***239. Station-Type Switchgear.*** The term "station-type switchgear" indicates a design in which the major component parts of a circuit, such as buses, circuit breakers, disconnecting switches, and current and voltage transformers, are in separate metal housings and the circuit breakers are of the stationary type (see Fig. 10-92). Phase segregation in metal-enclosed switch-

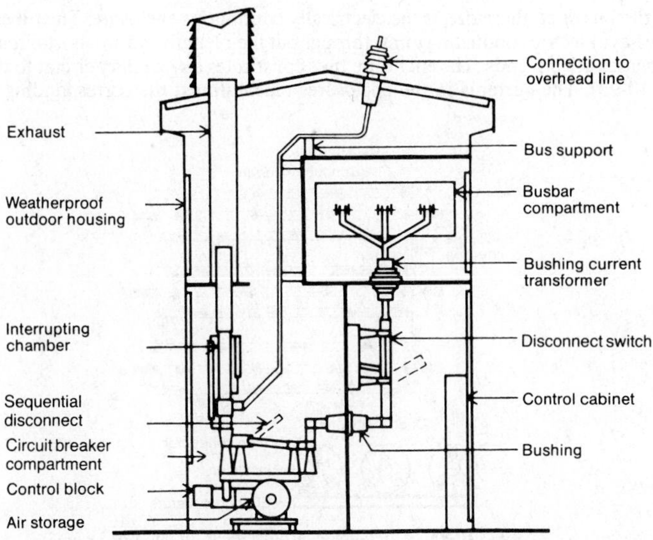


**FIG. 10-92** Metal-enclosed station-type switchgear cubicle for outdoor installation; equipped with a heavy-duty, air-blast circuit breaker, 14.4 kV, 3000 A, 50 kA. *(Brown Boveri.)*

gear is a type of design in which a 3-phase metal housing is divided into three single-phase compartments by means of single metal barriers.

Metal-enclosed station-type switchgear is used in industrial, commercial, and utility installations, generally for voltages of 14.4 to 69 kV, and ratings of the heavy-duty range up to some 5000 A continuous current.

***240. Isolated-Phase Metal-Enclosed Bus.*** Isolated-phase metal-enclosed bus is a type of design in which each phase is enclosed in an individual metal housing, and an air space is provided between the housings. It is considered to be the safest, most practical, and most economical way of preventing phase-to-phase short circuits by means of construction methods. The bus may be self-cooled or forced-cooled by circulating air or liquid.

Briefly, the isolated-phase bus duct has the following features:

Proof against contact; locked electrical premises not necessary. Faults only in the form of ground faults; protection against fault spreading to more than one phase

Field forces, static and dynamic, only between enclosure and conductor, not between phases

Protected against contamination and moisture

No losses in surrounding conducting material (grilles, railings, concrete reinforcement, lines, etc.)

The range of bus ducts includes continuous current ratings from about 5 up to some 25 kA self-cooled, or 40 kA with forced cooling. The momentary current ratings have to match the rating of attached equipment. With high current ratings, more attention must be paid to:

Progressive rise of conductor temperature due to skin effect

Heating of surrounding conducting material by the magnetic field of conductors

High forces on main or component conductors in the event of a short circuit

In an enclosure with sections of tube insulation (sectional enclosure), eddy currents exist with values as large as the conductor current. These give rise to heat losses, and so the magnetic field of the main conductor is not always compensated for sufficiently. An important technical feature of the bus duct, therefore, is the electrically continuous enclosure. The tubes enclosing each phase have electric conducting joints throughout their length and are short-circuited across the three phases at both ends. The enclosure thus constitutes a secondary circuit to the conductors (Fig. 10-93). The currents in the enclosures reach almost the corresponding conductor

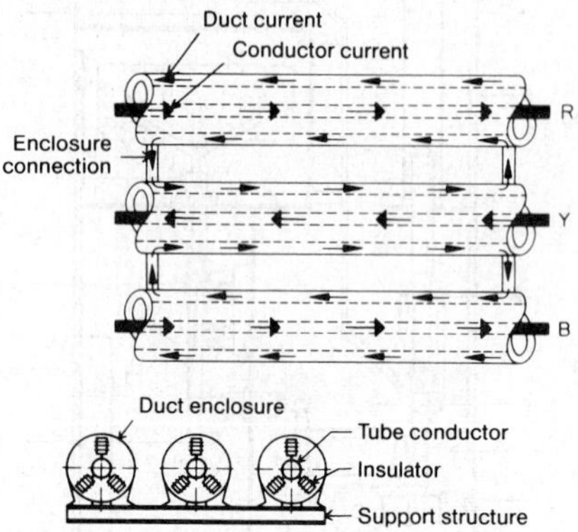


**FIG. 10-93** Three-phase arrangement of an isolated-phase bus-duct and principle of enclosure connection; according to Kirchhoff's law sum of conductor currents (+) and sum of duct currents (−) is zero.

currents, depending on the resistance of the duct, but are of the opposite direction. The magnetic field outside the enclosure is almost completely eliminated, and thus there are no external losses or field forces between the phases. Connections to machines and switchgear must be adaptable and removable.

Current transformers for measurement and protection are of the bushing type or are integrated into the bus duct at a suitable place. Voltage transformers can be contained in the bus duct or mounted in separate instrument boards. The same applies to protective capacitors. Care must be taken that branch lines are adequately dimensioned with regard to thermal short-circuit strength.

The reliability of generator bus ducts can be enhanced by employing means to maintain the air pressure in the duct. Although generally bus ducts are leakproof, the large number of dismantleable joints may cause slight leakage and might lead to moisture condensation during a plant shutdown. Supplying the bus duct with filtered, precompressed air at a slight pressure

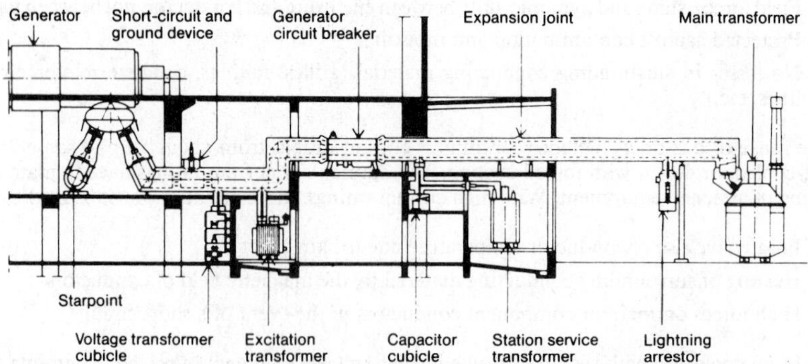


**FIG. 10-94** Generating-plant isolated-phase bus-duct arrangement with generator circuit breaker type DR. *(Brown Boveri.)*

ensures that the air flow is only outward; contamination of the conductors is not possible. Drying the air by precompressing prevents condensation.

Short-circuiting and grounding facilities are required in the bus duct to protect the generator and also for maintenance grounding purposes. Manually positioned links and straps are sufficient for small unit ratings; motor-operated grounding switches are recommended for higher capacities. A typical isolated-phase bus arrangement of a power station including generator circuit breaker is shown in Fig. 10-94.

***241. References***

Baltensperger, P.: Knowledge of switching phenomena; *Brown Boveri Rev.*, vol. 4/62, pp. 381–397.

Braun, A., Eidinger, E., and Ruoss, E.: Interruption of short circuit currents; *Brown Boveri Rev.*, vol. 4/79, pp. 240–253.

Kopainski, J., and Ruoss, E.: Switching of low inductive and capacitive currents; *Brown Boveri Rev.*, vol. 4/79, pp. 255–261.

Current Limiting Breaker, pp. 173–174. *Brown Boveri Switchgear Manual,* 6th. ed., 1977, W.F. Schmitt.

Petry, H.: Switching of High Currents under Vacuum; *Calor-Emag Rev.*, vol. 1/1983, pp. 3–11.

Köppl, G, and Vogt, R.: HV Outdoor Breakers; *Brown Boveri Rev.*, vol. 4/78, pp. 243–247.

Schaumann, R., and Poole, D.: $SF_6$-Medium-Voltage Switchgear, *Brown Boveri Rev.*, vol. 11/77, pp. 644–649.

Mauthe, G., and Szente-Varga, H. P.: GIS for 72.5 to 550 kV, *Brown Boveri Rev.*, vol. 4/78, pp. 220–230.

## CIRCUIT SWITCHERS

*By ARTHUR D. CRINO, DONALD L. LOTT, RONALD E. MITTELSTAEDT, and KELLY A. SHAW*

***242. Definitions.*** Circuit switchers are mechanical switching devices suitable for frequent operation; not necessarily capable of high-speed reclosing; capable of making, carrying, and breaking currents under normal circuit conditions; capable of making, and carrying for a specified time, currents under specified abnormal conditions; and capable of breaking currents under certain other specified abnormal circuit conditions. They may include an integral isolating device. Circuit switchers available today use sulfur hexafluoride as an interrupting medium and may be equipped with a trip device connected to a relay to open the circuit switcher automatically under specified abnormal conditions, such as overcurrent or faults.

A circuit switcher, like a circuit breaker, must carry normal load currents within a specified temperature range to prevent damage to key components such as contacts, linkage, terminals, and isolating device parts. Principal designating parameters of a circuit switcher are maximum operating voltage, basic insulation level (BIL), rated load current, interrupting current, whether an isolator is required, whether a trip device is required, and whether manual or motorized operation is required.

A circuit switcher essentially combines the functions of a circuit breaker (without reclosing capability) and a disconnecting switch (by providing visible isolation, but not necessarily meeting the safety requirements of all users). A circuit switcher provides a cost-effective alternative means of transformer protection and switching, line and loop switching, capacitor or reactor switching, and load management, with protection in most instances.

Evolution of the circuit switcher concept provides a more in-depth understanding of its application versatility and its limitations.

### History of Circuit-Switcher Development

***243. After World War II,*** the drive to electrify the remaining rural and sparsely populated areas of the United States was renewed. Providing fully rated circuit breakers for switching loaded circuits was frequently beyond budget limitations. This created a need for new transmission and subtransmission voltage circuit-switching devices. One such device could be described as a load interrupter. It appeared in a wide variety of forms. Most were attachments to disconnect switches (Fig. 10-95).

**FIG. 10-95** Example of 1950-era interrupter–disconnect switch combination.

Initially most of these devices used low-volume oil as an interrupting medium. Ablative gas-generating devices and later vacuum displaced oil. With rare exceptions, these devices had deficiencies.

In the mid-1950s, $SF_6$ (sulfur hexafluoride) was first employed as an interrupting medium. The application was an interrupter attachment for disconnect switches. One such installation is shown in Fig. 10-96.

**FIG. 10-96** Installation of mid-1950s disconnect switch with interrupter employing $SF_6$ as interrupting medium.

Whereas ablative devices and vacuum bottles are limited to approximately 30-kV recovery voltage per gap, this single-gap $SF_6$ device was readily applied on 138-kV systems for up to 600 A load switching. Most of these vacuum, ablative, and $SF_6$ devices were shunted into the circuit during the disconnect switch opening process. As the 1960s approached, the circuit switcher was born. It appeared as an in-line device. While the first version employed a number of ablative devices in series, it soon evolved into the use of $SF_6$ as a medium. Because of the unfavorable experience with the earlier devices, the general acceptance of the circuit switcher took much effort and considerable time. A typical installation is shown in Fig. 10-97.

Applications for circuit switchers have been primarily for transformer protection. The circuit switcher provides load-switching capability and low-fault protection. For applications where the available short-circuit current exceeds the device's capability, blocking relays can be used. However, in most applications this is not necessary.

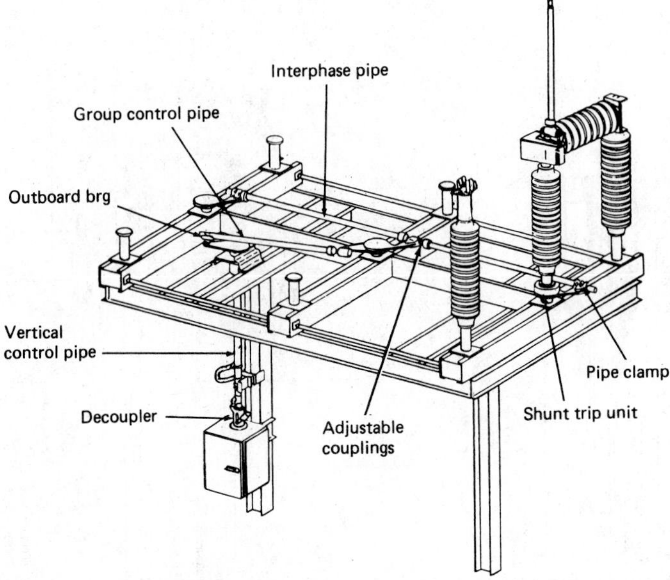


**FIG. 10-97** Schematic of typical 3-pole arrangement. Two poles have been deleted to clarify mechanical drive-train arrangement.

## General Construction

***244. Live-tank $SF_6$ gas "puffer"-type interrupters*** are utilized by most circuit switchers today. A typical cross section of an $SF_6$ puffer-type interrupter is shown in Fig. 10-98*c*. In the closed position the contacts are surrounded by a flow guide and piston assembly which is ready to mechanically generate a "puff" of $SF_6$ to cool and deionize the arc that is established prior to circuit interruption (see Fig. 10-98*a*). The moving cylinder attached to the contact assembly is driven by the main opening spring, causing the gas to be pressurized by the stationary piston. The stationary contact "follows" the moving contact as the piston assembly achieves the prepressurized gas condition (Fig. 10-98*b*). When the contacts (which are hollow tubes) part, an arc is established and the gas flow divides into two parts and flows down the stationary and moving contact tubes. The alternating nature of the arc current waveform results in two current zeros every cycle. As long as the arc is sufficiently "hot" or conductive through the $SF_6$ dielectric medium, the current will reestablish. At the first current zero where the $SF_6$ density is sufficient to stop the arc from reestablishing itself and to provide necessary dielectric strength, the arc is interrupted. This entire process from trip signal initiation to current interruption requires up to 8 Hz or 133 ms in modern circuit switchers.

Figure 10-99 illustrates a typical "blade-disconnect model" circuit switcher with the interrupter and blade connected in series. For opening, the trip device, called a "shunt trip," receives a trip signal when the relay system detects an abnormal condition within the specified range or when the operator desires a high-speed circuit opening. By discharging its operating spring, the shunt trip rotates the insulator above it at high speed, thus tripping and discharging the opening spring in the driver mechanism. This actuates the interrupter to open the circuit. If the insulator above the shunt trip continues to rotate, by motor or manual actuation of the drive-train controls, the blade opens to achieve visible isolation. The blade hinge mechanism is actuated directly by the rotating insulator through the driver mechanism. Continued rotation of the insulator after the blade is open will "toggle" the drive-train controls to lock the blade in its open position.

For closing, the reverse rotation of the insulator first releases the drive-train toggle and allows the blade to begin closing. The shunt trip units have already recharged during the opening

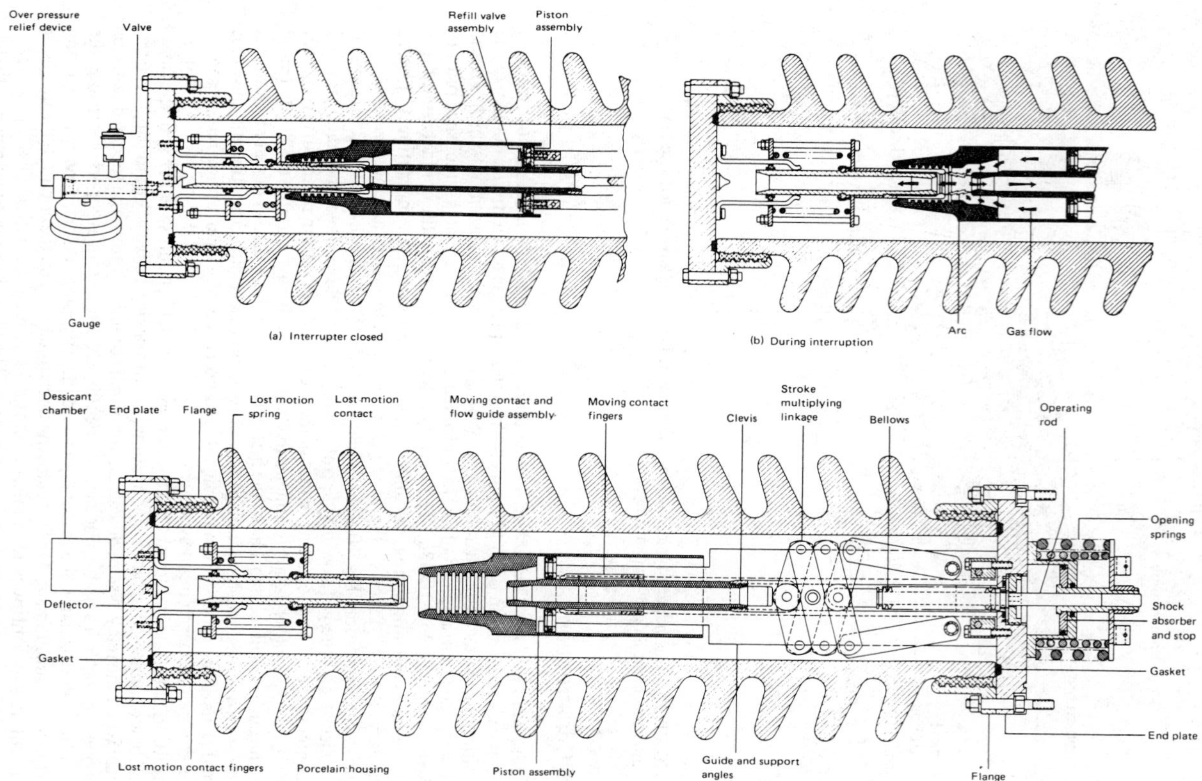


**FIG. 10-98** Cross section of typical $SF_6$ puffer interrupter. *(a)* Closed position; *(b)* opening process; *(c)* fully open position.

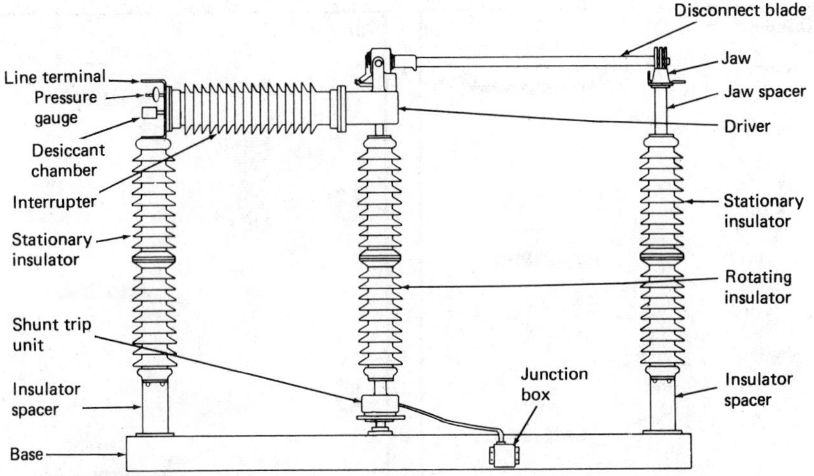


**FIG. 10-99** Single pole of blade-type circuit switcher.

operation. As the blade closes, the closing springs are charged in the driver. The last few degrees of closing rotation lock the blade in position and release the closing springs in the driver, thus closing the interrupter. The opening springs are charged as the closing springs discharge. If the unit has closed into a circuit condition that provides a trip signal to the shunt trip units, the opening process may immediately proceed, since all springs are charged and all controls are ready.

The closing operation may be achieved in other designs by closing the interrupter during the opening stroke of the blade. When a close operation is called for, all that is necessary is to close the blade, because the interrupter is already closed. Because of the arc established in air for this type of closing, high-speed operation of the blade is necessary to minimize damage to contacts and prevent flashovers. Both methods of closing are proven over many years of field use.

Bladeless circuit switchers operate exactly the same as blade models, except that on opening, the insulator rotation is used only for driver and interrupter actuation. Models that depend on high-speed blade operation for closing are available in bladeless nondisconnect configuration, but circuit closing must be accomplished by other means.

For models without shunt trip, opening is accomplished by rotating the insulator to the point where the driver opening spring would normally be tripped by the shunt trip's rotation. This configuration is used where protection duty is not a function of the circuit switcher.

The above sequence of operation is illustrated in Fig. 10-100 for blade units that use the interrupter for closing the circuit.

## Ratings

***245. Short-Circuit Current.*** Short-circuit or interrupting ratings of circuit switchers are less than those of circuit breakers. Circuit switchers are "limited-duty" or "medium-fault" interrupting devices and generally fulfill a requirement between high-power fuses and circuit breakers in a transmission or distribution system. Circuit-switcher designs to date do not encompass instantaneous reclosing capability since their most common application of transformer protection precludes this requirement.

Typical short-circuit current ratings are from 4 to 16 kA on a symmetrical current basis, depending on type of interrupting duty and voltage rating. The symmetrical current is the highest value of the ac component in rms amperes measured at the instant of contact part.

When interruption occurs, a transient recovery voltage (TRV) is imposed in microseconds across the interrupting device as a result of system adjustments to the new state before a steady-state condition is achieved known as the normal-frequency recovery voltage. TRV

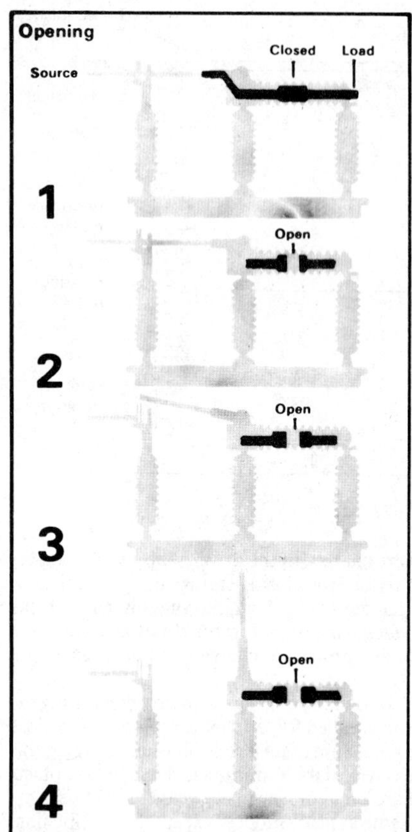


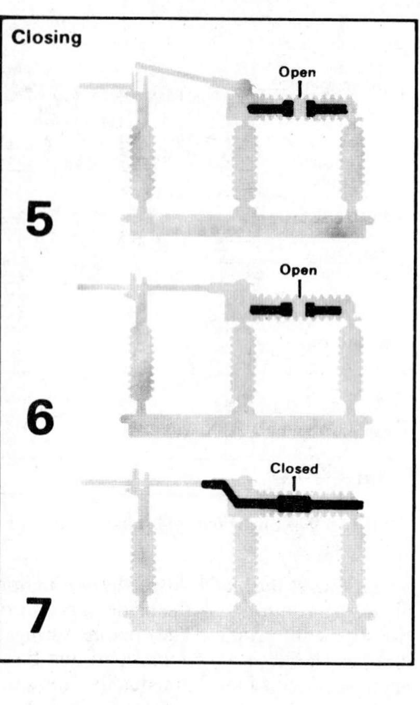


**FIG. 10-100** Sequence of operation for blade-type circuit switcher.

attempts to reestablish the arc by either thermal reignitions or dielectric breakdown of the interrupting gap. Refer to ANSI C37.04 and C37.06 for TRV definitions and ratings.

*Continuous Current.* A continuous current rating is the designated limit of current in rms amperes that can be carried continuously under usual service conditions and in an ambient temperature not in excess of 40°C without exceeding temperature limits assigned to the various materials comprising the current-carrying parts or that are in contact with these parts. For further information refer to ANSI C37.04, C37.30, and IEC 694.

Circuit-switcher continuous current ratings are 1200, 1600, and 2000 A.

*Short-Time Current.* This is a dual rating to verify the capability of the circuit switcher to carry abnormally high currents in the closed position for short periods of time:

1. Rated momentary current is the total current in asymmetrical rms amperes (ac and dc components) carried for 10 cycles on a 60-Hz basis. These ratings can approach 50 times the continuous current ratings and assure the device will withstand the high electromagnetic forces from initial transient conditions of a short circuit tending to bring about mechanical damage or contact separation.
2. Rated 3-s current is 62.5% of the rated momentary current and verifies that the current-carrying parts will withstand the heating effect without excessive annealing or contact welding. Common circuit switcher momentary ratings are 61, 70, and 80 kA.

*Close-and-Latch Current.* For those circuit switchers that make the circuit in the interrupting device, a close-and-latch rating is required as is a close rating for the type that closes the circuit on a disconnect switch blade. This rating verifies the capability of the contact structure and operating mechanism to withstand the forces developed by closing in on a fault. Close-and-latch ratings are 30 and 40 kA. Limited-duty close ratings for the blade type are of the same magnitudes.

*Rated Voltage.* Rated voltages for circuit switchers are generally stated in terms of maximum system design voltages that indicate the upper limit at which the device is designed to operate. Circuit switchers are most commonly applied at 72.5 through 242 kV but are also available for special applications at 362 and 550 kV.

In order to provide safe insulation levels to ground and across the device when in the open position, the circuit switcher must withstand certain specified magnitudes and waveshapes of test voltages without flashover or puncture of any of its insulation systems. This is the rated dielectric strength and consists of

1. One-minute dry and 10-s wet low-frequency (60-Hz) withstand voltage.
2. Dry lightning-impulse ($1.2 \times 50$ $\mu$s waveshape) withstand voltage. It is the positive-polarity withstand level that is the basic-insulation-level rating (BIL in kV).
3. Dry and wet switching-impulse ($250 \times 2500$ $\mu$s) withstand voltage. This rating only applies at 362 kV and above.
4. Dry chopped-wave impulse withstand voltage. The requirement for this rating has yet to be determined for a circuit switcher but is a standard rating for a circuit breaker. If applicable, this rating would apply to a bladeless model circuit switcher having graded gaps in series per pole.

*Interrupting Time.* The rated interrupting time is the maximum permissible interval between energizing the shunt trip coils at rated control voltage and the interruption of the main circuit in all poles when interrupting a current within the required interrupting capabilities. This interrupting time may be modified under certain asymmetrical current conditions as detailed in ANSI C37.04. Circuit switchers are available from 5- to 8-cycle interrupting times.

*Duty Cycle.* Operating duty is the short-circuit current to be interrupted, closed upon, etc. The duty cycle is a stated sequence of closing and opening operations. Various types of circuit switchers will have different duty cycles. An example is O-17s-CO where the first operation is an open followed by 17 s to permit the operating mechanism and blade to recycle in order to perform the close-open cycle.

*Corona-RIV.* A corona-free rating is commonly established to prevent radio and television interference. This is referred to as the radio influence voltage (RIV) and should be less than 500 $\mu$V as determined by a test circuit per NEMA Publication No. 107. As a general rule, a circuit switcher that produces no visible corona under dark conditions at 105% of rated voltage will have a negligible RIV level.

## Selection and Application

***246. The versatility of circuit switchers*** and related circuit-interrupting devices requires careful selection of the ratings, components, and accessories to be specified for a given protection and/or switching duty. The following criteria must be considered:

1. System data
   - *a.* Nominal service voltage
   - *b.* Maximum continuous current
   - *c.* Through fault current withstand
   - *d.* Basic impulse (insulation) level
2. Circuit protection duty
   - *a.* Present and future available fault currents
   - *b.* Length of overhead lines or underground cables

*c.* Transformer ratings, impedance, and connections
*d.* Transient recovery voltage
*e.* Interrupting time, maximum

3. Switching duty
   *a.* Inductive currents (unloaded transformers)
   *b.* Small capacitive currents (unloaded lines and cables)
   *c.* Inductive/capacitive currents (choke coils, capacitors, grounded or ungrounded)
   *d.* Closing on fault
   *e.* Closing of long lines

Approximately 80% of all circuit switchers are applied on the primary or high-voltage side of a transformer for switching and fault-protection duty. By means of protective relaying, faults which occur within the protection zone of the transformer can be detected to bring about tripping of the primary-side circuit switcher. Should the fault occur on the secondary side, which is most frequently the case, the magnitude of fault current at the circuit-switcher location is limited by the transformer impedance and is less than the available primary-fault current. Hence the use of the term "inherent secondary-fault current," and for some circuit switchers, a dual short-circuit current rating for primary or secondary faults.

Associated with these reflected secondary faults are the inherent capacitance and inductance of the transformer windings, which generally produce higher TRVs than is the case with a primary fault. These higher TRVs can increase the difficulty of the interruption process, while conversely, the impedance-limited fault current is easier to interrupt.

Primary-fault currents in excess of the switcher rating can often be cleared by source-side circuit breakers having interrupting times shorter than those of a circuit switcher. It is also common to block tripping of the circuit switcher when the fault current is in excess of its rating and pass this duty on to the source-side breaker.

Circuit switchers and interrupters are generally available in mounting arrangements similar to isolating air-disconnect switches. Since they are not freestanding, their application and that of the supporting structure design and/or optional freestanding pedestal is influenced by phase spacings desired, terminal height, altitude, atmospheric elements, seismic conditions, and wind, ice, and terminal pad loadings.

The phase spacing recommended in the ANSI C37.32 differentiates between the minimums for vertical-break disconnect switches (also applying to bus supports) and the increased spacings required for horn gap switches (switching devices utilizing arcing in air as an operating mode). From a safety viewpoint, it is important to consider the method of operation in addition to the duty applications of the particular device as a determination in selecting the phase spacing.

Elevation of the installation site, in addition to the temperature, humidity, and air-contamination characteristics, significantly affects the dielectric strength and coordination of these circuit-interrupting devices. The guidelines established in ANSI C37.30 for site elevations coupled with a further analysis of these other factors may dictate the use of higher-voltage-rated units or extra-leakage-distance porcelain standoff insulators.

Seismic, wind, ice, and terminal pad loadings must also be considered in the design of the supporting structure.

***247. Testing.*** Industry standards specifically for circuit switchers have yet to be developed. Accordingly, applicable sections of existing standards for circuit breakers and disconnect switches are used as a guide in establishing definitions, ratings, capabilities, and test procedures for circuit switchers (see ANSI C37.09 and C37.34).

Tests can be considered to consist of four different groupings

1. Type or design tests are for the purpose of proving the capability of the switcher to meet the specified ratings.
2. Reliability tests are part of a quality assurance program to demonstrate that the circuit-switcher design has achieved a specified mechanical reliability. They consist of growth or prototype tests to improve upon the established design reliability; demonstration or pilot run tests to assure the reliability requirements have been met; and acceptance or production tests to verify that production units comply with the demonstrated design reliability.

3. Routine or production tests are done to ensure that the production is in accordance with established procedures. This includes testing of individual components and subassemblies.
4. Installation testing to assure that the circuit switcher will perform its intended function in the system.

***248. Accessories and Maintenance.*** As discussed in a preceding section, circuit-switcher manufacturers offer optional freestanding support pedestals in addition to many other accessories, some of which require direct mounting to the supporting structures. These accessories include field-adjustable auxiliary switches both internal and external to the operating mechanism, grounding switches mounted on either the source or load end of the circuit interrupters, key interlocks, and hookstick-operated bypass switches which permit exercising the circuit switcher or interrupter without opening the high-voltage power circuit.

Present-day circuit switchers or interrupters are designed, manufactured, and tested in accordance with applicable ANSI circuit-breaker and disconnect-switch standards to assure reliability with minimum maintenance. Manufacturer's maintenance interval recommendations vary and are usually defined for the most rigorous service and operating conditions. Since circuit switchers are usually three group-controlled single-phase devices, some maintenance tests are recommended in addition to those generally required for circuit breakers. These periodic recommendations include, but are not limited to

1. Checking continuity of trip, control, and heater circuits
2. Checking $SF_6$ interrupting-medium gas pressures and correlating these observations with the ambient temperature
3. Checking operating simultaneity of all three single-pole units
4. Checking air-disconnect-switch contacts for wear, proper contact pressure, and alignment

## VOLTAGE REGULATORS

*By JAMES H. HARLOW and KELLY A. SHAW*

***249. Voltage Regulation.*** A primary objective of any electric system is to provide power users with a supply voltage compatible with their utilization equipment. Every electrical device is designed to operate at a certain rated voltage for optimum efficiency and maximum length of service. An ideal electric supply system provides constant voltage to all users under all loading conditions. No system is ideal; it is economically impractical to attempt an "ideal system" design approach. Today's ideal system is that system providing a voltage supply satisfactory to all utilization equipment, with the most economical use of available regulation equipment.

Several methods of improving the voltage profile on electric transmission or distribution systems are in use. These include transformer load-tap changers, switched and fixed capacitors, and step-voltage regulators. Induction-voltage regulators and synchronous condensers were common at one time, but they have been superseded by other, less-expensive alternatives. A relatively new approach, static var control, employing thyristor phase-angle firing control of fixed capacitors is being applied experimentally on several high-voltage transmission systems. Application of single-phase, step-voltage regulators dominates the distribution market to a great extent.

***250. An economical means of voltage regulation*** is desirable to provide power users with a voltage supply compatible with utilization equipment. The following list describes the effects of unregulated voltage:

I. Resistance loads (electric stoves, irons, water heaters, toasters, etc.)
   *A.* Low voltage
      1. Longer heating time.
   *B.* High voltage
      1. Shorter life of heating elements.

**II.** Motor loads (vacuum cleaners, washing machines, fans, refrigerators, etc.)

*A.* Low voltage

1. Increased slip-increased line current under excited motor. Results: *(a)* decreased efficiency and *(b)* motor overheating.

*B.* High voltage

1. Overexcited motor-increased torque. Result: possible damage to coupling or appliance.

**III.** Electronic loads (radio and television)

*A.* Low voltage

1. Poorer quality of reception on television sets: *(a)* picture not distinct and *(b)* decreased picture size.

*B.* High voltage

1. Shortens life of electron tubes.

**IV.** Illumination loads (incandescent and fluorescent lighting)

*A.* Low voltage

1. Decreased incandescent lamp efficiency; a 10% decrease in voltage results in 70% normal illumination output.
2. If voltage is too low, fluorescent lamps will be inoperative.

*B.* High voltage

1. Life expectancy of bulbs decreases.

Of course, many benefits of regulated voltage provided to consumers also benefit suppliers by virtue of decreased investment per kVA distributed, increased efficiency of distribution equipment, and increased revenue. The intangible benefit of customer satisfaction must not be overlooked.

## Methods of Regulation

***251. The first regulators were induction-type machines.*** These appeared very early in the development of the electric power industry and were used extensively for a number of years. An induction regulator consists of a rotor and a stator much like a motor. Like step regulators, induction regulators take voltage from the source and add to or subtract from it to hold the load voltage steady. Output voltage is changed by mechanically adjusting the position of the rotor relative to the stator. The rotor does not rotate continuously but has its position changed as required by a small, internal self-contained motor. This motor responds to a signal from a control circuit. At present, almost all new installations of induction-type voltage regulators are for industrial applications.

*Step-type voltage regulators,* the precursors of modern design, were introduced in the early 1930s. The first step-type regulators were developed from the autobooster concept. A 2400/120-V distribution transformer connected as an autotransformer gives a 5% boost. Adding a center tap to the secondary (series) winding gives two 2½% steps. Adding two more taps gives four 1¼% steps. A preventive autotransformer (or bridging reactor) divides these steps in half to give the modern ⅝% steps. In a ±10% regulator, thirty-two ⅝% steps, 16 above and 16 below neutral, provide regulation to all types of loads. Step-type voltage regulators are actually tapped autotransformers. An autotransformer is a transformer in which part of one winding is common to both the primary and the secondary circuits associated with that winding. In other words, the primary (exciter) winding is both electrically and magnetically connected to the secondary (series) winding. The exciter winding is common to both primary and secondary; the series winding is connected in series with load (output) current. Step-type voltage regulators are commonly provided in both single-phase and 3-phase styles.

*Transformer load-tap changers* (commonly referred to as LTC transformers) are actually combination transformers and step-type regulators. The tap-changing mechanism is mounted in an oil-filled compartment which is commonly sealed from the core and coil. Tap changing is accomplished by a gang-operated, 3-phase oil switch providing simultaneous regulation of the three phase voltages.

*Fixed and switched capacitors* are not voltage regulators in the electrical definition of the device. Capacitors are used to improve the line-voltage profile by connecting a "bank" of capacitors in shunt. These shunt capacitors first improve an otherwise lagging power factor "seen" by a source. The effect of the improved power factor will be reduced line losses and improved regulation. The use of capacitors beyond that required to attain unity power factor will result in a leading component of current in the line inductance, causing voltage on the line to rise. As long as total line current remains lagging, capacitors provide an improved voltage profile. When total current is leading, however, shunt capacitors increase line current (and line losses) and may cause a large voltage rise, resulting in excessively high voltage.

*Static var systems* (SVS), presently still in proof-testing stages, involve effectively regulating (or fine tuning) the reactive compensation afforded by the shunt capacitors by virtue of phase-angle firing control of thyristors. Basically, the thyristor conduction period will establish the capacitive (leading) current. The control system used is very rapid relative to system fluctuations, permitting optimization of the application. Other features of SVS, especially its use in improving system stability, have made it feasible for exploration on transmission systems. Thus far, SVS application at distribution voltages has been justified only for very large bulk power supplies.

***252. Method of Step-Voltage-Regulator Operation.*** A typical step-voltage regulator (Figs. 10-101 and 10-102) involves a shunt winding, a series winding, and a bridging reactor or preventive autotransformer. Taps, each representing essentially 1¼% voltage, are affixed to the series winding and brought to a specially designed dial switch.

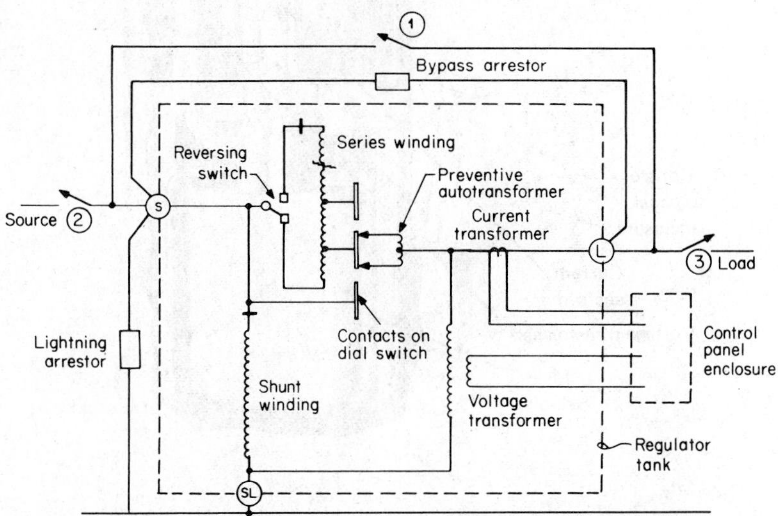


**FIG. 10-101** Wiring diagram of a typical distribution step-voltage regulator showing both external and internal connections. Preventive autotransformer shown on a nonbridging position. (1) Bypass switch; (2) source switch; (3) load switch.

The *main transformer* comprises the *shunt winding* and the *series winding.* The series winding most often is rated at 10% voltage of the shunt winding. Usually eight taps on the series winding are brought to a *dial-switch* assembly as individual contacts, with the voltage difference between contacts being 1¼% voltage.

The terminals of a center-tapped *preventive autotransformer* are able to transition (slide) between the dial switch contacts in a manner which avoids momentary loss of load. As one finger advances to the next step, arcing results because of the inductive nature of the reactor, but

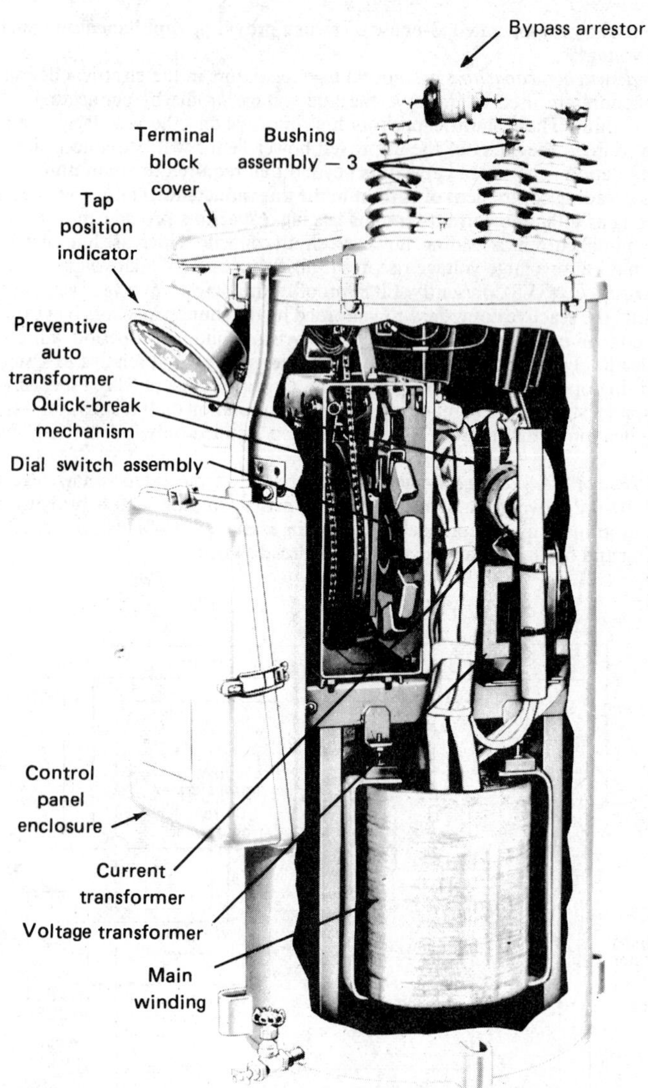


**FIG. 10-102** Cutaway view of a distribution regulator. *(Siemens-Allis, Inc.)*

load current is maintained (Fig. 10-103*b*) through the finger which does not part contact. When the parting finger remakes on the adjacent contact, a bridging condition is established (Fig. 10-103*c*). A circulating current is established through the preventive auto transformer (PA), or a center tapped bridging reactor, and load potential is seen to be the average potential of the taps being bridged. To minimize the arc duration, a *quick-break mechanism* accelerates the moving contacts. The rapid separation of the contacts and the use of contact tips composed of a tungsten-carbon alloy mitigate the attendant ablation of contact material.

A *reversing switch* permits the polarity of the series winding to be reversed relative to the shunt winding, thereby accommodating plus and minus regulation with the same series winding.

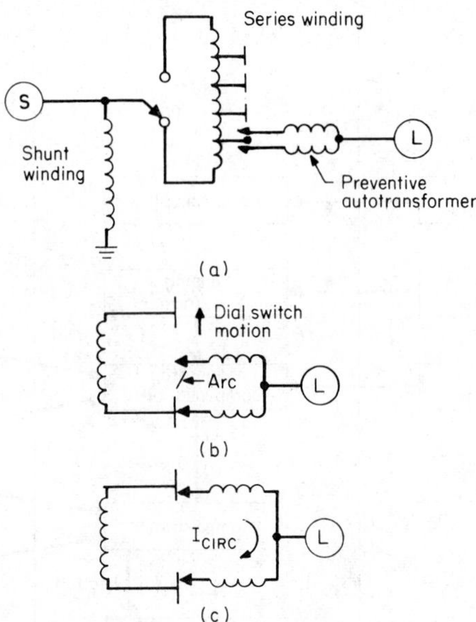


**FIG. 10-103** Operations of the internal mechanisms of the regulator.

A *voltage transformer* and *current transformer* are used to provide signals necessary for the control to perform its function.

Dielectric protection of the series winding is afforded by the *bypass arrestor.* A surge propagated on the line will be shunted past the regulator. *Lightning arrestors,* often provided at both the source and the load terminals, similarly protect the regulator from overvoltage surge conditions.

Other external appurtenances will include the three bushings (source, load, and SL or common), a *tap position indicator* display of the present operating position of the regulator and a *control panel enclosure.*

***253. Regulator control circuitry*** has progressed from a mechanically driven regulating relay beam to highly sophisticated, microprocessor-based digital control circuitry. All regulator controls comprise three major parts:

1. A *voltage-sensing device* which monitors the output voltage of the regulator and sends a signal to the control circuit.
2. An *amplifying* or *switching section,* with or without time delay, which delays and/or transmits the signal.
3. A *motor drive circuit* which responds to the signal by closing relay contacts or actuating electronic switches that cause the tap-changing motor to operate to correct the voltage.

The important functions of the typical regulator control are shown in the block diagram, Fig. 10-104.

An *auxiliary voltage source,* such as a nominal 120-V tertiary winding on the main core and/or a voltage transformer, will be provided to supply power and the signal for deriving the regulator output voltage. This 120 V is scaled down in a *sensing transformer* which may itself be tapped to facilitate ratio adjustment if the voltage source is not exactly as required. The output of the sensing transformer will, by operator selection, be altered to reflect the voltage drop between the regulator and the load. This drop, which is a function of the load current and the line impedance, is modeled by knowing the current from a *current transformer* and established *line resistive compensation* and *line reactive compensation* parameters.

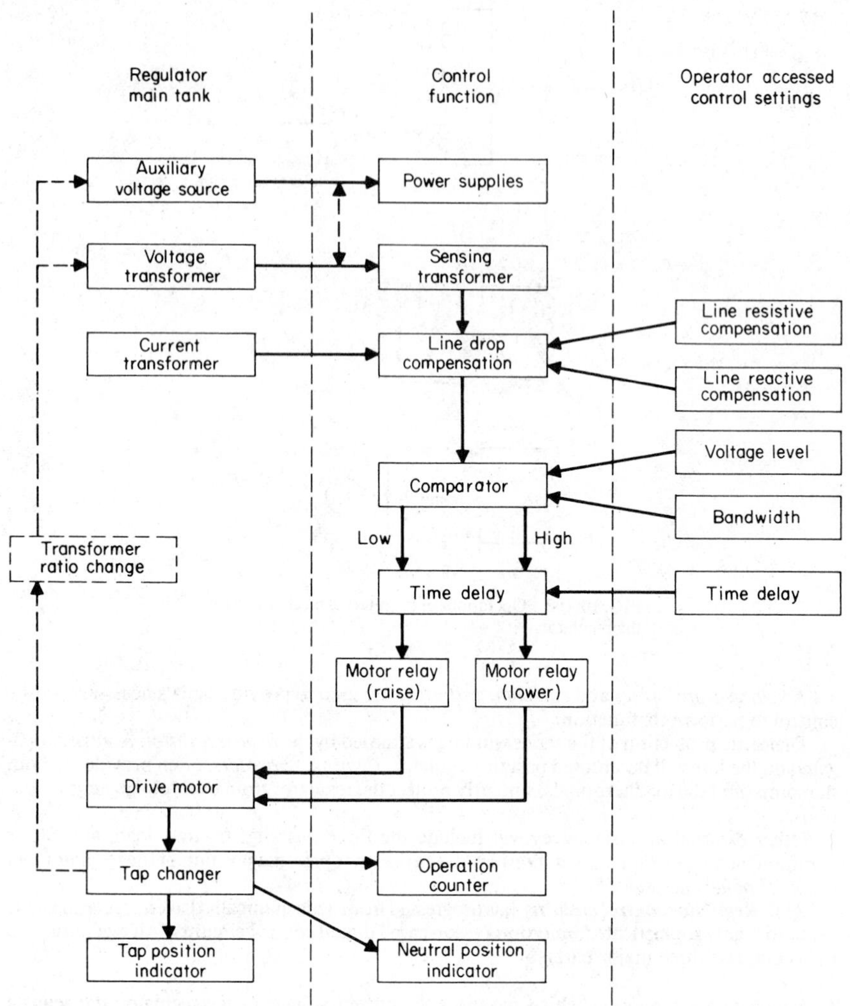


**FIG. 10-104** Regulator control functions.

The voltage, now compensated for line drop, is related in a *comparator* to the desired voltage level as established by a *voltage-level* setting and a *bandwidth* or tolerable voltage range limits setting.

If the compensated voltage exceeds the high limit (or is less than the low limit) a *time delay* is activated. Requiring the voltage to remain out-of-band for a period of typically 30 to 60 s assures that the voltage-level change is of sufficient duration to warrant regulator action.

After the preset delay is satisfied, a *drive motor* is powered and runs the *tap changer* to the next position. Tap-changer motion advances the *operation counter,* moves the *tap position indicator* pointer, and causes illumination of a special *neutral position indicator* light, if appropriate.

Of course the new tap position in turn causes a *transformer ratio change,* altering the input to the sensing transformer and closing the control function loop.

## Application of Regulators

***254. The rating of single-phase step-voltage regulators*** is a simple function of the rated range of regulation and the current. For example a 100-A regulator with a ±10% range of regulation for a 7620-V system will be rated

$$100\ \text{A} \times 0.10 \times 7620\ \text{V} = 76.2\ \text{kVA}$$

Early recognition that this is 10% of circuit kVA will avoid later difficulties. Modern single-phase feeder-voltage regulators are available in the following ratings:

2500 V, 50 to 416.3 kVA
5000 V, 50 to 416.3 kVA
7620 V, 38.1 to 889 kVA
13,800 V, 69 to 414 kVA
14,400 V, 72 to 833 kVA
19,920 V, 100 to 833 kVA

The ampere rating of many smaller regulators can be increased by limiting the range of regulation to accommodate higher line currents. By limiting the range of regulation, the portion of series winding through which line current passes is decreased. This decrease in the portion of series winding "seeing" load current decreases the internal regulator losses. By decreasing losses, heating inside the regulator is decreased. Consequently, the regulator may carry more load current at the same temperature rise. For example, a 100-A regulator, ±10%, may be operated at ±5% regulation at 160 A. The same regulator may be operated at +5%, 160 A, and −10%, 100 A. Limit-switched taps and overload capabilities are, by ANSI standards, 8¾% (110%), 7½% (120%), 6¼% (135%), 5% (160%), to a limit of 668 A.

Oil-immersed step-type regulators, by ANSI standards, are rated at 55°C top oil temperature rise. A 65°C hot-spot winding temperature is also specified in regulator standards. The short-circuit withstand capacity of standard regulators is 25 times normal full-load current for 2 s, and 40 times normal full-load current for 0.8 s.

***255. Single-phase regulators*** in three-phase installations is a very common application. Configuring the regulators in wye, closed delta, and open delta are acceptable alternatives, depending on system conditions.

The wye connection is invariably used on grounded-neutral 4-wire systems. Regulator sizing is a straightforward adaptation of single-phase principles.

The use of three single-phase regulators in closed delta requires that recognition be made of how current will flow in the regulator. Especially important is the fact that system line current and regulator series winding current is not the same. The application will be specified based on load current, but the current in the regulator series winding will establish the regulator size required. Also, the closed-delta application results in a ±15% range of regulation in the line voltage for a ±10% range on the individual single-phase regulators. Another point which is significant when modeling for line-drop compensation or sensing a power reversal is to note that the current and voltage signals from the regulators will be displaced by 30° (lead or lag) at unity power factor.

An open-delta connection is sometimes used to save expense. This system is like a wye connection in that line current and series winding current are identical, but like a delta in that the 30° phase shift occurs; in fact, in this case the shift is leading in one regulator and lagging in the other.

The various regulator connections utilize the installed kVA of regulators with differing efficiency. For each 1000 kVA of system capacity, the use of ±10% regulators will require total regulator capacity of

| | |
|---|---|
| Grounded wye | 100 kVA |
| Closed delta | 123 kVA |
| Open delta | 115 kVA |

Another important consideration when applying single-phase regulators in 3-phase installations is that certain single-phase regulator connections may be unsafe. In deciding on the proper and safe connection for a given application, three basic phenomena should be considered: (1) third harmonics, (2) system line surges, and (3) line faults.

For instance, when single-phase regulators are grounded on an ungrounded system, a resonant circuit is possible between the third-harmonic magnetizing reactance of the regulators and line capacitance. This resonance can intensify third-harmonic voltages to dangerous levels. A similar situation can occur when regulators are ungrounded on either a grounded system or an ungrounded system. To protect from line surges, regulator connection is unimportant as long as a bypass arrester and line-to-ground arresters are used. Line-to-ground fault problems may be serious when regulators are ungrounded on a grounded system or grounded on an ungrounded system. A problem may also occur when regulators are connected in either open- or closed-delta configuration on a grounded system. Table 10-12 summarizes safe and dangerous single-phase regulator connections.

**256. *A feature of regulators*** is that they may be bypassed in service such that load interruption is not necessary during installation procedures. It is critical that proper switching procedures be followed to avoid the extremely high circulating current which will appear in the series winding if bypassing occurs at other than the neutral tap position. When placing a regulator on the line (see Fig. 10-101), three switches are commonly used: a source switch, load switch, and bypass switch. To place the regulator in service, first the source switch is closed. The regulator is checked out by running the tap changer in the raise and lower directions. The regulator is returned to the neutral position, and the load switch is closed. Caution must be taken to make sure that the regulator will not make an automatic tap change when the load switch is closed. After the load switch is closed, the bypass switch may be opened. To take the regulator out of service, the procedure is reversed. First the regulator is run to the neutral position, the bypass switch is closed, the load switch is opened, and then finally the source switch is opened. Once again, caution must be taken to prevent the regulator from making an automatic tap change during this switching procedure by assuring isolation of the control power.

**257. *Regulators may be paralleled*** if and only if sufficient loop impedance exists to limit circulating currents, based on the maximum difference in the voltages of the two circuits $V_1$ and $V_2$. Equally important, the percent impedances of the two circuits must be equal or very close. The common practice is that the two impedances must be close enough so that circulating current does not exceed 10% of rated current. For impedance $Z_1$ of circuit 1 and $Z_2$ of circuit 2, percent circulating current is given approximately by

$$\% \text{ circulating current} = \frac{Z_1 - Z_2}{Z_1 + Z_2} \times 100$$

**TABLE 10-12** Safe and Dangerous Regulator Connections

| Regulator Connection | System Connection | Effect: Third Harmonic | Effect: Line Surges | Effect: Line Ground | Conclusion |
|---|---|---|---|---|---|
| Grounded Y | Grounded | S | $S_1$ | S | S |
| Ungrounded Y | Grounded | D | $S_1$ | D | D |
| Delta | Grounded | S | $S_1$ | $S_2$ | $S_2$ |
| Open Delta | Grounded | S | $S_1$ | $S_2$ | $S_2$ |
| Grounded Y | Ungrounded | D | $S_1$ | D | D |
| Ungrounded Y | Ungrounded | D | $S_1$ | S | D |
| Delta | Ungrounded | S | $S_1$ | S | S |
| Open Delta | Ungrounded | S | $S_1$ | S | S |

S = safe.
$S_1$ = safe if suitable bypass series winding protection is supplied.
$S_2$ = conditionally safe. Overexcitation of regulators may lead to their failure if fault allowed to persist.
D = dangerous.

When regulators are paralleled, it is necessary that the tap-changing mechanisms be on as nearly the same tap position as possible. If they are not, a circulating current of magnitude $I_c = (V_1 - V_2)/(Z_1 + Z_2)$ will flow in the loop. The most widely used method for paralleling regulators is the "current-balance" method. Circulating current is separated from load current by means of auxiliary current transformers. This current is fed into the voltage reference circuit to cause the unit to change taps to reduce the circulating current.

***258. Regulators are often applied in series,*** or cascaded, on the same feeder. In this application, two or more regulators operate to control the voltage profile along with a distribution line. The most important consideration in such applications is the time-delay settings on the regulator controls. The modern solid-state control is continuously adjustable from at least 15 to 90 s. When regulators are cascaded, the first regulator (at or closest to the substation) should be set with the shortest time delay, with progessively longer delays farther away from the station. If the settings are reversed, the farthest regulator would attempt to correct all voltage fluctuations first. As the load change appeared at the other regulators on the line, each would attempt to compensate for it in order. When finally the substation unit reacted, it would raise the overall voltage level, making it necessary for each successive unit down the line to back down accordingly.

***259. Setting of the regulator control*** must be accomplished with care to assure proper regulator operation and proper supply voltage for the users.

The *voltage-level setting* is the voltage which the regulator is to maintain at its output, expressed on a 120-V base. (Note that this will be the voltage at the "load center" if line-drop compensation is used, as explained later.)

To avoid a hunting condition of the regulator, a *bandwidth* is set to define the limits of acceptable voltage about the voltage-level setting. The stated bandwidth is the total range, such that a regulator set for 120 V with a 2-V bandwidth will be "in-band" if the output is in the range of 119 to 121 V. The bandwidth setting must be larger than the voltage change expected from a single tap change or a hunting situation will occur. Beyond this, it is a qualitative judgment; a lower setting will maintain a closer output voltage tolerance, a higher setting will reduce tap-changer operations, extending the regulator life.

The *time delay* is the time duration outside of the prescribed band required before tap-changer actuation. As noted earlier, this is very important in cascaded operations. Otherwise, it will be set at typically 30 to 60 s to avoid unduly quick responses to line-voltage fluctuations.

The use of *line-drop compensation* will cause the regulator to hold the voltage-level setting at a point remote from the regulator, rather than at the regulator location. The classic illustration of the application of line-drop compensation involves a "load center" some miles from the substation. It is required to hold a given voltage at the load. Given that the line is inductive in nature, this implies holding a higher voltage at the substation, the incremental voltage increase being a function of the line impedance (resistive and reactive) and the line current. Thus, the two line-drop compensation settings are the resistive and reactive models of the line, calibrated in a 0- to 24-V basis, reflecting the drop to be anticipated (on a 120-V base) between the regulator and the load when the system is carrying rated regulator current.

Tables are provided with the regulators to use in determining the proper settings. A simplified example demonstrates the procedure.

*Example*

1. The wye-connected regulators are rated 76.2 kVA, 7620 V (= 100 A).
2. The load center is 4 mi from the regulators.
3. The feeder to the load center is 2/0 ACSR on 36 in center spacing.

*Solution*

1. Tables provided with the regulator will show
   *a.* Line resistance ≃ 0.90 Ω/mi
   *b.* Line reactance ≃ 0.77 Ω/mi
2. Calculate compensation as
   *a.* $\dfrac{\text{CT primary rating}}{\text{Voltage transformer ratio}} \times R/\text{mi} \times \text{miles} = R \text{ comp set}$

*b.* $$\frac{\text{CT primary rating}}{\text{Voltage transformer ratio}} \times X/\text{mi} \times \text{miles} = X \text{ comp set}$$

$$R_{\text{comp set}} = 5.7 \text{ or } 6 \text{ V}$$
$$X_{\text{comp set}} = 4.9 \text{ or } 5 \text{ V}$$

Thus the required 120 V will be held at the load. Changes in load current are automatically compensated.

Each line-drop compensation application needs individual consideration. System conditions which would invalidate this example are single-phase or delta connection or the use of shunt capacitors on the feeder. Reference should be made to the user instructions.

***260. Several control accessories*** are available to provide more sophisticated feeder loading control. These devices have served to make modern feeder regulators the "nerve center" of the distribution system. A *voltage-limit control* device provides automatic limit of regulator output voltage. Settings for both upper and lower limits protect against extremely heavy, light, or unusual loading conditions and against regulator or control malfunction. A review of the use of line-drop compensation may show that at the highest anticipated loading, the voltage at the regulator is too high for proper customer utilization. Then an upper-limit voltage-limit control may be required to protect a load close to the regulator by overriding the line-drop compensation function. A *reverse power flow detector* can monitor source-side voltage with an internal source-side potential supply, detect a reversal of power flow direction, and "turn the regulator around" electrically so that it regulates in the proper direction. A *voltage-reduction control,* which can be operated locally or from a remote control, provides automatic voltage reduction in preselected percentages. This is particularly useful where system capacity is close to peak load; studies have shown that a 5% voltage reduction can reduce system load by almost 5%.

## Regulator Developments

***261. Innovation in the design of step-voltage regulators*** has been toward reduction (or effective elimination) of the arc commonly produced during each tap change.

Viewed as a black box, a regulator in which there is no arcing in oil would be judged the preferable alternative based on two facts.

1. There is no need for venting since no combustible gases are produced by arcing.
2. Routine periodic maintenance is virtually eliminated because the question of contact material ablation caused by the arc no longer exists, and the oil, being exposed to neither oxidation from the air nor the effects of arcing, would be expected to exhibit a life commensurate with a sealed distribution transformer.

Techniques to accomplish this performance have used both sealed vacuum interrupters and power thyristors in configurations which commutate the current to the special interruption medium during the tap change. Current cessation occurs in the vacuum bottle or thyristor such that current interruption never occurs in a contact parting under oil.

Regulators in this category may require special protection measures which confirm proper internal operation. Otherwise the control and interaction with the power system are identical to the conventional regulators.

***262. References on Voltage Regulators***

1. American National Standard C57.15, Requirements, Terminology and Test Code for Step-Voltage and Induction-Voltage Regulators.
2. Massara, J. M., and Tupper, D. L.: Choosing Safe Single-Phase Regulator Connections; *Allis-Chalmers Engineering Review,* 1968, vol. 33, no. 1.
3. Instruction manuals, features bulletins, and product catalogs published by General Electric Company, McGraw-Edison Company, and Siemens Energy and Automation, Inc.
4. Harlow, J. H., and Stich, F. A.: An Arcless Approach to Step-Voltage Regulation. *IEEE* Trans. *Power App. Syst.,* July 1982, vol. PAS-101, pp. 2096–2102.

# POWER CAPACITORS

By J. LAPP and J. R. WILLY

## Power-Factor Correction

***263. General.*** The efficiency of power generation, transmission, and distribution equipment is improved when it is operated near unity power factor. The least-expensive way to achieve near unity power factor is with the application of power-factor-correction capacitors (Fig. 10-105). Capacitors provide a static source of leading reactive current and can be installed

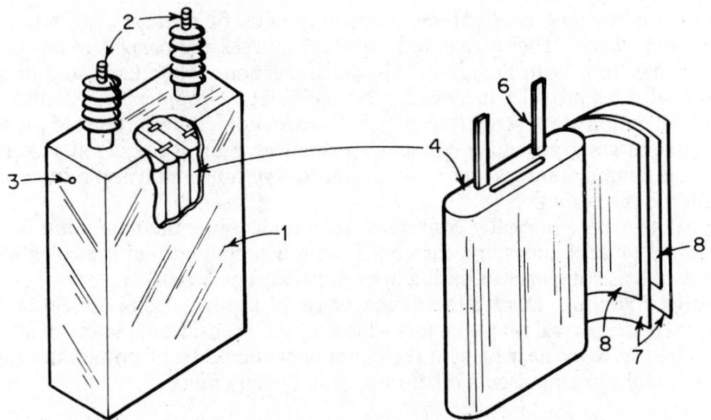


**FIG. 10-105** Typical power capacitor. (1) Capacitor tank; (2) porcelain bushings; (3) fill hole for liquid impregnant; (4) capacitor pack; (5 and 6) electrodes; (7) polypropylene film sheets; (8) aluminum foil sheets.

close to the load. Thus the maximum efficiency may be realized by reducing the magnetizing (lagging) current requirements throughout the system.

## Primary and Secondary Capacitors

***264. Capacitor installations*** are usually shunt-connected across the power lines and are either energized continuously or switched on and off during load cycles. There are two types of capacitors; secondary (low voltage) and primary (high voltage). Of the two types, the primary capacitor is more common.

Secondary, or low-voltage, capacitors are generally available in voltage ratings from 240 to 600 V over the range of 2.5 to 50 kilovars (kvar). When low-voltage capacitors are connected to the secondary lines, they are usually physically located nearer to the lagging reactive loads. This reduces the kVA requirements of the immediate lines and transformers or, conversely, allows a larger kilowatt load with the same-size lines and transformers.

Primary power-factor-correction capacitors are connected to high-voltage lines and are generally available in voltage ratings from 2.4 to 21.6 kV over the range of 50 to 400 kvar. Higher voltage and kvar ratings are achieved by connecting capacitor units in series and parallel arrangements. The cost of high-voltage capacitors is lower per kvar than low-voltage capacitors because of the basic difference in dielectric materials which allows high-voltage capacitors to be operated more efficiently. Also, present-day high-voltage capacitors operate at a lower watts loss per kvar than do low-voltage capacitors. For example, capacitors that utilize an *all-film dielectric* may operate with losses of less than 0.2 W per kvar. *Film-paper dielectric* capacitors may operate with losses of 0.5 W per kvar. Low-voltage capacitors using *kraft-paper dielectric* may experience losses of near 3 W per kvar.

## Capacitor Applications

***265. Typical High-Voltage Capacitor Applications.*** Capacitors for overhead distribution systems can be pole-mounted in banks of 300 to 1800 kvar at nearly any primary voltage up to 34.5 kV phase-to-phase. Pad-mounted capacitor equipments are used for underground distribution systems in the same range of sizes and voltage ratings.

The number of capacitors to install to raise the power factor from one value to another is given in Table 10-13.

***266. Common Capacitor Connections.*** Figure 10-106 shows four of the most common capacitor connections: 3-phase grounded wye, 3-phase ungrounded wye, 3-phase delta, and single-phase. Grounded or ungrounded wye connections are usually made on primary circuits whereas delta and single-phase connections are usually made on low-voltage circuits.

The majority of the power capacitor equipment installed on primary distribution feeders is connected grounded-wye. There are a number of advantages and benefits to be derived from this type of connection. With the grounded-wye connection, switch tanks and frames are at ground potential. This provides increased personnel safety. Grounded-wye connections provide for faster operation of the series fuse in case of a capacitor failure. Grounded capacitors can bypass some line surges to ground and therefore exhibit a certain degree of self-protection from transient voltages and lightning surges. The grounded-wye connection also provides a low-impedance path for harmonics.

If the capacitors are electrically connected ungrounded-wye, the maximum fault current would be limited to three times line current. If too much fault current is available, generally above 5000 A, the use of current-limiting fuses must be considered.

***267. Design Criteria.*** During the design phase of a power-factor-correction capacitor application, there are several major factors which must be considered. Safety of all personnel who are required to work near or with the equipment should be of prime importance. The following paragraphs briefly discuss other important design criteria.

## Protection Principles

***268. Fundamental Protection Principles.*** Several fundamental principles must be observed in the selection of fuses for capacitor application. These are:

1. The fuse link must be capable of continuously carrying 135% of the rated capacitor current.
2. The fuse cutout must have sufficient interrupting capacity to handle successfully the available fault current, clearing voltage, and available energy before the capacitor tank ruptures.
3. The fuse link must withstand, without damage, the normal transient current during bank energization or deenergization. Similarly, it must withstand the capacitor unit's discharge current during a terminal-to-terminal short.
4. For ungrounded-wye banks, maximum fault current is usually limited to three times normal line current. The fuse link must clear within 5 min at 95% of available fault current.
5. For effective capacitor protection, maximum asymmetric rms fault current should not exceed the current value at the intercept point of the capacitor tank-rupture time-current characteristic (TCC) curve and the minimum time shown on the fuse maximum-clearing time-current characteristic curve.
6. The maximum-clearing TCC curve of the fuse link must coordinate with the tank-rupture TCC curve of the capacitor.

For any capacitor installation, principle 1 establishes the continuous-current capacity of the required link, whereas principles 2 and 3 define the region in which the high-current end of the fuse must fall. Principle 5, defining the maximum permissible current through a shorted capacitor, is based upon the design limitations of a capacitor unit and the recognized minimum clearing time of an expulsion fuse link.

Tank-rupture curves are essential to the correct selection of fuse links for overcurrent protection of any capacitor installation. Fuse selections should be based upon the coordination of the fuse link maximum-clearing curve (Fig. 10-106) and capacitor tank-rupture curves (Figs.

Fuse
Gnd.
Grounded wye
Ungrounded wye
Delta
Single phase grounded to neutral

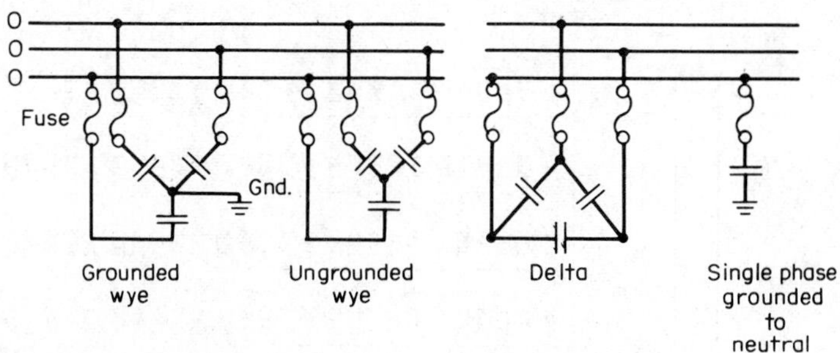

**FIG. 10-106*a*** Common methods for connecting power-factor capacitors.

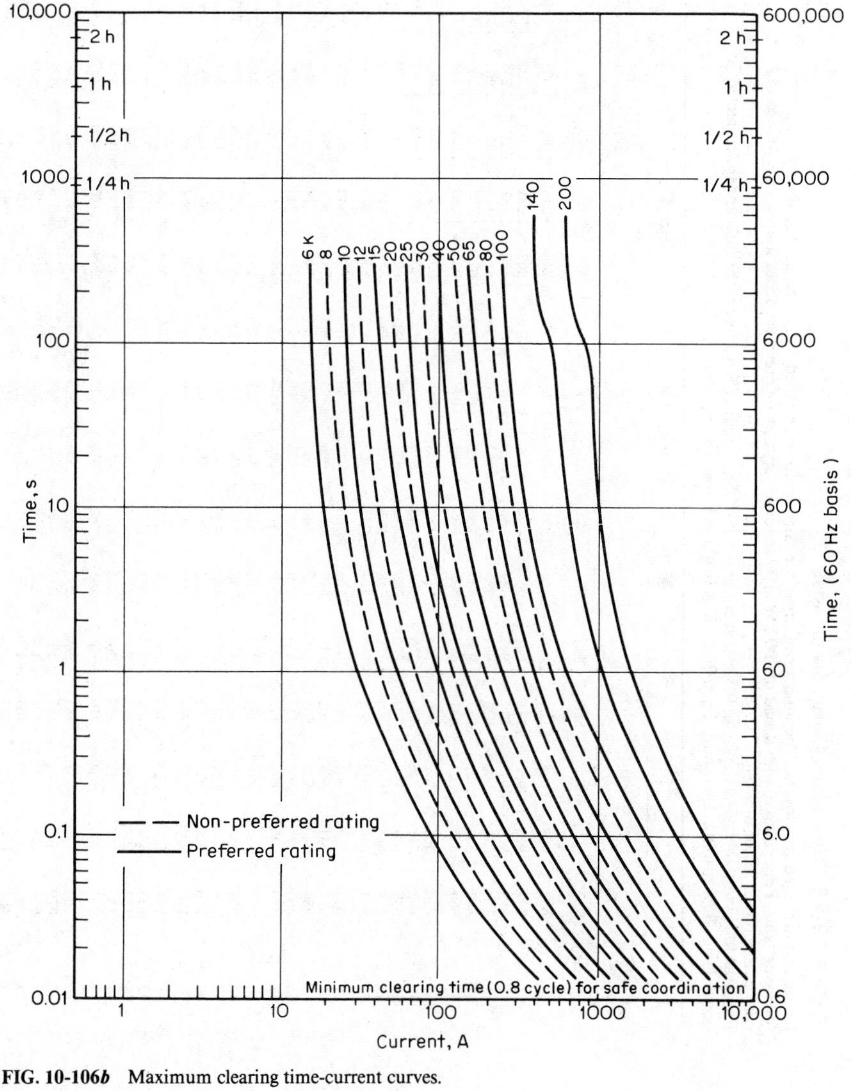

Time, s
Time, (60 Hz basis)
Current, A
2 h
1 h
1/2 h
1/4 h
6 K
8
10
12
15
20
25
30
40
50
65
80
100
140
200
Non-preferred rating
Preferred rating
Minimum clearing time (0.8 cycle) for safe coordination

**FIG. 10-106*b*** Maximum clearing time-current curves.

**TABLE 10-13** How Many Capacitors to Install

Capacitors are rated in kilovars. The number of capacitor kilovars to be installed can be computed simply from the following table. For example, with a load of 200 kW at 77% power factor, how many capacitor kilovars are needed to correct to a power factor of 95%?
From the following table select the factor 0.500 that corresponds to the present 77% reading to the right and the corrected power factor 95% reading downward. Then 200 kW × 0.500 = 100 kvars required.

| Present power-factor percentage | Corrected power-factor percentage | | | | | | | | | | | | | | | | | | | |
|---|---|---|---|---|---|---|---|---|---|---|---|---|---|---|---|---|---|---|---|---|
| | 80 | 81 | 82 | 83 | 84 | 85 | 86 | 87 | 88 | 89 | 90 | 91 | 92 | 93 | 94 | 95 | 96 | 97 | 98 | 99 |
| 50 | 0.982 | 1.008 | 1.034 | 1.060 | 1.086 | 1.112 | 1.139 | 1.165 | 1.192 | 1.220 | 1.248 | 1.276 | 1.306 | 1.337 | 1.369 | 1.403 | 1.442 | 1.481 | 1.529 | 1.590 |
| 51 | .937 | .962 | .989 | 1.015 | 1.041 | 1.067 | 1.094 | 1.120 | 1.147 | 1.175 | 1.203 | 1.231 | 1.261 | 1.292 | 1.324 | 1.358 | 1.395 | 1.436 | 1.484 | 1.544 |
| 52 | .893 | .919 | .945 | .971 | .997 | 1.023 | 1.050 | 1.076 | 1.103 | 1.131 | 1.159 | 1.187 | 1.217 | 1.248 | 1.280 | 1.314 | 1.351 | 1.392 | 1.440 | 1.500 |
| 53 | .850 | .876 | .902 | .928 | .954 | .980 | 1.007 | 1.033 | 1.060 | 1.088 | 1.116 | 1.144 | 1.174 | 1.205 | 1.237 | 1.271 | 1.308 | 1.349 | 1.397 | 1.457 |
| 54 | .809 | .835 | .861 | .887 | .913 | .939 | .966 | .992 | 1.019 | 1.047 | 1.075 | 1.103 | 1.133 | 1.164 | 1.196 | 1.230 | 1.267 | 1.308 | 1.356 | 1.416 |
| 55 | .769 | .795 | .821 | .847 | .873 | .899 | .926 | .952 | .979 | 1.007 | 1.035 | 1.063 | 1.090 | 1.124 | 1.156 | 1.190 | 1.228 | 1.268 | 1.316 | 1.377 |
| 56 | .730 | .756 | .782 | .808 | .834 | .860 | .887 | .913 | .940 | .968 | .996 | 1.024 | 1.051 | 1.085 | 1.117 | 1.151 | 1.189 | 1.229 | 1.277 | 1.338 |
| 57 | .692 | .718 | .744 | .770 | .796 | .822 | .849 | .875 | .902 | .930 | .958 | .986 | 1.013 | 1.047 | 1.079 | 1.113 | 1.151 | 1.191 | 1.239 | 1.300 |
| 58 | .655 | .681 | .707 | .733 | .759 | .785 | .812 | .838 | .865 | .893 | .921 | .949 | .976 | 1.010 | 1.042 | 1.076 | 1.114 | 1.154 | 1.202 | 1.263 |
| 59 | .618 | .644 | .670 | .696 | .722 | .748 | .775 | .801 | .828 | .856 | .884 | .912 | .939 | .973 | 1.005 | 1.039 | 1.077 | 1.117 | 1.165 | 1.226 |
| 60 | .584 | .610 | .636 | .662 | .688 | .714 | .741 | .767 | .794 | .822 | .850 | .878 | .905 | .939 | .971 | 1.005 | 1.043 | 1.083 | 1.131 | 1.192 |
| 61 | .549 | .575 | .601 | .627 | .653 | .679 | .706 | .732 | .759 | .787 | .815 | .843 | .870 | .904 | .936 | .970 | 1.008 | 1.048 | 1.096 | 1.157 |
| 62 | .515 | .541 | .567 | .593 | .619 | .645 | .672 | .698 | .725 | .753 | .781 | .809 | .836 | .870 | .902 | .936 | .974 | 1.014 | 1.062 | 1.123 |
| 63 | .483 | .509 | .535 | .561 | .587 | .613 | .640 | .666 | .693 | .721 | .749 | .777 | .804 | .838 | .870 | .904 | .942 | .982 | 1.030 | 1.091 |
| 64 | .450 | .476 | .502 | .528 | .554 | .580 | .607 | .633 | .660 | .688 | .716 | .744 | .771 | .805 | .837 | .871 | .909 | .949 | .997 | 1.058 |
| 65 | .419 | .445 | .471 | .497 | .523 | .549 | .576 | .602 | .629 | .657 | .685 | .713 | .740 | .774 | .806 | .840 | .878 | .918 | .966 | 1.027 |
| 66 | .388 | .414 | .440 | .466 | .492 | .518 | .545 | .571 | .598 | .626 | .554 | .682 | .709 | .743 | .775 | .809 | .847 | .887 | .935 | .996 |
| 67 | .358 | .384 | .410 | .436 | .462 | .488 | .515 | .541 | .568 | .596 | .624 | .652 | .679 | .713 | .745 | .779 | .817 | .857 | .905 | .966 |
| 68 | .329 | .355 | .381 | .407 | .433 | .459 | .486 | .512 | .539 | .567 | .595 | .623 | .650 | .684 | .716 | .750 | .788 | .828 | .876 | .937 |
| 69 | .299 | .325 | .351 | .377 | .403 | .429 | .456 | .482 | .509 | .537 | .565 | .593 | .620 | .654 | .686 | .720 | .758 | .798 | .840 | .907 |
| 70 | .270 | .296 | .322 | .348 | .374 | .400 | .427 | .453 | .480 | .508 | .536 | .564 | .591 | .625 | .657 | .691 | .729 | .769 | .811 | .878 |
| 71 | .242 | .268 | .294 | .320 | .346 | .372 | .399 | .425 | .452 | .480 | .508 | .536 | .563 | .597 | .629 | .663 | .701 | .741 | .783 | .850 |
| 72 | .213 | .239 | .265 | .291 | .317 | .343 | .370 | .396 | .423 | .451 | .479 | .507 | .534 | .568 | .600 | .634 | .672 | .712 | .754 | .821 |
| 73 | .186 | .212 | .238 | .264 | .290 | .316 | .343 | .369 | .396 | .424 | .452 | .480 | .507 | .541 | .573 | .607 | .645 | .685 | .727 | .794 |
| 74 | .159 | .185 | .211 | .237 | .263 | .289 | .316 | .342 | .369 | .397 | .425 | .453 | .480 | .514 | .546 | .580 | .618 | .658 | .700 | .767 |
| 75 | .132 | .158 | .184 | .210 | .236 | .262 | .289 | .315 | .342 | .370 | .398 | .426 | .453 | .487 | .519 | .553 | .591 | .631 | .673 | .740 |
| 76 | .105 | .131 | .157 | .183 | .209 | .235 | .262 | .288 | .315 | .343 | .371 | .399 | .426 | .460 | .492 | .526 | .564 | .604 | .652 | .713 |
| 77 | .079 | .105 | .131 | .157 | .183 | .209 | .236 | .262 | .289 | .317 | .345 | .373 | .400 | .434 | .466 | .500 | .538 | .578 | .620 | .687 |
| 78 | .053 | .079 | .105 | .131 | .157 | .182 | .210 | .236 | .263 | .291 | .319 | .347 | .374 | .408 | .440 | .474 | .512 | .552 | .594 | .661 |
| 79 | .026 | .052 | .078 | .104 | .130 | .156 | .183 | .209 | .236 | .264 | .292 | .320 | .347 | .381 | .413 | .447 | .485 | .525 | .567 | .634 |
| 80 | .000 | .026 | .052 | .078 | .104 | .130 | .157 | .183 | .210 | .238 | .266 | .294 | .321 | .355 | .387 | .421 | .459 | .499 | .541 | .608 |

| | | | | | | | | | | | | | | | | | | | |
|---|---|---|---|---|---|---|---|---|---|---|---|---|---|---|---|---|---|---|---|
| 81 | .000 | .026 | .052 | .078 | .104 | .131 | .157 | .184 | .212 | .240 | .268 | .295 | .329 | .361 | .395 | .433 | .473 | .515 | .582 |
| 82 | | .000 | .026 | .052 | .078 | .105 | .131 | .158 | .186 | .214 | .242 | .269 | .303 | .335 | .369 | .407 | .447 | .489 | .556 |
| 83 | | | .000 | .026 | .052 | .079 | .105 | .132 | .160 | .188 | .216 | .243 | .277 | .309 | .343 | .381 | .421 | .463 | .530 |
| 84 | | | | .000 | .026 | .053 | .079 | .106 | .134 | .162 | .190 | .217 | .251 | .283 | .317 | .355 | .395 | .437 | .504 |
| 85 | | | | | .000 | .027 | .053 | .080 | .108 | .136 | .164 | .191 | .225 | .257 | .291 | .329 | .369 | .417 | .478 |
| 86 | | | | | | | .026 | .053 | .081 | .109 | .137 | .167 | .198 | .230 | .265 | .301 | .343 | .390 | .451 |
| 87 | | | | | | | | .027 | .055 | .082 | .111 | .141 | .172 | .204 | .238 | .275 | .317 | .364 | .425 |
| 88 | | | | | | | | | .028 | .056 | .084 | .114 | .145 | .177 | .211 | .248 | .290 | .337 | .398 |
| 89 | | | | | | | | | | .028 | .056 | .086 | .117 | .149 | .183 | .220 | .262 | .309 | .370 |
| 90 | | | | | | | | | | | .028 | .058 | .089 | .121 | .155 | .192 | .234 | .281 | .342 |
| 91 | | | | | | | | | | | | .030 | .061 | .093 | .127 | .164 | .206 | .253 | .314 |
| 92 | | | | | | | | | | | | | .031 | .063 | .097 | .134 | .176 | .223 | .284 |
| 93 | | | | | | | | | | | | | | .032 | .066 | .103 | .145 | .192 | .253 |
| 94 | | | | | | | | | | | | | | | .034 | .071 | .113 | .160 | .221 |
| 95 | | | | | | | | | | | | | | | | .037 | .079 | .126 | .187 |
| 96 | | | | | | | | | | | | | | | | | .042 | .089 | .150 |
| 97 | | | | | | | | | | | | | | | | | | .047 | .108 |
| 98 | | | | | | | | | | | | | | | | | | | .061 |

10-107 and 10-108).This fundamental principle has been evolved to assume safe operation of capacitor equipments.

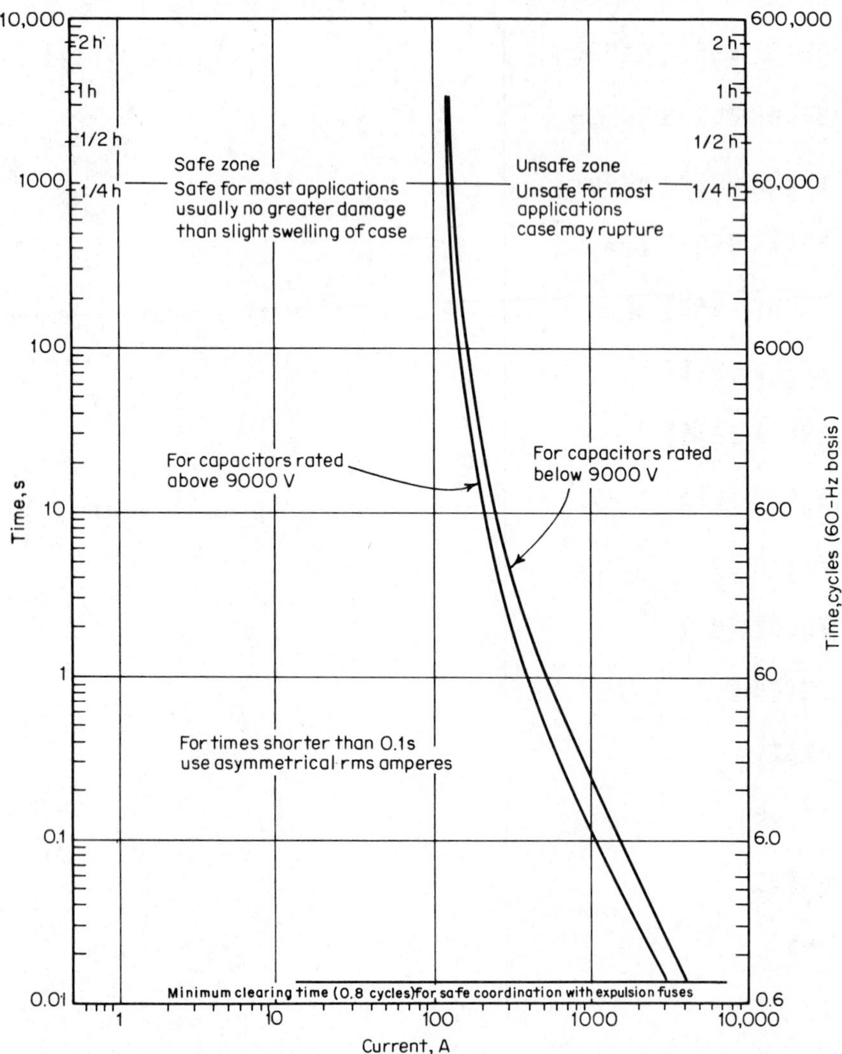


**FIG. 10-107*a*** Tank-rupture time-current curves for 100-kvar all-film single-phase capacitors.

***269. Capacitor Tank Rupture.*** Capacitor tank rupture will occur if the total energy applied to the capacitor under failure conditions is greater than the ability of the capacitor tank to withstand such energy. This energy application could occur under a wide variety of current-time conditions ranging from currents modestly in excess of normal ratings for periods of days, weeks, months, or longer, to very high currents at very short times, such as would occur during the dumping of energy stored in parallel capacitors in large shunt banks into a failed unit.

The cause of tank rupture is an internal pressure, either isobaric or localized, sufficient to

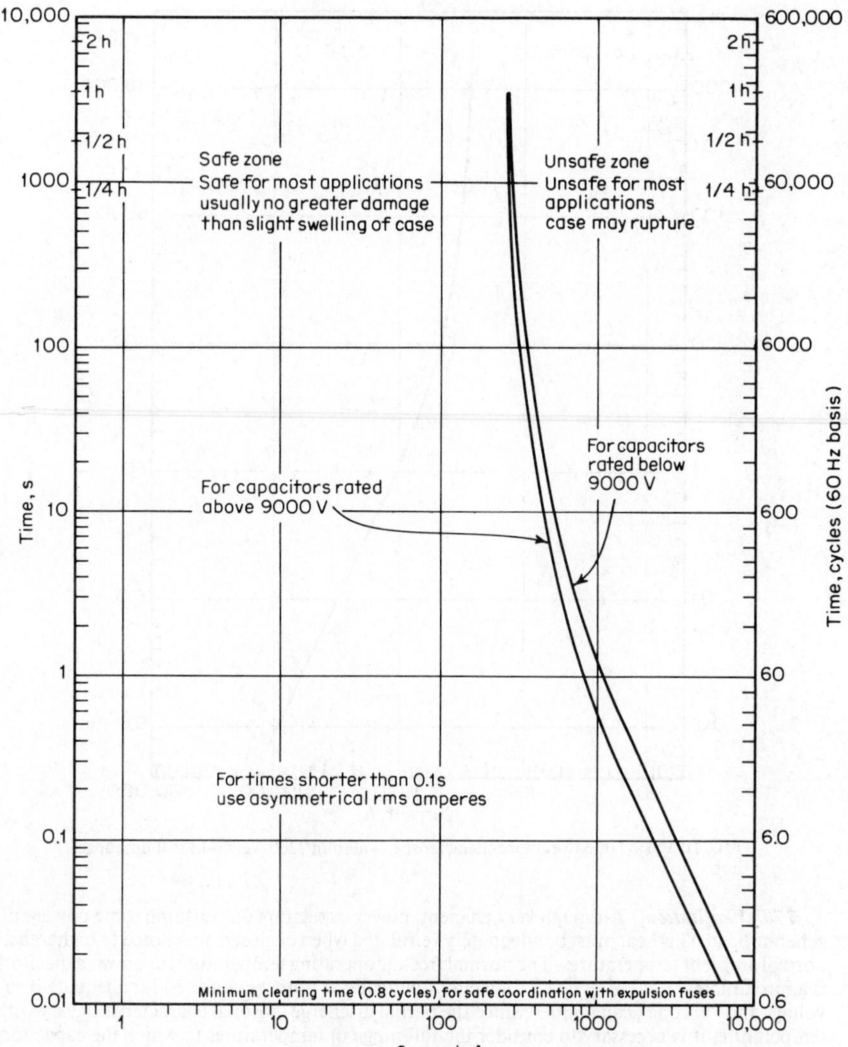


**FIG. 10-107*b*** Tank-rupture time-current curves for 150-, 200-, and 300-kvar all-film single-phase capacitors.

stress the capacitor tank beyond its ultimate strength. The increase in internal pressure is caused by self-generated heat losses and by arcing, with rupture times less than 10 min being caused primarily by the internal arcing. The amount of arcing in turn is determined by the location of the failure as modified by the design of the capacitor and by the available fault current.

The TCC (tank-rupture) curve in Fig. 10-108 shows that for fault currents from 1600 to 12,000 A, the arc energy required to produce tank rupture is essentially constant. This constant-energy relationship was expected at the higher currents. However, no claim can be made now as to how far this curve may be extrapolated toward still higher currents. As the lower current range is approached from the 1600-A level, increasingly longer periods of nonarcing time occur, until in the 300-A range and less there may be no arcing.

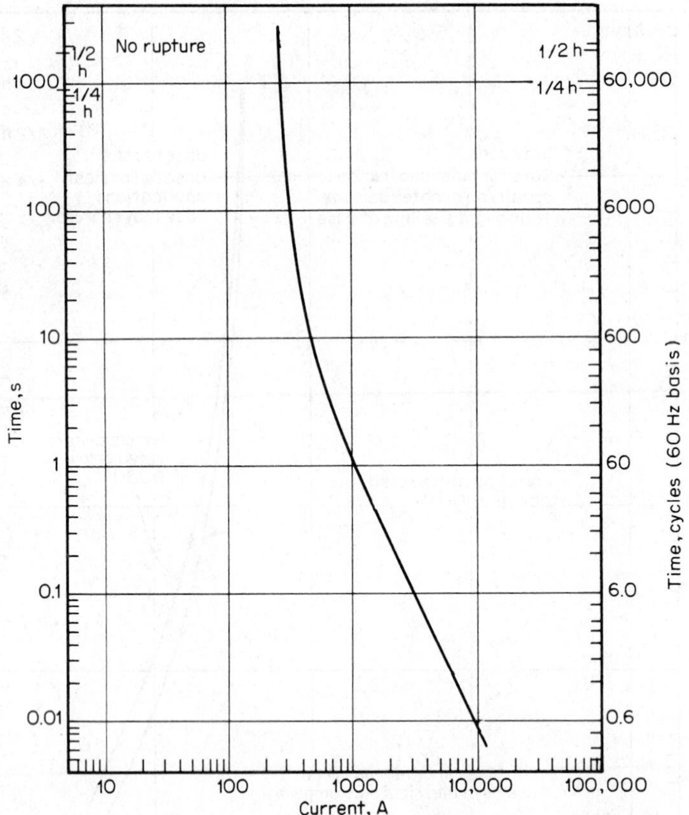


**FIG. 10-108** Time-to-rupture characteristic—all-film 7200-V, 200-kvar shunt capacitor.

***270. Ventilation.*** Although very efficient, power capacitors do consume some power and generate heat. This heat must be adequately ventilated when enclosed or exposed to higher than normal ambient temperatures. The normal free-air operating temperature for power capacitors is approximately −40 to +50°C. Capacitor applications must be designed for adequate overvoltage and "corona" capabilities. Since the partial discharge (corona) characteristics vary with temperature, it is necessary to consider the full range of temperatures to which the capacitors will be exposed, in both the energized and deenergized modes.

***271. System Voltage.*** Capacitors are designed for operation on 60-Hz sine-wave power lines at a specific voltage which is listed on the unit nameplate. However, they are designed to operate at overvoltages of 10% without damage to the capacitor. The kvar output of the capacitor increases as the square of the applied voltage.

$$\text{kvar}_{E_2} = \text{kvar}_{E_1} \frac{(E_2)^2}{(E_1)^2} \tag{10-76}$$

For example, a 200-kvar, 7200-V capacitor will supply 242 kvar at 7920 V.

$$\text{kvar}_{E_2} = 200 \frac{(7920)^2}{(7200)^2}$$

$$\text{kvar}_{E_2} = 242 \tag{10-77}$$

***272. Harmonic Distortion.*** Capacitors are designed to operate on sine-wave current with limited amounts of harmonic distortion. Typical applications that may cause harmonic current problems are arc furnaces, saturable reactors, and rectifiers. Normally, the capacitor manufacturer is aware of these conditions, when they exist, for proper consideration in designing the capacitor equipment.

***273. Discharge Resistors.*** When line voltage is removed from a power capacitor, the danger exists that, even days later, under certain conditions, the unit would retain an extremely high charge. This characteristic of retaining such a charge is demonstrated by the high-efficiency and low-loss operation of a power capacitor. To eliminate this hazard, all power capacitors contain internal-discharge resistors. This resistor assembly will reduce the terminal voltage from line voltage to 50 V within 5 min of deenergization for a capacitor rated higher than 1200 V ac, and within 1 min for capacitors rated less than 1200 V ac.

***274. High-Frequency Charging Currents.*** When the installation of a switched rack or bank is contemplated, the nearness of other capacitor equipment must be evaluated. High-frequency charging currents can result in blown fuses. The use of series reactors and special switches is sometimes required to reduce these currents to safe levels. Proper installation of lightning arresters will ensure the protection of capacitor equipment from lightning surges.

***275. References on Power Capacitors***

*NEMA Publication* CP 1-1973; revised Dec. 19, 1975.

*Standards Bulletin* CP, p. 113.

*IEEE Publication* 18/ANSI C55.1-1968.

## *FUSES AND SWITCHES*

*By C. A. POPECK*

### Fuses

***276. General Classifications.*** Fuses can be classified into two categories. The first is low-voltage fuses operating up through 600 V ac. Most devices are tested and approved by Underwriters' Laboratories, Inc., and are marketed in a wide variety of characteristics and physical configurations. The second classification is high-voltage fuses. These fuses are used by electric utilities to protect distribution-class equipment and by large industrial complexes which have their own electrical distribution systems.

***277. Low-Voltage Fuses.*** Low-voltage fuse standards are established by ANSI, NEMA, or Underwriters' Laboratories (see Table 10-14). Their characteristics include a voltage class, an ampere rating, and an interrupting rating and, for some classes of fuses, a current-limiting rating. Cartridge fuses are classified in the following voltage classes: not over 250-V ac, not over 300-V ac, and not over 600 V ac. Fuses should not be used for dc applications unless recom-

**TABLE 10-14** Class of Low-Voltage Fuses and Applicable Standards

| Fuse class | Standard |
|---|---|
| Plug fuses | UL 198F |
| Radio and appliances | UL 1417 |
| Mine duty | UL 198M |
| DC fuses | UL 198L |
| H | UL 198B and ANSI C97.1 |
| G, J, and L | UL 198C and ANSI C97.1 |
| K | UL 198D and ANSI C97.1 |
| R | UL 198E |
| T | UL 198H |

mended by the manufacturer. Most low-voltage fuses must be used in accordance with the National Electrical Code. Exceptions include those used in ships, railways, aircraft, and automotive vehicles other than mobile homes and recreational vehicles.

The standard lines of low-voltage fuses are available in several steps of ampere capacity, each of which is a different physical size (see Table 10-15).

**TABLE 10-15** Typical Dimension Grouping of Fuses

| Class | Volts | Amperes |
|---|---|---|
| G | 300 | 0–15 |
| G | 300 | 16–20 |
| G | 300 | 21–30 |
| G | 300 | 31–60 |
| H and K | 250 and 600 | 0–30 |
| H and K | 250 and 600 | 31–60 |
| H and K | 250 and 600 | 61–100 |
| H and K | 250 and 600 | 101–200 |
| H and K | 250 and 600 | 201–400 |
| H and K | 250 and 600 | 401–600 |
| L | 600 | 601–800 |
| L | 600 | 801–1200 |
| L | 600 | 1201–1600 |
| L | 600 | 1601–2000 |
| L | 600 | 2001–2500 |
| L | 600 | 2501–3000 |
| L | 600 | 3001–4000 |
| L | 600 | 4001–5000 |
| L | 600 | 5001–6000 |

A minimum of 10,000-A interrupting capacity is typical in low-voltage fuses, but some sizes and types are capable of interrupting up to 200,000-A ac or 100,000-A dc. The interrupting rating is the highest rms symmetrical alternating current which the fuse can interrupt at rated voltage.

Low-voltage current-limiting fuses are designed so that non-current-limiting fuses cannot be inserted within the fuse holder as a direct replacement. Thus Class K fuses which are interchangeable with Class H fuses are not permitted the "current-limiting" label.

A low-voltage current-limiting fuse successfully and safely interrupts all available currents within its specified interrupting rating, and within its current-limiting range limits the clearing time at rated voltage to an interval equal to or less than the first major current loop. These fuses also limit peak let-through current to a value less than the normal peak current that would be possible without current-limiting availability. The current-limiting characteristics of current-limiting fuses are expressed by two electrical measurements: *(a)* maximum peak let-through current, the maximum instantaneous value of current passed by the fuse during time of operation; and *(b)* maximum clearing $I^2t$ (amperes-squared-seconds), an expression of the energy available as a result of current flow during the clearing time of operation.

In recent years combinations of low-voltage circuit breakers and low-voltage current-limiting fuses have been devised to provide improved short-circuit protection, especially on motor branch circuits where contactors have low interrupting ratings.

A device which has been labeled a permanent power fuse is actually not a fuse, but a current limiter. A low-melting-temperature metal such as sodium is sealed in an insulated pressure vessel. Fault currents cause the metal to rapidly change its state to a gas. Pressures can reach 40,000 psi. The resistance increases several orders of magnitude, thus limiting the current, which is then cleared by a circuit breaker of relatively smaller capacity. Upon interruption of the current, the fuse reverts to its original state.

***278. High-Voltage Fuses.*** High-voltage fuses are defined as any fuse (above 600 V) or fuse devices used for the purpose of isolating an electric short circuit from a high-voltage electrical distribution system. Specific classes of fuses or fused devices are

Enclosed cutouts and fuses
Open cutouts and fuses
Open-link cutouts and fuses
Current-limiting fuses
Power fuses
Oil-immersed protective links

See Table 10-16.

**TABLE 10-16** Classes of High-Voltage Fuses or Fused Devices and Applicable Standards

| Class device | Standard |
|---|---|
| Distribution cutouts and fuse links | ANSI C37.42 |
| Distribution oil cutouts and fuse links | ANSI C37.44 |
| Power fuses | ANSI C37.46 |
| Current-limiting fuses | ANSI C37.47 |

These fuses are used to protect potential transformers, distribution or medium-sized power transformers, and lateral taps from main distribution feeder circuits. They are often used as sectionalizing devices on main feeder circuits. The ampacity and interrupting rating of these devices range up to 720 A, 500 mVA at 14,400 V, and 250 A, 2000 mVA at 138,000 V. Fuses are generally used in electrical series with other fuses or circuit-protective devices. Care must be taken in coordinating the time-current characteristics for proper isolation of the electric circuit during fault and overload conditions.

High-voltage fuse links for use with expulsion cutouts are available with many different time-current characteristics. Figure 10-109 shows the minimum melting time-current characteristics for ANSI Type K fuse links.

In an effort to standardize fuse-link characteristics, ANSI has adopted time-current characteristics for two basic fuse-link types: Type K (fast) and Type T (slow). These fuse links are designed so as to have the same time-current characteristics regardless of manufacturer. A wide variety of nonstandardized fuse-link characteristics are also available. Fuse-link characteristics are usually based on tests starting cold in an 18 to 32°C ambient temperature. For characteristics at other ambients or for preloading variations, consult the individual manufacturer. Figure 10-110 shows a typical open-type cutout.

## Performance Characteristics

***279. High-Voltage Fuse Performance Characteristics.*** The characteristics of various high-voltage protective devices used on distribution power systems are

1. **Expulsion cutouts**
   a. *Open type:* 20-kA asymmetrical maximum interrupting current (IC), 5 to 35 kV, violent in operation at high faults, low cost, both initial and refusing. Maximum continuous-current rating of 200-A; can be fitted with a solid blade for conversion to a disconnect switch with a rating of 300 A.
   b. *Enclosed type:* 8- to 10-kA asymmetrical maximum interrupting current, used primarily where safety codes dictate use.

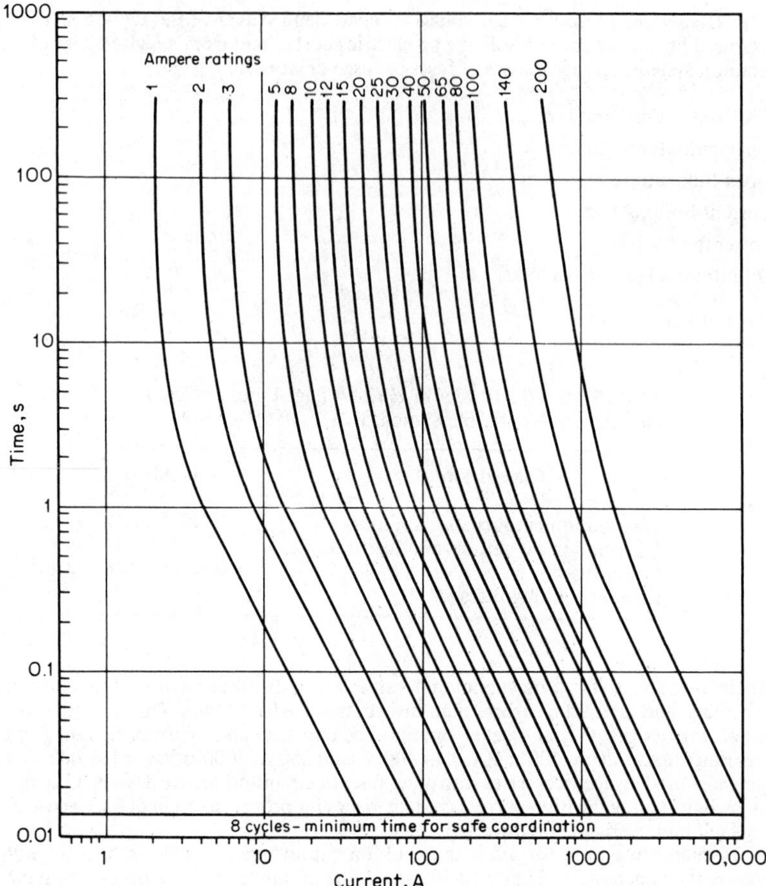


**FIG. 10-109** Minimum melting time-current characteristics of a NEMA Type K fuse. *(A. B. Chance Co.)*

*c. Open link:* 1200-A maximum interrupting current, 50-A maximum continuous current, applied on rural lines and/or small transformers.

2. **Oil cutouts:** Considerable application in the past, especially in underground vaults; however, low interrupting current now poses serious underrating problems.
3. **Liquid fuses:** Nonviolent, low interrupting current (8 to 10 kA maximum).
4. **Power fuses:** Reduced arc energy, somewhat less violent than cutouts on high faults, rated to 20-kA interrupting current; both initial purchase and replacement expensive.
5. **Under-oil protective link:** 4000-A asymmetrical maximum interrupting current, violent in operation, low cost, contaminates insulating oil.
6. **General-purpose current-limiting fuses:** Nonviolent, current-limiting, high interrupting current (50 kA), requires coordination study, generates peak arc voltage, not affected by system transient recovery voltage, both initial purchase and replacement expensive.
7. **High-range backup current-limiting fuses:** Current-limiting, high interrupting current (50 kA), requires a low-current interrupting device in series, operates only at high currents,

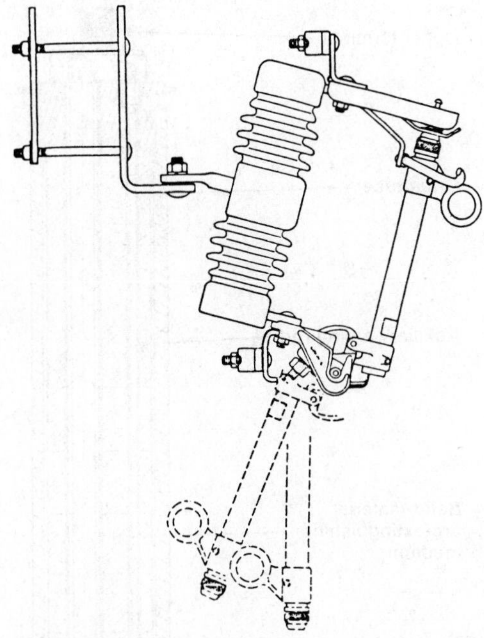

**FIG. 10-110** Open-type distribution fuse cutout. *(A. B. Chance Co.)*

does not affect existing system coordination, low re-fusing cost on majority of outages because of only blowing expulsion link, not affected by system transient recovery voltage.

8. **Vacuum fuses:** Function with no external arcing or violence. They are nonrenewable and the associated cost is high. They have not found widespread application.

All expulsion-principle fuses depend on arc-quenching material, either bone fiber, liquid solutions, or boric acid powder to develop water vapor and/or other gases to cool the arc from the melted fuse link. These fuses have no energy-limiting ability and require a natural current-zero crossing to successfully interrupt a short-circuit current. Figure 10-111*b* shows a cross section of a power-type fuse.

***280. Current-Limiting Fuses.*** High-voltage current-limiting fuses for use on distribution systems have two distinct classes: *(a) General-purpose current-limiting fuses* are devices which will successfully interrupt currents which will melt the fusible element in 1 h. These are some-

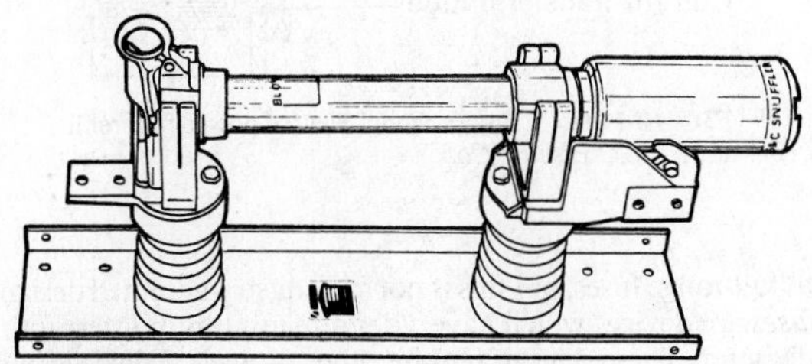

**FIG. 10-111*a*** Indoor-type power fuse.

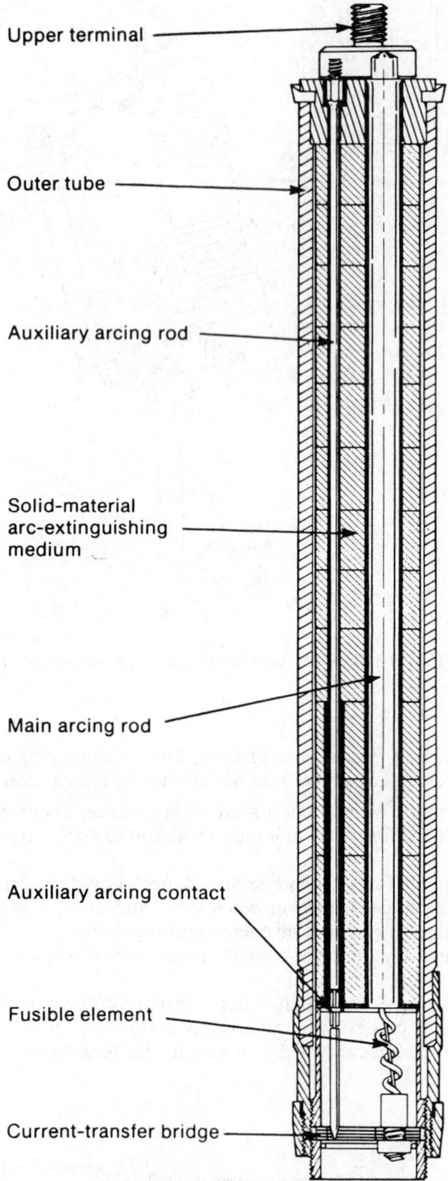


**FIG. 10-111*b*** Cross-sectional view of power-fuse refill unit. *(S&C Electric Co.)*

times referred to as full-range fuses, but this is not an industry-accepted designation. *(b) Backup current-limiting fuses* are devices which have a definite minimum interrupting rating as specified by the manufacturer. These devices require other protective devices in electrical series to protect the backup current-limiting fuse from currents below its minimum interrupting rating.

Three important parameters should be known about high-voltage current-limiting fuses:

1. *Continuous current rating:* The maximum current that the fuse is designed to carry continuously.
2. *Peak arc voltage:* Maximum voltage generated by the current-limiting fuse. If wire-wound, the voltage value is a function of fault current. If a ribbon-element fuse, then the voltage is a function of applied voltage across the fuse. See Fig. 10-112.
3. *$I^2t$ clearing:* Maximum allowed by the current-limiting fuse. This measures the energy-limiting effect of the fuse.

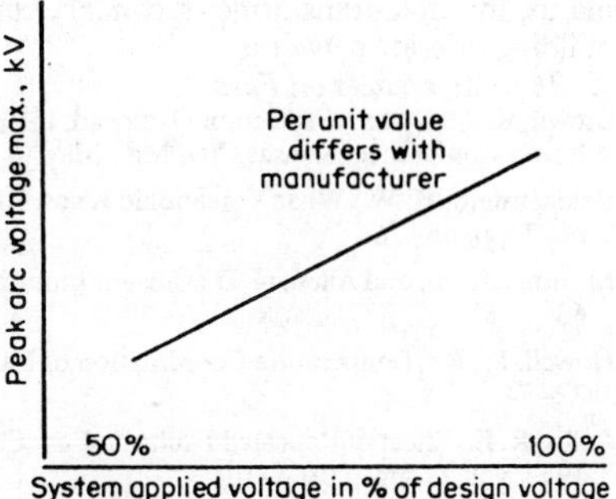


**FIG. 10-112** Peak arc voltage of ribbon-element current-limiting fuses versus system applied voltage.

Care in application of current-limiting fuses according to voltage rating must be maintained. In general these fuses should not be applied to circuits with a voltage less than 50% of the fuse-voltage rating to avoid excessive peak arc voltages. Figure 10-113 shows a typical backup current-limiting fuse to be used in series with overhead expulsion fuses.

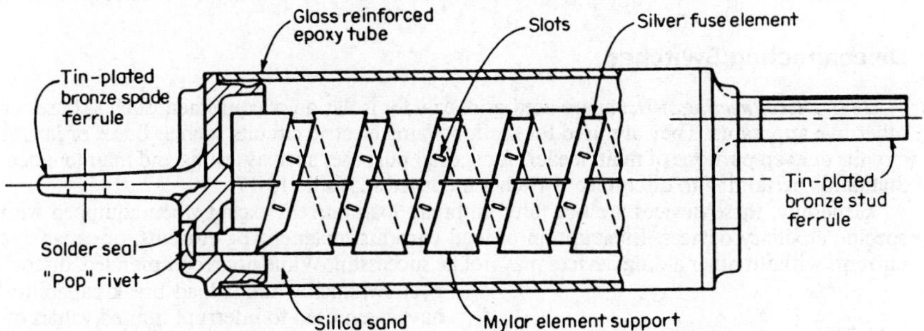


**FIG. 10-113** Backup current-limiting fuse for a distribution system. *(A. B. Chance Co., K-Mate 12.)*

It is equally important that fuses not be exposed to system recovery voltages in excess of their rating.

***281. Fuses in Enclosures.*** High-voltage fuses may be mounted in enclosures for several applications: industrial service entrance switchgear, pad-mounted switchgear or transformers for underground circuits, or in enclosures for subsurface applications. Most fuses will require special adaptation. Power fuses are fitted with a muffler to reduce the intensity of the exhaust gases when used in enclosures. Current-limiting fuses are supplied with special seals to prevent the ingress of fluid when applied under oil such as in transformers. Fuse cutouts are generally not recommended for use in enclosures or vaults.

Derating of fuses in enclosures may have to be considered because of restricted heat transfer. Consult the manufacturer.

***282. Limiters*** are time-delay fusible connectors designed to be installed in low-voltage network-mains cable at street-junction points. They are rated in cable sizes and have time-current characteristics to *(a)* allow the cable fault to burn itself clear if it does so promptly, without blowing the limiter; *(b)* blow before the cable insulation away from the fault is roasted and prevent the failure from spreading beyond the junction; and *(c)* obtain adequate selectivity so that, when installed in a network, only those limiters connected to the faulted cable blow (see Fig. 10-114 for limiter characteristics).

The usual applications of limiters are at the street-corner junctions of low-voltage network mains, multiple-transformer secondary cables, multiple-service cables, and intervault ties of building interior networks.

***283. References on Fuses***


Brown, R. A.: System Protection Overhead, IEEE C77352-81A, Presented at the IEEE 1977 Rural Electric Power Conference, Kansas City, Mo., May 15–17, 1977.

Reichenstein, H. W.: What You Should Know About Fuses; *Electr. Constr. Maint.*, February 1976, vol. 75, no. 2, pp. 65–76.

Bruning, A. M., and Allen, G. D.: Current Limiting Fusing; *Trans. Distrib.*, October 1974, vol. 26, no. 10, p. 60.

Howell, M. R.: Temperature Coordination of Relays and Fuses; *Electr. World*, July 11, 1942, vol. 118, pp. 72–74.

Reibs, R. E.: Effect of Repeated Faults on Fuse Characteristics; *Trans. AIEE Power Appar. Syst.*, December 1952, vol. 71, pp. 1101–1108.

Mathews, W. A.: Increasing Interrupting Capacity of Low Voltage Circuit Breaker Systems; *Electr. Eng.*, October 1963, vol. 82, pp. 613–617.

Cameron, F. L.: Application of High Voltage Power Fuses; *Westinghouse Eng.*, May 1963, vol. 23, pp. 90–93.

Peach, N.: Low Voltage Fuses Can be Adapted to Varied Jobs; *Power*, September 1964, vol. 108, pp. 166–167.

Mikulecky, H. W.: Current Limiting Fuse with Full Range Clearing Ability, *Trans. AIEE Power Appar. Syst.*, vol. 84, pp. 1107–1112.

Schultz, N. R., Hopkinson, R. H., and Easley, J. H.: Single-Phase Switching and Fusing in Three-Phase Circuits; 27th Annual Power Distribution Conference, University of Texas at Austin, October, 1974.


## Disconnecting Switches

***284. Disconnecting switches*** are used primarily for isolation of equipment such as buses or other live apparatus. They are used for sectionalizing electric circuits such as buses or lateral circuits or even portions of main feeders for special purposes such as testing and maintenance. Standards pertaining to disconnect switches are listed in Table 10-17.

Generally, these devices are not rated to break load current except when equipped with specific auxiliary devices. Interruption of load currents, magnetizing currents, or capacitive currents without other aiding devices may not be successful. While not recommended, disconnect switches without load break capability have been used to interrupt limited values of load, charging, and magnetizing current. Their performance depends not only upon the magnitude of the current but also the speed of operation, wind direction, and velocity and clearance to other energized or grounded components. The principal concern is that the arc which is created may establish a high-current fault. Arc reach may approach tens of feet or more depending upon system voltage and the current being interrupted. However, these switches must be designed to carry expected load currents and remain closed for momentary current flow such as fault currents. Fault currents in excess of a specific rating may cause the switch to be blown open by the magnetic forces due to the short-circuit current.

There are three classes of disconnect switches:

*a.* Station

*b.* Transmission

*c.* Distribution

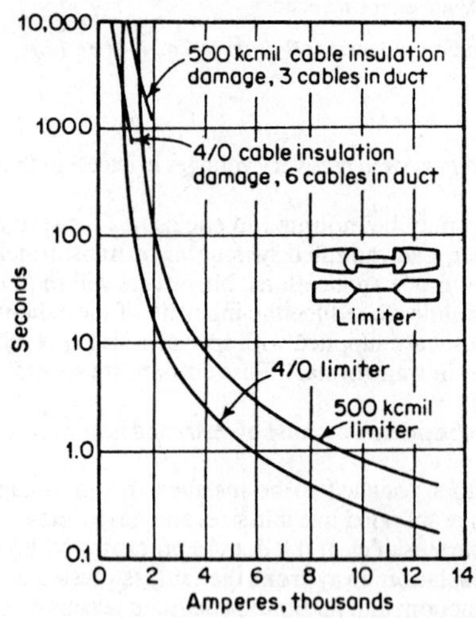


**FIG. 10-114** Time-current characteristics of low-voltage limiters for No. 4/0 and 500 Mcmil network-mains cables.

**TABLE 10-17** Standards Related to Disconnect Switches

| | |
|---|---|
| Ratings and Application Guide | ANSI C37.32 |
| Rated Control Voltages | ANSI C37.33 |
| Test Code | ANSI C37.34 |
| Operation and Maintenance | ANSI C37.35 |
| Loading Guide | ANSI C37.37 |

Switches can be further categorized as gang-operated or hookstick-operated and loadbreak or nonloadbreak types.

The basic insulation level (BIL) of station-class equipment is normally higher than for transmission or distribution equipment. Station equipment ranges from 2.4 to 765 kV at present. Disconnect switches rated up through 3000 A are available. Manual or automatic switching can be provided. Figure 10-115 shows a typical hookstick switch. Figure 10-116 shows one pole of a high-voltage gang-operated switch.

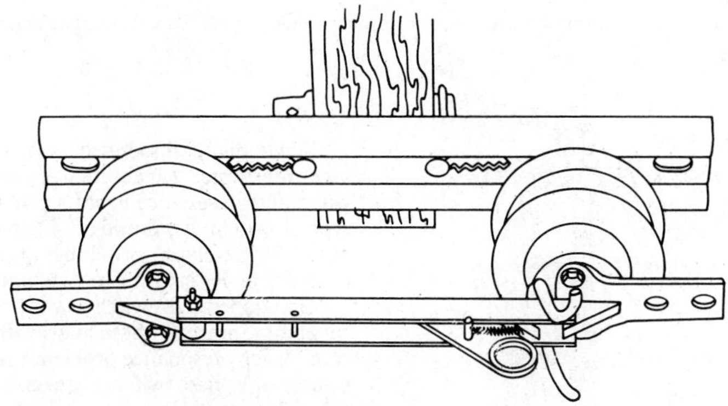

**FIG. 10-115** Distribution class hookstick-disconnecting switch for outdoor service. *(A. B. Chance Co.)*

The design of disconnecting switches demands considerable attention to the contact surfaces. Consideration must be given to the rigors of extreme environments.

High-pressure contacts are generally the form used to provide the current transfer. Current densities of 100,000 A/in$^2$ are not uncommon when using silver for contact points. Contact pressures as high as 500,000 lb/in$^2$ ensure that good cleaning action is achieved and keeps the current transfer points free from contamination.

Transmission disconnecting switches are generally used as load-management tools. Increasing needs for transmission lines and decreasing availability of right-of-way makes automatic switching of transmission load desirable.

Load management is achieved during "dead time" by switching the proper disconnects automatically through sensing loss of voltage. These systems are available through 161 kV, 1200 A. Figure 10-117 shows a typical gang-operated distribution disconnect switch.

The objective of load management is to minimize outage time and allow for more efficient utilization of substation capacity at the distribution level. There is a growing interest in automating distribution class switches to achieve load-management objectives.

Distribution disconnecting switches are becoming the method of providing for both single-phase and 3-phase sectionalizing. As distribution voltage grows to higher levels, the utility must provide more sectionalizing capability or switching capability or suffer large and longer outages during faults. Hence at 34.5 kV one may find some switching capability every 2000 to 3000 ft of overhead conductor.

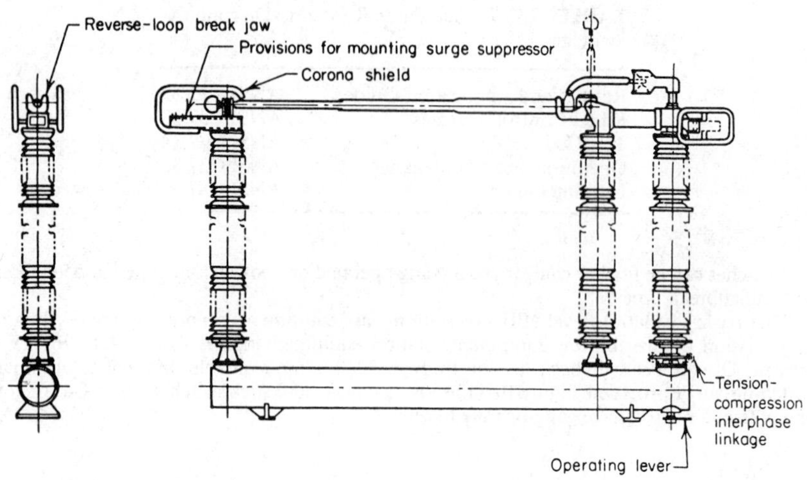


**FIG. 10-116** Single, vertical-break disconnect switch, outdoor type, 500 kV. *(Westinghouse Electric Corp.)*

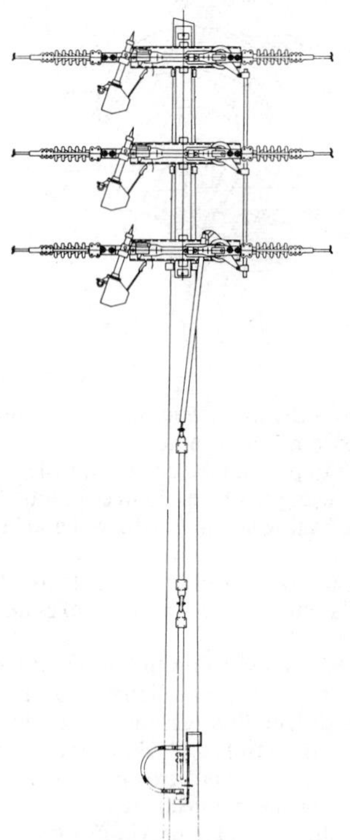

**FIG. 10-117** Gang-operated phase-over-phase load break switch for distribution load management. *(A. B. Chance Co.)*

Single-phase disconnect switching with load-interrupter capability can be applied where ferroresonance is not a problem. This type of switching is found on single-phase circuits. Single-phase switching of a heavily loaded 3-phase circuit is not desirable.

Gang-operated switches with loadbreak capability interrupt these loads without concern for ferroresonance problems. In application these 3-phase switches can be mounted in either horizontal or vertical configurations. To ensure proper operation, the mounting should be as rigid as possible. Care must be exercised in proper alignment of blades and clips. Attention to these matters allows proper operation without overstressing porcelain insulators or distortion of contact surfaces and operational handles. Figure 10-118 shows a unitized-type distribution disconnecting switch. Switches should be serviced in accordance with ANSI C37.35 and the manufacturer's recommendations.

Load-interrupter devices, when combined with disconnecting switches, provide the desired capability of switching load currents without circuit breakers. Generally these interrupters are auxiliary devices and are not continuous-duty in terms of carrying load. This load interruption can be achieved by:

***a.*** Insertion of a resistor in the circuit following opening of the main switch contact and interruption of the current drawn between arcing horns in the air.

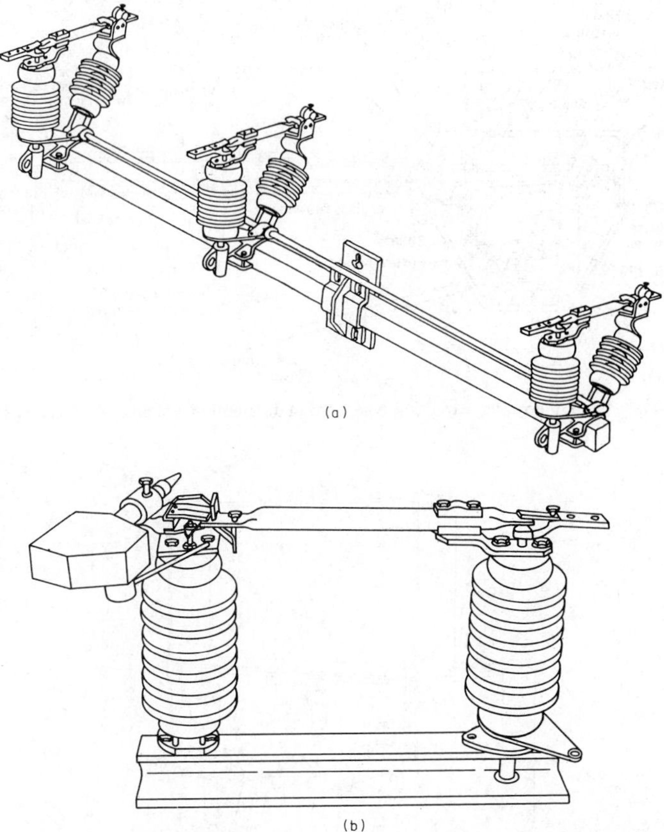


**FIG. 10-118** *(a)* Unitized-type, three-phase horizontal distribution disconnecting switch. *(b)* Single pole equipped with load interrupter. *(A. B. Chance Co.)*

*b.* Use of a blast of air or other gas to effectively lengthen the arc resulting from the main contacts opening.

*c.* Use of an interrupter paralleling the main contacts just prior to opening and interrupting in this auxiliary chamber after the main contacts open. This is typically accomplished with an expulsion-type device or vacuum switch.

Shown in Figure 10-119 is an expulsion-type load interrupter used on distribution disconnecting switches to assist in interrupting load current.

***285. Switches for Underground Circuits.*** The trend toward underground distribution circuits has created the need for a new class of pad-mounted switches. These are available in both live-front and dead-front configurations with the latter growing in acceptance. Live-front switches are typically air-insulated and utilize expulsion load-break interrupters and power fuses.

Dead-front switches are typically oil- or gas-insulated. These use vacuum switches and sometimes $SF_6$ devices for load-break operation. Oil-immersible current-limiting fuses are used almost exclusively. External connections are made by means of separable connectors which can be removed to provide a visible break when working on cables. A typical dead-front switch and fuse enclosure is shown in Figure 10-120.

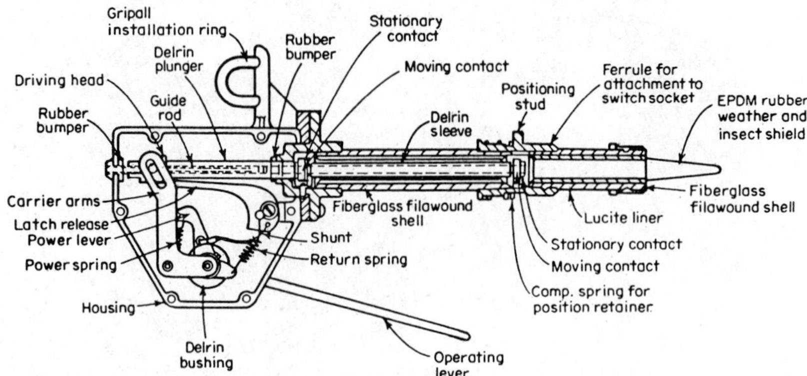


**FIG. 10-119** Load interrupter, expulsion type, used in distribution systems. *(A. B. Chance Co.)*

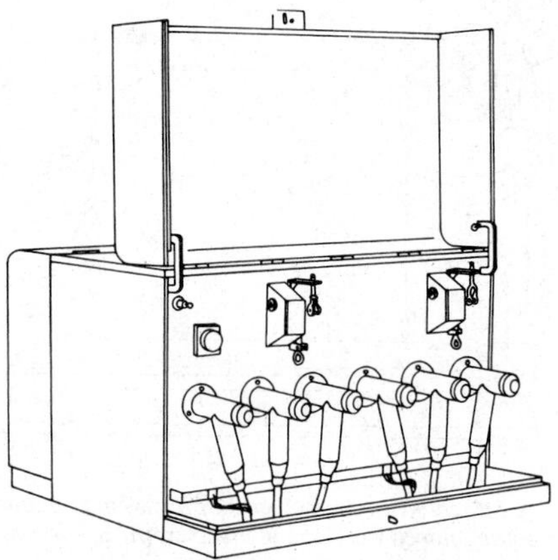

**FIG. 10-120** Dead-front pad-mounted switch including current-limiting fuses. *(A. B. Chance Co.)*

Subsurface switchgear is used extensively for isolating underground circuits in large metropolitan areas. Oil-break-type switches have an advantage of high momentary ratings up to 40,000 A. They have the disadvantage of contaminating the insulating oil when they are operated to break load. Vacuum switches either in an oil dielectric or $SF_6$ gas are gaining increased acceptance for this application. They have the drawback of momentary ratings only up to 20,000 A.

***286. Circuit switchers*** are a hybrid device combining the features of a disconnect switch and circuit breaker. Typically they combine an $SF_6$ fault interrupter with an air-break disconnect switch. The fault interrupter is actuated by appropriate relays. Fault interrupting ratings up to 10,000 A are available. Voltage classes range from 38 to 169 kV. The disconnect provides visual isolation. The device also functions as a load interrupter switch. Within its duty range, the device is more economical than most circuit breakers.

***287. References on Switches***

Beard, L. R., and Speas, T. P. Jr.: The Advantage of Three Phase Switching to Eliminate Ferroresonance Problems on Electrical Distribution Circuits; Pacific Coast Electrical Association Meeting, March 1983.

Shah, K. R., and Ward, W. W., Jr.: Switching Severity of Horn Gap Switches—A New Consideration; *RA&S*, July/August 1972, vol. PAS-91, no. 4, pp. 1602–1605.

IEEE Committee Report: New IEEE Temperature Limitations for Disconnecting Switches; *PA&S*, September 1969, vol. PAS-88, no. 9, pp. 1412–1423.

McNerney, A. M.: Results of High-Current Tests on 161 kV Disconnecting Switches; *AIEE Trans.*, April 1965, vol. 74, part 111, Power Apparatus and Systems, pp. 104–108.

IEEE Committee Report: Results of Survey on Interrupting Ability of Air Break Switches; *IEEE Trans. Power Appar. Syst.*, September 1966, vol. PAS-85, pp. 1008–1019.

Luehring, E. L., and Fitzgerald, J. P.: Switching the Magnetizing Current of Large 345 kV Transformers with Double-Break Air Switches; *IEEE Trans. Power Appar. Syst.*, October 1965, vol. PAS-84, pp. 902–906.

Flugum, R. W., and Hillesland, G. G.: Load Breaking on Distribution Circuits; *IEEE Trans. Power Appar. Syst.*, December 1965, vol. PAS-84, no. 12.

Rankin, E. C.: Experience with Methods of Extending the Capability of High-Voltage Air Break Switches; *AIEE Trans.*, February 1960, vol. 78, part 111-B, *Power Apparatus and Systems*, pp. 1634–1637.

Amchin, H. K., and Curto, R. T.: Switching Surge Voltages Due to the Interruption of Transformer Magnetizing Current; *AIEE Trans.*, December 1959, vol. 78, part 111-B, *Power Apparatus and Systems*, pp. 1443–1449.

Gostin, B. F.: High-Current Testing of Air Disconnect Switches; *AIEE Trans.*, April 1955, vol. 74, part 111, *Power Apparatus and Systems.*

Andrews, F. E.; Janes, L. R.; and Anderson, M. A.: Interrupting Ability of Horn Gap Switches; *AIEE Trans.*, 1950, vol. 69, part 11, pp. 1016–1027.

# SECTION 11
# ALTERNATE SOURCES AND CONVERTERS OF POWER


**Jerald D. Parker**
*Professor of Mechanical Engineering, Oklahoma State University*

**Michael F. McNitt-Gray**
*Instructor, Pennsylvania State University, Altoona, Pa.; Member, IEEE*

**Allen L. Clapp**
*Managing Director, Clapp Research Associates; Member, IEEE, PES/IAS*

**Robert W. Rex**
*Chairman, Republic Geothermal Company*

**R. Ramakumar**
*Professor of Electrical Engineering, Oklahoma State University; Senior Member, IEEE*

**James R. Birk**
*Director, Advanced Conversion and Storage Department, Electric Power Research Institute*

**George H. Miley**
*Chairman, Nuclear Engineering Program, University of Illinois at Urbana-Champaign*

**David Linden**
*Consultant; former Chief, Power Sources Technical Area, Electronic Technology and Devices Laboratory, U.S. Army Electronics Command; Fellow, American Institute of Chemists; Member, American Chemical Society*

**Robert D. Weaver**
*Project Manager, Battery Technology, Electric Power Research Institute*

**Arnold P. Fickett**
*Director, Energy Utilization Department, Electric Power Research Institute*

**Roland W. Ure, Jr.**
*Former Professor of Electrical Engineering and Professor of Materials Science, University of Utah; Senior Member, IEEE*

**Fred N. Huffman**
*Manager, Thermionic Conversion Department, Thermo Electron Corporation*

**John C. Cutting**
*Director of Power Systems, Autodynamics, Inc.*

**CONTENTS**

*Numbers refer to paragraphs*



## INTRODUCTION

***1. Definition of Alternate Sources.*** The threatened limitations on conventional sources of electric power, posed by future shortages of fossil and nuclear fission fuels and the reduced availability of hydroelectric power sites, has focused the attention of electrical engineers on "unconventional" sources, that is, those that are not now in large-scale use. These means of generating power have recently come under the general heading of "alternate sources." This classification (Par. 2) is a loose one, including some of the oldest techniques (e.g., batteries) as well as the more recent (e.g., the generation of power by collection of solar heat).

The present emphasis is not only on discovering new means of generating power but on improving the efficiency and lowering the cost of established techniques. This section brings together descriptions and data on a number of these alternate sources, currently under development to employ the presently untapped energy sources of the sun and the earth, as well as physical and chemical means of deriving electric power from new or improved materials, reactions, and conversion schemes.

*2. **Classification of Alternate Sources and Conversions.*** The material in this section follows the following classifications:

a. *Power from the sun:*
Solar power source and collectors, Par. **3.**
Photovoltaic converters, Par. **12.**

b. *Power from earth-bound sources:*
Subsurface (geothermal) sources, Par. **14.**
Surface (wind) power, Par. **21.**

c. *Application of solar and terrestrial sources to existing power systems.*
Storage, Par. **32.**

d. *Alternate conversion by physical means:*
Magnetohydrodynamics, Par. **135.**
Nuclear fusion, Par. **43.**
Thermoelectricity, Par. **125.**
Thermionics, Par. **130.**

e. *Alternate conversion by chemical means:*
Batteries, Par. **71.**
Fuel cells, Par. **97.**

## SOLAR POWER

*By JERALD D. PARKER, MICHAEL F. MCNITT-GRAY, AND ALLEN L. CLAPP*

*3. **Solar Constant.*** The sun irradiates the earth continuously with abundant energy. The mean distance between the sun and the earth, $1.5 \times 10^8$ km, varies about $\pm 3\%$ each year as the earth moves in an elliptical orbit about the sun. The *solar constant* $I_{SC}$ is the rate at which the sun's energy irradiates a unit area of surface normal to the sun's rays, in the space outside the earth's atmosphere at the earth's mean distance from the sun. The most recent estimates[1]* of the solar constant in various units are

428 Btu/(ft$^2$)(h)
1353 W/m$^2$
4871 kJ/(m$^2$)(h)
1.940 cal/(cm$^2$)(min)

The effective surface temperature of the sun is 5762 K, and the radiation emitted is about 7% ultraviolet (wavelength $< 0.38$ $\mu$m), 47% visible (wavelength between 0.38 and 0.78 $\mu$m), and about 46% infrared (wavelength $> 0.78$ $\mu$m). Thus the sun's radiation tends to be primarily of a much shorter wavelength than the thermal radiation emitted by most objects on the earth.

*4. **Radiation Received at Earth's Surface.*** The earth's atmosphere, with its water vapor, carbon dioxide, dust, smoke, and clouds, diminishes the sun's rays by absorption and scattering. In addition, the earth's rotation causes any spot on the earth to receive sunlight only part of the time. Because of the tilt of the earth's axis and the variation in the earth-sun distance, the amount of radiation received at any fixed location on earth varies with the seasons. Figure 11-1 shows the variation in the average daily solar insolation on a horizontal surface in the United States. The average for the 48 contiguous states is about 1500 Btu/ft$^2$/day. This value is useful for making estimates as to surface area needed for a solar-energy system of a specified size. For example, the average daily solar irradiation on 1 square mile

*Superior numbers refer to references, Par. **13.**

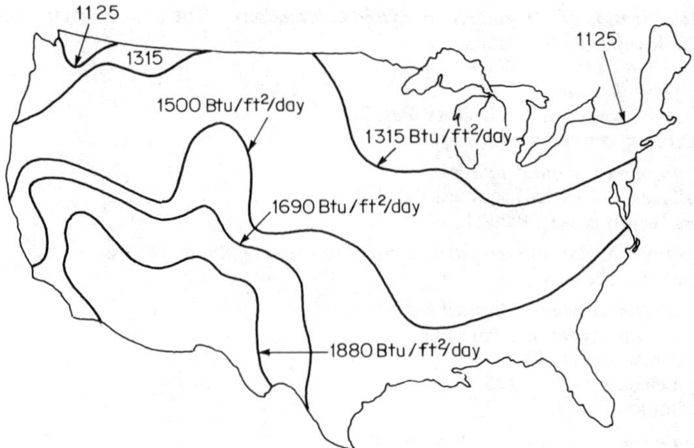


**FIG. 11-1** Variations in solar-energy rate (insolation) over the 48 contiguous states.

of the earth's surface would be

$$1500 \frac{\text{Btu}}{\text{ft}^2 \cdot \text{day}} (5280)^2 \text{ ft}^2 = 4.18 \times 10^{10} \text{ Btu/day}$$

The total yearly energy used in the United States is about $80 \times 10^{15}$ Btu/year, which is equivalent to the solar radiation on about 5200 square miles. Not all this radiation can be captured and utilized, of course, and the intermittent nature of solar radiation requires that the collected energy be stored or that there be some alternative standby system.

***5. Flat-Plate Collector.*** The simplest way to gather solar energy is to use a flat-plate collector to heat either a liquid or a gas. A simple flat-plate collector is shown in Fig. 11-2*a*. A blackened or specially treated metal surface called the absorber plate is the heart of the collector. The sun's rays striking the absorber plate are absorbed and raise the plate temperature. Fluid, in intimate contact with the plate, is heated as it passes through the collector. Heat losses to the surroundings are kept small by the use of insulation behind the absorber plate and by one or more transparent covers, usually glass or plastic. The covers

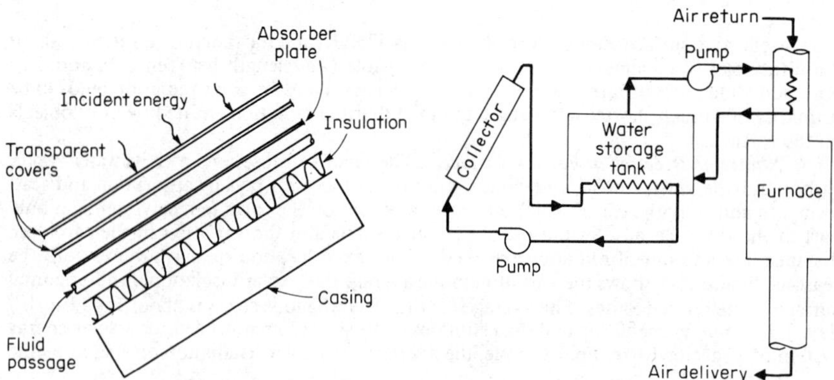


**FIG. 11-2a** Elements of the basic flat-plate collector.

**FIG. 11-2b** Solar-heating system with water storage.

permit the relatively short wavelength radiation from the sun to enter the collector, but absorb the relatively longer thermal radiation emitted by the absorber plate. This is sometimes referred to as the *greenhouse effect.*

The transparent covers also reduce the heat loss by convection to the surrounding atmosphere. The number of transparent covers is determined by the desired operating temperature of the collector, with two covers usually used for temperatures above about 82.2°C (180°F) and one cover for lesser temperatures. In some cases, such as with swimming-pool heaters where collector temperatures are quite low, the cover may be omitted.

***6. Collector Efficiency.*** The collector performance is evaluated in terms of *collector efficiency,* the ratio of the energy collected to that incident on the collector, usually expressed as a percent. At a fixed rate of solar insolation, the collector efficiency of a given collector decreases with temperature difference between the collector and the surrounding air. Thus there is a trade-off between temperature of collection and amount of energy collected. If high collection temperatures are desired, a larger amount of collector surface is needed than would be required to gather the same amount of energy at a lower collection temperature. Since the major cost of most solar-energy systems is in the collectors, it is important to keep both the unit cost of the collectors and the total amount of collector surface as small as possible.

Ordinarily a surface that is a good absorber is also a good emitter. Collector efficiency can be improved by the use of a *selective surface,* one which has a high absorptance for sunlight but a low radiation emittance. Selective surfaces are usually prepared by a plating on deposition process. At a fixed collector temperature, collector efficiency may be more than doubled over that of an ordinary surface when selective surfaces are utilized. Selective surfaces also permit the collection of energy at a higher than normal temperature without as large a decrease in collector efficiency as would occur for a nonselective absorber surface.

***7. Heating with Solar Energy.*** A typical solar-energy heating system using water as the collection and storage medium is shown in Fig. 11-2*b*. The main components of such a system are the collectors, the storage tank, a fan and coil control, and an auxiliary heat source. Pumps are also necessary to keep the fluid moving through the collectors and through the coil. The control system causes hot water to be circulated through the coil and turns on the fan when the house requires heat. If the temperature of the water in the storage tank is too low for heating, the control turns on the auxiliary system. The control should also cause water to circulate through the solar collector when sufficient sunlight is available to increase the storage temperature, and it should be capable of turning off the collection system should storage temperatures get too high. The heating of domestic hot water is a bonus that every solar-heating system should take advantage of.[2]

The storage tank provides energy during the night and for brief periods when clouds obscure the sun. Storage of energy sufficient to meet the requirements of several days without sunlight would be prohibitively expensive. An auxiliary heat source is usually necessary, owing to the extreme variability of sunlight.

An alternative solar-heating system employs a solar collector that heats air instead of water and uses a bed of rocks as the storage medium. In air systems there is no concern with freezing and leaks can usually be tolerated. The disadvantage can be that electricity consumption in the air blower can be much larger than the cost of pumping water through the collector.

***8. Solar Thermal-Conversion Plants.*** The thermal energy collected from a solar collector can be converted into work or mechanical energy by the use of a heat engine, which can then be used to generate electricity. The three types of thermodynamic cycles which seem to be practical for use with solar systems are the Rankine cycle, the Brayton cycle, and the Stirling cycle. The Rankine cycle is a vapor power cycle that is used with modifications in most large-sized electric generating plants. The Brayton cycle is a gas power cycle used as the basis of most gas-turbine power plants. The Stirling cycle is a high-efficiency cycle used as the basis for an external-combustion gas engine with relatively low pollution and noise characteristics. A schematic flow diagram for a solar power plant operating on the Rankine cycle is shown in Fig. 11-3.

The limitations imposed by the second law of thermodynamics apply to any solar thermal power cycle. This limitation is best expressed in terms of the Carnot principle, which says that it is impossible to construct an engine that operates between two given heat reservoirs and which has a higher thermal efficiency than a Carnot engine operating between

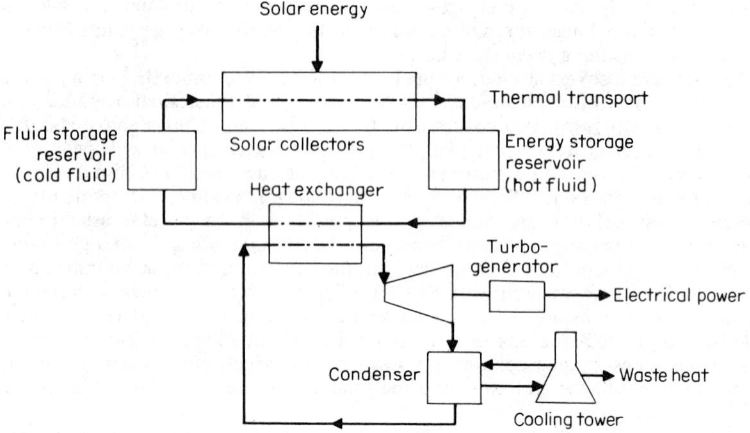


**FIG. 11-3** Schematic diagram of a solar power plant, operating on the Rankine cycle.

the same reservoirs. The Carnot engine is an idealized heat engine which is thermodynamically reversible and which receives heat from a high-temperature reservoir (the source) and rejects heat to a lower-temperature reservoir (the sink). The useful work done per cycle is the difference between the heat added and the heat rejected during the cycle.

The *thermal efficiency,* the ratio of useful work done to the heat supplied, is expressed for the Carnot cycle in terms of the temperature of the reservoirs with which it is exchanging heat.

$$\eta = 1 - \frac{T_L}{T_H}$$

where $\eta$ = thermal efficiency of the Carnot cycle
$T_L$ = absolute temperature (degrees Celsius + 273.15°) of sink
$T_H$ = absolute temperature of source

For a solar-energy system collecting heat at 121.1°C (250°F), the maximum thermal efficiency for any heat engine using that heat and rejecting heat to the atmosphere at 10°C (50°F) would be

$$\eta = 1 - \frac{273.15 + 10}{273.15 + 121.1} = 0.282 \tag{11-1}$$

In other words, even an ideal heat engine would convert only 28% of the solar energy collected if the collector exit temperature were 121.1°C (250°F). A real engine would convert considerably less.

***9. Concentrating Collectors.*** Flat-plate collectors have very poor collector efficiencies at temperatures above 93.3°C (200°F) and therefore are not suitable for use in solar thermal power plants. Collectors which concentrate the sun's rays can give higher collector efficiencies and much higher outlet temperatures than flat-plate collectors. The disadvantage lies in the fact that concentrating collectors rely primarily on direct solar radiation and thus must be rotated continually to track the sun. This tracking involves considerable expense in motors, mechanisms, and controls, and particular care must be taken to keep one collector from shading another. In addition, concentrating collector surfaces must be accurately shaped and highly polished and kept free of dust and other substances that might impair the reflection of the solar radiation.

The effect of concentration ratio on solar-collector efficiency and the combined effect of this efficiency and the Carnot principle on the overall plant efficiency[3] is shown in Fig. 11-4. This is for an assumed solar irradiation of 646 W/m$^2$ and a paraboloidal concentrator

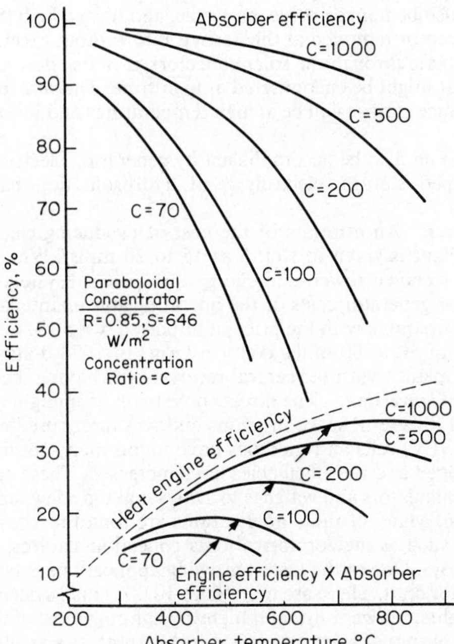


**FIG. 11-4** Absorber efficiency *(top)* and engine efficiency as functions of the concentration ratio *C*.

collector. The results presented are ideal; actual efficiencies would be much lower. The lower curve shows that for a given concentration ratio there is an optimum operating temperature for the absorber, above which the overall efficiency of the plant decreases with temperature.

***10. Central and Distributed Systems.*** Solar thermal power plants may be classified as *central receiver systems* or as *distributed systems.* In the central receiver system, solar energy is transferred optically from the individual collectors to a single receiver, for example, a boiler for a Rankine-cycle-type power plant. The most common approach for this type of plant is to locate the boiler at the top of a tall tower and to surround the tower with hundreds of mirrors which can reflect the sun's rays to the top of the tower. Systems have been proposed which would generate superheated steam at about 537.7°C (1000°F) to reach thermal efficiencies comparable with conventional fossil-fueled power plants. The turbine generator would be located on the ground near the base of the tower.

In a distributed system energy is transported from individual solar collectors by heated fluid flowing through pipes to a central boiler. The collectors are normally of the concentrating type such as parabolic troughs or paraboloidal dishes. Flat-plate collectors could be utilized, but the relatively low temperatures which they can produce lead to low thermal efficiencies. The resulting low overall plant efficiencies would require some rather large areas if the flat-plate collector were used. This is balanced to some degree by the lower cost per unit area of flat-plate collectors as compared with concentrating collectors.[4]

Because of the intermittent nature of the sun, a solar power plant should be operated as an energy-displacement system, in connection with a conventional power system, which is employed whenever sufficient solar energy is not available because of clouds or night. Another approach is to have fossil fuel available to the solar system to furnish the required heat as needed. A third approach is to store thermal energy collected by the solar system for use during the periods of inadequate solar insolation. In a solar power plant this would involve the storage of a liquid at high temperature. A suitable storage medium should be low in cost and have high heat capacity, high temperature capability, and a lower vapor pressure. It should also be noncorrosive, nontoxic, and have a high thermal conductivity so that heat can be stored or removed at the desired rate without excess heat-transfer surface. If the fluid is to circulate through the solar collectors, it is also desirable that it not freeze at the temperatures that might be encountered at nighttime. Thermal insulation of the storage tanks is necessary, since storage will be at high temperatures and for reasonably long periods of time.

Energy storage could also be accomplished by generating electricity and then using the electricity for pumped storage, electrolysis of hydrogen, the charging of batteries, or flywheels.

***11. Cost Estimates.*** An estimate of the cost of producing electricity from a central-solar-tower power plant is given in Ref. 5 as 15 to 20 mills/kWh. This is the cost of the heliostat (reflector), receiver, tower, and energy-transport subsystems but does not include the boiler, turbine, or generator costs or the operating and maintenance and parasitic load costs. The cost is comparable with the price of imported oil in 1974. Reference 6 gives total busbar energy costs of 30 to 60 mills/kWh (in terms of 1974 dollars) for several types of solar thermal power plants, with the central-receiver type having the lowest cost.

***12. Photovoltaic Conversion.*** The direct conversion of sunlight into electric power was primarily developed for use in space missions and was rarely used elsewhere until the late 1970s. Photovoltaic (PV) cells and modules have found many commercial applications as the prices have dropped and the efficiencies have increased. These applications range from a few small cells in calculators and watches to arrays making a few megawatts of peak power for utility generation. Many of these applications are found in sites remote from existing electric power grids such as meteorological data collection stations, radio repeater towers, and navigational buoys. One of the fastest-growing applications is battery charging for sailboats and small motor craft. There are more than 10,000 small systems in use in Africa and Asia for powering lights, refrigerators, and highway lighting; most of these installations have battery storage. The output of solar arrays is direct current; as a result, it must be converted and conditioned if it is intended to feed power synchronously into existing alternating-current networks.

In the mid-1980s, there were several commercially available and competing PV technologies: (1) single-crystal silicon, which uses a uniform chemical structure; (2) polycrystalline silicon, which uses a different manufacturing method to create a chemical structure that is a series of crystalline structures within one photovoltaic cell; (3) amorphous silicon, which uses another manufacturing method to give an almost random atomic chemical structure. In general, as the atomic structure becomes more random, less energy input and manufacturing complexity is required. However, more uniform structure means increased current collection and increased efficiency. In the mid-1980s, the typical solar cell consisted of one of these types of silicon treated with a thin layer of cadmium sulfide, gallium arsenide, indium phosphide, or similar compound. This produces a semiconductor junction between *n*-type and *p*-type materials. Sunlight impinging on the cell creates electron-hole pairs, the electrons being attracted to the positively charged *n*-type material and holes to the *p*-type (Fig. 11-5*a* and *b*). Research and development work in the mid-1980s focused on cost reduc-

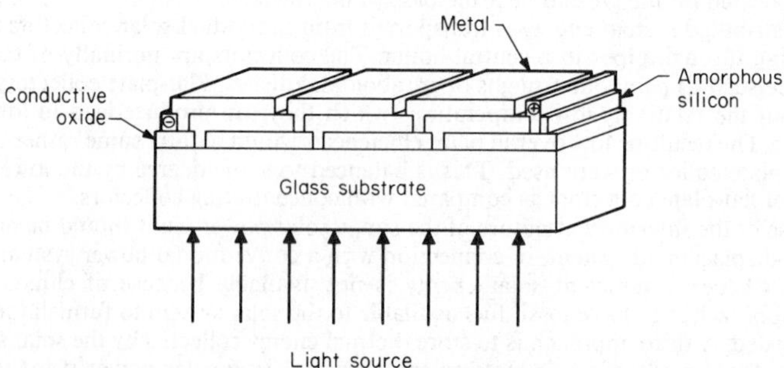


**FIG. 11-5a** Diagram of an amorphous-silicon module with three series-connected cells.

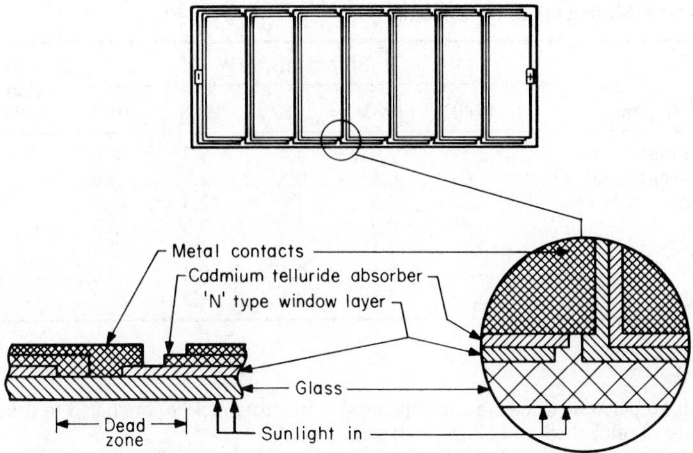


**FIG. 11-5b** Monolithic array pattern geometry of cadmium telluride system.

tion for silicon-based materials, development of other photovoltaic materials (such as gallium arsenide and cadmium sulfide), and the reduction of balance-of-system costs.

In the mid-1980s, the conversion efficiencies for the commercially available technologies ranged from 11.5% for single-crystal silicon to 7% for polycrystalline silicon to 5% for amorphous silicon. Laboratory efficiencies for amorphous silicon have already exceeded 11% in several locations; such systems should be in production before 1990. By that time, laboratory efficiencies are expected to be at 15%.

At present the performance of the crystalline and semicrystalline forms of silicon PV cells is greater than that of the amorphous silicon cells. The crystalline technology is the basis of the swiftly changing semiconductor industry, and PV benefits from this evolving knowledge. By the same token, the amorphous-silicon PV technologies benefit from other applications of amorphous silicon in copiers, laser printers, vidicons in television cameras, etc.

Although the predominant type of collector has been the flat-plate system, PV systems which concentrated the sun's rays on the PV cells with fresnel lenses or parabolic reflectors were also available commercially in the mid-1980s. These concentrator systems can increase the efficiency of the overall system and can use less-photosensitive material to achieve the same power output. Most of these systems are used with a tracking support system that effectively follows the sun's movement, keeping the photovoltaic array aligned for maximum power production. Some of the flat-plate systems also use tracking support systems to increase their overall production and extend the effective operating hours into the shoulder hours that are often coincident with utility peaks. See Table 11-1 for growth in shipments of PV systems.

According to data reported by the Federal Energy Administration in 1974, the cost of solar-cell arrays decreased sharply during the development of the space program, from about \$400 per watt in 1958 to about \$20 per watt in 1974. This report projected a cost of about \$1.50 per watt by 1985, provided that a substantial commitment was made to the development of cost-reduction methods. By 1985 these projections had not come true but were on the verge of doing so. Modules of photovoltaic cells could be purchased (in large quantities) at prices of slightly more than \$5 per watt in 1984. Balance-of-system costs for utility generation, which include wiring, support structure, and power conditioning costs, ranged from \$4 to \$8 per peak watt in the mid-1980s. These prices led to the installation of more than 5 MW of photovoltaic generation by 1985. By mid-1985, at least one manufacturer offered a 100-MW tracking concentrator system at installed costs significantly less than \$2 per watt (no quotes were available for a 1-MW system). New manufacturing facilities were installed in 1985 producing thin-film amorphous-silicon PV systems at less than \$2 per watt

**TABLE 11-1** Shipments of PV Systems

| Region | Shipments, MW | | | | | Percent change from 1983–1984 |
|---|---|---|---|---|---|---|
| | 1980 | 1981 | 1982 | 1983 | 1984 | |
| U.S. (flat plate) | 2.5 | 2.9 | 4.4 | 9.7 | 8.6 | |
| U.S. (concentrated) | 0 | 0.6 | 0.5 | 2.8 | 3.0 | |
| Total U.S. | 2.5 | 3.5 | 4.9 | 12.5 | 11.6 | −7 |
| Europe | 0.4 | 0.9 | 1.7 | 3.3 | 3.3 | 0 |
| Japan | 0.5 | 1.1 | 1.7 | 5.3 | 7.7 | 45 |
| Other | 0.1 | 0.1 | 0.1 | 0.4 | 0.6 | 33 |
| Total | 3.5 | 5.6 | 8.4 | 21.5 | 23.2 | |

(selling at $6 per watt to recoup startup costs), with costs in 1990 predicted to drop to 50¢ per watt to manufacture and $2 per watt sales price.

There are two main reasons why the thin-film technologies offer the promise of significantly reduced costs. First, the thin-film cells use only a few microns of direct material, instead of the tens of mils of silicon used by the crystalline, polycrystalline, or ribbon silicon modules. Cadmium telluride can absorb 99% of the sun's energy in less than 0.5 $\mu$m thickness, as opposed to the 8-mil requirement for crystalline silicon. The crystalline silicon requires 400 times as much material (if it is made from a ribbon) to 1000 times as much material (if it is sawed from a grown crystal). In addition, monolithic construction of thin-film modules can be done at the same time that the cells are formed, thus eliminating most of the cost of module fabrication from cells. In the conventional technologies, cells are cut into individual parts and then circuited back together like discrete silicon cells. Monolithic interconnection during cell fabrication eliminates labor and, as an added plus, produces a superior-looking product because of its uniform finish.

***13. References on Solar Power Sources***

1. Duffie, John A., and Beckman, William A.: *Solar Energy Thermal Processes;* New York, Wiley-Interscience, 1974.
2. Solar Energy Utilization for Heating and Cooling; *ASHRAE Applications Handbook,* 1974.
3. Oman, H., and Bishop, C. J.: A Look at Solar Power for Seattle; 8th Intersociety Energy Conversion Engineering Conference Proceedings, Philadelphia, Aug. 13–16, 1973.
4. Mahefkey, E. J., Jr.: The Solar Collector Thermal Power System, Its Potential and Development Status; 7th Intersociety Energy Conversion Engineering Conference Proceedings, San Diego, September 1972.
5. Easton, C. R., et al.: Evaluation of Central Solar Tower Power Plants; 9th Intersociety Energy Conversion Engineering Conference Proceedings, San Francisco, 1974.
6. Gervais, Robert L., and Bos, Piet B.: Solar Thermal Electric Power; *Astronaut. Aeronaut.,* November 1975.
7. Definition Report, National Solar Energy Research Development and Demonstration Program, ERDA 49, June 1975.
8. Pollard, W. G.: The Long-Range Prospects for Solar Energy; *American Scientist,* July/August 1976, Vol. 64, no. 4, p. 424.
9. Pope, C. S.: Dreams Come True; *Photovoltaics International,* June/July 1985.
10. Carlson, D. E.: Progress and Projections in Thin-Films; *Photovoltaics International,* June/July 1985.
11. Yerkes, J. W.: Considering Cost, *Photovoltaics International,* June/July 1985.
12. Watts, R., Smith, S. A., and Dirks, J. A.: *Photovoltaic Industry Progress through 1984,* Battelle Pacific Northwest Lab, 1985.
13. Buresch, M.: *Photovoltaic Energy System Design and Installation;* New York, McGraw-Hill Book Company, 1983.
14. Maycock, P. D., and Stirewalt, E. N.: *Photovoltaics—Sunlight to Electricity in One Step;* Andover, Mass., Brick House Publishing Co., 1981.

## GEOTHERMAL POWER

*By ROBERT W. REX*

***14. Introduction.*** The outer crust of the earth contains a very large reservoir of energy present as sensible heat. This resource is between one and two orders of magnitude larger than the recoverable energy from uranium and thorium in the same volume of rocks. This assumes the use of the 60 to 70% efficient breeder reactor technology which has yet to be developed. It also assumes appropriate inefficiencies for converting the thermal energy of the rocks of the earth's crust into electricity. The only natural fuel system that represents a larger energy resource on the earth is the fusion energy of deuterium.

***15. Origin of Geothermal Energy.*** Geothermal energy is present over the entire extent of the earth's surface. It varies only in its ease and cost of extraction. The need for developing the least-expensive resource first has resulted in the exploration of geothermal resources primarily in those areas which have associated volcanic activity. However, by no means is the resource restricted only to volcanic areas.

The primary processes of heat transfer from the earth's interior are

**a.** Direct heat conduction

**b.** Rapid injection of basaltic magma along rifts that penetrate deep into the mantle

**c.** Large plutons or bodies of magma which rise buoyantly toward the surface, have bubblelike geometry, often with volumes of thousands of cubic kilometers, and serve as feeders for volcanoes

A large magma body or pluton typically contains 100,000 to 300,000 MW-centuries of energy, allowing for the inefficiencies of the conversion from thermal to electrical energy. This is comparable to the amount of electricity that could be generated if one utilized all the oil in Kuwait to generate electricity.

The Greek word "batholith" describes large subsurface molten rock bodies. Batholiths typically originate about 70 to 100 km below the surface and rise buoyantly, undergoing fractional crystallization and chemical change by the assimilation of host rock and a loss of heavy minerals by gravity segregation. These large bodies typically rise close to the surface and occasionally break through to create large geothermal areas such as Yellowstone National Park, the Geysers dry steam field in California, the volcanoes that ring the Pacific Ocean, and the many other island arcs of the world's oceans.

In the United States this type of energy resource is restricted to the western third of the country. Rift geothermal areas appear to be associated with major sedimentary basins that undergo repeated injections of magma, often in small amounts but for such long periods of time that they accumulate massive amounts of hot water. Examples are the Imperial Valley of California and the Rift Valley of Africa.

The processes of natural heat conduction from the interior to the surface are strongly influenced by the conductivity and the permeability to water of earth materials. In sedimentary basins of recent origin, the impermeability of clay deposits inhibits escape of water. This trapped water acts as a heat reservoir to accumulate geothermal energy. In addition, natural gas appears to be highly abundant in these hot trapped waters of young sedimentary basins.

The weight of the overburden in these sedimentary basins compresses the trapped hot water, giving rise to what is called the geopressured geothermal resource. A well drilled into sands trapped within a geopressured system often encounters pressures that range from 3000 to 10,000 lb/in$^2$ above hydrostatic. These high pressures can serve to increase the productivity of hot-water wells. Experimental work is directed in the United States to recover this type of geothermal energy. The present focus for energy recovery from geopressured geothermal resources is in Texas and Louisiana. However, the resource is abundant in many other parts of the country, including Mississippi, Alabama, Oklahoma, California, Oregon, Washington, Alaska, and possibly parts of the Atlantic coast of the United States. Exploration for this resource is in its earliest stages and it will be more widely spread than is presently recognized.

***16. Utilization of Geothermal Energy.*** Geothermal energy has been recovered for electricity production since 1904 in Italy, and since that time there has been a steady expansion of this resource on a worldwide basis with over 4000 MW on line and 6000 MW expected to be on line by 1990. Currently over 2000 MW of electricity capacity is on line in California, and a doubling of this capacity is expected by 2000 with substantial additional expansion coming throughout the western United States.

The first resource to be developed in the United States was the dry steam field at the Geysers, about 90 miles north of San Francisco. The original technology employed here was a direct transfer of Italian and Japanese technology.

Commercial production of electricity from hot-water geothermal resources has been carried out for 35 years in New Zealand and for over 20 years in Japan. Electricity production utilizing hot-water geothermal resources is under way in California, Nevada, Hawaii, and Utah. Power production is currently under way from hot-water resource areas in Iceland, Mexico, El Salvador, the Philippines, the United States, Italy, New Zealand, Indonesia, Kenya, China, the Azores, Turkey, Nicaragua, Guadeloupe, and the Soviet Union, and experimental programs for further utilization of the resources are under development in many nations around the world.

Initial production of electricity from the hot-water resource is based on the natural flow of hot underground water into the wells where pressure release causes boiling. The steam and water mixture quickly rises in the well to the surface and passes into steam separators where the steam is extracted. This separated steam then is used to drive low-pressure steam turbines.

A new technology which has been employed for energy recovery in the United States, the Soviet Union, and in Japan is the use of heat exchangers and a secondary working fluid for recovery of thermal energy from hot water. Hydrocarbons such as propane or isobutane are commonly used as the secondary working fluid. Freon and ammonia are considered for some applications. A 40-MW heat-exchange power plant is in operation for use in the Heber area of the Imperial Valley, California. In addition, heat-exchange plants energized by low-pressure steam containing abundant noncondensable gases are under development. A 10-MW binary fluid power facility is in operation in the East Mesa geothermal field in California with additional capacity under construction.

The rapid development of the dry-steam and wet-steam or hot-water resource has been a consequence of favorable economics. Geothermal power plants are similar to conventional low-pressure steam power plants like those commonly utilized in the 1920s. However, modern geothermal plants utilize advanced metallurgy to resist the corrosive constituents sometimes found in geothermal fluids.

The principal advantage of a geothermal plant lies in lower capital cost, because no boiler or nuclear reactor is needed to generate steam. The net effect is that approximately 40 to 60% of the capital investment needed for most power plants is not needed in a geothermal plant. Geothermal energy appears to be producible at costs below those of competitive fossil fuels and uranium oxide, making an overall cost of electricity from the geothermal energy resource often less expensive than either coal, fuel oil, nuclear, or gas-produced electricity. Geothermal plants have proved most useful for base-load power plants and not for peaking service. This means that geothermal plants are primarily entering the market to fill a niche where modest-sized plants are needed, where low capital cost is important, where long-life fuel contracts are important, and where short construction periods are advantageous.

***17. Exploration for Geothermal Energy.*** Geothermal deposits were originally found by locating hot springs and geysers. Frequently these are associated with volcanoes. The majority of geothermal fields presently known have surface manifestations that led to the original drilling program that resulted in field discovery.

A new class of geothermal field has been found by geophysical exploration techniques. Examples include the East Mesa, Heber, and North, South, and East Brawley geothermal fields in the Imperial Valley of California, which were found entirely by the use of geophysical technology. These techniques include heat-flow studies, measurement of minute variations in the earth's gravitational and magnetic fields, and geochemical techniques which give indications of high-temperature fluids at depth. Among these techniques are the use of chemical thermometers that result from temperature-dependent chemical reactions between groundwater and the minerals of the subsurface.

Aerial infrared surveys have proved ineffective because their skin depth rarely exceeds more than a few millimeters. The propagation of solar energy into the ground ordinarily reaches a depth of 10 to 15 m. The seasonal fluctuation in solar energy introduces a periodic temperature signal in the upper 15 m of the earth's surface which almost always swamps out the steady-state component of natural heat flow from the earth's interior. Consequently, most geothermal exploration techniques need to have the ability to penetrate a kilometer or more below the surface. This restriction is probably the most important single factor limiting exploration technology.

The most widely used techniques for geothermal exploration include measurement of heat flow at depths below 100 m or more; electrical sounding techniques which measure the earth's resistivity to depths of several kilometers; telluric and magnetotelluric techniques that measure resistivity at substantial depths; gravity and magnetic techniques; and to a lesser extent seismic technology, including both active and passive methods. There is a substantial amount of research taking place directed toward the development of new exploration technology, but it has yet to play a major role in the discovery of new geothermal fields.

***18. Research and Development Programs.*** In addition to the dry- and wet-steam geothermal resource development around the world, there is a massive research and development effort directed toward developing new types of geothermal resources. These include the low-temperature water resource from 60 to 120°C which appears attractive for municipal heating, industrial, and agricultural purposes, and electricity production from high-temperature dry rocks. There are two types of hot-rock systems. One is molten lava or magma, which is a technologically difficult but energetically attractive target. The second hot dry-rock resource is the crystalline rock of the earth's interior with sufficient heat content to justify recovery.

Normal conduction of heat from the earth's interior and natural generation of heat from the decay of natural radioactive elements in crystalline rocks such as granite gives rise to temperatures in the range of 150 to 200°C within 20,000 ft of the surface in granite areas of the United States. The technology proposed for the recovery of this hot dry rock thermal energy is based on drilling deep wells into areas of hot granite and hydraulically fracturing the rock. Then cold water is circulated through these fractures, extracting sufficient heat via recovery wells so that electricity can be generated upon return to the surface.

The hot dry rock technology is being investigated by the Los Alamos Scientific Laboratory of the University of California at the Jemez volcano in New Mexico. The investigators in New Mexico have been successful in drilling wells to depths of 9000 ft and more into hot granite, hydraulically fracturing the granite, and establishing interconnection between two wells, generating steam, and producing small amounts of electricity. The maintenance of sustained flow between the two wells has been demonstrated successfully. The most important result of the work to date is that it is possible to drill into hot granite at costs that are no greater than those of drilling into sedimentary rocks of the midcontinent of the United States. Furthermore, it has been discovered that fracturing deep granite requires only about 10% more pressure than the total overburden load.

It appears that the basic concept is technically feasible and its development will depend on economically competitive technology.

***19. Volcanoes as Energy Resources.*** Current geothermal developments center on dry-steam and hot-water systems, but probably most of the enormous heat content of the earth's interior that is within reach of the drill is present in dry rock. As a basis for carrying out an order-of-magnitude appraisal of the geothermal resource potential, let us take as a case study the volcanic area surrounding the Valles Caldera in the Jemez Mountains in northern New Mexico. Working from the basis that the energy is only in the hot rock of the system, we have utilized currently available temperature gradient data, known lithology, and geophysical data to construct a simplified model. The measured temperature gradient to 700 m is 180°C/km. Projecting this gradient and using the known topography and geology to construct a temperature model, we assumed a small increase in conductivity with depth, a heat capacity of 0.20 cal/(°C)(g), an average rock density of 2.75 g/cm$^3$, and a lower working temperature of 100°C. The seasonal air temperature varies about 0 to 10°, making air cooling attractive for disposal of waste heat. Table 11-2 shows the results of this model for the Jemez Mountains case using the measured temperature gradient of 180°C/km.

**TABLE 11-2** Heat Characteristics of the Jemez Volcano Model Shallow $\Delta T$ = 180°C/km

| Depth, km | Volume available, $km^3$ | Mass, $10^{18}$/g | $T$, °C | Useful $\Delta T$, °C | Useful $\Delta H$, cal/g | Total available enthalpy, $10^{20}$/ g·cal |
|---|---|---|---|---|---|---|
| 0–1 | 1600 | 4.40 | 90 | 0 | 0 | 0 |
| 1–2 | 1200 | 3.30 | 290 | 190 | 38 | 1.25 |
| 2–3 | 1000 | 2.75 | 460 | 360 | 72 | 1.98 |
| 3–4 | 800 | 2.20 | 610 | 510 | 102 | 2.24 |
| 4–5 | 600 | 1.15 | 740 | 640 | 128 | 1.47 |
| 5–6 | 400 | 1.10 | 800 | 700 | 140 | 1.54 |
| Total | | | | | | 8.48 |

If we further assume in this model that the usable thermal energy in the rock system is completely recovered over a time period of 100 years, then the recoverable thermal energy in this volcano is 8.48 × $10^{20}$ g·cal. The conversion efficiency for producing electrical energy from geothermal energy is about 14%. Therefore, 1.19 × $10^{20}$ g·cal could be converted to electricity which would equal 158,000 MW-centuries (MW·cen) since

$$1 \text{ g}\cdot\text{cal} = 1.163 \times 10^{-9} \text{ MWh} \quad 1.19 \times 10^{20} \text{ g}\cdot\text{cal} \times 1.163 \times 10^{-9} \text{ MWh/g}\cdot\text{cal}$$

$$= 1.38 \times 10^{11} \text{ MWh} = 158{,}000 \text{ MW}\cdot\text{cen}$$

There appear to be at least 10 volcanic areas of this magnitude in Alaska, at least another 10 in the contiguous 48 states, and at least 5 in Hawaii. Consequently, volcanic energy alone constitutes a reserve of energy of about 4 × $10^6$ MW·cen, probably adequate to meet U.S. electrical energy needs for several centuries.

If the volcanic rock is in a fractured state and naturally water-filled, its water carries about 1.8 times the energy that the hot rock carries. Dry steam in the fractures is harder to appraise with respect to its energy content because of the uncertain density of the fluid. Although the net energy difference per unit volume between hot rock, hot water, and dry steam is not large enough to affect an order-of-magnitude resource size calculation, steam and water are significantly less costly than hot dry rock as sources of geothermal heat. The intense shattering of the Geysers geothermal field over a broad areal extent suggests that natural thermal stresses may play a major role in producing steam- or water-filled porosity in areas of major igneous intrusions. The widespread occurrence of deep-fracture porosity is becoming evident from deep drilling into crystalline rocks around the world. This suggests that wet-steam and not hot dry rock systems will become the major sources of future geothermal energy.

***20. Economics of Geothermal Energy.*** U.S. electricity economics are very complex and electricity prices are sensitive to alternative avoided costs. This means that the mix of hydropower, gas, oil, geothermal energy, and coal will be sensitive to fuel switching, availability of dump power, imported oil prices, and seasonal demand fluctuations. The decade of the 1980s is starting with substantial regional excess capacity. Consequently, new geothermal plants must have an economic competitive advantage in order to be built. In the western U.S., the Geysers field appears to be highly competitive with oil, coal, and gas. The wet-steam resource at Roosevelt Hot Springs in Utah also appears to be fully competitive with coal, oil, and gas and should see substantial development in the 1980s. The wet-steam resource of the Imperial Valley of California is currently being developed with three 10-MW pilot plants running in three different fields. Its long-term development will depend on cost trends for air-pollution-free fuel systems. Coal with pregasification and scrubbing will produce electricity at costs that should allow substantial geothermal development in the Imperial Valley.[1]

In summary, geothermal technology in the U.S. and worldwide has reached the point

[1]Based on early figures for the Southern California Edison Company Cool Water Plant.

where the industry is well established. Geothermal economics permits successful competition with fuel oil at expected long-term prices. Competition with coal depends on the degree of air pollution allowed to the two systems. High air-quality standards can easily be attained by geothermal plants at minimal incremental cost giving geothermal plants a competitive edge over clean coal-fueled plants. However, coal-fired power plants burning high sulfur coal and not controlling particulate emissions are less costly than most, but not all, wet-steam geothermal plants.

## WIND POWER

*By R. RAMAKUMAR AND MICHAEL F. McNITT-GRAY*

***21. Introduction.*** Wind energy, a manifestation of solar energy, is merely air in motion, set up and continually regenerated by a small fraction of the insolation reaching the outer atmosphere. It is estimated that, over the land area, nature is generating wind energy at the rate of $1.67 \times 10^{15}$ kWh annually. Obviously, only a small fraction of this can be harnessed for use in other forms. Energy available in winds over the globe is at least 10 times the figure given above.

In the early and mid-1980s, wind energy experienced an explosive growth spurt. Fueled by availability of sites with very attractive winds, escalating conventional power costs, attractive federal and state tax credits, and by the Public Utility Regulatory Policies Act (which required utilities to purchase power from private generating facilities at marginal costs), wind turbines went from research tools to commercial reality. By the end of 1984, over 8000 wind turbines, yielding 600 MW in nameplate capacity, were installed in the United States. Most of this activity took place in California. These installations were privately owned and were made in groups called wind farms, which consisted of 50 to 100 wind turbines. These wind farms showed load factors of 0.15 to 0.35 (on an annual basis) and some had operating availabilities of 95% or more.

While the majority of these machines were in the 25 to 65 kW range, later-model wind turbines ranged from 75 to 150 kW range. In addition, several federally funded demonstration wind turbines were constructed which ranged from 250 to 4000 kW. Trends in the wind turbine industry in the mid-1980s were toward wind turbines in the 200 to 2000 kW range.

The power density in moving air is given by

$$P_w = KV^3 \qquad \text{watts per unit area}$$

where $V$ is the wind speed and $K$ is a constant which depends on the units used. Values of $K$ for different combinations of units used to measure area and wind speed are given in Table 11-3.

Theoretically, only the fraction $^{16}\!/_{27} = 0.5926$ (called Glauert's limit or Betz coefficient) of the above power is recoverable by horizontal-axis machines. In practice, it is reasonable to recover 70% (called aerodynamic efficiency) of this theoretical maximum. Therefore, the overall conversion efficiency, from the total energy available in wind to the mechanical energy available in the rotating shaft, is limited to about 40%. Practical values range from 20 to 40%, depending on the aerodynamics of the aeroturbine.

**TABLE 11-3** Value of $K$ in Different Units

| | Wind-speed units | | | | |
|---|---|---|---|---|---|
| | Miles per hour | Kilometers per hour | Feet per second | Meters per second | Knots |
| Area units: | | | | | |
| Square feet | $5.3 \times 10^{-3}$ | $1.272 \times 10^{-3}$ | $1.68 \times 10^{-3}$ | $5.934 \times 10^{-2}$ | $8.08 \times 10^{-3}$ |
| Square meters | $5.7 \times 10^{-2}$ | $1.3687 \times 10^{-2}$ | $1.807 \times 10^{-2}$ | 0.6386 | $8.70 \times 10^{-2}$ |

The estimated total energy available in the winds over the land area of the United States is $1.15 \times 10^{14}$ kWh/year. The NSF/NASA Solar Energy Panel's study in 1972 reported that, by locating wind-energy systems in strategic areas (both inland and offshore), it is possible to generate $1.536 \times 10^{12}$ kWh of electrical energy in the United States annually by the year 2000. To generate the same amount of electrical energy by conventional means from crude oil with an overall efficiency of 40% will require the consumption of 6 million barrels of oil per day.

Wind energy does not represent a panacea but can play an important role in the global energy strategy to fulfill the needs of the future.

***22. Wind Characteristics and Wind Measurement.*** Wind is highly variable and site-specific. Since it varies with seasonal, synoptic, diurnal, and short-term scales, success of a wind-energy system requires careful evaluation of wind characteristics and selection of a proper site. This involves extended measurement and evaluation of wind data. Weather bureau and airport wind statistics normally available are imprecise for characterization of the wind-energy resource in space and time.

Wind speed increases roughly with the one-seventh power of height. At low speeds, there is usually a larger increase, but at the useful high wind speeds, this one-seventh power law is a good engineering approximation.

Ideally, an integrating instrument whose driving torque is proportional to the cube of wind speed would give an accurate indication of the annual energy output of a wind-energy system during the study period. This procedure is not always practical, and the designer usually has to make judgments from such data as hourly mean values of wind speeds, speed-duration curves, and speed-frequency data obtained from long-period measurements. Vertical and horizontal distribution of wind speed, required for the mechanical aspects of the design, is obtained at selected sites from medium-period measurements. Studies on the dynamic performance (both electrical and mechanical) of wind-energy systems require very short period wind measurements.

A factor that is commonly considered in the study of wind characteristics is the "energy-pattern factor." It is the ratio of actual energy in the varying wind to the energy calculated from the cube of the mean wind speed. This factor is always greater than unity, indicating that estimates of energy based on mean (hourly) wind speeds are pessimistic.

***23. The Weibull Distribution.*** Considering the wind speed $v$ at a site as a random variable, it can be described by a density function $f(v)$. The Weibull distribution given by the density function

$$f(v) = \beta \frac{v^{\beta-1}}{\alpha^{\beta}} \exp\left[-\left(\frac{v}{\alpha}\right)^{\beta}\right] \tag{11-2}$$

in which $\alpha$ and $\beta$ are two parameters, has found considerable acceptance for modeling wind, primarily because of the possibility of adjusting $\alpha$ and $\beta$ to fit available data. Even at one particular site, different models may be needed to model the wind during different seasons. The parameters $\alpha$ and $\beta$ can be calculated from the sample mean $m_v$ and variance $\sigma_v^2$ using the following relationships:

$$m_v = \alpha\Gamma\left(1 + \frac{1}{\beta}\right) \tag{11-3}$$

$$\left(\frac{\sigma_v}{m_v}\right)^2 = \frac{\Gamma(1 + 2/\beta)}{\Gamma^2(1 + 1/\beta)} - 1 \tag{11-4}$$

If the sample mean is the only information available, then a compromise value between 3 and 4 is chosen for $\beta$, which is then used along with the known $m_v$ to compute $\alpha$.

***24. Utilization Aspects.*** Wind-energy-utilization techniques fall under three broad categories:

**a.** Isolated continuous-duty systems, used in conjunction with suitable energy-storage and reconversion systems

**b.** Fuel-supplement systems, used in conjunction with conventional utility systems or isolated conventional generating units

c. Small rural systems for applications which are not critical and can use the energy in the wind when available

The small farm windmills and battery-storage systems that dotted the midwestern countryside in the United States during the thirties and forties belong to category *a*. Privately owned wind turbines that generate electricity and sell it back to electric utilities under PURPA (Public Utilities Regulatory Policies Act) as well as utility-owned wind generation belong to category *b;* large-scale entry of wind energy is expected to be in a fuel-sparing mode (category *b*), and systems that belong to category *c* are envisaged for use in developing countries.

Because of the ease with which wind energy can be converted to rotary mechanical energy, generation of electrical energy and pumping of water are the two principal applications considered for wind-energy utilization.

***25. Aeroturbine Types and Characteristics.*** Horizontal-axis aeroturbines (or rotors) proposed for harnessing wind power range from bamboo poles and cloth stuck to bullock-cart wheels, to sophisticated space-age design blades and hub arrangements. All have common features. By far the single most important parameter to be considered is the "tip-speed ratio" (or specific speed), that is, the ratio of its peripheral speed to wind speed. This ratio ranges from 2 to 10 for practical turbines. In general, smaller ratios (less than 4) require several blades (more than 3) and result in high starting torques and low rotational speeds. Higher ratios (4 through 10) require fewer blades and have lower starting torque and high rotational speed. High-tip-speed-ratio rotors inherently are less efficient because of increased frictional losses.

The *power coefficient* $C_p$, defined as the fraction of the power in the wind converted to mechanical shaft power, is a function of the tip-speed ratio and has a maximum value dependent upon the pitch angle of the blades. There appears to be no single aerodynamic solution to the selection of the aeroturbine, and the final choice is usually strongly influenced by economics rather than aerodynamics.

***26. Blade Arrangements.*** Single-blade rotors require a dead-weight counterbalance and do not appear to be attractive for large installations. Most of the aeroturbines contemplated for large-scale harnessing of wind power have two or three blades. The two-blade configuration is more cost-effective than the three-blade arrangement but suffers from vibration during orientation, which disappears with three or more blades. This is not considered to be a serious defect. It can be overcome using the advances in material technology and aerodynamics. The argument between proponents of two-blade and three-blade designs has not yet been completely settled.

For small high-speed machines, wood, with its excellent strength-weight ratio, is an adequate material and is still used to make blades. Larger machines require metal blades, fabricated using the technology developed by the aircraft industry. Glass-reinforced plastic has also been successfully used in blade building.

***27. Vertical-Axis Aeroturbines.*** Vertical-axis aeroturbines possess symmetry in the sense of being able to operate with wind from any direction. They also have the added advantage of delivering mechanical power at ground level, less weight aloft, simpler construction, and economical operation. At the present state of technology, efficiencies achieved with vertical-axis rotors have matched those of horizontal-axis machines.

Some of the vertical-axis machines that have been proposed are: (*a*) Savonius (or S) rotor, patented in 1929; (*b*) Darrieus rotor, patented in 1931; (*c*) several modifications and combinations of (*a*) and (*b*); (*d*) giromill (or cyclogiro rotor); and (*e*) tornado-type wind-energy system.

The *Savonius rotor* consists of an S-shaped metal airfoil, supported between two circular end plates. Wind impinging on the concave side is circulated through the center of the rotor to the back of the convex side, thus decreasing a high-negative-pressure region which would otherwise result. Tip-speed ratios typically range from 1 to 2, and efficiencies are around 15%. These rotors are self-starting and perform somewhat like a two-stage turbine. Ratio of height to diameter (aspect ratio) is usually less than 3.

The *Darrieus rotor* has two or more curved airfoil blades, held together at top and bottom and positioned such that it can accept wind from any direction. Physically, it resembles the lower section of an eggbeater. These rotors are not self-starting, operate at tip-speed ratios of 6 to 8, and have efficiencies around 35 to 40%. Several companies now manufacture vertical-axis wind turbines (VAWTs) and they are commercially available.

A *giromill* consists of a set of vertical blades attached to the axis by means of support arms at the top, bottom, and middle (if necessary). As the rotor rotates these arms in a circular path, the orientation of the blades is changed to achieve maximum force from the wind. The blades must be flipped from a positive to a negative orientation twice each revolution, at diametrically opposite points. The rotor easily adapts to a change in wind direction and can convert over 60% of the kinetic energy in a wind-stream tube having a cross-sectional area equal to rotor diameter times the span (height). This high efficiency is the result of the giromill affecting not only this wind-stream area, but also an additional area due to lift action of the blades.

In the *tornado* system, wind energy is collected by a stationary tower and a vortex (or tornado) is formed in the center of the tower by properly directing the wind by opening vanes in the windward side and closing in the back side. This vortex creates a low-pressure core directly above a horizontal turbine located at the throat of an inlet that is open at the bottom with a bellmouth shape. With a large vortex strength, a significant pressure difference can be maintained across the turbine. This large pressure difference results in high air velocities and high power densities.

***28. Wind to Electrical Energy Conversion.*** Three basic factors must be considered in selecting the proper system for generation of electrical energy from wind. They are

**a.** *Type of output:* dc, variable-frequency ac, or constant-frequency ac

**b.** *Aeroturbine rotational speed:* constant speed with variable-pitch blades, nearly constant speed with simpler pitch-changing mechanisms, or variable speed with fixed-pitch blades

**c.** *Utilization of the electrical-energy output:* battery storage, other forms of storage, or interconnection with conventional utility grid

Initial large-scale generation of electrical energy from wind is expected to be in constant-frequency ac form, to be fed synchronously into an existing utility grid. This does not involve energy storage (except in the form of fuel saved), and present economics heavily favors this approach. For this application, the choice of electrical subsystem boils down to either a constant-speed constant-frequency (CSCF) system or a variable-speed constant-frequency (VSCF) system. The historic Smith-Putnam 1.25-MW unit (built on Grandpa's Knob near Rutland, Vt., from 1941 to 1945) and the U.S. DOE's MOD-0A, MOD-1, and MOD-2 units employed constant-speed turbines and conventional synchronous machines.

Wind turbines with an electrical rating of greater than 100 kW typically use constant-speed turbines and conventional synchronous generators. On the other hand, most wind turbines rated less than 100 kW use fairly constant speed turbines with induction machines generating constant-frequency power.

Recent advances in solid-state power-switching technology and the availability of economical high-power devices such as diodes and thyristors have initiated an interest in the possibility of allowing the aeroturbine speed to vary optimally with wind and employing variable-speed constant-frequency (VSCF) generating systems to obtain electrical energy at the required fixed frequency to be pumped into existing utility mains.

The advantages of this approach are (*a*) simpler and more economical mechanical arrangement for the aeroturbine because of the lack of complex pitch-changing mechanisms, (*b*) operation of the aeroturbine always at its maximum efficiency point (constant tip-speed ratio), (*c*) the possibility of extracting a larger fraction of the energy available in the wind, and (*d*) significant reductions in the aerodynamic stresses associated with constant-speed operation.

These advantages must be weighted against the added cost of the electrical subsystem. The choice between these two approaches is not completely clear at present and should await the results of the research programs underway. It is believed that both approaches will be employed in wind-energy systems of the future, depending on the wind regime and size of the unit. The latest megawatt-scale units (MOD-5) of the U.S. DOE are being designed for variable-speed operation.

Several mechanical and electrical methods are available to obtain constant-frequency output from nonsynchronous prime movers (such as variable-speed aeroturbines). However, only three approaches are being considered at present. They are (*a*) field-modulated generator systems being developed at Oklahoma State University, (*b*) asynchronous ac-dc-ac systems, and (*c*) rotary devices such as ac commutator generators, cycloconverter-type frequency changers, and doubly fed machines.

***29. Operational and Control Aspects of Wind-Electric Systems.*** Sensors and servomechanisms are required to provide directional orientation of the rotor (yaw control) into the wind in the case of horizontal-axis machines. In addition, output power of the unit must be properly controlled to match wind variations and aeroturbine characteristics.

Typical wind-speed-duration and power-duration curves are shown in Fig. 11-6. The wind power plant (Fig. 11-6*a*) starts delivering power at a wind speed called the cut-in speed $V_C$, and the plant is shut down for wind speeds above a safe maximum speed called the furling speed $V_F$. In between, the output power is determined by the power coefficient $C_p$ of the aeroturbine and the efficiencies of the generator $\eta_g$ and of the mechanical-drive system $\eta_m$.

Conventional synchronous machines driven by aeroturbines and operating in parallel with power grids will lock into synchronism and maintain a constant rotational speed irrespective of the wind speed. Suitable controllers must be used to sense the mode (generating or motoring) of operation and duration in that mode to make the necessary pitch control and other changes needed for proper operation.

Once the generator output reaches its rated value, it is maintained constant at that value for further increases in wind speeds. This process "spills" valuable energy in the wind at high wind speeds as shown in Fig. 11-6*b*. Induction generators also maintain the aeroturbine speed at very nearly a constant value. Though they are simpler than synchronous generators, they draw their excitation voltamperes from the grid and require power-factor-correcting variable capacitors or some kind of variable-frequency rotor-feeding scheme. These two belong to the category of constant-speed constant-frequency (CSCF) systems discussed earlier.

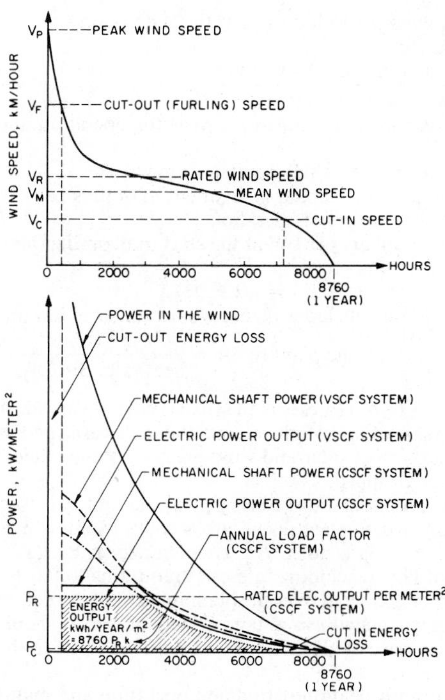


**FIG. 11-6** Typical wind-speed-duration (*a*) and power-duration (*b*) curves of wind-driven generator.

With variable-speed constant-frequency (VSCF) systems, proper output power control can be installed to maintain a constant tip-speed ratio to maximize $C_p$, and this will result in slightly higher power outputs throughout the operating range as compared with CSCF systems. In addition, certain unique features of VSCF systems (for example, the nonmotoring nature of field-modulated generator systems and of systems employing alternators with rectified outputs) may allow the extraction of part of the energy spilled by CSCF systems (as shown in Fig. 11-6*b*). VSCF systems for wind-energy utilization are still in the research and development stage and therefore are yet to be fully proved in the practical sense.

***30. Economic Aspects.*** Because of its dilute nature and unpredictable irregularity, wind-energy utilization requires cost-intensive hardware for collection and conversion to usable forms. Until economical energy-storage and reconversion systems are developed, large-scale use of wind energy will be limited to systems employed as supplementary energy sources in conjunction with existing conventional power systems. The cost of energy obtained from such "energy displacers" should then be compared with only the fuel cost incurred in conventional power plants.

Neglecting taxes and tax-related incentives and credits, the cost of generating electrical energy can be expressed as

$$c = \frac{r(1+r)^n}{(1+r)^n - 1}\frac{P}{8.76k} + \frac{mP}{8.76k} + 3.413\frac{f}{\eta} \tag{11-5}$$

in which

$$P = \frac{(1+r_1)^{(t+1)} - 1}{r_1}\frac{(1+e)^t}{t+1}P_1 \tag{11-6}$$

where $c$ = generation cost, mills/kWh

$e$ = average monthly inflation (escalation) rate per unit (p.u.) during planning and construction periods

$f$ = fuel cost, dollars per million Btu at the generation site

$k$ = annual load factor

$= \dfrac{\text{kWh generated per year by the unit}}{8760 \times \text{unit capacity}}$

$m$ = fraction of capital cost needed per year for operation and maintenance of the unit

$n$ = number of years over which capital is amortized

$P$ = adjusted value of the capital investment in dollars per kW of installed capacity at the year of commercial operation

$P_1$ = capital cost in dollars per kW at the start of planning (base-year cost)

$r$ = annual interest rate, p.u.

$r_1$ = monthly interest rate (p.u.) = $r/12$

$t$ = lead time, months; includes planning and construction time

$\eta$ = overall efficiency of the plant (p.u.) = $\dfrac{3413}{\text{heat rate, Btu/kWh}}$

In Eq. (11-5), the first term represents the (fixed) cost of capital, the second term represents the generation's share of operation and maintenance costs, and the third term accounts for the fuel costs. In the case of solar and wind-energy systems, there are no fuel costs, and the third term should be dropped.

The effect of inflation (escalation) during planning and construction periods and the interest costs during construction are taken into account by adjusting the capital-cost figure from its base-year value. This factor, $P/P_1$, is contained in Eq. (11-6) and is obtained by assuming the total cash flow (excluding interest) amounting to $[(1 + e)^t P_1]$ to be in $(t + 1)$ equal monthly installments, spread over the planning and construction periods. The interest during this lead time is calculated using standard techniques. It should be remembered that $c$ differs from consumer price of electricity by an amount equal to the cost of energy transmission and distribution.

Assuming a 20-year amortization period and operation and maintenance cost to be 5% of capital per year, generation costs for wind-energy systems are plotted in Fig. 11-7 for different capital costs, interest rates, and load factors. By comparing the cost of wind-gen-

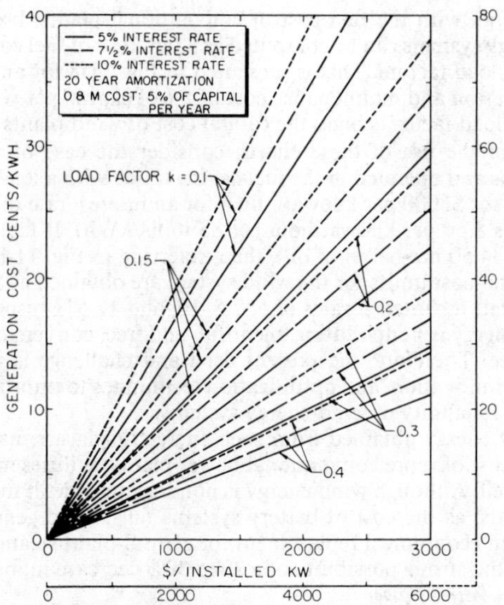


**FIG. 11-7** Generation costs of wind-energy systems.

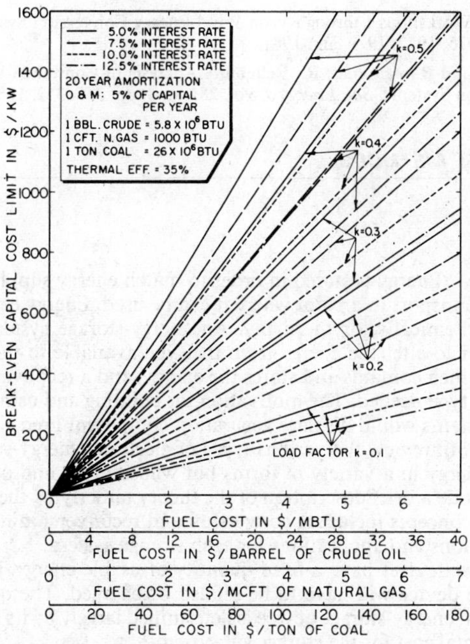


**FIG. 11-8** Break-even capital-cost limits for solar and wind-energy systems.

erated electrical energy with the fuel cost of conventional plants, break-even capital-cost limits for wind-energy systems can be computed as a function of fuel cost for different values of interest rates and load factors. This is presented in Fig. 11-8 for an amortization period of 20 years and operation and maintenance cost of 5% of capital per year. From these plots, it is clear that if the load factor is high, the capital cost of wind plants can also be high.

As an example of the use of these charts, consider the case of wind-energy systems located in wind farms and operated in the fuel-saving mode with a load factor of 0.4. If such systems can be built for $2000 per kilowatt, then for an interest rate of 7.5%, the generation cost (see Fig. 11-7) is 8.5¢ per kilowatthour (or 85 mills/kWh). If fuel costs $6 per million Btu (equivalent to $34.50 per barrel of oil), then, referring to Fig. 11-8, for a load factor of 0.4, break-even capital-cost limits for the wind system are obtained as $1572, $1386, $1230, and $1092 per kilowatt for interest rates of 5, 7.5, 10, and 12.5%, respectively.

Although wind energy is nonpolluting, plentiful, and free, conversion to useful forms for utilization is not free. Therefore, the present need and challenge is in the application of modern technology, innovation, and optimization techniques to improve the economic viability and long-term reliability of wind-energy systems.

While the cost of energy obtained from such energy displacers may primarily be compared with the fuel cost of more conventional power plants, utilities may realize some savings in capacity as well. Although wind energy is nondispatchable, it may nonetheless occur at utility peak periods; as the cost of battery systems falls, wind generation may become more important as an economical replacement for central-plant steam generation. Various projections indicate the strong possibility of a 20 to 30% decrease in installed costs by 1990.

***31. References on Wind Power***

1. Putnam, P. C.: *Power from the Wind,* Van Nostrand, New York, 1948.
2. Golding, E. W.: *The Generation of Electricity by Wind Power,* Philosophical Library, New York, 1955.
3. Ramakumar, R., Allison, H. J., and Hughes, W. L.: Prospects for Tapping Solar Energy on a Large Scale, *Sol. Energy J,* vol. 16, no. 2, pp. 107–115, October 1974.
4. Ramakumar, R.: Harnessing Wind Power in Developing Countries, Record of the Tenth Intersociety Energy Conversion Engineering Conference, IEEE Catalog 75 CHO 983-7 TAB, pp. 966–973, August 1975.
5. Proceedings of the Workshops I through V on Wind Energy Conversion Systems, sponsored by the U.S. DOE, 1973, 1975, 1977, 1979, and 1981.
6. Deshmukh, R. G., and Ramakumar, R.: Reliability Analysis of Combined Wind-Electric and Conventional Generation Systems, *Sol. Energy J,* vol. 28, no. 4, pp. 345–352, 1982.

## *ENERGY-STORAGE METHODS*

*By JAMES R. BIRK*

***32. Introduction.*** Energy is stored in order to match energy supply with demand and/or to contain it for transport to a point where it can be used. Energy can be stored mechanically, electrically, chemically, or thermally. An energy-storage system is generally comprised of a converter to alter the energy from the type available to the type best stored, a storage subsystem which contains and stores the energy, and a reconverter to transform the stored energy to the type needed. The motivation for building and using storage is economics, since storage systems would displace generation equipment fired with premium fuels.

This subsection will review the various types of electrical energy-storage systems—systems which store energy in a variety of forms but whose input and output are electricity. The review will include a brief description of the theory underlying the operation, a discussion of engineering concepts including conversion and reconversion methods, and an outline of the specifications and applications for each storage system.

All energy-storage devices have a fixed quantity of usable energy. Once the energy has been consumed, the device will have to be refilled or charged. The quantity of energy or hours of storage of an energy-storage device is determined largely by its intended application and by the incremental cost for the storage subsystem.

***33. Lead-Acid Battery.*** One of the oldest and best-known devices used to store energy is the lead-acid battery. The technology is based upon the reduction of lead dioxide to lead sulfate at the positive electrode and the simultaneous oxidation of lead to lead sulfate at the negative electrode. The electrolyte, sulfuric acid, is consumed, and energy is discharged during this process. Energy is stored by reversing these reactions; that is, charging the battery. The energy stored is proportional to the voltage (2.08 V per cell) and to the amount of lead. Lead has a high molecular weight, is an inherently inefficient chemical for battery energy storage, and is used for carrying current in the cell. As a result, the amount of lead required per kilowatthour of storage is 50 to 100 lb which unfavorably impacts both weight and cost of the storage system. A serious drawback of this battery is, therefore, its low energy density (that is, energy stored per unit weight of battery) which is about 10 to 25 Wh/lb depending upon battery design, operation, and desired life. Round-trip energy efficiency is 65 to 80%, depending also upon design and operation. Under suitable operating conditions, a well-designed and manufactured lead-acid battery can achieve 2000 cycles and last 10 years. However, life is substantially sacrificed to achieve higher energy densities required for certain applications, such as electric vehicles.

A lead-acid battery system is comprised of individual cells that have a capacity often ranging from a few tenths of a kilowatthour to a few kilowatthours. A large submarine cell can have a capacity of 10 kWh and weigh a thousand pounds. Lead-acid cells are often arranged in series strings to achieve the desired dc voltage for conversion or for the application intended. The number of strings will depend upon the energy requirements. Energy conversion is often achieved by solid-state converter/rectifier systems. Lead-acid batteries are used in submarines, forklift trucks, uninterruptible power supplies, electric vehicles, and short-term emergency power systems for several applications including telephones, computers, and nuclear power stations. Lead-acid batteries are also being considered for electric utility application. In fact, West Berlin's utility (BEWAG) is planning to install a 8.6 MW/9.3 MWh lead-acid battery system in 1987 to meet a part of their system regulation (maintaining frequency stability) and spinning reserve needs (supplying short-term emergency power).

***34. Nickel-Cadmium and Other Commercial Batteries.*** The nickel-cadmium battery is the only other commonly employed rechargeable battery besides lead-acid. The chemical reaction involves reduction of nickel oxide to nickel hydroxide and the oxidation of cadmium to cadmium hydroxide in an alkaline electrolyte (20 to 35% potassium hydroxide). The voltage of this couple (1.3 V) is less than that of lead-acid and the energy density (about 20 Wh/lb) is slightly better. Due to the high costs for nickel and cadmium, this battery is expensive and is only used when extremely long life or lighter weight is required. Major applications are satellites, portable equipment and tools, and various military applications. Other secondary (rechargeable) batteries considered for these and other specialized applications are silver-zinc, nickel-iron, and nickel-hydrogen. These systems are all expensive but they are durable and have very long lives.

***35. Advanced Batteries.*** Over the past two decades, a number of advanced batteries have been investigated. The aim of the research programs has been to lower costs while maintaining the desirable specifications (durability and performance) of the lead-acid and nickel-cadmium batteries. A secondary objective is to achieve improved energy densities. The advanced battery systems closest to commercialization are zinc-chloride, zinc-bromide, and beta (sodium-sulfur) batteries. These are compared in Table 11-4. In contrast to the lead-acid and nickel-cadmium systems, these advanced batteries use low-cost readily-available active materials and enjoy simple electrochemical reactions which should lead to excellent durability.

**TABLE 11-4** Advanced Battery Systems

| System | Positive electrode | Negative electrode | Cell voltage | Temperature, °C | Achievable energy density, Wh/lb |
|---|---|---|---|---|---|
| Na–S | S | Na | 2.1 | 300–350 | 45–55 |
| Zn–$Cl_2$ | $Cl_2$ | Zn | 2.1 | 20–50 | 30–40 |
| Zn–$Br_2$ | $Br_2$ | Zn | 1.8–1.9 | 20–50 | 20–30 |

These advanced batteries either operate at higher temperature (Na–S) or employ flowing electrolytes (zinc-halogen). The additional subsystems for such batteries to maintain flow or temperature add to overall system complexity which, in turn, impact their reliability. However, these subsystems will normally be built within the battery modules at the factory and the battery system will have the same modular character of the lead-acid battery. As a result of these additional subsystems, optimum economics will not occur until module sizes are in the 100- to 400-kWh range, a factor of 10 larger than the largest lead-acid battery modules. This size of module will have application only where there are fairly sizable energy-storage requirements, such as electric utility and commercial or industrial applications which require a capacity of a few hundred kilowatts to several megawatts. While there is hope that these advanced batteries can be used for electric-vehicle application, their economic optimum size and complexity suggest that the economic goals for this application will be extremely difficult to achieve. An electric-vehicle battery has to be inherently simple from both the design and engineering standpoint to meet owner expectations.

The advanced batteries are not yet commercially available at affordable costs. Engineering prototypes of the zinc-chloride system at a size of 500 kWh began independent testing in early 1984. However, similar sized prototypes of the other two advanced batteries are not scheduled for such testing until 1991–1992. Limited commercial availability of these systems can occur within about a year of successful independent testing; however, acceptable costs may not occur for yet an additional 3–5 years, and only then with a reasonable market.

***36. Pumped Hydro.*** Pumped hydro or pumped storage, as it is often called, has been used in electric utility systems for over 50 years. Electricity is used to operate a motor-pump combination to pump water to an elevated reservoir. When energy is required, water is allowed to flow down to a lower reservoir through a turbine-generator combination, much like the turbine-generators in conventional hydroelectric plants. The energy stored is proportional to the head (height differential between the upper and lower reservoir) times the stored volume of water. For a head of 1200 ft, 36 $ft^3$ of water is required for generation of 1 kWh. For a head of 120 ft, 360 $ft^3$ of water is required to generate 1 kWh.

In areas where topographic or ecological considerations rule out pumped-hydro plants, utilities now may choose to go underground. A modification of pumped hydro, underground pumped hydro (UPH), would have its lower reservoir located about a mile below the earth's surface in a competent hard-rock cavern. A UPH plant can, therefore, be sited in flat terrain and should have only a minimal effect on the surrounding environment. The most significant drawback of UPH is that the plants must be built in huge sizes (2000 MW/20,000 MWh) to be economically attractive.

Both pumped hydro and UPH, like conventional hydro, have very fast response characteristics (emergency full-power capability is 10s) and very high efficiencies (72 to 75%). These factors, along with low capital costs, have led to the construction and operation of 18,000 MW of pumped hydro in the United States. Another 16,000 MW is planned or under construction. Pumped hydro will typically be the storage alternative of choice for electric utilities having favorable topography.

***37. Compressed Air Energy Storage (CAES).*** CAES involves the compression of air into an underground cavern or reservoir (excavated rock, solution-mined salt, or aquifer) where it is stored. Favorable geologies exist in about 75% of the United States. Geologic considerations generally dictate that most of the heat of compression be removed (and used or stored, if possible) before the air is stored underground. During generation the air is expanded through a turbine to drive a generator. Economics dictates that heat be added to the air from a thermal store or through combustion of a fuel prior to the expansion-generation phase.

CAES technology has recently become commercialized for electric utility application. The commercial units incorporate modified combustion turbines wherein the compressor and expander are physically separated by clutches and a motor-generator. For the expansion-generation phase, the technology uses oil or gas, although other fuels can be used. When a heat exchanger is incorporated to use exhaust heat to raise the temperature of the pressurized air, the overall energy balance involves approximately 4000 Btu of oil or gas (during the generation phase) and 0.75 kWh of electricity (for the compression step) to produce 1 kWh of plant output. Typical energy-storage efficiencies cannot be calculated because a mixture of fuels is used. However, overall energy use is about 11,500 Btu/kWh, assuming the electricity for compression was generated at 10,000 Btu/kWh. A storage device using only electricity would have to have an efficiency of 87% to achieve an overall fuel use of 11,500

Btu/kWh, again assuming charging electricity is generated at 10,000 Btu/kWh. CAES plants for U.S. application using available compressors and expanders come in sizes of 25, 50, and 220 MW capacity.

A 290-MW CAES plant using two underground salt caverns is currently operating in the NWK utility system in Huntorf, West Germany. The salt caverns have a total volume of 390,000 $yd^3$ operating between 650 and 1060 $lb/in^2$. This plant was commissioned in 1978 and has a 3½ h storage capacity. It has been operating with a 99% starting reliability and 90% availability and can be fully operational in 6 min from a cold start. The second plant is currently being built by the Italian utility ENEL, using an aquifer store. The 25-MW plant is located in the Larderello area near Sesta, Italy.

***38. Flywheels.*** Flywheels involve the storage of kinetic energy in a rotating object. The stored energy is proportional to the object's moment of inertia times the square of its angular velocity. For a flat disk the moment of inertia is proportional to its mass times the square of its radius. Therefore, the energy stored is proportional to the mass and the square of the angular velocity. Two approaches to flywheel energy storage are generally pursued. The commonly used approach is to use a heavy material, like steel, and operate at a moderate speed. The other approach, the so-called advanced flywheel, uses lighter high-strength materials like glass or carbon fibers and operates at high speed.

Flywheels have smoothed our energy sources in such historic applications as the potter's wheel, the grain mill, and the water wheel. Today, its best-known use is in engines of various types. For large-scale electric energy storage, economics would dictate the use of lighter-weight high-speed (30,000 r/min) wheels operating in a vacuum and using magnetic bearings to reduce friction. Under these conditions energy efficiencies of about 80% are achievable. Since the speed of the wheel will change as energy is added or withdrawn, the motor-generator will have to accommodate variable speeds. Several approaches to variable-speed generation have been investigated for wind application, although today's wind machines use fixed-speed generators. Conservation measures have encouraged the use of variable-speed motors and several manufacturers now have this technology available in large sizes, for such applications as thermal power plant fans and pumps. The flywheel itself is an expensive way to store energy with costs probably twice that of the lead-acid battery. However, for applications such as utility system regulation where there are high power and low energy requirements, flywheels would be suitable if integrity and reliability can be demonstrated. The ideal use for today's flywheels is still to absorb and dissipate mechanical energy to smooth the operation of rotating machinery.

***39. Magnetic Storage.*** Electricity (dc) can be stored in magnetic fields wherein the energy stored is proportional to the inductance times the square of the current. An energy storage system based upon magnetics would involve an ac-dc-ac converter system (like batteries) and a large coil of a super-conductor (for example, a niobium-titanium alloy). To achieve superconductivity the conductor is maintained in a bath of liquid helium at about 1.8°K. Since there are no moving parts in the coil and electrical resistance is near zero, efficiencies can be above 90%. Energy losses occur in the converter and refrigeration system, although both can be designed to achieve high efficiency.

The cost per unit of stored energy of superconducting magnetic energy storage (SMES) varies with the minus one third power of the stored energy. In other words, cost per unit of stored energy decreases by 21% for each doubling of the storage capacity. In practice, therefore, very large sizes of several thousand megawatt hours are required for SMES to be cost-competitive.

The conductor for a large 5000 MWh SMES plant would be approximately a mile in diameter, carry 765,000 A through a 112 turn coil, and produce a 42,000 Gauss magnetic field. Such a conductor coil with its helium cooling system and support structure would be placed in a trench in competent rock that would be 24 m deep and 4.5 m wide. During operation, voltage across the converter would cycle between 4,500 and 1,500 V dc, and during idle periods the converter would be bypassed. Unfortunately, while such a system appears technically feasible, costs are high compared to alternative storage schemes.

SMES is, however, attractive for applications where high power is required for very short periods. For example, a small 10 MW 3-second, experimental system was built to inhibit subsynchronous resonance in a high voltage ac line from the state of Washington to California. Most well known, however, is the superconducting magnet at the Fermi National Accelerator Laboratory near Chicago used to focus particles in the accelerator.

Through R&D SMES costs might be sufficiently reduced to become competitive for bulk storage application. Ongoing research is directed to improve conductor current density and develop innovative design concepts that lower conductor support requirements. Even with successful research, issues involving the effects of magnetic fields, land use, and quality control suggest that practical systems would not be commercially available until well into the 21st century.

***40. Thermal.*** While thermal and cool storage have many applications, including thermal storage of off-peak energy in many European homes, it is not economically viable with electricity being both the energy source and the final product. The poor efficiency of converting heat to electricity is prohibitive. Thermal storage is generally only economically attractive when the ultimate energy required is heat. Site-specific applications where heat is stored as steam and hot water at power plants have proven economic. However, retrofit of such technology into an existing power plant would be difficult and costly. Since thermal and cool storage do not involve electricity as both the input and output, these storage concepts are beyond the scope of this section.

***41. Hydrogen.*** Many technical references include hydrogen and other chemicals among the various types of energy-storage systems. While hydrogen might ultimately be a desirable way to transport energy, economic studies suggest that hydrogen for storage is not economically viable. Processes of forming and using hydrogen (i.e., the conversion and reconversion steps) are inefficient and expensive. In addition, storage of hydrogen, as a gas, liquid, or hydride, is also likely to be expensive. As a result, hydrogen should not, in itself, be considered as an energy-storage alternative unless it is part of an integrated system that involves conversion, transportation, and reconversion and use, and where such an integrated system has favorable economics over alternative methods of converting, transporting, and reconverting and using energy.

***42. Summary.*** There is a diversity of storage technologies to store electrical energy. They are somewhat more complex to analyze than for conventional generation systems because the number of hours of operation will dictate capital costs *and* because use, particularly in electric utilities, is critically dependent upon existing and planned electric energy generation. It is important to note that the total cost of a storage system ($/kW) is the sum of the power related costs expressed in $/kW and the storage-related costs ($kWh) times the required hours of storage. Thus, if long periods of storage are required, it is critical to have an inexpensive storage subsystem. On the other hand, for high-power applications with short discharge periods, low conversion cost and fast response are the key requirements. Table 11-5 outlines relative economics for the storage systems discussed. For electric utility systems, a combination of storage systems might be optimal because of the relative costs. For example, batteries could be used for peaking (less than 5 h of storage) and CAES for intermediate duty cycle (more than 8 h of storage).

Besides economics there are several other considerations that should be addressed before a storage system is selected for a particular application: siting restrictions, environmental impact, response time, unit size, lead time, commercial availability, and operational experience. Only after all these factors are considered can a storage system be selected for any particular application. Once selected, storage systems will often be very competitive with generation technologies provided low-cost charging energy is available. Most storage systems have greater size and operational flexibility than the competing generation schemes and should therefore be seriously considered in any utility expansion plan and for certain utility customers.

**TABLE 11-5** Relative Economics for Energy Storage Systems

| Storage technology | Conversion cost, $/kW | Storage cost, $/kWh |
|---|---|---|
| Lead-acid battery | Low | High |
| Nickel-cadmium battery | Low | Very high |
| Advanced battery | Low | Moderate |
| Pumped hydro | Moderate | Low |
| Underground pumped hydro | Moderate | Low to Moderate |
| Compressed air | Moderate | Low |
| Superconducting magnet | Low | Very high |
| Flywheel | Low to Moderate | Very high |

## NUCLEAR FUSION

By GEORGE H. MILEY

***43. Introduction.***[1,2]* The substantial research effort now in progress aimed at controlling thermonuclear reactions to produce electric power at gigawatt levels is stimulated by the fact that the basic fuel consumed in fusion, deuterium, is present to the extent of one atom for every 500 atoms of hydrogen. This represents a virtually inexhaustible fuel source, whereas oil and fission fuels are sufficient to contribute to the world's power needs, at the present escalating rates of use, for less than a century. The production of fission fuels in breeder reactors and the employment of various forms of solar energy and coal, with nuclear fusion, provide the prospective means of avoiding severe shortages of electricity and other forms of power in the next century.

We anticipate an experimental demonstration of energy break-even from a fusion plasma within a few years. However, after this demonstration of "scientific feasibility," many years will be required to establish engineering and economic feasibility. Thus, current estimates for the introduction of commercial fusion plants are generally for sometime well after the year 2000. In fiscal year 1983 the federal outlay in support of magnetic fusion research in the United States exceeded $500 million annually. A somewhat lower, but still very significant, budget, largely provided by the Department of Defense, is devoted to inertial confinement fusion. The following pages are devoted to a summary of the techniques and processes of nuclear fusion research as of 1983.

***44. Thermonuclear Reactions.***[2] The reactions of interest for thermonuclear power occur when nuclei (ions) of elements of low atomic number are brought together in a plasma, at such temperature (i.e., velocity) as to cause the nuclei to collide with sufficient force to fuse, and thereby to transform a part of their mass into kinetic energy. The reaction most likely to be employed in first-generation plants for production of large amounts of electric power involves deuterium, tritium, and lithium, in the so-called D–T–Li cycle.

This cycle involves two types of reaction. In the first, deuterium and tritium react to produce an alpha particle (helium-4 nucleus) and a high-energy neutron:

$$\mathrm{D} + \mathrm{T} \rightarrow {}^4\mathrm{He} + n$$

The reaction endows the helium-4 ion with an energy of 3.5 MeV, and the neutron with 14.1 MeV.

The second reaction involves the production of tritium by lithium, which is contained in a "blanket" surrounding the plasma. Two neutron-induced reactions are involved in the production of tritium:

$${}^6\mathrm{Li} + n \rightarrow \mathrm{T} + {}^4\mathrm{He}$$

and

$${}^7\mathrm{Li} + n \rightarrow \mathrm{T} + {}^4\mathrm{He} + n$$

The tritium used in the D + T reaction is thereby regenerated by the lithium and is not required as a primary fuel. The power density produced by this cycle reaches a maximum when the kinetic temperature of the plasma is about 15 keV.

***45. Advanced Fuels.***[2,3] A number of fusion reactions are possible at temperatures of potential interest for fusion (<1 MeV) that involve various light elements through boron. The principal advantages of seeking such "advanced fuels" include: elimination of the need to breed tritium, reduced neutron fluxes (to reduce both materials damage and induced radioactivity in vessel and blanket structure), and an increased fraction of fusion energy carried by charged particles (making direct energy conversion attractive). Two general classes of fuels are of interest: deuterium-based and proton-based. Examples of the former include deuterium-deuterium (D–D) and deuterium–helium 3 (D–$^3$He) fusion while hydrogen(proton)–boron 11 ($p$–$^{11}$B) is a frequently cited proton-based reaction. The deuterium-based reactions all involve neutron production to some degree due to D–D reactions,

*Superior numbers refer to references, Par. **70**.

whereas $p$–$^{11}$B represents a fairly ideal fuel since the primary product is helium ($p + {}^{11}\text{B} \rightarrow 3\alpha$). The supply of hydrogen and boron 11 (from borax) is virtually as extensive as for deuterium; $^3$He, however, must be bred, either via D–D reactions or through the decay of tritium (12-year half-life).

Because of the relatively small fusion cross sections involved and the higher energies (plasma temperatures) required, all of the advanced fuels pose more stringent confinement requirements and offer lower power densities compared to D–T–Li fusion. Consequently, these fuels appear to be candidates for later-generation power plants.

***46. Power Production.***[2,4] The major part (about 80%) of the energy released by the D–T–Li cycle is carried by the high-energy neutrons. These, being chargeless, cannot interact with electric or magnetic fields, and hence cannot produce electric power directly. Rather, their power is converted to heat, which is extracted from the lithium blanket by thermal transfer. This heat (like that in fission- and fossil-fueled systems) must be converted to electricity by a thermal system, for example, one employing vapor (e.g., steam) or a working liquid (e.g., potassium).

The remaining 20% of the power generated involves charged alpha particles (helium-4 nuclei). This power may be abstracted either by thermal conversion or directly by causing the charged particles to interact with electrostatic or electromagnetic fields. With a $p$–$^{11}$B cycle, the charged particle energy could be increased to nearly 100%, making direct conversion a more important avenue for energy extraction.

***47. Energy Break-Even.***[2,5] In nuclear fission the reaction is neutron-induced and self-sustaining when criticality is reached. In nuclear fusion, on the other hand, the strong mutual repulsion between the positively charged deuterium and tritium nuclei must be overcome to cause them to fuse. To do so, power must be supplied to start and sustain the reaction and the power derived from the reaction must, of course, exceed the input power. Energy break-even is the condition at which the fusion energy released just equals the energy input required to heat the plasma to the fusion temperature.

To achieve energy break-even, the product $n\tau$ of the plasma density $n$ and the time $\tau$ during which the plasma is confined must lie within a narrow range for a specific reaction. In the D + T reaction this product (i.e., $n\tau$) must be roughly $10^{14}$ ions/cm$^3\cdot$s (frequently called the "Lawson criterion"). Thus, if confinement is achieved for a few tenths to several seconds (typical of magnetic confinement), the plasma density required lies in the range of $10^{14}$ to $10^{16}$ ions/cm$^3$. If the confinement time is of the order of a thousandth of a microsecond (typical of inertial systems), much higher densities, of the order of $10^{26}$ ions/cm$^3$, must be achieved. In all cases, the plasma temperatures must be in the range of 10 to 15 keV.

***48. Power Balance.***[2,4,5] To obtain useful amounts of power from fusion, not only must the conditions for power break-even be met and all the other power inputs to the reactor be supplied, but the total input power must be exceeded by the fusion-released power by a large amount, the difference being equal to the useful output. The power input to the auxiliaries of the fusion reactor, which heat and confine the plasma, is typically relatively large. In fact the power required by the magnet coils in magnetic confinement is so great that superconductive coils, operating at cryogenic temperatures, are used in most designs in order to reduce power requirements.

Since useful power output from a fusion reactor can be obtained only when substantial input power is provided, the reactor must be viewed as a power-amplifying device, the degree of power amplification differing according to the plasma-confinement scheme and the design of the lithium blanket.

***49. Nonelectrical Applications.***[6] In addition to electrical production, fusion reactors can potentially be used for a variety of other important purposes. For example, neutrons from D–T (and other D-based fuels) can be employed very effectively in a fusion-fission hybrid to breed fissile fuel in the blanket region of the fusion device. A unique feature of these hybrids is that very high "support ratios" are conceivable, e.g., with some concepts, as much as 20 times the energy production of the fusion plant can be supported in light-water fission plants. Fusion neutrons may also be used in conjunction with fission processes in hybrid systems and to accelerate the decay of fission wastes, i.e., providing a fission waste burner. Several methods for chemical production by fission plants have also been studied. These generally use the penetrating power of the high-energy neutron to create a high-temperature region in the blanket. This then appears well suited to various processes, e.g., hydrogen production from water, via either high-temperature electrolysis or thermochemical processes. The use of radiation, neutrons, or the plasma itself for chemical processing has also been considered, but this approach is not as straightforward (or as developed) as thermal processing.

***50. Plasma Confinement.***[7] The high temperatures and pressures of the thermonuclear reaction preclude the use of a material vessel to confine the plasma during the reaction. Two types of nonmaterial confinement are now under investigation, namely, magnetic and inertial confinement.

In the magnetic confinement scheme, intense magnetic fields are generated in the reactor, so disposed with respect to the reaction space that the plasma ions and electrons experience an inward pressure that resists the outward pressure of the hot plasma.

In the inertial confinement scheme, D–T-fueled targets (e.g., small glass ampules) are bombarded by laser or charged-particle laser beams. Several such beams are arranged to impinge symmetrically on the target from different directions; that is, beams are aimed at a common point at which the target is positioned. Each laser beam is pulsed and heats the surface of the target. Very high energy, exceeding a megajoule, is required in each pulse. Typically, each pulse lasts about $10^{-9}$ s and the pulse rate is about 1 per second. The symmetrical heating at the surface of the pellet causes its interior to be highly compressed, so that fusion will occur at the center and the reaction will burn outward. High-energy light ions and also heavy ions are also considered to be potentially attractive for inertial confinement fusion, the key advantages being more efficient coupling of the beam energy into the target than is possible with photons (lasers). Also, the potential for the development of economical, efficient, high-repetition-rate accelerators seems good based on developing accelerator technology.

***51. Magnetic Confinement.***[7,8] Two principal geometries are currently under development for magnetic confinement: the tokamak and the mirror reactor. There are also various alternative approaches, for example, the reversed field pinch, the spheromak, and the stellarator.

The performance of the magnetic confinement methods can be expressed by a parameter $\beta$, the ratio of the outward plasma kinetic pressure to the inward confining of the magnetic field. This parameter is a direct measure of the efficiency of the magnetic confinement; that is, high-$\beta$ systems make better use of the confining field than do the low-$\beta$ systems. The parameter $\beta$ is defined as

$$\beta = \text{constant} \times \frac{n(T_i + T_e)}{B^2}$$

where $n$ is the plasma density, $T_i$ and $T_e$ the ion and electron kinetic temperatures, and $B$ the confining magnetic field strength. High-$\beta$ reactors are expected to operate with $\beta$ greater than 0.8, and low-$\beta$ designs at values below 0.2. The fusion power density varies as the square of $\beta$.

***52. Tokamak Reactors.***[8] In tokamak reactors, the plasma-confinement space has closed, toroidal geometry, as shown in Fig. 11-9. A steady magnetizing current in the toroidal field coil produces a toroidal field of the order of 100 kG, which confines the plasma within the toroidal cavity. A second magnetic field, the poloidal field, is produced by external magnetic cores so that this field falls at right angles to the toroidal field everywhere within the cavity. The poloidal field is pulsed. The vector sum of the two fields (dashed line in Fig. 11-9) causes the plasma ions and electrons not only to be confined but to move through the torus, producing an axial current of the order of 10 to 20 MA. The ohmic heating associated with this current is an important source of power input for heating the plasma. However, in a tokamak ohmic heating is insufficient to produce fusion (ignition) temperatures alone. Additional plasma heating must be supplied by one of various methods, or combinations of them, such as radio-frequency (rf) heating, adiabatic compression, and the injection of high-energy (tens of kilo-electron-volts) beams of neutral fuel particles.

A third, transverse, pulsed magnetic field is also needed to provide control of the position of the plasma column within the cavity. This is needed to deal with the pressing technical problems of the contamination of the plasma by impurity ions that results when the plasma impinges on the metal wall of the cavity. Various schemes of avoiding this contamination, involving divertor coils and similar devices, are now under intensive development.

When operating in the burn cycle, the kinetic temperature of the plasma is expected to

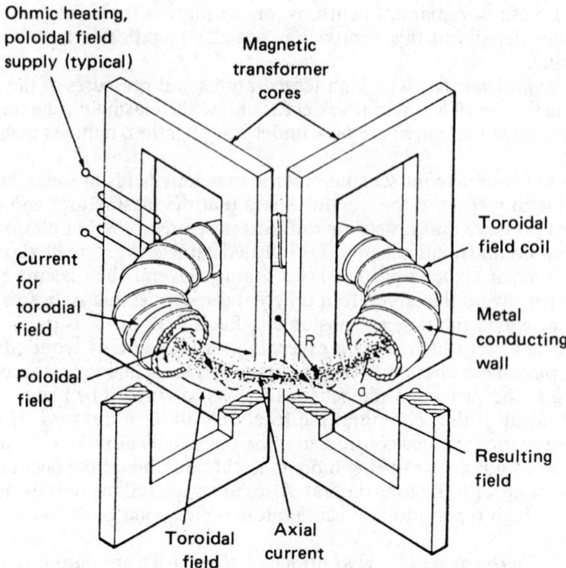


**FIG. 11-9** Elements of the tokamak reactor. Two magnetic fields (toroidal and poloidal) combine to produce a resulting field shown in the dashed line. This field both confines the plasma and causes its ions and electrons to move through the cavity. The ohmic heating of this axial current is one component of the heat required to achieve fusion of the plasma. Additional heating must also be supplied from an external source to maintain the fusion temperature in excess of 100 million degrees absolute. *(From Ref. 9.)*

range from 10 to 30 keV, the latter figure being above 300 million degrees absolute. In tokamak research in the 1980s, however, more modest objectives are in view. The projected operating temperature of the TFTR tokamak at Princeton University in 1984 is slightly less than the 7 to 10 keV required for ignition, but energy break-even should still be obtained by "beam-target" reactions created as injected high-energy ions slow down in the background plasma. In pulsed operation of the tokamak, successive periods of plasma confinement will last several seconds or tens of seconds. The total burn time, prior to purging and reloading the reactor, will be of the order of several hundreds or thousands of seconds. Auxiliary heating power of 10 to 100 MW is required to produce a fusion power output of 1 to 5 GW. Recent studies have concentrated on the possibility of using rf or other techniques to provide a current drive which can ultimately permit steady-state rather than pulsed operation. Deuterium and tritium fuel would be injected at a rate of about 2 to 4 $\times$ $10^{22}$ atoms per second. The typical tokamak is a low-$\beta$ (3 to 10%) and low-density ($10^{14}$ ions/cm$^3$) design.

Several conceptual studies of the prospective operating characteristics of tokamak reactors (when and if they are successfully built) have been made. Figure 11-10 and Table 11-6 are the results of such a study done by the staff at the Argonne National Laboratory along with industrial and university collaborators. The operating parameters of tokamak research devices, supported by DOE funds, are described in Ref. 8. Figure 11-11 shows a schematic drawing of the TFTR which is expected to demonstrate energy break-even. Built at the Princeton Plasma Physics Laboratory (PPPL) at a cost of over $500 million, this device began initial operation in December 1982, but several years of testing and experiments are necessary before optimum operation is expected.

The TFTR will employ ~30 MW of neutral beam heating in a device with 2.48-m major radius ($R$) and a 0.85-m minor radius ($a$) and an on-axis toroidal field of 5.2 T. The device

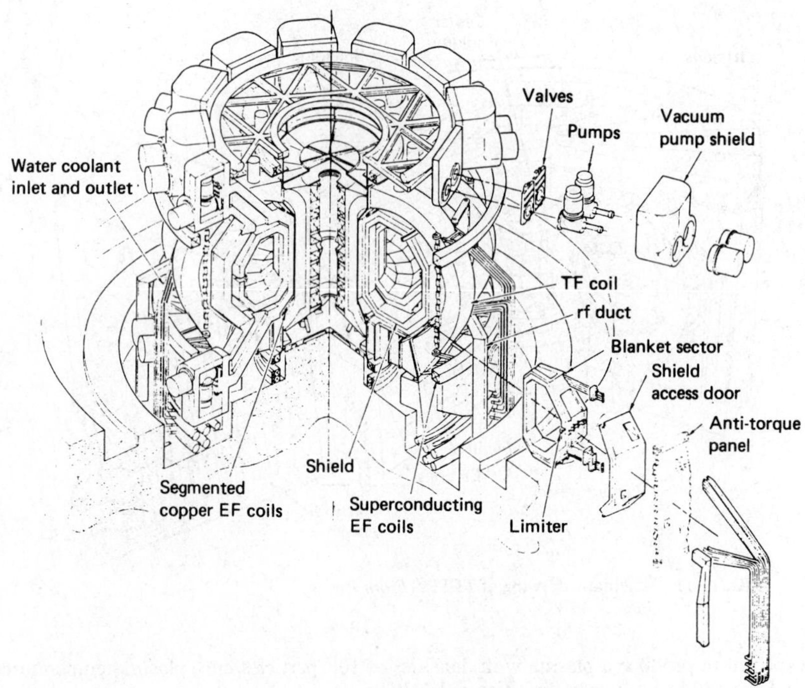


**FIG. 11-10** STARFIRE reference design—isometric view. *(From Ref. 8.)*

**TABLE 11-6** Operating Characteristics of the STARFIRE Tokamak

| Parameter | Value |
|---|---|
| Plasma radius | 1.94 m |
| Major radius of torus | 7.0 m |
| Plasma *3* | 6.7% |
| Ion density | $0.8 \times 10^{14}/\mathrm{cm}^3$ |
| Ion temperature | 11 keV |
| Maximum toroidal field | 110 kG |
| Plasma current | 10 MA |
| Fusion power density | $0.7\ \mathrm{MW/m^3}$ |
| Fusion power output | 4000 MW |
| Electrical power output | 1200 MW |
| Tritium burnup | 42% |
| Burn time | Continuous |
| Wall loading | $3.6\ \mathrm{MW/m^2}$ |
| Material | Modified 316 stainless steel |
| Coolant | Pressurized $H_2O$ |
| Breeder | $LiAlO_2$ |
| Power cycle | $H_2O$/steam |
| Impurity control | Limiter vacuum |
| Heating | rf |

SOURCE: Ref. 8.

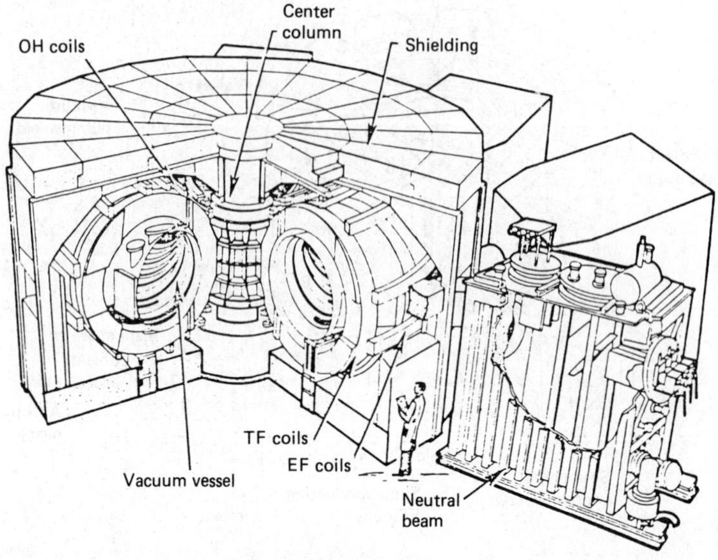


**FIG. 11-11** Schematic drawing of TFTR. *(From Ref. 8.)*

is expected to produce a plasma with densities of $10^{13}$ particles/cm$^3$, plasma temperatures of 10 keV, and fusion power densities of 1 MW/m$^3$.

The TFTR is representative of a class of tokamak devices that will begin operation within the next few years. Examples include the Joint European Torus, which is under construction at Culham Laboratory in the United Kingdom under the auspices of the European Economic Community and Atomic Energy Research Institute. The Soviet Union is considering building a tokamak device and Japan has a major experiment underway.

The achievements to date in tokamak experiments have been quite encouraging and have provided a basis for projecting with confidence the successful performance of the TFTR. For example, ISX-B at the Oak Ridge National Laboratory (ORNL) has demonstrated the achievement of $\beta$ values which are apparently about twice the theoretical prediction. The Princeton Large Torus (PLT) at PPPL has achieved central ion temperatures greater than 7 keV with the injection of 2.4 MW of neutral beam power. Also, rf heating experiments on PLT with ion cyclotron resonance waves (25 MHz) have demonstrated heating of tokamak plasmas with efficiencies comparable to achieved $n\tau$ values of $\sim 10^{14}$ s/cm$^{-3}$, i.e., equivalent to the Lawson requirement for break-even (but at lower temperatures). Fueling of tokamak devices via injection of small pellets ($\sim$1 mm in diameter) at velocities of $10^3$ m/s has been demonstrated on the ISX-B device. In addition, experiments on ISX-B have shown the desirability of rf-assisted breakdown and startup in reducing the initial voltage to be provided by the ohmic heating (OH) coil. This significantly reduces the design requirements for superconducting OH coils for reactor applications. The poloidal diverter experiment at PPPL has demonstrated that 70% of the plasma energy can be removed via diverter action. In addition, Doublet III at GA Technologies has produced plasma currents greater than 2 MA. Taken together, these experiments have provided a laboratory proof-of-concept demonstration for the tokamak.

Recently two design efforts, one national and one international in scope, have addressed the next step in the tokamak development program after TFTR. Within the United States, an activity is underway to establish a design concept of a tokamak engineering test facility (ETF) which is being carried out under the direction of the Fusion Engineering Design Center located at ORNL. Major parameters for the ETF are: major radius $R = 5.4$ m, minor radius $a = 1.3$ m, and an on-axis toroidal field of 5.5 T. With a $\beta < 5\%$, it is expected that the ETF would produce an ignited D–T plasma with burn times of $>100$ s. It would employ superconducting magnets and be capable of demonstrating the capability to breed tritium and remove the heat generated by the fusion neutrons in the blanket.

On the international front, the International Atomic Energy Agency has organized a worldwide effort to assess the feasibility of undertaking a major fusion reactor development project. The project is termed the International Tokamak Reactor (INTOR) Workshop and includes representatives for Euratom, Japan, the United States, and the Soviet Union.[10] The initial results of this assessment, which developed some conceptual designs and reviewed, in depth, the technical issues of such a project, indicate that such a project is feasible and that the tokamak should be selected because of its advanced stage of development. The workshop is continuing with emphasis on the identification of issues critical to the development of such a device.

The various steps required to take the tokamak from scientific feasibility to a commercial reactor concept are complex. These steps include TFTR, an ETF/INTOR-type device to demonstrate long D–T burn physics, and the integration of major reactor technologies, a demonstration reactor (DEMO) represented by a design developed by ORNL, and a commercial-size tokamak reactor represented by the STARFIRE design developed at Argonne National Laboratory. One notes that there is a factor of ~2 scaleup in physical dimensions from TFTR to ETF, but that there is only a small increase in dimensions to the DEMO and a commercial reactor. There are, of course, substantial differences between the ETF and DEMO with respect to reliability and lifetime of components.

***53. Mirror Reactors.***[8] Mirror-type fusion reactors are characterized by open magnetic field geometries, that is, magnetic flux passes through a mirror-type device and intersects material walls outside the reaction chamber. In order for such a device to be an adequate container of fusion plasma, it is essential that the end leakage of the plasma be strongly inhibited.

The earliest magnetic mirror configuration used a solenoid with increased magnetic field strength near its ends (Fig. 11-12). In this "simple" mirror, a charged particle travels in a helical orbit around an axially directed magnetic field line. When traveling into a region of increasing magnetic field strength, most particles are reflected, that is, their axial motion is stopped and reversed before they can penetrate to the high point of the mirror field. (Some

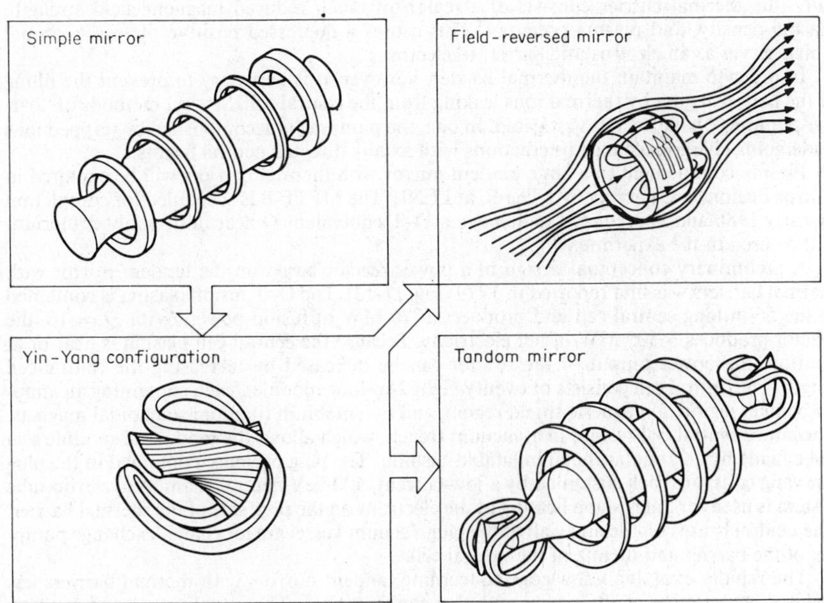


**FIG. 11-12** Evolution of mirror confinement concepts. *(From Ref. 8.)*

particles with highly directed velocity may still escape. Strong magnets can reduce this leakage, but it cannot be completely eliminated.)

Unfortunately, plasma in a simple mirror is grossly unstable to sideways motion because the magnetic field weakens in directions perpendicular to the coil axis. (The radial weakening of field is evidenced by the concave-inward surface of the magnetic flux bundle). To solve this gross, or MHD, stability problem, the simple mirror configuration was replaced by the minimum-$B$ mirror. From the center of a minimum-$B$ magnetic field—produced by a pair of solenoids and Ioffe bars, a baseball coil (shown in Fig. 11-12), or yin-yang coil—the field strength increases in all directions. However, by 1975, it was concluded from conceptual design studies of mirror reactors based on the minimum-$B$ geometry that end losses from such a reactor would severely limit its plasma $Q$ (fusion power divided by trapped injected power). The subsequent search for enhanced-$Q$ mirror machines led to two new concepts: the field-reversed mirror and the tandem mirror.

In a field-reversed mirror (FRM) the confinement of plasma occurs through a toroidal region of closed magnetic field lines generated by plasma currents in a nearly uniform background field (Fig. 11-12). The FRM offers the exciting possibility of fusion electric power reactors in small sizes. For example, conceptual design studies were carried out, for both a multicell reactor producing 75 MW (electric) of net electric power and a single-cell pilot reactor producing $\sim$11 MW (electric). However, following a series of initial experiments at the Lawrence Livermore National Laboratory (LLNL) that failed to build up sufficient plasma current for reversal, most of the effort shifted to the tandem mirror.

The tandem mirror confinement concept, invented in 1976, is now the mainline effort of the mirror fusion program. The basic concept entails the improved axial confinement of a long cylindrical fusion plasma within a solenoid by means of strong electrostatic potentials at the ends, produced by mirror-confined, end-plug plasmas (Fig. 11-12).

A major new invention for tandem mirrors—the thermal-barrier concept—was reported in 1979. This concept followed from the realization that enhancement of the confining electrostatic potential is possible by the establishment of a hotter electron population in the plugs than in the central cell. Further, colder central-cell electrons allow a higher fusion power density for a given magnetic field strength. However, in the normal operation, electron flow between the plugs and central cell presents large temperature differences. Thus a thermal barrier is necessary to reduce the passing of central-cell electrons into the plug. Basically, the thermal barrier consists of a region of much reduced magnetic field strength, plasma density, and plasma potential. This causes a depressed positive plasma potential which serves as an electrostatic barrier to electrons.

In order to maintain the thermal barrier, however, it is necessary to prevent the filling of the barrier region by thermal ions leaking from the central cell. Several methods of "barrier pumping" have been investigated. In one, the pumping is accomplished by trapped ions undergoing charge exchange interactions with axially directed neutral beams.

Plasma confinement in a large tandem mirror with thermal barriers will be explored in Mirror Fusion Test Facility (MFTF)-B, at LLNL. The MFTF-B is scheduled for completion by early 1986 and is predicted to achieve a D–T equivalent $Q$ near unity (only deuterium will be used in the experiment).

A preliminary conceptual design of a power reactor based on the tandem mirror with thermal barriers was first reported in 1979 (Fig. 11-13). The D–T fusion plasma is contained in the 56-m-long central cell and produces 1770 MW of fusion power. With $Q \simeq 10$, the reactor produces $\sim$500 MW of net electricity. Because the central-cell plasma is near or at ignition, the power output of the reactor can be increased by increasing the central-cell length. The central cell consists of twenty-eight 2-m-long modules, each containing an annular blanket region, a magnetic shield region, and two niobium titanium solenoidal magnets. The entire central cell resides in a vacuum trench, which allows the module-to-module seal to be made by an annular metal inflatable cushion. The plug plasmas, contained in the plug yin-yang coils, are each sustained by a low-current, 400-keV neutral beam. A gyrotron tube system is used for microwave heating of the electrons on the plug side of the thermal barrier. The neutral beams on the end wall of the plug vacuum vessel are for charge-exchange pumping of the barrier and fueling of the central cell.

The rapidly evolving knowledge concerning tandem mirrors with thermal barriers has resulted in a number of alternative end-plug configurations. The investigation and comparison of these different end plugs is a key focus of present reactor design studies.

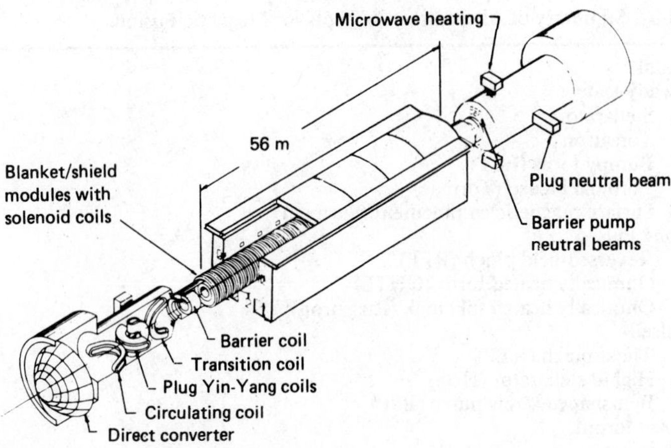


**FIG. 11-13** The TMR with thermal barriers. *(From Ref. 8.)*

***54. Alternative Magnetic Fusion Concepts.***[8,11] Table 11-7 summarizes a representative cross section of alternative fusion concepts (AFCs) that, in one way or another, have been or are being considered for the production of electrical power, chemical process heat, and/or fissile material. Depending on the confinement scheme considered, systems studies of AFCs range from a simple physics analysis, based on Lawson-like criteria, to detailed conceptual designs. With few exceptions, most reactor studies of alternative concepts fall into the less formalized part of this spectrum. For this reason, a quantitative intercomparison and ranking is not possible.

The toroidal AFCs summarized in Table 11-7 are classified as steady-state, long-pulsed (10 to 100 s) and pulsed ($\sim$1 s). A sampling from each category is given.

***55. Steady-State Toroidal Systems.***[8,11] The stellarator-Torsatron design is treated as a generic steady-state concept while both Tormac and Surmac are described separately.

*Stellarator-Torsatron Reactors.* Unlike the tokamak, the nonaxisymmetric stellarator achieves equilibrium in a toroidal geometry by externally inducing a rotational transform in the confining magnetic field (Fig. 11-14); ideally, no axial currents need be supported by the toroidal plasma column, as is required in a tokamak, although until very recently all stellarator experiments utilized such currents for ohmic heating. Implementation of a deformation (twist) into a simple toroidal field coil set allows the Torsatron magnetic geometry to be produced while eliminating the helical coil set in favor of a highly modular device (Fig. 11-15). In addition, more optimally oriented coil forces and lower stresses are anticipated for this modular Torsatron approach. These new advances have renewed interest in the reactor extrapolation of the stellarator-Torsatron concept, a renaissance that coincides with experimental success in heating a low-ohmic-current device, the latter being a prerequisite for eventual steady-state reactor operation.

Qualitative advantages that can be invoked for the Torsatron reactor concept include: steady-state magnetic fields and burn; operation at ignition or high $Q$ for low recirculating power; impurity and ash removal by means of a magnetic limiter and helical poloidal diverter that occur as a natural consequence of the topology; and no major plasma disruptions that could lead to an intense, local energy dump and no auxiliary position or field-shaping coils and moderate aspect ratio ($>$10), both of which ease maintenance access.

*Elmo Bumpy Torus (EBT) Reactor.*[8] The EBT concept is a toroidal array of simple magnetic mirrors. The promise of a steady-state, high-$\beta$ reactor that operates at or near D–T ignition emerges from this combination of simple mirrors and toroidal geometry. The creation of an rf-generated, low-density, and energetic electron ring at each position between mirror coils (i.e., midplane location) is needed to stabilize the bulk, toroidal plasma against well-known instabilities associated with simple mirror confinement. Figure 11-16 gives a schematic diagram of the EBT configuration and the rf-sustained electron rings.

**TABLE 11-7** Summary of Alternative Concepts for Magnetic Fusion

1. Toroidal
   *a.* Steady-state
      (1) Stellarator
      (2) Torsatron
      (3) Bumpy torus (EBT)
      (4) Toroidal bicusp (Tormac)
      (5) Surface magnetic confinement (Surmac)
   *b.* Long Pulsed
      (1) Reversed-field pinch (RFP)
      (2) Ohmically heated torus (OHTE)
      (3) Ohmically heated tokamak (Riggatron)
   *c.* Pulsed
      (1) Theta-pinch (RTP)
      (2) High-$\beta$ stellarator (HBS)
      (3) Belt-shaped screw pinch (BSP)
2. Compact toroid
   *a.* Stationary
      (1) Spheromak
      (2) Field-reversed mirror (FRM)
      (3) Triggered-reconnected adiabatically compressed torus (TRACT)
      (4) Electron-layer field-reversed mirror (Astron)
      (5) Slowly imploding liner (LINUS)
   *b.* Translating
      (1) Spheromak
      (2) Field-reversed theta pinch (CTOR)
      (3) Moving-ring field-reversed mirror (MRFRM)
      (4) Ion-ring compressor
3. Linear
   *a.* Steady-state
      (1) Multiple-mirror solenoid
   *b.* Pulsed
      (1) Linear theta pinch (LTP)
      (2) Laser-heated solenoid (LHS)
      (3) Electron-beam heated solenoid (EBHS)
4. Very dense (fast-pulsed, linear) systems
   *a.* Fast-imploding liner (FLR)
   *b.* Dense plasma focus (DPF)
   *c.* Wall-confined shock-heated reactor (SHR)
   *d.* Dense Z-pinch (DZP)
   *e.* Passive liners

SOURCE: Ref. 8.

The EBT was first examined as a reactor in 1976. Interim revisions and reassessments of this design have been made during the intervening years, and typical parameters, taken from a 1980 study at Los Alamos National Laboratory, are given in Table 11-8. An EBT configuration that is stabilized by energetic electron rings combines a number of unique features that describe a fusion reactor with the following attractions: steady-state operation in an ignited or a high-$Q$ mode; a potential for high-beta operation with the attendant efficient utilization of magnetic field energy; large aspect ratio to give an open and accessible geometry; an engineering assembly that is comprised of relatively simple and compact modules; ease of maintenance, modular construction, and a relatively simple magnet system. Although the earliest EBT reactor designs predicted relatively large power plants, the attainment of high magnetic aspect ratios in systems with lower physical aspect ratios through the use of aspect-ratio-enhancement (ARE) coils indicates that smaller reactors may be possible while simultaneously maintaining the above-mentioned reactor features.

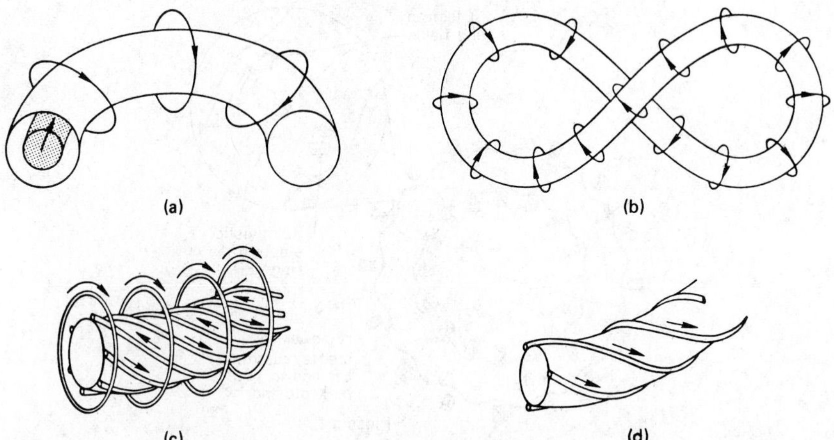


**FIG. 11-14** Toroidal configurations: (*a*) Tokamak. The main magnetic field is provided by poloidal coils with the poloidal field due mainly to a toroidal current in the plasma. Several external toroidal coils (which are not shown) provide a vertical magnetic field as well as drive the plasma current. (*b*) Figure-eight stellarator. Only one set of solenoidal coils is needed. The rotational transform is generated by the torsion of the magnetic axis. (*c*) Classical stellarator. A set of $2l$ helical coils with current flowing in opposite directions inside the toroidal field coils provides the rotational transform. (*d*) Torsatron. A single set of $l$ helical windings provides both the toroidal and poloidal fields. Usually an additional set of toroidal coils (which are not shown) is necessary to provide an additional vertical field. The standard heliotron configuration is similar to this, but has an additional set of toroidal field coils.

*Toroidal Bicusp (Tormac).*[8] Like the tokamak and the stellarator and Torsatron designs, the Tormac is a toroidal device that confines plasma on combined poloidal and toroidal magnetic fields. This configuration is illustrated in Fig. 11-17. By opening the outer poloidal field regions, however, the Tormac creates an absolute minimum-$B$ configuration that is MHD-stable for large aspect ratio and plasma $\beta$. The resulting toroidal line cusps support plasma on both closed field lines (i.e., high-$\beta$ bulk plasma) and open field lines, confinement of the latter plasma being enhanced by mirroring effects in the sheath region that separates regions of open and closed field lines.

**FIG. 11-15** Coils in a typical modular stellarator.

*Surface Magnetically Confined Systems (Surmac).*[8] The Surmac concept represents one example of a general class of multipole configurations in which electrical conductors are arrayed in either a linear or toroidal geometry to create a surface magnetic configuration with low magnetic field in the bulk plasma volume. Figure 11-18 illustrates a toroidal version of this high-$\beta$, steady-state system. The Surmac can operate with considerably reduced synchrotron radiation emanating from the bulk plasma, and, therefore, this concept appears to be particularly suitable for confining the high-temperature advanced-fuel plasmas.

***56. Long-Pulsed Toroidal Systems.***[8] In terms of power density, relative simplicity and symbiosis with the basic confinement scheme, ohmic dissipation of toroidal plasma currents represents a very efficient heating scheme. Two long-pulsed toroidal concepts are described that propose ohmic heating as the sole means to obtain an ignited thermonuclear plasma for reactor application: the Riggatron and the reversed-field pinch reactor (RFPR). A vari-

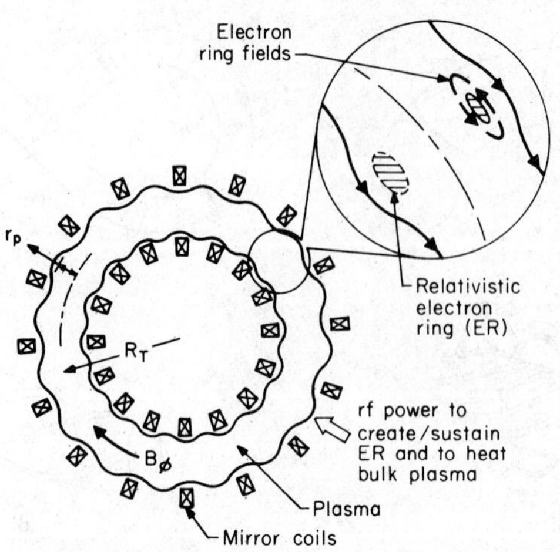


**FIG. 11-16** Simplified schematic drawing of the EBT concept. *(From Ref. 8.)*

**TABLE 11-8** Typical Parameters for TRACT

| | |
|---|---|
| Minor radius, m | 0.14 |
| Major radius, m | 0.36 |
| Length, m | 1.88 |
| Plasma density, $10^{20}/m^3$ | 28 |
| Average $\beta$ | 0.77 |
| Magnetic field, T | 7 |
| Pulsed energy, MJ | 570 |
| Burn time, s | 0.5 |
| Thermal power, MW | 520 |
| Net power, MWe | 130 |

SOURCE: Ref. 8.

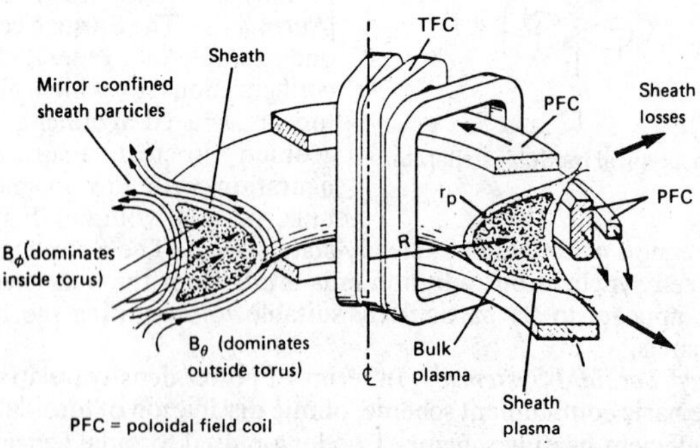


**FIG. 11-17** Schematic representation of the Tormac plasma-field-coil configuration. *(From Ref. 8.)*

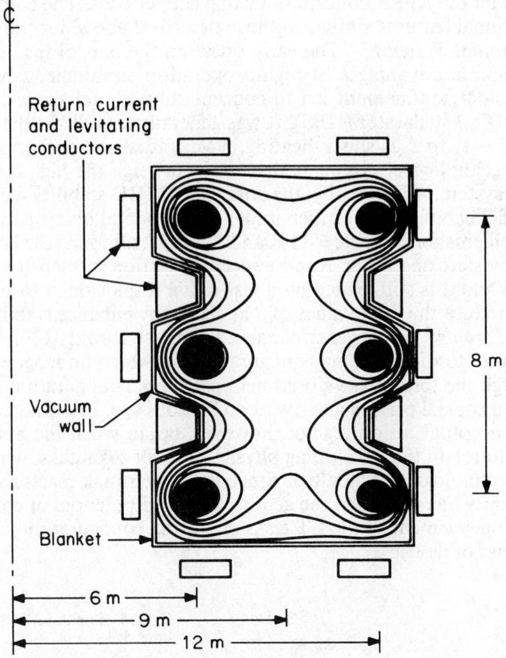


**FIG. 11-18** Schematic representation of the toroidal Surmac plasma and coil configuration. *(From Ref. 8.)*

ation of the RFPR has recently been proposed that would use an external helical winding to achieve a more controllable rotational transform in a reversed-field state; this concept is called the Ohmically Heated Toroidal Experiment (OHTE).

*The High-Field Ohmically Heated Tokamak Reactor (Riggatron).*[8,12] Although in principal a tokamak, the Riggatron represents a sufficient change in engineering approach and "conventional" tokamak physics to warrant classification as an alternate concept. The combined use of high toroidal current density (8 $MA/m^2$) and high toroidal field (16 to 20 T) copper coils positioned near the first wall allows net energy production in a relatively short burn period from a high-$\beta$, ohmically heated system. The severe thermal-mechanical environment necessarily dictates an engineered short life. The plasma chamber and the $D_2O$-cooled copper magnets would be small because of the increased plasma density (2 to 3 $\times$ $10^{21}$ $m^{-3}$) and high $\beta$ (0.15 to 0.25). The 6- to 10-MA Riggatron would generate 1 to 2 GW (thermal), the fusion neutron power being recovered in a fixed lithium blanket located outside the magnet system. Recovery of joule and neutron heating in the copper coils is also an essential element of the overall power balance. The short-lived, disposable reactor would operate in clusters four to six fusion modules, with two additional standby modules and a rapid "plug-in" capability providing high plant reliability and availability without in situ remote maintenance.

*Reversed-Field Pinch Reactor (RFPR).*[8,12] The RFPR is similar to a tokamak in that a toroidal axisymmetrical configuration is used to confine a plasma with toroidal current by a combination of poloidal and toroidal magnetic fields. Using a passive conduction shell, the RFPR relaxes inherent constraints on the magnetic field profiles for a tokamak such that the variation of the magnetic shear need not exhibit a minimum in a region enclosed by a first-wall conducting shell. By removing this constraint, the RFPR can operate with a current density that is sufficient for ignition by ohmic heating, an unrestricted aspect ratio, a higher $\beta$ value, and an appreciably lower magnetic field at the superconducting windings. Recently, considerable attention has also been given to the possibility of developing a compact reactor based on the RFPR concept but using copper coils. The compact RFPR would have many operational features similar to those described above for the Riggatron.

***57. Pulsed Toroidal Systems.***[8] The early quest on the part of fusion reactor designers to attain the economic advantages of high-$\beta$ operation simultaneously with the physics advantages of toroidal confinement led to concepts like the reference theta-pinch reactor (RTPR) and the high-$\beta$ stellarator (HBS). It was generally found that the fast-pulsed nature of the RTPR (i.e., ~1- to 2-$\mu$s shock heating, 30-ms adiabatic compression, ~0.5-s burn time) resulted in technological problems that may outweigh the high-$\beta$ (>0.8) advantages for that particular system. Additionally, the absence of MHD stability without fast feedback for the particular field configurations then under experimental investigation indicated other reactor-related problems for both the RTPR and the HBS, although the latter was eventually proposed for steady-state operation. A more recent variation is the belt-shaped screw pinch reactor (BSPR). While it is still heated by a fast radial implosion, a toroidal bias magnetic field is applied to reduce the final values of $\beta$ and thereby enhance stability.

***58. Compact Toroids.***[8] The generic name compact toroid (CT) has recently been applied to the class of toroidal plasma configurations in which no magnetic coil or material walls extend through the torus. This closed-field plasmoid configuration is not new, having been generated by a coaxial plasma gun over two decades ago. Interest in this configuration, as applied to a conceptual fusion reactor, however, began when the Spheromak was proposed as a means to retain the developing physics base for tokamaks, while simultaneously shedding certain technological difficulties. Since the Spheromak reactor was first proposed, the fusion community has identified the general area and potential of compact toroids, the Spheromak being one element of the CT class of plasma configurations. Figure 11-19 summarizes the CT class of devices.

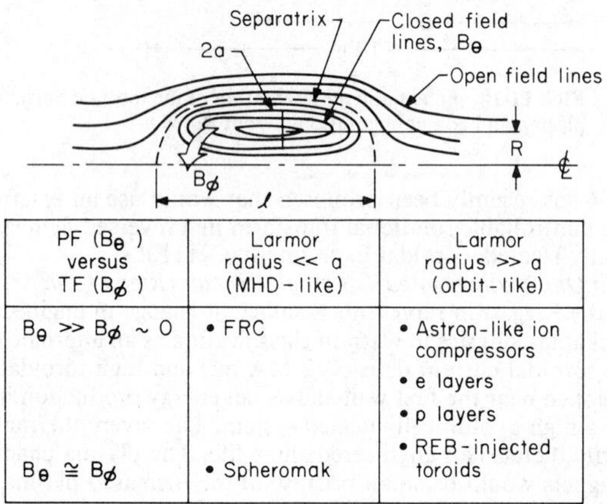


| PF ($B_\theta$) versus TF ($B_\phi$) | Larmor radius << a (MHD-like) | Larmor radius >> a (orbit-like) |
|---|---|---|
| $B_\theta \gg B_\phi \sim 0$ | • FRC | • Astron-like ion compressors<br>• e layers<br>• p layers |
| $B_\theta \cong B_\phi$ | • Spheromak | • REB-injected toroids |

**FIG. 11-19** Diagram and classification of CT configuration where $a$ is the minor radius, $R$ is the major radius, $B_\theta$ is the poloidal field, $B_\phi$ is the toroidal field, and l is the separatrax length. *(From Ref. 8.)*

Although representing a great diversity of plasmoid formation, heating, and confinement schemes, the fundamental physics of particle and energy transport and stability and equilibrium are not well known for many subsets of the CT class. Two approaches that involve field reversal with smaller-orbit ions driving the plasma currents, namely the Spheromak and the field-reversed theta pinch, are currently receiving the most study. These concepts, along with the field-reversed mirror, are normally classified as field-reversed configurations (FRCs).

Various reactor studies have been performed to examine the potential advantages of FRCs. Both stationary designs and concepts where the plasma is passed through a linear burn chamber have been considered. The TRACT (triggered-reconnection adiabatically compressed toroid) reactor design (Fig. 11-20) for the stationary burn of a field-reversed theta pinch is a good example of these concepts.

Table 11-9 also gives typical prototype reactor parameters for this ~1-Hz batch-burn system. Utilizing a longer burn period (~0.5 s) and modest magnetic fields (5.3 T), a hybrid superconducting (dc)/normal (ac) coil system would provide the required flux compression to achieve ignition in a plasmoid with an initial radius of 0.72 m. A first-wall copper coil

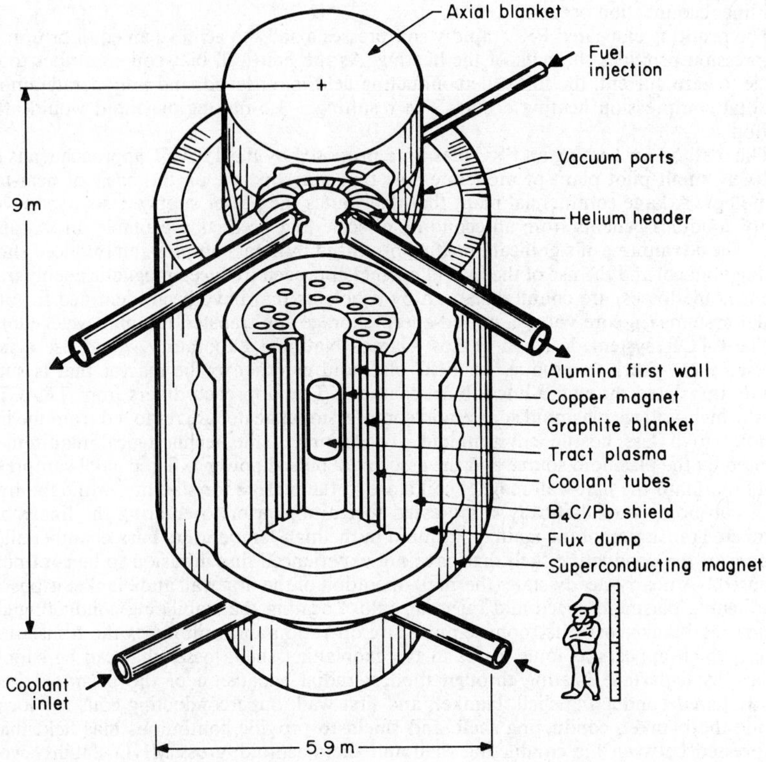


**FIG. 11-20** Conceptual layout of the TRACT reactor that would form and compress in situ an FRC to ignition and burn. *(From Ref. 8.)*

**TABLE 11-9** Typical Parameters for TRACT

| | |
|---|---|
| Minor radius, m | 0.14 |
| Major radius, m | 0.36 |
| Length, m | 1.88 |
| Plasma density, $10^{20}/m^3$ | 28 |
| Average $\beta$ | 0.77 |
| Magnetic field, T | 7 |
| Pulsed energy, MJ | 570 |
| Burn time, s | 0.5 |
| Thermal power, MW | 520 |
| Net power, MWe | 130 |

SOURCE: Ref. 8.

cancels and subsequently reverses for a few milliseconds the field generated by an exoblanket superconducting coil, during which time a low-temperature plasma is created. The superconducting flux is reestablished in two stages: a fast (shock) stage and a slower (adiabatic compression) stage. During the shock phase the plasma column and trapped (reversed) bias flux are radially compressed, providing significant shock heating. Cusp coils at each end of the 8-m-long plasma chamber reinforce the trapped flux while expanding the forward flux, leading to a delay in the reconnection of field lines. When the external field induced by the fast shock-heating power supply reaches a peak value, trigger coils are activated and field-line reconnection occurs.

The resulting elongated FRC rapidly compresses axially to achieve an equilibrium; this compression provides the bulk of the heating. As the first-wall bias coil continues to discharge to zero current, the full superconducting field is retrieved, and a moderate amount of radial compression heating occurs. The resulting ~1.5-m-long plasmoid would attain ignition.

The method for heating an FRC plasmoid proposed by the TRACT approach leads to a relatively small pilot plant of moderate cost that may operate on the basis of near-term technology. A large commercial plant that distributes the power supply costs over several reactor modules benefits from an economy of scale that predicts acceptable direct capital costs. The advantages of significant heating promised by axial compression (reduced shock-heating voltage) and the use of the hybrid magnet approach (reduced magnetic energy transfer and joule losses) are counterbalanced by problems that have been identified for other similar systems (in-core voltage, pulsed-energy storage and transfer, thermal cycle, etc.).

The CTOR system designed at Los Alamos National Laboratory would use a field-reversed theta pinch to produce an FRC plasmoid external to the reactor that is subsequently translated through a linear burn chamber. This approach differs from TRACT in that the high-voltage plasmoid source and compressional heater are removed from the burn chamber to a less hostile environment. To minimize the technological requirements imposed by the plasmoid source and the associated pulsed power, a flared axial compressor would maintain the first-wall magnet coil close to the plasma for stability, while the translating plasmoid is adiabatically compressed to ignition prior to entering the linear burn chamber. Translation of the ignited plasmoid in the high-temperature burn chamber allows portions of the conducting shell that have not experienced flux diffusion to be continually "exposed." A nearly steady-state (thermal) operation of the first wall and blanket is possible by adjusting plasmoid speed and injection rate. Locating the stabilizing conducting shell outside the blanket permits room-temperature operation and minimizes the translational power, which appears as joule losses in the exoblanket shell, losses that can be supplied directly by $\alpha$-particle heating through modest radial expansion of the plasmoid inside slightly flared conducting shell, blanket, and first wall. Superconducting coils are located outside the blanket, conducting shell, and shield to provide continuous bias field that is compressed between the conducting shell and the plasmoid; gross MHD stability would thereby be provided throughout the burn without requiring feedback stabilization. The plasmoid motion terminates in an end region where expansion directly converts internal plasma energy to electrical energy. Figure 11-21 gives a schematic view of the above described operation.

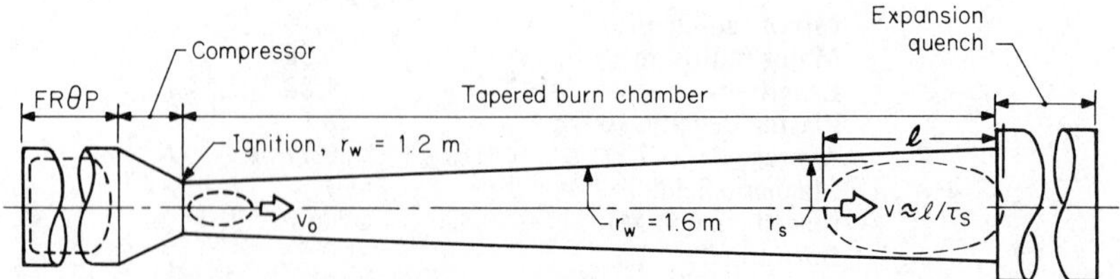


**FIG. 11-21** Schematic layout of a CTOR that would translate at an initial velocity $v_0$ an FRC down a linear burn chamber of length $L$. The FRC would be formed externally to the reactor by an FRθP plasmoid source.

The initial plasmoid ($T$ = 1.6 keV, $r_s$ = 2.5 m, $l$ = 9.7 m) is compressed to 8 keV ($r_s$ = 0.85 m, $l$ = 5.0 m) in 0.1 s. Here, $r_s$ is the plasmoid radius and $l$ is its length. This ignited plasmoid enters the burn chamber with an initial velocity equal to two to five times $l\tau_s$, where the electrical skin time, $\tau_s$, describes the decay of flux within the area between the first wall and plasma separatrix. The plasmoid velocity is subsequently reduced by tailoring the flare of the shell to maintain a constant first-wall neutron loading along the length of the burn chamber. Plasmoid motion proceeds until the velocity falls below $l/\tau_s$, at which point the reactor length is defined.

***59. Inertial Confinement Reactors.***[13–16] In the inertial confinement scheme, the surface of a small target of a solid material containing deuterium-tritium fuel is symmetrically illuminated by very high energy laser or charged particle beams. The density of the nuclei in the target prior to illumination is typically $10^{22}$ ions/cm$^3$. A thermal compression, caused by ablation as the beam energy impinges on the target, causes a further increase in density of up to $10^4$ times, producing a density at the center of the pellet of $10^{26}$ ions/cm$^3$. At this density, a density-confinement time product of $10^{16}$ can be achieved in a confinement time of 1 ns ($10^{-9}$ s) or less. (This is adequate for energy break-even, see Par. **47**.) Thus, the process is essentially one of a thermally induced implosion. The total laser energy which must be applied to the target to achieve useful fusion gains is estimated to be of the order of 5 MJ.

In terms of reactor design, the most important feature of inertial fusion is that the confinement of the plasma is decoupled from the functions of the reaction chamber, wall, and blanket. This is possible because the physics of energy absorption, implosion, ignition, and burn of this plasma are physically separable from the chamber conditions and structural requirements.

In inertial fusion reaction chamber design, the degree of freedom allowed by separability is exploited by placing fluids (gases or liquids) or fields (magnetic deflection) inside the chamber to protect the wall from ablative material loss due to the energy deposition of the short-range fusion radiation (25 to 35% in soft x-rays and debris) and, in the case of thick liquid-metal walls, from radiation damage due to deeply penetrating x-rays and neutrons. Similarly, the final optics of the driver can be protected by placing the final mirrors tens of meters away and protecting them by a gas, or by closing a rotating shutter in front of the ion beam ports to block the plasma debris.

The use of protective fluids, combined with the pulsed nature of the energy release, leads to new engineering constraints and techniques that are radically different from those of magnetic fusion. Understanding and manipulating the dynamic response of fluids and structures becomes of primary importance, and such factors as new materials development take less emphasis.

***60. The HYLIFE System.***[13] Over the last few years, the most detailed calculations and conceptual design for an inertial fusion power plant have been performed at LLNL, in conjunction with a team of university and industrial contractors. The high-yield lithium injection fusion energy (HYLIFE) chamber shown in Fig. 11-22 is designed to operate with yields of a few thousand megajoules at rates of a few hertz. A lithium energy-conversion blanket, consisting of a dense array of 20-cm-diameter jets, is continuously injected into the chamber. This provides an effective blanket thickness of 1 m between the fusion pellet and the inner steel wall. The 14-MeV neutron flux is reduced by a factor of 200 by this lithium blanket, allowing a chamber with a 5-m-radius to operate for 21 full-power years (30 years at 70% availability), at an integrated neutron energy flux of 0.3 MW/m$^2$. Since the flux in the wall without lithium protection would be 5.7 MW/m$^2$ (including pellet effects), the power density within the reactor vessel is very high, approaching that of a fission reactor. A common low-alloy ferritic steel (2.25 Cr–1 Mo) is used throughout. The tritium-breeding ratio is controllable between 0.1 and 1.7.

The chamber contains a hexagonal array (50% packing fraction) of 20-cm-diameter lithium jets, which gives an effective shock isolation from the effects of the fusion pulse. Since the hot gas merely blows through the array of jets, this configuration minimizes the wall stress due to the impact of lithium accelerated by high-pressure blowoff gas, caused by x-ray and debris energy deposition. A substantial fraction of the kinetic energy of expansion of fluid, resulting from the neutron absorption, is dissipated in the liquid-liquid interactions among colliding jets. Finally, the enormous surface area of the fluid acts as an effective condensation pump for the lithium vapor.

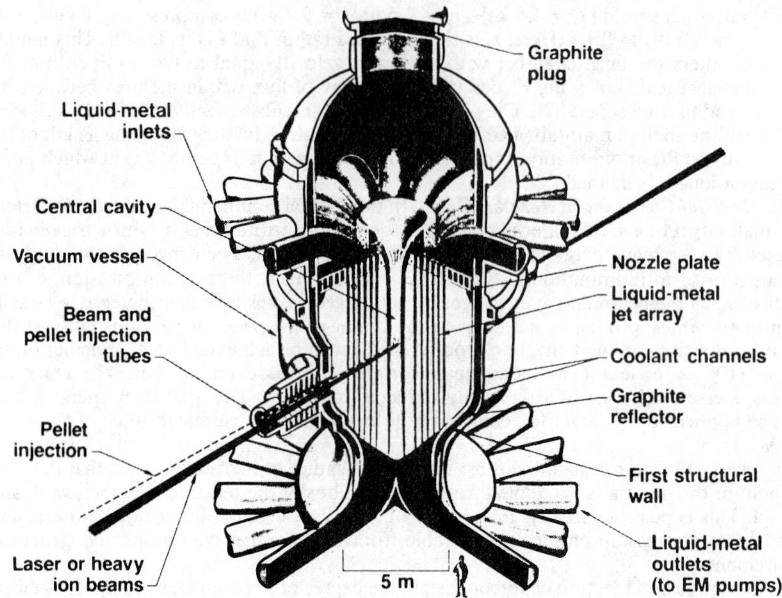


**FIG. 11-22** Artist's concept of the HYLIFE reactor chamber, showing the array of liquid-metal jets that form the reestablishable first wall and blanket. *(From Ref. 12.)*

The HYLIFE power plant requires 16 pumps to inject the liquid lithium into the 5-m-radius chamber. The power necessary for the lithium recirculation pumping is 20 MW (electric), only 1.6% of the gross electric power, if mechanical pumps are used. The laser mirrors are located at the far ends of the containment building, 60 m from the microexplosion. They are protected from the soft x-rays and debris by 0.25 torr of xenon in a 30-m section of the passageway. The neutron flux on the mirrors is 0.042 MW/m$^2$, low enough to assure substrate lifetimes of over a year. The power plant characteristics are summarized in Table 11-10.

***61. Fusion Breeder.*** The physical basis of the fusion breeder (also called the fusion-fission hybrid) is the prolific production of high-energy neutrons in fusion reactions. These neutrons can be used to breed fertile fuel (plutonium and $^{233}$U) for use in conventional fission reactors. Each D–T fusion reaction produces 14-MeV of total energy, or approximately four times as many neutrons per unit of energy as a fission event (~3 neutrons per 200 MeV). The 14-MeV fusion neutrons produce additional neutrons in the breeder blanket by neutron-multiplying nuclear reactions, either by fast fissions in fertile materials and/or nonfissioning reactions, such as $(n, 2n)$ reactions in beryllium.

***62. Breeder Types.*** Two different fusion breeder types based on these two methods of neutron multiplication have emerged. One uses a fast-fission blanket, and the other uses a suppressed-fission blanket. In the former, the D–T fusion source is surrounded by a blanket of fertile material ($^{238}$U and/or $^{232}$Th) and a lithium compound for tritium breeding as shown in Fig. 11-23. Fast fissions in the fertile material multiply both the fusion energy (3 to 10 times) and the neutrons (approximately two to four neutrons per fusion neutron). One of the neutrons is needed to breed tritium from lithium and the remainder are available for breeding fissile fuel.[17]

In the suppressed-fission blanket (see Fig. 11-23), a nonfissioning, neutron-multiplying material such as beryllium replaces most of the fertile fuel in the blanket. Consequently, the fast-fission source of neutron multiplication is replaced with a nonfissioning source of neutrons. To further suppress the fissioning of the bred $^{233}$U, its concentration is limited to about 1% or less of the fertile fuel. Thus, the breeding blanket must be designed to allow for

**TABLE 11-10** Summary of HYLIFE Power Plant Characteristics

| | |
|---|---|
| **System parameters** | |
| Fusion power, MW | 2700 |
| Net electric power, MW (electric) | 1004 |
| Net system efficiency, % | 32 |
| Tritium breeding ratio | 1.0 to 1.7 |
| **Laser and pellet parameters** | |
| Beam energy, MJ | 4.5 |
| Pellet gain (energy multiplication) | 400 |
| Yield, MJ | 1800 |
| Repetition rate, Hz | 1.5 |
| Laser efficiency, % | 5 |
| Laser power consumption, MW (electric) | 135 |
| Fusion energy gain | 20 |
| **Fusion chamber** | |
| Radius, m | 5 |
| Height, m | 8 |
| Material | 2¼ Cr–1 Mo |
| Midplane neutron flux, MW/m² (with lithium) | 0.68 |
| **Lithium array geometry** | |
| Number of jets | 175 |
| Midplane jet diameter, m | 0.2 |
| Midplane packing fraction | 0.57 |
| Effective array thickness, m | 0.47 |
| **Flow parameters** | |
| Midplane jet velocity, m/s | 13.3 |
| Array flow rate, m³/s | 72.2 |
| Total lithium pumping power, MW (electric) | 26.0 |
| Total lithium inventory, m³ | 1850 |
| Lithium outlet temperature, °C | 500 |
| Lithium temperature rise per pulse, °C | 18 |

SOURCE: Ref. 13.

on-line, or quick, low-cost removal and recovery of the bred fissile fuel. The low fission rate results in superior overall reactor safety characteristics. Also, a much lower fission product inventory and a lower decay afterheat results.

***63. Fuel Cycles.*** The suppressed-fission blanket can produce plutonium rather than $^{233}$U by substituting uranium carbide for the thorium, but then the number of client light-water reactors (LWRs) that can be fueled is lower. For example, for a recent design with a 2100-MW fusion driver (4000 $MW_{nuclear}$), the suppressed fission design could support 14 light-water reactors (LWRs) or 4000 $MW_{nuclear}$ each when producing $^{233}$U but only about nine when producing plutonium. The advantages of higher LWR support motivate the desire for new fuel-reprocessing technologies for $^{233}$U and thorium. Two technologies being considered are the conventional, aqueous THOREX reprocessing technology and a pyrochemical process that uses magnesium dissolution of the thorium leaving the $^{233}$U as a precipitate.

***64. Breeder Role.*** Fusion breeders are best understood in the context of a symbiotic fusion-breeder–fission-burner system that generates electricity. The fusion breeder would be incorporated into a fuel-cycle complex along with fuel-reprocessing plants, fuel-fabrication facilities, and possibly a waste-disposal facility—all in a safeguarded area. The fissile fuel produced in the fusion breeder is recovered by reprocessing, mixed with fertile fuel, fabricated into fuel rods, and shipped to the fission burner reactors. The spent fuel from the burner reactors is returned to the safeguarded fuel-cycle complex where the remaining fissile fuel is separated from the radioactive waste material.

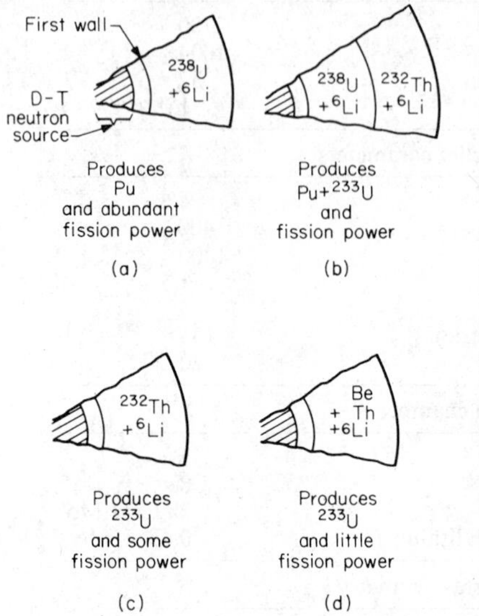

**FIG. 11-23** Methods of neutron multiplication. (*a*) $^{238}$U fast-fission blanket; (*b*) $^{238}$U + $^{235}$Th fast-fission blanket; (*c*) $^{232}$Th fast-fission blanket; (*d*) $^{9}$Be fission-suppressed blanket.

***65. Support Ratio.*** The thermal support ratio is defined as the nuclear (or thermal) power of the client fission reactor (e.g., LWR) divided by the nuclear power of the fusion breeder. A high thermal support ratio is advantageous because such a breeder could fuel a large number of fission reactors, thus having a large commercial impact. Also, for large support ratios, only a small fraction of the symbiotic system's cost and electricity generation is attributed to the addition of a fusion breeder into the existing electricity generation system (typically 15% of the overall capital cost and about 5% of the overall electricity generation). Because relatively little power would be produced within the safeguarded fuel-cycle park itself, the breeders could be in a remote location.

Thermal support ratios for fusion breeders range from 4 to 45, depending on the choice of fusion blanket and client thermal converter (e.g., LWRs or advanced converter reactors). The following support ratio estimates are typical: uranium fast-fission blankets produce enough plutonium to support 4 to 6 LWRs; thorium fast-fission blankets produce enough $^{233}$U to support 8 to 12 LWRs, or 14 to 28 advanced converters; uranium suppressed-fission blankets produce enough plutonium to support 9 to 11 LWRs; thorium suppressed-fission blankets produce enough $^{233}$U to support 12 to 16 LWRs, or 35 to 45 advanced converters.

The variations in these support-ratio estimates depend on the specifics of the fusion-blanket designs, the type of client fission reactor, and fuel-cycle choices.

These support ratios can be put in perspective by comparing them to the support ratios of a liquid-metal fast-breeder reactor (LMFBR). A typical LMFBR does not produce enough excess fissile fuel to support even one LWR of equivalent nuclear power. Furthermore, the LMFBR must also produce fissile fuel to satisfy the fissile inventory requirement of additional LMFBRs. Consequently, LWR support is not an effective mode of LMFBR operation. The fusion breeder, on the other hand, requires no initial fissile inventory, and its tritium inventory is quite low.

***66. Performance Requirements.*** Fusion breeders can operate economically with significantly lower fusion performance and higher cost than fusion electric power plants. We consider two performance indicators for magnetic fusion devices: plasma energy gain $Q$ and first-wall fusion neutron wall loading $\Gamma$. $Q$ is defined as the ratio of the fusion energy produced to the input energy required to heat and sustain the plasma (supplied by relatively expensive beams of energetic neutral atoms and/or rf heating systems). The fusion neutron first-wall loading in megawatts per square meter is indicative of the blanket power density.

Several conceptual design studies of fusion reactors have shown that pure fusion power plants will require $Q > 20$ and $\Gamma > 3$ MW/m$^2$ to produce competitive electricity. Although high $Q$ is preferred, fusion breeders will only require $Q$ values from 2 to 6 and $\Gamma > 1$ MW/m$^2$ to economically produce fuel for LWRs. Fusion breeders with fast-fission blankets can operate in the lower-$Q$ regime (i.e., $Q$ from 2 to 4); with suppressed-fission blankets, $Q$ values above 6 are required.

These performance requirements also apply to inertial confinement fusion (ICF) systems. For example, ICF hybrid studies have shown that a driver-efficiency target gain product ($\eta_D G$) of about 6 is acceptable while an $\eta_D G$ product greater than 20 is required for fusion electric power generation. These lower performance requirements can allow relaxation of technology requirements.

***67. Progress toward Attainment of Controlled Fusion.***[9] Figure 11-24 illustrates the progress toward the goal of controlled nuclear fusion to date. In this figure, the kinetic temperature of a pure deuterium-tritium plasma (ions and electrons taken to have the same

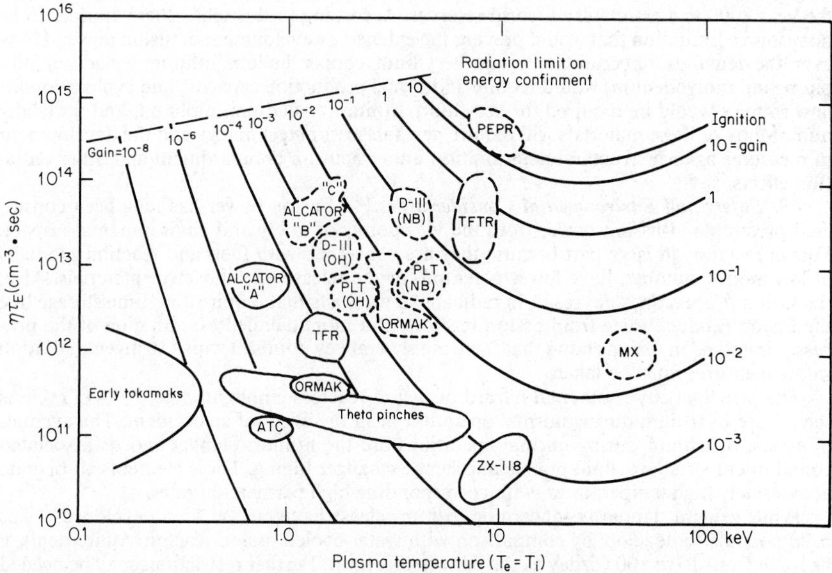


**FIG. 11-24** Progress toward the attainment of controlled nuclear fusion, in terms of plasma temperature and density-confinement-time product. Areas enclosed in solid lines indicate experimental systems that were operational prior to April 1976; those in dashed lines were planned for operation at later dates. The abbreviations have the following meanings:

ALCATOR—high-field tokamak, MIT: (A) 1974, (B) 100 kG, April 1978, (C) 140 kG, October 1978.

ATC—adiabatic toroidal compressor tokamak, Princeton: initial startup 1974, shutdown 1976.

D-III—Doublet tokamak, GA Technologies, Inc., (OH) with ohmic heating, 1978; (NB) upgrade with neutral beam injection 1979.

MX—Large mirror experiment, proposed by Lawrence Livermore Laboratory—experiment to start with 80-kV injection energy.

ORMAK—Oak Ridge National Laboratory tokamak: initial operation (lower left) in 1975, upgrade (center) in 1979.

PEPR—Prototypical experiment power reactor, conceptual design at Oak Ridge National Laboratory assisted by Westinghouse Corporation and GA Technologies, Inc., with assistance of Argonne National Laboratory, to follow TFTR; operation expected by mid-1980s.

PLT—Princeton Plasma Physics Laboratory large-torus tokamak: (OH) with ohmic heating in 1975, (NB) with neutral beam injection in 1977.

TFT—Tokamak Fontenay-aux-Roses (France); initial operation in 1972, shutdown for upgrade in 1976.

TFTR—Tokamak fusion test reactor, Princeton; hydrogen plasma operation in 1981.

2Z-IIB—Mirror reactor, Lawrence Livermore Laboratory; achieved physics goals in 1975, shutdown in 1976.

*(Courtesy U.S. Energy Research and Development Administration, Division of Magnetic Fusion Energy.)*

temperature) is plotted against the density-confinement time product $n\tau$ (Par. **47**). In these coordinates, ignition occurs at any point on or above the locus (top of the diagram) at which the thermonuclear energy gain exceeds 30.

Loci of constant gain (fusion power density at 17.6 MeV per fusion, divided by the external power density required to maintain the plasma temperature against losses) are plotted over a wide range (from $10^{-8}$ to 10) over which fusion confinement experiments have operated, or are projected to operate. The values of the operating areas of specific experiments are identified by the abbreviated symbols for several confinement devices. The highest attained kinetic temperature achieved is in mirror and certain pinch experiments where about 30 keV, equivalent to 300 million degrees absolute, has been achieved.

***68. Unique Resource Requirements.***[9] Table 11-11 lists a number of materials needed to support a nuclear-fusion power economy capable of delivering a total power of a million megawatts (third column) compared with the forecast U.S. demand for these materials in the year 2000 and estimates of world reserves. According to this table, there appears to be no resource limitation that would prevent the extensive development of fusion power. However, the demands for certain materials (beryllium, copper, helium, lithium, vanadium, niobium, and molybdenum) would require additional production capacity, and exploration for new sources would be required for beryllium, lithium, vanadium, niobium, and molybdenum. Many of these materials will require new fabrication technology and the development of measures against structural deterioration under neutron bombardment and other radiation effects.[18]

***69. Safety and Environmental Considerations.***[4,9] Fusion power has long been considered preferable to fission power from the viewpoints of safety and environmental impact. Fusion reactors, in large part because they are concerned with fuels and reaction products of low atomic number, have lower potential for the release of radioactive materials. Also, the fusion process does not result in radioactive products that require long-time storage like the fission-product waste from fission reactors. But more detailed examination of the processes involved in fusion shows that there are several key points at which stringent radiation safety measures must be taken.

The principal environmental hazard of fusion reactors employing the D–T–Li cycle is the release of tritium during normal operation or in the event of an accident. The avenues of escape of tritium during normal operations are the lithium blanket and its associated containment structure, fluid piping, and heat-exchanger tubing. These elements all operate at extremely high temperatures, with corresponding high permeation rates.

While general standards concerning tritium release have not yet been developed, it has been possible to develop, by comparison with water-cooled fission-reactor requirements, a figure of from 10 to 100 Ci/day as the permissible limit. Further restrictions may be needed to protect against the effects of tritium oxides, which are considered to be several hundred times more hazardous than the element itself.

In a 1000-MW fusion plant, the internal throughput of tritium of about $4 \times 10^6$ Ci/day would require a containment factor of 99.9999% to limit the escape to 10 Ci/day. While such a factor is feasible, its attainment at acceptable costs must be demonstrated before large-scale fusion plants employing tritium can be demonstrated to be cost-effective.

Fusion neutrons will also cause activation of the first-wall and blanket materials. This radioactivity can be minimized by judicious selection of materials. The overall biological-hazard potential of a typical fusion blanket system is lower than that of a liquid-metal fast-breeder fission reactor by at least one order of magnitude. While this fact does not in itself demonstrate that fusion reactors are safer than fission reactors, it does indicate that the technical problems of ensuring fusion safety probably will be simpler. Both types of reactor require stringent design constraints in this respect.

The problems of an accidental or forced shutdown involve two major factors: (1) the afterheat associated with loss of coolant of the lithium blanket and its structural elements, and (2) the energy storage of the magnets in the confinement system. The afterheat depends markedly on the structural materials (e.g., silicon carbide, sintered aluminum, vanadium, iron, or nickel) associated with the blanket. The duration of the cooling period may vary from hours to years. These materials have biological-hazard potentials lower than plutonium, but not negligibly so.

The forces associated with the large magnet structures at the high fields required for

**TABLE 11-11** Estimated Quantities of the Unique Resource Requirements Associated with Fusion Power*

| Material | Application | Inventory for $10^6$ MW capacity, metric megatons | Forecast U.S. demand for year 2000, metric megatons | World reserves, metric megatons | Quantity contained in upper 10 m of earth's crust, metric megatons | Ratio of quantity in earth's crust to world reserves |
|---|---|---|---|---|---|---|
| Beryllium | Neutron multiplication | 0.046 | 0.002 | 0.09 | 1.5(4)† | 1.6(5) |
| Aluminum | Coil conductor | 1.0–2.6 | 33 | 1000 | 5.1(8) | 5.1(5) |
| Copper | Coil conductor | 3.2–8.6 | 7 | 300 | 2.3(5) | 7.6(2) |
| Helium | Refrigerant | 0.04–1.1 | 0.02 | 1 | 4(3)‡ | 4(3) |
| Lithium | Fuel, coolant | 0.95–1.5 | 0.014 | 0.8 | 2(5)§ | 2.5(5) |
| Titanium | Structure, S.C.¶ | 0.5 | 2 | 150 | 2.8(7) | 1.9(5) |
| Vanadium | Structure, S.C. | 2.4 | 0.03 | 10 | 6.6(5) | 6.6(4) |
| Niobium | Structure, S.C. | 3.3 | 0.01 | 10 | 9.4(4) | 9.4(3) |
| Molybdenum | Structure | 2.8 | 0.09 | 5.4 | 6.2(3) | 1.1(3) |
| Tin | S.C. | 0.3 | 0.1 | 7.2 | 1.4(4) | 2(3) |
| Lead | Shielding | 11 | 2.5 | 95 | 9.4(4) | 1(3) |

*See Ref. 9 for basic assumptions used in estimates.
†To be read as $1.5 \times 10^4$.
‡In the atmosphere.
§There is approximately the same quantity of lithium in seawater.
¶S.C. = Superconductive materials.
SOURCE: Ref. 9.

fusion are of epic proportions. During normal operation, for example, in a typical tokamak reactor the force tending to push the toroidal field coils toward the center is 100,000,000 lb per coil at a field of 150 kG.

The stored energy in the system can be as high as $10^{11}$ J. Uncontrolled quenching of the field can produce extremely high voltages, so the insulators associated with the magnetic system must have a high safety factor. Even in normal operation, when the field energy is purposely dissipated by passage through a resistance bank, a residual amount of the energy, of the order of 1%, may remain. This residue ($10^{9}$ J of energy) is still enormous by conventional standards.

The foregoing examples illustrate the nature of extent of the engineering problems that will remain to be solved after controlled fusion itself is demonstrated and reduced to practice.

*Acknowledgments.* This article represents an update of the article by Donald G. Fink in the eleventh edition of the *Standard Handbook for Electrical Engineers.* In addition to material from this earlier article, considerable reliance has been placed on material in the excellent review articles by Baker et al.,[8] and by Monsler et al.[13] The reader interested in more details should not only consult the general articles referenced here, but also the more extensive references cited in them.

***70. References***

1. Gough, W. C.: Why Fusion; in Kristiansen, M. and M. O. Hagler (eds.); *Proc. Symposium on Thermonuclear Fusion Reactor Design,* Texas Tech. University, 1970, ORO-3778-3, p. 256, Clearinghouse for Federal, Scientific and Technical Information, Springfield, Va., 1971.
2. Miley, G. H.: *Fusion Energy Conversion,* American Nuclear Society, LaGrange, Ill., 1976.
3. Miley, G. H.: Advanced Fuel Concepts and Applications; *Fusion Reactor Design and Technology,* vol. II, IAEA-TC-392136, International Atomic Energy Agency, Vienna, Austria, 1983, pp. 15–28.
4. Conn, R. W.: The Engineering of Magnetic Fusion Reactors; *Scientific American,* vol. 249, no. 4, pp. 60–71, October 1983.
5. Lawson, J. D.: Some Criteria for a Power Producing Thermonuclear Reactor; *Proc. Royal Society (London),* vol. B70, p. 6, 1957.
6. Miley, G. H.: Overview of Nonelectrical Applications of Fusion; *Proc. 2nd Miami Int. Conf. Alternate Energy Source,* vol. 6, no. 19, p. 2585, 1980.
7. Dolan, T. J.: *Fusion Research;* New York, Pergamon Press, 1982.
8. Baker, C., Carlson, G. A., and Krakowski, R. A.: Trends and Developments in Magnetic Confinement Fusion Reactor Concepts; *Nuclear Technology/Fusion,* vol. 1, no. 5, 1981.
9. Fink, D. G.: Nuclear Fusion; *Standard Handbook for Electrical Engineers,* 11th ed; New York, McGraw-Hill, 1978, p. 11-58.
10. Stacey, W. M., Abdou, M. A., Montgomery, D. B., Rawls, J. M., Schmidt, J. A., Shannon, T. E., and Thome, R. J.: The FED-INTOR Activity; *Nuclear Technology/Fusion,* vol. 4, p. 202, 1983.
11. Johnson, J. L.: The Stellarator Approach to Toroidal Plasma Confinement; *Nuclear Technology/Fusion,* vol. 2, p. 340, 1982.
12. Krakowski, R. A., Glancy, G. E., and Dabiri, Ali E.: The Technology of Compact Fusion Reactor Concepts; *Nuclear Technology/Fusion,* vol. 4, p. 342, 1983.
13. Monsler, M. J., Hovingh, J., Cook, D. L., Frank, T. G., and Moses, G. A.: An Overview of Inertial Fusion Reactor Design; *Nuclear Technology/Fusion,* vol. 1, p. 302, 1981.
14. Brueckner, K. A., and Jorna, S.: Laser-Driven Fusion; *Rev. Modern Physics,* vol. 46, p. 325, 1974.
15. Moses, G. A., and Duderstat, J. J.: *Inertial Confinement Fusion;* New York, Wiley-Interscience, 1982.
16. Gibbs, G. W.: Special Report: The Status of Short-Wavelength Laser Fusion; *Laser Focus,* October 1983.
17. Maniscalco, J. A., Berwald, D. H., Moir, R. W., Lee, J. D., and Teller, Edward: The Fusion Breeder—An Early Application of Nuclear Fusion; UCRL-87801, Lawrence Livermore National Laboratory, June 1983.
18. Gold, R. E., Bloom, E. E., Clinard, F. W., Smith, D. L., Stevenson, R. D., and Wolfer, W. G.: Materials Technology for Fusion: Current Status and Future Requirements; *Nuclear Technology/Fusion,* vol. 1, 1981.

## BATTERIES

*By DAVID LINDEN and ROBERT D. WEAVER*

### Principles of Operation

*71. **Electrochemical Principles and Reactions.*** A battery is a device that converts the chemical energy contained in its active materials directly into electrical energy by means of an oxidation-reduction electrochemical reaction. This type of reaction involves the transfer of electrons from one material to another. In a nonelectrochemical reaction, this transfer of electrons occurs directly and only heat is involved. In a battery (Fig. 11-25) the negative electrode or anode is the component capable of giving up electrons, being oxidized during the reaction. It is separated from the oxidizing material, which is the positive electrode or cathode, the component capable of accepting electrons. The transfer of electrons takes place in the external electric circuit, connecting the two materials. Transfer of charge is completed within the electrolyte by movement of ions, not by electron flow.

The operation of the fuel cell (Par. **99**) is similar to that of a battery except that one or both of the reactants are not permanently contained in the electrochemical cell, but are fed into it from an external supply when power is desired. The fuels are usually gaseous or liquid (compared with the metal anodes generally used in batteries), and oxygen or air is the oxidant.

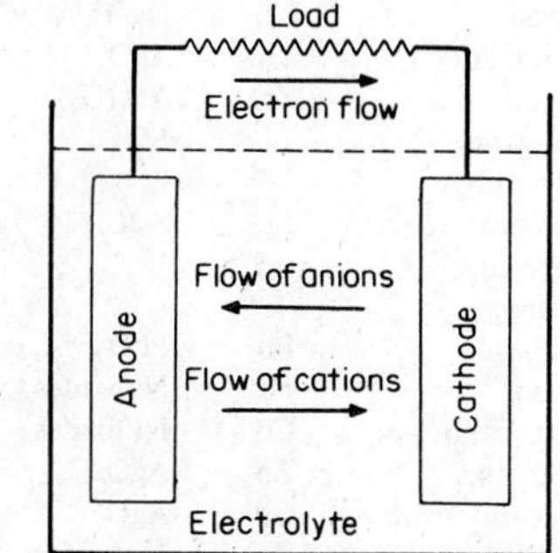


**FIG. 11-25** Electrochemical operation of a battery.

*72. **Components of Batteries.*** The basic unit of the battery is the cell. A battery consists of one or more cells, connected in series or parallel depending on the desired output voltage and capacity. The cell consists of three major components: the anode (the reducing material or fuel), the cathode or oxidizing agent, and the electrolyte which provides the necessary internal ionic conductivity. These electrolytes are usually liquid, but some batteries employ solid electrolytes which are ionic conductors at their operating temperatures. In addition, practical cell design requires a separator material (which serves to separate the anode and cathode electrodes mechanically), electrically conducting grid structures or materials added to each electrode to reduce internal resistance, and suitable containers.

*73. **Theoretical Cell Voltage and Capacity.*** The theoretical capacity (ampere-hours) of a battery system is determined by its active materials. The maximum electrical energy (watt-hours) corresponds to the free-energy change of the reaction. The theoretical voltage and ampere-hour capacities of a number of electrochemical systems are given in Table 11-12. The voltage is determined by the active materials selected, while the ampere-hour capacity is determined by the amount (weight) of available reactants. One gram-equivalent weight of material will supply 96,480 coulombs, or 26.805 Ah of electrical charge.

*74. **Factors Influencing Battery Voltage and Capacity.*** In practice, only a small fraction of the theoretical capacity is realized. This is due not only to the presence of nonreactive components (containers, separators, electrolyte) that add to the weight and volume of the battery, but to many other factors that prevent the battery from performing at its theoretical level.

Factors influencing the voltage and capacity of a battery are as follows:

*Voltage Level.* When a battery is discharged in use, its voltage is lower than the theoretical voltage. The difference is caused by *IR* losses due to cell resistance and by polarization of the active materials during discharge. This is illustrated in Fig. 11-26. The theoretical discharge curve of a battery is shown as curve 1. In this case, the discharge of the battery proceeds at the theoretical voltage until the active materials are consumed and the capacity

**TABLE 11-12** Characteristics of Batteries

| System | Anode | Cathode | Theoretical battery: Voltage, V | Theoretical battery: Capacity, Ah/kg | Practical battery: Typical voltage, V | Practical battery: Capacity[a] Wh/kg | Practical battery: Capacity[a] Wh/dm$^3$ |
|---|---|---|---|---|---|---|---|
| Primary: | | | | | | | |
| Leclanche | Zn | $MnO_2$ | 1.6 | 230 | 1.2 | 65 | 175 |
| Magnesium | Mg | $MnO_2$ | 2.0 | 270 | 1.5 | 100 | 195 |
| Organic cathode | Mg | *m*-DNB | 1.8 | 1400 | 1.15 | 130 | 180 |
| Alkaline $MnO_2$ | Zn | $MnO_2$ | 1.5 | 230 | 1.15 | 65 | 200 |
| Mercury | Zn | HgO | 1.34 | 185 | 1.2 | 80 | 370 |
| Mercad | Cd | HgO | 0.9 | 165 | 0.85 | 45 | 175 |
| Silver oxide | Zn | AgO | 1.85 | 285 | 1.5 | 130 | 310 |
| Zinc-air | Zn | Air ($O_2$) | 1.6 | 815 | 1.1 | 200 | 190 |
| Li-organic electrolyte | Li | $SO_2$[b] | 2.1–5.4 | 130–660 | 1.8–3.2 | 250 | 400 |
| Li-thionyl chloride | Li | $SOCl_2$ | 3.57 | 400 | 3.3 | 400 | 1000 |
| Secondary: | | | | | | | |
| Lead acid | Pb | $PbO_2$ | 2.1 | 55 | 2.0 | 37 | 70 |
| Edison | Fe | Ni oxides | 1.5 | 195 | 1.2 | 29 | 65 |
| Nickel-cadmium | Cd | Ni oxides | 1.35 | 165 | 1.2 | 33 | 60 |
| Silver-zinc | Zn | AgO | 1.85 | 285 | 1.5 | 100 | 170 |
| Silver-cadmium | Cd | AgO | 1.4 | 230 | 1.05 | 55 | 120 |
| Zinc-nickel oxide | Zn | Ni oxides | 1.75 | 185 | 1.6 | 55 | 110 |
| Zinc-air[c] | Zn | Air ($O_2$) | 1.6 | 815 | 1.1 | 150 | 155 |
| Cadmium-air[c] | Cd | Air ($O_2$) | 1.2 | 475 | 0.8 | 90 | 90 |
| Zinc-$O_2$ | Zn | $O_2$ | 1.6 | 610 | 1.1 | 130 | 120 |
| $H_2$-$O_2$ | $H_2$ | $O_2$ | 1.23 | 3000 | 0.8 | 45 | 65 |
| Zn-$Cl^2$[j] | Zn | $Cl_2$ | 2.12 | 396 | 2.1 | 65 | 90 |
| Reserve: | | | | | | | |
| Cuprous chloride | Mg | CuCl | 1.6 | 240 | 1.4 | 45 | 65[d] |
| Zinc-chloride | Mg | AgCl | 1.8 | 170 | 1.5 | 60 | 95[d] |
| Zinc-silver oxide | Zn | AgO | 1.85 | 285 | 1.5 | 30 | 75[e] |
| Thermal | Ca | K[b] | 2.8 | 240 | 2.6 | 10 | 20[f] |
| Ammonia-activated | Mg | *m*-DNB | 2.2 | 1400 | 1.7 | 22 | 60[g] |
| High-temperature | Na | S | 2.08 | 685 | 1.8 | 200 | [h] |
| | Li | S | 2.2 | 1150 | 1.8 | 200 | [i] |

[a]Delivered capacity when discharged at normal temperatures (20°C) at normal discharge rates.
[b]Based on fuel consumption only.
[c]Weight of air not considered in computation of watthours.
[d]Water-activated.
[e]Automatically activated; high rate discharge; 2- to 20-min rate.
[f]Fused salt; heat activated; high rate discharge; 2- to 10-min rate.
[g]Four-minute discharge rate.
[h]$\beta$-alumina electrolyte, 300°C operation.
[i]Fused salt; 350°C operation.
[j]Data for electric vehicle configuration.

fully utilized. The voltage then drops to zero. Under actual conditions, the discharge curve is similar to curve 2. The initial voltage is lower than theoretical, and it drops off as the discharge progresses.

*The Current Drain of the Discharge.* As the current drain of the battery is increased, the *IR* loss increases, the discharge is at a lower voltage, and the service life of the battery is reduced (curve 5). At extremely low current drains it is possible to approach the theoretical capacities (in the direction of curve 3). In a very long discharge period, the chemical self-deterioration during the discharge becomes a factor and causes a reduction of capacity.

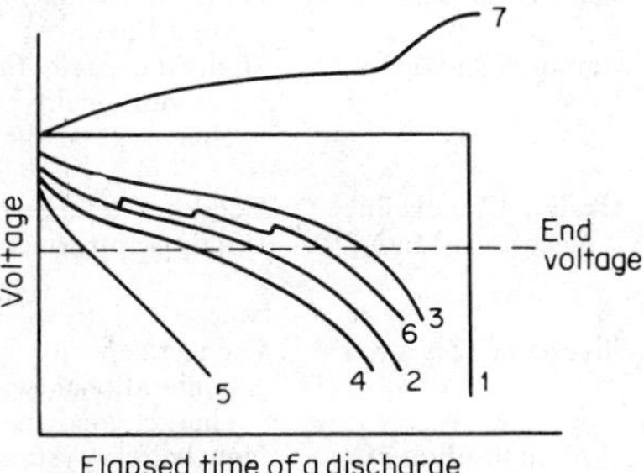


**FIG. 11-26** Battery-discharge characteristics.

*Voltage Regulation.* The voltage regulation required by the equipment is most important. As is apparent by the curves in Fig. 11-26, design of equipment to operate to the lowest possible end voltage results in the highest capacity and longest service life. Similarly, the upper voltage limit of the equipment should be established to take full advantage of the battery characteristics. In some applications, where only a narrow voltage range can be tolerated, voltage regulators may have to be employed to take full advantage of the battery's capacity. If a secondary battery is used in conjunction with another energy source which is permanently connected in the operating circuit, allowances must be made for the voltage required to charge the battery, as illustrated in curve 7, Fig. 11-26. The maximum voltage available from the charger must exceed the maximum battery charge voltage.

## Primary Batteries

***75. General.*** A number of different types of primary batteries are used widely in civilian, industrial, and military applications. They are a convenient, usually relatively inexpensive, lightweight source of power for portable electric devices. The general advantages of primary batteries are reasonably good shelf life, high energy densities at low to moderate rates, little, if any, maintenance, and ease of use. Typical characteristics, applications, or uses of these batteries are shown in Tables 11-12 and 11-13.

***76. Leclanche Cell ($Zn-MnO_2$).*** The Leclanche or carbon-zinc dry cell, known for over a hundred years, is still the most widely used of all the dry-cell batteries because of its low cost, reliable performance, and ready availability. Cells and batteries of many sizes and characteristics have been manufactured to meet the requirements of a wide variety of applications. Characteristics of typical cells are given in Table 11-14.

*Composition.* The Leclanche cell uses a zinc anode, a manganese dioxide cathode, and an electrolyte of ammonium chloride and zinc chloride dissolved in water. Powdered carbon (acetylene black) is mixed with the depolarizer to improve conductivity and retain moisture. As the cell is discharged, the zinc is oxidized and the manganese dioxide reduced. The overall cell reaction is

$$Zn + MnO_2 \rightarrow ZnO + Mn_2O_3$$

*Construction.* The Leclanche cell is made in many shapes and designs but in two basic constructions: cylindrical and flat. Similar chemical ingredients are used in both constructions.

In the common cylindrical cell (Fig. 11-27) a zinc can serves as the cell container and anode. The manganese dioxide is mixed with acetylene black and solid ammonium chloride, wet with a zinc chloride-ammonium chloride electrolyte, and made in the form of a bobbin. A carbon rod is inserted into the bobbin. The rod serves as a current collector and is porous enough to permit the escape of gases, which accumulate in the cell, without allowing leakage of electrolyte. The separator is a cereal paste, also wet with electrolyte, which

**TABLE 11-13** Major Characteristics and Applications of Primary Batteries

| System | Characteristics | Applications |
|---|---|---|
| Zinc-carbon (Leclanche) (zinc-$MnO_2$) | Popular common low-cost primary battery, available in variety of sizes | Flashlight, portable radios and electronics, toys, novelties, instruments, etc. |
| Magnesium (Mg-$MnO_2$) | High-capacity primary battery, long shelf life | Military receiver-transmitters, aircraft emergency transmitters |
| Mercury (Zn-HgO) | Highest capacity (by volume) of conventional types, flat discharge, good shelf life | Hearing aids, medical (heart pacers), photography, detectors, receiver-transmitters, military sensor and detection equipment |
| Alkaline (Zn-alkaline electrolyte-$MnO_2$) | Good low-temperature and high-rate performance, moderate cost | Cassettes and tape recorders, calculators, radio and TV—popular for high-drain primary-battery application |
| Silver-zinc (Zn-AgO) | Highest capacity (by weight) of conventional types, flat discharge, good shelf life | Hearing aids, photography, electric watches, missiles and space application (larger sizes) |
| Lithium (lithium-$SO_2$) | New battery system—recent development; highest-performance primary battery, excellent low-temperature performance, long shelf life | Will have wide, general-purpose application when available. First uses will be military and special civilian applications needing high-capacity and low-temperature performance |
| Solid electrolyte | Extremely long shelf life, low-power battery | Medical electronics, memory circuits, fusing |

**TABLE 11-14** Approximate Service Capacity of American National Standards Institute Sizes of Cylindrical and Flat Zinc-Carbon Cells at Various Current Drains

| USASI cell size | Starting drain, mA | Service capacity, h | USASI cell size | Starting drain, mA | Service capacity, h | USASI cell size | Starting drain, mA | Service capacity, h |
|---|---|---|---|---|---|---|---|---|
| N | 1.5 | 275 | G | 15 | 820 | F24 | 1 | 475 |
| | 7.5 | 52 | | 75 | 150 | | 5 | 150 |
| | 15 | 24 | | 150 | 65 | | 10 | 72 |
| AAA | 2 | 290 | 6 | 50 | 700 | F30 | 1.3 | 275 |
| | 10 | 45 | | 250 | 150 | | 6.5 | 40 |
| | 20 | 17 | | 500 | 70 | | 13 | 16 |
| AA | 3 | 350 | F15 | 0.4 | 210 | F40 | 1.3 | 450 |
| | 15 | 40 | | 2 | 30 | | 6.5 | 108 |
| | 30 | 15 | | 4 | 8 | | 13 | 52 |
| B | 5 | 420 | F12 | 0.5 | 435 | F60 | 2 | 190 |
| | 25 | 65 | | 2.5 | 103 | | 10 | 40 |
| | 50 | 25 | | 5 | 51 | | 20 | 15 |
| C | 5 | 430 | F17 | 0.6 | 710 | F70 | 3 | 550 |
| | 25 | 100 | | 3 | 155 | | 15 | 150 |
| | 50 | 40 | | 6 | 75 | | 30 | 65 |
| D | 10 | 500 | F20 | 0.7 | 210 | F80 | 3 | 600 |
| | 50 | 105 | | 3.5 | 35 | | 15 | 165 |
| | 100 | 45 | | 7 | 12 | | 30 | 72 |
| E | 15 | 400 | F22 | 0.8 | 475 | F90 | 3 | 770 |
| | 75 | 70 | | 4 | 98 | | 15 | 200 |
| | 150 | 30 | | 8 | 49 | | 30 | 90 |
| F | 15 | 520 | F25 | 1 | 500 | F100 | 5 | 1,000 |
| | 75 | 105 | | 5 | 105 | | 25 | 260 |
| | 150 | 45 | | 10 | 45 | | 50 | 110 |

physically separates the two electrodes and provides the means for ion transfer through the electrode. In the newer "paper-lined" cell, an absorbent kraft paper is used as the separator. This provides thinner separator spacing and lower internal resistance.

Another cylindrical cell is the "inside-out" construction shown in Fig. 11-28. In this cell, an injection-molded impervious inert carbon wall serves as the container of the cell and as the current collector. The zinc anode, in the shape of vanes to increase its surface area, is located inside the cell and surrounded by the cathode mix. This ensures efficient zinc consumption and, since zinc is not used as a container, a high degree of leakage resistance.

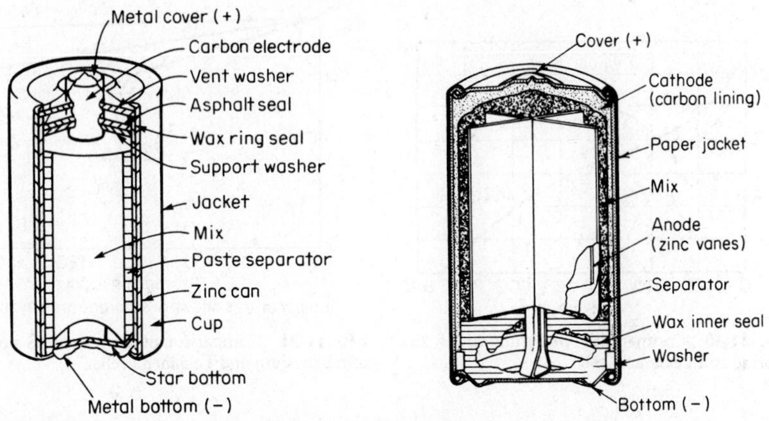


**FIG. 11-27** Cross section of a Leclanche cylindrical cell.

**FIG. 11-28** Cross section of a Leclanche "inside-out" cell.

The flat-cell construction is illustrated in Fig. 11-29. In this cell, carbon is coated on a zinc plate to form a duplex electrode—a combination of the zinc of one cell and the carbon of the adjacent one. The flat cell has a higher energy-to-volume ratio, as the rectangular shape utilizes the available volume better than the cylindrical cell.

*77. **Zinc Chloride Cell.*** A recent modification of the Leclanche cell is the zinc chloride electrolyte cell. The construction is similar to the conventional carbon-zinc cell but the electrolyte contains only zinc chloride, without the saturated solution of ammonium chloride. The zinc chloride cell is a high-performance cell with improved high-rate and low-temper-

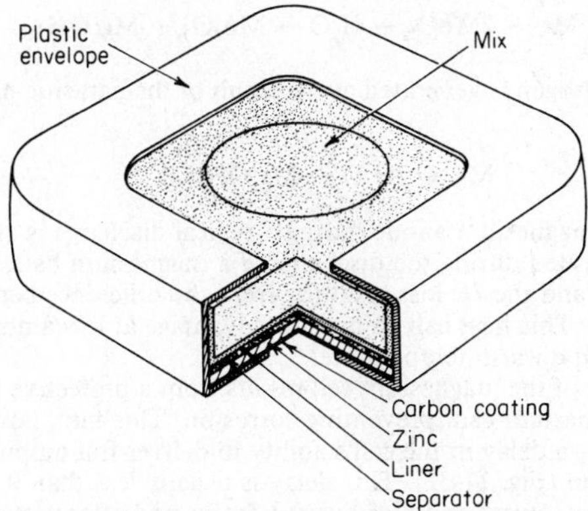


**FIG. 11-29** Leclanche flat cell.

ature performance and a reduced incidence of leakage. A comparison of the performance of the zinc chloride cell with the conventional cell is presented in Fig. 11-30.

***78. Magnesium Dry Cells (Mg–MnO$_2$).*** The magnesium dry cell has two main advantages over the zinc-carbon cell: twice the capacity or service life of an equivalently sized zinc cell and the ability to retain this capacity during storage, even at elevated temperatures (Fig. 11-31). The construction of the magnesium dry cell is similar to the cylindrical zinc-carbon cell except that a magnesium alloy is used instead of zinc. The cathode consists of an

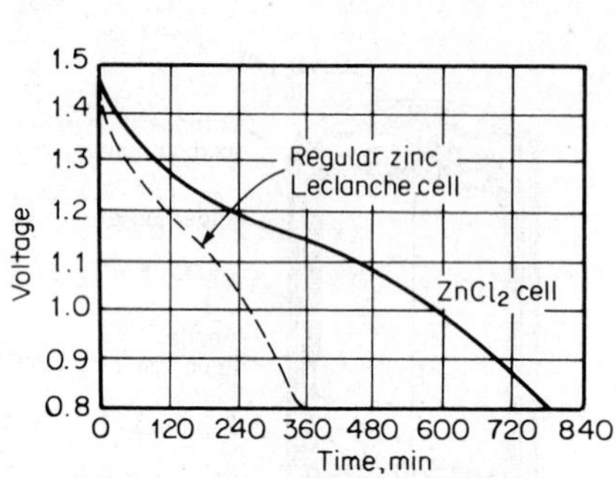


**FIG. 11-30** Comparative performance of zinc chloride and Leclanche cells.

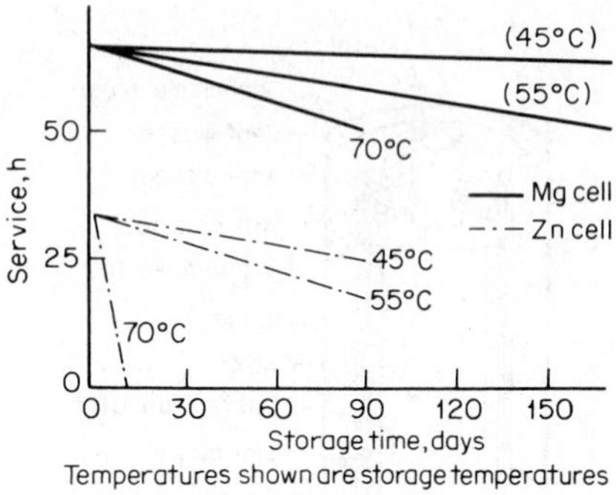

**FIG. 11-31** Comparison of service versus storage of magnesium and Leclanche cells.

extruded mix of manganese dioxide, acetylene black (to provide conductivity and moisture absorbency), magnesium perchlorate electrolyte, barium and lithium chromate as corrosion inhibitors, and magnesium hydroxide as a buffering agent to improve storageability. The degree of "wetness" or amount of water is critical as water participates in the anode reaction and is consumed during the discharge. A carbon rod serves as the cathode current collector. The separator is an absorbent kraft paper as in the "paper-lined" structure. Sealing of the magnesium cell is critical, as it must be tight to retain cell moisture during storage, but provide a means for the escape of hydrogen gas which forms as the result of a parasitic reaction during the discharge of the battery. This is accomplished by a mechanical vent—a small hole in the plastic top seal washer under a retainer ring which is deformed under pressure, releasing the excess gas. Magnesium batteries have not been fabricated successfully in flat-cell designs.

The overall reaction of the Mg–MnO$_2$ cell is

$$Mg + 2MnO_2 + H_2O \rightarrow Mn_2O_3 + Mg(OH)_2$$

At the same time, hydrogen is generated as the result of the parasitic magnesium corrosion reaction:

$$Mg + 2H_2O \rightarrow Mg(OH)_2 + H_2$$

The efficiency of the magnesium anode during a typical discharge is about 70%. Considerable heat thus is generated during the discharge of a magnesium battery owing to the exothermic side reaction and the *IR* loss resulting from the difference between the theoretical and operating voltage. This heat can be used to advantage at low ambient temperatures to maintain the battery at a warm temperature.

The good shelf life of the magnesium cell results from a protective film which forms on the inside of the magnesium can, preventing corrosion. This film, however, is responsible for a voltage "delay"—a delay in the cell's ability to deliver full output voltage after it has been placed under load (Fig. 11-32). This delay is usually less than 0.3 s but is longer for discharges at low temperatures and high current drains and after prolonged storage at high temperatures.

Typical discharge curves are given in Fig. 11-33. The magnesium battery is less sensitive to discharge rate than the zinc-carbon cell. The performance of the magnesium cell for various discharge rates and temperatures is shown in Fig. 11-34.

While successful in military use, the magnesium battery has not found wide commercial use. This is due to the fact that the magnesium battery loses its excellent storageability after being partially discharged and hence is unsatisfactory for long-term intermittent use. Other influencing factors are the higher unit cell voltage and the evolution of hydrogen and heat on discharge, which presents a potential safety hazard.

***79. Zinc-Mercuric Oxide Cells (Zn-HgO).*** The zinc-alkaline–mercuric oxide battery is noted for its high capacity per unit volume, a relatively constant output voltage during its discharge, and good storage characteristics.

*Composition.* The zinc-mercuric oxide cell uses amalgamated zinc as the anode, mercuric oxide (mixed with 5 to 10% graphite) as the cathode, and potassium hydroxide as the electrolyte. A saturated solution of zinc oxide is added to the electrolyte to retard the corrosion of zinc, minimize the production of hydrogen, and improve the stability of the cell.

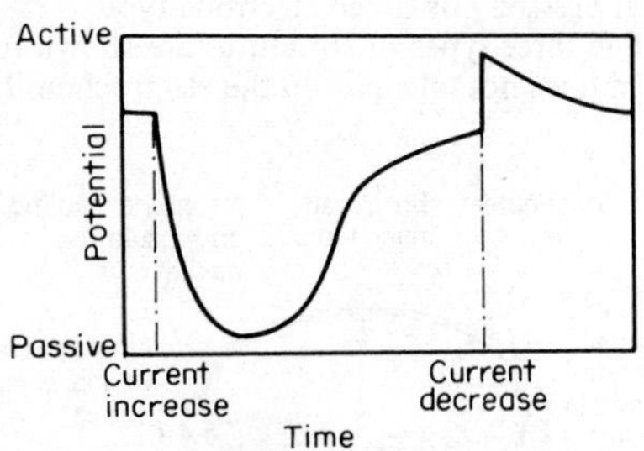


**FIG. 11-32** Voltage-delay characteristic of magnesium cell.

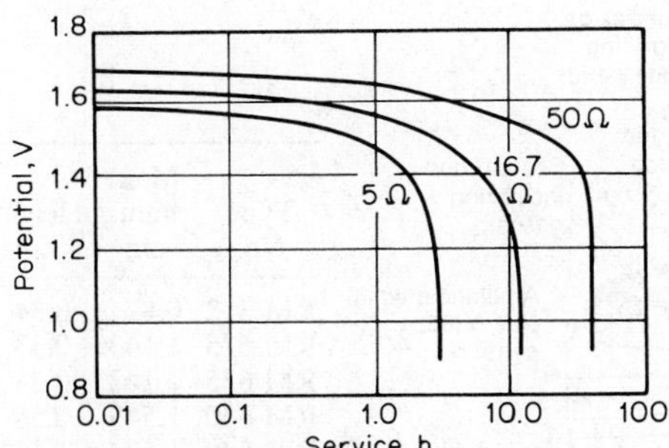


**FIG. 11-33** Discharge curves of magnesium cells (A size, 20°C).

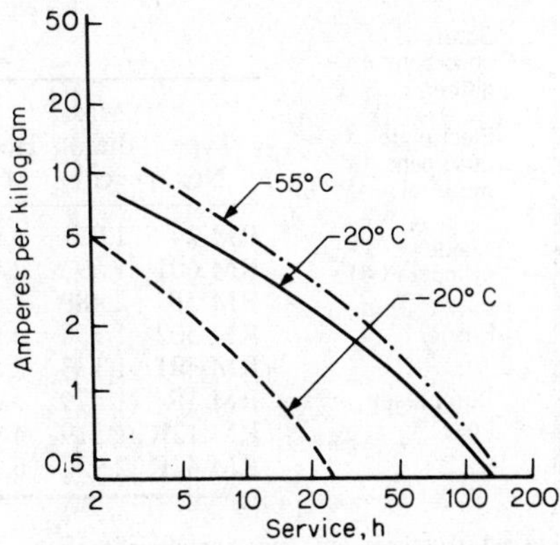


**FIG. 11-34** Service capacity of magnesium cell.

The overall chemical reaction during discharge is

$$Zn + HgO \rightarrow ZnO + Hg$$

The ampere-hour capacity of the mercuric oxide cathode and the zinc anode are equalized or balanced, and on completion of the cell's discharge, no residual unoxidized zinc remains. Without this feature, the zinc would continue to react, generating hydrogen gas in the cell.

*Construction.* The zinc-mercuric oxide cell is manufactured in three basic structures: the wound-anode type, the flat-pressed powdered cathode and anode type, and the cylindrical pressed powdered electrode type.

The three types of structures are shown in Fig. 11-35. All constructions use a steel can, which does not take part in the electrochemical reaction and is not consumable, for the cell

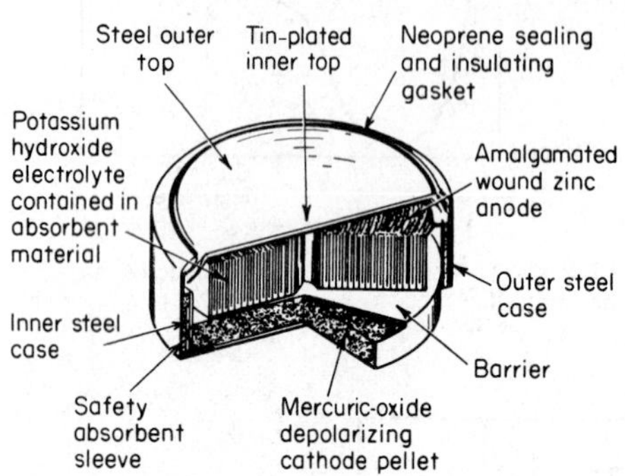


| Type No. | Max. diam., cm | Height, cm | Weight, g | Rated capacity, mAh |
|---|---|---|---|---|
| RM 640 | 1.587 | 0.965 | 9.68 | 360 |
| RM 3 | 2.498 | 1.37 | 22.56 | 1,500 |
| RM 1438 | 3.71 | 1.003 | 36.22 | 2,700 |
| RM 1450 | 3.71 | 136 | 51.80 | 4,500 |
| RM 2550 | 6.58 | 1.394 | 165.20 | 13,000 |

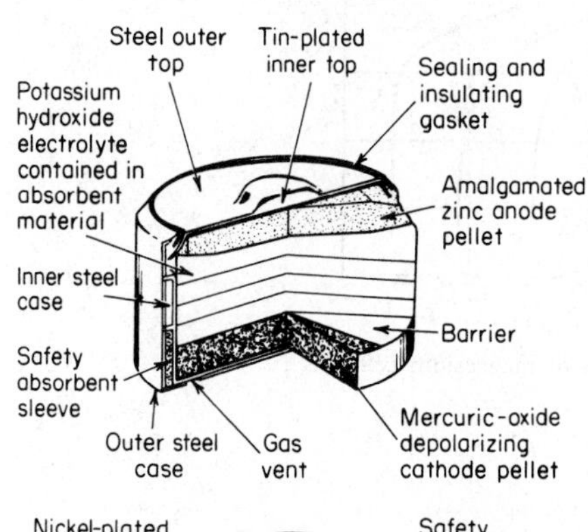


| Type No. | Max. diam., cm | Height, cm | Weight, g | Rated capacity, mAh |
|---|---|---|---|---|
| RM 312 | 0.87 | 0.34 | 0.56 | 36 |
| RM 575 | 1.143 | 0.33 | 1.4 | 100 |
| RM 675 | 1.143 | 0.54 | 2.24 | 160 |
| RM 630 | 1.549 | 0.58 | 4.76 | 350 |
| RM 640 | 1.574 | 1.104 | 7.84 | 500 |
| RM 4R | 3.02 | 1.66 | 40.88 | 3,400 |

| Type No. | Max. diam., cm | Height, cm | Weight, g | Rated capacity, mAh |
|---|---|---|---|---|
| RM 24 | 1.0 | 4.396 | 14.0 | 900 |
| RM 601 | 1.59 | 2.857 | 34.16 | 1,800 |
| RM 3R | 2.489 | 1.651 | 28.56 | 2,200 |
| RM 502 | 1.358 | 4.90 | 30.44 | 2,400 |
| RM 401 | 1.133 | 2.844 | 11.20 | 800 |
| RM 1R | 1.579 | 1.638 | 12.04 | 1,000 |
| RM 12R | 1.519 | 4.959 | 30.88 | 3,600 |
| RM 42R | 2.922 | 6.032 | 148.33 | 14,000 |

**FIG. 11-35** Zinc–mercuric oxide cell structures (left) and characteristics (right).

container. In the wound-anode type, the anode is composed of a corrugated zinc strip with an absorbent paper wound in an offset manner so that it protrudes at one end and the zinc protrudes at the other end. The zinc is amalgamated with 10% mercury and the paper impregnated with the electrolyte, which causes it to swell and produce a positive contact pressure. The cathode is a pellet made of powdered mercuric oxide and graphite which is pressed into the steel can. An absorbent KOH-resistant separator is placed between the two electrodes. The cell is crimped sealed; the can is separated from the top by an insulating neoprene or plastic grommet.

In the pressed-powder cells, the zinc powder is amalgamated and pressed into a pellet with sufficient porosity to allow electrolyte impregnation. A double-can structure is used in the larger-sized cells as a safeguard. Under excessive gas pressures, the compression of the upper part of the grommet by internal pressure allows the gas to escape into the space between the two cases. A paper tube surrounds the inner can so that any liquid carried by the discharging gas will be absorbed, maintaining a leak-resistant structure. Release of the excess gas pressure reseals the cell.

The cylindrical and button cell are variations of the pellet design. The button cell uses a gelled electrolyte to reduce electrolyte leakage further. The mechanical and electrical specifications of representative cells of each of the three constructions are given in Fig. 11-35.

*Voltage.* The open-circuit voltage of the mercury cell is 1.35 V and is reproducible within 0.001 V. On discharge, the cell is characterized by a very flat discharge as shown in Fig. 11-36. The cutoff voltage usually is 0.9 to 1.0 V per cell.

***80. Cadmium-Mercuric Oxide Cell (Cd-HgO).*** The substitution of cadmium for zinc results in a very stable system, with a predicted shelf life of up to 10 years, as well as performance at low temperatures (Fig. 11-37). The watthour capacity of this cell, because of its lower voltage, is about 60% of zinc-mercuric oxide cell capacity. In design, the cadmium-mercuric oxide cell is similar to the zinc-mercuric oxide cell.

***81. Zinc-Silver Oxide Cell (Zn-AgO).*** The primary zinc-alkaline-silver oxide cell also is similar in design to the small zinc-mercuric oxide button cell but uses silver oxide in place of mercuric oxide. Cells range in capacity up to 200 mAh for use at 50-h and lighter loads. The silver oxide cell has an open-circuit voltage of 1.6 V and operates about 0.2 V higher than the mercuric oxide cell. Typical discharge curves for this cell are given in Fig.

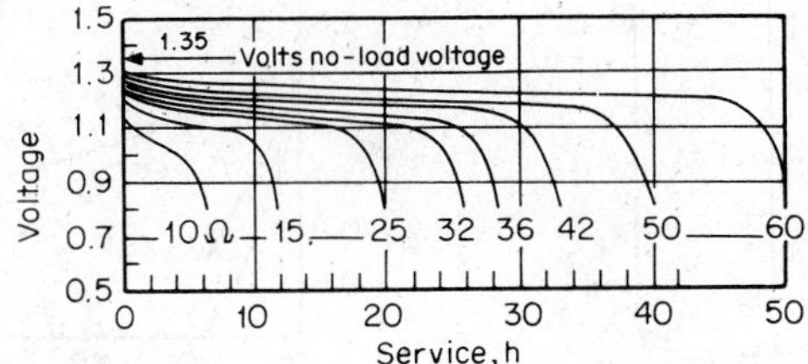


**FIG. 11-36** Discharge curves of zinc-mercuric oxide cells.

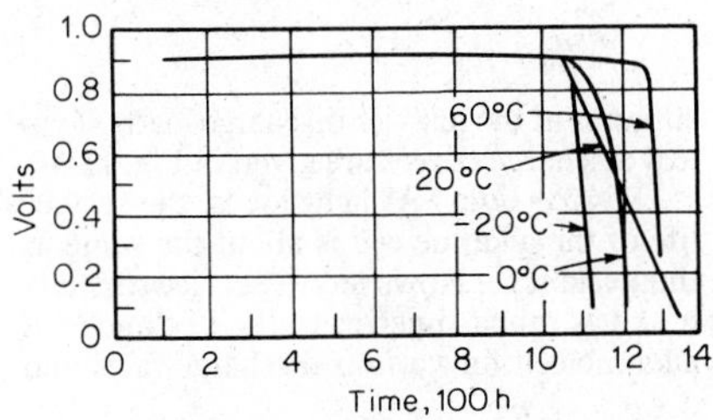


**FIG. 11-37** Discharge curves of zinc-mercuric oxide cell. *(R. Thorton, 26th PSS.)*

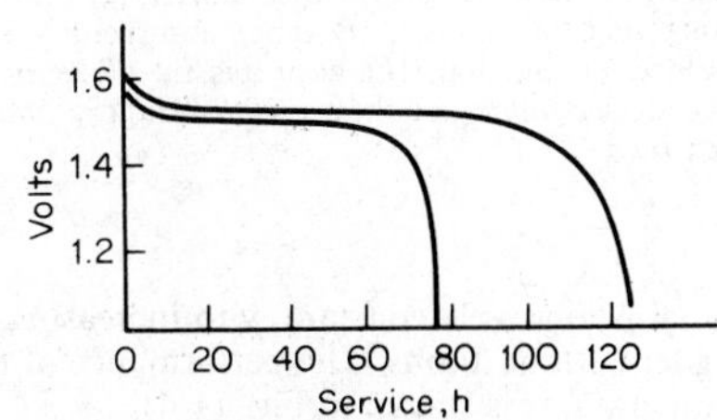


**FIG. 11-38** Discharge curves of zinc-silver oxide cell.

11-38. The silver oxide cell has a higher energy density (on a weight basis) and is less sensitive to a reduction in ambient temperature than the mercuric oxide cell. At the design loads, the cell will deliver about 70% of its 20°C performance at 0°C and 35% at −20°C. These characteristics make this battery desirable for use in hearing aids, photographic applications, and electronic watches.

***82. Alkaline-$MnO_2$ Cell (Zn–$MnO_2$).*** The zinc-alkaline-$MnO_2$ cell uses the same electrochemically active materials, zinc and manganese dioxide, as the Leclanche cell but differs in construction and in the use of highly conductive potassium hydroxide electrolyte which result in a lower internal resistance. The advantage on low-rate or intermittent discharge is marginal, but on high and continuous drain conditions, the alkaline cell can deliver from two to ten times the ampere-hour capacity of the Leclanche cell. Its performance at low temperatures is superior to other commercially available dry batteries.

$$Zn + 2MnO_2 \rightarrow Mn_2O_3 + ZnO$$

The electrolyte undergoes no change during the discharge, maintaining its high conductivity throughout the cell's life. It thus differs from the Leclanche cell, whose resistance increases during the discharge.

*Construction.* The principal features of the Zn–$MnO_2$ cell are the manganese dioxide cathode of high density, a zinc anode of high surface area, and the highly conductive potassium hydroxide electrolyte. As illustrated in Fig. 11-39, the $MnO_2$, mixed with graphite or carbon black, is pressed against the inner surface of the can which serves as the cathode current collector. The anode is centrally located and consists of a mixture of granular or powdered zinc, which is amalgamated to reduce hydrogen evolution, and the electrolyte. In some designs, a gelling agent is used to immobilize the electrolyte and minimize leakage. A highly absorbent chemically inert separator separates the electrodes.

*Voltage.* The open-circuit voltage of the alkaline-$MnO_2$ cell is 1.5 V. Its discharge also is similar to the Leclanche cell, but it is superior at the heavier discharge loads. Typical discharge curves are given in Fig. 11-40.

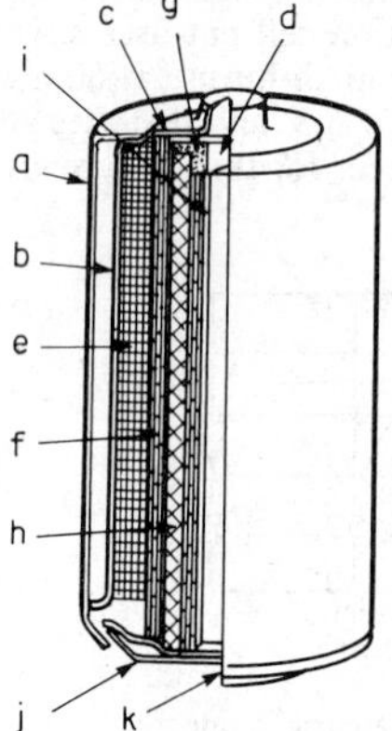


**FIG. 11-39** Construction of alkaline-$MnO_2$ cells. (*a*) Outer nickel-plated can; (*b*) tube adapter; (*c*) inner gold-plated can; (*d*) insulator disk; (*e*) depolarizer; (*f*) outer absorbent with barrier; (*g*) insulating ring; (*h*) anodes; (*i*) inner absorbent; (*j*) molded double top; (*k*) clear plastic dielectric jacket. Electrolyte not shown. *(P.R. Mallory and Co., Inc.)*

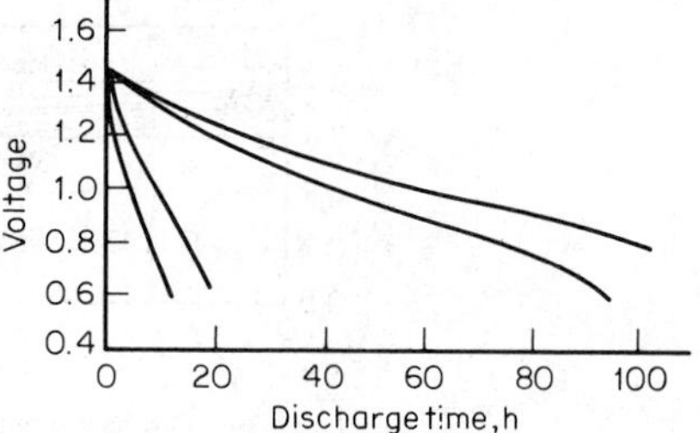


**FIG. 11-40** Discharge curves of an alkaline-$MnO_2$ cell at 20°C.

*Service Life.* At light loads, the service life of the alkaline cell is about the same as the Leclanche. However, its service capacity remains relatively constant with increasing load and is much superior to the Leclanche at higher current drains. The performance of the alkaline cell for various discharge rates and temperatures is shown in Fig. 11-41.

*Effect of Temperature.* The alkaline-$MnO_2$ cell performs well at low temperatures, excelling over the best Leclanche cells. The cell operates to temperatures as low as −40°C.

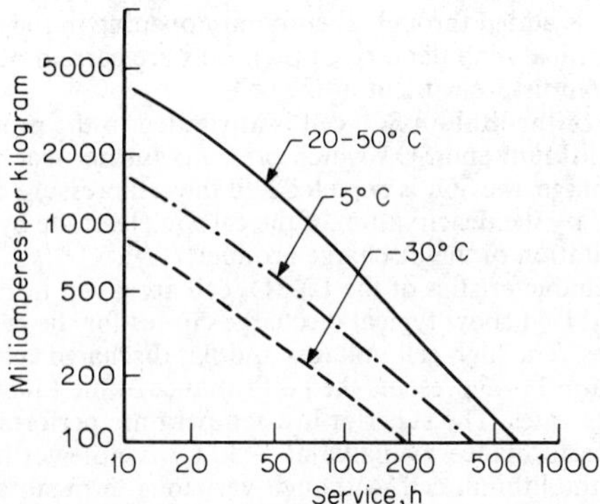


**FIG. 11-41** Service capacity of alkaline-$MnO_2$ cell.

*Shelf Life.* The shelf life of the alkaline-$MnO_2$ cell is moderately superior to the Leclanche cell. Capacity retention is about 90% after 1 year of storage at 20°C.

***83. Lithium Primary Batteries.*** The lithium battery is a recent development which has a number of advantages over other primary-battery systems. Lithium is an attractive anode because of its reactivity, light weight, and high voltage; cell voltages range between 2 and 3.6 V, depending on the cathode material.

The advantages of the lithium battery include: high energy density, in the order of 250 Wh/kg and 400 Wh/$dm^3$; high power density; flat discharge characteristics; excellent service over a wide temperature range, down to −40°C or below; good shelf life; up to 5 years without refrigeration is anticipated.

Nonaqueous solvents must be used as the electrolyte because of the solubility of lithium in aqueous solutions. Organic solvents, such as acetonitrile and propylene carbonate, and inorganic solvents, such as thionyl chloride, are typical. A compatible solute is added to provide the necessary electrolyte conductivity. A number of different materials—sulfur dioxide, carbon monofluoride, vanadium pentoxide, copper sulfide—are used as the active cathode material.

***84. Lithium-Sulfur Dioxide Cell.*** An advanced lithium primary cell uses sulfur dioxide for the cathode material and an electrolyte consisting of acetonitrile and lithium bromide. The cell reaction mechanism is

$$2Li + 2SO_2 \rightarrow Li_2S_2O_4 \text{ (lithium dithionite)}$$

The cell is typically fabricated in a cylindrical structure as shown in Fig. 11-42. A "jellyroll" construction is used, made by spirally winding strips of lithium ribbon, a polypropylene separator, and the cathode electrode (a Teflon-carbon mix pressed on an aluminum screen). This design provides the high surface area and low cell resistance which are necessary to obtain high-current and low-temperature performance. The roll is inserted in a steel container which is electrically connected to the anode. The cathode, in turn, is connected to its terminal and the cell hermetically sealed. The electrolyte-depolar-

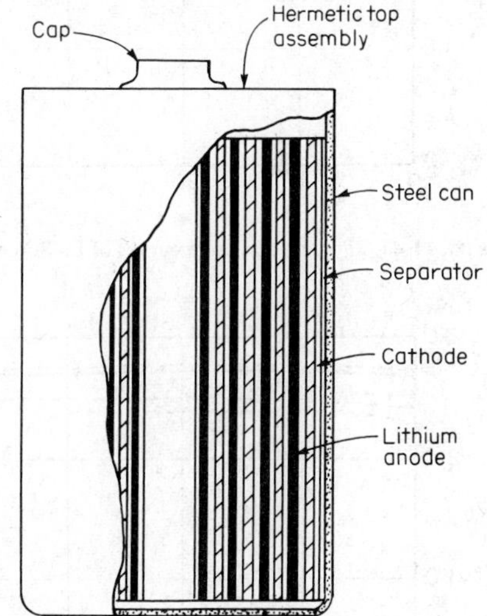


**FIG. 11-42** Cross section of cylindrically wound Li–$SO_2$ cell.

izer mixture (70% $So_2$) is added through a temporary opening in the cell to an $SO_2$ pressure of about 4 atm. The critical manufacturing operations are carried out in dry rooms or dry boxes to minimize the moisture content of the cell.

The good shelf life of the lithium-$SO_2$ cell is attributed to the protective film formed by the initial reaction of lithium and $SO_2$, which prevents further reaction or loss of capacity on stand. During discharge, the $SO_2$ is depleted and the cell pressure reduced. The discharge is generally terminated by the deactivation of the carbon electrode by blocking of the active area due to the precipitation of the discharge product.

The performance characteristics of the Li–$SO_2$ cell are given in Figs. 11-43 through 11-46. Figures 11-43 and 11-44 show typical discharge curves for the cell at various discharge loads and temperatures. The high cell voltages and flat discharge curves shown are characteristic of this cell. Figure 11-45 presents the performance of the Li–$SO_2$ cell at various temperatures and discharge rates. The superior low-temperature performance (over 60% of the normal-temperature performance available at −30°C) is noteworthy. Figure 11-46 illustrates the shelf life of the lithium cell. Although very long term storage has not been demonstrated experimentally, the data suggest a long-shelf-life capability even at high temperatures, particularly with the newer cell designs.

Special attention must be given to the design and use of the lithium battery, as it contains materials that are potentially flammable and toxic. Properly designed cells are equipped with safety vents which release $SO_2$ when the cells reach high temperatures and pressures, thus preventing explosive damage.

The lithium battery can deliver unusually high current. As high internal temperatures can develop from continuous high current drain, short circuiting, or inadvertent internal

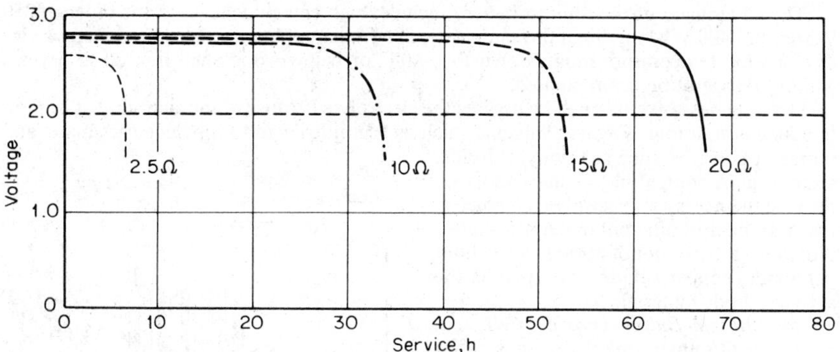


**FIG. 11-43** Typical discharge curves of Li–$SO_2$ cell at 20°C, at various loads.

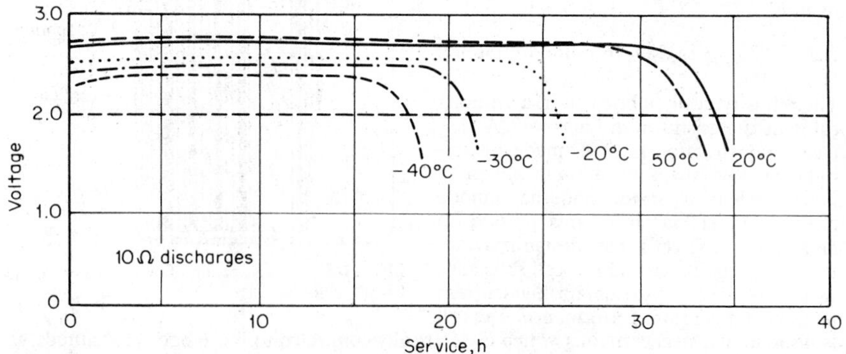


**FIG. 11-44** Effect of temperature on performance of Li–$SO_2$ cells (D size).

cell shorts, such use must be avoided. It is advisable to equip batteries with fuses to protect against short circuiting.

Charging lithium-$SO_2$ cells may result in explosion of cells (even those that are equipped with safety vents) and should not be attempted. Similarly, cells or groups of cells should not be connected in parallel without diode protection to prevent one group of cells from charging the other. Forced discharging, which could occur with cells which are connected in series or to an external power supply, may also result in venting and/or explosion. Currently, special procedures govern the transportation, shipment, and disposal of lithium batteries. While it requires special handling and design, the many advantages of the lithium cell predict an increasing use of this battery in military and civilian areas.

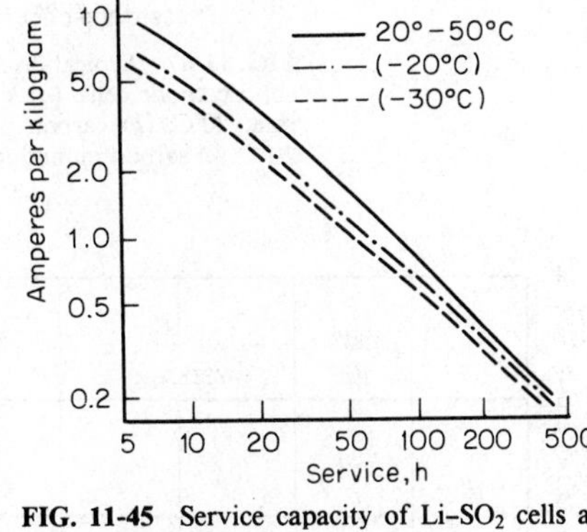


**FIG. 11-45** Service capacity of Li–$SO_2$ cells at various temperatures.

***85. Other Lithium-Organic Electrolyte Cells.*** A number of other cathode materials, mostly solid, are under the development for lithium primary cells, including reserve-type batteries. Cells using these solid cathodes have the advantage of being nonpressurized but do not have the high current capability of the $SO_2$ system. Some cells are specifically designed for low-rate applications, using "bobbin"-type constructions. These cells deliver higher energy outputs and should provide safe operation, as they are self-limiting in energy and current output. Typical discharge curves for the carbon monofluoride and vanadium pentoxide cells are shown in Fig. 11-47.

***86. Inorganic-Electrolyte Cells.*** Certain nonaqueous inorganic liquids, such as thionyl chloride ($SOCl_2$) and sulfuryl chloride ($SO_2Cl_2$), also are capable of forming a passivating film on the lithium anode, similar to that in $SO_2$, and can be used both as the electrolyte solvent and as the active cathode material. A commonly used solute is $LiAlCl_4$. Cells made with these components are similar in construction to the Li–$SO_2$ cell but are not pressurized at room temperature because of the lower vapor pressure of the thionyl chloride. The Li–$SOCl_2$ cells also operate at approximately 0.5 V higher than the comparable Li–$SO_2$ cell. Figure 11-48 compares the discharge curves of the two batteries and illustrates the

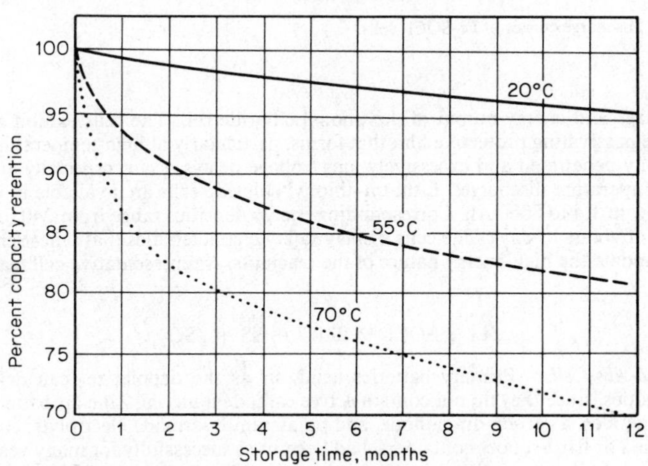


**FIG. 11-46** Shelf life of Li–$SO_2$ cell at different storage temperatures.

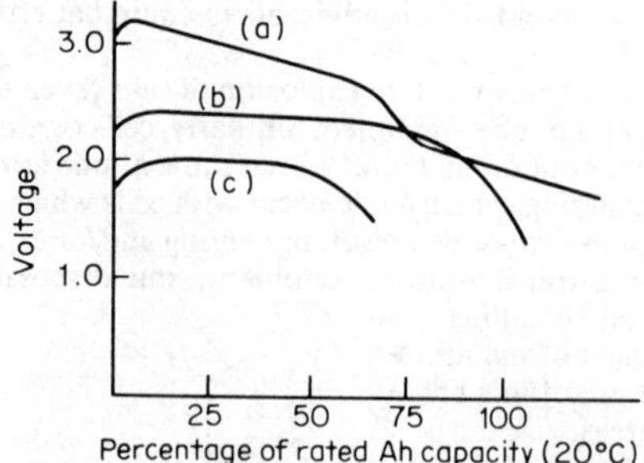


**FIG. 11-47** Typical discharge curves of lithium–solid cathode cells. (*a*) Vanadium pentoxide, 8-h rate, 20°C; (*b*) carbon monofluoride, 20-h rate, 20°C; (*c*) carbon monofluoride, 20-h rate, −20°C.

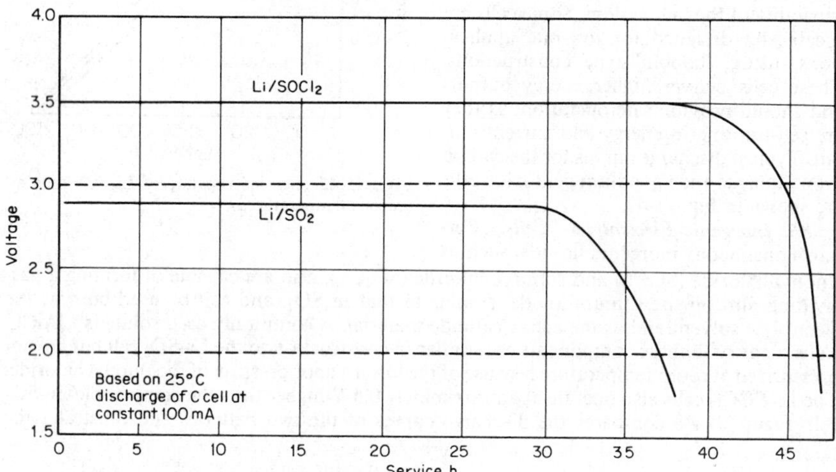


**FIG. 11-48** Discharge curves of Li–$SOCl_2$ cells.

higher voltage and energy output of the thionyl chloride cell. The cells exhibit good shelf life, but the passivating protective film that forms, particularly at high-temperature storage, is not readily penetrated and excessively long voltage delays occur, especially at high-rate and low-temperature discharges. Lithium–thionyl chloride cells are available in a range of capacities from 1.4 to 8000 Ah. Corresponding energy densities range from 240 to 450 Wh/kg at rates of about 10 days. The cells display safety characteristics that reflect engineering to accommodate the high-energy nature of the reactants. A representative cell reaction is

$$2Li + SOCl_2 \rightarrow 2LiCl + \tfrac{1}{2}S + \tfrac{1}{2}SO_2$$

***87. Zinc-Air Cells.*** Primary batteries using air as the depolarizer can deliver high energy densities, since they do not contain active cathode material. Zinc-air batteries, using bulky zinc anodes, a carbon-air cathode, and potassium hydroxide electrolyte, constructed in heavy-glass or hard-rubber containers, had been used successfully for many years in railway signals and similar applications. They were noted for their high energy densities, but low power output capability. Lower-capacity zinc-air batteries, up to the 20-Ah size, are now being developed, using thin Teflon-coated fuel-cell-type electrodes. These new structures have high energy densities, in the order of 220 Wh/kg, and are capable of moderately high current drains.

*Construction.* A typical construction of the primary zinc-air cell is shown in Fig. 11-49. It consists of a high-surface-area gelled zinc anode, fabricated by pressing zinc granules with a gelled potassium hydroxide electrolyte, and a teflonated air cathode made of low-cost non-noble-metal catalyst. Cylindrical-shaped and button cells have also been designed. Effective sealing is essential in this construction to prevent electrolyte leakage. The cells are stacked to form a battery; space must be left between each cell for air circulation.

*Performance Characteristics.* The zinc-air battery is best suited for continuous moderate drain discharges at temperatures between −20 and +35°C. Intermittent operation usually results in a loss of capacity due to the drying out of the cell. As the cell depends on oxygen from the air for its operation, differences in performance occur with variation in air circulation. Figure 11-50 shows typical curves for continuous discharge at 20°C. The flat discharge is charac-

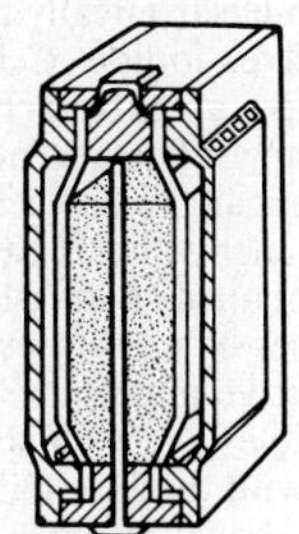

**FIG. 11-49** Construction of primary zinc-air cell. *(Gould, Inc.)*

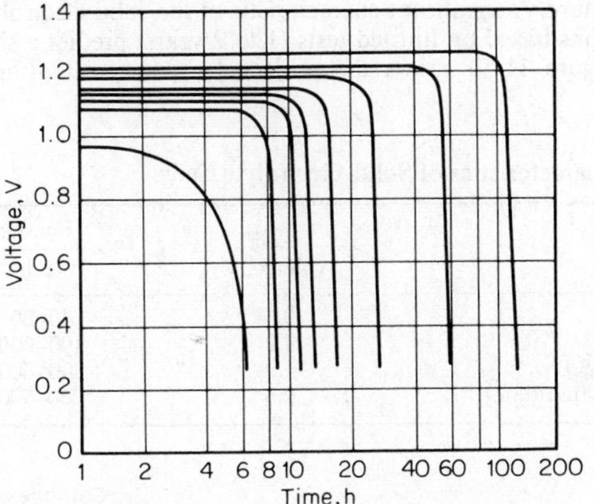


**FIG. 11-50** Typical discharge curves of zinc-air primary cells. *(Gould, Inc.)*

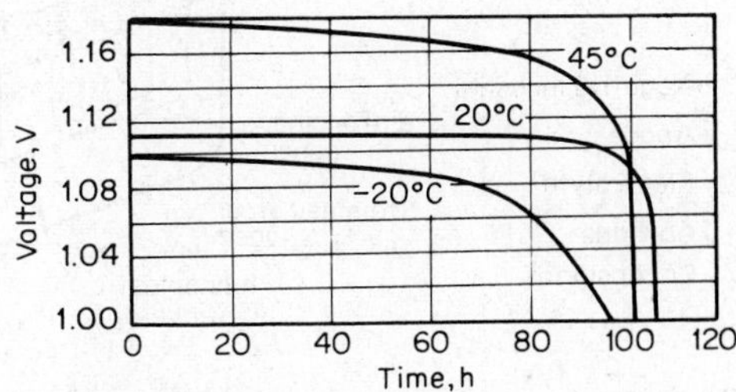


**FIG. 11-51** Voltage versus time curve of zinc-air cell at various temperatures. *(Gould, Inc.)*

teristic of this cell. Figure 11-51 shows the energy output of the zinc-air battery at various temperatures. In addition to the reduction of service life, the average voltage decreases with decreasing discharge temperature.

Primary zinc-air cells must be stored in sealed bags after manufacture to maximize storage life. Limits of storage as well as the integrity and leak resistance of the cell seal have yet to be determined. Once the packaging has been removed, the battery should be put into service shortly thereafter, since moisture loss results in dry-out and reduced capacity.

***88. Solid-Electrolyte Batteries.*** Most batteries depend on the ionic conductivity of liquid electrolytes for their operation. The solid-electrolyte batteries depend on the ionic conductivity of an electronically nonconductive salt in the solid state as, for example, $Ag^+$ ion mobility in silver iodide. Cells using these solid electrolytes are low-power (microwatt) devices but have an extremely long shelf life and the capability of operating over a wide temperature range. The absence of liquid eliminates corrosion and gassing and permits the use of a hermetically sealed cell. The solid-electrolyte batteries are used in medical electronics (in devices such as heart pacemakers), for memory circuits, and for fusing and other such applications requiring a long-life, low-power battery.

Several types of solid-electrolyte batteries are being marketed using different solid electrolytes and active materials. The characteristics of several of the available types are summarized in Table 11-15. Of special significance are the high energy densities (5 to 10 Wh/in$^3$) achieved with the Li-anode solid-electrolyte battery. Typical construction of solid-electrolyte cells is shown in Fig. 11-52. The design features a sealed structure to exclude moisture and maintain a high-density, void-free package.

The discharge curves for various solid-electrolyte cells at 25°C are given in Fig. 11-53. As the batteries are designed primarily for low current drain, continuous discharge at high rates is not practical. The energy density and power density also are dependent upon the operating temperature. A significant characteristic of the solid-electrolyte battery is its long shelf life. Projections based on limited tests (1 to 2 years) predict a shelf life exceeding 15 years at 20°C. Figure 11-54 shows the projected capacity retention at various storage temperatures.

**TABLE 11-15** Characteristics of Solid-Electrolyte Cells

| System | Cell voltage, V | Energy density at 100-h rate | |
|---|---|---|---|
| | | WH/dm$^3$ | Wh/kg |
| $Ag/RbAg_4I_5/I_2$ | 0.66 | 40–80 | 15–25 |
| $Li/LiI(Al_2O_3)/PbI_2$, Pb | 1.9 | 100–200 | 35–70 |
| $Li/LiI(Al_2O_3)PbI_2$,PbS,Pb | 1.9 | 300–500 | 75–150 |
| $Li/LiI/I_2$ (poly-2-vinylpyridine) | 2.8 | 250–500 | 120–180 |

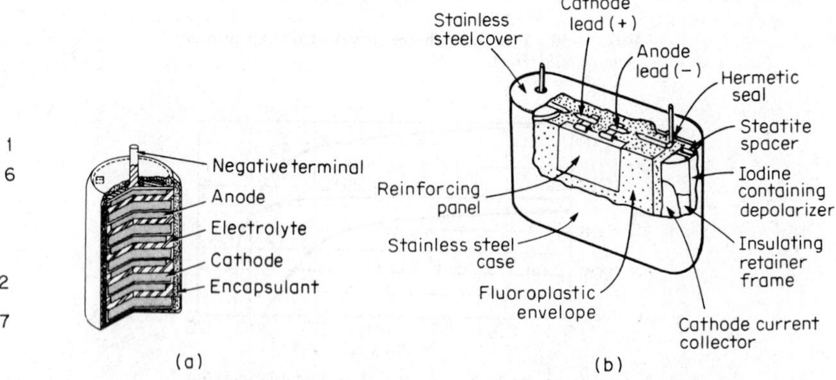


**FIG. 11-52** Construction of solid-electrolyte cells. (*a*) Silver iodide cell; (*b*) lithium iodide cell.

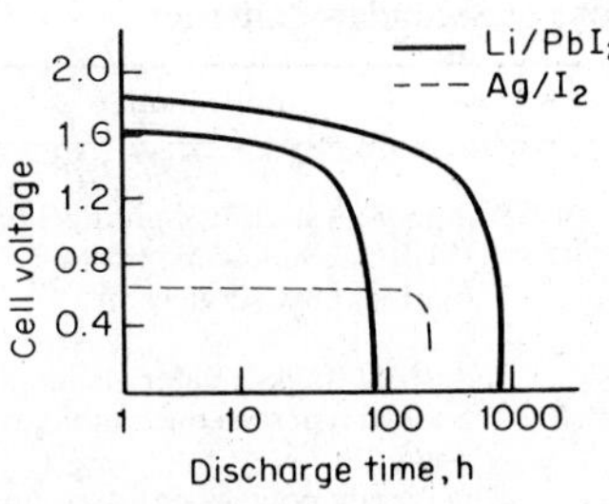


FIG. 11-53 Typical discharge curves of solid-dielectric-electrolyte cells.

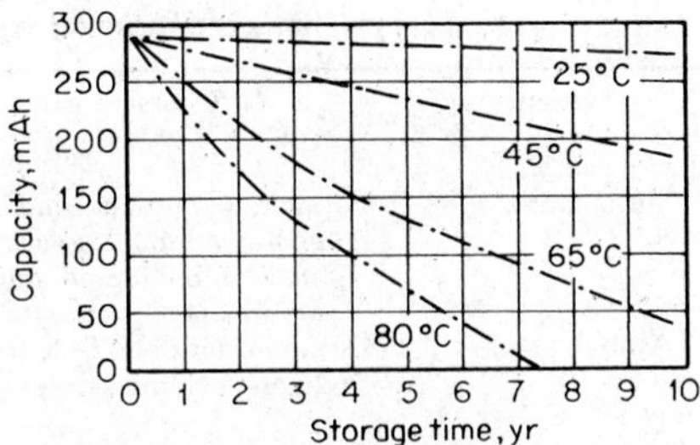


FIG. 11-54 Capacity retention versus temperature of solid-electrolyte cells. *(Gould, Inc.)*

***89. Other Primary Batteries.*** Many other electrochemical systems have been used as primary batteries to obtain special performance characteristics. The more prominent ones are listed in Table 11-13.

***90. Hybrid Configurations.*** In the hybrid configuration a high-energy-density primary battery is combined with a high-power-density secondary battery to improve the overall battery performance. The secondary battery handles the high-power and low-temperature performance requirements (where the primary battery is least efficient) and is recharged by the primary battery. The primary battery is discharged at a relatively light load to maximize the energy output. This configuration provides a means for obtaining high efficiency under adverse environmental and load conditions.

***91. Recharging Primary Batteries.*** Recharging primary batteries is a practice that should generally be avoided, as the cells are not designed for such use. In most instances it is impractical and it could be hazardous in cells that are tightly sealed and not vented to permit the release of gases that form during charge and discharge. Most primary cells and batteries are labeled with a cautionary notice advising that they should not be recharged.

Some Leclanche zinc-carbon cells can be recharged for several cycles under carefully controlled conditions. For successful recharging, the cell should be placed on charge soon after removal from discharge and at a low rate (about 16 h charge time). The cell voltage on discharge should not be below 1.0V when it is removed for charging. The cells must be returned to service soon after recharging, as the shelf life after recharge is poor.

## Secondary Batteries

***92. General.*** Secondary batteries are widely used in many applications. They are characterized, in addition to their ability to be recharged, by high power density, high discharge rate, flat discharge curves, and good low-temperature performance. Their energy densities are usually lower than those of primary batteries. Table 11-16 lists the characteristics of secondary batteries.

The applications of secondary batteries fall into two major categories:

1. Those applications where the secondary battery is used essentially as a primary battery, but recharged after use. Secondary batteries are used in this manner for convenience (as in hand-held calculators or electronic flash units), for cost savings (as they can be recharged rather than replaced) or for power drains beyond the level of primary batteries.
2. Those applications where the secondary battery is used as an energy-storage device, being charged by a prime energy source and delivering its energy to the load on demand. Examples are automotive and aircraft systems, emergency no-fail and standby power sources, and hybrid applications.

A summary of some of the major applications of the various types of secondary batteries is given in Table 11-16.

**TABLE 11-16** Major Characteristics and Applications of Secondary Batteries

| System | Characteristics | Applications |
|---|---|---|
| Lead-acid: | | |
| Automotive | Popular, low-cost secondary battery—moderate capacity, high-rate and low-temperature performance | Automobile starting, lighting, ignition (SLI); lawnmowers, tractors, marine, float service |
| Motive power | Designed for deep 6- to 9-h discharge, cycling service | Industrial trucks, materials handling. Special types used for submarine power |
| Stationary | Designed for standby float service, long stand life | Emergency power—utilities, no-break systems |
| Sealed | Sealed, maintenance-free, low cost, good float capability | TV, portable tools, lights and appliances, radios and cassettes and tape players |
| Nickel-cadmium: | | |
| Vented | Good high-rate, low-temperature capability; flat voltage, excellent cycle life | Aircraft batteries, industrial and emergency-power applications, communication equipment |
| Sealed | Good high-rate, low-temperature performance, excellent cycle life, maintenance-free | Photography, portable tools, appliances, standby power |
| Zinc–silver oxide | Highest energy density, good high-rate capability, low cycle life | Lightweight portable radio, TV, and communication equipment; torpedo propulsion, drones, submarines, and other military applications |

***93. Lead-Acid Battery ($Pb-PbO_2$).*** The lead-acid battery is the most widely used secondary battery. Its low cost, reliability, and generally favorable performance characteristics account for its acceptance in many different applications. This type of battery is manufactured in many sizes, ranging in capacity from less than 1 Ah (small plastic-encased or sealed portable cells) to several thousand ampere-hours for stationary and vehicle-propulsion types. Characteristics of typical cells are summarized in Table 11-16.

*Composition.* The lead-acid battery uses highly reactive sponge lead for the negative electrode, lead dioxide as the active positive material, and a sulfuric acid solution for the electrolyte. As the cell discharges, the active materials of both electrodes are converted into lead sulfate. The sulfuric acid electrolyte also takes part in the reaction producing water. On charge, the reverse actions take place. The state of charge of the battery can be determined by measuring the specific gravity, which decreases on discharge, increases on charge. The discharge and charge reactions of the battery are

$$Pb + PbO_2 + 2H_2SO_4 \underset{\text{charge}}{\overset{\text{discharge}}{\rightleftharpoons}} 2PbSO_4 + 2H_2O$$

At the end of the charge, electrolysis of water also occurs, producing hydrogen at the negative and oxygen at the positive electrode.

*Construction.* The most common construction for the lead-acid cell is the pasted-plate design. A cross section of an automotive-type battery using this construction is shown in Fig. 11-55. The active material for each electrode is prepared as a paste by mixing finely divided lead oxides and suitable expander materials with sulfuric acid. The paste is spread onto a lead-alloy grid which provides the necessary electrical conductivity and structure to hold the active materials. The resultant plates are soldered to connecting straps to form positive and negative groups which are interleaved. Separators are placed between the electrodes, and the completed element is placed in a container. The container is designed with a sediment space under the element to collect safely any of the active material that dislodges. Sufficient headroom is provided above the plates to hold excess electrolyte. Conventional lead-acid batteries employ antimonial-lead grids. The alloying is necessary to provide adequate strength for the thin grid structure to facilitate casting.

The maintenance-free lead-acid batteries, which became popular in the mid-1970s, use calcium-lead grids which are more resistant to corrosion and self-discharge. Water loss from gassing during charge has been minimized in this new battery. The battery is filled with an excess of electrolyte and sealed to prevent contamination damage. Other design features are improved separator materials which reduce the possibility of internal shorting, enclosed internal connectors, and lightweight, high-impact-strength polypropylene cases.

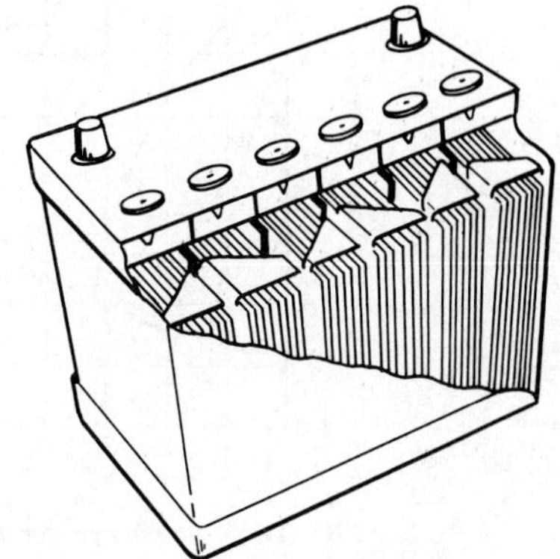

**FIG. 11-55** Cross section of lead-acid automotive battery. *(Gould, Inc.)*

Other lead-acid batteries are similar in design to the automotive battery but vary in lead-alloy composition, plate thickness, separators, container, etc., to optimize the performance characteristics for the particular application.

***94. General Performance Characteristics.*** The general performance characteristics of the lead-acid battery are given in Fig. 11-56.

*Voltage.* The nominal voltage of the lead-acid cell is 2 V; the voltage on open circuit is a direct function of the specific gravity, ranging from 2.12 V for a cell with 1.28 specific gravity to 2.05 V at 1.21. Figure 11-57 presents typical discharge curves for the lead-acid cell. The end voltage is usually about 1.75 V but can be as low as 1.0 V at extremely high rates, as in automotive starting service.

*Specific Gravity.* The selection of specific gravity used for the electrolyte depends on the service requirements. The electrolyte concentration must be high enough for good electrical conductivity and to fulfill electrochemical requirements, but not so high as to cause separator deterioration or corrosion of other parts of the cell, which would shorten life and increase self-discharge. A specific gravity of 1.26 to 1.28 is usually used in automotive and high-performance batteries and as low as 1.21 for stationary standby batteries. The specific gravity should be reduced in high-temperature climates.

During discharge (Fig. 11-56) the specific gravity decreases about 0.125 to 0.150 points from a fully charged to a fully discharged condition. The change is proportional to the

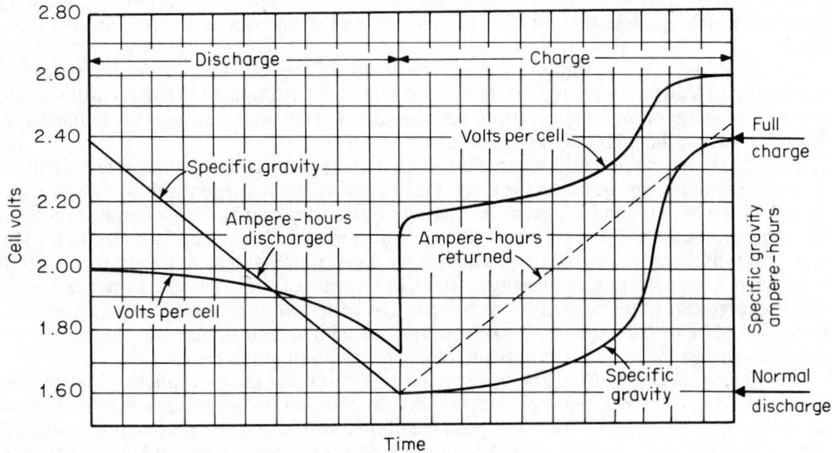


**FIG. 11-56** Performance characteristics of lead-acid batteries.

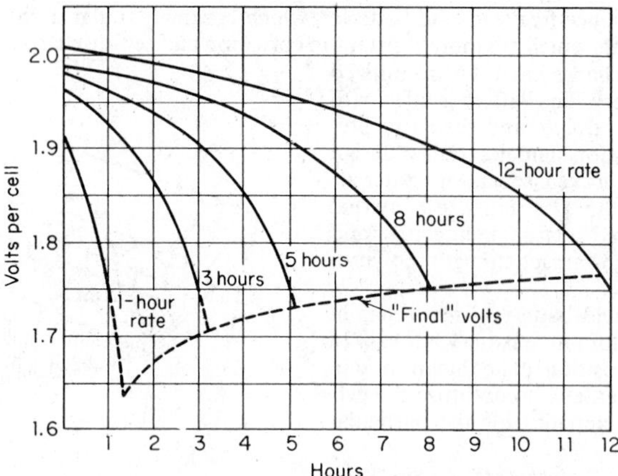


**FIG. 11-57** Discharge curves of lead-acid batteries at different hour rates.

ampere-hours discharged. The specific gravity is, thus, an excellent means for checking the state of charge of the battery. A short period of time should be allowed prior to measurement after completion of the discharge for equalization of the concentration throughout the cell. On charge, the change in specific gravity should similarly be proportional to the ampere-hour charge accepted by the cell, but there is a lag, as complete mixing of the electrolyte does not occur until gassing commences near the end of the charge.

*Service Life.* The service life of a typical automotive-type lead-acid cell is shown in Fig. 11-58 for different discharge rates and temperatures. These curves have been normalized at the 20-h rate at 25°C. A 100-Ah cell, for example, will deliver 20 h of service when discharged at 5 A at 25°C or 1 h of service when discharged at −40°C at 20 A. Typically, higher service capacity is obtained at lower discharge rates and higher temperatures. In general, a battery may be discharged without harm at any rate of current it will deliver, but the discharge should not be continued beyond the point where the cell approaches exhaustion or where the voltage falls below a useful value.

Automotive cells are made with thinner plates, have a larger surface area, and use a higher concentration of electrolyte than motive-power and stationary batteries, so higher currents can be drawn at higher voltage levels. Hence the electrical output of stationary batteries, which are designed for long-life service, will be inferior to that indicated in Fig. 11-58.

*Temperature.* The variation of the performance of the lead-acid cell at different temperatures and loads is given in another form in Fig. 11-59. Although the battery will operate over a wide temperature range, continuous operation at high temperatures may reduce cycle life as a result of the increase in the rate of corrosion.

*Self-Discharge.* The self-discharge rate (loss of battery capacity on storage) is dependent on a number of factors, inluding the type of lead alloy used, the concentration of electrolyte, the age of the battery, and temperature. Self-discharge is caused by local chemical reaction between components of the plate and occurs almost entirely in the negative electrode. The rate of self-discharge is about 15% per month for the antimonial-lead batteries at 25°C. Batteries using purer lead grids have substantially lower rates of self-discharge. Capacity lost by self-discharge can be recovered by recharging the battery. For best practice, a battery on stand should be recharged every 3 to 6 months, since prolonged storage can cause irreversible damage and make recharge difficult, owing to sulfation of the plates.

***95. Lead-Calcium Cells.*** The lead-calcium cell uses a small percentage of calcium to give the lead grid the necessary physical rigidity in place of the larger amount of antimony normally used. Thus it is closer to a pure-lead grid and has considerably reduced self-discharge due to local chemical action. The calcium alloy cells are best suited for standby float or open-circuit service rather than a cycling type of use. Frequent recharging in cycle use

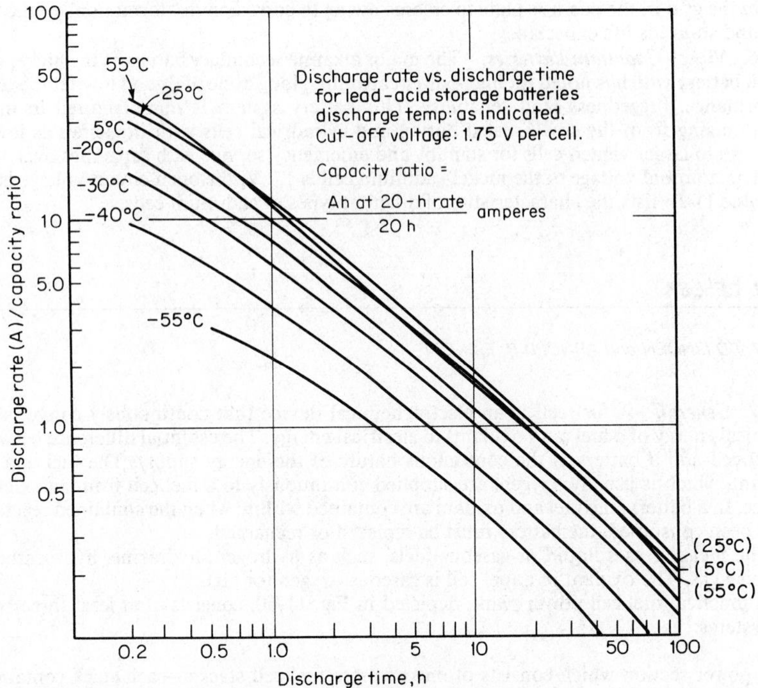


**FIG. 11-58** Service-capacity curves of lead-acid batteries.

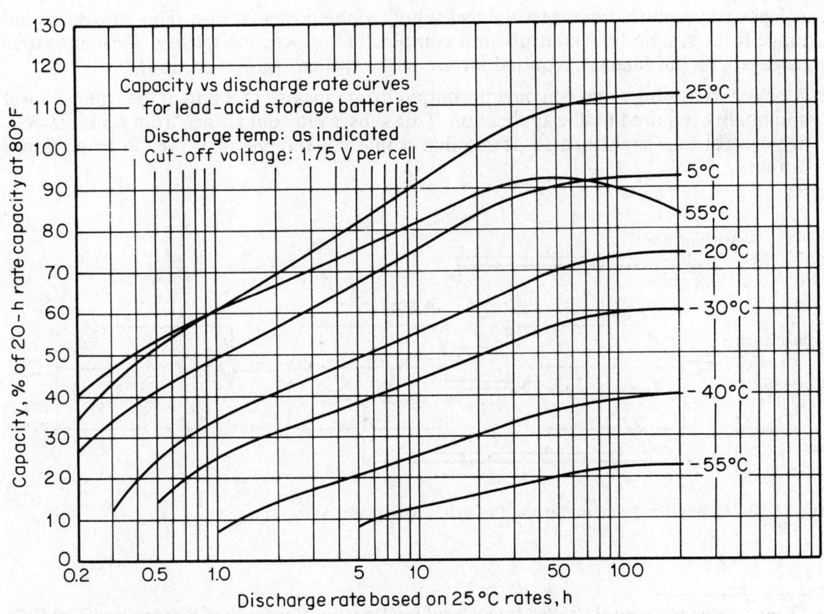


**FIG. 11-59** Performance of lead-acid batteries at various temperatures.

causes the grid in the positive plate to enlarge owing to corrosion or "formation" of the lead grid and shortens life expectancy.

***96. Nickel-Cadmium Batteries.*** The major alkaline secondary battery is the nickel-cadmium battery which is noted for high power capability, long cycle life, good low-temperature performance, ruggedness, and reliability. This battery system is manufactured in many sizes, ranging from the small sealed button and cylindrical cells with capacities as low as 0.020 Ah to larger vented cells for standby and emergency service with capacities over 1000 Ah. The nominal voltage of the nickel-cadmium cell is 1.2 V; the open-circuit voltage is 1.4 V. Table 11-16 lists the characteristics of different types of cadmium cells.

## FUEL CELLS*

*By DAVID LINDEN and ARNOLD P. FICKETT*

***97. General.*** A fuel cell is an electrochemical device that continuously converts the chemical energy of a fuel (and oxidant) to electrical energy. The essential difference between a fuel cell and a battery is the continuous nature of the energy supply. The fuel and the oxidant, which is usually oxygen, are supplied continuously to a fuel cell from an external source. In a battery, the fuel and oxidant are contained within; when the contained reactants have been consumed, the battery must be replaced or recharged.

The fuel cell uses liquid or gaseous fuels, such as hydrogen, hydrazine, hydrocarbons, and coal gas. The oxidant in a fuel cell is gaseous oxygen (or air).

A practical fuel cell power plant, depicted in Fig. 11-60, consists of at least three basic subsystems:

**a.** A power section which consists of one or more fuel cell stacks—each stack containing many individual fuel cells usually connected in series to produce a stack output ranging from a few to several hundred volts (direct current). This section converts processed fuel and the oxidant into dc power.

**b.** A fuel subsystem that manages the fuel supply to the power section. This subsystem can range from simple flow controls to a complex fuel-processing facility. This subsystem processes fuel to the type required for use in the fuel cell (power section).

**c.** A power conditioner that converts the output from the power section to the type of power and quality required by the application. This subsystem could range from a simple voltage control to a sophisticated device that would convert the dc power to an ac power output.

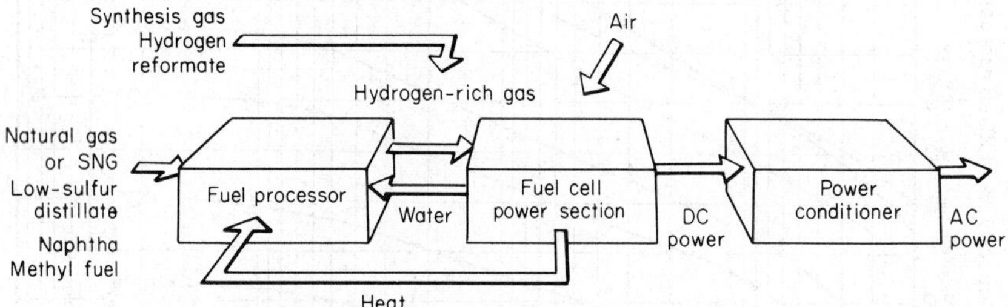


**FIG. 11-60** Generalized schematic of a fuel cell power plant.

*This material has been abstracted from David Linden (ed.), *Handbook of Batteries and Fuel Cells*, New York, McGraw-Hill Book Co., 1984.

In addition, a fuel cell power plant, depending on size, type, and sophistication, may require an oxidant subsystem as well as thermal and fluid management subsystems.

***98. Operation.*** A simple fuel cell is illustrated in Fig. 11-61. Two catalyzed carbon electrodes are immersed in an electrolyte (acid in this illustration) and separated by a gas barrier. The fuel, in this case hydrogen, is bubbled across the surface of one electrode while the oxidant, in this case oxygen from ambient air, is bubbled across the other electrode. When the electrodes are electrically connected through an external load, the following events occur:

1. The hydrogen dissociates on the catalytic surface of the fuel electrode, forming hydrogen ions and electrons.
2. The hydrogen ions migrate through the electrolyte (and a gas barrier) to the catalytic surface of the oxygen electrode.
3. Simultaneously, the electrons move through the external circuit to the same catalytic surface.
4. The oxygen, hydrogen ions, and electrons combine on the oxygen electrode's catalytic surface to form water.

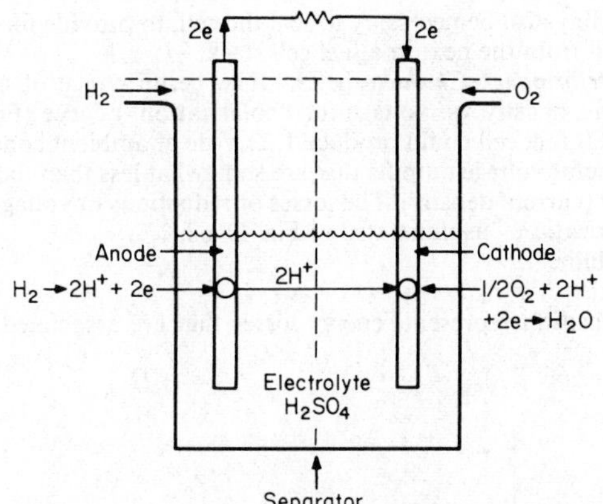


**FIG. 11-61** Operation (reaction mechanism) of the fuel cell.

The reaction mechanisms of this fuel cell, in acid and alkaline electrolytes, are shown in Table 11-17. The major differences, electrochemically, are that the ionic conductor in the acid electrolyte is the hydrogen ion (or, more correctly, the hydronium ion, $H_3O^+$) and the $OH^-$ or hydroxyl ion in the alkaline electrolyte. Further, in the acid electrolyte the product, water, is produced at the cathode and in the alkaline electrolyte fuel cell at the anode.

The net reaction is that of hydrogen and oxygen producing water and electrical energy. As in the case of batteries, the reaction of one electrochemical equivalent of fuel will theoretically produce 26.8 Ah of dc electricity at a voltage that is a function of the free energy

**TABLE 11-17** Reaction Mechanisms of the $H_2$–$O_2$ Fuel Cell

| | Acid electrolyte | Alkaline electrolyte |
|---|---|---|
| Anode | $H_2 \rightarrow 2H^+ + 2e$ | $H_2 + 2OH^- \rightarrow 2H_2O + 2e$ |
| Cathode | $\frac{1}{2}O_2 + 2H^+ + 2e \rightarrow H_2O$ | $\frac{1}{2}O_2 + 2e + H_2O \rightarrow 2OH$ |
| Overall | $H_2 + \frac{1}{2}O_2 \rightarrow H_2O$ | $H_2 + \frac{1}{2}O_2 \rightarrow H_2O$ |

of fuel-oxidant reactions. At ambient conditions, this potential is ideally 1.23 V dc for a hydrogen-oxygen fuel cell.

***99. Major Components of the Fuel Cell.*** The important components of the individual fuel cell are:

1. The *anode* (fuel electrode) must provide a common interface for the fuel and electrolyte, catalyze the fuel oxidation reaction, and conduct electrons from the reaction site to the external circuit (or to a current collector that, in turn, conducts the electrons to the external circuit).
2. The *cathode* (oxygen electrode) must provide a common interface for the oxygen and the electrolyte, catalyze the oxygen reduction reaction, and conduct electrons from the external circuit to the oxygen electrode reaction site.
3. The *electrolyte* must transport one of the ionic species involved in the fuel and oxygen electrode reactions while preventing the conduction of electrons (electron conduction in the electrolyte causes a short circuit). In addition, in practical cells, the role of gas separation is usually provided by the electrolyte system. This is often accomplished by retaining the electrolyte in the pores of a matrix (or inert blotter). The capillary forces of the electrolyte within the pores allow the matrix to separate the gases, even under some pressure differential.

Other components may also be necessary to seal the cell, to provide for gas compartments, and separate one cell from the next in a fuel cell stack.

***100. General Performance Characteristics.*** The performance of a fuel cell is represented by the current density vs. voltage (or "polarization") curve (Fig. 11-62). Whereas ideally a single $H_2$–$O_2$ fuel cell could produce 1.23 V dc at ambient conditions, in practice, fuel cells produce useful voltage outputs that are somewhat less than the ideal and decrease with increasing load (current density). The losses or reductions in voltage from the ideal are referred to as "polarization," as illustrated in Fig. 11-62.

These losses include:

1. Activation polarization represents energy losses that are associated with the electrode reactions.

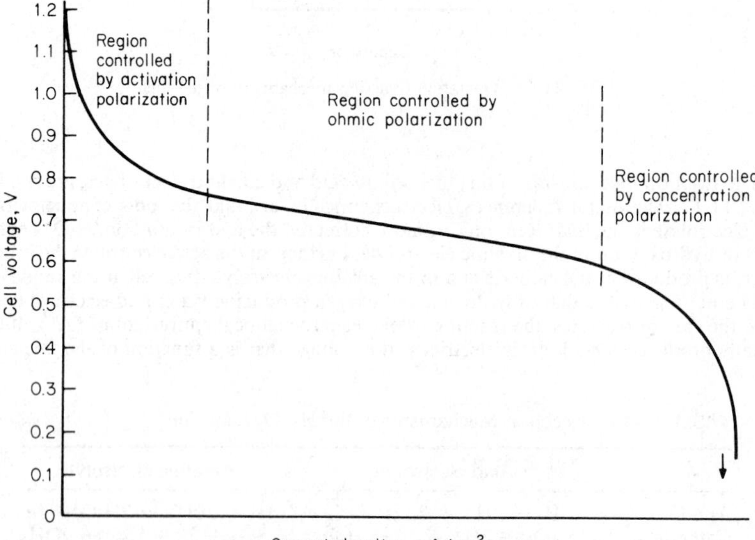


**FIG. 11-62** Fuel cell polarization curve.

2. Ohmic polarization represents the summation of all the ohmic losses within the cell, including electronic impedances through electrodes, contacts, and current collectors and ionic impedance through the electrolyte. These losses follow Ohm's law.
3. Concentration polarization represents the energy losses associated with mass transport effects. For instance, the performance of an electrode reaction may be inhibited by the inability of reactants to diffuse to or products to diffuse away from the reaction site.

The net result of these polarizations is that practical fuel cells produce between 0.5 and 0.9 V dc at currents of 100 to 400 mA/cm$^2$ of cell area. Fuel cell performances can be increased by increasing cell temperature and reactant partial pressure. For any fuel cell, the trade-off always exists between achieving higher performance by operating at higher temperature or pressure and confronting the materials and hardware problems imposed at the more severe conditions.

## Fuel Cell Systems

***101. Classification and Types.*** Fuel cell systems can take a number of different configurations, depending on the combination of type of fuel and oxidant, whether the fueling is direct or indirect, the type of electrolyte, the temperature of operation, etc., although in actual practice, the number of combinations is limited. A listing of the practical fuel cell systems is given in Table 11-18.

***102. Acid Fuel Cells.*** Table 11-18 lists two types of acid fuel cells (solid polymer electrolyte and phosphoric acid), aqueous alkaline fuel cells, molten carbonate fuel cells, and solid oxide fuel cells.

Acid fuel cells are characterized by:

Ionic conduction is provided by hydrogen ions [or by hydronium ions ($H_3O^+$)].

Platinum or platinum alloys (in very small quantity) are the active electrocatalysts.

Carbon (graphite) is an acceptable material of construction for current collectors, gas separators, etc., and is commonly used.

*Solid Polymer Electrolyte System.* The solid polymer electrolyte (SPE) system uses an ion exchange membrane as the electrolyte. The advantages of the SPE fuel cell are (1) the electrolyte, being a solid, cannot change, move about, or vaporize from the system. (2) The only liquid in the fuel cell is water, minimizing corrosion. The disadvantages are (1) the SPE must be hydrated (water-saturated) to perform; consequently, operation must be under conditions where the by-product water does not vaporize into the reaction air stream faster than it is produced; this constrains cell operation to under 60°C at ambient pressure and about 120°C at elevated pressures. (2) The SPE freezes at about 0°C and undergoes a freeze-drying phenomenon. This constrains applications to those where low-temperature capability is not a requirement.

Due to their inability to operate much above 120°C, SPE fuel cells are best suited for use with hydrogen-rich gases that contain little or no carbon monoxide. Carbon monoxide inhibits the fuel cell anode reaction, the degree of inhibition decreasing with increasing temperature. Consequently, SPE fuel cells have found their important applications in the space program operating on pure hydrogen or in military applications operating on hydrogen obtained by the decomposition of a hydride.

*Phosphoric Acid Electrolyte System.* The phosphoric acid electrolyte system operates at 150 to 220°C. At lower temperatures, phosphoric acid is a poor ionic conductor. At higher temperatures, material stability (carbon and platinum) becomes limiting. The advantages of phosphoric acid fuel cells are (1) the electrolyte is very stable, (2) the phosphoric acid can be highly concentrated (~100%) where the water vapor pressure is very low and steady-state water removal by the reactant gases will always equal product water rate, and (3) at 150 to 220°C, the anode performance is very good even on fuels containing up to 5% carbon monoxide. The disadvantage of phosphoric acid fuel cells is that the cathode performance is sluggish. In fact, the major technology thrusts in phosphoric acid are toward improvement of the cathode. The phosphoric acid fuel cell is a preferred system for use with fuels containing carbon oxides.

**TABLE 11-18** Classification of Practical Fuel Cells

| Application | Fuel | Oxidant | Electrolyte | Temperature |
|---|---|---|---|---|
| Remote: | | | | |
| Space | Direct $H_2$ | Liquid $O_2$ | Aqueous alkaline | Low, intermediate |
| | | | Solid polymer | Low |
| Undersea | Direct $H_2$ | Liquid $O_2$ | Aqueous alkaline | Low |
| | Direct hydrazine | Hydrogen peroxide | | |
| Military: | | | | |
| Low power, $\leq$ 100 W | Indirect hydride | Air | Aqueous alkaline | Low |
| | | | Solid polymer | |
| High power, $\geq$ 500 W | Indirect hydrocarbon | Air | Phosphoric acid | Intermediate |
| | Indirect methanol | | | |
| Commercial power: | | | | |
| Dispersed (or on-site) | Indirect hydrocarbon | | Phosphoric acid | Intermediate |
| | Indirect methanol, ethanol | Air | Molten carbonate | High |
| | Direct coal gas | | | |
| Central station | Indirect coal | Air | Phosphoric acid | Intermediate |
| | Direct coal gas | | Molten carbonate | High |
| | | | Solid oxide | Very high |
| Vehicle | Hydrogen | | | |
| | Hydride | Air | Phosphoric acid | Intermediate |
| | Indirect methanol | | | |
| | Indirect hydrocarbon | | | |

***103. Alkaline Fuel Cells.*** Although early alkaline fuel cells operated at relatively high temperature (~250°C) with concentrated (85 wt %) potassium hydroxide, systems developed more recently operate at much lower temperatures (<120°C) using less concentrated (35 to 50 wt %) potassium hydroxide. The lower temperature enables the use of matrices to retain the electrolyte and increases the life of other components.

Alkaline fuel cells are characterized by:

Ionic conduction is provided by hydroxyl ($OH^-$) ions.

A wide range of electrocatalysts can be used including nickel, silver, metal oxides, spinels, and noble metals—although the truly high performance systems use at least small amounts of noble metal.

Construction materials include carbon, nickel, and stainless steel.

The advantages of alkaline fuel cells are (1) cathode performance is much better than for acid fuel cells and (2) materials of construction tend to be low in cost. The primary disadvantage is that the electrolyte reacts with carbon oxides to form potassium carbonate. This severely limits the cells' performance. Thus, alkaline fuel cells have only limited application where carbonaceous fuels or air are used as reactants. The important applications (space and underseas) involve pure hydrogen and oxygen.

***104. Molten Carbonate Fuel Cells.*** Molten carbonate fuel cells use an alkali metal (Li, K, Na) carbonate as the electrolyte. Since these salts can function as electrolytes only when in the liquid phase, the cells operate at 600 to 700°C, which is above the melting points of the respective carbonates. Molten carbonate cells are characterized by:

Ionic conduction is by the carbonate ion; thus the carbonate ion must be involved in the two electrode reactions

$$\tfrac{1}{2}O_2 + CO_2 + 2e \rightarrow CO_3^{2-} \qquad \text{(cathode)}$$

$$\underline{CO_3^{2-} + H_2 \rightarrow CO_2 + H_2O + 2e \qquad \text{(anode)}}$$

$$\tfrac{1}{2}O_2 + H_2 \rightarrow H_2O \qquad \text{(net)}$$

A consequence of this is that $CO_2$ must be recycled from the anode to the cathode.

At 600 to 700°C, electrode reactions proceed without highly specific catalysts. Nickel and nickel oxide work quite well; noble metals are not used.

Construction materials include nickel and ceramics.

The advantages of molten carbonate fuel cells are (1) cell performance is good, activation polarization is small; and (2) at 600 to 700°C, any carbon monoxide in the fuel converts to hydrogen on the anode via the water gas shift reaction

$$CO + H_2O \rightarrow CO_2 + H_2$$

(as a result, fuel gases high in carbon monoxide are readily used), and (3) waste heat from the fuel cell can be available at a relatively high temperature (>500°C), enabling its use in bottoming or industrial heating cycles.

Disadvantages are (1) the high temperature imposes severe constraints on materials suitable for long lifetimes and (2) a source of carbon dioxide is required to complete the cathode reaction (this is provided by recycling $CO_2$ from the anode exhaust to the cathode inlet). As a result, molten carbonate fuel cells are best suited for applications that integrate the fuel cell with a carbonaceous fuel processor, i.e., a reformer or coal gasifier.

***105. Solid Oxide Fuel Cells.*** As the name implies, solid oxide fuel cells employ a solid, nonporous metal oxide electrolyte which allows ionic conductivity by the migration of oxide ions through the lattice of the crystal. Stabilized zirconia is commonly used as the electrolyte. The cells operate at 900 to 1000°C. Whereas practical cells of the technologies previously discussed are normally packaged into "filter press" or "plate and frame" stack assemblies, solid oxide fuel cells are configured into tubular cell stacks.

Characteristics of the solid oxide fuel cell include:

Ionic conduction is provided by oxide ions.

The cathode employs metal oxides, such as praseodymium oxide or indium oxide; the anode uses nickel or nickel cermet.

Because of the high temperature, materials of construction will likely be confined to be ceramics or metal oxides.

Solid oxide fuel cells offer advantages similar to those of molten carbonate cells, that is, good performance on fuels containing hydrogen or hydrogen and carbon monoxide, the elimination of noble-metal catalysts, and the availability of high-grade reject heat. In addition, they do not suffer the constraint of molten carbonate cells which require a carbon dioxide recycle to the cathode. The primary disadvantages are the very high temperature of operation and the severe material constraints imposed by the ~1000°C temperature.

## Low-Power Fuel Cell Systems

The advantageous characteristics of fuel cells led to the development of a number of different systems ranging in size from the portable units, 5 W or smaller to kilowatt power levels (where ease of operation, low maintenance, and silence are important), to large stationary plants delivering megawatts of power (where the high efficiency over the range from full to partial load and reduced pollution are significant). The lower-power fuel cells were designed mainly for military or special applications such as the space program. For the space applications and for forward-area military use, the fuel cell offers high energy densities that exceed the performance of batteries when operated over long periods of time.

***106. Space Power Systems.*** The fuel-cell systems built between 1960 and 1970 for the Gemini and Apollo space programs and in 1980 for the space shuttle orbiter are still among the most successful demonstrated to date. They used hydrogen as the fuel and oxygen as the oxidizer. The Gemini fuel cell, designed as a 1-kW system, used an ion exchange membrane separator-electrolyte, known as a solid polymer electrolyte (SPE). The larger 1.5-kW Apollo and 7-kW space shuttle orbiter units were based on a potassium hydroxide electrolyte system, but the latter system incorporated a number of significant advances in design. The space shuttle fuel cells are 20 kg lighter and deliver 6 to 8 times as much power as those of the Apollo system. The space shuttle system is made up of three fuel cell power plants and normally supplies 14 kW of power, with peak loads of 36 kW. The fuel cell power plant is illustrated in Fig. 11-63. Each power plant is 35 cm high, 38 cm wide, and 101 cm long and weighs 91 kg.

Each fuel cell power section contains two parallel stacks of 32 cells in series. Each is capable of supplying 12 kW at peak (27.5 V and 436 A), 7 kW average power; at 2 kW, each

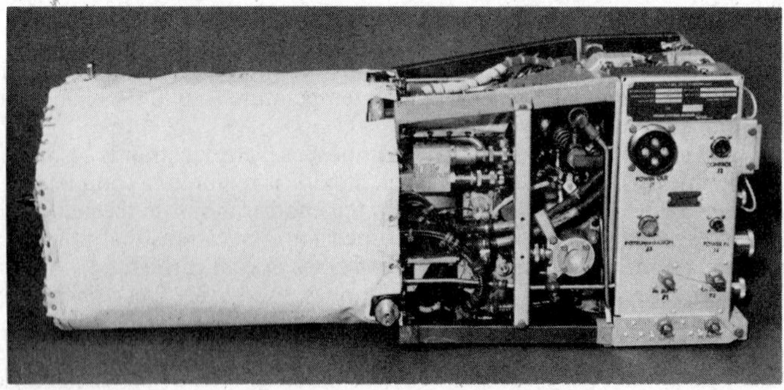

**FIG. 11-63** Space shuttle orbiter fuel cell power plant. *(Courtesy of United Technologies Corp.)*

power plant provides 32.5 V and 61.5 A. Cryogenically stored hydrogen and oxygen are delivered to each fuel. On a 3-day mission, the fuel cells utilize about 450 kg of $H_2$ and $O_2$; about 600 L of water is produced. The power plants will be serviced between flights and reused until each has accumulated 2000 h of on-line service.

***107. Portable Fuel Cells.*** Fuel cells in the power range of up to about 200 W can be an attractive alternative to batteries for long-term operation, with potential reduction in the weight and cost. The size limitation of these portable fuel cell systems is generally too small to permit the use of elaborate fuel conditioning. Hence, easily handled and readily oxidized fuels (e.g., liquid or gaseous fuels such as hydrazine, methanol, and ammonia) and fuels that provide hydrogen through a simple physical or chemical reaction have been used in most fuel cell systems of this size and type. For these ground applications, air-breathing, rather than pure oxygen, systems are used.

While the fuel cell systems in this size and power range potentially have advantageous characteristics, system complexities and the development of new battery systems with higher energy densities have limited the interest and successful use of the low-power fuel cell.

***108. Direct Fuel Cell Systems.*** Direct-type fuel cells, in which the fuel can be introduced into the fuel cell without requiring conversion to hydrogen, were considered for small fuel cell systems because they eliminated the need for a fuel-conditioning unit, thus saving important space and weight. Methanol ($CH_3OH$) and hydrazine ($N_2H_4$) were the main liquid fuels used. Methanol is directly oxidizable, but removal of the carbonate, one of the reaction products of the dissolved methanol fuel cell in alkaline electrolytes, from the electrolyte is extremely difficult. Efforts then shifted to hydrazine. Hydrazine decomposes easily into hydrogen and nitrogen at the electrode surface; in fact, the voltage observed is that of hydrogen.

Major effort was directed toward a silent power source for forward-area military use. A 60-W, 24-V hydrazine-air fuel cell was developed in a configuration similar to the one used later for the metal hydride cell. The fuel cell used a 35% potassium hydroxide electrolyte and a 64% hydrazine monohydrate fuel and operated between 55 and 70°C with a fuel utilization of 600 Wh/kg. A larger 300-W, 24-V power source was also developed for forward-area use. This system weighed 20 kg, with electrolyte and 4 L of fuel, and had a volume of 35 $dm^3$. The fuel was sufficient for 12 h of operation at 300 W. Field tests confirmed the successful electrochemical functioning of the cell, but mechanical deficiencies caused early failure of the system.

***109. Metal Hydride Fuel Cells.*** The majority of current portable fuel cell developments use a metal hydride as the source of hydrogen fuel. Metal hydrides are attractive because they can store large amounts of hydrogen more conveniently and with a higher energy density (total equivalents of hydrogen per total weight of hydrogen source and container) than hydrogen in a pressurized or liquefied form.

One type of metal hydride produces hydrogen by the reaction with water, e.g., calcium hydride ($CaH_2$)

$$CaH_2 + 2H_2O - Ca(OH)_2 + H_2$$

A second type of metal hydride, a reversible hydride, is based on the principle that certain metals or alloys (e.g., iron titanium, lanthanum nickel, and various other rare-earth metal and nickel alloys) have the ability to take up large amounts of hydrogen gas within their crystal structure. A reduction in pressure or an increase in temperature release the hydrogen. These hydrides can deliver hydrogen at about 500 Wh/kg.

A 60-W, 28-V hydrogen-air fuel cell is illustrated in Fig. 11-64. A similar package was also used for a 30-W fuel cell. The system consists of three sections: the hydrogen generator, the fuel cell stack, and the power conditioner. The overall unit weighs 7 kg and has a volume of 10 $dm^3$.

Hydrogen is supplied by a Kipp generator using the reaction between sodium aluminum hydride ($NaAlH_4$) and water

$$NaAlH_4 + H_2O\ (\text{excess}) \rightarrow Al(OH)_3 + NaOH + 4H_2$$

The hydride, in the form of a solid pellet, delivers in excess of 2000 Wh/kg. Hydrogen for 4 h of operation is supplied to the fuel cell with a single 120-g charge. The Kipp generator

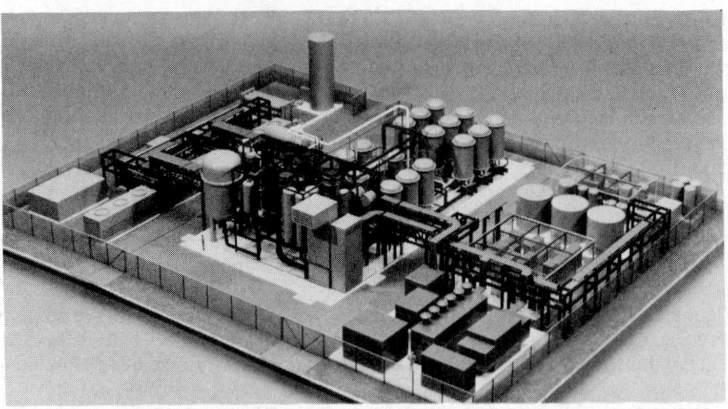

**FIG. 11-64** Hydrogen-air fuel cell system (60 W).

delivers hydrogen to the fuel cell on demand; when no hydrogen demand exists, the pressure builds up in the generator, forcing water away from the fuel pellet and stopping the reaction.

A later design used a reversible metal hydride, lanthanum pentanickel hydride, as the source of hydrogen

$$LaNi_5 + 3H_2 \rightleftarrows LaNi_5H_4$$

This change, replacing the exothermic sodium aluminum hydride generator with the reversible metal hydride source which absorbs heat on the release of hydrogen, could reduce the total heat output of the system by 65%. The system was found to operate satisfactorily at 20°C, but higher ambient temperatures still caused problems because of the much higher operating temperature of the fuel cell stack.

*SPE Fuel Cell.* There is renewed interest in the SPE fuel cell for applications in which mobility is important and power requirements are low. The advantages of the SPE cell are ease of product water removal, simple construction, stable electrode-electrolyte interface, and favorable life characteristics.

The SPE cell is being considered for power levels from a few watts to 500 W, operating in the −40 to 50°C range with no restrictions on humidity. Special designs, including insulation for low temperatures and methods for waste heat disposal, will probably be required to achieve this performance. The cell operates on hydrogen and ambient air. Preferred fuel sources for hydrogen generation are magnesium and aluminum, which are reacted with salt water. Bottled hydrogen, hydride-stored hydrogen, or hydrides may also be used. With magnesium, it is expected to obtain energy densities (fuel consumption) of about 220 Wh/kg on a wet basis and 1300 Wh/kg on a dry basis. For longer missions, where a larger system weight can be tolerated, reformed methanol combined with carbon monoxide absorption, is being considered.

***110. Indirect Methanol Fuel Cell Power Plants.*** A family of fuel cells, with output ratings of 500 W to 5 kW, is being developed for forward-area military use to provide a silent, lightweight source of electrical energy. Originally, conventional hydrocarbon fuels were considered for these power plants, but the changing availability in recent years of traditional petroleum-based fuels prompted a consideration of methanol as an alternative fuel. Methanol is attractive for fuel cell use; it is more convenient to store and handle than either hydrogen or ammonia, and it can be more readily reformed to hydrogen than the long-chain hydrocarbons. Methanol cannot be used successfully in alkaline electrolyte fuel cells because the carbon oxides that form during reforming cannot be tolerated. The development of the phosphoric acid electrolyte fuel cell, which could tolerate these carbon oxides, reactivated interest in the methanol fuel cell, even though the acid electrolyte presented more problems with corrosion and requires the use of noble-metal catalysts.

Three sizes of fuel cell power plants are being developed. The characteristics of these units are summarized in Table 11-19. Major development emphasis is on the 1.5-kW size.

**TABLE 11-19** Characteristics of Methanol Fuel Cell Power Plants

| Power rating, kW | Size (volume), $dm^3$ | Dry weight, kg | Fuel consumption, g/kWh |
|---|---|---|---|
| 1.5 | 200 | 70 | 1000 |
| 3.0 | 340 | 135 | 800 |
| 5.0 | 510 | 225 | 800 |

The methanol fuel cell power plant is based on a low-temperature steam reformer and a phosphoric acid fuel cell stack. The power plant consists of three subassemblies as shown in the schematic in Fig. 11-65, the fuel conditioner converting methanol to hydrogen, the fuel cell assembly converting the hydrogen to electric power (direct current), and the power conditioner which converts the fuel cell output power to regulated direct current or to an ac output.

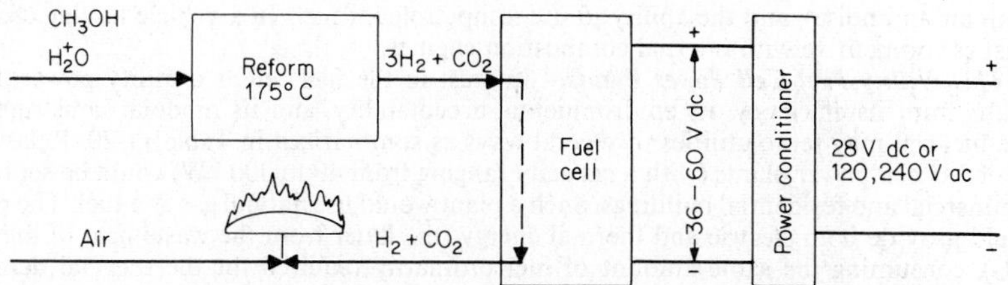


**FIG. 11-65** Simplified schematic of a 15-kW indirect methanol-air fuel cell system.

The 1.5-kW methanol fuel cell power plant is illustrated in Fig. 11-66. In the fuel-conditioning subsystem, an aqueous methanol feed (58% by weight of methanol) is vaporized and superheated to a temperature of 160°C and passed through the catalyst bed where the water and methanol react at a temperature of 250 to 300°C. Hydrogen and carbon oxides (approximately 75% $H_2$, 25% $CO_2$, 2% CO) are formed with residuals of water and methanol. The gaseous products are cooled with ambient air to about 50°C and routed to the fuel cell. The fuel cell subassembly consists of 80 phosphoric acid electrolyte cells and the ancillary

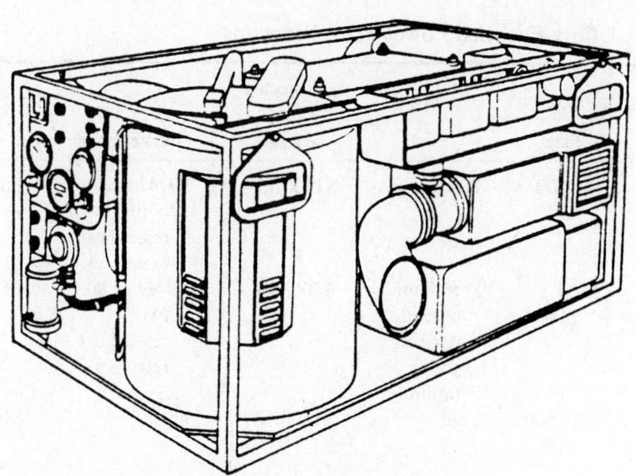

**FIG. 11-66** Methanol fuel cell power plant (1.5 kW).

equipment for thermal control. The cells are clamped between honeycomb end plates to ensure good electrical contact. Air flows in a single pass across the short dimension of the cell and fuel in a double pass across the long dimension. The fuel cell operates at about 170°C.

The power plant can have either a dc or ac output, depending on the power conditioner that is used. The inverter converts the 36- to 60-V dc to either 120- or 240-V ac. The dc-dc converter takes the same input voltage and delivers an adjustable output voltage in a range of 26- to 36-V dc.

This indirect methanol fuel cell system uses few moving parts and operates at relatively low temperatures; it should have a long and reliable life, with minimum maintenance (MTBF: 750 h; overhaul: 6000 h) and operate over the specified temperature range (−50 to +50°C) and environmental conditions.

A similar fuel cell system, sized at about 20 kW, is also being considered for electric vehicle use in a hybrid fuel cell–secondary battery power system. The most advanced approach uses methanol for the fuel, a re-former, and the phosphoric acid fuel cell. Also under consideration are fuel cell systems based on the solid polymer electrolyte (SPE) technology and the use of superacid electrolytes, such as trifluoromethanesulfonic acid (TFMSA). The aim of this program is to use the fuel cell's high efficiency, low pollution (both air and noise), and the ability to use nonpetroleum fuels in a vehicle having characteristics competitive with internal combustion engines.

***111. Utility Fuel Cell Power Plants.*** Interest in the fuel cell as a utility power plant results from its efficiency, its environmental acceptability, and its modular construction. The fuel cell may serve utilities in several ways as summarized in Table 11-20. Relatively small fuel cell power plants (with a capacity ranging from 40 to 300 kW) could be set up in commercial and residential buildings. Such a plant would use natural gas as a fuel. The plant would provide both electric and thermal energy (the latter from the waste heat of the fuel cells), consuming the same amount of fuel ordinarily required for the thermal demand alone. Overall efficiencies approaching 90% have been projected for fuel cell power plants of this type. An advantage of the fuel cell is that the waste heat can be used without altering the power production characteristics. Larger plants, ranging in capacity from 5 to 25 MW, could be dispersed throughout an electric utility system to perform load-following duty efficiently, taking advantage of the higher efficiency of the fuel cell even at reduced loads. In the future, fuel cells could be integrated with coal gasifiers to provide large, central-station, base-load power plants that utilize coal directly. The capacity of such plants would range from 150 to 1000 MW. A plant of this kind is projected to be more than 45% efficient, on the basis of the heating value of the coal consumed.

*Three major programs* aimed at hastening the development of the utility fuel cell as a commercial technology are in progress. These are the on-site program, the FCG-1 program,

**TABLE 11-20** Utility Fuel Cell Power Plant Programs

| Role | Size | Fuel | Projected cost, 1980 $ | Efficiency, % | Electrolyte |
|---|---|---|---|---|---|
| On-site power plants | 40–300 kW | Pipeline gas | $1000/kW | 39–42 or 90 (with reject-heat recovery) | Phosphoric acid |
| Dispersed (substation) power plants | 5–25 MW | Petroleum- or coal-derived gas or liquid | $800/kW | 41–47 or 80 (with reject-heat recovery) | Phosphoric acid |
| Central station power plants | 150–600 MW | Coal | $1500/kW* | 45–50 | Molten carbonate or phosphoric acid |

*Includes cost of coal gasifier.

and the molten carbonate fuel cell program. In addition to these three major programs, the high-temperature solid-oxide fuel cell has received intermittent support as a utility fuel cell. The on-site program is an extension of a project named TARGET (Team to Advance Research for Gas Energy Transformation). The project was originally sponsored jointly by the United Technologies Corporation and a consortium of gas and gas and electric utilities. Starting in 1967, the group supported the development of phosphoric acid fuel cells for on-site residential and commercial applications.

The *on-site power plant* concept utilizes gas (natural or synthetic) as the fuel, converting it into electricity and into thermal energy for heating and cooling. The efficiency is quite high, since virtually all the waste heat is directed toward the heating and cooling part of the operation. Now, in an extension of the program with the support of the Department of Energy and the Gas Research Institute, the objective is the testing of 50 plants of 40-kW capacity and the attainment of commercial availability of the technology by the end of the decade.

The second program originated with the electric utility industry in 1971, when a group of utility companies joined the Edison Electric Institute and United Technologies in an assessment of the potential benefit of fuel cells to the industry. The venture led in 1972 to the *FCG-1 (for fuel cell generator-1) program,* an effort sponsored by United Technologies and nine utility companies to develop a 26-MW phosphoric acid power plant for commercial service by 1980. The FCG-1 program developed and demonstrated (in 1976 and 1977) a 1-MW pilot plant. The demonstration showed that a power plant fueled by naphtha could provide large amounts of energy to a utility system while satisfactorily meeting the utilities' operational requirements for heat rate, emissions, and load-following capability.

In 1976, seeking to expedite the commercial application of fuel cells, the Electric Power Research Institute and the Energy Research and Development Administration (now the Department of Energy) became involved in the FCG-1 program. One result was a project to design, build, and test a 4.5-MW module of the FCG-1 power plant in a utility system. Consolidated Edison Company of New York was chosen as the host utility for the demonstration.

The third major program grew out of efforts undertaken in 1971 by the electric and gas utility industries to advance fuel cell technology in order to expand its applicability to a wider range of fuels, including coal, by the direct integration of the fuel cell to a coal gasifier. The focus of this program is the *molten carbonate fuel cell power plant* expected to be available for commercial service in about a decade. General Electric, the Institute of Gas Technology, Oak Ridge National Laboratory, Energy Research Corporation, and United Technologies Corp. are involved in the program. Argonne National Laboratory is managing the Department of Energy effort as well as conducting supporting research and development efforts.

Serious interest in the high-temperature solid oxide (HTSO) fuel cell developed in the early 1960s as a result of several anticipated advantages offered by this technology. However, by 1970 much of the enthusiasm toward HTSO fuel cells had waned, and by 1972 the major programs had been terminated. The loss of interest was principally due to problems of material stability at 1000°C; problems with cathode and intercell-connection materials were especially severe.

More recently, interest in the HTSO fuel cell has been renewed and concentrated in two areas: first, between 1973 an 1976, HTSO electrolytes were investigated that could operate at lower temperatures (700 to 850°C), and second, since 1976 HTSO fuel cells using a zirconia electrolyte operating at 1000°C have been explored.

***112. Phosphoric Acid Fuel Cells.*** Despite the increasing number of developers and the seeming wide range of applications (kilowatts to megawatts), the basic phosphoric acid fuel cell (PAFC) technology that is being developed is virtually the same in all cases.

The basic *cell structure* is shown in Fig. 11-67. It consists of:

1. *A carbon or graphite separator-current collector plate* that separates hydrogen from the air of the adjacent cell (in a multicell stack) and also provides the electrical series connection between cells.
2. *Anode current collector ribs* that conduct the electrons from the anode to the separator plate. The ribbed configuration provides gas passages for hydrogen distribution to the anode.

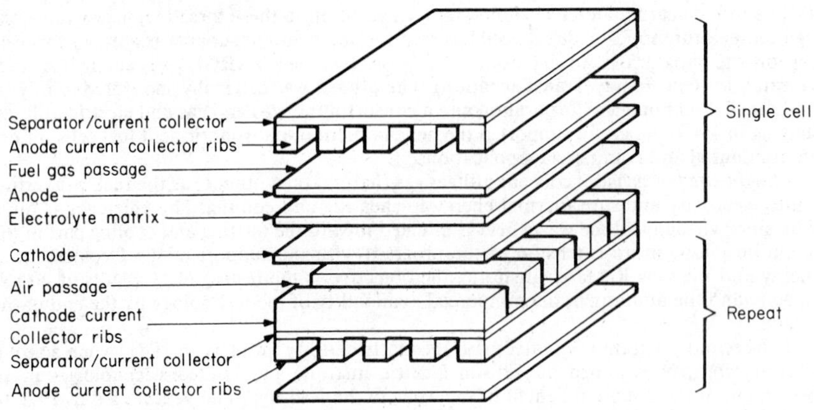


**FIG. 11-67** Basic phosphoric acid electrolyte fuel cell.

3. *An anode* that consists of a porous graphitic substrate with the surface adjacent to the electrolyte treated with a platinum or platinum alloy catalyst.
4. *An electrolyte matrix* that retains the concentrated phosphoric acid.
5. *A cathode* that is similar to the anode but uses a modified noble-metal catalyst and an increased catalyst loading (usually 0.5 mg/cm$^2$) to enhance the oxygen reduction kinetics.
6. *Cathode current collector ribs* that are also virtually identical to the anode ribs.

These single cells are stacked in series to produce the desired output power and voltage. For instance, the 4.5-MW power plant being installed in New York contains 20 stacks, each having nearly 500 cells of 3400 cm$^2$ area each. Fuel and air supply and exhaust manifolds are then connected along the respective sides of the stacks.

***113. Performance Characteristics.*** Typical performance for a PAFC is illustrated in Fig. 11-68. Present PAFC performance is almost entirely determined by the cathode. That is, concentration polarization, anode activation polarization, and ohmic losses are small compared with the activation losses exhibited at the cathode.

Figure 11-69 shows typical PAFC performance decay rates as a function of operating time at a constant current density. The interesting feature of this figure is the linear relationship of voltage decay with log (time) following the initial thousand hours of operation. This decay results from the tendency of the platinum catalyst to lose surface area and hence intrinsic activity as a function of operating time.

***114. Phosphoric Acid Power Plants.*** The characteristics targeted by the utility industry for commercial first-generation phosphoric acid power plants under development by United Technologies Corp. are summarized in Table 11-21. The similarity in the 40-kW on-site and the 10-MW dispersed power plants is apparent. The significant differences between the two systems (other than size) are the operational pressure-temperature and current densities. The 40-kW unit operates at ambient pressure, 200°C, and 190°C, and between 100 and 200 mA/cm$^2$; the 10-MW unit will operate at 6.5 to 8 $\times$ 10$^5$ Pa, 200°C, and between 200 and 300 mA/cm$^2$.

A prototype 40-kW on-site system design is shown in Fig. 11-70. This design is a modification of that fabricated and tested as a pilot 40-kW fuel cell power plant. The pilot power plant exceeded 18,000 h of operation, more than 8000 h with one power section and with a continuous run of more than 3000 h.

The prototype subsystem design and engineering development are complete, and testing has been initiated to verify all operating characteristics. A coordinated effort is presently being implemented by the Department of Energy, Gas Research Institute, United Technologies Corp., and several utilities that will place up to fifty 40-kW prototypes into the field for evaluation. The primary objective of this operational feasibility program is to establish the utilities', manufacturers', and sponsors' acceptance of and commitment to on-site fuel cell systems.

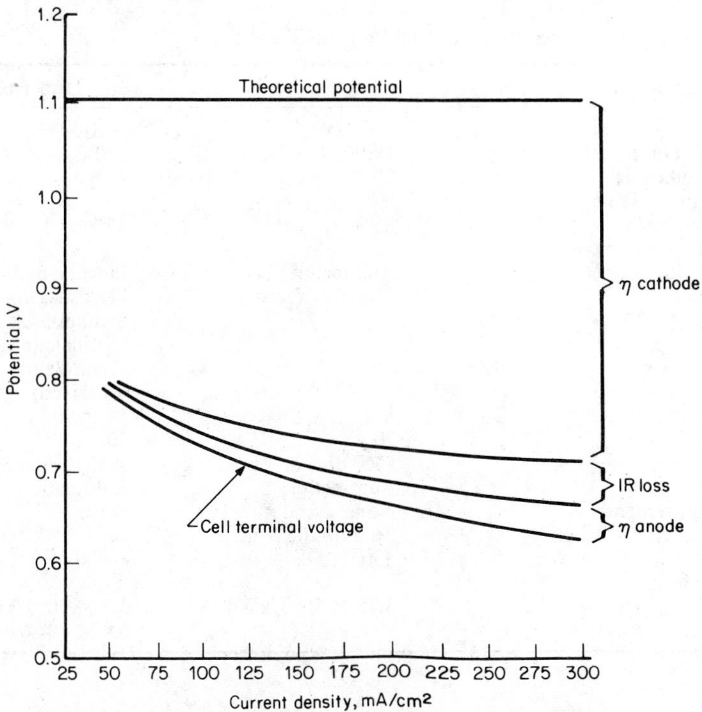


**FIG. 11-68** Typical performance of a phosphoric acid electrolyte fuel cell (190°C, 3 × $10^5$ Pa, air, $H_2$ fuel, platinum loading: 0.75 mg/cm$^2$).

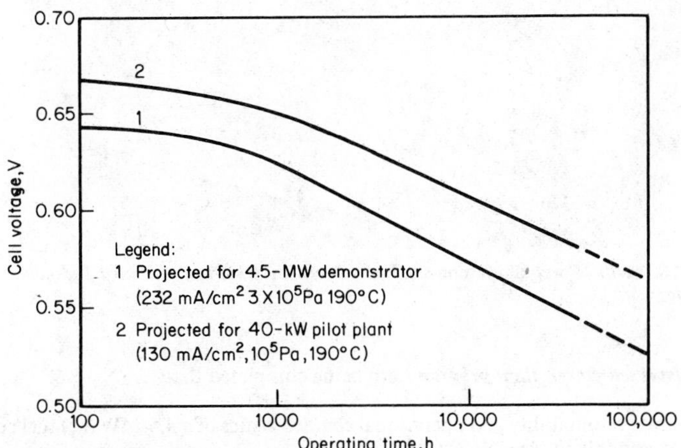


**FIG. 11-69** Typical performance decay for phosphoric acid fuel cells.

**TABLE 11-21** First-Generation Fuel Cell Program Targets

| Characteristic | On-site | Dispersed |
|---|---|---|
| Modular size | ~40 kW | ~10 MW |
| Stack temperature and pressure | 190°C, $1 \times 10^5$ Pa | 210°C, $6 \times 10^5$ Pa |
| Electrical efficiency (based on HHV) | 39% | 41% |
| Total efficiency (including heat recovery) | 85% | 85% |
| Fuel capability | All pipeline-quality fuels | Pipeline natural gas<br>Peak shaving gas<br>Synthesis gas<br>Distillate (coal or petroleum derived)<br>Methanol |
| Design life | 20 y | 20 y |
| Noise | <60 dbA at 3 m | <55 dbA at 30 m |
| Operation | Automatic | Automatic |
| Water requirements | None | None |
| Output | 3$\phi$, 120/208 V ac | 3$\phi$, 4–69 kV ac |
| Footprint | 4 $m^2$ | ~900 $m^2$ |
| Available heat: | | |
| at 80°C | $3.75 \times 10^6$ J/kWh | $3.3 \times 10^6$ J/kWh |
| at 150°C | | $0.4 \times 10^6$ J/kWh |

**FIG. 11-70** Forty-kilowatt on-site fuel cell system. *(Courtesy of United Technologies Corp.)*

***115. Dispersed power plant programs*** are being completed that:

1. Demonstrate the sitability and operational characteristics of a 4.5-MW (ac) fuel cell demonstration module in New York City
2. Demonstrate the characteristics of a similar 4.5-MW fuel cell in Tokyo, Japan

3. Develop a ~10-MW commercial prototype design which could be available for utility evaluation by the mid-1980s

An artist's drawing of the 4.8-MW dc (4.5-MW ac) module (fuel processing and power sections) is shown in Fig. 11-71. The power plant is currently undergoing installation at the downtown New York City site as shown in Fig. 11-72. The purposes of this demonstration are similar to those of the 40-kW field test, with a focus on establishing the power plant's sitability and operating characteristics.

A parallel effort is being implemented that will result in the design, development, and component-level verification of a commercial prototype power plant, nominally rated at 10 MW. This effort will lead to a prospectus for commercial prototype power plants with the expectation that the industry will subsequently deploy up to 500 MW of such units as part of a commercial feasibility program, leading to commercial service in the 1980s.

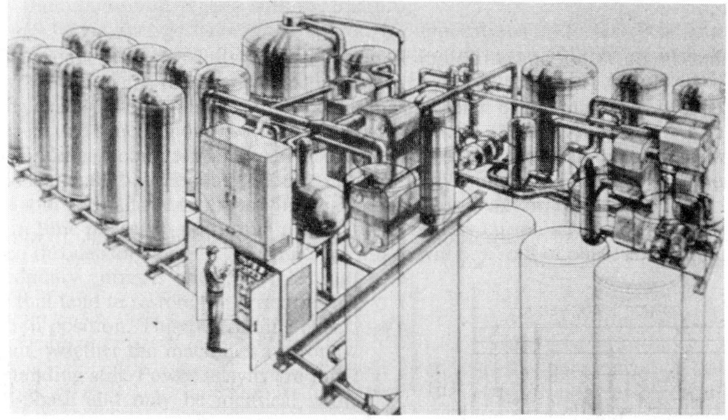

**FIG. 11-71** Artist's drawing of a 4.8-MW (dc) module.

**FIG. 11-72** Site of 4.5-MW fuel cell (New York, N.Y.).

Phosphoric acid fuel cell power plants are also being considered for use in central stations. These power plants would utilize integrated coal gasifiers to supply the hydrogen-rich fuel and would provide base load power. Such a power plant, based upon 1977 PAFC technology, projects an efficiency for coal to ac power of 34%. By use of current PAFC technology, this efficiency would approach 40%. Thus, such a power plant is not an unreasonable alternative.

***116. Molten Carbonate Fuel Cells.*** Interest in the molten carbonate fuel cell (MCFC) technology stems from its high efficiency, excellent environmental character, potential for competitive costs, and modularity. In contrast to the PAFC, MCFC development efforts are presently focused on large, multimegawatt (~600 MW) central-station power plants integrated with a coal gasifier. This focus is, in part, due to their high operating temperature and the problems inherent in starting (heating) and stopping (cooling) MCFC power plants.

***117. Cell structure for the MCFC*** is geometrically very similar to that of the PAFC, as shown in Fig. 11-67. The materials that are used are, however, very different from those used in the PAFC. As reference to Fig. 11-67 shows, the MCFC consists of:

1. *A separator and current-collector plate* that separates the fuel gas from the air of the adjacent cell in a multicell stack and also provides the electrical connection between cells. Like its PAFC counterpart, it must be impermeable to hydrogen and oxygen, a good electronic conductor, and stable to fuel and air environments in the presence of 650°C carbonate salts.
2. *An anode current collector* that conducts the electrons from the anode to the separator plate. This current collector must also provide passage for fuel flow. In some configurations, this function is provided by ribbing or folding the separator plate.
3. *An anode that* consists of a porous nickel treated with a refractory oxide to reduce sintering. At the 650°C temperature, no other catalyst is required.
4. *An electrolyte system* comprising a mixture of lithium-potassium carbonate and inert powder (presently a lithium aluminate). This mixture forms a paste when molten and freezes to form a "tile" when cooled. This electrolyte system presents major challenges to MCFC development. It must have minimum ionic resistance, while separating the fuel and oxidant gases at pressure differentials in excess of $0.07 \times 10^5$ Pa. In addition, it must be electronically insulating.
5. *A cathode* that is similar to the anode except that it uses nickel oxide (doped with lithium to impart electronic conductivity).
6. *A cathode current collector* that has similar requirements and configurational operations as the anode current collector. Since nickel is thermodynamically unstable, material options include lithium-doped nickel oxide and corrosion-resistant stainless steel.

As with the PAFC, individual cells are stacked in series to result in a cell stack of the required power and voltage output.

***118. Performance characteristics.*** Figure 11-73 illustrates a typical MCFC performance at the start of test. The performance of MCFCs tends to be better than that of PAFCs primarily due to the improved behavior of the air cathode. The performance losses in practical MCFCs are distributed among the various polarizations with $\eta_{ohm}$ and $\eta_{act}$ (cathode) being the largest contributors. Since *IR* losses due to the electrolyte and contact resistance can be severe, attention must be given to developing thin electrolyte structures (<0.05 cm) and maintaining good pressure contacts.

Figure 11-74 shows the best life test data for small single cells. The decrease in performance with time tends to be associated with a corresponding *IR* increase. The longest endurance demonstrated by a multicell stack is 2000 h. This 20-unit, 930-$cm^2$ cell stack also completed five thermal cycles without failure, confirming at least limited success in resolving this problem.

***119. Molten Carbonate Fuel Cell Power Plants.*** The thrust of the MCFC program is the development of a central-station power plant, comprising a coal gasifier, gas clean-up system, MCFC (topping cycle), and gas or steam turbine (bottoming cycle). A typical system energy flow for a central-station MCFC power plant is shown in Fig. 11-75. Design requirements and goals for a 675-MW power plant are described in Table 11-22. Key features are listed in Table 11-23. A possible power plant configuration would include:

600 MCFC stacks (having five hundred 1-$m^2$ cells each) for a total 450-MW output

15 coal gasifiers capable of handling $3 \times 10^{11}$ J/h each

10 heat recovery steam generators

Five 15-MW gas turbines

One 150-MW steam turbine

An artist's concept of a 675-MW central station MCFC power plant is shown in Fig. 11-76.

Although the ultimate application of the MCFC will likely be as a large coal-fueled, central-station power plant, other applications such as that of dispersed or on-site generators with and without reject heat recovery are also being considered. Because of the high operating temperature, the quality of the MCFCs waste heat could be compatible with a variety of industrial heating applicatons. Also, at 600°C, in situ reforming of methane or methanol is possible; this could result in a very efficient small power plant.

***120. High-Temperature Solid Oxide (HTSO) Fuel Cells.*** HTSO fuel cells employ a "tubular" rather than a "filter press" stack configuration. This tubular stack, shown in Fig. 11-77, avoids the problem of sealing the edges of the individual cells that is necessary in filter press stacks. Such edge seals are considered impractical due to the lack of nonporous, insulating, gasket materials for use at 1000°C. The state-of-the-art HTSO fuel cell stacks

**TABLE 11-22** Design Requirements and Goals for a Central Station MCFC Power Plant

| Requirements | |
|---|---|
| Central station plant | |
| Power level | −675 MW(e) |
| Fuel specification | Illinois no. 6 coal |
| Modular construction | |
| Environmental | Projected 1985 federal requirements |
| Site characteristics | "Middletown" except for cooling tower heat rejection |
| **Goals** | |
| Base load duty with daily load following capability | |
| Heat rate | $6.8 \times 10^6$ J/kWh |
| Capital installed cost (1982 dollars) | $1500/kW(e) |
| Plant availability | 85%. |
| Life goals (75% capacity factor) | |
| Fuel cell stacks | 6 years |
| Balance of plant | 30 years |
| Startup/shutdown: | |
| Startup: Cold startup in 4 to 6 h | |
| Shutdown: 100% to zero load in 3 h | |
| Daily load following: | |
| Large-load-change response time of 2 h | |
| Small-load-change response rate up to 2%/min | |
| Abnormal conditions: | |
| Complete-load rejection (breakers opening) | |
| Partial-load rejection (from power system breakup) | |
| Sustained abnormal voltage or frequency operation | |
| Limit fault current to 1.1 per unit current (rms basis) | |
| Other: | |
| Independent var control | |

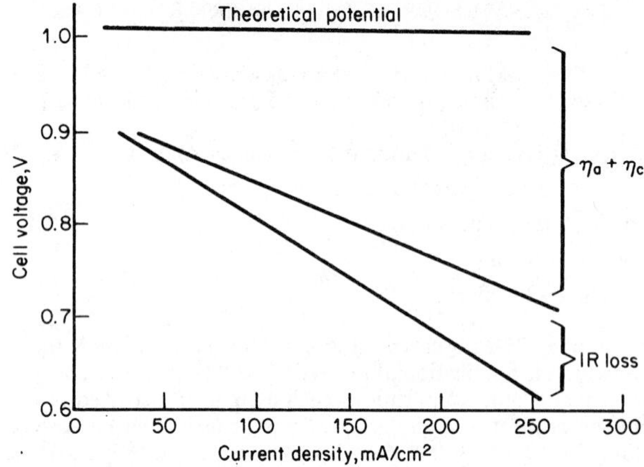


**FIG. 11-73** Typical molten carbonate fuel cell performance, 650°C, 1 $\times$ $10^5$ Pa, air and reformate fuel.

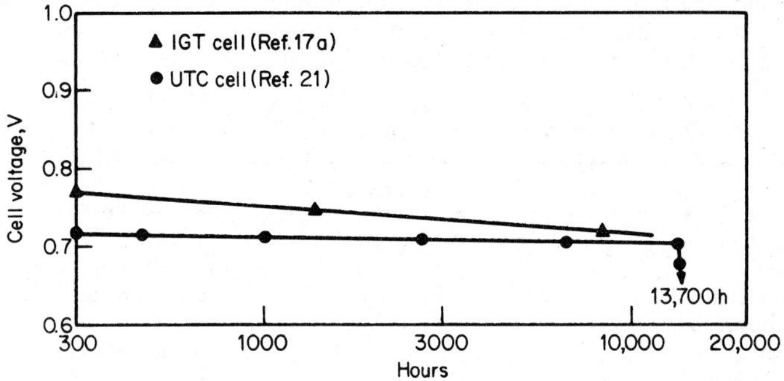


**FIG. 11-74** Performance vs. time for long-lived MCFC (data in both cells based on simulated reformate fuel).

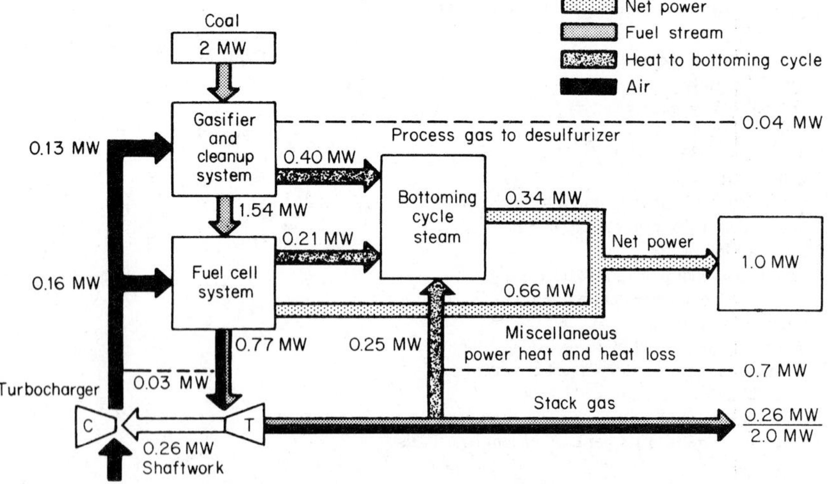


**FIG. 11-75** Typical integrated fuel cell coal gasifier energy flow.

**TABLE 11-23** Key Features of MCFC Power Plant*

| | |
|---|---|
| Rated power plant output | 675 MW(e) |
| Individual cell voltage* | 0.75 V dc |
| Power density* | 125 $MW/cm^2$ |
| Cell area/power plant | $3 \times 10^5\ m^2$ |
| Fuel cell output | ~365 MW(e) |
| Bottoming cycle output | 310 MW(e) |
| Overall power plant efficiency (coal to ac power) | 50% (~$6.8 \times 10^6$ J/kWh heat rate) |
| Gasifier | Texaco (oxygen blown) |
| Clean-up | Selexol + ZnO |

*At rated power.

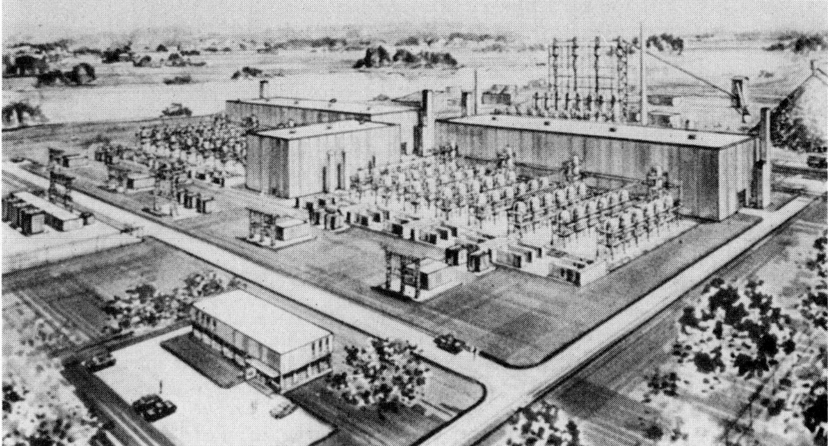

**FIG. 11-76 Artist's conception of a 675-MW MCFC fuel cell power plant.** *(Courtesy of United Technologies Corp.)*

consist of five basic components: a porous support tube, a fuel electrode, a solid electrolyte, an air electrode, and an electronically conducting interconnection.

The porous tube provides mechanical support for the other components. It is nominally 1 to 2 cm in diameter with a wall thickness of 0.1 cm. The fuel electrode is typically a porous nickel-zirconia cermet that serves as both the electrocatalyst and current collector. The electrolyte is a dense film of yttria-stabilized zirconia. The air electrode consists of a discontinuous catalyst layer coated with a porous, doped indium oxide current collector. The fuel electrode, electrolyte, and air electrode are each approximately 50 $\mu$m thick. Individual cells are series-connected by electronically shorting the fuel electrode of one cell to the air electrode of the adjacent cell. This interconnection must pass through the electrolyte via a gas-tight seal. The interconnection material has historically been a major concern because of the severe environment.

Practical considerations in developing the individual components are:

Porous support tubes are required to provide a mechanically strong structure. The support tube must allow access of the fuel gas to the fuel electrode. Thus, a proper trade-off between strength and porosity is important. It is also necessary that the thermal expansion characteristics of the support tube match those of the other components. These con-

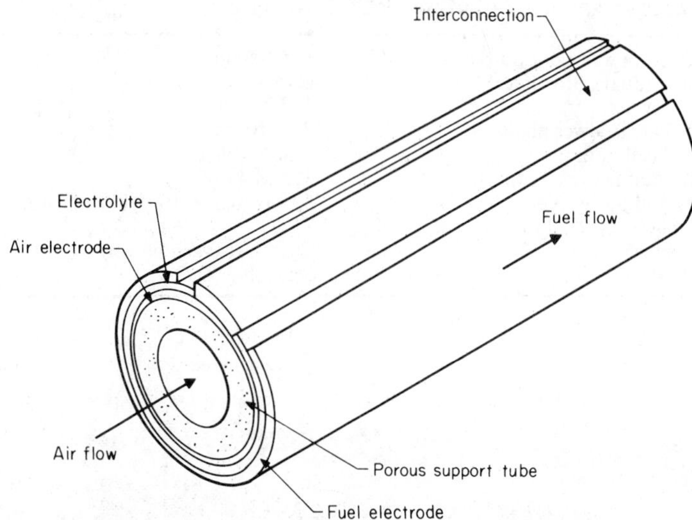


**FIG. 11-77** Cross section of Westinghouse thin-film, high-temperature, solid-electrolyte fuel cell stack. *(Courtesy of Westinghouse Electric Corp.)*

siderations coupled with cost have led to the selection of calcia-stabilized zirconia as the material presently in use in the Westinghouse program.[6] The Westinghouse requirements include 25 vol % open porosity, $3 \times 10^7$ Pa tensile strength, and $10 \times 10^{-6}$ m/m°C thermal expansion.

The fuel electrode must be electronically conductive, not crack or otherwise lose conductivity after a thermal cycle, allow fuel gas to reach the electrolyte interface, and catalyze the fuel oxidation reaction. The use of a porous nickel-zirconia cermet ensures a thermal expansion match and reasonable conductivity. The zirconia also provides a bond to the porous support and mitigates the sintering of the nickel catalyst.

The solid electrolyte must be an absolute gas barrier, provide good ionic conductivity, but be an electronic insulator. The HTSO electrolyte must tolerate thermal cycling without loss of integrity.

The air electrode must conduct electrons, withstand the oxidizing environment, allow oxygen to reach the electrolyte interface, adhere to the electrolyte thermal cycle without cracking, and catalyze the reduction of oxygen to oxide ions. Only electron-conducting metal oxides are able to satisfy these requirements. Candidate materials include tin-doped indium oxide and doped lanthanum manganite.

The interconnection between cells in a stack represents one of the major challenges in HTSO fuel cell development. The interconnection must have high electron conductivity (and negligible ionic conductivity), gastightness, low volatility, a thermal expansion match with other components, and chemical compatibility with both oxidizing and reducing environments. Compatibility with an oxidizing environment dictates the use of an oxide (nothing else is stable in air at 1000°C); yet most oxides are thermodynamically unstable in a reducing atmosphere, and only a few offer electronic conductivity. Modified lanthanum chromites have been identified as suitable cell interconnection materials.

***121. Performance Characteristics.*** Figure 11-78 shows the typical performance of a HTSO stack operating on hydrogen and air as well as the voltage loss contribution of the various components.

Endurance capability of HTSO fuel cells has steadily improved. Single-cell lifetimes of 34,000 h have been achieved at a constant output of 120 mA/cm$^2$ and 0.7 V dc.

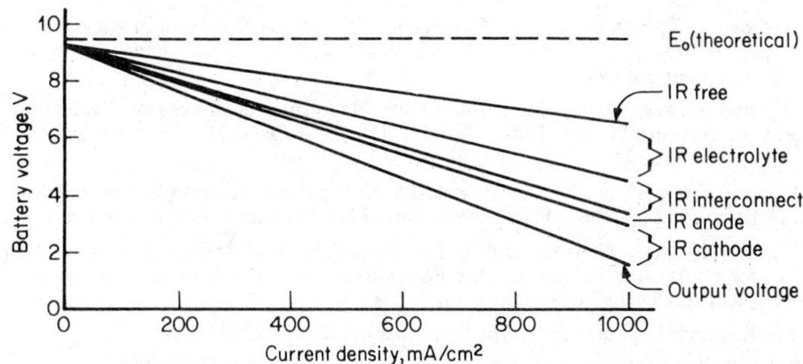


**FIG. 11-78** Typical voltage losses for a 10-cell battery of HTSO cells at 1000°C with hydrogen and air.

More recently, over 3000 h of operation have been achieved on a 10-cell stack. After 2000 h of operation at 400 mA/cm$^2$ on hydrogen-air, stack voltage had dropped 6%. This stack subsequently operated stably for an additional 3000 h at 150 mA/cm$^2$ on a simulated coal gas. Both tests involved 11 cool-down and reheat cycles, providing some confidence that cells and stacks can tolerate thermal cycling without performance deterioration.

***122. Power Plant Considerations.*** Under a NASA contract, Westinghouse analyzed power plant concepts involving the HTSO fuel cell integrated with a coal gasifier. The preferred power plant was similar to that previously shown (Fig. 11-76) for molten carbonate fuel cells and utilized a steam bottoming cycle to convert waste heat from the gasifier and the stack to additional power. An overall efficiency of 40 to 44% was calculated. This efficiency assumed a fuel cell operating point of 400 mA/cm$^2$ at 0.66 V dc. It is likely that efficiencies approaching 48% would have been achieved if lower current densities (200 mA/cm$^2$) had been used. That study also assumed a 10,000-h fuel cell life, and as a result the HTSO power plant was not economically competitive with molten carbonate fuel cell power plants. If a 40,000-h lifetime assumption is made, the resulting economics would likely have been competitive.

***123. Direct Fuel Cell Systems.*** Methanol ($CH_3OH$) and hydrazine ($N_2H_4$) are the liquid fuels that have been used in direct fuel cells, as they are more readily oxidized than the fossil fuels. Work is still in the development stage, and no systems were commercially available in the mid-1970s. Effort in recent years has been deemphasized.

*Hydrazine* has been used in a 60-W and larger configurations. The 60-W hydrazine unit is similar to the hydrogen fuel cell illustrated in Fig. 11-78 and delivers over 600 Wh/kg but was abandoned in favor of the hydrogen system. Similar experience was obtained with a larger 120-kg, 1.5-kW unit which degraded rapidly after 300 h of service.

A relatively small effort continues on direct hydrocarbon fuel cells, particularly with high-temperature (1000 to 1200°C) solid-electrolyte fuel cells. The higher temperatures should allow direct reaction of the hydrocarbons with improved kinetics, although the potential benefits may be offset by the problems of high-temperature operation.

### *124. Bibliography on Batteries and Fuel Cells*

Abens, S. G., Hofbauer, J. A., and Marchetto, P. G.: 3 and 5 kW Methanol Power Plant Program, *Abstr. Natl. Fuel Cell Semin.,* Courtesy Associates, Washington, D.C., 1981.

Adlhart, O. J.: An Assessment of the Air Breathing, Hydrogen Fueled SPE Cell, *Proc. 28th Power Sources Symp.,* Redbank, N.J., 1978.

Bacon, F. T.: Fuel Cells, *BEAMA J., 61,* 1954.

Barthelemy, R.: Defense Applications of Fuel Cells, *Abstr. Natl. Fuel Semin.,* Courtesy Associates, Washington, D.C., 1981.

Benjamin, T. G., Camara, E. H., and Marianowski, L. G.: *Handbook of Fuel Cell Performance,* prepared for the U.S. Department of Energy, Contract No. EC77C03-1545, Institute of Gas Technology, 1980.

Berger, Carl: *Handbook of Fuel Cell Technology,* Prentice-Hall, Englewood Cliffs, N.J., 1968.

Bett, J. S., Bushnell, C. L., and Buswell, R. F.: Advanced Technology Fuel Cell Program, Electric Power Research Institute Project 114-2, Annual Report No. EM1328, Power Systems Division of United Technologies Corporation, 1980.

Bonds, T. L., and Dawes, M. H.: Fuel Cell Power Plant Integrated Systems Evaluation, Electric Power Research Institute Project 1085-1, Interim Report No. EM1097, General Electric Company, 1979.

Bonds, T. L., and Dawes, M. H.: Fuel Cell Power Plant Integrated Systems Evaluation, Electric Power Research Institute Project 1085-1, Final Report No. EM1670, General Electric Company, 1981.

Cronin, P. G., Murphy, A. J., Newton, R. J., and Wagner, E. S.: Assessment of a Coal Gasification Fuel Cell System for Utility Application, Electric Power Research Institute Project 1041-8, Final Report No. EM2387, Kinetics Technology International, 1982.

Electric Power Research Institute: Technical Assessment Guide, 1982.

"Eveready" Battery Applications and Engineering Data, Union Carbide Corp., 1971.

Falk, S. Uno, and Salkind, Alvin J.: *Alkaline Storage Batteries,* Wiley, New York, 1967.

Fickett, A. P.: Fuel Cell Electrocatalysts—Where Have We Failed?, Paper presented at Spring Meeting Electrochem. Soc., Philadelphia, Pa., 1977.

Fleischer, Arthur, and Lander, John J.: *Zinc-Silver Oxide Batteries,* Wiley, New York, 1971.

Fuel Cell Users Group: Report to the Management Committee of the Electric Utility Fuel Cell Users Group, prepared by Fuels Subcommittee, 1980.

George, M., and Scozzofava, J.: Reversible Metal Hydride-Air Fuel Cell, ECOM Report 77-2644-F, Ft. Monmouth, N.J., 1978.

*Gould Battery Handbook,* Gould, Inc., 1973.

Grove, W. R.: On Voltaic Series in Combinations of Gases by Platinum, *Philos. Mag., 14,* 127–130, 1839.

Grove, W. R.: On a Gaseous Voltaic Battery, *Philos. Mag., 21,* 287–293, 1842.

Grove, W. R.: *The Correlation of Physical Forces,* 6th ed., Longmans, Green, New York, 1874.

Grubb, W. T.: Ion Exchange Batteries, *Proc. 11th Annu. Battery Res. Dev. Conf.,* pp. 5–8, Atlantic City, N.J., 1957.

Handley, L. M., and Cohen, R.: Specification for Dispersed Fuel Cell Generator, Electric Power Research Institute Project 1777-1, Interim Report No. EM2123, Power Systems Division of United Technologies Corporation, 1981.

Heise, George W., and Cahoon, N. Corey: *The Primary Battery,* vols. I, II, Wiley, New York, 1971, 1975.

Holtberg, P. D., Woods, T. J., Hill, R. H., and Rasmussen, J. J.: 1982 GRI Baseline Projection of U.S. Energy Supply and Demand, 1981–2000, Gas Research Insights, 1982.

Houghtby, W. E., King, J. M., and Thompson, R. A.: Advanced Technology Fuel Cell Program, Electric Power Research Institute Project 114-2, Annual Report No. EM956, Power Systems Division of United Technologies Corporation, 1978.

Intersociety Energy Conversion Engineering Conference 1967–1977, Institute of Electrical and Electronics Engineers.

Isenberg, A. O.: Processing and Performance of High Temperature Solid Oxide Fuel Cells, *Abstr. Natl. Fuel Cell Semin.,* Courtesy Associates, Washington, D.C., 1980.

Isenberg, A. O.: Recent Advancements in Solid Electrolyte Fuel Cell Technology, *Abstr. Natl. Fuel Cell Semin.,* Courtesy Associates, Washington, D.C., 1982.

Jackson, S. B.: Performance and Cost Impacts of Using Coal-Derived Fuel in Phosphoric Acid Fuel Cell Power Plants, *Abstr. Natl. Fuel Cell Semin.,* Courtesy Associates, Washington, D.C., 1981.

Jasinski, Raymond: *High Energy Batteries,* Plenum, New York, 1967.

King, J. M., Reiser, C. A., and Schroll, C. R.: Molten Carbonate Fuel Cell Systems Verification and Scale-Up, Electric Power Research Institute Project 1273-1, Interim Report No. EM2502, United States Technologies Corporation, 1982.

Kordesch, Karl V.: *Batteries,* Marcel Dekker, New York, 1974.

Krumpelt, M., Ackerman, J., Herceg, J., Zwick, S., Slack, C., and Lwin, Y.: Gas Systems, *Abstr. Natl. Fuel Cell Semin.,* Courtesy Associates, Washington, D.C., 1982.

Kudo, T., and Obayashi, H.: Ion-Electron Mixed Conduction in the Fluorite Type $Ce_{1-x}Gd_xO_{2-x/2}$, *J. Electrochem. Soc., 123,* 415, 1976.

Kunz, J. R., and Gruver, G. A.: The Catalytic Activity of Platinum Supported on Carbon for Electrochemical Oxygen Reduction in Phosphoric Acid, *J. Electrochem. Soc., 122,* 1279, 1975.

Lebhafsky, H. A., and Cairns, E. J.: *Fuel Cells and Fuel Cell Batteries,* pp. 18–47, Wiley, New York, 1968.

Lindstrom, O.: Fuel Cells, *ASEA J., 37,* 3, 1964.

Lorton, G. P.: Sulfur Removal Processes for Advanced Fuel Cell Systems, Electric Power Research Institute Project 1041-5, Final Report EM1333, C. F. Braun, 1980.

McBryar, H.: NASA Fuel Cell Program Plan, *Abstr. Natl. Fuel Cell Semin.,* Courtesy Associates, Washington, D.C., 1979.

McCormick, J. B., Huff, J., Srinivasan, S., and Bobbett, R.: Application Scenario for Fuel Cells in Transportation, Report LA 7634-MS, Los Alamos National Laboratory, 1979.

Mientek, A. P.: On Site Fuel Cell Power Plant Technology Development Program, prepared for GRI by United Technology Corporation, South Windsor, Conn., 1982.

*Nickel-Cadmium-Battery Application Engineering Handbook,* General Electric, 1975.

Rohr, F. S.: High Temperature Solid Oxide Fuel Cells—Present State and Problems of Development, *Ext. Abstr. Workshop High Temp. Solid Oxide Fuel Cells,* Brookhaven National Laboratory, New York, 1977.

Ross, P. N., Jr., and Benjamin, T. G.: Thermal Efficiency of Solid Electrolyte Fuel Cells with Mixed Conduction, *J. Power Sources,* January, 311–321, 1977.

Ross, P. N., Jr.: Oxygen Reduction on Supported Pt Alloys, and Intermetallic Compounds in Phosphoric Acid, Electric Power Research Institute Project 1200-5, Final Report No. EM-1553, Lawrence Berkeley Laboratories, Berkeley, Ca., 1980.

Stonehart, P., and Baris, J.: Preparation and Evaluation of Advanced Electrocatalysts for Phosphoric Acid Fuel Cells, *First Q. Rep. NASA CR 159843,* 1980.

Stonehart, P., and MacDonald, J. P.: Stability of Acid Fuel Cell Cathode Materials, Electric Power Research Institute Project 1200-2, Interim Report No. EM1664, Stonehart Associates, 1981.

Strasser, J.: Development of a 7 kW $H_2/O_2$ Fuel Cell Assembly with Circulatory Electrolyte in a Compact Modular Design, *Proc. Electrochem. Soc.,* Boston, Mass., 1979.

Vielstich, Wolf: *Fuel Cells,* Wiley-Interscience, New York, 1970.

## THERMOELECTRIC AND THERMIONIC CONVERSION

*By ROLAND W. URE, JR. and FRED N. HUFFMAN*

### Thermoelectric Devices

***125. General Description.*** Thermoelectric devices are solid-state devices which either generate electricity from a heat source, or pump heat when supplied with an electric current. A schematic diagram of these devices is shown on Fig. 11-79. The active elements are marked $p$ and $n$ on this figure. In the generator, an electric current will flow in the load $R$ if the opposite ends of the active elements have different temperatures. In the cooling device, heat is absorbed at one end and given off at the opposite end of the active elements if an electric current is passed through the elements as shown. A refrigerator can be made by surrounding the cold end of the elements with an insulated box.

Since these are solid-state devices with no moving parts, they require very little maintenance and have long life if properly made.

*Assumptions.* The following assumptions are used in the initial treatment here:

**a.** Problems of heat transfer at the ends of the arms are not considered, but $T_h$ and $T_c$ are taken as the actual temperatures at the two ends of the arms.

**b.** Perfect thermal insulation of the arms and heat reservoirs $T_h$ and $T_c$ is assumed.

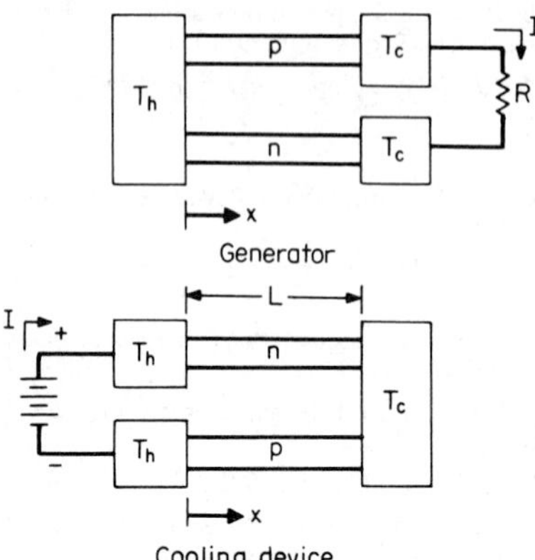


**FIG. 11-79** Schematic drawing of a Seebeck-effect thermoelectric generator and a Peltier-effect refrigerator. The elements marked $T_h$ and $T_c$ are heat reservoirs and are assumed to have zero electrical resistance. The elements marked $p$ and $n$ are called arms and usually have positive and negative absolute Seebeck coefficients, respectively.

**c.** Effects caused by electrical resistance at the junctions at the ends of the arms are neglected.

**d.** The properties of the arms such as electrical resistivity, thermal conductivity, and Seebeck coefficient are assumed independent of temperature.

The effects of these assumptions are considered later.

*Parameters.* The properties of the materials which are important are the electrical resistivity $\rho$, the thermal conductivity $\kappa$, and the Seebeck coefficient $\alpha$. The thermoelectric figure of merit Z is a parameter of the active materials; for a couple, it is defined as

$$Z = \alpha_{pn}^2/[(\rho_n\kappa_n)^{1/2} + (\rho_p\kappa_p)^{1/2}]^2$$

where the subscripts refer to the two materials making up the couple. In discussion of particular materials, it is convenient to define a figure of merit for a single material as

$$Z_i = \alpha_i^2/\rho_i\kappa_i$$

where $\alpha_i$ is the absolute Seebeck coefficient of the single material. The Seebeck coefficient of the couple $\alpha_{pn}$ is

$$\alpha_{pn} = \alpha_p - \alpha_n$$

If the two materials are similar except that their Seebeck coefficients have opposite signs, the figure of merit of the couple is approximately equal to the average of the figures of merit of the two individual materials.

The electrical resistance of the $n$ and $p$ arms in series is

$$R = \rho_n/\gamma_n + \rho_p/\gamma_p \qquad (11\text{-}7)$$

where

$$\gamma_n = A_n/L_n \quad \text{and} \quad \gamma_p = A_p/L_p$$

and $A$ is the cross-sectional area of the arm and $L$ is the length of the arm.

The thermal conductance of the $n$ and $p$ arms in parallel is

$$K = \kappa_n\gamma_n + \kappa_p\gamma_p \qquad (11\text{-}8)$$

**Power Generation—Single Couple**

***126. General.*** Consider first the design of a single couple. Devices are usually made with a number of couples; the design of such a device will be considered in a later section. The important design parameters for a power generator are the efficiency, power output, and heat input. The efficiency $\eta$ is defined as the ratio of the electrical power dissipated in the load $P_o$ to the thermal power input $q_h$ to the hot junction,

$$\eta = P_0/q_h \tag{11-9}$$

The reduced efficiency $\eta_r$ is defined as the ratio of the efficiency to the Carnot efficiency,

$$\eta_r = \eta T_h/\Delta T \tag{11-10}$$

where $\Delta T = T_h - T_c$. The thermal power input to the hot junction is given by

$$q_h = \alpha T_h I - \tfrac{1}{2} I^2 R + K\,\Delta T \tag{11-11}$$

where $R$ and $K$ are given by Eqs. (11-7) and (11-8). In a power generator, the positive direction for current is from the $p$ arm to the $n$ arm at the cold junction, and the current is given by

$$I = \alpha\,\Delta T/(R + R_L) = \alpha\,\Delta T/[R(s + 1)] \tag{11-12}$$

where $R_L$ is the load resistance and $s = R_L/R$. The electrical power output is

$$P_0 = I^2 R_L = (\alpha\,\Delta T)^2 s/[R(s + 1)^2] \tag{11-13}$$

The output voltage is

$$V = IR_L = \alpha\,\Delta T s/(s + 1) \tag{11-14}$$

and the efficiency is

$$\eta = \frac{2s\,\Delta T}{T_h(2s + 1) + T_c + 2(1 + s)^2 RK\alpha^{-2}} \tag{11-15}$$

There are five parameters which must be selected to optimize the generator design: $T_h$, $T_c$, $\gamma_n$, $\gamma_p$, and $s$. [Once $\gamma_n$ and $\gamma_p$ are determined, $R$ is given by Eq. (11-7). Thus adjustment of $s$ is done by varying the load resistance.] Three design criteria will be considered in the next three sections. In some applications other criteria may be important, and designs different from those considered here may be most suitable.

***127. Maximum Efficiency between Fixed Temperatures.*** The parameters of a couple which has been designed to maximize the efficiency when operating with fixed hot- and cold-junction temperatures will be given in this section. This type of operation assumes that $T_c$ and $T_h$ do not change as $R_L$, $\gamma_n$, and $\gamma_p$ are varied. The shape ratio which maximizes the efficiency is

$$\gamma_n/\gamma_p = (\kappa_p\rho_n/\kappa_n\rho_p)^{1/2} \tag{11-16}$$

With this shape ratio, $RK\alpha^{-2} = Z^{-1}$. The load resistance which gives maximum efficiency is

$$R_L = Rs_e \tag{11-17}$$

where $s_e = (1 + Z\overline{T})^{1/2}$ and $\overline{T} = (T_h + T_c)/2$. Values for $s_e$ are given in Fig. 11-80.

The efficiency with both the geometry and the load resistance optimized is

$$\eta = \frac{\Delta T}{T_h}\,\frac{s_e - 1}{s_e + (T_c/T_h)} \tag{11-18}$$

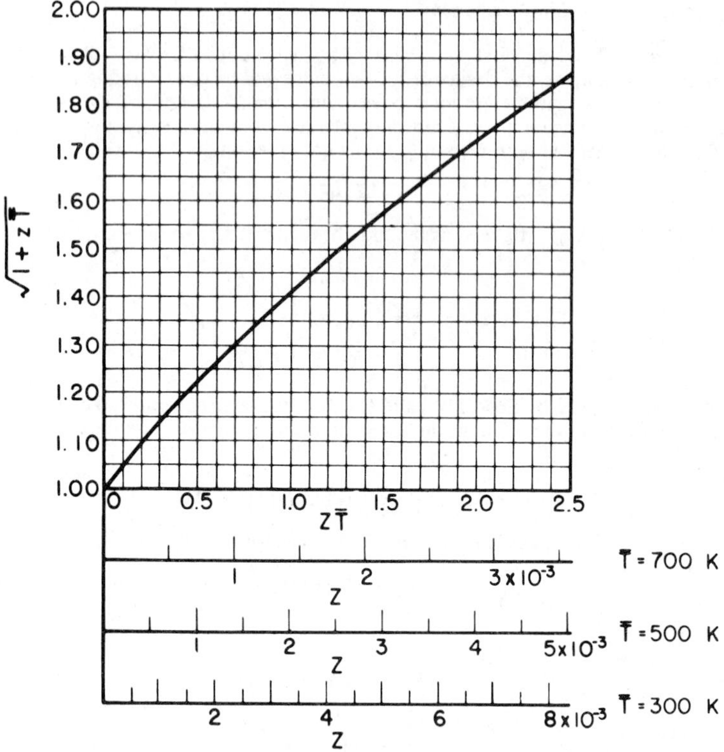


**FIG. 11-80** Values of $(1 + Z\overline{T})^{1/2}$ as a function of $Z\overline{T}$, or of $Z$, for three values of $\overline{T}$.

The optimum efficiency is shown in Fig. 11-81. With optimum load and geometry, the shape factor required to give a power output $P_o$ is

$$\gamma_n = \frac{A_n}{L_n} = \frac{P_0(1 + s_e)^2}{(\Delta T)^2 s_e \alpha} \left(\frac{\rho_n}{Z\kappa_n}\right)^{1/2} \tag{11-19}$$

and the internal resistance is

$$R = \frac{\alpha}{\gamma_n} \left(\frac{\rho_n}{Z\kappa_n}\right)^{1/2} = \frac{(\alpha\, \Delta T)^2 s_e}{P_0(1 + s_e)^2} \tag{11-20}$$

The output voltage is given by Eq. (11-14), with $s_e$ substituted for $s$.

For quick calculations using materials in which $\rho_n$ and $\rho_p$ are not too different and $\kappa_n$ and $\kappa_p$ are similar, the following expressions can be used,

$$\gamma_n \approx \gamma_p \approx 2P_0(1 + s_e)^2\bar{\rho}/[s_e(\alpha\, \Delta T)^2] \tag{11-21}$$

$$R \approx 2\rho/\gamma_n \tag{11-22}$$

$$Z \approx \alpha^2/4\overline{\rho\kappa} \tag{11-23}$$

where $\bar{\rho} = (\rho_n + \rho_p)/2$ and $\bar{\kappa} = (\kappa_n + \kappa_p)/2$ (11-24)

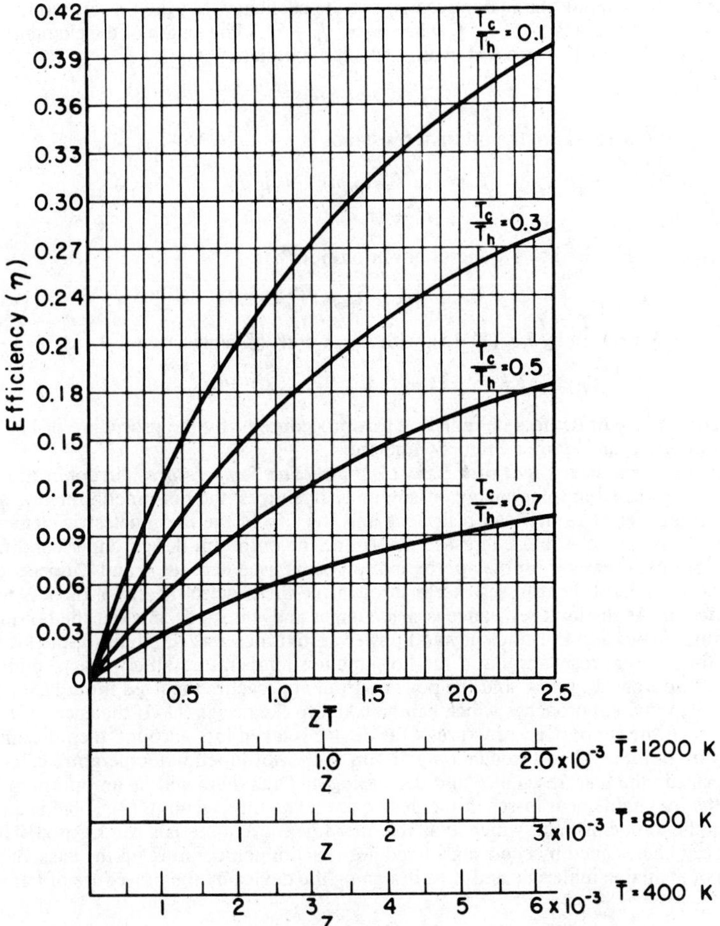


**FIG. 11-81** Efficiency of a thermoelectric couple as a function of $Z\overline{T}$, or of $Z$ for three values of $\overline{T}$. Designed for maximum efficiency.

***128. Maximum Power Output between Fixed Temperatures.*** A generator operating with fixed hot- and cold-junction temperatures can also be designed to maximize the power output.This maximum power output corresponds to $R_L = R$ or $s = 1$. With this load resistance, the current, power output, and load voltage are given by Eqs. (11-12), (11-13), and (11-14) with $s = 1$. The size factors which maximize the efficiency are again given by Eq. (11-16), and the efficiency is

$$\eta = \frac{2\,\Delta T}{3T_h + T_c + 8Z^{-1}} \tag{11-25}$$

For $Z\overline{T} = 3$, the efficiencies given by Eqs. (11-18) nd (11-25) differ by about 10%, and this difference becomes smaller for smaller $ZT$. Since the best materials available today have a $ZT$ less than 1.5, these two designs are practically identical.

However, the usual objective in designing for maximum power output is to minimize the size or weight of a generator or to minimize the thermoelectric material required. Thus the factor which should be maximized is, not the efficiency, but the power output divided by the sum of the size factors, $P_o/\gamma_t$, where $\gamma_t = \gamma_n + \gamma_p$. (The length of the elements should also be minimized.) The size-factor ratio which maximizes $P_o/\gamma_t$ is

$$\gamma_n/\gamma_p = (\rho_n/\rho_p)^{1/2} \tag{11-26}$$

With this size-factor ratio, the internal resistance is

$$\begin{aligned} R &= [\rho_n + 2(\rho_n\rho_p)^{1/2} + \rho_p]/\gamma_t \\ &= [\rho_n + (\rho_n\rho_p)^{1/2}]/\gamma_n \end{aligned} \tag{11-27}$$

The size factor required to produce a power output $P_o$ is

$$\gamma_n = 4P_o[\rho_n + (\rho_n\rho_p)^{1/2}]/(\alpha\,\Delta T)^2 \tag{11-28}$$

The efficiency is given by Eq. (11-15), with $s = 1$ and $RK$ replaced by

$$RK = \rho_n\kappa_n + (\rho_n\rho_p)^{1/2}(\kappa_n + \kappa_p) + \rho_p\kappa_p \tag{11-29}$$

The performance of devices designed by this procedure or by the procedure of Par. **127** is similar unless $\kappa_n$ and $\kappa_p$ are extremely different.

***129. Constant Heat Input with Fixed Cold-Junction Temperature.*** In applications such as generators utilizing solar energy or isotope heat sources, the hot-junction temperature is not fixed but varies as the load resistance and the size of the thermoelectric elements are changed. It is thus of some interest to determine the optimum design under conditions of fixed heat flux. However, in the approximation being used here ($\alpha$, $\rho$, and $\kappa$ independent of temperature, and no heat losses except through the thermoelectric arms), there is no optimum design. As the load resistance is made larger and $\gamma$ is made smaller, the hot-junction temperature rises and the efficiency and power output increase. In the limit, the load resistance, the internal resistance, and the hot-junction temperature all approach infinity, the efficiency approaches unity, and the power output approaches the fixed input heat flux.

There are two approaches which can be taken to the design. (1) If the effect of the heat loss through the supporting structure of the device is taken into account, then the efficiency and power output will go through a maximum as the hot-junction temperature is increased by increasing the load resistance and decreasing $\gamma$. Thus there will be an optimum design when the inevitable heat losses in the device are taken into account. (2) There is an upper limit to the temperature at which available thermoelectric materials can be used. Thus the device can be designed by choosing a fixed hot-junction temperature on the basis of performance of available materials and then designing the device by the procedure of Par. **127** or **128.**

## Thermionic Devices[1]

***130. Definition of Thermionic Converter.*** A thermionic converter is a vacuum or vapor-filled device with a hot electrode to emit electrons and a cold electrode to collect the electrons. It is a high-temperature static heat engine that obeys the Carnot cycle. Electrons are emitted from the emitter (see Fig. 11-82). The heat put into the emitter lifts the electrons over the work-function barrier of the surface of the metal, $\phi_E$. An electron just outside the metal has kinetic energy of motion, and potential energy relative to the electrons in the metal equal to $\phi_E$ or in some cases $V_m$, which is the greatest negative potential in the system. As the electrons are collected, they may fall through a potential $V_D$ to the surface of the collector. They then fall through an additional potential $\phi_c$ to the Fermi level of the collector. Any additional potential $V_0$ is available to drive the electrons through an external circuit. Thus, $V_0$ is the output voltage of this generator. For an ideal converter with no heat loss and with $V_D$ equal to 0, the converter would obey the Carnot efficiency

$$\eta = \frac{T_E - T_C}{T_E} = \frac{\phi_E - \phi_C}{\phi_E} \tag{11-30}$$

[1]A revision of the material by V. C. Wilson in the tenth edition.

***131. Types of Thermionic Converters***

*Vacuum Converters.* A fundamental problem of thermionic conversion is that of the negative space charge barrier which is a consequence of the finite transit time of the electrons crossing from the emitter to the collector. The electrons which are emitted are repelled by the electric field from the electrons in transit between the electrodes and will be reflected back to the emitter unless they have sufficient kinetic energy to overcome the repulsion and reach the collector. In principle, the space charge problem can be overcome by spacing the electrodes quite closely. The potential diagram of a vacuum converter approximates that of Figure 11-82*a*. Calculations by Webster[1] given in Fig. 11-83 indicate that interelectrode spacings less than 0.0005 in are required for practical output power densities (i.e., greater than a watt per square centimeter). J. E. Beggs[2] has constructed small-area vacuum converters whose performances agree well with Webster's calculations. In practice, it has not been possible to build large-area vacuum converters with the required close spacings using conventional designs.

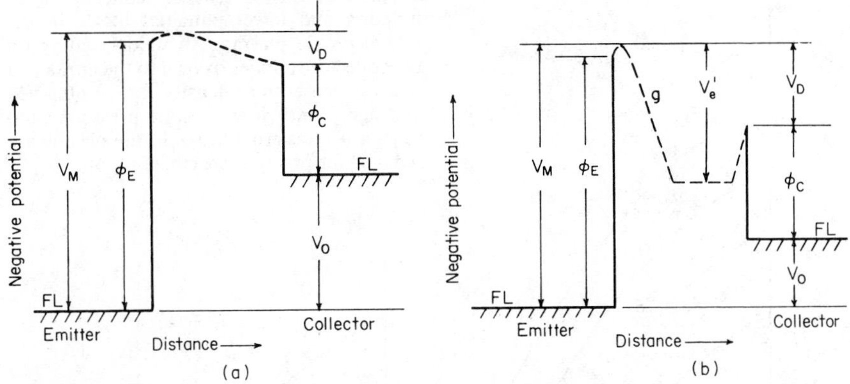


**FIG. 11-82** Different types of thermionic converter potential diagrams. (*a*) Unignited mode; (*b*) ignited mode.

*Cesium Vapor Converters.* Addition of cesium vapor between the electrodes of a thermionic converter has two beneficial results. First, the cesium adsorbs on the surfaces of the emitter and collector to provide favorable electrode work functions. Second, the cesium can provide positive ions to neutralize the electron space charge by surface contact ionization at the emitter and/or electron impact ionization in the interelectrode space. The cesium pressure is regulated by the temperature of a liquid cesium reservoir.

1. *Unignited Mode Operation.* At sufficiently high emitter temperature and emitter work function, positive Cs ions will be generated as the Cs atoms strike the hot emitter. Because of its large mass, a single Cs ion can allow up to 500 electrons to cross from the emitter to the collector. The electrode spacing can be increased to around 0.040 in since the Cs ions neutralize the electron space charge. For larger spacings, current loss due to electron scattering by the Cs atoms (typically, Cs pressure less than 0.1 torr) will significantly reduce the output power. Such a converter is said to be operating in the "unignited," "Knudsen," or "extinguished" mode. Unfortunately, this desirable operational mode is only possible for emitter temperatures above 2200 K. Such extremely high temperatures preclude most materials and heat sources.

[1]Webster, H. F.: *J. Appl. Phys.*, 1959, vol. 30, p. 488.
[2]Beggs, J. E.: *Adv. Energy Convers.*, 1963, vol. 3, p. 447.

2. *Ignited Mode Operation.* Virtually all thermionic converters operate in the "ignited" mode. This mode is achieved by increasing the cesium pressure up to 10 torr and reducing the interelectrode spacing to around 0.010 in. The high cesium pressure permits current densities of tens of amperes per square centimeter to be obtained from the emitter at temperatures above 1600 K. Since ions supplied by surface contact ionization are insufficient to neutralize the electron space charge under these conditions, ions must be supplied by electron impact ionization in the interelectrode space. The arc drop $V_D$ required to ignite the plasma between the electrodes is approximately 0.5 V and subtracts from the output voltage $V_o$ (see Figure 11-82*b*). Ignited mode operation results in practical thermionic converter power densities and efficiencies at temperatures at which heat sources and materials are available.

*Additive Oxygen Converters.* Controlled additions of oxygen into thermionic converters can yield substantial improvement in power density, especially at spacings greater than 0.020 in. The improved performance is due to a combination of reduced collector work function and lower potential losses in the interelectrode plasma. An oxide coating on the collector supplies oxygen to the emitter so that a given current density can be obtained at a significantly lower cesium pressure. Consequently, scattering losses in the plasma are reduced for practical spacings.

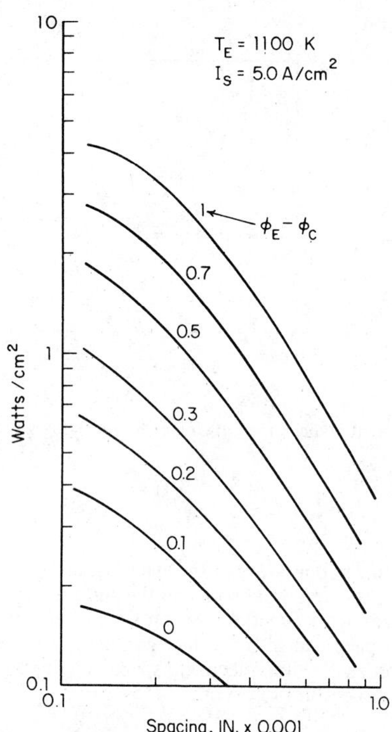


**FIG. 11-83** Maximum-output power density from a vacuum thermionic converter versus emitter-collector spacing for seven different values of work-function difference $\phi_E - \phi_C$ in electron-volts.

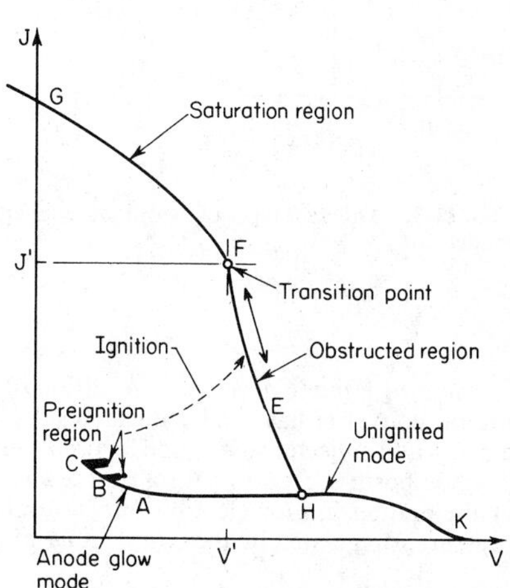


**FIG. 11-84** Identification of the various modes of operation for a high-cesium pressure diode thermionic converter.

*Advanced Converters.* The efficiency of the conventional thermionic converter would be doubled if the arc drop could be reduced to zero. Many technical approaches have been suggested for decreasing the arc drop. They include auxiliary electrodes, pulsed operation, structured electrodes, and electropositive additives. The most promising approaches require an emitter that provides high current densities ($>$10 A/cm$^2$) at low cesium pressures ($<$0.1 torr) and xenon gas in the interelectrode space. Thus far, progress has been stymied by the lack of a suitable emitter.

***132. Cesium Vapor Diode Converters.*** Although a diode thermionic converter is an extremely simple device, the physics involved is surprisingly complicated. The electricl potential distribution between the electrodes and the generation of ions depends upon the way in which the thermionic converter is operated. Figure 11-84 is a plot of the voltage generated versus the current density. At open circuit, point $K$, the electrons reaching the collector just equal the leakage currents from the collector. If one starts to draw current from the converter, the voltage drops rapidly along the line $K$ to $A$. In the region from $H$ to $A$ the current is limited by the number of ions that are generated at the hot emitter.

At point $B$, $V_D$ of Fig. 11-82*a* has become so large that some electrons arriving at the collector have obtained sufficient energy to generate Cs ions adjacent to the collector. This creates a glow adjacent to the collector. As one tries to draw more current, the converter "ignites" and the potential distribution at this point shifts from that of Fig. 11-82*a* to that of Fig. 11-82*b*. In this obstructed region there is a double sheath adjacent to the emitter. Closest to the emitter is an excess of electrons such that the negative potential increases as one moves away from the emitter. Next, there is a high density of positive ions creating a potential well of magnitude $V_e'$. Electrons are accelerated into this well and then rethermalize to a temperature much higher than the temperature of the emitter. The Cs atoms and Cs ions are at a temperature intermediate between the temperature of the two electrodes.

The high-temperature electrons ionize some of the Cs. At the transition point $F$, all the electrons emitted from the emitter enter the plasma, and the negative space charge next to the emitter disappears. One might think that this would produce a sharp limit to the current density. However, as the output voltage decreases, $V_D$ increases and forces the positive space charge closer to the emitter, causing increased emission by the Schottky effect.

***133. Applications.*** Thermionic converters may be coupled to any heat source which has a high enough temperature. Indeed, they have been coupled to nuclear reactor, radioisotope, solar, and fossil fuel heat sources.

*Space Applications.* Historically, thermionic conversion was first developed for space applications where it is especially attractive because of lack of moving parts, good efficiency, and high heat rejection temperature (which minimize radiator size and mass). The U.S.S.R. has reported on the tests of four thermionic (TOPAZ) reactors. Thermionic technology is being developed in the United States with a view to megawatt space reactor applications. Solar thermionic generators have been less attractive than photovoltaic arrays because of their more stringent tracking requirements. Radioisotope thermionic generators have not been launched because radioisotope thermoelectric generators present significantly reduced radiological hazards on launch and reentry.

*Terrestrial Applications.* Applications of thermionic converters to terrestrial applications required that a means of protecting the high-temperature refractory metals be found. A trilayer structure (tungsten emitter, silicon carbide layer, and intermediate graphite substrate) made by chemical vapor deposition has given good results in combustion atmospheres. Stable operation of combustion-heated thermionic converters using such a trilayer structure has been demonstrated for periods up to 12,500 h at emitter temperatures of around 1730 K. Potential terrestrial applications include power-plant topping, cogeneration with industrial processes, and solar heating systems. Although the primary barrier to commercialization has been the cost of fabricating the thermionic converters, manufacturing studies indicate that it should be possible to mass produce converters at a competitive cost.

***134. References***

*Thermoelectric Devices*

Heikes, R. R., and Ure, R. W., Jr.: *Thermoelectricity: Science and Engineering;* New York, Interscience Publishers, Inc., 1961.

Goldsmid, H. J.: *Thermoelectric Refrigeration;* New York, Plenum Press, 1964.

*Proc. IEEE,* Special Issue, May 1963, vol. 51, no. 5, pp. 699–724.

*Elec. Eng.,* Special Issues, May and June 1960, vol. 79, nos. 5 and 6, pp. 353–493.

*Proc. Symp. Thermoelec. Energy Conversion;* in Klein, P. H. (ed.), *Adv. Energy Convers.;* 1961, vol. 1, nos. 1–4, pp. 1–365; January–June 1962, vol. 2, pp. 1–312.

Cadoff, I. B., and Miller, E. (eds.): *Thermoelectric Materials and Devices;* New York, Reinhold Publishing Corporation, 1960.

Egli, P. H. (ed.): *Thermoelectricity;* New York, John Wiley & Sons, Inc., 1960.

Harman, T. C., and Honig, J. M.: *Thermoelectric and Thermomagnetic Effects and Applications;* New York, McGraw-Hill Book Company, 1967.

*Thermionic Devices*

Baksht, F. G., et al.: Thermionic Converters and Low-Temperature Plasmas; Technical Information Center/U.S. Department of Energy, 1978.

Blue, E., and Ingold, J. H.: Chapter 5 in Sutton, G. W. (ed.), *Direct Energy Conversion;* New York, McGraw-Hill Book Company, 1966.

Hatsopoulos, G. N., and Gyftopoulos, E. P.: *Thermionic Energy Conversion;* Cambridge, Mass., The MIT Press, vol. 1, 1973; vol. 2, 1979.

Rasor, N. S.: Chapter 5 in *Applied Atomic Collision Physics;* vol. 5, New York, Academic Press, 1982.

## MAGNETOHYDRODYNAMICS

*By JOHN C. CUTTING**

***135. Introduction.*** Recognition that the magnetic force on a gaseous or liquid conductor can produce desirable engineering results is not a new concept in electrical engineering. Designs for arc-expulsion gaps, which use self-induced magnetic forces to extinguish the arc, and for liquid-metal pumps are old in the art. These are, however, power-*consuming* rather than power-*generating* applications.

Since 1959, substantial effort has been devoted to exploring the conditions under which a conducting fluid, specifically a gaseous plasma or liquid metal, moving through a magnetic field might generate useful electrical power (Refs. 17 and 18, Par. **148**), or, inversely, might convert electrical energy into thrust for rocket propulsion in space (Refs. 10 and 16). Only the former is discussed here. The terms magnetoplasmadynamics (MPD), magnetofluiddynamics (MFD), magnetogasdynamics (MGD), and magnetohydrodynamics (MHD), as well as other similar combinations have been used to describe the combination of disciplines required to treat the phenomena. Common usage now favors magnetohydrodynamics as the generic term in power generation applications.

The motivation for the development and use of MHD generators in central-station application is the promise of a significant improvement in plant thermal efficiency because much higher temperatures are allowable, and required, in the thermodynamic cycle of MHD generators in which there are no rigid moving parts with close tolerances.

As originally envisaged, the MHD generator is a "topping" unit on an otherwise conventional turbine-generator station. Electric power is generated in the MHD unit, and its exhaust heat, with temperature as high as 2200 K, is used to create steam for the turbine generators. The limiting Carnot efficiency for a station might by such a scheme be raised from the present maximum of about 65% ($T_1$ = 850 K, $T_2$ = 300 K) upward toward 85% ($T_1$ = 2600 K, $T_2$ = 420 K).

The net thermal overall efficiency of the combined cycle can be expressed as $\eta_1 + \eta_2 (1 - \eta_1)$, where $\eta_1$ is the efficiency of the MHD generator and $\eta_2$ is the efficiency of the "bottom" steam plant. Typical values are 0.25 and 0.40 for $\eta_1$ and $\eta_2$, respectively, for an overall efficiency of 0.55.

Currently, a variety of MHD system conceptions, as summarized in Table 11-24, are subjects of continuing theoretical and experimental effort. Generally, with all these systems except some of the liquid-metal designs, the practical design configurations proposed have dc outputs taken from electrodes at the sides of the MHD channel, and it is envisaged that electronic inverters will be used to convert MHD power from dc to power-system-frequency ac.

*The author and editor acknowledge the important contribution of the late Clifford Mannal and Norman W. Mather to the treatment of this subject in the preceding edition of this handbook, which serves as the basis for the present revised version.

**TABLE 11-24** Principal MHD Power-Generating Systems

| Cycle | Heat source | Working fluid | Magnetic field by | Electrical output | Rejects heat to | Application | Features |
|---|---|---|---|---|---|---|---|
| Open | Coal *or* manufactured or natural gas *or* $H_2—O_2$ (produced using coal) *or* fuel oil | Potassium seeded combustion products (temperatures up to ~2500°C) | DC superconducting magnet (4–6 T) | DC converted to AC (with external inverter) | Bottom steam plant *or* gas turbine *or* directly to atmosphere (in peaking applications) | Base-load power plant *or* peak-load power plant | Good thermal efficiency if used with bottoming plant<br>Competitive economically<br>Low pollution potential |
| Closed | Nuclear high-temperature GCR *or* coal *or* natural gas *or* fuel oil | Cesium seeded helium (temperatures up to ~1400°C) | DC superconducting magnet (4–6T) | DC converted to AC (with external inverter) | Bottom steam plant *or* directly to atmosphere | Base-load power plant *or* space-mission power plant | Good thermal efficiency if used with bottoming plant<br>Low pollution potential |
| | | Two-phase closed-cycle, liquid metal in MHD channel (temperatures about 870°C max.) | Conventional or superconducting magnet (1–2 T) | DC converted to AC (with external inverter) | Bottom steam plant *or* directly to atmosphere | Base-load power plant *or* space-mission power supply | Fairly good thermal efficiency with bottom plant<br>Low pollution potential |
| | | | AC field coils | AC induction directly | | | |

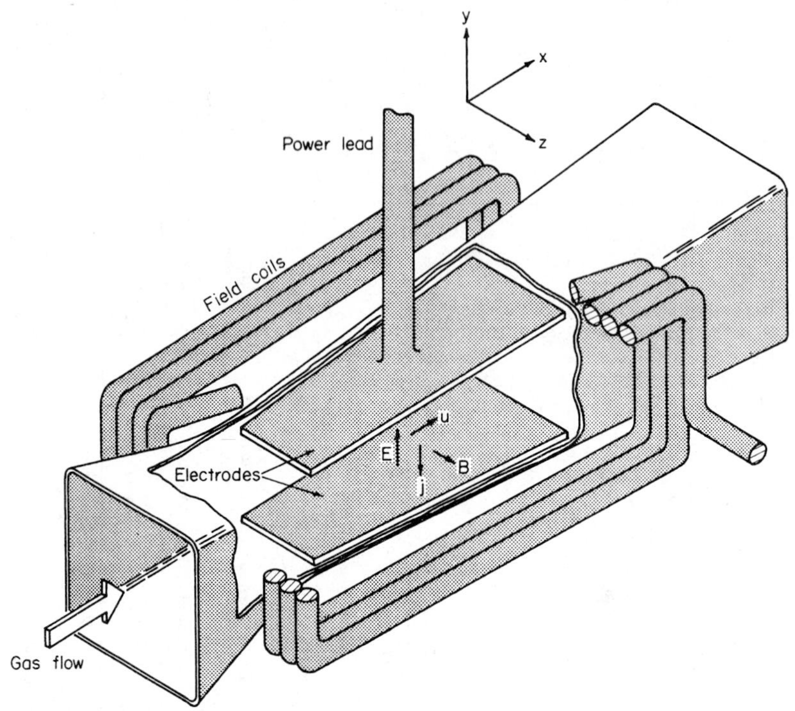


**FIG. 11-85** Basic elements of a magnetohydrodynamic generator.

***136. Fundamental Equations.*** Figure 11-85 shows the basic scheme for an MHD generator having quasi-one-dimensional flow of ionized gas (i.e., plasma) channeled through a perpendicular, static magnetic field. The charged particles experience a transverse force which is equivalent to an electric field $\mathbf{u} \times \mathbf{B}$ (see Table 11-25, Glossary) due to their velocity $\mathbf{u}$ through the magnetic field. This generates, between electrodes spaced a distance $d$ apart, an open-circuit voltage of

$$V_{oc} = \int_0^d (\mathbf{u} \times \mathbf{B}) \cdot d\mathbf{l} \tag{11-31}$$

An electric field $\varepsilon$ is therefore present between the electrodes such that $\varepsilon + \mathbf{u} \times \mathbf{B} = 0$, corresponding to the zero-current condition. When current $\mathbf{j}$ flows, owing to connecting an external load $R_L$, the electric field $\varepsilon$ and the electrode voltage $V$ are reduced because of the electrical resistance of the fluid. In general,

$$V = -\int_0^d \varepsilon \cdot d\mathbf{l} = R_L \int_{Ae} \mathbf{j} \cdot d\mathbf{s} \tag{11-32}$$

and the power delivered at a pair of electrodes per unit electrode area is $jV$, that is, electrode current density times electrode voltage. But, if uniform conditions can be assumed to apply, it is often more convenient to use the power density in the flow channel $\mathbf{j} \cdot \varepsilon$ (watts per cubic meter).

In the MHD channel both electrons and ions are affected by the $\varepsilon$ field, but the electrons, having much greater mobility, conduct almost all the current. Since the current carriers are predominantly electrons, another electric field, the *Hall field,* is created,

$$\varepsilon_H = \beta \mathbf{j} \times \mathbf{B} \tag{11-33}$$

**TABLE 11-25** Glossary

$A_c$ = cross-sectional area of flow channel, square meters
$A_e$ = electrode area, square meters
$a$ = speed of sound in fluids, meters per second = $(\gamma p/\rho)^{1/2}$ = $(\gamma RT/W_m)^{1/2}$
$B$ = magnetic induction field, webers per square meter = teslas
$C$ = electron rms thermal velocity, meters per second
= $(3\kappa T_e/m_e)^{1/2}$ for Maxwellian distribution
$c_p$ = specific heat at constant pressure, J-m³/kg-K
$c_v$ = specific heat at constant volume, J-m³/kg-K
$d$ = channel width, meters
$\varepsilon$ = electric field, volts per meter
$\varepsilon_H$ = Hall-effect electric field, volts per meter
$\varepsilon^*$ = electric field in moving frame, volts per meter
$\epsilon$ = internal energy of fluid, J-m³/kg
$e$ = electronic charge, $1.60 \times 10^{-19}$, coulombs
$H$ = total enthalpy of fluid, J-m³/kg
$h$ = static enthalpy, J-m³/kg = $H - u^2/2$
$I$ = output current, amperes
$j$ = current density, amperes per square meter
$K$ = Faraday generator coefficient or electrical efficiency = $\epsilon_y/uB$
$K_H$ = Hall generator coefficient = $-\varepsilon_x/\omega\tau uB$
$k$ = Boltzmann's constant, $1.38 \times 10^{-23}$ joules per degree Kelvin
$l$ = length, meters
$L$ = length of MHD channel, meters
$L_i$ = interaction length, meters
$M$ = Mach number = $u/a$
$m_a$ = atom mass, kilograms
$m_e$ = electron mass, $9.11 \times 10^{-31}$ kilograms
$m_i$ = ion mass, kilograms
$n_e$ = electron particle density, $m^{-3}$
$P$ = delivered power density, watts per cubic meter
$P_i$ = internally developed electrical power density, watts per cubic meter
$p$ = pressure, newtons per square meter; also, atmospheres, where 1 atm = $1.01 \times 10^5$ N/m²
$\hat{\mathbf{p}}$ = pressure tensor, newtons per square meter
$Q_{en}$ = electron-neutral collision cross section, square meters
$R$ = universal gas constant, $8.31 \times 10^3$ J/(kg mole)(K)
$R_i$ = internal resistance, ohms
$R_L$ = external load resistance, ohms
$s$ = surface or area variable, square meters
$T$ = temperature, kelvins, also, degrees Celsius = K − 273, and degrees Fahrenheit = (9/5)K − 459, are used
$T_a$ = temperature of background gas, kelvins
$T_e$ = electron temperature, kelvins
$u$ = fluid velocity, meters per second
$V$ = electrode voltage
$V_i$ = ionization voltage = ionization energy, electronvolts (since $e$ = 1 in electronvolt units)
$V_{oc}$ = open-circuit electrode voltage
$v$ = electron-drift velocity, meters per second
$W_m$ = molecular weight of gas, atomic units
$\mathbf{x}_1$ = unit vector in $x$ direction
$\mathbf{z}_1$ = unit vector in $z$ direction
$\beta$ = $1/en_e$ = Hall constant, cubic meters per coulomb
$\gamma$ = ratio of specific heats = $c_p/c_r$
$\theta$ = $-\tan^{-1}\varepsilon_y/\varepsilon_x$ = cross-connection angle in a diagonally connected generator
$\lambda$ = electron mean-free path, meters
$\mu_e$ = electron mobility, m²/V-s = $u/\epsilon$ = $v/\epsilon \approx 3.6 \times 10^{-16} T_e^{1/2}/(pQ_{en})$
$\mu_i$ = ion mobility, m²/V-s
$\rho$ = mass density, kilograms per cubic meter; $1/\rho$ = specific volume, m³/kg
$\sigma$, $\sigma_o$ = $nev$ = $ne\mu_e\varepsilon$ = fluid conductivity in absence of magnetic field, siemens per meter
$\hat{\boldsymbol{\sigma}}$ = tensor conductivity, including effect of magnetic field, siemens per meter
$\tau$ = electron mean free time, seconds = $\lambda/C$
$\tau_i$ = ion mean free time, seconds
$\omega$ = electron cyclotron frequency, $s^{-1}$ = $eB/m_e$ = $\mu_e B/\tau$
$\omega_i$ = ion cyclotron frequency, $s^{-1}$ = $eB/m_i$ (if singly ionized)

NOTE: Except where explicitly noted otherwise, mksa units used throughout.

Here $\beta = 1/n_e e$ is the Hall constant. The current component which is created by this field, the *Hall current*, is given by $-\mu_e \mathbf{j} \times \mathbf{B}$, in which $\mu_e = \omega\tau/B$ is the electron mobility. (The sign is negative since this is an electron current.)

Segmented electrode structures are used to minimize the Hall current. In the ideal limit of zero Hall current, the effect of the Hall field cancels the perpendicular force on the electrons as they conduct current in the transverse magnetic field. In that case the current flow in the ionized gas is described by *Ohm's law*,

$$\mathbf{j} = \sigma_0(\varepsilon + \mathbf{u} \times \mathbf{B}) \tag{11-34}$$

in which $\sigma_0$ is the conductivity. When there is Hall current, it is added into the right-hand side of this expression to give a form of *generalized Ohm's law*,

$$\mathbf{j} = \sigma_0(\varepsilon + \mathbf{u} \times \mathbf{B}) - \mu_e \mathbf{j} \times \mathbf{B} \tag{11-35}$$

A vector diagram of this relation (after dividing through by $\sigma_0$) is shown in Fig. 11-86. A theoretical discussion of the generalized Ohm's law is given in Refs. 2, 3, 4, and 9.

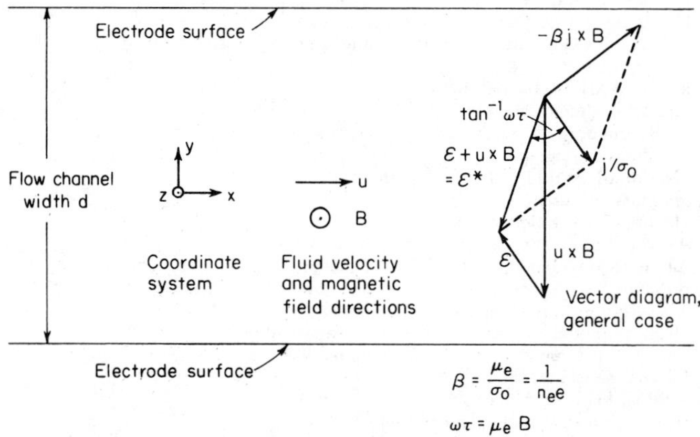


**FIG. 11-86** Vector quantities involved in magnetohydrodynamic conversion.

Various forms of generalized Ohm's law are obtained depending upon the approximations made in deriving the current-field relationship from the equations of motion of the constituent parts of the fluid and their interactions. [These are frequently referred to as the "transport" or "momentum" equations, which are derived from the Boltzmann equation of the kinetic theory of gases (see, e.g., Ref. 2).] The form given here, Eq. (11-35), is appropriate for weakly ionized gases in thermal equilibrium at moderate temperatures. It also has the equivalent form, neglecting ion current, given by

$$\mathbf{j} = \hat{\sigma} \cdot (\varepsilon + \mathbf{u} \times \mathbf{B}) = \hat{\sigma} \cdot \varepsilon^* \tag{11-36}$$

or

$$j_\nu = \sum_k \sigma_{\nu k}\varepsilon^*_k \qquad (\nu,k = x,y,z) \tag{11-37}$$

where, when $\mathbf{B} = z_1 B_2$,

$$\hat{\sigma} = |\sigma_{\nu k}| = \sigma_0 \begin{Vmatrix} \dfrac{1}{1+\omega^2\tau^2} & \dfrac{-\omega\tau}{1+\omega^2\tau^2} & 0 \\ \dfrac{\omega\tau}{1+\omega^2\tau^2} & \dfrac{1}{1+\omega^2\tau^2} & 0 \\ 0 & 0 & 1 \end{Vmatrix}$$

where $\omega = \dfrac{eB}{m_e}$ electron cyclotron angular frequency

$\tau = \dfrac{\lambda}{C}$ electron mean free time

Also, $\omega\tau = \mu_e B$ Hall parameter

Here, owing to the Hall component, the conductivity is anisotropic as indicated by the conductivity sensor $\hat{\sigma}$, unless $\omega\tau \ll 1$.

The electrical power developed internally in the MHD channel is proportional to the conductor velocity (i.e., the fluid velocity) times the force exerted on it via the perpendicular current flow in the magnetic field. Using vector field quantities, the internally developed power density is

$$P_i = \mathbf{u} \cdot (\mathbf{j} \times \mathbf{B}) \tag{11-38}$$

The fraction of this that is delivered to the load connected to the terminals is

$$P = \mathbf{j} \cdot \boldsymbol{\varepsilon} \tag{11-39}$$

Hence the local electrical (or isentropic) efficiency is

$$\eta_e = \frac{P}{P_i} = \frac{\mathbf{j} \cdot \boldsymbol{\varepsilon}}{\mathbf{u} \cdot (\mathbf{j} \times \mathbf{B})} = \frac{\mathbf{j} \cdot \boldsymbol{\varepsilon}}{\mu_x j_y B_z} \tag{11-40}$$

where the last form has utilized the geometry indicated in Figs. 11-85 and 11-86; however, the numerator has been left in general form since it varies depending on the generator configuration, as described in the next section.

The dimensionless product $\omega\tau$, often called the Hall parameter, which appears in the conductivity tensor in Eqs. (11-36) and (11-37), is an important characteristic number in MHD design, ordinarily ranging from about 1 to 5. On the microscopic scale this number indicates the relative average angular travel of electrons between collisions as they tend to move helically about the magnetic field lines because of the perpendicular forces that act on moving charges in a transverse magnetic field. Typically, $\lambda \sim 10^{-7}$ m, $C \sim 10^5$ m/s, so that $\tau \sim 10^{-12}$ s. Also, $\omega = 1.76 \times 10^{11}\,B$ or $\sim 10^{12}$ for $B \sim 6T$. These values give $\omega\tau \sim 1$. The mean free path $\lambda$ is inversely proportional to pressure. Low pressures and high values of $B$ give large values of $\omega\tau$.

On a macroscopic scale, the value of $\omega\tau$ indicates the relative importance of the Hall field and Hall current. When $\omega\tau = 1$, the total current vector is directed 45° to the left of the $\varepsilon^*$ vector, and for large values of $\omega\tau$ the current vector is nearly perpendicular to $\varepsilon^*$ (predominantly Hall current). In weakly ionized gases, if both the equivalent characteristic number for ions, $\omega_i\tau_i$, and $\omega\tau$ are large simultaneously, the angle is reduced. The conductivity then is lowered owing to a phenomenon called "ion slip." The ion mobility $\mu_i = \omega_i\tau_i/B$ is appreciable in this case, and the generalized Ohm's law of Eq. (11-35) has the additional term $\mu_e\mu_i(\mathbf{j} \times \mathbf{B}) \times \mathbf{B}$ added to the right-hand side (see Ref. 9 or Ref. 11). Since $\omega_i$ is much smaller than $\omega$ (because of much larger ion mass) even though $\tau_i$ is larger than $\tau$ owing to the smaller ion thermal velocity, the $\omega_i\tau_i$ product is ordinarily negligible. In any case, appreciable ion slip requires both $\omega_i\tau_i$ and $\omega\tau$ or, equivalently, the product $\mu_e\mu_i B^2$, to be large.

***137. Generator Configurations.*** If the geometry of Fig. 11-85 is used so that $\mathbf{u} = \mathbf{x}_1 u_x$ and $\mathbf{B} = z_1 B_z$, and if segmented, instead of continuous, electrodes are used so that the condition $j_x = 0$ can be established by electrically isolating each pair of electrodes along the flow channel, the generalized Ohm's law relation, Eq. (11-35), has the following components:

$$j_x = \sigma_0\varepsilon_x - \mu_e j_y B_z = 0 \quad \text{or} \quad \varepsilon_x = \frac{\mu_e B_z j_y}{\sigma_0} = \omega\tau\frac{j_y}{\sigma_0} \quad \text{Hall field}$$

$$j_y = -\sigma_0(u_x B_z - \varepsilon_y)$$

$$j_z = 0$$

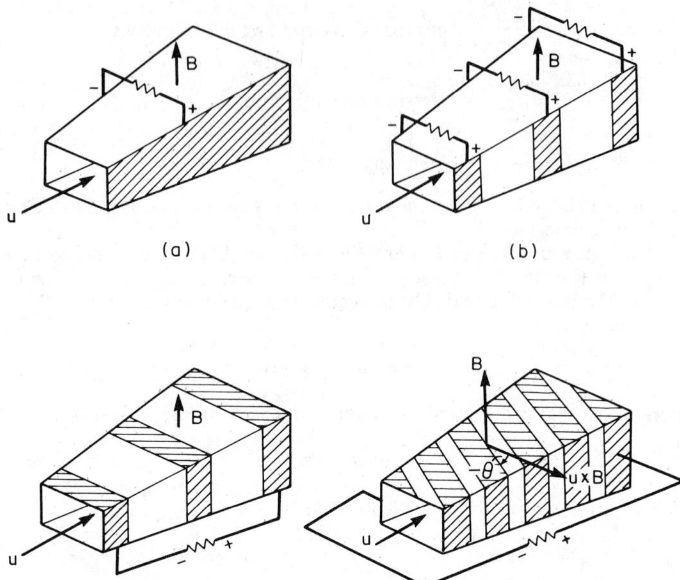


**FIG. 11-87** Alternative configurations. (*a*) Continuous-electrode Faraday generator; (*b*) segmented-electrode Faraday generator; (*c*) Hall generator; (*d*) diagonally connected generator.

If the circuit for $j_y$ is completed through external loads, the arrangement is known as the *Faraday generator configuration* (Fig. 11-87*b*). The open-circuit voltage and power density in this case, with uniform conditions over the cross section of the channel assumed, are

$$V_{oc|F} = -\int_0^d \varepsilon_y \, dy = -\int_0^d u_x B_z \, dy = -u_x B_z d \qquad (11\text{-}41)$$

$$P = \mathbf{j} \cdot \varepsilon = j_y \varepsilon_y = -\sigma_0 (u_x B_z - \varepsilon_y)\varepsilon_y = -\sigma_0 u_x^2 B_z^2 (1 - K)K \qquad (11\text{-}42)$$

where $K = \varepsilon_y / u_x B_z$ is a dimensionless loading parameter (see Par. **137**). The negative sign in Eq. (11-42) corresponds to power generation. Also, the local electrical efficiency $\eta_e$ is just equal to $K$.

A variation of the Faraday generator is the configuration having continuous, instead of segmented, electrodes along the sides of the channel (Fig. 11-87*a*). In this case $\varepsilon_x = 0$ and both $j_x$ and $j_y$ components of current are present. The $j_x$ component is in the direction of the fluid flow and has its circuit completed through the electrode walls. The $j_y$ current component, the power density, and the local electrical efficiency are as follows:

$$j_y = \frac{\sigma_0}{1 + \omega^2\tau^2}(\varepsilon_y - u_x B_z) = \frac{\sigma_0 u_x B_z}{1 + \omega^2\tau^2}(1 - K) \qquad (11\text{-}43)$$

$$P = \mathbf{j} \cdot \varepsilon = j_y \varepsilon_y = -\frac{\sigma_0}{1 + \omega^2\tau^2}(u_x B_z - \varepsilon_y)\varepsilon_y \qquad (11\text{-}44)$$

$$= -\frac{\sigma_0 u_x^2 B_z^2}{1 + \omega^2\tau^2}(1 - K)K \qquad (11\text{-}45)$$

$$\eta_e = K$$

The open-circuit terminal voltage is the same for both Faraday-generator configurations, as is also the local electrical efficiency, but the power density is reduced when the electrodes are continuous. Because of this and problems with electrodes that must conduct the circulating Hall current in addition to the load current, this configuration is never used.

Another scheme with segmented electrodes has opposite electrode pairs short-circuited so that the condition $\varepsilon_y = 0$ is achieved. The load is then connected between the first and the last electrode pairs, as shown in Fig. 11-87*c*. This is known as the *Hall-generator configuration.* The current components [using Eq. (11-37)] are

$$j_x = \frac{\sigma_0}{1 + \omega^2\tau^2}(\omega\tau u_x B_z + \varepsilon_x)$$

$$j_y = \frac{\sigma_0}{1 + \omega^2\tau^2}(u_x B_z - \omega\tau\varepsilon_x)$$

$$j_z = 0$$

The open-circuit voltage, the power density, and the local electrical efficiency for this case, assuming uniform conditions throughout the length $L$ of the channel between load-circuit connections (somewhat of an oversimplification), are

$$V_{oc}/_H = -\int_0^L \varepsilon_x\,dx = \int_0^L \omega\tau u_x B_z\,dx = \omega\tau u_x B_z L \tag{11-46}$$

$$P = \mathbf{j}\cdot\varepsilon = j_x\varepsilon_x = \frac{\sigma_0}{1 + \omega^2\tau^2}(\omega\tau u_x B_z + \varepsilon_x)\varepsilon_x = \frac{\sigma_0\omega^2\tau^2u_x^2B_z^2}{1 + \omega^2\tau^2}(1 - K_H)K_H \tag{11-47}$$

$$\eta_e = \frac{P}{u_x\varepsilon_y B_z} = \frac{(1 - K_H)K_H}{K_H + 1/\omega^2\tau^2} \tag{11-48}$$

where $K_H = -\varepsilon_{x/\omega\tau}u_xB_z$ is the Hall-generator loading parameter. A comparison between the Faraday- and Hall-generator local electrical efficiencies is given in Fig. 11-88. It is clear that good efficiency in a Hall generator requires high $\omega\tau$ and a small loading parameter value while the Faraday-generator efficiency is not dependent upon $\omega\tau$ and improves with large values of the loading parameter.

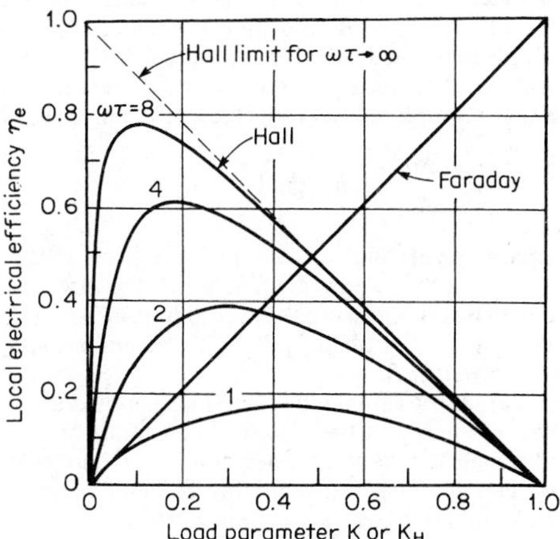


**FIG. 11-88** Comparison of local electrical efficiencies of Faraday and Hall generators.

Another scheme that has been favored recently for experimental rigs is the diagonal connection of segmented electrodes as shown in Fig. 11-87*d*. In this case the diagonal connections are made along equipotential surfaces at the angle $\theta = -\tan^{-1}\varepsilon_y/\varepsilon_x$ with respect to the vector $\mathbf{u} \times \mathbf{B}$. Since they are coincident with equipotentials, the diagonal connections can be made part of the MHD channel wall that is exposed to the hot plasma. The whole channel may then be made of metal sections consisting of electrodes on the sidewalls and their continuation as diagonal connections across the top and bottom (assuming that the magnetic-field direction is vertical).

Individual sections require insulation from each other, but such electrical insulation can be relatively thin and partially shielded by the metal members from the plasma. Ideally, with this connection, the Hall current is zero and the electrode voltage between opposite pairs is the same as for the Faraday generator. With the diagonal connections, the overall circuit is a series connection of Faraday-generator electrodes and the output, at comparatively high voltage, is taken between the first and last electrodes in the series. (More than one series string may be necessary to satisfy the end conditions.)

The characteristics of the diagonally connected generator are intermediate between those of the Faraday and Hall generators if the zero Hall current is not a constraint; however, the most desirable operation is with zero Hall current, that is, with $j_x = 0$. The current components are, in general,

$$j_x = \frac{\sigma_0}{1 + \omega^2\tau^2}[\varepsilon_x - \omega\tau(\varepsilon_y - u_x B_z)]$$

$$= \frac{\sigma_0 u_x B_z}{1 + \omega^2\tau^2}\left[\omega\tau(1 - K) - \frac{K}{\tan\theta}\right]$$

$$j_y = \frac{\sigma_0}{1 + \omega^2\tau^2}(\omega\tau\varepsilon_x + \varepsilon_y - u_x B_z)$$

$$= \frac{\sigma_0 u_x B_z}{1 + \omega^2\tau^2}\left(1 - K + \frac{\omega\tau K}{\tan\theta}\right)$$

The condition $j_x = 0$ requires a particular load-parameter value, namely,

$$K = \frac{\omega\tau \tan\theta}{1 + \omega\tau \tan\theta} \qquad (11\text{-}49)$$

Ordinarily, operating conditions are adjusted to maintain this value as closely as possible.

***138. Electric Power Relations.*** The electrical power density generated internally in the fluid is, as noted in Par. **136**, $\mathbf{u} \cdot (\mathbf{j} \times \mathbf{B})$ and involves the conversion of fluid power into electrical power. The fluid power is derived from the velocity times the rate of change of momentum (i.e., velocity × force) and velocity times the pressure gradient. These quantities must balance, and that requirement leads to the "momentum equation"

$$\rho\mathbf{u} \cdot \frac{d\mathbf{u}}{dt} = -\mathbf{u} \cdot \text{grad}\, p + \mathbf{u} \cdot (\mathbf{j} \times \mathbf{B}) \qquad (11\text{-}50)$$

The quantity $\rho$ is the mass density and $p$ is the pressure, here taken as a scalar quantity. (More generally, the pressure term is $-\mathbf{u} \cdot \textbf{div}\,\hat{\mathbf{p}}$ in which the viscous forces in the fluid are included as well as the hydrostatic pressure. Also in the more general form a term $\rho_e\mathbf{u} \cdot \varepsilon$, usually negligible, is added to the right-hand side.) The second term on the right is the internally developed electrical power $juB$.

The density of electrical power actually delivered is, as noted above, $j\epsilon$, and is less than $juB$ owing to dissipation of energy in the electrical resistance of the fluid. The difference remains in the fluid, although the result is a degradation of energy since kinetic energy is converted into thermal energy. The fluid enthalpy $H$ (sum of internal energy $\epsilon$, kinetic energy per unit mass $u^2/2$, and the pressure-specific volume product $p/\rho$) is therefore affected only by the power actually delivered, which yields the "energy equation"

$$\rho\frac{dH}{dt} = \mathbf{j} \cdot \varepsilon \qquad (11\text{-}51)$$

The simplest way to represent the external loading of a generator is through the use of a "loading parameter" which represents the ratio of the actual load-voltage component in the generator to the internally generated voltage component. It can be variously represented by the following forms:

$$K = \frac{\varepsilon_y}{u_x B_z} = \frac{V}{V_{oc}} = \frac{R_L}{R_i + R_L} \qquad \text{Faraday generator} \tag{11-52}$$

$$K_H = \frac{\varepsilon_x}{\omega\tau u_x B_z} = \frac{V}{V_{oc}} = \frac{R_L}{R_i + R_L} \qquad \text{Hall generator} \tag{11-53}$$

The last two forms on the right-hand side of each of these expressions are based on the existence of uniform conditions throughout the channel. Here $R_i$ is the internal resistance and $R_L$ is the load resistance. The condition for maximum power output is $K$ and $K_H$ equal to 0.5 (that is, $R_L = R_i$). In that case $\varepsilon = uB/2$ for the Faraday generator and $\varepsilon = \omega\tau uB/2$ for the Hall generator, and the power densities in the fluid are (with subscripts dropped)

$$P_{max}|_F = \frac{\sigma_0 u^2 B^2}{4} \qquad \text{W/m}^3 \tag{11-54}$$

$$P_{max}|_H = \frac{\sigma_0 \omega^2 \tau^2 u^2 B^2}{4(1 + \omega^2\tau^2)} \qquad \text{W/m}^3 \tag{11-55}$$

If $K$ is different from 0.5, the power density is

$$P = P_{max}(1 - K)K \tag{11-56}$$

for the Faraday generator and similarly for the Hall generator.

Target values for the various quantities in these expressions might be $\sigma_0 = 10$ S/m, $u = 950$ m/s, $B = 6$ T, $\omega\tau = 3$, giving a maximum available power density of $8 \times 10^7$ W/m$^3$ or 80 W/cm$^3$ for the Faraday generator and slightly less for the Hall generator. In actual generators these values may not be achievable. Furthermore, considerations of efficiency, secondary effects, etc., will probably make $K = 0.75$ and $K_H = 0.25$ nearer to overall optimum values. (See Table 11-26 for values as determined in typical design studies.)

***139. Flow Relations.*** In general the equations governing the power generation and flow cannot be solved in closed form; so numerical computation of individual cases is necessary. However, a few simple cases are soluble and, although highly approximate, are useful for the insight they give. The simplest is obtained for quasi-one-dimensional constant-velocity flow with ideal gas flow assumed. Here, the Faraday-generator configuration with segmented electrodes is also assumed. In this case the current component due to the Hall field is zero, and the *generalized Ohm's law* reduces to

$$j = \sigma(\varepsilon - uB) = -\sigma uB(1 - K) \tag{11-57}$$

where $\sigma$ is the scalar conductivity, previously denoted as $\sigma_0$. (Since, in the Faraday configuration, the current is just $j_y$, the velocity $u_x$, and the magnetic field $B_z$, it is the usual practice to drop the subscripts.)

The *momentum equation* for steady-state one-dimensional flow is obtained by setting equal to zero the $\partial u/\partial t$ part of the expansion of $du/dt$ in Eq. (11-50), giving

$$\rho u \frac{du}{dx} + \frac{dp}{dx} = jB \tag{11-58}$$

With constant-velocity flow as assumed here, the $du/dx$ term vanishes.

The *energy equation,* Eq. (11-51), becomes

$$\rho u \frac{dh}{dx} = j\varepsilon \tag{11-59}$$

since $H = u^2/2 + h$ and $u$ is constant. For an ideal gas $h = c_p T$, giving

$$\rho u c_p \frac{dT}{dx} = j\varepsilon \tag{11-60}$$

**TABLE 11-26** Conceptual Design Studies of Central-Station MHD-Steam Plants

| Item | Advanced energy conversion systems (ECAS) (Ref. 81) | Conceptual study of potential early commercial MHD power plants (CSPEC) (Ref. 82) | Parametric study of potential early commercial MHD power plants (PSPEC) (Ref. 83) | | Conceptual design of 200 MWe MHD engineering test facility (ETF) (Ref. 84) | Feasibility study of MHD retrofit (Ref. 85) | Parametric analysis of closed cycle MHD power plants (Ref. 86) |
|---|---|---|---|---|---|---|---|
| Date of report | 1975 | 1980 | 1979 | 1979 | 1981 | 1979 | 1981 |
| Combustor | 1 stage | 2 stage | 1 stage | 1 stage | 2 stage | | 2 stage |
| Fuel | Coal | Coal | Coal | Coal | Coal | Coal | Coal |
| Air preheat temp., K | 1644.0 | 922.0 | 866.3 | 866.3 | 866.3 | 1644.0 | 601.2 |
| Oxygen enrichment % (mole) | 21 | 37.6 | 42.0 | 34.0 | 30 | 21 | none |
| Pressure, atm | 9.0 | — | 9.30 | 8.30 | 4.51 | 3.32 | 10.0 |
| Temperature, K | 2830 | 2860 | 2924 | 2854 | 2690 | 2678 | 1977 |
| MHD Generator | | | | | | | |
| Cycle | Open | Open | Open | Open | Open | Open | Closed |
| Fluid | Comb. gas | Comb. gas | Comb. gas | Comb. gas | Comb. gas | Comb. gas | Argon |
| Magnetic field, Tesla | 5 | 6 | 6 | 6 | 6 | 6 | 6.0 |
| Velocity, Mach no. | 0.90 | Subsonic | Subsonic | Subsonic | 0.88 | 0.91 | 1.2 |
| Channel length, m | 28.0 | 14.0 | 18.0 | 25.0 | 15.2 | 9.0 | 11.8 |
| Generator coefficient $K$ | 0.80 | 0.78 | 0.78 | 0.79 | 0.78 | 0.70 | 0.8 |
| Power, MWe (net) | 1420 | 620.0 | 551.0 | 523.7 | 85.0 | 13.7 | 507.8 |
| Steam turbine | | | | | | | |
| Gross output, MWe | 587 | — | 814.0 | 636.0 | 168.6 | 860.1 | 554.4 |
| Net output, MWe | 512 | 470 | 549.0 | 407.0 | 117.2 | 780.9 | 493.0 |
| Plant output (net) | | | | | | | |
| Rating, MWe | 1932.0 | 1090.0 | 1100.0 | 930.7 | 202.2 | 794.6 | 1000.8 |
| Efficiency, % | 48.3 | 42.7 | 42.5 | 43.4 | 38.0 | 33.2 | 43.2 |
| Heat rate, Btu/kWh | 7066 | 7992 | 8030 | 7864 | 8979 | 10279 | 7900 |
| Cost 1983 $ | | | | | | | |
| Capital $/kWe | 695 | 1216.0 | 1145.7 | 865.6 | 918.2 | 307.3 | 1390.4 |
| COE mills/kWhr | 38.5 | 58.0 | 55.7 | 45.9 | — | 45.9 | 63.9 |

A *continuity*, or *mass-conservation*, law also applies to the fluid flow. For quasi-one-dimensional steady flow this is $\rho u A_c$ = mass flow rate = constant, where $A_c$ is the cross-sectional area of the channel. In this case the velocity is constant so that the relation can be written as

$$\frac{d}{dx}(\rho A_c) = 0 \tag{11-61}$$

where $A_c$ is the cross-sectional area of the channel.

From these equations the electrical efficiency $K = \epsilon/uB$ becomes

$$K = \rho c_p \frac{dT}{dp} \tag{11-62}$$

For an ideal gas

$$c_p T = \frac{\gamma}{\gamma - 1}\frac{p}{\rho} \tag{11-63}$$

which can be combined with Eq. (11-34) to give

$$K = \frac{\gamma}{\gamma - 1}\frac{d \ln T}{d \ln p} \tag{11-64}$$

This is integrated to give the relation between $p$ and $T$ in the constant-velocity channel,

$$\frac{T}{T_0} = \left(\frac{p}{p_0}\right)^{K(\gamma-1)/\gamma} \tag{11-65}$$

where $T_0$ and $p_0$ are the temperature and pressure at the entrance to the generator channel.

The momentum equation and Ohm's law when combined and integrated under the same conditions yield

$$p = p_0 - \sigma u B^2(1 - K)x \tag{11-66}$$

in which $x = 0$ at the entrance, where $p = p_0$. This may be written in the dimensionless form

$$\frac{p}{p_0} = 1 - \frac{x}{L_i} \qquad L_i = \frac{1}{1 - K}\frac{p_0}{\sigma u B^2} \tag{11-67}$$

where $L_i$ may be termed the "interaction length" (Ref. 22) and is an approximate measure of the channel length required in an actual design to extract an appreciable part of the gas energy. (Cf. G. W. Sutton in Ref. 13*a*, pp. 123–126; also Ref. 36. However, in Ref. 11 only the $p_0/\sigma u B^2$ part is called "the interaction length.") Only 0.6 to 0.8 of $p_0$ would be utilized in an actual design, but the average value of $\sigma$ is likely to be about the same fraction of the inlet valve so that taking $L \approx L_i$ is approximately correct.

The pressure relation, and other quantities found in a similar manner, are shown in Fig. 11-89. Also shown is the Mach number variation. The Mach number is

$$M \equiv \frac{u}{a} = \frac{u}{\sqrt{\gamma p/\rho}} = \frac{u}{\sqrt{\gamma R T/W_m}} \tag{11-68}$$

where $a$ is the local sound speed, $R$ is the universal gas constant, and $W_m$ is the molecular weight of the gas. For constant velocity $u$,

$$\frac{M}{M_0} = \left(\frac{T}{T_0}\right)^{-1/2} = \left(\frac{p}{p_0}\right)^{-(1/2)K(\gamma-1)/\gamma} \tag{11-69}$$

The quasi-one-dimensional flow and power-generation relations used in this example are often the starting point for a numerical calculation by computer. Typically in such a case,

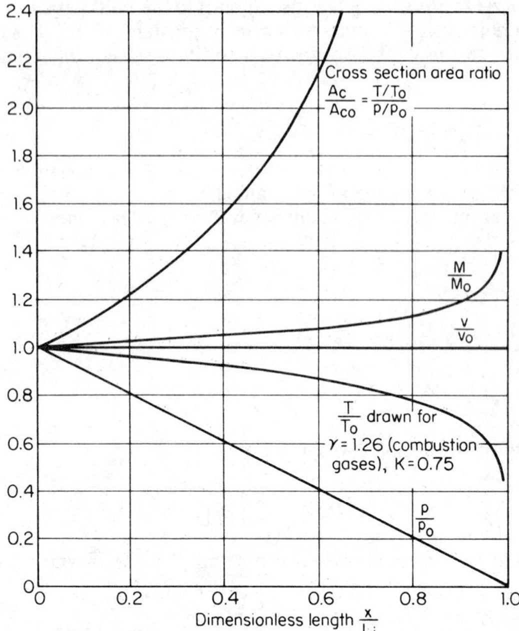


**FIG. 11-89** Values of dimensionless ratios for normalized length $x/L_i$ in an ideal constant-velocity channel.

quantities to account for electrode voltage drop, wall friction, and heat loss through the walls are attached, somewhat empirically, to the basic one-dimensional equations. The electrode voltage drop is taken into account by multiplying the electric field by a factor $(1 - \Delta)$ where $\Delta$ is a small dimensionless quantity representing the average effect of the electrode drop. The wall friction is accounted for by adding a term $-F$ on the right-hand side of the momentum equation, Eq. (11-58). Again, this is an averaged quantity. The heat loss is accounted for by adding a term $-G$ on the right-hand side of the energy equation, Eq. (11-59), and this, too, is an averaged quantity (e.g., see Refs. 12 and 55).

Detailed two- and three-dimensional channel-flow models have been generated by STD Research Corp. (e.g., see Refs. 87 and 88). These models numerically solve the governing coupled-fluid, electrical, and transport equations. The complex three-dimensional solutions have been effectively utilized in the analysis of flow instabilities capable of causing electrothermal damage in large channels with magnetic interaction.

***140. Duct Design.*** Design of an MHD duct for a full-scale power-producing generator requires the detailed numerical solution of the fundamental magnetohydrodynamic equations and the thermodynamical relations in the fluid. The choice of maximum cycle pressure and temperature is basic to everything else in the design. This will be strongly influenced by components outside the channel itself, namely, the combustor, air preheat equipment, the magnet, the various heat exchangers, and the temperature and pressure limitations of gas-cooled reactors if used in closed-cycle loops. Figure 11-90 shows in a general way the relation between pressure and temperature in the channel itself for various constant generator lengths and constant total power output (Refs. 11 and 22). High pressure reduces the size of heat exchangers, but it also reduces the gas conductivity and must be compensated by increased temperature or increased generator length.

Then a choice of input-flow velocity is necessary. Here a clear optimum in terms of performance exists (Ref. 19; also Ref. 11). Both $\sigma u^2$ and $\sigma u$ have maximums for a given total enthalpy at the channel entrance since increased $u$ means decreased temperature and, on assuming the electrons are in thermal equilibrium with the gas, almost exponentially

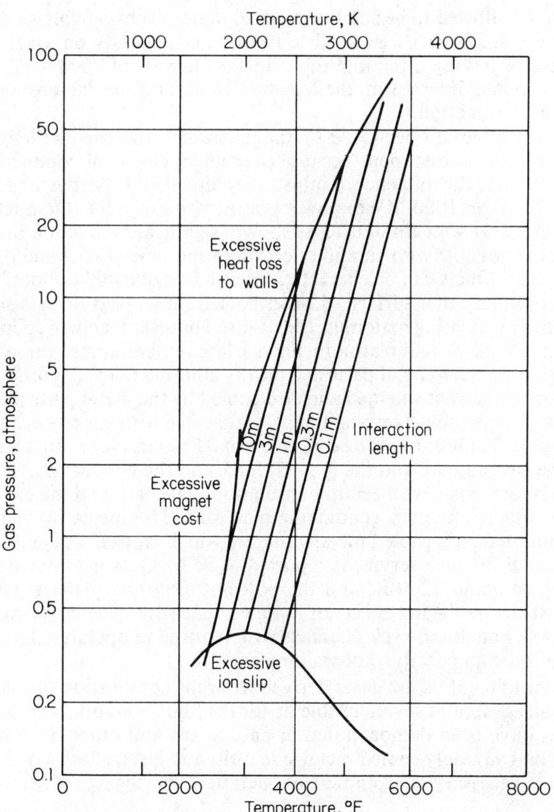


**FIG. 11-90** Gas pressure versus temperature for various interaction lengths. Curves are for a 500-MW closed-cycle MHD channel using cesium-seeded helium and a magnetic field of 6 T (Ref. 22). Limits imposed by heat loss, magnet cost, and ion slip are shown. *(Adapted from Ref. 22.)*

decreased $\sigma$. Also, the length of the channel is inversely proportional to $u$, as indicated in the expression for $L_i$ given above [Eq. (11-67)]. In the open-cycle optimization study reported in Ref. 44 the optimum velocity including the effect of wall friction was found to be 750 m/s ($M \approx 0.8$).

A majority of MHD channels tested to date have been supersonic. Although the optimum channel velocity may in some cases be in the high subsonic range, the decoupling between downstream pressure perturbation and the combustor has simplified testing. Although unplanned (Ref. 89), several channels operated successfully with super- or subsonic flow in the channel. The standing shock in the channel was observed to have a minor influence on overall performance. Recent analyses indicate that the optimum Mach number distribution for MHD channels in large fossil-fuel-fired units is 0.85 inlet and 1.0 exit (Ref. 90).

*Proportions and Size.* Beyond the nozzle the required cross-sectional area of a channel increases throughout the length of the generating section, as is illustrated for the constant-velocity subsonic case in Fig. 11-89. Usually the increase required can be closely approximated with a second-degree equation indicating that a linear taper of both the sides and the top and bottom can be used. The average cross section should be approximately square in order to minimize friction and heat loss at the walls. However, the width between electrode walls should not be allowed to be small enough to make electrode voltage drop a significant fraction of the voltage between electrodes. In one design study on a subsonic diagonally connected generator having a thermal input to the channel of 2000 MW, the initial cross section is $Y = 2$ m and $Z = 0.8$ m, the length is 16 m, and the exit cross section is square about 3.5 m in each direction.

In the case of coal-fired open-cycle MHD generators, the channel length is dependent upon coal source and pretreatment because they affect electrical conductivity. Figure 11-91*a,b* (Ref. 91) indicate the influence of moisture removal and combustor pressure on MHD channel length for large (1000-MWe) power plants with air preheat temperature of 2500°F and peak magnetic field strength of 6 T. As shown, significant reduction in generator length and magnet cost is possible with careful selection of the type of coal and its preparation.

*Wall Materials.* Ducts can be characterized as having "cold" or "hot" walls. The usual cold-wall design has an inner surface of water-cooled metal pegs or bars separated by thin layers of electrically insulating materials. Metals used or tested include copper, nonmagnetic stainless steel, super alloys, and platinum, and surface temperatures range from 500 to 1000 K or so. Insulation between metal parts is typically alumina ($Al_2O_3$), boron nitride, or mica. The insulators are somewhat shielded and also cooled by the metal parts they separate. Gas-tight integrity is provided by clamping the parts together with gaskets or O rings, or by use of an external glass-filament-reinforced plastic box. The structure must also be capable of withstanding the gas pressure and the ($\mathbf{j} \times \mathbf{B}$) forces on the conductors.

Hot-wall ducts are made with ceramic insulation materials, and the electrodes are made with ceramic or other refractory conducting materials. The insulating portions may have ceramic pegs mounted on a cooled metal structure which is itself segmented. The best non-conducting material for such service is reported to be MgO, which may be used at surface temperatures up to about 2250 K, and the best hot electrode material is zirconia ($ZrO_2$) stabilized against the damaging effects of ionic current with calcium or rare-earth-element additives (Ref. 48). For closed-cycle channels with reduced temperature levels, boron nitride has been shown to be an effective hot-wall material.

In the presence of combustion gases derived from the combustion of coal, the formation of slag on the walls generates severe problems for the "hot" ceramic wall designs. Although ceramic surfaces have been demonstrated in natural gas and other clean-fuel-fired experiments (Refs. 92 and 93), only cooled metal side walls and electrodes have demonstrated the potential lifetimes necessary for commercial operation in slagging environments.

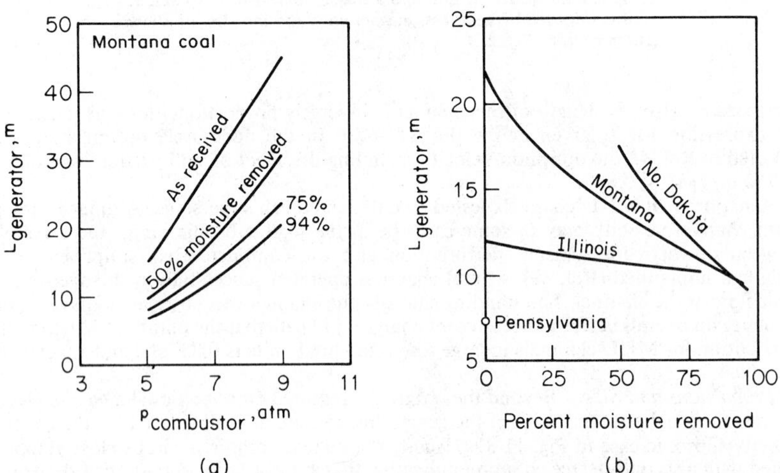


**FIG. 11-91** Generator length as a function of combustor pressure and coal pretreatment. *(From Ref. 91.)*

***141. Properties of High-Temperature Air.*** Since a proposed application of MHD power generation is as a topping unit in a fossil-fueled plant, designs which use the combustor gas directly as the working fluid have obvious advantages. Temperatures for fossil-fuel combustion (with preheated or oxygen-enriched air) are in the range of 4200 to 5000°F. The conductivity of high-temperature air is therefore of interest and is given by the lowest of the curves in Fig. 11-92.

Over a limited range, the conductivity is proportional to $1/p^m$ where $p$ is the pressure and $m$ is ~0.5 at 2760°C (5000°F) and ~0.3 at 5500°C (10,000°F). At atmospheric pressure ($p = 1$) the conductivity of normal air is too low to serve as the working fluid, and the conductivities of various combustion-product gases from the burning of fossil fuels are all similarly low. Increasing the conductivity by decreasing the density by, say, $10^{-1}$ would yield almost adequate conductivity at the cost, however, of a prohibitive demand on the size of the duct and the heat-exchanging surfaces.

***142. Properties of Seeded Combustion Products and Inert Gases.*** Conductivity is given by summing $en\mu$ (charge, density, and mobility) for all charged species present. But electrons, because of their high mobility, contribute most of the conductivity in an ionized gas even though, to a high degree of approximation, there is an equal number of ions in the gas. At low degrees of ionization, the mobility of electrons is inversely related to the collision frequency for momentum transfer with neutral particles. As the degree of ionization increases, electron-ion and electron-electron collisions cause the electron mobility to be reduced. This effect is so strong that at less than 1% ionization the conductivity is no longer increasing linearly with electron density, as at first, but instead increases more slowly toward the fully ionized value owing to the progressive reduction of electron mobility. A high degree of ionization is therefore not necessary for the achieving of an appreciable fraction of the fully ionized conductivity limit. However, at temperatures that are practicable in MHD

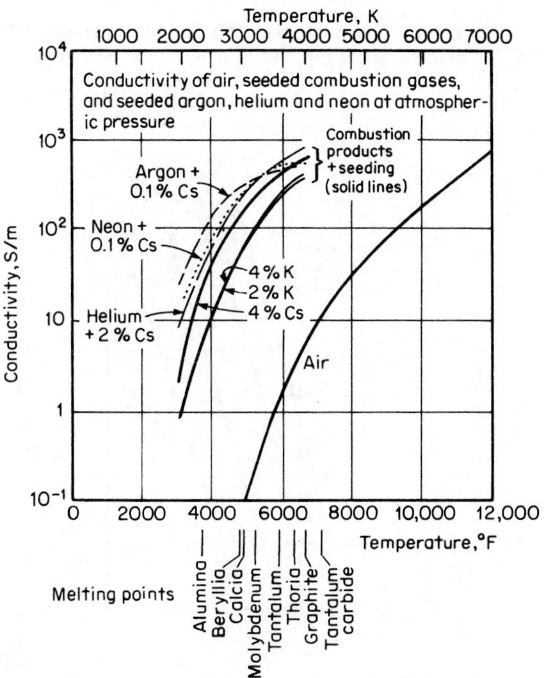


**FIG. 11-92** Conductivity of air, seeded combustion gases, and seeded argon, helium, and neon at atmospheric pressure. *(All curves, except air, from Ref. 31.)*

applications, the thermal ionization of air, combustion-product gases, and the inert gases is so low that the electron density is orders of magnitude below that necessary to obtain suitable conductivities.

A large increase in conductivity is obtained by seeding the gas with a small percentage of materials which have much lower ionization potentials than air, combustion products, and the inert gases. First ionization potentials of the air components are ~14 V, and those of the inert gases are even higher, while those of the alkali metals range from 3.89 V for cesium and 4.34 V for potassium to 5.4 V for lithium. Some calculated conductivities of seeded gases are shown as a function of temperature and pressure in Fig. 11-93 (Ref. 36). In general, these curves are in close agreement with measured values (see, e.g., Ref. 21).

Over the temperature range of interest, the conductivity of a seeded gas can be approximated by the relation

$$\sigma = n_e e / m_e \nu$$

Where $n_e$ is the electron number density, $\nu$ is the total electron-atom collision frequency, $e$ and $m_e$ are the electronic charge and mass, respectively. The electron number density can be estimated using the Saha equation

$$n_e = AT^a e \frac{-eV_i}{2kT}$$

where $A$ is an empirical factor, $k$ and $a$ are constants, and $V_i$ is the ionization voltage of the alkali metal seed. For potassium- or cesuim-seeded combustion products and noble gases $a$ is approximately 0.75.

The empirical factor $A$ accounts for the pressure of electron-negative components in the gas such as, for example, OH and Cl, which can significantly reduce the electron density. The seeding compound and the fuel should be chosen to minimize the formation of these negative ions.

Somewhat unfortunately, the collision cross sections of the alkali metals are large compared with those of the other gases of interest so that, as the percentage of seed atoms is

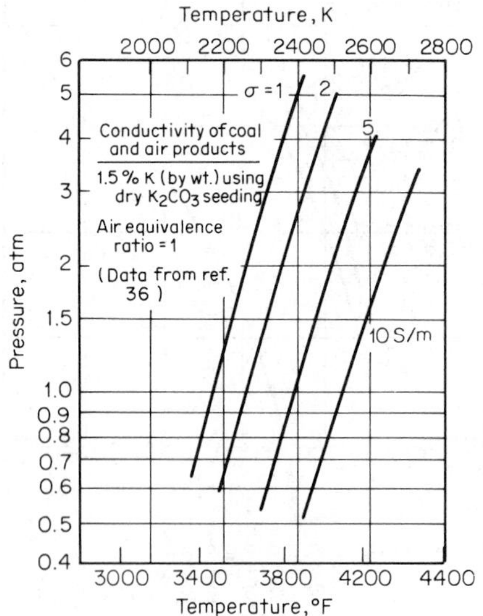


**FIG. 11-93** Pressure versus temperature for various conductivities of coal and air products plus 1.5% K by weight, using dry $K_2CO_3$. Air equivalence ratio = 1. *(Adapted from Ref. 36.)*

increased, the electron density increases as desired but the mobility begins to be markedly decreased, and an optimum seeding percentage exists. This is about 0.1% for Cs and K in argon and about 0.3% for these same alkali metals in neon (Ref. 31). At high temperatures, the electron-ion and electron-electron collisions, discussed above, dominate the collisions and limit the mobility and conductivity still further.

***143. Electrodes.*** Electrical contact with the working fluid occurs at the electrodes. When ionized gas is the working fluid, the requirements are rigorous: (1) Ability to withstand the corrosive and erosive actions, at high temperature, of the fluid, which usually has an alkali metal content and, in some cases, a slag and ash content as the result of burning coal. (2) Ability to emit copious quantities of electrons at the cathode side (cathode terminals are positive when generating power) either by diffuse arcing or by thermionic emission, or both, without excessive erosion. (3) Ability to withstand the arcing that occurs on the anode side without excessive erosion. (4) Good thermal conductivity so that either direct or indirect liquid cooling is practicable. Both metal and refractory ceramic electrodes have had extensive use in experimental rigs.

Electron emission occurs at the cathode side by surface arcing, which is relatively diffuse if the current density does not become too high. Contact between the plasma and the anode is made somewhat similar by means of arcs at the surface through the boundary layer into the main body of the plasma. The erosion caused by these arcs is the chief electrode problem, but with suitable choice of materials, adequate cooling, and minimization of the sulfur content of the fuel, seed additive operating lifetimes of several thousand hours can be expected in a full-sized power-station generator.

The Avco Corporation has over the years developed a metal-capped copper electrode design. Sketches of two different anode designs (Ref. 94) are shown in Fig. 11-94. These electrodes operate with a slag coating and have operated as an anode surface for over 1300 h to date (Ref. 95) and appear capable of operating for 5000 to 8000 h (Ref. 96). Configuration 1, which is entirely platinum clad, is used in the channel areas of higher electrical stress, while configuration 2, which employs both platinum and stainless steel caps, is used in lower stress areas. Both configurations employ an upstream platinum bar to protect the electrode from Hall-effect-induced current concentration.

Cathodes are not as prone to electrochemical corrosion as are anodes. Reacting ionic species are released in anode slag layers but are not released in the cathode slag layer. The cooled copper cathode surface, however, still requires caps to resist electrical and mechanical erosion by slag. Avco (Ref. 95) has found the lowest erosion rates in a cathode capped with ⅛-in of a tungsten-copper alloy. This design has operated for 250 h, with a much longer lifetime expected.

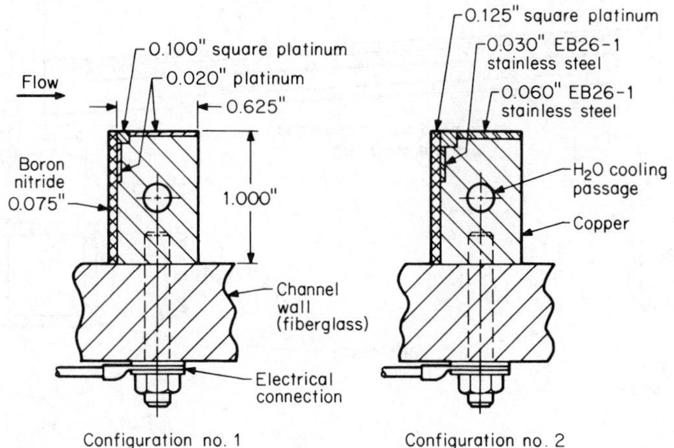


**FIG. 11-94** Typical anode configuration. *(From Ref. 94.)*

***144. Magnets.*** For MHD generators the comparative economics easily favors the use of superconducting magnets in preference to water-cooled or cryogenically cooled magnets with coils having normal conductivity (Ref. 12). The latter not only consume an inordinate amount of power (estimated at 40 MW in the water-cooled magnet of a 500-MW generator and about a tenth of that in a cryogenically cooled magnet of the same size) but all of that power is converted into heat in the windings, making necessary a large investment in cooling arrangements (Ref. 8).

Superconducting magnets create no joule heat requiring removal except in portions that are temporarily or continually not superconducting, particularly the leads. Consequently, if properly insulated, only a modest amount of heat removal is required. The coils, in typical designs, are mounted in a dewar made in the shape of a hollow circular cylinder or cone (see Fig. 11-95). The dewar itself consists of three nested vessels: the innermost is a liquid-helium vessel in which the coils with their supporting structure are mounted; next is a radiation shield cooled by helium vapor; finally, all this is enclosed in an evacuated room-temperature vessel to provide vacuum insulation (Refs. 63 and 65a).

Suitable magnetic fields for rectangular channels can be obtained with a pair of coils having either racetrack geometry (called "pseudo Helmholz arrangement" in Ref. 12) or saddle geometry, as illustrated in Fig. 11-96. Saddle coils are better from the standpoint of

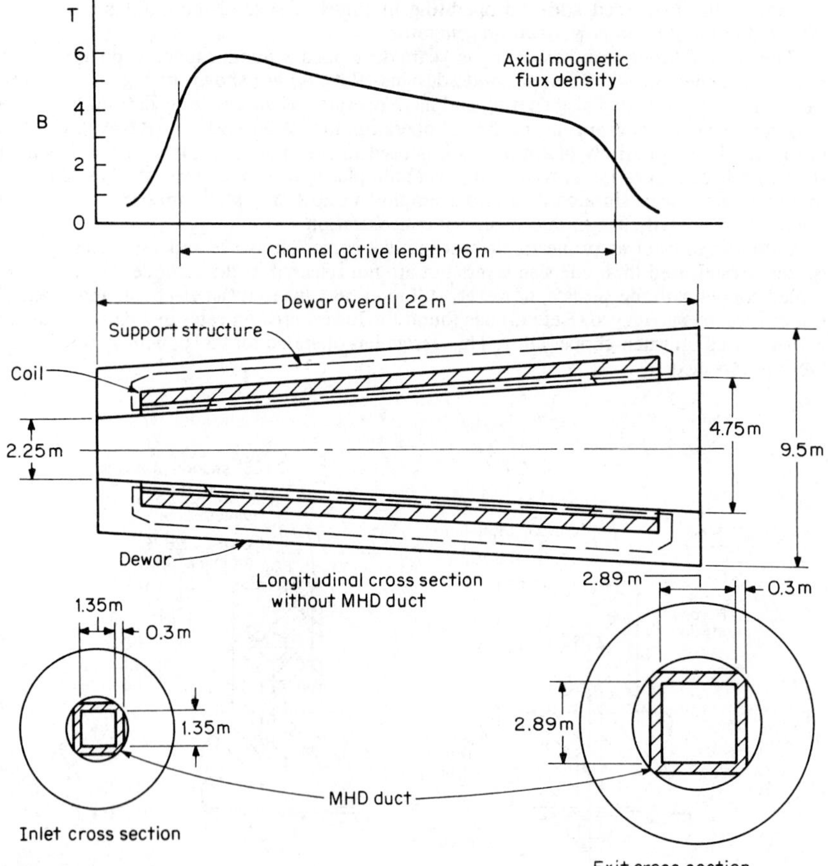


**FIG. 11-95** Saddle-type superconducting magnet and dewar designed for a 600-MW MHD generator. *(Adapted from Refs. 71 and 74.)*

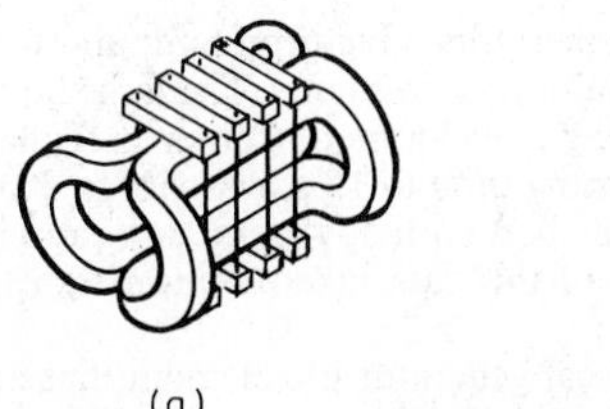


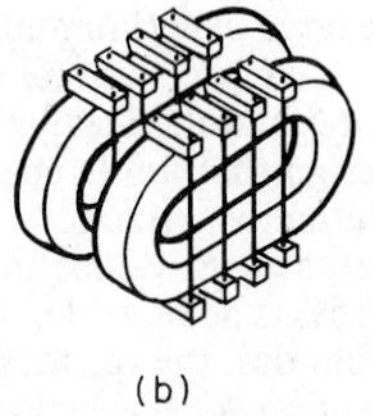


**FIG. 11-96** Superconducting magnet coil geometries. (*a*) Saddle coils; (*b*) racetrack coils. *(Adapted from Ref. 63.)*

winding effectiveness, requiring only about two-thirds as much conductor material as racetrack coils, but they are more difficult to fabricate and to support rigidly (Refs. 63 and 74). The coil and dewar shown in Fig. 11-95 are conceptual designs for a 600-MW generator (Refs. 71 and 74).

Superconductors for large magnets are made as composites of, usually, fine NbTi wires embedded in heavy copper wires. The ratio of copper to superconductor volume is typically about 12 to 1. Such a composite, if run at low enough current denisty and with sufficient cooling capacity available, is able to operate and dissipate the heat generated even if superconductivity is lost. Such a conductor is called "fully stable" (Ref. 12), and this condition is ordinarily specified in the design of large magnets. (For a general review see Ref. 27.)

***145. Liquid-Metal Generators.*** The very high temperatures required to maintain a plasma in the conducting state can be avoided if a liquid metal is used as the working fluid. Since first proposed in 1962 (Ref. 34), a large effort has been made on extending the basic theoretical concept and experimentally verifying the various schemes conceived. (See Table I of Ref. 26, listing the facilities which have been engaged in liquid-metal MHD research.) Working fluids investigated include mercury, mercury-potassium mixture, potassium, cessium, sodium, and sodium-potassium mixture in single-component systems, and Ar–Na, He–Na, He–Li, and others in two-component systems. The details of this effort are largely reported in the proceedings of the six international conferences listed in Ref. 14 and to some extent in the symposia listed in Ref. 13.

Besides operating in a lower-temperature regime than the gaseous plasma generators, the higher conductivity of the liquid metals makes possible high power density with moderate magnetic fields so that relatively small size generators are practicable. Direct ac induction generation is also proposed and has had a limited amount of testing. These characteristics have led to proposals of their use as space-mission power supplies, and thus have generated considerable government agency support for the research (Ref. 26).

The problem areas relate to efficiency of two-phase nozzles and other schemes for generating a high-speed liquid flow, separator efficiency, stability of the two-phase flow, and overall thermodynamic cycle efficiency. However, after a decade of effort, it is believed that in central-station applications a maximum cycle efficiency of 50% can be achieved with a maximum temperature of 870°C (1600°F) (Ref. 51).

***146. Experimental Status.*** Experiments with high-temperature seeded gases in thermal equilibrium are in general agreement with theoretical predictions, provided that the theory used adequately models the major physical phenomena and the experimental conditions are carefully controlled. The kinds of difficulties encountered include nonuniform condition in the plasma resulting in low conductivity, short electrode life, various boundary-layer problems including separation of the boundary layer in supersonic ducts, high electrical leakage currents, and frequent arcing. In addition, materials failures due to high temperature or thermal shock, erosion, chemical incompatibility with the alkali seed material, thermal deflection, water-cooling-system leaks, and gas leaks have plagued most experiments but have been largely overcome by sound development efforts. The vast general sources of information on developments as they have occurred are the preprint volumes of the Symposia on Engineering Aspects of Magnetohydrodynamics (Ref. 13) and the proceedings of the International Conferences on Magnetohydrodynamic Electric Power Generation (Ref. 14).

Several notable large experimental generators have been built and tested to date. In the United States there have been several major achievements. The large liquid-fueled LORHO Hall Generator at the Arnold Engineering Development Center in Tullahoma, Tennessee, with a heat-sink design that limits operating time to 15 s, was run to demonstrate achievability of effective enthalpy extraction. The best enthalpy extraction ratio reached was 11.6% at 35.5 MWe. Analytical extrapolation of the data taken demonstrates that the enthalpy extraction goal of 15% is achievable.

A single set of anodes, the most critical generator life element, has achieved over 1300 h (noncontinuous) of power generation testing on the Avco Mark VII generator at electrical potential conditions the same as may be expected commercially. Between 5000 and 8000 h of useful life are projected for these electrodes. A 50-MWt MHD generator was recently tested with ash injection to simulate coal firing at the Coal Fired Flow Facility in Butte, Montana. The generator is currently being used to demonstrate operability of an inverter designed and built by Westinghouse Corporation.

Two-stage slagging combustors are being developed for coal-fired MHD systems. Over 400 test firings on 10–20 $MW_t$ combustors have been performed by TRW (Ref. 97) at typical MHD conditions and with slag rejection rates of over 90%. In addition, a 50-$MW_t$ coal combustor is presently being tested at the Mountain States Energy facility in Butte, Montana (Ref 98). Scaleup of these designs to the larger units required for economic MHD units are planned.

The TRW 20-$MW_t$ slagging coal combustor and the existing supersonic Avco channel were integrated and installed in the Avco Mark VI facility with a 4-T iron-core magnet and tested for the express purposes of (1) obtaining definition of combustor and channel interface issues, (2) providing characterization data of a coal-fired MHD power train, and (3) establishing a basis for comparison with ash-injected combustor channel operation.

Components for the bottoming cycle portion of an MHD power plant have been under development at the University of Tennessee Space Institute in Tullahoma, Tennessee. Initial results form tests at 27-$MW_t$ coal-fired input show that $NO_x$ and $SO_x$ emission levels are controllable below NSPS standards. Tests are currently being set up to test material life and control of the very fine (1-$\mu$m mean size) particles exiting the MHD power train.

The Soviet National Program has been the most ambitious overall program to date. Construction of the first clean fuel-fired MHD generating unit having an electrical capacity of 500 MW is reported to have begun at the site of the Ryazan central station power plant. This unit is expected to be commissioned in the mid-eighties. Plans to introduce the first commercial coal-fired MHD plant in the United States must be preceded by a demonstration of a complete power plant of small scale, possibly 50 to 100 MWe. Methods of funding this facility are presently being considered.

***147. Economics and Environmental Impact.*** The primary target of the MHD-steam plant is the production of power at lower cost through the exchange of added capital investment and operating costs to obtain lower fuel costs per kilowatthour of power produced. Furthermore, this must be done within the environmental restraints that are now imposed by governmental authority. It is, however, this latter consideration that makes the coal-fired open-cycle MHD-steam plant seem to be the clear choice over all competitors from the standpoint of both economics and environmental impact (Ref. 56). Its higher efficiency of 50 to 55%, compared with 40% for the best conventional steam plant, means that the heat discharge from the plant is reduced by 15 to 25%, thereby reducing its potential for thermal pollution. The comparison is even more favorable over nuclear plants, which have thermal efficiencies of only about 30% (Ref. 28).

But the greatest importance of the MHD-steam plant, as now envisaged, is its potential of very low air pollution while burning high-sulfur coal (Ref. 56). The $SO_2$, $NO_2$, and particulate emissions are all reduced to levels well below stringent environmental requirements. In pilot-plant tests, 2.2 wt % sulfur coal was burned in a cyclone furnace at 2200°C (4000°F) with seed concentration of 1 g·mol $K_2CO_3$/kg coal with 99.8% removal of $SO_2$, leaving only 5 ppm $SO_2$ in the gaseous effluent. This occurs because of an affinity that the potassium seed material has for $SO_2$, so that seed recovery in the MHD system, which is necessary for economic reasons, also removes the $SO_2$. The removal costs are calculated as approximately a fifth of the $SO_2$ removal costs in a conventional coal-fired plant. In the same tests, through use of two-stage combustion, $NO_x$ emissions were reduced below 150 ppm, or 0.12 lb $NO_2$/million Btu, well below EPA regulations, and complete combustion of carbon monoxide was achieved in all these tests (Ref. 56; see also Ref. 54). For a quite different scheme of coal gasification with the MHD generator utilizing $H_2$–$O_2$ for air-pollution control see Refs. 66 and 79.

The economic analysis is also favorable. The capital and operating costs of a full-scale coal-burning MHD-steam plant are estimated to be less than those of a conventional steam plant using the same fuel and equipped for $SO_2$ removal. Table 11-27 shows cost calculations, based on 1983 prices, of two coal-burning MHD-steam systems as compared with a conventional coal-burning plant equipped for $SO_2$ removal.

**TABLE 11-27** Comparative Costs of Coal-Burning MHD-Steam Plants and Conventional Steam Plants*

| | MHD-steam plant | Conventional coal-burning steam plant |
|---|---|---|
| Total installed cost, $/kW | 950 | 920 |
| Plant efficiency, % | 45.4 | 34.9 |
| Operation and maintenance, mills/kWh | 9.3 | 6.0 |
| Fuel costs, mills/kWh | 20.6 | 27.1 |
| Fixed charges, mills/kWh | 29.9 | 28.9 |
| Cost of electricity, mills/kWh | 59.8 | 62.0 |

*1983 prices.

NOTE: Costs are based on plants with thermal input of 2000 mW, 1.57 $/ton coal, 5700 h/yr operation (0.65 load factor), fixed charge rate of 15%, and 30-year plant life.

### *148. References*


*Books Dealing Mainly with the Physical Theory*

1. Cowling, T. G.: *Magnetohydrodynamics;* New York, Interscience, 1957.
2. Spitzer, L., Jr.: *Physics of Fully Ionized Gases,* 2d ed.; New York, Interscience, 1962.
3. Pai, S-I: *Magnetogasdynamics and Plasma Dynamics;* Vienna, Springer-Verlag, Englewood Cliffs, N.J., Prentice-Hall, 1962.
4. Kulikovskiy, A. G., and Lyubimov, G. A.: *Magnetohydrodynamics;* Reading, Mass., Addison-Wesley, 1965.
5. Cramer, K. R., and Pai, S-I.: *Magnetofluid Dynamics for Engineers and Applied Physicists;* Washington, D.C., Scripta Publishing Co. (McGraw-Hill), 1973.
6. Mitchner, M., and Kruger, C. H.: *Partially Ionized Gases;* New York, Wiley-Interscience, 1973.

*Books Dealing with MHD Applications.*

7. Coombe, R. A. (ed.): *Magnetohydrodynamic Generation of Electrical Power;* London, Chapman & Hall, 1964.
8. Spring, K. H. (ed.): *Direct Generation of Electricity;* Chapter 3 by Swift-Hook, D. T., *MHD Generation;* London and New York, Academic, 1965.
9. Sutton, G. W., and Sherman, A.: *Engineering Magnetohydrodynamics;* New York, McGraw-Hill, 1965.
10. Jahn, R. G.: *Physics of Electric Propulsion;* New York, McGraw-Hill, 1968.
11. Rosa, R. J.: *Magnetohydrodynamic Energy Conversion;* New York, McGraw-Hill, 1968.
12. Heywood, J. B., and Womack, G. J. (eds.): *Open Cycle MHD Power Generation;* Oxford and New York, Pergamon, 1969.

*Conference Proceedings*

13. Symposia on Engineering Aspects of Magnetohydrodynamics (although no preprints on proceedings were issued for the First Symposium, held in Philadelphia on Feb. 18–19, 1960, the program is printed as Appendix I in *a* below. Preprints for other symposia were issued and may be purchased, insofar as supply remains, through Professor J. A. Fox, Department of Mechanical Engineering, University of Mississippi, University, Miss. 38677. Proceedings have been published only for the second and third symposia).

*a.* Second, 1961, University of Pennsylvania, Philadelphia, Pa. *Proceedings of Second Symposium;* Mannal, C., and Mather, N. W. (eds.), *Engineering Aspects of Magnetohydrodynamics;* New York, Columbia University Press, 1962.

*b.* Third, 1962, University of Rochester, Rochester, N.Y. *Proceedings of Third Symposium:* Mather, N. W., and Sutton, G. W. (eds.), *Engineering Aspects of Magnetohydrodynamics;* New York, Gordon and Breach, 1964.

*c.* Fourth, 1963, University of California, Berkeley, Calif.

*d.* Fifth, 1964, Massachusetts Institute of Technology, Cambridge, Mass.

*e.* Sixth, 1965, University of Pittsburgh and Carnegie Institute of Technology, Pittsburgh, Pa.

*f.* Seventh, 1966, Princeton University, Princeton, N.J.

*g.* Eighth, 1967, Stanford University, Stanford, Calif.

*h.* Ninth, 1968, The University of Tennessee Space Institute, Tullahoma, Tenn.

*i.* Tenth, 1969, Massachusetts Institute of Technology, Cambridge, Mass.

*j.* 11th, 1970, California Institute of Technology, Pasadena, Calif.

*k.* 12th, 1972, Argonne National Laboratory, Argonne, Ill.

*l.* 13th, 1973, Stanford University, Stanford, Calif.

*m.* 14th, 1974, The University of Tennessee Space Institute, Tullahoma, Tenn.

*n.* 15th, 1976, University of Pennsylvania, Philadelphia, Pa.

*o.* 16th, 1977, University of Pittsburgh, Pittsburgh, Pa.

*p.* 17th, 1978, Stanford University, Stanford, Calif.

*q.* 18th, 1979, Montana School of Mines, Butte, Mt.

*r.* 19th, 1981, University of Tennessee Space Institute, Tullahoma, Tn.

*s.* 20th, 1982, University of California, Irvine, Calif.

*t.* 21st, 1983, Argonne National Laboratory, Argonne, Ill.

*u.* 22nd, 1984, Mississippi State University, Starkville, Miss.

*v.* 23rd, 1985, Hidden Valley Conference Center, Somerset, Pa.

14. International Conferences on Magnetohydrodynamic Electric Power Generation (formerly, Symposium on Magnetoplasmadynamic Electric Power Generation (1st), or International Symposium on Magnetohydrodynamic Electric Power Generation (2d and 3d)).

*a.* First, 1962, Newcastle-upon-Tyne; *Proceedings:* I.E.E. Conference Report Series No. 4, The Institution of Electrical Engineers, London, 1963.

*b.* Second, 1964, Paris; *Proceedings:* European Nuclear Agency and Organization for Economic Cooperation, Paris, 1964.

*c.* Third, 1966, Salzburg, Austria; *Proceedings:* International Atomic Energy Agency and European Nuclear Energy Agency, Salzburg, 1966.

*d.* Fourth, 1968, Warsaw, Poland; *Proceedings: Electricity from MHD,* 6 vols., International Atomic Energy Agency, Vienna, 1968.

*e.* Fifth, 1971, Munich, West Germany; *Proceedings:* 6 vols., International Atomic Energy Agency, Munich, 1971.

*f.* Sixth, 1975, Washington, D.C.; *Proceedings:* 4 vols., Energy Research and Development Administration, Washington, D.C., 1975.

15. McGrath, I. A., Siddall, M. W., and Thring, M. W. (eds.) Advances in Magnetohydrodynamics; *Proc. Colloq. Sheffield Univ.,* October 1961; New York, Macmillan, 1963.

### *Review Articles*

16. Resler, E. L., Jr., and Sears, W. R.: The Prospects for Magneto-aerodynamics, *J. Aeron. Sci.,* April 1958, vol. 25, pp. 235–245, 258.

17. Sporn, P., and Kantrowitz, A.: Magnetohydrodynamics: Future Power Process? *Power,* November 1959, vol. 103, no. 11, pp. 62–65.

18. Steg, L., and Sutton, G. W.: Prospects of MHD Power Generation, *Astronautics,* August 1960, vol. 5, pp. 22–25.

19. Rosa, R. J.: Physical Principles of Magnetohydrodynamic Power Generation, *Phys. Fluids,* February 1961, vol. 4, pp. 182–194.

20. Gunson, W. E., Smith, E. E., Tsu, T. C., and Wright, J. H.: MHD Power Conversion, *Nucleonics,* July 1963, vol. 21, no. 7, pp. 43–47.
21. Brogan, T. R.: MHD Power Generation, *IEEE Spectrum,* February 1964, vol. 1, pp. 58–65.
22. Rosa, R., and Kantrowitz, A.: MHD *Power; Int. Sci. Technol.,* September 1964, no. 33, pp. 80–86, 89, 90, 92.
23. Petrick, M.: Liquid Metal Magnetohydrodynamics, *IEEE Spectrum,* March 1965, vol. 2, pp. 137–151.
24. Kerrebrock, J.: Magnetohydrodynamic Generators with Nonequilibrium Ionization, *J. AIAA,* April 1965, vol. 3, pp. 591–601.
25. Tsu, T. C.: MHD Power Generators in Central Stations, *IEEE Spectrum,* June 1967, vol. 4, no. 6, pp. 59–65.
26. Morse, F. H.: Survey of Liquid Metal Magnetohydrodynamic Energy Conversion Cycles, *Energy Conversion* (Pergamon Press), July 1970, vol. 10, pp. 155–176.
27. Livingston, J. D.: Superconductivity and Superconducting Materials: General Electric Co., Corporate Research and Development, September 1971, Report 71-C-275.
28. Kantrowitz, A., and Rosa, R. J.: MHD Power Generation, in *Physics and the Energy Problem—1974,* AIP Conference Proceedings 19, Fiske, M. D., and Havens, W. W., Jr. (eds.), American Institute of Physics, 1974, pp. 357–378.

*Journal and Conference Papers*

29. Hurwitz, H., Jr., Kilb, R. W., and Sutton, G. W.: Influence of Tensor Conductivity on Current Distribution in an MHD Generator, *J. Appl. Phys.,* 1961, vol. 32, pp. 205–216.
30. Kerrebrock, J. L.: Conduction in Gases with Elevated Electron Temperature, in Ref. 13*a*, pp. 325–346.
31. Frost, L. S.: Conductivity of Seeded Atmospheric Pressure Plasmas, *J. Appl. Pbhys.,* October 1961, vol. 32, pp. 2029–2036.
32. Brown, J. J. W.: Some Aspects of MHD Power Plant Economics, in Ref. 13*b*, pp. 223–241.
33. Hurwitz, H., Jr., Sutton, G. W., and Tamor, S.: Electron Heating in Magnetohydrodynamic Generators, *J. Am. Rocket Soc.,* 1962, vol. 32, pp. 1237–1243.
34. Jackson, W. D., and Pierson, E. S.: Operating Characteristics of the M.P.D. Induction Generator, in Ref. 14*a*, pp. 38–42.
35. Prem, L. L., and Perkins, W. E.: A New Method of MHD Power Conversion Employing a Fluid Metal, in Ref. 14*b*, vol. 2, pp. 971–984.
36. Way, S., and Young, W. E.: The Feasibility of Large-Scale MHD Power Generation, in Ref. 14*b*, vol. 3, pp. 1483–1495.
37. Kerrebrock, J. L.: Nonequilibrium Ionization Due to Electron Heating: I. Theory, *J. AIAA,* 1964, vol. 2, pp. 1072–1080.
38. Kerrebrock, J. L., and Hoffmann, M. A.: Nonequilibrium Heating: II. Experiments, *J. AIAA,* 1964, vol. 2, pp. 1080–1087.
39. Fishman, F. J.: Effect of Electrode Nonuniformities along the Magnetic Field in MHD Generators, *Ad. Energy Convers.,* December 1964, vol. 4, pp. 223–236.
40. Gallant, H.: Development of a Combustion Chamber for MHD Generators, *Brown Boveri Rev.,* December 1964, vol. 51, pp. 817–820.
41. Klepeis, J., and Rosa, R. J.: Experimental Studies of Strong Hall Effects and U × B Induced Ionization, *J. AIAA,* September 1965, vol. 3, pp. 1659–1666.
42. Novack, M. E., and Brogan, T. R.: Water Cooled Insulating Walls, *Ad. Energy Convers.,* July 1965, vol. 5 no. 2, pp. 95–102.
43. Rosner, M.: The Oil Fired MHD Power Plant in Ref. 14*c*, vol. III, pp. 771–784.
44. Tsu, T. C., Young, W. E., and Way, S.: Optimization Studies of MHD-Steam Plants, in Ref. 14*c*, vol. III, pp. 889–911.
45. Mattsson, A. C. J., et al.: Performance of a Self Excited Generator, *Mech. Eng.,* November 1966, vol. 88, pp. 38–41.
46. Wright, L. E., and Carson, J. E.: Operation of a 20 MW Hall Generator, in Ref. 13*h*, pp. 78–84.
47. Decker, R., Hoffman, M. A., and Kerrebrock, J. L.: Behavior of a Large Non-Equilibrium MHD Generator; *J. AIAA,* March 1971, vol. 9, pp. 357–364.

48. Nidolaieva, V. A.: Rapporteur Session on Electrodes and Insulating Materials, in Ref. 14*e*, General volume, pp. 112–118.

49. Sonju, O. K., and Teno, J.: An Experimental and Theoreitcal Investigation of a High Interaction Combustion-Driven MHD Generator, in Ref. 14*e*, General volume, pp. 71–84.

50. Solbes, A.: Instabilities in Non-Equilibrium M.H.D. Plasmas, A Review, AIAA Paper 70-40, AIAA 8th Aerospace Sciences Meeting, New York, Jan. 19–21, 1970.

51. Amend, W. E., Petrick, M., and Cutting, J. C.: Analysis of Liquid Metal MHD Power Cycles for Central Station Power Generation, in Ref. 13*k*, pp. IV. 1.1–.13.

52. Bergamn, P. D., Plants, K. D., Demeter, J. J., and Bienstock, D.: An Appraisal of Coal-Gasification Schemes for MHD Power Generation, in Ref. 13*k*, pp. VI.3.1–.7.

53. Gasparotto, M.: Electrical Insulation Problems in Design and Tests of Closed Cycle MHD Generators, in Ref. 13*k*, pp. I.2.1.–.8.

54. Hals, F. A., and Lewis, P. F.: Control Techniques for Nitrogen Oxides in MHD Power Plants, in Ref. 13*k*, pp. VI.5.1–.10.

55. Wu, Y. C. L., et al.: Theoretical and Experimental Studies of Magnetohydrodynamic Power Generation with Char; in Ref. 13*k*, pp. II.1.1.–.14.

56. Bienstock, D., et al.: Air Pollution Aspects of MHD Power Generation, in Ref. 13*l*, pp. VII.1.1.–.10.

57. Brederlow, G., and Wittke, K. J.: Effective Electrical Conductivity and Electron Temperature Measurements in a Nonequilibrium MHD Generator Plasma at Atmospheric and Higher Pressures, in Ref. 13*l*, pp. I.7.1.–.6.

58. Ring, L. E., Garrison, G. W., Brogan, T. R., and Schmidt, H. J.: Design of an MHD Performance Demonstration Experiment, in Ref. 13*l*, pp. V.8.1–.7.

59. Shanklin, R. V., et al.: The UTSI Coal Burning MHD Program, in Ref. 13*l*, pp. II.8.1–.8.

60. Kirillin, V. A., and Sheindlin, A. E.: Some Results on the U-25 MHD Power Plant; March–April 1974, *High Temperature* (translation of *Teplofiz, Vys. Temp.*), vol. 12, no. 2, pp. 325–327.

61. Bowen, H. K., et al.: Chemical Stability and Degradation of MHD Electrodes, in Ref. 13*m*, pp. IV.1.1–14.

62. Chang, K. Y., Jackson, W. C., and Wu, Y. C. L.: A Two-Dimensional Dynamic Programming Method for the Optimization of Diagonal Conducting Wall Magnetohydrodynamic Generators, in Ref. 13*m*, pp. III.1.1–.7.

63. Hatch, A. M., et al.: A Superconducting Magnet for Long Duration Testing of MHD Channels, in Ref. 13*m*, pp. II.4.1–.6.

64. Koester, J. K., and Unkel, W.: Performance and Hall Field Limitation Studies in the Stanford M-8 MHD Generator, in Ref. 13*m*, pp. I.6.1–.6.

65*a*. Purcell, J. R., and Wang, S. T.: Design Study for a High Field Superconducting Magnet for MHD, in Ref. 13*m*, pp. II.5.1–.5.

65*b*. Rosa, R. J., Petty, S. W., and Enos, G. R.: Long Duration Testing in the Mark VI Facility, in Ref. 13*m*, pp. I.5.1–.10.

66. Smith, J. M., Nichols, L. D. and Seikel, G. R.: NASA Lewis $H_2$-$O_2$ MHD Program, in Ref. 13*m*, pp. III.7.1–.5.

67. Wu, Y. C. L., et al.: On Direct Coal Fired MHD Generator, in Ref. 13*m*, pp. I.2.1–.7.

68. Zankl, G., et al.: Results of the IPP Test Generator and the Design of a 10 $MW_{el}$ Short-Time Combustion MHD Generator, in Ref. 13*m*, pp. I.1.1–.18.

69. Zauderer, B., Tate, E., and Marston, C. H.: Electrode Studies and Recent Results of Nonequilibrium MHD Generator Experiments, in Ref. 13*m*, pp. VII.5.1–.4.

70. Daily, J. W., Kruger, C. H., Self, S. A., and Eustis, R. H.: Boundary Layer Profile Measurements in a Combustion Driven Generator, in Ref. 14*f*, vol. I, pp. 451–470.

71. Hatch, A. M.: Characteristics of Superconducting Magnets for Large Generators, in Ref. 14*f*, vol. IV, pp. 131–141.

72. Hirabayashi, M., et al.: Design and Performance of a Testing Channel for Semi-Hot Operation, in Ref. 14*f*, vol. I, pp. 435–450.

73. Mason, T. O., et al.: Properties and Thermochemical Stability of Ceramics and Metals in an Open-Cycle Coal-Fired MHD System, in Ref. 14*f*, vol. II, pp. 77–103.

74. Montgomery, D. B., et al.: Superconducting Magnets for Base Load MHD Generators, in Ref. 14*f*, vol. IV, pp. 115–130.

75. Petty, S., Rosa, R., and Enos, G.: Devleopments with the Mark VI Long-Duration MHD Generator, in Ref. 14*f*, vol. I, pp. 231–249.

76. Ring, L. E., et al.: A Status Report on the MHD Demonstration Experiment, in Ref. 14*f*, vol. I, pp. 215–230.

77. Stickler, D. B., and DeSaro, R.: Replenishment Analysis and Technology Development, Ref. 14*f*, vol. II, pp. 31–49.

78. Wu, Y. C. L., et al.: Experimental and Theoretical Investigation on a Direct-Coal-Fired MHD Generator, in Ref. 14*f*, vol. I, pp. 199–213.

79. Wilson, D. R.: Development of a Theoretical Method for Predicting the Performance of Hydrogen-Oxygen MHD Generators; 9th Intersociety Energy Conversion Engineering Conference, San Francisco, Aug. 26–30, 1974, *Proceedings,* The American Society of Mechanical Engineers, New York.

80. Williams, J. R., Rosa, R. J., Yang, Y. Y., and Clement, J. D.: Exploratory Study of Several Advanced Nuclear-MHD Power Plant Systems; 8th Intersociety Energy Conversion Engineering Conference, Aug. 13–17, 1973, *Proceedings,* American Institute of Aeronautics and Astronautics, New York.

81. Energy Conversion Alternative Study, General Electric Co., Dec. 1976, NASA CR-134949.

82. Parametric Study of Potential Early Commercial MHD Power Plants, 1980, U.S. Department of Energy DEN 3-52.

83. Staiger, P. J. and Abbott, J. M.: Summary and Evaluation of the Parametric Study of Potential Early Commercial MHD Power Plants, June 1980, NASA TM-81497.

84. Magnetohydrodynamics (MHD) Engineering Test Facility (ETF) 200 MWe Power Plant—Conceptual Design Engineering Report, Sept. 1981, NASA DEN 3-224.

85. Feasibility Study—MHD Retrofit of Steam Power Plants, July 1979, U.S. Department of Energy Report ER-79-12.

86. Parametric Analysis of Closed Cycle Magnetohydrodynamic (MHD) Power Plants, September 1981, NASA CR-165472.

87. Demetriades, S. T., Maxwell, C. D., and Oliver, D. A.: Progress in Analytical Modeling of MHD Power Generators, in Ref. 13*t*.

88. Ishikawa, M. and Wu, Y. C. L.: Three Dimensional Current Distribution and Slagging Effects in Coal-Fired MHD Channels, in Ref. 13*r*.

89. Iserov, A. D., et al.: Study of U25B MHD Generator System in Strong Electric and Magnetic Fields, in Ref. 13*q*.

90. Hals, F., et. al.: Results from Study of Potential Early Commercial MHD Power Plants; Proc. 7th International Conference on MHD Electrical Power Generator, 1980, Cambridge, Mass.

91. Cutting, et al.: The Influence of Coal Properties of MHD Generator Parameters: in Ref. 13*o*.

92. Cook. C. S. and Dickinson, K. M.: Argon Contamination Associated with Eeramic Regenerative Heat Exchangers for Closed Cycle MHD, in Ref 13*o*.

93. Veefkind, A., et al.: Noble Gas MHD Generator Experiments at Low Stagnation Temperatures, in Ref. 13*p*.

94. Kessler, R.: MHD Generator Channel Design, Proc. Australian International MHD Symposium, 1983, Sydney, Australia

95. Hruby, V., Kessler R., et al.: 1000 Hour MHD Anode Test, in Ref. 13*s*.

96. Magnetohydrodynamics (MHD) Program Evaluation, Gilbert/Commonwealth, May 1983, prepared under contract DE-AC01-81FE-15618.

97. Bauer, M., et al.: Development Status of the TRW 20 $MW_t$ MHD Coal Combustor, in Ref. 13*t*.

98. Braswell, R., et al.: Combined Stage Performance characterization of the TRW 50 $MW_t$ Coal-Fired Combustor, in Ref. 13*v*.

# SECTION 12

# ECONOMICS OF BULK ELECTRIC POWER SUPPLY

**R. X. French**

*Manager, Electrical Department, Sargent & Lundy, Engineers; Chairman, Power System Engineering Committee, IEEE Power Engineering Society; Senior Member, IEEE*

**CONTENTS**

*Numbers refer to paragraphs*

***1. Introduction.*** The long-time trend in the cost of electric power* is shown in Table 12-1. As the electric power industry developed, technological improvements reduced the cost of electricity, even through the 1960s when all other costs were increasing rapidly.

The cost of electric energy charged to the customer consists of the total cost of the three categories of utility operations: generation, transmission, and distribution. This total cost can also be broken down into three major components: fuel, equipment, and wages. The relative magnitude of these various components tends to change over the years in response to changes in technological, economic, or environmental factors. Table 12-2 provides a breakdown of the cost of electric energy, reflecting the relative magnitude of each of the cost components for the U.S. electric power system in 1968.[2] It can be seen that at that time generation costs (fuel and generation equipment and wages) made up half of the total. These costs have continued to increase during the seventies and early eighties, becoming an even greater part of the total.

*Fuel costs* represented 16% of the total energy cost in 1968. These costs have significantly increased compared to other cost components during the seventies because of limitations in fuel supplies. By the early eighties, the share of the total electric energy cost allocated to fuel costs has increased to 37%.[1] This is equal to the share representing all equipment costs (i.e., generation, transmission, and distribution) at that time.[1] This trend, however, is not expected to continue because of reduced utility dependence on oil as a primary fuel source and the fuel economy of the nuclear plants expected to be added to the system.

*Generating equipment costs* have increased more than other equipment costs. In the late sixties, annual expenditures on construction of generating equipment represented 50% of all utility construction expenditures. This share increased to 75% by the early 1980s.[1] Generating equipment costs are expected to continue to increase more than other equipment costs because

* Superior numbers refer to numbered references, Par. **44**.

**TABLE 12-1** Average Revenue per Kilowatthour Sold by the Total Electric-Utility Industry

| | | Commercial and industrial | |
|---|---|---|---|
| Year | Residential, ¢/kWh | Small users, ¢/kWh | Large users, ¢/kWh |
| 1930 | 6.03 | 4.13 | 1.41 |
| 1940 | 3.84 | 3.08 | 1.06 |
| 1950 | 2.88 | 2.64 | 1.01 |
| 1960 | 2.47 | 2.46 | 0.97 |
| 1961 | 2.45 | 2.35 | 0.97 |
| 1962 | 2.41 | 2.37 | 0.96 |
| 1963 | 2.37 | 2.28 | 0.93 |
| 1964 | 2.31 | 2.19 | 0.91 |
| 1965 | 2.25 | 2.13 | 0.90 |
| 1966 | 2.20 | 2.06 | 0.89 |
| 1967 | 2.17 | 2.04 | 0.90 |
| 1968 | 2.12 | 2.00 | 0.90 |
| 1969 | 2.09 | 1.99 | 0.91 |
| 1970 | 2.10 | 2.01 | 0.95 |
| 1971 | 2.19 | 2.12 | 1.03 |
| 1972 | 2.29 | 2.22 | 1.09 |
| 1973 | 2.38 | 2.30 | 1.17 |
| 1974 | 2.83 | 2.85 | 1.55 |
| 1975 | 3.21 | 3.23 | 1.92 |
| 1976 | 3.45 | 3.46 | 2.07 |
| 1977 | 3.78 | 3.84 | 2.33 |
| 1978 | 4.03 | 4.10 | 2.59 |
| 1979 | 4.33 | 4.40 | 2.85 |
| 1980 | 4.93 | 5.10 | 3.41 |

SOURCE: Ref. 1.

of the additional costs added to power plants to accommodate environmental and other regulatory requirements.

*Wages* represented 26% of the total cost of electric energy in 1968. These costs have decreased compared to other costs during the seventies. The share represented by the cost attributed to wages was about 18% in the early 1980s.[1]

*Distribution costs* are determined principally by the geographic density of the load being served and are not related to generation and transmission costs.

*Transmission costs* are determined mainly by the need to transport power from the generating stations to delivery points in the distribution system and the need to interconnect the

**TABLE 12-2** Total Cost of Power in the U.S. in 1968*

| | Generation, % | Transmission, % | Distribution, % | Total, % |
|---|---|---|---|---|
| Fuel | 16 | | | 16 |
| Equipment | 24 | 11 | 23 | 58 |
| Wages | 10 | 2 | 14 | 26 |
| Total | 50 | 13 | 37 | 100 |

* Breakdown into major components (Ref. 2).

generating stations for reliability. Sometimes transmission costs enter into economic comparisons of alternative generation options (such as when comparing locating a generating plant near a fuel supply and transmitting the power, versus locating it near the load and transporting the fuel). However, they generally are not a factor when considering alternative generation options unless a large concentrated power source is involved.

The two principal factors involved in the economics of the bulk power supply are the cost of the generating equipment and the cost of the fuel. There are many trade-offs between these two items that can have significant effects on the cost of power.

To make valid comparisons and decisions between the existing available power sources and to evaluate potential new sources, it is necessary to combine the equipment (capital) costs, which are one-time costs, with the fuel costs and wages which occur year after year. Engineering economic calculations are used to make the comparisons. Recreation, aesthetics, and health values cannot be readily stated in dollars, so they should not be included in an economic evaluation. They must be judged separately. However, the economic comparisons inherently include these items to some degree. Additional cost represents additional utilization of concrete, steel, aluminum, copper, plastics, wood, coal, oil, uranium, etc. Therefore, a minimum-cost project will consume the minimum amount of resources and probably have a reduced effect on the environmental factors.

## PRIMARY SOURCES OF POWER

***2. General.*** Throughout the history of the electric power industry, the primary energy sources have been the combustion of fossil fuels (coal, oil, and natural gas) to produce steam to drive steam turbines, and the impoundment of rivers to provide water to drive hydraulic turbines. In the sixties, a third principal source was developed to the point of commercial viability, that is, nuclear reaction of uranium to produce steam to drive steam turbines. Many other sources are currently under investigation; however, much additional research and pilot-plant development will be required before it will be possible to evaluate their utilization in the overall bulk power supply.

***3. Fossil-Fuel Combustion.*** Prior to the advent of electric power, most power was generated by burning wood or coal in a boiler to produce steam to drive reciprocating steam engines which, in turn, drove machinery by a system of belts and pulleys. Early electric power generation followed the same procedure except that a generator was driven to produce electricity. Later, steam turbines replaced the reciprocating engines and the generator was connected directly to the turbine shaft to form the most commonly used device to generate electric power, the steam turbine-generator. This, along with its boiler, is referred to as a generating unit.

Several units are generally located in one plant in order to utilize more economically features that can be used in common by more than one unit such as fuel- and ash-handling equipment, buildings and equipment, operating and maintenance staff, and transmission-line substations.

Internal combustion engines are used to drive electric generators. However, owing to their limited size, they do not represent a significant part of the total power generation.

Gaseous and liquid fossil fuels can be burned and the hot gases used to drive a turbine directly. These combustion turbines eliminate the cost of the steam portion of the power plant and thus have lower first cost. However, they are much less efficient and require more expensive fuels and more maintenance. Therefore, they incur higher operating costs.

The inefficiency of the combustion turbine is a result of the large heat content of the exhaust gas. This gas is at a high temperature and contains a considerable amount of unburned oxygen; therefore, it is possible to utilize it to generate steam either directly in a waste-heat boiler, or as preheated combustion air in a conventional boiler. The steam produced can then drive a steam turbine-generator. This arrangement is called a combined-cycle plant.

***4. Nuclear-Fuel Reaction.*** In the 1960s nuclear reactors were developed to the point of practical, economical electric power production. The reactors produce either steam (boiler-water reactor, BWR), hot water under pressure (pressurized-water reactor, PWR), or hot gas (high-temperature gas reactor, HTGR). The two latter types use heat exchangers to generate steam. All three utilize the steam to drive turbine-generators.

The water used in the BWR and PWR is an intrinsic part of the nuclear reaction in the reactor core. It acts as a moderator which slows down the free neutrons so that they are present in the core longer and thus have more collisions with uranium nuclei and produce more free neutrons. In the United States and some other countries nuclear-power technology has concentrated on reactors that use ordinary ("light") water as a moderator, whereas Canadian technology has produced a reactor that uses "heavy" water (the hydrogen atoms of the heavy-water molecules contain a neutron as well as a proton). The heavy-water reactor can sustain the necessary chain reaction with natural uranium fuel, whereas the light-water reactors must use enriched uranium, which has had the concentration of fissionable uranium atoms increased.

The enrichment process consists of concentrating the amount of fissionable atoms that occurs in natural uranium. This process leaves a large amount of the uranium in a useless state. Another way of providing enriched uranium is to store natural uranium close around the core of a special reactor called a breeder. Some of the free neutrons in this reactor enter the natural uranium blanket and convert some of the nonfissionable nuclei into fissionable nuclei. Thus the breeder reactor enriches uranium for use as fuel in other reactors as well as producing power itself. Successful development of breeder-reactor technology will allow a manyfold increase in the energy that can be extracted from a given amount of uranium.

All the foregoing reactors obtain heat from the fissioning (breaking apart) of uranium nuclei. It is also possible to obtain heat by fusing hydrogen nuclei together. This process, called fusion, uses heavy hydrogen atoms (deuterium) as fuel. This material is present in very small amounts in all water. It is hoped that fusion reactors can produce power more economically and with less hazard to health and environment than fission reactors; however, several decades of research and development will be required to demonstrate their practicality and costs.

*5. **Hydroelectric Power.*** Natural precipitation provides a continuous source of water at elevations higher than sea level. The flow of water back to lower elevations has provided a source of power by means of waterwheels from early times. Impoundment by means of dams has provided a steady source and a higher water head to improve operation of the wheel. The natural elevation differential at Niagara Falls was used for the motive power for the first commercial alternating-current central station.

Hydropower has the advantage of requiring no fuel. However, its use is complicated by the need to dedicate a significant part of a river course to form a lake large enough to provide a steady water source. Initial costs for the dam and other construction work are higher than for other types of generation. This higher first cost must be offset by long-time fuel cost savings. Therefore, the justification of hydropower is very sensitive to the economic data and procedures. Often the cost of a project is divided between multiple uses of the water, such as power, navigation, irrigation, recreation, and flood control. These competing uses greatly restrict the availability (and thus the relative cost) of the power.

The use of tidal movement of water to generate power has been proposed in some coastal locations where there are large tides. Because of the relatively low water head provided by tidal action, it is necessary to impound huge quantities of water. The cost of the impounding structures has been found to be prohibitive. The structures also would probably have a significant environmental impact owing to their great size.

Efforts are being made to develop power from ocean-wave action, but these are too embryonic to evaluate at present.

*6. **Geothermal Steam.*** At a few locations in the world natural steam is close enough to the surface of the earth that it is accessible by using conventional drilling methods to pipe it to the surface. These locations are too few to be of any overall significance. The expansion of the use of geothermal steam to areas where the heat is not near the surface will require major progress in the development of very deep well-drilling technology.

*7. **Fuel Cells.*** Fuel cells generate low-level direct-current power as a result of a chemical reaction between a hydrocarbon fuel and oxygen. Development has progressed to the point where practical devices are available. However, the costs have not yet been reduced to the point where fuel cells can be considered as competitive with other conventional power sources except in special applications where small amounts of power are required.

*8. **Primary Batteries.*** Primary batteries utilize a chemical reaction between two components of the battery to produce a small amount of direct-current power. The battery components are used up in the process, so the cost is prohibitive for large-scale applications.

***9. Solar Electric Power.*** Electric power can be developed from the sun's rays in two ways: solar cells which produce low levels of direct-current power as a result of the sun's rays striking certain materials, and solar boilers, which consist of a system of mirrors which concentrate the rays from a large area onto a vessel containing water.

Practical use of solar electric power must overcome three fundamental problems: first, the sun's energy is so diffuse that very large earth surface areas must be covered by the mechanism used to collect and convert the energy; second, practical energy output is limited to part of the daylight hours on cloudless days; and third, because of the first two points, the only practical locations in the United States are in the southwestern deserts, which are so far from power-consuming areas as to require major transmission lines to deliver the power.

It will be necessary to justify solar electric power solely on the basis of the fuel that it might save. It will be necessary to have alternative power sources to supply the load when the sun is not shining.

***10. Wind Power.*** It is practical to generate power from propeller-driven generators. However, because of the limited size of the equipment and the variable nature of the wind, costs are high. Reliability of this equipment has been poor.

## ENERGY-STORAGE SYSTEMS

***11. General Aspects.*** Electric power is a highly perishable commodity. There is no means of storing it directly in electrical form. Thus sufficient generating capacity must be constructed to meet the peak load. This expensive capacity is underutilized during off-peak periods. Energy-storage systems can reduce the overall cost of power by reducing the amount of generating capacity required. The storage system absorbs energy during off-peak periods and delivers it to the load during peak periods. To be economically effective, the storage system's construction cost must be low and its efficiency high.

***12. Intermittent Sources.*** Another use for a storage system is to provide power to the load when power from intermittent sources is not available (hydropower when water is not available, solar power when there is no sun, etc.).

***13. Pumped-Storage Hydro.*** Storing energy can be accomplished by using an electric-motor-driven pump to raise water from a lower pool to an upper reservoir when the electric load demand is low (at night or on weekends) or when excess generating capacity is available. Later, the same motor pump can be operated in reverse as a turbine-generator using the water in the upper reservoir as an energy source.

For a pumped-storage system to be economically justified, the power-source fuel cost must be very low (hydro, nuclear, high-efficiency fossil, solar) and the construction cost of the pumped-storage plant must be lower than alternative generating capacity. Low construction costs per unit of power require a very large capacity plant, and a large elevation difference between the upper and lower pools (doubling the water head cuts the required storage volume in half). There are not many locations where the topography is suitable for this type of installation.

***14. Storage Batteries.*** Practical storage-battery systems are available to store surplus electrical energy in chemical form for use at a later time. However, at the present time overall cost benefits have not been sufficiently apparent to justify large-scale trial installations that are needed to verify costs and reliability. Research is being conducted on different types of batteries.

***15. Cryogenic Storage Magnets.*** Research is currently in progress on large cryogenic (supercold) magnets that have the capability of storing large amounts of energy in their magnetic field for long periods of time because of the very low electrical losses in the magnet conductors. Much additional research and development is required before the relative economics of this device can be determined.

***16. Hydrogen Fuel Cycle.*** A scheme that has been proposed to store the energy output of low-fuel-cost plants when they are not required to supply load is the hydrogen fuel cycle. The surplus generating capacity would be used to obtain hydrogen from water by electrolysis. The hydrogen would then be stored or transported for use as a fuel in another generating unit. Much development work will be required to determine the overall costs for this system.

*17. Flywheels.* The use of mechanical flywheels has been proposed for energy storage. Major development of strong materials will have to be made and pilot plants built to demonstrate the reliability and costs for this type of storage before it can be justified.

## BY-PRODUCT POWER

*18. By-product Power.* There are industrial processes which require large amounts of heat at temperatures and pressures well below those at which boilers can generate steam. Where this situation exists, it is possible to obtain very low cost power by generating steam at a higher temperature and pressure and running it through a turbine, exhausting the steam in the condition required by the industrial process. This arrangement (called a topping unit) provides very economical power for two reasons:

**a.** The additional construction cost for the higher-temperature and -pressure boiler plant is lower than the total cost of a boiler plant built to supply electric generation only.

**b.** The additional fuel required to generate the higher-temperature and -pressure steam is less than the average fuel cost for generating steam for electric generation only. The principal reason for this is that for a conventional generating unit, the steam must be condensed back into water in order to obtain good efficiency. The condenser used for this purpose must be supplied with cooling water which absorbs most of the heat in the steam exhausted from the turbine. This heat is lost. In a topping unit, there is no condenser heat loss because the exhaust steam is used for process heat.

Another method of producing by-product power is the use of an extraction turbine which has openings at one or more points to allow steam to be removed after it has passed part way through the turbine. This steam is at lower temperature and pressure than the inlet steam and can be used as process steam. As with a topping unit, the extraction steam does not lose heat to a condenser; therefore, its generation efficiency is very high.

The generation of by-product power is restricted to locations where large amounts of process heat are required. At such locations the cost of power can be reduced significantly.

## FUELS

*19. Coal.* The principal fuel for power production is coal. Bituminous, subbituminous, and lignite are classifications given to coals to indicate the amount of heat contained in a given amount of the material. Transportation costs are high for lignite, which has the lowest heat content. Therefore, this material is burned only in plants located at the fuel source.

The sulfur found in coal is converted to sulfur oxides which are discharged into the atmosphere, so the amount of sulfur in the coal is an important factor. Most of the eastern and all of the midwestern coals have high sulfur content which requires some form of sulfur-removal equipment. This significantly increases the plant cost and reduces the plant efficiency. Coals with much lower sulfur content are located in some western states. Transportation costs to bring this coal to the east and middle west add significantly to its cost.

Boilers and precipitators are designed for the specific heat content, sulfur content, and other physical and chemical properties of the coal to be used; therefore, it is often not possible to change the type of coal used without expensive modifications to the plant equipment.

Many projects are currently in progress to determine the practical feasibility and costs for converting coal to gaseous or liquid fuel, but it is too early to assess the results. These procedures are attractive because they offer the possibilities of sulfur removal before combustion and of providing fuel for combustion turbines as well as steam boilers.

*20. Residual Fuel Oil.* Residual fuel oil is a significant source of energy for power production. This oil contains the heavier components of crude oil that remain after gasoline and other light hydrocarbons have been removed. Oil-fired steam power plants are less expensive to build and operate than coal-fired plants. Combustion turbines utilize lighter oils as a fuel.

*21. Natural Gas.* Natural gas has been used as a fuel for steam power plants located near the oil fields where the gas is produced. Some gas has been burned in coal or oil boilers in other parts of the country when surplus pipeline capacity was available. However, owing to the high value of natural gas for chemical and space-heating uses, its future use as an energy source for power is questionable. Combustion turbines readily utilize natural gas as a fuel.

*22. Uranium.* Natural uranium is the basic fuel for all light-water fission reactors. It must be enriched (the content of fissionable uranium increased from the natural value of about 0.7% to about 3%) to be usable. This increase is accomplished by passing the natural product through filters or centrifuges which increase the concentration of fissionable material in part of the output while reducing it in the remainder, which is then unusable as fuel. A breeder reactor converts this depleted uranium back into usable fuel, thereby greatly extending the amount of usable uranium.

*23. Plutonium.* Plutonium is a fissionable by-product of nuclear-reactor operation. It can be mixed into natural or enriched uranium to recover the energy available in the plutonium.

*24. Deuterium.* Fusion reactors will utilize deuterium as fuel. This material exists in large quantities in water but would have to be extracted and converted into a usable form.

*25. Solid Waste.* Solid waste is currently being used as fuel by being burned in combination with coal in conventional boilers. The waste must be processed to remove metals and glass and to form the necessary consistency. This process has been found to be economical partly because of the value of the recovered materials and the elimination of waste-disposal costs; however, the amount of heat available is not large compared with the power needs of the area over which the waste is generated. Therefore, waste probably will not be a significant source of energy for bulk power production.

## DEVELOPMENT OF OVERALL COST OF ELECTRIC POWER

*26. Need for Power-Load Forecast.* The time required to obtain sites for design and to construct modern power plants varies from 12 years for nuclear plants to 8 years for large coal-fired plants and 3 years for combustion turbines. These long lead times require that decisions be made based on long-term forecasts of the load. The health, welfare, and progress of a modern society depend so heavily on the availability of electricity that a utility must make certain that sufficient generating capacity is available when required by its customers. The forecast of future load takes into account population increase, trends in the per capita use of electricity in the home, industrial expansion, and greater utilization of electricity by industrial processes and agriculture. The national load increase was about 7% per year for many years but dropped to less than half this rate following the 1973 oil crisis.

Recently there have been efforts to reduce the rate of growth by means of voluntary and compulsory curtailment and by adjusting rates to discourage use. These measures appear to have success in reducing the near-term rate of load growth. It is too early to tell whether the long-term rate of load increase would be affected. The conversion of transportation and heating loads from oil and gas to electric loads could cause an increase in load growth. Table 12-3 shows the division of the national electrical load by categories.[2]

**TABLE 12-3** Categories of Electric Power Use

| Year | Industrial, % | Residential, % | Commercial, % | Other,* % |
|---|---|---|---|---|
| 1965 | 41 | 24 | 18 | 17 |
| 1970 | 40 | 25 | 18 | 17 |
| 1990 (projected) | 41 | 24 | 20 | 15 |

* The "other" category includes streetlights, government use, railroads, and utilities, including energy losses.

SOURCE: Ref. 2.

The amount of generating capacity required is also affected by the amount of reserve capacity that is needed to provide a reliable power supply (see Par. **37**).

***27. Basic Concepts of Engineering Economics.*** The basic purpose of engineering economics is to determine the best allocation of resources. The common denominator used to compare the many different resources is their cost. Cost inherently includes a measure of the amount of natural resources required (steel, coal, oil, cement, copper, aluminum, plastics, wood, etc.). Therefore, cost is a rough measure of the environmental impact of a project. More detailed analyses of environmental aspects are made by determining the additional costs incurred to reduce specific impacts. These are called cost-benefit analyses.

Engineering economics, as applied to bulk power supply, is used to compare alternative methods of providing for the required generation, transmission, or distribution of electric power. For generation the following items are considered:

**a.** Type (fossil, nuclear, combustion turbines, etc.).

**b.** Size (few large units versus more smaller units).

**c.** Location (near load or near fuel supply).

**d.** Timing (defer installation by substituting short-term purchase of power).

In assessing the above points, it is necessary to combine costs associated with one-time investments with annually recurring costs such as fuel and operating labor. Thus it is necessary to include the time-value-of-money (Par. **36**) concepts of economics.[3] It is also necessary to include estimated effects of inflation, because the initial investment will not only be subjected to escalation in cost up to the time the facility is built but also fuel costs will continue to escalate over the lifetime of the project.

Even though sophisticated procedures are used in these analyses, final decisions must be made based on judgment of the accuracy of all the assumed future costs. Sensitivity checks must be made to determine how results would be affected if various future conditions do not occur in accordance with the assumptions used.

***28. Economic Philosophy of Revenue Requirements.*** The basic criterion generally used in electric power economic analyses is that the alternative requiring least revenue from the customer is the proper economic choice.[4] It is thus necessary to determine all costs incurred by an alternative. They are grouped into three classes:

**a.** *Ownership costs,* which include all responsibilities assumed when an investment is made (i.e., recovering the invested capital, paying a return on the borrowed funds while they are being used, and paying any taxes associated with the investment and with the return being earned on it). While the borrowed funds are spent at the time the project is built, the revenue required to repay them (plus return and taxes) is collected from the customers over the life of the project. This component of the required revenue is commonly called the "fixed charge." It is fixed in the sense that once the investment is made, there is an ongoing responsibility to obtain the revenue no matter how much power the plant produces. Insurance and local taxes are usually included in the fixed charges.

**b.** *Fuel costs,* including transportation to the power plant.

**c.** *Operator and maintenance cost.* This includes plant operating and maintenance personnel and supplies. The larger part of this cost is constant each year while some of it varies with the amount of power produced.

***29. Ownership Costs.*** To determine total fixed charges associated with an investment, it is necessary to determine the total cost incurred by the owner. Labor and material cost, including overhead and profit of the supplier (direct construction cost), is obvious; however, the following items must also be included to obtain the true total cost to the owner:

**a.** *Indirect construction costs.* Design and engineering costs, construction management costs, owner's overhead and administration costs. These are generally stated as a percentage of the direct construction cost.

**b.** *Allowance for funds used during construction (AFUDC).* Since the fixed charges (and income from the investment) start when the plant goes into service, the return on the funds spent prior to operation must be accounted for. For a construction period of several years, this figure is a significant part of the total investment.

**c.** *Escalation.* Since the estimate of direct construction costs is generally based on present-day actual costs, the labor and material cost escalation up to the time of completion of the plant must be included.

***30. Cost Breakdown.*** Figure 12-1 shows a breakdown of the various cost components for

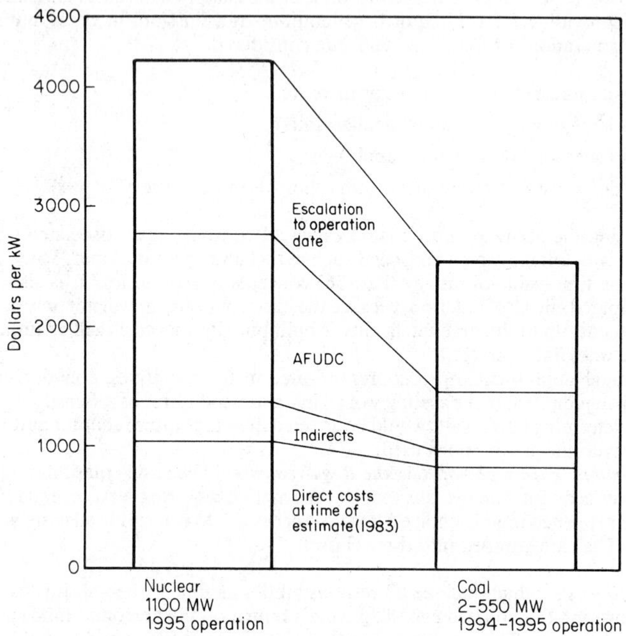


**FIG. 12-1** Breakdown of elements of total power-plant cost, coal-fired versus nuclear plants (Ref. 5).

large coal-fired and nuclear units.[5] This figure shows the general relationship between the various costs but should not be considered an accurate representation of the actual cost of a particular plant or of the differences between the cost of coal and nuclear capacity. Many items must be considered to determine an accurate construction cost:

**a.** Unit size and type.

**b.** Construction time—affects amount of AFUDC.

**c.** Operating date—affects escalation.

**d.** Inflation rate—affects escalation.

**e.** Geographical location—affects construction labor rates.

**f.** Environmental-control regulations—determine cost of air-cleaning equipment and cooling-water system.

**g.** First or later unit at a site.

**h.** Cost of money—affects AFUDC.

**i.** Efficiency (heat rate)—affects amount and cost of equipment.

The change in cost with unit size can be seen in Figs. 12-2 and 12-3. The radical difference in costs between the two figures is mainly a result of about 30 years of inflation in the construction industry.

*31. Construction-Cost Comparisons.* For construction-cost comparison purposes the cost of power-production facilities is generally stated in terms of dollars per kilowatt. When making a comparison on this basis, it is necessary to make certain that both the dollars and the kilowatts of the alternatives are equivalent. All indirect costs and AFUDC should be included in the dollars and they should be at the same point in time. With regard to the kilowatts, some types of generating capacity can produce full output continuously under all weather conditions while others cannot.

For instance, combustion turbines have wide ranges of output depending on air density, which is related to air temperature and site elevation. Also their output must be kept well below maximum capability so as not to cause excessive maintenance expenses and failures. Comparisons on the basis of dollars per kilowatt are not valid unless they include the same components of cost at the same point in time and the power is of the same availability and reliability.

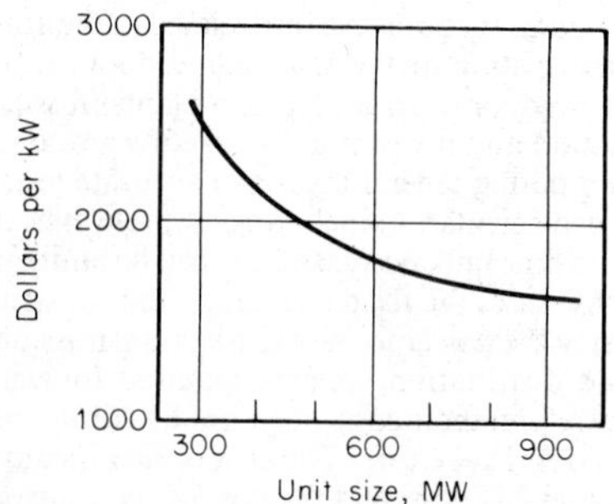


**FIG. 12-2** Estimated construction costs for different size coal-fired units (two-unit plants for 1992 operation).

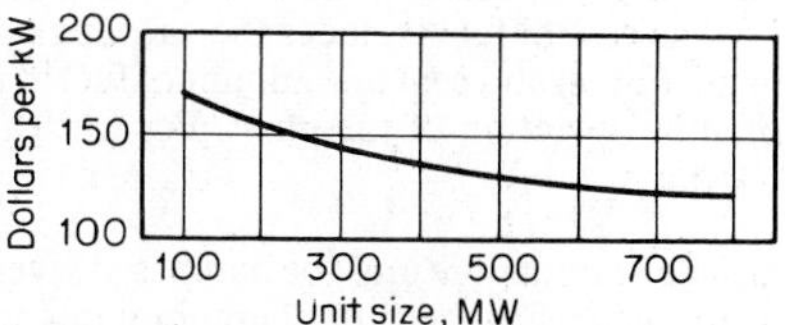


**FIG. 12-3** Cost trend for coal-fired units, two-unit plants, in 1964 (Ref. 6).

*32. Fixed Charges.* Fixed charges are stated as a percent of the total ownership cost (direct construction cost plus indirect costs plus AFUDC plus escalation to time of completion). They consist of the following components:

| | Typical levelized values for investor-owned utility |
|---|---|
| Recovery of capital (depreciation) | 4% |
| Return on borrowed funds: interest on debt (bonds) | 4% |
| Return on equity: | |
| Common stock | 3% |
| Preferred stock | 1% |
| Federal and state income taxes | 4% |
| Property taxes and insurance | 1% |
| | 17%/year |

Governmental agencies do not include taxes or return on equity and use a lower interest rate on their borrowed funds, resulting in a much lower fixed-charge rate. Sometimes governmental agencies add a factor to cover the business-venture nature of capital investments made by government agencies and to account for lack of taxes on government projects. This increases their fixed-charge rate to a value near that used by investor-owned utilities. Industrial firms would use a higher fixed-charge rate to reflect a shorter depreciation period, a higher expected return, and correspondingly higher taxes. The return used in the fixed-charge-rate calculation is the minimum acceptable return needed to attract new capital.

The fixed-charge rate (stated as a percentage of the initial investment) actually changes each year of the life of a plant because, as the capital is recovered, the return is based on the remaining amount. Income taxes are related to the amount of return, so they change also. Several different depreciation methods are allowed for income-tax purposes. Different depreciation periods are allowed for different types of plants (fossil, nuclear, combustion turbines). Accelerated depreciation and investment tax credits are sometimes applicable. State taxes are deductible when computing federal taxes. An accurate year-by-year fixed-charge determination requires a detailed calculation including all these effects.

Some bulk power sources can be built only at locations that are remote from the load area. In these cases the fixed charges on the construction cost of the transmission facilities necessary to bring the power to the existing transmission network should be included. The subtransmission and distribution systems required for various bulk power sources generally are the same; therefore, their costs need not be considered.

***33. Fuel Cost.*** The cost of fuel includes the price of the raw material at the mine or well, processing cost, and the cost of the transportation system (railroad, pipeline, port facility for imported fuel, etc.). Usually the cost of fuel is reduced to a single figure expressed in cents per million Btu delivered at the power plant. Escalation to time of delivery must be taken into account. This can include increase in market price due to value of a particular fuel as influenced by the price of alternative energy sources, increases in cost of production due to environmental restrictions, increases in labor costs, and transportation costs.

The cost of fuel includes the carrying charges on the supply of fuel that is maintained on hand. For oil and coal the amount of fuel in inventory is usually the amount required for 2 or 3 months operation of the plant. For nuclear fuel the carrying charges are very high for two reasons:

**a.** The uranium ore must be purchased several years before it is used. This lead time is required for processing the natural uranium into usable form and manufacture of the fuel assemblies that are placed into the reactor.

**b.** The entire core must be placed in the reactor initially even though the energy will be extracted over a 3- or 4-year period.

There are several distinct items of cost which make up the nuclear-fuel cost (mining and milling, conversion to uranium hexafluoride gas, enrichment of fissionable isotope content, reconversion to solid form and fabrication of fuel elements, reprocessing of spent fuel, credit for recovered plutonium). Each occurs at a different time before and after the insertion of the fuel into the reactor. A fraction (one-third to one-fourth) of the fuel in the reactor is removed at definite time intervals (12 to 18 months) and is replaced by fresh fuel. Thus, keeping track of the total cost of all the components of the various fractions, each with different ages and annual escalation rates, is a mammoth task requiring use of a large computer-system program.

For comparison purposes, fuel costs are generally stated in cents per million Btu of heat value in the fuel. The radical increases in all fuel costs since the 1973 oil crisis have made forecasts of future fuel costs rather meaningless. Up until 1969 the average cost of natural gas and coal was about 25¢/million Btu and that of oil was about 35¢/million Btu. The cost of uranium fuel for nuclear reactors was 15 to 20¢/million Btu. By the early 1980s, coal costs and nuclear-fuel costs tripled and oil cost rose by a factor of 8. Natural gas is no longer available as a utility fuel for new units.

***34. Heat Rates.*** The efficiency of generating units is generally stated in terms of the number of Btu required from the fuel to produce 1 kilowatthour of electric energy. This quantity (Btu/kWh) is called the heat rate. The net cost of fuel (mils/kWh) is then the product of the raw-fuel cost (¢/million Btu) and the unit heat rate (Btu/kWh) times $10^{-5}$ for unit conversion.

| | Typical heat rates at maximum output, Btu/kWh |
|---|---|
| Fossil-fired steam units | 9,000–10,000 |
| Nuclear units | 10,000–11,000 |
| Combustion turbines | 12,000–14,000 |

The unit heat rate can be altered, within limits, with a corresponding change in the construction cost. The heat rate for which a unit is designed is a function of the fixed charges on the incremental construction cost required to improve the heat rate, the assumed lifetime fuel cost, and the assumed lifetime capacity factor.

Actual unit heat rates will be greater than the design value at maximum output because units are always operated at outputs below their rating (see Pars. **37** to **40**). Some units are designed for maximum efficiency at outputs below their maximum rating. This allows for normal operation at a point of high efficiency with the ability to increase output (at lower efficiency) for peak-load periods.

***35. Operator and Maintenance Costs.*** These costs consist of the salary expenses, including overheads, for the plant operating and maintenance personnel, and maintenance and operating materials (other than fuel). They consist of a fixed component (dollars per year) and a component that varies with the amount of power produced (mils/kWh). Since the total of the two is a very small figure compared with ownership and fuel costs, it is desirable to combine the two components into a single figure. Reference 7 (Par. **44**) provides actual values for 1980, which vary over a wide range even for similar sizes and types of generating units. A figure of about \$12/kW/year is the median value, with many plants within ±50% of this value and a few beyond that range. Within this wide range there is no correlation with unit size, type (coal, oil, nuclear), or number of units in the plant.

***36. Methods of Economic Analysis.*** Comparison of alternative means of providing power requires combining costs that occur at different times. The cost for constructing a plant occurs over a several-year period prior to the initial operation of the facility. The same is true for the initial core of nuclear fuel. Fuel costs for fossil fuels occur close to the time the fuel is used and, therefore, are spread over the lifetime of the project. Costs of future possible power sources, such as solar and geothermal, will be all construction costs with essentially no future fuel costs.

Costs that occur at different times can be combined by utilizing the concept of the *time value of money.* This concept essentially states that money (resources) required at one time can be moved to another time by applying an interest rate (and financial algebra) to revalue the money at the new time. The total cost of a project can then be reduced to a single figure by adding all its lifetime costs, each moved to a common point in time. The present time or the time of initial operation is often used as the common time point. The summation of all the lifetime of costs moved to a common time point is called the *present worth* of the project. When comparing the present worth of two projects, it is necessary that they produce the same total result, such as supplying the same amount of power at the times and places it is needed and with the same reliability.

Computing the average production cost (mils/kWh) by dividing all costs by the total energy produced is another common comparison figure.

Comparisons using the total costs reduced to a single present worth (or average production cost) number have a disadvantage in that they obscure the fact that a considerable time may be required for the future fuel-cost savings of a project to offset the early high construction cost of that project. Use of present worth properly accounts for the time value of money, but the use of a single number removes the time factor from view.

Since the accuracy of the results depends on how closely the actual far-distant fuel costs agree with the values assumed in the calculations, a better procedure is to plot the results year by year. The graph should show the accumulation of savings of one alternative over another so that the time to "break-even" can be seen. If this break-even time is long, the risk of decision error due to changing conditions is high. Figure 12-4 shows the results of a typical economic comparison presented in this manner. Another advantage of this procedure is that it shows how large the early losses become before future savings start to reduce them.

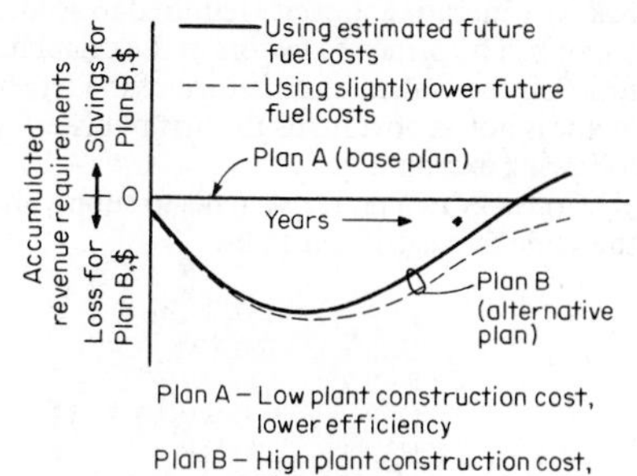


**FIG. 12-4** Method of presenting economic comparison results.

Since some of the values used in the comparisons are based on estimates of distant future costs, it is always advisable to check results for variations of the estimated values. These sensitivity checks provide more insight into the risks involved in any decision.

## RESERVE GENERATING CAPACITY, RELIABILITY, AVAILABILITY, CAPACITY FACTOR, AND LOAD FACTOR

*37. **Reserve Generating Capacity.*** A significant part of the cost of producing electric power is the cost of providing the extra generating capacity needed to serve the load when some generating units are out of service. Modern power plants are highly complex combinations of many devices and systems which often have been recently redesigned to make them more efficient or more economical. They are often operating at extremes of material limits of temperature and stress, again in an attempt to reduce overall power costs. These plants require significant amounts of downtime for regular preventative maintenance. They also experience a considerable amount of downtime owing to unexpected occurrences resulting from equipment breakdown and material failures. To prevent shutting off power to some customers when these scheduled and unscheduled outages occur, reserve generating capacity is installed on all power systems.

Reserve generating capacity is required to cover the following items:

**a.** Scheduled downtime for routine preventative maintenance.

**b.** Scheduled downtime for refueling of nuclear units.

**c.** Inadvertent forced outages caused by equipment malfunctions.

**d.** Extremely adverse weather conditions (load is very sensitive to weather conditions; sometimes the load forecast itself contains a margin for extremely adverse weather).

**e.** Actual load exceeding forecast (which was made several years earlier).

**f.** Unexpected reduction in output capability of generating units due to equipment malfunctions, poor fuel condition, higher cooling-water temperature, regulatory restrictions, etc.).

**g.** Delay of completion of new generating units.

The amount of reserve capacity that is required is a subjective judgment. Excessive reserve greatly increases the cost of power. Inadequate reserve causes great inconvenience and economic loss to customers. Utility managers must constantly assess a balance between these two extremes.

*38. **Reliability.*** The reliability of a power system is measured by its ability to serve all power demands made by all customers without failure over long periods of time. For instance, a common criterion is to install sufficient capacity to prevent interrupting power to customers oftener than one day in a 5-year period on a long-time average basis. The specific amount of reserve generating capacity required to accomplish this goal is determined by probability mathematics. The principal factors in the calculation are the average outage times for the generating units on the system and the size of the larger units relative to the size of the load. The second factor is not as obvious as the first but is equally important. The principle can be seen from the following example.

Consider two power systems to supply the same load, 700 MW. Assume that all units have the same average downtimes.

| | *System A* | | *System B* |
|---|---|---|---|
| Unit 1 | 100 MW | Unit 1 | 100 MW |
| Unit 2 | 100 MW | Unit 2 | 100 MW |
| Unit 3 | 200 MW | Unit 3 | 100 MW |
| Unit 4 | 200 MW | Unit 4 | 100 MW |
| Unit 5 | 300 MW | Unit 5 | 300 MW |
| Unit 6 | 300 MW | Unit 6 | 500 MW |
| | 1200 MW capacity | | 1200 MW capacity |
| | 700 MW load | | 700 MW load |
| | 500 MW reserve | | 500 MW reserve |

For each of these systems, loss of the largest unit at time of the peak load would not cause a loss of load because there is sufficient reserve to cover any one unit being out of service. The most likely event that would cause a loss of load would be two units being down at the same time. The following combinations of pairs of units would cause a loss of load because the total outage would be greater than the reserve:

| *System A* | | *System B* | | | |
|---|---|---|---|---|---|
| Unit 6 | 300 MW | Unit 6 | 500 MW | Unit 6 | 500 MW |
| Unit 5 | 300 MW | Unit 5 | 300 MW | Unit 2 | 100 MW |
| | 600 MW | | 800 MW | | 600 MW |
| | | Unit 6 | 500 MW | Unit 6 | 500 MW |
| | | Unit 4 | 100 MW | Unit 1 | 100 MW |
| | | | 600 MW | | 600 MW |
| | | Unit 6 | 500 MW | | |
| | | Unit 3 | 100 MW | | |
| | | | 600 MW | | |

Thus it is much more likely that system *B* will suffer a loss of load because there are many more combinations of equally likely events that can cause the outage to exceed the reserve.

Actually, system *B* is not five times more likely to suffer a loss of load, as might be assumed from the above analysis, because there are other combinations (trios of units off, quartets of units off, etc.) which will be somewhat more likely to cause a loss of load on system *A*. The reason for this is that units 3 and 4 on system *A* are larger than the same units on system *B*. This will bring the reliability of the two systems closer together, but system *A* will remain more reliable because the likelihood of two units being out at the same time is much greater than for three or more units being off at the same time.

The above analysis was made on the assumption that all units were equally reliable. Experience has been that, as unit size increases, the downtime, both forced and scheduled, increases. This is caused by the increased size and complexity of the equipment in the larger plants.

***39. Availability.*** Figure 12-5 shows the operating availability of various sizes of fossil units and nuclear units.[8] "Operating availability" is the percent of the time that a unit is available to

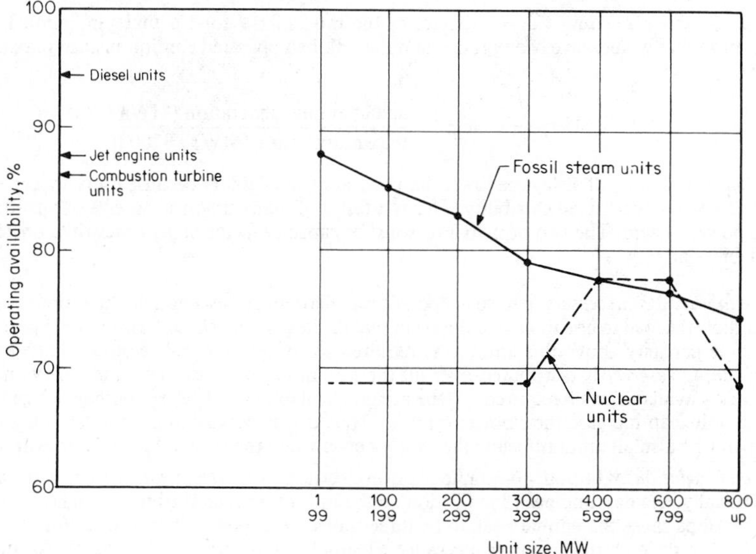


**FIG. 12-5** Operating availabilities of various types of power sources. The curve at the right shows the variation with megawatt ratings for fossil and nuclear steam-turbine plants (Ref. 8).

produce power whether needed by the system or not. It is a measure of overall unit reliability. It is equal to 100% less percent of time on forced outage less percent of time on scheduled outage. It should not be confused with "capacity factor," which is a measure of the energy actually produced compared with the energy that could have been produced if the unit were operated at its rated output continuously. See Fig. 12-6 for typical values.

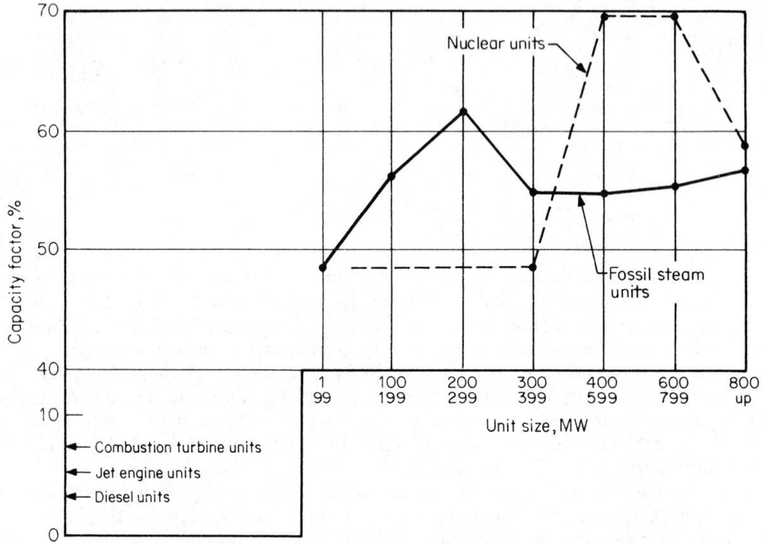


**FIG. 12-6** Capacity factors of various power sources. The curve at the right shows the variation with megawatt ratings of fossil and nuclear steam-turbine plants (Ref. 8).

***40. Capacity Factor.*** Capacity factor is the ratio of the total actual generation to the generation that would have been produced if the unit had operated continuously at maximum rating.

$$\text{Annual capacity factor} = \frac{\text{actual annual generation (MWh)}}{\text{maximum rating (MW)} \cdot 8760\ \text{h}}$$

Capacity factor will always be lower than operating availability because units are generally operated below their rated capability. This is a result of many practical aspects of operating a large power system. The two principal reasons for capacity factor being lower than operating availability are:

**a.** *Reliability.* It is necessary to have in operation at all times at least enough generating capacity to allow the load to be supplied in the event that the largest unit should trip out unexpectedly. (Some capacity above this amount is required for other practical reasons.) This reserve *(spinning reserve)* is distributed over all the generating units to make the reserve power quickly available when required. (There are practical limits to the speed with which generating units can increase their power output.) Thus it is necessary to have most of the units operating a small amount below their rating even when the system load is at maximum.

**b.** *Load variations.* The hour-to-hour load on an electric power system varies over a wide range. Typically, the daytime peak load is double the minimum load during the night. Seasonal variations cause the annual peak to be three times the annual minimum. A plot of all the 8760 hourly loads during the year is called a load-duration curve (see Fig. 12-7). The hourly loads during a typical week are shown in Fig. 12-8.

It is not practical to bring large generating units into and out of service every day (or every week in some cases). Therefore, it is necessary to have enough capacity on line all week long to carry the load of the peak hour of the week (see Par. **42** for more details). At other times it is necessary to reduce the output of some of the units (usually the older, smaller, less-efficient ones).

For technical reasons in the area of boiler flame stability and automatic control systems, it is not practical to reduce the output of fossil steam-turbine units below about 40% of their rating. Therefore, as the system load decreases, it is eventually necessary to reduce output from the efficient units also, so that the less-efficient units will not have to be removed from service. This reduction in output results in a lowering of the capacity factor on all units, the reduction being greater for the less-efficient units.

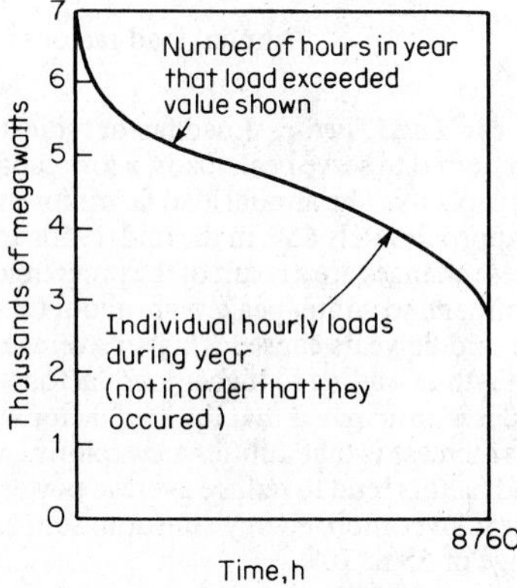


**FIG. 12-7** Annual load-duration curve.

*Generating unit capacity factor* should not be confused with *system load factor,* which is the ratio of the average system load to the peak system load.

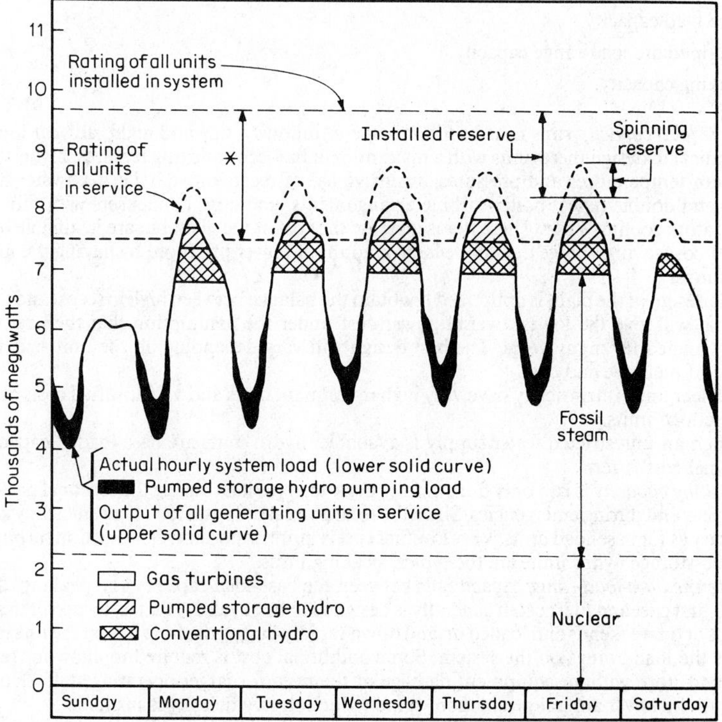


**FIG. 12-8** Weekly load and generation curves.

$$\text{Annual load factor} = \frac{\text{total annual load (MWh)}}{\text{annual peak load (MW)} \cdot 8760\ \text{h}}$$

***41. Load Factor.*** Load factor indicates, in a rough way, the excess generating capacity that is required to serve peak loads, a low-capacity factor indicating a larger need for excess generating capacity. The annual load factor for the entire country has varied from 60% in the late 1940s to approximately 65% in the mid-1950s and 1960s. In the 1970s it has fallen again to about 60%. These changes are a result of the progression from a high winter peak, through a period when the summer and winter peaks were about the same level, to a high summer peak. The dual peaks in the middle years caused a higher average load without a corresponding increase in the annual peak, thus leading to higher load factors. By the early 1980s, the national load factor was 62%, and it is anticipated that the load factor will increase to 64% by the early 1990s.[9] The reason for this increase is that utilities are exploring various methods to improve their load factors. Higher load factors tend to reduce average power costs because the investment costs for equipment are spread over more energy consumption. Load factors on individual systems generally are in the range of 55 to 70%.

## Generating-Capacity Mix

***42. Generating-Capacity Mix.*** The most economical means of supplying the cyclical load on an electric power system is to have three basic types of generating capacity available:

**a.** Base-load capacity.

**b.** Intermediate-load-range capacity.

**c.** Peaking capacity.

*Base-load capacity* runs near its full rating continuously, day and night, all year long. It is economical to design these units with a maximum of fuel-economizing features, highest practical steam temperatures and pressures, extensive use of regenerative boiler-feedwater heaters, reheat and double-reheat boiler-turbine arrangements, and large condensers with minimum-temperature cooling water. These items increase the cost of the plant but are justifiable because the fuel-cost saving is large due to the large amount of power produced by having the unit run continuously.

The design of the plant is optimized to obtain the balance between high first cost and low fuel cost that will give the lowest overall power cost under the assumption that the unit will be heavily loaded for many years. The best design will vary depending on the unit size, money costs, and fuel type and cost.

Nuclear units intrinsically have very high investment costs and very low fuel costs and thus are base-load units.

When an unrestricted water supply is available, hydro units are base-loaded capacity because fuel cost is zero.

*Peaking capacity* is run only during daily peak-load periods during the seasonal peak times of the year and during emergencies. Since the total annual output is low, high efficiency is not as necessary as for base-load units. Very low first cost is more important. Combustion turbines and pumped-storage hydro units are the typical peaking units.

*Intermediate-load-range* capacity fits between the base-load capacity and peaking capacity in both first cost and fuel cost. It generally is designed to be "cycled," that is, turned off regularly at night or on weekends and loaded up and down rapidly during the time it is on the line in order to take the load swings on the system. Some additional cost is required to allow for repeated starts and stops without equipment damage or the need for larger operating staffs. However, owing to the lower annual production, some reduction in efficiency is justified.

Older small-base-load units and hydro units with restrictions on water use are sometimes used for intermediate and peaking service.

The mix of the three types of capacity that will provide the most economical power for a particular power system over a long period of time can be determined only by detailed long-range system-planning studies.

| Designation | First cost | Fuel cost | Typical annual capacity factor, % |
|---|---|---|---|
| Base-load capacity | High | Low | 65–75 |
| Intermediate-load-range capacity | Intermediate | Intermediate | 30–40 |
| Peaking capacity | Low | High | 5–15 |

## Interconnections and Pooling

***43. Interconnections and Pooling.*** The two main factors affecting the cost of generating electric power are ownership cost and fuel cost. These are both affected greatly by the concepts of interconnection and pooling.

All utilities are interconnected with others to some degree. When there are strong interconnections capable of transporting large amounts of power, it is possible for utilities to build larger units while retaining good system reliability because the reserve generating capacity of all the interconnected companies is available when needed by any member. Larger units are inherently less expensive per kilowatt because many of the relatively fixed costs are spread over a larger number of kilowatts of capacity. This effect is termed *economy of scale.*

Interconnections also allow savings in fuel costs by allowing companies with surplus efficient generation to sell energy to others at rates lower than the cost that the others might incur if they generated on their own units. This is called *economy interchange.*

Interconnections also allow companies to exchange power when the times of their peak loads differ, such as between winter and summer. This is called *diversity interchange.* This allows each party to install less generating capacity while maintaining adequate system reliability.

The above benefits can be attained only by means of strong transmission interconnections. There is also an economy of scale for transmission lines; however, it is related to the voltage level. In general, longer transmission distances and higher loads justify higher transmission voltages. However, there are many complex technical matters that affect practical and economical transmission-system design. Figure 12-9 shows a very simplified relationship for one transmission distance.

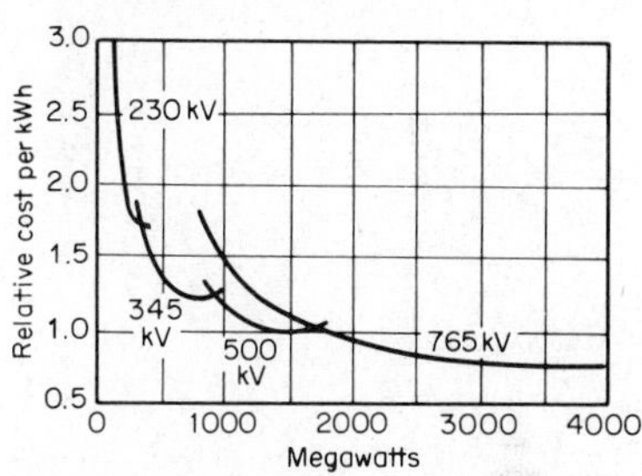


**FIG. 12-9** Relative cost of electric-energy transmission over a 200-mile distance, as a function of the line voltage (Ref. 2).

***44. References***

1. Statistical Yearbook of the Electric Utility Industry—1981, Edison Electric Institute.
2. The 1970 National Power Survey, Part I, Federal Power Commission.
3. Grant, E. L. *Principles of Engineering Economy,* New York, Ronald Press, 1970.
4. Jeynes, P. H. *Profitability and Economic Choice,* Ames, Iowa State University Press, 1968.
5. Brandfon, W. W. A Comparison of Future Nuclear and Coal-Fired Electric Generation in the Midwestern United States; presented at the American Nuclear Society 1982 winter meeting.
6. National Power Survey—1964, Part I; Federal Power Commission.
7. Thermal-Electric Plant Construction Cost and Annual Production Expenses—1980; Energy Information Administration.
8. Equipment Availability Report 1972–1981, Generating Availability Data System, North American Electric Reliability Council.
9. Electric Power Supply & Demand, 1983–1992, North American Electric Reliability Council, July 1983.

# SECTION 13
# PROJECT ECONOMICS

**Allen L. Clapp**
*Managing Director, Clapp Research Associates; Member, IEEE, AEA, AFA, ASA*

## CONTENTS

*Numbers refer to paragraphs*

***1. Bottom Line Economic Measurements.*** This primer is intended to give a quick introduction to the financial considerations which drive the decisions to start or abandon a project. The "bottom line" on any project is that it is either better or worse than alternative investments. Money is the usual medium for measuring "better" because all of the other factors, like risk, reputation, and enjoyment, can often be translated into a monetary equivalent.

The decision to start a project, and the selection of the method to finance it, may involve many interrelated factors. Chief among these factors are the values of project costs and receipts, interest rates, possible returns from other projects, tax regulations, and available financing. The remainder of this primer briefly discusses these items and illustrates the economic differences resulting from three different methods of financing a project: (1) 100% financing by the owner, (2) 50% owner's equity and 50% borrowed debt, and (3) leasing from another owner.

The illustrations below are intended to convey the certain knowledge that taking shortcuts on economic analysis may lead to an inappropriate decision. This is particularly true when a long-term project, like a new energy production system, is being evaluated against a short-term project, like purchasing specialty machinery for producing a product which has a limited sales life. The correct decision is the one which yields the greatest total value to the owner.

***2. The Value of Money.*** Money has no value of its own; its value is proportional only to the goods and services it provides. The amount of goods and services money can provide in a given year relates directly to the relative value of money at that one point in time. If inflation did not reduce the value of money over time, a specified amount of dollars could buy the same set of goods and services in one time as in another. Because of inflation, however, the value of that specific amount of dollars decreases over time; the same amount of money is worth fewer goods and services in later periods. As a result, the decision to start a project should consider both the *amounts* of expenditures and receipts associated with the project and the *timing* of those cash flows.

The terms used to express the effect of time on the value of money are (1) *real dollars* and (2) *nominal-year dollars* or *nominal dollars.* Nominal dollars refer to the amount of dollars received or spent in a given year. Because of inflation, a dollar received in year X will be worth more or less than a dollar received in year Y. In order to compare the two, the real purchasing power of a year X dollar must be compared against the real purchasing power of a year Y dollar. It makes no difference whether (1) year X dollars are converted into the number of year Y dollars that have the same real purchasing power, or (2) year Y dollars are converted into year X dollars. If more convenient, both can be converted into equivalent dollars of some other nominal year.

In order to consider the effects of inflation on a project, all cash flows from each of the various years of the project should be expressed on a directly comparable, common year basis so that their relative values can be considered. To accomplish this, the *nominal year dollars* of cash flow in each future year are converted to *constant year dollars* by *discounting* their value back to that of one common year.

If inflation is running at 10% per year, the relative value of $100 in hand in 1984 will be $110 in 1985 or $121 in 1986, etc. Likewise, future values must be discounted to obtain their value today. In other words, the *real value* of $146.41 (nominal year dollars) received 4 years away is only $100.00 in 1984 dollars. The illustration in Table 13-1 uses such a 10% *discount rate* to calculate the real value (in constant 1984 dollars) of future nominal year dollars for each year. The first two rows show the decline in real value (the ability to purchase goods and services) of a

**TABLE 13-1** Relationship of Nominal Dollars to Real Dollars

| | Dollars in year of receipt | | | | | |
|---|---|---|---|---|---|---|
| | 1984 | 1985 | 1986 | 1987 | 1988 | Total |
| Nominal value (actual dollars) | 100.00 | 100.00 | 100.00 | 100.00 | 100.00 | 500.00 |
| Real value (1984 buying power) | 100.00 | 90.91 | 82.64 | 75.13 | 68.30 | 416.98 |
| Nominal value (actual dollars) | 100.00 | 110.00 | 121.00 | 133.10 | 146.41 | 610.51 |
| Real value (1984 buying power) | 100.00 | 100.00 | 100.00 | 100.00 | 100.00 | 500.00 |

stream of $100 annual receipts. The second two rows show the increase in annual dollar receipts required to maintain the same real income in each future year.

If a project is to be a success, the sum of its real costs and real returns must be positive enough to overcome any uncertainty about the occurrence of future costs and returns. The *present value* of a future income stream (or cost stream) is the sum of the *real* values of the individual future receipts (or costs). The *net present value* of a project is calculated by subtracting the present value of project costs from the present value of expected project returns. The example in Table 13-2 illustrates both the time value of money and the process of calculating the net present value of a project. The nominal dollar values of cash stream A are identical, but in reverse order, to those in cash stream B.

In this illustration, if B is a revenue stream and A is a cost stream, the project makes some money in 5 years; the *net present value* (NPV) is a positive value of $30. If only the nominal dollar flows are considered, the project appears to break even; the *nominal return* over the life of the project is zero. However, that is *not* the case in real terms. Because of the time difference in the cash flows, the project earns a net positive real spendable return.

In this example, the project begins to lose money in 1986. Obviously, if the project can be stopped at the appropriate time, more income will be retained by the owner. If not, the project may still be the best alternative, especially if the scenario of Table 13-2 is the worst expected case and the "best guess" case would return significant profits. Whether this particular project would be started depends upon such factors as the relative returns that can be earned from alternative projects, the relative risk of each project, the availability of financing, and the type and usefulness of tax advantages.

***3. Decision Criteria.*** There are two measures of the relative worth of projects—the net spendable *amount* of the return (the net present value), and the rate of return on the investment required. The latter measure is the *internal rate of return.* Mathematically, the internal rate of return is the discount rate at which the present value of the cost stream (including both original investments and subsequent costs) equals the present value of the revenue stream. The internal rate of return of the above project is obviously greater than 10%, since the NPV is positive at a 10% discount rate. If the NPV had been negative, then it would have been obvious that the internal rate of return was less than 10%.

A decision criterion often used to discriminate between projects is the *payback period,* or *payback.* Mathematically, the payback period is the *cost of the improvement* divided by the *average annual savings.* Although first-year savings are sometimes used as the divisor, the average savings should be used and should include escalations over the life of the project. Using only the first-year savings can yield an incorrect payback.

The following discussion demonstrates that a payback criterion can often lead to the wrong conclusion. If cash stream A and cash stream B of Table 13-2 were both "savings" streams resulting from the investment of $400 in projects A and B, respectively, the payback would mathematically be the same for each project because they have the same total savings. The average savings (income) is $650 divided by 4 years, or $162.50 per year. The payback for each project is the $400 investment divided by the average annual savings of $162.50, or almost 2.5 years. However, the NPV of each is not equal. The NPV of project A is $100 ($500 − $400); project B's NPV is $130 ($530 − $400). The time value of money causes project B to clearly be the better project; the payback criterion fails to differentiate between the two.

**TABLE 13-2** Calculation of Net Present Value

| | Yearly cash flow, $ | | | | | | | | Total cash flow, $ | |
|---|---|---|---|---|---|---|---|---|---|---|
| | 1984 | | 1985 | | 1986 | | 1987 | | | |
| Cash stream | Nom.* | NPV† | Nom. | NPV | Nom. | NPV | Nom. | NPV | Nominal value | Present value @ 10% discount rate |
| A | 100 | 91 | 150 | 124 | 180 | 135 | 220 | 150 | 650 | 500 |
| B | 220 | 200 | 180 | 149 | 150 | 113 | 100 | 68 | 650 | 530 |
| B − A | 120 | 109 | 30 | 25 | −30 | −22 | −120 | −82 | 0 | 30 |

* Nom. = nominal value.
† NPV = net present value.

Because of the time-value-of-money problem, a payback criterion can actually indicate that a lesser project is better. For example, if the 1987 savings of project A increased from $220 to $240, the NPV of the project would increase from $100 to $114. Clearly project B with an NPV of $130 is still better, if the discount rate is 10%. However, the payback period for project A would now decrease from 2.5 to 2.4 years; as a result the wrong project would be picked if a payback criterion is used.

The type of payback discussed above is called a *simple payback* because it uses nominal year dollars in the calculations. If real (constant year) dollars are used, it is called the *discounted payback period* or the "breakeven period." In the above example, using a discounted payback criterion would have indicated the correct choice in both cases. In Table 13-2, the average discounted savings for projects A and B would be $125 [$500 present value (PV)/4 years] and $132.50, respectively; the discounted paybacks would be 3.2 years ($400/$125/year) and 3 years, respectively. Project B would be chosen because of its shorter payback period.

If the 1987 savings of project A increased to $240, the PV of savings would only increase to $514. Since this would still be less than the PV of $530 for the savings from project B, project A would have lower average discounted savings and a longer discounted payback than project B; the correct relative choice would be made. It is clear that, if paybacks are used at all, the discounted payback should be used.

Although the above illustration shows the possible folly in looking only at nominal numbers, Table 13-3 and Fig. 13-1 show that folly even better. Both project X and project Y require a $1000 initial investment. It should be clear from Table 13-3 that project Y is the better of the two investments. It would be chosen whether the decision criterion was NPV, internal rate of return, calculated discounted paybacks or calculated simple paybacks. However, if the first-year savings is used in the payback calculation, or if actual payback time (see the graph) is used, project X would be chosen. This shows the problem with using first-year savings instead of average savings; it also brings up another important point. It is *cash flows* which dominate business decisions; both the level and the timing of those flows can be critical. Project X could very well be the appropriate project to choose if the timing of its cash flows allowed other projects to be undertaken such that the aggregate benefit of all projects was increased. The final decisions on projects should be made on an overall benefit basis.

Another useful tool for comparing projects is the *benefit-cost ratio,* which is the present value of the benefits (savings) divided by the initial cost. For projects X and Y of Table 13-3, the

**TABLE 13-3** Net Present Value Versus Payback ($1000 original investment, 10% discount rate)

| | Net cash returns by year | | | | | Project return | | | | |
|---|---|---|---|---|---|---|---|---|---|---|
| | | | | | | Nominal $ | | Discounted $ | | |
| Project | 1 | 2 | 3 | 4 | 5 | Total | Net | PV | NPV | %IRR* |
| X | 500 | 500 | 250 | 100 | 50 | 1400 | 400 | 1155 | 155 | 18.5 |
| Y | 200 | 300 | 400 | 500 | 600 | 2000 | 1000 | 1444 | 444 | 23.3 |

| | Simple payback, yr | | | Discounted paybacks, yr | |
|---|---|---|---|---|---|
| | Calculated using | | | | |
| Project | 1st-year savings | Average savings | Actual payback | Calculated | Actual |
| X | 2.0 | 3.6 | 2.0 | 4.3 | 3.5 |
| Y | 5.0 | 2.5 | 3.2 | 3.5 | 3.8 |

* IRR = internal rate of return.

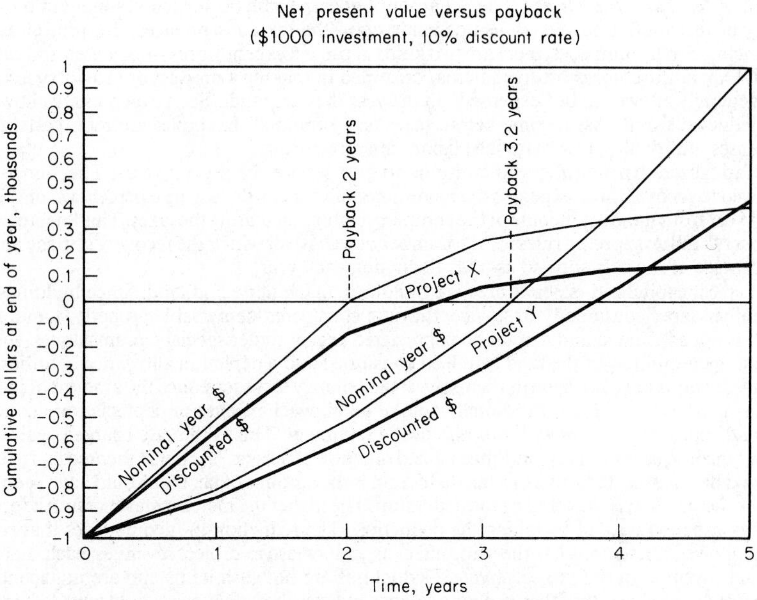

**FIG. 13-1** Graph of net present value versus payback (values from Table 13-3).

benefit-cost ratios are 1.155 and 1.444, respectively. When the appropriate discount rate is used, any benefit-cost ratio greater than unity (1.0) indicates that the project is profitable.

Calculating the NPV and the internal rate of return from each alternative project is a rational method of discriminating between projects and ranking them in an investment priority. First, the projects can be ranked in descending order by the internal rates of return. With an unlimited amount of money and management time, a company would be expected to start all projects with an internal rate of return greater than the cost of money to the company. However, in the "real world" that is usually not the case. The firm is usually limited in capital, or in management capability, and must choose a subset of the complete menu of alternative projects. The NPVs of the projects can be used to help match available resources to achieve the greatest total real return.

In addition to the consideration of the real income and the real rates of return from the various projects, the nominal dollar flows of each project must be considered to assure that the cash flow of the company will be great enough to provide the capital needed in each time period.

If the total cash outlay required for all projects is greater than the total income during any period, the company must either borrow the shortfall or pay it out of available cash. For many companies, available cash is tight, and expected business conditions are not good or are uncertain. These companies will rarely invest in a set of projects that may put them in financial jeopardy—even if the expected long-term returns are great. It is not unusual for a low-return project to be substituted for a high-return project when the cash requirements of the high-return project coincide with other cash demands and the company cannot economically provide the required funds at that time.

The example in Table 13-3 is simplistic. It incorrectly assumes that (1) the project costs and returns are certain and (2) all proceeds of the project can be retained by the owner. Uncertainty of cash flows should be considered by using "sensitivity analysis" and comparing expected results under both optimistic and pessimistic conditions. The tax consequences of the manner in which a project is financed are discussed in the next sections. Further comments on the characteristics affecting the type and amount of an investment are provided at the end of this primer.

***4. After-Tax Cash Flows.*** The net amount of cash available for reinvestment in the company or distribution to the owners depends upon the tax consequences of a project and its financing. For tax purposes, there are two kinds of project expenditures—expensed and capitalized. Expenditures for short-lived items consumed in making a product or providing a service are generally allowed to be "expensed" in the year they are made. Such expenses are allowed to be deducted from gross income before taxes are computed. Examples are rent, parts, travel expenses, utility bills, raw materials, labor, and advertising.

Capitalized expenditures will continue to give service for several years. The company is allowed to recover those expenses over a number of years by deducting a percentage of the cost each year from the gross income of the company before calculating the taxes. This "depreciation recovery" follows specific rules for the number of years over which the recovery is made and the percentage of the cost allowed as a tax deduction each year.

Typical capitalized expenditures are buildings, machinery, and land. Since buildings and machinery are "consumed" in service, they are considered depreciable property. Land, however, is not consumed and cannot be depreciated except under special circumstances, such as where the usefulness of the land is indeed consumed and a depletion allowance is authorized.

Deductions have no value in themselves; they merely serve to reduce the amount of income that is taxable. As a result, the actual value of an allowed expense or depreciation deduction depends upon the *incremental tax rate* of the company. This is the rate charged against the "last" income earned in a year. Since a tax deduction offsets or "shelters" income by reducing the taxable income, the value of a tax deduction is the amount of tax that would have been paid on the income that is sheltered by the deduction. The higher the incremental tax rate, the greater the tax expense avoided by taking the deduction. The reduction in income taxes that *results* from allowed deductions has the same effect as an increase in project revenues; each increases the net revenues of the project. (*Note:* Deductions are not cash items and are not spendable income; their value is that they generate savings in taxes that otherwise would have to be paid.)

Most projects qualify for one or more special tax subsidies called tax credits. A tax credit can offset a tax otherwise owed to the government; the actual cash required for paying taxes is thus reduced. Tax credits are usually in the form of a stated percentage of the capitalized project investment and are usually allowed only in the year of the investment. Unlike allowed depreciation, the effect on the company from a tax credit is independent of the incremental tax rate. The tax credit is a direct reduction in the tax liability of the company. If the tax credit is greater than the tax liability in that year, the unused portion can be applied in other years.

The net cash flow in spendable dollars yielded by a project depends upon the gross income and the cash expenditures which must be made as a result of the project. The tax effects of the investment and the method of financing the investment can sometimes "make or break" a project. Table 13-4 shows the items that must be considered when calculating tax liabilities.

**TABLE 13-4** Tax Calculation

| |
|---|
| Gross income |
| − interest payments |
| − operating expenses |
| − allowable amortization and depreciation on equipment |
| − other tax-deductible expenses |
| = taxable income (+ or −) |
| × incremental tax rate |
| = initial tax liability (+ indicates due, − indicates saved) |
| − total tax credits (only if tax liability is positive) |
| = actual tax (+ indicates due, − indicates saved) |

NOTE: Taxable income is the net difference between gross income and allowed deductions. Since taxable income determines the actual tax liability, it is easy to see the effect on after-tax income of increasing or decreasing the allowed deductions.

**TABLE 13-5** Cash Flow Calculations

| Method 1 |
|---|
| Taxable income |
| − principal payments on debt |
| + allowable amortization and depreciation (these are noncash-deductible expenses and, as such, are not spent but available) |
| = cash available for taxes |
| − tax due (or + tax savings) |
| = after-tax cash income |
| **Method 2** |
| Gross income |
| − interest payments |
| − principal payments |
| − other cash expenses |
| = cash available for taxes |
| − tax due (or + tax savings) |
| = after-tax cash income |

NOTE: These calculations assume that the total income of this project and other projects is great enough for the owner to use all of the benefits earned in this year. Otherwise, some of the benefits may be carried into another tax year—but they will be worth less because of the time value of money.

Table 13-5 shows two methods of calculating the effect of taxes on cash flow; both yield the same answer. These methods are presented here to aid in understanding the effect of nondeductible expenses and noncash tax deductions on the cash flow of a given year. Principal payments on loans are not allowed as a tax deduction, but they are cash payments that must be made during the year. On the other hand, depreciation on depreciable assets is allowed as a tax deduction, and therefore reduces taxes, but it is not an out-of-pocket cash expenditure.

***5. Financing Effects.*** The examples in Table 13-6 show the tax benefits that result from changing the method of financing a project. The project requires an initial investment of $4000. If as in line 1, the owner finances the whole project with personal equity funds, without borrowing any funds and going into debt, the only tax deduction allowed over the life of the project is the depreciation expense. Since both the tax credits and the depreciation expense are related only to the cost of the depreciable assets, and not to the method of financing, both are the same in all cases. If the owner has a 50% incremental tax rate, the allowed deductions generate $2000 in tax savings if the project is 100% equity-financed. The resulting tax benefits total 60% of the original equity investment.

If the owner borrows $1000 and invests $3000 of his or her own money, i.e., finances the project in a 25:75 debt-equity ratio, the allowed tax deductions rise by the $492 interest deduction, and the tax benefits increase. Financing part of a project with debt funds is called *leveraging* the equity investment. All of the benefits of the project continue to flow to the owner, and the tax benefits themselves increase. As a result of the increased benefits and the decreased equity investment, the ratio of tax benefits to equity increases; the rate of return on the investment thus increases, even though the project itself is bringing in the same gross income.

If the project is financed with a 75:25 debt-equity ratio, the tax benefits which accrue to the owner amount to over three times its original equity investment. There are no free lunches, however. If the project fails to reach its income objectives, or costs run higher than expected, the owner will still be liable for payment of the principal and interest payments on the money borrowed for the project. The higher the leverage of the investment in the project, the higher is the business risk the owner faces.

**TABLE 13-6** Examples of Tax Benefits
Total tax benefits received by owner of a $4000 project

| Percent equity financing | Owner equity investment, $ | Amount borrowed, $ | Depreciation expense deduction, $ | Interest expense deduction @ 15%, $ | Total deductions, $ | Taxes saved @ 50%, $ | Inv. tax credits, $ | Total cash benefits, $ | Ratio tax benefits-equity |
|---|---|---|---|---|---|---|---|---|---|
| 100 | 4000 | 0 | 4000 | 0 | 4000 | 2000 | 400 | 2400 | 0.60 |
| 75 | 3000 | 1000 | 4000 | 492 | 4492 | 2246 | 400 | 2646 | 0.88 |
| 50 | 2000 | 2000 | 4000 | 983 | 4984 | 2492 | 400 | 2892 | 1.45 |
| 25 | 1000 | 3000 | 4000 | 1476 | 5476 | 2738 | 400 | 3138 | 3.14 |

**TABLE 13-7** Data for Illustrations in Tables 13-8 through 13-14

If $1000 is borrowed for 5 years at 15% interest, the payments will be those shown in columns 2–7

| | Annual cash payments by year, $ | | | | | |
|---|---|---|---|---|---|---|
| | 1 | 2 | 3 | 4 | 5 | Total |
| Interest (tax-deductible) | 150.00 | 127.75 | 102.17 | 72.75 | 38.91 | 491.58 |
| Principal (not deductible) | 148.32 | 170.57 | 196.15 | 225.57 | 259.41 | 1,000.02 |
| Payment | 298.32 | 298.32 | 298.32 | 298.32 | 298.32 | 1,491.60 |

| | Year | | | | |
|---|---|---|---|---|---|
| | 1 | 2 | 3 | 4 | 5 |
| ACRS depreciation rates, % | 15 | 22 | 21 | 21 | 21 |
| Allowed depreciation deduction on $2000 | 300 | 440 | 420 | 420 | 420 |
| 10% investment tax *credit* (not deduction) | | | | | |
| First year only | 200 | | | | |

NOTE: Assumed combined federal and state tax rate = 50%.

Table 13-7 contains the data for the illustrations of financing effects in the remaining tables. The payments for principal and interest are shown for a debt of $1000 to be repaid over 5 years at 15% interest. The depreciation rates allowed under the accelerated cost recovery system (ACRS) during 1982 are shown along with the annual depreciation and the investment tax credit allowed on a $2000 depreciable investment. (*Note:* 1983 regulations do not allow the full 10% investment tax credit to be taken unless the depreciable basis of the property is reduced by half of the credit. The tables in this text were prepared using the 1982 regulations and have been retained for simplicity of illustration. Since tax credits and tax deductions change frequently, care should be taken to use the correct allowances.)

Table 13-8 shows the calculations of tax effects and cash flows for a $2000 project which the owner finances completely with equity investment. There are no interest deductions included in the tax calculations, since there is no debt to repay. Likewise, there are no principal payments included in the cash flow calculations. The incremental income tax rate of the owner is assumed to be 50%. This method of financing the project yields a nominal return of $4434 over 5 years from an original investment of $2000. The internal rate of return is 32.6%.

Table 13-9 shows the same project, except that it is now financed with 50% equity and 50% debt, with the debt cost assumed at a rate of 15% per year. A 50:50 debt-equity ratio increases the cash outflow required to service the debt; it reduces the overall nominal return over the 5 years to $3187. However, since the owner invested only $1000.00, the internal rate of return of the project increases to the 55% level. This indicates that, if the owner had $2000 to invest, it would be better (other things being equal) to invest $1000 each in two such projects. The yield would then be $6374 for a $2000 investment, as compared to $4434 if only one project is completely owner-financed.

Table 13-10 shows that the same project, with a 30% owner tax rate, yields $3982 in income for the $1000 initial investment. (*Note:* All of the tax credit could not be used in the first year because the tax liability was reduced by the lower tax rate.)

***6. Leasing.*** When a lease arrangement is worked out between two parties, the lessor party owns the installation and the lessee party pays for its use. Since the lessee must pay enough profit to the lessor for the lessor to be willing to install the property for the lessee's use, this arrangement might not appear advantageous to the lessee. However, leasing can be a great advantage in

**TABLE 13-8** 100% Owner Financing of a $2000 Project

| | Year | | | | | |
|---|---|---|---|---|---|---|
| | 1 | 2 | 3 | 4 | 5 | Total |
| **Tax calculations** | | | | | | |
| Revenues | 1500 | 1680 | 1880 | 2100 | 2360 | 9520 |
| − interest | 0 | 0 | 0 | 0 | 0 | 0 |
| − O&M expenses* | 500 | 550 | 605 | 666 | 732 | 3053 |
| − depreciation | 300 | 440 | 420 | 420 | 420 | 2000 |
| = taxable income | 700 | 690 | 855 | 1014 | 1208 | 4467 |
| × tax rate | 0.50 | 0.50 | 0.50 | 0.50 | 0.50 | 0.50 |
| = initial tax due | 350 | 345 | 427 | 507 | 604 | 2233 |
| − tax credits | 200 | 0 | 0 | 0 | 0 | 200 |
| = actual tax due | 150 | 345 | 427 | 507 | 604 | 2033 |
| **Cash flow** | | | | | | |
| Taxable income | 700 | 690 | 855 | 1014 | 1208 | 4467 |
| − principal payments | 0 | 0 | 0 | 0 | 0 | 0 |
| + depreciation | 300 | 440 | 420 | 420 | 420 | 2000 |
| = cash available | 1000 | 1130 | 1275 | 1434 | 1628 | 6467 |
| − tax due | 150 | 345 | 427 | 507 | 604 | 2033 |
| = after-tax cash income | 850 | 785 | 848 | 927 | 1024 | 4434 |

* Operation and maintenance.

NOTE: The original owner investment of $2000 returns over $4000 in 5 years for an internal rate of return of 32.6%.

several situations, particularly when the lessee does not want to, or cannot, borrow the initial money required. The tax advantages of a lease often make a project *go* with lease financing when it cannot go otherwise. Tables 13-11, 13-12, 13-13, and 13-14 examine the cash flows that occur in a leasing situation.

Table 13-11 calculates the revenue required for the lessor to recover its expenses and investment *without* any return, i.e., to break even, if it installs the project and leases it to a lessee. In this particular case, the lessor would make no profit and there would be no incentive to install the project. This case is shown only for the purpose of having a clean example to use as a base for leading into the following examples. Table 13-11 shows the effect of the tax deductions on the lessor; it also shows the out-of-pocket expenses of the lessor that must be covered by the lessee if the lessor breaks even. This is essentially the same set of calculations shown in Table 13-9, except that Table 13-11 calculates the breakeven point.

If the lessor breaks even, Table 13-12 shows the return to the lessee from leasing the project from the lessor. In this case, the lessee invests no money in the project and still reaps a handsome reward. One of the mechanisms that makes leasing work is that the lessee can take the entire cost of the lease as a deduction before taxes, *including the cost of the principal payments of the lessor.* If the lessee were to put the project in on its own, as in Table 13-9, it could deduct only depreciation and interest payments. By leasing, the lessee gets, in effect, two bites at the apple; it gets to deduct the entire lease payment before taxes. Since the lease payment includes both the principal payments and the tax effects of depreciation allowances, the lessee, in effect, gets to write the project off twice, once at the lessor's incremental tax rate and once at the lessee's incremental tax rate.

Table 13-13 is the same as Table 13-11, except that Table 13-13 calculates the revenue required to produce a 15% return on investment for the lessor, rather than a breakeven return.

**TABLE 13-9** 50% Debt and 50% Owner Financing of a $2000 Project

| | Year | | | | | |
|---|---|---|---|---|---|---|
| | 1 | 2 | 3 | 4 | 5 | Total |
| **Tax calculations** | | | | | | |
| Revenues | 1500 | 1680 | 1880 | 2100 | 2360 | 9520 |
| − interest | 150 | 128 | 102 | 73 | 39 | 492 |
| − O&M expenses | 500 | 550 | 605 | 666 | 732 | 3053 |
| − depreciation | 300 | 440 | 420 | 420 | 420 | 2000 |
| = taxable income | 550 | 562 | 753 | 941 | 1169 | 3975 |
| × tax rate | 0.50 | 0.50 | 0.50 | 0.50 | 0.50 | 0.50 |
| = initial tax due | 275 | 281 | 376 | 471 | 585 | 1988 |
| − tax credits | 200 | 0 | 0 | 0 | 0 | 200 |
| = actual tax due | 75 | 281 | 376 | 471 | 585 | 1788 |
| **Cash flow** | | | | | | |
| Taxable income | 550 | 562 | 753 | 941 | 1169 | 3975 |
| − principal payments | 148 | 171 | 196 | 226 | 359 | 1000 |
| + depreciation | 300 | 440 | 420 | 420 | 420 | 2000 |
| = cash available | 702 | 831 | 977 | 1135 | 1330 | 4975 |
| − tax due | 75 | 281 | 376 | 471 | 585 | 1788 |
| = after-tax cash income | 627 | 550 | 601 | 664 | 745 | 3187 |

NOTE: The original *owner* investment of $1000 returns over $3000 in 5 years for an *internal rate of return* of 54.5%.

*Leveraging* the owner's equity investment 1 : 1 with debt causes the internal rate of return on the owner's equity investment to rise because the owner invests only half the money but still receives the full tax benefits.

**TABLE 13-10** 50% Debt and 50% Owner Financing of a $2000 Project, with an Owner Tax Rate of 30%

| | Year | | | | | |
|---|---|---|---|---|---|---|
| | 1 | 2 | 3 | 4 | 5 | Total |
| **Tax calculations** | | | | | | |
| Taxable income | 550 | 562 | 753 | 941 | 1169 | 3975 |
| × tax rate | 0.30 | 0.30 | 0.30 | 0.30 | 0.30 | 0.30 |
| = initial tax due | 165 | 169 | 226 | 282 | 351 | 1193 |
| − tax credits | 165 | 35 | 0 | 0 | 0 | 200 |
| = actual tax due | 0 | 134 | 226 | 282 | 351 | 993 |
| **Cash flow** | | | | | | |
| Cash available | 702 | 831 | 977 | 1135 | 1330 | 4975 |
| − tax due | 0 | 35 | 226 | 282 | 351 | 993 |
| = after-tax cash income | 702 | 697 | 751 | 853 | 979 | 3982 |

**TABLE 13-11** Required Breakeven Revenue for Lessor for $2000 Project with 50:50 Debt-Equity Ratio

| | Year | | | | | |
|---|---|---|---|---|---|---|
| | 1 | 2 | 3 | 4 | 5 | Total |
| **Tax calculations** | | | | | | |
| Interest | 150 | 128 | 102 | 73 | 39 | 492 |
| + O&M expenses | 500 | 550 | 605 | 666 | 732 | 3053 |
| + depreciation | 300 | 440 | 420 | 420 | 420 | 2000 |
| = deductible expenses | 950 | 1118 | 1127 | 1159 | 1191 | 5545 |
| × tax rate | 0.50 | 0.50 | 0.50 | 0.50 | 0.50 | 0.50 |
| = initial taxes saved | 475 | 559 | 563 | 580 | 595 | 2772 |
| + tax credits | 200 | 0 | 0 | 0 | 0 | 200 |
| = actual taxes saved | 675 | 559 | 563 | 580 | 595 | 2972 |
| **Cash flow** | | | | | | |
| Interest & principal | 298 | 299 | 298 | 299 | 298 | 1492 |
| + O&M expenses | 500 | 550 | 605 | 666 | 732 | 3053 |
| − tax savings | 675 | 559 | 563 | 580 | 595 | 2972 |
| = operating cash outlay | 123 | 290 | 340 | 385 | 435 | 1573 |
| + investment recovery | 200 | 200 | 200 | 200 | 200 | 200 |
| = required cash | 323 | 490 | 540 | 585 | 635 | 2573 |
| × 2 (tax factor) | | | | | | |
| = required revenue | 646 | 980 | 1080 | 1170 | 1270 | 5146 |

**TABLE 13-12** Income of Lessee if Lessor Breaks Even

| | Year | | | | | |
|---|---|---|---|---|---|---|
| | 1 | 2 | 3 | 4 | 5 | Total |
| **Tax calculations** | | | | | | |
| Revenues | 1500 | 1680 | 1880 | 2100 | 2360 | 9520 |
| − lease payment | 646 | 980 | 1080 | 1170 | 1270 | 5146 |
| = taxable income | 854 | 700 | 800 | 930 | 1090 | 4374 |
| × tax rate | 0.50 | 0.50 | 0.50 | 0.50 | 0.50 | 0.50 |
| = tax due | 427 | 350 | 400 | 465 | 545 | 2187 |
| **Cash flow** | | | | | | |
| Revenues | 1500 | 1680 | 1880 | 2100 | 2360 | 9520 |
| − income taxes | 427 | 350 | 400 | 465 | 545 | 2187 |
| − lease payment | 646 | 980 | 1080 | 1170 | 1270 | 5146 |
| = after-tax cash income | 427 | 350 | 400 | 465 | 545 | 2187 |

NOTE: The long-run economics of leasing would depend upon the terms of the lease and the residual ownership and use of equipment after initial payoff.

**TABLE 13-13** Required Lessor Revenue if Lessor Makes 15% on Investment on a $2000 Project with 50:50 Debt-Equity Ratio

| | Year 1 | 2 | 3 | 4 | 5 | Total |
|---|---|---|---|---|---|---|
| Tax calculations | | | | | | |
| Interest | 150 | 128 | 102 | 73 | 39 | 492 |
| + O&M expenses | 500 | 550 | 605 | 666 | 732 | 3053 |
| + depreciation | 300 | 440 | 420 | 420 | 420 | 2000 |
| = deductible expenses | 950 | 1118 | 1127 | 1159 | 1191 | 5545 |
| × tax rate | 0.50 | 0.50 | 0.50 | 0.50 | 0.50 | 0.50 |
| = initial taxes saved | 475 | 559 | 563 | 580 | 595 | 2772 |
| + tax credits | 200 | 0 | 0 | 0 | 0 | 200 |
| = actual taxes saved | 675 | 559 | 563 | 580 | 595 | 2972 |
| Cash flow | | | | | | |
| Loan payment (i + p) | 298 | 299 | 298 | 299 | 298 | 1492 |
| + O&M expenses | 500 | 500 | 605 | 666 | 732 | 3053 |
| − tax savings | 675 | 559 | 563 | 580 | 595 | 2972 |
| = operating cash outlay | 123 | 290 | 340 | 385 | 435 | 1573 |
| + recovery of initial investment @ 15% return | 298 | 299 | 298 | 299 | 298 | 1492 |
| = required cash | 421 | 589 | 638 | 684 | 733 | 3065 |
| × 2 (the tax factor) | | | | | | |
| = required income | 842 | 1178 | 1276 | 1368 | 1466 | 6130 |

Required income almost doubles, primarily because of the income taxes that have to be paid on taxable income before the net cash is available to the lessor.

Table 13-14 shows that the effect of allowing the lessor to earn a 15% rate of return is to cut the lessee's after-tax income roughly in half. However, since the lessee still hasn't invested any money in the project, the rate of return of the lessee is infinitely large. When the return of $1695 from leasing is compared with the return of Table 13-9, where an initial investment of $1000 is required, the attractiveness of many leasing schemes is immediately seen. When such schemes are combined with provisions for the lessee to be able to buy the project from the lessor in the future at a reasonable price and at lessee's option, the package can be especially attractive.

In some cases, leasing is used to protect the lessee from buying a set of equipment that may not work well for its application. By leasing, the lessee gets a chance to work with the equipment and see if it performs as expected—before spending large amounts of investment capital on the installation.

*7. **Rate of Return Requirements.*** There are three components of *interest rates.* The first is the *liquidity factor.* There is a value in having cash available to use for whatever investment opportunity may appear in the future. Before one person will lend money to another, the interest earned must compensate the lender for the unavailability of its money while the borrower still has it, i.e., for the lack of liquidity. Second, just like one neighbor lending another a lawn mower, the lender of money expects to get it back in just as valuable a condition as when it was borrowed. In the case of money, the borrower must increase the interest rate paid to the lender enough to include the expected rate of *inflation.* This allows the lender to recover the same *value* as originally lent, albeit a greater number of dollars. The third item that must be included in the interest rate, before a lender is willing to part with the money, is enough

**TABLE 13-14** Income of Lessee If Lessor Makes 15%

| | Year | | | | | |
|---|---|---|---|---|---|---|
| | 1 | 2 | 3 | 4 | 5 | Total |
| Tax calculations | | | | | | |
| Revenues | 1500 | 1680 | 1880 | 2100 | 2360 | 9520 |
| − lease payment | 842 | 1178 | 1276 | 1368 | 1466 | 6130 |
| = taxable income | 658 | 502 | 604 | 732 | 894 | 3390 |
| × tax rate | 0.50 | 0.50 | 0.50 | 0.50 | 0.50 | 0.50 |
| = tax due | 329 | 251 | 302 | 366 | 447 | 1695 |
| Cash flow | | | | | | |
| Revenues | 1500 | 1680 | 1880 | 2100 | 2360 | 9520 |
| − income taxes | 329 | 251 | 302 | 366 | 447 | 1695 |
| − lease payment | 842 | 1178 | 1276 | 1368 | 1466 | 6130 |
| = after-tax cash income | 329 | 251 | 302 | 366 | 447 | 1695 |

NOTE: With a zero investment by the lessee, the lessor makes a 15% return and the lessee still makes $1695, an infinite return. The long-run economics of leasing would depend upon the terms of the lease and the residual ownership and use of equipment after the initial payoff.

additional interest to offset any *risks* associated with the loan. Obviously, the riskier a loan appears, the higher the interest rate required by the lender will be. All of these factors entail *uncertainty*. The lender is uncertain about what opportunities may come along later, the devaluation of the loan from inflation, the ability of the lender to repay the loan, changes in government regulations, and other factors.

These same factors influence the *minimum expected rate of return*, or the *hurdle rate*, that a company requires a project to meet or exceed before giving it full consideration. If the company must borrow money to finance the project, it will be concerned about its ability to repay the loan without jeopardizing the company. The very financing methods which leverage a company's investment and produce such high possible returns also leverage the company's financial risk. Usually, the more stable the expected earnings from projects, the more leverage the company is willing to risk.

If a hurdle rate is used to screen potential projects, the hurdle rate should appropriately reflect the weighted cost of capital to the firm. Using a hurdle rate that is significantly different from the weighted cost of capital incorrectly rejects and accepts projects.

It is not correct to use either the opportunity cost of using retained earnings or the interest rate on borrowed debt solely as the hurdle rate. If retained earnings are used in one project, they are unavailable for use in others. The opportunity cost of using those funds is the rate of return that could be earned by investing those funds in routine company business opportunities. As such, they are generally both higher cost and less extensive than available debt funds. Using that rate can deny worthwhile projects and choke the expansion of the firm.

Considering a project to be financed entirely by debt is also unrealistic. If funds are borrowed without a complementary equity investment, the debt-equity ratio rises, the debt coverage ratio falls, and the ability to borrow more funds decreases.

As a result of the above and related factors, the appropriate hurdle rate is the weighted cost of capital to the firm.

Hurdle rates are often used both as a threshold of profitability that projects must meet and as a method of discriminating *between* projects. As stated earlier, using a hurdle rate that is significantly different from the actual cost of capital to the firm will undercommit or overcommit the firm. As shown below, it may also lead to an incorrect choice of projects.

If all projects under consideration have positive cash flows in later years, almost any hurdle rate can be used to determine the "best" projects on a relative NPV basis. The higher the hurdle rate used to discount future cash flows, the lower the resulting NPV. The result may be the wrong NPV, but the relative ranking will not change. However, that is *not* the case where one or more of the projects have some later years with negative cash flows, such as when significant investments in maintenance or replacement are required; relative rankings may change.

In effect, the discount rate used as a hurdle is assumed to be a rate that can continue to be earned in other areas by the dollars returned from a project each year. It can be used to pay off debt and "earn" the avoided interest or it can be put into another income-producing project. In addition, the higher the discount rate, the less value are later revenues. It is these effects which require the hurdle rate to be close to the actual cost of capital. If a firm cannot actually earn the hurdle rate by reinvesting each year's proceeds from a project, the wrong project may be chosen.

***8. Characteristics Affecting Investments.*** The above discussions have briefly covered some of the factors that drive the decisions people make about new projects and affect the amounts and types of investments. The following is a summary of items that must be considered when any major project is examined:

Ability to borrow money

Cash on hand

Relative risk of the project

Ability to use tax benefits

Existence of tax credits or unusual benefits or constraints

Relative tax rates

Timing of the costs and revenues

Relative permanence of the investment

Ability to shift to another investment if one becomes more attractive

Ability to maintain and operate the project equipment

The ability of a party with money and a party with a project need to find a satisfactory arrangement for (1) financing the project and (2) appropriately sharing the risks and the benefits is almost limitless. Both parties (they may be the same party if the project is primarily owner-financed) must find an acceptable level of risk and reward.

***9. Risk and Reward.*** To many people, taking a risk is its own reward; to others, very little risk is worth taking. The successful manager will analyze alternative projects and will adjust project parameters and financing methods to yield combinations of risk and expected reward appropriate for all parties.

The successful project analyst will be guided by the TANSTAAFL principle: *There ain't no such thing as a free lunch.* Someone, somewhere pays for everything. The questions are who?, how much?, and when? Answering these provides the basis for sound decisions.

***10. Bibliography***

Childs, J. F.: *Encyclopedia of Long-Term Financing and Capital Management;* Englewood Cliffs, NJ, Prentice-Hall, Inc., 1976.

Clapp, A. L.: *Primer on Project Economics;* Research Triangle Park, NC, North Carolina Alternative Energy Corporation, 1984.

Schall, L. D., and Haley, C. W.: *Introduction to Financial Management;* New York, McGraw-Hill Book Company, 1977.

Weston, J. F., and Brigham, E. F.: *Managerial Finance,* 4th ed; Hinsdale, IL, The Aryden Press, 1972.

# SECTION 14
# TRANSMISSION SYSTEMS

**L. O. Barthold**
*Chairman, Power Technologies, Inc.; Fellow, IEEE*

**A. M. DiGioia, Jr.**
*President, GAI Consultants, Inc.; Fellow, ASCE; Member, IEEE*

**D. A. Douglass**
*Senior Engineer, Power Delivery, Power Technologies, Inc.; Senior Member, IEEE*

**I. S. Grant**
*Manager, Software Products Department, Power Technologies, Inc.; Senior Member, IEEE*

**J. A. Moran, Jr.**
*U. S. Technical Representative, Nokia Engineering; Senior Member, IEEE (formerly Manager, Underground Cable Systems, Power Technologies, Inc.)*

**J. D. Mozer**
*Technical Supervisor, GAI Consultants, Inc.; Member, ASCE*

**J. R. Stewart**
*Senior Consultant, Consulting Services Department, Power Technologies, Inc.; Senior Member, IEEE*

**J. A. Williams**
*Manager, Underground Cable Systems, Power Technologies, Inc.; Senior Member, IEEE*

**D. D. Wilson**
*President, Power Technologies, Inc.; Fellow, IEEE*

**CONTENTS**

*Numbers refer to paragraphs*

## *OVERHEAD AC POWER TRANSMISSION*

*Revised by L. O. BARTHOLD, A. M. DiGIOIA, JR., D. A. DOUGLASS, I. S. GRANT, J. D. MOZER, J. R. STEWART, and D. D. WILSON*

Overhead transmission of electric power remains one of the most important elements of today's electric power system. Transmission systems deliver power from generating plants to industrial sites and to substations from which distribution systems supply residential and commercial service. Those transmission systems also interconnect electric utilities, permitting power exchange when it is of economic advantage and to assist one another when generating plants are out of service because of damage or routine repairs.

Since the beginning of the electrical industry, research has been directed toward higher and higher voltages for transmission. As systems have grown, higher-voltage systems have rarely displaced existing systems, but have instead overlayed them. Economics normally dictate that an overlay voltage should be between two and three times the voltage of the system it is reinforcing. Thus it is common to see, for example, one system using lines rated 115, 230, and 500 kilovolts (kV). The highest ac voltage in commercial use is 765 kV. Research and test lines have explored voltages as high as 1500 kV, but there is growing doubt that, in the foreseeable future, much use will be made of voltages higher than those already in service. This plateau in growth is due to a corresponding plateau in the size of generators and power plants, more homogeneity in the geographic pattern of power plants and loads, and, to some extent, adverse public reaction to the visual impact of large towers. Recognizing this plateau, some research attention has shifted to making intermediate voltage lines more compact. The past decade has also seen important advances in design of transmission structures as well as in the components used in line construction, particularly insulators. The pace of current research promises continued improvements in lines of existing voltage as well as the prospect of some fundamentally different ac alternatives.

## Classification of Systems

***1. Transmission Systems.*** The fundamental purpose of the electric utility transmission system is to transmit power from generating units to the distribution system which ultimately supplies the loads. This objective is served by transmission lines that connect the generators into the transmission network, interconnect various areas of the transmission network, interconnect one electric utility with another, or deliver the electrical power from various areas within the transmission network to the distribution substations. Transmission system design is the selection of the necessary lines and equipment which will deliver the required power and quality of service for the lowest overall average cost over the service life. The system must also be capable of expansion with minimum changes to existing facilities.

Electrical design of ac systems involves (1) power flow requirements, (2) system stability and dynamic performance, (3) selection of voltage level, (4) voltage and reactive power flow control, (5) conductor selection, (6) losses, (7) corona-related performance (radio, audible, and television noise), (8) electromagnetic field effects, (9) insulation and overvoltage design, (10) switching arrangements, (11) circuit breaker duties, and (12) protective relaying.

Mechanical design includes (1) sag and tension calculations, (2) conductor composition, (3) conductor spacing (minimum spacing to be determined under electrical design), (4) types of insulators, and (5) selection of conductor hardware.

Structural design includes (1) selection of the type of structures to be used, (2) mechanical loading calculations, (3) foundations, and (4) guys and anchors.

Miscellaneous features of transmission-line design are (1) line location, (2) acquisition of right-of-way, (3) profiling, (4) locating structures, (5) inductive coordination (considers line location and electrical calculations), (6) means of communication, and (7) seismic factors.

## Economics

***2. Choice of Voltage Level.*** Standard transmission voltages are established in the United States by the American National Standards Institute (ANSI). There is no clear delineation between distribution, subtransmission, and transmission voltage levels. In some systems 34.5 kV may be a transmission voltage while in other systems it is classified as distribution, depending on function. Table 14-1 shows the standard voltages listed in ANSI Standards C84 and C92.2, all of which are in use at present.

**TABLE 14-1** Standard System Voltages, kV

| Rating | | Rating | |
|---|---|---|---|
| Nominal | Maximum | Nominal | Maximum |
| 34.5 | 36.5 | 161 | 169 |
| 46 | 48.3 | 230 | 242 |
| 69 | 72.5 | 345 | 362 |
| 115 | 121 | 500 | 550 |
| 138 | 145 | 765 | 800 |
| | | 1100 | 1200 |

The nominal system voltages of 345, 500, and 765 kV from Table 14-1 are classed as extrahigh voltages (EHV). They are used extensively in the United States and in certain other parts of the world. In addition, 400-kV EHV transmission is used, principally in Europe. EHV is used for the transmission of large blocks of power and for longer distances than would be economically feasible at the lower voltages. EHV may be used also for interconnections between systems or superimposed upon large power-system networks to transfer large blocks of power from one area to another.

One voltage level above 800 kV, namely, 1100 kV nominal (1200 kV maximum), is presently standardized. This level has not as yet been installed commercially, although sufficient research and development have been completed to prove technical practicability.*[1-3]

## Conductor Selection

***3. Considerations in Selection.*** The choice of a conductor for a transmission line, as with structure type, depends on the specific application. Once the mechanical strength requirement of the conductor is satisfied, the conductor choice considers the total costs associated with the conductor and also the corona-related electrical environmental effects of radio and audible noise. Corona also causes power loss, particularly during wet weather.

The electrical stress on the surface of a conductor is a function of the voltage on the conductor, the size (i.e., surface area) of a conductor, and the spacing between conductors and/or grounded objects.

The equivalent size of a conductor can be increased either by using a larger conductor or several smaller conductors electrically and physically connected together (bundled conductors). While a single, very large conductor would be electrically adequate, several smaller conductors offer practicality of manufacturing and transporting, ease of construction, and minimizing material usage and mechanical stresses on the supporting structures during high winds and/or ice on the conductors.

At voltages of 345 kV and above, the minimum conductor size or the minimum number of conductors and the individual conductor size in a bundle, are, in addition to cost considerations, normally determined by the corona-related electrical environmental effects. At voltages below 345 kV (e.g., 69 through 230 kV), the minimum size is normally based only on conductor economics.

The conductor sag in the span between structures will depend on conductor materials, conductor weight, conductor strength, conductor tension, conductor temperature, and ice accumulation on the conductor. Strong conductors can be installed at higher tensions and will sag less.

As the current in a conductor increases, the losses increase with a resultant increase in conductor temperature, causing the sag to increase. If the conductor is carrying heavy electrical load on a hot day, very significant increases in sag can occur. Short spans of 150 to 300 ft may have sags of 2 to 5 ft. Very long spans, for example, 1000 to 1500 ft, may experience sags of 20 ft or more.

Since a limiting design criterion is minimum conductor height above ground (for safety reasons), the maximum sags during operation can determine structure heights and span lengths. Similarly, in certain areas ice can form on the conductors of sufficient weight to limit the structure heights and span lengths to maintain ground clearance.

***4. Economics.*** Conductor economic analyses normally use the present worth of revenue required (PWRR) method. This considers the sum of the present worth of levelized annual fixed charges on the total line capital investment, plus annual expenses for line losses:

$$\text{PWRR} = \sum_{n=1}^{\text{NYE}} \left(1 + \frac{i}{100}\right)^{-n} \times \left(\text{CI} \times \frac{F_L}{100} + \text{ADC}_n + \text{AEC}_n\right) \qquad (14\text{-}1)$$

where PWRR = present worth of revenue required
NYE = number of years to be studied
$n$ = $n$th year
$i$ = annual discount rate in percent
CI = total per mile capital investment
$F_L$ = line fixed-charge rate in percent
$\text{ADC}_n$ = per mile demand charge for line losses for year $n$
$\text{AEC}_n$ = per mile energy charge for line losses for year $n$

*Superior numbers refer to references, Par. **121**.

The cost of line losses is based on the cost of generating the losses. Annual demand and energy charges are calculated as shown in the following equations.

Annual demand charge for line losses for year $n$

$$\mathrm{ADC}_n = \frac{C_{\mathrm{kW}} \times \mathrm{ESC}_n}{10^3} \times \frac{F_g}{100} \times \left[1 + \frac{\mathrm{RES}}{100} \times I_L^2 \times \frac{R}{N_c} \times N_{\mathrm{ckt}} \times N_p\right] \tag{14-2}$$

where $\mathrm{ADC}_n$ = annual demand charge for year $n$
$C_{\mathrm{kW}}$ = installed generation cost in dollars per kilowatt
$\mathrm{ESC}_n$ = escalation cost factor for year $n$
$F_g$ = generation fixed-charge rate in percent
RES = required generation reserve in percent
$I_L$ = demand phase current in amperes per circuit
$R$ = single conductor resistance in ohms per mile
$N_c$ = number of conductors per phase
$N_{\mathrm{ckt}}$ = number of circuits
$N_p$ = number of phases

Annual energy charge for line losses for year $n$

$$\mathrm{AEC}_n = \frac{C_{\mathrm{MWh}} \times \mathrm{ESC}_n}{10^6} \times 8760 \times \frac{L_f}{100} \times I_L^2 \times \frac{R}{N_c} \times N_{\mathrm{ckt}} \times N_p \tag{14-3}$$

where $\mathrm{AEC}_n$ = annual energy charges for year $n$
$C_{\mathrm{MWh}}$ = cost of generating energy in dollars per megawatthour
$\mathrm{ESC}_n$ = escalation cost factor for year $n$
$L_f$ = loss factor for determining energy losses in percent
$I_L$ = demand phase current in amperes per circuit
$R$ = single conductor resistance in ohms per mile
$N_c$ = number of conductors per phase
$N_{\mathrm{ckt}}$ = number of circuits
$N_p$ = number of phases

As the conductor size increases, the installed cost increases, both because of the increased conductor cost and the stronger structures necessary to support the larger, heavier conductor and the attendant mechanical loading. The larger conductor cross section, however, results in lower resistance and therefore lower losses. If corona losses are considered, these are also reduced for larger conductors, assuming other dimensions (e.g., phase spacing) remain constant. Therefore, there will be an overall minimum cost at a specific conductor size, where installed cost forces the PWRR higher for large conductors and the cost of losses forces the PWRR higher for smaller conductors. This is conceptualized in Fig. 14-1. In most

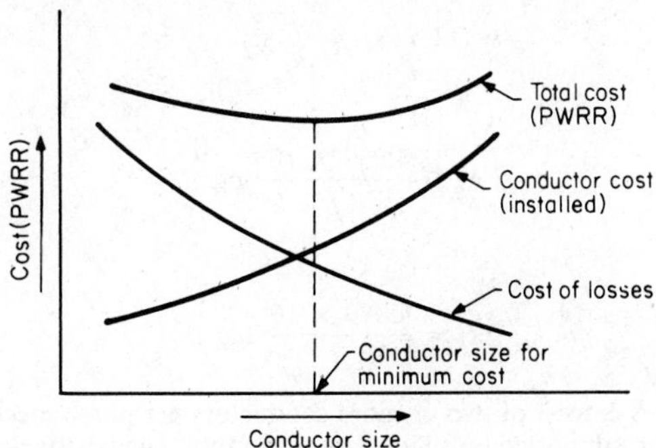


**FIG. 14-1** Conductor economic concept.

practical analyses, there is a relatively flat "minimum" total cost (PWRR) region such that the line designer can temper the economic choice with other factors. Various conductor designs and configurations, such as number of conductors per bundle and size of conductors in a bundle, are examples of areas of designer preference. The higher cost of energy, primarily due to increased fuel costs, has increased the significance of cost of losses in the economic analysis, skewing the economics toward larger conductors with lower losses.

Methods are presently available and are being continuously improved to interactively involve structure design and costing into the overall line economics, with conductor economics being a primary variable.

## Electrical Properties of Conductors

*5. **Positive-Sequence Resistance and Reactance.*** The conductors most commonly used for transmission lines are ACSR (aluminum conductor steel-reinforced), AAC (all-aluminum conductor), AAAC (all-aluminum alloy conductor), and ACAR (aluminum conductor alloy-reinforced). Tables of the electrical characteristics of the most commonly used ACSR conductors are in Sec. 4. Characteristics of other conductors can be found in conductor handbooks or manufacturers' literature.

The per mile resistance, reactance, and capacitance can be determined from the data in the tables of Sec. 4 and the spacing factors $X_d$ and $X'_d$.

The positive-sequence resistance is listed as the 60-Hz value at 50°C. The expression for inductive reactance per mile is

$$X_L = 0.004657f \log \frac{D}{\text{GMR}} \tag{14-4}$$

where $D$ = equivalent spacing in feet, GMR = geometric mean radius in feet as given in the conductor tables of Sec. 4, and $f$ = frequency in hertz. GMR for ACSR conductor is given at 60 Hz. However, 60-Hz values of GMR can be used at other commercial power-system frequencies with small error. $X_L$ also can be expressed as

$$X_L = X_a + X_d = 0.004657f \log \frac{1}{\text{GMR}} + 0.004657f \log D \tag{14-5}$$

When the spacing is 1 ft, $X_d$ becomes zero. Thus $X_d$ is frequently called the "one-foot" reactance. The expression for capacitive shunt reactance per mile,

$$X_c = \frac{4.099 \times 10^6}{f} \log \frac{D}{r_c} \tag{14-6}$$

where $r_c$ = conductor radius in feet, can also be expressed as

$$X_c = X'_a + X_d$$

where

$$X'_a = \frac{4.099 \times 10^6}{f} \log \frac{1}{r_c} \tag{14-7}$$

and

$$X'_d = \frac{4.099 \times 10^6}{f} \log D \tag{14-8}$$

*Bundle conductors* consist of two or more conductors per phase mechanically and electrically connected and supported by an insulator assembly. The positive-sequence resistance is, to a first approximation, the 60-Hz, 50°C values in the Sec. 4 tables divided by the number of conductors per phase. General formulas for the inductance and capacitance of bundle conductors are

$$L_\phi = \frac{1}{n}\left[0.08047 + 0.74113 \log \frac{24(S_{gm})^n}{d(M_{gm})^{n-1}}\right] \quad \text{mH/mi} \tag{14-9}$$

From Eq. (12-9) inductive reactance is found to be

$$X = \frac{1}{n}\left[K + 0.2794 \log \frac{24(S_{gm})^n}{d(M_{gm})^{n-1}}\right] \quad \Omega\text{/mi at 60 Hz} \tag{14-10}$$

and the capacitance is

$$C_\phi = \frac{0.03883n}{\log[24(S_{gm})^n/d(M_{gm})^{n-1}]} \quad \mu F\text{/mi} \tag{14-11}$$

In the above, $n$ = number of conductors per phase (bundle); $d$ = diameter of conductor in inches; $S_{gm}$ = geometric mean distance between conductors of different phases in feet, found by taking the mean distance from all conductors of one phase to all conductors of the other phases; $M_{gm}$ = geometric mean distance in feet between the $n$ conductors of one phase; $K$ = internal conductor reactance defined as

$$K = 0.004657f \log \frac{r_c}{\text{GMR}} \quad \Omega\text{/mi} \tag{14-12}$$

The reactive and capacitive shunt reactances for bundled conductors can also be found by using the $X_a + X_d$ method, by determining the equivalent $X_a$ and $X'_a$ of the conductor bundle. The expressions for the equivalents are given in Table 14-2. These expressions are for three-conductor bundles on equilateral spacing and for four-conductor bundles on

**TABLE 14-2** Equivalent Reactances

| Bundle | $X_{aeq}$ | $X'_{aeq}$ |
|---|---|---|
| 2 conductors | $\frac{1}{2}(X_a - X_s)$ | $\frac{1}{2}(X'_a - X'_s)$ |
| 3 conductors | $\frac{1}{3}(X_a - 2X_s)$ | $\frac{1}{3}(X'_a - 2X'_s)$ |
| 4 conductors | $\frac{1}{4}(X_a - 3X_s)$ | $\frac{1}{4}(X'_a - 3X'_s)$ |

square spacing. The subscript $s$ indicates the spacing of the conductors within the bundle in feet. Values for $X_a$ and $X'_a$ are in the conductor tables in Sec. 4. Values for $X_s$ and $X'_s$ are from the same formulas as $X_d$ and $X'_d$.

$$X_d = 0.004657f \log s \tag{14-13}$$

$$X'_d = \frac{4.099 \times 10^8}{f} \log s \tag{14-14}$$

where $s$ is in feet and $f$ is frequency in hertz. Equation (14-14) is correct for a ratio of spacing $s$ to conductor radius $r$ of 5 or more.

The value of $X_{aeq}$ is added to $X_d$ (the spacing factor which is determined for the mean spacing between the conductors of the different phases). $X'_{aeq}$ and $X'_d$ are handled in a like manner.

***6. Zero-Sequence Impedances.*** When earth-return currents due to faults or other causes are to be calculated, negative-sequence and zero-sequence impedances must be determined in addition to positive-sequence quantities. Negative-sequence quantities are the same as the positive-sequence values for transmission lines. Precise determination of the zero-sequence quantities is difficult because of the variability of the earth-return path.

Calculation of zero-sequence impedance parameters is far more complex than for positive-sequence quantities, being a function of conductor size, spacing, relative position of conductors with respect to overhead ground wires, electrical characteristics of overhead ground wires, and the resistivity of the earth-return circuit. Reference 4 includes a detailed analysis of zero-sequence parameters which are normally calculated using digital computer programs.

Table 14-3 lists representative values of positive- and zero-sequence impedances for different voltage transmission lines with shield wires. Zero-sequence reactance increases for unshielded lines.

**TABLE 14-3** Typical Transmission Line Impedance*

| Voltage, kV | $R_1$ | $X_{L1}$ | $X_{C1}$ | $R_0$ | $X_{L0}$ | $X_{C0}$ | $X_0/X_1$ |
|---|---|---|---|---|---|---|---|
| 69 | 0.280 | 0.709 | 0.166 | 0.687 | 2.74 | 0.315 | 3.86 |
| 115 | 0.119 | 0.723 | 0.169 | 0.625 | 2.45 | 0.265 | 3.39 |
| 230 | 0.100 | 0.777 | 0.182 | 0.591 | 2.26 | 0.275 | 2.91 |
| 345 | 0.060 | 0.590 | 0.138 | 0.551 | 1.99 | 0.208 | 3.37 |
| 500 | 0.028 | 0.543 | 0.127 | 0.463 | 1.90 | 0.198 | 3.50 |
| 765 | 0.019 | 0.548 | 0.128 | 0.428 | 1.77 | 0.185 | 3.23 |

NOTE: 1 mi = 1.61 km.

*$R_1$, $X_{L1}$, $R_0$, $X_{L0}$ are in ohms per mile; $X_{C0}$, $X_{C1}$ are in megohm-miles.

***7. Nominal-π Representation.*** Transmission lines can be represented by the nominal $\pi$ as in Fig. 14-2, in which half the capacitive susceptance, in siemens, is connected at each end of the line. The nominal-$\pi$ representation is used in digital computer studies involving lines of moderate length (usually under 100 mi).

***8. The nominal-T representation*** of a transmission line is shown in Fig. 14-3. The total line susceptance $b$, in siemens, is concentrated at $A$, the midpoint of the line.

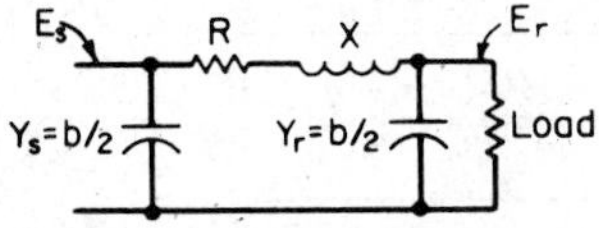


**FIG. 14-2** Nominal $\pi$ line.

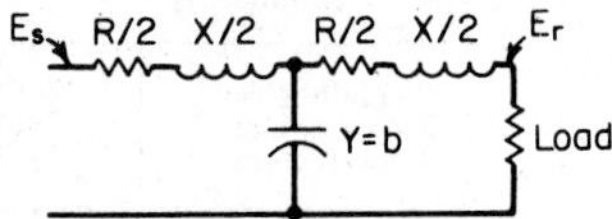


**FIG. 14-3** Nominal-$T$ line.

***9. ABCD parameters*** (general circuit constants) of a line are defined by the equations

$$E_s = AE_r - BI_r \tag{14-15}$$

$$I_s = CE_r - DI_r \tag{14-16}$$

For a short line (under 100 mi) if $Z_1 = R + j\omega L$ and $Z_2 = 2/jb$ (refer to the nominal-$\pi$ line of Fig. 14-2).

$$A = D = \frac{Z_1 + Z_2}{Z_2} \tag{14-17}$$

$$B = Z_1 \tag{14-18}$$

$$C = \frac{Z_1 + 2Z_2}{Z_2^2} \tag{14-19}$$

For longer lines

$$A = D = \cosh \gamma \tag{14-20}$$

$$B = Z_c \sinh \gamma \tag{14-21}$$

$$C = \sinh \gamma / Z_c \tag{14-22}$$

where

$$\gamma = \sqrt{(R + j\omega L)(j\omega C)} \tag{14-23}$$

and

$$Z_c = \sqrt{\frac{R + j\omega L}{j\omega C}} \tag{14-24}$$

and *R, L,* and *C* are line resistance, inductance, and capacitance per mile.

Formulas for *ABCD* constants for various circuit configurations are given in Table 14-4.

***10. Surge Impedance Loading.*** The surge impedance of a transmission line is the characteristic impedance with resistance set equal to zero (i.e., *R* is assumed small compared to $j\omega L$ of Eq. 14-24).

$$Z_c = \sqrt{L/C} \tag{14-25}$$

The power which flows in a lossless transmission line terminated in a resistive load equal to the line's surge impedance is denoted as the surge impedance loading (SIL) of the line.

**TABLE 14-4** Formulas for Generalized Circuit Constants

| No. | Type of network | | Equivalent constants | | | |
|---|---|---|---|---|---|---|
| | | | $A_E$ | $B_E$ | $C_E$ | $D_E$ |
| 1 | Series impedance | $E_s$ Z $E_r$ | 1 | Z | O | 1 |
| 2 | Shunt admittance | $E_{sn}$ Y $E_{rn}$ | 1 | O | Y | 1 |
| 3 | Uniform line | $E_s$ ABC $E_r$ | A | B | C | A |
| 4 | Two uniform lines | $E_s$ $A_2B_2C_2$ $A_1B_1C_1$ $E_r$ | $A_1A_2 + C_1B_2$ | $B_1A_2 + A_1B_2$ | $A_1C_2 + A_2C_1$ | $A_1A_2 + B_1C_2$ |
| 5 | Two nonuniform lines or networks | $E_s$ $A_2B_2C_2D_2$ $A_1B_1C_1D_1$ $E_r$ | $A_1A_2 + C_1B_2$ | $B_1A_2 + D_1B_2$ | $A_1C_2 + D_2C_1$ | $D_1D_2 + B_1C_2$ |
| 6 | General network and sending transformer impedance | $E_s$ $Z_{TS}$ ABCD $E_r$ | $A + CZ_{TS}$ | $B + DZ_{TS}$ | C | D |
| 7 | General network and receiving transformer impedance | $E_s$ ABCD $Z_{TR}$ $E_r$ | A | $B + AZ_{TR}$ | C | $D + CZ_{TR}$ |
| 8 | Two networks in parallel | $E_s$ $A_1B_1C_1D_1$ $A_2B_2C_2D_2$ $E_r$ | $\frac{A_1B_2 + A_2B_1}{B_1 + B_2}$ | $\frac{B_1B_2}{B_1 + B_2}$ | $C_1 + C_2 + \frac{(A_1 - A_2)(D_2 - D_1)}{B_1 + B_2}$ | $\frac{D_1B_2 + D_2B_1}{B_1 + B_2}$ |

NOTE: All constants in this table are complex quantities. $A = a_1 + ja_2$ and $D = d_1 + jd_2$ are numerical values. $B = b_1 + jb_2$ = ohms. $C = c_1 + jc_2$ = siemens. As a check on calculations of *ABCD* constants, note that $AD - BC = 1$.

Under these conditions, the receiving end voltage $E_R$ equals the sending end voltage $E_S$ in the magnitude, but lags $E_S$ by an angle $\delta$ corresponding to the travel time of the line. For a three-phase line

$$\text{SIL} = (E_{L-L})^2/Z_L \tag{14-26}$$

Since $Z_C$ has no reactive component, there is no reactive power in the line, $Q_S = Q_R = 0$. This indicates that for SIL the reactive losses in the line inductance are exactly offset by reactive power supplied by the shunt capacitance or $I^2\omega L = E^2/\omega C$.

SIL is a useful measure of transmission line capability even for practical lines with resistance, as it indicates a loading where the line's reactive requirements are small. For power transfer significantly above SIL, shunt capacitors may be needed to minimize voltage drop along the line, while for transfer significantly below SIL, shunt reactors may be needed.

SILs for typical transmission lines are given in Table 14-5. Cables normally have current ratings (ampacity) considerably below SIL, while overhead line current ratings may be either greater than or less than SIL. Figure 14-4 presents illustrative overhead line loadability as a function of line length and SIL.

**TABLE 14-5** SIL of Typical Transmission Lines

| System kV | $Z_c$, ohms | SIL, MW |
|---|---|---|
| | Overhead lines | |
| 230 | 367 | 144 |
| 345 | 300 | 400 |
| 500 | 285 | 880 |
| 765 | 280 | 2090 |
| 1200 | 250 | 5760 |
| | Cables | |
| 230 | 38 | 1390 |
| 345 | 25 | 4760 |

Although Fig. 14-4 is only illustrative of loading limits, it is a useful estimating tool. Long lines tend to be stability-limited and have a lower loading limit than shorter lines which tend to be voltage-drop- or conductor-ampacity-limited.

## Electrical Environmental Effects

*11. Corona and Field Effects.* There are two categories of electrical environmental effects of power transmission lines. Corona effects are those caused by electrical stresses at the conductor surface which result in air ionization ("corona") and include radio, television, and audible noise. Field effects are those caused by induction to objects in proximity to the line. While the generic term is electromagnetic effects, within the electric power industry the fields are divided into two types: electric field effects and magnetic field effects. Electric fields, related to the voltage of the line, are the primary cause of induction to vehicles, buildings, and objects of comparable size. Magnetic fields, related to the currents in the line, are the primary cause of induction to long objects, such as fences and pipelines.

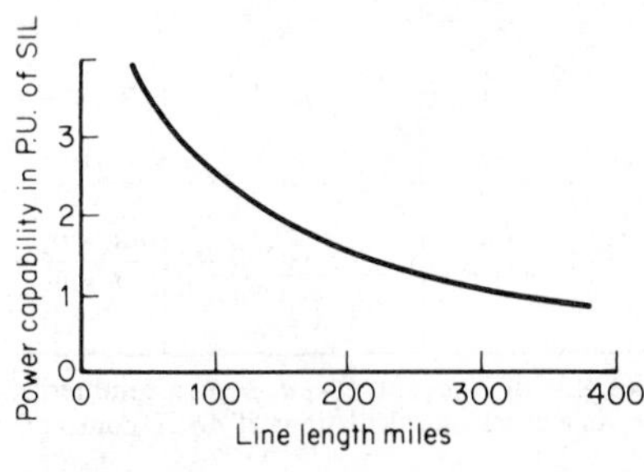


**FIG. 14-4** Overhead line loading in terms of SIL.

***12. Assessment Criteria.*** In an electrical environmental analysis, it is important to determine the proper criteria for assessment of the impact. For example, the audible noise criterion in a commercial or industrial area would be inappropriate in a quiet residential neighborhood.[5] Likewise, ground-level electric field criteria on a parking lot would be different from that in terrain inaccessible by motor vehicles. For audible noise, the only concern is annoyance, but for electric fields, safety, annoyance, and perception levels all may have to be considered.

Probability of exposure is also an important criterion. The impact of radio noise in arid locations is different from places with considerable rainfall. Since different people have different perception and annoyance thresholds, statistical evaluations are necessary, recognizing that some percentage of people will find a generally accepted noise level annoying. Because of the combination of worst-case events which are normally assumed in an electrical environmental analysis, the overall probability of annoyance is usually considerably smaller than initially presumed.

A predictive model is necessary to calculate the expected effect. Depending on the specific effect, it may be an empirical formula or may be quite sophisticated. However, it is only by calculating the effect and comparing it with specified criteria that the overall impact can be assessed. This is illustrated by Fig. 14-5,[6] which is a flowchart of the analysis procedure for an example case of electric-field-induced shock.

***13. Audible Noise.*** Corona-produced audible noise during foul weather, particularly during or following rain, can be an important design parameter for high-voltage ac transmission lines. Audible noise has two components, a random noise component and a low-frequency hum, each produced by different physical mechanisms. While the hum component is closely correlated with corona loss on the line, the random noise is not. Of these two, the most frequent cause of annoyance is the random noise, and it is this which is calculated and compared with acceptance criteria.

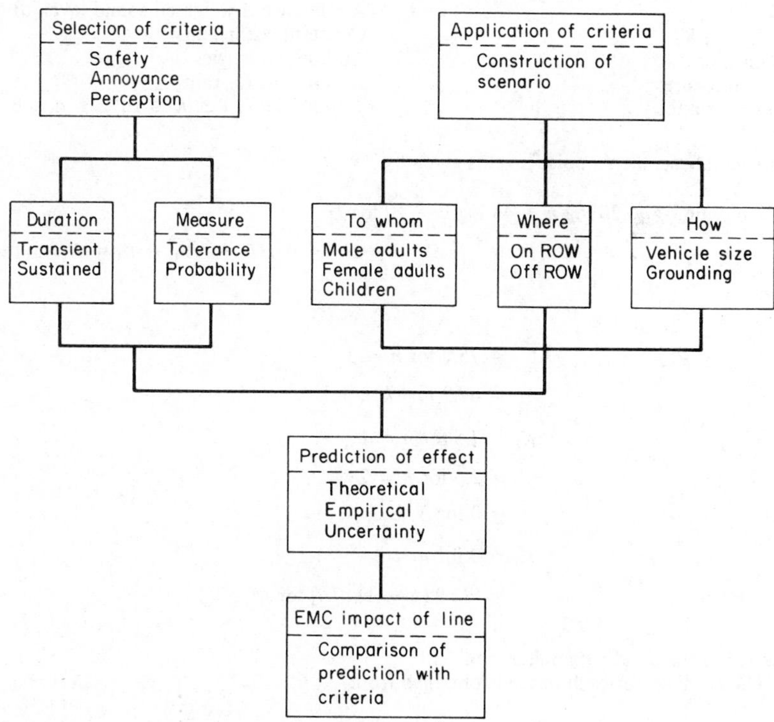


**FIG. 14-5** Factors affecting transmission line EMC for shock effects.

Analyses to predict levels of audible noise consider $A$-weighted sound level, [dB($A$)] during rain including:

$L_{50}$, which is the level exceeded 50% of the time during rain (considering all rain storms over a period of time, usually 1 year)

$L_5$, which is the level exceeded 5% of the time during rain

Average, which is the average level of noise expected during rain. (This is usually close to the $L_{50}$ value and is sometimes called "wet-conductor" noise.)

Heavy rain, which is the level expected during heavy rain. (This usually is representative of laboratory artificial rain tests but is assumed representative of the $L_5$ level.)

Reference 7 compares audible noise formulas which have been developed throughout the world. One formula for both $L_5$ and $L_{50}$ values was developed at Project UHV and is given by:

| | |
|---|---|
| $g$ | Average-maximum surface gradient of conductor or conductor bundle, kV/cm |
| $n$ | Number of subconductors in a phase (or pole) bundle |
| $d$ | Diameter of subconductors, cm |
| $D$ | Distance from line to point at which noise level is to be calculated, m |
| SL | $A$-weighted sound level of the noise produced by the line, dB($A$) |
| AN | $A$-weighted sound level of the noise produced by one phase of the line, dB($A$) |
| $AN_0$ | A reference $A$-weighted sound level, dB($A$) |
| $K_1, K_2, K_3, K_4$ | Constant coefficients |
| Application: | All line geometries |
| Noise measure: | $L_5$ rain and $L_{50}$ rain |
| Range of validity: | 230–1500 kV, $1 \leq n \leq 16$, $2 \leq d \leq 6$ |

For each phase, the $L_5$ noise level is given by

$$AN_5 = -665/g + 20 \log n + 44 \log d - 10 \log D - 0.02D + AN_0 + K_1 + K_2 \tag{14-27}$$

with

$$\begin{aligned} AN_0 &= 75.2 \text{ for } n < 3 \\ &= 67.9 \text{ for } N \geq 3 \\ K_1 &= 7.5 \text{ for } n = 1 \\ &= 2.6 \text{ for } n = 2 \\ &= 0 \text{ for } N \geq 3 \\ K_2 &= 0 \text{ for } n < 3 \\ &= [22.9\,(n - 1)\,d/B] \text{ for } n \geq 3 \end{aligned}$$

where $B$ is the bundle diameter, cm.

The $L_{50}$ level for each phase is obtained from

$$AN_{50} = AN_5 - \Delta A \tag{14-28}$$

where $\Delta A = 14.2\, g_c/g - 8.2 \qquad \text{for } n < 3$

$= 14.2\, g_c/g - 10.4 - 8\,[(n-1)\, d/B] \qquad \text{for } n \geq 3$

and $g_c = 24.4(d^{-0.24}) \qquad \text{for n} \leq 8$

$= 24.4(d^{-0.24}) - 0.25(n-8) \qquad \text{for } n > 8$

$$\text{SL} = 10 \log \sum_{i=1}^{P} 10^{\text{AN}i/10} \tag{14-29}$$

Figure 14-6 illustrates a typical presentation of audible noise calculations. The profile, in this case for a representative 500-kV line and wet conductors, quantifies the level of noise in dB($A$) greater than 0.002 μbar as a function of distance from the centerline of the structure. From this method of presentation, analysis of maximum levels as well as effect on width of right-of-way can be analyzed. Similarly, design variables such as conductor size, spacing, and configuration; height of conductors; weather variations; etc., can be considered.

Figure 14-7[3,8] quantifies experience with transmission line audible noise complaints. These mostly occur during wet-conductor conditions and low ambient noise, such as after rain or during fog. During heavy rain conditions, the noise of the rain masks the line noise. Other factors during heavy rain, such as closed windows, combine to make this condition less likely to result in complaints even though the noise is louder. In the absence of local noise regulations, comparison of calculated $L_{50}$ or average audible noise with Fig. 14-7 gives a reasonable preliminary evaluation of the possibility of audible noise annoyance. When measurements are to be taken to confirm ambient noise or line noise, care must be taken to follow proper procedures.[9]

***14. Radio and Television Noise.*** Electromagnetic interference from overhead power lines is caused by two phenomena: complete electrical discharges across small gaps (microsparks) and partial electrical discharges (corona). Gap-type sources occur at insulators, line hardware, and defective equipment and are a construction and maintenance problem rather than a design consideration. They are responsible for about 90% of noise complaints and can be located and eliminated as they occur.[10] Conductor and hardware corona is considered

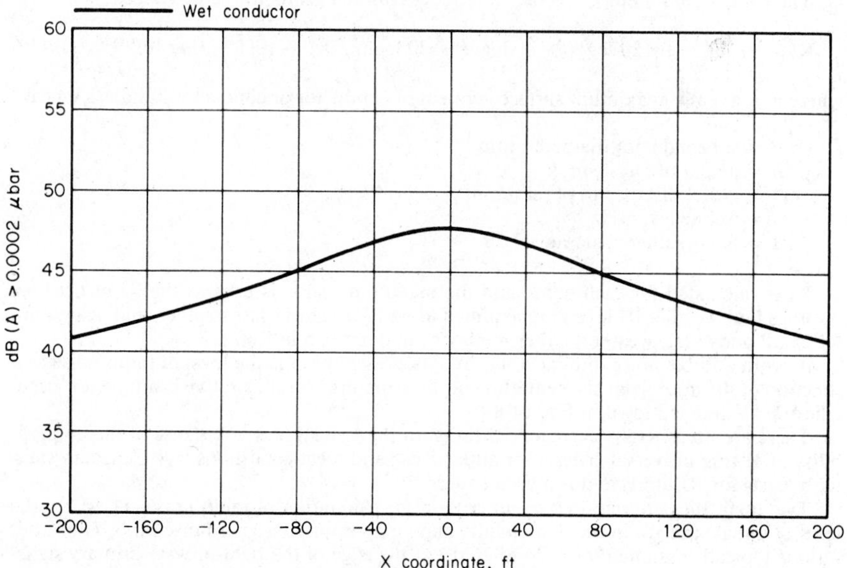


**FIG. 14-6** Audible noise profile for transmission line.

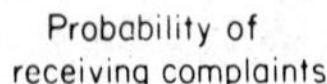


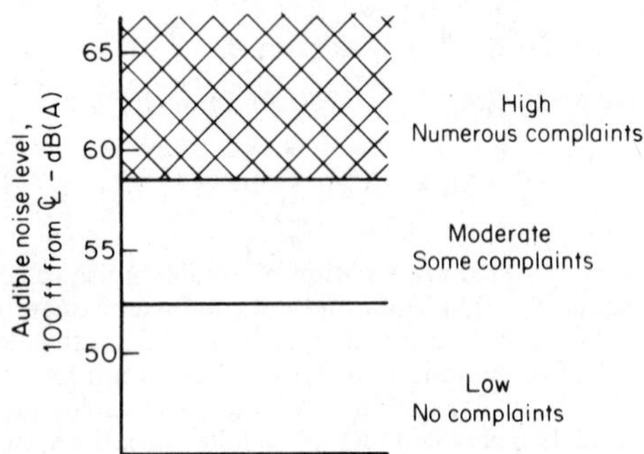


**FIG. 14-7** Audible noise complaint guidelines.

during the design phase. On a properly designed line, conductor corona noise rarely results in complaints except perhaps in weak signal fringe areas.

The specification of "corona-free" hardware is important to eliminate that source of electromagnetic interference and is especially important as lines are constructed with closer spacings and resulting higher electric fields on the hardware. Conductor clamps and other fittings which were formerly acceptable at traditional phase spacings may not be adequate for compact lines.

For ac lines, radio and television noise are functions of the weather. Fair-weather noise may be significant and varies with the season, wind velocity, and barometric pressure.

Two families of computation methods are available for radio noise: those based on conductor laboratory tests and analytical propagation theory (semianalytical methods) and those based on an empirical formula using data from long-term tests on operating lines (comparative methods).

The comparison method[11] is useful for conventional geometries and designs:

$$RI = -150.4 + 120 \log g + 40 \log d + 20 \log (h/D^2) + 10[1 - (\log 10f)^2] \quad (14\text{-}30)$$

where $g$ = average maximum surface gradient of conductor or conductor-bundle, kV peak/cm
$d$ = subconductor diameter, mm
$h$ = height of phase, m
$D$ = radial distance to observer, m
$f$ = frequency, mHz
RI = fair-weather radio noise, dB

RI is calculated for each phase and the maximum value is used as the RI of the line. Average foul-weather RI levels are assumed to be 17 dB above fair weather, and heavy-rain RI 24 dB above fair weather. Other methods are described in Ref. 3.

As with audible noise, the most useful data presentation is the level of radio noise as a function of distance from the centerline of the structure. An illustrative example for a specific 500 kV line is shown in Fig. 14-8.

There are no generally accepted RI limits in the United States, because of the impossibility of setting universal criteria for all land use and local conditions.[12] A Canadian standard exists for RI limits and is a useful guide.[13]

Two quantities are required to set criteria for evaluation of radio noise. These are the level of signal strength in the line vicinity and an appropriate signal/noise ratio. This latter ratio is typically assumed to be 24 to 26 dB at the edge of the right-of-way. Primary signal strengths may be 54 dB above 1 $\mu$V (0.5 mV/m) in rural areas to 88 dB or more in cities.

Prediction of television noise is not as advanced as that of radio noise, primarily because

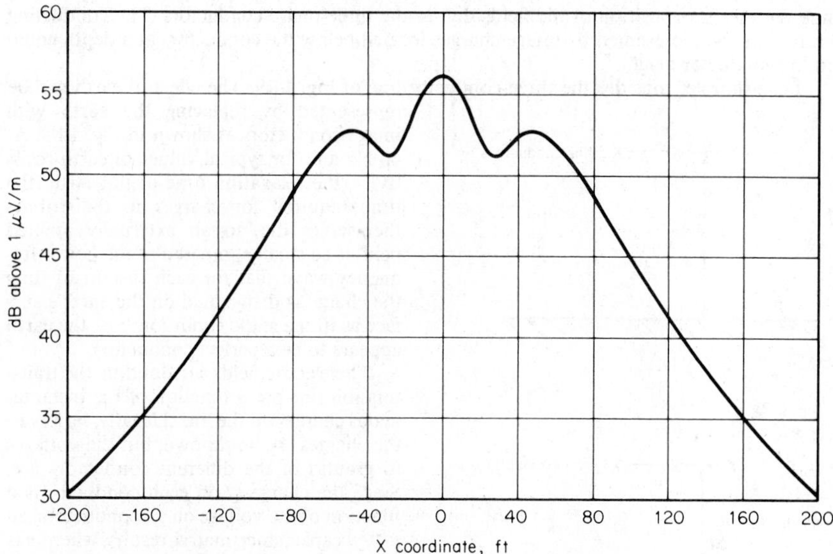


**FIG. 14-8** Radio noise profile for transmission line.

of the limited number of actual cases of conductor corona television interference. As with radio noise, most television interference complaints result from microsparks which can be located and eliminated as they occur. These are not generally a design consideration. In the few cases where corona-caused television noise has occurred in foul weather, it has often been possible to remedy the situation by an improvement in the receiving antennas rather than changes to the transmission line design. References 3 and 14 contain recent work on prediction and evaluation of TVI.

***15. Gaseous Oxidants.*** Gaseous oxidants can be produced by corona activity in air and, in sufficient concentrations, may produce adverse effects on flora and fauna. The most important oxidants are ozone ($O_3$) and oxides of nitrogen (mainly NO and $NO_2$), where ozone is the major constituent.

Federal standards limit photochemical oxidants to 0.12 parts per million for a maximum of one-hour concentration not to be exceeded more than once per year. Some states have more restrictive regulation; for example, the Minnesota Pollution Control Agency standards are for 0.07 ppm by volume (130 $\mu g/m^3$). Ozone can be detected by smell at minimum concentrations of 0.01 to 0.15 ppm.

Analytical studies and field measurements have been conducted on both operating and test lines[15–22]. The highest calculated value for 1 mi/h wind parallel to the line was 0.019 ppm maximum ground-level concentration. Measurements have indicated that transmission line contribution to gaseous oxidants cannot be detected within statistical limits of significance and accuracy. With instrumentation capable of detecting 0.002 ppm, the transmission line contribution was indistinguishable from ambient.

Thus, gaseous oxidants are not a concern with respect to electric power transmission lines.

***16. Ground-Level Electric Fields.*** Ground-level electric field effects of overhead power transmission lines relate to the possibility of exposure to electric discharges from objects in the field of the line. These may be steady currents or spark discharges. Other areas which have received attention are the possibility of fuel ignition and interference with wearers of prosthetic devices (e.g., pacemakers).[23]

It is appropriate to consider unlikely conditions when setting and applying electric field safety criteria because of possible consequences, thus statistical considerations are necessary. Annoyance criteria need not be as stringent and mitigating factors can be considered.

*Electric Field Calculations.* The resultant electric fields in proximity to a transmission line are the superposition of the fields due to the three-phase conductors. The conducting earth must be represented by image charges located below the conductors at a depth equal to the conductor height.

For example, consider the three-conductor line of Fig. 14-9. The effect of earth can be represented by replacing the earth with image conductors as shown in Fig. 14-9. At 60 Hz and for typical values of earth resistivity, the relaxation time of the earth (the time required for charges to redistribute themselves due to an externally applied field) is so small compared to the power frequency wave that for each instant of time the charge is distributed on the earth's surface as in the static condition (i.e., the earth appears to be a perfect conductor).

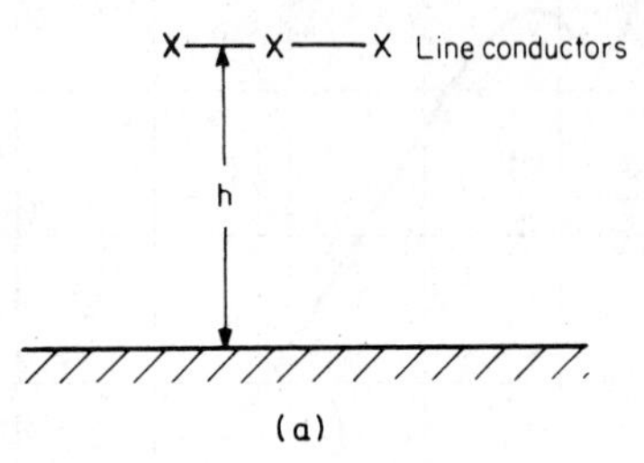


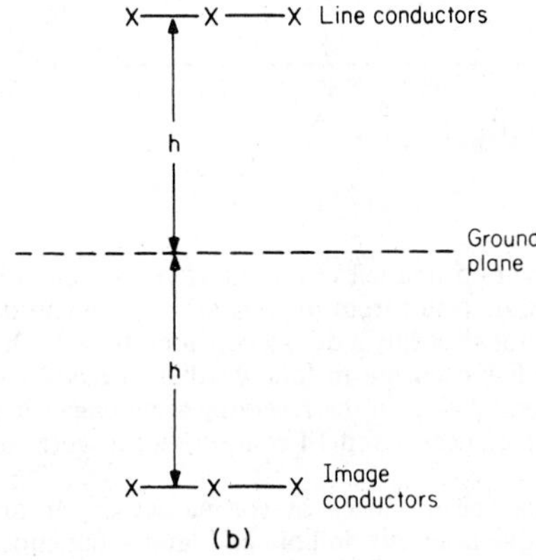


**FIG. 14-9** Representation of conducting earth: (*a*) earth; (*b*) image.

The electric fields surrounding the transmission line are a function of the instantaneous charges on the line. Usually, however, the charges are not known, but the voltages to ground of the different conductors are. Since the charge $Q$ on each conductor is a function of the voltage on all conductors, an $n \times n$ capacitance matrix results, where $n$ is the number of conductors, according to the formula:

$$[Q] = [C]\,[V] \tag{14-31}$$

which, for a three-conductor configuration (ignoring shield wires), is

$$Q_1 = C_{11}V_1 + C_{12}V_2 + C_{13}V_3 \tag{14-32}$$

$$Q_2 = C_{21}V_1 + C_{22}V_2 + C_{23}V_3 \tag{14-33}$$

$$Q_3 = C_{31}V_1 + C_{32}V_2 + C_{33}V_3 \tag{14-34}$$

The off-diagonal (mutual) capacitance terms significantly affect the final result. The individual terms of the capacitance matrix are computed by:

$$C_{nm} = \frac{Q_n}{V_m}\bigg|\ \text{all other voltages} = 0 \tag{14-35}$$

where $n$ and $m$ are conductors.

The potential coefficient matrix is, however, more amenable to computation and is defined by:

$$[V] = [P]\,[Q] \tag{14-36}$$

whose individual terms are given by:

$$P_{nm} = \frac{V_n}{Q_m}\bigg|\ \text{all other charges} = 0 \tag{14-37}$$

This is an open-circuit matrix where the individual terms can be computed by assuming a charge at one conductor and calculating the voltage at the prescribed location assuming all the other conductors nonexistent (open-circuited). For a single conductor of radius $r$ and a height $h$ above the earth, the self-potential coefficient is given by:

$$P_{nn} = \frac{1}{2\pi\varepsilon}\ln\frac{2h}{r} \tag{14-38}$$

For two conductors $n$ and $m$ where $d_{nm}$ is the distance between them, and $d_{nm'}$ is the distance between conductor $n$ and the image of conductor $m$, the mutual potential coefficient is given by:

$$P_{nm} = \frac{1}{2\pi\varepsilon} \ln \frac{d_{nm'}}{d_{nm}} \tag{14-39}$$

This potential coefficient matrix can be calculated and inverted to yield the capacitance matrix:

$$[C] = [P]^{-1} \tag{14-40}$$

This capacitance matrix allows the calculation of the charges on the individual conductors for the given initial voltage distribution according to Eqs. (14-32) through (14-34). Once these charges are obtained, the desired electric fields can be determined.

For the single conductor and observer location of Fig. 14-10, the ground-level electric

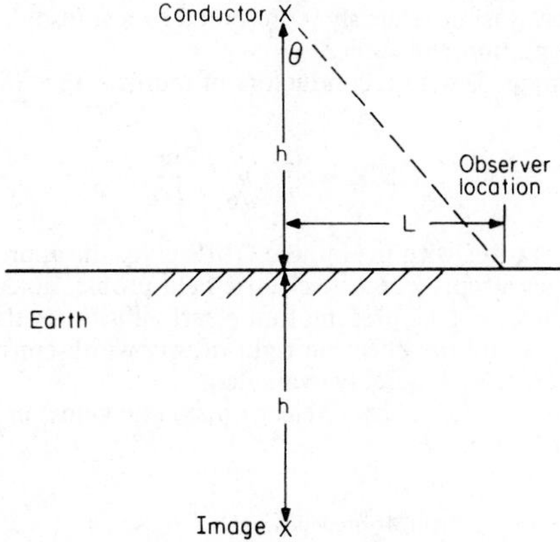


**FIG. 14-10** Single conductor.

field is determined from

$$E = \frac{Q_l}{2\pi\varepsilon r} \tag{14-41}$$

The distance from the conductor to the observer is

$$r = \sqrt{h^2 + L^2} \tag{14-42}$$

Thus

$$E = \frac{Q_l}{2\pi\varepsilon\sqrt{h^2 + L^2}} \tag{14-43}$$

$Q$ must be determined from $[Q] = [C]\,[V]$. For a single conductor this equation reduces to:

$$Q_l = P^{-1}V = \frac{1}{(1/2\pi\varepsilon)\ln(2h/r)}\,V \tag{14-44}$$

For a multiconductor configuration, $Q$ would come from the full matrix calculation.

$E$ is radially directed from the line charge. The vertical component is

$$|E| \cos\theta = \frac{Q_l}{2\pi\varepsilon\sqrt{h^2 + L^2}} \frac{h}{\sqrt{h^2 + L^2}} = \frac{Q_l}{2\pi\varepsilon} \frac{h}{h^2 + L^2} \tag{14-45}$$

The vertical component of the electric field at ground level because of the image is equal to the field from the conductor, since the image is the geometric mirror image and has the opposite sign charge. Thus, the total ground-level field is given by

$$E = \frac{Q_l}{\pi\varepsilon} \frac{h}{h^2 + L^2} \tag{14-46}$$

At ground level, the horizontal components of the electric fields of the conductor and its image cancel and the resultant field is purely vertical.

For a three-phase line, the fields of the three conductors and their images are computed separately and added.

For fields extremely close to the line conductors, care must be taken to represent the local efforts properly. For example, the surface field around the conductor is not uniform. For a bundled conductor, it is more nearly represented by a sinusoid. Farther from the conductors, a GMR representation will suffice.

For a bundle of diameter $D$ with $n$ conductors of radius $r$, the GMR is given by

$$\text{GMR} = \frac{D}{2} \sqrt[n]{\frac{2nr}{D}} \tag{14-47}$$

Replacing the conductor radius with the bundle GMR gives the appropriate representation.

Figure 14-11 illustrates a representative electric field profile, in kV rms per meter, from the centerline of the structure. This presentation clearly illustrates the maximum field, the location of the maximum, and the effect on right-of-way width considerations. Sensitivity to various parameters can also be quickly evaluated.

*Criteria for Evaluation.* The effects of electromagnetic fields on humans is due to dis-

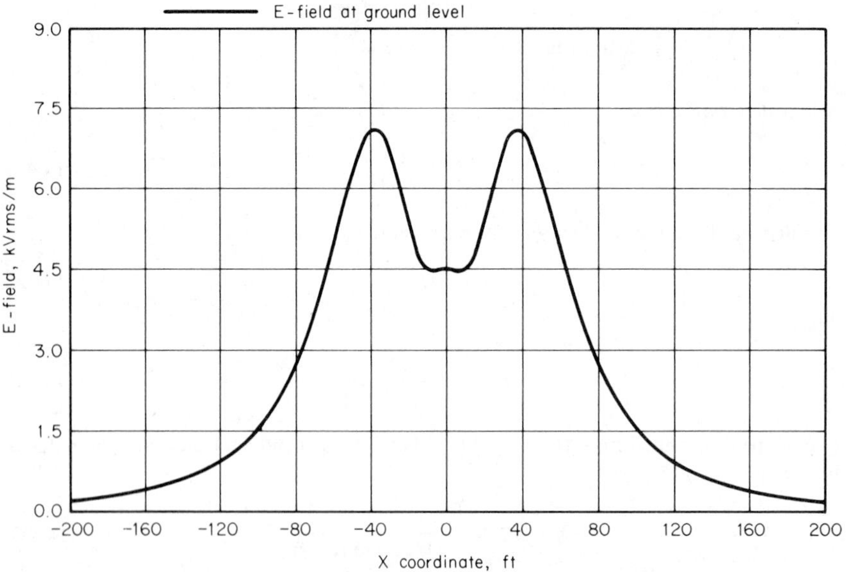


**FIG. 14-11** Electric field profile for transmission line.

charges from objects insulated from ground; typically vehicles, buildings, and fences which become electrically charged by induction from the line. Table 14-6 summarizes effects on humans, ranging from no perception through severe shock and possible ventricular fibrillation.[24]

Criteria for spark discharges are expressed in terms of stored charge or stored energy on the charged object. Levels for perception in adult males are of the order of 0.12 mJ, while experience indicates that approximately 2 mJ results in an annoying spark. Safety is seldom of concern, since approximately 25 J is required for injury, a value beyond that expected on objects beneath transmission lines.

Deno's work, using test data, relates short-circuit current to the undisturbed electric field

**TABLE 14-6** Threshold Levels for 60-Hz Contact Currents

| rms current, mA | Threshold reaction and/or sensation |
|---|---|
| | Perception |
| 0.09 | Touch perception for 1% of women |
| 0.13 | Touch perception for 1% of men |
| 0.24 | Touch perception for 50% of women |
| 0.33 | Grip perception for 1% of women |
| 0.36 | Touch perception for 50% of men |
| 0.49 | Grip perception for 1% of men |
| 0.73 | Grip perception for 50% of women |
| 1.10 | Grip perception for 50% of men |
| | Startle |
| 2.2 | Estimated borderline hazardous reaction, 50% probability for women (arm contact) |
| 3.2 | Estimated borderline hazardous reaction, 50% probability for women (pinched contacts) |
| | Let-go |
| 4.5 | Estimated let-go for 0.5% of children |
| 6.0 | Let-go for 0.5% of women |
| 9.0 | Let-go for 0.5% of men |
| 10.5 | Let-go for 50% of women |
| 16.0 | Let-go for 50% of men |
| | Respiratory tetanus |
| 15 | Breathing difficult for 50% of women |
| 23 | Breathing difficult for 50% of men |
| | Fibrillation |
| 35 | Estimated 3-s fibrillating current for 0.5% of 20-kg (44-lb) children |
| 100 | Estimated 3-s fibrillating current for 0.5% of 70-kg (150-lb) adults |
| | Established standards |
| 0.50 | ANSI standard for maximum leakage (portable appliance) |
| 0.75 | ANSI standard for maximum leakage (installed appliance) |
| 5.0 | NESC recommended limit for induced current under transmission line |

for objects insulated from ground.[23] Initial calculations assume the worst possible combination of circumstances: no leakage path to ground exists for the object, complete grounding of the person involved, steady contact, and orientation of the vehicle parallel to the line. Table 14-7 lists sample criteria and electric fields needed to meet them for three sample vehicles.

**TABLE 14-7** Limiting Electric Field for Given Criteria, kV/m

| | | Sample vehicles | | |
|---|---|---|---|---|
| | | Autos, pickups | Farm vehicles | Buses, trailer trucks |
| | Sample criteria | A | B | C |
| Safety | 5 mA | 22.32 | 10.86 | 6.33 |
| | 25 J | 259.00 | 159.00 | 106.50 |
| Annoyance | 2 mA | 8.92 | 4.35 | 2.50 |
| | 2 mJ | 2.37 | 1.41 | 0.95 |
| Perception | 1.1 mA | 4.91 | 2.39 | 1.39 |
| | 0.12 mJ | 0.58 | 0.35 | 0.23 |

High voltages may develop due to electric field coupling, but the available short-circuit current is small (i.e., high-impedance source); thus calculations are based on a Norton equivalent and the short-circuit current. A relatively high resistance ground is sufficient to reduce electric-field-coupled voltage.

Table 14-8 lists maximum electric fields on the right of way under lines of different voltage classes. The fields attenuate rapidly with distance from the line and are usually much lower at the right-of-way edge.

**TABLE 14-8** Likely Range of Maximum Vertical Electric Field for Various Voltage Transmission Lines

| Line voltage, kV | Near-ground vertical electric field, kV/m |
|---|---|
| 765 | 8–13 |
| 500 | 5–9 |
| 345 | 4–6 |
| 230 | 2–3.5 |
| 161 | 2–3 |
| 138 | 2–3 |
| 115 | 1–2 |
| 69 | 1–1.5 |

***17. Fuel Ignition.*** Theoretical calculations indicate that if several unlikely conditions exist simultaneously, a spark could release sufficient energy to ignite gasoline vapors. These conditions include a perfectly grounded person refueling a car perfectly insulated from ground with a metal can while the car is parked directly under a line. The spark would have to occur in the precise location of optimum fuel-air mixture. Recent research[3,25] confirms the low probability of accidental fuel ignition under actual conditions.

No confirmed cases of accidental ignition under transmission lines exist, confirming the low probability of these factors occurring simultaneously. Because of the consequences of a gasoline fire, some electric utilities advise that gasoline-fueled vehicles not be refueled near a line of 500 kV or above. If refueling were necessary, the vehicle could be grounded or the can connected to the vehicle to prevent sparks.

***18. Ground-Level Magnetic Fields.*** Magnetic field coupling affects objects which parallel the line for a distance, such as fences and pipelines, and is generally negligible for vehicle- or building-sized objects. As opposed to electric field coupling, magnetic field coupling is a low-voltage, low-impedance source with relatively high short-circuit currents. Single grounds are ineffective in preventing magnetically coupled voltages and multiple low-resistance grounds are needed. The resistance of the person touching a fence or pipeline is the dominant current-limiting impedance in the equivalent electrical circuit.[26] Calculations are based on a "longitudinal-electric-field" approach and are described in Refs. 27 to 29.

A consideration in the calculation of magnetic fields which is different from the electric field calculation concerns the images. A perfectly conducting earth can be assumed for the electric field problem, even for realistic values of earth resistivity. The assumption of a transmission line in free space (no earth at all) gives a closer approximation to the ground-level magnetic fields than does the assumption of perfectly conducting earth. This effect is frequently treated by use of an image conductor located at a greater depth in the earth than the conductors are above the earth. Distances of several hundred meters are commonly used for this image depth, according to the relation $D = 660\sqrt{\rho/f}$ meters where $\rho$ is the earth resistivity in ohm-meters and $f$ is the frequency. Magnetic field calculations are given in Ref. 9, including the use of Carson's terms to evaluate the effects of imperfectly conducting earth.

It is normally adequate to consider conductors in free space without images. For the conductor of Fig. 14-8 without its image:

$$B = \frac{\mu I}{2\pi r} = \frac{\mu I}{2\pi\sqrt{h^2 + L^2}} \tag{14-48}$$

This is then separated into vertical and horizontal components by multiplying by $\sin\theta$ and $\cos\theta$. In general, both components must be retained. For a three-phase line, all conductors must be computed. Horizontal and vertical components of $B$ from the three conductors must then be combined individually as phasors, considering the angles of the different currents. The combined horizontal and vertical components in general have different angles, causing their resultant to trace an ellipse in time.

In this same manner, image currents at some assumed depth can be computed and their fields included. The use of matrix calculations allows inclusion of ground wires and bundled conductors as is the case of electric fields.

With both electric and magnetic fields it is essential to follow proper measurement procedures[30] for comparison with calculations.

## Line Insulation

***19. Requirements.*** The electrical operating performance of a transmission line depends primarily on the insulation. An insulator not only must have sufficient mechanical strength to support the greatest loads of ice and wind that may be reasonably expected, with an ample margin, but must be so designed as to withstand severe mechanical abuse, lightning, and power arcs without mechanically failing. It must prevent a flashover for practically any power-frequency operating condition and many transient voltage conditions, under any conditions of humidity, temperature, rain, or snow, and with such accumulations of dirt, salt, and other contaminants which are not periodically washed off by rains.[31]

***20. Insulator Materials.*** The majority of present insulators are made of glazed porcelain. Porcelain is a ceramic product obtained by the high-temperature vitrification of clay, finely ground feldspar, and silica. Insulators of high-grade electrical porcelain of the proper chemical composition free from laminations, holes, and cooling stresses have been available for many years.

The insulator glaze seals the porcelain surface and is usually dark brown, but other colors such as gray and blue are used. Porcelain insulators for transmission may be disks, posts, or long-rod types.

Porcelain insulators have been used at all transmission line voltages from 115 through 765 kV and up to UHV and, if correctly manufactured and applied, have high reliability.

A typical porcelain disk insulator is shown in Fig. 14-12.

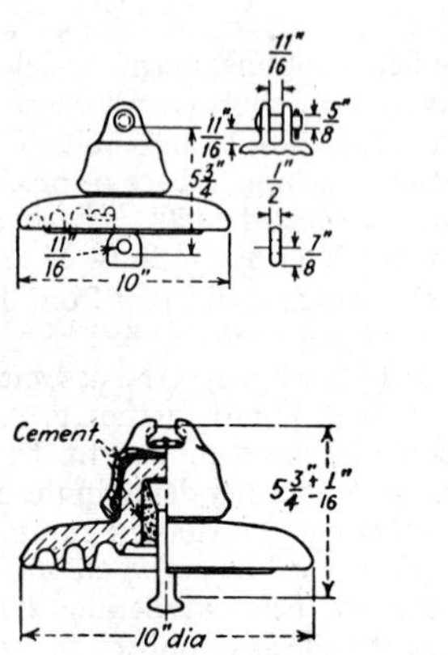


**FIG. 14-12** Typical porcelain disk insulator. (a) Clevis type; (b) Ball-and-socket type. *(Locke Insulators Inc.)*

Glass insulators have been used on a significant proportion of transmission lines. These are made from toughened glass, and are usually clear and colorless or light green. For transmission voltages they are available only as disk types. Most glass disk insulators are designed so the skirts shatter completely when damaged, but without mechanically releasing the conductor. This provides a simple method of inspection.

Synthetic insulators, originally pioneered by the General Electric Company in 1963 for high-voltage transmission lines,[32] and more recently introduced by several manufacturers, are finding increasing acceptance. Most consist of a fiberglass rod covered by weather sheds or skirts of polymer (silicon rubber, polytetrafluoroethylene, cycloaliphatic resin, etc.)[33] as shown in Fig. 14-13. Other types include a cast polymer concrete called Polysil R[34] and a coreless type with alternating metal and insulating sections.[35]

Improvements in design and manufacture in recent years have made synthetic insulators increasingly attractive since their strength-to-weight ratio is significantly higher than that of porcelain and can result in reduced tower costs, especially on EHV and UHV transmission lines.

These insulators are usually manufactured as long-rod or post types. The light weight of most designs and resistance to damage aids construction. In addition, their performance under contaminated conditions may be significantly better than that of porcelain.[36] An IEEE application guide will shortly be available.

Use of synthetic insulators on transmission lines is relatively recent and a few questions

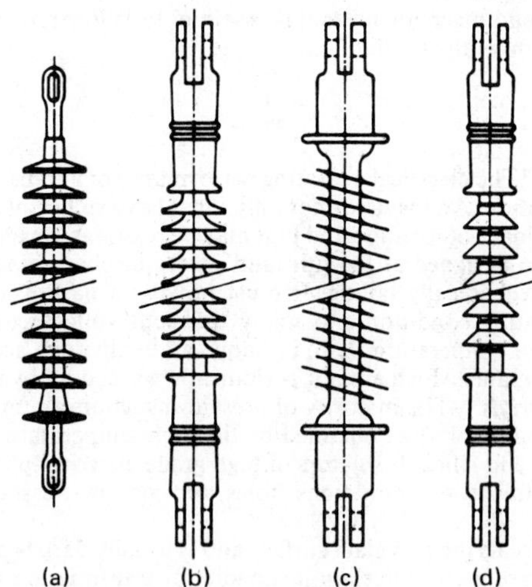


**FIG. 14-13** Typical nonceramic insulators.

are still under study, in particular the lifetime behavior of insulating shed materials under contaminated conditions. It has been found necessary to use grading rings on some types at higher voltages to prevent damage to the sheds, and a very small number of insulators have experienced "brittle fractures," in which the fiberglass core breaks close to an end fitting. Despite these problems it appears that reliable synthetic insulators are presently available.

***21. Insulator Design.*** Transmission insulators may be strings of disks (either cup and pin or ball and socket), long-rods, or line posts. Posts are only infrequently applied above 230 kV.

Present suspension insulators conform to ANSI Standard C29.2, and standards have been established for 15,000, 25,000, 36,000 and 50,000-lb ratings. It is common practice to use a factor of safety of 2 for the maximum stress applied to porcelain or glass insulators. For fiberglass-core insulators it is more common for the manufacturer to supply a recommended maximum working load.

Each manufacturer supplies catalogs which provide a physical description of the insulator's mechanical characteristics, wet and dry 60-Hz flashover strength, and positive and negative impulse ($1.2 \times 50$ $\mu$s) critical (50%) flashover strength. Switching surge performance ($250 \times 3000$ $\mu$s) is usually not supplied. In clean conditions most insulators of equivalent dimensions have very similar performance.

Suspension insulator strings, i.e., insulators used to support the conductor weight at a suspension or tangent structure, may be in I (vertical) or V configurations. The V configuration is used to prevent conductor movement and resultant clearance reductions at the structure. At dead-end or tension structures the insulators must also support the conductor tension, and it is not uncommon for these tension strings to be given a slightly higher flashover strength (e.g., by adding disks) to reduce the likelihood of a flashover that might lead to insulator string mechanical failure. Two or more strings of insulators in parallel can be used on suspension and tension strings to provide higher mechanical strength if required.

The electrical strength of line insulation may be determined by power frequency, switching surge, or lightning performance requirements. At different line voltages, different parameters tend to dominate. Table 14-9 shows typical line insulation levels and the controlling parameter. In compacted or uprated designs, considerably fewer insulators than these have been successfully used.[37,38]

Detailed descriptions of insulation design for electrical performance for different conditions, line voltages, and line types are available[3,39–41] from a number of studies.

***22. Insulator Standards.*** NEMA Publication, "High Voltage Insulator Standards," and AIEE Standard 41 have been combined in ANSI C29.1 through C29.9. Standard C29.1 covers all electrical and mechanical tests for all types of insulators. The standards for the various insulators covering flashover voltages; wet, dry, and impulse; radio influence; leakage distance; standard dimensions; and mechanical-strength characteristics are as follows: C29.2, suspension; C29.3, spool; C29.4, strain; C29.5, low- and medium-voltage pin; C29.6, high-voltage pin; C29.7, high-voltage line post; C29.8, apparatus pin; C29.9, apparatus post. These standards should be consulted when specifying or purchasing insulators.

***23. Line Insulation Design***

*Power-Frequency Design.* The criteria for power-frequency design is usually that flashover shall not occur for normal operating conditions, including reduced clearances to the

**TABLE 14-9** Typical Line Insulation

| Line voltage, kV | No. of standard disks | Controlling parameter (typical) |
|---|---|---|
| 115 | 7–9 | Lightning or contamination |
| 138 | 7–10 | Lightning or contamination |
| 230 | 11–12 | Lightning or contamination |
| 345 | 16–18 | Lightning, switching surge, or contamination |
| 500 | 24–26 | Switching surge or contamination |
| 765 | 30–37 | Switching surge or contamination |

structure from high wind. A typical wind-design limit is the 50- or 100-year return period wind, i.e., a wind velocity which occurs only once in 50 or 100 years. This velocity is obtained from local wind records and may be typically 80 to 100 mi/h. Maximum operating voltages are defined by ANSI C84 and C92 standards and are 5 or 10% above the nominal value.

In clean conditions, power-frequency voltage is not a controlling parameter for insulator design (as distinct from air gap clearance). However, even in quite lightly contaminated conditions it may become so.

Design for contamination is usually expressed as inches of creepage per kilovolt, where the creepage distance is the length of the shortest path for a current over the insulator surface and ranges up to 2 in per kilovolt or more for heavy contamination. Standard insulator disks (10 × 5¾ in) have a typical creepage length of 11.5 in per disk. To avoid very long insulator strings for contamination, disks with additional creepage distance are made. The creepage can be extended by use of lengthened skirts and deeper grooves in the underside. Fog-type disks have up to 21.5 in of creepage per 13½ × 8 in units. A typical fog-type insulator is illustrated in Fig. 14-14.

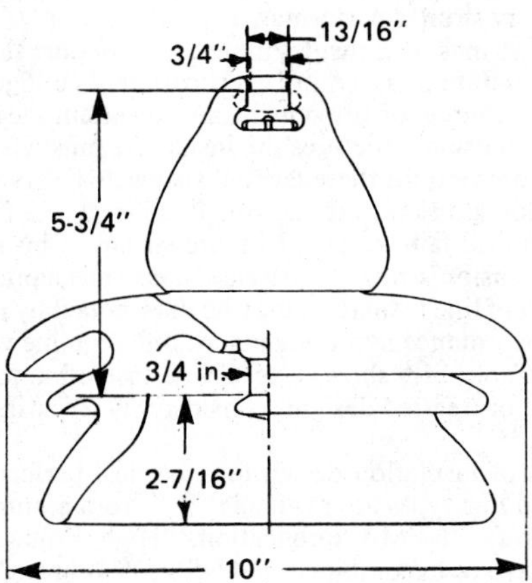


**FIG. 14-14** Typical fog-type disk insulator.

In extremely contaminated conditions, insulation with extended creepage may not be enough. In these cases insulator washing or the use of a silicone grease coating (replaced at regular intervals) may be used.

Table 14-10 provides a simplified indication of creepage distance as a function of contamination,[39] and Fig. 14-15 shows guidelines from the IEEE application guide.[40]

For nonceramic insulation the same approach is used, except that subject to manufacturer's recommendations, a reduction in creepage distance up to 30% may be possible. This is due to the physical behavior of the nonceramic insulating material in moist conditions.

Another approach which has been used to combat contamination effects is the semiconductive glaze insulator. The semiconducting glaze allows a small but definite power-frequency current to flow over the surface. The insulator does not improve the standard test values, such as wet and dry power-frequency flashover and short-time impulse flashover, although it may have some value under switching surge conditions.

The glaze has a surface resistivity of about 10 MΩ per square. This is achieved by special formulations of materials involving, at the present stage of development, the use of tin-antimony additive to a more normal glaze composition. The presence of this small leakage current, of the order of 1 to 2 mA for suspension insulators, but which can be several times that value for large porcelains (such as are used in high-voltage bushings) has three effects:

1. Linearization of the voltage distribution over the insulator or string of insulators. This aids greatly in improving the performance of the insulator with respect to corona disturbance and RIV performance, plus having some benefits under dry and clean conditions.

**TABLE 14-10** Insulation Requirements for Contamination: Provisional EHV Line Insulation Design Table for Various Contamination Conditions

Standard 5¾- by 10-in Vertical Insulator Units

| Contamination | | Equivalent amount NaCl, mg/cm² | Provisional design values | | |
|---|---|---|---|---|---|
| | | | Leakage distance, in/kV rms line-to-ground | Average kV rms | |
| Class | Types | | | Per in axial length | Per unit |
| A | Clean atmosphere—rural and forest regions. No industrial contamination | 0–0.03 | Insulation requirements not set by contamination | | |
| B | Slight atmospheric contamination. Suburbs of large industrial regions. Railways. Frequent washing rains | 0.04 | 1.04 | 2.0 | 11.5 |
| C | Moderate contamination containing soluble salts up to 5%. Furnaces, dust from metallurgical plants, mine dust, fly ash, fertilizer dust in small quantities | 0.06 | 1.31 | 1.6 | 9.1 |
| D | Severe contamination containing 15% or more of soluble salts; dust from aluminum and chemical works, cement plants, heavy agricultural fertilizing, fly ash with high salt or sulfur content | 0.12 | 1.74 | 1.2 | 6.9 |
| E | Salt precipitation—seaside regions, salt marshes | 0.30 | 2.11 | 1.0 | 5.7 |

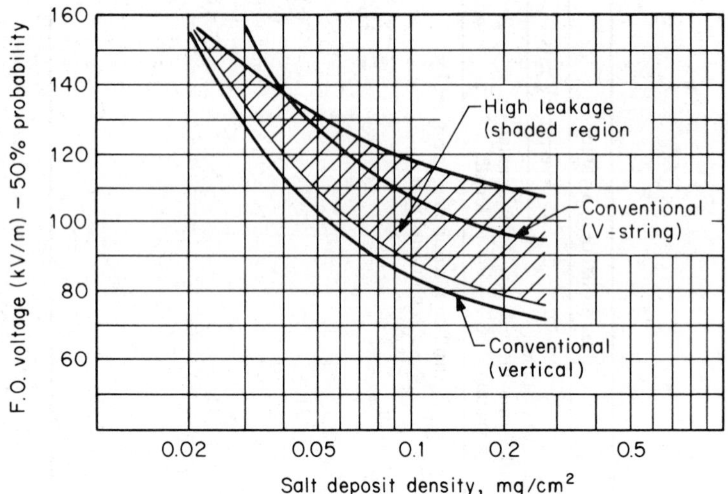


**FIG. 14-15** Power-frequency withstand voltage of contaminated suspension insulators in fog expressed in kV/m of connection length (spacing).

2. Heating of the insulator. This occurs because of the power loss associated with the leakage current flow to a temperature which is usually about 5°C over the ambient air conditions. The heating effect enables the insulator to remain dry during conditions of fog or mist. This eliminates the majority of contaminated-insulator flashovers which occur when accumulated contamination becomes damp. This damp contamination condition is the most usual cause for contaminated-insulator flashover because most contaminants are more electrically conducting when damp or wet.
3. The elimination of "dry banding," which is recognized as another major cause of flashover of standard insulators when contaminated. This occurs when the insulator has been thoroughly wetted, such as in a rain storm which wets but does not thoroughly clean the contamination from the insulator's surface. Under these conditions dry bands will form as the standard insulator dries, and arcs strike across the dry-band area. These arcs can progress until flashover of the entire insulator occurs. With a semiconducting insulator, the relatively low resistance of the glaze shunts the dry-band area as the insulator dries and prevents the striking of the small power-frequency arcs.

The improved performance possible with semiconducting insulators has been proved in the laboratory and field,[42-46] but, because of the energy losses associated with the inherent leakage current, they are not widely used.

In some severe contamination areas, the problem has been effectively attacked by the use of silicone grease coatings. The unique amoebic action of a thick layer of silicone grease on an insulating surface is such as to envelop conducting solid particles which are said to "load" up the silicone grease to the saturation point, at which time the "used" silicone grease is removed and replaced with new silicone grease. In severe contamination areas, the greasing and degreasing cycles may be required every few months; in less severe contamination areas the cycle may be a year or more, depending on experience acquired. In this manner, the time between insulator cleanings can be greatly extended, thus making for substantial savings. Once the silicone coating is used, the coatings must usually be wiped off and replaced manually, as and when necessary. Among the manufacturers of silicone grease are the General Electric Company and the Dow-Corning Corporation.

For the cleaning operation to remove contamination from the insulator surface, many contaminants such as salt deposits and water-soluble conducting liquids can be successfully removed by hot-line washing, using high-pressure water and insulated nozzles and hoses. Another method is "dry cleaning" by the use of an abrasive powder such as a limestone mixture or biodegradable plastic pellets, discharged at high pressure through hose and nozzle on the insulating surface. In many cases either hot-line washing or dry cleaning alone is sufficient to cope with the rate of accumulation encountered with the particular contaminant. An exception is substantially conducting materials, which take a chemical "set" after exposure to water, such as cement dust, some forms of gypsum, or asbestos, which often must first be manually chipped off or scrubbed off the insulating surface and then covered with silicone grease as previously described.

It should be emphasized that these problems may be very severe or even nonexistent, due to the variability of contamination exposure, which in turn depends on the chemical and electrical nature of the contaminant, prevailing wind direction, persistence of fog, smog, or other weather factors.

To monitor buildup of contaminants, some utilities collect data at the site to warn operating departments of an impending flashover, so as to promptly implement contamination-combative procedures.

*Switching Surge Design.* Operation of a circuit breaker on a transmission line can cause transient overvoltages. If the breaker is opening, this may be due to restrikes across the breaker contacts as they separate, although restriking has been nearly eliminated with present breaker technology. If the breaker is closing, the cause may be unequal voltages on each side of the breaker, including the effect of residual charge on the line from a recent deenergization. The crest magnitudes of switching surges are normally defined in per unit of maximum power-frequency-crest phase-to-ground voltage. For example, on a 138-kV line (145 kV maximum), the per unit value is 118 kV. Typical switching surges range from 1 to as high as 4 or 5 per unit, and the varying characteristics of breaker operations provide a distribution of surge magnitudes which is often modeled as a truncated gaussian distribution.

The criterion for switching surge design is usually that flashover shall not occur for most or all switching events. Several design methods have been used, including:

**a.** The maximum expected surge is determined, e.g., from a transient network analyzer (TNA) or digital study, and the line insulation is designed to withstand that surge.

**b.** Rather than the maximum surge, a surge value corresponding to a statistical level is used, typically the 2% value (i.e., the crest value determined from the statistical distribution of surge crests, such that the level will be exceeded by only 2% of all surges).

**c.** Rather than design insulation to withstand a maximum surge, a statistical approach is used to design for a low number of flashovers per switching event. Typical levels are one flashover per 100 or 1000 breaker operations. This often results in a more economical design than either of the withstand approaches above.

**d.** By modeling the statistical distribution of switching surge crests, the distribution of insulator flashover with voltage, and the statistical distribution of weather that can be obtained from local weather stations, a probabilistic design can be prepared using a relatively simple computer program based on the allowable flashover rate. Typical procedures, data, and examples for such calculations are provided in several publications.[47,48]

*Impulse Surge Design.* Impulse surges on a line are caused by lightning strokes to or near the line. At transmission insulation levels, only strokes that directly intercept the line are capable of causing flashovers.

A number of methods of calculating transmission line lightning performance have been published, and are summarized in the references to Chap. 12 of Ref. 31, together with a simplified calculation method. A computer program for this simplified calculation method is available from the IEEE WG on Transmission Line Lightning Performance, and a more sophisticated program for evaluation of multicircuit lines will shortly be available from EPRI.

It is unusual for line insulation to be determined by lightning performance. More typically, insulation is determined by other requirements and the lightning performance is then verified. If this performance is unsatisfactory, it is often more efficient to change other design parameters such as shield wires or grounding than to add insulation.

Other methods of improving lightning performance have included addition of surge arresters at relatively frequent intervals along a line, and on double-circuit lines the use of unbalanced insulation so one circuit will flash over first and protect the other. Use of arresters is presently quite costly and is rare, although new arresters may change this. Use of unbalanced insulation appears to have little, if any, advantage.

*Phase-to-Phase Insulation.* The controlling paths for flashovers on most presently installed transmission lines are phase-to-ground, since there are grounded structure components between phases. However, for some new designs, such as the Chainette,[49] and compact lines the controlling path may be phase-to-phase air gaps or even phase-to-phase insulators.

Design methods for phase-to-phase insulation are essentially the same as for phase-to-ground insulation. Until recently, there was lack of knowledge of conductor clearance at midspan under various dynamic loading conditions, and lack of phase-to-phase switching surge data. Research studies sponsored by EPRI have now provided adequate design information on both topics.[41,47,48]

***24. Protective and Grading Devices.*** Damage to insulators from heavy arcs was a serious maintenance problem in the past, and several devices were developed to ensure that an arc would stay clear of the insulator string. Subsequent improvements in the use of overhead ground wires and fast relaying have reduced the likelihood of insulator damage to the point that arc protection devices are now rarely used in the United States.

Earlier protective measures consisted of attaching small horns to the clamp, but it was found that horns with a large spread both at the top of the insulator and at the clamp were required to be effective. Under lightning impulse the arc tends to cascade the string, and tests show that the gap between horns should be considerably less than the length of the insulator string. Protection by arcing horns thus resulted in either a reduced flashover voltage or an increase in the number of units and length of the string. In any event, flashover persisted as a power arc until the line tripped out. For these reasons arcing horns have not been used in the United States for many years, although they are fairly common in Europe.

The arcing ring or grading shield is mainly for the purpose of improving the voltage distribution over the insulator string, and its effectiveness is due to the more uniform field. Protection of the insulator is not, therefore, dependent on simply providing a shorter arcing path, as is the case with horns. Efficient rings are rather large in diameter and, for suspension strings, clearances to the structure should be at least as great as from ring to ring. These considerations have made this device generally unattractive for modern construction. Grading rings are now used only at very high voltages for special applications, or with nonceramic insulators.

## Line and Structure Location

***25. Preparation for Construction.*** The cost of preparing for transmission-line construction is a considerable part of the total costs—under some conditions as much as 25%. Right-of-way and clearing are more or less fixed by local conditions, but the cost of surveys, accompanying maps, profiles, and engineering layout is to some extent governed by judgment. Many times in the past the overall costs have been increased by right-of-way difficulties and by delays in receiving proper materials because of inadequate preparations. The engineering work, properly carried out, makes it possible to obtain the right-of-way and complete the clearing well in advance of construction and to purchase every item of material and deliver it to the correct location.

The work of locating and laying out a line does not require great refinement, but careful planning is essential. With inexperienced surveyors or drafters, it must be assumed that errors will be made, and every possible device must be used to discover these errors before construction is started.

***26. Location.*** The general character of the line location should be determined because it has a definite bearing on the type of design. In extreme cases, such as difficult mountainous sections or in highly developed areas near cities, this may be a determining factor in the selection of the conductor and type of structures.

On heavy trunk lines, minor repairs and replacements are not an important item, and accessibility may often be rightly sacrificed to obtain the economy of a more direct route. Light wood lines must, however, be readily accessible for inspection and repairs. Line location is a matter of judgment and requires a person of wide general experience capable of correctly weighing the divergent requirements for inexpensive and available right-of-way, low construction costs, and convenience in maintenance. In mountainous country or in thickly populated areas, it is generally not advisable to attempt a direct route or try to locate on long tangents. Small angles of a few degrees cost little more and add little to the length of line. Most designs provide suspension structures for line angles of 5 to 15° which are not excessively costly. High, exposed ridges should be avoided, to afford protection against both wind and lightning.

Following a general reconnaissance by ground and air, for which 10 to 20 days per 100 mi should be allowed, and the assembling of all available maps and information, control points can be established for a general route or areas selected for more detailed study which may prove to be determining factors in the location of the line.

With this preliminary work completed, the major difficulties should have been determined. The policy as to such matters as right-of-way condemnation, electrical environmental assessments, telephone coordination, navigable-stream crossing, air routes, airports, and crossings with other utilities must be decided as definitely as possible.

Preliminary specifications should be issued before the final survey is started. These should include (1) outline drawings of the various structures with the important dimensions, (2) conductor sag curves and a sag template, (3) the maximum spans and angles for each type of structure, and (4) the requirements for right-of-way and clearing. Estimated costs are valuable, especially comparative costs of the various types of structure. With this information the field engineer can often, in a difficult section, choose the location best suited to the design.

Aerial maps can often be secured at much less cost than preliminary surveys, and in highly developed areas may be used to advantage for completely laying out the line without sending surveyors into the area until after the right-of-way has been secured.

Photographs taken at approximately ½ mi to the inch give sufficient detail for most work. Such maps can be photographically enlarged about four times for special detail. With a ½-mi-to-the-inch scale, the route of the line can be determined within a width of about 3 mi and sufficient landmarks located on a fairly accurate map to serve as a guide for flying the line.

*Location Survey.* The actual survey party can typically be divided into four divisions, each of which can complete at least a mile a day in average weather and country. Their operations may be carried out separately or nearly concurrently by allowing a full week's separation between successive operations and transferring personnel as needed.

The work falls naturally into the following: (1) an alignment party, choosing the exact location and cutting out the line; (2) a staking party, driving stakes at 100-ft stations and locating all obstructions; (3) a level party, taking elevations and side slopes; (4) a property and topography party, locating property lines.

A field drafting force located at a convenient point for receiving field notes can complete the final plan and profile drawings as fast as the survey can be made.

The method of procedure and size of survey organization depend upon the character of the country, the length and type of line, the experienced personnel available, and the schedule which must be maintained. In level, sparsely populated country, satisfactory, but incomplete, property surveys and profiles have been made during an open dry winter for a wood H-frame line 50 mi in length in approximately 4 months' time, with the personnel averaging a crew of eight and an engineer.

On a development involving the construction of several hundred miles of steel-tower line, the survey for a 65-mi line in rather difficult country, including 25 mi of inaccessible mountainous country, was completed with property maps and profiles in the form for permanent records in 2 months' time with a crew of about 20 and a locating engineer.

*Purchase.* Generally, right-of-way is not purchased in fee, but a perpetual easement is secured in which the owner grants the necessary rights to construct and operate the line but retains ownership and use of the land. The width of the right-of-way may be stated as a definite width or in general terms, but the easement must provide for (1) a means of access to each structure; (2) permission to erect all structures and guys; (3) all trees and brush to be cleared over a specified width for erection; (4) the removal of trees which would not safely clear the conductor if the conductor were to swing out under maximum wind or which would not safely clear the conductor if they were to fall; (5) the removal of buildings, lumber piles, haystacks, etc., which constitute a fire hazard. One of the major causes of serious line outages is the neglect to adhere strictly to conservative rules for clearing.

***27. Tower Spotting.*** The efficient location of structures on the profile is an important component of line design. Structures of appropriate height and strength must be located to provide adequate conductor ground clearance and minimum cost. In the past, most tower spotting has been done manually, using templates, but several computer programs have been available for a number of years for the same purpose.

*Manual Tower Spotting.* A celluloid template, shaped to the form of the suspended conductor, is used to scale the distance from the conductor to the ground and to adjust structure locations and heights to (1) provide proper clearance to the ground, (2) equalize spans, and (3) grade the line (Fig. 14-16).

The template is cut as a parabola on the maximum sag (usually at 49°C) of the ruling span (see Par. **42**) and should be extended by computing the sag as proportional to the square of the span for spans both shorter and longer than the ruling span. By extending the template to a span of several thousand feet, clearances may be scaled on steep hillsides. The form of the template is based on the fact that, at the time when the conductor is erected, the horizontal tensions must be equal in all spans of every length, both level and inclined, if the insulators hang plumb. This is still very nearly true at the maximum temperature. The template, therefore, must be cut to a catenary or, approximately, a parabola. The parabola is accurate to within about one-half of 1% for sags up to 5% of the span, which is well within the necessary refinement.

Since vertical ground clearances are being established, the 49°C no-wind curve is used in the template. Special conditions may call for clearance checks. For example, if it is known that a line will have high temperature rise because of load current, conductor clearance should be checked for the estimated maximum conductor temperature. One crossing over a navigable stream was designed for 88°C at high water. Ice and wet snow many times cause weights several times that of the ½-in radial ice loading, and conductors have been known to sag to within reach of the ground. Such occurrences are not normally considered in line design, and when they occur, the line is taken out of service until the ice or snow drops. Checks made afterward have nearly always shown no permanent deformation.

The template must be used subject to a "creep" correction for aluminum conductors. Creep is a nonelastic conductor stretch which continues for the life of the line, with the rate of elongation decreasing with time. For example, the creep elongation during the first 6 months is equal to that of the next 9½ years. All conductors of all materials are subject to creep, but to date only aluminum conductors have had intensive study. Creep is not substantial in other conductors, but the conductor manufacturers should be consulted. The IEEE Committee Report, Limitations on Stringing and Sagging Conductors, in the Decem-

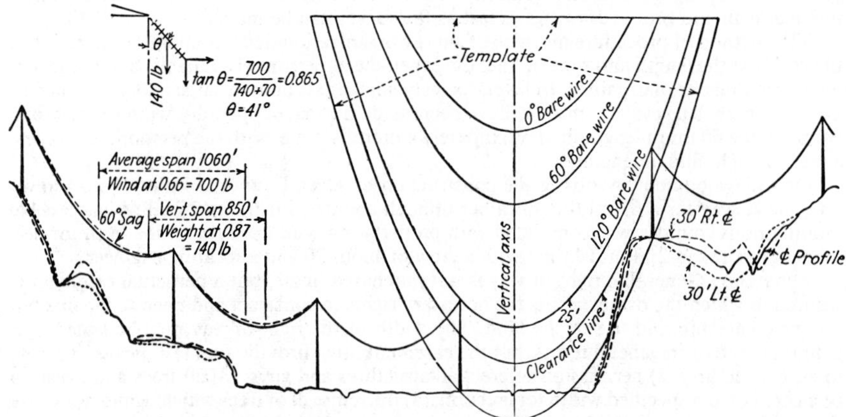


**FIG. 14-16** Sag template determines clearance of a suspended conductor from the ground.

ber 1964 *Transactions of the IEEE Power Group* discusses creep, and the reader should examine that report.[50]

Creep causes a continuous slow increase in the sag of the line which must be estimated and allowed for. The aluminum-conductor manufacturers will furnish creep-estimating curves, and most sag-tension computer programs now available are capable of calculating sags with and without creep. These curves are at approximately constant temperatures, around 15.5 to 21°C, and plot stress against elongation, one curve for each period of time, 1 h, 1 day, 1 month, 1 year, 10 years, etc. The values are integrated values for the period and are considered to be reasonable estimates. The temperature used is a reasonable average of the year's temperature across the center of the United States.

Precise values for creep are impossible to determine, since they vary with both temperature and tension, which are continuously varying during the life of the line. From Fig. 3 of the above committee report, it is found that a 1000-ft span of 954,000-cmil 48/7 ACSR when subjected to a constant tension of approximately 18% of its ultimate strength at a temperature of 15.5°C will have a sag increase in 1 day of approximately 5.5 in; in 10 days, 13 in; in 1 year, 27 in; in 10 years, 44 in; and in 30 years, 52 in.

Unless it is known that the line will have a life of less than 10 years, not less than 10 years creep should be allowed for. Creep has come into consideration in transmission-line design only during the past 30 years, and to date no standards have been established for handling it. Probably the simplest approach is to check all close clearance points on the profile with a template made with no creep allowance and to specify higher structures at these points if the addition of liberal creep sag infringes on the required clearances. It is possible to prestress the creep out of small conductors, but for large conductors this requires time and special tensioning facilities not normally available. Also the time lost in constructing an EHV line will more than pay for the extra structure height required to compensate for the creep. Prestressing changes the modulus of elasticity, and this new modulus should be used in the design.

The vertical weight supported at any structure is the weight of the length of conductor between low points of the sag in the two adjacent spans. For bare-conductor weights, this distance between low points can be scaled by using a template of the sag at any desired temperature. The maximum weight under loaded conditions should be scaled from a template made for the loaded sags. For most problems, the horizontal distance may be taken as equal to the conductor length. Distances to the low point of the sag may be computed by Eq. (14-68) (Par. **47**).

*Uplift.* On steep inclined spans the low point may fall beyond the lower support; this indicates that the conductor in the uphill span exerts a negative or upward pull on the lower tower. The amount of this upward pull is equal to the weight of the conductor from the lower tower to the low point in the sag. Should the upward pull of the uphill span be greater than the downward load of the next adjacent span, actual uplift would be caused, and the conductor would tend to swing clear of the tower.

It is important that abrupt changes in elevation of the structures should not occur, so that the conductor will not tend to swing clear of any structure even at low temperatures. This condition would be indicated if the 0°C curve of the template can be adjusted to hang free of the center support and just touch the adjacent supports on either side. In northern states it would be well to add a curve to the template for the below-zero temperatures experienced.

*Insulator Swing.* The uplift condition should not even be approached in laying out suspension insulator construction; that is, each tower should carry a considerable weight of conductor. The minimum weight that should be allowed on any structure may be logically determined by finding the transverse angle to which the insulator string may swing without reducing the clearance from the conductor to the structure too greatly. Also, the ratio of vertical weight to horizontal wind load should be kept such as not to allow the insulator to swing beyond this angle. The maximum wind is usually assumed at a temperature of 60°F. The wind pressure, measure in pounds per square foot, to be used in swing calculations is a matter of judgment and depends upon local conditions. Under high-wind conditions it is reasonable to require somewhat less than normal clearances. Generally a clearance corresponding to about 75% of the flashover value of the insulator is adequate. The insulator will swing in the direction of the resultant of the vertical and horizontal forces acting on the insulator string as shown in Fig. 14-16.

*Long Spans.* Rough country may necessitate spans considerably longer than contemplated in the design and may involve a number of factors including (1) proper clearance between conductors, (2) excessive tensions under maximum load, and (3) structures adequate to carry the additional loads.

Safe clearance between conductors is often based on the National Electrical Safety Code (NESC) formula, in which the spacing $a$ in inches is given as proportional to the square root of sag; $d$ is in inches.

$$a = 0.3 \text{ in/kV} + 8\sqrt{d/12} \tag{14-49}$$

This relation was developed for, and is useful on, comparatively short span lines of the smaller conductors and for voltages up to 69 kV; but for very long spans and heavy conductors, the formula results in spacings considerably larger than have proved satisfactory. It also results in spacings that are questionably small for very light conductors on long spans. Percy H. Thomas proposed an empirical formula which takes into account the weight of the conductor and its diameter, requiring less spacing for heavy conductors and a greater spacing for small conductors by the ratio of diameter $D$ in inches to weight $w$ in pounds per foot ($D/w$) (discussed in Par. **51**) as a means of determining the required conductor spacing for the average span of the line. The factor $C$ Eq. (14-73) includes an allowance to permit the standard spacing to be used on somewhat longer spans than average construction. The same formula, however, may be used to examine the spacings which have been successfully used on maximum spans and a value for $C$ selected from experience for determining the safe spacing required for an occasional unusually long span.

Excessive tensions on very long spans may be avoided by dead-ending at both ends and computing such a stringing sag as will result in the same maximum tension as elsewhere in the line. Such a span will be found to have considerably greater stringing sag and lower stringing tension than the normal span. Sag curves or charts are often prepared giving the sag for dead-end spans of various lengths such that the maximum tension under loaded conditions will be the same.

Dead-end construction is costly, and consideration should be given to avoiding this additional expense. It is common practice to permit spans up to double the average span without dead ends, although spans of this length may require additional spacing between wires. A careful examination of some trial figures on the sags and tensions developed in a long span will often indicate how great a span may be carried on suspension structures. The maximum loaded tension which would occur in a long span, if this span were dead-ended and sagged to the same stringing tension as the rest of the line, compared with the maximum tension for normal span lengths, is a good indication of the necessity for dead-end construction.

In case a number of long spans are encountered in a line or section of line, it may prove more economical to reduce the tension in the entire section to the long-span values and accept an increase in sag and corresponding reduction in span length in order to avoid dead ends.

*Computer Tower Spotting.* In a line of any significant length there are a very large number of possible permutations of tower spotting. Considerable experience is required to determine the least expensive, and since even the most experienced designer cannot examine all the possibilities, the least-cost design may often be missed by manual methods.

Several computer programs were developed for tower spotting with the more widespread use of computers in power engineering design since the 1950s. These programs are of several different types, including the "look ahead,"[51] "dynamic,"[52] "reiterative dynamic,"[53] and, more recently, the computer-aided design (CAD). This last type can be either a manual design method, but using a computer and CRT display as aids to facilitate design, or a more sophisticated version.

Until recently, computer tower spotting has not been widely used. Reasons for this have included the effort required for data preparation; the size of the computer and cost of computer time; constraints on the number of alternatives that the program can consider, as a control of computer costs; and limits of the programs themselves. Dramatic improvements in computing power and cost are changing this picture, together with modern methods of measuring the terrain plan and profile. EPRI project RP2151 includes the development of a tower-spotting program on the EPRI work station (a dedicated computer for engineering line designers).

If properly designed and applied, a tower-spotting program can provide designs equivalent or better than manual methods,[53] together with the advantages of speed, reduced labor, and convenience of redesign.

The availability of these programs in a convenient-to-use form and improved computers may significantly increase their use.

## Mechanical Design of Overhead Spans

***28. Conductor Loads.*** The span design consists of determining the sag at which the conductor shall be erected so that heavy winds, accumulations of ice or snow, and low temperatures, even if sustained for several days, will not stress the conductor beyond the elastic limit, cause a serious permanent stretch, or result in fatigue failures from continued vibrations.

The dead weight of the conductor and the weight of the accumulated ice or snow act vertically; the wind load is assumed to act horizontally and at right angles to the span; the resultant is the vectorial sum. Under combined vertical and horizontal loading, the conductor swings out into an inclined plane whose angle with the vertical is the angle between the direction of the vertical force and the resultant force. The resulting deflection is measured in this inclined plane.

The wind pressure, $p_z$, at height $z$ above ground level, in pounds per square foot, is given by the following formula:[54]

$$p_z = 0.00256V^2C_fK_zG_z \tag{14-50}$$

where $V$ = the basic wind speed, in miles per hour, determined from the wind speed contour map in Fig. 14-17

$C_f$ = the force coefficient given in Table 14-11 or 14-12

$K_z$ = the velocity pressure exposure coefficient given in Table 14-13

$G_z$ = the gust response factor given in Fig. 14-18 or Table 14-14

The exposure categories required for the determination of $p_z$ are defined in Table 14-15. These exposure categories and the basic wind speed map in Fig. 14-17 are not applicable to sections of transmission lines that cross high mountain ridges, large river valleys, or other topographic features where localized wind speed-up effects may occur. In these cases, special meteorological studies should be conducted to establish the appropriate wind loadings.

The gust response factors in Fig. 14-18 and Table 14-14 account for the dynamic response of the conductors and towers, respectively, to wind gusts. However, in computing the wind loading on the conductors for sag-tension analysis, a gust response factor of 1.0 should be used in Eq. (14-50).

***29. Safety Code Loadings.*** The NESC[55] district loadings have generally been accepted as a guide in determining the thickness of ice, wind velocity, and temperature which may be expected in any section of the country (Fig. 14-19). These loading assumptions, which are given in Table 14-16, are convenient as a basis of design, in that the loads caused by ice, wind, and low temperatures are assumed to occur simultaneously; however, consideration should be given to past experience and local conditions. For instance, accumulation of ice and snow on the conductors is rare in Minnesota, but extremely low temperatures are common; ice loads considerably greater than heavy loading but without extreme winds have occurred on several occasions from Maryland to New England, as well as in many other locations.

Unit wind and ice loadings for conductors are found by the following formulas:

$$\text{Wind load (lb/ft)} = \frac{p}{12}D \tag{14-51}$$

$$\text{Ice load (lb/ft)} = 1.244\,(Dr + r^2) \tag{14-52}$$

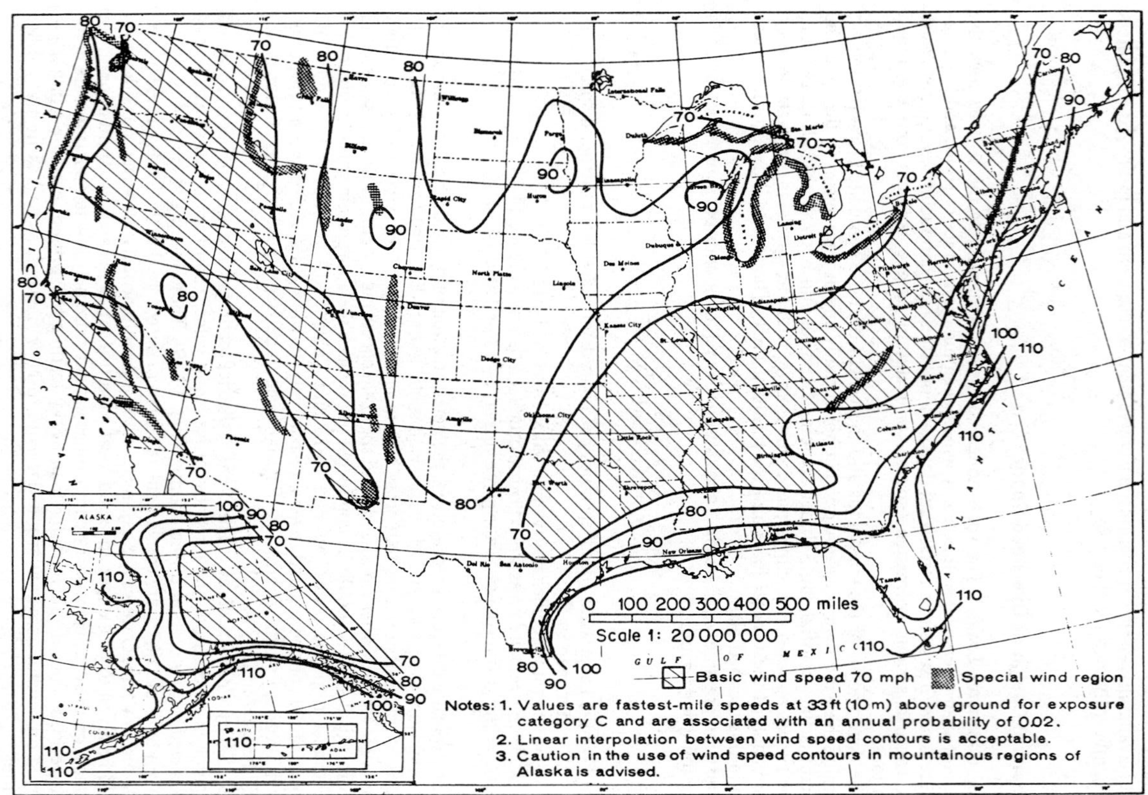

0 100 200 300 400 500 miles
Scale 1: 20 000 000
GULF OF MEXICO
Basic wind speed 70 mph
Special wind region
ALASKA
Notes: 1. Values are fastest-mile speeds at 33 ft (10 m) above ground for exposure category C and are associated with an annual probability of 0.02.
2. Linear interpolation between wind speed contours is acceptable.
3. Caution in the use of wind speed contours in mountainous regions of Alaska is advised.


**FIG. 14-17** Basic wind speed (mi/h).

**TABLE 14-11** Force Coefficients for Cylindrical Surfaces

| Description of surface | $C_f$ |
|---|---|
| Stranded cables (conductors, ground wires, guy wires) | 1.0 |
| Smooth circular cylinder | 1.0 |
| Rough circular cylinder | 1.2 |
| 16-sided polygon | 0.8 |
| 12-sided polygon | 1.0 |
| Octagon | 1.4 |

**TABLE 14-12** Force Coefficients for Lattice Towers, $C_f$

| | $C_f$ | |
|---|---|---|
| | Square towers | Triangular towers |
| <0.025 | 4.0 | 3.6 |
| 0.025–0.44 | $4.1 - 5.2\varepsilon$ | $3.7 - 4.5\varepsilon$ |
| 0.45 –0.69 | 1.8 | 1.7 |
| 0.7 –1.0 | $1.3 + 0.7\varepsilon$ | $1.0 + \varepsilon$ |

NOTES : $\varepsilon$ is the ratio of solid area to gross area of tower face.

Force coefficients are given for towers with structural angles or similar flat-sided members.

For towers with rounded members, the design wind force shall be determined using the values in the above table multiplied by the following factors:

| | |
|---|---|
| $\varepsilon \leq 0.29$ | factor = 0.67 |
| $0.3 \leq \varepsilon \leq 0.79$ | factor = $0.67\varepsilon + 0.47$ |
| $0.8 \leq \varepsilon \leq 1.0$ | factor = 1.0 |

For triangular-section towers, the design wind forces shall be assumed to act normal to a tower face.

For square-section towers, the design wind forces shall be assumed to act normal to a tower face. To allow for the maximum horizontal wind load, which occurs when the wind is oblique to the faces, the wind load acting normal to a tower face shall be multiplied by the factor $1.0 + 0.75\varepsilon$ for $\varepsilon < 0.5$ and shall be assumed to act along a diagonal.

in which $p$ = wind pressure in pounds per square foot, $D$ = diameter of conductor in inches, $r$ = radial thickness of ice. Ice is taken at 57 lb/ft$^3$.

The NESC also specifies an extreme wind-loading condition which is to be applied to the bare conductor (no ice) at a temperature of +60°F and constant $K = 0$. For this case, the extreme wind pressure is determined from a wind pressure contour map based on fastest mile of wind data from an earlier edition of ANSI A58.1. In lieu of this procedure, the basic wind speed map in Fig. 14-17 can be used in conjunction with Eq. (14-50) to obtain extreme wind loadings.

***30. Stresses in a Span.*** The high-tension stress in the conductor is the result of attempting to support a vertical load by a member, that is, the conductor, extending in a very nearly horizontal direction or nearly at right angles to the direction of the load. The slope of the conductor at the support is generally only a few degrees below the horizontal, which causes a stress in the conductor many times the weight supported. From mechanics, the horizontal tension in the wire, $t$, is equal to the weight supported $V$ (which is the length of wire $l/2$ times the weight per foot $w$) divided by the tangent of the angle of slope $\theta$ (Fig. 14-20).

***31. The Parabola and Catenary Curves.*** The slope of the conductor at a support, which is the factor determining the conductor tension, indicates the form of the curve assumed by

**TABLE 14-13** Velocity Pressure Exposure Coefficient, $K_z$

| Height above ground level, z, ft | $K_z$ | | |
|---|---|---|---|
| | Exposure *A* | Exposure *B* | Exposure *C* |
| 0–15 | 1.20 | 0.80 | 0.12 |
| 20 | 1.27 | 0.87 | 0.15 |
| 25 | 1.32 | 0.93 | 0.17 |
| 30 | 1.37 | 0.98 | 0.19 |
| 40 | 1.46 | 1.06 | 0.23 |
| 50 | 1.52 | 1.13 | 0.27 |
| 60 | 1.58 | 1.19 | 0.30 |
| 70 | 1.63 | 1.24 | 0.33 |
| 80 | 1.67 | 1.29 | 0.37 |
| 90 | 1.71 | 1.34 | 0.40 |
| 100 | 1.75 | 1.38 | 0.42 |
| 120 | 1.81 | 1.45 | 0.48 |
| 140 | 1.87 | 1.52 | 0.53 |
| 160 | 1.92 | 1.58 | 0.58 |
| 180 | 1.97 | 1.63 | 0.63 |
| 200 | 2.01 | 1.68 | 0.67 |

NOTES: Linear interpolation for intermediate values of height *z* is acceptable.
Exposure categories are defined in Table 14-14.
1 ft = 0.3048 m.

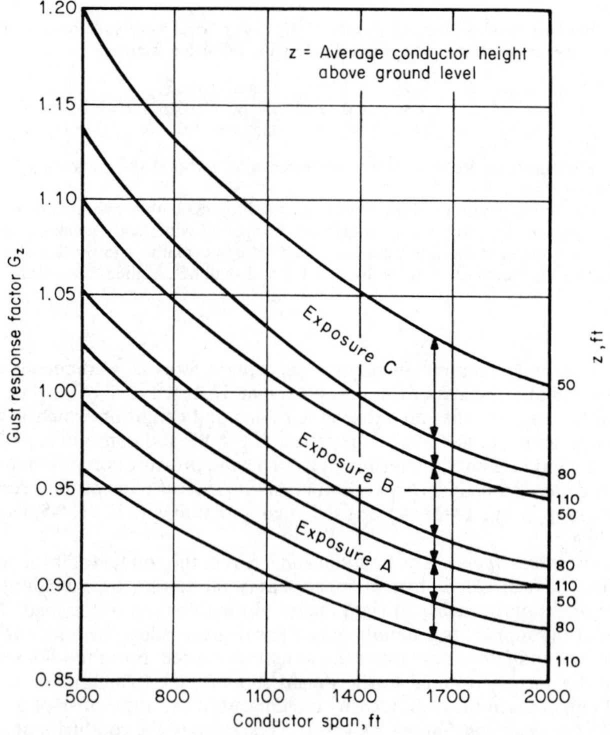


**FIG. 14-18** Conductor gust response factors.

**TABLE 14-14** Tower Gust Response Factors

| Height above ground level, $z$, ft† | $G_z$* | | |
|---|---|---|---|
| | Exposure $A$ | Exposure $B$ | Exposure $C$ |
| 50 | 1.19 | 1.31 | 1.48 |
| 100 | 1.12 | 1.22 | 1.35 |
| 150 | 1.08 | 1.16 | 1.27 |

*Applicable only for self-supporting lattice towers. Special analysis required for flexible-pole and guyed structures.

†$z$ is height to center of pressure of mean wind loading on structure.

NOTE: 1 ft = 0.3048 m.

**TABLE 14-15** Description of Exposure Categories

| Exposure category | Description |
|---|---|
| $A$ | Flat, unobstructed coastal areas directly exposed to wind flowing over large bodies of water |
| $B$ | Open terrain with scattered obstructions having heights generally less than 30 ft; e.g., cultivated fields and grasslands |
| $C$ | Suburban areas, wooded areas, or other terrain with numerous closely spaced obstructions having the size of single-family dwellings or larger |

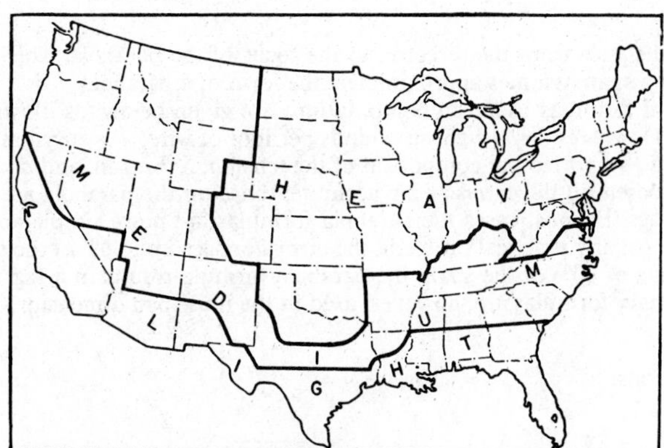


**FIG. 14-19** NESC district loading map of United States for mechanical loading of overhead lines (from the 1981 ed.).

the conductor. In Fig. 14-20 it will be seen that the cosine of the angle of slope $\theta$ at the support is the ratio of the horizontal tension $t$ to the resultant tension $T$ in the conductor, or $\cos\theta = t/T$. The mathematical form of the slope of the tangent to a catenary is $\cos\theta = c/y$, in which $c$ = ordinate of the low point of the curve and $y$ = ordinate of the point of tangency. It is thus evident that the span must form a catenary and that the ordinate $y$ should be considered as a dimension of a span corresponding to the tension at the point of support; or, more definitely, this imaginary dimension $y$ is a length of wire whose weight is

**TABLE 14-16** NESC District Loading Requirements

| | Loading districts | | |
|---|---|---|---|
| | Heavy | Medium | Light |
| Radial thickness of ice, in | 0.50 | 0.25 | 0 |
| Horizontal wind pressure, ($lb/ft^2$) | 4 | 4 | 9 |
| Temperature, °F | 0 | +15 | +30 |
| Constant *K*, (lb/ft)* | 0.30 | 0.20 | 0.05 |

*To be added to the resultant of the horizontal and vertical forces in the direction of the resultant.

NOTE: 1 in = 25.4 mm; 1 lb/ft = 14.59 N/m; 1 $lb/ft^2$ = 47.88 Pa; $t_{°C} = t_{°F} - 32)/1.8$.

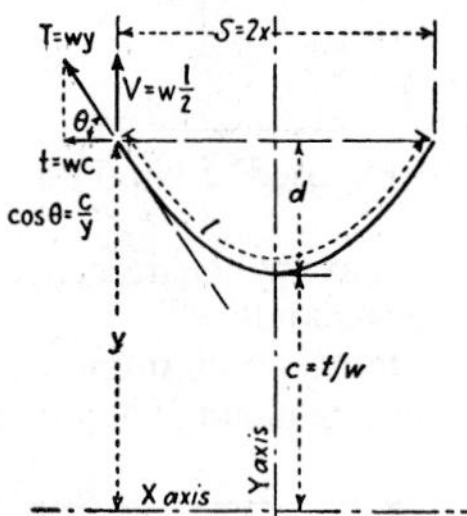


**FIG. 14-20** The catenary.

equal to the tension $T$. Similarly, the ordinate $c$ corresponds to the horizontal tension existing at the low point of the sag and should also be considered a dimension of the span.

In Fig. 14-21 the weight of wire in the half span is approximately equal to $wx$, since $2x$ approximately equals $l$, if the sag is not too great. The tangent of the slope $\theta = wx/t$, which is the mathematical form of the tangent to a parabola; that is,

$$\tan \theta = x/p$$

in which $p$ = distance from the directrix to the focus of the parabola. This demonstrates that the wire in a span assumes approximately the form of a parabola.

The principal formulas useful in computations are given below (as illustrated in Figs. 14-20 and 14-21), where $w$ = weight in pounds per foot of wire, $T$ = tension in pounds at the support, and $t$ = horizontal component of the tension; $S$ = span, and $d$ = sag, both in feet. Formulas based on the catenary are accurate; those on the parabola are approximate for very large sags. In most cases the parabola formulas are more simple and sufficiently accurate for almost any practical problem; the error, for sags up to 6% of the span, is about ½%; and for a sag of 10% of the span, the parabola formula results in a sag about 2% too small. The catenary formulas are, however, used in the unit-span dimensions.

Catenary equations:

$$y = c \cosh \frac{x}{c} \tag{14-53}$$

$$\frac{l}{2} = c \sinh \frac{x}{c} = \sqrt{y^2 - c^2} \tag{14-54}$$

$$d = y - c = c\left(\cosh \frac{x}{c} - 1\right) \tag{14-55}$$

Parabola equations:

$$x^2 = 2py \tag{14-56}$$

$$d = \frac{wS^2}{8t} \tag{14-57}$$

$$l = S\left(1 + \frac{8}{3}\frac{d^2}{S^2}\right) \text{ approx.} \tag{14-58}$$

$$T = t + wd \tag{14-59}$$

The dimensions $y$ and $c$ of the catenary give actual dimensions of the curve corresponding to tension in the wire; that is, $y = T/w$, $c = t/w$. The relation $T = t + wd$ is very useful.

***32. The Unit Span.*** A considerable saving in tedious arithmetic and a gain in accuracy are made by using the unit-span dimensions as devised by Percy H. Thomas.[56] On the unit-span basis the catenary corresponding to the actual span of wire is not considered to have the large physical dimensions of the span itself but to be reduced, in proportion, to a curve of unit length.

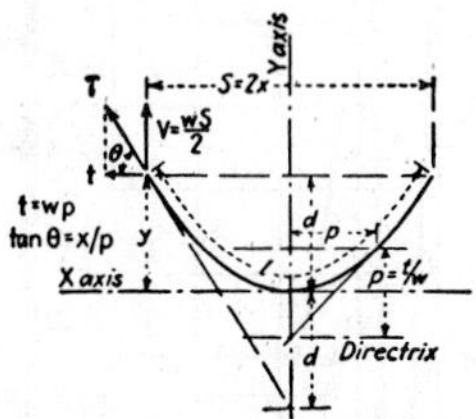


FIG. 14-21 The parabola.

Each dimension of the unit curve is therefore in the ratio of $1/S$ to the dimensions of the actual span, including the dimensions $y$ and $c$ corresponding to tension. In the following discussion the unit dimensions $y_1$ will be referred to as the tension factor $T_1$. In Figs. 14-20 and 14-22, the unit dimensions are as follows:

Unit sag: $$d_1 = d/S = y_1 - c_1 = c_1[\cosh (1/2c_1) - 1] \qquad (14\text{-}60)$$

Unit length: $$l_1 = l/S = 2c_1 \sinh (1/2c_1) \qquad (14\text{-}61)$$

Tension factor: $$T_1 = y_1 = y/S = T/wS = c_1 \cosh (1/2c_1) \qquad (14\text{-}62)$$

$$t_1 = c_1 = c/S = t/wS \qquad (14\text{-}63)$$

All spans having the same ratio of sag to span or the same ratio of length to span, or the same tension factor, are represented by the same unit curve; and thus a comparatively brief tabulation listing the unit dimensions may be prepared by assuming values of $c_1$ and calculating $d_1$ $l_1$, and $y_1$. If, for any span, the sag, the length, or the tension is known, the corresponding unit dimension is readily figured, and the unknown unit dimensions are read from the tabulation. Interpolations are best made by plotting these values in a curve as shown in Fig. 14-23. Three sets of curves are shown, each plotted to a different scale. Working charts should be plotted to convenient scale from data given in Par. **37**. A very elaborate tabulation of unit-span dimensions was prepared by Martin[57] and published by the Copperweld Bimetallics Division, Copperweld Corp. In this tabulation, interpolations may be made directly from the table (see Par. **41**).

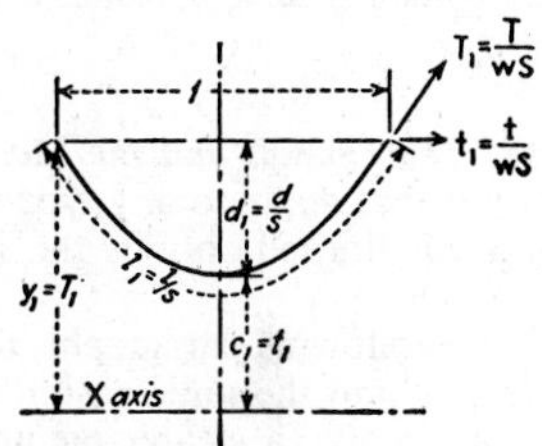


FIG. 14-22 Unit span.

***33. Span Calculations.*** The problems of span design may be divided into three classes: (*a*) The sag in a span resulting from a conductor of given weight at a given tension (or the reverse). These problems are direct solutions of Eqs. (14-56) to (14-59) or may be read from the Thomas chart as described in Par. **32**. (*b*) The sags or tensions resulting from unequal spans or differences in elevation of supports. These effects are generally of minor importance, and it is usually necessary only to investigate a few limiting conditions by graphic methods. Special cases may be solved by the methods outlined in Par. **47**. (*c*) The sags resulting from changes in loading or temperature. These problems are complicated, but the two simultaneous equations involved are so readily solved as two intersecting curves that, in common with most design problems, nothing is gained by attempts at algebraic solutions. The following paragraphs describe manual methods of sag-tension calculations. These are now rarely used, since most conductor manufacturers provide convenient computer programs for this purpose that include allowance for temperature, creep, wind and ice loads, and span length. A program which permits calculations up to high conductor temperatures was recently developed by Ontario Hydro.[58]

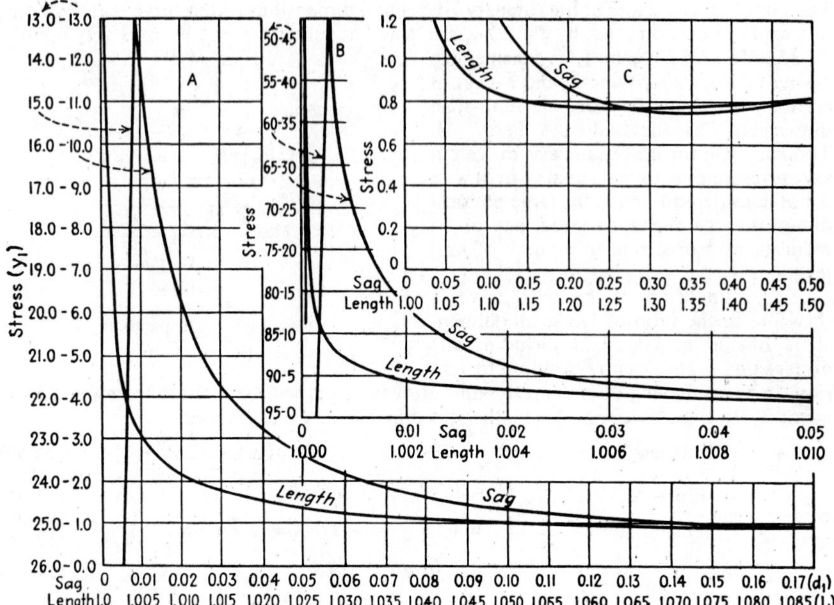


**FIG. 14-23** Thomas chart. Curve *A* is good for sags from 2 to 15%; curve *B* for sags less than 2%; curve *C* for very large sags, being especially useful for spans on steeply inclined slopes. Since the chart is too small for accurate work, the data from which the curves are plotted are given in Table 14-17. Figure 14-24 is a portion of curve *B*, plotted to a larger scale.

***34. The Thomas and the Martin methods*** are the simplest methods of making sag calculations that apply to any type of conductor. No approximations are introduced, and the accuracy is limited only by the scale to which the charts are drawn (Fig. 14-23 and Table 14-17).

In the following paragraphs, the unit sag is often referred to as the sag, the unit length as the length, and the unit tension factor as the tension. In discussing the catenary, the unit dimensions give a clearer picture, whereas consideration of the stretch and temperature expansion seems to apply more naturally to actual dimensions. There is some advantage in not attempting to distinguish too closely between the actual and unit dimensions, as the two are, for many purposes, interchangeable. Likewise, in the following paragraphs, it will be noted that the "length" is considered alternatively as the length of the catenary arc and as the length of the physical conductor. Some confusion will be avoided if it is realized that, in the span, these two must be the same, and it is only for the purpose of visualizing the process of computation that the two lengths are separated.

***35. Length of the Conductor in the Span.*** The process of erecting the conductor and adjusting the sag, usually considered as adjusting the tension, is actually a matter of adjusting the length of wire in the span. A little wire is taken out if the sag is too great or a little added in the span if the sag is too small. Changes in the length of wire due to elastic stretch and to expansion and contraction from temperature, with corresponding changes in sag, produce similar adjustments. In this adjustment two sets of simultaneous conditions must be satisfied: (1) the length of catenary arc determined entirely by the form of the curve dependent on the sag and reflected in the tension must be equal to (2) the length of the wire determined by the stress and elastic stretch.

***36. The Length Curve.*** All values of the first of these two simultaneous conditions, that is, the length of the conductor as determined by the length of the catenary arc and the $y$ dimension of many possible forms of the catenary, are plotted as the length curve of the Thomas chart (Fig. 14-23). Any particular point on this length curve represents the length and tension for a particular unit catenary.

**TABLE 14-17** Stress and Length in Terms of Ratio of Sag to Span

Unit-span dimensions*

| $c_1$ | $y_1$ | $d_1$ | $l_1$ | $c_1$ | $y_1$ | $d_1$ | $l_1$ |
|---|---|---|---|---|---|---|---|
| Horizontal stress factor | Stress factor | Unit sag | Unit length | Horizontal stress factor | Stress factor | Unit sag | Unit length |
| 100.0000 | 100.001 3 | .001 250 | 1.000 004 2 | 6.2500 | 6.270 0 | .020 01 | 1.001 066 |
| 90.9091 | 90.910 5 | .001 375 | 1.000 005 1 | 5.8824 | 5.903 6 | .021 26 | 1.001 205 |
| 83.3333 | 83.334 8 | .001 500 | 1.000 006 1 | 5.5555 | 5.578.1 | .022 52 | 1.001 351 |
| 76.9231 | 76.924 7 | .001 625 | 1.000 007 1 | 5.2632 | 5.286 9 | .023 77 | 1.001 503 |
| 71.4286 | 71.430 3 | .001 750 | 1.000 008 2 | 5.0000 | 5.025 0 | .025 02 | 1.001 668 |
| 66.6667 | 66.668 5 | .001 875 | 1.000 009 4 | 4.7619 | 4.788 2 | .026 27 | 1.001 839 |
| 62.5000 | 62.502 0 | .002 000 | 1.000 010 7 | 4.5455 | 4.573 0 | .027 53 | 1.002 017 |
| 58.8235 | 58.825 7 | .002 125 | 1.000 012 0 | 4.3478 | 4.376 6 | .028 78 | 1.002 205 |
| 55.5555 | 55.557 8 | .002 250 | 1.000 013 5 | 4.1667 | 4.196 7 | .030 04 | 1.002 402 |
| 52.6316 | 52.633 9 | .002 375 | 1.000 015 0 | 4.0000 | 4.031 3 | .031 29 | 1.002 606 |
| 50.0000 | 50.002 5 | .002 50 | 1.000 017 | 3.8462 | 3.878 7 | .032 55 | 1.002 819 |
| 45.4545 | 45.457 3 | .002 75 | 1.000 020 | 3.7037 | 3.734 2 | .033 80 | 1.003 040 |
| 41.6667 | 41.669 7 | .003 00 | 1.000 025 | 3.5714 | 3.606 5 | .035 06 | 1.003 270 |
| 40.0000 | 40.003 1 | .003 13 | 1.000 026 | 3.4483 | 3.484 6 | .036 31 | 1.003 508 |
| 38.4615 | 38.464 8 | .003 25 | 1.000 028 | 3.3333 | 3.370 9 | .037 57 | 1.003 754 |
| 35.7143 | 35.717 8 | .003 50 | 1.000 033 | 2.9412 | 2.983 8 | .042 60 | 1.004 825 |
| 33.3333 | 33.337 1 | .003 75 | 1.000 037 | 2.5000 | 2.550 2 | .050 17 | 1.006 680 |
| 31.2500 | 31.254 0 | .004 00 | 1.000 043 | 2.2727 | 2.328 0 | .055 22 | 1.008 086 |
| 29.4118 | 29.416 0 | .004 25 | 1.000 048 | 2.0000 | 2.062 8 | .062 83 | 1.010 444 |
| 28.5714 | 28.575 8 | .004 38 | 1.000 051 | 1.8519 | 1.919 8 | .067 91 | 1.012 194 |
| 27.7777 | 27.782 3 | .004 50 | 1.000 054 | 1.6667 | 1.742 2 | .075 56 | 1.015 068 |
| 26.3158 | 26.320 5 | .004 75 | 1.000 060 | 1.5625 | 1.643 2 | .080 68 | 1.017 154 |
| 25.0000 | 25.005 0 | .005 00 | 1.000 067 | 1.4286 | 1.517 0 | .088 40 | 1.020 542 |
| 22.7273 | 22.732 8 | .005 50 | 1.000 081 | 1.3514 | 1.444 9 | .093 56 | 1.022 973 |
| 20.8333 | 20.839 3 | .006 00 | 1.000 096 | 1.2500 | 1.351 3 | .101 34 | 1.026 881 |
| 20.0000 | 20.006 3 | .006 25 | 1.000 104 | 1.1905 | 1.297 0 | .106 55 | 1.029 660 |
| 19.2308 | 19.237 3 | .006 50 | 1.000 113 | 1.1111 | 1.225 5 | .114 41 | 1.034 093 |
| 17.8571 | 17.864 1 | .007 00 | 1.000 131 | 1.0638 | 1.183 5 | .119 68 | 1.037 224 |
| 16.6667 | 16.674 2 | .007 50 | 1.000 150 | 1.0000 | 1.127 6 | .127 63 | 1.042 19 |
| 15.6250 | 15.633 0 | .008 00 | 1.000 171 | 0.9091 | 1.050 1 | .141 00 | 1.051 19 |
| 14.7059 | 14.714 4 | .008 50 | 1.000 193 | 0.8333 | 0.987 9 | .154 55 | 1.061 09 |
| 13.8889 | 13.897 9 | .009 00 | 1.000 216 | 0.7143 | 0.896 5 | .182 26 | 1.083 69 |
| 13.1579 | 13.167 4 | .009 50 | 1.000 241 | 0.6250 | 0.835 8 | .210 83 | 1.110 13 |
| 12.5000 | 12.510 0 | .010 00 | 1.000 267 | 0.5555 | 0.796 2 | .240 61 | 1.140 57 |
| 11.6279 | 11.638 7 | .010 75 | 1.000 308 | 0.5000 | 0.771 54 | .271 54 | 1.175 20 |
| 10.6383 | 10.650 1 | .011 75 | 1.000 368 | 0.4545 | 0.758 42 | .303 87 | 1.214 23 |
| 10.0000 | 10.012 5 | .012 50 | 1.000 417 | 0.4167 | 0.754 44 | .337 77 | 1.257 88 |
| 9.0909 | 9.104 7 | .013 75 | 1.000 504 | 0.3846 | 0.758 04 | .373 43 | 1.306 45 |
| 8.3333 | 8.348 3 | .015 00 | 1.000 600 | 0.3571 | 0.768 18 | .411 04 | 1.360 21 |
| 7.6923 | 7.708 4 | .016 26 | 1.000 704 | 0.3333 | 0.784 14 | .450 80 | 1.419 52 |
| 7.1428 | 7.160 4 | .017 51 | 1.000 817 | 0.3125 | 0.805 46 | .492 96 | 1.484 73 |
| 6.6667 | 6.685 4 | .018 76 | 1.000 938 | | | | |

*Percy H. Thomas, Sag Calculations for a Suspended Wire, *Trans. AIEE,* 1911, vol. 30, p. 2229, Refer to Fig. 14-22.

The length of the unit catenary, corresponding to the actual span under consideration, is the basis from which the sags and tensions are determined for any temperature or loading condition. This length of the catenary is not used directly; instead, the "unstressed length" of the conductor is more convenient. The unstressed length is found by subtracting the elastic stretch of the conductor from the catenary length.

$$\text{Stretch/ft} = P/E \tag{14-64}$$

in which $P$ = stress in pounds per square inch and $E$ = modulus of elasticity.

***37. Temperature Change.*** The effect of change in temperature is considered to take place after the conductor is relieved of all stress, thus eliminating the complicated adjustments between change in length due to temperature and change in stretch from the resulting change in tension. The process is as if the conductor were removed from the span and laid on the ground before the rise in temperature takes place.

$$(\text{Change in length})/\text{ft} = \alpha t \tag{14-65}$$

where $\alpha$ = linear coefficient of expansion and $t$ = change in temperature. This gives a new unstressed length $P_1$ at the new temperature. This is one point on the curve of the second of the two simultaneous conditions described in Par. **33**; viz., the curve of the length of the wire as determined by the elastic stretch. Points on this curve may be computed from the modulus of elasticity [Eq. (14-64)] or read from a stress-strain curve of the conductor. If the modulus is constant, this will be a straight line. The intersection of the elastic-stretch curve with the catenary-length curve is the solution of these two simultaneous conditions and the tension at the new temperature. It is as if the same conductor lying on the ground at the new temperature were lifted back into the span and stretched until the ends just reached the supports. The intersection is the point at which the stretch is just sufficient to allow the conductor to hang freely.

Points on the chart (Fig. 14-23) are unit dimensions; the stretch is the stretch per unit length; and the tension is expressed as the unit tension factor $T/wS$.

***38. Change in loading*** changes the value of $w$ in the equation $T_1 = T/wS$. Thus, in plotting the curve of the stretch of the conductor on the Thomas chart, the slope of the stretch curve is quite different if the ice and wind are removed and the value of $w$ correspondingly changed.

***39. Example Calculation*** (see Fig. 14-24). Required: the sag at which a 600-ft span of 4/0 19-strand hard-drawn bare-copper conductor shall be erected at 15.5°C so that, under

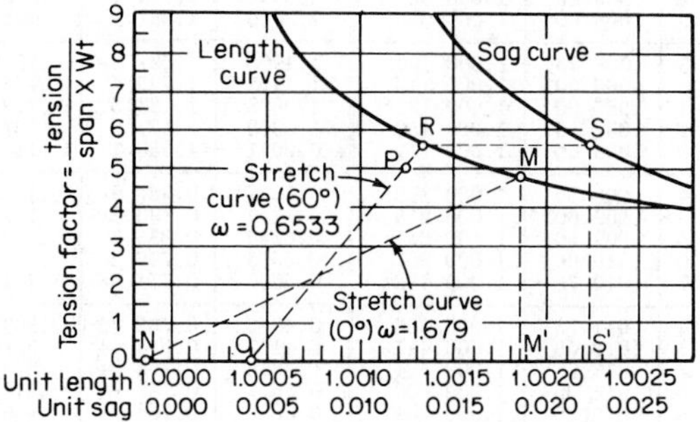


**FIG. 14-24** Sag and tension determination for copper conductor.

heavy loading, ½-in ice, 4-lb wind at −18°C, the tension shall not exceed one-half the ultimate strength. The following data are taken from Table 14-18.

| | |
|---|---|
| Ultimate strength | 9,617 lb |
| Maximum tension $T$ | 4,808 lb |
| Conductor weight bare $w$ | 0.6533 lb/ft |
| Resultant weight $w_r$ | 1.679 lb/ft |
| Area | 0.1662 in² |
| Modulus of elasticity $E$ | 14,500,000 (initial) |
| Coefficient of expansion $X$ | 0.0000094 |

Dimension $y$ = 4808/1679 = 2864 ft, and the unit-span dimensions corresponding to a catenary of $y$ = 2864 ft for a 600-ft span are:

Tension factor $T_1$ = 2864/600 = 4.77 (point $M$, Fig. 14-24).

**TABLE 14-18** Physical Properties of Conductor Materials

**This table lists the physical properties of the more common conductor and overhead ground-wire materials. National Electrical Safety Code recommended initial and final moduli of elasticity are given where available and are denoted by an asterisk (*). (See also Sec. 4.)**

| Material | Ultimate strength, lb/in² | Modulus of elasticity, initial | Modulus of elasticity, final | Coefficient of expansion/°F |
|---|---|---|---|---|
| Copper, stranded, hard-drawn | 57,000– 59,000 | | | |
| 3–12 strand | .............. | 14,000,000* | 17,000,000* | .000,009,4* |
| 7–19 strand | .............. | 14,500,000* | 17,000,000* | .000,009,4* |
| 37 strand | .............. | 14,500,000 | 17,000,000 | .000,009,4 |
| Solid, hard drawn | 49,000– 63,000 | 14,500,000* | 17,000,000* | .000,009,4* |
| All-aluminum cable, stranded, hard-drawn | 23,000– 30,000 | .......... | 10,000,000 | .000,012,8 |
| ACSR all sizes | 40,000– 56,000 | See Par 45 | $E_aH_a + E_sH_s$ See Par 45 for symbols | $(E_aH_a/E_{as})\theta_a + (E_sH_s/E_{as})\theta_s$ |
| Steel | 190,000 | .......... | 29,000,000 | .000,006,4 |
| Aluminum | 23,000– 30,000 | .......... | 10,000,000 | .000,012,8 |
| No. 2 and No. 4 7/1 | 54,000– 55,000 | 11,600,000* | 12,600,000* | .000,010,1* |
| No. 6 to No. 4/0 6/1 | 43,000– 49,000 | 10,250,000* | 11,500,000* | .000,010,5* |
| Copperweld, solid | 80,000–170,000 | 22,000,000* | 24,000,000* | .000,007,2* |
| Stranded | 100,000–170,000 | 20,500,000* | 23,000,000* | .000,007,2* |
| Copperweld—copper: | | | | |
| No. 4 to 350,000 cmil Cu eq | 65,000– 96,000 | | | |
| No. 6A to No. 2A | 86,000– 96,000 | 16,500,000* | 19,000,000* | .000,008,5* |
| No. 2F to No. 1/0F | 71,000– 73,000 | 15,500,000* | 18,000,000* | .000,009,0* |
| Galvanized-steel strand, all strengths | 45,000–190,000 | .......... | 27,000,000 | .000,006,7 |
| Stainless-steel strand, Page type 301 | 235,000 | .......... | 26,000,000 | .000,009,2 |
| Bronze, stranded 8.5 to 85 cond. ASTM B105 alloys | 67,000–135,000 | .......... | 16,000,000 | .000,009,4 |
| Alumoweld 7-strand | 180,000 | .......... | 23,000,000 | .000,007,2 |

NOTE: 1 lb/in² = 6.895 kPa; $t_{°C} = (t_{°F} - 32)/1.8$.

Unit length stressed $l_1$ = 1.00186 (read from chart, point *M'*).

Owing to the tension of 4808 lb, or a stress of 4808/0.1662 = 28,929 lb/in², the conductor is stretched 28,929/14,500,000 = 0.001995 ft/ft, or, for the unit length of 1.00186 ft, a total of 1.00186 × 0.001995 = 0.001999 ft.

The unstressed length at −18°C ($l_0$ at − 18°C) = 1.00186 − 0.001999 = 0.99986 (point *N*).

A change in temperature from −18°C to 15.5°C results in an increase in length of 60 × 0.0000094 = 0.000564 ft/ft or a total increase of 0.000564 $<$ 0.99986 = 0.000564 ft.

The unstressed length at 15.5°C ($l_0$ at 15.5°C) = 0.99986 + 0.000564 = 1.000424 (point *O*).

It is now assumed that with the load removed and starting with the unstressed length computed above, the conductor is stretched until the stretch curve intersects the length curve. Since the modulus is constant, the stretch curve is a straight line determined by any two points, for example, the unstressed length and the length corresponding to any tension factor such as 5.00, or a tension, for the bare cable, of

$$5.00 \times 600 \times 0.6533 = 1960 \text{ lb}$$

This will give a stretch of

$$1.000424[1960/(0.1662 \times 14{,}500{,}000)] = 0.000813$$

The unit length for tension factor 5.00 at 60°F = $l_0$ at 60°F = 1.000424 + 0.000813 = 1.001237 (point *P*).

The line *OP* extended intersects the length curve at *R* or at a tension factor of 5.60, which, for the 600-ft span of 4/0 copper, corresponds to

$$5.60 \times 600 \times 0.6533 = 2195 \text{ lb}$$

The unit sag for a tension factor of 5.60 is read from the sag curve (point *S'*) as 0.02242, or an actual sag for the 600-ft span of 0.02242 × 600 = 13.45 ft, which is the sag at which the conductor should be erected at 15.5°C.

The intersection of the stretch curve with the length curve may be made very accurately by trial from the values given in Martin's tables, as discribed in the introduction to those tables.

The computations are repeated in condensed form below and include the sags for 120°F.

| | | | Point |
|---|---|---|---|
| $T_1$ = | $4{,}808/(1.679 \times 600)$ | = 4.77 | *M* |
| $l_1$ = | | = 1.00186 | *M* |
| Stretch = | $1.00186[4{,}808/(0.1662 \times 14{,}500{,}000)]$ | = <u>0.001999</u> | |
| $l_0$ at −18°C = | | = 0.99986 | *N* |
| Change 15.5°C = | $0.99986(60 \times 0.0000094)$ | = <u>0.000564</u> | |
| $l_0$ at 15.5°C = | | = 1.00042 | *O* |
| $l_0$ at 49°C = | | = 1.000988 | |
| Stretch at $T_1$ = | $5.00 = 1.000424[5.00 \times 600 \times 0.6533/(0.1662 \times 14{,}500{,}000)]$ | = 0.000813 | |
| $l_1$ at 15.5°C when $T_1$ = | 5.00 | = 1.001237 | *P* |
| $l_1$ at 49°C when $T_1$ = | 5.00 | = 1.00180 | |
| $T_1$ at 15.5°C = | 5.60 | | |
| Tension = | $5.60 \times 600 \times 0.6533$ | = 2,195 lb | |
| $d_1$ at 15.5°C = | 0.02242 | | |
| Sag = | $0.02242 \times 600$ | = 13.45 ft | |
| $T_1$ at 49°C = | 4.88 | | |
| Tension = | $4.88 \times 600 \times 0.6533$ | = 1,913 lb | |
| $d_1$ at 49°C = | 0.0257 | | |
| Sag = | $0.0257 \times 600$ | = 15.42 ft | |

***40. Conductor Materials.*** See Table 14-18.

***41. Martin's Tables.*** A very complete tabulation of unit-span values has been prepared by James S. Martin and published by the Copperweld Bimetallics Division, Copperweld Corp. The unit values are given in such small steps that these tables may be used in exactly the same manner as trigonometric functions. The stress factor as given in Martin's tables is the reciprocal of the stress factor in the Thomas chart, that is, $WS/T$. Martin's tables also include an additional dimension, which may be described as the average stress factor between the stresses at the support and at the low point. In computing the stretch it is evident that this average should be used rather than the maximum at the support. This may have an appreciable effect on extraordinarily large sags.

***42. Stresses Due to Unequal Spans.*** It is impractical to dead-end and erect each span separately; therefore, each span in a level line must, as erected, have approximately the same tension. Any change in temperature has a much greater effect on short spans than on long, but changes in loading have a greater effect on long spans than on short. However, high tensions on short spans are equalized by a very slight movement of supports, whereas this is not the case on long spans. Computations for sagging are usually made on a somewhat longer span than the average, often designated as the *ruling span.* A formula commonly used to compute the ruling span is as follows: average span + ⅔(longest span − average span). The movement of suspension insulators and deflection of poles may generally be relied upon to equalize the tensions in occasional long or short spans. Sag charts for erecting conductor should usually not cover a range greater than one-half to double the average span, as extreme cases should be investigated.

The effect of insulators swinging longitudinally and increasing or reducing the tension may be computed with the Thomas chart method by taking the horizontal movement of the insulator as an increase or decrease in the length of conductor in the span, exactly as in the case of changes in length due to temperature (see Par. **36**). If the insulator swings toward the span, the effect is to increase the length of conductor by the amount of the swing. The small change in span length resulting from the swing is negligible.

***43. Sagging Charts for Erecting Conductors.*** From the computations outlined in preceding paragraphs, sag and tension curves or tables are prepared for erecting the conductors giving the tension for a range of temperatures in 5.5°C intervals and the sag for a range of spans in 20-ft intervals. Actually, even though the work of preparing these tables has been carried out with great accuracy and the work of erection is done with great care, the irregular profile and unequal span lengths usually encountered in a line result in conductor tensions, especially under maximum loading, slightly different from the computed result (also see Par. **80**). It is impossible to avoid these irregularities, and the differences are of little importance on the average line if (1) the calculations are based on a span length representative of the particular line, (2) some allowance is made in the maximum tension in case of particularly rough country, and (3) correction measures are applied for very long spans or great differences in elevation.

***44. Initial and Final Sags.*** The modulus of elasticity of a conductor for usual computations is assumed to be constant and of a definite value for any given material. Actually, however, individual tests to determine the modulus of elasticity show a considerable variation. The first loading of a conductor gives a slightly curved stress-strain diagram; that is, the modulus is not strictly constant. If the test is continued to a load approaching the elastic limit and then backed down to no load, the conductor will return along a straight stress-strain curve. Subsequent loading of the conductor to the same maximum value will follow this return stress-strain curve on a constant modulus. From a large number of careful tests the average initial stress-strain curve and the average final modulus may be determined.

A conductor erected in a span and not previously stressed to the maximum design tension will stretch, under maximum load, along the initial stress-strain curve; upon release of the load, the conductor will contract along the final modulus and will not return to the initial length by the amount of the permanent set. This results in a slightly greater sag than that at which the conductor was originally installed. Also, the conductor will never quite reach the same maximum tension if the same maximum load is applied a second time. Often sag computations do not warrant consideration of these refinements. However, for conductors stressed beyond the theoretical elastic limit and for composite conductors such as ACSR at high maximum tensions, the final sags may be considerably larger than the initial, and these conditions must be considered.

Such computations are made on the Thomas chart by considering that the maximum tension is the result of the conductor stretching on the initial stress-strain curve from the tension at which it is erected to the maximum loaded tension. By reversing this process on the chart, the initial unstressed length of the conductor is computed by subtracting the initial stretch from the loaded length, and the final unstressed length is found by subtracting the stretch as determined by the final modulus from the loaded length. The initial and final bare-wire sags at stringing temperatures are computed as usual from these initial and final no-load lengths, respectively, the initial curve being used for computing the initial sag and the final curve for computing the final sag. A typical stress-strain curve for copper is shown in Fig. 14-25. The stretch under various conditions is best read directly from the chart rather than attempting to compute from initial and final values of the modulus of elasticity. Thus, from Fig. 14-25, *OA′* is the initial stretch for a load of half the ultimate, which is subtracted from the length of the conductor under maximum load to give the initial unstressed length: and *PA′* is the final stretch, which subtracted from the length under maximum load gives the final unstressed length.

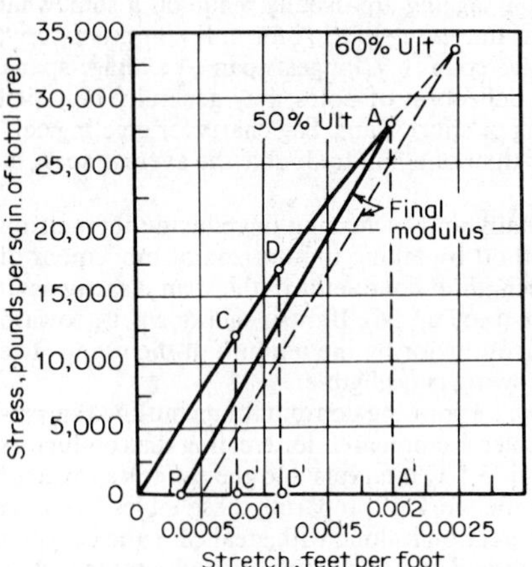


**FIG. 14-25** Stress-strain curve for hard-drawn seven- and nine-strand copper cable.

In drawing the stretch line to intersect the length curve on the Thomas chart it is not necessary to draw the entire stretch curve *OCD* but is more convenient to plot any two points on the Thomas chart corresponding, for instance, to the points *C* and *D*, which have a stretch of *OC′* and *OD′*, respectively. A line drawn through these two points to intersect the length curve will give the initial bare-conductor tension. In making such computations, various short cuts will suggest themselves.

***45. Sag computations for ACSR*** are made exactly as described in Pars. **39** and **34** except for the complications introduced by the composite stress-strain diagram. Most of these difficulties may be eliminated by reading the stretch for any condition or assumed tension directly from the diagram, as described for copper conductors in Par. **44**, and not attempting to compute the stretch from the modulus. The stress-strain diagram must be obtained from tests on actual conductors.

A typical stress-strain diagram for ACSR is shown in Fig. 14-26. It will be noted that the final modulus curve shows a distinct break at *P*. This is due to the permanent set in the aluminum strands, which throws all the load on the steel strands at low tensions. For loads from *M* to *P* the aluminum strands are slightly loose on the steel, and the conductor stretches as if it consisted of the steel core alone; for loads greater than *P*, the aluminum and steel strands work together as a composite conductor.

Changes in temperature change the relative length of aluminum and steel strands and thus change the point at which the steel and aluminum begin to work together. It is only occasionally that conductor tensions fall below the point *P*; often, therefore, computations need not take this effect into account. This is the case in the computation given below.

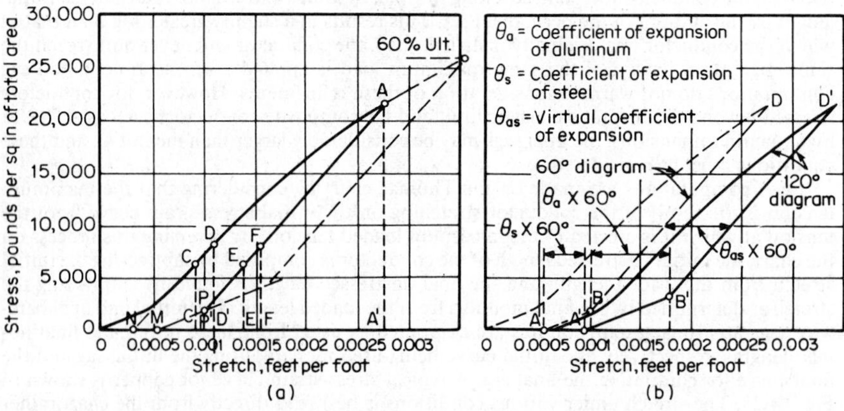


**FIG. 14-26** Stress-strain diagram, No. 4/0 ACSR.

Changes in temperature actually occur while the conductor is under tension and not, as assumed in computations, in the unstressed condition. Such temperature changes change the relative amount of stress in the steel and aluminum even though the total tension is kept constant. Such a temperature change therefore produces a different expansion or contraction in the composite conductor, depending on the conductor tension existing at the time the temperature change occurs. The coefficient of expansion is therefore not the same at all tensions as it is in a conductor of only one material. A very accurate method of computing sags, taking into account all changes in modulus and the coefficient of expansion, is described in a pamphlet published by the Alcoa Conductor Products Co. of the Aluminum Company of America.[59] This pamphlet also contains catenary tables and creep curves for some conductors. Creep is discussed in considerable detail.

However, fairly accurate computations may be made by use of the "virtual coefficient of expansion," which assumes the temperature change to occur under no load. The coefficient of expansion thus obtained is slightly higher than the actual. The virtual coefficient of expansion is used in the following example and is obtained thus: Where $E_a$ and $E_s$ are the modulus of the aluminum and steel strands, respectively, and $E_{as}$ is the modulus of the composite cable, $H_a$ and $H_s$ are the percent area of aluminum and steel, respectively, and $\theta_a$, $\theta_s$, and $\theta_{as}$ are the coefficients of expansion of the aluminum, steel, and composite cable. Average values of $E_a$, $E_s$, $\theta_s$ are given in Table 14-18.

$$\theta_{as} = \theta_a \frac{E_a H_a}{E_{as}} + \theta_s \frac{E_s H_s}{E_{as}} \tag{14-66}$$

$$E_{as} = E_a H_a + E_s H_s \tag{14-67}$$

The coefficient of expansion for 4/0 ACSR, in which $H_a = 0.857$ and $H_s = 0.143$, is 0.0000107.

*Example Calculation.* Required the sag at which a 600-ft span of 4/0 ACSR shall be erected at 15.5°C so that, under heavy loading, ½-in ice, 4-lb wind at −18°C, the tension shall not exceed one-half the ultimate strength; also the final sag at 15.5°C which will occur after the conductors have been loaded to the maximum tension. Symbols are as given in Par. **39.** The following data are taken from Table 14-18.

| | | | |
|---|---|---|---|
| Ultimate strength | | 8420 lb | |
| Maximum tension | | 4200 lb | |
| Conductor weight bare | | 0.292 lb/ft | |
| Resultant weight | | 1396 lb/ft | |
| Area | | 0.1939 in$^2$ | |
| Modulus of elasticity | | Read stretch from Fig. 14-26 | |
| Coefficient of expansion | | 0.0000107 | |
| $T_1 = 4{,}200/(1{,}396 \times 600)$ | | 5026 (Fig. 14-27) (point *M*) | |
| $l_0$ at −18°C | | 1.00166 (point *M'*) | |
| Initial | | | Point |
| Stress | = 4200/0.1939 | = 21,712 lb/in$^2$ (Fig. 14-26*a*) | (*A*) |
| Stretch | | = 0.00280 (Fig. 14-26*a*) | *O-A'* |
| $l_0$ at −18°C | = 1.00166 − 0.00280 | = 0.99886 | |
| Change 15.5°C | 60 × 0.0000107 | = 0.00064 | |
| $l_0$ at 15.5°C | = 0.99886 + 0.00064 | = 0.99950 | |

It is not necessary to plot the foregoing points on the Thomas chart. As described in Par. **44,** it is more convenient to plot two points on the stretch curve such as *C* and *D* (Fig. 14-27), corresponding to points *C* and *D* (Fig. 14-26*a*), than to draw in the entire stretch curve

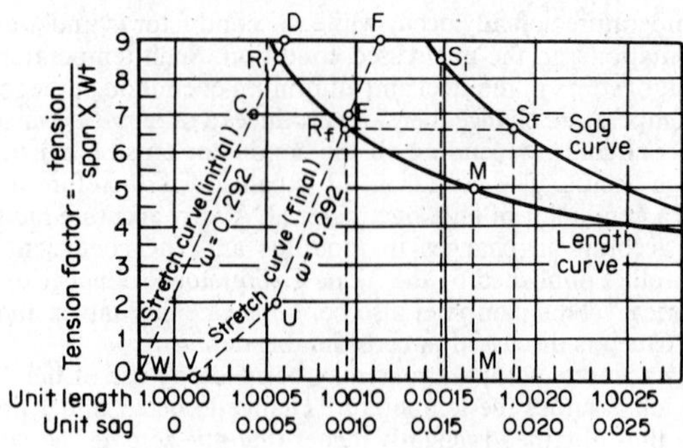


**FIG. 14-27** Sag and tension determination for ACSR.

*OCD* (Fig. 14-26*a*). The intersection of a line through these two points with the length curve gives the tension factor at 15.5°C bare conductor.

| | | | Point |
|---|---|---|---|
| Stress when $T_1$ = 9.00 | = 9.00 × 0.292 × 600/0.1939 | = 8130 lb/in² | |
| Stress when $T_1$ = 7.00 | = 7.00 × 0.292 × 600/0.1939 | = 6320 lb/in² | |
| Initial stretch at 8130 lb/in² | | = 0.00112 (Fig. 14-26*a*) | *O-D'* |
| Initial stretch at 6320 lb/in² | | = 0.00094 | *O-C'* |
| Length at $T_1$ = 9.00 | = 0.99950 + 0.00112 | = 1.00062 (Fig. 14-27) | |
| Length at $T_1$ = 7.00 | = 0.99950 + 0.00094 | = 1.00044 | |
| $T_1$ at 60°F | | = 8.44 | $R_i$ |
| Initial tension | = 8.50 × 600 × 0.292 | = 1489 lb | |
| $d_1$ at 60°F | | = 0.01475 | $S_i$ |
| Initial sag | = 0.01475 × 600 | = 8.85 | |
| Final | | | |
| Stress | = 4200/0.1939 | = 21,650 lb/in² (Fig. 14-26*a*) | *A* |
| Stretch | | = 0.00222 | *M-A'* |
| $l_0$ at 0°F | = 1.00166 − 0.00222 | = 0.99944 | |
| $l_0$ at 60°F | = 0.99947 + 0.00064 | = 1.00011 (Fig. 14-27) | *V* |
| Stress when $T_1$ = 9.00 | = 9.00 × 0.292 × 600/0.1939 | = 8130 lb/in² | |
| Stress when $T_1$ = 7.00 | = 7.00 × 0.292 × 600/0.1939 | = 6320 lb/in² | |
| Final stretch at 8130 lb/in² | | = 0.00102 (Fig. 14-26*a*) | *M-F'* |
| Final stretch at 6320 lb/in² | | = 0.00086 | *M-E'* |
| Length at $T_1$ = 9.00 | = 1.00011 + 0.00102 | = 1.00113 (Fig. 14-27) | |
| Length at $T_1$ = 7.00 | = 1.00011 + 0.00086 | = 1.00097 | |
| $T_1$ at 60°F | = | = 6.67 | $R_f$ |
| Final tension | = 6.67 × 600 × 0.292 | = 1168 lb | |
| $d_1$ at 60°F | = | = 0.0188 | $S_f$ |
| Final sag | = 0.0188 × 600 | = 11.28 ft | |

It will be noted in Fig. 14-26*a* that in the above example the point *P*, where the steel starts to take the entire stress, is at 1800 lb/in$^2$, or a tension of 350 lb. This corresponds to a tension factor for this conductor of 350/(600 $\times$ 0.292), or 1.99. This stress of 1800 lb/in$^2$ causes a stretch of 0.00046 (distance *MP* as measured on Fig. 14-26*a*) and would result in a length of 1.00013 + 0.00046 = 1.00059.

This point of unit tension factor 1.99 and length 1.00059 is plotted on the Thomas chart (Fig. 14-27), point *U*, to illustrate that this break in the final stress-strain curve does not in this case have any effect on the solution of the problem.

As a matter of interest, the point *U* is plotted on the Thomas chart, which corresponds to the point *P* on the stress-strain chart (Fig. 14-26*a*). The point *W* on the Thomas chart corresponds to the point *N* on the stress-strain diagram. This point *N* is actually a more convenient reference point than point *O*.

The effect of change in temperature is accurately accounted for in the method described in the Alcoa Company's pamphlet. However, fairly accurate results can be obtained by assuming that temperature expansion and contraction take place at no load. It is not necessary to redraw the stress-strain diagram for each temperature, as by this assumption the stress-strain diagrams for all temperatures are identical except for the location of the point at which the aluminum begins to pick up load, that is, point *B*. The method of computing this change is illustrated in Fig. 14-26*b*. The 120° stress-strain diagram is indicated as being offset from the 60° diagram by the virtual coefficient of expansion, or 0.00064 ft, showing that the composite conductor increases this amount in length. However, between points *A* and *B*, where the steel alone is acting, the conductor changed in length only by the expansion of the steel, or 60° $\times$ $\theta_s$ = 0.0004 ft, shown as $AA_1$. The aluminum strands expanded 0.00077 ft (60° $\times$ $\theta_a$) at the same time; and, although this change in the length of the aluminum strands did not change the length of the conductor, it did move the point at which the aluminum begins to take up stress from *B* to *B'*.

***46. Unbalanced Ice Loads.***[60,61] The jump of the conductor resulting from ice dropping off one span of an ice-covered line has been the cause of many serious outages on long-span lines where conductors are arranged in the same vertical plane. The vertical spacing required to prevent "ice-jump" trouble may be estimated by static calculations of the differential sag of two vertically adjacent conductors assuming one conductor has ice and the other has no ice. Normally, sufficient clearance is provided to accommodate this sag difference, including a factor for sag error with a margin for switching surge withstand.

Utility practice, based on historically satisfactory field performance, is typically based on the following criteria for vertical phase-to-phase clearance due to ice.

A maximum sag error of 6 in is assumed.

The upper conductor is assumed to be subject to maximum ice load, typically 1 in for short ruling spans, 0.5 in for unusually long spans.

The lower conductor is assumed to be completely free of ice.

Clearance sufficient to withstand the maximum switching surge is to be provided.

The calculation is performed using normal sag-tension procedures. For certain line designs (e.g., compact lines with post insulators), dynamic techniques are possible.[41,47] However, the trouble has been practically cleared up by horizontally offsetting the conductors from 18 in to 3 ft on medium-voltage lines. The conductor jumps in practically a vertical plane, and this should be true if no wind is blowing, since then all forces and reactions are in a vertical direction.

***47. Supports at Different Elevations.*** For the usual cases encountered in a line, the slight local variations from the calculated sags and tensions, due to the difference in elevation of supports at the ends of a span, are of no importance. The differences in tension due to this and to the method of supporting the conductor in sagging are discussed under Conductor Installation (Pars. **79** to **81**).

The sag *d* (Fig. 14-28), measured vertically to a tangent to the conductor which is parallel to a line through the supports, will be very nearly equal to the sag in a level span of a length equal to the slope distance $S_1$. For very large sags on inclined spans, Martin suggests a slight correction to the slope distance $S_1$ which gives very accurate results. The sag in the inclined span is more accurately equal to the sag in a level span of a length equal to $S_1 + (S_1 - S)$. Generally, however, the difference between the horizontal and slope distance is negligible, and the sag is taken as the same as the sag in a level span of the same horizontal length. This latter statement would be theoretically correct if the conductor hung in the form of a parabola, that is, if the weight of wire were measured in feet of horizontal distance instead of along the length of the wire.

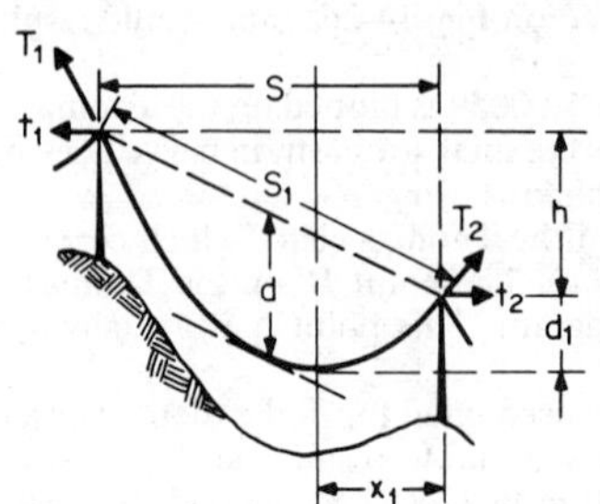


**FIG. 14-28** Span with support at different elevations.

The dimensions to the low point of the sag $d_1$ and $x_1$ shown in Fig. 14-28 are obtained from the parabola and are as follows:

$$d_1 = d\left(1 - \frac{h}{4d}\right)^2 \qquad (14\text{-}68)$$

$$x_1 = \frac{S}{2}\left(1 - \frac{h}{4d}\right) \qquad (14\text{-}69)$$

On steep hillsides the low point of the sag may fall outside the low support; in such a case $h/4d$ would be greater than 1.

The horizontal components of the conductor tensions $t_2$ and $t_2$ (Fig. 14-28) must be equal, and as the vertical component is greater on the uphill side, the resultant tension $T_1$ must be greater than $T_2$. The catenary formula applied to this case is

$$T_1 = T_2 + wh \qquad (14\text{-}70)$$

in which $w$ = weight per foot of the wire and $T_1$ and $T_2$ = resultant tension in the uphill and downhill support, respectively.

***48. Inclined-Span Calculations.*** The mathematics of the inclined catenary is quite complicated and will not be analyzed here. However, workable methods[62] have been devised and may be used when necessary. On the other hand, the equations of the inclined parabola are very simple, and calculations for any usual span may be made on the parabola formulas with sufficient accuracy for almost any practical purpose.

The equation of the inclined parabola is exactly the same form as for the parabola with level supports. This is stated in mathematical language as follows: The equation of a parabola, referred to a tangent and the diameter through the point of contact, is

$$x^2 = 2py$$

Thus, if in Fig. 14-29 the $y$ axis is vertical and the $x$ axis inclined, $y$ is measured vertically and $x$ is measured parallel to the $x$ axis. The sag is

$$d = \frac{wS_1^2}{8t_1} \qquad (14\text{-}71)$$

The sag $d$ is measured vertically, and $S_1$ = slope distance between supports, $t_1$ = tension in the direction of the $x$ axis, and $w$ = weight per foot, also as measured parallel with the $x$ axis. In other words, the total weight of conductor in the span is approximately $wS_1$.

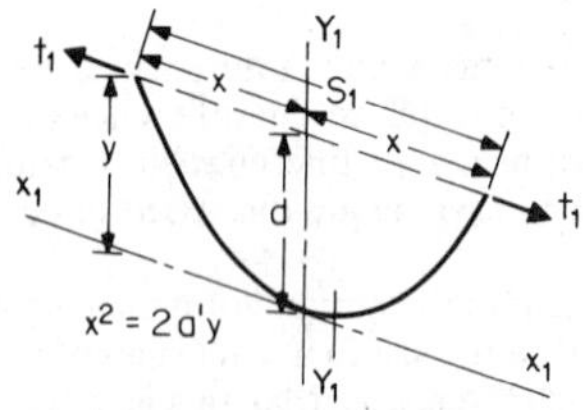

**FIG. 14-29** Inclined parabola referred to nonrectangular coordinates.

***49. Wind-Induced Motion of Overhead Conductors.*** In addition to ordinary "blowout" of overhead conductors (i.e., the swinging motion of the conductor due to the normal component of wind), there are three types of cyclic wind-induced motions that can be a source of damage to structures or conductors or that can result in sufficient reduction in electrical clearance to cause flashover. The categories of wind-induced cyclic motion are aeolian vibration, galloping, and wake-induced oscillation.

Aeolian vibration can occur when conductors are exposed to a steady low-velocity wind. If the amplitude of such vibration is sufficient, it can result in strand fatigue and/or fatigue of conductor accessories. The amplitude of vibration can be reduced by reducing the conductor tension, adding damping by using dampers (or clamps with damping characteristics), or by the use of special conductors which either provide more damping than standard conductors or are shaped so as to prevent resonance between the tensioned conductor span and the wind-induced vibration force.

Galloping is normally confined to conductors with a coating of glaze ice over at least part of their circumference and thus is not a problem in those areas where ice storms do not occur. It may be controlled by the use of various accessories attached to the conductor in the span to change mechanical and/or damping characteristics. Specially shaped conductors or conductor accessories which alter the iced conductor's aerodynamic characteristics, particularly those that increase aerodynamic damping, are also effective. The amplitude of galloping motions can be reduced by the use of higher conductor tensions and evidence suggests higher tensions can also reduce the possibility of occurrence. Galloping and aeolian vibration occur in both single and bundled conductors.

Wake-induced oscillation is limited to lines having bundled conductors and results from aerodynamic forces on the downstream conductor of the bundle as it moves in and out of the wake of the upstream conductor. Wake-induced oscillation is controlled by maintaining sufficiently large conductor spacing in the bundle, unequal subspan lengths, and tilting the bundles.

*Aeolian Vibration.* As wind blows across a conductor, vortices are shed from the top and bottom of the conductor. The vortex shedding is accompanied by a varying pressure on the top and bottom of the conductor that encourages cyclic vibration of the conductor perpendicular to the direction of wind flow. The frequency at which this alternating pressure occurs is given by the expression

$$f = 3.26 \times U/d \tag{14-72}$$

where $U$ = wind speed, mi/h
$d$ = conductor diameter, in
$f$ = frequency, Hz

For a 1.0-in-diameter conductor exposed to a 10-mi/h wind, the vortex shedding force oscillates at 32.6 Hz. To develop significant amplitudes, there must be a resonance between this oscillating wind force and the vibrating catenary (conductor). The fundamental frequency of vibration of the suspended conductor is in the range of 0.1 to 1.0 Hz. Therefore, the aeolian vibration force will be unlikely to excite a fundamental span mode. This is verified by actual conductor performance where significant amplitudes are usually observed for frequencies in the range of 10 to 100 Hz. Practical wind speeds cause vortex shedding forces of greater than 10 Hz, eliminating frequencies below this level, and frequencies above 100 Hz are not present because of the rapid increase in conductor self-damping for these higher frequencies.

The maximum alternating stress resulting in strand fatigue normally occurs at the conductor clamp. The stress is related to the amplitude of conductor vibration and is the amplitude normally measured by field recording devices. Stress and amplitude of vibration can be related by analytical means such as the Poffenberger-Swart formula.[63] The amplitude of aeolian vibration is fixed by the balance of energy input from the wind-induced vortex shedding forces and the energy loss due to conductor, accessory, and structure damping. The addition of dampers (see Par. **75**) to the conductor has been established as an effective means of control.[64] Special conductors such as SDC and SSAC[65] have also been shown effective in reducing the strand stress levels.

Another effective means of limiting vibration fatigue problems is to increase the self-damping of standard conductors by reducing tension. As a practical approximation, stringing conductors to a final unloaded tension of 15% or less at the minimum seasonal average temperature (usually 0 to 30°F) will prevent vibration fatigue problems. Higher tensions are routinely used in areas where the line is parallel to existing lines and the higher tension on the existing line has not resulted in problems. The use of vibration dampers or special antivibration conductors can also allow the use of higher tension levels.

As with single conductors, bundled conductors are subjected to aeolian vibration. However, the interaction of conductors in the bundle due to slightly different tensions and increased damping from spacers results in lower vibration levels for bundles than for single conductors in the same wind exposure.

*Galloping.* Both bundled and single conductors are subject to galloping during or after glaze ice storms. Galloping oscillations occur at frequencies near the fundamental span mode or its second or third harmonic (0.1 to 1.0 Hz) and exhibit maximum amplitudes as large as the conductor sag. While there has been extensive debate concerning the galloping mechanism, and considerable experimental and analytical study, it appears that there presently exist a number of control methods that are effective in reducing the amplitude and incidence of galloping motion. In-span hardware, such as the "detuning pendulum" developed by EPRI,[66] and the "winddamper" developed by A.S. Richardson,[67] are effective for existing spans where galloping occurs. The T2 conductor,[68] developed by Kaiser, and several other hardware devices are available to control galloping in new lines.

In contrast, control methods such as sleet melting by use of high current levels appear to be almost totally ineffective in stopping galloping and can result in annealing damage to the conductor.

*Wake-Induced Oscillation.* Bundled conductors are subject to wake-induced oscillations with amplitudes and frequencies typically between that of aeolian vibration and galloping. The frequencies of oscillation are normally in the range of 1 to 10 Hz and the amplitudes are in the range of 10 conductor diameters. The modes in which such vibration occurs are considerably more complex than the modes exhibited during either galloping or the almost invisible aeolioan vibrations. The source of wind energy for wake-induced oscillation is, as the name suggests, the wake from the windward conductor of the bundle which causes the motion of the downwind conductor.

There are three basic approaches to the control of wake-induced oscillation[69]: two involve reducing the input of wind energy and the third involves detuning the mechanical bundle system to prevent resonance. The methods based on reducing wind energy input to the bundle are bundle tilting and bundle sizing. By tilting the bundle to angles of 20° or more, the downwind conductors are moved to the edge of the upwind conductor's wake and the energy input is reduced. By keeping the subconductor spacing to the order of 20 times the conductor diameter, the wind energy input to the windward conductor is reduced by being moved to a wake region of reduced intensity. The third commonly used method to control or eliminate wake-induced oscillations is to stagger the length or simply to shorten the average subspan length. This method does not control those oscillations where the bundle moves as a rigid body and is somewhat dependent on the mechanical characteristics of the spacers.

In comparison to the damage that can result from aeolian vibration or galloping, field reports of wake-induced oscillation damage are usually of a minor nature being primarily conductor abrasion from clashing and spacer breakage, neither of which normally results in system outages.

## Supporting Structures

***50. Types of Supporting Structure.*** Numerous types of structure are used for supporting transmission-line conductors, for example, self-supporting steel towers, guyed steel towers, self-supporting aluminum towers, guyed aluminum towers, self-supporting steel poles, flexible and semiflexible steel towers and poles, rope suspension, wood poles, wood H frames, and concrete poles. The type of supporting structure to use depends upon such factors as the location of the line, importance of the line, desired life of the line, money available for initial investment, cost of maintenance, and availability of material. Because of the wide conductor spacing required for electrical clearances and insulation, the high tensile stresses used in conductors and ground cables to pull these cables up to a sag which will keep the heights of the structures within reason, the long spans necessary for crossing ravines in mountainous country, and the reliance to be placed on a major trunk line, lines exceeding 345 kV are frequently built of self-supporting steel towers although guyed and rope-suspension structures are increasingly applied. A line built with self-supporting steel towers is very satisfactory in all respects, as it requires less inspection and has a maximum life with minimum maintenance costs. However, high-strength aluminum-alloy towers are available, and their use is on the increase. They have the advantage of better resistance to corrosive atmospheres than steel.[70] The structural configurations and design details are the same as with steel, with the added problem of greater deflections when stresses are applied owing to the lower modulus of elasticity of aluminum. The effect of long-time creep of aluminum is yet to be determined. Self-supporting steel poles are frequently used in congested districts where right-of-way is limited and short spans are necessary. The advent of EHV has brought a great variety of new structural configurations. Details of some of these have been published. *Electrical World,* Nov. 15, 1965, pp. 95–118, contains outline drawings of 35 towers and six wood-pole H-frame structures as applied to EHV, as well as a tabulation of specification items of EHV lines in the United States and Canada. The *Transmission Line Reference Book, 345 kV and Above,* 2d ed., 1982, published by EPRI,[3] contains details of a broad spectrum of 345- through 800-kV structures. Some examples are illustrated and described in Par. **56**.

*Wood poles* are used extensively where they are readily available. Medium- and lower-voltage lines can be built economically with such poles fitted with either steel or wood crossarms. *Wood H-frames* composed of two poles tied together at the top with wood or steel crossarms have been successfully used for the higher-voltage lines up to 345 kV. To take full advantage of the transverse strength, such poles can be braced internally for at least a portion of their height with wood X bracing.

*Concrete poles* have been used in some parts of the world where timber is scarce and where the ingredients for making concrete are readily obtainable. Another advantage is that they are impervious to insect damage and other forms of decay prevalent with wood structures in tropical or subtropical climates. They are generally cast in units, by using standard forms, and transported to the site, although they may be manufactured where used. Concrete poles should always have sufficient prestressed steel reinforcement to take care of the bending stresses due to wind loads, pulls from cables, etc., in addition to being designed as columns under vertical loads. In all structures conductor configuration and the effect of various forces which may act upon them must be taken into account.

***51. Conductor Spacing and Clearances***

*Horizontal Configuration.* The minimum spacing of conductors on structures where post-type or V-string insulators are used on medium-length spans will generally depend upon the least separation that can be used at midspan without the conductors approaching too closely under adverse wind- or ice-loading conditions.

With suspension insulators a different problem exists, as the swing of the insulator string has to be considered and clearances to the structure determined. This will generally give conductors a spacing at the supports which will be greater than the required midspan separation. One typical rule is to calculate the swing of the insulator string, both with the wind on the bare conductor and the wind on the ice-coated conductor with the corresponding vertical loads acting at the point of conductor suspension, to determine which condition gives the maximum deflection. The vertical loads should be taken on a length of span which is two-thirds the span for the horizontal loads. This will allow a certain amount of leeway in using a standard height structure at a location where the ground is lower than at the two adjacent structures. After the length of the insulator string has been determined electrically and the angle of insulator swing calculated, a normal electrical clearance is established to the structure from the deflected position of the conductor, which, when applied to the three conductors in their relative positions, will determine the necessary horizontal separation of the phases at the supports. This separation should then be checked to see whether or not it is sufficient for the midspan separation required. Midspan separations that will not be subject to flashover if the conductors begin to swing out of step are usually inherent on high-voltage lines owing to the clearances required at the structures. On very long spans and on the longer spans of low-voltage lines, these spacings may be insufficient. Thomas[71] proposed a horizontal-spacing formula for the determination of safe midspan spacings in windy territory where gusts and strong eddies might cause wires to start swinging at different periods,

$$\delta = CdD/w + A + L/2 \qquad (14\text{-}73)$$

in which $\delta$ = horizontal spacing in feet; $C$ = an experience factor discussed later; $d$ = percent sag of the condition to be studied; $D$ = overall diameter of the conductor; $w$ = conductor weight, in pounds per foot, used in calculating $d$; $A$ = arcing distance of the line voltage (1 ft/110 kV); $L$ = length, in feet, of the swinging portion of the insulator string. Thomas proposed an experience factor of 4 for copper and 3.5 for ACSR. It has since been found that, in areas not subject to frequent violent winds, values of $C$ as low as 1 will provide safe midspan spacings. Thomas was doubtful whether or not the added $L/2$ distance is necessary, since insulators seldom swing out of step. This doubt seems to have been justified. Spacing is further discussed in Par. **68.**

*Vertical Configuration.* Where the conductors are arranged in vertical configuration, the same electrical clearances will apply for the same voltage as for horizontal configuration; but it may be necessary to increase the vertical separation somewhat to prevent the conductors from coming together or approaching too closely at the center of the span when unequal ice-loading conditions occur or the ice falls off a lower conductor first (see Par. **46**).

In Fig. 14-30, $\theta$ = angle of insulator swing from vertical, $H$ = horizontal span, $V$ = vertical span, $w$ = weight of conductor with or without ice load per lineal foot, $w_e$ = weight of insulator string including hardware. Then

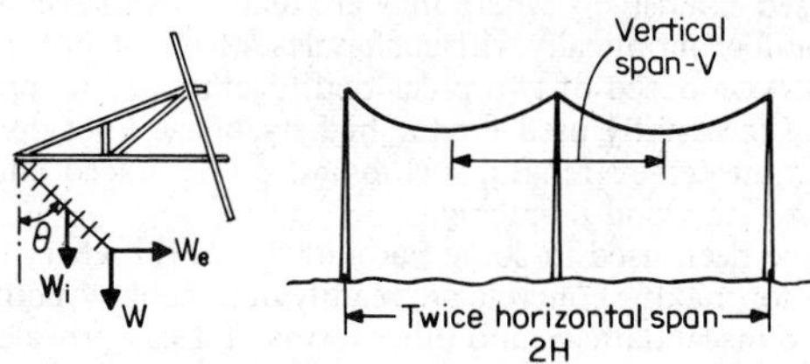


**FIG. 14-30** Determination of suspension insulator swing.

$$\tan \theta = Hw_e/(Vw + w_i/2) \qquad (14\text{-}74)$$

Ground wires, if used, are located above the conductors for lightning protection and in such a position that there is no danger of contact with the conductors at midspan. As ground wires are generally strung with less sag than the conductor cables, ample clearance at midspan is readily obtainable.

The above considerations taken together with the maximum vertical sag to be used and the height required for the conductors above the ground level will determine the height and width of the supporting structure. Extensions can be used where the terrain requires a higher structure than normal.

**52.** ***Transverse Forces on Support Structures.*** Transverse forces acting on towers or poles are due to wind on the conductors and ground cables (and ice coating if in ice districts), wind on the structures, and horizontal components of the tensions in the cables at angle turns in the line (Fig. 14-31).

The stress due to an angle in the line is computed by finding the resultant force produced by the wires in the two adjacent spans. For example, in Fig. 14-31, if the change in the direction of the line is the angle $a$ and the stresses $t$ in the adjacent spans are equal to each other, the resultant force

$$F = 2t \sin (a/2) \qquad (14\text{-}75)$$

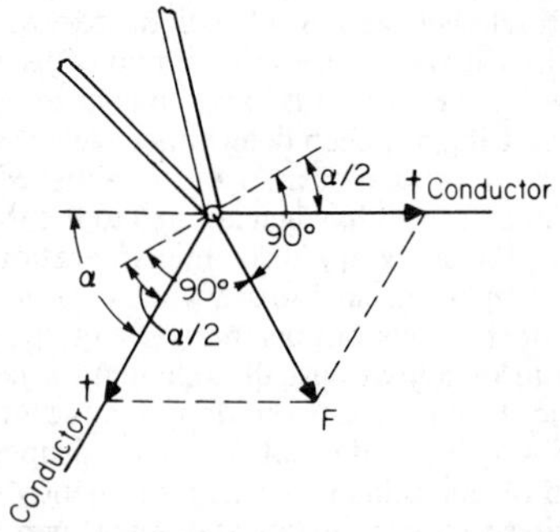

**FIG. 14-31** Determination of transverse forces.

Table 14-19 gives the resultant force $F$ due to a tension $t$ of 1000 lb in each conductor of two adjacent spans. The resultant force due to each conductor may be thus computed and the moments about the ground line determined. These moments may be added to those produced by the wind pressure to find the maximum stress.

In applying wind loads to the structure, the appropriate force coefficients, exposure coefficients, and gust reponse factors given in Par. **28** should be used.

***53. Longitudinal Forces on Support Structures.*** Longitudinal forces acting on towers or poles are due mainly to the maximum tension which is assumed to exist in the conductor and ground wires if broken. Ordinarily, especially with suspension insulator strings, these tensions are balanced in the adjacent spans; but if a conductor breaks, a distinct force is produced along the line due to unbalanced tension. If the break occurs on a conductor at the end of a crossarm, there is, in addition to the longitudinal force, a torsional force introduced which must be resisted by the structure. Wind acting in the direction of the line is not ordinarily a factor, as the maximum tension in the conductor is produced when the wind is blowing transverse to the line. As to the reduced stress which occurs in a span from the breaking of a conductor with the suspension insulator string deflecting in the direction of the line, the best practice is to ignore this reduction in tension, as the force due to breaking may cause an impact which more than offsets the reduction in tension. Special release clamps were devised for use on the Plymouth Meeting–Siegfried line of the Philadelphia Electric Company so that, if an insulator string deflected to an angle of 20° in the direction of the line, the clamping mechanism would release the pressure with only the friction in the saddle holding the conductor. This reduced the tension in the conductor considerably; and by assuming a low value for the tension in the conductor due to a break, a more economical structure was obtained.

***54. Vertical Forces on Support Structures.*** Vertical forces acting on towers or poles are those caused by the weight of that portion of the conductors, plus ice loading if any, which is supported by the structure in question. In addition, there are the weights due to insulators and accessories and the weight of the structure itself. If a structure is located in a valley, there may actually be uplift on it, if the vertical components of the tensions in the conductors exceed the downward loads.

***55. Combined Forces on Support Structures.*** In determining the maximum forces acting on towers or poles, it is necessary to combine the transverse forces, longitudinal forces (including torsion), and vertical forces so that they act simultaneously. Several different combinations of loading conditions may be desirable, as follows:

**a.** A condition with all conductors intact and the full transverse and vertical forces acting. These forces should correspond to the appropriate NESC district loadings and extreme wind loadings described in Par. **29.** The NESC also specifies overload capacity factors which must be applied to the transverse, vertical, and longitudinal loadings to provide adequate strength of the support structures. These factors depend on the grade of construction and on the type of structure.

**TABLE 14-19** Resultant Force Due to Equal Tensions of 1000 Lb in Adjacent Spans

| Angle | | Resultant $F$, lb | Angle | | Resultant $F$, lb |
|---|---|---|---|---|---|
| $a$ deg | $a/2$ deg | | $a$ deg | $a/2$ deg | |
| 10 | 5 | 174.4 | 70 | 35 | 1147.2 |
| 20 | 10 | 347.2 | 80 | 40 | 1285.6 |
| 30 | 15 | 517.6 | 90 | 45 | 1414.2 |
| 40 | 20 | 684.0 | 100 | 50 | 1532.0 |
| 50 | 25 | 845.2 | 110 | 55 | 1638.4 |
| 60 | 30 | 1000.0 | 120 | 60 | 1732.0 |

NOTE: 1 lb = 4.448 N.

**b.** A condition with all conductors intact, except the number it is desired to assume broken, with the transverse and vertical forces computed for each particular conductor, according to whether or not it is assumed broken. The longitudinal forces due to broken conductors must be combined with the transverse and vertical forces at all points of support where the conductors are assumed broken. It is customary, when more than one conductor is assumed broken, to consider all breaks in the same span and at the suports which will produce the maximum overturning moment, the maximum torque, or a combination of both.

**c.** A condition in some localities where extraheavy vertical loads, caused by an unusually large formation of ice on the conductors, may occur. These loads are combined with the weight of the structure.

**d.** A condition with vertical loads acting upward at the conductor supports.

NOTE. It is not customary to combine transverse and longitudinal loads with the loads specified under *c* and *d*.

Other factors may enter into the determination of the maximum forces acting on supporting structures in special cases, such as the horizontal and vertical components of tensions in guys and the addition of pole-top transformers, switches, and working platforms.[72]

The proper number of conductors to assume broken is a debatable question and depends upon what margin of safety is desired and the amount of money it is desired to invest for this security. Generally speaking, the minimum number of conductors to assume broken for tangent suspension single-circuit towers should be either one ground wire or any one conductor, and for double-circuit towers either one ground wire and one conductor or any two conductors on the same side of the tower and in the same span, by using the different cable supports for application of the forces to determine the maximum stress in each member of the tower. Anchor or dead-end towers should be able to withstand all or any number of conductors and ground wires broken. Generally, the condition of broken conductors and ground wires on one side of the tower will produce greater stresses in the web members than if all the conductors and ground wires are considered broken, owing to the unbalanced torsional forces existing when only the conductors and ground wires on one side of the tower are broken.

***56. Types of Metal Structures.*** Structures may support single, double, or multiple circuits. The first two types are generally used for transmission lines except in congested areas where right-of-way is very expensive and it is desired to transmit large blocks of power over one line. In such a case three or more circuits may be supported by the structures.

*Self-Supporting or Rigid Structures.* On both single- and double-circuit tower lines of any considerable length, at least three kinds of towers are required for economic reasons:

1. A tangent suspension tower which can be used for normal spans where no angles in the line occur (Figs. 14-32 and 14-33).
2. An angle suspension tower which can be used for normal spans with a small angle turn in the line or with longer spans on tangents.
3. An angle tower which can be used for normal spans with a large angle turn in the line, with extralong spans on tangent, or as a full dead-end tower for anchoring. Insulators may be either suspended or in the strain position.

Very often it is desirable to introduce a fourth kind of tower with insulators always in the strain position to take care of exceptionally large angle turns in the line; in extremely long spans on tangent; and also, where required, as a full dead-end tower. When this type of tower is provided, the tower listed in item 3 may be of lighter construction and not used for dead-end purposes.

Double-circuit towers with the vertical configuration of conductors, as used on different lines, are very much alike in appearance, generally being square in cross section. It is customary to locate the middle conductors outside the upper and lower conductors, for reasons explained in Par. **46**.

With single-circuit towers and the conductors arranged in horizontal configuration, a different problem arises which has resulted in the design of special patented structures for

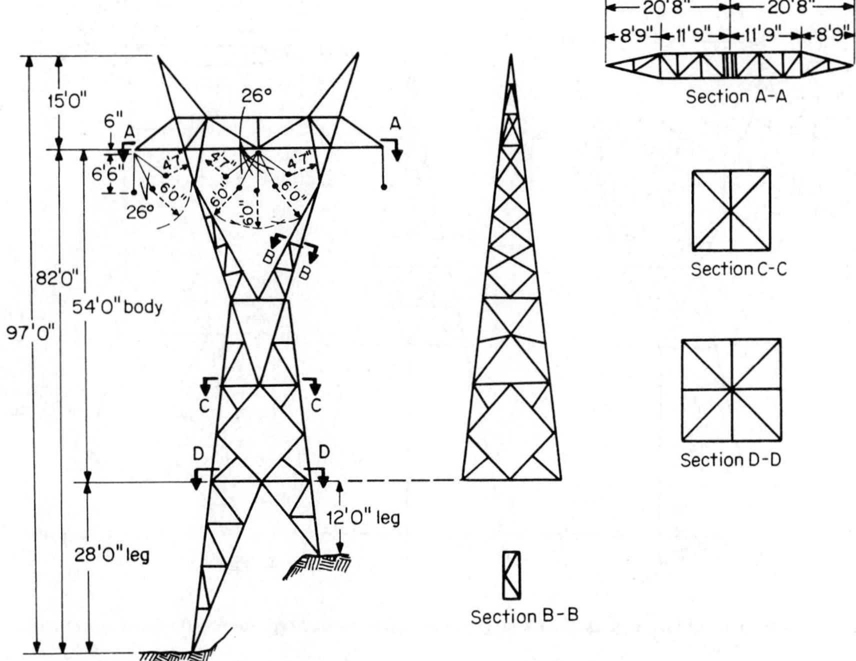


**FIG. 14-32** Tennessee Valley Authority 161-kV single-circuit tangent suspension corset-type tower. *(Designed by Blaw-Knox Co.)*

the higher-voltage lines with wide conductor spacing. The shape of these towers has been developed with a view to minimizing the weight of steel required in the superstructure and also reducing the size of footings by minimizing the effect of torsion. The more common types are the Blaw-Knox tower (Fig. 14-32), or corseted type, as originally used on the Plymouth Meeting–Conowingo line of the Philadelphia Electric Company; and the American Bridge Company's rotated tower (Fig. 14-33), used on the first Hoover Dam–Los Angeles line (this line has since been uprated to 500 kV) and also by the Bonneville system and on lines of the Tennessee Valley Authority. Either of these types serves the purpose for which it was intended. The theory behind the rotated tower is that the greatest overturning moment is caused by a combination of the transverse forces and longitudinal forces, due to broken conductors, acting simultaneously, which produces a resultant force acting at an angle of approximately 45° with the direction of the line. In this case the whole four tower legs are resisting the overturning moment, thereby reducing foundation loads and consequently costs. Under normal conditions of loading, with only the transverse forces acting, two legs on the diagonal separation will take care of the overturning moment. Obviously the greatest advantage of the rotated type over the nonrotated type is on tangent towers and towers used for small-angle turns in the line when the transverse and longitudinal forces are approximately equal.

Figure 14-34*a* shows a TVA 500-kV conventional-design tangent self-supporting tower for a bundle-conductor line having three 971,600-cmil ACSR conductors per phase. The overhead ground-wire clamps are suspended and insulated from the tower by means of distribution-type guy strain insulators. The overhead ground wires are composed of seven strands of No. 9 Alumoweld and are used for carrier-current communication channels.[73] Each ground-wire insulator is provided with a spill-over gap to protect it during lightning discharges.

It is interesting to compare Fig. 14-34*a* with Fig. 14-34*b*. Both show 500-kV towers, but

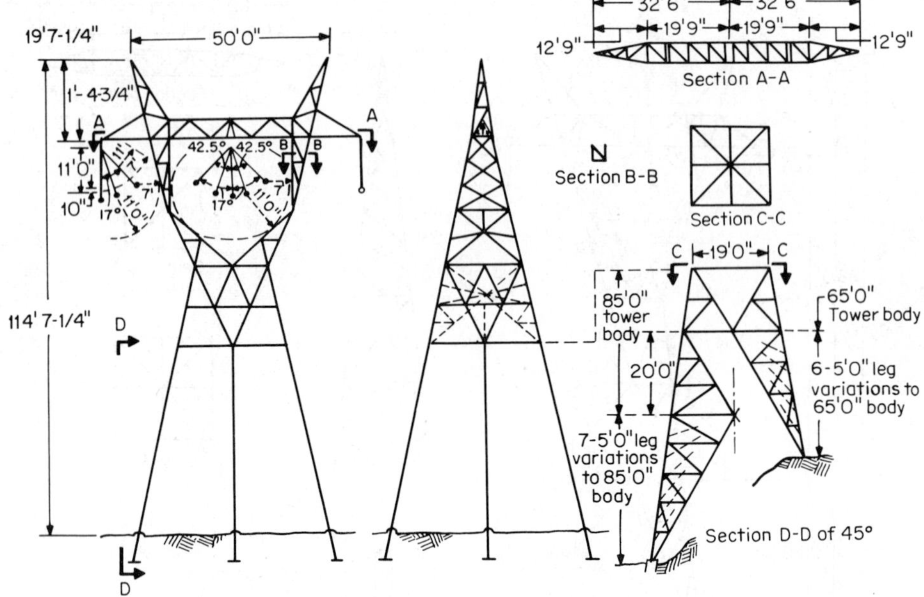


**FIG. 14-33** City of Los Angeles 287-kV tangent suspension rotated-type tower. *(Designed by American Bridge Company.)*

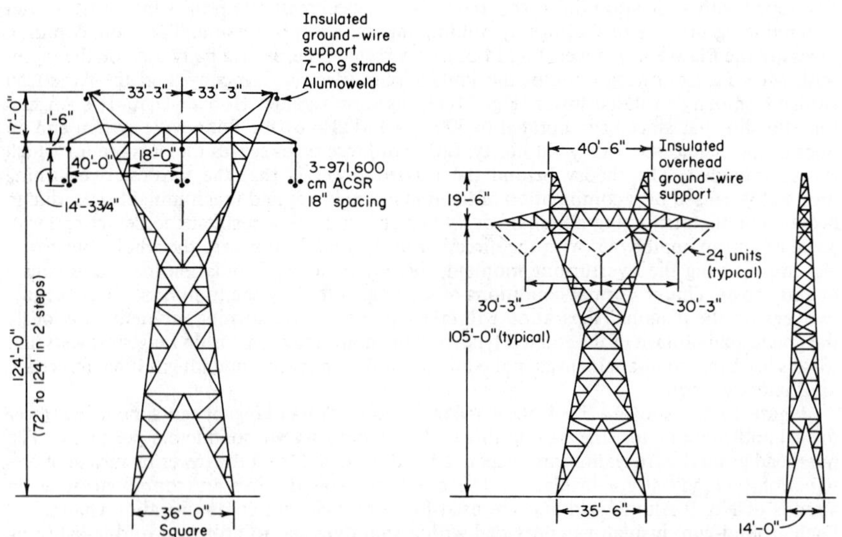


**FIG. 14-34a** 500-kV tangent self-supporting tower. *(Tennessee Valley Authority.)*

**FIG. 14-34b** 500-kV semiflexible steel tangent tower of Arkansas Power and Light Company.

Fig. 14-34*b* is designed for a narrower right-of-way. The wind side swing of the conductors in a span is half the sag at 30° side swing, and this is common to both towers. Therefore, the saving in right-of-way for Fig. 14-34*b* is 40 ft plus 7 ft, 7 in, less 30 ft, 3 in, or 17 ft, 4 in on each side, or a total of approximately 35 ft.

Figure 14-35 shows a light-suspension 500-kV single-circuit tower typically used by the Bonneville Power Administration (BPA), supporting three 1,192,500-cmil ACSR Bunting conductors per phase and two 7-strand No. 8 Alumoweld overhead ground wires. BPA uses continuous overhead ground wires throughout its entire 500-kV network except on single-circuit lines west of the Cascade Mountains. In the latter case, overhead ground wires extend

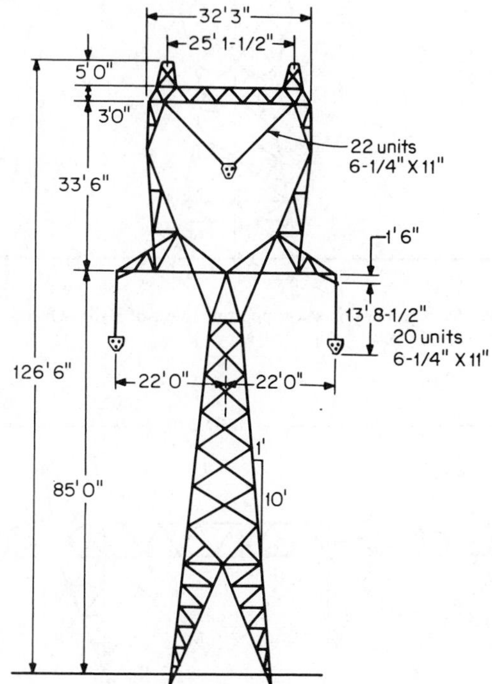


**FIG. 14-35** Suspension-type 500-kV tower. *(Bonneville Power Administration.)*

1 mi out from the substations. Typically, BPA 500-kV lines are designed to withstand 100-mi/h winds and solid ice coatings up to 1½ in.

A steel suspension self-supporting tower used by Hydro-Quebec for 735-kV Manicouagan lines is shown in Fig. 14-36. Line conductors consist of a four conductor bundle per phase, each conductor being 1028-kcmil ACSR insulated with 33 insulator units (5¾ × 10 in) per phase. This type of tower was used on the first stages of the Manicouagan project since September, 1965, and in subsequent stages of the same project.

A unique structure is the 765-kV self-supporting steel tower used by American Electric Power Company (Fig. 14-37). This tower, weighing from 44,000 to 66,500 lb, including grillage foundation, was designed by American Bridge Company for erection in parts, if desired, by a Skycrane helicopter. Like AEP's 765-kV V tower shown in Fig. 14-34, there are 30 insulator disks (5¾ × 10 in) per leg of V strings in the outside phases and 32 insulator units in the middle phase. Also, like the tower shown in Fig. 14-34, this tower is designed to meet the same special AEP loading criteria already described. Two overhead ground wires provide a 15° shielding angle to the outside four-conductor bundles.

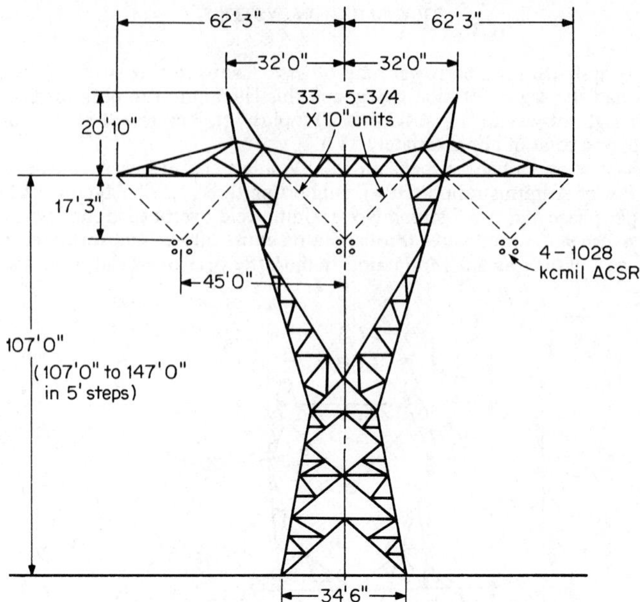


**FIG. 14-36** Steel suspension self-supporting tower of Hydro-Quebec for 735-kV Manicouagan lines.

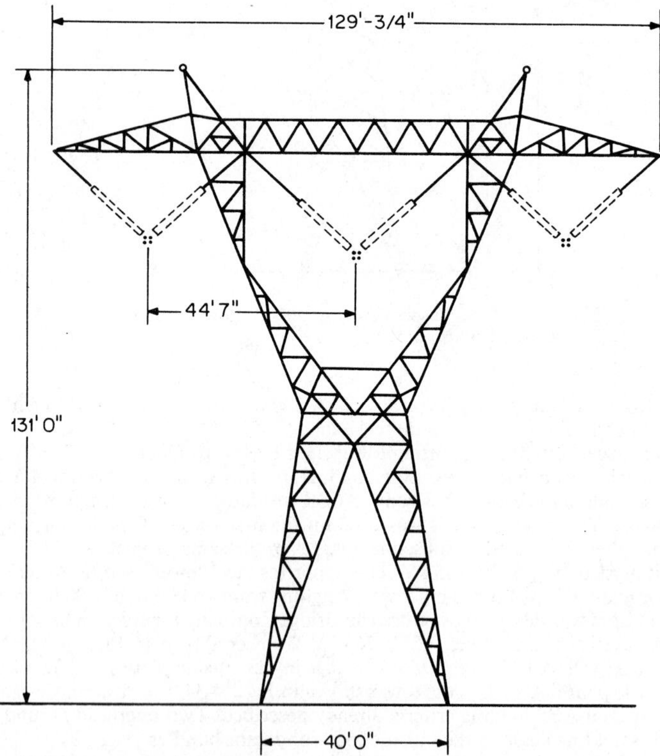


**FIG. 14-37** Steel 765-kV suspension tower, self-supporting design. *(American Electric Power Company.)*

*Semiflexible Structures.* Such structures have been used to some extent for the voltages under EHV. This type of tower has a narrow base in the direction of the line. The ground wires are strung tightly to take up unbalanced loads due to broken conductors and form part of the structural system. In case a conductor breaks, the unbalanced load will be taken up by the ground wires and transmitted by them to the next anchor tower.

With the advent of EHV and bundle conductors, semiflexible self-supporting towers are receiving more consideration, and some are being used. With the heavy bundle conductors, the breaking of one conductor is not serious, and the breaking of all conductors of a phase is practically nonexistent. Possible causes are airplanes and tornadoes, which no practical tower could withstand. Figure 14-34*b* shows such a tower as used on the 500-kV system of the Arkansas Power and Light Company. The overhead ground wires are insulated from the towers as they are in Fig. 14-34*a* for communication purposes. Figure 14-38*a* shows a steel-saving semiflexible tower used by the Pacific Gas and Electric Company. Note the X guying used between tower legs to obtain the required lateral strength.

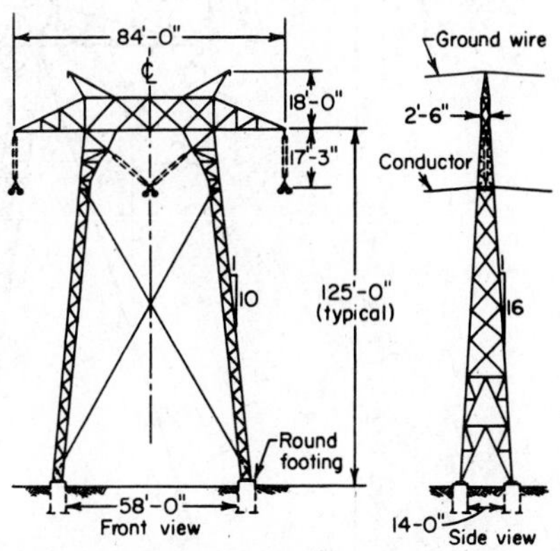


**FIG. 14-38a** 500-kV steel tower used in valley areas by Pacific Gas and Electric Company.

*Guyed Towers.* Such towers overcome the weakness of semiflexible towers in line with the line. They can be used for single-conducter lines or for any other service. Figure 14-38*b* shows a guyed steel tower used by the Pacific Gas and Electric Company in mountainous country. A feature of the tower is that the legs do not have to be of equal length. This tower has the same internal X guying as the tower of Fig. 14-38*a*. The self-supporting feature of the tower of Fig. 14-38*a* is replaced by four guys in the direction of the line and with an increase in strength.

Figure 14-39 shows a Kaiser aluminum guyed-V 345-kV tower used by the American Electric Power Company. Weighing from 3350 to 5400 lb (1510 to 2450 kg), this tower was erected by using a helicopter to "tilt up" the assembled tower by pivoting about a special hinge at the center foundation. This allows the use of a helicopter with a lifting capacity

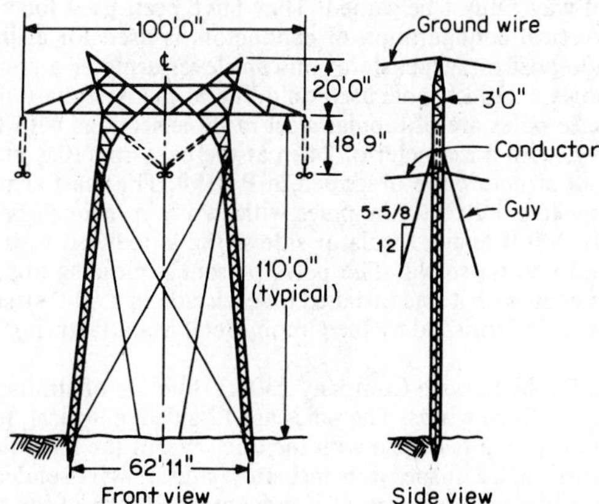


**FIG. 14-38b** 500-kV steel tower used in mountainous areas by Pacific Gas and Electric Company.

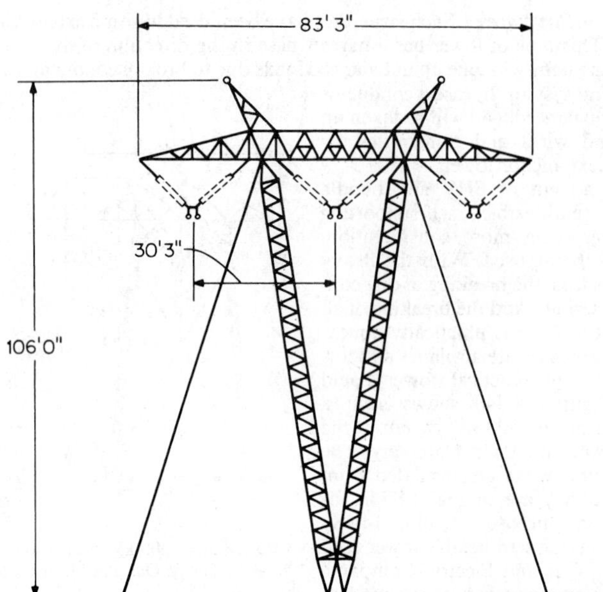


**FIG. 14-39** 345-kV guyed V aluminum suspension tower. *(American Electric Power Company.)*

smaller than the weight of the tower since most of the tower weight is supported by the foundation while it is being tilted up. There are 15 insulator units (5¾ × 10 in) per leg of the V strings on this tower.

Figure 14-40 shows a Kaiser aluminum guyed-V 765-kV tower also used by American Electric Power Company. There are 30 insulator units (5¾ × 10 in) per leg of the V strings in the outside phases and 32 insulator units in the middle phase.

Each of these towers has been designed to withstand special AEP loading criteria which include 100-mi/h winds with no ice, 50-mi/h winds with 1 in of ice, and 1¼ in of ice with no wind, in addition to the NESC loading requirements.

*Tubular Steel Poles.* These poles are being used on city streets and in congested areas where a wide right-of-way cannot be gained. They have been used for voltages up to and including 345 kV. Vertical configuration of conductors is used for all high-voltage lines. Insulators may be side post or suspension[74] on cantilever arms or a combination[75] of the two. Figure 14-41 shows a 230-kV pole used on a line of the Arizona Public Service Company in Phoenix. These poles are of tubular steel in three sections with telescoping joints. The poles are tapering, with a diameter of 24 in at the base and 10.8 in at the top. Foundations for this type of structure are described in Par. **90.** The mast arms are 8 ft long, of tubular steel, with brackets bolted to the poles with two ¾-in through bolts. The poles are spaced approximately 300 ft apart. Insulator side swing is reduced by a 200-lb combined hold-down weight and corona shield. The poles present a pleasing appearance and have elicited no objections even with a line installed on each side of a 60-ft street. The poles, side arms, and accessories were furnished by the Union Metal Manufacturing Company of Canton, Ohio.

The New Orleans Public Service Company 230-kV line[75] is of similar construction but is designed for hurricane-force winds. The poles are 12-sided, elliptical, high-strength steel, with the short diameter, which is in line with the line, 75% of the long diameter. The insulators are a combination of 12 suspension insulators and a swivel-ended strut (side-post) insulator equal to 12 suspension insulators, to prevent side swing of the suspension insulators. Some poles have side-post 230-kV insulators only. The poles have no base for bolting

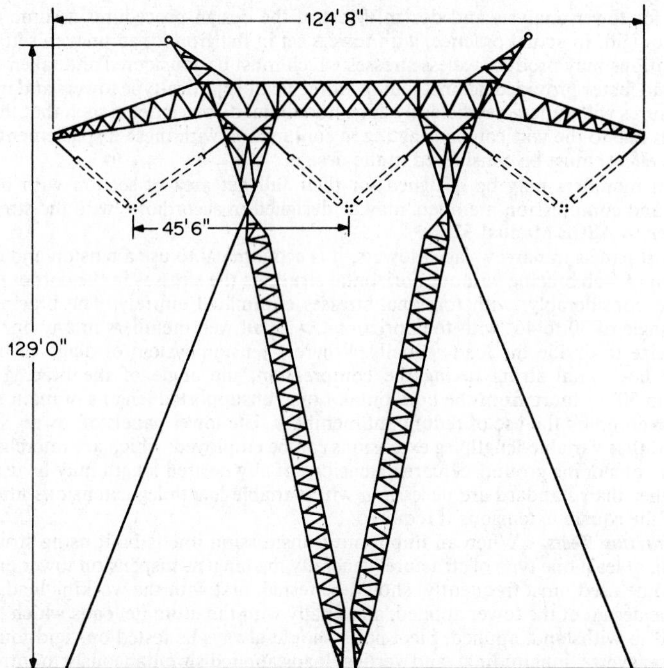


**FIG. 14-40** 765-kV guyed V aluminum suspension tower. *(American Electric Power Company.)*

to a foundation but do have baseplates and are set in concrete in holes 25 ft deep. The holes are made by driving 32-in-diameter steel casings to a depth somewhat deeper than 25 ft and cleaning them out.

*Special Structures.* Transposition towers require special structures where it is not expedient to make the transpositions on a standard tower by the use of special crossarms. Long spans over rivers and bays and crossings over important highways and trunk-line railroads frequently require towers which either are much higher than normal or must have a larger factor of safety against collapse. Anchor towers near substations, towers for mounting switches, and towers for turning 90° angles also may come under this classification. Such special structures are designed to suit local requirements and are subject to regulations of the U.S. Army Engineer Corps in the case of navigable-river crossings, to state public-utility commissions or other bodies for highway crossings, and to the particular regulations of railroads which are concerned.

***57. Stresses in Structures—Design.*** Stresses in towers can be computed analytically by the historic graphic methods or by use of one of the available computer

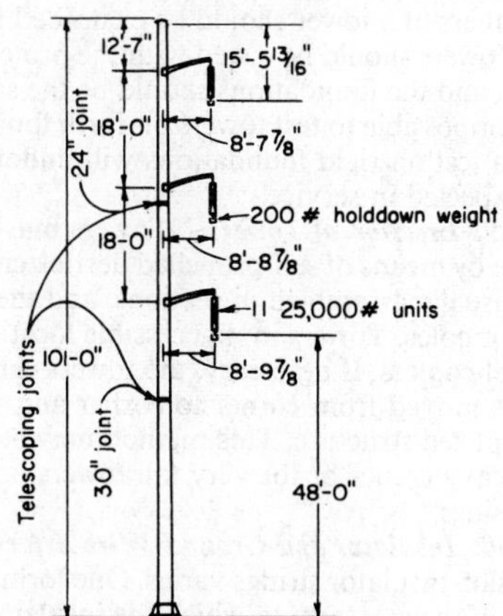


**FIG. 14-41** Steel pole for 230 kV. *(Arizona Public Service Company.)*

programs for tower analysis and design. Most of the design procedures assume the foundations are rigid. In actual practice, with towers set in the ground, an uneven settlement of the foundations may produce excess stresses which must be considered and taken care of in the overload factor provided against failure. Some flexible lattice-type towers and most pole-type structures will undergo sufficiently large horizontal displacements such that the nonlinear stresses due to the vertical loads acting in conjunction with these displacements (the so-called $P$-$\Delta$ effect) must be considered in the design.

Tension members may be designed for their full net area of section with bolt holes deducted, and compression members may be designed in accordance with the strength formulas given in ASCE Manual 52.[76]

For short panels in narrow-faced towers, it is economical to use a tension and compression system of web bracing without horizontal struts, as the stresses in the corner posts will be reduced considerably, with torsional stresses eliminated entirely. The bracing should make an angle of 30 to 45° with the horizontal, with all web members in any one panel of the same size to divide the loads equally. Where a tension system of diagonal bracing is used with horizontal struts taking the compression, the angle of the bracing may be increased to 50° or more from the horizontal. Long, unsupported lengths of main members can be broken up by the use of redundant members. The lower panels of towers should be so designed that variable-length leg extensions can be employed, which are interchangeable, to take care of sloping ground. Square extensions of any desired length may be used where towers higher than standard are necessary, with variable-length leg extensions fitted to the bottom of the square extensions if required.

***58. Structure Tests.*** When an important transmission line is built using structures of new design, at least one type of structure (generally the tangent suspension tower or the type which is to be used most frequently) should be tested, first with the working loads as specified for the design of the tower applied, and finally with the ultimate loads which the tower is expected to withstand, applied. Steel poles should always be tested on rigid foundations with the transverse, longitudinal, and vertical loads applied simultaneously to introduce all direct bending and torsional stresses. Crossarms should be tested for the additional torsional stresses introduced, where pin insulators are used, and combined with the longitudinal loads and the heaviest vertical loads specified. Equivalent concentrated loads may be used in some cases to avoid applying a multiplicity of small loads at different points, which would cause delay in shifting loads, but care should be taken to see that all combinations of loads or individual loads which will produce the maximum stress in each member are applied. After the structure has successfully withstood all the specified loads, a destruction test is desirable to determine the overload factor. This can generally be made with the test loads which cause the maximum stresses in the greatest number of members on a tower in place, by increasing the transverse loads indefinitely until failure occurs. After a test is completed, members of a tower should be examined for elongation of bolt holes, straightness, etc.

Towers should be tested with the protective coating which is to be used in service on the steel, and the foundations should be the same as those for which the towers are designed. If it is impossible to test towers on earth foundations, they may be tested on rigid foundations, but a test on rigid foundations will undoubtedly show a greater overload factor than may be expected in service.

***59. Erection of Towers.*** Towers may be assembled on the ground and then lifted into place by means of self-propelled derrick cranes or latticed-steel gin poles.[77] Very large towers are usually assembled in sections, and the sections are lifted into place by means of cranes or gin poles. Towers in inaccessible locations may be transported and assembed by the use of helicopters. If necessary, the towers can be erected from the ground up by the use of gin poles moved from corner to corner and with the erection crew climbing up on the partly completed structure. This method may be used for small jobs which do not warrant the use of heavy cranes or for very tall towers beyond the reach of cranes, such as those for river crossings.

***60. Insulator and Ground-Wire Attachments*** (Fig. 14-42). The method of attaching suspension insulator strings varies. One form of attachment is the U bolt fastened on the underside of the crossarm to which the insulator hardware is attached. This device will give flexibility both longitudinally and transversely. Another attachment is in the form of a bent plate or angle fastened to the underside of the crossarm with a fairly large hole to receive a hook or shackle at the top of the insulator string.

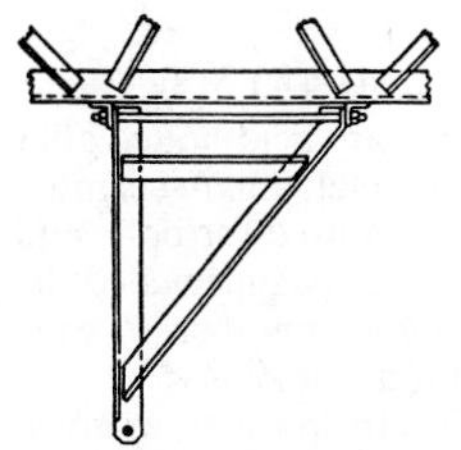


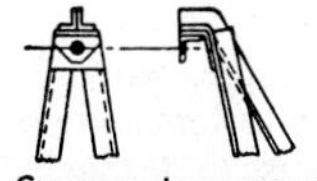


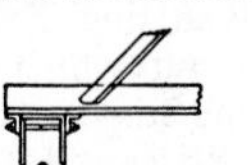


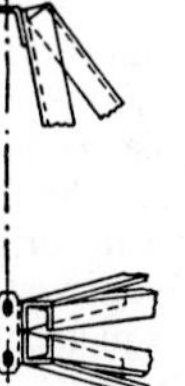


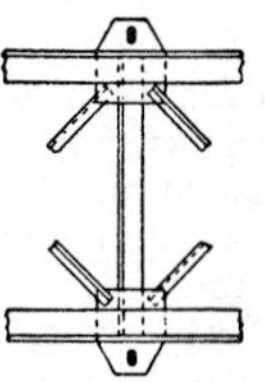

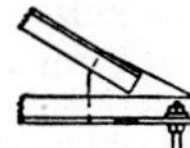

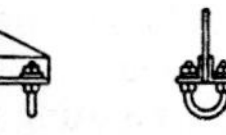


**FIG. 14-42** Insulator and ground-wire attachments.

In order to keep the conductor spacing to a minimum on high-voltage lines and take advantage of clearances to steelwork, the point of attachment of the insulator string may be dropped several inches below the crossarm, in which case flexible hangers are required, that is, hangers which are hinged at the crossarm and free to pivot in the direction of the line. These hangers may be made of plates, shapes, or bent round rods, with suitable connection at the bottom for receiving the insulator string. With suspension-type insulators in the strain position, horizontal pull-off plates are required.

To minimize failure of ground wires due to vibration, they are preferably suspended, and the attachment at the tower consists of an angle or bent plate with a hole to receive the suspension clamp. Patented rigid ground-wire clamps may be obtained if desired, which can be bolted directly to the steel structure. These are generally of the V-groove type.

***61. Ladders and Step Bolts.*** Ordinary steel towers are provided with step bolts on one corner post for climbing the tower. Special high structures, such as river-crossing towers, should have ladders extending up to the level of the top crossarm. Such ladders should be provided with guard cages supported on the sides of the ladder.

***62. Seismic Effects.*** This field is complex,[78] involving the various disciplines of the seismologist; the special dynamicist, who understands both structures and equipment and their differences; the vibration test engineer; and the designer who is experienced in the practical aspects of the lines and equipment and their function. There are many subtleties which are rarely appreciated by any one person and, consequently, many authorities must be consulted.

As contrasted with substation equipment, transmission-line structures built to withstand above-ground weather also will generally survive moderate earthquake tremors without noticeable distress. As studies of seismic effects are becoming more sophisticated, investigators have established seismic design criteria in the United States by dividing the country into four zones of seismic probability, the most severe being on the Pacific coast.[79] In such areas periodic review of existing foundations from a seismic standpoint is recommended. Experience in the San Fernando, California, earthquake of February 9, 1971, showed that over the years the foundation strength of a number of towers had been sufficiently reduced by erosion or adjacent excavation that their slopes failed during the earthquake.

A recent study[80] analyzes the effect of earthquake ground motion on both wood and steel transmission structures. It concludes that, except where damage to foundations and anchors because actual earth fissures or slippage develop, seismic disturbances produce no overstress in transmission structures designed in conformity with the National Electrical Safety Code.

***63. Protective Coatings for Steel Structures***

*Galvanizing.* For important transmission lines where long life is desired, it is almost universal practice to galvanize fabricated-steel towers and poles which are field-bolted after the other shop work has been completed, because under such conditions galvanizing is more economical than painting. The method of "hot-dip" galvanizing is also used for bolts and nuts, with the threads rerun for nuts after galvanizing. This has practically superseded the sherardizing, or "dry galvanizing," of bolts and nuts which was in use for a number of years. All galvanizing should be in accordance with the Standard Specifications for Zinc (Hot Galvanized) Coatings on Structural Steel Shapes, Plates and Bars and Their Products, as given by ASTM Designation A123, which calls for an average coating of 2 oz of zinc per square foot of surface.

To test the uniformity in the thickness of the galvanized coating, the Preece test is used. This is described in the Standard Methods of Determining Weight and Uniformity of Coating in Zinc-Coated (Galvanized) Iron or Steel Articles, as given by ASTM Designation A90.

Structures located near industrial plants, if subjected to sulfuric acid and fumes, should not be galvanized.

*Painting.* Painting is sometimes resorted to for fabricated-steel towers and poles, and generally shop-riveted, welded, or special steel poles are painted. Towers located near industrial plants in a smoky atmosphere should be painted. The base coat should be of a mixture of red lead and raw linseed oil in something like the proportion of 33 lb of dry red lead to a gallon of oil with a little dryer added. The outer coats may be of any good all-weather paint. To keep structures in good condition, painting is necessary every 2 or 3 years. With structures having a larger number of small members, this may make the cost of maintenance very high. On structures having few pieces and large, flat surfaces, painting may be economical.

Where steel structures are buried in the ground, a special problem sometimes arises at the ground level where moisture occurs. At this point, galvanizing may deteriorate after a short interval of time, especially if the soil has a sulfur or acid content. A paint made from asphaltum compounds will often prove useful for protection at these points.

Newly galvanized steel should not be painted until it has a chance to weather for a period of 6 months, and the galvanized surfaces should then be clean. Aluminum paint is ordinarily used to paint galvanized towers when the galvanizing has deteriorated.

*Weathering, or self-painting, steels* have been developed by the major steel companies. These are steels which are not treated with any kind of protective coating, but the chemical composition of which is such that they may be said to "paint" themselves. They are installed completely uncoated but thoroughly cleaned of all mill scale and foreign matter. In a few years, a dense dark-brown oxide with a purplish cast forms which becomes a permanent protective coating to all surfaces exposed to the weather. A slight loss of thickness occurs, which eventually stops, as the corrosion rate is nonprogressive.

***64. Wood Poles.*** Wood poles are considerably cheaper than steel for many types of construction. The lower cost is due, in part, to the more conservative basis of design normally adopted for steel. Generally, steel structures are designed to support safely one or more broken conductors, whereas wood structures are often not so designed. It is logical that the reasons for choosing the more expensive steel construction should require conservative design throughout and that conditions justifying the cheaper and shorter-lived wood structures would warrant accepting some of the more theoretical hazards.

For voltages of 69 kV and lower, wood is quite generally used.

Wood-pole construction for many years has been used for all voltages up to and including 345 kV. H frames with various modifications have been designed, the most popular using the main crossarm as the bottom member of a truss (see Par. **68**).

Butt-treated cedar and full-treated pine are used almost exclusively in transmission-line construction; the use of untreated poles has been practically abandoned as uneconomical since the supply of chestnut and northern cedar poles has been exhausted. Treated fir has also been supplied in some quantity from the Northwest.

Cedar poles resist decay, but satisfactory life is not secured unless the butt is treated. The pole is usually treated from the butt to about 2 ft above the ground line. The balance of the pole is not treated. Pine and fir require complete treatment of practicially all the sapwood. This treatment is applied under pressure.

No universally effective protection has been devised against woodpecker damage. Some localities are often subject to serious epidemics of woodpecker trouble.

*Preservative Treatment.* Pole decay is due to a fungus which requires air, moisture, warmth, and food for its subsistence; the wood of the pole constitutes its food. The conditions most favorable to the growth of the fungus are found at the ground line. The preservative has toxic or antiseptic properties which make the wood either poisonous or unfit food for the fungus.

Preservatives and preserving methods conforming to the standards of the American Wood Preservers Association (AWPA)[81] should be used in the treatment of poles. There are many wood preservatives, including those using poisonous salts such as copper, mercury, zinc, and arsenic compounds. However, there are only two included in AWPA recommendations for poles, Standard C-4-74-C, and they are

1. Coal-tar creosote, AWPA Standard P1-65
2. A 5% solution of pentachlorophenol in a petroleum distillate, AWPA Standard P8 (commonly called "penta")

By AWPA Standard M1-70, pentachlorophenol is not recommended for use in coastal waters. Coastal waters are defined as salty waters. One other preservative is increasing in popularity. This is AWPA Standard P11-70, a creosote-pentachlorophenol mixture in which pentachlorophenol is not less than 2% of the mixture. All of these preservatives are applied by the following methods.

*a. The open-tank method,* applied to cedar poles, consists in boiling the butts of the poles in a tank of creosote oil, after which the oil is allowed to cool or the poles are transferred to a cold tank of oil. The duration of the hot and cold treatment, usually 8 h or more, depends on several factors, the most important of which is the degree of seasoning. The treatment is based on the fact that the wood cells expand with heat and on cooling draw the creosote into the wood under atmospheric pressure. The sapwood of unseasoned poles has annular rings of a nearly impervious fiber which prevent penetration of the oil. In seasoning, this fiber dries and breaks open. To ensure penetration of the greater part of the sapwood, which is usually less than 1 in in depth, an incision process has been developed and is almost universally used. Narrow cuts, parallel with the wood fibers, are made to a depth of about ½ in at frequent intervals around the circumference of the pole for a distance above and below the ground line. Complete penetration is obtained to a depth somewhat greater than the depth of the incisions even on unseasoned poles.

*b. Pressure treatment* is applied to pine and fir. The poles, on a truck, are run into a steel cylinder and subjected to a steam treatment for a period of several hours at a temperature which will not damage the wood cells, usually specified at not more than 259°F (126°C). The pressure is then removed and a vacuum applied. The steam treatment opens up the wood cells and allows the preservative to penetrate. The length of time required for the steam and vacuum treatment depends on the condition of the wood, the amount of oil that is to be injected, and the depth of penetration desired. From this point in the process, one of two methods may be followed. The *full-cell,* or *Bethel, process* allows all the preservative injected to remain in the wood. This process is generally used for piling and underwater work when it is desired to exclude water from the wood and to resist the attack of marine borers. The *empty-cell process* draws off excess oil and secures protection from decay by the coating of oil left on the walls of the wood cells. The empty-cell process is adequate and preferable for usual structures and is almost exclusively used for poles and arms.

The empty-cell treatment is obtained by either the Rueping or the Lowry process. The Rueping process seems to be in more general use, although the Lowry process is equally successful.

In the *Rueping process,* following the steam treatment, an air pressure is applied. While still under pressure, hot oil is forced into the cylinder. The oil is held under this pressure and maintained at a temperature of about 200°F (93.5°C) by steam coils within the cylinder, for a period of several hours. Upon removing the oil and reducing the pressure, the compressed air within the wood cells forces out the surplus oil. The amount of oil retained depends on the pressures applied and the time of treatment, although it is possible to remove only a part of the oil that has been injected.

The *Lowry process* is similar to the Rueping process except that no compressed air is used. After the preservative has been forced into the wood under pressure, a high vacuum is quickly created, causing a sudden expansion of the air within the wood cells and thus driving out surplus preservative (see also Sec. 4).

*Strength Calculations.* As used in a line, the pole is a cantelever beam, fixed in the earth at the butt and supporting the transverse wind load from the conductors of a length equal to half the sum of adjacent spans. Computation of the safe load that may be carried is a matter of simple mechanics outlined in Fig. 14-43. Some slight approximations have been introduced for simplicity.

The fact that if the pole were a part of a perfect cone, the maximum fiber stress might occur at a point above the ground line is of more theoretical interest than practical use. The difference between the load carried at the critical section and at the ground line is less than may readily be caused by irregularities in the pole.

Poles are almost universally classified according to the ANSI dimensions (see Sec. 4), which have been arranged so that the nominal ultimate strength is the same for all lengths and species of the same class. Poles are classified as Class 1, 2, 3, etc., and H1, H2, H3, etc., and the minimum circumference 6 ft from the butt is specified for each class and each species to give the desired nominal strength. The nominal ultimate was computed from conservative average ultimate fiber stresses from a very large number of tests.

The top diamaters are specified but are given only as a minimum and are the same for all species. Actually, the taper of various kinds of timber, although fairly uniform, is quite different for different species, and the average top diameter of ANSI-class poles will be considerably larger than this minimum.

*a.* ***Pole tests*** give very erratic results, and tests on a few poles should never be given great consideration. Designs should if possible be based on accepted average unit fiber stresses rather than test results unless a considerable number of duplicate tests can be made and averaged.

*b.* ***Factor of Safety.*** It has been found from experience with heavy transmission-line construction that a factor of safety of 2 on the accepted average ultimate is conservative. On light construction, this is sometimes slightly reduced, but a material reduction is usually not justified in view of the deterioration of wood with age. On sustained loads, such as heavy angles, a liberal additional factor of safety is desirable to prevent the pole's warping and giving the appearance of being overloaded. When possible, guys should be attached close to the load to eliminate heavy continued bending.

*Setting Depth.* The strength of the pole foundation is difficult to reduce to figures and is not of such primary importance to the safety of the structure as in the design of a tower. Failure of the foundation, in the sense that failure is used in the design of steel towers, that is, a considerable movement of the pole in the ground, is of little consequence except for

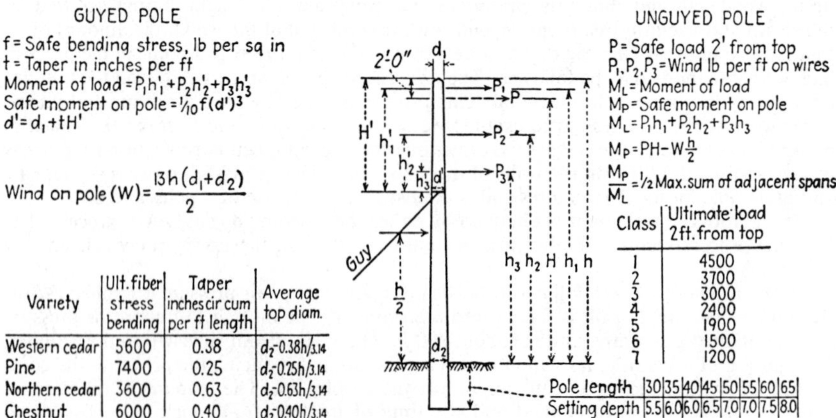


**FIG. 14-43** General pole-strength calculations.

the inconvenience and expense of straightening up the line and retamping the poles. The setting depth for poles of various lengths has been pretty well established by general practice and is almost universally used (see Fig. 14-43). These depths seem somewhat illogical in that no account is taken of the strength of the pole or of the quality of the soil; however, this appears more reasonable when it is considered that the desired result is not to obtain a rigid foundation but to prevent the pole from "kicking" out of the ground.

***65. Wood Crossarms.*** These are generally manufactured of creosoted yellow pine or untreated Douglas fir. Untreated pine arms of the timber commercially available are not satisfactory. Untreated fir arms are widely used and are apparently giving a life comparable with that of the poles. Arms should be of the highest-quality timber. The smaller arms, up to 5 by 6 in and 10 or 12 ft in length, can generally be supplied on standard crossarm specifications, although structural specifications give very satisfactory arms. Heavy arms for H frames, that is, 6- by 8- and 6- by 10-in timbers and 3- by 8- and 3- by 10-in planks, 20 to 35 ft long, are best purchased as high-grade structural timbers. Structural timbers are furnished under the rigid specifications and inspection of the large timber manufacturers' associations.

*Plank Arms.* The eccentric connection of large arms to the pole, especially when carrying heavy conductors, is not desirable, and a number of designs make use of two planks, one on each side of the pole attached together at the ends. Generally two 3- by 8-in planks are used in place of a 6- by 8-in timber arm. The plank-arm construction has several advantages in addition to the better connection to the pole, although the end hardware is somewhat complicated, and in many designs the strength of the crossarm against longitudinal loads is somewhat reduced.

***66. Wood versus Steel Arms.*** Wood crossarms are lower in cost than steel arms of the same strength and, aside from the shorter life and possibility of being shattered by lightning, are satisfactory. On wood-pole construction the longer life of steel arms is of little value, and the possibility of lightning damage is the price paid for the lightning insulation of the wood. Lightning damage to arms is usually not a major operating problem; and on lines thoroughly shielded with overhead ground wires, crossarm and pole damage is practically eliminated.

***67. Design of arms*** must provide for carrying the vertical load with an ample margin of safety, but often neither the arm nor the connecting hardware is well suited for carrying the full load of a broken conductor as is required of steel towers. Crossarms on single-pole construction have practically no resistance to longitudinal loads. If a heavy conductor breaks, the arm will swing around to very nearly a longitudinal position, restrained only by the attachment of the unbroken conductors. This would be likely to result in badly bent hardware and probably a split and disfigured arm but little serious damage; the major damage would be the broken conductor and not the effects of the break. H-frame construction (see Fig. 14-47*c*) is better adapted to such loads, but the effect of a break is much the same, in that the deflection of the poles and movement of the poles in the ground relieve the greater part of the load and usually result only in some minor damage to the arms and hardware, which is easily repaired.

Double-arm construction can be considered very little, if any, stronger than twice the sufficient bolts and keys to develop the shear. The shear is several times the applied load and makes a very heavy joint necessary.

The common sizes used, 5 by 6 in for lighter single-pole construction and 6 by 8 in for H-frame, allow ample vertical strength for ordinary spans. Conservative unit stresses should be used in vertical load on the arms.

The connecting hardware, as generally used, is not designed as would be necessary in a framed structure, such as a truss, where movement in a joint would cause serious secondary stresses in the main members. Only one ¾-in bolt is ordinarily employed, even in heavy H-frame construction in types of construction carrying wire heavier than has been general practice, and in very long spans the use of these connections, based entirely on experience, should not be followed without a careful check. The same applies to designs carrying heavy angles on crossarm construction where the entire angle load must be transmitted through the bolts to the pole. For such angles, the 3-pole structure is a more positive arrangement.

***68. Conductor Arrangement and Spacing.*** In wood construction with short spans over comparatively level terrain, these parameters are determined largely by the line voltage. A wide variety of conductor arrangements is found in past practice. However, with the use of larger conductors and longer spans, the conductor configuration and separation are often a matter of providing the safest arrangement with ample spacing, especially for the occasional longer-than-normal spans encountered in rolling country. The conductor arrangement should provide spacing for these occasional long spans, as it is generally more economical to design a standard structure with spacings suitable for a span about 50% longer than normal rather than to use too many special structures.

The H-frame design gives one of the best conductor arrangements and mechanical strength for long-span construction and may be used as a special structure for especially long spans in almost any type of line.

Wood-pole H-frame structures are used on EHV lines at an appreciable saving over metal towers. Figure 14-44 shows an H-frame structure with trussed crossarm as used on the Kansas Gas and Electric Company 345-kV lines. This line uses two 795-kcmil ACSR

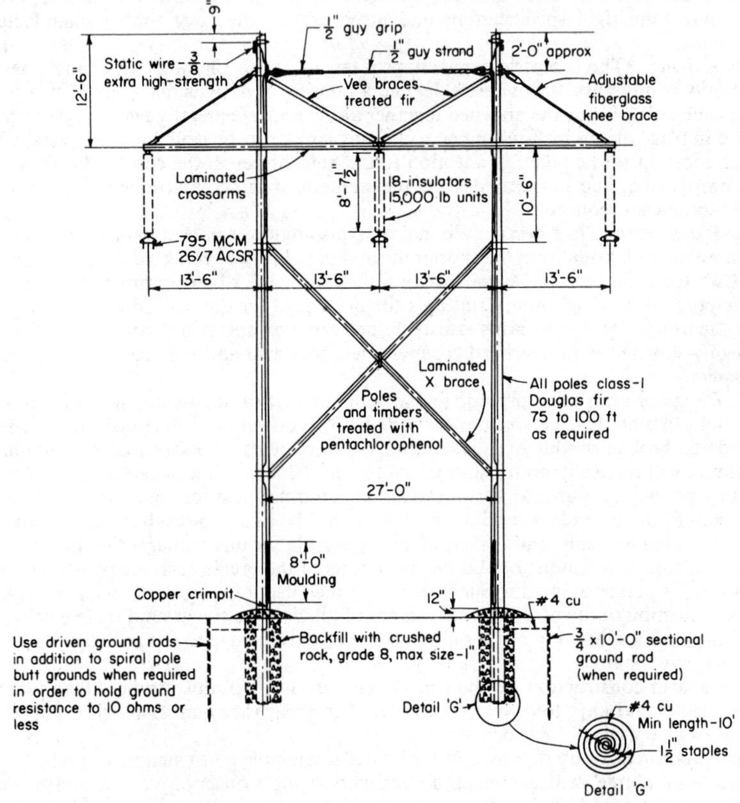


**FIG. 14-44** 345-kV wood H-frame structure of Kansas Gas and Electric Company.

conductors per phase on 18-in bundle spacing and 27-ft phase spacing. The insulator suspension hardware is not grounded, and full advantage is taken of the impulse insulation of the crossarm. Lightning flashovers would be expected to take place between the conductors and the ground wires on the poles and not to follow the insulator string and crossarm. All poles and timbers are penta-treated fir.

Figure 14-45 shows a modified H-frame wood-pole structure, designated a K frame, as used by the Northern States Power Company on its 345-kV system.[82] This structure also is designed to carry two 795-kcmil ACSR conductors per phase on 18-in bundle spacing and

27-ft horizontal phase spacing, but the center conductor is approximately 6 ft higher than the outside phases. This structure also takes full advantage of the impulse insulation of the crossarm.

It should be noted that the conductor spacing is primarily a function of the sag and that a conductor arrangement entirely satisfactory for a large or high-strength conductor would be hazardous for a small conductor, with correspondingly greater sags, in the same length of span. Also, a conductor spacing safe for a light loading district should not be used with the heavy sags required for heavy loading conditions. A small or lightweight conductor would require more spacing than a larger or heavier conductor in the same span and with the same sag.

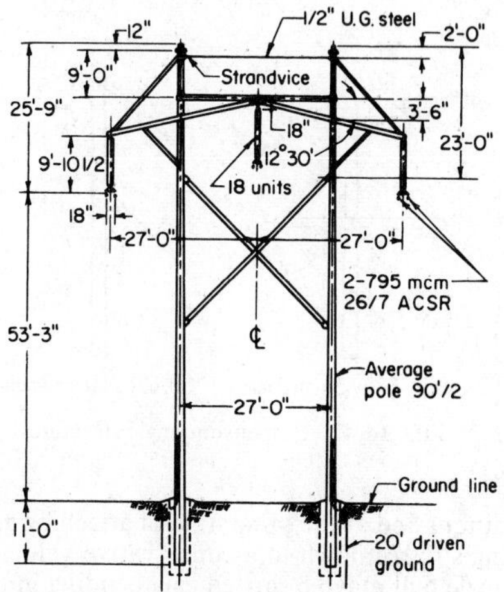


**FIG. 14-45** 345-kV wood K-frame structure of Northern States Power Company.

On suspension construction the spacings are usually determined by the clearance required for swing of the suspension insulators as discussed under steel tower design (see Par. **51**). A detailed layout is required for each conductor, as the size and material have a marked effect on the swing characteristics (Fig. 14-46*c*). A fairly conservative assumption, which results in reasonable design, requires that the clearance from the conductor to a grounded structure shall be at least 0.75 the dry-flashover distance of the insulator or the "tight-string" distance under an 8-lb wind on the bare conductor at a temperature of 60°F. This may be modified in details, and it is common practice to allow somewhat reduced clearances to wood members. Typical layouts are shown in Fig. 14-46.

The swing should be taken for a somewhat more unfavorable case than level spans. The

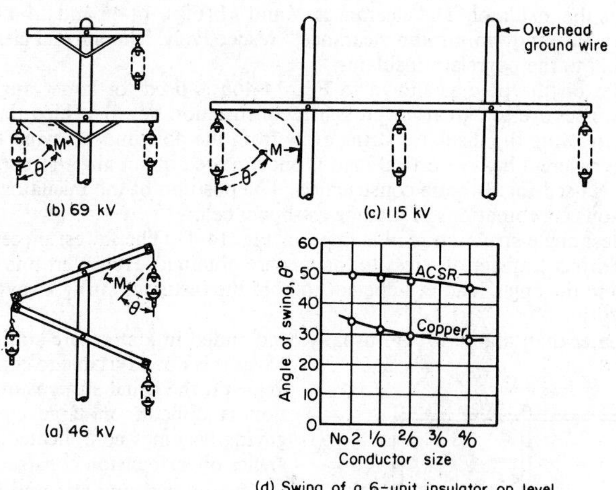


**FIG. 14-46** Typical suspension-insulator arrangements on wood construction.

usual range of conditions encountered would be fairly well covered if a vertical span of three-fourths to two-thirds the horizontal span is assumed; that is, the clearances provide for cases where it is necessary to locate a structure somewhat below the elevation of the adjacent supports. The insulator-swing calculations are discussed in Par. **51.**

Angle structures in general use are shown in Figs. 14-47 and 14-48. The design is a matter of providing clearance from the conductor to the structure and to the guys under all conditions and at the same time of attaching guys as close to the load as possible to keep bending stresses down to a conservative value. On small angles where the loads are small, the angle pull may be carried as a bending in the pole and arm; but on larger angles, the loads should be carried directly by the guys, insofar as possible.

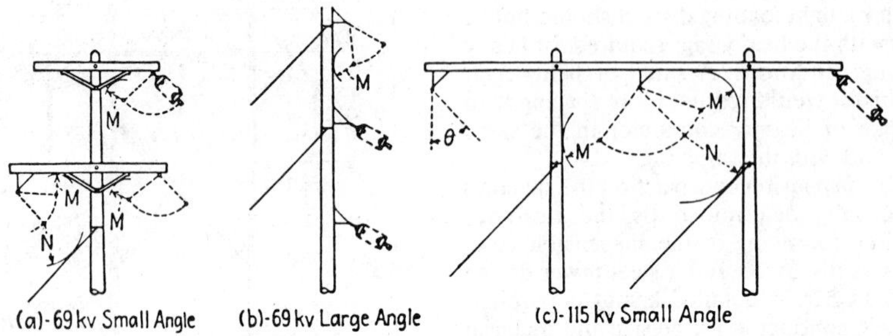


**FIG. 14-47** Suspension-angle structures.

Figure 14-47*a* shows the usual small-angle construction, illustrating how it may be necessary to offset the arms to give clearance to the inside conductor. Figure 14-47*c* illustrates a similar design for small angles on heavy H-frame construction where the angle is so small that, if the maximum wind should blow from right to left in the illustration, it would cause the insulator to swing somewhat to the left of vertical. Therefore, clearance *M* must be provided, not only to the pole on the right, in the illustration, but also to the pole on the left. In the above designs the guy is attached some distance below the crossarm in order to give a clearance *N* from the conductor to the guy, which is somewhat greater than the flashover distance across the insulator. The clearances *N* and *M* (Figs. 14-46 and 14-47) indicate the "normal clearance" and "minimum clearance," respectively. The normal clearance should be at least equal to the porcelain insulator.

The bracket on the pole as shown in Fig. 14-46*b* is used for larger angles where the mechanical stresses are too great for crossarm construction but for which the angle pull is not sufficient to swing the insulator string away from the pole under a wind from the right or at locations where a heavy vertical load is encountered on an angle structure. A similar 3-pole design is used for H-frame construction. The position of the insulator may be computed for various combinations of loading as shown below.

The simplest angle structure is illustrated in Fig. 14-47. The fewest pieces of hardware and the most direct transfer of stress to the guy are obtained. However, this design can be used only where the angle load is sufficient to hold the insulator string away from the pole under all conditions.

Angles greater than about 50° are usually dead-ended in a structure similar to Fig. 14-48, as it is not advisable to carry too large an angle on the usual suspension clamp. Erection is difficult on large conductors, and guying becomes complicated for very large angles on suspension construction.

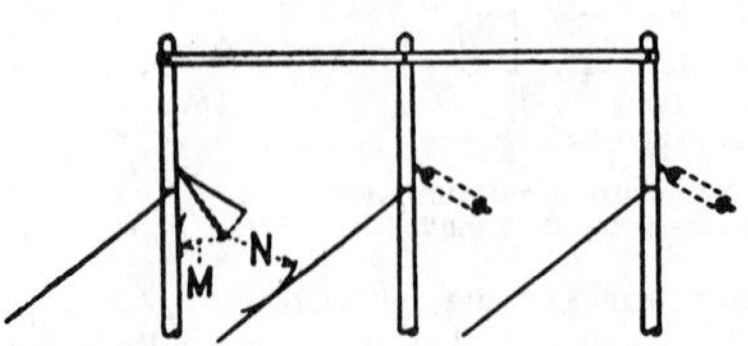


**FIG. 14-48** 115-kV large-angle structure.

If grounded guys are used, a ground wire should be carried up the pole; and when the guy is attached close to the insulator, contact should be made with the insulator hardware to avoid the possibility of burning the pole from leakage or splintering from lightning. It is common practice to use one or two additional insulators on such angle structures.

Clearances on angles should be the same as on tangent construction, that is, under normal conditions must be somewhat greater than the flashover distance over the insulator but under maximum wind conditions may be reduced to 0.75 of normal, with some slight further reduction if this clearance is to ungrounded wood (see Table 14-20).

**TABLE 14-20** Clearances for Various Lengths of Insulator String

| Insulator, 5¾″ units | Normal clearance, in | Min. clearance 0.75 normal, in |
|---|---|---|
| 4 | 25¼ | 19 |
| 6 | 36¾ | 27 |
| 8 | 48¼ | 36 |
| 12 | 71¼ | 50 |
| 16 | 94¼ | 70 |

NOTE: 1 in = 25.4 mm.

The greatest swing, that is, angle load and wind in same direction, may occur with wind on the bare wire at 0°F, but usually the combined ice and wind load is limiting because of the larger conductor tension. In the case of the wind blowing against the angle, clearances must be computed for a high temperature and resulting low conductor tension. Under normal conditions, for example, 60°F, full clearance should be maintained, equal at least to the dry-flashover distance of the insulator.

The angle load, that is, the transverse component of the conductor tension $t$ at an angle $a$, is found as follows:

$$\text{Angle load} = 2t \sin (a/2) \tag{14-76}$$

$$\tan \theta = \frac{(\text{angle load}) \pm (\text{wind load})}{(\text{vertical load}) + (\tfrac{1}{2}\ \text{weight of insulator})} \tag{14-77}$$

in which $\theta$ = swing of the insulator from the vertical; the vertical load is the weight of the conductor supported by the insulator, or the weight per foot times the distance between the low points of the sag in the adjacent spans; and the horizontal load is the wind load on the spans supported by the insulator.

***69. Pole Ground Wires.*** Ground wires should be installed on all poles, at all voltages, in lightning areas:

1. To prevent splitting of poles by lightning.
2. To provide a direct connection to ground and prevent pole burning if an insulator breaks down. Since the ground wire on these lines has relatively high resistance to ground, the wire can be small as No. 6 galvanized iron and the ground connection can be several wraps of the wire around the butt of the pole.

## Line Accessories (Lines under EHV)

***70. Suspension Clamps.*** These designs are fairly well standardized for the usual conductors. Simple, light, well-designed clamps in both malleable iron and forged steel are available for almost any conductor. The seat and clamping surfaces should be smooth, without any projections or sharp bends, and should be formed to support the conductor on long, easy curves and at the comparatively sharp bends formed at horizontal and vertical angles. Heavy, complicated clamps, unless very carefully designed, are generally avoided to allow as much freedom as possible at the support. For the same reason care is exercised to avoid rigid connections of any kind.

***71. Trunnion-Type Clamps.*** These are designed to give an almost completely flexible connection by supporting the clamp on a pivot, approximately on the axis of the conductor (Fig. 14-49). Thus any vibration of the conductor tends to be transmitted through the clamp, eliminating much of the heavy binding stresses caused by a fixed support.

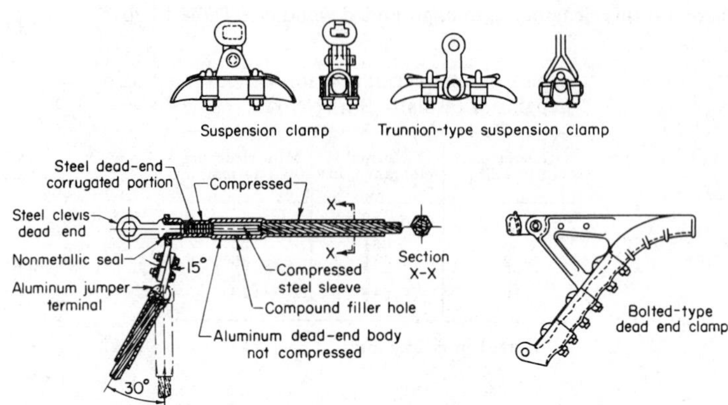


**FIG. 14-49** Conductor clamps.

The suspension clamp is intended primarily to support the weight of the conductor and to prevent any longitudinal movement from accidental unequal tensions in adjacent spans. It is generally considered desirable but not always essential that the suspension clamp hold the conductor in case of a break. For large conductors under heavy tensions it is difficult to design a light, flexible connection that will not slip under such a contingency.

***72. Slip, or Releasing, Clamps.*** Several especially heavy lines have been designed on the proposition that, since suspension clamps could not reasonably be secured that would positively hold the conductor, a clamp would be used that would hold under all ordinary conditions but would slip at something like one-half the maximum conductor tension in case of a break. This arrangement justified a considerable reduction in the exceedingly large longitudinal design loads on the towers and resulted in a considerable saving in tower and foundation costs. Several designs of slip clamps and releasing clamps have been used.

***73. Dead-End Clamps.*** These clamps are of the bolted type and are available for practically all copper and aluminum conductors. However, for the larger ACSR conductors the compression-type dead-end clamp is generally used (see Fig. 14-49). This is very similar to the compression splice used on ACSR.

The dead end for the steel core, which may have a clevis or an eye-type end, is pressed on after the aluminum sleeve has been slipped out over the conductor. The aluminum sleeve is then slipped back over the steel sleeve until the aluminum body makes contact with the shoulder of the steel sleeve. The electrical connection tongue on the aluminum body is aligned with the clevis or eye of the steel-core dead end as required, after which the aluminum body is filled with the nonoxidizing compound furnished with the body and the body is compressed. Similar pressed-on dead ends are available for copper, Copperweld, and other conductors. Several manufacturers furnish ACSR dead ends in all sizes required.

***74. Armor Rods.*** These rods are quite generally used on ACSR lines as a protection against fatigue of the aluminum strands from vibration. Armor rods consist of a bundle of aluminum rods, somewhat larger in diameter than the strands of the conductor, laid parallel to the length of the conductor and arranged to form a complete covering. These are spirally twisted by a tool to lie approximately parallel with the lay of the strands in the cable and are clamped in place at each end. The suspension clamp is attached at the center, with the armor extending 2 or 3 ft on each side. The bending stresses caused by vibration are reduced by the increased diameter and area of metal and distributed over a longer section of conductor.

***75. Vibration Dampers.*** The Stockbridge damper, as well as several other designs, are devices for damping vibration out of the entire span. Such dampers have been used on ACSR, copper, and steel conductors and ground wires as illustrated in Fig. 14-50. The cause of conductor vibration and the action of the Stockbridge damper are outlined in Par. **49.**

*Overhead Ground-Wire Vibration.* Overhead ground wires are especially subject to vibration; in fact, most steel ground wires will often be found in rather irregular vibration of small amplitude which generally does not appear to have any ill effects. Ground-wire attachments should be made with at least as great care as given the conductor clamps. Rigid clamps have been almost entirely abandoned in favor of a suspension clamp similar to that used on the conductor and attached by links or shackles so as to give a perfectly flexible connection.

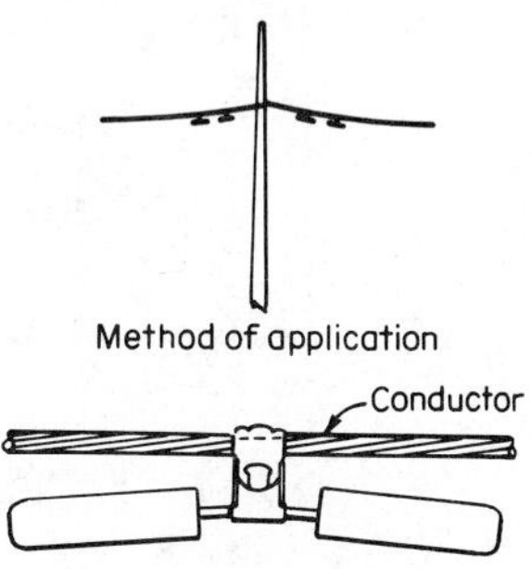


**FIG. 14-50** Vibration damper.

***76. Hardware.*** Many items of hardware have become fairly well standardized. The dimensions of the eye of eye bolts, the length of thread on various-length bolts, end links, and hardware for suspension insulators are quite uniform. It is usually possible to obtain about identical stock material from a number of manufacturers. Many other items such as shackles, guy clamps, and crossarm braces are furnished in such a wide variety that considerable care is required to choose the most commonly used but suitable stock items. Much expense and confusion in both construction and maintenance are saved by limiting the number of hardware items.

***77. Insulating Braces and Guys.*** With the use of wood to increase the impulse insulation to decrease the line's sensitivity to lightning flashovers, steel crossarm braces have been replaced with wood on a number of lines. Connections are made by pressed-steel fittings. The use of a 48-in wood brace in place of steel, for additional wood insulation, is roughly the equivalent in lightning-flashover strength of adding one suspension unit to the insulation. The effect on 60-Hz flashover is, however, negligible.

To obtain equal wood insulation at guyed structures to what may be obtained on unguyed construction requires long wood insulators in the guys. These guy insulators are quite efficient because of the high tensile strength of clear wood; an ultimate strength of 6000 lb/in$^2$ on the net section is conservative. A 2- by 2-in fir insulator will develop the full strength of a ⅜-in Siemens-Martin guy strand. The design of the connection to the pressed-steel fitting requires only that several bolts of insufficient diameter be used to give the necessary bearing area between the wood and the shank of the bolt. The bolts should be placed alternately through the face and side of the stick to prevent splitting.

Reinforced fiberglass is receiving increased favor as guy-wire insulators in place of wood, and as crossarm braces in place of wood or steel. Impulse flashover voltages of fiberglass line hardware can be supplied by the manufacturers.

***78. Guys.*** The various grades of guy strand are almost universally furnished in accordance with ASTM specifications. The ultimate strength for each size and grade is given in Sec. 4. The so-called double-galvanized is generally used. Common guy strand is not ordinarily employed in transmission construction, as the best-quality galvanizing is not furnished in this grade. Siemens-Martin strand is most commonly used for the lighter lines and high-strength strands for heavy construction.

More than one size of guy strand is not economical for a line, and often the same size may be used for several designs. The ⅜-in size, in either Siemens-Martin or high-strength grade, is most generally used both for guys and for overhead ground wires.

In the usual wood-pole construction great refinement is not required in designing guys. Usually it is sufficient to determine the number of guys, of the size and quality to be used on the line, required to support the load, an additional guy being employed for any fractional part. In transmission construction a factor of safety of 2 is general for guys, although this may be somewhat reduced.

A common problem in guy design is illustrated in Fig. 14-51. The ratio of the guy load $L$ to the conductor load $T$ is the same as the ratio of the length of the guy $B$ to the distance $A$. The length $B$ is readily determined from a sketch drawn to scale.

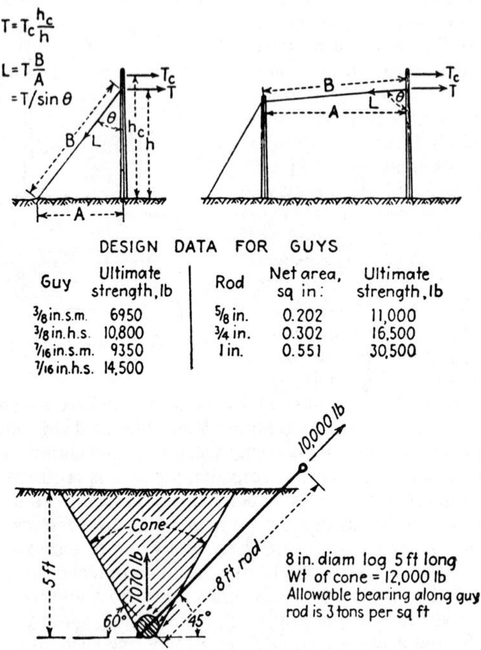


DESIGN DATA FOR GUYS

| Guy | Ultimate strength, lb |
|---|---|
| 3/8 in. s.m. | 6950 |
| 3/8 in. h.s. | 10,800 |
| 7/16 in. s.m. | 9350 |
| 7/16 in. h.s. | 14,500 |

| Rod | Net area, sq in. | Ultimate strength, lb |
|---|---|---|
| 5/8 in. | 0.202 | 11,000 |
| 3/4 in. | 0.302 | 16,500 |
| 1 in. | 0.551 | 30,500 |

**FIG. 14-51** General guy and log anchor-strength calculations.

$$L = T\frac{B}{A} \quad \text{or} \quad L = \frac{T}{\sin\theta} \tag{14-78}$$

If the conductor load $T_c$ is above the point of the attachment of the guy,

$$T = T_c h_c / h \tag{14-79}$$

## Conductor and Overhead Ground-Wire Installation

***79. Conductor Stringing.*** This operation requires an experienced crew, not only to prevent damage to the conductor and overhead ground wire but also to maintain the sags and tensions specified in the design. Correct sags are essential to give the required mechanical safety, but it is equally important that the actual sag in the line correspond to that used in the design, to ensure proper clearances to the ground. More detailed procedures and guidelines are available in an IEEE publication.[83]

***80. Stringing Equipment.*** All transmission conductors and overhead ground wires should be strung over free-running snatch blocks or rollers made for this purpose. Both conductors and ground wires of any material are easily damaged, and with the long spans and heavy conductors used in modern construction, satisfactory sags cannot be obtained at reasonable cost except by eliminating all possible friction at the supports. Dynamometers for measuring the tension are of value as a means of knowing the tension at all times, but they cannot be relied upon to set the sag. The final sag should be adjusted by sighting. Walkie-talkie radios are widely used as standard erection equipment; better and more efficient work is obtained by having direct communication between the reel crew, the pulling crew, and the workers doing the sagging.

Sags are measured by setting sights on the structures at each end of the span at a vertical distance below the conductor support equal to the sag. This method is convenient and well within the necessary accuracy, even for inclined spans. For average inclined spans, the sag is taken as the sag for a level span of the same horizontal length, although the sag for a level span equal to the slope distance is more nearly correct. Except in extreme cases the horizontal and slope distances are practically the same. On long, inclined spans, when the low point of the sag falls below the ground level of the lower tower, it is more convenient to measure the vertical sag below the lower support, as given by Eq. (14-68) (Par. **47**).

***81. Accuracy of Sagging.*** Friction in sagging blocks prevents the wire from reaching exactly the same tension at all points. As the wire is pulled up, the sag tends to be greater in spans from the pulling point; and when slacked back somewhat, more sag is thrown into the nearer spans. These effects are usually fairly well eliminated by allowing some time for the tension to equalize and by skillful handling.

Curves of "span versus sag" and "span versus tension" for possible stringing temperatures are used in sagging. The actual conductor temperature in sagging is very important. The IEEE Committee Report on Stringing and Sagging Conductors, *Trans. IEEE Power Group,* December, 1964, p. 1235, recommends that the temperatures be obtained by direct measurement at the time of sagging by means of a thermometer placed inside a short length of the conductor and suspended at least 15 ft above the ground. Accurate temperature also is important in making allowance for creep during stringing. Creep elongation starts as soon as the conductor is pulled into the air, and it is important that this elongation be allowed to take place and not to pull the conductor back up to the calculated sag. One way to do this is to use a temperature curve which will indicate the calculated sag plus the additional sag due to creep up to the time of clipping in. The creep sag must be estimated from the manufacturer's curves.

In spans of varying length a greater sag tends to form in the long spans; and on steep grades the sags at the higher elevations tend to be less than at the bottom of the hill. These effects are not of importance except in extreme cases and are due to the fact that the wire is, and must be, supported on rollers in such a way as to be entirely free to travel. Thus the tension in the wire on each side of the roller must be equal irrespective of the slope of the wire away from this support, and the resultant on the support is not vertical but in the direction of the bisector of the angle between the slopes of the wires as they leave the roller on each side. At a support between a short and long span (Fig. 14-52*a*) the wire on the short-span side is more nearly horizontal, *OA,* whereas on the long-span side the wire may have a considerable slope *OB.* The tension on each side must be equal, $OA = OB$, but the horizontal component of the tension is therefore less in the long span *BD* than in the short span *AE.* The resultant *OC* is inclined.

It is theoretically possible, although very difficult practically, to clamp the conductor at the correct position so that the resultant will be vertical and the horizontal tensions equal as in Fig. 14-52*b*. This is the condition assumed in the computations and office location and, for all except extreme cases, is the reasonable assumption.

Similarly, the different slopes of the wire leaving the roller on hillsides with spans of equal length but at different elevations cause the horizontal component of the tension to be less in the wire with the greatest slope (Fig. 14-53). With a series of spans on a slope this effect tends to accumulate, for the horizontal tension $t_2$ at the upper support must be the same as the horizontal tension $t_2$ at the lower support, whereas the resultant tension $T_2$ is less than the tension $T_1$ because of its smaller slope. The relation $T_1 = t_2 + wD$ and $T_2 = t_2 + w\delta$ (see Fig. 14-53) is discussed in Par. **47**.

The differences in sag are not usually carried from one conductor pull to the next, but each pull is sagged to approximately the correct tension independent of the other; thus when the snubs between pulling sections are removed, differences in tension tend to equalize. For this reason it is best not to clamp in the conductors too close behind the sagging crew. Often skillful sagging reduces these effects by using the friction in the blocks to prevent the conductor from "collecting in the low spots."

These irregularities are of little consequence generally, especially when it is realized that the important consideration is to have equal tensions under maximum load rather than

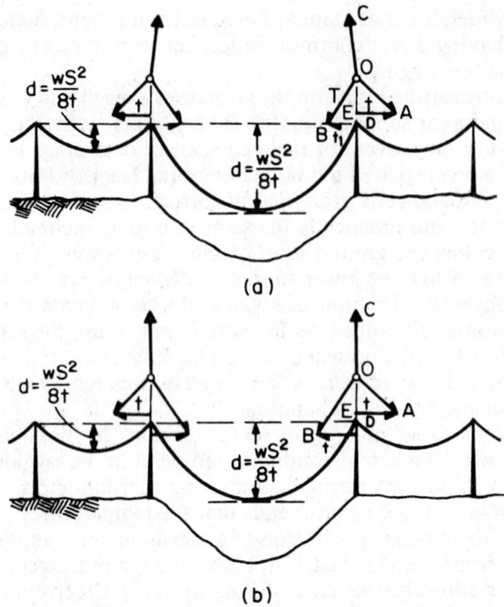


**FIG. 14-52** Illustrating the change in tension in long spans. (*a*) On rollers; (*b*) clamped in theoretically correct position.

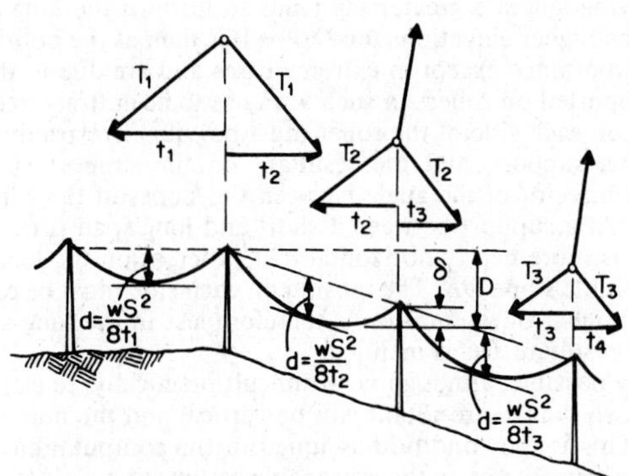


**FIG. 14-53** Illustrating the change in tension on hillsides. (This diagram is much exaggerated.)

under bare-wire conditions. In extreme cases, provision for the above conditions may be made by special sags, allowing somewhat higher tensions in spans above the normal level of the line and providing extra clearance in low sections so that slightly larger than normal sags may be used. Occasionally special methods must be devised.

***82. The McIntyre Joint*** (Fig. 14-54). This splice is used chiefly on small sizes of conductor. It consists essentially of seamless copper or aluminum tubing, oval in section, into which each conductor is pushed from opposite ends, until the conductors project about 2 in beyond the ends of the sleeve. The tube is then twisted the required number of turns by special tools. The joint shown in Fig. 14-54 is used for 1/0 to 4/0 ACSR and steel ground wires. For steel-reinforced aluminum (ACSR), the simple twisted-sleeve joint can be used in sizes up to 4/0 and develops about 80% of the strength of the cable.

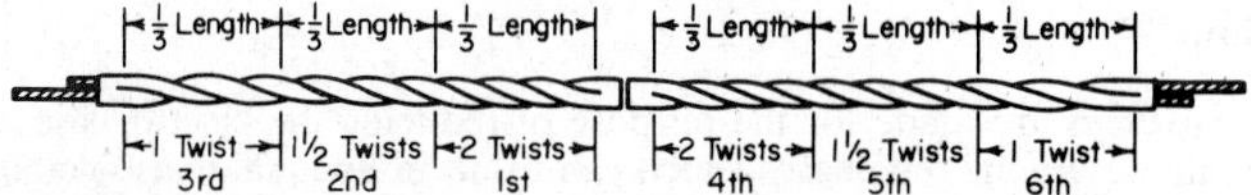


**FIG. 14-54** McIntyre or twisting joint.

***83. Compression Joints.*** These joints are used with the large ACSR conductors and the large "all-aluminum" conductors, including the high-strength types. As with the compression dead ends, the most widely used joints are those made by Somerset Products Co., subsidiary of Thomas & Betts Corp., and the Alcoa Conductor Products Co. of the Aluminum Company of America. Cutaway drawings of these joints are shown in Fig. 14-55. They consist of aluminum sleeves and steel sleeves for ACSR. Installation procedures call for the aluminum cable and the insides of the aluminum sleeves to be thoroughly cleaned.

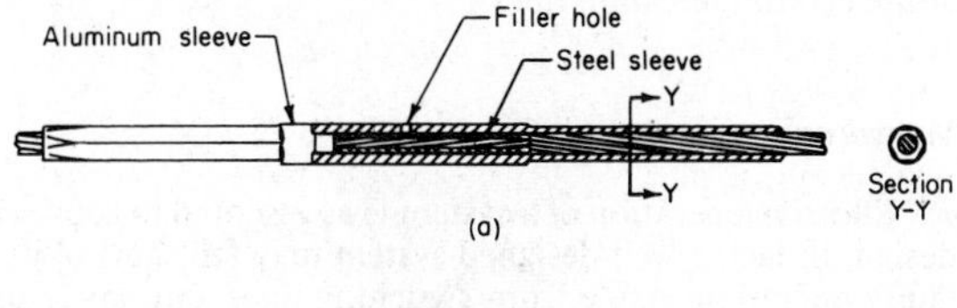


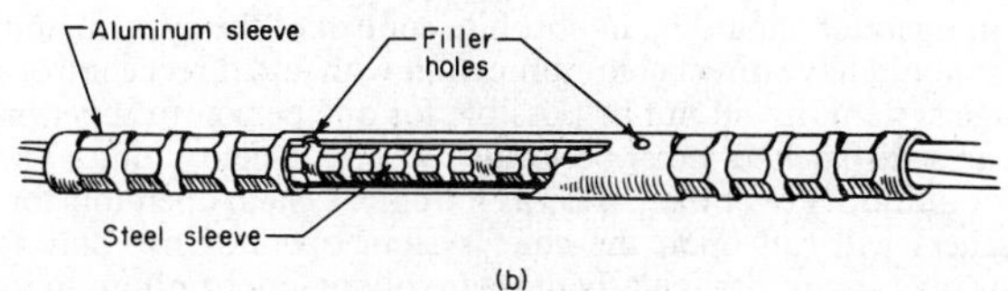


**FIG. 14-55** Compression joints. (*a*) Aluminum Company of America. (*b*) Thomas & Betts.

If the conductor is weathered, the strands should be unlayed and all scale removed. The aluminum sleeve is then slipped on the cable and backed out of the way. The aluminum on each cable end is next carefully cut back, care being taken not to nick the steel core, for a distance equal to one-half the length of the steel sleeve plus a distance of ½ in or more, depending upon the size of the conductor, so that the elongation of the steel sleeve on compression will not interfere with the free lay of the aluminum strands. The conductor ends are then marked by tape or other suitable means to center the sleeve. The steel sleeve is put in place and compressed, working from the center out. The aluminum sleeve is next slipped into place and filled with the nonoxidizing compound furnished with it, the filler holes are plugged, and the joint is ready for compression. The sleeves are compressed by working from the center out. The center section of the aluminum sleeve over the steel sleeve is not compressed. When the compression is completed, the Alcoa sleeve is hexagonal and smooth from overlapping compressions, and the Thomas & Betts sleeve, also hexagonal, has uncompressed ribs between the compressions as shown in Fig. 14-55*b*.

***84. Overhead Ground-Wire Installation.*** Overhead ground wires should receive no less care in erecting than the conductors, for the usual zinc or copper protective coating is very easily destroyed. Ground wires should be sagged in the same way as the conductors except that the important factor in ground-wire sags is to maintain ample clearance to the conductors. Generally, ground wires are sagged to about 80% of the conductor sags, thus ensuring proper clearance even under ice loads. McIntyre joints are frequently used for splicing, but a much more efficient joint can be made with the pressed-steel joint similar to the ACSR joint illustrated in Fig. 14-55.

## Transpositions[84,85]

***85. Transpositions*** are made for the purpose of reducing the electrostatic and electromagnetic unbalance among the phases which can result in unequal phase voltages for long lines. Untransposed lines also can cause inductive interference with paralleling wire communication lines. However, communication interference in the past has been largely with overhead long-distance telephone and telegraph lines. Most of these lines are now going underground, and other overhead lines are being replaced by microwave radio.

For some time, transpositions were little used. With the large power-system networks comprising most of the country's transmission lines, the unbalance of an untransposed line is largely smoothed out by the phase-balancing effect of the rotating equipment scattered over the system. The transpositions, however, do enhance the reliability and efficiency of the line in the following respects: (1) They restrain the amount of current which one line will induce in the other, thereby enhancing the reliability of fault arc extinction in the event of a faulty circuit. (2) They serve to reduce transmission power losses. (3) Depending on their location, transpositions can serve to reduce the electromagnetic coupling of power-line currents in adjacent telecommunication lines.

## Operation and Maintenance

***86. Operation.*** Effective operation of a system is as essential to good service as is excellent engineering design. In fact, a well-designed system may fall short of its service requirements owing to faulty operation. Aside from switching lines and power units to meet the load conditions of the system, operation consists not only in restoring service promptly after an interruption but also in detecting and removing faulty apparatus, thus actually preventing the development of faults.

A chief system operator should be in absolute control of the system, and if it is a small system, he or she should have direct communication with and direct control over every part of it. If it is a large system, it will not be possible for one person to supervise all switching operations, and area dispatchers must be located at convenient points. These dispatchers will have the same authority over their areas as the chief system operators for small systems. The area dispatchers will call upon the chief system operator not only for approval of unusual switching operations, particularly those involving interruptions to important loads. They will, however, make reports each shift, as convenient, on routine switching operations and will report major interruptions as soon as possible, with cause if known. Dummy boards are useful at dispatching centers. These boards should show the one-line diagrams of the circuits at all stations under the dispatcher's supervision, and provision should be made to show whether switches are open or closed. These boards must be kept correct up to the minute, even if it is necessary to do so by temporary means. It is best to anticipate system changes so that the dummy board can show the changes as soon as they are made. During normal operating conditions no switching should be done, including that of generators, without the dispatcher's permission. All dispatching orders should be reported back in order to prevent misunderstandings and should be recorded in log books both by the dispatcher and by the operator who will do the switching. The logs should show a record of all transactions, with particular care about times of receiving orders, of opening and closing switches, and of cases of trouble.

Emergency routines should be set up for all stations and should be followed at times of catastrophic storms when all means of communication with the dispatcher are interrupted. These routines will list the sequence for doing emergency switching on the operator's responsibility in an effort to restore service.

Supervisory control systems make it possible for operators at one transmission substation to operate several nearby substations as well as their own and thereby reduce operating personnel. Supervisory control also makes it possible for one central dispatching office to operate all the transmission substations serving a large metropolitan area. The supervisor may utilize carrier-current, microwave radio, or telephone channels for transmitting information and operating switches.

Sleet or glaze formation on lines is highly undesirable, and some companies prevent it by raising the temperature of the conductors with current. Ice will form on conductors over a small temperature range which is on the order of −3 to +2°C. The current required to prevent ice formation may be found according to Clem[86] from

$$I^2 = \frac{\theta\sqrt{dv}}{8.18 \times R} \times 10^4 \qquad (14\text{-}80)$$

in which $I$ = current, $\theta$ = temperature rise in degrees Celsius above surrounding air, $R$ = conductor resistance in ohms per mile at 20°C, $d$ = diameter of conductor, $v$ = wind velocity in miles per hour.

With the lines in service, it is usually difficult to obtain sufficient current. However, the necessary current sometimes may be obtained by transferring load from other lines to the line in trouble. Dead lines may be heated by short-circuiting them at one end and sending the necessary current from the other end. The approximate voltage to neutral is $E = I\sqrt{R^2 + X^2}$, where $R$ = resistance per wire and $X$ = reactance per wire.

Melting the ice after it has formed is considerably more difficult and requires more current than is required to prevent formation. Clem's article gives the various formulas required to calculate the melting current.

***87. Maintenance.*** Periodic inspection is normally maintained over all lines, with the frequency of inspection depending on the country traversed and the importance of the lines. In some densely settled areas, patrols of once a week may be considered necessary, whereas important lines in areas subject to heavy storms or other hazards may not require inspections more than once in 2 months. Patrollers may cover the line on foot, on horseback, by automobile, or by helicopter, depending on the characteristics of the right-of-way. Close and accurate patrolling is not obtained by one person in an automobile, in general, even when the line follows a highway. Helicopter patrol is by far the best over mountainous and sparsely settled country, and 200 mi a day can be covered readily. The helicopter can fly as slowly and as close to the line as is necessary, and it has the great advantage that the patroller is looking down on the line instead of up against the bright sky as a background. Tower and wood-pole structure numbers should be fastened to the tops of the towers or structures in such a manner that they can be read without trouble by the person in the helicopter. Helicopter patrol cannot be used over urban areas or congested industrial areas because of governmental restrictions on height of flight. Patrols on foot are best in such areas. Horseback patrol is best in cattle-range country, if aerial patrol is not available.

Landslides, washouts, danger timber, or anything else that is a potential danger to the line, such as piles of brush or straw which if burned could cause hot gases to short circuit the line, should be reported. Of course the person on patrol must also be on a close lookout for damaged conductors, insulators, and structures. A pair of field glasses is usually considered indispensable.

Personnel on ground patrols should keep the dispatcher informed as to their whereabouts and should call in from all patrol telephone stations and from substations as they reach them. They should call in not less often than morning, noon, and night. If a storm comes up while patrollers are out on a line, they should call the dispatcher as soon as possible, telling where they can be reached. Patrol cars and helicopters are radio-equipped.

Tree gangs whose sole duty is to remove brush, trim trees, and remove danger timber have been found to be advantageous by large companies. The use of chemical sprays to kill brush along rights-of-way is satisfactory from the standpoint of killing the brush if legally permissible but leaves a potential fire hazard. Care must be taken in spraying that wind does not blow the chemicals over growing crops. This danger has been found to be a disadvantage in helicopter spraying.

Emergency crews are stationed at locations always available by telephone or radio so that every important section of the transmission system can be reached by a crew within a reasonably short time. A light truck, provided with two-way radio and with the necessary tools and materials for making immediate repairs, is used by many companies. In addition to spotlights on the truck, a spotlight operated from a portable storage battery is frequently very useful. Small houses containing spare parts, such as insulators, lengths of cables, and clamps, should be located at intermediate and accessible points along the line in sparsely settled country. Such houses should be kept locked. Some companies employ concrete construction with iron doors. A routine inspection and checking of materials in such houses is advisable.

Line repairs and replacements are accomplished either when the line is energized or deenergized. The line crew should notify the dispatching office when a particular line or section is desired. If deenergized maintenance is desired, the line should then be not merely cleared through the circuit breakers but opened by disconnecting switches as well. If it is to be out of service for an extended period and there is danger of lightning, the line should be grounded out at its terminals to prevent the possibility of flashing over switches or insulators at the terminals. If the line is not equipped with grounding switches, it may be grounded out by equipment such as is used by line crews. Before the line crew is allowed to work, the line should be short-circuited and well grounded on each side of the location where the crew will work, with the grounding equipment in full sight. The grounding equipment should consist of heavy extraflexible copper cables, which should be attached by means of "hot-line" tools and clamps, the line being considered to be "hot" until the grounding equipment is applied. Ground chains are not safe and should not be used. Reliance should not be placed in grounding switches or grounding cables at the ends of the line.

In order to make repairs and replacements without interrupting the service, special "live-line" tools have been devised whereby insulators may be replaced, conductors spliced, etc., on lines of all voltages while the line is hot. Live-line maintenance methods are described in an IEEE Task Group Report[87] of the IEEE Transmission and Distribution Committee (1973). It covers methods and equipment for live-line maintenance. A foundation is presented from which working clearances and methods can be developed for specific needs in particular applications. Safety in live-line maintenance procedures was presented in a previous report,[88] in 1967.

Damaged insulators and insulator sections may be detected while they are in service, but suitable precautions should be followed. In general, the methods employed for faulty-insulator detection are based upon the measurement of the voltage gradient across the individual units of a string of suspension insulators or across the parts of multipart pin-type insulators. For safety reasons, none of the test methods should be used in wet weather.

Faulty insulators may also be detected by special radio interference locators consisting essentially of a sensitive battery-operated receiver coupled to either a directional loop antenna or a "whip" antenna. The latter type may be attached to a hot-line stick to enable close investigation of the insulator under test. Infrared techniques are also employed.

The Doble method[89] is, in effect, a spark-gap voltmeter which is safe to use and which gives high accuracy in measuring potentials in the field on live transmission-line insulators.

There are two general types of Doble safety tester: the type A single-prong tester for multipart pin-type insulators on either wood or steel construction, and the type B two-prong tester for multiunit suspension-type insulators on either wood or steel construction. The type B is most applicable to transmission lines.

The equipment consists of a micrometer spark gap in series with a capacitor and a special telephone-type headset with which to listen to gap sparkover. The telephone receiver is heard through a rubber hose connected to a highly insulated hollow tube which is long enough so that there is no danger to the operator.

Both sides of the electric circuit of the tester are terminated by exposed metal tips (Fig. 14-56) arranged so as to bridge readily a single insulator disk or section. In the circuit between the two contact tips, a protective insulating capacitor is built into the tester in such a manner as to make the impedance between its terminals greater than the impedance of a single good disk. Thus, in operation, the tester *does not short-circuit* the disk under test. The tester may be considered as a voltmeter which indicates the voltage between the points on the insulator touched by the tips of the tester.

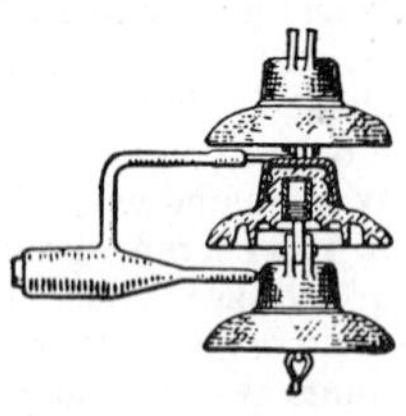

**FIG. 14-56** Doble insulator tester using the principle of a spark-gap voltmeter for suspension insulators.

The degree of defect in the disk under test is indicated by the size of the maximum gap at which a noise is heard in the tester, as compared with the size of a maximum gap for a good disk in the same position in a string; a totally dead disk gives no sound, irrespective of the gap setting. In practice, one gap length is fixed in advance and used for all units of the string. The operator judges if a unit is defective from the noise heard in the headset.

This apparatus is in use for testing insulators on lines at voltages from 11 kV through the medium transmission ranges.

The I-T-E Imperial Corp. live-line insulator tester detects defective insulators, either multipart pin-type or suspension, by comparing the measured voltage distribution over insulators or insulator strings while in service with characteristic curves plotted for good insulators of the same type. It is suitable for use in testing insulator integrity on transmission systems with nominal voltages through 230 kV. The tester employs a single-prong head for multipart pin-type insulators, a small two-prong head (Fig. 14-57) for suspension strings or small one-piece insulators, and a large two-prong head for multipart pedestal insulators. Visual indication is given by means of a meter which shows a deflection in proportions to the voltage gradient. Since relative indications only are required, the meter is calibrated simply in units of deflection. Tests may be made of all shells of multipart pin-type insulators and all units of suspension strings with equal facility. As with the Doble tester, tests should be made only on perfectly dry insulators.

**FIG. 14-57** The I-T-E Imperial Corp. two-prong live-line insulator tester.

Doble field power-factor test is used (in contrast to that previously described) for testing the insulation of electrical power apparatus *with the apparatus out of service.* Dielectric watts loss and charging current are measured at selected test voltages up to 20 kV, from which power factor, capacitance, ac resistance, and the presence of ionization (corona) can be determined. The specimens to be tested may be in the substation or in the service building.

The test equipment is capable of determining the condition of electrical insulation of bushings, bus supports, cables, capacitors, circuit breakers, insulators, surge arresters, liquid insulation, potheads, rotating machinery, transformers, and voltage regulators. Power-factor measurements with this equipment have been adopted by many companies as a criterion for servicing power-apparatus insulation.

High power factors or sudden increases in power factor from a previous test indicate contaminated or deteriorated insulation which may be an operating hazard. Changes in the watts loss, charging current, ac resistance, and capacitance between tests are also used for indicating operating hazards in apparatus insulation.

A variety of other makes of portable insulation testers, both ac and dc, are available for use in testing line insulation, if desired, when the line is out of service.

## Foundations

***88. Lattice-Tower Foundation Loads and Displacement Criteria.*** Lattice-tower foundation loads consist of vertical tension (uplift) or compression forces and horizontal shear forces. For tangent and small line angle towers, the vertical loads on a foundation may be either uplift or compression. For terminal and line angle towers, the foundations on one side may always be loaded in uplift while the other side may always be loaded in compression. The distribution of horizontal forces between the foundations of a lattice tower vary with the bracing of the structure. A typical free-body diagram of foundation loads is shown in Fig. 14-58.

When the foundations of a tower displace and the geometric relationship of the tower to its foundations remains the same, any increase in load due to this displacement will have a minimal effect on the tower and its foundation. However, foundation movements which change the geometric relationship between the tower and its foundations will redistribute

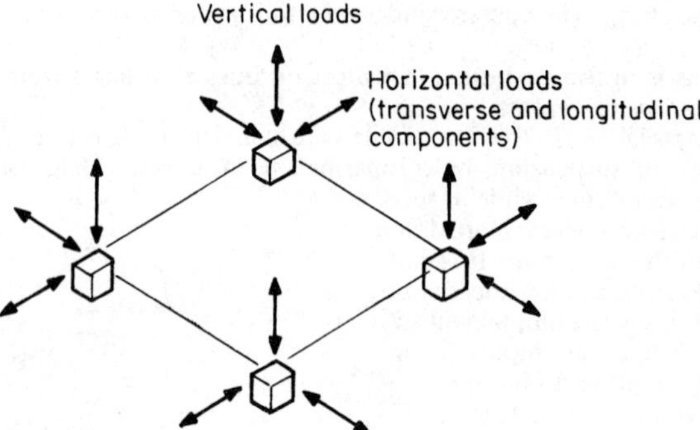


FIG. 14-58 Typical loads acting on lattice-tower foundations.

the loads in the tower members and foundations. This will usually cause greater reactions on the foundation that moves least relative to the tower, which in turn will tend to equalize this differential displacement.

Presently, the effects of differential foundation movements are normally not included in tower design. Several options are available should the engineer decide to consider differential foundation displacements in the tower design. These options include designing the foundations to satisfy performance criteria which will not cause significant secondary loads on the tower, or design the tower to withstand specified differential foundation movements.

***89. Single-Shaft Foundation Loads and Displacement Criteria.*** These structures have one foundation so that differential foundation movement is precluded. The foundation reactions consist of a large overturning moment and usually relatively small horizontal, vertical, or torsional loads. Figure 14-59 presents a free body diagram of the loads.

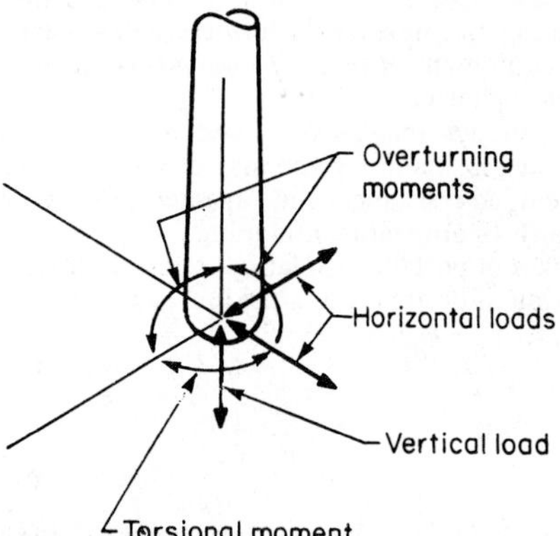


FIG. 14-59 Typical loads acting on foundations for single-shaft structures.

For single-shaft structures, the foundation movement of concern is the angular rotation of the shaft in the vertical plane and horizontal displacement of the top of the foundation. When these displacements have been determined, the displacement of the conductors can be computed. Under high wind loading, a corresponding deflection of the conductors perpendicular to the transmission line can be permitted. Accordingly, a large ground-line displacement of the foundation could also be permitted. Due to foundation rotation, the clearance between the conductors and the structure would only be decreased for structures with single string insulators. The midspan ground clearance and the change in line angle would also decrease a negligible amount.

In establishing displacement criteria for single-shaft-structure foundations, consideration should be given to how much total, as well as permanent, displacement can be permitted. In some cases, large permanent displacements might be aesthetically unacceptable and replumbing of the structures and/or their foundations may be required. In establishing displacement criteria, the cost of replumbing should be compared to the cost of a foundation that is more resistant to displacement.

For terminal and large line angle structures, large foundation deflections parallel to the conductor may be intolerable. For these structures, excessive deflections may reduce the conductor-to-ground clearance or affect the load capacity of adjacent structures. There are also problems in the stringing and sagging of conductors if the deflections are excessive. This problem is usually resolved by construction methods or use of permanent guys.

***90. Framed Structure Foundation Loads and Displacement Criteria.*** These structures are dependent in part for their stability on one or more of their joints resisting moment. The foundation reactions are dependent on which joints can resist moment and the relative stiffness of the members. Figures 14-60 and 14-61 present free body diagrams of four- and

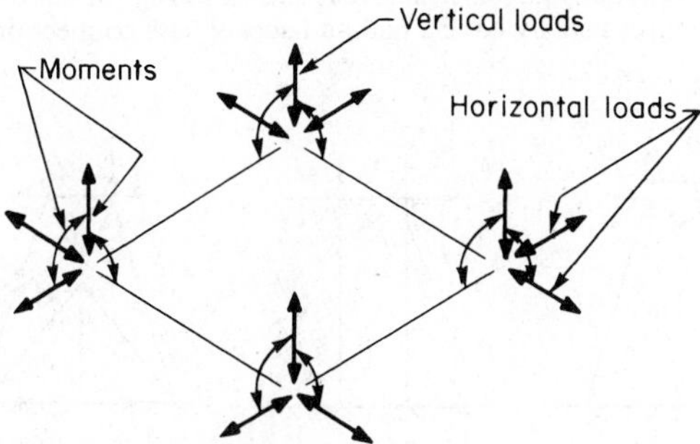


**FIG. 14-60** Typical loads acting on foundations for four-legged framed structures.

two-legged framed structures. If the bases of structures are designed with pins or universal joints, then the moments acting on the foundations will theoretically be zero.

Many different types of two-legged, H-framed structures are in use in transmission lines. This has been particularly true in recent years since visual impact has become of greater concern.

The H-frame structure is particularly applicable for wood, tubular steel, and concrete poles. The crossarm may be pin-connected to the poles. These structures may be unbraced, braced, or internally guyed as shown in Fig. 14-62.

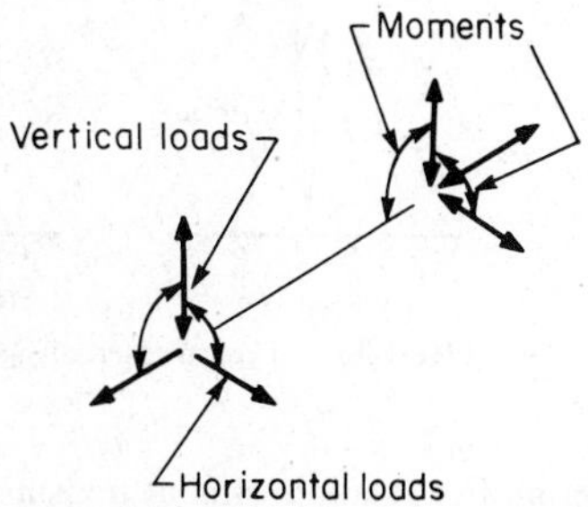

**FIG. 14-61** Typical loads acting on foundations for two-legged framed structures.

As with lattice towers, past practice has not usually included the influence of foundation displacement and rotation in H-frame structure design. Significant foundation movements will redistribute the frame and foundation loads. The foundations can be designed to experience movements which will not produce significant secondary stresses or the structure can be designed to a predetermined maximum allowable displacement and rotation.

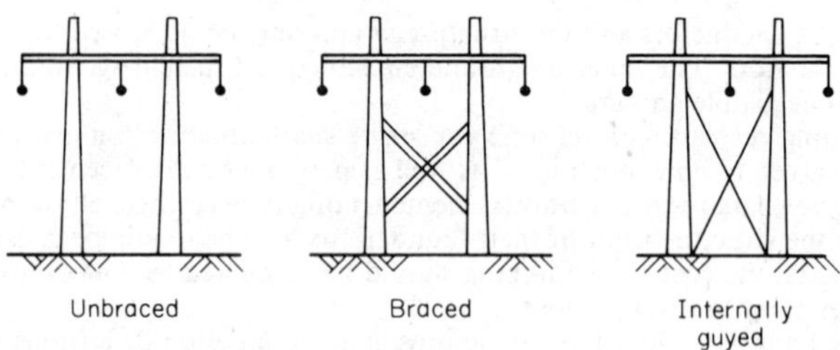


**FIG. 14-62** Typical H-frame structures.

***91. Externally Guyed Structures and Displacement Criteria.*** There are three general types of externally guyed structures. For all types, the guys produce uplift loads on the guy foundations and compression loads on the structure foundation. The guys are generally adjustable in length to permit plumbing of the structure during construction and to account for creep in the guy and movement of the uplift anchor.

The first type of externally guyed structure is shown in Fig. 14-63. In this case, the shaft or shafts of the structures usually have a ball-and-socket base connection to the foundation

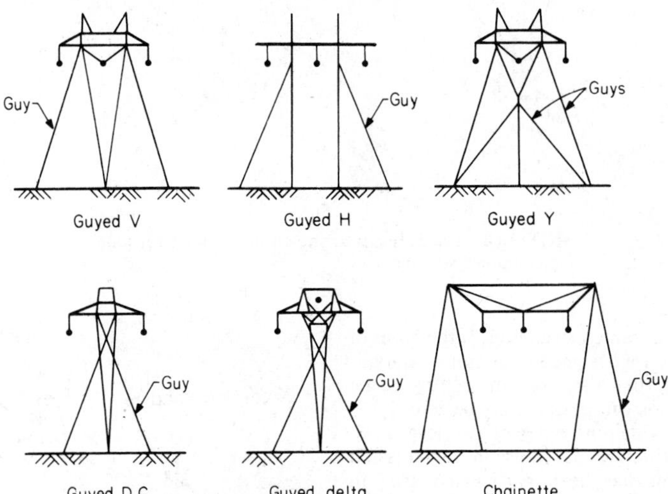


**FIG. 14-63** Typical externally guyed structures.

to permit free rotation without transmitting moment to the foundation. This will produce compression loading with a small shear load. This type of guyed structure can generally tolerate large foundation movements. Considerations in establishing displacement criteria are similar to those discussed for single-shaft structures.

The second type of guyed structure consists of a single shaft as shown in Fig. 14-64. This type of structure is often used as a terminal and large line angle structure and is quite flexible, allowing most of the load to be resisted by tension in the guys and compression in the main shaft. This type of guyed structure can generally tolerate significant foundation movement as far as its own structural integrity is concerned; but, like the terminal and large line angle poles discussed previously, if excessive guy anchor slippage occurs, conductor-to-ground clearance, security of adjacent structures, and the stringing and sagging of conductors can become a problem.

The third type of externally guyed structure is a conventional lattice tower that is guyed to reduce its leg loads and foundation reactions. This approach is often used to upgrade existing towers. The flexibility of the guy, together with the flexibility of the tower, are needed to compute the foundation reactions and anchor loads. The maximum amount of anchor slippage can be selected and the tower and anchors designed accordingly. The initial and final modulus of elasticity of the guys together with the creep of the guys should be considered. The amount of pre-tension in the guys should be specified. Load testing of the guy anchors is recommended to ensure against excessive slippage. Figure 14-65 shows a typical installation.

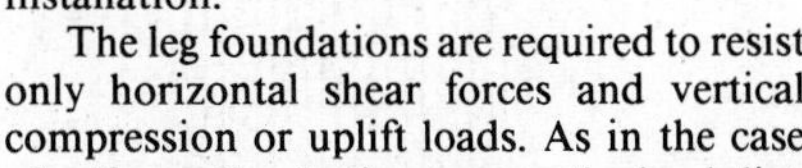

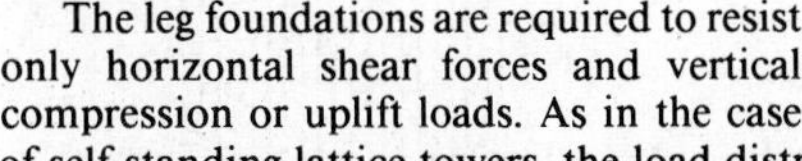

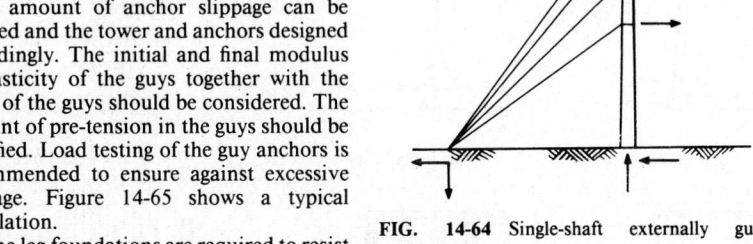

**FIG. 14-64** Single-shaft externally guyed structure.

The leg foundations are required to resist only horizontal shear forces and vertical compression or uplift loads. As in the case of self-standing lattice towers, the load distribution among the members of the structure is sensitive to the foundation performance. Differential displacements of the legs of the tower will result in load redistribution and may affect the integrity of the tower.

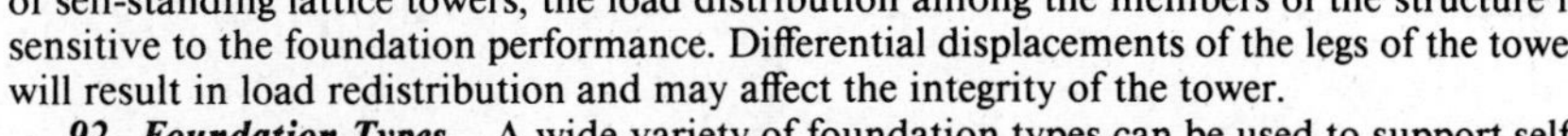

***92. Foundation Types.*** A wide variety of foundation types can be used to support self-standing or guyed lattice tower, framed, and single-shaft structures. A summary for each structure type is given below:

| *Lattice Tower* | *Framed and Single-Shaft Structures* |
|---|---|
| Steel grillages | Poured-concrete spread foundations |
| Poured-concrete spread foundations | Drilled shafts |
| Rock foundations | Direct embedment |
| Drilled shafts | |

*Steel Grillages.* Figure 14-66 shows three typical types of steel grillages. Figure 14-66*a* is a pyramid arrangement in which the leg stub is connected to four smaller stubs which in turn are connected to the grillage at the base. The advantage of this type of construction is

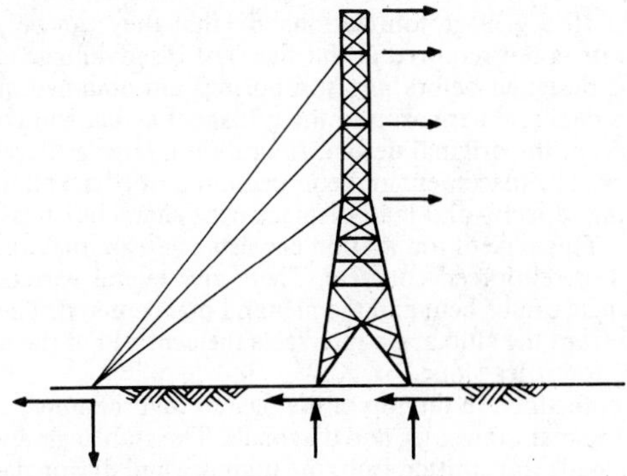

**FIG. 14-65** Externally guyed lattice tower.

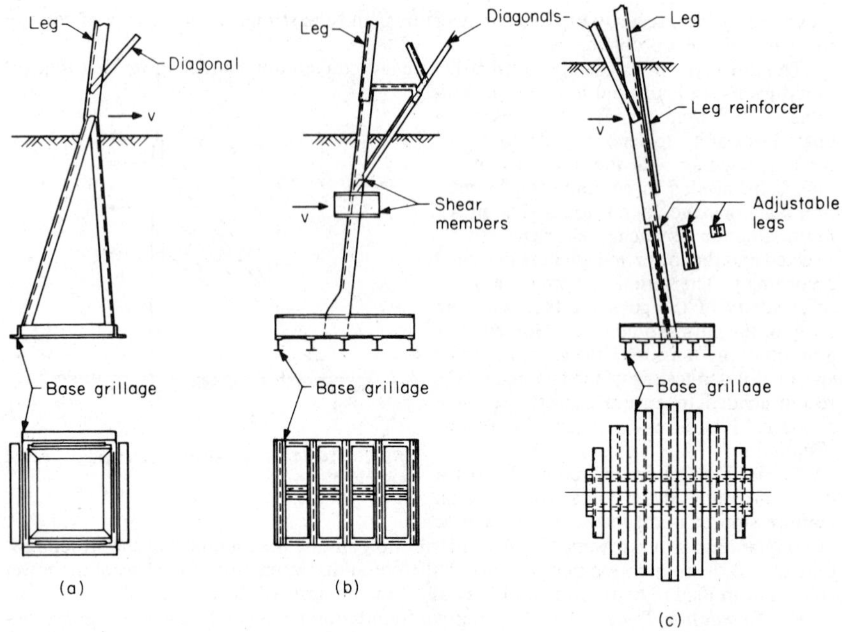


**FIG. 14-66** Typical steel grillage foundations.

that the pyramid can transfer the horizontal shear load down to the grillage base by truss action. However, the pyramid arrangement does not permit much flexibility for adjusting the assembly, if needed. In addition, it is difficult to compact the backfill inside the pyramid.

Figure 14-66*b* shows a grillage foundation which has the single leg stub carried directly to the grillage base. The horizontal shear is transferred through shear members that engage the passive lateral resistance of the adjacent compacted soil.

Figure 14-66*c* also has the single leg stub carried directly to the grillage base. This type of grillage foundation has a leg reinforcer which increases the area for mobilizing passive soil pressure as well as increasing the leg strength. The shear is transferred to the soil via the leg and reinforcer and resisted by passive soil pressure. The base grillage of these three typical foundations consists of steel beams, angles, or channels which transfer the bearing or uplift load to the soil.

The advantages of steel grillage foundations are that they can be purchased with the tower steel and concrete is not required at the site. The disadvantage is that these foundations usually must be designed before any soil borings are obtained and may have to be enlarged by pouring a concrete base around the grillage if actual soil conditions are not as good as those assumed in the original design. In addition, large grillages are difficult to set with required accuracy. The placement and compaction of the backfill material is critical to the actual load-carrying capacity and load-displacement characteristics of the foundation.

*Poured Concrete.* This type of foundation consists of a base mat and a square or round pier. It is constructed of reinforced concrete. There are several variations as indicated in Fig. 14-67. The stub angle can be bent and the pier and mat centered. The mat can be located so that the projection from the stub angle intersects the centroid of the mat or the pier itself can be battered to the tower leg slope.

The stub angle is embedded in the top of the pier so that the upper exposed section can be spliced directly to the main tower leg and diagonals. The stub angle should be of adequate size to resist the axial loads transmitted from the main leg and diagonals plus any secondary bending moment from the horizontal shear, if applicable. The stub angle must be embedded in the concrete to a sufficient length to transmit the load to the concrete. Bolted clip angles

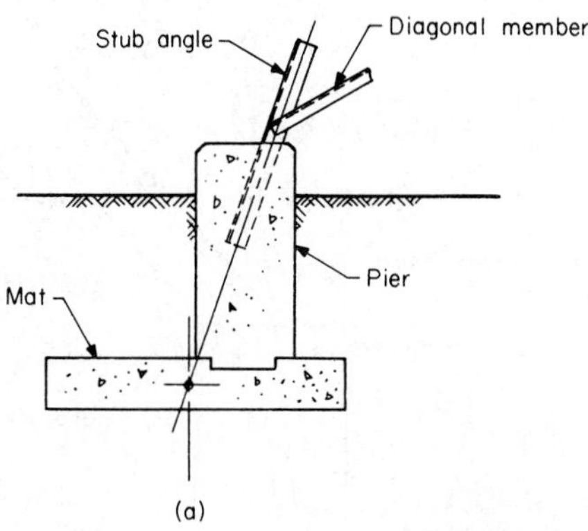


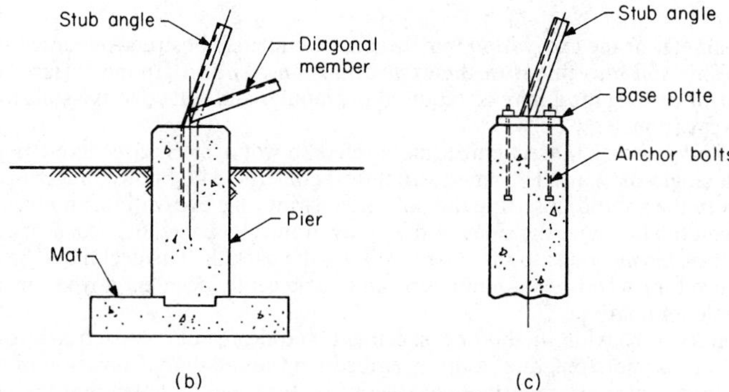


**FIG. 14-67** Typical poured concrete foundation (reinforcing not shown).

or welded stud shear connectors may be added on the end of the stub to reduce this length. Anchor bolts can also be used in lieu of the direct embedment stub angle as shown in Fig. 14-67*c*.

*Rock Foundations.* Many areas of the United States have bedrock either exposed at the ground surface or covered with a thin mantle of soil. Relatively simple, economical, and efficient rock foundations may be installed where this type of terrain is encountered. A rock foundation can be designed to resist both uplift and compression loads plus horizontal shear and, in some structure applications, bending moments. Where suitable bedrock is encountered at the surface or close to the surface, a rock foundation, as shown in Fig. 14-68, can be installed.

*Drilled Concrete Shaft Foundations.* The drilled concrete shaft is the most common type of foundation presently being used to support lattice towers, framed structures, and single shafts. Drilled concrete shafts are constructed by power augering a circular excavation, placing the reinforcing steel and pouring concrete to form a shaft foundation. Lattice towers are attached by embedment of a stub angle or through the use of base plates and anchor bolts. Framed structures and single shafts are attached through the use of base plates and anchor bolts.

Drilled shafts can be constructed in a wide variety of soil and rock types. However, the construction of drilled concrete shafts may encounter problems under certain soil conditions. For example, granular soils may collapse into the excavation before concrete can be poured. In soft, cohesive soils, squeezing or shear failure of the soil can occur producing a

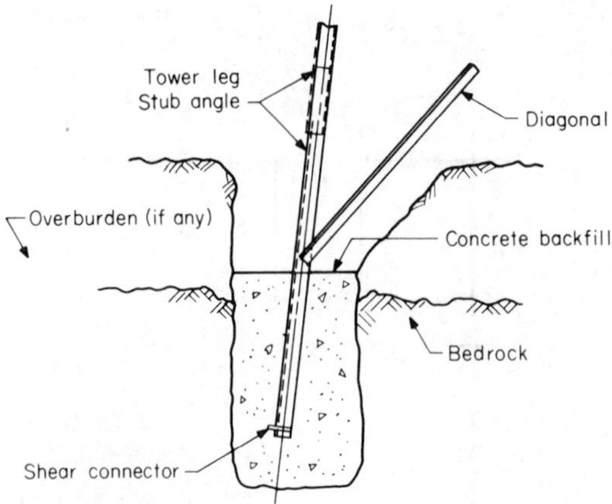


**FIG. 14-68** Rock foundation.

reduced diameter or the excavation may become completely obstructed before the concrete is placed. This soil movement in the excavation can result in ground-surface settlement. Casing and/or drilling mud may be required in granular and soft cohesive soils to maintain an open excavation.

*Direct Embedment.* Direct embedment refers to wood, steel, or concrete pole foundations (both single-shaft and H-framed structures) constructed by power augering a circular excavation in the ground, inserting the pole directly into the excavation, and backfilling the void between the pole and the sides of the excavation. Thus, the pole acts as its own foundation by transferring loads to the in-situ soil via the backfill. This technique has been traditionally used for wood-pole foundations and has recently been employed for metal- and concrete-pole foundations.

The quality of backfill, method of placement, and degree of compaction strongly influence the stiffness and strength of a direct embedment foundation. Corrosion of an embedded metal pole is also an important consideration. It should be noted that the presence of granular or soft, cohesive soils may cause the same construction problems for direct embedment foundations as for drilled concrete-shaft foundations.

***93. Subsurface Investigations.*** The technical requirement of assuring a safe and cost-effective foundation design for transmission structures requires a thorough knowledge of the subsurface considerations along the right-of-way (ROW). The intent of this section is to provide a guide for performing an adequate subsurface investigation for the design of transmission-line structure foundations.

When designing a foundation, the engineer should be concerned with three factors: (1) the ultimate load-bearing capacity of the subsurface material, (2) the quality of available backfill materials, and (3) the allowable displacements of the foundation. Hence, the objectives of a subsurface investigation are to determine the stratigraphy, physical characteristics, and engineering properties (particularly the strength and deformation characteristics) of the soil or rock underlying a given site.

To determine the most cost-effective foundation, it is necessary to consider the engineering and physical properties of the subsurface materials; construction costs; the construction aspects of a particular foundation type and how they are influenced by such factors as groundwater elevation, safety requirements, contractor capability and experience, and environmental constraints.

The scope of a subsurface investigation will vary depending on foundation loads, type of structure, and probably foundation types, types of subsurface materials, and previous knowledge of subsurface conditions along the line route. It is necessary to use engineering judgment when considering the scope of the subsurface investigation. A detailed outline for developing a cost-effective investigation is given in an EPRI report.[90]

***94. Design of Spread Foundations.*** The design of spread foundations for transmission towers must consider both the direction (uplift or compression) and orientation (inclination and eccentricity) of the applied loads. The foundation must be designed to prevent excessive displacement or shear (bearing capacity or uplift) failure of the support soil. A detailed presentation of estimating the uplift and compression capacity of spread foundations is given in an EPRI report.[90]

***95. Design of Drilled Shaft Foundations.*** Drilled shaft foundations are used to support lattice-tower, framed, and single-shaft structures. This type of foundation supports vertical compression loads through a combination of side shear and end bearing and supports vertical uplift loads by side shear. Lateral loads and overturning moments are supported by lateral resistance of the soil and/or rock in which the shaft is embedded plus the vertical shearing resistance on the perimeter of the shaft, and the horizontal shear on the base and the base moment.

*Compression and Uplift Capacity.* Methods for computing the compression and uplift capacity of drilled shaft foundations are given in an EPRI report.[90]

*Lateral Load Capacity.* The response of a drilled shaft to lateral loads is the result of complex interactions between the shaft and the soil and/or rock in which it is embedded. A common method of modeling this interaction is called the subgrade modulus approach. Reference 91 provides a detailed explanation of a method for determining the lateral capacity of drilled shafts. The computer program, PADLL,[91] which was developed as part of the EPRI research project is available for design and eliminates the simplifying assumptions associated with prior models.

*Direct Embedment.* The response of direct embedment foundations in compression, uplift, and lateral loads is similar to that of drilled concrete shafts. Most of the analytical techniques used in drilled shaft design are relevant to direct embedment design. The principal differences between direct embedment foundations and drilled concrete shaft foundations are: (1) the backfill which intervenes between the pole and the in-situ soil, and (2) the stiffness of the embedded structure shaft relative to that of a drilled concrete shaft. Drilled shafts transfer loads directly to the in-situ soil.

*Construction Considerations.* Factors affecting, and problems associated with, the construction of drilled shaft foundations can be divided into two general areas: (1) geotechnical factors influencing construction, and (2) construction-related problems.

***96. Design of Anchors and Anchor Foundations.*** An anchor is a device which will provide resistance to an upward (tensile) force transferred to the anchor by a guy wire or structure leg member. An anchor may be a steel plate, wooden log, or concrete slab buried in the ground, a deformed bar or a steel cable grouted into a hole drilled into either soil or rock, or one of several manufactured anchors which are either drilled or rotated into the ground. Anchorage may also be provided by vertical or battered drilled shafts or piles. Typical types of anchors are shown in Fig. 14-69.

Anchors may be classified either as deadman or prestressed. Deadman anchors are defined as those anchors which are not loaded until the structure is loaded. Prestressed anchors are loaded to specified load levels during installation of the anchor. An advantage of a prestressed anchor is that most of the initial strains of the anchorage system have been removed before the structural load is applied. Therefore, the full capacity of the anchor can be attained at very small deformation (movements in soil of less than ¼ in are typical). Another advantage of prestressed anchors is that they are proof-loaded to their design load at the time of installation. Disadvantages of prestressed anchors are that they are generally more expensive than deadman anchors and they should not be used in compressible soils.

Another advantage of prestressed anchors is that shallow anchors may obtain additional strength by the increased effective stress created by the influence of the bearing plate on the soils adjacent to the anchors.

Deadman anchors may include any of the systems shown in Fig. 14-69. Initial strains in deadman anchors may be reduced by as much as 50% by prestressing them to their design load at the time of installation.

*Anchor Application.* Anchors are used to permanently support guyed structures, as well as to temporarily support other structure types during erection and stringing. The legs of lattice towers can be anchored directly by rock anchors or helix-type anchors. The uplift capacity of spread foundations may be increased through the use of anchors as shown in Fig. 14-70. Guys and anchors are also extensively used to terminate wire loads on wood structures and to increase wood structure capacity for high transverse loading. At interme-

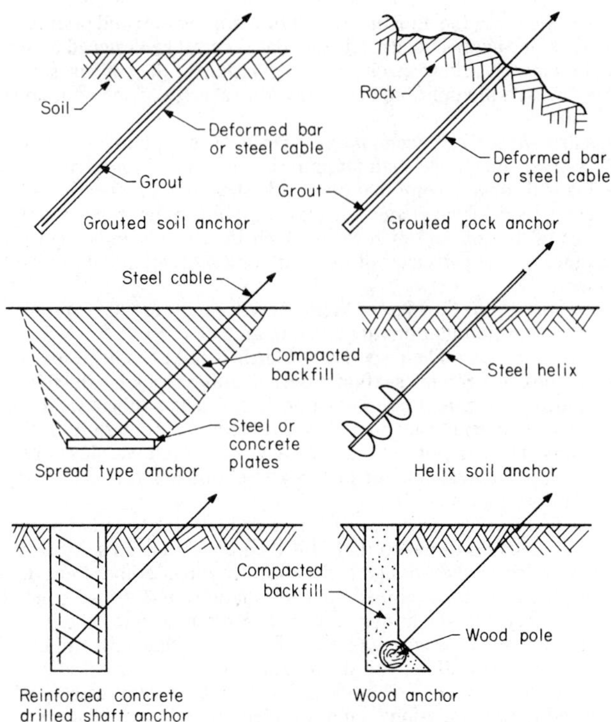


**FIG. 14-69** Typical anchors.

diate structure locations, guys and anchors may be utilized to provide additional longitudinal strength. Anchors can be used to increase the load capacity of existing foundations.

*Design.* The design of an anchor depends upon a knowledge of the peak and residual shear strength properties of the soil or rock in which it is embedded. In rock, it is also important to know the degree and depth of any weathering which may have occurred, together with the orientation and spacing of joints and foliation. In addition, an understanding of the load characteristics and the structure deflection tolerance combined with the guy cable elongation is important in selecting and designing the type of anchor. Anchor pullout tests are often conducted to confirm design assumptions where prior experience is lacking.

References 90 and 92 provide detailed information on the design of anchors.

## Recent Technology Options

***97. Voltages above 765 kV.*** Historically, ac transmission has been characterized by increasing voltage levels, since the amount of power transfer for a consistent impedance between generation and load is mathematically proportional to the square of the system voltage.

Figure 14-71 illustrates the historical voltage trend.[3] With each succeeding voltage increase there has been increased research, development, and design efforts previous to application. This is illustrated by the ratio of phase spacing to flashover spacing for conventional transmission line designs as illustrated in Fig. 14-72. Research into voltage levels up to 1500 kV has been conducted at Project UHV at Pittsfield, Massachusetts, funded by the Electric Power Research Institute; the American Electric Power, Ohio Brass, ASEA, IREQ Project on UHV; and the more recent Bonneville Power Administration 1100-kV project to assess both electrical and mechanical characteristics.

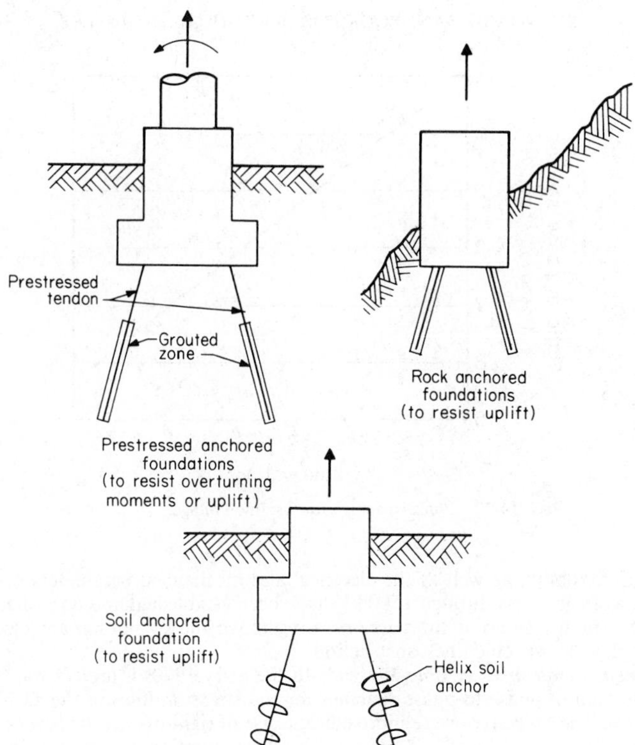


**FIG. 14-70** Typical anchored spread foundations.

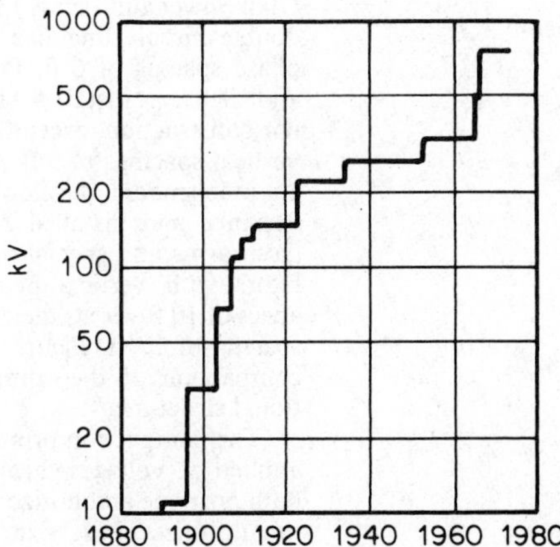


**FIG. 14-71** Maximum transmission voltage in North America.

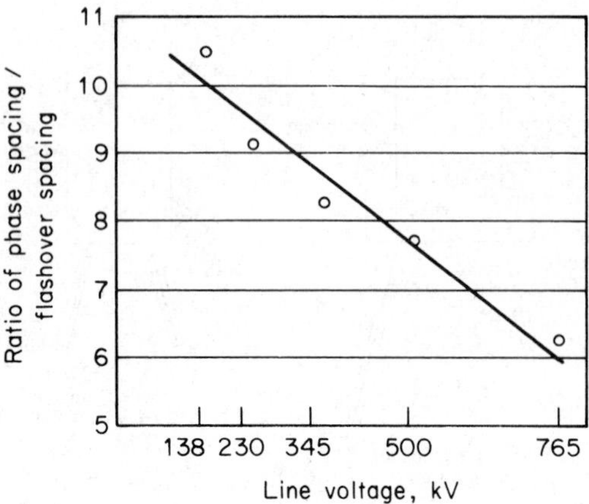


**FIG. 14-72** Phase-spacing ratio vs. line voltage.

Technical feasibility as well as the electrical and mechanical parameters necessary for design of ac voltage levels through 1500 kV have been established and published.

At present, no transmission lines are operating above 765 kV nor has any electric utility announced plans for specific line construction.

***98. Compact Transmission Line Design.*** In the early 1970s a project was initiated to explore reduction of phase-to-phase spacings for transmission lines in the 115- to 138-kV voltage ranges. The objectives were more efficient use of right-of-way for new construction, voltage uprating of existing lower-voltage lines, and possible economies. Methods of design and supporting data were developed for practical designs of very compact lines with reduced clearances. Because of the need to restrain conductor motion at the structure, most applications use posttype insulators. Figure 14-73 illustrates one of the first commercial applications on the Utah Power and Lights 138-kV system. This double circuit line has vertical phase-to-phase spacing of 6 ft. Puget Sound Power and Light has installed 115-kV lines of similar construction, except single circuit, with vertical spacings of 5 ft. Ontario Hydro, in an independent development project, has designed and installed 230-kV lines using davit arms and special resistive-glaze insulators with vertical phase-to-phase clearances of 10 ft versus their previous standard spacing of 27 ft. Figure 14-74 illustrates a comparison of the compact and conventional structures.[93]

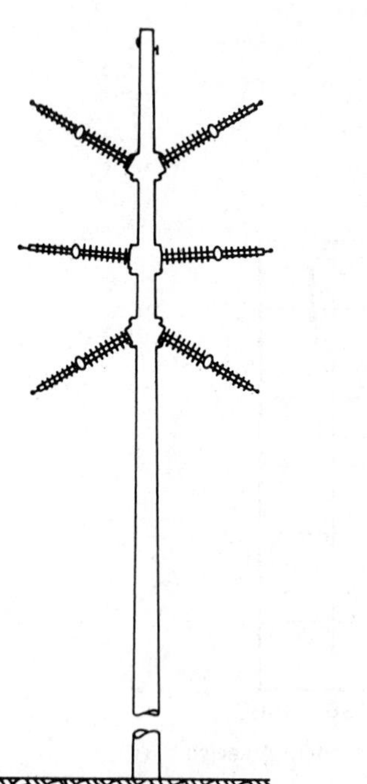

**FIG. 14-73** Representative double-circuit 138-kV compact line.

The compaction principle has also been applied at voltages through 500 kV using both posttype and horizontal-Vee insulators to minimize phase spacings. Figure 14-75 illustrates a 500-kV compact design using horizontal-Vee insulators. This line, on the Pennsylvania Power and Light Company system, is constructed on a 140-ft right-of-way.

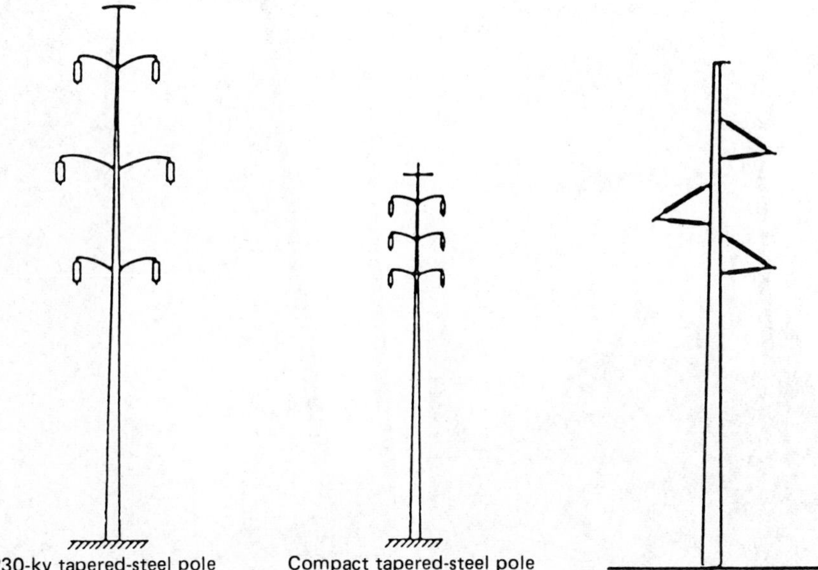


**FIG. 14-74** Comparison of conventional and compact 230-kV lines.

**FIG. 14-75** 500-kV compact line.

Compact line design, a relatively recent development, is becoming widely accepted by the electric utility industry.

***99. High-Phase-Order Transmission.*** The use of more than three phases for electric power transmission has been studied intensively for several years.[94–97]

High-phase-order transmission is basically the utilization of more than one 3-phase system to employ the inherent field or insulation characteristics of a conductor system. As an example, Fig. 14-76*a* illustrates a 3-phase system *A-B-C* and Fig. 14-76*b* a second 3-phase system interposed at 60° from the original 3-phase system, thus producing a 6-phase set. It can be seen that the total space required for conductors defined by a circumscribed circle has not been increased by addition of the second set of conductors. Additional 3-phase groups, appropriately phase-shifted, can be integrated on the circle's periphery without increasing space requirements. With the increased number of phases, the voltage between phases decreases in proportion to the reduction in distance maintaining essentially constant kilovoltage per foot. For example, Fig. 14-76*c* illustrates a 12-phase set.

The concept can be applied at all voltage levels. The thermal capacity of a higher-phase-order line is directly proportional to the number of phases while, because of mutual effects, surge impedance loading increases at a lesser rate. Thermally, the capacity of a 12-phase line

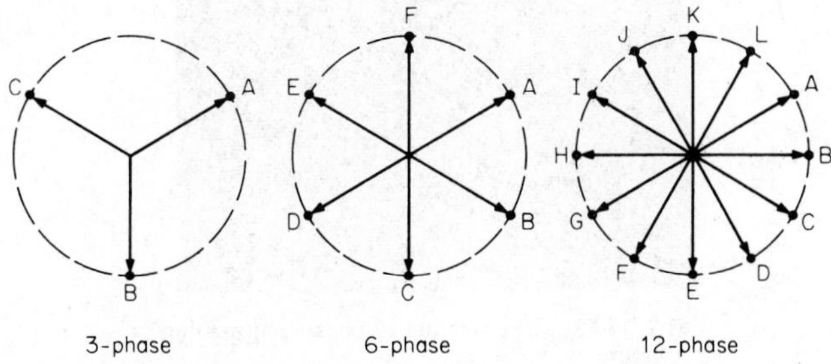


**FIG. 14-76** Illustration of high-phase-order phasor diagrams.

**FIG. 14-77** Prototype 6-phase transmission structures.

**FIG. 14-78** Experimental 12-phase transmission structure.

would be four times that of a normal 3-phase line, resulting in essentially a four-fold increase in right-of-way power density.

Analytical studies have shown the feasibility of high-phase-order or polyphase transmission, and the economics appear to be favorable for a variety of applications. These include uprating of existing double circuit lines, either to reduce losses or to increase the power transfer capability, and new construction in areas where right-of-way is limited or impossible to obtain. The acceptance of a high-phase-order system requires construction and testing an experimental line. Analytical results were also needed, especially for switching surges. Whereas phase-ground surges have historically predominated in line design with six or more phases—as with compact 3-phase lines—phase-phase surges have assumed increased importance.

In a recently completed project sponsored by the U.S. Department of Energy, construction of experimental 6- and 12-phase lines and associated electrical and mechanical testing were completed.[98] Comprehensive studies of electrical environmental effects, overvoltages, insulation design, and economic evaluations have been conducted.

Figure 14-77 illustrates two 80-kV phase-ground 6-phase structures on the test line, and Fig. 14-78 illustrates an 80-kV 12-phase structure.

The dimensions of the phase conductor arrays resulted in a 4-m window for the 6-phase line and a 6-m window for the 12-phase line. The towers were designed for NESC heavy loading requirements for a 370-m span of Drake (795-kcmil ACSR) conductor.

Six-phase and twelve-phase transmission lines are feasible and can be attractive alternatives to conventional three-phase transmission at system voltage levels from 115 through 1200 kV. Studies have shown both economic equivalence and electrical environmental advantage (e.g., audible noise and electric field effects) for comparable power capacities to three phase.

Several utilities are presently evaluating high-phase-order alternatives.

## UNDERGROUND POWER TRANSMISSION

### General

***100. Cable Applications.*** Underground cable systems are used for electric-energy transport where overhead construction is impracticable, unsafe, costly, or environmentally unacceptable. The principal applications have been in heavily urbanized areas where overhead rights-of-way are unavailable or prohibitively costly, or where local or state ordinances mandate undergrounding. There are also many spot locations where undergrounding may be practical for reasons of safety, security, reliability, or aesthetics; these include airport approaches, station and substation exits, long water crossings, and areas of unusual scenic value or with extreme vulnerability to damage by natural forces or vandalism.

Although several alternative systems are available for selection—depending on voltage, power requirements, length, cost, and reliability considerations—the oil-impregnated paper-insulated pipe-type cable system has strongly dominated United States' practice. There are over 3000 circuit miles of underground transmission cables rated 69 to 345 kV in operation in the United States, over 90% of which employ the pipe-type design.

Pipe-type designs rated to 550 kV have been proved feasible and are available but have only been used commercially in Japan. In the United States, there are two relatively short cable installations at the 550-kV level; one utilizes self-contained cable for generator leads from a high-head hydroelectric station to the overhead spreader yard; the other uses an $SF_6$-insulated "dip" in a high-capacity overhead line crossing another transmission corridor.

### Cable System Types

***101. Cable Characteristics.*** Cables for operation at transmission voltages are highly miniaturized devices as compared with most other electrical apparatus. Conductor sizes and insulation thicknesses must be held to the minimum consistent with reliable operation because of manufacturing, shipping, and installation constraints as well as costs. Design dielectric stresses therefore tend to be high, imposing the need for scrupulous control over all aspects of a project regardless of the cable system selected. All transmission cables must be free of destructive ionization at the highest voltages to be experienced in service, and it is largely the technique used to achieve ionization freedom which differentiates the system types.

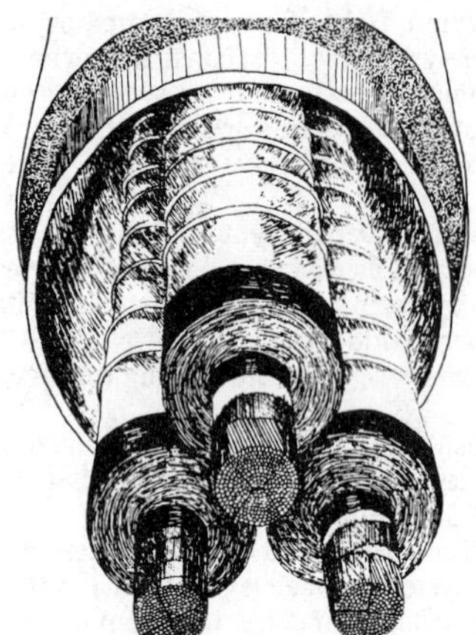

**FIG. 14-79** Pipe-type cable.

***102. High-Pressure Oil-Filled (HPOF) System.*** The HPOF pipe-type cable system comprises a welded, coated, and usually cathodically protected steel pipe (typically 8⅝ in at the 138- and 230-kV levels, 10 or 12 in at the 345- and 550-kV levels) into which, after testing and vacuum drying, are drawn three paper-insulated mass-impregnated cables (Fig. 14-79) shipped to the job site on sealed reels, typically in lengths of ½ mi or greater. Ability to install cable in long lengths subsequent to burial and backfilling of the pipe has significant advantage for construction in crowded urban areas.

After splicing, terminating, and final evacuation, the HPOF system is filled with oil and pressurized to about 200 lb/in$^2$ to suppress ionization. In effect, the small voids which form during handling, subsequent to manufacture, are reimpregnated during this final treatment.

Because of the large volumes of oil in these systems (typically 2 gal/ft), large storage tanks (5% of the total volume) and oil-pressure control units with pumps and relief valves are needed to accommodate oil contraction and expansion. Longitudinal pressure drops are usually low since the pipe is less than half full of cable. This gives the important advantage of providing the ability to circulate the oil either for temperature averaging of hot spots or for forced cooling to increase rating.

***103. Low-Pressure Oil-Filled (LPOF) Systems.*** The LPOF cable system (the original self-contained cables), developed in the mid-1920s, also operates in the ionization-free mode by virtue of pressurization. A very low viscosity oil-filling medium is continuously maintained gas-free at positive pressure to preclude ionization, starting from time of manufacture and throughout all the installation phases, by means of rather exacting installation techniques. The conductors of these cables are stranded with a hollow core which provides the longitudinal feeding path and maintains the factory metal-sheathed cable internally pressurized—thus the name "self-contained" (see Fig. 14-80), as no further pressure enclosure or pipe is required. The single-conductor direct-buried design which permits spacing between the cables has significantly better heat-dissipation characteristics than the HPOF pipe cable with its three conductors enclosed in a common pipe.

It should be noted that the hollow core directly increases the volume of materials required for all the cable components over the expanded conductor (paper, oil, sheath, and jacket). The core is therefore kept as small as practicable consistent with hydraulic pressure drop, which is inversely proportional to the fourth power of the core diameter and which strongly influences the spacing between feeding locations.

Normal pressures of 15 to 50 lb/in$^2$ gage are maintained by low-pressure bellows-type expansion tanks distributed along the route at spacings dependent upon the profile, the hydraulic resistance, ambient temperature, and normal and transient ratings. Hydraulic design of LPOF cables has a significant effect on system cost, since each intermediate feeding location entails a joint in the cable system, usually incorporating isolation of oil sections (stop joint).

Reinforced lead-sheathed designs or aluminum-sheathed designs (usually corrugated for flexibility) afford the opportunity for significant accessory-cost savings by operation at higher nominal pressure, permitting increased spacing between pressurizing stations. Pres-

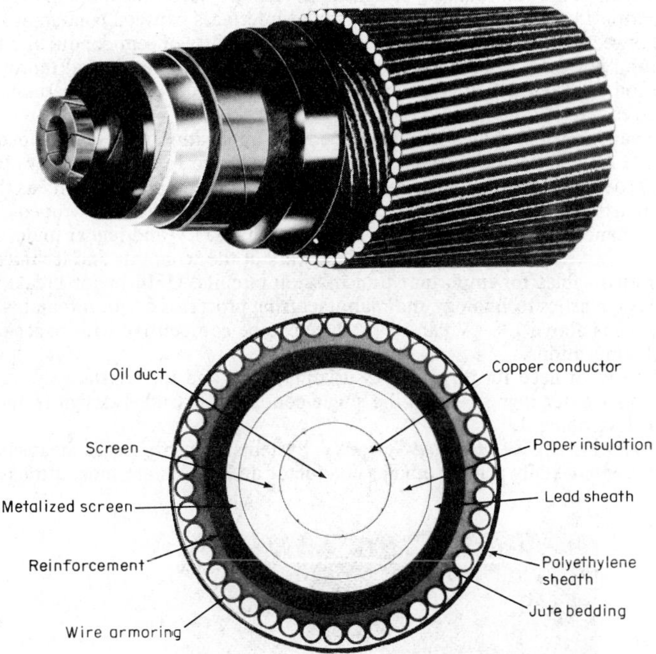


**FIG. 14-80** Self-contained cable. *(Photo courtesy Pirelli, Italy.)*

sures as high as 300 lb/in$^2$ have been specified for installations involving large elevation differences.

Long submarine lengths have no accessibility for intermediate feeding joints and must be fed from the terminals only. This makes hydraulic design still more onerous. Pumped pressurizing stations and very low viscosity impregnants are frequently indicated; otherwise very large diameter cores must be used with resulting increased cable manufacturing difficulty and cost.

The reliability of LPOF systems is primarily dependent upon the integrity of the sheath, requiring thorough protection from corrosion. Synthetic jackets are usually provided and must be protected against mechanical damage. In the case of open-circuited or cross-bonded sheath operation (to reduce sheath losses), the jacket must be protected against transient overvoltage puncture by using surge arresters, which further complicates system design.

Because of the single-conductor design, a spare fourth conductor (or seventh in the case of double circuits) can be employed in special cases to reduce outage time in event of a failure. It is customary in submarine-cable circuits to separate the individual conductors several hundreds of feet in an attempt to afford security against anchor or trawler damage to more than one conductor. Since virtually all failures of submarine circuits result from external damages, it has become standard practice to embed the cables below the sea bed in areas of shipping or bottom-fishing activity.

Because of long lengths involved for submarine cables, the sheath, its reinforcement, and wire armor must be carefully bonded to prevent galvanic corrosion from small differences in induced potentials.

Two other types of high-voltage systems have evolved over the last decade.

***104. Extruded-Dielectric Cable.*** These cable systems require no pressurizing equipment and are generally easier to install than are oil and paper cables. Extruded polymeric insulations (polyethylene, cross-linked polyethylene, and EPR—ethlyene-propylene-rubber) have volumetric expansion coefficients of the same order of magnitude as oil, posing corollary problems of accommodating dimensional changes. Here the problem is not so much void formation *in* the dielectric but at the radial interfaces between conductor and insulation and between the insulation and shield. The availability of semiconducting versions of the insulation polymers makes it possible to simultaneously extrude and thoroughly bond the insulation and conducting layers inside and outside to effect sound, permanently void-free interfaces.

Such processing techniques have extended the application area of extruded-dielectric cables (Fig. 14-81) to the lower transmission voltages (up to 138 kV), but they are generally not yet regarded by U.S. utility users as being of the same order of reliability as their paper-oil counterparts. This is in contrast to other areas of the world where extruded cables have gained acceptance for voltages of 220 to 275 kV with 400 kV and higher under consideration. There is mixed feeling in the U.S. industry about the economic and technical suitability of extruded cables for important transmission circuits. Until major breakthroughs in both the basic plastics technology and manufacturing processes occur, it remains to be seen whether designs above 138 kV can be produced to be competitive with paper-oil systems for critical applications.

The absence of need for oil and pressurizing plants and the resistance to catastrophic failure through water ingress, keeps the single-conductor extruded design in the forefront for further development.

***105. Compressed-Gas-Insulated Systems.*** Systems using $SF_6$ gas as the insulation (Fig. 14-82) between coaxially spaced tubular conductor and sheath are most attractive at very

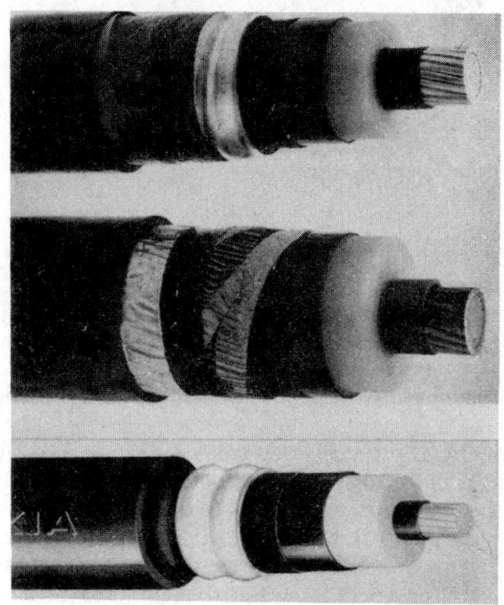

**FIG. 14-81** Extruded-dielectric cables. *(Photo courtesy of Nokia Engineering.)*

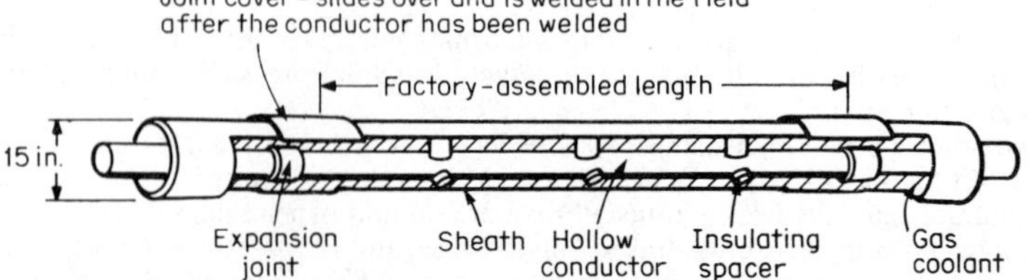


**FIG. 14-82** Gas-insulated transmission line.

high power levels at extra high voltages. Experience to date is limited to short lengths (300 m) of which there are several in operation at 345 kV. Two installations at 500 kV are operating in North America and one at 400 kV in Europe. Designs to date are rigid, requiring a joint every 13 to 18 m, and are large in diameter.

Reduced-diameter flexible systems in continuous lengths and 3-conductor designs in a common enclosure are under development, each of which would improve applicability of compressed-gas-insulated systems for major underground circuits.

## Cable Capacity Ratings

***106. Ampacity.*** Ampacity of paper-insulated and extruded-dielectric cables is governed by the maximum allowable conductor temperature. The conductor temperature in turn is determined by the losses ($I^2R$ and dielectric) in the system and the thermal resistance from heat-producing components to ambient temperature. A detailed ampacity calculation is beyond the scope of this handbook; Refs. 99 to 102 give calculation procedures plus empirical numbers for various system parameters.

*Pipe-Type Circuits.* An indication of the major components that enter into an ampacity calculation for a pipe-type cable is shown in Fig. 14-83. They include:

*Conductor ohmic loss,* which is the $I^2R$ loss in the conductor, taking into account skin

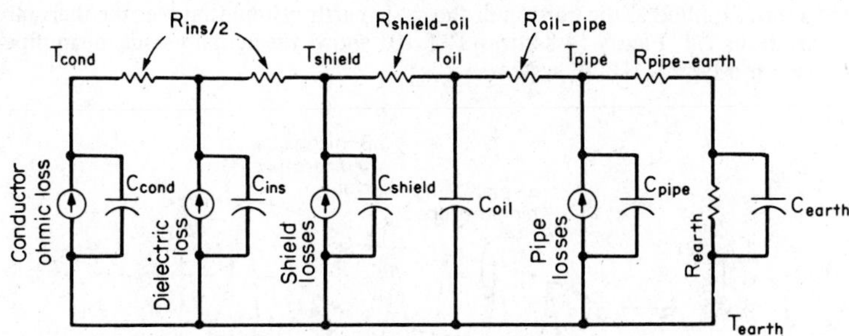


FIG. 14-83 Electrical analog of HPOF cable thermal circuit.

and proximity effects (see Sec. 4, Conductors). Conductor resistance is temperature-dependent.

*Conductor temperature,* set by AEIC specifications (85°C, reduced to 75°C if complete knowledge of the earth thermal parameters for the complete cable route is not available; 85°C is generally permissible if oil circulation is employed; see Par. **56**), and determined by acceptable aging rate of the insulation.

*Insulation thermal resistance,* which is expressed in thermal ohms per foot (a 1°C temperature rise occurs for the flow of one watt per longitudinal foot of conductor through a resistance of one thermal ohm per foot). The thermal resistance is obtained from the insulation volume thermal resistivity $\rho_i$, in °C·cm/W, by the relationship

$$R_{\text{ins}} = 0.012\rho_i \log\left(1 + \frac{2t}{d}\right) \tag{14-81}$$

where $t$ is the insulation thickness and $d$ is the conductor diameter.

*Dielectric loss,* which is voltage- and temperature-dependent and can be calculated from Ref. 99.

*Shield losses,* which are the current-dependent losses in the cable shield due to circulating and eddy currents.

*Pipe losses,* which are current-dependent losses from hysteresis and eddy currents due to incomplete magnetic-field cancellation from the three conductors. Pipe losses can account for as much as 40% of the total heat generation. Stainless-steel pipes show promise of significantly reducing pipe losses where the much higher pipe costs can be tolerated.

*Earth thermal resistance,* which depends upon the thermal resistivity $\rho_e$ of the backfill material, the depth of burial of the pipe, and the pipe diameter in the relationship

$$R_{\text{earth}} = 0.012\rho_e \log \frac{4L}{D} \tag{14-82}$$

where $L$ is the depth of burial and $D$ is the pipe diameter. The earth thermal resistivity varies widely, from 40°C·cm/W for quartz sand to 400°C·cm/W for soil with a relatively high organic content. Typically the earth portion accounts for 75% of the thermal resistance from conductor to ambient. In many cases a special low-resistivity backfill is placed in the trench, and a composite thermal resistivity is used in Eq. (14-82).

Soil resistivity is strongly moisture-dependent, and moisture tends to migrate away from high-temperature regions. In some cases, therefore, heat flux density and earth interface temperature may govern cable ampacity.

The *interface resistances* $R_{\text{shield-oil}}$, $R_{\text{oil-pipe}}$, and $R_{\text{pipe-earth}}$, which are typically small and calculable.

The *thermal capacitances* are important for calculation of short-term ratings and for evaluation of effects of daily load cycling.

A cable system will normally not carry a constant load; the current will experience daily, weekly, and seasonal fluctuations. All losses except dielectric losses vary as the square of the current. Cable rating calculations use the "loss factor" to account for daily load variations. The loss factor is applied to the portion or the cable-earth system that sees the thermal effect of daily variations.[99,100] Figure 14-84 from Ref. 101 shows the actual versus mean pipe-temperature rise for a typical HPOF system.

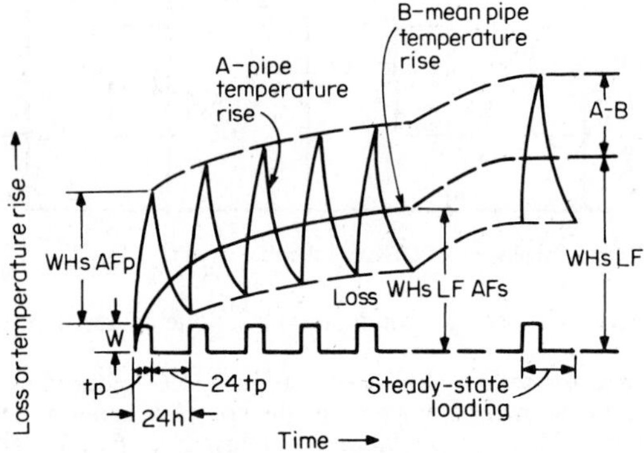


**FIG. 14-84** Pipe temperature rise for rectangular loss cycle for a single-pipe system.

The procedure for calculating ampacity is somewhat cumbersome but is amenable to straightforward computer programming. Ratings for many standard configurations and conditions are tabulated in Ref. 101.

*Self-Contained and Extruded-Dielectric Cables.* Ampacity calculations for self-contained and extruded-dielectric cables follow the general procedures given above for pipe-type circuits,[99] although many engineers use the IEC format for self-contained cables.[100] Since oil circulation is uncommon for self-contained cables, and the total thermal profile of the route is seldom known, conductor temperatures are often limited to 75°C. Normal conductor temperature is 85°C for polyethylene and 90°C for cross-linked polyethylene. For a given voltage class and conductor size, direct-buried single-conductor cables will have a higher ampacity than pipe-type because of decreased thermal resistance to ambient earth. Wider spacing of the individual cables in the trench will reduce mutual heating effects and increase the current rating. Placing self-contained or extruded cables in a duct reduces the rating because of dead air space in the duct. For single-conductor installations the method of sheath bonding and associated sheath currents affects the allowable conductor current. AEIC-ICEA tables give ratings for many standard configurations.[103]

*Gas-Insulated Transmission Lines (GITL).* Again, the rating procedure does not differ greatly from that for paper-insulated cables; the procedures of Ref. 99 or Ref. 100 may be employed. Temperature constraints are generally imposed by the earth-interface temperature rather than conductor temperature.

Both extruded-dielectric and gas-insulated lines have dielectric losses substantially lower than those for paper-insulated cables because of lower dielectric constants and dissipation factors.

*Transient and Emergency Ratings.* Transient ratings are those in which the heat-storage capacity of the cable system permits short-duration overcurrents without exceeding normal temperatures. Emergency ratings permit a higher conductor temperature for a specified length of time, recognizing that a reduction in insulation life results. AEIC specifications dictate the maximum number of occurrences per year that will still permit an acceptable loss of life. AEIC-specified-conductor temperatures are given in Table 14-21.

The transient rating increase of pipe-type systems is greater than that of self-contained, extruded, or GITL systems because of the greater mass of material. The equivalence for emergency ratings must be calculated for the individual systems. Reference 104 gives details of transient rating procedures.

*Forced Cooling.* The capacity of a cable circuit can be increased by "forced cooling" to provide a lower-resistance path for the losses to flow from the cable. For HPOF circuits, the procedure is to periodically remove the pressurizing oil from the cable pipe, pass it through a heat exchanger, and return it to the pipe at a remote location. Figure 14-85 shows the hydraulic circuit for a typical forced-cooled HPOF system.

**TABLE 14-21** AEIC Conductor Temperatures

| | Allowable maximum temperature, °C | | | |
|---|---|---|---|---|
| System | Normal* | 100-h emergency* | No specific time limit | 300-h emergency |
| HPOF | 85 | 105 | | 100 |
| Self-contained | 85 | 105 | | 100 |
| Polyethylene | 75 | | 90 | |
| Cross-linked polyethylene | 90 | | 130† | |

*These temperatures must be reduced by 10°C if the thermal environment is not defined over the total length of the circuit. Oil circulation for HPOF cable eliminates this requirement and installation of controlled backfill eliminates this requirement for all cable types.

†This limit is currently under review owing to thermomechanical effects in cables and accessories.

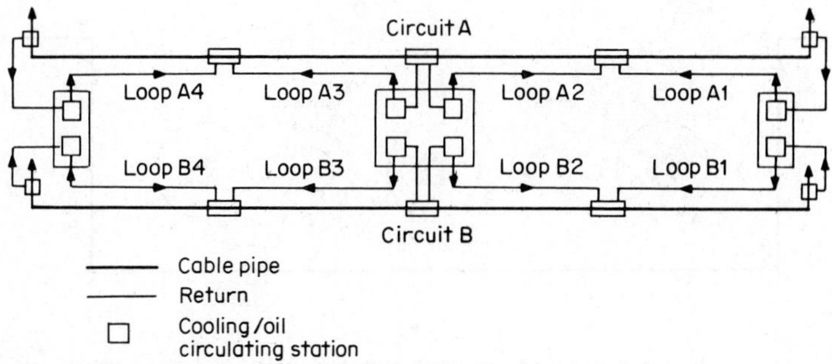


**FIG. 14-85** Hydraulic circuit, typical forced-cooled HPOF cable system.

The major additional components to a self-cooled HPOF system include return pipe for the oil, heat exchangers, circulating pump (200 to 400 gal/min), lower-viscosity oil, and modifications to the cable pipe to prevent cable damage from impinging oil. Reference 105 describes a typical 345-kV forced-cooled system and Ref. 106 describes a procedure for making thermal calculations.

The added cost and complexity of forced-cooling equipment, along with potentially high energy changes for operation of the equipment, have tended to limit forced-cooled installations to areas of very high installation costs, or to retrofitting on existing HPOF cable lines.

Self-contained and extruded cables can also be forced-cooled.[107] Three possible methods are depicted in Fig. 14-86.

Internal cooling has the advantage of removing oil with a high heat content from the core of the cable itself. This system, which has not seen commercial use, requires special joints to periodically remove the line-potential oil, pass it through heat exchangers, and return it to the cable core.

Lateral and integral cooling have a somewhat lower efficiency, but the simplicity of the

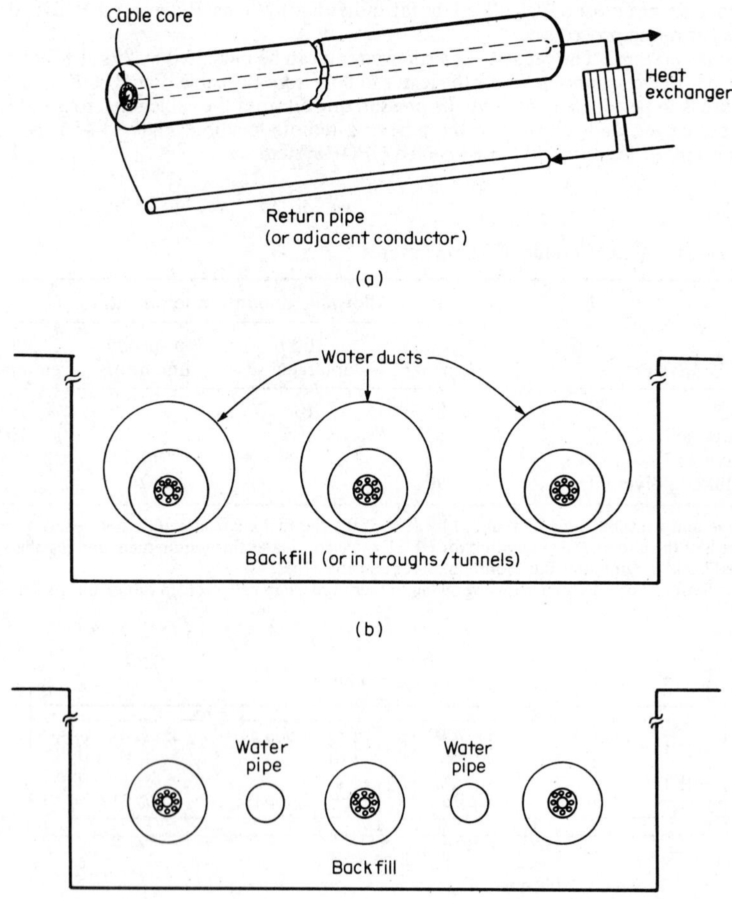


**FIG. 14-86** Alternate methods of cooling self-contained cables. (*a*) Internal (core) cooling; (*b*) integral cooling; (*c*) lateral cooling.

cooling system has led to several commercial installations. Extruded-dielectric cables can make use of either integral or lateral cooling. It is also possible to cool gas-insulated transmission lines in the same manner, although there have been no such installations to date. Researchers overseas have reported attempts to remove and cool the dielectric gas itself.

### Electrical Characteristics

***107. Calculation of Electrical Parameters.*** The electrical characteristics of cables (resistance, capacitance, and inductance) can be calculated for other than pipe-type cables with reasonable accuracy.

Calculation of resistance and capacitance for HPOF cables is straightforward, but because of the magnetic steel pipe and the change in position of the cables within the pipe due to load fluctuations, there is no precise method of calculating the inductance of this type of cable system. Several papers present approximate formulas and empirical results for a limited range of conductor sizes.[108–111]

The high capacitance of a cable circuit makes it necessary to consider shunt-reactor compensation for higher-voltage systems in excess of 15 mi. Surge arrester duty is more severe for cable circuits than for overhead lines.

### Installation

***108. Pipe-Type Cables.*** Installation of high-pressure oil-filled (pipe-type) circuits is described in Ref. 112. The procedure is an involved one; major steps include:

1. Obtain preliminary engineering design, survey, and permits, and final engineering design.
2. Break pavement if street installation; dig trench (see Fig. 14-87.)

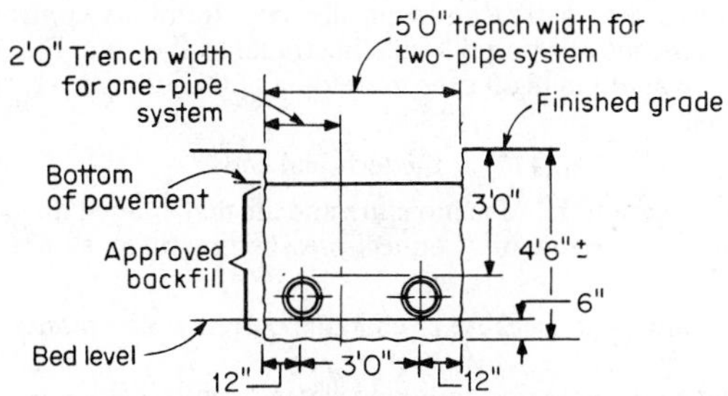


**FIG. 14-87** Typical trench cross section for 345-kV HPOF cable.

3. Install 40- to 80-ft sections of precoated pipe, weld, test, and apply corrosion coating at weld areas.
4. Install splicing manholes at spacings dictated by pulling tension calculations—typically 2000 to 5000 ft, depending on conductor and cable size, route, etc.[113]
5. Backfill as pipe installation proceeds. A backfill of controlled thermal resistivity may be used if thermal conditions warrant. Local regulations usually limit the length of open trench to a few hundred feet. Restore pavement.
6. Evacuate length of pipe between manholes; perform a pressure-rise test to detect leaks and moisture; fill with dry nitrogen to 10 to 25 lb/in$^2$; leave until cable pulling begins.
7. Specify exact lengths between manholes to the cable manufacturer, adding perhaps 20 ft plus 1%. Cables are shipped on sealed reels to the utility, which in turn mounts three

reels on a trailer such as that shown in Fig. 14-88, or on other suitable supports. A winch at the other end of the pipe section pulls the cables into the pipe with a previously installed wire rope (Fig. 14-89); the reels are unsealed shortly before the pull.

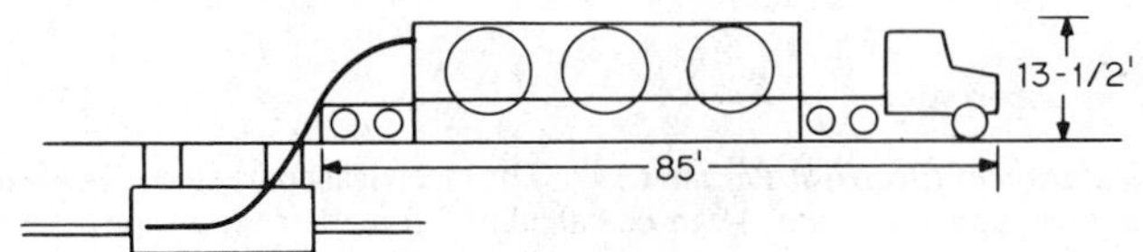


**FIG. 14-88** Reel trailer.

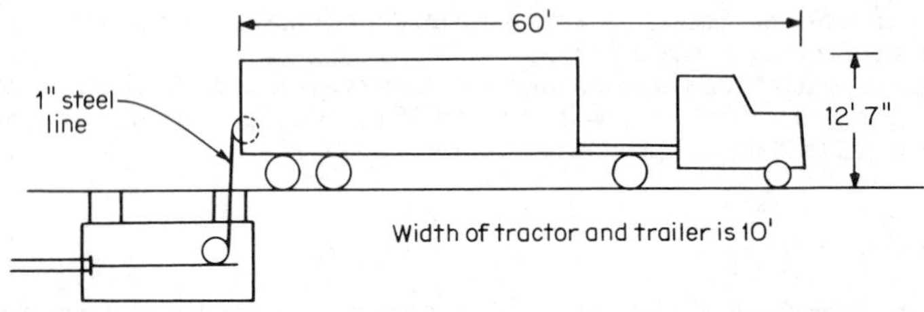


**FIG. 14-89** Pulling-winch trailer.

8. Once the cable is pulled, install temporary sealing ends ("night caps") and pressurize the section with dry nitrogen to protect the cable until splicing begins. Some utility specifications require evacuation prior to nitrogen pressurization; others require evacuation if unforeseen exposure to the elements has occurred.
9. Splicing operations are carried out typically in a humidity-controlled environment above 138 kV by means of a portable splicing trailer as shown in Fig. 14-90. A complete splice, from removal of the night caps to welding of the joint sleeve, could take 2 weeks for a 345-kV line.
10. Install "potheads" (see Par. **113**) at the terminal ends.
11. Evacuate the line again to remove moisture and air that entered during splicing, and fill the line with the pressurizing oil. Connect pressurizing plant, slowly bring pressure up to operating value.
12. Perform a high-voltage (dc because of charging current) "acceptance test."

Crossings of major highways and railroads where open trenching is not permitted may be accomplished by boring and installing a sleeve, or driving a casing, and pushing or pulling the prewelded pipe sections through the sleeve. Attention should be paid to thermal conditions for cable operation.

***109. Self-Contained and Extruded-Dielectric Cables.*** Self-contained and extruded-dielectric cables require different installation procedures. The individual cables are preferably directly buried in a trench with perhaps 1-ft spacing between conductors, and the length between splices is governed by the reelable length of cable, by the length of trench allowed to remain open, or by maximum sheath voltage permitted. If later access is required, or if length of open trench is limited, the cables may be pulled into previously installed ducts. The section length between splices is then severely limited by friction between cable covering and duct wall during pulling. Oil reservoirs for self-contained cables are spaced according to elevation changes or maximum transient hydraulic pressures during load cycling. Generally each substation end has a set of reservoirs, and they may be placed in manholes as needed. Reference 114 gives details of a typical self-contained cable installation.

Trial installations show that plowing-in transmission extruded-dielectric cables may offer advantages in rural and suburban areas where subsurface conditions permit.

***110. Gas-Insulated Cables.*** Gas-insulated transmission lines require a conductor and

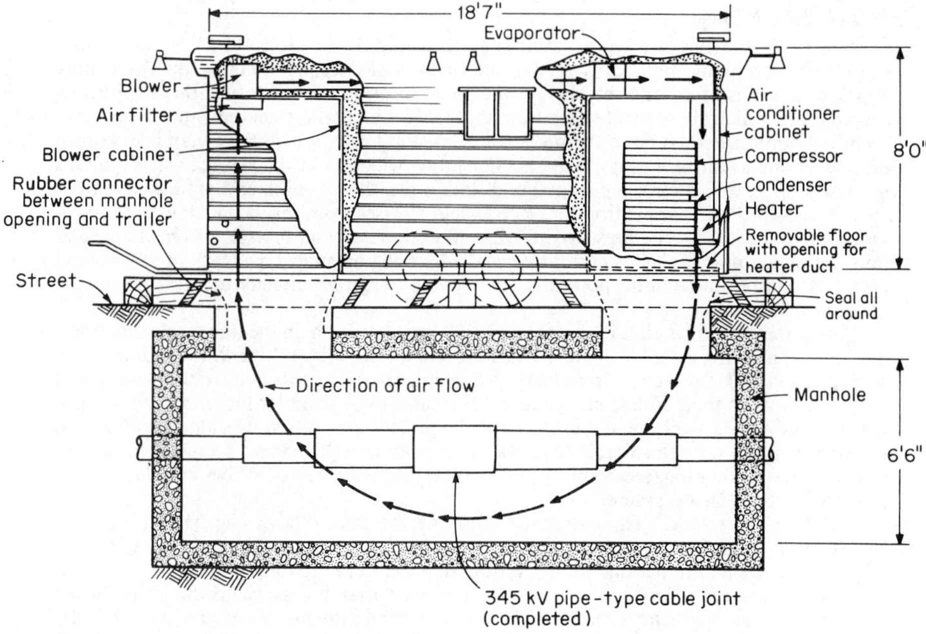


**FIG. 14-90** Manhole and splicing trailer.

enclosure joint for each phase every 40 to 60 ft. Humidity control is not required, but the joint area must be kept clean. The jointing is a straightforward procedure, requiring only the welding (or plugging in) of adjacent conductors, cleaning the joint area and spacers thoroughly, and welding an enclosure sleeve into place, with application of corrosion coating in the weld area. The inflexibility of commercial GITL systems makes their use in city streets awkward and dictates a thorough survey in other installations so that factory-formed bends may be specified. It is desirable not to backfill the line until high-voltage acceptance tests are completed, to aid in location and repair of trouble spots.[115]

***111. Water Crossings.*** Water crossings can be accomplished in several ways. Any of the systems described may be installed in tunnels, and special methods to account for the vertical drop in tunnel shafts must be investigated for each. For direct water crossings, pipe-type cables are limited to the length that can either (1) be manufactured (impregnating-tank limitations), (2) be shipped, or (3) be pulled. A 7000-ft 345-kV water crossing is the longest to date at that voltage. The setting up of pulling and splicing barges, with the decision to have submerged joints, has permitted longer HPOF water crossings in special instances.

Self-contained cables laid either directly onto the bottom or into trenches are currently the preferred means for long water crossings in the United States and abroad. Reference 116 describes a 12-mi 138-kV installation under Long Island Sound.

***112. Conductor Materials.*** Copper conductors are presently used for a majority of transmission cable installations, although aluminum is gaining wider application for HPOF and extruded-dielectric cables. Use of aluminum presents no undue installation difficulty except for a more elaborate conductor splice requiring both pressing and welding of connectors for oil/paper cables.

## Accessories

***113. General.*** None of the presently used underground transmission systems requires particularly elaborate accessories. The cost, except for joints and forced-cooling plants, accounts for a small percentage of the project cost for systems more than a couple of miles long.

### *114. HPOF Cable*

*Joints.* A drawing of a typical joint is given in Fig. 14-91 for a 345-kV system. Total time (from removing night caps to welding of joint sleeves) for making this joint may approach 2 weeks of around-the-clock work. In addition to normal joints, there are trifurcating joints, which allow a transition from three cables in a mild-steel line pipe to one cable each in a nonmagnetic pipe to the potheads; semistop joints, which have glands to reduce oil flow in the event of a major leak; anchor joints which hold the stainless-steel ribbons used to support cables for steep slopes; and skid joints, the downhill end of anchor joints.

Potheads provide a transition for electric-field stresses from insulated cable to air, provide a pressure seal for the pressurizing oil, and exclude air and moisture from the insulation. Although most potheads currently installed are the open-air type (Fig. 14-92), several HPOF circuits employ "mini potheads" that give a transition directly to an $SF_6$-insulated metal-clad substation.

Pressurizing plants maintain the nominal 200-lb/in$^2$ pressure in the line. A plant consists of pumps (including installed spares), controls, and a reservoir tank sized to accommodate oil expansion and contraction from load cycling and seasonal earth-temperature changes. It is not uncommon for a tank feeding a long EHV cable to be sized for more than 50,000 gal. It is common for a two-pipe circuit to have the pressurizing pump operate continually at perhaps 10 gal/min (with suitable relief valves to maintain the correct pressure) to accomplish thermal smoothing (see Ampacity, Par. **106**). Reference 117 describes calculation procedures for the cable oil system.

*Cathodic Protection.* The external coating of the steel pipe is the basic protection against corrosion. Cathodic protection by means of a resistor-rectifier set or sacrificial anodes provides additional margin in event of damage to the coating.

*Temperature Monitoring.* A small insulated wire may be placed in the space in the center of the four conductor segments. The wire is carried through splices to a joint typically three to four cable sections away from the pothead, where the insulation is removed and the wire is joined to the cable conductor. A monitor and transmitter are installed in the pothead corona shield, and a receiver is located at ground potential. Thermistors may also be installed at discrete locations on the cable.

Forced cooling of HPOF cables can be done in several ways. Common to all systems are a circulating pump to move the oil through the line and heat exchanger, a return pipe to send cooled oil to the cable pipe, and an oil-to-cooling-medium heat exchanger. Oil-to-air heat exchangers have been in use since the late 1940s.[118] Oil-to-water evaporative-cooling-tower systems have a higher capacity, and oil-to-water heat exchangers may be used. Oil-to-

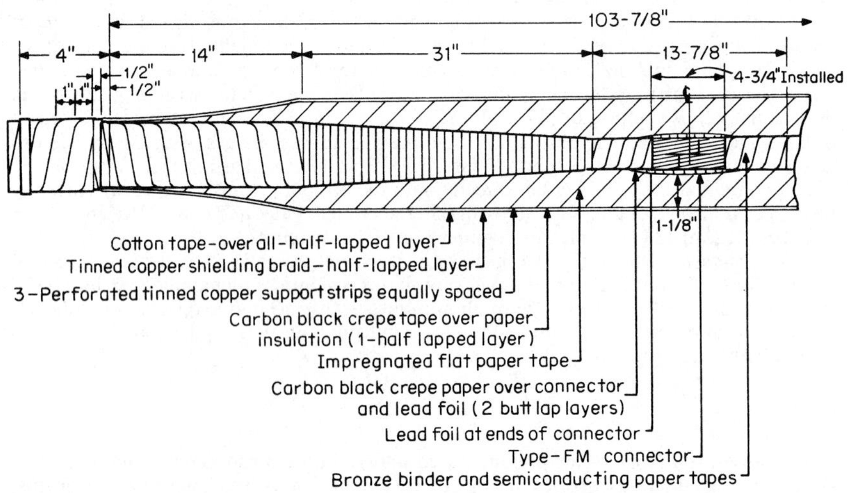


**FIG. 14-91** Typical 345-kV joint.

refrigerant exchangers are preferred on intensively cooled lines, but heat rejection from the refrigerant must still be to air or water. Extensive forced-cooling systems were installed in the 1970s. There have been far fewer installations in the 1980s, reflecting low usage of transmission cables in general, plus high energy and maintenance costs for refrigeration plants. A schematic diagram of a typical forced-cooling plant, employing oil-to-air precoolers and main oil-to-refrigerant exchangers, is shown in Fig. 14-93.

Lateral or integral cooling of self-contained cables does not require oil-to-water heat exchangers, and pressure constraints are less severe on the water-handling equipment. It is uncommon to use refrigerant-type units for these systems.

***115. Self-Contained Cables.*** Normal joints are similar to HPOF, except that a bore must be left in the conductor for continuation of the oil duct, and a nonmagnetic (lead, copper, or aluminum) sleeve is placed over the completed splice. Oil-feeding joints use insulating barriers and the oil itself for insulation, since oil must flow freely to the conductor bore.

Potheads are similar in appearance to HPOF. No trifurcating joint is required since most self-contained cables are single-core.

Oil reservoirs at locations along the feeder are determined by elevation changes and the allowable pressure on the cable sheath.

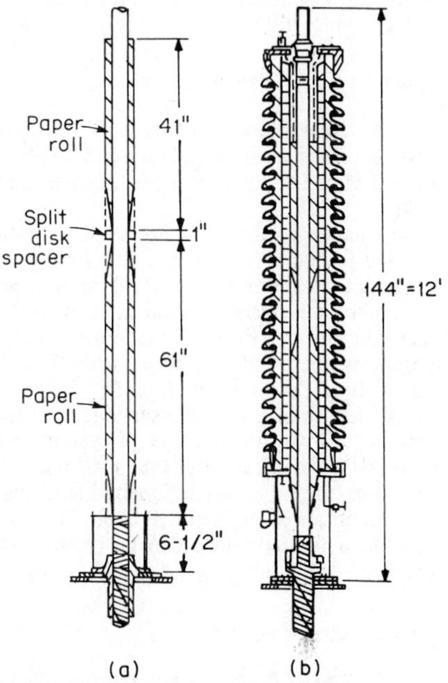

**FIG. 14-92** 345-kV pothead. *(G&W Electric Company.)*

***116. Extruded-Dielectric Cables.*** These cables have the smallest requirement for accessory equipment. Joints are taped and often also fused. Potheads are commonly the "slip-on" type, which require less preparation than oil-filled ones. No pressurizing plant or reservoir is required, nor is cathodic protection. The system is amenable to temperature monitoring. Research continues in the 1980s to develop more reliable accessories for this type of cable.

***117. Gas-Insulated Transmission Lines.*** Again, accessory requirements are small. Joints occur every 40 to 60 ft (see Installation, Par. **110**) but are simple and straightforward. Potheads are factory-made and attach to the rest of the system in the same manner as a normal joint. Expansion and contraction of the insulating gas is accommodated by allowing the pipe pressure to increase or decrease. Pressure-monitoring devices must be installed to

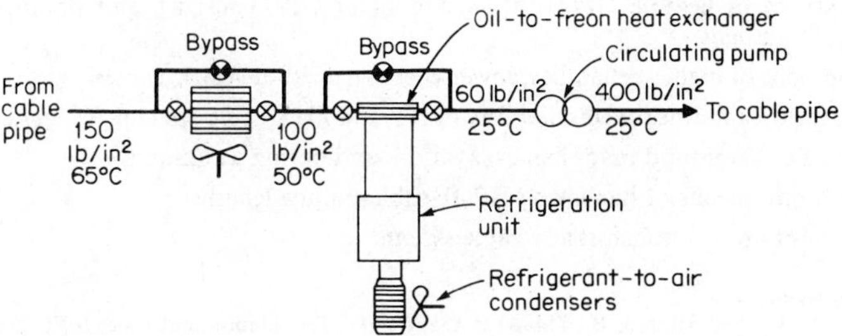


**FIG. 14-93** Diagram of a typical forced-cooling plant.

warn of a gas leak. Since the outer pipe is aluminum, proper attention to corrosion coating and cathodic protection is vital.

***118. Direct-Current Cables.*** DC cables are preferred for long circuits where ac charging currents are excessive. Of 16 installations in service in the mid-1980s, all but one involved major submarine crossings. Voltages are as high as 280 kV for these water crossings, but 600-kV cables are well-proven in the laboratory.[119]

## Operation and Maintenance

***119. Requirements.*** Operation and maintenance requirements of underground transmission cables are minimal compared with most electrical power equipment and are essentially confined to the accessories for pressure maintenance, cathodic protection, and forced cooling.

For paper-insulated cables, a dc insulation-resistance test every year or two is recommended to detect increases in moisture level or other degradation, and joints in locations susceptible to thermomechanical bending should be x-rayed every few years. Semiannual, or perhaps quarterly, dew-point readings should be taken on gas-insulated transmission lines. Oil samples should be taken for analysis on HPOF and self-contained cables. For all these tests, the trend with time rather than the individual reading is of interest for monitoring any degradation of the line.

The length of time necessary for fault location and repair varies widely for the various systems, or even for one type of system, depending upon the location of the failure. Generally, HPOF cable failures take the longest to repair because of the necessity of freezing the oil so that the pipe may be opened, and the length of time to make a repair splice. Except for self-contained or extruded cable in duct, it is uncommon to replace a complete section of cable; a faulted splice can be rebuilt, or a new splice and a short "piece-out" can be installed.

## Future Developments

***120. R&D Efforts.*** Since the middle 1960s there has been a trend toward broadened utility and governmental support to further research and development programs in underground transmission systems. These can be generally characterized as directed toward (*a*) lowering the energy transport costs of state-of-art systems both by improvement of power ratings and by streamlining installation methods, and (*b*) development of advanced systems for future transport of bulk power in the multigigawatt ($10^9$ W) range over increased distances.[120]

It is beyond the scope of this section to discuss these programs in detail. Listings and reports covering these activities are available from the Electric Power Research Institute (EPRI)[121] and the U.S. Department of Energy,[122] the two principal funding agencies.

Recent or currently active programs include:

1. Development of low-loss synthetic-taped insulations to replace cellulosic paper.
2. Research into forced cooling of HPOF cables to increase ratings
3. Development of EHV dc cables for $\pm 600$ and $\pm 1000$ kV operation
4. Development of flexible GITLs, three-conductor GITL designs, and optimization of rigid GITL designs
5. Development of higher-reliability, lower-cost extruded-dielectric cables
6. Simplification and improved reliability of HV and EHV cable-splicing design techniques
7. Advanced underground mapping, excavation, and boring techniques
8. Research into means of increasing HPOF cable-pulling lengths
9. Research into novel transmission cable sytems

***121. References***

1. Barnes, H. C., and Thoren, B.: The AEP-ASEA UHV Test Station and Line; IEEE Conf., Paper C73-319-1, presented at IEEE Summer Power Meeting, 1973.

2. Development of Ultra-high Voltage Transmission; Bonneville Power Administration, Portland, Ore., July 1974.

3. *Transmission Line Reference Book, 345 kV and Above,* 2nd ed., Electric Power Research Institute, Palo Alto, Calif., 1982.

4. Clarke, Edith: *Circuit Analysis of A-C Power Systems;* John Wiley & Sons, Inc., New York, 1943.

5. Keast, D. N.: Assessing the Impact of Audible Noise from AC Transmission Lines: A Proposed Method; *IEEE Trans. Power Appar. Syst.*, May/June 1980, vol. PAS-99, no. 3, p. 1021.

6. Clayton, R. E., and Stewart, J. R.: Transmission Line Electromagnetic Compatibility, 1975 IEEE Electromagnetic Compatibility Symposium Record, IEEE publication 75CH1002-5EMC.

7. A Comparison of Methods for Calculating Audible Noise of High Voltage Transmission Lines, IEEE Task Force Report, *IEEE Trans. Power Appar. Syst.*, October, 1982, vol. PAS-101, no. 10, p. 4290.

8. Perry, D. E.: An Analysis of Transmission Line Audible Noise Based upon Field and Three-Phase Test Line Measurements; *IEEE Trans. Power Appar. Syst.*, May/June, 1972, vol. PAS-91, p. 857.

9. Measurement of Audible Noise from Transmission Lines; IEEE Task Force Report, *IEEE Trans. Power Appar. Syst.*, March, 1981, vol. PAS-100, no. 3, p. 1442.

10. *The Location, Correction and Prevention of RI and TVI Sources from Overhead Power Lines,* IEEE Tutorial Publication 76 CH1163-5-PWR, 1976.

11. Chartier, V. L., et al.: Investigation of Corona and Field Effects of AC/DC Hybrid Transmission Lines; *IEEE Trans. Power Appar. Syst.*, January, 1981, vol. PAS-100, no. 1, p. 72.

12. Review of Technical Considerations on Limits to Interference from Power Lines and Stations; IEEE Committee Report, *IEEE Trans. Power Appar. Syst.*, January/February, 1980, vol. PAS-99, no. 1, p. 365.

13. *Tolerable Limits and Methods of Measurement of Electromagnetic Interference from Alternating Current High Voltage Power Systems,* CSA Standard C108.3.1-1975, Canadian Standards Association.

14. *Human Repsonse to Interference with TV Picture Quality,* Report EL-1587, Project 68-4, Electric Power Research Institute, Palo Alto, CA, 1980.

15. Scherer, H. N., Jr., Ware, B. J., and Shih, C. H.: Gaseous Effluents Due to EHV Transmission Line Corona; Paper T 72 550-2, *IEEE Trans. Power Appar. Syst.*, May/June, 1973, vol. PAS-92, no. 3.

16. Frydman, M., Levy, A., and Miller, S. E.: Oxidant Measurements in the Vicinity of Energized 765 kV Lines; IEEE Paper T 72 551-0, *IEEE Trans. Power Appar. Syst.*, May/June, 1973, vol. PAS-92, no. 3.

17. Fern, W. J., and Brabets, R. I.: Field Investigation of Ozone Adjacent to High Voltage Transmission Lines; IEEE Paper T 74 057-6, *IEEE Trans. Power Appar. Syst.*, Sept/Oct, 1974, vol. PAS-93, No. 5.

18. Frydman, M., and Shih, C. H.: Effects of the Environment on Oxidants Production in AC Corona; IEEE Paper T 73 407-4, *IEEE Trans. Power Appar. Syst.*, January/February, 1974, vol. PAS-93, no. 1.

19. Roach, J. F., Chartier, V. L., and Dietrich, F. M.: Experimental Oxidant Production Rates for EHV Transmission Lines and Theoretical Estimates of Ozone Concentrations Near Operating LInes; IEEE Paper T 73 414-0, *IEEE Trans. Power Appar. Syst.*, March/April, 1974, vol. PAS-93, no. 2.

20. Abel, W. A.: Comparison of Ozone Instrumentation; IEEE Paper A 78 166-7, abstract in *IEEE Trans. Power Appar. Syst.*, July/August, 1978, vol. PAS-97, no. 4, p. 1009.

21. Roach, J. F., et al.: Ozone Concentration Measurements on the C-Line at the Apple Grove 750 kV Project and Theoretical Estimates of Ozone Concentrations Near 765 kV Lines of Normal Design; *IEEE Trans. Power Appar. Syst.*, July/August, 1978, vol. PAS-97, no. 4, p. 1392.

22. Sebo, S. A., et al.: Examination of Ozone Emanating from EHV Transmission Line Corona Discharges; *IEEE Trans. Power Appar. Syst.*, March/April, 1976, vol. PAS-95, no. 2.

23. *The Electrostatic and Electromagnetic Effects of AC Transmission Lines,* IEEE Tutorial 79 EH 0145-3 PWR, 1979.

24. Electric and Magnetic Field Coupling from High Voltage AC Power Transmission Lines—Classification of Short-Term Effects on People; IEEE Committee Paper, *IEEE Trans. Power Appar. Syst.*, November/December, 1978, vol. PAS-97, no. 6, p. 2243.

25. Chiu, M. C.: Fuel Ignition by High Voltage Capacitive Discharges; report JHU PPSE T-18, March 1983, John Hopkins University Applied Physics Laboratory, Laurel, Md.

26. Hamaam, M. S., and Baishiki, R. S.: A Range of Body Impedance Values for Low Voltage, Low

Source Impedance Systems of 60 Hz; *IEEE Trans. Power Appar. Syst.*, May, 1983, vol. PAS-102, no. 5, p. 1097.

27. Jaffa, K. C.: Magnetic Field Induction from Overhead Transmission and Distribution Power Lines on Parallel Fences; *IEEE Trans. Power Appar. Syst.*, April, 1981, vol. PAS-100, no. 4, p. 1624.
28. Dabkowski, J.: The Calculation of Magnetic Coupling from Overhead Transmission Lines; *IEEE Trans. Power Appar. Syst.*, August, 1981, vol. PAS-100, no. 8, p. 3850.
29. Taylor, R. J.: Hazard Analysis for Magnetic Induction from Electric Transmission Lines; report JHU PPSE T-23, March 1982, John Hopkins University Applied Physics Laboratory, Laurel, Md.
30. *IEEE Recommended Practices for Measurement of Electric and Magnetic Fields from AC Power Lines,* IEEE Standard 644-1979, IEEE, New York, NY.
31. Kaminski, J., Jr.: Long Time Mechanical and Electrical Strength in Suspended Insulators; *Trans. AIEE,* August, 1963, p. 446.
32. Nicholas, F. S., and Vose, F. C.: A Polymer Insulator for High Voltage Transmission Lines; *Elec. Eng.,* vol. 82, 1963.
33. Abilgaard, E. H., Bauer, E. A., et al.: Composite Longrod Insulators and Their Influence on the Design of Overhead Lines; CIGRE paper 22-03, 1976.
34. Development of Polymer Bonded Silica, (Polysil) for Electrical Applications; EPRI Report EL488, May, 1977.
35. The Metapol Insulator: Dulmison (Australia) Inc. Catalog.
36. Karady, G., and Lamontagne, G.: Electrical and Contamination Performance of Synthetic Insulators for 735 kV Transmission Lines, IEEE Paper A76 502-5, presented at IEEE PES Summer Meeting, Portland, Ore., July 18–23, 1976.
37. Broschat, M.: Transmission Line Uprating 115 kV to 230 kV, Report on Operating Performance; *IEEE Trans.*, March/April 1972, pp. 545–548.
38. Update Line to 345 kV on same ROW; *Electr. World,* Nov. 15, 1973, pp. 66–67.
39. *EHV Transmission Line Reference Book,* EEI 1968.
40. IEEE Working Group on Insulator Contamination: Application Guide for Insulators in a Contaminated Environment; *IEEE Trans. Power Appar. Syst.*, September/October 1979, pp. 1676–90.
41. *Transmission Line Reference Book, 115–138 kV Compact Line Design,* EPRI Publication, 1978.
42. Moran, J. H.: The Effect of Cold Switch-on on Semi-conducting Glazed Insulators; IEEE Paper C74 071-7 presented at IEEE Power Meeting, New York, January, 1974.
43. Moran, J. H., and Powell, D. G.: A Possible Solution to the Insulator Contamination Problem; IEEE Paper 71CP41 presented at IEEE Power Meeting, New York, January, 1971.
44. Falter, S. L., and Powell, D. G.: Radio Influence Voltage Characteristics of Transmission Line Assemblies, Using Semi-conducting Glazed Insulators; IEEE paper C73 416-5 presented at IEEE Summer Power Meeting, 1973, Vancouver, B.C.
45. Fukui, Hiroshi, Naito, Katsuhiko, Irie, Takashi, and Komoto, Iwas: A Practical Study on Application of Semi-conducting Glaze Insulators to Transmission Lines; IEEE Paper T74 073-3 presented at IEEE Winter Power Meeting, New York, January, 1974.
46. Nigol, O., Reichman, J., and Rosenblatt, G.: Development of New Semi-conductive Glaze Insulators; Paper T73 420-7, *IEEE Trans. Power Appar. Syst.*, March/April, 1974, vol. PAS-93, pp. 614–622.
47. Bundled Circuit Design for 115-138 kV Compact Transmission Lines; EPRI Report EL 1314 (2 vols.), February, 1980.
48. Phase to Phase Switching Surge Design; EPRI Report EL 1550, September, 1980.
49. Souchereau, et al.: Validation of a Chainette Tower for a 735 kV Line; *CIGRE Paper* 22-04, 1978.
50. IEEE Committee Report: Limitations on Stringing and Sagging Conductors; *IEEE Trans. Power Appar. Syst.*, December, 1964, vol. 83, no. 12, pp. 1230–1235.
51. Converti, V., Hyland, E. J., Tickle, D. E.: Optimized Transmission Tower Spotting on Digital Computer; *AIEE CP60-1201, October 1960.*
52. Stagg, G. W., Watson, M.: Dynamic Programming—Transmission Line Design; *IEEE International Conversion Record,* part 3, *Power,* 1964, pp. 55–61.
53. Glasson, G. T.: A Dynamic Programming Method for the Optimum Positioning and Selection of Transmission Towers; *IE Australia, EE Trans.*, March, 1968, pp. 127–134.
54. ANSI A58.1-1982, Minimum Design Loads for Buildings and Other Structures; American National Standards Instutute, New York, 1982.

55. ANSI C2-1981, *National Electrical Safety Code,* New York, Institute of Electrical and Electronics Engineers, 1981.

56. Thomas, P. H.: Sag Calculations for Suspended Wires; *Trans. AIEE,* 1911, vol. 30, p. 2229.

57. Martin, James P.: Sag Calculations by the Use of Martin's Tables; Copperweld Corp., 1931, rev. 1961.

58. Barrett, J. S., Dutta, S. and Nigol, O.: A New Computer Model of ACSR Conductors; *IEEE Trans. Power Appar. Syst.,* March, 1983, vol. PAS-102, no. 3, pp. 614–623.

59. Alcoa Conductor Products Co.: *ACSR Graphic Method for Sag-Tension Calculations;* 1961 ed.

60. Greisser, V. H.: Effects of Ice Loading on Transmission Lines; *Trans. AIEE,* 1913, vol. 32, p. 1829.

61. Healy, E. S. and Wright, J. A.: Unbalanced Conductor Tensions; *Trans. AIEE,* 1926, p. 1064.

62. Nash, John F., and Nash, John F., Jr.: Calculations for Cable and Wire Spans; *Trans. AIEE,* 1945, vol. 65, p. 685, and discussion, p. 984.

63. Poffenberger, J. C., and Swart, R. L.: Differential Displacement and Dynamic Conductor Strain; *IEEE Trans.,* vol. PAS-84, 1965, pp. 281–289.

64. *Transmission Line Reference Book—Wind Induced Conductor Motion,* Electric Power Research Institute, Palo Alto, Calif., 1979, chap. 3.4.

65. Ibid., chap. 3.5.

66. Nigol, O., and Havard, D. G.: Control of Torsionally-Induced Conductor Galloping with Detuning Pendulums; IEEE Paper A78 125-7, January 1978.

67. Richardson, A. S.: Design and Performance of an Aerodynamic Anti-Galloping Device; IEEE Conf. Paper C68 670-PWR, June 1968.

68. Douglass, D. A., and Roche, J. B.: Anti-Galloping Potential of a New Twisted Conductor Design, *Proceedings of the Canadian Electrical Association International Symposium on Overhead Conductor Dynamics,* June 1981, pp. 83–98.

69. *Transmission Line Reference Book—Wind Induced Conductor Motion,* chap. 5.

70. Sellers, A. H., and Williams, J. E.: All-Aluminum Transmission Tower Line; *Trans. AIEE,* June 1961, p. 169.

71. Thomas, Percy H.: Formula for Minimum Horizontal Spacing; *Trans. AIEE,* 1928, vol. 47, p. 1323.

72. Farr, F. W., Ferguson, C. M., McMurtrie, N. J., Steiner, J. R., White, H. B., and Zobel, E. S.: A Guide to Transmission Structure Design Loadings, *Trans. IEEE Power Group,* November 1964, p. 1073.

73. Farmer, G. E.: The Use of Insulated Ground Wires on a Transmission Line for Communication Purposes; *IEEE Trans. Power Appar. Syst.,* December, 1963, p. 884.

74. Ramthun, M. K., Pitzel, B. H., and Campbell, D. W.: Stream-Lined 230-kV Transmission Passes Overhead in City's Streets; *Electr. World,* June 29, 1964, p. 94.

75. Stumpf, M. W., and Mouton, R. A.: 12 Sided Single Poles Carry 760 MVA Capacity Line (New Orleans, La); *Electr. World,* Nov. 16, 1964, p. 94.

76. Guide for the Design of Steel Transmission Towers; *Manual 52,* American Society of Civil Engineers, New York, 1971.

77. Richardson, W. B., New Techniques Speed Construction of 500 kV Lines; *Electr. World,* January 15, 1965, p. 27.

78. Newmark, N. M., and Rosenblueth, E. H.: *Fundamentals of Earthquake Engineering,* Englewood Cliffs, N.J., Prentice-Hall, Inc., 1971.

79. Klopfenstein, A., McDonald, J. F., Pecknold, D., A. W., and Walker, W. H.: Seismic Test and Analysis of Capacitor Banks; *IEEE Trans. Paper* T74 406-5, January/February 1975, vol. PAS-94, no. 1, p. 81.

80. Long, L. W.: Analysis of Seismic Effects on Transmission Structures; *IEEE Trans. Paper* T-73-326-6, January/February 1974, vol. PAS-93, no. 1, pp. 248–254.

81. *Standard of Recommended Practices,* American Wood Preservers Association, Washington, D.C.

82. Weber, L. C., Glass, E. C., and Alexander, G. W.: Application of Statistical Methods in the Design and Uprating of Wood-pole Transmission Lines; *Trans. IEEE Power Group,* August, 1965, p. 725.

83. *Installation of Overhead Line Conductors,* IEEE 524, 1980.

84. Fowle, Frank F.: The Transposition of Electrical Conductors; *Trans. AIEE,* 1904, vol. 23, p. 659.

85. Von Voigtlander, F.: Transposition Practices; *Elec. Eng.,* January, 1943.

86. Clem, J. E.: Currents Required to Remove Conductor 'Sleet'; *Electr. World,* December 6, 1930, p. 1053, and January 31, 1931, p. 245.

87. Live Line Maintenance Methods; *IEEE Trans. Paper* T 73-157-5, *IEEE Trans. Power Appar. Syst.,* September/October, 1973, vol. PAS 92, no. 5, pp. 1642–1648.

88. Recommendations for Safety in Live-Line Maintenance; *IEEE Trans. Paper* 31 TP67-96.

89. Doble, F. C.: Progress in Field Testing of Insulators; *Electr. World,* 1923, vol. 81, p. 1397.

90. Cornell University: *Transmission Line Structure Foundations for Uplift/Compression Loadings;* Electric Power Research Institute Report EL-2870, Palo Alto, Calif., February, 1983.

91. GAI Consultants, Inc.: *Laterally Loaded Drilled Pier Research,* vols. 1 & 2; Electric Power Research Institute, Report EL2197, Palo Alto, Calif., January, 1982.

92. Goldberg, D. T., Jaworski, W. E., and Gordon, M. D.: *Lateral Support System and Underpinning,* vol. I, *Design and Construction,* prepared for Federal Highway Administration, U.S. Department of Commerce Publication PB-257 210, April 1, 1976.

93. *Electr. World,* July 15, 1976 p. 63.

94. Barthold, L. O., and Barnes, H. C.: High Phase Order Power Transmission; *Electra* 1973, no. 24, pp. 139–153.

95. Venkata, S. S., Guyker, W. C., et al.: 138 kV Six Phase Transmission System: Fault Analysis; IEEE Paper 81SM485-2, Summer Power Meeting, July 1981.

96. Stewart, J. R., and Wilson, D. D.: High Phase Order Transmission—A Feasibility Analysis, Part 1, Steady State Considerations; *IEEE Trans. Power Appar. Syst.,* November/December, 1978, vol. PAS-97, no. 6, p. 2300.

97. Stewart, J. R., and Wilson, D. D.: High Phase Order Transmission—A Feasiblity Analysis, Part II, Overvoltages and Insulation Requirements, *IEEE Trans. Power Appar. Syst.,* November/December, 1978, vol. PAS-97, no. 6, p. 2308.

98. Technical and Economic Characteristics of High Phase Order Power Transmission; DOE Report DOE/ET/29297-2, March 1983.

99. Neher, J. H., and McGrath, M. H.: The Calculation of Temperature Rise and Load Capability of Cable Systems; *Trans. AIEE, Power Appar. Syst.,* October, 1957, vol. 76, pp. 752–772.

100. Calculation of the Continuous Current Rating of Cables (100% Load Factor); International Electrotechnical Commission, Publication 287, 1969.

101. Morris, M., and Burrell, R. W.: Current-Carrying Capacity of Pipe-Cable Systems Under Steady State and Transient Loading Conditions; *Trans. AIEE, Power Appar. Syst.,* June, 1954, pp. 650–660.

102. Katz, C., Eager, G. S., Jr., Seman, G. W., Garcia, F. G., Smith, W. G., McCourt, J. W.: Progress in the Determination of AC/DC Resistance Ratios of Pipe-Type Cable Systems; *IEEE Trans. Power Appar. Syst.,* November/December, 1978, vol. PAS-97, no. 6.

103. Insulated Power Cable Engineers Association, American Institute of Electrical Engineers: Power Cable Ampacities; New York, 1962.

104. Neher, J. H.: A Simplified Mathematical Procedure for Determining the Transient Temperature Rise of Cable Systems; *Trans. AIEE Power Appar. Syst.,* August, 1953, p. 712.

105. Buckweitz, M. D., and Pennell, D. B.: Forced Cooling of UG Lines; *Transm. Distrib.,* April, 1976, vol. 28, no. 4, pp. 51–58.

106. The Calculation of Continuous Ratings for Forced Cooled Cables; Working Group 08 (Forced Cooled Cables) of CIGRE Study Committee No. 21 (HVDC Cables), *Electra,* no. 66, pp. 59–84.

107. Ball, E. H., Endacott, J. D., Skipper, D. J.: UK Requirements for Future Prospects for Forced-Cooled Cable Systems; *Proc. IEE,* March, 1977, vol. 124, no. 3, pp. 334–338.

108. Neher, J. H.: The Phase Sequence Impedance of Pipe-Type Cables; *IEEE Trans. Power Appar. Syst.,* August, 1964, vol. 83, pp. 795–804.

109. Del Mar, W. A.: Reactance of Large Cables in Steel Pipe or Conduit; *Trans. AIEE Power Appar. Syst.,* 1948, vol. 67, pt. 2, pp. 1409–1415.

110. Thomas, E. R., and Kershaw, R. A.: Impedance of Pipe-Type Cable; *IEEE Trans. Power Appar. Syst.,* October, 1965, vol. 84, no. 10, pp. 953–964.

111. Insulated Power Cable Engineers Association: Committee Report on AC/DC Resistance Ratios at 60 Cycles; June, 1958 (booklet).

112. Hatcher, C. T., Gillette, R. W., and Burrell, R. W.: 345-kV Underground Transmission on the Consolidated Edison Company of New York Systems; *IEEE Trans. Power Appar. Syst.,* April, 1966, vol. PAS-85, no. 4, pp. 353–360.

113. Rifenburg, R. C.: Pipe Line Design for Pipe-Type Feeders; *Trans. AIEE,* December, 1953, pp. 1275–1288.

114. Kozak, S., Corbett, J. T., and Bender, F. J.: Features of the New 138-kV Self-Contained Oil-Filled Cable System for Detroit Edison; *IEEE Trans. Power Appar. Syst.,* May/June, 1975, vol. PAS-94, pp. 949–958.

115. Supplee, G. W., Kyle, R. J., Snow, R. V., and Bolin, P. C.: Installation of 230-kV Compressed Gas Insulated Bus; *IEEE Trans. Power Appar. Syst.,* January/February, 1974, vol. PAS-93, pp. 349–353.

116. Gazzana-Priaroggia, P., Piscioneri, J. H., and Margolin, S. W.: Long Island Sound Submarine Cable Interconnection; *IEEE Trans. Power Appar. Syst.,* July/August, 1971, vol. PAS-90.

117. Oil Flow and Pressure Calculations for Pipe-Type Cable Systems, AIEE Committee Report; *Trans. AIEE,* pt. 3, April, 1955, vol. 74, pp. 251–261.

118. Burrell, R. W.: Application of Oil-Cooling in High Pressure Oil-Filled Pipe-Cable Circuits; *IEEE Trans. Power Appar. Syst.,* September, 1965, vol. PAS-84, no. 9.

119. Moran, J. A., Jr., and Williams, J. A.: HVDC Cables—An Overview of Design Practices and Experiences; presented at the Symposium on Urban Applications of HVDC Power Transmission, Philadelphia, Pa., October, 1983, sponsored by U.S. Department of Energy.

120. Underground Power Transmisson; A study prepared by Arthur D. Little, Inc. for the Electric Research Council, ERC Publ. 1-72, October, 1971.

121. Electric Power Research Institute (EPRI), 3412 Hillview Avenue; Box 10412; Palo Alto, Calif. 94303.

122. U.S. Department of Energy, Electric Energy Systems Divison, 1000 Independence Avenue, S.W., Washington, D.C. 20585.

123. *Underground Systems Reference Book;* prepared by an Editorial Staff of the Edison Electric Institute Transmission and Distribution Committee, published by the Edison Electric Institute; New York, 1957.

# SECTION 15
# DIRECT-CURRENT POWER TRANSMISSION

**P. F. Albrecht**
*Senior Applications Engineer, Electric Utility System Engineering Department (EUSED), General Electric Co.*

**G. D. Breuer**
*Consultant, EUSED, General Electric Co.*

**K. Clark**
*Applications Engineer, EUSED, General Electric Co.*

**R. C. Degeneff**
*Manager, HVDC Transmission Engineering, EUSED, General Electric Co.*

**H. J. Fiedler**
*Senior Applications Engineer, EUSED, General Electric Co.*

**C. W. Flairty**
*Manager, Control Development, HVDC Control Engineering, General Electric Co.*

**D. W. Houghtaling**
*Senior Applications Engineer, EUSED, General Electric Co.*

**E. T. Jauch**
*Senior Applications Engineer, EUSED, General Electric Co.*

**J. J. LaForest**
*Senior Applications Engineer, EUSED, General Electric Co.*

**E. V. Larsen**
*Senior Applications Engineer, EUSED, General Electric Co.*

**J. C. McIver**
*Applications Engineer, EUSED, General Electric Co.*

**F. Nozari**
*Senior Applications Engineer, EUSED, General Electric Co.*

**R. L. Rofini**
*Senior Valve Design Engineer, HVDC Systems Operation, General Electric Co.*

**H. M. Schneider**
*High Voltage Transmission Research Facility, General Electric Co.*

**J. D. Stickler**
*Project Manager, HVDC System Operation, General Electric Co.*

**J. Urbanek**
*Manager, HVDC Valve and Auxiliary Engineering at Skeats High Power Laboratory, General Electric Co.*

**L. E. Zaffanella**
*Manager, High Voltage Transmission Research Facility, General Electric Co.*

## CONTENTS

*Numbers refer to paragraphs*

## INTRODUCTION

***1. General.*** From the beginning of electric power history, dc lines and cables have been less expensive than those for 3-phase ac transmission. Alternating current, however, is more advantageous than direct current for generation, low-voltage distribution, and electric power consumption. To utilize the savings dc offers, generated ac power must be converted to dc power at a converter station, then transmitted over a dc line to another converter station where it is converted back to ac. The lack of reliable high-voltage power-conversion equipment made the application of dc systems impractical until the mid-1950s, when the development of the high-voltage mercury-arc valve resulted in a commercially competitive position for dc transmission.

***2. History.*** Over the years, many attempts have been made to develop converters for high-voltage direct-current (HVDC) transmission. The best-known was developed by the Swiss engineer Thury in 1889. Thury's system consisted of dc generators and motors connected in series on the dc side and was used in Europe from 1890 to 1937. Converters based on mechanical switches were tested in England and Sweden in the 1920s and 1930s. In the United States, the General Electric Company built converters for dc lines during the 1930s. These converters used mercury-arc valves with relatively low ratings and were in operation from 1937 to 1945. The first commercial dc installation, which remains in operation, was the Gotland transmission system in Sweden, commissioned in 1954.

Toward the end of the 1960s, solid-state semiconductor technology was introduced to HVDC converter systems. The first thyristor converters were commissioned around 1970 in the Gotland scheme as a commercial extension and in Sakuma, Japan, as an experimental back-to-back installation. In 1972, General Electric commissioned the world's first all-solid-state HVDC system at Eel River in New Brunswick, Canada.

***3. High-Voltage DC Applications.*** During the past 15 years, there has been a significant increase in the interest in HVDC transmission. The application of HVDC transmission can be attributed to one or more of the following reasons:

*Economical.* DC systems often provide a more economical alternative than ac. For systems with long overhead transmission lines, the lower cost of dc transmission lines offsets the higher converter terminal costs. There is no universally correct break-even distance, since the economical comparison between dc and ac alternatives depends largely on local conditions, such as requirements placed on line performance and properties of connecting ac systems. Studies show that under normal conditions, however, it is advantageous to consider dc for overhead lines when the transmission distance is 500 km or more. In areas with a high cost for right of way, HVDC becomes feasible at a shorter distance.

For underground cables, the extra cost of converter stations for a dc system would be paid for by savings in cable and associated costs. The considerable difference between cable costs for dc and ac transmission is more pronounced than for overhead line costs. Break-even distances for dc cable average 30 km. Distances of 60 km or more which are not feasible for uncompensated ac transmission may be eligible for an HVDC application. There is a practical limit to the possible noninterrupted length of ac cables. This is due to capacitive charging currents, which are especially pronounced in high-voltage cables. Underground cables can be compensated for by intermediate shunt reactors. However, this method is not practical for sea cables. Shunt reactors add substantially to transmission costs, required land, and intermediate terminals. DC cables have no steady-state capacitive charging currents to influence the design.

Largely due to technological advances, HVDC costs per kilowatt of transmitted power have not increased at the same rate as comparable costs for ac systems.

*Functional.* HVDC systems also offer functional characteristics and performance not achievable with ac systems. These include nonsynchronous interconnections, control of power flow, and modulation to increase power flow. Interconnection between power systems is justified where sufficient production or load diversity exists, and by limiting required spinning reserve to enable an increase of the maximum power-production unit size. Although diversity is normally small when compared with power-network size, such an interconnection often represents a large-size intertie. Maintaining a stable weak ac intertie between two independently controlled power systems is both costly and technically difficult, and often impossible. In such cases, an asynchronous dc intertie provides an appropriate solution. DC transmissions can supply small, remote systems, such as islands, where the power flow through the link also controls the local-system frequency. The asynchronous nature of a dc link can allow a system to run pump storage stations at different speeds for pumping and generating operations. Such a mode of operation provides maximum efficiency without any switching in the main circuits.

When interconnecting two power systems with different nominal frequencies, HVDC offers technical advantages even where the length of the dc line is small.

*Environmental.* Environmentally, HVDC systems are often more compatible and provide less operational difficulties than a comparable ac system.

For these reasons, coupled with the dramatic increase in fuel costs in the late 1970s and early 1980s, the utility industry has seen a tremendous increase in the number of HVDC systems under study and construction. Figure 15-1 shows the installed, committed, and studied HVDC capacity of systems worldwide.

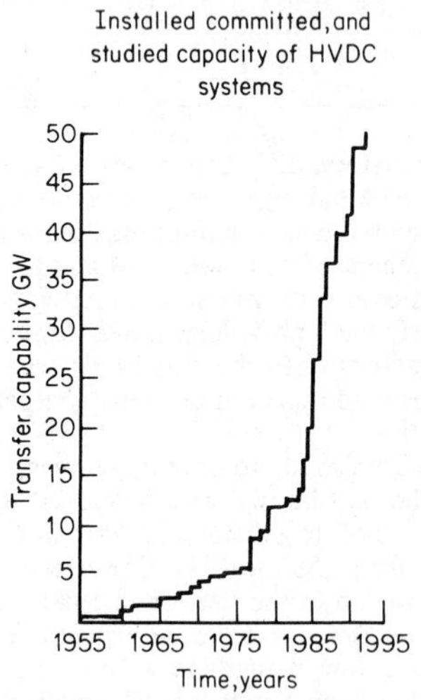


**FIG. 15-1** The economical, functional, and environmental advantages offered by HVDC have resulted in increased utility interest in HVDC transmission. Since 1983, 12 GW of HVDC systems have been installed and are operational, 20 GW are committed (orders placed and/or under construction), and 18 to 50 GW of HVDC systems are actively being studied.

## FUNDAMENTALS

### Converter Behavior and Mathematical Relationships

***4. Symbols and Definitions.*** The following symbols and definitions are used in this subsection:

$U_d$ = direct voltage
$U_{di0}$ = ideal no-load direct voltage
$I_d$ = direct current, average value
$U_L$ = phase-to-phase fundamental voltage on ac bus, rms value
$I_L$ = current from ac network, rms value
$E_L$ = ac network internal emf
$P$ = active power passing through converter
= $U_d I_d$
$Q$ = reactive power
$\cos \phi$ = power factor
$Q_L$ = ac network short-circuit capacity = $U_{LN}^2/X_L$
$Q_c$ = reactive power generation of ac filters and capacitor banks at rated voltage
$s$ = number of 6-pulse bridges series-connected on the dc side
$m$ = transformer ratio
$k$ = proportional constant = $3\sqrt{2}/\pi \cong 1.35sm$
$\alpha$ = rectifier delay angle
$\gamma$ = inverter extinction angle (commutation margin)
$R_c$ = equivalent commutating resistance
$d_N$ = direct-voltage drop, percentage of $U_{di0N}$

The subscript $N$ is used to denote the rated value of the corresponding symbol.

***5. Basic Relationships between AC and DC Quantities for a Rectifier Station.*** A simplified schematic diagram of a converter station and the feeding ac network is shown in Fig. 15-2. The ac network is represented by a Thévenin equivalent as an emf $E_L$ connected in series with an impedance. In the following, it is assumed to be lossless and is denoted by the reactance $X_L$. Due to the connection of an ac filter at the ac bus, the voltage at this point is basically sinusoidal. The rms value of the phase-to-phase fundamental voltage at the ac

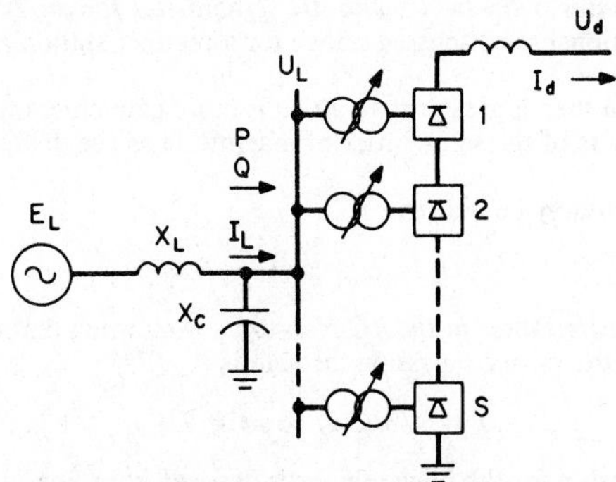


**FIG. 15-2** A simplified schematic diagram of a converter station (rectifier) and the ac feeding network is shown.

bus is denoted $U_L$ and its rated value $U_{LN}$. The total generation of reactive power in the ac filter and shunt capacitor banks at rated voltage is denoted $Q_c$.

The converter station consists of a number of 6-pulse converter bridges connected in series and parallel on the dc side. As only the number of bridges connected in series influences further discussion, the case without parallel connection has been illustrated in Fig. 15-2. The number of series-connected bridges is denoted as $s$, the most common arrangement being $s = 2$.

Because of the dc reactor, current on the dc side is almost completely smoothed. The dc component is denoted $I_d$. The designation for the corresponding rated value is $I_{dN}$ and for the direct voltage is $U_d$.

As the ac bus voltage has been assumed to be basically sinusoidal, the relationships between ac quantities $U_L$ and $I_L$ and dc quantities $U_d$ and $I_d$ are the same as for those 6-pulse converters fed by a stiff sinusoidal emf. Accordingly, the ideal no-load direct voltage is

$$U_{di0} = \frac{3\sqrt{2}}{\pi} smU_L = 1.35smU_L = kU_L \qquad (15\text{-}1)$$

$U_{di0}$ is then directly proportional to the ac bus voltage $U_L$, the number of bridges connected in series $s$, and the transformer ratio $m$.

The no-load voltage can be further decreased by phase control, giving for the rectifier

$$U_{d0} = kU_L \cos\alpha = U_{di0} \cos\alpha \qquad (15\text{-}2)$$

where $\alpha$ is the delay angle.

Reactances and losses in the transformer and valves will give additional voltage drop. The basic equation for determining direct voltage is

$$U_d = U_{di0} \cos\alpha - R_c I_d$$

with

$$R_c = \frac{3}{\pi}\omega L_c \qquad (15\text{-}3)$$

where $R_c$ is the equivalent commutating resistance and $L_c$ is transformer short-circuit reactance.

It can be shown that the ratio between ac and dc can be very well approximated by the equation

$$I_L/I_d = k/\sqrt{3} \qquad (15\text{-}4)$$

The converter thus gives a fixed transformation between the ac $I_L$ and the dc $I_d$ determined only by the number of bridges connected in series $s$, the transformer ratio $m$, and a proportional constant.

***6. Basic Relationships between AC and DC Quantities for an Inverter Station.*** The assumptions and relationships discussed above for a rectifier station are basically valid for an inverter station.

As the delay angle $\alpha$ then is greater than 90°, it is more convenient to define an extinction angle $\gamma$, which usually is of the same order of magnitude as the delay angle in the rectifier (15 to 18°).

Equation (15-3) can now be written

$$U_d = U_{di0} \cos\gamma - R_c I_d \qquad (15\text{-}5)$$

***7. Active and Reactive Power in the AC Network.*** Assuming that there are no losses in the converter station, the power $P$ passing through is

$$P = \sqrt{3}\, U_L I_L \cos\phi = U_d I_d \qquad (15\text{-}6)$$

The following relationship for the power factor is derived from Eqs. (15-1), (15-4), and (15-6):

$$\cos\phi = \frac{U_d}{kU_L} = \frac{U_d}{U_{di0}} = \frac{U_{di0}\cos\alpha - R_c I_d}{U_{di0}} \qquad (15\text{-}7)$$

The power factor is thus determined only by the ratio between the actual direct voltage $U_d$ and the ideal no-load voltage $U_{di0}$, independent of whether the reduction in voltage is due to increased $\alpha$ or increased inductive voltage drop in the transformer.

Active losses in the transformer and valves can be taken into account by increasing $U_d$ in the rectifier and decreasing $U_d$ in the inverter by an amount corresponding to the losses.

From Eqs. (15-1) and (15-4), the following relationship can be derived for the fundamental power from the ac bus:

$$S = \sqrt{3}\, U_L I_L = U_{di0} I_d \tag{15-8}$$

and for the reactive power

$$Q = \sqrt{S^2 - P^2} = U_d I_d \sqrt{\left(\frac{U_{di0}}{U_d}\right)^2 - 1} \tag{15-9}$$

Solving Eq. (15-3) for $U_d/U_{di0}$ and substituting into Eq. (15-9), it can be seen that $Q$ varies nonlinearly with delay angle $\alpha$ and transformer reactance drop. This relationship is illustrated by Fig. 15-3.

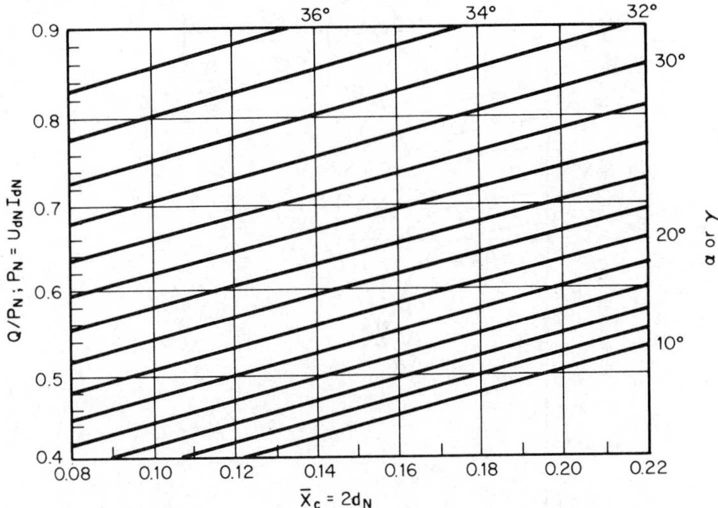


**FIG. 15-3** Reactive power (per unit of rated active power $P_N$) varies with per unit transformer reactance $x_c$ and delay angle $\alpha$ (rectifier) or extinction angle $\gamma$ (inverter).

***8. AC Network Loaded by a Converter.*** The basic relationships presented in the previous paragraphs are sufficient to represent the converter in computer programs for load flow and stability analysis. However, simplified studies without the use of extensive computer programs can be performed using a representation of the ac network as shown in Fig. 15-2. The ac network is represented by an emf $E_L$ behind a lossless impedance having a reactance $X_L$. The transient reactance of synchronous machines should be used in most cases. The reactance $X_L$ is preferably expressed as a short-circuit capacity defined by

$$X_L = U_{Ln}^2 / Q_L \tag{15-10}$$

To simplify the calculation further, the ac network, with ac filter and shunt capacitor bank, and transformed by using Thévenin's theorem, is shown in Fig. 15-4. The ac network equivalents are

$$X_c = \frac{U_L^2}{Q_c}; \qquad X_q = \frac{U_L^2}{Q}; \qquad X_p = \frac{U_L^2}{P} \tag{15-11}$$

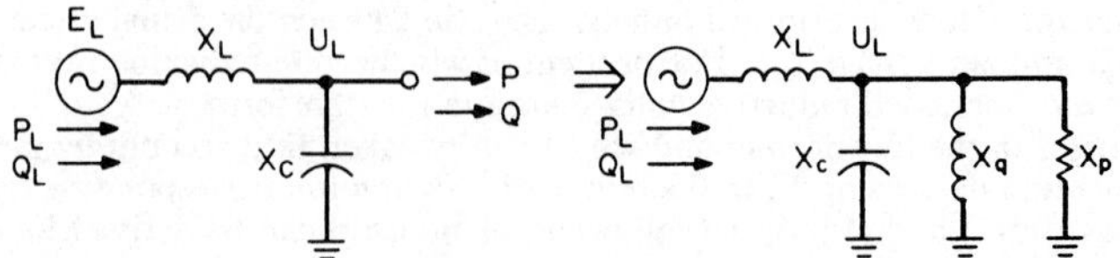


**FIG. 15-4** The ac network is transformed using Thévenin's theorum.

The relationship between $E_L$, $U_L$, $P$, and $Q$ is

$$E_L = U_L\left(1 + \frac{X_L}{X_p} + \frac{X_L}{X_q} + \frac{X_L}{X_c}\right) = U_L\left(\frac{P + Q_L + Q_c + Q}{Q_L}\right) \tag{15-12}$$

Assuming a lossless ac network, $P_L = P = \sqrt{3}\, U_L I_L \cos\phi$. Figure 15-5 graphically displays Eq. (15-12).

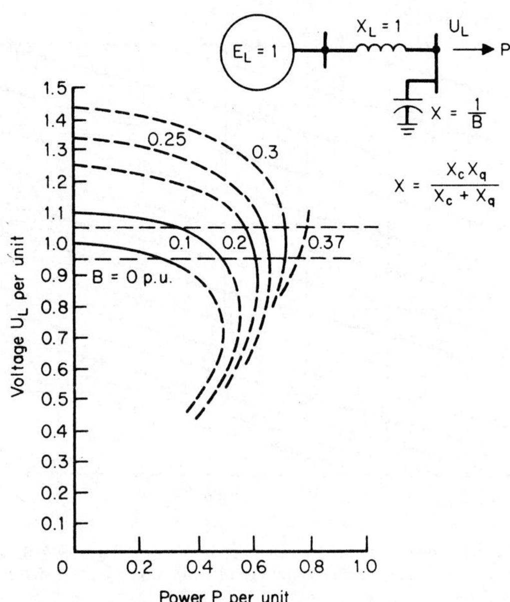


**FIG. 15-5** **AC bus voltage $U_L$ as a function of converter load $P$ for various susceptances $B$.**

***9. Transient Change of Direct Current.*** The equations of the converter and the feeding ac network represented in the previous section are suitable for studying performance during a sudden change in direct current. The magnitude of the internal emf $E_L$ behind the transient network impedance may then be considered constant. Therefore, $E_L$ before the current change is a function of $I_{dN}$, $U_{dN}$, $U_{LN}$, etc., and $E_L$ after the current change is a function of $I_d$, $U_d$, $U_L$, etc. The following equation can be written where the left-hand side represents $E_L$ prior to the current change and the right-hand $E_L$ side after the current change:

$$U_{LN}\left(U_{dN} I_{dN}\left\{1 + \left[\left(\frac{U_{di0N}}{U_{dN}}\right)^2 - 1\right]^{1/2}\right\}\frac{X_L}{U_{LN}^2} + \frac{X_L}{X_c} + 1\right)$$

$$= U_L\left(U_d I_d\left\{1+\left[\left(\frac{U_{di0}}{U_d}\right)^2-1\right]^{1/2}\right\}\frac{X_L}{U_L^2}+\frac{X_L}{X_c}+1\right) \qquad (15\text{-}13)$$

***10. Transient Voltage Increase at Load Rejection.*** The transient voltage increase at complete load rejection from rated conditions is determined from Eq. (15-13) by setting $I_d = 0$. This yields

$$U_{LN}\left(U_{dN} I_{dN}\left\{1+\left[\left(\frac{U_{di0N}}{U_{dN}}\right)^2-1\right]^{1/2}\right\}\frac{X_L}{U_{LN}^2}+\frac{X_L}{X_c}+1\right) = U_L\left(\frac{X_L}{X_c}+1\right) \qquad (15\text{-}14)$$

It should be noted that the phase angle of network impedance has been assumed to equal 90°, that is, a pure inductive network. In a real network, there are always some active losses and loads which are of importance for a very low ratio between the short-circuit capacity in the network and the transmitted power on the dc system. This resistive component in the network causes an increased transient voltage in the rectifier and a decreased transient voltage in the inverter.

***11. Maximum Transient Power Increase.*** The maximum fast power increase on a converter is limited by the fact that a sudden increase in direct current causes a decrease in direct voltage. The ac bus voltage $U_{L0}$ with $\alpha = 0$, thus also $U_{di0}$, will decrease because of the voltage drop in the ac network impedance. Owing to the transformer reactance, the internal voltage drop in the converter will increase. The variation in the ac bus voltage $U_L$ for the rectifier is studied first.

The maximum power is transmitted when the delay angle $\alpha$ is equal to zero, giving maximum direct voltage. A ratio of power at $\alpha = 0$ and $\alpha_N = 15°$ gives

$$1 = \frac{U_{LN}\left\{I_{d0}(kU_{LN} - R_c I_{d0})\left[1+\sqrt{\left(\frac{1}{1 - R_c I_{dN}/kU_{L0}}\right)^2 - 1}\right] + \frac{U_{L0}^2}{X_c} + \frac{U_{L0}^2}{X_L}\right\}}{U_{L0}\left\{I_{dN}(kU_{LN}\cos\alpha_N - R_c I_{dN})\left[1+\sqrt{\left(\frac{1}{\cos\alpha_N - R_c I_{dN}/kU_{LN}}\right)^2 - 1}\right] + \frac{U_{LN}^2}{X_c} + \frac{U_{LN}^2}{X_L}\right\}} \qquad (15\text{-}15)$$

From this equation, $U_{di0}/U_{di0N} = U_L/U_{LN}$ may be found.

Figure 15-6 shows $U_L/U_{LN}$ as a function of $I_d/I_{dN}$ with $I_{dN}U_{di0N}/Q_t$ and $d_N$ as parameters.

The ratio between transient power and rated power, assuming constant emf $E_L$, can be written:

$$\frac{P_{\alpha=0}}{P_N} = \frac{U_{di0}}{U_{di0N}}\frac{I_d}{I_{dN}}\frac{(1 - R_c I_d/U_{di0})}{(\cos\alpha_N - R_c I_{dN}/U_{di0N})} \qquad (15\text{-}16)$$

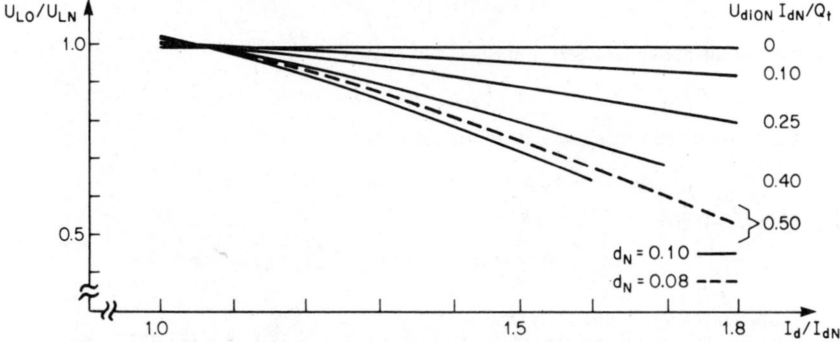


**FIG. 15-6** Transient voltage decreases at sudden current increases.

For this equation it was assumed that the emf $E_L$ behind the ac network is unchanged. After a short period of time, $E_L$ will increase as the voltage regulation in the ac network tries to restore the voltage $U_L$. This will enable a further increase in transmitted power.

***12. Current Limitation.*** It is obvious from the curves given in Fig. 15-7 that the current must be limited such that an attempt to increase the transmitted power will not result in an uncontrolled current increase. This is particularly important when ac network short-circuit capacity is low. Normally, the current should be limited to a value considerably lower than the value corresponding to maximum power, since the power maximum is rather flat.

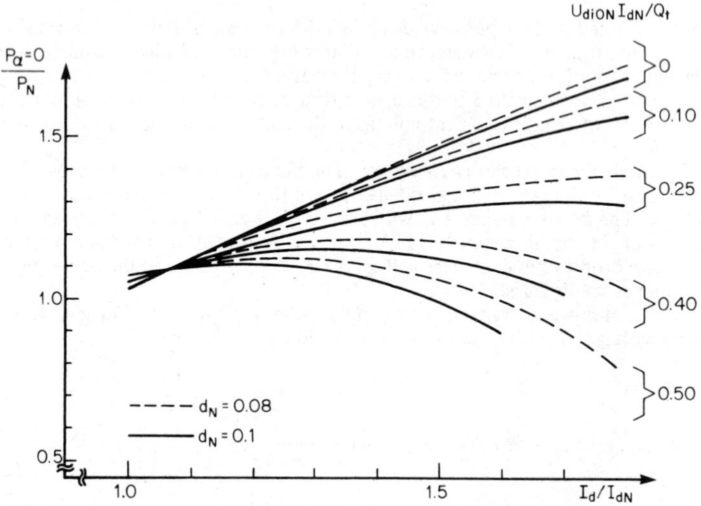


**FIG. 15-7** Transient transmitted power changes at a sudden current increase.

It is, of course, also necessary to have a current limit to ensure that the converter will not be overloaded.

***13. Example of Transient Power Maximum Calculation.*** For a rectifier and the ac network to which it is connected, the following data are assumed to be valid:

| | |
|---|---|
| $U_{LN}$ = 350 kV | $U_{dN}$ = 400 kV |
| $d_N$ = 10% | $I_{dN}$ = 1000 A |
| $Q_t$ = 1850 MVA | $P_N$ = 400 MW |
| | $\alpha_N$ = 15° |

Insertion of the above numbers into Eq. (15-3) gives

$$U_{di0N} = 461 \text{ kV}$$

The transformer ratio $m$ is calculated from Eq. (15-1):

$$461 = 1.35 \times 1 \times m \times 350 \qquad m = 0.976$$

Equation (15-4) gives

$$I_{LN} = 1.35 \times 0.976 \times 1000/\sqrt{3} = 761 \text{ A}$$

The ratio $U_{di0N}I_{dN}/Q_t = 461 \times 1/1850 = 0.25$

From Fig. 15-7 it is found that the maximum value of $P_{\alpha=0}/P_N$ then is 1.323, corresponding to 528 MW and a direct current $I_d$ of around 1700 A. It should be noted that a 70% increase in current will result in only a 30% increase in transmitted power. From Fig. 15-6 it is found that the ac bus voltage has dropped to 84% of rated voltage 350 kV, that is, 294 kV. If, on the other hand, the maximum current is limited to 1200 A, an overcurrent of 20%, the transmitted power is increased to 472 MW, an increase of 18%, almost the same percentage as the increase in current.

***14. Influence on Power Maximum of DC Line Losses.*** In point-to-point transmission, the lowest direct voltage occurs at the inverter terminal. The calculations above have been performed only for the rectifier terminal with $\alpha = 0$. Similar calculations apply for the inverter, although the minimum value of $\gamma$, $\gamma_{min}$, is fairly close to the normal value of 15 to 18°. Because of this, the transient maximum transmitted power is slightly reduced when compared with the rectifier, assuming the other conditions to be similar.

Again, it has been assumed that the ac network impedance has a phase angle of 90°. As active power is transmitted into the ac network, a resistive component in the impedance will increase the inverter ac bus voltage when the current is increased. The inductive component lowers the voltage as for the rectifier and causes the transient maximum power limitation.

If the network impedances of the two ac networks are similar, about the same maximum power can be transmitted through the rectifier as through the inverter. Since the power losses in the dc line also have to pass the rectifier, however, it will usually be the rectifier which limits received power.

Owing to dc line resistance, the power received by the inverter is lower than that transmitted by the rectifier. Losses also occur in the inverter and in the rectifier. If these losses, together with line resistance, are represented by a resistance $R_t$, the ratio between the maximum power received by the inverter ac network and its corresponding rated power can be written as

$$\frac{P_{inv}}{P_{N,inv}} = \frac{U_{di0}I_{dN}}{U_{di0N}I_{dN}} \frac{\{1 - [(R_c + R_t)I_d/U_{di0}]\}}{\{\cos \gamma_N - [(R_c + R_t)I_{dN}/U_{di0N}]\}} \tag{15-17}$$

In accordance with Eq. (15-17), the curves in Fig. 15-7 will be altered somewhat to represent the relationship for received power. For long dc lines, the influence of line losses may be fairly large.

***15. Calculated Example Including DC Line Losses.*** In the Par. **13** example, the maximum transmitted power is 528 MW. The rectifier is now assumed to be connected to an inverter by a transmission line which is 700 km long and has a dc resistance of 15 Ω. The inverter ac network impedance is the same as for the rectifier network.

Power received by the inverter at rated current is $P_{Ninv} = 400 - 15\text{ MW} = 385\text{ MW}$. This power can be increased transiently to $P_{inv} = P_{\alpha=0} - RI^2 = 528 - 15(1.7)^2\text{ MW} = 484\text{ MW}$ by increasing the current from 1000 to 1700 A. This is a current increase of 70%, which gives a power increase of only 25.9%.

A transient current increase to 1200 A gives a transmitted power of 472 MW in the rectifier station. The received power in the inverter will then be $P_{inv} = 472 - 15 \times (1.2)^2 = 450$ MW. A transient power increase will thus be about 17% for a current increase of 20%.

***16. Power Transfer by a DC Link.*** The power in the sending and receiving end of a dc line as illustrated in Fig. 15-8 may be expressed as

$$P = \frac{U_{d1}^2 - U_{d2}^2}{2R} \pm \frac{(U_{d1} - U_{d2})^2}{2R} \tag{15-18}$$

where index 1 refers to the sending end and index 2 to the receiving end. The first term gives the power in the middle of the line, and the second term gives half the line losses. The plus sign gives the power at the sending end and the minus sign at the receiving end.

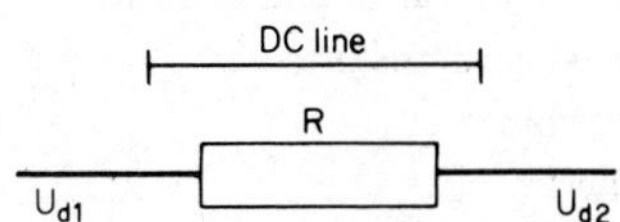


**FIG. 15-8** Equivalent dc line used for the calculation of power at the sending and receiving ends.

The active and reactive power on the ac buses as illustrated in Figure 15-9 may be written as follows:
Active power at the sending end:

$$P_1 = k_1 U_{L1} \cos \phi_1 \frac{k_1 U_{L1} \cos \phi_1 - k_2 U_{L2} \cos \phi_2}{R} \tag{15-19}$$

Reactive power at the sending end:

$$Q_1 = k_1 U_{L1} \sin \phi_1 \frac{k_1 U_{L1} \cos \phi_1 - k_2 U_{L2} \cos \phi_2}{R} \tag{15-20}$$

Active power at the receiving end:

$$P_2 = k_2 U_{L2} \cos \phi_2 \frac{k_1 U_{L1} \cos \phi_1 - k_2 U_{L2} \cos \phi_2}{R} \tag{15-21}$$

Reactive power at the receiving end:

$$Q_2 = k_2 U_{L2} \cos \phi_2 \frac{k_1 U_{L1} \cos \phi_1 - k_2 U_{L2} \cos \phi_2}{R} \tag{15-22}$$

The factors $k_1$ and $k_2$ are as defined in Par. **5.**

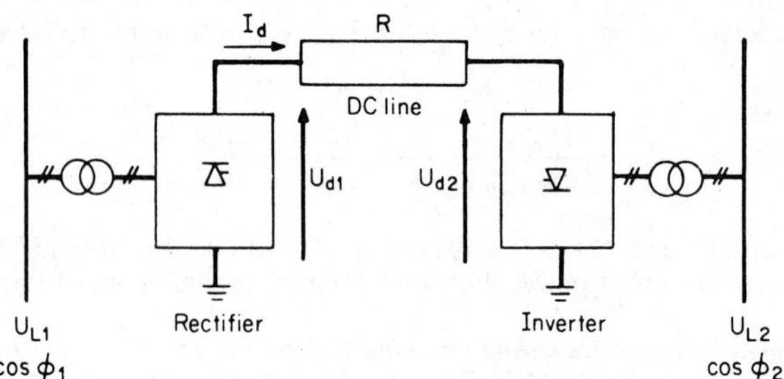


**FIG. 15-9** A one-line diagram of a dc link or between ac systems.

## Terminal Design and System Configuration

***17. Terminal Design.*** The HVDC terminal is an integral part of the HVDC system. It provides the basic system function—conversion of ac to dc or vice versa. When a terminal converts ac to dc, it is referred to as a *rectifier* terminal; when it converts dc to ac, it is referred to as an *inverter* terminal. For simplicity, and since most terminals today are designed for both modes of operation, either terminal can be referred to as a *converter* terminal.

The converter terminal (Fig. 15-10) can therefore be defined as an operative unit comprised of the following major components:

Solid-state valves and controls
Converter transformers
Reactors
Filters
Reactive power supplies
Protective, monitoring, measuring, communication, and auxiliary equipment

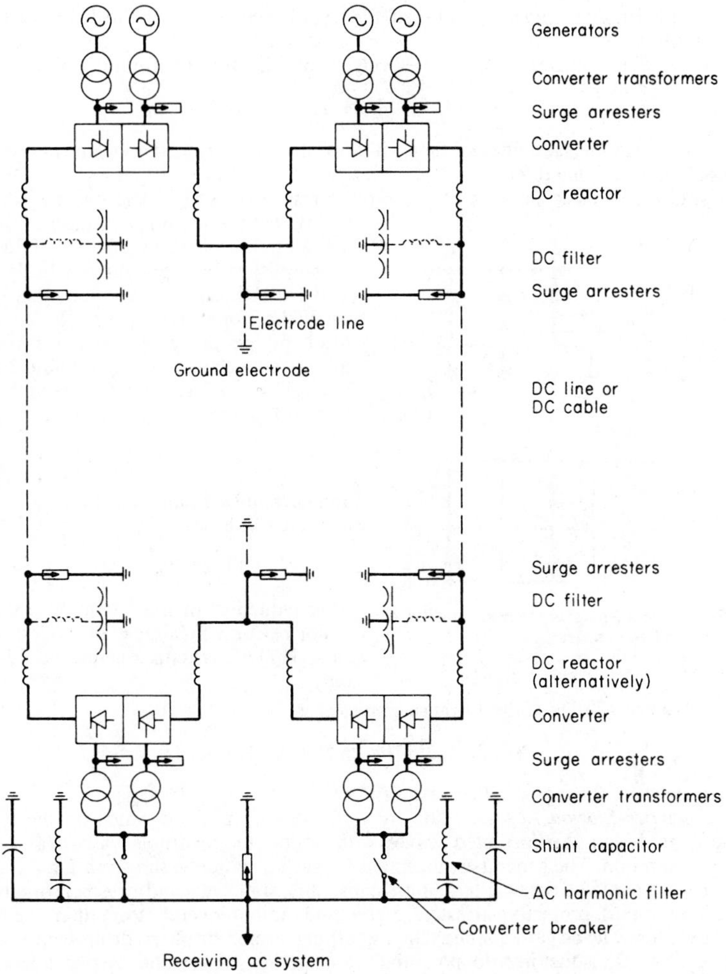

Generators
Converter transformers
Surge arresters
Converter
DC reactor
DC filter
Surge arresters
Electrode line
Ground electrode
DC line or
DC cable
Surge arresters
DC filter
DC reactor
(alternatively)
Converter
Surge arresters
Converter transformers
Shunt capacitor
AC harmonic filter
Converter breaker
Receiving ac system

**FIG. 15-10** A transmission arrangement with the sending end at point of generation, the receiving end connected to an ac system via a bipolar dc line, and ground return used as a spare conductor.

The 6-pulse Graetz bridge (Fig. 15-11) is the basic building block for the ac/dc/ac conversion process. Each 6-pulse bridge consists of six elements. Each element represents an optimized number of solid-state thyristors connected in series, and sometimes in parallel, to form a solid-state valve.

The dc output voltage of a 6-pulse bridge contains voltage harmonics of the order

$$6, 12, 18, \ldots, 6n \qquad n = 1, 2, 3, \ldots$$

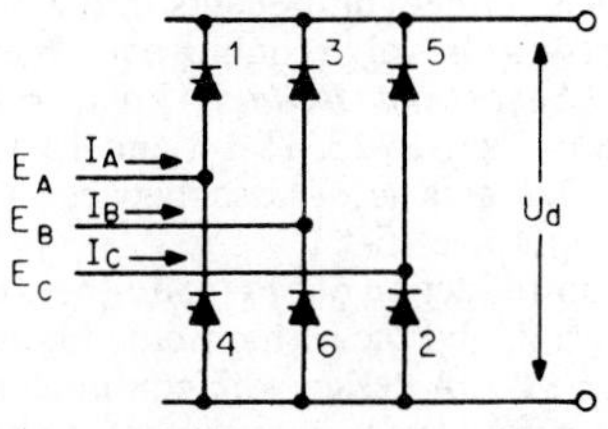


**FIG. 15-11** The six-pulse Graetz bridge is the basic building block for the ac/dc/ac conversion process.

Unless these harmonic voltages are filtered, the resulting current down the dc line can cause telephone interference.

The valve-side ac current of a 6-pulse bridge contains current harmonics of order:

$$5, 7, 11, 13, 17, 19, 23, 25, \ldots, 6n \pm 1$$

Unless these harmonics are filtered, they will flow into the ac system and can cause voltage distortion and telephone interference.

A 12-pulse converter consists of two 6-pulse bridges in series. The two are identical except that the ac supply voltages of the two are shifted in phase by 30°. This is usually accomplished by supplying one bridge with a wye-wye transformer and the other with a wye-delta connection (Fig. 15-12). As a result of this 30° phase shift, certain harmonics above are canceled out, thus leaving voltage harmonics on the dc side of the converter of order:

$$12, 24, 36, \ldots, 12n$$

and current harmonics on the ac side of the converter of order:

$$11, 13, 23, 25, \ldots, 12n \pm 1$$

This reduction in filter requirements is a major reason why almost all modern solid-state HVDC systems operate as 12-pulse only.

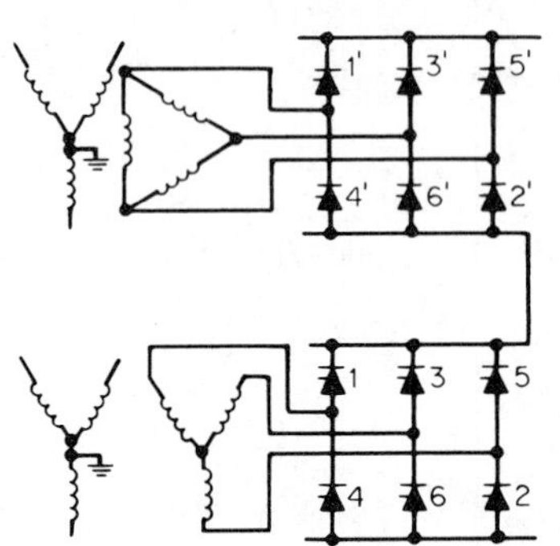


**FIG. 15-12** The 12-pulse converter consists of two 6-pulse bridges in series.

The valve firing order of the 12-pulse converter is

$$1, 1', 2, 2', 3, 3', 4, 4', 5, 5', 6, 6', 1, 1', \text{etc.}$$

In the steady state, there are 30° between firings.

***18. Converter-Station Layout.*** The converter-station layout depends on the type of valve and its design. Air-insulated valves with double or quadruple valves (as defined below) are common. The general arrangement for such a station is shown in Fig. 15-13.

The ac yard contains the ac filters, shunt capacitors, static vars and/or synchronous condensers (if required), protective arresters, carrier and radio frequency noise filters, ac breakers, and switches. The dc yard contains the smoothing reactor, dc filters, dc protective arresters, dc current transductors, dc potential transductors, power-line carrier filters, and disconnect switches.

To obtain a compact design, converter transformers and smoothing reactors are located close to building walls with the valve side bushings pointing into the valve hall. The layout in Figs. 15-14 and 15-15 shows quadruple valves, that is, each physical valve structure contains four valve functions. Fans for the closed-loop air-cooling system are indicated in the basement. Other components in the building include valve protective surge arresters, converter controls, valve cooling equipment, and auxiliaries.

***19. Space and Building-Volume Requirements.*** An air-insulated station layout as shown in Figs. 15-13, 15-14, and 15-15 requires a minimum ground area of some 20 $m^2$/MW. This area is divided between the main substation components approximately as shown in Table 15-1.

By introduction of $SF_6$ (sulfur hexafluoride) insulation for the major parts, the ac and dc switchyard, and the ac harmonic filters, it is possible to improve the ground area required to 2 $m^2$/MW. A design with equipment in several levels and including $SF_6$-insulated valves offers reduced space requirements to below 1 $m^2$/MW.

To minimize building size, it is important to consider geometrical configurations of buses and equipment and the shape of the ground plane. This makes possible substantial reductions in the clearance distances between live parts and ground compared with what

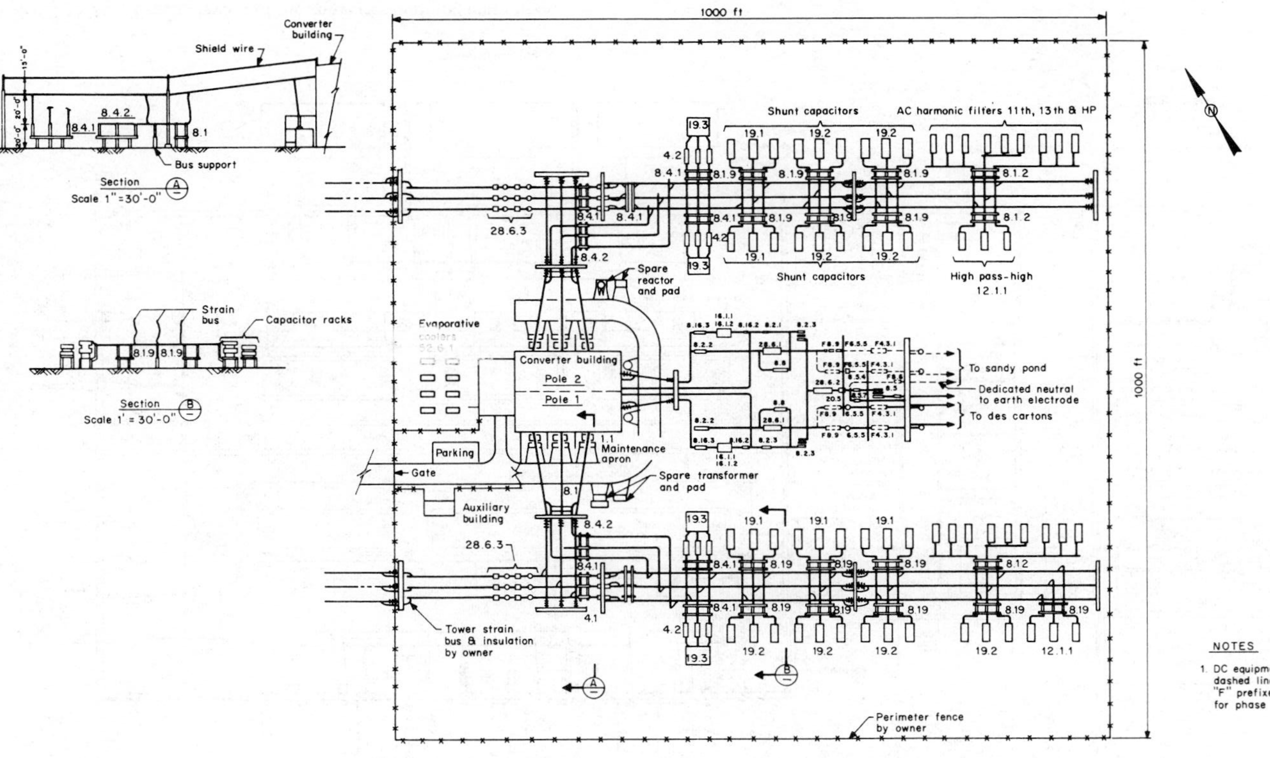


FIG. 15-13 The general arrangement of a converter station with air-insulated double or quadruple valves.

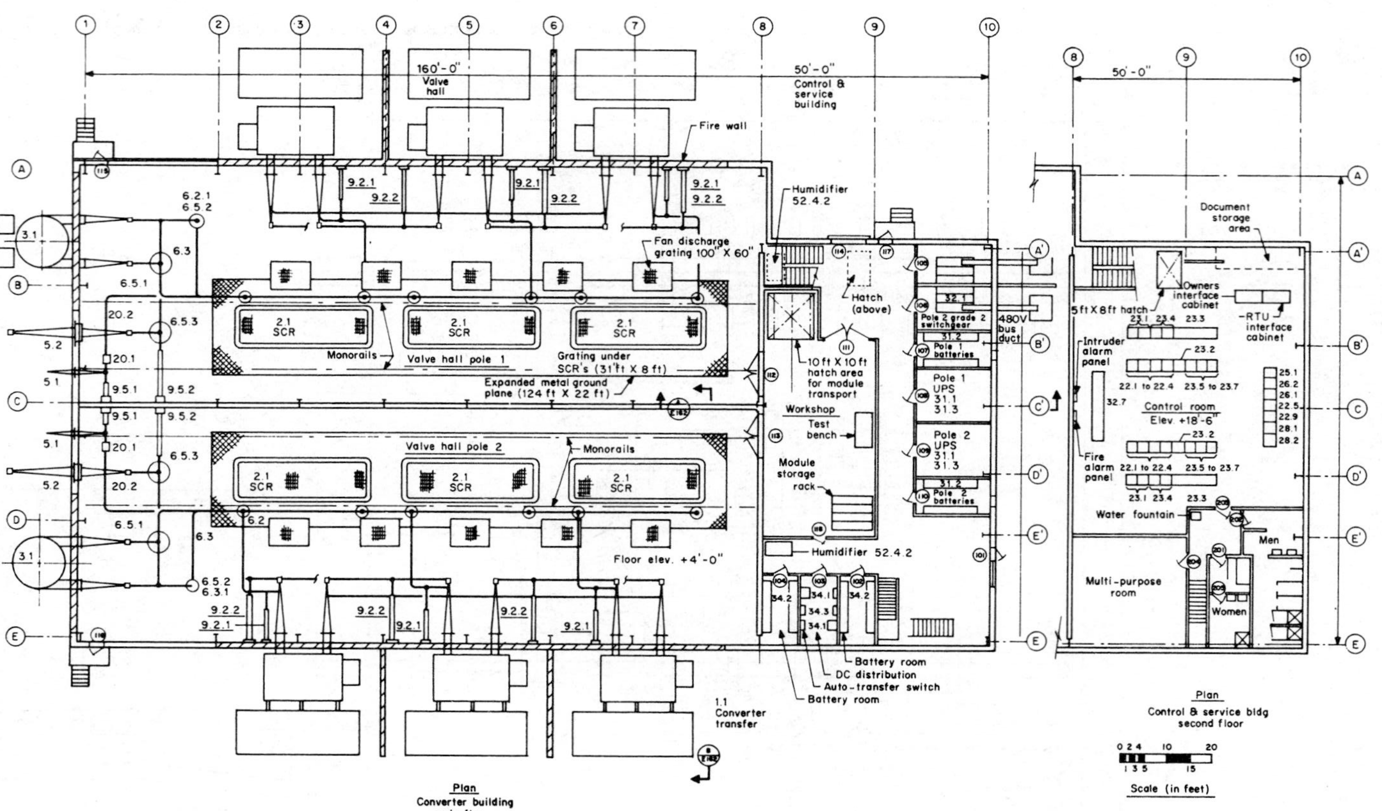


FIG. 15-14 Typical layout of the converter building main floor.

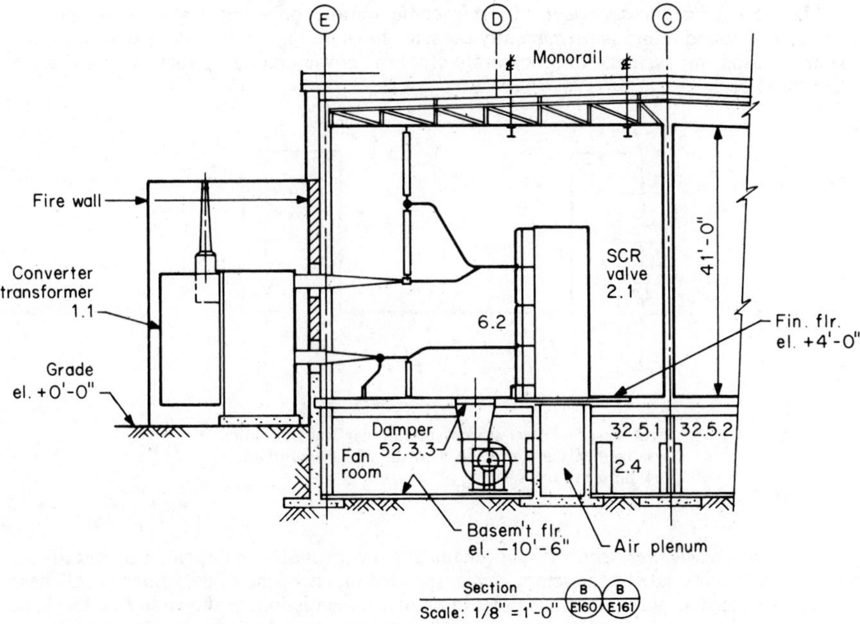


**FIG. 15-15** Side view of a typical converter building.

**TABLE 15-1** Space-Requirement Analysis

| Item | Ground area occupied, % |
|---|---|
| Valve building | 8 |
| AC filters | 35 |
| AC buswork and transformers | 45 |
| DC yard | 11 |

would normally be required in an outdoor switchyard. The building volume primarily depends on the direct voltage of the converter, and the volume increases more than linearly with voltage (see Fig. 15-16).

***20. Transmission Arrangements.*** In transmission, ground return can be used as one conductor. That is, each separately insulated transmission conductor, together with the ground-return path, forms a separate electric circuit. Based on this fundamental principle, the following basic circuit arrangements can be considered.

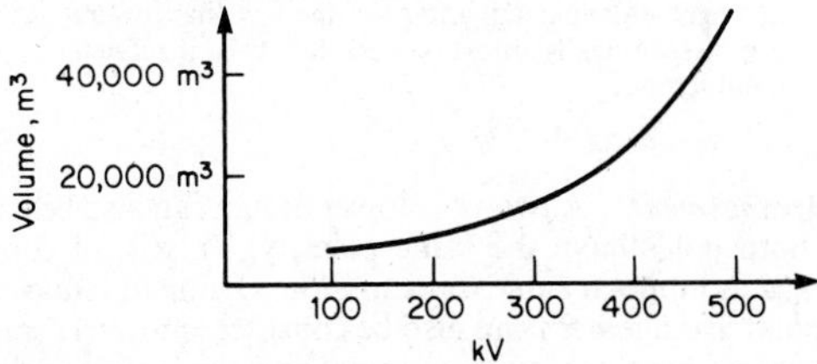


**FIG. 15-16** Converter building volume increases more than linearly with direct voltage.

***21. Monopolar Arrangement.*** In this configuration, only one transmission pole is installed. Ground return is permanently used, as shown in Fig. 15-17. Monopolar transmission is used in systems of comparatively low power rating, primarily with cable transmission.

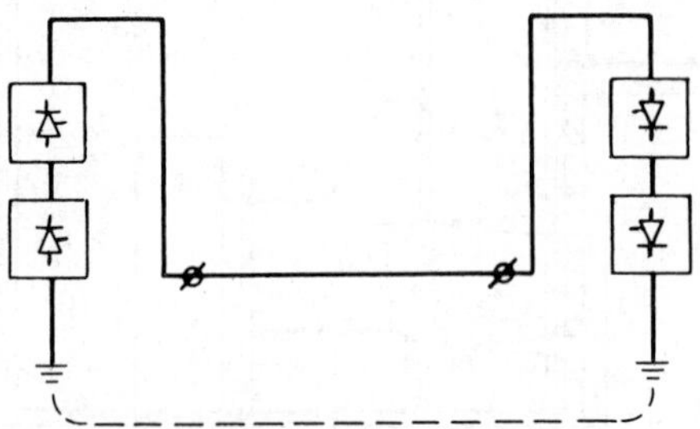

**FIG. 15-17** Direct-current monopolar transmission arrangements are used in systems of comparatively low power rating.

***22. Bipolar Arrangement.*** It is mechanically most suitable to design an overhead-line tower for two insulated conductors, one suspended on each side of the center post. These can be arranged as plus and minus poles in bipolar transmission, as shown in Fig. 15-18. In bipolar transmission, ground return is not necessarily used but is normally provided to increase the transmission availability in case of pole failure.

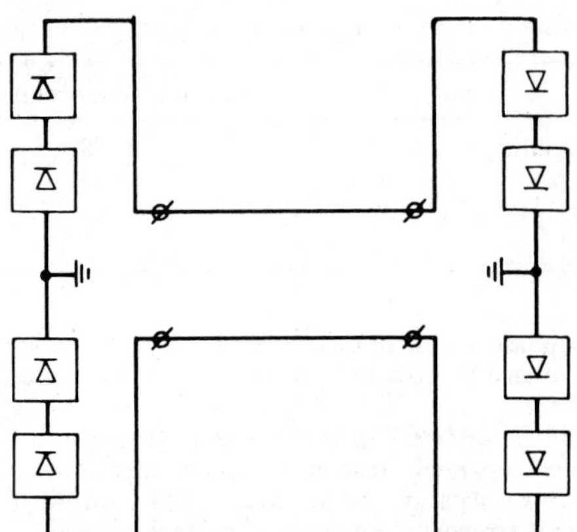

**FIG. 15-18** Direct-current bipolar transmission arrangements take advantage of the fact that towers are mechanically most suited for two insulated conductors.

***23. Homopolar Arrangement.*** A two-pole tower design can also be used in a homopolar transmission where both poles have the same polarity. This is, of course, applicable for overhead lines feeding a monopolar cable transmission. Owing to reduced corona, however, the use of the homopolar arrangement can also be considered for very large overhead transmission systems where two bipolar circuits are used and the insulator chains of each tower are carrying two separate conductors with the same polarity (shown in Fig. 15-19).

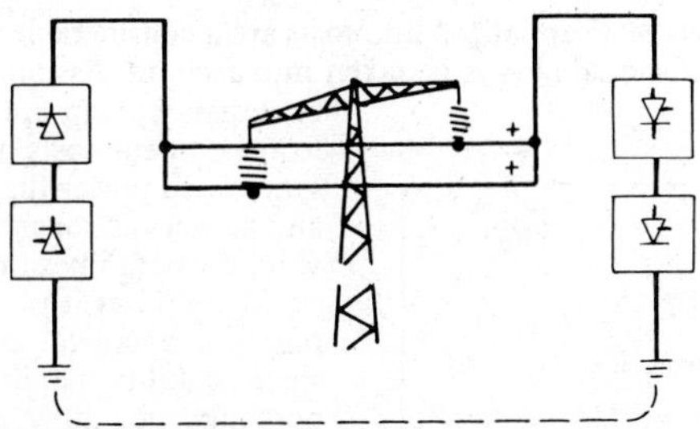

**FIG. 15-19** Direct-current homopolar transmission arrangements can be considered for very large overhead transmission systems where two bipolar circuits are used.

## Economics and Efficiency

***24. Choice of DC Line Voltage.*** In a dc link, transmission-line voltage can be chosen freely to meet the economical optimum of the dc transmission system. For a given rated power to be transmitted, the optimum line voltage must be calculated considering converter-station costs, line costs, and total loss costs. The calculations follow the same rules as for ac lines, but the limitations set by corona phenomena for ac lines are much less pronounced for dc lines.

It is practical to study cost relations for the line and the terminals separately. A total optimization should consider transmission distance and other factors which may be of importance, such as environmental requirements.

***25. DC Line Costs.*** DC line costs vary according to Fig. 15-20, where each voltage level corresponds to a specified creepage distance and flashover clearance. For a direct voltage,

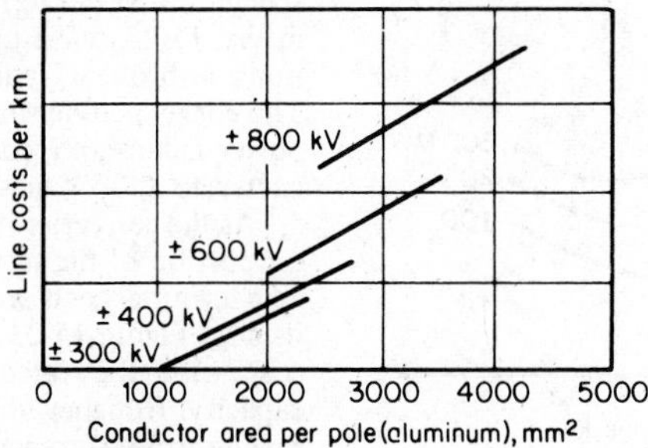


**FIG. 15-20** HVDC line cost as a function of total pole conductor area (aluminum) for different system voltage levels.

below 400 kV, the voltage level has only a minor influence on costs. For higher voltages, however, the cost increases rapidly due to the nonlinear withstand characteristics of switching surges.

From these curves, line costs versus line voltage-to-ground for a given rated load, including capitalized loss costs, are calculated according to Fig. 15-21. The lowest curve gives the capitalized loss cost separately, which is the same for all curves, and is included in the total loss curves. The voltage optimum curve indicates the highest voltage which should be considered at all. In practice, a value of about 75 to 80% of the optimum voltage is found reasonable, considering line length, station voltage-dependent costs, load utilization of the line, etc.

It should be realized that capitalized line costs are a considerable portion (30 to 35%) of the total line costs and should always be taken into account. Assuming that an ac line can be designed in an optimal way with respect to investment costs and loss costs, for the same rated power the cost ratio between dc and ac lines is roughly 2:3. This relation is valid for both investment and loss costs.

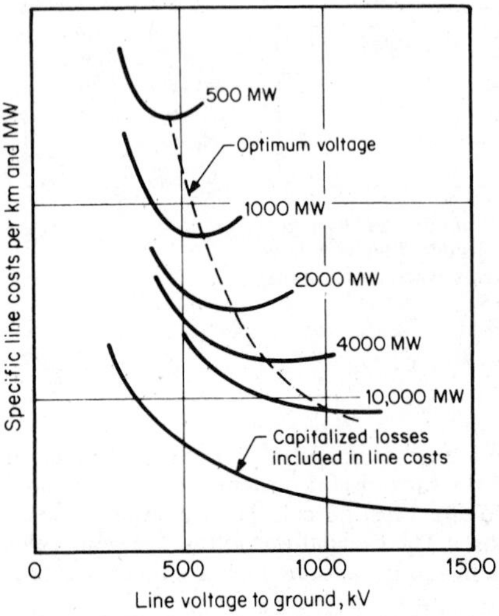

**FIG. 15-21** Bipolar HVDC line costs for different transmitted power ratings including capitalized loss costs. The position and shape of the cost curves depend upon the actual loss costs in relation to investment costs.

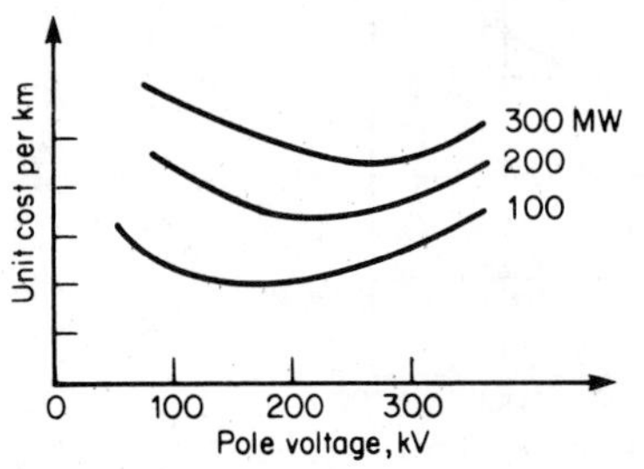

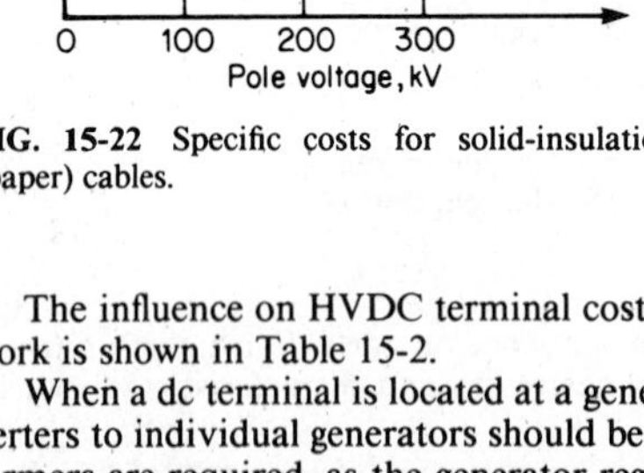

**FIG. 15-22** Specific costs for solid-insulation (paper) cables.

Many times it is not possible to make a purely economical design for an ac line, since stability requirements, corona phenomena, etc., may require a more rigid design. It may also be necessary to provide some type of line compensation. In such cases, the cost ratio is still lower, sometimes as low as 1:2 or less.

***26. Cable Costs.*** DC cables are considerably less costly than ac cables for the same rating. A resulting cable cost ratio of 1:3, comparing dc and ac alternatives for a given project, is not unusual. The relative cost of solid-insulation cables as a function of power rating has a characteristic similar to that shown in Fig. 15-22. Because conductor temperature must not exceed a certain value, current density has to be decreased rapidly above a certain current rating. Consequently, the unit cost strongly depends on the choice of cable parameter ratings in the low-voltage range, as illustrated by the curves.

***27. DC Converter Terminal Costs.*** The specific converter terminal cost curve, given in Fig. 15-23, refers to bipolar converter stations with one 12-pulse converter per pole. The curve includes the turnkey cost for fully power-factor-compensated terminals connected to 230-kV ac systems.

As the converter itself makes up the largest portion of the terminal equipment, specific terminal costs are mainly dependent on its size. Figure 15-23 shows the variation of cost with the rated power transmission capacity. If higher voltage levels are chosen (owing to total transmission-cost optimization), the per unit cost will normally increase. Specific terminal costs for a monopolar transmission of the same power rating are slightly higher.

The influence on HVDC terminal costs of the alternating voltage in the connected network is shown in Table 15-2.

When a dc terminal is located at a generating station, direct connection of separate converters to individual generators should be considered. No tap changers on converter transformers are required, as the generator regulators can give voltage control. No phase-compensating shunt capacitor banks are needed, because the generators can supply reactive power. Furthermore, with the converters as the sole generator load, no harmonic filters are required, since generators can be designed to withstand the harmonics. Typical HVDC system component costs are shown in Table 15-3.

***28. Line Losses.*** The losses on a dc line are resistive conductor losses and corona losses. The latter are, in most cases, relatively small. There is no skin effect to consider, and

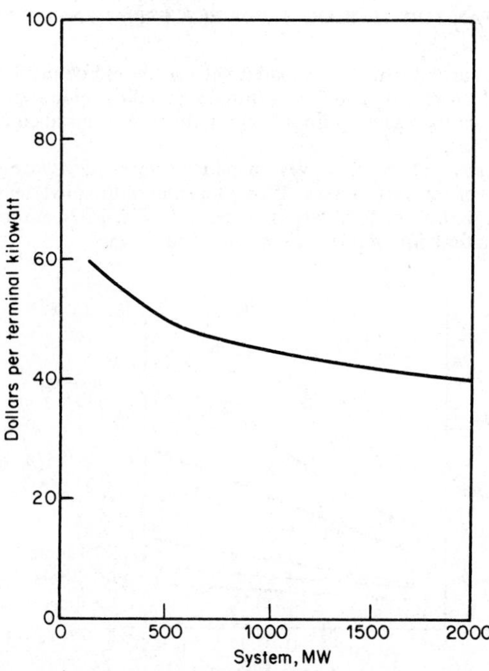


**FIG. 15-23** DC turnkey converter terminal cost (1985 dollars) versus bipolar rating.

**TABLE 15-2** Influence of Alternating Voltage on HVDC Terminal Costs

| Voltage, kV | Percentage of cost, % |
|---|---|
| 130 | 98 |
| 230 | 100 |
| 345 | 101 |
| 500 | 103.5 |
| 765 | 107.0 |

**TABLE 15-3** HVDC System Component Costs as a Percentage of Total Project Costs

| Component | Percentage of cost, % |
|---|---|
| Equipment: | |
| Converter transformer | 20–25 |
| DC valve (including controls and cooling) | 20–30 |
| Filter and var supply | 5–20 |
| Miscellaneous (communications, dc reactor, arresters, relaying, etc.) | 5–15 |
| Engineering (system studies, project management) | 2–5 |
| Civil work and site installation | 15–30 |

a somewhat higher current density compared with ac should be justified. A normal value for aluminum conductors is 0.75 to 0.90 A/mm$^2$, depending upon capitalized loss costs. It is not critical to reach the exact optimum, as a reduction in installed cost is counteracted by increased line losses.

***29. Station Losses.*** Thyristor valves introduce a forward voltage drop which constitutes the main part of the valve losses. These, together with converter transformer losses, form the dominant part of the total terminal losses. In Fig. 15-24 approximate component losses are given as a function of relative converter load.

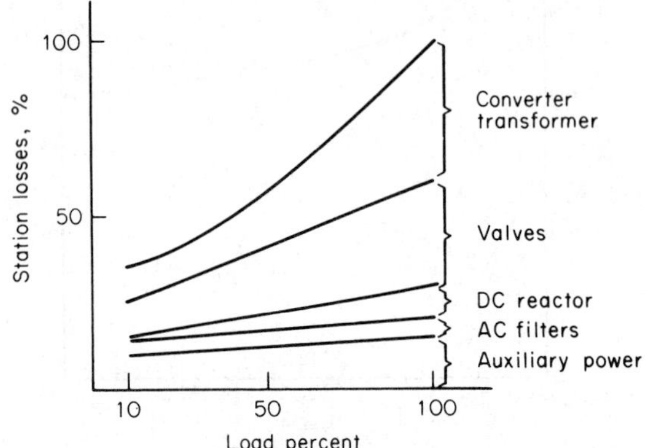


**FIG. 15-24** Distributed converter station losses as a function of load.

For a fully compensated converter station, efficiency can, to a large extent, be influenced by equipment design. Design, therefore, is a function of the loss evaluation. For modern HVDC terminals, 0.65 to 1.0% losses per terminal are typical.

## *STUDIES LEADING TO TERMINAL EQUIPMENT SPECIFICATIONS*

### Overview

***30. System Study Requirements.*** The design of an HVDC system requires coordinating performance characteristics of a great number of components as shown in Fig. 15-25. These components are characterized by complex electrical, mechanical and civil interactions which must be coordinated to meet the technical specifications, performance requirements, and specified availability and reliability characteristics. The system must operate within prescribed loss levels and at a reasonable cost. To achieve these objectives, a number of highly interactive studies are performed using sophisticated digital and analogue computer tools. The major system study elements that must be accomplished to assure a successful HVDC system design are listed in Table 15-4.

The following paragraphs discuss each major study element and pertinent relationships.

***31. Power System Studies.*** Power system studies are necessary to ensure that a dc system will interact with the existing ac system as anticipated.

DC power transfer studies determine the amount of power that can be transferred under normal and contingency operations. Var requirements and switching and control strategies are examined during these studies. AC system loading is examined under a variety of load and contingency situations. Reactive power studies determine the system var requirements in light of the total ac system, dc converter and specifications. The major parameters are

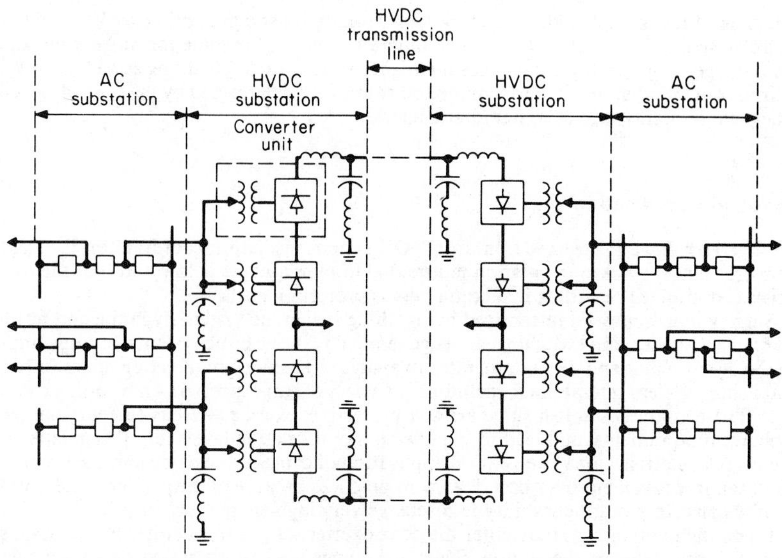


**FIG. 15-25** HVDC system components are characterized by complex electrical, mechanical, and civil interaction.

transformer impedance, range of converter angle operation, power transfer, and ac system characteristics. Reactive power requirements are met as required by switched static reactive compensation supplemented by dynamic reactive compensation. Static reactive compensation includes ac filters, shunt capacitors and shunt reactors. Dynamic reactive compensation may be supplied by static var control systems, synchronous condensers, and dc system control action.

Possible fundamental frequency overvoltage estimates are also determined. This information is used in establishing protection schemes, control action and insulation coordination during load rejection and faults. The ac stability limit can often be increased with the appropriate use of dc control because of the high relative speed at which dc power flow can be controlled.

HVDC controls are typically designed to meet specific operational and performance

**TABLE 15-4** Major System Studies

- Availability/reliability
- Power system studies:
  - DC power transfer
  - Reactive power requirements
  - Fundamental frequency overvoltage
  - Modulation
  - Control requirements
- Subsynchronous torsional interaction:
  - Torsional stability screening
  - Subsynchronous damping controller
- Insulation coordination:
  - Transient overvoltage
  - Station lightning protection
  - DC terminal arrester selection
- Filter design:
  - AC filter design
  - DC filter design
- Power line carrier and radio interference
- Electric field, audible noise
- Circuit-breaker and switch requirements
- Protective relaying
- Equipment specification
- Efficiency

objectives of the overall ac-dc system. System availability and protection are key to defining control systems. Basic converter control functions remain the same for most applications, and the type of application influences the higher-level control functions as well as any specialized control subsystems. Performance objectives are determined by the type of link, customer requirements, and ac system parameters.

## Insulation Coordination

***32. Overvoltage Protection.*** In an HVDC system insulation coordination is essential. It ensures that the risk of insulation failure due to overvoltages is low and that the overall design is optimized regarding insulation versus protection costs.

Surge voltages may be introduced by lightning strikes, ac system dynamic and ferroresonant overvoltages, and switching surges generated by the ac or dc systems. Studies employing the dc simulator and digital programs to analyze transient circuit action are essential for establishing system interactions, including the effect of surge arresters on resulting overvoltages. These studies establish surge arrester protective levels, as well as current and energy discharge requirements, and lead to the selection of insulation levels for all elements of the dc converter station. Converter station application of gapless zinc oxide surge arresters represents an improvement over past dc system practices because of the greater protective levels of the arresters and their ability to discharge very high energy surges.

Lightning overvoltages may enter the dc converter station from either the ac or dc side from strikes to lines or the station. Switching surges may be generated on the ac side by circuit-breaker action or from line faults, and on the dc side by pole-to-ground faults. Dynamic overvoltages on the ac side result from loss of load on a converter and consequent reduction of var consumption. Ferroresonant overvoltages result from transformer saturation following faults and the subsequent action of harmonic currents on the ac system parameters. Certain converter conditions can generate overvoltages, and all foreseeable contingencies of control and circuit action must be considered.

A typical protective scheme for a back-to-back converter station is shown in Fig. 15-26. The major elements of equipment are indicated, and all surge arresters are identified (i.e., 6.2, 6.3, etc.). The basis of surge arrester selection is that all circuit locations are protected from overvoltages from any cause. Surge arrester discharge capabilities are established based on contingencies and resulting circuit behavior.

Typical minimum criteria for insulation coordination margins are 15% for switching surges and 20% for impulse surges.

***33. Creepage Distance.*** A summary of current practice regarding creepage distances in HVDC stations is available.[1] Table 15-5 summarizes the range of values that have been utilized. Those ratings in parentheses are values that are generally used. For indoor and outdoor creepage distances, the lower values of the range have generally provided satisfactory performance and have been selected for current projects. The higher values were used in some indoor cases where there was particular concern with dust and related issues of filtering valve hall air. In the case of outdoor creepage distances, the higher values were used due to the incidence of salt deposits for installations near sea coasts.

## AC Harmonic Filters

***34. Generation of Harmonic Currents.*** Converters generate harmonic currents and force them into the ac system. AC harmonic filters reduce the harmonics flowing into the ac system by providing a low harmonic impedance to ground. Additionally, they provide part of the reactive power consumed by the converter.

AC harmonic filter design involves calculating the harmonic current generated, devel-

---

[1]Survey of Creepage Distances and Clearances as in HVDC Converter Stations, *IEEE Working Group* 79.5, IEEE Paper 85 WM 164-9.

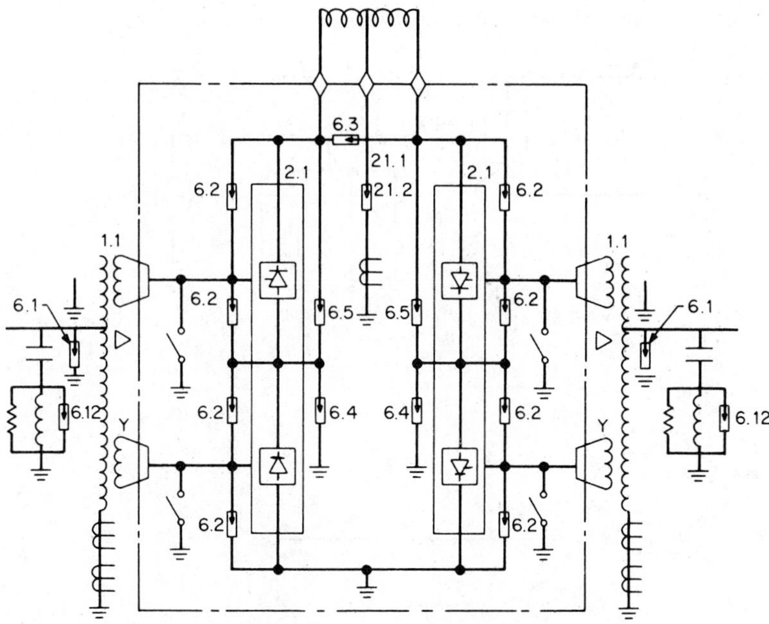


**FIG. 15-26** This line diagram illustrates the components required for an HVDC back-to-back converter station.

**TABLE 15-5** Creepage Distance in HVDC Stations, cm/kV

| | Insulator | Bushings transformer | Wall |
|---|---|---|---|
| Indoor | 1.4–2.3<br>(1.5) | 1.4–3.0<br>(1.4) | 1.4–3.0<br>(1.4) |
| Outdoor | 2.3–5.1<br>(2.6) | 1.85–3.9<br>(2.3) | 1.83–6.3<br>(2.5) |

oping filter configuration, establishing switching strategy for var banks, estimating harmonic impedance characteristics of the ac system, and calculating filter performance and rating.

An HVDC converter can be modeled as a harmonic current source, as shown in Fig. 15-27. The major harmonic currents generated by the 12-pulse converter under balanced conditions are for characteristic harmonics of order:

$$h = 12n \pm 1 \qquad n = 1, 2, 3, 4, \ldots$$

The harmonic current of order $h$ has a magnitude of less than $I_1/h$ where $I_1$ is the fundamental current amplitude. The magnitude of harmonic current also varies with the firing angle and the dc load level.

Under unbalanced conditions, noncharacteristic harmonic currents are also generated, although their magnitudes are typically smaller (0.1 to 0.3%). Odd-order harmonic currents are generated if the ac bus voltage or the converter transformer are unbalanced, and even-

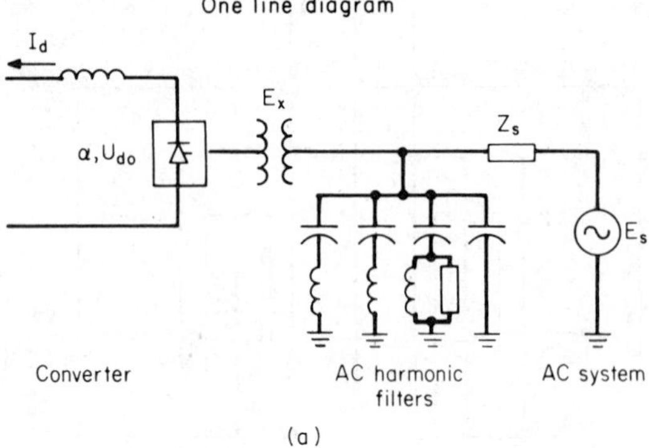


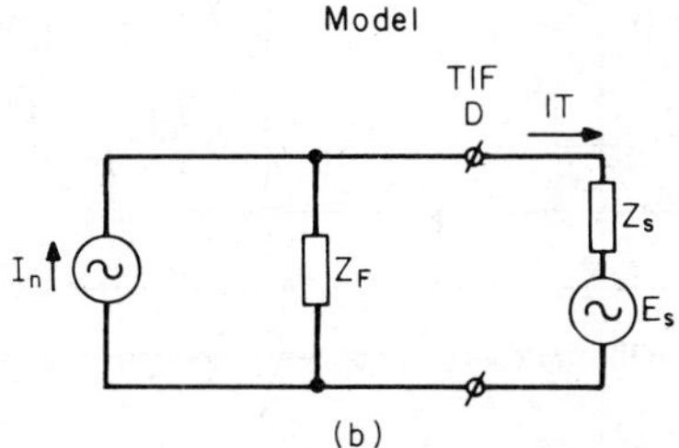


**FIG. 15-27** (*a*). One-line diagram of an HVDC converter, ac harmonic filters, and ac system. (*b*). Model of the one-line diagram with the HVDC converter modelled as a harmonic current source.

order harmonic currents are generated if the valve firing pulses jitter. Typical ranges of the parameters and variations encountered are ac bus voltage unbalance, 1 to 2%; valve firing angle, 10 to 80°; valve firing jitter, 0.1 to 0.2°; transformer impedance, 0.1 to 0.2 per unit; and transformer impedance unbalance, 1 to 3%.

Figure 15-28 shows characteristic harmonics as functions of the dc load level for a fixed firing angle.

***35. Types of Filters.*** Bandpass filters, double bandpass filters, and high-pass filters are configurations normally encountered in HVDC systems. Bandpass filter (Fig. 15-29) characteristics are simplicity, good performance, and low losses (i.e. high impedance) at other frequencies.

The impedance of a double bandpass filter (Fig. 15-30) is low at two tuned frequencies, and the impedance characteristics are similar to the net characteristics of two single bandpass filters. These filters can offer a practical and economic solution when specified performance measures and harmonic characteristics of the ac system necessitate filters tuned to low-order harmonics.

A high-pass filter (Fig. 15-31) is insensitive to the detuning effects of temperature and source frequency variations. Associated losses are high, especially for low-order harmonics. Normal use would be a low-impedance path for higher-order harmonics, in conjunction with bandpass filters for each of the larger low-order harmonics (e.g., characteristic harmonics at the 11th and 13th).

AC filters normally comprise 40 to 50% of the required converter compensation. These filter banks, shunt capacitor banks, and reactors are switched depending on the ac system reactive power demand and the converter's var consumption. The maximum Mvar sizes of

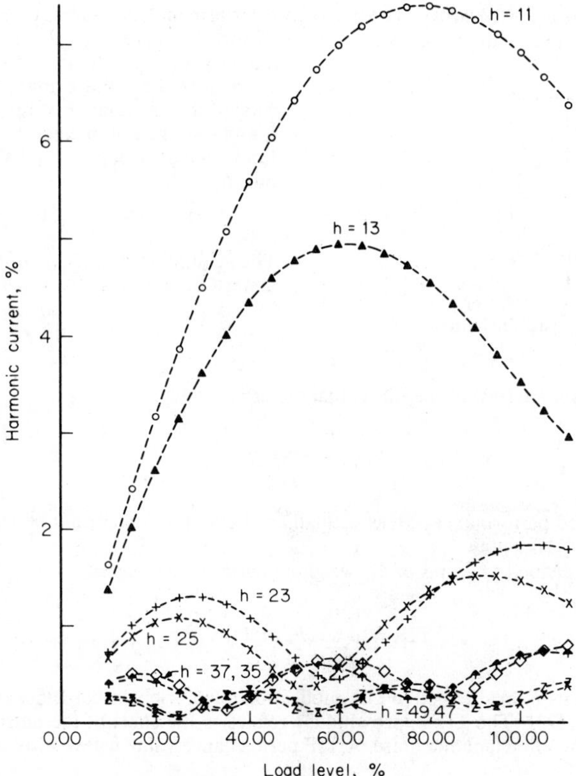


**FIG. 15-28** Characteristic harmonics as functions of the ac load level for a fixed firing angle.

**FIG. 15-29** Bandpass filter circuit.

**FIG. 15-30** Double bandpass filter circuit.

switched filter banks, capacitor banks, and reactors are limited by the largest permissible converter bus voltage variation.

Proper evaluation of system harmonic impedance is important since it affects the size of the filter required to achieve a desired performance level. AC system harmonic impedance can be determined by actual measurements on the system, measurements on a scale model, calculations based on a system model, or realistic assumptions.

***36. Performance Measures.*** Three types of performance measures have been used in HVDC system specifications. They are voltage distortion, telephone influence factor (TIF), and IT product. Individual distortion ($D_h$) is defined as the voltage magnitude ($V_h$, kV rms) of the harmonic $h$ at the converter bus (or a power line of interest) in percent of the fundamental voltage ($V_1$, kV rms) of the bus, or

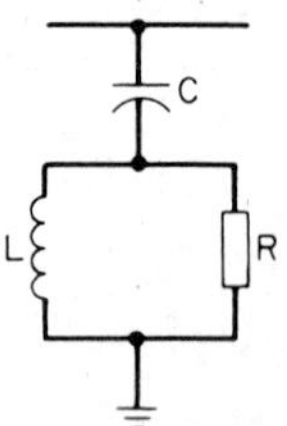


**FIG. 15-31** High-pass filter circuit.

$$D_h = 100 \times V_h/V_1 \quad \% \qquad (15\text{-}23)$$

The arithmetic sum ($D_{\text{sum}}$) of the individual distortion values is

$$D_{\text{sum}} = \sum_{h=2}^{49} D_h \qquad (15\text{-}24)$$

The root sum square (rss) of the individual distortion values is

$$D_{\text{rss}} = \left( \sum_{h=2}^{49} D_h^2 \right)^{1/2} \qquad (15\text{-}25)$$

Typical specified performance values with all ac filters in service would be $D_h = 1.0\%$, $D_{\text{rss}} = 2.0\%$, and $D_{\text{sum}} = 4.0\%$.

The TIF is defined as an rss of the weighted harmonic voltages:

$$\text{TIF} = \left[ \sum_{h=2}^{49} (F_h \times V_h/V_1)^2 \right]^{1/2} \qquad (15\text{-}26)$$

where the weights $F_h$ are defined in EEI publication 60–63, September 1960, and graphically shown in Fig. 15-32. The TIF approximately relates to the effect of the harmonic voltages of the power line on telephone noise. A TIF performance value below 35 is acceptable.

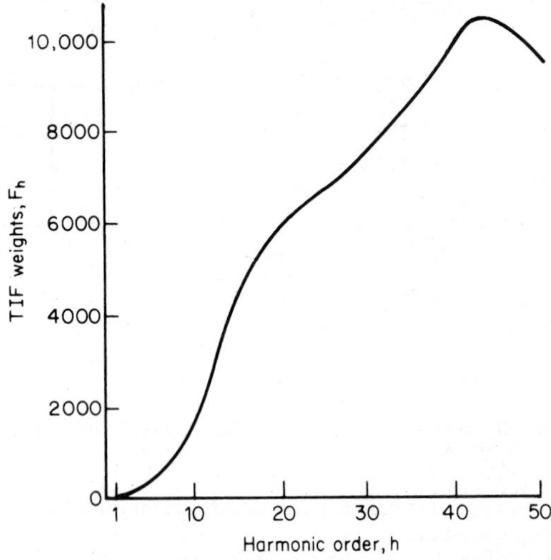


**FIG. 15-32** TIF weights. (*EEI Publication 60-63, September 1960.*)

The IT product is defined as an rss of the weighted harmonic currents:

$$\mathrm{IT} = \left[ \sum_{h=2}^{49} (F_h I_h)^2 \right]^{1/2} \tag{15-27}$$

where $I_h$ is the rms value of the $h$th harmonic current, and the same weights $F_h$ are used as for the TIF. Sometimes, kIT (= IT/1000) is used. The IT product approximately relates to the effect of the harmonic currents flowing in the power circuit on telephone noise.

## DC Harmonic Filters

***37. Design Criterion.*** Filters are required on the dc side of converters to limit interference with communication circuits in the neighborhood of the dc transmission line. This section focuses on filters applied to limit interference in the voice-frequency spectrum (e.g., 100 Hz to 5 kHz). Such equipment is usually referred to as harmonic filters, as opposed to the power-line carrier (PLC) and radio interference (RI) filters discussed in other sections.

The design criterion for dc harmonic filters is a function relating the flow of harmonic currents at any point along the dc line to the interference with adjacent telephone lines. Significant parameters are the location of telephone lines with respect to the power line, their shielding, the presence of any ground wires on the dc tower, and the earth's resistivity. These parameters should be taken into account when planning an HVDC system to determine the criterion for dc harmonic filter design. Such a criterion is typically expressed as equivalent disturbing current or induced voltage in a test line.[1,2] Since interference is inherently lower in the balanced bipolar mode of operation than in the monopolar mode and since the monopolar mode is atypical, a higher level is usually specified for the monopolar operating modes.

To design the dc harmonic filter, the entire dc network must be modeled to calculate the parameters for the criterion.

The harmonic generation of the converter must be determined, considering the full range of operating conditions and all sources of unbalance. The unbalances in ac voltage, transformer impedances, and operating conditions between poles create noncharacteristic harmonics. These noncharacteristic harmonics, particularly those arising from unbalances between poles, can contribute substantially to interference and hence to required filter size.

It is important to coordinate the dc harmonic filters with the smoothing reactor size for all converters on the dc system. In addition to the voice-frequency interference criterion, dc system resonance near low-order harmonics must be avoided.

## System Protection

***38. AC Protection.*** The philosophy of protective relaying on equipment associated with HVDC stations is similar to that of any important HVAC station. It is essential that each piece of equipment be protected with high-speed primary protection. The highest speed protection is usually a closed-zone differential-type protection, since time delay for coordination with other system protection is then unnecessary. This protection must detect all types of faults in the equipment and isolate the smallest portion of the system or station possible, consistent with available breakers and switching equipment.

Additionally, each piece of equipment must be included in a second zone of protection to provide backup. It is imperative that no less than two protective relays detect any given

---

[1]Fletcher, D. E., and Patterson, N. A.: The Equivalent Disturbing Current Method for DC Transmission Line Inductive Coordination Studies and DC Filter Performance Specification, *Proceedings of International Conference on DC Power Transmission,* June 4–8, 1984, Montreal, Quebec, Canada, pp. 198–204.

[2]Lasseter, R. H., et al.: DC Filter Design Methods for HVDC Systems, *IEEE Trans.,* March/April 1977, vol. PAS-96, pp. 571–578.

fault. It is not necessary to have one specific backup relay for each primary, however, since one may provide backup for several primary relays. The timing of the local backup protection must be consistent and coordinated with any other system relays which may also detect a fault (Fig. 15-33).

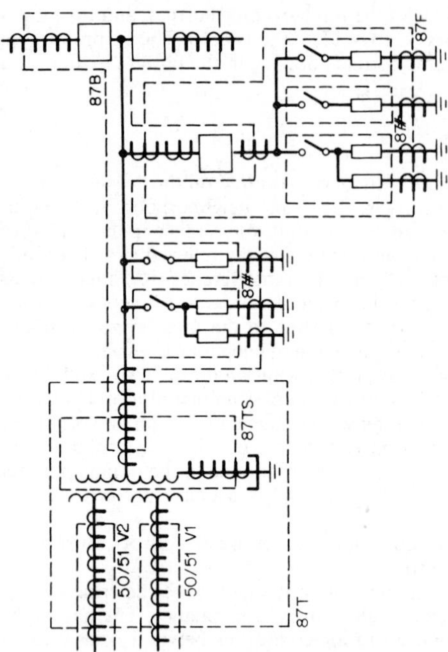


**FIG. 15-33** Primary zones for protection in an HVDC system are the converter zone, ac line zone, neutral zone, and, if used, a metallic return zone.

The studies required to determine protective relay types, ranges, protective levels, and operating times include the following:

**a.** Short-circuit analysis for maximum and minimum generation conditions including both phase and ground faults.

**b.** Coordination studies on adjacent lines or nearby stations to assure proper operating times.

**c.** Relay–current transformer–potential transformer circuit analysis to determine relay settings.

**d.** Voltage distribution studies for capacitor and filter banks.

***39. DC Protection.*** Protection of the dc link, from the valve side of the converter transformers on, is largely contained in the converter control. The protective functions may be broadly categorized as dc power circuit protection for faults or valve protection for malfunctions or conditions which might overstress the valve.

The purpose of protection devices is to detect and take appropriate action for faults and other abnormalities in the circuit. Protection is necessary to safeguard the equipment and circuit, and should encompass reasonable contingencies while minimizing the operational impact of abnormalities. Primary and backup protective circuits and redundant trip relaying are employed to protect against any single contingency failures. Redundant ac and dc current transformers are provided for critical protection zones. Protection is on a per-pole basis to minimize the probability of a single fault or failure affecting both poles. Protection zones are defined to take selective action based on the location of a fault.

Communications between terminals may be used for some of the protective functions to enhance performance, but they are not relied upon for system protection.

DC power circuit protection encompasses the entire dc circuit bounded by ac power transformer secondaries at each terminal. Each terminal protection region is divided into three major zones, plus a fourth, if metallic return is used:

Converter zone

DC line zone

Neutral zone

Metallic return zone

All protection sequences involve only one pole. It is possible to clear and isolate any fault limited to one pole by stopping only that pole. Depending on location, faults in the neutral zone may involve both poles, but immediate shutdown is not normally required.

In addition to circuit fault protection, specific valve protection features are normally incorporated in the control as follows: ac undervoltage, ac overvoltage, snubber thermal protection, loss of control power, unordered power reversal, loss of control current (current exceeds order), prolonged gating without current, time-overcurrent protection, valve thermal protection, and smoothing reactor protection.

### Communication

***40. Telecontrol Considerations.*** Telecontrol for converter terminal operation generally consists of channels for supervisory control and data acquisition (SCADA), emergency voice communication between converter terminals and other remote locations, and communication for control and protection purposes. SCADA equipment facilitates remote control of the converter terminal equipment. Voice communication is required between converter terminals and the remote dispatch office for testing purposes and occasional emergency manual load control operation. Control communication is provided for the damping signal control function and requires high signal security and dependability.

A typical overall telecontrol one-line diagram is illustrated in Fig. 15-34. SCADA is shown between the system dispatch center and the converter terminals. Interterminal communication, between converter terminals, provides the control and protection signaling requirements. SCADA may also be used between converter terminals.

The telecontrol system is generally configured to communicate simultaneously over alternate communication paths. The telecontrol system is installed at the pole level in a hierarchical structure, as shown in Figure 15-35.

Telecontrol system terminal equipment provides the encoding, decoding, and modulation of signals. Security requires a very high probability that the equipment *never* give a false output without a signal, whereas dependability requires that the equipment *always* give an output when a desired signal is sent. Likewise, equipment availability (mean time between failure), redundancy, and signaling speed are critical factors to evaluate. Signaling speed is a function of the modem bandwidth employed by the telecontrol system terminal equipment, which, in turn, directly affects the number of voice channels and baud rates required for HVDC telecontrol needs.

The communication channel equipment may be a single-function, single-sideband power-line carrier, microwave, or leased telephone. The trade-off factors to evaluate for each alternative include speed, signal-to-noise ratio (S/N), number of voice channels required, signal update time, signal security and dependability, and cost.

Voice communication between converter terminals usually requires only one full-duplex voice channel. SCADA for remote control, metering, and alarms also normally requires only one full-duplex voice channel, since these functional requirements can be slow (in the order

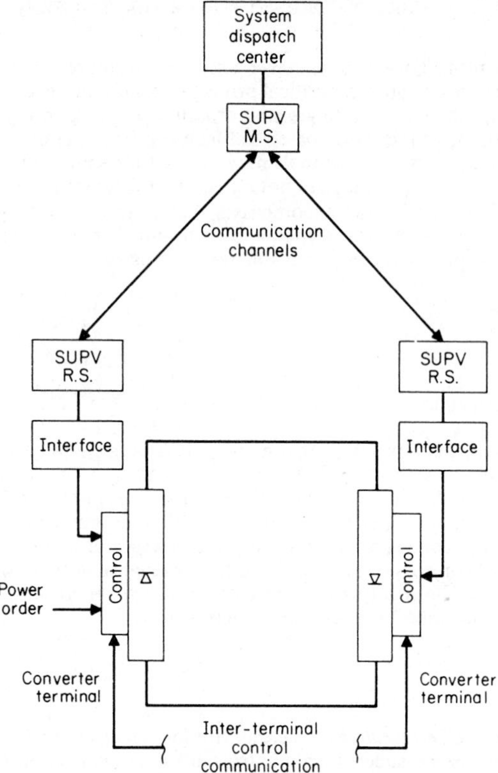


**FIG. 15-34** A typical overall HVDC telecontrol system covers communication between the dispatch center and the converter terminals as well as interterminal communication.

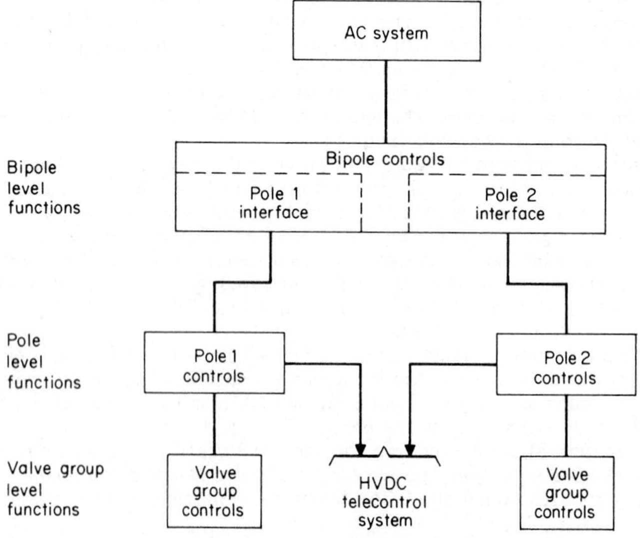


**FIG. 15-35** Telecontrol systems are installed in a hierarchical structure.

of seconds). HVDC telecontrol requirements include control and protection signals, with update times varying between 20 to 50 ms. These telecontrol signals are provided for each pole.

## Switching

***41. Overvoltage and Transient Current Studies.*** Switching studies for equipment in an HVDC station are necessary due to the potential for high voltages and high transient currents. Extra stress on switches can be caused when station equipment such as capacitor and filter banks is energized, deenergized, or left connected when HVDC load is dropped. Back-to-back switching of capacitive elements in the station and faults on the station bus can create exceptionally high currents in the equipment. Studies include examples of overvoltage, to ensure switch and breaker capability to interrupt at the highest voltages possible, and transient current, to ensure the latching and withstand capabilities of these switches.

## Reliability

***42. Performance Measure.*** Energy availability, the key reliability performance measure for HVDC systems, measures the fraction of continuous rated capacity available, averaged over the time period being considered. Energy availability is computed by expressing outage time in terms of equivalent hours of full outage, as shown in the following equations:

$$\begin{matrix}\text{Equivalent} \\ \text{outage time} \\ \text{(EOT)}\end{matrix} = \begin{matrix}\text{actual} \\ \text{outage} \\ \text{time}\end{matrix} \times \begin{matrix}\text{fraction of} \\ \text{capacity on} \\ \text{outage}\end{matrix} \tag{15-28}$$

$$\begin{matrix}\text{Energy} \\ \text{unavailability} \\ \text{(EU)}\end{matrix} = \frac{\text{equivalent outage hours}}{\text{total time}} \times 100\% \tag{15-29}$$

$$\begin{matrix}\text{Energy} \\ \text{availability} \\ \text{(EA)}\end{matrix} = (100 - \text{energy unavailability})\ \% \tag{15-30}$$

The equivalent outage hours for the period is the sum of the equivalent outage times for individual outages.

***43. Performance Data.*** Reliability performance data on HVDC systems have been collected by CIGRE (International Conference on Large High Voltage Electric Systems). Table 15-6 lists the complete record of CIGRE reports which contain data for solid-state HVDC systems.

The average annual energy availability by system for the 11 systems reporting to CIGRE through 1982 ranges widely from 88.4 to 98.8%. In a particular system, energy availability

**TABLE 15-6** Solid-State HVDC Reliability Data Base

| CIGRE report | Operating years covered | No. of systems reporting | Cumulative experience system-years |
|---|---|---|---|
| 14-09(1974) | 1972 | 1 | 1 |
| 14-08(1976) | 1973–1974 | 1 | 3 |
| 14-08(1980) | 1975–1978 | 8 | 20 |
| 14-06(1982) | 1979–1980 | 10 | 39 |
| 14-06(1984) | 1981–1982 | 11 | 60 |

depends on operation and maintenance practices as well as equipment design reliability. If systems are excluded for which reported statistics appear to be largely influenced by operation and maintenance, the range of average system energy availability is 96.2 to 97.6%, and the average annual energy availability is 97.2%, corresponding to an energy unavailability of 2.8%. These results include both terminals, but transmission line outages have been excluded. Subsequent discussion is based on these selected data.

Outages are classified by CIGRE as forced or scheduled. Forced outages are unplanned maintenance events which cannot be deferred at least 1 week after the need for maintenance arises. Scheduled outages comprise all other maintenance. Thus, scheduled outages include both planned maintenance (e.g., annual shutdown) and unplanned maintenance which can be deferred at least 1 week.

Figure 15-36 shows the average energy unavailability by year for 1977 through 1982 and indicates the breakdown of energy unavailability between forced and scheduled outages. For

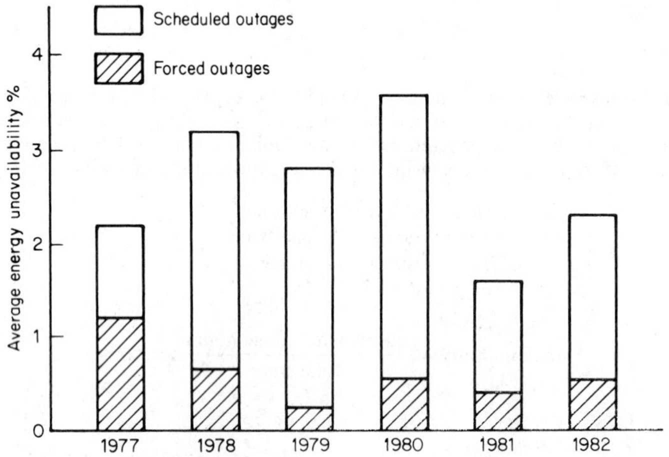


**FIG. 15-36** Average energy unavailability by year is based on CIGRE data for 1977–1982. Systems with large operation and maintenance effects are excluded.

the period covered by Fig. 15-36, forced outages contribute 23% of the total energy unavailability, while scheduled outages contribute 77%.

Forced outages are further classified by CIGRE into five categories as follows:

| | |
|---|---|
| AC-E | AC and auxiliary equipment |
| V | Valves, including valve auxiliaries |
| C + P | Control and protection equipment |
| DC-E | Other primary dc equipment |
| O | Unknown cause and operator error |

Figure 15-37 shows the percentage contribution of each category to forced energy unavailability for each documented year. The final bar shows the actual percentage for each category for the entire 6-year period.

Reliability is particularly important to the success of HVDC systems because outages can have a very large economic impact. For example, if replacement energy cost is \$30 per megawatthour, a 1% increment in energy availability on a 1000-MW system has an annual cost of

$$30(1000)(24)(3.65) = \$2,628,000 \text{ per year}$$

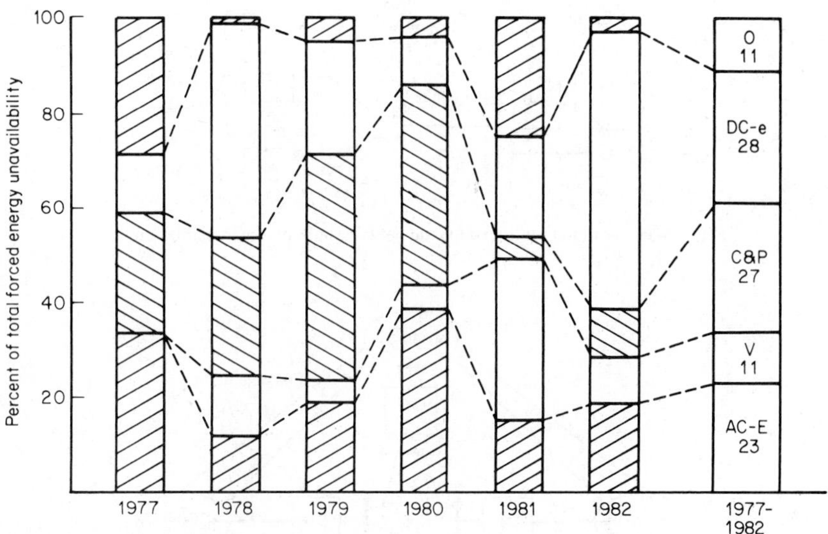


**FIG. 15-37** Forced energy unavailability by equipment category is based on CIGRE data for 1977–1982. Systems with large operation and maintenance effects are excluded.

Dividing the annual cost by the fixed charge rate (carrying charge) gives the capitalized value of the annual cost. Assuming a fixed charge rate of 0.20 (20%),

$$\frac{2,628,000}{0.20} = \$13,140,000$$

This is the capital investment that can be justified by a 1% increase in energy availability.

## Multiterminal Considerations

***44. Multiterminal DC Systems.*** Successful applications of point-to-point dc links worldwide have shown power system planners that three-terminal dc (3TDC) and multiterminal dc (MTDC) schemes may be the vehicles to fully utilize economic and technical advantages of HVDC technology in the future.

MTDC systems may be planned in advance or evolve from expansion of operating two-terminal dc (2TDC) links. In the latter case, advanced planning can often be advantageous in minimizing required modification, thereby minimizing outage of the link during expansion.

***45. Series or Parallel Schemes.*** There are two possible connection schemes for MTDC systems: a constant voltage parallel scheme and a constant current series scheme. In the parallel scheme, converters are connected in parallel and operate at a common voltage. Here, the dc network may be of either the radial (Fig. 15-38) or mesh type (Fig. 15-39). In the series scheme, converters are connected in series with a common direct current flowing through all terminals. The parallel connected scheme is discussed here, since it is widely accepted as the most practical configuration with the fewest operational problems.

***46. Potential Applications and Benefits.*** Potential applications for MTDC systems are similar to those for 2TDC systems and can be classified under three basic categories:

Bulk power transmission—where low-cost energy from several power plants is transmitted over a long distance to different ac systems (Fig. 15-38).

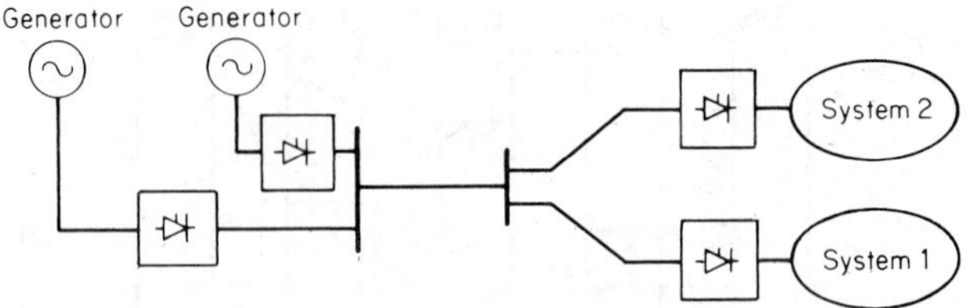


**FIG. 15-38** A radial multiterminal dc network bulk power transmission.

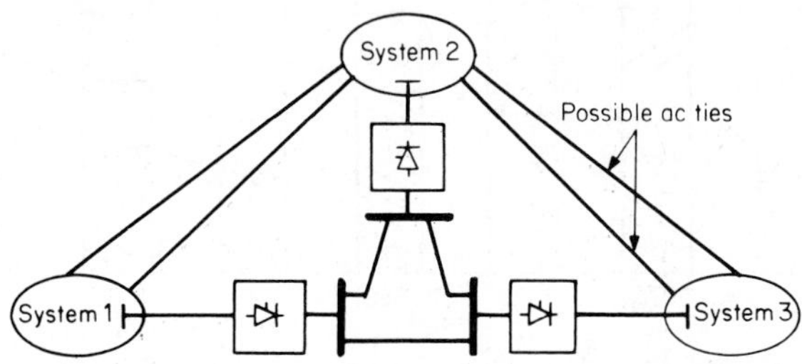


**FIG. 15-39** A mesh dc network used as an ac network interconnection.

AC network interconnection over a long or medium distance—where generation-load balancing and sharing of spinning reserves are of primary concern (Fig. 15-39).

Reinforcement of an ac network—where limited ac expansion possibilities exist. Energy from a new power plant is fed to different locations of an ac system, usually metropolitan (Fig. 15-40).

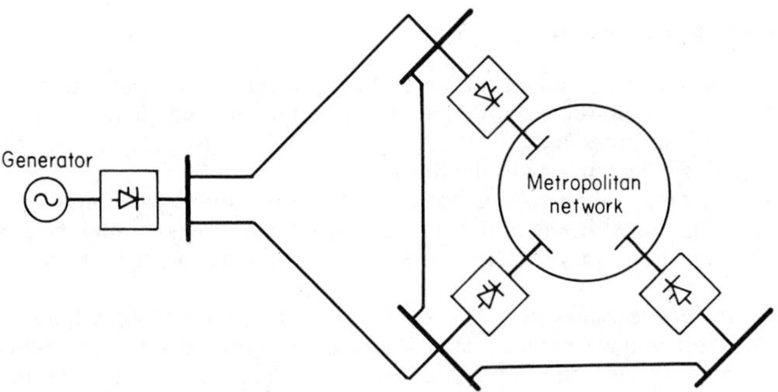


**FIG. 15-40** A multiterminal dc system reinforces an urban network.

MTDC systems offer both economical and technical advantages over several equivalent 2TDC systems. The primary economic advantages are as follows:

The total installed converter rating in an MTDC system is usually less than that of several equivalent 2TDC systems.

MTDC systems offer lower-cost transmission lines and/or cables.

The inherent overload capability of MTDC transmission lines can increase the capacity of transmission corridors.

Technical advantages of MTDC systems include:

MTDC systems provide greater flexibility in dispatching transmitted power. In mesh dc networks, the inherent overload capability of transmission lines allows for more flexible dc power transfer patterns.

In a larger interconnected power system, MTDC systems can provide a powerful control action to damp out troublesome electromechanical oscillations.

In conjunction with phase shifting transformers and generation shifting, MTDC systems may be used to enforce desired power flow patterns in a large interconnected power system.

***47. Control of MTDC Systems.*** The fundamental control principle for MTDC systems is a natural generalization of that for existing 2TDC systems. That is, each converter is controlled according to control characteristics virtually identical to those for 2TDC systems (see Pars. **104** to **110**). As with 2TDC systems, there is a current margin requirement in MTDC systems: the sum of the rectifier current orders must exceed the sum of inverter current orders by a value known as current margin.

In normal operational mode for a given power transfer dispatch, a selected terminal establishes a normal direct voltage profile throughout the system. This voltage setting terminal (VST) operates on either constant voltage control or on constant angle control, exhibiting a nearly horizontal characteristic. Other terminals in the MTDC system operate on constant current, i.e., the vertical characteristic. Generally, the horizontal segment of the VST characteristic is at a relatively lower voltage than those of the current controlling terminals. Also, the current at the VST differs from its respective reference current by the current margin. If the VST is an inverter, the actual current is greater than the respective reference current by the current margin. If the VST is a rectifier, the actual current is less than the reference current.

These relationships are shown in Fig. 15-41 for a four terminal dc system (4TDC). In Fig. 15-41*a*, inverter terminal four is the VST, while the other converters are on current

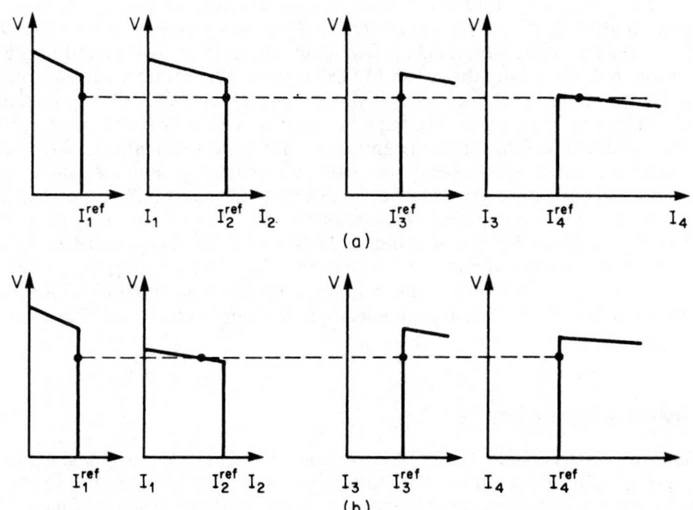


**FIG. 15-41** Control characteristics for a 4TDC system. The voltage setting terminal (VST) is the inverter terminal 4 for (*a*) and the rectifier 2 for (*b*).

control. The current at terminal four is therefore more than the reference current, as previously described. In Figure 15-41*b*, it is assumed that the alternating voltage for rectifier terminal two has dropped and, therefore, terminal two becomes the voltage-setting terminal. As a result, its operating current is decreased by the current margin. In this backup mode of operation, the previous VST (inverter terminal four) operates on constant current control.

Unlike 2TDC systems, the direct voltage polarity in MTDC systems may not be reversed unless it is desired to change power flow direction at every terminal in the system. Therefore, should any of the converters be changed from an inverter to a rectifier, or vice versa, its current must be reduced to zero, a polarity-reversing switch must be operated there, and, finally, the converter current must be increased to the desired value. Coordination of current orders to ensure availability of sufficient current margin in the dc system is an essential requirement for such maneuverings.

***48. Fault Clearing and Switching.*** In case of a fault in the dc transmission network, three methods are available for clearing the fault. The methods are based on the kind of switching device used to isolate the faulted line. The most versatile and flexible method is the use of dc breakers capable of interrupting the maximum fault current that the system can produce.

The breakers should be designed to react rapidly to a relay signal and interrupt the fault current in a matter of milliseconds. A rapid and accurate fault-sensing or relay system is usually required as well. Fault sensing is local and structured similarly to one of the schemes presently used in ac systems.

If the faulted line radially connects a converter to the remainder of the MTDC system, its disconnection would result in isolation of the converter from the system. Consequently, coordinating current orders of the remaining converters in the MTDC system is necessary and should be done centrally, via high-speed communication, to achieve the best performance. By using the breaker scheme, a fault in the dc transmission network may be cleared in approximately 10 ms.

Another method of clearing a fault is by using load break switches in conjunction with the converter controls. The switch is rated to interrupt direct currents as high as normal operating current—not the fault current. When a fault occurs on the dc transmission network and no other control action is implemented, the converter controls, by use of current regulators, reduce the fault current to less than normal values. The faulted line can then be isolated from the system by activating load break switches, in about 100 ms, and MTDC system operation may then be resumed. The requirements for balancing current orders in this case are similar to those for fault clearing by a dc breaker.

The third method forces all the converters to temporarily operate as inverters as soon as the fault is detected. The stored energy in the dc system is then delivered to the ac systems, resulting in deenergizing the entire MTDC system. When direct currents are driven to zero, the faulted line is disconnected by high-speed isolators. The MTDC system can then be restarted and resume operation. The requirements for current order balancing are similar to those discussed earlier. This protective measure may be implemented in 200 to 300 ms.

For MTDC systems that are energy ties, the third method of fault clearing is generally acceptable. The scheme, however, may not be desirable for an MTDC system with a relatively large number of converters and transmission lines. The scheme may even be unacceptable when the 200- to 300-ms period of MTDC shutdown may jeopardize rotor-swing stability of generators in the vicinity of the dc terminals. Disconnecting the faulted line by load break switches or dc breakers is more appropriate for these situations. Generally, the appropriate method of fault clearing is application-dependent and should be selected after sufficient system studies.

## Subsynchronous Oscillations

***49. Torsional Interaction.*** HVDC transmission systems and turbine generator shaft torsional systems can interact in an unfavorable manner. This interaction, which appears as subsynchronous oscillations in the electrical systems, can lead to shaft damage. The first step in studying the torsional interaction phenomenon is to determine which, if any, units in the electrical vicinity of an HVDC terminal have the potential for significant interaction.

The greatest amount of interaction occurs with a turbine-generator unit feeding an HVDC line radially, i.e., with no parallel ac connections. Conversely, a unit connected to an ac transmission system several hundred miles away from an HVDC terminal is unlikely to experience any interaction. An approximate relationship between the magnitude of the interaction and the ac system strength has been developed. This relationship can be used as a quantitative screening tool to identify units and system contingencies requiring detailed studies:

$$UIF_i = \frac{MVA_{HVDC}}{MVA_i}\left(1 - \frac{SC_i}{SC_{tot}}\right)^2 \qquad (15\text{-}31)$$

where $UIF_i$ = unit interaction factor of $i$th unit
$MVA$ = rating as per subscript (HVDC or $i$th unit)
$SC_{tot}$ = short-circuit capability at HVDC commutating bus including $i$th unit
$SC_i$ = short-circuit capability at HVDC commutating bus excluding $i$th unit

The results of extensive sensitivity studies suggest that a unit with an interaction factor less than about 0.1 will not have significant interaction and can be neglected in further studies.

***50. Subsynchronous Damping Controller.*** It has also been determined that designing an HVDC control to ensure positive damping for the unit with the greatest interaction also ensures positive damping for all other units. This is because the magnitude of the interaction varies with the strength of the ac transmission system between a given unit and the HVDC line, while the phase remains relatively fixed. Thus, designing a control for the most extreme case of radial HVDC operation represents a complete solution with respect to changes in the ac transmission network.

If it is found from the screening test that the HVDC system and generator interactions must be evaluated in more detail, studies such as those in EPRI report EL-2708 should be conducted. In this report, methods have been developed for designing a subsynchronous damping controller (SSDC), and its performance and robustness demonstrated on both digital computer programs and an analogue scale model HVDC simulator.

An SSDC can change the inherent negative damping characteristic of the HVDC system to one of positive damping in the subsynchronous torsional frequency range. In fact, an HVDC system with an SSDC can provide more positive damping to turbine-generator rotor torsionals than a conventional ac transmission system. In some situations, an SSDC can be designed to provide several times the amount of damping required to maintain stability, thereby increasing the rate of decay of transient torsional oscillations.

***51. Influence on Planning.*** Planning for the location and size of major transmission and generation facilities should not be influenced by concerns over the HVDC torsional interaction phenomenon. Major equipment can be chosen and sited for optimum economics and enhancement of steady-state power flows and transient stability by HVDC control. The influence of torsional interaction should be viewed as a potential side benefit by reducing fatigue duty on the shaft. This contrasts with the problems associated with series-compensated ac transmission systems, where the subsynchronous resonance (SSR) interaction can limit the amount of power transfer through an ac line or require additional equipment to allow higher levels of compensation.

## PLC and RI Filters

***52. Noise Sources and Noise Level Calculations.*** Electrical noise is produced by HVDC converter stations in the power-line carrier frequency band from 30 to 500 kHz and in the radio frequency band from 500 kHz to 300 MHz. This noise may be conducted on ac or dc power lines connected to the converter station or may be radiated from the converter station and connecting transmission lines. In determining the need and configuration of a power-line carrier–radio interference (PLC-RI) filter, the following must be considered: noise developed by the converter, permissible noise on the system, additional noise sources, and methods for reducing converter-conducted and -radiated noise.

Noise developed by the converter depends on the magnitude of the commutation transient and in turn on the magnitude of the dc voltage. Noise voltages on the line side of the converter are influenced by the internal resonances of the converter transformer on the ac side and the smoothing reactor on the dc side. If valve noise level in the PLC and RI range are known for an operating system, approximations of noise levels at the converter terminals for a new system with different dc voltage can be determined using the following relationships:

$$\text{dB carrier correction} = 20 \log \left(\frac{\text{BW}}{\text{BW}_b}\right)^{1/2} + 20 \log \frac{V_{\text{dc}}}{V_{\text{dc}b}} \tag{15-32}$$

$$\text{dB RI correction} = 20 \log \left(\frac{\text{BW}}{\text{BW}_b}\right)^{1/2} + 20 \log \frac{V_{\text{dc}}}{V_{\text{dc}b}} + 10 \text{ dB} \tag{15-33}$$

where BW = bandwidth of measurement for system of interest, kHz
$\text{BW}_b$ = bandwidth for base case, kHz
$V_{\text{dc}}$ = dc voltage of system of interest, kV
$V_{\text{dc}b}$ = base system's dc voltage, kV

The implementation of carrier noise filters is a practical method for reducing HVDC converter-conducted noise. Series noise filters are effective in reducing the carrier frequency noise conducted, induced, or radiated from ac and dc transmission lines. This method works well down to carrier frequencies as low as 30 kHz. Noise filters should be installed in each phase on the ac line side, and in each pole and electrode line on the dc line side. An example of typical PLC noise limits for 500-kV ac and dc lines and 230-kV ac lines is illustrated in Fig. 15-42.

There are three aspects of radio interference to consider: noise levels along the transmission lines, noise levels around the converter premises as a result of the conversion process, and background noise due to corona and atmospheric conditions on the line. The elec-

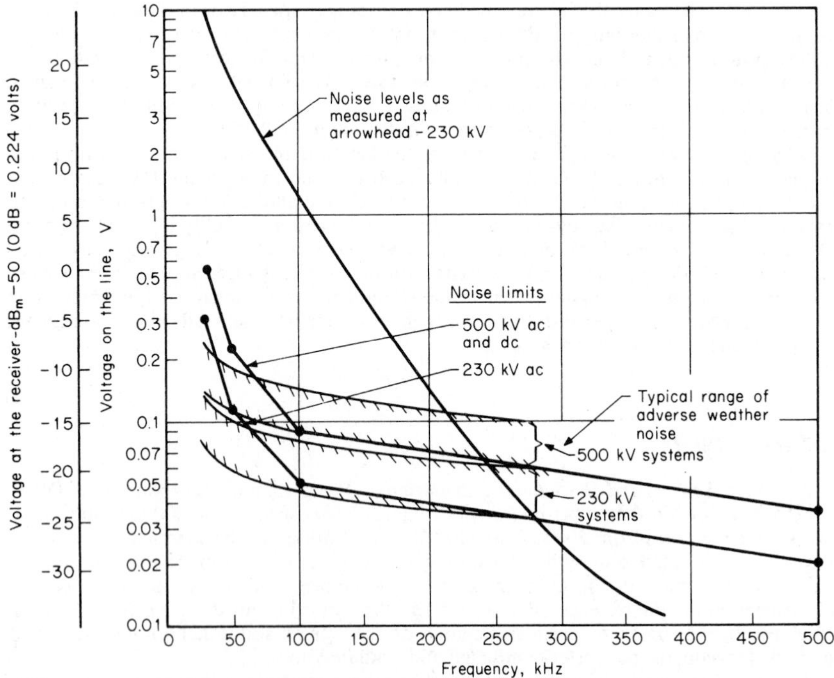


**FIG. 15-42** Typical PLC noise levels and limits for 500-kV ac and dc lines and 230-kV lines.

tric field at the edge of the right of way is a function of the voltage on the line (Fig. 15-43). That phenomenon is explained by the following equation

$$E = \frac{2h}{h^2 + s^2} \cdot \frac{V}{\ln \frac{2h}{r}} \tag{15-34}$$

where $h$ = height of conductor to ground, m
$s$ = perpendicular distance from conductor to point of interest, m
$r$ = equivalent conductor radius, m
$V$ = noise voltage on the conductor

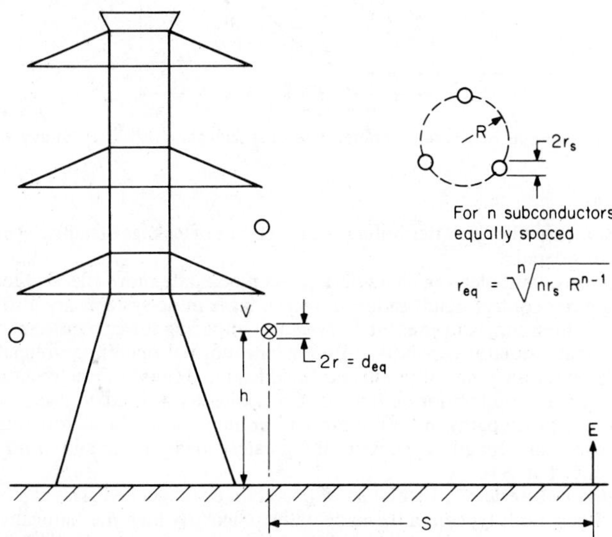


**FIG. 15-43** The electric field at the edge of the right of way is a function of the voltage on the line and the geometry.

A commonly specified radio interference limit (RIL) is 100 $\mu$V/m at given locations and contour lines (Fig. 15-44). From the allowable line noise voltage, the required noise filtering level can be determined. The RI level around the converter premises is due in part to the line and in part to the converter facilities. If series noise filters are used, the electric field from the lines is negligible at points 1500 ft from the lines. Therefore, all noise must come from other parts of the converter facilities. Generally, the construction of dc converters inherently makes direct radiation from converter premises low, since the valves are contained in a metal-shielded building which provides inherently good electromagnetic shielding and effectively eliminates any direct radiation from the valves.

The RI from the converter will coexist on the lines with RI from corona, and that total will be what is measured at a distance from the outside conductor.

## *OVERHEAD LINES AND CABLES*

### Line Insulation

***53. Requirements.*** The insulation in HVDC overhead transmission lines (as in HVAC) is stressed by lightning overvoltages, fault-induced overvoltages, and normal operating voltage. The first two types of stress impose requirements primarily on insulation string length

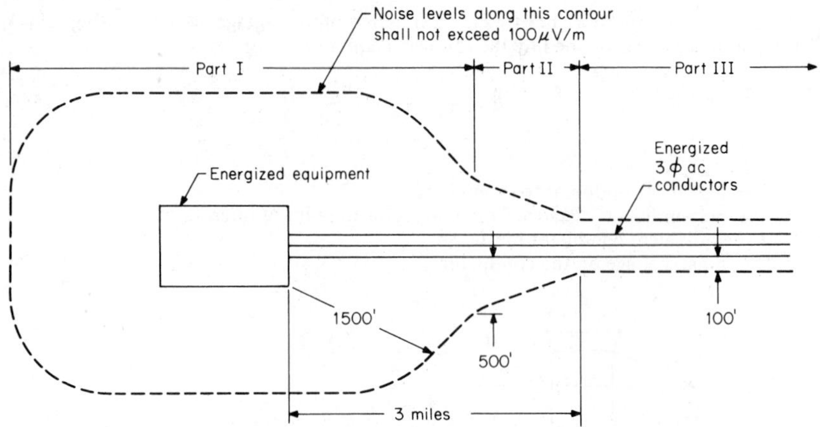


**FIG. 15-44** An example of a radio interference limit specification of 100 μV/m at given locations and contours.

and strike distance, while the latter influences the choice of leakage distance, especially when pollution is considered.

For dc systems, fault-clearing characteristics can be made more effective than for ac by means of converter control. Fault-induced overvoltages in dc systems are also appreciably lower than switching surges in present ac systems. Imposing lower requirements on string length means that insulator capability to withstand normal operating voltage will be the factor considered for all zones of pollution. In addition, because of electrostatic attraction, pollution is a greater problem for dc than ac. Consequently, it is advantageous to use insulators with long leakage paths in both clean and polluted areas. Insulators must also have sufficient mechanical strength to support the greatest loads of ice and wind that can be expected (see also Par. **54**).

***54. Insulator Materials.*** Insulators of porcelain or toughened glass can be used in dc transmission lines. Both types can be made with sufficiently high mechanical strength. For dc, it is important to choose materials with high internal resistance. Sodium content must be kept low to prevent internal leakage current at normal temperature and normal service voltage from exceeding a few microamperes. Insulators of organic materials have not been widely used for dc or ac, but their use is under active consideration for dc.

***55. Atmospheric Pollution.*** The severity of insulator pollution depends on atmospheric pollution, including the amount of insoluble material and dissociable matter, such as salt. The frequency and intensity of rain are of vital importance, since rain washes insulators periodically and, in effect, causes an equilibrium state of insulator pollution. Table 15-7 gives characteristic data for atmospheric pollution in equilibrium state for various districts.

The actual pollution on insulators is usually referred to as the amount of dissociable matter present. It is generally given in terms of equivalent sodium chloride (NaCl) deposit density. This is defined as the amount of NaCl which gives the same electrical conductivity

**TABLE 15-7** Characteristic Data for Atmospheric Pollution in Equilibrium State for Various Districts

| District | Equivalent NaCl deposit density on line insulators, mg/cm$^2$ |
|---|---|
| Clean areas | 0.02–0.05 |
| Industrial areas | 0.05–0.10 |
| Heavily polluted and coastal areas | 0.10–0.40 |

as the pollutant when dissolved in a given quantity of distilled water divided by the area of the washed surface.[1]

Because of the different mechanisms of insulator particle attraction and differences in washing properties for various parts of the insulator, pollutants become unevenly distributed over the insulator surface. The equivalent NaCl-deposit density is, in most cases, greatest on the undersurface, as shown in Fig. 15-45. Pollutants are also unevenly distributed along an insulation string, as shown in Fig. 15-46. There is a pronounced tendency for the elements at the ends of the string to be more polluted than others, especially for dc.

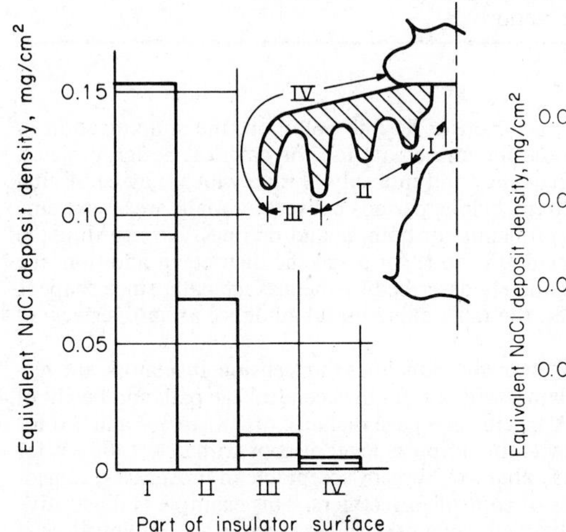


**FIG. 15-45** Equivalent NaCl deposits on different parts of a dc insulator at live end of +250-kV chain. Solid line (———) = after 32 months' operation in a ±250-kV bipolar long-term test. Broken line (---) = after 9 months' operation on the ± 250-kV bipolar Konti-Skan line.

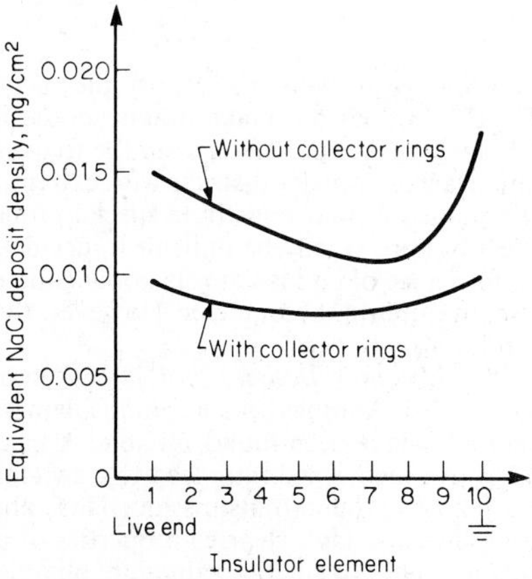

Equivalent NaCl deposit density, mg/cm²

0.020 0.015 0.010 0.005 0

Without collector rings

With collector rings

1 2 3 4 5 6 7 8 9 10

Live end

Insulator element

**FIG. 15-46** Measured deposit density on undersurfaces of two insulator chains, showing the influence of grading rings at both ends, after 4 months' operation at +250 kV.

***56. Necessary Leakage Paths.*** Usually, the leakage properties of insulators are characterized by the length of their leakage paths. It is sometimes convenient to distinguish between protected and unprotected leakage paths, or to introduce some form factor. In the following section, the necessary leakage path for a dc line will first be discussed under the assumption that insulators of "good design" are used. The factors which contribute to this good design will be discussed as well.

Table 15-8 gives the recommended specific creepage distance in different zones of pollution. The figures in Table 15-8 are based on ac experience, where the ac rms voltage to ground is comparable with the dc voltage to ground. Note that, for zone 1, the figure 2.3 has been recommended instead of 2.8. Although the requirement of withstanding internal overvoltages in ac systems automatically leads to leakage paths of 2.8 cm/kV phase-to-ground or more, there is little doubt that smaller leakage paths are sufficient to sustain continous voltage stress. Cases have been reported where leakage paths as low as 2.0 cm/kV have given satisfactory service. When data for specific existing ac lines are available, they should be used as a basis for determining necessary creepage distance for passage of the dc line through that district. However, it should be recognized that dc lines tend to accumulate more pollutant than ac lines in the same locations.

[1]Forrest, J. S., et al.: Research on the Performance of High Voltage Line Insulators in Polluted Atmospheres, *J. IEE,* 1960, vol. 107/A.

**TABLE 15-8** An Overview of Recommended DC Specific Creepage Distance in Different Zones of Pollution

| Zone no. | Description | DC specific creepage distance, cm/kV |
|---|---|---|
| 1 | Agricultural area, woodlands, | 2.3 |
| 2 | Outskirts of industrial areas | 4.0 |
| 3 | Industrial area—direct vicinity of the sea | 5.0 |
| 4 | Direct vicinity of extremely dirty industries such as certain chemical industries, power stations | 7.0 |

Although of value to identify the spectrum of dc line pollution, the simplification of Table 15-8 does not make it appropriate for immediate use for practical design projects. The table makes no reference to the frequency and intensity of rain, which may be of vital importance. Very dry districts with extremely long periods between rainfalls are an extreme exception and may constitute special problems for both ac and dc lines. Disregarding the effect of rain, it may be difficult to decide the zone for a specific district. In addition, the performance of an insulator is not adequately described by the leakage path, since shape is also an important parameter. However, the table offers useful guidance as an overview of conditions.

***57. Insulator Design.*** For HVDC transmission lines, suspension insulators are recommended. Antifog cap-and-pin insulators with a ratio between leakage path and height of 2.8 to 3.2 have been found suitable. A specific creepage distance of 2.8 cm/kV and 3.0 for the above-mentioned ratio lead to a switching impulse level of approximately 1400 kV for a 400 kV-to-ground transmission. This value would seem acceptable in most cases, considering the good fault-clearing properties of controlled rectifiers. This example is illustrative only as higher or smaller values are obtained when other leakage paths are employed, or if the insulators have a different ratio of leakage path to height.

Mechanical requirements do not differ from those of ac insulators (see Par. **54**). Electrical requirements, on the other hand, are different from those for ac. In dc, voltage distribution is resistive, instead of capacitive and resistive as in the case of ac. Even under dry conditions, air humidity will wet the pollutant. Consequently, voltage distribution will be affected by the distribution of pollutant on the string. Drying processes must also be taken into account. The leakage current density, for instance, will be highest near the pin of a cap-and-pin suspension insulator and cause effective drying. The surface voltage gradient may become very high there, although the equivalent NaCl-deposit density is high. Voltage distribution is also affected by the shape of the petticoats.

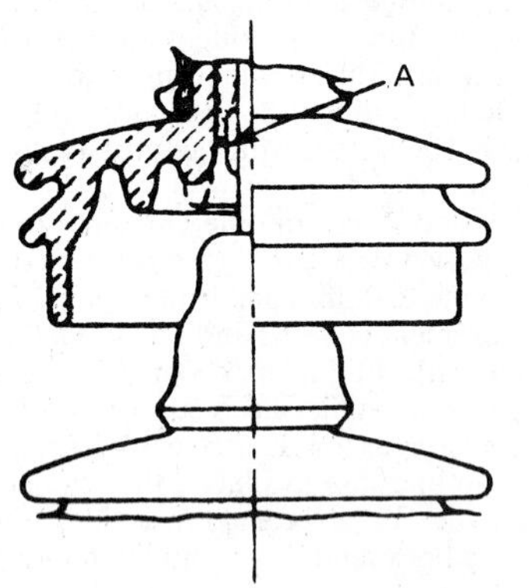


**FIG. 15-47** Antifog cap-and-pin suspension insulator, showing good design for direct current (solid lines) and alternating current (broken lines). *A* indicates deformable material.

Self-cleaning properties of a dc insulator are very important. Therefore, the surface of a good dc insulator must be very smooth. The protected leakage path should be about 60 to 70% of the total path. This means that the sheds of an insulator should be relatively large. The petticoats on the undersurface can be made deep, but if so, the distance between them must be large. The broken lines in Fig. 15-47 show a good ac insulator which did not behave well under dc conditions. By reducing the height of the inner petticoat, shown by the solid line, performance was improved considerably. This latter insulator is used in New Zealand on a ±250-kV, 575-km-long overhead transmission line. In clean districts, 12 elements are used in each insulator string, giving a specific creepage distance of 2.3 cm/kV.

***58. Electrolytic Phenomena.*** Overall, the effects of internal leakage currents are of no consequence to dc insulators, especially if precautions regarding material are followed. The effects of external electrolysis are more easily detected.

Surface currents are transported by anions and cations. Anions are neutralized at the cathode, where metals are deposited as a visible cathode growth while hydrogen is liberated. Cations, which are neutralized at the anode, may escape as a gas or chemically attack the anode. At the anode, secondary processes lead to the formation of hydroxides and carbonates. As these compounds require more space than the pure metal, a swelling of the anode results. The swelling is most pronounced if the anode has small geometric dimensions, for example, if the pin in a cap-and-pin insulator is the anode.

To avoid insulator cracking due to pin growth, a sleeve of a weak deformable material can be placed around the pin, as shown in Fig. 15-47 near point *A*. In principle, the insulator shape should be such that mechanical tension forces are avoided. This problem can also be alleviated with the application of a zinc sleeve to the pin as a sacrificial electrode.

***59. Collector Rings to Reduce Pollution.*** The nonlinear deposit distribution along the string is caused by the forces acting on the particles in the air surrounding the string. These are forces due to wind, gravitation, electric forces on charged particles, and electric forces on uncharged particles in a nonuniform field. Therefore, under the influence of a unidirectional electric field, both charged and uncharged particles in the air surrounding a dc string are set into motion and attracted especially to the energized and grounded end of the string where the electric field strength is highest. The nonlinear deposit distribution causes a nonlinear voltage distribution along the string, which, to a certain extent, reduces the breakdown voltage.

To reduce the insulator pollution and to improve deposit distribution along the string, collector rings can be mounted at the top and bottom of the insulator string. The electrostatic field around the insulators will then change in that the gradient concentration now occurs on the surface of the ring itself. Consequently, the particles will move to the collector ring instead of the string.

The reduction in insulator pollution by the use of collector rings is greater at positive than at negative polarity. Also, the influence of collector rings on pollution is greater on the undersurface of the insulator elements than on the upper. Figure 15-46 shows the influence of collector rings on the pollutant distribution along a dc insulator string.

***60. Greasing to Reduce Influence of Pollution.*** Another means for reducing the effect of pollutant accumulation on flashover voltage is to apply coatings on the insulator. Silicone grease applied in a 0.5- to 2-mm surface layer has been used with good results. Different kinds of petroleum grease are also available. These coatings reduce the external leakage current considerably by encapsulating the particles in the grease and isolating them from each other. However, when the grease has absorbed a certain quantity of pollutant, its effectiveness is reduced suddenly and flashover may occur without warning. If the grease is replaced regularly, coatings of this kind can be helpful in high-pollution districts. The time between repeated coatings of the insulators can range from a couple months in an extremely dirty area to a couple years in cleaner areas.

***61. Insulator Tests.*** As a consequence of the requirements for adequate leakage path in a dc insulator, there is a need for other pollution tests in addition to those outlined for ac insulators. Investigations have shown that the results of dc rain tests can differ very much from those of dc pollution tests. Pollution tests can be carried out indoors or outdoors. No standards exist for indoor tests, but test methods have been proposed[1] which give reproducible results and offer the same ranking to insulators as do the long-term outdoor tests. When polluting insulators artificially, it may be important to do so under voltage to duplicate flied pollutant deposit distribution.

[1]Annestrand, and Schel: *Direct Current,* 1967, vol. 12, no. 1, pp. 1–8.

## Environmental Considerations

***62. Design Criteria.*** Comparison of dc and ac transmission systems requires an evaluation of different design criteria. Design height is determined by a complex analysis of many factors including maximum electric field strength, field strength at the edge of the right of way, strength and weight of conductors, span length, National Electric Safety Code (NESC) clearances, and thermal rating and its effect on sag.

Conductor sizes are usually not a direct function of the thermal rating of the line. Environmental factors such as maximum electric fields, TV interference (TVI), radio interference (RI), and audible noise determine conductor size and the number of conductors in a bundle. All of these criteria are site-specific. Local regulations as well as local environmental factors must be included in an evaluation.

A major difference between ac and dc transmission systems is the reduced right-of-way (ROW) requirements of dc overhead lines. Since dc systems use two poles, less ROW width is needed for dc overhead lines when they are installed in a horizontal configuration. Estimates of typical ROW width are difficult since there are many factors to be considered, e.g., electric fields, audible noise, radio interference, NESC clearances, and space charge. These factors are a function of voltage span length, pole spacing, conductor height, conductor diameter, bundle diameter, and number of conductors per bundle. Additionally, dc line design must take into account the effects of dc space charge and ion drift, which are presently being studied in an attempt to further understand these phenomena.

Typical ROW widths for dc lines are 75 ft for ±250 kV with a 600-ft span (115 ft for a 1000-ft span) and 140 ft for 400 kV. These values can vary widely, depending on line designs and local conditions.

Another environmental effect from a dc system is harmonics on both the ac and dc side of the converter. Their effects on the power system and adjacent communication systems are of concern.

Hybrid ac and dc lines on the same ROW or the same structures are being considered. Coupling between systems, reliability, and all of the previously discussed environmental effects must be considered.

## Conversion of Transmission Lines from AC to DC

***63. Necessary Changes.*** New transmission line construction has become increasingly expensive and time-consuming. Acquisition of new ROW is difficult, particularly with increasing environmental concerns. Conversion of existing ac lines to HVDC can provide a feasible option for increasing transmission capacity at relatively low cost and with comparatively short lead times.

HVDC transmission allows operation at higher voltage and current than possible with ac transmission over the same line, thus increasing power transfer potential over a given ROW. The economic advantages of the power interchange capability often result in a prompt payback of the initial converter and line modification expenses.

Temporary overvoltages are usually not significant in HVDC line insulation, as HVDC system voltages are limited by the converter controls. Transient overvoltages on HVDC lines are almost exclusively caused by the surge induced on a healthy pole by a ground fault on the other pole of a bipolar line. This type of surge is moderate in magnitude (approximately 1.7 pu), much less than a typical ac circuit switching surge. Consequently, an ac line converted to HVDC operation can typically operate at a pole-to-ground voltage in excess of the phase-to-phase (rms) ac voltage parameters of the initial design.

AC transmission line loading is generally limited to levels considerably less than the thermal limit by voltage regulation, stability, and phase angle considerations. This is particularly true of EHV lines where minimum conductor diameter is often established by corona criteria. HVDC conversion allows controlled loading of a line up to its thermal limit, assuming adequate provision for voltage drop.

Conversion of a typical ac transmission line to dc operation at the thermal limit increases power transfer capacity three to five times the ac circuit's surge impedance loading. This assumes that two phases are used as dc poles, and one phase is used as a metallic return conductor. Other configurations are possible which allow more power to be transmitted.

Little change is required in the transmission line to permit HVDC operation. The existing tower structures and conductors can be used without change. Ordinary ac insulators, however, do not have adequate contamination performance to exploit the maximum HVDC voltage permitted by the line clearances (see Par. **55**). Fog-type ac insulators or special HVDC insulators allow dc operating voltage equal to or greater than the ac line-to-line voltage for which the line was designed. Both insulator types provide the additional creepage distance to resist contamination effects.

In some cases, conductor changes, and the necessary tower structural reinforcements, can be made to increase the capacity and reduce the losses of the converted line.

***64. Various Line Configurations.*** A number of line configuration options exist for converting three-phase ac lines to a bipolar HVDC line. Figure 15-48 shows a single-circuit line

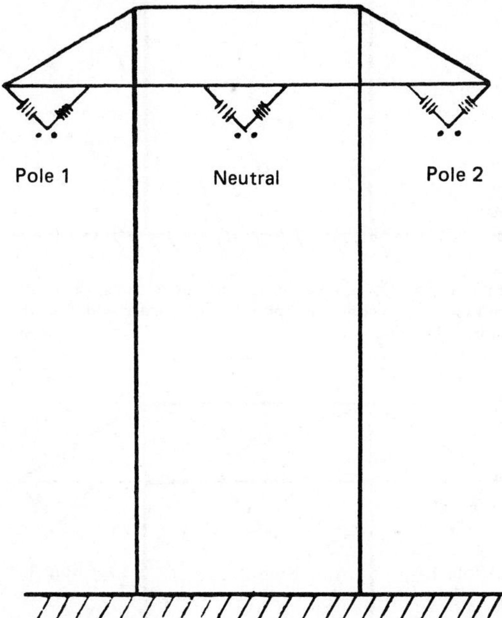


**FIG. 15-48** Some single-circuit line conversions require only a line insulator change with the center phase used as a permanent metallic neutral.

conversion requiring only a change of line insulators with the center phase providing a permanent metallic neutral. Costs and lead time associated with siting, designing, and constructing an earth electrode can be eliminated by using the metallic neutral.

On some lines, it may be possible to eliminate the center conductors and add conductors to the outside bundles as shown in Fig. 15-49. The same conductor weight is supported, but the changed mechanical force distribution possibly requires structural reinforcement. This option increases the line thermal capacity by 50%.

Another option, illustrated by Fig. 15-50, doubles the conductors in the center bundle. The outer bundles are connected in parallel to form one pole and the upgraded center bundle forms the other pole. The total conductor weight supported increases by 33%, and structural changes are required. The additional conductors double the line thermal capacity of the configuration in Fig. 15-48.

Availability of a double circuit ac line for conversion allows the simple configuration shown in Fig. 15-51.

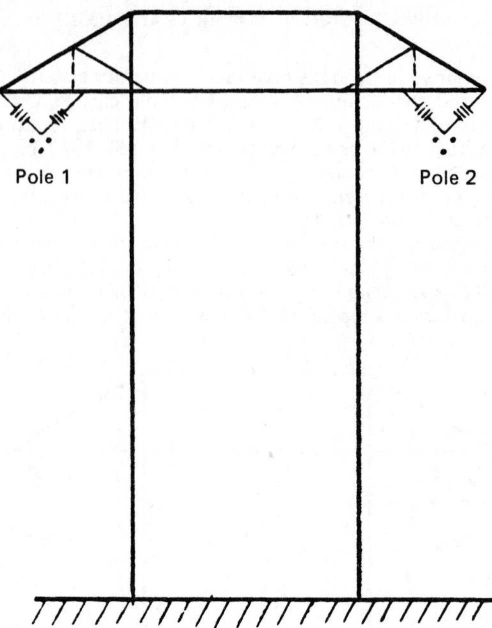


**FIG. 15-49** On some ac to dc line conversions, the center conductors are eliminated and conductors are added to the outside bundles.

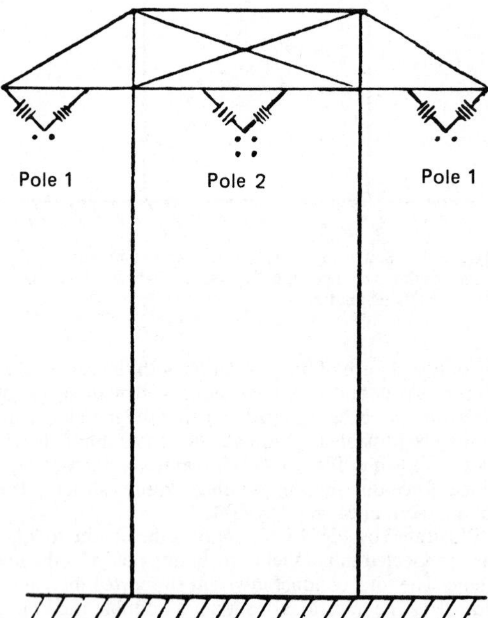


**FIG. 15-50** Doubling the number of conductors in the center bundle is another option in converting ac lines to dc.

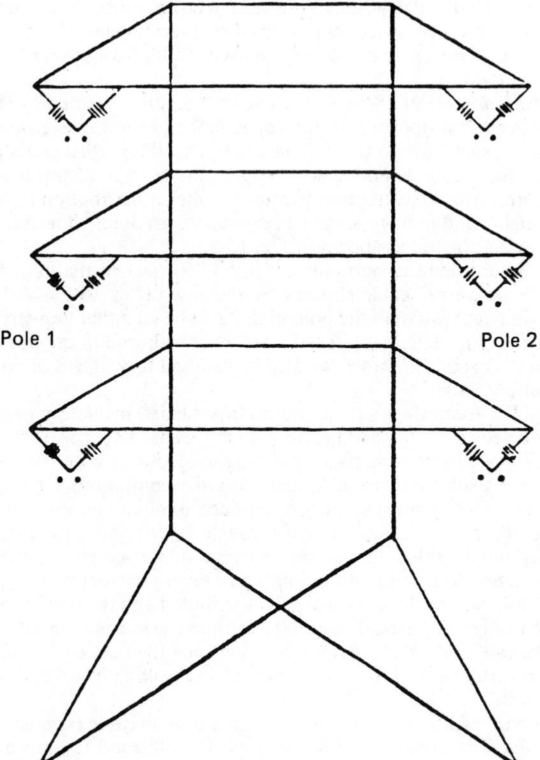


**FIG. 15-51** The availability of a double-circuit ac line for conversion to dc allows this configuration.

## Cables

***65. Types of HVDC Cables.*** Power transmission cables used for ac transmission can also be used for dc. Through better utilization of insulation and with the absence of capacitive currents, these cables are normally capable of transmitting more dc than ac power. A cable especially designed for HVDC transmission will, however, have somewhat different design from that of an ac cable.

There are four different types of cable to consider: paper-insulated solid-type, oil-filled, gas-pressure, and plastic-insulated. The solid-type insulation cable is probably the most economical for voltages up to about 300 kV. At present, conductor temperature has to be limited to about 50°C to prevent migration of impregnating oil in the insulating material.

Oil-filled cable always operates with oil pressure and can be used for the highest voltages. Oil supply has to be arranged, and this requires high pressure on long submarine cables. A variety of the oil-filled cable—the flat cable—for which no external oil supply is required has also been used for dc transmission.

Gas-pressure cable has, as its name implies, gas instead of oil as the pressure medium. This type of cable has been used for direct voltage up to 250 kV to ground.

Extruded-plastic insulation is widely used for ac cables up to 150 kV. For HVDC, however, this type of insulation has not yet been used. Tests are under way for evaluation of its suitability for this application.

***66. DC Cable Performance.*** A power cable used for dc transmission has no capacitive leakage currents, and power transmission is limited by conductor losses only.

How much the cable insulation can withstand in ac operation is normally limited by the maximum voltage stress in service and at impulse overvoltages. For dc cables a stress of three to five times that for an ac cable may be used. In the Konti-Skan cable (solid insulation) 25 kV/mm is used.

***67. Cable Insulation at DC Stresses.*** In ac cables, stress created by the electrical field is distributed in inverse proportion to the capacitance of the cable dielectric. This always gives the highest stresses close to the conductor. In dc cables, voltage distribution is determined by insulation resistance and space charges, and is dependent on temperature. In a cold cable (with uniform dielectric temperature), voltage distribution is the same as in an ac cable. At a high conductor-to-sheath temperature gradient, however, the stress may become highest at the sheath, as shown in Fig. 15-52.

In dc transmission, there is normally a demand for fast regulation, including polarity reversals. At such a reversal, cable stresses are increased (Fig. 15-53). As the stress is proportional to the time derivative of the potential, it can be seen that unit stress becomes high. Overvoltages from connected dc overhead lines must be borne in mind, even though overvoltage protection is used. The cable can also be exposed to voltages of power frequency in case of valve malfunctions.

***68. Design of DC Cables.*** As no dielectric losses exist, insulation paper can be chosen to give the best dielectric strength. Therefore, highly dense paper is used. To refine insulation performance even more, this paper can be used close to both the conductor and the outside surface. For a solid-type cable, a high-viscosity compound is needed to prevent oil migration, but the viscosity must still be low enough to obtain good flexibility.

In a dc cable, there is a leakage current through the insulation between conductor and sheath. In underground cables, this leakage current should be given a special path to the armoring and designed to ground at the joints to prevent corrosion. For sea cable without corrosion protection, current leakage is evenly distributed and the risk of corrosion is minor. Where corrosion protection between the lead sheath and armoring is used, it should be interconnected at intervals to prevent high voltages between the two. By locating the earth electrode at sufficient distance from the cable, the current leaking in is kept low and corrosion is prevented as well.

Normal steel wire can be used for the armoring of a dc cable because of the absence of eddy-current losses. The armor must be designed to withstand the mechanical stresses of laying and repair (around $3 \times 10^4$N for the Konti-Skan cable at 70 m depth). Figure 15-54 shows the cable design used for the Swedish part of the Konti-Skan transmission.

The auxiliaries for dc cables are the same as for ac cables. It must be kept in mind, however, that a long creepage distance is required for the pothead at the terminals.

***69. Cable Laying.*** HVDC cables are laid, jointed, and connected in the same way as equivalent ac cables. For submarine purposes, flexible joints have to be used. These should have the same properties—electrically and mechanically—as the cable itself.

***70. Cables for Large Depth.*** The dc cable used for the Skagerrak HVDC transmission between Norway and Denmark is of special interest owing to the great depth (maximum

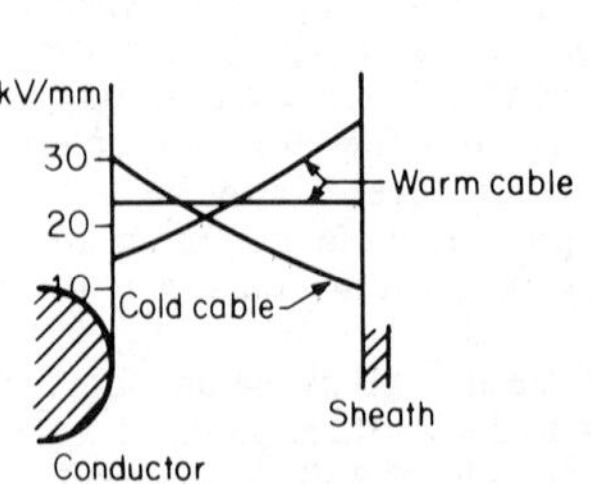


**FIG. 15-52** Stress distribution in an HVDC cable during cold and warm (load) conditions shows that stress on warm cable may be highest at the sheath.

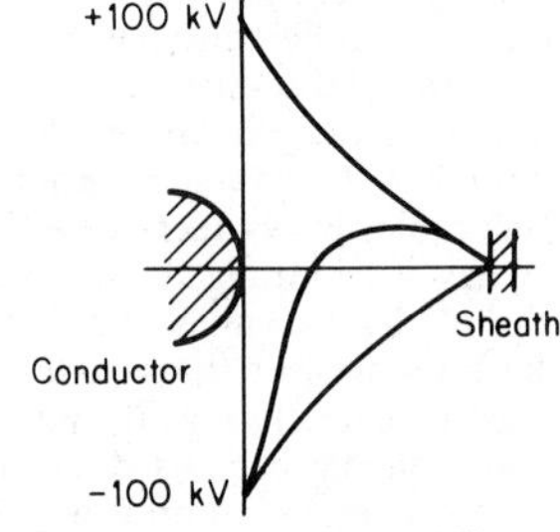


**FIG. 15-53** Stress is increased during the reversal in polarity of a warm dc cable.

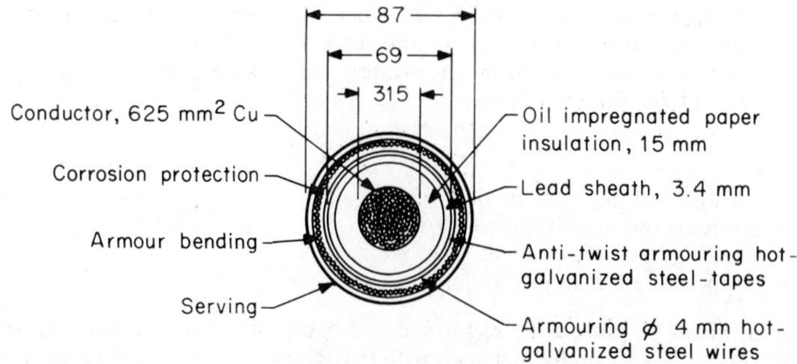


**FIG. 15-54** The cable used in the Konti-Skan 250-kV dc transmission from Gothenburg, Sweden, to Alborg, Denmark, was designed to withstand the mechanical stresses of laying and repair. A submarine-cable cross section is shown. Weight in air = 24.5 kg/m; in water = 19.6 kg/m.

530 m) along its route. This places extra-high stresses on the cable during laying which could be critical, especially if the cable has to be spliced in this area. As a consequence, special caution was taken regarding design, laying techniques, and repair.

The cable, designed for 250-kV, 1000-A dc, has a 800-cm$^2$ copper conductor, 16-mm insulation, and a 3.5-mm lead sheet. Besides a steel strap, the cable is also provided with two heavy steel wire armors, wound in opposite directions. The weight is 48 kg/m. The cable was delivered in one piece for the total length, 130 km.

A special cable ship was built, with a horizontal rotating drum (diameter, 30 m), which could load the whole cable (6500 tons). The ship was completely equipped not only for cable laying and repair, but for pipeline operations as well.

The cable was dug into the bottom soil as protection against damage. Repairs must be made, however, and cable handling in such instances uses a remotely controlled submarine with manipulator and TV cameras.

## Ground Return

***71. General Aspects.*** It has proved possible to use the ground as the return path for direct current in HVDC transmissions, thus decreasing losses as well as installation costs. Ground return may be used in single-pole transmission or as a spare "conductor" in case of a fault on one pole of a bipolar transmission.

Voltage drop in the ground-return path is concentrated at the ground or sea electrodes because the direct current will follow good conducting layers several miles under the earth's surface between the electrodes. For an actual case, the voltage drop per unit length may be seen in Fig. 15-55. The voltage gradient $E$ created at the ground surface by a horizontal electrode in a simplified case is given by

$$E = \frac{\rho I}{\pi l} \frac{x}{x^2 + h^2} \quad \text{V/m} \qquad (15\text{-}35)$$

where $\rho$ = electric resistivity of soil, Ω·m
$I$ = rated current of electrode, A
$x$ = lateral distance from electrode, m
$h$ = depth to center of electrode, m
$l$ = length of electrode, m

The electrode may be placed in the ground, in the sea, or on the seashore.

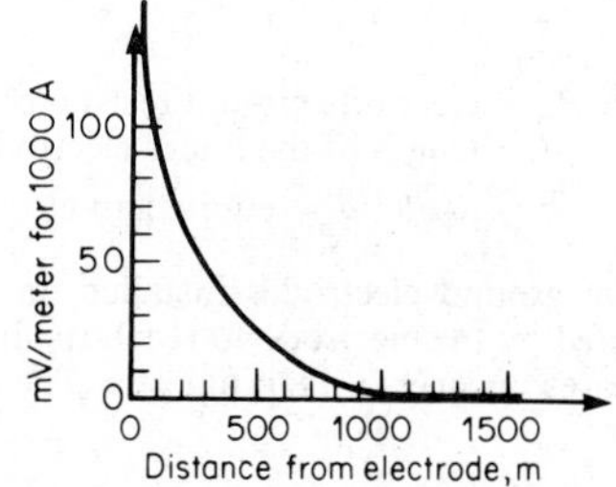


**FIG. 15-55** Voltage gradient in the ground return path indicates voltage drop per unit length.

*72. **Ground Electrodes.*** The electrical and thermal resistivity of the ground are both strongly dependent on its moisture content and increase greatly when the water content is below 15%. The temperature rise of the electrode must be kept below the boiling point of water. The following relationship exists:

$$V_e = (2\lambda\rho\theta)^{1/2} \tag{15-36}$$

where $V_e$ = allowable voltage rise of electrode, V
$\lambda$ = thermal conductivity of soil, J/m³·°C
$\rho$ = electrical resistance of soil, Ω·m
$\theta$ = temperature rise of electrode, °C

The ground must be thoroughly examined and constants $\lambda$ and $\rho$ measured before the electrode site is fixed. $\rho$ normally increases with particle size. Typical values are $\rho = 10$ for clays and $\rho = 10^4$ for certain kinds of sand. The thermal conductivity does not vary that much; typical values are 0.5 to 2.5 W/(m)(°C).

Moisture around the electrode decreases because of influence from the electrical (electroosmosis) and thermal fields (thermoosmosis) but is compensated for by hydrostatic pressure. A ground electrode should not be placed in very fine grained ground, to avoid its drying out because of electroosmosis.

A suitable area for the ground electrode is found from thermal and electrical properties and groundwater level. Experiments should be conducted with small electrodes on site to check computer results. The electrode arrangement depends upon economic and environmental factors. Figure 15-56 shows a possible electrode arrangement. The coke, with a resistivity of approximately 0.3 Ω·m, is the actual electrode. The graphite electrode (standard cathodic protection electrode) simply transmits the direct current from the electrode line to the coke. This arrangement protects the graphite electrode from anode corrosion.

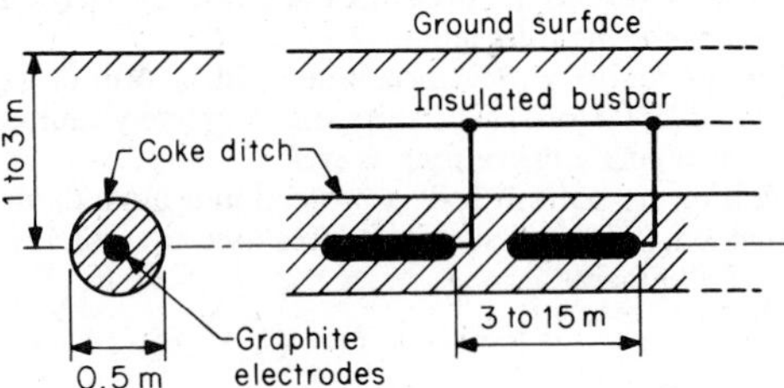


**FIG. 15-56** Typical arrangement of a ground electrode uses the coke for the actual electrode. The graphite electrode transmits the direct current from the electrode line to the coke.

The classic design equation for a linear electrode is

$$R_e = \frac{\rho}{\pi l}\left(\ln\frac{2l}{b} - 1\right) \tag{15-37}$$

where $R_e$ = electrode ground resistance, Ω
$l$ = length of the linear electrode, m
$b = \sqrt{dh}$; $d$ = equivalent electrode diameter, m; $h$ = depth of burial, m

For ground electrodes installed on land, the time constant $T$ is defined by the time required by the electrode to reach its final steady-state temperature $\theta_{max}$ if the temperature increases linearly at its initial rate. The expression for the time constant is

$$T = \frac{\gamma}{2\lambda}\frac{V_e^2}{\rho J} \quad \text{s} \tag{15-38}$$

where $\gamma$ = heat capacity of the soil, J/m³C, and $J$ = current density, A/m².

***73. Sea Electrodes.*** A negative electrode may consist of simple copper conductors laid on the sea bottom. A positive electrode should be protected so that fish do not get within about 10 m of the electrode. Graphite or magnetic electrodes are preferable. For the Gotland scheme (200 A), 12 magnetic electrodes were used, mounted inside an area protected from the open sea by a stone wall. In cases where special attention must be paid to the risk of corrosion on underground metallic structures and pipelines, the electrodes should be located in the open sea on the sea bottom at a depth to reduce the leakage currents (see Par. **75**). The electrode for the south end of the HVDC Pacific intertie is a good example.[1] For the critical Los Angeles area, which has a comprehensive network of pipelines and cables, leakage currents must be kept extremely low. The electrode was, therefore, placed on the ocean bottom some 2000 m out to sea at a depth of about 15 m. It consists of 24 electrode elements, mutually series-connected and fed in parallel by six conductors. Each element is built up by two parallel rods of a special high-silicon-iron alloy placed inside a concrete enclosure serving as a shield. The design is similar to that shown in Fig. 15-57. Very low field strength

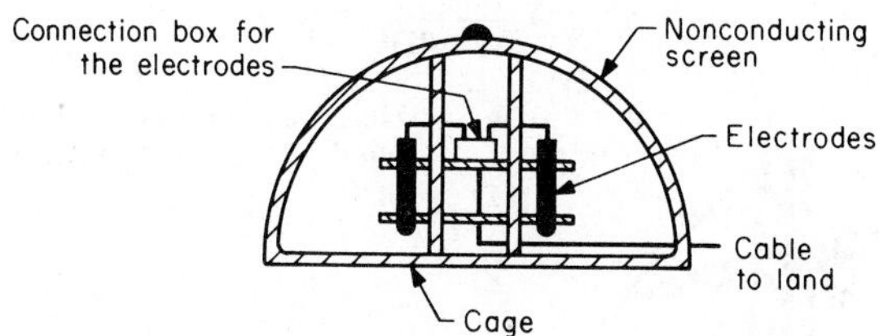


**FIG. 15-57** Arrangement for a sea electrode featuring a nonconducting protective cage.

and leaking current levels are achieved with the sea electrode. The effective ground resistance is on the order of 0.01 to 0.02 Ω.

***74. Electrode Line.*** In order to avoid corrosion (see Par. **75**), the electrode should be placed at least 3 to 5 km away from the converter station. Either a cable or an overhead line can be used for the connection. In the station, a transient ground via a capacitor (>10 mF) or a surge arrester for overvoltage protection should be arranged.

***75. Corrosion.*** To avoid corrosion on metallic structures, cables, or pipelines, the electrode should be placed several hundred yards from small structures or several miles from cables or pipelines. Protection could be achieved by giving the metallic structure a small (about 1 V) negative potential (so-called cathodic protection) as well.

For cables and pipelines, the leaking dc density must be kept below specified values, normally 0.1 to 1 $\mu A/cm^2$. An approximate formula for such currents occurring in cables or pipelines at a distance $D$ meters from a shore or sea electrode (Fig. 15-58) is given below:

$$i_{c,\max} = \frac{1}{2\pi aR} \frac{I_0}{2[\alpha/\rho_1 + (\pi - \alpha)/\rho_2]D^3} \quad \text{A/m}^2 \tag{15-39}$$

where $I_0$ = direct current, A
$\rho_1$ = sea resistivity, Ω·m
$\rho_2$ = earth resistivity, Ω·m
$R$ = resistance of cable armoring or pipeline, Ω/m
$a$ = cable (pipeline) radius, m
$\alpha$ = angle between ground and sea bottom, rad

At low values of $D$, Eq. (15-39) gives values that are far too high for the current leaking into cable, because of the cable's influence on the potential field. The expression in that equation must then be multiplied by a correction factor. For submarine cables, however, the approximation gives reasonable values.

[1]Elder, G. F., and Whitney, D. B.: The Los Angeles HVDC Ocean Electrode, 1962 Conference, Cleveland, Ohio, National Association of Corrosion Engineers.

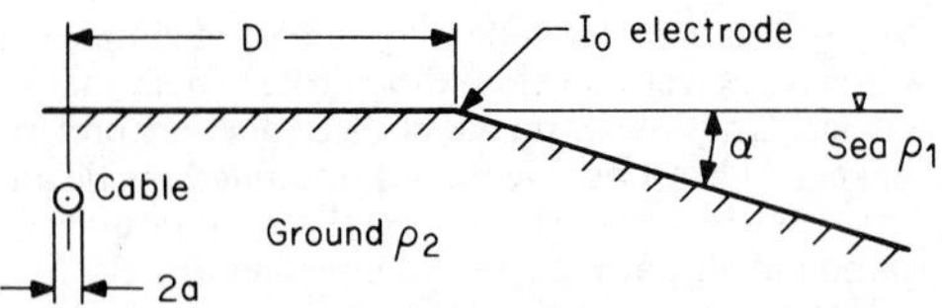


**FIG. 15-58** For cables and pipelines, the leaking dc density must be kept below specified values. An arrangement of a short electrode shows the quantities involved in figuring current leakages.

For a ground electrode, the corresponding formula would be

$$i_{c,\max} = \frac{I_0\rho_2}{4\pi^2 aRD^3} \quad \text{A/m}^2 \tag{15-40}$$

Here, the expression should also be multiplied by a correction factor at low values of $D$.

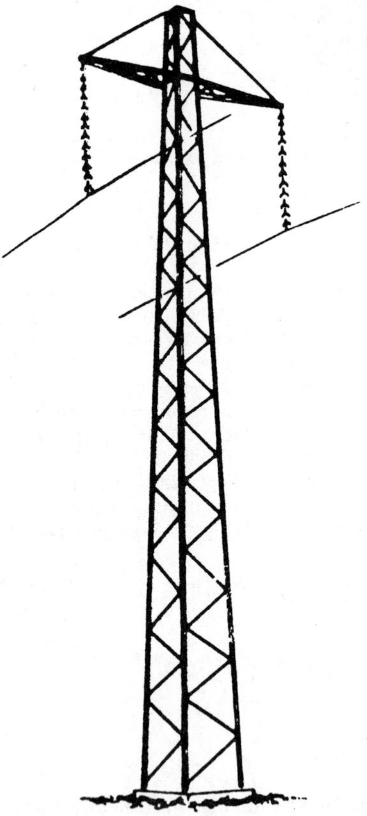

**FIG. 15-59** A self-supporting tangent tower was used on the Danish side of the Konti-Skan transmission system.

## Towers

***76. Mechanical Tower Design.*** The principle for mechanical design of transmission towers and calculation of conductor forces is similar to that for ac transmission. As the dc tower normally carries only two insulated conductors, the natural arrangement is a center mast and crossarm which holds the two-pole insulator strings. When an overhead ground wire is used, this arrangement is most conveniently supported by an extension of the center mast.

***77. Two-Pole Towers.*** As an example of self-supporting or rigid two-pole towers, Fig. 15-59 shows the design for the Danish overhead line in the Konti-Skan 250-kV transmission. The arrangement employs a steel tower which can be equipped with two fully insulated poles. On the Swedish side of the Konti-Skan transmission system, shown in Fig. 15-60, guyed aluminum towers are used.[1] The tower is equipped with a center-torque crossbeam which prevents the tower from turning in case of conductor breakage.

The guyed aluminum tower for dc transmission is especially suitable for helicopter erection, as its weight is low. The weight of the Swedish Konti-Skan tower, designed for a transmission rating of ±250 kV at 500 MW, is around 700 kg.

***78. Single-Pole Tower.*** Such a design is, of course, used mainly for monopolar

[1] "The Konti-Skan HVDC Project," CIGRE Report 408, June 1964.

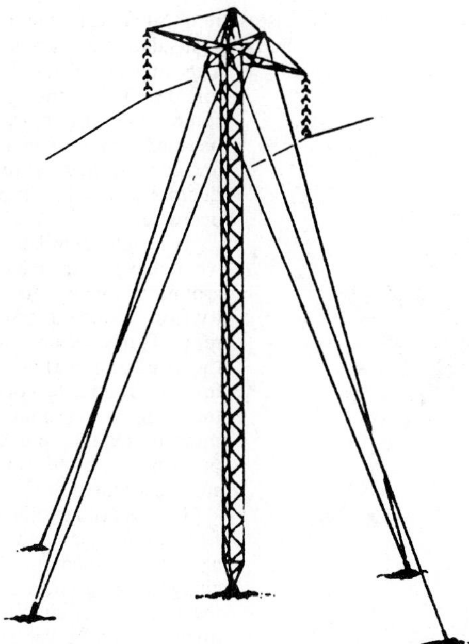

**FIG. 15-60** A guyed aluminum tangent tower was used on the Swedish side of the Konti-Skan transmission system.

transmissions. In cases of high-security requirements, it has sometimes been considered necessary to separate the poles of a bipole transmission by using two monopolar lines. Figure 15-61 shows a single-pole tower for the Inga-Shaba transmission (±500 kV).

## Corona

***79. Corona Phenomena.*** Corona is associated with the ionization phenomena in the vicinity of conductors. It occurs when the electric field strength is high enough to cause a breakdown of the surrounding air. The discharges give rise to current pulses which cause power losses, audible noise, and radio interference. In addition, corona causes the formation of a space charge which modifies the ambient electrical environment, electric field, and air in the proximity of dc lines. At negative polarity, corona is characterized by frequently repeated pulses of a few picocoulombs, the so-called Trichel pulses. At positive polarity the pulses are much less frequent but contain, instead, charges up to thousands of picocoulombs. The current pulses have a duration of only a fraction of a microsecond (shorter for negative polarity than for positive).

Although the physics of corona discharges is well known, it is almost impossible to express the losses and interference in an exact mathematical form, because corona starts at local points on the conductor where electric field strength can be considerably higher than the nominal average field strength. Therefore, not only the conductor roughness, but also the accumulation of insects, dirt particles, water droplets, etc., is of vital importance.

Since the conductor surface conditions change with time, from washing by rain and other natural occurrences, the losses, as well as the interference, can be treated only in a statistical manner.

***80. Corona Losses.*** In principle, corona loss is not as serious for dc as for ac lines. For an overhead transmission line, corona losses are dominated by the discharges on the conductors, while corona on the metal elements and insulators is a small part of the total loss. The intensity of the discharges on conductors depends on their surface conditions. In some cases, losses are high for new lines because of intensive corona from scratches and foreign material on the cables. As conductors age, the average corona losses decrease.

**FIG. 15-61** A monopolar, self-supporting tangent tower was used in the Inga-Shaba transmission system.

It is characteristic of dc corona that when released, the charge must be carried to ground or to a conductor of opposite polarity (this is not the case for alternating current). Thus, a space charge, which reduces the conductor surface voltage gradient, is formed around the conductor. The effect of the charge is so pronounced that it is usually more reasonable to characterize the corona performance by the line voltage than by the surface gradient.

Because of large dc space charges, losses from monopolar and bipolar transmission are quite different. In the bipolar case, charge mixing takes place as negative ions reach the direct vicinity of the positive conductor and vice versa, thus increasing the gradient on both conductors. Therefore, losses from bipolar lines are larger in comparison to those from monopolar lines than would be expected from nominal surface gradients.

***81. Bipolar Corona Loss.*** Corona loss for a bipolar line has been expressed by equations of the following type:

$$P = P_0 + a(\rho_{max} - \rho_0)^2 \qquad (15\text{-}41)$$

where $\rho_{max}$ = maximum conductor surface gradient, kV/cm

$\rho_0$ = corona onset conductor surface gradient, kV/cm

$P_0$ = corona loss at onset gradient, kW/km

Measurements on a 2 × 46 mm conductor at 18.3-m pole spacing show that

$$P_0 = 0.3 \text{ kW/km} \qquad a = 0.02 \qquad \rho_0 = 14 \text{ kV/cm}$$

Measurements on a 4 × 30.5 mm conductor at 18.3-m pole spacing show that

$$P_0 = 0.3 \text{ kW/km} \qquad a = 0.03 \qquad \rho_0 = 14 \text{ kV/cm}$$

The loss is divided about equally between positive and negative conductors.

***82. Monopolar and Homopolar Corona Loss.*** Although the corona discharge mechanism is very different for positive and negative polarity, average corona loss is about the same for both polarities. For a given conductor, the loss decreases approximately with the square of the height of the conductors. The loss of a monopolar line is less than one-half (one-fifth to one-third) of the loss of a bipolar line operating at the same voltage. The average corona loss of a homopolar line is slightly less than the sum of the losses of the corresponding monopolar lines operating at the same voltage.

***83. Weather Influence.*** DC corona losses increase with bad weather, during rain and snow, and with hoarfrost. Compared to ac, however, the increase is small, and the ratio between bad- and fair-weather losses is usually between 2 and 4 and very seldom exceeds 10. The increase during bad weather is less if the conductors have a high surface gradient. This is a significant advantage of HVDC systems over ac.

***84. Annual Mean Losses.*** As in the case of ac corona, average annual corona losses can be calculated if the corona-loss characteristic and the range of weather conditions during an average year are known. Table 15-9 compares a 500-kV 3-phase ac and a ±400-kV bipolar dc overhead transmission line with the same capability. The following assumptions regarding the weather conditions have been made:

5880 h, fair weather

2000 h, fog, dew, and hoarfrost

880 h, rain

As indicated, fair-weather corona losses are roughly the same for both systems, while the annual mean losses and maximum losses are considerably lower for dc than for ac. For an ac system, the maximum corona losses may amount to the same magnitude as the load losses. They will be considerably less for a dc system, however.

In effect, these quantitative differences will cause dc corona losses to be of minor economic importance for many practical cases. Less corona loss also means that the choice of conductor arrangement and conductor type in dc systems generally can be made without special regard to the corona-loss problem.

***85. Radio Interference from HVDC Overhead Lines.*** HVDC lines generate radio interference (RI) principally in two different ways:

By pulses occurring at the ignition of the valves, then transmitted via the switchyard to the line

By corona on the line

The first source can be reduced or eliminated by taking precautions in the terminal station. The latter is dealt with in the following manner.

***86. Characteristics of Measuring Instruments.*** RI measurements are usually carried out with an instrument specified by NEMA, CISPR, or other agency. In principle, such apparatus consists of a quasi-peak voltmeter connected to a bandpass filter via a weighting circuit. The response characteristic of the instrument varies with the pulse repetition frequency, pulse duration, bandwidth of the apparatus, etc. For random pulses, the reading becomes roughly proportional to the rms values of the entrance voltage multiplied by the

**TABLE 15-9** Comparison for Corona Power Losses between a Bipolar ±400-kV DC and a 3-Phase 500-kV AC Line

DC line conduction 2 × 46 mm; pole separation 10.5 m; average height 21.7 m.
AC line conduction 3 × 36 mm; phase spacing 11.2 m; average height 18.6 m.

| | Corona losses kW/km | |
|---|---|---|
| | ±400 kV dc | 500 kV ac |
| Average losses in fair weather | 1.3 | 1.3 |
| Minimum losses in fair weather | 0.6 | 0.1 |
| Maximum losses in a short line section under worst weather conditions | 10 | 130 |
| Maximum losses for the whole line under worst weather conditions | 6 | 20 |
| Annual mean losses for the whole line | 2.3 | 5.6 |

square root of the apparatus bandwidth. Instrument reading, therefore, is a function of input power within the apparatus passband.

Since the characteristics of occurrence of dc and ac corona pulses are different, the same instrument reading will not give the same subjective disturbance (on radio receivers tuned for acceptable listening conditions). A signal-to-noise ratio up to three times lower can be allowed for dc than for ac for achieving the same quality reception.

***87. RI from Bipolar Lines.*** Only the positive conductor will significantly contribute to RI. The median level of RI in fair weather may be estimated by

$$\mathrm{RI} = 51\ \mathrm{dB} + 1.5\,(g - 20.9) + 10\ \mathrm{lag}\,(n/2) + 40\ \mathrm{lag}\,(d/4.58) - 40\ \mathrm{lag}\,(R/20) - 2.6 \qquad (15\text{-}42)$$

where RI = dB above 1 $\mu$V/m at 1 MHz
$g$ = nominal maximum surface gradient, kV/cm
$n$ = number of subconductors in a bundle
$d$ = subconductor diameter, cm
$R$ = distance from positive conductors to point at which RI is calculated, m

***88. RI from Monopolar and Homopolar DC Lines.*** Only lines with positive voltages will cause significant RI. Fair-weather RI is calculated by the same equations used for bipolar lines. The absence of a conductor of opposite polarity decreases the nominal surface gradient, causing a lowering of the RI by a few decibels.

***89. RI Frequency Spectrum.*** A typical RI frequency spectrum, normalized to a 1-MHz bandwidth, is shown in Fig. 15-62.

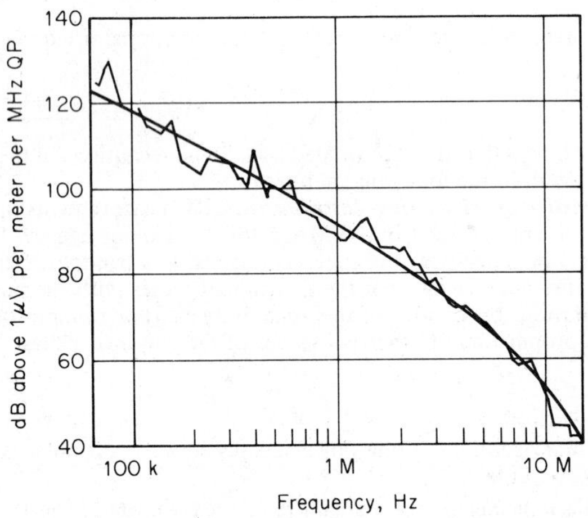


**FIG. 15-62** Radio interference frequency spectrum normalized to 1-MHz bandwidth.

***90. Influence of Conductor Bundling on RI.*** For ac lines, the use of bundled conductors with the same total cross-sectional area as a single conductor can reduce the RI level. This is possible because the interference strongly depends on the electric field strength at the conductor surface, which is reduced for a bundled conductor.

For dc lines, however, the presence of a space charge around the conductor counterbalances the effect of an increased number of bundled conductors. Although an increased number of bundled conductors for a constant total pole area slightly decreases the RI also for dc lines, RI considerations normally should not be decisive when choosing the number of bundled conductors.

***91. RI during Bad-Weather Conditions.*** With ac, the RI during rain is higher than during fair weather, the increase depending on both the rain intensity and the conductor surface voltage gradient. With dc, just the opposite occurs, as illustrated by Fig. 15-63. For dc, this reduction also occurs during snow. In most cases, wind increases the interference level, but in a very irregular manner.

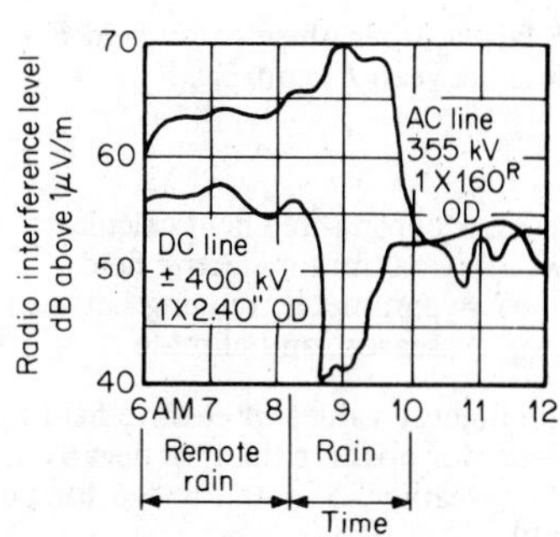


**FIG. 15-63** Measured radio-interference level at 0.83 MHz and 30 m distance from a bipolar dc and a 3-phase ac line, before, during, and after rain.

***92. Television Interference.*** The noise in the television frequency range, i.e., above 50 MHz, is dominated by insulator corona. For this phenomenon, the interference during fair-weather conditions is very small but increases during bad-weather conditions. Generally, the noise level in this frequency range is very low and can be ignored for most practical cases.

***93. Audible Noise from DC Lines.*** Only positive dc conductors significantly contribute to audible noise (AN). The uncertain level of AN in fair weather may be estimated by

$$\text{AN} = 57\text{ dB} + 126 \text{ lag}(\rho/25) + 18 \text{ lag}(n/2) + 25 \text{ lag}(d/4.45) - 10 \text{ lag } R - 0.02R + k_n \qquad (15\text{-}43)$$

where AN = dB(A), above 20 μPa
$g$ = nominal maximum surface gradient, kV/cm
$n$ = number of subconductors
$d$ = subconductor diameter, cm
$R$ = distance from positive conductor to point at which AN is calculated, m
$k_n = 0$ for $n \geq 3$, $k_n = 2.6$ for $n = 2$, $k_n = 7.5$ for $n = 1$

In bad-weather conditions, audible noise is less than during fair weather.

***94. Electric Field and Ion Density at Ground.*** The ionization of air in proximity to the conductor surface results in the formation of a space charge in the entire region between conductors and ground. In the dc line right of way, air ions predominantly contribute to space charge. The electric field at ground level is affected by the presence of a space charge and may be significantly greater than the electric field calculated for corona-free conditions. For monopolar and homopolar lines, the density of air ions and the electric field at ground level reach their highest values directly underneath the conductors. For bipolar lines, that condition occurs slightly outside the conductors.

***95. Electric Field at Ground at the Worst Location in the Right of Way.*** The electric field at ground ($E_e$) in corona-free conditions is given by

$$E_e = \frac{2Vh}{\ln(4h/d_{\text{eq}}) - \frac{1}{2}\ln(4h^2 + p^2/p^2)} \cdot \left[\frac{1}{h^2 + (x - p/2)^2} - \frac{1}{h^2 + (x + p/2)^2}\right] \qquad (15\text{-}44)$$

where $V$ = voltage to ground
$h$ = height above ground
$p$ = pole spacing (infinitely large for monopolar lines)
$d_{\text{eq}}$ = equivalent bundle diameter = $D^n \sqrt{nd/D}$
$n$ = number of subconductors in a bundle
$d$ = subconductor diameter
$D$ = bundle diameter
$x$ = horizontal distance from center of a bipolar line

The electric field of a line in corona may reach values much higher than $E_e$. Maximum values are needed in areas where there may be heavy rain or ice on the conductors. These values may reach a value $E_{max}$ at the worst location within the right of way, given by

$$E_{max} = 1.31(1 - e^{-1.7p/h}) \qquad (15\text{-}45)$$

In fair weather, the electric field $E$ at the worst location within the right of way is intermediate between $E_e$ and $E_{max}$:

$$E = E_e + S(E_{max} - E_e) \qquad (15\text{-}46)$$

where $E_e$ = corona-free field calculated at location $x$
$E_{max}$ = maximum electric field
$S$ = parameter, ranging between 0 and 1, dependent on surface gradient, weather, season, and climate

The highest values of electric field in fair weather are reached during the summer and are higher for positive than for negative conductors. At normal conductor gradients in summer fair weather, $S = 0.2$ to 0.6 for positive polarity and $S = 0.05$ to 0.35 for negative polarity.

***96. Ion Density at Ground at the Worst Location in the Right of Way.*** The ion density may vary from ambient values in conditions of no corona to maximum values given by

$$N_{max} = 68.8 \frac{(1 - e^{-0.7p/h})}{(1 - e^{-1.7p/h})} \quad \text{V/h}^2 \qquad (15\text{-}47)$$

where $N$ = number of ions/cm$^3$
$h$ = height of conductors above ground, m
$p$ = pole spacing (infinitely large for monopolar lines), m
$V$ = voltage to ground, kV

The value is reached underneath the conductors only in extreme foul-weather conditions. In fair weather, the highest ion densities occur during the summer and are higher under the positive than under the negative pole. In summer fair weather at the worst location in the right of way, positive ion densities are 40 to 80% of $N_{max}$ and the negative ion densities are 10 to 60% of $N_{max}$.

## EQUIPMENT CHARACTERISTICS

### HVDC Valves

***97. Requirements.*** The main requirements of an HVDC valve for power transmission are

Low forward voltage drop during conduction
Ability to withstand high negative and positive voltages without breaking down
Controllable firing instant
Overcurrent capability to withstand internal and external faults

Different types of equipment to fulfill these requirements have been used in commercial installations. Before 1970, mercury-arc multigap technology was used in valve construction. In 1972, the first converter station composed entirely of thyristor valves went into operation. Since 1975, new converter stations have used only thyristor valves.

***98. Mercury-Arc Valves.*** Although mercury-arc valves are not being used in new construction, there are converter stations in operation using mercury-arc valves. The design of a mercury-arc valve with a high number of grading electrodes is shown in Fig. 15-64. The valve consists of a water-cooled stainless-steel evacuated tank on which several anode assemblies are mounted. The anode, grading electrodes, and control grid are mounted in a porcelain cylinder. The grading electrodes (about 20 per 125 kV valve) are connected to an

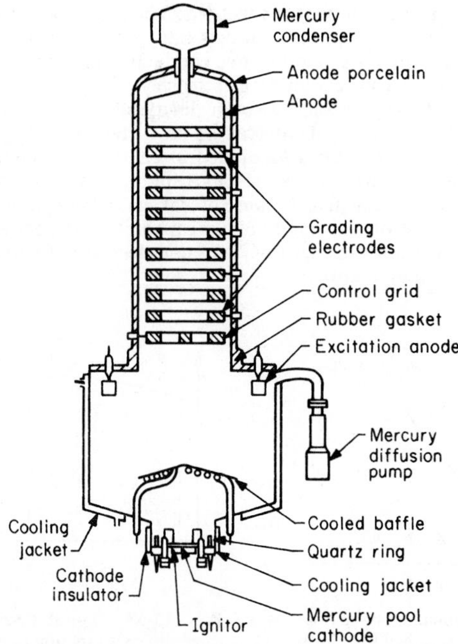


**FIG. 15-64** Principal design of the multigap mercury-arc valve for voltages above 100 kV.

external capacitive-resistive voltage divider, which gives the most favorable voltage control. The anode porcelain is sealed to the tank by a rubber gasket secured by a mercury seal. All other seals are of the permanent type (kovar-glass-porcelain). The anodes are connected in parallel via an external current divider. The arc voltage drop is about 50 V. A typical rating of a 6-pulse bridge of six anode valves of this type is direct voltage, 120 to 150 kV; direct current, 1800 A.

***99. Thyristor Characteristics.*** A thyristor (IEC and IEEE description: "reverse blocking triode thyristor") is a three-terminal (anode, cathode, gate) semiconductor device. It is functionally a unidirectional current flow switch. Over the years, thyristor voltage and current ratings have increased to a present rating of about 6 kV and a single thyristor current capability of about 3000 A.

The typical thyristor package consists of the silicon semiconductor wafer encapsulated in a disc-type porcelain package. The metal faces of the package are clamped under pressure to heat dissipators (either air- or liquid-cooled) (Fig. 15-65). The metal faces are also the

**FIG. 15-65** Typical thyristor package for HVDC application.

thyristor's power (anode and cathode) connections. The major power dissipation of a thyristor occurs when it is conducting current (functionally a closed switch). When the thyristor first turns on, only a portion of its junction area conducts. The area of conduction increases with time until maximum conduction area is reached. This phenomenon is more pronounced in larger area thyristors and has been described as a "spreading" loss in recent literature. Consequently, thyristor manufacturers have started to describe the conduction loss of a thyristor as a function of the *di/dt* of the front of the conduction current waveform. This typical relationship is shown in Fig. 15-66. A minor power loss occurs when the thyristor is functionally an "open switch" because of leakage currents from the impressed voltage on the thyristor. Thyristor withstand voltage capability is dependent on the thyristor's junction temperature, as shown in Fig. 15-67. A rapid decrease of voltage capability begins at about 125°C junction temperature.

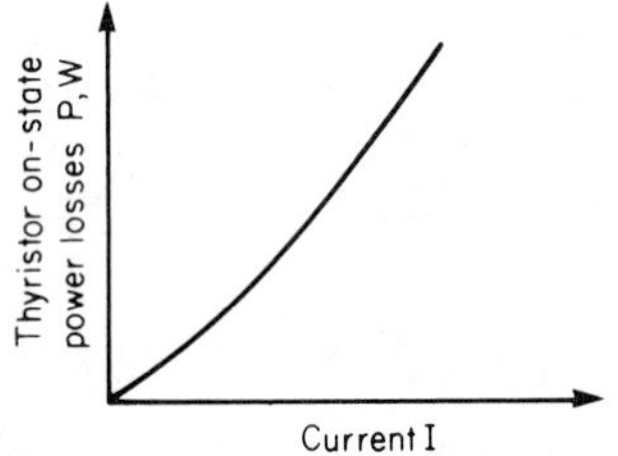


**FIG. 15-66** Thyristor on-state power losses as a function of current for a specified waveform.

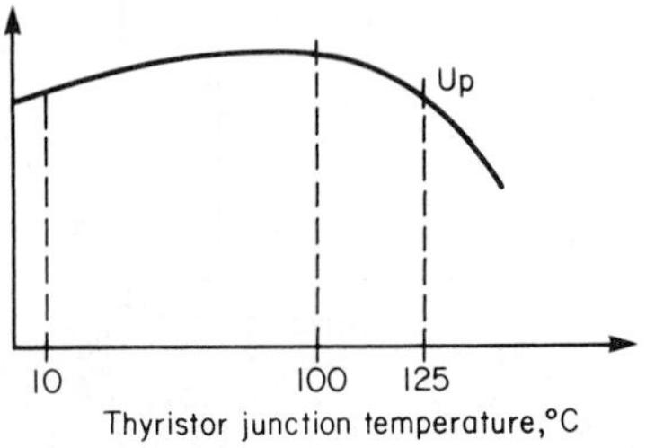

**FIG. 15-67** Typical thyristor blocking voltage capability as a function of junction temperature.

***100. Thyristor Valve Circuits.*** Thyristor valves consist of a series connection of thyristors and other circuit components. The number of series-connected thyristors varies from about 20 to 150 thyristor levels depending on the voltage rating of the valve and thyristor. Parallel connection of thyristors may be used, depending on the valve current rating.

Circuits vary between valve manufacturers. All valve circuits contain saturable reactors connected in series with the thyristors. The reactor must control the initial *di/dt* of the thyristor current and minimize the effect of time variations between each thyristor's turn-on. Valves also contain snubber circuits (series connection of resistors and capacitors) to control voltage distribution within the valve and the valve voltage transient that occurs when the thyristors commutate off (end current conduction). Valve protective circuits protect the thyristor levels from excessive applied voltages (auxiliary gating and voltage breakover circuits).

A status monitoring circuit in the valve indicates the thyristor levels that have become short-circuited during operation (typical maintenance interval for replacement of short-circuited levels is 1 year or more). Fiberoptic light guides provide the communication link between the valve and ground potential for status monitoring and initiating valve gating.

Thyristors and auxiliary circuits are usually arranged in modules and are connected in series to make up a valve. Valve modules have contained 2 to 12 series-connected thyristor levels. The valve module has the same electrical properties as the complete valve, except at a reduced voltage rating. The valve module concept permits testing of a prorated stress encountered by the total valve in normal and abnormal operation. It is also easy to handle during installation and maintenance.

The valve modules are connected in series in a valve structure to form a valve. The valve structure can contain as many as 12 valves, depending on the operating voltage. A common arrangement is a group of four valves in one valve structure with three valve structures forming a series connection of two Graetz bridge circuits as shown in Fig. 15-68. The advantage of such an arrangement is that the valve operating at the highest dc voltage can be physically located the greatest distance from the ground plane, thus simplifying the electrical grading.

Combinations of valve insulation and cooling mediums are

Air for both insulation and cooling

Oil for both insulation and cooling

Air for insulation and water, oil, Freon, etc., for cooling

$SF_6$ for insulation and oil, Freon, etc., for cooling

Air insulation with air or water cooling is the dominant design approach. Air insulation and cooling, however, is usually the simplest design. The alternatives may offer a more compact valve design, although valve size is not a strong contributor to converter station size. In a building containing valve structures for a 500-kV dc rating, spacing for electrical clearances has more influence on the floor plan.

***101. Valve Firing.*** Valve firing is the function of delivering a gating signal to the valve thyristors to initiate valve conduction. The typical approach is to provide a light pulse by a fiberoptic light guide to the valve. At the valve, a light-to-electrical interface provides the electrical signal to the gates of the individual thyristors. The firing system often includes redundant firing circuits for improved reliability. The complexity and energy requirements of the valve's gating system can be reduced by the use of light-triggered thyristors.

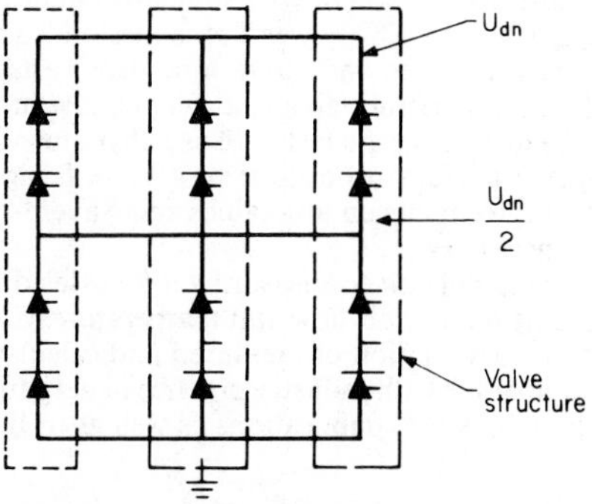

**FIG. 15-68** A common arrangement of three valve structures, each containing four valves, to form a 12-pulse converter.

***102. Valve Cooling.*** The valve cooling system is an integral part of the valve design. The thyristor's current capability is dependent on the cooling system. The combination of thyristor current and cooling system determine the thyristor's junction temperature.

A main part of the thyristor valve cooling is the thyristor heat dissipator, which is physically clamped on each side of the thyristor. A typical relationship of the junction temperature to the cooling medium flow is shown in Fig. 15-69. A satisfactory valve cooling design is shown in Figs. 15-66, 15-67, and 15-69. The current rating requirement defines the thyristor losses (see Fig. 15-66). The cooling characteristic (Fig. 15-69) and cooling media temperature determine the junction temperature, which should be compatible with the desired withstand voltage capability of Fig. 15-67.

The closed-loop system which simplifies control of contaminants in the loop can be

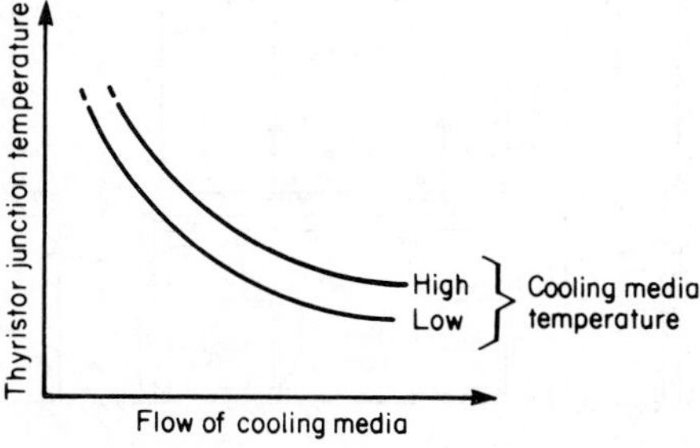


**FIG. 15-69** Thyristor junction temperature as a function of cooling medium flow.

either air- or liquid-cooled. A typical arrangement for a closed-loop air-cooled valve is shown in Fig. 15-70. The air-to-liquid heat exchanger (coil) is part of the primary cooling system. The primary cooling system can reject its heat to the ambient air, a cooling tower, or a body of water, such as a lake or river. The cooling system must be able to operate at the maximum ambient temperature. Therefore, at this maximum, an optimum valve design would not have a continuous overload capability. As ambient temperature or load changes during operation, the cooling flow rates can be controlled to obtain the lowest possible operating losses.

***103. Testing Thyristor Valves.*** Since thyristor valves are a relatively new type of electrical apparatus, appropriate testing procedures are still evolving. The generally recognized test description was issued by the International Electrotechnical Commission (IEC) in 1981 as the first edition of IEC Standard Publication 700, "Testing of Semiconductor Valves for High-Voltage DC Power Transmission." This document describes the type and production tests for valves and valve modules. Tests include dielectric, fault current, and operating tests. A thyristor valve is a unique, low-impedance device compared to other devices subjected to dielectric tests. Also, a thyristor valve with protective firing will not allow a specified waveform to exist if protective firing occurs. The IEC recognizes that compromises and agreements on test values and waveforms between the manufacturer and customer may be necessary.

It is difficult to measure valve losses directly in the field because of tolerances in measuring fluid or air flow and temperature. Consequently, valve losses are usually determined by a combination of measured and calculated losses of various valve components. This is compatible with industry practice in determining losses of large rectifier circuits as described in various IEC publications as well as in IEC 700.

## Control and Operation of HVDC Links

***104. General Requirements and Control Principles.*** The fundamental objectives of an HVDC control system are

To control a system quantity such as dc line current, transmitted power, or frequency of either of the two connected ac networks with sufficient accuracy and speed of response.

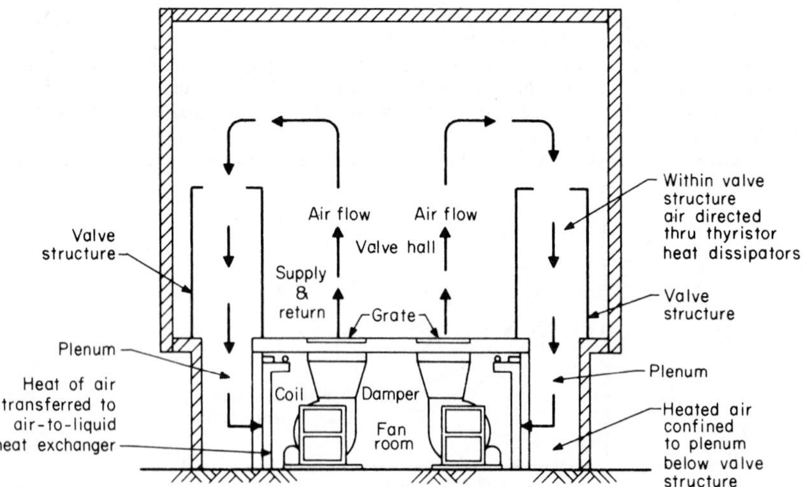


**FIG. 15-70** The closed-loop system, which simplifies control of contaminants in the loop, can be either air- or liquid-cooled.

To ensure stable operation, with reliable commutation in the presence of system disturbances.

To fulfill the above objectives at a minimum reactive power consumption.

In addition to these items the control system shall, if possible, ensure correct operation at large disturbances, or at least minimize the consequences when the fault is cleared, during normal operation.

The principle of direct-current control is illustrated by Fig. 15-9, from which it is clear that the transmitted current is

$$I_d = \frac{U_{d1} - U_{d2}}{R} \tag{15-48}$$

Because of the rather small value of $R$, the current will vary rapidly with changes in the differences between terminal direct voltages, $U_{d1} - U_{d2}$. Two methods are available for the control of $U_{d1}$ and $U_{d2}$. One is to change the converter transformer ratio by tap-changer control. The other is to change the ratio between direct voltage and alternating voltage by phase control, that is, change of the delay angle $\alpha$ (see Par. **5**).

The first method is slow because of the highly limited speed of the transformer tap changer. The delay angle $\alpha$, on the other hand, may be changed very rapidly by the phase control system. However, this is accomplished at the cost of an increased amount of reactive power being consumed by the converter (see Par. 7).

To minimize the amount of reactive power consumed, it is usual to operate the rectifier at a minimum nominal delay angle $\alpha$ and the inverter at a minimum nominal margin of commutation $\gamma$. The mode of operation for the inverter is either margin angle regulation or voltage regulation, and the reference values of margin angle $\gamma$ or dc voltage $U_{d2}$ are determined by the requirement for secure inverter operation without excessive commutation failures, even in the presence of minor ac system disturbances. A value between 15 and 18° is usually chosen. A larger value would give a decreased risk for commutation failures but would simultaneously give increased stresses on the valves and increased demand for reactive power.

If the margin angle regulation mode is used, the dc line voltage $U_{d2}$ may then be adjusted by tap-changer control at the inverter. If the voltage regulation mode is used as the primary mode, the margin angle mode becomes a backup for disturbances that would lead to commutation failures as a result of fast reduction of the ac voltage. Margin angle regulation continues at the expense of a reduced dc voltage until the transformer load tap changer has raised converter ac voltage sufficiently to again permit dc voltage regulation. The current and thus also the transmitted power may then be rapidly controlled by grid control in the rectifier; i.e., control of the rectifier direct voltage $U_{d1}$ in Fig. 15-9 is accomplished by means of delay angle control. At steady-state operation it is suitable to operate with a delay angle in the range 12 to 18°. A smaller delay angle would give less demand for reactive power but also would limit the ability to rapidly increase the rectifier voltage by decreasing the delay angle.

As the direct voltage is determined by the inverter, the suitable delay angle range for the rectifier may be obtained by tap-changer control in the rectifier.

***105. Static Characteristics of a Converter.*** Assume a converter with a basic constant-current-control system with a current order $I_{d1}$. The converter is also provided with a constant voltage control mode and/or a constant $\gamma$ control mode. The latter prevents the margin of commutation from decreasing below a set reference value in steady state.

In Fig. 15-71 are shown the characteristics for such a converter when the load is increased from zero. From point $A$ to point $B$ (to the left in Fig. 15-71) the load could be assumed to be a resistance decreasing from infinity at point $A$. The load resistance is too high and the converter voltage is too low for the converter to be able to deliver the ordered current. The converter operates at the minimum limit value of the delay angle, which usually is set between 5 and 8°.

At point $B$, the resistance of the load has decreased to such a value that the desired current is achieved if the converter voltage has its maximum value. At a further decrease of

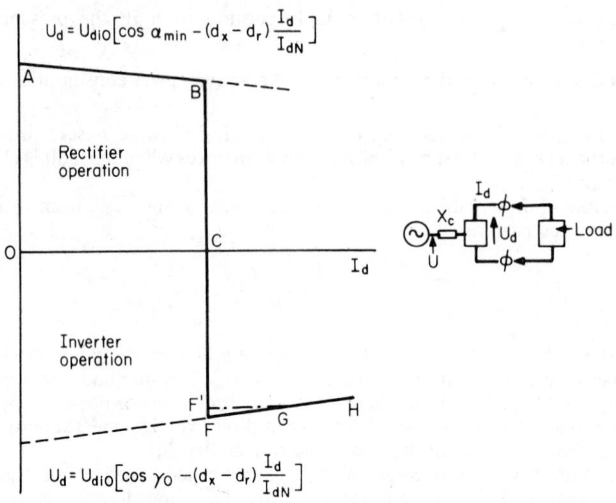


**FIG. 15-71** Voltage current characteristics for a current-controlled converter are shown when the load is increased from zero.

resistance, the converter must increase its delay angle to keep the direct current at the value $I_{d1}$. This occurs in the region *BF* in which the converter operates in current-control mode.

At point *C*, the converter voltage changes polarity and, if the overlap angle is neglected, has a delay angle equal to 90°. From this point the converter works as an inverter and the load must be active, for example, a direct voltage source in series with a resistor.

At point *F*, the delay angle $\alpha$ has increased to a value where the margin of commutation has reached its minimum value, $\gamma_0$, and the $\gamma$ control system takes over. *F* represents the point at which the inverter has its maximum voltage.

From *F* through *H* the margin of commutation is kept constant. Thus when the voltage of the load, and thus the direct current and the overlap angle $\mu$ of the converter, is increased, the converter direct voltage is decreased. This gives the converter a characteristic in this region corresponding to a negative resistance.

When constant voltage control mode is used, the inverter operating point is at *F′*. As load is increased for a constant ac system voltage, the operating point moves from *F′* to *G*, the overlap angle of the converter is increased, and the converter margin angle is decreased to maintain constant dc voltage. At point *G* the operating mode switches to constant margin angle $\gamma$ and follows operating *G* to *H* for further load increases.

From this discussion, the following modes of operation for a converter may be summarized:

Rectifier operation against the minimum limit in the delay angle (from *A* to *B*). This occurs when the converter voltage is not high enough to generate the desired current.

Constant current control (between *B* and *F*, or *B* to *F′*). There are no principal differences between rectifier and inverter operation.

Control on constant margin of commutation (from *F* and further) or control on constant voltage mode from *F′* to *G* and with constant margin mode from *G* to *H*.

***106. Cooperation Between Sending and Receiving Terminals.*** Consider an HVDC transmission with the principle shown in the configuration of Fig. 15-9. Each station may consist of a series connection of a number of 6-pulse converters. The terminal with index 1 is operating as rectifier and the other as inverter.

The rectifier has the characteristic *ABC* of Fig. 15-71, and the inverter is principally represented by the curve *CFH* or *CF′GH*. The inverter is, however, given a current reference $I_{r2}$ which is slightly less than the reference of the rectifier $I_{ri}$. The difference

$$\Delta I = I_{ri} - I_{r2} \tag{15-49}$$

is called the current margin and is normally in the order of 10 to 15% of rated current.

The cooperation between the rectifier and inverter terminals with characteristics as above is illustrated in Figs. 15-72 and 15-73. The dc line voltage drop can either be regarded as neglected or included in any of the characteristics for the terminals.

In contrast to Fig. 15-71, where the line voltage has been defined as positive for rectifier operation, the inverter characteristic has been defined as positive for inverter operation in Figs. 15-72 and 15-73.

Figures 15-72 and 15-73 also indicate that the inverter characteristic has been slightly modified by cutting the sharp corner (*Aa* in Fig. 15-72*a*). This measure may be taken to

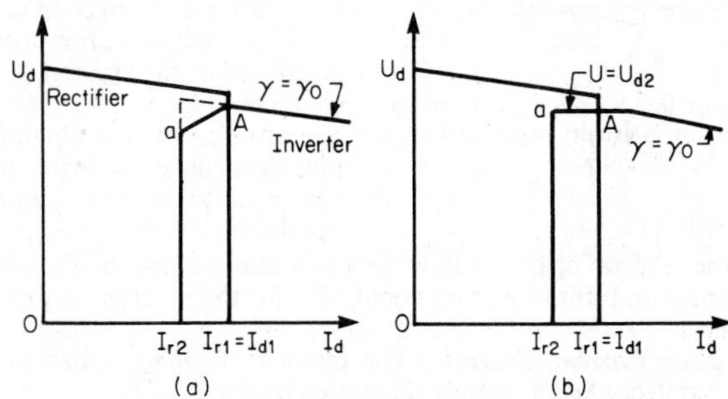


**FIG. 15-72** Direct voltages are plotted against the direct current with higher maximum voltages on the rectifier side.

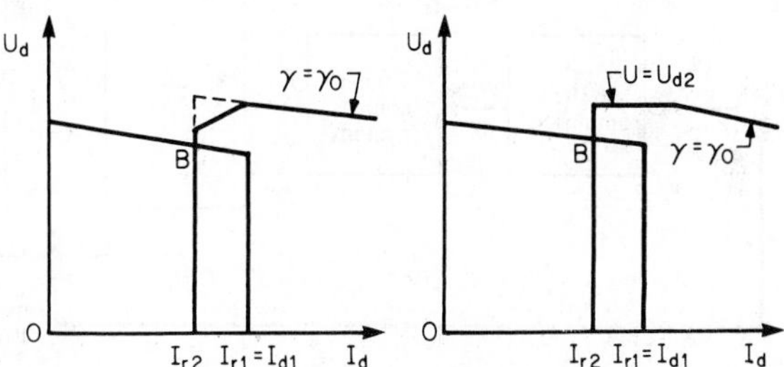


**FIG. 15-73** Direct voltages are plotted against the direct current with higher maximum voltages on the inverter side.

avoid an undefined intersection between the rectifier and inverter characteristics if they coincide or nearly coincide at voltage deviations in either or both of the connected ac networks. The inverter characteristic for the voltage control mode is shown in Fig. 15-72*b*. In this case the undefined intersection between rectifier and inverter is prevented by a small increase in the voltage reference $U_{d2}$ upon first detection of current mode at the inverter.

Figure 15-72*a* and *b* depicts the operation mode with the minimum delay angle in the rectifier giving a higher direct voltage than either the minimum extinction angle or the voltage references $U_{d2}$ in the inverter. The interconnection point *A* between the two characteristics defines a stable operating point. The direct voltage is determined by the inverter and the current by the rectifier. If the alternating voltage is decreased in the rectifier terminal or increased in the inverter terminal, the condition may change in such a way that the inverter takes over the current control as depicted in Fig. 15-73*a* and *b* at operation point *B*. With more than one converter connected in series, such a transition also occurs when one rectifier is disconnected (blocked) or when another inverter is connected in series (deblocked).

The reason for introducing a margin between current orders in the rectifier and inverter becomes obvious from the two figures, as zero margin or a negative margin would not have given any stable operating point.

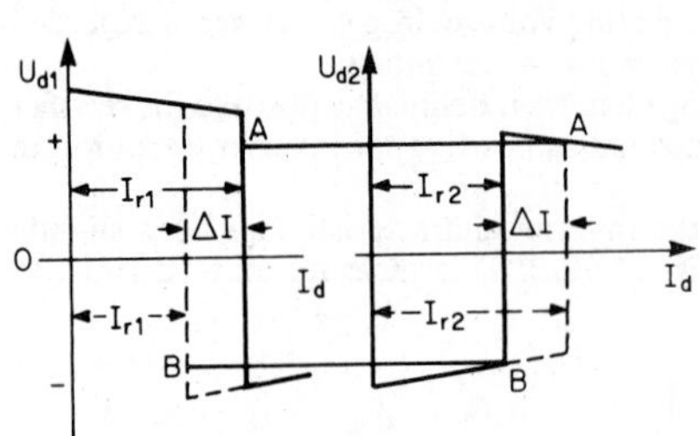


**FIG. 15-74** A change in power-flow direction for an HVDC transmission is usually performed by a polarity change of the line voltage.

A change of power-flow direction for an HVDC transmission is usually performed by a polarity change of the line voltage. This is illustrated in Fig. 15-74 where the complete converter characteristics have been drawn for both stations. The initial operating point is *A* and the current margin is $\Delta I$. If the latter is removed in station 2 (initially inverter) and the current order in station 1 is reduced by the amount $\Delta I$, the broken line characteristics will be valid and the new operating point *B* is obtained. Thus the dc line has changed polarity, and the two converter stations have changed mode of operation.

In practice, the reversal of power-flow direction also requires other operations, such as switching the minimum limit of $\alpha$ from about 100° for the inverter to about 5° for the rectifier and vice versa.

***107. Closed-Loop Current Control.*** The rectifier normally is provided with a basic feedback current control system, simply illustrated by Fig. 15-75.

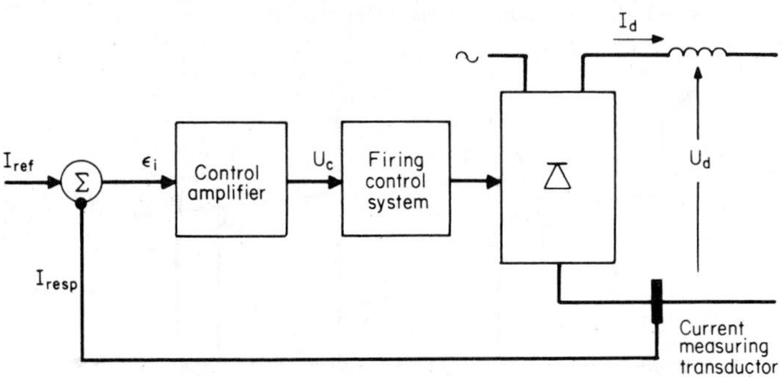


**FIG. 15-75** Simplified block diagram for a constant-current control system.

The control amplifier is used to give suitable static gain and dynamic compensation for stabilization of the control loop. Control pulses are generated by the firing control system and transmitted to valve potential by the valve control system (assumed included in the valve bridge symbol in Fig. 15-75).

Firing control system designs may differ considerably, but they belong mostly to one of two types, here called the individual phase control system and the equidistant firing control system, respectively.

In the individual phase control system the valves connected to the same phase are controlled separately from other valves. Such a system may be so designed that the output voltage from the control amplifier (in this case approximately integrating) is compared to a monotonically increasing voltage, one for each valve, which starts at the zero crossing of the commutation voltage which corresponds to $\alpha = 0$. At equality between the two voltages a control pulse is generated.

Individual phase control is the alternative for HVDC. It was used prior to the Pacific Intertie transmission, which was placed into operation in the late 1960s. In later applications, however, it has been replaced by the equidistant control principle.

The basic part of an equidistant firing control system is a voltage-controlled oscillator. Its frequency in steady state is equal to the product of the pulse number of the converter and the fundamental frequency of the ac network. The output frequency is controlled by the output voltage $U_c$ from the current control amplifier as shown in Fig. 15-76.

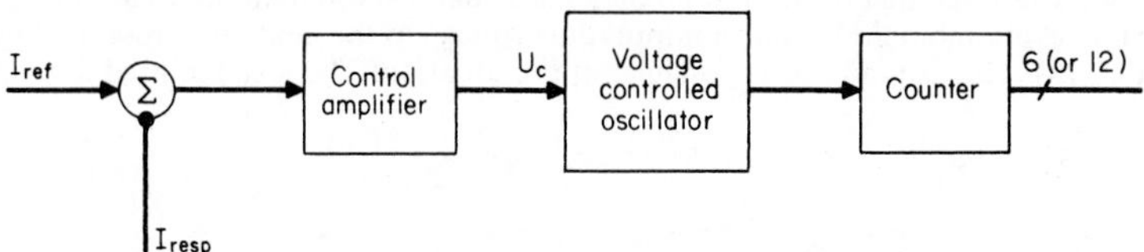


**FIG. 15-76** A block diagram showing the generation of control pulses in an equidistant firing control system.

When transiently $I_{ref} > I_{resp}$,[1] $U_c$ is given a suitable value by the control amplifier to decrease the frequency of the oscillator. On the contrary, when $I_{ref} < I_{resp}$, $U_c$ increases the same frequency. A decreased frequency means a steadily decreasing $\alpha$ and increasing voltage. In this way the direct current is restored again to its nominal value.

The output signal from the controlled oscillator is a pulse train which contains all triggering pulses for a valve bridge. A separation must be done to obtain individual control pulses for each individual valve. This is performed by a digital counter of suitable design. $U_c$ in Fig. 15-76 is obviously proportional to $d\alpha/dt$, as the oscillator is frequency-controlled by its input voltage. It is also possible to use a phase-controlled oscillator by which $U_c$ corresponds directly to $\alpha$.

For a practical system the block diagram in Fig. 15-76 must be completed with special functions, the objective of which is to prevent the oscillator from falling out of phase. Thus the control pulses must always be generated within a time interval defined by the limits $\alpha_{min}$ and $\gamma_{min}$.

***108. Margin of Commutation Control.*** By analogy to current control, the control on a minimum margin of commutation may be of the individual phase or equidistant firing type. Two further alternatives exist for both types, namely, a predictive system or a feedback control system.

A predictive system is a feedback type of control system which, based on instantaneous measurements of direct current and commutation voltage, calculates the correct instant for firing. This type of system is, of course, inherently individual for each valve or phase. However, if the most critical valve is chosen and allowed to determine the firing for all valves in an equidistant way, an equidistant control system is obtained.

A typical feedback control system measures the resulting $\alpha$, compares it to a reference value $\gamma_r$, and uses the difference $U_\gamma = \gamma_r - \gamma$ to control a voltage-controlled pulse oscillator as described for the current control system. In this case an equidistant firing control system is obtained. However, an individual phase control system using the feedback principle is also possible. Both types of system, predictive and feedback, have commutation failures at disturbances, but the feedback principle gives better precision of control and may be simpler, especially when combined with a voltage-controlled oscillator system for current control. The block diagram in Fig. 15-76 can also be used to represent a feedback control system if $I_{ref}$ and $I_{resp}$ are replaced by $\gamma_{ref}$ and $\gamma_{resp}$, which represent the smallest measured $\gamma$ for the valves within the bridge.

A predictive margin of a commutation control system may also work; the margin of commutation is then defined as the remaining voltage-time area of the commutation voltage from the end of the overlap to the succeeding zero crossing of the same voltage. A voltage-time area or turnoff time is needed for the commutation.

[1]Subscripts stand for "reference" and "response," respectively.

The prediction process is characterized as a continuous calculation of the resulting margin of commutation which would result if the firing occurred in the relevant moment. When the predicted margin has decreased to a reference value, a firing pulse is generated. When one valve has fired, a special selection unit connects the commutation voltage for the next firing valve to a computer circuit, the predictor. This circuit is also fed with a voltage proportional to the half-period time of the ac system frequency, as well as a voltage proportional to the elapsed time from the preceding zero crossing of the relevant commutation voltage.

Using these three quantities, the predictor calculates the remaining total voltage-time area $A(t)$ at every moment for the commutation voltage to the next zero crossing (Figs. 15-77 and 15-78). This is made as a triangular approximation. The area required for commutation $A$ is proportional to the direct current. When this area is subtracted for the total area, we get

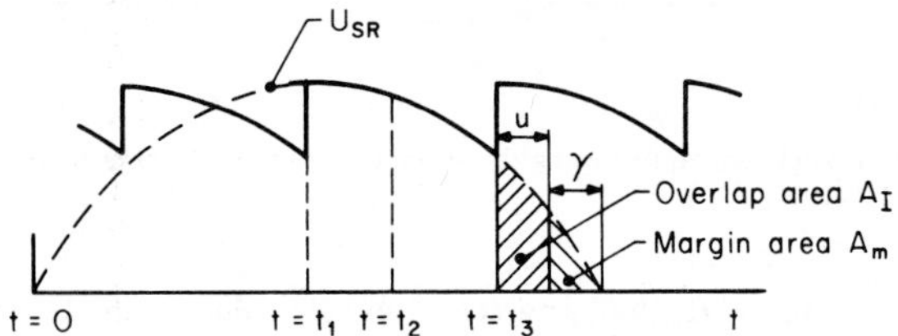


**FIG. 15-77** Prediction of the remaining voltage-time area is calculated for the commutation voltage.

$$A_p(t_2) = A(t_2) - A_I \tag{15-50}$$

$A_p(t_2)$ being the commutation voltage-time area margin when firing occurs at the arbitrary time $t_2$. The value corresponding to $A_p(t)$ is compared with a reference value $A_{m,\mathrm{ref}}$. When $A_p(t) = A_{m,\mathrm{ref}}$, a firing pulse is released at the time $t = t_3$. After the commutation, a commutation margin area $A_m$ is available.

The prediction process as described above is not exact. To correct for this, a feedback loop may be included. The difference between the predicted margin $A_p$ at firing and the actually obtained margin $A_m$ may then be calculated. The difference

$$\Delta A = A_p - A_m \tag{15-51}$$

is used as a correction at a later firing (for example, one period later) and added to $A_{m,\mathrm{ref}}$.

The firing control principles discussed above result in a form of individual phase control and thereby nonequidistant firing at ac network asymmetries. It is suitable to include special functions to attain firing symmetry even at asymmetric alternating voltage.

The valve which requires the earliest firing (the valve with the smallest margin of commutation) will, in steady-state operation, determine the firing for the other valves with equal distances (60° in a 6-pulse converter) between firings.

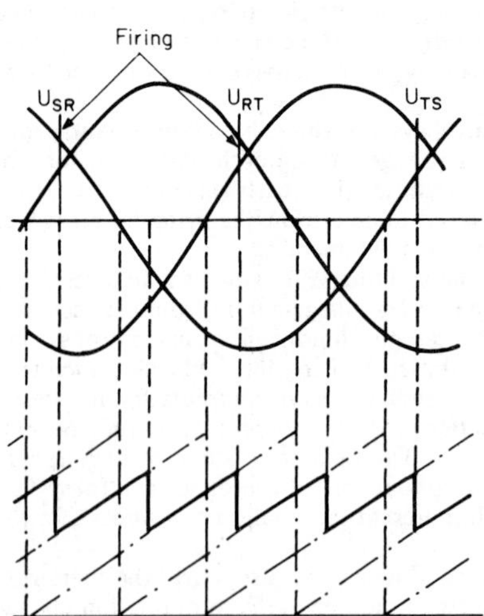


**FIG. 15-78** Generation of a voltage is proportional to the elapsed time after the previous passage through zero of the commutation voltage for the valve in turn to fire.

The firing control system may be designed for either 6- or 12-pulse operation with regard to both current control and margin of commutation control. A 12-pulse firing control system is here understood to mean one equipment common to both 6-pulse valve bridges within the 12-pulse converter and the voltage-controlled oscillator described earlier working with a frequency equal to 12 times the fundamental frequency of the ac network.

***109. Master Control.*** The basic current control system was described in the preceding paragraph. Since the dc line voltage is almost constant as long as the number of converters connected in series in each terminal is not changed, the direct current control is almost equivalent to power control. This is especially true when the inverter is operating in the constant voltage control mode. However, series converters may be connected or disconnected or short-term ac system voltage depressions may occur which result in considerable change of direct voltage. It might then be favorable to control the current in such a way that the transmitted power is kept constant. The traditional way of doing this is to provide each terminal with a current-order calculator which supplies the reference to the current control system. This reference is calculated from the desired transmitted power by dividing it by the measured dc line voltage.

As identical current orders have to be supplied to both terminals, a telecommunication link must be used for system control. This may transmit either a current order or a power order, or both. The common order is either set manually in any of the terminals or set from a control system of higher order, which may also consider the situation in the sending or receiving ac networks as described in the next paragraph.

Another type of master controller is shown in Figs. 15-79 and 15-80. In this case, all important signal conditioning of the orders is performed in common for both stations in a lead master controller which is placed in either of the two stations. Figure 15-79 shows this equipment, from which the current order is transmitted to a slave master controller, shown in Fig. 15-80, in the other station. The order transmission is performed in a synchronous manner by which changes in the order settings produce simultaneous changes in the effective current order in the two stations.

Manual power-order setting and rate-of-change setting can be performed in either of the two stations. Orders are transmitted by the telecommunication link to each station with the designated lead master controller being transferable between stations.

When the dc link is used for stabilization of an ac network, a power-order contribution is generated in a damping controller and added to the normally set power order as shown in Fig. 15-79. In this case, it is suitable (but not necessary) to place the lead master control equipment in the station which is connected to the network to be stabilized to minimize the total telecommunication delay.

In the lead master controller, the power order and rate-of-change order are converted to a ramp by which the power on the dc transmission is changed when an execution button is pushed. A current order is calculated as the quotient between the power order and dc line voltage, and the order is limited by a master load limiter before it is transmitted to the other station (Fig. 15-79).

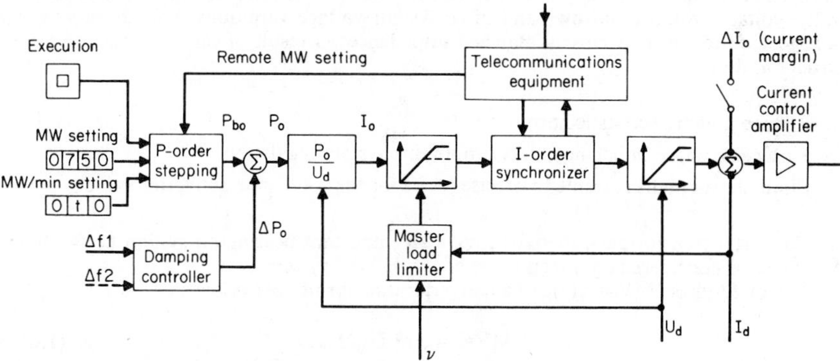


**FIG. 15-79** The master controller in the lead station transmits the current order.

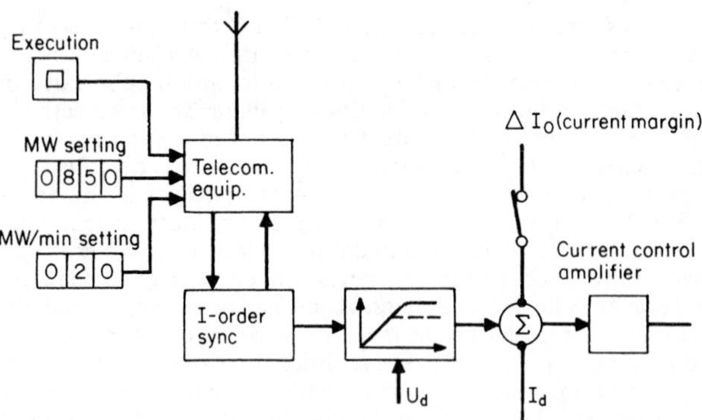


**FIG. 15-80** The master controller in the trail station receives the current order.

The master load limiter performs an optional limitation with regard to the current through the valves and possibly the cooling-medium temperature $\nu$.

A voltage-dependent current-order limiter is present in both stations.

***110. Interconnection of Nonsynchronous and Synchronous AC Networks.*** The interconnection between the sending and receiving ac networks is either nonsynchronous or synchronous, that is, there may or may not be a parallel ac connection. In the first case the frequency might also be different, for example, 50 and 60 Hz. From the point of view of control this is, however, of minor interest.

In the nonsynchronous case the transmitted power on the dc link is sometimes made dependent upon the frequency in any of the networks either to control this frequency completely or just to improve the stability in any of the networks during certain transient conditions.

In the synchronous case the transmitted power may be partly controlled from the phase difference or a time derivative of the phase difference (frequency difference) which might have a stabilizing effect at disturbances as described in Par. **113.**

## Converter Transformers

***111. Function and Design.*** Converter transformers are the link between the ac and dc systems. They provide a natural barrier between systems, preventing dc voltage and current from reaching the ac system. They also provide the necessary phase shifts required for 12-pulse valve operation through wye and delta secondary connections, and they maintain the valve voltage within a narrow band of ac system voltage variations through on-load tap changers. Through their closely matched impedances, a result of careful design and construction, they

Reduce noncharacteristic harmonics

Reduce ac phase imbalance, thus simplifying system regulation

Limit short-circuit currents to precise levels for the solid-state converter valve

Converter transformers, therefore, have an important bearing on system performance and must be considered an integral part of any HVDC system.

The three-phase MVA rating of a converter transformer per bridge is

$$\text{MVA} = \pi/3\ U_{d0N} I_{dN} \qquad (15\text{-}52)$$

$I_{dN}$ is by IEC convention.

$$E_{d0N} = \frac{E_{dN}}{\cos\alpha - \overline{X}\left(\frac{1}{2} + \frac{\pi}{3}\frac{\overline{R}}{\overline{X}}\right)} \quad \text{rectifier} \tag{15-53}$$

$$E_{d0N} = \frac{E_{dN}}{\cos\gamma - \overline{X}\left(\frac{1}{2} - \frac{\pi}{3}\frac{\overline{R}}{\overline{X}}\right)} \quad \text{inverter} \tag{15-54}$$

where $R$ = resistance at specific tap distance
$\overline{R}$ = average transformer resistance
$\overline{X}$ = average transformer reactance
$X$ = reactance at specific tap positions

A load tap changer (LTC) maintains rated dc voltage with variation in dc load and ac system voltage. The tap range on a converter transformer ac winding may be as large as 30% of the ac winding turns. The basic equation for LTC compuation is

$$E_{d0} = \frac{E_d + \left(0.5 \pm \frac{\pi}{3}\frac{R}{X}\right)\overline{X}I_d E_{d0N}}{\cos\alpha/\gamma} \tag{15-55}$$

Under normal conditions, losses in a converter transformer are composed of the same losses in an ac transformer plus losses due to harmonic currents produced during conversion. An approximate formula for converter transformers loss is

$$\text{Total losses} = I^2R + (1 + k)\,(\text{stray losses} + \text{eddy losses}) \tag{15-56}$$

The factor $k$ ranges from 0.75 to 2.5 and depends upon specific transformer design materials and configurations.

Converter transformers have a unique set of dielectric design characteristics since the windings are subjected to direct voltage stress superimposed upon the customary ac voltage stress. The ac stress distributes as it would in a capacitive network, and the steady-state dc stress distributes as it would in a resistive network.

### Smoothing Reactors

***112. Smoothing Reactor Function.*** A smoothing reactor is placed in service with dc converters to smooth dc ripple current and reduce current transients during system contingencies (see Fig. 15-10). It serves a secondary purpose of protecting the converter valve from voltage surges coming from the dc line. Special filtering circuits may be installed to reduce telephone interference from the dc line. These must be coordinated with the smoothing reactor to avoid a resonance problem. For a 6-pulse group, the difference between the average and minimum direct current is given by

$$I_d = \frac{V_{d0}}{\omega L_d}\,(0.0931 \sin\alpha) \tag{15-57}$$

where $L_d$ = smoothing reactor inductance.

### AC Breakers

***113. Breaking and Reclosing of Parallel AC Ties.*** Transmitted dc power can be controlled to stabilize one or more associated ac systems during the "dead" time between opening the parallel ac tie and its subsequent reclosure. The controls generally continue to act after the reclosure until both ac systems are returned to equilibrium. The following alternatives for control of the dc link are assumed:

**a.** Constant power.

**b.** Constant phase angle between the ac sides. The objective of this method of control is to vary the dc power to maintain a constant difference between the phase angles of the two ac systems. Since measurement of phase angle difference across long distances is difficult, the control of angle acceleration $d^2\phi/dt^2$ or the time derivative of frequency difference $\Delta f$ is often used. This type of control is equivalent to controlling the acceleration of synchronous machines in the two ac systems.

**c.** As (a), with the addition of a term proportional to the frequency difference.

**d.** As (b), with the addition of a term proportional to the frequency difference.

It is also assumed that during the dead time no controls in the ac network act, and that the dc link can carry the necessary overload.

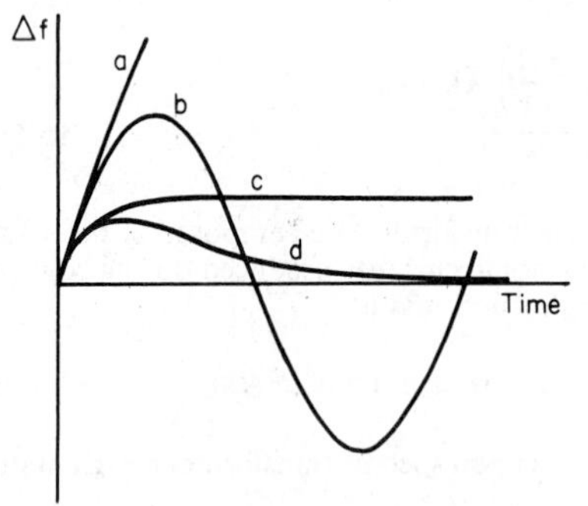


**FIG. 15-81** The function of frequency deviation $\Delta f$ in a parallel dc link stabilizing an ac link.

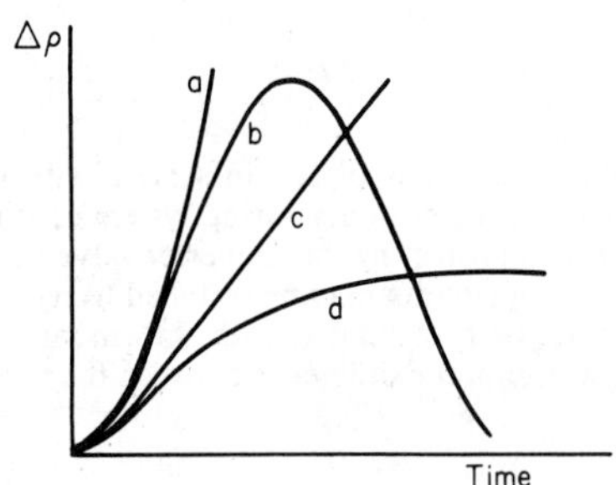


**FIG. 15-82** The function of phase-angle deviation $\Delta\phi$ in a parallel dc link stabilizing an ac link.

*Case a.* After the opening of the ac line, the frequency difference $\Delta f$ between the two ac networks increases continuously (curve *a*, Fig. 15-81), along with the phase angle $\Delta \phi$ (curve *a*, Fig. 15-82). The ac line must be reclosed before $\Delta f$ and $\Delta\phi$ have exceeded allowable values.

*Case b.* The dc link responds like an ac link. The oscillations of $\Delta f$ and $\Delta\phi$ are shown as curve *b* in the above-mentioned figures.

*Case c.* This situation is shown by the *c* curves. The frequency deviation is limited, and the phase-angle difference increases more slowly than in case *a*, giving more time for reclosing. It should even be possible to wait one or more revolutions before the ac line is reclosed.

*Case d.* This situation is shown by the *d* curve. The phase-angle deviations are limited, and the frequency difference brought down to zero after some time. The ac line can be reclosed at any time.

## HVDC Breakers

***114. Basic Concepts.*** In contrast to ac systems, the fundamental problem of breaker operation in dc systems is the absence of current zeros. DC transmission circuits usually have large inductances in the form of smoothing reactors and long transmission lines. Therefore, short of reversing driving voltages, direct current can be brought to zero only by applying a counter voltage in excess of the driving voltage. A dc breaker attempts to insert this countervoltage or breaker voltage in one manner or another. While bringing the direct current to zero, the breaker has to absorb significant amounts of energy. In a simple representation of a dc circuit shown in Fig. 15-83, the breaker is shown inserting a countervoltage $V_b$. The energy $E_b$ absorbed by the breaker in bringing the current to zero by applying this countervoltage is

$$E_b = \frac{1}{2}LI_d^2\frac{V_b}{V_b - V_d} \tag{15-58}$$

***115. Breaker Capabilities.*** There are four dc breaker capabilities of special interest: voltage capability, current capability, energy absorption capability, and switching time. Voltage capability of the breaker is of two types: maximum breaker voltage and voltage withstand capability of the breaker. The term voltage capability of the breaker is intended to reflect the voltage withstand capability of the interruption part, i.e., the maximum breaker voltage. In general, the highest permissible breaker voltage gives the best performance in terms of lower energy absorption demand and shorter switching time.

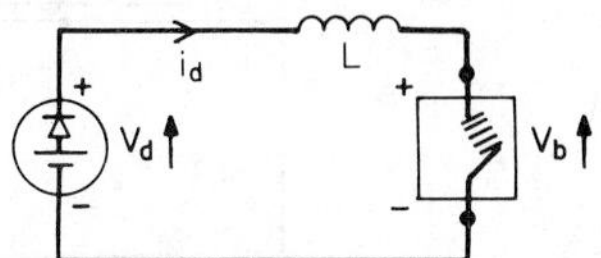


**FIG. 15-83** A simple representation of a dc circuit showing the breaker inserting a counter voltage $V_b$.

The interrupting current capability of the breaker is of special interest and a dominant consideration in the design. In dc systems, fault currents can reach high values compared to the steady state rating of different parts of the dc circuit. Automatic constant current control response of the converters should bring this fault current within the limits of the converter steady-state current ratings in a short time, depending on characteristics of the faulted dc circuit.

The energy capability of the breaker depends on the ac circuit parameters, inductance, converter voltage, interrupting current, breaker, voltage, and duty cycle. In high-voltage breakers with series-connected modules, the commutation process spread of different modules may also affect the energy rating of the breaker.

Switching time has four major components: time for fault detection, time between trip signal and energy absorber insertion, time to bring the current to zero, and time to bring the system to a new steady-state condition. Time for fault detection may be on the order of 2 ms for station faults and 10 to 30 ms for line faults. If the breaker trip signal is to be delayed to bring fault current to breaker interrupting current capability, additonal time delay might be involved, Providing for this, the maximum time for this phase is not likely to exceed about 40 ms. The time between the trip signal and the insertion of energy absorbers depends primarily on the opening time of the current-carrying breaker and possibly on the commutation process. The time required to bring the current to zero after insertion of the energy absorber decreases as the breaker voltage increases. After current interruption, some time will be needed to bring the dc system back to its approximate postdisturbance steady-state condition.

Based on the above considerations and recognizing the variations in breakers and the dependence of interruption on dc circuit and fault conditions, an approximate range of switching time for fault clearing would be 30 to 200 ms.

There are opportunities to improve system performance by the application of HVDC breakers in multiterminal systems.

## Protective Relay Equipment

***116. Relays.*** Protective relay equipment found in HVDC converter stations and associated ac stations is the same as that used for individual equipment protection in other ac stations and is detailed in other sections of this handbook.

## DC Current and Potential Transductors

***117. DC Current Transductor.*** For an HVDC power system, measurements of direct current are required as an input to the converter controller and for metering and instrumentation.

The dc current transductor (DCCT) measures the dc current and provides a signal, in the form of a dc voltage proportional to the dc current. The DCCT also provides isolation between the HVDC line, where the current is measured, and the control system ground.

A regular (unipolarity) DCCT is comprised of a transductor unit and a sensor unit (Fig. 15-84). The transductor unit consists of two toroidal saturable cores, each with primary and secondary windings. In this case, the primary winding is merely the dc line conductor or

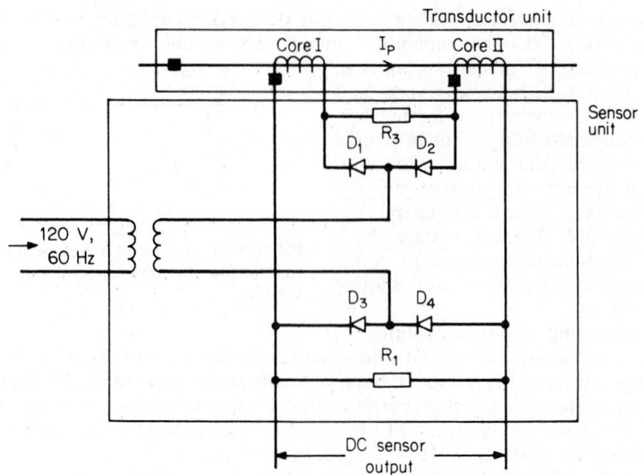


**FIG. 15-84** A regular or unipolarity dc current transductor (DCCT) is comprised of a transductor unit and a sensor unit.

cable passing through the windows of the toroidal cores. The secondary windings are connected in series and are excited by ac voltage. The sensor unit of each DCCT processes the secondary currents of the saturable cores and provides the output voltage.

The DCCTs have a nominal rating of 1000 A dc and require a 120-V, 60-Hz auxiliary voltage supply to give a dc output signal. The output voltage is 10 V dc across a 10-Ω resistor for rated primary dc current.

***118. DC Potential Transductor.*** As previously indicated, measurements of dc line voltage are required as an input to the converter controller and for metering and instrumentation of an HVDC system.

The dc potential transductor (DCPT) measures dc line voltage with isolation between the power system and control system ground. The DCPT consists of the following main parts:

Current-limiting precision resistor

Auxiliary unit

Transductor cores

Distribution and sensor box

The precision resistor, auxiliary unit, and transductor cores are located in the valve hall. The auxiliary unit is located at the base of one of the resistor columns.

Figure 15-85 shows a schematic arrangement of the equipment forming the DCPT circuit for one pole. One DCPT is necessary for each pole.

AC control power is required to obtain dc output voltage and is obtained from a first-grade power source.

## Arresters: Metal-Oxide Arresters, Valve Element Characteristics

***119. Disk Characteristics.*** Metal-oxide arrester elements, comprised primarily of zinc oxide, but containing a number of other metal oxides as well, have an extremely nonlinear voltage current characteristic. As illustrated in Fig. 15-86, a typical disk that conducts less than 1 μA of current at normal operating voltage exhibits a voltage only slightly more than twice normal operating voltage at a current of 10,000 A. Because of this degree of nonlinearity, it is possible to design both ac and dc arresters without series gaps. This is a simpli-

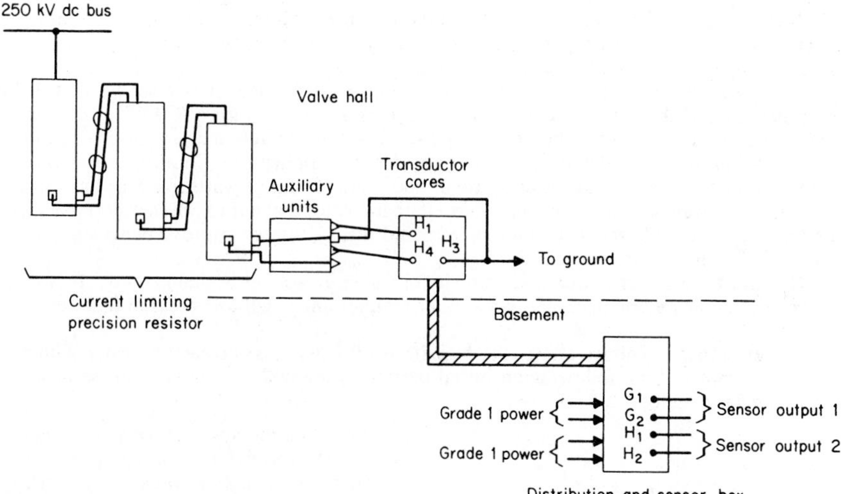


**FIG. 15-85** This schematic for a dc potential transductor (DCPT) indicates locations for the main parts. One DCPT is necessary for each pole.

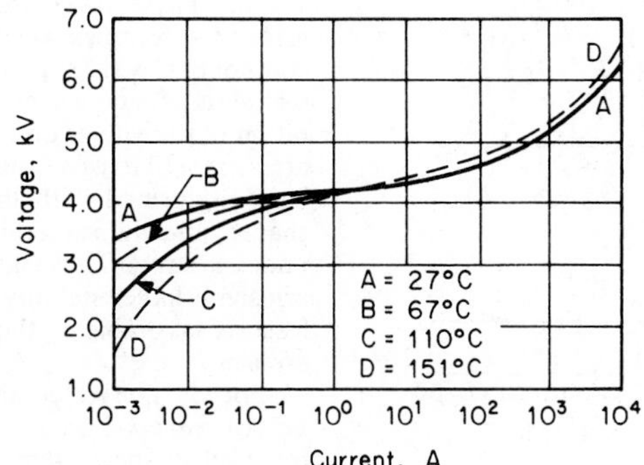


**FIG. 15-86** These voltampere curves for a typical zinc-oxide disk show operating characteristics for a temperature range of 27 to 151°C.

fication which is especially important for dc arresters because it is difficult to make a gap capable of clearing against dc voltage, particularly after the gap has been subjected to a high-energy surge.

When the current is drawn at its maximum continuous operating voltage, a typical zinc-oxide disk has a negative temperature coefficient, as indicated by Fig. 15-86. That is, as the temperature increases, the leakage current and resulting watts loss increase, and this increase must be considered in the thermal design of the arrester. However, the temperature coefficient becomes very slightly positive at currents above a few amperes. This slight positive temperature coefficient at higher current, in contrast to the strong negative coefficient of silicon-carbide nonlinear elements, makes it possible to operate zinc-oxide elements in parallel to discharge high-energy surges.

***120. Long-Term Stability.*** Elements used in either ac or dc arresters designed without series gaps must be able to withstand the effects of continuously applied voltage for a reasonable lifetime without significant change in characteristics. Several investigators report a gradual increase in leakage current and watts loss of metal-oxide disks under continuous operating stress.[1,2,3] It has also been shown that the rate of increase can be influenced considerably by disk composition and processing.[4]

Long-term stability studies of ac disks have been performed. It was shown that disks manufactured at that time would require 200 years at a temperature of 45°C to double in watts loss. This degree of stability, even allowing a considerable margin for error in data extrapolation, was considered adequate for ac arrester application. Continued tests on the same disks have shown that the estimate of 200 years to double watts loss was too conservative, and extrapolation of the longer-term data indicates that approximately 400 years will be required to double the watts loss in those old disks. Process improvements since 1976 have still further improved stability.

It is not generally true that disks which are stable on ac will be stable on dc. It is true, however, that ac and dc disks act in a similar manner under voltage stress in that

> Under constant voltage stress, the logarithm of watts loss increases linearly with the square root of time following an initial period during which a slight decrease in watts loss may be noted.

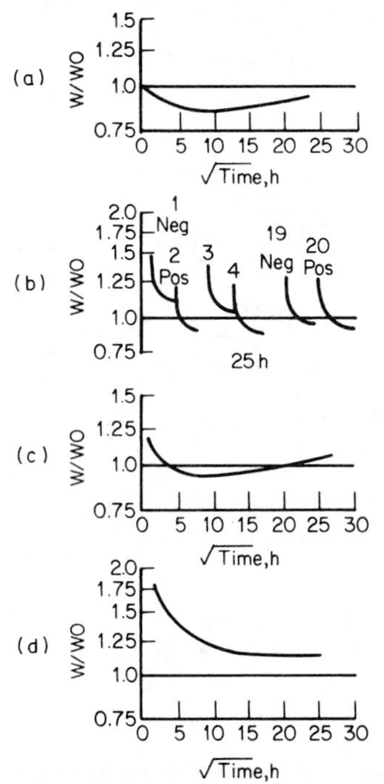


**FIG. 15-87** These curves for watts loss versus time relationship at 115°C for dc zinc-oxide elements show the effect of polarity reversal.

The phenomenon is thermally activated, and the logarithm of time required for a specified change in watts loss is linearly related to the reciprocal of the absolute temperature.

In HVDC converter station applications, arresters are subjected to various waveshapes. These range from dc on the line arrester to complex waves consisting of a dc component with superimposed steps and segments of sine waves caused by commutation of the valves on the valve and bridge arresters. Extensive stability testing has been performed with these various waveshapes, and the same phenomena occur as under ac stress. The disks used for dc system arresters have stability characteristics at least as favorable as those achieved for ac arresters.

For dc converter arresters, there is a voltage reversal each time power flow is reversed in the system. The effect of such polarity reversals was determined during accelerated life tests with results shown in Fig. 15-87. For these tests, dc leakage (or watts loss) was first determined with one polarity over an extended time to establish a trend for the actual disks being tested. After approximately 600 h, data were taken over a period of 28 days, with the polarity being reversed once each day except on weekends. After the 21st reversal, long-term

---

[1]Sakshaug, E. C., Kresge, J. S., and Miske, S. A., Jr.: A New Concept in Station Arrester Design, Paper No. F76-393-9, *Pow. Appar. Syst.*, March/April 1977, Vol. PAS-96, No. 2.

[2]Tominaga, A., Azume, K., Nitta, T., Najai, T., Imataki, M., and Kuwabara, M.: Reliability and Application of Metal Oxide Surge Arresters for Power Systems, Paper No. F78-703-1, *Pow. Appar. Syst.*, May/June 1979, Vol. PAS-98, No. 3.

[3]Kobayaski, M., Mizuno, M., Aizawa, T., Hayashi, M., and Mitani, K.: Development of Zinc Oxide Non-Linear Resistors and Their Application to Gapless Surge Arresters, Paper No. F77-682-8, *Pow. Appar. Syst.*, May/June 1979, Vol. PAS-98, No. 3.

[4]Written discussion of previous reference. *Pow. Appar. Syst.*, May/June 1979, Vol. PAS-98, No. 3.

data were again taken without polarity reversal, but at a polarity opposite to that of the initial data. Finally, the polarity was again reversed and long-term data taken for a polarity the same as that of the initial data. As Fig. 15-87 indicates, the basic effect of polarity reversal is to cause a temporary increase of up to 75% in leakage current. This is of no consequence in the arrester design, and the current will return to the previous level within a day or two. Thereafter, as shown by the final two long-term tests, the drift rate is the same as or perhaps less than, that at the original polarity. The slope of the increasing portion of the curves in Fig. 15-87 is such that a time of approximately 1 year would be required to double the watts loss if the operation were at 115°C. Field tests reported previously indicate that even in the worst of circumstances, the weighted average operating temperature of an arrester will not exceed 45°C.[1] The drift rate at 45°C will be approximately 800 times slower than at 115°C, confirming that long-term stability drift is of no consequence for these arresters.

Conclusions from extensive dc stability tests are that

With proper processing, entirely satisfactory zinc-oxide arrester stability is obtained for all waveshapes within the dc converter.

Polarity reversal will have no adverse effects on long-term arrester stability.

***121. Transient Stability and Heat Transfer Considerations.*** As discussed earlier and shown in Fig. 15-86, metal-oxide disks exhibit a negative temperature coefficient of resistance at normal operating voltage. Therefore, when the disks in an arrester become hot due to energy absorption during a surge discharge, leakage current and watts loss are increased. The arrester design must incorporate adequate heat transfer means to dissipate this increased loss while allowing the disks to cool from the transient condition. Because the heat transfer requirements are similar, the dc arrester utilizes the same heat transfer design as ac arresters with heat transferred to the porcelain container by means of silicone rubber molded around the disk periphery.

Figure 15-88 graphically illustrates the operation of the heat transfer design as determined for a dc arrester operating at a high ambient temperature of 60°C. Curve 1 of Fig. 15-88 shows the watts loss in a zinc-oxide element as a function of disk temperature. Curve 2 shows the watts that can be transferred from a hot disk to the 60°C porcelain housing as a function of disk temperature.

Under normal conditions, the operating point is the lower crossover point of curves 1 and 2, with the disk a degree or two warmer than the procelain. When disk temperature rises as a result of high-energy absorption during a discharge, watts loss in the disk increases in accordance with curve 1, and the disk cools at a rate determined by the excess dissipation capability, which is the difference between curves 2 and 1. If sufficient energy were absorbed to drive the disk temperature to the upper crossover of curves 1 and 2 (185°C), there would be no excess dissipation capability, the disk could not cool, and thermal runaway would result. Curve 3 of Fig. 15-88 represents a pessimistic estimate of increased leakage characteristics which might result after long-term use. It is seen that even then, the ultimate disk temperature could reach 155°C before there is a risk of thermal runaway.

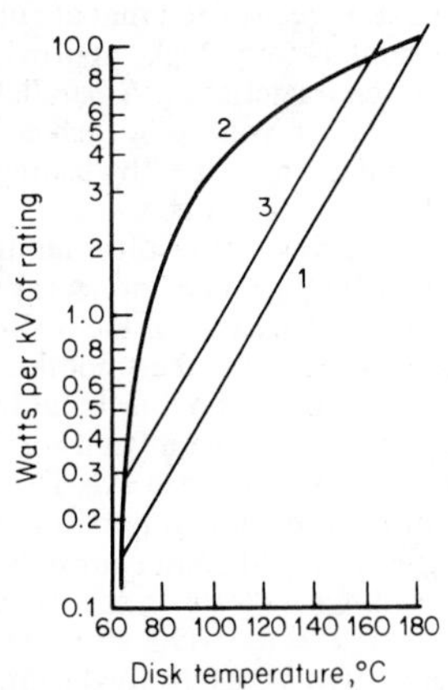


**FIG. 15-88** These watts loss and dissipation characteristic curves show the components of thermal design for a surge arrester.

***122. Energy Capability—Multiple Column Designs.*** The ultimate limit on the energy a disk can absorb is that which

[1]Sakshaug, E. C., Kresge, J. S., and Miske, S. A., Jr.: A New Concept in Station Arrester Design, Paper No. F76-393-9, *Pow. Appar. Syst.*, March/April 1977, Vol. PAS-96, No. 2.

causes disk cracking from thermal shock. This limit is reached if the sudden temperature rise exceeds 55°C, which requires energy absorption of approximately 6 kJ per kilovolt of maximum continuous operating voltage (MCOV) for the arrester. A sudden rise of disk temperature to only 115°C could result in disk cracking if the ambient procelain temperature were as high as 60°C. Consequently, it is obvious from Fig. 15-89 that the heat transfer system does not limit the amount of energy that an arrester can absorb in a single operation.

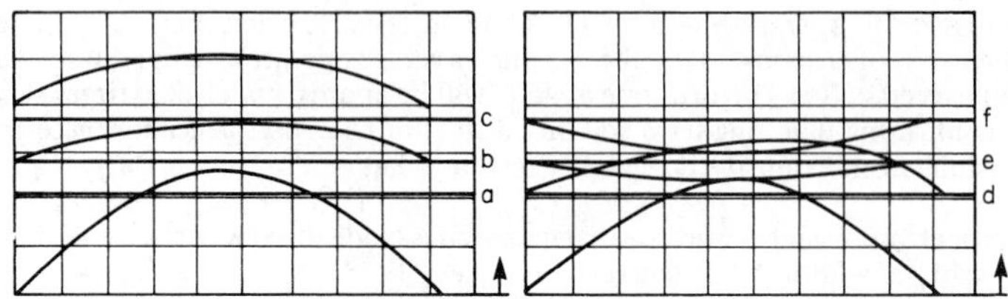


I = Total arrester current, 2600 a crest (650 a/col)
a = column 1-column 2 current (−0.5 a crest)
b = column 1-column 3 current (+2.1 a crest)
c = column 4-column 3 current (+3.5 a crest)
d = column 2-column 3 current (+3.3 a crest)
e = column 2-column 4 current (−0.9 a crest)
f = column 1-column 4 current (−1.4 a crest)

**FIG. 15-89** The results of current-sharing tests on valve arresters supplied for the Brazilian Itaipu HVDC project are shown. A four-column model was used. Therefore, six different currents were determined for a total arrester current crest of 2600 A.

In many dc applications, the energy capability of a single column of disks would be inadequate, and multiple column designs are used. The temperature coefficient of resistance for currents greater than a few amperes is slightly positive (Fig. 15-86), so there is no inherent tendency for columns to be driven out of parallel during high-energy discharges. However, it must be recognized that at current levels of a few hundred amperes per column, the disk exponent is very high (approximately 30) and the columns must be very carefully matched in characteristics. A parallel column arrester is constructed by selecting disks so the total discharge voltage of each column is the same at a preselected current. Proper current sharing is then verified by testing the columns in parallel and measuring the difference currents between columns.

Current-sharing test results obtained during design tests on valve arresters supplied for the Brazilian Itaipu HVDC project are shown in Fig. 15-89. A four-column model was tested to simulate the discharge of the 800-mi dc transmission line, plus dc filter capacitors, by a single valve arrester via the smoothing reactor. There were six values of difference currents to be determined for the four-column model. By taking the measurements on two consecutive tests with the current transformer connections changed between tests, all six values were measured as shown in Fig. 15-89. The total arrester current was 2600 A crest (650 A per column) and the maximum difference current was 3.5 A (0.5%). There were no discernible changes in the difference currents, either in magnitude or waveshape, through the extensive series of design tests on the various samples tested.

***123. Design Tests.*** DC arresters are usually designed to fulfill the specific needs of a particular system as determined from system studies. A number of design tests such as discharge voltage, high-current withstand, insulation withstand, radio interference voltage (RIV), and pressure relief are performed essentially in accordance with ANSI C6.2.1 as they would be for an ac arrester. However, a particularly significant test is a high-energy operating duty test done in place of the duty cycle and transmission line discharge tests of ANSI C6.2.1. An arrester model, including the correct heat transfer representation, is subjected to full-energy discharges starting at a maximum operating temperature (usually 60°C). Generally, two discharges are applied, spaced 1-min apart. Immediately following the second discharge, a dynamic overvoltage of duration and magnitude defined by system studies is applied, followed by the application of normal maximum system operating voltage. It is required that the arrester disks absorb the energy without damage and that the test sample cool to ambient temperature while operating voltage is applied.

It is common practice to require tests consisting of five operations at full energy and 20 operations at 80% of full energy. The difference currents between columns are monitored during the discharges to assure that current sharing is within design tolerances and to show that current sharing does not deteriorate as a result of the discharge duty. The oscillograms of Fig. 15-89 were obtained during such a test on the valve arrester design for Itaipu.

Experience indicates that it is not difficult to match columns to share current to within a 10% tolerance and that the balance between columns is not degraded even slightly as a result of the number of operations.

The application of zinc-oxide arresters with improved and demonstrated features has permitted the achievement of performance and reliability levels required by HVDC terminals and systems.

### Capacitors

***124. Capacitors*** in an HVDC converter station must provide some of the reactive power required by the HVDC converter. Capacitor banks are typically outdoor, rack-mounted installations, containing many individual capacitor units in multiple series- and parallel-connected groups. The bank is usually designed to operate at the incoming ac network's transmission-level voltage, although individual capacitor units are rated to operate at a maximum of 10 to 20 kV. The number of series groups in the bank is therefore determined by the bank's operating voltage. The number of parallel groups is dependent on the Mvar size of the bank—more Mvar of reactive power means more groups of parallel capacitors.

Capacitor bank and filter switching is done with loadbreak switches and/or circuit breakers, depending on switching duty. Current-limiting reactors are often incorporated to limit the magnitude and frequency of inrush currents for fault conditions and back-to-back capacitor bank switching. Capacitor units require individual fusing to (1) protect against can rupture in the case of an internal dielectric faliure, and (2) indicate the presence of failed units.

Although either internal or external fusing may be employed, the use of internal fusing requires an additional means of failure detection.

When an individual capacitor unit fails, the impedance of the affected series section increases and the voltage redistributes among the series sections. Capacitor units are rated to operate continuously at $>=$ 110% of their normal operating voltage. This accounts for the overvoltages seen in a capacitor bank prior to the replacement of a failed unit.

Capacitor units are given high-potential production tests to screen manufacturing defects and reduce the field failure rate. AC units typically get dc potential tests at 4.3 or 6.25 times rated ac voltage, or an ac potential test at 200% of rating. DC units are generally tested at three times rated dc voltage.

An additional requirement for dc capacitors is the provision for grading resistors within the capacitor bank. These are often required to fix dc voltage distribution across the bank in cases of environmental contamination. Without them, contaminant deposits on the capacitor bushings would establish voltage distribution in an arbitrary and nonuniform pattern.

## *PROJECT CONSIDERATIONS*

### Installation

***125. Scheduling and Coordination.*** The installation of an HVDC transmission terminal typically involves activity spanning the final one-half to two-thirds of an overall project schedule. A representative installation schedule is given in Fig. 15-90. Major installation-related activities include

Site preparation
Building erection
Yard foundations
Receiving materials and equipment at the site
Equipment erection
Equipment, subsystem, and system testing

These activities involve the coordination of varied electrical and civil engineering functions, as well as the scheduling of material and equipment shipments to the site to coincide with the installation sequences. The key to this phase of the project is strong on-site technical management by personnel familiar with the technical requirements of the equipment being installed, as well as labor relations and project scheduling and coordination.

HVDC systems may be constructed under various forms of management. The most prevalent is the turnkey approach, which places all responsibility on the prime contractor for constructing a system that achieves customer goals. HVDC systems may be constructed through a supplier-customer partnership, with the customer assuming (as an example) site responsibility. An engineer-constructor might also have turnkey responsibility under a licensing agreement. HVDC projects have been completed in as little time as 17 months and have required as much as 54 months. Thirty to thirty-six months would normally be adequate to completely install an HVDC system. An HVDC project of this kind could be broken into three parts:

System studies to define the required equipment
Equipment construction in the factory
Installation at the HVDC project site

## Testing and Commissioning

***126. Factory and Field Testing.*** Commissioning of the station occurs after the user has confirmed that the specified performance criteria are achieved at the converter station. Overall receipt of equipment, site preparation, erection of the building, installation, and testing may involve a time period of approximately 1 to 2 years, depending on system size and complexity.

There are two basic areas of testing for an HVDC project: factory and field testing. Fac-

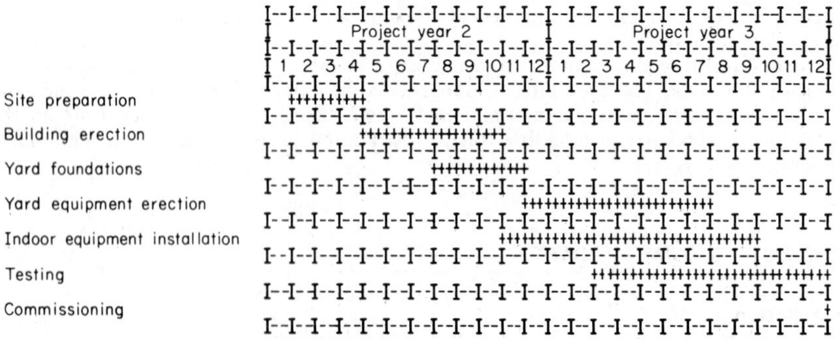

```
                                I--I--I--I--I--I--I--I--I--I--I--I--I--I--I--I--I--I--I--I--I--I--I--I--I
                                I          Project year 2             I          Project year 3           I
                                I--I--I--I--I--I--I--I--I--I--I--I--I--I--I--I--I--I--I--I--I--I--I--I--I
                                I 1  2  3  4  5  6  7  8  9 10 11 12 I 1  2  3  4  5  6  7  8  9 10 11 12 I
                                I--I--I--I--I--I--I--I--I--I--I--I--I--I--I--I--I--I--I--I--I--I--I--I--I
Site preparation                 ++++++++++
                                I--I--I--I--I--I--I--I--I--I--I--I--I--I--I--I--I--I--I--I--I--I--I--I--I
Building erection                         ++++++++++++++++++
                                I--I--I--I--I--I--I--I--I--I--I--I--I--I--I--I--I--I--I--I--I--I--I--I--I
Yard foundations                                 ++++++++++++
                                I--I--I--I--I--I--I--I--I--I--I--I--I--I--I--I--I--I--I--I--I--I--I--I--I
Yard equipment erection                                    ++++++++++++++++++++++++
                                I--I--I--I--I--I--I--I--I--I--I--I--I--I--I--I--I--I--I--I--I--I--I--I--I
Indoor equipment installation                             +++++++++++++++++++++++++++++++++++
                                I--I--I--I--I--I--I--I--I--I--I--I--I--I--I--I--I--I--I--I--I--I--I--I--I
Testing                                                             ++++++++++++++++++++++++++++++
                                I--I--I--I--I--I--I--I--I--I--I--I--I--I--I--I--I--I--I--I--I--I--I--I--I
Commissioning                                                                                         +
                                I--I--I--I--I--I--I--I--I--I--I--I--I--I--I--I--I--I--I--I--I--I--I--I--I
```

**FIG. 15-90** Installation of an HVDC transmission system typically involves activity spanning the final one-half to two-thirds of an overall project schedule.

tory testing involves production testing of each piece of electrical or mechanical equipment. In some cases, design or prototype testing may be required for newly developed equipment. Factory or production testing is performed to ensure that each piece of equipment has been manufactured properly and meets all NEMA, ANSI, and other applicable standards. These tests range from functional operation to checking alignment and fit for mechanical equipment, to power testing for electrical equipment to confirm guaranteed performance criteria such as losses, efficiency, and other production quality measurements. Installation or field testing involves testing each piece of equipment after it is installed to confirm that it is functioning in accordance with the agreed upon requirements. Subsystem testing is performed on all systems prior to the application of high voltage to the system. Table 15-10 is an itemization of installation and testing for a typical HVDC project.

Energization of high-voltage switchyard equipment is performed after subsystem control and protection tests have been completed. Station system testing involves initial testing of the dc valves with high voltage applied to the converter transformer terminals. In addition to preliminary subsystem testing, prerequisite subsystem testing prior to converter operation includes the following: thyristor monitoring, dc controls, light guides, valve cooling, valve firing, valve phasing, valve high voltage, and dc short circuit. Typical performance tests are conducted during operation of the converter where the transmission system starts at low power levels and progresses to rated power and voltage.

## Operation and Maintenance

***127. Operation.*** The operation of an HVDC converter station is similar to that of an ac generator station where remote control is possible. Unlike ac transmission, dc transmits power in a controlled manner. The system dispatcher, or HVDC station operator, determines the desired power level, and the dc system delivers that amount. Normal disturbances are automatically accommodated by the converter control and resolved within the converter station.

One of the converter stations may have an operator, and from this station the power, frequency, and current can be manually set. The operator receives essential information about other stations via a communication channel. This information includes fault signals and position of circuit breakers. The operator can stop and start the transmission, if necessary. The degree of automation will be determined from system to system.

The communication channel is also used to control changes in the transmitted power.

**TABLE 15-10** Subsystem Installation and Testing Sequence

| |
|---|
| AC auxiliary power |
| Station lighting and ac services |
| DC station battery |
| Uninterruptable power supply |
| Converter transformer and dc smoothing reactor |
| High-voltage switches and breakers |
| Reactive supply |
| Potential and current transformers |
| Fire protection and valve hall smoke detection |
| Air conditioning/heating |
| Valve cooling equipment |
| AC control and protection |
| Sequence and events/transient fault recorder |
| Thyristor monitoring |
| Valves |
| DC valve control |
| Round power (if applicable) |
| Through power |
| Acceptance and commissioning |

In case this communication link is interrupted, power on the transmission line remains at the value existing before the interruption.

On the assumption that conductor dc equals rms ac current, and the conductor dc voltage-to-ground equals the peak ac voltage-to-ground, the power transmitted per phase is

$$P_d = V_d I_d \quad \text{for the single-line dc system} \tag{15-59}$$

$$P_a = V_a I_a \cos \Phi \quad \text{for the single-line ac system} \tag{15-60}$$

and the ratio is

$$\frac{P_d}{P_a} = \frac{\sqrt{2}}{\cos \Phi} \quad \text{for a single-phase system} \tag{15-61}$$

For a bipolar dc system and a three-phase ac system, the ratio is

$$\frac{P_d}{P_A} = \frac{2\sqrt{2}}{3 \cos \Phi} \tag{15-62}$$

Because the power limit of overhead ac lines is often determined by factors other than conductor heating, the ratio of dc power to ac power per conductor may be very high (1.5 to 4).

***128. Maintenance.*** At certain intervals, normally once a year, all converter equipment including the valves should be subjected to a checking procedure. For control equipment, special measuring points are available that permit voltages, currents, and oscillograph traces to be compared with reference values and oscillograms with the station in operation. Valves should be checked regarding all fundamental functions of each thyristor circuit as well as the common equipment in the valve.

For a bipolar station with one air-insulated 12-pulse converter per pole, approximately 250 worker-hours of maintenance per year will normally be required by trained electricians.

## Bibliography

***129. References for Terminal Design***

Engström, Mutanda: Power by HVDC to Refine Copper in Africa, *IEEE Spec.*, December 1975.

Eriksson, G., and Haglöf, L.: HVDC Station Design, *ASEA J.*, 1975, Vol. 3, No. 3, pp. 61–65.

Klein, et al.: HVDC to Illuminate Darkest Africa, *IEEE Spec.*, October 1974.

Stairs, C. M.: Development of an HVDC Solid-State Back-to-Back Asynchronous Tie, Manitoba Power Conference, June 1971.

***130. References for Economics and Efficiency***

Danfors, P., and Hammarlund, B.: Must HVDC Go to UHV Levels?, *Electr. World*, April 1, 1973, p. 60.

Danfors, P.: The Role of HVDC in the Future, Transmission System Design, *Energy* 73, International Energy Exhibition, September 1973.

Kaiser, et al.: Freileitungen fur die floch spannungs-Gleich-stroms-Ubertragung, *ETZ-A*, 1968, Vol. 89, No. 9, Berlin.

***131. References for Insulation Coordination***

Flisberg, G., and Uhlmann, E.: Insulation Levels for HVDC Terminals, *Electra* (CIGRE), 1974, No. 34, pp. 43–61.

Survey of Creepage Distances and Clearances as in HVDC Converter Stations, IEEE Working Group 79.5, IEEE Paper 85 WM 164-9.

***132. References for DC Filters***

Fletcher, D. E., and Patterson, N. A.: The Equivalent Disturbing Current Method for DC Transmission Line Inductive Coordination Studies and DC Filter Performance Specification, *Proceedings of International Conference on DC Power Transmission*, June 4–8, 1984, Montreal, Quebec, Canada, pp. 198–204.

Lasseter, R. H., et al.: DC Filter Design Methods for HVDC Systems, *IEEE Trans.*, Vol. PAS-96, March/April 1977, pp. 571–578.

***133. References for Communication***

Bacon, G. H., Fiedler, H. J., Lindh, C. B., Molnar, A. J.: Power Line Carrier for High Voltage DC Systems, April 21–23, American Power Conference, Chicago, Illinois.

Hammarlund, B.: Telecommunications for HVDC Transmission, *IEEE Trans. Power Appar. Sys.*, March 1968, Vol. PAS-87, No. 3, pp. 690–694.

Picot, Y., and Wong, L. K.: Telecommunication Design Requirements for HVDC Systems, *CIGRE*, 1972, Paper 35-01.

### *134. References for Multiterminal Systems*

Kanngiesser, K. W., Bowles, J. P., Ekström, A., Reeve, J., and Rumpf, E.: HVDC Multiterminal Systems, *CIGRE*, 1974, Paper 14-08.

Lamm, U., Uhlmann, E., and Danfors, P.: Some Aspects on Tapping of HVDC Transmission, *Proc. Am. Power Conf.*, 1963, Vol. 25, pp. 736–744; also published in *Direct Current*, 1963, Vol. 8, No. 5, pp. 124–129.

Long, W. F., Reeve, J., McNichol, J. R., Harrison, R. E., and Bowles, J. P.: Considerations for Implementing Multiterminal DC Systems, *IEEE Trans. Power Appar. Syst.* September 1985, Vol. 104, pp. 2521–2530.

Nozari, F., Grund, C. E., and Hauth, R. L.: Current Order Coordination in Multi-terminal HVDC Systems, *IEEE Trans. Power Appar. Syst.*, Vol. 100, November 1981, pp. 4628–4635.

### *135. References for Subsynchronous Oscillations*

Bahrman, M. P., Larsen, E. V., Patel, H. S., and Piwko, R. J.: Experience with HVDC-Turbine-Generator Torsional Interaction at Square Butte, *IEEE Trans. Power Appar. Syst.*, May/June 1980, Vol. PAS-99, pp. 966–975.

Bowler, C. E. J., Demcko, J. A., Mankoff, L., Kotheimer, W. C., and Cordray, D.: The Navajo SMF Type Subsynchronous Resonance Relay, *IEEE Trans. Power Appar. Syst.*, September/October 1978, Vol. PAS-97, No. 5, pp. 1489–1495.

IEEE SSR Working Group, "Countermeasures to Subsynchronous Resonance Problems," *IEEE Trans. Power Appar. Syst.*, September/October 1980, Vol. PAS-99, No. 5, pp. 1810–1818.

Mortensen, K., Larsen, E. V., and Piwko, R. J.: Field Tests and Analysis of Torsional Interaction Between the Coal Creek Turbine-Generators and the CU HVDC System, *IEEE Trans. Power Appar. Syst.*, January 1981, Vol. PAS-100, No. 1, p. 336.

Piwko, R. J., and Larsen, E. V.: HVDC System Control for Damping of Subsynchronous Oscillations, presented at the IEEE Transmission and Distribution Conference, Minneapolis, Minnesota, September 20–25, 1981, paper no. 81 TD 660-0.

### *136. References for PLC/RI*

Fisher, F. A., Lindh, C. B., Lasseter, R. H., and Reeve, J.: Measurements of the Electrical Noise Produced in the Carrier Band, PES Paper 81TD627-9, IEEE PES 1981 T&D Conference, Minneapolis, Minnesota, September 20–25, 1981.

*IEEE Guide for Power Line Carrier Applications,* IEEE Std. 643-1980.

### *137. References for Line Insulation*

Annestrand, and Schel: *Direct Current,* 1967, Vol. 12, No. 1, pp. 1–8.

Bailey, B. M.: Test Line Experience with HVDC Overhead Transmission, *Trans IEEE Power Appar. Syst.*, January 1970, Paper 70 TP74-PWR.

Clark, F. M.: *Insulating Materials for Design and Engineering Practice,* New York, John Wiley & Sons, Inc., 1962.

DC Conductor Development, *EPRI* Report EL-2257, February 1982.

Harrison, R. E.: Insulation Requirements for DC Transmission Lines, June 1973, *CIGRE* Working Group 33.

*HVDC Transmission Line Research,* EPRI Report EL-2419, May 1982.

Knudsen, N., and Iliceto, F.: Contribution to the Electrical Design of EHV DC Overhead Lines, IEEE January/February 1974, Vol. PAS-93, No. 1, pp. 233–239.

Knudsen, M., and Iliceto, F.: Contribution to the Electrical Design of EHVDC Overhead Lines, *Trans. IEEE,* July 1973, Paper T73 413-2.

*Transmission Line Reference Book, 345 kV and Above,* 2d ed., EPRI, 1982.

*Transmission Line Reference Book, HVDC to +/-600 kV,* EPRI 1977.

### *138. Reference for Cables*

Nyberg, B. R., Hersted, K., Bjoerloev-Larsen, K.: Numerical Methods for Calculation of Electrical Stresses in HVDC Cables with Special Application to the Skagerrak Cable, *IEEE Trans.*, March/April 1975, Vol. PAS-94, No. 2, Paper T74 461-0.

### *139. References for Ground Return*

Dell, D. G.: The North Island Sea Electrode for the Benmore-Haywards HVDC Transmission Scheme, *New Zealand Eng.*, 1965, Vol. 20, No. 6.

Dell, D. G.: The Benmore Land Electrode for the Benmore-Haywards HVDC Transmission Scheme, *New Zealand Eng.*, 1965, Vol. 20, No. 5.

Elder, G. F., and Whitney, D. B.: The Los Angeles HVDC Ocean Electrode, 1962 Conference, Cleveland, Ohio, National Association of Corrosion Engineers.

*HVDC Ground Electrode Design,* EPRI Research Project 1467-1, Report EL-2020, International Engineering Co., Inc.

Köhler, A.: HVDC Transmission: Earth Electrode Arrangements, *Direct Current,* 1965, Vol. 10, No. 1, pp. 18–24.

Landholm, R.: DC Transmission with Return Current through the Earth for HVDC Transmission, *Direct Current,* 1953, Vol. 1, No. 4, pp. 79–86.

Rusck, S.: HVDC Power Transmission: Problems Relating to Earth Return, *Direct Current,* 1962, Vol. 7, No. 11, pp. 290–298, 300.

Schemie, R., and Simons, D. S.: Ground Current Return Electrode Design, Manitoba Power Conference EHV-DC.

### *140. Reference for Tower Design*

The Konti-Skan HVDC Project, *CIGRE,* Report 408, June 1964.

### *141. References for Corona*

Gehrig, E. H., et al.: BPA's 1100-kV DC Test Project. Part II. Radio Interference and Corona Loss, *Trans. IEE Power Appar. Syst.,* March 1967, Vol. 86, No. 3, pp. 278–290.

Hylten-Cavallius, N., et al.: Insulation Requirements, Corona Losses, and Radio Interference for High Voltage Direct Current Lines, *Trans. IEEE,* Paper 63-998.

Hylten-Cavallius, N., et al.: Corona Losses, Radio Interference and Insulator Requirements for HVDC Lines. Studies Regarding Insulator Interference for Frequencies between 30 and 1500 MHz, *CIGRE,* Paper 407, 1964.

IEEE Committee Report: *A Comparison of Methods for Calculating Audible Noise of High Voltage Transmission Lines,* IEEE Vol. PAS-101, October 1982, pp. 4090–4099.

Kovalskaja, O. T., et al.: Investigation of Corona Losses on Experimental Section DC Electrical Transmission Line, *Direct Current Sci. Rev. Inst. Bull.,* 1960, Vol. 5.

*Methods of Measuring Radio Noise,* NEMA Publ. 107, 1960.

Morris, R. M., and Morse, A. R.: Radio Interference and Corona Loss Measurements on "Nelson River" Conductors, Manitoba Power Conference, June 1971, National Research Council of Canada.

Morris, R. M., and Rakoshdas, B.: An Investigaion of Corona Loss and Radio Interference from Transmission Line Conductors at High Direct Voltages, *Trans. IEEE Power Appar. Syst.,* January 1964.

Reiner, G. L., and Gehrig, E. H.: Celilo/Sylmar ±400 kV Line RI Correlation with Short Test Line, IEEE Vol. PAS-96, No. 3, May/June 1977, pp. 955–961.

Specification for CISPR Radio Interference Measuring Apparatus for the Frequency Range 0.15 MHz to 30 MHz, IEC Publ. 1, 1961.

Witt, H.: Insulation Levels and Corona Phenomena on HVDC Transmission Lines, Techn. Dr. Thesis, Chalmers University, Gothenburg, Sweden, 1961.

### *142. References for HVDC Thyristor Valves*

Final report of the Task Force on a Guide for Test Procedures for HVDC Thyristor Valves sponsored by DC Transmission Subcommittee of the IEE T&D Committee and DC Converter Stations Subcommittee of the IEEE Substations Committee, February 28, 1984, IEEE Std. Project P-857.

Hingorani, N. G.: Special Report for Group 14 (DC Links), International Conference on Large High Voltage Electric Systems, 112, Boulevard Haussmann, 75008 Paris 1984 Session, 29th August–6th September.

International Electrotechnical Commission, IEC Standard Publ. 700, 1st ed., 1981, Testing of Semiconductor Valves for High-Voltage D.C. Power Transmission.

Krishnayya, P. C. S.: Important Characteristics of Thyristors of Valves for HVDC Transmission and Static VAR Compensators, International Conference on Large High Voltage Electric Systems, 112, Boulevard Haussmann, 75008 Paris 1984 Session, 29th August–6th September.

### *143. References for Control and Operation of HVDC Links*

Ainsworth, J. D.: Harmonic Instability between Controlled Static Converters and AC Networks, *Proc. IEE,* July 1967, Vol. 114, No. 7.

Ainsworth, J. D.: The Phase Locked Oscillator: A New Control System for Controlled Static Converters, *IEEE Trans. Power Appar. Syst.,* 1968, Vol. 87, pp. 858–865.

Ekstrom, A., and Liss, G.: A Refined HVDC Control System, *IEEE Trans. Power Appar. Syst.,* 1970, Vol. 89, pp. 723–732.

Engstrom, G.: Operation and Control of HVDC Transmission, *Trans. IEEE,* Paper.

Foswell, H.: The Gotland DC Link, The Grid Control and Regulation Equipment, *Direct Current,* June 1955, Vol. 2, No. 5, December , No. 7.

Hauth, R. L., Patel, H. S., and Piwko, R. J.: HVDC Control Developments—Addressing System Requirements, *Electric Forum Magazine,* 1984, Vol. 9, No. 3, pp. 25–31.

Kimbark, E. W.: *Direct Current Transmission,* Wiley-Interscience, Vol. 1, pp. 157–183.

Persson, E.: Calculation of Transfer Functions on Grid-Controlled Converter Systems, *Proc. IEE,* May 1970, Vol. 117, No. 5.

Rumpf, E., and Ranade, S.: Comparison of Suitable Control Systems for HVDC Stations Connected to Weak AC Systems, Part 1. New Control Systems, Part II. Operational Behavior of the HVDC Transmission, *IEEE Trans. Power Appar. Syst.,* 1972, Vol. 91, pp. 549–564.

Uhlmann, E.: Stabilization of an AC Link by a Parallel DC Link, *Direct Current,* August 1964, Vol. 2, No. 3.

Uhlmann, E.: AC Network Stabilization by DC Links, *CIGRE,* 32-01, 1970 Session.

Uhlmann, E.: Static Converters Connected to AC Networks with Unsymmetrical Voltages, *Elteknik,* 1967, Vol. 10, No. 8, pp. 149–153.

Breuer, G. D., Grund, C. E., Hauth, R. L., and Pohl, R. V.: The Dynamic Performance of AC Systems with HVDC Transmission Lines, Paper 14-02, *CIGRE,* 1976 session, August/September 1976, Paris, France.

Engström, G.: Operational and Control of HVDC Transmission, *Trans. IEEE,* Paper 1963–83.

Persson, E.: Calculation of Transfer Functions in Grid-Controlled Converter Systems, *Proc. IEE,* May 1970, Vol. 117, No. 5.

Rumpf, E., and Ranade, S.: Comparison of Suitable Control Systems for HVDC Stations Connected to Weak AC Systems, Part I: New Control Systems; Part II: Operational Behavior of the HVDC Transmission, *IEEE Trans. Power Appar. Syst.,* 1972, Vol. 91, pp. 549–564.

Uhlmann, E.: Stabilization of an AC Link by a Parallel DC Link, *Direct Current,* August 1964.

Uhlmann, E.: AC Network Stabilization by DC Links, *CIGRE,* 32-01, 1970 session.

Uhlmann, E.: Static Converters Connected to AC Networks with Unsymmetrical Voltages, *Elteknik,* 1967, Vol. 10, No. 8, pp. 149–153.

### *144. Reference for HVDC Breakers*

Vithayathil, J. J.: HVDC Breakers and Its Applications, Sharing the Brazilian Experience, March 1983, Rio de Janeiro.

### *145. References for Arresters*

Kobayashi, M., Mizuno, M., Aizawa, T., Hayashi, M., and Mitani, K.: Development of Zinc Oxide Non-Linear Resistors and Their Application to Gapless Surge Arresters, Paper No. F77-682-8.

Sakshaug, E. C., Kresge, J. S., and Miske, S. A., Jr.: A New Concept in Station Arrester Design, Paper No. F76-393-9, *Power Appar. Syst.,* March/April 1977, Vol. PAS-96, No. 2.

Sakshaug, E. C., Kresge, J. S. and Miske, S. A., Jr.: Arrester Protection of High Voltage DC Transmission System Converter Terminals, Paper No. 71TP47, PWR, *Power Appar. Syst.,* July/August 1971, Vol.PAS-90, No. 4.

Tominaga, A., Azume, K., Nitta, T., Najai, T., Imataki, M., and Kuwabara, M.: Reliabiilty and Application of Metal Oxide Surge Arresters for Power Systems, Paper No. F78-703-1, *Power Appar. Syst.,* May/June 1979, Vol. PAS-98, No. 3.

Written Discussion of Previous Reference. *Power Appar. Syst.,* May/June 1979, Vol. PAS-98, No. 3.

### *146. General References*

Adamson, C., and Hingorani, N. G.: *High Voltage Direct Current Power Transmission,* London, Garraway, Ltd., 1960.

*Direct Current Periodical Journal:* 1952–1967, London, Garraway, Ltd.; 1969–1971, Oxford, Pergamon Press.

Engström, G.: Operation and Control of DC Power Transmission, *Trans. IEEE,* Paper 1963–83.

Haglöf, L., et al.: Transmision por continua en alta tension (CCAT), *Rev. Electrotecnica,* 1975, Vol. 61, No. 1 (Spanish).

High Voltage DC and/or AC Power Transmission: IEE Conference Publication 107 with discussion record, London, November 1973.

HVDC Symposium, Canadian Electrical Association, Vancouver, March 1971.

Kimbark, E. W.: *Direct Current Transmission,* Vol. 1; New York, Wiley Interscience, 1971.

Lamm, U.: What Is the Place of HVDC Transmission in Today's Power Systems? *Electr. World,* 1963, Vol. 159, No. 20, pp. 98–99, 129–130, 132.

Pucher, W., Kanngiesser, K. W., Koetzold, B., and Schultz, W.: HVDC Circuit Breaker in Intermeshed Multiterminal HVDC Systems, *CIGRE,* 1972, Report 31–08.

*Proceedings; Manitoba Power Conference EHV-DC,* University of Manitoba, June 1971.

Uhlmann, E.: AC Network Stabilization by DC Links, *CIGRE,* 1970, Report 32-01.

Uhlamnn, E.: *Power Transmission by Direct Current,* Berlin, New York, Springer-Verlag, 1975.

1974 Minnesota Power Systems Conference, IEEE Twin Cities Section, Minneapolis, October 1974.

# SECTION 16

# POWER-SYSTEM INTERCONNECTIONS

**Nathan Cohn**
*General Partner and Senior Associate, Network Systems Development Associates; Fellow, IEEE*

**Bruce F. Wollenberg**
*Principal Consultant, Power System Security, Control Data Corporation*

**W. A. Elmore**
*Consulting Engineer and Section Manager, Relay and Telecommunications Division, Westinghouse Electric Corporation*

**Jalal Gohari**
*Principal Electrical Engineer, Gibbs & Hill, Inc.*

## CONTENTS

*Numbers refer to paragraphs*

## CONTROL OF GENERATION AND POWER FLOW

*By Nathan Cohn*

***1. Growth of Power Production.*** The world's capacity for producing electric energy has grown rapidly in recent years.

The United States accounts for more than 35% of the world's electric-energy production. The six countries with the largest generating capacity for 1982 are shown in Fig. 16-1 (Ref. 1). Data for the same year on energy production and consumption per capita for these and the next nine countries are shown in Table 16-1. Growth of electric power in the United States had been at an annual compound rate of about 7%, resulting in generating capability doubling about every 10 years. Growth continues, but at a rate of about 3%. Growth of installed capacity from 1920 to 1983 is shown in Fig. 16-2. Use of electric energy in the United States has kept pace with generating capability, until recently. Generation in the United States from 1920 to 1983 is shown in Fig. 16-3. In 1983, 75% of United States generation was from fossil fuels, 14% was hydro, and 13% was nuclear.

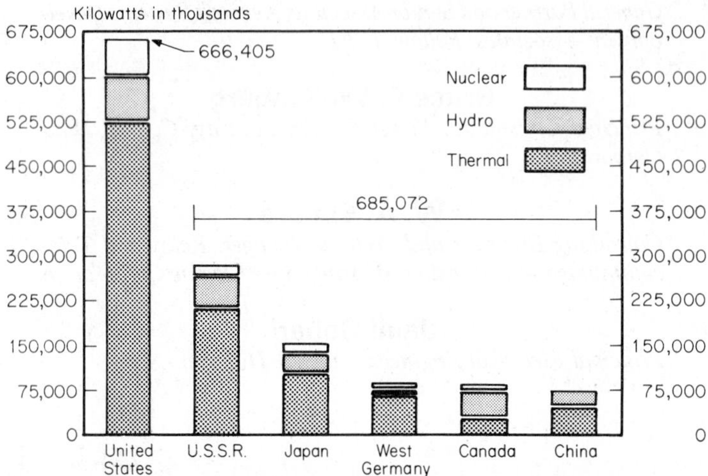


**FIG. 16-1** World generating capacity for the six largest countries—1982.

***2. Interconnected Systems.*** Adjacent electric power systems have found it advantageous, for reasons of reliability and economy, to interconnect and to operate in parallel.

*Advantages of Interconnection.* In periods of need, interconnected companies can draw on one another's rotating reserves, increasing reliability of system operation and contributing to continuity of service to customers. In normal operating periods, adjacent companies can schedule bulk power transfers over intercompany ties to take advantage of energy-cost differentials in the respective areas. Load diversity, seasonal conditions, time-zone differences, or shared investment in larger, and hence more efficient, generating units may make excess generation available in one area at a cost lower than energy could, at that time, be generated in adjacent areas. Bulk transfers are correspondingly scheduled. Resultant savings are equally shared by participating companies.

*NERC.* An organization that contributes importantly to the successful operation of interconnected systems in the United States and Canada is the North American Electric Reliability Council (NERC) and The North American Power Systems Interconnection Committee (NAPSIC). Formed voluntarily by the electric utility industry in 1968, NERC directs its efforts to augment the reliability and adequacy of bulk power supply of the electric utility systems in North America. NERC consists of nine regional reliability councils whose memberships comprise essentially all the electric power systems in the United States and Canada.

**TABLE 16-1** World Power Data

Fifteen Countries with Greatest Installed Generating Capacity, 1982

| Country | Installed capacity, kW in thousands | | | | Energy production*, kWh in millions | | | | Population (thousands) | kWh per capita |
|---|---|---|---|---|---|---|---|---|---|---|
| | Hydro | Nuclear | Thermal | Total | Hydro | Nuclear | Thermal | Total | | |
| United States | 78,428 | 63,042 | 524,935 | 666,405 | 310,788 | 282,773 | 1,710,650 | 2,304,211 | 231,786† | 9,941 |
| U.S.S.R. | 55,889 | 18,000 | 211,603 | 285,492 | 175,277 | 80,000 | 1,111,823 | 1,367,100 | 268,800 | 5,086 |
| Japan | 33,311 | 17,342 | 104,158 | 154,811 | 84,039 | 102,430 | 394,678 | 581,147 | 118,600 | 4,900 |
| Germany (West) | 6,509 | 9,826 | 69,434 | 85,769 | 19,646 | 63,577 | 283,654 | 366,877 | 61,700 | 5,946 |
| Canada | 46,400 | 5,600 | 31,000 | 83,000 | 261,055 | 35,312 | 91,084 | 387,460 | 24,400 | 15,880 |
| China (mainland) | 26,000 | — | 50,000 | 76,000 | 74,400 | — | 253,280 | 327,680 | 1,008,175 | 325 |
| France | 21,200 | 23,284 | 29,500 | 73,984 | 70,900 | 103,000 | 92,000 | 265,900 | 54,200 | 4,906 |
| United Kingdom | 2,451 | 6,490 | 60,250 | 69,191 | 5,637 | 43,972 | 222,553 | 272,162 | 56,100 | 4,851 |
| Italy | 16,877 | 1,273 | 31,873 | 50,023 | 44,080 | 6,804 | 133,560 | 184,444 | 57,400 | 3,213 |
| Brazil | 32,892 | — | 6,012 | 38,904 | 141,224 | — | 10,865 | 152,089 | 127,700 | 1,191 |
| India | 13,058 | 860 | 24,890 | 38,808 | 52,675 | 3,210 | 82,792 | 138,677 | 713,000 | 194 |
| Spain | 12,700 | 1,500 | 15,700 | 29,900 | 28,036 | 8,772 | 79,952 | 116,760 | 37,900 | 3,081 |
| Sweden | 15,215 | 6,440 | 8,029 | 29,684 | 55,624 | 39,090 | 5,410 | 100,124 | 8,310 | 12,049 |
| Australia | 6,288 | — | 21,255 | 27,543 | 15,000 | — | 89,890 | 104,890 | 15,000 | 6,993 |
| Poland | 1,807 | — | 24,181 | 25,988 | 2,605 | — | 114,975 | 117,580 | 36,300 | 3,239 |

* Data presented on a gross generation basis.

† U.S. population July 1, 1982; excludes armed forces abroad. Dept. of Commerce, Bureau of Census.

SOURCE: Ref. 1.

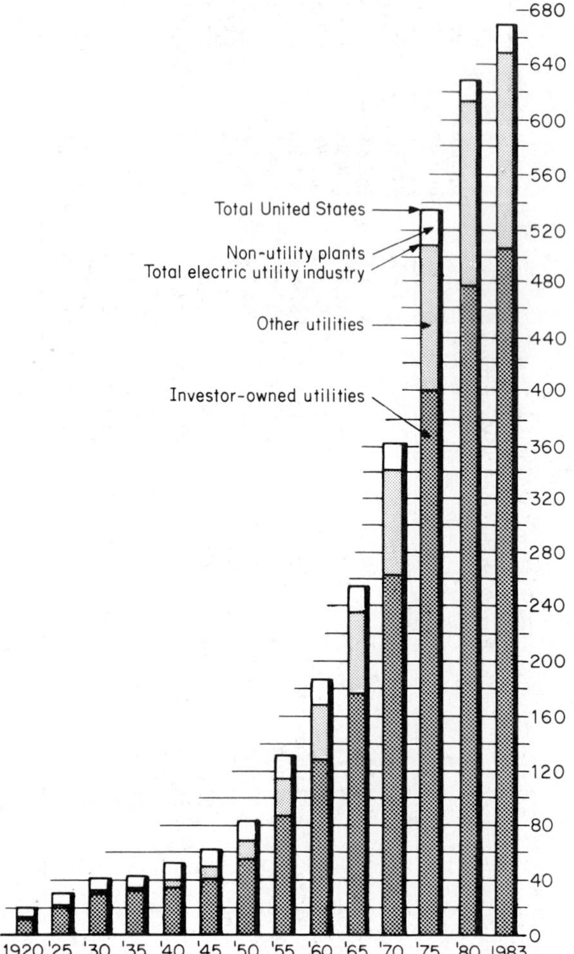


**FIG. 16-2** Growth of United States installed generating capacity, 1920–1983.

A map of the nine NERC regions is shown in Fig. 16-4. A summary of 1983 generating capacities and peak loads for the NERC regions is tabulated in Table 16-2. NERC projections of generating capacity and peaks for 1988 and 1993 are shown in Table 16-3.

For additional information on NERC operations and related studies see Refs. 2 and 3.

*NAPSIC.* The North America Power Systems Interconnection Committee served for many years as the operating organization responsible for coordinating operating matters, especially those factors that promote reliability of operation among the interconnected systems in North America. NAPSIC developed operating guides, recommendations, criteria, and standards to cover operating matters that required coordination.

NAPSIC was conceived in 1962 and formalized in January 1963, although its predecessor organizations date back as far as 1933. From an independent organization in the beginning, NAPSIC now serves as the operating committee of NERC. NERC continually maintains and updates the original NAPSIC operating guides and minimum criteria for operating reliability, now included in the NERC Operating Manual (Ref. 4).

*Interconnections in the United States and Canada.* Interconnections between adjacent

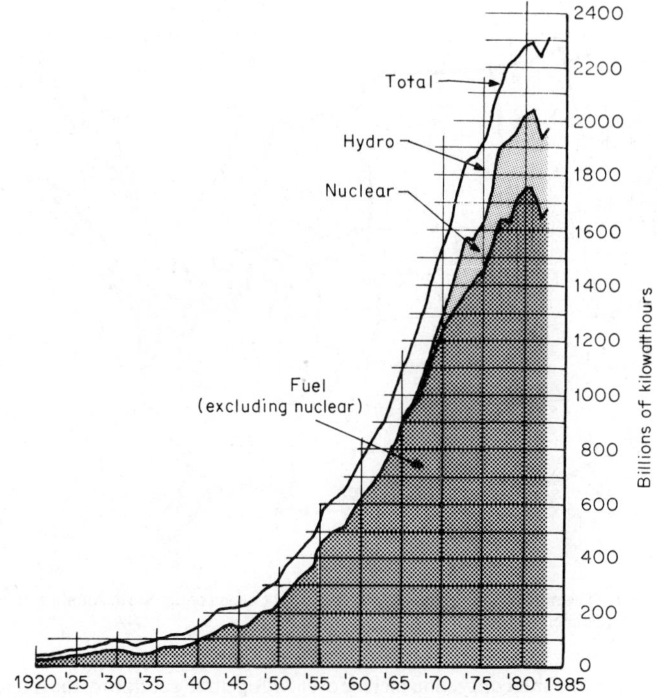


**FIG. 16-3** Growth of United States generation, 1920–1983.

operating utilities have been steadily expanded in the United States and Canada. Major interties in the United States are shown in Fig. 16-5. There are, as shown by the segmented map of Fig. 16-6, three major synchronized interconnected areas in the United States and Canada.

The largest of these interconnections extends to the east from the Rocky Mountains and includes the Middle West, the Gulf Coast, the eastern seaboard, and eastern Canada. Constitu-

**TABLE 16-2** NERC Regions

Year-end Capacities and Summer Peaks, 1983

| Code | Region | 1983 Capacity, MW | 1983 Summer peak, MW |
|---|---|---|---|
| ECAR | East Central Area Reliability Coordination Agreement | 90,914 | 65,959 |
| ERCOT | Electric Reliability Council of Texas | 43,800 | 34,972 |
| MAAC | Mid-Atlantic Area Council | 46,734 | 34,783 |
| MAIN | Mid-America Interpool Network | 40,797 | 34,784 |
| MAPP* | Mid-Continent Area Power Pool | 34,322 | 32,333 |
| NPCC* | Southeastern Electric Reliability Council | 98,522 | 66,285 |
| SERC | Southeastern Electric Reliability Council | 124,999 | 97,235 |
| SPP | Southeast Power Pool | 59,294 | 45,881 |
| WSCC* | Western Systems Coordinating Group | 126,754 | 84,396 |

* Includes Canadian areas.

SOURCE: Ref. 2.

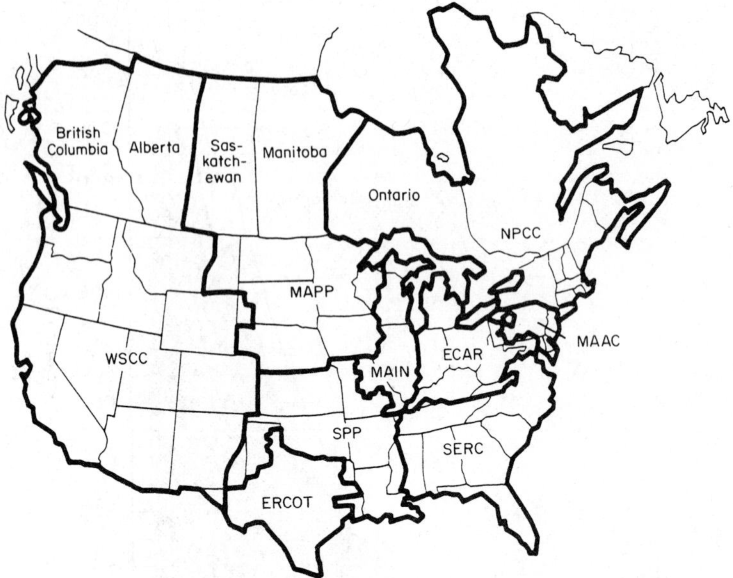


**FIG. 16-4** The nine regions in the United States and Canada of the North American Electric Reliability Council (NERC).

ent groups within this interconnection include over 200 operating utilities of the Interconnected Systems Group (ISG), the 12 operating companies of the Pennsylvania–New Jersey–Maryland pool (PJM), and the more than 50 operating utilities of the Canadian–Eastern United States group (CANUSE). This interconnection encompasses the ECAR, MAAC, MAIN, MAPP, NPCC, SERC, and SPP regions of NERC. The many utilities of this interconnection, some publicly owned, some investor-owned (large and small), having a 1983 noncoincident summer peak load about 338 million kW, operate continuously in parallel.

Similar parallel operation is achieved in each of the other two interconnections.

The Western Interconnected Systems includes the Northwest Power Pool, the Rocky Mountain Pool, the California-Nevada Systems, and the Arizona–New Mexico Systems. It encompasses the WSCC region of NERC. It has an aggregate noncoincident summer peak of

**TABLE 16-3** NERC Projected Year-End Capacities and Summer Peaks, 1988 and 1993

| | Projected capacity, MW | | Projected peak, MW | |
|---|---|---|---|---|
| Region | 1988 | 1993 | 1988 | 1993 |
| ECAR | 102,451 | 103,997 | 71,438 | 79,110 |
| ERCOT | 51,158 | 61,346 | 43,061 | 52,407 |
| MAAC | 48,498 | 51,169 | 36,690 | 38,150 |
| MAIN | 50,037 | 49,954 | 36,874 | 40,165 |
| MAPP | 35,370 | 38,725 | 28,816 | 31,372 |
| NPCC* | 113,175 | 119,037 | 73,023 | 80,922 |
| SERC | 141,596 | 152,347 | 106,741 | 121,873 |
| SPP | 66,752 | 71,940 | 51,486 | 58,970 |
| WSCC* | 131,931 | 150,938 | 97,498 | 109,901 |

* Includes Canadian areas.

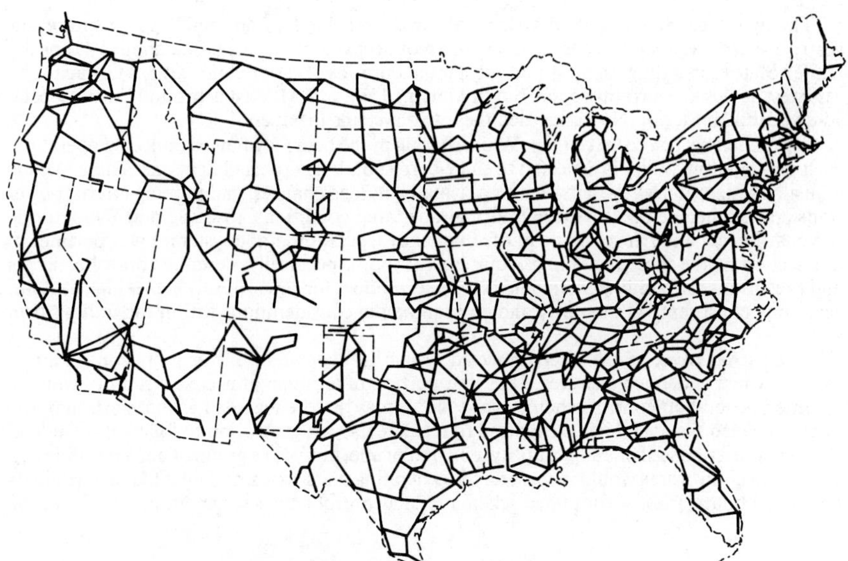

**FIG. 16-5** Major interconnections in the United States.

about 84 million kW. WSCC supplements Ref. 4 with Minimum Operating Reliability Criteria (Ref. 5). The Northwest Power Pool (NWPP) has a further operating manual for its members (Ref. 6).

The Texas Interconnected System encompasses substantially the ERCOT region of Texas and has a noncoincident peak of close to 24 million kW.

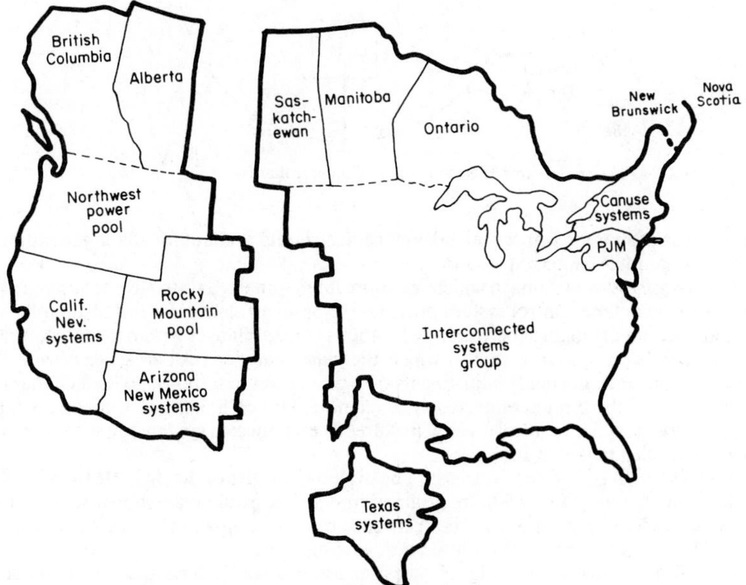


**FIG. 16-6** The three major United States and Canadian interconnected systems.

The two large interconnected systems of Fig. 16-6 have, for appreciable periods in recent years, operated as a single coast-to-coast interconnected system, including about 94% of the United States generating capacity and substantial portions of eastern and western Canada. The east-west ac ties are currently open, but in March 1977 a 110-MW asynchronous back-to-back ad-dc-ac link at Stegall, Nebraska, tied these two systems together.

*Obligations of Interconnection.* While sharing in the benefits of interconnected operation, each participating utility is expected to share correspondingly in its obligations. These include adjusting generation levels in its own area to match its load changes, maintaining intertie power transfers at scheduled levels during normal operating conditions, assisting neighbors during periods of need, and participating in the frequency regulation of the system. Later portions of this section define these regulating requirements more specifically, describe control concepts and executions for fulfilling them, and discuss techniques for concurrently achieving optimum economy, consistent with security and environmental considerations within the utility's own system.

***3. Control Areas.*** Each interconnected system is made up of one or more control areas, each of which is defined (see Ref. 7) as that portion of an interconnected system to which a common generation control scheme is applied. It may also be regarded as that portion of the interconnected system which is expected to regulate its own generation to follow its own load changes. It may consist of a single utility, or a part of one, or a whole group of pooled utilities. In each case a control area would include all the generating units, loads, and lines that fall within its prescribed boundaries. A simplified schematic of a control area is shown in Fig. 16-7. All the

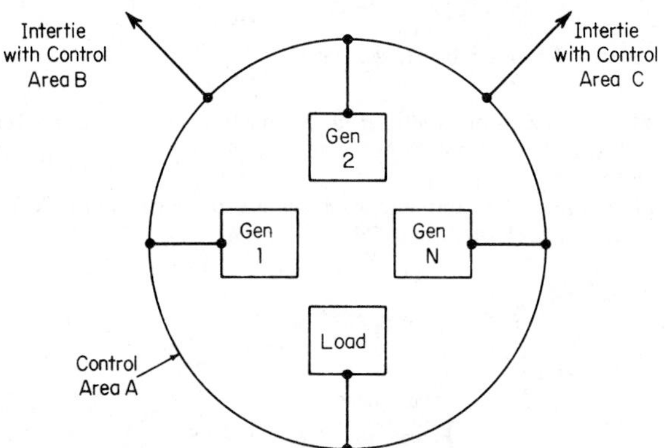


**FIG. 16-7** Simplified schematic of a control area.

control areas of an interconnection, taken together, should account for all the generation, load, and ties of the interconnected system.

*A single-area system* is one in which the entire interconnected system is encompassed within one control area. One control system provides the basic regulation for the entire interconnection and does not distinguish between the locations of load changes within the interconnection.

*A multiple-area system* is one in which there are many control areas, each with its own control system, each normally adjusting its own generation in response to load changes within its own area. All three interconnected systems in the United States and Canada operate on a multiple-area basis. The control areas of the three interconnected systems are shown in Fig. 16-8 and are identified in Table 16-4.

*ISG-PJM-CANUSE Control Areas.* The over 200 utilities of the ISG-PJM-CANUSE eastern interconnection of Fig. 16-6 are grouped into 105 control areas, shown schematically as connected circles in Fig. 16-8. Interties between adjacent companies are shown as single lines, though in most cases there are multiple ties between areas.

Each of the control areas of Fig. 16-8 has generation, load, and ties after the manner of Fig. 16-7 and with varying degrees of size, complexity, and geographical extent. Illustrative of the extent of a single control area is circle 68 of Fig. 16-8, which represents the 13-company New

England Power Exchange Pool (NEPEX) shown in Fig. 16-9. In earlier years these companies constituted several control areas. At present, despite varying ownership, they have operationally been integrated into a single control area (see Ref. 8). Note the many ties (tabulated in parentheses in each circle) each of the companies has with its neighbors.

Circle 72 of Fig. 16-8 represents the New York Power Pool (NYPP). It includes a group of eight independently owned utilities operating as a single control area as tabulated in Fig. 16-10. The number of ties with neighboring utilities is shown in parentheses.

Another illustration of an extensive single control area is the seven-state American Electric Power System, circle 1 in Fig. 16-8. Its five operating companies, which have common ownership, the ties between them, and its 144 interties with 24 other control areas are shown schematically in Fig. 16-11.

A large group of independently owned utilities that operate as a single control area is the PJM pool, shown as circle 83 in Fig. 16-8. Its 12 participating companies, the ties between them, and its 30 major interties with other control areas are illustrated schematically in Fig. 16-12. Prior to its interconnection with ISG and CANUSE, the PJM pool operated as a single-area interconnection.

*Western Control Areas.* Control areas for the western interconnected system are included in Fig. 16-8 and are identified in Table 16-4.

*Texas Control Areas.* The control areas of the Texas interconnection and their interties are shown schematically and identified in Fig. 16-8 and Table 16-4.

**TABLE 16-4** Controlling Organization and Center Location of Control Areas of Fig. 16-8 (listed alphabetically)

| | | | |
|---|---|---|---|
| 1. | AEP | AMERICAN ELECTRIC POWER SERVICE CORP. | CANTON, OH |
| 2. | APS | ALLEGHENY POWER SERVICE CORP. | CHARLEROI, PA |
| 3. | ARPS | ARIZONA PUBLIC SERVICE CO. | PHOENIX, AZ |
| 4. | ASEC | ASSOCIATED ELECTRIC COOPERATIVE, INC. | SPRINGFIELD, MO |
| 5. | AUST | AUSTIN ELECTRIC DEPARTMENT | AUSTIN, TX |
| 6. | BIRI | BIG RIVERS RURAL ELECTRIC COOP. CORP. | HENDERSON, KY |
| 7. | BPA | BONNEVILLE POWER ADMINISTRATION | VANCOUVER, WA |
| 8. | BROV | BROWNSVILLE PUBLIC UTILITY BOARD | BROWNSVILLE, TX |
| 9. | CAJN | CAJUN ELECTRIC POWER COOPERATIVE | NEW ROADS, LA |
| 10. | CAPO | CAROLINA POWER & LIGHT CO. | RALEIGH, NC |
| 11. | CEIL | CENTRAL ILLINOIS LIGHT CO. | PEORIA, IL |
| 12. | CEIP | CENTRAL ILLINOIS PUBLIC SERVICE | SPRINGFIELD, IL |
| 13. | CEKP | CENTRAL KANSAS POWER CO. | HAYS, KA |
| 14. | CELE | CENTRAL LOUISIANA ELECTRIC CO. | ALEXANDRIA, LA |
| 15. | CEPL | CENTRAL POWER & LIGHT CO. | CORPUS CHRISTI, TX |
| 16. | CTKS | CENTRAL TELEPHONE & UTILITIES CORP./ WESTERN POWER DIV. | GREAT BEND, KA |
| 17. | CHPU | CHELAN COUNTY PUD #1 | WENATCHEE, WA |
| 18. | CIGE | CINCINNATI GAS & ELECTRIC CO. | CINCINNATI, OH |
| 19. | CLEI | CLEVELAND ELECTRIC ILLUMINATING CO. | BRECKSVILLE, OH |
| 20. | COLM | COLUMBIA WATER & LIGHT DEPT. | COLUMBIA, MO |
| 21. | COEC | COMMONWEALTH EDISON CO. | LOMBARD, IL |
| 22. | DAPC | DAIRYLAND POWER COOPERATIVE | LA CROSSE, WI |
| 23. | DAPL | DALLAS POWER & LIGHT CO. | DALLAS, TX |
| 24. | DAPO | DAYTON POWER & LIGHT CO. | DAYTON, OH |
| 25. | DOPU | DOUGLAS COUNTY PUD #1 | EAST WENATCHEE, WA |
| 26. | DUPC | DUKE POWER CO. | CHARLOTTE, NC |
| 27. | DULC | DUQUESNE LIGHT CO. | PITTSBURGH, PA |
| 28. | EAIL | EASTERN IOWA LIGHT & POWER COOP. | WILTON, IA |
| 29. | ELPE | EL PASO ELECTRIC CO. | EL PASO, TX |
| 30. | ELEN | ELECTRIC ENERGY INC. | JOPPA, IL |
| 31. | EMDE | EMPIRE DISTRICT ELECTRIC CO. | JOPLIN, MO |
| 32. | FLPC | FLORIDA POWER CORP. | ST. PETERSBURG, FL |
| 33. | FLPL | FLORIDA POWER & LIGHT CO. | MIAMI, FL |
| 34. | GAMW | GAINSVILLE REGIONAL UTILITIES | GAINSVILLE, FL |
| 35. | GRCP | GRANT COUNTY PUD #2 | EPHRATA, WA |
| 36. | GUSU | GULF STATES UTILITIES CO. | BEAUMONT, TX |
| 37. | HOLP | HOUSTON LIGHTING & POWER CO. | HOUSTON, TX |
| 38. | IDPC | IDAHO POWER CO. | BOISE, ID |
| 39. | ILPC | ILLINOIS POWER CO. | DECATUR, IL |
| 40 | IMID | IMPERIAL IRRIGATION DISTRICT | IMPERIAL, CA |
| 41. | INSR | INDIANA STATEWIDE REC., INC.-HOOSIER ENERGY DIV. | BLOOMINGTON, IN |
| 42. | INPL | INDIANAPOLIS POWER & LIGHT CO. | INDIANAPOLIS, IN |
| 43. | INPD | INTERSTATE POWER CO. | DUBUQUE, IA |
| 44. | IOEL | IOWA ELECTRIC LIGHT & POWER CO. | CEDAR RAPIDS, IA |
| 45. | IOIG | IOWA-ILLINOIS GAS & ELECTRIC CO. | DAVENPORT, IA |
| 46. | IOPL | IOWA POWER & LIGHT CO. | DES MOINES, IA |
| 47. | IOPS | IOWA PUBLIC SERVICE CO. | SOUIX CITY, IA |
| 48. | IOSU | IOWA SOUTHERN UTILITIES CO. | CENTERVILLE, IA |
| 49. | JACO | JACKSONVILLE ELECTRIC AUTHORITY | JACKSONVILLE, FL |
| 50. | KACP | KANSAS CITY POWER & LIGHT CO. | KANSAS CITY, MO |
| 51. | KAGE | KANSAS GAS & ELECTRIC CO. | WICHITA, KS |
| 52. | KAPL | KANSAS POWER & LIGHT CO. | TOPEKA, KS |
| 53. | KEUC | KENTUCKY UTILITIES CO. | LEXINGTON, KY |
| 54. | LAFA | LAFAYETTE UTILITIES SYSTEM | LAFAYETTE, LA |
| 55. | LALW | LAKELAND DEPT. OF ELECTRIC & WATER UTILITIES | LAKELAND, FL |
| 56. | LASD | LAKE SUPERIOR DISTRICT POWER CO. | ASHLAND, WI |
| 57. | LOAN | LOS ANGELES DEPT. OF WATER & POWER | LOS ANGELES, CA |
| 58. | LOCR | LOWER COLORADO RIVER AUTHORITY | AUSTIN, TX |
| 59. | LOGE | LOUISVILLE GAS & ELECTRIC CO. | LOUISVILLE, KY |
| 60. | MAGE | MADISON GAS & ELECTRIC CO. | MADISON, WI |
| 61. | MECS | MICHIGAN ELECTRIC COORDINATED SYSTEM | ANN ARBOR, MI |
| 62. | MIPL | MINNESOTA POWER & LIGHT CO. | DULUTH, MN |
| 63. | MIPU | MISSOURI PUBLIC SERVICE CO. | JEFFERSON CITY, MO |
| 64. | MOPO | MONTANA POWER CO. | BUTTE, MT |
| 65. | MSS | MIDDLE SOUTH SERVICES, INC. | PINE BLUFF, AR |
| 66. | MUSC | MUSCATINE WATER & ELECTRIC PLANTS | MUSCATINE, IA |
| 67. | NEPC | NEVADA POWER CO. | LAS VEGAS, NV |
| 68. | NEPEX | NEW ENGLAND POWER EXCHANGE | W. SPRINGFIELD, MA |
| 69. | NEPP | NEBRASKA PUBLIC POWER DISTRICT | HASTINGS, NE |
| 70. | NIPS | NORTHERN INDIANA PUBLIC SERVICE CO. | HAMMOND, IN |
| 71. | NOSM | NORTHERN STATES POWER CO. | MINNEAPOLIS, MN |
| 72. | NYPP | NEW YORK POWER POOL | GUILDERLAND, NY |
| 73. | OHEC | OHIO EDISON CO. | AKRON, OH |
| 74. | OHVE | OHIO VALLEY ELECTRIC CORP. | PIKETON, OH |
| 75. | OKGE | OKLAHOMA GAS & ELECTRIC CO. | OKLAHOMA CITY, OK |
| 76. | OMPB | OMAHA PUBLIC POWER DISTRICT | OMAHA, NE |
| 77. | ORLA | ORLANDO UTILITIES COMMISSION | ORLANDO, FL |
| 78. | OTTP | OTTER TAIL POWER CO. | FERGUS FALLS, MN |
| 79. | PAGE | PACIFIC GAS & ELECTRIC CO. | SAN FRANCISCO, CA |
| 80. | PAPL | PACIFIC POWER & LIGHT CO. | PORTLAND, OR |
| 81. | PAPL/WY | PACIFIC POWER & CO.-WYOMING | GLENROCK, WY |
| 82. | PASA | PASADENA LIGHT & POWER DEPT. | PASADENA, CA |
| 83. | PJM | PENNSYLVANIA-NEW JERSEY-MARYLAND INTERCONNECTION | VALLEY FORGE, PA |
| 84. | PLAQ | PLAQUEMINE LIGHT DEPT. | PLAQUEMINE, LA |
| 85. | POGE | PORTLAND GENERAL ELECTRIC CO. | PORTLAND, OR |
| 86. | PSCO | PUBLIC SERVICE CO. OF COLORADO | DENVER, CO |
| 87. | PSIN | PUBLIC SERVICE CO. OF INDIANA | PLAINFIELD, IN |
| 88. | PSNM | PUBLIC SERVICE CO. OF NEW MEXICO | ALBUQUERQUE, NM |
| 89. | PSOK | PUBLIC SERVICE CO. OF OKLAHOMA | TULSA, OK |
| 90. | PSPL | PUGET SOUND POWER & LIGHT CO. | REDMOND, WA |
| 91. | SAJL | SAINT JOSEPH LIGHT & POWER CO. | SAINT JOSEPH, MO |
| 92. | SARV | SALT RIVER PROJECT | PHOENIX, AZ |
| 93. | SAAN | CITY PUBLIC SERVICE BOARD OF SAN ANTONIO | SAN ANTONIO, TX |
| 94. | SADG | SAN DIEGO GAS & ELECTRIC CO. | SAN DIEGO, CA |
| 95. | SEAT | SEATTLE CITY LIGHT | SEATTLE, WA |
| 96. | SIPP | SIERRA PACIFIC POWER CO. | RENO, NV |
| 97. | SMEA | SOUTH MISSISSIPPI ELECTRIC POWER ASSN. | HATTIESBURG, MS |
| 98. | SOCA | SOUTH CAROLINA PUBLIC SERVICE AUTH. | MONCKS CORNER, SC |
| 99. | SOCE | SOUTHERN CALIFORNIA EDISON CO. | ROSEMEAD, CA |
| 100. | SOCG | SOUTH CAROLINA ELECTRIC & GAS CO. | COLUMBIA, SC |
| 101. | SOCO | SOUTHERN SERVICES, INC. | BIRMINGHAM, AL |
| 102. | SOEP | SOUTHWESTERN ELECTRIC POWER CO. | SHREVEPORT, LA |
| 103. | SOIG | SOUTHERN INDIANA GAS & ELECTRIC CO. | EVANSVILLE, IN |
| 104. | SOIP | SOUTHERN ILLINOIS POWER COOP. | MARION, IL |
| 105. | SOTE | SOUTH TEXAS ELECTRIC COOP. INC. | VICTORIA, TX |
| 106. | SUCO | SUNFLOWER ELECTRIC COOP. | HAYS, KS |
| 107. | SWLP | SPRINGFIELD WATER, POWER & LIGHT | SPRINGFIELD, IL |
| 108. | SWPS | SOUTHWESTERN PUBLIC SERVICE CO. | AMARILLO, TX |
| 109. | SWPA | SOUTHWESTERN POWER ADMINISTRATION | SPRINGFIELD, MO |
| 110. | TACO | TACOMA CITY LIGHT | TACOMA, WA |
| 111. | TAEC | TAMPA ELECTRIC CO. | TAMPA, FL |
| 112. | TEES | TEXAS ELECTRIC SERVICES CO. | FORT WORTH, TX |
| 113. | TEPL | TEXAS POWER & LIGHT CO. | DALLAS, TX |
| 114. | TOEC | TOLEDO EDISON CO. | TOLEDO, OH |
| 115. | TUEP | TUCSON ELECTRIC POWER CO. | TUCSON, AZ |
| 116. | TVA | TENNESSEE VALLEY AUTHORITY | CHATTANOOGA, TN |
| 117. | UNEC | UNION ELECTRIC CO. | ST. LOUIS, MO |
| 118. | UNPA | UNITED POWER ASSOCIATION | ELK RIVER, MN |
| 119. | UPPP | UPPER PENINSULA POWER CO. | HOUGHTON, MI |
| 120. | UTPL | UTAH POWER & LIGHT CO. | SALT LAKE CITY, UT |
| 121. | VEBM | VERO BEACH MUNICIPAL UTILITIES | VERO BEACH, FL |
| 122. | VIEP | VIRGINIA ELECTRIC POWER CO. | RICHMOND, VA |
| 123. | WAPA/LC | WESTERN AREA PWR. ADM., LOWER COLORADO | PHOENIX, AZ |
| 124. | WAPA/LM | WESTERN AREA PWR. ADM., LOWER MISSOURI | LOVELAND, CO |
| 125. | WAPA/UC | WESTERN AREA PWR. ADM., UPPER COLORADO | MONTROSE, CO |
| 126. | WAPA/UM | WESTERN AREA PWR. ADM., UPPER MISSOURI | WATERTOWN, SD |
| 127. | WAMP | WASHINGTON WATER POWER CO. | SPOKANE, WA |
| 128. | WEFA | WESTERN FARMERS ELECTRIC COOP. | ANADARKO, OK |
| 129. | WIEP | WISCONSIN ELECTRIC POWER CO. | MILWAUKEE, WI |
| 130. | WIPC | WESTERN ILLINOIS POWER COOPERATIVE | JACKSONVILLE, IL |
| 131. | WIPL | WISCONSIN POWER & LIGHT CO. | MADISON, WI |
| 132. | WIPS | WISCONSIN PUBLIC SERVICE CO. | GREEN BAY, WI |
| 133. | YADI | YADKIN, INC. | BADIN, NC |

**CANADIAN CONTROL AREAS**

| | | | |
|---|---|---|---|
| 1. | BCHA | B. C. HYDRO & POWER AUTHORITY | VANCOUVER, BC |
| 2. | CPL | CALGARY POWER LTD. | ALBERTA |
| 3. | GRLA | GREAT LAKES POWER COMPANY, LTD. | SAULT ST. MARIE, ONT. |
| 4. | HQ | QUEBEC HYDRO-ELECTRIC COMMISSION | MONTREAL, QUEBEC |
| 5. | MHEB | MANITOBA HYDRO-ELECTRIC BOARD | WINNIPEG, MANITOBA |
| 6. | NB | NEW BRUNSWICK ELECTRIC POWER COMM. | FREDERICTION, NB |
| 7. | NSPC | NOVA SCOTIA POWER CORP. | HALIFAX, NS |
| 8. | OH | ONTARIO HYDRO | TORONTO, ONT. |
| 9. | SAGU | SAQUENAY POWER SYSTEM | QUEBEC |
| 10. | SPC | SASKATCHEWAN POWER CORP. | REGINA, SASK. |

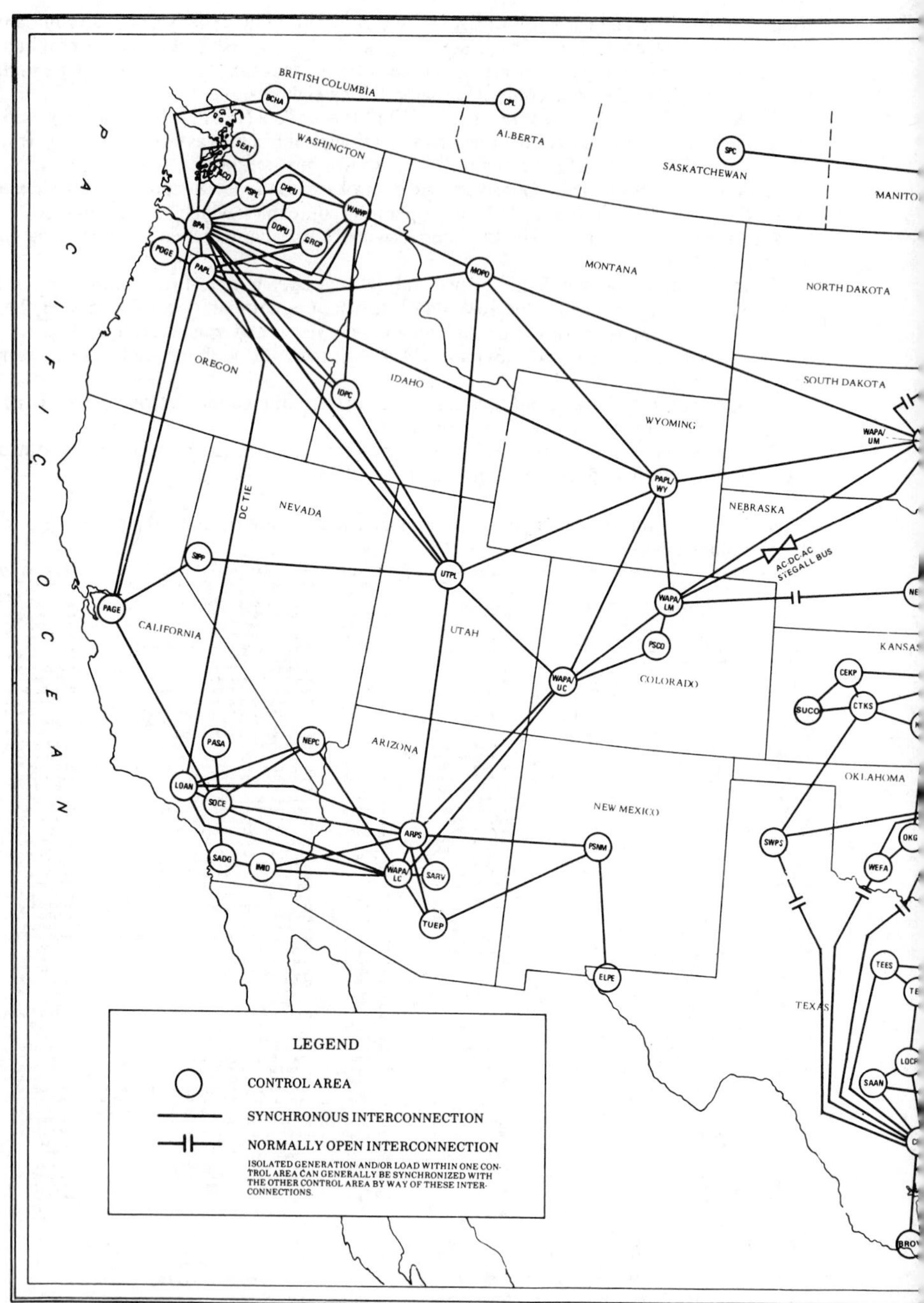


**FIG. 16-8** North American Interconnected Control areas. See Table 16-4 for controlling organization and center locations. *(Map and table from U.S. Dept. of Energy, Federal Energy Regulatory Commission, February 1981.)*

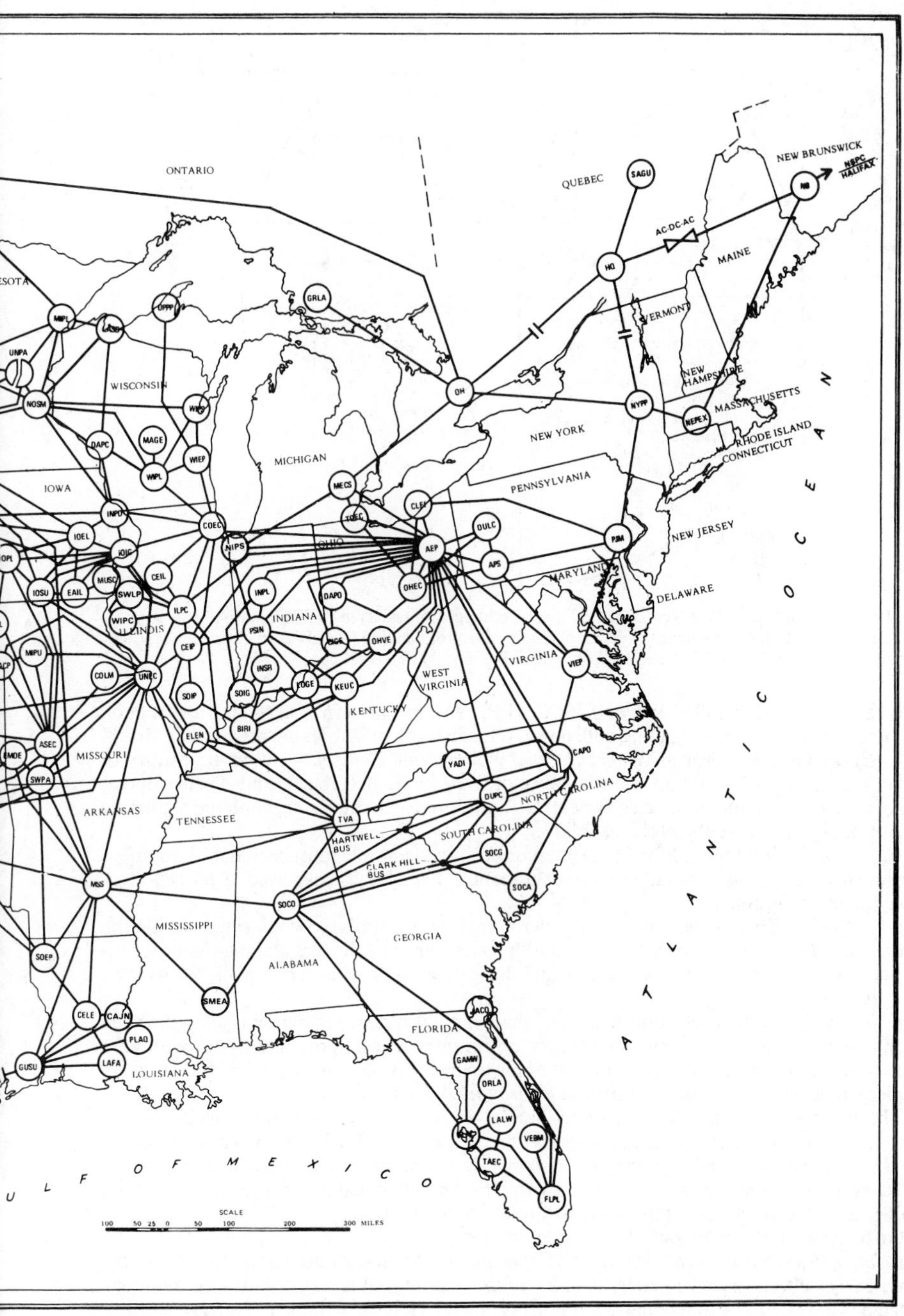

ONTARIO
QUEBEC
NEW BRUNSWICK
NSPC HALIFAX
AC-DC-AC
MAINE
VERMONT
NEW HAMPSHIRE
MASSACHUSETTS
RHODE ISLAND
CONNECTICUT
NEW YORK
PENNSYLVANIA
NEW JERSEY
DELAWARE
MARYLAND
VIRGINIA
WEST VIRGINIA
KENTUCKY
NORTH CAROLINA
SOUTH CAROLINA
TENNESSEE
GEORGIA
ALABAMA
MISSISSIPPI
FLORIDA
LOUISIANA
ARKANSAS
MISSOURI
ILLINOIS
INDIANA
OHIO
MICHIGAN
WISCONSIN
IOWA
ESOTA
HARTWELL BUS
CLARK HILL BUS
A T L A N T I C O C E A N
U L F O F M E X I C O
SCALE
100 50 25 0 50 100 200 300 MILES

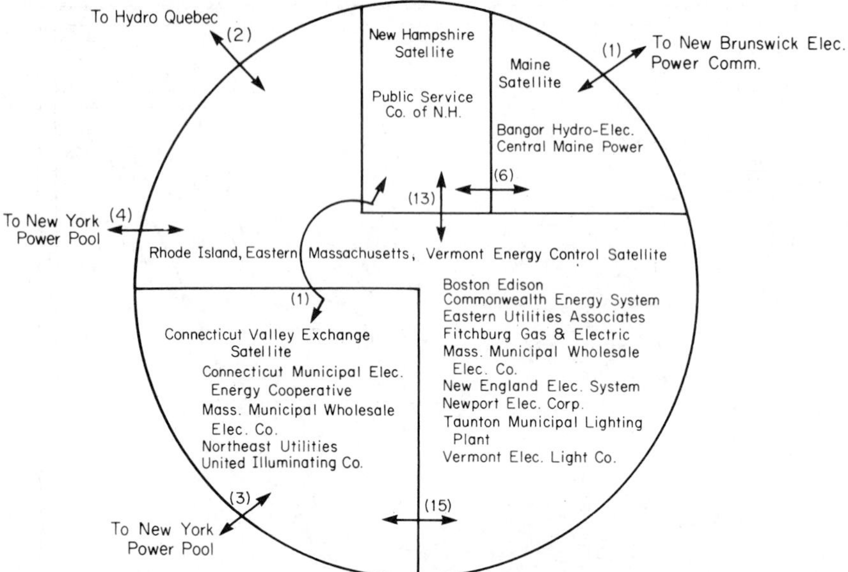


**FIG. 16-9** Composition of the New England Power Exchange Control Area (NEPEX). Member groups are Rhode Island, Eastern Massachusetts, Vermont Energy Control Satellite (REMVEC), Connecticut Valley Exchange Satellite (CONVEX), New Hampshire Satellite, and Maine Satellite.

Operating practices followed by NEPEX, AEP, PJM, the western systems, the Texas pool, and others in scheduling and billing bulk power transfers with neighbors are discussed in Ref. 8.

***4. Operating Objectives of Generation and Power-Flow Control.*** Automatic control of generation and power flow is an essential need for the smooth, neighborly, and effective operation of a widespread interconnected system. On a multiple-area interconnection the regulating or control objectives are threefold, as follows:

*Objective 1.* Total generation of the interconnection as a whole must be matched, moment to moment, to the total prevailing customer demand. This in itself is achieved by the self-regulating forces of the system.

*Objective 2.* Total generation of the interconnected system is to be allocated among the participating control areas so that each area follows its own load changes and maintains scheduled power flows over its interties with neighboring areas. This objective is achieved by *area regulation.*

*Objective 3.* Within each control area, its share of total system generation is to be allocated among available area generating sources for optimum area economy, consistent with area security and environmental considerations. This objective is achieved by *economic dispatch,* supplemented as required by *security* and *environmental dispatch.*

The area means of achieving objectives 2 and 3 are referred to as *supplementary control,* or currently more generally as *automatic generation control* (AGC). Such control may be regarded as a *reallocation control* redistributing the systemwide governing responses that had occurred to load changes in various areas, to generators within the areas that had the change. Each area then follows its own load change, with scheduled internal distribution.

On a single-area system, objective 2 does not apply.

*Relative Priority of Control Objectives.* For all practical purposes, and except for short-term stored-energy effects discussed later, electric energy as produced and distributed on interconnected power systems cannot be stored. It must be made as it is used. Matching total generation of the interconnection to total load of the interconnection is therefore a first and paramount objective if continuity of service to customers is to be maintained. Within the limits of intertie transfer capabilities and any other applicable factors of area safety or security, objective 1 as listed above therefore takes precedence over the other two objectives.

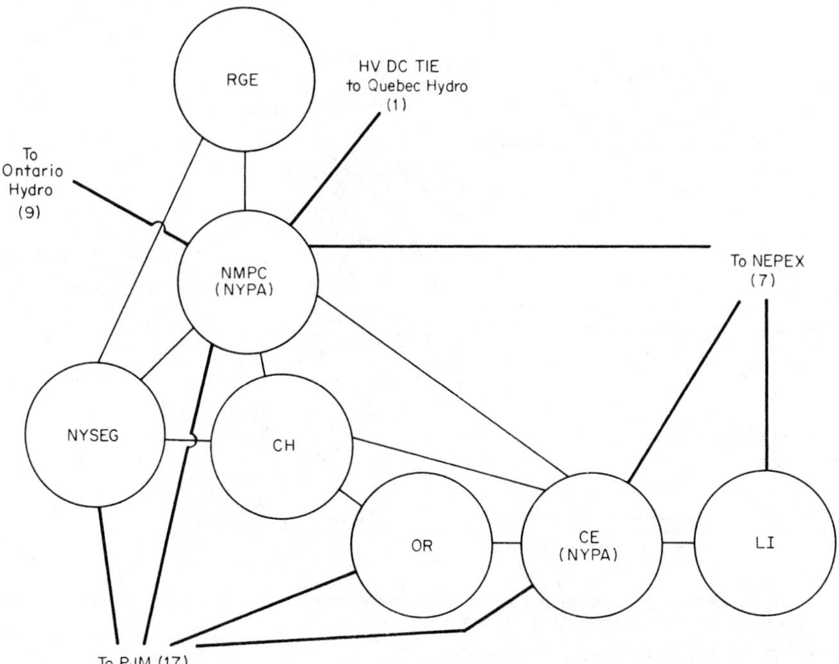


**FIG. 16-10** Operating companies and tie lines of the New York Power Pool (NYPP):

CE: Consolidated Edison Company
CH: Central Hudson Gas and Electric Corporation
LI: Long Island Lighting Company
NYSE&G: New York State Electric and Gas Corporation
NMPC: Niagara Mohawk Power Corporation
OR: Orange and Rockland Utilities, Inc.
RGE: Rochester Gas and Electric Company
NYPA: Power Authority of the State of New York

As between objectives 2 and 3, it is generally felt that each area has an obligation to the interconnection to place higher priority on good execution of area regulation than on optimum area economic dispatch. In other words, if an operating conflict develops in an area, making it difficult for the control simultaneously to achieve objectives 2 and 3, each area will generally be expected to subordinate its desire for optimum economic dispatch to its responsibility to provide area regulation that coordinates effectively with the needs of the interconnection as a whole. Each interconnected system itself establishes criteria on the degree of departure from good area regulation that it considers acceptable for its constituent control areas (Ref. 4–6).

*Evaluation of Control Performance.* Relative control performance (Ref. 9) of an area control system in achieving objectives 2 and 3 is illustrated in the curves of Fig. 16-13. The figure includes four sets of hypothetical curves. Each pair plots, on a common time axis, departures from fully effective area regulation (on the left) and departures from optimum area economic dispatch (on the right). Departures from fully effective area regulation are identified as an *area control error* above zero (area overgenerating) or below zero (area undergenerating).

The curves at *(a)* show excellent control performance for both regulation and economic dispatch. At *(b)* area regulation is again excellent, but this time it is achieved at the expense of internal area economy. At *(c)* area regulation has been sacrificed to achieve internal area

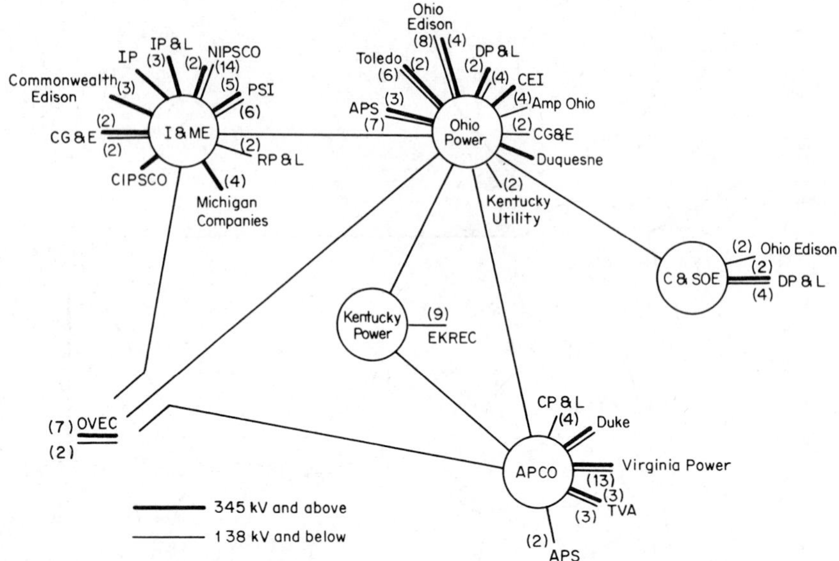


**FIG. 16-11** Operating companies and major tie lines of the American Electric Power Company. Numbers in parentheses indicate number of ties.

AMP OHIO: American Municipal Power of Ohio
APCO: Appalachian Power Company
APS: Allegheny Power System
CEI: Cleveland Electric Illuminating Company
CG&E: Cincinnati Gas & Electric Company
CIPSCO: Central Illinois Public Service
C&SOE: Columbus & Southern Ohio Electric Company
CP&L: Carolina Power & Light Company
DP&L: Dayton Power & Light Company
EKREC: East Kentucky Rural Electric Cooperative
I&ME: Indiana & Michigan Electric Company
IP&L: Indianapolis Power & Light Company
NIPSCO: Northern Indiana Public Service Company
OVEC: Ohio Valley Electric Company
PSI: Public Service Company of Indiana
RP&L: Richmond Power & Light Company

economy. At *(d)* control fails to achieve either good area regulation or internal economic dispatch.

From the viewpoint of the interconnection as a whole, control as reflected in the curves at *(a)* and *(b)* would be regarded as excellent, the control area in both instances completely fulfilling its obligation to the interconnection. The area regulation at *(c)* and *(d)* might be regarded by the interconnection as less than satisfactory, depending upon the magnitude and duration of the departures of area control error from zero, as compared with performance criteria that the interconnection would regard as reasonable.

Additional factors which must be considered in judging area control performance are area security and environmental considerations. These factors may at times dictate significant

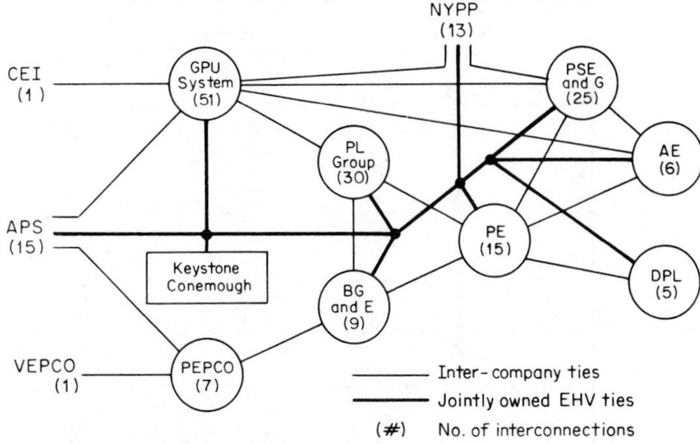


**FIG. 16-12** Operating companies and major tie lines of the Pennsylvania–New Jersey–Maryland Pool.

APS: Allegheny Power System

BG&E: Baltimore Gas & Electric Company

CEI: Cleveland Electric Illuminating Company

GPU: General Public Utilities System
Jersey Central Power & Light Company
Pennsylvania Electric Company
Metropolitan Edison Company

NYPP: New York Power Pool

PEPCO: Potomac Electric Power Company

PSE&G: Public Service Electric & Gas Company

PE: Philadelphia Electric Company

AE: Atlantic City Electric Company

DPL: Delmarva Power & Light Company

PL Group: Pennsylvania Power & Light Company
UGI Corporation

departures from good area regulation and desired economy, yielding results as reflected in the curves at *(d)* of Fig. 16-13. For additional discussion of area control performance see Par. **15**.

## Governing

***5. System Governing.*** There are two components in the governing action that matches total system generation to total system load. One derives from the speed-governing systems (Ref. 10) with which most generating units are equipped. The other results from the frequency coefficient of connected load. In addition, there is a short-term transient effect from the spinning stored energy of the system when load changes occur.

The sequence of steps by which system load changes are accommodated by corresponding changes in system generation is, on assuming, for example, an *addition* of load to the system, as in the following list:

1. *Change in stored energy.* The additional increment of load applied to the system is intially satisfied from the stored spinning energy of the system, resulting in a corresponding decrease of system speed or frequency.

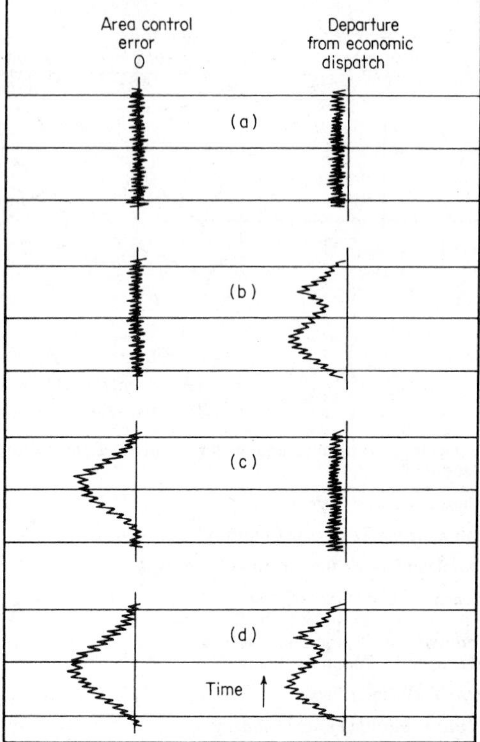


**FIG. 16-13** Relative control performance of an area control system. (*a*) Excellent area regulation and economic dispatch; (*b*) excellent regulation, but at the expense of internal area economy; (*c*) area regulation sacrificed to achieve internal area economy; (*d*) poor regulation and poor internal area dispatch.

2. *Change in effective load.* The total system load, because of its frequency coefficient, will effectively decrease because of the frequency decrease, correspondingly releasing already available generation to serve the newly added load increment.
3. *Change in generation.* The reduced frequency will also cause generating-unit–speed-governing systems to increase input to their prime movers, increasing generation and arresting further change in frequency.

For a *drop* in connected-system load, system stored energy and system frequency would both *increase,* the remaining system load would effectively *increase,* and total system generation would be *decreased.*

*Effect of System Frequency.* The above steps of system governing result in sustained frequency deviations following system load changes, causing a decrease in system frequency when load is added and an increase when load is dropped. On a large interconnected system, a sudden load change of about 2% of system spinning capacity would typically result in a frequency change of about 0.1 Hz. Similarly, the effect on system frequency when a large block of generation is lost is of this same approximate magnitude. These figures assume that the impact of the sudden load or generation change will not disturb stable system operation and that the complete interconnection remains in synchronism. Unless specifically stated to the contrary, this same assumption applies throughout this section.

*Index to Mismatch of Generation and Load.* A changing system frequency is the index of a mismatch between total system generation and total system load. Governing action is in process while such acceleration or deceleration of system speed is taking place. It has been effectively completed and the match between generation and load restored when frequency is steady, even though it is at a value other than the normal frequency schedule, 60 Hz in the United States.

*Importance of Constant Frequency.* A constant frequency at its normal scheduled value is not necessarily, of itself, a major system operating objective (Ref. 11). The deviations from normal value that result from governing action might, therefore, not be regarded as objectionable. Synchronous time would be affected but could readily be corrected periodically. Much more significant, however, is the fact that the automatic control equipment which carries out the functions of area regulation and economic dispatch utilizes prevailing frequency as a parameter in the control execution. As is discussed later, this is the factor that permits the automatic controls in the many control areas of an interconnection to operate simultaneously without interaction or hunting between them. More specifically, proper allocation of total system generation to each of the control areas, under this type of area regulation, occurs only when system frequency is at its normal value. For this reason, maintaining system frequency constant, which is to say restoring it to its normal value after departure owing to system governing action, becomes a system control objective.

On the large ISG-PJM-CANUSE interconnection, frequency typically has a moment-to-moment bandwidth of about 0.01 to 0.02 Hz, with superimposed deviations of the band average, on a time scale of minutes, usually well within ±0.01 Hz of nominal 60 Hz. A section of frequency chart from this interconnection, made on an exceptionally high resolution measuring instrument, is shown in Fig. 16-14.

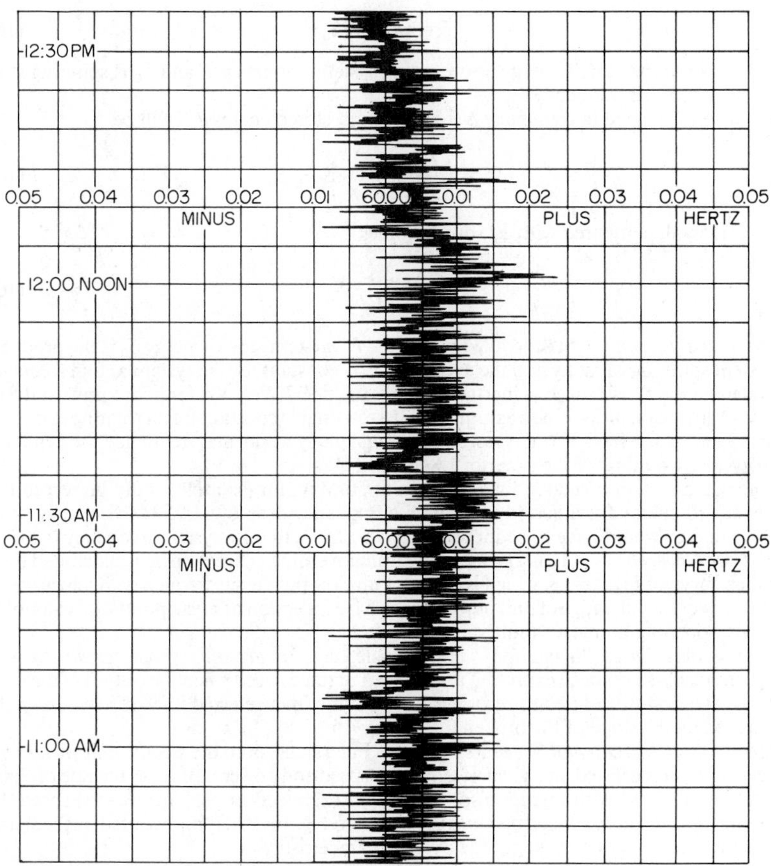


**FIG. 16-14** Frequency-deviation chart of the ISG–PJM–CANUSE interconnection.

*Natural Regulation and Supplementary Control.* System governing actions are sometimes referred to as *natural regulation.* Subsequent regulating steps, executed manually or automatically through the speed changers of one or more system generator governors to correct frequency deviations, area generation allocations, tie-line loadings, economic dispatch, or combinations of these conditions, are defined as *supplementary regulation* or *supplementary control* and more generally as *automatic generation control* (AGC). In older literature this was referred to as *load-frequency control.*

***6. Stored-Energy Considerations.*** Restoration of system frequency to its nominal value, for a given connected load, is a matter of restoring system stored spinning energy to the value that corresponds to normal frequency. Such restoration is achieved by supplementary control, creating overgeneration if frequency is to be raised or undergeneration if frequency is to be lowered. This in effect creates a new mismatch between total generation and total load, system governing action follows, and a new frequency level is established corresponding to the new level of stored spinning energy in the system. Such supplementary control is continued until normal frequency is restored.

It should be noted that supplementary control action, because of the mismatch between generation and load it creates, is always followed by governing action. Within the limits of governor sensitivities and dead band, there will correspondingly be reallocation or shifting of generation between units which have remained on governing control only and those to which supplementary control has been applied.

*Stored-Energy Equations.* Stored spinning energy may be written as a function of the square of system frequency as follows:

$$S = (F/F_0)^2 S_0 \tag{16-1}$$

where $S$ is spinning stored energy at prevailing system frequency $F$ and $S_0$ is spinning stored energy at scheduled frequency $F_0$.

For a small change in frequency $\Delta F$, a change in stored energy $\Delta S$ will be

$$\Delta S = \left(\frac{F_0 + \Delta F}{F_0}\right)^2 S_0 - S_0 \tag{16-2}$$

When $F$ is small compared with $F_0$, this becomes

$$\Delta S = \frac{2\Delta F}{F_0} S_0 \tag{16-3}$$

*Numerical Example.* At scheduled frequency $F_0$, system stored energy $S_0$ is the product of the system spinning capacity and the average inertia constant for the system at that frequency. One reference (Ref. 12) suggests inertia constants of 2 to 6 kWs/kVA for hydro units and 5 to 9 kWs/kVA for steam units. Upon assuming that a reasonable constant for all rotating equipment of the system is 6 kWs/kVA, the stored spinning energy at normal frequency for a 300,000-MVA system would be of the order of 1,800,000 MWs.

On such a system a sudden load increase of 50 MW, for example, could be served from system stored energy for a period of 6 s with a frequency drop of 0.005 Hz. Similarly, stored spinning energy would have to be increased by 300 MWs to raise system frequency 0.005 Hz.

***7. Unit Governors.*** A typical generating-unit governing characteristic is illustrated by the solid-line curve of Fig. 16-15, which is a plot of unit output versus frequency. Such curves are not linear over the full range of unit output. They have inflection or break points, for example, at inlet valve-opening points on multivalve turbines.

*Steady-State Speed Regulation.* For a single unit, *steady-state speed regulation* is the change in steady-state speed expressed as a percent of rated speed when the output of the unit is gradually reduced from rated to zero power. It is sometimes referred to as *percent droop.* It is represented by the slope of the broken line in Fig. 16-15.

*Steady-State Incremental Speed Regulation.* For a single unit, the *steady-state incremental speed regulation* is defined, at a given steady-state speed and power output, as the rate of change of steady-state speed with respect to power output expressed in percent of rated speed. It is sometimes referred to as *percent incremental droop.* In Fig. 16-15 it is represented by the slope of each of the segments that make up the overall characteristic.

*Dead Band.* The *dead band* of a speed-governing system is the measure of its insensitivity to changes in system speed and is expressed in percent of rated speed.

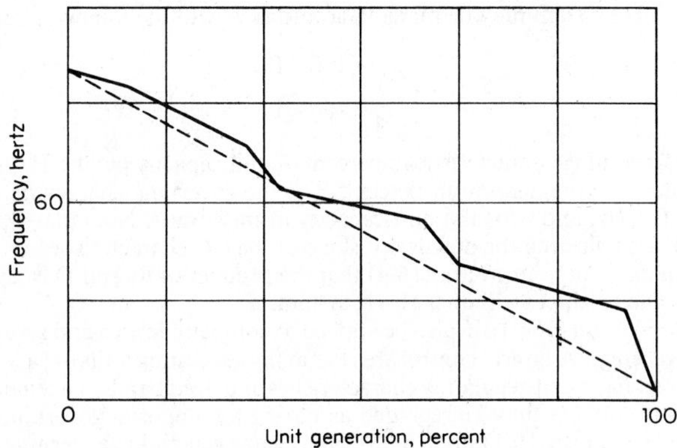


**FIG. 16-15** Typical generating-unit governing characteristic.

*Hypothetical Governing Characteristic.* For many aspects of power systems' operations and control analysis it is convenient and acceptable to consider that unit governing characteristics are linear over the small range of system speed being considered. Such a characteristic, solid line *GG*, is shown in Fig. 16-16.

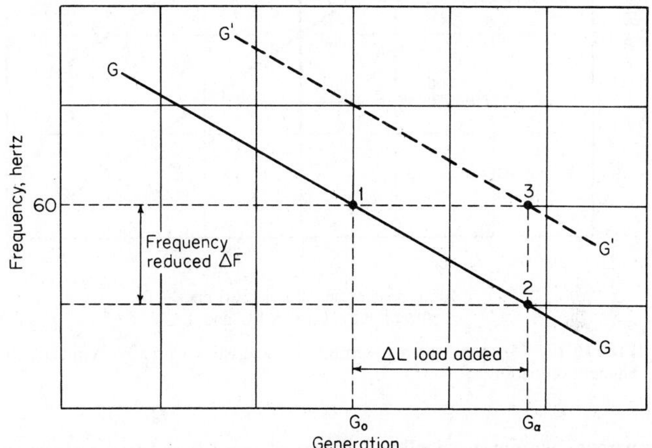


**FIG. 16-16** Unit governing characteristics (considered linear over a small range of system speed).

*Shifting the Governing Characteristic.* Applying supplementary control to the speed changer of a speed-governing system has the effect of shifting the unit characteristic parallel to itself, illustrated by the broken line *G′G′* in Fig. 16-16. A complete cycle of governing action and supplementary control for an isolated unit with a governing characteristic as shown in Fig. 16-16 and with a connected load having zero frequency coefficient is as follows: Point 1 on *GG* defines initial conditions of generation and load $G_0$ and 60 Hz. Load is increased by $\Delta L$, and governing action achieves balance at point 2. Supplementary control shifts the generating characteristics to *G′G′*, on which frequency is restored and the new load accommodated at point 3.

*Alternative Expressions for Governing Characteristic.* In addition to stating speed regulation or governing characteristics in terms of percent of rated speed, it is sometimes desirable to express these parameters either in percent capacity per 0.1 Hz or megawatts per 0.1 Hz. Conversions for a 60-Hz system may be made in accordance with the following relations:

$$\beta_1' = 100/6D \tag{16-4}$$

$$\beta_1 = M/6D \tag{16-5}$$

where $\beta_1'$ is the governing characteristic in percent of unit capacity per 0.1 Hz, $D$ is the steady-state incremental speed regulation in percent, $\beta_1$ is the governing characteristic expressed in megawatts per 0.1 Hz, and $M$ is the unit capacity in megawatts. Note that $D$, $\beta_1$, and $\beta_1'$ are negative quantities reflecting the downward slope of the *GG* characteristic.

It is convenient to note from Eq. (16-4) that the product of $\beta_1'$ and $D$ is always 16⅔ on a 60-Hz system. The product is 20 on a 50-Hz system.

Equations (16-4) and (16-5) can also be applied to comparable area and system parameters.

***8. Area Governing.*** A typical control area has many generating units of varying types, sizes, and ages, with varying speed-regulating characteristics and dead bands. Taken in the aggregate, operating units of an area may be regarded as having a composite governing characteristic similar to curve *GG* in Figs. 16-17 and 16-18. Typical area generation governing characteristics

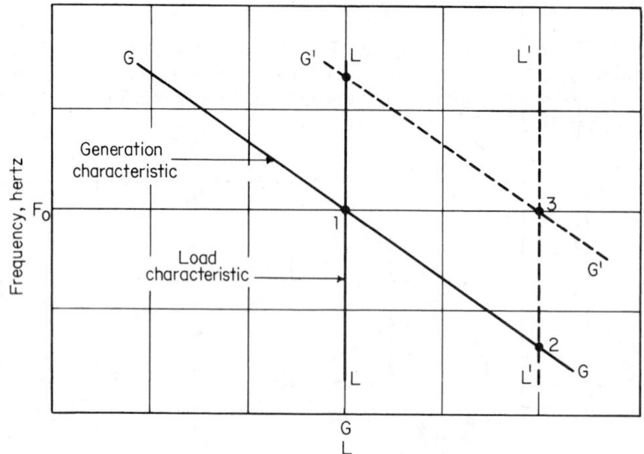


**FIG. 16-17** Composite area-governing characteristics, with zero frequency coefficient of load.

expressed as percent of rated speed fall in the range of about 16 to 5%, corresponding according to Eq. (16-4) to a range for $\beta_1'$ of about 1 to 3.5% of capacity per 0.1 Hz.

*Area-Load Frequency Characteristic.* The frequency coefficient of connected load in a control area defines its *area-load frequency characteristic.* The latter is the change in total area load that results from a change in system frequency. It is expressed in percent of connected load per 0.1 Hz and designated $\beta_2'$, or in megawatts per 0.1 Hz and designated $\beta_2$. Both $\beta_2'$ and $\beta_2$ are positive quantities. This characteristic is reported (Ref. 13) to cover a fairly broad span in various areas. Typical averages are in the range of 0.3 to 0.5%/0.1 Hz. In Fig. 16-17 the load-frequency characteristic *LL* is drawn to illustrate zero frequency coefficient. In Fig. 16-18, the characteristic *LL* is shown with a typical value of about one-third the generation governing characteristic.

*Matching Area Generation and Load—Isolated Operation.* Consider an area with governing characteristics first as in Fig. 16-17 and then as in Fig. 16-18, operating isolated. In both graphical representations, balance between area generation and area load exists where the load-frequency characteristic *LL* intersects the generation governing characteristic *GG*. In both figures, initial conditions are as at point 1. Load is then increased to the level defined by *L'L'*. A

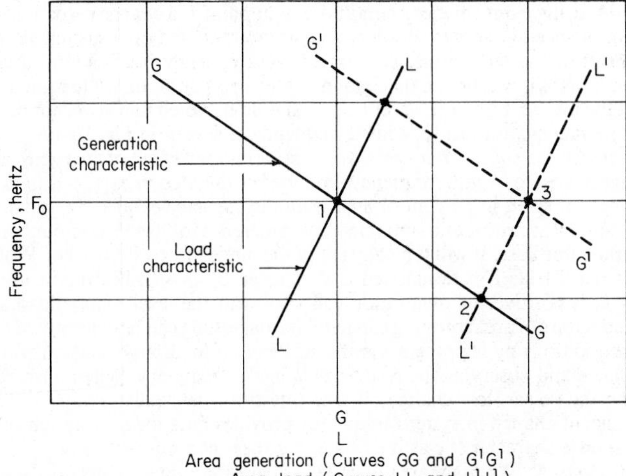


**FIG. 16-18** Composite area-governing characteristics, with load-frequency characteristic at one-third the generation governing characteristic.

new point of balance is achieved at point 2, by generation governing only in Fig. 16-17, and by both generation governing and the effect of area-load frequency characteristic in Fig. 16-18. In each case supplementary control is then applied, shifting *GG* to a new position *G′G′*, and, as shown in each of the figures, balanced conditions with frequency returned to normal are achieved at point 3.

*Matching within Dead-Band—Isolated Operation.* Upon assuming an appreciable overall dead band for the aggregate of its unit governors, the area generation governing characteristic for small load changes that fall within the limits of the dead band may be considered to be a vertical line, such as *GG* in Fig. 16-19. The load-frequency characteristic is as shown by *LL*. On isolated

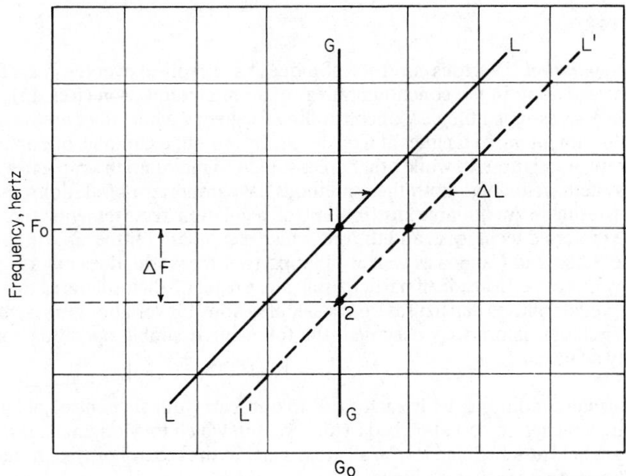


**FIG. 16-19** Area generation-governing characteristic for very small load changes (*GG* vertical line; governing is by load-frequency characteristic only).

operation and starting from balanced conditions as at point 1, a change in load $\Delta L$ too small to initiate generator speed-governor action can be accommodated on a sustained basis after the initial response from stored spinning energy by the frequency coefficient of area load. Area generation remains as it was before the load increase. This condition is illustrated by point 2 in Fig. 16-19, at the intersection of $GG$ and $L'L'$. It should be noted that governor dead band may occur at any frequency, not just scheduled frequency as shown in Fig. 16-19.

*Area Frequency-Response Characteristic.* The combined effect of area generation governing and change in area load with frequency is defined as the *area frequency-response characteristic.* It may be expressed in percent of area spinning capacity per 0.1 Hz, in which case it is designated $\beta'$ and is the arithmetic sum (algebraic difference) of its two components, $\beta'_1$ and $\beta'_2$, expressed to the same base. It will typically fall in the range of 1 to 4%/0.1 Hz. When expressed in megawatts per 0.1 Hz it is designated $\beta$. It is given by $\beta_1 - \beta_2$, bearing in mind that $\beta_1$ is negative and $\beta_2$ is positive. Its magnitude will vary with the level of area load and with the magnitude and nature of area spinning capacity. Its magnitude can be determined for a given set of operating conditions by tripping a significant block of load or generation when an area is operating isolated and observing the resulting change in frequency. When operating interconnected, noting the power-flow change on area interties when a disturbance outside the area causes a significant change in system frequency provides data for a computation of the area frequency-response characteristic as it exists under these conditions.

*System Responses.* When an area operates as part of an interconnection, it responds not only to its own load changes but also to the load changes that occur in other areas. The governing characteristic curves of Figs. 16-17 to 16-19, though referred to earlier as applicable to an isolated area, may be regarded as applicable to a complete interconnected system, which in itself is really a large isolated area. The $GG$ curves of these figures represent the composite characteristic of all the generation governors in operation within the system. The $LL$ curves are the aggregate of all area load-frequency characteristics. The algebraic difference of the $GG$ and $LL$ characteristics, $\beta'_1 - \beta'_2$, represents the *system frequency-response characteristic* and is typically of the order of 2% of spinning capacity per 0.1 Hz for large interconnections. It is a negative quantity.

For an analysis of the dynamic relationships of system self-regulating forces, see Ref. 14.

Because governors cannot recognize the origin or location of a load change, all areas of an interconnection share in system load changes in proportion to their respective frequency-response characteristics. Supplementary control is required to allocate generation changes to the control areas in which the load changes occur.

## Bias Control

***9. Area Regulation.*** Various control techniques have evolved over the years for achieving effective area regulation in the constituent areas of an interconnection (Ref. 15). In one technique, an area was assigned the task of controlling frequency while other areas endeavored to hold power flow on interarea tie lines at fixed levels. In another technique, one area still had the frequency-control assignment, while other areas sought to maintain interarea ties at levels that varied with system frequency. Both these methods have important limitations, as analyzed in Ref. 16, and result in inequitable distribution of regulation requirements and in interarea hunting. The preferred technique, and the one which is standard for the large interconnections in the United States and Canada as well as other parts of the world, does not assign frequency control to any one area. Instead, all participating areas regulate their interarea ties in a manner that permits predetermined departures from scheduled flows for variations in system frequency. When this regulation is properly executed, the following desirable operating conditions are automatically achieved:

1. Under normal conditions, with each area able to carry out its control obligations, load changes are assigned to and absorbed by the areas in which they originate, interarea power transfers are held at scheduled levels, all areas share in the control of system frequency, and normal system frequency is obtained.
2. Under abnormal or emergency conditions, when one or more areas are unable to carry out their control assignments, and assuming the interconnection remains in synchronism, all other areas automatically assist the areas that are in need by permitting interarea power

transfers to depart from normal schedules in the direction that will help the troubled areas, and frequency is permitted to depart from normal to the extent required to provide the necessary assistance over the interarea ties to the troubled areas.

This type of automatic control is identified as *frequency-biased net interchange control.* It is also referred to as *net interchange bias control, frequency bias control,* or just *bias control.*

*Interchange Flow Paths.* When an area has bulk transfer ties with more than one additional area, scheduled interchange transfers between them, though maintained at scheduled levels, do not necessarily flow directly between them. Flows are likely to split over parallel paths through other areas. In some cases, flow on an individual tie line may be in the direction opposite to its schedule, though in the aggregate each area is achieving the correct net of all its prevailing schedules (Ref. 17). This can be seen by considering area A of Fig. 16-7 tied to area B and C as shown in Fig. 16-20. Scheduled interarea transfers are as shown by the broken-line arrows. Actual flows may be as shown by the solid-line arrows. Scheduled and actual flows on individual ties are summarized in Table 16-5.

It will be noted in Fig. 16-20 that the actual net interchange of each area with the interconnection as a whole does match the algebraic sum of its schedules with adjacent areas. This is summarized in Table 16-6.

In applying automatic control, and as suggested by Tables 16-5 and 16-6, an area having ties

**TABLE 16-5** Scheduled and Actual Tie-Line Power Flows as Shown in Fig. 16-20

| Interarea tie | Scheduled flow | Actual flow |
|---|---|---|
| AB | 20 to B | 20 to A |
| AC | 80 to C | 120 to C |
| BC | 60 to C | 20 to C |

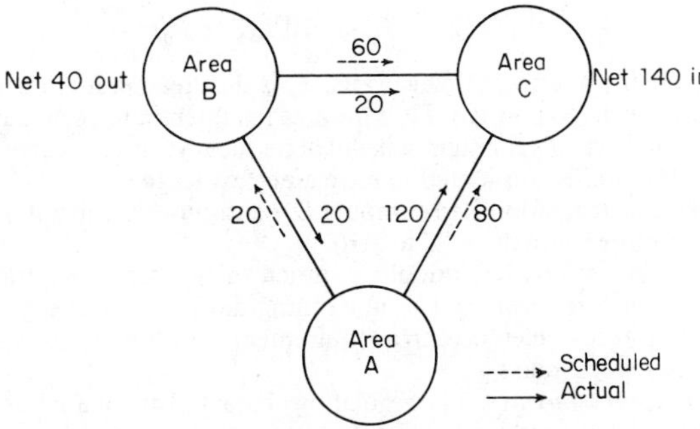


**FIG. 16-20** Power flow in individual tie lines at different levels and in a direction opposite to schedules.

**TABLE 16-6** Scheduled and Actual Net Interchange Flows as Shown in Fig. 16-20

| Area | Net of area interchange schedules | Actual area net interchange |
|---|---|---|
| A | 100 out | 100 out |
| B | 40 out | 40 out |
| C | 140 in | 140 in |

with more than one other area cannot regulate just one interarea schedule. It must regulate to the net of all its interarea schedules.

*Net Interchange as a Measure of Area Balance.* An objective of area regulation is to match area generation changes to area load changes while interarea transfers are maintained on schedule. The aggregate of area load changes or the extent to which they have been matched by area generation changes cannot be measured directly. However, deviations of net interchange from schedule are an index of the extent to which area load changes have not been matched by area generation changes and become a direct measure of the generation required in the area to match generation to load. For an interconnected system of $n$ areas, the following relations apply for each area:

$$G_{nr} = L_n + T_{0n} \tag{16-6}$$

$$G_n = L_n + T_n \tag{16-7}$$

where $G_{nr}$ is the area generation required to match prevailing area load $L_n$ and maintain area net interchange schedule $T_{0n}$, interchange "out" is +, interchange "in" is −, and $G_n$ is the prevailing area generation, with prevailing area load $L_n$ and actual net interchange $T_n$. Subtracting Eq. (16-6) from Eq. (16-7),

$$G_n - G_{nr} = T_n - T_{0n} \tag{16-8}$$

Equation (16-8) states that the change required in area generation to follow its own load changes, and at the same time restore net interchange to schedule, is given by the generation error, which is a measure of the deviation of area net interchange from schedule.

***10. Net Interchange Tie-Line Bias Control.*** Area regulation based on Eq. (16-8) would represent *constant net interchange control,* sometimes referred to as *flat tie-line control,* and would have the limitations (Ref. 16) already referred to. Expanding Eq. (16-8) to include a frequency bias factor provides the effective and cooperative type of area regulation identified as *frequency-biased net interchange tie-line bias control* (Refs. 16 and 18).

*Area Control Error.* In operating on net interchange tie-line bias control, deviation from desired area generation is defined as the *area control error* and is given for area $n$ by

$$E_n = (T_n - T_{0n}) - 10B_n(F - F_0) \tag{16-9}$$

where $E_n$ is the area control error, in megawatts; $T_n$ is the area net interchange in megawatts, power "out" being considered positive; $T_{0n}$ is the area net interchange schedule in megawatts; $F$ is system frequency, in hertz; $F_0$ is system scheduled frequency; and $B_n$ is area frequency bias, in megawatts per 0.1 Hz, and is considered to have a negative sign.

The objective of area regulation is to compute $E_n$ continuously and automatically to adjust area generation as required to reduce $E_n$ to zero.

Although no longer preferred terminology, much of the basic literature on tie-line bias control utilizes the term *area requirement* in defining deviation of area generation from the desired value. For ready cross reference, area requirement is arithmetically equal to area control error but is of opposite algebraic sign.

*Bias Regulating Characteristic.* The regulating characteristic of a net interchange tie-line bias control which operates to reduce $E_n$ of Eq. (16-9) to zero is shown as curve *CC* in Fig. 16-21. Point 1 on *CC* defines the scheduled net interchange $T_{0n}$ at normal frequency $F_0$. The slope of curve *CC* is the reciprocal of $10B_n$.

At each point in time, a plot of prevailing system frequency at prevailing net interchange will define a point on the coordinates of Fig. 16-21. When this point falls on *CC*, the area control error will be zero. The controller will be in balance, and it will not change area generation. When a point such as $p$ does not fall on *CC*, its horizontal displacement from curve *CC* is the control error $E_n$. When it falls above the curve, $E_n$ is plus, indicating a need to decrease area generation. When it falls below the curve, $E_n$ is minus, reflecting a need to increase area generation. In each of the latter cases, the automatic control would adjust area generation until $E_n$ is reduced to zero, and the point defining the resultant net interchange and frequency falls on curve *CC*.

Rotating characteristic *CC* clockwise to a vertical position would yield the control curve for *constant net interchange control,* that is, $B_n = 0$. Rotating the characteristic *CC* counterclockwise to a horizontal position would provide the control curve for *constant frequency control,* that is, $B_n = \infty$.

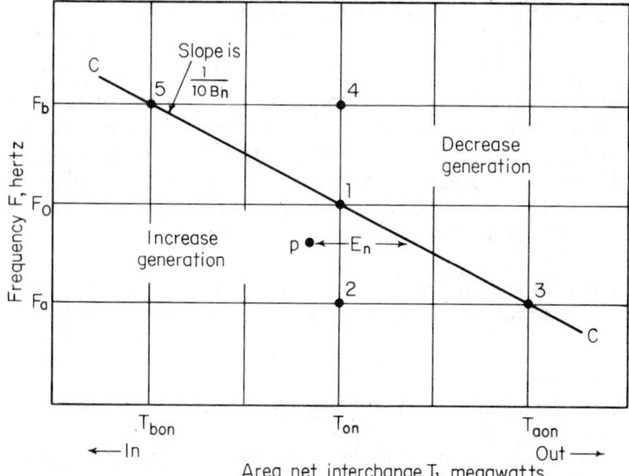


**FIG. 16-21** Area tie-line bias zero-control-error curve. $T_{0n}$ is the net interchange schedule at normal frequency $F_0$; $T_{a0n}$ is the schedule at $F_a$; $T_{b0n}$ is the schedule at $F_b$.

*Computation of Area Control Error.* A schematic diagram of the computation of area control error in accordance with Eq. (16-9) for the control area of Fig. 16-7 is shown in Fig. 16-22. Although this area has only two interties that need to be summated to obtain area net interchange $T_n$, most areas have many ties. All must be included in the net interchange summation.

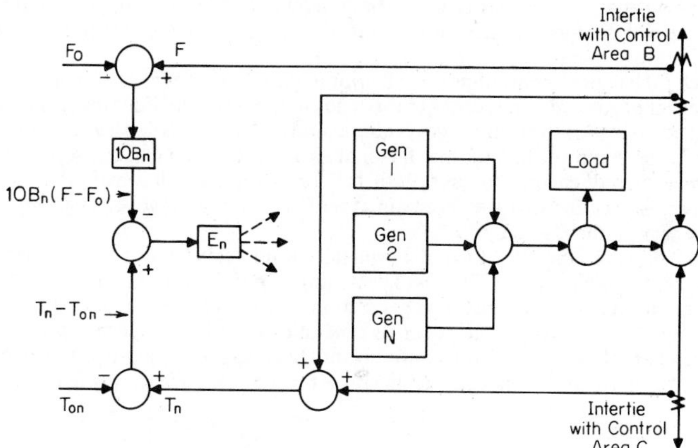


**FIG. 16-22** Computation of an area control error.

Power flow on tie lines normally swings or oscillates at a period of a few seconds. To obtain an accurate measurement of net interchange, and to ensure that any power that is being wheeled through the area be excluded from the measurement, tie-line telemetering equipment should provide simultaneity of measurement of all ties that enter into the computation.

The broken lines leaving block $E_n$ in Fig. 16-22 represent the application of control to area generators to reduce $E_n$ to zero (see Pars. **20–28** for a summary of pertinent techniques and considerations applicable to this facet of generation control).

*Functions of the Bias Factor.* The inclusion of the frequency bias factor $10B_n(F - F_0)$ in Eq. (16-9) fulfills two functions.

1. It coordinates area regulation with area governing responses to load changes that occur in other control areas. Depending on the bias setting, the bias factor will permit such area governing responses to persist, will add to them, or will diminish them.
2. It assigns to each control area a share of the frequency-control burden. For example, if net interchange is on schedule for all control areas, but frequency is too low, then for each area the factor $T_n = T_{0n}$ is zero, but the bias factors are not. All areas, therefore, have an area control error $E_n$ proportionate to the frequency deviation $F - F_0$ and to the respective bias settings. All areas will therefore participate in overgenerating to restore system stored energy and system frequency to their normal values.

*Bias as a Schedule Shift.* The bias factor of Eq. (16-9) may be regarded as causing a shift in area net interchange schedule with frequency. Equation (16-9) may be rewritten

$$E_n = T_n - T_{x0n} \tag{16-10}$$

where $T_{x0n}$ is the biased scheduled interchange for the area at frequency $F$ and is equal to $T_{0n} + 10B_n(F - F_0)$.

In Fig. 16-21, for example, point 2 defines normal interchange schedule $T_{0n}$ and frequency $F_a$. The point does not fall on *CC*, however, and the new biased schedule for this frequency is $T_{a0n}$, defined by point 3 on *CC*. Similarly, point 4 is not on *CC*, and for frequency $F_b$ the new biased schedule is $T_{b0n}$, defined by point 5 on *CC*.

*An Illustration of Bias Action.* The simplified sketches in Fig. 16-23 illustrate how the bias factor coordinates area regulation with area governing responses to remote load changes for the ideal case where the bias setting matches the area governing characteristic. They also illustrate the concept of schedule shift with frequency. In this example (Ref. 19), area *A*, in which a load change of five units in magnitude occurs, is assumed to have one-fifth the interconnection capacity. The remaining four-fifths of the interconnection capacity is shown as another single area *B*. Governing characteristics and bias settings are assumed to be of the same percentage magnitude in each area and hence, when expressed in megawatts per 0.1 Hz, are proportional to respective area capacity. Both areas are on net interchange bias control, and the interchange schedule is zero. Also, for simplicity, it is assumed that $\beta_2$, the frequency coefficient of load in each area, is zero.

At sketch *(a)* balanced conditions prevail prior to the addition of five units of load at area *A*. At *(b)*, the effect of governing responses to this load change is shown. Four units of the new load at *A* are picked up by *B* as the result of governing action and appear as departure from normal on the *B-A* tie. Upon assuming that the bias factor at area *B* is set to match its governing characteristic $\beta$, the new biased schedule for area *B* will call for four units of flow from *B* to *A*. Since this flow already exists, $E_B$ for area *B* will remain at zero, permitting the governing contribution of area *B* to persist on the *B-A* tie.

At area *A*, however, the plot of *F* and $T_n$ will fall considerably below and to the left of its *CC* curve, like point *p* of Fig. 16-21. There will be a minus control error $E_A$, and control will act to increase generation correspondingly. As generation at *A* is increased, frequency will be raised and governing action at area *B* will reduce its generation contribution correspondingly. In this hypothetical example, these steps will continue until fully balanced conditions are restored as at sketch *(c)*, with all five units of added load in area *A* being accommodated by five units of added generation in that area.

## Bias Settings

***11. Influence of Bias Settings on Control.*** The magnitude of the bias setting, $B_n$ in Eq. (16-9), has relatively little effect on the ability of a tie-line bias controller to respond to load changes that occur within its own area. Even with zero bias the controller could fulfill this function. The bias setting has considerable influence, however, on the response of the controller to load changes in other areas and on the coordination of the controller with area governing and with controller actions in other areas (see Refs. 20 to 22).

*Response to Remote Load Change.* Load changes that occur outside a given control area are considered from the viewpoint of that area as *remote* changes.

Figure 16-23 has demonstrated the usefulness of ideally matching area *B* bias setting to its

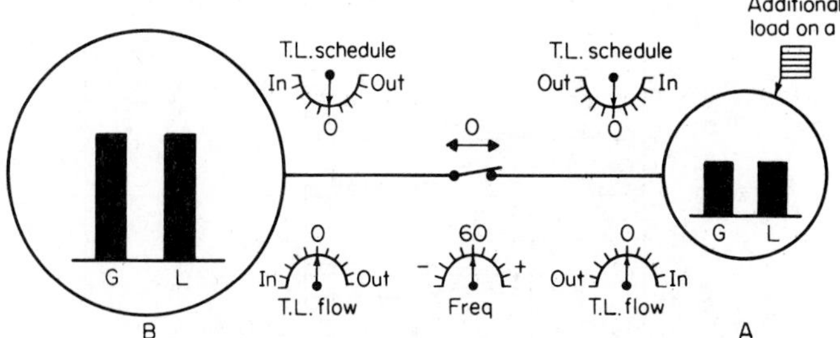

(a) Balanced conditions prior to load change at A

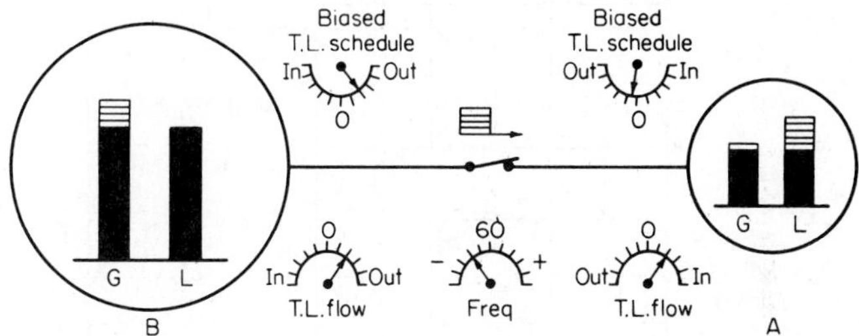

(b) Conditions following governing action in response to load increase at A

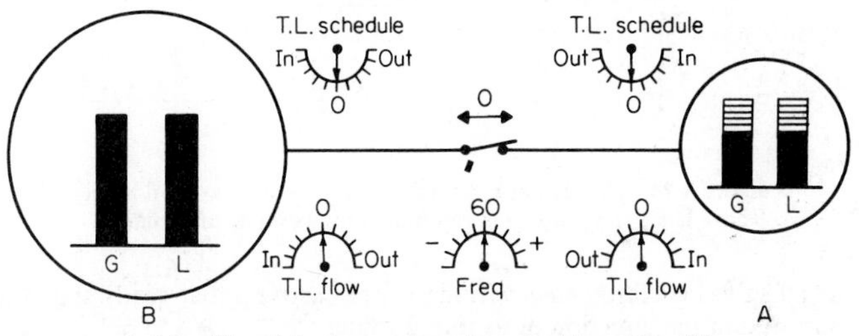

(c) Conditions following supplementary regulator action at A

**FIG. 16-23** Governing and supplementary regulation responses to a remote load change for the ideal case of area bias setting matching area governing characteristic. Also illustrates net interchange schedule shifts with frequency.

frequency-response characteristic. For this condition, area $B$ bias control simply sustains the area's self-regulating contribution to remote area $A$, neither adding to it nor reducing it. Such ideal settings cannot always be achieved. Conditions resulting when it is not will now be examined.

The nature of the changes in system frequency $B$, tie-line power flow $T_B$, effective load $L_B$ at area $B$, and generation $G_B$ at area $B$ in response to a load change at remote area $A$, for varying bias settings at area $B$, is shown in Fig. 16-24.

Initial steady-state conditions apply from $t_0$ to $t_1$. The step-function load change, an increase in this example, occurs at area $A$ at $t_1$. From $t_1$ to $t_2$, bias control acts at area $B$, resulting in the new steady-state conditions shown from $t_2$ to $t_3$.

*Effect of Bias Ratio.* The action from $t_1$ to $t_2$ (Fig. 16-24) depends on the bias ratio at area $B$, which is the ratio $R_B$ of the bias setting $B_B$ in area $B$ to its combined frequency-response characteristic $\beta_B$. When $R_B = 1$, as it was in Fig. 16-23, the several parameters do not change from $t_1$ to $t_2$. There is full coordination of the bias control with governing responses at area $B$.

When $R_B < 1$, bias control imposes additional changes on all four parameters, moving

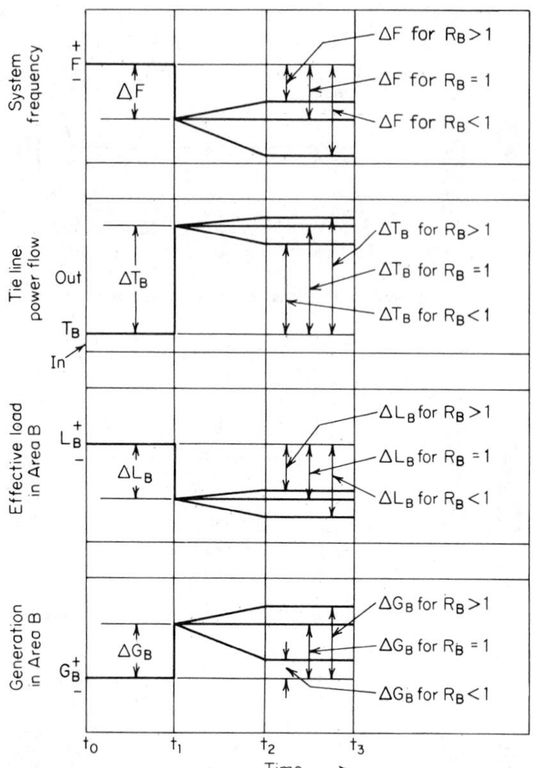


**FIG. 16-24** Changes in system frequency, tie-line power flow, effective load, and generation resulting from a remote load change.

frequency and effective load farther away from their respective initial steady-state values, while generation is moved in the direction of its initial value.

When $R_B > 1$, bias control imposes additional changes on frequency and load in the direction toward their respective initial steady-state values, while generation and tie-line flow are moved farther away from their initial values providing greater assistance to area $A$.

The curves drawn in Fig. 16-24 are for the condition that $R_B$ is the same percentage below unity when $R_B < 1$ as it is above unity when $R_B > 1$. It will be noted that the imposed effects are smaller where $R_B > 1$ than they are for $R_B < 1$.

*Other Factors Affecting Bias-Control Responses.* In addition to the bias ratio, other factors which influence the magnitude of one or more of the imposed effects of bias control on the parameters of Fig. 16-24 when $R_B$ is not unity are

*Size ratio.* This is the ratio of the size of the *disturbance area* to the size of the total system, based on bias magnitude. It is designated $Y_A$ and is given by

$$Y_A = B_A/B_S \tag{16-11}$$

where $B_A$ is the bias setting in disturbance area $A$, and $B_S$ is the summation of bias settings in all control areas of the interconnected system. For this example,

$$B_S = B_A + B_B \tag{16-12}$$

*Governing ratio.* This is the ratio in the *nondisturbance* area of its generation governing characteristic to its frequency-response characteristics. It is designated $P_B$ and is given by

$$P_B = \beta_{1B}/\beta_B \tag{16-13}$$

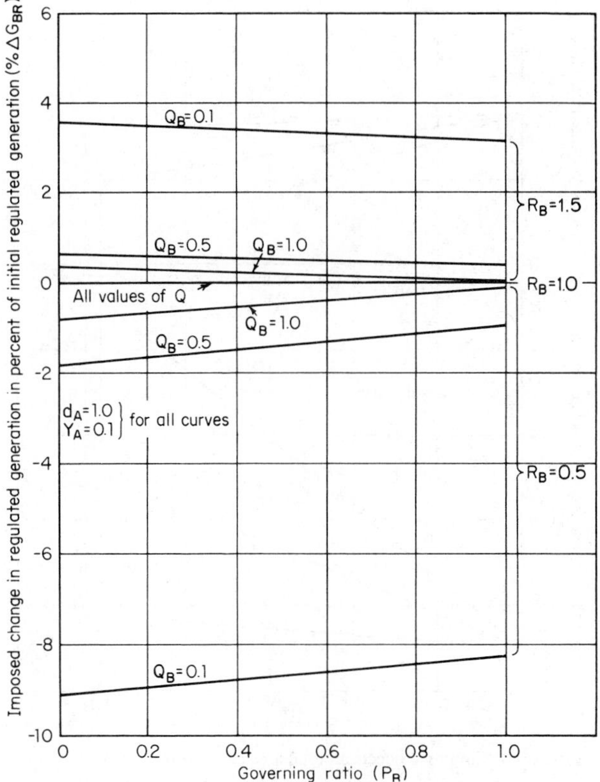


**FIG. 16-27** Imposed effects on initial regulated generation.

**6.** The larger the amount of local spinning generation not subject to bias control, the greater are the changes in area generation caused by bias settings that differ from the frequency-response characteristic.

***12. Bias Control on an Isolated Area.*** An important question related to tie-line bias control is what the nature of its performance will be if the area becomes disconnected from the interconnected system. Under such conditions, if the bias controller remains in operation and can reduce area control error to zero, it will perform like a *constant frequency control.* The frequency it will seek to maintain is a function of the normal frequency schedule $F_0$, the bias setting $B_n$, and the tie-line schedule $T_{0n}$ for which the control was set. It is given as follows:

$$F_{0n} = F_0 - \frac{T_{0n}}{10B_n} \tag{16-19}$$

where $F_{0n}$ is the frequency to which the area will be forced by the tie-line bias controller action. Standard algebraic convention is for $B_n$ to be negative, and $T_{0n}$ is positive for outgoing, negative for incoming, power schedules.

This relation may also be written

$$F_{0n} = F_0 - \frac{W_n}{10B'_n} \tag{16-20}$$

where $W_n$ is the tie-line schedule $T_0$ expressed as a percentage of the area spinning capacity. It is

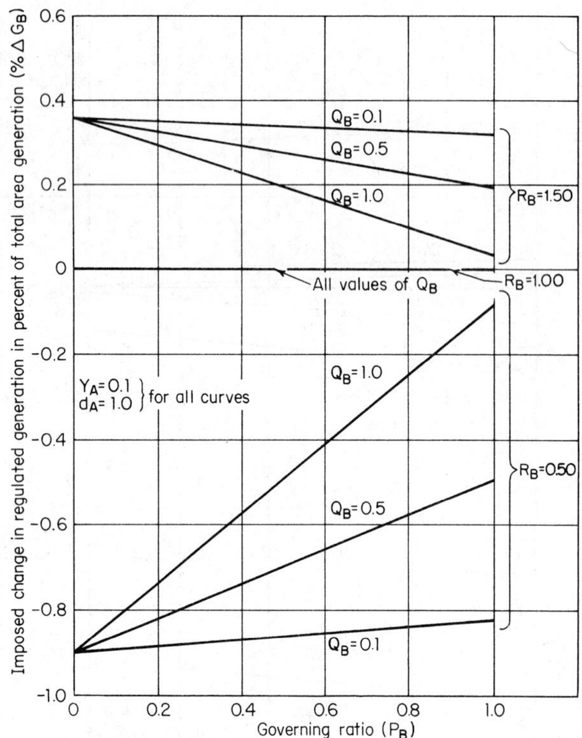


**FIG. 16-28** Imposed effects on total area generation.

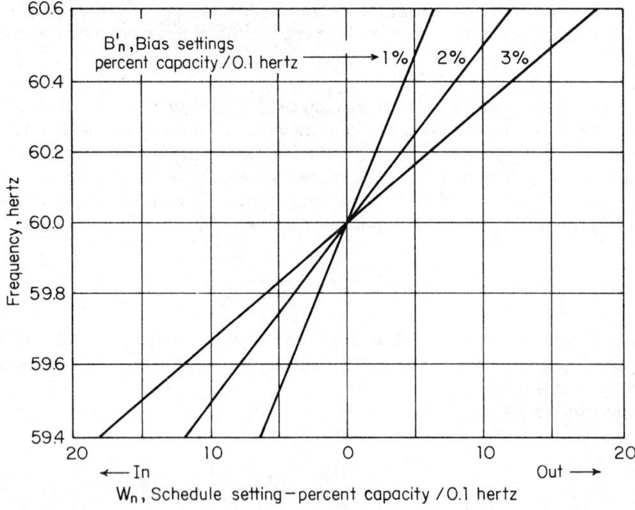


**FIG. 16-29** Isolated bias controller as a frequency regulator.

positive for outgoing, negative for incoming, power schedules. $B'_n$ is the bias $B_n$ expressed in percent of area spinning capacity per 0.1 Hz and is negative.

The relationships of Eq. (16-20) are illustrated in the curves of Fig. 16-29. With a net interchange schedule of zero, frequency would be regulated to normal 60 Hz. For other interchange schedules, the frequency that is held on isolation is in the direction to minimize the regulating burden on the isolated area. If there were incoming interchange before isolation occurred, the frequency would be held at a lower level than normal, minimizing the effect of the loss of incoming power. If there were outgoing interchange before isolation occurred, the frequency would be held at a level higher than normal, utilizing the local load frequency characteristic and increased stored spinning energy to help absorb the excess generation in the area. The controller can be shifted to regulation of frequency at its normal value as local conditions permit.

## Schedule Deviations

***13. Nature and Causes of Deviations.*** When certain criteria are fulfilled, net interchange tie-line bias control in all areas of an interconnected system, each operating in accordance with Eq. (16-9), will maintain system frequency at its scheduled value, and all area net interchanges at their respective area schedules.

*Swings versus Trends.* The parameters of system frequency, net interchange power flows, and area control errors vary or swing continually, generally within a period of a few seconds, normally within a band that is superimposed on slower sustained variations or trends. (See Fig. 16-14 for an example of a frequency record.) Thus when commenting in this section on matching system frequency to its schedule, and net interchange flows to their schedules, and reducing area control error to zero, the intent is to identify the short-term average or base value of the swings, generally referred to as the *steady-state values.*

*Criteria to Achieve Schedules.* Frequency and net interchange schedules are achieved when: (1) all portions of the interconnected system are included in one of the control areas, (2) there are no metering or schedule setting errors in any area, (3) the algebraic sum of all net interchange schedules equals zero, (4) the algebraic sum of all measured interchange flows is zero, (5) the frequency schedule is the same in all areas, (6) a common frequency is measured in all areas, and (7) all areas regulate to reduce respective area control errors to zero.

*Errors Causing Schedule Deviations.* When the foregoing criteria are not fulfilled, deviations of system frequency from schedule and area net interchanges from respective schedules will occur. The specific causes for such deviations are threefold (see Ref. 23).

First, there is the failure of one or more control areas to regulate effectively, that is, one or more areas fail to reduce respective area control errors $E_n$ to zero. Such errors have been identified as *E errors.* Second, there may be errors in the measurement of area net interchange, or errors or offsets in the setting of bulk power transfer schedules, defined as *tau* ($\tau$) *errors.* Finally, in one or more areas, there may be errors in the measurement of system frequency or errors or offsets in the setting of frequency schedules, defined as *phi* ($\phi$) *errors.*

*Control Equations Incorporating Errors.* When tau or phi errors are present in a given area, the basic equation for area control errors, Eq. (16-9), becomes

$$E_n = (T_n + \tau_{1n} - T_{0n} - \tau_{0n}) - 10B_n(F + \phi_{1n} - F_0 - \phi_{0n}) \tag{16-21}$$

where $\tau_{1n}$ is any error in measurement of $T_n$, $\tau_{0n}$ is any error or offset in setting $T_{0n}$, $\phi_{1n}$ is any error in measurement of $F$, and $\phi_{0n}$ is any error or offset in setting $F_0$.

Combining the individual tau factors and the individual phi factors, respectively, Eq. (16-21) becomes

$$E_n = (T_n - T_{0n} - \tau_n) - 10B_n(F - F_0 - \phi_n) \tag{16-22}$$

$$\tau_n = \tau_{0n} - \tau_{1n} \tag{16-23}$$

$$\phi_n = \phi_{0n} - \phi_{1n} \tag{16-24}$$

For practical purposes, if parameters as actually measured or set are substituted for the true parameters and the errors or offsets, Eqs. (16-21) and (16-22) can be written

$$E_n = (T'_n - T'_{0n}) - 10B_n(F'_n - F'_{0n}) \tag{16-25}$$

where

$$T'_n = T_n + \tau_{1n} \tag{16-26}$$

$$T'_{0n} = T_{0n} + \tau_{0n} \tag{16-27}$$

$$F'_n = F + \phi_{1n} \tag{16-28}$$

$$F'_{on} = F_0 + \phi_{0n} \tag{16-29}$$

*Effect of Errors on Frequency.* The cumulative effect (Refs. 23 and 24) on system frequency of $E$, $\tau$, and $\phi$ errors is given by

$$\Delta F = -\sum_{n=1}^{N} (1/10B_s)(E_n + \tau_n - 10B_n\phi_n) \tag{16-30}$$

where

$$\Delta F = F - F_0 \tag{16-31}$$

$$B_S = \sum_{n=1}^{N} B_n \tag{16-32}$$

*System Time Deviation.* System synchronous time is the time integral of system frequency. Deviation of system time from true time is given by

$$\epsilon = \frac{3600}{F_r}\int_0^t (F - F_r)\, dt \tag{16-33}$$

where $\epsilon$ is time deviation in seconds, $F_r$ is the system reference or rated frequency, and $t$ is the time in hours over which the integration is made.

For a 60-Hz system, Eq. (16-33) becomes

$$\epsilon = 60\int_0^t (F - 60)\, dt \tag{16-34}$$

A useful, simplified relationship, derived from Eq. (16-34), is

$$\epsilon = fm \tag{16-35}$$

where $f$ is the average frequency deviation over a time period $m$, in minutes. Thus, a frequency deviation of 0.1 Hz sustained for 10 min would cause a time deviation of 1 s.

*Effect of Errors on Time Deviation.* For a 60-Hz system, combining Eq. (16-30) with Eq. (16-34) yields, for the effect of errors on system time deviation,

$$\epsilon = -6\sum_{n=1}^{N} 1/B_s\left(\int_0^t E_n\, dt + \int_0^t \tau_n\, dt - 10B_n\int_0^t \phi_n\, dt\right) \tag{16-36}$$

*Effect of Errors on Net Interchange.* The cumulative effect (Refs. 23 and 24) on area net interchange of $E$, $\tau$, and $\phi$ errors in its own area, and $E$, $\tau$, and $\phi$ errors in all other areas, is given by

$$\Delta T_n = (1 - Y_n)(E_n + \tau_n - 10B_n\phi_n) - Y_n\sum_{\substack{i=1\\ i\neq n}}^{N}(E_i + \tau_i) + 10B_n\sum_{\substack{i=1\\ i\neq n}}^{N} Y_i\phi_i \tag{16-37}$$

where

$$\Delta T_n = T_n - T_{0n} \tag{16-38}$$

$$Y_n = B_n/B_s \tag{16-39}$$

and $i$ designates areas other than area $n$.

The first term of Eq. (16-37) defines the deviations in area net interchange that result from errors or offsets in its own area. The last two terms define deviations that result from errors or offsets in other areas, emphasizing the interdependence between the control areas of an interconnected system.

*Area Inadvertent Interchange.* The time integral of true area net interchange minus the time integral of true scheduled net interchange, which is to say the time integral of the term $T_n - T_0$ in Eq. (16-9), is defined as the *inadvertent interchange* of the area. Reference 7 notes that it develops in two ways: One results from the *intentional* deviation from normal interchange schedule due to frequency bias action, caused by errors or offsets in other areas. The other is the *unscheduled* deviation from interchange schedule that results from its own metering errors, schedule setting errors, or failure of its control to reduce its own area control error to zero. (See Par. **15** for suggested new terms which avoid possible ambiguity and define inadvertent interchange sources more specifically.)

Area inadvertent interchange $I_n$ is given by

$$I_n = \int_0^t (T_n - T_{0n})\,dt \tag{16-40}$$

where $t$ is the time in hours over which the integration is to apply.

*Effect of Errors on Inadvertent Interchange.* Combining Eqs. (16-37) and (16-40) yields, for the cumulative effect of errors in all areas, on the accumulation of inadvertent interchange in a given area:

$$I_n = (1 - Y_n)\left(\int_0^t E_n\,dt + \int_0^t \tau_n - 10B_n \int_0^t \phi_n\,dt\right) - Y_n \sum_{\substack{i=1\\ i\neq n}}^{N} \left(\int_0^t E_i\,dt + \int_0^t \tau_i\,dt\right) + 10B_n \sum_{\substack{i=1\\ i\neq n}}^{N} Y_i \int_0^t \phi_i\,dt \tag{16-41}$$

The first term of Eq. (16-41) defines the inadvertent interchange in an area that results from its own $E$, tau, and phi errors or offsets. The second and third terms define the effects on an area's inadvertent interchange by $E$, tau, and phi errors or offsets in other areas.

If an area has no $E$, tau, or phi errors of its own, the first term of Eq. (16-41) would be zero, and all the area inadvertent interchange, if any, would be so-called intentional, derived from its frequency-bias assistance to other areas having errors or needs.

*Relationship of Inadvertent Interchange to Time Deviation.* The time integrals of an area's $E$, tau, or phi errors are factors in defining the relationship of the area's inadvertent accumulation to the system time deviation, as follows (see Refs. 23 and 24):

$$I_n = \int_0^t E_n\,dt + \int_0^t \tau_n\,dt - 10B_n \int_0^t \phi_n\,dt + \frac{B_n}{6}(\epsilon) \tag{16-42}$$

When the time integrals of an area's $E$, tau, and phi errors are equal, in the aggregate, to zero, Eq. (16-42) becomes

$$I_n = \frac{B_n}{6}(\epsilon) \tag{16-43}$$

***14. Energy Balancing and Time Correction.*** Because interconnected systems wish to maintain system synchronous time within reasonable limits, time-deviation correction procedures have been adopted. Also, because interarea billings for bulk energy transfers are based on scheduled transfers rather than actual transfers, payback for departures from energy transfer schedules, which is to say reduction of inadvertent interchange to zero, is required. General practice for energy balancing is to consider a day divided into "on-peak" and "off-peak" periods which are common to all participating areas of the system, with the expectation that unscheduled energy received by an area during on-peak hours will be repaid during on-peak hours, whereas off-peak unscheduled energy will be repaid during off-peak hours. On the eastern interconnected system, on-peak hours for weekdays and Saturdays start at 0600 and end at 2200 hours Central Standard (Daylight) Time. On the western interconnected system, it is the same hours Pacific Standard (Daylight) Time. Other hours, all day Sundays, and certain holidays are off-peak hours.

*Time-Deviation Correction.* As part of their time-deviation correction procedure, it is the practice of the United States and Canadian interconnected systems to assign to one area the maintenance of a system time standard. For the eastern interconnected system, the standard is maintained by the American Electric Power Company at Columbus, Ohio. Through designated communications channels, information on the status of system time deviation is relayed to all control areas, and, as required, certain periods are designated as time correction periods. During such periods, all areas are expected to simultaneously offset their frequency schedules by an amount related to accumulated system time deviation, yielding in each area a computation of area control error in accordance with Eq. (16-44) which is Eq. (16-25) augmented with the frequency schedule offset, as follows:

$$E_n = (T'_n - T'_{0n}) - 10B_n(F'_n - F'_{0n} - b\epsilon) \tag{16-44}$$

where $b$ is a time-deviation modifier, Hz/s, common to all areas, and is a minus quantity. In conventional practice $b$ equals $-0.02$ Hz per 2-s deviation. A schematic of such system time correction is shown in Fig. 16-30, with switch $S$ in the (b) position.

If all areas use the same $b\epsilon$ frequency schedule offset, and if there is effective regulation in all areas, system frequency will be shifted by the amount of the offset, as reflected by Eq. (16-30). Referring to Eq. (16-35) it will be seen that, with $b = -0.02$ Hz/s, such control would correct a 1-s error in 50 min.

Also, as reflected by Eq. (16-37), such a uniformly applied and effectively executed systemwide correction procedure would not cause any shifts in net interchange power flows.

For the western system, similar time correction is continuously applied automatically by almost all areas of the interconnection. The reference time standard is maintained by the Southern California Edison Company.

*Energy Balancing.* As outlined in Ref. 4, present-day practice gives areas two choices for correcting inadvertent interchange accumulations.

One technique provides for bilateral correction. Here an area with inadvertent interchange in one direction arranges with another area having inadvertent interchange in the opposite direction for both to offset their interchange schedules simultaneously and in the same amount, but of opposite sign. A corresponding change in net interchange flow between the two areas is created, reducing the inadvertent interchange accumulation of each, without—when properly executed—affecting the system frequency or time, or net interchange of other areas, as can be seen from Eqs. (16-30), (16-36), and (16-37).

Recognizing the difficulty of one area always finding another with whom to achieve energy balancing, an alternative procedure is for an area to make a unilateral shift in net interchange schedule to correct for its inadvertent interchange accumulation, provided the resultant control action will be in a direction to contribute toward the correction of prevailing time deviation. This technique finds frequent use.

The area control-error computation for an area utilizing either the bilateral or the unilateral technique for inadvertent interchange correction would be made in accordance with Eq. (16-45), which is Eq. (16-25) augmented with a net interchange schedule offset as follows:

$$E_n = (T'_n - T'_{0n} + k_n I_n) - 10B_n(F'_n - F'_{0n}) \tag{16-45}$$

where $K_n$ is a constant, common to both areas practicing bilateral inadvertent interchange corrections. A schematic of such area inadvertent interchange correction appears in Fig. 16-30 with switch $S$ in the (c) position.

*Synchronized Systemwide Energy Balancing and Time Correction.* The techniques that have been described and are currently in use consider time correction as a system matter, in which all areas are expected simultaneously to participate on a system-instructed basis, whereas inadvertent interchange has been regarded as a separate area matter, to be taken care of by each area unilaterally or two areas in concert. A decade ago, a new approach was suggested (Refs. 25 to 27), based on the concept that inadvertent interchange and time deviation result from common causes and should be corrected by common, fully correlated action. The proposed technique, an extension of an earlier proposal (see Refs. 23, 24, and 28), claims advantages of simultaneous correction of inadvertent interchange and time deviation, assignment of corrective action only to areas that created the schedule deviations, and compensation for sustained tau and phi errors in constituent areas.

For a 60-Hz system, the computation of area control error for the proposed synchronized

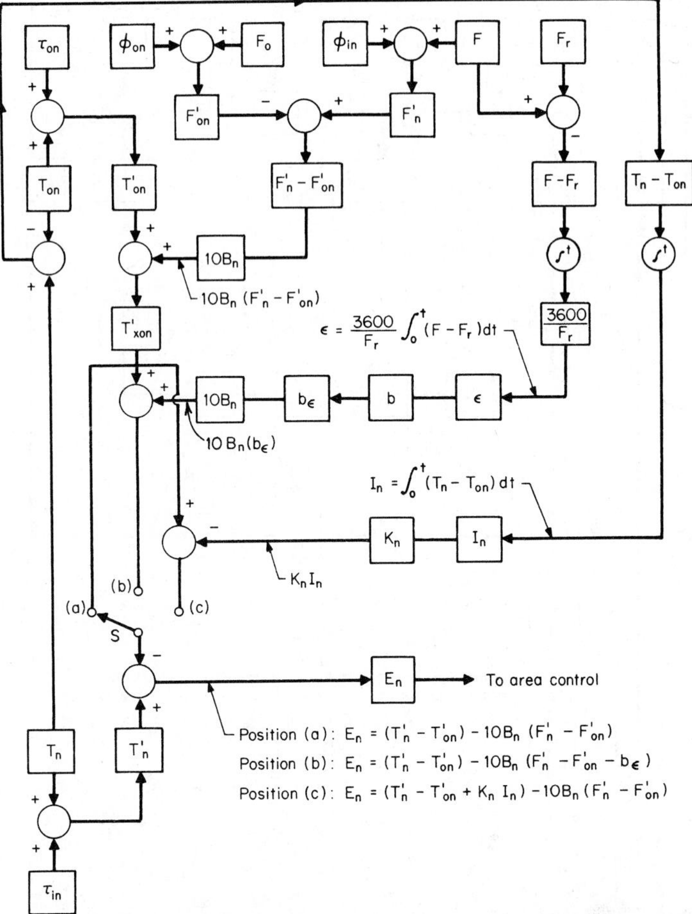


**FIG. 16-30** Schematic of conventional control and correction computing circuits. Switch position: *(a)* conventional frequency biased net interchange control [Eq. (16-25)]; *(b)* conventional universal offset of frequency schedule for time-deviation correction [Eq. (16-44)]; *(c)* conventional bilateral (or unilateral) offset of net interchange schedule for inadvertent interchange correction [Eq. (16-45)].

technique would be as follows:

$$E_n = (T'_n - T'_{0n} + I_n/H) - 10B_n(F'_n - F'_{0n} + \epsilon/60H) \qquad (16\text{-}46)$$

where $H$ is common to all areas and is the time, in hours, within which $I_n$ is to be corrected.

A stated limitation of the method was that all areas of the interconnected system had to simultaneously participate. A schematic (see Ref. 26) of the method is shown in Fig. 16-31.

*Single-Step Area Energy Balance and Area Component Time Correction.* Another, more recent proposal provides single-step correction of area-caused components of system time deviation and area inadvertent interchange. It is based on the *components concept* of Refs. 29 and 30. It has the advantage over the synchronized systemwide technique of Refs. 24 to 26, and Fig. 16-31, by being applicable in a single area, regardless of whether other areas do or do not use it. A description of the components concept and its applications, including this one, appears in Par. **15**.

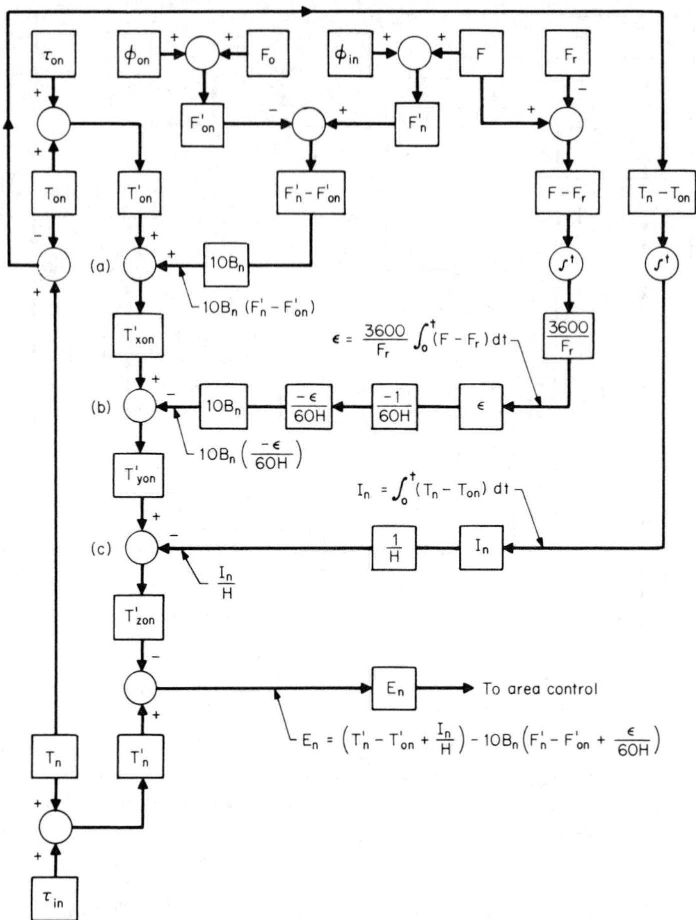


**FIG. 16-31** Schematic of a proposed technique for universal synchronized correction of inadvertent interchange and time deviation on a coordinated systemwide basis (60-Hz system). Schedules for area net interchange, $T'_{0n}$, are adaptively shifted to achieve at *(a)* a revised schedule $T'_{x0n}$ by shifting $T'_{0n}$ from frequency deviation; at *(b)* a revised schedule $T'_{y0n}$ by shifting $T'_{x0n}$ from time deviation; and at *(c)* a revised schedule $T'_{z0n}$ by shifting $T'_{y0n}$ from inadvertent interchange.

## The Components Concept

***15. The Concept.*** A new concept introduced and described in Refs. 29 and 30 holds that system time deviation and area inadvertent interchange are each made up of individual components, each caused by the regulating deficiencies in each of the systems areas, and that each of these components can be separated and its magnitude determined by a defined *decomposition technique* that utilizes only known or measurable parameters.

The component of system time deviation that is caused by an area's regulating deficiencies is defined as the *area component of time deviation.* The component of an area's inadvertent interchange that is caused by its own regulating deficiencies is defined as its *primary inadvertent.* The components of an area's inadvertent interchange that is caused by the regulating deficiencies in other areas are defined as its *secondary inadvertent.*

Such components have the following uses:

1. By a single area to determine quantitatively its *control performance,* that is, its effect on system time, on its own inadvertent interchange, and on the inadvertent interchange of other areas; and the net effect of other areas on its inadvertent interchange.
2. By two or more areas to determine additionally the reciprocal effects of each area on the other's inadvertent interchange.
3. By a single area for single-step simultaneous corrective control of its own component of time deviation, its component of primary inadvertent, and the components of secondary inadvertent it has caused in other areas.
4. By all areas to introduce a dollar payback technique for unscheduled interchange.

*System Time Deviation Components.* The component of system time deviation for each area, $\epsilon_n$, accumulated in a specified time period is given by:

$$\epsilon_n = -\frac{6}{B_s}\left(I_n - \frac{B_n\epsilon}{6}\right) \tag{16-47}$$

where $B_s$ is the sum of the bias settings of all areas and is a minus quantity and $I_n$ and $\epsilon$ are accumulated in the same time period.

The right-hand parenthetical term, which appears also in the other decomposition equations, represents the summation of regulating deficiencies and offsets in the area during the same time period. It is a summation of the $\epsilon$, tau and phi errors of the area, including corrective schedule offsets, and for simplicity is referred to in the aggregate as the *area regulating deficiencies.* When multiplied by appropriate factors, its *effect* on the respective left-hand parameter of Eqs. (16-47), (16-48), and (16-49), and comparably for Eq. (16-50), is shown.

*Primary Inadvertent.* The primary component of area inadvertent interchange, $I_{nn}$, accumulated in a specified time period, is given by

$$I_{nn} = (1 - Y_n)\left(I_n - \frac{B_n\epsilon}{6}\right) \tag{16-48}$$

where $Y_n$ is the ratio of $B_n$ to $B_s$.

*Secondary Inadvertents.* Each of the secondary components caused by area $n$ in a remote area $i$, $I_{in}$, in a specified time period, is given by

$$I_{in} = -Y_i\left(I_n - \frac{B_n\epsilon}{6}\right) \tag{16-49}$$

where $Y_i$ is the ratio of $B_i$ to $B_s$. Also,

$$I_{in} = -\frac{Y_i}{1 - Y_n} I_{nn} \tag{16-50}$$

Similarly, each component of secondary inadvertent in area $n$ caused by a remote area $i$, $I_{ni}$, is given by

$$I_{ni} = -Y_n\left(I_i \frac{-B_i\epsilon}{6}\right) \tag{16-51}$$

and

$$I_{ni} = -\frac{Y_n}{1 - Y_i} I_{ii} \tag{16-52}$$

Where double subscripts are used for inadvertent interchange, as in primary and secondary inadvertents, the first subscript identifies the area that has the inadvertent, the second identifies the area that caused it.

*Component Relationships.* The time deviation, primary inadvertent, and secondary inadvertent components caused by the technical deficiencies of an area result from a common cause and are hence interrelated, as shown by the following equations:

$$I_{nn} = -(1 - Y_n)\frac{B_s \epsilon_n}{6} \tag{16-53}$$

and

$$I_{in} = \frac{(B_i)}{6}\epsilon_n \tag{16-54}$$

*Corrective Control.* From Eqs. (16-53) and (16-54) it is seen that these area-caused components are linearly related, and when one is zero, all are zero. This is the basis of single-area single-step simultaneous corrective control of area-caused components. It is achieved by offsetting the area frequency schedule or net interchange schedule automatically to introduce area regulating deficiencies of magnitude in megawatthours equal to those that originally created the components, but of opposite sign, and setting a time in hours, $H_n$, in which the correction is to occur.

The frequency schedule offset, $\phi_n$, for such control is a function of the area component of system time deviation, as given by

$$\hat{\phi}_n = \frac{-\epsilon_n}{60 Y_n H_n} \tag{16-55}$$

The net interchange schedule offset, $\tau_n$, is a function of primary inadvertent, as given by

$$\hat{\tau}_n = \frac{-I_{nn}}{(1 - Y_n)H_n} \tag{16-56}$$

Either offset provides the same results, namely, the single-step simultaneous correction in time period $H_n$ of all previously accumulated area-caused components.

The control arrangement for corrective operations are shown in Fig. 16-32.

*Dollar Payback.* Current practice among control areas provides for energy payback in kind for unscheduled transfers. The components concept provides a means of supplementing energy payback with *dollar paybacks.* For each hour, or other specified period, a higher-than-schedule dollar value is assigned to *primary import* and *secondary export.* A lower-than-schedule dollar value is assigned to *primary export* and *secondary import.* A given area pays the high charge for its unscheduled primary import and the low charge for its secondary import. The area receives the low payment for its primary export and the high payment for its secondary export. Equitable penalties and rewards for unscheduled interchange are thus achieved.

## Economic Dispatch

***16. Objectives of Economic Dispatch.*** The objective of area economic dispatch is to allocate the total generation required of the area to alternative sources in order to achieve best possible area economy consistent with safe, effective operation. The nature of this problem, for the simplified area of Fig. 16-7, is shown in schematic form in Fig. 16-33. Area load can be satisfied, alternatively and at different costs, by adjusting outputs at generators 1, 2, . . . , *N* or by scheduling new levels of bulk power transfer over interties with area *A* and area *C.*

*Factors Influencing Economic Dispatch.* Generators will be of different sizes and efficiencies, fuel costs will vary, and there will be variations in transmission losses from the various generating sources to load centers. Each of these factors influences the cost at which power can be generated and delivered to users within the area. In addition, each intertie represents opportunity for possible purchase of power at prices that may be more attractive than local area delivered costs.

*Factors Overriding Economy.* There are, at the same time, factors that limit achievement of highest possible area economy. For example, the operating range of generators may be restricted to certain high or low limits, system security may require changes in network flows or location and amounts of spinning reserve that differ from the most economic allocation, there may be

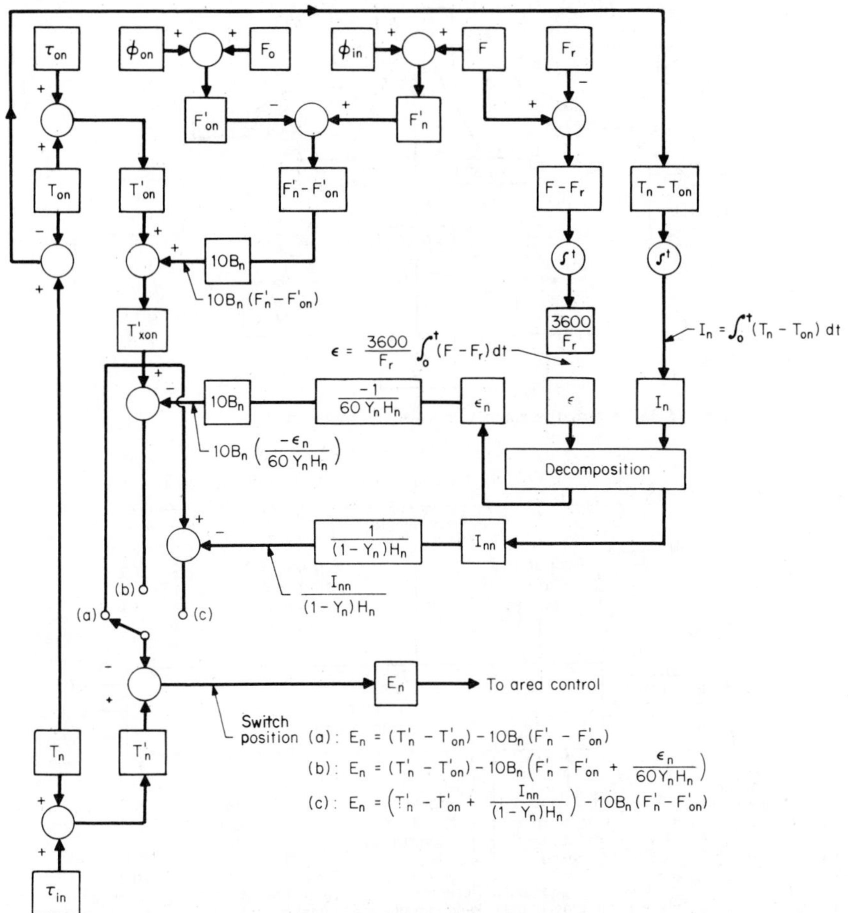


**FIG. 16-32** Schematic of control and correction computing circuits using "components concept." Switch position *(a)* provides conventional frequency biased net interchange control; position *(b)* offsets the frequency schedule of the area with the $\hat{\phi}_n$ factor of Eq. (16-55) and provides single-area single-step simultaneous correction of the area component of system time deviation, the primary component of its own inadvertent interchange, and the secondary components of inadvertent interchange it has caused in all other areas. Optionally, position *(c)* provides the same single-step corrective control by an equivalent offset of area net interchange schedule with the $\hat{\tau}_n$ factor of Eq. (16-56).

transmission-line limitations, there may be environmental limitations, or there may be operating requirements related to stream flow or storage at hydrounit locations.

Finally, and of particular importance, the demand for generation change from the area controller may exceed the permissible rate of generation change of the sources next in line to change generation for optimum economy. In this case, the control will bypass the economic allocation schedules and assign generation changes, in what is termed *area assist action,* to faster-responding units in order more effectively to reduce area control error to zero. Subsequently, as system conditions permit, a reallocation will automatically be made to achieve optimum internal area economy.

*Economic Dispatch Control.* Factors which are used for computing area economic dispatch and those which sometimes make it necessary to override economic allocations are shown in

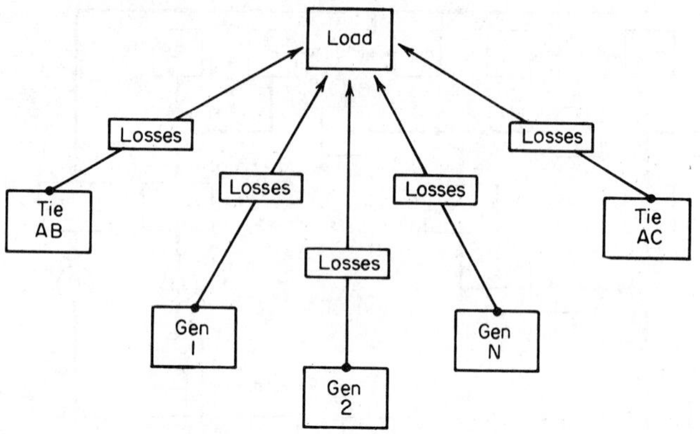


FIG. 16-33 Alternative sources for satisfying area load.

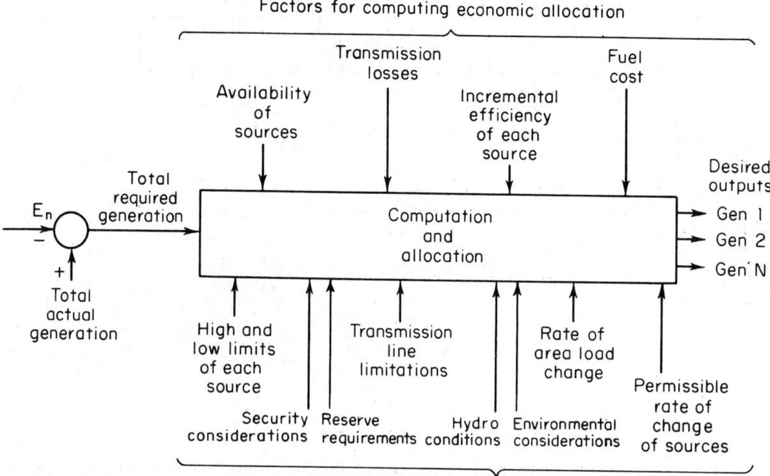


FIG. 16-34 Allocating total required generation to individual area sources.

Fig. 16-34. This schematic also illustrates how a single control system combines the objectives of economic dispatch with the requirements of area regulation.

Area control error $E_n$, computed in accordance with Eq. (16-9) and Fig. 16-22, or in accordance with augmented Eqs. (16-25), (16-44), (16-45), or (16-46), combined as a feed-forward parameter with total actual generation as a feedback defines the total generation required of the area. By appropriate computation and allocation, this total requirement is divided among available area sources. When control causes these individual assignments to be achieved, area control error will have been reduced to zero and optimum available area economic dispatch will simultaneously have been established.

***17. Coordination Equation.*** For a steam turbogenerator unit the equation that coordinates delivered cost considerations with parameters related to unit heat rate, fuel cost, and transmission losses is as follows:

$$\lambda_n = \frac{(dH_n/dP_n)f_n}{1 - \partial P_L/\partial P_n} \tag{16-57}$$

where $\lambda_n$ is the *incremental cost* of *delivered power* for source $n$; $dH_n/dP_n$ is the *incremental heat rate* for source $n$; $f_n$ is the cost of incremental fuel for source $n$, adjusted to include other varying costs such as maintenance cost at source $n$; and $\partial P_L/\partial P_n$ is the *incremental transmission loss* for source $n$.

Equation (16-47) may be rewritten

$$\lambda_n = \frac{dF_n/dP_n}{1 - \partial P_L/\partial P_n} \tag{16-58}$$

where $dF_n/dP_n$ is the *incremental generating cost* at source $n$ and is equal to the product of $dH_n/dP_n$ and $f_n$.

The individual terms of Eqs. (16-57) and (16-58) are more fully defined in the paragraphs that follow. For detailed derivation and discussion of these equations and their parameters see Refs. 31 to 38.

*Incremental Heat Rate.* The incremental heat rate of a steam turbogenerator at any particular output is the ratio of a small change in heat input per unit time to the corresponding change in power output. At each unit output it is given by the first derivative of the unit input-output curve. It is usually expressed in Btu per kilowatthour. It appears as $dH_n/dP_n$ in the numerator of Eq. (16-47).

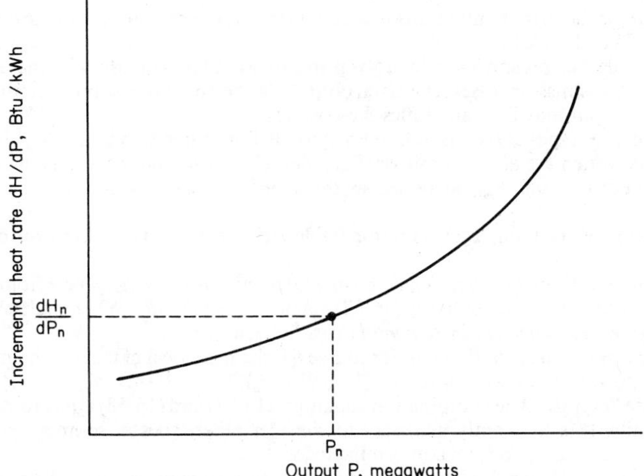


**FIG. 16-35** Hypothetical curve of incremental heat rate versus unit output.

A hypothetical curve of incremental heat rate versus unit output is shown in Fig. 16-35. In practice, curves are not this smooth. Discontinuities and slope reversals are frequently encountered. For such curves, approximations of the actual curves are used, drawn so that the curves are continuous and have no slope reversals. Such approximations are in general well within the basic accuracy of the method.

*Cost of Incremental Fuel.* For a given generating source, the cost of incremental fuel is defined as the ultimate replacement cost of the fuel that would be consumed to supply an additional increment of power. In Eq. (16-57) it is combined with other incremental costs, such as maintenance, and the combination is designated $f_n$.

*Incremental Generating Cost.* The incremental cost of generated power is the product of incremental heat rate and the cost of incremental fuel adjusted to include all variable costs, including maintenance. At any particular value of generation it is the ratio of the additional cost incurred in producing an increment of generation to the magnitude of that increment of generation. It is usually expressed in mills per kilowatthour. It appears as $dF_n/dP_n$ in the numerator of Eq. (16-58).

*Total Transmission Losses.* Total transmission losses do not enter directly in the coordination equation, but they provide the basis for deriving the incremental loss factors that do appear in the equation.

The preferred equation for total transmission loss, derived in Ref. 36, is

$$P_L = \sum_m \sum_n P_m B_{mn} P_n + \sum_n B_{n0} P_n + K_{L0} \tag{16-59}$$

where $P_L$ is the total transmission loss; $P_m$ is the power output of source $m$; $P_n$ is the power output of source $n$; $B_{mn}$ are the constants related to the nature and characteristics of the area, it being noted that $B_{mn}$ is not necessarily equal to $B_{nm}$; $B_{n0}$ is a constant related to source $n$; and $K_{L0}$ is a constant that may be regarded as representing total system losses under the imaginary condition of zero system power supply.

*Incremental Transmission Losses.* The incremental transmission loss for a source is the fraction of power loss incurred by transmitting a small increment of power from that source to another point, in this case the hypothetical load center of the area. It is defined for a given source as the partial derivative of Eq. (16-59) with respect to that source. For source $n$ this may be written

$$\frac{\partial P_L}{\partial P_n} = \sum_m 2B_{nm} P_m + B_{n0} \tag{16-60}$$

where $\partial P_L/\partial P_n$ is the incremental transmission loss for source $n$. It appears in the denominator of Eqs. (16-57) and (16-58).

Incremental transmission losses for a given area may rise fairly rapidly with increased power output. Total transmission losses for a given output may be only 5 or 6%, but incremental losses at that same output may be many times these values.

There are three types of $B$ constants in Eq. (16-60). For source $n$, these are $B_{nm}$, identified as *self-constants,* which are always positive; $B_{mn}$, identified as *mutual constants,* which may be positive or negative; and $B_{n0}$, identified as the added constants, which may be positive or negative.

Methods for determining $B$ constants for a wide variety of system conditions are discussed in Ref. 36.

*Incremental Delivered Power.* The *incremental fraction of delivered power* from source $n$ is represented by the denominator of Eq. (16-57), namely, $1 - \partial P_L/\partial P_n$. When multiplied by 100, it may be expressed as *percent incremental delivered power.*

*Penalty Factor.* The *penalty factor* for source $n$ is the reciprocal of the incremental fraction of delivered power and hence is given by $(1 - \partial P_L/\partial P_n)^{-1}$.

***18. Other Sources.*** The coordination equations (16-57) and (16-58) apply to steam turbogenerator units. It is frequently necessary to consider other types of sources, including hydrounits, intertie points, and nonconforming loads.

*Hydro Plants.* Hydro plants may be handled by assigning an equivalent incremental generation cost to each plant (see Ref. 39).

*Tie Points.* Interties with adjacent areas may be considered as generating sources, with the numerator of Eq. (16-58) representing the incremental cost of purchased power at the intertie points. This is a particularly significant computation in determining the possible economic advantages of bulk transfer over interties between adjacent areas.

*Large Nonconforming Loads.* A *nonconforming load* is one that does not vary linearly with total area load. When the magnitude of such a load is large compared with the generation at or near the substation that serves this load, it should receive special consideration in the computation of incremental transmission losses. A nonconforming load is telemetered to the computing network and is treated as a negative power source. In this way, appropriate factors for such a load and its $B$ constants can be introduced into the computation (see Ref. 36 for additional details).

***19. Equal Incremental Costs.*** Optimum operating economy for a group of sources is achieved when they are loaded to equal incremental costs. The incremental costs that are to be compared and equalized to achieve such economic dispatch depend on whether there are significant differences in incremental transmission losses for the several sources.

*General Case.* For the general case, applicable to control areas whose sources have significantly differing incremental transmission losses, economic dispatch is achieved when sources are loaded to equal incremental costs of delivered power. The corresponding relationship for $n$

sources of an area is

$$\lambda = \lambda_1 = \lambda_2 = \cdots = \lambda_n \tag{16-61}$$

where $\lambda$ is the incremental cost of delivered power for area as a whole, and $\lambda_1, \lambda_2, \ldots, \lambda_n$ are the incremental costs of delivered power for sources 1, 2, . . . , $n$, respectively.

*Cases Where Incremental Transmission Losses Can Be Ignored.* Where the incremental transmission losses for the several sources of an area, or the differences between them, are small, the denominators of Eq. (16-58) as applied to the several sources become essentially equal. For such cases typically encountered in small, closely knit areas, economic dispatch can be achieved by loading the sources to equal incremental generating costs. Economic dispatch is thus achieved when

$$\frac{dF_1}{dP_1} = \frac{dF_2}{dP_2} = \cdots = \frac{dF_n}{dP_n} \tag{16-62}$$

*Within Stations.* For a given station having $n$ sources tied to a common bus, any applicable incremental transmission losses are the same for all $n$ units. Thus, *within* such a station optimum economy is achieved when its units are loaded to equal incremental generating costs in accordance with Eq. (16-62).

When the $f_n$ factor of Eq. (16-57), the cost of incremental fuel and other varying costs, is the same for all units within such a station, optimum economy within the station is achieved by loading its units to equal incremental heat rates, as follows:

$$\frac{dH_1}{dP_1} = \frac{dH_2}{dP_2} = \cdots = \frac{dH_n}{dP_n} \tag{16-63}$$

*Sources Out of Range.* In applying the principle of loading to equal incremental costs to achieve economic dispatch it should be noted that some of the stations or units of the area then in operation may be *out of range* of the incremental cost being used at that particular time as the loading reference. Units with *lower* incremental costs over their full range will have been fully loaded by earlier incremental cost allocations. Units with *higher* incremental costs over their full range may be in operation for reserve purposes or in anticipation of imminent increases in demand.

## Control Application

***20 General Practice.*** Modern practice is to apply automatic control to most or all of the generating units of a control area, thereby achieving economic, security, and environmental dispatch while simultaneously fulfilling the obligations of area regulation. Figure 16-36 shows in

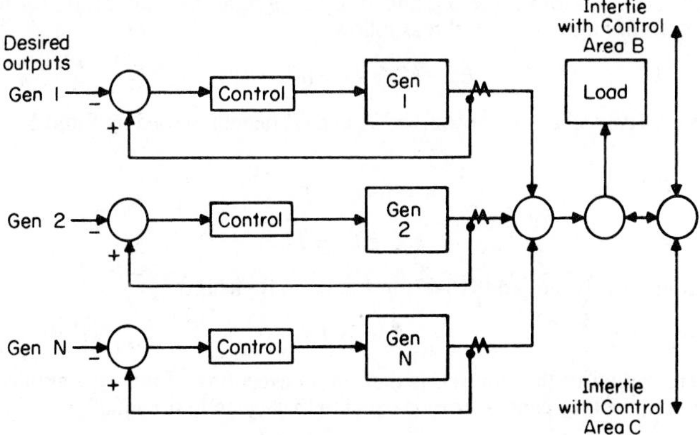


**FIG. 16-36** Method of achieving computed allocations on area sources.

schematic form how the desired output for each of the sources of an area, computed in accordance with the schematic of Fig. 16-34, is achieved by a closed-loop execution. The computed desired output is compared with a feedback from actual output from each source, and control action is applied to the source until the two are matched.

The combination of Figs. 16-22, 16-34, and 16-36 provides a complete schematic representation of how all sources of the area of Fig. 16-6 are automatically controlled to maintain area control error at zero while simultaneously achieving area economic dispatch, tempered as required by security and environmental considerations or other prevailing area conditions.

***21. Control Classification.*** Control systems have been classified (Ref. 40) by the portions of the area control problem which they seek to solve and by the nature of the programming techniques used for achieving economic dispatch. Classifications are as follows:

*Class I.* In this category, control is applied to area regulation only. Modern controls rarely have this limited objective.

*Class II.* In this category, control is applied to both area regulation and economic dispatch. Generation allocation programs are manually preset. Such controls are generally of the analog type.

*Class III.* In this category, control is applied to both area regulation and economic dispatch. Generation allocation programs are based on stored incremental generating cost data and are automatically computed and applied in real time. Such controls are increasingly of the digital computer type and frequently include extensive programs for state estimation, security dispatch, and environmental constraints.

***22. Manually Set Programming.*** Class II controls are sometimes referred to as being of the *flexible-programming* type, or the *base and participation* type. Allocation programs are manually set and periodically reset as required to meet changed system conditions. For example, programs must be reset when units which are subject to control are brought on or taken off the line. The base and participation technique is used as a part of a digital-computer assembly of the Class III type.

*Example.* A control execution (Ref. 41) of Class II which utilizes as the programming reference a feed-forward from area control error algebraically combined with a feedback from actual area generation is shown schematically in Fig. 16-37. For simplicity, this schematic applies to an area having two generating sources $G_A$ and $G'_A$ and two interties *AB* and *AC*. For each source, desired generation as a function of total-area required generation is programmed with manually set dials to approximate the loading curves shown at the upper right of Fig. 16-37. The allocation curve for each source is made up of straight-line segments by applying suitable settings to *base point* and *slope* (or *participation*) dials. At all points, the curves are set so that the sum of source generation allocations is equal to the total generation required of the area. Thus,

$$(G + G') - E_n = P + P' \qquad (16\text{-}64)$$

where $G$ is the output of source $G_A$; $G'$ is the output of source $G'_A$; $E_n$ is the area control error, and hence $-E_n$ is the generation change required to reduce the area control error to zero; $P$ is the programmed allocation for source $G_A$; and $P'$ is the programmed allocation for source $G'_A$.

Equation (16-64) may be rewritten as follows:

$$E_n = (G - P) + (G' - P') \qquad (16\text{-}65)$$

Let the *source control errors* for sources $G_A$ and $G'_A$ be represented by $S$ and $S'$, respectively; then

$$S = G - P \qquad (16\text{-}66)$$

and

$$S' = G' - P' \qquad (16\text{-}67)$$

Substituting Eqs. (16-66) and (16-67) in Eq. (16-65) yields

$$E_n = S + S' \qquad (16\text{-}68)$$

The generalization of this analysis is that, in an execution of this type applied to an area having $n$ sources, the area control error is equal to the algebraic sum of all of the source errors,

$$E_n = \Sigma S_n \qquad (16\text{-}69)$$

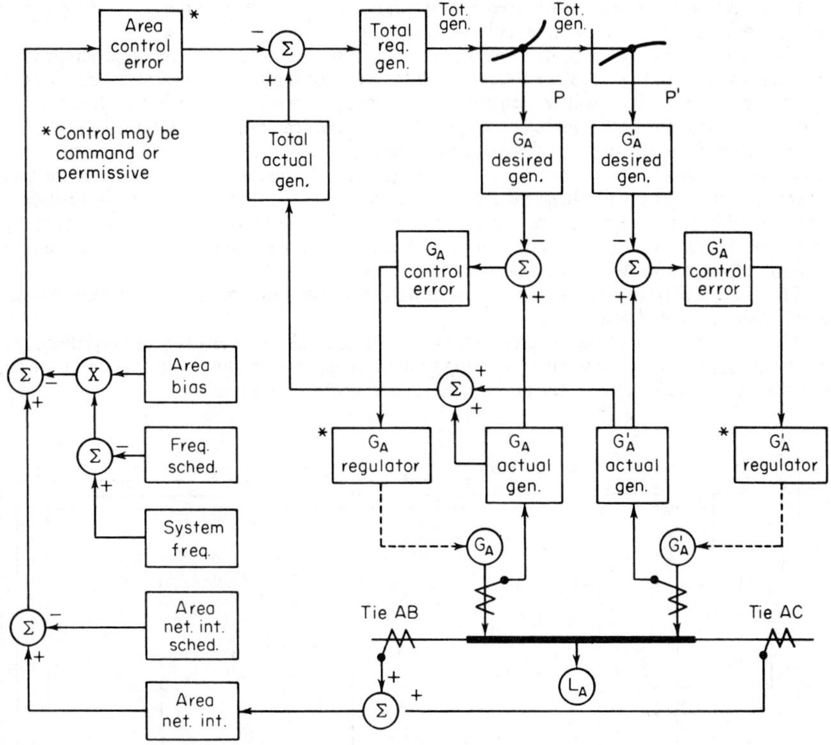


FIG. 16-37 Predictive flexible-programmed analog control system (Class II).

An important characteristic of such a control execution is that it predictably computes the generation assignment to each source. That is to say that the programmed allocation for each source is independent of the rate at which other controlled sources respond to their allocations. Hunting between units is thereby minimized.

Many control systems of this type have been used (Ref. 41) in the United States and elsewhere in the world.

***23. Stored Cost Programming.*** Class III controls utilize a *lambda reference* for allocating total required area generation to individual sources. Generation at each source is automatically adjusted until its incremental cost of delivered power, source lambda, matches a common area lambda, thereby achieving economic dispatch in accordance with Eq. (16-61). This execution is sometimes referred to as the *fixed-programming type.* It has the advantage that program allocations automatically adapt to the number of sources in service and require no reprogramming as do the Class II executions when the number of units on the line is changed.

*Computing Lambda Reference.* The common lambda to which each source lambda is matched can be established by automatic adjustment until the resultant control action reduces area control error to zero; or it can be automatically computed to satisfy the criterion that total source allocations shall equal the algebraic sum of area control error and existing total area generation. The latter criterion is the same as that expressed for one type of Class II control in Eq. (16-64).

Another way of applying the criterion for area lambda computation that derives from Eq. (16-64) is automatically to adjust the lambda reference until the area control error is equal to the algebraic sum of source control errors, in accordance with Eq. (16-69).

For computation techniques based on either Eq. (16-64) or Eq. (16-69), the area reference lambda is established predictively, without waiting for area control error to be reduced to zero.

In these executions, the allocation to each source is again independent of the rate at which other control sources respond to their respective allocations.

*Example.* A control execution of this type for a simplified two-source area having two interties is shown in Fig. 16-38. The area lambda reference is adjusted until area control error equals the summation of source control errors in accordance with Eq. (16-69). Incremental transmission loss for each source is computed in accordance with Eq. (16-60) from stored $B$ constants and from real-time telemetered measurements of source outputs and intertie power flows. By using stored data for the cost of incremental fuel, the incremental heat rate for each source is computed on the basis of Eqs. (16-58) and (16-57). From stored relationships of incremental heat rate versus source output, represented by the curves in the upper right of Fig. 16-38, the desired outputs for each source are obtained. Control is applied to each source until desired output is matched by actual output.

***24. Control Executions.*** Table 16-7 summarizes the hierarchical levels that constitute an interconnected power system.

There are unique loading objectives for each of these levels, but there is normally, discounting the availability and use of phase-shifting transformers, only one level where generation can actually be adjusted, namely, the energy conversion units themselves.

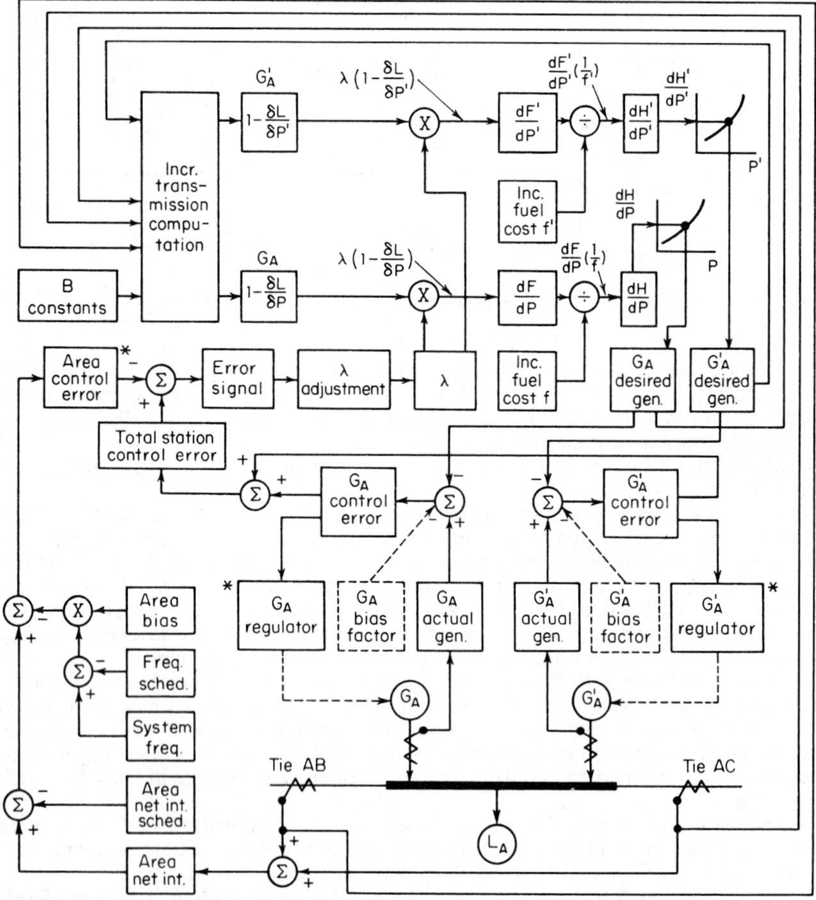


**FIG. 16-38** Predictive lambda reference computer-control system (Class III).

**TABLE 16-7** Hierarchical Levels of a Power System

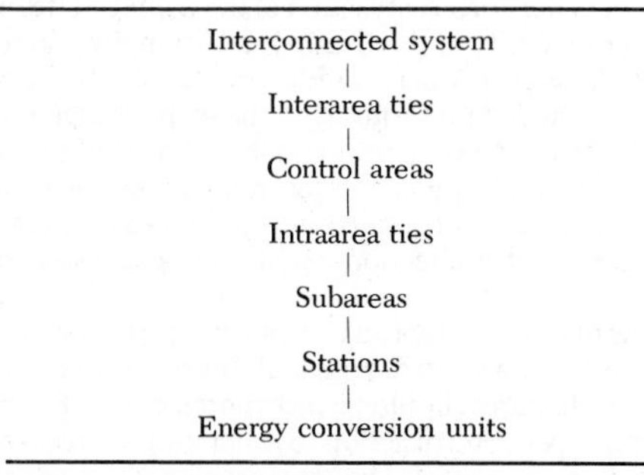


Two general techniques are available for applying control signals to individual sources. One is termed *permissive control,* the other *command control.* The former puts a priority, when a choice must be made, on area regulation as against source regulation or loading objectives at intermediate levels.

*Permissive Control.* In the permissive-control execution, source generation is permitted to change only when, in correcting a source control error, it will also help reduce area control error and achieve intermediate loading objectives.

*Command Control.* In the command-control execution, source generation is changed whenever a source control error exists, regardless of whether the action will reduce or augment area control error or achieve intermediate loading objectives. This type of control has also been referred to as *mandatory control.*

***25. Unit Bias.*** In applying command control to the individual generator it is helpful to include a unit frequency bias (Ref. 40) in the computation of unit control error. This will coordinate the control with governor responses to remote load changes and avoid opposition to such responses by the command control. A unit bias may also be used to apply a specific governing characteristic to a generating unit, overriding nonlinearities or dead band in the unit governor.

Such unit-bias factors are shown as broken-line blocks in Fig. 16-38.

***26. Analog and Digital Techniques***

*Analog Executions.* For about 40 years following the first installation, in 1927, of automatic generation control on an interconnected system, all AGC assemblies utilized analog executions. Starting with pioneering experimental concepts in the then new field, successive advances followed (Ref. 15). By 1937 tie-line bias control had been introduced. By about 1948, fully distributed frequency-biased net interchange control had become the accepted standard. In the following decade, advances in economic dispatch included the predictive base-participation technique of Fig. 16-37, and the lambda reference computer-control systems of Fig. 16-38. These and related techniques developed and utilized in the analog period remain as basic operating practices in present-day comprehensive digital-computer control systems. Analog execution reached its zenith in about 1960.

*Digitally Directed Analog Execution.* For a brief period following the analog era, digitally directed analog assemblies were installed. These included a flexibly programmed analog console of the type shown in Fig. 16-37 supplemented by a digital computer. The latter was programmed to compute an area economic dispatch every few minutes and was arranged to automatically set the allocation dials of the analog console to correspond to these economic-dispatch computations. The digital computer was available for other services, such as determining when units should be put on or taken off the line, computing advantageous interchanges with neighboring areas, collecting data for bulk-transfer billings, and monitoring system security and environmental constraints. When the digital computer is in use for such other services, or was down for repair or maintenance, the analog console continued in service for area regulation and source allocation control.

*Direct Digital Control.* The capabilities of digital computers, their flexibility, speed, and data storage, the continuing improvements in these characteristics and in their reliability and their performance/price ratios, led in the 1960s and thereafter to the general adoption of direct digital control without an auxiliary analog console. Initially they repeated the well-established concepts of bias control and economic dispatch, with system monitoring for pertinent operating data. The massive blackout in the northeast section of the United States in 1965 stimulated great interest in improved system security and reliability, additionally and formally emphasized by the formation in 1968 of the North American Electric Reliability Council (NERC). Software and peripheral developments added importantly to the scope of digital systems. New advanced applications, beyond those of analog systems, include on-line studies of load flow, state estimation, contingency analysis, external equivalents for monitoring, and real-time system optimization. Many such applications are incorporated into present-day energy management systems (EMS). For discussions on current applications techniques and system security see Ref. 48, and Pars. **30–35**.

***27. Protection.*** Components of a typical computer control system may be spread out over the thousands of square miles of a control area and linked together by long telemetering and control channels. Protective features should be incorporated into the computer control system so that it will not apply improper control action to generating sources when components fail or do not function properly. There should also be comparable protective action when system, area, or station conditions are beyond the corrective scope or capability of the control system.

Typical conditions which may be utilized to interrupt or suspend control automatically, partly, or completely are:

1. System frequency swings to abnormally high or low value.
2. Voltages of power supplies are outside their normal ranges.
3. A telemetering or control channel is lost.
4. Normal telemetering or control signals are not received.
5. Abnormal telemetering or control signals are received.
6. A high or low limit is reached on an individual generator.
7. A high limit is reached on a tie line.

Suitable equipment is incorporated into the computer control system and takes programmed protective action, with suitable alarm warnings to operators, when one or more abnormal or emergency conditions are detected.

***28. Installation Data and Future Outlook.*** Reference 43 includes data and bibliography of the state of power systems control as the analog era ended and the digital era began. Reference 44 reviews on-line applications. A study of contemporary systems planning (Ref. 47), suggests

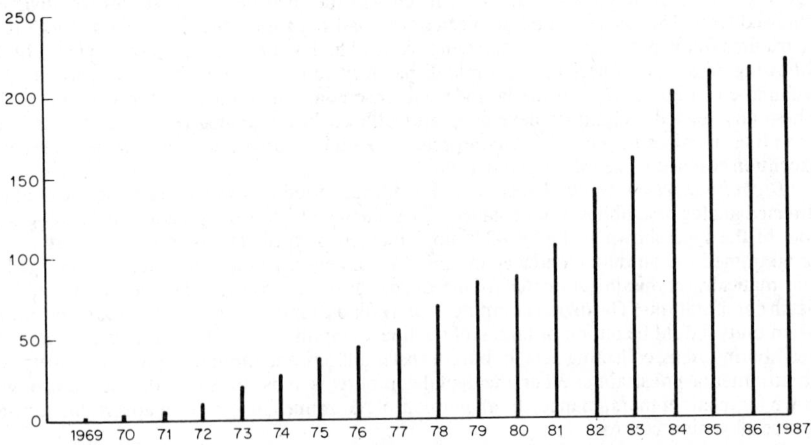


**FIG. 16-39** Growth of modern digital computer control centers, 1969–1987. By 1988, there will be 268 centers in 58 countries. (Ref. 51.)

the need for future research activity. Ref. 49 describes one of the most recently commissioned installations as the handbook goes to press. A summary of developments in computer control for interconnected systems, by decades, over its 60-year history, which started in 1927, covering both the analog and digital eras, is contained in Ref. 50. Fig. 16-39, based on Ref. 51, summarizes the growth of digital-computer control center installations from 1969 to 1987, by 1988 there will be 268 known or contracted such centers in 58 countries.

### *29. References*

*Control of Generation and Power Flow*


1. Statistical Year Book of the Electric Utility Industry for 1985; *Publ.* 84–51, Washington, D.C., Edison Electric Institute.
2. Annual Report; North American Electric Reliability Council, Princeton, N.J., 1984.
3. Review of Overall Reliability and Adequacy of the North American Bulk Power Systems, 14th Annual Review; Report of the North American Electric Reliability Council, Princeton, N.J., July 1975.
4. Operating Manual; North American Electric Reliability Council, Princeton, N.J.
5. Minimum Operating Reliability Criteria, Western Systems Coordinating Council, Salt Lake City, Utah.
6. Operating Manual, Northwest Power Pool, Portland, Oregon.
7. Definitions of Terminology for Automatic Generation Control on Electric Power Systems; *IEEE Publ.* 94, November, 1965.
8. Cohn, N., Mochon, H. H., Jr., Kleinbach, W. S., Welch, T. P., McNulty, M. B., Brabston, J. E., Jr., Timme, E. F., and Scarth, E. D.: Symposium on Scheduling and Billing of Bulk Power Transfers; *Proc. Am. Power Conf.,* Chicago, Ill., 1972, vol. 34, pp. 904–967.
9. Cohn, Nathan: Considerations in the Regulation of Interconnected Areas; *IEEE Trans.,* 1967, vol. PAS-86, no. 12, pp. 1527–1538.
10. Recommended Specification for Speed-Governing of Steam Turbines Intended to Drive Electric Generators Rated 500 kW and Larger; *IEEE Publ.* 122 (AIEE 600), December, 1959.
11. Cohn, Nathan: Common Denominators in the Control of Generation on Interconnected Power Systems; presented before Systems Operation Committee, Pennsylvania Electric Association, May, 1957. *Reprint* 461–5(8), Leeds & Northrup Company.
12. *Westinghouse Electrical Transmission Reference Handbook,* 4th ed.; Westinghouse Electric Corporation, chap. 13, pt X, p. 486.
13. The Effect of Frequency and Voltage on Power System Load; IEEE Winter Power Meeting, Jan. 30–Feb. 4, 1966, *Paper* 31-CP 66-64.
14. Cohn, Nathan: Power Systems Control Practice; *Proc. Ninth Annu. Allerton Conf. on Circuit and Systems Theory,* 1971, pp. 719–730.
15. Cohn, Nathan: Recollections of the Evolution of Realtime Control Applications; *Automatica,* 1984, vol. 20, no. 2, pp. 145–162.
16. Cohn, Nathan: Power Flow Control-Basic Concepts for Interconnected Systems; *Proc. Midwest Power Conf.,* Chicago, Ill., 1950, vol. 12, pp. 159–175. Also *Electr. Light Power,* 1950, vol. 28, no. 8, pp. 82–94, and no. 9, pp. 100–107.
17. Cohn, Nathan: Control of Power Flow in System Interconnections; presented before Electric Section, Wisconsin Utilities Association, November, 1954. *Reprint* ND4-56-461(9), Leeds & Northrup Company.
18. Cohn, Nathan: Automatic Control of Power Systems, Symposium on Reliability of Bulk Power Supply in Large Interconnected Power Systems; *Proc. IEEE Int. Conv.,* March, 1966, Pt. 12, *Paper* 50.1.
19. Cohn, Nathan: Principles and Applications of Tie Line Bias Control and Economic Loading; *Tenth Annu. Am. Public Power Assoc. Eng. Operations Workshop,* New Orleans, La., January, 1966. *Reprint* E7-3111 RP, Leeds & Northrup Company.
20. Cohn, Nathan: Some Aspects of Tie-line Bias Control on Interconnected Power Systems; *Trans. AIEE,* 1956, vol. 75, pt. III, pp. 1415–1428.
21. Cohn, Nathan: A Step-by-step Analysis of Load-frequency Control Showing the System Regulating Responses Associated with Frequency Bias; presented before the 1956 Meeting of the Interconnected Systems Committee; Des Moines, Iowa, March, 1956. *Reprint* 461-5(6), Leeds & Northrup Company.
22. Cohn, Nathan: Bias Revisited; presented before the Spring 1970 Meeting of the East Central Systems Group of the North American Power Systems Interconnection Committee, St. Joseph, Mich., April, 1970. *Reprint* E7.0014, Leeds & Northrup Company.

23. Cohn, Nathan: Techniques for Improving the Control of Bulk Power Transfers on Interconnected Systems; *IEEE Trans.*, 1971, vol. PAS-90, no. 6, pp. 2409–2419.

24. Cohn, Nathan: Energy Balancing; *Proc. Am. Power Conf.*, Chicago, Ill., 1973, vol. 34, pp. 995–1024.

25. Cohn, Nathan: Methods of the Systems for Synchronized Coordination of Energy Balancing and System Time in the Control of Bulk Power Transfers; U.S. Patent 3,898,442, Aug. 5, 1975.

26. Cohn, Nathan: Some New Thoughts on Energy Balancing and Time Correction on Interconnected Systems; *Proc. IEEE Region 5 Conf. on Control of Power Systems*, 1976, Oklahoma City, Oklahoma, *IEEE Publ.* 76CH1057-9REG5.

27. Cohn, Nathan: Research Opportunities in the Control of Bulk Power and Energy Transfers on Interconnected Systems; A paper of Ref. 43.

28. Cohn, Nathan: Methods of and Systems for Coordinated System-wide Energy Balancing in the Control of Bulk Power Transfers; U.S. Patent 3,701,891, Oct. 31, 1972. Assigned to Leeds & Northrup Company.

29. Cohn, Nathan: Decomposition of Time Deviation and Inadvertent Interchange on Interconnected Systems. I: Identification, Separation and Measurement of Components. II: Utilization of Components for Performance Evaluation and Corrective Control, *IEEE Trans. Power Apparatus and Systems*, 1982, PAS-101, no. 5, May, p. 1144, and no. 8, August, p. 2711.

30. Cohn, Nathan: Energy Conservation by Improved Control of Bulk Power on Interconnected Systems, U.S. Patent 4,267,571, May 12, 1981; Canadian Patent 1,142,589, Mar. 8, 1983.

31. Kaufmann, P. G.: Load Distribution between Interconnected Power Stations; *J. IEE*, 1943, vol. 90, pt. II, no. 14, pp. 119–130.

32. George, E. E.: Intrasystem Transmission Losses; *Trans. AIEE*, 1943, vol. 62, pp. 153–158.

33. Ward, J. B., Eaton, J. R., and Hale, H. W.: Total and Incremental Losses in Power Transmission Networks; *Trans. AIEE*, 1950, vol. 69, pp. 626–632.

34. Kirchmayer, L. K., and Stagg, G. W.: Analysis of Total and Incremental Losses in Transmission Systems; *Trans. AIEE*, 1951, vol. 70, pp. 1197–1204.

35. Harder, E. L., Ferguson, R. W., Jacobs, W. E., and Harker, D. C.: Loss Evaluation, Part II. Current-power-form Loss Formulas; *Trans. AIEE*, 1954, vol. 73, pp. 716–731.

36. Early, E. D., Watson, R. E., and Smith, G. L.: A General Transmission Loss Equation; *Trans. AIEE*, 1955, vol. 74, pt. III, pp. 510–520.

37. Brownlee, W. R.: Coordination of Incremental Fuel Costs and Incremental Transmission Losses by Functions of Voltage Phase Angles; *Trans. AIEE*, 1954, vol. 73, pt. III, pp. 529–541.

38. Kirchmayer, L. K., and Stagg, G. W.: Evaluation of Methods of Coordinating Incremental Fuel Costs and Incremental Transmission Losses; *Trans. AIEE*, 1952, vol. 71, pt. III, pp. 513–521.

39. Fereshetian, H., Liechty, M. D., and Brown, N. E.: Coordination of Desired Generation Computer with Area Control; *Proc. Am. Power Conf.*, Chicago, Ill., 1959, vol. 21, pp. 554–563.

40. Cohn, Nathan: Methods of Controlling Generation of Interconnected Power Systems; *Trans. AIEE*, 1962, vol. 80, pt. III, pp. 270–282.

41. Cohn, Nathan: Area-wide Generation Control—A New Method for Interconnected Systems; *Proc. Am. Power Conf.*, Chicago, Ill., 1953, vol. 15, pp. 316–344. Also *Electr. Light Power*, June, 1953, vol. 31, no. 7, pp. 167–175; July, 1953, no. 8, pp. 97–108; August, 1953, no. 9, pp. 77–83.

42. Control Centres, *International Journal of Electrical Power Systems*, 1983, special issue, vol. 5, no. 4.

43. Cohn, Nathan: State of the Automatic Control Art in the Electric Power Industry of the United States; *Proc. Sixth Joint Automatic Control Conf.*, Troy, N.Y., June, 1965. Also *IEEE Spectrum*, 1956, vol. 2, no. 11, pp. 67–77.

44. Cohn, N., Biddle, S. B., Jr., Lex, R. G., Jr., Preston, E. H., Ross, C. W., and Whitten, D. R.: Online Computer Applications in the Electric Power Industry; *Proc. IEEE*, 1970, vol. 58, no. 1, pp. 78–87.

45. Special Issue on Computers in the Power Industry; *Proc. IEEE*, 1974, vol. 62, no. 7.

46. Energy Control Centers; U.S. and Canada, Session 17; U.S.S.R., U.K., Sweden, Japan, and France, Session 24; *Proc. 6th Triennial World Congr., Int. Fed. Automatic Control*, Part II, Applications, 1975, Boston/Cambridge, Mass., Instrument Society of America, Pittsburgh, Pa.

47. Power Systems Planning and Operations—Future Problems and Research Needs; *Proc. Eng. Foundation Conf.*, sponsored by EPRI with cooperation of ERDA, 1976, Henniker, N. H., EPRI, SR-EL-377 Palo Alto, Calif.

48. Energy Management Systems, several authors on recent installations, *Proceedings of the IEEE PICA Conference*, Nov. 1983.

49. Dopazo, J. F. et al.: The New AEP System Control Center; *Proc. Am. Power Conf.*, 1983 Chicago, Ill., vol. 45, pp. 603–609.

50. Cohn, Nathan: Developments in Computer Control of Interconnected Power Systems—Exercises in Co-operation and Co-ordination Among Independent Entities; *Elektron, J.*, South African IEE, vol. 2, no. 7, June–July, 1985, Johannesburg, South Africa.

51. Dy Liacco, T. E. and Rosa, D. L.: Survey of System Control Centers for Generation-Transmission Systems; privately printed, Cleveland, Ohio, January 1984.

*Additional Bibliography*

Cohn, Nathan: *Control of Generation and Power Flow on Interconnected Systems;* New York, John Wiley & Sons, Inc., 1966; 2 ed., 1971. Distributed by Leeds & Northrup Company, North Wales, Pa.

Wood, A. J. and Wollenberg, B. F.: *Power Generation, Operation and Control;* New York, John Wiley & Sons, Inc., 1984.

## POWER SYSTEM SECURITY

*By BRUCE F. WOLLENBERG*

***30. Introduction.*** Power systems, like all physical devices and systems, are subject to failure. Failures may be caused by internal breakdowns (insulation failure, plant auxiliary equipment failure, etc.) or by events outside the power system itself (lightning strikes, objects falling on lines, etc.). When a failure of a piece of equipment occurs on a power system it must be taken out of service and the remaining equipment must be able to make up for the loss without further equipment overloading.

Many equipment failures result in overloads that must be eliminated quickly, and for this reason automatic relay equipment is usually installed to remove any piece of electrical apparatus that is overloaded. However, this does not mean that protection of the system is totally under the control of the protective relays. The way in which the system is operated has a primary influence on the security of the system and this responsibility falls on system operators.

Two different examples can be presented to show how operator's decisions affect the outcome of equipment failures.

1. When a generator is taken out of service, the remaining generators must take up the loss in generation. However, if there is insufficient generation remaining on the system the frequency drop can be so severe that recovery is impossible. Operators must therefore be sure to commit enough generation so that the loss of any generating units will still leave enough capacity to safely restore frequency.
2. When a transmission line fails, the current flowing on it will redistribute on the remaining circuits. If the new flows on the remaining circuits cause one or more of them to be overloaded, the relays protecting them will open, resulting in still more redistributions of currents. This process, if it continues, is called a cascading failure and can result in a blackout on parts or all of the system.

Keeping the system operating so that failures do not lead to cascading system breakdowns is referred to as maintaining *power system security* and is a primary reason for the installation of computerized control and information-gathering systems by electric utilities (Ref. 1). These "energy management systems" usually consist of central digital computers connected to remote terminal units (RTU's) via communications channels. The computers have a variety of programs to gather status (open/close, on/off, etc.) and analog (megawatts flows, megavar flows, bus kilovolts, etc.) from throughout the power system as well the ability to send control commands such as circuit breaker open/close and generator megawatt set point to substations and power plants. Most of the operator actions we will discuss below are exercised through such a computer system via color cathode-ray-tube displays and keyboards. The other means the operations personnel have at their disposal are the telephone as well as written orders issued to utility personnel.

Operators can take a variety of actions to see that system security is adequate. These actions are summarized in Table 16-8.

***31. System Monitoring***

*Data Acquisition.* The first job performed by the system operators in maintaining system security is simply to monitor the system. This means watching the values read by transducers in

**TABLE 16-8** Actions Taken by System Operators to Maintain System Security

| Operator action | Variables adjusted |
|---|---|
| Generator commitment | Generator on/off status |
| Generator dispatch | Generator megawatt output schedule |
| Generator bus voltage | Unit exciter setting |
| Load tap changing transformer | Tap position |
| Phase-shift transformer | Tap position |
| Neighboring system interchange | Interchange schedule |
| Network configuration | Substation breaker open/close status |
| Load shedding | Distribution feeder breaker status |

substations and power plants that are transmitted to the energy management system through the RTUs and taking note of any out-of-limit or other unusual conditions. However, with tens of thousands of status and analog values, this is a humanly impossible task and is therefore usually handled by the computer system, which checks each value as it comes into the system. Changes in status and out-of-limit analog values are brought to the operator's immediate attention through alarm messages on the console displays.

*Alarm Processor.* Alarm messages are generated for a variety of system conditions and are usually logged in permanent files and printed for archival study if needed. Some of the system conditions that create alarm messages when out of limit or not in normal state are shown in Table 16-9.

Because of the experience of the system operators, alarm messages are often enough to determine the source of trouble and allow actions to be taken to reduce the chances of serious failures. However, knowledge of single values by themselves is often insufficient, and the operator must be able to draw conclusions from knowing the status and values of many variables. In the case of breaker status values, the operators are provided with one-line display diagrams of each monitored substation in the power system. These one-line diagrams have graphic indications of breakers, busbars, switches, transformers, etc. Further, the breaker positions are shown in color so that a quick accurate assessment of a switching action can be obtained by looking at the display.

*Status Processor.* When a switching action affects more than one substation, the computer system can analyze the transmission system network using a status processing program (Ref. 2). This program requires a complete description of the transmission system stored in the computer. When supplied with the telemetered status values, the status processor analyzes the

**TABLE 16-9** Power System Alarm Conditions

| Quantity monitored | Alarm indication |
|---|---|
| Bus voltage | Bus kV above/below limit |
| Transformer/transmission line flow | Megawatt, megavar, MVA, amperes above or below limit |
| Transformer/cable equipment | Temperature or pressure above or below limit |
| Communications channel to RTU | Normal/failure |
| Protective relay communication channel | Normal/failure |
| Breaker status | Open/close |
| Breaker $SF_6$ | Normal/low pressure |
| System frequency | Above/below limit |
| Generator status | On/off line |
| Generator unit | Megawatts, megavar, kilovolts above limit |
| Load | Megawatts, megavar, above limit |
| Load shedding equipment | Load shed/load restore |

topology of each substation and then the entire network to see which buses are connected together and which lines are in service and whether the system is connected or has been switched into electrical islands. Often the output of the status processor is sent to indicators on large graphic diagrams of the electrical system that take up one wall of the operations room.

*Reserve Monitor.* Another program that monitors the information that is being transmitted to the energy management system has the responsibility of calculating the generation reserves and comparing this amount to established reliability criteria. The reserve monitoring program provides the operator with a display of current reserves as well as generating alarm messages when insufficient reserves are seen.

*State Estimator.* A common problem that operators often have to contend with is poor-quality data. Transducers may drift out of calibration or their output connected incorrectly, and data communications channels may fail, making some data unavailable. Operators must make their decisions within this context, and it is useful if the computer system can indicate which measured values are bad and estimate those values not available because of telemetery failure.

The program aimed at this function is called a state estimator (see Refs. 3 and 4). The state estimator takes a mathematical model of the power system that has been built using the output of the status processor plus measured analog values and calculates a best estimate of the state variables for the system. In the case of an electric power system, the state variables are the voltage magnitude and phase angle at each bus in the network. If there are many more measured values than states, the quality of the estimate improves. Once the voltage magnitude and phase angle are available, the state estimator can calculate the flows in all transmission lines and transformers in the network. In addition, if sufficient redundant measurements are available, the presence of a bad measurement can be detected and identified so that it can be repaired (Ref. 5).

Finally, the state estimator, by virtue of its need for a mathematical model that matches the system, provides an ideal base for modeling contingencies.

Table 16-10 summarizes the functions most commonly used in the monitoring of power systems.

***32. Contingency Analysis.*** Many of the problems that arise on a power system cannot be corrected quickly enough to prevent cascading system failures. In order to help the operators guarantee that the system can withstand equipment failures, modern energy management systems often provide predictive programs to study the power system before trouble happens. Using mathematical models of the power system, operators can predict the outcome of hypothetical failures before they happen and take evasive action to protect the system if necessary.

The process of studying the power system for hypothetical failures is often referred to as contingency analysis or security analysis (Ref. 6). In a contingency analysis program a list of

**TABLE 16-10** System Monitoring Functions

| Function name | Function performed |
|---|---|
| Data acquisition | Process messages from RTUs<br>Check analog measurements against limits<br>Check status values against normal value<br>Send alarm conditions to alarm processor |
| Alarm processor | Format alarm messages<br>Transmit messages according to priority |
| Status processor | Determine status of each substation for proper connection<br>Determine states of network for equipment out of service, islanding, etc.<br>Send alarm messages to alarm processor |
| Reserve monitor | Check generator megawatt output on all units against unit limits<br>Alarm if insufficient reserves |
| State estimator | Determine system state variables using measured values and network model<br>Detect presence of bad measured values<br>Identify location of bad measurements<br>Initialize network model for other application programs |

possible system failures is modeled one after the other and a record kept of those failures that could cause overloads or voltage problems on the transmission system. To model the network under transmission line, transformer, and generating unit outages requires a mathematical calculation that can predict network quantities given a complete description of the network as it exists at the present time.

The model of the network consists of its admittance matrix, which is built from the information produced by the status processor. The real and reactive power at each bus come from the state estimator. The most widely accepted method of calculating the flows and voltages under an outage is the use of an ac load-flow program. The ac load-flow must solve two simultaneous quadratic equations (one for the real power and the other for reactive power) for each bus in the network. By keeping the bus loading conditions as determined by the state estimator and then altering the admittance matrix to reflect the outage, the contingency analysis program can calculate the effects of the outage and alarm the operator. The basic flow of information from the power system to the contingency analysis program is shown in Fig. 16-40.

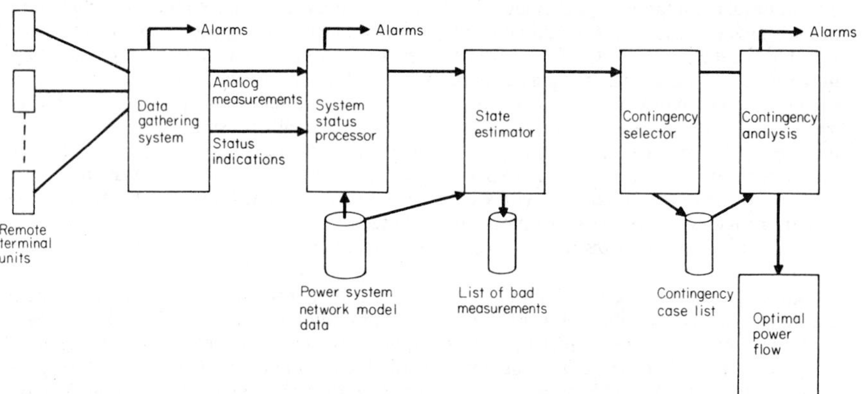


**FIG. 16-40** Power-system security program sequence.

Several problems arise in the execution of a contingency analysis program as shown in Fig. 16-40. The main difficulty is the fact that there are many more credible failures than time available to carry out the study. In the normal operation of a power system nothing is static for very long. This means that loadings change, units must be added or taken off, interchange with neighboring systems changes, and the network itself can change as equipment is switched. Therefore the study of which failures are important must be repeated often to be useful. Unfortunately, however, the ac load-flow calculation can take a great deal of time to perform, especially if the network being modeled is large. Typical contingency analysis models can run to several thousand buses in modern energy management systems, and this may take anywhere in the range of 15 s to several minutes to solve depending on the capability of the computers installed in the energy management system. If, for example, the execution time of the ac load flow took 1 min for each outage and several hundred were to be tested, the operators might not know all the results for a few hours—which is much too long. In order to solve the contingency analysis in a reasonable time, two approaches have been taken: contingency selection and faster load-flow calculation techniques.

*Contingency Selection.* Contingency selection is based on the idea that of all the possible failure events to be studied only a few will give problems. Therefore, in contingency selection techniques a calculation is made which tries to tell which failure cases will cause trouble and then only those cases are run through a full ac load flow.

Two methods of contingency selection are currently in use. The first (Ref. 7) uses a single calculation based upon a performance index of system loading conditions and obtains an approximate ordering of the performance index for each possible network outage. The second uses a screening technique (Ref. 8) which executes a partial load-flow calculation for each outage and then completes the calculation for those cases that appear heading for trouble. These two methods are compared in Table 16-11.

**TABLE 16-11** Comparison of Contingency Selection Methods

| Contingency selection method | Characteristics of method |
|---|---|
| Single calculation | Uses performance index of network loading<br>Calculates approximate ordered list of contingency event (ordered with most severe contingency at top of list)<br>Full ac load flows are run on top members of list and limits checked and alarmed<br>Time to calculate list equal to run time for one to two full ac load flows<br>Present technology does not include ordering based on out-of-limit bus voltages |
| Contingency screening | Based on calculation of partial ac load flow<br>If partial load flow indicates possible limit violations, then remainder of load flow is run and limits checked and alarmed<br>Time to perform screening is equal to one-quarter to one-fifth ac load-flow time for each possible contingency event. Total time is therefore much longer than single calculation method.<br>Can select contingencies for bus voltage and line/transformer flow-limit violations |

*Fast Load-Flow Methods.* Here the object is to carry out the ac load flow or an approximation to it as quickly as possible. Two methods are generally in use today. The first (Refs. 9 and 10), which is somewhat dated and is being used less and less, uses factors to predict line and transformer loading when other network equipment fails or when generating units fail. This method is very fast, but it is only an approximation and cannot predict voltage or reactive flows on the network. The second (Ref. 11) involves use of a fast ac load-flow method known as the "decoupled load flow" algorithm, in which the real and reactive aspects of the power system network are separated. These two methods are compared in Table 16-12.

***33. Correcting Network and Generation Problems.*** Whenever the contingency analysis programs report that the failure of a piece of equipment will cause overloads or voltage limit

**TABLE 16-12** Fast Load Flow Methods

| Method of load calculation | Characteristics |
|---|---|
| Distribution factors | Uses one set of factors to determine network loading for generator outage contingencies and another set to determine network loading for outage of transmission lines and transformers<br>Must calculate and store factors ahead of use; when network is altered by switching factors must be recalculated<br>Calculation of contingency loading of network is extremely fast<br>Can only predict megawatt transmission line and transformer loading |
| Decoupled ac load flow | Decouples real and reactive calculations<br>Must iterate ac load-flow solution<br>Sometimes has trouble when network branch reactance to resistance ratio is low (will not converge to a solution)<br>Can simulate generator var limit control which may be important in some cases<br>Can calculate megawatt and megavar flows as well as bus voltage magnitudes for limit checking |

**TABLE 16-13** Objectives and Constraints in Optimal Power Flow Calculations

| | |
|---|---|
| Objectives (only one in use at one time) | Minimize system operating cost ($/h)<br>Minimize system megawatt losses<br>Minimize overloads |
| Constraints | Generator unit megawatts within limit<br>Generator unit megavars within limit<br>Transformer tap position within limit<br>Bus voltage magnitude within limit<br>Transmission line megawatts, MVA, or amperes within limit<br>Transformer megawatts, MVA within limit<br>Area interchange within limit<br>Flow over groups of circuits within limit<br>AC load-flow conditions met at all buses |

violations, the operators must decide how to react. If the operators decide that the failure would cause too severe an overload or voltage-limit violation, then they must make adjustments to the power system to prevent such an overload.

Often the operators will know what adjustments to make so that no problems result if the failure studied does in fact occur. When the adjustments require changes to many variables simultaneously, however, the operators must rely on programs to calculate them. These adjustments can involve taking the generating units off their economic dispatch set points, changing the commitment schedule of the units, or adjusting interchange schedules. In such cases there is an economic penalty for adjustments and the utility must set policies that guide the operators as to whether to incur the economic penalty or the risk of system failure.

In many modern energy management systems the real power losses on the transmission lines and transformers is minimized as a further way to economize. Here again, when the contingency analysis indicates a system problem, it may need to be corrected by an adjustment to a generator voltage schedule or to a transformer tap position, and this may also cause an economic penalty, although not nearly as much as with an adjustment to generation megawatt schedules.

*Optimal Power Flow.* Two methods are used to calculate the proper adjustments to a power system to relieve overloads and voltage violations. The first makes use of linear programming and solves a decoupled model much as in the decoupled load-flow calculation (Refs. 12 and 13). In the linear-programming routine, the objective is to minimize operating cost for any given supplied load while meeting a variety of constraints such as circuit flows and bus voltages. The second method solves a completely coupled ac load-flow model and uses megawatt losses or operating cost as the objective function (Refs. 14, 15, and 16). Generally the same constraints are used as in the first method. These methods both go under the name optimal power flow or optimal load flow, since they involve solving the minimum of some power system objective while meeting the full ac load-flow conditions as constraints.

The objectives and constraints of an optimal power flow are summarized in Table 16-13.

***34. Conclusions.*** Modern power system energy management systems are justified by electric utilities on the basis of their ability to save the utility money through better optimization of the operation and their ability to help operators prevent large system failures. The power system security programs prove their economic worth many times in preventing costly system blackouts. In addition, the power system security programs also allow operators to operate the system closer to its limits, thus giving better economic operation (Ref. 17).

***35. References***

1. Dy Liacco, T. E.: The Adaptive Reliability Control System; *IEEE Trans.*, 1967, vol. PAS-86, pp. 517–531.
2. Sasson, A. M., Ehrman, S. T., Lynch, P., and VanSlyck, L. S.: Automatic Power System Network Topology Determination; *IEEE Trans.*, 1973, vol. PAS-92, pp. 610–618.
3. Schweppe, F. C., and Wildes, J.: Power System Static State Estimation, Part I: Exact Model, *IEEE Trans.*, 1970, vol. PAS-89, pp. 120–125.
4. Schweppe, F. C., and Handschin, E.: Static State Estimation in Power Systems; *IEEE Proc.*, 1974, vol. 62.

5. Dopazo, J. F., Klitin, O. A., and Sasson, A. M.: State Estimation for Power Systems: Detection and Identification of Gross Measurement Errors; *Proc. 8th Power Industry Computer Applications Conference,* Minneapolis, June 1973.

6. Debs, A. S., and Benson, A. R.: Security Assessment of Power Systems; *Systems Engineering for Power: Status and Prospects,* US Government Document CONF-750867, 1967, pp. 1–29.

7. Wollenberg, B. F., and Ejebe, G. C.: Automatic Contingency Selection; *IEEE Trans.,* 1979, vol. PAS-99, pp. 97–109.

8. Albuyeh, F., Bose, A., and Heath, B.: Reactive Power Considerations in Automatic Contingency Selection; *IEEE Trans.,* 1981, vol. PAS-100, pp. 107–112.

9. El-Abiad, A. H., and Stagg, G. W.: Automatic Evaluation of Power System Performance—Effects of Line and Transformer Outages; *AIEE Trans.,* 1963, vol. PAS-81, pp. 712–716.

10. Brown, H. E.: Contingencies Evaluated by a Z Matrix Method; *IEEE Trans.,* 1969, vol. PAS-88, pp. 409–412.

11. Stott, B., and Alsac, O.: Fast Decoupled Load Flow; *IEEE Trans.,* 1974, vol. PAS-93, pp. 859–869.

12. Wollenberg, B. F., and Stadlin, W. O.: A Real Time Optimizer for Security Dispatch; *IEEE Trans.,* 1974, vol. PAS-93, pp. 1640–1644.

13. Stott, B., Marinho, J., and Alsac, O.: Review of Linear Programming Applied to Power System Rescheduling, *Proceedings 1979 Power Industry Computer Applications Conference,* IEEE Document No. 79CH1381-3-PWR, pp. 142–154.

14. Dommel, H. W., and Tinney, W. F.: Optimal Power Flow Solutions; *IEEE Trans.,* 1968, vol. PAS-87, pp. 1866–1876.

15. Burchett, R. C., Happ, H. H., and Vierath, D. R.: Quadratically Convergent Optimal Power Flow, *IEEE Trans.,* 1984, vol. PAS-103, No. 11, pp. 3267–3276.

16. Sun, D., Ashley, B., Brewer, B., Hughes, A., and Tinney, W. F.: Optimal Power Flow by Newton Approach; *IEEE Trans.,* 1984, vol. PAS-103, No. 10, pp. 2864–2880.

17. Masiello, R. D., and Wollenberg, B. F.: Cost Benefit Justification of an Energy Control Center; *IEEE Trans.,* 1981, vol. PAS-100, pp. 2568–2574.

*Additional Bibliography*

Wood, A. J., and Wollenberg, B. F.: *Power Generation, Operation and Control,* John Wiley & Sons, New York, 1984.

## RELAYING AND PROTECTION

*By W. A. ELMORE*

***36. General.*** The fundamental concept of protective relaying is to detect and isolate faults and other destructive phenomena in the shortest possible time consistent with economics and security. The principles vary at different points in the power system because of differing constraints. Distribution system relaying must coordinate with fuses and reclosers for faults while ignoring "cold-load pickup," capacitor bank switching, and transformer energization.

Transmission line relaying, on the other hand, must be sufficiently discriminating to locate and isolate any type of fault and do so with sufficient speed to preserve stability, to reduce fault damage, and to minimize the impact on the power system. This dictates the use of one or more pilot relaying systems.

Subtransmission relaying varies from complete pilot relaying to simple directional overcurrent relaying depending on the importance and general nature of the subtransmission system.

***37. Distribution System Relaying.*** Typical distribution circuit relaying is shown in Fig. 16-41. Only one set of feeder relays is shown. This complement would be repeated for each feeder. The time-delayed phase and ground relays 51 and 51N usually have a high degree of inverseness in their current-time characteristic to coordinate with the fuses and reclosers that are farther out on the circuit. The instantaneous units 50 and 50N are typically set to trip the feeder breaker and protect the fuses when a temporary fault occurs beyond the fuse and are removed from service by a reclosing relay to allow the fuse to blow when reclosing into a permanent fault.

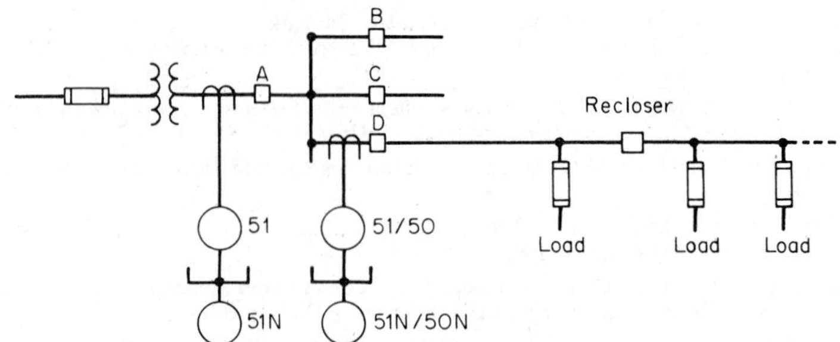


**FIG. 16-41** Typical distribution circuit relaying.

The 51N relay must be set with care to avoid its operation on loss of single-phase lateral load when a fuse blows. The "normal" load unbalance can be controlled to a reasonable degree by carefully supervising the balance of load connected to each individual phase (usually a 4-wire circuit with line-to-neutral connected loads). The opening of a fuse to clear a fault, and thereby drop load associated with one phase, will produce a much higher than normal load unbalance. This must not be allowed to cause operation of the ground relay. Its sensitivity is largely regulated by this consideration.

Cold load pickup is the phenomenon whereby a feeder being reenergized after a long outage will experience a load appreciably in excess of maximum steady-state load (as a result of loss of diversity by thermostatically controlled devices). The feeder relays must ignore this if sectionalized reenergization is to be avoided. The relays on breaker *A* in Fig. 16-41 provide primary protection for the bus and backup protection for the feeder relays and breakers. In general they are time-delayed and coordinate with the feeder relays with the accepted sacrifice of clearing speed for bus faults. These phase relays provide some measure of thermal protection for the supply transformer.

***38. Subtransmission Relaying.*** Loops and multiple power sources used in feeding loads from the subtransmission system usually dictate the use of directional overcurrent relaying, distance relaying, or pilot relaying. In general a subtransmission system is not intended to transmit bulk power from one location to another. Multiple sources are used purely in the interests of continuity of service.

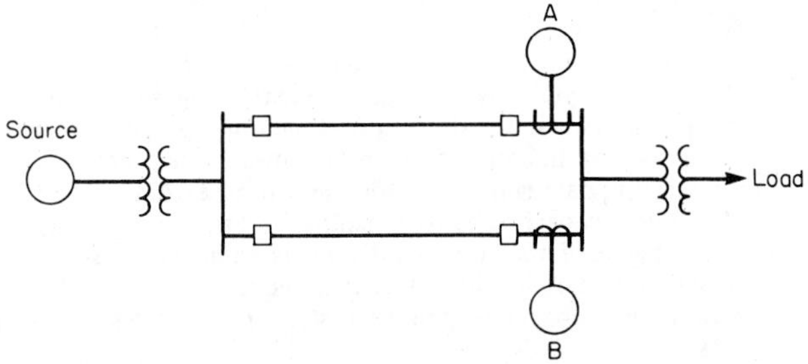


**FIG. 16-42** Partial one-line diagram of typical subtransmission system showing locations where directional relays are required.

Figure 16-42 shows an example requiring directional overcurrent relaying. A fault on the upper line would cause equal currents to flow in relays *A* and *B*. For this fault case, it is desired that relay *A* trip and *B* restrain. A fault on the lower line also causes equal current to flow in relays *A* and *B*. For this case, it is desired that relay *B* operate and relay *A* restrain.

These two cases produce requirements that are mutually exclusive using simple overcurrent relays. The requirements can be met with directional overcurrent relays. If directional, the *A*

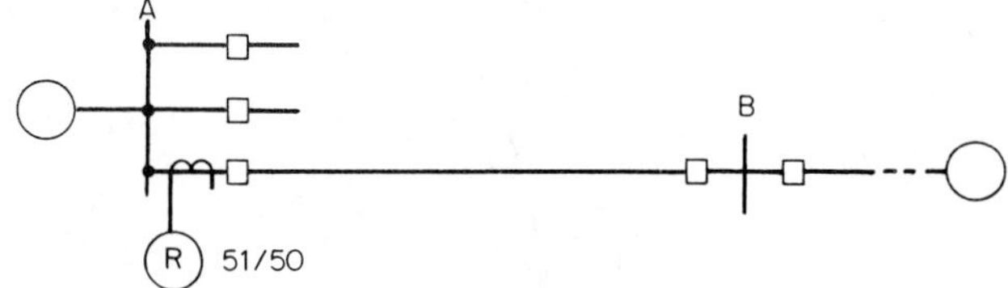


**FIG. 16-43** Directional relaying criterion.

relays would respond only to faults on the upper line and the *B* relays only to faults on the lower line. Coordination between *A* and *B* then becomes unnecessary.

Figure 16-43 defines in the simplest form a criterion for establishing where directional overcurrent relays are desirable.

Relay *R* in Fig. 16-43 requires consideration of distinctly different criteria, depending upon whether instantaneous or intentional time delay tripping is involved. An instantaneous device at *R* must be set in such a way that it will never respond to a fault beyond bus *B*. The setting will be dictated by the maximum fault contribution (phase-fault contribution for phase relays or gound-fault contribution for ground relays *or phase relays*) for a fault at *B* and by the influence on the measuring unit of the dc component in the fault current. For example, a maximum fault at *B*, producing 20 A in relay *R* would require a setting in excess of 20 A. If the maximum overreach factor for the particular instantaneous unit in use were 1.3 and a 10% margin were desired, a setting of 1.3 (1.1) (20) = 28.6 A would be required.

If a reverse fault such as a fault near bus *A* on other circuits could cause current in relay *R* to exceed 20 A (symmetrical), a higher setting would be required for this instantaneous unit than 28.6 A because the same overreach and margin factors would apply.

Since the extent of line coverage is dependent on the setting of the device as well as the source-line impedance ratio, a reverse fault which dictated a higher setting would cause the extent of line coverage to be smaller. By using directional control, no consideration need be given to reverse faults.

If the magnitude of relay current for this maximum magnitude reverse fault were less than 20 A, no consideration need be given to the inclusion of directional control for the instantaneous unit. A nondirectional relay will be satisfactory in this application because the relative fault currents make the relay inherently directional.

Time-delay overcurrent relays differ in their criteria from those of the instantaneous unit. In the interests of backup protection, relay *R* should always be able to detect the minimum fault on and beyond bus *B*. Further, in any time-delay relay applications, this minimum case should produce an adequate multiple of pickup current in the relay to assure a clearly predictable operating time.

If, for example, the minimum fault at *B* produced 14 A in relay *R*, a setting of 7 A would be required (to give a multiple of pickup of 2 for this minimum fault case). If a reverse fault could deliver current sufficiently large to cause operation of a relay set at this level, consideration should be given to the use of directional control of the time unit. A frequently used conservative summary of this concept is: If the maximum reverse fault current can exceed 25% of the minimum fault current at the next bus, use directional control.

The combined criterion for these concepts is: Use directional control if a reverse fault could influence the sensitivity of relaying used to detect forward faults or if selectivity would not otherwise be possible. If source variations restrict instantaneous coverage to less than 50% of the protected line, or if the tripping times realizable for time-delay relays become undesirably long, distance relays should be used.

Distance relays respond to the voltage and current applied to them and are usually more highly responsive at some lagging current angle. Figure 16-44 shows a typical *R-X* diagram that describes the behavior of these devices. Most distance relays in current use, phase and ground, have a characteristic similar to curve 1 or curve 2. Faults producing an apparent impedance at the relay location that falls inside the characteristic circle will cause the relay to operate. Since a distance relay has a distinct "reach" irrespective of source impedance and is directional, it is said to protect a "zone." Zone 1 relays cover a portion such as 80 to 90% of a subtransmission or transmission line. Zone 2 relays respond to faults at all locations on the line and also to others in

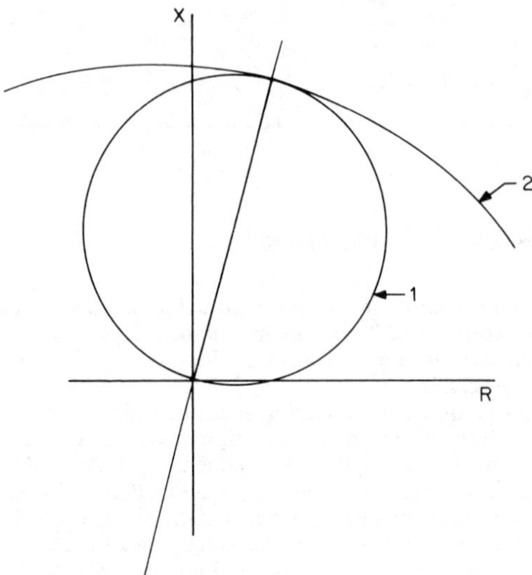


**FIG. 16-44** Resistance-reactance plot of distance relay characteristics.

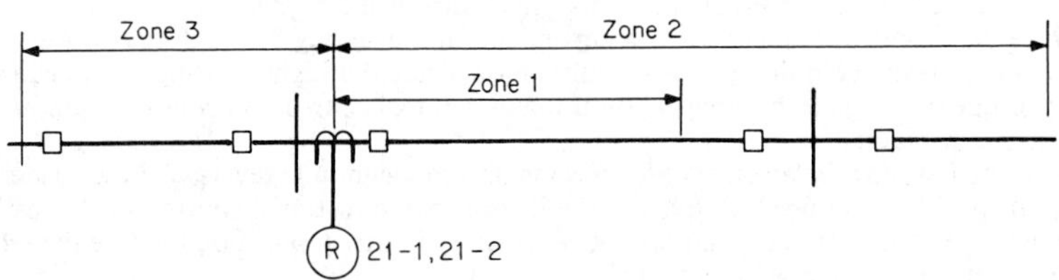


**FIG. 16-45** One-line diagram showing concept of distance relay zones.

proximity of the line end. This is shown in Fig. 16-45. Zone 2 relays are typically set to cover 125 to 175% of the protected line. Since they overreach the next bus at the end of the protected line, they must have a time delay or be associated with a pilot relaying system in order to preserve selectivity with other relays.

A zone 3 relay is also often used and may be directional in the same sense or opposite sense as the zone 1 and zone 2 relays, or in some applications may be nondirectional. Figure 16-45 shows a one-line diagram with a "reverse-looking" zone 3 relay.

Simplified distance-relaying schemes are sometimes used in the interests of economy. One type uses a complete complement of relays for one zone, which is initially set for a zone 1 function. A "starting" unit (overcurrent or distance) senses the presence of a fault. After a time delay, the setting (reach) of the relay is extended to zone 2 and still later to zone 3 (forward). A further abbreviation of this scheme allows the "starting" units to identify the type of fault and to connect the appropriate voltages and currents to a single distance unit.

These systems vary substantially in complexity, redundancy, dependability, and cost. The choice of one system over the others is dictated by the relative importance that is placed on each of these factors and the significance of the compromises involved in making such a choice.

***39. Transmission Line Relaying.*** High-speed clearing of faults is universally required on transmission systems in the interests of maintaining stability, minimizing shock to wide areas of the power system, and decreasing fault damage. Pilot relaying is an important ingredient in this process. Pilot relaying entails the use of information obtained from one or more remote terminals as well as local information to establish the need to trip (or refrain from tripping) a local breaker. The remote information is transmitted by power-line carrier, microwave, tones, pilot wires, optical fiber, or some combination thereof. An abundance of pilot-relaying systems are in use, each having their individual strengths and marginal weaknesses, and each having varying degrees of dependence on the integrity of the channel.

*Pilot Channels.* Figure 16-46 shows one of the many types of pilot channels in use. This

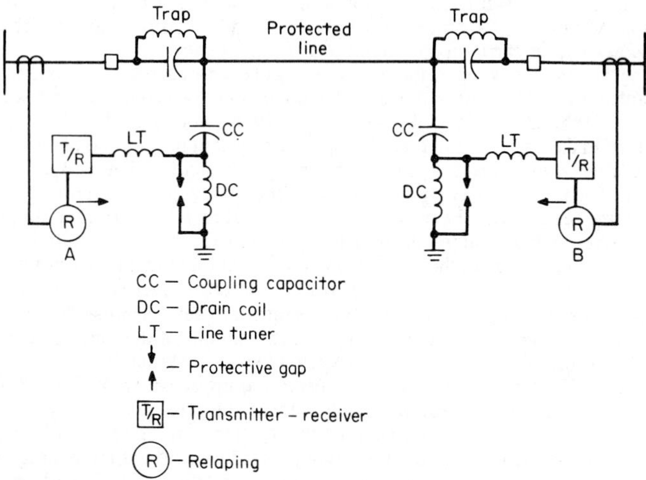


**FIG. 16-46** Representative channel for pilot relaying.

particular arrangement uses "power-line carrier." The pilot channel is chosen sufficiently higher than the power frequency to allow separation to be achieved easily, generally 30 to 300-kHz.

***40. Types of Protective Relaying Systems.*** Two basic systems form the nucleus for the families of relaying systems in predominant use. They are the directional-comparison and the phase-comparison systems.

*Directional-Comparison Relaying.* The fundamental concept of the directional-comparison system is shown in Fig. 16-46. A directional relay at *A* responds to faults to its right as shown by the directional arrow in the figure. A similar relay at *B* responds to faults to the left of *B*. Both relays respond simultaneously only to faults on the protected line. The communication channel informs *A* about the state of *B* and another informs *B* about the state of *A*.

One- to one-and-a-half-cycle initiation of tripping is commonly achieved at both terminals following the occurrence of a fault on such a protected line. No tripping of these relays occurs for faults on other line sections. Abbreviated descriptions of the commonly used directional comparison schemes follow.

*Directional-Comparison Blocking.* In this system, each of the terminals is equipped with tripping and carrier-starting relays. The tripping relays are directional toward the protected line and are set to respond to all faults on the protected line and 25 to 50% beyond. This is called an overreaching setting. The carrier signal is required to prevent tripping for faults in that 25 to 50% area. Tripping at *A* is blocked by a signal transmitted from *B* and received at *A*. Transmission of the signal is initiated by a carrier-starting relay that operates for faults outside the protected line section. Internal faults are cleared by the tripping relays at all terminals which have overriding control to stop all carrier transmission. A single-frequency on-off carrier may be used for both directions of transmission (*A* to *B* and *B* to *A*) because all carrier is turned off for an internal fault.

*Underreach Blocking.* This system uses a zone-extension scheme to limit, in the interests of economy, the number of distance units required. A relay set to cover zone 1 (the area from the relay location out to 80 or 90% of the protected transmission line length) is stepped, after a coordinating delay such as 4 ms to zone 2 reach (covers the entire line) provided blocking carrier is not received from other terminals. If carrier is received, zone extension is still carried out, but at a much later time (often 15 cycles), to provide backup coverage for remote bus line sections and apparatus. Different carrier frequencies are required for the two carrier channels. Station *A* carrier cannot be allowed to block station *A* tripping.

*Acceleration.* Zone-extension is again used with this system. A frequency-shift carrier channel is preferred because transmission through a fault on the protected line may be required. A guard frequency is transmitted during nonfault conditions. The protective relays are given a zone 1 setting. All faults on the protected line are seen by one or both of the relays at the two ends of the line. Each causes carrier to be shifted to a trip frequency.

Receiving trip frequency causes the zone 1 setting of each local relay to be extended to zone 2 distance immediately. All faults in the area of overlap of the two zone 1 settings will be cleared without regard to the carrier signal. End-zone faults (faults not covered by the zone 1 relays at one of the terminals) will be cleared at high speed and essentially simultaneously.

*Overreaching Transfer Trip.* A transfer-trip relaying system is identified as an "overreaching" or an "underreaching" system, depending on the setting of the directional distance relay that keys the frequency shift tone or carrier transmitter at each line terminal. If it has a setting that causes it to respond to faults on the protected line, and additionally to faults beyond the end of the protected line, it overreaches and the system is identified as an "overreaching-transfer-trip" system. This system is also called a "permissive-intertrip-overreach" system.

Tripping occurs when the distance relay operates at each terminal and a trip signal is received at that terminal. The distance relays at the two ends of the line cooperate to clearly identify a fault as being "internal" to the protected line or "external."

*Underreaching Transfer-Trip.* These schemes are equipped with directional distance relays set to respond to faults within approximately 80% of the protected line length. When they operate, they key the frequency-shift channel transmitter from "guard" to "trip" as well as immediately tripping the local breaker(s) without regard to action at the remote terminal.

The two categories of these systems are identified as "direct" and "permissive." In the "direct-underreaching-transfer-trip" system, receiving the channel trip causes tripping of the terminal breaker(s). No local fault-detector relay operation is required. Strictly speaking, the direct scheme is not a directional-comparison system, because operation of the zone 1 relay issues a command to trip all breakers associated with the protected line and no comparison takes place.

In the permissive scheme, a local directional distance relay supervises tripping. These schemes are also called "permissive-intertrip-underreach" systems. Note that the overreaching-transfer-trip systems require that a signal be received by the channel equipment in order for tripping to take place.

These systems are usually committed to channels that are not dependent upon the integrity of the protected power line itself such as pilot wires and microwave.

*Unblock System.* The unblock pilot relaying system is virtually identical to the overreaching-transfer-trip system but contains provision for allowing short time (100 to 150 ms usually) tripping when the channel fails, provided a local overreaching distance relay operates. Tripping of the transmission line prevents "loss of channel" from occurring on external faults. Loss of channel not accompanied by operation of a distance relay merely sounds an alarm to indicate that condition.

Each of the schemes above represents varying layers of complexity imposed on the basic concept of allowing one or more distance relays at each terminal to identify the existence of and the direction to a fault. Use of the pilot channel allows the two terminals to share this information and to initiate the appropriate action based upon the comparison. While the description is in terms of two-terminal applications, they may in general be applied to the protection of three terminal lines.

These systems incorporate subtle differences and small variations in their levels of security and dependability. They do differ in cost and capability and their choice is greatly influenced by personal choice and individual previous experience.

*Phase-Comparison Relaying.* This form of pilot relaying compares, over a communication channel, the instantaneous direction of current at the two ends of the transmission line. To allow the use of a single channel, some such systems use a combination of the individual phase currents to generate a single phase quantity for comparison. Others use a combination of the symmetrical components (positive, negative, and zero sequence) of the phase currents, and by applying appropriate weighting factors to each and adding the combination, a single-phase sinusoidal voltage is produced and converted to a square wave for comparison at the two terminals.

The concept of the scheme is that external faults will cause the local and received remote quantities to be essentially equal in magnitude but opposite in direction, while internal faults will cause them to be possibly different in magnitude but essentially in phase. In the comparison, the local quantity is delayed by an amount equal to the inherent channel delay, providing near-perfect coincidence for external faults.

The "segregated-phase-comparison" system compares the instantaneous direction of current at the two ends of the transmission line for each phase rather than utilizing some weighted combination of the currents or their symmetrical components. Each phase requires a separate channel. A local sinusoidal voltage proportional to phase current is converted, for each phase, to a square wave delayed by an amount dependent on channel delay and compared to the received remote quantity for the corresponding phase. Internal faults will produce essentially in-phase comparisons. External faults will produce comparisons essentially 180° out of phase. Considerable angular variation in these comparisons will still provide precise information regarding fault location.

Three channels are required for individual phase comparisons for a three-phase system. To ease the complexity of the channel requirement, it is possible to combine currents and use the "delta-ground" configuration. The delta comparison uses a voltage proportional to the difference of two currents (e.g., $I_A - I_B$). The ground comparison uses $3I_0$ current at the two ends of the transmission line.

***41. Generator Relaying.*** Generators are a vital part of a power system, and their protection deserves critical consideration. For the larger machines, 50,000 kW and above, a consistent pattern of protection has evolved. For the smaller machines, economics usually dictates that greater risks be accepted.

***42. Large-Machine Protection***

*Hazards.* The hazards against which protective devices guard are faults, unbalanced currents, loss of field, field ground, instability, and other miscellaneous phenomena that will be described later.

*Phase Faults.* Phase-fault protection is invariably provided by differential relays as shown in Fig. 16-47. By using identical ratio and accuracy-class current transformers, any "through"

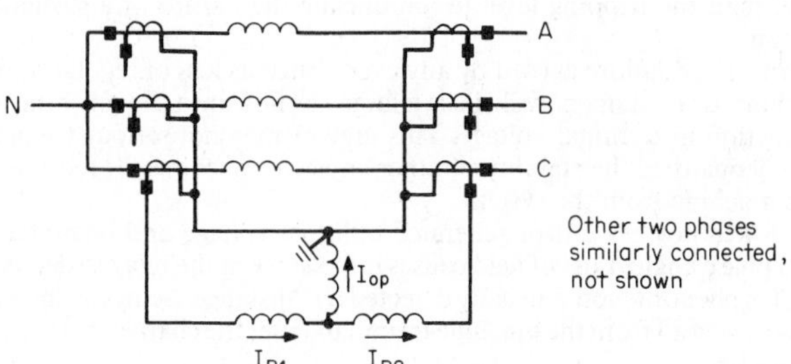


FIG. 16-47 Typical differential protection for generator.

phenomenon such as load, external faults, or power swings will produce essentially equal restraint currents $I_{R1}$ and $I_{R2}$. Operating current, $I_{OP}$, will be the difference of the two ct (current-transformer) error currents, or zero in the case of equal or negligible errors.

Internal faults will generally cause $I_{R2}$ to reverse with respect to $I_{R1}$ and $I_{OP}$ to equal the transformed total fault current. The relays that are usually applied here have a sensitivity that is dependent upon the restraint. For high through current, restraint is high and the required $I_{OP}$ is high, thereby restraining properly for possible high differences in error currents. For low internal fault current, restraint is much lower and the $I_{OP}$ required is much lower, allowing sensitive detection of the fault. With this concept, large differential current can be ignored where this is the proper response, and small differential current can be recognized where tripping is the desired response.

*Ground Faults.* Stator faults involving conductor contact with grounded elements may cause essentially no current flow or curent comparable to phase-fault levels, depending on the system neutral grounding. Most large machines are unit-connected, meaning the turbine, the generator and the transformer are treated as a unit, with no fault switching at generator voltage level. The low-voltage winding of the unit transformer is delta-connected, providing zero-sequence isolation from all other segments of the power system. The generator neutral is grounded through a high-impedance circuit, usually a distribution transformer loaded with a secondary resistor. This combination limits ground-fault current to a few amperes, *which is undetectable* by the generator differential relay. With this widely used grounding method, the generator neutral shift is dependent upon fault location. A ground fault at a generator terminal will cause full line-to-neutral voltage to exist between neutral and ground. The closer the fault to the neutral, the lower the magnitude of this voltage.

A relay connected across the secondary terminals of the distribution transformer will be able to detect this voltage. It can be given sufficient sensitivity to detect faults from the line terminal down to approximately 4% of the neutral. It must ignore the normal $3^D$ harmonic voltage, neutral to ground, to achieve this sensitivity.

Recent occurrences of neutral-to-ground faults which short out the ground-fault protection described above have led to consideration of several kinds of equipment to complement or replace it. These schemes use the $3^D$ harmonic voltage neutral to ground and sense its absence for a neutral-to-ground fault or they interject a current at another frequency and supervise its level. Neutral-to-ground faults rarely occur, and, in themselves, are of no consequence. A second ground fault will not only go undetected with neutral-to-ground fundamental-voltage-detection but also may destroy the generator.

*Unbalanced Faults.* Inherent in unbalanced faults is the fact that negative-sequence current is present. Flux associated with negative sequence rotates in a direction opposite to rotor rotation. This causes appreciable current flow in rotor structural parts that are not designed for such current and excessive heating occurs. A relay designed to respond in a similar way to the machine is applied for this protective function. It is $I_2^2t$ responsive, where $I_2$ is per-unit negative-sequence current (on the machine full-load current base) and $t$ is time in seconds. Generators vary in capability from $I_2^2t$ of 5 to 40 for negative-sequence currents in excess of full load, depending upon the type and size of machine.

The negative-sequence current relay protects the generator against a prolonged contribution to an unbalanced fault beyond the generator breaker. It often contains provision for "alarming" at a lower level than the tripping level to annunciate the hazard of a sustained unbalanced current condition.

*Loss of Field.* Field failure caused by any event, such as loss of regulator, opening of field breaker, field short, or field open, will cause a large var flow into the machine and generally a substantial reduction in terminal voltage. This may or may not seriously jeopardize the machine or it may jeopardize the stability of other adjacent machines. It requires detection and removal of the machine from the system.

Most loss-of-field devices utilize generator terminal voltage and phase current to obtain impedance and phase angle. Loss of field causes impedance at the relay to decrease and current to lead more. This phenomenon is usually detected by "distance" relays as shown in Fig. 16-48. Apparent ohms as viewed from the machine terminals enter the characteristic circle of the relay, causing it to operate.

All such relays are equipped with time delay to avoid undesired tripping on power swings. Some contain directional and undervoltage units to permit additional sensitivity to partial loss of field and allow coordination with regulator minimum excitation units, the machine capability curve, and the steady-state stability curve.

*Field Ground.* A single field ground causes no machine distress. Allowed to go uncorrected until a second field ground occurs, it can cause sufficient magnetic unbalance to produce *catastrophic* vibration. For "brush-type" machines, detection of the first ground is usually accomplished by detecting current flow in a high-impedance dc-measuring circuit to ground. AC is also used in other devices through the introduction of an ac voltage between the dc field circuit and ground and monitoring the low-magnitude normal current that is allowed to flow.

Where a "brushless" arrangement is used, no normal access exists to the field circuit because there are no nonrotating parts at field voltage level as there are in brush-type machines. Monitoring for grounds is achieved by periodically dropping, manually or automatically, pilot brushes onto collector rings provided for the purpose. The rings are connected to each side of the field. Measurements are then made of voltage and current to ground.

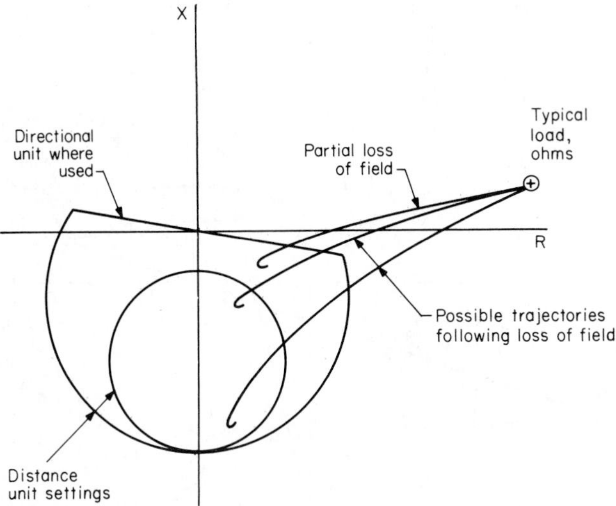


**FIG. 16-48** Detection of generator loss of field by measurement of impedance.

*Instability.* When the electrical center appears to be in the transmission system, distance relays applied to protect the transmission lines can be used to detect instability and to separate the two system parts. This can usually be done discriminatingly with out-of-step blocking at some locations and tripping at others: all done in the interests of maintaining as nearly as possible a generation-load match after the separation.

On the other hand, when the electrical center falls in the unit transformer or in the machine, the normal complement of relays applied to generator or transformer protection either will not detect the out-of-step condition or will be time-delayed to the point of being unreliable for this function. In these cases, out-of-step relaying is applied.

Figure 16-49 demonstrates the system behavior for a fault condition and for an out-of-step condition as viewed from the machine terminals and plotted in terms of a resistance-reactance diagram. Advantage is taken of the fact that emergence from the area between the blinder lines is on the same side as entry for normal fault clearing and on the opposite side from entry for an out-of-step condition. A blinder-type out-of-step relay trips for the latter case.

*Other Protection.* For the large important units, relaying is included to detect: motoring of the generator, inadvertent energization when the machine is at standstill, excessive volts per hertz that in turn causes excessive transformer and generator iron heating, stator and field overcurrent, and any malfunction not detected by the first line relaying (i.e. backup must be included to prevent catastrophic failure in the event of protective device malfunction).

***43. Small-Machine Protection.*** Much individual preference goes into the choice of protective equipment for small machines. For the very small, only voltage-restrained or supervised overcurrent relays may be used. In some cases only over-or undervoltage and frequency detection is applied. In other cases, protection approaching that for larger machines is used.

In some cases, compromises with the more elaborate protection are used. For very small machines, time-delayed overcurrent relays with insensitive settings are used in the differential configuration. Specially connected watt relays are used for a combination loss-of-field and out-of-step detection function.

***44. Motor Protection.*** Both synchronous and induction motors have protective requirements similar to those of generators. One important difference is that motors are accelerated by applying full or reduced voltage to their terminals, while generators are brought up to speed by their prime mover before being connected to the power system. Large starting current, then, is a normal expected phenomenon associated with motors that generators do not experience. Both types of devices contribute to external phase faults. Motor neutrals are not generally grounded, so no ground current will flow in an unfaulted motor.

Any protective device applied to protect a motor must ignore those conditions of starting

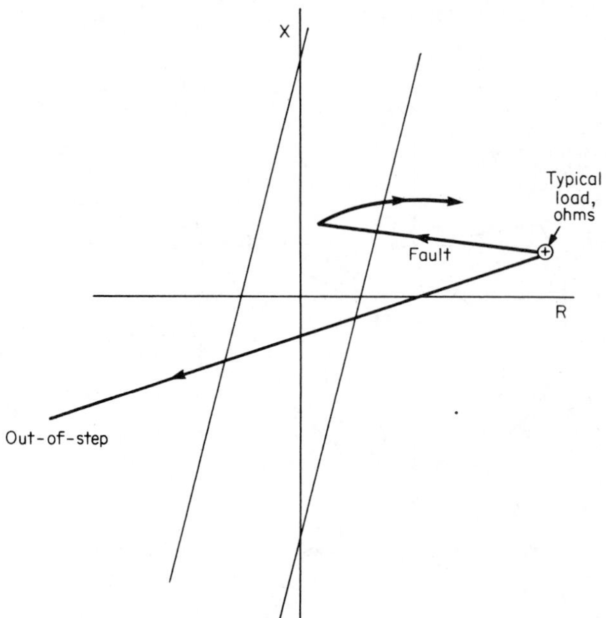


**FIG. 16-49** Blinder scheme for generator out-of-step detection.

current, load, and "through-fault" current, at the same time being able to sense low-magnitude internal-fault current. Differential relays perform this function well, often using a through-type current transformer with the two leads associated with each phase physically inserted through the ct window. Equal in-and-out currents generate no secondary voltage, so no operation of the relay connected to the ct secondary occurs. Internal faults cause unequal currents which generate a secondary voltage to cause instantaneous relay tripping. For the larger motors, differential relaying schemes identical to those used for generators are used for phase-fault detection.

A ground-relaying variation of the "through-type ct" scheme requires that all three phase conductors be inserted through the ct window. Only ground faults on the motor side of the ct can cause the relay to operate. This is a widely used scheme. Another important element for detecting a fault in a motor is an instantaneous-trip phase device. It must, of course, be set above motor starting current, but available phase-fault current magnitude will usually greatly exceed the starting current magnitude and very effective use can be made of this inexpensive and simple device.

*Thermal Protection.* Motors are usually equipped with devices that detect and relieve motor overloading. These are either devices that experience a heating effect comparable to that of the motor itself and act accordingly, or are relays that detect the temperature of a resistance-temperature detector (RTD) (through a measurement of its resistance) embedded between conductors in the stator slot. As the motor temperature increases beyond the allowable level, the RTD resistance rises and tripping of the controller takes place.

*Locked-Rotor Protection.* Neither of the relays used for thermal protection will, in general, protect a motor with a locked rotor. A time overcurrent relay receiving one phase current will normally perform this locked-rotor protective function adequately. In some special large-motor applications where permissible locked-rotor time is less than the required starting time, distance relays have been used successfully to run timers to protect for the locked-rotor condition based on a measurement of a combination of motor impedance and phase angle.

*Unbalance Protection.* Any degree of voltage unbalance at the motor terminals will manifest itself in the form of increased heating in the motor, well beyond that which could be predicted from the increase in stator current. This can be sensed by a relay which measures voltage unbalance or negative-sequence voltage. Buses that supply a large number of motors are usually equipped with this kind of protection. Phase-current-magnitude comparison has also been used very successfully on circuits supplying a single large motor.

*Synchronous-Motor Protection.* Because of the unique characteristics of synchronous motors, they are usually equipped with loss-of-field and out-of-step protection. This is often provided by a relay responsive to volt-amperes at an angle representative of the var flow into the motor on loss of field. It will also respond on loss of synchronism if the rate of pole slippage is compatible with the relay operating time or if the relay has a delayed resetting characteristic.

***45. Transformer Relaying.*** Protection of large transformers generally consists of differential protection, gas space or oil rate-of-rise of pressure, or gas accumulation detection plus time overcurrent relays for backup.

*Differential Relaying.* The differential-relaying concept (Fig. 16-50) is applicable to trans-

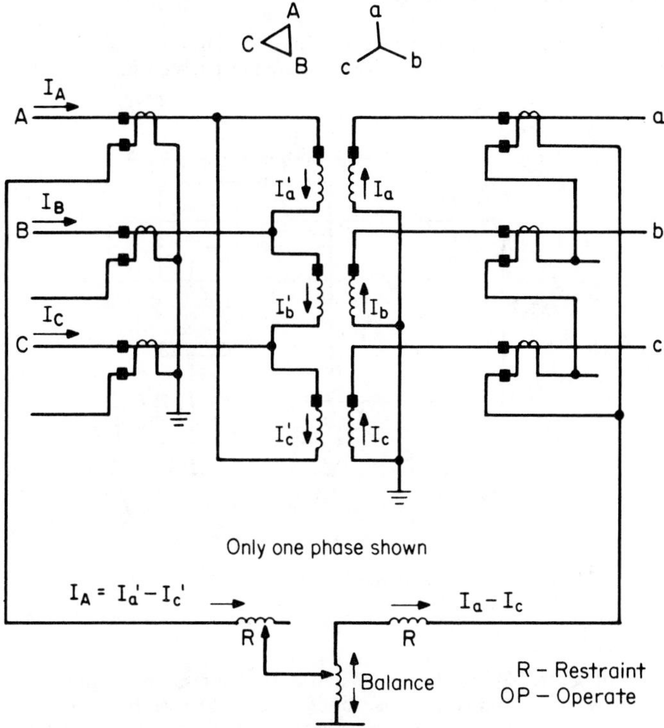


**FIG. 16-50** Concept of transformer differential protection.

former protection in a manner similar to that for generator protection, but distinct differences exist. While current transformers having essentially identical ratios and characteristics are obtainable in generator protection, no such identity is possible with the ct's used in transformer protection. Inherently, they *must* have different ratios and probably will have quite different characteristics. Also, inrush current on initial energization and following external fault removal is a very real phenomenon that must be accommodated by the transformer differential relay. These two circumstances, different ct's and inrush, force the use of a less-sensitive differential relay for the protection of transformers.

In addition to the fact that "through" conditions such as load or external faults produce different currents on the two sides of the transformer (to cause equal ampere turns in the windings), for a wye-delta or delta-wye transformer there is also a phase shift between the line currents on the two sides. Further, the standard ratios of ct's (such as 1200:5, 600:5, 100:5) used on the two sides of the transformer do not generally produce equal secondary currents for comparison by the differential relays for through conditions. As a result of these considerations:

1. Delta-side ct's are connected in wye.
2. Wye-side ct's are connected in delta.
3. Balance of input currents in the ratio of as much as 3:1 may be done inside the relay.
4. Inrush current is distinguished from internal-fault current in most transformer differential relays by using all harmonics, a combination of harmonics, or second harmonic only for inrush restraint.
5. Restraint is produced in proportion to the magnitude of the through current causing the relay to be sensitive at low current where ct error is likely to be low and to be insensitive at high current where ct error will be higher.

A widely used scheme for protecting a wye winding of a transformer against ground faults is shown in Fig. 16-51. The auxiliary transformer is carefully chosen with a ratio that will mini-

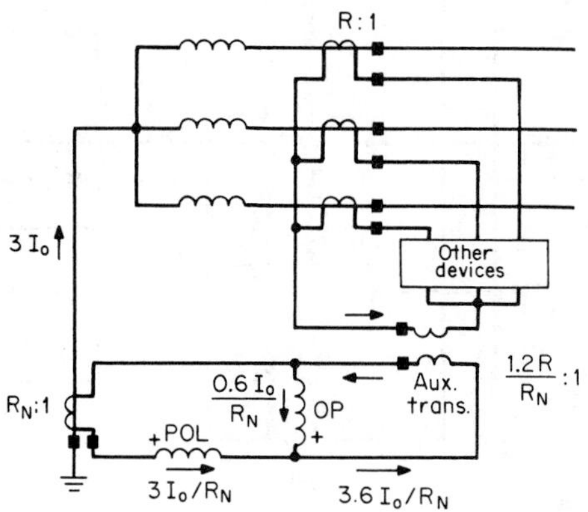


**FIG. 16-51** Transformer wye-winding differential protection.

mize the effect of ct error for external faults and to force a restraint condition (currents not flowing into the winding polarity markers simultaneously) to exist. Internal faults produce a reversal in the operating current direction with respect to the polarizing (reference) current direction causing the relay to operate. Another common application uses a time overcurrent relay supplied by a neutral ct connected in a wye-winding ground connection. It must be time-coordinated with other ground relays on the power system connected to the wye winding. Where differential relays are used, the primary function of this neutral ground relay is to back up these other devices.

A netural-ground relay may accomplish a primary (or first line) relaying function where low-resistance grounding is used and high-voltage fuses are used. The typical fuse size required for full load capability will not detect a low-voltage winding failure to ground in such a case. The ground relay will, depending on fault-current level. Remote tripping of a breaker feeding the fused transformer will be required. Tripping of a low-voltage breaker will not clear this type of fault.

*Rate of Rise of Pressure or Gas Accumulation.* Depending on whether a transformer is designed to have a nitrogen space above oil or to have a "conservator tank" and be completely filled with oil, use will be made of a rate of rise of gas pressure or a rate of rise of oil pressure device in larger transformers. Normal load cycling causes pressure change, but the rate of change is moderate. Faults under oil cause a much higher rate of change, and this distinction allows this type of device to distinguish between load change and faults.

Gas-accumulation relays collect any gas generated under oil by arcing or excessive temperature and base their fault detection on the extent of this collection.

## POWER-LINE CARRIER

*By J. GOHARI*

### Carrier Current

***46. Carrier current*** provides a means of conveying speech, metering indications, control impulses, etc., from one station to another by existing transmission lines without interfering with their normal function of transmitting power.

*Fundamental justification* lies in utilizing existing power lines of sturdy construction to gain reliability, economy, or special characteristics. High-voltage lines are so constructed that they are less affected by the elements than any other circuits except underground cable.

***47. Elements of a carrier channel*** are the sending terminal assembly including line matching and tuning, a coupling means, receiving-station coupling and terminal assembly, and the power line.

***48. Coupling capacitors*** are the most widely used and effective coupling means, the paper capacitor being the standard carrier coupling device. Made up of a large number of sections connected in series to obtain the proper voltage, they are stacked mechanically to provide units in the higher-voltage insulation classes. Typical units range in voltage and capacitance from 46 kV, 0.015 μF, to 765 kV, 0.004 μF. The capacitor is mounted on a metal base to provide convenient installation and space for connection to its lower terminal. The base contains a 60-Hz drain coil and may also contain protective gaps, grounding switches, and part or all of the coupling network.

Coupling capacitors may be provided with a tap or an auxiliary capacitance in their base, which is used as a voltage source for a potential device. This tap is so arranged that it will be maintained at a fixed nominal voltage above ground regardless of the voltage class of the unit.

***49. Line tuning*** is required to tune out the reactance of a capacitor with a suitable inductance. The simplest application involves the coupling of a single frequency between a single line conductor and ground using a single coupling capacitor. The capacitor is series-resonated with a variable inductor at the operating frequency of the carrier terminal equipment, thus providing an efficient path for coupling the carrier signal to the line conductor (Fig. 16-52).

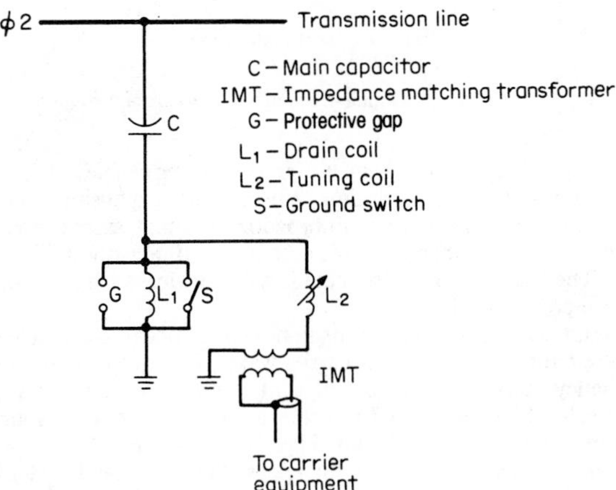


**FIG. 16-52** Single-frequency line-to-ground coupling.

***50. The carrier terminal*** with its coaxial cable is matched to the impedance of the power line by an adjustable impedance-matching transformer. Impedance matching of the transmitter, receiver, and the transmission line is required to allow maximum transfer of energy. Transmission line characteristic is defined in its simplest form, for a lossless circuit, to be

$$Z_0 = \sqrt{l/c}$$

where $l$ and $c$ represent the distributed inductance and capacitance of the line. Characteristic impedance for a single overhead conductor is $Z_0 = 138 \log (2\, h/r)$ where $h$ is the height of the conductor above ground and $r$ is the conductor radius. Impedance mismatch will result in undesired attenuation because of signal reflection. The magnitude of the loss is defined as

$$\text{Mismatch loss} = 20 \log \frac{Z_0 + Z_1}{2Z_0Z_1} \quad db$$

where $Z_0$ = characteristic impedance
$Z_1$ = termination impedance

The coupling capacitor may be used for more than one frequency if more tuning elements are added.

***51. Transmitters*** generally require a resistive load to be capable of modulation without excessive distortion. Therefore, where there is more than one transmitter coupled to the line at a common point, there should be a resonant path for each frequency. To permit independent tuning, antiresonant traps must be inserted in each path to reject all except the resonant frequency (Fig. 16-53).

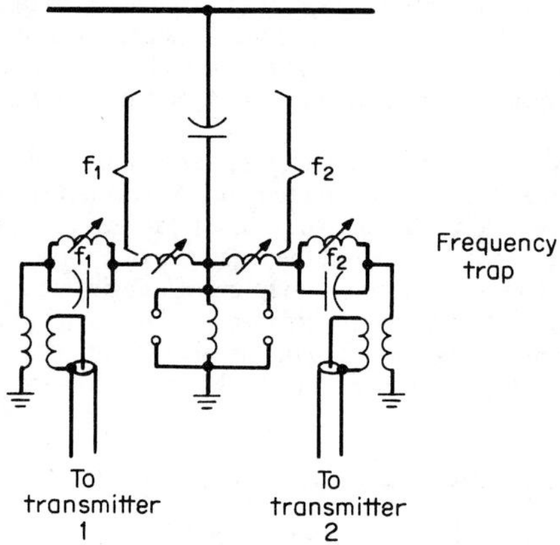


**FIG. 16-53** Double-frequency line-to-ground coupling.

*The practical upper limit for resonant tuning* is two frequencies. Above this, broadband tuning provides a most satisfactory solution. Where future expansion is expected, it may be preferable initially to design the circuit with broadband tuning, since a reasonable number of additional carrier circuits may then be introduced at any time without disturbing the existing circuit or circuits. The line tuner may also include safety devices such as a protective gap, 60-Hz drain coil, and grounding switch.

***52. Line traps*** are used to make the transmission line appear as a simple two-terminal line used in telephone circuits. They direct the carrier wave over a given circuit, increase efficiency, smooth out frequency characteristics, minimize interference, prevent interruption of the communication channels when the ground switches are closed, block off spur lines, and allow transmission during a nearby external fault (Figs. 16-54 and 16-55).

*Commercial traps* are available with inductances of approximately 180 μH to 1.8 mH, but there is no theoretical limit to the inductance which can be provided. Size, weight and cost limit

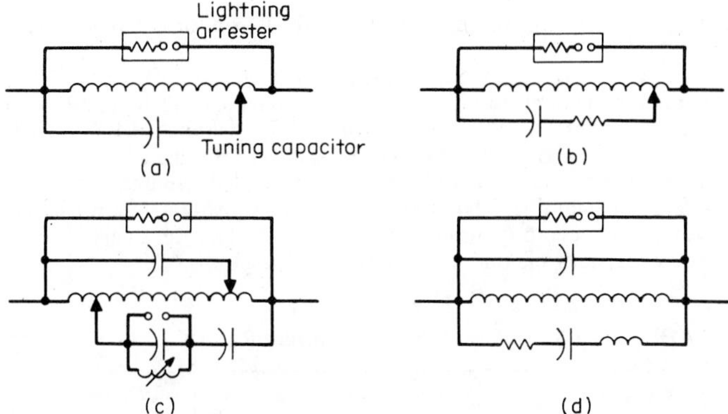


**FIG. 16-54** Line-trap schematics. *(a)* Single frequency, high $Q$; *(b)* single frequency, low $Q$; *(c)* double frequency; *(d)* wide band.

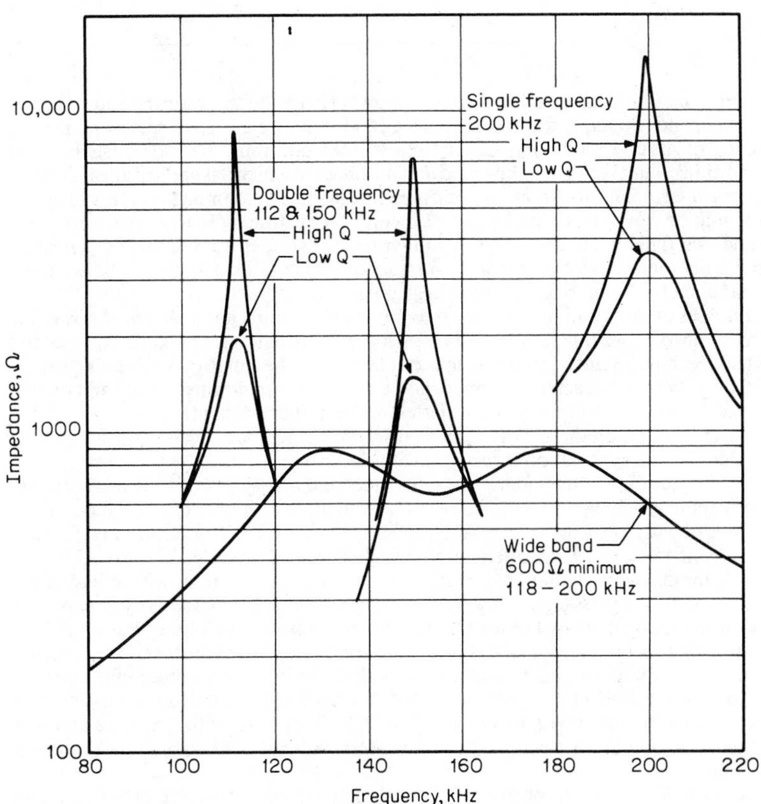


**FIG. 16-55** Typical impedance characteristics for single-frequency, double-frequency, and wide-band traps. Main coil inductance is 0.265 mH.

production to an approximate 2-mH range. They can be tuned to one or two frequencies or may be broadbanded.

Since a wave trap is connected directly to the power line, it must be rated for the full operating current and should be able to withstand the same dynamic and thermal-limit current as the other station equipment. Operating currents are within the range of 100 to 3000 A. In some instances, coils rated for above 2 mH have been installed to eliminate the need for tuning capacitors. This eliminates trouble caused by the possibility of the tuning capacitors being ruined by overvoltages or lightning. The present cost of these extremely large inductances precludes their use at higher operating currents. Table 16-14 is a short summary of continuous rating and short-circuit capacity of commercially available line traps.

**TABLE 16-14** Typical Line Trap Current Ratings

| Continuous current rating, A | Symmetrical short-circuit current, 2 s | Asymmetrical fully offset current |
|---|---|---|
| 400 | 15,000 | 42,000 |
| 800 | 20,000 | 56,000 |
| 1200 | 36,000 | 102,000 |
| 1600 | 44,000 | 125,000 |
| 2000 | 63,000 | 178,000 |
| 3000 | 63,000 | 178,000 |
| 4000 | 63,000 | 178,000 |

***53. Attenuation*** is a measure of the loss of energy between the transmitting and receiving terminal and depends on many factors: frequency, conductor size and spacing, line configuration, presence of ground wires or parallel circuits, transpositions, ground resistivity, and weather conditions. The type of coupling used and the phase to which it is applied affect the total attenuation from terminal to terminal and must be considered in planning the carrier channel.

Line attenuation can be predicted by use of a theoretical treatment of carrier channel known as modal analysis. This method is in effect similar to the symmetrical component generally utilized to predict the behavior of transmission lines at 60 Hz. Calculations can be accurate, provided that end effects and reflections are negligible.

Natural modes of propagation are applied to the phase currents and voltages of powerline carrier signals between two line terminals. For each mode, there is a set of phase voltages and currents which bear a constant relationship to each other along the length of the line. Each mode is represented by its own characteristic impedance, a constant specific attenuation, and its own phase velocity. Practically no energy is transferred between individual modes.

*Mode 1.* The current in the middle phase is twice that in the outer phase and flows in the opposite direction. The attenuation is the least with this mode.

*Mode 2.* The currents in the outer phases are equal and opposed, while the middle phase does not contribute anything to the transmission. Attenuation is greater than for mode 1.

*Mode 3.* The currents flow in the same direction in all three phases and are equal in magnitude. Owing to earth return, attenuation is highest with this mode.

Please note the change in definition of mode 1 and mode 3 in this text. When modal analysis is studied in references, it is essential to ascertain the definition of each mode as used in the text.

Various forms adopted in practice for transmission over this type of line generally represent a combination of two or three modes. To achieve the lowest possible losses, the available transmission power should be fed to the line by mode 1. This becomes more important the longer the line and the higher the carrier frequency. At a certain distance from the transmitter, mode 2 and 3 components are attenuated to a low level. Therefore, with long lines and high carrier frequencies, only proper coupling using mode 1 will cause the signal to arrive at the receiving point.

***54. Line reflection*** occurs at the end of a line, at the junction of two or more lines, and at substations. Spur lines having attenuation of 10 dB or more do not generally produce either a serious loss or a serious distortion of the frequency characteristics. Short spur lines of low total attenuation produce severe losses over a wide frequency range.

*55. **Transmission-line loop circuits*** may cause a partial cancellation of carrier energy arriving over two sides of the loop or, at least, distortion of a carrier telephone channel. Corrective measures consist in the judicious use of line traps or line terminations, changing the coupling mode, or shifting the carrier frequencies.

*56. **Carrier-frequency noise*** on the lower-voltage transmission lines is caused chiefly by defective insulators and loose hardware and on the higher-voltage lines by corona discharge. All circuits are subject to the noise effects of atmospherics, line faults, switching surges, and faulty apparatus. Noise increases substantially in bad weather as the combined result of lightning, corona from drops of water or particles of snow or ice on the conductor, and leakage over insulators. Background noise is generally greater at the lower frequencies, but disturbances and lightning storms may produce high noise levels throughout the entire spectrum. Noise during thunderstorms can produce a value 10 times or more the fair-weather noise. All these factors must be considered for best utilization of the transmission line.

*The proper characteristic of the noise* must be considered for each application, and receiver bandwidth as well, since the noise response is usually a function of bandwidth.

It is not ordinarily possible to reduce appreciably the noise level present at a given receiving point in a carrier system. Therefore, the only practical way to improve signal-to-noise ratio is to raise the signal level at the receiving point. It is not usually feasible to raise signal levels by increasing the transmitting power, because appreciable gains in terms of decibels require inordinately large increases in power. Although it is practical to raise the signal level from a 1-W transmitter to 100 W or from a ⅒-W transmitter to 10 W, raising the level of a 10-W transmitter by 20 dB requires an increase to 1000 W, or 100 times the original power. A much more practical solution is to reduce the channel attenuation by using every means available.

## Carrier Communication

*57. **Two-frequency duplex telephony*** uses different frequencies for the two directions. This is a simple scheme in which transmitters and receivers at both ends operate continuously during a conversation. This system is readily adaptable to extension over wire-line telephone lines through private branch exchanges (PBXs), but it is not fundamentally party-line, nor does it conserve frequencies. Hence, each length requires an additional two frequencies within the carrier spectrum unless widely separated by many line sections.

*58. **Single-frequency simplex telephony*** uses one frequency for both directions, conserves frequencies, provides party-line service, and is accomplished with equipment similar to two-frequency duplex, except that a relay is added to transfer from send to receive. A pushbutton in the handset controls this relay. The requirement for manual control prevents operation through a PBX.

*59. **In single-frequency automatic simplex,*** the carrier transmitter is started and stopped automatically by the presence or absence of sound at the microphone. Voice-operated relays (either electronic or mechanical) start and stop the transmitter as well as block and unblock the receiver in correct sequence and with the proper rate of change to accomplish a smooth transition and also to hold the transmitter inoperative on local noise while signals are being received.

*60. **Amplitude modulation*** is used for the majority of carrier telephone installations at the present time, utilizing both double-sideband and single-sideband transmission. With double-sideband transmission, both the carrier and two sidebands are transmitted. Therefore, it will generally occupy a frequency band of 10 kHz.

*61. **With single-sideband transmission,*** the frequency band assigned to a channel will be only half as wide. By using both upper and lower sidebands in the single-sideband system, two telephone-quality voice circuits can be obtained in the same bandwidth as one double-sideband channel. The carrier itself need not be transmitted but may be reinserted at the receiving end. The reduction in bandwidth and improvement in signal-to-noise ratio are important advantages.

*62. **Frequency modulation*** is also applicable to power-line carrier. Because of the limited frequency bands available, some of the advantages of this method must be sacrificed by restricting the modulation index. To avoid distortion, the frequency characteristic of the power system must be relatively flat throughout the width of both sidebands.

*63. **Carrier transmitters*** are generally arranged for mounting in relay racks or rack-type cabinets. Attention is given to harmonic suppression by using high-capacitance tank circuits; to frequency stability by using crystal-controlled or stabilized oscillators; and to high modulation percentages by using automatic gain control circuits. Output power requirements range from about 1 W for short and medium hauls to 10 W for most applications. In two-channel single-sideband systems, power as high as 100 to 200 W is common. Most sets today are designed to operate from a standard 48- or 125-V dc source but have an optional alternative power pack, enabling them to be operated from either 115-V 60-Hz or 250-V dc supplies.

*64.* ***Receivers*** may be simple for short hauls where maximum frequency utilization is not required. For general use, selectivity should be as high as obtainable, consistent with commercial fidelity. Automatic gain control circuits which hold constant output over at least a 40-dB range are usually incorporated. Many receivers for long-range work are equipped with special squelch circuits to block the receiver during periods of no received carrier.

*65.* ***Calling methods*** include the use alone, or in combination, of loudspeakers, code ringing, and selective ringing. Code ringing normally utilizes an audible tone with audio filter and time-delay relay at the receiver to prevent false operation from voice or from line noise. Selective ringing is obtained similarly by the addition of standard selectors, in which case a distinctive "ring back," or revertive ring, indicates that selective equipment at the call station functions properly to actuate the ringer.

## Carrier Relaying

*66.* ***Directional-Comparison Relaying.*** Carrier-current channels used in place of pilot wires in directional-comparison relay schemes have proved effective, fast, and economical. The resulting system is selective as to line sections protected, and the function delegated to the carrier is only to close a relay. A typical arrangement consists of directional relays at the two ends of a line section to determine whether fault power flows into or out of the section. If fault current flows into one end and out the other, the line section is sound and the fault is external. If fault current flows into the line section at both ends, an internal fault must be present. Carrier transmitters at the two ends of the line section are each arranged to transmit only when fault power flows out of the section, in which case the line is sound and carrier receiving relays operate to prevent tripping of the circuit breakers. If fault current flows inward at each end, neither set transmits and the circuit breakers at both ends are allowed to trip from fault detector relays. Each line section so protected is independent of all other line sections. The carrier system is required to transmit and receive over its own line section during an overload or fault outside its section. Therefore, line traps at the ends of the section are necessary to prevent an external fault occurring a short distance outside the section from short-circuiting the carrier path. Receivers must not respond to the high-frequency arc component of the fault current. Consequently, they are made relatively insensitive. Transmission over a line section is not required in both directions at the same time, which permits a single-frequency two-way channel. Total elapsed time from the occurrence of the fault to energizing the trip coil is usually 1 to 3 c, but even greater speeds are required as the transmission system becomes more sophisticated.

*67.* The requirement for ever greater speeds has fostered the development of *static relaying,* in which all relays are replaced by their equivalent electronic circuits. Overall operating time is ⅜ to 1 c.

*68.* ***Phase-comparison relaying,*** by balancing the current phase angles at the near and far ends of the transmission line, is the counterpart of differential relaying.

*69.* ***Transfer trip relaying*** provides a means for high-speed control of a remote circuit breaker. It is especially applicable for limiting transformer damage, in those cases where a high side circuit breaker is used, by fast operation of the far-end circuit breaker. Although the carrier equipment is basically the same as used for telemetering, special precautions must be taken in the design of the entire system to assure positive action when needed with maximum protection against false tripping. A continuously transmitted guard channel is usually employed with line traps at both ends and equipment designed for maximum rejection of noise and spurious signals. For tripping the remote circuit breaker, the channel frequency is shifted a small amount, which causes the remote trip relay to operate.

*70.* ***Frequency-shift blocking*** may be used where power-line sections have a high attenuation or other irregularities. Where the attenuation is between 40 and 60 dB, frequency-shift equipment often provides the only satisfactory channel. Because of the narrow band used, the response time of such channels is somewhat longer than for the assemblies previously described. This additional delay is usually secondary to reliability in importance, but the delay does limit the application of frequency-shift blocking in the relay system. Wide-band frequency-shift equipment with as little as 4-ms operate time is commercially available for special applications. A continuous carrier-frequency signal is transmitted over the high-voltage line. Upon occurrence of a fault, frequency is changed slightly under the control of the protective relays. This slight change in frequency actuates a relay at the receiving end, which allows blocking or tripping as required. This system has the advantage of being relatively insusceptible to interference. Also, continuous channel supervision is readily provided.

*The receiver limiter and discriminator circuits* which are employed ensure that the receiver output will be actuated only by a carrier signal of exactly the right frequency and also having more power than any other signal having only slightly different frequencies. Therefore, this equipment can be used on power circuits where other types of carrier would be entirely inadequate.

## Other Uses for Power-Line Carrier

***71. Carrier may be used for telemetering and supervisory control.*** These channels cover distances ranging from one line section of a few miles in length up to many sections totaling several hundred miles as required for power-interchange readings between large interconnected systems. Choice of the carrier method depends on channel length, available frequency band, speed of response, accuracy required, and whether or not full-time operation is essential. Carrier channels for remote control, supervisory control, automatic load control, or telemetering generally use a narrow bandwidth, compared with telephone channels. For short-haul work, simple on-off keying is adequate and may sometimes be accomplished as a secondary function of telephone or relaying equipment. For long-haul work, where both attenuation and noise are important considerations, better performance results from using some form of continuous carrier.

***72. The straight continuous-wave system*** consists of an unmodulated transmitter keyed by supervisory transmitting contacts or by the sending meter.

***73. In the frequency-shift system,*** transmission on one frequency or the other is continuous, which minimizes the effect of interference from noise or other carrier channels and permits continuous compensation by automatic-volume-control circuits for a change in signal level. A typical system using a 25-W transmitter is capable of producing reliable indications through attenuations up to about 60 dB, representing several hundred miles of average transmission line with intermediate stations. When several quantities are to be transmitted between two points, economy in both equipment and spectrum results from multiplexing with audio tones, one of which is used for each quantity. The tones are shifted or frequency-modulated as required for each individual telemetering indication, and the entire group is used to modulate the carrier.

***74. Supervisory control of unattended stations*** requires transmission of signals of the same general type as for telemetering. Since it is generally desired to control many operations, some form of multiplexing is indicated. Tones, impulses, or combinations of both have been successfully employed. In general, a similar type of system is used for conveying the indications back to the control point.

***75. Selected References***

IEEE Guide for Power-Line Carrier Applications, IEEE Std. 643-1980.

1960–1974 Index; *AIEE/IEEE Trans. Power Appar. Syst.*, 75-CH-1048-8-PWR.

PLC Bibliography; *IEEE Paper* 31-TP-65-25.

Application Guide for Power Line Carrier; *IEEE Tech. Paper* 54-12.

Natural Modes of PLC on Horizontal 3-Phase Line; *IEEE Paper* 63-936.

*Communications System Engineering Handbook;* New York, McGraw-Hill Book Company, 1967.

Power Line Carrier Application Guide; General Electric Co. (Available to power company personnel on request.)

Applied Protective Relaying; Westinghouse Electric Corporation, 1976.

# SECTION 17
# SUBSTATION DESIGN

**Joseph Basilesco**
*Consulting Engineer; President, Basilesco Consultants, Wakefield, MA; Member, IEEE; Member, Power Systems Communications Committee; Past Chairman, Microwave Subcommittee of Power Systems Communications Committee; Member, CIGRE; Convener, CIGRE Working Group WG 23-06, Power Station Auxiliaries; Author of IEEE papers on Protective Relaying, Control, and Communications.*

**CONTENTS**

*Numbers refer to paragraphs*

***1. System Components.*** In large, modern ac power systems, the transmission and distribution system functions to deliver bulk power from generating sources to the users at the load centers. Transmission facilities generally include generating stations, step-up transformers, interconnecting transmission lines, switching stations, and step-down transformers. The distribution system comprises primary distribution lines or networks, service transformer banks, and secondary lines or networks, all of which service the load areas.

***2. Design Objective.*** As an integral part of the transmission system, the substation or switching station functions as a connection and switching point for transmission lines, subtransmission feeders, generating circuits, and step-up and step-down transformers. The substation design objective is to provide maximum reliability, flexibility, and continuity of service and to meet these objectives with the lowest investment costs that satisfy system requirements.

***3. Voltage Levels.*** System requirements include the selection of optimum voltage levels, which depend on the load requirements and transmission-line distances involved. Many large thermal and nuclear power plants are located at great distances from the load centers in order to capitalize on low site costs, ample cooling-water supply, economical fuel supply, and less stringent environmental considerations. For these reasons the use of transmission voltages as high as 765 kV is becoming more common.

The substations used in distribution systems operate at voltage classes from 13.8 to 69 kV. Transmission substations serving bulk power sources operate at voltages from 69 to 765 kV. Voltage classes used in the United States for major substations include 69, 115, 138, 161, 230, and 287 kV (considered high voltage or HV class) and 345, 500, and 765 kV (considered "extra high voltage" or EHV class). Even higher voltages now in various stages of planning or construction include 1100 and 1500 kV and are referred to as "ultra high voltage" or the UHV class.

***4. Design Considerations.*** Many factors influence the proper selection of the type of substation for a given application. The most appropriate type of station depends on such factors as voltage level, load capacity, environmental considerations, site space limitations, and transmission-line right-of-way requirements. In addition, design criteria can vary among systems.

With the continuing general increase in the cost of equipment, labor, land, and site treatment, every effort must be made to select criteria that represent the best compromise to satisfy system requirements at minimum costs. Since the major substation costs are reflected in the power transformers, circuit breakers, and disconnecting switches, the bus layout and switching arrangement selected will determine the number of switches and power circuit breakers required. The choice of insulation levels and coordination practices affects cost considerably, especially at EHV. A drop of one level in BIL (basic insulation level) can reduce the cost of major electrical equipment by thousands of dollars. A careful analysis of alternative switching schemes is essential, especially at EHV levels, and can also result in considerable savings by choosing the minimum equipment to satisfy system requirements.

A number of factors must be considered in the selection of bus layouts and switching arrangements for a substation to meet system and station requirements. A substation must be reliable, economical, safe, and as simple in design as possible. The design should provide a high level of service continuity. The design should also provide for further expansion, flexibility of operation, and low initial and ultimate costs. Means should be provided for maintaining lines, breaker, and switches with no service interruptions or hazard to personnel.

The physical orientation of the transmission-line routes often dictates the substations' location and bus arrangement. The selected site should be such that a convenient arrangement of the lines can be accomplished.

For reliability, the substation design should prevent total substation shutdown caused by breaker failure or bus faults and should be such as to permit rapid restoration of service after a fault occurs. Planned arrangement of lines with sources connected opposite to loads improves reliability. The layout must permit future additions and extensions without interrupting service.

***5. Main Bus Connections.*** The substation scheme selected determines the electrical and physical arrangement of the switching equipment. Different bus schemes come about as emphasis is shifted between the factors of reliability, economy, safety, and simplicity as warranted by the function and importance of the substation.

The substation bus schemes most often used are:

**a.** Single bus
**b.** Double bus, double breaker
**c.** Main and transfer bus
**d.** Double bus, single breaker
**e.** Ring bus
**f.** Breaker and a half

Some of these schemes may be modified by the addition of bus-tie breakers, bus sectionalizing devices, breaker bypass facilities, and extra transfer buses. Figures 17-1 to 17-6 show one-line diagrams for some of the typical schemes listed above.

***6. Single Bus.*** The single-bus scheme (Fig. 17-1) is not normally used for major substations. Dependence on one main bus can cause a serious outage in the event of breaker or bus failure. The station must be deenergized in order to carry out bus maintenance or add bus extensions. Although the protective relaying is relatively simple, the single-bus scheme is considered inflexible and subject to complete outage.

***7. Double Bus, Double Breaker.*** The double-bus–double-breaker scheme (Fig. 17-2) requires two circuit breakers for each feeder circuit. Normally each circuit is connected to both buses. In some cases, half of the circuits could operate on each bus. For these cases bus or breaker failure would cause the loss of half of the circuits. The location of the main buses must be such as to prevent faults spreading to both buses. The use of two breakers per circuit makes this scheme expensive. However, it represents a high order of reliability when all circuits are connected to operate on both buses.

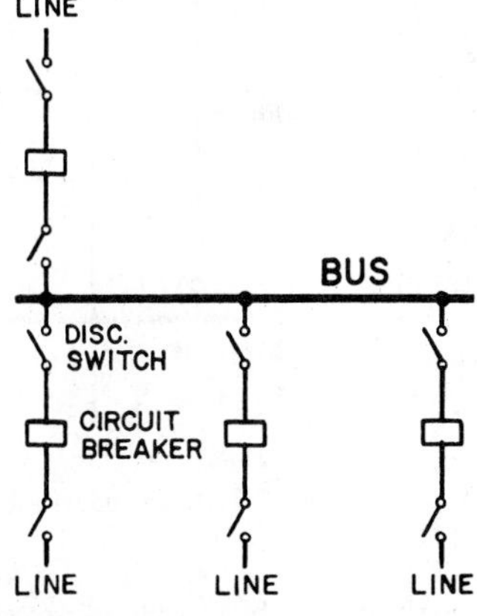


**FIG. 17-1** Single bus.

***8. Main and Transfer Bus.*** The main- and transfer-bus scheme (Fig. 17-3) adds a transfer bus to the single-bus scheme. An extra bus-tie circuit breaker is provided to tie the main and transfer buses together.

When a circuit breaker is removed from service for maintenance, the bus-tie circuit breaker is used to keep that circuit energized. Unless the protective relays are also transferred, the bus-tie

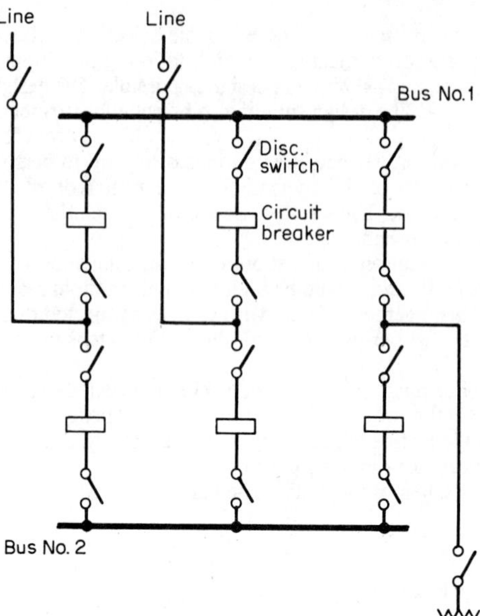


**FIG. 17-2** Double bus, double breaker.

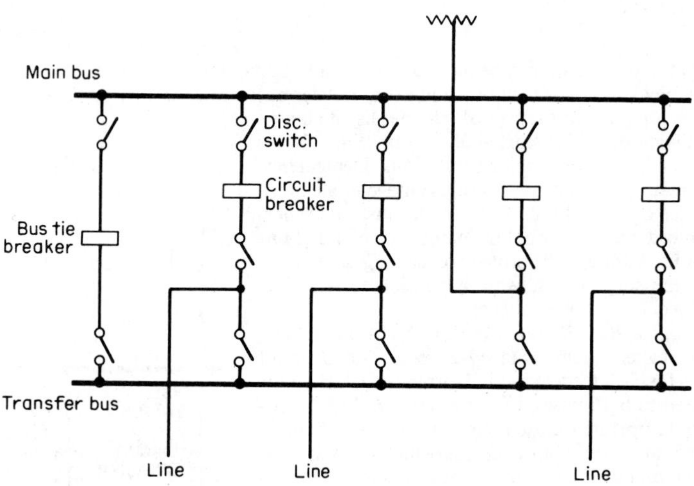


**FIG. 17-3** Main and transfer bus.

relaying must be capable of protecting transmission lines or generators. This is considered rather unsatisfactory since relaying selectivity is poor.

A satisfactory alternative consists of connecting the line and bus relaying to current transformers located on the lines rather than on the breakers. For this arrangement line and bus relaying need not be transferred when a circuit breaker is taken out of service for maintenance, with the bus-tie breaker used to keep the circuit energized.

If the main bus is ever taken out of service for maintenance, no circuit breakers remain to protect any of the feeder circuits. Failure of any breaker or failure of the main bus can cause complete loss of service of the station.

Disconnect switch operation with the main- and transfer-bus scheme can lead to operator error, injury, and possible shutdown. Although this scheme is low in cost and enjoys great popularity in many countries, it does not provide the high degree of reliability and flexibility required by power utilities in the United States.

***9. Double Bus, Single Breaker.*** This scheme uses two main buses, and each circuit includes two bus selector disconnect switches. A bus-tie circuit (Fig. 17-4) connects to the two main buses

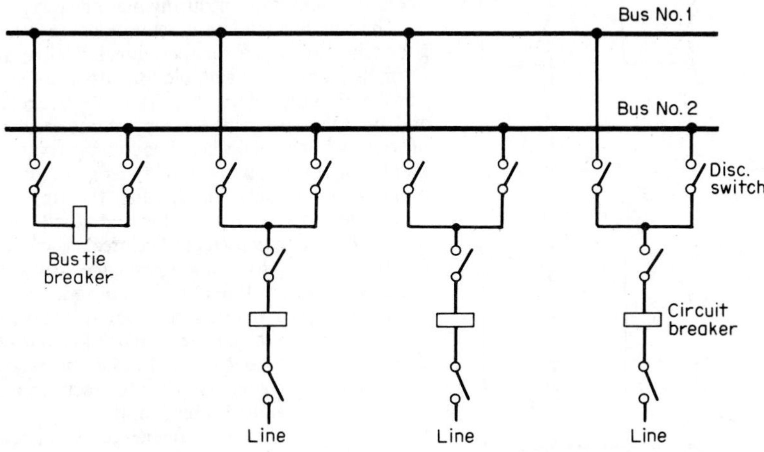


**FIG. 17-4** Double bus, single breaker.

and, when closed, allows transfer of a feeder from one bus to the other bus without deenergizing the feeder circuit by operating the bus selector disconnect switches. The circuits may all operate from the No. 1 main bus, or half of the circuits may be operated off either bus. In the first case, the station will be out of service for bus or breaker failure. In the second case, half of the circuits would be lost for bus or breaker failure.

In some cases circuits operate from both the No. 1 bus and No. 2 bus and the bus-tie breaker is normally operated closed. For this type of operation a very selective bus protective relaying scheme is required in order to prevent complete loss of the station for a fault on either bus.

Disconnect switch operation becomes quite involved, with the possibility of operator error, injury, and possible shutdown. The double-bus–single-breaker scheme is poor in reliability and is not normally used for important substations.

***10. Ring Bus.*** In the ring-bus scheme (Fig. 17-5) the breakers are arranged in a ring with circuits connected between breakers. There are the same number of circuits as there are breakers. During normal operation, all breakers are closed. For a circuit fault, two breakers are tripped, and in the event one of the breakers fails to operate to clear the

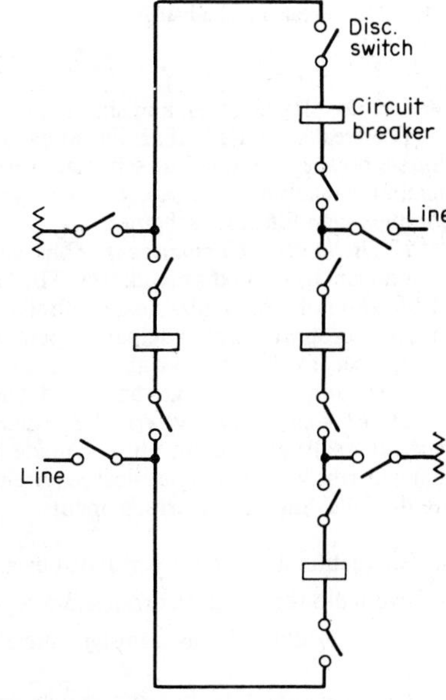


**FIG. 17-5** Ring bus.

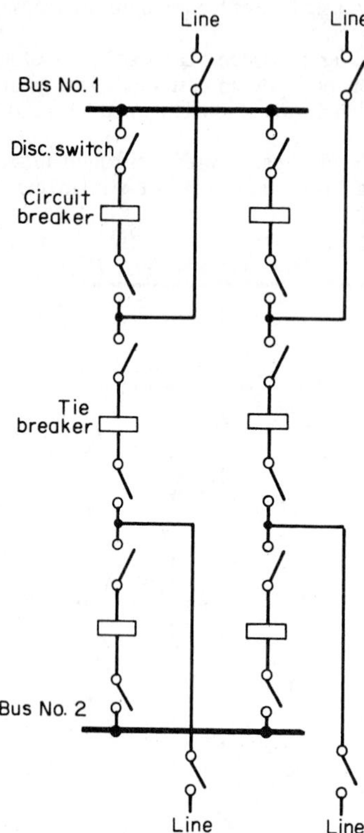


**FIG. 17-6** Breaker and a half.

fault, an additional circuit will be tripped by operation of breaker-failure backup relays. During breaker maintenance, the ring is broken, but all lines remain in service.

The circuits connected to the ring are arranged so that sources are alternated with loads. For an extended circuit outage, the line disconnect switch may be opened and the ring can be closed. No changes to protective relays are required for any of the various operating conditions or during maintenance.

The ring-bus scheme is economical in cost, has good reliability, is safe for operation, is flexible, and is normally considered suitable for important substations up to a limit of five circuits. Protective relaying and automatic reclosing are more complex than for previous schemes described. It is common practice to build major substations initially as a ring bus; for more than five outgoing circuits, the ring bus is usually developed to the breaker-and-a-half scheme.

***11. Breaker and a Half.*** The breaker-and-a-half scheme (Fig. 17-6), sometimes called the three-switch scheme, has three breakers in series between the main buses. Two circuits are connected between the three breakers, hence the term *breaker and a half.* This pattern is repeated along the main buses so that one and a half breakers are used for each circuit.

Under normal operating conditions all breakers are closed and both buses are energized. A circuit is tripped by opening the two associated circuit breakers. Tie breaker failure will trip one additional circuit, but no additional circuit is lost if a line trip involves failure of a bus breaker. Either bus may be taken out of service at any time with no loss of service. With sources connected opposite loads, it is possible to operate with both buses out of service. Breaker maintenance can be done with no loss of service, no relay changes, and simple operation of the breaker disconnects.

The breaker-and-a-half arrangement is more expensive than the other schemes, except the double-breaker–double-bus scheme. However, the breaker-and-a-half scheme is superior in flexibility, reliability, and safety. Protective relaying and automatic reclosing schemes are more complex than for other schemes.

***12. Reliability Comparisons.*** The various schemes have been compared to emphasize their advantages and disadvantages. The basis of comparison to be employed is the economic justification of a particular degree of reliability. The determination of the degree of reliability involves an appraisal of anticipated operating conditions and the continuity of service required by the load to be served. Table 17-1 contains a summary of the comparison of switching schemes to show advantages and disadvantages.

***13. Physical Arrangements.*** Once determination of the switching scheme best suited for a particular substation application is made, it is necessary to consider the station arrangement that will satisfy the many physical requirements of the design. Available to the design engineer are the following station arrangements:

**a.** Conventional outdoor open-type bus-and-switch arrangements

**b.** Inverted-bus substation arrangements

**c.** Sulfur hexafluoride gas mini-type metal-clad substations

Outdoor open-type bus-and-switch arrangements are used generally in connection with

**TABLE 17-1** Summary of Comparison of Switching Schemes

| Switching scheme | Advantages | Disadvantages |
|---|---|---|
| 1. Single bus | 1. Lowest cost. | 1. Failure of bus or any circuit breaker results in shutdown of entire substation.<br>2. Difficult to do any maintenance.<br>3. Bus cannot be extended without completely deenergizing substation.<br>4. Can be used only where loads can be interrupted or have other supply arrangements. |
| 2. Double bus, double breaker | 1. Each circuit has two dedicated breakers.<br>2. Has flexibility in permitting feeder circuits to be connected to either bus.<br>3. Any breaker can be taken out of service for maintenance.<br>4. High reliability. | 1. Most expensive.<br>2. Would lose half of the circuits for breaker failure if circuits are not connected to both buses. |
| 3. Main and transfer | 1. Low initial and ultimate cost.<br>2. Any breaker can be taken out of service for maintenance.<br>3. Potential devices may be used on the main bus for relaying. | 1. Requires one extra breaker for the bus tie.<br>2. Switching is somewhat complicated when maintaining a breaker.<br>3. Failure of bus or any circuit breaker results in shutdown of entire substation. |
| 4. Double bus, single breaker | 1. Permits some flexibility with two operating buses.<br>2. Either main bus may be isolated for maintenance.<br>3. Circuit can be transferred readily from one bus to the other by use of bus-tie breaker and bus selector disconnect switches. | 1. One extra breaker is required for the bus tie.<br>2. Four switches are required per circuit.<br>3. Bus protection scheme may cause loss of substation when it operates if all circuits are connected to that bus.<br>4. High exposure to bus faults.<br>5. Line breaker failure takes all circuits connected to that bus out of service.<br>6. Bus-tie breaker failure takes entire substation out of service. |
| 5. Ring bus | 1. Low initial and ultimate cost.<br>2. Flexible operation for breaker maintenance.<br>3. Any breaker can be removed for maintenance without interrupting load.<br>4. Requires only one breaker per circuit.<br>5. Does not use main bus.<br>6. Each circuit is fed by two breakers.<br>7. All switching is done with breakers. | 1. If a fault occurs during a breaker maintenance period, the ring can be separated into two sections.<br>2. Automatic reclosing and protective relaying circuitry rather complex.<br>3. If a single set of relays are used, the circuit must be taken out of service to maintain the relays. (Common on all schemes.)<br>4. Requires potential devices on all circuits since there is no definite potential reference point. These devices may be required in all cases for synchronizing, live line, or voltage indication.<br>5. Breaker failure during a fault on one of the circuits causes loss of one additional circuit owing to operation of breaker-failure relaying. |
| 6. Breaker and a half | 1. Most flexible operation.<br>2. High reliability.<br>3. Breaker failure of bus side breakers removes only one circuit from service.<br>4. All switching is done with breakers.<br>5. Simple operation; no disconnect switching required for normal operation.<br>6. Either main bus can be taken out of service at any time for maintenance.<br>7. Bus failure does not remove any feeder circuits from service. | 1. 1½ breakers per circuit.<br>2. Relaying and automatic reclosing are somewhat involved since the middle breaker must be responsive to either of its associated circuits. |

generating stations and substations. Arrangements and general design characteristics of outdoor switching structures are influenced by the function and type of the installation and by its capacity, voltage, and ground-area limitations.

***14. Substation Components.*** The electrical equipment in a typical substation can include the following:

**a.** Circuit breakers

**b.** Disconnecting switches

c. Grounding switches
d. Current transformers
e. Potential transformers or capacitor voltage transformers
f. Coupling capacitors
g. Line traps
h. Lightning arresters and/or gaps
i. Power transformers
j. Shunt reactors
k. Current-limiting reactors
l. Station buses and insulators
m. Grounding system
n. Series capacitors
o. Shunt capacitors

***15. Support Structures.*** In order to properly support, mount, and install the electrical equipment, structures made of steel, aluminum, or wood, and concrete foundations are required. The typical open-type substation requires strain structures to support the transmission-line conductors; support structures for disconnecting switches, current transformers, potential transformers, lightning arresters, line traps, capacitor voltage transformers; and structures and supports for the strain and rigid buses in the station.

When the structures are made of steel or aluminum they require concrete foundations; however, when they are made of wood, concrete foundations are not required. Additional work is required to design concrete foundations for supporting circuit breakers, reactors, transformers, capacitors, and any other heavy electrical equipment.

Substation-equipment support structures can be fabricated of aluminum or steel and may consist of single wide-flange or tubular-type columns; rigid-frame structures composed of wide flanges or tubular sections; or lattice structures composed of angle members. Substation strain structures can be wood-pole structures; aluminum or steel lattice-type structures; or steel A-frame structures. Aluminum and weathering-steel structures can be used in their natural unfinished state. Normal carbon-steel structures should have galvanized or painted finishes. Wood structures should have a thermal- or pressure-process-applied preservative finish.

Aluminum structures are lightweight, have an excellent strength-to-weight ratio, and require little maintenance but have a greater initial cost than steel structures. Weathering-steel structures can be field-welded without the special surface preparation and touch-up work required on galvanized or painted steel structures, and the self-forming protective corrosion oxide eliminates maintenance. In addition, the weathering-steel color blends well in natural surroundings. Galvanized or painted steel structures have a slightly lower initial cost than weathering-steel structures; however, they require special treatment before and after field welding and require more maintenance.

Lattice-type structures are light in weight, have small wind-load areas, and are low in cost. Single-column support structures and rigid-frame structures require little maintenance, are more aesthetically pleasing, and can be inspected more quickly than lattice structures, but they have a greater initial cost. In order to reduce erection costs, rigid-frame structures should be designed with bolted field connections.

The design of supporting structures is affected by the phase spacings and ground clearances required, by the types of insulators, by the length and weight of buses and other equipment, and by the wind and ice loading. For data on wind and ice loadings, see National Electric Safety Code and ANSI Standard C2.2-1974. For recommended clearances and phase spacings see Pars. **26** and **27**.

Other structural and concrete work required in the substation includes site selection and preparation, roads, control houses, manholes, conduits, ducts, drainage facilities, catch basins, and fences.

***16. Site Selection.*** Civil-engineering work associated with the substation should be initiated as early as possible in order to ensure that the best available site is selected. This work includes a study of the topography and drainage patterns of the area together with a subsurface

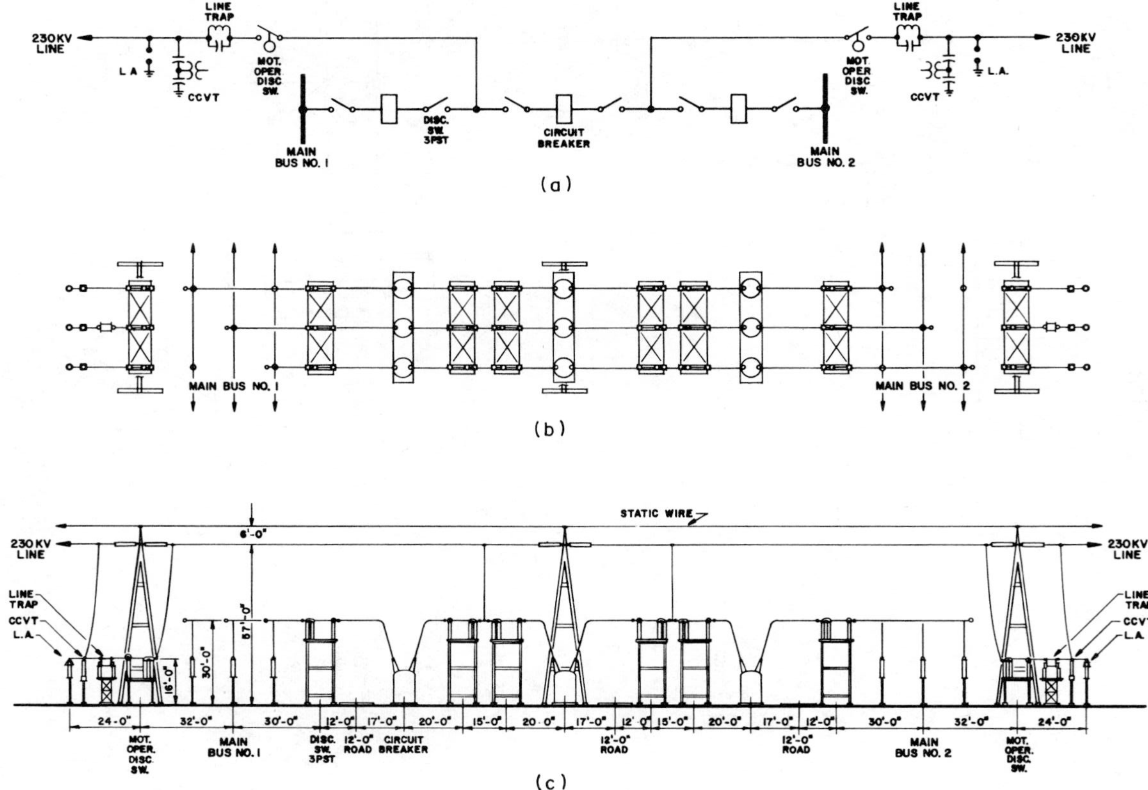


FIG. 17-7 Typical conventional substation layout, breaker-and-a-half scheme. *(a)* Main one-line diagram; *(b)* plan; *(c)* elevation.

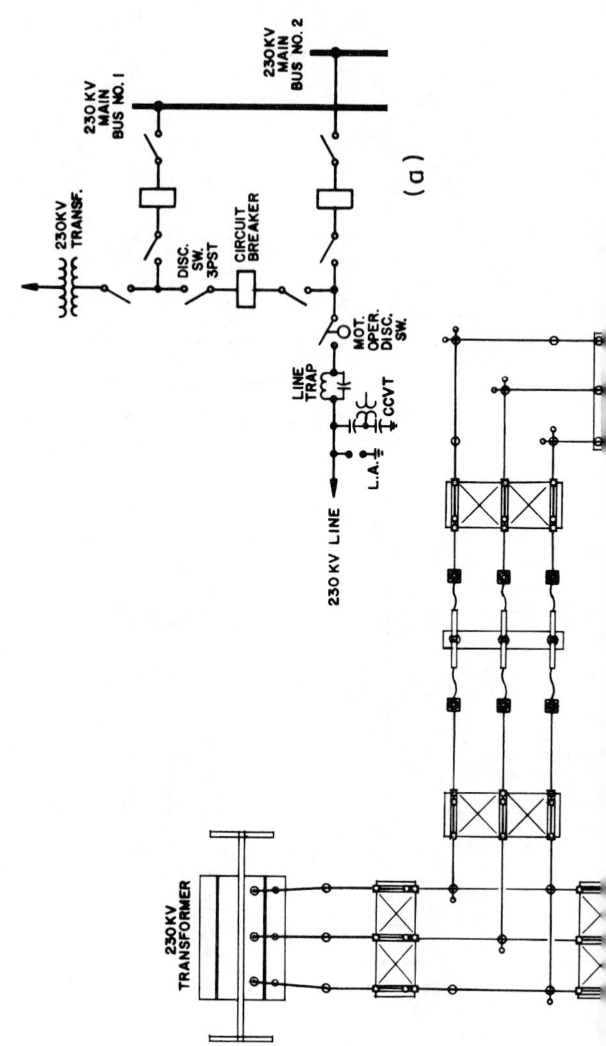
230KV TRANSF.
230KV MAIN BUS NO. 1
230KV MAIN BUS NO. 2
DISC. SW. 3PST
CIRCUIT BREAKER
MOT. OPER. DISC. SW.
LINE TRAP
CCVT
L.A.
230 KV LINE
(a)
230KV TRANSFORMER

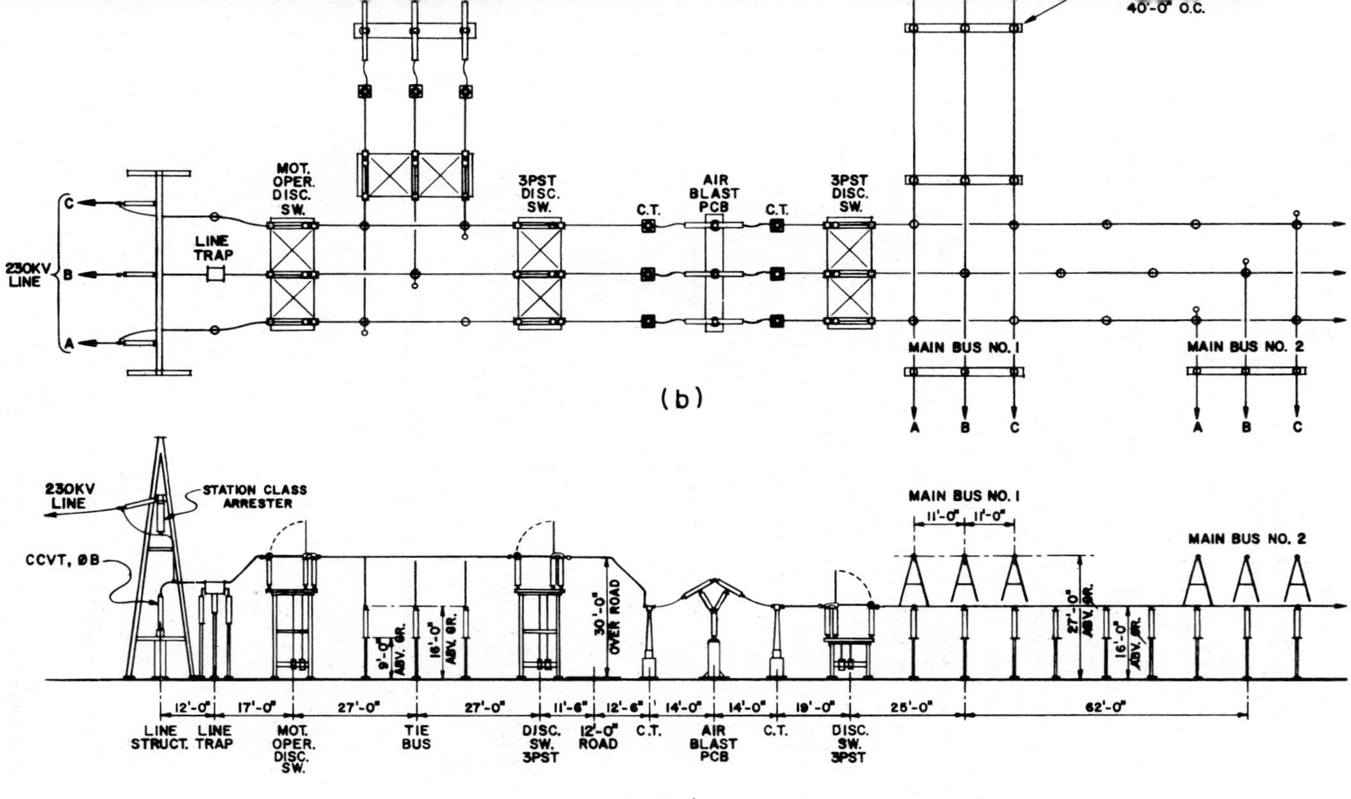


**FIG. 17-8** Typical 230-kV inverted-bus substation. (*a*) One-line diagram; (*b*) plan; (*c*) elevation.

soil investigation. The information obtained from the subsurface soil investigation will also be used to determine the design of the substation foundations. For large substations or substations located in areas with poor soils, it may be necessary to obtain additional subsurface soil tests after the final selection of the substation site has been made. The additional information should fully describe the quality of the soil at the site since the data will be used to design equipment foundations.

***17. Open-Bus Arrangement.*** A typical conventional open-bus substation arrangement consists essentially of open-bus construction using either rigid- or strain-bus design or combinations of rigid and strain bus. In schemes using double-bus arrangements such as the breaker-and-a-half arrangement, shown in Fig. 17-7, the buses are arranged to run the length of the station and are located toward the outside of the station. The transmission-line exits cross over the main bus and are dead-ended on takeoff tower structures. The line drops into the bay in the station and connects to the disconnecting switches and circuit breakers.

Use of the conventional arrangement requires three distinct levels of bus to make the necessary crossovers and connections to each substation bay. Typical dimensions of these levels at 230 kV, for example, are 16 ft for the first level above ground, 30 ft high for the main bus location, and 57 ft for the highest level of bus (see Fig. 17-7).

The conventional arrangement, in use since the mid-1920s, has been widely used by many electric utilities in the United States, has the advantages of requiring a minimum of land area per bay and relative ease of maintenance, and is ideally suited to a transmission-line through connection where a substation must be cut into a line right-of-way.

***18. Inverted Bus.*** An arrangement gaining great popularity is the inverted-bus, breaker-and-a-half scheme for EHV substations. A typical layout is outlined in Fig. 17-8. A one-line diagram of a station showing many variations of the inverted-bus scheme is indicated in Fig. 17-9. With this arrangement all outgoing circuit takeoff towers are located in the outer perime-

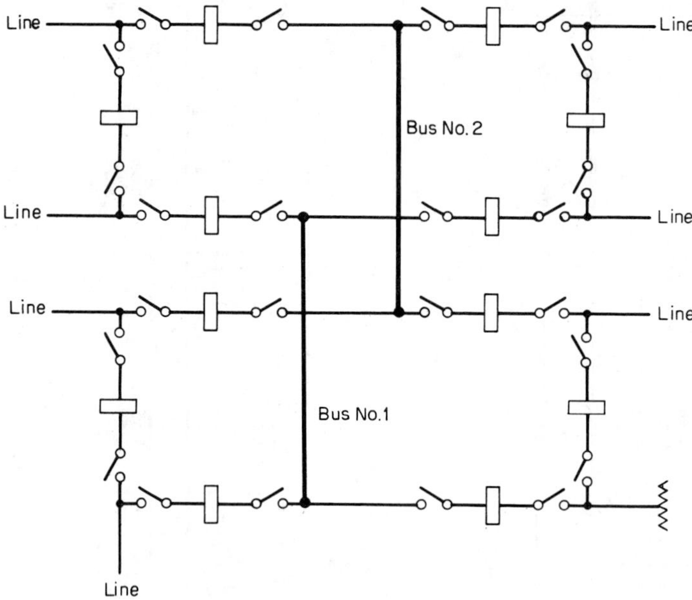


**FIG. 17-9** EHV substation, low-profile, inverted breaker-and-a-half.

ter of the substation, eliminating the crossover of line or exit facilities. Main buses are located in the middle of the substation, with all disconnecting switches, circuit breakers, and all bay equipment located outboard of the main buses. The end result of the inverted-bus arrangement presents a very low profile station with many advantages in areas where beauty and aesthetic qualities are a necessity for good public relations. The overall height of the highest bus in the 230-kV station indicated above reduces from a height of 57 ft above ground in the conventional arrangement to a height of only 30 ft above ground for the inverted-bus low-profile scheme.

***19. Gas-Insulated Substations.*** For those difficult applications where space requirements become a problem in using the conventional or inverted open-type bus arrangements and for those stations along the seacoast where trouble can occur because of salt contamination of the high-voltage insulators, $SF_6$-gas-insulated substations are available at voltage classes up to 765 kV.

A typical 230-kV conventional-type five-breaker ring-bus substation using the conventional open-bus type of construction shown in Fig. 17-7 occupies an overall area of 300 by 120 ft. An $SF_6$-gas-insulated substation for the same five-breaker ring arrangement occupies a space of no more than 26 by 72 ft, a reduction in site area of one-twentieth of the area required by the conventional substation. The volume reduction can be as much as one-sixtieth of the conventional type of station.

Although the $SF_6$ gas, sulfur hexafluoride, was synthesized as early as 1900, it was not until 1947 that it was produced on an industrial scale in the United States. Sulfur hexafluoride has two properties that make it important to the switchgear designer:

**a.** The gas exhibits excellent dielectric strength. Flashover curves indicate that at a gage pressure of 10 lb/in² the voltage withstand is higher than that of oil and many times that for air. (See Fig. 17-10.)

**b.** The $SF_6$ gas has high arc-extinguishing characteristics due to its low thermal time constant, its high free-electron absorption, and its high chemical stability,which allow a rapid recombination of the products resulting from the action of the arc. For plain break arcs the arc-quenching ability of $SF_6$ is of the order of 100 times that of air.

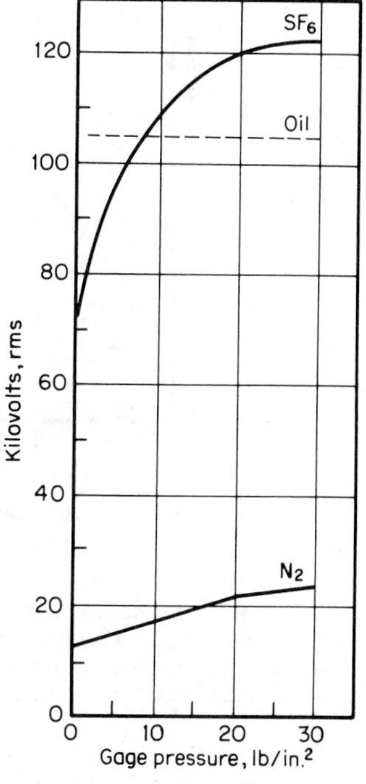


**FIG. 17-10** 60-Hz dielectric strength of gas.

The physical properties of sulfur hexafluoride are shown in Fig. 17-11. Sulfur hexafluoride is an odorless, colorless, inert, nontoxic, and incombustible gas. The curve of its vapor pressure versus temperature shows that its liquefaction temperature is low.

At densities up to 1.5 lb/ft³ [38 lb/in² (gage) at 70°F], $SF_6$ remains a gas at temperatures as low as −50°C (see Fig. 17-11). Thus the need for a heating system to control the bus temperature is eliminated when using low pressures.

With regard to general design considerations of Gas-Insulated Substations (GIS), manufacturers employ $SF_6$ gas almost exclusively today as the insulating material. Since no aging effects or other unknown shortcomings have been detected to date, it is likely that $SF_6$ will remain the predominant insulating gas for GIS.

The choice of rated pressure of $SF_6$ gas is based on a search for an optimum value. The pressure determines the diameter of the gas enclosures. The diameter must be sufficient to ensure proper dielectric performance and limit temperature rise, but it must be as small as practicable since it affects the size of the substation. As shown on the curve of Fig. 17-11, the use of a high pressure would require that the gas be heated in order to prevent liquefaction. A consideration of the factors has resulted in a choice of a rated pressure of the order of 50 lb/in². The small pressure prevents the condensation of any humidity which is present inside metal-clad switchgear in spite of precautions taken during filling.

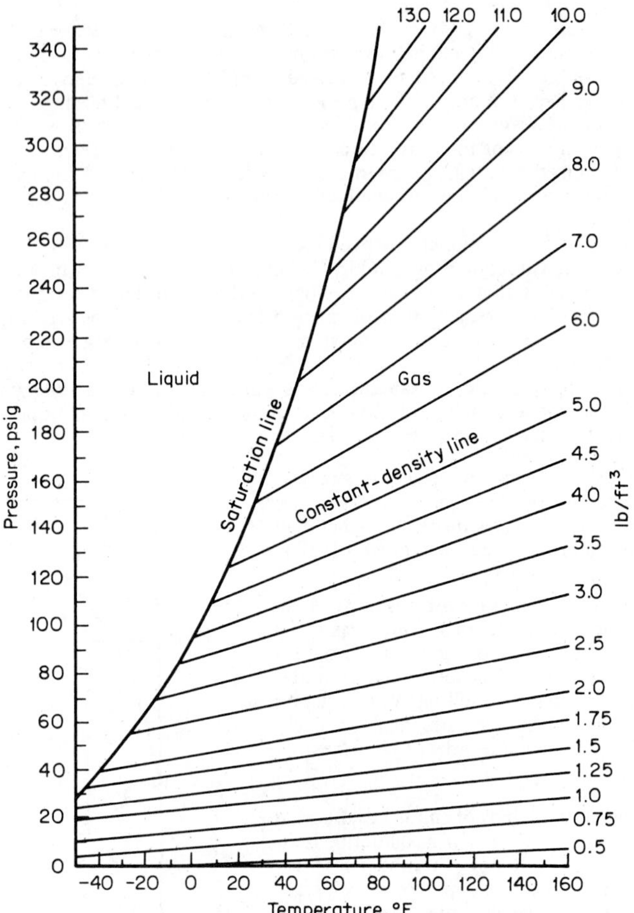


**FIG. 17-11** Physical properties of $SF_6$.

GISs using $SF_6$ as the insulating gas were introduced in 1965, with their use expanding dramatically as the many advantages of GIS were realized. In addition to the advantages already mentioned such as compact size and superior performance against contamination, there are several additional factors to consider:

**a.** Indoor installations always possible.

**b.** Higher reliability and longer maintenance intervals than with conventional air-insulated equipment.

**c.** Low visual impact on the environment.

**d.** Lower costs for site clearance and buildings.

**e.** Lower erection costs.

**f.** Superior performance in areas with severe seismic conditions.

Both single-phase and three-phase enclosures using steel and fabricated and cast aluminum are used throughout the world with equally good service records. For applications at EHV for continuous current-carrying ratings of 3000 A or above, aluminum enclosures are preferred over steel to minimize losses. While three-phase busbars are installed at voltage levels as high as 420 kV, the main applications for GIS using three-phase enclosures are for voltage levels up to 170 kV with single-phase enclosures normally used for voltages above 170 kV.

GISs are constructed with segregated gas compartments in order to limit the extent of the equipment that is taken out of service for maintenance when the gas has to be removed, to limit gas loss in the event of a leak, to localize and minimize the effect of an accident due to internal arcing, and to allow easier supervision of gas and leakage detection. In general there is a trend to combine more enclosures to form larger gas sections.

The gas density of $SF_6$ gas affects the breakdown voltage. Supervision of the gas density consists of gas-pressure-measuring systems or density monitors of the different gas compartments.

Pressure-relief devices or external bursting disks are widely used to prevent bursting of an enclosure as a result of overpressure created by internal arcing. When the enclosure contains large volumes of gas such as in main busbar sections or gas-insulated transmission lines (GITL), pressure-relief devices serve no useful purpose since overpressure will not reach an unsafe level in the presence of a large amount of gas. It is possible to avoid danger to personnel by directing the exhaust away from locations where people are likely to be working.

To avoid the use of external bursting disks, it is possible to use internal bursting disks, e.g., weak spacers between compartments, and thus enlarge the volume of gas to reduce the pressure rise. When spacers are used as internal bursting disks, the enclosure must withstand a pressure high enough to ensure that the spacer ruptures before the enclosure.

GISs have been installed outdoors in industrial areas, along the seashore, and in snowy climates. However, most GISs are installed indoors in buildings. Due to the compact size of the switchgear, relatively inexpensive buildings are needed. The indoor arrangement allows easier installation and maintenance, especially for those countries with severe climatic conditions.

The circuit breakers used in the GIS are almost exclusively single-pressure $SF_6$ puffer-type breakers. Disconnecting switches are generally of the no-load type but can operate successfully with the normal inherent capacitive currents (0.3 to 1.0 A at 145 to 800 kV). Earthing switches only intended for maintenance are provided with slow action manual or electrical drive mechanisms and have no making capabilities. To cope with the possibility of induced line voltages and currents, earthing switches with making capacity of up to 80 kA at 550 kV are available and may be added to ensure personnel safety.

Severe damage or burn-through can occur in the event a moving arc remains stationary at one point such as at a spacer insulator. Burn-through for a stationary arc using aluminum enclosure material has been measured to occur in the order of a few cycles for 50- to 63-kA arc currents. To prevent the possibility of severe damage to the equipment and any possibility of injury to personnel, every effort is made to prevent the occurrence of these internal arcs by careful design and testing of the equipment, careful application of insulation coordination, and by continuous supervision of the gas density. If a flashover occurs, even after taking the above precautions, it is very important to provide high-speed busbar protective relaying and high-speed transmission-line protective relays to ensure rapid clearing of the arc to prevent the possibility of burn-through.

The mini-substation is available in all types of bus arrangements including double-busbar, single-breaker, breaker-and-a-half, ring-bus, and single-bus schemes. The main elements built into the substation and completely enclosed in the $SF_6$ installation include circuit breakers, load break switches, disconnecting switches, ground switches, current and potential transformers, busbars, coupling capacitors, and $SF_6$ leadout bushing insulators for connection to overhead lines, transformer, or other external equipment. Equipment such as transformers, shunt reactors, shunt capacitors, power-line carrier line traps, etc., are not manufactured as part of the $SF_6$-insulated system. However, long runs of bus using $SF_6$ insulation may extend some distance from the substation in order to make the connection to the other conventional equipment. For a typical GIS used in a 230-kV scheme, refer to Fig. 17-12.

***20. Substation Buses.*** Substation buses are a most important part of the station structure since they carry high amounts of energy in a confined space. They must be carefully designed in order that the construction will provide adequately and economically for the utilization of the electric energy generated and at the same time have sufficient structural strength to withstand the maximum stresses that may be imposed on the conductors, and in turn on the structure, by heavy currents under short-circuit conditions.

In their early development substations in the HV class were usually of the strain-bus design. The strain bus is similar to a transmission line and consists of a conductor such as ACSR (aluminum cable steel reinforced), copper, or high-strength aluminum alloy strung between substation structures. EHV substations normally use the rigid-bus approach and enjoy the advantage of low station profile and ease of maintenance and operation (see Fig. 17-8). Mixing of rigid- and strain-bus construction is normally employed in the conventional arrangement as shown in Fig. 17-7. Here the main buses use rigid-bus design and the upper buses between transmission towers are of strain-bus design. A typical design at 765 kV uses a combination of both rigid and strain bus (see Fig. 17-13).

A comparison of rigid and strain buses indicates that careful consideration should be given to the selection of the proper type of bus to use. The rigid bus has a number of advantages such as:

**a.** Less steel is used and structures are simple.

**b.** Rigid conductors are not under constant strain.

**c.** Individual pedestal-mounted insulators are more accessible for cleaning.

**d.** The rigid bus is lower in height, has a distinct layout, and can be definitely segregated for maintenance.

**e.** Low profile with the rigid bus provides good visibility of the conductors and apparatus and gives a good appearance to the substation.

Some disadvantages to using the rigid bus are:

**a.** More insulators and supports are usually needed for rigid-bus design, thus requiring more insulators to clean.

**b.** The rigid bus is more sensitive to structural deflections causing misalignment problems and possible damage to the bus.

**c.** The rigid bus usually requires more land area than the strain bus.

**d.** Rigid-bus designs are comparatively expensive.

Strain-bus installations have the following advantages:

**a.** Comparatively lower cost than the rigid bus.

**b.** Substations employing the strain bus may occupy less land area than stations using the rigid bus.

**c.** Fewer structures are required.

Some disadvantages of using strain-bus construction are as follows:

**a.** Strain structures require larger structures and foundations.

**b.** Insulators are not conveniently accessible for cleaning.

**c.** Painting of high steel structures is costly and hazardous.

**d.** Emergency conductor repairs are more difficult.

The design of station buses depends on a number of elements which may be classified as follows:

**a.** Current-carrying capacity

**b.** Short-circuit stresses

**c.** Establishing minimum electrical clearances

The current-carrying capacity of a bus is limited by the heating effects produced by the

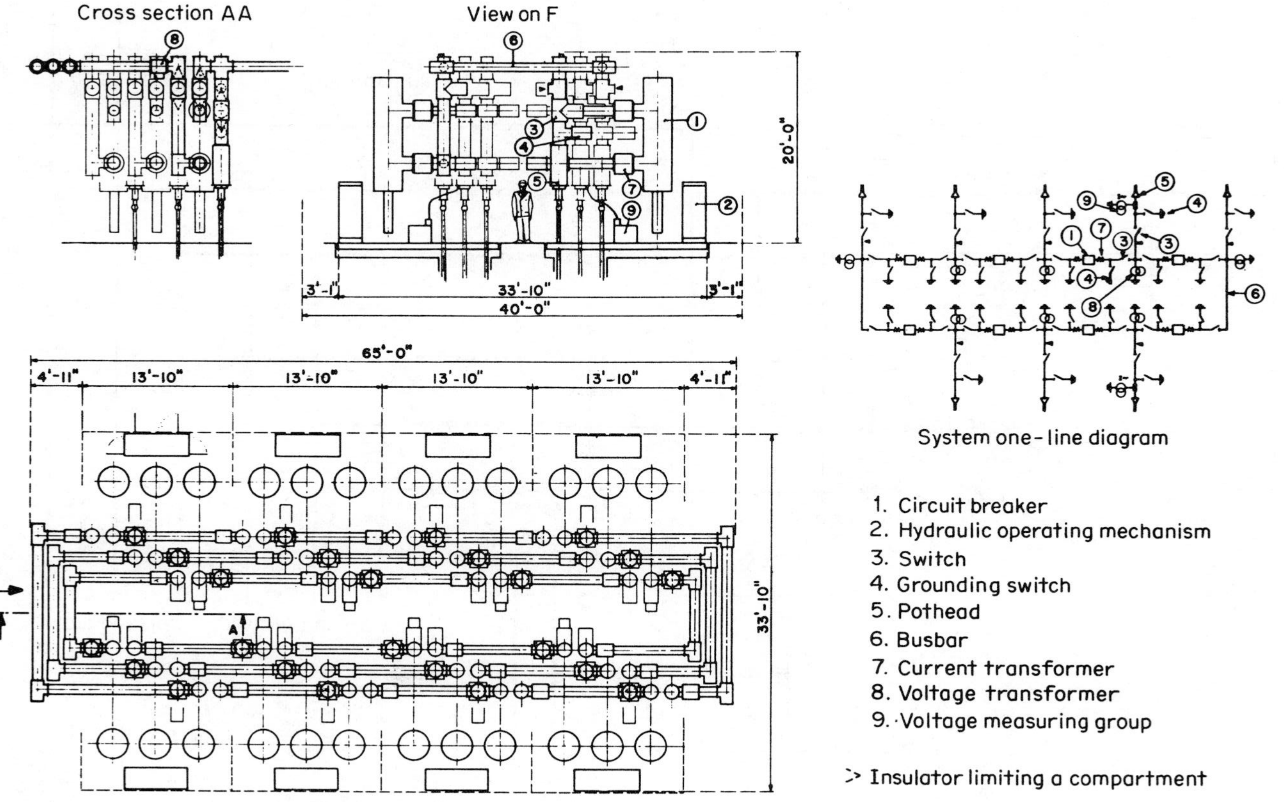


FIG. 17-12 A 230-kV $SF_6$ Gas-Insulated Substation.

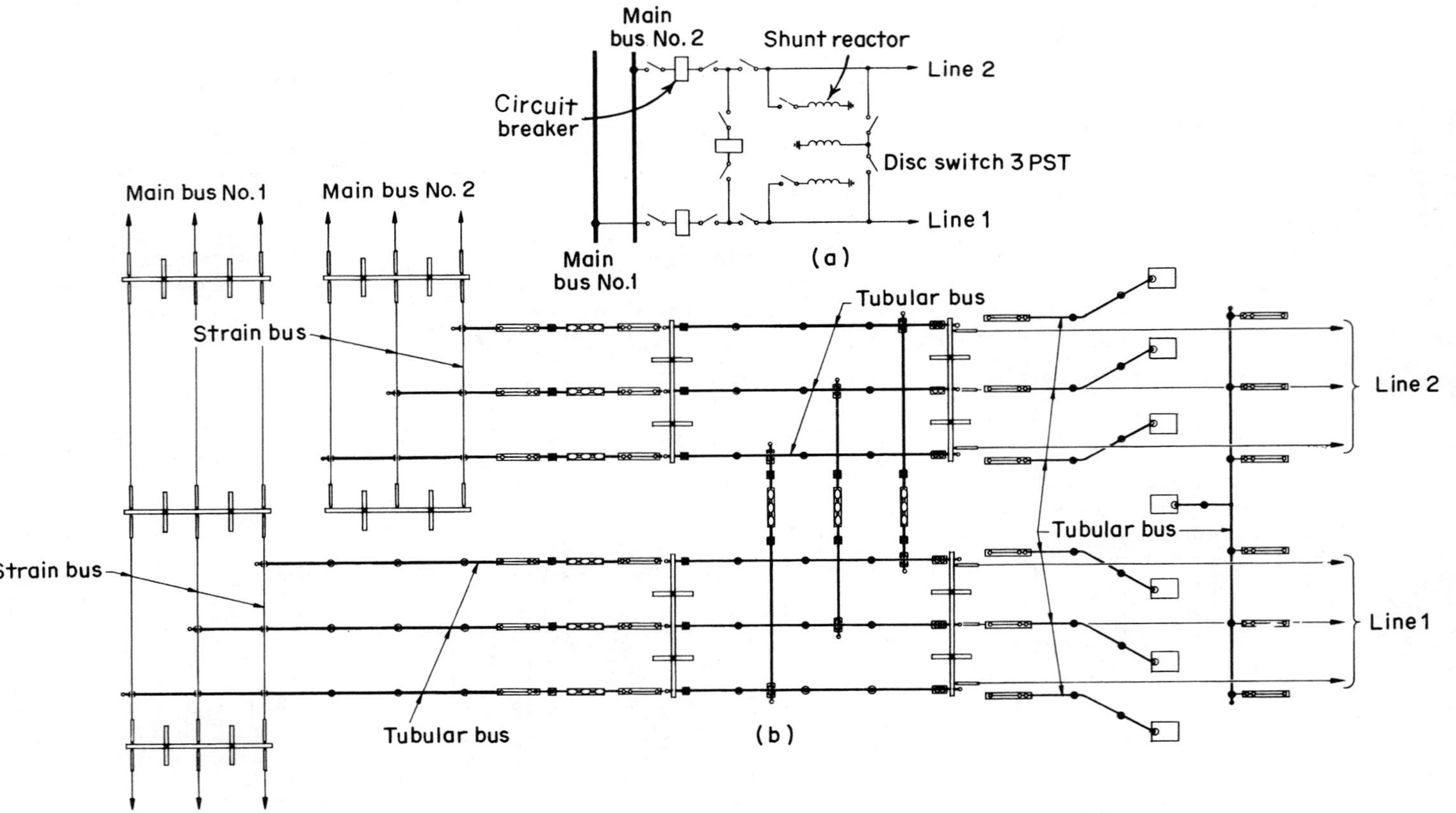

Main bus No. 2
Shunt reactor
Line 2
Circuit breaker
Disc switch 3 PST
Line 1
Main bus No.1
(a)
Main bus No.1
Main bus No. 2
Strain bus
Tubular bus
Line 2
Tubular bus
Strain bus
Line 1
Tubular bus
(b)

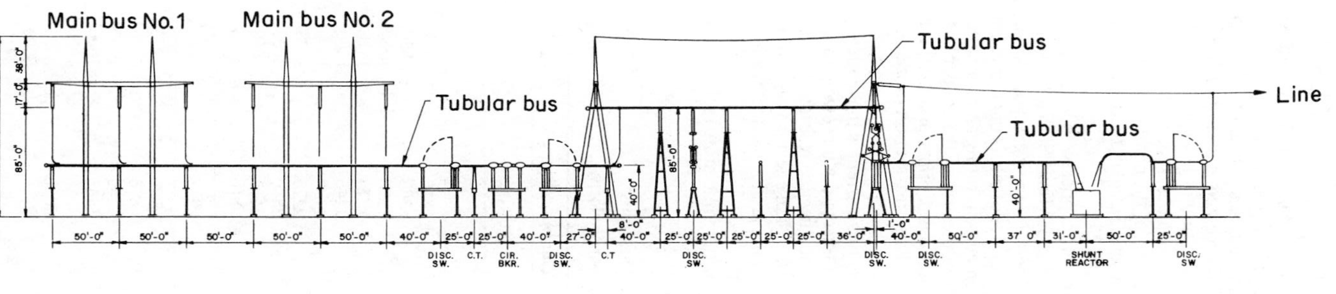


(c)

**FIG. 17-13** A 765-kV EHV substation using both rigid- and strain-bus design. *(a)* Main one-line diagram; *(b)* plan; *(c)* elevation.

current. Buses are generally rated on the basis of the temperature rise which can be permitted without danger of overheating equipment terminals, bus connections, and joints.

The permissible temperature rise for plain copper and aluminum buses is usually limited to 30°C above an ambient temperature of 40°C. This value is the accepted standard of IEEE, NEMA, and ANSI. This is an average temperature rise, and a maximum or hot-spot temperature rise of 35°C is permissible. Many factors enter into the heating of a bus, such as the type of material used, the size and shape of the conductor, the surface area of the conductor and its condition, skin effect, proximity effect, conductor reactance, ventilation, and inductive heating caused by the proximity of magnetic materials.

***21. Bus Material.*** Bus materials in general use are aluminum and copper, with hard-drawn aluminum, especially in the tubular shape, the most widely used in HV and EHV open-type outdoor stations. Aluminum has the advantage of being about one-third the weight of copper. In addition, the aluminum requires little maintenance, and proper use of alloys of aluminum provides the rigidity needed to serve as bus material. For a given current rating and for equal limiting temperatures, the area of aluminum bus is about 133% of the area of the copper bus. Aluminum tubing is used almost exclusively in HV and EHV stations. Copper or aluminum tubing as well as other special shapes is sometimes used for low-voltage distribution substation buses.

***22. Skin Effect.*** Skin effect in a conductor carrying an alternating current is the tendency toward crowding of the current into the outer layer, or "skin," of the conductor, owing to the self-inductance of the conductor. This results in an increase of the effective resistance of the conductor and in a lower current rating for a given temperature rise. Skin effect is very important in heavy-current buses where a number of conductors are used in parallel, because it affects not only each conductor but each group of conductors as a unit.

Tubes have less skin-effect resistance than flat conductors of the same cross section, and

**TABLE 17-2** Current Ratings for Bare Copper Tubular Bus, Outdoors*

(40°C ambient temperature, 98% conductivity copper, frequency 60 Hz, wind velocity 2 ft/s at 90° angle)

| Nominal size | Outside diameter, in | Inside diameter, in | Current ratings, A | | |
|---|---|---|---|---|---|
| | | | 30°C rise | 40°C rise | 50°C rise |
| | | Standard Pipe Sizes | | | |
| ½ | 0.840 | 0.625 | 545 | 615 | 675 |
| ¾ | 1.050 | 0.822 | 675 | 765 | 850 |
| 1 | 1.315 | 1.062 | 850 | 975 | 1080 |
| 1¼ | 1.660 | 1.368 | 1120 | 1275 | 1415 |
| 1½ | 1.900 | 1.600 | 1270 | 1445 | 1600 |
| 2 | 2.375 | 2.062 | 1570 | 1780 | 1980 |
| 2½ | 2.875 | 2.500 | 1990 | 2275 | 2525 |
| 3 | 3.500 | 3.062 | 2540 | 2870 | 3225 |
| 3½ | 4.000 | 3.500 | 3020 | 3465 | 3860 |
| 4 | 4.500 | 4.000 | 3365 | 3810 | 4305 |
| | | Extra-heavy Pipe Sizes | | | |
| ½ | 0.840 | 0.542 | 615 | 705 | 775 |
| ¾ | 1.050 | 0.736 | 760 | 875 | 970 |
| 1 | 1.315 | 0.951 | 1000 | 1140 | 1255 |
| 1¼ | 1.660 | 1.272 | 1255 | 1445 | 1600 |
| 1½ | 1.900 | 1.494 | 1445 | 1650 | 1830 |
| 2 | 2.375 | 1.933 | 1830 | 2080 | 2325 |
| 2½ | 2.875 | 2.315 | 2365 | 2720 | 3020 |
| 3 | 3.500 | 2.892 | 2970 | 3365 | 3710 |
| 3½ | 4.000 | 3.358 | 3380 | 3860 | 4255 |
| 4 | 4.500 | 3.818 | 3840 | 4350 | 4850 |

* From Anderson Electric Technical Data, Table 13.

NOTE: 1 in = 25.4 mm; 1 ft/s = 0.3048 m/s.

tubes with thin walls are affected the least by skin effect. Aluminum conductors are affected less by skin effect than copper conductors of similar cross section because of the greater resistance of aluminum.

***23. Proximity Effect.*** Proximity effect in a bus is the distortion of the current distribution caused by induction between the out and return conductors, which causes a concentration of current in the parts of the buses nearest each other, thus increasing their effective resistance. The proximity effect must be taken into account for buses carrying alternating current; the effect on three-phase buses is less than on single-phase buses.

Tubular conductors used on alternating current have a better current distribution than any other shape of conductor of similar cross-sectional area, but they also have a relatively small surface area for dissipating heat losses. These two factors must be properly balanced in the design of a tubular bus.

Tubing provides a relatively large cross-sectional area in minimum space and has the maximum structural strength for equivalent cross-sectional area, permitting longer spaces between supports. In outdoor substations spans up to 40 and 50 ft with copper or aluminum tubes up to 6-in diameter are considered practicable, the tendency in design being to use long spans and thus reduce the number of insulator posts to a minimum.

Current-carrying capacities of copper and aluminum tubular buses of different dimensions are shown in Tables 17-2 and 17-3.

***24. Thermal Expansion.*** Thermal expansion and contraction of bus conductors is an

**TABLE 17-3** Current Ratings for Bare Aluminum Tubular Bus, Outdoors*

(Ratings based on 30°C over 40°C ambient, frequency 60 Hz, wind velocity 2 ft/s cross-wind)

| Nominal size | Outside diameter, in | Inside diameter, in | Current ratings, A | |
|---|---|---|---|---|
| | | | † | ‡ |
| ASA Schedule 40 (Standard Pipe Size) | | | | |
| ½ | 0.840 | 0.622 | 405 | 355 |
| ¾ | 1.050 | 0.824 | 495 | 440 |
| 1 | 1.315 | 1.049 | 650 | 575 |
| 1¼ | 1.660 | 1.380 | 810 | 720 |
| 1½ | 1.900 | 1.610 | 925 | 820 |
| 2 | 2.375 | 2.067 | 1150 | 1020 |
| 2½ | 2.875 | 2.469 | 1550 | 1370 |
| 3 | 3.500 | 3.068 | 1890 | 1670 |
| 3½ | 4.000 | 3.548 | 2170 | 1920 |
| 4 | 4.500 | 4.026 | 2460 | 2180 |
| 5 | 5.563 | 5.047 | 3080 | 2730 |
| ASA Schedule 80 (Extra-heavy Pipe Size) | | | | |
| ½ | 0.840 | 0.546 | 455 | 400 |
| ¾ | 1.050 | 0.742 | 565 | 500 |
| 1 | 1.315 | 0.957 | 740 | 655 |
| 1¼ | 1.660 | 1.278 | 930 | 825 |
| 1½ | 1.900 | 1.500 | 1070 | 945 |
| 2 | 2.375 | 1.939 | 1350 | 1200 |
| 2½ | 2.875 | 2.323 | 1780 | 1580 |
| 3 | 3.500 | 2.900 | 2190 | 1940 |
| 3½ | 4.000 | 3.364 | 2530 | 2240 |
| 4 | 4.500 | 3.826 | 2880 | 2560 |
| 5 | 5.563 | 4.813 | 3640 | 3230 |

* Data from Aluminum Company of America.
†6063-T6 = 53% IACS typical.
‡6061-T6 = 40% IACS typical.
NOTE: 1 in = 25.4 mm.

important factor in bus design, particularly where heavy-current buses or buses of long lengths are involved. An aluminum bus will expand 0.0105 in/ft of length for a temperature rise of 38°C (100°F). In order to protect insulator supports, disconnecting switches, and equipment terminals from the stresses set up by the thermal expansion of the conductors, provision should be made for expansion by means of expansion joints and clamps permitting tubing to slide.

Vibration of long tubular-bus spans has been experienced to some extent but can be eliminated by inserting a piece of cable inside the tubular bus.

***25. Bus Spacing.*** The spacing of buses in substations is largely a matter of design experience. However, in an attempt to arrive at some standardization of practices, minimum electrical clearances for standard basic insulation levels were established and published by the National AIEE Committee on Substations. The data are summarized in AIEE *Paper* 54-80 which appeared in the June, 1954, *Transactions,* p. 636. This guide, shown in Table 17-4, provides minimum-clearance recommendations for electric transmission systems designed for impulse withstand levels up to and including 1175 kV BIL.

Ongoing studies attempt to extend the clearance recommendations to include the EHV range. The data published in 1954 are satisfactory to withstand anticipated switching-surge requirements of electric systems rated 161 kV and below. For systems rated 230 kV and above, more accurate determination of the switching-surge characteristics of insulation systems was required before final clearance recommendations could be made.

***26. Clearances.*** In 1972 the Substations Committee of the IEEE published *Trans. Paper*

**TABLE 17-4** Minimum Electrical Clearances for Standard BIL Outdoor Alternating Current

| kV class (1) | BIL level, kV withstand (2) | Minimum clearance to ground for rigid parts, in, (3) | Minimum clearance between phases (or live parts) for rigid parts, in, metal to metal (4) | Minimum clearance between overhead conductors and grade for personnel safety inside substation, ft (5) | Minimum clearance between wires and roadways, inside substation enclosure, ft (6) |
|---|---|---|---|---|---|
| 7.5 | 95 | 6 | 7 | 8 | 20 |
| 15 | 110 | 7 | 12 | 9 | 20 |
| 23 | 150 | 10 | 15 | 10 | 22 |
| 34.5 | 200 | 13 | 18 | 10 | 22 |
| 46 | 250 | 17 | 21 | 10 | 22 |
| 69 | 350 | 25 | 31 | 11 | 23 |
| 115 | 550 | 42 | 53 | 12 | 25 |
| 138 | 650 | 50 | 62 | 13 | 25 |
| 161 | 750 | 58 | 72 | 14 | 26 |
| 230 | 825 | 65 | 80 | 15 | 27 |
| 230 | 900 | 71 | 89 | 15 | 27 |
| | 1,050 | 83 | 105 | 16 | 28 |
| | 1,175 | 94 | 113 | 17 | 29 |

NOTES:

1. Coordinate kV class and BIL when choosing minimum clearances.
2. The values above are recommended minimums but may be decreased in line with good practice, depending on local conditions, procedures, etc.
3. The values above apply to 3300 ft above sea level. Above this elevation, the above values should be increased according to Par. 22-4 of AIEE Standard No. 22A, Air Switches, Insulator Units and Bus Supports, October, 1949.
4. These recommended minimum clearances are for rigid conductors. Any structural tolerances, or allowances for conductor movement, or possible reduction in spacing by foreign objects should be added to the minimum values.
5. These minimum clearances are intended as a guide for the installation of equipment in the field only, and not for the design of electric devices or apparatus such as circuit breakers, transformers, etc.

1 in = 25.4 mm; 1 ft = 0.3048 m.

T72 131-6, which established recommendations for minimum line-to-ground electrical clearances for EHV substations, based on switching-surge requirements. The recommendations are based on a study of actual test data of the switching-surge strength characteristics of air gaps with various electrode configurations as reported by many investigators. The results are shown in Table 17-5 and include minimum line-to-ground clearances for EHV system voltage ratings of 345, 500, and 765 kV.

The Substations Committee of the IEEE has an ongoing effort to study test data to determine phase-to-phase air clearances and will publish revised reports, particularly for the EHV levels of 345, 500, and 765 kV. The clearances given in Table 17-4 are considered adequate for both line-to-ground and phase-to-phase values for the voltage classes up through 230 kV nominal system voltage where air-gap distances are dictated by impulse (BIL) withstand characteristics.

**TABLE 17-5** Minimum Electrical Clearances for EHV Substations Based on Switching Surge and Lightning Impulse Requirements
(Line-to-ground)

| System voltage, kV | | Transient voltage | | SS clearances, in | | BIL clearances, in | |
|---|---|---|---|---|---|---|---|
| Nom. | Max. | PU SS | Withstand SS crest, kV | Equivalent SS CFO, kV | Line to ground | Withstand BIL, kV | Line to ground |
| 345 | 362 | 2.2 | 650 | 785 | 84 | 1050 | 84 |
| | | 2.3 | 680 | 821 | 90 | | |
| | | 2.4 | 709 | 857 | 96 | | |
| | | 2.5 | 739 | 893 | 104 | 1300 | 104 |
| | | 2.6 | 768 | 928 | 111 | | |
| | | 2.7 | 798 | 964 | 118 | | |
| | | 2.8 | 828 | 1000 | 125 | | |
| | | 2.9 | 857 | 1035 | 133 | | |
| | | 3.0 | 887 | 1071 | 140 | | |
| 500 | 550 | 1.8 | 808 | 976 | 124 | 1550 | 124 |
| | | 1.9 | 853 | 1031 | 132 | | |
| | | 2.0 | 898 | 1085 | 144 | 1800 | 144 |
| | | 2.1 | 943 | 1139 | 156 | | |
| | | 2.2 | 988 | 1193 | 168 | | |
| | | 2.3 | 1033 | 1248 | 181 | | |
| | | 2.4 | 1078 | 1302 | 194 | | |
| | | 2.5 | 1123 | 1356 | 208 | | |
| | | 2.6 | 1167 | 1410 | 222 | | |
| | | 2.7 | 1212 | 1464 | 238 | | |
| | | 2.8 | 1257 | 1519 | 251 | | |
| 765 | 800 | 1.5 | 982 | 1186 | 166 | 2050 | 167 |
| | | 1.6 | 1047 | 1265 | 185 | | |
| | | 1.7 | 1113 | 1344 | 205 | | |
| | | 1.8 | 1178 | 1423 | 225 | | |
| | | 1.9 | 1244 | 1502 | 246 | | |
| | | 2.0 | 1309 | 1581 | 268 | | |
| | | 2.1 | 1375 | 1660 | 291 | | |
| | | 2.2 | 1440 | 1739 | 314 | | |
| | | 2.3 | 1505 | 1818 | 339 | | |
| | | 2.4 | 1571 | 1897 | 363 | | |
| | | 2.5 | 1636 | 1976 | 389 | | |
| | | 2.6 | 1702 | 2055 | 415 | | |

NOTES:

*A.* Minimum clearances should satisfy either maximum switching-surge or BIL duty requirement, whichever dictates the larger dimension.

*B.* For installations at altitudes in excess of 3300 ft elevation, it is suggested that correction factors, as provided in ANSI C37.30-1970, be applied to withstand voltages as given above.

SS: switching surge
CFO: critical flashover
1 in = 25.4 mm.

Considerable data have been published by CIGRE relative to establishing phase-to-phase air clearances in EHV substations as required by switching surges. The CIGRE method is based on nearly simultaneous and equal opposite-polarity surge overvoltages in adjacent phases. The phase-to-ground surge overvoltage is multiplied by a factor of up to 1.8. (The theoretical maximum phase-to-phase would be twice the phase-to-ground surge overvoltage.) The estimated value of phase-to-phase overvoltage is then compared with test data to obtain clearances. Refer to an article in CIGRE, Electra No. 29, 1973, Phase-to-Ground and Phase-to-Phase Air Clearances in Substations, by L. Paris and A. Taschini.

Suggested values of phase-to-phase clearances for EHV substations based on the CIGRE method are shown in Table 17-6. The table was formulated by choosing various phase-to-ground transient voltage values such as are used in Table 17-5, Minimum Electrical Clearance for EHV Substations Line-to-Ground. These values of phase-to-ground overvoltage were multiplied by a factor of 1.8 to arrive at a value of estimated phase-to-phase transient overvoltages. An equivalent phase-to-phase critical flashover value of voltage is next assumed by multiplying the switching-surge phase-to-phase voltage by 1.21. Finally this value is compared with data in the CIGRE article prepared by Paris and Taschini to arrive at air-clearance values based on switching-surge impulse voltages.

EHV substation bus phase spacing is normally based on the clearance required for switching-surge impulse values plus an allowance for energized equipment projections and corona rings. This total distance may be further increased to facilitate substation maintenance as illustrated in Table 17-6.

***27. Mechanical and Electrical Forces.*** A station bus must have sufficient mechanical strength to withstand short-circuit stresses. Two factors are involved: *(a)* the strength of the insulators and their supporting structure and *(b)* the strength of the bus conductor.

A simple guide for the calculation of electromagnetic forces exerted on buses during short-circuit conditions is stated in ANSI Standard C37.32-1972, American National Standard Schedules of Preferred Ratings, Manufacturing Specifications, and Application Guide for High Voltage Air Switches, Bus Supports, and Switch Accessories.

**TABLE 17-6** Suggested Electrical Clearances for EHV Substations Based on Switching Surge Requirements and Including U.S. Utility Practice
(Phase-to-phase)

| System voltage, kV | | Transient voltage | | SS clearances, in* | | Present practice |
|---|---|---|---|---|---|---|
| Nom. | Max. | SS L-G PU | Withstand SS crest, kV | Equivalent L-L SS CFO, kV | Rod to rod withstand* | U.S. utility phase spacing, ft |
| 345 | 362 | 2.2 | 650 | 1405 | 103 | 15 to 18 |
| | | 2.6 | 768 | 1660 | 128 | |
| | | 3.0 | 887 | 1915 | 159 | |
| 500 | 550 | 1.8 | 808 | 1745 | 138 | 20 to 35 |
| | | 2.2 | 988 | 2135 | 190 | |
| | | 2.5 | 1123 | 2425 | 239 | |
| | | 2.8 | 1257 | 2715 | 294 | |
| 765 | 800 | 1.5 | 982 | 2120 | 189 | 45 to 50 |
| | | 1.8 | 1178 | 2545 | 261 | |
| | | 2.1 | 1375 | 2970 | 356 | |
| | | 2.4 | 1571 | 3395 | 480 | |

* The values of L-L switching-surge clearances are based on the use of SS L-G crest voltages multiplied by 1.8. This value of L-L SS voltage is then multiplied by 1.21 to indicate an SS CFO value of voltage used to determine the clearances. For a description of method used, refer to CIGRE report by L. Paris and A. Taschini, Phase-to-Ground and Phase-to-Phase Air Clearances in Substations, CIGRE, *Electra* no. 29, 1973, pp. 29–44.

L-G: line-to-ground
L-L: line-to-line
SS: switching surge
CFO: critical flashover
NOTE: 1 in = 25.4 mm; 1 ft = 0.3048 m.

The electromagnetic force exerted between two current-carrying conductors is a function of the current, its decrement rate, the shape and arrangement of conductors, and the natural frequencies of the complete assembly, including mounting structure, insulators, and conductors. Obviously, it is not feasible to cover each and every case with one simple equation, even if some approximations are made, because of the large number of variables involved, including the wide range of constants for support structures.

The force calculated by the following equation is that produced by the maximum peak current. In most cases, the calculated force is higher than that which actually occurs, due to inertia and flexibility of the systems, and this fact tends to compensate for the neglect of resonant forces. The equation, therefore, is sufficiently accurate for usual practical conditions.

$$F = M\frac{5.4 \times I^2}{S \times 10^7} \tag{17-1}$$

where $F$ = pounds per foot of conductor
$M$ = multiplying factor
$I$ = short-circuit current, A (defined in Table 17-7)
$S$ = spacing between center lines of conductors, in

After determining the value of $I$, select the corresponding $M$ factor from Table 17-7.

**TABLE 17-7** Multiplying Factor ($M$) for Calculation of Electromagnetic Forces

| Circuit | Amperes ($I$) expressed as | Multiplying factor ($M$) |
|---|---|---|
| dc | Max. peak | 1.0 |
| ac, 3-phase | Max. peak | 0.866 |
| ac, 3-phase | rms asymmetrical | $(0.866 \times 1.63^2) = 2.3$ |
| ac, 3-phase | rms symmetrical | $(0.866 \times 2.82^2) = 6.9$ |
| 1 phase of 3 phase or 1 phase | Max peak | 1.0 |
| 1 phase of 3 phase or 1 phase | rms asymmetrical | $(1.63^2) = 2.66$ |
| 1 phase of 3 phase or 1 phase | rms symmetrical | $(2.82^2) = 8.0$ |

Structures with long spans held in tension by strain insulators cannot be calculated for stresses by the above procedure, but approximate estimates can be made by following the procedure generally used for calculating mechanical stresses in transmission-line conductors.

The total stress in an outdoor bus is the resultant of the stresses due to the short-circuit load, together with the dead, ice, and wind loads.

**a.** *Buses up to 161 kV.* The distance between phases and the character of the bus supports and their spacing are such that wind loading may usually be neglected. Ice load of ½ in is usually considered.

**b.** *Buses for 230 kV and higher voltages.* The spacing between phases is usually so large that the mechanical effects of short-circuit currents may not be the determining factor, and such buses, when properly designed for the mechanical loads only, may be found to satisfy the electrical requirements, such as current-carrying capacity. However, short-circuit duties on modern systems have been increasing rapidly, and the electrical forces should be checked by formula (17-1).

Deflections and stresses on aluminum bus can be determined by referring to Tables 17-8 and 17-9. All loads are assumed to be uniformly distributed. Loading includes the dead load of the bus and, in addition, includes ice loadings of ½- and 1-in coating on the bus. Wind loads are

**TABLE 17-8** Aluminum Round Tubular Busbar Deflections and Stresses*
(Standard iron pipe sizes)

| IPS size, in | Loading | Span, ft: 20 Deflection, in | 20 Stress, lb/in² | 25 Deflection, in | 25 Stress, lb/in² | 30 Deflection, in | 30 Stress, lb/in² | 35 Deflection, in | 35 Stress, lb/in² | 40 Deflection, in | 40 Stress, lb/in² | 45 Deflection, in | 45 Stress, lb/in² | 50 Deflection, in | 50 Stress, lb/in² |
|---|---|---|---|---|---|---|---|---|---|---|---|---|---|---|---|
| 1¼ | Bare | 1.45 | 2010 | 3.54 | 3135 | | | | | | | | | | |
| | ½" ice | 3.94 | 5445 | 9.61 | 8510 | | | | | | | | | | |
| | ½" ice + 8 lb wind | 5.12 | 7090 | 12.51 | 11075 | | | | | | | | | | |
| | 1" ice | 7.57 | 10470 | 18.48 | 16360 | | | | | | | | | | |
| 1½ | Bare | 1.09 | 1725 | 2.66 | 2700 | | | | | | | | | | |
| | ½" ice | 2.83 | 4475 | 6.90 | 6990 | | | | | | | | | | |
| | ½" ice + 8 lb wind | 3.61 | 5715 | 8.81 | 8930 | | | | | | | | | | |
| | 1" ice | 5.28 | 8365 | 12.90 | 13070 | | | | | | | | | | |
| 2 | Bare | 0.68 | 1350 | 1.67 | 2110 | 3.45 | 3040 | | | | | | | | |
| | ½" ice | 1.65 | 3265 | 4.03 | 5100 | 8.35 | 7345 | | | | | | | | |
| | ½" ice + 8 lb wind | 2.05 | 4055 | 5.00 | 6340 | 10.38 | 9125 | | | | | | | | |
| | 1" ice | 2.95 | 5845 | 7.21 | 9135 | 14.95 | 13150 | | | | | | | | |
| 2½ | Bare | 0.47 | 1130 | 1.15 | 1765 | 2.38 | 2540 | 4.42 | 3455 | | | | | | |
| | ½" ice | 0.96 | 2310 | 2.36 | 3610 | 4.89 | 5200 | 9.05 | 7080 | | | | | | |
| | ½" ice + 8 lb wind | 1.14 | 2730 | 2.78 | 4270 | 5.77 | 6150 | 10.70 | 8370 | | | | | | |
| | 1" ice | 1.61 | 3845 | 3.92 | 6010 | 8.13 | 8655 | 15.06 | 11780 | | | | | | |
| 3 | Bare | 0.31 | 910 | 0.76 | 1425 | 1.58 | 2050 | 2.93 | 2790 | 5.00 | 3640 | | | | |
| | ½" ice | 0.61 | 1775 | 1.49 | 2775 | 3.08 | 3995 | 5.71 | 5440 | 9.74 | 7105 | | | | |
| | ½" ice + 8 lb wind | 0.71 | 2060 | 1.72 | 3220 | 3.58 | 4635 | 6.62 | 6310 | 11.30 | 8240 | | | | |
| | 1" ice | 0.98 | 2860 | 2.39 | 4465 | 4.96 | 6430 | 9.19 | 8755 | 15.68 | 11435 | | | | |

| | | | | | | | | | | | | | | | |
|---|---|---|---|---|---|---|---|---|---|---|---|---|---|---|---|
| 3½ | Bare | 0.24 | 790 | 0.58 | 1230 | 1.20 | 1775 | 2.22 | 2415 | 3.79 | 3155 | 6.06 | 3995 | | |
| | ½" ice | 0.45 | 1490 | 1.09 | 2330 | 2.26 | 3355 | 4.19 | 4565 | 7.15 | 5960 | 11.46 | 7545 | | |
| | ½" ice + 8 lb wind | 0.51 | 1710 | 1.25 | 2670 | 2.59 | 3845 | 4.81 | 5230 | 8.20 | 6835 | 13.14 | 8650 | | |
| | 1" ice | 0.70 | 2350 | 1.72 | 3670 | 3.57 | 5280 | 6.61 | 7190 | 11.27 | 9390 | 18.05 | 11885 | | |
| 4 | Bare | 0.19 | 695 | 0.45 | 1090 | 0.94 | 1565 | 1.74 | 2130 | 2.97 | 2785 | 4.76 | 3525 | 7.25 | 4350 |
| | ½" ice | 0.34 | 1275 | 0.83 | 1995 | 1.72 | 2870 | 3.19 | 3910 | 5.45 | 5105 | 8.72 | 6465 | 13.30 | 7980 |
| | ½" ice + 8 lb wind | 0.39 | 1450 | 0.94 | 2265 | 1.96 | 3260 | 3.62 | 4435 | 6.18 | 5795 | 9.90 | 7335 | 15.09 | 9055 |
| | 1" ice | 0.53 | 1975 | 1.28 | 3085 | 2.66 | 4440 | 4.93 | 6045 | 8.42 | 7895 | 13.49 | 9990 | 20.55 | 12330 |
| 4½ | Bare | 0.15 | 620 | 0.36 | 970 | 0.76 | 1400 | 1.40 | 1905 | 2.39 | 2490 | 3.83 | 3150 | 5.83 | 3890 |
| | ½" ice | 0.27 | 1115 | 0.65 | 1740 | 1.35 | 2505 | 2.51 | 3410 | 4.28 | 4455 | 6.85 | 5640 | 10.44 | 6960 |
| | ½" ice + 8 lb wind | 0.30 | 1255 | 0.73 | 1960 | 1.52 | 2820 | 2.82 | 3840 | 4.81 | 5015 | 7.71 | 6345 | 11.75 | 7835 |
| | 1" ice | 0.41 | 1695 | 0.99 | 2650 | 2.06 | 3810 | 3.81 | 5190 | 6.51 | 6780 | 10.42 | 8580 | 15.89 | 10590 |
| 5 | Bare | 0.12 | 555 | 0.29 | 870 | 0.61 | 1250 | 1.12 | 1705 | 1.92 | 2225 | 3.07 | 2815 | 4.69 | 3475 |
| | ½" ice | 0.21 | 970 | 0.51 | 1520 | 1.06 | 2185 | 1.96 | 2975 | 3.35 | 3885 | 5.37 | 4920 | 8.18 | 6070 |
| | ½" ice + 8 lb wind | 0.23 | 1085 | 0.57 | 1695 | 1.18 | 2440 | 2.19 | 3320 | 3.74 | 4335 | 5.99 | 5490 | 9.13 | 6775 |
| | 1" ice | 0.31 | 1455 | 0.77 | 2275 | 1.59 | 3275 | 2.94 | 4455 | 5.02 | 5820 | 8.04 | 7365 | 12.26 | 9095 |
| 6 | Bare | 0.08 | 465 | 0.20 | 725 | 0.42 | 1040 | 0.79 | 1420 | 1.34 | 1850 | 2.15 | 2345 | 3.28 | 2895 |
| | ½" ice | 0.14 | 775 | 0.34 | 1210 | 0.71 | 1745 | 1.32 | 2375 | 2.25 | 3105 | 3.60 | 3930 | 5.49 | 4850 |
| | ½" ice + 8 lb wind | 0.15 | 855 | 0.38 | 1335 | 0.78 | 1925 | 1.45 | 2615 | 2.48 | 3420 | 3.97 | 4325 | 6.05 | 5340 |
| | 1" ice | 0.21 | 1135 | 0.50 | 1770 | 1.04 | 2550 | 1.92 | 3470 | 3.28 | 4530 | 5.26 | 5735 | 8.02 | 7080 |

* From Kaiser Aluminum Electrical Conductor Technical Manual.

The tabulated deflections are for single-span, simply supported buses. Deflections for fixed-end buses are one-fifth of the values given above, and the deflections for continuous buses for the center spans are also one-fifth of the values above. The deflections for the end spans are two-fifths of the values given. The stresses given in the above table are the stresses in the outer fibers as calculated for simply supported beams with a uniformly distributed load.

NOTE: 1 in = 25.4 mm; 1 ft = 0.3048 m; 1 lb = 0.4536 kg; 1 lb/in$^2$ = 6.895 kPa.

**TABLE 17-9** Aluminum Round Tubular Busbar Deflections and Stresses*

(Extra-heavy pipe sizes)

| IPS size, in | Loading | Span, ft 20 Deflection, in | 20 Stress, lb/in² | 25 Deflection, in | 25 Stress, lb/in² | 30 Deflection, in | 30 Stress, lb/in² | 35 Deflection, in | 35 Stress, lb/in² | 40 Deflection, in | 40 Stress, lb/in² | 45 Deflection, in | 45 Stress, lb/in² | 50 Deflection, in | 50 Stress, lb/in² |
|---|---|---|---|---|---|---|---|---|---|---|---|---|---|---|---|
| 1¼ | Bare | 1.54 | 2130 | | | | | | | | | | | | |
| | ½" ice | 3.54 | 4900 | | | | | | | | | | | | |
| | ½" ice + 8 lb wind | 4.42 | 6110 | | | | | | | | | | | | |
| | 1" ice | 6.47 | 8945 | | | | | | | | | | | | |
| 1½ | Bare | 1.15 | 1825 | 2.82 | 2855 | | | | | | | | | | |
| | ½" ice | 2.53 | 4005 | 6.17 | 6255 | | | | | | | | | | |
| | ½" ice + 8 lb wind | 3.09 | 4895 | 7.55 | 7645 | | | | | | | | | | |
| | 1" ice | 4.47 | 7085 | 10.92 | 11070 | | | | | | | | | | |
| 2 | Bare | 0.72 | 1425 | 1.76 | 2225 | | | | | | | | | | |
| | ½" ice | 1.46 | 2890 | 3.57 | 4520 | | | | | | | | | | |
| | ½" ice + 8 lb wind | 1.73 | 3430 | 4.23 | 5360 | | | | | | | | | | |
| | 1" ice | 2.46 | 4870 | 6.01 | 7610 | | | | | | | | | | |
| 2½ | Bare | 0.49 | 1185 | 1.21 | 1850 | 2.50 | 2665 | | | | | | | | |
| | ½" ice | 0.89 | 2125 | 2.17 | 3320 | 4.49 | 4780 | | | | | | | | |
| | ½" ice + 8 lb wind | 1.01 | 2420 | 2.46 | 3780 | 5.11 | 5440 | | | | | | | | |
| | 1" ice | 1.40 | 3345 | 3.41 | 5225 | 7.07 | 7525 | | | | | | | | |
| 3 | Bare | 0.33 | 955 | 0.80 | 1495 | 1.66 | 2150 | 3.07 | 2925 | | | | | | |
| | ½" ice | 0.56 | 1625 | 1.36 | 2540 | 2.82 | 3660 | 5.23 | 4980 | | | | | | |
| | ½" ice + 8 lb wind | 0.62 | 1815 | 1.52 | 2840 | 3.15 | 4085 | 5.84 | 5560 | | | | | | |
| | 1" ice | 0.85 | 2465 | 2.06 | 3850 | 4.28 | 5545 | 7.93 | 7550 | | | | | | |

| | | | | | | | | | | | | | | | |
|---|---|---|---|---|---|---|---|---|---|---|---|---|---|---|---|
| 3½ | Bare | 0.25 | 825 | 0.60 | 1290 | 1.25 | 1860 | 2.32 | 2530 | 3.96 | 3305 | | | | |
| | ½" ice | 0.41 | 1360 | 1.00 | 2125 | 2.07 | 3060 | 3.83 | 4165 | 6.53 | 5440 | | | | |
| | ½" ice + 8 lb wind | 0.45 | 1500 | 1.10 | 2345 | 2.28 | 3380 | 4.23 | 4600 | 7.21 | 6010 | | | | |
| | 1" ice | 0.60 | 2015 | 1.48 | 3145 | 3.06 | 4530 | 5.67 | 6170 | 9.67 | 8055 | | | | |
| 4 | Bare | 0.19 | 725 | 0.47 | 1135 | 0.98 | 1635 | 1.82 | 2230 | 3.10 | 2910 | 4.97 | 3680 | 7.58 | 4545 |
| | ½" ice | 0.31 | 1165 | 0.76 | 1820 | 1.57 | 2620 | 2.91 | 3565 | 4.97 | 4655 | 7.96 | 5895 | 12.13 | 7275 |
| | ½" ice + 8 lb wind | 0.34 | 1270 | 0.83 | 1990 | 1.72 | 2865 | 3.18 | 3900 | 5.43 | 5095 | 8.70 | 6445 | 13.26 | 7955 |
| | 1" ice | 0.45 | 1690 | 1.10 | 2640 | 2.28 | 3800 | 4.22 | 5170 | 7.20 | 6755 | 11.54 | 8550 | 17.59 | 10555 |
| 4½ | Bare | 0.16 | 650 | 0.38 | 1015 | 0.79 | 1460 | 1.46 | 1990 | 2.49 | 2600 | 4.00 | 3290 | 6.09 | 4060 |
| | ½" ice | 0.24 | 1015 | 0.59 | 1585 | 1.23 | 2285 | 2.28 | 3110 | 3.90 | 4060 | 6.24 | 5135 | 9.51 | 6340 |
| | ½" ice + 8 lb wind | 0.26 | 1100 | 0.65 | 1720 | 1.34 | 2475 | 2.48 | 3370 | 4.23 | 4405 | 6.77 | 5575 | 10.32 | 6880 |
| | 1" ice | 0.35 | 1445 | 0.85 | 2260 | 1.76 | 3255 | 3.26 | 4430 | 5.55 | 5785 | 8.90 | 7320 | 13.56 | 9040 |
| 5 | Bare | 0.13 | 580 | 0.31 | 905 | 0.63 | 1305 | 1.17 | 1775 | 2.00 | 2320 | 3.20 | 2935 | 4.88 | 3625 |
| | ½" ice | 0.19 | 885 | 0.47 | 1380 | 0.97 | 1990 | 1.79 | 2710 | 3.05 | 3535 | 4.89 | 4475 | 7.45 | 5525 |
| | ½" ice + 8 lb wind | 0.21 | 950 | 0.50 | 1490 | 1.04 | 2140 | 1.93 | 2915 | 3.29 | 3810 | 5.26 | 4820 | 8.02 | 5950 |
| | 1" ice | 0.27 | 1240 | 0.65 | 1935 | 1.35 | 2785 | 2.51 | 3795 | 4.28 | 4955 | 6.85 | 6270 | 10.44 | 7745 |
| 6 | Bare | 0.09 | 485 | 0.21 | 755 | 0.44 | 1090 | 0.82 | 1485 | 1.40 | 1940 | 2.25 | 2455 | 3.43 | 3030 |
| | ½" ice | 0.13 | 700 | 0.31 | 1095 | 0.64 | 1580 | 1.19 | 2150 | 2.04 | 2810 | 3.26 | 3555 | 4.97 | 4390 |
| | ½" ice + 8 lb wind | 0.13 | 745 | 0.33 | 1165 | 0.68 | 1675 | 1.27 | 2280 | 2.16 | 2980 | 3.46 | 3775 | 5.27 | 4660 |
| | 1" ice | 0.17 | 950 | 0.42 | 1485 | 0.87 | 2140 | 1.61 | 2910 | 2.75 | 3800 | 4.41 | 4810 | 6.72 | 5940 |

* From Kaiser Aluminum Electrical Conductor Technical Manual.

The tabulated deflections are for single-span, simply supported buses. Deflections for fixed-end buses are one-fifth of the values given above, and the deflections for continuous buses for the center spans are also one-fifth of the values above. The deflections for the end spans are two-fifths of the values given. The stresses given in the above table are the stresses in the outer fibers as calculated for simply supported beams with a uniformly distributed load.

NOTE: 1 in = 25.4 mm; 1 ft = 0.3048 m; 1 lb = 0.4536 kg; 1 lb/in² = 6.895 kPa.

assumed to be 8 lb/ft² of the projected area of tubing including ½ in of ice. Large deflections should be avoided even if the maximum bending stress is found to be within safe limits. It is generally satisfactory, in approximation of bus diameter, to allow 1 in of bus outside diameter for every 10 ft of bus span. Refer to the note below Tables 17-8 and 17-9 for the method of support and number of spans.

Stresses on disconnecting switches under short-circuit conditions may be sufficient to open them, with disastrous results; therefore, modern switch designs embody locks, or overcenter mechanisms, to prevent this from occurring. The force on the switch blade varies as the square of the current. This force will be increased if the return circuit passes behind the switch and will vary inversely as the distance from the center of the switch blade to the center of the return conductor.

Bus supports are designed for definite cantilever strength, expressed in inch-pounds and measured at the cap supporting the conductor clamp. Ample margin of safety with regard to insulation and structural strength should be provided, manufacturers' data should be carefully checked, and units so selected that allowable values for the particular units are not exceeded. Good practice recommends that the working load must not exceed 40% of the published rating, and short-circuit loads must not exceed the insulator published rating. These loads should include forces for ultimate short-circuit growth and worst mechanical loading.

***28. Protective Relaying.*** The substation employs many protective relaying systems in order to protect the equipment associated with the station, the most important of which are:

**a.** Transmission lines emanating from the station

**b.** Step-up and step-down transformers

**c.** Station buses

**d.** Breaker failure

**e.** Shunt reactors

**f.** Shunt and series capacitors

Substations serving bulk transmission systems for HV, EHV, and UHV circuits must provide a high order of reliability and security in order to provide continuity of service to the electric system. More and more emphasis is being placed on very sophisticated relaying systems which must function reliably at high speeds to clear line and station faults with maximum security and freedom from false tripping.

Many EHV and UHV systems now use two sets of protective relays for lines, buses, and transformers. Many utilities use one set of electromechanical relays for transmission-line protection, with a completely separate, redundant set of solid-state relays to provide a second protective relaying package. Use of two separate sets of relays operating from separate potential and current transformers and from separate station batteries allows testing of relays without the necessity of removing the protected line or bus from service. For more difficult relaying applications, such as EHV lines using series capacitors in the line, some companies use two sets of solid-state relays to provide the protection systems.

Transmission-line relay terminals are located at the substation and comprise many different types of relaying schemes as follows:

**a.** Direct underreaching

**b.** Permissive underreaching

**c.** Permissive overreaching

**d.** Directional comparison

**e.** Phase comparison

**f.** Pilot wire

These schemes comprise pilot relaying systems applicable for the protection of bulk-power transmission lines and are briefly described below. Pilot relaying is an adaptation of the principle of differential relaying to line protection and functions to provide high-speed clearing of the line for faults anywhere on the line. Pilots include wire pilot, using a two-wire pair between the ends of the line; carrier current pilot; microwave pilot; fiber-optics pilot; and the use of audio tone equipment over wire, carrier, fiber optics, or microwave. The transmission lines may have two or more terminals each with circuit breakers for disconnecting the line from the rest of the power system. All the relaying systems described can be used on two-terminal or multiterminal lines. The relaying systems program the automatic operation of the circuit breakers during power-system faults.

***29. Direct Underreaching Fault Relays.*** These relays (Fig. 17-14) at each terminal of the

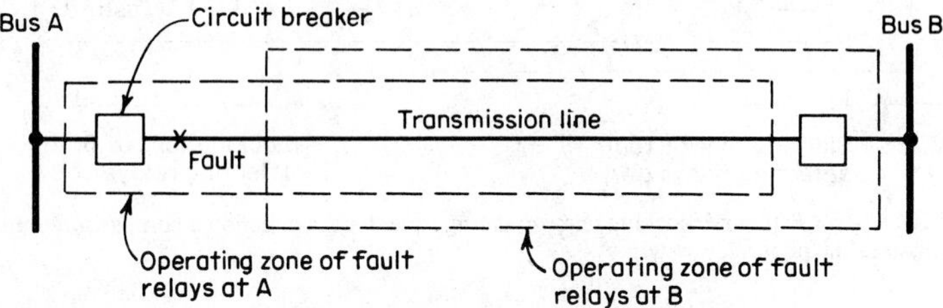


**FIG. 17-14** Fault relay operating zones for the underreaching transfer trip transmission-line pilot relaying system.

protected line sense fault power flow into the line. Their zones of operation must overlap but not overreach any remote terminals. The operation of the relays at any terminal initiates both the opening of the local breaker and the transmission of a continuous remote tripping signal to effect instantaneous operation of all remote breakers. For example, in Fig. 17-14, for a line fault near bus *A*, the fault relays at *A* open (trip) breaker *A* directly and send a transfer trip signal to *B*. The reception of this trip signal at *B* trips breaker *B*.

***30. Permissive Underreaching Relays.*** The operation and equipment for this system are the same as that of the direct underreaching system, with the addition of fault detector units at each terminal. The fault detectors must overreach all remote terminals. They are used to provide added security by supervising remote tripping. Thus, the fault relays operate as shown in Fig. 17-14, and the fault detectors as shown in Fig. 17-15. As an example, for a fault near *A* in Fig. 17-14, the fault relays at *A* trip breaker *A* directly and send a transfer trip signal to *B*. The reception of the trip signal plus the operation of the fault detector relays at *B* (Fig. 17-15) trips breaker *B*.

***31. Permissive Overreaching Relays.*** Fault relays at each terminal of the protected line sense fault power flow into the line, with their zones of operation overreaching all remote terminals. Both the operation of the local fault relays and a transfer trip signal from all of the remote terminals are required to trip any breaker. Thus, in the example of Fig. 17-15 for the line

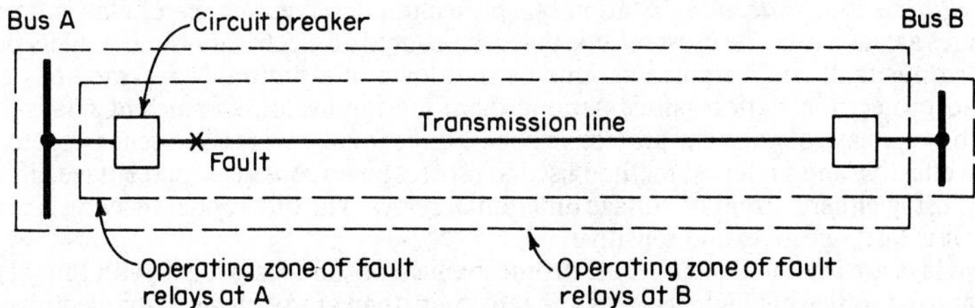


**FIG. 17-15** Fault relay operating zones for the overreaching transmission-line pilot relaying system.

fault near *A*, fault relays at *A* operate and transmit a trip signal to *B*. Similarly the relays at *B* operate and transmit a trip signal to *A*. Breaker *A* is tripped by the operation of the fault relay *A* plus the remote trip signal from *B*. Likewise, breaker *B* is tripped by the operation of fault relay *B* plus the remote trip signal from *A*.

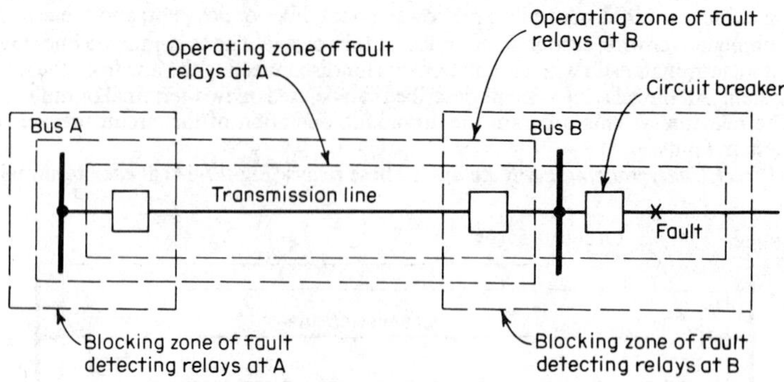


**FIG. 17-16** Fault and blocking relay operating zones for the directional-comparison transmission-line pilot relay system.

***32. Directional-Comparison Relays.*** The channel signal in these systems (Fig. 17-16) is used to block tripping in contrast to its use to initiate tripping in the previous three systems. Fault relays at each terminal of the protected line section sense fault power flow into the line. Their zones of operation must overreach all remote terminals. Additional fault detecting units are required at each terminal to initiate the channel blocking signal. Their operating zones must extend further or be set more sensitively than the fault relays at the far terminals. For example, in Fig. 17-15, the blocking zone at *B* must extend further behind breaker *B* (to the right) than the operating zone of the fault relays at *A*. Correspondingly, the blocking zone at *A* must extend further out into the system (to the left) than the operating zone of the fault relays at *B*.

For an internal fault on line *AB*, no channel signal is transmitted (or if transmitted, it is cut off by the fault relays) from any terminal. In this absence of any channel signal, fault relays at *A* instantly trip breaker *A*, and fault relays at *B* instantly trip breaker *B*. For the external fault to the right of *B* as shown in Fig. 17-15, the blocking zone relays at *B* transmit a blocking channel signal to prevent the fault relays at *A* from tripping breaker *A*. Breaker *B* is not tripped because the *B* operating zone does not see this fault.

***33. Phase-Comparison Relays.*** The three line currents at each end of the protected line are converted into a proportional single-phase voltage. The phase angles of the voltages are compared by permitting the positive half-cycle of the voltage to transmit a half-wave signal block over the pilot channel. For external faults these blocks are out of phase so that alternately the local and then the remote signal provide essentially a continuous signal to block or prevent tripping. On internal faults the local and remote signals are essentially in phase so that approximately a half cycle of no channel signal exists. This is used to permit the fault relays at each terminal to trip their respective breakers.

***34. Station Bus Protection.*** Station bus protection deserves very careful attention, since bus failures are, as a rule, the most serious that can occur to an electrical system. Unless properly isolated, a bus fault could result in complete shutdown of a station. Many methods are employed to protect the station buses. Among them are the use of overcurrent relays, backup protection by relays of adjacent protective zones, directional-comparison schemes, etc. By far the most effective and preferred method used to protect buses consists of percentage differential relaying, using either current or voltage differential relays. The differential relaying is preferred because it is fast, selective, and sensitive.

The relays are available in either electromechanical or solid-state form, with the solid-state units featuring somewhat higher speeds and sensitivity than are available in the electromechanical models. Operating times of 5 to 8 ms can be achieved with solid-state bus differential relays.

Because of the high magnitude of currents encountered during bus faults, current transformers may saturate and thus cause false tripping during external faults. The possibility of ac and dc saturation during faults makes it mandatory that current transformers used for bus differential protection be accurate and of the best quality possible. Also, current transformers should be matched to provide similar ratios and characteristics.

Some bus differential relays developed in solid-state form in Europe have been designed to function correctly even when using current transformers of inferior quality and different ratios. However, it is considered good practice to provide the best possible current transformers for use in bus differential relay applications.

For a sensitive bus differential scheme using current percentage differential relays, refer to Fig. 17-17. For a percentage differential scheme using high impedance voltage differential relays refer to Fig. 17-18.

Because the effective resistance of the voltage relay coil circuit is so high, of the order of 3000 Ω, a voltage limiting element must be connected in parallel with the rectifier branch in order to prevent the CT secondary voltage from being excessive. The overcurrent relay in series with the voltage limiter provides high-

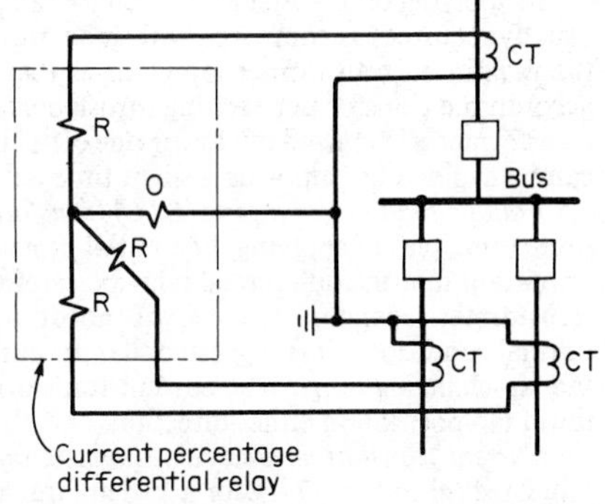


**FIG. 17-17** Bus differential protection using current percentage differential relays. CT: current transformer; O: operating coil; R: restraining coil.

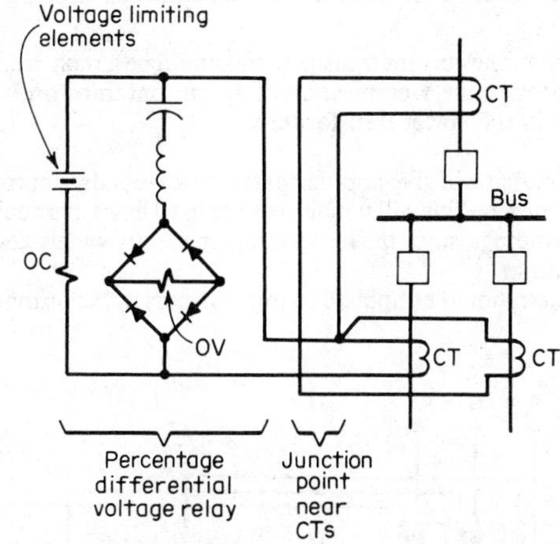


**FIG. 17-18** Bus differential protection using voltage percentage differential relays. OC: high set overcurrent relays; OV: voltage element; CT: current transformer.

speed operation for bus faults of high currents. All current transformer leads are paralleled at a junction point in the substation near the circuit breakers, and only one set of leads is required to be run into the control house where the relay is normally located.

***35. Transformer Protection.*** Transformers may be subjected to short circuits between phase and ground, open circuits, turn-to-turn short circuits, and overheating. Interphase short circuits are rare and seldom develop as such initially, since the phase windings are usually well separated in a three-phase transformer. Faults usually begin as turn-to-turn failures and frequently develop into faults involving ground.

It is highly desirable to isolate transformers with faulty windings as quickly as possible to reduce the possibility of oil fires, with the attendant destruction and the resulting cost for replacements. Differential protection is the preferred type of transformer protection due to its simplicity, sensitivity, selectivity, and speed of operation. If the current-transformer ratios are not perfectly matched, taking into account the voltage ratios of the transformer, autotransformers or auxiliary current transformers are required in the current-transformer secondary circuits to match the units properly so that no appreciable current will flow in the relay operating coil, except for internal fault conditions.

In applying differential protection to transformers, somewhat less sensitivity in the relays is usually required, as compared with generator relays, since they must remain nonoperative for the maximum transformer tap changes that might be used. It is also necessary to take into account the transformer exciting inrush current which may flow in only one circuit when the transformer is energized by closing one of its circuit breakers. As a rule, incorrect relay operation can be avoided by imposing a slight time delay for this condition.

*Voltage-load tap-changing (LTC) transformers* may be protected by differential relays. The same principles of applying differential protection to other transformers hold here as well. It is important that the differential relay be carefully selected so that the unbalance in the current-transformer secondary circuits will not in any case be sufficient to operate the relay under normal conditions. It is suggested that the current transformers be matched at the midpoint of the tap-changing range. The current-transformer error will then be a minimum for the maximum tap position in either direction.

*Current-transformer and relay connections* for various types of differential protection are indicated *(a)* in Fig. 17-19 for a Y-delta transformer and *(b)* in Fig. 17-20 for a three-winding Y-delta-Y transformer. Two rules, frequently used in laying out the wiring for differential protection of transformers whose main windings are connected in Y and delta, are:

1. The current transformers in the leads to the Y-connected winding should be connected in delta; current transformers in the leads to a delta-connected winding should be connected in Y.
2. The delta connection of the current transformers should be a replica of the delta connection of the power transformers; the Y connection of the current transformers should be a replica of the Y connection of the power transformers.

*Current transformers* that will give approximately 5-A secondary current at full load on the transformer should be chosen. This will not be possible in all cases, particularly for transformers having three or more windings, since the kVA ratings may vary widely and may not be proportional to the voltage ratings.

*Overcurrent protection* should be applied to transformers as the primary protection where a

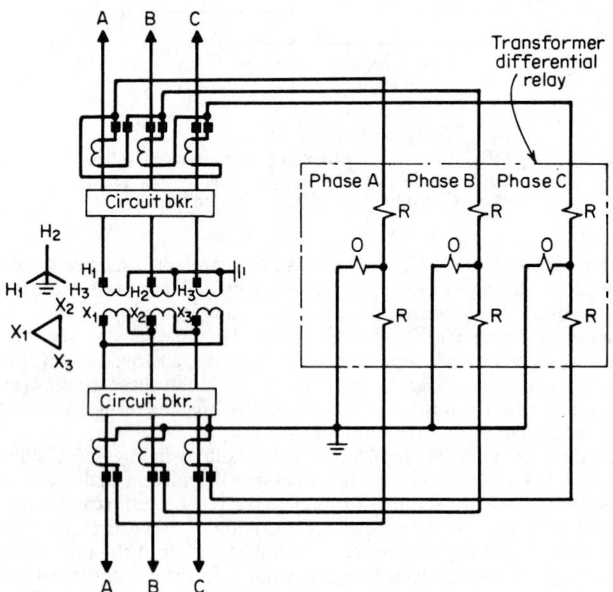


**FIG. 17-19** Transformer differential protection for a Y-delta transformer.

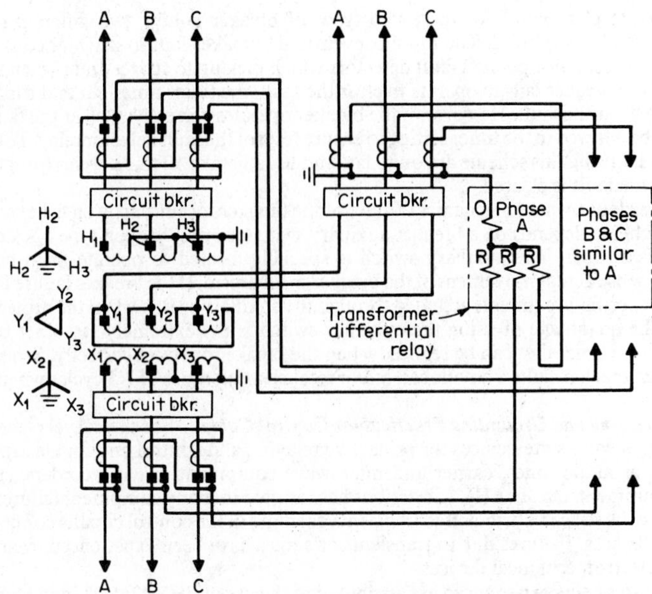


**FIG. 17-20** Transformer differential protection for a Y-delta-Y transformer.

differential scheme cannot be justified or as "backup" protection if differential is used. Frequently faster relaying may be obtained for power flow from one direction by the use of power-directional relays.

*Transformer overheating protection* is sometimes provided to give an indication of overtemperature, rarely to trip automatically. Overload relays of the replica type may be connected in the current-transformer circuits to detect overloading of the unit. Others operate on top-oil temperature, and still others operate on top-oil temperature supplemented with heat from an adjacent resistor connected to a current transformer in the circuit. A recently developed sensor using a glass chip sensitive to temperature changes employs fiber-optics techniques to measure winding hot-spot temperatures.

Gas- or oil-pressure relays are available for attachment to the top or side of transformer tanks to indicate winding faults which produce gas or sudden pressure waves in the oil. Rapid collection of gas or pressure waves in the oil, due to short circuits in the winding, will produce fast operation. New more sophisticated methods to detect incipient failures by frequent monitoring of gas samples are being developed.

***36. Circuit-Breaker Protection.*** In recent years great emphasis has been placed on the need to provide backup protection in the event of failure of a circuit breaker to clear a fault following receipt of a trip command from protective relays. For any fault the protective relays operate to trip the necessary circuit breakers. In addition, these same protective relays, together with breaker-failure fault detector relays, will energize a timer to start the breaker-failure backup scheme. If any breaker fails to clear the fault, the protective relays will remain picked up, permitting the timer to time out and trip the necessary other breakers to clear the fault.

Circuit-breaker failure can be caused by loss of dc trip supply, blown trip fuses, trip coil failure, failure of breaker trip linkages, or failure of the breaker current-interrupting mechanism. The two basic types of failures are (1) mechanical failure or (2) electrical failure of the breaker to clear the fault.

Mechanical failure occurs when the breaker does not move following receipt of a trip command because of loss of dc trip supply, trip coil failure, or trip linkage failure.

Electrical failure occurs when the breaker moves in an attempt to clear a fault upon receipt of the trip command but fails to break the fault current because of misoperation of the current interrupter itself.

In order to clear faults for these two types of breaker failure, two different schemes of protection can be employed. The more conventional breaker-failure schemes consist of using instantaneous current-operated fault detectors which pick up to start a timer when fault relays operate. If the breaker fails to operate to clear the fault, the timer times out and trips necessary breakers to clear the fault. However, if the breaker operates correctly to clear the fault, enough time must be allowed in the timer setting to ensure reset of the fault detector relay. Total clearing times at EHV using this scheme are quite fast and usually take 10 to 12 cycles from the time of fault until the fault is cleared.

For those faults where mechanical failure of the breakers occurs, an even faster scheme is in use. This scheme depends on a breaker auxiliary switch (normally open type 52-A contact) to initiate a fast timer. The auxiliary switch is specially located to operate from breaker trip linkages to sense actual movement of the breaker mechanism. If the breaker failure is mechanical, the breaker-failure timer is actuated through the auxiliary switch when the protective relays operate. The advantage of using the auxiliary switch is the extremely fast reset time of the breaker-failure timer that can be realized when the breaker operates correctly. Schemes in use with the fast breaker-failure circuit can attain total clearing times of 7.5 cycles when a breaker failure occurs.

***37. Shielding and Grounding Practices for Control Cables.*** For several years the increased application of solid-state devices for protective relaying and control and for electronic equipment such as audio tones, carrier and microwave equipment, event recorders, supervisory control equipment, etc., in EHV substations has resulted in many equipment failures. Many of these failures have been attributed to transients or surges in the control circuits connected to the solid-state devices. Failures due to transients or surges have been experienced even with conventional electromechanical devices.

The failures being experienced are attributed to the use of EHV (345 kV and higher voltage levels) as well as the presence of unusually high short-circuit currents. One of the major sources of transient voltages is the switching of capacitances, for example, the operation of a disconnect switch which generates high-magnitude, high-frequency oscillatory surge currents. The transient magnetic fields associated with these high-frequency surge currents are both electrostatically and magnetically coupled to cables in the area. Induced voltages have been reported to be as high as 10 kV in cables without shielding, and the frequencies of these induced voltages have been reported to be as high as 3 MHz.

In order to avoid insulation breakdown at 10-kV crest and possible false operation of relays, it is important that station design include necessary precautions to limit the undesirable surges and control circuit transients to an acceptable minimum.

In any station design there are several precautions that can be taken. All cable circuits that are used in a substation should be run radially, with each circuit separated from any other circuit and with both supply and return conductors contained within the same cable. If a conductor is routed from the control house to a point in the switchyard with the return circuits following different paths, loops may be formed that are inductive and are subject to magnetically induced voltages. However, when the two conductors involved are both affected by the same field, the voltage appearing between them at the open end should be essentially zero.

Because of ground-mat potential differences and longitudinally induced voltages in the radial circuits, proper cable shielding is necessary to maintain lowest possible voltages on the cable leads. The cables which require shielding include control, current, and potential transformer circuits. The shield should be as low resistance as possible, and it should be connected to the ground grid at least at both ends. To reduce penetration of a magnetic flux through the nonferrous shield (lead, copper, bronze, etc.) a current must flow in the shield to produce a counter flux which opposes the applied flux. Ground-grid conductors should be placed in parallel to and in close proximity to the shield to maintain as low a resistance between the ends as possible, and also to form a small loop to reduce the reactance between ground and the shield. Without close coupling of the conductor and ground shield, the propagation time of the two paths could differ so that a voltage impulse could arrive at the receiving end with a time difference, hence causing an unwanted voltage difference.

All control, potential transformer, and current transformer cables should be shielded, with the shield grounded at the switchyard end and at the control-house end. In addition, each group or run of conduits and cables should be installed with a separate No. 4/0 bare stranded copper cable buried directly in the ground and grounded and bonded to the control cable shield at each end of each cable. The bare copper cable should run as closely as possible to the cable run. The heavy cable functions to provide a low-resistance path in an attempt to prevent heavy fault currents from flowing in the shield and to reduce reactance between ground and shield.

In order to limit induced voltages, the control cable runs should be installed, where possible, at right angles to high-voltage buses. Where it is necessary to run parallel to a high-voltage bus for any appreciable distance, the spacing between cables and high-voltage buses should be made as great as possible. Distances of at least 50 ft should be maintained.

It is further considered good practice to have both current-transformer and potential-transformer leads installed with the ground for the secondary wye neutral made at the control-house end rather than at the switchyard end. Any rise due to induced voltages will be concentrated at the switchyard and will ensure operator safety at the control switchboard in the control house.

The shield can be grounded by using a flexible tinned copper braid of from ½ to 1 in wide. The shielded-cable outer insulation is peeled back, exposing the sheath. The 1-in braid is wrapped around the sheath and soldered carefully to it. The other end of the braid is connected to a lug, and solder should be run over the lug to the braid connection. The lug is then bolted securely to the ground busbar. The flexible copper braid circuits should be kept as short as possible and should be run directly to the ground bus without any bends, if possible.

It should be pointed out that the shields should be grounded at multiple points, rather than at a single point, because of the tendency to lose any advantage from single-point grounding at 50 kHz and above. As an example: Assume that one input and one output terminal of a system are grounded, each at different points on a common ground plane. A small noise voltage will usually exist across these ground points because of currents flowing in the finitely conductive ground plane. If either the load or source ground is lifted, a ground loop is no longer formed and coupling of unwanted signals is minimized. This is the advantage of having one physical ground.

Removal of one of the ground connections achieves a single-point ground only for dc and low-frequency signals. At higher frequencies, ground loops will be created by capacitance coupling. Frequencies below 50 kHz are considered the arbitrary crossover point for single-point grounding. At EHV the transient voltages above 50 kHz represent the more serious problem; for this reason all cable shields should be grounded at least at two points. It should be noted that shielding of control cables is normally provided for substations operating at voltage levels of 138 kV and above.

***38. Substation Grounding.*** Grounding at substations is highly important. The functions of a grounding system are listed below:

**a.** Provide the ground connection for the grounded neutral for transformers, reactors, and capacitors.

**b.** Provide the discharge path for lightning rods, arresters, gaps, and similar devices.

**c.** Ensure safety to operating personnel by limiting potential differences which can exist in a substation.

**d.** Provide a means of discharging and deenergizing equipment in order to proceed with maintenance on the equipment.

**e.** Provide a sufficiently low resistance path to ground to minimize rise in ground potential with respect to remote ground.

Substation safety requirements call for the grounding of all exposed metal parts of switches, structures, transformer tanks, metal walkways, fences, steelwork of buildings, switchboards, instrument-transformer secondaries, etc., so that a person touching or near any of this equipment cannot receive a dangerous shock if a high-tension conductor flashes to or comes in contact with any of the equipment listed. This function in general is satisfied if all metalwork between which a person can complete contact, or which a person can touch when standing on the ground, is so bonded and grounded that dangerous potentials cannot exist. This means that each individual piece of equipment, each structural column, etc., must have its own connection to the station grounding mat.

A most useful source of information concerning substation grounding is contained in the comprehensive guide IEEE Standard 80-1976, IEEE Guide for Safety in Substation Grounding, published in June, 1976. Much of the following information is based on recommendations stated in the IEEE Standard *80*.

*The basic substation ground system* used by most utilities takes the form of a grid of horizontally buried conductors. The reason that the grid or mat is so effective is attributed to the following:

**a.** In systems where the maximum ground current may be very high, it is seldom possible to obtain a ground resistance so low as to ensure that the total rise of the grounding system potential will not reach values unsafe for human contact. This being the case, the hazard can be corrected only by control of local potentials. A grid is usually the most practical way to do this.

**b.** In HV and EHV substations no ordinary single electrode is adequate to provide needed conductivity and current-carrying capacity. However, when several are connected to each other, and to structures, equipment frames, and circuit neutrals which are to be grounded, the result is necessarily a grid, regardless of original objectives. If this grounding network is buried in soil of reasonably good conductivity, this network provides an excellent grounding system.

*The first step in the practical design of a grid or mat* consists of inspecting the layout plan of equipment and structures. A continuous cable should surround the grid perimeter to enclose as much ground as practical and to avoid current concentration and hence high gradients at projecting ground cable ends. Within the grid, cables should be laid in parallel lines, and at reasonably uniform spacing. They should be located, where practical, along rows of structures or equipment to facilitate the making of ground connections. The preliminary design should be adjusted so that the total length of buried conductor, including cross connections and rods, is at least equal to that required to keep local potential differences within acceptable limits.

*A typical grid system* for a substation might comprise 4/0 bare stranded copper cable buried 12 to 18 in below grade and spaced in a grid pattern of about 10 by 20 ft. (Other conductor sizes, burial depths, and grid conductor spacings, however, are frequently used.) At each junction of 4/0 cable, the cables would be securely bonded together, and there might also be connected a driven copper-covered steel rod approximately ⅝ in in diameter and approximately 8 ft long. In very high resistance soils it might be desirable to drive the rods deeper. (Lengths approaching 100 ft are recorded.) A typical grid system usually extends over the entire substation yard, and sometimes a few feet beyond the fence which surrounds the building and equipment. Figure 17-21 shows a grounding plan for a typical EHV substation operating at 345 kV.

In order to ensure that all ground potentials around the station are equalized, the various ground cables or buses in the yard and in the substation building should be bonded together by heavy multiple connections and tied into the main station ground. This is necessary in order that appreciable voltage differences to ground may not exist between the ends of cables which may run from the switchyard to the substation building.

Heavy ground currents such as those which may flow in a transformer neutral during ground faults should not be localized in ground connections (mats or groups of rods) of small area in order to minimize potential gradients in the area around the ground connections. Such areas should have reinforced wire sizes where necessary to handle adequately the most severe condition of fault-current magnitude and duration.

*Copper cables or straps* are usually employed for equipment-frame ground connections. However, transformer tanks are sometimes used as part of the ground path for lightning arresters mounted thereon. Similarly, steel structures may be used as part of the path to ground if it can be established that the conductivity, including that of any joints, is and can be maintained as equivalent to the copper conductor that would otherwise be required. Studies by some utilities have led to their successful use of steel structures as part of the path to the ground mat from overhead ground wires, lightning arresters, etc. Where this practice is followed, any paint films which might otherwise introduce a high-resistance joint should be removed and a suitable joint compound applied or other effective means taken to prevent subsequent deterioration of the joint from oxidation.

Connections between the various ground leads and the cable grid and connections within the cable grid are usually clamped, welded, or brazed. Ordinary soldered connections are to be avoided because of possible failure under high fault currents or because of galvanic corrosion.

Each element of the ground system (including grid proper, connecting ground leads, and electrodes) should be so designed that it will:

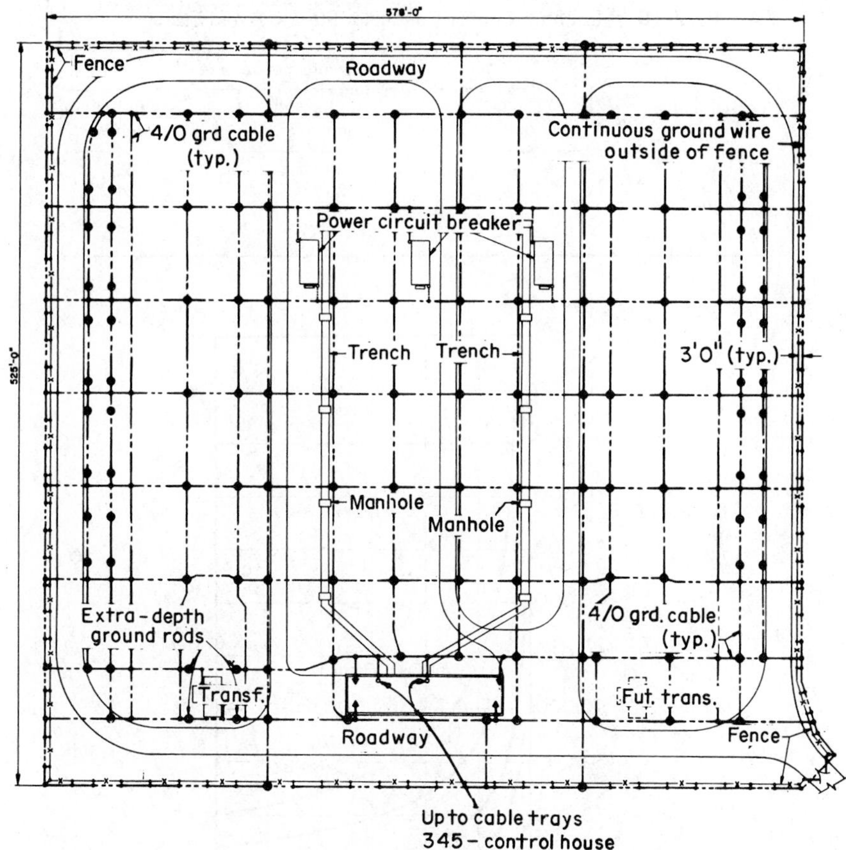


**FIG. 17-21** Grounding plan for a 345-kV substation.

**a.** Resist fusing and deterioration of electric joints under the most adverse combination of fault-current magnitude and fault duration to which it might be subjected.

**b.** Be mechanically rugged to a high degree, especially in locations exposed to physical damage.

**c.** Have sufficient conductivity so that it will not contribute substantially to dangerous local potential differences.

Adequacy of a copper conductor and its joints against fusing can be determined from Table 17-10 and by referring to Fig. 17-22.

If the switchyard is on soil of high resistivity so that it is impossible to obtain suitably low resistance from rods driven within the station, it is possible to reduce the resistance by extending the main ground grid outside the enclosed substation area to a secondary ground mat located

**TABLE 17-10** Minimum Copper Conductor Sizes to Avoid Fusing

| Time duration of fault, s | Circular mils per ampere | | |
|---|---|---|---|
| | Cable only | With brazed joints | With bolted joints |
| 30 | 40 | 50 | 65 |
| 4 | 14 | 20 | 24 |
| 1 | 7 | 10 | 12 |
| 0.5 | 5 | 6.5 | 8.5 |

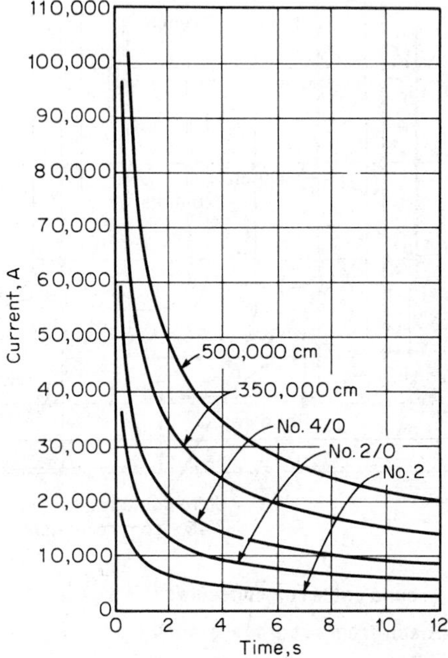


**FIG. 17-22** Short-time fusing curves for copper cable.

adjacent to the substation. The effective resistance of the complete grounding system can be lowered appreciably by the use of a more extensive grid area and of additional grid conductor length. An important reason for trying to lower the grid resistance is to minimize ground-potential rise with respect to remote ground during ground faults.

*Ground-potential rise* depends upon fault-current magnitude, system voltage, and ground-system resistance. The current through the ground system multiplied by its resistance measured from a point remote from the substation determines the ground-potential rise with respect to remote ground. The current through the grid is usually considered to be the maximum available line-to-ground fault current. For example, a ground fault of 15,000 A flowing into a ground grid with a value of 0.5-Ω resistance to absolute earth would cause an *IR* drop of 7500 V. The 7500-V *IR* drop due to the fault current could cause serious trouble to communications lines entering the station if the communications facilities are not properly insulated or neutralized.

*Low-resistance station grounds* are frequently difficult to obtain. In such cases, the use of driven grounds will provide the most convenient means of obtaining a suitable ground connection. The arrangement and number of driven grounds will depend upon the station size and the nature of the soil. The ground mat of Fig. 17-21 has a value measured to be of the order of

0.5 Ω. The best soils for ground mats are wet and marshy, with clay or clay loam as the next best. Sand and sandy soils are of higher resistance, making it difficult to obtain low-resistance ground connections.

The size of the rods used is determined mainly by the depth to which they must be driven, although small rods can be driven to considerable depths by the use of driving collars. Figure 17-23 shows the relationship between rod size and resistance obtained. Driving more rods in a

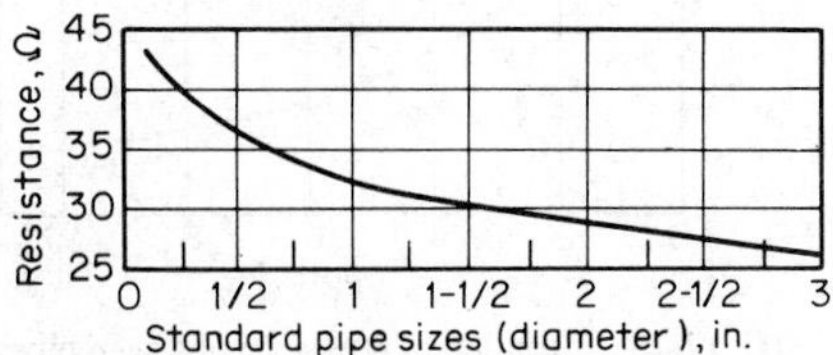


**FIG. 17-23** Relation between pipe diameter and ground resistance. *(NBS Technologic Paper 108, June 1918.)*

given space will help reduce resistance, but the reduced resistance is not a function of the number of rods. Figure 17-24 shows the effect on resistance of spacing and number of rods in

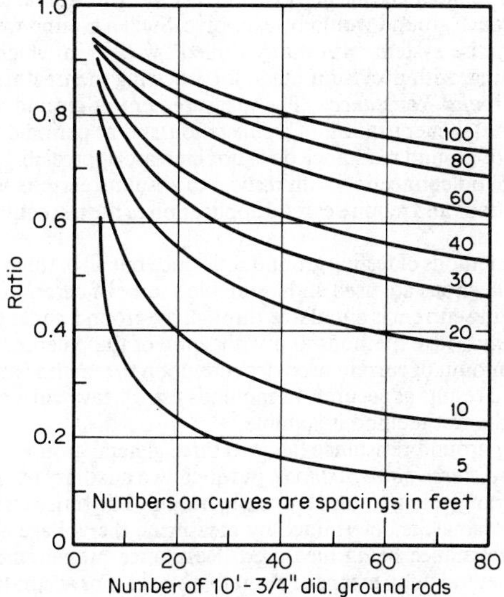


**FIG. 17-24** Ratio of conductivity of ground rods in parallel on an area to that of isolated rods. *(H. B. Dwight, Trans. AIEE, vol. 55, p. 1936.)*

square areas. These curves apply to ¾-in by 10-ft rods. The rods or pipes for permanent stations should be of noncorroding materials. Figure 17-25 shows the effect of increased length of rods in uniform soil. Usually the improvement is much greater than indicated because the rods penetrate into better-conducting earth as they are driven deeper. In addition, where the ground can become frozen, rods must be driven below the frost line to obtain low resistance.

In general, it is advisable to obtain reduced ground resistance by the use of a more extensive mat and more ground rods rather than by treating the earth around the rods with salt, because of

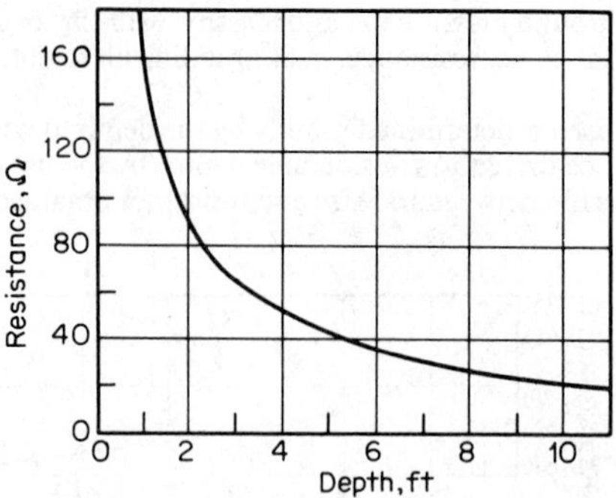


**FIG. 17-25** Variation of resistance of driven pipes with depth. Soil fairly wet. External diameter of pipe is 1.02 in. *(NBS Technologic Paper No. 108, June 1918.)*

the impermanence of the treatment. However, treatment of the soil is sometimes the only means whereby suitable resistance can be obtained.

It is not possible to describe all methods of obtaining ground connections of suitably low resistance. The problem sometimes presents great difficulties and calls for considerable extra expense. Substations should not be located on solid rock with little or no topsoil, since the cost of obtaining a low-resistance ground would be excessive. Such a ground would require the use of an extensive counterpoise system with many drilled "wells," in which electrodes would be inserted in treated filling, with provision made for renewing the treatment.

***39. Measuring Ground Resistance.*** The measurement of ground resistance is necessary both at the time of initial energization of a substation and at periodic intervals thereafter to ensure that the value of ground resistance does not increase appreciably. The measurement of the resistance of a ground connection with respect to absolute earth is somewhat difficult. All results are approximations and require careful application of the test equipment and selection of reference ground points.

There are several methods of testing ground resistance but all of them are similar in that two reference ground connections are used and a suitable source of current is required for the test. Some form of alternating current is circulated through the ground under test in amounts from a few milliamperes, as in bridge methods and with some of the patented ground testers, up to 100 A or more. The amount of current used depends upon the method and methods using very small currents will give results as accurate as methods using heavy currents if the ground under test is one for which the test method is suitable.

Methods of testing ground resistance fall into three general groups:

1. *Triangulation or three-point methods,* in which two auxiliary test grounds and the point to be measured are arranged in a triangular configuration. The series resistance of each pair of ground points in the triangle is determined by measuring the voltage across and the current through the ground resistance being measured. Resistance measurements are made by the voltmeter-ammeter method or by means of a suitable bridge. For accurate results the resistance of the auxiliary grounds and the ground under test should be of the same order of magnitude, and results may be meaningless if the test grounds have more than 10 times the resistance of the ground under test. This method is suitable for measuring the resistance of tower footings, isolated ground rods, or small grounding installations. It is not suitable for measurement of low-resistance grounds such as the ground grid at large substations.

2. *Ratio methods,* in which the series resistance $R$ of the ground under test and a test probe is measured by means of a bridge which operates on the null-balance principle. A calibrated slide-wire potentiometer is connected to the two ground connections, with the slider of the potentiometer connected to a second test probe. The potential of the slider to ground is adjusted to zero or null. If $D$ is the total slide-wire resistance and $d_1$ is the resistance from the slider to the ground under test, the resistance $R$ of the ground under test is $(d_1/D) \times R$. The Vibrometer and the Groundometer, self-contained test instruments, make use of this principle. This method is much more satisfactory than triangulation methods, since ratios of test-probe resistance to the

resistance of the ground under test run as high as 300:1 with test instruments such as the Groundometer. Although this method has its limitations in testing low-resistance grounds of large areas, suitable readings can be obtained by locating the test probes in a straight line, in a direction 90° from the substation fence and with the distance of the furthest probe twice the width of the substation. Best accuracy can be attained by taking measurements at the greatest possible distance from the ground grid being measured.

3. *Fall-of-potential methods,* which include methods using close-in reference grounds, usually less than 1000 ft from the ground under test. The principle of the fall-of-potential method using close-in reference grounds is illustrated by Fig. 17-26. A fixed probe is driven in

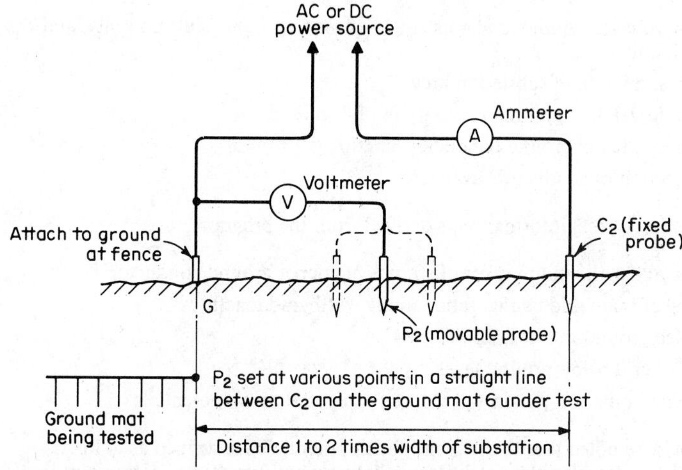


**FIG. 17-26** Field setup for making ground-resistance tests by means of the fall-of-potential method.

the ground at point $C_2$ with a movable probe $P_2$ set at various points in a straight line between $C_2$ and the ground mat $G$ under test. Either alternating current or direct current is circulated through ground $G$ and fixed test probe $C_2$. A voltmeter is connected between point $G$ and probe $P_2$ and an ammeter is connected to observe current flow through probe $C_2$. Voltmeter readings $E$ are taken simultaneously with ammeter readings $I$. The reading $E/I$, which equals the resistance in ohms, is plotted in Fig. 17-27. The resistance shown on the flat part of the curve or

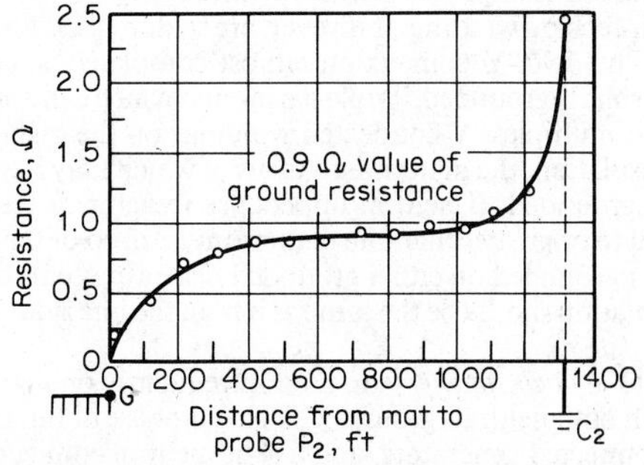


**FIG. 17-27** Ground-resistance curve for a substation ground mat.

at the point of inflection is taken as the resistance of the ground. This method may be subject to considerable error if stray currents are present. It is normally applied by using several test-probe readings at 10% intervals of the distance from $G$ to $C_2$, with the test probe located midway between $G$ and $C_2$. Self-contained test instruments which make use of this method are available; among them are the Ground Ohmer and Megger ground tester. These instruments give better results than the voltmeter-ammeter method, since they are designed to eliminate the effects of stray currents.

In recent years considerable emphasis has been placed on the use of computer programs to calculate the design parameters of substation ground systems. These programs normally employ methods detailed in IEEE Guide 80. Normal input data required to run a typical program consist of the following:

**a.** System voltage, symmetrical rms single-phase-to-ground fault currents, and the clearing time of faults

**b.** Length and width of substation area

**c.** Estimated value of ground resistivity in ohm-meters

**d.** Assumed value of ground conductor length

**e.** Cross section of conductors available

The following typical information is derived from the program:

**a.** Size, total length, and number of strands of copper ground conductor

**b.** Spacing of main grid configuration along width and length

**c.** Expected ground-mat resistance

**d.** Depth of grid below ground level

**e.** Tolerable limits and maximum values of step and touch potentials

It should be noted that the step and touch potentials are defined as follows:

$E_{\text{step}}$ is the tolerable potential difference between any two points on the ground surface which can be touched simultaneously by the feet.

$E_{\text{touch}}$ is the tolerable potential difference between any point on the ground where a person may stand and any point which can be touched simultaneously with either hand.

***40. Transformer Connections.*** *Delta-delta connected* transformers are used mainly on the lower transmission voltages. This is due to the fact that the complete winding must be insulated for full line-to-line voltage; for voltages above 73 kV the cost increase is appreciable over Y-connected transformers with graded insulation. The delta-delta-connected transformers have one advantage in that the bank can be operated open delta at 86.6% of the capacity of the two remaining transformers.

*The delta-star connection* is in common use for both step-up and step-down purposes. When used as a step-up transformer, the high-tension winding is Y-connected; when used for step-down purposes, the low-tension winding is usually Y-connected in order to provide a grounded neutral for secondary transmission or for primary distribution.

*Delta-connected* high-tension windings, however, are seldom used for transmission voltages of 138 kV and above. The delta-Y connection almost completely suppresses the triple harmonics with the neutral solidly grounded. Triple harmonics which can appear on power systems are the third and its odd multiples. Y-connected windings on the higher voltages are usually provided with graded insulation, the neutral-end turns of which may have very little insulation if the neutral is solidly grounded. If neutral impedance (reactor or resistor) is used, neutral insulation must be equal to or greater than the maximum $IZ$ drop of the neutral impedance. If the neutral is to be left ungrounded on either grounded neutral systems or ungrounded neutral systems, the neutral insulation should be the same as it is on the line side to avoid traveling-wave troubles.

*Star-star-connected (Y-Y) transformers* are used infrequently on high-voltage transmission systems. When used with both neutrals grounded, if single-phase or three-phase shell type, they must be used with Y-connected generators, and a solid neutral connection must be provided between the generator, or generators, and the low-tension transformer neutral in order to minimize triple-harmonic troubles. The various types of Y-Y-connected transformers can be used with both neutrals ungrounded with satisfactory results or with neutrals grounded if of the three-phase core type. The triple harmonics are nearly suppressed in three-phase core-type transformers.

Star-star-connected transformers with a delta-connected third winding (tertiary) overcome the difficulties of the simple Y-Y connection. The tertiary winding may be for the suppression of harmonics only, in which case no connections are brought out with three-phase transformers. Y-delta-Y transformers are frequently used to supply two distribution voltages or a distribution voltage and a secondary-transmission voltage. If the service supplied from the delta-connected winding is four-wire three-phase, the neutral must be obtained from a separate grounding transformer. A common use for the tertiary winding is to provide substation station-service power to operate station auxiliary equipment. Three-winding transformers, all windings of which are used, are frequently rated with two outputs: (1) the individual output of each secondary winding alone with the other secondary winding carrying no load and (2) a simultaneous loading rating in which each secondary winding is given a rated loading with the primary-winding loading the resultant of the two secondary loadings.

*Autotransformers* are generally used for transforming from one transmission voltage to another when the ratio is 3 : 1 or less. Such transformers are normally connected in Y with the neutral solidly grounded and when so connected should be provided with a closed delta tertiary winding of adequate capacity for the suppression of harmonics, for ground-fault duty, and to provide station-service power. The tertiary is frequently used to provide a supply of distribution voltage. Autotransformers are superior to separate-winding transformers owing to their lower cost, greater efficiency, smaller size and weight, and better regulation. Autotransformers may also be obtained with zigzag-connected windings or with delta-connected windings. Both these types are free from triple-harmonic troubles but in general are more expensive.

*Delta-connected autotransformers* have a possible disadvantage in that they insert a phase shift into the transformation, which means that the system being served must be radial or else it must be served by similar transformations at other points.

***41. Transformer Loading Practice.*** Because of the varying load cycle of most transformers, it is customary to permit loading considerably in excess of the transformer nameplate rating. There may be limitations on the transformer imposed by bushings, leads, tap changers, cables, disconnecting switches, circuit breakers, etc. Good engineering design, however, will permit operation without these limitations.

The increase in transformer loading is limited by the effect of temperature on insulation life. High temperature decreases the mechanical strength and increases the brittleness of fibrous insulation and makes transformer failure increasingly likely even though the dielectric strength of the insulation may not be seriously decreased. Overloading should be limited then by giving consideration to the effect on insulation life and transformer life. For recurring loads, such as the daily load cycles, the transformer would be operated for normal life expectancy. For emergencies, either planned or accidental, loading would be based on some percentage loss of life.

In a typical case for a failure of part of the electrical system, a 2.5% loss of life per day for a transformer may be acceptable. Loading recommendations based on the evaluation of the loss

**TABLE 17-11** Percent Daily Peak Load for Normal Life Expectancy with 30°C Cooling Air

| Duration of peak load, h | Self-cooled with % load before peak of | | | Forced-air-cooled up to 133% of self-cooled rating, with % load before peak of | | | Forced-air-cooled over 133% of self-cooled rating, or forced-oil-cooled, with % load before peak of | | |
|---|---|---|---|---|---|---|---|---|---|
| | 50% | 70% | 90% | 50% | 70% | 90% | 50% | 70% | 90% |
| 0.5 | 189 | 178 | 164 | 182 | 174 | 161 | 165 | 158 | 150 |
| 1 | 158 | 149 | 139 | 150 | 143 | 135 | 138 | 133 | 128 |
| 2 | 137 | 132 | 124 | 129 | 126 | 121 | 122 | 119 | 117 |
| 4 | 119 | 117 | 113 | 115 | 113 | 111 | 111 | 110 | 109 |
| 8 | 108 | 107 | 106 | 107 | 107 | 106 | 106 | 106 | 105 |

of insulation life as affected by temperature are contained in USAS C57.92-1962, Guide for Loading Oil-Immersed Power Transformers with 55°C Average Winding Rise Insulation Systems. *NEMA Publ.* TR98-1964 contains corresponding recommendations for loading power transformers with 65°C average winding rise insulation systems. USAS C57.92-1962 Guide states that an average loss of life of 1% per year or 5% in any one emergency operation is considered reasonable.

Daily overload cycles consistent with normal life expectancy for air-cooled power transformers at 30°C ambient temperature are given in Table 17-11, which is a condensation of data taken from USAS C57.92-1962, Section 92-01.250. For a listing of transformer loading above normal with some sacrifice of life expectancy, data given in *NEMA Publ.* TR 98-1964, Part 3, are condensed in Table 17-12.

Ambient temperature affects load capacity by an amount dependent on the type of cooling as shown in Tables 17-11 and 17-12. For changes from this average ambient temperature, transformer ratings may be adjusted as shown in Table 17-13. The table applies to both the 55°C and the 65°C average winding-temperature-rise transformers. For the ambient temperature of air-cooled transformers, use the average value over a 24-h period or 10°C under the maximum during the 24-h period, whichever is higher.

The following temperatures and load limitations are generally applied to transformers. The temperaure of the top oil should never exceed 100°C. The maximum hot-spot winding temperature should not exceed 150°C for 55°C rise transformers or 180°C for 65°C rise transformers. Short-time peak loading for ½ h or more should not exceed 200% rating. At abnormally high temperatures it may be necessary to remove some oil in order to avoid overflow or excessive pressure.

***42. Surge Protection.*** A substation should be designed to include safeguards against the hazards of abnormally high voltage surges that can appear across the insulation of electrical equipment in the station. The most severe overvoltages are caused by lightning strokes and by switching surges.

The main methods to prevent these overvoltages from causing insulation failures include:

a. Use of surge arresters
b. Equipment neutral grounding
c. Proper selection of equipment impulse insulation level
d. Proper selection and coordination of equipment basic insulation levels
e. Careful study of switching-surge levels that can appear in the substation

The main device used to prevent dangerous overvoltages, flashovers, and serious damage to equipment is the surge arrester. The surge arrester conducts high surge currents, such as can be caused by a lightning stroke, harmlessly to ground and thus prevents excessive overvoltages from appearing across equipment insulation. For a detailed description of the characteristics and application of arresters, refer to Sec. **27**.

The important consideration in applying surge arresters and in selecting equipment insulation levels depends greatly on the method of grounding used. Systems are considered to be effectively grounded when the coefficient of grounding does not exceed 80%. Similarly, systems are noneffectively grounded or ungrounded when the coefficient of grounding exceeds 80%.

A value not exceeding 80% is obtained approximately when, for all system conditions, the ratio of zero sequence reactance to positive sequence reactance ($X_0/X_1$) is positive and less than 3, and the ratio of zero sequence resistance to positive sequence reactance ($R_0/X_1$) is positive and less than 1. What this says in effect is that if neutrals are grounded solidly everywhere and if a ground occurs on one of the conductors, then the voltage that can appear on the healthy phases cannot exceed 80% of normal phase-to-phase voltage.

Thus the coefficient of grounding is defined as the ratio of maximum sustained line-to-ground voltage during faults to the maximum operating line-to-line voltage. On many HV and EHV systems the coefficient of grounding may be as low as 70%.

Surge arrester ratings are normally selected on the basis of the coefficient of grounding; thus, for effectively grounded systems the 80% arrester is selected when using the conventional gap-type arrester. When using the gapless metal oxide arrester, a lower-value arrester may be selected based on the maximum continuous operating voltage (MCOV) equal to the maximum

**TABLE 17-12** Allowable Peak Loads (in Multiples of Maximum Nameplate Rating) for Moderate Sacrifice of Life Expectancy with 30°C Cooling Air*

| Duration of peak load, h | Hottest-spot temperature reached, °C | Life loss in percent not more than | Self-cooled (OA) with % load before peak of | | | | Forced-air-cooled (OA/FA) up to 133% of self-cooled rating with % load before peak of | | | | Forced-air-cooled (OA/FA/FA) over 133% of self-cooled rating or forced-oil-cooled (FOA or OA/FOA/FOA) with % load before peak of | | | |
|---|---|---|---|---|---|---|---|---|---|---|---|---|---|---|
| | | | 50% | 70% | 90% | 100% | 50% | 70% | 90% | 100% | 50% | 70% | 90% | 100% |
| ½ | 171 | 0.25 | 2.00 | 2.00 | 2.00 | 1.96 | 2.00 | 1.95 | 1.85 | 1.80 | 1.64 | 1.60 | 1.54 | 1.51 |
| | 180 | 0.50 | 2.00 | 2.00 | 2.00 | 2.00 | 2.00 | 2.00 | 1.95 | 1.90 | 1.69 | 1.66 | 1.60 | 1.57 |
| 1 | 163 | 0.25 | 1.96 | 1.89 | 1.80 | 1.74 | 1.77 | 1.72 | 1.65 | 1.61 | 1.47 | 1.45 | 1.49 | 1.39 |
| | 180 | 1.00 | 2.00 | 2.00 | 1.99 | 1.94 | 1.93 | 1.88 | 1.81 | 1.78 | 1.57 | 1.55 | 1.52 | 1.50 |
| 2 | 155 | 0.25 | 1.68 | 1.63 | 1.57 | 1.53 | 1.53 | 1.50 | 1.47 | 1.44 | 1.33 | 1.32 | 1.31 | 1.30 |
| | 171 | 1.00 | 1.83 | 1.79 | 1.71 | 1.64 | 1.66 | 1.64 | 1.60 | 1.58 | 1.42 | 1.41 | 1.39 | 1.39 |
| | 180 | 2.00 | 1.91 | 1.83 | 1.71 | 1.64 | 1.74 | 1.71 | 1.65 | 1.61 | 1.47 | 1.46 | 1.44 | 1.43 |
| 4 | 147 | 0.25 | 1.44 | 1.41 | 1.39 | 1.37 | 1.35 | 1.34 | 1.33 | 1.32 | 1.24 | 1.23 | 1.23 | 1.23 |
| | 163 | 1.00 | 1.55 | 1.52 | 1.47 | 1.44 | 1.47 | 1.46 | 1.45 | 1.45 | 1.32 | 1.32 | 1.32 | 1.32 |
| | 180 | 4.00 | 1.55 | 1.52 | 1.47 | 1.44 | 1.51 | 1.50 | 1.47 | 1.46 | 1.40 | 1.40 | 1.39 | 1.39 |
| 8 | 139 | 0.25 | 1.28 | 1.27 | 1.27 | 1.26 | 1.24 | 1.24 | 1.24 | 1.24 | 1.18 | 1.18 | 1.18 | 1.18 |
| | 155 | 1.00 | 1.38 | 1.37 | 1.36 | 1.36 | 1.36 | 1.36 | 1.36 | 1.36 | 1.27 | 1.27 | 1.27 | 1.27 |
| | 171 | 4.00 | 1.38 | 1.37 | 1.36 | 1.36 | 1.42 | 1.42 | 1.41 | 1.41 | 1.35 | 1.35 | 1.35 | 1.35 |

* Based on capability tables in *NEMA Publ.* TR98, Part 3.

For forced-air-cooled transformers, the peak loads are calculated on the basis of all cooling being in use during the period preceding the peak load. When operating without fans, use the tables for OA transformers.

Differences in cooling methods used with forced-oil-cooled transformers result in differences in peak-load-carrying ability. Consult the manufacturer before applying loads above the values given in the table.

**TABLE 17-13** Effect of Ambient Temperature on kVA Capacity

| Type of cooling | % of rated kVA decrease in capacity for each °C increase over 30°C air | % of rated kVA increase in capacity for each °C decrease under 30°C |
|---|---|---|
| Self-cooled—OA | 1.5 | 1.0 |
| Forced-air-cooled—OA/FA, OA/FA/FA | 1.0 | 0.75 |
| Forced-air-cooled—FOA, OA/FOA/FOA | 1.0 | 0.75 |

normal line-to-neutral voltage. For example, a 115-kV system (maximum operating voltage equals 121 kV) can use a 97-kV conventional arrester, that is, 80% of 121 kV, when operating on a solidly grounded system, and can use a gapless-type metal oxide arrester rated 70 kV. It should be noted that other factors such as resonant conditions, system switching, etc., could increase the value of the coefficient of grounding and thus should be studied in each individual system.

*The impulse insulation level* of a piece of equipment is a measure of its ability to withstand impulse voltage. It is the crest value, in kilovolts, of the wave of impulse voltage that the equipment must withstand. However, at EHV the switching-surge insulation level may be lower than the corresponding impulse level and thus the switching-surge level becomes the dominant factor in establishing insulation levels.

Basically the coordination of insulation in a substation means the use of no higher-rated arrester than required to withstand the 60-Hz voltage and the choice of equipment insulation levels that can be protected by the arrester. Careful study of switching-surge levels that can occur at the substation as determined, for example, by transient network analyzer studies can also be used to determine and coordinate proper impulse insulation and switching-surge strength required in a substation electrical equipment.

### *43. References*

*Books*

Mason, C. R.: *The Art and Science of Protective Relaying;* New York, John Wiley & Sons, Inc., 1967.

*Electrical Transmission and Distribution Reference Book;* East Pittsburgh, Pa., Westinghouse Electric Corporation, 1950.

Blume, L. F., Boyajian, A., Camilli, G., Lennox, T. S., Minneci, S., and Montsinger, V. M.: *Transformer Engineering;* New York, John Wiley & Sons, Inc., 1951.

Van C. Warrington, A.R.: *Protective Relays, Their Theory and Practice,* vol. 1; London, Chapman & Hall, Ltd., 1971.

Van C. Warrington, A.R.: *Protective Relays, Their Theory and Practice,* vol. 2; London, Chapman & Hall, Ltd., 1974.

*Applied Protective Relaying;* Newark, N.J., Westinghouse Electric Corporation.

*Alcoa Aluminum Bus Conductor Handbook;* Pittsburgh, Pa., Aluminum Company of America, 1957.

*$SF_6$ Gas-Insulated Substations*

IEEE Committee Report, Bibliography of Gas Insulated Substations, *IEEE Trans. Paper* 82-WM-169-1, January 1982.

Troger, H., Boeck, W., Högg, P., Larrue, H., Lightle, D., Mazza, G., Pettersson, K., Tutein, P., and Vigreux, J.: The State of International Development and Experience with $SF_6$ Gas Insulated High Voltage Switchgear; *CIGRE Rep.* 23-01, 1982.

Chu, F. Y., and Tahiliani, V.: Gas Insulated Substations Fault Survey, *IEEE Trans., Power Appar. Syst.,* Nov/Dec 1980, Vol. 99, pp. 2351–2356.

Kawamura, T., Ishii, T., Satoh, K., Hashimoto, Y., Tokoro, K., and Harumoto, Y.: Operating Experience of Gas Insulated Switchgear (GIS) and Its Influence on the Future Substation Design; *CIGRE Rep.* 23-04, 1982.

Boggs, S. A., Chu, F. Y., Hick, M. A., Rishworth, A. B., Trolliet, B., and Vigreux, J.: Prospects for Improving the Reliability and Maintainability of EHV Gas Insulated Substations; *CIGRE Rep.* 23-10, 1982.

Graybill, H. W., Cronin, J. C., and Field, E. J.: Testing of Gas Insulated Substations and Transmission Systems; IEEE Summer Meeting Power Engineering and Systems, 1973, *IEEE Trans. Paper* T73 366-2.

IEEE Committee Report, Safety and Reliability Considerations in Compact Substations; Substation Committee Working Group 70.1, Compact Substations, American Power Conference, April 1972.

Maury, E., Pariselle, R., and Vigreux, I.: Commissioning and Initial Operating Results of 225 kV $SF_6$-insulated Metal-clad Substations; *CIGRE Rep.* 23-06, 1970.

Reimers, T. D., and Owens, J. B.: 345 kV Metal-clad Pressurized $SF_6$ Substation Design; *CIGRE Rep.* 23-04, 1970.

Graybill, H. W., and Williams, J. A.: Underground Power Transmission with Isolated Phase Gas-insulated Conductors; *IEEE Trans. Power Appar. Syst.*, 1970, vol. 89, pp. 17–23.

Takagi, T., Saba, S., and Kataoka, K.: Miniaturization of Substation in Urban District; Paper 1.3-159, Eighth World Energy Conference, Bucharest, 1971.

### *Station Design*

AIEE Committee Report, Basic Structural Design for Transmission Substations Including Light Metals; *Electr. Eng.*, April 1952, vol. 71, pp. 344–350.

AIEE Committee Report, A Guide for Minimum Electrical Clearances for Standard Basic Insulation Levels; *Trans. AIEE, Power Appar. Syst.*, June 1954, vol. 73, pp. 636–641.

Hertig, G. E.: High- and Extra-high-voltage Substation Design and Economic Comparisons; *Trans. AIEE*, 1963, vol. 81, pp. 832–840.

Committee Report, Design Standardization Methods and Techniques for Substation Facilities (Bibliogr.); *Trans. AIEE, Power Appar. Syst.*, October, 1964, vol. 83, pp. 1029–1034.

Dolan, P. R., and Peat, A. J.: Design of the First 500 kV Substations on the Southern California Edison Company System; *IEEE Trans., Power Appar. Syst.*, 1967, vol. 86, pp. 531–539.

Committee Report, 500 kV AC Substation Design Criteria, Summary of Industry Practices; *IEEE Trans., Power Appar. Syst.*, 1969, vol. 88, pp. 854–861.

Scherer, H. N.: 765 kV Station Design; *IEEE Trans. Power Appar. Syst.*, 1969, vol. 88, pp. 1372–1376.

Committee Report, 700/765 kV AC Substation Design Criteria, a Summary of Industry Practices; *IEEE Trans., Power Appar. Syst.*, 1970, vol. 89, pp. 1521–1524.

Colombo, A., Sartorio, G., and Taschini, A.: Phase to Phase Air Clearances in EHV Substations as Required by Switching Surges; *CIGRE Paper* 33-11, 1972.

Committee Report, Minimum Line-to-Ground Electrical Clearances for EHV Substations Based on Switching Surge Requirement; *IEEE Trans., Power Appar. Syst.*, 1972, vol. 91, pp. 1924–1930.

Paris, L., and Taschini, A.: Phase-to-Ground and Phase-to-Phase Air Clearances in Substations; CIGRE, *Electra*, 1973, no. 29. (Recommended by CIGRE S.C. 23 and CIGRE S.C. 33.)

ANSI Standard C2-1984, National Electrical Safety Code.

### *Bus Construction*

Rayleigh, J. W. S.: *Aeolian Tones;* New York, Cambridge University Press, 1920.

Dwight, H. B.: Skin Effect and Proximity Effect in Tubular Conductors; *Trans. AIEE*, February 1922, pp. 189–198.

Wagner, C. F.: Current Distribution in Multi Conductor Single Phase Buses; *Electr. World*, March 18, 1922, vol. 79, no. 11.

Schurig, O. R., and Sayre, M. F.: Mechanical Stresses in Bus Bar Supports During Short Circuits; *J., AIEE*, April 1925, vol. 44, pp. 365–372.

Temple, G., and Brickley, W. G.: *Rayleigh's Principle;* New York, Oxford University Press, 1933.

Higgins, T. J.: Formulas for Calculating Short Circuit Stresses for Bus Supports for Rectangular Tubular Conductors; *Trans. AIEE*, August 1942, vol. 61, pp. 578–580.

Higgins, T. J.: Formulas for Calculating Short Circuit Forces between Conductors of Structural Shape; *Trans. AIEE*, October 1943, vol. 62, pp. 659–663.

Milton, R. M., and Chambers, F.: Behavior of High-voltage Busses and Insulators during Short Circuits; *Trans. AIEE*, August 1955, vol. 74, pp. 742–749.

Taylor, D. W., and Stuehler, C. M.: Short Circuit Tests on 138 kV Busses; *Trans. AIEE*, August 1956, vol. 75, pp. 739–747.

Committee Report, Use of Aluminum for Substation Busses; *IEEE Trans., Power Appar. Syst.*, 1963, vol. 82, pp. 72–102.

Foti, A.: Design and Application of EHV Disconnecting Switches; *Trans. AIEE, Power Appar. Syst.*, October 1965, vol. 84, pp. 868–876.

Attri, N. S., and Edgar, J. N.: Response of Bus Bars on Elastic Supports Subjected to a Suddenly Applied Force; *IEEE Trans., Power Appar. Syst.*, 1967, vol. 86, pp. 636–650.

Fischer, E. G.: Seismic Design of Bus Runs and Supports; *IEEE Trans., Power Appar. Syst.*, 1973, vol. 92, pp. 1493–1500.

ANSI Standard C37.30-1971 (IEEE Standard 324-1971), Definitions and Requirements for High Voltage Air Switches, Insulators and Bus Supports.

ANSI Standard C37.32-1972, Schedules of Preferred Ratings, Manufacturing Specifications, and Application Guide for High Voltage Air Switches, Bus Supports and Switch Accessories.

NEMA Standard SG6-1974, Power Switching Equipment.

*System Protection*

Blackburn, J. L.: Future Automatic Switching of EHV Transmission Lines—Development and Application of Solid-state Relays; *Proc. Am. Power Conf.*, 1965, vol. 27, pp. 998–1008.

IEEE Committee Report, Ground Relaying Practices and Problems: A Power System Relaying Committee Survey; *IEEE Trans., Power Appar. Syst.*, May 1966, vol. PAS-85, no. 5, pp. 524–532.

Sutton, H. J.: The Application of Relaying on an EHV System; *IEEE Trans.*, April 1967, vol. 86, no. 4, pp. 408–415.

Vanderleck, J. M.: Measurement of Composite Error of Relay-type Current Transformers; *Ontario Hydro Res. Q.*, 1967, vol. 19, no. 3, pp. 15–18.

Ungrad, H.: Back-up Protection; *Brown Boveri Rev.*, June 1968, vol. 55, no. 6, pp. 297–305.

Korponay, N., and Ungrad, H.: The Requirements Made of Current Transformers by High-speed Protective Relays; *Brown Boveri Rev.*, June 1968, vol. 55, no. 6, pp. 289–297.

Committee Report, Relaying the Keystone 500 kV System; *IEEE Trans.*, June 1968, vol. 87, no. 5, no. 6, pp. 1434–1439.

Horowitz, S. H., and Seeley, H. T.: Relaying the AEP 765 kV System; *IEEE Trans.*, September 1969, vol. PAS-88, no. 9, pp. 1382–1389.

ANSI Standard C37.90-1978 (IEEE Standard 313-1971), Relays and Relay Systems Associated with Electric Power Apparatus.

ANSI Standard C37.90a-1974 (IEEE Standard 472-1974), Guide for Surge Withstand Capability (SWC) Tests.

Schumm, G. P.: The Philosophy of Protective Relaying in the United States and Europe; *Proc. Am. Power Conf.*, 1971, vol. 33, pp. 1105–1115.

Chadwick, J. W., and Goff, L. E.: Development of a Static Single-pole Relaying Scheme for the TVA 500-kV System; *Proc. Am. Power Conf.*, 1971, vol. 33, pp. 1127–1133.

Boyaris, E., and Guyot, W. S.: Experience with Fault Pressure Relaying and Combustible Gas Detection in Power Transformers; *Proc. Am. Power Conf.*, 1971, vol. 33, pp. 1116–1126.

Narayan, V.: Distance Protection of H. V. and E. H. V. Transmission Lines; *Brown Boveri Rev.*, July 1971, vol. 58, no. 7, pp. 276–286.

Ungrad, H.: Distance Relays with Signal Transmission for Main and Backup Protection; *Brown Boveri Rev.*, July 1971, vol. 58, no. 7, pp. 293–304.

ANSI Standard C37.91-1985 (IEEE Standard 273-1967), Guide for Protective Relay Applications to Power Transformers.

Elmore, W. A.: Some Guidelines for Selecting a Solid-state Transmission Line Relaying System; *Westinghouse Eng.*, March 1972, vol. 32, no. 2, pp. 50–59.

Rockefeller, G. D.: What are the Prospects for Substation-Computer Relaying?; *Westinghouse Eng.*, September 1972, vol. 32, no. 5, pp. 152–156.

Sykes, J. A., and Morrison, I. F.: A Proposed Method of Harmonic Restraint Differential Protection of Transformers by Digital Computer; *IEEE Trans.*, May/June 1972, vol. 91, no. 3, pp. 1266–1272.

Forford, T., and Linders, J. R.: A Half Cycle Bus Differential Relay and Its Applications; *IEEE Trans.*, July/August 1974, vol. 93, no. 4, pp. 1110–1120.

Emanuel, A. E., and Vora, J. P.: Sensor Coil for Internal Fault Protection of Shunt Reactors; *IEEE Trans.*, November/December 1974, vol. 93, no. 6, pp. 1917–1926.

IEEE Committee Report, Bibliography of Relay Literature 1972–1973.

*Shielding of Control Cables*

Sutton, H. J.: Transients Induced in Control Cables Located in EHV Substation, *IEEE Trans.*, July/August 1970, vol. 89, no. 6, p. 1069.

Dietrick, R. E., Ramberg, H. C., and Barber, J. C.: BPA Experience with EMI Measurements and Shielding in EHV Substations; *Proc. Am. Power Conf.*, 1970, vol. 32, pp. 1054–1061.

Kotheimer, W. C.: Control Circuit Transients, Pt. 1; *Power Eng.*, January 1969, vol. 73, no. 1, pp. 42–46.

Kotheimer, W. C.: Control Circuit Transients, Pt. 2; *Power Eng.*, February 1969, vol. 73, no. 2, pp. 54–56.

*Grounding*

Committee Report, Principles and Practices in Grounding; *Edison Electr. Inst. Ser. Rep.* D9, October 1936.

Bellasi, P. L.: Impulse and 60-cycle Characteristics of Driven Grounds; *Trans. AIEE*, March 1941, vol. 60, pp. 123–128.

Eaton, J. R.: Grounding Electric Circuits Effectively, I, II, III; *Gen. Electr. Rev.*, June, July, and August 1941.

AIEE Committee Report, Application Guide on Methods of Substation Grounding; *Trans. AIEE, Power Appar. Syst.*, April, 1954, vol. 73, pp. 271–275.

IEEE Standard 80-1976, IEEE Guide for Safety in Substation Grounding.

Kinyon, A. L.: Earth Resistivity Measurements for Grounding Grids; *Trans. AIEE, Power Appar. Syst.*, December, 1961, vol. 80, pp. 795–800.

IEEE Standard 81-1983, Recommended Guide for Measuring Ground Resistance and Potential Gradients in the Earth.

ANSI C114.1-1973 (IEEE Standard 142-1982), IEEE Recommended Practice for Grounding of Industrial and Commerical Power Systems.

EPRI Final Report EL-2682, Analysis Techniques for Power Substation Grounding Systems, vol. 1, Design Methodology and Tests.

# SECTION 18

# POWER DISTRIBUTION

**Norman R. Schultz**

*Deceased. Formerly Manager, Power Distribution Systems Engineering Operation, General Electric Company; Fellow, IEEE; Executive Chairman, 1971 IEEE Underground Distribution Conference*

**Harold E. Campbell**

*Retired. Formerly Senior Engineer, Power Distribution Systems Engineering Operation, General Electric Company; Fellow, IEEE; Distribution Editor, Electric Forum Magazine; Registered Professional Engineer, New York*

**Jack H. Easley**

*Deceased. Formerly Senior Engineer, Power Distribution Systems Engineering Operation, General Electric Company; Chairman, IEEE Distribution Subcommittee; Distribution Editor, Electric Forum Magazine; Registered Professional Engineer, New York*

**Allen L. Clapp, P.E.**

*Managing Director, Clapp Research Associates; Member IEEE-PES/IAS; Chairman, National Electrical Safety Code Committee*

**Walter J. Ros**

*Manager, Transmission and Distribution Product Application Engineering, Systems Development and Engineering Department, General Electric Company*

## CONTENTS

*Numbers refer to paragraphs*

## ELEMENTS OF DISTRIBUTION PRACTICE

***1. Distribution Defined.*** Broadly speaking, "distribution" includes all parts of an electric utility system between bulk power sources and the consumers' service-entrance equipments. Some electric utility distribution engineers, however, use a more limited definition of distribution as that portion of the utility system between the distribution substations and the consumers' service-entrance equipment. In general, a typical distribution system consists of (1) subtransmission circuits with voltage ratings usually between 12.47 and 245 kV which deliver energy to the distribution substations; (2) distribution substations which convert the energy to a lower "primary system" voltage for local distribution and usually include facilities for voltage regulation of the primary voltage; (3) primary circuits or "feeders," usually operating in the range of 4.16 to 34.5 kV and supplying the load in a well-defined geographical area; (4) distribution transformers in ratings from 10 to 2500 kVA which may be installed on poles, grade-level pads, or in underground vaults near the consumers and transform the primary voltages to utilization voltages; (5) secondary circuits at utilization voltage which carry the energy from the distribution transformer along the street or rear-lot lines; and (6) service drops which deliver the energy from the secondary to the user's service-entrance equipment. Figures 18-1 and 18-2 depict the component parts of a typical distribution system.

***2. Distribution investment*** in the past constituted 35 to 50% of the capital investment of a typical electric utility system. In recent years, the annual capital expenditure for distribution has averaged around 16% of the total electric power system expenditures.

***3. The function of distribution*** is to receive electric power from large, bulk sources and to distribute it to consumers at voltage levels and with degrees of reliability that are appropriate to the various types of users.

For single-phase residential users. American National Standard Institute (ANSI) C84.1-1982 defines "Voltage Range A" as 114/228 V to 126/252 V at the user's service entrance and 110/220 V to 126/252 V at the point of utilization. This allows for voltage drop in the consumer's system. Nominal voltage is 120/240 V. Within Range A utilization voltage, utilization equipment is designed and rated to give fully satisfactory performance.

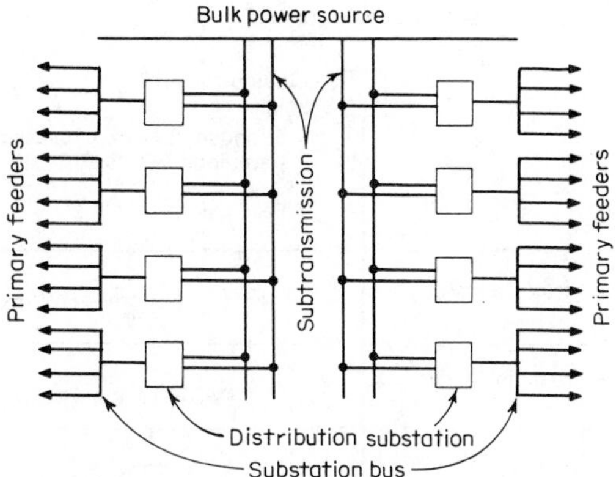


**FIG. 18-1** Typical distribution system.

As a practical matter, voltages above and below Range A do occur occasionally; however, ANSI C84.1 specifies that these conditions shall be limited in extent, frequency, and duration. When they do occur, corrective measures shall be undertaken within a reasonable time to improve voltages to meet Range A requirements.

Rapid dips in voltage which cause incandescent-lamp "flicker" should be limited to 4 or 6% when they occur infrequently and 3 or 4% when they occur several times per hour. Frequent dips, such as those caused by elevators and industrial equipment, should be limited to 1½ or 2%.

Reliability of service is impossible to define in a quantitative manner; it can be described by factors such as frequency and duration of service interruptions. While short and infrequent interruptions may be tolerated by residential and small commercial users, even a short interruption can be costly in the case of many industrial processes and can be dangerous in the case of hospitals and public buildings. For such sensitive loads, special measures are often taken to ensure an especially high level of reliability, such as redundancy in supply circuits and/or supply equipment. Certain computer loads may be sensitive not only to interruptions but even to severe voltage dips and may require special power-supply systems which are virtually uninterruptible.

***4. From a system-planning*** and design point of view, the optimum choice of subtransmission voltage and system arrangement is closely interrelated with distribution substation size and with the primary distribution voltage level. At any given time, the most economical arrangement is achieved when the sum of the subtransmission, substation, and primary feeder costs to serve an area is a minimum over the life of the facilities. In practice, the number, size, and availability of bulk supply sources for feeding the subtransmission may be significant factors as well.

A distribution system should be designed so that anticipated load growth can be served at minimum expense. This flexibility is needed to handle load growth in existing areas, as well as load growth in new areas of development.

***5. Overhead and underground*** distribution systems are both used in large metropolitan areas. In the past in smaller towns and in the less-congested areas of larger cities, overhead distribution was almost universally used; the cost of underground distribution for residential areas was several times that of overhead. During the past 15 to 20 years, the cost of underground residential distribution (URD) has been reduced drastically through the development of low-cost, solid-dielectric cables suitable for direct-burial, mass production of pad-mounted distribution transformers and accessories, mechanized cable-installation methods, etc. The cost of a typical URD system is about 50% greater than that of an overhead system in many areas; in others, there is little or no differential due to local land conditions. As a result, some utilities will justifiably have some type of extra charge for underground. With the increased public interest in improving the appearance of residential areas and the declining cost of URD, the growth of URD has been extremely rapid. Today, perhaps as much as 70% of new residential construction is served underground. A number of states have enacted legislation making underground distribution mandatory for new residential subdivisions.

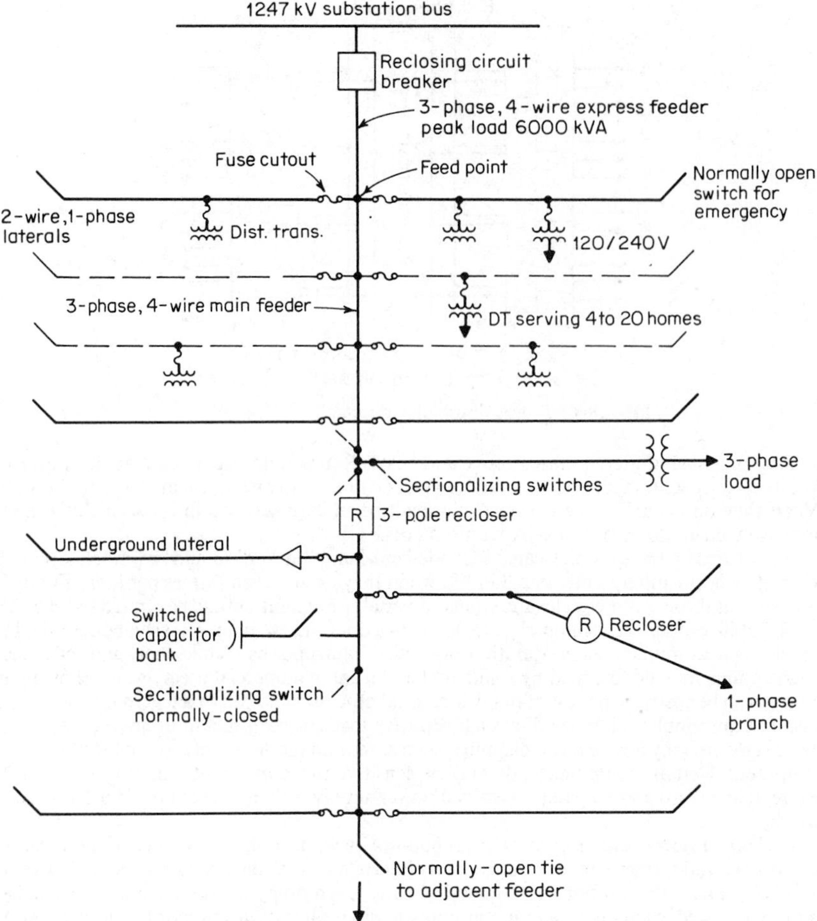


**FIG. 18-2** One-line diagram of typical primary distribution primary feeder.

As load densities increase, overhead construction becomes unwieldy because of the larger transformers and conductors required. For this reason, in the "downtown" commercial areas of most cities, underground distribution is customarily used. However, where unexpected growth is experienced, it may be next to impossible to physically, much less economically, add capacity to existing underground systems unless duct work with sufficient spare capacity was provided in the original installation. In residential areas, unexpected growth is a particular problem because direct burial, rather than ducted conduit, is the predominant underground construction type.

In recent years, the electronics age has added uncertainty to the process of distribution planning. On the one hand, customer-owned microprocessors and utility-controlled switches are increasingly controlling loads at the time of utility peak loads. Sometimes, however, this is at the expense of individual peak loads during system off-peak hours, such as when thermal-storage systems are employed to move heating or cooling loads entirely off peak. In addition, electrical loads from computers, larger color television sets, increased air-conditioning and heat-pump use, and a myriad of new electrical appliances confound the planning process by adding new load, often *after* underground facilities are constructed to serve them.

On the other hand, some appliances, like microwave ovens, are more efficient and tend to reduce load. The one sure bet is that it will be at least a decade before the effects of the electronics and electrical revolution are known well enough for underground planners to operate efficiently.

***6. Rural service*** has been extended to most farmers and rural dwellers through the efforts of utilities, cooperatives, and government agencies. Rural construction must be of the least-expensive type consistent with durability and reliability because there may be only a few users per mile of line. Historically, rural construction has been overhead, but the advent of cable-plowing techniques has made underground economically competitive with overhead in some parts of the country, and a growing amount of rural distribution is being installed underground.

***7. Higher primary voltages*** of 24.9Y/14.4 and 34.5Y/19.92 kV are gaining rapid acceptance, although primary voltages in the "15-kV class" predominate. The 5-kV class continues to decline in usage.

Surveys indicate that in recent years approximately 78% of the overhead and underground line additions are at 15 kV, 11% are at 25 kV, and 7.5% are at 35 kV.

Generally, when a higher distribution voltage is initiated, it is built in new, rapidly growing load areas. The economic advantage of the higher voltages usually is not great enough to justify massive conversions of existing lower-voltage facilities to the higher level. The lower-voltage areas are contained and gradually compressed over a period of years as determined by economics, obsolescence, and convenience.

Virtually all modern primary systems serving residential and small commercial and small industrial loads are 4-wire, multigrounded, common-neutral systems.

***8. Distribution-System Automation.*** With increasing emphasis on reliability of service coupled with the increasing use of underground systems and higher primary voltages, a definite trend is under way to make greater use of protective and sectionalizing equipment in the primary system in order to minimize the number of customers involved in an outage and to reduce the outage time. Proposed schemes run the gamut from manually operated devices to automatic devices remotely controlled from distribution centers. The remote-controlled schemes vary from some type of supervisory control to distributed microprocessor-based computer-controlled systems with built-in logic to cope quickly with the various system problems which may arise.

Automation practices will become more standardized in the future as the necessary equipment and communication channels are developed.

## CLASSIFICATION AND APPLICATION OF DISTRIBUTION SYSTEMS

***9. Distribution systems may be classified in various ways:***

**a.** As to current—alternating or direct.

**b.** As to voltage—120-V, 12,470-V, 34,500-V, etc.

**c.** As to scheme of connection—radial, loop, network, multiple, and series.

**d.** As to loads—residential, small light and power, large light and power, street lighting railways, etc.

**e.** As to number of conductors—2-wire, 3-wire, 4-wire, etc.

**f.** As to type of construction—overhead or underground.

Alternating-current circuits may be further classified as to phases: single-phase, 2-phase, or 3-phase; and as to frequency: 25-Hz, 60-Hz, etc.

***10. Application of Systems.*** In American practice, alternating-current (ac) 60-Hz systems are almost universally used for electric power distribution. These systems comprise the most economical method of power distribution, owing in large measure to the ease of transforming voltages to levels appropriate to the various parts of the system. These transformations are accomplished by means of reliable and economic transformers, with no need for rotating machinery. By proper system design and the application of overvoltage and overcurrent protective equipment, voltage levels and service reliability can be matched to almost any consumer requirement.

Single-phase residential loads generally are supplied by simple radial systems at 120/240 V. The ultimate in service reliability is provided in densely loaded business/commercial areas by means of grid-type secondary-network systems at 208Y/120 V or by "spot" networks, usually at 480Y/277 V. Secondary-network systems are used in about 90% of the cities in this country having a population of 100,000 or more, and in more than one-third of all cities with populations between 25,000 and 100,000.

Where secondary-network systems do not supply sufficiently reliable service for critical loads, emergency generators and/or batteries are sometimes provided together with automatic switching equipment so that service can be maintained to the critical loads in the event that the normal utility supply is interrupted. Such loads are found in hospitals, computer centers, key industrial processes, etc.

*11. **Single-phase*** residential loads are almost universally supplied through 120/240-V, 3-wire, single-phase services. Large appliances such as ranges, water heaters, and clothes dryers are served at 240 V. Lighting, small appliances, and convenience outlets are supplied at 120 V.

An exception to the above comments occurs when the dwelling unit is in a distributed secondary-network area served at 208Y/120 V. In this case, large appliances are supplied at 208 V and small appliances at 120 V.

*12. **Three-phase, 4-wire,*** multigrounded, common-neutral primary systems, such as 12.47Y/7.2 kV, 24.9Y/14.4 kV, and 34.5Y/19.92 kV, are used almost exclusively. The fourth wire of these Y-connected systems is the neutral for both the primary and the secondary systems. It is grounded at many locations. Single-phase loads are served by distribution transformers, the primary windings of which are connected between a phase conductor and the neutral. Three-phase loads can be supplied by 3-phase distribution transformers or by single-phase transformers connected to form a 3-phase bank.

Primary systems in the 15-kV class are most commonly used, but the higher voltages are gaining rapid acceptance. Figure 18-2 illustrates a typical radial primary feeder.

The 4-wire system is particularly economic for URD systems since each primary lateral or branch circuit consists of only one insulated phase conductor and the bare, uninsulated neutral, rather than two insulated conductors. Also, only one primary fuse is required at each transformer and one surge arrester in overhead installations.

*13. **Three-phase, 3-wire primary*** systems are not widely used for public distribution, except in California. They can be used to supply single-phase loads by means of distribution transformers having primary winding connected between two phase conductors. Single-phase primary laterals consist of two insulated phase conductors; each single-phase distribution transformer requires two fuses and two surge arresters (where used).

Three-phase loads are served through 3-phase distribution transformers or appropriate 3-phase banks.

*14. **Two-phase systems*** are rarely used today.

*15. **Direct-Current Systems.*** Today, only a few vestiges remain of the old Edison dc systems established in the formative days of electric utilities. They have been almost entirely replaced by the more flexible and economic ac systems.

*16. **Series systems*** have been used for street lighting since the earliest days of electric lighting. They are inherently high-voltage systems, well suited to street lighting where simultaneous control is desired for all lamps on a circuit. Originally they were dc systems, but later they were supplanted by ac series systems for incandescent lighting. Although many series systems remain in use today, they have largely been replaced by multiple systems, often supplied from general-service secondary systems, with photoelectric cells or other devices to control lighting.

## CALCULATION OF VOLTAGE REGULATION AND $I^2R$ LOSS

*17. **General.*** When a circuit supplies current to a load, it experiences a drop in voltage and a dissipation of energy in the form of heat. In dc circuits voltage drop is equal to current in amperes multiplied by ohmic resistance of the conductors, $V = IR$. In ac circuits voltage drop is a function of load current and power factor and the resistance and reactance of the conductors. Heating is caused by conductor losses, traditionally called "copper losses"; for both dc and ac circuits they are computed as the square of current multiplied by conductor resistance in ohms. Watts $= I^2R$, or kW $= I^2R/1000$. Capacitance can be neglected for calculation of the usual steady-state quantities in distribution circuits because its effect on voltage drop is negligible for the circuit lengths and operating voltages used. In circuit design, a conductor size should be selected so that it will carry the required load within specified voltage-drop limits and will have an optimized value of installed cost and cost of losses. Today, a conductor size meeting these criteria will operate well within safe operating temperature limits. In some cases, short-circuit current requirements will dictate minimum conductor size.

***18. Percent voltage drop or percent regulation*** is the ratio of voltage drop in a circuit to voltage delivered by the circuit, multiplied by 100 to convert to percent. For example, if the drop between a transformer and the last customer is 10 V and the voltage delivered to the customer is 230, the percent voltage drop is $10/230 \times 100 = 4.35\%$. Often the nominal or rated voltage is used as the denominator because the exact value of delivered voltage is seldom known.

***19. Percent $I^2R$*** or percent conductor loss of a circuit is the ratio of the circuit $I^2R$ or conductor loss, in kilowatts, to the kilowatts delivered by the circuit (multiplied by 100 to convert to percent). For example, assume a 240-V single-phase circuit consisting of 1000 ft of two No. 4/0 copper cables supplies a load of 100 A at unity power factor.

$$I^2R = 100^2 \times 2 \times 0.0512 = 1024 \text{ W} = 1.024 \text{ kW}$$

$$\text{Load delivered} = 240 \times 100 = 24{,}000 \text{ W} = 24 \text{ kW}$$

$$\%\ I^2R \text{ loss} = 1.024/24 \times 100 = 4.26\%$$

***20. Direct-current voltage drop*** is easily calculated by multiplying load amperes $I$ by ohmic resistance $R$ of the conductors through which the current flows (see Sec. **4** for ohmic resistance of various conductors). *Example:* A 500-ft dc circuit of two 4/0 copper cables carries 200 A. What is the voltage drop? Resistance of 1000 ft of 4/0 copper cable is 0.0512 Ω.

$$\text{Drop} = IR = 200 \times 0.0512 = 10.24 \text{ V}$$

If 240 is the delivered voltage,

$$\%\ \text{regulation} = 10.24/240 \times 100 = 4.26\%$$

***21. Voltage-drop calculations for 3-wire dc circuits*** require separate computations for each conductor if the load is appreciably unbalanced. Drop in the neutral that carries the unbalanced current must be added to drop in the more heavily loaded leg and *subtracted* from drop in the more lightly loaded leg to obtain voltage drops from leg to neutral.

***22. $I^2R$ or conductor loss in dc or ac circuits*** is calculated by multiplying the square of the current in amperes by ohmic resistance of the conductors through which the current flows. The result is in watts.

***23. In dc circuits percent voltage drop and percent conductor loss*** are identical, and their ratio is unity.

$$\text{Percent voltage drop} = IR/V \times 100$$

$$\text{Percent } I^2R = I^2R/VI \times 100 = IR/V \times 100$$

In ac circuits the ratio of percent conductor loss to percent voltage regulation is given approximately by the following formula:

$$\frac{\%\ I^2R \text{ loss}}{\%\ \text{voltage drop}} = \frac{\cos \phi}{\cos \theta \cos (\phi - \theta)} \tag{18-1}$$

where $\theta$ = power-factor angle and $\phi$ = impedance angle; that is, $\tan \phi = X/R$.

***24. Table 18-1*** gives voltage drop in volts per 100,000 A · ft for 2-wire dc circuits for a number of sizes of conductor. Ampere-feet is the product of the number of amperes of current flowing and the distance in feet between the sending and receiving terminals multiplied by 2 to

**TABLE 18-1** Voltage Drop in Volts per 100,000 A · ft, 2-Wire DC Circuits (Loop)

| Conductor size, AWG or kcmil | | Volts drop per 100,000 A·ft, 90° copper temp |
|---|---|---|
| Copper | Approx. equivalent aluminum | |
| 6 | 4 | 102.8 |
| 4 | 2 | 64.6 |
| 2 | 1/0 | 40.7 |
| 1/0 | 3/0 | 25.6 |
| 2/0 | 4/0 | 20.3 |
| 4/0 | 336 | 12.8 |
| 350 | 556 | 7.71 |
| 500 | 795 | 5.39 |
| 1000 | | 2.70 |
| 1500 | | 1.80 |
| 2000 | | 1.35 |

NOTE: 1 ft = 0.3048 m.

take into account the drop in both the outgoing and return conductors. Or the feet can be considered to be the total number of conductor feet, outgoing and return.

Table 18-1 also gives the voltage drop for 3-wire circuits when serving balanced loads, where the term "feet" is taken to mean twice the number of feet between sending and receiving terminals.

***25. Use of Table 18-1.*** *Example 1.* What is the voltage drop and percent voltage drop when 200 A dc flows 1500 ft one way through a 2-wire, 120-V, 556-kcmil aluminum circuit? First determine ampere-feet factor as 100 × 1500/100,000 = 1.5. From Table 18-1, the voltage drop is 7.71 V per 100,000 A · ft. This value multiplied by the 1.5 factor gives the total voltage drop = 1.5 × 7.71 = 11.6 V. The percent voltage drop = 11.6 × 100/120 = 9.64%. From Par. **23**, the percent conductor loss also is 9.64%, which is equivalent to 120 × 100 × 0.0954 = 1.16 kW.

*Example 2.* A mine 1 mile from a motor-generator station must receive 100 kW dc at not less than 575 V. Maximum voltage of the generator is 600 V. What conductor size should be used?

$$\text{Max. current} = \frac{100{,}000\ \text{W}}{575\ \text{V}} = 173.9\ \text{A}$$

$$\text{Loop ft} = 2 \times 5280 = 10{,}560\ \text{ft}$$

$$\frac{\text{A} \cdot \text{ft}}{100{,}000} = \frac{173.9 \times 10{,}560}{100{,}000} = 18.36$$

18.36 × voltage drop per 100,000 A · ft from Table 18-1 = 25 V

Therefore, voltage drop per 100,000 A · ft = 25/18.36 = 1.36.

From Table 18-1, the copper conductor size corresponding to 1.36 V/100,000 A · ft is 2000 kcmil copper.

***26. Calculating Voltage Drop in AC Circuits.*** The voltage drop per mile in each round wire of 3-phase 60-Hz line with equilateral spacing $D$ inches between centers or in each wire of a single-phase line $D$ inches between centers is

$$\underset{\cdot}{V}\ \text{drop} = \underset{\cdot}{I}R + j\underset{\cdot}{I}\left(0.2794 \log \frac{D}{r} + 0.03034\,\mu\right) \quad \text{volts in phasor form} \qquad (18\text{-}2)$$

where $\underset{\cdot}{I}$ is in phasor amperes; $R$ = 60-Hz resistance of the wire per mile, Ω; log- = log to base 10; $r$ = radius of round wire, in; and $\mu$ = permeability of the wire (unity for nonmagnetic materials such as copper or aluminum). $j$ in Eq. (18-2) denotes an angle of 90°; $+j$ means 90° leading, $-j$

means 90° lagging. Thus the expression for phasor current lagging the reference voltage is $\underset{\cdot}{I} = I_x - jI_y = I\sqrt{\theta°}$ with reference to a conveniently chosen horizontal axis of reference—usually sending- or receiving-end voltage. The dot beneath $\underset{\cdot}{I}$ or $\underset{\cdot}{V}$ indicates phasor values. Voltage drops determined in this manner are also phasors and are with respect to the reference axis.

When wire is stranded, an equivalent radius must be used for $r$ in Eq. (18-2). $r = 0.528\sqrt{A}$ for 7 strands, $r = 0.5585\sqrt{A}$ for 19 strands, $r = 0.5675\sqrt{A}$ for 37 strands, where $r$ = equivalent radius, in, and $A$ = area of metal, in².

Frequency is 60 Hz for the constants in parentheses in Eq. (18-2), which gives reactance $X$ in ohms per mile. For 25 Hz, multiply by 25/60. The equation is sometimes written

$$\underset{\cdot}{V} \text{ drop per mile} = \underset{\cdot}{I}(R + jX) = \underset{\cdot}{I}\underset{\cdot}{Z} \qquad \text{volts in phasor form} \tag{18-3}$$

where $I$ is in phasor amperes and $Z = Z/\phi°$ Ω/mi at 60 Hz.

***27. Three unsymmetrically spaced wires*** *a, b,* and *c* of a 3-phase circuit with correct transpositions can have voltage drop in each wire calculated by Eq. (18-2) by substituting for $D$ the geometric mean of the three interaxial distances:

$$D = \sqrt[3]{D_{ab}D_{bc}D_{ca}}$$

***28. The Phasor Method.*** In Eq. (18-3), $I$ is in vector amperes,

$$\underset{\cdot}{I} = I_x - jI_y = I\sqrt{\theta°}$$

where $\theta$ = angle that the current lags (or leads) the voltage. The sending-end voltage is usually chosen as the axis, or phasor, of reference in drawing the phasor diagram. For example, consider Fig. 18-3, where sending voltage $V_s = V_s\underline{/0°}$, load current $I = I\underline{/\theta°}$, circuit impedance $\underset{\cdot}{Z}$ =

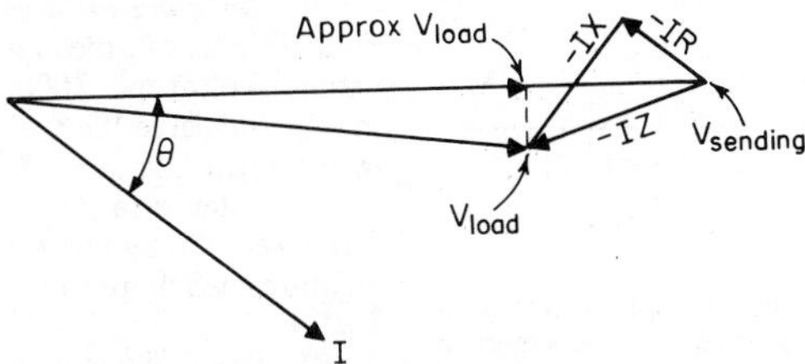


**FIG. 18-3** Phasor diagram showing voltage relationships.

$Z\underline{/\phi°} = R + jX$, and load voltage $\underset{\cdot}{V}_L = \underset{\cdot}{V}_s - \underset{\cdot}{I}\underset{\cdot}{Z}$ (all phasors). The symbol $\overline{/}$ is used for negative angles, assuming that the counterclockwise direction from the phasor or reference is positive and the clockwise directions negative. Assume: $V_s = 230/0°$, $\underset{\cdot}{I} = 50\,\overline{/36.87°}$, $\underset{\cdot}{Z} = 0.2\underline{/71.57°}$, and $\underset{\cdot}{Z} = R + jX$. Thus:

$$\begin{aligned}\underset{\cdot}{V}_L &= 230/0° - 50\,\overline{/36.87°} \times 0.2\underline{/71.57°} \\ &= 230\underline{/0°} - 10\underline{/34.70°} \\ &= 230 - 10\cos 34.70° - j\,10\sin 34.70° \\ &= 230 - 8.22 - j\,5.69 = 221.78 - j\,5.69 \\ &= 221.78 \text{ (very nearly)}\end{aligned}$$

Neglecting the term $-j\,5.69$ simplifies the final calculation and gives the load voltage within a fraction of 1% of the precise result. This method is sufficiently accurate for practically all distribution engineering calculations and can be thought of as

$$V \text{ drop} = IR\cos\theta + IX\sin\theta = IZ\cos(\phi - \theta) \tag{18-4}$$

where $I$ and $Z$ are absolute magnitudes, not phasor quantities; $\phi$ = impedance angle; and $\theta$ = power-factor angle by which the current lags (or leads) the voltage. Calculating the drop in the above example by this method:

$$V \text{ drop} = 50 \times 0.2 \times \cos 71.57° \times \cos 36.87° + 50 \times 0.2 \times \sin 71.57° \times \sin 36.87° = 2.53 + 5.69 = 8.22 \text{ V}$$

or

$$V \text{ drop} = IZ \cos(\phi - \theta) = 50 \times 0.2 \times \cos(71.57° - 36.87°) = 10 \cos(34.7°) = 10 \times 0.822 = 8.22 \text{ V}$$

***29. Impedance Z*** can be visualized as the hypotenuse of a right triangle in which the base is the resistance $R$ and the altitude is the reactance $X$. In phasor form $\underset{\cdot}{Z} = R \pm jX$, where the positive sign is used for inductive reactance, and the negative sign for capacitive reactance. Impedance can also be expressed as $\underset{\cdot}{Z} = Z\underline{/\phi}$, where $Z$ is the absolute magnitude and $\phi$ is the angle between $\underset{\cdot}{Z}$ and $R$ in Fig. 18-4. This angle is an absolute value in that it has no relationship to the axis of reference in a phasor diagram, as do voltage and current. Alternating current causes a voltage drop in resistance which is in time phase with the current and in inductive reactance a drop which leads the current by 90 electrical degrees, assuming the positive direction for measurement of angles is counterclockwise. Or conversely, the current in an inductive reactance lags the voltage drop by 90°.

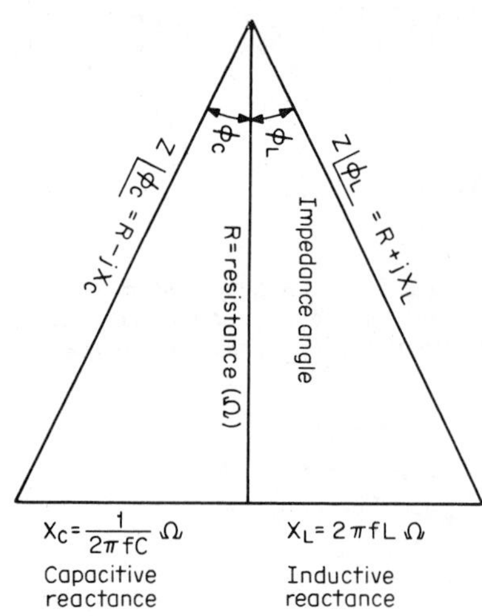


**FIG. 18-4** Impedance diagrams for series connection of resistance and reactance. ($L$ = inductance in henrys, $C$ = capacitance in farads, $F$ = frequency in hertz.)

***30. Impedance Values.*** Tables are available (see Sec. 4) which give 60-Hz impedance values in ohms per 1000 ft for common sizes of wire and cable. The values can be expressed in the form $\underset{\cdot}{Z} = R + jX$, which can be converted to the form $Z\underline{/\phi°}$ if desired. The latter form is convenient to use in voltage-drop calculations when the current is expressed as $I\underline{/\theta°}$.

***31. Power Factor.*** In typical distribution loads, the current lags the voltage, as shown in Fig. 18-3, where $\theta$ is shown as the angle between current and sending voltage; and $\cos\theta$ is referred to as the "power factor" of the circuit. In a purely resistive circuit, the current and voltage are in phase; consequently the power factor is 1.0 or unity. In a purely inductive circuit, the voltage and current are out of phase by 90 electrical degrees, resulting in a power factor of zero. In a circuit consisting of a resistance in series with a reactance of equal ohmic value ($\phi = 45°$), $\theta \pm 45°$, also. Thus the power factor is $\cos 45° = 0.707$, or 70.7%.

In a single-phase ac circuit the load in kW can be expressed as

$$\text{kW} = EI \cos\theta \qquad (18\text{-}5)$$

where $E$ = magnitude of rms line-to-line voltage, kV
$I$ = magnitude of current, rms amperes
$\theta$ = electrical angle between phasor voltage and current

From Eq. (18-5) it is obvious that the magnitude of the current for a given voltage and kilowatt load depends on the power factor.

$$I = \text{kW}/E \cos\theta \qquad (18\text{-}6)$$

The corresponding equations for balanced 3-phase circuits are

$$\text{kW} = \sqrt{3}\, EI \cos\theta \qquad (18\text{-}7)$$

and

$$I = \text{kW}/\sqrt{3}\, E \cos\theta \qquad (18\text{-}8)$$

where the symbols are as specified above, and $\theta$ is measured as the angle between the line-to-neutral voltage of a given phase and the current in that phase.

***32. Example of Calculation of Single-Phase Line Drop.*** Given a load of 500 kW at 80% power factor (lagging), 7.2 kV circuit voltage, 60-Hz, single-phase circuit using 1/0 aluminum conductor spaced 30 in on centers. The load is located 1 mi from the substation. What is the voltage drop? From tables on conductor characteristics,

$$r = 0.185\ \Omega/1000\ \text{ft}$$

$$x = 0.124\ \Omega/1000\ \text{ft}$$

Therefore,

$$R + jX = 5.28\ (0.185 + j\,0.124)$$
$$= 0.9769 + j\,0.6547\ \Omega$$

From Eq. (18-6),

$$I = \frac{\text{kW}}{E \cos \theta} = \frac{500}{7.2 \times 0.8} = 86.81\ \text{A}$$

$$E = 7.2\underline{/0^\circ}$$

$$\cos \theta = 0.80$$

$$\theta = 36.87^\circ$$

and

$$\sin \theta = 0.60$$

From Eq. (18-4),[1]

$$\text{Voltage drop} = 2(IR \cos \theta + IX \sin \theta) = (86.81 \times 0.9769 \times 0.8 + 86.81 \times 0.6547 \times 0.6)$$
$$= 2(67.84 + 34.10) = 203.88\ \text{V}$$

***33. Calculation of 3-Phase Line Drops with Balanced Loads.*** In 3-phase circuits with balanced loads on each phase, the line-to-neutral voltage drop is merely the product of the phase current and the conductor impedance as determined from standard tables. There is no return current with balanced 3-phase loads. Thus the line-to-line voltage drop is $\sqrt{3}$ times the line-to-neutral drop or

$$V_{\text{drop}\ L-L} = \sqrt{3}(IR \cos \theta + IX \sin \theta) \qquad (18\text{-}9)$$

For example, assume that the circuit of Par. **32** now is a 3-phase 12.47-kV circuit 1 mi long with the same 1/0 aluminum conductors at an equivalent spacing of 30 in, and a load of $3 \times 500 =$ 1500 kW at 0.8 pf lagging. What is the line-to-line voltage drop? $R$ and $X$ are the same values as previously; that is, $R + jX = 0.9769 + j\,0.6547\ \Omega$.

The current per phase from Eq. (18-7) is

$$I = \frac{\text{kW}}{\sqrt{3}\,E \cos \theta} = \frac{1500}{\sqrt{3} \times 12.47 \times 0.8} = 86.81\ \text{A}$$

as before.

$$V_{\text{drop}\ L-L} = \sqrt{3}\,(IR \cos \theta + IX \sin \theta)$$
$$= \sqrt{3}\,(86.81 \times 0.9769 \times 0.8 + 86.81 \times 0.6547 \times 0.6)$$
$$= 117.51 + 59.06 = 176.57\ \text{V (approx.)}$$

***34. Calculation of Voltage Drop in Unbalanced Unsymmetrical Circuits.*** If there are $n$ different wires $a, b, c, d, \ldots, n$ carrying currents $I_a, I_b, I_c, \ldots, I_n$, respectively, whether 2-,

[1] The factor of 2 is used for a single-phase system to represent the impedance of the outgoing conductor and the return conductor.

3-, or 6-phase, the voltage drop in wire $a$ per mile at 60 Hz is

$$I_aR_a + j\left[0.2794\left(I_a \log\frac{1}{r} + I_b \log\frac{1}{D_{ab}} + I_c \log\frac{1}{D_{ac}} + \ldots + I_n \log\frac{1}{D_{an}}\right) + 0.03034\,\mu\, I_a\right] \quad \text{volts in phasor form} \qquad (18\text{-}10)$$

where currents are in phasor amperes; $R_a$ = 60-Hz ohmic resistance of conductor $a$ per mile; $r$ = equivalent radius, in inches, of conductor $a$; $D_{ab}$, $D_{ac}$, $D_{an}$ = distances, in inches, between centers of conductors $a$ and $b$, $a$ and $c$, and $a$ and $n$; and $\mu$ is the permeability of conductor $a$ (unity for nonmagnetic material). To get the drop in $b$, replace all $a$'s by $b$'s and all $b$'s by $a$'s in Eq. (18-10); similarly, to get the drop in $c$, interchange $a$'s and $c$'s; likewise for $n$. For 25 Hz, multiply that part of Eq. (18-10) which is in brackets by 25/60. Equation (18-10) gives voltage drop for any degree of load unbalance, power factor, or conductor arrangements. In using this formula calculations are made easier by choosing voltage to neutral as the reference axis.

***35. Approximate Method of Calculating Voltage Drop in Unbalanced, Unsymmetrical Circuits.*** Equation (18-10) requires laborious calculations and is used only when exact results are necessary. Voltage drops sufficiently accurate for engineering purposes can be made by using an equivalent impedance for each conductor. The reactance component of the equivalent impedance is computed from a spacing $D$ equal to the geometric means of the interaxial distances of the other conductors to the conductor being considered. For instance, if there are four conductors $a$, $b$, $c$, and $n$ for conductor $a$, $D = \sqrt[3]{D_{ab}, D_{ac}, D_{an}}$; for conductor $b$, $D = \sqrt[3]{D_{ab}, D_{bc}, D_{bn}}$.

***36. Phasor and connection diagrams*** are drawn in computing voltage drops in unbalanced circuits. Figure 18-5 shows an unbalanced 4-wire 3-phase 4160Y/2400-V circuit with assumed loads, power factors, and equivalent line impedances. Phase-to-neutral drops between source and load are given by the following using one of the many possible voltage-notation conventions:

$$\begin{aligned} V_{na} - V_{n'a'} &= I_aZ_a + I_nZ_n \\ V_{nb} - V_{n'b'} &= I_bZ_b + I_nZ_n \\ V_{nc} - V_{n'c'} &= I_cZ_c + I_nZ_n \end{aligned} \qquad (18\text{-}11)$$

Phase-to-phase drops between source and load are given by the following:

$$\begin{aligned} V_{ba} - V_{b'a'} &= I_aZ_a - I_bZ_b \\ V_{ac} - V_{a'c'} &= I_cZ_c - I_aZ_a \\ V_{cb} - V_{c'b'} &= I_bZ_b - I_cZ_c \end{aligned} \qquad (18\text{-}12)$$

In computing line-to-neutral drop in phase $a$, it is convenient to choose $V_{na}$ as the axis of reference.

$$\begin{aligned} V_{na} - V_{n'a'} &= I_aZ_a + I_nZ_n = (100\,/\underline{20^\circ})(1.2/\underline{49^\circ}) + (43.2\,/\underline{32.2^\circ})(0.5/\underline{40^\circ}) \\ &= 120/\underline{29^\circ} + 21.6/\underline{7.8^\circ} = 126.4 + j61.9 \end{aligned}$$

$$\text{Load voltage } V_{n'a'} = 2400 - 126.4 - j61.9 = 2273.6 \text{ V (very nearly)}$$

Likewise, in computing line-to-neutral drop in phase $b$, it is convenient to choose $V_{nb}$ as the axis of reference. The phasor diagram of Fig. 18-5 must be rotated in a counterclockwise direction 120°; then $I_b = 90\,/\underline{10^\circ}$ and $I_n = 43.2/\underline{87.8^\circ}$.

$$\begin{aligned} V_{nb} - V_{n'b'} &= I_bZ_b + I_nZ_n = (90\,/\underline{10^\circ})(1.1/\underline{47^\circ}) + (43.2/\underline{87.8^\circ})(0.5/\underline{40^\circ}) \\ &= 65.8 + j76.6 \end{aligned}$$

$$\text{Load voltage } V_{n'b'} = 2400 - 65.8 - j76.6 = 2334.2 \text{ V (very nearly)}$$

***37. Drop in the neutral conductor*** of a 4-wire 3-phase circuit or a 5-wire 2-phase circuit makes resultant drop on the more heavily loaded phases greater than it would be for the same

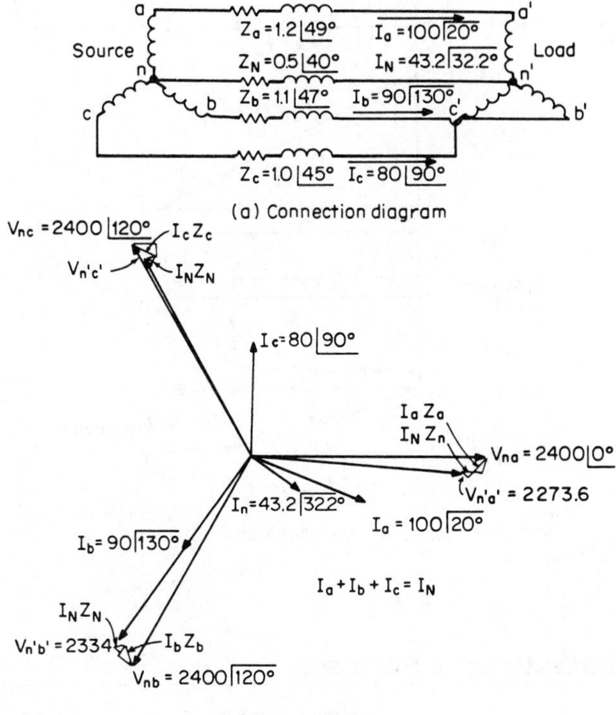


**FIG. 18-5** Connections and phasor diagrams for unbalanced loads and unsymmetrical circuit.

current under balanced conditions. Likewise, net drop is less on more lightly loaded phases than for the same current when balanced.

***38. Distributed Loads; Voltage Drop and $I^2R$ Loss.*** Voltage drop and conductor power losses resulting from a concentrated load on a distribution line can be easily calculated as shown in earlier parts of this section. However, distribution circuit loads are generally considered to be distributed—often, but not always, uniformly. Distributed load may be considered as effectively concentrated at one point along the circuit to calculate total voltage drop and at another point to calculate conductor $I^2R$ losses in the conductor. If the load is uniformly distributed along the feeder, the total voltage drop can be calculated by assuming that the entire load is concentrated at the midpoint of the circuit, and the total $I^2R$ losses can be calculated by assuming that the load is concentrated at a point one-third the total distance from the source.

However, if there is a superimposed through load beyond the given feeder section, this method of calculation becomes cumbersome. It is possible to develop a single precise equivalent circuit for both the voltage-drop and loss calculations. Figure 18-6 shows the load representation and equivalent for uniformly distributed loads. Equivalents also can be developed for other types of distribution.

Figure 18-6 shows the equivalent circuit of two-thirds of the total load concentrated at three-quarters of the total distance from the source.

***39. Annual conductor-loss factor*** is the ratio of average yearly conductor loss to conductor loss at peak load. It is used to determine the total annual energy dissipated in conductor losses by multiplying annual loss factor by kilowatt loss at peak load by the number of hours in a year. The loss factor is related to, but not a direct function of, annual load factor, which is the ratio of average yearly load to annual peak load.

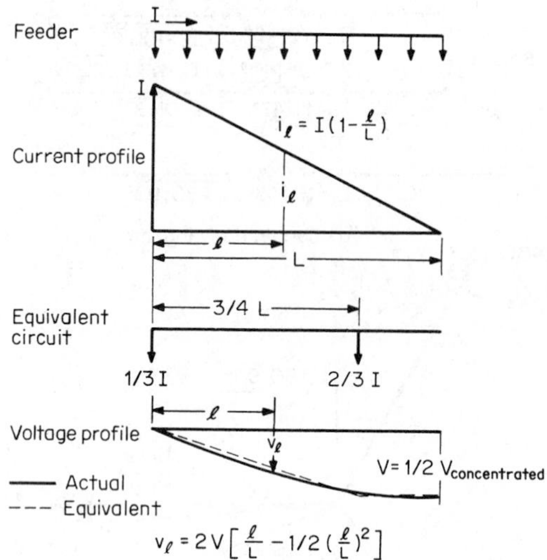


**FIG. 18-6** Uniformly distributed loads.

## THE SUBTRANSMISSION SYSTEM

***40. Definition.*** Subtransmission is that part of the utility system which supplies distribution substations from bulk power sources, such as large transmission substations or generating stations. In turn, the distribution substations supply primary distribution systems. Subtransmission has many of the characteristics of both transmission and distribution in that it moves relatively large amounts of power from one point to another, like transmission, and at the same time it provides area coverage, like distribution.

In some utility systems transmission and subtransmission voltages are identical; in other systems, subtransmission is a separate and distinct voltage level (or levels). This is easy to account for because in the evolutionary development of utility systems, today's transmission voltage naturally tends to become tomorrow's subtransmission voltage, just as today's subtransmission voltage tends to become tomorrow's primary distribution voltage.

Because of the wide range of voltages used in subtransmission and because of the wide variation in geographical conditions and local ordinances, subtransmission circuits are sometimes built on pole lines on city streets, or on tower lines on private rights-of-way, or in underground cables.

***41. Voltages of subtransmission circuits*** range from 12 to 245 kV, but today the levels of 69, 115, and 138 kV are most common. The use of the higher voltages is expanding rapidly as higher primary voltages are receiving increased usage. Current practice as indicated by a recent informal utility survey is shown in Fig. 18-7: 115 and 138 kV together comprise about half the usage, 69 kV about 20%; 230 kV usage is becoming substantial, reflecting the growing use of 25- and 34.5-kV primary distribution.

***42. Conductors*** of ACSR or aluminum have generally supplanted copper in overhead construction, and aluminum conductors are being used increasingly in cables.

***43. Voltage Regulation of Subtransmission.*** The size of conductors used in subtransmission systems is determined by *(a)* magnitude and power factor of the load, *(b)* emergency loading requirements, *(c)* distance that the load must be carried, *(d)* operating voltage, *(e)* permissible voltage drop under normal and emergency loading, *(f)* optimum economic balance between installed cost of the conductor and cost of losses. Table 18-2 gives the line-to-neutral voltage drops per 100,000 A · ft for common cable and overhead conductor sizes and representative

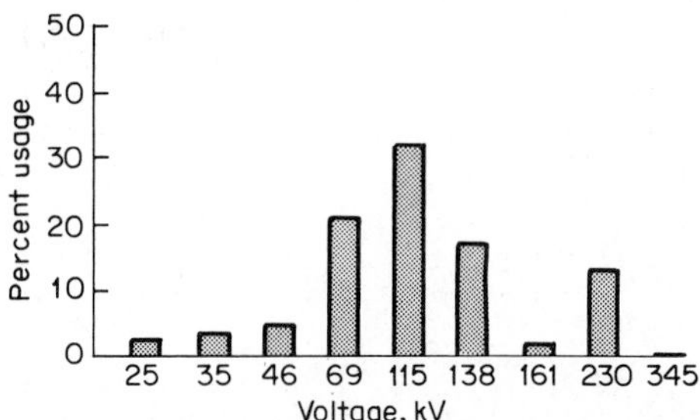


**FIG. 18-7** Usage of distribution substation high-voltage rating.

power factors for 34.5- and 69-kV subtransmission. Values in the table are based on the approximate formula (18-4).

$$V_{\text{drop}} = IR \cos \theta + IX \sin \theta = IZ \cos (\phi - \theta)$$

where $R$, $X$, and $Z$ are 60-Hz resistance, reactance, and impedance in ohms per 1000 ft of a single conductor, $\theta$ is the power-factor angle in electrical degrees; $\phi$ is the impedance angle, tan − 1($X/R$).

***44. Examples of How to Use Table 18-2.*** Determine the voltage drop when a 3-phase 20,000-kVA load at 95% power factor is carried 10 miles over an overhead 69-kV circuit with No. 2/0 ACSR conductor. Assuming the receiving-end voltage to be 69 kV, the current is

$$I = \frac{\text{kVA}}{\sqrt{3}\,E} = \frac{20{,}000}{\sqrt{3} \times 69} = 167.35 \text{ A}$$

Circuit feet are

$$10 \times 5{,}280 = 52{,}800 \text{ ft}$$

Thus,

$$\frac{\text{A} \cdot \text{ft}}{100{,}000} = \frac{167.35 \times 52{,}800}{100{,}000} = 88.36$$

From the overhead portion of Table 18-2, the voltage drop per 100,000 A · ft at 95% power factor for a No. 2/0 ACSR conductor is 19.1 V. Therefore, the total voltage drop for the example is 88.36 × 19.1 = 1687.68 V line-to-neutral. Since normal line-to-neutral voltage is $69/\sqrt{3}$ = 39.838 kV, or 39,838 V, the percent voltage drop is 1687.68 × 100/39,838 = 4.24%.

Assuming that permissible voltage drop is the limiting factor, what overhead ACSR conductor size should be used to supply a load of 40,000 kVA at 95% power factor and receiving-end voltage of 69 kV with a permissible drop of 5% and 8 mi between sending and receiving ends?

$$\text{Current} = \frac{40{,}000}{\sqrt{3} \times 69} = 334.71 \text{ A}$$

$$\text{Circuit feet} = 8 \times 5280 = 42{,}240 \text{ ft}$$

$$\frac{\text{A} \cdot \text{ft}}{100{,}000} = \frac{334.71 \times 42{,}240}{100{,}000} = 141.38$$

The permissible voltage drop is $0.05 \times 69{,}000/\sqrt{3} = 1991.92$ V line-to-neutral. The corresponding permissible voltage drop per 100,000 A · ft is

$$\frac{1991.92}{141.38} = 14.1 \text{ V}/100{,}000 \text{ A} \cdot \text{ft}$$

From Table 18-2, it is seen that this corresponds approximately to No. 4/0 ACSR.

**TABLE 18-2** Voltage Drops per 100,000 A · ft* for 3-Phase, 60-Hz, 34.5- and 69-kV Subtransmission

| | Voltage class | | | | | | | | | | |
|---|---|---|---|---|---|---|---|---|---|---|---|
| | 34.5 kV | | | | | 69 kV | | | | | Approx. amp. capacity for air moving at 2 ft/s |
| | Lagging power factor | | | | | | | | | | |
| Conductor size | 0.7 | 0.8 | 0.9 | 0.95 | 1.00 | 0.7 | 0.8 | 0.9 | 0.95 | 1.00 | |
| | Underground subtransmission† | | | | | | | | | | |
| Aluminum: | | | | | | | | | | | |
| No. 1/0 | 18.3 | 19.9 | 21.1 | 21.5 | 21.0 | | | | | | |
| No. 2/0 | 15.4 | 16.5 | 17.4 | 17.6 | 16.9 | | | | | | |
| No. 4/0 | 10.7 | 11.2 | 11.5 | 11.4 | 10.5 | | | | | | |
| 350 kcmil | 7.69 | 7.84 | 7.77 | 7.55 | 6.50 | 8.04 | 8.10 | 7.92 | 7.62 | 6.38 | |
| 500 kcmil | 6.15 | 6.12 | 5.88 | 5.59 | 4.50 | 6.53 | 6.43 | 6.10 | 5.74 | 4.48 | |
| 750 kcmil | 4.96 | 4.80 | 4.44 | 4.10 | 3.00 | 5.25 | 5.05 | 4.63 | 4.23 | 3.01 | |
| 1000 kcmil | 4.32 | 4.12 | 3.73 | 3.37 | 2.30 | 4.69 | 4.44 | 3.96 | 3.55 | 2.32 | |
| | Overhead subtransmission‡ | | | | | | | | | | |
| ACSR: | | | | | | | | | | | |
| No. 4 | 42.9 | 45.5 | 47.3 | 47.5 | 44.7 | 43.6 | 46.1 | 47.7 | 47.8 | 44.7 | 120 |
| No. 2 | 31.5 | 32.5 | 32.7 | 32.1 | 28.4 | 32.2 | 33.1 | 33.1 | 32.4 | 28.4 | 165 |
| No. 1/0 | 24.1 | 24.1 | 23.2 | 22.1 | 18.0 | 24.8 | 24.7 | 23.7 | 22.4 | 18.0 | 225 |
| No. 2/0 | 21.6 | 21.2 | 20.1 | 18.8 | 14.6 | 22.3 | 21.8 | 20.5 | 19.1 | 14.6 | 260 |
| No. 4/0 | 17.3 | 16.6 | 15.1 | 13.8 | 9.66 | 18.0 | 17.2 | 15.5 | 14.1 | 9.66 | 355 |
| 336.4 kcmil | 12.7 | 11.8 | 10.4 | 9.13 | 5.57 | 13.4 | 12.4 | 10.8 | 9.44 | 5.57 | 480 |
| 477 kcmil | 11.2 | 10.3 | 8.72 | 7.44 | 3.92 | 12.0 | 10.9 | 9.15 | 7.75 | 3.92 | 605 |
| 795 kcmil | 9.73 | 8.68 | 7.06 | 5.78 | 2.37 | 10.4 | 9.28 | 7.49 | 6.09 | 2.37 | 850 |

Regulation of copper conductors can be estimated with reasonable accuracy as that of aluminum conductors two sizes larger.

For ampacities of cables, see Tables 18-24 and 18-25.

* Values in the table give the difference in absolute value between sending-end and receiving-end line-to-neutral voltages of a balanced 3-phase circuit.

† Underground cable impedances are based on 90°C conductor temperature with close triangular spacing of cables using typical solid-dielectric insulation, 100% insulation level, single conductor, shielded and jacketed.

‡ Overhead conductor impedances are based on 50°C conductor temperature, ACSR construction, 600 A/in$^2$ density with 60-in equivalent spacing for 35 kV and 90 in for 69 kV.

NOTE: 1 in = 25.4 mm; 1 in$^2$ = 645 mm$^2$; 1 ft = 0.3048 m.

***45. Subtransmission-System Patterns.*** A wide variety of subtransmission-system designs are in use, varying from simple radial systems to systems similar to networks. The radial system is not generally used because most utilities today plan their subtransmission-distribution substation systems so that one major contingency such as outage of a subtransmission circuit or failure of a distribution substation transformer will not result in loss of load—or at least the loss of load will be of short duration while automatic switching operations take place. Thus, loop and multiple circuit patterns predominate. Figures 18-8 and 18-9 illustrate the basic nature of these two patterns. The loop pattern implies that a single circuit originating at one bulk power source "loops" through several substations before terminating at another bulk source, or even at the original source. Reinforcing ties, as indicated by the dotted connection, are used when the number of substations exceeds some predetermined level.

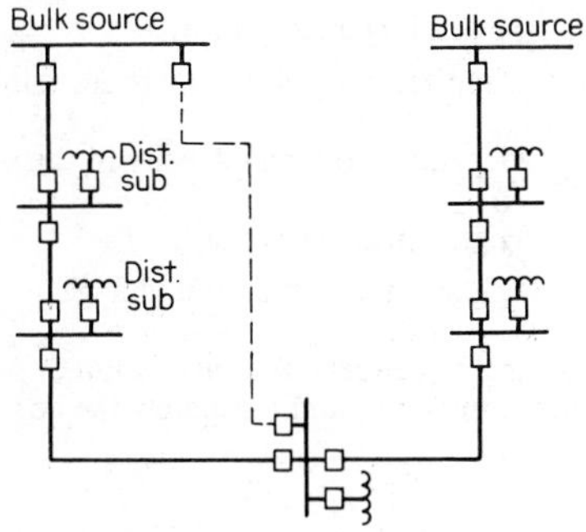

Multiple circuit pattern implies the use of two or more circuits which are tapped at each substation as illustrated in Fig. 18-9. The circuits may be radial or may terminate in a second bulk power source. Many variations of the two basic patterns are found. From a recent informal survey of approximately 50 major utilities, it appears that the two patterns are about equally used.

**FIG. 18-8** Loop pattern.

A vast majority of today's subtransmission is of overhead construction, much of it built on city streets as contrasted to private right-of-way. However, appearance and environmental considerations, difficulty in obtaining substation sites and right-of-way, and rapid growth of underground distribution are certain to exert continuing pressure on the undergrounding of subtransmission. Even with the use of direct-buried, solid-dielectric cables, the cost of under-

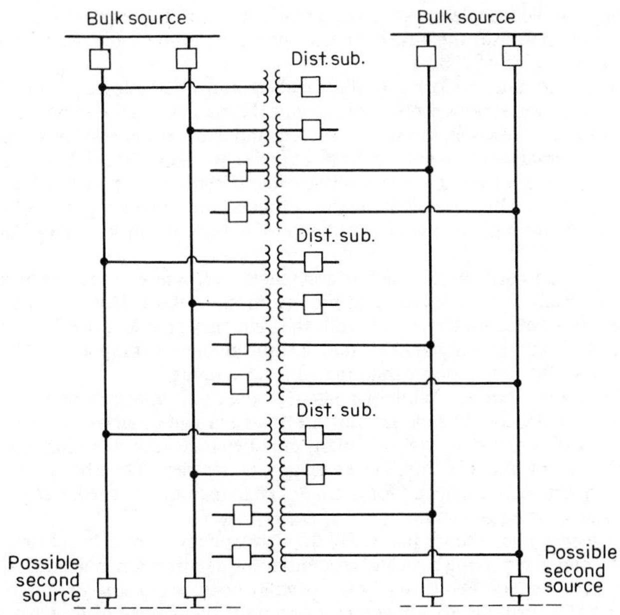


**FIG. 18-9** Multiple pattern.

ground subtransmission is many times the cost of overhead circuits, particularly where the overhead subtransmission can be built on city streets.

Thus, a requirement to build future subtransmission underground would have major impact on the balance of overall subtransmission-substation-primary distribution costs. It undoubtedly would focus attention on minimizing the amount of subtransmission circuitry needed to cover the load area, which in turn would favor:

Fewer, larger substations.

Loop subtransmission pattern, rather than multiple parallel circuits.

Depending on load density in this area, it *could* favor:

Higher primary voltage.

Higher subtransmission voltage.

Changes in either subtransmission or primary voltage levels are major decisions which require study in depth, and ultimately the commitment of large financial resources.

## PRIMARY DISTRIBUTION SYSTEMS

***46. Primary Systems.*** The primary distribution system takes energy from the low-voltage bus of distribution substations and delivers it to the primary windings of distribution transformers.

***47. Overhead Primary Systems.*** Typically, overhead primary distribution systems have been operated as radial circuits from the substation outward. Figure 18-2 shows schematically a typical primary feeder in a predominantly residential area; an overhead 12.47Y/7.2-kV system is used for illustrative and functional purposes, but underground systems will be discussed later.

The main feeder backbone usually is a 3-phase 4-wire circuit from which the single-phase lateral or branch circuits are tapped through fuse cutouts to protect the system from faults on the lateral circuits. The single-phase lateral circuits consist of one phase conductor and the neutral. Distribution transformers are connected between the phase and the neutral; in this case they would have a rating of 7200 V.

Utilities use automatic reclosing feeder breakers and line reclosers to minimize service interruptions. However, serious problems involving the main will cause an outage to some or all of the feeder until line crews can locate the problem and manually operate pole-top disconnecting switches appropriately to isolate the problem and to pick up as much load as possible from adjacent feeders. Switches of this kind usually are found in both the main and lateral circuits as indicated in Fig. 18-2. Also, it is often possible to make and to remove connections while the system is energized through the use of hot-line tools, hot-line clamps, insulated bucket trucks, etc.

Generally this approach has provided an acceptable level of service because overhead system troubles are relatively easy to locate, and repair times are short. However, when the entire primary system is installed underground, while the frequency of serious trouble is expected to be lower than in overhead systems, it is likely that the time involved in pinpointing the location and making repairs will be much longer than in overhead systems.

***48. Underground System.*** While a relatively small percentage of new general-purpose feeders is being installed totally underground, the trend is growing and is expected to continue to grow. Since it is difficult to accomplish many maintenance and operating functions on an underground system while it is "hot," or energized, in contrast to overhead-system practices, specific provisions must be made in the system design to incorporate needed sectionalizing and overcurrent protective equipment.

The main feeder plan shown in Fig. 18-10 is reasonably typical of present practice on underground systems supplying basically residential and small commercial loads. Note that the main feeders are operated radially, but with normally open ties to adjacent main feeders. The main feeder switches usually are 3-phase, 600-A, manually operated load-break switches. The single-phase and 3-phase lateral circuits also are operated as normally open loops.

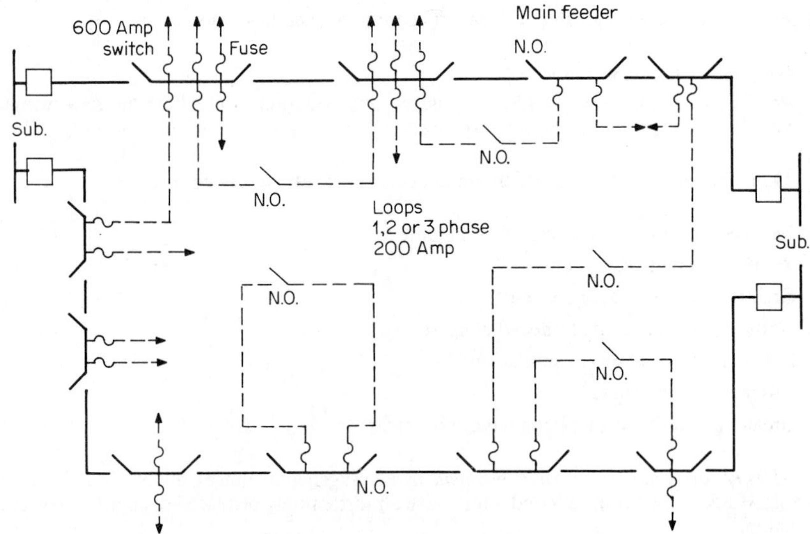


**FIG. 18-10** Typical main-feeder underground circuit. (All switches closed unless shown otherwise.)

Switching in the 200-A circuits can be accomplished by means of either load-break switches or separable, insulated cable connectors. Usually, two main feeder switches are grouped along with the lateral circuit switching and protective equipment into one piece of pad-mounted equipment.

The primary feeders supplying secondary-network systems in metropolitan areas usually are radial 3-wire circuits consisting of 3/c cables in underground duct lines. The 3-phase network transformers are T-tapped to the primary feeders.

***49. Automation.*** With increasing emphasis on reliability of service, a definite trend is under way to make greater use of protective and sectionalizing equipment in the primary system in order to minimize the number of customers involved in an outage and to reduce the outage time. Proposed schemes run the gamut from manually operated devices to automatic devices remotely controlled from distribution centers. The remote-controlled schemes vary from some type of supervisory control to computer-controlled systems with built-in logic to cope quickly with the various problems which may arise.

***50. Primary-Distribution-System Voltage Levels.*** Since World War II, the 15-kV distribution class has become firmly entrenched and today represents 60 to 80% of all primary distribution activity. Very little expansion of lower-voltage systems is taking place. There is a trend, however, toward increasing usage of primary voltage levels above the 15-kV class. This trend has an impact on substation and subtransmission practices as well because higher primary voltages almost axiomatically lead to larger substations and higher subtransmission voltages.

The two principal voltages above 15 kV are 24.49Y/14.4 kV and 34.5Y/19.92 kV. New line additions at these voltage levels now average more than 20% of those at 15 kV.

To achieve economy, the higher primary voltages also require heavier feeder loadings which could imply reduced service reliability because more customers are affected by primary faults. Greater use of automatic switching and protective equipment can do much toward preserving a level of reliability to which the public has become accustomed. This is another reason that most observers believe that an increased amount of automation is inevitable in our distribution systems.

For example, a typical 12.47-kV feeder serves a normal peak load in the order of 6000 to 7000 kVA. On this basis, the probable peak loading of a fully developed 34.5-kV feeder would be expected to be in the neighborhood of 18,000 to 20,000 kVA.

Why go to high-voltage distribution (HVD)? Most of today's systems in the 15-kV class are not voltage-drop-limited, and cost of higher-voltage laterals and associated equipment needed to cover the load area is greater. The major economic advantages are

Larger (and fewer) substations.

Possibility of eliminating a system voltage-transformation level where the new primary voltage is the former subtransmission level.

Other advantages of HVD which are difficult to evaluate in dollars are

**a.** Reduced losses in early stages of development.
**b.** Reduced voltage regulation.
**c.** Greater distance or area coverage.
**d.** Fewer circuits per route (reduced congestion).
**e.** Fewer circuit positions at substations.
**f.** Fewer substation sites.
**g.** Greater flexibility in supplying large spot loads.

***51. The conductor sizes used in overhead primaries*** generally range from No. 2 AWG to 795 kcmil. ACSR and aluminum conductors have almost entirely displaced copper for new construction.

Aerial cable is used occasionally for primary conductors in special situations where clearances are too close for open-wire construction or where adequate tree trimming is not practicable. One type of aerial cable is made up of three fully insulated conductors (shielded cables) lashed to a bare conductor which serves both as mechanical support and as the neutral. This form of aerial cable is not frequently used. Another type of construction more frequently used consists of covered conductors (nonshielded) supported from the messenger by insulating spacers of plastic or ceramic material. The conductor insulation, usually a solid dielectric such as polyethylene, has a thickness of about 150 mils for a 15-kV class circuit and is capable of supporting momentary contacts with tree branches, birds, and animals without puncturing. This type of construction is commonly referred to as "spacer cable."

***52. The conductor sizes most commonly used in underground primary distribution*** vary from No. 4 AWG to 1000 kcmil. Four-wire main feeders may employ 3- or 4-conductor cables, but single-conductor concentric-neutral cables are becoming more and more popular for this purpose. The latter usually employ polyethylene insulation, often have a concentric neutral of one-half or one-third of the main conductor cross-sectional area.

The smaller-sized cables used in lateral circuits of URD systems are nearly always single-conductor, concentric-neutral, polyethylene-insulated, and usually directly buried in the earth. Insulation thickness is on the order of 175 mils for 15-kV-class cables and 345 mils for 35-kV class with 100% insulation level.

Stranded or solid aluminum conductors have virtually supplanted copper for new construction, except where existing duct sizes are restrictive. With the solid-dielectric construction, in order to limit voltage gradient at the surface of the conductor within acceptable limits, a minimum conductor size of No. 2 AWG is common for 15-kV-class cables, and No. 1/0 AWG for 35-kV class.

***53. Voltage Regulation of Primary Distribution.*** Table 18-3 can be used to determine the voltage drop of an existing circuit when the load data are known or to determine minimum conductor size required to meet a given voltage-drop limit. Data are given for various underground-cable and overhead-conductor configurations for 12.47 and 34.5 kV.

*Example.* What is the voltage drop for a 34.5-kV overhead circuit 3 mi long using 4/0 aluminum conductor and carrying a balanced 3-phase load of 15,000 kVA at 90% power factor: The current is $15{,}000/\sqrt{3} \times 34.5 = 251$ A. The circuit feet are $3 \times 5280 = 15{,}840$ ft. Thus $\text{A} \cdot \text{ft}/100{,}000 = 251 \times 15{,}840/100{,}000 = 39.758$. From Table 18-3, the appropriate voltage drop per 100,000 A · ft is 14.0 V line-to-neutral. Therefore, the total voltage drop for the example is

$$39.758 \times 14.0 = 556.6 \text{ V line-to-neutral}$$

**TABLE 18-3** Line-to-Neutral Voltage Drops per 100,000 A · ft* for 12.47Y/7.2 and 34.5Y/19.92 kV and Balanced 3-Phase Loads

| | Voltage class | | | | | | | | | | |
|---|---|---|---|---|---|---|---|---|---|---|---|
| | 12.47Y/7.2 kV | | | | | 34.5Y/19.92 kV | | | | | Approx. amp. capacity for air moving at 2 ft/s |
| | Lagging power factor | | | | | | | | | | |
| Conductor size | 0.7 | 0.8 | 0.9 | 0.95 | 1.00 | 0.7 | 0.8 | 0.9 | 0.95 | 1.00 | |
| | Underground primary | | | | | | | | | | |
| Aluminum: Concentric neutral—direct buried, cross-linked polyethylene, conductor 70°C, neutral 60°C, earth resistivity 90 Ω·cm³, triplex configuration, full installation | | | | | | | | | | | |
| No. 1/0 | 17.1 | 18.5 | 19.8 | 20.2 | 19.8 | 17.6 | 19.0 | 20.1 | 20.4 | 19.8 | |
| No. 2/0 | 14.1 | 15.1 | 16.0 | 16.3 | 15.7 | 14.6 | 15.6 | 16.3 | 16.5 | 15.7 | |
| No. 4/0 | 9.82 | 10.4 | 10.7 | 10.7 | 9.96 | 10.3 | 10.8 | 11.0 | 10.9 | 9.95 | |
| 350 kcmil | 7.01 | 7.19 | 7.17 | 7.00 | 6.11 | 7.37 | 7.49 | 7.39 | 7.16 | 6.11 | |
| 500 kcmil | 5.66 | 5.69 | 5.55 | 5.31 | 4.40 | 6.04 | 6.00 | 5.76 | 5.47 | 4.40 | |
| 750 kcmil | 4.63 | 4.55 | 4.30 | 4.03 | 3.12 | 4.95 | 4.82 | 4.49 | 4.16 | 3.11 | |
| 1000 kcmil | 4.10 | 3.98 | 3.69 | 3.41 | 2.52 | 4.37 | 4.20 | 3.85 | 3.52 | 2.51 | |
| Single conductor shielded and jacked, cross-lined polyethylene, conductor 70°C, unigrounded shield, triplex configuration, full insulation | | | | | | | | | | | |
| 350 kcmil | 7.29 | 7.49 | 7.51 | 7.35 | 6.47 | 7.55 | 7.72 | 7.67 | 7.47 | 6.47 | |
| 500 kcmil | 5.78 | 5.82 | 5.67 | 5.45 | 4.54 | 6.08 | 6.07 | 5.86 | 5.58 | 4.54 | |
| 750 kcmil | 4.64 | 4.54 | 4.26 | 3.97 | 3.02 | 4.88 | 4.74 | 4.41 | 4.08 | 3.02 | |
| 1000 kcmil | 4.02 | 3.85 | 3.52 | 3.21 | 2.26 | 4.23 | 4.03 | 3.65 | 3.31 | 2.26 | |
| | Overhead primary † | | | | | | | | | | |
| No. 4 | 42.3 | 45.4 | 47.8 | 48.5 | 46.6 | 43.4 | 46.3 | 48.5 | 49.0 | 46.6 | 115 |
| No. 2 | 29.8 | 31.2 | 32.0 | 31.9 | 29.3 | 30.9 | 32.2 | 32.7 | 32.4 | 29.3 | 160 |
| No. 1/0 | 21.8 | 22.2 | 22.1 | 21.5 | 18.5 | 23.0 | 23.2 | 22.8 | 22.0 | 18.5 | 215 |
| No. 2/0 | 19.0 | 19.1 | 18.6 | 17.8 | 14.7 | 20.1 | 20.0 | 19.3 | 18.3 | 14.7 | 250 |
| No. 4/0 | 14.7 | 14.3 | 13.3 | 12.4 | 9.20 | 15.9 | 15.3 | 14.0 | 12.7 | 9.20 | 340 |
| 336.4 kcmil | 11.8 | 11.2 | 9.97 | 8.91 | 5.80 | 13.0 | 12.1 | 10.7 | 9.41 | 5.80 | 465 |
| 477 kcmil | 10.4 | 9.58 | 8.27 | 7.18 | 4.10 | 11.5 | 10.5 | 8.97 | 7.68 | 4.10 | 590 |
| 795 kcmil | 8.22 | 7.92 | 6.52 | 5.40 | 2.40 | 9.96 | 8.88 | 7.22 | 5.90 | 2.40 | 820 |

For ampacities of cables, see Tables 18-25 and 18-26.

Regulation of copper for overhead conductors can be estimated with reasonable accuracy the same as that of aluminum conductors two sizes larger.

For single-phase overhead primaries, the voltage drop is approximately two times the 3-phase values given in the table. For underground single-phase primaries in concentric-neutral, direct-buried cables; see section on URD systems.

Cables are 15- and 35-kV classes, respectively.

* Values in the table give the difference in absolute value between sending-end and receiving-end line-to-neutral voltages of a balanced 3-phase circuit, in volts.

† Overhead conductor impedances are based on 50°C conductor temperature, aluminum conductor with 30-in equivalent spacing for 12.47Y kV and 60-in for 34.5Y kV.

NOTE: 1 in = 25.4 mm; 1 ft = 0.3048 m.

Since normal line-to-neutral voltage is $34{,}500/\sqrt{3} = 19{,}920$ V, the percent voltage drop

$$556.6 \times 100/19{,}920 = 2.79\%$$

*Example.* What is the minimum aluminum conductor size to carry 6000 kVA at 90% power factor of balanced 3-phase load over a 2-mi, 12.47Y/7.2-kV feeder with no more than a 3% voltage drop? Load current is $6000/\sqrt{3} \times 12.47 = 277.8$ A. Circuit feet $= 2 \times 5280 = 10{,}560$ ft. Thus

$$\frac{\text{A} \cdot \text{ft}}{100{,}000} = \frac{277.8 \times 10{,}560}{100{,}000} = 29.34$$

$$\text{Permissible voltage drop} = 0.03 \times \frac{12{,}470}{\sqrt{3}} = 216 \text{ V}$$

The corresponding drop per 100,000 A · ft is 216/29.34 = 7.36 V, line-to-neutral. From Table 18-3, this value falls between 477 and 795 kcmil, so that the latter size would be chosen.

***54. Loading of primary feeders*** varies greatly depending on primary voltage, load density, emergency loading requirements, etc. Typical peak loads on 15-kV class feeders are 6 to 7000 kVA. Peak loads on 25- and 35-kV class, fully developed feeders probably will be proportionally greater in the future, assuming that appropriate measures can be taken to maintain acceptable reliability of service.

***55. Voltage drop in the primary feeder*** is an important factor in system design; however, it is only one of the many voltage-drop considerations involved in determining the range of voltages delivered to the customers' service entrances. American National Standard "Voltage Ratings for Electric Power Systems and Equipment (60-Hz)," ANSI C84.1-1982, defines in detail the voltage ranges which should be observed. Outside of the distribution substation voltage drops occur in the primary system, the distribution transformer, the secondary system, the service drop, and in the users' wiring systems, as well. Remedial measures, such as voltage regulators and shunt capacitor banks, can be used to counteract or reduce the voltage drop due to load flow.

A traditional rough rule of thumb has been to allow a voltage drop of about 3% in the primary of urban and suburban systems at time of peak load. Actually, with typical load densities and primary systems of 15-kV class or higher, it is very probable that economic system designs have a primary voltage drop smaller than 3%.

In rural systems which are typified by long lines and light load densities, primary voltage drops may be somewhat larger. This is offset somewhat by the absence of secondaries in serving individual farms; however, the service drops often are longer than in urban systems. The design objective, of course, is to keep delivered voltage to all customers in an acceptable and satisfactory range.

## THE COMMON-NEUTRAL SYSTEM

***56. The 4-wire, multigrounded, common-neutral distribution system*** now is used almost exclusively because of the economic and operating advantages it offers. Usually the windings of the substation transformers serving the primary system are wye-connected and the neutral point is solidly grounded. Occasionally a small amount of impedance is connected between the transformer neutral and ground in order to limit line-to-ground short-circuit currents on the primary system to a predetermined value. The neutral circuit must be a continuous metallic path along the primary routes of the feeder and to every user location. Where primary and secondary systems are both present, the same conductor is used as the "common" neutral for both systems. The neutral is grounded at each distribution transformer, at frequent intervals where no transformers are connected, and to metallic water pipes or driven grounds at each user's service entrance. The neutral carries a portion of the unbalanced or residual load currents for both the primary and secondary systems. The remainder of this current flows in the earth and/or the water system. For typical conditions, it is estimated that about one-half of the return current flows in the neutral conductor, although the division can vary widely depending on

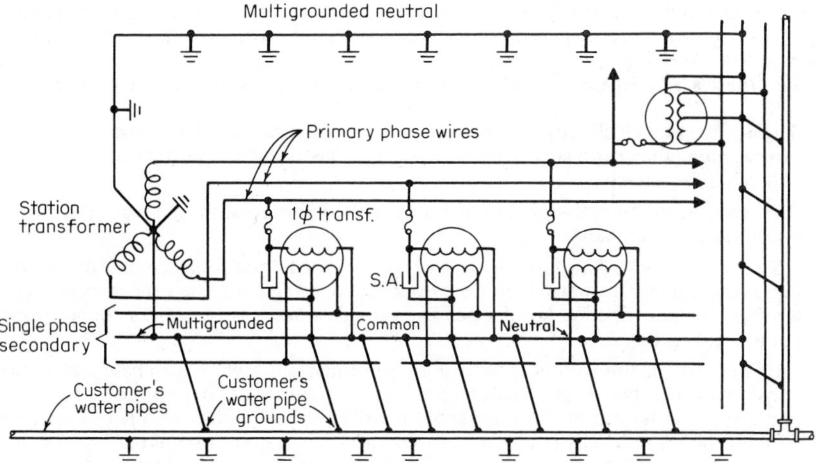


**FIG. 18-11** Common-neutral methods of distribution.

earth resistivity and the relative routing of the electric and water systems. Figure 18-11 is a schematic representation of a common-neutral system.

***57. Grounding of Neutral.*** Rules related to grounding on the utility system neutral are given in the National Electrical Safety Code, ANSI C2-1984, and regulations governing the grounding of the neutral on users' premises are stated in the National Electrical Code (NESC), ANSI C1-1984. In brief, the secondary neutral is grounded at every service through a metallic water-piping system and through "made electrode grounds" such as other underground metal systems, building steel, or driven ground electrodes. The increasing use of nonmetallic water piping and insulating couplings on metal water systems is requiring the use of other grounding means. The secondary neutral also is grounded at the distribution transformer, usually by means of driven grounds. Although it is often general practice to install a metal butt plate or a wire butt wrap on poles to help in grounding the system neutral and other equipment, the NESC requires two such devices to equal one *made electrode;* as a result, neither can be used to satisfy the NESC requirement for a direct earth ground with a made electrode at each transformer or other arrester location.

The resistance to ground of a typical metallic water-piping system usually is less than 3 Ω. When made electrode grounds are used, they should have a resistance of not more than 25 Ω. Many utilities strive for lower values such as 5, 10, or 15 Ω.

Where there is no secondary neutral, as such, and no distribution transformers, the primary neutral should be grounded at intervals of not less than 1000 ft. Many utilities require grounding at smaller spacing, such as 500 ft, to meet the NESC requirements for a multigrounded neutral, there must be a minimum of the equivalent of four made electrodes in *each* mile.

In URD systems, the primary circuits usually are in direct-buried, concentric neutral cable, so that excellent grounding is obtained.

***58. The neutral*** must have a continuous metallic path between the substation and users' services. No disconnecting devices should be installed in the common neutral. In no case should the earth or buried metallic-piping systems be used as the only path for the return of normal load current.

***59. Size of Primary Neutral.*** On single-phase primary circuits (phase and neutral) the neutral conductor should be large enough to carry almost as much current as the phase conductor. Often the same neutral conductor size is used for both, or the neutral has "100%" conductivity.

In 3-phase primary circuits carrying reasonably balanced load, the neutral conductor can be considerably smaller than the phase conductors; 50% conductivity is not uncommon; some utilities specify a specific size of neutral conductor, such as No. 1/0 aluminum, regardless of the size of the phase wires.

Secondary-system neutral conductors are often the same size as the phase conductors where open-wire construction is used. Where triplexed construction is used, the neutral frequently has a reduced cross section.

***60. The 4-wire, common-neutral primary system*** has many advantages over 3-wire systems:

- **a.** Single-phase branch circuits, or laterals, consist of one insulated phase conductor and the neutral, rather than two insulated phase conductors. The economic advantage is very great in underground systems.
- **b.** On overhead systems, only one lightning arrester is required at each single-phase distribution transformer, rather than two.
- **c.** Only one primary bushing or cable termination is needed on each single-phase distribution transformer, rather than two. In the case of underground systems where the primary "loops through" each distribution transformer, two primary cable terminations or connectors are needed, rather than four.
- **d.** Only one fuse or fuse cutout is needed in the primary of each single-phase distribution transformer. Not only is this a substantial economic advantage, but a short circuit in the primary of the transformer is interrupted positively by the action of a single fuse; and primary voltage is thereby removed from the transformer. In the case of the 3-wire system with the distribution transformer connected phase-to-phase, a second fuse must operate to remove primary voltage and the fault. There may be appreciable time between operation of the two fuses during which fault current continues to flow and abnormal voltages may be experienced by the user.
- **e.** Single-phase primary lateral circuits can be protected by a single fuse cutout, rather than two. Line-to-ground short circuits are promptly cleared by operation of one fuse and voltage removed from the branch circuit. In a 3-wire system (assumed grounded at the substation) single-phase lateral protection, if used, would require two fuse cutouts; a line-to-ground fault would blow only one fuse, leaving all the distribution transformers on that circuit excited at only 58% of normal as long as the faulted phase remains grounded. Under these conditions users' equipment would be exposed to abnormally low voltage. The ability to fuse lateral circuits contributes substantially to reliability of service, since a major amount of the total circuit exposure is comprised of the primary laterals in residential areas.

***61. Common-Neutral and Telephone Circuits.*** Usually, no problems are encountered in the joint use of poles for overhead distribution circuits and telephone circuits, particularly when the telephone circuits are in cable, as is now common practice. Also, in underground residential circuits, power cables and telephone cables often are installed in the same trench with no intentional physical separation of the power and communication facilities, that is, "random lay." Where separate grounding electrodes are employed for supply and communication facilities at customer's premises, the electrodes shall be bonded together with not less than No. 6 AWG copper wire.

## VOLTAGE CONTROL

***62. System Voltage Levels and Voltage Ranges.*** Since about 1900, there have been several recommendations for certain voltages as standard or preferred for primary and secondary distribution systems, as well as for higher-voltage systems. The latest listing of standard system voltages is American National Standards Institute (ANSI) Standard C84.1-1982, "Voltage Ratings for Electric Power Systems and Equipment (60 Hz)." This standard was formulated by both utilities and manufacturers, and its recommendations are followed by both segments of the industry. Observance of this standard enables the utilities and manufacturers to work in harmony. In many states, ANSI C84 is the basis for rulings of the regulatory commission as far as voltage requirements are concerned.

This standard designates certain specific numerical values as standard nominal voltages including 120/240 V single-phase, 480Y/277 V, 12,470Y/7200 V, as well as the higher primary voltages, 24,940Y/14,400 V, and 34,500Y/19,920 V, and others.

Using the nominal 120/240-V system as an example, the standard designates two different ranges of voltage, Range A and Range B. Range A service voltage specifies that utility supply system be so designed and operated that most service voltages are within the limits specified, for example, 114/228 and 126/252 V. The occurrence of service voltages outside these limits is to be infrequent.

With the typical voltage drops between the service entrance and the points of utilization, the utilization equipment is designed and rated to give fully satisfactory performance within Range A.

Range B service voltage includes voltages above and below Range A that necessarily result from practical design and operating conditions on supply or user systems. These conditions are limited in extent, frequency, and duration. When they occur, corrective measures should be undertaken within a reasonable time to improve voltages to meet Range A requirements.

Insofar as practicable, utilization equipment is designed to give acceptable performance within Range B.

The design and operating bogey of the utilities is to provide service voltage to all customers at all times within Range A limits.

***63. Voltage Profiles.*** It is usually convenient to discuss distribution-feeder-voltage regulation in terms of voltage "profiles" of the feeder, because the voltages are everywhere different on the feeder. A profile is simply a graph of feeder-voltage magnitude versus location on the feeder.

For a simple case of one load at the end of the feeder (assuming uniform conductor), the one-line diagram and profile are as shown in Fig. 18-12.

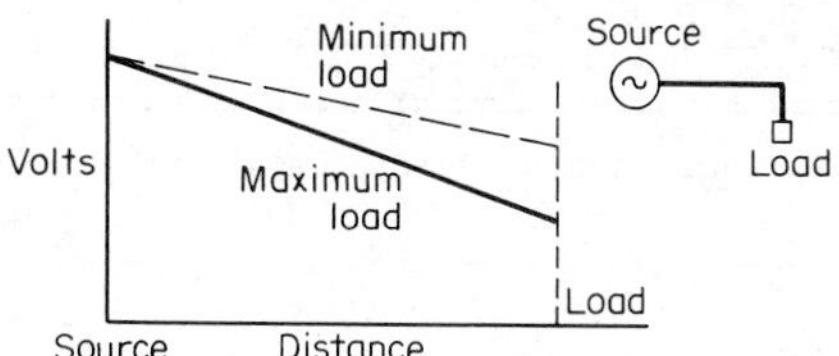


**FIG. 18-12** Voltage profile for concentrated load.

The profile is a straight line between source and the load, and the voltage regulation at any point between is proportional to the distance from the source. It may be, as shown by the dashed-line profile, that minimum load is not zero, in which case the voltage variation is less than the calculated regulation, since regulation is usually calculated on the voltage difference between no-load and full-load conditions.

If additional loads are distributed along the feeder, the profile becomes a broken line, and if the load is uniformly distributed, the profile becomes a smooth curve, as shown in Fig. 18-13.

The shape of the profile is of less consequence than knowing the extremes, because there are generally customers connected at all points on the feeder, and no customer's voltage should be too high or too low. Since most feeders neither supply a single load nor are uniformly loaded, it usually is necessary to calculate the voltage profile on a piece-by-piece basis, representing the loads and feeder configurations as accurately as the situation warrants.

In addition to the distribution-feeder-voltage profile, there is additional regulation in the distribution transformer and its secondaries and services. This additional regulation can be added to the profile as shown in Fig. 18-14. For protection of the first customer on the feeder

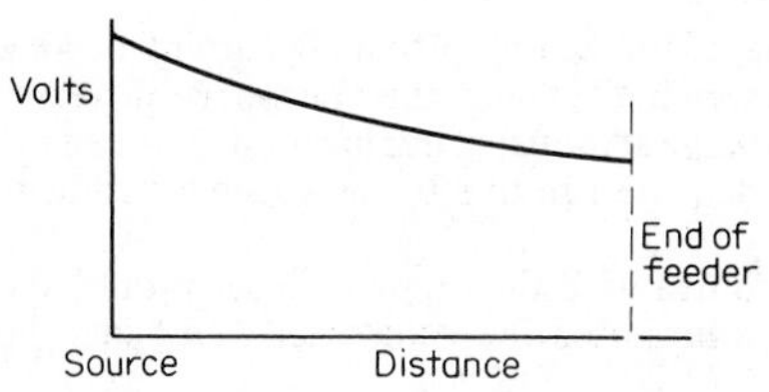


**FIG. 18-13** Voltage profile for distributed load.

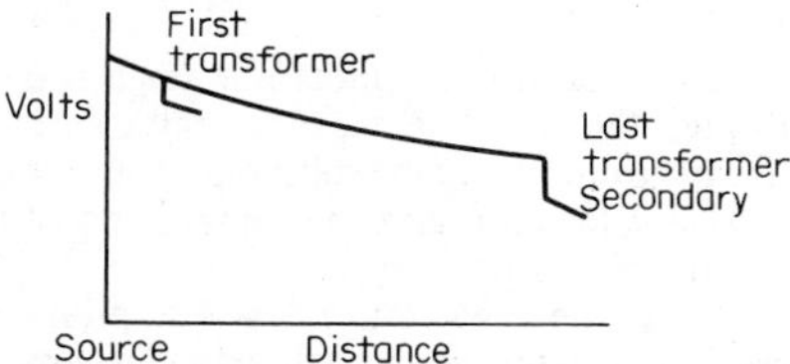


**FIG. 18-14** Additional regulation due to transformer and secondary.

from possible overvoltage it is usual to assume only a partially loaded transformer rather than one at full load.

It is now possible to establish a limiting band of voltage within which all customers must lie for satisfactory service, usually Range A. In turn, this will also establish the maximum permissible difference between the full-load and light-load primary voltage.

The problem of holding the right voltage at each customer location at all times may be visualized by referring to Fig. 18-15.

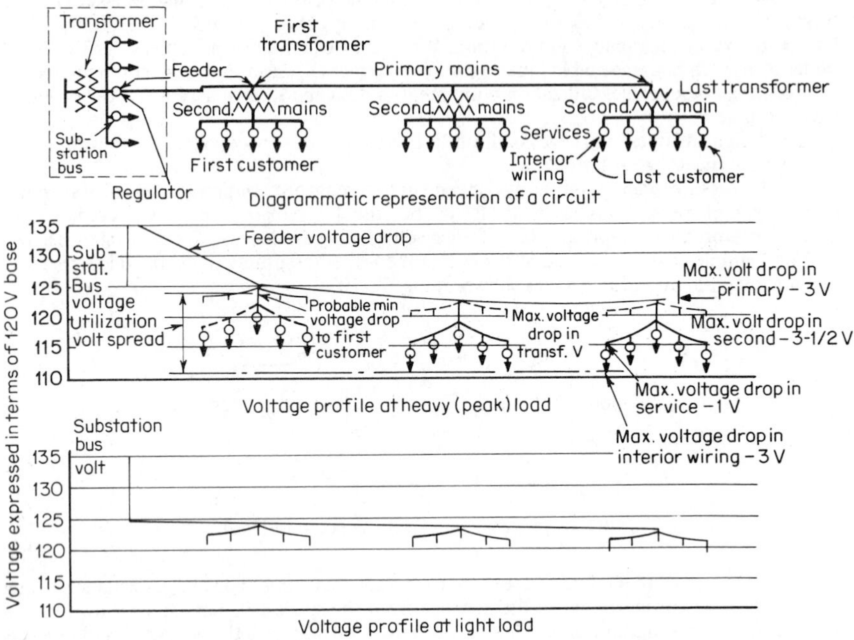


**FIG. 18-15** Distribution circuit with voltage profiles at heavy and light loads.

***64. Voltage Control.*** As implied in Fig. 18-15, usually there is voltage-control equipment in the substation consisting of load tap changer on the power transformers or bus or feeder voltage regulators. This regulating equipment can control only the voltage *level* of the primary system. It can have no effect on the voltage *spread* between the first and last customers on the feeder.

There are several procedures which can be taken to correct for increasing voltage drops as the load on the feeders grows, among them are capacitors and supplementary feeder-voltage-regulator installations.

The effect of capacitor application is illustrated in Fig. 18-16, where the load is assumed to be uniformly distributed along the feeder, and a capacitor bank is installed as indicated in Fig. 18-16.

The capacitor produces a voltage rise because of its leading picofarad current flowing through the inductive reactance of the feeder. As is seen in the figure, this voltage rise increases linearly from zero at the substation to its maximum value at the capacitor location. Between the capacitor location and the remote end of the feeder, the rise due to the capacitor is at its maximum value.

When the capacitor voltage-rise profile is combined with the original feeder profile, the resulting net profile is obtained. The capacitor has increased the voltage *level* all along the feeder, resulting also in a reduced voltage *spread.*

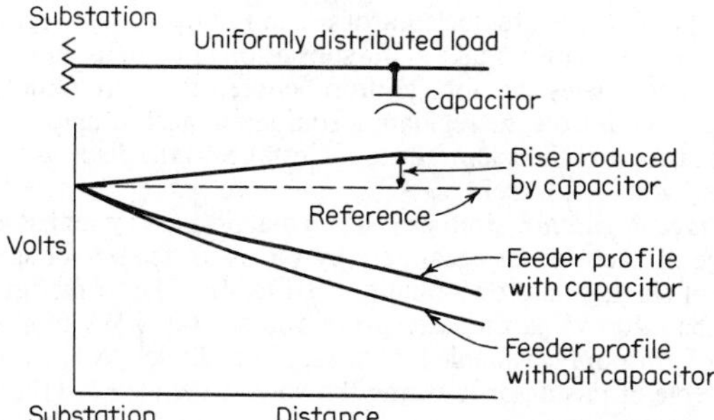


**FIG. 18-16** Effect of shunt-capacitor application.

In practical applications the capacitor bank can be a permanently connected or "fixed" bank as shown, or an automatically switched bank. The fixed bank is limited in size by the allowable voltage rise during light-load conditions and therefore may not produce sufficient voltage rise during heavy-load conditions. It can then be supplemented by additional switched capacitors which automatically switch on at heavy-load conditions and off again as the load decreases.

The effect of applying a supplementary feeder-voltage regulator is shown in Fig. 18-17.

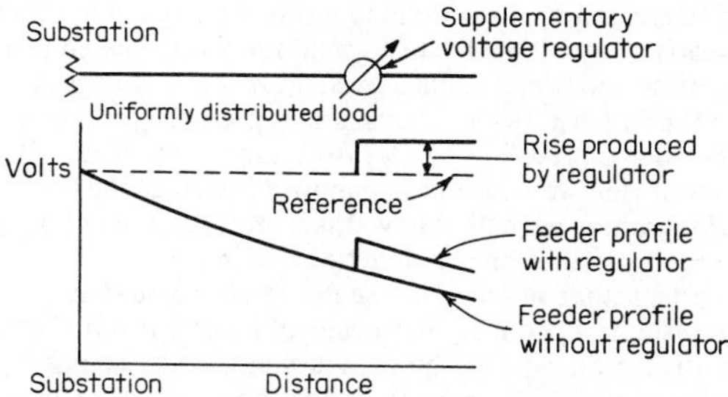


**FIG. 18-17** Effect of supplementary voltage regulator.

Note that the regulator produces no voltage effect between the source and the regulator location and its entire boost effect between the regulator location and the remote end of the feeder.

A typical primary feeder serves distributed loads, as well as concentrated loads, and may also have shunt capacitors and supplementary voltage regulation, such that all these previous concepts must be employed in studying voltage conditions.

***65. Voltage regulation*** in distribution substations usually is accomplished by individual feeder-voltage regulators or by automatic load-tap-changing equipment in the substation transformers. Individual feeder-voltage regulators are advantageous where feeders of differing lengths and diverse load characteristics are supplied from the same substation bus. Automatic load-tap-changing equipment in the power transformer provides voltage control on the substation bus, or group regulation, when feeder lengths and load characteristics are reasonably homogeneous.

Voltage control is needed to compensate not only for the voltage regulation in the subtransmission system and substation transformer, which is measurable at the substation, but also for the voltage regulation which occurs in the distribution transformers and in the primary and secondary systems beyond the substation. The latter portion of the overall system voltage regulation is a function of the load flow and system impedances and cannot be measured directly at the substation. Therefore, the control systems of the voltage regulators or tap-changing equipment not only sense the voltage at the substation but also usually contain a "line-drop compensator" which simulates the voltage drop between the station and some point in the distribution system and controls the regulating equipment accordingly.

Switched shunt capacitor banks sometimes are installed at the distribution substation as part of the overall system voltage control.

***66. Feeder-Voltage Regulator.*** In the typical radial primary system, it is often necessary to regulate the voltage of each feeder separately by means of feeder-voltage regulators. These regulators may be of single-phase or 3-phase construction. The former are available in sizes from 25 to more than 400 kVA, the latter from 500 to 2000 kVA. For distribution-system application they are commonly available for voltages from 2.5 kV to 34.5 grd Y kV. Regulators commonly are capable of raising or lowering the voltage delivered to the feeder by 10% and normally are rated on this basis.

Modern voltage regulators all are of the step-voltage type, which has completely supplanted the earlier induction voltage regulators. The step-voltage regulator basically is an autotransformer which has numerous taps in the series winding. Taps are charged automatically under load by a switching mechanism which responds to a voltage-sensing control in order to maintain voltage as close as practicable to a predetermined level. The voltage-sensing control receives its inputs from potential and current transformers and provides control of system voltage level and bandwidth. In addition, it permits selection of line-drop compensation and provides features such as operation counter, time-delay selection, test terminals, and control switch.

Most feeder-voltage regulators are of the 32-step design. Since they usually operate over a range of voltage of 20%, the voltage change per step is ⅝%. If the full range of regulation of $\pm 10\%$ is not required, the regulators can carry more than rated current. For example, operating with a range of $\pm 5\%$, 160% of rated current can be carried.

***67. Line-Drop Compensator.*** In simplified terms, the regulator voltage (local voltage) is stepped down by means of a potential transformer and fed to the control system where it is compared with the desired and preset voltage level. If the actual voltage deviates from the preset level by more than $\pm \frac{1}{2}$ of the bandwidth, which also is preset by the operator, the tap-changing mechanism operates, after a preset time delay, to return the voltage within the preset band. From a practical point of view, the minimum bandwidth is twice the size of the voltage step, or $2 \times \frac{5}{8}\% = 1.25\%$. Maintaining a small bandwidth is important in reducing voltage variations and in making full use of the allowable system voltage drop.

The line-drop compensator consists of adjustable resistance and reactance components and is preset to simulate system impedance. By means of a current transformer, current proportional to load current is circulated through the resistance and reactance, producing a voltage signal which is combined with the signal from the local voltage. The net result is that the line-drop compensator causes a higher voltage to be held at the voltage regulator during periods of heavy load. In this way, a constant voltage is held at some point in the system, as determined by the compensator setting. This helps to achieve the goal of minimizing the voltage change with varying loads at any location.

***68. Supplementary Voltage Regulation.*** In some long primary circuits such as rural feeders, it is often necessary to provide voltage regulation in addition to that incorporated in substation equipment because of large voltage drops in the system. This supplementary voltage regulation usually is improved by single-phase automatic step regulators in the smaller ratings. These regulators are suitable for pole mounting.

***69. Bus regulation*** at the distribution substation usually is provided by automatic load-tap-changing equipment built into the substation transformer or by large step-voltage regulators.

***70. Switched shunt capacitors*** are often applied at distribution substations or out on the primary feeders to accomplish a portion of the overall voltage-regulation job. Most utilities apply shunt capacitors primarily as a tool in economic system design. Usually fixed (unswitched) shunt capacitors are applied to bring the light-load power factor to more or less 100%. Then, additional automatically switched shunt capacitor banks are added to achieve an economic full-load power factor, which is usually in the order of 95 to 100%.

These capacitors, in addition to their economic functions, such as reducing losses and releasing system capacity, improve system conditions substantially. Usually additional voltage control is needed, however, and this is most economically accomplished with voltage-regulating equipment.

## *OVERCURRENT PROTECTION*

*71. **General Principles of Overcurrent Coordination.*** Coordination of overcurrent protection devices means their proper arrangement in series along a distribution circuit so that they function to clear faults from the lines and equipment in accordance with a prearranged sequence of operation. Fuse cutouts, automatic circuit reclosers, sectionalizers, and relayed circuit breakers are the overcurrent protective devices most commonly used. Ratings and characteristics can be obtained from appropriate product bulletins of the manufacturers.

When the protective devices are properly applied and coordinated:

They can eliminate service outages resulting from temporary faults.

They reduce the extent of outages, that is, the number of users affected.

They are helpful in locating the fault, thereby reducing the duration of interruptions.

*72. **Main-Line Sectionalizing.*** Usually the first protective device on a primary feeder is a circuit breaker or a power-class recloser located in the substation. If the circuit is overhead, the circuit breaker often is provided with reclosing relays so that it operates in much the same manner as a recloser. If the circuit is primarily underground, reclosing is not generally used.

If portions of the main feeder and long branches extend beyond the zone of protection of the relayed breaker or recloser at the substation, additional overcurrent protective equipment usually will be installed out on the main feeder.

Manually operated sectionalizing equipment such as pole-top disconnecting switches or solid blade cutouts also are installed at strategic locations along the main feeder to:

Provide a convenient means of isolating faults so that repairs can be made after other parts of the feeder are restored to service.

Provide means of connecting the feeder to adjacent feeders so that service can be maintained to most customers while repair or maintenance operations are taking place.

On underground feeders, this sectionalizing equipment is often in the form of 3-phase, manually operated, load-break switches.

*73. **Branch-Circuit Protection.*** It is exceedingly important to isolate faults on branch and subbranch lines, even short ones, in order to maintain service on the rest of the feeder. Not only does the branch-circuit protection protect the rest of the feeder, but it helps to pinpoint the location of the fault.

Also, usually there is much more mileage and much more exposure in the branch circuit or laterals than in the feeder main. The simple expulsion-fuse cutout is almost universally used for branch and subbranch overcurrent protection. It may be used in combination with reclosers.

On underground feeders, the lateral circuits usually are fused at the point where the main feeder is tapped to establish the lateral. Often the fuses for several lateral circuits are grouped into a sectionalizing equipment which may also incorporate main-feeder and load-break sectionalizing switches.

*74. **Temporary Fault Protection.*** On overhead distribution circuits, a large portion of the faults are of a temporary nature or are potentially of a temporary nature. For example, some types of transitory faults include momentary contacts with tree limbs, and lightning sparkover of insulators or crossarms where no sustained 60-Hz follow current is established and no protective devices operate. Other types of faults which result in 60-Hz follow current can be of a transient nature if the circuit voltage can be removed quickly for a short period of time and then restored after the fault path has recovered adequate dielectric strength. Such faults can result from lightning flashovers, bird or animal contacts, conductors swinging together, etc. Reclosers and reclosing breakers provide the function of fault deenergization, pause for deionization of the arc path, and reestablishment of voltage.

If the fault has disappeared during the "dead time," the reclosure is successful. If not, one or more additional reclosing cycles may be attempted. If the fault persists after the prescribed number of reclosing operations, the breaker or recloser will lock open; or the fault will be removed by operation of a fuse or sectionalizer.

It should be recognized that the reclosing function is provided to eliminate the effects of *temporary* faults only. If all faults were of a permanent nature, reclosing would be pointless.

To provide effective protection against temporary faults, all parts of the feeder should be within the zone of a reclosing device. That is, if the station recloser or relayed circuit-breaker sensing does not reach to the remote ends of the circuit, they should be supplemented with reclosers out on the line. (The term "reach" here is used with the meaning of "sense" faults or "sense and operate" for faults.)

***75. Permanent Fault Protection.*** Permanent faults are those which require repairs, maintenance, or replacement of equipment by the utility operating department before voltage can be restored at the point of fault. System overcurrent protection is provided to disconnect the faulted portion of the system automatically so that an outage is experienced by a minimum number of consumers. Isolation of permanent faults is usually accomplished by the operation of fuse cutouts. It is also achieved in some cases by operation (to lock out) of reclosers, circuit breakers, or sectionalizers.

***76. Combination of Permanent and Temporary Fault Protection.*** If all faults were of a permanent nature, low-cost fuse cutouts would be the best solution for primary line protection. If all faults were temporary, automatic reclosing devices capable of covering the entire circuit would be the best solution.

In actual practice both kinds of faults occur, and the problem becomes one of selecting the type of device or combination of devices to provide best overall results. For selection of a system of overcurrent protection, it is necessary to give proper consideration to many factors such as importance of service, total number of faults per year, ratio of temporary to permanent faults, cost to utility of service interruptions, and annual charge on investment.

***77. Selection of Overcurrent Protective Equipment—General.*** The one-line diagram of a distribution circuit, as shown in Fig. 18-18, will show how a well-coordinated installation of overcurrent protective equipment can be made.

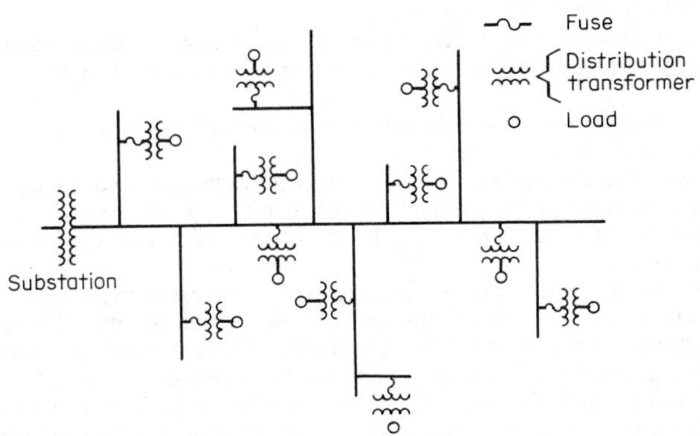


**FIG. 18-18** Distribution feeder.

At the left is the substation, which steps down the voltage from high-voltage subtransmission level to primary-distribution voltage level. It is at this point that the distribution system starts. A distribution substation usually has a number of radial 3-phase feeders radiating from it. However, for the purposes of illustration, only a single feeder will be considered, and it is shown extending to the right from the substation. At various points along the feeder, branch lines or laterals are tapped off and in some cases subbranches are tapped from these branches. There are, of course, loads (residences, stores, garages, etc.) all along the feeder, branches, and subbranches. Only a few of these loads are shown, for the sake of clarity of the diagram.

It is general practice to install a fuse on the primary (incoming) line side of each distribution transformer, as shown in Fig. 18-18. This may be a transformer internal fuse or an external fuse installed in a cutout. Transformer fusing will be discussed later in Pars. **83** to **87**. Figure 18-18 shows the basic system to which additional overcurrent protective equipment must be added to assure good service continuity.

In order to apply overcurrent protective equipment to this system properly, it will be necessary to know the highest and lowest (maximum 3-phase and minimum line-to-ground or line-to-line) values of short-circuit currents which can flow if a fault should occur where the feeder leaves the substation, at each branch junction point, and at each subbranch junction point, as well as the minimum line-to-ground short-circuit current which could flow if a fault should occur at the end of any of the branches or subbranches. These short-circuit currents may be calculated easily by conventional methods.

***78. Clearing Nonpersistent or Temporary Faults.*** Operating records, as well as numerous studies, indicate that a reduction of 75 to 90% in the number of total outages on an overhead system can be attained by the installation of automatic reclosing devices (automatic circuit recloser or reclosing circuit breaker). The recloser or breaker will open the circuit "instantaneously" when a fault occurs, and reclose it after a short period of time. The recloser will repeat this operation if the fault is still on the line after the first reclosure. These two instantaneous openings are intended to clear nonpersistent faults (lightning flashover of the line insulators, momentary contact of two conductors, momentary tree contact, etc.) from the circuit.

Referring to Fig. 18-18, automatic circuit reclosers will be applied to protect the entire system against temporary faults. To achieve this sort of protection, the first recloser should be installed on the main feeder at the substation or the power circuit breaker at the substation should be equipped with overcurrent and reclosing relays.

In applying reclosers to do this job, certain factors must be considered: *(a)* The voltage rating of the recloser must be high enough to meet the requirements of the system. *(b)* Load current, or the amount of current which flows at the point of installation of the recloser under full-load conditions, should not exceed the amount of current which the manufacturer has rated the recloser to carry continuously (continuous-current rating). Recloser ratings are usually selected to be 140% of the peak load current of the circuit. This allows for normal load growth. *(c)* The highest value of short-circuit current which will flow through the recloser and which the recloser must interrupt. This value should not be greater than the highest value of current which the recloser is rated to interrupt (interrupting rating).

Referring to Fig. 18-19, a recloser or breaker with reclosing relays will be located at A to meet the three application principles mentioned above. This device will be depended upon to clear nonpersistent faults which occur in the feeder, branches, or subbranches, anywhere within its protective orbit Zone *A* (shown by dotted line in Fig. 18-19). This protective orbit extends to the

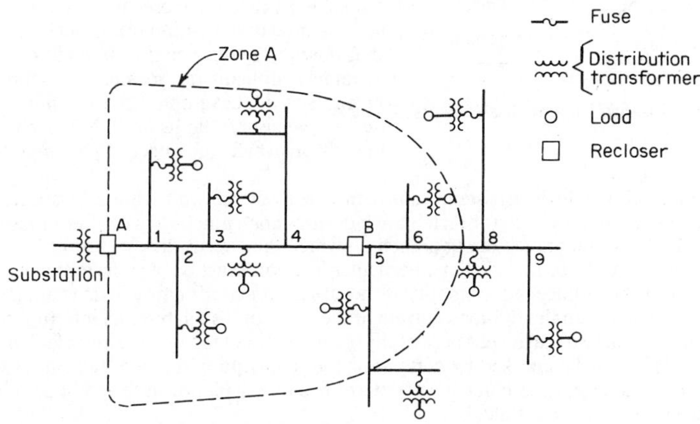


**FIG. 18-19** Distribution feeder with automatic reclosers.

point where the minimum available short-circuit current, as determined by calculation, is equal to the smallest value of current which will cause the device to operate. This value of current required to operate the recloser or breaker is called minimum pickup current. For a recloser it is usually equal to *twice the continuous current rating of the recloser.* A fault beyond this zone may not cause the recloser or breaker *A* to operate; and, therefore, another recloser, *B*, with a lower minimum pickup current rating, should be installed just inside of Zone *A*, thus resulting in so-called overlapping protection.

This second recloser, *B* in Fig. 18-19, is placed on the source side (side nearest source of power) of branch 5 so that it can protect the end of this branch from nonpersistent faults which may not cause recloser *A* to operate. It is applied according to the same considerations as was the recloser at *A*. It will be assumed that a fault on the feeder or any branch or subbranch beyond (to the right of) *B* will cause enough current to flow to operate the recloser at *B*. Every point on the entire circuit is now protected against nonpersistent faults because every point is within the protective orbit of some reclosing device. Obviously, if every point were not within the protective orbit of some reclosing device another recloser would have to be installed still farther out on the line.

***79. Clearing Persistent Faults.*** The first requirement of protecting the circuit against nonpersistent or transient faults has been taken care of by recloser application. It is necessary now to concentrate on the second and third requirements, that is, confining persistent faults to the shortest practical section of line and making persistent faults easy to locate.

If a permanent fault occurs anywhere on the system beyond a recloser, the recloser will operate once, twice, or three times instantaneously, depending on adjustment, in an attempt to clear the fault. However, since a persistent fault will still be on the line at the end of these operations, it must be cleared by some means other than the instantaneous recloser operations. For this reason the recloser is provided with one, two, or three time-delay operations, depending on adjustment. These additional operations are purposely slower (time-delay operations) to provide coordination with fuses or to allow the fault to "self-clear." If the fault is still on the line after the fourth opening, the recloser will not close in but lock open.

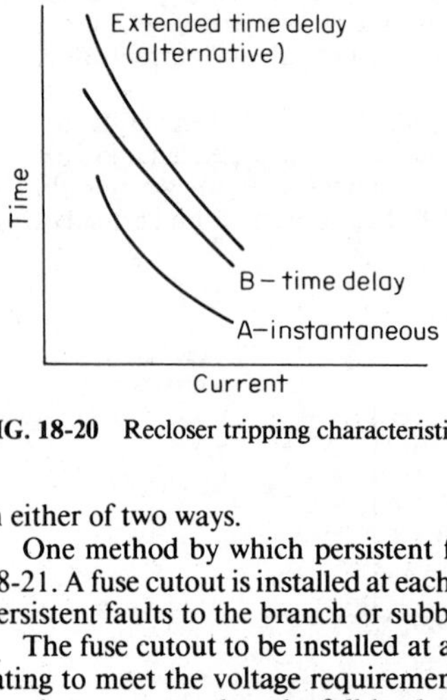

**FIG. 18-20** Recloser tripping characteristics.

Referring to Fig. 18-20, Curve *A* repesents the instantaneous tripping characteristic with respect to time for the first and second opening of a conventional automatic circuit recloser. Curve *B* represents the tripping characteristics for the third and fourth openings. Following the fourth trip on time delay, the recloser will lock out and must be manually reclosed after the cause of the fault has been remedied.

A persistent fault on a branch or subbranch line *should not* cause a recloser to lock open, since a fault on a relatively unimportant subbranch could shut down the entire circuit, in addition to being extremely difficult to locate. Therefore, some means should be employed to confine outages due to persistent faults to the branch or subbranch on which they occur. This may be done in either of two ways.

One method by which persistent faults can effectively be dealt with is illustrated in Fig. 18-21. A fuse cutout is installed at each branch or subbranch junction to confine outages due to persistent faults to the branch or subbranch on which they occur, that is, fuses 1, 2, 3, 4, etc.

The fuse cutout to be installed at a particular location must be of sufficiently high voltage rating to meet the voltage requirements of the circuit. Its continuous current rating must be equal to or greater than the full-load current at the point of installation. Its interrupting rating must be high enough so that it will successfully open the circuit for any persistent fault occurring beyond it. This may be checked by comparing the interrupting rating of the cutout with the maximum available short-circuit current determined by calculation at the point on the system where the cutout is to be installed.

When the correct ratings of fuse links are used throughout the system, *no fuse will be blown or even damaged by a temporary fault beyond it; that is, the recloser will open the circuit one, two, or three times on instantaneous operations without the fuse link being damaged.* On a permanent fault, the first fuse link on the source side of the fault will be blown, and the circuit thus will be opened by the blowing of the fuse during the third or fourth (time-delay) operation of the recloser, before the recloser will lock open. Hence the fault will be isolated by the fuse and the recloser will reset automatically, restoring service everywhere except beyond the blown fuse. The recloser should never lock open on a permanent fault beyond the fuse if it has been properly

**TABLE 18-4** Automatic Recloser and Fuse Range of Coordination*

| Recloser rating, rms A (continuous) | | Fuse link ratings, rms A | | | | | | | |
|---|---|---|---|---|---|---|---|---|---|
| | | 25T | 30T | 40T | 50T | 65T | 80T | 100T | 140T |
| | | Range of coordination, rms A | | | | | | | |
| 50 | Min | 190 | 480 | 830 | 1200 | 1730 | 2380 | | |
| | Max | 620 | 860 | 1145 | 1510 | 2000 | 2525 | | |
| 70 | Min | 140 | 180 | 365 | 910 | 1400 | 2000 | 2750 | |
| | Max | 550 | 775 | 1055 | 1400 | 1850 | 2400 | 3200 | |
| 100 | Min | 200 | 200 | 200 | 415 | 940 | 1550 | 2280 | |
| | Max | 445 | 675 | 950 | 1300 | 1700 | 2225 | 3050 | |
| 140 | Min | | 280 | 280 | 280 | 720 | 710 | 1750 | |
| | Max | | 485 | 810 | 1150 | 1565 | 2075 | 2875 | |
| 200 | Min | | | | 400 | 400 | 400 | 880 | 3200 |
| | Max | | | | 960 | 1380 | 1850 | 2600 | 4000 |
| 280 | Min | | | | | | 620 | 620 | 1350 |
| | Max | | | | | | 1500 | 2200 | 4000 |

* Recloser sequence: two instantaneous plus two standard time-delay operations.

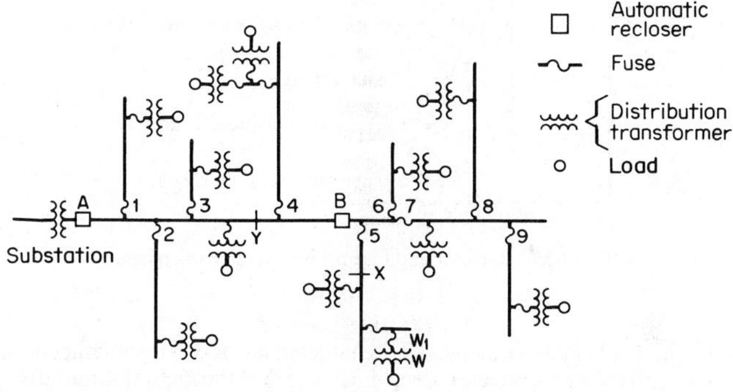


**FIG. 18-21** Distribution feeder with automatic reclosers and fuse cutouts.

coordinated with the recloser. Extensive coordination tables are available, as illustrated in Table 18-4, to simplify and facilitate the job of coordinating reclosers with fuse links.

***80. Recloser-Fuse Coordination.*** Figure 18-22 shows the time-current characteristic curves of the automatic circuit recloser similar to those shown in Fig. 18-20. On these curves, the time-current (TC) characteristics of a fuse *C* are superimposed. It will be noted that fuse curve *C* is made up of two parts; that is, the upper portion of the curve (low current range) represents the total clearing-time TC curve, and the lower portion (high current range) represents the melting TC curve for the fuse. The intersection points of the fuse curves *C* with the recloser curves *A* and *B* illustrate the limits between which coordination will be expected. Basically, this is correct within the interest of simplicity. However, to establish intersection points *a* and *b* accurately and to prepare coordination charts, it is necessary that the characteristic curves of both recloser and fuse be shifted, or modified, to take into account alternate heating and cooling of the fusible element as the recloser goes through its sequence of operations. For example, if the fuse is to be protected for two instantaneous openings, it is necessary to compute the heat input to the fuse during these two instantaneous recloser operations.

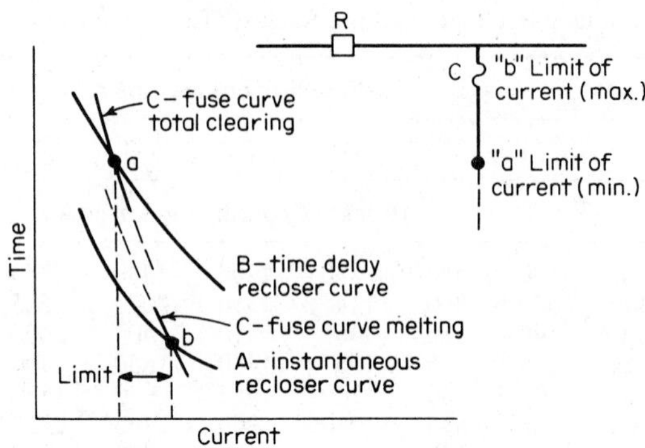


FIG. 18-22 Recloser and fuse time-current characteristics.

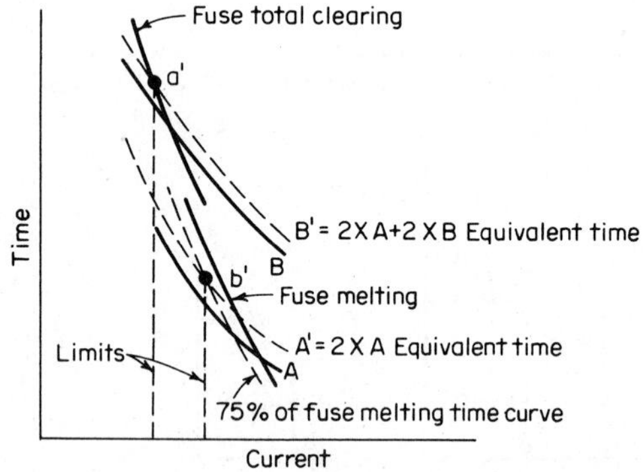


FIG. 18-23 Recloser and fuse time-current characteristics.

Curve $A'$ in Fig. 18-23 is the equivalent TC characteristic of two instantaneous openings *(A)* and is compared with the fuse-damage curve, which is 75% of the melting-time curve of the fuse. This will establish the high current limit of satisfactory coordination indicated by intersection point $b'$. To establish the low current limit of successful coordination, compare the total heat input to the fuse represented by curve $B'$, which is equal to the sum of two instantaneous *(A)* plus two time-delay *(B)* operations, with the total clearing-time curve of the fuse. The point of intersection is indicated by $a'$.

On the basis of all corrections added, the fuse will coordinate successfully with recloser between the current limits of $a'$ and $b'$.

To further clarify what is meant by coordination within prescribed limits, refer to Fig. 18-21—branch 5 and recloser *B*, and also Fig. 18-23 to establish how coordination is achieved, between the limits of $a'$ and $b'$. Assume that fuse 5 beyond recloser *B* is to be protected against blowing or being damaged during two instantaneous operations of the recloser *in the event of a transient fault at X*. If the maximum calculated short-circuit current at the fuse location does not exceed the magnitude of current indicated by $b'$, the fuse will be protected against blowing during all transient faults. By observation of the characteristics in Fig. 18-23, for any magnitude of short-circuit current less than $b'$, but greater than $a'$, the recloser will trip on its instantaneous characteristic once or twice to clear the fault before the fuse-melting characteristic is approached. On the other hand, *if the fault at X is persistent,* the fuse at 5 should blow before the recloser *B* locks out. If the minimum (line-to-ground) calculated short-circuit current available *at the end of branch 5* is substantially greater than the current indicated by $a'$, the fuse will blow

(Fig. 18-23) in accordance with the total clearing characteristic, probably before the first time-delay characteristic of the recloser is approached.

The correct fuse link for any application may be selected by comparing its TC characteristics curve with those of the recloser and making certain allowances and corrections as shown. However, tables have been prepared similar to Table 18-4 to simplify greatly the job of coordinating reclosers with fuse links.

This table shows the maximum and minimum currents at which certain ratings of fuse links will coordinate with certain ratings of reclosers. The only requirement in their use is a knowledge of the available short-circuit currents and load currents, on the system.

Other sequences of recloser operation can be employed, but two instantaneous and two time-delay operations is the combination most widely used. In some cases, it is necessary to coordinate recloser operation with a relayed breaker at the substation. The principles of coordination are similar to the previous discussions, but a detailed study is beyond the scope of this handbook. This is also true of the application requirements for power-class reclosers for substation and line protection.

***81. Fuse-to-Fuse Coordination.*** It may be desirable to use more than two fuses in series beyond a recloser in order to reduce the number of consumers affected by an outage. An example of this would be the fuses at points 7, 8 and at transformers on branch 8, Fig. 18-21. The coordination of these fuses in series beyond the recloser *B* may be accomplished by coordinating adjacent fuses first with each other and then with the recloser in the manner just outlined.

Figure 18-24 illustrates the general principle of coordinating fuses in series. Fuse 7 is called the protected fuse, and fuse 8 is called the protecting fuse. For perfect coordination, fuse 8 must clear the circuit during a fault anywhere beyond it, such as at *X*, before use 7 is damaged or partially melted. From this can be seen the requirement for melting-time–current curves plotted to minimum values and total-clearing-time–current curves plotted to maximum values for each fuse-link rating. *Total-clearing-time* curves represent the total time, including melting time and arcing time, plus manufacturing tolerance, for the fusible elements to clear the circuit. *Melting-time* curves represent the minimum time, based upon factory test, at which the fusible element melts for various currents. From the melting-time curves, "damaging"-time curves can be determined by applying a factor of safety. It usually is suggested that the damaging-time curve be made by taking 75% of the melting time (in seconds) of a particular size at various current values.

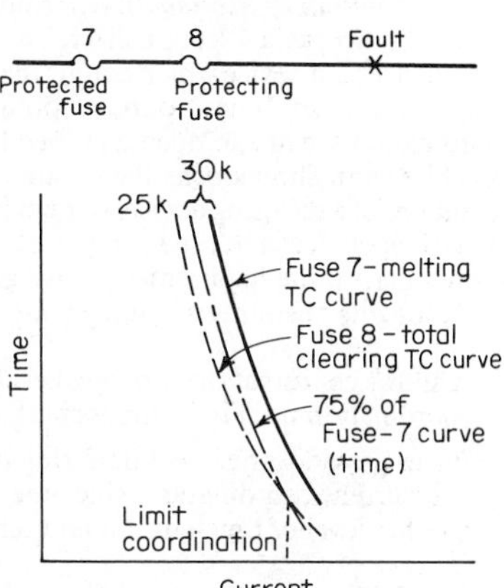


**FIG. 18-24** Fuse time-current characteristics.

To establish coordination of two fuses in series, it is necessary to compare the total-clearing-time–current curve of the "protecting" fuse with the damage-time–current curve of the "protected" fuse. If there is no intersection of these two curves throughout their entire current range, coordination or selectivity can be expected. Where there is an intersection of the curves, the current value indicated by the point of intersection will establish the limit of selectivity.

Because of the inherent characteristics of fuses, the maximum available short-circuit current in that section (determined by calculation) controlled by the protecting link (8 in Fig. 18-24) is the determining current which establishes coordination possibilities.

Most fuse-link manufacturers publish tables which make coordination very simple. These tables eliminate the necessity of comparing actual fuse-characteristic curves.

Table 18-5 is illustrative of tables used for fuse-to-fuse coordination. The values in the left-hand column are the protecting fuse ratings and the values across the top are the protected fuse ratings. The numerical values in the table show the magnitude of current or curve intersection points at which, or below which, fuse 7 will be protected by fuse 8. These current magnitudes are maximum values; in other words, for any short-circuit current greater than that shown, fuse 7 will be damaged. Hence, a larger-rated fuse will have to be selected for location 7 or else its position must be changed.

**TABLE 18-5** Fuse Ratings

Source 7 8

| Type K EEI-NEMA ratings, A, of the protecting fuse links (8 in diagram) | Type K EEI-NEMA ratings, A, of the protected fuse links (7 in diagram) | | | | | | | | |
|---|---|---|---|---|---|---|---|---|---|
| | 6K | 8K | 10K | 12K | 15K | 20K | 25K | 30K | 40K |
| | Max short-circuit rms A to which fuse links will be protected | | | | | | | | |
| 1K | 135 | 215 | 300 | 395 | 530 | 660 | 820 | 1100 | 1370 |
| 2K | 110 | 195 | 300 | 395 | 530 | 660 | 820 | 1100 | 1370 |
| 3K | 80 | 165 | 290 | 395 | 530 | 660 | 820 | 1100 | 1370 |
| 5-A series Hi-surge | 14 | 133 | 270 | 395 | 530 | 660 | 820 | 1100 | 1370 |
| 6K | | 37 | 145 | 270 | 460 | 620 | 820 | 1100 | 1370 |
| 8K | | | 133 | 170 | 390 | 560 | 820 | 1100 | 1370 |
| 10-A series Hi-surge | | 16 | 24 | 260 | 530 | 660 | 820 | 1100 | 1370 |
| 10K | | | | 38 | 285 | 470 | 720 | 1100 | 1370 |
| 12K | | | | | 140 | 360 | 660 | 1100 | 1370 |
| 15K | | | | | | 95 | 410 | 960 | 1370 |
| 20K | | | | | | | 70 | 700 | 1200 |
| 25K | | | | | | | | 140 | 580 |

SOURCE: General Electric Company.

***82. Isolation by Sectionalizer.*** Another method of isolating persistent faults is to install a device, known as a "sectionalizer," at locations where a fuse might otherwise be used. A sectionalizer is a device which counts the operations of a backup automatic-interrupting device such as a recloser. It has no interrupting capacity of its own but operates in a predetermined coordination scheme to open a faulted lateral before the backup device locks out.

The sectionalizer opens the circuit after a predetermined number (usually two or three) operations of a reclosing device. Its opening operation occurs during a period when the reclosing device is open. It can be used to replace a lateral sectionalizing fuse or to replace a lateral recloser where interrupting requirements have grown beyond the capability of the recloser.

Among its operating advantages are

It allows coordination with breakers or reclosers where fault current is above 5000 A. Such coordination usually is impossible with expulsion fuses.

It can provide a new sectionalizing point on an existing circuit without upsetting existing overcurrent coordination, since the device operates as a counter and does not introduce another level of time-current coordination.

***83. Equipment Protection.*** *General.* It is necessary to provide overcurrent protection for distribution equipment such as capacitors and distribution transformers:

To protect the system from the effects of equipment failures.

To reduce the probability of violent failures.

To indicate the location of the fault.

A detailed discussion of all aspects of overcurrent protection of equipment is beyond the scope of this handbook. However, because of its importance, a few comments will be included regarding the overcurrent protection of distribution transformers.

*Self-Protected Transformers.* The term "self-protected" distribution transformer is applied to units which incorporate an internal primary expulsion fuse, a direct-mounted arrester, and an internal secondary circuit breaker. The low-voltage circuit breaker protects the transformer from excessive overload and from some of the faults originating on the secondary system. The expulsion fuse has the sole function of removing a failed transformer from the system.

The rating of the internal expulsion fuse usually is quite large compared with the continuous current rating of the transformer, perhaps 10 to 14 times. This is done

To ensure that the fuse is not damaged by the maximum tripping current of the circuit breaker.

To minimize the possibility of extraneous fuse blowing because of lightning current effects.

Another reason is that fuse removal and replacement may require that the transformer be taken to a shop facility.

Transformer internal expulsion fuses are installed at the factory and are given a designating number rather than an ampere rating for coordination purposes. For a 7200-V transformer, the internal expulsion fuse, often called "weak link," has an interrupting capacity of about 3000 A. Weak links for higher-voltage transformers have somewhat lower interrupting capacity.

Despite the fact that self-protected transformers often are installed at locations on the system where the interrupting capacity of the weak link may be exceeded for a solid fault, experience over the years has been excellent, probably because most transformer failures begin as relatively low fault-current turn-to-turn failures. As the fault current progressively becomes larger, the fuse will operate well before its interrupting capacity is exceeded. Thus, while high-current transformer faults can occur, their frequency of occurrence is very small.

However, there is growing concern among utility companies regarding the occasional violent failures of transformers, and many users are using, or are considering the use of, current-limiting fuses as one method to minimize the energy input into a failed transformer.

The secondary circuit breaker is depended upon to provide protection against excessive transformer loads and secondary system faults that occur within its zone of protection, or "reach." Its TC characteristic should be such that it will always operate before the primary fuse suffers any damage, as illustrated in Fig. 18-25. On the other hand, the breaker should not operate for faults beyond the customer's service-entrance-protective equipment.

Likewise, the internal primary fuse should operate to clear transformer faults before damage occurs to the line sectionalizing fuses back toward the source.

*Conventional Transformers.* Conventional distribution transformers usually are protected by separately mounted expulsion fuse cutouts in series with the primary winding. No secondary

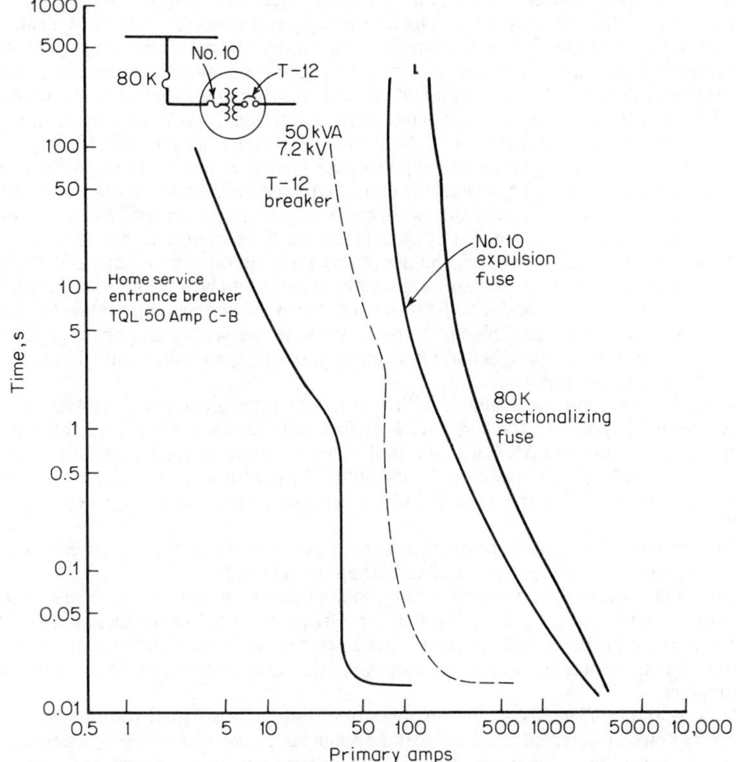


**FIG. 18-25** Overcurrent coordination for self-protected distribution transformer.

overcurrent protection is provided, so that protection against extreme overloads or secondary faults, if any, must come from the primary fuse. Therefore, the size of the primary fuse is relatively much smaller than for the self-protected transformer, usually being chosen in the range of two to three times the full-load current of the transformer.

It is desirable to keep the fuse rating as low as possible consistent with certain application limitations:

**a.** When a transformer is energized by closing of its cutout or operation of a recloser or other switch, a large "magnetizing inrush" current can occur. Initially, this current can be as much as 20 or more times normal, rapidly decaying to normal in a short time—perhaps ½ to 1 s or more. The primary fuse link must be large enough to avoid damage by the magnetizing inrush current; so it usually is selected at least large enough to carry 12 times rated transformer current for 0.1 s without damage.

**b.** The primary fuse should not be damaged by lightning currents or arrester discharge currents (depending on connection used) or large magnetizing currents which can result from saturation of the core due to lightning currents. Many utilities assign an arbitrary minimum fuse size which they will employ. With expulsion fuses, 10- or 15-A rating is often designated as the minimum size.

With a fuse *rating* of two to three times rated transformer current, the minimum melting current under long-time conditions will be in the range of four to six times transformer rating. Consequently, little overload protection is obtained.

Many utilities do not expect to obtain effective overload protection from the primary fuses but count on a transformer load-management program or seasonal load-survey techniques to keep their "burnouts" at an acceptable level. Also, the primary fuse has a limited "reach" as far as secondary faults are concerned; therefore, secondary faults can occur which cannot be "seen" by the fuse. Often these faults—especially on underground systems—will burn clear.

***84. Distribution expulsion cutouts*** are by far the most common type of protective device used on overhead primary-distribution systems. The open-type cutout has generally supplanted the porcelain-enclosed style. The cutout consists of an insulating structure and a hinged fuse tube of hollow cylindrical construction which contains the fuse link. When the fuse link melts, the ensuing arc impinges on the wall of the fibrous tube holder (and usually a small auxiliary tube), generating gas which provides the expulsion action needed to extinguish the fault current. Separation of the fuse link also releases the cutout latching mechanism so that the fuse holder falls to the open position and can readily be located by operating personnel. The fuse holder also can be switched manually with a switch stick, much like a disconnect switch. In some cases, a solid blade is used in place of the fuse holder to provide a disconnecting function. The cutout also can be provided with load-breaking accessories so that it can be used as a load-break switch.

Generally cutouts are available in 100- and 200-A continuous current ratings for fuses, and 300- or even 400-A with solid blades. Cutouts are available with voltage ratings for all the common primary system voltages and interrupting capacities generally from 1200 up through 16,000 A symmetrical, and more.

Fuse links for cutouts are available with a variety of TC characteristics. However, the two most widely used types are the Type K (fast) fuse links and the Type T (slow) links with characteristics as defined in ANSI C37.43. Both types have certain application advantages and disadvantages which must be evaluated by the utility. Ordinarily a given utility uses one type or the other, not both. Use of the Type K links is believed to be somewhat greater than use of Type T.

Other less common types of primary fuses employ fusible elements immersed in oil, or in chemical compounds such as boric acid or carbon tetrachloride.

Fuses are not widely used in electric utility secondary systems, with the notable exception of secondary network systems, where "limiters" are frequently used in the secondary cable circuits. Limiters are fusible elements whose TC characteristics are coordinated with the cable size and insulation characteristics to prevent damage to the cable when faults do not burn clear or self-extinguish.

***85. Current-Limiting Fuses.*** The use of current-limiting fuses in distribution systems has been growing rapidly in recent years. Since these applications are relatively new, a description of the fuses and their characteristics is in order. The fuse generally is constructed of silver wire or ribbon fusible elements—often several in parallel—spirally wound on a core or spider and packed in a quartz-sand filler in a sealed cylindrical glass or epoxy-glass container. Provisions for suitable electrical connections are provided at the ends. When operation takes place under high-fault-current conditions, the fusible element melts almost instantaneously at a series of reduced sections all along its length. The resulting arc dissipates its heat rapidly into the surrounding sand, melting the sand around the arc into a glasslike structure called a fulgorite. This action builds up the apparent resistance of the fuse extremely rapidly, resulting in a "back voltage" greater than system voltage. Thus the fault current is limited to a value much less than the available system fault current.

Current-limiting fuses are characterized by

**a.** High-current interrupting ability. Interrupting ratings of 50,000 A symmetrical or greater are commonly available.

**b.** Operation is noiseless, and there is no expulsion of the arc or arc products. Thus the fuse can be "packaged" into relatively confined space in transformers and protective equipment, making it extremely attractive for use on underground systems.

**c.** In the current-limiting mode of operation, the interrupting time is very fast, one-half cycle or less.

**d.** Current-limiting action and fast operation reduce the amount of $I^2t$ (or fault energy) let through into failed equipment, thereby reducing resultant damage. In the case of distribution transformers applied on systems of high available fault current, protection by current-limiting fuses can virtually eliminate violent failures due to high fault current.

***86. General-purpose current-limiting fuses*** are designed to clear fault currents over a broad range. They are defined by ANSI Standard C37.40-3.2.2.2 as fuses capable of interrupting all currents from the maximum interrupting current down to the current causing melting of the fusible element in one hour. Current-limiting fuses inherently are excellent fault-current interrupters in the high current range. Typical general-purpose fuses operate in the current-limiting mode at fault currents equal to *approximately* 25 times rated current or larger.

Special design and construction techniques are required to obtain clearing of low-fault-current values. For operating times greater than about 0.01 s, the fuses have TC characteristics which are plotted on loglog coordination paper in the same manner as expulsion fuse characteristics.

***87. Backup current-limiting fuses*** are defined by ANSI Standard 37.40-3.2.2.1 as fuses capable of interrupting all currents from the rated maximum interrupting current down to the

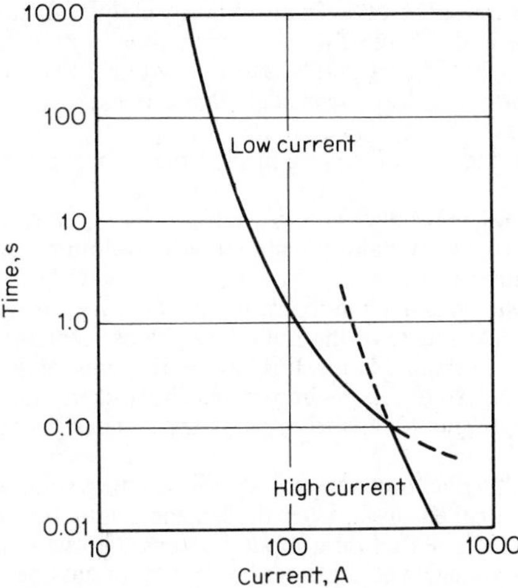


**FIG. 18-26** Time-current characteristic for backup current-limiting fuse in series with expulsion fuse.

rated minimum interrupting current as given by the manufacturer. The low current clearing must be accomplished by an auxiliary device, most commonly an expulsion fuse. In this case, the TC characteristics are a composite of the two fuses as shown in Fig. 18-26. The backup current-limiting fuse can be retrofitted into existing pole-type distribution transformer installations which have expulsion fuse protection only.

## OVERVOLTAGE PROTECTION

***88. Causes of Overvoltage.*** There are many varied causes for the occurrence of overvoltages on distribution systems. The most prominent ones are

**a.** Lightning.
**b.** Neutral displacement during line-to-ground faults.
**c.** Operation of switching and overcurrent protective equipment.
**d.** Resonance effects associated with inductance and capacitance in series.
**e.** Accidental contacts with higher-voltage systems.
**f.** Forced current-zero interruptions.

*Lightning.* Lightning is the most frequent cause of overvoltages on distribution systems. Basically, lightning is a gigantic spark resulting from the development of millions of volts between clouds or between a cloud and the earth. It is akin to the dielectric breakdown of a huge capacitor.

The voltage of a lighning stroke may start at hundreds of millions of volts between the cloud and earth. Although these values do not reach the earth, millions of volts can be delivered to the building, tree, or distribution line struck. In the case of overhead distribution lines, it is not necessary that a stroke contact the line to produce overvoltages dangerous to equipment. This is because "induced voltages" caused by the collapse of the electrostatic field with a nearby stroke may reach values as high as 500 kV.

The amount of current in a stroke is a statistical quantity, depending on the energy in the cloud and the voltage difference between the cloud and the earth at the start of the stroke. A few stroke currents in excess of 200,000 A have been measured; however, 50% of all stroke currents are less than 15,000 A.

The time duration of the current flow in the majority of the high-current strokes is only tens or hundreds of microseconds. Typically, the current rises to its maximum in 1.5 to 10 $\mu$s, decreases to half value in 20 to 50 $\mu$s, and falls to zero within 100 to 200 $\mu$s. On a 60-Hz basis, these are extremely short times if one considers that one-half cycle is equivalent to $1/120$ s or $1{,}000{,}000/120 = 8333$ $\mu$s.

Numerous field investigations have established the numerical statistics which apply to lightning.

In summary, lightning can produce voltages dangerous to the distribution system and all its component equipment. It poses a major threat to service continuity and must be coped with by means of distribution surge arresters.

*Neutral Displacement.* When a line-to-ground fault occurs, even on a "solidly grounded" system, the line-to-ground voltage of the unfaulted phases increases because of the neutral displacement during line-to-ground faults. It is beyond the scope of this handbook to present a detailed mathematical analysis of this phenomenon. The student who is interested in pursuing this subject in greater depth should refer to any standard text dealing with symmetrical components analysis.

The importance of this phenomenon is that a distribution surge arrester must be able to successfully "perform" if called upon, even though there may be a line-to-ground fault on another phase. Suffice it to say that on a modern overhead 4-wire, multigrounded primary distribution system, this voltage will not exceed 1.25 × nominal line-to-ground voltage.

Since the arrester rating basically is the maximum voltage against which it will reseal, the standard arrester rating equal to or greater than 1.25 × nominal line-to-ground voltage usually is selected. For example, on a 12.47Y/7.2-kV system, 1.25 × 7.2 = 9 kV; and a 9-kV arrester would be applied.

*Operation of Switches and Overcurrent Protective Devices.* The operation of switches and overcurrent protective devices produces a short-time transient overvoltage or "switching surge." The *normal* operation of these devices will not produce voltages exceeding twice normal which would not be expected to result in falure of equipment or operation of arresters.

However, operation of improperly applied or inadequate equipment can produce switching surges of greater magnitude. For example, the phenomenon of "restriking" in a switch being used to disconnect capacitors can result in voltages theoretically three times normal or greater. Voltages in the order of 2.5 times normal are more typical of field measurements.

*Resonance Effects.* Ferroresonance is another potential source of transient overvoltages. In 3-phase circuits, single-phase switching (by cutouts or separable insulated cable connectors, for example), fuse blowing, or a broken conductor can result in overvoltages when ferroresonance occurs between the magnetizing impedance of a transformer and the system capacitance of the isolated phase or phases. The term "ferro" is used because of the steel core of the transformer.

A myriad of practical circuit situations can occur which may result in ferroresonant phenomena. Basically, the necessary conditions can arise when one or two open phases result in capacitance being energized in series with the nonlinear magnetizing impedance of a transformer, as in Fig. 18-27, where the switches could be fuse cutouts at a cable riser pole, the

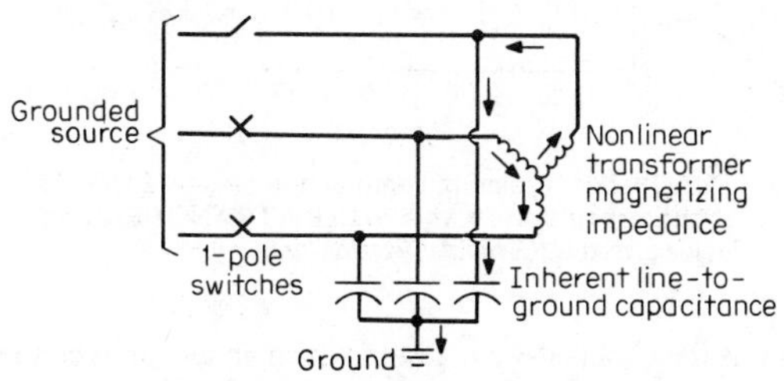


**FIG. 18-27** Single-phase switching in 3-phase circuit.

capacitance could be that of a length of cable connecting to the ungrounded winding of a pad-mounted transformer.

Ferroresonance cannot always be entirely avoided, but steps can be taken to reduce the probability of occurrence such as location of fuses and switches electrically close to the transformer, use of grounded-wye, grounded-wye transformer connection, etc.

*Accidental Contacts.* Often overhead primary distribution circuits are built underneath higher-voltage circuits on the same pole line. Broken high-voltage conductors can fall upon the lower-voltage circuit, possibly causing considerable damage to equipment.

*Forced Current-Zero Interruptions.* Most fault-current-interrupting devices, such as expulsion fuses, reclosers, and power circuit breakers accomplish their arc interruption at a normal 60-Hz current zero. The resulting transient voltages usually are twice normal or less. However, it is possible under some conditions for circuit-opening devices to "force" a current zero rather than "waiting" for a 60-Hz current zero. This can lead to very high transient voltage, depending on the amount of current "chopped," the rate of chopping, and the system configuration. Current-limiting fuses, for example, force current to zero by a rapid increase in their effective resistance when they operate in the current-limiting mode. They must produce an arc voltage greater than system voltage in order to "force" the current zero.

Care is taken in the design of current-limiting fuses to limit the *maximum* arc voltage to a value of approximately 2.2 times normal 60-Hz *peak* voltage. When employed in accordance with proper application guides, they do not expose systems to excessive overvoltages.

***89. Insulation Characteristics.*** *Distribution Transformers.* The low-frequency insulation strength of distribution transformers is confirmed by the 60-Hz high-potential test and the induced voltage test. Low-frequency insulation strength, however, is not necessarily indicative of the impulse strength of a transformer, and industry standards have been established which designate basic impulse insulation levels (BILs) for various voltage classes and define the impulse tests which are made to demonstrate the impulse insulation withstand characteristics.

The standard insulation impulse test wave is defined as a 1.2 × 50 μs voltage wave which increases to its crest value in 1.2 μs and decreases to one-half of the crest value in 50 μs. The waveshape is reasonably typical of natural lightning.

Two impulse withstand values are defined for distribution transformers. The first is the full-wave value or BIL. For a 7200-V transformer which will operate in a 15-kV class distribution system, the BIL is 95 kV. Transformers commonly used on 34.5-kV systems have a BIL of 125 or 150 kV.

The 95-kV BIL, for example, means that the transformer must withstand without failure a 1.2 × 50 μs wave having a crest value of 95 kV. On a time-voltage plot such as Fig. 18-28, the 95-kV point usually is plotted in the 7 to 10 μs range.

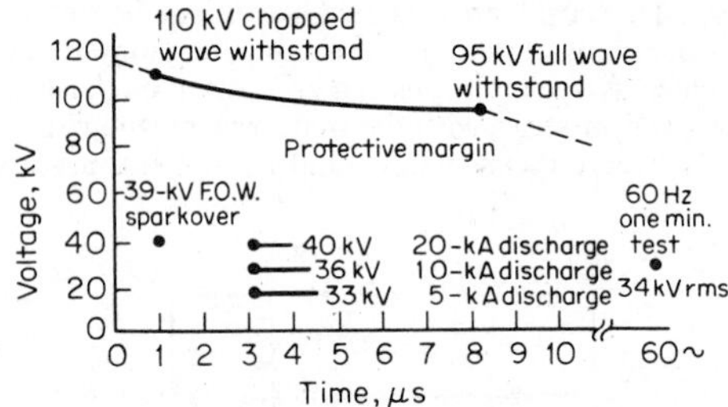


**FIG. 18-28** Insulation coordination plot—7200-V distribution transformer with 9/10 kV TRANQUELL zinc oxide arrester. *(General Electric Company.)*

The second point is the chopped-wave test in which an air gap is connected at the bushing terminal so that it will chop the 1.2 × 50 μs wave at about 3 μs. The transformer chopped-wave test is at 115% of the full-wave value, or 110 kV for 95-kV BIL. It usually is plotted in the 1½- to 3-μs range as shown in the coordination plot of Fig. 18-28.

*Other Equipment.* Other equipment also is built to specified BILs. So-called distribution-class equipment for use on 15-kV systems has a 95-kV BIL; some "power-class" equipment (power circuit breakers, power transformers, etc.) and some overcurrent protective and switching equipment have a higher BIL, for example, 110 kV for the 15-kV class. Distribution-class equipment for use on 25 kV has a 125-kV BIL; for 34.5-kV distribution systems, both 125- and 150-kV BILs are being used at present.

*Cable.* The insulation characteristics for insulated cable usually are determined by the 60-Hz requirements rather than the impulse level. The BIL of modern solid-dielectric power cables used in underground distribution systems is specified by AEIC CS5 to be the same as the corresponding power-class equipment. The cable BIL therefore is slightly higher than the BIL of the distribution-class equipment. The BIL of 15-kV concentric neutral cable, for example, is 110 kV.

*Overhead Lines.* Obviously, the BIL of overhead circuits can vary widely depending on the type of construction. Wood pole lines in the 15-kV category may have a BIL as great as 400 to 500 kV, whereas the BIL of armless construction on steel poles would be equivalent only to that of the insulators used.

***90. Distribution Surge-Arrester Fundamentals.*** Two types of surge arresters are in use on distribution systems, the silicon carbide gapped arrester and the zinc oxide gapless arrester. The silicon carbide arrester, which has been in use for many years, consists of a controlled, enclosed spark-gap assembly in series with silicon carbide nonlinear resistance or valve elements assembled into a porcelain housing.

The gap structure is carefully designed to assure a low and consistent impulse sparkover. The impulse current flows through the gaps and valve element to ground. The gap structure also is carefully designed so that it can efficiently interrupt the ensuing 60-Hz current, or "follow current," thus resealing against normal system voltage at the end of the half cycle (of 60-Hz current) in which the surge occurred, ready for additional operations. The maximum 60-Hz rms voltage for which the arrester is designed to reseal is the arrester rating.

The nonlinear resistance of the valve element exhibits relatively high resistance at low current levels so that it effectively controls the follow current and relatively low resistance at high impulse discharge currents so that the arrester discharge voltage (*IR* voltage) is well below the BIL of the protected equipment.

The newer zinc oxide arrester uses a zinc oxide valve material which is so electrically nonlinear that a change in discharge current of 100,000 to 1 results in a change in discharge voltage of only 56%. Because of this exceptional characteristic, gaps are not required. The arrester consists of zinc oxide valve elements assembled into a porcelain housing. The zinc oxide arrester is designed to draw only milliamperes at normal system voltage. When a surge reaches the arrester, only that current is conducted which is necessary to limit the overvoltage. There is no power follow current. As a result, the arrester absorbs the minimum amount of energy required to provide the desired protection.

The arrester protective level is established by a front-of-wave sparkover voltage at 0.5 to 1.0 μs and the ensuing *IR* discharge voltage which is dependent on the amount of impulse current and is usually plotted in the area of 3 to 5 μs as shown in Fig. 18-28.

The zinc oxide arrester has no front-of-wave sparkover voltage because it has no gaps. However, it is customary to use an "equivalent front-of-wave protective level" for zinc oxide arresters which is the discharge voltage for a 10-kA impulse current wave which produces an arrester voltage crest occurring in 1 μs.

Numerical values of front-of-wave sparkover and *IR* discharge voltages can be obtained from manufacturers' product bulletins or handbooks for various arrester ratings. Usually the discharge voltages are given at several levels of current from 1500 to 20,000 A. These values are based on a standard impulse-current waveshape of 8 × 20 μs. A 10,000-A discharge voltage is commonly used as the basis for insulation coordination studies on distribution systems.

For a 9/10 kV GE TRANQUELL zinc oxide distribution surge arrester, which is normally applied on a 12.47Y/7.2-kV system, the equivalent front-of-wave protective level is 39.2 kV crest. Values of maximum discharge voltage (kV crest) for 8 × 20 μs current waves are tabulated below:

| | 1500 A | 5000 A | 10,000 A | 20,000 A |
|---|---|---|---|---|
| Crest kV | 29.2 | 32.5 | 35 | 40 |

***91. Insulation Coordination.*** The impulse withstand strength of distribution transformers has been discussed as well as the front-of-wave sparkover and *IR* discharge voltage levels of arresters. The protective margin is computed as the percentage by which apparatus withstand capability exceeds arrester protective level.

As shown in Fig. 18-28, the 95-kV BIL distribution transformer (15-kV class) has a chopped-wave test of 100 kV, a full-wave test of 95-kV crest, and a 60-Hz 1-min high-potential test of 34 kV rms.

Current practice is to suggest a 20% margin in the impulse time area which provides a factor of safety to allow for some deterioration of transformer insulation strength with age, etc. To protect 95-kV BIL equipment, this would call for the maximum arrester front-of-wave sparkover of 92 kV or (110/1.2) and a maximum *IR* at 10 kA of 79 kV or (95/1.2) on the basis of coordination at 10 kA. Note that arrester front-of-wave sparkover is compared with equipment chopped-wave withstand because both voltages have steep fronts. Similarly, arrester *IR* voltage is compared with BIL of equipment.

Note from Fig. 18-28, that the protective margin for 15-kV class equipment is more than 100% where the arrester is located at the transformer terminals. It should be pointed out, however, that protective margins are much smaller for 25- and 34.5-kV equipment because the equipment BIL is increased only to 125 kV, whereas the applicable arrester ratings must increase in proportion to system voltage, that is, 18- and 27-kV ratings, respectively.

It is also important to note that the typical insulation coordination plot implies that the arrester is connected directly to the transformer terminals. In actual practice there is some length of leads connecting the arrester to the transformer and to a ground connection. Flow of impulse current through the inductance of the leads produces a voltage which must be added to the arrester *IR* voltage. A commonly used figure is that each foot of lead length can result in about 2 kV of impulse voltage. Since this voltage adds to the arrester voltage, it reduces the margin of protection. Therefore, the length of the arrester connections to the insulation being protected should be kept as short as possible.

***92. Protection of Overhead Lines.*** Most overhead primary-distribution lines do not have any specific lightning protection other than that provided by the arresters on distribution transformers. In a few areas of the country where the incidence of lightning is very high, some utilities have used shield wires or overhead ground wires as a protective measure. The efficacy of shield wires for protection of primary circuits is debatable because low ground footing resistances must be achieved.

Several years ago, a joint laboratory and theoretical study by the General Electric Company and a group of 10 utilities showed that the use of arresters on all three phases of an overhead line is the most effective means of line protection. The arresters are spaced at intervals of several hundred feet, depending on the insulation level of the line and other factors. This method of protection now is being used by a number of utilities, and the members of the original study group are accumulating and comparing experience data from their installations.

***93. Protection of Underground Systems.*** Many underground circuits are derived from overhead systems, so that overvoltage protection is required. In overhead systems arresters can easily be installed at equipment terminals, whereas in UD systems it is not always possible to install arresters at the equipment terminals. In fact, they may be located several hundred cable feet away, usually at the junction pole between the overhead and underground systems. Also, most underground systems have a normally open tie point where traveling waves of voltage can be doubled by reflection.

Thus, with an arrester at the cable riser pole, whatever voltage it permits to enter the UD system can be doubled, or nearly so. The question arises as to whether this arrester can provide adequate protection for the entire UD system.

In summary, for system voltages up through the 15-kV class, a distribution-type surge arrester at the riser pole generally will provide protection because of the large margin between the arrester performance and the equipment BILs. However, for 25- and 34.5-kV systems, it may become necessary to apply an arrester at the normally open tie point as well, or to use an intermediate or station-type arrester at the riser pole.

Consideration should also be given to the surge protection of the cable. In the past, it was felt that if reasonable protection was afforded the transformers, the cable was adequately protected because of its higher BIL rating. Field failures of aging underground cable, particularly occurring during and after the time of year when lightning incidence is highest, suggests that everything practical should be done to minimize the level of impulse voltages on the cable.

***94. Footing Resistance.*** In the installation of pole-type, single-phase transformers, it is common practice to interconnect the transformer tank, neutral (primary and secondary), and surge arrester ground terminal at the top of the pole with a single grounding lead carried down the pole to a driven ground. In such installations, the surge arrester is in close shunt relation to the primary insulation of the transformer, and low footing resistance is not required for protection of the primary insulation. However, it is important that low values of footing resistance be attained in order to reduce the possibility of high voltages being generated on the secondary system during lightning discharges.

Low values of footing resistance are not required for line protection where surge arresters on all three phases are applied.

The National Electrical Safety Code, ANSI C2, specifies that the grounding lead shall have a conductivity equivalent to a copper conductor not smaller than No. 6 AWG. It also suggests that the resistance of the driven ground should not exceed 25 Ω. Many utilities strive for footing resistances much lower than 25 Ω. Where the system neutral is interconnected to continuous metallic water-piping systems, ground resistances are often less than 3 Ω.

## DISTRIBUTION TRANSFORMERS

***95. Distribution transformers*** convert electrical energy from primary voltages (2.4 to 34.5 kV) to utilization voltages (120 to 600 V). Momentary drops in lighting voltage caused by the starting current of motors often necessitate use of separate transformers where 3-phase motors 20 hp and larger and 1-phase motors 6.5 hp and larger must be served from radial circuits.

***96. Standard Ratings of Single-Phase Distribution Transformers.*** By agreement between users and manufacturers, certain features of line-transformer design have been standardized for sizes up to 500 kVA and for voltages up to 67,000 V. Capacities are 10, 15, 25, 37½, 50, 75, 100, 167, 250, 333, and 500 kVA.

Voltage rating on primary windings are 2400/4160Y, 4800/8320Y, 7200/12,470Y, 12,470GrdY/7200, 7620/13,200Y, 13,200GrdY/7620, 12,000, 13.200/22,860GrdY, 13,200, 13,800GrdY/7970, 13,800/23,900GrdY, 13,800, 14,400/24,940GrdY, 16,340, 24,940GrdY/14,400, 19,920/34,500GrdY, 34,500GrdY/19,920, 22,900, 34,400, 43,800, and 67,000. On the secondary, side windings are built for 3-wire operation at voltages of 120/240 or for 240/480. For some of the larger kVA sizes, secondary side windings are available at voltages from 2400 to 7970 V. Bushings for both primary and secondary terminals are equipped for solderless connectors. They are located on the side of the case, except that primary bushings for 7200 V and higher are cover-mounted. Supporting lugs are arranged to permit mounting either by bolting to the pole or by hanging on crossarms. Where necessary, provision is made for a grounding connection to the case or from the secondary neutral terminal to the case.

Similar standards have been promulgated by ANSI for 3-phase pole-type transformers up to 500 kVA.

***97. Electrical characteristics*** typical of single- and 3-phase transformers of the 12470Y/7200V class are given in Table 18-6. Distribution transformers with different primary voltages will have values only slightly different from those shown in Table 18-6. Transformer regulation for a kVA load of power factor cos $\theta$, at rated voltage, can be calculated from the formula:

Percent regulation

$$= \frac{\text{kVA load}}{\text{kVA transformer}} \left[ \% \, IR \cos\theta + \% \, IX \sin\theta + \frac{(\% \, IX \cos\theta - \% \, IR \sin\theta)^2}{200} \right]$$

**TABLE 18-6** Typical Electrical Characteristics of Single-Phase and 3-Phase 60-Hz Distribution Transformers (Loss factors can vary according to evaluation factors.)

| Size, kVA | Percent *IR* | Percent *IX* | Percent *IZ* | Percent no-load loss | Percent load loss |
|---|---|---|---|---|---|
| Pole-type single-phase transformers—voltage rating 7200/12,470Y to 120/240 V | | | | | |
| 10 | 1.6 | 1.4 | 2.1 | 0.59 | 1.65 |
| 15 | 1.3 | 1.0 | 1.6 | 0.51 | 1.28 |
| 25 | 1.2 | 1.7 | 2.1 | 0.38 | 1.26 |
| 37½ | 1.3 | 1.9 | 2.3 | 0.37 | 1.31 |
| 50 | 1.1 | 1.8 | 2.1 | 0.36 | 1.10 |
| 75 | 1.0 | 2.1 | 2.3 | 0.34 | 1.03 |
| 100 | 1.0 | 2.1 | 2.3 | 0.32 | 1.02 |
| 167 | 1.0 | 2.0 | 2.2 | 0.29 | 0.96 |
| 250 | 1.0 | 2.3 | 2.5 | 0.23 | 0.99 |
| 333 | 0.9 | 2.4 | 2.6 | 0.21 | 0.90 |
| 500 | 0.8 | 2.5 | 2.6 | 0.20 | 0.82 |
| Pad-mounted 3-phase transformers—voltage ratings 12,470Y/7200 to 208Y/120 V | | | | | |
| 75 | 1.0 | 3.0 | 3.2 | 0.52 | 0.95 |
| 112.5 | 1.1 | 3.2 | 3.4 | 0.40 | 1.15 |
| 150 | 1.0 | 3.4 | 3.5 | 0.39 | 0.96 |
| 225 | 1.0 | 3.4 | 3.5 | 0.36 | 0.98 |
| 300 | 1.0 | 3.8 | 3.9 | 0.33 | 0.97 |
| 500 | 1.0 | 3.9 | 4.0 | 0.27 | 0.97 |

***98. Transformers are installed on poles*** in the following ways: transformers 100 kVA and smaller are bolted directly to the pole and sizes 167 to 500 kVA have support lugs attached to the transformer and intended for bolting to adapter plates for direct pole mounting or hung on crossarms by means of steel hangers attached securely to the transformer.

Banks of three single-phase transformers are hung side by side on heavy double arms, usually located low on the pole, or on a "cluster" bracket which spaces them around the pole.

Three or more transformers 167 kVA and larger are installed on a platform supported by two poles set 10 to 15 ft apart. The transformer-platform structure is often placed on the customer's premises to reduce the distance that secondaries must be run and to avoid pole congestion on public thoroughfares.

***99. Transformers are installed*** in street vaults, in manholes, on pads at ground level, subsurface, within buildings, or are direct-buried when underground construction is employed.

When installed within buildings where the possibility of submersion is remote, the overhead or inside types of transformer and cutout are used. Transformer vaults within a building are of fireproof construction, except when transformers are dry type or filled with nonflammable liquid.

***100. Booster Transformers.*** Where it is desired to raise the voltage by a fixed percentage, as when line drop is excessive, this may be accomplished by a transformer used as a booster. This is a transformer so connected that the secondary is in series and in phase with the main line, and thus the primary voltage is raised by the amount of the secondary voltage, 5% in *(a)* and 10% in *(b)* of Fig. 18-29, where the ratios of transformation are 20:1 and 10:1, respectively.

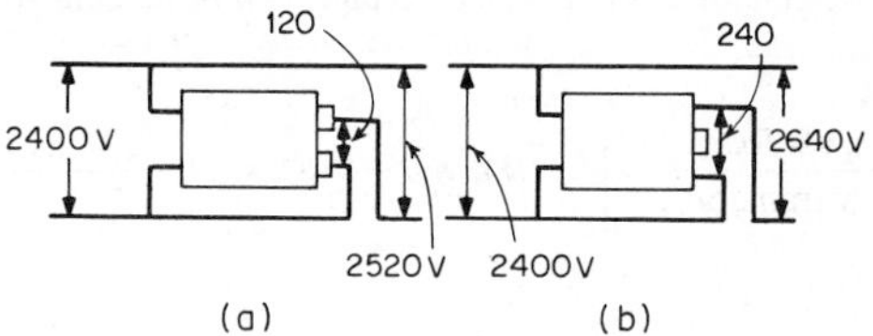


**FIG. 18-29** Booster-transformer connections.

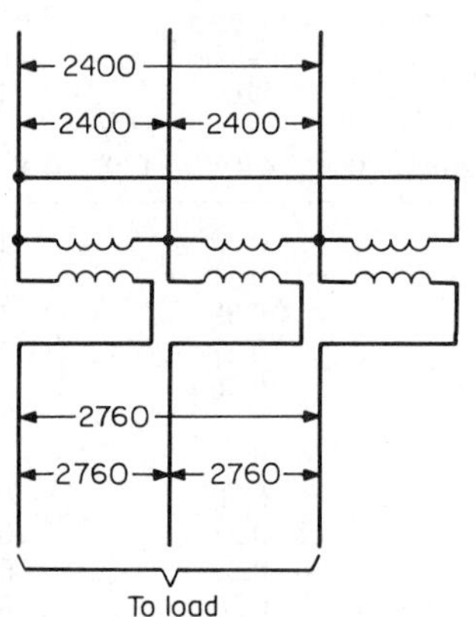


**FIG. 18-30** The connection of boosters in 3-phase circuit.

In installing boosters care must be taken not to open the primary of the booster while the secondary is carrying the line current. Dangerous voltages would be induced in the primary coils in such a case. The safest way of connecting or disconnecting a booster is to have the main line open.

***101. The use of boosters*** in a Δ-connected 3-phase system is not so simple as in single-phase circuits. The booster secondary is looped into the line, and voltage is taken for the primary from an adjoining phase, as in Fig. 18-30. Insertion of a booster on one phase affects the voltage on two phases, as shown diagrammatically in Fig. 18-31. The effect of a booster in each phase is shown in Fig. 18-32. Three boosters are required, therefore, to balance a 3-phase 3-wire circuit.

***102. Automatic Boosters or Pole-Mounted Regulators.*** Fixed boosters are seldom used today, except for occasional 208Y/120-V customers whose motors demand a full 230 V to handle their loads. For voltage boost on primary circuits, their place was first taken by automatic boosters which contactors cut in and out of service under control of a voltage relay. Later improvements provided two steps, then four steps, until today 16- or 32-step regulators built for pole mounting cover the customary ±10% range. Contact-making voltmeters and

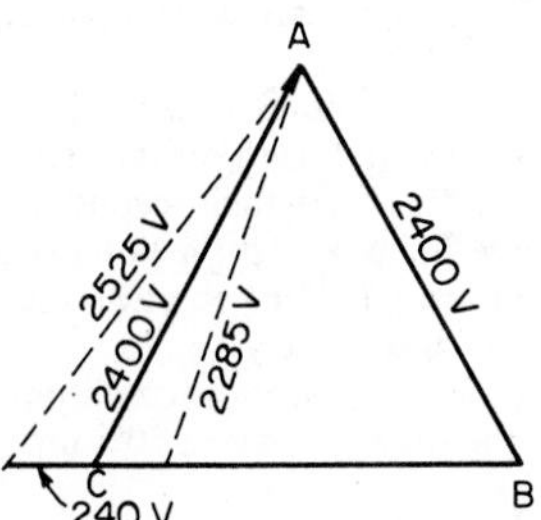


FIG. 18-31 The effect of a booster in one phase of a delta 3-phase circuit.

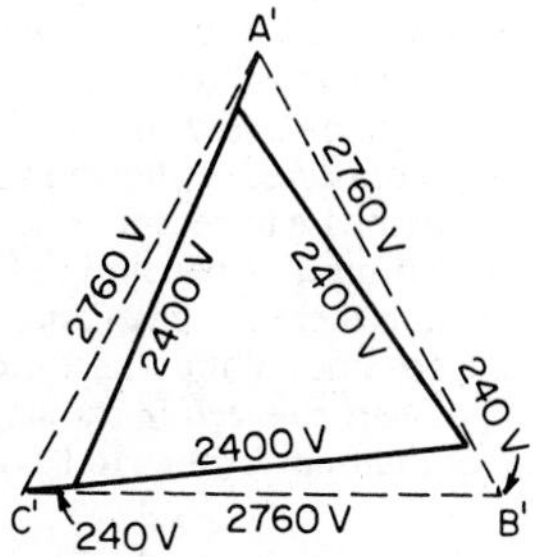


FIG. 18-32 The effect of a booster in each phase of a delta 3-phase circuit.

line-drop compensators give them essentially the same characteristics as the larger station regulators.

***103. Open-Δ connection*** enables small power customers to receive 3-phase service from two transformers connected to a 3-phase circuit, thus reducing the investment in transformers.

Open-Δ from a 3-wire system is the usual Δ connection with one transformer omitted. The connection from a 4-wire system is shown in Fig. 18-33, 2-phase wires and neutral being used on the primary side of the

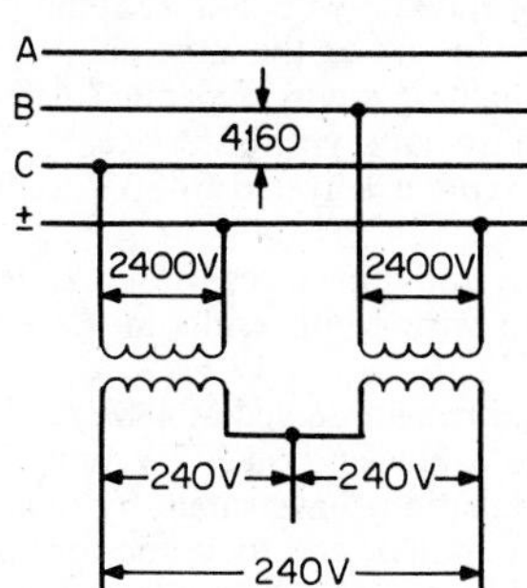


FIG. 18-33 Open Y connection from 4-wire, 3-phase system.

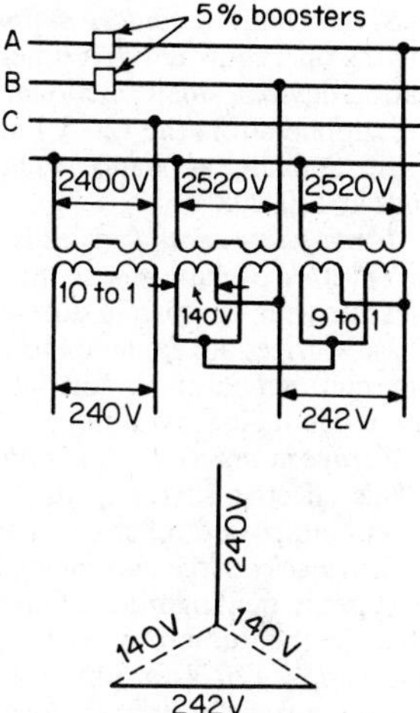


FIG. 18-34 Two-phase service from 4-wire, 3-phase system with three transformers.

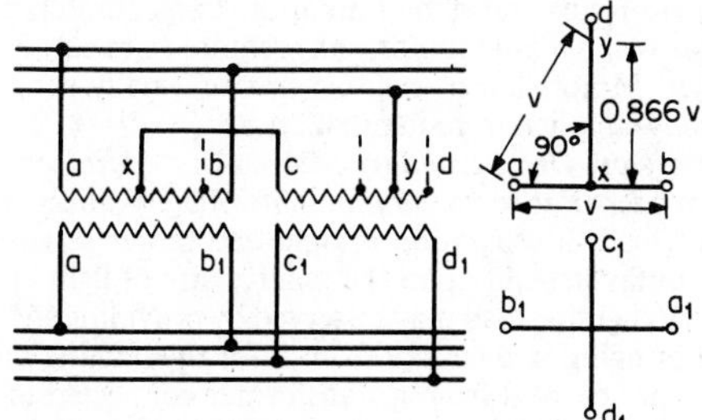


FIG. 18-35 Balanced T- or Scott-transformer connection for 3- to 2-phase transformation.

transformers. Current in each of two single-phase transformers connected in open Δ is 73% greater than in each of three transformers connected in closed Δ.

***104. The 2-phase 240-V service*** from a 3-phase 4-wire system may be secured with three transformers, connected as shown in Fig. 18-34. The unit at the left has a ratio of 10 : 1 and is connected from phase to neutral. The other two have ratios of 9 : 1, with their secondary coils in multiple and arranged as two limbs of a Y connection to give 242 V across the outer wires. The 3-phase system is therefore unbalanced by this arrangement, since half the energy is taken from one phase. Capacities of the transformers should be selected accordingly.

***105. The Scott connection*** shown in Fig. 18-35. gives an accurate transformation but requires one of the transformers to have an 86.6% tap and the other to have a 50% tap.

## SECONDARY RADIAL DISTRIBUTION

***106. Secondary mains*** operate at utilization voltage and serve as the local distributing main. In early commercial radial systems secondary mains that supply general lighting and small power are usually separate from mains that supply 3-phase power because of the dip in voltage caused by starting motors. This dip in voltage, if sufficiently large, causes an objectionable lamp flicker.

***107. Single-phase secondary mains supplying general lighting*** and small power are usually 3-wire mains operating at 120 V line-to-neutral and 240 V line-to-line. Incandescent lamps, fans, heating devices, small fractional-horsepower motors, and other appliances rated 115 or 120 V are supplied from the 120-V line and neutral. Electric ranges, larger single-phase motors up to 6.5 hp, and large appliances rated 230 or 240 V are supplied 240 V. Some utilities supply these loads at 120/208 V.

***108. Three-phase secondary mains*** are commonly operated 3-wire 240 V. Some utilities offer 208Y/120-V 3-phase 4-wire service. The 3-phase mains are on the same poles or in the same duct line (but in separate ducts) with single-phase lighting mains. Separate single-phase and 3-phase services are extended to customers who require both types of service.

In large commercial and industrial installations power is often delivered at 480 V to effect an economy in conductor investment.

***109. European practice*** is to supply 220 V to lighting and appliances from a 220/380-V system. This effects a saving in distribution and interior wiring but results in less efficient incandescent lamps and other small appliances.

***110. In America*** large commercial buildings and factories are served at 480Y/277 V because most permanent lighting is fluorescent, which operates efficiently at 277 V, and 480 V is well suited for the numerous 3-phase motors. Such installations have small dry-type transformers to supply 120 V for portable lights, convenience outlets, and tools and for business machines; these transformers are located near the 120-V loads and supplied from the 480-V system.

***111. Fractional-horsepower motors*** up to about ¾ hp are regularly supplied by single-phase 120-V mains. Industry committees, sparked by sudden acceptance of home air conditioning, several years ago agreed to permit starting currents not to exceed 50 A for 115-V motors. Special design enabled motors up to ¾ hp to meet this limitation. Larger motors up to 6.5 hp are usually served at 240 V, although 3- and 6.5-hp motors may require extra care in distribution design to avoid troublesome flicker. Motors larger than 6.5 hp are usually connected 3-phase. Three-phase service is not usually supplied in residential areas.

***112. Light and Power from One Secondary Main.*** In a *radial* system, 3-phase service is sometimes supplied from a separate secondary main if voltage is affected by elevator motors or other intermittently used load. If separation of light and power service is not necessary, the nature of the connection may depend upon the relative size of light and power loads. When power load is predominant, lighting load may be served by providing additional capacity in one of the transformers and bringing in a neutral from it for the lighting service. The neutral for lighting service is sometimes derived from a transformer connected to one phase of 240- or 480-V power circuits giving 120/240 V for lighting. This is the usual procedure where power is served at 480 V. When the lighting load is predominant, service is often provided at 208Y/120 V, 4-wire.

***113. Transformer and Secondary-Main Economy, Overhead Distribution.*** Several independent studies have been made to determine the proper combination of transformer and radial secondary main that provides satisfactory voltage regulation and costs a minimum per kVA of load served. All these studies indicate that, for 120/240-V single-phase distribution, overhead secondary mains should be three No. 1/0 to three No. 4/0 aluminum, the latter being preferred when air conditioning or heating is to be served.

***114. Permissible length*** of the three No. 1/0 aluminum secondary mains depends on the load density. On the assumption of evenly distributed loads and 3% drop in the mains, for 15 kW/1000 ft, the permissible length is 600 ft; and for 30 kW/1000 ft, 400 ft. Widespread use of ranges and motor-driven appliances establishes an additional limit for flicker at 200 to 300 ft.

***115. Transformer size*** should be such that the initial peak load is between 75 and 100% of rated capacity. In medium-load densities 25- and 50-kVA transformers will fulfill this requirement. Transformers should be allowed to remain in service until their winter peak load reaches at least 150 to 180% of rated capacity. When this occurs the "hot-spot" winding temperature is approaching 110°C—the maximum safe temperature.

***116. Load growth*** should be taken care of by *installing additional transformers* and cutting radial secondaries or by *increasing the size* of the existing transformers where secondary-main regulation permits.

***117. The three No. 1/0 to 4/0 aluminum single-phase secondary mains*** should not be replaced by larger conductors to improve secondary-main regulation. Additional transformers should be installed and parts of the existing mains transferred to the new transformers.

***118. Underground systems*** should also be designed initially with capacity for growth. In order to accomplish this, many utilities in underground residential distribution (URD) work do not use secondary mains. Rather, one transformer is used to supply four to six homes by installing service drops large enough for future loads from the transformer to each home. With this system design it is relatively easy to change out the transformer to a larger size when the load grows.

***119. Pad-mounted transformers*** can be sized and operated the same as overhead-type transformers. Advantage can be taken of the short-time overload capability given by ANSI C57.91, "Guide for Loading Mineral Oil Immersed Overhead-Type Distribution Transformers with 55C or 65C Average Winding Rise." Subsurface transformers in close-fitting cylindrical vaults require special baffles and chimney specified by the manufacturer in order that they might be loaded the same as an overhead-type transformer.

***120. Subway-type transformers*** should not be replaced or relieved of load until the calculated hot-spot winding temperature exceeds 110°C, provided, of course, that voltage at the ends of the secondary is satisfactory. To calculate hot-spot winding temperature, the maximum load and top-oil (or case) temperatures must be measured. Maximum case temperature has been found to be within 3°C of top-oil temperature. It is assumed in making the calculation that the difference between hot-spot-winding and top-oil temperature is 20°C at full load and that this difference varies as the square of load. This is a conservative assumption. For example, assume maximum case temperature 67°C when 130% load is on the transformer. Then the calculated winding hot-spot temperature is given by

$$67°C + 3°C + 20°C(1.30)^2 = 114°C$$

Fans to supplement natural air movement have been used to boost safe capability of vault transformers.

***121. Table 18-7*** gives the voltage drop per 10,000 A · ft for single-phase and 3-phase secondaries for a variety of load power factors. The underground portion of the table can be used for underground systems and also overhead systems where triplex cable construction is employed. The overhead part of the table gives the voltage-drop information for overhead aluminum conductors on racks. The table can be used to determine voltage drop quickly on any secondary circuit if load, curcuit length, and conductor size are known.

All values in the table are for aluminum conductors at 50°C temperature. Values for copper can be determined with satisfactory accuracy by using the table for a conductor of equivalent resistance; that is, use an aluminum conductor two sizes larger than the copper conductor.

In using the table, the first thing required is the number of ampere-feet involved in the problem. This is obtained by multiplying the amperes per phase by length of circuit in feet. (For single-phase use number of feet between source and load; impedance of return circuit is in-

**TABLE 18-7** Voltage Drops per 10,000 A · ft* for Single-Phase and 3-Phase Secondaries, 60 Hz

| | Voltage | | | | | | | | | |
|---|---|---|---|---|---|---|---|---|---|---|
| | 120/240-V single-phase | | | | | 208Y/120 V, 240 V, 480Y/277 V, and 480-V 3-phase | | | | |
| | Lagging power factor | | | | | | | | | |
| Conductor size | 0.7 | 0.8 | 0.9 | 0.95 | 1.00 | 0.7 | 0.8 | 0.9 | 0.95 | 1.00 |
| | Underground or triplex secondary† | | | | | | | | | |
| Aluminum: | | | | | | | | | | |
| No. 2 | 4.524 | 5.042 | 5.530 | 5.752 | 5.858 | 2.262 | 2.521 | 2.765 | 2.876 | 2.929 |
| No. 1 | 3.690 | 4.084 | 4.450 | 4.606 | 4.646 | 1.845 | 2.042 | 2.225 | 2.303 | 2.323 |
| No. 1/0 | 3.002 | 3.304 | 3.574 | 3.686 | 3.684 | 1.501 | 1.652 | 1.787 | 1.843 | 1.842 |
| No. 2/0 | 2.458 | 2.684 | 2.880 | 2.954 | 2.920 | 1.229 | 1.342 | 1.440 | 1.477 | 1.460 |
| No. 3/0 | 2.028 | 2.194 | 2.334 | 2.380 | 2.318 | 1.014 | 1.097 | 1.167 | 1.190 | 1.159 |
| No. 4/0 | 1.684 | 1.804 | 1.898 | 1.920 | 1.840 | 0.842 | 0.902 | 0.949 | 0.960 | 0.920 |
| 350 kcmil | 1.166 | 1.218 | 1.238 | 1.228 | 1.114 | 0.583 | 0.609 | 0.619 | 0.614 | 0.557 |
| | Overhead secondary‡ | | | | | | | | | |
| Aluminum: | | | | | | | | | | |
| No. 2 | 5.530 | 5.888 | 6.146 | 6.192 | 5.860 | 2.801 | 2.974 | 3.095 | 3.112 | 2.930 |
| No. 1/0 | 3.932 | 4.088 | 4.150 | 4.102 | 3.700 | 2.002 | 2.074 | 2.097 | 2.067 | 1.850 |
| No. 2/0 | 3.372 | 3.456 | 3.448 | 3.368 | 2.940 | 1.722 | 1.758 | 1.746 | 1.700 | 1.470 |
| No. 4/0 | 2.516 | 2.504 | 2.406 | 2.284 | 1.840 | 1.294 | 1.282 | 1.225 | 1.158 | 0.920 |
| 336.4 kcmil | 1.940 | 1.876 | 1.792 | 1.594 | 1.160 | 1.006 | 0.968 | 0.918 | 0.813 | 0.580 |
| 477 kcmil | 1.646 | 1.556 | 1.392 | 1.248 | 0.820 | 0.858 | 0.808 | 0.718 | 0.640 | 0.410 |
| 795 kcmil | 1.336 | 1.224 | 1.042 | 0.892 | 0.480 | 0.704 | 0.642 | 0.543 | 0.462 | 0.240 |

Regulation of copper conductors can be estimated with reasonable accuracy the same as that of aluminum conductors two sizes larger.

* Values in the table give the difference in absolute value between sending-end and receiving-end line-to-neutral voltages of balanced 3-phase circuit and phase-to-phase or phase-to-neutral voltages of single-phase circuit.

† Underground cable impedances are based on 50°C conductor temperature with close triangular spacing of cable using typical solid-dielectric insulation, 100% insulation level, single conductor.

‡ Overhead conductor impedances are based on 50°C conductor temperature with 8-in equivalent spacing for single-phase and 10-in spacing for 3-phase.

NOTE: 1 in = 25.4 mm; 1 ft = 0.3048 m.

cluded in table.) Divide this ampere-feet by 10,000 to determine the multiplier to be used with values in the table. For the proper voltage, conductor size, and power factor find the voltage-drop factor in the table and multiply by the multiplier determined previously. This will be the absolute line-to-neutral volts difference (drop) between the sending and receiving ends of the circuit. Dividing by the line-to-neutral voltage of sending end or receiving end and multiplying by 100 will express this as a percentage of sending- or receiving-end voltage, respectively.

*Example.* Given a 3-phase 60-Hz secondary 500 ft in length, which consists of No. 4/0 aluminum conductor cable; conductor temperature 50°C; receiving-end load 100 kVA at 0.8 power factor lagging; receiving-end line-to-line voltage 480.

$$A \cdot ft = \frac{100}{\sqrt{3} \times 0.48} \times 500 \text{ ft} = 60{,}142, \text{ or } 6.014 \text{ times tabular value}$$

From Table 18-7 for No. 4/0 cable, 0.8 pf the value is 0.902. Line-to-neutral voltage drop is $0.902 \times 6.014 = 5.425$. This is $5.425/277 \times 100 = 1.96\%$ voltage drop on basis of receiving end.

## BANKING OF DISTRIBUTION TRANSFORMERS

***122. Banking.*** Tying together the secondary mains of adjacent transformers supplied by the same primary feeder is known as *banking.* The practice of banking, when used, is usually applied to the secondaries of single-phase transformers, and all transformers in a bank must be supplied from the same phase of the primary circuit. The use of banking is not as prevalent as it was formerly. Banked distribution transformers differ from the low-voltage ac network in that one circuit supplies all transformers where secondaries are banked together, whereas different circuits supply adjacent transformers in an ac low-voltage network. Only a few companies operate their transformers banked.

***123. Advantages claimed for banking*** compared with secondary radial distribution are (1) reduction in lamp flicker caused by starting motors, (2) less transformer capacity required because of greater load diversity among a larger group of customers, (3) better average voltage along the secondary, and (4) greater flexibility for load growth.

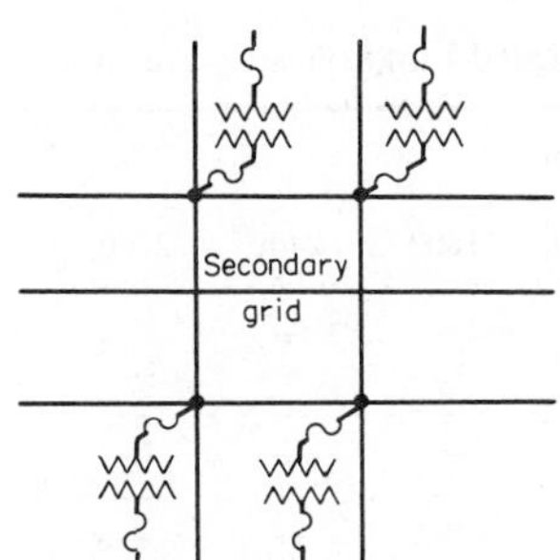

**FIG. 18-36** Fuse application in grid systems.

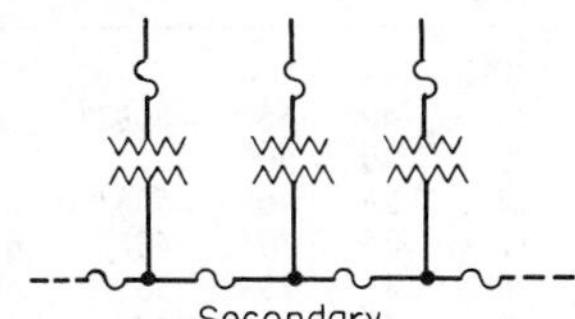

**FIG. 18-37** Fuse application in straight-line systems.

There are two general types of secondary banking: the grid type and the straight-line type, as shown in Figs. 18-36 and 18-37.

## APPLICATION OF CAPACITORS

***124. Correction of Power Factor.*** It is desirable to add shunt capacitors in the load area to supply the lagging component of current. The cost is frequently justified by the value of circuit and substation capacity released and/or reduction in losses. Installed cost of shunt capacitors is usually least on primary distribution systems and in distribution substations.

The application of a shunt capacitor to a distribution feeder produces a uniform voltage boost per unit of length of line, out to its point of application. Therefore, it should be located as far out on the distribution system as practical, close to the loads requiring the kilovars. There are some cases, particularly in underground distribution, where secondary capacitors are economically justified despite their higher cost per kilovar.

Development of low-cost switching equipment for capacitors has made it possible to correct the power factor to a high value during peak-load conditions without overcorrection during light-load periods. This makes it possible for switched capacitors to be used for supplementary voltage control. Time clocks, temperature, voltage, current, and kilovar controls are common actuators for capacitor switching. See Par. **64.**

***125. Capacitor Installations.*** Capacitors for primary systems are available in 50- to 300-kvar single-phase units suitable for pole mounting in banks of 3 to 12 units. Capacitors should be connected to the system through fuses so that a capacitor failure will not jeopardize system reliability or result in violent case rupture. To assure that the proper fuse protection is provided, the installed capacitor fuse ratings are listed in Tables 18-8 and 18-9.

***126. Effect of Shunt Capacitors on Voltage.*** Proposed permanently connected capacitor applications should be checked to make sure that the voltage to some customers will not rise too high during light-load periods. Switched capacitor applications should be checked to determine that switching the capacitor bank on or off will not cause objectionable voltage flicker. The curves in Fig. 18-38 can be used to compute voltage rise.

***127. Effect of Shunt Capacitors on Losses.*** The maximum loss reduction on a feeder with distributed load is obtained by locating capacitor banks on the feeder where the capacitor kilovars is equal to twice the load kilovars. This principle holds whether one or more than one capacitor bank is applied to a feeder.

Capacitor kilovars up to 70% of the total kilovar load on the feeder can be applied as one bank with little sacrifice in the maximum feeder-loss reduction possible with several capacitor banks. A rule of thumb for locating a single capacitor bank on a feeder with uniformly distributed loads is that the maximum loss reduction can be obtained when the capacitor kilovars of

**TABLE 18-8** Recommended Group Fusing, K- or T-Rated Links (Floating-Y Banks)

| | 3-phase kilovar | | | | | | | | | | |
|---|---|---|---|---|---|---|---|---|---|---|---|
| Volts | 150 | 300 | 450 | 600 | 900 | 1200 | 1350 | 1800 | 2400 | 2700 | 3600 |
| 2,400 | 40K | — | — | — | — | — | — | — | — | — | — |
| 4,160 | 25 | 40 | 65*[a] | 80K*[c] | — | — | — | — | — | — | — |
| 4,800 | 20 | 40 | 50[b] | — | — | — | — | — | — | — | — |
| 7,200 | 12 | 25 | 40 | 50K[b] | 80*[d] | — | — | — | — | — | — |
| 8,320 | 12 | 25 | 30 | 40 | 65[e] | 80K[g] | — | — | — | — | — |
| 12,470 | 8 | 15 | 25 | 30 | 50[f] | 65[h] | 65* | 80K[i] | — | — | — |
| 13,200 | 8 | 15 | 20 | 25 | 40 | 50 | 65* | 80K*[i] | 100K*[j] | — | — |
| 13,800 | 6 | 12 | 20 | 25 | 40 | 50 | 65* | 80*[j] | 100K*[j] | — | — |
| 14,400 | 6 | 12 | 20 | 25 | 40 | 50K | 65* | 80*[j] | — | — | — |
| 20,800 | — | 8 | 12 | 20 | 25 | 40 | 40 | 50 | 65 | 80* | 100K* |
| 21,600 | — | 8 | 12 | 15 | 25 | 30 | 40 | 50 | 65 | 80* | — |
| 23,000 | — | 8 | 12 | 15 | 25 | 30 | 40 | 50 | 65 | 80T* | — |
| 23,900 | — | 8 | 12 | 15 | 25 | 30 | 30 | 50 | 65 | 65 | 80K |
| 24,900 | — | 8 | 12 | 15 | 25 | 30 | 30 | 50 | 65[i] | 65 | 80K |
| 34,500 | — | — | 8 | 10 | 15 | 20 | 25 | 30 | 40 | 50 | 65 |

NOTES:
Fusing is in safe zone unless otherwise shown.
Max bank size for 50 kvar units is 600 kvar.
Max bank size for 100 kvar units is 1200 kvar.
Max bank size for 150 kvar units is 1800 kvar.
Max bank size for 200 kvar units is 2400 kvar.
* Zone 1
[a] 150-kvar units only.
[b] Zone 1 for 50-kvar units.
[c] 200-kvar units only.
[d] 300-kvar units only.
[e] Zone 1 for 100- or 150-kvar units.
[f] Zone 1 for 100-kvar units.
[g] Zone 1 with 200-kvar units. Not suitable for 100-kvar units.
[h] Zone 1 for 100- and 200-kvar units.
[i] For 200-kvar and larger only, zone 1 for 200 kvar units.
[j] For 300-kvar and larger only.

**TABLE 18-9** Recommended Group Fusing, K- or T-Rated Links
Grounded-Y- and Δ- Connected Banks

| | 3-phase kilovar | | | | | | | | | | |
|---|---|---|---|---|---|---|---|---|---|---|---|
| Volts | 150 | 300 | 450 | 600 | 900 | 1200 | 1350 | 1800 | 2400 | 2700 | 3600 |
| 2,400 | 40 | 80 | — | — | — | — | — | — | — | — | — |
| 4,160 | 25 | 50 | 80 | 100 | — | — | — | — | — | — | — |
| 4,800 | 20 | 40 | 65 | 80 | 140 | — | — | — | — | — | — |
| 7,200 | 15 | 30 | 40 | 65 | 80 | — | — | — | — | — | — |
| 8,320 | 12 | 25 | 40 | 50 | 80 | 100 | — | — | — | — | — |
| 12,470 | 8 | 15 | 25 | 40 | 50 | 65 | 80 | 100 | 140 | — | — |
| 13,200 | 8 | 15 | 25 | 30 | 50 | 65 | 80 | 100 | 140 | — | — |
| 13,800 | 8 | 15 | 25 | 30 | 50 | 65 | 65 | 100 | 140 | 140 | — |
| 14,400 | 8 | 15 | 20 | 30 | 40 | 65 | 65 | 80 | 140 | 140 | — |
| 20,800 | 6 | 10 | 15 | 20 | 30 | 40 | 50 | 65 | 80 | 100 | 140 |
| 21,600 | 6 | 10 | 15 | 20 | 30 | 40 | 40 | 65 | 80 | 80 | 140 |
| 23,000 | 6 | 10 | 15 | 20 | 25 | 40 | 40 | 50 | 80 | 80 | 100 |
| 23,900 | 6 | 8 | 12 | 20 | 25 | 40 | 40 | 50 | 80 | 80 | 100 |
| 24,900 | 6 | 8 | 12 | 15 | 25 | 40 | 40 | 50 | 65 | 80 | 100 |
| 34,500 | 6 | 6 | 10 | 12 | 20 | 25 | 25 | 40 | 50 | 50 | 80 |

NOTES:
1. Refer to Table 18-9A for fuse sizes within fault current limits.
2. Maximum link size for each unit—check Table 18-9A for all:

| | | | |
|---|---|---|---|
| 50 kvar | 65K, 30T | 200 kvar | 100K, 65T |
| 100 kvar | 80K, 50T | 300 kvar and up | 140K, 80T |
| 150 kvar | 100K, 50T | | |

3. Ratio of fuse continuous current rating to nominal capacitor current is 1.65 minimum.

the bank is equal to two-thirds of the kilovar load on the feeder. This bank should be located two-thirds of the distance out on the distributed feeder portion. Deviation of the capacitor bank location from the point of maximum loss reduction by as much as 10% of the total feeder length does not appreciably affect the loss benefit. Therefore, in practice, in order to make the most out of the capacitor's loss reduction and voltage benefits, it is best to apply the capacitor bank just beyond the optimum loss-reduction location.

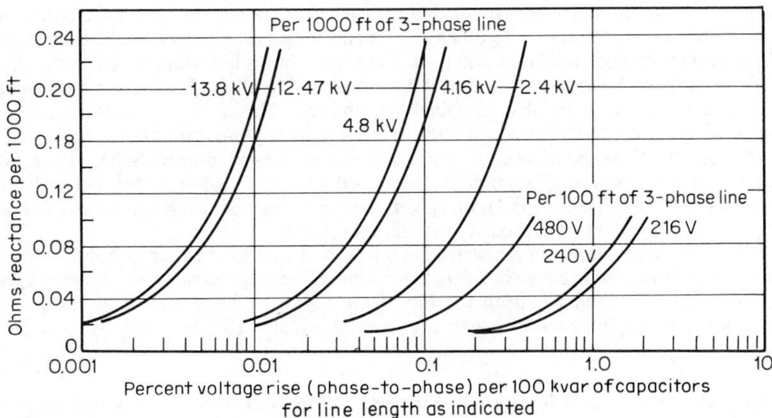


**FIG. 18-38** Curves of voltage rise caused by capacitor application.

**TABLE 18-9A** Coordination Table: Grounded-Y and Δ Connected Banks (Maximum fault current for zone indicated)

| | 50 kvar unit | | 100 kvar unit | | 150 kvar unit | | 200 kvar unit | | 300 and 400 kvar unit | |
|---|---|---|---|---|---|---|---|---|---|---|
| Fuse link | Safe zone | Zone 1 | Safe zone | Zone 1 | Safe zone | Zone 1 | Safe zone | Zone 1 | Safe zone | Zone 1 |
| 30 K & lower | 2900 | 3900 | 4000 | 5300 | 4600 | 6300 | 5400 | 7000 | 5800 | 7000 |
| 40 K | 2700 | 3900 | 4000 | 5300 | 4600 | 6300 | 5400 | 7000 | 5800 | 7000 |
| 50 K | 2000 | 3700 | 3900 | 5300 | 4600 | 6300 | 5400 | 7000 | 5800 | 7000 |
| 65 K | — | 2400 | 2800 | 5300 | 4000 | 6300 | 5400 | 7000 | 5800 | 7000 |
| 80 K | — | — | 700 | 3500 | 2200 | 5500 | 4100 | 7000 | 5000 | 7000 |
| 100 K | — | — | — | — | — | 2800 | 1700 | 6300 | 2800 | 7000 |
| 140 K | — | — | — | — | — | — | — | 1800 | — | 3500 |
| 20 T & lower | 2900 | 3900 | 4000 | 5300 | 4600 | 6300 | 5400 | 7000 | 5800 | 7000 |
| 25 T | 2200 | 3900 | 4000 | 5300 | 4600 | 6300 | 5400 | 7000 | 5800 | 7000 |
| 30 T | 800 | 2800 | 3200 | 5300 | 4200 | 6300 | 5400 | 7000 | 5800 | 7000 |
| 40 T | 220 | 1000 | 1700 | 4300 | 3000 | 6300 | 4500 | 7000 | 5600 | 7000 |
| 50 T | — | 200 | 400 | 2500 | 1100 | 4000 | 2800 | 7000 | 4200 | 7000 |
| 65 T | — | — | — | 500 | — | 2100 | 1600 | 5500 | 2500 | 6800 |
| 80 T | — | — | — | — | — | — | — | 3500 | 1000 | 5000 |
| 100 T | — | — | — | — | — | — | — | — | — | 2200 |

Safe zone—Rupture probability less than 10%. Zone 1—Rupture probability 10 to 50%.

## POLES AND STRUCTURES[1]

***128. Overhead Construction.*** Although overhead distribution construction remains less costly than underground, the vast majority of *new* residential developments are being served by underground systems. However, most new main feeder circuits and rural and semirural systems are being constructed overhead. Also, because of the tremendous amount of overhead distribution plant already in place, it is important to have a good grasp of overhead construction.

***129. Appearance Considerations.*** For many years, the traditional overhead distribution construction embodied one or more crossarms on each pole for mounting primary insulators, distribution transformers, surge arresters, cutouts, etc. However, back in the early 1960s when public interest in appearance made itself manifest, many new concepts in overhead construction were introduced in the interest of obtaining better-appearing systems. The principal change was the introduction and wide use of "armless" construction wherein insulators and equipment are directly mounted to the poles. Other ideas also were introduced, such as use of shorter poles, cabled secondaries and services, strategic use of steel poles to reduce or eliminate need for guy wires, and fewer circuits per pole. It should be noted, however, that cabled secondaries employ a "covered conductor" which can withstand momentary contact between conductors but which is appropriately considered by the National Electrical Safety Code ANSI C2-1984 to be the same as bare conductor for clearances to other objects and for all other purposes. In addition, the black covering adds both thickness and contrast to the cables when viewed against the sky.

***130. Wood poles*** have been used almost universally for overhead distribution lines because of the abundance of the material, ease of handling, and cost. The life of wood poles is materially extended with wood preservatives and cedar, pine, and fir are most commonly used.

***131. Specifications and dimensions for wood poles*** are presented in ANSI 05.1 1979. Poles meeting the requirements of this standard are grouped into classes based on their circumference at a location 6 ft from the butt. Poles of a given class and length are designed to have approximately the same load-carrying capacity regardless of species.

The minimum circumferences specified at 6 ft from the butt have been calculated in order for each species in a given class to develop, at the groundline, appropriate stresses when a given horizontal load is applied 2 ft from the top of the pole. The horizontal loads used in the calculations for identifying the 15 classes are as given in Table 18-10.

[1] To avoid inappropriate duplication, certain discussions common to transmission and distribution may be found in Sec. 14.

**TABLE 18-10** Loads Used to Identify Pole Classes

| Pole class | Horizontal load, lb |
|---|---|
| H6 | 11,400 |
| H5 | 10,000 |
| H4 | 8,700 |
| H3 | 7,500 |
| H2 | 6,400 |
| H1 | 5,400 |
| 1 | 4,500 |
| 2 | 3,700 |
| 3 | 3,000 |
| 4 | 2,400 |
| 5 | 1,900 |
| 6 | 1,500 |
| 7 | 1,200 |
| 8 | 740 |
| 10 | 370 |

NOTE: 1 lb = 0.4536 kg.

In making the calculations, it was assumed that the pole is used as a simple cantilever and that the maximum fiber stress in the pole due to the bending moment will occur at the ground level. The circumference at the ground line was calculated using standard engineering formulas. This circumference value was then used to calculate the circumference 6 ft from the butt using typical tapers per foot of length between the ground line and the point 6 ft from the butt.

The assumption of maximum fiber stress at the ground line is theoretically correct if the taper of the pole is such that the circumference at the ground line is not more than 1½ times the circumference at the point of loading. If the circumference at ground line is more than 1½ times the circumference at the point of loading, the maximum fiber stress theoretically occurs at a location above the ground line where the circumference is 1½ times the circumference at the point of loading. This makes it necessary to calculate specific loads supported.

Typical tapers used in determining the required circumference 6 ft from the butt circumference range from 0.38 for western cedar to 0.20 for western hemlock. These numerical values are in change in circumference per foot of length.

***132. Concrete poles*** reinforced with steel were originally employed chiefly for street-lighting standards, where a neat appearance is demanded. But some concrete poles have been used for general distribution as well, usually with a minimum of attached wires and apparatus.

With the increased manufacturing and quality control capability for prestressed concrete poles, and the need for tall structures for narrow rights-of-way or aesthetic requirements, has come the more frequent use of prestressed concrete poles in transmission lines. As demands for transmission facilities in urban narrow rights-of-way have increased, they have been most often met with steel single-pole structures, especially where two circuits are required.

***133. Steel poles,*** ordinarily set in concrete, have long been used to support street lights. More recently, in a more ornamental form and bolted to concrete foundations, they have been used for parkway lighting and, to a limited extent, for distribution where appearance demands.

***134. Aluminum poles*** also are employed for parkway lighting standards and for certain other locations. They are bolted to concrete foundations to avoid the attack of fresh cement on aluminum. Although use of aluminum on transmission lines has increased, it has been with latticed structures, not generally with poles.

***135. Types of Loading.*** Poles carrying overhead distribution lines are subject to vertical and horizontal forces, of which some are continuous and others are applied only under abnormal or occasional conditions. Normal vertical forces are the weight of wires, transformers, and other equipment, and these are less than normal horizontal forces in many cases. Abnormal vertical load is imposed when wires are coated with ice, which may increase their normal weight 200 to 400%. For example, the weight of six covered No. 6 copper wires 100 ft long is normally 67 lb; but ice to a radial thickness of 0.5 in increases their weight to about 370 lb.

Normal horizontal forces acting upon a pole are the unbalanced component of wire tension at turns and corners, the side pull of service drops, and the horizontal component of weight when the pole is not vertical. Abnormal horizontal stresses are imposed by wind pressure, by breakage of conductors, or by failure of supporting guys.

***136. Application of Loading.*** Vertical loading of wires and equipment is applied through crossarms and other attachments to the pole. These forces are amply sustained by poles chosen to meet requirements of transverse forces, except that, for line transformers, poles having 1-in greater diameter than line poles may be chosen. Transverse forces from unbalanced conductor tension at corners and bends are normally the greatest forces acting on the line. These are usually carried by guy cables secured to suitable anchorages, which relieve the pole itself of the stress. In some cases, the pole is underbraced and carries the entire load.

***137. Ice Loading.*** When wires are loaded with ice, conductor tension is increased in direct proportion to the added weight of ice and may become two to four times as great as normal. This stress is borne by the conductors and, through them, communicated to the pole and the guying system. Where ice loading occurs, the guying system must have a suitable factor of reserve to meet abnormal loadings. The tension of conductors being increased with ice loading, elasticity in the wire permits a slight increase in length which makes tension less than the calculated amount for nonelastic conductors and supports.

***138. Wind Loading.*** Loading due to wind pressure becomes appreciable in the design of poles and structures when wind velocities of over 40 mi/h are prevalent. Such forces are most noticeable on overhead lines when the direction of wind is at right angles to the direction of wires, both because the area exposed is greatest at that angle and because the force exerted is sustained by the pole without the aid of guying.

The area of conductors exposed to wind is much increased by a coating of ice, and the combination of ice with high wind is often the most severe loading condition to which a line is subjected. In many parts of the United States such a condition is never experienced, and it is very rare even where ice coatings occur almost every winter, since the conductor movement due to wind tends to break off the ice coating.

However, with the introduction of large-diameter conductors and bundled-conductor configurations, high winds in summer or other storm periods not involving ice may cause the greater problem. As a result, the National Electrical Safety Code, ANSI C2-1984 requires that severe wind conditions be checked if any portion of the structure or conductors exceed 60 ft above grade.

***139. Strength of Wood Poles.*** The strength of a wood pole must be sufficient to withstand transverse forces, such as wind pressure, on the pole and conductors; unbalanced pull on conductors when they are broken; and side pull on curves and corners where guys cannot be provided. These forces place the fiber on the wood under tension, and the load which a pole will carry is determined by the inherent strength of its wood fiber under tension and the moment of forces. The moment is

$$M = PL + P_1L_1 + P_2L_2 + \cdots \tag{18-13}$$

in which $P$ = force, lb, acting at one crossarm; $L$ = height, ft, at which the arm is attached; $P_1$, $P_2$, etc. = forces acting on other arms; and $L_1$, $L_2$, etc. = respective heights. If $s$ = fiber stress, lb/in², and $c$ = circumference at the ground, in, the allowable moment of a pole of given size is

$$M = 0.00026386sc^3 \tag{18-14}$$

Thus for a pole having a ground-line circumference of 40 in and an allowable fiber stress in an emergency of 2500 lb/in², the maximum allowable moment is

$$M = 0.00026386 \times 2500 \times (40)^3 = 42{,}218 \text{ lb} \cdot \text{ft}$$

If the average height of attachments is 30 ft, total force is 42,200/30 = 1407 lb.

The ultimate fiber stresses for various species of wood poles are listed in ANSI 05.1 1979. In practice, the actual pole stresses are limited to some allowable percentages of ultimate stress. The National Electrical Safety Code, ANSI C2-1984, provides guidelines for allowable stresses under vertical and transverse loading. These guidelines suggest 25% for Class B and 37.5 to 50% for Class C construction at time of installation. Some utilities use larger factors of safety in their designs; different factors of safety may be used for normal unbalanced forces and for abnormal forces of a temporary nature.

When the maximum fiber stress is above the ground line because of taper of the pole, the allowable moment can be calculated in a manner similar to Eq. (18-14), using the circumference of the pole at the point of load application and the taper of the pole to determine the location where the circumference is 1½ times that at the location at the load as described beginning with Par. **128**. Usually decay is greatest at the ground line, and as the pole ages, its ground-line circumference is reduced to make it the point of greatest fiber stress.

***140. Example of Calculation of Pole Size.*** ANSI 05.1-1979 specifies a value of ultimate fiber stress for western cedar of 6000 lb/in². Assume a 40-ft pole and a depth of setting of 6 ft.

Assume the pole will carry a transverse pull of 400 lb at a height of 32 ft and another of 90 lb at 30 ft. What class of pole is required and what are its circumferences at ground line and at top?

$$M = PL + P_1L_1 = 400 \times 32 + 90 \times 30 = 15{,}500 \text{ ft} \cdot \text{lb}$$

If a factor of safety of 4 : 1 is used with respect to the ultimate stress, the allowable fiber stress is 6000/4 = 1500 lb/in$^2$.

From Eq. (18-14),

$$c = \sqrt[3]{\frac{M}{0.00026386s}} = \sqrt[3]{\frac{15{,}500}{0.00026386 \times 1500}} = \sqrt[3]{39{,}162} = 33.96 \text{ in at the ground line}$$

From ANSI 05.1-1979, Table 5, this corresponds very closely to a Class 5 pole, which has a circumference of 34.0 in (10.82 in diameter) at 6 ft from the butt. Minimum circumference of this class of pole at the top is 19 in (6.05 in diamter).

If the allowable fiber stress were increased to one-half the ultimate (factor of safety = 2), the moment could be increased to

$$M = 0.00026386 \times 3000 \times (34)^3 = 31{,}112 \text{ ft} \cdot \text{lb}$$

If this were a northern white cedar pole, having an ultimate fiber stress of 4000 lb/in$^2$, for a safety factor of 4, the ground-line circumference would be

$$c^3 = \frac{15{,}500}{0.00026386 \times 1000} = 58{,}743 \text{ ft} \cdot \text{lb}$$

$$c = \sqrt[3]{58{,}743} = 38.87 \text{ in}$$

From Table 3 of ANSI 05.1-1979, this is very close to the Class 5 pole, which has a specified circumference of 39 in 6 ft from the butt. This also illustrates that the pole classification, in effect, defines the loading capability of the pole regardless of the species.

***141. Wind Pressure.*** Wind pressure must be taken into account when designing a pole line. For purposes of calculation, the following formulas are often used to calculate pressures due to wind:

$$P = 0.004\ V^2 \tag{18-15}$$

where $P$ = pressure, lb/in$^2$, on flat surfaces normal to the wind, and $V$ = wind velocity, mi/h.

$$P = 0.0025\ V^2 \tag{18-16}$$

for cylindrical surfaces such as wires and poles. Values of $V$ can be obtained from weather bureau records for the particular locality.

Wind pressure on a 40-ft pole, of which 34 ft is above ground, with 7 in top diameter and 14 in butt diameter, can be calculated as follows:

$$\text{Projected area} = [(7 + 14) \times \tfrac{1}{2}] \times 34 \times 12 = 4284 \text{ in}^2$$
$$= 4284/144 = 29.75 \text{ ft}^2$$

A wind of 60 mi/h would cause a force calculated by Eq. (18-16):

$$P = 0.0025 \times (60)^2 = 9.0 \text{ lb/ft}^2$$

Since the uniform wind pressure is applied to a long, slender trapezoidal area whose center of gravity is 15.11 ft above the ground line, the resulting moment about the ground line is

$$9 \times 29.75 \times 15.11 = 4046 \text{ ft} \cdot \text{lb}$$

With the diameter of a typical distribution conductor taken as about 0.35 in, the total wind force on a 150-ft span assuming 60 mi/h would be

$$\left(150 \times \frac{0.35}{12}\right) \times 0.0025 \times (60)^2 = 39.375 \text{ lb}$$

On six conductors it would be 236.25 lb. Assuming that the conductors have an effective height of 31 ft at the pole, the resultant moment is 236.25 × 31 = 7324 ft · lb. The sum of the moments from wind pressure on pole and wires is

$$4046 + 7324 = 11{,}370 \text{ ft} \cdot \text{lb}$$

When distribution transformers, capacitors, voltage regulators, or other equipment are mounted on the pole, resulting wind forces should be taken into consideration.

In some areas where (1) overhead facilities are subject to high winds, (2) the soil is such as to provide relatively poor overturning resistance, and (3) underground facilities are inappropriate, such as barrier islands subject to hurricane and tidal forces, a unique system is employed to provide for minimum storm damage and quick return to service. In these cases, the structures are deliberately overdesigned relative to their natural foundations so that they will lean, rather than break, under excessive wind loading. Power is turned off as winds reach hurricane speed. After the storm, the lines are inspected, leaning poles are straightened, and service is restored, all with a minimum outage time and expense.

***142. Ice- and wind-loading*** requirements vary widely throughout the United States, and each utility has adopted design practices suitable for its own conditions of terrain and climate. The National Electrical Safety Code, ANSI C2-1984, presents some guidelines on conductor loading.

The actual loading on conductors is equal to the resultant loading per foot due to the vertical load on the conductor, assumed ice-covered where appropriate, and the transverse loading due to horizontal wind pressure upon the projected area of the conductor, again assumed ice-covered where appropriate. Usually a design constant is added to the loading so calculated.

To establish the guidelines for the loading of overhead lines, the National Electrical Safety Code has divided the United States into three loading districts, heavy, medium, and light. These are roughly: *(a)* the northeastern section extending east to west from the Atlantic through the Dakotas, and from the Canadian border southward into Texas and the Ohio Valley; *(b)* the Pacific Northwest and a narrow belt eastward to the Atlantic including parts of Arizona, Texas, Louisiana, Alabama, Georgia, and the Carolinas; *(c)* the remaining narrow belt extending from coast to coast and bordered on the south by Mexico and the Gulf of Mexico. These are, respectively, the heavy, medium, and light loading districts.

Table 18-11 illustrates suggested loadings.

***143. Equipment Loading.*** Poles are subject to heavy loads when used to support equipment such as distribution transformers, voltage regulators, and capacitor banks. Such loads are chiefly vertical but do have a transverse component when the pole is bent or drawn away from a vertical position. The equipment also presents additional transverse loading due to wind pres-

**TABLE 18-11** Conductor Loadings Due to Ice and Wind

| | Loading district | | | |
|---|---|---|---|---|
| | Heavy | Medium | Light | Extreme wind loading |
| Radial thickness of ice, in | 0.50 | 0.25 | 0.00 | 0.00 |
| Horizontal wind pressure, lb/ft² | 4 | 4 | 9 | 9 to 31 |
| Temperaure, °F | 0 | +15 | +30 | +60 |
| Constant to be added to the resultant, lb/ft | 0.30 | 0.20 | 0.05 | 0.00 |

Selected data from the National Electrical Safety Code, ANSI C2-1984.

Since heavy ice does not often form on conductors in a heavy wind, the transverse loading assumed is deemed sufficient for the purpose but is not sufficient to represent the vertical (or combined) load which is imposed on conductors by heavy deposits of ice which frequently form in comparatively still air. In order to apply a total loading to conductors representing more nearly the conditions encountered in practice, constants are added to the conductor loading.

NOTE: 1 lb/ft = 1.488 kg/m; 1 lb/ft² = 4.882 kg/m²; $t_{°C} = (t_{°F} - 32)/1.8$.

sure. Therefore, poles to be used for supporting equipment usually are specified to be of a better class than those supporting conductors only.

Three-unit installations of distribution transformers normally employ the "cluster-mounting" arrangement in which the transformers are supported directly on the pole by suitable brackets. Three-phase banks as large as 500 and 750 kVA are installed in this manner. The former practice of supporting the transformers on a platform carried by two or more poles has nearly disappeared because of appearance and cost considerations. For 3-phase transformer installations of 300 to 500 kVA and larger, it is often more economical to install a pad-mounted unit near the user's service entrance with relatively short primary-cable connections to the overhead system and with cable or bus-duct connections on the secondary.

***144. Unbalanced Loads.*** Transverse forces are imposed on a pole where there is a change in direction of the line, that is, where the conductors on either side of the pole form an angle. These forces usually are offset by guys where practicable. The loading on the structure, including guys, is considered to be the resultant load equal to the transverse wind load and the load imposed by the conductors due to the change in direction. If it is not practical to guy, the pole and its setting must have sufficient strength to withstand the stresses imposed. The force applied to the pole at the point of attachment of the conductors can be calculated as in Fig. 18-39. Divergence from a straight line can be determined by joining two points, each 100 ft from the corner pole, by a straight line. The distance $A$ from the corner pole perpendicular to this line can then be used to calculate the transverse force applied by the line wires on the pole.

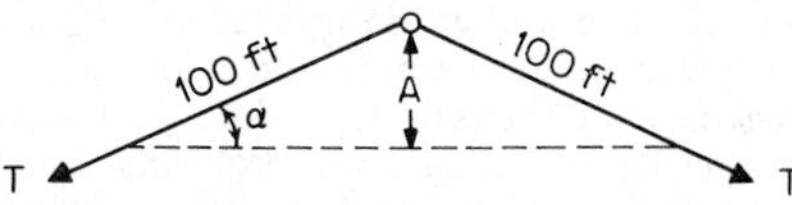


**FIG. 18-39** Determining transverse force on a corner pole.

Assume $T$ to be the total tension, in pounds, of all the wires carried by the pole.

$$\text{Transverse force} = 2T \sin \alpha = 2T \frac{A}{100} \tag{18-17}$$

For example, if $A$ is 10 ft and if there are six wires having a total tension of

$$T = 6 \times 250 = 1500 \text{ lb}$$

the transverse force due to conductor tension to be sustained normally by the pole is

$$2 \times 1500 \times \frac{10}{100} = 300 \text{ lb}$$

For right-angle turns, $\alpha$ is 45°, $\sin \alpha = 0.707$, and $A = 70.7$ ft. This is the most severe condition ordinarily encountered. In such cases tensions should be reduced as much as practicable at the corner pole by shortening spans and by transferring a part of the stress to guy wires. Stresses at dead-end poles are treated similarly.

***145. Safety-Code Requirements.*** The National Electrical Safety Code, in Part 2, sets up minimum safety requirements for loading, strength, and clearance for those parts of supply lines which are involved in crossings of railroad or communication circuits or which come into such proximity to these as to create a "conflict." It also includes joint use of lines and separate use on any public way. The code recognizes differences in degree of estimated hazard, which are assumed to depend upon voltage of the supply line, importance of railroad and other communication systems, classification of loading district, types of crossings, etc. The NESC allows reduced loadings where utility facilities are sheltered from the wind, but it requires full loadings to be met if those shelter mechanisms do not exist. Caution is advised in depending upon the continued existence of buildings and other artificial objects as a sheltering influence—especially in the case of buildings; they can either (1) be removed or (2) be joined by others and produce channeling effects that would adversely affect the line loading. Similarly, trees are not generally considered to be an effective shelter because of the possibility of clear-cutting or other removal.

## STRUCTURAL DESIGN OF POLE LINES

***146. Pole Location.*** In residential areas, poles generally are spaced from about 100 to 150 ft apart depending on the size of the lots. The span usually is an integral number of lot widths in length, and poles are set to provide convenient points for connection of services. Longer spans up to 300 ft or longer are used in rural areas. In either case, pole spacing should consider expected future growth and service requirements in the area. The choice of pole spacing is also a function of the load to be carried and the relative economics of longer spans. Pole spacing and line location both are a function of the obstacles in the area. The National Electrical Safety Code, ANSI C2-1984, has specific requirements about placement of poles near other poles, roads, buildings, fire hydrants, etc.

Poles are required to be at least 3 ft away from fire hydrants and should be at least 4 ft away. Although these distances are required to allow fire fighters access in emergencies, they also give room for personnel working on the pole to move around on the ground. No fire hydrant, telephone pedestal, or other like object should be located on the climbing side of the pole; the climbing side of the pole should be kept clear of protruding obstacles to provide a clear drop zone for line workers if they chip out of the pole and fall.

There are two hazards with chipping on a pole; the first is the possibility of damage due to picking up splinters from previous gaff marks, and the second is the possibility of damage from the fall. Because of the severe problems caused from splinter damage, some communication line workers (who do not climb as high as supply workers) are taught to climb a pole without belting off until they reach the work height; this allows them to push away from the pole as they go down and allows them to roll when they hit the ground. In such cases, no obstruction should be allowed on the climbing side within 10 ft.

Poles and their supported equipment up to 15 ft above a roadway are required to be located far enough from roadways that an ordinary vehicle that is using and located on the traveled way will not contact the utility facilities.

Poles are required to be at least 12 ft from railroad rails, except in some limited sidings where the clearance must be 7 ft, and room must be left to unload cars. Where some other facility is the controlling obstruction, the clearance may be reduced from 12 ft but may not be less than 7 ft.

***147. Selection of Poles.*** The height of poles is determined by required clearances over obstructions, streets, and crossings; the span lengths; and the number and character of conductors or circuits to be carried. The most usual lengths for poles used in distribution construction have been 30, 35, and 40 ft. Armless construction and joint use tend to favor the 40-ft length. In recent years, however, as space for CATV and other facilities is provided on joint-use poles, 45-ft poles are even more commonly used, especially in hilly terrain. The Class 5 and 6 poles are very popular, although larger poles are often required where heavy equipment is to be mounted or longer spans with greater wind loadings are used. The class of pole is determined by stress requirements for the grade of construction being employed.

Where primary conductors are to be carried with joint use, the 35-ft pole usually is the minimum used, and 40 ft is quite common. Without primary conductors and with joint use, 30-ft poles are used occasionally. Joint use of pole lines by two or more public service companies is encouraged where feasible because (1) it makes maximum use of capital investment and avoids undesirable duplication of pole lines, and (2) it provides fewer opportunities for one circuit to fall into another during a storm. When a pole or section of poles is broken as a result of the action of falling trees, flying debris, or errant vehicles, the tendency is for all circuits of a joint-use line to be promptly and automatically deenergized. However, if a conflicting line falls into another line, facilities or personnel working on the other line may be damaged. Thus conflicting line locations are discouraged where joint-use locations are practical. In either case, the National Electrical Safety Code, ANSI C2-1984, requires greater strengths than for single-circuit installations.

***148. Pole-Setting Depths.*** Table 18-12 lists typical depths of setting for wood poles of various lengths.

***149. Guying Longitudinal, Angle, and Dead-end Forces with Tensile Guys and Anchors.*** When the horizontal loads to be carried by poles are greater than can be safely supported by the poles, guys or braces are required to provide additional support. Guys and anchors are commonly used wherever conductor tensions are not balanced, as at dead ends, corners, or where the direction of the line changes substantially. Down guys transmit force from the overhead

**TABLE 18-12** Depth of Wood-Pole Settings

| Length of pole, ft | Depth of setting, ft |
|---|---|
| 30 | 5.5 |
| 35 | 6 |
| 40 | 6 |
| 45 | 6.5 |
| 50 | 7 |
| 55 | 7.5 |
| 60 | 8 |
| 70 | 9 |

NOTE: 1 ft = 0.3048 m.

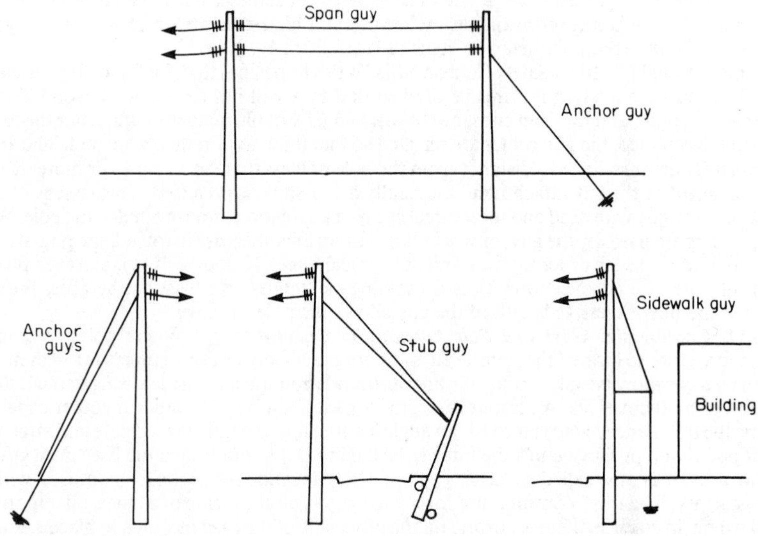


**FIG. 18-40** Tensile guys.

structure system to a buried anchor system (see Fig. 18-40). They are located opposite the forces and use materials in tension to balance the forces. Where it is not practical to place these facilities to continue the imbalanced forces to ground, a compressive guy or pole brace must be used.

Where traffic ways or other obstacles do not allow direct anchoring of the forces at the imbalanced pole, an overhead span guy is used to transmit the force to another pole in line with those forces and in a place that allows a down guy to be used. The best down guy is a straight anchor guy composed of appropriate wire, fastenings, and anchor; it is the most economical to install in most locations, is the least trouble to maintain, deteriorates the least over time, and is the most reliable type of down guy.

The National Electrical Safety Code, ANSI C2-1984 requires guys to have an 18-ft clearance over roadways and 8 ft over spaces or ways accessible to pedestrians only. Where an anchor guy cannot be used because of interference with pedestrian or vehicular traffic, and where there is no practical location for an auxiliary pole and anchor guy that would allow a span guy to be used, a stub guy may have to be used. A stub guy consists of an angled span guy from the imbalanced pole to a short stub pole that is (1) high enough to provide the clearance required for the guy over the affected area, and (2) low enough that the resulting moment arm is short enough that the stub can take the forces involved. Often the stub pole will have to be separately braced below ground to take the forces. Obviously, these conditions are not desired, and they are not practical for balancing large forces.

Another down guy used in constrained locations is the sidewalk guy. With the sidewalk guy, a horizontal strut is attached to the pole midway down the pole; the guy wire runs angled from the top of the pole to the strut and straight down to an anchor from there. These guys are most often used to guy small taps off a main line where there is not room for a full-length anchor-guy lead. The strut is usually just long enough to place the anchor back of a sidewalk and high enough to allow pedestrian clearance, hence the name. This guy should not be confused with the compressive guy discussed below; both are sometimes called strut guys. Sidewalk guys cannot be used for large forces because they will bow the pole due to their horizontal force; this, in turn, causes serious vertical-loading and bending-moment problems if the forces are great enough.

Where unbalanced tension is experienced by crossarms, such as in sidearm construction or dead ends for heavy conductors, crossarms are often guyed. Often a span guy will be attached to the back of the crossarm at each conductor location and run to another pole in line with the forces; otherwise, excess vertical loading on the crossarms can result.

***150. Guy wires*** generally are made of stranded steel cable, usually galvanized for weather resistance. Several grades are available, including extra-high strength, which is commonly used. It is available in several diameters in steps of 1/16 in from 3/16 in up.

The National Electrical Safety Code, ANSI C2-1984, specifies that, for Grade B construction and for transverse loads, guy wires be used so that they will not be stressed beyond 37.5% of ultimate strength, and for dead ends, not beyond 66.67% of ultimate strength. After the tension has been calculated, the guy cables are selected so that these requirements are met. The anchor guys usually are installed at a distance from the pole not less than one-quarter or more than 1 to 1½ the height of the guy attachment. Generally a 1 : 1 slope is preferred. This lessens both the tension in the guy wire itself and the vertical component of forces transmitted to the pole. Where steep angles are used for the guy, or where large forces are otherwise involved, the pole size may need to be increased in order to withstand the vertical forces. In poor soils, this can even push the pole into the ground over time, thus decreasing clearances and twisting the pole, the latter because the now-excessive length of the guy allows the pole to move.

***151. Compressive Guys and Pole-Foundation Underbracing.*** Where poles with imbalanced forces are so located that tensile guys cannot practically be used, either the forces must be taken by a compressive guy or the pole and its foundation must be made strong enough to take the forces by themselves. A compressive guy consists of a pole (or other member capable of taking the imbalanced load) placed at an angle leaning into the imbalanced pole and attached at the top so that it pushes against the force to be balanced. Its common name is a "push guy" for this reason; it is also called a "strut guy" in a few locations—not to be confused with the sidewalk guy. The most common use for a push guy is on the inside of a curve on a mountain road where, in essence, there is no land on the other side of the road in which to place a pole and anchor to use a span guy; span guys are preferred where they are practical because of their relatively simple installation, longer life, and ease of maintenance compared to the push guy. Push guys are difficult to attach to the supported poles, often can only be attached significantly lower than the imbalanced conductors, especially on sharp angles, and require the imbalanced pole to be oversized as a result. It is generally difficult to position the push guy in the ground, or cut it to the right length, so that the imbalanced pole will remain upright under load.

Where poles supporting unbalanced stresses are so situated that it is not practical to support them by either tensile or compressive guying, they must be underbraced to withstand the force imposed with as little deviation from original position as possible. This normally requires that the pole have more than usual diameter and be no taller than absolutely necessary for clearance.

For very long spans or heavy conductors, steel poles are often required. If wood poles are used, top diameters of 8 to 10 in are required for such positions to avoid bending. In addition, the pole is underbraced by timbers bolted to it below ground line and at the butt, as shown in Fig. 18-41. An alternative method, using concrete, is also good. Small boulders or crushed stone may be used in backfill to advantage.

In the use of plank or concrete, the pole is set at a slight angle, or rake, in a direction opposite that from which stress is to be applied, to allow for compacting of soil when wires are pulled up. The area of plank or concrete should be about 4 ft$^2$, both top and bottom. Where steel poles are used to secure strength, they are usually set in a concrete base of such dimensions as to bear the stresses imposed.

***152. Vertical clearances*** given in Table 18-13 are taken from the National Electrical Safety Code, ANSI C2-1984. They apply to crossings where span lengths do not exceed 175 ft in

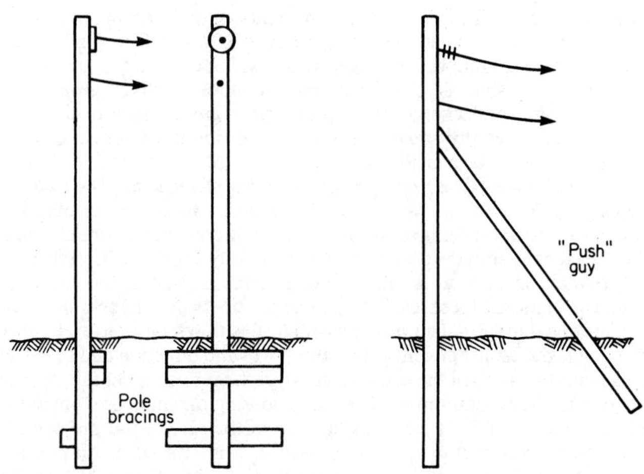


FIG. 18-41 Compressive guy and pole bracing.

**TABLE 18-13** Minimum Vertical Clearances above Grade or Rails, in Feet

| Nature of surface beneath conductors | Guys, communication, most neutrals, shielded supply cables, and cables below 750 V | Supply conductors: Nonshielded cables alone 750 V; open wires 0–750 V | Supply conductors: Open 750 V to 22 kV | Supply conductors: Open 750 V to 50 kV | Trolley and railroad contact conductors: 0 to 750 V | Trolley and railroad contact conductors: 750 V to 50 kV |
|---|---|---|---|---|---|---|
| Track rails | 27 | 27 | 28 | 29 | 22 | 22 |
| Roads, streets and alleys | 18 | 18 | 20 | 21 | 18 | 20 |
| Nonresidential driveways, parking lots, etc., *not* subject to truck traffic | 18 | 18 | 20 | 21 | 18 | 20 |
| Residential driveways | 12 | 15 | 20 | 21 | 18 | 20 |
| Commercial areas not subject to truck traffic | 12 | 15 | 20 | 21 | 18 | 20 |
| Orchards, forests, grazing lands, and other lands | 18 | 18 | 20 | 21 | — | — |
| Spaces or ways accessible to pedestrians only | 15 | 15 | 15 | 16 | 16 | 18 |
| Water areas *not* suitable for sailboating | 15 | 15 | 17 | 17 | — | — |
| Water areas suitable for sailboating (add 5 ft for rigging and launching areas) | 18–38 | 18–38 | 20–40 | 21–41 | — | — |

NOTE: 1 ft = 0.3048 m.

heavy-loading districts, 250 ft in medium-loading districts, or 350 ft in light-loading districts. These clearances are based on a conductor temperature of 60°F, with no wind displacement and voltages not over 50,000 to ground. For longer spans and higher voltages, greater clearances are required. All voltages in Table 18-13 are phase-to-ground for effectively grounded circuits and other circuits where prompt deenergization occurs after a ground fault or subsequent breaker operation. Other circuits require greater clearances. Exceptions are provided for many of the stated clearances for low-voltage facilities.

***153. Clearances between Crossing or Adjacent Wires, Conductors, or Cables Carried on Different Supporting Structures.*** Where any two wires, conductors, or cables cross or are adjacent in open span without being supported on a common structure, the clearances between them must be sufficient to limit the possibility of contact between them. The effect of movement under wind, thermal, or ice loading must be considered as well as the required clearances between them. The National Electrical Safety Code considers that the two suspended facilities must first be considered to be in their most proximate position relative to each other, assuming that both experience the same ambient conditions (i.e., wind direction and speed, air temperature, and icing conditions); then the distance between them must meet the clearance requirements for the two facilities. Caution is advised in choosing the most proximate position. The two facilities may move different amounts because of the same wind pressure; one may be unloaded at ambient temperature or fully ice loaded, while the other may be at maximum operating temperature. One could be at initial sag while the other is at final sag. Detailed examination of Rule 233 of the National Electrical Safety Code is recommended.

When the suspended facilities are at their closest proximity, both the vertical clearance and the horizontal clearance should be checked to see which is controlling. The horizontal clearance must be at least 5 ft. Where the voltage potential between them exceeds 129 kV, an additional clearance of 0.4 in/kV is required. The vertical clearances shown in Table 18-14 are required for spans not exceeding 175, 250, or 350 ft in heavy-, medium-, or light-loading districts, respectively. For longer spans, operating temperatures above 120°F, and voltages higher than 50 kV, additional clearances are required.

***154. Clearances from Buildings and Other Installations.*** Wires, conductors, and cables which pass by buildings, signs, supporting structures of a second line, pools, bridges, tanks, etc., are required to have clearances from those structures, when not attached to them, that allow normal use of those facilities. Required clearances are given in the National Electrical Safety Code, ANSI C2-1984. In general, clearances above portions of structures which are accessible to pedestrians or vehicles are the same or similar to those above ground for the same activity. Where pedestrian access is restricted, and the area is normally accessible only to workers, lesser clearances are allowed. Clearances allow normal maintenance of the structures being passed by.

The minimum horizontal clearance to supporting structures of a second line, lighting support, or traffic signal support are 5 ft; the minimum vertical clearance is 6 ft (7 ft above 15 kV).

The vertical clearance required above building roofs or projections not accessible to pedestrians is 10 ft; this allows repair workers with hand tools the room to work on the roof. Above signs, this clearance drops to 8 ft for open conductors. Both vertical clearances are reduced for communication, guys, and most neutrals and cables. The horizontal clearance to buildings, signs, etc., is 3 ft for guys, neutrals, cables, etc.; 5 ft for open conductors of 0 to 8700 V and higher-voltage cables; 6 ft for 8701 to 22,000 V; and 7 ft for 22,001 to 50,000 V.

Bridges have lesser clearances where workers are allowed and generally the same clearances as buildings where pedestrians are allowed. Greater clearances are required around pools and associated structures to accommodate pool skimmer poles, rescue poles, and diving.

***155. Clearances for Facilities Suspended from the Same Structure.*** The clearances required by the National Electrical Safety Code, ANSI C2-1984, between wires, conductors, and cables carried on the same supporting structure provide adequate room on the structure for workers to operate and maintain the lines, adequate room out in the span for communication workers to work under supply facilities, and adequate separation between suspended facilities to limit contact by them during operation. Because some communication cables on longer spans will "gallop" in high winds, and because such cables may gallop as high as a straight line between their points of attachment at supports, supply conductors above 750 V generally are prohibited from sagging lower than such a straight line.

The vertical clearance at supports for primary-supply conductors above other facilities is required to be 16 in above neutrals and 40 in above communication (60 in if above 8700 V). For

**TABLE 18-14** Vertical Clearances, in Feet

| | | | Supply | | | | |
|---|---|---|---|---|---|---|---|
| | | | Cable | | Open Conductors | | |
| Nature of facilities crossed over | Communication conductors, cables and messengers | Guys, neutrals, surge protection wires | Shielded; nonshielded 0–750 V | Nonshielded above 750 V | 0–750 V | 751 V–22 kV | 22–50 kV |
| Communication conductors, cables, and messengers | 2 | 2 | 2 | 4 | 4 | 6 | 6 |
| Guys, neutrals, surge protection wires | 2 | 2 | 2 | 2 | 2 | 4 | 4 |
| Shielded cable; nonshielded cable 0–750 V | 2 | 2 | 2 | 2 | 2 | 2 | 4 |
| Nonshielded cable above 750 V | 4 | 2 | 4 | 2 | 2 | 2 | 4 |
| Open conductor | | | | | | | |
| 0–750 V | 4 | 2 | 4 | 2 | 2 | 2 | 4 |
| 751 V–22 kV | 6 | 4 | 4 | 4 | 4 | 2 | 4 |
| 22–50 kV | 6 | 4 | 6 | 6 | 6 | 4 | 4 |
| Trolley and railway contact conductors and span/messenger wires | 4 | 4 | 4 | 4 | 4 | 6 | 6 |

NOTE: 1 ft = 0.3048 m.

supply conductors, at voltages up to 8700 between conductors, the minimum horizontal clearance provided by the National Electrical Safety Code ANSI C2-1984 is 12 in. For higher voltages it is required that 0.4 in be added for each 1000 V above 8700. For sags of more than 24 in, separations greater than 12 in are required and are determined, for wires of No. 2 AWG or larger, as

$$\text{Separation} = 0.3 \text{ kV} + 8\sqrt{S/12} \quad \text{in} \tag{18-18}$$

where $S$ = sag, in. For wires smaller than No. 2 AWG the rule is

$$\text{Separation} = 0.3 \text{ kV} + 7\sqrt{(S/3)} - 8 \quad \text{in} \tag{18-19}$$

*Multiconductor, spacer, and other cabled types* of supply-circuit construction are exempt from the above phase-spacing requirements.

***156. Climbing Space.*** Climbing space must be provided on poles for workers to move up and through facilities to reach each of the facilities on the structure unless it is the unvarying rule that workers do not climb the poles. If workers climb some portions of the structure and not others, for example, communication but not supply, and the remainder are worked from insulated bucket trucks, only the portions climbed must have climbing space. The climbing space may move around the pole to allow access to other parts or to avoid traveling through certain locations, as long as full space is allowed to facilitate the turns. These dimensions are intended to provide a clear climbing space of 24 in when the conductors bounding the space are covered with appropriate temporary protective covering. The vertical dimension is 40 in above and below the conductors (60 in if above 8700 V). The National Electrical Safety Code ANSI C2-1984 specifies the horizontal climbing-space dimensions given in Table 18-15. Where conductors of the same voltage classification are on the same crossarm, the horizontal dimensions are projected vertically not less than 40 in above and below the limiting conductors.

**TABLE 18-15** Horizontal Climbing-Space Dimensions (Voltage-to-Ground)

| Supply conductors | Horizontal climbing space, in |
|---|---|
| 0–750 V | 24 |
| 750 V–15 kV | 30 |
| 15–28 kV | 36 |
| 28–38 kV | 40 |
| 38–50 kV | 46 |

NOTE: 1 in = 25.4 mm.

Equipment such as transformers, regulators, capacitors, surge arresters, and switches when located below the conductors should be mounted outside the climbing space.

***157. Working Space.*** Working spaces are required by the National Electrical Safety Code, ANSI C2-1984, at each side of the climbing space and extending to the outermost conductor positions. The size of the working space is linked to the vertical clearances required between conductors at the support and the size of the climbing space, both vertical and horizontal.

***158. Clearances between Supply and Communication Equipment.*** The National Electrical Safety Code, ANSI C2-1984, generally requires a minimum of 40-in clearance between supply equipment and communication equipment and between the conductors of each to the equipment of the other. This provides for footroom for the supply workers and headroom for the communication workers. Special provisions are made to allow luminaires to be placed between these facilities.

***159. Vertical and Lateral Conductors.*** Vertical and lateral conductors may be run within the normal supply space and communication space on a pole if they are so located as to meet the requirements of the National Electrical Safety Code, ANSI C2-1984. Generally such conductors are required to be insulated and placed out of the climbing footroom area or held away from the pole on the opposite side.

## *LINE CONDUCTORS*

***160. Conductor Factors.*** Copper and aluminum are the metals most used as conductors in distribution systems. Proportions are fixed by the combined effect of conductivity, weight, strength, and cost. Recent years have seen such a shift in availability and cost that aluminum has gained almost universal use in distribution, supplanting copper, which was preferred for many years. A number of experimental installations were made using sodium as the conductor in underground cables. While no major problems were encountered, sodium has not at this time received any appreciable acceptance as an electrical conductor.

***161. Conductor Materials.*** Aluminum has the advantage of about 70% less weight for a given size, but its conductivity is only about 61% that of annealed copper. For distribution, it is commonly rated as equivalent to a copper conductor two AWG sizes smaller, which has almost identical resistance. Its tensile strength is less than copper, and it is commonly used, particularly in the smaller sizes, by stranding aluminum around a steel core of proper size to give the desired tensile strength. In larger sizes, the tensile-strength requirements of distribution are satisfied by stranded aluminum without the reinforcing steel. Another way of obtaining high tensile strength is to combine steel with copper or aluminum wires. Steel is combined with copper in a high-strength strand known as Copperweld, which has 30 to 40% of the conductivity of a copper conductor of equal size. In a similar manner aluminum and steel conductor can be combined into what is known as Alumoweld. (For further data about properties of conductors, see Sec. **4**.)

Both copper and aluminum are suitable for use as substation buses, being available in flat bars, tubes, and rods. For very heavy currents, channel shapes are used to make up box-type buses, which are the most economical for such applications.

***162. Use of Copper.*** Where copper is used for overhead circuits with span lengths of 200 ft or more, it is commonly used in the "hard-drawn" form because of its greater tensile strength. For common types of local distribution circuit where spans are shorter and flexibility is desirable, "medium-hard-drawn," or annealed, copper is used. Mechanical connectors are extensively used for joints and taps on overhead copper.

*Underground copper cables* are usually made of standard soft copper because of its greater flexibility. The smaller size of copper conductors helps to offset unfavorable price levels because of savings in insulating and sheathing material as well as the ability to put maximum carrying capability in a given size of duct.

***163. Use of Aluminum.*** In rural line work, where long spans and conductors of high tensile strength are an economic necessity, the combined requirements of conductivity and strength have been met with aluminum stranded around a steel core sized to give the required strength. Such a cable is known as "aluminum cable steel-reinforced" and is commonly designated as ACSR. Development of high-strength aluminum alloys has led to such alternative cables as "aluminum conductor alloy-reinforced" (ACAR) and "all-aluminum-alloy conductor" (AAAC), which also combine conductivity with tensile strength.

Urban distribution uses ACSR and all-aluminum conductors. Stranded aluminum is common where large conductors are required.

*Underground Aluminum Cables.* The development of such synthetic insulations as polyethylene has made aluminum almost universally used for underground distribution. In the smaller sizes for URD, a solid conductor is often applied rather than stranded construction.

*Jointing of aluminum* requires special care to secure good contact and to guard against corrosion. Jointing is often done with compression devices, although mechanical connectors packed with corrosion-inhibiting compound can be used.

***164. Use of Steel.*** Steel conductors are rarely used for distribution circuits because of their high resistance. But steel with a heavy covering of copper, known as Copperweld*, or with a heavy covering of aluminum, known as Alumoweld*, has conductivity approaching 40% that of copper and can be used in some applications. Such coated conductors are also very attractive as high-strength strands or reinforcements for composite cables, which get improved conductivity from strands of hard-drawn copper over the Copperweld or hard-drawn aluminum over the Alumoweld.

Conductors reinforced with steel have impedances which increase somewhat as current

---

* Copperweld and Alumoweld are registered trademarks of the Copperweld Bimetallic Group.

density increases. Voltage drops are correspondingly higher than those of copper or aluminum conductors of equal conductivity.

Copperweld and Alumoweld are generally more durable than galvanized-steel cables. They have therefore been used to some extent for guy cables. They are also used widely for shield wires.

## OPEN-WIRE LINES

***165. Crossarms.*** Southern pine and Douglas fir are the best woods for crossarms because of thin, straight grain, high tensile strength, and durability. Experience indicates that a cross section 3½ in wide by 4½ in high is ample for the average distribution line. Main lines are commonly built with six-pin arms, and smaller lines use four-pin arms. Minimum spacing of pins is 12 in, and spacings of 14 to 16 in are commonly used. Minimum spacing of pole pins is 30 in to provide climbing space.

Crossarms also are used for supporting transformers and other equipment.

There is increasing use of armless construction.

***166. Double crossarms*** are installed on poles at corners, at terminals, and at other points where unusual loads are to be supported.

***167. Vertical racks*** are installed on poles to support secondary and multiple street-lighting wires. They are available with two-, three-, or four-spool insulators. Rack construction is less expensive than crossarms and has supplanted them to a large extent. When services run to houses on both sides of the street, two racks are required, one on each side of the pole. In addition, several pole-top designs mount insulators directly on the pole, eliminating the use of crossarms.

***168. Wire Stringing.*** In erecting wire, the *tension* should be sufficient to prevent too much sag in the spans and yet not so great as to stress the wire unduly. For practical purposes the approximate formula given by Rankine may be used:

$$t = S^2w/8d \quad \text{lb} \tag{18-20}$$

in which $t$ = tension, lb; $S$ = span length, ft; $w$ = resultant load, including weight of wire, lb/ft of conductor; and $d$ = sag, ft, at the center of a horizontal span. If span length is doubled, tension must be quadrupled in order to keep sag the same. If tension is the same on several spans of different lengths, sag is different in each span. The sag of any span when tension is known is found by changing Eq. (18-20) to the form

$$d = S^2w/8t$$

***169. Sag Tables.*** Maximum tension in a span is limited by the strength of the wire and supports. Conductor sags under the assumed loading conditions for the particular loading district (see National Electrical Safety Code, ANSI C2-1984 edition) should be such that the tension of the conductor should not exceed 60% of its ultimate strength. Also, the tension at 60°F, without external load, should not exceed 35% of the conductor ultimate strength under its initial unloaded condition.

It is not unusual to design so that the tension of the conductor will not exceed 50% of its ultimate strength under loaded conditions or a 2000-lb limitation. In some cases, the same sag values are employed for a range of wire sizes so that the appearance of a line carrying different conductor sizes will be improved.

The sags given in Table 18-16 are selected from standard sheets of a large utility.

***170. Expansion and Contraction of Conductors with Temperature Change.*** Changes in sag due to expansion and contraction of conductors under varying temperature conditions are important in the stringing of conductors. Lines erected in winter months are likely to be too slack during the summer unless allowances are made. The length of wire in a span, elastic stretching due to mechanical loading being disregarded, varies as determined by the coefficient of expansion of the conductor and the temperature range,

$$L_t = L_0(1 + \alpha_0 t) \tag{18-21}$$

**TABLE 18-16** Sags for Typical Distribution Conductors (Heavy-Loading District—60°F)

| Size, AWG or M cmils | Conductor material | Sags (in) for span lengths (ft) of | | | | | | | |
|---|---|---|---|---|---|---|---|---|---|
| | | 80 | 100 | 125 | 150 | 175 | 200 | 250 | 300 |
| Open wire: | | | | | | | | | |
| No. 0 | Al. alloy, bare | .. | 10 | 16 | 23 | 31 | 40 | 27* | 38* |
| No. 3/0 | Al. alloy, bare | .. | 10 | 16 | 23 | 31 | 40 | 32* | 46* |
| 336.4 | Aluminum, bare | .. | 10 | 16 | 23 | 31 | 40 | 75 | 108 |
| No. 0 | Al. alloy, polyeth. | .. | 10 | 16 | 23 | 31 | 40 | 59* | 90* |
| No. 3/0 | Al. alloy, polyeth. | .. | 10 | 16 | 23 | 31 | 40 | 70 | 101 |
| 336.4 | Aluminum, polyeth. | .. | 18 | 27 | 38 | 51 | 66 | | |
| Cabled secondaries: | | | | | | | | | |
| 3 No. 0 | Al. alloy, insul. | .. | 10 | 16 | 23 | 31 | 40 | | |
| 3 No. 3/0 | Al. alloy, insul. | .. | 18 | 27 | 38 | 51 | 66 | | |
| 4 No. 3/0 | Al. alloy, insul. | .. | 18 | 29 | 43 | 60 | 80 | | |
| Cabled service drops: | | | | | | | | | |
| 3 No. 4 | Aluminum, insul. | 32 | 52 | 73† | | | | | |
| 3 No. 0 | Aluminum, insul. | 51 | 79 | 116† | | | | | |
| 4 No. 3/0 | Aluminum, insul. | 68 | 108 | 171† | | | | | |

* Taken up to 2000-lb tension limit for spans over 200 ft.
† For 120-ft service drops, 450-lb limit.
NOTE: 1 in = 25.4 mm; 1 ft = 0.3048 m; 1 lb = 0.4536 kg.

where $\alpha_0$ = coefficient of expansion, ft/ft of length/°F above 10°F
$t$ = temperature, °F (above 0°)
$L_0$ = length of wire at 0°F

For aluminum: $\alpha_0 = 0.000024/°C$ or $0.0000133/°F$
For ACSR: $\alpha_0 = 0.0000112/°C$ or $0.0000062/°F$ (nearly that of steel)
For copper: $\alpha_0 = 0.000017/°C$ or $0.0000094/°F$

Ampere loadings on distribution circuits often require that the temperature rise of the conductor due to resistance losses must be taken into account, also.

## JOINT-LINE CONSTRUCTION

***171. Joint-line construction*** is used where two or more utilities would otherwise maintain separate pole lines, such as where both power and communication lines are routed along the same street.

***172. Basis of Joint Use.*** Poles are used jointly under a *joint-ownership agreement or under a lease agreement.* Under joint ownership, the cost of providing the pole is borne jointly by the companies which share in its ownership. Division of expense is, in general, made in proportion to the space allotted to respective users. Clearance space, required between power and communication circuits and between the lowest attachment and ground, is disregarded in determining percentage of ownership. Clearance between higher-voltage and lower-voltage power circuits is chargeable to the higher-voltage circuits.

In case of leased space, the lessee acquires only the right to occupy a specified space. The owning company installs and maintains the pole and includes all charges in the rental price. Lessees usually install their own attachments and maintain them, though pin space is sometimes leased where space for only a few wires is required.

***173. Construction Specifications.*** The types of construction, clearances, and relative levels of different classes of circuit should be provided for by a suitable specification, forming part of the agreement under which joint use is entered into. The general purpose is that construction of all parties be such as not to jeopardize the service or equipment of any of the other parties to the agreement. Construction requirements are set forth in the National Electrical Safety Code, ANSI C2-1984 edition. Some of the most important parts of such a specification are discussed in the following paragraphs.

***174. Relative Levels of Supply and Communication Conductors.*** When supply and communication conductors are located on the same poles, it is generally desirable that the supply conductors be located at the higher levels. This places the higher voltages near the pole top and the communication conductors at the lower levels. This relative location of facilities provides a lower probability of contact between the two systems since the supply conductors are usually larger than the communication lines. This also provides easier access to the lower-voltage or communication circuits by the line crews and avoids the need to climb through the supply conductors to work on the lower-voltage or communication systems. Where 600-V trolley feeders are carried on joint poles, the feeders are located for convenience at the approximate level of the trolley contact conductor.

***175. Vertical Clearances.*** Spacing of conductor attachments must be appropriate with the requirements of safety in operation and maintenance. Clearance requirements are spelled out in detail in the National Electrical Safety Code, ANSI C2-1984. Generally a minimum vertical clearance of 40 in is used between communication conductors (or open wires of 0 to 750 V) and supply conductors operating between 750 and 8700 V to ground. Separations of 60 in are required for supply conductors operating above 8700 V. If the communication circuits belong to the utility for use in operating supply lines, reduced spacings of 16 and 40 in, respectively, are permitted. See Pars. **152** to **159**.

***176. Grades of Construction.*** Strength of poles must be such as to withstand ice and wind loadings normally experienced in the locality where the line is built, for all of the conductors to be carried. These conditions vary greatly in different parts of the United States, there being no ice in some parts of the country and a greater prevalence of wind in others. The material presented in Pars. **135** to **145** is applicable.

***177. The size of conductors*** should be such that they do not experience a tension more than 60% of their rated breaking strength under conditions of maximum loading and not more than 25% of this value for final unloaded tensions at 60°F. Very little use is made in new construction of wire sizes smaller than No. 1/0 stranded aluminum or No. 2 ACSR.

***178. Inductive Coordination.*** A vast majority of the newer telephone circuits on joint-use distribution lines are in cable, rather than open-wire, and telephone interference is rarely encountered. Where long exposures of open-wire circuits do exist, it may be necessary to make suitable transpositions in both power and communication circuits to eliminate electrical unbalances as much as practicable.

***179. Aerial-Cable Construction.*** Insulated aerial cables carried by steel messenger cables have been used occasionally in primary distribution circuits where undergrounding is not practicable and special conditions, such as reduced clearances or unusually severe tree problems, exist. This type of cable is fastened to a galvanized-steel cable, or "messenger," by means of brackets or lashings in a manner similar to large communication cables. Usually it consists of one, two, or three insulated conductors spiraled around the messenger which supports the assembly mechanically and usually serves as a neutral as well.

This type of construction is finding very little usage in new system extensions.

***180. Spacer-cable construction*** provides many of the advantages of aerial cable at lower cost. It consists of primary conductors having less insulation than the cable insulation which is customary for the circuit voltage, supported on a messenger and separated from each other by insulating spacers installed at suitable intervals along the line. See Par. **51**.

## UNDERGROUND RESIDENTIAL DISTRIBUTION

***181. Underground Residential Distribution.*** During the past 10 years, the evolution of underground distribution systems, particularly single-phase systems to serve residential areas (URD), has proceeded at a rapid rate. For a so-called mature industry, the rate of change has been phenomenal. Costs have been steadily reduced through the introduction of new system concepts, improved installation practices, and the development of specialized equipment.

Nearly every utility in the United States now has a policy covering the installation of URD in new residential tracts. Conditions vary all the way from a substantial differential payment by the developer to a no-charge basis by the utility, although the developer usually is requested to assist with excavation. In addition, a number of states have established legal requirements mandating that all new residential developments in excess of a given number of homes be served by an underground distribution system. As a result, perhaps as many as two-thirds of new residential dwelling units are being served underground.

Based on data generated by the continuing series of IEEE Conferences on Underground Distribution, Fig. 18-42 shows the growth in cable installation, which indicates that the quantity is doubling every 2.2 years.

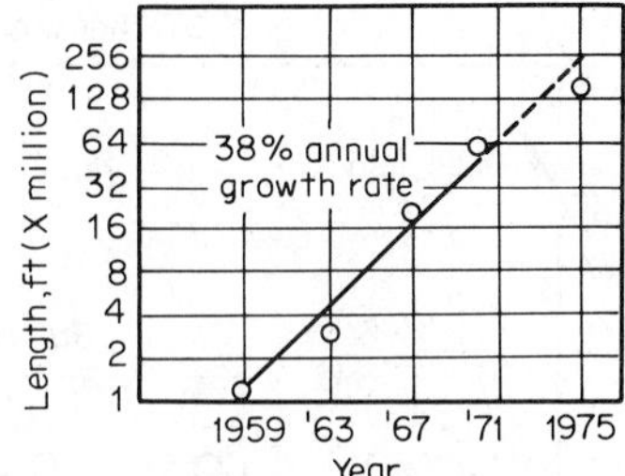


**FIG. 18-42** Total primary URD cable installed by 28 utilities.

***182. Cost.*** Underground distribution systems often cost more than comparable overhead systems. What are the principal factors contributing to the rapid growth of URD? These include:

**a.** Greater public interest in the aesthetic appearance of residential communities.

**b.** Reduced cost of underground equipment and installations brought about by:

Solid dielectric insulated cables—lower-cost—suitable for direct burial without duct systems.

Factory-built cable terminations and splices of low cost easily prepared in the field by ordinary line crews without the aid of highly trained cable splicers.

Mass production of specialized equipment such as pad-mounted transformers and accessories.

Improved installation technique and equipment.

As a result, substantial progress has been made in reducing the relative cost of underground distribution as shown in Fig. 18-43. These data, also compiled from IEEE Underground Distribution Conferences, indicate that a leveling off is occurring. In many locations, large lots would require an intermediate service pole for overhead wires and therefore it may be economical to put them underground.

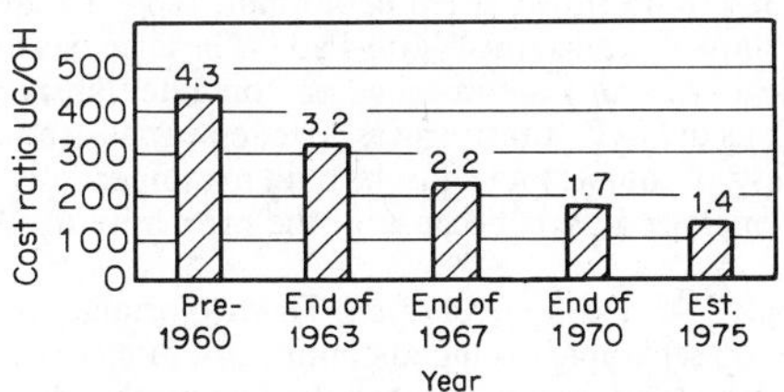


**FIG. 18-43** Average cost ratio of underground to overhead for 26 utilities.

***183. Performance.*** Most observers are of the opinion that the frequency of faults is lower on underground systems than on overhead systems and that the faults are not so likely to "bunch up" because of storm conditions. However, faults are much more difficult and time-consuming to find, to isolate, and to repair on underground systems. This, coupled with the fact that many operating procedures cannot be performed on an underground system while it is energized and that it is impossible to make many of the temporary improvisations on underground circuits that can be accomplished on overhead systems, has led to the development of protective and sectionalizing equipment such as switches and separable cable connectors which often are physically integrated as accessory devices in the underground distribution transformers.

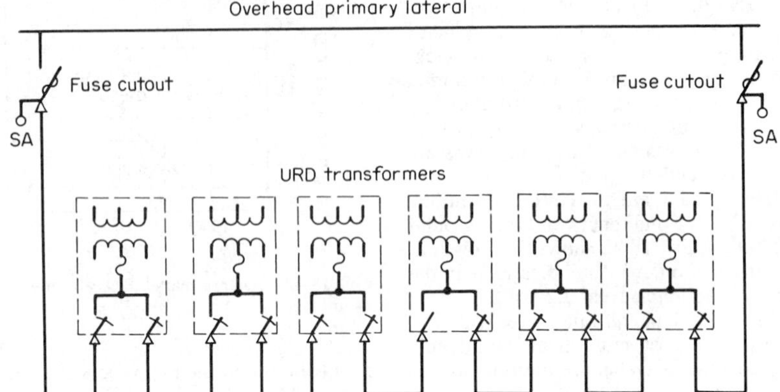


**FIG. 18-44** URD system derived from existing overhead circuits.

***184. Service-restoration*** requirements have also resulted in primary-system designs which operate as a normally open loop as shown in Fig. 18-44. In the case of a cable fault, this facilitates location and isolation of the failure and more rapid service restoration to all customers on the unfaulted portion of the primary loop. It is estimated that about 85% of primary URD systems are being operated as loops, the remainder being radial. Where radial laterals are used, many utilities provide portable, aboveground cables so that faulted cables can be bypassed temporarily and service restored while repairs are being made.

***185. Transformers.*** The heart of the URD system is the single-phase distribution transformer because the primary cable terminations, switching and sectionalizing equipment, and overcurrent protective equipment usually are housed in the transformer enclosure. Thus most operating procedures require access to one or more distribution transformers. Three general types of single-phase transformers are in use.

*Pad-Mounted.* Figure 18-45 shows the predominant type of transformer being used for URD. The transformer shown is called the "Mini-Pad." The term pad derives from the fact that transformers in this category usually are installed on concrete slabs, or pads.

The electrical functions of URD transformers cover essentially the same range as pole-type units. For reasons of safety, of course, they must be built in tamper-resistant configurations with no exposed electrically energized parts because of the proximity of such transformers to the general public.

The Mini-Pad in Fig. 18-45 has its cover open. The two primary bushings at the upper left are for use with load-break, separable insulated connectors, or elbows. This results in a "dead-front" configuration which is required to achieve the low-height Mini-Pad construction. The three 120/240-V bushings are at the right-hand side.

Many other combinations of pad-mounted construction and accessory equipment are available, including "live-front" primary connections with stress cones for the cables, internal or external primary fuses and switches, secondary circuit breakers, etc. Refer to appropriate product bulletins of the manufacturers or handbooks for further equipment details. Generally the loadability of pad-mounted transformers is comparable with that of pole types.

*Residential Subsurface Transformers (RST).* Although usage of pad-mounted transformers predominates, a number of residential subsurface transformers are used. RSTs are installed in relatively tight-fitting vaults with the cover grating of the vault at ground level.

Cooling is accomplished by natural convection of the air, although some users increase the efficiency of circulation by means of special chimneys to direct and control the circulation. With properly designed and installed chimneys, the loadability of RSTs is equal to that of pole types.

RSTs must be submersible and therefore utilize dead-front primary cable terminations, usually the separable insulated connectors or "elbows." Provisions for operation of accessories such as switches, fuses, and circuit breakers are located on the cover of the transformer so that

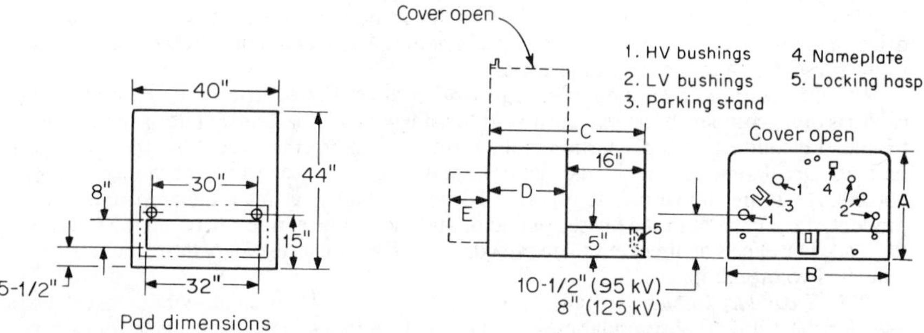


95- and 125-kV BIL (35 kV GrdY and below)

| kVA | Dimensions, in | | | | |
|---|---|---|---|---|---|
| | A | B | C | D | E |
| 15 | 24 | 36 | 33.9 | 17.9 | |
| 25 | 24 | 36 | 33.9 | 17.9 | |
| 37.5 | 24 | 36 | 33.9 | 17.9 | |
| 50 | 24 | 36 | 33.9 | 17.9 | 8.0 |
| 75 | 24 | 36 | 40.0 | 24.0 | 8.0 |
| 100 | 26 | 36 | 40.0 | 24.0 | 8.0 |
| 167 | 32.5 | 36 | 40.0 | 24.0 | 8.0 |

**FIG. 18-45** Mini-Pad distribution transformer. *(General Electric Company.)*

they can be operated by a member of the line crew standing on the surface of the ground. Usually the vault is too small for a person to enter.

***186. Primary Cables.*** Primary URD cables are almost universally of the single-conductor concentric-neutral type employing polyethylene or cross-linked polyethylene insulation. Ordinary polyethylene is a thermoplastic which melts at temperatures in the order of 110°C. The process of "cross-linking" polyethylene converts it into a thermosetting material which does not have a melting point, per se.

Figure 18-46 shows a section of primary URD cable. The central conductor is the energized phase conductor, and the external concentric wires serve as the neutral.

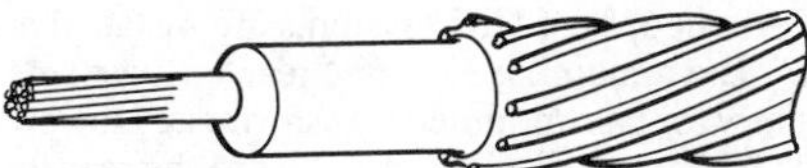

**FIG. 18-46** Concentric-neutral type of primary URD cable.

Corrosion of the tinned-copper concentric-neutral wires of primary URD cables results in reduced cross-sectional area of the wires, increasing their resistance. In some instances, the continuity of the wires is destroyed. Neutral corrosion may cause safety and operating problems on the URD circuit. Corrosion occurs when the neutral wires become anodic, which results in loss of metal. The wires may become anodic due to nearby dissimilar metals or to variations in soil characteristics along the cable route. Determining the location and extent of corrosion damage is a complex procedure which may involve surveys, testing, and, in some cases, excavation. Corrective actions for existing cables include replacing portions of the cable, reestablishing the neutral circuit, and installing sacrificial anodes for cathodic protection. Corrosion in new installations can be controlled by the proper selection of materials, cable construction, type of installation, and cathodic protection.

Most utilities directly bury the primary cables, although some duct is used for street and driveway crossings or where excavation problems are severe. Often the URD cables are placed in the same trench as the telephone cables.

***187. The precise calculation of voltage drop*** in direct-buried, concentric-neutral primary cables is quite complex because a portion of the single-phase load current flows in the concentric-neutral conductors and a portion in the earth surrounding the cable. Also, there may be an induced circulating current in the neutral conductors. Typical values of voltage drop per 100,000 A · ft are shown in Table 18-17. To use the table, calculate the ampere-feet as the product of the current in the phase conductor and the distance in feet between the source and the load. The effects of direct burial on impedance of the return current path are included in the tabulated voltage drops.

***188. Secondary Cables.*** Usually three polyethylene-insulated, single-conductor cables are used for the 120/240-V secondaries and services. These may be separate cables or of triplex construction. In some cases a bare copper neutral conductor is used. The secondary and service cables are usually directly buried.

***189. Cable Terminations.*** A major advantage of the polyethylene-insulated primary cable, in addition to low cost, is the ease with which it can be terminated in contrast with earlier traditional paper and lead cable, that is, cable insulated with oil-saturated paper with an outer lead sheath or jacket (PILC). Termination and splicing of a PILC cable requires a skilled cable splicer working with paper tapes and equipment for soldering and "wiping" the lead cover to potheads or to another section of PILC cable. Several hours are needed to prepare a 15-kV PILC cable termination.

The polyethylene concentric-neutral cable can easily be prepared for termination by means of either a factory-made stress cone or a separable insulated connector. This preparation can be done by an ordinary lineworker in a hour or less using cutting jigs and tools to prepare the cable for the installation of the factory-made termination.

When the URD primary cable is terminated by means of a simple stress cone, the electrical connection to the terminal of the connected device usually is an exposed or "live-front" connection. When a separable insulated connection is used for termination, a "dead-front" construction is obtained; that is, all exposed surfaces of the cable and its termination are essentially at ground potential and thus present less of a hazard to operating personnel. In some cases, dead-front configuration allows a reduction in dimensions of the equipment.

Insulated connector modules are available in a great variety of configurations such as the elbow and bushing, multitaps, stand-off bushings for temporary use on parking stands, load-break modules, and T taps. Figure 18-47 illustrates a cutaway view of a switch (load-break) module, and Fig. 18-48 is a similar illustration of an elbow connector and module.

The basic separable connector system used with URD distribution transformers usually is rated 200 A continuous and is available to either load-break or non-load-break form. See appropriate product bulletins or handbooks of the manufacturers for further product description.

***190. System Types.*** About 65% of URD systems are installed along the streets in front of the houses, or "front-lot." The remaining 35% are "rear-lot" systems. Many utilities prefer the rear-lot location, particularly for pad-mounted transformers. However, there are obvious operating and maintenance problems associated with access to the rear-lot location. At the moment there does not appear to be any strong trend toward either option.

An extremely large number of combinations of transformer equipment are being used, and a detailed discussion of them is beyond the scope of this handbook. However, the following listing is reasonably representative of "typical" practice:

**a.** Pad-mounted transformer of the Mini-Pad configuration.

**b.** Primary laterals operated as normally open loops.

**c.** 12.47 grounded Y/7.2-kV primary voltage.

**d.** The primary lateral loops through each distribution transformer; that is, there are two primary cable connections to each transformer (see Fig. 18-44).

**e.** Front-lot construction.

**f.** Four to eight homes served by each transformer.

**g.** Transformers of dead-front construction with load-break separable insulated connectors.

**TABLE 18-17** Single-Phase Voltage Drops per 100,000 A · Ft* for 15- and 35-kV Direct-Buried Concentric-Neutral Cables (Loop Values)

| | Voltage class | | | | | | | | | |
|---|---|---|---|---|---|---|---|---|---|---|
| | 15 kV | | | | | 35 kV | | | | |
| | Lagging power factor | | | | | | | | | |
| Conductor Size | 0.7 | 0.8 | 0.9 | 0.95 | 1.00 | 0.7 | 0.8 | 0.9 | 0.95 | 1.00 |
| Underground primary | | | | | | | | | | |
| Aluminum: Concentric-neutral—direct-buried, Cross-linked polyethylene, conductor 70°C, neutral 60°C, earth resistivity 90 Ω-cm³, full insulation | | | | | | | | | | |
| No. 2 | 44.1 | 46.6 | 48.1 | 48.1 | 44.8 | | | | | |
| 1/0 | 30.9 | 32.8 | 34.0 | 34.1 | 32.0 | 31.3 | 33.0 | 34.1 | 34.1 | 31.7 |
| 2/0 | 25.3 | 27.0 | 28.1 | 28.3 | 26.8 | 25.8 | 27.3 | 28.3 | 28.4 | 26.6 |
| 4/0 | 17.0 | 18.2 | 19.1 | 19.3 | 18.5 | 17.5 | 18.6 | 19.3 | 19.5 | 18.4 |

* Values in the table give the difference in absolute value between sending-end and receiving-end line-to-neutral voltages, in volts.

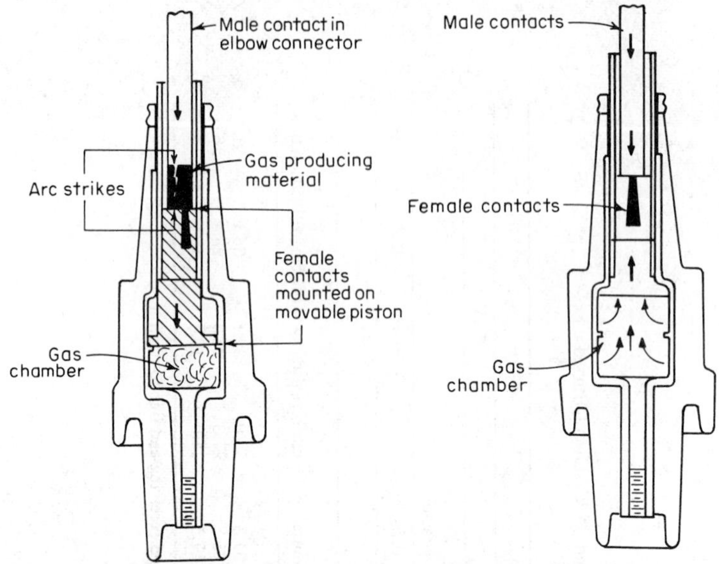


FIG. 18-47 Piston-action 25-kV connector. *(General Electric Company.)*

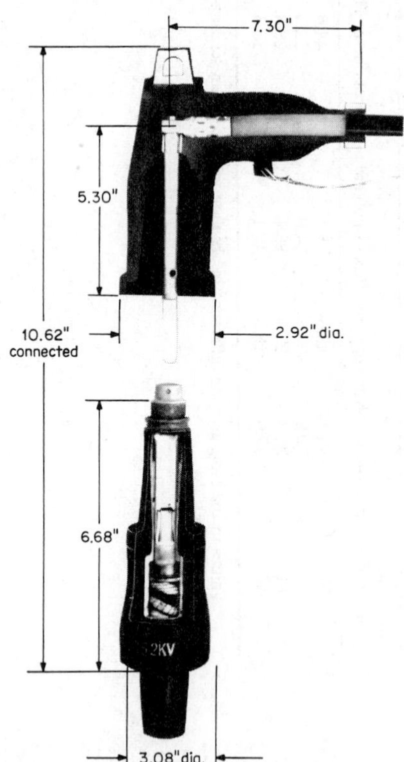


**FIG. 18-48** Cutaway view of elbow connector and switch module. *(General Electric Company.)*

**h.** Internal fusing for each transformer.

**i.** Direct-buried, polyethylene and/or cross-linked polyethylene insulated cables.

***191. Homes Served per Transformer.*** There is an optimum number of homes to serve from each transformer depending on the load per home, size of lots, and type of system to be used. For a given load per home and lot size, the cost per kVA of transformer decreases as the number of homes increases. This is because increasingly larger transformers would be used.

However, as the number of homes per transformer increases, the cost of the secondary and service system increases because of the larger secondary cable required. Since the total cost is the sum of those costs, an optimum number of homes per transformer will exist.

In making such an economic study, it is necessary to examine the secondary-service-system voltage drop. A detailed study also should evaluate transformer and cable losses for the various arrangements.

Four to eight homes served per transformer seems to be reasonably typical of present practice. Larger loads per home and larger lot sizes favor a smaller number of homes per transformer. Conversely, smaller loads per home and smaller lot sizes favor serving more homes from each transformer.

***192. Lightning Protection of URD Systems.*** See Par. **93** for information concerning lightning protection of URD systems.

## *UNDERGROUND SERVICE TO LARGE COMMERCIAL LOADS*

***193. Large commercial loads*** constitute one of the major segments of utility distribution systems, especially in built-up areas where underground supply systems are a requirement. Demands range from a few hundred to many thousands of kVA per customer, and the engineering and design time to provide adequate service facilities is substantial. Each job is special, requiring selection of appropriate and correctly sized equipment, negotiation of space and layout with building owners or their consultants, and frequently a detailed discussion of facilities, charges, rates, and contracts. The best tool the distribution engineer has is an adequate knowledge of the systems and components which are available, together with guidelines on their cost and reliability. Beyond this, engineering common sense and reasonable operating practices must be combined with the other factors in order to arrive at a decision on the method of service.

***194. Characteristics of Large Commercial Loads.*** All large commercial loads generally involve the following factors:

**a.** *Loads* are in the range of 300 to 4000 kVA or more. The larger loads (even up to values of 50 or 75 MVA) are normally supplied by multiples of lower-capacity services.

**b.** *Utilization voltage* is 480Y/277, although smaller loads may be 208Y/120 and some of the larger institutional loads may be 4160Y/2400 (with the customer providing further step-down).

**c.** Individual *service size* is limited to about 4000 A by availability of service entrance switching, maximum fuse or breaker sizes, largest commercial wiring systems, and a growing "gut feeling" that this represents enough eggs in any one basket. Providing adequate interrupting capacity is also a definite factor, and single transformers above 4000 A may be priced as specials.

**d.** *Installation space* is limited and has a high value to the owner. Utility equipment must be as compact as possible and should not require exceptional customer requirements for auxiliaries.

**e.** Each job is one-of-a-kind and requires much *custom engineering* as well as detailed coordination with the owner of the building facilities. Complex commercial considerations are also involved, covering rates, ownership of facilities, contracts, and future maintenance responsibilities.

**f.** Service quality must be high, as to both voltage regulation and continuity. Frequent interruptions are not tolerable, and long planned interruptions are not feasible. Service complaints when expressed are long and loud.

***195. Service Arrangements.*** Several basic service arrangements can be considered for these loads:

1. Radial
2. Primary loop
3. Primary selective
4. Secondary selective
5. Spot network

If radial service were adequate, there would be no need for the succeeding systems because the radial system is the least complex and the least expensive. Unfortunately, when the supply system is underground, it also is the least reliable and generally is unsatisfactory except in special cases. The principal drawback of the radial system is its exposure to long interruptions due to component failure and the necessity for repeated planned interruptions for routine maintenance or new construction.

These five basic service systems are illustrated in Fig. 18-49, which also shows a basic main feeder system of two similar underground feeders.

***196. Radial System.*** The radial system is exposed to many interruption possibilities, the most important of which are those due to primary cable failure or transformer failure. Either

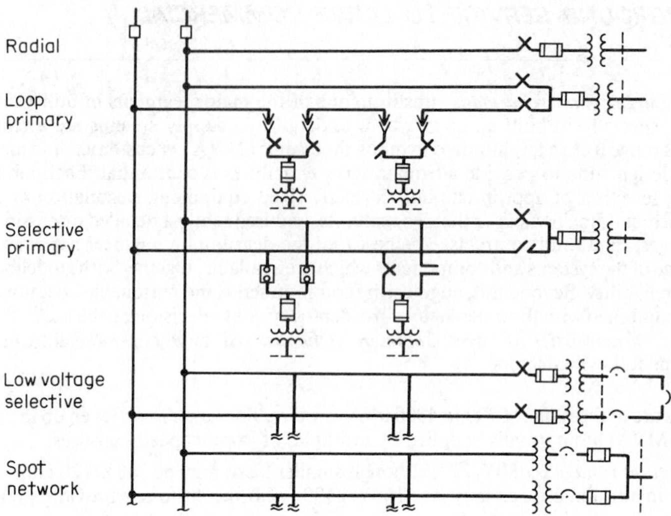


FIG. 18-49 Five basic service systems.

event will be accompanied by a long interruption, reported nominally by utilities as 10 to 12 h. Both components have finite failure rates, and such interruptions are expected and statistically predictable. The system will be satisfactory *only* if the interruption frequency is very low and if there are ways to operate the system without planned outages.

***197. Primary Loop.*** A great improvement is obtained by arranging a primary loop, which provides two-way feed at each transformer. In this manner, any section of the primary can be isolated, without interruption, and primary faults are reduced in duration to the time required to locate a fault and do the necessary switching to restore service. The cable in each half of the loop must have capacity enough to carry all the load. The additional cable exposure will tend to increase the frequency of faults, but not necessarily the faults per customer. The addition of a loop tie switch at the open point also introduces the possibility of a single equipment fault causing an interruption to both halves of the loop. Murphy's law generally applies to these situations.

Automatic loop switching to reduce interruption duration further is very difficult to arrange and is not normally applied to these systems.

***198. Primary Selective.*** This system uses the same basic components as in the primary loop, but arranged in a dual or main/alternate scheme. Each transformer can "select" its source, and automatic switching is frequently used. When automatic, the interruption duration can be limited to 2 or 3 s. Each service represents a potential two-feeder outage (if the open switch fails), but under normal contingencies, service restoration is rapid and there is no need to locate the fault (as with the loop) prior to doing the switching. This scheme is in popular use on many underground systems.

***199. Secondary Selective.*** This service system uses two transformers and low-voltage switching. It is not in popular use by utilities for 480-V service, but is common in industrial plants and on institutional properties. Primary operational switching is eliminated and with it some causes of difficulty. Duplicate transformers virtually eliminate the possibility of a long interruption due to failure. Load is divided between the two units, and automatic transfer is employed on loss of voltage to either load. There must be close coordination of utility and customer during planned transfers, and the split responsibility is probably the principal reason for its limited use as a service system.

***200. Secondary Spot Network.*** Maximum service reliability and operating flexibility are gained by a spot network using two or more transformer/protector units in parallel. The low-voltage bus is continuously energized by all units, and automatic disconnection of any unit is obtained by sensitive reverse power relays in the protector. Maintenance switching of primary feeders can be done without customer interruption or involvement. Spot networks are common in downtown, high-density areas and have been applied frequently in outlying areas for large commercial services where the supply feeders can be made available. This system also represents the most compact and reliable arrangement of components for service in underground systems.

## LOW-VOLTAGE SECONDARY-NETWORK SYSTEMS

***201. Distributed or grid-type secondary network systems*** have been used for many years by electric utility companies to serve high-density load areas in the downtown section of cities. Secondary networks are used in about 90% of the cities in this country having a population of 100,000 or more, and in more than one-third of all cities with population between 25,000 and 100,000.

The service voltage is 208Y/120 V supplying light and power loads in stores, hotels, restaurants, office buildings, apartment houses, and in some cases individual residences. The systems and equipment are entirely underground, and the 208Y/120-V portion consists of grids of interconnected cables supplied at numerous points by network transformers which feed the grid through network protectors.

A given secondary network is supplied by several primary feeders suitably interlaced through the area in order to achieve acceptable loading of the transformers under emergency conditions and to provide a system of extremely high service reliability. Primary voltages are found in the range of 5 to 34.5 kV, with the 15-kV class predominating. See Fig. 18-50.

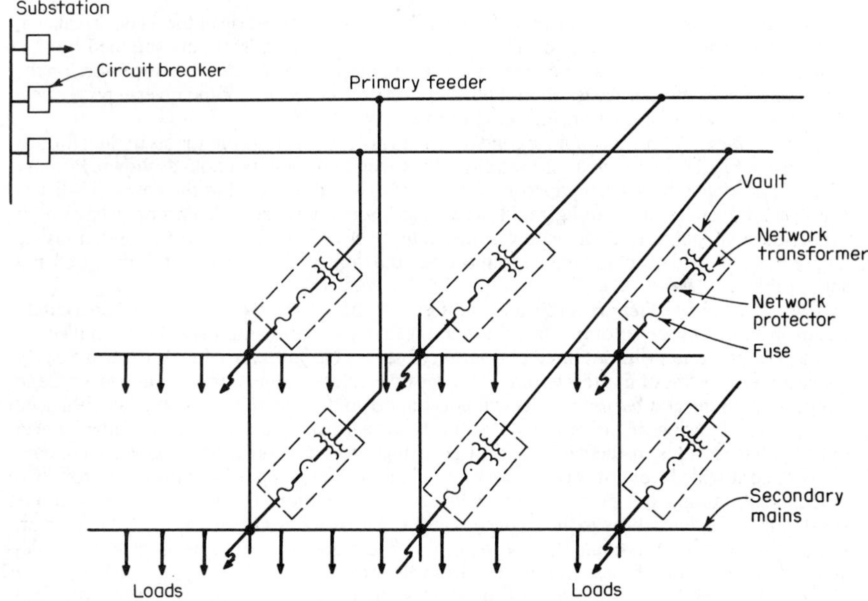


**FIG. 18-50** Schematic diagram of small segment of a secondary network.

***202. The number and routing of the primary feeders*** is usually based on the assumption that the loss of one or two feeders will not cause a service interruption. For example, the design bogey may be such that the network can operate satisfactorily during the forced outage of one feeder when another feeder is out of service for repairs or maintenance (single contingency).

***203. The secondary cable system*** is designed so that the loss of one transformer will not cause low voltage or a service interruption. Secondary cable faults in 208Y/120-V networks are allowed to burn clear or are cleared by means of limiters, which essentially are fuses having characteristics proportioned to protect the cable and to coordinate with other protective devices. Secondary faults usually will *not* burn clear on 480Y/277-V networks, so limiters are used extensively in these systems. Usually the secondary mains consist of two or more cables in parallel so that failure of one cable does not result in a service interruption.

***204. Network Transformers.*** The network protector is mounted on one end of a network transformer and the primary disconnecting and grounding switch on the other end. The typical network transformer is 3-phase, 216Y/125 V, oil-cooled, in a heavy corrosion-resistant tank suitable for installation in subsurface vaults under streets or sidewalks. Occasionally there are installations in dry locations where submersible construction is not required. Five hundred kVA is a very common rating, although 750- and 1000-kVA units are available and in use. (For *spot* networks of 480 Grd Y/277 V, transformer ratings are available through 2500 kVA.) There are only two or three distributed street networks at 480Y/277 V in the United States.

***205. Network Protectors.*** The network transformer is connected to the secondary network through a network protector (NWP) as shown in Fig. 18-50. The NWP is an air circuit breaker with relays and auxiliary devices and backup fuses, all enclosed in a metal case, which usually is physically mounted on the secondary side of the transformer. The functions of the relays are

**a.** To open the NWP on power-flow reversal, or in case of a fault in the transformer or in the primary feeder.

**b.** To reclose the NWP when the voltage of the primary feeder is of the correct magnitude and phase relation with respect to the network voltage so that when the NWP closes, power will flow *from* the feeder *into* the network.

Thus, if there is a fault on a primary feeder, it is cleared by operation of the feeder breaker at the substation and the opening of all network protectors on transformers supplied by that feeder. Also, if a feeder breaker is opened manually in preparation for maintenance work on the feeder, all NWPs on that feeder should open automatically because of the reverse power flow caused by excitation of the transformers from the low-voltage side.

***206. Cables.*** All primary and secondary cables are routed along the streets in duct lines as indicated in Fig. 18-50. Loads are served along the streets and at intersections as shown. Primary cables traditionally have been paper-insulated, lead-covered (PILC), but the new solid-dielectric insulated cables are gaining rapid acceptance. Secondary cables commonly have been rubber-covered, but polyethylene now is used extensively. Manholes at street intersections are large enough to hold numerous cable connections and limiters and to allow workers to pull and splice cables.

***207. Continuity of service*** is the outstanding advantage of a network system. When a failure occurs in a primary feeder or in a transformer, the faulty feeder is automatically disconnected, and service continues without interruption. Secondary cable faults are allowed to burn clear or are cleared by means of limiters without loss of service. Substations supplying networks are so designed that typical substation faults will not shut down the network; this is further enhanced by careful interlacing of the primary feeders through the load area and their connection to different bus sections in the substation. It is strongly recommended that a given secondary network be supplied from only one substation. If a network is supplied, for instance, from two different substations, it is possible under some system conditions for power to flow from one substation to the other through the secondary grid and network transformers. Should this occur, some network protectors could "see" reverse power flow and open, thus resulting in the undesirable disconnection of these tranformers from the network.

***208. Network Size.*** A secondary network supplied by five or more feeders will keep transformer loadings at 125% or less during the outage of one primary feeder. If the feeders are in the 15-kV class, each feeder could easily supply six 1000-kVA or twelve 500-kVA network units. Thus five feeders interlaced could supply a 30,000-kVA network under idealized conditions. With 500-ft-square blocks, 1000 kVA per block corresponds to 112 MVA of load per square mile.

Some utilities plan for the emergency outage of one feeder while a second is out of service for maintenance.

In general, 208Y/120-V networks are in the order of 30,000 to 40,000 kVA in size. There are many networks smaller than this range and some larger. One important limitation to the size of a secondary network is the ability to restore service after the network has been shut down.

***209. Spot Networks.*** New commercial buildings in existing 208Y/120-V network areas usually have very large electric loads. These loads frequently are supplied by 480Y/277-V *spot* networks, since it is impractical to handle individual loads much larger than about 200 kVA from the 208Y/120-V street networks, and 480Y/277 V is an excellent voltage for supplying large commercial buildings. In some cases the spot networks are supplied from primary feeders which also serve a distributed network.

***210. High-Rise Buildings.*** Primary-voltage feeders are being used as the riser feeder in high-rise buildings. A rule of thumb is that if an apartment building is 10 floors or more, it is more economical from an overall point of view to use the primary voltage rather than utilization voltage for the riser feeder. Similarly, for a commercial building with 480Y/277-V utilization, a building of 50 floors or more usually justifies the use of primary voltage feeders as risers in the building.

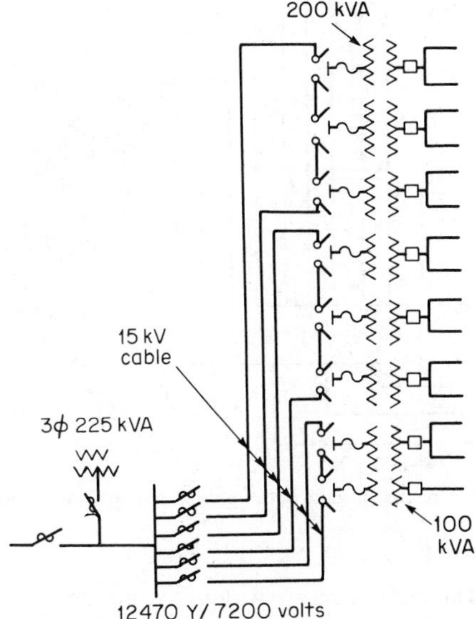


**FIG. 18-51** Schematic diagram of a looped primary system.

The primary system pattern within a high-rise building is usually a loop as shown in Fig. 18-51 or multiple as shown in Fig. 18-52. This will allow cable faults to be manually isolated by proper switching so that cable faults will cause only short interruptions to customers. Customers fed radially through a transformer will be without service if the transformer fails until a temporary connection can be made to an adjacent transformer. For more important loads like elevators, hall lighting, and fire pumps, better reliability is often obtained by using a spot network or low-voltage selective system.

***211. Dry type*** is the insulating medium of the transformers most desired for high-rise buildings. This is because no special provisions have to be made in the transformer room, such as fireproofing for oil-filled or venting the transformer to the outside of the building for nonflammable liquid-filled transformers. Many of the transformer rooms for apartment buildings and some commercial buildings are in the core of the building, which makes it difficult to use the liquid-filled transformer. However, transformers for supplying heavy loads, such as air conditioning in commercial buildings, can usually be located against an outside wall on machinery

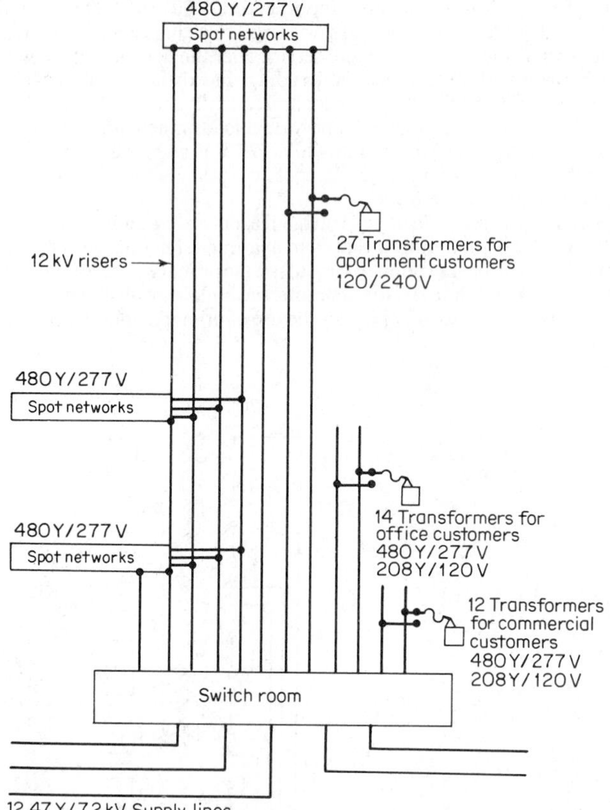


**FIG. 18-52** One-line diagram of the multiple primary system for John Hancock Center.

floors of the building. This makes it relatively easy to vent a nonflammable liquid-filled transformer to the outside of the building. Hence network transformers are often used for this application.

The primary load-break switch or load-break connector with a fuse can be arranged for either the multiple or loop type of feed where the short-circuit current available is within their rating. Usually the primary short-circuit current available is in the 8000- to 10,000-A range. The current-limiting type of fuse is often used for this "inside-the-building" application because it does not discharge ionized gases or noise during interruption.

## CONSTRUCTION OF UNDERGROUND SYSTEMS FOR DOWNTOWN AREAS

*212.* ***Underground construction*** is required in the built-up downtown areas of cities where the distribution system serves a multiplicity of concentrated commercial loads. Usually the distribution circuits are installed in conduits or duct lines along the city streets. Equipment such as switches and transformers is installed in vaults under the streets or sidewalk or in rooms located within the buildings.

*213.* ***Inflexible conduit systems*** have not been widely used in the United States except for the early Edison systems and are now completely obsolete. In this system, the conductors were insulated copper rods which were placed in an iron pipe which was then filled with a melted asphaltic compound which solidified on cooling. The tube sections were laid in a trench, joined together, and directly buried.

For many years, inflexible conduit systems were used in Europe; however, in the United States flexible systems gained preference because of the expense and inconvenience of digging through street surfaces in order to make repairs or replacements.

***214. The flexible underground system*** consists of ducts or pipes extending between manholes. This type of system has the advantage of minimum disturbance of street pavement and interference with traffic. Cables can be drawn or withdrawn from manhole locations for repairs or changes.

Manholes are placed at all junction points, corners, and as needed for secondary and service cables. The spacing of manholes depends on the types of circuits installed, varying considerably between through-type and local distribution circuits. In straight runs, the intervals may be as great as 500 to 700 ft, depending on the allowable cable-pulling tensions and utility practice.

***215. Duct Materials and Practices.*** Many types of suitable materials have been used for cable ducts, such as fiber-clay tile, concrete, plastic, fiberglass, and soapstone. Preference varies from one utility to another. In general the material should be impervious to water and not degraded by chemical action or electrolysis. The bore usually is round and should be smooth to avoid damage to the cable sheath or jacket. The diameter of the bore should be adequate to accept the largest-diameter cable envisioned for the foreseeable future. Several types of single duct are shown in Fig. 18-53. Diameters ranging from 3½ to 5 in are common.

For underground distribution systems, a duct line usually is built up of a number of single ducts, often arranged in a rectangular array and encased in concrete as shown in Fig. 18-54. For secondary-network systems, duct lines containing 6 to 12 ducts are frequently used.

***216. The number of ducts*** in a given duct line should be sufficient to accommodate anticipated load growth for a number of years in the future.

Theoretically, the most economic form of duct structure would be two ducts wide. However, when more than six or eight ducts are required, this design may lead to excessive depth. As a result, usually a rectangular construction, three or four ducts wide and three or four ducts deep, is used as shown in Fig. 18-54. The maximum number of ducts to be put into a duct line is

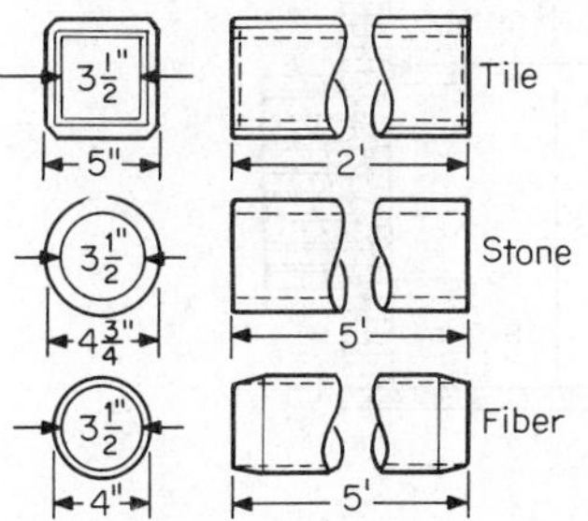


**FIG. 18-53** Types of single duct for underground conduits.

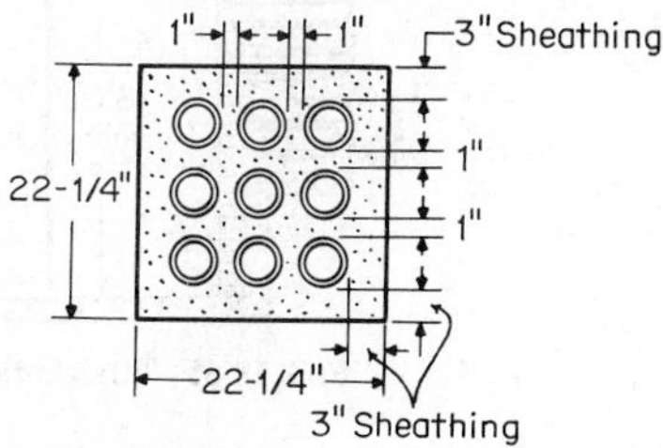


**FIG. 18-54** Arrangement of conduit with concrete sheath.

governed chiefly by thermal limitations. It is desirable to have as many ducts as possible on the outside of the bank in order to facilitate heat transfer to the surrounding earth. Insofar as possible, the inner ducts should be used for cables which produce little heat.

Since space for training cables in manholes is limited, it becomes more difficult to rack and train cables when the incoming duct line is several ducts wide and several ducts deep.

***217. Location of Manholes.*** Manholes provide protected and accessible space in which cables and associated equipment can be operated efficiently. They must be provided in sufficient number to permit pulling in cable without excessive tension, to house necessary tranformers and switching equipment, and to provide for splices and service connections.

On long runs the manhole spacing usually is not more than 500 or 600 ft. Where local distribution circuits are involved with numerous service connections, manholes may be located about 100 ft apart. Manholes or vaults to house transformers must be large enough to provide working room and proper ventilation. Location of transformer vaults in the sidewalk area is preferred, and sidewalk gratings are commonly provided to improve ventilation. For locations in the street or in areas accessible to vehicular traffic, the vault roof and gratings must be designed to withstand anticipated loadings.

***218. Many sizes and shapes*** of manholes are used. The design used for a particular installation may well be governed by the presence of local obstructions such as gas lines, water pipes, or conduit lines of other utilities. For cable manholes in the streets, many utilities have standardized on a rectangular or coffin-shaped design with the long axis parallel to the duct line. The manhole should be deep enough to allow the lowest duct to enter about a foot above the floor and should have 5 to 6 ft of clear headroom for workers. Also, the bottom of the manhole should be higher than the adjacent sewer system so that a drain can be effective in keeping it dry. Figures 18-55 and 18-56 illustrate two types of cable manholes.

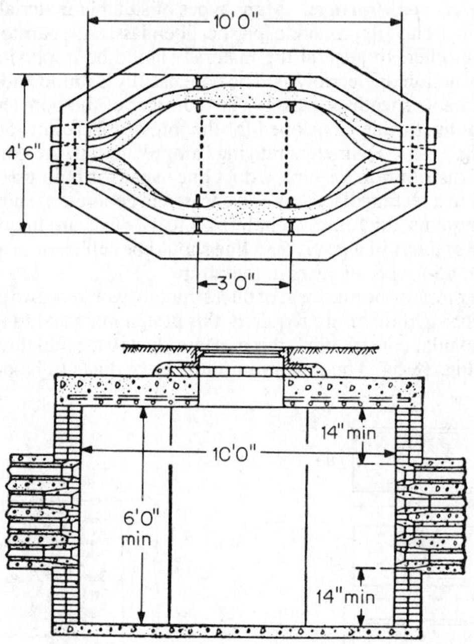


**FIG. 18-55** Straight-type manhole.

***219. Handholes.*** In some cases, a shallow form known as a handhole is used for local-distribution-circuit connections. These are usually built above the conduit line so that only the top row of ducts enters the handhole. Secondary-distribution circuits are thus accessible for service taps and do not interfere with through lines in the lower ducts.

***220. Installation of Conduit System.*** Typically there is considerable congestion of underground structures under city streets. Therefore, when a new duct line is planned, a survey should be made to select a location which will present as few obstructions as possible. This is done by noting the position of manholes of existing systems and by consulting whatever map records are available.

The exact final depth of the ditch often cannot be determined until the depths of pipes and conduits crossing it have been disclosed by excavation. Alignment and grades should be established with surveying instruments to ensure proper drainage and avoid pockets. The line may be curved slightly to avoid obstructions, but dips should be avoided where water may accumulate and freeze.

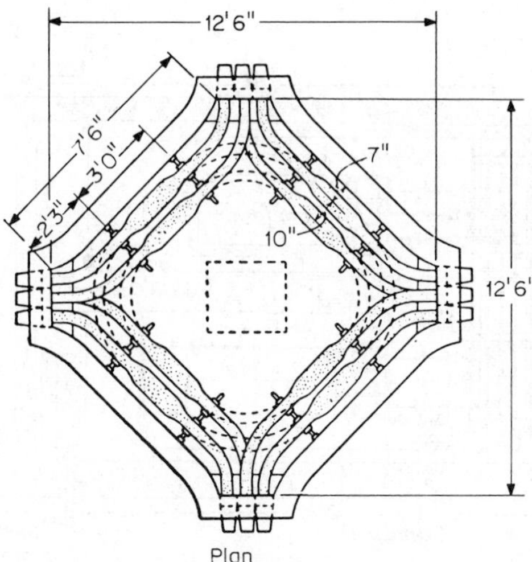


FIG. 18-56 Nine-duct X-type manhole.

Where conditions are suitable, the ditch can be dug so that the earth can be used as the form for the bottom and sides of the concrete encasement.

*221. Structural Design of Cable Manholes.* The walls of manholes generally are constructed of brick, nonreinforced concrete, or reinforced concrete. The reinforced-concrete manholes may be poured in the field or may be precast. Many variations in detailed structural design are found with different utilities. The roof must have sufficient strength to support the heaviest street traffic passing over it, which often necessitates use of steel reinforcement or structural steel beams.

To provide access of personnel and the installation of equipment, manhole frames and covers are provided and are supported on the roof of the manhole. The covers for the use of personnel usually are round and made of cast iron or steel. While some square or rectangular covers are used, they usually are heavier for the same effective opening and can fall into the manhole during replacement.

Transformer vaults are described in Par. **223.**

*222. Cable supports or hangers* usually are mounted on the walls of the manhole to support the cables in their trained position and to maintain an orderly arrangement of the cables.

*223. Transformer Vaults.* When transformers or other equipment are installed beneath streets, sidewalks, or alleys, manholes or vaults are provided. Usually enough room is provided around the equipment so that workers can operate or maintain it. In some cases involving nonnetwork service, commercial loads are supplied from "commercial subsurface transformers" where access to accessory equipment, such as fuses, internal switches, and separable cable terminations, is available from ground level; the vault may be close-fitting since it is not necessary for workers to enter the vault.

The most prevalent types of transformer vaults are found in secondary-network systems. They may be located under sidewalks or streets or partly or entirely within buildings. The arrangement, size, and shape of a network vault are determined by the number and rating of network transformers to be installed and the nature of accessory equipment, such as primary switches. Figure 18-57 shows the general arrangement of a sidewalk vault arranged for two network units with network protectors and primary switches. The roof consists of removable slabs of sidewalk concrete (not shown), and access is available at either end by iron steps going into space provided for circulation of air. Under normal conditions, the entrances are covered with suitable metal gratings or grills which allow for circulation of air.

When a nearby sewer is readily available, a drainage connection is often used in street vaults. Sidewalk vaults often do not accumulate enough water to justify a sewer connection and usually are provided with a small sump to facilitate pumping out with portable equipment, if necessary.

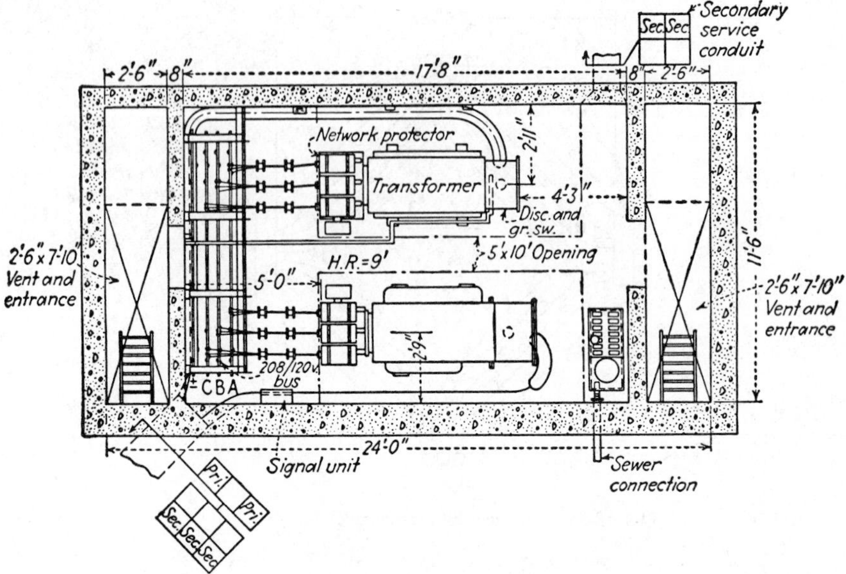


FIG. 18-57 General arrangement of a network vault under a sidewalk (plan view).

Secondary connections are made from the network protector terminals to a suitable bus and then to street mains if a distributed network is involved. Generally the secondary voltage is 208Y/120 V for distributed street networks and 480Y/277 V for spot networks.

## UNDERGROUND CABLES

***224. Types of Cables.*** Underground distribution systems have been in use for many years in the downtown built-up areas of American cities. In most instances these are secondary-network systems with facilities installed beneath streets and sidewalks, and the cables are usually installed in conduit or duct systems. For primary-voltage circuits from 5 to 35 kV, paper-insulated, lead-covered (PILC), 3-conductor cable has been used extensively. Single-conductor secondary cables with rubber insulation and neoprene jacket are common. More recently single-conductor polyethylene-insulated cables are being used for both primary and secondary. Copper conductor predominated in the past, but aluminum has nearly displaced copper in new installations, except where existing duct space is limiting.

In residential and suburban areas, new underground distribution systems to serve commercial loads, such as shopping centers and commercial and industrial parks, often employ direct-buried cables; conduits may be provided in locations where subsequent excavation would be excessively expensive or inconvenient. Aluminum conductors are almost universal. For primary cables, solid-dielectric insulation is used almost exclusively, with cross-linked polyethylene and EPR insulations predominating. Concentric-neutral wires are common. Secondary cables in these systems generally have aluminum conductors and solid-dielectric insulation, with cross-linked polyethylene being the most common. The secondary neutral is usually an insulated conductor, although there is some use of bare copper neutrals.

For most distribution circuits in the 5-kV class or higher, the cables employ a shielded construction. Shielding is the use of a conducting or semiconducting material on the surface of insulating material to confine the electric field to the insulation proper and to avoid undesired concentrations of electric stress. Shielding is used on the outer surface of the cable insulation or directly over the main conductor, or both. Outside shielding, often in the form of metallic tapes, metallic sheaths, or concentric wires, must be effectively grounded. This shielding also provides a return path for short-circuit current in the event of cable failure and protects workmen from the shock of charging current.

***225. Number of Conductors.*** Cables can be classified as single-conductor, 2-conductor, 3-conductor, etc., according to the number of separately insulated conductors enclosed by a single sheath or jacket (see Fig. 18-58).

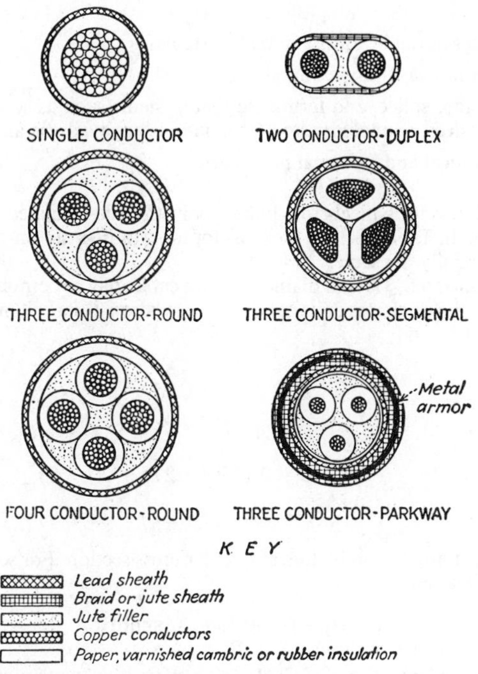


**FIG. 18-58** Cross sections of typical cables.

***226. Cable Insulation.*** Electric supply cables are insulated with a wide variety of insulating materials depending on voltage ratings, type of service, installation conditions, etc. In the past, the following have been commonly used:

**a.** Rubber and rubberlike for 0 to 35 kV

**b.** Varnished cambric for 0 to 28 kV

**c.** Impregnated paper of the solid type for voltages up to 69 kV and with pressurized gas or oil up to 345 kV or higher

These insulation systems usually require a sheath or suitable jacket to prevent infiltration of moisture; loss of oil, gas, or impregnant; and to provide protection against corrosion and electrolysis. In some cases, an armor overlay is used to provide mechanical protection.

With impregnated-paper insulation of the solid type, a lead sheath is usually provided.

A wide variety of joints, splices, and terminations are used, depending on the voltage, cable insulation, number of conductors, jacketing or sheathing material, and method of shielding. Joints, splices, and terminations are discussed in more detail in Pars. **236** to **241**.

Single-conductor cables are used, of course, in single-phase primary system for residential service and normally are used in single-phase or 3-phase secondary systems where many taps and connections are involved. Single-conductor cables also are frequently used in direct-buried, 3-phase primary systems.

Three-conductor primary cables are often used in duct systems where they have the advantage of occupying only one duct. Several typical cables are shown in Fig. 18-58.

At the present time, solid-dielectric insulating materials such as polyethylene, cross-linked polyethylene, and EPR are receiving the widest application in underground distribution systems, both direct-buried and duct systems. Principal reasons for the wide usage of these insulations are

**a.** Low cost

**b.** Suitable for direct burial or for use in duct systems.

**c.** Sheath or jacket not generally required.

**d.** Much easier to tap, splice, and terminate than systems such as solid impregnated paper. Factory-made splices, connectors, and terminations are available and widely used.

**e.** Excellent mechanical and electrical properties.

Insulation thickness for typical cross-linked, polyethylene-insulated, nonjacketed distribution cables are given in Table 18-18. Thickness for most ratings of non-cross-linked polyethylene cables are essentially the same.

***227. Cable Diameters.*** Overall diameter $D$ of a cable may be computed from the diameter of its conductors $d$, the thickness of its conductor insulation $T$, its belt insulation $t$, and its lead sheath $S$, as follows:

Single-conductor: $$D = d + 2T + 2S \qquad (18\text{-}22)$$

2-conductor: $$D = 2(d + 2T + t + S) \qquad (18\text{-}23)$$

3-conductor: $$D = 2.155(d + 2T) + 2(t + S) \qquad (18\text{-}24)$$

4-conductor: $$D = 2.414(d + 2T) + 2(t + S) \qquad (18\text{-}25)$$

These formulas apply to conductors of circular cross section. For sector-type 3-conductor cables, the overall diameter

$$D_3 = D - 0.35d \quad \text{approx.} \qquad (18\text{-}26)$$

***228. Electrical Characteristics of Cable.*** Skin effect is an ac phenomenon whereby alternating current tends to flow more densely near the outer surface of a conductor than near the center. This is the magnetic-flux linkages of current near the center of the conductor are relatively greater than the linkages of current flowing near the surface of the conductor. The net effect is that the effective resistance of the cable is greater for alternating current than for direct current. This effect increases as the conductor size increases and as the frequency increases. It is also a function of the relative resistance of the conductor material, being less for materials of higher resistance; for example, the skin effect for a given diameter of cable is great if the material is copper rather than aluminum. Because of skin effect, large cables are sometimes built up over a central core of nonconducting material.

The nonuniform distribution of alternating current across the cross section of the cable also has the effect of reducing the effective internal inductance of the cable. Usually, this effect is extremely small in distribution circuits and is neglected.

It should be noted that magnetic flux linking the cable because of nearby current also can affect the cross-sectional distribution of current and can significantly change the effective ac resistance of the cable for multiconductor cables or cables in the same duct. This is known as the "proximity effect." Most tables of conductor characteristics list factors which combine the results of the skin effect and proximity effect.

If an insulated cable has an outer metallic wrapping such as sheaths, metal pipes, or concentric-neutral conductors installed in such a manner that induced circulating currents can flow normally in these external conductors, losses will occur in these circuits, reducing the ampacity of the cable.

***229. Skin-Effect Coefficients.*** Skin-effect and proximity-effect coefficients are given in Table 18-19 for copper and aluminum conductors at 25°C.

**TABLE 18-18** Insulation Thickness for Cross-Linked, Thermosetting, Polyethylene-Insulated Cable*

| Rated circuit voltage, phase-to-phase volts | Conductor size, AWG or kcmil | Insulation thickness for 100 and 133% insulation levels | |
|---|---|---|---|
| | | mils | mm |
| 0–600 | 14–9 | 45 | 1.14 |
| | 18–2 | 60 | 1.52 |
| | 1–4/0 | 80 | 2.03 |
| | 225–500 | 95 | 2.41 |
| | 525–1000 | 110 | 2.79 |
| 601–2,000 | 4–9 | 60 | 1.52 |
| | 8–2 | 70 | 1.78 |
| | 1–4/0 | 90 | 2.29 |
| | 225–500 | 105 | 2.67 |
| | 525–1000 | 120 | 3.05 |
| 2,001–5,000 | 8–1000 | 90 | 2.29 |
| 5,001–8,000 | 6–1000 | 115 | 2.92 |
| 8,001–15,000 | 2–1000 | 175 | 4.45 |
| 15,001–25,000 | 1–1000 | 260 | 6.60 |
| 25,001–28,000 | 1–1000 | 280 | 7.11 |
| 28,001–35,000 | 1/0–1000 | 345 | 8.76 |

* Adapted from IPCEA Pub. S-66-524, NEMA Pub. WC-7-1471, Revision No. 3, September 1974.

*100% level* applied where system overcurrent protection is such that ground faults are cleared within 1 min. Applies to the great majority of distribution systems.

*133% level* applied where clearing time of 100% level cannot be met, but there is assurance of fault clearing within 1 h.

Minimum-size conductors should be in accordance with above values to limit maximum voltage stress on the insulation at the conductor to a safe value.

To determine the skin effect on the effective resistance of a single-conductor 1000-kcmil copper cable operating at 25°C, refer to Table 18-19, where the dc resistance is 0.01079 Ω/1000 ft and the skin-effect coefficient is 1.067. The effective resistance at 60 Hz is 1.067 × 0.01079 = 0.0115 Ω/1000 ft, 6.7% greater than for direct current. For a similar 2000-kcmil, the increase in resistance for alternating current is 23.3%; the ampacity of the cable is reduced to 100/1.233 = 81.1%.

The last two columns of Table 18-19 give the coefficients for multiconductor cables or cables in the same duct. They are used in the same manner as in the previous examples. For the larger conductors, the derating factor is substantial.

***230. Electrostatic Capacitance.*** The capacitance of a shielded or concentric-neutral single-conductor cable is

$$C = \frac{0.00736K}{10^6 \log_1 (D/d)} \tag{18-27}$$

where $C$ = capacitance, farads/1000 ft
$K$ = dielectric constant of insulation
$D$ = diameter over the insulation
$d$ = diameter over the conductor shield

***231. Charging Current.*** The charging current of a single-conductor cable is

$$I = \frac{0.0463\ EfK}{1000 \log_1 (D/d)} \tag{18-28}$$

where $E$ = voltage to neutral, kV
$f$ = frequency, Hz
$I$ = amperes per 1000 ft, charging current

**TABLE 18-19** DC Resistance and Correction Factors for AC Resistance

| Conductor size, AWG or kcmil | DC resistance, Ω/1000 ft at 25°C* | | AC resistance multiplier | | | |
|---|---|---|---|---|---|---|
| | | | Single-conductor cables† | | Multiconductor cables‡ | |
| | Copper | Aluminum | Copper | Aluminum | Copper | Aluminum |
| 8 | 0.6532 | 1.071 | 1.000 | 1.000 | 1.00 | 1.00 |
| 6 | 0.4110 | 0.6741 | 1.000 | 1.000 | 1.00 | 1.00 |
| 4 | 0.2584 | 0.4239 | 1.000 | 1.000 | 1.00 | 1.00 |
| 2 | 0.1626 | 0.2666 | 1.000 | 1.000 | 1.01 | 1.00 |
| 1 | 0.1289 | 0.2114 | 1.000 | 1.000 | 1.01 | 1.00 |
| 1/0 | 0.1022 | 0.1676 | 1.000 | 1.000 | 1.02 | 1.00 |
| 2/0 | 0.08105 | 0.1329 | 1.000 | 1.001 | 1.03 | 1.00 |
| 3/0 | 0.06429 | 0.1054 | 1.000 | 1.001 | 1.04 | 1.01 |
| 4/0 | 0.05098 | 0.08361 | 1.000 | 1.001 | 1.05 | 1.01 |
| 250 | 0.04315 | 0.07077 | 1.005 | 1.002 | 1.06 | 1.02 |
| 300 | 0.03595 | 0.05897 | 1.006 | 1.003 | 1.07 | 1.02 |
| 350 | 0.03082 | 0.05055 | 1.009 | 1.004 | 1.08 | 1.03 |
| 500 | 0.02157 | 0.03538 | 1.018 | 1.007 | 1.13 | 1.06 |
| 750 | 0.01438 | 0.02359 | 1.039 | 1.015 | 1.21 | 1.12 |
| 1000 | 0.01079 | 0.01796 | 1.067 | 1.026 | 1.30 | 1.19 |
| 1500 | 0.00719 | 0.01179 | 1.142 | 1.058 | 1.53 | 1.36 |
| 2000 | 0.00539 | 0.00885 | 1.233 | 1.100 | 1.82 | 1.56 |

* To correct to other temperatures, use the following:

For copper: $R_T = R_{25}\,[(234.5 + T)/259.5]$

For aluminum: $R_T = R_{25}\,[(228 + T)/253]$

where $R_T$ is the new resistance at temperature $T$ and $R_{25}$ is the tabulated resistance.

† Includes only skin effect (use for cables in separate ducts).

‡ Includes skin effect and proximity effect (use for triplex, multiconductor, or cables in the same duct).

NOTE: 1 ft = 0.3048 m.

For overhead circuits at distribution voltages and power frequencies, the charging current usually is negligible. It may become significant in high-voltage transmission circuits as discussed in Sec. 14.

For insulated cables, the charging current is relatively greater than in overhead circuits because of close spacing and the higher dielectric constant of the cable insulation; $K = 1$ for air and 3.3 for impregnated paper. For unfilled polyethylene $K = 2.3$, and it may run as high as 2.9 for filled, cross-linked polyethylene.

***232. Geometric Factors.*** Charging current of 3-phase three-core cable is affected by arrangement of conductors (round or sector) and by relative thicknesses of conductor insulation $T$ and belt insulation $t$. These relations have been put into usable form by working out logarithmic denominators of the equation for various ratios of thickness of insulation to diameter of conductor. This has been termed the *geometric factor.* Charging current of a three-core 3-phase cable is

$$I = \frac{3 \times 0.106 EfK}{1000 G_2} \quad \text{A/1000 ft} \tag{18-29}$$

For impregnated-paper cable, $K$ is 3.3, and the equation for 60-Hz circuits becomes

$$I = \frac{3 \times 3.3 \times 0.106 \times 60E}{1000 G_2} = 0.063 \frac{E}{G_2} \quad \text{amperes} \tag{18-30}$$

Values of $G$ for single-conductor and $G_2$ for 3-conductor cable may be taken from Table 18-20.

***233. Geometric Factors of Cables.*** See Table 18-20. Intermediate values may be found by interpolation.

***234. Example.*** Find 60-Hz charging kVA for 33-kV cable having three 350,000-cmil sector-type conductors each with 10/32 in of paper and a 5/32-in belt.

$$T = 0.313 \text{ in} \qquad t = 0.156 \text{ in} \qquad d = 0.681 \text{ in} \qquad t/T = 0.5$$

$$(T + t)/d = (0.313 + 0.156)/0.681 = 0.69; \ E = 33/1.73 = 19 \text{ kV}$$

Interpolating in Table 18-20, we find $G_2 = 2.78$.

***235. For sector-type cable,*** $G_2$ must be multiplied by the sector factor for 0.69, which is seen to be 0.86 in the sector-factor column in Table 18-20. For such a cable,

$$G_2 = 0.86 \times 2.78 = 2.39 \qquad \text{and} \qquad I = (0.063 \times 19)/2.39 = 0.5\text{A}/1000 \text{ ft}$$

Charging kVA $= 3IE = 3 \times 0.5 \times 19 = 28.5$ kVA/1000 ft, and for a cable having a length of 20 mi it would be $20 \times 5.28 \times 28.5 = 3010$ kVA. For single-conductor cables, $t = 0$ and $(T + t)/d = T/d$, which is used to get the value of $G$ from the values for single-conductor cable in Table 18-20.

***236. Cable Terminations.*** A cable termination must perform several functions:

**a.** Provide means for electrical connection of the cable to an equipment or circuit.

**b.** Control the electrostatic stresses so that there is no electrical-discharge activity in the termination at design voltage levels. One important consideration is to control the voltage stresses where the change is from a uniform radial field within the (shielded) cable to a new configuration beyond the termination of the shield. Other considerations are to provide adequate flashover and creepage strength to nearby grounds.

**c.** Prevent loss of gas or liquid insulation impregnant from the cable, where needed, or from the termination.

**d.** Provide suitable mechanical and/or hermetic termination of the sheath, where used.

**e.** Serve as a load-break switch or separable connection, where needed.

Many types of terminations are in use, ranging from those made by hand in the field to factory-made types requiring very little work in the field. Two broad classifications are "live-

**TABLE 18-20** Table of Geometric Factors of Cables

| Ratio $\frac{T+t}{d}$ | $G$ Single conductor | Sector factor | Three-conductor cables | | | | | |
|---|---|---|---|---|---|---|---|---|
| | | | $G_1$ at ratio $t/T$ | | | $G_2$ at ratio $t/T$ | | |
| | | | 0 | 0.5 | 1.0 | 0 | 0.5 | 1.0 |
| 0.2 | 0.34 | ..... | 0.85 | 0.85 | 0.85 | 1.2 | 1.28 | 1.4 |
| 0.3 | 0.47 | 0.690 | 1.07 | 1.075 | 1.08 | 1.5 | 1.65 | 1.85 |
| 0.4 | 0.59 | 0.770 | 1.24 | 1.27 | 1.29 | 1.85 | 2.00 | 2.25 |
| 0.5 | 0.69 | 0.815 | 1.39 | 1.43 | 1.46 | 2.10 | 2.30 | 2.60 |
| 0.6 | 0.79 | 0.845 | 1.51 | 1.57 | 1.61 | 2.32 | 2.55 | 2.95 |
| 0.7 | 0.88 | 0.865 | 1.62 | 1.69 | 1.74 | 2.55 | 2.80 | 3.20 |
| 0.8 | 0.96 | 0.880 | 1.72 | 1.80 | 1.86 | 2.75 | 3.05 | 3.45 |
| 0.9 | 1.03 | 0.895 | 1.80 | 1.89 | 1.97 | 2.96 | 3.25 | 3.70 |
| 1.0 | 1.10 | 0.905 | 1.88 | 1.98 | 2.07 | 3.13 | 3.44 | 3.87 |
| 1.1 | 1.16 | 0.915 | 1.95 | 2.06 | 2.15 | 3.30 | 3.60 | 4.05 |
| 1.2 | 1.22 | 0.921 | 2.02 | 2.13 | 2.23 | 3.45 | 3.80 | 4.25 |
| 1.3 | 1.28 | 0.928 | 2.08 | 2.19 | 2.29 | 3.60 | 3.95 | 4.40 |
| 1.4 | 1.33 | 0.935 | 2.14 | 2.26 | 2.36 | 3.75 | 4.10 | 4.60 |
| 1.5 | 1.39 | 0.938 | 2.20 | 2.32 | 2.43 | 3.90 | 4.25 | 4.75 |
| 1.6 | 1.44 | 0.941 | 2.26 | 2.38 | 2.49 | 4.05 | 4.40 | 4.90 |
| 1.7 | 1.48 | 0.944 | 2.30 | 2.43 | 2.55 | 4.17 | 4.52 | 5.05 |
| 1.8 | 1.52 | 0.946 | 2.35 | 2.49 | 2.61 | 4.29 | 4.65 | 5.17 |
| 1.9 | 1.57 | 0.949 | 2.40 | 2.54 | 2.67 | 4.40 | 4.76 | 5.30 |
| 2.0 | 1.61 | 0.952 | 2.45 | 2.59 | 2.72 | 4.53 | 4.88 | 5.42 |

front" and "dead-front." The former involves exposed, bare electrical connections and possibly lengths of unshielded cable. The latter type of termination is completely enclosed in a semiconducting or metallic structure essentially at ground potential, such that it can be touched without hazardous shock while the equipment is energized.

*237. Elementary Stress-Cone Termination.* Figure 18-59 shows a single-conductor shielded cable prepared for termination. Figure 18-60 shows an elementary stress-cone termination.

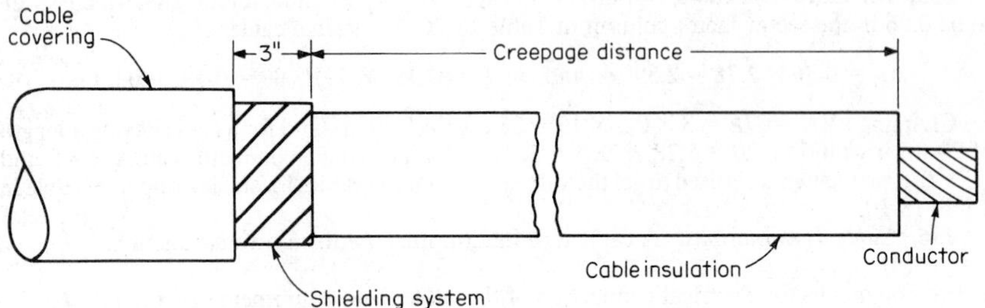


**FIG. 18-59** Elementary plain-shield termination.

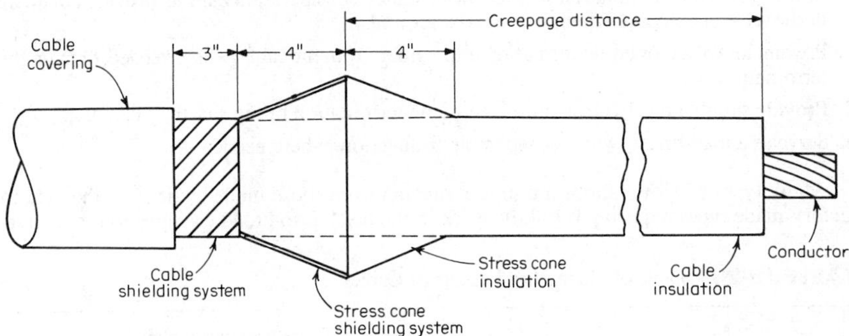


**FIG. 18-60** Elementary stress-cone termination.

The stress cone may be built up by using tapes of a material compatible with the cable insulation, or it may be a factory-molded stress cone which is slipped over the (solid-dielectric) cable insulation system after the end of the cable has been properly dressed.

The stress cone serves to keep dielectric stresses at acceptable values. Without the stress cone, the plain termination of Fig. 18-59 would experience excessive stresses at normal voltage near the end of the shielding system, leading quickly to failure.

Occasionally, an overall cover tape may be provided. Usually a compression-type connector is installed on the cable conductor to provide a means of electrical connection to the equipment.

*238. Separable Insulated Connectors.* Accompanying the rapid growth in the use of single-conductor, concentric-neutral (or shielded cables employing solid-dielectric insulation) has been the development of separable insulated connectors of the "dead-front" classification. A typical termination consists of:

**a.** An elbow, as shown in Fig. 18-48, which is physically and electrically connected to the end of a properly dressed cable.

**b.** A load-break switch module where the load-break function is specified (Fig. 18-47).

**c.** An apparatus bushing. When a load-break module is not used, the elbow is mated directly to the apparatus bushings.

The external shield and the insulation of the elbow and switch module are made of synthetic rubber. The stress-relief function is designed into the molded configuration.

When the elbow is in its normal connected position, the termination is dead-front and submersible.

A wide variety of accessory components are available, including multitaps, insulated bushings, feed-through bushings, insulating caps, etc., so that wide flexibility in operating procedures can be obtained for differing system and equipment configurations.

Separable insulated connectors are available in 200-A ratings for use on grounded-wye systems of the 15-, 25-, and 35-kV classes. Ratings of 600 A also are available, but these usually cannot be opened under load or while the cable is energized.

To install the elbow, the end of the cable is dressed according to the manufacturer's specifications; often a jig is used so that proper removal of semiconducting shield, correct dressing dimensions, exposure of the conductor, etc., are easily obtained. A crimped connector is then installed on the cable conductor, and the elbow is then slipped over the cable, internal male contact installed, and the cable or concentric neutral is connected to the semiconducting outer shield of the elbow.

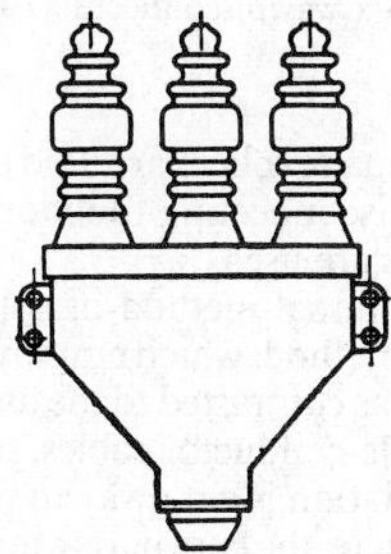

**FIG. 18-61** Three-conductor disconnecting-type pothead.

Obviously, a termination of this kind can be installed much more quickly and with a lower level of skill needed than in terminating the traditional paper-insulated, lead-sheathed cables.

***239. Pothead Terminations.*** Where cables are connected to overhead systems or to switchgear equipment, they often are terminated by means of potheads, such as that shown in Fig. 18-61. An extremely wide variety of types is in use, depending on

**a.** System voltage.

**b.** Type of cable insulation.

**c.** Single- or multiconductor cable.

**d.** Type of jacket or sheath.

**e.** Whether "live" side of pothead is outdoor or within an equipment. Some potheads are of the disconnecting type as indicated in Fig. 18-62.

There is an increasing trend toward the use of factory-made molded-rubber terminations in place of the traditional potheads, particularly with cables having solid-dielectric insulation.

***240. Subway junction boxes*** are sometimes used in cable systems to interconnect distribution circuits at points where it is desired that the connection be opened, at times, for construction or operating purposes and where overhead disconnecting facilities are not available. In low-voltage systems, such as 208Y/120-V secondary networks, such boxes may include sectionalizing fuses or connecting links. They are used in a number of different circuit configurations, a six-way junction box being shown in Fig. 18-63.

***241. Splicing.*** Cable splices to a great extent resemble "back-to-back" portions of cable terminations. The completed splice provides:

**a.** Electrical connection between the cable conductors, usually by means of crimped connectors.

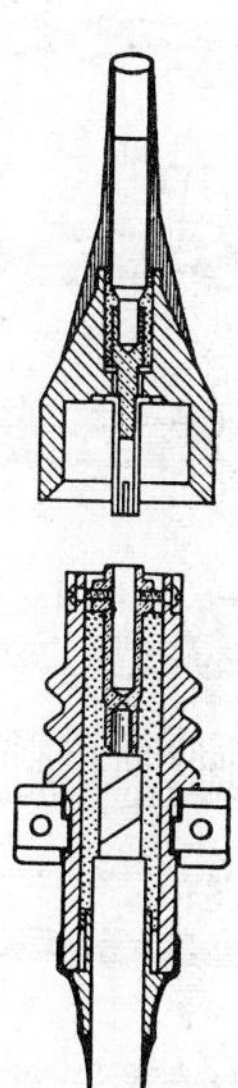

**FIG. 18-62** Disconnecting-type pothead.

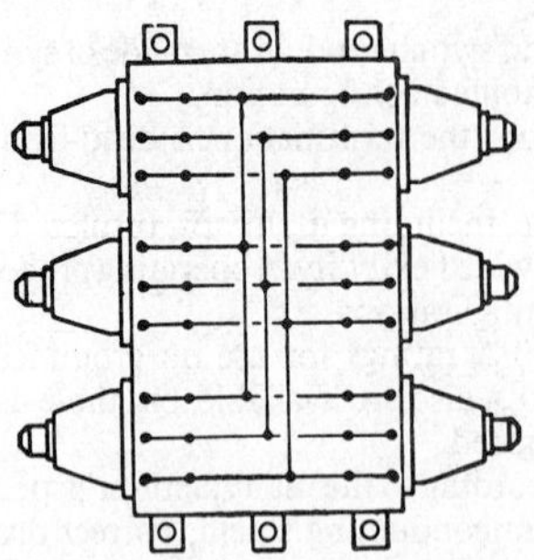

**FIG. 18-63** Six-way disconnecting cable-junction box.

**b.** Insulation over the exposed conductors and connector.

**c.** Jointing of the shielding or concentric neutral systems of the two cable sections so that electrical stresses are properly controlled.

**d.** Jointing of the jacketing systems or sheaths.

**e.** A "stop" function where the two cable insulation systems are different, for example, oil-impregnated paper on one side and solid dielectric on the other. Here it is necessary to contain the oil in the paper insulation and to exclude its penetration into the solid dielectric.

**f.** A disconnecting function, where required.

Factory-made splices are used extensively for splicing cables with solid-dielectric insulation. Where the disconnecting function is required, various combinations of multitaps and elbow terminations are used.

The traditional method of splicing paper-insulated, lead-sheathed cable is shown in Fig. 18-64. This method, which employs hand-wrapped insulating tapes, requires a high level of skill and training as contrasted to the use of factory-made splices for joining solid-dielectric cables. In jointing single-conductor cables, the lead sheath is removed about 6 in back from the end, and enough insulation is cut away to permit a soldered connection to be made. When the connection is complete, the bare parts are wrapped with tape until the equivalent of cable insulation has been applied. A lead sleeve which has previously been slipped over one of the cables is now wiped on the two cable sheaths so as to enclose the joint. Air space around the joint is then filled with hot insulating compound poured into a small hole in one end of the sleeve; a similar hole is left in the other end to allow air to escape. These holes are then closed by soldering. The joint should be allowed to cool before it is moved, so that the compound will hold the parts rigidly in place.

In jointing 3-conductor cables, the lead must be removed about 10 in to facilitate taping the conductors (see Fig. 18-64). In making joints for 6600 V and higher, it is important that as little air remain in taping as possible. If paper tape is used, each layer should have compound poured over it before the next is applied.

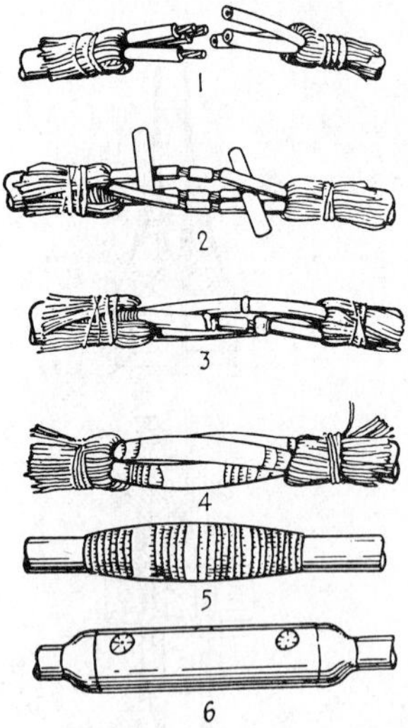

**FIG. 18-64** Successive stages in cable splicing.

***242. Installation of Cable.*** Generally, *direct-buried cable* in underground residential distribution (URD) systems is buried in a trench, usually 36 in or more deep. Often, random lay of the power and telephone cables is employed, with no intentional separation. When soil conditions permit, the trench is backfilled with the original material. In some cases a selected backfill and/or protective covering over the cables may be necessary.

Where URD circuits must be routed under streets or other paved surfaces, or in locations such that it would be impractical to dig in order to repair a faulted cable, duct installation often is used.

Where soil conditions and circuit configurations are favorable, it is possible to directly plow in the cable by means of a special plow which breaks the earth ahead of the cables and guides them into the furrow.

*In duct installations,* the most common method of preparation has been to use wood or metal rods which are pushed into the duct section by section as they are connected together. When the opposite end of the duct is reached, a wire is attached to the end of the last rod and is pulled into the duct as the rods are withdrawn.

When the duct is airtight, a piston with attached flexible wire can be blown through the duct by means of compressed air. This method is quicker and less laborious than the rodding process.

Cables are pulled through the duct by means of a pulling line, usually wire rope, and a power-driven cable-pulling winch. Cables of moderate size and length usually are pulled by means of cable grip, which is a type of woven-wire basket designed to increase its grip on the cable as the tension increases.

Often a flexible pull-in tube is used in the manhole to prevent damage to the cable being pulled and to other exposed cables in the manhole.

With long sections or larger cables, it may be necessary to use a pulling eye rather than a cable grip because the pulling tension may exceed the capability of the cable grip. The pulling eye is a steel eye which usually is fastened directly to the cable conductors.

When a new cable is to be installed in an existing duct, it is generally desirable that the diameter of the duct be at least ¾ in greater than that of the cable. Where the duct section is exceptionally long or contains relatively sharp bends, a clearance of 1 in may be needed. On short, straight sections, ½-in clearance may be acceptable.

Where several single-conductor cables are to be drawn into the same duct, the cable reels are set up in tandem and all cables are pulled into the duct simultaneously.

***243. Cable Training.*** The location of cables in manholes is determined primarily by the ducts which they occupy. The cables should be fanned out as they leave their ducts so that they do not cross other cables or ducts. Sufficient length must be left in manholes to permit training on racks around the manhole walls, as shown in Figs. 18-55 and 18-56, and for joining. Radius of bends should be greater than the minimum safe bending radius for the cable, and cable movement, also, must be considered. The safe bending radius varies with the size, type of sheath or armor, etc., and generally is in the order of 8 to 12 times the overall diameter of the cable for power cables.

With large cables, the periodic load cycles cause repeated flexing and movement in the manholes, and duct-month protection is often used to prevent cracking of lead sheaths. This may consist of a piece of galvanized metal inserted under the cable and arranged to prevent the sheath from being pressed against the sharp edges of the duct mouth.

In order to limit damage resulting from cable faults to the failed cable, it is quite common to fireproof the cables in a vault or manhole. Such fireproofing usually is done with asbestos tapes and asbestos or mortar cements.

***244. Sheath Bonding.*** With cables having a lead or other metallic sheath, it is common practice to bond together the different sheaths in a manhole or vault. Bonding consists of electrically connecting together the various sheaths. Various types of bonding systems are in use to maintain the sheaths at a common potential near ground potential, thus reducing the danger to workers who may be in the manhole when a cable fault occurs. This also eliminates the possibility of serious arcing occurring between the sheaths of the faulted and unfaulted cables.

***245. Selection of Duct Position for Cable.*** Cables used in local distribution should be installed in the top row of ducts so that manholes for service connections and lateral circuits can be of a relatively shallow construction. In distribution manholes, the higher-voltage cables and cables of through lines should be placed in the lower and outside ducts, where possible, and they should be trained with the least possible interlacing with other cables.

***246. Ambient earth temperatures*** vary with geographical location, season, and depth. Average daily air temperature at a given location follows a more or less sinusoidal curve over the seasons of the year. As a result, the earth ambient temperatures also exhibit an annual sinusoidal variation. At depths greater than 1 to 2 ft, there is essentially no daily variation in earth ambient, but there is a seasonal variation which is greater at shallow depths, decreasing with depth. In addition, as the depth increases, the variation in ambient temperature increasingly lags behind the daily ambient air temperature curve; at a depth of 6 ft, this lag may be as great as 6 to 8 weeks. At depths of 20 to 30 ft, the earth temperature remains practically constant at about the mean annual air temperature. As a result the earth temperature tends to increase with depth in the winter and decrease with depth in the summer. At a depth of 3½ ft, typical temperatures are as follows:

| | Temperature, °C | |
|---|---|---|
| | Summer | Winter |
| Northern U.S. | 20–25 | 2–15 |
| Southern U.S. | 25–30 | 10–20 |

***247. Calculation of Ampacities of Cables.*** The precise calculation of ampacities of cables is extremely complex and has been the subject of numerous technical papers. This complexity is due not only to the characteristics of the thermal circuit, such as heat transfer through the cable insulation and sheath, transfer to duct or earth, and transfer from duct bank to earth, but also to the fact that losses in the cable conductor are subject to skin and proximity effects. Also, additional losses can occur in the cable-shielding system depending on the nature of the installation.

Presently accepted methods of calculation, empirical data, and numerous references are treated in the following references:

**a.** "Power Cable Ampacities"; vol. I, Copper Conductors (1962); AIEE (now IEEE) Publication S-135-1, IPCEA Publication P-46-426.

**b.** "Power Cable Ampacities"; vol. II, Aluminum Conductors (1962); AIEE Publication S-135-2, IPCEA Publication P-46-426.

**c.** IPCEA-NEMA Standards Publication, "Ampacities Including Effect of Shield Loss for Single-Conductor Solid-Dielectric Power Cable 15 kV through 35 kV (Copper and Aluminum Conductors)" (1972); IPCEA Publication P-53-426, NEMA Publication WC50-1976.

***248. Maximum Allowable Conductor Temperature.*** The IPCEA temperature ratings for polyethylene (thermoplastic) and cross-linked-polyethylene-insulated power cables are

| | Max. conductor temperature, °C | |
|---|---|---|
| Insulation | Normal operation | Emergency overload |
| Polyethylene | 75 | 90 |
| Cross-linked polyethylene | 90 | 130 |

Maximum conductor temperatures for impregnated paper-insulated cables are given in Table 18-21, as adapted from Publication P-46-426 of the IPCEA.

***249. Ampacity of Cables.*** There is a growing usage of single-conductor, solid-dielectric power cables for important 3-phase distribution circuits in the 15- to 35-kV class. Various shielding systems are in use including concentric wires, ribbons, and tapes. In many cases, on 4-wire primary-distribution circuits, the concentric wires are used as neutral conductors. When the shields are bonded together and grounded at multiple locations, circulating currents can flow in the shields, resulting in $I^2R$ losses and appreciable heating effect. Such losses may be significant when the cables are spaced.

The AIEE-IPCEA ampacity tables mentioned in Par. **247** do not include the effects of circulating-current losses, but these effects are included in the ampacity tables of IPCEA Publication P-53-426; NEMA Publication WC50-1976, "Ampacities Including Effect of Shield Losses for Single-Conductor Solid-Dielectric Power Cable 15 kV through 35 kV (Copper and Aluminum Conductors)." The ampacity data in Table 18-22 have been taken from this publication and apply to directly buried, solid-dielectric power cable.

Table 18-23 has been adapted from the IPCEA Publication P-53-426 (and NEMA Publication WC50-1976) to illustrate typical ampacities of single-conductor, solid-dielectric power cables installed in underground ducts. The type of installation is assumed to be directly buried fiber or plastic duct of inside diameter nominal pipe size to provide a minimum diametral clearance of 0.75 in between cable outside diameter and inside diameter of the duct.

The assumed arrangement of the ducts is shown in Fig. 18-65.

**TABLE 18-21** Maximum Conductor Temperatures for Impregnated-Paper-Insulated Cable*

| Rated voltage, kV | Conductor temperature, °C | | |
|---|---|---|---|
| | Normal operation | Emergency operation | |
| **Solid-type multiple conductor belted** | | | |
| 1 | 85 | 105 | |
| 2–9 | 80 | 100 | |
| 10–15 | 75 | 95 | |
| **Solid-type multiple conductor shielded and single conductor** | | | |
| 1–9 | 85 | 105 | |
| 10–17 | 80 | 100 | |
| 18–29 | 75 | 95 | |
| 30–39 | 70 | 90 | |
| 40–49 | 65 | 85 | |
| 50–59 | 60 | 75 | |
| 60–69 | 55 | 70 | |
| **Low-pressure gas-filled** | | | |
| 8–17 | 80 | 100 | |
| 18–29 | 75 | 95 | |
| 30–39 | 70 | 90 | |
| 40–46 | 65 | 85 | |
| **Low-pressure oil-filled and high-pressure pipe type** | | | |
| | | 100 h | 300 h |
| 15–17 | 85 | 105 | 100 |
| 18–39 | 80 | 100 | 95 |
| 40–162 | 75 | 95 | 90 |
| 163–230 | 70 | 90 | 85 |



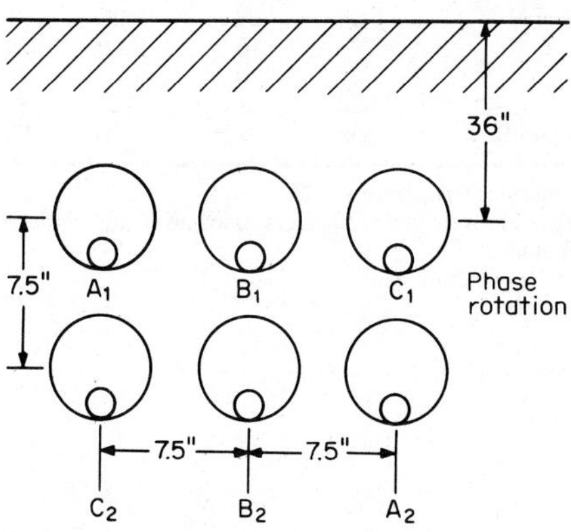


**FIG. 18-65** Arrangement of ducts.

**TABLE 18-22** Ampacity of Single-Conductor Solid-Dielectric Power Cable Installed Direct-Buried

**Single circuit, balanced 3-phase load, flat spacing, 7.5-in separation, 36-in depth of burial, 90 *p* soil resistivity, 15 kV, aluminum conductors, earth ambient temp. 25°C, 100% load factor**

| | | Max. conductor temp | | | | Max. earth interface temp | |
|---|---|---|---|---|---|---|---|
| | | 75°C | | 90°C | | | |
| Cond. size | Shield resistance at 25°C, μΩ/ft, | A | * | A | * | 50°C, A | 60°C, A |
| 4/0 | Open | 295 | 63 | 328 | 74 | 245 | 284 |
| 4/0 | 1505 | 292 | 63 | 326 | 75 | 242 | 280 |
| 4/0 | 1003 | 291 | 63 | 324 | 75 | 240 | 279 |
| 4/0 | 502(1/6 neut) | 287 | 63 | 321 | 75 | 236 | 274 |
| 4/0 | 251(1/3 neut) | 281 | 64 | 314 | 76 | 229 | 266 |
| 350 | Open | 387 | 64 | 431 | 76 | 318 | 368 |
| 350 | 910 | 379 | 64 | 422 | 76 | 308 | 358 |
| 350 | 607 | 375 | 65 | 418 | 76 | 304 | 353 |
| 350 | 303(1/6 neut) | 364 | 65 | 407 | 77 | 292 | 341 |
| 350 | 152(1/3 neut) | 348 | 65 | 390 | 78 | 275 | 322 |
| 500 | Open | 470 | 65 | 524 | 77 | 382 | 443 |
| 500 | 637 | 451 | 66 | 504 | 78 | 361 | 420 |
| 500 | 425 | 443 | 66 | 495 | 78 | 352 | 410 |
| 500 | 212(1/6 neut) | 421 | 67 | 473 | 79 | 330 | 386 |
| 500 | 106(1/3 neut) | 395 | 68 | 445 | 81 | 305 | 358 |
| 750 | Open | 586 | 66 | 652 | 78 | 469 | 544 |
| 750 | 425 | 539 | 67 | 605 | 80 | 421 | 492 |
| 750 | 283 | 521 | 68 | 585 | 81 | 403 | 472 |
| 750 | 142(1/6 neut) | 481 | 69 | 543 | 82 | 366 | 430 |
| 750 | 71(1/3 neut) | 446 | 70 | 503 | 83 | 337 | 396 |
| 1000 | Open | 680 | 66 | 757 | 79 | 543 | 630 |
| 1000 | 637 | 635 | 67 | 712 | 80 | 497 | 580 |
| 1000 | 425 | 617 | 68 | 692 | 81 | 478 | 559 |
| 1000 | 212 | 572 | 69 | 645 | 82 | 435 | 511 |
| 1000 | 106(1/6 neut) | 518 | 70 | 587 | 83 | 388 | 457 |
| 1250 | Open | 761 | 67 | 849 | 79 | 605 | 702 |
| 1250 | 509 | 692 | 68 | 778 | 81 | 535 | 626 |
| 1250 | 340 | 665 | 69 | 749 | 82 | 508 | 597 |
| 1250 | 170 | 604 | 70 | 684 | 83 | 453 | 534 |
| 1250 | 85(1/6 neut) | 543 | 71 | 616 | 84 | 402 | 475 |
| 1500 | Open | 834 | 67 | 930 | 80 | 659 | 765 |
| 1500 | 425 | 736 | 69 | 829 | 82 | 562 | 660 |
| 1500 | 283 | 700 | 69 | 791 | 83 | 528 | 622 |
| 1500 | 142 | 626 | 70 | 711 | 84 | 463 | 548 |
| 1500 | 71(1/6 neut) | 560 | 71 | 637 | 85 | 412 | 486 |
| 2000 | Open | 955 | 68 | 1068 | 81 | 747 | 870 |
| 2000 | 318 | 795 | 70 | 900 | 83 | 595 | 701 |
| 2000 | 212 | 744 | 70 | 845 | 84 | 550 | 650 |
| 2000 | 106 | 650 | 71 | 742 | 85 | 472 | 560 |
| 2000 | 53(1/6 neut) | 589 | 72 | 668 | 86 | 430 | 507 |

* Corresponding earth interface temperature, °C.



NOTE: 1 in = 25.4 mm; 1 ft = 0.3048 m.

**TABLE 18-23** Ampacity of Single-Conductor Solid Dielectric Power Cable Installed in Underground Ducts

**Double circuit, balanced 3-phase load, 90 *p*, soil resistivity 15 kV, aluminum conductors, earth ambient temp. 25°C, 100% load factor.**

| | | Max conductor temp | | | | Max. earth interface temp | |
|---|---|---|---|---|---|---|---|
| | | 75°C | | 90°C | | | |
| Cond. size | Shield resistance at 25°C, μΩ/ft | A | * | A | * | 50°C, A | 60°C, A |
| 4/0 | Open | 205 | 54 | 231 | 64 | 192 | 221† |
| 4/0 | 1505 | 204 | 54 | 229 | 64 | 190 | 219† |
| 4/0 | 1003 | 203 | 54 | 228 | 64 | 189 | 218† |
| 4/0 | 502(1/6 neut) | 200 | 54 | 226 | 64 | 186 | 215† |
| 4/0 | 251(1/3 neut) | 196 | 54 | 221 | 64 | 181 | 210† |
| 350 | Open | 270 | 55 | 304 | 65 | 247 | 285† |
| 350 | 910 | 265 | 55 | 298 | 66 | 241 | 279† |
| 350 | 607 | 262 | 56 | 296 | 66 | 238 | 276† |
| 350 | 303(1/6 neut) | 255 | 56 | 288 | 66 | 231 | 268† |
| 350 | 152(1/3 neut) | 243 | 56 | 276 | 66 | 219 | 255† |
| 500 | Open | 328 | 56 | 369 | 67 | 296 | 341† |
| 500 | 637 | 316 | 57 | 356 | 67 | 283 | 328† |
| 500 | 425 | 310 | 57 | 250 | 67 | 277 | 322† |
| 500 | 212(1/6 neut) | 295 | 57 | 335 | 67 | 262 | 306† |
| 500 | 106(1/3 neut) | 276 | 57 | 314 | 68 | 243 | 285† |
| 750 | Open | 408 | 58 | 459 | 68 | 361 | 418† |
| 750 | 425 | 378 | 58 | 428 | 69 | 331 | 386† |
| 750 | 283 | 366 | 58 | 415 | 69 | 319 | 373† |
| 750 | 142(1/6 neut) | 339 | 58 | 386 | 69 | 293 | 343† |
| 750 | 71(1/3 neut) | 311 | 59 | 355 | 70 | 268 | 314† |
| 1000 | Open | 476 | 59 | 535 | 70 | 416 | 481† |
| 1000 | 637 | 448 | 59 | 506 | 70 | 387 | 451† |
| 1000 | 425 | 436 | 59 | 493 | 70 | 376 | 438† |
| 1000 | 212 | 406 | 59 | 462 | 71 | 347 | 407† |
| 1000 | 106(1/6 neut) | 368 | 60 | 420 | 71 | 311 | 367 |
| 1250 | Open | 533 | 59 | 600 | 71 | 461 | 535† |
| 1250 | 509 | 490 | 60 | 555 | 71 | 419 | 489 |
| 1250 | 340 | 473 | 60 | 537 | 71 | 402 | 471 |
| 1250 | 170 | 432 | 60 | 493 | 72 | 364 | 428 |
| 1250 | 85(1/6 neut) | 386 | 61 | 442 | 72 | 323 | 381 |
| 1500 | Open | 584 | 60 | 657 | 71 | 501 | 581 |
| 1500 | 425 | 524 | 60 | 594 | 72 | 442 | 518 |
| 1500 | 283 | 500 | 61 | 569 | 72 | 420 | 494 |
| 1500 | 142 | 450 | 61 | 514 | 73 | 374 | 441 |
| 1500 | 71(1/6 neut) | 399 | 61 | 457 | 73 | 330 | 390 |
| 2000 | Open | 668 | 61 | 753 | 72 | 566 | 658 |
| 2000 | 318 | 570 | 61 | 650 | 73 | 473 | 556 |
| 2000 | 212 | 537 | 62 | 613 | 73 | 442 | 522 |
| 2000 | 106 | 471 | 62 | 540 | 74 | 384 | 455 |
| 2000 | 53(1/6 neut) | 419 | 62 | 480 | 74 | 342 | 404 |

**Caution: These ampacities apply only to the phase sequence of Fig. 18-65.**

* Corresponding earth interface temperature, °C.

† Conductor temperature exceeds 75°C.



NOTE: 1 ft = 0.3048 m.

## FEEDERS FOR RURAL SERVICE

***250. Basic Conditions.*** Rural distribution differs from urban in that consumers are farther apart and load units are generally small. Since distances are great, the primary-system voltage should be of the 15-kV class or higher, and the load per mile being low requires the cost of feeder construction to be as low as is consistent with a reasonable degree of permanence and reliability. One transformer per customer is required in many cases. Rural construction since the late 1930s has made electric service available to practically every farm. Efforts are directed now to bolstering capacity to serve the growing loads.

***251. Poles and Spans.*** Design of overhead lines for rural service differs from that of urban lines in several respects. Costs are reduced by using longer spans and as few accessories as possible. Longer spans mean greater sag and higher poles to get proper clearance at the low point of the span. The increase in sag may, however, be reduced by use of higher tensile stresses in conductors. This is possible when steel is employed in conjunction with copper or aluminum wires. Steel is combined with copper in a high-strength strand known as Copperweld, which has 30 or 40% of the conductivity of a copper conductor of equal size, or in a similar aluminum and steel conductor known as Alumoweld. When greater conductivity is needed, one or more strands of hard copper are stranded with or around the Copperweld or hard aluminum strands around Alumoweld. Steel is also stranded with aluminum wires into ACSR conductor. Such types of conductor have ample conductivity for rural lines, and they have been used widely.

In level country, spans of 400 to 600 ft are practical, while in hilly country spans of 800 to 900 ft are occasionally possible.

***252. Cable.*** Design of underground circuits for rural service is similar to that of urban underground circuits. Concentric-neutral cables are likely to be used in both types of systems. Because the rural circuits have longer uninterrupted runs of cable, there is a better opportunity to plow in the cable rather than digging trenches. Plowing results in lower installed costs per unit length of cable. In fact, some electric suppliers report that the total cost of a rural underground system is less than the cost of an overhead system to serve the same load.

***253. Location of Circuits.*** Rural-service circuits are run along main highways, where the largest number of users may be reached. Branches along intersecting roads are extended as may be warranted by service requirements. In some cases, private rights-of-way, maintained for transmission lines, may be utilized.

***254. Voltage.*** Rural circuits may be extended 5 to 50 mi from the point of supply, and voltage used for primary distribution must be chosen accordingly. Loadings are often so small that the minimum size of conductor required for dependable strength for overhead or cable insulation for underground is sufficient to meet requirements of voltage drop and line loss. This is particularly true when the higher voltages are used.

The most common primary voltage used in rural areas is 12,470Y/7200 V, 4-wire for normal load densities and 24,940Y/14,400 V for very light load densities. There is a trend toward using both 24,940Y/14,400 V and 34,500Y/19,900 V for all types of rural areas.

Single-phase circuits are most economical for the usual light loads found in rural areas and where power units do not exceed 10 hp.

Vee-phase circuits consisting of two phase conductors and the neutral are an economical method of supplying 3-phase loads using open-wye–open-delta transformer banks. Full 3-phase, 4-wire construction will be desirable for many areas.

In some cases there may be relatively small 3-phase loads in a single-phase area. Often these loads can be supplied economically from a single-phase system by means of a phase converter, the output of which is 3-phase voltage.

***255. Limitations of voltage and distance*** are illustrated by the following figures showing kilowatt-miles corresponding to a 5% line drop at 80% power factor for a circuit of 1/0 ACSR, or its equivalent in other metals.

Kilowatts × miles for 5% voltage drop, power factor 80%

| System | 4.16 kV | 12.47 kV | 24.94 kV | 34.5 kV |
|---|---|---|---|---|
| Single-phase | 82 | 737 | 2949 | 5646 |
| 3-phase | 488 | 4375 | 17919 | 33815 |

Values for other sizes are approximately in proportion to relative cross section. In order to determine specific voltage-drop values for 3-phase overhead and underground circuits, refer to Table 18-3.

***256. Conductors and Spans.*** Because of the economy of using long spans, the choice of span lengths and conductor strength is of much importance in planning rural lines. Single-phase lines are commonly taken from a 3-phase system with neutral grounded. The grounded conductor is carried on a bracket about 2 ft below the phase wire, which rests on an insulator carried on the top of the pole. No crossarm is required, except on a main line of more than one phase.

While the conductivity of No. 4 ACSR or Copperweld may be thermally adequate for the greater part of a rural system, system economics can dictate the use of larger conductors. The strength of No. 4 ACSR or Copperweld is usually ample for spans of 350 to 600 ft, depending upon design-loading conditions. Conductors should be sagged in accordance with the conductor manufacturer's recommendations.

***257. Poles.*** The strength of poles should be determined for the height required by the methods described in Pars. **131** to **144**.

The length of poles required in any situation must be such as to allow for depth of setting and height of wire supports needed to give proper clearance above ground at the low point of the span. Such clearances should be not less than value shown in the National Electrical Safety Code (ANSI C2-1984) (see Par. **157**). In the case of road and railroad crossings, the necessary clearance may sometimes be more readily had by placing one end of the span near the crossing, thus avoiding having the low part of the span over the crossing. In rolling or hilly areas, it is desirable to locate poles on higher elevations to permit use of longer spans and greater sags.

Where no ice loading is likely, 30-ft poles can be used for two conductors of a single-phase branch on level ground or on long even slopes with span lengths to 400 ft. This will often preclude, however, the addition of communication circuits to the poles. Where ice and wind loading is expected with some regularity, it is necessary to use 35-ft poles for spans exceeding 300 ft. At corners or angles, poles should be supported by guying or bracing to support unbalanced longitudinal stresses.

***258. Crossarms or equivalent equipment*** are required for the main 3-phase circuits and for lines supplying any user taking 3-phase service for power. A two-pin arm is often used with the third phase on the pole top. The grounded neutral is carried on the side of the pole about 2 ft below the arm.

***259. Transformer Installations.*** Transformers usually supply not more than one or two customers, and sizes, therefore, are small compared with the average used in urban work, 10 to 15 kVA being average for single-phase installations. Where points of use are more than about 500 ft apart, it is usually most economical to provide separate transformers. When two users are within this distance, they can be served by placing a transformer between them and constructing a secondary. Rural loads on some systems have grown to the point where 15- and 25-kVA transformers are required.

Transformer capacity may usually be selected on the basis of loading the transformer to 150% of nameplate rating for peak loads lasting for 1 to 2 h. Pumping for drainage or irrigation is likely to require rated capacity more nearly equal to load.

***260. Stray Voltages.*** Stray voltages on dairy farms may cause lowered milk production and increased mastitis in dairy cattle. Dairy cattle are particularly sensitive to low magnitudes of voltage. Voltages on the order of 1.0 V occurring between metal stanchions or metal drinking cups and the concrete floor may be troublesome. It should be noted that the same symptoms may be due to other causes and that stray voltages are not always to blame.

One characteristic of the common-neutral distribution system, in which the neutral conductor is common to both the primary and secondary systems, is the multiplicity of ground connections between the neutral conductor and earth. Unbalanced load conditions on either the farm secondary system or the utility primary system result in current flow in the neutral conductor. Due to the multiplicity of ground connections, some portion of the neutral current flows in earth. These earth currents cause stray voltages to appear in the earth. Deteriorated insulation on farm wiring and machinery can also cause earth currents, and these causes should be eliminated first.

One way to minimize stray voltages in the dairy barn is to cast wire mesh into the concrete floor and to bond the mesh and all metal structures together to establish equal potentials. In cases where current flow in the utility primary neutral is identified as a cause of stray voltages, it may be necessary to isolate the primary and secondary neutrals of the distribution transformer serving the farm. The National Electrical Safety Code, ANSI C2-1984, addresses this situation in Section 97D and requires a spark gap to interconnect the two neutrals.

## DEMAND AND DIVERSITY FACTORS

***261. Demand Factor.*** The ratio of maximum demand to total load connected, expressed as a percentage, is termed the demand factor of an installation. For example, if a residence having equipment connected with a total rating of 6000 W has a maximum demand of 3300 W, it has a demand factor of 55%. Demand factors of various types of large loads are helpful in designing systems, particularly those in buildings. As an example, a single household electric clothes dryer, of course, has a demand factor of 100%, but 25 dryers in a group have a demand factor of 33%. Similarly, three to five all-electric apartments in a multifamily dwelling have a demand factor of 45%. The lower the demand factor, the less system capacity required to serve the connected load. However, summer air conditioning and winter electric heating are loads that make for high demand factors.

***262. Coincidence or Diversity Factor.*** The coincidence factor is defined as the ratio of the maximum demand of the load as a whole, measured at its supply point, to the sum of the maximum demands of the component parts of a load. The diversity factor is the reciprocal of the coincidence factor. Coincidence factors can be applied to known consumer demands for estimating the loading of distribution transformers, lines, and other facilities. Coincidence factors for residential consumers can vary over a wide range for different types of consumers. The coincidence factor for a large group of consumers with no major appliance might be as low as 30%, whereas a group of electric-heating consumers might be as high as 90%.

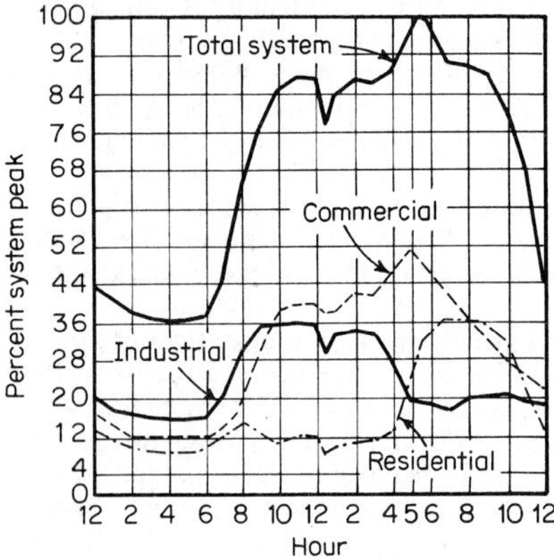


**FIG. 18-66** Characteristic metropolitan load pattern.

***263. Diversity between Classes of Users.*** The daily-load curve of a utility is a composite of demands made by various classes of users. The load curve on the day of maximum total system peak occurs when class loads gang up to create this maximum demand for the year. This is not necessarily the day, and usually is not the day, of any particular class peak. Class load curves on the day of system peak are illustrated in Fig. 18-66.

Air-conditioning loads have shifted these curves for many systems to cause daytime peaks during hot weather in the summer. Electric house heating builds heavy morning and evening loads during cold weather in the winter.

**TABLE 18-24** Diversity Factors

| Elements of system between which diversity factors are stated: | Diversity factors for | | | |
|---|---|---|---|---|
| | Residence lighting | Commercial lighting | General power | Large users |
| Between individual users | 2.0 | 1.46 | 1.45 | |
| Between transformers | 1.3 | 1.3 | 1.35 | 1.05 |
| Between feeders | 1.15 | 1.15 | 1.15 | 1.05 |
| Between substations | 1.1 | 1.10 | 1.1 | 1.1 |
| From users to transformer | 2.0 | 1.46 | 1.44 | |
| From users to feeder | 2.6 | 1.90 | 1.95 | 1.15 |
| From users to substation | 3.0 | 2.18 | 2.24 | 1.32 |
| From users to generating station | 3.29 | 2.40 | 2.46 | 1.45 |

***264. Diversity in the Feeder System.*** The diversity of demands by transformers on a radial feeder makes the maximum load on the feeder less than the sum of the transformer loads. The diversity factors of a feeder vary greatly depending upon load conditions. Some typical diversity factors are given in Table 18-24. The diversity factor of lighting feeders ranges from 1.1 to 1.5, while that of mixed light-and-power feeders is likely to be 1.5 to 2 or more. At the substation there is also a diversity factor of 1.05 to 1.25 between the sum of feeder maxima and the substation maximum. A large system has a further diversity factor between substations of 1.05 to 1.25.

***265. Total diversity factors*** in a large system are somewhat as in Table 18-24.

## DISTRIBUTION ECONOMICS[1]

***266. Economic Comparisons.*** The most straightforward and generally applicable technique to use in distribution-system investment problems is that of making economic comparisons on the basis of the present value of all future annual costs. That is, the economic choice is the one with the lowest present value of all future costs. With this as a criterion, the procedure for making an economic comparison between alternatives is a simple two-step operation, that is,

**a.** Estimate for each alternative the annual costs for each year.

**b.** If annual costs are not uniform, calculate their present value.

***267. Time Value of Money.*** Money does have time value, and rent or interest on its use has to be paid. It is obvious that an alternative which requires the least expenditure immediately would be best, everything else being equal.

The process of taking money and finding its equivalent value at some future time is called a "future worth" or "future value" calculation. This calculation is the same as that used in determining the effect of compound interest.

If 8% is the established interest rate, then \$100 today is equivalent to \$100(1 + 0.08) or \$108 a year from now, and $100(1 + 0.08) + 100(1 + 0.08) \times 0.08 = 100(1 + 0.08)^2$ 2 years from now, and $100(1 + 0.08)^{10}$ 10 years from now. The expression $(1 + i)^n$ is called the single payment compound amount factor, where $i$ is the interest rate and $n$ is the number of years. These factors and others are readily available for various interest rates and number of years in economic books such as *Principles of Engineering Economy* by Eugene L. Grant.

It should be noted that the use of 8% for the interest rate is for illustrative purposes only. The actual interest rate to use will be determined by the economic conditions at the time.

Hence, to find the future worth of \$100, 10 years later in the above example, first, look up the compound amount factor in the 8% interest table for year 10, then multiply it by 100. The compound amount factor for this case is 2.159 and the future worth calculates to be 100(2.159) = \$215.90.

The process of finding the equivalent value of money at some earlier time is called a "present worth" or "present value" operation.

The present worth calculation is the reverse of the future worth calculation. If \$100 today has a future worth a year from now of \$108, then we can also say that \$108 a year from now has a present worth of \$100 today. The present worth factor is the inverse of the future worth factor, and it also may be found in interest tables. Since the future worth factor for $n$ years is $(1 + i)^n$, where $i$ is the interest rate, the present worth factor is $1/(1 + i)^n$.

To determine the present worth, as of today, of a \$100 cost anticipated to be incurred 2 years from now where the interest rate is 8%, first the present worth factor of 0.8573 is obtained from interest tables. Then multiplying this factor by \$100 gives the present worth of \$100(0.8573) = \$85.73.

Formulas for calculating the compound interest factors and a graphical interpretation of these factors are shown in Fig. 18-67.

***268. Annual Charges.*** It is desirable to have a convenient method of calculating the annual costs of capital investments made in an alternative scheme. Fortunately, this can be done by using a level carrying charge which is expressed as a percentage of the original investment.

[1] Also see Sec. 13.

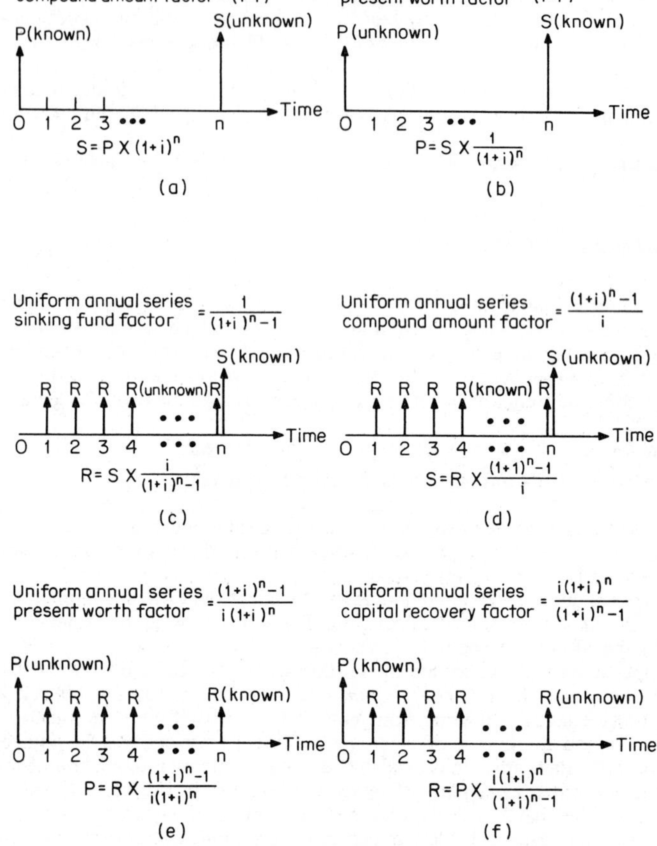


**FIG. 18-67** Graphical interpretation of compound interest factors.

The total revenue requirements of a piece of equipment are the sum of the annual charges for:

**a.** Return on investment

**b.** Depreciation

**c.** Income tax

**d.** Property taxes

**e.** Insurance

**f.** Operating and maintenance expenses

The first five of these charges can conveniently be estimated as a percentage of original investment. The operating and maintenance charges should be separately estimated for each project because they do not relate to capital investment as a percentage.

***269. Level Annual Carrying Charges.*** The level annual carrying charge is the percentage by which the capital investment can be multiplied to determine its annual cost on a uniform basis. The value of this carrying charge is very much dependent upon the expected life of the piece of equipment because depreciation varies in accordance with life expectancy. A method of obtaining the level annual carrying charge is as follows: (1) calculate the sum of the annual charges for return on investment, depreciation, income tax, property tax, and insurance for each year of the expected life of the piece of equipment; (2) use the appropriate present worth factor with each annual cost to convert the annual cost to a present worth value; (3) sum up these values to obtain the total present worth of the annual carrying charges, and (4) multiply the total present worth by the capital recovery factor (see Fig. 18-67) to get the equivalent uniform annual charge. Figure 18-68 shows graphically the actual and equivalent carrying charges for a capital investment of a piece of equipment with a 5-year life and an assumed 8% cost of money.

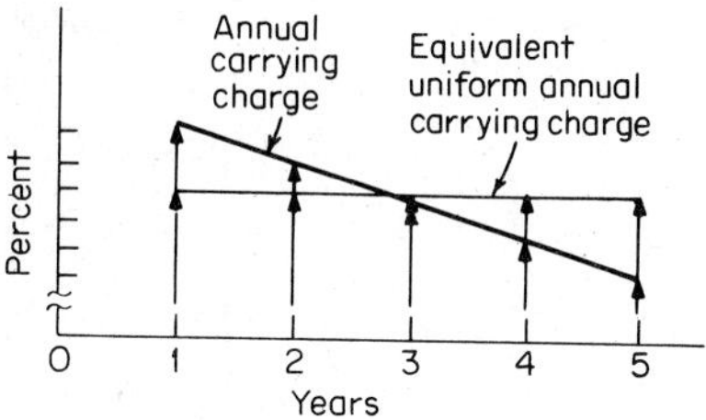


**FIG. 18-68** Representation of carrying charges.

The total carrying charges with 8% cost of money for various service life are estimated as follows:

| Years of life | Level annual total carrying charge in % |
|---|---|
| 5 | 30.82 |
| 10 | 20.59 |
| 15 | 17.44 |
| 20 | 16.04 |
| 25 | 15.34 |
| 30 | 14.96 |
| 35 | 14.76 |
| 40 | 14.67 |
| 45 | 14.63 |
| 50 | 14.63 |

*270. **Operating and Maintenance Expenses.*** This cost component varies with the nature of the project. It is usually not a direct function of the capital invested and may have an inverse tendency. That is, alternatives often exist for higher capital expenditures to reduce operating costs. Therefore, it is *not* expressed as a percent of capital investment in most cases. Nevertheless, it should be included in annual costs.

*271. **Study Period.*** When determining the economic comparison of alternates by comparing the present worth of annual costs, the study period should be taken to the point that the alternates are equivalent in capability. If this is not practical, the study should be taken so far into the future that the difference in present worth would be insignificant.

*272. **Economic Evaluations.*** A simple example will show a comparison between two alternatives. Let *CC* represent the capital investment multiplied by the level annual carrying charge. *O&M* represent annual operation and maintenance, and *RR* represent the total revenue requirement necessary annually to carry the project. A pad-mounted sectionalizing switch is needed for an underground circuit. The choice is between two manufacturers who can supply the switch but with different characteristics as follows:

| | Mfr. A | Mfr. B |
|---|---|---|
| Installed cost | $3600 | $3300 |
| Operating and maintenance | 50/year | 100/year |
| Expected life | 30 years | 20 years |

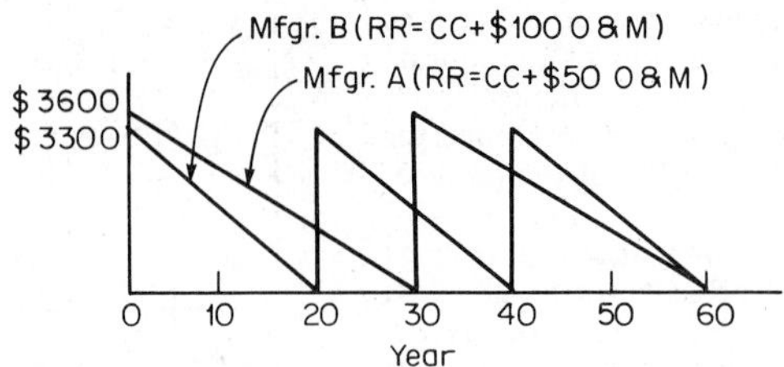


**FIG. 18-69** Time diagram.

There is no salvage value at end of life. Determine which alternative is less expensive.

The first step is to draw a time diagram like Fig. 18-69. The common point in time for the two alternatives is 60 years, so two cycles of $A$ should be compared with three cycles of $B$.

Present worth analysis:

$$\begin{aligned}\text{PW Mfr. } A \text{ alternative} &= 3600 \times 0.1496 \times 11.258 + 50 \times 11.258 + \\ &\quad (3600 \times 0.1496 \times 11.258 + 50 \times 11.258)\, 0.0994 \\ &= 6063.11 + 562.90 + 658.63 \\ &= \underline{\underline{7284.64}}\end{aligned}$$

$$\begin{aligned}\text{PW Mfr. } B \text{ alternative} &= 3300 \times 0.1604 \times 9.818 + 100 \times 9.818 + \\ &\quad (3300 \times 0.1604 \times 9.818 + 100 \times 9.818)\, 0.2145 + \\ &\quad (3300 \times 0.1604 \times 9.818 + 100 \times 9.818)\, 0.0460 \\ &= 5196.86 + 981.80 + 1325.32 + 284.22 \\ &= \underline{\underline{7788.20}}\end{aligned}$$

where 3600 = installed cost of Mfr. $A$ switch
0.1496 = level annual carrying charge for 30-year $A$ switch
11.258 = 8%, 30-year uniform annual series present worth factor
50 = *O&M* of $A$ switch
0.0994 = 8%, 30-year single-payment present worth factor
3300 = installed cost of Mfr. $B$ switch
0.1604 = level annual carrying charge for 20-year $B$ switch
9.818 = 8%, 20-year uniform annual series present worth factor
100 = *O&M* of $B$ switch
0.2145 = 8%, 20-year single payment present worth factor
0.0460 = 8%, 40-year single payment present worth factor

Manufacturer $A$ switch would be the overall lowest cost and would be the better deal provided the capability and reliability of the two switches are equivalent.

## DISTRIBUTION-SYSTEM LOSSES

***273. General.*** About 8% of the total output of a large power system is lost or unaccounted for. Much of this loss is in the distribution system. Since investment must be made in facilities to supply these losses, they should be an important consideration in the engineering design of the system. A knowledge of their magnitude is essential and they should not be omitted from overall comparisons of alternative facilities without a study of each specific situation.

***274. Line Losses.*** The line losses, which are the sum of the $I^2R$, or resistance losses, can be easily found when the currents at peak load are known. Simplifying assumptions can often be made in making these calculations. For instance, if the load can be considered as being uniformly distributed, the losses are the same as if the total load were concentrated at a point one-third of the way out on the feeder.

***275. Transformer Losses.*** Transformers have a no-load loss as well as a load loss. The transformer no-load loss is independent of load, whereas the load loss will vary as the square of the current. These losses for distribution transformers are usually published as no-load and total loss when the transformer is operating at rated voltage and rated kVA. The load loss at full-load current is the difference between total and no-load losses.

***276. Working Principles.*** The problem of converting kWh of lost energy to dollars and cents has resulted in considerable controversy among system operators, because of the difficulty of determining the value of the energy. It is not the purpose of this handbook to take sides in the controversy, but rather to show the principles involved so that engineers will be able to evaluate losses using appropriate system costs.

The cost of supplying losses can be broken down into two major parts:

**a.** Energy component, or production cost to generate kWh losses.

**b.** Demand component, or annual costs associated with system investment required to supply the peak kW of loss.

The two components of cost are usually combined into a single figure either in terms of cents per kilowatthour of total energy loss or as dollars per kilowatt of peak loss. Expressing losses in terms of dollars per kilowatt is usually called "capitalized" cost of losses, and it has some advantage in that it shows directly the amount of money that could be economically spent to save 1 kW of loss. However, the expression of cost of losses in cents per kilowatthour is usually a more convenient form to use in most engineering studies.

The cost of losses depends on the point in the system at which they occur. The farther out on the system the greater value losses have. One kilowatt of loss saved on the secondary system is worth more than 1 kW loss at generation because of the cumulative effect of increments of losses as they pass through various elements of the system.

In calculating loss, present-day or future cost of system investment should be used. The primary interest is to find the incremental investment, in dollars, required to supply an incremental load in kilowatts.

Opinions differ widely as to the degree to which the demand component of losses shall be evaluated. This ranges all the way from the dollar cost per kilowatt for future system expansion to no value at all for this component.

The great majority of utility engineers prefer to assign full value to the demand component of losses.

***277. Responsibility Factor.*** Owing to diversity between classes of loads (i.e., residential, industrial, etc.) on a distribution system, peak loads on distribution, transmission, and generation do not usually occur at the same time. Therefore, a loss which contributes 1 kW to the distribution-system peak might contribute less than this to transmission- and production-plant peak because its maximum does not occur at the same time as the transmission or generation peak. This introduces "peak responsibility factors" used for evaluating cost of losses in various parts of the system.

***278. Loss Factor.*** If the peak conductor losses of line or transformer have been calculated, it will still be necessary to know the loss factor or percent equivalent hours before it is possible to calculate the actual losses over a period of time.

Loss factor is usually defined as the ratio of the average power loss, over a designated period of time, to the maximum loss occurring in that period. The term can refer to any part or all of the electric system. It is sometimes referred to as the "load factor of the losses."

A corollary to loss factor is the term "equivalent hours." This is defined as the number of hours per day, week, month, or year of peak load necessary to give the same total kilowatthours of loss as that produced by the actual variable load over the selected period of time. The period of time for distribution studies is usually 1 year, and it is obvious that "percent equivalent hours" has the same meaning as the term "percent loss factor."

***279. Relation between Loss Factor and Load Factor.*** Definitions of "loss factor" and "load factor" are quite similar. Care should be taken that the latter is not used in place of loss factor when considering system losses. There is a relationship between the two factors which is dependent upon the shape of the load curve. Because resistance losses vary as the square of the

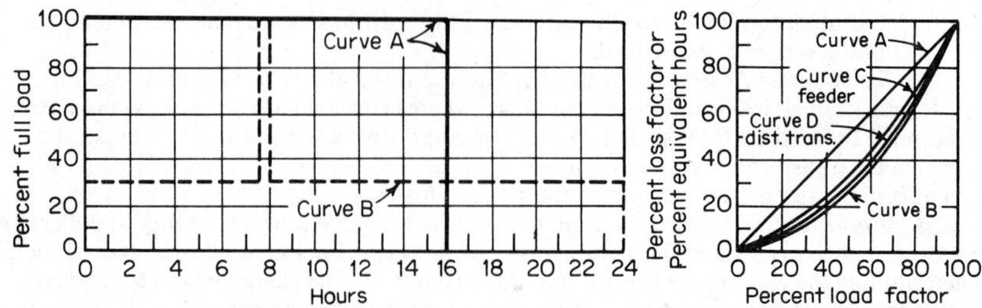


**FIG. 18-70** Relationship between load factor and loss factor or equivalent hours.

load, it can be shown that the value of loss factor can vary between the extreme limits of load factor and load factor squared. A number of typical load curves have been studied to determine this relationship for distribution feeders and distribution transformers. The relation is shown in Fig. 18-70. Note that loss factor is always less than load factor except where they are both unity, as would be the case for transformer core losses. The relationship between load factor and loss factor at the distribution transformer can be expressed by the empirical formula:

$$\text{Loss factor} = 0.15 \text{ load factor} + 0.85 \text{ (load factor)}^2$$

It should be noted that when the shape of the load curve is known or can be reasonably estimated, the loss factor should be calculated directly and not determined by the empirical formula.

***280. Cost of Losses.*** The two parts of the cost to supply losses are as follows:

$$\text{Energy component} = 8760 F_L E$$

$$\text{Demand component} = F_S P$$

where $F_L$ = loss factor of load
$E$ = cost of energy, dollars/kWh
$F_S$ = responsibility factor
$P$ = annual cost of system capacity, dollars/kW-year

Annual cost of losses can be combined into one value, in terms of either dollars per kilowatt-hour or dollars per kilowatt-year of peak loss with the following formulas:

$$\text{Cost of losses, \$/kWh} = \frac{F_S P}{8760\, F_L} + E$$

$$\text{Capitalized cost of losses, \$/kW-year} = F_S P + 8760\, F_L E$$

## STREET-LIGHTING SYSTEMS

***281. Characteristics.*** The lighting of streets, parkways, and other roadways is about the only service in which the electric utility is often responsible for the utilization equipment. This involves the complete service of installation, maintenance, and operation of lighting systems during the hours of darkness (approximately 4000 h) when they are required. Series (constant-current) circuits, historically a common supply for street lighting, have become obsolete.

***282. Multiple Circuits.*** Street-lighting units today are normally supplied directly from the local distribution (120/240 V). The high-intensity-discharge lamps used have compatible ballasts for all common voltages 120, 208, 240, 277, 480 V, designed to strike an arc within the light source and provide stable operating conditions. Ballasts may be high-power-factor or normal-power-factor types. Photoelectric controls are most frequently used integrally with individual lights but may also be used to switch contactors controlling circuits used for lighting only. An example of this is highway lighting where extended systems from a power-supply point are normally designed with 480 V.

***283. Lamps and Luminaires.*** Present street-lighting systems are designed using high-intensity-discharge sources. The three principal types are clear or phosphor-coated mercury, metal halide, and high-pressure sodium. These lamps are available in several sizes ranging from less than 100 to 1000 W. Metal halide is not widely used because of its short life and poor lumen maintenance. Mercury lighting was the most popular type, since it was extensively used to replace older incandescent and fluorescent systems in the recent past. However, high-pressure sodium is the newest and most efficient lamp available. Its efficiency is over twice as high as mercury. The compact arc also allows for better control of the light distribution by the luminaire. The higher lamp efficiency and better control reduce the street-lighting power requirements to less than half that required for mercury. High-pressure sodium is taking over as the leading lighting system because of its economics.

Luminaires are sealed and also can be filtered. This minimizes the light loss due to dirt collecting in the luminaire. This is done to match the luminaire dirt depreciation so it matches the 4-year lamp life and minimizes the cleaning required during relamping.

There is a large variety of street-lighting equipment. This is required to fit different mounting heights, street widths, and lamp wattage. There are also differences in daytime appearance that are needed to fit the needs of the environment.

***284. Underground Systems.*** While most utility-owned lighting systems were originally attached to existing wood-pole overhead distribution lines, these convenient supports are increasingly on rear lot lines or nonexistent (underground distribution). This means that public lighting systems must be designed with underground supply run from the nearest transformer, joint trench, or handhole. Integrating the street-lighting circuits properly with the overall underground system from the outset is essential for proper economics and cost control.

## Bibliography

### *285. Books for General Reference*

Lewis, Walter W.: *Protection of Transmission Systems against Lightning;* New York, John Wiley & Sons, Inc., 1950.

Edison Electric Institute: *Underground Systems Reference Book;* New York, Edison Electric Institute, 1957.

National Electric Light Association: *Overhead Systems Reference Book;* New York, National Electric Light Association, 1927.

*Electrical Transmission and Distribution Reference Book;* East Pittsburgh, Pa., Westinghouse Electric Corporation, 1964.

*Electric Utility Engineering Reference Book—Distribution Systems;* East Pittsburgh, Pa., Westinghouse Electric Corporation, vol. 3, 1965.

Skrotzki, B. G. A.: *Electric Transmission and Distribution;* New York, McGraw-Hill Book Company, 1954.

*Distribution Data Book;* GET-1008M; Schenectady, N.Y., General Electric Company, 1980.

*Overcurrent Protection for Distribution Systems;* GET-1751A; Schenectady, N.Y., General Electric Company, 1962.

### *286. References to Transactions of IEEE (formerly AIEE)*

*System*

Coleman, J. O., and Davis, R. F.: Inductive Co-ordination of Common-Neutral Power Distribution Systems and Telephone Circuits; 1937, p. 17.

Bankus, H. M., and Gerngross, J. E.: Unbalanced Loading and Voltage Unbalance on Three-Phase Distribution Transformer Banks; 1954, vol. 73, pt. III, p. 367.

Bankus, H. M., and Gerngross, J. E.: Combined Single-Phase and Three-Phase Loading of Open-Delta Transformer Banks; *Power Apparatus and Systems,* February 1958, pp. 1337–1343.

Brieger, L., Xenis, C. P., Bisson, A. J., and DeLellis, J.: Distribution Equipment Used on 265/460-Volt Networks and Its Operating Features; 1954, vol. 73, pt. III, p. 1525.

Easley, J. H., and Shula, W. E.: Cost and Reliability Evaluation of Four Underground Primary Distribution Feeder Plans; Transactions Paper, Conference Record—1974 Underground Transmission and Distribution Conference, 74-CH0832-6-PWR, pp. 436–443 (for future publication in *Power Apparatus and Systems*).

Mitchell, C. F., Sweeney, J. O., and Cantwell, J. L.: An Economic Analysis of Distribution Transformer Application; *Power Apparatus and Systems,* December 1959, pp. 1196–1202.

### System Planning

Campbell, H. E., Ender, R. C., Gangel, M. W., and Talley, V. C.: Economic Analysis of Distribution Systems; *Power Apparatus and Systems,* August 1960, pp. 423–443.

Anderson, A. S., and Thiemann, V. A.: Distribution Secondary Conductor Economics; *Power Apparatus and Systems,* February 1960, pp. 1839–1843.

Van Wormer, F. C.: Some Aspects of Distribution Load Area Geometry; *Power Apparatus and Systems,* December 1954, pp. 1343–1349.

Blake, D. K.: Some Observations on the Economic Benefits in Going from One System Voltage Level to a Higher System Voltage Level; *Power Apparatus and Systems,* vol. 71, pt. III, pp. 585–592.

Smith, J. A.: Determination of Economical Distribution Substation Size; *Power Apparatus and Systems,* October 1961, pp. 663–670.

Smith, J. A.: Economics of Primary Distribution Voltages of 4.16 through 34.5 kV; *Power Apparatus and Systems,* October 1961, pp. 670–683.

Webler, R. M., Gangel, M. W., Carter, G. K., Zeman, A. L., and Ender, R. C.: Secondary Distribution System Planning for Load Growth; *Power Apparatus and Systems,* December 1963, pp. 908–927.

Sarikas, R. H., and Thacker, H. B.: Distribution System Load Characteristics and Their Use in Planning and Design; *Power Apparatus and Systems,* August 1957, pp. 564–573.

### Overvoltage and Overvoltage Protection

Task Force Report—Investigation and Evaluation of Lightning Protective Methods for Distribution Circuits—Part I, Model Study and Analysis; Part II—Application and Evaluation; *Power Apparatus and Systems,* August 1969, vol. PAS-88, no. 8, pp. 1232–1247.

Hopkinson, R. H.: Ferroresonant Overvoltage Control Based on TNA Tests on Three-Phase Wye-Delta Transformer Banks; *Power Apparatus and Systems,* February 1968, vol. PAS-87, no. 2, pp. 352–361.

Hopkinson, R. H.: Ferroresonant Overvoltage Control Based on TNA Tests on Three-Phase Delta-Wye Transformer Banks; *Power Apparatus and Systems,* October 1967, vol. PAS-86, no. 10, pp. 1258–1265.

Hopkinson, R. H.: Ferroresonance during Single-Phase Switching of Three-Phase Distribution Transformer Banks; *Power Apparatus and Systems,* April 1965, vol. PAS-4, pp. 289–293. Discussion, June 1965, pp. 514–517.

Hendrickson, P. E., Johnson, I. B., and Schultz, N. R.: Abnormal Voltage Conditions Produced by Open Conductors on Three-Phase Circuits Using Shunt Capacitors; *Power Apparatus and Systems,* 1953, vol. 72, pt. III, p. 1183.

Clayton, J. M., and Hileman, A. R.: A Method of Estimating Lightning Performance of Distribution Lines; *Power Apparatus and Systems,* 1954, vol. 73, pt. III, p. 953.

Powell, R. W.: Lightning Protection of Underground Residential Distribution Circuits; *Power Apparatus and Systems,* September 1967; vol. PAS-86, no. 9, pp. 1052–1056.

### Overcurrent and Overcurrent Protection

Slepian, J., and Strom, A. P.: Arcs in Low-Voltage A-C Networks; *Power Apparatus and Systems,* 1931, p. 847.

Xenis, C. P.: Short-Circuit Protection of Networks by the Use of Limiters; *Power Apparatus and Systems,* 1937, p. 1191.

Linder, F. W.: Graphical Method of Calculating Fault Currents on Rural Distribution Systems; *Power Apparatus and Systems,* 1945, p. 16.

Auer, G. G., Ender, R. C., and Wylie, R. A.: Digital Calculation of Sequence Impedances and Fault Currents for Radial Primary Distribution Circuits; *Power Apparatus and Systems,* February 1961, pp. 1264–1277.

Johnson, W. H., and Meler, T. J.: Distribution Circuit Protection, American Electric Power Company; *Power Apparatus and Systems,* February 1960, pp. 1833–1839.

Gray, D. M.: Internal Fault Tests on Distribution Transformers; Transactions Paper T74-480-0, *Power Apparatus and Systems* (to be published in a future issue).

Harner, R. H.: Distribution System Recovery Voltage Characteristics. Part I—Distribution Transformer Secondary-Fault Recovery Voltage Investigation; *Power Apparatus and Systems,* February 1968, vol. PAS-87, no. 2, pp. 463–487.

Harner, R. H., and Rodriguez, J.: Transient Recovery Voltages Associated with Power-System, Three-Phase Transformer Secondary Faults; *Power Apparatus and Systems,* September/October 1972, vol. PAS-91, no. 5, pp. 1887–1896.

Arndt, R. H., Koch, R. E., and Schultz, N. R.: Concept Alternatives and Application Considerations in the

Use of Current-Limiting Fuses for Transformer Protection; Transactions Paper, Conference Record-1974 Underground Transmission and Distribution Conference, 74-CH0832-6-PWR, pp. 259–267 (for future publication in *Power Apparatus and Systems*).

*Voltage Regulation and Kilovar Supply*

Johnson, I. B., Schultz, A. J., Schultz, N. R., and Shores, R. B.: Some Fundamentals on Capacitance Switching; *Power Apparatus and Systems,* August 1955, pp. 727–736.

Darrow, K. G., Phillips, V. E., Schultz, A. J., and Were, A. E.: Test Circuits for Capacitance Switching Devices; August 1955, vol. 71, pt. III, pp. 624–635.

Neagle, N. M., and Samson, D. R.: Loss Reduction from Capacitors Installed on Primary Feeders; *Power Apparatus and Systems,* October 1956, pp. 950–959.

*Underground Systems*

Barger, J. V., and Smith, D. R.: Impedance and Circulating Current Calculations for UD Multi-Wire Neutral Circuits; *Power Apparatus and Systems,* May–June 1972, vol. PAS-91, no. 3, pp. 992–1006.

Schultz, N. R., and Thomas, J. F.: Underground Distribution Thermal Tests in the Phoenix Area; *Power Apparatus and Systems,* September 1967, vol. PAS-86, no. 9, pp. 1042–1051.

*Grounding*

Application Guide on Methods of Substation Grounding; AIEE Group on Substation Grounding Practices, 1954, vol. 73, pt, III, p. 271.

***287. Standards and Standards Publications***

American National Standards Institute (ANSI) C1-1984, National Electric Code.

American National Standards Institute (ANSI) C2-1984, National Electrical Safety Code.

ANSI C84.1-1982, Voltage Ratings for Electric Power Systems and Equipment (60-Hz).

ANSI 05.1-1979, Specifications and Dimensions for Wood Poles.

AIEE-IPCEA (Insulated Power Cable Engineers Association); AIEE Pub. S-135-1; IPCEA Pub. P-46-426; "Power Cable Ampacities, Vol. I—Copper Conductors," 1962.

AIEE-IPCEA; AIEE Pub. S-135-2; IPCEA Pub. P-46-426, "Power Cable Ampacities, Vol. II—Aluminum Conductors," 1962.

IPCEA-NEMA (National Electrical Manufacturers Association); IPCEA Pub. P-53-426; NEMA Pub. WC50-1972, "IPCEA-NEMA Standards Publication—Ampacities Including Effect of Shield Losses for Single-Conductor Solid-Dielectric Power Cable, 15 kV through 35 kV (Copper and Aluminum Conductors)," 1972.

IPCEA Pub. S-66-524, NEMA Pub. WC7-1971, "IPCEA-NEMA Standards Publication-Cross-Linked-Thermosetting-Polyethylene-Insulated Wire and Cable for the Transmission and Distribution of Electric Energy," 1970.

Clapp, A. L.: *National Electrical Safety Code Handbook—1984,* New York, IEEE, 1984.

***288. References to Periodicals***

*System*

Blake, D. K.: Low-Voltage A-C Networks; *Gen. Electr. Rev.,* February, March, April 1928, pp. 82, 140, 186.

Urge Utilities Study Flicker Limits; *Electr. World,* Nov. 3, 1958, p. 49.

Shuptar, Raymond: Raise Voltage Flicker Limits 20–60%; *Electr. World,* June 26, 1961, p. 52.

Gangel, M. W., and Propst, R. F.: Investigating Distribution Transformer Load Characteristics; *Distribution Magazine,* July 1961, p. 6.

Robinson, P. B., and Spencer, J. H.: New Advances in Automatic Metering; *Distribution Magazine,* fall 1973, p. 12.

Frey, L. T. III, and Waldon, P. L.: Cyclic Loading of Direct-Buried Transformers; *Distribution Magazine,* January 1970, p. 3.

Klein, K. W.: Computer Gives Economic Loading for Transformer and Secondary; *Electr. World,* Nov. 30, 1959, p. 83.

*System Planning*

Van Wormer, F. C.: Design and Operation of Spot Networks; *Distribution Magazine;* 2d/3d Quarter 1966, p. 5; 4th Quarter, 1966, p. 19.

Campbell, H. E.: Serving Critical Loads; *Distribution Magazine,* 4th Quarter, 1966, p. 9.

Brown, P. G., Propst, H. R., and Tice, J. B.: Unity Power Factor Is Essential to Emergency Kilowatt Transportation; *Electric Forum Magazine,* fall 1975, p. 10.

Hayes, R. H., and Hill, O. L.: Progress in Remote Line Switch Control; *Transmission and Distribution,* June 1975, p. 52.

*Overvoltage and Overvoltage Protection*

Auer, G. G.: Basic Considerations in Lighting Protection of URD Systems; *Distribution Magazine,* April 1968, p. 16.

*Overcurrent and Overcurrent Protection*

Howard, S. B. and Stroebel, R. W.: Can Single-Phase Cutouts Be Applied to Three-Phase Circuits; *Distribution Magazine,* 2d Quarter 1964, p. 4.

Lasseter, J. A.: Burndown Tests on Bare Conductors; *Electr. Light and Power,* Dec. 15, 1956, p. 94.

*Voltage Regulation and Kilovar Supply*

Gangel, M. W.: Compensator Settings Made Easier; *Distribution Magazine;* pt. 1, April 1960, p. 22; pt. 2, July 1960, p. 18.

Schultz, N. R.: Calculating Voltage Drop and Power Loss; *Distribution Magazine,* January 1969, p. 11.

*Underground Systems*

Kemnitz, L. A. and Smith, J. C.: Power and Telephone in Common Trench Cuts Underground Cost; *Electr. World,* Jan. 15, 1962, p. 48.

Thomas, J. F.: Soil Heat Tests Yield Useful Thermal Data; *Electr. World,* Feb. 22, 1965, p. 90.

Chiles, R. E., and Smith, B. E.: 34.5 kV Cable Used for URD Primary; *Electr. World,* Mar. 22, 1965, p. 111.

Medek, J. D. and Steeve, E. J.: Devices Help Find URD Cable Faults; *Electr. World,* Oct. 4, 1965, p. 102.

Xenis, C. P.: Aluminum Conductor Use for Networks Widens; *Electr. World,* Jan. 10, 1966, p. 50.

Van Wormer, F. C.: Underground Distribution Systems for Residential Areas; *Distribution Magazine;* pt. 1, January 1959, p. 3; pt. 2, April 1959, p. 12; pt. 3, April 1960, p. 16; pt. 4, April 1962, p. 3; pt. 5, April 1963, p. 22.

*Grounding*

Jensen, Claude: Grounding Principles and Practice; Establishing Grounds; *Electr. Eng.,* February 1945, p. 68.

### *289. References to Miscellaneous Publications*

*System*

Gangel, M. W., and Propst, R. F.: Transformer Characteristics Correlated to Loading; Power Distribution Conference, University of Texas, October 1963.

"Specifications and Drawing for 7.2/12.5 kV Line Construction," REA Form 802, Rural Electrification Administration, U.S. Department of Agriculture, revised August 1962. (For sale by the Superintendent of Documents, Government Printing Office, Washington, D.C. 20402.)

"Description of Units, Specifications, and Drawings for 14.4/24.9 kV Line Construction," REA Form 803, Rural Electrification Administration, U.S. Department of Agriculture, revised December 1956. (For sale by the Superintendent of Documents, Government Printing Office, Washington, D.C. 20402.)

Beaty, H. Wayne: 10th Annual T&D Construction Survey; *Electr. World,* Sept. 1, 1975, pp. 35–42.

*System Planning*

*Electr. World,* Load Growth Forces Higher Voltages; June 1, 1974, pp. 154–163.

Campbell, H. E.: Today and Tomorrow; Underground Distribution to High Rise Buildings; IEEE Conference Record—Special Technical Conference on Underground Distribution, 31C35, September 1966, pp. 223–239.

Crawford, J. W. and Hamner, F. G.: "Demand and Diversity Characteristics of Residential Loads"; Southeastern Electric Exchange, Engineering and Operating Conference, April 1963.

*Overvoltage and Overvoltage Protection*

"Distribution Surge Arrester Seminar Text"; Publication PEP-51, General Electric Company, Pittsfield, Mass., 1973.

### *Overcurrent and Overcurrent Protection*

Beaty, H. Wayne: Special Report—Switching and Overcurrent Protection for Distribution Systems; *Electrical World,* Apr. 1, 1974, pp. 41–56.

Campbell, H. E.: "Implication of Increased Short-Circuit Duty on Residential Distribution Systems"; American Power Conference, vol. 35, 1973, pp. 1098–1104.

### *Underground Systems*

"Specifications and Drawings for Underground Electric Distribution"; REA Form 806, Rural Electrification Administration, U.S. Department of Agriculture, January 1975.

Lewis, S. M.: URD Survey Report; *Transmission and Distribution,* July 1973, pp. 88–95.

"IEEE Conference Record—1974 Underground Transmission and Distribution"; 74CH0832-6-PWR and 74CH0832-6-PWR (SUP.), Apr. 1–5, 1974.

"Underground Corrosion Control Guide," NRECA Research Project, August, 1982.

# SECTION 19

# WIRING DESIGN FOR COMMERCIAL AND INDUSTRIAL BUILDINGS

**Allen L. Clapp, P.E.**
*Managing Director, Clapp Research Associates*

**Wendell G. Leisinger**
*Manager, Circuit Breaker Marketing Section, Square D Company*

**R. H. (Bob) Wood**
*Manager, Busway Marketing, Square D Company*

## CONTENTS

*Numbers refer to paragraphs*

## BASIC INSTALLATION RULES AND INSPECTION

*1.* ***Codes.*** *The National Electrical Code (NEC)* establishes the minimum standards of wiring design and installation practice for consumer-owned wiring and equipment in the United States. Its rules are written to protect the public from fire and life hazards. It is revised periodically by a committee drawn from industry associations, insurance groups, organized labor, and representatives of municipalities. It is sponsored by the National Fire Protection Association, and approved by the American National Standards Institute as ANSI C1. It forms the basis of the vast majority of municipal electrical wiring ordinances, which adopt successive editions of the Code as issued.

The National Electrical Safety Code (NESC) establishes the minimum standards of electric supply system design and installation for utility-owned conductors and equipment in the United States. It is also revised periodically by a committee drawn from utility groups, industries, state and federal regulators, insurance groups, organized labor and other interested parties. Its secretariat is the Institute of Electrical and Electronics Engineers; the NESC is American National Standard ANSI C2. The NEC oversees supply and communication wiring that are in and on consumer-owned buildings but not an integral part of a generating plant, substation, or control center. The NEC does not cover communication utility wiring, nor does it cover electric utility generation, transmission, or distribution system wiring. The NESC covers the latter systems. The NESC also covers similar systems under the control of qualified persons, such as those associated with large industrial complexes. In recent years, the provisions of the NESC relating to underground wiring have become increasingly applicable in commercial complexes as extremely large commercial complexes have become more frequent. Some of the latter systems are not unlike those utility systems found in small towns or compact subdivisions.

Lists of Inspected Electrical Equipment and Appliances are issued yearly by the *Underwriters' Laboratories, Inc. Electrical Testing Laboratories, Inc.* and *Factory Mutual Engineer ing Corp.* are other testing laboratories that function as third-party certifiers of the basic safety of manufactured products used in electrical work. One function of the laboratories is to examine and pass upon electrical materials, fittings, and appliances in order to determine if they comply with the standard-test specifications set up by these laboratories.

*2.* ***Legal Status of Code.*** The rules in the National Electrical Code are enforced by being incorporated in ordinances passed by various cities and towns, covering the installation of electric wiring. The Occupational Safety and Health Administration (OSHA) requires that all new electrical installations must conform to all the rules of the NEC. The NESC is adopted by state utilities commissions and is referred to by the NEC for some high-voltage applications.

When installing any electrical equipment, first ascertain whether local installation rules in the form of ordinances are enforced in the community. If so, follow such rules; if none exists, follow the requirements of NEC.

*3.* ***1984 Editions.*** Where reference is made in this section to installation rules, the 1984 edition of NEC or NESC is used as a basis.

*4.* ***Code Not a Design Manual.*** Design of an installation in accordance with the Code minimizes fire and accident hazards but does not guarantee satisfactory or efficient operation of the system. Other design standards are necessary to accomplish the latter purposes.

*5.* ***License.*** In many areas the installation of electric wiring is controlled by city, county, or state license often combined with installation rules.

*6.* ***Rules of Electric Service Companies.*** Electric lighting and power companies generally issue certain rules of their own, based to a large extent on peculiar requirements which are necessary in order to give the best possible service to the greatest number of customers and on NESC requirements.

These rules are concerned mostly with matters of distribution engineering. They relate to locations and details of service entrance, provision for meters, the kind of electricity furnished by the company, its frequency and voltage, the types and sizes of motors, rules in connection with starting characteristics of such motors, and similar matters.

The electric-service company usually supplies copies of its rules at no charge.

*7.* ***Inspection.*** Every electrical installation should be inspected wherever an experienced inspector is available to ensure that it complies with local and NEC rules. Such inspection is usually mandatory in cities having electrical ordinances. In some areas the fire underwriters maintain inspectors who check electrical wiring, while in others the municipality makes a check through its electrical inspectors. Where inspection is not mandatory, it is always advisable to request the most convenient fire underwriters' bureau to make the necessary inspection.

Federal and state buildings usually require inspection by authorized federal or state inspectors. In these instances inspection includes not only safety considerations but the requirements of the particular job specifications. Other inspection may be required but it is often waived. OSHA compliance officers may and do make inspections of existing electrical systems at any time.

## METHODS OF WIRING

***8. Wiring Methods Classified.*** The discussion of wiring methods in this section relates to *interior circuits for light, heat, and power* and does not cover signaling or communication systems.

Numerous methods of wiring are authorized by NEC, most of them used to a greater or lesser extent in commercial and industrial buildings. Those of interest can be grouped as follows:

**a.** Raceways for general use.
1. Rigid metal conduit.
2. Intermediate metal conduit (IMC).
3. Electric metallic tubing (EMT).
4. Nonmetallic conduit.
5. Surface raceways.

**b.** Cable-assembly systems for general use.
1. Nonmetallic sheathed cable.
2. Underground feeder and branch-circuit cable.
3. Metal-clad cable (armored cable).
4. Mineral-insulated metal-sheathed cable (MI).
5. Aluminum-sheathed cable (ALS).

**c.** Conductor systems for general use.
1. Open wiring on insulators.
2. Concealed knob-and-tube wiring (only as permitted in NEC Sec. 324-3).

**d.** Cable-assembly systems for limited use.
1. Service-entrance cable.
2. Nonmetallic extensions.
3. Underplaster extensions.

**e.** Raceway systems for limited use.
1. Flexible-metal conduit.
2. Liquid-tight-flexible-metal conduit.
3. Underfloor raceway.
4. Cellular-metal-floor or cellular-concrete-floor raceway.
5. Wireways.
6. Cable trays.

**f.** Special systems.
1. Busways.
2. Cable bus.
3. Multioutlet assemblies.
4. Electrical floor assemblies.

***9. Installation Methods.*** Requirements to be met in installing each of the foregoing systems are found in the current edition of the NEC. The requirements are specific and detailed and change somewhat as the art progresses; hence reference should be made to the Code for the exact circumstances under which each system is permitted or prohibited, together with the precise rules to be followed in installation.

The discussion in the following paragraphs compares the systems generally and indicates the major limitations on use of each.

## TYPES OF CONDUCTORS

***10. General Provisions Applying to All Wiring Systems.*** The types of wiring discussed may be used for voltages up to 600 V unless otherwise indicated. Each type of insulated conductor is approved for certain uses and has a maximum operating temperature. If this is exceeded, the insulation is subject to deterioration. In recent years, modified ethylene tetrafluoroethylene (Z and ZW) and perfluoroalkoxy (PFA and PFAH) cables have been allowed for high-temperature operations. See Table 19-1. Each conductor size has a maximum current-carrying capacity, depending on type of insulation and conditions of use. These ratings should not be exceeded (see Tables 19-2A through 19-2L and Fig. 19-1 for ratings and underground conduit systems). Conductors may be used in multiple usually in large sizes only (sizes 1/0 and larger, see Sec. 310-4, NEC).

Conductors of more than 600 V should not occupy the same enclosure as conductors carrying less than 600 V, but conductors of different light and power systems of less than 600 V may be grouped together in one enclosure if all are insulated for the maximum voltage encoun-

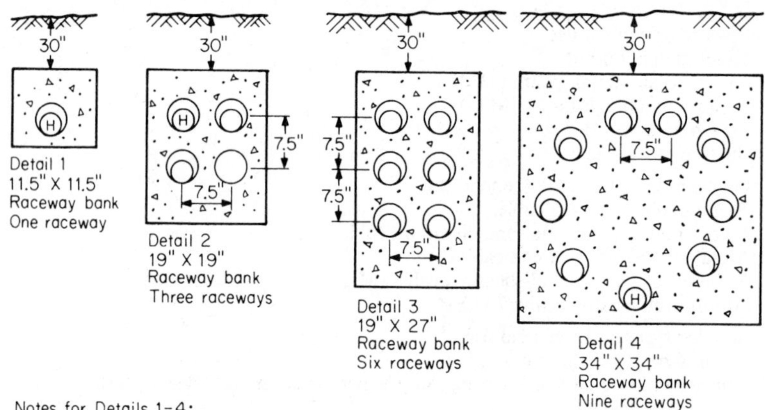


Notes for Details 1–4:

1. Larger circles represent raceways.
2. Smaller circles represent the installed conductor(s) or cables

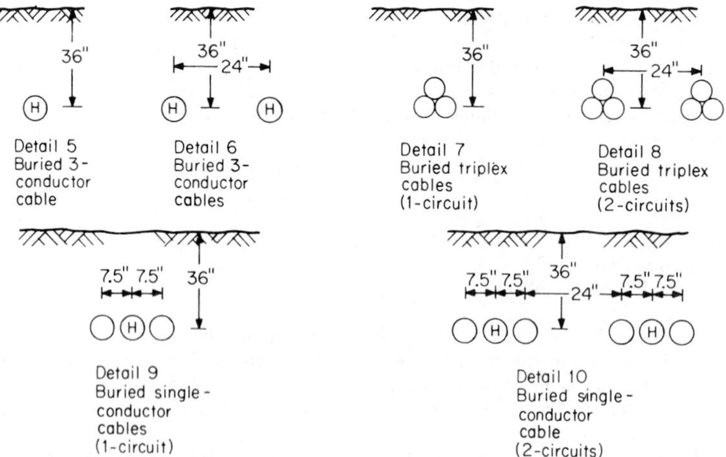


**FIG. 19-1** Configurations for buried systems using conductors in Tables 19-2A through 19-2L.

tered. In general, communication circuits should not occupy the same enclosure with light and power wiring.

Boxes or fittings must be installed at all outlets, at switch or junction points of raceway or cable systems, and at each outlet and switch point of concealed knob-and-tube work.

***11. Provisions Applying to All Raceway Systems.*** The number of conductors permitted in each size and type of raceway is definitely limited to provide ready installation and withdrawal. For conduit and electrical metallic tubing see Tables 19-3A, 19-3B, and 19-3C. Raceways, except surface-metal molding, must be installed as complete empty systems, the conductors being drawn in later. Conductors must be continuous from outlet to outlet without splice, except in auxiliary gutters and wireways.

Conductors of No. 8 AWG and larger must be stranded. Raceways must be continuous from outlet to outlet and from fitting to fitting and shall be securely fastened in place.

All conductors of a circuit operating on alternating current, if in metallic raceway, should be run in one enclosure to avoid inductive overheating. If, owing to capacity, not all conductors can be installed in one enclosure, each raceway used should contain a complete circuit (one conductor from each phase).

***12. Rigid-metal conduit, intermediate metal conduit, and electrical metallic tubing*** are the systems generally employed where wires are to be installed in raceways. Both conduit and tubing may be buried in concrete fills or may be installed exposed. Wiring installed in conduit is approved for practically all classes of buildings and for voltages both above and below 600 V. Certain restrictions are placed on the use of tubing.

Conduit consists of standard-weight steel pipe (preferably either galvanized or cadmium-plated, although it may be black-enameled for use indoors and where not subject to severe corrosive influences) or of aluminum. Electrical metallic tubing (EMT) has the same internal diameter as conduit but a thinner wall of higher-quality steel.

*Note on Tables 19-2A through 19-2L. Use of Conductors with Higher Operating Temperatures.* Where the room temperature is within 10°C of the maximum allowable operating temperature of the insulation, it is desirable to use an insulation with a higher maximum allowable operating temperature; although insulation can be used in a room temperature approaching its maximum allowable operating temperature limit if the current is reduced in accordance with the correction factors for different room temperatures as shown in the following correction factor table.

Correction Factors for Ambient Temperatures Other than 40°C

| °C | °F | 60°C (140°F) | 75°C (167°F) | 85°C (185°F) | 90°C (194°F) |
|---|---|---|---|---|---|
| 25 | 77 | 1.32 | 1.20 | 1.15 | 1.14 |
| 30 | 86 | 1.22 | 1.13 | 1.11 | 1.10 |
| 35 | 95 | 1.12 | 1.07 | 1.05 | 1.05 |
| 40 | 104 | 1.00 | 1.00 | 1.00 | 1.00 |
| 45 | 113 | | 0.93 | 0.94 | 0.95 |
| 50 | 122 | | 0.85 | 0.88 | 0.89 |
| 55 | 131 | | 0.76 | 0.82 | 0.84 |
| 60 | 140 | | 0.65 | 0.75 | 0.77 |
| 70 | 158 | | 0.38 | 0.58 | 0.63 |
| 80 | 176 | | | 0.33 | 0.44 |

Fittings and connectors used with conduit may be threaded or threadless. EMT fittings are usually threadless.

Sizes of EMT above 2 in have the same external diameter as the equivalent rigid-conduit size.

Nonmetallic rigid conduits, in approximately the same dimensions as rigid-metal conduits,

*(text continues on p. 19-26)*

**TABLE 19-1** Conductor Application and Insulations

| Trade name | Type letter | Max. operating temp. | Application provisions | Insulation | Thickness of insulation: AWG or MCM | Thickness of insulation: Mils | Outer covering |
|---|---|---|---|---|---|---|---|
| Asbestos | A | 200°C (392°F) | Dry locations only. Only for leads within apparatus or within raceways connected to apparatus. Limited to 300 V. | Asbestos | 14<br>12–8 | 30<br>40 | Without asbestos braid |
| Asbestos | AA | 200°C (392°F) | Dry locations only. Only for leads within apparatus or within raceways connected to apparatus or as open wiring. Limited to 300 V. | Asbestos | 14<br>12–8<br>6–2<br>1–4/0 | 30<br>30<br>40<br>60 | With asbestos braid or glass |
| Asbestos | AI | 125°C (257°F) | Dry locations only. Only for leads within apparatus or within raceways connected to apparatus. Limited to 300 V. | Impregnated asbestos | 14<br>12–8 | 30<br>40 | Without asbestos braid |
| Asbestos | AIA | 125°C (257°F) | Dry locations only. Only for leads within apparatus or within raceways connected to apparatus or as open wiring. | Impregnated asbestos | <br>14<br>12–8<br>6–2<br>1–4/0<br>213–500<br>501–1000 | Sol. / Str.<br>30 / 30<br>30 / 40<br>40 / 60<br>60 / 75<br>— / 90<br>— / 105 | With asbestos braid or glass |

| | | | | | | | | | | |
|---|---|---|---|---|---|---|---|---|---|---|
| Asbestos and varnished cambric | AVA | 110°C (230°F) | Dry locations only. | Impregnated asbestos and varnished cambric | | 1st Asb. | VC | AVA 2d Asb. | AVL 2d Asb. | AVA-asbestos braid or glass |
| | | | | | 14–8 (solid only) | — | 30 | 20 | 25 | |
| | | | | | 14–8 | 10 | 30 | 15 | 25 | |
| | | | | | 6–2 | 15 | 30 | 20 | 25 | |
| | | | | | 1–4/0 | 20 | 30 | 30 | 30 | |
| | | | | | 213–500 | 25 | 40 | 40 | 40 | |
| | | | | | 501–1000 | 30 | 40 | 40 | 40 | |
| | | | | | 1001–2000 | 30 | 50 | 50 | 50 | |
| Asbestos and varnished cambric | AVB | 90°C (194°F) | Dry locations only. | Impregnated asbestos and varnished cambric | | | | VC | Asb. | Flame-retardant, cotton braid (switchboard wiring) |
| | | | | | 18–8 | | | 30 | 20 | |
| | | | | | 6–2 | | | 40 | 30 | |
| | | | | | 1–4/0 | | | 40 | 40 | |
| | | | | | | | Asb. | VC | 2d Asb. | Flame-retardant, cotton braid |
| | | | | | 14–8 | | 10 | 30 | 15 | |
| | | | | | 6–2 | | 15 | 30 | 20 | |
| | | | | | 1–4/0 | | 20 | 30 | 30 | |
| | | | | | 213–500 | | 25 | 40 | 40 | |
| | | | | | 501–1000 | | 30 | 40 | 40 | |
| | | | | | 1001–2000 | | 30 | 50 | 50 | |

**TABLE 19-1** Conductor Application and Insulations *(Continued)*

| Trade name | Type letter | Max. operating temp. | Application provisions | Insulation | Thickness of insulation: AWG or MCM | | | | Mils | Outer covering |
|---|---|---|---|---|---|---|---|---|---|---|
| | | | | | | 1st Asb. | VC | AVA 2d Asb. | AVL 2d Asb. | |
| Asbestos and varnished cambric | AVL | 110°F (230°F) | Dry and wet locations. | Impregnated asbestos and varnished cambric | 14–8 (solid only) | — | 30 | 20 | 25 | AVL-lead sheath |
| | | | | | 14–8 | 10 | 30 | 15 | 25 | |
| | | | | | 6–2 | 15 | 30 | 20 | 25 | |
| | | | | | 1–4/0 | 20 | 30 | 30 | 30 | |
| | | | | | 213–500 | 25 | 40 | 40 | 40 | |
| | | | | | 501–1000 | 30 | 40 | 40 | 40 | |
| | | | | | 1001–2000 | 30 | 50 | 50 | 50 | |
| Fluorinated ethylene propylene | FEP or FEPB | 90°C (194°F) | Dry locations | Fluorinated ethylene propylene | 14–10 | | | | 20 | None |
| | | 200°C (392°F) | Dry locations—special applications.* | | 8–2 | | | | 30 | |
| | | | | Fluorinated ethylene propylene | 14–8 | | | | 14 | Glass braid |
| | | | | | 6–2 | | | | 14 | Asbestos braid |
| Mineral insulation (metal sheathed) | MI | 85°C (185°F) | Dry and wet locations. | Magnesium oxide | 16–10 | | | | 36 | Copper |
| | | | | | 9–4 | | | | 50 | |
| | | 250°C (482°F) | For special application.* | | 3–250 | | | | 55 | |

* Where environmental conditions require maximum conductor operating temperatures above 90°C.

| | | | | | | | | |
|---|---|---|---|---|---|---|---|---|
| Moisture-, heat- and oil-resistant thermoplastic | MTW | 60°C (140°F) | Machine tool wiring in wet locations as permitted in NFPA Standard No. 79. (See Article 670.) | Flame-retardant, moisture-, heat- and oil-resistant thermoplastic | | A | B | A. None |
| | | 90°C (194°F) | Machine tool wiring in dry locations as permitted in NFPA Standard No. 79. (See Article 670.) | | 22–12 | 30 | 15 | B. Nylon jacket or equivalent |
| | | | | | 10 | 30 | 20 | |
| | | | | | 8 | 45 | 30 | |
| | | | | | 6 | 60 | 30 | |
| | | | | | 4–2 | 60 | 40 | |
| | | | | | 1–4/0 | 80 | 50 | |
| | | | | | 213–500 | 95 | 60 | |
| | | | | | 501–1000 | 110 | 70 | |
| Paper | | 85°C (185°F) | For underground service conductors, or by special permission. | Paper | | | | Lead sheath |
| Perfluoroalkoxy† | PFA | 90°C (194°F) | Dry locations. | Perfluoro-alkoxy | 14–10 | | 20 | None |
| | | 200°C (392°F) | Dry locations—special applications.* | | 8–2 | | 30 | |
| | | | | | 1–4/0 | | 45 | |
| Perfluoroalkoxy† | PFAH | 250°C (482°F) | Dry locations only. Only for leads within apparatus or within raceways connected to apparatus. (Nickel or nickel-coated copper only.) | Perfluoro-alkoxy | 14–10 | | 20 | None |
| | | | | | 8–2 | | 30 | |
| | | | | | 1–4/0 | | 45 | |
| Heat-resistant rubber | RH | 75°C (167°F) | Dry locations. | Heat-resistant rubber | 14–12‡ | | 30 | Moisture-resistant, flame-retardant, nonmetallic covering§ |
| Heat-resistant rubber | RHH | 90°C (194°F) | Dry locations. | | 10 | | 45 | |
| | | | | | 8–2 | | 60 | |
| | | | | | 1–4/0 | | 80 | |
| | | | | | 213–500 | | 95 | |
| | | | | | 501–1000 | | 110 | |
| | | | | | 1001–2000 | | 125 | |
| Moisture- and heat-resistant rubber | RHW | 75°C (167°F) | Dry and wet locations. For over 2000-V insulation shall be ozone-resistant. | Moisture- and heat-resistant rubber | 14–10 | | 45 | Moisture-resistant, flame-retardant, nonmetallic covering§ |
| | | | | | 8–2 | | 60 | |
| | | | | | 1–4/0 | | 80 | |
| | | | | | 213–500 | | 95 | |
| | | | | | 501–1000 | | 110 | |
| | | | | | 1001–2000 | | 125 | |

* A new addition to the NEC.
† Some rubber insulations do not require an outer covering.
‡ For 14–12 sizes RHH shall be 45 mils thickness insulation.
§ Where environmental conditions require maximum conductor operating temperatures above 90°C.

**TABLE 19-1** Conductor Application and Insulations *(Continued)*

| Trade name | Type letter | Max. operating temp. | Application provisions | Insulation | Thickness of insulation: AWG or MCM | Thickness of insulation: Mils | Outer covering |
|---|---|---|---|---|---|---|---|
| Heat-resistant latex rubber | RUH | 75°C (167°F) | Dry locations. | 90% unmilled, grainless rubber | 14–10<br>8–2 | 18<br>25 | Moisture-resistant, flame-retardant, nonmetallic covering |
| Moisture-resistant latex rubber | RUW | 60°C (140°F) | Dry and wet locations. | 90% unmilled, grainless rubber | 14–10<br>8–2 | 18<br>25 | Moisture-resistant, flame-retardant, nonmetallic covering |
| Silicone-asbestos | SA | 90°C (194°F)<br>125°C (257°F) | Dry locations.<br>For special application.* | Silicone rubber | 14–10<br>8–2<br>1–4/0<br>213–500<br>501–1000<br>1001–2000 | 45<br>60<br>80<br>95<br>110<br>125 | Asbestos, glass or other suitable braid material |
| Synthetic heat-resistant | SIS | 90°C (194°F) | Switchboard wiring only. | Heat-resistant rubber | 14–10<br>8<br>6–2<br>1–4/0 | 30<br>45<br>60<br>80 | None |

* Where environmental conditions require maximum conductor operating temperatures above 90°C.

| | | | | | | | | |
|---|---|---|---|---|---|---|---|---|
| Thermoplastic | T | 60°C (140°F) | Dry locations. | Flame-retardant, thermo-plastic compound | 14–10<br>8<br>6–2<br>1–4/0<br>213–500<br>501–1000<br>1001–2000 | | 30<br>45<br>60<br>80<br>95<br>110<br>125 | None |
| Thermoplastic and asbestos | TA | 90°C (194°F) | Switchboard wiring only. | Thermo-plastic and asbestos | 14–8<br>6–2<br>1–4/0 | Th'pl'.<br>20<br>30<br>40 | Asb.<br>20<br>25<br>30 | Flame-retardant, nonmetallic covering |
| Thermoplastic and fibrous outer braid | TBS | 90°C (194°F) | Switchboard wiring only. | Thermo-plastic | 14–10<br>8<br>6–2<br>1–4/0 | | 30<br>45<br>60<br>80 | Flame-retardant, nonmetallic covering |
| Extruded polytetrafluoro-ethylene | TFE | 250°C (482°F) | Dry location only. Only for leads within apparatus or within raceways connected to apparatus, or as open wiring. (Nickel or nickel-coated copper only.) | Extruded polytetra-fluoro-ethylene | 14–10<br>8–2<br>1–4/0 | | 20<br>30<br>45 | None |
| Heat-resistant thermoplastic | THHN | 90°C (194F) | Dry locations. | Flame-retardant, heat-resistant thermo-plastic | 14–12<br>10<br>8–6<br>4–2<br>1–4/0<br>250–500<br>501–1000 | | 15<br>20<br>30<br>40<br>50<br>60<br>70 | Nylon jacket or equivalent |

**TABLE 19-1** Conductor Application and Insulations *(Continued)*

| Trade name | Type letter | Max. operating temp. | Application provisions | Insulation | Thickness of insulation: AWG or MCM | Thickness of insulation: Mils | Outer covering |
|---|---|---|---|---|---|---|---|
| Moisture and heat-resistant thermoplastic | THW | 75°C (167°F)<br>90°C (194°F) | Dry and wet locations.<br><br>Special applications within electric discharge lighting equipment. Limited to 1000 open-circuit volts or less. (Size 14–8 only as permitted in Section 410-31.) | Flame-retardant, moisture- and heat-resistant thermo-plastic | 14–10<br>8–2<br>1–4/0<br>213–500<br>501–1000<br>1001–2000 | 45<br>60<br>80<br>95<br>110<br>125 | None |
| Moisture- and heat-resistant thermoplastic | THWN | 75°C (167°F) | Dry and wet locations. | Flame-retardant, moisture- and heat-resistant thermo-plastic | 14–12<br>10<br>8–6<br>4–2<br>1–4/0<br>250–500<br>501–1000 | 15<br>20<br>30<br>40<br>50<br>60<br>70 | Nylon jacket or equivalent |
| Moisture-resistant thermoplastic | TW | 60°C (140°F) | Dry and wet locations. | Flame-retardant, moisture-resistant thermo-plastic | 14–10<br>8<br>6–2<br>1–4/0<br>213–500<br>501–1000<br>1001–2000 | 30<br>45<br>60<br>80<br>95<br>110<br>125 | None |

| | | | | | | | |
|---|---|---|---|---|---|---|---|
| Underground feeder & branch-circuit cable-single conductor. (For Type UF cable employing more than one conductor, see Article 339.) | UF | 60°C (140°F) | See Article 339. | Moisture-resistant | 14–10<br>8–2<br>1–4/0 | 60*<br>80*<br>95* | Integral with insulation |
| | | 75°C† (167°F) | | Moisture- and heat-resistant | | | |
| Underground service-entrance cable-single conductor. (For Type USE cable employing more than one conductor, see Article 338.) | USE | 75°C (167°F) | See Article 338. | Heat- and moisture-resistant | 12–10<br>8–2<br>1–4/0<br>213–500<br>501–1000<br>1001–2000 | 45<br>60<br>80<br>95‡<br>110<br>125 | Moisture-resistant nonmetallic covering [See 338-1(2).] |
| Varnished cambric | V | 85°C (185°F) | Dry locations only. Smaller than No. 6 by special permission. | Varnished cambric | 14–8<br>6–2<br>1–4/0<br>213–500<br>500–1000<br>1001–2000 | 45<br>60<br>80<br>95<br>110<br>125 | Nonmetallic covering or lead sheath. |

* Includes integral jacket.

† For ampacity limitation, see Section 339-1(a).

‡ Insulation thickness shall be permitted to be 80 mils for listed Type USE conductors that have been subjected to special investigations.

NOTE: The nonmetallic covering over individual rubber-covered conductors of aluminum-sheathed cable and of lead-sheathed or multiconductor cable is not required to be flame-retardant.

**TABLE 19-1** Conductor Application and Insulations *(Continued)*

| Trade name | Type letter | Max. operating temp. | Application provisions | Insulation | Thickness of insulation: AWG or MCM | Thickness of insulation: Mils | Outer covering |
|---|---|---|---|---|---|---|---|
| Moisture- and heat-resistant cross-linked synthetic polymer | XHHW | 90°C (194°F) | Dry locations. | Flame-retardant cross-linked synthetic polymer | 14–10 | 30 | None |
| | | | | | 8–2 | 45 | |
| | | | | | 1–4/0 | 55 | |
| | | 75°C (167°F) | Wet locations. | | 213–500 | 65 | |
| | | | | | 501–1000 | 80 | |
| | | | | | 1001–2000 | 95 | |
| Modified ethylene tetrafluoro-ethylene* | Z | 90°C (194°F) | Dry locations. | Modified ethylene tetra-fluoro-ethylene | 14–12 | 15 | None |
| | | | | | 10 | 20 | |
| | | 150°C (302°F) | Dry locations—special applications.† | | 8–4 | 25 | |
| | | | | | 3–1 | 35 | |
| | | | | | 1/0–4/0 | 45 | |
| Modified ethylene tetrafluoro-ethylene | ZW | 75°C (167°F) | Wet locations. | Modified ethylene tetra-fluoro-ethylene | 14–10 | 30 | None |
| | | | | | 8–2 | 45 | |
| | | 90°C (194°F) | Dry locations. | | | | |
| | | 150°C (302°F) | Dry locations—special applications.† | | | | |

* A new addition to the NEC.

† Where environmental conditions require maximum conductor operating temperatures above 90°C.

SOURCE: NEC Table 310-13.

**TABLE 19-2A** Ampacities for Bare and Covered Linewire*

Based on 40°C ambient, 80°C total conductor temperature, 2 ft per second wind velocity

| Bare copper conductors | | Weatherproofed copper linewire | |
|---|---|---|---|
| AWG, MCM | A | AWG, MCM | A |
| 8 | 98 | 8 | 103 |
| 6 | 124 | 6 | 130 |
| 4 | 155 | 4 | 163 |
| 2 | 209 | 2 | 219 |
| 1/0 | 282 | 1/0 | 297 |
| 2/0 | 329 | 2/0 | 344 |
| 3/0 | 382 | 3/0 | 401 |
| 4/0 | 444 | 4/0 | 466 |
| 250 | 494 | 250 | 519 |
| 300 | 556 | 300 | 584 |
| 500 | 773 | 500 | 812 |
| 750 | 1000 | 750 | 1050 |
| 1000 | 1193 | 1000 | 1253 |
| **Bare AAC aluminum conductor** | | **Weatherproofed AAC aluminum** | |
| AWG, MCM | A | AWG, MCM | A |
| 8 | 76 | 8 | 80 |
| 6 | 96 | 6 | 101 |
| 4 | 121 | 4 | 127 |
| 2 | 163 | 2 | 171 |
| 1/0 | 220 | 1/0 | 231 |
| 2/0 | 255 | 2/0 | 268 |
| 3/0 | 297 | 3/0 | 312 |
| 4/0 | 346 | 4/0 | 364 |
| 266.8 | 403 | 266.8 | 423 |
| 336.4 | 468 | 336.4 | 492 |
| 397.5 | 522 | 397.5 | 548 |
| 477.0 | 588 | 477.0 | 617 |
| 556.5 | 650 | 556.5 | 682 |
| 636.0 | 709 | 636.0 | 744 |
| 795.0 | 819 | 795.0 | 860 |
| 954.0 | 920 | 1033.5 | 1017 |
| 1033.5 | 968 | 1272 | 1201 |
| 1272 | 1103 | 1590 | 1381 |
| 1590 | 1267 | 2000 | 1527 |
| 2000 | 1454 | | |

* Effective January 1, 1987.

**TABLE 19-2B** Ampacities of Multiconductor Cables with Not More Than Three Insulated Conductors, Rated 0–2000 V in Free Air*

**Based on ambient air temperature of 40°C (For NM, NMC, AC, TC, MC, MI and SNM cables)†**

| Size: AWG, MCM | Temperature rating of conductor (see Table 19-1) | | | | | | | |
|---|---|---|---|---|---|---|---|---|
| | 60°C | 75°C | 85°C | 90°C | 60°C | 75°C | 85°C | 90°C |
| | Copper | | | | Aluminum or copper-clad aluminum | | | |
| 18 | | | | 11† | | | | |
| 16 | | | | 16† | | | | |
| 14 | 18† | 21† | 24† | 25† | — | — | — | — |
| 12 | 21† | 28† | 30† | 32† | 18† | 21† | 24† | 25† |
| 10 | 28† | 36† | 41† | 43† | 21† | 28† | 30† | 32† |
| 8 | 39 | 50 | 56 | 59 | 30 | 39 | 44 | 46 |
| 6 | 52 | 68 | 75 | 79 | 41 | 53 | 59 | 61 |
| 4 | 69 | 89 | 100 | 104 | 54 | 70 | 78 | 81 |
| 3 | 81 | 104 | 116 | 121 | 63 | 81 | 91 | 95 |
| 2 | 92 | 118 | 132 | 138 | 72 | 92 | 103 | 108 |
| 1 | 107 | 138 | 154 | 161 | 84 | 108 | 120 | 126 |
| 0 | 124 | 160 | 178 | 186 | 97 | 125 | 139 | 145 |
| 00 | 143 | 184 | 206 | 215 | 111 | 144 | 160 | 168 |
| 000 | 165 | 213 | 238 | 249 | 129 | 166 | 185 | 194 |
| 0000 | 190 | 245 | 274 | 287 | 149 | 192 | 214 | 224 |
| 250 | 212 | 274 | 305 | 320 | 166 | 214 | 239 | 250 |
| 300 | 237 | 306 | 341 | 357 | 186 | 240 | 268 | 280 |
| 350 | 261 | 337 | 377 | 394 | 205 | 265 | 296 | 309 |
| 400 | 281 | 363 | 406 | 425 | 222 | 287 | 317 | 334 |
| 500 | 321 | 416 | 465 | 487 | 255 | 330 | 368 | 385 |
| 600 | 354 | 459 | 513 | 538 | 284 | 368 | 410 | 429 |
| 700 | 387 | 502 | 562 | 589 | 306 | 405 | 462 | 473 |
| 750 | 404 | 523 | 586 | 615 | 328 | 424 | 473 | 495 |
| 800 | 415 | 539 | 604 | 633 | 339 | 439 | 490 | 513 |
| 900 | 438 | 570 | 639 | 670 | 362 | 469 | 514 | 548 |
| 1000 | 461 | 601 | 674 | 707 | 385 | 499 | 558 | 584 |
| Ambient temp. °C | For ambient temperatures other than 40°C multiply the ampacities shown above by the appropriate factor shown below. | | | | | | | |
| 21–25 | 1.32 | 1.20 | 1.15 | 1.14 | 1.32 | 1.20 | 1.15 | 1.14 |
| 26–30 | 1.22 | 1.13 | 1.11 | 1.10 | 1.22 | 1.13 | 1.11 | 1.10 |
| 31–35 | 1.12 | 1.07 | 1.05 | 1.05 | 1.12 | 1.07 | 1.05 | 1.05 |
| 36–40 | 1.00 | 1.00 | 1.00 | 1.00 | 1.00 | 1.00 | 1.00 | 1.00 |
| 41–45 | 0.87 | 0.93 | 0.94 | 0.95 | 0.87 | 0.93 | 0.94 | 0.95 |
| 46–50 | 0.71 | 0.85 | 0.88 | 0.89 | 0.71 | 0.85 | 0.88 | 0.89 |
| 51–55 | 0.50 | 0.76 | 0.82 | 0.84 | 0.50 | 0.76 | 0.82 | 0.84 |
| 56–60 | — | 0.65 | 0.75 | 0.77 | — | 0.65 | 0.75 | 0.77 |
| 61–70 | — | 0.38 | 0.58 | 0.63 | — | 0.38 | 0.58 | 0.63 |
| 71–80 | — | — | 0.33 | 0.44 | — | — | 0.33 | 0.44 |

* Effective January 1, 1987.

NOTE: The overcurrent protection for conductor types marked with an obelisk (†) shall not exceed 7 A for 18 AWG, 10 A for 16 AWG, and 15 A for 14 AWG, 20 A for 12 AWG, and 30 A for 10 AWG copper; or 15 A for 12 AWG and 25 A for 10 AWG aluminum and copper-clad aluminum.

**TABLE 19-2C** Ampacities of Single-Insulated Conductors, Rated 0–2000 V in Free Air*

Based on ambient air temperature of 40°C

| Size: AWG, MCM | Temperature rating of conductor (see Table 19-1) | | | | | |
|---|---|---|---|---|---|---|
| | 60°C | 75°C | 90°C | 60°C | 75°C | 90°C |
| | Types †T, †TW, †RUW | Types †RH, †RHW, †RUH, †THW, †THWN, †XHHW, †ZW | Types TA, TBS, SA, AVB, SIS, FEP, †FEPB, †RHH, †THHN, †XHHW** | Types †T, †TW, †RUW | Types TA, TBS, †RH, †RHW, †RUH, †THW, †THWN, †XHHW | Types TA, TBS, SA, AVB, SIS, †RHH, †THHN, †XHHW** |
| | Copper | | | Aluminum or copper-clad aluminum | | |
| 18 | | | 16† | | | |
| 16 | | | 22† | | | |
| 14 | 24† | 30† | 35† | — | — | — |
| 12 | 30† | 39† | 45† | 24† | 29† | 36† |
| 10 | 41† | 51† | 61† | 30† | 39† | 45† |
| 8 | 55 | 71 | 83 | 43 | 55 | 64 |
| 6 | 73 | 94 | 109 | 57 | 73 | 85 |
| 4 | 96 | 124 | 145 | 75 | 97 | 113 |
| 3 | 112 | 145 | 169 | 88 | 113 | 132 |
| 2 | 128 | 165 | 192 | 100 | 128 | 150 |
| 1 | 148 | 191 | 223 | 115 | 149 | 174 |
| 0 | 171 | 221 | 258 | 133 | 172 | 201 |
| 00 | 198 | 255 | 298 | 154 | 199 | 232 |
| 000 | 229 | 295 | 345 | 178 | 230 | 269 |
| 0000 | 266 | 343 | 400 | 207 | 268 | 312 |
| 250 | 295 | 381 | 445 | 230 | 297 | 347 |
| 300 | 331 | 427 | 499 | 259 | 334 | 389 |
| 350 | 366 | 473 | 552 | 287 | 370 | 431 |
| 400 | 397 | 514 | 500 | 312 | 402 | 469 |
| 500 | 460 | 595 | 695 | 361 | 466 | 544 |
| 600 | 514 | 664 | 776 | 404 | 522 | 609 |
| 700 | 567 | 733 | 857 | 447 | 578 | 674 |
| 750 | 594 | 768 | 898 | 469 | 606 | 707 |
| 800 | 617 | 798 | 934 | 488 | 631 | 736 |
| 900 | 664 | 859 | 1005 | 527 | 680 | 795 |
| 1000 | 711 | 920 | 1076 | 566 | 730 | 853 |
| 1250 | 809 | 1048 | 1228 | 650 | 840 | 982 |
| 1500 | 898 | 1166 | 1367 | 730 | 944 | 1103 |
| 1750 | 978 | 1271 | 1493 | 803 | 1039 | 1216 |
| 2000 | 1051 | 1367 | 1606 | 871 | 1128 | 1321 |
| Ambient temp. °C | For ambient temperatures other than 40°C multiply the ampacities shown above by the appropriate factor shown below. | | | | | |
| 21–25 | 1.32 | 1.20 | 1.14 | 1.32 | 1.20 | 1.14 |
| 26–30 | 1.22 | 1.13 | 1.10 | 1.22 | 1.13 | 1.10 |
| 31–35 | 1.12 | 1.07 | 1.05 | 1.12 | 1.07 | 1.05 |
| 36–40 | 1.00 | 1.00 | 1.00 | 1.00 | 1.00 | 1.00 |
| 41–45 | 0.87 | 0.93 | 0.95 | 0.87 | 0.93 | 0.95 |
| 46–50 | 0.71 | 0.85 | 0.89 | 0.71 | 0.85 | 0.89 |
| 51–55 | 0.50 | 0.76 | 0.84 | 0.50 | 0.76 | 0.84 |
| 56–60 | — | 0.65 | 0.77 | — | 0.65 | 0.77 |
| 61–70 | — | 0.38 | 0.63 | — | 0.38 | 0.63 |
| 71–80 | — | — | 0.45 | — | — | 0.45 |

* Effective January 1, 1987.

** For dry locations only. See 75° column for wet locations.

NOTE: The overcurrent protection for conductor types marked with an obelisk (†) shall not exceed 7 A for 18 AWG, 10 A for 16 AWG, 15 A for 14 AWG, 20 A for 12 AWG, and 30 A for 10 AWG copper; or 15 A for 12 AWG and 25 A for 10 AWG aluminum and copper-clad aluminum.

**TABLE 19-2D** Ampacities of Three Single-Insulated Conductors, Rated 0–2000 V, Triplexed on a Messenger*

Based on ambient air temperature of 40°C

| | Temperature rating of conductor (see Table 19-1) | | | |
|---|---|---|---|---|
| | 75°C | 90°C | 75°C | 90°C |
| Size: AWG, MCM | Types RH, RHW RUH, THW, THWN, XHHW, ZW | Types THHN, RHH, XHHW** | Types RH, RHW, RUH, THW, THWN, XHHW | Types THHN, RHH, XHHW** |
| | Copper | | Aluminum or copper-clad aluminum | |
| 8 | 57 | 66 | 44 | 51 |
| 6 | 76 | 89 | 59 | 69 |
| 4 | 101 | 117 | 78 | 91 |
| 3 | 118 | 138 | 92 | 107 |
| 2 | 135 | 158 | 106 | 123 |
| 1 | 158 | 185 | 123 | 144 |
| 0 | 183 | 214 | 143 | 167 |
| 00 | 212 | 247 | 165 | 193 |
| 000 | 245 | 287 | 192 | 224 |
| 0000 | 287 | 335 | 224 | 262 |
| 250 | 320 | 374 | 251 | 292 |
| 300 | 359 | 419 | 282 | 328 |
| 350 | 397 | 464 | 312 | 364 |
| 400 | 430 | 503 | 339 | 395 |
| 500 | 496 | 580 | 392 | 458 |
| 600 | 553 | 647 | 440 | 514 |
| 700 | 610 | 714 | 488 | 570 |
| 750 | 638 | 747 | 512 | 598 |
| 800 | 660 | 773 | 532 | 622 |
| 900 | 704 | 826 | 572 | 669 |
| 1000 | 748 | 879 | 612 | 716 |
| Ambient temp. °C | For ambient temperatures other than 40°C multiply the ampacities shown above by the appropriate factor shown below. | | | |
| 21–25 | 1.20 | 1.14 | 1.20 | 1.14 |
| 26–30 | 1.13 | 1.10 | 1.13 | 1.10 |
| 31–35 | 1.07 | 1.05 | 1.07 | 1.05 |
| 36–40 | 1.00 | 1.00 | 1.00 | 1.00 |
| 41–45 | 0.93 | 0.95 | 0.93 | 0.95 |
| 46–50 | 0.85 | 0.89 | 0.85 | 0.89 |
| 51–55 | 0.76 | 0.84 | 0.76 | 0.84 |
| 56–60 | 0.65 | 0.77 | 0.65 | 0.77 |
| 61–70 | 0.38 | 0.63 | 0.38 | 0.63 |
| 71–80 | — | 0.45 | — | 0.45 |

* Effective January 1, 1987.

** For dry locations only. See 75° column for wet locations.

**TABLE 19-2E** Ampacities of Three Single-Insulated Conductors, Rated 0–2000 V in Conduit in Free Air*

Based on ambient air temperature of 40°C

| Size: AWG, MCM | Temperature rating of conductor (see Table 19-1) | | | | | |
|---|---|---|---|---|---|---|
| | 60°C | 75°C | 90°C | 60°C | 75°C | 90°C |
| | Types †RUW, †T, †TW, †UF | Types †RH, †RHW, †RUH, †THW, †THWN, †XHHW, †USE, †ZW | Types SA, AVB, †FEP, †FEPB, †THHN, †RHH, †XHHW** | Types †RUW, †T †TW, †UF | Types †RH, †RHW, †RUH, †THW, †THWN, †XHHW, †USE | Types SA, AVB, †THHN, †RHH, †XHHW** |
| | Copper | | | Aluminum or copper-clad aluminum | | |
| 14 | 18† | 22† | 25† | — | — | — |
| 12 | 23† | 28† | 32† | 18† | 22† | 26† |
| 10 | 29† | 37† | 42† | 23† | 29† | 34† |
| 8 | 36 | 48 | 55 | 28 | 37 | 43 |
| 6 | 50 | 64 | 75 | 37 | 50 | 58 |
| 4 | 65 | 83 | 97 | 50 | 65 | 76 |
| 3 | 76 | 98 | 114 | 59 | 76 | 89 |
| 2 | 87 | 112 | 130 | 68 | 87 | 102 |
| 1 | 104 | 134 | 156 | 81 | 104 | 122 |
| 0 | 119 | 153 | 179 | 93 | 119 | 139 |
| 00 | 135 | 175 | 204 | 106 | 137 | 159 |
| 000 | 160 | 207 | 242 | 125 | 162 | 189 |
| 0000 | 184 | 238 | 278 | 144 | 186 | 217 |
| 250 | 210 | 271 | 317 | 165 | 213 | 249 |
| 300 | 232 | 300 | 351 | 183 | 236 | 276 |
| 350 | 254 | 328 | 384 | 201 | 259 | 303 |
| 400 | 274 | 354 | 415 | 218 | 281 | 329 |
| 500 | 314 | 407 | 477 | 252 | 326 | 381 |
| 600 | 345 | 448 | 525 | 280 | 362 | 424 |
| 700 | 376 | 489 | 574 | 308 | 399 | 467 |
| 750 | 392 | 509 | 598 | 322 | 417 | 488 |
| 800 | 403 | 524 | 616 | 334 | 432 | 506 |
| 900 | 426 | 555 | 653 | 357 | 463 | 542 |
| 1000 | 499 | 585 | 689 | 380 | 493 | 578 |
| Ambient temp. °C | For ambient temperatures other than 40°C multiply the ampacities shown above by the appropriate factor shown below. | | | | | |
| 21–25 | 1.32 | 1.20 | 1.14 | 1.32 | 1.20 | 1.14 |
| 26–30 | 1.22 | 1.13 | 1.10 | 1.22 | 1.13 | 1.10 |
| 31–35 | 1.12 | 1.07 | 1.05 | 1.12 | 1.07 | 1.05 |
| 36–40 | 1.00 | 1.00 | 1.00 | 1.00 | 1.00 | 1.00 |
| 41–45 | 0.87 | 0.93 | 0.95 | 0.87 | 0.93 | 0.95 |
| 46–50 | 0.71 | 0.85 | 0.89 | 0.71 | 0.85 | 0.89 |
| 51–55 | 0.50 | 0.76 | 0.84 | 0.50 | 0.76 | 0.84 |
| 56–60 | — | 0.65 | 0.77 | — | 0.65 | 0.77 |
| 61–70 | — | 0.38 | 0.63 | — | 0.38 | 0.63 |
| 71–80 | — | — | 0.45 | — | — | 0.45 |

* Effective January 1, 1987.

** For dry locations only. See 75° column for wet locations.

NOTE: The overcurrent protection for conductor types marked with an obelisk (†) shall not exceed 15 A for 14 AWG, 20 A for 12 AWG, and 30 A for 10 AWG copper; or 15 A for 12 AWG and 25 A for 10 AWG aluminum and copper-clad aluminum.

**TABLE 19-2F** Ampacities of Three Insulated Conductors, Rated 0–2000 V, within an Overall Covering (Three-Conductor Cable), in Conduit in Free Air*

**Based on ambient air temperature of 40°C**

| Size: AWG, MCM | Temperature rating of conductor (see Table 19-1) | | | | | |
|---|---|---|---|---|---|---|
| | 60°C | 75°C | 90°C | 60°C | 75°C | 90°C |
| | Types †RRW, †T, †TW | Types †RH, †RHW, †RUH, †THW, †THWN, †XHHW,** †ZW | Types †THHN, †RHH, †XHHW** | Types †RHH, †T, †TW | Types †RH, †RHW, †RUH, †THW, †THWN, †XHHW | Types †THHN, †RHH, †XHHW |
| | Copper | | | Aluminum or copper-clad aluminum | | |
| 14 | 16† | 21† | 24† | — | — | — |
| 12 | 21† | 27† | 31† | 17† | 21† | 24† |
| 10 | 28† | 35† | 40† | 21† | 27† | 30† |
| 8 | 35 | 45 | 52 | 27 | 35 | 41 |
| 6 | 46 | 59 | 69 | 36 | 46 | 53 |
| 4 | 61 | 78 | 91 | 47 | 61 | 71 |
| 3 | 71 | 92 | 107 | 56 | 72 | 84 |
| 2 | 81 | 105 | 123 | 64 | 82 | 96 |
| 1 | 94 | 121 | 141 | 73 | 94 | 110 |
| 0 | 110 | 143 | 166 | 86 | 111 | 130 |
| 00 | 126 | 163 | 190 | 99 | 127 | 149 |
| 000 | 144 | 186 | 218 | 113 | 146 | 170 |
| 0000 | 169 | 219 | 255 | 133 | 171 | 200 |
| 250 | 187 | 241 | 282 | 147 | 189 | 221 |
| 300 | 209 | 269 | 315 | 165 | 212 | 248 |
| 350 | 230 | 297 | 348 | 182 | 235 | 274 |
| 400 | 247 | 319 | 374 | 197 | 254 | 296 |
| 500 | 280 | 363 | 425 | 226 | 291 | 341 |
| 600 | 305 | 396 | 465 | 250 | 322 | 377 |
| 700 | 330 | 429 | 504 | 273 | 353 | 414 |
| 750 | 342 | 445 | 524 | 285 | 369 | 432 |
| 800 | 350 | 456 | 537 | 294 | 381 | 446 |
| 900 | 365 | 477 | 564 | 313 | 405 | 475 |
| 1000 | 381 | 499 | 590 | 331 | 429 | 504 |
| Ambient temp. °C | For ambient temperatures other than 40°C multiply the ampacities shown above by the appropriate factor shown below. | | | | | |
| 21–25 | 1.32 | 1.20 | 1.14 | 1.32 | 1.20 | 1.14 |
| 26–30 | 1.22 | 1.13 | 1.10 | 1.22 | 1.13 | 1.10 |
| 31–35 | 1.12 | 1.07 | 1.05 | 1.12 | 1.07 | 1.05 |
| 36–40 | 1.00 | 1.00 | 1.00 | 1.00 | 1.00 | 1.00 |
| 41–45 | 0.87 | 0.93 | 0.95 | 0.87 | 0.93 | 0.95 |
| 46–50 | 0.71 | 0.85 | 0.89 | 0.71 | 0.85 | 0.89 |
| 51–55 | 0.50 | 0.76 | 0.84 | 0.50 | 0.76 | 0.84 |
| 56–60 | — | 0.65 | 0.77 | — | 0.65 | 0.77 |
| 61–70 | — | 0.38 | 0.63 | — | 0.38 | 0.63 |
| 71–80 | — | — | 0.45 | — | — | 0.45 |

* Effective January 1, 1987.

** For dry locations only. See 75° column for wet locations.

NOTE: The overcurrent protection for conductor types marked with an obelisk (†) shall not exceed 15 A for 14 AWG, 20 A for 12 AWG, and 30 A for 10 AWG copper; or 15 A for 12 AWG and 25 A for 10 AWG aluminum and copper-clad aluminum.

**TABLE 19-2G** Ampacities of Single-Insulated Conductors, Rated 0–2000 V in Nonmagnetic Underground Raceways (One Conductor per Raceway)*

**Based on ambient earth temperature of 20°C, raceway arrangement as per Fig. 19-1, 100% load factor, thermal resistance (RHO) of 90, conductor temperature 75°C**

| Size: MCM | 3 Raceways (Fig. 19-1 Detail 2) | 6 Raceways (Fig. 19-1 Detail 3) | 9 Raceways (Fig. 19-1 Detail 4) | 3 Raceways (Fig. 19-1 Detail 2) | 6 Raceways (Fig. 19-1 Detail 3) | 9 Raceways (Fig. 19-1 Detail 4) |
|---|---|---|---|---|---|---|
| | Types RHW, THW, THWN, XHHW, USE | Types RHW, THW, THWN, XHHW, USE | Types RHW, THW, THWN, XHHW, USE | Types RHW, THW, THWN, XHHW, USE | Types RHW, THW, THWN, XHHW, USE | Types RHW, THW, THWN, XHHW, USE |
| | Copper | | | Aluminum or copper-clad aluminum | | |
| 250 | 344 | 295 | 273 | 269 | 230 | 213 |
| 350 | 418 | 355 | 328 | 327 | 277 | 256 |
| 500 | 511 | 431 | 397 | 401 | 337 | 311 |
| 750 | 640 | 534 | 490 | 505 | 421 | 387 |
| 1000 | 745 | 617 | 566 | 593 | 491 | 450 |
| 1250 | 832 | 686 | 628 | 668 | 551 | 504 |
| 1500 | 907 | 744 | 680 | 736 | 604 | 552 |
| 1750 | 970 | 793 | 723 | 796 | 651 | 594 |
| 2000 | 1027 | 836 | 762 | 850 | 693 | 631 |
| Ambient temp. °C | For ambient temperatures other than 20°C multiply the ampacities shown above by the appropriate factor shown below. | | | | | |
| 6—10 | 1.09 | 1.09 | 1.09 | 1.09 | 1.09 | 1.09 |
| 11–15 | 1.04 | 1.04 | 1.04 | 1.04 | 1.04 | 1.04 |
| 16–20 | 1.00 | 1.00 | 1.00 | 1.00 | 1.00 | 1.00 |
| 21–25 | 0.95 | 0.95 | 0.95 | 0.95 | 0.95 | 0.95 |
| 26–30 | 0.90 | 0.90 | 0.90 | 0.90 | 0.90 | 0.90 |

* Effective January 1, 1987.

**TABLE 19-2H** Ampacities of Three Insulated Conductors, Rated 0–2000 V, within an Overall Covering (Three-Conductor Cable) in Underground Raceways (One Cable per Raceway)*

**Based on ambient earth temperature of 20°C, raceway arrangement as per Fig. 19-1, 100% load factor, thermal resistance (RHO) of 90, conductor temperature 75°C**

| Size: AWG, MCM | 1 Raceway (Fig. 19-1 Detail 1) | 3 Raceways (Fig. 19-1 Detail 2) | 6 Raceways (Fig. 19-1 Detail 3) | 1 Raceway (Fig. 19-1 Detail 1) | 3 Raceways (Fig. 19-1 Detail 2) | 6 Raceways (Fig. 19-1 Detail 3) |
|---|---|---|---|---|---|---|
| | Types RHW, THW, THWN, XHHW, USE | Types RHW, THW, THWN, XHHW, USE | Types RHW, THW, THWN, XHHW, USE | Types RHW, THW, THWN. XHHW, USE | Types RHW, THW, THWN, XHHW, USE | Types RHW, THW, THWN, XHHW, USE |
| | Copper | | | Aluminum or copper-clad aluminum | | |
| 8 | 54 | 48 | 42 | 42 | 37 | 32 |
| 6 | 71 | 63 | 54 | 55 | 49 | 42 |
| 4 | 93 | 81 | 69 | 72 | 63 | 54 |
| 2 | 121 | 105 | 89 | 94 | 82 | 70 |
| 1 | 140 | 121 | 102 | 109 | 94 | 79 |
| 0 | 160 | 137 | 116 | 125 | 107 | 90 |
| 00 | 183 | 156 | 131 | 143 | 122 | 102 |
| 000 | 210 | 178 | 148 | 164 | 139 | 116 |
| 0000 | 240 | 202 | 168 | 187 | 158 | 131 |
| 250 | 265 | 222 | 184 | 207 | 174 | 144 |
| 350 | 321 | 267 | 219 | 252 | 209 | 172 |
| 500 | 389 | 320 | 261 | 308 | 254 | 207 |
| 750 | 478 | 388 | 314 | 386 | 314 | 254 |
| 1000 | 539 | 435 | 351 | 447 | 361 | 291 |
| Ambient temp. °C | For ambient temperatures other than 20°C multiply the ampacities shown above by the appropriate factor shown below. | | | | | |
| 6–10 | 1.09 | 1.09 | 1.09 | 1.09 | 1.09 | 1.09 |
| 11–15 | 1.04 | 1.04 | 1.04 | 1.04 | 1.04 | 1.04 |
| 16–20 | 1.00 | 1.00 | 1.00 | 1.00 | 1.00 | 1.00 |
| 21–25 | 0.95 | 0.95 | 0.95 | 0.95 | 0.95 | 0.95 |
| 26–30 | 0.90 | 0.90 | 0.90 | 0.90 | 0.90 | 0.90 |

* Effective January 1, 1987.

**TABLE 19-21** Ampacities of Three Single-Insulated Conductors, Rated 0–2000 V, in Underground Raceways (Three Conductors per Raceway)*

Based on ambient earth temperature of 20°C, raceway arrangement per Fig. 19-1, 100% load factor, thermal resistance (RHO) of 90, conductor temperature 75°C

| Size: AWG, MCM | 1 Raceway (Fig. 19-1 Detail 1) | 3 Raceways (Fig. 19-1 Detail 2) | 6 Raceways (Fig. 19-1 Detail 3) | 1 Raceway (Fig. 19-1 Detail 1) | 3 Raceways (Fig. 19-1 Detail 2) | 6 Raceways (Fig. 19-1 Detail 3) |
|---|---|---|---|---|---|---|
| | Types †RHW, †THW, †THWN, †XHHW, †USE | Types †RHW, †THW, †THWN, †XHHW, †USE | Types †RHW, †THW, †THWN, †XHHW, †USE | Types †RHW, †THW, †THWN. †XHHW, †USE | Types †RHW, †THW, †THWN, †XHHW, †USE | Types †RHW, †THW, †THWN, †XHHW, †USE |
| | Copper | | | Aluminum or copper-clad aluminum | | |
| 12 | 36† | 31† | 24† | 28† | 22† | 18† |
| 10 | 46† | 41† | 32† | 36† | 31† | 25† |
| 8 | 58 | 51 | 44 | 45 | 40 | 34 |
| 6 | 77 | 67 | 56 | 60 | 52 | 44 |
| 4 | 100 | 86 | 73 | 78 | 67 | 57 |
| 3 | 116 | 99 | 83 | 91 | 77 | 65 |
| 2 | 132 | 112 | 93 | 103 | 87 | 73 |
| 1 | 153 | 128 | 106 | 119 | 100 | 83 |
| 0 | 175 | 146 | 121 | 136 | 114 | 94 |
| 00 | 200 | 166 | 136 | 156 | 130 | 106 |
| 000 | 228 | 189 | 154 | 178 | 147 | 121 |
| 0000 | 263 | 215 | 175 | 205 | 168 | 137 |
| 250 | 290 | 236 | 192 | 227 | 185 | 150 |
| 300 | 321 | 260 | 210 | 252 | 204 | 165 |
| 350 | 351 | 283 | 228 | 276 | 222 | 179 |
| 400 | 376 | 302 | 243 | 297 | 238 | 191 |
| 500 | 427 | 341 | 273 | 338 | 270 | 216 |
| 600 | 468 | 371 | 296 | 373 | 296 | 236 |
| 700 | 509 | 402 | 319 | 408 | 321 | 255 |
| 750 | 529 | 417 | 330 | 425 | 334 | 265 |
| 800 | 544 | 428 | 338 | 439 | 344 | 273 |
| 900 | 575 | 450 | 355 | 466 | 365 | 288 |
| 1000 | 605 | 472 | 372 | 494 | 385 | 304 |
| Ambient temp. °C | For ambient temperatures other than 20°C multiply the ampacities shown above by the appropriate factor shown below. | | | | | |
| 6–10 | 1.09 | 1.09 | 1.09 | 1.09 | 1.09 | 1.09 |
| 11–15 | 1.04 | 1.04 | 1.04 | 1.04 | 1.04 | 1.04 |
| 16–20 | 1.00 | 1.00 | 1.00 | 1.00 | 1.00 | 1.00 |
| 21–25 | 0.95 | 0.95 | 0.95 | 0.95 | 0.95 | 0.95 |
| 26–30 | 0.90 | 0.90 | 0.90 | 0.90 | 0.90 | 0.90 |

* Effective January 1, 1987.

NOTE: The overcurrent protection for conductor types marked with an obelisk (†) shall not exceed 20 A for 12 AWG and 30 A for 10 AWG copper; or 15 A for 12 AWG and 25 A for 10 AWG aluminum and copper-clad aluminum.

**TABLE 19-2J** Ampacities of Two or Three Insulated Conductors, Rated 0–2000 V, Cabled within an Overall (Two- or Three-Conductor) Covering Directly Buried in Earth*

Based on ambient earth temperature of 20°C, arrangement per Fig. 19-1, 100% load factor, thermal resistance (RHO) of 90

| | 1 Cable (Fig. 19-1 Detail 5) | | 2 Cables (Fig. 19-1 Detail 6) | | 1 Cable (Fig. 19-1 Detail 5) | | 2 Cables (Fig. 19-1 Detail 3) | |
|---|---|---|---|---|---|---|---|---|
| | 60°C | 75°C | 60°C | 75°C | 60°C | 75°C | 60°C | 75°C |
| | Types | | Types | | Types | | Types | |
| Size: AWG, MCM | †UF | †RHW, †THW, †THWN, †XHHW, †USE | †UF | †RHW, †THW, †THWN, †XHHW, †USE | †UF | †RHW, †THW, †THWN, †USE | †UF | †RHW, †THW, †THWN, †XHHW, †USE |
| | Copper | | | | Aluminum or copper-clad aluminum | | | |
| 12 | 38† | 43† | 34† | 41 | 30† | 34† | 26† | 32† |
| 10 | 47† | 56† | 43† | 52† | 38† | 45† | 34† | 41† |
| 8 | 64 | 75 | 60 | 70 | 51 | 59 | 47 | 55 |
| 6 | 85 | 100 | 81 | 95 | 68 | 75 | 60 | 70 |
| 4 | 107 | 125 | 100 | 117 | 83 | 97 | 78 | 91 |
| 2 | 137 | 161 | 128 | 150 | 107 | 126 | 110 | 117 |
| 1 | 155 | 182 | 145 | 170 | 121 | 142 | 113 | 132 |
| 00 | 177 | 208 | 165 | 193 | 138 | 162 | 129 | 151 |
| 00 | 201 | 236 | 188 | 220 | 157 | 184 | 146 | 171 |
| 000 | 229 | 269 | 213 | 250 | 179 | 210 | 166 | 195 |
| 0000 | 259 | 304 | 241 | 282 | 203 | 238 | 188 | 220 |
| 250 | | 333 | | 308 | | 261 | | 241 |
| 350 | | 401 | | 370 | | 315 | | 290 |
| 500 | | 481 | | 442 | | 381 | | 350 |
| 750 | | 585 | | 535 | | 473 | | 433 |
| 1000 | | 657 | | 600 | | 545 | | 497 |
| Ambient temp. °C | For ambient temperatures other than 20°C multiply the ampacities shown above by the appropriate factor shown below. | | | | | | | |
| 6–10 | 1.12 | 1.09 | 1.12 | 1.09 | 1.12 | 1.09 | 1.12 | 1.09 |
| 11–15 | 1.06 | 1.04 | 1.06 | 1.04 | 1.06 | 1.04 | 1.06 | 1.04 |
| 16–20 | 1.00 | 1.00 | 1.00 | 1.00 | 1.00 | 1.00 | 1.00 | 1.00 |
| 21–25 | 0.94 | 0.95 | 0.94 | 0.95 | 0.94 | 0.95 | 0.94 | 0.95 |
| 26–30 | 0.87 | 0.90 | 0.87 | 0.90 | 0.87 | 0.90 | 0.87 | 0.90 |

* Effective January 1, 1987.

NOTE: The overcurrent protection for conductor types marked with an obelisk (†) shall not exceed 20 A for 12 AWG and 30 A for 10 AWG copper; or 15 A for 12 AWG and 25 A for 10 AWG aluminum and copper-clad aluminum.

**TABLE 19-2K** Ampacities of Three Triplexed Single-Insulated Conductors, Rated 0–2000 V, Directly Buried in Earth*

**Based on ambient earth temperature of 20°C, arrangement per Fig. 19-1, 100% load factor, thermal resistance (RHO) of 90**

| Size: AWG, MCM | See Fig. 19-1 Detail 7 | | See Fig. 19-1 Detail 8 | | See Fig. 19-1 Detail 7 | | See Fig. 19-1 Detail 8 | |
|---|---|---|---|---|---|---|---|---|
| | 60°C | 75°C | 60°C | 75°C | 60°C | 75°C | 60°C | 75°C |
| | Types | | Types | | Types | | Types | |
| | †UF | †USE | †UF | †USE | †UF | †USE | †UF | †USE |
| | Copper | | | | Aluminum or copper-clad aluminum | | | |
| 12 | 41† | 48† | 39† | 46† | 32† | 38† | 31† | 36† |
| 10 | 54† | 63† | 50† | 59† | 42† | 49† | 39† | 46† |
| 8 | 72 | 84 | 66 | 77 | 55 | 65 | 51 | 60 |
| 6 | 91 | 107 | 84 | 99 | 72 | 84 | 66 | 77 |
| 4 | 119 | 139 | 109 | 128 | 92 | 108 | 85 | 100 |
| 2 | 153 | 179 | 140 | 164 | 119 | 139 | 109 | 128 |
| 1 | 173 | 203 | 159 | 186 | 135 | 158 | 124 | 145 |
| 0 | 197 | 231 | 181 | 212 | 154 | 180 | 141 | 165 |
| 00 | 223 | 262 | 205 | 240 | 175 | 205 | 159 | 187 |
| 000 | 254 | 298 | 232 | 272 | 199 | 233 | 181 | 212 |
| 0000 | 289 | 339 | 263 | 308 | 226 | 265 | 206 | 241 |
| 250 | | 370 | | 336 | | 289 | | 263 |
| 350 | | 445 | | 403 | | 349 | | 316 |
| 500 | | 536 | | 483 | | 424 | | 382 |
| 750 | | 654 | | 587 | | 525 | | 471 |
| 1000 | | 744 | | 665 | | 608 | | 544 |
| Ambient temp. °C | For ambient temperatures other than 20°C multiply the ampacities shown above by the appropriate factor shown below. | | | | | | | |
| 6–10 | 1.12 | 1.09 | 1.12 | 1.09 | 1.12 | 1.09 | 1.12 | 1.09 |
| 11–15 | 1.06 | 1.04 | 1.06 | 1.04 | 1.06 | 1.04 | 1.06 | 1.04 |
| 16–20 | 1.00 | 1.00 | 1.00 | 1.00 | 1.00 | 1.00 | 1.00 | 1.00 |
| 21–25 | 0.94 | 0.95 | 0.94 | 0.95 | 0.94 | 0.95 | 0.94 | 0.95 |
| 26–30 | 0.87 | 0.90 | 0.87 | 0.90 | 0.87 | 0.90 | 0.87 | 0.90 |

* Effective January 1, 1987.

NOTE: The overcurrent protection for conductor types marked with an obelisk (†) shall not exceed 20 A for 12 AWG and 30 A for 10 AWG copper; or 15 A for 12 AWG and 25 A for 10 AWG aluminum and copper-clad aluminum.

**TABLE 19-2L** Ampacities of Three Single-Insulated Conductors, Rated 0–2000 V, Directly Buried in Earth*

**Based on ambient earth temperature of 20°C, arrangement per Fig. 19-1, 100% load factor, thermal resistance (RHO) of 90**

| Size: AWG, MCM | See Fig. 19-1 Detail 9 | | See Fig. 19-1 Detail 10 | | See Fig. 19-1 Detail 9 | | See Fig. 19-1 Detail 10 | |
|---|---|---|---|---|---|---|---|---|
| | 60°C | 75°C | 60°C | 75°C | 60°C | 75°C | 60°C | 75°C |
| | Types | | Types | | Types | | Types | |
| | UF | USE | UF | USE | UF | USE | UF | USE |
| | Copper | | | | Aluminum or copper-clad aluminum | | | |
| 8 | 84 | | 78 | | 66 | | 61 | |
| 6 | 107 | | 101 | | 84 | | 78 | |
| 4 | 139 | | 130 | | 108 | | 101 | |
| 2 | 178 | | 165 | | 139 | | 129 | |
| 1 | 201 | | 187 | | 157 | | 146 | |
| 0 | 230 | | 212 | | 179 | | 165 | |
| 00 | 261 | | 241 | | 204 | | 188 | |
| 000 | 297 | | 274 | | 232 | | 213 | |
| 0000 | 336 | | 309 | | 262 | | 241 | |
| 250 | | 429 | | 394 | | 335 | | 308 |
| 350 | | 516 | | 474 | | 403 | | 370 |
| 500 | | 626 | | 572 | | 490 | | 448 |
| 750 | | 767 | | 700 | | 605 | | 552 |
| 1000 | | 887 | | 808 | | 706 | | 642 |
| 1250 | | 979 | | 891 | | 787 | | 716 |
| 1500 | | 1063 | | 965 | | 862 | | 783 |
| 1750 | | 1133 | | 1027 | | 930 | | 843 |
| 2000 | | 1195 | | 1082 | | 990 | | 897 |
| Ambient temp. °C | For ambient temperatures other than 20°C multiply the ampacities shown above by the appropriate factor shown below. | | | | | | | |
| 6–10 | 1.12 | 1.09 | 1.12 | 1.09 | 1.12 | 1.09 | 1.12 | 1.09 |
| 11–15 | 1.06 | 1.04 | 1.06 | 1.04 | 1.06 | 1.04 | 1.06 | 1.04 |
| 16–20 | 1.00 | 1.00 | 1.00 | 1.00 | 1.00 | 1.00 | 1.00 | 1.00 |
| 21–25 | 0.94 | 0.95 | 0.94 | 0.95 | 0.94 | 0.95 | 0.94 | 0.95 |
| 26–30 | 0.87 | 0.90 | 0.87 | 0.90 | 0.87 | 0.90 | 0.87 | 0.90 |

* Effective January 1, 1987.

are also a general-use raceway. Some restrictions are imposed, affecting particularly installations exposed to possible mechanical injury. Grounding continuity is provided by an additional grounding conductor pulled into the raceway with the circuit conductors or as part of a cable assembly.

Nonmetallic PVC rigid conduits are commonly assembled with matching fittings by adhesives. Field bends are made by softening the plastic in a hot air stream of several hundred degrees from an electric heater-blower.

Nonmetallic PVC raceways of relatively flexible construction and with conductor already drawn in are used for direct burial in airport, highway, parkway, and similar installations.

Asbestos-cement, PVC, and fiber conduits are extensively used in underground distribution. They may be installed directly in earth or encased in concrete envelopes.

***13. Cable-assembly systems*** are used extensively for concealed wiring not embedded in masonry or concrete. They may also be installed exposed in dry locations and, depending upon the particular construction and ratings, in wet locations. Branch-circuit sizes are conventionally 600-V-rated. Cables rated for 5 through 15 kV are frequently used for primary distribution feeders in large commercial and industrial electrical systems.

In industrial plants and commercial utility areas cable assemblies are often installed in expanded metal trays, ladder racks, or other approved cable-support systems.

Nonmetallic-sheathed cables are almost universally used in single-family house wiring in the United States and in many multifamily occupancies, although armored cable is still required in some multifamily dwellings. Armored cable is extensively used in commercial applications. (See Fig. 19-2.) Armored cable is used in extending branch circuits from outlet boxes on rigid conduit or EMT systems to lighting fixtures in suspended ceiling work.

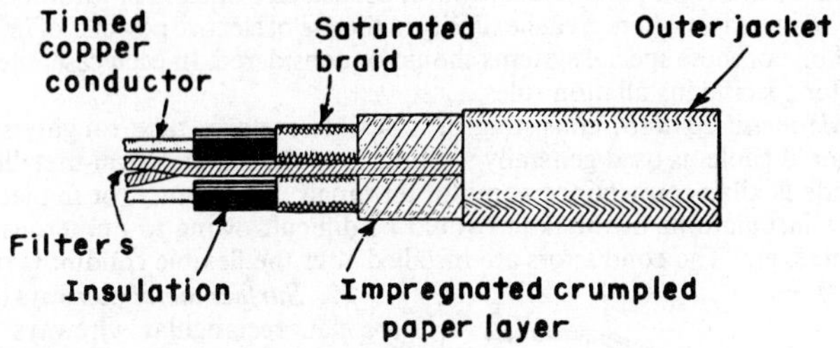


**FIG. 19-2** Nonmetallic-sheathed cable.

Metal-clad type MC cable applies to constructions using interlocked armor, close fittings, or flexible corrugated tube over No. 14 copper, No. 12 aluminum, or larger conductors.

***14. Two other metal-sheathed cables*** of special construction are recognized by the Code. Mineral-insulated metal-sheathed cable (MI) is copper-clad, containing one or more conductors and insulated with highly compressed refractory mineral insulation. It is widely used in industrial power and control wiring and in either wet or dry locations.

Aluminum-sheathed cable (ALS) consists of one or more insulated conductors in an impervious, continuous, closely fitting tube of aluminum and may also be used in wet or dry locations. Copper-sheathed cable (type CS) is a copper-clad version.

Both MI and ALS must be terminated and connected by means of fittings designed and approved for the purpose.

***15. Open wiring on knobs and cleats*** is rarely encountered in current work. Open feeders are still used in some industrial construction where low cost is a consideration, where no safety hazard is involved, and appearance is unimportant. (See Fig. 19-3.)

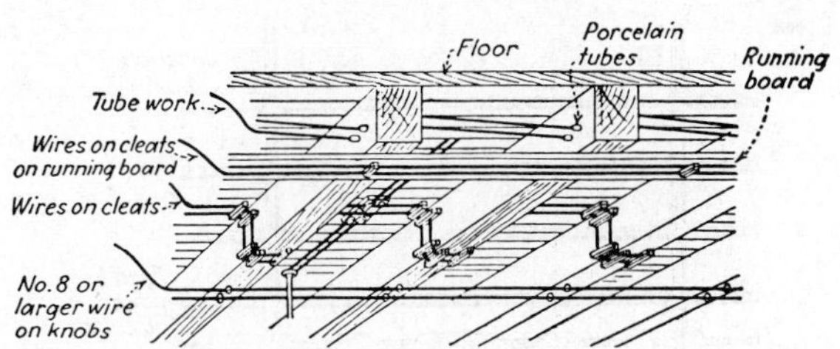


**FIG. 19-3** Methods of supporting open wiring.

***16. Cable-Assembly Systems for Limited Use.*** Several cable assemblies have been developed for particular uses, rather than for complete wiring systems for a building. The NEC should be consulted for specific requirements in each case.

***17. Service-entrance cable*** is a form of armored or nonmetallic-sheathed cable specifically approved for service-entrance use. It is available in four types: ASE, with interlocked metallic-armor protection; SE, without armor; SD, for service drops, construction similar to SE; USE, underground service-entrance cable suitable for direct burial in the ground.

***18. Nonmetallic surface extensions*** are 2-wire assemblies limited to exposed work in office (or residence) occupancies, where additional outlets are to be installed in the same room with the outlet from which the extension originates. The location must be dry and not subject to corrosive vapors. The voltage should not exceed 150 V between conductors.

*19. **Underplaster extensions*** may be used as a concealed-wiring method to install additional outlets on an existing branch circuit. They may be buried in the plaster finish of walls or ceilings in buildings of fire-resistive construction.

*20. **Raceway Systems for Limited Use.*** In general, the raceway systems developed for special purposes and discussed in Pars. 22 to 27 are of more commercial importance and find a more varied use than do the special cable-assembly systems previously discussed. This is particularly true of underfloor and cellular raceways for concealed work and of wireways and busways for exposed work. In cases where great flexibility in the use of electric power is of importance, the application of one of these special systems should be considered. In each case, the NEC should be consulted for specific installation rules.

*21. **Flexible-metal conduit,*** consisting of a flexible metallic tube roughly similar to the armor of armored cable, is used generally with rigid-conduit or electrical-metallic-tubing systems, to provide flexible connections at motor terminals, for instance, or in place of the rigid product where installations of the latter would be difficult owing to numerous bends, close working quarters, etc. The conductors are installed after the flexible conduit is in place.

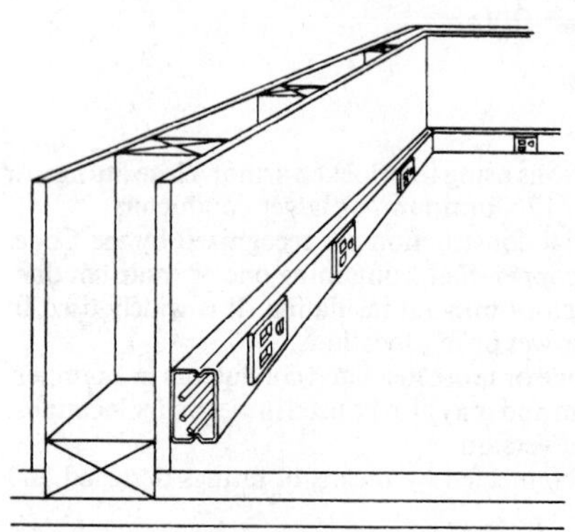

**FIG. 19-4** Typical surface raceway with plug receptacles.

*22. **Surface metal raceways*** (see Fig. 19-4) are flat, rectangular wireways used for exposed work in dry locations. They are frequently used to install additional outlets in a building already wired, where concealment of conductors is difficult, and are also used for special purposes, e.g., installation of cove lighting and for show-window reflectors. Unless made of a metal at least 0.040 in thick, they are limited to use on circuits not exceeding 300 V.

*23. **Liquid-tight flexible-metal conduit*** is, as the name suggests, a type of flexible-metal conduit having an outer jacket impervious to liquids and terminated in liquid-tight fitting. It is most widely used for connecting motors to rigid-conduit systems or fixed-equipment enclosures.

*24. **Underfloor raceways*** (Fig. 19-5) are

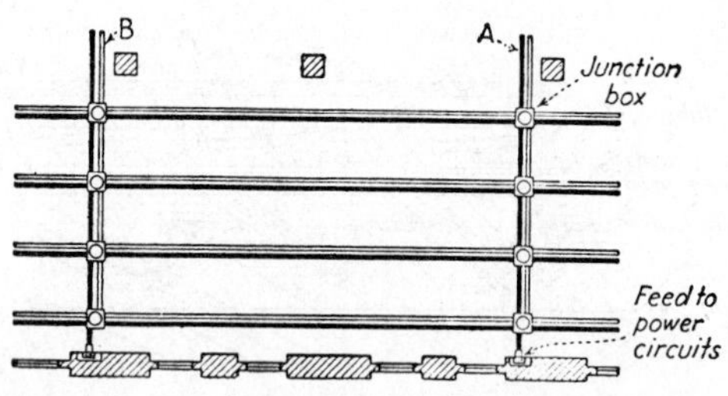


**FIG. 19-5** Layout of double underflow duct system. *(A)* For power circuits; *(B)* for signal and telephone circuits.

employed in buildings of fire-resistant construction to provide readily accessible raceways in the floor slab for light and power, telephone, and signal circuits. One, two, or three ducts are installed, depending on the desired uses. Junction boxes which mark each end of a run of raceway, and the tops of which are flush with the floor covering, make it possible to locate accurately the run of duct and, hence, to install additional outlets with the special tools provided by the manufacturer. Owing to its flexibility, this type of construction is particularly suitable for large office areas or where outlet locations are subject to change.

***25. The cellular-metal-floor raceway*** involves a cellular-steel floor (Fig. 19-6*a*), which is a structural load-carrying element whose hollow cells form the wire raceway and a system of transverse headers, together with the necessary fittings and adapters. The headers are also wire raceways, providing electrical access from distribution points to any predetermined number of cells. The system can be designed to provide overall floor and ceiling electrical service for conductors not larger than No. 0 AWG, not only for light and power but also for telephone and

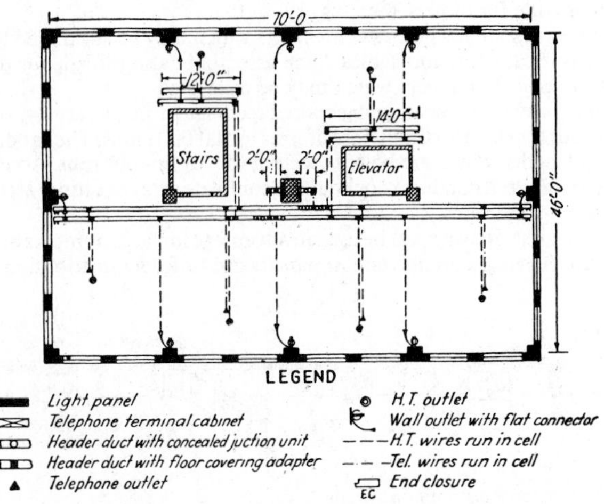


**FIG. 19-6a** Cellular-flow wiring layout.

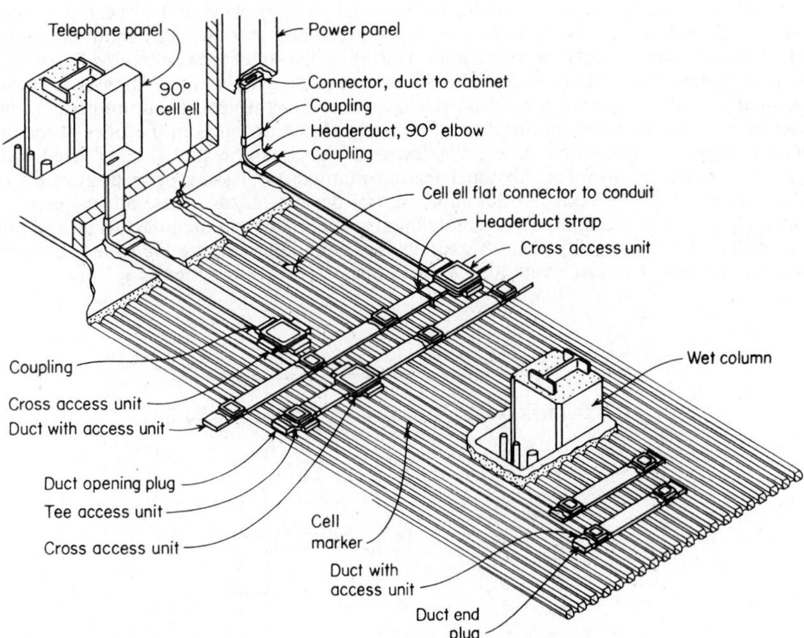


**FIG. 19-6b** Floor ducts and access units.

signal circuits. The large internal-cell areas (normally on 6-in centers) afford adequate conductor space, while the complete floor and ceiling coverage provides for great flexibility in use during the building life, since access can be had to headers and cells at any time for additional outlets, new or rerouted circuits, etc.

Cellular-concrete-floor raceways are precast slabs with tubular "cells" designed to line up in a continuous raceway. Cells terminate in metallic header ducts and other special fittings for connection to other parts of the electrical systems. Fittings approved for the purpose are inserted into the cell to provide for outlets. (See Fig. 19-6*b*.)

Structural raceways are formed-steel members which may be assembled to provide for the installation of electrical wires and cables. Such assemblies also provide for the installation of wiring devices in vertical members which may be concealed.

***26. Wireways*** provide a convenient, exposed rectangular metal raceway or trough (usually 4 by 4 in in section) for electric conductors of sizes to 500,000 cmils. The product is available in several standard lengths, which are bolted together for continuous runs. Access at any point is through hinged covers and conduit knockouts. A complete array of fittings assures flexibility for various installation conditions.

Owing to their size, wireways can be used to advantage for large numbers of conductors, e.g., a group of circuits leaving a branch circuit panelboard or feeder distribution board.

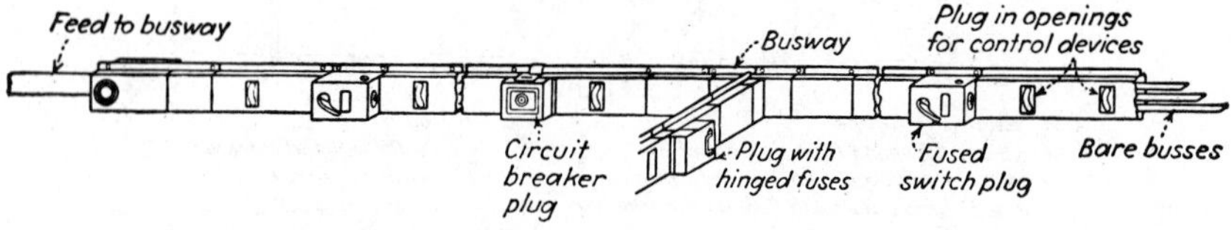


**FIG. 19-7** Units of busway distribution system.

***27. Busways*** (Fig. 19-7) are one of the more important recent developments for exposed heavy-capacity feeder and circuit wiring in industrial plants because of their flexibility in use, which makes them readily adaptable to future needs and to changing conditions such as relocation or revamping of production lines. The initial investment can be confined to immediate requirements and additions made at any time as requirements increase. The system consists essentially of interconnected prefabricated lengths or sections of steel or aluminum duct which enclose bus bars mounted on insulators. Regularly spaced openings in the sides of the duct permit plugging in branch-circuit control devices of the circuit-breaker, fuse, or fused-switch type, for convenient control of individual or group motor drives, lighting or heating circuits, etc. The ease of relocating both the duct and control devices makes its use advantageous for supplying power to machines on assembly lines, mass production manufacturing, and other applications where flexibility of electric supply is essential. Busways are available in capacities ranging from about 125 to about 3000 A, for 3-phase 3- or 4-wire systems.

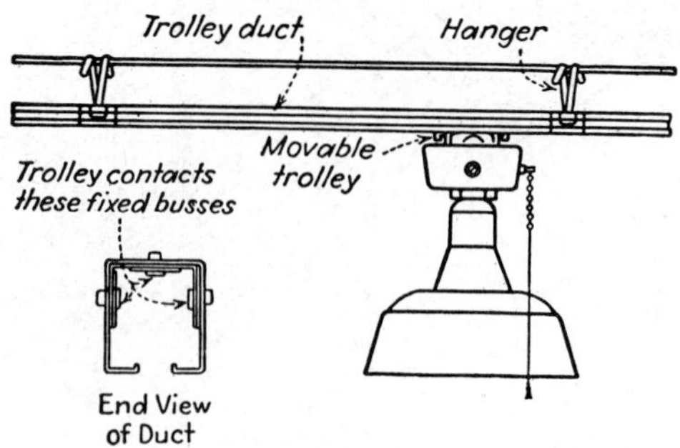


**FIG. 19-8** Trolley duct used for movable lighting fixture.

The so-called trolley duct (Fig. 19-8) is a variation of the busway in which the metal duct and electrical buses (either single-phase or 3-phase) are so arranged that access is had to the buses at any point in the run. Current is collected from the buses by movable trolleys to which are wired portable or movable electrical devices. In industrial plants the system is used to supply power to cranes and hoists, to portable tools on assembly lines and benches, etc. It has found some application in drafting rooms, stock departments, and similar locations, where ability to move lighting units quickly is of advantage.

***28. Multioutlet assemblies*** are surface-mounted raceways of metal or plastic with plug receptacle outlets at spaced intervals or provisions for the insertion of receptacles as desired. Multioutlet assemblies are widely used where a number of cord-connected appliances must be served (as along the back of a workbench or laboratory table). They are also used to provide greater convenience for the attachment of portable cords. In this application they are usually installed along the top of the baseboard (as around the perimeter of a private office).

***29. Conductors for Building Wiring.*** The various types of conductor available for interior wiring, together with their sizes, insulations, and uses, are indicated in Tables 19-1 to 19-3. Rubber and thermoplastic insulations are available in a number of compounds and constructions for resistance to heat, moisture, or other environmental conditions.

Other insulations used in building wiring include magnesium oxide, fluorinated ethylene propylene, silicone rubber, and the long-familiar varnished-cambric and asbestos constructions.

Various connector types are shown in Fig. 19-9.

***30. Dimensions of rubber-covered and thermoplastic-covered conductors*** are given in Table 19-4.

***31. Current-Carrying Capacity (Ampacity).*** As the conductors of an electrical wiring system offer some resistance, a current-carrying conductor dissipates heat. Under practical

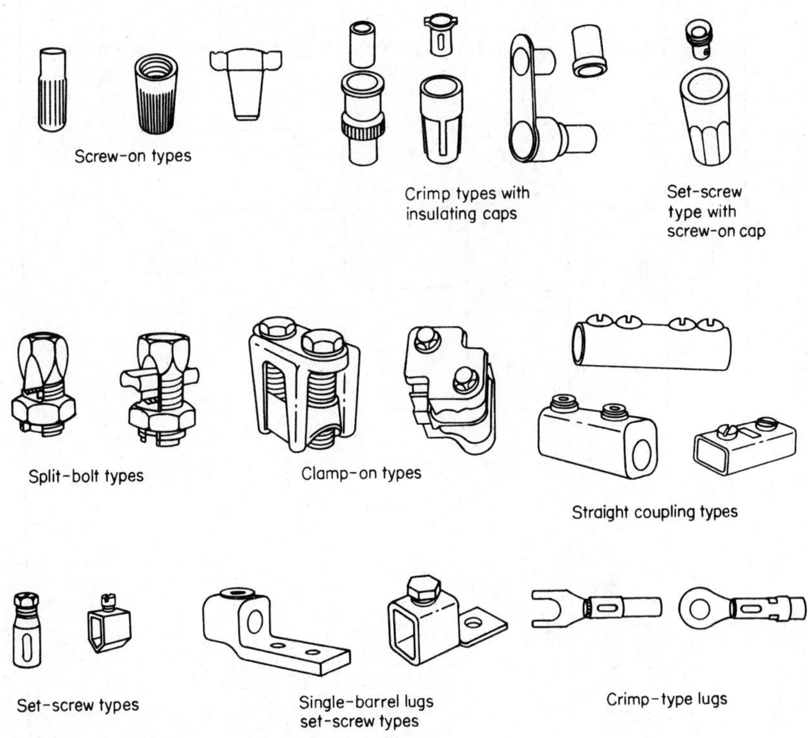


**FIG. 19-9** Types of wire connectors.

**TABLE 19-3A** Maximum Number of Conductors in Trade Sizes of Conduit or Tubing
(Based on Table 1, NEC Chapter 9)

| Type letters | Conductor size, AWG or kcmil | Conduit trade size, in | | | | | | | | | | | | |
|---|---|---|---|---|---|---|---|---|---|---|---|---|---|---|
| | | ½ | ¾ | 1 | 1¼ | 1½ | 2 | 2½ | 3 | 3½ | 4 | 4½ | 5 | 6 |
| TW, T, RUH, | 14 | 9 | 15 | 25 | 44 | 60 | 99 | 142 | | | | | | |
| RUW, | 12 | 7 | 12 | 19 | 35 | 47 | 78 | 111 | 171 | | | | | |
| XHHW (14 thru 8) | 10 | 5 | 9 | 15 | 26 | 36 | 60 | 85 | 131 | 176 | | | | |
| | 8 | 2 | 4 | 7 | 12 | 17 | 28 | 40 | 62 | 84 | 108 | | | |
| RHW and RHH | 14 | 6 | 10 | 16 | 29 | 40 | 65 | 93 | 143 | 192 | | | | |
| (without outer | 12 | 4 | 8 | 13 | 24 | 32 | 53 | 76 | 117 | 157 | | | | |
| covering), | 10 | 4 | 6 | 11 | 19 | 26 | 43 | 61 | 95 | 127 | 163 | | | |
| THW | 8 | 1 | 3 | 5 | 10 | 13 | 22 | 32 | 49 | 66 | 85 | 106 | 133 | |
| TW, | 6 | 1 | 2 | 4 | 7 | 10 | 16 | 23 | 36 | 48 | 62 | 78 | 97 | 141 |
| T, | 4 | 1 | 1 | 3 | 5 | 7 | 12 | 17 | 27 | 36 | 47 | 58 | 73 | 106 |
| THW, | 3 | 1 | 1 | 2 | 4 | 6 | 10 | 15 | 23 | 31 | 40 | 50 | 63 | 91 |
| RUH (6 thru 2), | 2 | 1 | 1 | 2 | 4 | 5 | 9 | 13 | 20 | 27 | 34 | 43 | 54 | 78 |
| RUW (6 thru 2), | 1 | | 1 | 1 | 3 | 4 | 6 | 9 | 14 | 19 | 25 | 31 | 39 | 57 |
| FEPB (6 thru 2), | 0 | | 1 | 1 | 2 | 3 | 5 | 8 | 12 | 16 | 21 | 27 | 33 | 49 |
| RHW and | 00 | | 1 | 1 | 1 | 3 | 5 | 7 | 10 | 14 | 18 | 23 | 29 | 41 |
| RHH (with- | 000 | | 1 | 1 | 1 | 2 | 4 | 6 | 9 | 12 | 15 | 19 | 24 | 35 |
| out outer covering) | 0000 | | | 1 | 1 | 1 | 3 | 5 | 7 | 10 | 13 | 16 | 20 | 29 |
| | 250 | | | 1 | 1 | 1 | 2 | 4 | 6 | 8 | 10 | 13 | 16 | 23 |
| | 300 | | | 1 | 1 | 1 | 2 | 3 | 5 | 7 | 9 | 11 | 14 | 20 |
| | 350 | | | | 1 | 1 | 1 | 3 | 4 | 6 | 8 | 10 | 12 | 18 |
| | 400 | | | | 1 | 1 | 1 | 2 | 4 | 5 | 7 | 9 | 11 | 16 |
| | 500 | | | | 1 | 1 | 1 | 1 | 3 | 4 | 6 | 7 | 9 | 14 |
| | 600 | | | | | 1 | 1 | 1 | 3 | 4 | 5 | 6 | 7 | 11 |
| | 700 | | | | | 1 | 1 | 1 | 2 | 3 | 4 | 5 | 7 | 10 |
| | 750 | | | | | 1 | 1 | 1 | 2 | 3 | 4 | 5 | 6 | 9 |

NOTE: 1 in = 25.4 mm.

**TABLE 19-3B** Maximum Number of Conductors in Trade Sizes of Conduit or Tubing
(Based on NEC Table 1, NEC Chapter 9)

| Type letters | Conductor size, AWG or kcmil | Conduit trade size, in | | | | | | | | | | | | |
|---|---|---|---|---|---|---|---|---|---|---|---|---|---|---|
| | | ½ | ¾ | 1 | 1¼ | 1½ | 2 | 2½ | 3 | 3½ | 4 | 4½ | 5 | 6 |
| | 14 | 13 | 24 | 39 | 69 | 94 | 154 | | | | | | | |
| | 12 | 10 | 18 | 29 | 51 | 70 | 114 | 164 | | | | | | |
| THWN, | 10 | 6 | 11 | 18 | 32 | 44 | 73 | 104 | 160 | | | | | |
| | 8 | 3 | 5 | 9 | 16 | 22 | 36 | 51 | 79 | 106 | 136 | | | |
| THHN, | 6 | 1 | 4 | 6 | 11 | 15 | 26 | 37 | 57 | 76 | 98 | 125 | 154 | |
| | 4 | 1 | 2 | 4 | 7 | 9 | 16 | 22 | 35 | 47 | 60 | 75 | 94 | 137 |
| FEP (14 thru 2), | 3 | 1 | 1 | 3 | 6 | 8 | 13 | 19 | 29 | 39 | 51 | 64 | 80 | 116 |
| FEPB (14 thru 8), | 2 | 1 | 1 | 3 | 5 | 7 | 11 | 16 | 25 | 33 | 43 | 54 | 67 | 97 |
| PFA (14 thru 4/0) | 1 | | 1 | 1 | 3 | 5 | 8 | 12 | 18 | 25 | 32 | 40 | 50 | 72 |
| PFAH (14 thru 4/0)<br>XHHW (4 thru | 0 | | 1 | 1 | 3 | 4 | 7 | 10 | 15 | 21 | 27 | 33 | 42 | 61 |
| 500 kcmil) | 00 | | 1 | 1 | 2 | 3 | 6 | 8 | 13 | 17 | 22 | 28 | 35 | 51 |
| | 000 | | 1 | 1 | 1 | 3 | 5 | 7 | 11 | 14 | 18 | 23 | 29 | 42 |
| Z (14 thru 4/0) | 0000 | | 1 | 1 | 1 | 2 | 4 | 6 | 9 | 12 | 15 | 19 | 24 | 35 |
| | 250 | | | 1 | 1 | 1 | 3 | 4 | 7 | 10 | 12 | 16 | 20 | 28 |
| | 300 | | | 1 | 1 | 1 | 3 | 4 | 6 | 8 | 11 | 13 | 17 | 24 |
| | 350 | | | 1 | 1 | 1 | 2 | 3 | 5 | 7 | 9 | 12 | 15 | 21 |
| | 400 | | | | 1 | 1 | 1 | 3 | 5 | 6 | 8 | 10 | 13 | 19 |
| | 500 | | | | 1 | 1 | 1 | 2 | 4 | 5 | 7 | 9 | 11 | 16 |
| | 600 | | | | 1 | 1 | 1 | 1 | 3 | 4 | 5 | 7 | 9 | 13 |
| | 700 | | | | | 1 | 1 | 1 | 3 | 4 | 5 | 6 | 8 | 11 |
| | 750 | | | | | 1 | 1 | 1 | 2 | 3 | 4 | 6 | 7 | 11 |
| XHHW | 6 | 1 | 3 | 5 | 9 | 13 | 21 | 30 | 47 | 63 | 81 | 102 | 128 | 185 |
| | 600 | | | | 1 | 1 | 1 | 1 | 3 | 4 | 5 | 7 | 9 | 13 |
| | 700 | | | | | 1 | 1 | 1 | 3 | 4 | 5 | 6 | 7 | 11 |
| | 750 | | | | | 1 | 1 | 1 | 2 | 3 | 4 | 6 | 7 | 10 |

NOTE: 1 in = 25.4 mm.

**TABLE 19-3C** Maximum Number of Conductors in Trade Sizes of Conduit or Tubing
(Based on NEC Table 1, NEC Chapter 9)

| Type letters | Conductor size, AWG or kcmil | Conduit trade size, in | | | | | | | | | | | | |
|---|---|---|---|---|---|---|---|---|---|---|---|---|---|---|
| | | ½ | ¾ | 1 | 1¼ | 1½ | 2 | 2½ | 3 | 3½ | 4 | 4½ | 5 | 6 |
| | 14 | 3 | 6 | 10 | 18 | 25 | 41 | 58 | 90 | 121 | 155 | | | |
| | 12 | 3 | 5 | 9 | 15 | 21 | 35 | 50 | 77 | 103 | 132 | | | |
| RHW, | 10 | 2 | 4 | 7 | 13 | 18 | 29 | 41 | 64 | 86 | 110 | 138 | | |
| | 8 | 1 | 2 | 4 | 7 | 9 | 16 | 22 | 35 | 47 | 60 | 75 | 94 | 137 |
| RHH | 6 | 1 | 1 | 2 | 5 | 6 | 11 | 15 | 24 | 32 | 41 | 51 | 64 | 93 |
| | 4 | 1 | 1 | 1 | 3 | 5 | 8 | 12 | 18 | 24 | 31 | 39 | 50 | 72 |
| (with | 3 | 1 | 1 | 1 | 3 | 4 | 7 | 10 | 16 | 22 | 28 | 35 | 44 | 63 |
| outer | 2 | | 1 | 1 | 3 | 4 | 6 | 9 | 14 | 19 | 24 | 31 | 38 | 56 |
| covering) | 1 | | 1 | 1 | 1 | 3 | 5 | 7 | 11 | 14 | 18 | 23 | 29 | 42 |
| | 0 | | 1 | 1 | 1 | 2 | 4 | 6 | 9 | 12 | 16 | 20 | 25 | 37 |
| | 00 | | | 1 | 1 | 1 | 3 | 5 | 8 | 11 | 14 | 18 | 22 | 32 |
| | 000 | | | 1 | 1 | 1 | 3 | 4 | 7 | 9 | 12 | 15 | 19 | 28 |
| | 0000 | | | 1 | 1 | 1 | 2 | 4 | 6 | 8 | 10 | 13 | 16 | 24 |
| | 250 | | | | 1 | 1 | 1 | 3 | 5 | 6 | 8 | 11 | 13 | 19 |
| | 300 | | | | 1 | 1 | 1 | 3 | 4 | 5 | 7 | 9 | 11 | 17 |
| | 350 | | | | 1 | 1 | 1 | 2 | 4 | 5 | 6 | 8 | 10 | 15 |
| | 400 | | | | 1 | 1 | 1 | 1 | 3 | 4 | 6 | 7 | 9 | 14 |
| | 500 | | | | 1 | 1 | 1 | 1 | 3 | 4 | 5 | 6 | 8 | 11 |
| | 600 | | | | | 1 | 1 | 1 | 2 | 3 | 4 | 5 | 6 | 9 |
| | 700 | | | | | 1 | 1 | 1 | 1 | 3 | 3 | 4 | 6 | 8 |
| | 750 | | | | | | 1 | 1 | 1 | 3 | 3 | 4 | 5 | 8 |

Notes to Tables 19-3A, B, and C:

1. Tables 19-3A to 19-3C apply only to complete conduit or tubing systems and are not intended to apply to short sections of conduit or tubing used to protect exposed wiring from physical damage.

2. Equipment grounding conductors, when installed, shall be included when calculating conduit or tubing fill. The actual dimensions of the equipment grounding conductor (insulated or bare) shall be used in the calculation.

3. When conduit nipples having a maximum length not to exceed 24 in are installed between boxes, cabinets, and similar enclosures, the nipple shall be permitted to be filled to 60 percent of its total cross-sectional area.

4. For conductors not included in NEC Chapter 9, the actual dimensions shall be used.

5. See the following table for the allowable percentage of conduit or tubing fill.

Percent of Cross Section of Conduit and Tubing for Conductors (See Table 2 of the NEC for Fixture Wires)

| | Number of conductors | | | | |
|---|---|---|---|---|---|
| | 1 | 2 | 3 | 4 | Over 4 |
| All conductor types except lead-covered (new or rewiring) | 53 | 31 | 40 | 40 | 40 |
| Lead-covered conductors | 55 | 30 | 40 | 38 | 35 |

NOTE 1: See Tables 19-3A, 19-3B, and 19-3C for number of conductors all of the same size in trade sizes of conduit ½ inch through 6 inch.

NOTE 2: For conductors larger than 750 MCM or for combinations of conductors of different sizes, use Tables 4 through 8, NEC Chapter 9, for dimensions of conductors, conduit and tubing.

NOTE 3: Where the calculated number of conductors, all of the same size, includes a decimal fraction, the next higher whole number shall be used where this decimal is 0.8 or larger.

NOTE 4: When bare conductors are permitted by other Sections of the NE Code, the dimensions for bare conductors in Table 8 of NEC Chapter 9 shall be permitted.

NOTE 5: A multiconductor cable of three or more conductors shall be treated as a single conductor cable for calculating percentage conduit fill area.

**TABLE 19-4** Dimensions of Rubber-Covered and Thermoplastic-Covered Conductors

| Size, AWG or kcmil (1) | Types RFH-2, RH, RHH,* RHW,* SF-2 | | Types TF, T, THW,† TW, RUH,‡ RUW‡ | | Types TFN, THHN, THWN | | Types §FEP, FEPB, FEPW, TFE, PF, PFA, PFAH, PGF, PTF, Z, ZF, ZFF | | Type XHHW | |
|---|---|---|---|---|---|---|---|---|---|---|
| | Approx. diam., in (2) | Approx. area, in² (3) | Approx. diam., in (4) | Approx. area, in² (5) | Approx. diam., in (6) | Approx. area, in² (7) | Approx. diam., in (8) | Approx area, in² (9) | Approx. diam., in (10) | Approx. area, in² (11) |
| 18 | .146 | .0167 | .106 | .0088 | .089 | .0064 | .081 | .0052 | ..... | ....... |
| 16 | .158 | .0196 | .118 | .0109 | .100 | .0079 | .092 | .0066 | ..... | ....... |
| 14 | 30 mils .171 | .0230 | .131 | .0135 | .105 | .0087 | .105 .105 | .0087 .0087 | | |
| 14 | 45 mils .204¶ | .0327¶ | .... | ..... | .... | ..... | .... .... | ..... ..... | | |
| 14 | ..... ..... | ..... | .162† | .0206† | .... | ..... | .... .... | ..... ..... | .129 | .0131 |
| 12 | 30 mils .188 | .0278 | .148 | .0172 | .122 | .0117 | .121 .121 | .0115 .0115 | | |
| 12 | 45 mils .221¶ | .0384¶ | .... | ..... | .... | ..... | .... .... | ..... ..... | | |
| 12 | ..... ..... | ..... | .179† | .0251† | .... | ..... | .... .... | ..... ..... | .146 | .0167 |
| 10 | ...... .242 | .0460 | .168 | .0224 | .153 | .0184 | .142 .142 | .0159 .0159 | | |
| 10 | ...... ..... | ..... | .199† | .0311† | .... | ..... | .... .... | ..... ..... | .166 | .0216 |
| 8 | ...... .328 | .0854 | .245 | .0471 | .218 | .0373 | .206 .186 | .0333 .0272 | | |
| 8 | ...... ..... | ..... | .276† | .0598† | .... | ..... | .... .... | ..... ..... | .241 | .0456 |
| 6 | .397 | .1238 | .323 | .0819 | .257 | .0519 | .244 .302 | .0467 .0716 | .282 | .0625 |
| 4 | .452 | .1605 | .372 | .1087 | .328 | .0845 | .292 .350 | .0669 .0962 | .328 | .0845 |
| 3 | .481 | .1817 | .401 | .1263 | .356 | .0995 | .320 .378 | .0803 .1122 | .356 | .0995 |
| 2 | .513 | .2067 | .433 | .1473 | .388 | .1182 | .352 .410 | .0973 .1316 | .388 | .1182 |
| 1 | .588 | .2715 | .508 | .2027 | .450 | .1590 | .420 .... | .1385 ..... | .450 | .1590 |
| 0 | .629 | .3107 | .549 | .2367 | .491 | .1893 | .462 .... | .1676 ..... | .491 | .1893 |
| 00 | .675 | .3578 | .595 | .2781 | .537 | .2265 | .498 .... | .1974 ..... | .537 | .2265 |
| 000 | .727 | .4151 | .647 | .3288 | .588 | .2715 | .560 .... | .2463 ..... | .588 | .2715 |
| 0000 | .785 | .4840 | .705 | .3904 | .646 | .3278 | .618 .... | .2999 ..... | .646 | .3278 |
| 250 | .868 | .5917 | .788 | .4877 | .716 | .4026 | | | .716 | .4026 |
| 300 | .933 | .6837 | .843 | .5581 | .771 | .4669 | | | .771 | .4669 |
| 350 | .985 | .7620 | .895 | .6291 | .822 | .5307 | | | .822 | .5307 |
| 400 | 1.032 | .8365 | .942 | .6969 | .869 | .5931 | | | .869 | .5931 |
| 500 | 1.119 | .9834 | 1.029 | .8316 | .955 | .7163 | | | .955 | .7163 |

| | | | | | | | | | | |
|---|---|---|---|---|---|---|---|---|---|---|
| 600 | 1.233 | 1.1940 | 1.143 | 1.0261 | 1.058 | .8792 | | | 1.073 | .9043 |
| 700 | 1.304 | 1.3355 | 1.214 | 1.1575 | 1.129 | 1.0011 | | | 1.145 | 1.0297 |
| 750 | 1.339 | 1.4082 | 1.249 | 1.2252 | 1.163 | 1.0623 | | | 1.180 | 1.0936 |
| 800 | 1.372 | 1.4784 | 1.282 | 1.2908 | 1.196 | 1.1234 | | | 1.210 | 1.1499 |
| 900 | 1.435 | 1.6173 | 1.345 | 1.4208 | 1.259 | 1.2449 | | | 1.270 | 1.2668 |
| 1000 | 1.494 | 1.7531 | 1.404 | 1.5482 | 1.317 | 1.3623 | | | 1.330 | 1.3893 |
| 1250 | 1.676 | 2.2062 | 1.577 | 1.9532 | | | | | 1.500 | 1.7672 |
| 1500 | 1.801 | 2.5475 | 1.702 | 2.2748 | | | | | 1.620 | 2.0612 |
| 1750 | 1.916 | 2.8895 | 1.817 | 2.5930 | | | | | 1.740 | 2.3779 |
| 2000 | 2.021 | 3.2079 | 1.922 | 2.9013 | | | | | 1.840 | 2.6590 |

*Dimensions of RHH and RHW without outer covering are the same as THW.

† Dimensions of THW in sizes 14 to 8. No. 6 THW and larger is the same dimension as T.

‡ No. 14 to No. 2.

§ In Columns 8 and 9 the values shown for sizes No. 1 through 0000 are for TFE only. The right-hand values in Columns 8 and 9 are for FEPB only.

¶ The dimensions of Types RHH and RHW.

**TABLE 19-5** Conductor Properties

| Size: AWG, MCM | Area, cmils | Stranding | | Overall | | DC resistance ohms/1000 ft | | | | | |
|---|---|---|---|---|---|---|---|---|---|---|---|
| | | | | | | Copper | | | | Aluminum | |
| | | | | | | 75°C (167°F) | | 25°C (77°F) | | | |
| | | Quantity | Diam., in | Diam., in | Area, $in^2$ | Uncoated | Coated | Uncoated | Coated | 75°C (167°F) | 25°C (77°F) |
| 18 | 1620 | 1 | — | 0.040 | 0.001 | 7.77 | 8.08 | 6.51 | 6.79 | 12.8 | 10.7 |
| 18 | 1620 | 7 | 0.015 | 0.046 | 0.002 | 7.95 | 8.45 | | | 13.1 | |
| 16 | 2580 | 1 | — | 0.051 | 0.002 | 4.89 | 5.08 | 4.10 | 4.26 | 8.05 | 6.72 |
| 16 | 2580 | 7 | 0.019 | 0.058 | 0.003 | 4.99 | 5.29 | | | 8.21 | |
| 14 | 4110 | 1 | — | 0.064 | 0.003 | 3.07 | 3.19 | 2.57 | 2.68 | 5.06 | 4.22 |
| 14 | 4110 | 7 | 0.024 | 0.073 | 0.004 | 3.14 | 3.26 | | | 5.17 | |
| 12 | 6530 | 1 | — | 0.081 | 0.005 | 1.93 | 2.01 | 1.62 | 1.68 | 3.18 | 2.66 |
| 12 | 6530 | 7 | 0.030 | 0.092 | 0.006 | 1.98 | 2.05 | | | 3.25 | |
| 10 | 10380 | 1 | — | 0.102 | 0.008 | 1.21 | 1.26 | 1.018 | 1.06 | 2.00 | 1.67 |
| 10 | 10380 | 7 | 0.038 | 0.116 | 0.011 | 1.24 | 1.29 | | | 2.04 | |
| 8 | 16510 | 1 | — | 0.128 | 0.013 | 0.764 | 0.786 | 0.6404 | 0.659 | 1.26 | 1.05 |
| 8 | 16510 | 7 | 0.049 | 0.146 | 0.017 | 0.778 | 0.809 | | | 1.28 | |
| 6 | 26240 | 7 | 0.061 | 0.184 | 0.027 | 0.491 | 0.510 | 0.410 | 0.427 | 0.808 | 0.674 |
| 4 | 41740 | 7 | 0.077 | 0.232 | 0.042 | 0.308 | 0.321 | 0.259 | 0.269 | 0.508 | 0.424 |
| 3 | 52620 | 7 | 0.087 | 0.260 | 0.053 | 0.245 | 0.254 | 0.205 | 0.213 | 0.403 | 0.336 |
| 2 | 66360 | 7 | 0.097 | 0.292 | 0.067 | 0.194 | 0.201 | 0.162 | 0.169 | 0.319 | 0.266 |
| 1 | 83690 | 19 | 0.066 | 0.332 | 0.087 | 0.154 | 0.160 | 0.129 | 0.134 | 0.253 | 0.211 |
| 1/0 | 105600 | 19 | 0.074 | 0.373 | 0.109 | 0.122 | 0.127 | 0.102 | 0.106 | 0.201 | 0.168 |

| | | | | | | | | | | | |
|---|---|---|---|---|---|---|---|---|---|---|---|
| 2/0 | 133100 | 19 | 0.084 | 0.419 | 0.138 | 0.967 | 0.101 | 0.0811 | 0.0843 | 0.159 | 0.133 |
| 3/0 | 167800 | 19 | 0.094 | 0.470 | 0.173 | 0.0766 | 0.0797 | 0.0642 | 0.0668 | 0.126 | 0.105 |
| 4/0 | 211600 | 19 | 0.106 | 0.528 | 0.219 | 0.0608 | 0.0626 | 0.0509 | 0.0525 | 0.100 | 0.0836 |
| 250 | — | 37 | 0.082 | 0.575 | 0.260 | 0.0515 | 0.0535 | 0.0431 | 0.0449 | 0.0847 | 0.0708 |
| 300 | — | 37 | 0.090 | 0.630 | 0.312 | 0.0429 | 0.0446 | 0.0360 | 0.0374 | 0.0707 | 0.0590 |
| 350 | — | 37 | 0.097 | 0.681 | 0.364 | 0.0367 | 0.0382 | 0.0308 | 0.0320 | 0.0605 | 0.0505 |
| 40 | — | 37 | 0.104 | 0.728 | 0.416 | 0.0321 | 0.0331 | 0.0270 | 0.0278 | 0.0529 | 0.0442 |
| 500 | — | 37 | 0.116 | 0.813 | 0.519 | 0.0258 | 0.0265 | 0.0216 | 0.0222 | 0.0424 | 0.0354 |
| 600 | — | 61 | 0.992 | 0.893 | 0.626 | 0.0214 | 0.0223 | 0.0180 | 0.0187 | 0.0353 | 0.0295 |
| 700 | — | 61 | 0.107 | 0.964 | 0.730 | 0.0184 | 0.0189 | 0.0154 | 0.0159 | 0.0303 | 0.0253 |
| 750 | — | 61 | 0.111 | 0.998 | 0.782 | 0.0171 | 0.0176 | 0.0144 | 0.0148 | 0.0282 | 0.0236 |
| 800 | — | 61 | 0.114 | 1.03 | 0.834 | 0.0161 | 0.0166 | 0.0135 | 0.0139 | 0.0265 | 0.0221 |
| 900 | — | 61 | 0.122 | 1.09 | 0.940 | 0.0143 | 0.0147 | 0.0120 | 0.0123 | 0.0235 | 0.0197 |
| 1000 | — | 61 | 0.128 | 1.15 | 1.04 | 0.0129 | 0.0132 | 0.0108 | 0.0111 | 0.0212 | 0.0177 |
| 1250 | — | 91 | 0.117 | 1.29 | 1.30 | 0.0103 | 0.0106 | 0.00863 | 0.00888 | 0.0169 | 0.0142 |
| 1500 | — | 91 | 0.128 | 1.41 | 1.57 | 0.00858 | 0.00883 | 0.00719 | 0.00740 | 0.0141 | 0.0118 |
| 1750 | — | 127 | 0.117 | 1.52 | 1.83 | 0.00735 | 0.00756 | 0.00616 | 0.00634 | 0.0121 | 0.0101 |
| 2000 | — | 127 | 0.126 | 1.63 | 2.09 | 0.00643 | 0.00662 | 0.00539 | 0.00555 | 0.0106 | 0.00885 |

NOTES: These resistance values are valid *only* for the parameters as given. Using conductors having coated strands, different stranding type, and especially, other temperatures, change the resistance.

Formula for temperature change: $R_2 = R_1 [1 + \alpha(T_2 - 20)]$ where: $\alpha_{cu} = 0.00393$, $\alpha_{AL} = 0.00403$.

Class B stranding is listed as well as solid for some sizes. Its overall diameter and area is that of its circumscribing circle. The construction information is per NEMA WC8-1976 (Rev 5-1980). The resistance is calculated per National Bureau of Standards Handbook 100, dated 1966, and Handbook 109, dated 1972.

Conductors with compact and compressed stranding have about 9 percent and 3 percent, respectively, smaller bare conductor diameters than those shown.

The IACS conductivities used: bare copper = 100%, aluminum = 61%.

1 in = 25.4 mm; 1 in$^2$ = 645 mm$^2$.

**TABLE 19-6** Dimensions and Percent Area of Conduit and of Tubing*

(Areas of Conduit or Tubing for the Combinations of Wires Permitted in Table 1, NEC Chapter 9)

| | | Area, in² | | | | | | | | |
|---|---|---|---|---|---|---|---|---|---|---|
| | | | Not lead covered | | | Lead covered | | | | |
| Trade size | Internal diameter, in | Total 100% | 2 Cond. 31% | Over 2 cond. 40% | 1 Cond. 53% | 1 Cond. 55% | 2 Cond. 30% | 3 Cond. 40% | 4 Cond. 38% | Over 4 cond. 35% |
| ½ | .622 | .30 | .09 | .12 | .16 | .17 | .09 | .12 | .11 | .11 |
| ¾ | .824 | .53 | .16 | .21 | .28 | .29 | .16 | .21 | .20 | .19 |
| 1 | 1.049 | .86 | .27 | .34 | .46 | .47 | .26 | .34 | .33 | .30 |
| 1¼ | 1.380 | 1.50 | .47 | .60 | .80 | .83 | .45 | .60 | .57 | .53 |
| 1½ | 1.610 | 2.04 | .63 | .82 | 1.08 | 1.12 | .61 | .82 | .78 | .71 |
| 2 | 2.067 | 3.36 | 1.04 | 1.34 | 1.78 | 1.85 | 1.01 | 1.34 | 1.28 | 1.18 |
| 2½ | 2.469 | 4.79 | 1.48 | 1.92 | 2.54 | 2.63 | 1.44 | 1.92 | 1.82 | 1.68 |
| 3 | 3.068 | 7.38 | 2.29 | 2.95 | 3.91 | 4.06 | 2.21 | 2.95 | 2.80 | 2.58 |
| 3½ | 3.548 | 9.90 | 3.07 | 3.96 | 5.25 | 5.44 | 2.97 | 3.96 | 3.76 | 3.47 |
| 4 | 4.026 | 12.72 | 3.94 | 5.09 | 6.74 | 7.00 | 3.82 | 5.09 | 4.83 | 4.45 |
| 4½ | 4.506 | 15.94 | 4.94 | 6.38 | 8.45 | 8.77 | 4.78 | 6.38 | 6.06 | 5.56 |
| 5 | 5.047 | 20.00 | 6.20 | 8.00 | 10.60 | 11.00 | 6.00 | 8.00 | 7.60 | 7.00 |
| 6 | 6.065 | 28.89 | 8.96 | 11.56 | 15.31 | 15.89 | 8.67 | 11.56 | 10.98 | 10.11 |

* This table gives the nominal size of conductors and conduit or tubing for use in computing size of conduit or tubing for various combinations of conductors. The dimensions represent average conditions only, and variations will be found in dimensions of conductors and conduits of different manufacture.

NOTE: 1 in = 25.4 mm; 1 in² = 645 mm².

SOURCE: Adapted from NEC Table 4, Chapter 9.

conditions of installation and operation, the temperatures reached must not result in the destruction of the insulation or risk to surrounding material.

Tables of maximum allowable current-carrying capacity are given in the NEC. Allowable ampacities for insulated conductors are based upon an allowable temperature rise above an ambient of 30°C (86°F). A list of temperature ratings for types of insulated conductors is given in Table 19-1.

***32. Allowable ampacities for copper conductors*** and aluminum conductors in accordance with the temperature rating of the insulation are given for installation in conduit and for installation in free air in Table 19-2A through 19-2F.

***33. Conductor and conduit diameters and areas*** are frequently necessary to calculate fill. Nominal values for conductors are given in Table 19-4; for conduit and tubing, in Tables 19-3A, B, and C. Table 19-5 gives resistances of conductors in ohms per 1000 ft.

***34. Permissible Percent Raceway Fill for Conductor Combinations.*** See Table 19-6.

***35. Number of Conductors in One Conduit or Tubing.*** The number of conductors of a certain size that may be installed in a given-sized conduit or electrical metallic tubing is limited to provide for ready installation and withdrawal without injury to conductor or insulating covering. Tables 19-3A, B, and C give these values for commonly used conductors.

***36. In considering the ampacity*** of conductors and conduit fill it is important to note that derating of ampacity applies for increased ambient temperatures and for more than 3 conductors, excluding neutrals, in a conduit.

***37. Flexible cords and fixture wire*** in some cases may be as small as No. 18 AWG; hence these are exceptions to the general rule that no conductor smaller than No. 14 should be used in light and power wiring. Owing to the heat generated in the lamp, heat-resistant wiring is required in fixtures.

## TYPES OF CIRCUIT

***38. Services and Feeders.*** No limit is placed on the electrical capacity of service conductors and service protection employed in bringing the electric supply into a building, since one supply only should be introduced whenever possible. Near the point of entrance of the supply, the heavy-service conductors are tapped by feeders which conduct the electricity to panelboards at various load centers in the building, where the final branch circuits which supply individual lighting, heating, and power outlets originate (see Fig. 19-10). No limits are placed on the

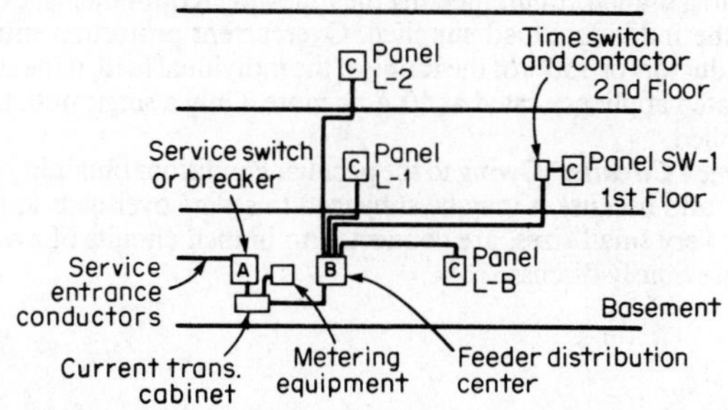


**FIG. 19-10** Riser diagram showing location of *(A)* service; *(B)* feeder, *(C)* branch circuit overcurrent protective devices.

electrical capacity of feeders, but for practical purposes they are limited in size by the difficulty of handling large conductors and raceways in restricted building spaces, by voltage drop, and by economic considerations.

Each lighting fixture, motor, heating device, or other item of utilization equipment must be supplied by one of the types of branch circuit of Pars. **39** to **41.**

**TABLE 19-7** Summary of Branch-Circuit Requirements
(Type FEP, FEPB, RUW, SA, T, TW, RH, RUH, RHW, RHH, THHN, THW, THWN, and XHHW conductors in raceway or cable)

| | Circuit rating | | | | |
|---|---|---|---|---|---|
| | 15 A | 20 A | 30 A | 40 A | 50 A |
| Conductors: | | | | | |
| (Min. size) | | | | | |
| Circuit wires* | 14 | 12 | 10 | 8 | 6 |
| Taps | 14 | 14 | 14 | 12 | 12 |
| Fixture wires and cords | Refer to NEC Section 240-4 | | | | |
| Overcurrent protection | 15 A | 20 A | 30 A | 40 A | 50 A |
| Outlet devices: | | | | | |
| Lampholders permitted | Any type | Any type | Heavy duty | Heavy duty | Heavy duty |
| Receptacle rating† | 15 A max. | 15 or 20 A | 30 A | 40 or 50 A | 50 A |
| Maximum load | 15 A | 20 A | 30 A | 40 A | 50 A |
| Permissible load | Refer to NEC Section 210-23(a) | Refer to NEC Section 210-23(a) | Refer to NEC Section 210-23(b) | Refer to NEC Section 210-23(c) | Refer to NEC Section 210-23(c) |

* These ampacities are for copper conductors where derating is not required. See Tables 19-2A to 19-2D.
† For receptacle rating of cord-connected electric-discharge lighting fixtures see NEC Section 410-14.
SOURCE: Adapted from NEC.

***39. Branch Circuits for Grouped Loads.*** The uses and limitations of the common types of branch circuit are outlined in Table 19-7. It will be noted that lighting branch circuits may carry loads as high as 50 A, although fluorescent lighting is limited to use on circuits of 15-A or 20-A rating. Such circuits are extensively employed in commercial and industrial occupancies. Branch circuits supplying convenience outlets for general use in other than manufacturing areas are usually limited to a maximum of 20 A, as the type of outlet required for heavier-capacity circuits usually will not accommodate the connection plug found on portable cords or lamps, motor-driven office machinery, etc.

***40. Individual Branch Circuits.*** Any individual piece of equipment (except motors) may also be connected to a branch circuit meeting the following requirements: Conductors must be large enough for the individual load supplied. Overcurrent protection must not exceed the capacity of the conductors or 150% of the rating of the individual load, if the single load device is a non-motor-operated appliance rated at 10 A or more. Only a single outlet or piece of equipment may be supplied.

***41. Motor Branch Circuits.*** Owing to the peculiar conditions obtaining during the starting period of a motor, and because it may be subjected to severe overloads at frequent intervals, motors, except for very small sizes, are connected to branch circuits of a somewhat different design from that previously discussed.

## PROTECTION

*By WENDELL LEISINGER*

***42. Molded Case Circuit Breakers.*** A circuit breaker is defined by the NEC as

> A device designed to open and close a circuit by nonautomatic means and to open the circuit automatically on a predetermined overcurrent without injury to itself when properly applied within its ratings.

**FIG. 19-11** Typical molded-case circuit breakers.

In addition, the purpose of a circuit breaker (Fig. 19-11), as stated in the NEC, is to open the circuit if the current reaches a value that will cause an excessive or dangerous temperature in a conductor or its insulation.

Underwriters' Laboratories (UL) Standard UL489, "Molded Case Circuit Breakers and Circuit Breaker Enclosures," further specifies test and construction requirements for molded-case circuit breakers specifically intended to provide service entrance and feeder and branch circuit protection in accordance with the NEC. These UL requirements cover molded-case circuit breakers rated through 600 V. UL specifies the applicable standards to which the circuit breaker is designed, whereas the NEC discusses proper applications of circuit breakers.

***43. Ratings.*** The ratings which apply to circuit breakers and their actual assigned numerical values reflect mechanical, electrical, and thermal capabilities of those circuit breakers that comply with industry standards published by the National Electrical Manufacturers Association and UL. These ratings, which appear on the breaker, include the following:

1. *Voltage.* Circuit breakers are designed and marked with the maximum voltage at which they can be applied. They can be used on any system where the voltage is lower than this breaker rating. This includes both ac and dc voltage systems.
2. *Frequency.* Circuit breakers are normally suitable for use on 50- and 60-Hz electrical distribution systems. Rerating of the circuit breaker may be required at other frequencies.
3. *Continuous current.* Standard molded-case circuit breakers are calibrated to carry 100% of their rated current in free air at 40°C ambient. In accordance with the NEC, when installed in their enclosures, these breakers should not be continuously loaded over 80% of their current rating. However, there are certain molded-case circuit breakers specifically approved for 100% continuous rating available, and these are so marked.
4. *Current interrupting.* The current-interrupting rating is expressed in rms symmetrical amperes. It may vary with the applied voltage and is the maximum current the breaker can be expected to safely interrupt. These current-interrupting ratings may vary from 1500 through 200,000 A, depending upon applied voltage. UL requires all circuit breakers to be so marked with their proper ratings.

***44. Thermal-Magnetic Circuit Breakers.*** Thermal-magnetic circuit breakers provide two forms of overcurrent protection. The first is overload protection, which is achieved by a bimetal providing an inverse time-current response. The second is overcurrent protection, which is achieved magnetically.

Overload tripping is obtained through deflection of the bimetal, which is heated by the load current. During an overload condition, the bimetal deflects, causing the breaker to trip or open mechanically. The larger the overload, the faster the tripping of the circuit breaker; the smaller the overload, the longer it takes the circuit breaker to trip. This is commonly called an *inverse time principle.*

Overcurrent (short-circuit) protection is obtained through electromagnetic tripping action without any intentional time delay. An overcurrent condition must be interrupted rapidly (usually less than 20 ms) in order to protect downstream equipment. During overcurrent conditions, an armature is moved by electromagnetic force and initiates tripping action.

All circuit breakers are rated for continuous current. This continuous-current rating is dependent upon the maximum current the load is expected to draw for 3 h or more continuously. Ratings may be fixed or adjustable, depending on the circuit breaker. Some constructions may require replacing all or part of the trip unit to change the continuous current rating. Overcurrent trip characteristics are a function, multiple, or percentage of the continuous-current rating.

***45. Current-Limiting Circuit Breakers.*** The NEC defines a current-limiting overcurrent protective device as

> a device which, when interrupting current in its current limiting range, will reduce the current flow in the faulted current to a magnitude substantially less than that obtainable in the same circuit if the device were replaced with a solid conductor having comparable impedance.

Additionally, UL has the following current-limiting circuit breaker definition:

> A circuit breaker that does not employ a fusible element and that when operating within its current-limiting range, limits the let-through $I^2$ to a value less than the $I^2$ of a ½ cycle wave of the symmetrical prospective current.

Current-limiting circuit breakers (Fig. 19-12) not only provide high interrupting capability but also limit let-through current and energy to downstream devices. $I^2t$ is an expression related to the energy resulting from current flow. Specific manufacturer's literature should be consulted for information on the current-limiting and energy-limiting characteristics of their particular circuit breakers.

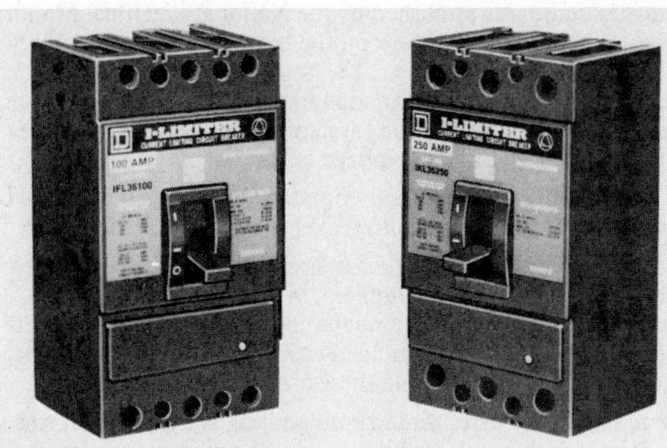


**FIG. 19-12 Current-limiting circuit breaker.**

Current-limiting circuit breakers provide the system designer with a means of reducing fault current and energy levels at downstream system components while still retaining the advantages, such as common trip and reusability, of circuit breaker protection. Current-limiting breakers can be reset and service restored in the same manner as conventional thermal-magnetic circuit breakers. There is nothing to replace even after clearing maximum fault currents. Figure 19-13 illustrates the current waveform resulting from current-limiting operation. At

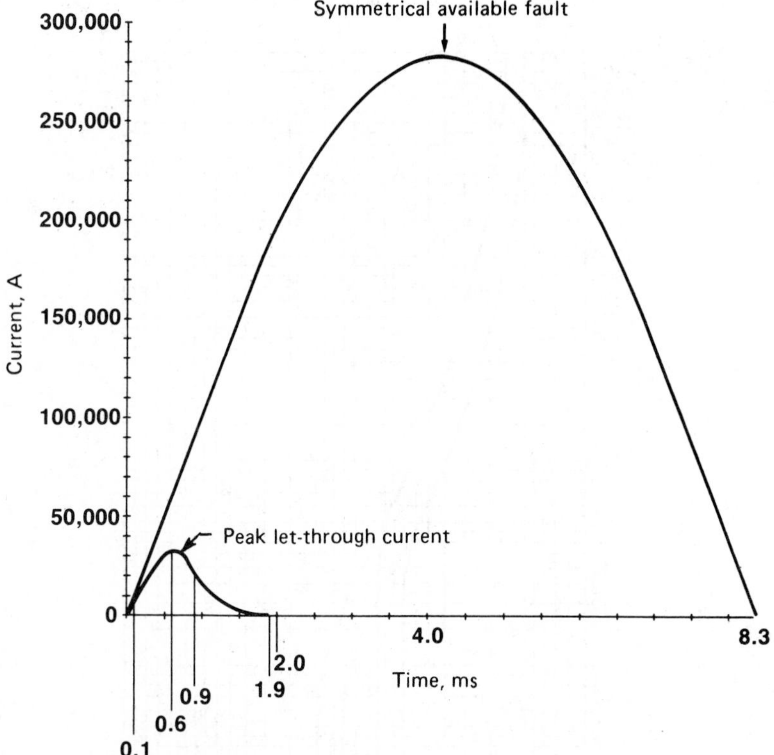


FIG. 19-13 Time-current curve for a current-limiting breaker.

current levels below the threshold of current limitation, conventional noncurrent limiting operation is the same as described for thermal-magnetic circuit breakers.

***46. Solid-State Trip Circuit Breakers.*** Solid-state trip circuit breakers utilize solid-state electronic components to measure and time the output from current transformers. The solid-state circuitry then initiates tripping action based on predetermined settings. These components are housed in the trip unit section of the breaker.

The basic function of the trip unit is to provide long-time current-time delay and instantaneous current response characteristics necessary for proper circuit protection. Combinations of these characteristics provide time delay to override transient overload, delayed tripping for sustained overloads, and instantaneous tripping for high-level short circuits.

Solid-state trip units may also be equipped with short-time pickup and delay functions. These characteristics are useful for providing coordination with downstream or upstream circuit breakers and/or fuses. The resulting combination of long-time delay and short-time delay characteristics provide delayed tripping for all levels of overcurrent below the instantaneous response. This provides time for downstream breakers to operate and clear the fault.

Solid-state trip units may also provide time-current characteristics for equipment ground fault protection in accordance with the NEC Section 230-95 requirements. The maximum current setting is 1200 A per NEC requirements, and time delay settings are usually adjustable to permit coordination with other protective devices.

***47. Time-Current Characteristic Curves.*** Circuit breaker time-current characteristic curves are a function of the type of trip function and its associated settings. A typical time-current characteristic curve is shown in Fig. 19-14 for a 400-A frame thermal-magnetic molded-case circuit breaker.

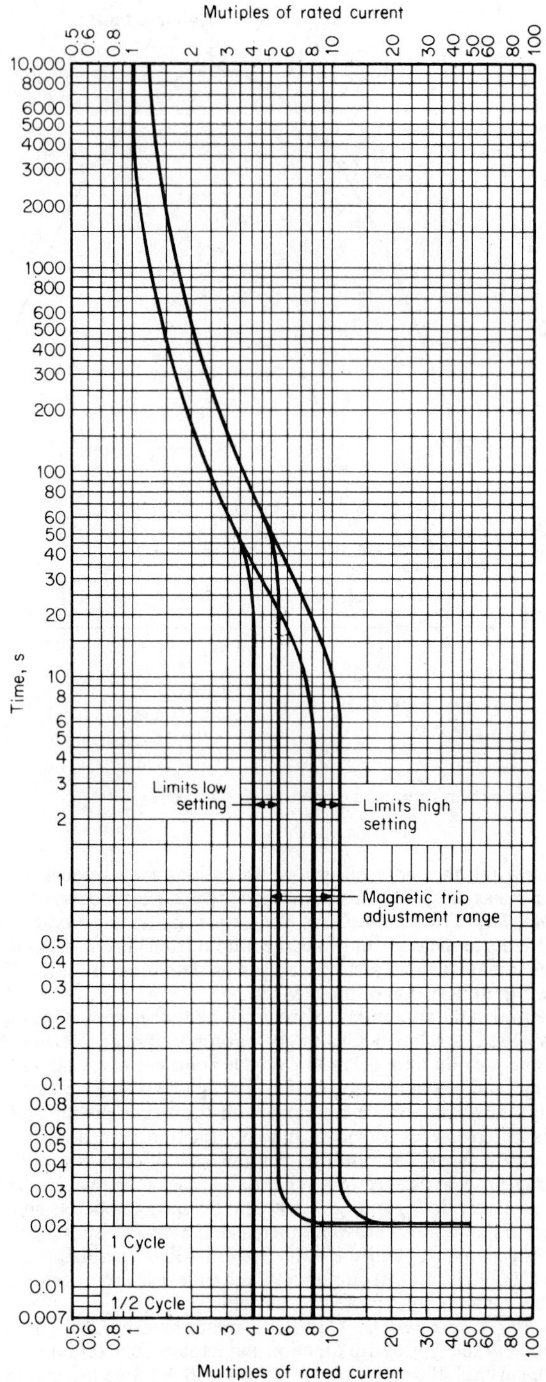


**FIG. 19-14** Characteristics tripping curve for molded-case breaker.

This particular curve applies for the continuous-current rating of 400 A and indicates total operational time from fault current initiation to clearing. This is a typical characteristic curve for a molded-case circuit breaker where overload sensing is achieved through a thermal element (a bimetal) and where instantaneous tripping is achieved magnetically.

In circuit breaker frame sizes larger than 100 A instantaneous operation is adjustable within approximately 5 to 10 times the continuous ampere rating.

Molded-case circuit breakers with more complex time-current characteristics are usually equipped with solid-state trip units. The continuous current rating may be adjustable using the long-time pickup function. In addition, long-time delay and short-time pickup and delay functions may be adjustable. The instantaneous response may also be adjustable.

Circuit breakers with solid-state trip units may also include integral equipment ground fault protection with current pickup and time delay adjustments. Figure 19-15 illustrates a combined time-current characteristic curve for a solid state electronic trip circuit breaker with all of these adjustments.

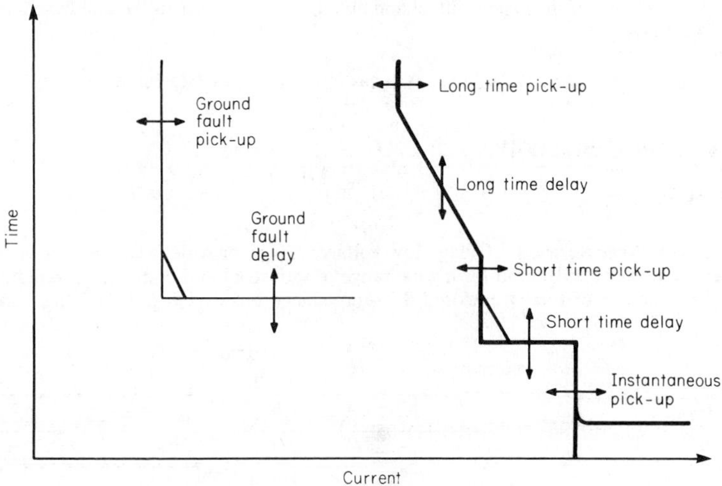


**FIG. 19-15** Time-current curve for solid-state electronic breaker.

Further refinement of the solid-state trip unit short-time delay characteristics for overcurrent protection or the ground fault time delay characteristics may be obtained by making part of the response curve a function of the product of time and the square of current ($I^2t$) illustrated by the sloping portions in Fig. 19-16. This sloping response curve often more readily coordinates with upstream or downstream thermal-magnetic circuit breakers and/or fuses.

It should be noted that solid-state trip units are available with varying degrees of complexity, and not all functions are provided for most applications. The specific manufacturer's literature should be consulted for guidance on individual units.

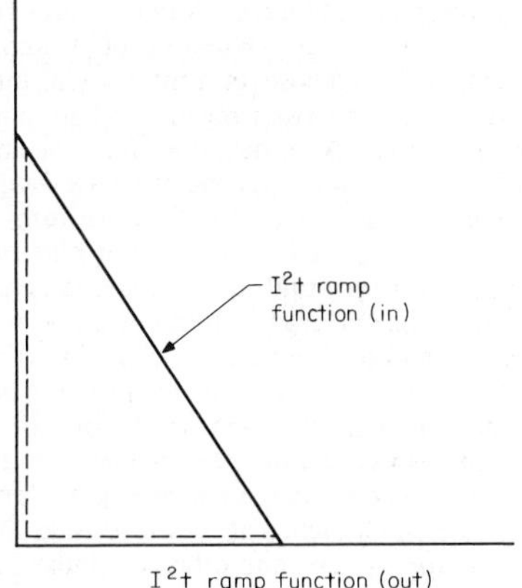


**FIG. 19-16** Sloping response curve for coordinating with thermal magnetic breakers or fuses.

***48. Series-Connected Ratings.*** *Series connection* of molded-case circuit breakers is a viable protection scheme, provided testing has verified performance. UL presently recognizes series-connected short circuit ratings and specifies test procedures to verify performance. Series-connected ratings must be based on tests and are only valid for specific circuit breaker types listed in the UL test reports. Individual manufacturer's series-connected ratings may be found in the UL Recognized Component Directory. Fuse-breaker coordinated combinations are also tested by UL and are applicable within their established guidelines.

***49. Coordination.*** When protection is being considered, the performance of a circuit breaker with respect to the connected conductors and load is a primary concern. Consideration is also given to the performance of circuit breakers with respect to other upstream and downstream circuit breakers and other protective devices. The object in coordinating protective devices is to make them selective in their operation with respect to each other. In doing so the effects of short circuits on systems are reduced to a minimum by disconnecting only the affected parts of a system. Stated in another way, only the circuit breaker nearest the short circuit should open, leaving the rest of the system intact and able to supply power to the unaffected areas.

***50. References***

1. National Electric Code, 1984.

2. IEC Publication 157-1, *Low Voltage Switchgear and Controlgear Part 1:* Circuit Breakers, 1973.

## LOW-VOLTAGE BUSWAY

*By Bob Wood*

***51. Low-Voltage Busways.*** Today's low-voltage busway provides a convenient, economical means of power distribution for a wide range of industrial and commercial buildings. In recent years the use of totally enclosed, low-impedance busway (Fig. 19-17) has increased

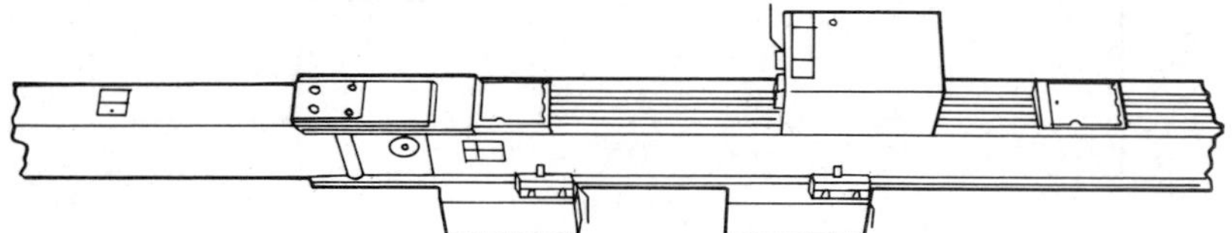

**FIG. 19-17** Low-voltage busway.

dramatically. This rapid growth is the result of technical and economical improvements which offer the system designer, installer, and end-user significant advantages.

*Description.* Low-voltage (600 V and below) busways consist of a factory prefabricated housing which encloses current-carrying busbars. The housing protects the busbars from physical harm and provides mechanical support. Busbars, which are normally electroplated and insulated, may be either aluminum or copper.

*Types.* Busway uses encompass a wide range of applications. For this reason various types of busways have evolved. For medium- and high-current applications (100 to 5000 A) low-impedance *feeder* busway and *plug-in* busway are available. These types of busway are utilized to supply and/or distribute large blocks of power within a factory or office building. *Lighting* bus and *trolley* duct normally supply power to small lighting loads or moving loads such as hand tools on an assembly line. This type bus is rated 30 to 100 A.

Feeder busway is designed to provide maximum efficiency of power transfer from one point to another. If multiple-power tap-off points are required, then a plug-in busway is utilized. Plug-in busway is similar to a feeder busway except that plug-in openings are located at convenient intervals along the entire length. Various types of fusible or circuit breaker plug-in units may be easily attached at these openings to supply power to individual loads. Lighting contactors, motor starters, and other specialized devices are also available in plug-in unit form.

*Applications.* Busway uses are many and varied. However, they may be grouped into four major categories: (1) the main service busway, (2) the long feeder busway run, (3) the plug-in busway distribution, and (4) the electrical riser for tall buildings.

The *main service busway* is usually a fairly short run of high-ampere feeder busway which brings electricity from the power company supply point into the main switchboard in a building.

*Long feeder busway runs* are designed to carry large blocks of electrical power from one point to another with maximum efficiency. A typical example of this type of application would be a busway used to bring power from the main service switchboard to a remote distribution switchboard.

*Plug-in busway distribution* is the normal method for power distribution in most modern manufacturing areas today. The plug-in busway is positioned throughout the manufacturing area, and loads are connected to the busway using fusible or circuit breaker plug-in units. This wiring method is flexible and allows for the addition or movement of loads simply by relocating the plug-in units which attach to the busway.

Finally, *electrical busway risers* are used to form the backbone of the power distribution system in tall metropolitan buildings. The busway is positioned vertically within the building and tap-offs are used on each floor to provide the necessary power requirements. As load requirements increase in the future, the busway riser can be preplanned to allow for additional tap-offs, thus continuing to grow with the building.

*Ratings.* Low-voltage busway is available for applications up to 600 V ac and may be supplied for use on 3-phase, 3- or 4-wire systems. A wide range of feeder or plug-in busway ampere ratings may be selected (see Table 19-8). These ratings are based on actual heat rise

**TABLE 19-8** Low-Voltage Busway—Ampere Ratings

| Plug-in | | Feeder |
|---|---|---|
| ■ | 225 A | |
| ■ | 400 A | |
| ■ | 600 A | |
| ■ | 800 A | ■ |
| ■ | 1000 A | ■ |
| ■ | 1200 A | ■ |
| ■ | 1350 A | ■ |
| ■ | 1600 A | ■ |
| ■ | 2000 A | ■ |
| ■ | 2500 A | ■ |
| ■ | 3000 A | ■ |
| ■ | 4000 A | ■ |
| | 5000 A | ■ |

within the busway. Current standards limit the maximum "hot spot" temperature within the busway housing to 55°C above ambient.

*Ground Bus.* A busway with a separate, low-impedance ground conductor is also available to provide for (1) a low-level static ground path, and (2) a secure low-impedance ground return path for medium- and high-level ground faults. This ground conductor is generally bonded to the busway housing at intervals of 10 ft or less. It is designed to carry a continuous 50% ground fault current as well as high-level ground fault currents of a short duration.

*Conductors.* A busway can be supplied with either aluminum or copper conductors. Based on size considerations only, copper is a more efficient electrical conductor than aluminum. However, when actual cost is considered, aluminum conductors are favored even though they must be slightly larger than the copper conductor required for the same ampere rating. Both aluminum and copper conductors are electroplated using tin or silver to ensure good surface conductivity at all electrical joints and plug-in unit connection points.

*Available Fault Current Levels.* These have increased dramatically in recent years. In order to ensure that a busway can safely withstand the extreme physical forces which can occur during a high-level fault, tests are conducted and maximum short-circuit levels are assigned to all busway components. The industry-recommended minimum short-circuit ratings for feeder and plug-in busway are shown in Table 19-9. In some instances optional higher short-circuit levels are available using alternate construction methods.

The actual available fault current throughout the entire electrical distribution system should be determined to ensure that all system components will function safely if a fault should occur.

**TABLE 19-9** Low-Voltage Busway Short-Circuit Ratings

| Continuous-current rating of busway, A | | Minimum short-circuit current ratings, A |
|---|---|---|
| Plug-in | Feeder | Symmetrical |
| 100 | — | 10,000 |
| 225 | — | 14,000 |
| 400 | — | 22,000 |
| 600 | — | 22,000 |
| — | 600 | 42,000 |
| 800 | — | 22,000 |
| — | 800 | 42,000 |
| 1000 | — | 42,000 |
| — | 1000 | 75,000 |
| 1200 | — | 42,000 |
| 1350 | — | 42,000 |
| — | 1200 | 75,000 |
| — | 1350 | 75,000 |
| 1600 | — | 65,000 |
| — | 1600 | 100,000 |
| 2000 | — | 65,000 |
| — | 2000 | 100,000 |
| 2500 | — | 65,000 |
| — | 2500 | 150,000 |
| 3000 | — | 85,000 |
| — | 3000 | 150,000 |
| 4000 | — | 85,000 |
| — | 4000 | 200,000 |
| — | 5000 | 200,000 |

*Voltage Drop.* Excessively low voltage or wide voltage variations on an electrical system can create many operating problems for the user. Modern low-impedance busway is designed to introduce an absolute minimum amount of voltage drop into the electrical system, thus helping to provide a stable source of electrical power. Typical feeder and plug-in busway voltage drop curves are shown in Table 19-10. Individual busway manufacturers can provide exact voltage drop curves for the specific busway ratings and types involved in each individual application.

*Installation.* A low-voltage busway system is made up of individual, factory-prefabricated

**TABLE 19-10** Typical Feeder and Plug-in Busway Voltage Drop Curves

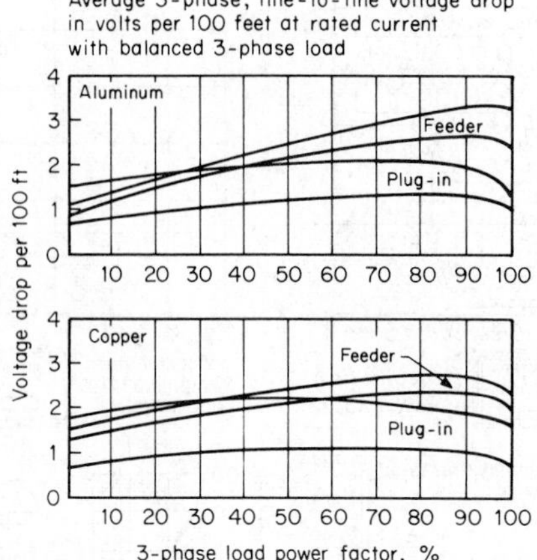


components such as 10-ft straight lengths, elbows, tees, flanged ends, cable tap boxes, etc. These individual components are designed so they can be quickly joined together at the job site with a minimum of field labor. Typically, one grade-5 bolt and two large-diameter Belleville washers are utilized to connect all the bars in an entire bus stack (see Fig. 19-18). Larger ampere ratings may utilize two or three busbar stacks, which may require slight additional assembly time.

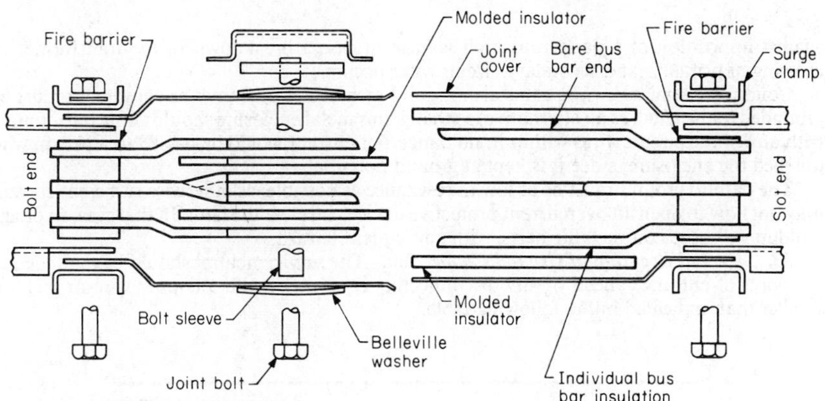


**FIG. 19-18** Busway one-bolt joint.

Prefabricated busway components reduce actual installation time to a minimum and result in a very low *installed cost* for the low-voltage busway. (The installed cost is the total material and labor cost necessary to provide a completed busway installation.)

## *PROTECTIVE GROUNDING*

***52. Purpose of Grounding.*** Secondary ac distribution systems should be grounded at the neutral conductor if the maximum voltage to ground does not exceed 150 V and may be grounded if this voltage is above 150 but does not exceed 300 V (Fig. 19-19). This is to guard

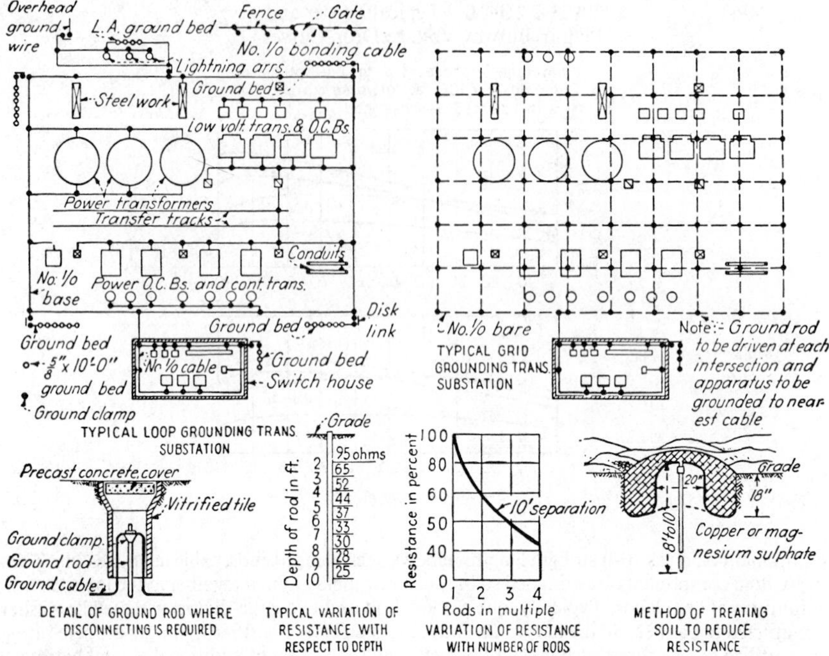


**FIG. 19-19** Complex equipment-grounding details showing connections, wire sizes, and grounds.

against imposition of a dangerous high voltage in case a breakdown in the transformer or crossing of primary- and secondary-circuit wires occurs.

Conduits, metal raceways, cable armors, and metal cases or frames of equipment must be grounded (or isolated, as an alternative) so that, if this metal enclosure should come into contact with any of the circuit wires within it, no dangerous current would be passed to a person who touched the enclosure, since it is kept at ground potential.

The path to ground must be as low in resistance as possible, as otherwise sufficient current may not flow to open an overcurrent protective device and clear the fault. In that case, a voltage buildup will occur on metallic parts, with consequent hazards.

***53. Size and Location of Ground Connection.*** The service neutral should be grounded at the point of entrance ahead of any disconnecting equipment with a copper wire or bus not smaller than indicated in the following table:

| Size of largest service conductor | AWG size of copper grounding conductor |
|---|---|
| 2 or smaller | 8 |
| 1 or 0 | 6 |
| 00 or 000 | 4 |
| Over 000–350,000 cir mils | 2 |
| Over 350,000–600,000 cir mils | 0 |
| Over 600,000–1,100,000 cir mils | 00 |
| Over 1,100,000 cir mils | 000 |

Metallic enclosures are commonly grounded through the grounding conductor used for the service neutral, although conduit or electrical metallic tubing may be used.

The grounding connection should, wherever possible, be made to a continuous underground water-piping system. Resistance of such a ground will usually be less than 0.1 Ω, thereby ensuring effectiveness.

***54. Polarization of Wiring.*** The conductor to be grounded must be continuously identified throughout the system to avoid errors in connections. For No. 6 AWG or smaller, this is accomplished by finishing the insulating covering with a white or natural-gray finish. Ends of conductors larger than No. 6 AWG are painted white or gray or wrapped with white tape, where exposed in outlets or panelboards.

## SYSTEMS OF INTERIOR DISTRIBUTION

***55. Standard Secondary-Voltage Types.*** Common interior distribution systems for buildings or plants having appreciable loads are

*a.* 3-phase 4-wire 120/208-V serving power and lighting.

*b.* Single-phase 3-wire 115/230-V serving lighting with 3-phase 3-wire 240- or 480-V for power.

*c.* 3-phase 4-wire 277/480-V serving power and fluorescent or mercury lighting with single-phase 115/230-V circuits for other utilization provided from the power system by means of local dry-type transformers.

For very large buildings and industrial plants distribution is often provided at higher voltage, notably 13.2, 4.1, or 2.3 kV, stepped down to utilization voltage at strategically load-centered substations.

***56. Three-Wire Single-Phase Systems.*** The 3-wire 115/230-V single-phase system (Fig. 19-20) is very commonly used for interior wiring for lighting and miscellaneous purposes but not for motor loads much in excess of 5 hp. The neutral wire is grounded, hence it should not be fused at any point. The branch circuits may be 2-wire 115- or 230-V or 3-wire 115/230-V. The neutral conductor carries only the unbalance in load between the two ungrounded conductors. For circuits up to 200 A, it should have the same capacity as the ungrounded conductors. A factor of 0.7 may be applied to unbalanced loads above 200 A in determining its size.

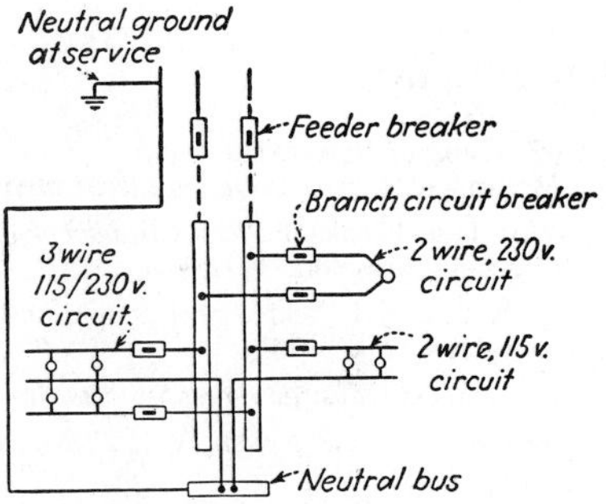

**FIG. 19-20** Three-wire, single-phase system showing types of circuits.

***57. Three-phase 3-wire system*** is usually employed where motors form a substantial load and where lighting is supplied from a separate single-phase system or by transformer. The usual voltage is 240 or 480 V. Branch circuits may be either 2-wire single-phase or 3-wire 3-phase.

***58. Three-Phase 4-Wire Systems.*** The 3-phase 4-wire system (Fig. 19-21) is widely used. Branch lighting circuits are connected between any one of the phase wires and the neutral wire. Power is taken from the 3-phase wires. The neutral wire is grounded. The voltage between phase wires is usually 208 V, and between any phase wire and neutral it is 120 V.

Circuits that may be supplied include 2-wire 120- or 208-V; 3-wire 120/208- or 208-V; and 4-wire 120/208-V. In the case of a 3-wire circuit consisting of 2-phase wires and the neutral, the "neutral" differs from that of a 3-wire single-phase circuit in that with both sides of the circuit evenly loaded it will carry equal current in the phase wires.

***59. Two-Wire Systems.*** The 2-wire system is used where current is supplied at 115 V, as when obtained from a 115-V generator, and on installations connected to utility companies' lines when the installation is of small capacity. It is limited in capacity; requires excessively large copper for heavy loads; and, in general, is not considered a modern complete distribution method.

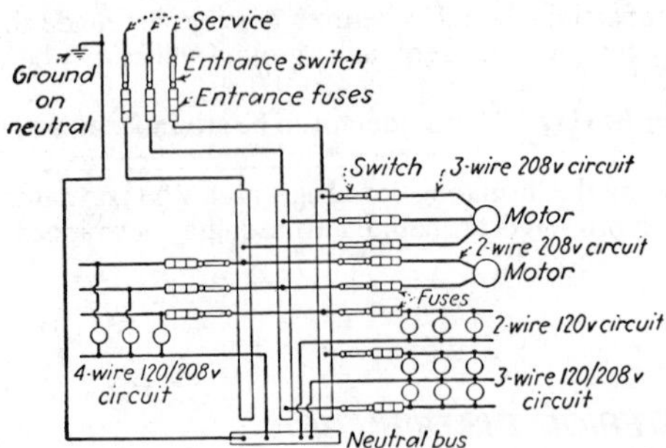


**FIG. 19-21** Four-wire, 3-phase system showing types of circuits.

***60. Two- and 3-wire dc systems*** are similar in connections (except grounding on the premises) and use to the 2- and 3-wire single-phase ac systems.

***61. Two-phase distribution*** may be effected with 4 or with 3 wires. In the former case there is a pair of wires for each phase, while in the latter there is 1 wire for each phase and a common wire for both phases. The circuits must be balanced on either side, just as in the case of a 3-wire single-phase system. Where 3 wires are used, the common wire should be 1.4 times as large as either of the other two, since it must carry 1.4 times as much current. Motors are connected to both phases and employ all 3 or all 4 wires, as the case may be. With 4 wires, the lamps are connected to each phase as though the supply were single-phase, and care should be exercised to balance the phases as nearly as possible.

## Bibliography

***62. General References***

1. McPartland, J.: *How to Design Electrical Systems,* McGraw-Hill Book Company, New York, 1968.
2. Watt, J., and Stetka, F.: *NFPA Handbook of the National Electrical Code,* 3d ed., McGraw-Hill Book Company, New York, 1972.
3. Croft, T., Carr, C., and Watt, J.: *American Electricians' Handbook,* 9th ed., McGraw-Hill Book Company, New York, 1970.
4. Beeman, D.: *Industrial Power Systems Handbook,* McGraw-Hill Book Company, New York, 1955.
5. McPartland, J., and Novak, W. J.: *Electrical Equipment Manual,* McGraw-Hill Book Company, New York, 1965.
6. McPartland, J., and Novak, W. J.: *Practical Electricity,* McGraw-Hill Book Company, New York, 1964.
7. National Electrical Code, National Board of Fire Underwriters, Boston, Mass., 1984.
8. Techniques of Electrical Construction and Design, *Electr. Constr. Maint.,* McGraw-Hill Book Company, New York, 1976.
9. McPartland, J.: Making Electrical Calculations, *Electr. Constr. Maint.,* New York, 1975.
10. McPartland, J.: Motor and Control Circuits, *Electr. Constr. Maint.,* New York, 1976.
11. McPartland, J.: Selecting Sizes and Ratings of Electrical Equipment, *Electr. Constr. Maint.,* New York, 1976.
12. McPartland, J.: What You Should Know About High-Voltage Electrical Work, *Electr. Constr. Maint.,* New York, 1976.
13. McPartland, J. F.: *McGraw-Hill's National Electrical Code Handbook,* McGraw-Hill Book Company, New York, 1984.
14. National Electrical Safety Code, Institute of Electrical and Electronics Engineers, New York, 1984.
15. Clapp, A. L.: National Electrical Safety Code Handbook, Institute of Electrical and Electronics Engineers, New York, 1984.

# SECTION 20
# MOTORS

**Alexander Kusko, Sc.D.**
*President, Alexander Kusko, Inc., Consulting Engineers; former Associate Professor, Department of Electrical Engineering, Massachusetts Institute of Technology; author of numerous technical papers and books; Fellow, IEEE*

**Syed M. Peeran, Ph.D.**
*Senior Engineer, Alexander Kusko, Inc., Consulting Engineers; Lecturer, Graduate School of Engineering, Northeastern University, Boston, MA; Member IEEE*

**CONTENTS**

*Numbers refer to paragraphs*

## GENERAL

***1. Types of Electric Motors.*** Electric motors provide motive power to a wide variety of domestic and industrial machinery. Their versatility, reliability, and economy cannot be equaled by any other form of drive. Successful motor application depends upon selecting a type of motor which satisfies the kinetic requirements of the driven machinery. There are several methods of classifying electric motors. First, based upon the electric power supply motors are classified as dc and ac motors. Figure 20-1 shows further classification of ac and dc motors based upon the stator and rotor construction.

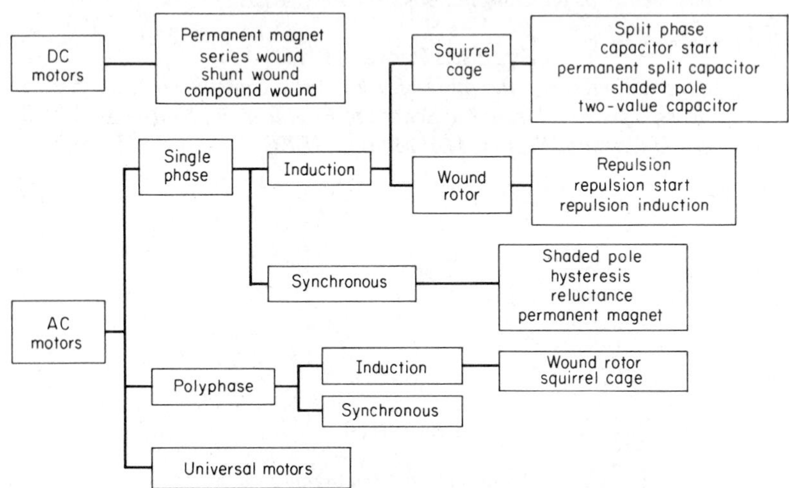


**FIG. 20-1** Classification of ac and dc motors.

Classifications based upon size and applications are fractional-horsepower, integral-horsepower, gear, torque, servo, and stepper motors. Various types of enclosures have been standardized by the National Electric Manufacturers' Association, U.S.A. (NEMA). The following are the standard enclosure types and their characteristics:

| Types | Characteristics |
|---|---|
| Open: | |
| Drip-proof | Operate with dripping liquids up to 15° from vertical. |
| Splash-proof | Operate with splashing liquids up to 100° from vertical. |
| Guarded | Guarded by limited size openings (less than ¾ in). |
| Semiguarded | Only top half of motor guarded. |
| Drip-proof fully guarded | Drip-proof motor with limited size openings. |
| Externally ventilated | Ventilated with separate motor-driven blower; can have other types of protection. |
| Pipe ventilated | Openings accept inlet ducts or pipe for air cooling. |
| Weather protected type 1 | Ventilating passages minimize entrance of rain, snow, and airborne particles. Passages are less than ¾ in in diam. |
| Weather protected type 2 | Motors have, in addition to type 1, passages to discharge high-velocity particles blown into the motor. |

| Types | Characteristics |
|---|---|
| Totally enclosed: | |
| Nonventilated (TENV) | Not equipped for external cooling. |
| Fan-cooled (TEFC) | Cooled by external integral fan. |
| Explosion-proof | Withstands internal gas explosion. Prevents ignition of external gas. |
| Dust-ignition-proof | Excludes ignitable amounts of dust and amounts of dust that would degrade performance. |
| Waterproof | Excludes leakage except around shaft. |
| Pipe-ventilated | Openings accept inlet ducts or pipe for air cooling. |
| Water-cooled | Cooled by circulating water. |
| Water-and-air-cooled | Cooled by water-cooled air. |
| Air-to-air-cooled | Cooled by air-cooled air. |
| Guarded TEFC | Fan cooled and guarded by limited-size openings. |
| Encapsulated | Has resin-filled windings for severe operating conditions. |

NEMA classification according to the variability of speed includes constant-speed motor such as ac synchronous motors, induction motors with small slip and dc short-wound motors, varying-speed motors such as dc series motors or repulsion motors, and variable-speed motors such as dc shunt-, series-, and compound-wound motors.

***2. Standards.*** Motors and generators are required to meet various industry and national standards and in some instances specific local codes and customer specifications. The more important of these standards may be briefly described as follows.

*a. NEMA Standards* are voluntary standards of the National Electrical Manufacturers Association and represent general practice in the industry. They define a product, process, or procedure with reference to nomenclature composition, construction, dimensions, tolerances, operating characteristics, performance, quality, rating, and testing. Specifically, they cover such matters as frame sizes, torque classifications, and basis of rating.

*b. IEEE Standards (AIEE).* IEEE Standards concern fundamentals such as basic standards for temperature rise, rating methods, classification of insulating materials, and test codes.

*c. USA Standards* are national standards established by the United States of America Standards Institute, which represents manufacturers, distributors, consumers, and others concerned. USA Standards may be sponsored by any responsible body and may become national standards only if a consensus of those having substantial interest is reached. Standards may cover a wide variety of subjects such as dimensions, specifications of materials, methods of test, performance, and definition of terms. USA Standards frequently are those previously adopted by and sponsored by NEMA, IEEE, etc. The chief motor and generator standard of USASI is C50, "Rotating Machinery," which is substantially in agreement with current NEMA Standards.

*d. National Electrical Code* is a USA Standard sponsored by the National Fire Protection Association for the purpose of safeguarding persons and buildings from electrical hazards arising from the use of electricity for light, heat, power, and other purposes. It covers wiring methods and materials, protection of branch circuits, motors and control, grounding, and recommendations regarding suitable equipment for each classification.

*e. Underwriters' Laboratories, Inc.,* is an independent testing organization which examines and tests devices, systems, and materials with particular reference to life, fire, and casualty hazards. It develops standards for motor and control for hazardous locations through cooperation with manufacturers. It has several different services by which a manufacturer can indicate compliance with Underwriters' Laboratories Standards. Such services are utilized on motors only in the case of explosion-proof and dust-ignition-proof motors where label service is used to indicate to code-enforcing authorities that motors have been inspected to determine their adherence to Underwriters' Laboratories Standards for motors for hazardous locations.

*f. Federal Specification CC-M-641* for integral-horsepower ac motors has been issued by the Federal government to cover standard motors for general government uses. Standard motors meet these specifications, but other Federal Specifications issued by various branches of the government for specific use may require special designs.

*g. World Standards.* Standards similar to our NEMA Standards have been established in other countries. The most significant are

1. IEC (International Electrotechnical Commission) Standard 72-1, Part 1
2. German Standard DIN 42673
3. British Standard BSI-2960, Part 2

These standards specify dimensions, classes of insulation, and in some cases horsepower ratings.

## DIRECT-CURRENT MOTORS

***3. Classes of DC Motors.*** Direct-current motors are used in a wide variety of industrial applications because of the ease with which the speed can be controlled. The speed-torque characteristic may be varied to almost any useful form. Continuous operation over a speed range of 8:1 is possible. While ac motors tend to stall, dc motors can deliver over five times the rated torque (power supply permitting). Reversal is possible without power switching. Permanent-magnet motors are available in fractional-horsepower ratings, while wound-field dc motors are classified as (1) shunt motor, in which the field winding is connected in parallel with the armature; (2) series motor, in which the field winding is connected in series with the armature, and (3) compound motor, which has a series-field and shunt-field winding. The shunt motor is used in constant speed applications such as drives for dc generators in dc motor-generator sets. The series motor is used in applications where a high starting torque is required, such as in electric traction, cranes, and hoists, etc. In compound motors, the droop of the speed-torque characteristic may be adjusted to suit the load.

***4. The construction of dc motors with a wound field*** is practically identical to that of dc generators; with minor adjustment the same dc machine may be operated either as a dc generator or as a motor. (See Sec. **8** of this handbook for construction, armature windings commutator, etc.).

Permanent-magnet dc motors have fields supplied by permanent magnets that create two or more poles in the armature by passing magnetic flux through it. The magnetic flux causes the current-carrying armature conductors to create a torque. This flux remains basically constant at all motor speeds—the speed-torque and current-torque curves are linear.

***5. Shunt Motors.*** DC shunt motors are suitable for application where constant speed is needed at any control setting or where appreciable speed range (by field control) is needed. The field circuit connection is shown in Fig. 20-2*a*.

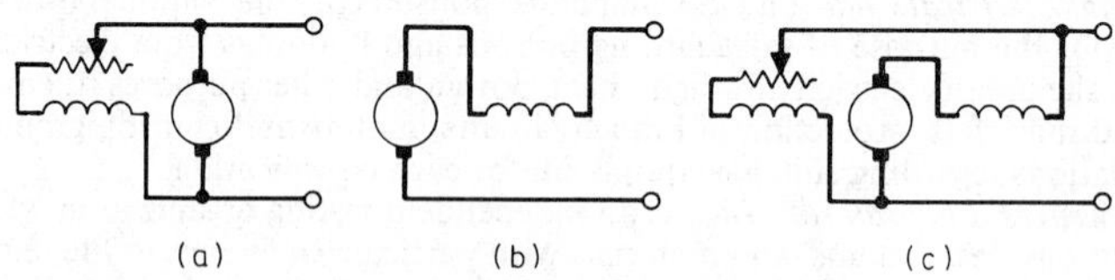


**FIG. 20-2** Field circuit connections of dc motor.

Since a motor armature revolves in a magnetic field, an emf is generated in the conductors which is opposed to the direction of the current and is called the counter emf. The applied emf must be large enough to overcome the counter emf and also to send the armature current $I_a$ through $R_m$, the resistance of the armature winding, the brushes; or

$$E_a = E_b + I_a R_m \quad \text{volts} \tag{20-1}$$

where $E_a$ = applied emf and $E_b$ = counter emf. Since the counter emf at zero speed, that is, at starting, is identically zero and since normally the armature resistance is small, it is obvious in view of Eq. (20-1) that, unless measures are taken to reduce the applied voltage, excessive current will circulate in the motor during starting. Normally, starting devices consisting of variable series resistors are used to limit the starting current of motors.

The torque of a motor is proportional to the number of conductors on the armature, the current per conductor, and the total flux in the machine. The formula for torque is

$$\text{Torque} = 0.1175 Z\phi I_a \frac{\text{poles}}{\text{paths}} \times 10^{-8} \quad \text{lb-ft} \tag{20-2}$$

where Z = total number of armature conductors, $\phi$ = total flux per pole, and $I_a$ = armature current taken from the line.

$$E_b = E_a - I_a R_m = Z\phi \frac{\text{r/min}}{60} \frac{\text{poles}}{\text{paths}} \times 10^{-8} \quad \text{volts} \tag{20-3}$$

or

$$\text{r/min} = 60 \frac{E_a - I_a R_m}{Z\phi} \frac{\text{paths}}{\text{poles}} \times 10^{8} \tag{20-4}$$

For a given motor the number of armature conductors Z, the number of poles, and the number of armature paths are constant. The torque can therefore be expressed as

$$\text{Torque} = \text{constant} \times \phi I_a \tag{20-5}$$

and the speed, likewise, is expressed as

$$\text{Speed} = \text{constant} \times (E_a - I_a R_m)/\phi \tag{20-6}$$

In the case of the shunt motor, $E_a$, $R_m$, and $\phi$ are constant, and the speed and torque curves are shown as curves 1 (Fig. 20-3); the effective torque is less than that generated by the torque required for the windage and the bearing and brush friction. The drop in speed from no load to full load seldom exceeds 5%; indeed, since $\phi$, the flux per pole, decreases with increase of load, owing to armature reaction, the speed may remain approximately constant up to full load.

***6. Speed Changes of Shunt Motors under Rapidly Fluctuating Loads.*** When the load on a shunt motor increases slowly, the flux per pole decreases as the result of armature reaction and the speed [Eq. (20-6)] remains approximately constant. If, however, the load changes rapidly, the flux per pole cannot change rapidly, owing to the self-inductance of the field coils; the machine then operates for the instant as a constant-flux machine, and the speed drops rapidly to allow the counter emf to decrease and the necessary current to flow.

The instantaneous speed drop is affected by the armature circuit resistance, the armature circuit inductance, the terminal voltage, the speed of the motor, and the moment of inertia of the motor. The armature circuit resistance is the most important factor, since the speed drop is directly proportional to the voltage drop in the armature circuit. Accurate calculation of the speed drop requires lengthy, precise calculation of all the factors involved.

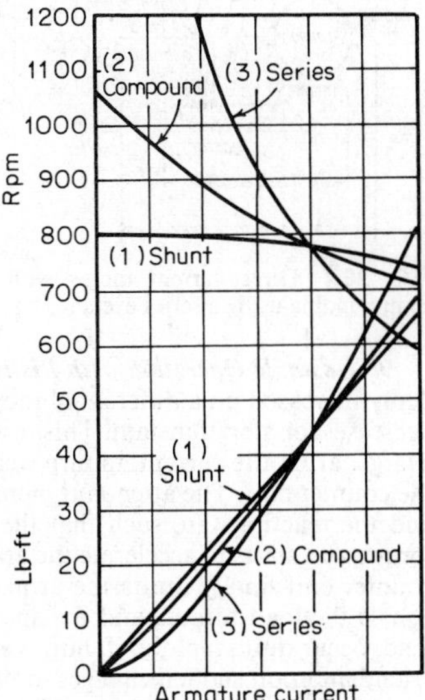


**FIG. 20-3** Motor characteristics.

A 1000-hp 600-V 200-r/min motor designed with a normal volume of core iron and with proportions chosen to give a minimum impact speed drop would have approximately 5.5% impact speed drop. This can be reduced to approximately one-half the above value by doubling the size of the motor and proportioning the machine parts properly. Machines of smaller sizes will approach 8% impact speed drop.

*7.* ***Speed and Torque of Series Motors.*** Equations (20-6) and (20-5) apply to motors of all continuous-current types. In the case of series motors the flux $\phi$ increases with the armature current $I_a$; the torque would be proportional to $I_a^2$ were it not that the magnetic circuit becomes saturated with increase of current. Since $\phi$ increases with load, the speed drops as the load increases. The speed and torque characteristics are shown in curves 3 (Fig. 20-3).

If the load on a series motor becomes small, the speed becomes very high, so that a series motor should always be geared or direct-connected to the load. If it were belted and the belt were to break, the motor would run away and would probably burst.

For a given load, and therefore for a given current, the speed of a series motor can be increased by shunting the series winding or by short-circuiting some of the series turns so as to reduce the flux. The speed can be decreased by inserting resistance in series with the armature.

*8.* ***The compound motor connections*** are shown in Fig. 20-2*c*. *The compound motor* is a compromise between the shunt and the series motors. Because of the series winding, which assists the shunt winding, the flux per pole increases with the load, so that the torque increases more rapidly and the speed decreases more rapidly than if the series winding were not connected; but the motor cannot run away under light loads, because of the shunt excitation. The speed and torque characteristics for such a machine are shown in curves 2 (Fig. 20-3).

The speed of a compound motor can be adjusted by armature and field rheostats, just as in the shunt machine.

*Indirect compounding* is used on some dc motors. In this case, the heavy strap-wound series field is replaced by a wire-wound field similar to a small shunt field. This field is excited by an unsaturated dc exciter, usually separately driven at constant speed. This exciter is excited by the line current of the motor for which it supplies the series excitation (see Fig. 20-4). The output voltage and the current from the exciter are proportional to the main motor current; so a given proportionality exists between the load current of the motor and its wire-wound series-field strength. The use of a reversing switch and rheostat in the armature circuit of the series exciter permits variations in strength and even polarity of the series field. This furnishes an easy method of changing the compounding of the motor if desired for various speeds, to maintain constant-speed regulation over a speed range. If desired, the series exciter rheostat can be mechanically connected to the shunt-field rheostat to accomplish this automatically.

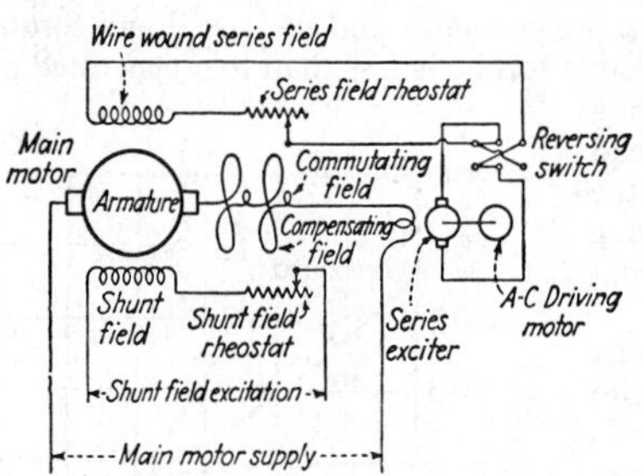

**FIG. 20-4** Direct-current motor with indirect compounding using a series exciter.

*9.* ***Unstable Operation with Rising-Speed Characteristic.*** When the load torque is suddenly increased on a differential motor designed for a rising-speed characteristic, the speed decreases for a brief instant. This results in a momentary drop of counter emf, which admits a larger armature current, in turn weakening the resultant field strength and further reducing the counter emf. The attendant increase in armature current increases the armature torque, and the reactions are such that the latter continues to increase until it exceeds the load torque and starts to accelerate the armature. The increase in speed continues until the rising counter emf finally limits the armature current to a value at which the armature torque equals the load torque, and the speed becomes constant. These reactions, with change of load, occur quite rapidly; if, however, the field cores are large and massive, changes in flux attendant upon sudden changes in mmf will lag by an appreciable time interval, on account of eddy currents in the cores. The presence of an appreciable flux lag, with very rapidly changing loads, results in unstable operation; for example, when the load is suddenly increased, the speed will drop appreciably before it begins to accelerate; and when the load is suddenly removed, the speed will rise appreciably before it begins to decrease. The effect of the armature inertia will accentuate these defects in speed regulation. Such defects are not found in motors whose speed decreases with increasing load, that is, those which have a drooping speed characteristic.

*10.* ***Power supplies*** to dc motors may be batteries, a dc generator, or rectifiers. The permanent-magnet and miniature motors use battery power supplies. Large integral-horsepower dc motors such as rolling-mill motors use dc generators as the power supply. Most fractional-horsepower and integral-horsepower dc motors operate with rectifier power supplies. Some of the types of rectifier power supplies are as follows:

1. Single-phase, half-wave
2. Single-phase, half-wave, back rectifier
3. Single-phase, half-wave, alternating-current voltage controlled
4. Single-phase, full-wave, firing angle controlled
5. Single-phase, full-wave, firing angle controlled, back rectifier
6. Three-phase, half-wave, voltage controlled
7. Three-phase, half-wave, firing angle controlled

The NEMA standard letter designations of dc motor test power supplies are as follows:

Power supply A—dc generator

Power supply C—3-phase 6-pulse controlled rectifier (230 V L-L, 60 Hz)

Power supply D—3-phase 6-pulse controlled rectifier (with three thyristers and three diodes) with free-wheeling diode (230/460 V L-L, 60 Hz)

Power supply E—3-phase 3-pulse controlled rectifier (460 V L-L, 60 Hz)

Power supply K—1-phase full-wave controlled rectifier with free-wheeling diode (230/115 V, 60 Hz)

When a direct-current integral-horsepower motor is operated from a rectified alternating-current supply, its performance may differ materially from that of the same motor when operated from a low-ripple direct-current source of supply, such as a generator or a battery. The pulsating voltage and current waveforms may increase temperature rise and noise and adversely affect commutation and efficiency. Because of these effects, direct-current motors must be designed or specially selected to operate on the particular type of rectified supply to be used. Armature-current form factor and ripple are two important parameters to be specified for motors which are required to operate with rectifier power supplies. The *form factor* is defined as the ratio of the rms value to the overage value of the armature currents. Recommended rated form factors vary from 2.0 for 1-phase half-wave rectifier supplies to 1.1 for 3-phase full-wave rectifier supplies (see NEMA MG1-14.60). Because the letters used to identify the power supplies in common use have been chosen in alphabetical order of increasing magnitude of ripple current, a motor rated on the basis of one of these power supplies may be used on any power supply designated by a lower letter of the alphabet. For example, a motor rated on the basis of an E power supply may be used on a C or D power supply.

*11.* ***DC Motor Ratings.*** NEMA standard ratings of industrial dc motors for 240-V and 500/550-V dc supply voltages are given in Tables 10-4 and 10-5 of NEMA standard MG1. The rating is continuous unless otherwise specified. All short-term load tests shall commence only when the windings and other parts of the machine are within 5°C of the ambient temperature at the time of starting the test. Continuous and short-term ratings are based upon maximum ambient temperature and insulation class. Except in engine and boiler rooms, the maximum ambient temperature is 40°C and the insulation classes are A, B, and F, rated for temperature rises of 70°C, 100°C, and 130°C, respectively.

*12.* ***Losses and Efficiency.*** Power losses in dc motors are due to bearing friction, brush friction, windage, eddy currents and hysteresis in the armature core and pole faces, brush contact-drop, $I^2R$ losses in the armature and field windings, and stray load losses. Typical values of total losses in industrial motors are 4 to 10% of the output. The bearing friction and brush friction losses are proportional to the speed of the motor, while the windage loss is proportional to the square of the speed. Eddy current loss in the armature teeth and in the armature core is proportional to the square of the speed and to the square of the air-gap flux density. Hysteresis loss in the armature teeth and core is proportional to the speed and the square of the flux density in the air gap. Brush contact drop is typically 1 V per brush arm for carbon-graphite brushes and 0.25 V for metal-graphite. Stray load losses are due to eddy currents in armature conductors, brush short-circuit losses in the commutator, and additional core loss arising from distortion of the magnetic field due to armature reaction. The efficiency of the dc motor is defined as

$$\eta = (\text{Input electric power} - \text{losses})/\text{input power} \times 100\%.$$

Typical efficiency variation with output is shown in Fig. 20-5.

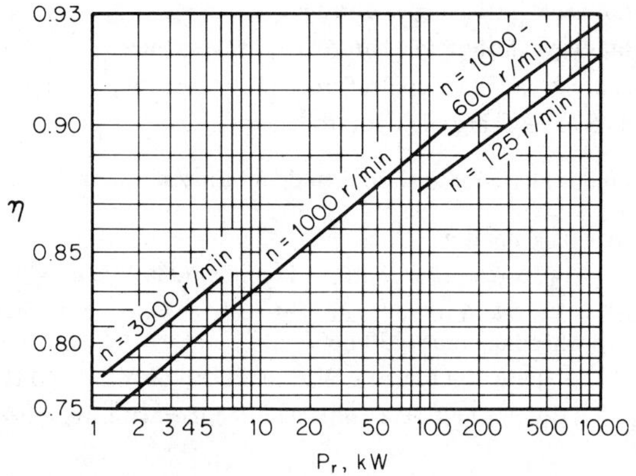


**FIG. 20-5** Typical efficiency curves of dc machines (Ref. 5, Par. 23).

***13. Short-Time Ratings.*** The effect of time and enclosure on motor rating may be seen from the following: A given frame will have a rating of 12 hp at 500 r/min as an enclosed machine on continuous duty, or 19 hp at 500 r/min as an open machine on continuous duty, or 31 hp at 500 r/min with a 1-h rating, or 40 hp at 500 r/min with a ½-h rating. The temperature rise on full load is 40°C as an open machine and 50°C as an enclosed machine. The horsepower is proportional to the speed over a range of 30% above or below the rated speed.

***14. Methods of Speed Control.*** Speed of a dc motor is controlled either by varying the voltage across the armature, the field winding, or both. Series-parallel combinations are an effective means of reducing armature voltage and motor speed. This method is applied in cam-controlled traction motors. Two identical motors are connected in parallel or in series. When in parallel, full voltage is applied across each motor, causing them to run at base speed. When in series, the motor speeds are essentially one-half of base speed. Field-series resistance in shunt motors weakens the field, which causes the motors to run above the base speed. Speed range as high as 8:1 may be obtained in special motors. Armature-series resistance used with shunt or series motors produces motor speed below the base speed. In the series motor the field winding is also affected by the armature-series resistance, producing greater effect on the speed-torque characteristic than for the short motor where the field is constant. Speed control by this method is usually limited to approximately 50% of the base speed. The above-speed control method results in power losses in the external resistors; solid-state dc motor control eliminates the power losses (see Par. **15** of this section).

***15. Permanent-Magnet DC Motors.*** Permanent-magnet (PM) motors are available in fractional and low integral-horsepower sizes. They have several advantages over field-wound types. Excitation power supplies and associated wiring are not needed. Reliability is improved, since there are no exciting field coils to fail, and there is no likelihood of overspeed due to loss of field. Efficiency and cooling are improved by elimination of power loss in an exciting field. And the torque-vs-current characteristic is more nearly linear. Finally a PM motor may be used where a totally enclosed motor is required for a continuous-excitation duty cycle.

Temperature effects depend on the kind of magnet material used. Integral-horsepower motors with Alnico-type magnets are affected less by temperature than those with ceramic magnets because flux is constant. Ceramic magnets ordinarily used in fractional-horsepower motors have characteristics that vary about as much with temperature as do the shunt fields of excited machines.

Disadvantages are the absence of field control and special speed-torque characteristics. Overloads may cause partial demagnetization that changes motor speed and torque characteristics until magnetization is fully restored. Generally, an integral-horsepower PM motor is somewhat larger and more expensive than an equivalent shunt-wound motor, but total system cost may be less.

A PM motor is a compromise between compound-wound and series-wound motors. It has better starting torque, but approximately half the no-load speed of a series motor. In applications where compound motors are traditionally used, the PM motor could be considered where slightly higher efficiency and greater overload capacity are needed. In series-motor applications, cost consideration may influence the decision to switch. For example, in frame sizes under 5 in in diameter the series motor is more economical. But in sizes larger than 5 in, the series motor costs more in high-volumes. And the PM motor in these larger sizes challenges the series motor with its high torques and low no-load speed.

Previously, Alnico and ceramic magnets have been used in PM motors. Recently, however, processing improvements have lowered the cost of rare earth-cobalt magnets so they are now feasible for dc motors.

Substituting rare earth-cobalt magnets for conventional Alnico and ceramic types can cut motor weight by one-third and reduce motor diameter by 1 in—without increasing cost or reducing torque capability. Also, the unusually high field strength of these magnets permits new motor designs such as "inside-out" configurations—with magnets on the rotor and windings on the stator—completely opposite from conventional motor design. The result is a smaller, lighter motor with less rotor inertia, greater heat dissipation and larger torque ratings.

Rare earth-cobalt magnets combine the advantages of high coercivity (resistance to demagnetization) and good magnetic induction at a reasonable cost. Coercivity is very important in a motor—especially at heavy load. High coercivity produces a torque response that is linear with motor current. Getting this kind of linearity with conventional magnets requires expensive electronic controls or complex mechanical gearing.

Although rare earth-cobalt magnets have distinct advantages for use in motors, they cannot be directly substituted for other types. Because of the extremely high energy product of these magnets, one-for-one substitution for conventional magnets generally results in an over-designed, uneconomical motor that is prone to malfunction. In fact, the motor might not work at all. Using rare earth-cobalt magnets in a motor requires the same degree of planning afforded to other magnet types.

***16. Brushless DC Motors.*** Brushless dc motors have a stationary armature and a rotating field structure, exactly opposite to how those elements are arranged in conventional dc motors. This construction speeds heat dissipation and reduces rotor inertia. Permanent magnets provide magnetic flux for the field. DC current to the armature is commutated with transistors rather than with the brushes and commutator bars of conventional dc motors.

Armatures of dc brushless motors typically contain 2 to 6 coils, whereas conventional dc motor armatures have from 10 to 50. Brushless motors have fewer coils because either two or four transistors are required to commutate each motor coil. This arrangement becomes increasingly costly and inefficient as the number of windings increases. A typical circuit of a brushless dc motor is shown in Fig. 20-6.

The transistors controlling each winding of a dc brushless motor are turned on and off at specific rotor angles. The transistors provide current pulses to the armature windings that are similar to those provided by a commutator. The switching sequence is arranged to produce a rotating magnetic flux in the air gap that stays at a fixed angle to the flux produced

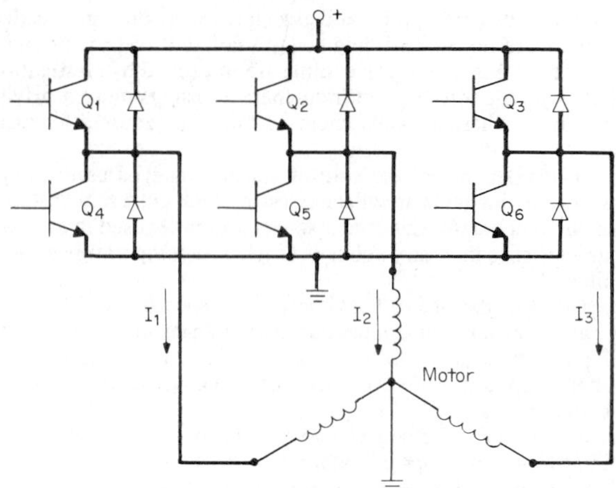


**FIG. 20-6** Typical circuit of a brushless dc motor.

by the permanent magnets on the rotor. Torque produced by a brushless dc motor is directly proportional to armature current.

***17. DC traction motors*** are dc series motors typically rated 140 hp, 310 V, 2500 r/min. Four motors are used in each transit car, two on each axle. The power supply is 600 to 1000 V dc from the third rail, which is powered by 2500- to 5000-kW rectifier sets in rectifier substations located along the track. Starting and speed control are either by a cam controller or by a chopper controller on board the transit car.

***18. DC Servomotors.*** DC servomotors are high-performance motors normally used as prime movers in computers, numerically controlled machinery, or other applications where starts and stops must be made quickly and accurately. Servomotors have lightweight, low-inertia armatures that respond quickly to excitation-voltage changes. In addition, very low armature inductance in these motors results in a low electrical time constant (typically 0.05 to 1.5 ms) that further sharpens motor response to command signals. Servomotors include permanent-magnet, printed-circuit, and moving-coil (or shell) motors. The rotor of a shell motor consists of a cylindrical shell of copper or aluminum wire coils. The wire rotates in a magnetic field in the annular space between magnetic pole pieces and a stationary iron core. The field is provided by cast Alnico magnets whose magnetic axis is radial. The motor may have two, four, or six poles.

Each of these basic types has its own characteristics, such as inertia, physical shape, costs, shaft resonance, shaft configuration, speed, and weight. Although these motors have similar torque ratings, their physical and electrical constants vary considerably. The choice of a motor may be as simple as fitting one into the space available. However, this is generally not the case since most servosystems are very complex.

***19. References for DC Motors***

Fitzgerald, A. E., Kingsley, C., Kusko, A.; *Electric Machinery;* New York, McGraw-Hill Book Company, 1971.

Nasar, S. A., and Unnewehr, L. E.; *Electromechanics and Electric Machines;* New York, Wiley, 1979.

Say, M. G., and Taylor, E. O.; *Direct Current Machines;* New York, Wiley, 1980.

Smeaton, R. W.; *Motor Applications and Maintenance Handbook;* New York, McGraw-Hill Book Company, 1969.

Kostenko, M., and Piotrovsky, L.; *Electrical Machines,* vol. 1; Moscow, MIR Publishers, 1974.

NEMA Standard MS1—*Motors and Generators.*

Kusko, A.; *Solid State—DC Motor Drives;* Cambridge, MA, MIT Press, 1969.

Dewan, S., Slemon, G. R., Straughen, A.; *Power Semi-Conductor Drives;* New York, Wiley, 1984.

Lightband, D. A., and Bicknell, D. A.; *The Direct Current Traction Motor;* London, Business Books Ltd., 1970.

## SYNCHRONOUS MOTORS

***20. Definition.*** A synchronous motor is a machine that transforms electric power into mechanical power. The average speed of normal operation is exactly proportional to the frequency of the system to which it is connected. Unless otherwise stated, it is generally understood that a synchronous motor has field poles excited with direct current.

***21. Types.*** The synchronous motor is built with one set of ac polyphase distributed windings, designated the *armature,* which is usually on the stator and is committed to the ac supply system. The configuration of the opposite member, usually the rotor, determines the type of synchronous motor. Motors with dc excited field windings on salient-pole or round rotors, rated 200 to 100,000 hp, are the dominant industrial type. In the brushless synchronous motor, the excitation (field current) is supplied through shaft-mounted rectifiers from an ac exciter. In the slip-ring synchronous motor, the excitation is supplied from a shaft-mounted exciter or a separate dc power supply. Synchronous-induction motors rated below 5 hp, usually supplied from adjustable-speed drive inverters, are designed with a different reluctance across the air gap in the direct and quadrature axis to develop reluctance torque. The motors have no excitation source for synchronous operation. Synchronous motors below 1 hp usually employ a permanent-magnetic type of motor. These motors are usually driven by a transistor inverter from a dc source; they are termed *brushless dc motors.*

***22. Standards.*** DC separately excited synchronous motors are covered by ANSI Standard C50.10-1965, Synchronous Machines, and C50.11-1965, Synchronous Motors. They are also covered by Part 21 of NEMA Standard MG-1 1972.

***23. Theory of Operation.*** The operation of the dc separately excited synchronous motor can be explained in terms of the air-gap magnetic-field model, the circuit model, or the phasor diagram model of Fig. 20-7.

In the magnetic-field model of Fig. 20-7*a*, the stator windings are assumed to be con-

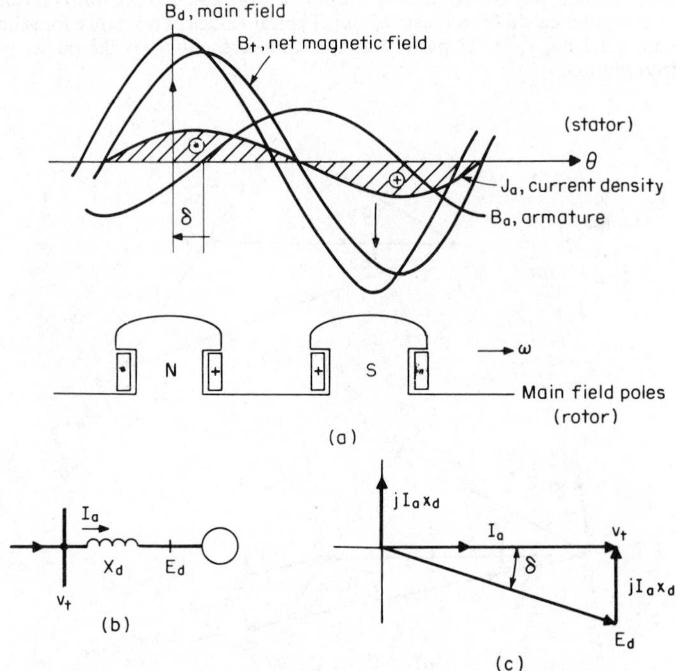


**FIG. 20-7** Operation of synchronous motor. (*a*) Air-gap magnetic-field model; (*b*) circuit model; (*c*) phasor-diagram model.

nected to a polyphase source, so that the winding currents produce a rotating wave of current density $J_a$ and radial armature reaction field $B_a$ as explained in Par. **41.** The rotor carrying the main field poles is rotating in synchronism with these waves. The excited field poles produce a rotating wave of field $B_d$. The net magnetic field $B_t$ is the spatial sum of $B_a$ and $B_d$; it induces an air-gap voltage $V_{ag}$ in the stator windings, nearly equal to the source voltage $V_t$. The current-density distribution $J_a$ is shown for the current $I_a$ in phase with the voltage $V_t$, and pf = 1. The electromagnetic torque acting between the rotor and the stator is produced by the interaction of the main field $B_d$ and the stator current density $J_a$, as a **J** × **B** force on each unit volume of stator conductor. The force on the conductors is to the left $(-\phi)$; the reaction force on the rotor is to the right $(+\phi)$, and in the direction of rotation.

The operation of the synchronous motor can be represented by the circuit model of Fig. 20-7*b*. The motor is characterized by its synchronous reactance $x_d$ and the excitation voltage $E_d$ behind $x_d$. The model neglects saliency (poles), saturation, and armature resistance, and is suitable for first-order analysis, but not for calculation of specific operating points, losses, field current, and starting.

The phasor diagram of Fig. 20-7*c* is drawn for the field model and circuit model previously described. The phasor diagram neglects saliency and armature resistance. The phasors correspond to the waves in the field model. The terminal voltage $V_t$ is generated by the field $B_t$; the excitation voltage $E_d$ is generated by the main field $B_d$; the voltage drop $jI_ax_d$ is generated by the armature reaction field $B_a$; and the current $I_a$ is the aggregate of the current-density wave $J_a$. The power angle $\delta$ is that between $V_t$ and $E_d$, or between $B_t$ and $B_d$. The excitation voltage $E_d$, in pu, is equal to the field current $I_{fd}$, in pu, on the air-gap line of the no-load (open-circuit) saturation curve of the machine.

***24. Two-Reaction Model.*** Two types of phasor diagrams can be used for analyzing the synchronous motor and calculating its performance, as shown in Fig. 20-8. The round-rotor diagram is suitable for synchronous motors without salient poles, and for a first-order representation of salient-pole motors where saliency is neglected. The consequence of neglecting saliency is that the phasor diagram yields too large a power angle $\delta$ and excitation voltage $E_d$. Neglecting saliency may be required to simplify a hand calculation, but is certainly not required if a computer calculation is carried out. Typical values of reactance for salient-pole motors are $x_d$ = 1.1 pu, $x_q$ = 0.7 pu for low-speed motors, and $x_d$ = 0.8 pu, $x_q$ = 0.5 pu for high-speed motors.

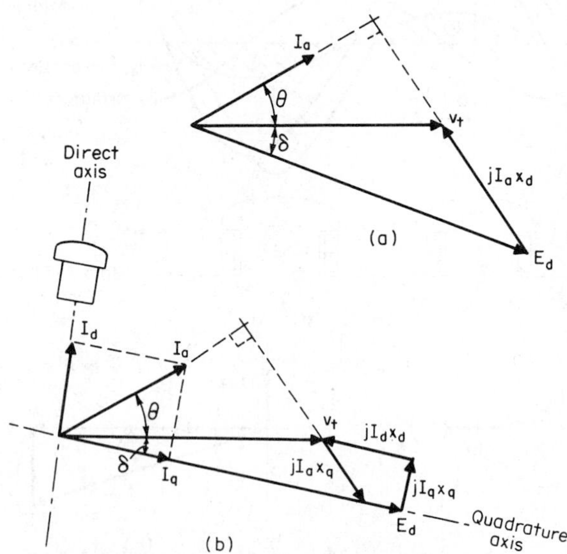


**FIG. 20-8** Phasor diagrams of synchronous motor operating at a leading pf of 0.8. (*a*) Round-rotor model; (*b*) two-reaction model.

The two-reaction diagram is shown in Fig. 20-8*b*. In accordance with the Blondel two-reaction theory the armature current is resolved into two components: one, the direct-axis component producing magnetization whose axis coincides with the center lines of the poles; and the other, the quadrature-axis component producing magnetization whose axis is midway between the field poles. The direct-axis component tends either to magnetize or to demagnetize the field poles; that is, it adds to or subtracts from the magnetization produced by the direct current in the field winding. The quadrature-axis component, on the other hand, produces magnetization midway between poles, or what is frequently called *cross magnetization*. The total voltage drop in the machine is the vector sum of the drops due to the direct- and quadrature-axis components of current separately, the direct-axis drop being the product of the direct-axis current by the direct-axis impedance, and the quadrature-axis drop being the product of the quadrature-axis current and the quadrature-axis impedance.

For small motors, the armature resistance can be included, but the pu resistance will always be relatively small compared to the pu values of $x_d$ and $x_q$. Under operating conditions of several times rated armature and field current, and close to the pull-out torque, the effect of saturation can be introduced by reducing $x_d$ from the saturated value by Kingsley's method.

***25. Power-Factor Correction.*** Synchronous motors were first used because they were capable of raising the power factor of systems having large induction-motor loads. Now they are also used because they can maintain the terminal voltage on a weak system (high source impedance), they have lower cost, and they are more efficient than corresponding induction motors, particularly the low-speed motors. Synchronous motors are built for operation at pf = 1.0, or pf = 0.8 lead, the latter being higher in cost and slightly less efficient at full load.

The selection of a synchronous motor to correct an existing power factor is merely a matter of bookkeeping of active and reactive power. The synchronous motor can be selected to correct the overall power factor to a given value, in which case it must also be large enough to accomplish its motoring functions; or it can be selected for its motoring function and required to provide the maximum correction that it can when operating at pf = 0.8 lead. In Fig. 20-9, a power diagram shows how the active and reactive power components $P_s$ and $Q_s$ of the synchronous motor are added to the components $P_i$ and $Q_i$ of an induction motor to obtain the total $P_t$ and $Q_t$ components, the kVA$_t$, and the power factor. The $Q_s$ of the synchronous motor is based on the rated kVA and pf = 0.8 lead, rather than the actual operating kVA.

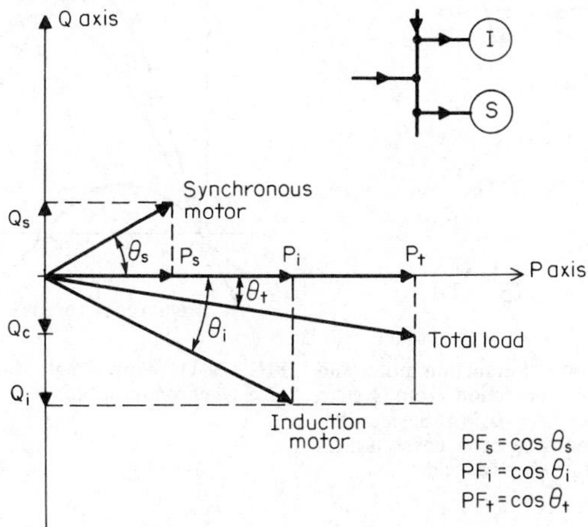


**FIG. 20-9** Power diagrams of induction motor and synchronous motor operating in parallel, showing component and net values of *P* and *Q*.

The synchronous motor can support the voltage of a weak system, so that a larger-rating synchronous motor can be installed than an induction motor for the same source impedance. With an induction motor, both the $P$ and $Q$ components produce voltage drops in the source impedance. With a synchronous motor operating at leading power factor, the $P$ component produces a voltage drop in the source resistance, but the $Q$ component produces a voltage rise in the source reactance that can offset the drop and allow the terminal voltage to be normal. If necessary, the field current of the synchronous motor can be controlled by a voltage regulator connected to the motor bus. Phasor diagrams of an induction motor and a synchronous motor connected to the same source are shown in Fig. 20-10. The leading current of the synchronous motor in Fig. 20-10*c* is able to develop a sufficient voltage rise through the source reactance $x_e$ to overcome the voltage drop in $r_e$ and maintain the motor voltage $V_t$ equal to the source voltage $V_s$.

***26. Power-Angle Diagrams.*** The power delivered to the terminals of a synchronous motor as it is loaded mechanically can be derived from the phasor diagram of Fig. 20-8*b* for the lossless machine:

$$P_c = \frac{V_t E_d}{x_d} \sin \delta + \frac{V_t^2}{2} \left( \frac{1}{x_q} - \frac{1}{x_d} \right) \sin 2\delta \tag{20-7}$$

Equation (20-7) is shown in per unit. The value of $P_c$ will be in watts, if the voltages are line-line and the reactances in ohms. The corresponding expression for torque can be obtained by dividing by the synchronous speed, $T_{em} = P_c/\omega_0$, where $\omega_0 = 1$ pu, or $2\pi\,(2f/\text{p})$ rad/s.

A plot of the torque is shown in Fig. 20-11. The total value is the *synchronous torque.* The peak value is the *pull-out torque.* The second term of Eq. (20-7) represents the *reluctance torque.* For round-rotor motors, $x_d = x_q$, the reluctance torque is zero, and the expression reduces to the first term. The curve gives the torque and power as a function of the power angle $\delta$. The peak value of the first term can be increased by raising the field current, which increases the excitation voltage $E_d$. The pull-out torque can be raised substantially above the rated pull-out torque by this means.

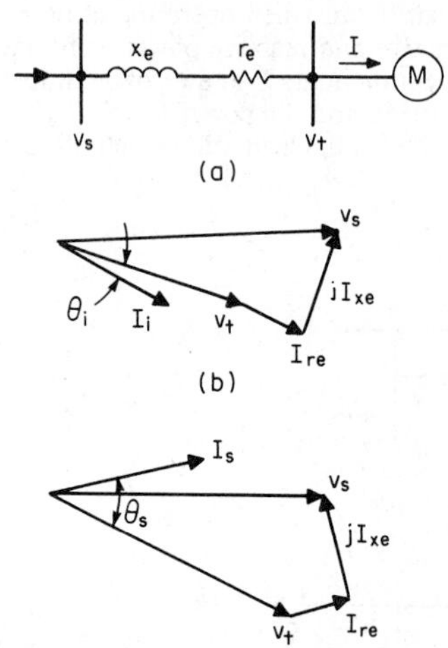


**FIG. 20-10** Comparison of induction motor and synchronous motor for operation from system with supply impedance $r_e + jx_e$. (*a*) System diagram; (*b*) induction motor at pf = cos $\theta_i$ lag; (*c*) synchronous motor at pf = cos $\theta_s$ lead.

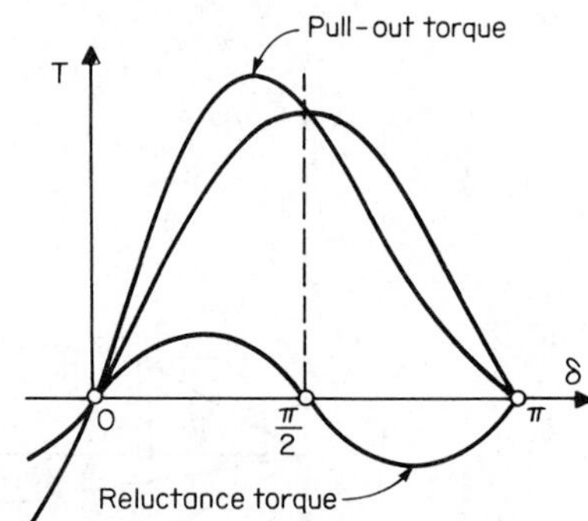


**FIG. 20-11** Power-angle diagram of a salient-pole synchronous motor.

***27. Starting.*** The interaction of the main field produced by the rotor and the armature current of the stator will produce a net average torque to drive the synchronous motor, only when the rotor is revolving at speed $n$ in synchronism with the line frequency $f$; $n = 120$ f/p, p = poles. The motor must be started by developing other than synchronous torques.

Practically, the motor is equipped with an induction-motor-type squirrel-cage winding on the rotor, in the form of a damper winding, in order to start the motor.

The motor is started on the damper windings with the field winding short-circuited, or terminated in a resistor, to attenuate the high "transformer" induced voltages. When the motor reaches the lowest slip speed, practically synchronous speed, the field current is applied to the field winding, and the rotor poles accelerate and pull into step with the synchronously rotating air-gap magnetic field. The damper windings see zero slip and carry no further current, unless the rotor oscillates with respect to the synchronous speed.

Starting curves for a synchronous motor are shown in Fig. 20-12. The damper winding is designed for high starting torque, as compared to an induction motor of the same rating. The closed field winding contributes to the starting torque in the manner of a 3-phase induction motor with a 1-phase rotor. The field winding produces positive torque to half speed, then negative torque to full speed, accounting for the anomaly at half speed. The maximum and minimum torque excursion at the anomaly is reduced by the resistance in the closed field winding circuit during starting. The effect is increased by the design of the damper winding.

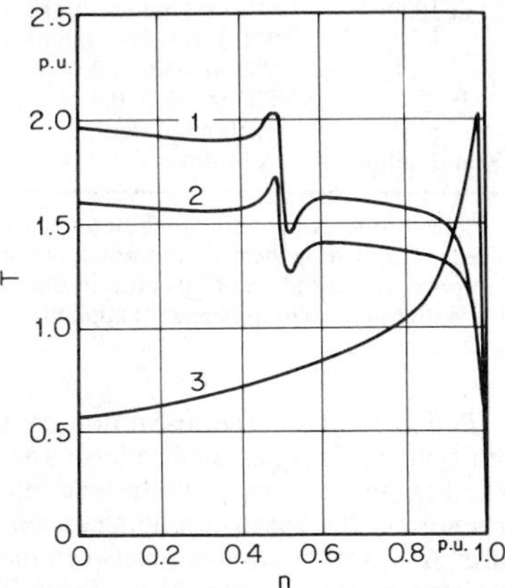


**FIG. 20-12** Characteristic torque curves for 5000-hp synchronous induction motor during run-up at full voltage. (1) Synchronous motor for pf = 1; (2) synchronous motor for pf = 0.8; (3) squirrel-cage induction motor.

The velocity of the rotor during the synchronizing phase, after field current is applied, is shown in Fig. 20-13. The rotor is assumed running at 0.05 pu slip on the damper winding. The undulation in speed, curve 1, is the effect of the poles attempting to synchronize the rotor just by reluctance torque. The added effect of the field current is shown by curve 2, and the resultant by curve 3. The effect of the reluctance torque of curve 1 is not dependent on pole polarity. The synchronizing torque of curve 2, with the field current applied, is pole polarity dependent; the poles want to match the air-gap field in the forward torque direction. Curve *a* shows a successful synchronization. Curve *b* shows the condition of too much load or inertia to synchronize.

***28. Torque Definitions.*** The torques described in the following paragraphs are listed in the Standards. The minimum values are given in Table 20-1.

*Locked-rotor torque* is the minimum torque which the synchronous motor will develop at rest for all angular positions of the rotor, with rated voltage at rated frequency applied.

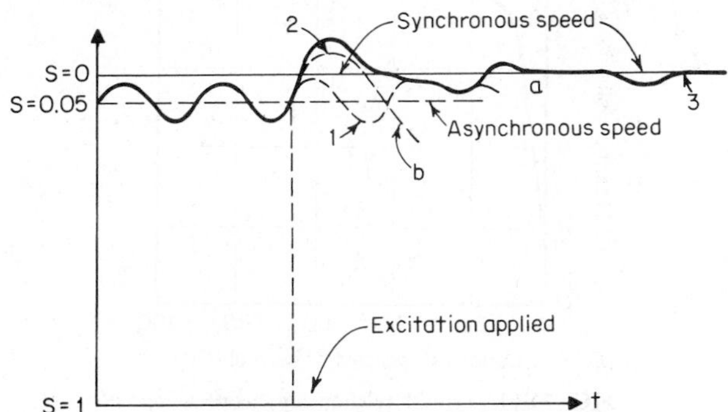


**FIG. 20-13** Relationship between slip and time for a synchronous motor pulling into synchronism. (*a*) Successful; (*b*) unsuccessful.

**TABLE 20-1** Locked-Rotor, Pull-in, and Pull-out Torques for Synchronous Motors (From ANSI C50.11-1965, Table 2)

| | | Percent of rated full-load torque* | | | |
|---|---|---|---|---|---|
| | | | Pull-in (based on normal $Wk^2$ of load)† | Pull-out† | |
| r/min | hp | Locked rotor | | 1.0 pf | 0.8 pf |
| 514 to 1800 | 200 and below; 1.0 pf<br>150 and below; 0.8 pf | 100 | 100 | 150 | 175 |
| | 250 to 1000; 1.0 pf<br>200 to 1000; 0.8 pf | 60 | 60 | 150 | 175 |
| | 1250 and larger | 40 | 60 | 150 | 175 |
| 450 and below | All ratings | 40 | 30 | 150 | 200 |

*The torque values with other than rated voltage applied are approximately equal to the rated voltage values multiplied by the ratio of the actual voltage to rated voltage in the case of the pull-out torque, and multiplied by the square of this ratio in the case of the locked-rotor and pull-in torque.

†With rated excitation current applied.

*Pull-in torque* is the maximum constant-load torque under which the motor will pull into synchronism, at rated voltage and frequency, when its rated field current is applied. Whether the motor can pull the load into step from the slip running on the damper windings depends on the speed-torque character of the load and the total inertia of the revolving parts. A typical relationship between maximum slip and percent of normal $Wk^2$ for pulling into step is shown in Fig. 20-14. Table 20-1 specifies minimum values of pull-in torque with the motor loaded with normal $Wk^2$; these values are given in Par. **31.** *Nominal pull-in torque* is the value at 95% of synchronous speed, with rated voltage at rated frequency applied, when the motor is running on the damper windings.

*Pull-out torque* is the maximum sustained torque which the motor will develop at synchronous speed for 1 min, with rated voltage at rated frequency applied, and with rated field current.

In addition, the *pull-up torque* is defined as the minimum torque developed between standstill and the pull-in point. This torque must exceed the load torque by a sufficient margin to assure satisfactory acceleration of the load during starting.

The *reluctance torque* is a component of the total torque when the motor is operating synchronously. It results from the saliency of the poles and is a manifestation of the poles

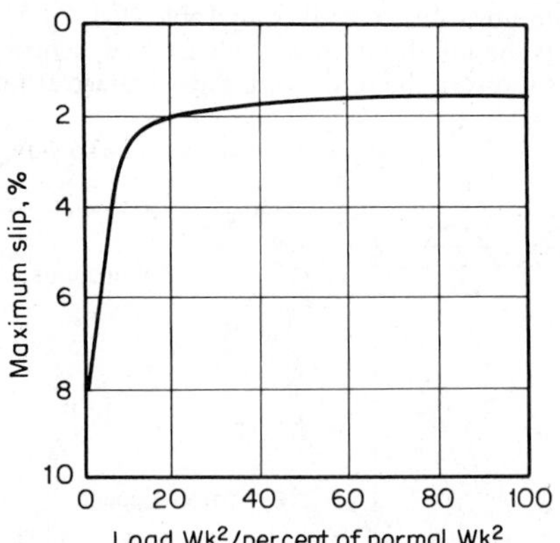


**FIG. 20-14** Typical relationship between load inertia and maximum slip for pulling synchronous motors into step.

attempting to align themselves with the air-gap magnetic field. It can account for up to 30% of the pull-out torque.

The *synchronous torque* is the total steady-state torque available, with field excitation applied, to drive the motor and the load at synchronous speed. The maximum value as the motor is loaded is the pull-out torque, developed as a power angle $\delta = 90°$.

***29. Synchronization.*** Synchronization is the process by which the synchronous motor "pulls into step" during the starting process, when the field current is applied to the field winding. Initially, the rotor is revolving at a slip with respect to the synchronous speed of the air-gap magnetic-field waves. The rotor torque, produced by the damper windings, is in equilibrium with the load torque at that slip. The ability of the rotor to accelerate and synchronize depends upon the total inertia ($Wk^2$), the initial slip, and the closing angle of the poles with respect to the field wave at the instant field current is applied.

Figure 20-15 shows the torque versus angle $\delta$ locus for the rotor during a successful synchronization. The rotor is subjected to the synchronous torque $T_s$, which is a function of $\delta$, and the damper torque $T_d$, which is a function of the slip velocity ($n_0 - n$). The torque $T_a$ available to accelerate the rotor is the residual of $T_a = T_s + T_d - T_l$. In the figure, the closing angle is assumed zero at point *a*. Furthermore, $T_d = T_l$, so that the residual torque $T_a$ is zero. The rotor has a finite slip, so that the power angle $\delta$ increases. As it does, the synchronous torque $T_s$ increases, $T_a$ increases and the rotor accelerates to point *b*, where $n = n_0$, $T_d = 0$. The slip goes negative, reverses the direction of the damper torque, but the rotor continues to accelerate to point *c*, where the speed is maximum and the accelerating torque is zero. The rotor falls back to points *d* and *e* at minimum speed, accelerates again, and finally synchronizes at point *f*.

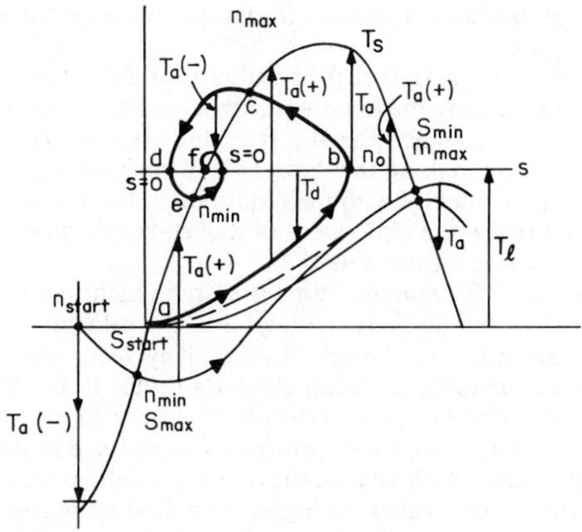


**FIG. 20-15** Locus of torque and speed vs. power angle $\delta$ for a synchronous motor during a successful and an unsuccessful attempt to synchronize.

If the initial slip is excessive, or if the inertia and/or load too great, the locus in Fig. 20-15 could follow the path *ab′*. The condition of $T_a = 0$ is reached below synchronous speed; the rotor never pulls into step, but oscillates around the initial slip velocity until the machine is tripped off.

***30. Damper Windings.*** Damper windings are placed on the rotors of synchronous motors for two purposes: for starting and for reducing the amplitude of power-angle oscillation. The damper windings consist of copper or brass bars inserted through holes in the pole shoes and connected at the ends to rings to form the equivalent of a squirrel-cage. The rings can extend between the poles to form a complete damper. Synchronous motors with solid pole shoes, or solid rotors, perform like motors with damper windings.

The design of the damper winding requires the selection of the bar and ring material to meet the torque and damping requirements. Figure 20-16 shows the effect on the starting

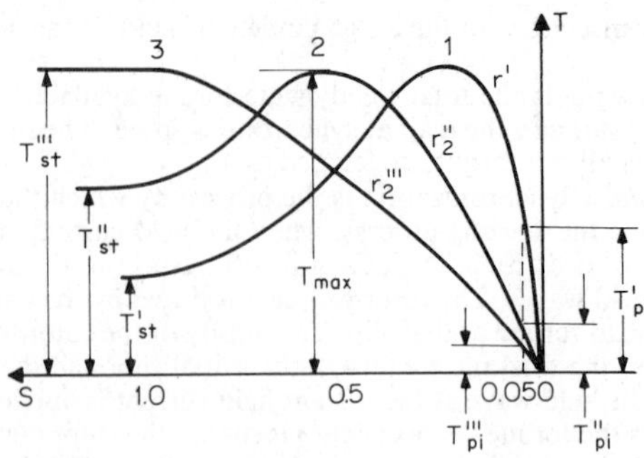


**FIG. 20-16** Effect of resistivity of damper material on the starting and pull-in torque of the synchronous motor. Damper winding 1, least resistance; damper winding 3, maximum resistance.

curves for the damper winding of varying the material from a low-resistance copper in curve 1, to a higher-resistance brass or aluminum-bronze alloy in curve 2. Curve 1 gives a starting torque of about 0.25 pu, and a pull-in torque of 1.0 pu, of the nominal synchronous torque. Curve 2 gives a higher starting torque of about 0.5 to 1.0 pu, but a pull-in torque of about 0.4 pu of the nominal value. The additional starting torque of the field winding is superimposed on the torque of the damper alone. The damper winding must be designed to meet the characteristics of the load.

To design the damper winding so that the amplitude of the natural-frequency oscillation is reduced, the bar currents during the low-frequency sweeping of the air-gap flux across the pole faces must be maximized. Since the slip frequency is low, the currents and damper effectiveness is maximized by making the dampers low resistance, corresponding to curve 1 in Fig. 20-16. This design coincides with the requirement for low-starting torque and high pull-in torque. In special cases, the equivalent of a deep-bar or double-bar damper can be used, if there is adequate space on the pole shoe.

***31. Methods of Starting.*** The method used to start a synchronous motor depends upon two factors: the required torque to start the load and the maximum starting current permitted from the line. Basically, the motor is started by using the damper windings to develop asynchronous torque or by using an auxiliary motor to bring the unloaded motor up to synchronous speed. Solid-state converters have also been used to bring up to speed large several-hundred-MVA synchronous motor/generators for pumped storage plants.

Techniques for asynchronous starting on the damper windings are the same as for squirrel-cage induction motors of equivalent rating. Across-the-line starting provides the maximum starting torque, but requires the maximum line current. The blocked-rotor kVA of synchronous motors as a function of pole number is shown in Fig. 20-17. If the ac line to the motor supplies other loads, the short-circuit kVA of the line must be at least 6 to 10 times the blocked rotor kVA of the motor to limit the line-voltage dip on starting. The starting and pull-in torques for three general classes of synchronous motors are shown in Fig. 20-18. The torques are shown for rated voltage; for across-the-line starting, the values will be reduced to $V_l^2$ (pu).

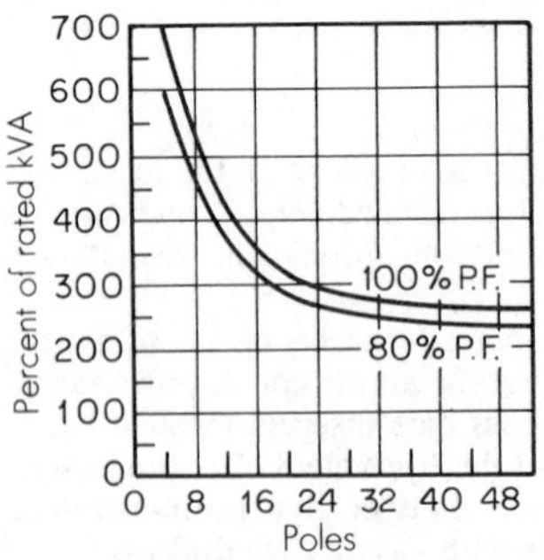


**FIG. 20-17** Approximate blocked-rotor kVA of synchronous motors.

Reduced-voltage starting is used where the full starting torque of the motor is not required and/or the ac line cannot tolerate the full starting current. The starter includes

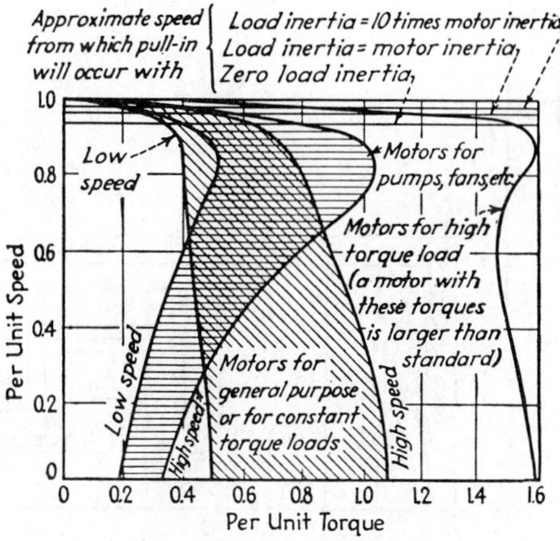


**FIG. 20-18** Approximate starting performance of synchronous motors.

a 3-phase open-delta or 3-winding autotransformer, which can be set to apply 50, 65, or 80% of line voltage to the motor on the first step. The corresponding torque is reduced to 25, 42, or 64%. The starter switches the motor to full voltage when it has reached nearly synchronous speed, and then applies the field excitation to synchronize the motor.

ANSI C50.11-1965 limits the number of starts for a synchronous motor, under its design conditions of $Wk^2$, load torque, nominal voltage, and starting method, to the following:

1. Two starts in succession, coasting to rest between starts, with the motor initially at ambient temperature, or
2. One start with the motor initially at a temperature not exceeding its rated load operating temperature.

If additional starts are required, it is recommended that none be made until all conditions affecting operation have been thoroughly investigated and the apparatus examined for evidence of excessive heating. It should be recognized that the number of starts should be kept to a minimum since the life of the motor is affected by the number of starts.

***32. Exciters.*** DC separately excited synchronous motors are provided with a shaft-driven exciter to supply the field power. Exciters are classified into slip-ring types and brushless types. The slip-ring type consists of a dc generator, whose output is fed into the motor field winding through slip rings and stationary brushes. The brushless type consists of an ac generator, with rotating armature and stationary field; the output is rectified by solid-state rectifier elements mounted on the rotating structure and fed directly to the motor field winding. In each type, the motor field current is controlled by the exciter field current. Typical kilowatt ratings for exciters for 60-Hz synchronous motors are given in MG1-21.16 as a function of hp rating, speed, and power factor. For a given hp rating, the exciter kW increases as the speed is reduced, and as the power factor is shifted from pf = 1.0 to pf = 0.8 lead.

During starting, the motor field winding must be disconnected from the exciter and loaded with a resistor, to limit the high induced voltage, to prevent damage to the rectifier elements of the brushless type, and to prevent the circulation of ac current through a slip-ring-type dc exciter. The switching is done with a contactor for the slip-ring type, and with thyristors on the rotating rectifier assembly for the brushless type. Except for the disconnection for starting, the synchronous-motor excitation system is practically the same as for an ac generator of the same rating.

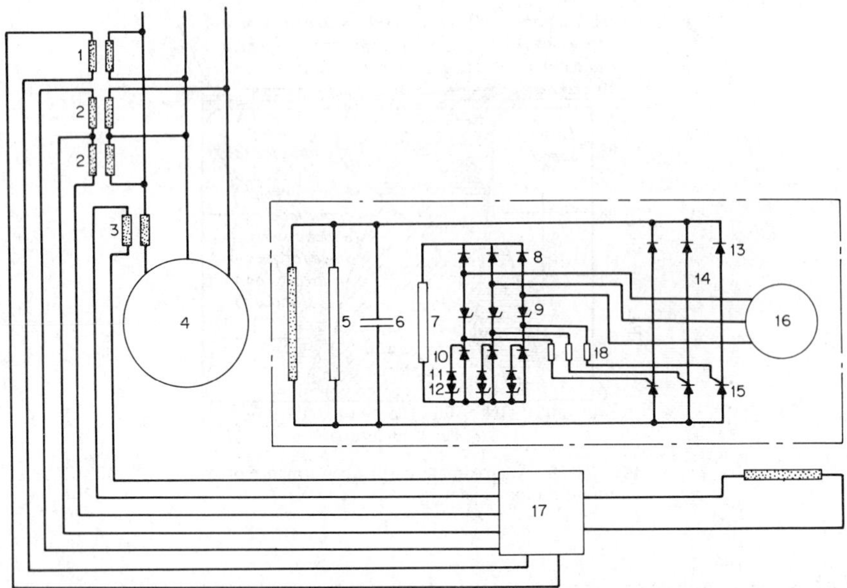


**FIG. 20-19** Brushless-type excitation system for a synchronous motor.

Brushless-type exciters are now used on all new high-speed synchronous motors (2 to 8 poles) that formerly were built with direct-drive dc exciters and slip rings. The brushless-type exciters require minimum maintenance and can be used in explosive atmospheres. The circuit of a typical brushless-type excitation system is shown in Fig. 20-19. The semicontrolled bridge with three diodes and three thyristors rectifies the output of the ac exciter generator and supplies the motor field winding. The thyristors act as a switch to open the rectifier during starting and to close it during running, whereas the ac exciter generator is excited with its own field current. The resistor is permanently connected across the motor field winding during starting and running. It improves the torque characteristics during starting, and protects the bridge elements against transient overvoltages during running. The capacitor protects the diodes and thyristors against commutation overvoltages caused by hole-storage phenomena in conjuntion with the inductances of the armature windings of the ac exciter generator.

The control system (Fig. 20-19) comprises a simple auxiliary rectifier arrangement connected in parallel with the main rectifier bridge and loaded with an auxiliary resistor 7. Each main thyristor has an auxiliary thyristor which provides the gate current and operates on the same phase of the ac exciter voltage. Consequently the trigger signal always occurs at the correct instant, that is, when the thyristors have a forward loading. No trigger signal is given during the blocking period. There is no excitation at the exciter during run-up, and therefore no trigger signal is applied to the gates of the thyristors and they remain blocking. The alternating current induced in the field winding flows in both directions through the protection resistor 5. When the machine has been run-up to normal speed, the field voltage is applied to the ac exciter. It then supplies the control current and the thyristors are fired. Control losses are only 0.1 to 0.2% of the exciter power and are therefore negligible. The auxiliary thyristor 10 together with the diode 11 and zener diode 12 prevents preignition of the thyristors during run-up due to high residual voltage in the ac exciter. On the other hand the gates of the other thyristors are protected against overload by zener diode 9 and resistor 18. If the voltage exceeds the zener voltage, the zener diode conducts the excess current.

***33. Standard Ratings.*** Standard ratings for dc separately excited synchronous motors are given in NEMA MG1-1978, Part 21. Standard horsepowers range from 20 to 100,000 hp. Speed ratings extend from 3600 r/min (2-pole) to 80 r/min (90-pole) for 60-Hz machines, and five-sixths of the values for 50-Hz machines. The power factor shall be unity or 0.8 leading. The voltage ratings for 60-Hz motors are 200, 230, 460, 575, 2300, 4000, 4600, 6600 and 13,200 V. It is not practical to build motors of all horsepower ratings at these speeds and voltages.

***34. Efficiency.*** Efficiency and losses shall be determined in accordance with IEEE test procedures for synchronous machines, Publication 115. The efficiency shall be determined at rated output, voltage, frequency, and power factor. The following losses shall be included in determining the efficiency: (1) $I^2R$ loss of armature and field; (2) core loss; (3) stray-load loss; (4) friction and windage loss; (5) exciter loss for shaft-driven exciter. The resistances should be corrected for temperature.

Typical synchronous motor efficiencies are shown in Fig. 20-20. The unity-power-factor

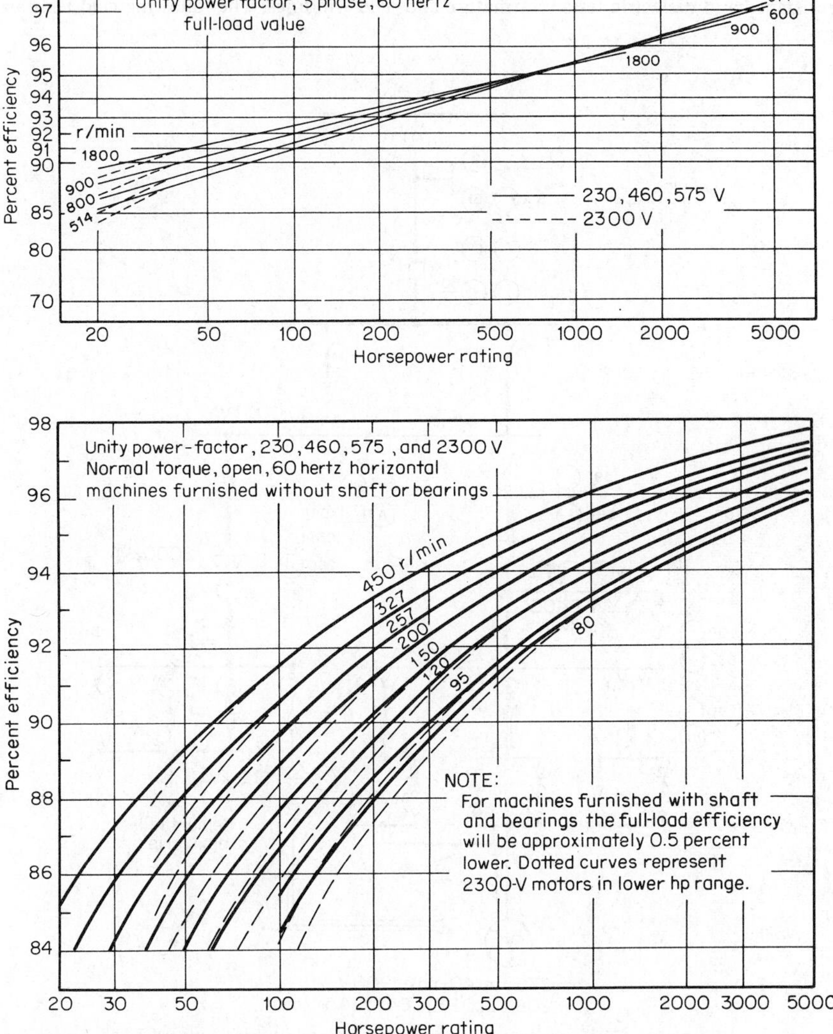


**FIG. 20-20** Full-load efficiencies of (*a*) high-speed general-purpose synchronous motors and (*b*) low-speed synchronous motors.

synchronous motor is generally 1 to 3% more efficient than the NEMA Design B induction motors. The 0.8 pf synchronous motor, because of the increased copper loss, is lower in efficiency; its efficiency is closer to that of the induction motor at high speed, but better at low speed.

***35. Standard Tests.*** Tests on synchronous motors shall be made in accordance with IEEE Test Procedure for Synchronous Machines, Publ. No. 115, and ANSI C50.10-1965. The following tests shall be made on motors completely assembled in the factory and furnished with shaft and complete set of bearings: resistance test of armature and field windings; dielectric test of armature and field windings; mechanical balance; current balance at no load; direction of rotation. The following tests may be specified on the same or duplicate motors: locked-rotor current; temperature rise; locked-rotor torque; overspeed; harmonic analysis and TIF; segregated losses; short-circuit tests at reduced voltage to determine reactances and time constants; field-winding impedance; speed-torque curve.

The following tests shall be made on all motors not completely assembled in the factory: resistance and dielectric tests of armature and field windings. The following field tests are

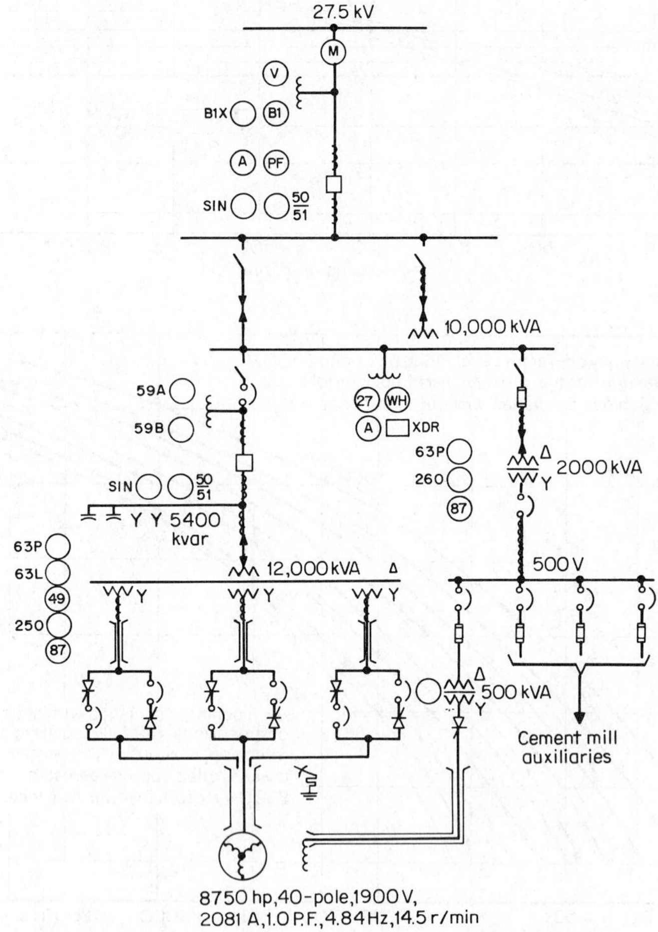


**FIG. 20-21** Cycloconverter-synchronous motor gearless drive system for ball mill.

recommended after installation: resistance and dielectric tests of armature and field windings not completely assembled in the factory; mechanical balance; bearing insulation; current balance at no load; direction of rotation. The following field tests may be specified on the same or duplicate motors: temperature rise; short-circuit tests at reduced voltage to determine reactances and time constants; field-winding impedance.

The dielectric test for the armature winding shall be conducted for 1 min, with an ac rms voltage of 1000 V plus twice the rated voltage. For machines rated 6 kVand above, the test may be conducted with a dc voltage of 1.7 times the ac rms test value. The dielectric test for the field winding depends upon the connection for starting. For a short-circuited field winding, the ac rms test voltage is 10 times the rated excitation voltage, but no less than 2500 V, nor more than 5000 V. For a field winding closed through a resistor, the ac rms test voltage is twice the rms value of the *IR* drop, but not less than 2500 V, where the current is the value that would circulate with a short-circuited winding. When a test is made on an assembled group of several pieces of new apparatus, each of which has passed a high-potential test, the test voltage shall not exceed 85% of the lowest test voltage for any part of the group. When a test is made after installation of a new machine which has passed its high-potential test at the factory and whose windings have not since been disturbed, the test voltages should be 75% of the original values.

***36. Cycloconverter Drive.*** A unique application for large low-speed synchronous motors is for gearless ball-mill drives for the cement industry. For a recently installed drive, the motor is rated 8750 hp, 1.0 pf, 6850 kVA, 14.5 r/min, 1900 V, 4.84 Hz, 40 poles, Class B. The power is provided by a cycloconverter over the range 0 to 4.84 Hz, as shown in Fig. 20-21. The cycloconverter consists of six thyristor rectifiers, each of which generates one polarity of the 3-phase ac voltage wave applied to the motor. The cycloconverter can be used effectively up to about one-third of the line frequency. The motor can be controlled in speed by the cycloconverter frequency, or in torque by the angle between the armature voltage and the field-pole position, approximately the power angle $\delta$.

***37. Inverter-Synchronous Motor Drive.*** Synchronous motors over about 1000 hp are being driven by machine-commutated inverters for adjustable-speed drives for large fans, pumps, and other loads. The machine-commutated inverter drive consists of two converters interconnected by a dc link as shown in Fig. 20-22*a*. The synchronous motor operates at constant volts per hertz, i.e., voltage proportional to frequency and speed. The converter characteristics are shown in Fig. 20-22*b* and 22*c*. The $\pm V_d$ values are 1.35 times the line-line voltage on the ac side of each converter. For a given motor speed, frequency, and voltage, the firing angle of the rectifier is set at $\alpha_r$ to yield the required dc voltage $V_l$ for the link. The firing angle of the inverter is set at $\alpha_i$ in the inverting quadrant of the converter so that the link voltage $V_l$ matches the internal ac voltage generated by the motor at the given speed. Power flows from the rectifier at $V_e I_d$ into the inverter and the motor. The inverter firing signals are synchronized to the motor voltage. For decelerating the motor, the rectifier and inverter functions are reversed by shifting the firing angles. Power flows from the motor into the dc link and to the supply line.

## Bibliography

***38. References for Synchronous Motors***


Fitzgerald, A. E., Kingsley, C., Jr., Kusko, A.; *Electric Machinery,* 3d ed.; New York, McGraw-Hill Book Company, 1971.

Sarma, Mulukutla S.; *Synchronous Machines;* New York, Gordon & Breach, 1979.

Concordia, C.; *Synchronous Machines;* New York, Wiley, 1951.

Say, M. G.; *Alternating Current Machines;* New York, Wiley, 1976.

NEMA Std. MS1—*Motors and Generators.*

IEEE Std. 115—1965 *Test Procedures for Synchronous Machines.*

IEEE Std. 421—1972 *Criteria and Definition for Excitation Systems for Synchronous Machines.*

Bose, B. K.; *Adjustable Speed AC Drives;* IEEE Press, 1980.

Kostenko, M., and Piotrovsky, L.; *Electric Machines,* vol. 2; Moscow, MIR Publications, 1974.

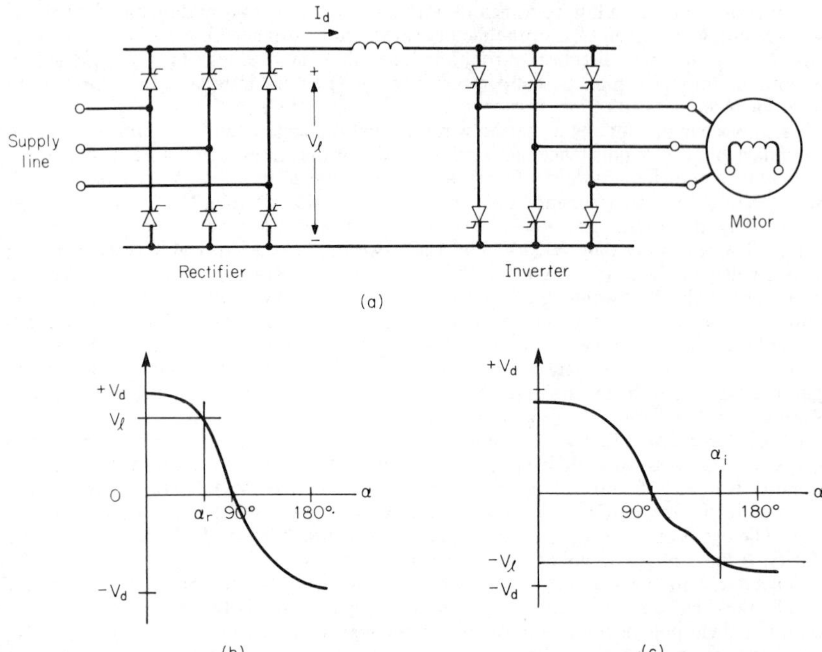


**FIG. 20-22** (*a*) Diagram of machine-commutated synchronous motor drive; (*b*) dc voltage vs. firing angle $\alpha_r$ characteristic of rectifier; (*c*) dc voltage vs. firing angle $\alpha_i$ characteristic of inverter.

## INDUCTION MACHINES

### Theory of the Polyphase Induction Motor[1]

***39. Principle of Operation.*** An induction motor is simply an electric transformer whose magnetic circuit is separated by an air gap into two relatively movable portions, one carrying the primary and the other the secondary winding. Alternating current supplied to the primary winding from an electric power system induces an opposing current in the secondary winding, when the latter is short-circuited or closed through an external impedance. Relative motion between the primary and secondary structures is produced by the electromagnetic forces corresponding to the power thus transferred across the air gap by induction. The essential feature which distinguishes the induction machine from other types of electric motors is that the secondary currents are created solely by induction, as in a transformer, instead of being supplied by a dc exciter or other external power source, as in synchronous and dc machines.

Induction motors are classified as squirrel-cage motors and wound-rotor motors. The secondary windings on the rotors of squirrel-cage motors are assembled from conductor bars short-circuited by end rings or are cast in place from aluminum or another conductive alloy. The secondary windings of wound-rotor motors are wound with discrete conductors with the same number of poles as the primary winding on the stator. The rotor windings are terminated on slip rings on the motor shaft. The windings can be short-circuited by brushes bearing on the slip rings, or they can be connected to resistors or solid-state converters for starting and speed control.

***40. Construction Features.*** The normal structure of an induction motor consists of a cylindrical rotor carrying the secondary winding in slots on its outer periphery and an encir-

[1] P. L. Alger; *The Nature of Polyphase Induction Machines;* New York, John Wiley & Sons, Inc., 1951.

cling annular core of laminated steel carrying the primary winding in slots on its inner periphery. The primary winding is commonly arranged for 3-phase power supply, with three sets of exactly similar multipolar coil groups spaced one-third of a pole pitch apart. The superposition of the three stationary, but alternating, magnetic fields produced by the 3-phase windings produces a sinusoidally distributed magnetic field revolving in synchronism with the power-supply frequency, the time of travel of the field crest from one phase winding to the next being fixed by the time interval between the reaching of their crest values by the corresponding phase currents. The direction of rotation is fixed by the time sequence of the currents in successive phase belts and so may be reversed by reversing the connections of one phase of a 2- or 3-phase motor.

Figure 20-23 shows the cross section of a typical polyphase induction motor, having in this case a 3-phase 4-pole primary winding with 36 stator and 28 rotor slots. The primary winding is composed of 36 identical coils, each spanning 8 teeth, one less than the 9 teeth in one pole pitch. The winding is therefore said to have 8/9 pitch. As there are three primary slots per pole per phase, phase A comprises four equally spaced "phase belts," each consisting of three consecutive coils connected in series. Owing to the short pitch, the top and bottom coil sides of each phase overlap the next phase on either side. The rotor, or secondary, winding consists merely of 28 identical copper or cast-aluminum bars solidly connected to conducting end rings on each end, thus forming a "squirrel-cage" structure. Both rotor and stator cores are usually built of silicon-steel laminations, with partly closed slots, to obtain the greatest possible peripheral area for carrying magnetic flux across the air gap.

***41. The Revolving Field.*** The key to understanding the induction motor is a thorough comprehension of the revolving magnetic field.

The rectangular wave in Fig. 20-24 represents the mmf, or field distribution, produced by a single full-pitch coil, carrying $H$ At. The air gap between stator and rotor is assumed to be uniform, and the effects of slot openings are neglected. To calculate the resultant field produced by the entire winding, it is most convenient to analyze the field of each single coil into its space-harmonic components, as indicated in Fig. 20-24 or expressed by the following equation:

$$H(x) = \frac{4H}{\pi}\left(\sin x + \frac{1}{3}\sin 3x + \frac{1}{5}\sin 5x + \frac{1}{7}\sin 7x + \cdots\right) \tag{20-8}$$

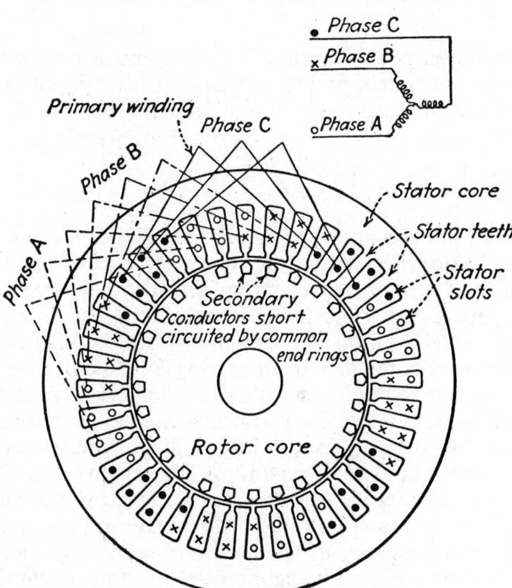


**FIG. 20-23** Section of squirrel-cage induction motor, 3-phase, 4-pole, 8/9-pitch stator winding.

When two such fields produced by coils in adjacent slots are superposed, the two fundamental sine-wave components will be displaced by the slot angle $\theta$, the third-harmonic components by the angle $3\theta$, the fifth harmonics by the angle $5\theta$, etc. Thus, the higher space-harmonic components in the resultant field are relatively much reduced as compared with the fundamental. By this effect of distributing the winding in several slots for each phase belt, and because of the further reductions due to fractional pitch and to phase connections, the space-harmonic fields in a normal motor are reduced to negligible values, leaving only the fundamental sine-wave components to be considered in determining the operating characteristics.

The alternating current flowing in the winding of each phase therefore produces a sine-wave distribution of magnetic flux around the periphery, stationary in space but varying sinusoidally in time in synchronism with the supply frequencies. Referring to Fig. 20-25, the field of phase A at an angular distance $x$ from the phase axis may be represented as an alternating phasor $I \cos x \cos \omega t$ but may equally well be considered as the resultant of two phasors constant in magnitude but revolving in opposite directions at synchronous speed:

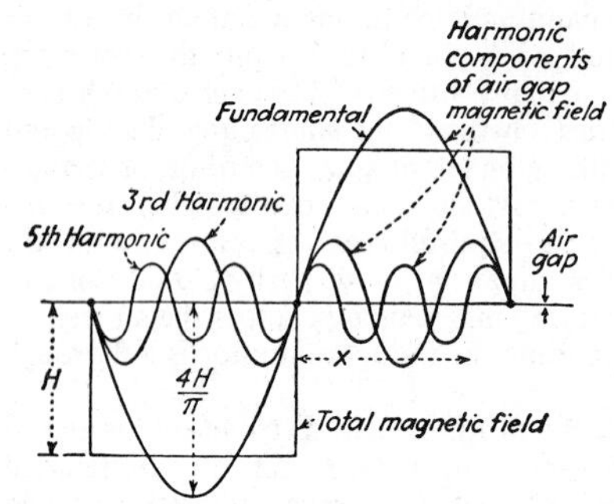

**FIG. 20-24** Magnetic field produced by a single coil.

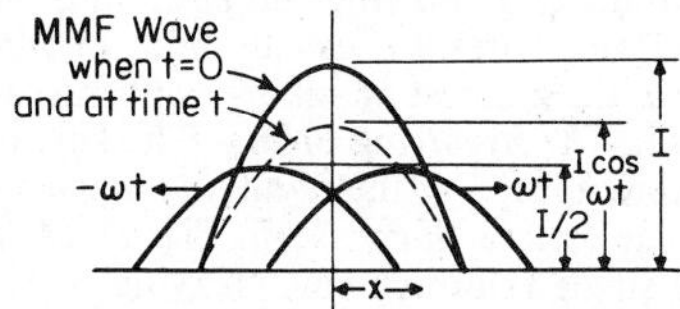


**FIG. 20-25** Resolution of alternating wave into two constant-magnitude waves revolving in opposite directions.

$$I \cos x \cos \omega t = \frac{I}{2} [\cos (x - \omega t) + \cos (x + \omega t)] \qquad (20\text{-}9)$$

Each of the right-hand terms in this equation represents a sine-wave field revolving at the uniform rate of one pole pitch, or 180 elec deg, in the time of each half cycle of the supply frequency. The synchronous speed $N_s$ of a motor is therefore given by

$$N_s = \frac{120f}{P} \quad \text{r/min} \qquad (20\text{-}10)$$

where $f$ = line frequency in hertz and $P$ = number of poles of the winding.

Considering phase $A$ alone (Fig. 20-26), two revolving fields will coincide along the phase center line at the instant its current is a maximum. One-third of a cycle later, each will have traveled 120 elec deg, one forward and the other backward, the former lining up with the axis of phase $B$ and the latter with the axis of phase $C$. But at this moment, the current in phase $B$ is a maximum, so that the forward-revolving $B$ field coincides with the forward $A$ field, and these two continue to revolve together. The backward $B$ field is 240° behind the backward $A$ field, and these two remain at this angle, as they continue to revolve. After another third of a cycle, the forward $A$ and $B$ fields will reach the phase $C$ axis, at the same moment that phase $C$ current becomes a maximum. Hence, the forward fields of all 3 phases are directly additive, and together they create a constant-magnitude sine-wave-shaped synchronously revolving field with a crest value 3⁄2 the maximum instantaneous value of the alternating field due to one phase alone. The backward-revolving fields of the 3 phases are separated by 120°, and their resultant is therefore zero so long as the 3-phase currents are balanced in both magnitude and phase.

If a 2-phase motor is considered, it will have two 90° phase belts per pole instead of three

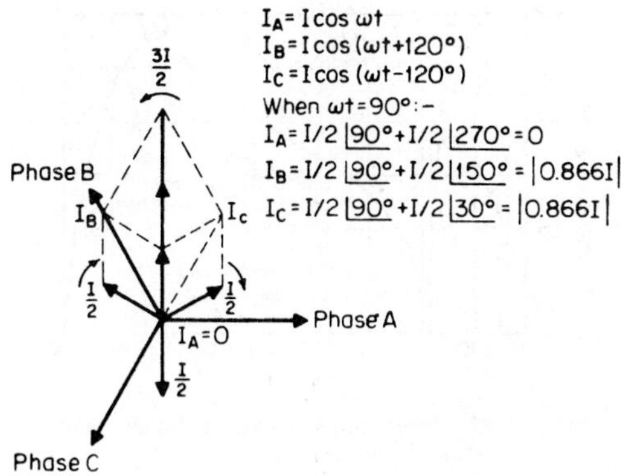


**FIG. 20-26** Resolution of alternating emf of each phase into oppositely revolving constant-magnitude components shown at instant when phase $A$ current is zero ($\omega\tau = 90°$).

60° phase belts, and a similar analysis shows that it will have a forward-revolving constant-magnitude field with a crest value equal to the peak value of 1 phase alone and will have zero backward-revolving fundamental field. A single-phase motor will have equal forward and backward fields and so will have no tendency to start unless one of the fields is suppressed or modified in some way.

While the space-harmonic-field components are usually negligible in standard motors, it is important to the designer to recognize that there will always be residual harmonic-field values which may cause torque irregularities and extra losses if they are not minimized by an adequate number of slots and correct winding distribution. An analysis similar to that given for the fundamental field shows that in all cases the harmonic fields corresponding to the number of primary slots (seventh and nineteenth in a nine-slot-per-pole motor) are important and that the fifth and seventh harmonics on 3-phase, or third and fifth on 2-phase, may also be important.

The third-harmonic fields and all multiples of the third are zero in a 3-phase motor, since the mmf's of the 3 phases are 120° apart for both backward and forward components of all of them. Finally, therefore, a 3-phase motor has the following distinct fields:

**a.** The fundamental field with $P$ poles revolving forward at speed $N_s$.

**b.** A fifth-harmonic field with $5P$ poles revolving backward at speed $N_s/5$.

**c.** A seventh-harmonic field with $7P$ poles revolving forward at speed $N_s/7$.

**d.** Similar thirteenth, nineteenth, twenty-fifth, etc., forward-revolving and eleventh, seventeenth, twenty-third, etc., backward revolving harmonic fields.

Figure 20-27 shows a test speed-torque curve obtained on a 2-phase squirrel-cage induction motor with straight (unspiraled) slots. The torque dips due to three of the forward-revolving fields are clearly indicated.

***42. Torque, Slip and Rotor Impedance.*** When the rotor is stationary, the revolving magnetic field cuts the short-circuited secondary conductors at synchronous speed and induces in them line-frequency currents. To supply the secondary $IR$ voltage drop, there must be a component of voltage in time phase with the secondary current, and the secondary current, therefore, must lag in space position behind the revolving air-gap field. A torque is then produced corresponding to the product of the air-gap field by the secondary current times the sine of the angle of their space-phase displacement.

At standstill, the secondary current is equal to the air-gap voltage divided by the secondary impedance at line frequency, or

$$I_2 = \frac{E_2}{Z_2} = \frac{E_2}{R_2 + jX_2} \tag{20-11}$$

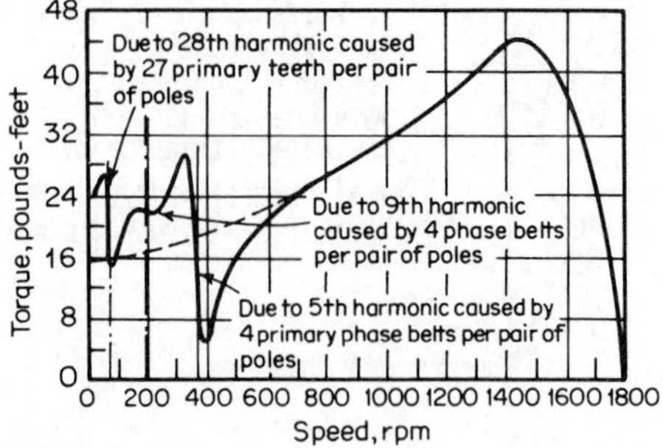


**FIG. 20-27** Speed-torque curve of 2-phase motor showing harmonic torque.

where $R_2$ = effective secondary resistance and $X_2$ = secondary leakage reactance at primary frequency.

The speed at which the magnetic field cuts the secondary conductors is equal to the difference between the synchronous speed and the actual rotor speed. The ratio of the speed of the field relative to the rotor, to synchronous speed, is called the slip $s$,

$$s = \frac{N_s - N}{N_s}$$

or

$$N = (1 - s)N_s \tag{20-12}$$

where $N$ = actual and $N_s$ = synchronous rotor speed.

As the rotor speeds up, with a given air-gap field, the secondary induced voltage and frequency both decrease in proportion to $s$. Thus, the secondary voltage becomes $sE_2$, and the secondary impedance $R_2 + jsX_2$, or

$$I_2 = \frac{sE_2}{R_2 + jsX_2} = \frac{E_2}{(R_2/s) + jX_2} \tag{20-13}$$

The only way that the primary is affected by a change in the rotor speed, therefore, is that the secondary resistance as viewed from the primary varies inversely with the slip.

In practice, the effective secondary resistance and reactance, or $R_2$ and $X_2$, change with the secondary frequency, owing to the varying "skin effect," or current shifting into the outer portion of the conductors, when the frequency is high. This effect is employed to make the resistance, and therefore the torque, higher at starting and low motor speeds, by providing a double cage, or deep-bar construction, as shown in Fig. 20-28. The leakage flux between the outer and inner bars makes the inner-bar reactance high, so that most of the current must flow in the outer bars or at the top of a deep-bar at standstill, when frequency is high.

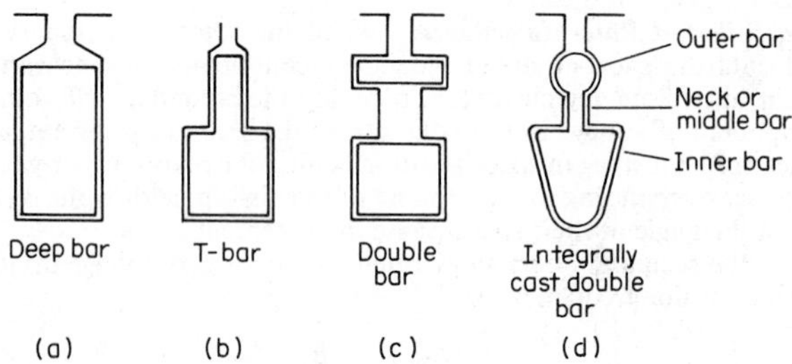


**FIG. 20-28** Alternative forms of squirrel-cage rotor bars.

At full speed, the secondary frequency is very low, and most of the current flows in the inner bars, or all over the cross section of a deep bar, owing to their lower resistance.

***43. Analysis of Induction Motors.*** Induction motors are analyzed by three methods: (1) circle diagram; (2) equivalent circuit; (3) coupled-circuit, generalized machine. The first two methods are used for steady-state conditions; the third method is used for transient conditions. The circle diagram is convenient for visualizing overall performance but is too inaccurate for detailed calculations and design. The magnetizing current is not constant, but decreases with load because of the primary impedance drop. All of the circuit constants vary over the operating range due to magnetic saturation and skin effect. The equivalent circuit method predominates for analysis and design under steady-state conditions. The impedances can be adjusted to fit the conditions at each calculation point.

***44. Circle Diagram.*** The voltage-current relations of the polyphase induction machine are roughly indicated by the circuit of Fig. 20-29. The magnetizing current $I_M$ proportional to the voltage and lagging 90° in phase is nearly constant over the operating range, while the load current varies inversely with the sum of primary and secondary impedances. As the slip $s$ increases, the load current and its angle of lag behind the voltage both increase, following a nearly circular locus. Thus, the circle diagram (Fig. 20-30) provides a picture of the motor behavior.

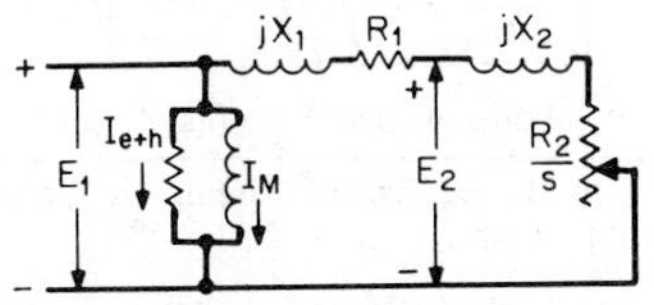


**FIG. 20-29** Equivalent circuit for circle diagram.

The data needed to construct the diagram are the magnitude of the no-load current $ON$ and of the blocked-rotor current $OS$ and their phase angles with reference to the line voltage $OE$. A circle with its center on the line $NU$ at right angles to $OE$ is drawn to pass through $N$ and $S$. Each line on the diagram can be measured directly in amperes, but it also represents voltamperes or power, when multiplied by the phase voltage times number of phases. The line $VS$ drawn parallel to $OE$ represents the total motor power input with blocked rotor, and on the same scale $VT$ represents the corresponding primary $I^2R$ loss. Then $ST$ represents the power input to the rotor at standstill, which, divided by the synchronous speed, gives the starting torque.

At any load point $A$, $OA$ is the primary current, $NA$ the secondary current, and $AF$ the motor power input. The motor output power is $AB$, the torque × (synchronous speed) is $AC$, the secondary $I^2R$ loss is $BC$, primary $I^2R$ loss $CD$, and no-load copper loss plus core loss $DF$. The maximum power-factor point is $P$, located by drawing a tangent to the circle from $O$. The maximum output and maximum torque points are similarly located at $Q$ and $R$ by tangent lines parallel to $NS$ and $NT$, respectively.

The diameter of the circle is equal to the voltage divided by the standstill reactance or to the blocked-rotor current value on the assumption of zero resistance in both windings.

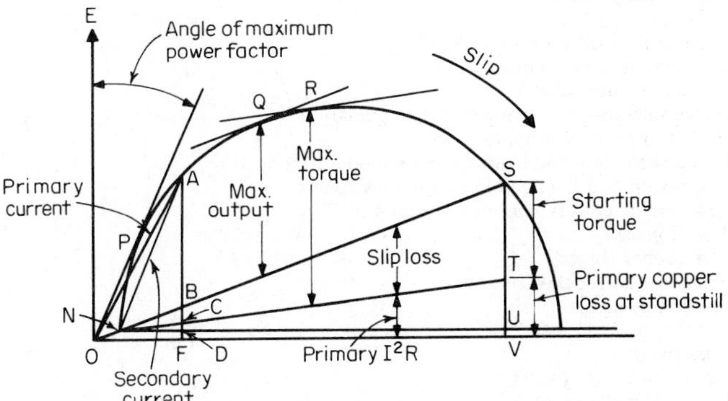


**FIG. 20-30** Circle diagram of polyphase induction motor.

The maximum torque of the motor, measured in kilowatts at synchronous speed, is equal to a little less than the radius of the circle multiplied by the voltage *OE*.

***45. Equivalent Circuit.*** Figure 20-31 shows the polyphase motor circuit usually employed for accurate work. The advantages of this circuit over the circle-diagram method are that it facilitates the derivation of simple formulas, charts, or computer programs for calculating torque, power factor, and other motor characteristics and that it enables impedance changes due to saturation or multiple squirrel cages to be readily taken into account.

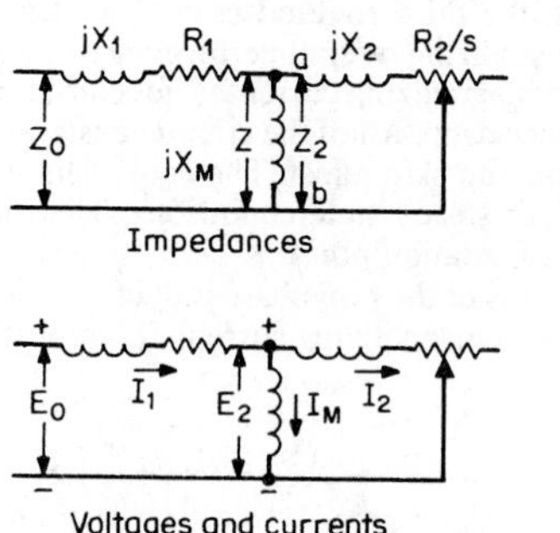


**FIG. 20-31** Equivalent circuit of polyphase induction motor.

Formulas for calculating the constants from test data are given in Table 20-3, Par. **53**, and their definitions are given in Table 20-2.

Inspection of the circuit reveals several simple relationships which are useful for estimating purposes. The maximum current occurs at standstill and is somewhat less than $E/X$. Maximum torque occurs when $s = R_2/X$, approximately, at which point the current is roughly 70% of the standstill current. Hence, the maximum torque is approximately equal to $E^2/2X$. This gives the basic rule that the percent maximum torque of a low-slip polyphase motor at a constant impressed voltage is about half the percent starting current.

By choosing the value of $R_2$, the slip at which maximum torque occurs can be fixed at any desired value. The maximum-torque value itself is affected, not by changes in $R_2$, but only by changes in $X$ and to a slight degree by changes in $X_M$.

**TABLE 20-2** Definitions of Equivalent-Circuit Constants

Unless otherwise noted, all quantities except watts, torque, and power output are per phase for 2-phase motors and per phase Y for 3-phase motors.

$E_0$ = impressed voltage (volts) = line voltage ÷ $\sqrt{3}$ for 3-phase motors
$I_1$ = primary current (amperes)
$I_2$ = secondary current in primary terms (amperes)
$I_M$ = magnetizing current (amperes)
$R_1$ = primary resistance (ohms)
$R_2$ = secondary resistance in primary terms (ohms)
$R_0$ = resistance at primary terminals (ohms)
$X_1$ = primary leakage reactance (ohms)
$X_2$ = secondary leakage reactance (ohms)
$X = X_1 + X_2$
$X_0$ = reactance at primary terminals (ohms)
$X_M$ = magnetizing reactance (ohms)
$Z_1$ = primary impedance (ohms)
$Z_2$ = secondary impedance in primary terms (ohms)
$Z_0$ = impedance at primary terminals (ohms)
$Z$ = combined secondary and magnetizing impedance (ohms)
$s$ = slip (expressed as a fraction of synchronous speed)
$N$ = synchronous speed (revolutions per minute)
$m$ = number of phases
$f$ = rated frequency (hertz)
$f_t$ = frequency used in locked-rotor test
$T$ = torque (foot-pounds)
$W_0$ = watts input
$W_H$ = core loss (watts)
$W_F$ = friction and windage (watts)
$W_{RL}$ = running light watts input
$W_s$ = stray-load loss (watts)

The magnetizing reactance $X_M$ is usually eight or more times as great as $X$, while $R_1$ and $R_2$ are usually much smaller than $X$, except in the case of special motors designed for frequent-starting service.

The equivalent circuit of Fig. 20-31 shows that the total power $P_{g1}$ transferred across the air gap from the stator is

$$P_{g1} = mI_2^2 \frac{R_2}{s} \tag{20-14}$$

The total rotor copper loss is evidently

$$\text{Rotor copper loss} = mI_2^2 R_2 \tag{20-15}$$

The internal mechanical power $P$ developed by the motor is therefore

$$P = P_{g1} - \text{rotor copper loss} = mI_2^2 \frac{R_2}{s} - mI_2^2 R_2 \tag{20-16}$$

$$= mI_2^2 R_2 \frac{1-s}{s}$$

$$= (1-s)P_{g1}$$

We see, then, that of the total power delivered to the rotor, the fraction $1 - s$ is converted to mechanical power and the fraction $s$ is dissipated as rotor-circuit copper loss. The internal mechanical power per stator phase is equal to the power absorbed by the resistance $R_2(1 - s)/s$. The internal electromagnetic torque $T$ corresponding to the internal power $P$ can be obtained by recalling that mechanical power equals torque times angular velocity. Thus, when $\omega_s$ is the synchronous angular velocity of the rotor in mechanical radians per second,

$$P = (1-s)\omega_s T \tag{20-17}$$

with $T$ in newton-meters. By use of Eq. 20-16,

$$T = \frac{1}{\omega_s} mI_2^2 \frac{R_2}{s} \tag{20-18}$$

For $T$ in foot-pounds and $N_s$ in revolutions per minute,

$$T = \frac{7.04}{N_s} mI_2^2 \frac{R_2}{s} \tag{20-19}$$

***46. Torque and Power.***[1] Considerable simplification results from application of Thévenin's network theorem to the induction-motor equivalent circuit. Thévenin's theorem permits the replacement of any network of linear circuit elements and constant phasor voltage sources, as viewed from two terminals by a single phasor voltage source $E$ in series with a single impedance $Z$. The voltage $E$ is that appearing across terminals $a$ and $b$ of the original network when these terminals are open-circuited; the impedance $Z$ is that viewed from the same terminals when all voltage sources within the network are short-circuited. For application to the induction-motor equivalent circuit, points $a$ and $b$ are taken as those so designated in Fig. 20-31. The equivalent circuit then assumes the forms given in Fig. 20-32. So far as phenomena to the right of points $a$ and $b$ are concerned, the circuits of Figs. 20-31 and 20-32 are identical when the voltage $V_{1a}$ and the impedance $R_1 + jX_1$ have the proper values. According to Thévenin's theorem, the equivalent source voltage $V_{1a}$ is the

[1] Fitzgerald, A. E., Kingsley, Jr., C., and Kusko, A.; *Electric Machinery* 3d ed.; New York, McGraw-Hill Book Company, 1971.

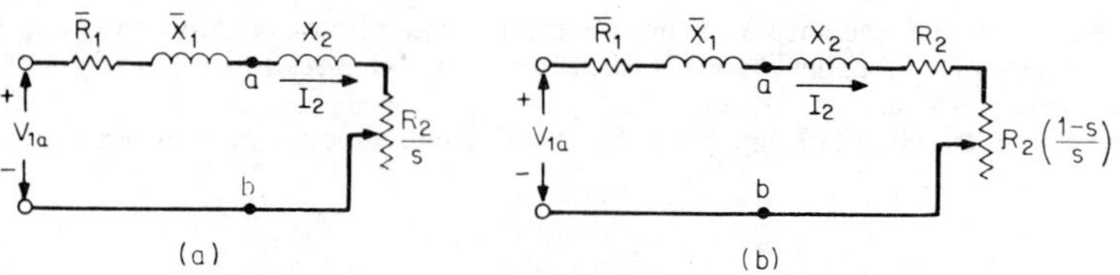


**FIG. 20-32** Induction-motor equivalent circuits simplified by Thévenin's theorem.

voltage that would appear across terminals *a* and *b* of Fig. 20-31 with the rotor circuits open and is

$$V_{1a} = E_0 - I_0(R_1 + jX_1) = E_0 \frac{jX_M}{R_1 + jX_{11}} \tag{20-20}$$

where $I_M$ is the zero-load exciting current and

$$X_{11} = X_1 + X_M$$

is the self-reactance of the stator per phase and very nearly equals the reactive component of the zero-load motor impedance. For most induction motors, negligible error results from neglecting the stator resistance in Eq. 20-20. The Thévenin equivalent stator impedance $R_1 + jX_1$ is the impedance between terminals *a* and *b* of Fig. 20-31, viewed toward the source with the source voltage short-circuited, and therefore is

$$\overline{R}_1 + j\overline{X}_1 = R_1 + jX_1 \qquad \text{in parallel with } jX_M$$

From the Thévenin equivalent circuit (Fig. 30-32) and the torque expression (Eq. 20-18), it can be seen that

$$T = \frac{1}{\omega_s} \frac{mV_{1a}^2(R_2/s)}{(\overline{R}_1 + R_2/s)^2 + (\overline{X}_1 + X_2)^2} \tag{20-21}$$

The slip at maximum torque, $s_{\max\,T}$, is obtained by differentiating Eq. (20-21) with respect to *s* and equating to zero:

$$s_{\max\,T} = \frac{R_2}{\sqrt{\overline{R}_1^2 + (\overline{X}_1 + X_2)^2}}$$

The corresponding maximum torque is

$$T_{\max} = \frac{1}{\omega_s} \frac{0.5mV_{1a}^2}{\overline{R}_1 + \sqrt{\overline{R}_1^2 + (\overline{X}_1 + X_2)^2}}$$

***47. Example of Application of Thévenin's Theorem.*** For the motor whose equivalent circuit parameters are $R_1 = 0.160\Omega$, $X_1 = 0.318\Omega$, $X_M = 12.4\Omega$, $R_2 = 0.188\Omega$, $X_2 = 0.318\Omega$, let it be required to find

**a.** The load component $I_2$ of the stator current, the internal torque *T*, and the internal power *P* for a slip $s = 0.03$

**b.** The maximum internal torque and the corresponding speed

**c.** The internal starting torque and the corresponding stator-load current $I_2$

The motor is rated 32 hp, 220 V ac, 3-phase, 6-pole, Y-connected, 60 Hz.

*Solution.* First, reduce the circuit to its Thévenin's equivalent. From Eq. (20-20) $V_{1a} = 123.8$ V and $\overline{R}_1 + j\overline{X}_1 = (0.152 + j0.312)\Omega$.

**a.** At $s = 0.03$, $R_2/s = 6.26$; then

$$I_2 = \frac{123.8}{\sqrt{(6.26 + 0.152)^2 + (0.63)^2}} = 19.2 \text{ A}$$

From Eq. (20.21)

$$T = \frac{3 \times (123.8)^2 \times 6.26}{2\pi(^{1200}/_{60})[(6.26 + 0.152)^2 + (0.63)^2]}$$

$$= 55.0 \text{ N}\cdot\text{m} = 40.5 \text{ lb-ft}$$

$$P = 3(19.2)^2 0.188 \frac{(1 - 0.03)}{0.03} = 6715 \text{ W}$$

**b.** At the maximum torque point

$$s_{\max T} = \frac{0.188}{\sqrt{(0.152)^2 + (0.63)^2}} = 0.29$$

$$\text{Speed at } T_{\max} = (1 - 0.29) \times 1200 = 851.9 \text{ r/min}$$

$$T_{\max} = \frac{1}{2\pi \times (^{1200}/_{60})} \frac{0.5 \times 3 \times (123.8)^2}{[0.152 + \sqrt{(0.152)^2 + (0.63)^2}]} = 228 \text{ N}\cdot\text{m} = 168 \text{ ft-lb}$$

**c.** At starting $s = 1$, $R_2$ will be assumed constant; therefore,

$$R_2/s = 0.188 \qquad R_1 + \frac{R_2}{s} = 0.34$$

$$I_{2 \text{ start}} = \frac{123.8}{\sqrt{(0.34)^2 + (0.63)^2}} = 172.9 \text{ A}$$

$$T_{\text{start}} = \frac{1}{2\pi \times (^{1200}/_{60})} \times (172.9)^2 \times 3 \times 0.188$$

$$= 134.3 \text{ N}\cdot\text{m} = 99.1 \text{ ft-lb}$$

## *TESTING OF POLYPHASE INDUCTION MACHINES*

***48. General.*** Proof of guaranteed performance, the determination of torque or efficiency of driven machines, and the evaluation of design changes are some of the purposes that require accurate tests of induction machines. Normally, running-light, locked-rotor, resistance, and dielectric tests only are made on standard motors. Input-output tests or segregated-loss tests are made when accurate efficiency determination is required. The inconvenience of making input-output tests and the inaccuracies inherent in any method which determines the losses as a small difference between two large quantities make the segregated-loss methods of test preferable in many cases. Such tests are especially necessary when actual performance under the varying conditions of service is to be determined from a limited number of factory or laboratory test runs. Experience has shown that the equivalent-circuit method of calculation enables accurate predictions of efficiency and other performance data to be made, provided the circuit "constants" are determined in advance by careful tests.

The AIEE Test Code for Induction Machines[1] gives authoritative procedures for conducting all usual tests, and many of the data contained in the following sections are derived from this source.

[1]American Standard Test Code for Polyphase Induction Motors and Generators; USASC50.20-1954.

***49. Running-Light Test.*** The motor is run at no load with normal frequency and voltage applied, until the watts input becomes constant. On slip-ring motors, the brushes are short-circuited. Readings of amperes and watts are taken at one or more values of impressed voltage, with rated frequency maintained. Accurately balanced phase voltages and a sine-wave form of voltage are necessary for good results, requiring operation of the test alternator and transformers well below magnetic saturation. The watts input at rated voltage will be the sum of the friction and windage, core loss, and no-load primary $I^2R$ loss. Subtracting the primary $I^2R$ loss at the temperature of test from the input gives the sum of the friction and windage and core loss. Segregation of the core loss from the windage and friction is not necessary for normal efficiency or other rated-voltage performance calculations. However, the segregation can be made, if desired, by taking amperes and watts input readings, at rated frequency, at different voltages varying from 125% of normal down to about 15% voltage, or the point of minimum current. Plotting the input watts, less primary $I^2R$, against the square of the voltage and extrapolating the lower part of the curve in a straight line to intercept the zero-voltage axis determines the friction and windage. Typical data of such a test are shown in Fig. 20-33.

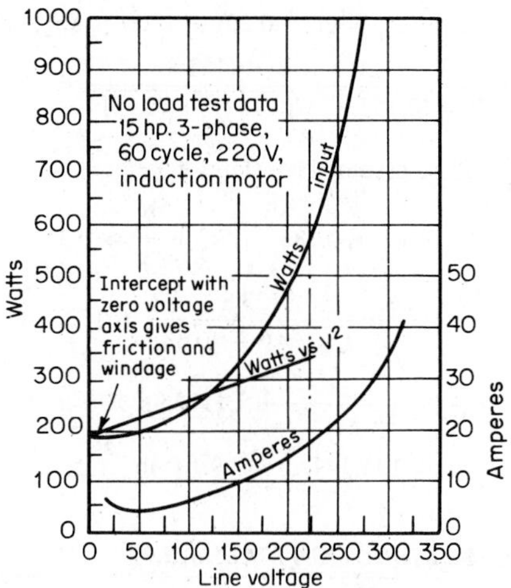


**FIG. 20-33** No-load excitation curves.

The value of the magnetizing reactance $X_M$ in Fig. 20-31 is determined from the no-load current at rated voltage $I_0$ by using the value of primary leakage $X_1$ determined from locked-rotor test data.

***50. Locked-Rotor Test.*** The motor is blocked so it cannot rotate; a reduced voltage of rated frequency is applied to the terminals; and readings of volts, watts, and amperes are taken. Readings should be taken quickly, and the temperature of the windings should be observed before and after the test to minimize errors due to changing resistance values. In the case of machines with closed-slot rotors or very small air gaps, magnetic saturation of the leakage paths will occur, and it may be desirable to take readings at half or full voltage to establish the actual value of starting current. Equivalent-circuit performance calculations, however, should be based on data taken at approximately rated current.

When only low-voltage test data are available, the locked-rotor current at higher voltages can be estimated by the formula

$$I = \frac{V - V_0}{V_t - V_0} I_t \tag{20-22}$$

where $V_t$, $I_t$ = test values of voltage and locked-rotor current; $V$, $I$ = corresponding values at a different voltage; and $V_0$ = intercept of test current-voltage curve with zero-voltage axis, obtained by extrapolating the test curve as a straight line through points in the approximate range of 50 to 200% current. $V_0$ represents the voltage due to flux of saturation density crossing closed slot bridges and similar leakage flux paths.

The motor impedance per phase is determined from the volts, amperes, and watts readings. The total resistance component for a 3-phase motor is

$$R = \frac{W}{3I^2} \quad \Omega\text{/phase Y} \qquad (20\text{-}23)$$

and the reactance component is

$$X = \sqrt{\frac{V^2}{3I^2} - R^2} \quad \Omega\text{/phase Y} \qquad (20\text{-}24)$$

where $W$ = watts input, $I$ = line current, and $V$ = voltage between lines.

Normally, the primary and secondary leakage-reactance values $X_1$ and $X_2$ are assumed equal, each having the value $X/2$.

The primary resistance is measured with direct current, a current about one-quarter of full-load value being preferably used, and readings being taken quickly to avoid errors due to temperature changes during the test. The primary resistance per phase Y is equal to one-half the resistance between any two terminals.

Subtracting the primary resistance at the temperature of test from the resistance component of the total impedance gives the effective secondary resistance at standstill. The starting torque may be calculated from this value by the equation

$$\text{Starting torque} = \frac{7.04KmI^2R_{2e}}{N_s} \quad \text{ft}\cdot\text{lb} \qquad (20\text{-}25)$$

where $I$ = amperes starting current per phase at specified voltage; $m$ = number of phases; $N_s$ = synchronous speed in r/min; $R_{2e}$ = resistance component of motor impedance, less primary resistance at temperature of test, in ohms per phase; $K$ = an empirical constant, usually approximately 0.9, which allows for nonfundamental secondary losses.

In practice, it is usual to measure the torque produced, by means of a lever arm and scale, in which case Eq. (20-25) provides a useful check on the accuracy of the measurements.

In the case of deep-bar or double squirrel-cage motors, the effective secondary reactance at speed is materially higher than at standstill, owing to the progressive shifting of the secondary current from the low-reactance high-resistance paths into the low-resistance high-reactance paths as the secondary, or slip, frequency decreases. Hence, for accurate performance calculations, it is necessary to determine the motor reactance at low secondary frequency. If a low-frequency supply is available, the locked-rotor test may be repeated at 15 Hz, or at most 25 Hz, for a 60-Hz motor. Calculation of the low-frequency reactance by Eq. (20-24) and multiplying this by the ratio of the rated to the test frequency will give the proper value to use in operating performance calculations.

Alternatively, the reactance value at speed may be obtained by adding an amount $\Delta X$ to the reactance determined by full-frequency locked-rotor test. The value of $\Delta X$ is approximately

$$\Delta X = R_{2e} - R_2 \qquad (20\text{-}26)$$

where $R_2$ = secondary resistance of full-load slip, determined by the slip test of Par. **51**.

***51. Slip Test.*** Whenever feasible, a current-slip curve should be taken under actual load conditions, with rated voltage and frequency maintained at the motor terminals. Measurements at a few points in the neighborhood of full-load current are usually sufficient; but for slip-ring motors a wider range should be covered, owing to the variable resistance of the brushes. The slip is normally too small to be determined by tachometer readings and should therefore be measured with a slip meter or stroboscopically. The slip-meter method makes use of a revolution counter differentially geared to the motor under test and to a small synchronous motor driven from the same power supply at the same synchronous speed. Care must be taken to correct the observed values of slip for the difference between the test temperature and the standard value of 75°C or the temperature attained in a full-load heat run with an ambient temperature of 25°C.

In practice, the value of current corresponding to an assumed value of $R_s2/s$ is calculated exactly by the equivalent circuit; the corresponding value of $s$ is read off the slip-current curve; and the true value of $R_2$ is obtained by multiplying $R_2/s$ by this value of $s$. However, $R_2$ may be approximately determined as follows:

Very roughly, the secondary resistance is equal to

$$R_2 = 1.1\frac{E \cdot s}{I_1} \quad \text{approx. } \Omega\text{/phase} \qquad (20\text{-}27)$$

where $E$ = terminal voltage per phase, $s$ = ratio of revolutions per minute of slip to synchronous speed, and $I_1$ = observed phase current.

The coefficient 1.1 varies over a range of about 1 to 1.2, depending on the motor characteristics and the value of the test load.

In case direct slip measurements are not practicable, the value of $R_2$ determined by Eq. (20-23) in a low-frequency locked-rotor test may be used. Or, in the case of a wound rotor, the actual resistance between slip rings may be measured and multiplied by the square of the ratio of primary to secondary volts to obtain the resistance referred to primary. The voltage ratio is obtained by measurement of primary and secondary voltages at standstill with the slip rings open-circuited. Averages of several rotor positions are taken to avoid errors due to possible unbalance.

***52. Stray-Load Loss Tests.*** Stray-load losses, $W_s$, are defined as the excess of the total measured losses above the sum of the friction and windage, core, and copper losses calculated for the conditions of load from the no-load tests described above. These extra losses are made up chiefly of high-frequency core losses and rotor $I^2R$ losses caused by the pulsations of the leakage-reactance fluxes produced by load currents. While the stray-load losses may be determined by direct input-output tests with a dynamometer or calibrated driving motor, the result is a small difference between two large quantities and so accuracy is very difficult to obtain. Whenever such tests are made, it is desirable to repeat them with the direction of power flow reversed, so the measurement errors may be substantially canceled out.

There are several ways of determining the stray-load loss by separate loss measurements, but the procedure is fairly complex and must be carefully done if accurate results are to be obtained. These are described in the AIEE Test Code for Polyphase Induction Machines.

***53. Performance Calculations.*** From the foregoing tests, all the circuit constants may be determined, enabling the equivalent-circuit calculations as outlined in Par. **45** to be carried out. To facilitate this, the formulas for calculating the constants as defined in Table 20-2 are collected in Table 20-3.

The procedure in making performance calculations based on test data is first to divide $E_0$ by the approximate expected value of normal current, an arbitrary value of $R_2/s$ being thus obtained. With this value and the known circuit constants, calculations are carried through for one point, determining the actual value of $I$. By entering the test slip-current

**TABLE 20-3** Formulas for Calculating Circuit Constants from Test Data for 3-Phase Motors

$$X = \frac{f}{f_t}\sqrt{\frac{V^2}{3I^2} - \left(\frac{W}{3I^2}\right)^2} \qquad \text{(Par. 50)}$$

$X_1 = X_2 = 0.5X$ for single squirrel-cage or wound-rotor motors

$X_1 = 0.4X$ and $X_2 = 0.6X$ for low-starting-current motors

$$W_H + W_F = W_{RL} - 3I_M^2R_1 \qquad \text{(Par. 49)}$$

$W_s$ from Par. **52**

$$X_M = \frac{E_0}{I_M} - X_1$$

curve (Par. **51**), the true value of $s$ is found, and from this and $R_2/s$, $R_2$ is calculated. All the circuit constants are then known, whence the efficiency, power factor, torque, etc., are determined. Additional points are calculated with different values of $s$, covering the desired range of loads, and the exact characteristics are taken off curves plotted from the calculated results.

If values of torque, current, etc., are desired for considerable overloads or throughout the accelerating range, the values of $R_2$ and $X$ should be modified to allow for magnetic saturation and eddy currents. Curves of reactance against current obtained by locked-rotor tests over the desired range of values and values of $R_{2e}$ and corresponding values of $\Delta X$ obtained by locked-rotor tests at different frequencies are desirable for this purpose, especially in cases of closed-slot or double squirrel-cage rotors.

***54. Temperature tests*** are made to determine the temperature rise of insulated windings under load conditions. ANSI Standards specify a limiting temperature for continuous-rated machines of 50°C by thermometer or 60°C by either the resistance- or the embedded-detector method for Class A insulating materials and corresponding values of 70°C by thermometer and 80°C by resistance or embedded detector for Class B insulation. Usually, the temperature is measured by mercury thermometers or thermocouples applied to the hottest accessible parts of the core and windings in several different locations. A small amount of putty is used to shield thermometer bulbs from the surrounding air, and care is taken to avoid external air currents, varying ambient temperatures, or other factors which may introduce errors.

The preferred method of making a full-load temperature test is to maintain nameplate voltage, current, and frequency until the temperature becomes constant, readings being taken every half hour. When constant temperature is reached, the motor is stopped as quickly as possible and additional thermometers are applied to the rotating parts as soon as these have come to rest. The maximum permissible time of stopping is 1 min for machines of less than 50 kW rating, 2 min for 50 to 200 kW ratings, and 3 min for machines larger than 200 kW. The winding temperatures usually increase after shutdown; so readings must be recorded at frequent intervals until definitely falling temperatures are observed. The highest temperature reached at any time during the test is taken as the correct value. If the temperatures fall continuously after shutdown, a curve should be plotted of temperature vs. time and extrapolated back to the moment of shutdown.

For protected-type or totally enclosed machines, it is often preferable to determine the temperature by the rise-of-resistance method. In this case, the "cold" resistance of the winding is measured at a known temperature, usually after the machine has been standing overnight at a uniform room temperature; and the "hot" resistance is measured immediately after shutdown. The hot resistance is taken as the highest value obtained after shutdown or is extrapolated back to the moment of shutdown if the resistance falls continuously.

The temperature is then calculated from the following formula:

$$T = \frac{R_T(234.5 + t)}{R_t} - 234.5 \qquad (20\text{-}28)$$

where $T$ = winding temperature when $R_T$ was measured, $R_T$ = hot resistance, $R_t$ = cold resistance, and $t$ = winding temperature when $R_t$ was measured.

## CHARACTERISTICS OF POLYPHASE INDUCTION MOTORS

***55. Types.*** All polyphase induction motors may be classified as squirrel-cage or wound-rotor, and may be of the single-speed or multispeed type. Squirrel-cage motors are further classified by NEMA[1] for torque-speed and current-speed curves as Designs A, B, C, and D, and by Code designations from A to V for locked-rotor kVA/hp. For all induction motors, the allowable temperature rises and insulation systems are designated by classes A, B, F, and H. Finally, the mechanical dimensions are designated by frame sizes, and in enclosures from drip-proof to totally enclosed with various types of ventilation. Both squirrel-cage and wound-rotor motors may be of the single-speed or multispeed type.

[1]NEMA, Motor and General Standards, Publ. MG1-1978.

Based upon efficiency, motors are also classified as standard and energy-efficient motors. Several manufacturers have developed product lines of energy-efficient motors under various trade names. Some of these trade names are E-plus (Gould), Energy Saver (GE), XE-Energy Efficient (Reliance Electric Co.), MacII High Efficiency (Westinghouse), Energy Efficient Corro-Duty (US Electric Motors), High Efficiency Pacemaker (Louis Allis Co.), Delco $E^2$ (GM) etc.

***56. Squirrel-Cage Motors.*** All integral-horsepower induction-motor design categories can mechanically withstand the magnetic stresses and locked-rotor torques of full-voltage line starting. The torque- and current-speed curves for Design A, B, C, and D squirrel-cage motors are shown in Figs. 20-34 to 20-36. Design B motors are most widely used; they have starting-torque and line-starting current characteristics suitable for most power systems. Design C and D motors have higher torque than Class B motors. For all design motors, the percentage torques tend to decline with increased hp rating.

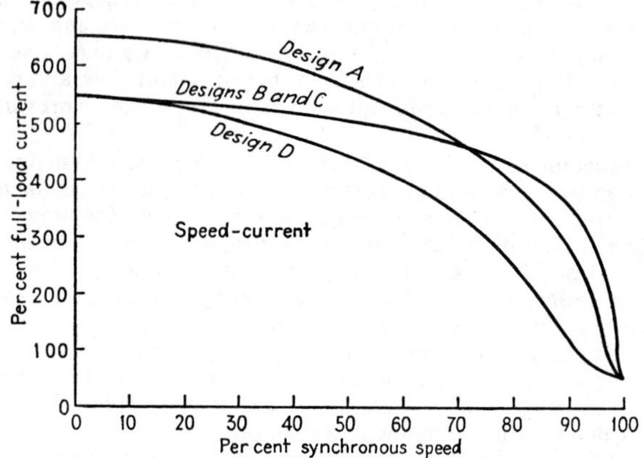


**FIG. 20-34** Typical speed-current curves for squirrel-cage induction motors.

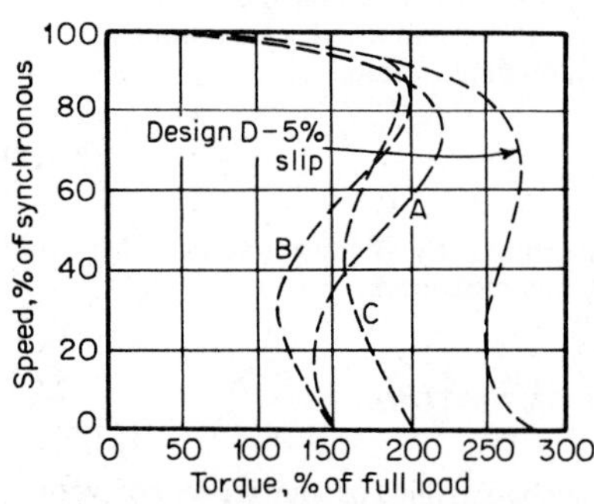


**FIG. 20-35** Speed-torque curves for typical Design A, B, C, and D squirrel-cage motors.

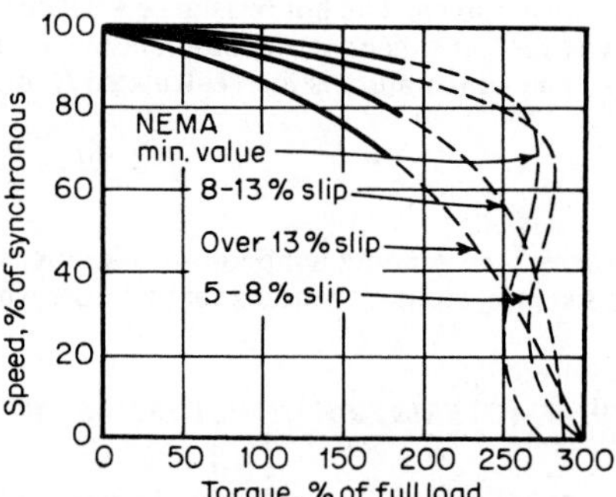


**FIG. 20-36** Speed-torque relationship for Design D squirrel-cage motors.

Design A motors are designed for the same locked-rotor torques and slips as Design B motors, but have higher breakdown torques and locked-rotor currents. These motors are suitable for loads with breakaway torque of 40 to 70%, accelerating torque of 20 to 50%, and peak torque of 130 to 175% of rated torque, where starting and stopping are infrequent. Reduced voltage starting may be required because of the high locked-rotor current.

Design B motors are usually line started at full voltage. These motors can accelerate to full speed any load that they can start. These motors are suitable for loads with breakaway torque less than 50%, accelerating torque less than 50%, and peak torque less than 125%. The low slip precludes loads of torque pulsation. The motor is suitable for continuous steady-load operation with infrequent starting and stopping.

Design D motors are designed for full-voltage starting and develop locked-rotor torques of 275% of rated torque. Locked-rotor currents are the same as Design B. These motors have more than 5% slip at rated torque and are designed for loads that are frequently applied and removed. These motors are divided into groups of 5 to 8% slip, 8 to 13% slip, and over 13% slip, as shown in Fig. 20-36.

***57. Application of Induction Motors.*** An induction motor should be selected to meet the starting and running requirements of the load, the constraints of the ambient conditions and supply line, and the duty cycle, with the minimum size, horsepower rating, and cost. Oversize motors can always be selected, but at the expense of starting current, running power factor, and cost. Typical load conditions and applications for Design A, B, C, and D motors are given in Table 20-4.

***58. Wound-Rotor Motors.*** An insulated winding, usually 3-phase, is provided on the rotor; the terminal of each phase is connected to a slip ring on the shaft. The stationary brushes, which bear on the slip rings, are connected to external adjustable resistance or solid-state converters by which power can be removed from, or injected into, the rotor to adjust the speed. Speed-torque and speed-current curves for a typical wound-rotor motor for various amounts of external resistance are shown in Fig. 20-37. The numbers on the curves refer to the percent external resistance; 100% resistance gives rated torque at standstill. The use of solid-state converters in a modified Krämer system is described in Par. **141.**

Wound-rotor motors are normally started with relatively high external resistance and this resistance is short-circuited in steps as the motor comes up to speed. Liquid rheostats are used in the higher ratings. This procedure allows the motor to deliver high-starting and accelerating torques, yet draw relatively light line current. Furthermore, most of the rotor-circuit losses during acceleration are dissipated in the external resistor rather than within the motor.

The curves of Fig. 20-37 indicate that the external resistance reduces the speed at which the motor will operate with a given load torque. For any one value of external resistance, the motor has varying speed characteristics, any change in load results in a considerable change in speed. The lower the operating speed, the more pronounced the effect, so that it is usually not feasible to operate at less than 50% of full speed by this method. Furthermore, because the power loss in the rotor and external resistor is proportional to the slip, the efficiency is reduced in direct proportion to the speed reduction.

Breakdown torque is given by NEMA in MG1-12.40. Secondary data, including open-circuit slip-ring voltage and short-circuit slip-ring current, at standstill, are given in MG1-10.34.

Slip-ring motors with external resistance are used as adjustable-speed motors from 50% to full speed for loads such as pumps and fans. They are used over the full speed range for hoists, elevators, and ski lifts. In addition, slip-ring motors are used to provide high starting and accelerating torque with low current for centrifuges, crushers, pulverizers, and other high-inertia loads. Solid-state ac and dc drives have replaced wound-rotor motors in many applications.

***59. Multiple-Speed Squirrel-Cage Motors.*** Multispeed squirrel-cage motors may be of the single-winding or two-winding type. The former have a stator winding which can be connected to give either one of two speeds having a ratio of 2:1. The method of connection is usually furnished by the controller manufacturer. The frame of the two-speed single-winding motor is about the same as that of the single-speed motor. The two-winding motor has two separate stator windings which can be wound for any number of poles so that any two synchronous speeds can be obtained. In addition one or both of the stator windings may be arranged for reconnection as in a single-winding motor, giving a total of three or four speeds, but the two speeds obtained on a single winding must have a ratio of 2:1. Thus a four-speed two-winding motor might have speeds of 1800, 900, 1200, and 600 r/min.

Multispeed motors are designed as (1) variable-torque motors, (2) constant-torque motors, and (3) constant-horsepower motors. The rated torque at four speed points for each

**TABLE 20-4** Induction-Motor Application Outline

| For this type of equipment | Requiring these torques: Starting | Requiring these torques: Max. running | With these load characteristics | Type and description |
|---|---|---|---|---|
| Petroleum and chemical<br>Industrial and chemical pumps<br>Cooling towers<br>Air-handling equipment<br>Compressors<br>Conveyors<br>Process machinery<br>Petroleum and chemical process equipment | 100–150% of full-load torque | 200–250% of full-load torque | Continuous operation, constant speed, high speed (over 720 rpm), easy starting; subject to short time overloads; good speed regulation | *Energy efficient:* NEMA design B, normal torques: normal starting current; can be used with variable-frequency/variable-voltage inverters; higher efficiency than standard design B motors |
| Centrifugal pumps<br>Blowers and fans<br>Drilling machines<br>Grinders<br>Lathes<br>Compressors<br>Conveyors | 100–150% of full-load torque | 200–250% of full-load torque | Variable load conditions, constant speed; subject to short time overloads; good speed regulation | *NEMA design B,* normal torques: normal starting current; can be used with variable-frequency/variable-voltage inverters |
| Reciprocating pumps<br>Stokers<br>Compressors<br>Crushers<br>Ball and rod mills | 200–300% of full-load torque | Not more than full-load torque | High starting torque due to high inertia, back pressure, standstill friction, or similar mechanical conditions; torque requirements decrease during acceleration to full-load torque; not subject to severe overloads; good speed regulation | *NEMA design C,* high torque: normal starting current; not recommended for use with variable-frequency inverters |

| | | | | |
|---|---|---|---|---|
| Punch presses<br>Cranes<br>Hoists<br>Press brakes<br>Shears<br>Oil well pumps<br>Centrifugals | Up to 300% of full-load torque | 200–300% of full-load torque; loss of speed during peak loads required | Intermittent loads; may require frequent start, stop, and reverse cycles; machine uses a flywheel to carry peak loads; poor speed regulation to smooth power peaks; may require acceleration of high-inertia load | *NEMA design D,* high torque: high slip; standard types have slip characteristics of 5–8% or 8–13% slip |
| Blowers<br>Fans<br>Machine tools<br>Mixing machines<br>Conveyors<br>Pumps | Some require low torque; others require several times full-load torque | 200% of full-load torque at each speed | Speed selection is desired, and two, three, or four fixed speeds are sufficient; starting torque can be low on blowers to high on conveyors; metal cutting machines are usually constant hp; friction loads (conveyors) are usually constant torque; fluid or air loads (blowers) are variable torque | *Multispeed:* general normal torque on dominant winding or speed; consequent pole windings or separate windings for each speed; based on load requirement, can be constant horsepower, constant torque, variable torque |
| Crushers<br>Conveyors<br>Bending rolls<br>Ball and rod mills<br>Centrifugal blowers<br>Pumps<br>Printing presses<br>Cranes and hoists<br>Centrifugals | Can provide torque up to maximum torque at standstill | 200–300% of full-load torque | Loads that require very high starting torque with low starting current; require speed adjustment over limited range (2 to 1); torque control during acceleration or controlled acceleration | *Wound rotor:* requires rotor control system to provide desired characteristic; control may be resistors or reactors or fixed-frequency inverter in the secondary (rotor) circuit; actual load speed depends on setting of rotor control |

SOURCE: Andreas, John C.; *Energy Efficient Electric Motors;* New York, Marcel Dekker, Inc., 1982.

type is shown in Fig. 20-38. Variable-torque motors have 1200/600 r/min, and are used on loads, such as in centrifugal pumps and fans whose horsepower requirement decreases more rapidly than the square of the reduction in speed. Constant-torque motors have horsepower ratings at each speed directly proportional to the speed, for example, 20/10 hp and 1200/600 r/min, and are used on conveyors, mixers, reciprocating compressors, printing presses, and other "constant-torque" loads. Constant-horsepower motors have the same horsepower rating at all speeds. They are used principally on machine tools, such as lathes, boring mills, planers, and radial drills. Multispeed motors of the constant-torque or variable-torque type are usually given a standard horsepower rating at the top speed but may have odd horsepower ratings at the lower speeds, since the latter are fixed by the speed ratios.

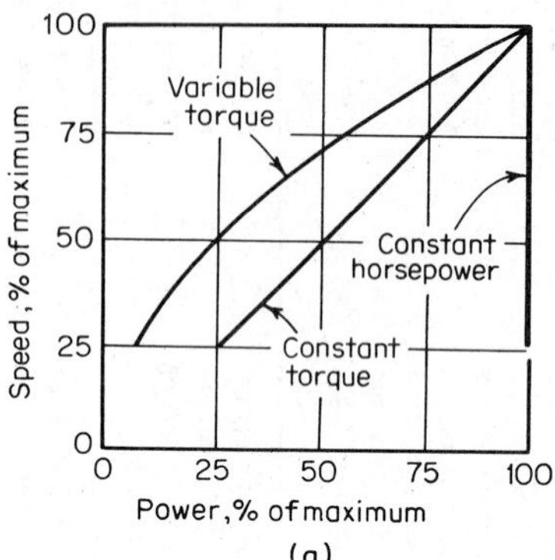


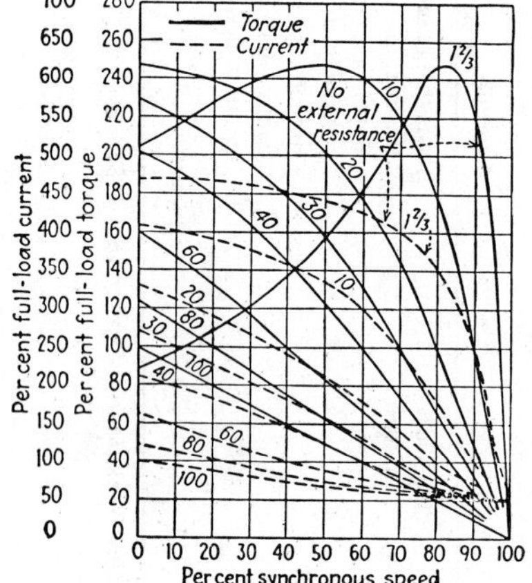


**FIG. 20-37** Speed-torque and speed-current curves of typical wound-rotor induction motor.

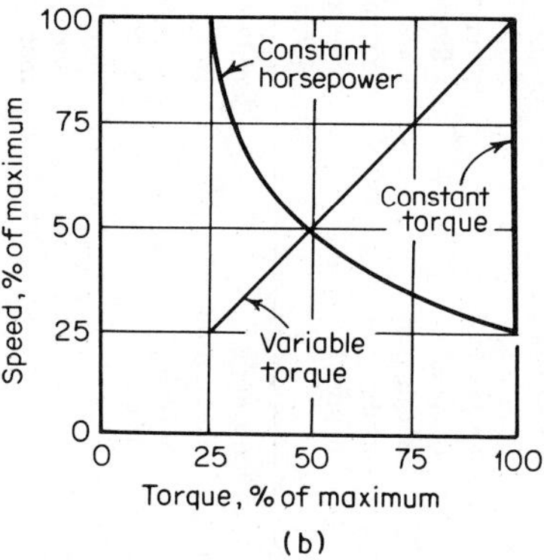


**FIG. 20-38** Basic load characteristics of multispeed motors having a 4:1 maximum speed ratio: (*a*) power; (*b*) torque.

***60. Temperature Rise.*** Temperature rise is no longer used as a rating method. Instead the manufacturer specifies the ambient temperature and the insulation class. The temperature rise will not exceed the limit for the insulation system when the motor is loaded to its rating or to its service factor load. The temperature rise limits are given in Table 20-5.

The temperature attained by squirrel-cage windings, cores, and mechanical parts shall not injure the machine in any respect. Temperatures shall be determined in accordance with the IEEE Test Procedures, Publication Nos. 112A and 114. For Class F and H insulation systems, special consideration shall be given to the bearings and lubrication.

The temperature rise for motors operating at any other ambient temperature $T_a$ than 40°C shall not exceed the values

For items a, b, e, f, i: Temperature rise $= 0.9(T_h - T_a)$

For items c, d, g, h: Temperature rise $= 0.965(T_h - T_a)$

**TABLE 20-5** Temperature Rise for Single-Phase and Polyphase Induction Motors

| | Class of insulation system | | | |
|---|---|---|---|---|
| | A | B | F | H |
| **Integral horsepower** | | | | |
| All motors with 1.15 service factor or higher | 70°C | 90°C | 115°C | |
| Totally-enclosed fan-cooled motors | 60°C | 80°C | 105°C | 125°C |
| Totally-enclosed non-ventilated motors | 65°C | 85°C | 110°C | 135°C |
| Motors with encapsulated windings, 1.0 service factor | 65°C | 85°C | 110°C | |
| All other motors | 60°C | 80°C | 105°C | 125°C |
| **Fractional horsepower** | | | | |
| Open motors with 1.15 service factor or higher | 70°C | 90°C | 115°C | |
| Totally-enclosed non-ventilated and fan-cooled | 65°C | 85°C | 110°C | 135°C |
| Any motor in frame smaller than 42 frame | 65°C | 85°C | 110°C | 135°C |
| All other open motors | 60°C | 80°C | 105°C | 125°C |

NOTE: Based on ambient temperature of 40°C, 3300-ft altitude. Temperature determined by the resistance method.

where $T_h$, the hot-spot temperature, is given by the following table:

| Class | Items *a* and *f* | All other items |
|---|---|---|
| A | 115°C | 105°C |
| B | 140°C | 130°C |
| F | 165°C | 155°C |
| H | | 180°C |

Preferred values of ambient temperature above 40°C are 50°C, 65°C, 90°C, 115°C.

The time ratings for single-phase and polyphase induction motors shall be 5, 15, 30, 60 min, and continuous. All short-time ratings are based upon a load test which shall commence when the windings and parts of the motor are within 5°C of the ambient temperature.

***61. Service Factor.*** General-purpose fractional- and integral-horsepower motors are given a "service factor," which allows the motor to deliver greater than rated horsepower, without damaging its insulation system. The motor is operated at rated voltage and frequency. The standard service factors are 1.4 for motors rated 1/20 to 1/8 hp; 1.35 for 1/6 to 1/3 hp; 1.25 for 1/2 hp to the frame size for 1 hp at 3600 r/min. For all larger motors through 200 hp, the service factor is 1.15. For 250 to 500 hp, the service factor is 1.0.

***62. Efficiency and Power Factor.*** Typical full-load efficiencies and power factors of standard Design B squirrel-cage induction motors are given in Figs. 20-39 and 20-40, respectively. The efficiencies of Design A motors are generally slightly lower, and those of Design D motors considerably lower. The power factors of Design A squirrel-cage induction motors are slightly higher, and those of Design C are slightly lower. Energy-efficient motors are those whose design is optimized to reduce losses. Comparative efficiencies of standard and energy-efficient motors of NEMA Design B are shown in Fig. 20-41.

***63. Full-Load Current.*** With the efficiency and power factor of a 3-phase motor known, its full-load current may be calculated from the formula

$$\text{Full-load current} = \frac{746 \times \text{hp rating}}{1.73 \times \text{efficiency} \times \text{pf} \times \text{voltage}} \tag{20-29}$$

where the efficiency and power factor are expressed as decimals.

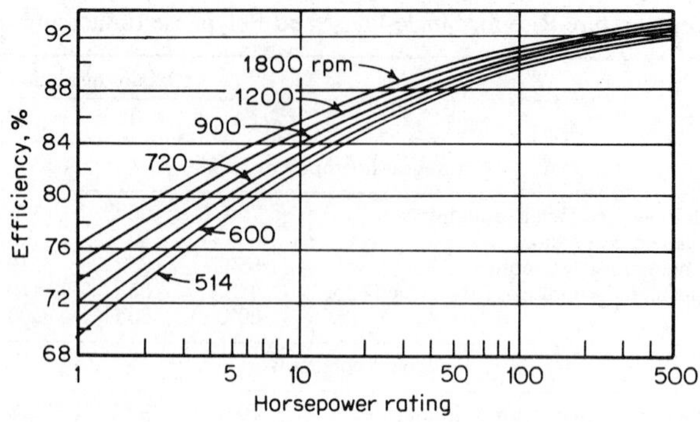

**FIG. 20-39** Typical full-load efficiencies of Design B squirrel-cage motors.

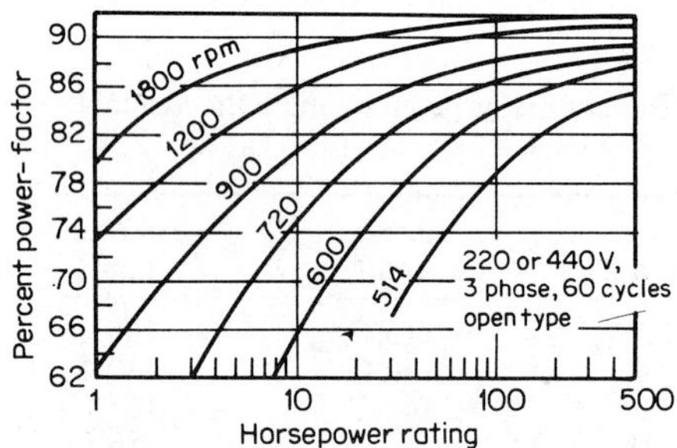

**FIG. 20-40** Typical full-load power factor of Design B squirrel-cage motors.

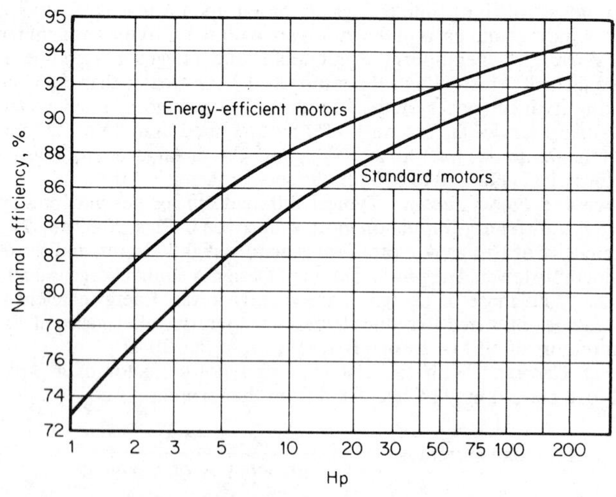

**FIG. 20-41** Nominal efficiencies for NEMA Design B 4-pole motors, 1800 r/min; standard vs. energy-efficient motors.

***64. Voltage and Frequency Variations; Voltage Unbalance.*** Polyphase induction motors are designed to operate successfully under the following conditions of voltage and frequency variation but not necessarily in accordance with standards established for operation at normal rating.

1. Where the voltage variation does not exceed 10% above or below normal.
2. Where the frequency variation does not exceed 5% above or below normal.
3. Where the sum of the voltage and frequency variation does not exceed 10% (provided the frequency variation does not exceed 5%) above or below normal. The effect of voltage and frequency variation on the characteristics of typical polyphase induction motors is given in Table 20-6.

These statements presuppose the same line voltage in each phase. When line voltages applied to a polyphase induction motor are not exactly the same, unbalanced currents will flow in the stator winding, the magnitude depending upon the amount of unbalance. A small amount of voltage unbalance may increase the current an excessive amount.

The percent voltage unbalance is defined as

$$\text{Percent voltage unbalance} = 100 \times \frac{\text{maximum voltage deviation from average}}{\text{average voltage}}$$

Temperature rise will be increased considerably, for example, 25% increase in rise for 3.5% unbalance. Torques and full-load speed will decrease slightly. The locked-rotor current will be unbalanced to the same degree as the voltage unbalance, but the full-speed current will be greatly unbalanced, on the order of approximately 6 to 10 times the percentage voltage unbalance, making selection of overload protective devices difficult.

**TABLE 20-6** General Effect of Voltage and Frequency Variation on Induction-Motor Characteristics

| Characteristic | Alternating-current (induction) motors | | | |
|---|---|---|---|---|
| | Voltage | | Frequency | |
| | 110 % | 90 % | 105 % | 95 % |
| Torque:* | | | | |
| Starting and maximum running | Increase 21 % | Decrease 19 % | Decrease 10 % | Increase 11 % |
| Speed:† | | | | |
| Synchronous | No change | No change | Increase 5 % | Decrease 5 % |
| Full load | Increase 1 % | Decrease 1.5 % | Increase 5 % | Decrease 5 % |
| Per cent slip | Decrease 17 % | Increase 23 % | Little change | Little change |
| Efficiency: | | | | |
| Full load | Increase 0.5 to 1 point | Decrease 2 points | Slight increase | Slight decrease |
| ¾ load | Little change | Little change | Slight increase | Slight decrease |
| ½ load | Decrease 1 to 2 points | Increase 1 to 2 points | Slight increase | Slight decrease |
| Power factor: | | | | |
| Full load | Decrease 3 points | Increase 1 point | Slight increase | Slight decrease |
| ¾ load | Decrease 4 points | Increase 2 to 3 points | Slight increase | Slight decrease |
| ½ load | Decrease 5 to 6 points | Increase 4 to 5 points | Slight increase | Slight decrease |
| Current: | | | | |
| Starting | Increase 10 to 12 % | Decrease 10 to 12 % | Decrease 5 to 6 % | Increase 5 to 6 % |
| Full load | Decrease 7 % | Increase 11 % | Slight decrease | Slight increase |
| Temperature rise | Decrease 3 to 4 C | Increase 6 to 7 C | Slight decrease | Slight increase |
| Maximum overload capacity | Increase 21 % | Decrease 19 % | Slight decrease | Slight increase |
| Magnetic noise | Slight increase | Slight decrease | Slight decrease | Slight increase |

*The starting and maximum running torque of ac induction motors will vary as the square of the voltage.

†The speed of ac induction motors will vary directly with the frequency.

***65. Torques and Starting Currents.*** Starting and breakdown torques of common Design A, B, and C squirrel-cage induction motors are given in Table 20-7. Relative values for other classes of squirrel-cage motors are indicated by the curves of Fig. 20-36. The minimum breakdown torque for wound-rotor motors is 200% of full-load torque. As indicated by the curves of Fig. 20-37, the starting torque and starting current of wound-rotor motors vary with the amount of external resistance in the secondary circuit.

The starting kVA of a squirrel-cage motor is indicated by a code letter stamped on the nameplate. Table 20-8 gives the corresponding kVA for each code letter, and the locked-rotor current can be determined from

$$\text{Locked-rotor current} = \frac{\text{kVA/hp} \times \text{hp} \times 1000}{k \times \text{line volts}} \tag{20-30}$$

where $k = 1$ for single-phase, and $k = 1.73$ for 3-phase.

Maximum locked-rotor current for Design B, C, and D 3-phase motors has been standardized as shown in Table 20-9 for 230 V. The starting current for motors designed for other voltages is inversely proportional to the voltage.

***66. Starting Methods.*** Wound-rotor motors are invariably started on full voltage but with external resistance in the secondary circuit. Ordinarily sufficient resistance is provided to give 100% torque at standstill, which means that 100% current will be drawn from the

**TABLE 20-7** Torques—Polyphase Induction Motors
(Percent of full-load torque)

| Rpm | 3,600 | | 1,800 | | | | 1,200 | | | | 900 | | | | 720 | |
|---|---|---|---|---|---|---|---|---|---|---|---|---|---|---|---|---|
| Torque | LR | BD | LR | LR | BD | BD | LR | LR | BD | BD | LR | LR | BD | BD | LR | BD |
| Design | AB | B | AB | C | B | C | AB | C | B | C | AB | C | B | C | AB | B |
| ½ hp | ... | ... | ... | ... | ... | ... | ... | ... | ... | ... | 150 | ... | 250 | ... | 150 | 200 |
| ¾ hp | ... | ... | ... | ... | ... | ... | 175 | ... | 275 | ... | 150 | ... | 250 | ... | 150 | 200 |
| 1 hp | ... | ... | 275 | ... | 300 | ... | 175 | ... | 275 | ... | 150 | ... | 250 | ... | 150 | 200 |
| 1½ hp | 175 | 275 | 265 | ... | 300 | ... | 175 | ... | 275 | ... | 150 | ... | 250 | ... | 150 | 200 |
| 2 hp | 175 | 250 | 250 | ... | 275 | ... | 175 | ... | 250 | ... | 150 | ... | 225 | ... | 145 | 200 |
| 3 hp | 175 | 250 | 250 | ... | 275 | ... | 175 | 250 | 250 | 225 | 150 | 225 | 225 | 200 | 135 | 200 |
| 5 hp | 150 | 225 | 185 | 250 | 225 | 200 | 160 | 250 | 225 | 200 | 130 | 225 | 225 | 200 | 130 | 200 |
| 7½ hp | 150 | 215 | 175 | 250 | 215 | 190 | 150 | 225 | 215 | 190 | 125 | 200 | 215 | 190 | 120 | 200 |
| 10 hp | 150 | 200 | 175 | 250 | 200 | 190 | 140 | 225 | 200 | 190 | 125 | 200 | 200 | 190 | 120 | 200 |
| 15 hp | 150 | 200 | 165 | 225 | 200 | 190 | 135 | 200 | 200 | 190 | 125 | 200 | 200 | 190 | 120 | 200 |
| 20 hp | 150 | 200 | 150 | 200 | 200 | 190 | 135 | 200 | 200 | 190 | 125 | 200 | 200 | 190 | 120 | 200 |
| 25 hp | 150 | 200 | 150 | 200 | 200 | 190 | 135 | 200 | 200 | 190 | 125 | 200 | 200 | 190 | 120 | 200 |
| 30 hp | 150 | 200 | 150 | 200 | 200 | 190 | 135 | 200 | 200 | 190 | 125 | 200 | 200 | 190 | 120 | 200 |
| 40–200 hp | * | 200 | * | 200 | 200 | 190 | * | 200 | 200 | 190 | 125 | 200 | 200 | 190 | 120 | 200 |

NOTE: LR = locked-rotor torque; BD = breakdown torque; A, B, and C refer to Design A, etc.
*Progressively lower values for these larger ratings.

**TABLE 20-8** Locked-Rotor kVA for Code-Letter Motors

| Code Letter* | Kva per Hp, with Locked Rotor | Code Letter* | Kva per Hp, with Locked Rotor |
|---|---|---|---|
| A | 0–3.14 | L | 9.0– 9.99 |
| B | 3.15–3.54 | M | 10.0–11.19 |
| C | 3.55–3.99 | N | 11.2–12.49 |
| D | 4.0 –4.49 | P | 12.5–13.99 |
| E | 4.5 –4.99 | R | 14.0–15.99 |
| F | 5.0 –5.59 | S | 16.0–17.99 |
| G | 5.6 –6.29 | T | 18.0–19.99 |
| H | 6.3 –7.09 | U | 20.0–22.39 |
| J | 7.1 –7.99 | V | 22.4 and up |
| K | 8.0 –8.99 | | |

*National Electrical Code.

**TABLE 20-9** Locked-Rotor Current for 3-Phase Motors at 230 V

| Rated horsepower | Classes B, C, D, amperes | Rated horsepower | Classes B, C, D, amperes | Rated horsepower | Classes B, C, D, amperes | Rated horsepower | Classes B, C, D, amperes | Rated horsepower | Class B amperes |
|---|---|---|---|---|---|---|---|---|---|
| 1 | 30 | 7½ | 127 | 30 | 435 | 100 | 1450 | 250 | 3650 |
| 1½ | 40 | 10 | 162 | 40 | 580 | 125 | 1815 | 300 | 4400 |
| 2 | 50 | 15 | 232 | 50 | 725 | 150 | 2170 | 350 | 5100 |
| 3 | 64 | 20 | 290 | 60 | 870 | 200 | 2900 | 400 | 5800 |
| 5 | 92 | 25 | 365 | 75 | 1085 | | | 450 | 6500 |
| | | | | | | | | 500 | 7250 |

line. If a higher torque is required to start the load, less external resistance must be used, and the current drawn is proportionately higher. As the motor accelerates, the external secondary resistance is short-circuited in one or more steps.

The locked-rotor values in Table 20-8 are generally recognized as the minimum needed by motor designers to obtain the required torque characteristics for general-purpose motors. Squirrel-cage motors with these values are usually acceptable for full-voltage starting on power lines and also on combined light and power secondaries of 208 or 230 V, if manually controlled (infrequently started). In the case of automatically controlled (frequently started) equipment, with 208- or 230-V motors supplied from combined light and power secondaries, current-reducing starters to reduce the current to about 65% of these values may be required, unless consultation with the power company indicates that the available system capacity will permit use of full-voltage starting. In any case consultations with the power company for motor applications above 25 hp are advisable.

Autotransformer starters (compensators) are the most popular of any reduced-voltage type. They have the advantage that the ratio of torque developed by the motor to the current drawn from the line remains substantially the same as for full-voltage starting. The motor torque and the current drawn from the line (neglecting the magnetizing current of the autotransformer) are both reduced in proportion to the square of the voltage impressed on the motor. The magnetizing current of the autotransformer generally does not exceed 25% of motor full-load current. Normally, the motor accelerates nearly to full speed on the reduced-voltage connection and is then transferred to full voltage. Since the circuit to the motor is opened and then immediately reclosed, a transient inrush of current occurs which may be of much greater magnitude than the current normally drawn by the motor at the speed at which the transfer is made. This transient inrush, however, is of such extremely short duration that it does not produce an objectionable voltage disturbance on the average power system. Standard autotransformer starters are provided with 65 and 80% voltage taps in sizes up to 50 hp and with 50, 65, and 80% voltage taps in the larger sizes.

"Part-winding" starting is being more widely used for reducing starting current. This involves arranging the stator winding so that, by use of adequate control devices, one part of the stator winding is first energized and subsequently the remainder of the winding is energized in one or more steps. The purpose is to reduce the initial values of the starting current drawn and/or the starting torque developed by the motor. The usual arrangement involves energizing one-half the stator winding on the first step, resulting in approximately 50% of normal locked-rotor torque and approximately 60% of normal locked-rotor current. While this torque may be insufficient to start the motor in some applications, it permits drawing full-winding starting current from the system in two increments. Another method is to connect two-thirds of the winding on the first step, by using a 4-pole contactor, in which case the motor should accelerate promptly to full speed. The remaining third of the winding is then connected by closing a second contactor with only two poles.[1]

Resistor-type reduced-voltage starters are sometimes used. They have the disadvantage that the current drawn from the line is reduced in direct ratio to the impressed voltage, while the torque developed by the motor is reduced as the square of this voltage. The resistor is short-circuited, either all at once or in steps, when the motor comes up to speed. The circuit for the motor is not broken in transferring to full voltage, as is the case with the autotransformer starter. These features make the resistor-type starter adapted for use where "increment-type" starting-current restrictions exist. With the resistor-type starter, the contactors which short-circuit the resistors, as well as the line contactors, must carry the full current of the motor, whereas in part-winding starting, the contactors for the two parts of the winding each carry only half the total current.

Reactor-type reduced-voltage starters are sometimes used on larger motors, most frequently on high-voltage motors (2300 V or above), where oil circuit breakers are necessary to provide sufficient current-interrupting capacity. In such cases, the reactor and starting

[1]P. L. Alger, H. C. Ward, Jr., and F. H. Wright; Split-winding Starting in 3-Phase Motors; *Trans. AIEE,* 1951, vol. 70, pt. 1, p. 867.

P. L. Alger and Lorraine Agacinsky; A New Method for Part-winding Starting of Polyphase Motors; *Trans. AIEE,* Power Apparatus and Systems, February 1956, no. 22, p. 1455.

P. L. Alger; Performance Calculations for Part-winding Starting of Three-Phase Motors; *AIEE Conf. Paper,* 56–515.

circuit breaker are placed in the neutral of the motor. The breaker can then be of low interrupting capacity, since the fault current at this point is limited by the reactance of the motor windings.

Wye-delta starting, though quite common abroad, is used in the United States primarily for refrigeration compressors. This starter consists of a switching arrangement which transfers the motor winding from Y for starting to Δ for running. The current drawn and the torque developed by the motor are thus reduced to only one-third their full voltage values. This very low torque, the extra contactors required, and the current inrush when the circuit is reclosed on Δ make this scheme less attractive than others.

Motors are frequently supplied from power systems consisting of complex networks for which calculation of the voltage drop would be difficult. The voltage drop may be estimated, however, if the short-circuit kVA is known at the point of power delivery.

When motor-starting kVA is drawn from a system, the voltage drop in percent of the initial voltage is approximately equal to 100 times the motor-starting kVA divided by the sum of this kVA and the short-circuit kVA. The motor-starting kVA used should be that drawn by the motor if the initial system voltage is maintained. For example, if a 1000-hp motor has a starting kVA of 5000 at the initial system voltage and the system short-circuit kVA is 50,000, the voltage drop will be approximately

$$\frac{5000 \times 100}{(5000 + 50{,}000)} = 9\% \text{ of initial voltage}$$

In many systems the short-circuit kVA varies over a wide range depending upon the number of parallel lines in service, system interconnections, etc. While the highest short-circuit kVA is of interest for circuit interruption, the minimum short-circuit kVA should be used for voltage-drop calculations since it gives the highest value.

***67. Shell-type motors*** consist of stators and rotors only, without shafts, end shields, bearings, or conventional frames (Fig. 20-42). The rotors are mounted directly on a shaft of the driven machine, which must also include a suitable support for the stator and a ventilating arrangement. The motors are built with relatively small outside diameters but may be slightly longer than standard machines. Furthermore, horsepower ratings over a rather wide range are built in each frame diameter, the ratings for the different diameters overlapping slightly. Although a great many of the motors used are for operation at standard commercial frequencies giving speeds up to 3600 r/min (on 60 Hz), they are frequently supplied for operation at higher frequencies and correspondingly higher speeds. Frequencies up to 2000 Hz with a corresponding 2-pole motor speed of 120,000 r/min have been used, but the more common "high" frequencies range from 60 to 240 Hz, giving 2-pole motor speeds up to 14,400 r/min.

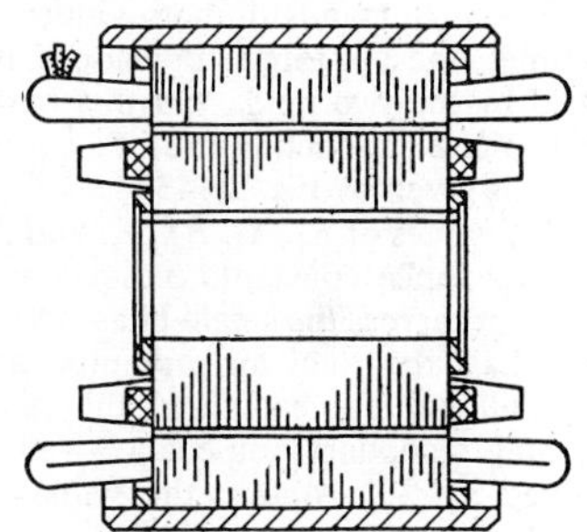

**FIG. 20-42** Cross section of shell-type motor.

Shell-type motors are used principally on machine tools and woodworking machinery. Their relatively small physical size facilitates a compact design with maximum flexibility in arrangement of machine parts. The small diameter of the motors is of particular value, since it allows close spacing of spindle shafts.

The wide range of ratings available in each diameter reduces the cost of providing suitable mountings for the motors.

Motors of similar mechanical construction but with special insulation are used in hermetically sealed refrigeration and air-conditioning compressors, where the motor runs in an atmosphere of refrigerating gas. The insulation must neither harm nor be harmed by the refrigerant and, so that the refrigerant may be kept clean and dry, must not trap moisture or dirt.

***68. Dimensions.*** NEMA has standardized mounting dimensions for various types of motors, those standardized for polyphase induction motors covering ratings from 1 to 125 hp (at 1800 r/min). For convenience each set of standardized dimensions has been assigned a frame number, and the various ratings of motors have been assigned frame numbers from the series. Any motor offered by a manufacturer having a frame number from this series will have the corresponding standardized mounting dimensions. These are listed in NEMA Motor and Generator Standard, Publ. MG1-1972.

These NEMA frame dimensions along with a close counterpart in the metric system are included in the *Rept. on Dimensions* 72 issued by The International Electrotechnical Commission covering progress toward an international standard of motor dimensions.[1]

## SINGLE-PHASE INDUCTION MOTORS

***69. General Theory.*** If one supply line to a polyphase induction motor is opened, the motor will not develop any starting torque, although if it is already operating, it will continue to run at slightly reduced speed, with a somewhat lower breakdown torque. The crux of the single-phase motor problem, therefore, is in providing auxiliary means for starting.

The magnetic field of a single-phase winding carrying alternating current may be represented as a phasor stationary in space but alternating in time, or as the sum of two equal and oppositely revolving field phasors, which are constant in magnitude. In a polyphase motor, the backward-revolving field phasors of the several phases cancel each other, and the forward-revolving ones add directly, giving a uniform revolving field. In the single-phase motor, means are provided to reduce the backward field, but this field has always some remaining magnitude (except at one particular load in the case of certain capacitor-run motors), and consequently a single-phase induction motor always has extra losses and a double-frequency pulsating torque not possessed by a polyphase motor.

A simple way to visualize the effects of this backward field is to consider that the forward- and backward-revolving fields are separately produced by the same stator current; that is, they are connected in series. Each field may then be treated as a separate polyphase induction motor, the forward field having a slip $s$ with respect to the rotor, and the other a slip $2 - s$. At standstill, both values of slip are unity, and the two circuits are identical. At all times, the net torque developed is equal to the difference of the separate torques produced by the two fields. On this basis, the single-phase induction-motor equivalent circuit is given by Fig. 20-43.

The values of $R_1$, $X_1$, $R_2$, $X_2$, and $X_M$ are the impedance constants derived by measurements across the single-phase terminals. Since half the total air-gap impedance at standstill is due to each field, the magnetizing and secondary impedance values are divided by 2 to obtain the values corresponding to the separate fields.

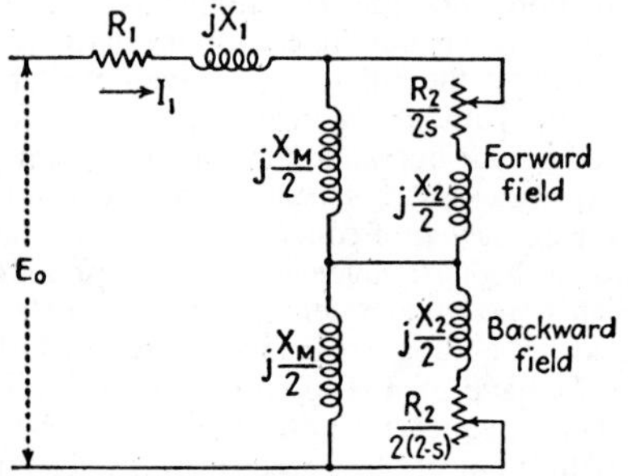


**FIG. 20-43** Equivalent circuit of single-phase induction motor.

Inspection of this circuit reveals several interesting properties of the motor. At full speed, $s$ is very small, and the backward field appears as an external series impedance of $R_2/4 + j(X_2/2)$. The corresponding loss $I^2R_2/4$ represents the power delivered to the rotor by the backward field. However, there is an equal loss due to the rotor's being driven forward at speed $1 - s$ against the backward-field torque; so the total loss caused by the backward field is $I^2R_2/2$, approximately. Since the backward-field rotor currents occur at double-line frequency, any double squirrel-cage or deep-bar rotor design which had an increased resistance at high frequency would greatly increase the power losses; and such designs, therefore, are seldom used for

[1]This report is available through American National Standards Institute, 1430 Broadway, New York, NY 10018.

single-phase motors. The breakdown torque of a single-phase motor may be approximately calculated for a polyphase induction motor, if the impedance of the backward-revolving field is considered as a series impedance added in the primary circuit of the polyphase motor (see Par. 47). Hence, any increase in the secondary resistance of a single-phase motor actually reduces the breakdown torque, as well as lowering the speed at which breakdown occurs.

Another interesting characteristic is the double-frequency torque pulsation. The double-frequency current in the rotor reacting upon the slip-frequency forward magnetic field evidently produces a torque pulsation, even at no load. Physically, the no-load part of the pulsating torque provides the means for supplying and removing the magnetic field twice each cycle in the axis at right angles to the stator winding, and the additional part under load corresponds to the double-frequency pulsation of the single-phase power input to the rotor. To prevent objectionable transmitted vibration and noise from this cause, it is usual to mount single-phase machines on supports with torsional elasticity of some type, often rubber rings encircling the bearing housings in the case of fractional-horsepower motors.

***70. Shaded-Pole Motor.*** The simplest way of providing a single-phase induction motor with starting torque is to place a permanently short-circuited winding of relatively high resistance in the stator at an electrical angle of 30 to 60° from the main winding. Usually this auxiliary winding, called a "shading coil," consists of an uninsulated copper strip encircling approximately one-third of a pole pitch. The current induced in the shading coil, by the portion of the main field linking it, reduces the magnitude of this flux and also causes it to lag in time phase. In consequence, the air-gap field has two components, an undamped alternating flux and a damped flux displaced both in space and in time. Shaded-pole motors are used only in very small sizes normally below 50 W output. Principal applications are for desk fans and air circulators, where their simplicity, low torque, and low cost are well suited to the requirements.

The inherently high slip of a shaded-pole motor makes it convenient to obtain speed variation on a fan load by reducing the impressed voltage. It is common practice to provide multispeed fan operation by employing a small switched-series reactor or a phase-controlled solid-state device to control the motor voltage.

***71. Resistance Split-Phase Motors.*** A considerably greater starting torque can be obtained by providing a separate starting winding, or auxiliary phase, 90° displaced in space from the main winding of a single-phase induction motor. This extra winding is normally wound with fewer turns of a much smaller size of wire, so that it has a considerably greater resistance to reactance ratio than the main winding, and it is connected directly across the power supply, in parallel with the main winding. Just as in the case of the shaded-pole motor, the field of the auxiliary winding is displaced in time and space, so that its vectorial combination with the main field gives a much larger forward than backward field component. The motor can be reversed by reversing either the main or the auxiliary winding.

Since the auxiliary winding is normally located 90° from the main winding, the two are mutually noninductive at standstill and the standstill characteristics may be calculated from two independent circuits each like that of Fig. 20-43. By a similar analysis to that of the preceding section, the starting torque of a split-phase motor is

$$T = \frac{14.1aK}{N_s} I_M I_A R_2 \sin\theta \qquad \text{ft}\cdot\text{lb} \tag{20-31}$$

where $I_M$ = main winding starting current in amperes; $I_A$ = auxiliary winding starting current in amperes; $N_s$ = synchronous speed; $R_2$ = resistance component of standstill impedance of main winding, less the primary resistance; $a$ = ratio of effective number of turns in auxiliary winding to main-winding effective turns; $K$ = an empirical coefficient which allows for nonfundamental rotor losses, usually equal to 0.9; and $\theta$ = angle of phase split between $I_M$ and $I_A$.

Design limitations usually prevent $\theta$ from being greater than 30°, so that the starting torque per voltampere cannot exceed half that of a 2-phase motor built in the same parts. Since, in addition, both $I_M$ and $I_A$ are drawn from a single phase of the power supply, the starting current is excessive, limiting the use of the resistance split-phase motor to sizes below ⅓ hp.

The auxiliary winding is opened automatically as the motor approaches full speed, as otherwise prohibitive losses would occur in it. Usually this is accomplished by means of a centrifugal switch or, in the case of hermetically sealed motors, by an electromagnetic relay. The high current density used to obtain an adequate resistance value makes the initial rate of temperature rise of the auxiliary winding very great, sometimes more than 50°C/s, so that these motors are not satisfactory for repeated starting or for inertia loads.

Earlier split-phase motor designs included motors with stationary external squirrel-cage members, with the primary windings on the rotor, receiving their power through slip rings. They are now normally built, however, with uniformly distributed partly closed stator slots, enameled-wire concentric stator windings, and a cast-aluminum or welded-copper squirrel cage on the rotor.

Typical characteristic curves for a ⅙-hp 60-Hz 1725-r/min, resistance split-phase motor are shown in Table 20-12.

***72. Repulsion-Start Induction-Run Motor.*** A common way of obtaining single-phase-motor starting torque is to provide a dc winding and commutator on the rotor, with a single pair of short-circuited brushes for starting and a centrifugal mechanism which short-circuits the entire commutator as the motor approaches full speed. This gives a pure repulsion-motor starting characteristic with very high torque per ampere and pure single-phase induction-motor operating characteristics. These motors were widely used in sizes up to about 5 hp. Typical characteristics of a 1-hp 60-Hz 1800-r/min motor of this type are shown in Table 20-12.

***73. Capacitor Motors.*** Low-cost low-voltage capacitors have proved extremely useful in improving the performance of split-phase motors. By inserting an external series capacitor in the auxiliary winding circuit and making this winding with many more turns of much lower resistance, the angle of phase split $\theta$ can be increased to 90°, or even more, and the coincident increase in the turn ratio $a$ permits a further decrease in the auxiliary winding current. Thus, the capacitor-start motor gives an adequate starting torque for a reasonable starting current and at the same time has so much greater thermal capacity than a resistance split-phase motor, by virtue of the reduced winding-current density, that it is satisfactory for nearly all industrial single-phase motor applications.

Figure 20-44 illustrates a convenient method of determining the best size of capacitor to use with a given motor. $I_M$ represents the locked-rotor current in the main winding and $I_A$ the current in the auxiliary winding. With no external capacitor, $X_C = 0$, and the motor becomes a plain resistance split type. As $X_C$ is increased, $I_A$ moves ahead in time phase, following a circular locus, increasing the torque and reducing the total current drawn from the line. Points of maximum starting torque, and maximum starting torque per ampere are indicated on the diagram.

Usually low-voltage electrolytic capacitors are used for starting purposes, since these are economical in 115-V intermittent ratings. However, in cases of severe starting duty, higher-voltage motors, or where capacitors are retained in the circuit during operation, paper or film-type, oil-filled ac capacitors are used.

For most applications, the auxiliary winding is opened by a centrifugal switch or relay, as the motor approaches full speed, just as in the case of the resistance split-phase motor.

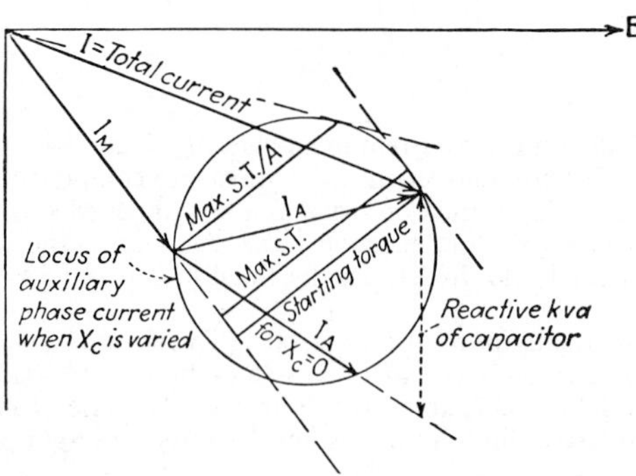


**FIG. 20-44** Capacitor-motor starting-torque diagram.

Such motors are called *capacitor-start motors.* In some cases of smaller-sized motors with low-starting-torque requirements, however, it is permissible to leave the capacitor permanently in circuit. These are called *permanent-split capacitor motors.* The limitations of starting torque and motor size on this type are the result of the inherent tendency of the auxiliary-winding current to increase in magnitude and shift backward in time phase as the motor accelerates, so that unless the capacitor impedance is very high, the motor will have objectionable losses and large torque pulsations at full speed. However, the power loss in the capacitor circuit at speed is very much less than in a shading coil for a given starting torque, and so the permanent-split capacitor motor is finding increasing use for fan drive in sizes up to ¼ hp.

For the larger capacitor motors, in sizes of ½ hp and up, it is frequently economical to retain the auxiliary winding in circuit with a reduced capacitor size, to improve the operating characteristics. This is usually accomplished by providing a large electrolytic or highly stressed capacitor in parallel with a small oil capacitor at starting and cutting the former out of circuit with a centrifugal switch or relay when the motor approaches full speed. Such motors are called *capacitor-run motors* and have winding connections as shown in Fig. 20-45. The analysis of the capacitor motor is done with an equivalent circuit of the type shown in Fig. 20-43 with the capacitors properly introduced.

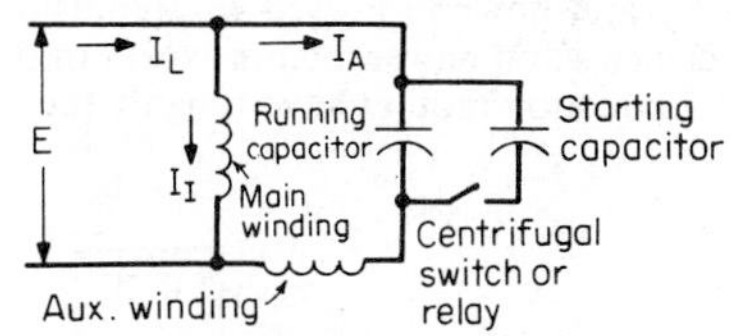


**FIG. 20-45** Capacitor-run motor winding connections.

***74. Horsepower, Speed, and Voltage Ratings.*** Standard horsepower and speed ratings of single-phase motors are given in Table 20-10. Motors built in frames having a continuous rating of less than 1 hp, open type, at 1700 to 1800 r/min are designated *fractional-horsepower* motors, and those built in larger frames are called *integral-horsepower* motors. Somewhat different standards of performance have been established for the two classes. ANSI C-50 and NEMA Motor and Generator Standards, Publ. MG1-1978, include basic standards for both fractional- and integral-horsepower motors which are normally followed in specifications and testing.

**TABLE 20-10** Standard Horsepower and Speed Ratings—Single-Phase Constant-Speed Motors

Standard horsepower ratings

| | | | | | |
|---|---|---|---|---|---|
| 1/20 | 1/6 | ½ | 1½ | 5 | 15 |
| 1/12 | ¼ | ¾ | 2 | 7½ | 20 |
| ⅛ | ⅓ | 1 | 3 | 10 | 25 |

Standard speed ratings

| Rpm 60 cycles | Fractional hp | Integral hp |
|---|---|---|
| 3,600 | 1/20–1 | 1½–25 |
| 1,800 | 1/20–¾ | 1 –25 |
| 1,200 | 1/20–½ | ¾–25 |
| 900 | 1/20–⅓ | ½–25 |
| **Rpm 50 cycles** | | |
| 3,000 | 1/20–1 | 1½–20 |
| 1,500 | 1/20–¾ | 1 –20 |
| 1,000 | 1/20–½ | ¾–20 |
| 750 | ....... | ½–20 |

Both capacitor and split-phase motors are available in the multispeed as well as the single-speed type. They are used principally for belt and direct drive of centrifugal and propeller fans and are of the variable-torque class. The multispeed motors for fan drive allow a change in fan speed without changing pulleys, which is essential where remote or automatic control of the rate of air delivery is required.

The standard voltage ratings for single-phase motors are 115 and 230 V for supply lines rated 120 and 240 V. Power companies place a limit on the size of motors that may be connected to single-phase lines. The limit usually falls between ½ and 1 hp for 120-V circuits and between 3 and 10 hp for 240-V circuits.

***75. Temperature Rise.*** The standard temperature rises and service factors of single-phase motors are the same as for polyphase motors.

***76. Efficiencies and Power Factors.*** Typical efficiencies and power factors of the various types of induction motor that might be used to fill the requirements of the different ratings are shown in Fig. 20-46. Repulsion-start induction-run motors have about the same efficiencies and power factors except in the 1½- to 3-hp range, where they are lower. Repulsion-induction motors have roughly the same efficiencies but higher power factors.

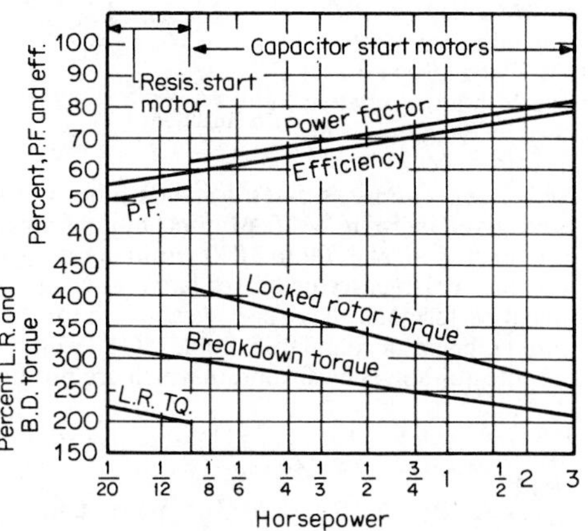


**FIG. 20-46** Typical operating characteristics of 1800-r/min single-phase motors.

***77. Full-load current*** of a single-phase motor is equal to

$$\frac{746 \times \text{hp}}{\text{Efficiency} \times \text{voltage}} \tag{20-32}$$

where the efficiency and power factor are expressed as decimals. Approximate values of full-load current are given in Table 20-11. These are used for selecting wire and fuse sizes if no more accurate data are available.

***78. Single-Phase Motor Characteristics.*** 60-Hz 4-pole 1800-r/min single-phase motor characteristics are shown in Fig. 20-46.

***79. Torques.*** The horsepower rating of a single-phase motor is defined by its breakdown torque. Thus, any 1800 r/min motor with a breakdown torque between 31.5 and 40.5 oz · ft is, by definition, a ⅙-hp motor. The value used for definition is the minimum of the range of manufacturing variation for that particular design.

***80. Starting Current.*** Maximum values of locked-rotor current are established by NEMA for 60-Hz motors as shown in Table 20-11. In integral-horsepower sizes, NEMA has established two sets of locked-rotor values. The Design L motors include those types having inherently higher locked-rotor values. The Design L motors include those types having inherently higher locked-rotor current than Design M motors.

**TABLE 20-11** Single-Phase Motor Characteristics

| | Hp | Approximate full load, amp 115 volts | Approximate full load, amp 230 volts | Locked rotor, amp 115 volts | | Locked rotor, amp 230 volts | | Breakdown torque (for defining hp ratings), oz-ft above line; lb-ft below line: 3,600 rpm | 1,800 rpm | 1,200 rpm | 900 rpm |
|---|---|---|---|---|---|---|---|---|---|---|---|
| Fractional hp | 1/6 | 4.4 | 2.2 | 20 | | 10 | | 8.7–11.5 | 16.5–21.5 | 24.1–31.5 | 31.5–40.5 |
| | 1/4 | 5.8 | 2.9 | 23 | | 11½ | | 11.5–16.5 | 21.5–31.5 | 31.5–44.0 | 40.5–58.0 |
| | 1/3 | 7.2 | 3.6 | 31 | | 15½ | | 16.5–21.5 | 31.5–40.5 | 44.0–58.0 | 58.0–77.0 |
| | 1/2 | 9.8 | 4.9 | 45 | | 22½ | | 21.5–31.5 | 40.5–58.0 | 58.0–82.5 | |
| | 3/4 | 13.8 | 6.9 | 61 | | 30½ | | 31.5–44.0 | 58.0–82.5 | 5.16–6.9 | |
| | | | | Design L | Design M | Design L | Design M | | | | |
| Integral hp | 1 | 16 | 8 | 70 | .. | 35 | ... | 44.0–58.0 | 5.16–6.8 | 6.9–9.2 | |
| | 1½ | 20 | 10 | .. | .. | 50 | 40 | 3.6–4.6 | 6.8–10.1 | 9.2–13.8 | |
| | 2 | 24 | 12 | .. | .. | 65 | 50 | 4.6–6.0 | 10.1–13.0 | 13.8–18.0 | |
| | 3 | 34 | 17 | .. | .. | 90 | 70 | 6.0–8.6 | 13.0–19.0 | 18.0–25.8 | |
| | 5 | 56 | 28 | .. | .. | 135 | 100 | 8.6–13.5 | 19.0–30.0 | 25.8–40.5 | |
| | 7½ | 80 | 40 | .. | .. | 200 | 150 | 13.5–20.0 | 30.0–45.0 | 40.5–60.0 | |
| | 10 | 100 | 50 | .. | .. | 260 | 200 | 20.0–27.0 | 45.0–60.0 | | |

NOTE: 1 oz-in = 0.00706 N·m; 1 lb-ft = 1.356 N·m.

## OTHER TYPES OF ELECTRIC MOTORS AND RELATED APPARATUS

***81. Induction Generators.*** Any induction motor, if driven above its synchronous speed when connected to an ac power source, will deliver power to the external circuit. The generator operation is easily visualized from the equivalent circuit of Fig. 20-31, corresponding to negative slip. The induction generator must always take reactive power from the load or the line for excitation and for the $I^2X$ losses. For this reason, the induction generator can only operate in parallel with an electric power system or independently with a load supplemented by capacitors. For independent operation, the speed must be increased with load to maintain constant frequency; the voltage is controlled with the capacitors.

An induction generator delivers an instantaneous 3-phase short-circuit current equal to the terminal voltage divided by its locked-rotor impedance. Its rate of decay is much faster than that of a synchronous generator of the same rating, corresponding to the subtransient time constant $T'_{do}$; sustained short-circuit current is zero.

The virtue of the induction generator is its ability to self-synchronize when the stator circuit is closed to a power system. At one time induction generators were used for small, unattended hydro stations. Today, induction generators are being used in a similar manner for wind turbines and cogeneration units. They have also been used for high-speed, high-frequency generators, because of their squirrel-cage rotor construction.

***82. Synchronous Induction Motors.*** There are three types of motors that can start and run as induction motors yet can lock into the supply frequency and run as synchronous motors as well. They are (1) the wound-rotor motor with dc exciter; (2) the permanent-magnet (PM) synchronous motor; (3) the reluctance-synchronous motor. The latter two types are used today primarily with adjustable-frequency inverter power supplies. In Europe, wound-rotor induction motors have often been provided with low-voltage dc exciters that supply direct current to the rotor, making them operate as synchronous machines. With secondary rheostats for starting, such a motor gives the low starting current and high torque of the wound-rotor induction motor and an improved power factor under load. Several different forms of these synchronous induction motors have been proposed, but they have not shown any net advantage over usual salient-pole synchronous or induction machines and are very seldom used in the United States. The PM synchronous motor is shown in Fig. 20-47*a*. The construction is the same as that of an ordinary squirrel-cage motor (either single or polyphase), except that the depth of rotor core below the squirrel-cage bars is very shallow, just enough to carry the rotor flux under locked-rotor conditions. Inside this shallow rotor core is placed a permanent magnet, fully magnetized. The rotor core serves as a keeper, so that the rotor is not demagnetized by removing it from the stator. In starting, the rotor flux is confined to the laminated core. As the speed rises, the rotor frequency decreases and the rotor flux builds up, creating a pulsating torque with the field of the magnet, as when a synchronous motor is being synchronized after the dc field has been applied. As the motor approaches full speed, therefore, the ac impressed field locks into step with the field of the magnet and the machine runs as a synchronous motor. The absence of rotor $I^2R$ loss, the synchronous speed operation, and the high efficiency and power factor make the motor very attractive for special applications, such as high-frequency spinning motors. When many such motors are supplied from a high-frequency source, the kVA requirements are reduced to perhaps 50% of those needed for usual induction motor types, with consequent large savings.

If the rotor surface of a $P$-pole squirrel-cage motor is cut away at symmetrically spaced points, forming $P$ salient poles, the motor will accelerate to full speed as an induction motor and then lock into step and operate as a synchronous motor. The synchronizing torque is due to the change in reluctance and, therefore, in stored magnetic energy, when the air-gap flux moves from the low- into the high-reluctance region. Such motors are often used in small-horsepower sizes, when synchronous operation is required, but they have inherently low pull-out torque and low power factor, and also poor efficiency, and therefore require larger frames than the same horsepower induction motor. The PM synchronous motor has superior performance in every way, except possibly cost. A cross section of the reluctance-synchronous motor is shown in Fig. 20-47*b*. These motors are available up to about 5 hp.

If the number of rotor salients is $nP$, instead of $P$, and if the $P$-pole motor winding is arranged to also produce a field of $(n - 1)P$ or $(n + 1)P$ poles, the motor may lock into step at a subsynchronous speed and run as a subsynchronous motor. For the $P$-pole fundamental mmf, acting on the varying rotor permeance, will create $(n + 1)P$ and $(n - 1)P$ pole fields from this case, and these will lock into step with the independently produced $(n - 1)P$- or $(n + 1)P$-pole field, when the rotor speed is such as to make the two harmonic fields turn at the same speed in the same direction.

It is difficult to provide much torque in such subsynchronous motors, and their use is therefore limited to very small sizes, such as may be used in small timer or instrument motors.

***83. Linear Motors.*** Linear induction motors (LIMs) have been built in fractional-horsepower ratings for such applications as moving drapes, and up to several thousand horsepower for driving tracked air-cushion transit vehicles on a guideway. Other applica-

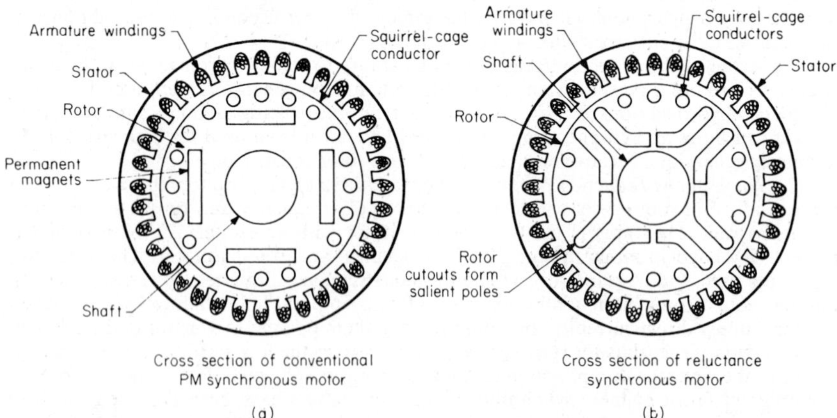


**FIG. 20-47** Cross section of (*a*) a PM synchronous motor and (*b*) a reluctance-synchronous motor.

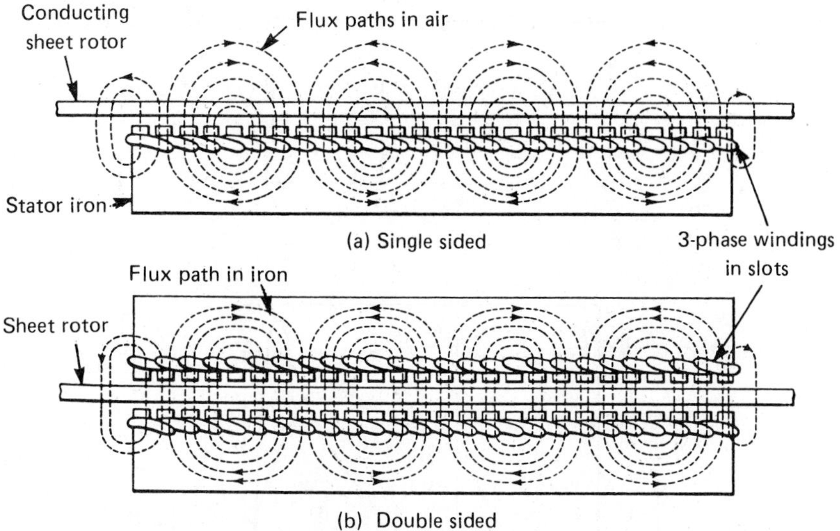


**FIG. 20-48** Single-sided and double-sided linear induction motors (LIMs) with sheet rotor.

tions include moving freight cars in yards, driving people-mover vehicles, and providing reciprocating motion for machine tools. LIMs are built like rotary induction motors with distributed multipole polyphase windings placed in the slots of a plane laminated status as shown in Fig. 20-48. When the windings are excited by a polyphase voltage of frequency $f$, an air-gap space flux wave is propagated along the length of the stator at a velocity of $v = 2fp$, where $p$ is the pole pitch. The rotor consists of an aluminum or copper sheet, which is propelled by the field with a slip velocity to provide the required thrust. LIMs are either double-sided, with two facing stators operating on a single rotor, or single-sided, with the rotor sheet backed by a moving or stationary magnetic return path. The magnetic force density normal to the stator surface is considerable compared to the tangential force density that moves the rotor, which requires that the stator be well-braced mechanically to maintain constant air-gap distances over the surface of the stator. The typical tangential force density is about 3 lb/in$^2$ for air-cooled windings, where the normal force density is about 30 lb/in$^2$.

The magnetic air gap of a double-sided LIM is the thickness of the sheet rotor plus the clearance between the rotor and the stators on either side. Whereas most rotary induction motors are built with an air gap of 0.025 to 0.1 mils, the air gap in the LIM is 0.25 to 1.5 in. For this reason, the magnetizing reactance of the LIM is lower than that of an equivalent rotary induction motor. Also the stator leakage reactance is higher. The equivalent circuit of the LIM is shown in Fig. 20-49*a*. Figure 20-49*b* shows the thrust-slip power factor and efficiency curves of a double-sided LIM[1]. This LIM has an air gap of 1.47 in, a rotor sheet thickness of 0.25 in, and a stator length of 9.8 in. The two 3-phase windings of the stator are excited at 173 Hz from inverters to produce a linear synchronous velocity of 395 ft/s. Speed control and breaking of LIMs is done in the same way as in the rotary induction motors.

***84. Eddy-Current Clutches.*** By constructing a squirrel-cage induction or synchronous motor with additional bearings, so that both primary and secondary members can revolve independently, a useful form of eddy-current clutch or coupling is obtained. Usually, such devices are designed with salient poles and dc excitation, to simplify the control. The dc-excited member is generally fixed on the load shaft, and the squirrel-cage member (or frequently a solid iron or steel cylinder in which eddy currents are induced) is attached to the rotor of the driving motor. The motor is started up with the clutch unexcited, the load

[1]M. G. Say; *Alternating Current Machines;* New York, John Wilcox & Sons, 1983.

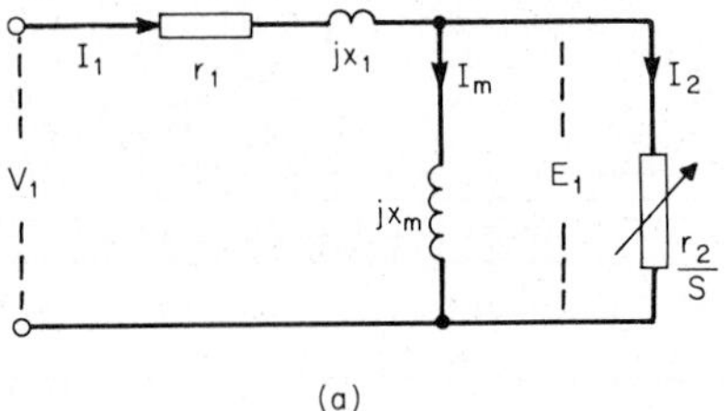


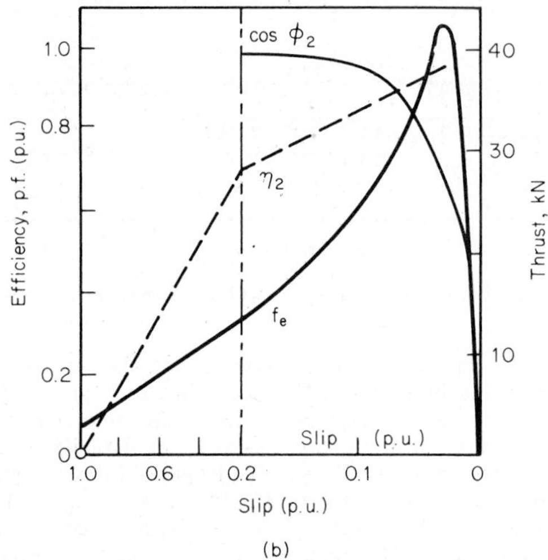


**FIG. 20-49** (*a*) Equivalent circuit of a double-sided LIM, and (*b*) the characteristic curves of a typical LIM.

remaining at rest. By then supplying a graduated amount of direct current to the clutch, a magnetic field is produced which induces squirrel-cage currents and a torque that brings the load up to any desired speed. Of course, the entire power represented by the driving-motor torque, multiplied by the difference between the speed of the driving motor and the speed of the load shaft, is a power loss dissipated in the clutch; so the efficiency is no better than with a rheostatically controlled wound-rotor motor. However, the smooth adjustment of the clutch torque and, therefore, the load speed by control of the dc excitation is an advantage that may justify the use of the clutch.

The speed-torque characteristic of the clutch is shown in Fig. 20-50. Eddy-current clutches are typically used for a load speed range of 3 to 1. The clutches are air-cooled and utilize liquid cooling with external heat exchangers in larger sizes.

***85. Selsyn Systems.*** The word *selsyn* is an abbreviation of the term *self-synchronous*, indicating machines which operate in space-phase alignment by virtue of electrical interconnection. If two wound-rotor induction motors are connected in parallel on their primary sides and to a common rheostat on their secondary sides, they will exert a synchronizing torque opposing any change in their relative speeds. The synchronizing torque falls off rapidly, however, as the external secondary resistance is reduced, and it is, of course, zero when both rotors are short-circuited. Hence, it is necessary to provide separate selsyn machines on the power-motor shafts if synchronism is to be maintained over normal ranges of speed and load. This characteristic is utilized in a variety of industrial applications, such as lift bridges and high-speed newspaper presses, where it is desired to have two or more motors operate in approximate synchronism during starting and load changes, without any mechanical tie between them.

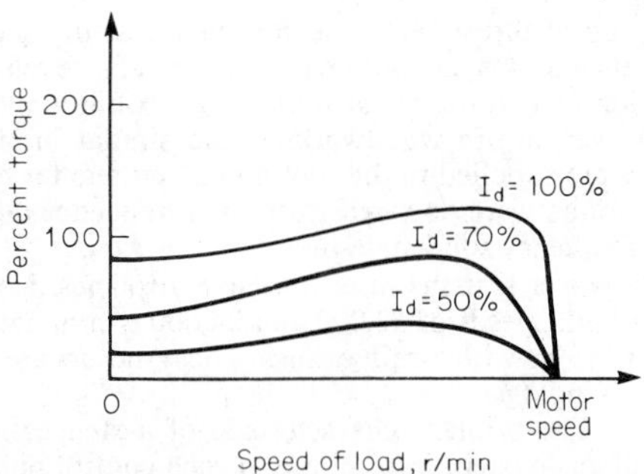


**FIG. 20-50** Torque-speed characteristic of eddy-current clutch as a function of direct clutch current $I_d$.

The principle is also usefully applied in remote-indicating and follow-up mechanisms of many types. Generally, there are a master selsyn and one or more receiver selsyns, the latter following exactly the angular position of the master. Each of the machines is essentially a small ac-excited synchronous motor, usually with a single-phase primary and a 3-phase secondary winding. A common source of ac excitation is connected to all the primary windings in parallel, and all the secondary windings are similarly connected in parallel, but not to any external source. If all the rotors are in the same angular position, all the secondary voltages will be in time phase; so no torque-producing secondary currents will flow. If one rotor is displaced through an angle $\theta$, its secondary voltage will be $\theta°$ out of phase with all the others and secondary currents will flow, creating torques that tend to restore the primaries to a common position. This positioning effect is present, whether the machines are rotating or standing still. Power selsyns are normally 3-phase and may be identical with wound-rotor induction motors.

Figure 20-51*a* to *c* shows the connection diagram, 3-phase standstill torque-angle characteristics, and 3-phase maximum torque vs. speed for two 25-hp 60-Hz 900-r/min, slip-ring motors connected in parallel on both sides.

Power-selsyn applications, therefore, consist of induction-driving motors, with separate direct-connected selsyns, often duplicates of the driving motors, all connected to the same power source. To bring the motors into alignment before starting the load, it is usual to apply single-phase voltage to the selsyn primaries for a brief period before applying 3-phase to all units.

***86. High-Frequency Motors.*** For high-speed tools and for spinning of rayon and other threads, a variety of interesting motor constructions have been developed. Normally these are two-pole 3-phase motors, with special high-frequency power supply of 90, 120, or 180 Hz, giving operating speeds between 500 and 10,500 r/min and up to 25 hp. In textile applications, the motors usually drive individual spinning buckets, which are subject to considerable unbalance

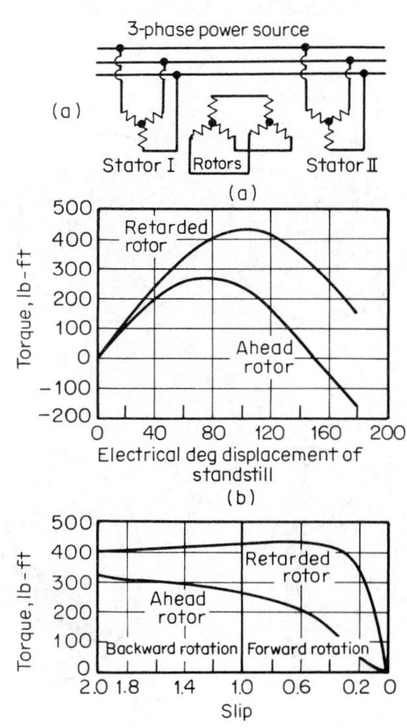


**FIG. 20-51** Selsyn connections and characteristics.

due to uneven building up of thread, etc. The continual starting and stopping for loading and unloading the buckets requires the motors to carry unbalance reliably through the entire speed range, necessitating careful design of mounting flexibility and shaft stiffness. Most usual applications, however, are in woodworking and similar industries, where separate motor stators and rotors are supplied to the tool manufacturers for building into their particular devices. These motors were powered from high-frequency alternators, but are now powered by adjustable-frequency solid-state inverters.

Three-phase 400-Hz power systems, used on large airplanes, have led to the development of 400-Hz motors with speeds of 12,000 and 24,000 r/min, having weights averaging 2 lb/hp for motors of 1 to 15 hp with 5-min ratings. These motors are open, with an external fan to force air over the windings.

***87. Stepper Motors.*** The primary characteristic of a stepper motor is its ability to rotate a prescribed small angle (step) in response to each control pulse applied to its windings. Below about 200 pulses per second, the motor rotates in discrete steps in synchrony with the pulses; at higher frequencies up to 16,000 pulses per second, the motor skews without stopping between pulses. Although motors are available for step angles of 90 to 0.180°, the common step is 1.8°. Stepper motors are categorized as permanent-magnet rotor (PM),

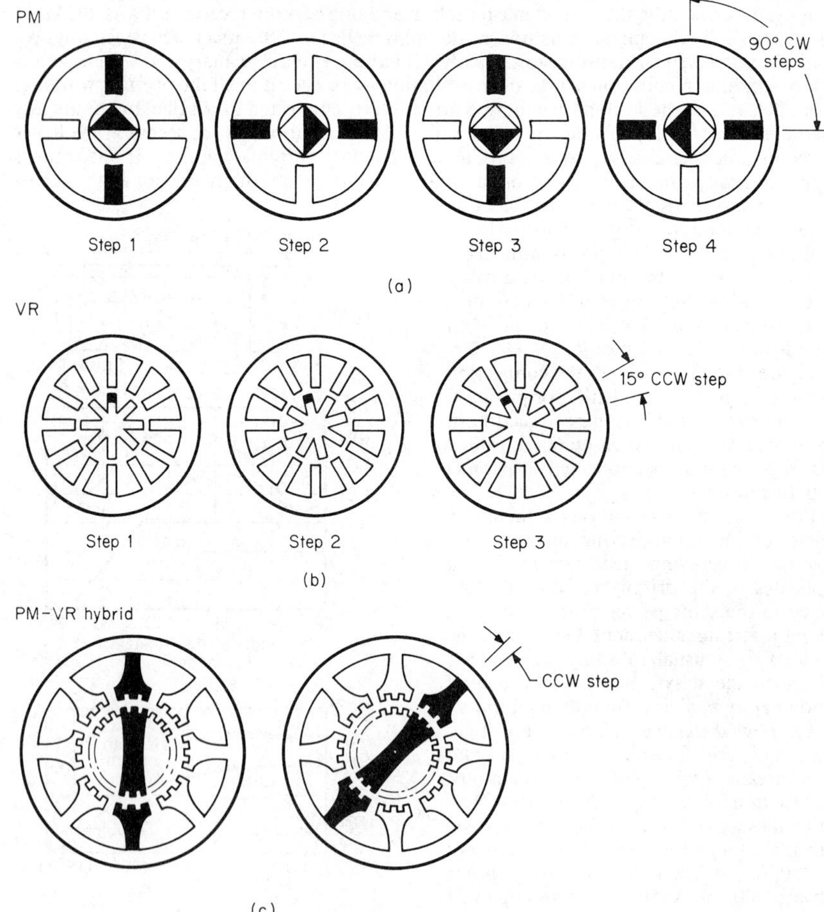


**FIG. 20-52** Three types of stepper motors.

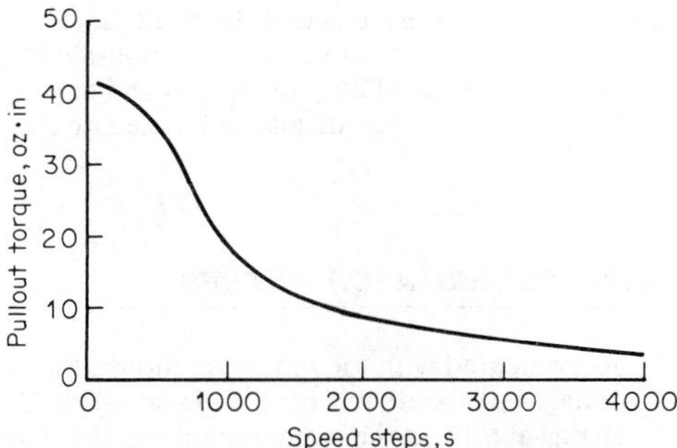


**FIG. 20-53** Pull-out torque vs. speed for a four-phase 5° step VR step motor running at half steps (2½°).

variable reluctance (VR), or hybrid (PM-VR). The rotor of the PM aligns itself with the energized stator poles as shown in Fig. 20-52*b*. The rotor turns until the poles are aligned at each step. The PM-VR hybrid shown in Fig. 20-52*c* has a high skew rate yet retains holding torque when the power is turned off. Motors can be made to rotate in half-steps to increase accuracy. Performance of stepper motors is described by two types of curves: the pull-out torque vs. speed curve, as shown in Fig. 20-53; and the holding torque angle curves, as shown in Fig. 20-54. Stepper motors are available with holding torques up to 4000 oz-in.

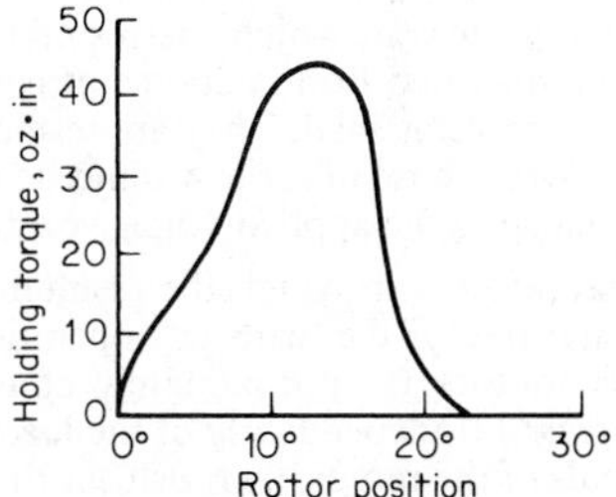


**FIG. 20-54** Holding torque vs. rotor position for a three-phase, VR step motor, 24 steps per revolution, bidirectional, 1600 steps/s.

***88. Hysteresis Motors.*** By constructing the secondary core of an induction motor of hardened magnet steel, in place of the usual annealed low-loss silicon-steel laminations, the secondary hysteresis loss can be greatly magnified, producing effective synchronous motor action. Such hysteresis motors, having smooth rotor surfaces without secondary teeth or windings, give extremely uniform torque, are practically noiseless, and give substantially the same torque from standstill all the way up to synchronous speed. A hysteresis motor is a true synchronous motor, with its load torque produced by an angular shift between the axis of rotating primary mmf and the axis of secondary magnetization. When the load torque exceeds the maximum hysteresis torque, the secondary magnetization axis slips on the rotor, giving the same effect as a friction brake set for a fixed torque.

Despite the interesting characteristics of this type of motor, it is limited to small sizes, because of the inherently small torque derivable from hysteresis losses. Only moderate flux densities are practicable, owing to the excessive excitation losses required to produce high densities in hard magnet steel, and, therefore, about 20 W/lb of rotor magnet steel represents the maximum useful synchronous power on 60 Hz. Hysteresis motors have found an important use for phonograph-motor drive, their synchronous speed enabling a governor to be dispensed with and freedom from tone waver to be secured.

The Telechron motor, which is so widely used for operating electric clocks, also operates on the hysteresis-motor principle. In the Telechron motor, a two-pole rotating field is produced in a cylindrical air space, and into this space is introduced a sealed thin-metal cylinder containing a shaft carrying one or more hardened magnet-steel disks, driving a gear train. The 60-Hz magnetic field causes the steel disks to revolve at 3600 r/min, driving through the gears a low-speed shaft, usually 1 r/min, which emerges from the sealed cylinder through a closely fitting bushing designed to minimize oil leakage. Although the magnetic field has to cross a very considerable air-gap length and pass through the tin walls of the metal cylinder, the power required to drive a well-designed clock is so small that ample output is obtained with only about 2 W input for ordinary household-clock sizes.

The hysteresis motor has been displaced for phonograph and tape-reel drives by the transistor-driven brushless dc motor. It has been displaced for electric clocks by solid-state circuits with digital readout.

## ALTERNATING-CURRENT COMMUTATOR MOTORS

***89. Classification.***[1] As compared with the induction motor, the ac commutator motor possesses two of the advantages of the dc motor: a wide speed range without sacrifice of efficiency, and superior starting ability. In the induction motor, the starting torque is limited by the small space-phase displacement between the air-gap flux and the induced secondary current and by magnetic saturation of the flux paths. In the ac commutator motor, on the other hand, the air-gap flux and current are held at the optimum space-phase displacement by proper location of the brush axis, and the secondary current is not limited by magnetic saturation, giving high torque per ampere at starting. Furthermore, the series commutator motor may be operated far above the induction-motor synchronous speed, giving high power output per unit of weight.

Alternating-current commutator motors may be grouped into two classes:

**a.** Those motors in which the resultant mmf providing the flux increases with the load. When operated from a source of constant voltage, the speed of such motors decreases with increasing load. They are termed *series motors* from the similarity of their characteristics to those of series-wound dc motors. The speed at any given load may be varied by changing the applied voltage or, in some cases, by shifting the brushes.

**b.** Those motors in which the resultant mmf providing the flux is substantially constant irrespective of the load. For operation from a source of constant voltage, the speed of such motors is approximately constant. The speed may, however, be increased or decreased (independently of the load) by increasing or decreasing the voltage at the terminals of the motor, by brush shifting, or by the provision of suitably disposed and connected auxiliary coils. Such motors are termed *shunt motors.*

Alternating-current commutator motors are either single-phase or polyphase. A unique characteristic of all single-phase motors is a double line-frequency pulsation of the torque produced, corresponding to the sinusoidal variation twice each cycle of the single-phase power supplied. This torque pulsation is partly transmitted to the load, causing small speed pulsations and necessitating special coupling and mounting designs to minimize vibration and fatigue stresses.

Polyphase commutator motors have the advantage of better inherent commutating ability, due in part to the need for shifting the rotor current only 60° in time phase at each brush stud for a 6-phase motor or 30° for 12 phases, as compared with 180° shift for a single-phase or dc machine. Single-phase motors are generally limited to sizes below about 10 hp, except for railway applications.

With the advent of solid-state devices, the ac commutator motor is being displaced by the thyristor-rectifier–powered dc motors and inverter-fed induction motors, at less cost and superior performance. The dc motor does not have the difficulties of commutation, the requirement for extra windings, and shifting brush arrangements, and can be built on an unlaminated frame. The induction motor has no commutator and can run at the high speeds of the commutator motor.

***90. Single-Phase Straight Series Motor.*** An ordinary dc series motor, if constructed with a well-laminated field circuit, will operate (although unsatisfactorily) if connected to a suitable source of single-phase alternating current. Since the armature is in series with the field, the periodic reversals of current in the armature will correspond with simultaneous

---

[1]C. W. Olliver; "The A-C Commutator Motor"; Princeton, N.J., D. Van Nostrand Company, Inc., 1927.

reversals in the direction of the flux, and consequently the torque will always be in the same direction. But the inductance of the motor will be so great that the current will lag far behind the voltage, and the motor will have a very low power factor. The entire amount of armature flux produced along the brush axis generates a reactive voltage in the armature, which must be overcome by the applied voltage, without performing any useful function whatever.

When the motor is first thrown in the circuit, and before the armature has moved from rest, the field constitutes the primary of a transformer and sends flux through the armature core. Those armature turns which at that instant are short-circuited under the brushes act as short-circuited secondary coils and are traversed by heavy currents which serve no useful purpose whatever and occasion serious heating. When the armature starts to revolve, these short-circuited turns are opened as they pass out from under the brushes and are replaced by other turns which are momentarily short-circuited and then opened. These interruptions of heavy currents are accompanied by serious sparking, since the heating is concentrated at the few segments on which the brushes rest. As soon, however, as a certain speed is acquired, the heating is distributed over all the segments and the conditions are ameliorated. This source of sparking is, then, most serious at the moment of starting. This difficulty has been minimized by operating at a lower frequency than 60 Hz and by the employment of leads of high resistance connecting the winding to the commutator segments.

The simple single-phase series motor has therefore two major faults, low power factor and poor commutation at low speeds, confining its use to fractional horsepower and very high speed applications.

***91. Single-Phase Compensated Series Motor.*** In all except the smallest sizes, it is usual to employ a compensating winding on the stator, in series with the armature and so arranged that its mmf as nearly as possible counteracts the armature mmf. A commutating winding is also frequently used, which somewhat overcompensates the armature reaction along the interpolar, or commutating-zone, axis and so provides a voltage to aid the current reversal, just as in a dc motor. By these means, the flux along the brush axis is reduced to a small fraction of its uncompensated value, and the power factor of the motor is greatly improved. Further improvement of the power factor is secured by using a smaller air gap and correspondingly fewer field ampere-turns than in an uncompensated motor, thus reducing the reactive voltage in the series field to a minimum.

***92. Universal Motors.*** Small series motors up to about ½-hp rating are commonly designed to operate on either direct current or alternating current and so are called *universal motors.* Universal motors may be either compensated or uncompensated, the latter type being used for the higher speeds and smaller ratings only. Owing to the reactance voltage drop, which is present on alternating current but absent on direct current, the motor speed is somewhat lower for the same load ac operation, especially at high loads. On alternating current, however, the increased saturation of the field magnetic circuit at the crest of the sine wave of current may materially reduce the flux below the dc value, and this tends to raise the ac speed. It is possible, therefore, to design small universal motors to have approximately the same speed-torque performance over the operating range, for all frequencies from 0 to 60 Hz. On a typical compensated-type ¼-hp motor, rated at 3400 r/min, the 60-Hz speed may be within 2% of the dc speed at full-load torque but 15% or more lower at twice normal torque, while on an uncompensated motor the speed drop will be materially greater.

The commutation on alternating current is much poorer than on direct current, owing to the current induced in the short-circuited armature coils, and this provides a definite limitation on their size and usefulness. If wide brushes are used, the short-circuit currents are excessive and the motor-starting torque is reduced, while if narrow brushes are used, there may be excessive brush chatter at high speeds, causing short brush life. Good design, therefore, requires careful proportioning of commutator and brush rigging to meet conflicting electrical, mechanical, and thermal requirements. Universal motors are generally used for vacuum cleaners, portable tools, food mixers, and similar small devices operating at maximum speeds of 3000 to 10,000 r/min.

The speed of the universal motor is controlled by means of a half-wave thyristor, or full-wave triac, as shown in Fig. 20-55.[1] The control device governs the half-wave average volt-

---

[1]A. Kusko; *Solid-State DC Motor Drives;* Cambridge, Mass., MIT Press, 1969.

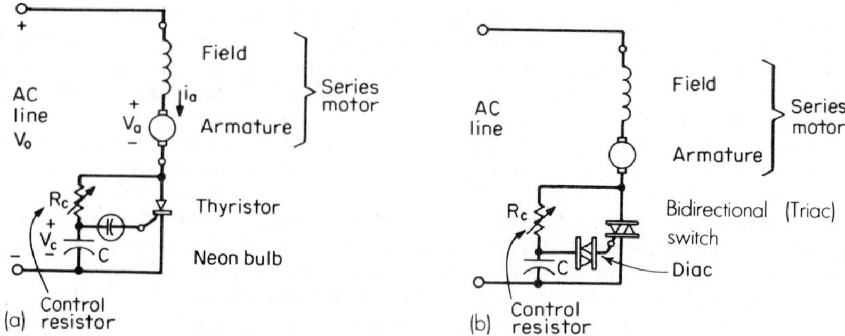


**FIG. 20-55** (*a*) Half-wave series universal motor circuit; (*b*) full-wave series universal motor circuit.

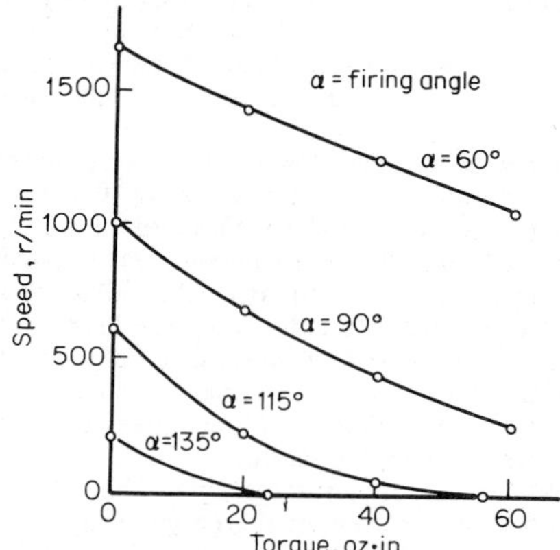


**FIG. 20-56** Measured speed-load torque characteristics of series motor and half-wave thyristor control.

age applied to the motor as a function of the firing angle. The firing circuits are usually relatively simple. The speed is controlled by changing a resistance value, such as $R_c$, in Fig. 20-55. The characteristics of a universal (series) motor with half-wave control are shown in Fig. 20-56.

***93. Single-Phase Railway Motor.*** The series-motor characteristic with its very high starting torque and high light-load speed is ideal for traction purposes, so that single-phase series motors were used in the majority of ac railway electrifications. The outstanding design problem in this case is to provide adequate commutating ability and thermal capacity to withstand the high circulating currents induced in the short-circuited armature coils at starting. The voltage induced in the short-circuited turn is proportional to the applied frequency. A frequency of 25 Hz, or commonly 16⅔ Hz in Europe, has been adopted for railway electrifications, because of the serious limitations on higher-frequency motors imposed by this relationship. However, the present trend in ac railway electrification is to use utility power at 60 Hz in the United States and 50 Hz in Europe. Rectifiers on the locomotive or self-propelled car convert the power to dc form for application to dc series or independent-field motors. The rectifiers are used to control motor voltage and speed by phase control. More recently, the dc power has been inverted to adjustable frequency ac power and applied to ac induction motors specially designed for traction duty, eliminating commutators altogether.

***94. Repulsion Motor.*** If, instead of connecting the field and armature in series, the supply voltage is connected directly across the field and the armature brushes are short-circuited, the simple repulsion motor is obtained, as illustrated in Fig. 20-57. Actually, the field and compensating functions are performed by a single stator winding, and the brush axis is placed at an angle $\alpha$ with respect to the axis of this winding.

If the brushes are shifted into line with the stator winding axis, the motor becomes a short-circuited transformer, drawing an excessive current and producing zero torque. As the brushes are shifted away from this position, called the "live neutral," a flux is produced in the cross axis and a torque is developed in the same direction as the shift. Further shift of the brushes results in a gradually decreasing current and a torque that passes through a maximum at an angle of 15 to 25°. Finally, at the false neutral, the induced armature current is zero, and the torque is also zero.

As the motor speeds up, at any brush angle, the speed voltage due to the armature conductors cutting the field flux opposes the transformer voltage due to the compensating flux and the induced armature current also decreases, allowing a gradual increase in the flux along the brush axis. At synchronous speed, the fluxes in the two axes are equal, giving a pure rotating magnetic field.

The repulsion motor has better commutation than the series motor at speeds up to synchronism and poorer commutation at very high speeds, owing to the greater values of currents induced in the short-circuited coils under the latter conditions. These induced currents provide a braking effect which serves to limit the high no-load speed.

The repulsion motor has two important advantages over the series motor for small-power industrial applications. The armature, being isolated from the line, may be designed for any convenient low voltage, to ease the problem of commutation, and no brush-yoke insulation is needed. Also, the pure rotating magnetic field produced at synchronous speed gives exact neutralization of the voltage induced in the short-circuited coils and, therefore, gives good commutation at this speed. Figure 20-58 illustrates the performance of the repulsion motor with brush shifting.

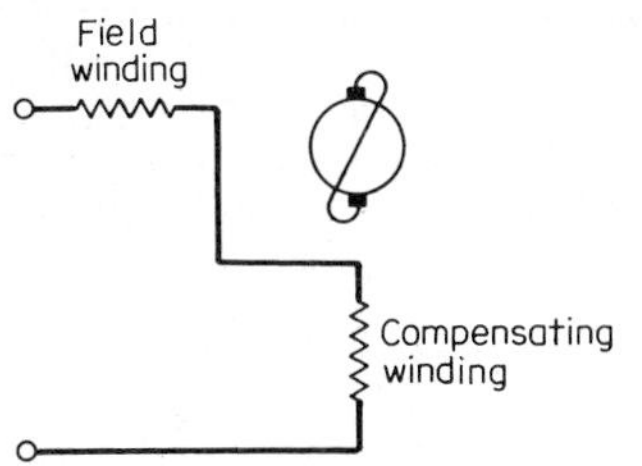


**FIG. 20-57** Repulsion-motor circuit diagram.

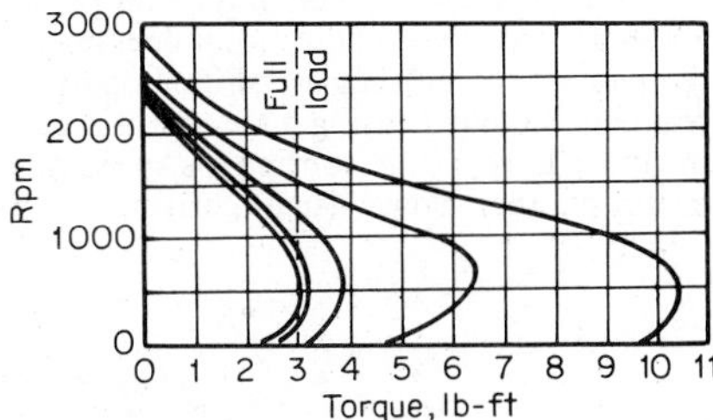


**FIG. 20-58** Speed-torque curves of repulsion motor with brush shifting.

***95. Repulsion-Start Induction Motor.*** The desirable combination of the high starting torque and low starting current of the repulsion motor with the constant-speed characteristic of the induction motor has led to numerous designs which automatically shift from the repulsion to the induction characteristic at about two-thirds of synchronous speed. The most usual type has some form of centrifugal mechanism which short-circuits the commutator bars at a predetermined speed. At all higher speeds, therefore, the armature winding effectively becomes a simple squirrel cage, and the motor has normal single-phase induction-motor characteristics. Many such motors combine a brush-lifting mechanism with the short-circuiting device, so that the brush noise and wear are absent during normal load operation. The usual range of sizes is ¼ to 5 hp at 1800 r/min, with starting torques of 300 to 600% of full-load value.

The capacitor-start single-phase induction motor using a centrifugal switch has replaced the repulsion-start induction motor for most applications.

***96. Brush-shifting Series Motors.*** Speed control of a series motor may be obtained by shifting the brushes, as in the repulsion motor, instead of varying the impressed voltage. Because of the need for low armature voltage, especially at industrial frequencies, to ensure commutation, it is usual to supply the rotor through a transformer in series with the stator, as shown in Fig. 20-59. Motors may be either single-phase or polyphase, the latter having the advantages of improved commutation and increased capacity. When the brushes are on live neutral, with the stator and rotor currents in exact opposition, the motor draws maximum current and produces zero torque. Shifting the brushes in either direction produces a torque in the same direction, the motor impedance rising with the angle of shift, until, at the dead neutral, 180°, the stator and rotor mmfs are additive, and maximum flux is produced with again zero torque.

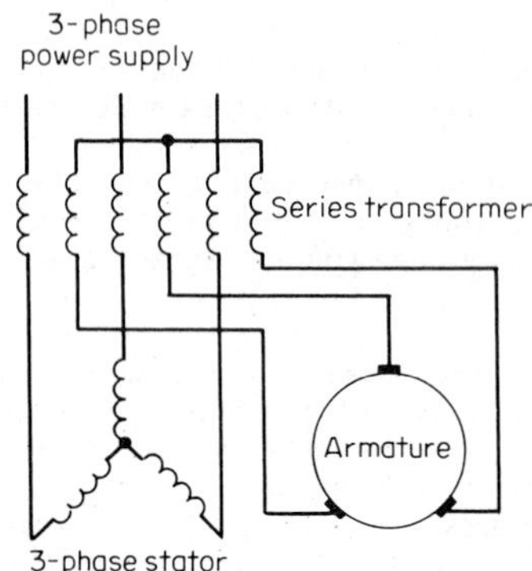


**FIG. 20-59** Brush-shifting series-motor circuit diagram.

The saturation limit of the transformer is conveniently used to hold the no-load speed down to a suitable value, usually about 150% of synchronous speed. This feature gives this type of motor a somewhat better speed regulation than other forms of series motors.

***97. Brush-shifting Polyphase Shunt Motor.*** The so-called Schräge motor gives an adjustable constant-speed characteristic with good torque and efficiency. This motor consists of a polyphase induction motor with the primary winding on the rotor and insulated secondary winding on the stator, the latter being connected across two independently movable sets of polyphase brushes on a commutator. The commutator is fed by an auxiliary adjusting winding, located in the outer parts of the rotor slots, as shown in Fig. 20-60. When the brushes connected to opposite ends of one phase of the stator winding are on the same commutator segment, they are short-circuited and the motor operates as a normal polyphase wound-rotor motor. When the two sets of brushes are shifted apart, a slip-frequency voltage is applied to the stator winding, thereby causing a change in speed. One direction of brush shift raises the speed; the other lowers it. A continuous speed range from full speed to standstill may be secured in this manner. Power-factor correction is also provided, by unsymmetrical spacing of the

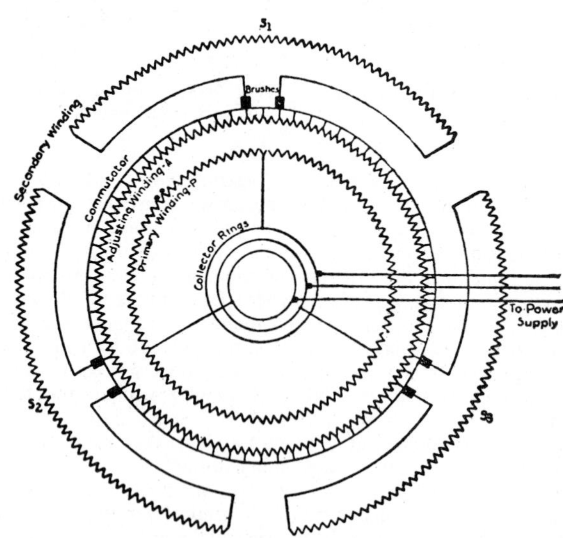


**FIG. 20-60** Brush-shifting polyphase shunt-motor circuit diagram.

brush positions, introducing a secondary voltage component at right angles to the speed voltage.

To minimize the difficulties of commutation, these motors have large numbers of commutator bars and brushes, which together with the special windings result in relatively high costs. However, each motor is complete in itself, without external transformer or regulator, and they are suitable for direct connection to usual industrial voltages of 550 V and below.

They are employed for a wide range of industrial services, driving fans, pumps, conveyors, packaging machinery, paper mills, etc. Sizes up to several hundred horsepower have been built. They are normally rated on a constant-torque basis, the horsepower output varying in direct proportion to the speed.

The no-load speeds of these machines range from 7 to 10% above their full-load speeds with the brushes in the maximum-speed position and from 20 to 40% above the full-load speeds with the brushes in the minimum-speed position. Their temperature rise is 40°C when operating from maximum to 50% of maximum speed and 50°C when operating at lower speeds. Maximum running torques range from 140 to 250% depending on the rating and the brush position.

## FRACTIONAL-HORSEPOWER-MOTOR APPLICATIONS

***98. Scope.*** A fractional-horsepower motor is defined by NEMA as either (1) a motor built in a frame with a NEMA two-digit frame number, or (2) a motor built in a frame smaller than the NEMA frame for a 1-hp, open-construction, 1700- to 1800-r/min, induction motor. The two-digit frame number is defined as 16D, where D is the height of the shaft centerline above the bottom of the mounting base. Fractional-horsepower motors include 1-phase and 3-phase induction and synchronous motors, 1-phase universal motors, and dc motors. Ratings, with minor exceptions, are $\frac{1}{20}$ to 1 hp, inclusive. Motors of smaller ratings are classified as subfractional or miniature.

***99. Purpose.*** General-purpose motors are of open construction, rated a 60°C temperature rise by resistance over a 40°C ambient temperature. They are designed according to standard ratings with standard operating characteristics and mechanical construction for rise under usual service conditions without restriction as to a particular application or type of application. A definite-purpose motor is any motor designed according to standard ratings with standard operating characteristics or mechanical construction for use under service conditions other than the usual or for on a particular type of application (see Par. **108**).

***100. Selection of Type.*** The principal characteristics of fractional-horsepower motors are shown in Table 20-12. For any application, the motor selected should meet the application and power supply requirements at the least cost. Numerous trade-offs are possible, for example, speed changing in finite steps compared to continuously adjustable-speed solid-state drives.

***101. Ratings.*** Standard voltage and frequency ratings are listed in Tables 20-13*a* and 20-13*b*.

***102. Service Conditions.*** General-purpose motors are designed to operate under the usual service conditions of 0 to 40°C ambient temperature, altitude to 3300 ft (1000 M), and installation on rigid mounting surfaces where there is no interference with the ventilation. Some general-purpose, definite-purpose, and special-purpose motors can operate under one or more unusual service conditions, which include exposure to dust, lint, fumes, radiation, steam, fungus, shock; operation where voltage, frequency, waveform, and form factor deviate from standards; and overspeed, overtemperature, and excess altitude operation. The manufacturer should be consulted for operation under unusual service conditions.

***103. Thermal Protection.*** Many single-phase motors are now available with a built-in thermal protector which affords complete protection from burnout due to any type of overload, even a stalled rotor. Most such devices are automatic-resetting, but some are manual-resetting. Motors so protected usually are marked externally in some way to indicate the fact.

***104. Reversibility.*** In general, standard motors of the types listed in the table can be arranged by the user to start from rest in either direction of rotation. There are exceptions, however. Shaded-pole motors, unless of a special design, can be operated in only one direc-

## TABLE 20-12 Characteristics of Fractional-Horsepower Motors

| | Alternating | | | | |
|---|---|---|---|---|---|
| | Single-phase | | | | |
| | Split-phase types | | | | |
| | General-purpose | High-torque | Two-speed, pole changing | Capacitor-start | Capacitor (1-value, or perm, split) |
| Schematic diagram of connections<br>Arrangements shown are typical or representative; most of the types illustrated have numerous other arrangements which are also used. | | | | | |
| Characteristic speed-torque curves<br>Ordinates are speed; 1 division = for all ac motors, 20% of syn. r/min; for universal motors, 1000 r/min; for dc motors, 20% of full-load rpm. Each abscissa division = 100% of full-load torque. | | | High<br>Low | | Fan Curve |
| Rotor construction | Squirrel-cage | Squirrel-cage | Squirrel-cage | Squirrel-cage | Squirrel-cage |
| Built-in automatic starting mechanism | Centrifugal switch | Centrifugal switch | Centrifugal switch | Centrifugal switch | None required |
| Horsepower ratings commonly available | 1/20–1/3 | 1/8–1/3 | 1/8–3/4 | 1/8–3/4 | 1/20–3/4 |
| Usual rated full-load speeds (for 60-Hz ac motors; also dc motors) | 3450, 1725, 1140, 865 | 1725 | 1725/1140<br>1725/865 | 3450, 1725, 1140, 865 | 1620, 1080, 820 |
| Speed classification | Constant | Constant | Two-speed | Constant | Constant, or adjustable varying |
| Means used for speed control | | | Two-speed switch | | 2-speed switch or auto-transformer |
| Comparative torques: Locked-rotor | Moderate | High | Moderate | Very high | Low |
| Comparative torques: Breakdown | Moderate | High | Moderate | High | Moderate |
| Radio interference, running | None | None | None | None | None |
| During acceleration | One click | One click | Two clicks | One click | None |
| Approximate comparative costs between type, for same horsepower rating: Below 1/20 hp | | | | | |
| 1/20–1/4 hp | 100 | 75 | 210 | 125 | 140 |
| 1/2–3/4 hp | 80 | 54 | 150 | 100 | 100–110 |
| General remarks | For constant-speed operation, even under varying load conditions, where moderate torques are desirable or mandatory, this type is often used in preference to the more costly capacitor-start motor. Meets NEMA starting currents. Typical applications: blowers; centrifugal pumps; duplicating machines; refrigerators; oil burners; unit heaters. | High locked-rotor currents (in excess of NEMA) limit the use of this type on lighting circuits to applications where the motor starts only very infrequently, because of a tendency to cause flickering of the lights. Principal applications: washing and ironing machines; cellar-drainer pumps; tools for a home workshop. | Used where two definite speeds independent of load are required. Ratings above 1/4 hp usually made capacitor-start. Motor shown always starts on high-speed connection; transfer to low speed made by starting switch. Common applications: belted blowers for warm-air furnaces or for other purposes; attic ventilators; air conditioning apparatus. | A general-purpose motor suitable for most applications requiring constant speed under varying loads, high starting and running torques, high overload capacity. Also available as two-speed pole-changing motor above 1/4 hp. A few important applications are: refrigeration and air conditioning compressors; air compressors; stokers; gasoline pumps. | Primarily used for unit heaters, or for other shaft-mounted fans. Essentially a constant-speed motor, but by means of a two-speed switch, or by means of an autotransformer, other speeds can be obtained, *with fan loads*, of horsepower rating selected closely matches the fan load. Can also be made in intermittent ratings for plug-reversing service. |

General remarks

Standard motors are ordinarily designed to operate in ambient temperatures from 10 to 40°C. Variations in line voltage of plus or minus 10%, or variations in frequency of plus or minus 5% are allowable.

Locked-rotor currents for single-phase motors, except split-phase high-torque and synchronous types, usually do not exceed the following limits established by NEMA:

| Rating, hp | Amperes at 115 volts | Amperes at 230 volts |
|---|---|---|
| 1/6 and smaller | 20 | 10 |
| 1/4 | 23 | 11½ |
| 1/3 | 31 | 15½ |
| 1/2 | 45 | 22½ |
| 3/4 | 61 | 30½ |

Fractional horsepower motors are built for across-the-line starting.

The standard direction of rotation is counterclockwise facing the end opposit the shaft extension.

| Current | | | | Dc or ac (60 Hz or less), universal types | | Direct current | |
|---|---|---|---|---|---|---|---|
| Motors | | 1, 2, or 3 phase | Polyphase | | | | |
| Repulsion-start | Shaded-pole | Nonexcited synchronous (reluctance) | Squirrel-cage induction | Without governor | With governor | Shunt or compound | Series |
| 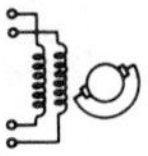 |  | Stator winding may be: split-phase, capacitor-start, capacitor, polyphase | 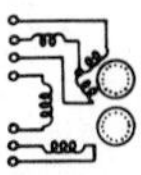 |  | 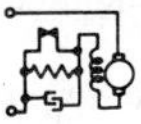 |  |  |
|  |  |  |  | 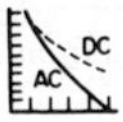 AC DC |  | 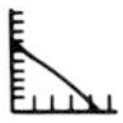 |  |
| Drum-wound; commutator | Squirrel-cage | Cage, with cutouts | Squirrel-cage | Drum-wound; commutator | Drum-wound; commutator | Drum-wound; commutator | Drum-wound; commutator |
| Short-circuit | None | Depends on stator winding | None | None | None | None | None |
| 1/8–3/4 | 1/2000–1/6 | 1/3000–1/3 | 1/8–3/4 | 1/150–1 | 1/50–1/20 | 1/20–3/4 | 1/25–1/30 |
| 3450, 1725, 1140, 865 | 1450–3000 | 3600,1800, 1200, 900 | 3450, 1725, 1140, 805 | 3000–11,000 | 2000–4000 | 3450, 1725, 1140, 865 | 900–2000 |
| Constant | Constant, or adjustable varying | Absolutely constant | Constant | Varying, or adjustable varying | Adjustable | Constant, or adjustable varying | Varying, or adjustable varying |
| | Choke or resistor | | | Choke or resistor | Adjustable governor | Armature resistance | Resistor |
| Very high<br>High | Low<br>Low | Low<br>Moderate | Very high<br>Very high | Very high | Very high | Very high | Very high |
| None<br>Continuous | None<br>None | None | None<br>None | Continuous<br>Continuous | Continuous<br>Continuous | Continuous<br>Continuous | Continuous<br>Continuous |
| 128<br>100 | 100 | 200–200<br>275 | 165–195<br>100 | 75<br>105–175 | 110<br>140–160 | 175–225<br>120–140 | 185 |
| A constant-speed motor suited to general-purpose applications requiring high starting torque, such as pumps and compressors. An associated type, the repulsion induction (buried cage) is used for door openers and other plug-reversing applications. Has been displaced for many applications by the capacitor-start motor. | For ratings below 1/20 hp, this is a general-purpose motor. For fan applications, speed control is affected by use of a series choke or resistor. Applications: fans, unit heaters, humidifiers, hair driers, damper controllers. | Cutouts in rotor result in synchronous-speed characteristics. Curve shown is for split-phase stator. Pull-in ability is affected by inertia of connected load. Used for teleprinters, facsimile-picture transmitters, graphic instruments, etc. Clocks and timing devices usually use shaded-pole hysteresis motors rated at a few millionths of a horsepower. | Companion motor to capacitor-start motor with comparable torques and generally suited to same applications if polyphase power is available. Inherently plug-reversible and suitable for door openers, hoists, etc. High-frequency motors used for high-speed applications, as for woodworking machinery, rayon spinning, and portable tools. | Light weight for a given output, high speeds, varying-speed and universal characteristics make this type very popular for hand tools of all kinds, vacuum cleaners, etc. Ratings above 1/4 hp usually compensated. Some speed control can be effected by a resistor or by use of a tapped field. Used with reduction gear for slower speed applications. | By means of a centrifugal governor, a constant-speed motor having the advantages of the universal motor is obtained. Governor may be single-speed or adjustable even while running. Speed is independent of applied voltage. Used in typewriters, calculating machines, food mixers, motion-picture cameras and projectors, etc. | A constant-speed companion motor for the capacitor-start or split-phase motor for use where only d-c power is available. For unit-heater service, armature resistance is used to obtain speed control. Not usually designed for field control. | Principally used as the dc companion motor to the shaded-pole motor for fan applications. Used in these small ratings in place of shunt motors to avoid using extremely small wire. |

**TABLE 20-13a** Voltage Ratings of AC Fractional-Horsepower Motors

| Motors | Frequency, Hz | Voltage |
|---|---|---|
| Single-phase | 60 | 115, 230 |
| | 50 | 110, 220 |
| Three-phase | 60 | 115, 200 |
| | | 230, (460) |
| | 50 | 220, 380 |
| Universal | 60* | 115, 230 |

*Can operate from dc to 60 Hz.

**TABLE 20-13b** Voltage Ratings of DC Fractional-Horsepower Motors

| Primary power source | Rating, hp | Armature voltage | Field voltage |
|---|---|---|---|
| Los-ripple dc | 1/20 to 1 | 115, 230 | 115, 230 |
| 1-phase rectifier | 1/20 to 1/2 | 75 | 50, 100 |
| | | 90 | 50, 100 |
| | | 150 | 100 |
| | 3/4 to 1 | 90 | 50, 100 |
| | | 180 | 100, 200 |
| 3-phase rectifier | 1/4 to 1 | 240 | 100, 150 |
| | | | 240 |

tion of rotation. Small dc and universal motors often have the brushes set off neutral, preventing satisfactory operation in the reverse direction. Single-phase motors which use a starting switch ordinarily cannot be reversed while running at normal operating speeds, because the starting winding, which determines the direction of rotation, is then open-circuited. By use of special relays this limitation of split-phase and capacitor-start motors can be overcome when necessary. Such motors are built for small hoists. High-torque intermittent-duty permanent-split capacitor motors; repulsion-induction (buried-cage) motors; and split-series dc or universal motors are often built for plug-reversing service. Standard polyphase induction motors can be reversed while running, as can the smaller ratings of dc motors; such applications should preferably be taken up with the motor manufacturer.

***105. Mechanical Features.*** Rigid and rubber-mounted motors are commonly available. Sleeve and ball bearings are both standard. Sleeve-bearing motors are designed for operation with the shaft horizontal, but ball-bearing motors can be operated with the shaft in any position. For operation with the shaft vertical, sleeve-bearing motors may require a special design. Rubber mounting is widely used for quiet operation, because all single-phase motors have an inherent double-frequency torque pulsation. An effective and common arrangement uses rubber rings concentric with the shaft and so arranged as to provide appreciable freedom of torsional movement but little other freedom. Sometimes the driven member picks up the double-frequency torque pulsation and amplifies it to an objectionable noise, for example, a fan with large blades mounted rigidly on the shaft. The cure for this difficulty is an elastic coupling between the shaft and the driven member; no amount of elastic suspension of the stator can help. Standard motors are generally open and of drip-proof construction. Splashproof and totally enclosed motors are easily available.

***106. Inputs of Small Single-Phase Motors.*** See Table 20-14. Full-load torque, in terms of horsepower and rated speed, is

$$\text{Full-load torque, oz}\cdot\text{ft} = \frac{84{,}000 \times \text{hp}}{\text{r/min}} \tag{20-33}$$

**TABLE 20-14** Approximate Starting and Full-Load Current for Single-Phase 115-V Motors

| Rating, hp | Max. locked-rotor current, A | | 3450 r/min | | 1725 r/min | | 1140 r/min | | 865 r/min | |
|---|---|---|---|---|---|---|---|---|---|---|
| | Des. O | Des. N | A | W | A | W | A | W | A | W |
| ⅛ | 50 | 20 | 2.9 | 207 | 2.7 | 176 | 3.9 | 207 | 5.4 | 245 |
| ⅙ | 50 | 20 | 3.2 | 254 | 3.0 | 214 | 4.3 | 254 | 6.0 | 296 |
| ¼ | 50 | 26 | 4.2 | 352 | 3.9 | 301 | 5.6 | 352 | 8.1 | 414 |
| ⅓ | 50 | 31 | 5.3 | 460 | 4.9 | 395 | 7.0 | 460 | 9.8 | 540 |
| ½ | 50 | 45 | 7.4 | 678 | 6.9 | 574 | 9.8 | 678 | — | — |
| ¾ | — | 61 | 10.6 | 981 | 9.9 | 835 | — | — | — | — |
| 1 | — | 80 | 13.3 | 1260 | — | — | — | — | — | — |

NOTE: A = amperes; W = watts; Des. = Designation.

***107. Application Tests.*** The primary object of any application test is to determine the power requirements of the appliance or device under various significant operating conditions. A convenient way of doing this is to use a motor of approximately the right horsepower rating and of predetermined efficiency at various outputs. Watts input are carefully measured under each condition. From the watts input observed (never use current as a measure of load except for dc motors) and the known efficiency, the load is readily determined. Care should be taken in measuring the watts input to correct for the meter losses.

A second, and equally important, object of the test is to determine the actual locked-rotor and pull-up torques required by the appliance. The locked-rotor and pull-up torques of the test motor should be known or measured at rated voltage and frequency. (Locked-rotor torque often varies with slight changes in rotor position.) Using a transformer or induction regulator to obtain a variable voltage (do not use a resistance or choke for this purpose), measure the minimum voltage at which the motor will start the appliance and also the minimum voltage at which it will pull it up through switch-operating speed. Assuming that the pull-up and locked-rotor torques each vary as the square of the applied voltage, it is then a simple matter to determine the actual locked-rotor and pull-up torques required by the device. After a motor has been selected, it should be determined whether or not it can operate the device at 10% above and below normal rated voltage of the motor or over a wider range of voltage, if desired. If exceptional load conditions may occasionally be encountered, use of a motor equipped with inherent-overheating protection is often desirable.

***108. Definite-Purpose Motors.*** For a number of important applications, involving large quantities of motors, NEMA has developed standards to meet these special requirements effectively and economically. Motors built to these standards are usually more readily obtainable and economical than special motors tailored to one application. Highlights and distinguishing features are given in Table 20-15. More details can be obtained in NEMA Standards.

***109. Small Synchronous Motors.*** Small synchronous motors in the 1.5- to 25-W range for timing, tape drives, small fans, and record players are available as brushless dc motors or as hysterisis motors. The brushless dc motor consists of a permanent-magnet field, 2-phase, synchronous motor driven by transistors from a dc source. The transistors are switched from a Hall-device signal which senses the rotor position. A regulator maintains constant speed.

*Shaded-pole hysteresis motors,* which operate at synchronous speed, are essentially the same as shaded-pole induction motors except that they use rotors of hardened-steel rings of a material having high hysteresis loss. Large quantities of such motors are built for clocks and timing devices. Clock motors have an input of 1.5 to 2 W and an output of a few millionths of a horsepower. Large motors with inputs up to 15 W are built for heavier duty applications. Rotor speeds are commonly 450, 600, and 3600 r/min. Most of these motors are furnished with built-in reduction gears to give output speeds of 60 r/min to 1 r/month.

*Reluctance motors,* both self-starting and manual-starting types, are available for similar

**TABLE 20-15** NEMA Standards for Definite-Purpose Motors

| Application | Principal types | Distinguishing features |
|---|---|---|
| Universal motor | Universal: Salient-pole and distributed field | Dimensional standards; common practices utilizing parts |
| Hermetic motors | Split-phase, capacitor-start, polyphase | Parts only for hermetic refrigeration condensing units |
| Belt-drive refrigeration compressors | Capacitor-start. Repulsion-start, polyphase | Open; sleeve bearings, extended rear oiler; automatic-reset thermal overload protection |
| Jet-pump motors | Split-phase, capacitor-start, repulsion-start polyphase | 3450 r/min; ball bearings; open; machined back end shield; automatic-reset overload protection |
| Motors for shaft-mounted fans and blowers | Split-phase, permanent-split capacitor, polyphase | Enclosed; horizontal; sleeve bearings; vertical, ball bearings; extended through bolts; capacitors on front end shield |
| Shaded-pole motors for shaft-mounted fans and blowers | Shaded-pole; two-speed, three-speed | Open or totally enclosed; sleeve bearings; high slips |
| Belted fans and blowers | Split-phase, capacitor-start, repulsion-start; two-speed split-phase and capacitor-start | Open; sleeve bearings; resilient mounting; automatic-reset overload protection; extended rear oiler |
| Stoker motors | Capacitor-start; repulsion start; polyphase | Totally enclosed recommended; automatic reset overload protection |
| Motors for cellar drainers and sump pumps | Split-phase | Vertical, dripproof, 50°C; two ball bearings, or one ball, one sleeve; mounts on support pipe; built-in float-operated line switch; overload protection |
| Gasoline-dispensing pumps | Capacitor-start, repulsion-start polyphase | Explosionproof; sleeve bearing; built-in line switch and capacitor; voltage-selector switch on single-phase |
| Oil-burner motors | Split-phase | Enclosed, face-mounted, round-frame; manual-reset overload protection; two line leads |
| Motors for home-laundry equipment | Split-phase | Low-cost, high starting current; open, 50°C; round-frame with ungrounded mounting rings; shaft extension with flat and hole for coupling |
| Motors for coolant pumps | Split-phase, capacitor-start, repulsion-start, polyphase | 3450- and 1725-r/min; totally enclosed; ball bearings; machined back end shield |
| Submersible motors for deep-well pumps | Split-phase, capacitor, polyphase | 3450 r/min; designed for operation totally submerged in water not over 25°C (77°F); use external relay for starting |

applications. Another type used is the synchronous-inductor motor, which is essentially an inductor alternator used as a motor; field excitation is furnished by a permanent magnet.

### Bibliography

***110. References on Fractional-Horsepower-Motor Applications***

Veinott, Cyril G.; *Fractional Horsepower Electric Motors,* 2d ed.; New York, McGraw-Hill Book Company, 1948.

Standards for Fractional Horsepower Motors; *NEMA Publ.* MG2, 1951, New York, National Electrical Manufacturers Association.

## MOTOR CONTROL

***111. Industrial motor control*** includes (1) motor-starting devices, (2) speed-control devices, (3) stopping devices, and (4) motor-protecting devices.

*Industrial motor control* is designed and built in accordance with rules and standards established by several organizations. Detailed design-construction and test information is contained in such publications as the National Electrical Code, National Board of Fire Underwriters; Standards for Industrial Control Apparatus, Institute of Electrical and Electronics Engineers; Industrial Control Standards, National Electrical Manufacturers Association; Standard for Industrial Control Equipment, Underwriters Laboratories; and Standard Rotation, Connections and Terminal Markings for Electric Power Apparatus, American National Standards Institute.

***112. The essential functions of motor control*** are the starting, speed regulating, stopping, and protecting of electric motors.

## MOTOR-STARTING DEVICES

***113. A contactor*** is a device, generally magnetically actuated, for repeatedly establishing and interrupting an electric power circuit.

Figure 20-61 illustrates a single-pole dc contactor. Contactors of this type are rated on a continuous-current-carrying-capacity basis and on an intermittent-duty basis, at values depending on the duty cycle. The NEMA Standard 8-h open ratings range from 25 to 2500 A. The intermittent ratings are 133⅓% of the open ratings. The shunt operating coil is designed to withstand 110% of rated voltage continuously and to close the contactor successfully at 80% of rated voltage.

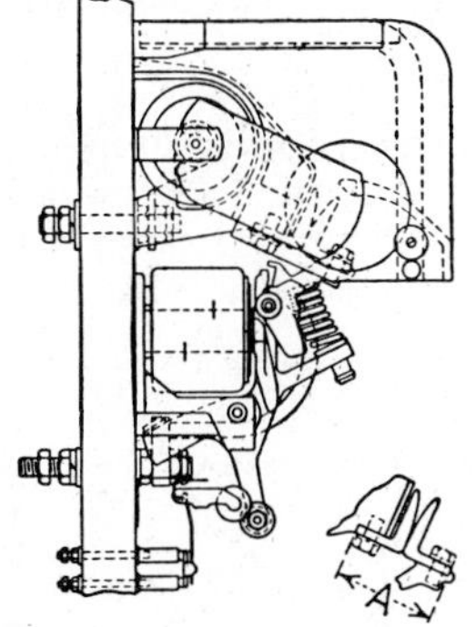

**FIG. 20-61** Direct-current contactor.

***114. The magnetic blowout*** consists of a coil wound on a steel core and mounted between steel pole pieces. The pole pieces are lined with refractory material. The assembly is enclosed in an insulated box, which is swung down over the contacts. The blowout coil is generally connected in series with the contactor and carries motor current with the contactor closed. The current sets up a magnetic field through the core and pole pieces of the blowout structure and across the contact tips. When an arc is formed, the magnetic field of the arc and the magnetic field of the blowout repel each other and the arc is forced upward and away from the contacts. The extinguishing action, due to the lengthening of the arc and the cooling of the refractory material, is extremely rapid and thereby greatly reduces the wear and burning of the contacts.

***115. Performance of Contactors.*** To obtain trouble-free service and maximum contact life, the following items should be in accordance with the manufacturer's specifications: initial and final contact pressures, magnetic gap, arc gap, and wear allowance. The contact pressures can be measured by means of a spring balance, initial pressure with the contactor open, and final pressure with the contactor closed. The magnetic gap is the distance from the center line of the core to a corresponding point on the armature lever, and the arc gap is the distance between the arcing tips. Contacts can be kept smooth by filing with a fine file. Contact-wear allowance is the total thickness of material that may be worn away before the contact between the two surfaces becomes ineffective. The contacts should be renewed when worn so that the distance A (Fig. 20-61) between the back edges of the contacts, with the contactor closed, becomes less than the specified amount. This usually corresponds to the condition where the contacts are worn to approximately half their original thickness.

Current-carrying contacts are usually made from copper, either left plain or plated with silver or cadmium. The contacts should not be lubricated. Hinge pins and bearings should be lubricated with a light machine oil. The surface of the core and the armature, which seal when the contactor is closed, should be kept clean.

***116. An ac contactor*** is similar in construction to a dc contactor, except that laminated iron structures are used. A shading coil is used at the core face to ensure continuous force of the armature and thus obtain quiet operation. AC contactors are available with two, three, or four main poles for interrupting all line circuits to single-phase, 3-phase, or 2-phase 4-wire motors (see Fig. 20-62 and Table 20-16). Standard ac contactors are designed to interrupt 10 times rated motor current, based on the contactor horsepower rating. The contactors are also designed to thermally withstand 15 times rated motor current for 1 s, to permit

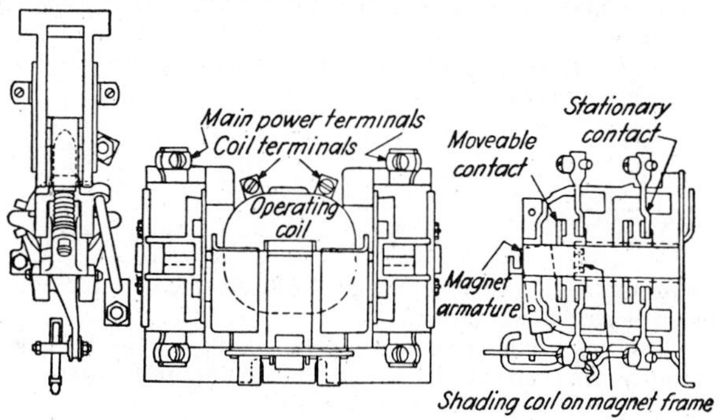


**FIG. 20-62** Alternating-current contactor.

**TABLE 20-16** Typical Ratings of AC Contactors Used as Across-the-Line Magnetic Starters with 3-Phase Motors

| Contactor size | Rating, A | Horsepower at 110 V | Horsepower at 220 V | Horsepower at 440–550 V |
|---|---|---|---|---|
| 00 | 9 | ¾ | 1½ | 2 |
| 0 | 18 | 2 | 3 | 5 |
| 1 | 27 | 3 | 7½ | 10 |
| 2 | 45 | | 15 | 25 |
| 3 | 90 | | 30 | 50 |
| 4 | 135 | | 50 | 100 |
| 5 | 270 | | 100 | 200 |
| 6 | 540 | | 200 | 400 |
| 7 | 810 | | 300 | 600 |
| 8 | 1215 | | 450 | 900 |
| 9 | 2250 | | 800 | 1600 |

protective devices such as circuit breakers and fuses to clear fault current carried by the contactor.

***117. A drum switch*** consists of stationary contact fingers held by spring pressure against contact segments on the periphery of a rotating cylinder or sector. Drum controllers have many advantages over the faceplate and multiple-switch types. The mechanical construction is better, heavy contact pressures can be maintained, parts can be well insulated, blowout magnets and arc shields can be used, and the structure can easily be completely enclosed. Less space is required by the drum control, and it is easier to operate. Drum controllers are built in 8-h ratings for dc motors up to 40 hp, 115 V, and 75 hp, 230 V.

***118. The design of starting-duty resistors*** requires the determination of the total ohms, the distribution of this resistance between the steps available, and the calculation of the current-carrying capacity, and the selection of the resistance material. Standard resistors to meet various classes of service are designated by class numbers in accordance with the NEMA Table of Classification of Resistors. NEMA also publishes a resistor application table intended as a guide in specifying and designing resistors (see NEMA Industrial Control Standards). This table lists typical machines with the corresponding NEMA resistor-classification number. For example, a lathe should have a dc starter with resistor classification No. 115. This resistor has sufficient total ohms to limit the current inrush on starting to 150% of full-load current. It will be designed to have current-carrying capacity for an average accelerating current (rms value) of 125% full-load current, on the basis of starting once during each 80-s period and with an accelerating time of 5 s.

Resistors must be available in a wide range of ohmic values and current capacity. Resistor units are stacked in parallel and series combinations to achieve the required values. For low ohms and high capacity, cast-iron or punched-steel grids are used. These steel grids range from 0.01-Ω, 160-A, to 0.40-Ω, 20-A continuous rating per grid. For high ohms and low capacity, wire-wound resistors are used with a unit resistance range from 4.0 to 6400 Ω and dissipation up to 900 W. Intermediate resistance requirements are fulfilled with edge-wound ribbon resistors with a unit resistance range of 0.05 to 8.6 Ω and dissipation up to 1320 W.

Resistors must be sized and arranged in assemblies so that the temperature rise for bare resistive elements does not exceed 375°C above 40°C ambient. The maximum current should not exceed the 10-s rating. Care must be taken to limit excess voltage on wire-wound resistors to avoid surface flashover. Resistors are generally insulated for 600 V rms to ground. Manufacturers will provide assembled resistor units for specific functions or provide the derating curves for the arrangement of the resistor units into assemblies, as a function of duty cycle, ambient temperature, and grouping.

## AC Motor Starting

***119. Selection of an ac starter*** is a compromise between requirements and cost. The primary requirements of the starter, obviously, are that the motor starting torque shall be adequate to start the load under worst-case line voltage and load conditions; also, that the line current shall not exceed limits set by the utility or plant voltage dip. A useful table from Millermaster is shown in Table 20-17. The available starters are listed in descending value of starting torque, based on a 60-hp, 440-V, 60-Hz, 900-r/min motor. Compared to the starters with series-connected elements, the autotransformer starter provides a means for reducing the line current below the motor current. The current and torque in Table 20-17 are shown for 100% line voltage. For reduced line voltage, the current is reduced in proportion; the torque is reduced as the square of the voltage. Limit on voltage dip is 15 to 20%.

The secondary requirements in starter selection include smoothness of acceleration, maintenance, power factor, reliability, and efficiency. The selection of a closed-transition starter depends upon whether the motor and the supply line can withstand the peak current at the time the starter transfers the motor to full voltage.

***120. Alternating-current across-the-line starters*** are simple in construction, easy to install and maintain, and inexpensive. A typical starter consists of a three-pole contactor with a thermal overload relay for protecting the motor. The starter connects the motor directly to the line, impressing full voltage to the motor terminals. It is particularly suitable for squirrel-cage motors. Since these starters connect the motor directly to the supply lines,

**TABLE 20-17** Comparison of Methods for Starting 3-Phase Motors
(Example of 60-hp, 900-r/min, 60-Hz motor)

| Method of starting | Starting current drawn from the line as a percentage of full-load current | Starting torque as a percentage of full-load torque |
|---|---|---|
| Connecting motor directly to the line full potential | 470 | 160 |
| Autotransformer 80% tap | 335 | 105 |
| Resistor starter to give 80% applied voltage | 375 | 105 |
| Part winding | 235 | 70 |
| Autotransformer 65% tap | 225 | 67 |
| Resistor starter to give 65% applied voltage | 305 | 67 |
| Solid-state starter | 300 | 65 |
| Star-delta starter | 158 | 54 |
| Resistance starter to give 58% applied voltage | 273 | 54 |
| Autotransformer 50% tap | 140 | 43 |
| Resistor starter to give 50% applied voltage | 233 | 43 |

the motor will draw an inrush current of 6 to 10 times running current. In the majority of installations this is not objectionable and will not damage the motor or the driven machinery. When the starting inrush must be lower, some form of reduced-voltage starting must be used. The common types of starters are autotransformer, primary-resistance, part winding, Y-Δ, and solid-state.

***121. Autotransformer starters*** have two autotransformers connected in open Δ to provide reduced-voltage starting. Three taps are supplied, as shown in Fig. 20-63, giving 50, 65, and 80% of full line voltage. The motor current varies directly as the voltage impressed on the motor terminals. The line current varies as the square of the impressed voltage and is therefore lower than with resistor-type starters. The torque also varies as the square of the impressed voltage. The 50% voltage tap will therefore provide 25% starting torque. Connections should be made to the lowest tap that will give the required starting torque.

Characteristics of this type of starter are low line current, low power from the line, and a low power factor. Acceleration is not continuous, because the torque developed by the motor remains practically constant during the starting period, on the first step; then changes to another value on the second step. The starter shown in Fig. 20-63 is an open-transition type. The motor is disconnected from the line during the transfer period. A closed-transition

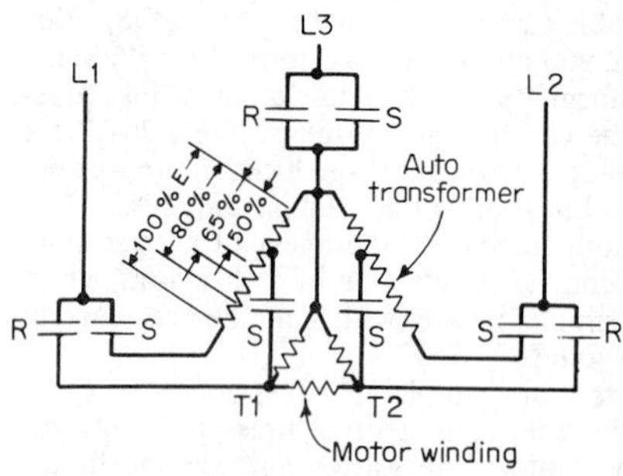


**FIG. 20-63** Connections for autotransformer starter.

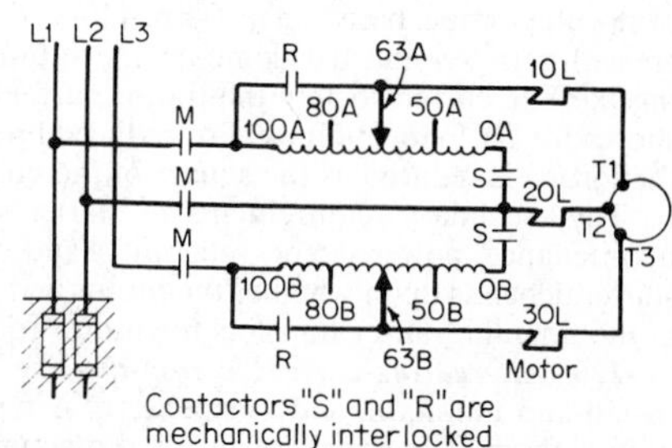


**FIG. 20-64** Connections for a closed-transition autotransformer motor starter.

starter which uses the Korndorfer connection is shown in Fig. 20-64. When the S contactor opens, the autotransformers act as series reactors until the R contactor closes.

Autotransformer starters are available in manual and automatic types. In the manual type the contacts are operated by means of a lever extending to one side of the enclosing case. The lever is equipped with a low-voltage release magnet. The automatic open-transition starter (Fig. 20-63) consists of a five-pole starting contactor S and a three-pole running contactor R. The closed-transition starter (Fig. 20-64) consists of a three-pole main contactor (M) and two-pole start and run contactors (S and R). When the "run" button is pressed, the starting contactor closes, connecting the transformer to the line and the motor to the reduced-voltage taps. A timing relay is operated by the starting contactor. The motor accelerates, and, after a specified number of seconds, the timer contacts close, deenergizing the starting contactor and energizing the running contactor. The transformer is disconnected and the motor connected to line voltage.

***122. Primary-resistor-type starters*** connect the motor to the line through a series resistor. Reduced voltage at the motor is obtained because of the voltage drop across the resistor. As the motor accelerates, the current drawn from the line declines, and consequently the voltage drop across the resistor is lowered, and the motor voltage at the motor terminals is increased. The torque delivered by the motor is therefore constantly increased as the motor speed increases. After a definite interval, a timing device operated by the main contactor energizes the accelerating contactor, which short-circuits the resistor. There is no transfer period during which the motor may lose speed, and therefore smooth acceleration is obtained. In comparison with the autotransformer-type starter, the primary-resistor type takes more power from the line on starting but provides smoother acceleration, faster acceleration with a given initial torque, and higher power factor. In the smaller sizes the primary-resistor starter costs less than the autotransformer starter.

***123. Part-Winding AC Starters.*** The motor winding must be in two parts, and at least six terminal leads must be provided on the motor. The method is therefore applicable to those motors which are designed for use on either of two voltages, the windings being in parallel on the lower voltage and in series on the higher voltage. For example, a 230/460-V motor could be used on 230 V with a part-winding controller. The controller would then be arranged to connect one section of the winding to the supply lines as soon as the starting button was pressed. Then, after a time delay provided by a timing relay, a second contactor would connect the other section of the motor winding to the supply lines, in parallel with the first section. In this way the starting current is reduced to approximatley one-half of what would be required if both winding sections were connected at the same time, as they would be with a standard 3-lead motor. The starting torque when the first winding section is connected will be less than half of the torque that would be obtained if both sections were connected at the same time. Contactors used for part-winding starters need capacity to handle only the circuit which they control, and so may be rated at one-half of the rating that would be required to handle the whole motor. Overload relays are provided for each section of the winding.

***124. Y-Δ ac starters*** are a form of reduced voltage starter used with six-lead motors in which 57% voltage is applied to the windings on the first step, full voltage on the second step. The starting current and starting torque are 33% of the full-voltage values. The Y-Δ starter is used for compressors and other loads that can be unloaded for starting, or can tolerate the 33% starting torque.

Y-Δ starters are built for open-transition, as shown in Fig. 20-65, or closed-transition operation, as shown in Fig. 20-66. The open-transition starter operates as follows: Relay contactor *S* is energized, connecting the motor windings in Y. A normally open auxiliary contact on contactor *S* closes, energizing contactor 1*M*. Its contacts close, energizing the motor windings in Y. After a predetermined interval, the contactor is deenergized, contacts on timer *TR* open, de-energizing contactor *S*, opening its contacts and thereby opening the Y-connected winding. The motor is now temporarily deenergized. A normally closed auxiliary contact on contactor *S* closes, energizing contactor 2*M* in addition to 1*M*. The motor is reenergized in Δ.

The closed-transition starter of Fig. 20-66 operates as follows: Contactor *S* is energized, connecting the motor windings in *Y*. A normally open auxiliary contact on contactor *S* closes, energizing contact 1*M*, closing its contacts, energizing the motor windings in Y. After a predetermined interval, contactor 1*A* is energized, connecting resistors RES in Y and par-

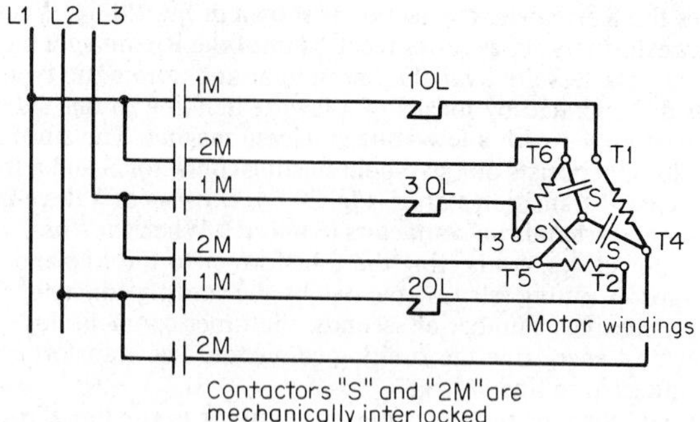


**FIG. 20-65** Open-transition Y-Δ starter.

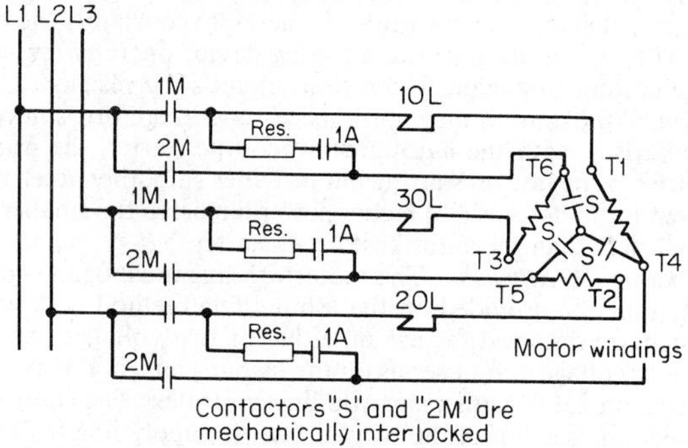


**FIG. 20-66** Closed-transition Y-Δ starter.

alleling them across the Y-connected motor winding. A normally closed auxiliary contact on contactor 1*A* opens, deenergizing contactor *S*, opening its contacts and placing resistors *RES* in series with the motor winding. The motor is now connected in Δ. A normally closed auxiliary contact on contactor *S* closes, energizing contactor 2*M*, closing its contacts and thereby shorting out the resistors *RES*. The Δ-connected compressor motor is now energized at full voltage.

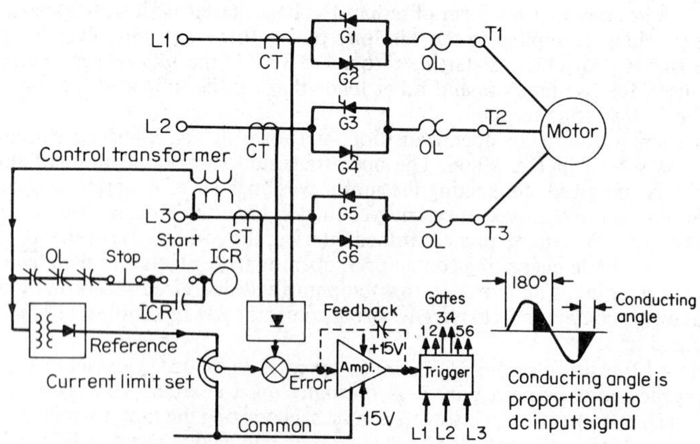


**FIG. 20-67** Diagram of solid-state ac motor starter.

***125. Solid-state ac starters*** employ back-to-back phase-controlled thyristors in two or three of the lines to the motor as shown in Fig. 20-67. The thyristors are controlled during the starting period to maintain about 300% line and motor current by gradually increasing the motor voltage from the initial value. Starting is smooth; the current and starting torque can be adjusted easily. The solid-state starter is applied where the line current is critical and where repetitive motor starting limits the life of electromagnetic contactors.

***126. Slip-ring ac motor starters*** consist of a contactor to connect the motor primary to the supply lines and a resistor and resistor switching means for the secondary circuit. The starting torque depends on the ohmic value of resistance used, maximum torque being obtained when the resistance is selected for an inrush of approximately three times full-load current. Sufficient resistance is generally used to limit the inrush current to 150 or 200%. The resistor is cut out step by step as the motor accelerates, until the slip rings are short-circuited. The commutating means may be a faceplate controller, a drum, or a series of magnetic contactors controlled by current or time relays. High starting torque and low running slip can be obtained with a slip-ring motor.

## DC Motor Starting

***127. Direct-current motors*** of small capacity may be started by connecting the motor directly to line voltage. Motors rated 2 hp or more generally require a reduced-voltage starter. The reduced voltage for starting is obtained by using resistance in series with the motor armature or by varying the armature supply voltage. Manual or magnetic control may be used.

DC motors in adjustable-voltage, adjustable-speed drives are started by turning the speed control up from zero to the desired speed, or by internal circuits that ramp the armature voltage to the desired value. Starting equipment, other than the armature-voltage rectifier or generator, is not required.

***128. Direct-current manual starters*** are satisfactory for applications that do not require frequent starting and stopping and where the starter can be mounted near the operator without requiring long motor leads. Across-the-line starters provide the simplest means of starting small dc motors. Manually operated switches for this service are available in sizes up to 1.5 hp at 115 V and 2 hp at 230 V. For larger motors resistance is connected in series with the motor armature to limit the current inrush on starting. A manually operated means is then provided for removing the resistor from the circuit in a series of steps. Starters are available in the faceplate type, the multiple-switch type, and the drum type. The faceplate type is built for motors up to 35 hp, 115 V, and 50 hp, 230 V. It consists of a movable lever and a series of stationary contact segments to which sections of resistor are connected. The resistor sections are short-circuited one at a time by moving the lever across the segments.

Manual starters have generally been replaced by push-button-operated magnetic control that incorporates overload protection and other safety features.

***129. Direct-current magnetic starters*** are used for applications where ease and convenience of operation are important; where the starter is operated frequently; where the motor is located at a distance from the operator; where automatic control by means of a pressure switch, limit switch, or similar device is desired; and for large motors which require the switching of heavy currents. Resistance is connected in series with the motor armature to limit the initial current and is then short-circuited in one or more steps.

For larger motors a series of magnetic contactors is used, each of which cuts out a step of armature resistance. The magnetic contactors are operated as the motor starts by one of two methods called *current-limit acceleration* and *time-limit acceleration;* the starting time is always matched to the burden of the load. Time-limit acceleration is advantageous where the starting time of the motor must be integrated into a timing sequence for an overall machine or process. Examples of each will be given.

***130. Time-Limit Acceleration.*** Figure 20-68 shows a type of *time-limit acceleration* where the operation of contactors, and therefore the rate of acceleration, is governed by a *magnetically operated definite time relay.* This time relay operates on the principle of discharging a capacitor, thus obtaining a definite time period which is unaffected by changes in temperature and load or by dust and dirt. With the motor at rest a circuit is obtained through a normally closed contact on *M* to energize the *CT* timing-relay coil and to charge capacitor *C*1. Contacts *CT*1 and *CT*2 on relay *CT* are open with the relay energized. Capacitor *C*2 is charged through the normally closed contact on the 2*A* contactor. Pressing the "start" button energizes the main contactor *M*, which maintains itself through a normally open interlock finger. Relay *FA* is energized, and its contact *FA*1 short-circuits the field rheostat. The motor accelerates from rest to a speed determined by the value of the *R*1-*R*3 resistor. The circuit to timing relay *CT* is opened by the interlock on *M*, and capacitor *C*1 discharges through the *CT* coil and the *AB* resistor. Contacts *CT*1 and *CT*2 can be individually adjusted to close at any time during the capacitor discharge period. Closing *CT*1 energizes the 1*A* contactor, which short-circuits the *R*1-*R*2 resistor. The motor then accelerates to a speed determined by the value of the *R*2-*R*3 resistor step. Closing *CT*2 energizes 2*A* and connects the motor across the line, permitting it to accelerate to normal speed. Relay *FA* is deenergized when 2*A* closes. Contacts *FA*1 and *FA*2 remain closed for a definite time because of the discharge of capacitor *C*2 through the *FA* relay coil. When contact *FA*1 opens, a resistance is inserted in the motor shunt field, equivalent to the field-rheostat resistance and *XY* resistance in parallel. Opening *FA*2 disconnects *XY*, and the motor runs at a speed determined by the setting of the field rheostat.

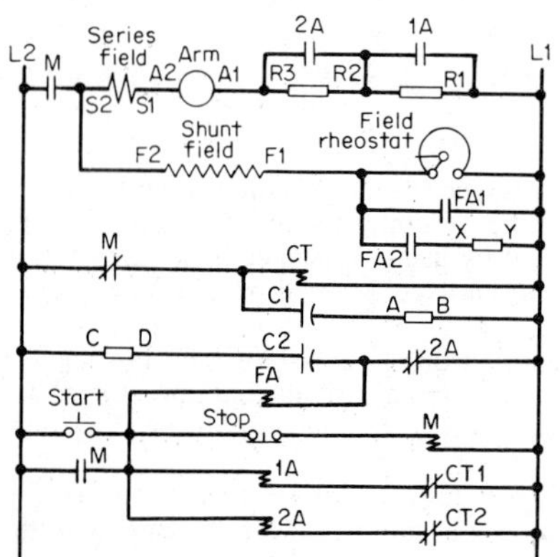


**FIG. 20-68** Time-limit acceleration with definite-time relay.

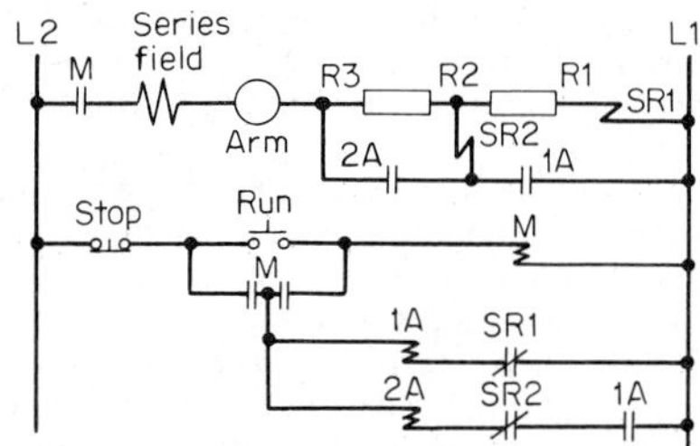


**FIG. 20-69** Direct-current series-relay accelerations.

***131. Direct-current starters with current-limit acceleration*** are designed to halt the starting operation whenever the required starting current exceeds an adjustable predetermined value, the starting operation being resumed when the current falls below this limit. With current-limit acceleration, the time required to accelerate will depend entirely upon the load. When the load is light, the motor will accelerate rapidly, and when it is heavy, the motor will require a longer time to accelerate. For this reason a current-limit starter is not so satisfactory as a time-limit starter for drives having varying loads. Time-limit starters are simpler in construction, accelerate a motor with lower current peaks, use less power during acceleration, and always accelerate the motor in the same time regardless of variations in load. Current-limit starters are desirable for motors driving high-inertia loads. A typical current-limit starter is shown in Fig. 20-69. The relays *SR*1-*SR*2 have normally closed contacts connected in series with the coils of the accelerating contactors. The coils of these relays are connected in the main motor circuit. The relays are provided with an adjustment so they can be set to close on a selected value of current.

Pressing the "run" button energizes the main contactor *M*, which closes and connects the motor to the line in series with the *R*1-*R*3 resistor. Motor current will flow through the *SR*1 coil, and its contacts will open rapidly and prevent 1*A* from closing. When the motor has accelerated enough to bring the line current down to the value for which *SR*1 is set, the relay contacts will close. A circuit is then provided for 1*A*, which closes, cutting out the first step of resistance and short-circuiting the *SR*1 coil. Current now flows through the *SR*2 coil and the 1*A* contacts. *SR*2 relay contacts open and prevent 2*A* from being energized. The motor accelerates again, and when the current falls to the value for which *SR*2 is set, its contacts close, energizing 2*A* and connecting the motor across the line.

***132. Magnetic controllers for larger dc motors*** are manufactured in forms to suit the application. The controllers are available in the following forms: (1) nonreversing, without and with dynamic braking; (2) nonreversing with speed regulation by field control, without and with dynamic braking; (3) reversing with dynamic braking, without and with speed regulation by field control.

## Synchronous Motor Starting

***133. Methods of Starting Synchronous Motors.*** The method used to start a synchronous motor depends upon two factors: the required torque to start the load, and the maximum starting current permitted from the line. Basically, the motor is started by using the damper windings to develop asynchronous torque, or by using an auxiliary motor to bring the unloaded motor up to synchronous speed. Recently, solid-state frequency converters have been designed to bring up to speed large several-hundred-MVA synchronous motor/generators for pumped-storage plants.

***134. Synchronous-motor starters*** of the full-voltage type connect the motor directly to the supply lines. The field winding is short-circuited through a discharge resistor during the starting period. The field is connected to the dc lines when the motor is at a speed near synchronism. Reduced-voltage starters connect the motor to a reduced voltage for starting and transfer to full voltage at a speed just below synchronism. This transfer may be controlled by a time relay or a frequency relay. The field is energized either immediately before or immediately after the full-voltage switch closes. Most modern synchronous motors obtain their field voltage from a brushless exciter on the shaft.

Figure 20-70 is a simplified diagram of a synchronous motor controller arranged for starting the motor directly across the line and synchronizing by a relay operating at a selected frequency. Pressing the start button energizes relay 1*CR* and contactor *M* to connect the motor stator windings to the ac lines. Current at line frequency is induced in the rotor field winding and flows through the discharge resistor *FD* and the coil of relay *FR*. A small portion of this current flows through the reactor *X*, but the amount is limited, as the frequency is high. Relay *FR* closes rapidly, and the contacts on *FR* open the circuit to the *FS* field contactor. As the motor accelerates, the frequency of the induced current in the field winding decreases and an increasing portion of the current flows through the reactor *X*. At a speed close to synchronism most of the current flows through *X*, and there will no longer be enough current flowing through the coil *FR* to keep the relay armature closed. Relay *FR* then opens, and the contacts on *FR* close to energize field contactor *FS*. Contactor *FS* connects the field to the dc lines and opens the field discharge circuit through resistor *FD*, and the motor pulls into synchronism.

Relay *FR* is polarized by a coil connected across the dc lines through interlock contacts $M_a$. Polarizing the synchronizing relay provides a means for energizing the field contactor at a point in the ac wave most favorable to synchronism. Brushless synchronous motors use electronic circuits on the rotating portion to control the switching time for the field.

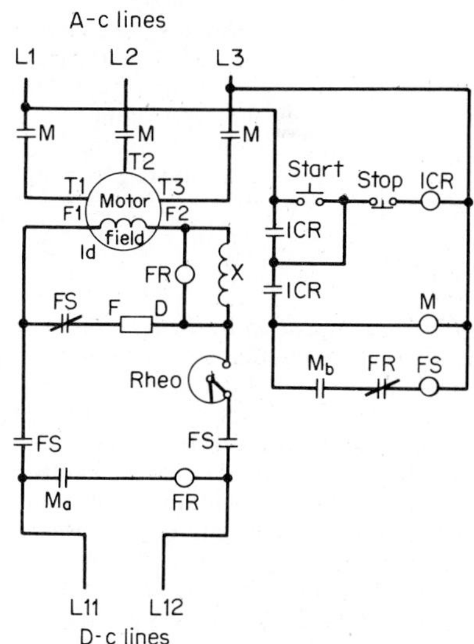


**FIG. 20-70** Full-voltage asynchronous motor controller with synchronization based on frequency.

***135. Speed-Control Devices.*** Speed control of electric motors may be obtained by various means. The design of a speed-regulating controller is determined by the type of motor with which it will be used. Table 20-18 lists the various types of motor in general use and the corresponding type of speed control for each.

**TABLE 20-18** Types of Motor Speed Control

| Type of motor | Type of speed control | Speed control range | Speed drops: No load to full load, % base speed |
|---|---|---|---|
| AC, squirrel | Multispeed<br>Pole changing by multiple windings or reconnectable single winding | Up to 4 initial speeds | Up to 5% (slip) |
| | Solid-state primary voltage control<br>NEMA Design D motor | 5 to 1 | 20% |
| | Stator frequency control, constant V/Hz<br>Solid-state inverter | 20 to 1 at constant torque, plus 3 to 1 at constant hp | 3% |
| DC, shunt wound | Stator frequency control<br>M-G set or solid-state frequency converter | 20 to 1 at constant torque, plus 3 to 1 at constant hp | Zero |
| AC, slip ring | Secondary resistors connected to slip rings | 3 to 1 | 3%, full speed; 50%, minimum speed |
| | Pumpback of slip-ring power<br>M-G set or solid-state converter | 20 to 1 | 3% |
| DC, shunt wound | Adjustable armature voltage<br>Solid-state converter or M-G set (Ward Leonard) plus field weakening | 20 to 1 at constant torque, plus 3 to 1 at constant hp | Up to 5% |
| DC, series wound | Series resistors<br>Solid-state dc chopper | 20 to 1<br>20 to 1 | Up to 100%<br>Up to 3% |

***136. Multispeed squirrel-cage motors*** are suitable for applications that require up to four operating speeds but that do not require speed control between these fixed speeds. However, a solid-state inverter plus a single-speed motor might be less costly than a multispeed motor.

Controllers for multispeed ac squirrel-cage motors may be of either the drum type or the magnetic type. Drum controllers are widely used, because the many changes in connections required to obtain different speeds can be readily accomplished. Drum controllers can be used with reconnected winding or separate-winding-type motors and with constant-torque, variable-torque, or constant-horsepower motors. Low-voltage and overload protection can be obtained by using a magnetic contactor and overload relay. When complete control by pushbuttons or other pilot devices is required, magnetic contactors are used to change the motor connections. Controllers of this type can be arranged to permit starting at any speed or to permit starting only at the slowest speed and changing to each higher speed in sequence.

***137. Primary Voltage Control.*** AC squirrel-cage motors are inherently constant-speed motors when supplied directly from utility lines. Narrow-speed-range control is obtained by adjusting the primary voltage on Design D motors using saturable reactors or solid-state phase-controlled thyristors in the stator circuits. Wide-speed-range control is obtained by adjusting the primary frequency and voltage on Design B motors using motor-alternator sets or solid-state frequency converters. The frequency of 60-Hz motors is typically adjusted from 3 to 120 Hz. From 3 to 60 Hz, the voltage is raised proportional to frequency so that the motor can deliver its full rated and breakdown torque. From 60 to 120 Hz, the voltage is kept constant so that the motor can deliver its rated horsepower.

Speed is controlled with thyristors in each of the lines to the stator of the induction motor as shown in Fig. 20-71*a*. Retarding the firing angles of the thyristors reduces the stator voltage of the motor. The torque at each speed is reduced as $V_2$, as shown in Fig. 20-71*b*. NEMA Design D motors ensure a sufficient range of descending torque in which the motor can stably drive its load. The power loss in the rotor is proportional to the torque × slip. With pump and fan loads, as shown in Fig. 20-71*b*, the torque is reduced as speed, so that the rotor power loss is acceptable at reduced speed. Typical ranges of pump and fan operation are 50 to 100% speed, 10 to 100% power.

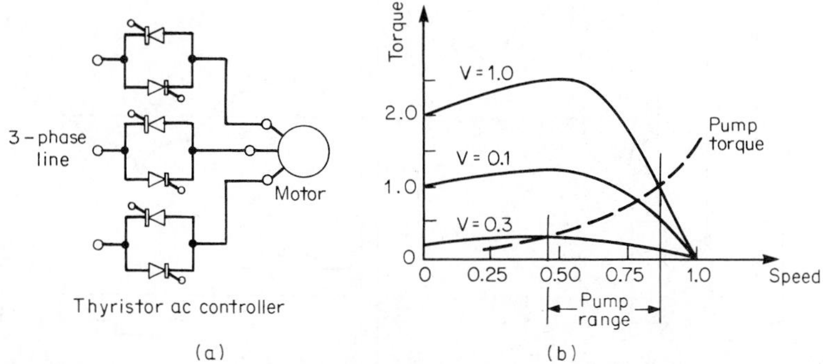


**FIG. 20-71** Primary voltage control. (*a*) Circuit of controller; (*b*) torque-speed characteristic at three-stator voltages; pump characteristic and range for 10 to 100% power.

***138. Adjustable-frequency ac induction motor drives*** consist of a solid-state rectifier, a solid-state inverter, the motor, and necessary controls. As shown in Fig. 20-72*a*, the rectifier converts power at 60-Hz line frequency to dc power; the inverter converts dc power to power for the motor which is adjustable in frequency and voltage. Inverters are classified by their output; they include six-step voltage, current source, and pulse-width modulated voltage. The six-step inverter in Fig. 20-72*a* causes the motor current to approximate a sine wave.

These ac drives operate in two modes with respect to base speed, as shown in Fig. 20-72*b*. From near zero to base speed, the inverter frequency and voltage are raised in proportion so that they both reach rated value for the motor at base speed. This is termed the *constant-torque mode* because the motor can deliver its rated torque anywhere in the speed ranges below base speed. From base speed to 200% or more of base speed, the inverter frequency is raised, but the voltage is maintained constant at the rated motor voltage. The consequence is that the magnetic field in the air gap of the motor decreases and the motor is able to deliver only one-seventh times its rated torque. However, the product of torque and speed is constant; the operation is termed the *constant horsepower mode*. The maximum speed depends on the mechanical capability of the motor to run above base speed and the maximum design frequency for the inverter.

***139. AC synchronous motors*** are speed-controlled in special applications where their self-excitation simplifies the frequency conversion equipment and where two or more motors must operate in synchronism with the supply. Adjustable-frequency, adjustable-voltage power is supplied from a motor-alternator set, or from a solid-state frequency converter. The solid-state converter is of the cycloconverter type for speeds up to 30% of the rated speed. For wide-speed range, the frequency converter is a forced-commutated inverter or a load-commutated inverter, which relies on the self-excitation of the motor to commutate the thyristors.

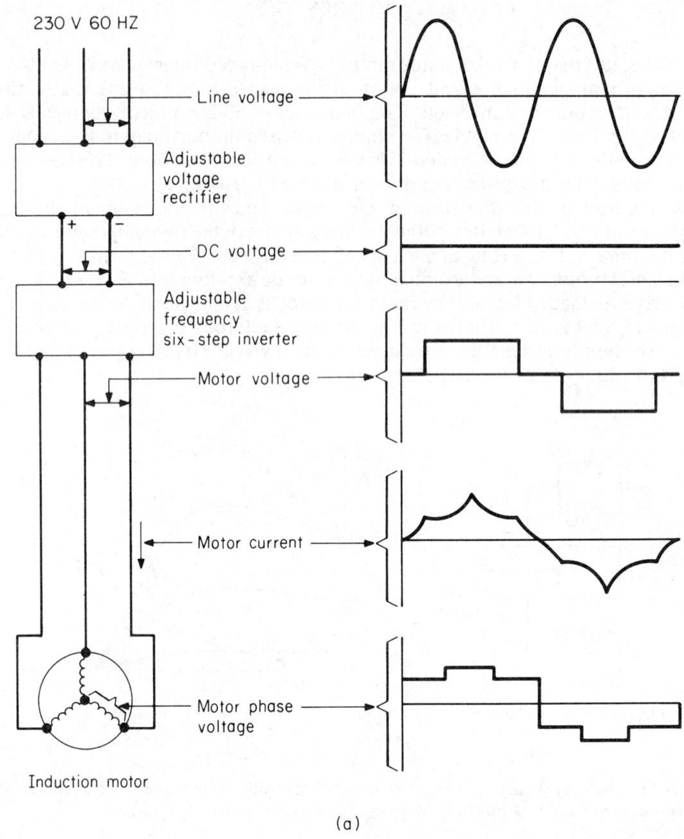


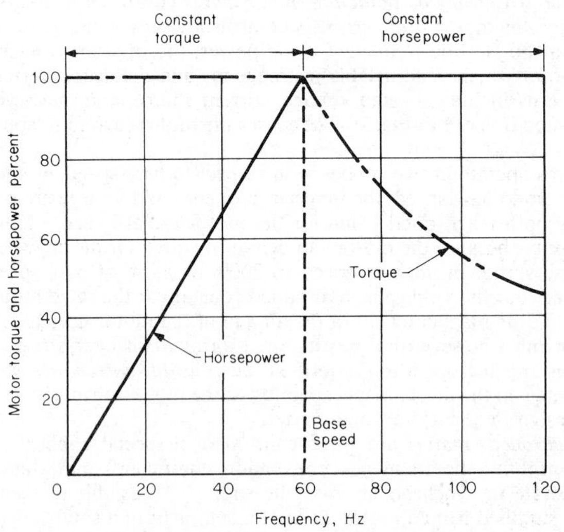


**FIG. 20-72** (*a*) Adjustable-frequency induction-motor drive using 6-step inverter; (*b*) torque and horsepower capabilities of adjustable-speed drive below and above base speed.

Synchronous motors with permanent-magnet fields rated up to about 30 hp are used in textile applications, where multiple motors are operated in synchrony from a common power supply. Synchronous motors in the range of 750 to 5000 hp and larger are used for driving large blowers, boiler fans in power plants, large pumps, and other applications requiring large motors. The block diagram for a load-commutated inverter drive is shown in Fig. 20-73. The two converters are phase-controlled thyristor bridges which can rectify or invert power, as necessary. The thyristors in the inverter operate in the firing-angle range from 90 to less than 180°, and they are commutated by the voltage generated in the stator windings of the motor by the rotating field.

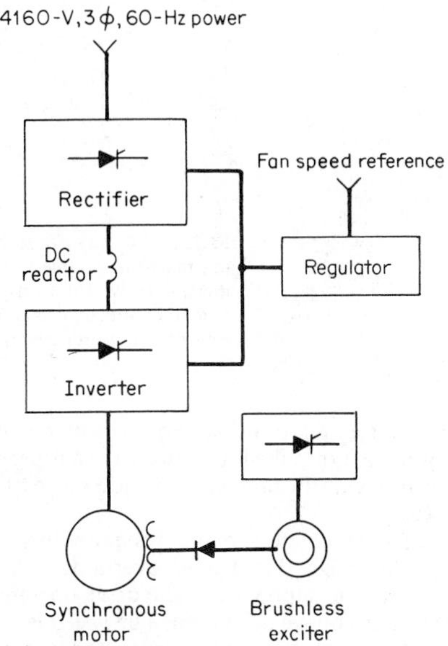


**FIG. 20-73** Block diagram of dual-converter synchronous-motor fan drive.

***140. AC slip-ring motor control*** requires that power be extracted from the rotor windings via the slip rings to reduce the motor speed, that is, increase the slip. Three methods are used: (1) secondary resistors; (2) rotor-power recovery by auxiliary rotating machines; (3) rotor-power recovery by auxiliary solid-state rectifier and converter. The auxiliary systems recover the electric energy that would be dissipated in the secondary resistors.

***141. AC slip-ring motor secondary-resistor speed regulators*** consist of a contactor to connect the primary of the motor to the supply lines and some form of resistance-switching device for the secondary circuit. The switching device may be a three-arm faceplate controller, a drum, or magnetic contactors. Regulating devices differ from starting devices in that the switching means can remain continuously on any one of the resistor steps. The motor will therefore operate continuously at a reduced speed, as determined by the amount of resistance remaining in the motor circuit. The use of secondary resistance for speed control is not an efficient method because of the power loss in the resistor. The amount of speed reduction obtained will vary directly with the load on the motor. Speed controllers of this type are usually designed for 50% speed reduction. Under favorable conditions, however, motors can be operated at 75% speed reduction.

The resistors are of the same type as the resistors used for armature regulation of dc motors (Par. **146**).

Rotor-power recovery drives are often classified as either constant horsepower or constant torque; the designation refers to the inherent limitation in power based on full current and flux in the main machine. In the first scheme (constant-horsepower drive), the slip energy is converted into mechanical power and then returned to the main motor shaft. Since horsepower is a function of the product of torque and speed, such motors have high torque at low speeds and lower torque at higher speeds. In drives using this arrangement the auxiliary machine is mounted on or mechanically geared to the main motor shaft (Fig. 20-74*a*).

In the second scheme (constant-torque drive), the slip energy is converted into electric power of the frequency and voltage of the supply circuit and is returned or fed back into the line. Since this power is not delivered to the main motor shaft, the auxiliary machine is not mechanically attached to the shaft but is separately driven. The limiting torque of the main motor being constant, the maximum horsepower output is proportional to the operating speed (Fig. 20-74*b*).

The classical recovery systems using auxiliary machines are termed the *Scherbius drive* and the *Krämer drive.* The former employs ac commutator machines; the latter relies on a dc link and rotary converters. A variation of the Krämer drive uses a synchronous motor

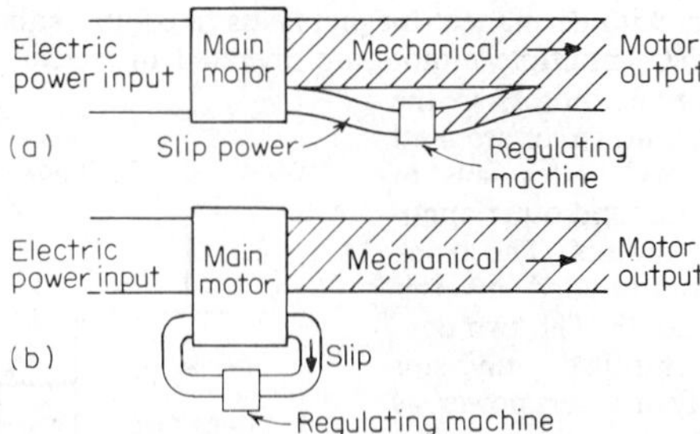


**FIG. 20-74** (*a*) Constant-horsepower drive; regulating machine, coupled to main motor, returns power mechanically. (*b*) Constant-torque drive; regulating machine, mechanically separate from main motor, returns slip power electrically.

and a dc generator in place of the rotary converter and a constant-speed set feeding the slip power back into the line. This drive has been used for a number of large wind-tunnel drives. It is particularly adapted to a wide range of speed control and to minimum disturbance on starting.

In a solid-state version of the modified Krämer drive, the slip-ring power is rectified in a diode bridge, then passed over a dc link to a line-commutated inverter that returns the power to the supply line. The drive operates with characteristics similar to the adjustable-armature-voltage dc system. The speed is controlled by the firing angle of the inverter. The recovery systems are economical for narrow speed ranges, such as for fans and pumps, or where the horsepower rating is so large that the costs of the controls are minimized.

***142. DC solid-state adjustable-voltage control*** utilizes phase-controlled 1-phase or 3-phase thyristor bridge rectifiers to provide armature and field voltage to independent-field dc motors. A typical armature-voltage supply is shown in Fig. 20-75*a*. As the firing angles of the thyristors are adjusted from 90 to 0°, the dc armature voltage rises from zero to rated value, and the speed follows in proportion.

The speed-torque characteristics of the dc drive are shown in Fig. 20-75*b*. In the range from zero to base speed, the field curent is at maximum, but the armature voltage is raised until it reaches rated value at base speed. At any one setting of armature voltage, the speed is relatively constant over the load torque range. The firing angles of the thyristor are adjusted by a regulator to maintain the speed constant at the set point. At any speed, the rated torque is available in this range. Above base speed, the armature voltage is held at rated, but the field current is reduced to obtain the higher value of speed. The maximum output of the motor is its rated horsepower in this range.

***143. The Ward Leonard adjustable armature-voltage system*** provides a flexible and efficient means for obtaining a wide range of adjustable speeds from shunt-field dc motors. The armature voltage is supplied from an M-G set in the Ward Leonard system, or from a solid-state converter. The motor speed is sensed from the armature voltage or from a tachometer, compared with a reference setting, and corrected to the required value. The M-G set will regenerate energy to the line, when the motor is decelerating, and will reverse the direction. A typical Ward-Leonard drive system is shown in Fig. 20-76. The dc generator can also be driven by a diesel engine, as in a strip-mine shovel or drilling platform.

A full-wave bridge rectifier is used for the motor field supply with a series resistor for adjusting the maximum motor field. The generator field current is supplied by the self-saturating magnetic amplifier *SR*1. A reference voltage is supplied for the speed-setting rheostat by a transformer and full-wave rectifier. A portion of this reference voltage, as determined by the position of the speed-setting rheostat, is matched through the magnetic-amplifier control winding with the generator armature voltage. If the generator voltage is low compared with the reference voltage set up by the rheostat, the current in the control winding will increase the output of the amplifier supplying the generator field. The increased generator field current will increase the armature voltage until it matches the reference volt-

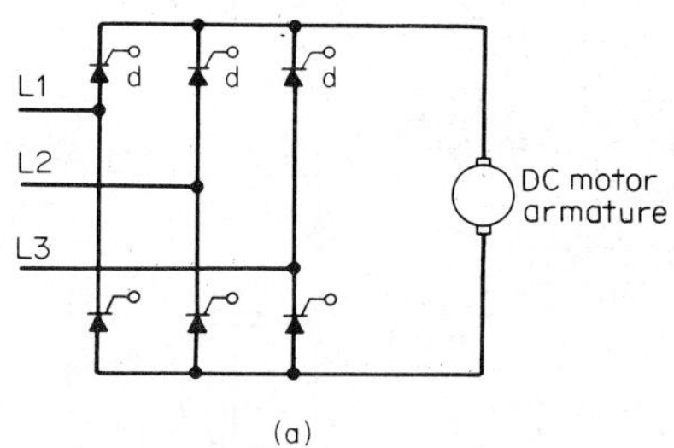


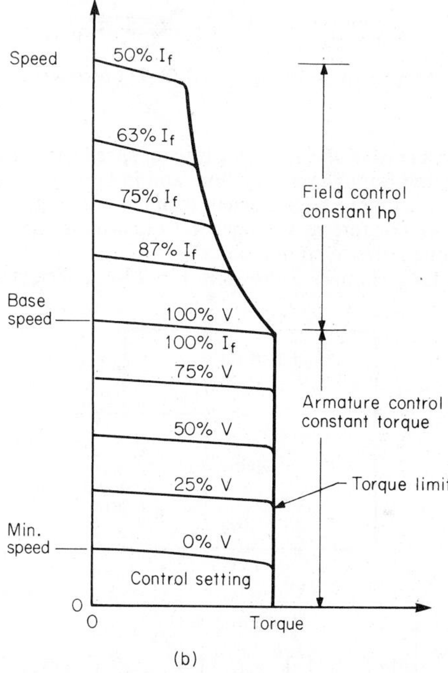


**FIG. 20-75** (*a*) Three-phase thyristor converter for armature-voltage supply. (*b*) Speed-torque characteristics of dc adjustable-speed drive.

age. The magnetic-amplifier control functions can be accomplished with transistor and thyristor controllers.

***144. DC series-wound motors*** are used for hoisting and electric traction applications where the speed is controlled either by series switched resistors or by a solid-state chopper (Par. **145**). Where full voltage is applied, the motor operates on a varying speed-torque characteristic. For example, the speed will vary over a 10 to 1 range, as the motor load is reduced from rated to minimum torque. The series-motor controller shown in Fig. 20-77 shifts the series-motor characteristic in four resistance steps, R1 through R4. Adding resistance forces the motor to run at a lower speed for the same torque. For each resistance step, the motor has a varying speed-torque characteristic. In electric traction applications, higher motor speed is obtained by shunting the series field with additional resistors. The series motor will regenerate energy to the dc line when the series field current is supplied from another source to force the generated armature voltage higher than the line voltage.

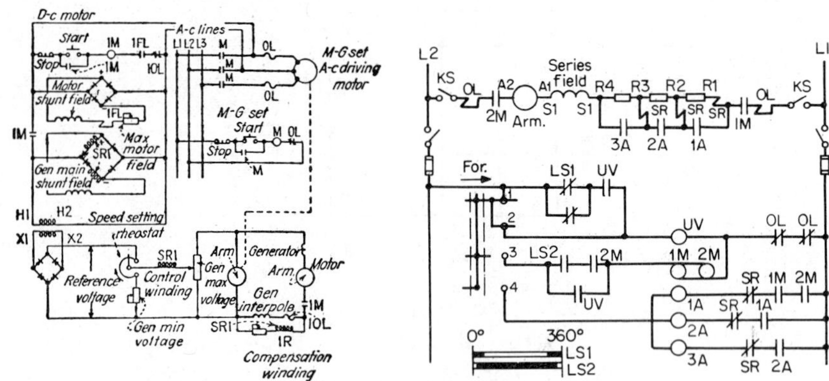


**FIG. 20-76** Adjustable-voltage Ward-Leonard drive system.

**FIG. 20-77** Series-motor controller.

***145. Solid-state chopper control*** is replacing resistor speed control of dc series motors in rapid-transit cars operating from wayside dc lines and in battery-powered electrical industrial trucks. The chopper is more energy efficient than resistor controllers and permits stepless control of motor speed or torque. Choppers for rapid-transit cars are rated up to 1000 kW at 1000 V dc, for four-series traction motors.

A simplified circuit for a chopper is shown in Fig. 20-78*a*. The chopper applies voltage

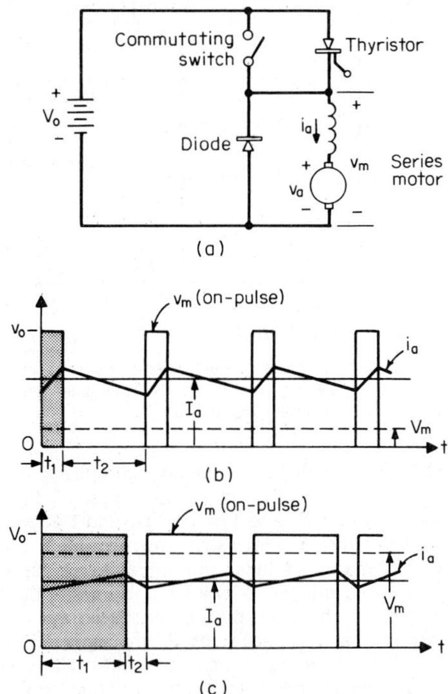


**FIG. 20-78** Chopper speed control for series motor. (*a*) Circuit; (*b*) waveforms at low speed; (*c*) waveforms at high speed.

to the motor in a series of pulses, as shown in Fig. 20-78*b*. The pulse frequency is typically in the 200 to 400 pulses per second range. The width of the pulse can be varied over about a 10:1 range in a system called time-ratio control. The widest pulse provides full voltage and maximum motor speed, as shown in Fig. 20-78*c*; the narrowest pulse provides minimum voltage and speed. Each pulse is ended by turning off the thyristor with the commutating switch, which is actually another thyristor circuit. When the thyristor is turned off, the motor current continues through the diode, until the thyristor is turned on to start the next pulse. The energy loss in the thyristor is small compared to the resistors of the controller shown in Fig. 20-77. The thyristor, diode, and motor can be reconfigured for the motor to regenerate to the dc line or the battery.

***146. Armature-regulating and field-regulating resistors*** are used for the speed control of dc motors fed from fixed-voltage dc supplies as shown in Fig. 20-79. Armature-regulating resistors provide starting and speed-control duty in multiple steps, usually for fan loads. The resistors are usually designated for continuous duty, in accordance with a NEMA classification number between 91 and 96, inclusive. The specific classification within this group is determined by the motor-starting current allowed, with all resistance in the circuit. The total ohms in the resistor are usually determined by the speed reduction required and the torque of the load. The *IR* drop across the resistor is the difference between the dc line voltage and the armature voltage, which is proportional to the speed. The current *I* is proportional to the torque. For example, at 50% speed a fan load requires 25 to 40% of its current at full speed. The total ohms are divided into steps to give equal speed changes per step.

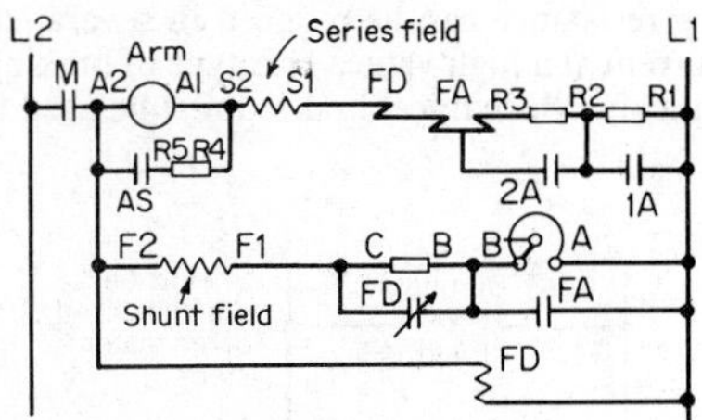


**FIG. 20-79** Armature speed-control resistors *R*1-*R*2 and *R*2-*R*3. Field speed-control resistors *A-B* and *B-C*.

***147. Field-regulating resistors*** should be designed on the basis of actual test data on the motor with which the control is to be used. The motor manufacturer will supply such data as the resistance of the shunt field or the maximum field amperes, the required rheostat resistance or the minimum field amperes. The total ohms in the rheostat will therefore be determined by the motor data. The number of steps in the rheostat must be large enough to limit the speed increase per step to an acceptable value. The character of the resistor material used may require a large number of steps, in order to keep the wattage per step on the rheostat within limits. The number of steps must be sufficiently large to keep the voltage drop at any step within specified limits to prevent arcing on the contacts. The arcing limits for sliding contacts vary from 200 V at 0.4 A to 50 V at 1 A and 25 V at 4 A. This voltage is the product of the step ohms and the amperes flowing before the step is inserted. The amount of resistor material required in the field rheostat is determined by the wattage which must be dissipated. Each step must be designed on the basis of the maximum current which it will have to carry. The total wattage capacity of the rheostat will be the summation of the step wattage. The distribution of the resistor between the various steps can be determined on the basis of obtaining either equal speed changes per step or equal percentage speed changes per step. Common practice at present is to provide equal speed changes per step, and a speed reduction to 20% of what is rated can be obtained with armature-regulating resistors. A speed increase up to 200% can be obtained by field resistors with a properly designed motor.

## STOPPING DEVICES

***148. Electric braking*** is used when the dc or ac motor is allowed to make several revolutions before coming to rest. In electric dynamic braking, the kinetic energy of the motor and the load is absorbed in the resistance of the motor windings, or in a resistor switched into the circuit. In electric regenerative braking, the energy is returned to the supply line. Magnetic brakes are used to obtain quick, accurate stopping and to hold the load after stopping. Most brakes are spring-set and electrically released, so that braking will be obtained even though an electrical failure occurs.

***149. Dynamic braking of a dc shunt-wound motor*** can be obtained by disconnecting the motor from the line and shunting the armature with a resistor. As shown in Fig. 20-80, when the motor is running, contacts *A*1 and *A*2 are closed, and *A*3 is open. When the stop button is pressed, contact *A*1 opens to disconnect the armature from the line; holding contact *A*2 opens; and contact *A*3 closes to put the braking resistor across the armature terminals. The field must remain connected to the line during braking. For constant-speed motors the ohmic value of the braking resistor is $R = (E - I_aR_a)/I$. The value of $I$ determines the amount of braking obtained and may vary from 150 to 300% of normal current. With shunt motors, having speed regulation by shunt-field control, the general practice is to strengthen the field during braking by short-circuiting the field rheostat. When this is done, the ohms in the braking resistor should be $R = (E - I_aR_a)/I \times$ speed range by field. The braking obtained decreases as the motor speed is reduced. The final stopping of the motor is due to friction, since at standstill no braking torque is obtained. For very quick stopping the braking resistance can be reduced in several steps as the speed decreases, thereby keeping the current at a high value. This type of braking is effective where it is necessary to stop a motor quickly. Mechanical brakes must be used to hold a load at a standstill.

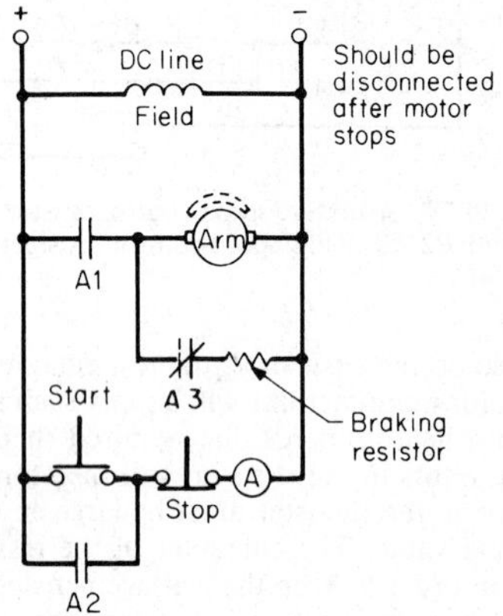


**FIG. 20-80** Dynamic braking circuit for shunt wound dc motor.

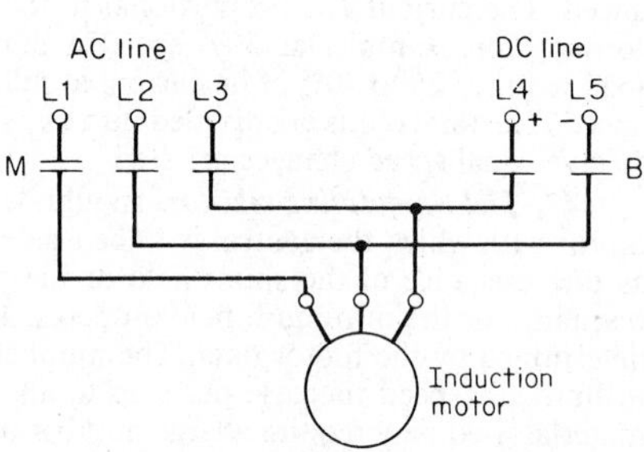


**FIG. 20-81** Dynamic braking circuit for 3-phase induction motor. Contacts *M* are for line-starting the motor. Contacts *B* are for applying direct current for braking.

***150. Dynamic braking with an ac induction motor*** can be obtained by disconnecting the motor from the power supply and applying direct current to one of the stator phases. As shown in Fig. 20-81, the kinetic energy of the rotor and the load is dissipated in the rotor circuit resistance.

***151. Plugging is used with dc motors*** to obtain very rapid reversing and is accomplished by connecting the motor to the line in the reverse direction while it is still rotating in the forward direction. The countervoltage of the armature is added to line voltage to force current through the armature and series resistor. The total resistance in circuit for plugging should be $R_t = 2(E - I_aR_a)/\text{inrush current}$. With the armature drop assumed as 10%, the formula may be written $R_t = 1.8E/1.5I = 1.2E/I$, where $E$ = line voltage and $I$ = normal current. The plugging resistance is determined by subtracting the armature resistance from $R_t$.

***152. Plugging is used with ac motors*** to obtain a very quick stop. Three-phase squirrel-cage and wound-rotor motors may be plugged by reversing the line connections to any two of the stator terminals while the motor is running in the forward direction. In order to use plugging as a stopping means, a zero-speed switch is necessary to open the reverse contactor and prevent reversing the motor. A common form of zero speed switch is the friction type, in which a contact is held closely by the friction of a small belt over a pulley driven by the motor. Any slight reversal of the motor will cause the contacts to open. Another type of zero-speed switch uses a disk which rotates in the magnetic field produced by Alnico magnets. Eddy currents are induced in the driven disk, and the magnetic reaction turns the magnet assembly to close the contacts. The contacts can be adjusted to open when the motor is near zero speed.

An electrical plugging relay is also used which remains closed until the line current decreases to the normal inrush current corresponding to the starting condition, or zero speed on the motor. This device has the advantage of not requiring a mechanical connection to the motor or machine.

***153. Regenerative braking*** is used with dc adjustable-armature-voltage controllers to obtain rapid stopping. For a solid-state armature converter, either the dc motor field current is reversed to reverse the armature voltage, or a second reverse polarity converter is switched in. In the first case, the direction of the armature current remains the same, but the armature voltage reverses. In the second case, the direction of the armature current reverses, but the armature voltage remains the same. The regulator controls the firing angles of the converters to maintain prescribed armature current during the braking period. For an ac-motor-driven dc generator supplying the dc motor in a Ward Leonard system, the generator voltage exceeds the countervoltage of the driven motor, and power is taken from the ac lines to keep the driven machine rotating. If the field strength of the generator is decreased, the generator voltage becomes less than the countervoltage of the motor and the motor feeds power back to the generator and to the ac lines. When the countervoltage exceeds the generator voltage, a heavy reverse current is obtained, as the value of this current is limited only by the low resistance of the loop circuit. A very rapid stop is obtained, since the voltage across the generator field can be reduced to zero in about 3 s.

***154. Direct current brakes*** are set by a spring and are released by means of a solenoid or a direct-operating magnet. The coil in the operating device may be for either series or shunt connection and for continuous or intermittent duty. Series-wound brakes are operated by motor current and require 80% full motor current to release with a continuous-duty coil and about 40% full-load current with an intermittent-duty coil. The brake will be held released on about 10% full-load motor current. Intermittent-duty series brakes are rated as either ½-h duty or 1-h duty, to correspond to the rating of intermittent-duty series motors. A series-brake coil will carry full motor current continuously, for a period corresponding to its rating, without exceeding a temperature rise of 75°C. Shunt brakes may be for either continuous or intermittent duty. Intermittent duty is defined as 1 min on and 1 min off or the equivalent, the longest time on not to exceed 1 h. Shunt brakes will release at 80% of normal voltage, when adjusted for rated torque. The larger-sized brakes use partial-voltage coils and protecting resistors. Series brakes have a heavy wire coil, which is less likely to give trouble, are faster in operation, and will set whenever the armature circuit is open. Data on a line of dc magnet-operated shoe brakes are given in Table 20-19.

**TABLE 20-19** Direct-Current Magnet-Operated Shoe Brakes

| Wheel diam., in | Maximum torque, lb-ft | | | |
|---|---|---|---|---|
| | Shunt wound | | Series wound | |
| | Int. duty | Cont. duty | ½-hr duty | 1-hr duty |
| 8 | 100 | 75 | 100 | 65 |
| 10 | 200 | 150 | 200 | 130 |
| 13 | 550 | 400 | 550 | 365 |
| 16 | 1000 | 750 | 1000 | 650 |
| 19 | 2000 | 1500 | 2000 | 1300 |
| 23 | 4000 | 3000 | 4000 | 2600 |

NOTE: Int. = intermittent; Cont. = continuous; 1 in = 25.4 mm; 1 lb-ft = 1.356 N·m.

**TABLE 20-20** Alternating-Current Torque-Motor-Operated Brakes

| Wheel diameter, in | Maximum torque, lb-ft | | Volt-amperes | | $WR^2$ of wheel | Safe maximum rpm | Weight of brake, lb |
|---|---|---|---|---|---|---|---|
| | Int. duty | Cont. duty | Int. duty | Cont. duty | | | |
| 10 | 160 | 125 | 160 | 105 | 3.1 | 2015 | 150 |
| 13 | 400 | 325 | 210 | 140 | 12 | 1550 | 240 |
| 16 | 800 | 600 | 300 | 240 | 25 | 1260 | 370 |
| 20 | 1600 | 1200 | 1000 | 470 | 75 | 1012 | 750 |
| 25 | 3200 | 2400 | 1500 | 550 | 220 | 806 | 1210 |

NOTE: 1 in = 25.4 mm; 1 lb = 0.4536 kg; 1 lb-ft = 1.356 N·m.

***155. Alternating-current brakes*** have three forms of operating mechanism: solenoid type, torque-motor type, and thrustor type. The smaller sizes of brakes are usually made in the solenoid type. On the larger sizes, a vertically mounted torque motor and antifriction ball jack provide a quiet, low-inrush-current operating means. Upon application of power to the motor, the rotary motion of the armature is transformed into straight-line motion through the antifriction jack. With the brake fully released, the torque motor is stalled across the line. When the motor is disconnected from the line, the spring in the brake overhauls the motor mechanism and applies the brake. Data on a line of brakes of this type are given in Table 20-20.

The thrustor-type operating mechanism consists of a self-contained motor-driven centrifugal pump, oil chamber, and a piston which produces a straight-line movement to release the brake.

***156. The brake size*** for most applications can be determined by using the formula $T = 5252 \times \text{hp/r/min}$, where $T$ = full-load motor torque in pound-feet, hp = motor horsepower, r/min = speed of shaft on which brake wheel is mounted. A brake should be selected with a torque rating equal to or greater than the full-load motor torque $T$. In some cases the braking torque is determined by extreme operating conditions against which the brake must hold, for example, heavy ice loads on bascule bridges or conditions of unbalance on skip hoists. In these cases the maximum load must be calculated and translated into pound-feet torque at the shaft on which the brake is mounted. Sufficient lining area is provided on all sizes of brake for the average application. However, a careful check as to lining area must be made when brakes are used for frequent stopping or for stopping high-inertia loads.

## MOTOR-PROTECTING DEVICES

***157. Fuses*** should be provided for motor circuits, in accordance with the NEC. The current rating of the fuse must be considerably higher than the current rating of the motor, or the fuse will blow when the motor is started. For that reason fuses do not provide adequate overload protection. They furnish protection for the motor only in case of a short circuit or a very heavy overload. Their primary purpose is to protect the circuit rather than the motor.

***158. Magnetic-type overload relays*** are operated by direct magnetic action of the motor current on a plunger. The relay consists of a series coil connected in the motor circuit and a plunger which is pulled up into the center of the coil when a certain value of current has been reached. When the plunger is lifted, a contact is tripped, opening the motor contactor-coil circuit and disconnecting the motor from the line. The tripping current can be varied by adjusting the initial position of the plunger with respect to the coil. Time delay in tripping is obtained by attaching a small oil dashpot to the plunger. The time delay can be adjusted so that the overload will not trip on the starting-current inrush but will trip on small sustained overloads.

***159. Thermal overload relays*** are available in the bimetallic type and the fusible-alloy type. The bimetallic type has two heaters in series with the circuit to be protected, and above these heaters are two strips of bimetallic material, which act as latches for the contact members. Bending of the bimetallic strips under heating of overload current will release the latches and allow the contacts to open. The fusible-alloy type has two heaters, each surrounding a thermal element consisting of a small tube, inside which is a loose-fitting shaft. The tube and shaft are rigidly joined by a special low-melting eutectic alloy. On overload, the increased current drawn melts the alloy, allowing the shaft to turn and the contacts to open.

Characteristics of a typical thermal overload are shown in Fig. 20-82. An inspection of these curves shows that the thermal overload adequately protects the wiring, that the fuse blows first on short-circuit current, and that the thermal relay allows the motor ample time to accelerate. A thermal overload has a tripping characteristic which corresponds closely to the heating characteristics of a motor and, therefore, provides an ideal protecting means. An overload coil should be selected so that the maximum permissible output can be obtained from the motor. A motor rated 40°C rise on the basis of 40°C ambient temperature will have a final safe temperature of approximately 95°C and will operate at 15% overload continuously without overheating. An overload coil should therefore be selected having an ultimate tripping current equivalent to 15% overload on the motor. A continuous overload of 15% would therefore ultimately trip the thermal relay. For overloads in excess of 15% the tripping time would be shorter than the time required for the motor to reach a dangerous temperature.

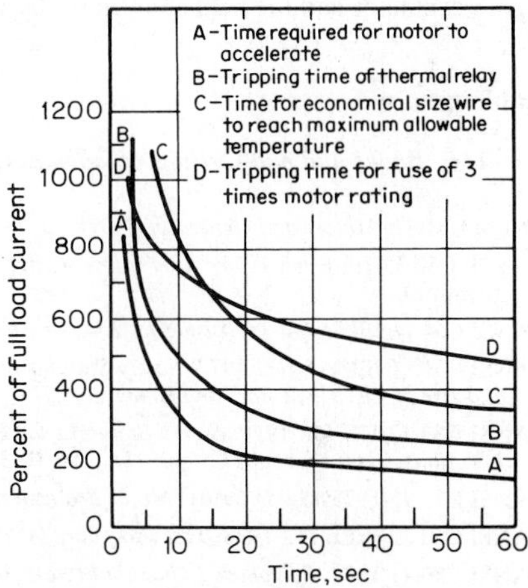


**FIG. 20-82** Characteristics of thermal overload relays.

***160. Low-voltage protection*** is the effect of a device, operative on the reduction or failure of voltage, to cause and maintain the interruption of power to the main circuit.

With magnetic controllers, this protection is obtained by using some form of 3-wire master switch. Should the line voltage drop to a low value or fail altogether, the main-line contactor will open and remain open, stopping the motor. To restart, it is necessary to push the "start" button. This type of control should always be used where the unexpected restarting of a motor after voltage failure may be dangerous to workers or equipment.

***161. Low-voltage release*** is the effect of a device, operative on the reduction or failure of voltage, to cause the interruption of power to the main circuit but not to prevent the reestablishment of the main circuit on return of voltage. Such protection is obtained when a 2-wire pilot device, for example, a snap switch, float switch, or pressure switch is used.

***162. Phase-failure protection*** is the effect of a device, operative upon the failure of power in one wire of a polyphase circuit, to cause and maintain the interruption of power in all the wires of the circuit.

***163. Phase-reversal protection*** is the effect of a device, operative on the reversal of the phase rotation in a polyphase circuit, to cause and maintain the interruption of power in all wires of the circuit. Protection of this type is necessary on elevators, where reversing of the phases would cause the car to start in a direction opposite to that in which the operator expects it to move.

***164. Field-failure protection*** is usually provided in controllers for dc shunt- and compound-wound motors. The coil of a relay is connected in series with the motor shunt field, and a normally open contact of the relay is connected in the stop circuit. If the field circuit is opened, the relay will be deenergized and the motor will be disconnected from the line. This prevents overspeeding the motor owing to an open circuit in the field. A *field protective relay* is used to insert resistance in series with the shunt field whenever the motor is not running. The coil of the relay is connected in parallel with the main switch coil, and a normally open relay contact is used to short-circuit a step of resistor in the field circuit. The resistor should be designed to reduce the voltage across the field to one-half line voltage. This reduces the field wattage to one-fourth the normal value and prevents overheating the field with the motor at standstill.

***165. A field-discharge resistor*** should be provided for 230-V motors rated 7½ hp or more and for 550-V motors rated 5 hp or more whenever the shunt-field circuit must be opened. The ohmic value of a discharge resistor should be one to three times the ohms in the field. If a resistance of three times the field ohms is used, the induced voltage, when the circuit is opened, will be four times normal line voltage. This voltage, caused by the inductance of the field, must be limited to prevent damage to the insulation of the field windings. On nonreversing controllers without dynamic braking, the shunt field can be connected behind the main contactor and the field allowed to discharge through the motor armature.

## Bibliography

### *166. Books and publications on ac motors*

*Standards*

NEMA MG1; *Motors and Generators,* 1978.

ANSI C50.35/IEEE 86-1975; *Alternating Current Electric Machines,* definition of basic per unit quantities.

ANSI C64.1/NEMA CB-1; *Brushes for Electric Machines.*

ANSI C51.1/NEMA MG2-1977; *Safety Standard for Construction and Guide for Selection, Installation, and Use of Electric Motors and Generators.*

ANSI C33.72/UL 674-1979; *Safety Standard for Electric Motors and Generators for Use in Hazardous Locations,* Class I Groups C1 and 1D, class II Groups E, F, and G.

ANSI C37.96-1975; *Guide for AC Motor Protection.*

ANSI C50.20/IEEE 112-1978; *Test Procedure for Polyphase Induction Motors and Generators.*

ANSI C50.41-1977; *Polyphase Induction Motors for Generating Stations.*

ANSI C50.21/IEEE 114-1969; *Test Procedure for Single-phase Induction Motors.*

ANSI C33.26/UL 547-1979; *Safety Standard for Thermal Protectors for Motors.*

ANSI C34.3/IEEE 444-1973; *Thyristor Converters for Motor Drives.* vol. 1. *Converters for DC Armature Motor Drives, Practices and Requirements.*

*Books*

Shoults, D. R., Rife, C. J., and Johnson, T. C.; *Electric Motors in Industry;* New York, Wiley, 1942.

Jones, R. W.; *Electric Control Systems,* 3d ed.; New York, Wiley, 1953.

Fitzgerald, A. E., Kingsly, C., Jr., and Kusko, A.; *Electric Machinery,* 3d ed.; New York, McGraw-Hill Book Company, 1971.

Kostenko, M., and Piotrovsky, L.; *Electrical Machines,* vol. 2; Moscow, MIR Publishers, 1974.

Andreas, J. C.; *Energy-Efficient Electric Motors;* New York, Marcel Dekker, 1982.

Lloyd, T. C.; *Electric Motors and Their Applications;* New York, Wiley-Interscience, 1969.

Say, M. G.; *Alternating Current Machines,* 5th ed; New York, Wiley, 1983.

Smeaton, R. W.; *Motor Application and Maintenance Handbook;* New York, McGraw-Hill Book Company, 1969.

Pearman, R. A.; *Power Electronics—Solid-State Motor Control;* Reston, Virginia, Reston Publishing Company, 1980.

Takenchi, T. J.; *Theory of SCR Circuits and Application to Motor Control;* Tokyo, Tokyo Electrical Engineering College Press, 1968.

Bose, B. K.; *Adjustable Speed AC Drive Systems;* New York, IEEE Press, 1980.

Murphy, J. M. D.; *Thyristor Control of AC Motors;* New York, Pergamon, 1973.

Laithwaite, E. R.; *Induction Machines for Speed Purposes;* New York, Chemical Publishing Company, 1966.

Nasar, S. A., and Boldea, I.; *Linear Motion Electric Machines;* New York, Wiley, 1976.

Poloujadoff, M.; *The Theory of Linear Induction Machines;* Oxford, England, Clarendon Press, 1980.

# SECTION 21

# INDUSTRIAL AND COMMERCIAL APPLICATIONS OF ELECTRIC POWER

**N. Richard Friedman**
*Chairman, Resource Dynamics Corporation*

**Joseph D'Auria**
*Project Manager, Resource Dynamics Corporation*

**Daniel Jackson, Jr.**
*Associate Editor, Coal Age*

**John Lusti**
*Vice President, Corporate Technical Development, Otis Elevator Co.*

## CONTENTS

*Numbers refer to paragraphs*

## INTRODUCTION

*By N. RICHARD FRIEDMAN*

***1. Trends toward Automation.*** In the industrial and commercial applications of electric power, since the last edition of this handbook, there has appeared an accelerated trend to use sophisticated automation systems, based largely on electronic computer control and processing. The dedicated process computer and minicomputer have been challenged (and, in many applications, replaced) by the *microprocessor* and microcomputer, which can perform the same functions at about one-tenth the cost of the minicomputer.

These developments have fostered the emergence of a new family of commercial and industrial process automation techniques, referred to as facility management systems (for commercial buildings) and flexible manufacturing systems (for industrial applications). Commercial facility management systems typically include sensors for input data, remote terminal units, the central processor, and human-machine interface devices. The size and complexity of these systems vary greatly, but functions can include heating, ventilating, and air conditioning (HVAC); fire management; security; access control; and energy management.

Industrial flexible manufacturing systems are assemblies of one or more machine tools and cutting tool and workpiece-handling devices which are employed to process a variety of finished parts. Flexible manufacturing can occur at different levels:

1. Flexible manufacturing *module:* Single machine with part-changing equipment and self-contained computer numerical control
2. Flexible manufacturing *cell:* Two or more numerical control machine tools linked to operate sequentially
3. Flexible manufacturing *system:* Combination of modules or cells linked via a supervisory computer

True flexible manufacturing systems (FMS) are combinations of one or more machine tools, inspection devices, part-washing equipment, and/or material-handling and storage equipment all operating in a coordinated manner under the control of a central computer. A schematic diagram of an FMS is shown in Fig. 21-1.

Among the areas in which rapid development of computer methods has occurred are agriculture, elevators, steel mills, and heating and cooling. These are treated at length in the

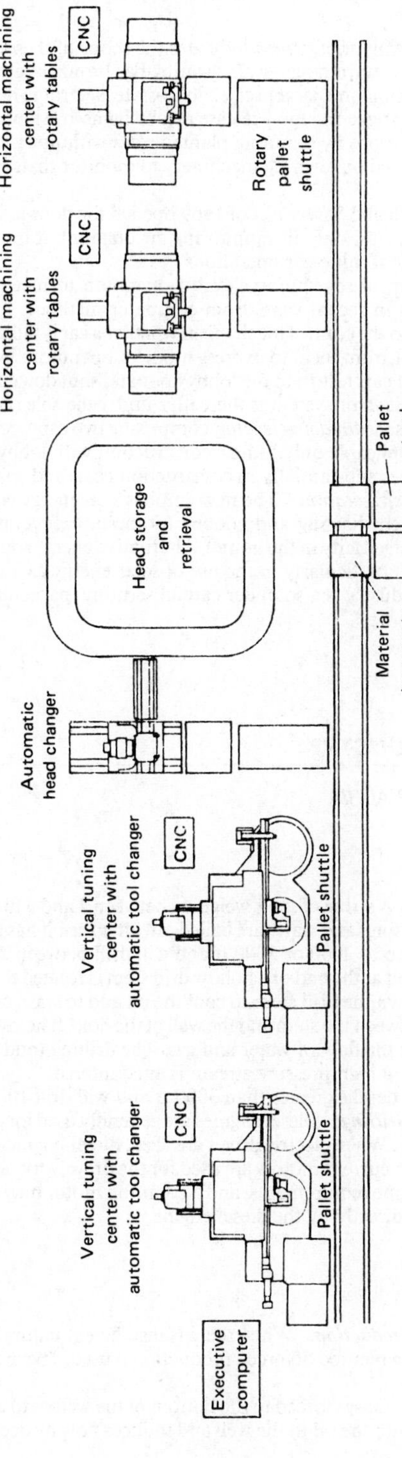


**FIG. 21-1** Schematic illustration of a flexible manufacturing system. *(Courtesy of Giddings and Lewis, Inc.)*

following pages. The following brief review indicates the scope of these developments.

*Agriculture.* The trend toward larger-scale farming has brought with it both the need and the means to apply automation on a larger scale. Solid-state electronic devices are now used to control livestock feeding by triggering food-release mechanisms at established intervals; phototransistors are used to thin crops by scanning planted rows with precision and at high speed; electronic sensors are also used on farming machinery to monitor shaft speeds, materials flow, and other parameters.

*Coal Mines.* The Health and Safety Act of 1969 opened the door to increased automation of the coal-mining industry, not only in monitoring air quality but in controlling the speed, frequency, and safety limits of mine-car operations.

*Elevators.* The problem of scheduling elevator operation to meet changing patterns of traffic has benefited greatly in recent years from computer methods. In addition, electronic sensors now communicate to the control mechanism whether a car is full or empty, and whether to park the car or convert it from local to express mode of operation. New fire regulations in some cities now require that cars return to the lobby floor and shut down if a fire signal registers in the control system; the elevator system is thereafter under the sole control of fire officials.

A major design change is the *tandem elevator,* comprising two cars occupying the same shaft and cables. This system, applicable only to new construction with lobby facilities suitable for this mode of operation, saves substantially in construction costs and space requirements.

*Air Conditioning and Refrigeration.* The new emphasis on energy conservation has led to engineering for more efficient heating and cooling (domestic refrigeration accounts for the major portion of the use of electricity in the home). Alternative energy sources are being actively investigated and developed, particularly in the use of solar energy as a supplement for house heating. Energy usage in industry is also under careful scrutiny, particularly in improving the efficiency of heat recovery.

## *THE PETROLEUM INDUSTRY*

*Revised by JOSEPH A. D'AURIA*

### Well Drilling

***2. Cable Tool Drilling.*** A string of tools weighing between 2 and 5 tons is repeatedly raised and dropped by a cable. Cuttings and water are bailed out after the bit has been withdrawn. This method is used for depths up to 2000 or 3000 ft and will drill between 20 and 150 ft/day.

***3. Rotary Drilling.*** A bit at the end of a hollow drill stem is rotated to cut or grind the rock. Drilling mud is pumped down the drill stem to cool the bit and to wash cuttings to the surface. The mud travels upward between the stem and the wall of the hole. It helps to prevent collapse of the wall and seals it against the flow of water and gas. The drilling mud also helps to prevent "blowing" of the well when a high-pressure stream is encountered.

Rotary rigs are used for depths greater than 3000 ft and will drill 100 to 300 ft/day.

***4. Power Required for Drilling.*** Diesel engines are generally used for drilling because of the requirements for portability. When electric drives are used, diesel-engine-driven generators are provided with the rig. Direct-current motors are used for the draw works and rotary table drives because of high starting-torque requirements, and a common motor may be used for both. The mud pump is often driven directly by the diesel engine.

### Oil-Well Pumping

***5. Methods of Forced Production.*** When there is insufficient natural pressure to force the crude oil to the surface, some method of forced production is used. The most common methods are:

*High-Pressure Gas Lift.* Gas is forced to the bottom of the well or to an intermediate point in the well. The gas mixes with the oil in the well and induces flow by decreasing the density of the fluid.

*Water Flooding.* Treated water is forced into the formation through nearby wells in order to increase the pressure in the formation and induce flow.

*Bottom-Hole Hydraulic Pump.* High-pressure crude oil is carried down the well in tubing, and it is used to drive a reciprocating pump located at the bottom of the well.

*Bottom-Hole Centrifugal Pump.* A special motor-driven multistage centrifugal pump is lowered to the bottom of the well. This method is used where large volumes of fluid must be pumped.

*Sucker-Rod Pump.* A reciprocating single-acting pump is installed at the bottom of the well on the end of a tube inside the well casing.

***6. Sucker-Rod Pump Drives.*** The plunger is operated by a sucker rod from the surface. Various methods are available to drive the sucker-rod string, but generally a walking beam is used to provide the desired vertical motion.

***7. Central Power Units.*** Central power units driven by electric motors are sometimes used to serve as many as 15 or 20 wells. Operating rods lead out to each pump to provide reciprocating motion to the walking beams.

***8. Individual Engine or Motor Drives.*** These are more commonly used than multiple drives because they can be started and stopped individually and the pumps can be operated at different speeds. Electric motor drives are preferred because they can be started and stopped by a timer, they provide consistent, trouble-free performance regardless of weather conditions, and maintenance and investment costs are low. Also, it is easy to measure power demand and energy consumption of an electric motor. The well may be counterbalanced readily by an ammeter.

***9. Motor Types.*** Torque requirements vary widely during the pumping cycle, and peaks occur when the sucker-rod string and fluid load are lifted and when the counterweight is lifted. NEMA design D motors, although relatively expensive, are well suited to this service, since they minimize current peaks and provide adequate torque under all service conditions, including automatic operation by time control. NEMA design C motors may be used where operating conditions are less severe. NEMA design B motors must be used with care in this service to avoid high cyclic current peaks, which may be objectionable on a small system, particularly if several wells should "get in step." The use of design B motors can also lead to oversizing of motors in an attempt to obtain sufficient starting torque. This results in the operation of the motor at a relatively low load factor, with consequent low power factor.

***10. Double- or Triple-Rated Motors.*** These are special motors developed for oil-well pumping. They are totally enclosed, fan-cooled NEMA design D motors that can be reconnected for 2- or 3-hp ratings at a common speed of 1200 r/min. Typical horsepower ratings are 20/15/10 and 50/40/30. They provide flexibility in the field since they permit the selection of the horsepower rating at which the motor may be operated most efficiently. They also permit changing the pumping speed by changing the motor pulley and reconnecting the motor.

***11. Single-Phase Operation.*** If single-phase power only is available, it is advisable to consider the use of single-phase to 3-phase converters and 3-phase motors. This avoids the use of large single-phase capacitor start motors, which are relatively expensive and contain a starting switch which could be a source of trouble due to failure or to the presence of flammable gas in the vicinity of the well.

***12. Oil-Well Control.*** A packaged control unit is available to control individual oil-well pumps. It contains, in a weatherproof enclosure, a combination magnetic starter, a time switch that can start and stop the motor according to a predetermined program, a timing relay that delays the start of the motor following a power failure, and lightning arresters. Pushbutton control is also provided.

***13. Power-Factor Correction.*** The induction motors used for oil-well pumping have high starting torques with relatively low power factors. Also, the average load on these motors is fairly low. Therefore, it is advisable to consider the installation of capacitors to avoid paying the penalty imposed by most power companies for low power factor. They will be installed at the individual motors and switched with them, if voltage drop in the distribution system is to be corrected as well as power factor. Otherwise they may be installed in larger banks at the distribution center, if it is more economical to do so.

## Gas-Processing Plants

***14. Natural Gas.*** Natural gas varies widely in composition and contains undesirable materials such as water and sulfur compounds, which must be removed before the gas enters the transmission pipeline. Various chemical processes are used, and plant capacity ranges from 5 to 1000 million $ft^3$ of gas processed/day. By-products such as propane, butane, pentanes, and elemental sulfur are produced and marketed.

***15. Power Supply.*** Purchased power is used where available. Local generation is by reciprocating gas engines or gas turbines.

Electrical installation practice in gas plants is similar to that followed in oil refineries, as described below.

## Oil Pipelines

***16. Gathering Systems.*** These collect crude oil from the individual wells, or from tanks located near them, and carry it to tankage, where shipments may be accumulated.

***17. Trunk Lines.*** These feed crude oil from gathering systems to the main crude pipeline pumping station.

***18. Crude Lines.*** Crude lines are generally operated as common carriers. Since they handle crude oil for several companies and because crude-oil shipments vary greatly in quality and composition, storage tanks are necessary at stations along the pipeline so that batch shipments may be handled.

***19. Products Lines.*** These convey products from a refinery to the market area. Some products lines are operated as common carriers, while others are privately owned and operated.

***20. Operation of Pipelines.*** Batches of crude oil or products are dispatched through a pipeline and are withdrawn to tankage at the end of the line or at intermediate points. If a batch is being drawn off at an intermediate point, the downstream stations will operate at reduced flow.

Little mixing occurs at the interface between different batches. By careful scheduling and a knowledge of the pipeline it is possible to predict fairly accurately when an interface will arrive at a station. Interface detectors are installed also.

***21. Pumping Stations.*** A schematic diagram of a typical pumping station is shown in Fig. 21-2.

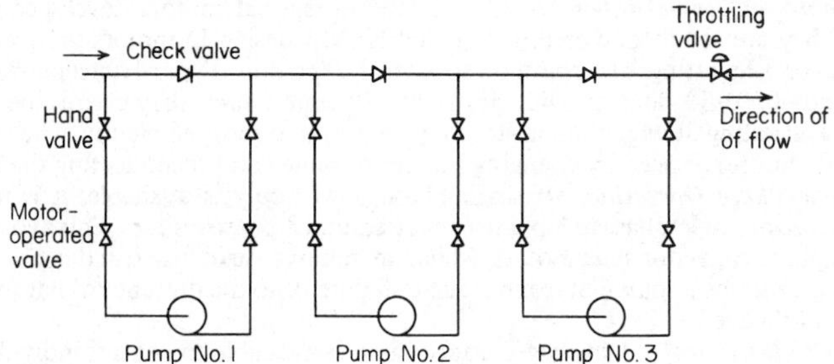


**FIG. 21-2** Pipeline pumping station.

***22. Arrangement of Pumping Stations.*** These are located at the head of the pipeline and at intervals along the line. Intermediate or booster stations must be capable of operating under varying conditions due to differences in liquid gravity, withdrawals at intermediate points, and the shutting down of other booster stations. Pumping stations often contain two or three pumps connected in series, with bypass arrangements using check valves across each pump. The pumps may all be of the same capacity, or one of them may be half size. By operating the pumps singly or together, a range of pumping capacities can be achieved.

Throttling of pump discharge may also be used to provide finer control and to permit operation when pump suction pressure may be inadequate for full flow operation.

***23. Control of Pumping Station.*** Pumping stations are often unattended and may be remotely controlled by radio or telephone circuits.

***24. Electrical System.*** Figure 21-3 is a typical electrical single-line diagram for a pumping station.

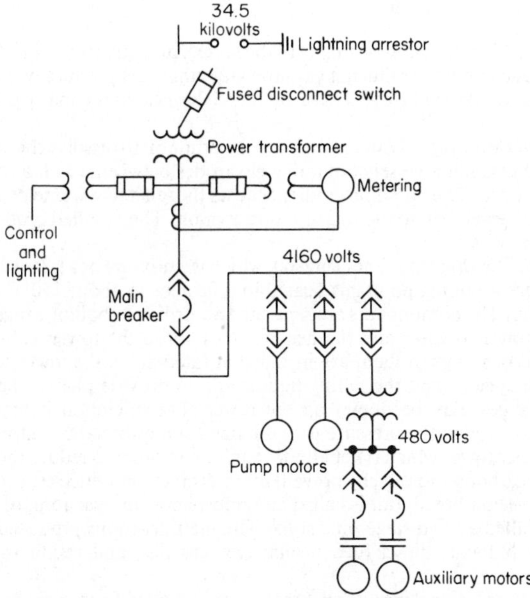


**FIG. 21-3** Single-line diagram of a pipeline pumping station.

***25. Motor Type for Main Pumps.*** The main pumps are driven by 3600-r/min induction motors having NEMA design B characteristics. Full-voltage starting is used.

***26. Motor Enclosure.*** Motor enclosures for outdoor use are NEMA weather-protected Type II, totally enclosed, fan-cooled, or dripproof with weather protection. Motors of the latter type are widely used. Not only are they less expensive than the other types, but they also have a service factor of 1.15. The above enclosure types are all suitable for the Class I, group D, division 2 classifications usually encountered.

If the pumps are located indoors, a division 1 classification is likely to apply. Motors must be Class I, group D, explosion-proof, or they may be separately ventilated with clean outside air brought to the motor by fans. Auxiliary devices such as alarm contacts on the motor must be suitable for the area classification. The installed costs, overall efficiencies, and service factors associated with the enclosures that are available will influence the selection.

## Natural-Gas Pipelines

***27. Natural-Gas Pipelines.*** The operation of gas pipelines is similar to that of oil pipelines except that a single gaseous product is handled instead of batches of liquids. Compressors, either reciprocating or centrifugal, are used instead of pumps, and the pipe diameters are much larger than those used in oil pipelines.

***28. Electrical Installation.*** The electrical installation and control systems are similar to those found in oil pipeline systems.

***29. Compressor Drivers.*** These are usually reciprocating gas engines or gas turbines, to make use of the energy available in the pipeline. Electric motor drives use slow-speed synchronous motors for reciprocating compressors and four- to six-pole induction motors with gear increasers for high-speed centrifugal compressors. Motor voltages, types, and enclosures are selected as for oil pipeline pumps. Motors used with centrifugal compressors must develop sufficient torque at the voltage available under inrush conditions to accelerate the high inertia load. They must also have adequate thermal capacity for the long starting time required, which may be 20 or 30 s.

## Oil Refineries

***30. General.*** Oil refineries vary greatly in the variety and quantity of products and crude oil throughput. A basic refinery producing gasoline and other fuel products would include operations such as those described below. These are merely typical of the many processes in current use.

***31. Crude-Oil Desalting.*** Water is added to the crude oil to dissolve the unwanted salt. The mixture is passed through a vessel containing electrodes between which a potential of several kilovolts is maintained. The potential gradient causes the salt and water to coalesce and settle to the bottom of the vessel, where the mixture is drawn off. The desalted crude oil is discharged near the top of the vessel.

***32. Crude-Oil Distillation.*** The crude oil, which is a mixture of a large variety of hydrocarbons having different boiling points, is heated in a furnace to about 750°F and then enters a fractionating tower. The components are separated according to boiling range, since the lighter ones rise in the tower as gases and the heavier ones fall in the tower as liquids. Trays with specially designed openings in them are installed at intervals in the tower to ensure intimate mixing of the rising gases and the falling liquids and to provide places where liquids having certain boiling ranges may be drawn off the tower. The operation is first performed in a distillation tower in which the pressure is maintained somewhat above atmospheric pressure and again in another tower which is kept under vacuum in order to reduce the boiling temperatures of the hydrocarbons and thereby prevent thermal cracking, which is decomposition due to excessive heat. The combined unit is called an atmospheric and vacuum, or A & V, unit. It is also sometimes called a "two-stage pipe still." The main fractions produced are condensable gas, gasoline components, diesel fuel, heaing oils, gas oils, and residuals. All of these are processed further.

***33. Fluid Catalytic Cracking.*** To increase gasoline yield from a crude oil, heavy gas oil from the distillation unit is processed in a cracking unit. The heavy molecules are brought in contact with a catalyst under proper conditions of temperature and pressure and are converted into lighter molecules. Thus lighter products are formed which are suitable for use as gasoline and distillate fuel components. The use of a catalyst promotes the cracking reaction at a lower temperature and pressure and produces larger quantities of products having more valuable qualities than is possible with straight thermal cracking.

The clay catalyst is in powder form, and it is handled as a fluid. The cracking reaction causes the formation of carbon deposits on the catalyst particles. These are removed by controlled burning in a regenerator vessel. The catalyst is continually being circulated through the reactor and the regenerator by means of gas and air flow, respectively.

***34. Combustion Air Blower.*** Combustion air for the regenerator is provided by a large centrifugal air blower driven by an induction motor and gear increaser or by a steam or gas turbine.

***35. Gas Compressor.*** Some of the products of the catalytic cracker are drawn off as gas. This gas is compressed, condensed, and fractionated to provide other fuel products and feedstocks for petrochemical processes. A centrifugal compressor is used, and it is driven by an induction motor through a gear increaser or by a directly coupled steam turbine or gas turbine.

***36. Catalytic Re-forming.*** In this process hydrocarbon molecules are rearranged and recombined to form molecules of higher octane rating which can be used as gasoline components. Hydrogen is produced in this process. Large reciprocating and centrifugal compressors are used to move the large volumes of gas involved in the process. Hydrogen is a by-product of this process.

***37. Hydrofining.*** This process uses hydrogen in the treatment of other products, such as jet fuels, distillates and lubricating oils, to improve quality and to remove sulfur.

***38. Cooling-Water Pumps.*** Large volumes of water are used for cooling-process streams and for condensers. Water may be conserved by the use of induced-draft cooling towers.

Cooling-water circulating pumps are driven by vertical motors. Standby pumps are driven by steam turbines.

***39. Power Supply.*** Refining operations are continuous processes, and uninterrupted runs of 1 or more years are expected between planned shutdowns for maintenance or turnarounds. Therefore, it is essential that the power supply be extremely reliable.

Often duplicate full-capacity feeders are installed to the refinery, and sometimes these are run from different substations, for increased security. Many refineries practice cogeneration to take advantage of their high steam loads.

***40. Distribution Systems.*** Because refinery loads are often concentrated to a large extent in fairly well-defined process areas, it is common to install unit substations in the major process areas. These substations contain power transformers to provide 4160- or 2400-V power, 480-V power and lighting transformers, or provision for feeding lighting transformers. They also contain all associated switchgear, motor-control, and emergency generators.

***41. Process-Unit Distribution.*** A typical refinery process-unit distribution system is shown in Fig. 21-4.

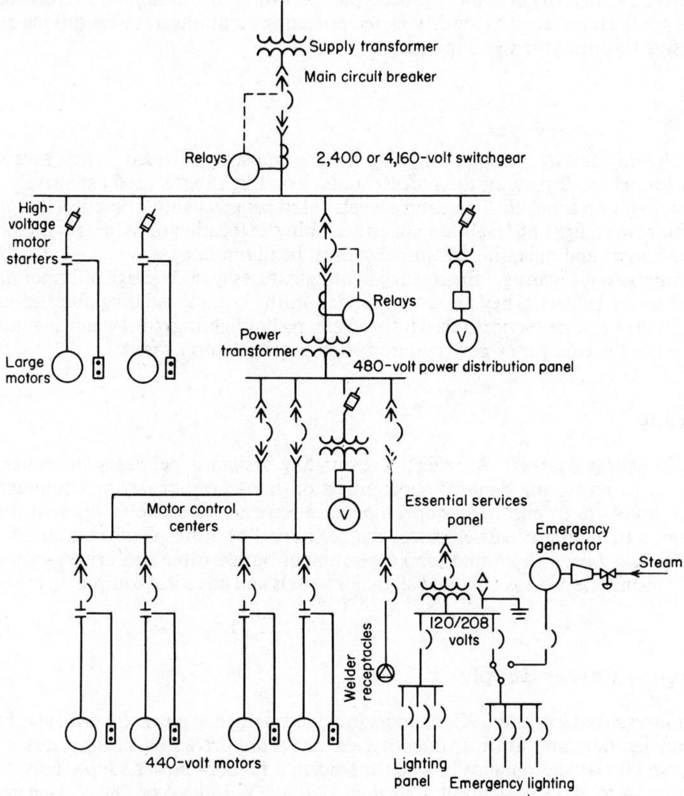


**FIG. 21-4** Typical refinery-process unit distribution system.

***42. Area Classification.*** Flammable gases and vapors are processed in oil refineries. Therefore, it is necessary to classify the various locations according to the material that is present and also according to the degree of hazard expected.

Reference should be made to the applicable electrical code for requirements governing installations in classified areas.

The actual classification of areas is usually made by the electrical design engineer in consultation with persons who are familiar with the operation of the process.

The most common classification for refinery process units is Class I, group D, the class being that of hazardous gases and vapors, and the group comprising gasoline and many of the petroleum products.

Current practice is to classify outdoor, freely ventilated process areas as division 2. Indoor process areas that are not freely ventilated and places below grade level are classified as division 1. Areas in which a permanent ignition source is located, such as around a furnace, are not classified. Pressure-ventilated unit substations and control buildings are not classified. However, some companies follow the practice of classifying control rooms as division 2.

## Electric Motors

***43. Type of Motor.*** Two-pole induction motors having NEMA design B characteristics are used to drive the majority of refinery-process pumps. Motors operating at slower speeds are used for some applications, such as for driving reciprocating compressors or for driving centrifugal compressors through gear-speed increasers.

## Lighting

***44. Lighting Levels.*** Adequate lighting coverage must be provided for safe access by operating personnel to all parts of the process units. Local lighting is used extensively on tower platforms, in pump areas, etc. Footcandle levels follow general lighting practice. In some places, such as on towers, lighting levels are not critical, but the location of fixtures is important. For example, ladders and operating instruments must be illuminated.

***45. Emergency Lighting.*** Emergency lighting fixtures give coverage of important accessways and tower ladders. They are also provided in the control building and the substation. Fixtures used in regular service are fed from an emergency lighting panel which is automatically transferred to the emergency power source when the main supply fails.

## Grounding

***46. Grounding System.*** An effective grounding system is necessary in refinery process units *(a)* to avoid having dangerous potentials on non-current-carrying equipment during electrical faults, *(b)* to ensure fast operation of electrical protective equipment under fault conditions, and *(c)* to dissipate electrical charges caused by atmospheric conditions.

***47. Ground Loop.*** A ground loop consisting of buried cable and driven ground rods is installed around the process unit, and all major vessels and electrical equipment are connected to it.

## Emergency Power Supply

***48. Emergency Generator.*** Generators provide emergency power for critical motor loads, emergency lighting, and for instruments. Generators are driven by steam turbines or diesel engines, which start automatically when the normal refinery power supply fails. Capacities range from 15 to 300 kW. Output voltage is 120/208 V 3-phase or 480 V 3-phase. Load is automatically transferred to the generator after it has reached normal operating speed.

***49. Batteries.*** Batteries are used when a continuous supply of power must be available for process instruments, emergency controls, and shutdown devices. They are also used for remote-control systems. Voltages of 24, 48 and 120 V are used. Inverters provide power for critical ac instruments and are sometimes used for computer power supply. Battery chargers are transferred to the emergency generator upon loss of normal power.

Batteries should have enough capacity to carry full load for half an hour with the charger off. Charger capacity must be sufficient to carry the full dc load and to recharge the battery in 8 h.

## Chemical Plants

***50. General.*** Electrical installations in chemical plants are similar to those found in oil refineries. However, some processes may involve materials that require special attention because of factors such as corrosivity and toxicity.

## *STEEL MILLS*

*Revised by N. RICHARD FRIEDMAN*

***51. Industry Description.*** The domestic steel industry includes blast-furnace integrated steel makers; nonintegrated minimills, and independent producers of wire, bar, and pipe made from raw steel. Blast-furnace operations use either an open hearth or basic oxygen furnace, with electric arc furnace used in the minimills. In 1984, approximately one-third of total raw steel was melted by electric furnaces, with the balance primarily melted by the basic oxygen furnace.

***52. Power-Distribution System.*** Integrated steel mills having blast furnaces and coke plants make use of the combustible gases from these processes by burning them to produce power and process steam. Many older steel plants produce and use power at 25 Hz, utilizing primary distribution voltages of 6.9 to 13.8 kV and secondary systems of 4160 or 2400 V and 480 V. Modern steel plants and modern parts of older plants utilize 60-Hz power exclusively, with primary distribution at 69 or 138 kV and secondary voltages of 13.8 kV, 4160 or 2400 V, and 480 V. Power can be supplied by public utilities, by in-plant generation, or by a combination of the two. Some plants having both 25- and 60-Hz systems have conversion equipment, typically large Scherbius sets or rectifier-inverter systems, for transfer of power from one frequency to the other. Where power is supplied from both a public utility and in-plant generation, there is usually some provision for controlling the maximum demand and improving the power factor of the portion of the load supplied by the utility to avoid high penalty charges.

The presence of numerous cranes and other equipment requiring dc motors with some control over speed has led to the extensive use of 250-V constant-potential dc shop circuits in most steel mills. In older plants the direct current is supplied by rotary converters, motor-generator sets, or mercury-arc rectifiers. In modern plants it is supplied by silicon diode or thyristor rectifiers. The trend is toward the elimination of dc shop circuits by the use of ac cranes and package thyristor power supplies for drives requiring variable speed.

***53. Primary Production.*** The basic steelmaking areas of an integrated steel plant consist of *coke-oven batteries* for conversion of coal to coke, *blast furnaces* for conversion of iron ore to molten iron, and *steel-producing units* for refining molten iron and other alloy ingredients to steel. Once this basic steel has been produced in ingot (block) form or "continuous cast" into semifinished bars, it is ready for subsequent rolling into a usable size and shape.

Power consumption per ton of steel produced is low in the basic steelmaking areas because the products are handled in molten or bulk form, compared with the rolling mills, where reheated or cold steel is literally squeezed and stretched to the desired size and shape. Much of the electric power consumption in these primary producing areas is associated with auxiliary drives involved in material handling, water, air, and by-product utilization and mobile equipment.

The processes of iron reduction and steel refining, with their many, sometimes elusive variables, do lend themselves to automatic and computer control. Open-loop computer systems have been applied to *blast-furnace* and *basic-oxygen-furnace* (BOF) operations. Raw-material handling and charging functions in *blast-furnace stock houses* and BOF have been automated extensively.

In a minimill, which includes specialty shops, the basic raw material is steel scrap which is melted and refined to produce raw steel and steel products. The steel is melted by an electric arc in which the current passes from one electrode through an arc to the scrap charge, and then from the charge to another electrode. The molten steel is then refined by rapid oxidation and the reaction of impurities with added slag materials.

***54. Rolling Mills.*** Rolling mills are classified either according to their construction or according to the material processed. Classified according to construction, mills are generally two-high or four-high, with a few existing three-high mills. Four-high mills consist of the usual two work rolls in contact with the product, with two additional "backup" rolls which are much larger and allow high rolling pressure without excessive deflection of the work rolls. A "universal" mill has vertical or edging rolls in tandem with the horizontal rolls. This permits a reduction of width or control of the edges of the product in the same stand where a reduction in thickness is taking place.

The principal types of mills, classified as to product rolled, are as follows:

*a. Blooming mills* roll ingots to blooms, or slabs. All material rolled in steel mills, except that which is direct continuous cast into slabs, blooms, or bars, first passes through this type of

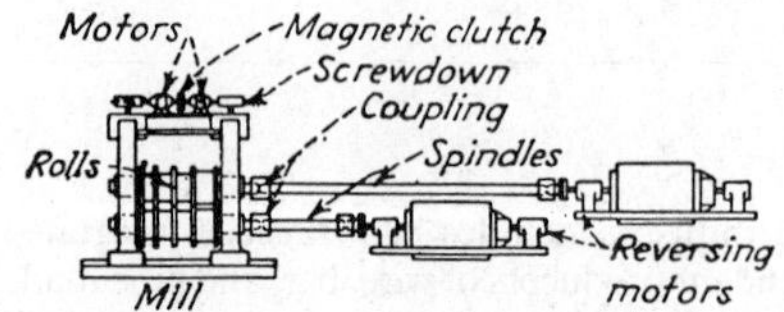


**FIG. 21-5** Two-high reversing mill with twin roll-drive motors.

mill, or its equivalent, to be reduced to proper dimensions for handling in the finishing mills. These are generally single-stand, two-high reversing. Slabbing mills are a modification of the blooming mill. They are usually universal mills, which permits convenient rolling of wide slabs by eliminating frequent turning of the ingot. (See Fig. 21-5.) Blooming and slabbing mills may have such automatic features as preset of roll openings and speed synchronization between main and edger rolls. The preset information is sometimes stored on business-machine cards and read into the mill control system by a card reader, or it can be stored in the computer memory in computer-controlled installations. Blooming and slabbing mills are powered by large low-speed dc motors operating from a variable-voltage system. A typical slabbing mill has a total of four 3000-hp motors driving the horizontal rolls and two 2000-hp motors driving the edger rolls at rated motor speeds of 40/80 r/min. The dc power for blooming mills has been traditionally supplied by generators using the Ilgner system, but the present trend is toward the use of thyristor power supplies connected in rectifier-inverter configurations.

*b. Hot-strip mills* roll sheets, strip, and plate from heated slabs. These mills can be placed in two categories: continuous and semicontinuous. (These terms designate the type of rougher which precedes the finishing train.) The *continuous mill* has two to six mills in line which reduce the slab to a predetermined thickness for subsequent rolling through the finishing train (see Fig. 21-6).

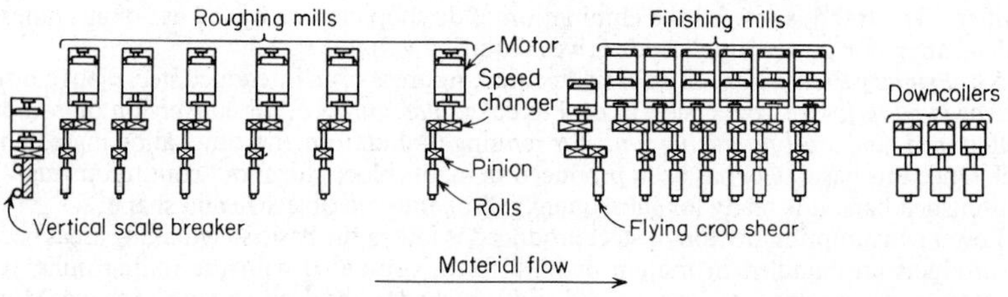


**FIG. 21-6** Hot-strip mill.

The *semicontinuous mill* has a reversing rougher on which reduction is made by running the piece back and forth through the mill, which reduces the slab to a predetermined thickness for subsequent rolling through the finishing train.

The finishing train consists of five to seven stands, closely coupled and synchronized in speed, in which the piece is reduced to the desired gage. As a general rule, the piece is in all finishing stands simultaneously.

Roughing stands of a continuous mill are usually driven by ac motors, since speed synchronization with an adjacent stand is not required. Many existing roughing mills utilize wound-rotor induction motors with flywheels and water-slip regulators, but today synchronous motors predominate. The roughers on semicontinuous mills are driven by dc motors.

Direct-current motors operating from a variable-voltage system are universally used for finishing stands. Tension between stands must be accurately controlled at a relatively low value, since the hot steel is in plastic form and excessive tension results in "necking," or breaking. Constant-tension "loopers" are used between stands for interstand tension regulation. The looper is pushed up by air, hydraulic pressure, or a torque motor, and the upward movement is restricted by the strip. The looper position is then fed back to the electrical control system, and the speed of individual finishing-stand motors is regulated to maintain the proper looper position.

Various measuring devices must always be in operation, measuring gage, screw position (roll opening), looper position, roll force, and width of strip. With the feedback from these devices, the positioning of the screw-downs, which control gage, and the edgers, which control width, can be regulated during the rolling of the strip. The newer mills make use of digital positioning to set up the mills prior to the entrance of the bar into the mill.

The thread-speed reference is set into the speed controller, and this speed is maintained by utilizing the pilot-generator feedback. When the strip enters the mill, the load increases sharply, bringing the current feedback to the armature controller into effect. By controlling the armature and field, rapid speed changes while complete control is maintained are possible, both with and without load. The acceleration control, when called upon, feeds into all the finishing-mill motors for uniform acceleration.

For setting up the mills in advance, digital position regulators are generally used; however, some of the newest mills have small digital computers controlling each stand. The principle of operation is basically the same in both systems. In Fig. 21-7, the desired position for the drive is

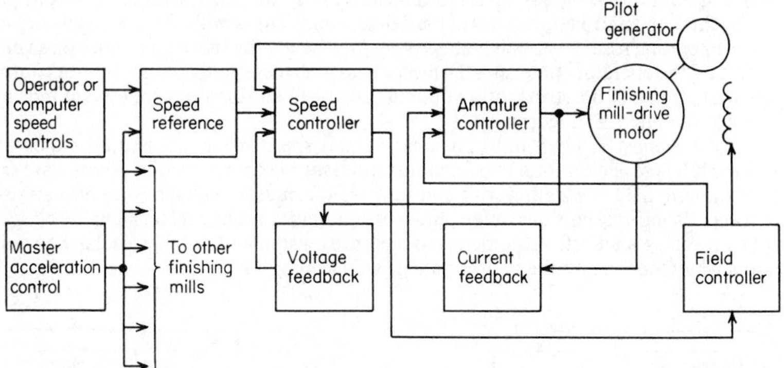


**FIG. 21-7** Position regulator.

set into the regulator, and then by taking the difference between this reference and the feedback signal from a position sensor, the drive is operated until this difference is zero. The thread speeds of the mills can be set up the same way except that, instead of moving to a preset position reference, the stands are accelerated or decelerated to a preset speed reference.

As the strip cools during rolling, it becomes harder and the output gage changes. Part of this change in gage can be controlled by accelerating the mill, but much of it must be controlled by changing the position of the screw-downs. The newer hot-strip mills use automatic gage control systems (AGC) to regulate screw-down position while the strip is in the finishing mill. There are two types of AGC in use, a constant-gage system and an absolute system. Both systems work basically the same, by taking a reference either from the head end of the strip in the constant-gage system or from preset switches in the absolute system and maintaining it throughout the length of the strip. This reference is compared with the actual gage of the strip, measured as a function of roll force and screw-down position, or by an x-ray gage, and the screw-downs are moved to bring both these signals to the same value. This is a continuous operation during the period of time the strip is in the finishing mills (see Fig. 21-8).

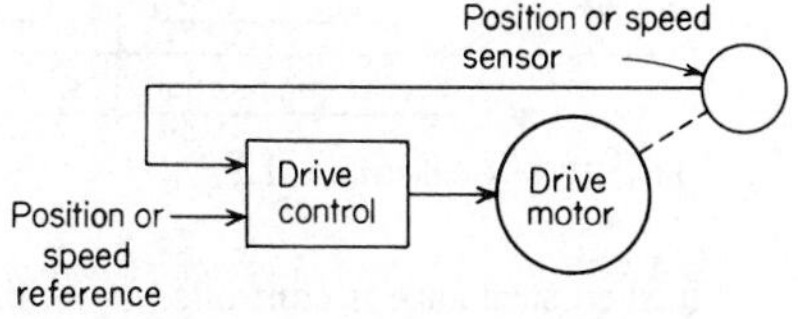


**FIG. 21-8** Automatic-gage control system.

Modern hot-strip mills are built with partial or complete computer control options. Rolling in a completely automatic or computer-controlled state, the mills function as previously explained, except that the computer provides the position and speed references instead of operators. With a computer-controlled mill, these references can be constantly changed as the computer receives new information from the strip.

The handling of steel between roughing stands and between the finishing stands and the downcoilers is accomplished by roll "tables," which are a series of motor-driven rolls upon which the steel lies. In some installations, one large table motor drives a number of table rolls through mechanical gearing and shafting ("line-shaft drive"); in others, each roll is driven by one small motor. Permanent-magnet field dc motors have been particularly successful for this arrangement. Table rolls are generally driven by dc motors operating from a variable-voltage system, but a number of variable-frequency induction-motor systems are in use. As thyristor adjustable-frequency power supplies have become more reliable, this system is becoming standard.

*c. Tandem cold-strip mills* are used to cold-reduce previously rolled hot-strip mill products down to thicknesses as low as 0.002 in. Special "foil" mills have been built which can roll even lighter gages. A cold-strip mill is similar to the finishing stands of a hot-strip mill, except that tension between stands plays a much more important role in reducing the thickness of the steel. Modern cold-strip mills are built for finishing speeds in excess of 5000 ft/min.

Cold-strip mills are generally three- to six-stand mills, four-high, with a coil box or payoff reel on the entry end and a tension reel at the delivery end. These mills are universally driven by dc motors operating from a variable-voltage system. Usually the individual stands are voltage-regulated, and the operator establishes the motor field and, thereby, the stand speed according to the gage-reduction and the strip tension desired. Load-cell tensiometers are used to indicate to the operator the interstand tensions.

Some of the most modern mills have each stand speed regulated, and all major drives incorporate full field acceleration to base motor speed for maximum torque. Above base motor speed, automatic field weakening with constant rated armature voltage is used to attain top motor speed. Payoff-reel tensions, interstand tensions, and winding-reel tensions are accurately controlled by using load-cell tensiometers to measure tension and provide feedback to tension regulators which operate on the appropriate drives (see Fig. 21-9).

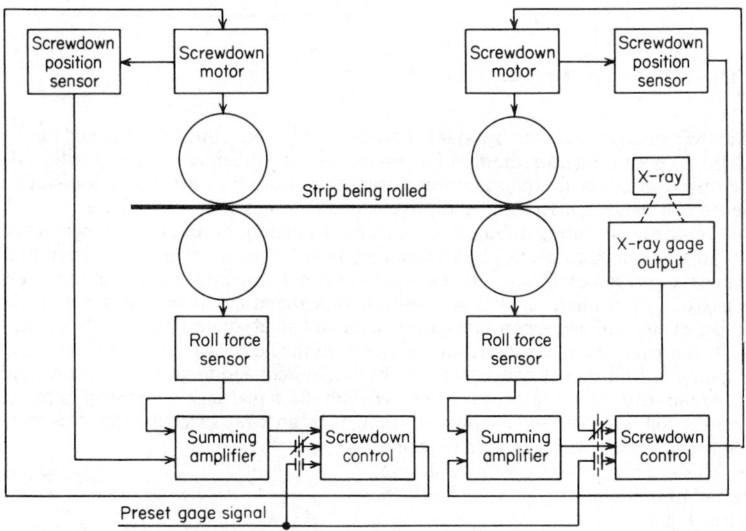


**FIG. 21-9** Cold-strip mill.

Finished steel gage is controlled by utilizing x-ray gages to provide information to AGC regulators. Gage control may be accomplished by operating on interstand tensions only or by a combination of interstand tensions and work-roll openings. Work-roll openings ("screw-down positions") are controlled by digital position regulators which hold the screw-down position constant to within 0.0001 in. To help further in producing constant finished gage, the rolling forces on each stand are held constant by using speed-programmed digital regulators to hold constant rolling force at all speeds. Rolling force is measured and indicated by load cells placed under each stand housing.

*d. Billet mills,* used to roll blooms into billets, are frequently of a continuous type with rolling stands in tandem with several sets of passes in the rolls so that different-sized billets can be produced from a given bloom without changing rolls. This type of mill is used for producing only a limited range of sizes, which are further reduced in finishing mills.

*e. Plate mills* produce plates from slabs previously rolled by a blooming or slabbing mill. These mills are generally single-stand mills, either two- or four-high reversing or three-high running continuously in one direction, though some mills are provided with several stands in tandem, one serving as a rougher or breakdown stand. They are sometimes universal mills, being provided with vertical rolls for finishing the edges of the plates.

*f. Structural mills* are used for rolling beams, heavy angles, channels, etc., from blooms or billets. Such mills rolling the smaller structural sections and miscellaneous shapes are sometimes called "bar mills." They are frequently three-high, with more than one stand in line, and frequently have a separate rougher or breakdown stand. Some also have two-high reversing finishing mills with edger and vertical, as well as horizontal, rolls, to produce wide-flange (H) beams.

*g. Rail mills* are special mills for rolling this product only from blooms or billets, though rails are sometimes rolled on structural mills.

*h. Merchant mills* are used for rolling small angles, channels, rounds, squares, etc., from billets. This classification is generally applied to mills rolling the smaller sections used for miscellaneous purposes. Most are two-high, in tandem; however, some are arranged for "cross-country" rolling.

*i. Rod mills* are a specialized type of merchant mill for rolling small rounds, usually from No. 5 BWG and upward, which are later drawn into wire. Modern rod mills are arranged to roll a multiplicity of strands simultaneously.

*j. Temper mills* are used to produce steel strip of the desired temper, flatness, surface, and luster by using rolling pressure and tension. Reduction in thickness is incidental in the process and is normally very slight. Temper mills generally have one or two four-high stands and are similar to cold-strip mills, except that they are of lighter construction, have less powerful motors, and have simpler electrical control systems. Temper mills have been built and operated at speeds in excess of 7000 ft/min.

*k. Tube mills* are used to produce tubes, for example, pipe and conduit, by either the seamless or the butt- or lap-weld process. Seamless mills pierce a solid billet and then form the pierced billet into tubes of the desired size and thickness. Butt- or lap-weld mills form and weld previously prepared pipe skelp into tubes.

***55. Process Lines.*** In most fully integrated steel mills, there are various types of process lines such as pickle lines, galvanizing and aluminizing lines, tin lines, continuous-annealing lines, and shear and slitting lines. Process lines are generally powered by dc motors operating from a variable-voltage system. Tensiometers are sometimes used to indicate and even control tension at strategic places. The most important process lines are described below.

*a. Pickle Lines.* After steel is hot-rolled and before it can be cold-rolled, it is processed through a pickle line. In this process, all scale and impurities that have built up on the surface of the strip are removed. This is accomplished by running the strip through a tank of acid, which "pickles" the steel.

To maintain constant material flow through the pickling tank, a welding machine is used to weld the tail end of one strip to the head end of the following strip. To avoid stopping the line every time a weld is to be made, slack in the strip is allowed to develop immediately following the welder in a pit called a looping pit. Then, as the tail end of the strip comes into the welder, it is held there, and the slack in the looping pit is taken up to maintain constant material flow in the pickle tanks while the weld is made. There is another looping pit at the end of the line preceding the recoiler that performs the same function while a full coil is being removed and another started.

*b. Galvanizing,* or coating with zinc, and aluminizing can be described as the same process since the same line is used for both processes, with only the coating changed.

The steel must first be pickled and cold-rolled before coming to the galvanizing line. There, just as in the pickle line, the first process is welding, head to tail, of one strip to the next. In place of looping pits, however, traveling looping cars are used to absorb or supply strip while the entry and delivery ends of the line are stopped. From the entry loop car, the strip goes through a controlled atmosphere to a temperature near the melting point of the plating material. The actual plating then takes place as the strip is "dipped" or run through a tank of molten plating material. After plating, the strip passes through the delivery loop car and is coiled or cut to length.

*c. Tin lines* are used for coating strip steel with a protective layer of tin after it has been cold-rolled, annealed, and temper-rolled.

A welder and a looping tower are used at the beginning of the line. The looping tower stores strip in numerous vertical passes, as opposed to the horizontal looping car in the galvanizing line. Cleaning and rinsing is the next step in the process. The tin is deposited on the steel strip electrolytically, by using pure tin anodes and a plating solution which may be acid or alkaline, depending upon the process. The low-voltage plating current is produced by plating generators or (on modern tin lines) by silicon-diode rectifiers. The current is applied directly to the tin anode and, after passing through the solution and the steel, is returned to its source through current-collector rolls in contact with the strip. After plating, the strip is put through a "reflow" process which heats it to the melting point of tin and gives it a shiny appearance and the proper iron-tin alloy bond. The reflow furnace may be gas-fired, or it may be a radio-frequency induction heating process.

After reflow the strip passes through the delivery looping tower and is then coiled or cut into predetermined lengths.

***56. Instrumentation.*** More than any other factors, the rapid advances in instrumentation technology and techniques have made possible the revolutionary strides in automated process control and increased productivity in the steel industry.

Typical examples of steel-mill instrumentation are as follows:

*a. X-ray and radioactive-isotope devices* for (1) continuous measurement of moving strips and slabs from strips as thin as 0.002 in to slabs as thick as 2 in by using absorption techniques, (2) continuous measurement of coating thicknesses on steel strips by using fluorescence or reflectance techniques, (3) analysis of the chemical properties of steel by using spectrographic techniques, and (4) detection of refractory erosion in blast furnaces by using absorption techniques.

*b. Infrared and ultraviolet devices* for (1) continuous width measurements of moving hot-steel strips and slabs, optical scanning, and tracking of the edges; (2) measurement of hot-steel temperatures by using optical pyrometric techniques; and (3) detection of pinholes down to 0.001 in or less in diameter in moving strips, usually by using visible or near-ultraviolet light.

*c. Ultrasonic devices* for the detection of flaws in slabs.

*d. Magnetic devices,* which produce inferred measurements of physical or Rockwell hardness of moving strips by measurement of magnetic hardness.

*e. Communications equipment,* which is employed for (1) two-way voice communications by using either frequencies in the radio-frequency spectrum or frequencies near 100 kHz for communications with overhead cranes by utilizing the hot rails as the information-transportation medium; (2) remote control of equipment such as overhead cranes; (3) closed-circuit television systems, which are especially useful for viewing high-temperature and remote areas; and (4) telemetry and supervisory control equipment, which allows operating personnel to remotely monitor and control the different processes involved in the production of steel.

*f. Digital devices* for (1) positioning of screw-downs; (2) control of machine tools; (3) measurement and control of elongation or stretching, length, and speed of steel strips; and (4) analog-to-digital conversion of process information for introduction into process control computers, data loggers, regulating systems, and telemetering systems.

***57. Computer Control.*** Digital computers are used both to monitor and to control processes in primary production and rolling-mill areas, including blast furnaces, basic oxygen furnaces, blooming and slabbing mills, hot-strip mills, plate mills, tandem cold mills, and temper mills. Since new steel facilities are larger, faster, and thus more demanding than older plant operations, most cannot operate without process computer systems. Efforts to increase computer reliability have also led to the application of hierarchical computer systems and redundant hardware in critical areas.

Modern process computer systems as applied to steel mills have some or all of the following capabilities:

*a. Mathematical modeling* for predictive control of the process by simulating process functions in the computer.

*b. Data acquisition* by instruments and sensors, which provide process feedback, as well as digital-control replacement of hard-wired controllers, wired from the primary sensor to the computer as an input. Output is wired directly to the control of the process variable.

*c. Alarm functions* performed by the computer, including first-out functions and subsequent sequence of start-up.

*d. Computer peripherals,* particularly CRT terminals, are used to chart long-term trending of process variables for storage in bulk memory or on magnetic tape for future recall and/or analysis.

## ELECTRICITY IN AGRICULTURE

*Revised by N. RICHARD FRIEDMAN*

### Field of Application

***58. Extent of Electric Service.*** By 1985, approximately 99% of the 2,285,000 farms in the United States were receiving central-station electric service. The 1983 nation's farm population was 7,029,000, a sharp decrease from the 8,864,000 farm residents in 1975, and an even more dramatic decline from the 1963 farm population of 13,367,000.

At the same time, the average farm size has increased from 420 acres in 1975 to 445 acres in 1985. Over the decade ending in 1985, farm output increased by 22%, primarily as a result of better farm management practices, the application of chemical fertilizers, and mechanization.

***59. Energy Use.*** The average annual use of electric energy on United States farms in 1980 was 16,301 kWh. As a result of automation and mechanization, this is expected to continue rising at an accelerated rate.

***60. Electronic Controls.*** Agricultural use of electronics is primarily for the control of energy-using systems, communications, and data processing.

Planting and harvesting machines are equipped with electronic monitoring devices indicating information such as shaft speeds, material flow rates, temperatures, and seeding malfunctions. Reed switches or hinged-plate switches are used to flash light signals or actuate buzzers and horns, and miniature electrical generators, the output of which is proportional to the speed, indicate shaft revolutions per minute relative to a desired value. This information is displayed on a console within the combine cab, or on the tractor.

Electronic crop thinners, using a phototransistor scanning system for each crop row, have resulted in higher vegetable crop yields than is possible with hand thinning, since earlier thinning and a more uniform plant population are possible. Other uses of electronics include automatic temperature and humidity controls for crop drying and storage, and automated surface sprinkler systems such as the solid-set, permanent-overhead, and center-pivot type (utilizing control consoles, operating through buried cables, microwave channels, VHF radio, pressure and temperature control switches) to provide round-the-clock irrigation control. Overriding time controls provide cooling when predetermined temperatures are reached.

Livestock-feeding systems use electronic controls for time-interval feeding of individual animals based upon current production level, weight, age, etc. A transmitter at the feed station sends a signal to a transponder unit on the animal, which upon activation switches on a relay controlling the feed unit. Automatic data recording can be accomplished for individual animals by means of a special neck band, thus enabling rapid and detailed collection of data which, aided by a computer, facilitate efficient management of larger enterprises. Current research and development indicate that automatic guidance of tractors and machines, and electric motor drives for farm tractors are possible in the near future.

### Farm Structures

***61. Water Systems.*** Water requirements for the farm household and farm enterprises, excluding irrigation, are frequently supplied by a single well. The water-supply equipment is usually an automatic hydropneumatic or air system having pumping capacity of 300 to 600 gal/h and using a ¼- to 1-hp motor, depending upon the total head in feet and the rate of pumping. Home water requirements average 50 gal/(person)(day). In addition, livestock requirements must be added: each horse, steer, or dry cow, 12 gal; each milk cow, 35 gal for drinking and washing equipment; each hog, 3 gal; each sheep, 2 gal; each 100 chickens, 8 gal. For yard fixtures, each ¾-in hose outlet requires 300 gal/h.

Where the source of supply is not more than 22 ft below the pump, a shallow-well system can be used. A jet-centrifugal pump has a practical lift limit of 80 to 100 ft, and piston-type pumps can go as deep as 800 ft with a suction lift below cylinder of 22 ft. This type is placed directly over the well and is generally recommended where pumping depths exceed 80 ft. Automatic pressure switches are usually set to start the pump when the pressure falls to 20 lb and stop it when 40 lb has been obtained. The energy requirement per 1000 gal of water pumped rarely exceeds 2 kWh.

***62. Heating Systems.*** Electrical heating of farmstead structures is generally confined to milk houses, individual pen-type areas for young livestock, and poultry brooders. Electric heating in the milk house is ideal, as it is odorless, conveniently controlled, and meets the high sanitary standards required. The milk-house temperature should not exceed 40°F. Several types of heaters have been successfully used: (1) the forced-air circulating type requiring 1500 to 3000 W, (2) batteries of 250-W infrared heat lamps directed toward working areas and water pipes, and (3) "heat-pump" systems, which utilize the heat removed in cooling the milk. In this type the ice-bank refrigeration system (either bulk or immersion coolers) extracts heat from the water in building up the ice, the heat thus being available for the milk house. Electricity used in this indirect manner produces about three times as much heat as it would if directly used in a resistance heater. Only coolers with ½-hp or larger motors are recommended for this.

In the colder regions, the milk house must be insulated for the most economical cost of installation and operation. In these areas an electrically heated milk house needs at least a 1500-W heater serviced by a 230-V line. Thermostats are usually attached to the heater unit, and operating consumption ranges from 1000 to 3000 kWh a season.

The need for infrared heat lamps during the first week of hog farrowing and sheep lambing has been proved. A 250-W lamp will heat an area 24 in in diameter when 3 ft above the floor. The lamps should be positioned at least 6 in above the animals and at least 30 in above the floor when bedding is used.

***63. Ventilation Systems.*** Electrically powered mechanical ventilation of livestock structures provides low-cost positive control for the removal of excess animal body heat, objectionable odors, condensation, and temperature and humidity control. A full-grown cow will give off 3000 Btu/h of body heat; 1000 chickens about 800 Btu/h. Accurately controlled tests with dairy cows at the University of Missouri showed that temperatures above 75°F and relative humidities over 75% resulted in sharp declines in milk production and body weight.

In general, summer ventilation should maintain inside temperatures equal to or below the outside temperature, while in winter the reverse is true. Thermostatically controlled motor-driven fans are installed as required, with adequate fresh-air intakes to prevent excessive energy costs. Two-speed fans, chosen to move the maximum air volumes required for various livestock, will permit airflow to be reduced in cold weather. Fan motors range from 1/20 to ½ hp and will consume 250 kWh/year and up, depending on usage. One kilowatthour of electricity will move about 1 million $ft^3$ of air.

## Dairy and Livestock Production

***64. Automatic Feeding.*** Mechanical and automatic feeding of livestock and poultry is of growing importance. The flow of materials proceeds from individual storages to blenders, grinders, conveyers, and finally to distributors, as shown in Fig. 21-10.

In a typical automatic system feed is processed at a low rate over a maximum time period. This allows the use of small, low-capacity electric motors at each stage and does not require overhaul of the 100- and 200-A service found on most farms. Capacitor or repulsion start motors, totally enclosed and with continuous-duty rating, are recommended. Hammer or burr grinders with 2- or 3-hp motors are required for small grains (up to 5-hp motors for ear corn). Other motors are generally ½- to 1½-hp for the movement of the materials. In proportionally blending feed components from storages, belt, auger, or electric-vibrator feed meters are used. After this mixture is ground, it is moved by auger or pneumatic blower to distributor conveyers in the feed lot or barn which are of the endless-chain, belt, or auger types. In cattle-feeding systems silage may be automatically removed from the silo. For feeder cattle the grain-supplement mixture is then metered onto the silage and the complete ration conveyed to the feed bunk for distribution. Dairy cows usually receive the grain ration separately from the silage, either in the barn or in the milking parlor.

***65. Controls*** are of the simple off-on switch type for manually controlled systems, but for

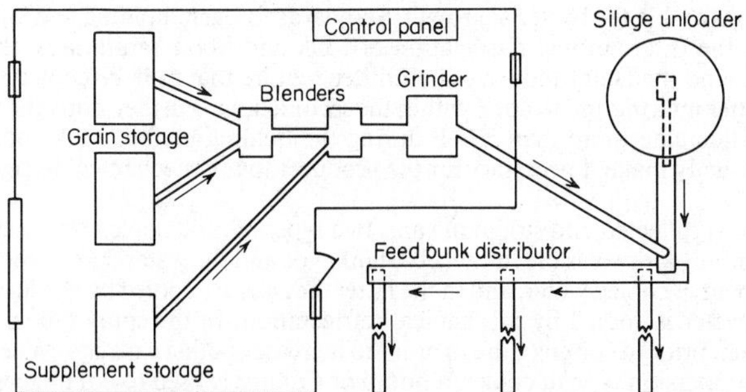


**FIG. 21-10** Cattle-feeding system designed to feed silage in addition to grain and supplement.

fully automatic feeding the switching is accomplished by electric timers or sensing elements, which measure temperature, level of material, or pressures, at each point in the movement of feed from storage to final distribution. These sensing devices are used with current-carrying relays such as a magnetic starter. This starter may have auxiliary switches which control circuits to other magnetic starters. Thus, interlocking of each stage of feed transmission is accomplished. This is necessary where a sequence of operations requires overload protection for each motor. Since these systems depend on continuance of electric power, it is advisable to provide an emergency generator to handle at least a part of the demand in the event of a power failure.

Automatic feeding equipment introduces new hazards for operators. Starting by switch or time clock may be unexpected. Open conveyers should be shielded. Power must be disconnected while repairs are made. Wiring should be of the underground-feeder type, and all electrical equipment must be grounded.

***66. Milk Handling.*** With the increase in numbers of cows being milked and the milk quality required to retain grade A fluid-milk markets, many producers have shifted to some degree of automation in milk handling. The completely automatic system is one whereby milk is taken from the cow, carried through a vacuum pipeline to a bulk cooling and storage tank, whence it is pumped into a bulk transporting truck for delivery to the processing plant, without having been lifted by, carried by, or exposed to human hands.

In the latest development, cows enter a central milking room, commonly called a milking parlor. Three or more are milked simultaneously by machine. The milk is moved by suction through glass or stainless-steel tubing to the bulk tank, located in a separate room. Bulk tanks are either of the vacuum or of the atmospheric type. The latter requires the addition of a releaser, since the tank itself is not under vacuum.

In conventional stanchion barns cows are milked by machine, the milk moving directly from the teat cups through a pipeline system to the bulk tank in a continuous flow. One variation in this system is where the milk is first collected in the milker pail, then poured into a centrally located receiver, to be moved through portable piping into the tank, either by vacuum or by a milk pump.

Where the above systems are not yet in use, the operator moves the machine from cow to cow, collecting the milk in cans, after which it is carried to the milk room for cooling.

***67. Milking machines*** are operated by intermittent suction and pressure produced by a vacuum pump. The degree of vacuum is controlled by a relief valve and is usually maintained at 13 to 17 in Hg. The pulsations range from 45 to 55/min and may be regulated by an electromagnet energized through a distributor from a small generator driven by the pump motor, by the operation of the vacuum upon a piston, by a rotary or slide valve, or by the timed strokes of the pump itself.

Use of a milking machine saves about one-half the labor or one-fourth the cost of milking 20 to 25 cows. For larger herds, the savings increase. Milking technique has so improved that the machine is left on the cow only 2 to 4 min, and very little stripping is required. The power requirements depend upon the type and size of the system but should not exceed ¼ hp/milking unit. Average use of energy is below 2 kWh/(cow)(month).

***68. Milk Cooling.*** Milk as drawn from the cow is at a temperature of approximately 98°F, a temperature favorable for bacterial growth. Standards for farm holding and/or cooling tanks formulated by the International Association of Milk and Food Sanitarians, the U.S. Public Health Service, and the Dairy Industry Committee require that milk be cooled to 50°F within the first hour after milking and to 40°F within the second hour. Furthermore, the milk temperature must not fluctuate more than ±2°F during the holding period. Milk-cooling standards such as these usually make it necessary for the producer to use mechanical refrigerated cooling equipment.

Where milk is collected and stored in cans, two types of milk coolers are used on the farm. They are commonly known as the immersion-tank type and the water-spray type. In the former the cans are stored in cooled water, and in the latter the cans are cooled by a cold-water spray. In each case the water is cooled by mechanical refrigeration. In the spray type a bank of ice is frequently frozen prior to cooling time in order to increase cooling capacity. Where the cans are cooled in the immersion type of cooler, a pump or agitator is used to circulate water for more rapid cooling.

***69. Bulk handling and cooling*** of milk eliminate the necessity for using the conventional 10-gal milk can because the milk is cooled and stored in stainless-steel vats. Tank trucks pick up the milk and deliver it to processing plants. Two methods are used for cooling vats. The first is commonly called the direct-expansion refrigeration type, where the Freon refrigerant is in direct contact with the milk-tank liner, and the second is a water-spray cooling type, using an ice bank to cool the water.

Recent tests indicate that direct-expansion air-cooled refrigeration units require a motor rating of 1 hp for each 50 gal of milk to be cooled per milking. For water-spray types using an ice bank, power requirements are sharply reduced and are estimated at ⅓ hp for each 50 gal of milk cooled. Direct-expansion units are higher in first cost but cool milk more rapidly and at a slightly lower energy rate than water-spray types. Electrical energy per 100 lb of milk cooled averaged 3.5 kWh for bulk coolers and 2.9 kWh for can coolers, according to University of Georgia reports.

***70. Water Heating.*** A source of hot water is required by nearly all dairy milksheds for such purposes as sanitizing milking equipment, massaging udders before machine milking, veterinarian and insemination services, and milk testing. Two types of electric water heaters for the dairy are in common use, namely, the nonpressure, or "pour-in," and the pressure type. The latter is increasing in use.

The 10-gal pour-in type has been popular in small dairies. No water system or piping is required. To get a gallon of hot water, a gallon of cold water is poured in to force the hot water out. The heaters are insulated and have heating units in sizes ranging from 250 to 1500 W.

The pressure type of electric water heater is the same as that used in the home. Where more than 15 gal of hot water is required daily, the pressure type of electric water heater is preferred by dairy operators. Immersion-type heaters are also available for heating water in open pails. They are relatively low in efficiency and range in size from 500 to 1500 W.

Warm water may be used for drinking by calves, other livestock, and poultry. Automatic, electrically heated stock waterers are used in areas where freezing is a possibility. Heating units of 150 to 200 W thermostatically controlled are usually adequate.

***71. Egg-Handling Equipment.*** Automatic egg graders, washers, or cleaners, and cooling and storage rooms are utilized in commercial egg production. Egg prices are based on quality and size. Mechanical, electrically driven graders have become standard equipment on large poultry farms because of possible labor savings. Some graders have a candler, which eliminates this separate operation. The larger units grade four sizes of eggs at the rate of 12 cases/h. Electric-energy requirement is about 1 kWh for each 100 cases of eggs graded. Dirty eggs are cleaned by automatic washing machines or by an abrasive wheel or buffer attached to the shaft of a small electric motor. Automatic machines are available that will dry-clean up to 1500 eggs/h. Electrical-energy use is nominal, varying from ¼ to ½ kWh/case of eggs dry-cleaned and 1 to 1½ kWh for each 10 cases for the washing and drying models. Gathering eggs two or three times per day and cooling to about 55°F within 24 h help the poultry producer secure premium prices. Eggs held in refrigerated rooms at temperatures below 60°F and 75% relative humidity are becoming increasingly popular.

***72. Brooding.*** Four different types of electrically heated chick brooding units are in common use. They are the (1) conventional hover type, (2) infrared lamp, (3) electrically heated slab brooder, and (4) underfloor hot water, heated electrically.

Conventional hover-type brooders are constructed for 300- to 500-chick capacity. The top and sides of these brooders are insulated in order to reduce heat loss. The resistance types of heating units, 750 to 1000 W capacity, are used as sources of heat. Temperature is controlled thermostatically. The initial starting temperature is 90 to 95°F, with a gradually reduced rate of 5°F/week until 65°F is reached. Brooding periods vary from 6 to 10 weeks, depending on temperature and climatic conditions. Space requirements for chicks vary from 7 to 10 $in^2$ for leghorns to 10 to 14 $in^2$/chick for large breeds. Energy consumption varies from 0.5 to 2 kWh/(chick)(brood), depending upon climatic conditions, condition of brooder house, and season of the year.

Brooding with infrared heat lamps in sizes from 125 to 375 W is common practice. A 250-W lamp provides heat for 50 to 100 chicks, depending upon brooding-room facilities and climatic conditions. Infrared heat lamps are used for farm-sized flocks and also for commercial chick brooding of 20,000 or more chicks in one house. In larger installations some of the lamps may be thermostatically controlled. Reports from a Virginia experiment station give energy consumption at 2 to 3 kWh/chick for winter brooding.

Concrete slabs heated with embedded and thermostatically controlled heating cable are also being used for chick brooding. The cable is customarily installed about ½ in below the surface of the slab. From 30 to 50 W heating capacity/$ft^2$ of slab is required. Space requirements for chicks are similar to the hover-type brooder. Slab brooders are customarily covered with a hover. In some cases a combination of concrete-slab infrared-heat-lamp brooding system is used. In the event of power failure slab brooders tend to retain heat, and temperatures will drop only a few degrees in a 6- to 8-h power-off period. Energy use is 5 to 6 kWh/chick.

## Plant Production

***73. Irrigation Pumping.*** More electrical energy is used for irrigation pumping than for any other field operation. Proper design of an irrigating system will depend upon the following factors: *(a)* the acreage and kind of crop to be irrigated, *(b)* the amount of water that must be supplied, *(c)* the amount of underground water available, *(d)* the depth at which it is found.

Except where the water requirements are small and the depth to water great, plunger pumps are rarely used. The more common type is the centrifugal turbine pump, but where the lift is not more than 15 ft, the horizontal centrifugal pump is also used. The bowl of the turbine pump should be set below any expected drawdown in the well, and this will depend upon the porosity of the surrounding strata as well as upon the rate of pumping.

Vertical turbine pumps require vertical motors with either solid or hollow shafts and thrust bearings capable of carrying the pump load. Horizontal pumps should be connected to their motors through flexible couplings to avoid the use of belts. With average allowance for evaporation, irrigating an acre 1 ft deep requires 340,000 gal. The soil can be wet to a depth of 4 ft by using 4 to 6 in of water. From 10 to 20 in is required to produce the ordinary crops. With an overall efficiency of 50% for pump and motor, each acre-foot of water will require about 2 kWh of electricity for each foot of lift.

***74. Methods of irrigation*** include overhead pipes, stationary spray plants, and portable sprinkler systems. In the overhead type the discharge pipes are supported on posts and are located about 50 ft apart in lengths up to 600 ft. The pipes are usually supported on rollers so that they can be oscillated by a type of water motor, and nozzles are spaced 2 ft or more apart. Sixty gal/min of water per acre at 30 lb pressure is satisfactory. Stationary spray plants can reduce spraying time in orchards by 50% or more compared with portable units. A central pumping station, mixing tanks, and symmetrically located discharge pipes complete the layout. The pumps are usually three- or four-cylinder, single-action, with capacities of 10 to 60 gal/min at pressures up to 600 lb or more, requiring motors of 5 to 30 hp. Outlets are located at regular intervals for attaching the spray hose. Spray nozzles discharge up to 8 gal/min depending on pressure and orifice size. Power required is usually under 10 kWh/(acre)(application). Portable systems utilize lightweight, quick-coupled pipes, with sprinklers attached. Laid on the ground, they require considerable labor to move, but the initial investment is less than with other types. Sprinklers operate at pressures of 20 to 50 lb/$in^2$ and cover circles 40 to 90 ft in diameter, delivering 3 to 30 gal/min. A motor as small as 2 hp will apply 1 in water to 3 acres of land per week, although larger outfits are commonly used.

***75. Grain Conditioning.*** Field harvesting and on-the-farm storage losses of small grains and ear corn can be materially reduced where mechanical crop-drying or conditioning equipment is utilized. Early harvest reduces field losses due to shattering or lodging of grain and shelling, which may occur during mechanical harvesting. Crops can be harvested when weather conditions are most favorable as soon as possible after they mature, thus reducing the chance of storm damage while the crop dries in the field.

***76. Heated-Air Crop Dryers.*** Equipment needed includes an oil burner, a power-driven fan, and a drying bin for the ear corn or small grain.

Most of the dryers are portable. Each unit consists of a power-driven fan, a heater, and safety controls. Such dryers have two characteristics that determine their performance in drying grain: (1) the rate at which heat is supplied (rate of fuel consumption per hour) and (2) the rate of air supply in cubic feet per minute. These dryers are normally equipped with oil burners that consume fuel at the rate of 3 to 14 gal/h and fans powered by 3- to 5-hp electric motors that deliver 9000 to 15,000 $ft^3$/min of air. Usually 9000 $ft^3$/min of 30°F air, with a relative humidity of 70%, can be heated to 70°F with an oil consumption of 3 gal/h used in a direct-heat dryer and 4.2 gal/h for the dryer if a heat exchanger is used. The U.S. Department of Agriculture reports that 1000 bu of ear corn was dried from 30 to 13% moisture in 167 h.

Shelled corn, wheat, and oats can also be dried with heated air. Depth of grain in drying bins is 4 to 5 ft. Airflow must be uniform through grain, and temperatures of heated air should not exceed 110°F for seed corn and 140°F for wet milling. Temperatures up to 200 have been used without affecting feed value.

***77. Unheated-Air Crop Dryers.*** Wheat, oats, and barley are harvested in the summer, when atmospheric conditions are relatively favorable for grain drying with unheated air. Wheat combined at a moisture content as high as 20% can be successfully dried with unheated air. Minimum airflow is 3 $ft^3$/(min)(bu) with grain up to depths of 4 ft. With 16% moisture content airflow may be as low as 1 $ft^3$/(min)(bu) with wheat up to depths of 8 to 10 ft.

***78. Forage Conditioning.*** Hay is finished in the mow by forcing air through it from ducts laid in the mow floor. The depth is limited to 6 to 8 ft for long hay or 4 to 6 ft for chopped or baled hay at one drying operation, but it is practicable to dry a total depth of 16 to 18 ft of long hay or 10 to 14 ft of chopped hay in the mow by repeated runs.

The hay should remain in the field until its moisture is reduced to 35 to 40%. Even then, 500 to 700 lb of water must be driven off to obtain 1 ton of finished hay containing 20% moisture or less. About 15 $ft^3$/(min)($ft^2$) of mow floor area is considered a desirable volume of air. Both centrifugal and propeller-type fans have been used satisfactorily, with motors of 3 to 7¼ hp. Depending upon moisture in the hay, humidity of the air, and efficiency of the system, 25 to 65 kWh is used per ton of finished hay.

## Materials Handling

***79. Conveyers and Elevators.*** Livestock and crop production requires much time and labor for loading, transporting, and unloading materials. Portable chain and flight conveyers, commonly called elevators, are available in lengths of 8 to 50 ft or more and in widths of 6 in to more than 20 in. They may be operated at angles up to 70°, depending on the material being handled, but care must be taken to prevent overturning or collapsing, particularly at the greater angles. The smaller sizes are generally used for moving loose, bulky materials such as small grains, chopped forage, and bedding and will require up to ¾-hp motors. Larger sizes are mounted on wheels and are used for baled, bagged, and packaged products, as well as other materials. Power requirements range from ¼ up to 5 hp, depending on the speed, angle of elevation, and weight of the material being handled. Vertical elevators for baled hay are mounted directly to the outside of barn walls. A 42-ft model will require a 2-hp motor.

***80. Auger conveyers*** requiring fractional-horsepower motors are used for the horizontal and vertical moving of grains. Automatic feeding arrangements may employ 10-in-diameter forage augers in multiples of 5- or 10-ft sections up to 100 ft in length. Three-horsepower motors are required for lengths up to 90 ft, and 5 hp is needed for longer units. *Pneumatic conveyance* of grains and feed is increasingly popular on farms where the distance between storage or processing areas and feeding areas is considerable. This method is safe, has few moving parts, and is dust-free. The pipe can be placed in almost any path, above- or belowground. A University of Illinois report states that an air velocity of 4000 ft/min is required for proper operation. A 5-in pipe will convey 4500 lb of grain/h. This will require 2¾ hp for each 100 ft of length, according to the report.

***81. Silo Unloaders.*** Mechanically operated silo unloaders remove the silage from the silo and deposit it at the foot. The operating mechanism of the top-unloading type is essentially a radial beam with scrapers or augers which collect the silage and bring it to the center of the silo, where it is picked up by a motor-driven air or mechanical device and delivered to the silo chute. Silage then falls down the silo chute, where it is collected for feeding.

There is also a bottom type of silage unloader. The operating mechanism consists of an endless chain mounted on a movable beam. The chain is equipped with scrapers which move the silage out of the silo as the chain revolves.

Unloaders eliminate the need for climbing the silo daily, reduce spoilage by removing silage at a uniform depth, and save up to 200 h/year of time. Results of Ohio State University tests indicate that top removal of grass silage at a rate of 1 ton/h requires 4.3 kWh and that 1.6 tons/h of corn silage requires 2.5 kWh. Three- to ten-hp motors operate the unloaders, and approximately 300 kWh is used annually.

***82. Barn Cleaners.*** Electrically operated mechanical devices remove manure from poultry, dairy, and livestock barns. In poultry houses the cleaners may be placed under a slatted floor or in a wire-covered pit under tiers of mechanical feeders and waterers. In dairy barns they are installed in the gutters behind the cows. The dragline type uses a motor-driven drum to pull a belt or chain conveyer, equipped with cross flights, to an inclined elevator at the end of the barn, depositing the manure in a field spreader or pit.

The endless-chain type is well adapted to the larger stable where two rows of cows are housed. A single chain with wood or steel paddles travels around the gutters and up a short elevator, discharging the manure outside the stable. In this type of installation connecting or cross gutters must be installed at each end of the two rows of existing gutters so that an endless chain can be installed. The oscillating type uses a reciprocating bar with hinged paddle or auger conveyer. Portable types generally use a scoop steered by the operator and drawn along the gutter by a cable attached to a motor-driven drum. Cleaners are operated by electric motors of 2- to 5-hp capacity. They can be set to operate automatically for a predetermined cleaning period or can be switched on as need arises. Electric-energy use ranges from ½ to 1 kWh a month for each cow housed in the stable.

***83. Feed Grinding.*** Two types of feed grinders are in common use for processing grains: the burr mill, in which grain is crushed between plates, and the hammer mill, in which rapidly revolving blades strike the grain, reducing its size and forcing it through a metal screen.

Feed-grinding and -handling units are used in conjunction with automatic feeding systems and may also be equipped with conveyers or blowers to move the processed feed to storage bins.

Portable grinder-mixers are of growing importance among livestock feeders. With these units mixtures of small grains, ear corn, hay, concentrates, and home-grown feeds of all types may be fed to cattle, hogs, sheep, and poultry at different locations on the farm. They are of the hammer-mill type, are equipped with loading and unloading augers, and are pulled and powered by 25- to 35-hp tractors.

Several manufacturers market small burr and hammer mills that operate efficiently when powered by 3- to 7½-hp electric motors. Storage bins should be overhead so that grain will flow to the mills by gravity feed.

The most efficient speed for hammer mills is 3300 to 4000 r/min. Capacity and energy requirements vary widely with the kind and condition of grain and the fineness of grinding. In general, 1 kWh of energy will produce 35 lb of finely ground oats to 400 lb of coarsely cracked corn.

## Maintenance

***84. Emergency Power.*** With increased dependency upon electric power for time-controlled mechanical feeding, pipeline milking systems, manure removal, etc., the added investment in emergency power units may be justified compared with the possible economic loss if regular power fails. Generators ranging from 3 to 15 kW and rated at 120/240 V are available in tractor power-takeoff (PTO) and engine-driven types. The latter may be manually or automatically started. Automatic generators must be of higher capacity, because peak-connected loads will be carried if power fails. Nonautomatic types should have a "power-off" alarm and must be PTO-equipped with an overload circuit breaker. The tractor PTO-driven generators are least expensive to purchase, as the tractor engine serves as the generator drive. Output is controlled by an engine tachometer and/or voltmeter in the generator unit. Manufacturers claim a voltage rating within 2% of the normal supply voltage. Required generator capacity is obtained by totaling the power needs of essential loads, plus allowances for future loads and high starting currents of the motors. Double-throw switches must be used at the point of connection into the wiring, to prevent generator damage and power feedback into the supply line.

***85. Arc Welders.*** A highly mechanized agriculture requires that many machinery and structural repairs be made by the farmers themselves. A survey by the Kansas Farm Electrification Council covering the period 1958 to 1962 indicates that the number of dollars invested in electric welders was greater than for any other item of electric farm equipment. The electric arc welder is inexpensive, efficient, and an almost indispensable tool on modern farms. The 180-A transformer-type ac welder is satisfactory for most farm shops. This machine can cut, hard-surface, and weld metals up to ½ in thick. It requires a line voltage of 220 to 240 V single-phase, 60 Hz. Current outputs from 30 to 180 A are possible. Duty cycle at maximum output is 20% with an open-circuit voltage of 25 V. A carbon-arc-torch attachment is used for brazing, soldering, and heating purposes. Larger generator-type units, either engine or tractor PTO-driven (hence portable), may be used as emergency power generators, supplying 5000 W of 230- or 115-V single-phase 60-Hz power.

***86. Phase Converters.*** Most farms have 100- or 200-A single-phase service, which limits them to the use of 7½- or 10-hp motors. Two types of phase converters are available which will convert single-phase to 3-phase current. By connecting the converters between the electric meter and the motor, they will permit the use of 3-phase motors up to 20 hp or more. In addition, the National Electric Code states that service entrances need be heavy enough to handle only the largest-power-demand equipment, plus a portion of all other equipment, rather than the total connected load as before. This is advantageous where irrigation pumps, grain dryers, large feed mills, etc., are in use.

***87. Other shop equipment*** includes electrically powered air compressors, drill presses, grinders, hoists, lathes, saws, and paint sprayers. These generally require ¼- to ½-hp motors. Battery chargers drawing approximately 2 kWh/charge are popular.

***88. Bibliography.*** Sources of additional information on the use of electricity in agriculture are as follows:

Agricultural experiment stations and cooperative extension services, state colleges.

Farm Electrification Section, Agricultural Research Service, U.S. Department of Agriculture, Beltsville, Md.

National Rural Electric Cooperative Association, 1800 Massachusetts Avenue N.W., Washington, D.C. 20036.

Richey, C. B., Jacobson, P., and Hall C. W. "Agricultural Engineers' Handbook"; New York, McGraw-Hill Book Company.

Rural Electrification Administration, Washington, D.C.

## COAL MINES

*By DANIEL JACKSON, JR.*

### Power Systems

***89. Electric Power Systems.*** Power consumed by coal companies is purchased from utilities. The standard practice is for the utility to supply power to a main or central substation, reduce voltage to the level required by the coal company, and then meter the power consumed. From this point, it is up to the coal company to distribute the power to all company-owned facilities, install transformer stations for voltage reduction as required, and transmit the power to points of use.

Although standards for distributing surface power by coal companies are established in part by the supplier, coal companies must operate under strict regulations set forth by the U.S. Bureau of Mines (USBM). These requirements begin at the point of purchase and end at the points of use in underground and surface mines. Even in working places of underground mines all electrical equipment must be approved by the USBM under Schedule 2G.

***90. Primary Power.*** Utility substations may reduce high voltage to the final levels of 220 or 440 V, or may drop it only to 13,000, 7200, 4160, or 2400 V transmission to load centers for final reduction. The number of voltage steps and the choice of voltage depend upon a coal company's system load, transmission distance, safety, and limitations imposed by mine laws.

Few coal mines today generate their own electric power. The principal exceptions are small- or medium-sized mines located in isolated areas where it is not economically possible to install power lines. Mines operating under these conditions install engine-generator sets in capacities up to 500 kW. Engines usually are diesel types, and the generators or alternators normally produce 250 V dc or 220 or 440 V ac, 60 Hz, with 110 V in some instances. When demand exceeds the capacities of the largest standard "off-the-shelf" unit, twin or triple units are installed.

Sources of primary coal-mine power, aside from privately owned public utilities, include, to a limited extent, REA-supported and local systems. Utilities make power available at 6600—rarely less—to 69,000 V. Most commonly the voltage is 33,000. With one-step transformation, primary mine distribution usually is 4160 or 7200 V, nominal, with a trend toward 13,000 V. The 2300-V level has been eliminated where it is economically feasible to do so.

In two-step transformation, the primary voltage usually is 13,000 V. Permanent transformer stations may employ either single- or 3-phase transformers, with a trend toward the latter.

***91. Metering.*** Both center metering and metering at several load centers are employed by mining companies. Each case must be decided on its merits. In the past several years a number of companies have increased the primary voltage level and changed to central metering to reduce power cost and improve efficiency.

The electricity required to produce a ton of coal is approximately 14.5 kWh. Power cost per ton of coal increased in the 1970s to about 19 cents, compared with 12 cents in 1955. This increased cost is reflected not only in the continued trend to mechanical mining but also in the use of higher-capacity mining equipment. Power consumption, for example, by one of the largest coal companies in the United States, producing some 50 million tons of coal annually, consumes more than 50,000,000 kWh of power per month. The total installed transformer capacity required to supply this power is in excess of 300,000 kVA. In addition, transmission facilities include approximately 300 mi of overhead line and 90 mi of high-voltage cable.

***92. Primary Distribution.*** Coal-mine power-distribution systems generally fall into two classifications, including pole-mounted high lines and cable systems, with many companies using a combination of the two. The number and length of pole lines depend upon mine size, terrain, and so on. The goal is to serve secondary substations at load centers on the surface or underground. The stations are moved as the workings advance.

***93. Strip Mines.*** At strip mines, pole-line practice is largely standardized with a main line. Pole-line laterals at intervals of 1200 to 1500 ft are run to the pit, terminating in switch houses which supply auxiliary transformers for low-voltage equipment and also supply the cables to the larger high-voltage equipment. As the pit moves across country, the laterals are shortened at intervals until the pit approaches the main line, which then is moved to restart the cycle. Cables on the equipment usually are 1000 ft long. Thus, with a lateral spacing of 1200 to 1500 ft, equipment can operate freely between laterals with enough cable to spare to permit terminating laterals some distance back when shortening is necessary.

A large number of strip mines are now using the "ground-cable system" instead of the pole-line and cable combination. Otherwise, the basic plan is the same. A complete system consists of the main cable and cable laterals, the cable being sectionalized in 1000- to 1500-ft lengths, as a rule, with connectors for termination in switch houses or for joining main cable lengths by junction boxes. Several types of cable may be employed, but the most common is type SHD with copper-shielding braid over each insulated conductor to equalize surface stresses and eliminate static discharge, and grounding conductors in interstices.

The latest improvement in this cable is the SHD-GC, with ground-check conductors for continuous cable monitoring. It is the safest and most widely used for high-voltage (up to 15,000 V) portable power applications at both strip and deep mines. Other SH cable types and also type W and G cables, without and with ground wires, respectively, may be used. A maximum rating of 5000 V for W and G, however, usually results in such cables being limited to 2300 V when employed in coal-mining work.

The largest power cable used in a strip operation to date is an SHD-GC type rated at 8000 V, 880 A. It is 5⅜ in in diameter and weighs 24½ lb/ft. It consists of six 500,000-cmil conductors—two for each phase—one 750,000-cmil center ground conductor, and three No. 8 insulated ground-check conductors.

Another version of the ground-cable system for distributing primary power is to provide two separate cable circuits. This reduces the size of the individual cables and permits continuous operation in the event one cable fails, at least on a limited basis.

***94. Deep Mines.*** Underground mines may use a combination of surface pole-mounted high lines and cables. Where depth of cover and terrain make surface lines costly or impracticable, cable systems are used for primary distribution at underground mines. In this case the cables enter the mine at the main opening or through boreholes. Armored cable is employed in some instances, but the usual, as in surface mines, is a synthetic-jacketed type, usually type SHD, although other SH types or W or G may be employed. Cable manufacturers have designed a mine-power feeder cable for this purpose.

Where depth of cover and terrain favor pole-mounted high lines, the primary distribution system will include pole lines which follow the mine workings on the surface. Boreholes are drilled at intervals and the power is taken into the mine, rated voltage being kept near the working areas. The pole lines may serve dc power-conversion units on the surface or underground or transformers near the working areas for reducing the voltage to 440- or 1000-V levels.

***95. Power Factor.*** Without correction, power factor at the average coal mine would be 75 to 95%—sometimes less but seldom more. Corrective steps are necessary. Power contracts include rates which are based on classification, diversity factors, load power factor, demand power factor, monthly power factor, and cost of producing power. All contracts do not contain power-factor clauses. Whether this clause is included or not, it is to the advantage of the coal company to maintain not less than 85% power factor. Most power-factor clauses require the consumer to maintain at least an 85% power factor.

Corrective methods and equipment used to improve power factor include:

**a.** Synchronous motors for equipment requiring, for example, 100 hp or more, pumps underground and in-surface processing plants, air compressors, ventilating fans, and crushers.

**b.** Synchronous motors, 0.8 leading, in the motor-generator sets in large excavating units in surface mining and for supplying direct current underground.

**c.** Capacitors supplement other types of correction and bring power factor up to 90 to 95%, the usual goal, which normally takes the mine out of the penalty area. Theoretically, correction should be installed with each motor of any size, but a more practicable system is capacity for a group of motors near the center of such a group. The farthest location back, and the less preferable, is in the substation.

***96. Primary-Voltage Regulation.*** Increasing load density makes it more difficult to maintain economically adequate voltage at all points of the system. Increasing sophistication of automation equipment also requires more precise voltage, not wider voltage swings. Though voltage drop can be reduced by increasing the size of the conductors, shortening circuits, or reducing reactance with busway or differing cable configurations, these means can become uneconomical if carried to the extent sometimes required for optimum performance of extremely expensive and critical equipment.

In those cases where precise voltage is required at the load, an automatic voltage regulator can provide accurately maintained voltage of the correct level at the point of utilization with minimum investment. The conductor in the feeder can be selected on the basis of load-carrying capability only, with the regulator compensating for any resultant voltage drops. The regulator can be selected to control a single feeder to a critical load, or it can be increased in size to handle the total output of a unit substation transformer.

The latter alternative is selected when all the loads in a particular area, if supplied with best-quality voltage, will produce at a significantly improved performance level. Even though the utility supply is closely regulated, the load voltages will vary. Thus it is rarely sufficient for an operator of a distribution system to rely solely upon primary power-supply regulation. Some of the newer automation equipment includes "built-in" voltage regulators, but in most cases the user must provide the well-regulated supply needed.

***97. Safety.*** Safety standards in all these areas of mine operation have changed considerably since Congress passed the Health and Safety Act of 1969, and changes are continually being made. To obtain the latest changes, consult the Department of the Interior, Bureau of Mine's publications "Mandatory Safety Standards, Surface Coal Mines and Surface Work Areas of Underground Coal Mines," Title 30, Part 77, and "Mandatory Safety Standards, Underground Coal Mines," Title 30, Part 75.

## Underground Power

*98. Power Supply.* Underground power systems have been in a state of change for several years. The "standard" voltage for underground coal-mine use was for many years 250 V dc. Approximately 15 years ago the industry began using 440 V ac at the working face. Practically all new mines installed today are using this voltage, plus the 4160- and 1000-V levels for high-horsepower machines, with many of the older mines converting to the higher voltage.

Higher horsepower per machine, plus the fact that coal-seam thickness and size of opening limit the physical size of mining equipment, is generating pressure for higher voltage ratings, however. The use of 440, 1000, and 4160 V at the face has provided many advantages, including safety, that were not available with 250 V dc. As loads continue to increase beyond the capabilities of 440 V, companies are switching to 1000 to 4160 V for large mining equipment such as continuous miners. State and federal mining laws have been or are being changed so that the higher voltage levels can be applied in the work areas.

Alternating-current and direct-current loads encountered in underground coal mining fall into two categories: locomotive and shuttle-car haulage, calling for direct current in most instances, and all other applications for which either direct or alternating current may be employed.

Because direct current is so well suited to locomotive and shuttle-car operation, it is still widely used in mining. Underground power systems may consist of a combination of alternating and direct current, especially where locomotives and shuttle cars are used. The advantages of alternating current have been proved to be great enough at the working face to justify dual voltage systems, ac and dc.

Exceptions to the general rule of 250 V dc and 440 V ac underground include 2300 and 4160 V ac for large slope hoists, pumps, compressors, and other equipment requiring 100-hp motors and larger. Some 110-V ac power is used underground, but normally only for lighting and certain special applications.

An underground ac power system consists of a surface substation, high-voltage cable couplers, portable underground switch houses, mine load centers (transformers), and distribution boxes. These circuits generally have a voltage rating of 4.16 or 13.9 kV, although in a few instances the primary underground distribution is handled at levels as low as 2.4 kV and as high as 13.8 kV.

Figure 21-11 is a one-line diagram of a typical power system for an ac mining operation. The equipment in the circles are the surface substation, where service voltage is stepped down to the primary distribution level of the mine, and the underground switch houses, which provide branch-circuit switching and relaying equipment for proper protection of the branch circuits and their connected apparatus.

All equipment for alternating current underground normally is portable to permit quick and easy transfer from one location to another. Transformers employed underground are of the dry type, along with sealed, nitrogen-filled units. A three-unit 150-kVA installation of this type, mounted on a skid, can be made with a total height of 25½ in and a total weight, including air circuit breaker and polarized plugs and receptacle potheads, of 2600 lb. Capacitors on separate skids may be used with such substations, for example, 45 kVA with a 225-kVA power center at another mine, which also provides skid-mounted 440-V distribution boxes with two circuit breakers, three feed receptacles, and ground-trip relays.

Underground voltage regulation is a major consideration in designing a power system for ac mining. The need for good voltage regulation has generally resulted in the use of oversized cables selected primarily on the basis of allowable voltage drop in the system. Since the cable represents 50 to 60% of the initial investment for the mine power system, it is important to look for alternatives to ensure adequate voltage at the face. An economic study was recently made of a proposed ac power system for a new mine, with the objective of comparing the cable cost of providing for voltage regulation in oversized cables or, as an alternative, through the use of a 10% buck-or-boost induction voltage regulator. Cable cost was $20,000 greater in the system employing cables selected primarily on the basis of voltage drop. These savings were based on the early operation of the mine when only one-third of the ultimate footage of high-voltage cable would be required. For the maximum distances expected after full development of the mine, total saving in first costs for high-voltage cable was approximately $80,000.

The voltage regulator should be installed in the surface substation. This will eliminate additional costs for a special low-height unit or construction of a vault for an underground unit. The surface-substation location is ideal for the added reasons that it can be applied to compen-

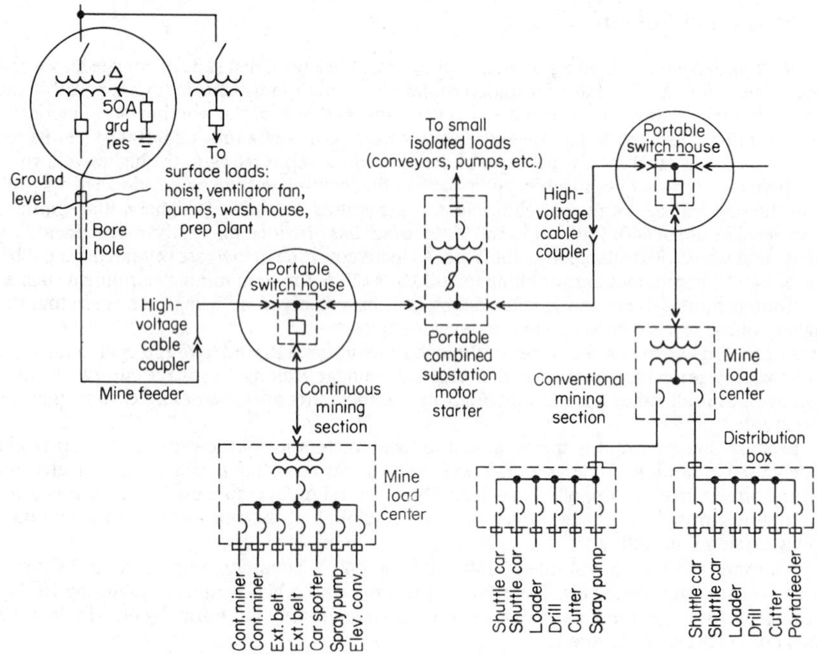


**FIG. 21-11** Alternating-current power system in a deep mine.

sate for voltage drop in the system ahead of the substation and that, through line-drop compensation, it will adjust for the voltage drop in the mine feeder circuits.

***99. Direct-Current Supply.*** Although considerable rotating equipment (motor-generator sets and rotary converters) and mercury-arc rectifiers are still in service, the silicon-diode rectifier is the standard conversion unit today. Mercury-arc rectifiers are being converted to silicon-diode units because the latter offer high efficiency, long life, great reliability, simplicity, and a minimum of maintenance. The total annual saving for a 500-kW silicon-diode unit compared with a mercury-arc rectifier, reportedly, is approximately $1800, based on a price of $12 per kilowatt for the silicon equipment. A 500-kW unit will pay for itself in less than 4 years.

Most silicon-diode rectifiers are portable, with rail wheels for movement over mine track systems. Others are mounted on skids, and still others are mounted on rubber tires. Stationary units also are available.

Automatic operation is the rule, and portability is essential to keep down the cost of the more frequent moves necessitated by the rapid extraction with concentrated mining, particularly in the thinner seams, from which more and more of the tonnage is coming. Among its other advantages, the rectifier has reasonably good efficiency, even at one-quarter load, and also can withstand heavy overloads for relatively long periods of time.

Like motor size and connected horsepower, the rectifier is increasing in size. Some 15 years ago, 150 or 200 kW was common. Now, a conversion unit under 300 kW is seldom installed, and with larger face machines, particularly with continuous miners, 500 kW and up is becoming common. One reason is the increase in peak and steady loads. Momentary peaks of up to 2000 A have been observed, along with average demands of 300 A or more.

***100. Basic Distribution.*** Where depth of overburden permits frequent use of boreholes, or where a hilltop area permits frequent drift openings, it may be more economical to use pole lines for primary ac service and distribute alternating or direct current from surface substations. Where underground stations are employed, the usual practice is to take primary alternating current in by cable—usually type SHD, carrying 4.16 to 13.8 kV. Type W and G cables, normally rated 5000 V maximum, also may be used, but usually only if the distribution voltage is not over 2300.

In dc mines, which normally rely on locomotive haulage, the basic distribution system is the trolley wire, with the track for return, supplemented by feeder and additional return circuits when the loads require them, which normally is the case. Some 4/0 trolley wire is employed, but the tendency is to use at least 6/0 and up to No. 9 section, with supplementary feeder, where several large locomotives are used. Big locomotives also bring up the problem of not only placement and capacity of substations but also the effect on the remainder of the dc system as the locomotive passes a given point. To reduce disturbances in face operation as a result of locomotive operation, substations may be equipped with automatic load distributors.

***101. Sectionalization.*** The normal distribution system naturally includes circuit breakers and other basic protection devices. Overcurrent settings should conform to good practice.

Aside from the basic facilities, a mark of a good distribution system is sectionalization with both manual and automatic equipment. Manual switches at intervals of 1000 ft in trolley lines and feeders permit isolating sections or parts of the system in case of trouble or to allow change or repair, frequently without interruption in sections not experiencing trouble. Automatic reclosing circuit breakers protect stub-end trolley and feeder lines and the sections they serve from fire, annealing of trolley, and other difficulties resulting from short circuits. The breakers automatically cut out sections experiencing trouble and leave the remainder of the system free to continue operations.

***102. Face Distribution.*** Mobile face equipment, whether alternating or direct current, must receive power through cables, with the exception of storage-battery-powered gathering locomotives and shuttle cars. Instead of connecting wire to wire or cable to cable, either solidly or by connectors, the practice today is to use junction and distribution boxes. Most such boxes, or circuit centers, also are designed to include circuit breakers and ground-protective equipment. Permissible types are available for use where permissible face equipment is required, usually where the Bureau of Mines or some other agency in authority finds an air sample containing 0.25% or more methane.

Main cables and, where the type of equipment, such as a stationary room conveyer, permits, distribution cables also are now being increasingly installed in short lengths, for example, 150 ft, for convenience in shortening, lengthening, and handling. Permissible connectors are available and may also be obtained with pilot pins, corresponding to pilot wires in the cable, to trip the circuit breaker before the main pins are disengaged, thus making it impossible to open the connection under power.

***103. Trailing Cables.*** The federal safety code and most state regulations require that frames of electrically operated equipment be physically grounded to protect persons in case of short circuits, either through the rail upon which the equipment rests, as a locomotive, or by a ground wire or cable. Where trailing cables serve equipment, the general practice is to include a grounding conductor or conductors in such cables.

For ac service, type W and G round, multiple-conductor cables normally are employed. One conductor in three-conductor type W may be used as a ground wire. Type G normally is made with three ground wires in the interstices between the cabled main conductors. These cables, plus heavy-duty drill cord, which is similar in general construction, are tough, flexible, and designed to take mining abuse.

In dc locomotive service, single-conductor cables are the common type. Concentric mining-machine cables find limited use in gathering locomotives as well as in other equipment, but while diameters are small, this type is not designed to take the abuse cables must withstand in modern high-capacity mining.

For practically all dc service underground, except cable-reel locomotives, the flat-twin cable (type W without ground and type G with ground) normally is preferred. Advantages include smaller size; lighter weight; ability to bend around small-diameter sheaves, cable drums, and guides; and good resistance to runovers and physical abuse.

Normally, the maximum circuit voltage with single-conductor, two-conductor concentric, two-conductor round, and flat-twin cables should be limited to 1000 V. Usual ratings for underground mining equipment are 600 to 5000 V.

One important new safety requirement for trailing cables is that no cable will contain more than one temporary splice, and it must be replaced with a permanent splice within 24 h. Mechanical and heat-shrink cable splices are now available to replace temporary cable splices quickly and economically.

***104. Protection.*** The solid wire or cable from machine frame to ground, if maintained, provides protection to persons from short circuit but, previously, no protection for the machine. With the new safety standards, this has been corrected. Mines employ equipment to open the circuit and thus reduce or eliminate damage as well as the hazard. In recent years, there has been a definite trend to circuit-breaker protection. With ac service, the secondaries of the transformers are Y-connected, with neutrals grounded, and current transformers are included to trip the circuit breaker when a ground occurs. With Y-connected and a full 440 V at the face, the maximum potential from any conductor to ground is limited. One new ac system includes doughnut-type current transformers around the three power conductors to detect imbalance in case of short circuit and to trip the circuit breaker in the same circuit center or distribution box.

In dc service, a three-pole circuit breaker may be employed, with one pole in the ground circuit with a 5- or 10-A current-limiting relay in series with it to trip the circuit breaker in case of a fault in the machine or cable. In such cables, the regular power conductors are supplemented by a grounding conductor or conductors. Ground trips also may be used with two-pole circuit breakers, but the relay may be damaged if heavy current flows and the grounding circuit is not opened.

Proposed alternatives, particularly for such mobile units as shuttle cars, include the polarized relay, the polarized short-circuiting device, and various solid-state devices.

***105. Battery Service.*** Aside from use in locomotives, the principal use of batteries is in powering tractors used to pull rubber-tired trail cars that haul coal at many of the small mines, and in tractors that haul supply cars and personnel carriers in larger, trackless operations. The use of battery-powered equipment is gaining in popularity owing mainly to improved batteries and more sophisticated equipment.

## Surface Mine Power

***106. Operating Voltage and Loads.*** Ten years ago 60- and 70-yd$^3$ shovels and draglines with connected horsepowers of approximately 4650 of main ac driving motors with peak loads of 7500 kW required 7200 V to drive these machines. One of the largest shovels in operation today includes a 140-yd$^3$ unit with more than 12,000 hp of main-drive ac motors and peak loads on the order of 17,500 kW. Certain smaller units, such as drills and pumps, operate on 220 or 440 V, supplied by small service transformers usually in or near the pit where the equipment is being used. Some older and smaller machines operate on 2300 or 4160 V nominal. The latest increase in voltage level has been to 13,800 V to power a 65-yd$^3$ shovel.

Trailing cables as a rule are type SHD with ground-check conductors to facilitate ground protection. Resistor values are fixed so that maximum voltage to ground is limited to a safe value. Quick tripping is necessary to reduce the time of possible shock and to limit damage to equipment.

Figure 21-12 shows the major items of electrical distribution apparatus for a surface mine, including the following:

1. The main substation, including equipment which transforms the incoming transmission-line voltage from the local utility to a primary distribution voltage for the mine.
2. A portable substation, including equipment which transforms the primary distribution voltage down to a lower voltage for the smaller machines not suitable for operation on the primary distribution voltage.
3. Portable switch houses, including equipment which automatically protects the distribution circuits at the distribution voltage level(s) and enables flexible isolation and control of mine feeders.
4. Portable power centers, including equipment which transforms power from the distribution voltage to utilization levels of 600 V and below for auxiliary equipment.
5. Transmission lines, cables, and cable couplers which carry power at the voltage used in the pit between the above electrical apparatus and eventually to the loads served by the distribution system.

The prime considerations in the choice of electrical apparatus used to perform these functions are safety, voltage regulation, reliability, and cost.

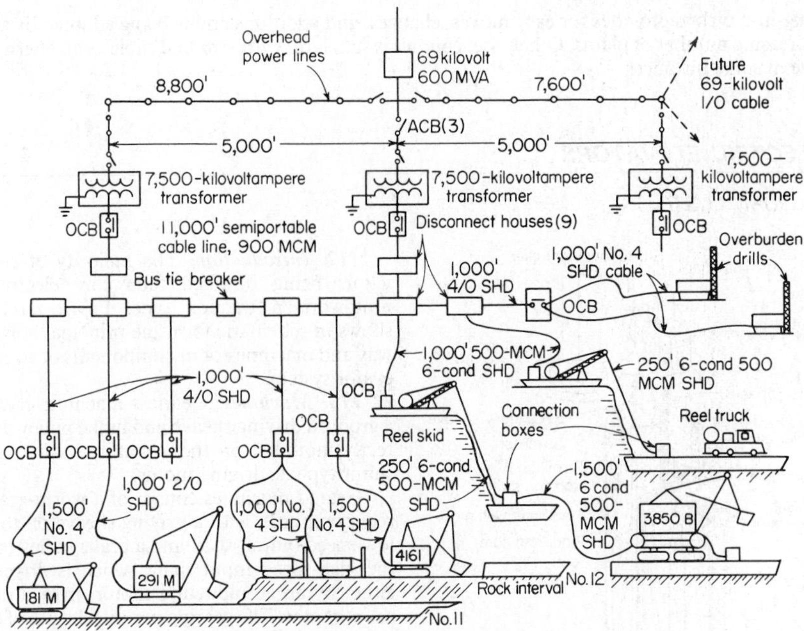


FIG. 21-12 Power system in a strip mine.

## Plant Power

***107. Voltage.*** Accepted voltage for most of the stationary motors in preparation plants is 440, leaving in most instances only the question of whether 2300 or 4160 V should be used for certain large units such as pumps and crushers. A rough rule is that motors of 100 hp and larger should be powered by the higher voltages.

***108. Transformer Location.*** Packaged substations with oil-filled transformers are available for outdoor service, with nonflammable units for indoor. If high-line voltage is over 10,000 V and the reduction to 440 V is made in one step, the substation should be of the outdoor type. If the supply is under 10,000 V, packaged indoor substations with nonflammable transformers are the general choice, principally because they can be placed closer to the load center.

***109. Controls.*** Starters grouped in factory-assembled central cabinets are now standard for coal-preparation plants. One central cabinet is satisfactory for a small plant, but a large plant normally requires cabinets at several locations to keep the starters reasonably close to the motors. Draw-type starters which can be pulled out for quick replacement by a spare constitute the probable pattern for the future. Dust is a major problem with starters. The best method of eliminating it is to house cabinets in enclosed rooms with forced filtered-air ventilation to keep the pressure slightly higher than outside.

***110. Capacitors.*** The induction-motor load of the preparation plant produces a low power factor, offset by capacitors to bring the lagging power factor up to or near 100%. Theoretically, an appropriate capacitor should be connected to the motor terminals, but cost, space, and maintenance considerations normally dictate grouping the capacitors in the control room.

***111. Motor Protection.*** The wide use of water and the difficulty of completely eliminating dust require some form of protection for many motors in preparation-plant service. This has led to growing use of special motors, such as dripproof, splashproof, and totally enclosed fan-cooled motors. The major type is the standard normal-starting-torque squirrel cage. Double-deck high-torque units are used where extra starting power is required, with wound-rotor units for extra-heavy starting. Control systems almost invariably provide for starting and stopping in sequence; some provide for locking out certain units when not needed or for maintenance.

Plug-in-type bus provides for easy moves, changes, and additions and is being adopted by an increasing number of plants. Otherwise, conduit is standard, with armored cable as an alternative in some instances.

## ELECTRIC ELEVATORS

*By JOHN LUSTI*

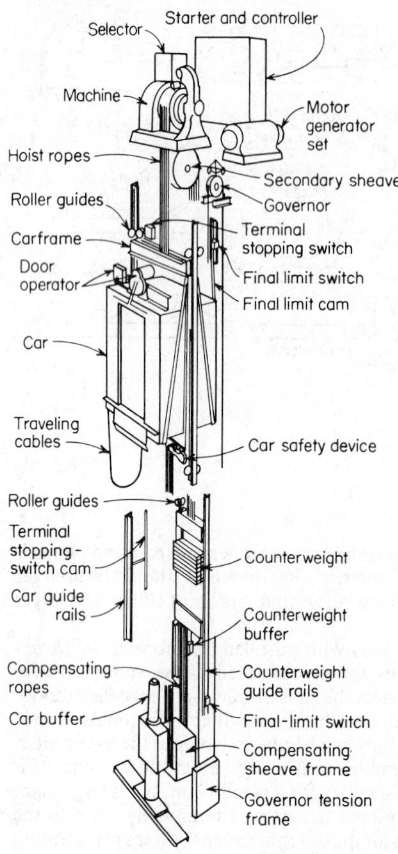


**FIG. 21-13** Elevator machine room and hoistway equipment.

***112. Introduction.*** The majority of elevators being installed today are electric-motor-driven traction type. Figure 21-13 shows in schematic form the principal hoistway and machine-room components of an elevator system.

***113. Machines.*** Gearless machines have a grooved driving sheave and brake pulley directly mounted on the armature shaft of a shunt-type dc driving motor.

Geared machines consist of a worm-gear reduction unit, with a driving sheave on the low-speed output shaft and a brake pulley on the high-speed input shaft, which is driven either by an ac induction motor or by a dc shunt motor. The gearing permits the use of a small high-speed drive motor such that the combination gear and motor cost less than a gearless machine.

To date, gearless machines are used almost exclusively for car speeds of 400 ft/min and up. Geared machines are rarely used beyond speeds of 350 ft/min because of excessive problems with noise and vibration and difficulties with gear wear and backlash brought about by frequent force reversals on the gearing during and due to the frequent starting and stopping of elevators.

***114. Brakes.*** Magnet-operated brakes are generally employed. They release electrically and apply through friction shoes held by springs against a cylindrical brake drum (pulley) on the machine driving shaft.

In the operation of gearless machines the brake function is to hold and not to slow down the elevator; therefore, its size is determined solely by torque requirements. Its function is the same on higher-speed geared machines; therefore, the brake mechanism, like the motor, is smaller on geared machines because of the torque reduction back through the gearing. On slow-speed ac motor-driven machines, where the stopping is done solely by the brake, the heat generated by the absorption of the system kinetic energy becomes a major factor in determining brake size.

Direct-current magnets are used wherever possible, because they are quiet and can be readily controlled to give quick but smooth application. Alternating-current brakes may be magnet- or motor-operated and are usually provided with dashpots to regulate their application.

***115. Traction.*** The elevator car is raised and lowered by hoisting ropes which pass over the machine drive sheave, the necessary traction being obtained by the friction between these ropes and the grooved surfaces of the sheave, and the pressure being applied by the weight of the elevator car and its load, the counterweight, and the weight of the ropes (and compensating chains or ropes, if used).

The traction-type elevator has an inherent safety feature in that, when either the car or the counterweight bottoms, the tension in the hoisting ropes is relieved and the driving sheave may rotate without moving the elevator owing to the loss in traction.

***116. Ropes.*** Sets of three to eight steel hoisting ropes with diameters ranging from ½ to 1 in are used in parallel. The rope diameter used for the set determines the minimum sheave diameter which may be used. Sheaves that are too small will introduce excessive stresses in the ropes while flexing as they go over the sheave, causing shortened rope life. The sheave diameter is usually 40 or more times the rope diameter.

***117. Roping Ratio*** (Figs. 21-14 and 21-15). Traction elevators usually have 1:1 or 2:1 ratios. The 2:1 rope gearing is employed on lower-speed gearless machines to reduce required machine size. The rope load on the drive sheave due to the suspended masses is halved, permitting lighter machine structure. The machine-torque requirements are halved owing to the ratio alone and are further reduced because the reduced load per rope permits smaller-diameter ropes and therefore smaller-diameter drive sheaves.

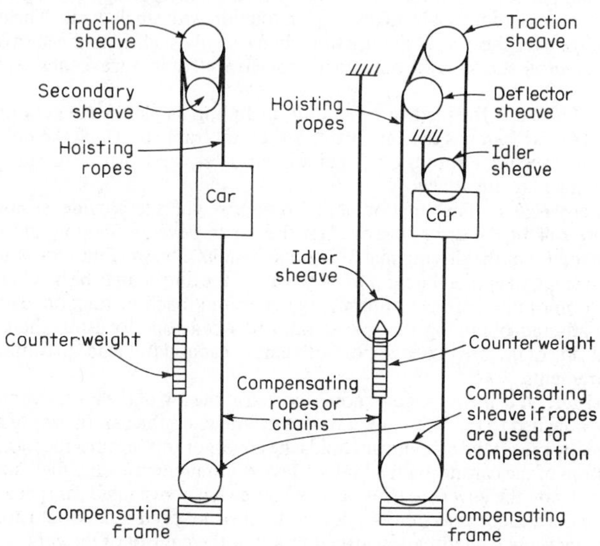


**FIG. 21-14** One-to-one traction elevator with double wrap.

**FIG. 21-15** Two-to-one rope-geared elevator with single wrap.

***118. Roping Wrap*** (Figs. 21-14 and 21-15). The hoist ropes may pass over the driving sheave only once, "single-wrap"; or they may be "double-wrap," wherein, after passing over the drive sheave, they go around a secondary idler sheave and over the driving sheave again in another set of grooves.

***119. Roping Compensation.*** When the weight of the hoist ropes becomes appreciable, compensating ropes or chains are provided to keep the traction load on the machine from changing because of the weight of the hoist ropes as they pass from one side of the system to the other. The compensating sheave in the pit (used with ropes only) can be so arranged that it can take up slack by traveling in the down direction but cannot travel in the up direction; thus the car and the counterweight are constrained to move together, eliminating the possibility of getting appreciably slack hoist ropes owing to jumping of the car (counterweight) as a result of rapid retardation of a down-traveling counterweight (car). This arrangement is provided on higher-speed elevators primarily to ensure that all the moving masses will act together in the event of a mechanical safety or buffer operation.

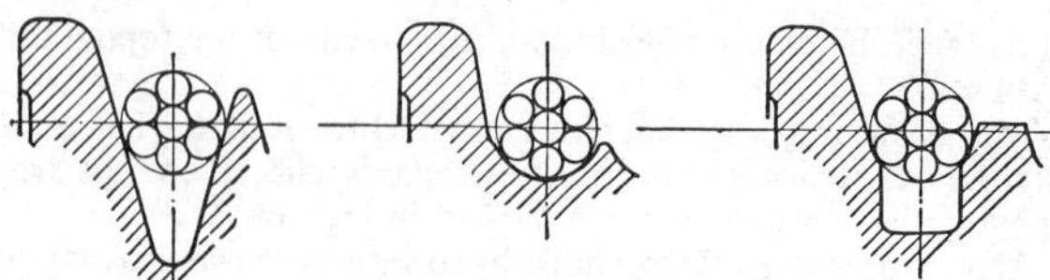

**FIG. 21-16** Methods of grooving traction sheaves, left to right: V groove, U groove, and undercut groove.

***120. Sheaves.*** Figure 21-16 shows the methods of grooving traction sheaves; the U groove allows more load per rope than the others but requires double wrapping to obtain sufficient traction. The V or undercut usually gives enough traction with single-wrap. The undercut type has the advantage over the V that the traction changes only slightly as the sheave wears.

The machine-drive sheave is the only traction sheave; all other sheaves always employ the U grooves.

***121. Car.*** The car is the load-carrying unit, including its platform, enclosure, car frame, and car door. The car frame is the supporting structural frame to which the hoisting ropes or hoisting rope sheaves (2:1 roping), car guides, car safety and platform, and generally the door-operator machine are attached.

***122. Counterweights.*** The function of the counterweight is to provide tension in the ropes on the opposite side of the drive sheave from the car to develop traction and to reduce the maximum net load that the driving machine must handle. The weight of the counterweight is generally equal to the weight of the car plus 40 to 45% of the duty load to be handled. This weight is generally a compromise between minimizing maximum machine traction load, minimizing average net load, and obtaining the lowest ratio between rope loads on the car versus the counterweight side of the drive sheave (for both empty car and full load) to minimize traction-relation requirements.

***123. Governor.*** A centrifugal governor is located at the top of the hoistway and driven by a governor rope attached to the safety operating mechanism on the car. In case of car overspeed (governor tripping speed) the mechanism holds the governor rope against the motion of the car, causing actuation of the car mechanical safety device. On moderate- and high-speed elevators, an electric contact on the governor trips open at somewhat above rated car speed and stops the car through the normal control circuits before it reaches the speed necessary to apply the safety. Additional contacts are sometimes furnished to aid in the control of the car.

***124. Car Safety Devices.*** The car safety device consists of at least one mechanism mounted on each side of the car frame between or below the lower members of the car frame (safety plank). They stop the car by clamping the guide rails. On low-speed cars (to 150 ft/min) knurled rollers or cams are jammed between the safety block and the guide rails to bring the car to an abrupt stop. On high-speed cars smooth safety jaws clamp the rails with a controlled force to bring the car to a sliding stop. Maximum permissible retardation, by code, is gravity with full load in the car.

Car safeties are permitted to function only in the down direction, since retardation may exceed gravity, which would cause persons to leave the floor of the car on an up application. Excessive retardation must be avoided because of possible injurious effects on passengers.

Where there is occupied space under the pit or a used passageway, the counterweight is also provided with a mechanical safety.

***125. Guides.*** The car and counterweight run on steel T-rail guides. Sliding guide shoes were formerly standard, but on many passenger elevators these are being replaced by roller guides, which give smoother riding and less friction. As they require no lubrication on the rail surfaces, it becomes easier to keep the hoistway clean and the fire hazard due to oil from the rails running down into the pit is eliminated.

***126. Buffers.*** Electric elevators are provided with buffers in the pit under the car and counterweight. Spring buffers are used for slow-speed and oil buffers for high-speed elevators. On the latter, the counterweight oil buffer is frequently made a part of the counterweight structure.

***127. Doors and Door Operators.*** Passenger-elevator car and hoistway doors are generally of the horizontal sliding type. The door panels have rollers at the top which ride on a steel track for support and guidance. The doors are guided at the bottom by shoes sliding in a machined self-cleaning slot. The door operator, mounted on supports to the car frame, is driven by an electric motor and coupled to the car door by belts, chains, or levers. The hoistway doors are automatically coupled to the car doors, when the elevator is at the landing, and operate in synchronism with them.

Freight elevators usually have a vertically sliding car gate and vertically sliding bi-parting hoistway doors. The car gate is counterweighted; the two halves of the bi-parting hoistway doors counterbalance each other. When the doors are power-operated, each door has an electrically powered door operator.

***128. Selector.*** The function of a selector is to give car-position information to the control and operating systems so that automatic stops may be made at prescribed landings. Selectors located in the machine room may be mechanically coupled to the car by steel tape or wire, in which case they may be used to give all required positions down to the small fraction of an inch required for floor-stopping accuracy. Some are coupled to the car by way of a sheave and the hoist ropes running over it. These are not sufficiently precise to permit accurate floor stops unless they are supplemented by devices, mounted on the car and interacting with cams fixed in relation to each landing, to take over position dictation when the car is within approximately 1 ft of the landing.

On slow-speed elevators switches in the hoistway handle the whole positioning, sometimes with an assist from a stepping selector in the machine room which is driven by pulses from the hoistway switches.

***129. Traveling Cables.*** Electric power and signals are carried to the car to operate the lights, door operators, etc., and call registration, door position, etc.; information is carried back to the machine room through traveling cables suspended from the bottom of the car at one end and attached to a junction box at the midpoint or top of the hoistway at the other end.

## Motors, Generators, and Controllers

***130. Special requirements*** of elevator motors are:

1. Utmost safety and reliability.
2. Automatic handling of negative or "overhauling" loads.
3. Ability to stand up and perform well under repeated starting and stopping. (A busy elevator may make up to 200 stops/h.)
4. Quietness and freedom from vibration.

***131. Alternating-Current Elevator Motors.*** Alternating-current elevator motors are almost exclusively squirrel-cage induction motors, either single-speed or two-speed. Squirrel-cage elevator motors generally have high-resistance rotors to provide high starting torque and limited starting current, because a large proportion of the duty cycle consists of starts from rest. Usually the full-load slip of these motors is about 20%; thus a motor having a synchronous speed of 900 r/min may run at 720 r/min when the elevator is carrying full load in the up direction. Figure 21-17 shows typical characteristics for a squirrel-cage hoist motor.

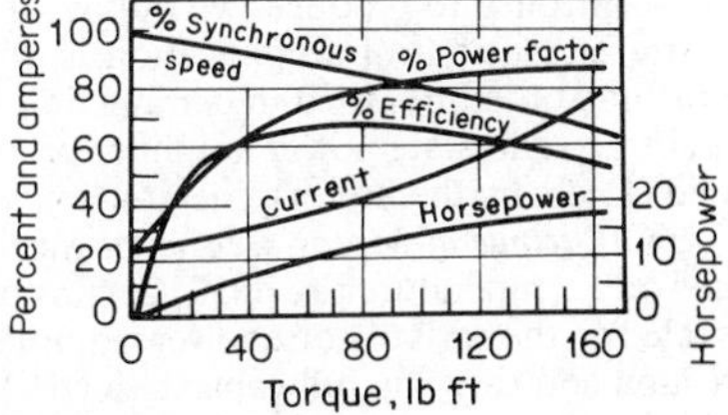

**FIG. 21-17** Typical characteristics of squirrel-cage induction hoist motor for geared elevator.

Double-squirrel rotors are coming into more common use for elevators. They enable maintaining the high-starting torque characteristic, but the full-load full-speed slip can be reduced to approximately 9%, giving better control and efficiency.

Frequent starting causes the outer high-resistance motor bars to heat more rapidly than the inner bars. Care must be taken in the rotor design to prevent stresses due to thermal expansion causing mechanical failure of the double cage where the bars are interconnected.

***132. Direct-Current Elevator Motors.*** Practically all dc motors now being built are the shunt type, with their speed controlled by varying the armature voltage.

Direct-current motors for worm-geared machines usually run from 10 to 75 hp, with speeds ranging from 1400 r/min in the smaller units to 600 r/min in the larger, the speeds being limited by the requirements of the gear unit. For gearless machines the range is 20 to 150 hp and 60 to 150 r/min. These motors are required to deliver torque far in excess of nominal during acceleration and slowdown of the hoistway masses. The high currents produced during these periods increase armature reaction and reactive voltage effects. Extra strong motor field excitation is used to overpower the armature reaction and prevent the loss of available torque due to its field-weakening effect. There is a greater need for interpoles to buck out excessive armature reactive voltage, which causes commutator sparking, than there is in the more common continuous-duty motors.

Figures 21-18 and 21-19 show typical characteristics of dc elevator motors.

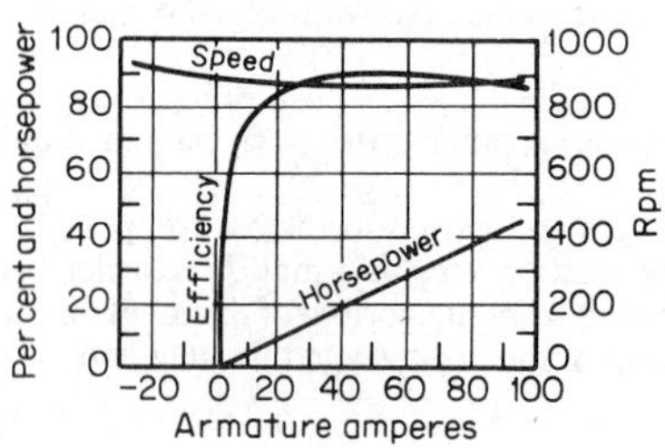


**FIG. 21-18** Typical characteristic curves of a dc hoist motor for geared elevator.

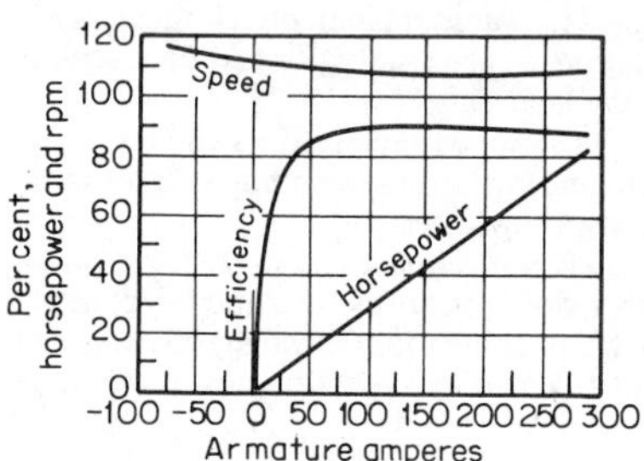


**FIG. 21-19** Typical characteristic curves of gearless traction elevator motor.

***133. Power Amplifiers***

*Motor-Generators.* Motor-generators are usually of the two-bearing type, but the larger sizes (above 75 kW) may be three- or four-bearing. It is important that they be as compact as possible, to limit the space occupied. Elevator motor-generator sets are usually of 10 to 75 kW capacity (continuous rating).

The generator is the power-amplifying element of the motor-generator set. The armature voltage, and thereby the power delivered to the dc elevator motor, is controlled by the much smaller power fed to the field of the generator.

On ac power supply, low-slip induction driving motors are used. Sixty-hertz synchronous speeds are usually 1800 r/min up to about 50 kW and 1200 r/min above 50 kW.

*Solid-state voltage control power amplifiers* noticeably entered the elevator field in the 1970s.

DC power amplification is obtained from thyristors, arranged in a full-wave bridge, and phase-controlled to produce two pulses (of the proper duration) per cycle per phase.

AC voltage-control amplification is obtained from various combinations of phase-fired thyristors, sometimes used in conjunction with power diodes and switches, in each phase.

The dc solid-state power amplifier provides the most power-efficient elevator system, followed closely by the motor-generator set, with the voltage-controlled ac amplifier far behind.

***134. Ratings of Motor-Generator Sets and Motors.*** Ratings of motor-generator sets are based on a temperature rise of 50°C with continuous run at nameplate rating. Elevator motors are rated on the basis of horsepower required when the elevator is traveling at full speed in the up direction and carrying full-capacity load. As this occurs infrequently in actual operation, the rating for elevator motors is intermittent (¼-, ½-, or 1-h time basis). As rated by the manufacturers, they will usually operate in service without exceeding a 50°C rise.

Motor-generator sets are rated lower than the elevator motors but not less than the rms elevator-motor horsepower for the service duty cycle.

Solid-state amplifiers have very low thermal capacity and therefore must be rated to handle the peak power demand of the elevator. Motors used with solid-state amplifiers must be specially designed and rated to compensate for the effects of the "chopped" input power provided by the thyristors.

***135. Alternating-Current Rheostatic Control.*** Squirrel-cage motors up to about 6 hp driving slow-speed elevators (usually under 80 ft/min) are often started "across the line." For higher speeds and ratings one or more steps of accelerating resistance or reactance are used in the stator circuit for squirrel-cage motors and the rotor circuit for slip-ring motors to control starting torque.

Stopping is accomplished by disconnecting the motor and applying the brake. When two-speed motors are used, they start on the fast-speed winding (as above), and stopping is accomplished by first transferring to the slow-speed winding, one or more resistance steps being used to smooth out the transition, and then dropping the brake with the car running at slow speed. Single-speed motors are rarely used above 150 ft/min, because of poor stopping accuracy due to excessive changes in slide through the brake because of changes in net load. Two-speed motors are rarely used above 250 ft/min.

The resistance or reactance steps are ordinarily controlled by timed magnet switches, as, for example, ac magnet switches timed by dashpots. It is more common to use capacitor-timed dc switches, the dc power for electromagnetic switch and brake operation being supplied by selenium rectifiers.

***136. Voltage Control***

*Solid-State Systems.* In this form of control, the voltage applied to the elevator motor is smoothly variable by control of the gain of a power amplifier which feeds the motor. By proper control of the voltage the speed, acceleration, and rate of change of acceleration (jerk) may be regulated to conform to specifications related to passenger comfort, peak power drain, and speed of service.

The ideal elevator run pattern consists of an initial period of constant rate of change of acceleration (jerk) until a specified acceleration is reached. This acceleration is maintained until the maximum velocity for the particular run is approached, at which time a constant rate of change of acceleration of the opposite polarity is applied to bring the car to the desired velocity at the moment the acceleration reaches zero.

If this run is a great enough distance, the velocity reached will be the maximum specified and the car will continue to run at this velocity until it must start to slow down to stop at a landing.

Slowdown is started by applying a rate of change of acceleration opposite to that used at the start, when it is computed that the car with its present acceleration and velocity can just be made to reach the stopping point (zero distance, velocity, and acceleration) without exceeding specified jerk and acceleration rates. How closely this pattern can be matched is dependent not only on the sophistication of the contol system but also on the size of the duty load, the maximum velocity, and the total distance the car must travel.

*Nonfeedback Systems.* Nonfeedback control systems can be used only with motor-generator power amplifiers.

The simplest form of voltage control employs a motor-generator set wherein the main generator field power is obtained by self-excitation, direction and slow-speed operation are controlled by a smaller separately excited field, and load compensation is provided by a cumulatively compounded series field. On start, the separately excited field is energized through resistances and switches to set the field flux in the proper magnitude and direction for self-excited buildup. The self-excited field is controlled by switches and resistances to provide desired rate of buildup of elevator speed. On slowdown, the distance to the stop is measured at discrete points. The self-excited resistance is cut back to provide the desired slowdown rates and the cutout near the floor, leaving the final slow-speed approach to the floor controlled by a separately excited field.

Self-excitation is used at lower speeds, to give smooth but rather imprecise slow performance with simple resistance-switching circuitry.

Resistance switching with separately excited generator field control has been rather highly developed. It requires rather complicated switching circuitry and specially designed motors and generators and careful adjustment to achieve top performance. On start, the acceleration is controlled by operation of control switches as a function of time. On slowdown, the switches are operated as a function of distance. During the run, switching circuitry compensates for load variations, brush drops, residual field voltages, generator field saturation, and armature reaction.

*Feedback Systems.* Feedback control systems measure the speed of the elevator, compare this speed with the speed desired, and if there is a difference, apply a signal to the power amplifier to bring this signal to zero.

When solid-state power amplifiers drive dc elevator motors, the input signal to maintain a given motor voltage is largely load-dependent, and feedback is therefore required. The speed feedback may be taken from the motor armature, but tachometric feedback is universally used as this eliminates the motor-speed-regulation error. In the case of ac induction motors, the amplifier can control the motor torque only and tachometric feedback is therefore mandatory.

With feedback systems, the speed is not inherently a function of the prime input signal as with nonfeedback systems; as a result, guarding against malfunctions which could cause hazardous conditions requires additional checking equipment and great care in circuit design.

On small elevator installations, the overall response to input signals is fast enough so that the desired elevator run pattern can be followed with relatively simple feedback control.

On large, high-rise, high-speed installations, there are a number of slow-response elements, resulting in the requirement for multiple feedback and filter circuits in order to permit sufficient feedback for proper response without putting the whole system into oscillation.

***137. Leveling Control and Operation.*** In essence, leveling is moving the car in the vicinity of the landing toward the landing, with the motion so constrained that it can be safely done with the car and hoistway doors open.

The car position in which leveling is permissible and the direction of leveling motion are controlled through mechanical engagement of cams in the hoistway with switches on the car, through a change in magnetic reluctance when electromagnetic "inductor" devices on the car pass iron vanes in the hoistway, or by similar leveling devices on positively driven selectors located in the elevator machine room. Special precautions are taken in the control systems of elevators whose normal running speed is greater than that permitted during leveling, to minimize the possibility of any faults causing excessive speed during leveling.

Automatic leveling is used on almost all moderate- and high-speed elevators, and to a large extent on slow-speed elevators, especially those used for freight. It permits the door-opening operation to start while the elevator is approaching the landing so that the doors will be open far enough to permit passenger transfer as soon as the car is at floor level, thereby speeding up service. It also permits correcting for initial stopping inaccuracies and maintaining the car at floor level in loading or unloading if stretching or contracting of the hoist ropes should cause the car to move.

***138. Continuous-pressure operation*** utilizes an up button and a down button in the car and at each landing which control the starting, stopping, and direction of travel of the elevator. The buttons must be held pressed to keep the car in motion. Automatic or manual means are provided to prevent interference from landing buttons when the car buttons are being used.

Continuous-pressure operation, owing to its simplified nature, is usually applied to freight elevators and limited to rheostatic control.

***139. Car switch operation*** means a master switch is used in the car, which gives a car attendant sole control of the starting, stopping, and direction of travel. An annunciator or single-flash signal is used to notify the operator of the location of landing calls. However, owing to automation, a lack of skilled attendants, and difficulties encountered in synchronizing waiting-passenger signals and power door operation with the position of the elevator, car switch operation is rarely used today and is confined mostly to low- and moderate-speed freight-service applications.

***140. Single automatic pushbutton operation (SAPO)*** is designed to serve only one person or group of persons at a time, and is used largely in small apartment houses or private residences. This service uses one button in the car for each floor served and one button at each landing. Momentary pressure of any button will cause the car to move toward, and automatically stop at, the corresponding floor. Once the car is in motion, all other commands will be ignored until the trip is completed.

***141. Selective collective operation*** employs one button in the car for each floor served and an up and down button on each landing and responds to several calls on one trip. Momentary pressure registers a call, which is stored in any sequence and remains registered until answered. Stops in response to the landing calls are made in the order in which the landings are reached, as the car travels up or down. However, uppermost or lowermost calls are answered when reached, irrespective of the direction of travel. In some cases, nonselective collective operation is used, where a single button is employed at each landing and all registered stops from landing or car are made in the order in which the landings are reached, regardless of the direction of car travel.

***142. Group automatic operation*** is applicable when two or more automatic passenger elevators are used, equipped with power-operated car and hoistway doors. Operation of the cars is coordinated by a control system. In its simplest form, the "selective collective" cars are operated from a common set of landing buttons; when there are no calls, one car parks at the lobby while the other car parks at a landing. The lobby car responds to calls from the lobby and the other car responds to calls from the various floors or landings. The system becomes a duplex collective system when the lobby car is permitted by the supervisory system to help the floor car respond to calls when there is an excess of calls and the floor car has not been able to clear up the calls within a certain period of time. All service of this nature is applicable to buildings where service demand is light; however, a more complex form of group automatic operation is used in all large, heavily trafficked buildings.

***143. Dispatching operations*** are thus employed in heavily trafficked, larger buildings to maximize the service capabilities of the elevators. Service is technically maximized when the average passenger loses the least amount of time getting from one floor to another. Service is optimized by programming the system for maximum handling capacity during peak loads (usually during morning and evening "rush" hours) and for minimum and equal waiting and traveling time for each passenger, individually, at other times.

Terminal dispatching basically requires that the elevators are stored in the lobby during traffic lulls and selected for dispatch, one by one, upon demand, one loading interval after the previous car, or sooner if 60 to 80% loaded as indicated by a load-measuring device. The cars go to the highest call and return to the lobby, answering any calls that may have been registered on the way. When traffic peaks are encountered, cars may be programmed to bypass landing calls or may be assigned to a limited number of floors to minimize stops. When the number of calls becomes low enough, the cars generally come to rest at the lobby and wait for the next call, each car time-spaced in sequential order, as discussed.

Multiple-zone dispatching is being increasingly used today, however. This system minimizes needless elevator travel, thereby saving money in terms of electricity used to run the system. Multiple-zone dispatching determines the distribution of cars among floors served by a group of elevators so that a passenger will find a car at or near his or her floor at all times. Floors are separated into zones, as many as there are cars available, with each car assigned to a specific zone. After answering the calls from the zone it covers, the car parks and shuts off in an "uncovered zone" until the next call. Multiple zoning better facilitates the spacing of the cars, since once spacing is disturbed, without reprogramming the cars have a tendency to get bunched: this is particularly true without adequate supervision in a terminal-dispatch situation.

***144. Control systems*** or dispatching and operation are, more often than not today, under supervision of a minicomputer. Usually there is one electronic computer controller per car, and the fully automatic systems use solid-state circuitry to control the motors driving the car, to program dispatching, to monitor car acceleration and deceleration, travel time between floors, door opening and closing times, and other constants, which are programmed into the computer's memory.

Sensors are used to report to the computer on system status at all times; for example, various types of electronic sensors can detect the location of a car at any given time and tell whether or not the car is loaded or empty. Typically, if a car is empty, the control aborts any command that may be remaining, such as may be the case with the prankster who pushes all the buttons in the car and then gets out. One important control under computer guidance is the elevator-capture system. In this case, if a fire is registered in the building, all controls are overridden by a master control which brings all cars to the lobby, where they are parked with doors open. Firefighters, equipped with a special key switch, can then use the elevators manually as they wish.

***145. Individual car features*** used in busy buildings, especially where supervisory dispatching or computer automatic control is used, include the use of hall lanterns with gongs to announce the arrival of a car at a particular floor. The hall lantern and gong indicate in which direction the car will be leaving the floor, and do so several seconds in advance of the arrival of the car on the floor, permitting passengers enough time to move into position in front of the car. Once the lantern is operated, it is important that the direction indicated does not change until all unidirectional requests are fulfilled; to obtain full utilization of the lanterns it is important that the passengers have complete confidence in them. It is especially important that the direction does not change while the doors are open, as a person might be entering the car at the time of the change, not be able to see the lantern change, and be carried in the wrong direction.

Load-measuring devices are used to bypass landing calls if there are too many people, approximately 75% of capacity, in the car to pick up further passengers efficiently. This has the effect of giving express service to a large number of people and reduces losses in passenger transfer time caused by shuffling around of people in the car to let people on and off at each stop, as well as eliminating stops when no further passengers could get on the car.

When the car stops for a call, the doors are opened and held open for a sufficient amount of time for normal passenger transfer. Various times are given as follows: for a stop to answer a car call only minimum time is allowed for passenger transfer; if a hall stop is answered, the time is increased because the prospective passenger is not quite as ready for the opening of the doors and generally hesitates, to allow for someone leaving the car; at a main landing where multiple transfers usually take place, the time is further increased; and when many people are in the car, allowance may be made for the additional time required as a function of load. Where it is felt that appreciable time should be allowed, light-ray detectors with the beams intersecting the door are sometimes used to cancel the remaining time after a person has been detected crossing the threshold.

Of course not all passengers can transfer within the door-open times allotted; so automatic door-reversal devices are used to reopen the doors if they are closing while passengers are in the way. The light-ray device may be used for this purpose, but since it will cause door reversal any time a person crosses the threshold whether or not the doors are near enough to hit the passenger, it will sometimes cause unnecessary reversal, with consequent unnecessary delay in car operation. Less delay is experienced with a lightweight movable mechanical shoe extending slightly ahead of the door while it is closing, which when contacted will cause the doors to reverse, or with an electronic field device which will cause reversal before the doors can touch the passenger. With the latter device doors can be made to back away just far enough to keep from touching a person, eliminating the necessity of a full reopening.

If people should insist on holding up the doors in any manner, thereby delaying service after a period of time, the door-reopening operations may be cut out, at which time the doors are closed at slow speed to urge people away without danger of injury.

***146. Elevatoring large buildings*** demands selection of the proper equipment based on a considerable number of facts, especially those which relate to peaks of traffic. The density of population in a single-purpose office building varies from one person per 90 ft$^2$ to one person per 120 ft$^2$ of net working area. In diversified-tenancy office buildings, the density may be one person per 100 to 150 ft$^2$.

Generally, the 5-min peak of morning arrival varies from one-fifth the population in a single-purpose building to one-ninth the population in a diversified-tenancy building. The 5-min peak traffic period during luncheon hours varies with the occupancy.

The evening 5-min peak, when the building is being emptied, usually exceeds the morning peak; however, there is generally a more efficient passenger transfer during the down peak, so that the number of elevators required is usually determined by the morning arrival peak.

The handling capacity of the cars is calculated on the basis of peak loading, the probable number of stops based on loading and total possible stops, the time of passenger transfer as a function of load, and the round-trip running time based on the number of stops and probable distance traveled. To forecast the proper number, speed, and size of the elevators that will be required in a building, it is thus necessary to predict accurately the peaks of traffic that will be experienced. This requires a large accumulation of data obtained from many different buildings of similar occupancy and purpose in similar locations. Often these data must be correlated and run through a computer to generate a proper program and eventual determination of the building's elevator needs.

***147. Grouping and Arrangement of Elevators.*** In large buildings, different banks of elevators are usually established to serve a set number of floors. Often, too, express and local banks are designated; however, since this particular arrangement reduces the number of elevators which serve a specific zone in the building, care must be taken to ensure that the arrangement results in a satisfactory interval between cars.

Car capacity in large diversified-tenancy office buildings is usually 3000 lb, and 3500 to 4000 lb in large single-purpose buildings. Groups generally consist of from 4 to 10 elevators in each bank, divided on either side of a large hallway. Car speeds are chosen in accordance with the distance traveled; typically 500 ft/min for 12 stories, 800 ft/min for 24 stories, and 1200 ft/min for 48 stories.

In newer and larger diversified-tenancy buildings with two lobby entrances, a current trend is to provide tandem elevator service. In this arrangement, two cars, placed one on top of the other, use a single elevator shaft on a single set of cables. One car serves even-numbered floors, while the other serves odd-numbered floors. Escalators link the two lobby levels, while signs or displays are used to indicate to passengers which floors are served by the tandem cars. This arrangement saves shaft space and cuts down construction costs.

## Performance

***148. Overall Performance.*** The final measures of performance are the number of passengers or amount of material carried per unit of time, the speed and dispatch with which they are handled, and for passengers, the waiting time for an elevator and the comfort of the ride. The performance depends on many factors such as speed, acceleration, and power-door operation. Even the shape of the elevator car and width of door opening influence the efficiency with which passengers can be loaded or unloaded.

An important performance index is the time required to run various distances, for example, one floor or two floors. An 800-ft/min elevator can make a 12-ft run in about 4 or 5 s, start to stop, not including door-operation time.

***149. Acceleration.*** This is limited by traction requirements and the comfort of the passengers.

To make the traction requirements reasonable, accelerations are generally limited to 5 ft/s$^2$ or below. However, where sufficient traction is available, acceleration as high as 10 ft/s$^2$ has been used.

Passenger comfort is affected by acceleration, rate of change of acceleration, and the length of time that they are experienced. Passenger feelings vary with the types and physical condition of the persons involved.

Accelerations of 5 ft/s$^2$ or less and rates of change of acceleration of 10 ft/s$^3$ are generally accepted as comfortable during the starting and stopping periods.

## Safety Features

***150. Door Interlocks.*** The American Standard Safety Code for Elevators (now ANSI) prohibits the starting of a car away from a floor unless the hoistway door is both closed and locked. This is usually accomplished by door interlock switches which open the control circuit when the door is either open or unlocked. Since these functions must be performed automatically on automatic elevators, a motor- or magnet-operated "retiring cam" is frequently furnished on the car. With power-operated doors, the interlock is usually combined with the operating mechanism.

***151. Gate Contact.*** This opens the control circuit if the car door or gate is open.

***152. Emergency Stop Switch.*** This is mounted in the car and when manually opened acts to cut off all power and apply the machine brake.

***153. Slack-Rope Switch.*** This is required for the hoist ropes of a drum-type elevator only. It is customary, however, to furnish a switch for the compensating sheave (Figs. 21-14 and 21-15) to stop the machine and apply the brake if the sheave moves above or below its normal position.

***154. Terminal Devices.*** Electric elevators are usually provided with two types of electrical devices for automatically limiting travel at either terminal. The "normal-terminal stopping device" slows down and stops the elevator when it is at or near the terminal landing and prevents further operation in the one particular direction. The "final-terminal stopping device" is arranged to function shortly after the normal device. It cuts off power to the elevator motor and brake and prevents further operation in either direction except by manual control from the machine room, which is necessary to return the elevator to its regular limits of travel.

Terminal stopping devices are usually mounted on the car or in the hoistway and are actuated by the movement of the car.

***155. Door Clearances.*** On automatic-operation elevators, the landing doors must be placed substantially flush with the edge of the landing sill, and the space between these doors and a car door or gate must be kept small to avoid trapping a person, particularly a child, when the doors or gate is closed (see American Standard Safety Code for Elevators).

***156. Runby and Clearance.*** The runby required for elevators of various speeds is usually fixed by law based on the American Standard Safety Code for Elevators. The clearance in the pit or between the top of the car and the overhead structure, when the car or counterweight bottoms, is ordinarily specified by law as 24 in minimum so that a person will be protected if lying flat either in the pit or on top of the car.

***157. Overloading.*** It requires extreme crowding with very tall people to overload seriously a passenger car conforming with the American Standard Safety Code for Elevators. This code prescribes the minimum hoisting capacity as a function of inside net platform area. Typical figures are for 30 ft$^2$, 2600 lb minimum; 5000 lb for 50 ft$^2$; 7600 lb for 70 ft$^2$.

Additionally, the code requires that the passenger elevator shall be designed to safely lower, stop, and hold loads up to 125% of rated load.

## Energy Demand and Feeders

***158. The kinetic energy*** of elevators is important, as energy must be supplied for every run to start the moving masses. With voltage control this is partly recovered in stopping, but with rheostatic it is practically all lost. For high-speed elevators operated by gearless hoisting machines, the principal kinetic energy is in the moving masses in the hoistway, and the kinetic energy of the motor and brake pulley is of comparatively small importance. For slow- and medium-speed elevators of the geared type, the rotor and the brake pulley have a large percentage of the kinetic energy of the elevator. When ac motor-geared hoists are used, the kinetic energy of the rotor is considerably higher than for direct current, for the reason that the rotors are larger in diameter than the armature of dc motors. For induction motors the kinetic energy per horsepower increases rapidly as the motors get larger (see Table 21-1 for computed kinetic energies of different parts of typical elevator equipment).

***159. Energy consumption*** is generally given in kilowatthours per car-mile for a certain load and number of stops per car-mile. Figure 21-20 shows how the consumption varies with load

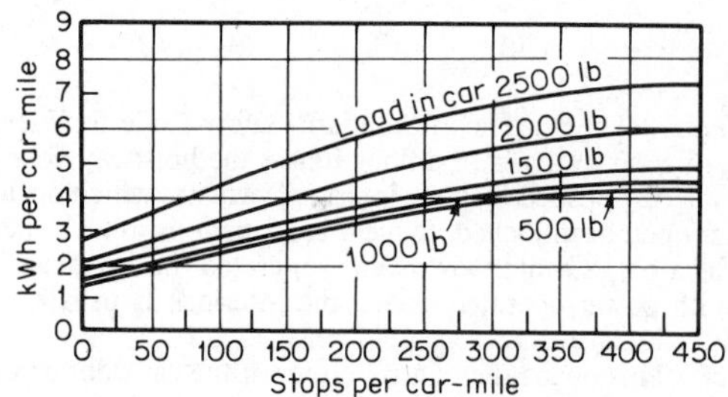


**FIG. 21-20** Typical energy-consumption curves of gearless traction elevator with voltage control, ac power, and duty of 2500 lb at 600 ft/min.

**TABLE 21-1** Analysis of Kinetic Energy of Typical Elevators

| Traction-machine type, hoist motor and control | Rated duty | Rise, ft | Kinetic energy ft-lb at rated load and speed in | | | | | |
|---|---|---|---|---|---|---|---|---|
| | | | Machine | Car | Counterweight | Ropes | Load | Total |
| Gearless 1:1, d-c motor, voltage control. | 2,500 lb at 800 fpm | 435 | 4180 | 15,250 | 18,200 | 10,000 | 6900 | 54,530 |
| Gearless 2:1, d-c motor, voltage control. | 3,000 lb at 500 fpm | 200 | 2900 | 6,750 | 8,050 | 3,900 | 3250 | 24,850 |
| Worm gear 1:1, d-c motor, voltage control. | 3,500 lb at 250 fpm | 125 | 6200 | 1,500 | 1,880 | 160 | 950 | 10,590 |
| Worm gear 1:1, a-c motor, 1 speed, rheostatic control. | 2,500 lb at 150 fpm | 100 | 2450 | 460 | 560 | 50 | 245 | 3,765 |

NOTE: 1 ft = 0.3048 m; 1 lb = 0.4536 kg; 1 ft-lb = 1.356 J.

**TABLE 21-2** Average Energy Consumption per Car-Mile in Typical Elevator Installations

| Traction-machine type and roping | Kind of supply and control | Rated duty | Rise, ft | Class of building | Kind of service | KWh per car-mile |
|---|---|---|---|---|---|---|
| Gearless 1:1..... | A-c voltage | 2,500 lb at 800 ft/min | 321 | 25-story office building | Passenger, express 1–25, no stops below 13 | (30 mi in 10-h day) 4.4 |
| Gearless 1:1..... | A-c voltage | 2,400 lb at 600 ft/min | 162 | 15-story office building | Passenger, local 1–15 | (20 mi in 10-h day) 4.1 |
| Gearless 2:1..... | A-c voltage | 4,500 lb at 450 ft/min | 135 | 12-story department store | Passenger, stop at each floor | (10 mi in 10-h day) 13.4 |
| Worm gear 1:1.. | A-c rheostatic | 1,800 lb at 100 ft/min | 61 | 7-story apartment house | Passenger | 4.1 |
| Gearless 1:1..... | A-c voltage | 2,500 lb at 600 ft/min | 267 | 18-story hotel | Passenger, express 1–18, no stops below 10 | (12 mi in 15-h day) 5.8 |

NOTE: 1 ft = 0.3048 m; 1 mi = 1.61 km; 1 lb = 0.4536 kg.

and stops. The curves are plotted on the basis of average consumption for up and down trips; so the one for 1000-lb load, which is about "balanced load," is the lowest of all. Usual values for stops per mile are 150 for local elevators, 75 or less for express elevators.

Note that curves such as those of Fig. 21-20 show only the energy consumed while the elevator is in motion. With voltage control the idling losses in the motor-generator, while the elevator is stopped at floors, must be added.

Considering all factors, with a moderate- or high-speed elevator that is reasonably busy, the total energy consumption with voltage control is usually less than with ac rheostatic control.

With car-switch or continuous-pressure operation some extra energy is used in inching to floors, especially if the control is rheostatic or no automatic leveling is provided.

From 80 to 90% of the power taken from the input power lines for the operation of the elevators is dissipated in the elevator machine room in the form of heat. With gearless machines and generator field control, approximately two-thirds of the heat released is from the motor-generator set, the remaining one-third being from the hoisting motor and control equipment. Since performance adjusted for normal temperatures becomes unsatisfactory at excessively high temperatures, it is the responsibility of the architect, consulting engineer, or owner to provide means, if necessary, to hold the ambient temperature of the machine room at a reasonable level. Spill air from the building air-conditioning system, separate air-conditioning units, or machine-room ventilating fans are often used for this purpose. The Btu released per hour by the elevators can readily be calculated from the kilowatthours per car-mile and the estimated car-miles per hour averaged on the basis of the working hours that the elevators remain in service.

***160. Average Energy Consumption for Various Types of Installation (see Table 21-2).*** The table is presented only to give an idea of the quantities. Widely different figures may prevail on different installations of similar type, owing to differences in operating conditions.

***161. Nature of Feeder Load.*** This is intermittent. Busy express elevators may actually run about 60% of the time they are in service. For other classes of service the time duty ranges down to 25% or even less. Figure 21-21 gives an idea of the duty cycle for a busy elevator. This is for voltage control and applies equally well to ac or dc supply.

The running current varies with the load, as does also the peak starting current with voltage control. With rheostatic control the starting peak is less dependent on the load.

The ratio of peak-starting kilovoltamperes (kVA), to running both with full load up, has the following typical values: high-speed passenger, voltage control, 1.75; low-speed heavy-duty freight, voltage control, 1.4 or less; ac rheostatic, 2.0 to 3.0 depending on refinement of control. The power factor of the starting peak is over 0.9 on voltage control and over 0.75 on rheostatic.

***162. Feeders for Single and Multiple Elevators.*** These should comply with the requirements of the National Electric Code (motor branch circuits) and any local regulations. The NEC bases feeder sizes on the nameplate current rating of the motor suitably modified for elevator service. In the case of motor-generator sets, the nameplate current is less than the full-load current owing to the continuous rating. However, the nameplate current rating is always equal to or greater than the rms of the current in the line during the duty cycle.

In addition to current capacity, voltage drop must be considered in the design of feeders. This should not exceed 3%.

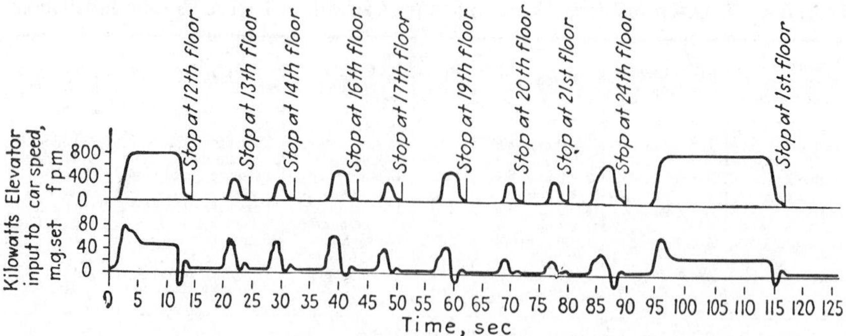


**FIG. 21-21** Typical round-trip record of express elevator in morning peak of filling an office building: duty, 2500 lb at 800 ft/min; travel—first to twelfth floor express, twelfth to twenty-fourth local; rise 261 ft.

When two or more elevators operate from the same set of feeders, it may happen that their current peaks coincide, in which case the rms line current becomes the arithmetic sum of the rms currents of all the cars. However, it is improbable that this will occur very frequently, so that advantage may be taken of the diversity of operation of the elevators in calculating the size of feeder.

For a group of $n$ similar cars, each with an rms current of $I$, the line current will be $nI$ when the peaks coincide and $\sqrt{n}I$ when the peaks are in the most random pattern. The line rms current falls somewhere between these two limits. The diversity factor (that is, the ratio of the most probable line current to $nI$) varies within considerable limits and depends not only on the number of elevators using one set of feeders but on the type and intensity of the elevator service.

***163. Feeder Protection.*** If fuses are used for protection, they should be of a "lag" type which will not blow on short-time overloads. Fuses are affected by repeated heating and cooling and by heat conducted from poor connections at the clips, and so the best protection for feeders is by delayed overload circuit breakers.

***164. Regeneration.*** The regenerative power of high-speed voltage-control elevators has been known to trip reverse power relays in the supply system. Such an elevator, fully loaded, running full speed down and stopping, may pump back about 20 and 90%, respectively, of its normal full-load-up running power. These figures are rough and subject to wide variation owing to differences in guide friction, etc. The problem of handling this negative power has required supplementary power-dissipating means when rectifiers are used to supply dc elevator motors from ac supply systems. Also, emergency power-supply systems have to be checked for the ability to handle the negative power.

***165. Cost of Operating Elevators.*** This varies greatly with the service. In general, the higher class of equipment is the most economical for intensive-traffic elevators, since it gives the greatest service return per unit of power used. Power-consumption cost is only a part of elevator operating costs and should be secondary to the necessity of providing a good quantity and quality of elevator service.

In addition to power, operating costs include building costs chargeable to the elevators, such as the space occupied by the equipment, also attendants' wages, welfare, uniforms, etc. The elimination of the attendant costs with automatic elevators effects a considerable saving.

***166. Hydroelectric elevators*** of the direct-lift plunger type are frequently installed for low-rise (up to 50 ft) low-speed (up to 150 ft/min) applications. The cars are generally not counterweighted, and safeties and governors are not required. The fluid (usually hydraulic oil) is pumped directly from a supply tank to the cylinder by an electric-motor-driven pump to move the car up, and gravity moves the car down.

Hydraulic motion-control valves operated by solenoid pilot valves control the motion of the car.

***167. Moving Stairways (Escalators).*** These are extensively used in department stores; railway, elevated, and subway stations; and airline buildings. They have also been applied to commercial buildings between the lower floors and to schools, racetracks, theaters, and other places of public assembly.

Moving stairways have the advantages that they can handle a large number of persons in a given time, there is no delay in getting on or off, the traffic is continuous, and even with very heavy traffic the transportation is comfortable.

Moving stairways are usually furnished in two sizes, 32 and 48 in wide between balustrades, with the nominal handling capacities in passengers per hour of 5000 and 8000, respectively, at a speed of 90 ft/min. A speed of 120 ft/min is sometimes used, which increases the handling capacity, although not necessarily in proportion to the ratio of 120:90.

A number of large office buildings provide moving-stairway service to six or seven floors above the ground floor, which is reasonably close to the limit of traveling time that persons will accept in busy buildings. The use of moving stairways versus elevators requires an economic consideration of such factors as space occupied, first cost, power consumption, maintenance, and operating personnel, if any.

Moving stairways are particularly desirable between the ground floor and basement or lower floors in office buildings. They enable basement and lower-floor space to be effectively used for merchandising, restaurants, banks, etc., which, under suitable conditions, return excellent rentals. With moving-stairway service, the lower-floor and basement space is almost as valuable as ground-floor space.

***168. Moving-Stairway Motors and Starters.*** On alternating current, squirrel-cage motors are generally used. They are designed for 6 to 10% slip, as the starting load is fairly heavy. A 32-in stairway with a rise of 20 ft requires about 10 hp when fully loaded.

On stairways that may run up or down, it is desirable to operate the ac motors on reduced excitation for down motion, thus saving power and improving power factor. Y-Δ starters are well adapted to such operation.

***169. Moving-Stairway Safety Devices.*** Moving stairways have magnetically released brakes to stop and hold them in case of failure of power. They also have switches to cut off power in case of slack chain and stop buttons at top and bottom landings.

## AIR CONDITIONING AND REFRIGERATION

### Air Conditioning

***170. Air Conditioning Defined.*** The definition of air conditioning as approved by the American Society of Heating and Ventilating Engineers is as follows: "The simultaneous control of all or at least the first three of those factors affecting both the physical and chemical conditions of the atmosphere within any structure. These factors include temperature, humidity, motion, distribution, dust, bacteria, odors, toxic gases, and ionization, most of which affect in greater or less degree human health or comfort."

Accepted trade definitions consider air conditioning to be the control of temperature, humidity, motion, and cleanliness of the air within any structure, adding the qualifying prefix "summer" to those systems designed only to maintain temperature and humidity at a point below the normal outside summer condition and "winter" to those systems designed only to maintain elevated temperatures and humidities during cold weather. Systems designed to afford the stipulated control during all seasons of the year are usually designated as "year-round" air-conditioning systems.

***171. Psychrometric Formula.*** A major step toward the development of air conditioning as a science was made by Willis H. Carrier with the derivation of the rational psychrometric formula (*Trans. ASME,* 1911) which established the relationship between dry-bulb and wet-bulb temperature, both easily ascertainable, and other important properties of the air such as density, relative and absolute humidity, vapor pressure, and total heat, the exact valuation of which is essential to the solution of problems involving the exchange of heat and humidity between air-conditioning apparatus and conditioned spaces and their contents. The rational psychrometric formula is the basis for numerous psychrometric charts.

***172. Psychrometric Terms***

*Dry-bulb temperature* is the commonly observed temperature of the air as indicated by a conventional thermometer not affected by the relative humidity of the air.

*Wet-bulb temperature* is the temperature of adiabatic saturation of air or the temperature at which air would normally saturate without any change in its heat content. Wet-bulb temperature is, therefore, also a measure of the total heat content of air; that is, at a given wet-bulb temperature the total heat is constant regardless of changes in the dry-bulb temperature.

*Relative humidity* within the normal range is the percent ratio of the weight of water vapor actually contained in a unit of space to the weight of water vapor that the same unit of space would contain if fully saturated at the same dry-bulb temperature.

*Absolute humidity* is the actual weight of water vapor in air at any given condition and is usually expressed in grains per cubic foot or grains per pound of dry air.

*Dew point* is the temperature of saturation of air, or the point beyond which any further reduction in temperature would result in condensation.

*Effective temperature* is an arbitrary index of the degree of warmth or cold felt by the human body in response to temperature, humidity, and air motion. Experimentally determined, the scale has been fixed by the temperature of saturated air which reproduces the sensation of warmth felt at other combinations of conditions.

*Sensible-heat effect* is the exchange of heat between air and its surroundings or apparatus that changes the dry-bulb temperature without affecting the moisture content. It is convenient to consider sensible heat as dry heat.

*Latent-heat effect* is the exchange of moisture between air and its surroundings or apparatus by evaporation or condensation resulting in a change in the moisture content of the air with consequent change in the total latent heat of vaporization represented by the moisture in the air. It is convenient to consider latent heat as humidity.

***173. Comfort Cooling.*** Investigations carried on under the auspices of the ASHVE indicate that effective temperatures varying from 70 to 73°F are acceptable inside conditions for peak summer weather depending upon length of occupancy and for normally clothed, normally active commercial workers. Table 21-3 gives approximate dry-bulb temperature and relative-humidity combinations within normal range which produce effective temperatures within the stated range with air motion 15 to 25 ft/min.

**TABLE 21-3** Inside Design Conditions*

| Effective temperature | Relative humidity 45 % | 50 % | 55 % |
|---|---|---|---|
| | Dry-bulb temperature, deg | | |
| 73 | 79½ | 79 | 78 |
| 72 | 78½ | 77½ | 77 |
| 71 | 77 | 76½ | 76 |
| 70 | 76 | 75 | 74½ |

* Data from ASHVE Comfort Chart; *Trans. ASHVE*, 1932, vol. 38, p. 410.
NOTE: $t_{°C} = (t_{°F} - 32)/1.8$.

Inside design conditions of 78 to 80°F dry bulb with 45 to 55% relative humidity, depending upon the character of occupancy, are accepted as standard inside design conditions against peak outside conditions for commercial comfort-cooling applications in almost all parts of the country.

It is seldom desirable to design for inside temperatures in excess of 80°F. It is generally impracticable to reduce the humidity sufficiently to produce effective temperatures within the desired range at dry-bulb temperatures much in excess of 80°F.

***174. Industrial-Process Air Conditioning.*** In industrial or process air conditioning, inside

**TABLE 21-4** Maximum Summer Design Conditions*

| City | D | W | City | D | W | City | D | W |
|---|---|---|---|---|---|---|---|---|
| Boston | 88 | 74 | Cleveland | 89 | 75 | Seattle | 80 | 65 |
| New York | 93 | 76 | Chicago | 91 | 76 | San Francisco | 77 | 62 |
| Philadelphia | 90 | 77 | St. Louis | 94 | 78 | Los Angeles | 90 | 70 |
| Washington, D.C | 92 | 77 | Atlanta | 92 | 77 | | | |
| Jacksonville | 94 | 79 | Birmingham | 94 | 78 | | | |
| New Orleans | 91 | 80 | Dallas | 99 | 78 | Denver | 90 | 64 |
| Houston | 94 | 80 | Phoenix | 106 | 76 | Salt Lake City | 94 | 66 |

* *ASHRAE Guide and Data Book,* 1956, Chap. 27, Table 1.

design conditions are frequently dictated entirely by the product requirements. In confectionery plants where chocolate is handled, it is usually not desirable to have temperatures in excess of 65°F (dry bulb), which is far below the comfort range. In chemical or pharmaceutical plants where deliquescent chemicals are handled, lower-than-average humidities may be required for satisfactory production. Many products can be satisfactorily processed at widely varying conditions as long as the condition is maintained constant and the process adjusted to it.

Where conditions are established by product requirement, design is usually directed toward maintenance of satisfactory working temperatures. Investigations to determine the limits for optimum health and efficiency of the industrial workers indicate that conditions should not exceed 80°F effective temperature, that is, approximately 82.8°F, d.b. at 80% RH, 84.4°F d.b. at 70% RH, 86.2°F d.b. at 60% RH, 88.2°F d.b. at 50% RH, and 90.3°F d.b. at 40% RH. Lower effective temperatures are required where perspiration must be prevented to avoid contamination or staining of products during handling.

There has been a marked increase in the use of complete air conditioning, including refrigeration, even in industrial operations where atmospheric control is not required by the process involved. An example is the trend toward larger manufacturing areas under single roofs, where windows cannot be depended upon for light or ventilation. In many cases it has proved less expensive in both first cost and operating cost to provide full air conditioning with refrigeration rather than ventilation.

***175. Winter Heating.*** Winter heating systems for normally occupied interiors are almost universally designed to maintain 70°F inside, with outside conditions assumed at 10°F above the lowest recorded temperature for the district in question.

***176. Maximum outside design conditions*** are usually taken at values which will not exceed 2.5% of the total hours during June, July, August, and September. Accepted maximum design dry-bulb (D) and wet-bulb (W) temperatures in degrees Fahrenheit for larger cities representative of various climatic divisions of the United States are given in Table 21-4.

***177. Calculation of the cooling load*** involves consideration of all possible internal as well as external heat gains not only for transmission and infiltration but also from sunlight radiation.

***178. Sunlight radiation*** for 40°N lat on July 21 attains maximum values in Btu per square foot per hour transmitted by unshaded glass for variously oriented surfaces as follows: East, 200 at 8 A.M.; South, 91 at 12 noon; West, 200 at 4 P.M.; horizontal 258 at 12 noon.[1] Major consideration must be given to sun-exposed glass. Of the total impinging radiation, the amount passing through the net area of the window is approximately as follows: unshaded, 100%; half covered by buff shade, 70%; fully covered by light-finish venetian blind, 70%; equipped with canvas awning, 30%.[2] Although sunlight-radiation effect upon walls and roofs can be calculated, allowance is usually made for this effect by calculating heat transmission at a temperature above the design outside temperature. A substantial temperature head (usually 50°F) should be allowed in calculating transmission through the roof surface due to long hours of exposure to the sun. Heat from sunlight contributes only to the sensible-heat load.

***179. Transmission of heat*** through the structural surfaces of buildings or spaces is directly proportional to the temperature difference between the two sides of the surface and to the

---

[1] *ASHRAE Guide and Data Book,* 1965, chap. 27, Table 12.

[2] Abstracted by permission from *Research Repts.* 975, *Trans. ASHVE,* 1934, vol. 40; and 1180, *Trans. ASHVE,* 1941, vol. 47.

**TABLE 21-5** Coefficients of Heat Transmission $U$*

| Surface composition | $U$ |
|---|---|
| Single window | 1.13 |
| Double window | 0.56 |
| 12-in brick wall, no interior finish | 0.35 |
| 12-in brick wall finished with ¾-in plaster on metal lath, furred | 0.25 |
| 10-in hollow-tile wall, stucco exterior, no interior finish | 0.33 |
| 10-in hollow-tile wall, stucco exterior finished with ¾-in plaster on metal lath, furred | 0.24 |
| 10-in concrete wall, no interior finish | 0.61 |
| 10-in concrete wall finished with ¾-in plaster on metal lath, furred | 0.36 |
| Wood siding, 1-in sheathing, studs, ¾-in plaster on metal lath | 0.26 |
| Wood siding, 1-in sheathing, studs, ¾-in plaster on metal lath, 3-in rockwool fill | 0.07 |
| Double partition, metal lath and ¾-in plaster on both sides of studding | 0.39 |
| 4-in hollow-tile partition, plastered on both sides | 0.37 |
| Yellow-pine flooring (25/32 in) on joists, no ceiling | 0.45 |
| Yellow-pine flooring (25/32 in), ¾-in plaster on metal-lath ceiling below joists | 0.31 |
| 6-in concrete slab floor, ¼-in asphalt tile, no ceiling | 0.60 |
| 6-in concrete slab floor, ¼-in asphalt tile, ¾-in plaster on metal lath hung or furred ceiling | 0.38 |
| 1-in wood flat roof, built-up roofing, no ceiling | 0.48 |
| 1-in wood flat roof, built-up roofing, ¾-in plaster on metal-lath ceiling below joists | 0.33 |
| 1-in wood flat roof, 1-in rigid insulation under built-up roofing, no ceiling | 0.21 |
| 1-in wood flat roof, 1-in rigid insulation under built-up roofing, ¾-in plaster on metal-lath ceiling below joists | 0.17 |
| 4-in concrete-slab flat roof, built-up roofing, no ceiling | 0.70 |
| 4-in concrete-slab flat roof, 1-in rigid insulation under built-up roofing, no ceiling | 0.24 |
| 4-in concrete-slab flat roof, built-up roofing, ¾-in plaster on metal-lath ceiling, hung or furred | 0.41 |
| 4-in concrete-slab flat roof, 1-in rigid insulation below built-up roofing, ¾-in plaster on metal-lath ceiling, hung or furred | 0.18 |

* *ASHRAE Guide and Data Book,* 1965, Chap. 24, Tables 5 to 18.
NOTE: 1 in = 25.4 mm.

thermal conductivity of the material or materials forming it. Overall coefficients for a number of common structural elements are given in Table 21-5. The coefficient $U$ is expressed in Btu per hour per square foot per degree Fahrenheit temperature difference across the surface and is based upon an outside-surface coefficient for 15-mi/h wind velocity.

***180. Total heat transmission*** through any exposed building surface may be computed by the formula

$$H = A \times U(t_o - t_i)$$

where $H$ = sensible heat in Btu per hour, $A$ = area of the surface element in square feet, $U$ = coefficient of heat transmission of the surface, and $t_o$ and $t_i$ =, respectively, the outside and inside temperatures in degrees Fahrenheit. This formula is used for computing heat loss from a space as well as heat gain to it.

***181. Heat from Occupants.*** Every occupant of a conditioned space contributes both sensible and latent heat to the atmosphere, the relative proportions being dependent upon the dry-bulb temperature and the total quantity upon the state of activity of the individual (see Table 21-6).

Only the sensible-heat fraction affects the temperature of the conditioned space, since latent heat from evaporation of body moisture appears as vapor at room temperature.

***182. Electrical Heat.*** The heat equivalent of the electrical input to all lights and electrical heating or power appliances used within a conditioned space contributes to the atmospheric heat which must be removed by the air-conditioning apparatus. In general practice, the approximate heat equivalent of 3400 Btu/(h)(kW) input to lights and appliances and 2500 Btu/(h)(hp) of motor capacity is used in calculating heat gain from this source. In considering heat from appliances and motors, load factor should be carefully investigated, since appliances or motors seldom run continuously at full load. Unless appliances are used for water heating or evaporation, the entire load from electrical sources will be sensible heat.

***183. Process Heat.*** Heat from gas appliances is the most common form of process heat. Average heat input to common gas appliances in Btu per hour is as follows: 14-in coffee urn, 7500; 12-in coffee urn, 5000.[1] Heat output will be approximately 75% sensible heat and 75% latent heat. Consideration should be given to load factor of appliances.

[1] *ASHRAE Guide and Data Book,* 1965, chap. 27, Table 27.

**TABLE 21-6** Heat Loss from Average Person, in Btu per Hour, as Sensible Heat *SH* and Latent Heat *LH**

| Condition of activity | Total heat loss, Btu/hr | Dry bulb temperature, deg F | | | | | |
|---|---|---|---|---|---|---|---|
| | | 70 | | 75 | | 80 | |
| | | Segregated heat loss, Btu/hr | | | | | |
| | | SH | LH | SH | LH | SH | LH |
| Seated at rest | 400 | 300 | 100 | 260 | 140 | 220 | 180 |
| Light work | 660 | 350 | 310 | 290 | 370 | 220 | 440 |
| Moderate work | 850 | 430 | 420 | 360 | 490 | 270 | 580 |

* Data from Charts of Research Report No. 908, *Trans. ASHVE,* 1931, vol. 37, p. 541.
NOTE: 1 Btu/hr = 0.293 W; $t_{°C} = (t_{°F} - 32)/1.8$.

***184. Ventilation and Infiltration.*** Air quantity for ventilation purposes is usually based upon the number of occupants of the conditioned space. Ventilation is required chiefly to overcome normal odors. Average accepted practice calls for the introduction of at least 10 ft³/min of outside air for each nonsmoking occupant and 20 ft³/min for each smoking occupant. For commercial interiors, such as restaurants and drugstores, where moderate smoking occurs, 15 ft³/min per person is usually allowed. In theaters, where ceiling heights are usually liberal, where there is no smoking, and peak occupancy seldom occurs for long periods, an allowance of 7½ ft³/min per peak occupant is considered ample. Air positively exhausted from a space must be deducted from the positive ventilation supply in considering possible infiltration. If the exhaust exceeds the supply, infiltration will naturally be induced. Air for ventilation should be first brought to the conditioning apparatus and conditioned before introduction to the conditioned space.

*Infiltration* varies with the construction, exposure, and size of a conditioned area as well as with door traffic and location of doors. The following quantities are suggested as a general guide for buildings of average construction: cubical contents 0 to 5000 ft³, one complete change of volume in 30 min; 5000 to 50,000 ft³, 40 min; 50,000 to 100,000 ft³, 60 min; 100,000 to 200,000 ft³, 90 min; over 200,000 ft³, 120 min. Infiltration in cubic feet per minute for load calculation and comparison with positive supply in cubic feet per minute may be determined by dividing cubical contents of the space by the time factor for that space magnitude.

The sensible-heat component of ventilation or infiltration may be computed approximately from the formula

$$H_S = Q \times 1.07(t_o - t_i) \qquad \text{(Par. \textbf{186})}$$

where $H_S$ = sensible heat in Btu per hour, $Q$ = cubic feet per minute of air, and $t_o$ and $t_i$ = outside and inside temperatures in degrees Fahrenheit, respectively.

The latent-heat component of ventilation or infiltration may be computed approximately from the formula

$$H_L = Q \times 0.675(h_o - h_i) \qquad \text{(Par. \textbf{187})}$$

where $H_L$ = latent heat in Btu per hour, $Q$ = cubic feet per minute of air, and $h_o$ and $h_i$ = outside and inside absolute humidities, respectively, expressed in grains per pound of dry air (see a psychrometric chart).

Only infiltration air entering the conditioned space directly contributes to the internal heat load. Ventilation air conditioned before introduction to the space contributes to the load upon the conditioner and cooling apparatus only.

***185. The internal sensible-heat load*** will be the summation of the sensible heat from transmission, sunlight radiation, occupants, lights, electrical devices, process heat, and infiltration. The *internal latent-heat load* will be the summation of the latent heat from occupants, process heat, and infiltration. A factor of safety of 10% is often added to these figures to allow for inaccuracies and, in the case of sensible-heat load, to cover the power input for air circulation.

The *total heat load* on the conditioner or cooling apparatus will be the sum of the internal sensible- and latent-heat loads plus the sensible- and latent-heat load of ventilation air. Grand-total heat is usually expressed in tons of refrigeration effect.

***186. Air Quantity for Sensible-Heat Absorption.*** The sensible-heat load usually serves as the first guide to the selection of proper air quantity. The specific heat of air within the normal air-conditioning range is approximately 0.241 Btu/(lb)(°F). One Btu will raise 4.16 lb of air 1°F, or approximately 56 ft³ of air 1°F. The air quantity required to absorb a given sensible-heat quantity may be determined by the formula

$$Q = \frac{H_S}{60} \times \frac{56}{t_i - t_d} = \frac{H_S}{1.07(t_i - t_d)}$$

where $Q$ = air quantity in cubic feet per minute, $H_S$ = sensible heat in Btu per hour, $t_i$ = space temperature in degrees Fahrenheit, $t_d$ = delivery temperature of air into the conditioned space in degrees Fahrenheit, 60 = minutes per hour, and 56 = cubic feet per Btu per degree Fahrenheit temperature rise. Modern diffusion-type outlets will permit satisfactory delivery of air into spaces of 10- to 12-ft ceiling at 60°F. With higher ceiling heights and good conditions, air may be introduced at 55 to 50°F.

***187. Latent-Heat Absorption (Dehumidification).*** For the quantity of air selected to meet the sensible-heat load by the method just suggested, the moisture content at which the air must be introduced into the space to absorb the internal latent-heat load may be approximately calculated by the following formulas:

$$h_r = h_L/Q \times 0.675 \qquad \text{and} \qquad h_d = h_i - h_r$$

where $h_r$ = humidity deficiency in grains per pound of dry air; $h_d$ and $h_i$ = absolute humidity in grains per pound of dry air at delivery condition and within the room, respectively; $H_L$ = internal latent heat in Btu per hour; $Q$ = air delivery in cubic feet per minute; and the factor 0.675 is calculated from the average density of air, the average latent heat of moisture in Btu per grain, and rate correction. The delivery-moisture condition thus calculated, together with the delivery temperature determined as in Par. **186**, will permit graphical determination of the required delivery-air conditions from a psychrometric chart. *Dehumidification* is usually accomplished by condensation of moisture in the process of cooling the air.

***188. Cooling-Dehumidifying Apparatus.*** Two types of apparatus are used for cooling and dehumidifying: (1) the spray-type dehumidifier and (2) the surface cooling coil. In either, cooling and dehumidification may be accomplished by cold fluid at the proper temperature from any source, for example, cold water from wells or ice tanks or water cooled directly by refrigeration apparatus or by the direct expansion of refrigerants in cooling coils.

*Spray Type.* In the spray-type dehumidifier, air to be conditioned is passed through a spray chamber into which a large volume of finely divided water is introduced. Spray-type dehumidifiers are of two general types: *(a)* those in which water cooled externally is introduced at a reduced temperature and *(b)* those in which water is sprayed over coils cooled by fluid from an external source.

Spray-type dehumidifiers have the advantage of permitting evaporative cooling (see Par. **191**), of cleaning the air of siliceous dust, and of absorbing some gases and odors.

*Surface Cooling Coils.* Surface cooling coils are widely used for cooling and dehumidification. Cooling of surface coils is usually accomplished by direct expansion of refrigerant though frequently cold water or brine is used.

Common designs include *(a)* pierced-plate types in which the tubes pass through closely spaced thin metallic plates, *(b)* helical extended surface in which a ribbon of metal crimped or slit to accommodate curvature is wound in a helix about the central tube, and *(c)* the type in which fins are extruded directly from the tube body by rolling.

***189. Heating Surface.*** Heating surface is generally similar to cooling surface in construction. Extended cast-iron surface is still used to some extent. Steam is an accepted heating medium, though use of hot water is increasing, as it facilitates control.

***190. Humidification and Humidifiers.*** In many comfort cooling applications and some industrial-process operations which have low winter heat loads and require only moderate humidification during the winter, this is accomplished by small sprays which introduce finely divided water droplets into the airsteam usually directly before or after the heating coil so that evaporation is stimulated. Pan-type humidifiers in which water is evaporated into the airsteam by steam coils or direct steam sprays may also be employed. It should be noted, however, that these procedures add to the loading within the space and therefore should be employed only where there is a deficiency in heat gain or a heating requirement.

In large commercial or industrial operations employing spray-type air washers for summer cooling and dehumidification, humidity control is usually achieved by maintaining a constant dew-point temperature and then allowing the air to rise to the dry-bulb temperature, resulting in the desired final relative humidity in absorption of heat gains within the space. It should be noted that the majority of industrial and many large commercial applications involve internal loading from power, lights, etc., that require the application of cooling during all seasons of the year when there is normal occupancy or operation. During winter this cooling may be obtained by outside air, as described subsequently for evaporative cooling, but there is no heating requirement during normal service.

The major shortcoming of the dew-point control method for humidity maintenance lies in the fact that there must be a fixed relationship between the maintained dew point and the final dry-bulb condition within the space to achieve the desired relative humidity. Thus there is a fixed relationship between the load and the temperature rise in supply air which may be taken to absorb this load, and this relationship may result in the requirement for excessively high circulating air quantities, particularly if high relative humidities must be maintained as is common in many textile-mill operations.

Accordingly, in such operations where high relative humidities must be maintained with high continuous internal heat loads, common practice is to employ a *split system* in which the dew point of the circulating air is maintained at an arbitrarily low level which would normally result in insufficient humidification after absorption of the heat load, but the humidity is increased within the space by the direct application of water droplets from compressed-air or motor-operated mechanical atomizers. The atomized moisture introduced directly into the space is in water form and in the process of evaporation not only increases the relative humidity but absorbs sensible heat from the space in the process of being converted from water to vapor at the room condition. The moisture thus introduced does not add anything to the heat load within the space and further, by the process noted above, increases the sensible-heat-absorbing capacity of the primary supplied air, thus making it possible to perform the required cooling with a substantially lower volume of circulating air than would be needed for the straight dew-point method.

Atomizers are also frequently used as prime means for humidification in areas such as textile mills. Employed in this fashion or with the central-station system they also afford an excellent facility for direct humidity control in zones within a major area.

***191. Evaporative Cooling.*** During intermediate seasons in normal climates or in climates such as prevail in dry mountainous country where temperatures are frequently excessive but humidities continuously low, evaporative cooling systems are very effective. Air is saturated adiabatically and thus assumes the wet-bulb temperature (see Par. **172**). Sensible cooling is thus done at the expense of increased humidity. Calculations for evaporative cooling systems are identical with those for other cooling loads except that 100% outdoor air is used.

Commercial evaporative coolers are similar to spray-type washers or humidifiers or spray-type dehumidifiers. The spray water is continuously recirculated by means of a pump.

***192. Fans for Air Conditioning.*** *Propeller-type fans* used in cooling or heating units and normal ventilation service are usually driven by direct-connected motors, since their operating speed is within a suitable range. Most propeller-type fans require only fractional-horsepower motors, and capacitor-type single-phase motors, single- or multispeed, are usually provided as standard by manufacturers.

*Axial-flow fans* offer good efficiency and pressure characteristics and are used extensively for *industrial* air-conditioning services in all capacity ranges. While their total sound output compares favorably with equivalent centrifugal fans, noise generation is at a high frequency and is relatively difficult to eliminate. Accordingly, they are not widely employed for *commercial* applications.

*Centrifugal-type fans* with volute housings are most commonly used in commercial air-conditioning work. Fan speeds are generally moderate to low, depending upon the air quantity to be circulated, and system resistance will vary, depending upon the design of ductwork, apparatus, etc. Thus centrifugal fans are almost universally belt-driven, since this permits the use of economical ac motor speeds, and since fan speed may be selected to fit the estimated requirement or readily changed on the job if necessary. Where single-phase ac motors are used for fan drive, as on residential or small commercial air-conditioning applications, the capacitor-type motor is favored because of its quietness. Polyphase squirrel-cage motors are widely used for fan drive, since the average-size air-conditioning installation involves sufficient total motor load to justify polyphase supply. Regular squirrel-cage motors perform satisfactorily in most fan service. Two-speed single-phase or polyphase fan motors are sometimes used for year-round air-conditioning systems where circulation of air need not be so high in winter as in summer.

***193. Filters.*** The most commonly used is the *unit-type viscous impingement filter,* which consists of a cell with a fibrous filler or filtering mat impregnated with an adhesive to which dust adheres upon impingement.

*Dry-type unit filters* are those in which the air is cleaned by straining action in passing through felt, cloth, or paper. Because of the density of the filtering media, large filtering area must be provided to keep resistance low.

Unless unit-type filters are carefully maintained, they may become so clogged as to restrict air circulation materially and affect adversely the operation of the system.

With the increase in size of systems in service and the increase in cost of maintenance labor, extensive development of automatic filtration equipment has occurred. One widely used design employs roll-type filter media which are fed clean onto a movable screen located across the airstream; are conveyed across the airstream, accumulating its dirt load; and then are taken up as a dirty roll on the opposite side of the filter elements. Since a substantial area of medium can be provided in the roll, only occasional replacement of clean rolls and removal of dirty rolls are required. For normal dust loads a deep fibrous mat impregnated with adhesive is normally employed. For lint loads and similar dirt loads which will not penetrate such a mat and therefore cannot be stored in it, thin paper or glass-fabric media are generally employed. Electric drive is used to convey the filter medium across the airstream, either operated on a fixed time cycle or arranged to automatically advance the filter medium as the pressure drop across the unit increases owing to the accumulation of dirt. Other automatic filters employing fixed-medium screens with electrically operated vacuum-cleaning devices which traverse the screen area are also available and find application, especially in areas where lint accumulation is the major problem.

The mechanical-filter types listed are efficient in removal of large dust particles which represent the greatest weight percentage of normal atmospheric dirt. Their inability to remove small particles which have little tendency to settle under gravitational action has resulted in the development of *electrostatic dust-precipitation equipment* which is efficient in removing even the smallest-sized particles such as those composing the visible elements of tobacco smoke.

The dirt accumulated on electrostatic filters must be periodically removed by cleaning. Large electrostatic units are often equipped with automatic washers to perform this operation, and one device employs collector plates mounted on conveying chains similar to the automatic filters described above which pass the plates continuously through an oil bath to clean them.

***194. Automatic Control.*** Most control devices depend upon physical, thermal, or hygrostatic force for their operation. For example, thermostats frequently are operated by the distortion of bimetallic strips caused by the unequal expansion of the metals with increase in temperature or by the increase in pressure with rise in temperature of a liquid or gas contained in a closed element. Hygrostats generally depend upon the expansion or contraction of human-hair strands with changes in humidity to perform the basic control operation, though silk, paper, or wood is used in certain applications. The thermostatic and hygrostatic elements usually operate other devices to perform the control operation.

***195. Air-conditioning control equipment*** may be divided into three separate classes: *(a)* self-contained, *(b)* pneumatic, and *(c)* electrical.

*Self-contained devices* are usually composed of thermostatic-fluid elements directly operating steam, water, or brine valves. These devices are not used where close control is necessary or where instantaneous tight shutoff is required. However, they are simple, cheap, and reliable and inherently give modulated control.

*Pneumatic-control* systems are those in which the thermostatic or hygrostatic element controls the rate of air leakage from a compressed-air system. The air pressure in the control system is used to operate diaphragm motors controlling valves and dampers. The spring loading of these motors can be varied so that within the given control-pressure range a whole sequence of separate valve and damper operations can be accomplished. Various types of relay and compensating arrangements are used to overcome inherent tendencies of such systems to override or "hunt" if set for high sensitivity or to give too wide a control range if set for low sensitivity.

The most common *electric thermostatic or hygrostatic controls* are of the *snap-action type;* that is, the condition change simply serves to cause the device to make or break an electric circuit, which in turn controls the major element or elements to be operated. Because of contact difficulties, line-voltage controllers generally use mercury-tube contactors, which limit the accuracy of control. All wiring to the control must be run in conduit. Low-voltage controls (24 V ac) which operate major motor drives or other elements through relays are more common; the light, open contacts permit accurate control, and open wiring (usually metal-clad) is allowed.

Electric controls employing thermocouples or resistance thermometers for sensing temperature conditions and various electrical leakage devices to determine humidity conditions are finding extensive application for air-conditioning control. However, for large-system work in the actuation of valves and dampers it is hard to find a satisfactory substitute for the pneumatic or hydraulic operator, so that electric or electronic control circuits usually act through other relays to actuate the major system elements by means of such operators. Nevertheless, there is an increase in the use of electric and electronic equipment because of the trend to centralizing main control observation, recording, and actuating points as a result of the complexity and extent of the systems being installed in large commercial and industrial operations.

***196. Air-Conditioning Systems.*** *Central systems* are usually assemblies of air-conditioning apparatus designed to afford the necessary atmosphere control and may be arranged with (1) a single fan and single apparatus to care for one space, (2) two or more fans and a single apparatus to care for two or more zones, (3) a single fan blowing through warm and cool air chambers with mixing dampers delivering properly proportioned air to two or more zones, (4) a single apparatus and central fan delivering highly conditioned air to individual recirculating fans in two or more zones, (5) a single central apparatus and fan with supplementary cooling or heating equipment in the ducts leading to two or more zones.

The wide application of air conditioning to multiroom, multistory commercial structures presents a number of problems which have resulted in the development of a number of new systems and techniques. Among systems in common use are the following:

**a.** The fan-coil system which employs individual units served with hot or cold water from a central circulating system or zone.

**b.** The high-pressure induction unit system where highly conditioned air at a high velocity is used to aspirate the flow of room air through localized heat-transfer coils supplied with hot or cold water from an external source, the primary air caring for ventilation and dehumidification while the secondary coil cares for the major sensible-heat gain or loss within the zone.

**c.** The low-pressure hot and cold plenum system in which each zone is supplied from a central unit or apparatus where warm and cold air are mixed as required to meet the zone demand.

**d.** The high-pressure double-duct system in which warm and cold air from central apparatus served through parallel trunk supply systems are mixed at a local distributing point in each zone through a sound-deadening mixing-valve device subject to control from the immediate zone.

**e.** The single-effect high-pressure interior system for substantial areas with relatively constant load characteristics with control for various sections by manual or automatic volume regulation, etc.

***197. The unit air conditioner*** is a factory assembly within a suitable casing of component apparatus necessary for atmosphere control which may include filters, sprays, cooling or heating coils, fan motor, dampers, etc. Unit air conditioners which include in the same casing the refrigerating system are defined as *self-contained unit air conditioners. Direct units* are those designed for location within the conditioned space. *Indirect or remote units* are those which are designed for location outside the conditioned space. *Free-delivery units* are those with fixed fan characteristics designed for use without air ducts.

Self-contained room air conditioners designed for mounting in a window or special opening through an outside wall have been highly developed as a plug-in appliance requiring only electrical connection. These units consist of filter and cooling coil, conditioned air fan, and outlet and inlet grilles, together with refrigeration compressor, motor and drive, condenser air fan, and air-cooled condenser, the latter elements located outside the building wall.

These units are available with total heat-removal capacities in Btu per hour ranging from approximately 5000 to 24,000 and with nominal power input approximately 1 hp/10,000 Btu capacity. Units up through 8000 Btu capacity are generally suitable for operation on 115-V single-phase service. Larger units usually require 230-V single-phase service, but this is available for such use in most areas. Units of this type almost universally employ air-cooled condensers.

Self-contained commercial units with the refrigeration system built in as part of the factory assembly are manufactured in various sizes with nominal total heat-removal capacities ranging from 2 tons to in excess of 30 tons. Units in capacities of 2 to 10 tons are frequently employed as direct units within the space, and many employ water-cooled condensers. However, the current trend is to air-cooled condensers located external to the space. A number of designs are available in capacities ranging from 5 to 30 tons designed for roof mounting and complete with air-cooled condensers and in many instances with electric-resistance heating coils or automatic gas or oil-fuel burner systems for heating. Such units are sometimes used in multiple to take care of fairly substantial commercial or industrial areas in single-story construction.

Self-contained and factory-assembled condensing units complete with either water- or air-cooled condensers and suitable for connection to remote air-conditioning units are currently available in capacities of 2 to 75 tons.

***198. Drives for Air-Conditioning Systems.*** There is no satisfactory substitute for electric-motor drive for the power auxiliaries of air-conditioning systems. Compression-refrigeration systems which are used in conjunction with the large majority of air-conditioning installations are primarily designed for electric-motor drive.

***199. Fuel and Energy Conservation.*** Because of the rising cost of fuels and energy, and the rapid rate at which fossil fuels are being used, methods of energy conservation are being emphasized. Some of these include:

**a.** *Use of computer programs* to determine the efficient energy use of the building or plant. Massive data relating to constants and variables, such as building occupancy, fuel consumption, and electrical characteristics of the building, are fed into the computer and compared with programs to conserve fuel. An analysis of the comparison shows which proposals are feasible and which are not.

**b.** *Automated monitoring* of key systems; for example, logic controllers can be set up on boilers to ensure efficient fuel utilization automatically.

**c.** *Heat-recovery systems.* Heat-recovery cooling units are growing in popularity; typical systems can be water-to-air or air-to-air types. Heat is recovered from areas of a building or plant that requires cooling and transferred to a portion of the plant that requires heating. Some systems recover flash steam for building heating, and still other systems can turn a dust-collection unit into a heating system by filtering the dust and particles through a hopper into drums for disposal, while redirecting the clean filtered air into the plant or building for heating.

**d.** *Solar air conditioning, heating, and cooling.* Solar collector arrays and reflector panels are located on a building's roof and are capable of regulating and storing the energy received from the sun for equal distribution during the day.

## Refrigeration

***200. Refrigeration Cycles.*** All practical refrigeration cycles produce heat removal by free evaporation in an enclosed chamber (evaporator) of a liquid (refrigerant) under pressure conditions that produce the desired evaporation temperature. The refrigerant liquid absorbs its latent heat of vaporization from the medium being cooled and in this process is converted to vapor at the same pressure and temperature. This vapor is conveyed to another chamber (condenser), in which the pressure is maintained at a level sufficiently high to permit condensation of the refrigerant by water or air at normal temperatures. The heat quantity abstracted in the condenser is the latent heat of condensation (or vaporization reversible) of the refrigerant fluid, together with the heat that has been added to the refrigerant in the process of conveying it from the evaporator pressure level to the condenser pressure level. The condensed refrigerant (liquid) is allowed to flow from the condenser, through suitable throttling valves, back to the evaporator to repeat the cycle.

***201. The standard unit of refrigeration*** is the ton, which is considered as heat removal at the rate of 200 Btu/min or 12,000 Btu/h or 228,000 Btu/24 h. This rate of cooling is about equivalent to the average cooling effect obtained by melting 1 ton of ice in 24 h at 43°F (latent heat of fusion of ice is 144 Btu/lb).

***202. Refrigeration Systems.*** Practical refrigeration systems differ only in the method used to convey the refrigerator vapor from the evaporator, or low side, to the condenser, or high side. In closed refrigeration systems, three methods are used to accomplish this transfer: *(a)* mechanical compression or pumping of the refrigeration vapor, *(b)* chemical absorption of the refrigerant vapor at low pressure with subsequent transfer to the high side as solute in a solution, and *(c)* physical absorption of the refrigerant vapor at low pressure with subsequent transfer to the high side as adsorbed vapor in a solid. Related to systems *(b)* and *(c)* are absorption and adsorption dehumidification systems used in air-conditioning processes requiring drying of air without cooling it.

***203. Compression-Refrigeration Cycle.*** In the compression-refrigeration cycle a mechanical compressor or pump is used to convey the refrigerant vapor from the (low) evaporator pressure to the (high) condenser pressure. Vapor is drawn from the evaporator, where it has been evaporated in performing the cooling work, and is compressed and delivered to the condenser where it is liquefied, its latent heat of vaporization being absorbed by the condenser-cooling medium, which is usually water or air at normal temperature. The liquefied refrigerant is collected in the bottom of the condenser or in a separate container called a "receiver" and from there is fed back to the evaporator through suitable throttling valves.

There are three types of mechanical compressor in common use: *(a)* the reciprocating compressor, *(b)* the rotary compressor, and *(c)* the centrifugal compressor. To these must be added the jet compressor, which, while not usually considered as a mechanical compressor, is quite similar in operation. The *reciprocating and rotary compressors* are positive-displacement compressors; that is, each cycle of the compressing member definitely opens a fixed volume into which fluid flows and then closes or occupies that same volume, forcibly expelling the fluid. The centrifugal or jet compressors depend for compression upon the kinetic or velocity energy imparted to the fluid.

The *reciprocating compressor* is suited to the compression of relatively small volumes of fluid through a high-pressure range. The *rotary compressor* is suited to the movement of moderate fluid volumes through moderate pressure ranges. The *centrifugal compressor* is suited to the movement of large volumes of fluid through small pressure ranges. Ammonia, carbon dioxide, Freon-12, methyl chloride, and sulfur dioxide are almost always used with reciprocating compressors. Freon-21 is used with rotary compressors, while Freon-11, methylene chloride, and water are used with centrifugal compressors.

***204. Refrigerants*** which are used commonly enough to merit classification by the National Board of Fire Underwriters are listed in Table 21-7. The Freon group is now among the most widely used of all refrigerants. These halogenated hydrocarbons are chiefly derived from methane ($CH_4$) by the replacement of hydrogen molecules with chlorine and fluorine molecules and include F-11, F-12, and F-22.

Those listed as inflammable or in toxicity classification groups 1 to 4 (mildly to seriously toxic depending upon concentration, exposure, etc.) include ammonia, butane, dichloroethylene, ethane, ethyl bromide, ethyl chloride, methyl bromide, methyl chloride, methyl formate, propane, and sulfur dioxide.

Those listed as nonflammable (or practically so) and less toxic than group 4 (nontoxic in normally possible concentrations) include carbon dioxide, dichlorodifluoromethane (F-12), monochlorodifluoromethane (F-22), dichlorotetrafluoromethane (F-114), monofluorotrichloromethane (F-11), and methylene chloride (dichloromethane). Water is also used as a refrigerant and is, of course, considered as nontoxic and nonflammable.

***205. Refrigerant Properties.*** Physical and thermodynamic properties of the nine most commonly used refrigerants and comparison of the performance of these refrigerants with the ideal (Carnot) cycle for standard conditions are listed in Table 21-7.

Ammonia is cheap and easily available; its efficiency is so high that it is the practical standard by which other refrigerants are rated. For these reasons it is widely used for ice manufacture and food processing, for example, ice-cream manufacture, meat packing, cold storage, and brewery refrigeration.

**TABLE 21-7** Comparison of Refrigerants

Refrigerating Data Book

| Refrigerants | Displacement cfm per ton | Work, hp per ton | Coefficient of performance | Pressure of saturated vapor, psia,* at saturation temperature | | | |
|---|---|---|---|---|---|---|---|
| | | | | 5 F | 40 F | 86 F | 100 F |
| 1. Anhydrous ammonia ($NH_3$) | 3.44 | 0.989 | 4.76 | 34.27 | 73.32 | 169.2 | 211.9 |
| 2. Carbon dioxide ($CO_2$) | 0.96 | 1.840 | 2.56 | 331.9 | 567.8 | 1043 | |
| 3. Dichlorodifluoromethane ($CCl_2F_2$)(F-12) | 5.81 | 1.002 | 4.70 | 26.51 | 51.68 | 107.9 | 131.6 |
| 4. Monochlorodifluoromethane ($CHClF_2$) (F-22) | 3.60 | 1.011 | 4.66 | 43.02 | 83.72 | 174.5 | 212.6 |
| 5. Monofluorotrichloromethane ($CCl_3F$)(F-11) | 36.32 | 0.927 | 5.09 | 2.93 | 7.02 | 18.30 | 23.60 |
| 6. Methyl chloride ($CH_3Cl$) | 5.95 | 0.962 | 4.90 | 20.80 | 42.60 | 95.50 | 119.0 |
| 7. Methylene chloride ($CH_2Cl_2$) | 74.30 | 0.963 | 4.90 | 1.17 | 3.38 | 10.60 | 13.25 |
| 8. Sulfur dioxide ($SO_2$) | 9.09 | 0.968 | 4.87 | 11.81 | 27.10 | 66.50 | 84.50 |
| 9. Water ($H_2O$) | 476.70 | 1.125 | 4.10 | ...... | 0.25† | ....... | 1.93† |

* To obtain gage pressures deduct 14.7 psi atmospheric pressure.

† These pressures given in inches of mercury absolute.

NOTE: Figures are based on standard conditions and dry compression and represent the theoretical performance of the refrigerants for these conditions based upon their thermodynamic properties. Figures for carbon dioxide and water are unfavorable under standard conditions, since 86°F condensing temperature is close to the critical temperature of $CO_2$ and 5°F suction temperature is below the freezing point of water. 1 psi = 6.895 kPa; $t_{°C} = (t_{°F} - 32)/1.8$; 1 ton = 907.2 kg.

Freon refrigerants, which have satisfactory characteristics with respect to toxicity and flammability as well as stable operation characteristics within the normal range, have largely replaced carbon dioxide for air conditioning, etc., where toxic refrigerants present public hazard.

Methyl chloride, sulfur dioxide, and Freon refrigerants find broad application in commercial and domestic refrigeration service.

***206. Effect of Operating Conditions on Compressor Capacity and Power Requirement.*** The capacity of any refrigerating compressor depends upon the mass of refrigerant vapor that it can convey from the low- to the high-pressure condition. The power requirement of the compressor depends upon the mass of vapor conveyed and the pressure differential between the evaporator and the condenser. For this reason the capacity and power requirement of any given compressor operating at constant displacement vary widely depending upon the temperature and pressure conditions of the suction and discharge vapor. The magnitude of this effect is illustrated in Table 21-8.

It should be noted that although the unit power requirement is much greater for low suction temperatures, the capacity of a constant-displacement machine is greatly reduced owing to reduced density of the vapor and reduced volumetric efficiency, so that the gross power requirement may be far below that at higher suction temperatures. For this reason it is general practice to operate belt-driven compressors at higher speeds when used at lower suction pressures to obtain greater displacement and capacity within the safe power rating of the machine. Care must be exercised in selecting equipment to be sure that safe power limitations will not be exceeded during starting-up periods when abnormally high temperatures may be encountered owing to high refrigerant temperatures. This is a very important consideration in the design of ultra-low-temperature systems, especially those employing multiple compressors in cascade arrangement. Within the normal range of condensing temperatures, the change in vapor density is not so pronounced, and the chief effect of departure above the design condensing temperature is a reduction in the output of the compressor.

***207. Evaporators.*** The largest classification of evaporators for direct air cooling is surface cooling coils. For cooling water or brine, simple arrangements of submerged pipe coils or plate surface are sometimes used for small applications, especially where refrigeration storage is desirable. More widely used with reciprocating compression equipment is the so-called dry-ex-

**TABLE 21-8** Characteristic Variation in Compressor Capacity with Changes in Suction and Discharge Condition*

Freon-12, Constant Displacement

| Varying suction condition (discharge constant, 105 F, 126.2 psig) | | | |
|---|---|---|---|
| Saturated suction gas | | Compressor capacity, Btu/hr/bhp | Gross capacity of compressor |
| Temp, deg F | Pressure, psig | | |
| 0 | 9.2 | 5,900 | Increases 375% through this range |
| 10 | 14.7 | 7,070 | |
| 20 | 21.1 | 8,460 | |
| 30 | 28.5 | 10,380 | |
| 40 | 37.0 | 12,760 | |
| 50 | 46.7 | 15,800 | |
| **Varying discharge condition (suction constant, 40 F, 37.0 psig)** | | | |
| Saturated discharge gas | | Compressor capacity, Btu/hr/bhp | Gross capacity of compressor |
| Temp, deg F | Pressure, psig | | |
| 82 | 87.0 | 22,400 | Decreases 33% through this range |
| 90 | 99.6 | 17,900 | |
| 98 | 113.3 | 14,700 | |
| 105 | 126.2 | 12,750 | |
| 112 | 140.1 | 11,050 | |

* Data calculated from typical compressor ratings.
NOTE: 1 psi = 6.895 kPa; 1 Btu/hr = 0.293 W; $t_{\cdot C} = (t_{\cdot F} - 32)/1.8$.

pansion liquid cooler in which refrigerant is expanded into the tube surface within a shell and water passed under forced circulation across the tube surface as directed by baffles. Some larger reciprocating compressor installations and all large-capacity centrifugal refrigeration systems employ a conventional shell-and-tube surface, with forced circulation of water through the tubes, which are mounted within a pool or evaporating refrigerant within the shell surface. The majority of such applications employ integrally finned tubing.

***208. Expansion Methods.*** Two general types of refrigerant expansion are employed: *(a)* dry expansion and *(b)* flooded expansion.

In the *dry-expansion system* the refrigerant is introduced into the evaporator through an expansion valve or pressure-reducing valve and makes one pass through the evaporating surface going to the compressor or absorber suction line. Since any liquid refrigerant which leaves the evaporator represents a loss of cooling effect, care is taken to assure only dry gas leaving the evaporator. The constant-pressure valve has been largely supplanted by the thermal-expansion valve, which through the use of the self-contained thermostatic-valve principle controls the flow of refrigerant to give a constant superheat at the outlet of the evaporator regardless of the load. Dry-expansion systems are widely used with expensive refrigerants, since only a minimum refrigerant quantity is required to feed the evaporator properly. The dry-expansion system also permits feeding evaporators from the top so that any oil carried in the refrigerant will flow through the evaporator and be conveyed back to the compressor.

*The flooded-expansion system* maintains through float-valve control a constant level of refrigerant in the evaporator. In coil-type evaporators this is accomplished by means of a surge chamber external to the coil, usually located above the top and bottom of the coil. Refrigerant is fed into the surge chamber, which is connected with the top and bottom of the coil and to the suction line. As liquid refrigerant passes up through the evaporator coil, its density is reduced owing to the formation of vapor bubbles, so that the refrigerant circulates rapidly through the coil to the surge chamber, where the gas bubbles are released to the suction line. The major disadvantages of the flooded system are the relatively large refrigerant charge required and the necessity of providing means for oil removal when refrigerants that mix readily with oil are used.

***209. Condensers.*** The function of the condenser is exactly the reverse of the evaporator, that is, to afford a rapid transfer of heat from the condensing refrigerant to the cooling medium. For this reason there is a great similarity between condensing and evaporating equipment. Air-cooled condensers are frequently used with small compression systems which are located in well-ventilated places and where the difficulty of obtaining water supply and drain connections is not commensurate with the operating economies realized through the use of water cooling. Air-cooled condensers are similar in construction to surface-cooling coils.

Because of the effect that high condensing pressures have upon the economy of operation of refrigeration plants, water-cooled condensers are widely used. It is customary to assume the condensing requirement of the average compression plant to be 15,000 Btu/(ton)(h), of which 12,000 Btu represents the cooling work, and 3000 Btu the energy applied in heat pumping. This amount of heat can be absorbed by 1 gal/min of water rising through 30°F, 2 gal/min through 15°F, or 3 gal/min through 10°F. The design of the condenser and the amount of water required will depend upon the available water temperature and the desired condensing temperature and pressure.

***210. Water Conservation.*** The increased use of tap water for condensing purposes in air-conditioning and refrigeration installations has seriously taxed the *water-supply* and *water-disposal* systems of many cities, with the result that restrictions in the form of increased water rates for condensing purposes, taxes upon water disposal, or outright prohibition of the use of water for condensing purposes have been effected by many municipalities. This has resulted in the development of equipment for the conservation of water. The familiar *cooling tower* has been redesigned to meet requirements of downtown buildings as to size and appearance, and the spray pond and atmospheric cooling tower find broader application in the industrial field. With spray ponds and atmospheric towers usually 5 gal/(min)(ton capacity) is allowed with water temperature entering the condenser at 10°F above the maximum outside wet-bulb temperature (see Par. **176**). With forced-draft towers usually 3 gal/(min)(ton) will suffice with the same temperature limits.

Thoroughly accepted in this field is the *evaporative condenser,* which consists of an extended-coil condenser over which air is driven by fans and which is continually bathed in a water spray. Efficient condensing effect is obtained by the combined effect of air cooling and evaporation of water by the hot condenser coils. Small sizes for capacities up to 5 tons use propeller fans and direct tap-water connection for spray. Larger sizes (condensers in excess of 250 tons capacity have been built) are quite similar to unit air conditioners and employ centrifugal fans and a recirculating-spray system with a centrifugal pump.

Maintenance difficulties attendant upon the use of either cooling towers or evaporative condensers have resulted in a vast increase in the application of air-cooled condensers for reciprocating refrigeration systems in capacities up to 200 tons and possibly higher. While the use of air-cooled condensers results in a considerable operating-power penalty under maximum temperature condition, the apparatus employed is simple and relatively maintenance-free, factors which frequently justify the power penalty resulting from their use.

***211. Refrigeration Compressors.*** The design of *reciprocating compressors* has been profoundly influenced by development of the production-line internal-combustion engine as offered today for automotive and aircraft service. Multicylinder in-line arrangements of 2 to 8 cylinders are thoroughly accepted, as are V and W arrangements of 4 to 16 cylinders and radial arrangements of 3, 5, and 7 cylinders. Practically all machines are single-acting with enclosed crankcases. With multicylinder design, large capacities are obtainable with good balance and relatively low piston speed, so that it is not unusual for machines of up to 120 tons capacity, with forced-feed lubrication, to operate at speeds as high as 1750 r/min. Speeds below 500 r/min (except with very large machines) are the exception rather than the rule. Capacity variation is obtained by unloading groups of cylinders or by bypass or clearance-pocket arrangements.

The overall application of refrigerated air conditioning to large industrial and commercial buildings requires capacities far beyond the range of the largest ice-making or storage refrigeration installations. These requirements have been met adequately by development in the *centrifugal refrigeration* system, which finds current application in single units from 200 to 1800 tons or over in capacity. Drive motors will vary from a low of 0.85 hp/ton for industrial applications to a high of 1.05 hp/ton for commercial installations.

The original designs for centrifugal systems employed multistage compressors with shaft seals and operating at speeds from 3500 to 7000 r/min. Standard four-pole or six-pole open motors with step-up gears were customarily used for compressor drives. Wound-rotor motors affording 25% speed regulation were frequently employed for capacity control. However, since 1955 design has largely been directed toward hermetic systems in which the drive motor is enclosed in the gas passage with the compressor, and with this arrangement the drive is naturally limited to the use of the squirrel-cage induction motor. Control of compressor capacity is afforded by regulating dampers in the compressor suction connection. Common designs employ either two-stage compressors direct-driven by two-pole motors at approximately 3500 r/min or single-stage compressors operating at 11,000 r/min or higher and driven through step-up gears by enclosed two-pole motors. One design employs an external frequency converter providing 300-Hz current for operation of a two-pole induction motor for direct compressor drive at approximately 17,500 r/min. Centrifugal compressors are still offered without the enclosed motor and with the shaft extended through the compressor housing with a suitable seal; as such they can employ the open-motor or turbine drive. However, the general design and control arrangement of the system is generally the same for either hermetic or open drives for manufacturing standardization. Even with open units induction-motor drives are generally selected, for though motor sizes are large, average experience indicates that even where power-factor penalties are involved, power-factor correction can be obtained more economically by the use of capacitors in connection with induction motors rather than through the use of synchronous motors. Where closed motors are employed for compressor drive, they are cooled either by water jacketing or by an arrangement employing a bypass of refrigerant or refrigerant gas for motor cooling.

***212. Compressor Drives.*** While conventional open motors and V-belt drives are generally employed for large-capacity, low-speed reciprocating compressors, which are still offered, the bulk of the production of reciprocating equipment is now of the direct-drive design generally employing four-pole 1750-r/min motors. This design permits an extremely compact arrangement which is well suited to the unit-system combinations now widely offered by the industry.

In both the reciprocating and centrifugal compressor fields there is contention as to whether the use of standard open motors or sealed hermetic arrangements is most acceptable. The use of standard open-motor drives necessitates the shaft seal, which offers some disadvantages. The use of the hermetic arrangement requires careful design of the electrical components and close control of the refrigerant atmosphere, which must be kept absolutely moisture-free if electrical problems are to be avoided. Obviously, if failure occurs, the repair of the hermetic motor is a far more serious problem than the repair of the open motor. On the other hand, if properly designed and installed the fully enclosed motor operates in a controlled atmosphere and may be less likely to give trouble. However, with either arrangement the motor is generally a standard two-pole or four-pole induction motor as far as the basic electrical design is concerned.

Reciprocating systems generally employ means of unloading by holding suction valves open to afford capacity control in operation and are similarly arranged to start unloaded. Centrifugal systems employ dampers for capacity control, which are kept closed during the starting cycle so that these units also start unloaded. However, the characteristics of the reciprocating system are such that the starting-torque requirement is fairly high. Accordingly, line start for drive motors is preferred wherever utility-company regulations will permit, and since these units are usually to approximately 125-hp maximum size and are frequently installed in multiple units of smaller capacity, line start is generally permissible. Where reduced voltage starting is required, autotransformer or resistance-type reduced-voltage starters are generally employed, though frequently these may be effective only as a means of reducing inrush, as in many instances full line voltage must be applied before the compressor will actually pick up the load.

Centrifugal refrigeration systems generally represent larger-capacity units than reciprocating systems, with motor sizes ranging from 75 to 2000 hp or occasionally higher. However, their starting-torque requirement is very moderate, and it is entirely practical to place the compressor in satisfactory operation by use of the Y-Δ starting system, in which the compressor drive motor is connected across the 3-phase source in a Y arrangement during the starting interval and switched to the Δ connection when the compressor has reached practically full speed and before the compressor damper is opened, permitting the unit to assume load. With this arrangement the current during the starting period can be limited to approximately 130% of the normal full-load running current, and this arrangement is generally entirely satisfactory to power companies. This starting arrangement is almost universally employed with centrifugal systems except for relatively small units, which may be within a range suitable for line start.

***213. Motor Voltage.*** All standard equipment is based on the assumption that 60-Hz current is available, though some manufacturers will still list a capacity rating of their units of

60-Hz design when employed on 50-Hz service. All systems involving motors up through 5 hp are generally designed for 230-V 60-Hz 1-phase service, but in sizes 3 hp and above are equally available for 208/220/440-V 60-Hz 3-phase service, which is generally available in most urban areas. Equipment for practically the entire available capacity range is designed for service on 208-V 60-Hz 3-phase current, which is commonly available as urban low-voltage distribution. However, it should be noted that because of the rigorous nature of compressor service it is usually inadvisable to attempt to use 220-V equipment for the compressor drive where only 208 V is supplied. Many feel that good design dictates the use of 200-V motors for such service. For larger commercial or industrial applications which are provided with primary service through independent substations, 440-V 60-Hz 3-phase equipment is generally preferred, as it permits a substantial reduction in wiring cost. Currently a number of systems employ 480/277-V 60-Hz 1-phase four-wire service, where the higher single-phase voltage may be satisfactorily used for fluorescent-lamp operation. This arrangement affords considerable economy in circuit design but presents complications with respect to the use of standard 115-V 60-Hz 1-phase conventional power or office equipment, for which separate transformers and circuiting must be provided. In the textile industry extensive use is made of 550-V 60-Hz 3-phase service, and equipment is generally available throughout the entire capacity range for this voltage characteristic. In larger plant operations employing plant primary service at either 2300 or 4160 V, these voltages are frequently used for large compressor drives, and most lines make equipment available for this voltage in the range from 200 to 2000 hp, though economics would generally indicate 500-hp drives as being the lowest acceptable size for this voltage.

***214. Reverse-Cycle or Heat Pump.*** The reverse-refrigeration cycle is for the purpose of heating interiors with electricity. Heat absorbed at a low-temperature level is pumped to a level sufficient to permit satisfactory heating with the expenditure of only the amount of energy necessary to perform the pumping work, which is also reclaimed at the high-temperature level. Under certain circumstances this cycle is entirely practical. Many installations have been made where well water at moderate temperatures (±50°F) is available as a low-temperature heat source, and these installations have shown coefficients of performance up to 400%, that is, 5 Btu of heating effect for 1 Btu of energy applied. However, the performance of the refrigeration machine is sharply limited as the temperature head between the evaporator and the condenser is increased, so that the practical possibilities of this cycle are affected not only by the cost of electrical energy compared with other fuels but by the level of the low-temperature heat source and the temperature that must be maintained within the interior.

Application of air conditioning to residences has accelerated activity in the heat-pump field. Since the basic refrigeration equipment is being installed for summer conditioning, its utility is increased by arranging it to handle at least a part of the winter heating requirements. Most new developments are directed toward the use of outside air as a low-temperature heat source as well as a means for providing condensing effect when the system is on the cooling cycle. Heating capacity of such a system will drop fairly rapidly as outside temperatures drop. However, there is a balance point between the requirement for summer cooling and available capacity for winter heating which will permit a rationally designed apparatus to care for the full heating requirements on the reverse-cycle basis for substantial periods during the heating season. Naturally, the more temperate the climate in which the apparatus is employed, the greater proportion of the total winter heat requirement can be handled entirely by the heat-pump principle. Since peak heating demand is experienced only for short periods, usually during early morning hours, the current procedure is to supplement the available output of the heat-pump device by direct resistance heating. This represents a substantial spot load, but its period of use is relatively short, and under most utility rate schedules the overall economies of the operation are reasonable. However, it presents problems in power distribution. A moderate-sized residence in the South Atlantic area with a basic 6-kW requirement for summer or heat-pump operation may be supplemented with as much as 12 kW of additional resistance heating load for use during peak heating periods. This imposes a maximum load of up to 18 kW per residential unit upon the lines of the utility, with little possible diversity.

On the other hand, this load occurs at a time when other urban loads are at a minimum and when outside ambient-temperature conditions are at a very low level, so that distribution circuits, transformers, etc., can handle a substantial overload with no difficulty. Accordingly, utility companies are finding that they can handle this load with no serious problems and feel that it is attractive.

A number of ingenious arrangements employing the heat-pump principle or other heat-recovery systems have been incorporated in recent designs of installations for large commercial buildings. It is evident that the core sections of these buildings are subject to cooling requirements throughout the entire year, as they are shielded from exterior exposure, but subject to heat gains from lighting and occupancy. A number of systems have been employed which use refrigeration for these interior zones throughout the entire year but during the wintertime take the effluent heat from the condenser of the refrigeration system and apply this as a means of heating the periphery of the building which is subject to normal winter exposure. Thus, in effect, the excess heat from the interior core of the building is transferred to the exterior for heating purposes while affording control of the heat load within the core of the building.

Other arrangements provide for direct cooling of lighting equipment either by circulating water or by exhaust ventilation and transfer the heat thus directly recovered to the periphery of the building for heating purposes. There are obviously many possibilities of this nature which must be studied for each individual project. However, where this means of heating is used, it is dependent upon a continued supply of electrical energy to the space as a primary heat source. In some instances arrangement is made to turn on interior lighting during cold periods for the purpose of heating the premises, or where lights are turned off, the equivalent amount of electrical energy may be applied through resistance heating, thus affording a heat source with no increase in electrical demand. These arrangements frequently make full electric heating entirely practical and economical for many large buildings.

***215. Refrigeration Storage: Moderate Temperature.*** Many refrigeration applications having high peak loads for short periods of time can be most economically handled by the use of relatively small refrigeration machines which are operated over long periods of time to store refrigeration effect to meet the high short-period requirement. Refrigeration effect may be stored in bodies of brine or water, though a relatively large storage volume will be required, since the water or brine can only absorb or give up its sensible heat. A more efficient storage method involves the formation of ice upon submerged evaporating coils. Every pound of ice thus formed absorbs and can release upon melting at 32°F, its latent heat of fusion, 144 Btu/lb.

This principle has been frequently used to improve the load factor of refrigeration systems used in conjunction with air-conditioning systems having relatively short use periods (restaurants serving only one meal a day, theaters with only evening showings, and churches or infrequently used auditoriums).

***216. Refrigeration Storage: Low Temperature.*** A solution of water and salt (or, in general, of any two substances) has a certain concentration which results in the lowest freezing temperature. A solution of this concentration is known as a *eutectic mixture.* A brine with a salt content lower than this concentration will start to freeze above this minimum temperature, and as cooling progresses, pure ice crystals will freeze out, thus increasing the concentration of the brine until the eutectic concentration is reached. At this point freezing will take place at a constant temperature until all latent heat is removed, the resulting crystals being a mechanical mixture of salt and frozen water. If the starting salt concentration of the mixture is greater than the eutectic concentration, salt crystals will first freeze out until the eutectic concentration is reached.

The use of eutectic mixture as a means of storing refrigerating effect, that is, latent heat of fusion, at a temperature below 32°F (the freezing point of water ice) has received wide attention. The eutectic mixture of water 76.7% and sodium chloride (common salt) 23.3% has a freezing temperature of −6°F, and a latent heat of fusion of 101.5 Btu/lb is available at this temperature. Plate evaporators charged with eutectic mixtures are frequently used in refrigerator trucks which are cooled at night to produce sufficient stored cooling effect to meet the next day's operation.

# SECTION 22
# ELECTRONICS

**P. Wood**
*Senior Member IEEE; Consulting Engineer, Westinghouse Research Laboratories*

**L. Gyugyi**
*Member, IEEE; Manager, Power Electronics Department, Westinghouse Research Laboratories*

**Jerome B. Brewster**
*Senior Member IEEE; Scientist, Westinghouse Research Laboratories*

**T. M. Heinrich**
*Senior Member IEEE; Manager, Motor Drive Systems, Westinghouse Research Laboratories*

**R. M. Oates**
*Member, IEEE; Manager, Advanced Systems Laboratory, Westinghouse Research Laboratories*

**B. R. Pelly**
*Member, IEEE; Vice President, Worldwide High Power Products, International Rectifier*

**Donald Galler**
*Member, IEEE; Senior Engineer, Alexander Kusko, Inc.*

**CONTENTS**

*Numbers refer to paragraphs*

## POWER ELECTRONICS

By P. WOOD

### Introduction

*1. **Power Electronics Defined.*** Power electronics deals with the application of electronic devices and associated components to the conversion, control, and conditioning of electric power. The primary characteristics of electric power which are subject to control include its basic form (ac or dc), its effective voltage or current (including the limiting cases of initiation and interruption of conduction), and its frequency and power factor (if ac). The control of electric power is frequently desired as a means for achieving control or regulation of one or more nonelectrical parameters, for example, the speed of a motor, the temperature of an oven, the rate of an electrochemical process, or the intensity of lighting.

*2. **Efficiency Requirements.*** Aside from the obvious difference in function, power electronics technology differs markedly from the technology of low-level electronics for information processing in that much greater emphasis is required on achieving high power efficiency. Few low-level circuits exceed a power efficiency of 15%, but few power circuits can tolerate a power efficiency less than 85%. High efficiency is vital, first, because of the economic value of the wasted power, and second, because of the cost of dissipating the heat it generates. This high efficiency cannot be achieved by simply scaling up low-level circuits; a different approach must be adopted.

*3. **Switching Power Converters.*** The approach that must be adopted is to use switching devices to implement the conversion functions needed. Since an ideal switch would dissipate no power when closed and carrying current or when open and withstanding (blocking) voltage, a switching power converter has a theoretical efficiency of 100%. Practical switching devices have power losses when conducting and blocking and when switching from one state to the other (opening or closing), but practical switching power converter efficiencies are rarely less than 80% and in some cases exceed 95%.

The basic characteristics of switching power converters do not depend on the type of switching devices used—electromechanical, vacuum or gas-filled tube, magnetic, or semiconductor. Most present-day converters use bipolar silicon semiconductor switches—diodes, transistors, thyristors, gate turn-off thyristors, etc. Some use metal-oxide semicon-

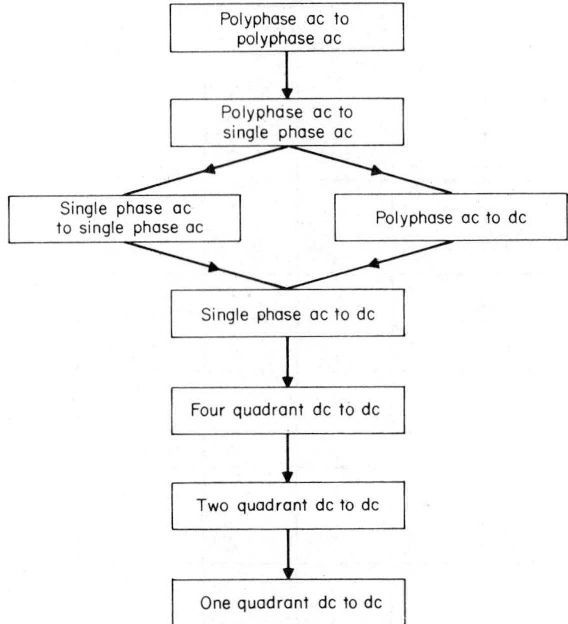


**FIG. 22-1** Hierarchy of power conversion functions.

ductor switches such as field-effect transistors and Schottky diodes. However, the particular devices used only influence the detail behavior of converters; fundamental converter properties are universal.

***4. General Theory.*** The general theory of switching power converters[1,2*] rests on their classification by function. There are only four possible power conversion functions, namely: ac to ac, ac to dc, dc to ac, dc to dc.

This observation leads to the ordered hierarchy or family of power conversion functions that is shown in Fig. 22-1. The "lower" members of the family are obtained by progessive degeneration (simplification) from the most complex, the conversion of polyphase ac with some voltage, frequency, and phase number to polyphase ac with other voltage, frequency, and phase number.

It follows that if a converter which can perform the most complex function is known, or can be postulated, converters capable of performing all other functions can be derived therefrom.

***5. Switching Matrices.*** All switching power converters can be considered as arrays (matrices) of switches connecting a source to a load (sink). The switches are opened and closed in repetitive patterns to accomplish the desired conversion function. For analytic purposes, the source and sink are presumed to be ideal voltage and current sources, respectively, or ideal current and voltage sources, respectively. These are called defined quantities.

Given these conditions, a polyphase ac to polyphase ac converter, $N$ phase to $M$ phase, is as depicted in Fig. 22-2. If the $N$ terminals are connected to an $N$-phase set of voltage sources and the $M$ terminals to an $M$-phase set of current sources, and the switches operated in the appropriate pattern, then the voltages impressed on the current sources will contain a constituent at their frequency and the currents drawn from the voltage sources will contain a constituent at their frequency. These impressed voltages and currents are called dependent quantities, since they depend on the defined quantities *and* the switching patterns.

The switching matrix of Fig. 22-2 could also be operated with voltage sources connected to the $M$ terminals, current sources connected to the $N$ terminals, and an appropriate trans-

*Superior numbers refer to Bibliography, Par. **80**.

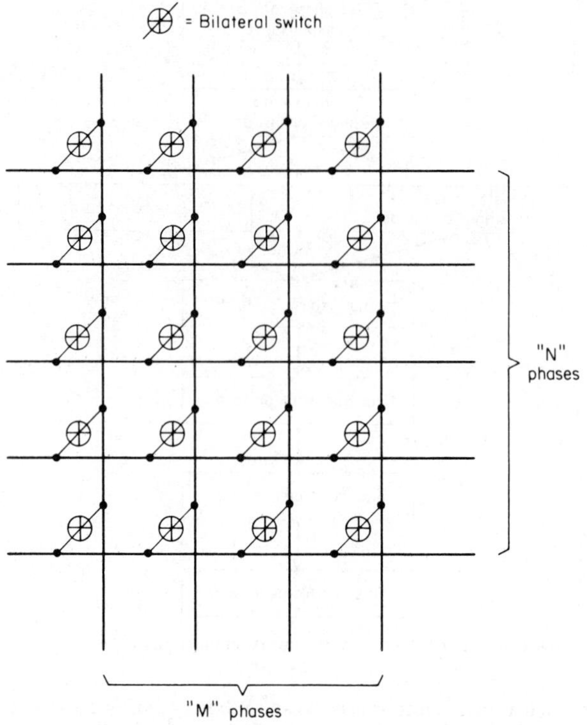


**FIG. 22-2** General switching matrix.

position of switching patterns. It then becomes the electrical reciprocal of that first described.

Power flow, in either case, can be in either direction—from the $N$ terminals to the $M$ terminals or vice versa. The direction of power flow is determined largely by the timing (phase) of switching patterns relative to source conditions.

The general switching matrix of Fig. 22-2 degenerates to the polyphase ac to single-phase ac matrix of Fig. 22-3 simply by reducing one set of terminals to two, and further to the

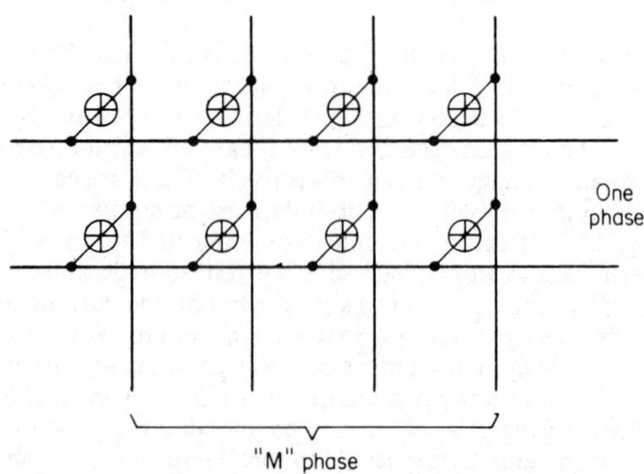


**FIG. 22-3** Polyphase to single-phase switching matrix.

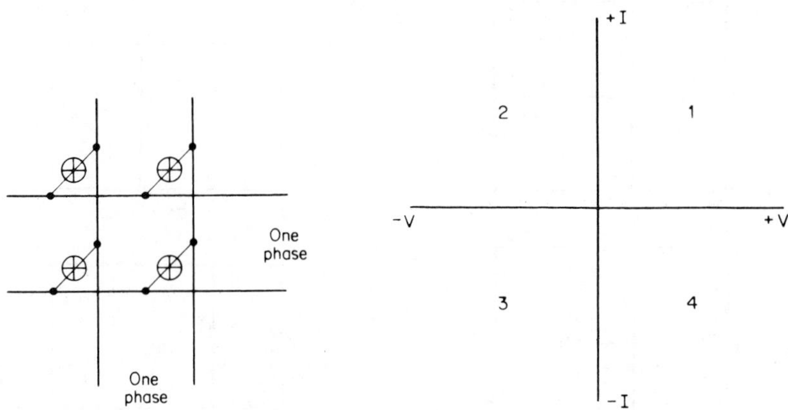


**FIG. 22-4** Single-phase to single-phase switching matrix.

**FIG. 22-5** *V-I* plane diagram.

single-phase to single-phase converter of Fig. 22-4 by reducing the other set of terminals to two. In the polyphase to single-phase case, the single-phase (two) terminals may be connected to a voltage or current source with the *N*-phase terminals connected to current or voltage sources, giving rise to reciprocal converters. In the single-phase to single-phase case, the reciprocal converters are identical. In all cases, power flow is reversible.

The switches in all these ac-to-ac converters are required to carry current in either direction when closed and block voltage of either polarity when open. They are said to require bidirectional current-carrying, bidirectional voltage-blocking, or fully bilateral capabilities.

The ac-to-dc and dc-to-ac conversion functions arise when the frequency of a defined single-phase quantity is reduced to zero. Thus the matrix of Fig. 22-3 becomes either an ac-to-dc voltage or an ac-to-dc current converter, depending on whether the single-phase defined quantity was originally voltage or current. Since power flow in both these converters is still reversible, they are also dc voltage to ac or dc current to ac converters, and are reciprocals.

The same applies to the single-phase versions of Fig. 22-4, which become single-phase ac-to-dc voltage or single-phase ac-to-dc current converters, also capable of performing the corresponding dc-to-ac conversions and mutual reciprocals. The reciprocity of these ac-to-dc–dc-to-ac converters is indicated by the terminology used to describe them. Those converting to or from a dc voltage are called voltage-sourced (or fed), and those converting to or from a dc current are called current-sourced (or fed).

The switch requirements for these converters also show reciprocity. For the voltage-sourced converters, the switches need bidirectional current-carrying capability, but only unidirectional voltage-blocking capability. For the current-sourced converters, the switches need bidirectional voltage-blocking capability, but only unidirectional current-carrying capability.

These ac-to-dc–dc-to-ac converters are said to have two-quadrant capability. Observing the *VI* plane diagrams of Fig. 22-5, the voltage-sourced converters operate as ac-to-dc converters in the first quadrant and as dc-to-ac converters in the fourth quadrant. Similarly, the current-sourced converters operate as ac-to-dc converters in the first quadrant but as dc-to-ac converters in the second quadrant. If switch capabilities in either case are reduced to unilateral, unidirectional current carrying, unidirectional voltage blocking, then they become single-quadrant converters. Although in principle the quadrant can be ac to dc or dc to ac in either case, depending on the current-carrying or voltage-blocking polarity selected for the switches, common semiconductor switch properties generally limit practical versions to single-quadrant ac-to-dc current-sourced converters or dc-to-ac voltage-sourced converters.

The dc-to-dc converters arise from progressive degradation of the single-phase matrix of

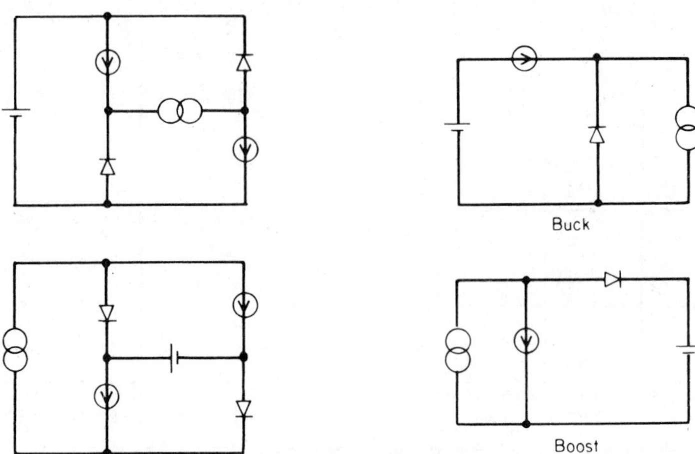


**FIG. 22-6** Two-quadrant dc-to-dc converters.

**FIG. 22-7** Single-quadrant dc-to-dc converters.

Fig. 22-4. As a single-phase ac-to-dc–dc-to-ac converter, it is also a dc to four-quadrant-dc converter, either current- or voltage-sourced. Reducing half of the switches to unilateral capability (diodes) gives the two reciprocal versions of the dc to two-quadrant-dc converter shown in Fig. 22-6. Then eliminating half of the switches yields the common single-quadrant dc-to-dc converters, the buck and boost converters, shown in Fig. 22-7. These are still mutual reciprocals.

***6. Existence (Switching) Functions.*** To define the performance of switching power converters, some means of describing switching patterns is needed. The description is best accomplished by the use of existence functions; these are simple repetitive pulse trains assigned the value 1 when the switch is closed and 0 when it is open. The simplest existence functions, like that depicted in Fig. 22-8, are unmodulated—all periods of unit value have equal duration and all periods of zero value have equal duration in steady-state operation. However, the times of unit and zero value may or may not be equal, depending on converter type and operating conditions.

For ac-to-dc converters and some ac-to-dc–dc-to-ac converters, modulated existence functions are needed. These do not have all equal unit-value or zero-value periods. As shown in Fig. 22-9, the periods vary (are modulated) in a prescribed manner. However, groups of pulses with identical periodic variations repeat to form a switch pattern used to generate a periodic function (ac-dependent quantity). Such functions, both unmodulated and modulated, are readily expressible in the time domain as Fourier expansions.

Using these functions to derive converter performance is easy, because the converter is now regarded as a modulator in which the defined quantities are modulated (multiplied) by the switch existence functions to produce the dependent quantities. Thus if the defined voltage vector for the matrix of Fig. 22-2 is $[V]_N$, a set of $N$ polyphase voltages, and the existence function matrix for the switches is $[H]_{NM}$, then

$$[H]_{NM}[V]_N = [V]_M$$

where $[V]_M$ is the dependent voltage vector, a set of $M$ polyphase voltages together with accompanying unwanted components that result from switching converter action. Similarly, then, if the defined current vector is $[I]_M$, then the dependent current vector $[I]_N$ is given by

**FIG. 22-8** Unmodulated existence function.

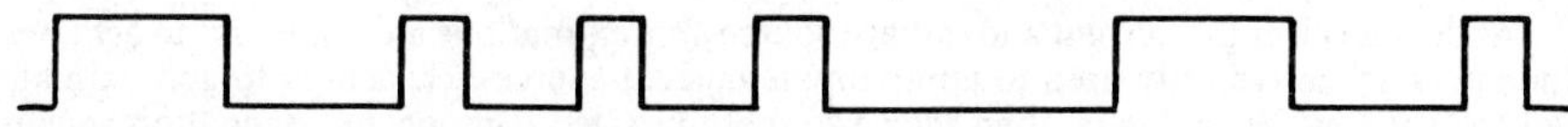

**FIG. 22-9** Modulated existence function.

$$[H]^T_{NM}\,[I]_M = [I]_N$$

where $[H]^T_{NM}$ is the transpose of $[H]_{NM}$.

These simple equations give the dependent quantities mathematically and graphically. Graphical construction is achieved by graphing defined quantities and existence functions on the same time scale and then graphing the dependent quantities as those segments of the defined quantities bounded by the unit-value periods of the existence functions, as shown in Fig. 22-10.

The switch stresses and waveforms within the converter are also derived from the equations and graphs, as shown in Fig. 22-10. Both equations and graphs show a primary disadvantage of switching power converters. Because of the way dependent quantities are fabricated, by successive connection to segments of defined quantities, they can rarely consist only of their wanted components. The wanted components are the dc or sinusoidal (usually) ac quantities that it is the converter's objective to produce so as to effect the power transfer desired. They are almost always accompanied by a set of unwanted components, harmonics of a supply or switching frequency, or sidebands, combinations of multiples of supply and switching frequencies.

*7. **Interfacing Converters.*** The defined voltage and current sources are convenient postulates for switching converter analyses. Although some sources and loads are reasonably valid approximations of voltage sources, few are good approximations of current sources. It is necessary to interface switching matrix terminals with the actual sources and loads so that they appear to the matrix to be reasonable approximations of the voltage and, particularly, current sources it needs to function. Usually, the interfacing fulfills a second function. It is used to attenuate the unwanted components in the converter's dependent quantities to levels tolerable by the actual sources and loads.

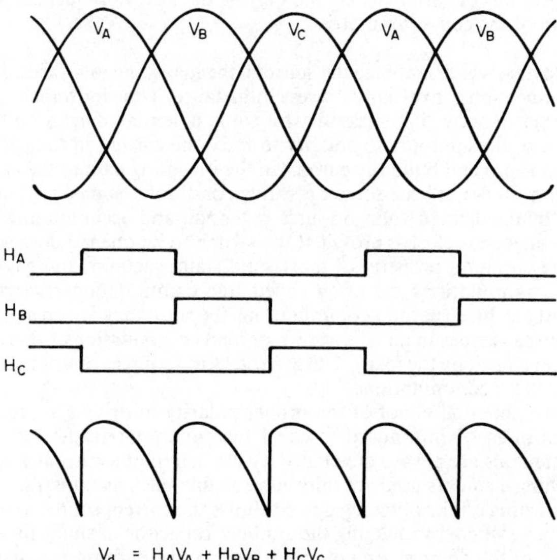


**FIG. 22-10** Waveform construction using existence functions.

At dc interfaces the current and voltage source approximations are quite easy to achieve. Inductors are universally used to approximate current sources, capacitors to approximate voltage sources. When, as is often the case, the unwanted component attenuation sought cannot easily be attained by the use of a single passive component, the direct interface component, inductor or capacitor, becomes part of a simple single-section *LC* filter placed between converter and load or between source and converter. Occasionally, when very stringent specifications apply, multisection filters are used.

AC interfaces pose more problems, especially since many converters produce or work from variable frequencies. For single-frequency applications, parallel-tuned circuits can be used to approximate voltage sources and series-tuned circuits can be used to approximate current sources. However, the transient response problems of the latter, coupled with the variable-frequency requirement in many cases, have led to the ac current source interface most often being implemented by means of a simple inductor. As in the dc case, this component is often made part of a single-section *LC* filter providing the unwanted component attenuation needed.

AC voltage sources, such as the electric utility supply, are often directly connected to the switching matrix. When the unwanted component burden is too high to tolerate, filtering is usually provided by some number of shunt-tuned traps where the frequency is fixed, or by an *LC* filter network when it is not. If no good voltage source approximation is present, parallel-tuned circuits (often with the capacitive branches realized as multiple series-tuned trap filters) are used for fixed-frequency applications; pure shunt capacitance, sometimes as part of an *LC* filter, is used when the frequency is variable.

The design of interfaces and the selection and construction of interface components are of paramount importance in most switching power converter applications. Various techniques, all involving multiple switches, higher switching frequencies, or both, are often used to improve or modify converter-dependent quantity spectra so as to minimize interfacing requirements.

***8. Commutation in Converters.*** Within any switching power converter, commutations occur whenever the switches open or close. A single commutation comprises the following actions:

1. Transfer of current from one conducting path to another
2. Transfer of voltage connection from one terminal to another
3. The opening of one converter switch and closing the next in sequence in that particular line or column of the converter matrix

The conducting paths, which embrace the sources, the loads, the interfaces, and the switches and their interconnections, invariably possess inductance. Thus for transfer of current from one path to another to occur, it is necessary for some potential to act for a finite time. This potential must be of the appropriate polarity to force the current in the path existing prior to commutation to zero and build the current in the incoming path to the appropriate level. In most converters, under at least some operating conditions, some or all of the commutations are driven by the defined voltage-source potentials and occur automatically when the next switch in sequence is closed, provided the switch to be opened does so automatically at current zero (a common property of most solid-state, vacuum, and gas-filled switching devices). These commutations are often called line commutations (because the voltage source is the utility ac line), natural commutations (because they are erroneously attributed to the natural current zeroes in an ac system), or load commutations (when the commutating potential is developed by the load). The generic term "source commutation" will be used here to delineate these commutations.

When a source potential is not of the proper polarity to drive a source commutation, another potential must be introduced to effect the commutation desired within the converter. These potentials are always developed by the action of a current source on a capacitor. If the switching elements used are fully controllable, such as transistors or gate turn-off thyristors, then turning off the outgoing device allows the current source to develop a potential across the open switch, by charging the snubber capacitor in shunt therewith, and also allows a potential of the appropriate polarity to reduce the current in that external path to zero and to build the current in the incoming path and switch to the desired level. With these switches, the potential can be allowed to assume the required polarity very rapidly;

however, if the inductances in the conducting paths are substantial, it is necessary to provide means for restraining the magnitude to safe levels.

If thyristors are used, there is a need to maintain reverse bias for a finite time before forward blocking capability is regained. This results in the use of a large "commutating" capacitor to slow up the rate of change of potential so that device requirements are met. This capacitor, and associated components together called the commutating circuit, are needed to make the thyristor(s) effective switches in a converter needing this type of commutation. Commutating circuits are parts of switches, not parts of switching matrices or converter circuits proper.

Such commutations are generally termed forced commutations, and converters which rely wholly or largely thereon are said to be force-commutated or self-commutated. A more apt term, deriving from the short time potential that must be generated within the converter, is impulse commutation. Converters using such commutations are then said to be impulse-commutated.

## SOLID-STATE SWITCHING DEVICES

*By JEROME B. BREWSTER*

***9. Introduction.*** Present-day power converters use a limited number but a large variety of solid-state devices. Power devices of different types are now used to perform similar functions as designers vie to produce the least costly, most efficient, smallest, or best-performing system. As an example, ac motor drives can be found using thyristors, gate turn-off switches (GTOs), or transistors. Economics and performance bear heavily upon the competitive usage.

In approaching the available solid-state power devices, it is appropriate first to look briefly at the different means by which devices may be cataloged and compared. In this examination, a primary division could be to discriminate between linear devices and those primarily used as switches. Most devices function as switches, but few types have real linear power capability. Devices may also be separated by structure, i.e., one junction, two junctions, or three junctions, thus giving unipolar devices, bipolar devices, two terminal, three terminal, nonregenerative, regenerative, high voltage, high current, high speed, etc. Besides the differences between generic classes, power devices are optimized for function within a class. High-voltage devices have performance characteristics different from high-current devices; high-frequency devices have characteristics different from their low-frequency counterparts. The array of characteristics for a single device type originates from a spread of characteristics obtained in the manufacturing process. In trying to choose a device for a circuit function, the design engineer must choose among available devices performing similar functions. Then, selections must be made from ratings of devices of a similar type. To do so properly with reasonable rationale requires understanding of the measurement of device characteristics and how ratings are applied. By understanding ratings, logical decisions can be made to discriminate among different device types. Once a device type is selected, ratings may be applied to achieve the desired performance or circuit function.

The rating of solid-state power devices follows the successful achievement of device manufacturability. Since competition exists, the existence of many manufacturers leads to the development of several independent ratings for similar devices. The maturity of a device determines how uniform device characteristics and ratings between manufacturers will be; new devices show the most rating variability. For mature devices, marketplace forces demand uniformity, and it is achieved to a degree, but never in totality. Manufacturers making devices of similar design with different equipment, slightly different processes, and different personnel will produce devices which show differences. Any batch of devices will show a spread of characteristics after processing. This spread will contain a percentage of devices of the originally intended design, called the yield, plus some devices not of the intended design, which may be usable or nonusable. The nonusable devices are scrapped, and the usable, out-of-specification units are reclassified as units of different voltage ratings, current ratings, etc.; when examining a sheet of ratings and characteristics, the spread of ratings generally results from a production distribution. The technique of selling the non-intended product distribution can result in some design sacrifice. As an example, fallout of yield from a high-voltage rectifier design will necessarily have less current capability than a low-voltage rectifier design of the same wafer or chip size. If a device rating is satisfied by a fallout of a high-voltage design, then it is appropriate and economically efficient to fill this need with such products.

The trend in modern semiconductor device manufacture is the use of rigidly controlled processing and high purity of materials so that production yields are "tight," requiring a minimum of device ratings to sell all production output. Characteristics are also tight due to product maturity, but the spread of characteristics never reduces to zero. Hence, characteristics are generally seen on a data sheet with minimums and maximums.

The presence of a registration number for a power device indicates some degree of device

**TABLE 22-1** Letter Symbols for Diodes and Thyristors

| Rating symbol | Meaning |
|---|---|
| $V_{DRM}$ | Repetitive peak forward off-state voltage (thyristor) |
| $V_{RRM}$ | Repetitive peak reverse voltage (thyristor, rectifier) |
| $V_{RSM}$ | Nonrepetitive peak reverse voltage (thyristor, rectifier) |
| $I_T$ rms | rms forward conducting current |
| $I_{T,\mathrm{av}}$ | Average on-state current (thyristor) |
| $I_{F,\mathrm{av}}$ | Average forward current (rectifier) |
| $I_{TMS}$ | Single-cycle (nonrepetitive) surge current (thyristor) |
| $I_{FSM}$ | Single-cycle (nonrepetitive) surge current (rectifier) |
| $I^2t$ | Nonrepetitive (rms) ampere$^2$-seconds overcurrent capability |
| $I_{GTM}$ | Peak forward gate current |
| $V_{GRM}$ | Peak reverse gate voltage |
| $I_{GRM}$ | Peak reverse gate current |
| $P_{GM}$ | Peak gate power |
| $P_{G,\mathrm{av}}$ | Average gate power |
| $T_J$ | Virtual junction temperature |
| $T_{\mathrm{stg}}$ | Storage temperature |
| $di/dt$ | Critical rate of rise of on-state current |
| $dv/dt$ | Critical rate of rise of off-state voltage |
| $V_{BO}$ | Breakover voltage |
| $V_{TM}$ | Maximum (steady-state) on-state voltage (thyristor) |
| $V_{FM}$ | Maximum (steady-state) on-state voltage (rectifier) |
| $V_{T,\mathrm{av}}$ | Average on-state voltage (thyristor) |
| $V_{TO}$ | Dynamic on-state voltage (during turn-on) |
| $I_{DRM}$ | Repetitive peak off-state current (thyristor) |
| $I_{RRM}$ | Repetitive peak reverse current (thyristor, rectifier) |
| $I_L$ | Latching current |
| $I_H$ | Holding current |
| $I_{DSM}$ | Nonrepetitive peak off-state current |
| $I_{RSM}$ | Nonrepetitive peak reverse current |
| $I_{RM,\mathrm{rec}}$ | Peak reverse recovery current |
| $V_{GT}$ | Gate trigger voltage |
| $I_{GT}$ | Gate trigger current |
| $V_{GD}$ | Gate nontrigger voltage |
| $I_{GD}$ | Gate nontrigger current |
| $t_d$ | Gate-controlled delay time |
| $t_r$ | Gate-controlled rise time |
| $t_{on}$ | Sum of $t_d$ plus $t_r$, turn-on time |
| $t_f$ | Fall time |
| $t_{rr}$ | Reverse recovery time |
| $t_q$ | Circuit-commutated turn-off time |
| $t_q$ (diode) | Circuit-commutated turn-off time with reverse-parallel rectifier |
| $\theta_{JC}$ | Thermal resistance, junction-to-case |
| $\theta_{JC}(t)$ | Transient thermal impedance, junction-to-case at time $t$ |

maturity, but with power devices of large size, the tendency is away from Joint Electron Device Engineering Council (JEDEC) registrations. A registered device implies a well-defined set of ratings and a well-defined set of characteristics, all guaranteed by expensive testing. With large power devices, the tendency for minor characteristic variations that disqualify a device from a rigid registration format is greater, and with this device goes a large loss of material and processing costs. By staying with "in-house numbers," maximum numbers of devices may reach the marketplace, with reasonable characteristics and reasonable cost. Rigid registration formats and tight characteristics can always be met by payment of the required testing cost. This is nowhere more evident than with military-type devices which in many cases only differ from standard industrial designs by extensive testing programs.

One of the most important reasons for variations in device characteristics and ratings is the intentional changes in process to optimize a particular characteristic. It can be stated that process optimization for one device characteristic usually results in the compromise of one or several other characteristics. This compromise effect is most apparent with junction devices such as transistors, thyristors, and rectifiers. The effects of such trade-offs should be understood and anticipated for intelligent interpretation of data sheet characteristics and ratings. The most important trade-off effects will be discussed for each power device type.

This section will cover solid-state devices and their ratings. Power devices are described by ratings and characteristics presented in power device data sheets. The data sheet presents samples and summaries of measurements made on devices, and these measured parameters are called characteristics. At other places, power device data sheets present limiting values of characteristics, which are guaranteed. Ratings are not measurable quantities because they are simply limitations of measured characteristics, as set by the manufacturer. Hence, blocking voltage of a thyristor should be greater than the rated value, and measured leakage current at rated blocking voltage should be less than maximum, etc. In some cases, limiting values of characteristics do not appear as ratings but only as an indication of characteristic spread. Further, only the most important characteristic limit may be presented, with some typical values of a particular characteristic. The absence of some characteristic limits means they lack importance and test dollars are being conserved.

Ratings of solid-state power devices will be discussed in a step-by-step fashion as related to selection and application. A listing of rectifier and thyristor rating and characteristic symbols is given in Table 22-1. A similar listing for transistors is given in Table 22-2.

**TABLE 22-2** Letter Symbols for Transistors

| Rating symbol | Meaning |
|---|---|
| $V_{EBO}$ | Emitter-to-base voltage (collector open) |
| $V_{CEO,\text{sus}}$ | Collector-to-emitter sustaining voltage (base open) |
| $V_{\text{CEO}}$ | Collector-to-emitter voltage (base open) |
| $V_{\text{CEV}}$ | Collector-to-emitter voltage (base at specified voltage) |
| $I_B$ | Base current (dc) |
| $I_C$ | Collector current (dc) |
| $P_T$ | Total power dissipation |
| $h_{FE}$ | DC current gain (common emitter) |
| $V_{\text{ce,sat}}$ | Collector-to-emitter saturation voltage |
| $V_{BE,\text{sat}}$ | Base-to-emitter saturation voltage |
| $I_{EBO}$ | Emitter-to-base cut-off current (collector open) |
| $t_s$ | Storage time, turnoff |
| $t_f$ | Fall time, turnoff |
| $f_T$ | Gain-bandwidth product |
| $C_{\text{ob}}$ | Output capacitance |
| $I_{SB}$ | Forward-biased, second-breakdown, collector current |
| $E_{s/b}$ | Reverse-biased, second-breakdown energy |
| $t_d$ | Delay time, turn-on |
| $t_r$ | Rise time, turn-on |
| $t_{\text{on}}$ | $t_d$ plus $t_r$ |
| $t_{\text{off}}$ | $t_s$ plus $t_f$ |

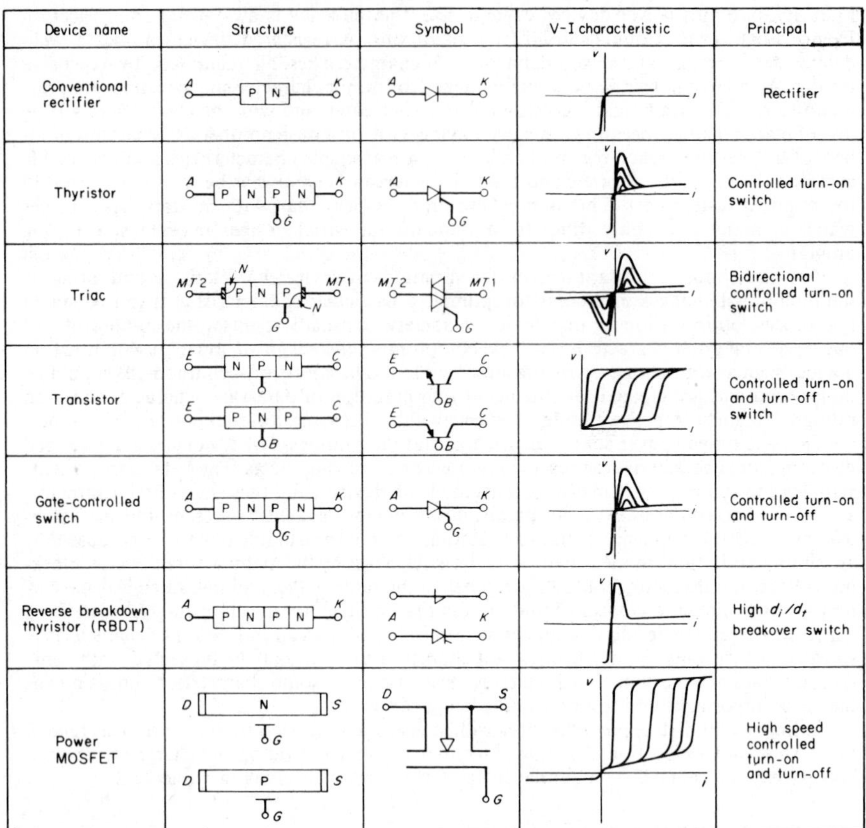


**FIG. 22-11** Semiconductor switching devices.

In addition, Fig. 22-11 summarizes the junction symbols, circuit symbols, and main terminal *VI* characteristics of the most common power devices.

## POWER RECTIFIERS

*By JEROME B. BREWSTER*

***10. Rectifier Average Current Rating.*** Signal diodes and power rectifiers conduct in one direction and block in the reverse direction. A rectifier conducts as soon as the potential in the forward direction exceeds the barrier potential, about 0.7 V. Power rectifiers are considered to include devices with current ratings from 1 A up.

Power rectifiers are characterized with an average current rating applicable for power frequencies of 60 to 400 Hz. This rating applies for a half-sinusoidal current waveform, as shown in Fig. 22-12.

For this waveform, the root-mean-square (rms) value of the current is 1.57 times the average value. The maximum average current rating of a rectifier is generally the first rating to be seen on a data sheet.

This rating reflects an rms current limit or may represent the thermal limit of a properly

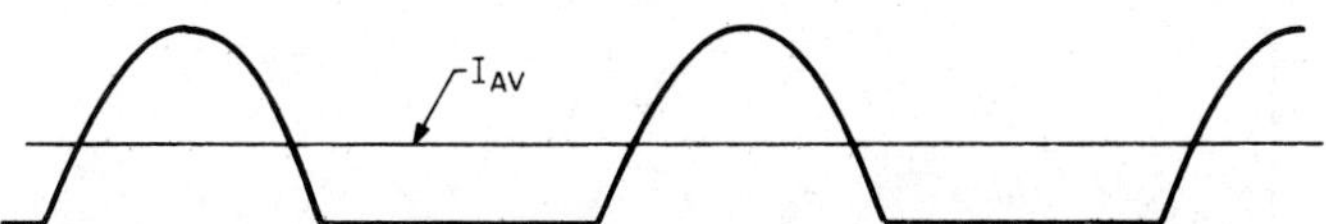


**FIG. 22-12** Half-sinusoidal current wave used for rectifier current rating.

cooled rectifier. An rms limit can exist for lead mount wires or wires bonded to the rectifier die. Average current ratings for lead-mounted rectifiers depend on ambient temperature. As ambient temperature rises, the current capability of the device falls because of its thermal limit.

RMS limiting also can apply to existing stud-mounted devices. Pigtail rectifier leads may limit rectifier current rating, and hence the rms limit for stud-mounted devices should be closely observed even if maximum junction temperature is held lower than rated maximum by effective cooling.

Flatpack, hockey puck, or press-pack rectifiers are thermally limited rather than rms-current-limited. The design rationale for a thermally limited rectifier is that as ambient temperature rises, rectifier current is progressively reduced to keep junction temperature within the device less than the rated $T_{J,\text{max}}$. Maximum junction temperatures range from 125 to 200°C for rectifiers.

The actual operating current rating of a rectifier is determined from its power losses as a function of average current. A sample loss curve is shown in Fig. 22-13.

The losses are shown for the most common waveforms found in rectifier systems. These are the 180° half sine wave, the 120° rectangular wave, and the 60° rectangular wave. They occur in resistively loaded half- or full-wave rectifiers, inductively filtered, 3-phase, full-wave rectifiers, and inductively filtered, 6-phase star rectifiers, respectively. The current waveforms are shown in Fig. 22-14.

From Fig. 22-13, it may be concluded that as the rms/avg ratio (form factor) of a rectifier current waveform increases, rectifier power dissipation increases. The information from Fig. 22-13 may be used to generate a rectifier rating curve from

$$T_{C,\text{max}} = T_{J,\text{max}} - P_{\text{av,max}} \times \theta_{JC}$$

The maximum allowable rectifier case temperature ($T_{c,\text{max}}$) is determined as the maximum junction temperature ($T_{J,\text{max}}$) minus the average junction power dissipation ($P_{\text{av,max}}$) times the thermal resistance, junction-to-case ($\theta_{JC}$). Maximum operating junction tempera-

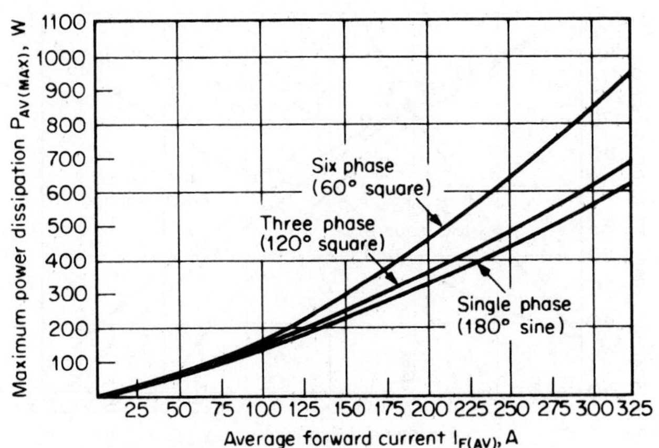


**FIG. 22-13** Typical rectifier conducting loss curves. *(Westinghouse Electric Corp.)*

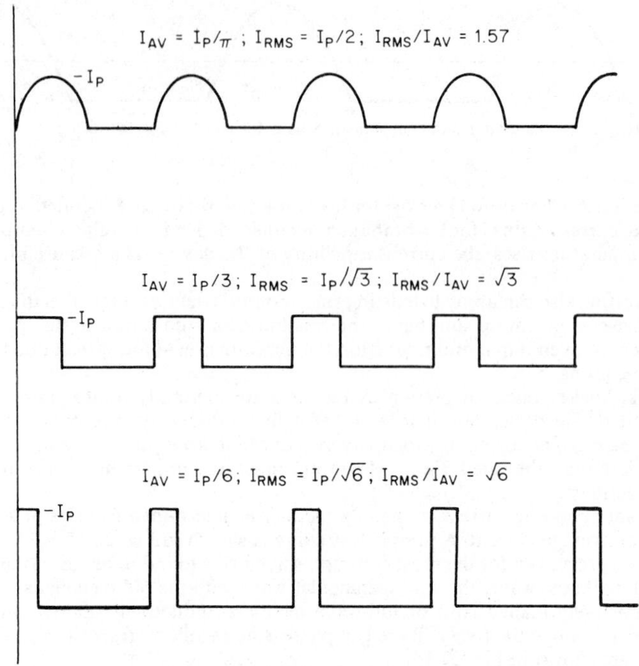


**FIG. 22-14** Some typical rectifier current waveforms—two-pulse with resistance load, three-pulse and six-pulse with infinite inductance load.

ture and steady-state junction-to-case thermal resistance are data sheet specifications. Maximum rectifier power dissipation as a function of average current is presented for any rectifier in curves such as Fig. 22-13. From the equation for $T_{C,\max}$, manufacturers generate rating curves for maximum case temperature as a function of average current. A typical curve is shown in Fig. 22-15. Manufacturers also publish guaranteed maximum rectifier

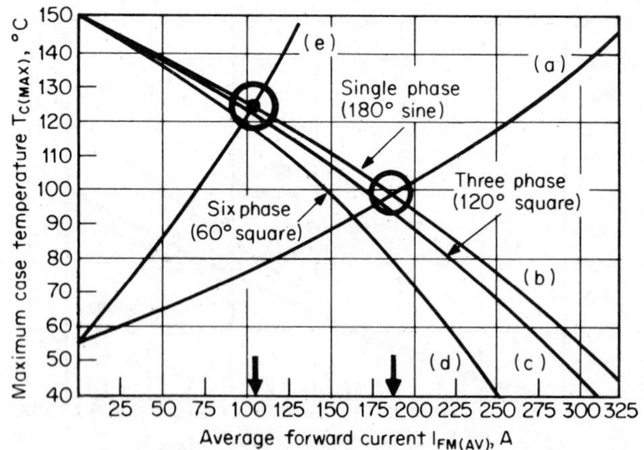


**FIG. 22-15** Permitted case temperature versus current for rectifier [(*b*), (*c*), and (*d*)], together with case temperatures versus current for two different heat sink conditions (*a*) and (*e*). *(Westinghouse Electric Corp.)*

conducting drop, $V_{FM}$, at a specified peak 60-Hz half-sine-wave current pulse ($I_{FM}$), and junction temperature. It is assumed that if a device is within the $V_{FM}$ limit, then the loss curve of Fig. 22-13 will be met. A proper rectifier $V_{FM}$ level indicates proper chip design, mounting, and bonding. These factors also affect thermal impedance and losses.

All points of Fig. 22-15 which lie below the appropriate curve are acceptable operating points. To determine where a particular cooling system will operate, the following equation must be considered.

$$T_C = T_A + P_{\text{av,max}} (\theta_{CS} + \theta_{SA})$$

where $T_C$ is the case temperature of the device in question, $T_A$ is the ambient temperature in degrees celsius, $P_{\text{av,max}}$ originates from Fig. 22-13, $\theta_{CS}$ is the device thermal impedance from rectifier package to sink (published in the device data sheet), and $\theta_{SA}$ is the thermal impedance of the heat sink chosen by the designer to remove heat from the device. Using this equation, the designer can achieve a desired current rating by choosing the appropriate heat sink to complement the rectifier. As an example, Figs. 22-13 and 22-15 are typical for a 250-A average-current-rated stud-mounted rectifier for which $\theta_{CS} = 0.10$°C/W. A heat sink suitable for this size device shows the following thermal resistances:

| Heat sink thermal resistance $\theta_{SA}$, °C/W | Airflow |
|---|---|
| 0.60 | Convection cooled |
| 0.20 | Forced air, 500 linear ft/min |
| 0.10 | Forced air, 1000 linear ft/min |

If this heat sink is used with $\theta_{SA} = 0.1$°C/W, curve *a* can be plotted on Fig. 22-15 for half-sine-wave current if $T_A$ is assumed to be 55°C. The intersection of the calculated curve and the manufacturer's curve, curve *b*, indicates that an average current of 187 A may be achieved with this system at a 55°C cooling ambient. A 105-A rating is obtained for a convection-cooled sink, shown as the intersection of curve *e* with curve *b*.

The calculations and their graphical presentation do not completely define a rectifier cooling system. After design, the calculated result must be checked by an actual heat run where the case temperature of the rectifier is measured using thermocouples. In rectifier systems, heating caused by other components or rectifiers can be influential in determining the final case temperature. The top rectifier in a convection- or fan-cooled stack should be carefully checked, because this device is subject to the greatest temperature rise. Measured data from a heat run may then be used to determine the intersection with the maximum permissible case temperature curve. Measured thermal checks need not be carried out at the maximum anticipated ambient temperature, since data can be adjusted by adding to each data point the temperature difference between the actual and maximum ambient temperatures.

The increase in rating obtained from better package configurations is shown in Table 22- 3.

**TABLE 22-3** Current Comparison for Stud and Disk Rectifiers

| | Stud-mounted rectifier | Disk-type rectifier |
|---|---|---|
| $I_{av}$ | 250 A | 400 A |
| $V_{RRM}$ | 1400 V | 1400 V |
| $I_{FSM}$ | 4500 A | 4500 A |
| $I^2t$ | 85,000 $A^2 \cdot s$ | 85,000 $A^2 \cdot s$ |
| $V_{FM}$ @ 800 A | 2.0 V | 2.0 V |
| $\theta_{JC}$ | 0.17°C/W | 0.095°C/W (double-sided cooling) |
| $\theta_{CS}$ | 0.10°C/W | 0.025°C/W (double-sided cooling) |

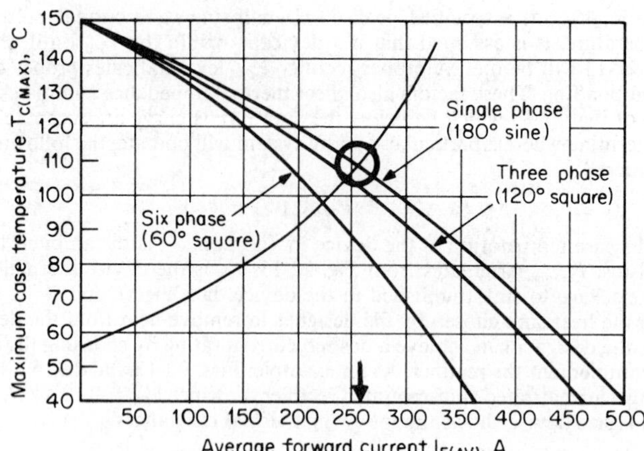


**FIG. 22-16** Another example of permitted case temperature and heat-sink-developed case temperature curves for $\theta_{SA}$ = 0.05°C/W and $\theta_{CS}$ = 0.03°C/W. *(Westinghouse Electric Corp.)*

For the disk type rectifier, assume that a heat sink is chosen such that $\theta_{CA}$ of 0.08°C/W is achieved. The disk device rating curves are shown in Fig. 22-16. Note that power dissipation curves for the stud- and disk-mounted rectifiers would be the same.

***11. Surge Current Rating.*** The surge current rating of a rectifier presumes the device to be operating at maximum rated junction temperature, carrying rated average current before the surge, and blocking rated reverse voltage in the interim half-cycle periods. After a repetitive reverse voltage pulse, the rated surge current $I_{FSM}$ is applied for half a cycle, followed by one half cycle of nonrepetitive surge voltage level, then followed by rated average current and voltage until the device cools to rated maximum junction temperature. Thereafter the surge current may be repeated, up to 100 times, in a similar fashion. A rectifier should not be designed to operate at the surge level, and the single-cycle surge is characterized only to define a fuse-blowing level after a short circuit develops and control of current has been lost. Although JEDEC defines the surge current test with the intent that a rectifier tolerate this at least 100 times, it is not good practice to intentionally exercise devices at this level. The single-cycle surge current defines a half cycle fusing $I^2t$ for 8.3 ms pulses as follows:

$$I^2t = I_{FSM}^2(T/2) \qquad \text{A}^2\cdot\text{s, where } T/2 = 8.3 \text{ ms}$$

A rectifier data sheet will publish both the rectifier single-cycle surge rating and the $I^2t$ corresponding to this 8.3-ms pulse. At this $I^2t$ level[3], the rectifier will survive a single-cycle surge. The rectifier will be adequately protected if it is serially connected with a fuse which is rated to clear with an $I^2t$ less than the rectifier $I^2t$ rating. The rectifier fuse need only be used if available fault current exceeds the single-cycle surge rating of the rectifier before a breaker or a "slow" fuse would open.

The multiple-cycle surge rating is the level of an integral number of current pulses ($n$) needed to produce the same junction temperature by the series of $n$ pulses as was produced by the single-cycle surge pulse.

***12. Transient Thermal Impedance.*** The main objective of power rectifier rating systems is to be certain that maximum operating junction temperature is not exceeded. For continuous trains of current conduction pulses, analysis using steady-state cooling parameters is appropriate. However, for current pulses of short duration, or short pulse trains, transient thermal impedances must be considered. Figure 22-17 shows a transient thermal impedance curve for a rectifier. The curve arises because of thermal capacity effects. When a rectifier experiences a step power pulse, the rise in junction temperature begins immediately, but the

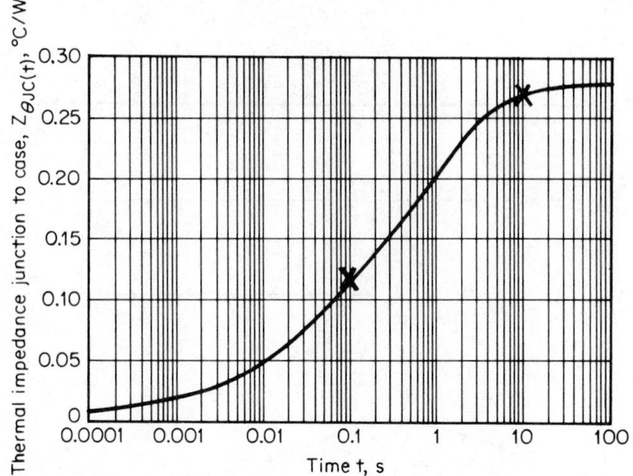


**FIG. 22-17** Typical rectifier transient thermal impedance curve. *(Westinghouse Electric Corp.)*

final temperature depends on the duration of the power pulse. The rise of junction temperature may be expressed quantitatively as

$$T_{jp} = T_{J,\text{av}} + P_0\theta(t_i)_{JC}$$

where $T_{J,\text{av}}$ is average junction temperature and $\theta(t_i)_{JC}$ is the transient thermal impedance, junction-to-case, at time $t_i$.

This assumes that all junction dissipation prior to a transient is replaced by an average dissipation, $P_{\text{av}}$, such that the junction temperature and case temperature at pulse initiation may be stated as

$$T_{C,\text{av}} = P_{\text{av}}(\theta_{CS} + \theta_{SA}) + T_{\text{amb}}$$

$$T_{J,\text{av}} = P_{\text{av}}R_{\theta,JC} + T_{C,\text{av}}$$

where all thermal resistances are steady-state.

As an example, assuming $P_0 = 300$ W and $T_{J,\text{av}} = 110$°C, where $\theta(t_i)_{JC}$ is the transient junction-to-case thermal impedance at time $t_i$, the junction temperature the rectifier achieves when subject to this power pulse $P_0$ after 0.1 s [$\theta(0.1\text{ s})_{JC} = 0.115$] is given by

$$T_{jp} = 110°\text{C} + 300(0.115°\text{C/W}) = 144.5°\text{C}$$

For a device rated 150°C maximum junction temperature, this temperature excursion is within limits, as long as the power pulse is removed after 0.1 s. If the same power pulse were applied for 10 s, such that the steady-state thermal impedance applied, then

$$T_{jp} = 110°\text{C} + 300(0.27°\text{C/W}) = 191°\text{C}$$

Maximum temperature (150°C) is exceeded and the design must be modified.

To calculate $T_{jP}$ after the power pulse is removed, a superposition technique is used whereby the effect of the positive power pulse is canceled by applying a negative power pulse when the actual power pulse terminates, as shown in Fig. 22-18. If the negative power pulse is applied after 0.1 s, and the temperature is calculated at 0.2 s, then

$$T_j = T_{J,\text{av}} + P_0\theta(0.2\text{ s})_{JC} - P_0\theta(0.2\text{ s} - 0.1\text{ s})_{JC}$$

$$= T_{J,\text{av}} + 300(0.135) - 300(0.115) = T_{J,av} + 6°\text{C}$$

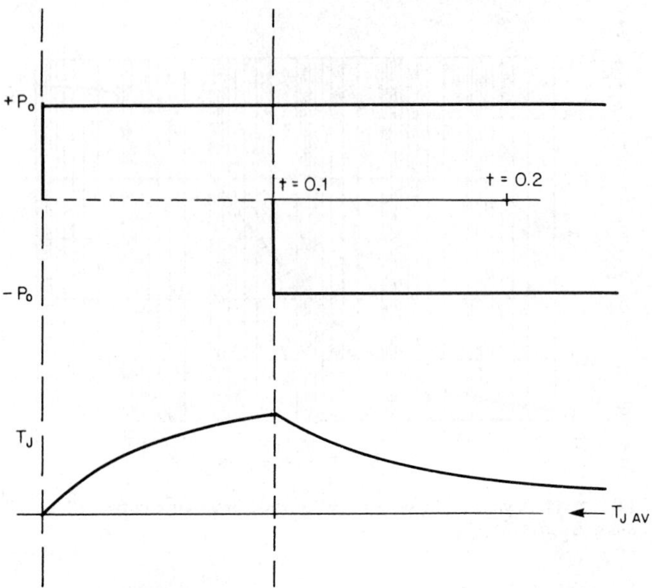


**FIG. 22-18** Superposition technique used to calculate transient excursions in junction temperature.

To compute the rise in junction temperature immediately after three different power pulses of a pulse train, as in Fig. 22-19,

$$T_j(t_5) = T_{J,av} + P_1\theta(t_5) - P_1\theta(t_5 - t_1) + P_2\theta(t_5 - t_2) - P_2\theta(t_5 - t_3) + P_3\theta(t_5 - t_4)$$

where $\theta(tn)$ = transient thermal impedance existing at time $t_n$
$\theta(t_n - t_j)$ = transient thermal impedance existing at time $t_n - t_j$
$T_j(tn)$ = junction temperature at time $t_n$

Modeling techniques may be used[4,5] to construct rectangular pulse approximations of complex pulses. The junction temperature at the end of a power pulse, whether simple or complex, must not exceed maximum $T_J$ as specified by the rectifier maker. Transient thermal impedance calculations generally assume that the power rectifier is mounted on a heat sink, and that no change in case temperature occurs during the time of the power pulse(s). This assumption is valid if the duration of the power pulse(s) is negligible compared to the thermal time constant of the heat sink. For longer power pulses, the case temperature change should be considered, and this is discussed in Ref. 5.

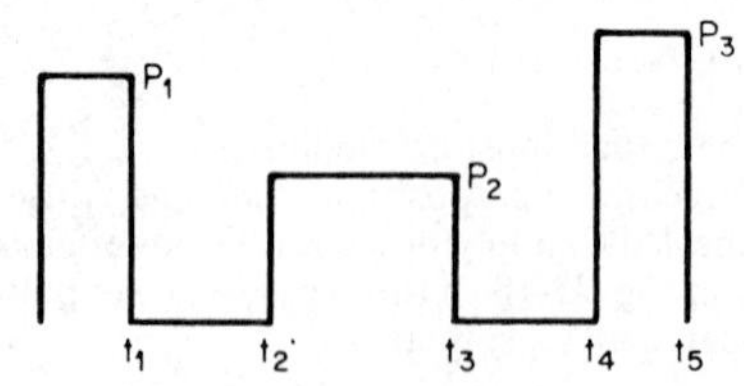


**FIG. 22-19** Example of complex power pulse train.

***13. Rectifier Reverse Recovery Phenomena.*** In normal operation, rectifiers are subjected to circuit conditions such that rectifier forward current is reduced to zero and thereafter the rectifying junction is required to block reverse voltage. Resulting device behavior is called reverse recovery. Typical rectifier current and voltage response are shown in Fig. 22-20. When rectifier current is forced to zero, conduction does not cease but continues in the reverse direction. The reverse current reaches a peak and then decays toward zero at a device-determined

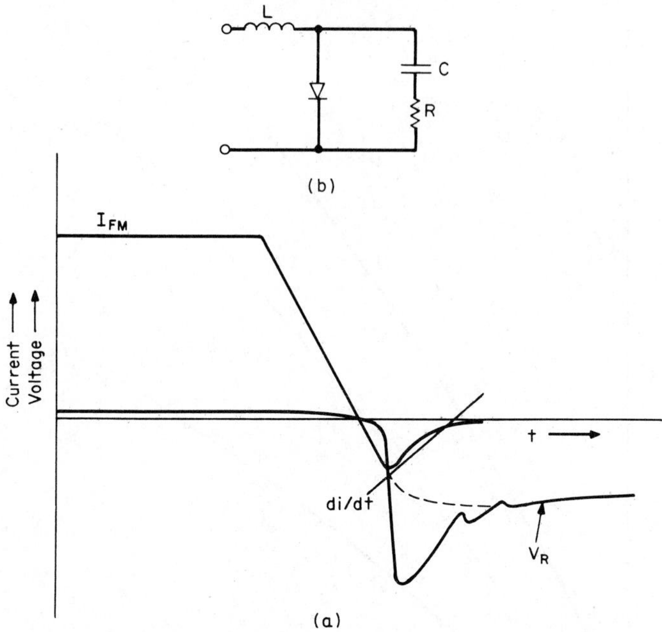


**FIG. 22-20** (*a*) Rectifier voltage and current waveforms during reverse recovery; (*b*) snubber network.

rate. The voltage transient produced by $L \times di/dt$ can exceed rectifier voltage rating, unless some means are provided to prevent this. The components used to prevent excessive rectifier reverse voltage are put together in what is called a snubber network. Usually a simple *RC* snubber is connected in shunt with the rectifier as shown in Fig. 22-20; its effect is also shown in the figure.

Rectifier reverse recovery response is defined in Fig. 22-21. The reverse recovery time ($t_{rr}$) is dependent upon junction temperature, current decay rate ($di_R/dt$), and the peak forward current from which the decay started ($I_{FM}$). Some typical data showing peak reverse

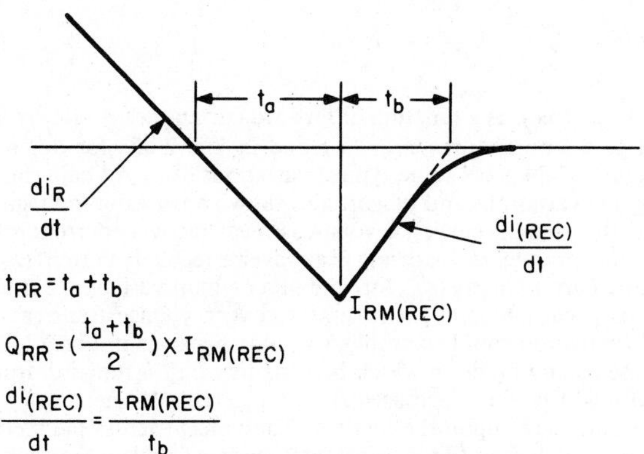


**FIG. 22-21** Typical data sheet definition of rectifier reverse recovery current waveform. *(Westinghouse Electric Corp.)*

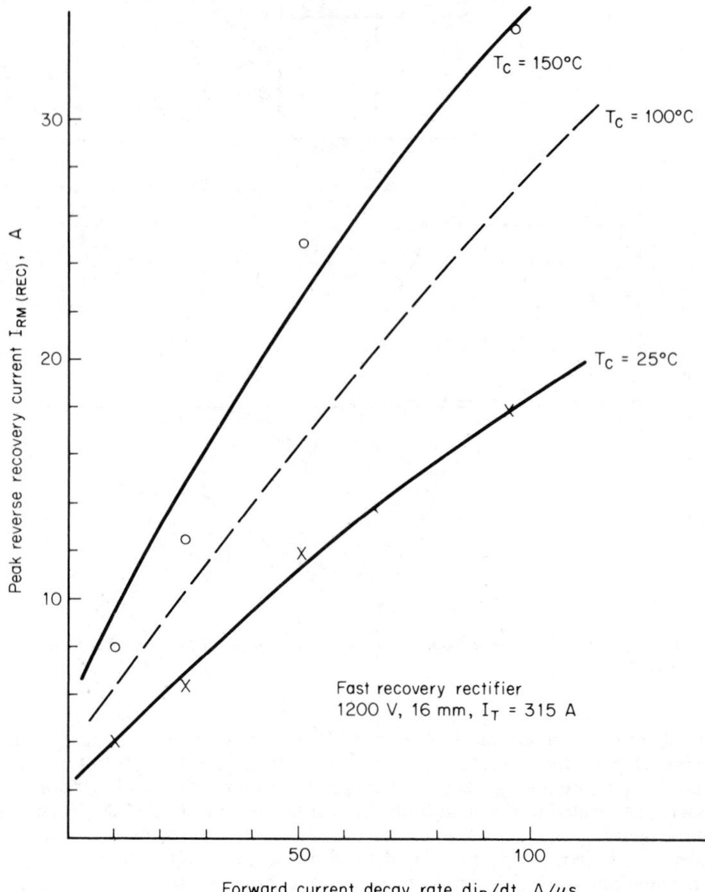


**FIG. 22-22** Typical variation of peak reverse recovery current with temperature and $di_R/dt$.

recovery current, $I_{RM}$,(rec), as a function of forward current decay rate are shown in Fig. 22-22 for different case temperatures $T_C = T_J$, and in Fig. 22-23 for different peak forward currents $I_{FM}$. Figure 22-24 gives some typical variations of $t_{rr}$. All data shown are for a fast-recovery device; the variations and magnitudes shown are greater for a standard rectifier of comparable size. By using the curves, a worst-case peak recovery current may be determined from known circuit parameters. Knowing peak reverse recovery current ($I_{RM}$, rec), the shape of the peak reverse current decay ($di_{rec}/dt$) and other circuit variables, a snubber resistor and capacitor may be chosen to prevent reverse recovery voltage transients from exceeding $V_{RRM,\max}$. The formulation snubber problem would appear as in Fig. 22-25. In this formulation, $i(t)$ represents any function which best fits the $di_{rec}/dt$ function from $I_{RM,rec}$ to zero. Such a function might be described as: $i(t) = I_{RM,rec}\,(e^{-t_r/\tau})$, where $I_{RM,rec}$ and $\tau$ are determined from measurements upon the rectifier. The voltage across the rectifier can be computed for various $RC$ values until the transient voltage is less than the peak repetitive device rating by a sufficient margin.

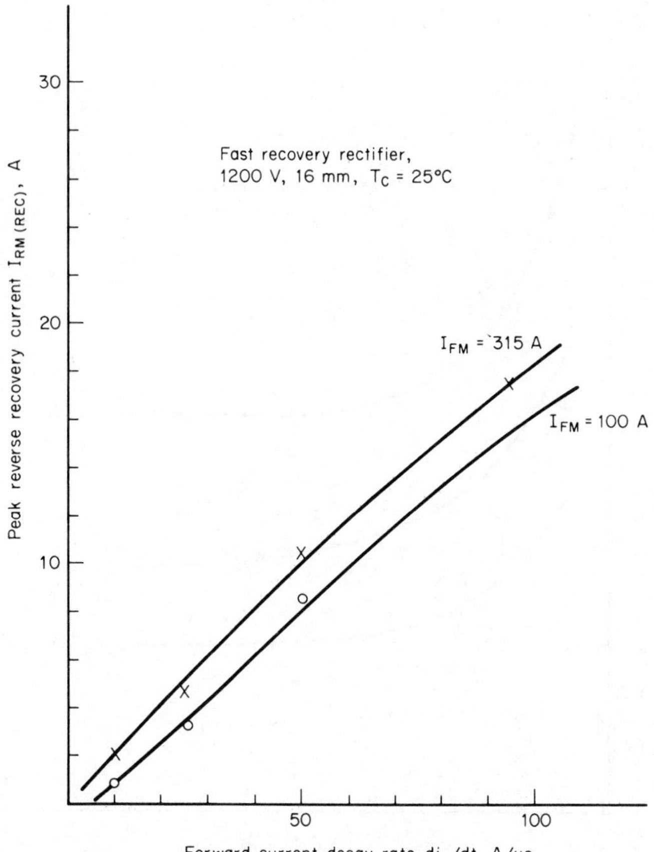


**FIG. 22-23** Typical variation of $I_{RR}$ with $I_{FM}$.

***14. Rectifier Voltage Ratings.*** Repetitive peak reverse voltage rating ($V_{RRM}$) is defined as the sustainable voltage level with repetitive 60 to 400-Hz voltage excitation which may be blocked by the rectifier at rated temperature. Under these conditions, the rectifier may produce rated leakage, but service life is guaranteed.

For nonrepetitive conditions, it is possible to exceed the repetitive rating, possibly producing higher-than-repetitive-rated leakage, and a junction temperature in excess of $T_{J,\max}$. This voltage is called the nonrepetitive transient peak reverse voltage, $V_{RSM}$. After a nonrepetitive surge, a rectifier must block repetitive reverse voltage on the next 50- to 60-Hz negative half cycle of voltage. During this following repetitive voltage application, reverse leakage current must be stable. If the nonrepetitive surge level is too high, leakage current will rapidly increase and will cause device destruction. It may be said that during the nonrepetitive surge, junction temperature exceeds $T_{J,\max}$, but by the following repetitive voltage application, $T_J$ must have cooled to a value lower than $T_{J,\max}$. Nonrepetitive blocking levels are selected by manufactueres' tests of devices at increasing levels of voltage for which leakage currents are monitored; a safe level of nonrepetitive voltage blocking may be chosen by this technique.

In high-current, high-voltage rectifiers, the blocking voltage limit of a rectifier may be set by factors other than avalanche breakdown in the bulk of the rectifier junction. Some

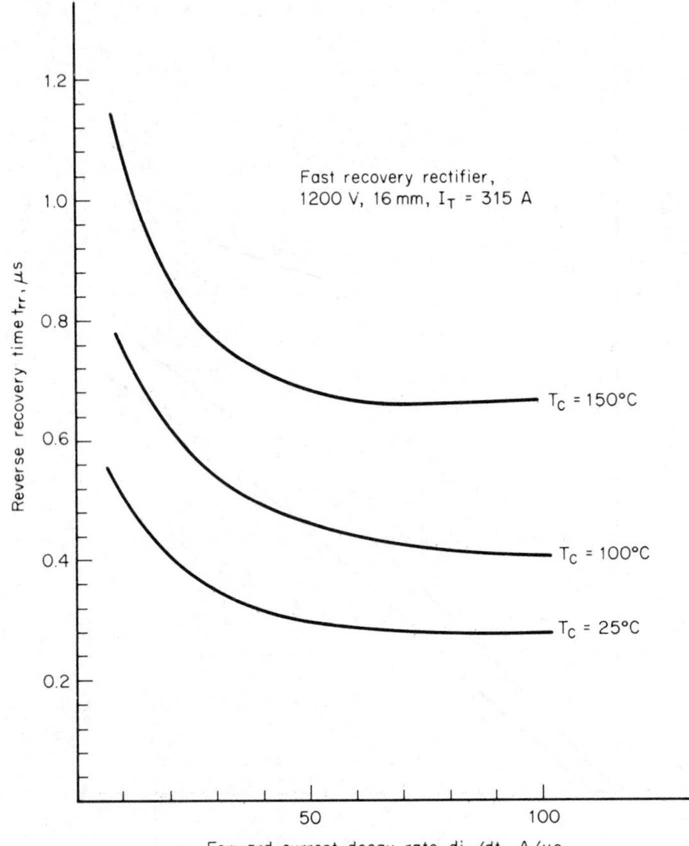


**FIG. 22-24** Typical variation of reverse recovery time with temperature and $di_R/dt$.

devices are limited by surface leakage along the rectifier beveled edge. At the edge, the junction reaches the surface and a bevel angle is created to reduce the field strength there. The rectifier may break down at the edge first. Because of the edge limitation, the rectifier may not tolerate rated leakage current at room temperature. Hence the nonrepetitive transient rated voltage is arrived at by some manufacturers in the same fashion as the normal repetitive rating, i.e., the nonrepetitive rating is in actuality a value arrived at by repetitive techniques.

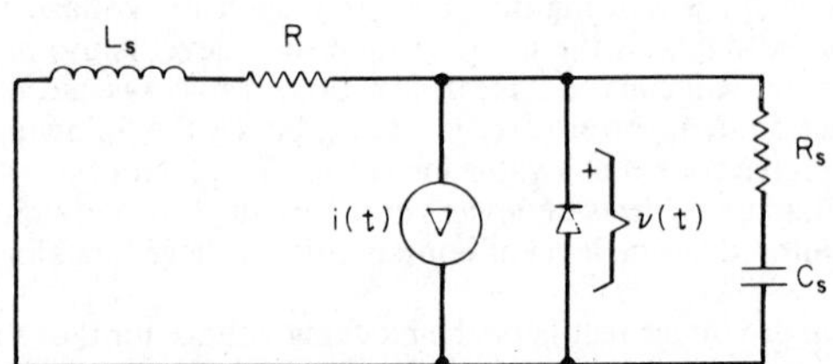


**FIG. 22-25** Circuit model used for snubber design.

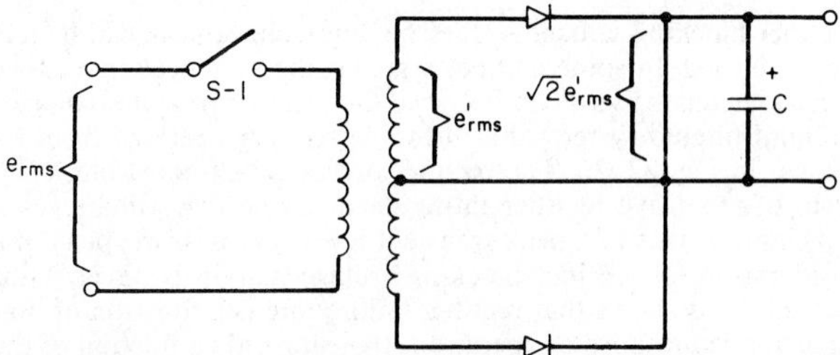


**FIG. 22-26** Two-pulse rectifier circuit with "capacitive input" filter at dc terminals.

By definition, a rectifier may produce rated leakage current at rated repetitive blocking voltage, but rated leakage should not be forced through the rectifier by increasing the applied voltage above the rated level. This can result in irreversible damage, especially at low temperatures. This applies especially to edge-limited devices.

A prime consideration in rectifier applications is that any voltage level in excess of the nonrepetitive rating, no matter how short the duration, may cause tracking along the rectifier bevel edge, resulting in catastrophic failure.

Voltage transients are always present in rectifier systems; the most common technique to limit or clip nonrepetitive transients involves the use of selenium or metal-oxide nonlinear devices. Selenium transient suppressors have been available since the evolution of power semiconductors, whereas metal-oxide devices are newer. These devices are available as unipolar or bipolar clippers and are placed either directly across each rectifier or from line to line in ac systems. Suppressors increase in size with energy absorption ability and rms withstand capability. Clamping voltage levels for selenium devices range from 2.5 to 2.8 times the ac rms voltage rating. For example, if a nonpolarized selenium suppressor has an rms rating of 480 V ac, then one grade of this device would clamp transients at 2.5 × 480 V = 1200 V. Since for many years selenium was the only means of protecting rectifiers, it was common to choose 1200-V repetitive-rated rectifiers when designing for 480 V ac, giving a rectifier blocking voltage to peak-of-line-voltage ratio of near 2:1. The safety margin

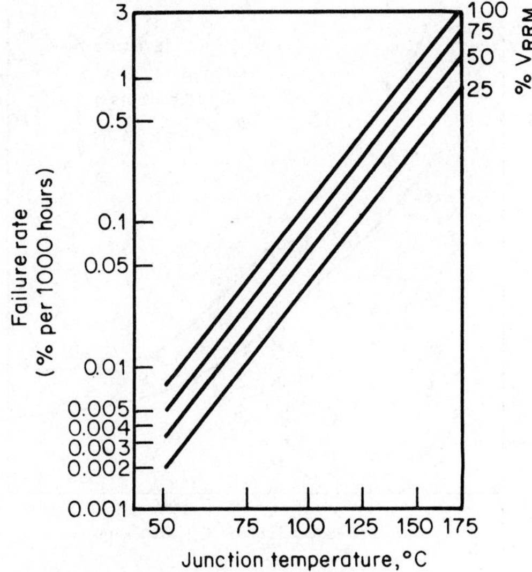


**FIG. 22-27** Typical failure rate curves.

of repetitive rectifier blocking voltage to the rms line-voltage peak can be reduced if better clamping devices are used. In some rectifier systems, the total peak inverse voltage seen by the individual rectifier repetitively can be twice the transformed line-voltage peak, such as in a capacitive-input-filtered system of a full-wave rectifier operated from a center-tapped transformer, shown in Fig. 22–26. The rectifier of Fig. 22–26 must block $2 \times e'$ rms, but the safety margin of repetitive rectifier rating need not be twice this level since the filter capacitor would clamp $e'$ rms to a peak transient level very near the peak of $e'$ rms.

A final consideration for rectifier blocking voltage margin is device failure rate, illustrated in Fig. 22–27. This shows that rectifier failure rate is a function of junction temperature and voltage applied in normal operation. By using only a fraction of the rectifier repetitive rating, and therefore a large margin, the failure rate of a rectifier can be decreased. Two-to-one rectifier repetitive blocking voltage margins are compatible with existing transient clippers and result in good reliability.

***15. High-Voltage Rectifiers.*** High voltage is relative. Steady progress in achieving ever-increasing voltage ratings has been made because of the needs of rectifier systems. When the needs cannot be met by individual devices, they are satisfied by series-connecting a number of devices to achieve the desired rating. High-voltage rectifier design requires that a very high resistivity silicon be used, with increased-base-width *n*-type silicon. With this design, rectifier conducting drop increases as a result of both the high-resistivity material and the increased thickness of silicon; the result is loss of current rating for a fixed-size device. To recover current rating lost by increased forward drop, the usual design technique is to increase conducting area. High-voltage rectifiers are larger than their low-voltage cousins of equal current rating.

Some rectifier design variations decrease rectifier forward drop. One of these variations is known as a pin-type rectifier design and results in 10 to 20% reduction of forward drop[4,5] compared to normal and standard avalanche-type design.

High-voltage rectifiers are often used in series strings for applications with very high blocking voltage requirements; here an *RC* network is required across each rectifier to assure reverse voltage transient sharing. *RC* networks around each rectifier in a series string maintain recovery current flowing around rectifiers which recover first, to prevent a buildup of

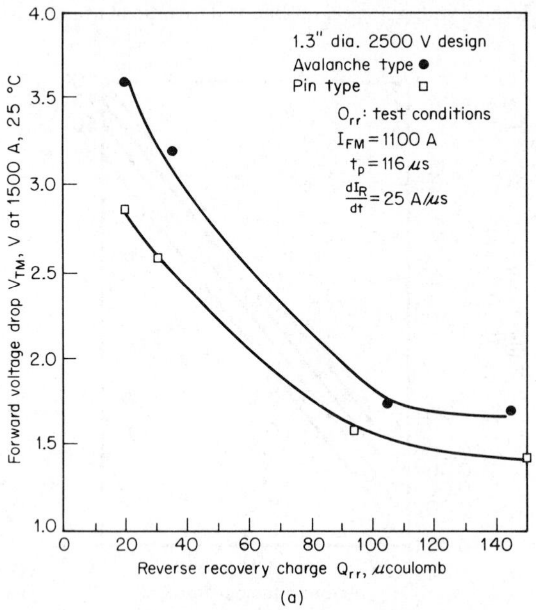


**FIG. 22-28*a*** Typical $Q_{rr}$ variations, 1100 A. *(Ref. 31.)*

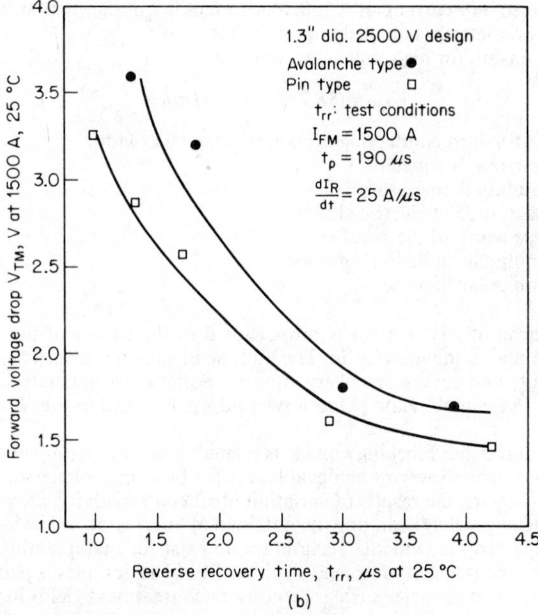


**FIG. 22-28*b*** Typical $Q_{rr}$ variations, 1500 A. *(Ref. 31.)*

voltage above the device rating because of recovery current still flowing in other rectifiers in the string. Design of such grading networks must consider device-to-device variations of the reverse recovery charge $Q_{rr}$, as well as its absolute magnitude.

***16. Rectifier Trade-offs.*** The three prime factors descriptive of rectifier performance are current rating, reverse blocking voltage rating, and reverse recovery speed, as evidenced

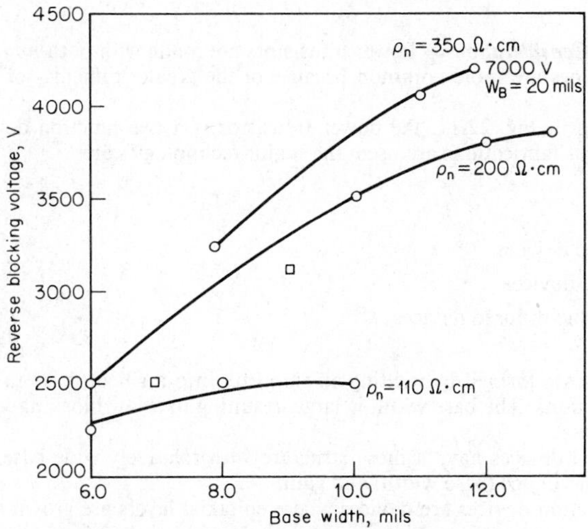


**FIG. 22-29** Reverse blocking voltage versus base width and resistivity. *(Ref. 31.)*

by a low reverse recovery current. It is inherent to rectifier design that an attempt to optimize one prime parameter has an effect upon the other two.

A design relationship for high-voltage rectifiers is[6]

$$V_{TM} = 3KT/2q \times w^2/DaKa$$

where $V_{TM}$ = the forward conducting drop across the base width
$K$ = Boltzman's constant
$T$ = absolute temperature
$q$ = magnitude of electric charge
$w$ = base width of the rectifier
$Da$ = ambipolar diffusion constant
$Ja$ = ambipolar lifetime

The forward drop of any rectifier is proportional to the square of the base width and inversely proportional to the lifetime $Ja$. The lifetime of silicon is directly related to reverse recovery charge $Q_{rr}$ and reverse recovery time $t_{rr}$. Some experimental design results are shown in Fig. 22-28*a*, which relate $Q_{rr}$ to forward drop $V_{FM}$ and in Fig. 22-28*b* to relate $t_{rr}$ to $V_{FM}$.

It is known that rectifier blocking voltage is related directly to base-width thickness $W_B$. Some experimental curves showing achievable rectifier blocking voltage for different design base widths, and showing the results of variation of silicon resistivity, are presented in Fig. 22-29. Higher blocking voltage can only be obtained by increasing base width; the effect on forward conducting drop is evident. Techniques do exist for manipulating device characteristics after chip processing is complete. Typical of these techniques is particle irradiation with electrons or heavier particles such as protons. Such treatment yields lifetime control of the finished crystal resulting in desired recovery time at an optimized conducting drop. By using this technique as the final processing step, rectifiers may be fine-tuned to achieve a desired balance of characteristics. Annealing at an elevated temperature stabilizes irradiation effects.

## POWER TRANSISTORS

*By JEROME B. BREWSTER*

**17. General Considerations.** Power transistors are made with both *npn* and *pnp* structures; the *npn* types are more common because of the greater difficulty of fabricating the *pnp*.

As diagramed in Fig. 22-11, the power transistor is a two-junction device with three terminals. Various fabrications are used; the major technologies are

**a.** Alloy devices
**b.** Single-diffused devices
**c.** Epitaxial base devices
**d.** Triple-or double-diffused devices

Alloy devices are formed by melting an impurity into a silicon base to form collector and emitter junctions. The base width is large, resulting in a low blocking voltage and low frequency response.

Single-diffused devices have a mesa structure with relatively wide base, resulting in a trade-off between voltage (base width) and gain.

Epitaxial junction devices are devices where epitaxial layers are grown as collectors or bases, or for both junctions, resulting in narrow base widths, good frequency response, and good collector sustaining voltage.

Triple- or double-diffused devices achieve reasonable switching speeds with high collector sustaining voltage and good gain ($h_{EE}$) profiles. Diffusion is preferable for processing large-area silicon slices with multifinger emitter structures.

All transistors are susceptible to forward bias second breakdown. The tendency is for the tolerance to decrease with increasing frequency response or decreasing base width. Hence, alloy devices are tolerant to forward bias second breakdown, making good linear audio amplifiers, but are poor as high-speed switches. The second breakdown characteristics of the other types are published by manufacturers who seek to produce good frequency response, high voltage, and adequate forward bias second breakdown tolerance.

**FIG. 22-30** Typical interdigitated transistor. *(Westinghouse Electric Corp.)*

High-current transistors are devices with large lateral dimensions and fingered or interdigitated emitters, as depicted in Fig. 22-30. Narrow-finger structures are required to produce good second breakdown tolerance. Current tends to crowd to the edges or center of the emitter for forward-biased or reverse-biased conditions, respectively. High-current transistor design provides for this crowding by providing large emitter length and periphery.

Emitter contacts are made by mesa structures; the emitter (enclosed dark area in Fig. 22-30) is elevated by using preforms that only contact the emitter, while other contacts connect to the base (light area in Fig. 22-30). The fusion is shown with preform in place. It should be noted that the transistor requires a long emitter edge (interdigitation of fingers) to achieve the required base current control. This same interdigitation is found in high-speed thyristors and GTOs to achieve fast turn-on and turnoff.

***18. Power Transistor Current Ratings.*** The continuous current rating for small power transistors is primarily an rms current limit of the emitter connecting wires. For large compression-bonded transistors, the continuous current rating is established by maximum power losses and gain falloff.

The parameters used to determine maximum current rating for a transistor operated in the linear or switching mode are

Continuous rated collector current, $I_C$

Thermal resistance, $\theta_{JC}$

Saturation voltage, $V_{CE,\mathrm{sat,max}}$

Maximum ambient temperature, $T_{A,\mathrm{max}}$

Maximum junction temperature, $T_{J,\mathrm{max}}$

*Example.* Consider a transistor for which $T_{J,\mathrm{max}} = 200°\mathrm{C}$. When operating as an inverter switch with 50% duty cycle, $V_{CE,\mathrm{sat,max}}$ is 1 V at $I_C = 50$ A. Then

$$P_{\mathrm{av}} = 0.5 \times 1\ \mathrm{V} \times 50\ \mathrm{A} = 25\ \mathrm{W}$$

If the transistor possesses a junction-to-case thermal impedance $\theta_{JC}$ of 0.09°C/W, a case-to-sink thermal impedance $\theta_{CS}$ of 0.05°C/W, a sink-to-ambient thermal impedance $\theta_{SA}$ of 0.16°C/W, then the maximum junction temperature is:

$$T_{J,\mathrm{max}} = T_A + P_{\mathrm{av}}(\theta_{JC} + \theta_{CS} + \theta_{SA})$$

With $T_A = 55°\mathrm{C}$,

$$T_{J,\mathrm{max}} = 55°\mathrm{C} + 25(0.09 + 0.05 + 0.16) = 62.5°\mathrm{C}$$

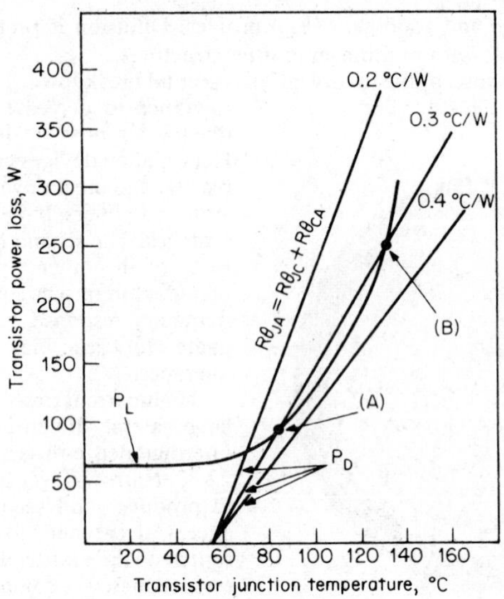

**FIG. 22-31** Power transistor thermal stability. *(Ref. 29.)*

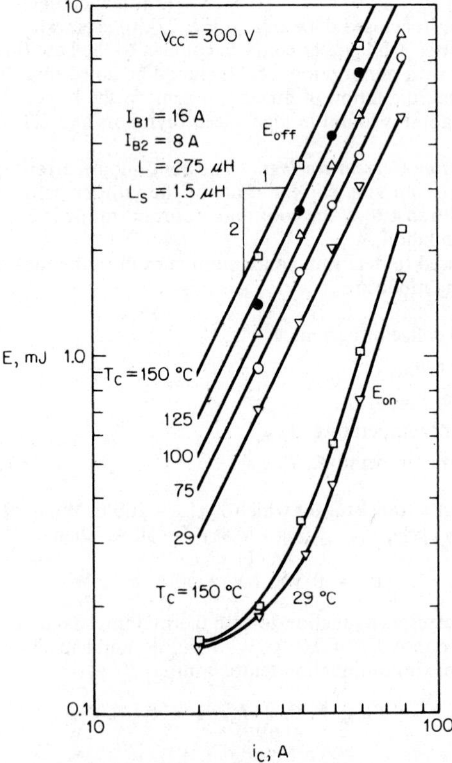

**FIG. 22-32** Typical switching losses for a transistor. *(Ref. 30.)*

This calculation gives a junction temperature well under $T_{J,\max}$. However, unlike rectifiers or thyristors, transistors have a positive temperature coefficient of power loss versus junction temperature; as the temperature of the transistor increases, transistor $P_L$ losses increase. The stability criterion is

$$\delta P_L/\delta T_J \leq 1/\theta_{JA} \qquad \text{and for safe operation } T_J \leq T_{J,\max}$$

The expression is better understood by consideration of Fig. 22-31. The curve illustrates two important transistor functions. The first is represented by the straight-line functions labeled $P_D$. The inverse slope of each curve is in units of degrees Celsius per watt and is the data sheet thermal resistance of the transistor plus the thermal resistance of the particular heat sink used, This curve increases with junction temperature, since both saturation voltage ($V_{CE,\text{sat}}$) and both turn-on and turnoff switching losses increase with temperature.

For stability, $\delta P_L/\delta T_J \leq 1/\theta_{JA}$ must be true at an intersection of the two functions. As an example, at point $A$ of Fig. 22-31, the slope of $P_L$ is less than that of $P_D$ and the transistor would be stable. For points on $P_L$ above $A$ but not $B$, the system's ability to remove heat is greater than the rate of heat generation, and the system will always fall back to the stable point $A$. For points above $B$, thermal instability would result.

To generate the information in Fig. 22-31, the curve $P_L$ must be determined. An example of how this can be done is given in Ref. 7.

Transistors also dissipate power because of switching losses. Some examples of measured switching losses for a large-current, high-voltage, triple-diffused transistor are given

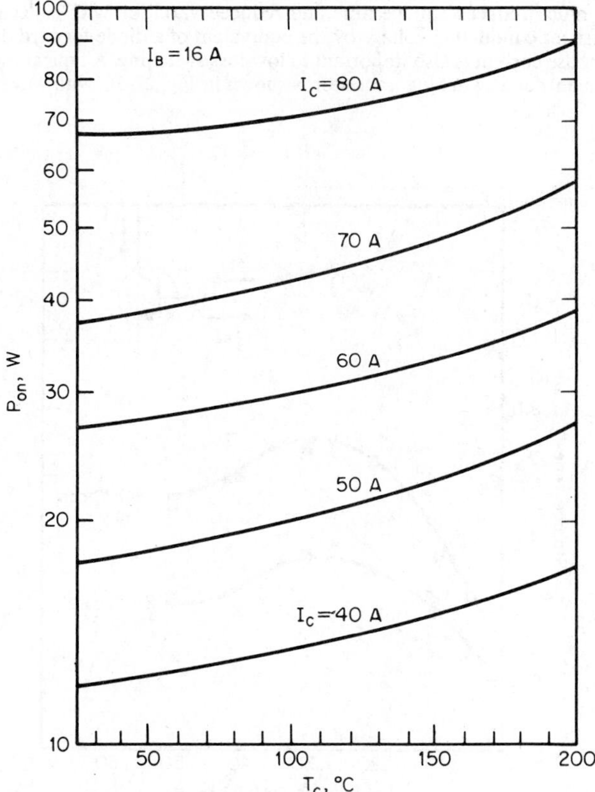


**FIG. 22-33** Typical conduction losses for a transistor. *(Ref. 30.)*

in Fig. 22-32. Turnoff losses are seen to increase substantially with temperature, but turn-on losses hold nearly constant.

Figure 22-33 shows calculated transistor conducting losses versus temperature, all determined from measured $V_{CE} - I_C$ transistor saturation data, measured at steadily increasing junction temperature.

***19. Base Drive Considerations.*** The first parameter to be considered regarding transistor base drive is the maximum continuous base current $I_{B,\max}$. This current is simply the rms current capability (with margin) of the base connecting wire and should not be exceeded.

An important base drive consideration related to power switching is the relationship of transistor base current to $V_{CE,\text{sat}}$. Low saturation voltage, and low conduction losses, requires overdrive of two to three times the minimum base current required for saturation. If the design approach is to provide base drive to a transistor always at a level that anticipates maximum collector current, then the device is overdriven at collector currents less than the assumed maximum. The effect of gross overdrive is a lenghtening of storage time $t_s$ just prior to turnoff; this is undesirable. For a constant load current, constant base drive is simple and adequate. For systems with variable load currents, more sophisticated drive systems that provide constant turn-on gain result in much less extension of storage time. Results for a transistor driven with constant gain ($I_C/I_B$) are shown in Fig. 22-34.

A second technique frequently used to alleviate storage-time problems at constant base drive is to utilize the antisaturation combination of diodes in the base and collector of a transistor. This effectively clamps a heavily driven transistor to a $V_{CE}$ level above deep saturation by effectively diverting excess base drive to the collector. The resultant storage-time effects, with and without a clamp, are shown in Fig. 22-35. Although the antisaturation circuit provides much reduced storage time and reduced variation with collector current, it increases transistor conducting voltage by the equivalent of a diode forward drop.

Transistor base current is also important to low-loss switching. A typical transistor base-drive arrangement for a switching transistor is shown in Fig. 22-36, with waveforms in Fig. 22-37.

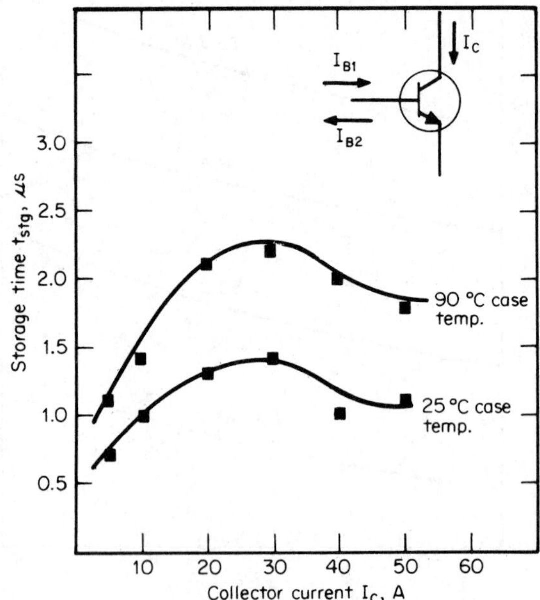


**FIG. 22-34** Transistor storage time. *(Ref. 29.)*

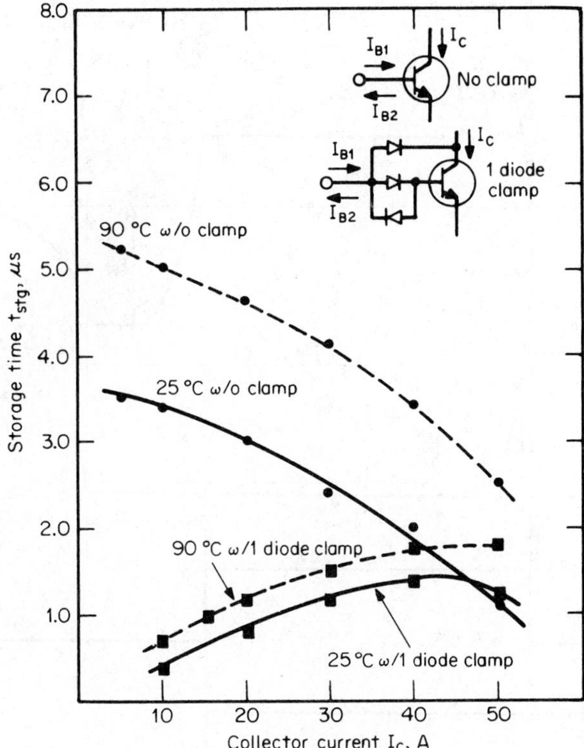


**FIG. 22-35** Influence of antisaturation diode on storage time. *(Ref. 29.)*

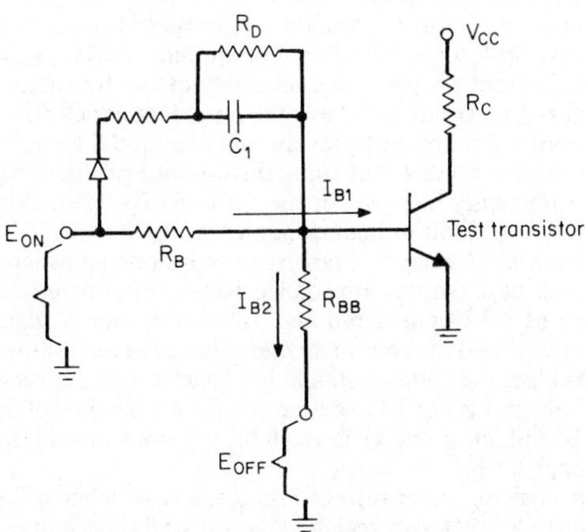


**FIG. 22-36** Typical transistor base drive.

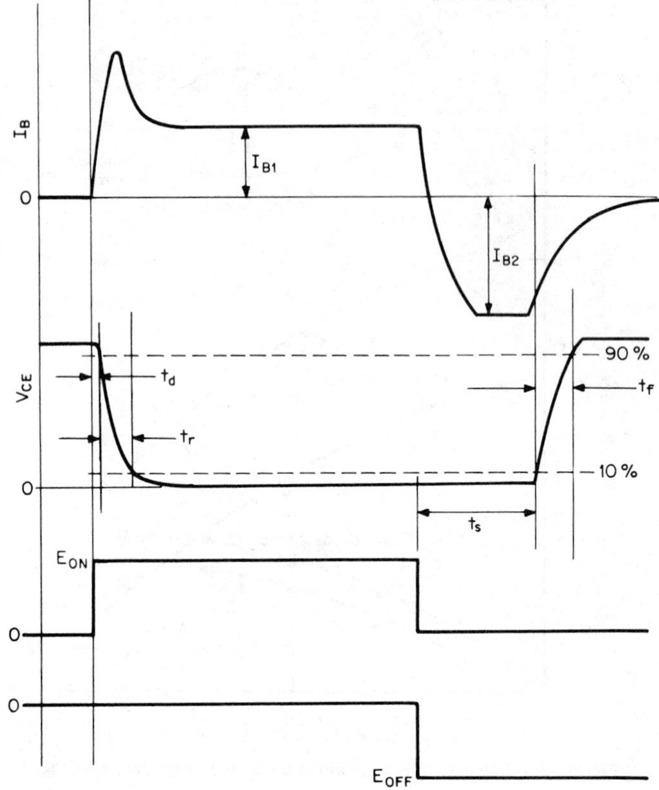


**FIG. 22-37** Typical base-current waveforms.

For turn-on and turnoff, $I_{B1}$ and $I_{B2}$ are the standard terms for dc-level turn-on and turn-off current, respectively. Almost no characterization is to be found for the rate of rise of current associated with a turn-on specification or of the peak base drive required to turn on a high-rate-of-rise current waveform. To achieve high-speed turn-on for high *di*/*dt* collector currents, base overdrive with high rate of rise is required. If drive is inadequate to turn on a high *di*/*dt* collector current properly, the tendency of the transistor is to limit *di*/*dt* by maintaining a collector-emitter voltage above the normal conducting value. Since this turn-on effect only lasts from a fraction of to several microseconds, forward-bias safe operating area (SOA) limits are not exceeded, but some disturbance of the current rise occurs, with higher turn-on loss. Sufficiently high rate-of-rise turn-on drive should be provided to avoid turn-on hang-up effects as a result of high *di*/*dt* switching.

***20. Transistor Blocking Voltage.*** Transistor collector-to-emitter voltage ratings are defined with a particular base-emitter impedance, base-emitter reverse voltage, or as a function of both. Because of SOA considerations, collector-emitter voltages are defined for a current-limited, energy-limited measurement condition to avoid failures during testing. For this reason, transistor blocking voltages should not be tested on a curve tracer, but only with the type of circuit shown in Fig. 22-38 to determine the avalanche voltage $V_{\mathrm{CEO,sus}}$. With this circuit, only the stored inductive energy is available to sweep current through the avalanche at the sustaining voltage level.

When the TUT (transistor under test) of Fig. 22-38 is switched off, the collector voltage sweeps through a collector breakover region ($V_{BRC,EO}$), and thence into a region of collector current. Along this portion of the curve, manufacturers define ($V_{CEO,\mathrm{sus}}$) as the voltage exist-

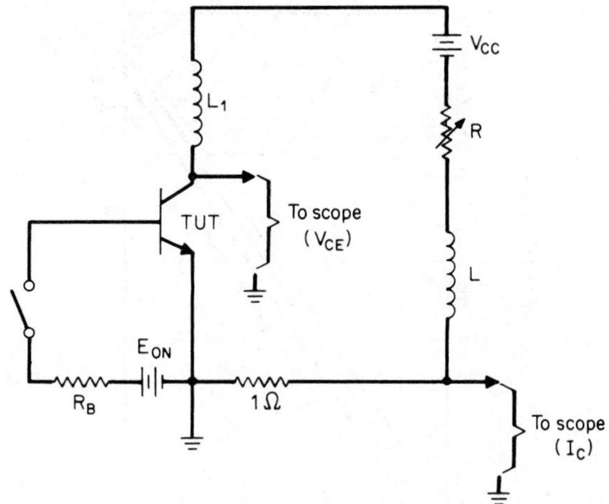


**FIG. 22-38** Measuring transistor blocking voltage.

ing at a measured current of several hundred milliamperes. Figure 22-39 shows a typical *V-I* trace from the circuit of Figure 22-38.

$V_{CEO,\text{sus}}$ is defined for the open base condition and is the lowest of all the transistor's voltage ratings; $V_{CER}$, $V_{CES}$, and $V_{CEV}$, specified on transistor data sheets, define transistor blocking capabilities under less stringent conditions.

In switching applications, transistors can operate at collector voltages less than $V_{CEO,\text{sus}}$ with little difficulty. For example, a device with a 450-V sustaining voltage can reliably switch in an inverter with a supply voltage of 300 V dc. The 150-V margin would be considered good design practice, but snubbers might be required to guarantee that transients do not exceed the $V_{CEO,\text{sus}}$ level. Figure 22-40 shows the relationship of transistor failure rate to

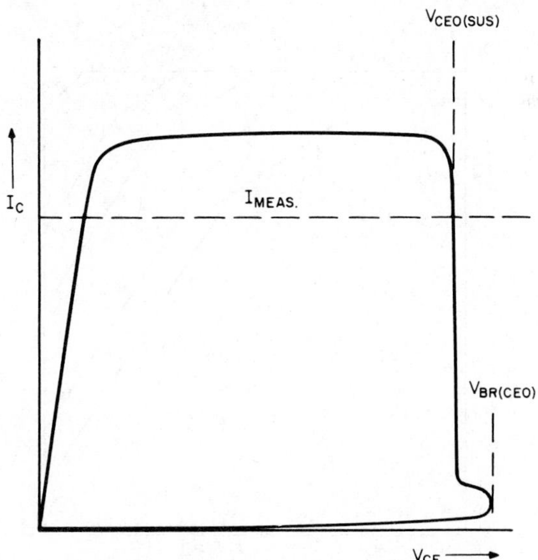


**FIG. 22-39** *V-I* trace of transistor while switching.

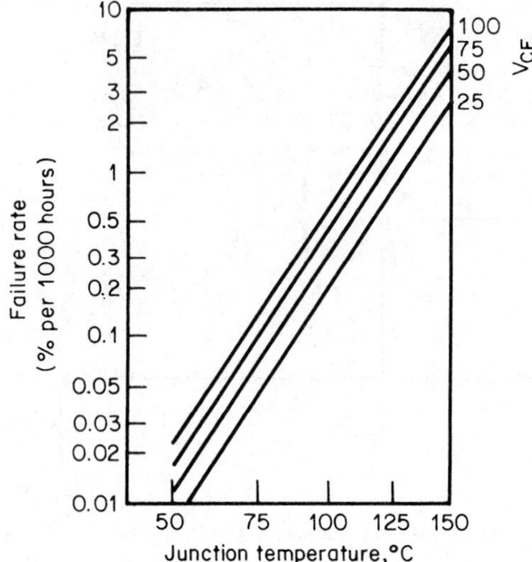


**FIG. 22-40** Transistor failure rates. *(Westinghouse Electric Corp.)*

blocking voltage and temperature. The effects of voltage and temperature safety margins are evident.

Transistor rating is also often specified in the reverse-biased, turnoff condition with clamped collector-emitter voltage. This is a nonavalanche condition at less than $V_{CEV,sus}$ for which a transistor can be switched with little suppression for voltage rate of rise. The condition approximates switching with an inductive load, except for the rate of rise of collector voltage, which may be faster than rate of current fall. Suppression may be necessary to con-

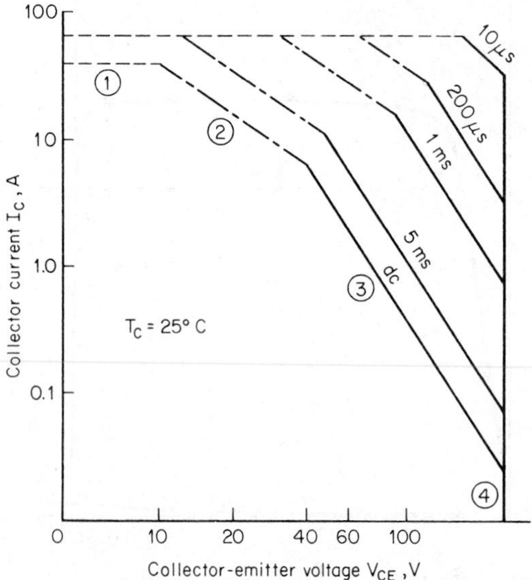


**FIG. 22-41** Forward bias SOA.

trol transient overshoot above the clamp level because of parasitic inductance. Permissible operation in this mode is defined by reverse bias SOA curves, which limit turnoff energy (nonavalanche) and vary in form from one manufacturer to the next. The key consideration regarding reverse bias SOA curves is to constrain transistor collector current and collector-emitter voltage within the area designated by the device manufacturer.

***21. Forward Bias SOA.*** A typical forward bias SOA curve is shown in Fig. 22-41. The four sections of the SOA boundaries are numbered and describe four different limiting conditions as follows:

*a.* Section 1 is the rms current limit boundary. In small devices, it is the rms limit of the current-carrying capacity of the wires bonding the outside transistor terminals to the transistor crystal. In large compression-bonded transistors, it derives from the rms limit of the base connection, which in turn determines the limit for the collector current.

*b.* Section 2 is the thermal limit of the transistor. Below this boundary, collector dissipation is low enough that $T_J$ is less than $T_{J,\max}$. For pulse duty, it has been computed by the manufacturer that pulse operation at the specified pulse width and collector-emitter voltages and currents will result in a $T_{J,\max}$ at the end of a single pulse, with starting case temperature as specified. The pulse may be repeated if $T_J$ is allowed to cool back to starting junction temperature ($T_C$ = 25°C in Fig. 22-41). Thermal ratings linearly derate to zero at 200°C case temperature.

*c.* Section 3 applies to the region of forward bias second breakdown. For dc bias points outside the limit, or a longer repetitive pulse at the prescribed $V_{CE}$-$I_C$ point, forward bias second breakdown will result.[8] Biasing a transistor outside the second breakdown limit will result in catastrophic failure.

*d.* Section 4 limit is that resulting from $V_{CEO,\text{sus}}$ voltage limitation, which should not be exceeded except for special devices and bias conditions guaranteed by the manufacturer to be safe.

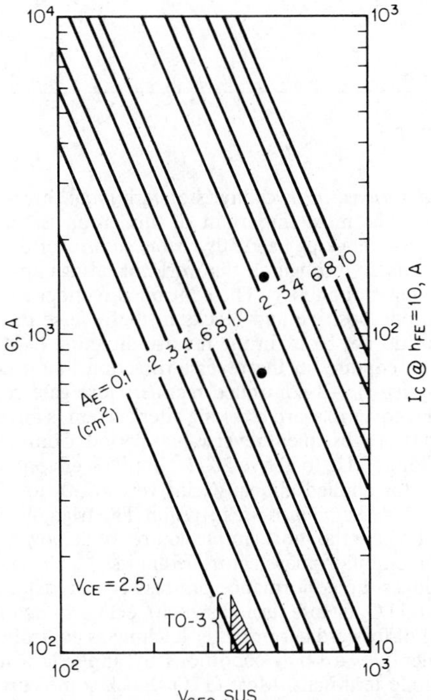


**FIG. 22-42** Voltage rating versus current capability for transistors. *(Ref. 28.)*

***22. Transistor Voltage-Current Trade-off.*** Transistors vary in size from TO-5 package units to those in flatpack or epoxy-type packages measuring several inches in diameter. For a constant semiconductor die size, it can be shown that transistor current rating drops with increasing collector-emitter voltage rating.[9]

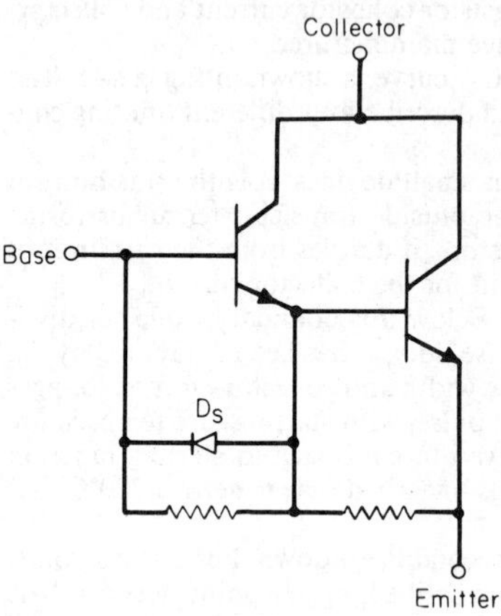


**FIG. 22-43** Darlington connection.

An example of this relationship for various emitter sizes ($A_E$) is shown in Fig. 22-42. As an example, a 23-mm transistor ($A_E \simeq 1$ cm$^2$), capable of 65-A current rating at a $V_{CEO,\text{sus}}$ voltage of 450 V would only be capable of a 10-A current rating at a sustaining voltage design level of 1000 V. The trade-off of voltage for current in junction transistors follows a square law relationship.

***23. Darlington Transistors.*** The Darlington transistor combination is shown in Fig. 22-43 and represents the union of two transistors on one chip. Physically, a smaller transistor supplies its emitter current to the base of the larger transistor, realizing a gain increase. Some manufacturers show a parasitic diode $D_P$ in reverse parallel across the main power transistor; this diode generally has very limited current capacity. Some Darlingtons also have a speed-up diode $D_s$ across the base emitter of the driver transistor. When the base of the Darlington is biased negative, the speed-up diode provides access to the output transistor base, enhancing the removal of carriers from the base of the output transistor and thus reducing storage time.

## POWER THYRISTORS

*By JEROME B. BREWSTER*

***24. General Considerations.*** Power thyristors are *pnpn* three-junction structures with two or three terminals. The most important of this group is the three-terminal silicon-controlled rectifier (reverse blocking triode thyristor) with reverse blocking capability. Controlled rectifiers are available throughout the world in ratings up to thousands of amperes and up to blocking voltages of 5000 V. The controlled rectifier switches in its defined forward direction with anode positive and blocks in the reverse direction. Some controlled rectifiers are made that do not block in the reverse direction (ASCR, or asymmetric controlled rectifier), or that conduct in the reverse direction like a power rectifier (RCT, or reverse conducting thyristor). All controlled rectifiers lose gate control when the turn-on action is complete and recover forward blocking after current is forced to zero by some other circuit means followed by a subsequent reverse-bias period. Controlled rectifiers are applied at power frequencies from 50 Hz to above 20 Hz, but 95% of applications are below 5 kHz.

An important thyristor applied almost exclusively at 50- to 60-Hz frequencies is the triac, or generically the *bidirectional triode thyristor.* The triac blocks and switches in both directions and is useful where the most simple control of ac power or current is necessary, either controlling in the on-off or phase-controlled mode.

A device which bridges the performance gap between thyristor and junction transistor is the turnoff thyristor (GTO, or gate turnoff switch). GTOs switch in the forward direction but do not lose control of forward current after latching as controlled rectifiers. GTOs turn off anode current under forward-bias conditions by applying a negative turnoff voltage between gate and cathode terminals. Most GTOs block some current only moderately in the reverse direction due to structural design.

Two other devices of similar structure are the reverse-blocking diode thyristor (RBDT) and the reverse-conducting diode thyristor (previously called the RSR, or reverse-switching rectifier). Both two-terminal devices are triggered by a high-rate *dv/dt* avalanche trigger, the one blocking and the other conducting in the reverse direction. Both devices offer a high-speed broad-area turn-on; they are used exclusively for radar modulators.

## SEMICONDUCTOR-CONTROLLED RECTIFIERS (SCRs)

*By JEROME B. BREWSTER*

***25. General Considerations.*** An early power SCR design is shown in Fig. 22-44. This is a four-layer *pnpn* device with connections to anode, cathode, and gate. Pole pieces applied to anode and cathode apply pressure to the silicon chip and make electrical contact. The gate signal of the design shown is applied by a third connection to the center metallization, connecting directly to the *p*-base gate terminal of the device. When triggered by a positive gate signal, conduction begins at the edge of the cathode nearest the gate and spreads at a finite velocity from the gate area to the outer periphery of the device, under the cathode. Similar structures were also made with the gate terminal at the edge of the disk.

Several deficiencies are encountered with this design. Although suitable for power frequencies, this design is not satisfactory at high switching frequencies because of the limited spreading speed of the plasma throughout the structure. When a gate signal is applied, a high-rate-of-rise load current can burn out the device near the gate. This is called *di/dt* failure. This structure will turn on when a forward positive-going transient voltage appears across the anode-cathode of the device. This is called turn-on by *dv/dt*. The effect can be alleviated by connecting a resistance from gate to cathode, but not enough to improve the transient capability sufficiently for reasonable performance, especially with large-area devices.

Two design modifications are shown in Fig. 22-45. The first improves *dv/dt* tolerance. By photoetching small pinhole wells into the SCR emitter through to the *p*-base area and then allowing cathode metallization to fill the holes, minute resistances can be placed throughout the cathode area. These act similarly to the placement of a resistance across the gate emitter but are spatially located to be effective over the complete area of the cathode. With this emitter shunt construction, *dv/dt* tolerance as high as 800 V/$\mu$s is now commonplace. The second modification of Fig. 22-45 is the use of an auxiliary SCR to drive the main gate. Names used for this structure are dynamic gate, amplifying gate, regenerative gate, or accelerated cathode excitation. The symbolic equivalent circuit is shown in Fig. 22-46. The structure in the middle of the device can be seen to form an auxiliary *pnpn* SCR whose output cathode, or floating *n* ring, is connected to a surface shorting ring through the lateral *p*-base resistance to the edge of the main SCR gate.

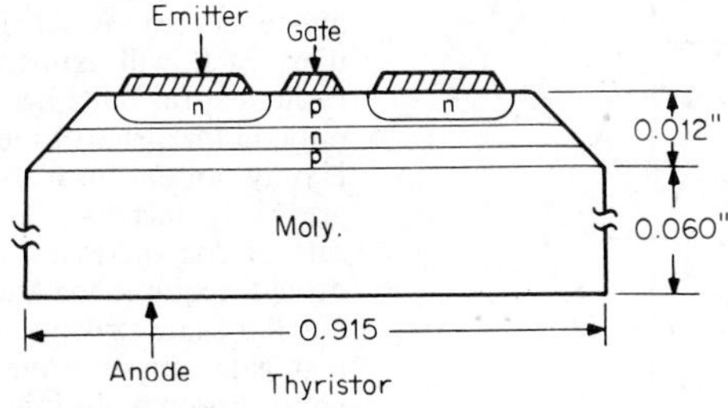


**FIG. 22-44** Power thyristor design.

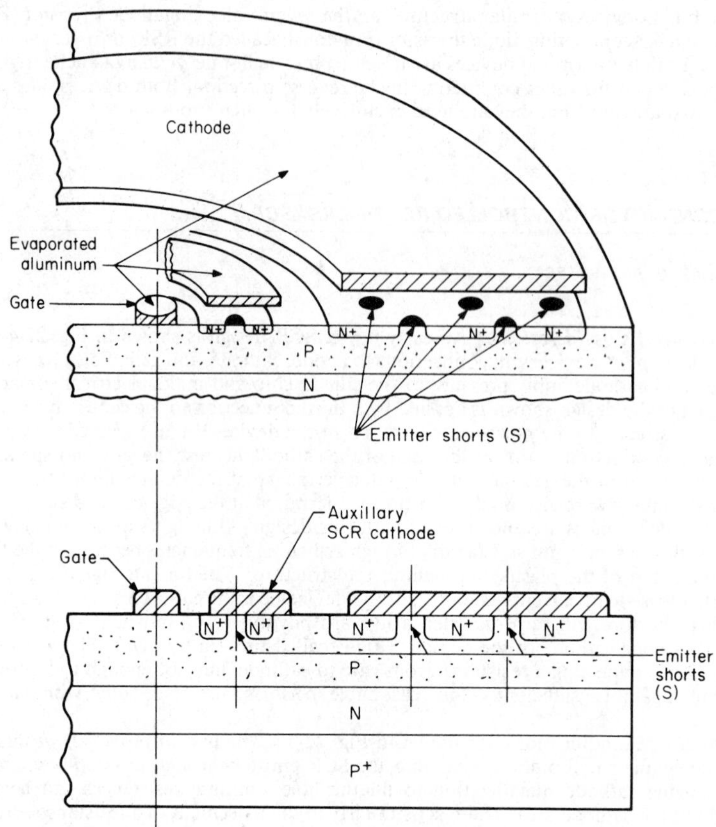


**FIG. 22-45** Modification to thyristor design for enhanced properties. *(Westinghouse Electric Corp.)*

When gate drive is applied to the device, current turn-on occurs in the auxiliary SCR first. Gate drive to the auxiliary flows through the main cathode gate shunts but is not enough to develop sufficient voltage to turn on the main cathode, because of the shorted cathode used for *dv/dt* protection. When the auxiliary turns on, its current provides high drive to the edge of the main cathode, if the anode current rate of rise is high. The auxiliary SCR will extinguish when the main cathode turns on. Since the auxiliary SCR is only on for a short time, the feedback effect is only present for a short time, but it progressively increases as the anode current rate of rise increases, i.e., when increased drive is required to handle higher *di/dt*. The auxiliary gate structure handles the need for high gate turn-on drive and results in complete turn-on of the inner cathode peripheral edge, allowing much higher *di/dt* ratings.

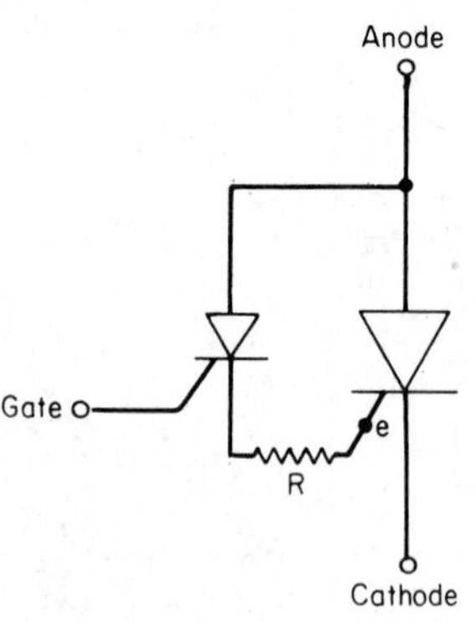


**FIG. 22-46** Equivalent circuit of dynamic gate thyristor.

A third advance in power SCRs has been required because of the finite spreading

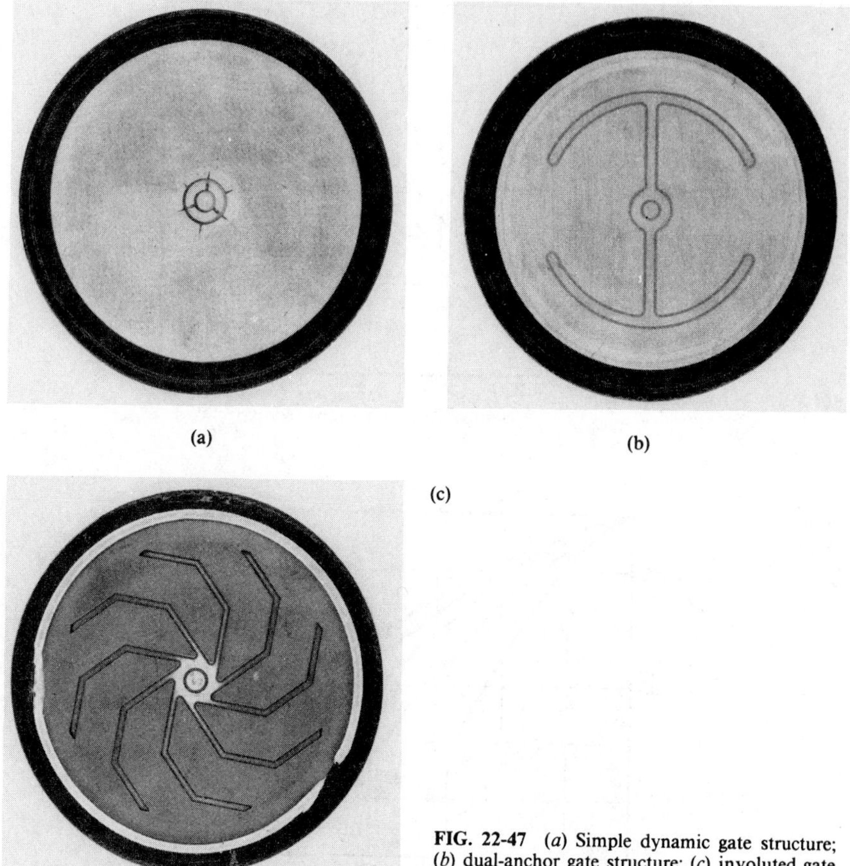


**FIG. 22-47** (*a*) Simple dynamic gate structure; (*b*) dual-anchor gate structure; (*c*) involuted gate structure.

speed of current conduction laterally beneath the cathode. If a current pulse is of such width that it is less than the spreading time of current throughout the cathode, then the effective emitter area is reduced, resulting in higher losses for such a high-frequency pulse. To alleviate this problem, device designs have been constructed to spread the amplified current throughout the emitter area. This has been done with various "distributed gate" designs, a progression of which is shown in Fig. 22-47. Each figure shows progressively increasing levels of interdigitation, resulting in a distribution of amplified gate current to the outermost area of the cathode. In Fig. 22-47, the dots in the middle area of each structure are the shorting metallization(s). The outermost area is the main cathode contact. To avoid shorting the gate (shorting region *S* and the cathode), chemically milled cathode contact preforms isolate the cathode area from the remaining areas. Increasing levels of interdigitation cause a progressively higher loss of cathode area, thus lowering average current rating at the expense of turn-on speed, but interdigitation allows low pulse loss performance up to 20 kHz.

***26. SCR Current Rating.*** Current ratings of controlled rectifiers are very similar to those of power rectifiers, with some differences created by device structure and capability. Maximum current ratings for SCRs are limited by a maximum junction temperature of 125°C. At higher junction temperatures, forward leakage currents would seriously degrade forward blocking voltage.

For power-line frequencies of 50 to 400 Hz, rating curves are provided showing average

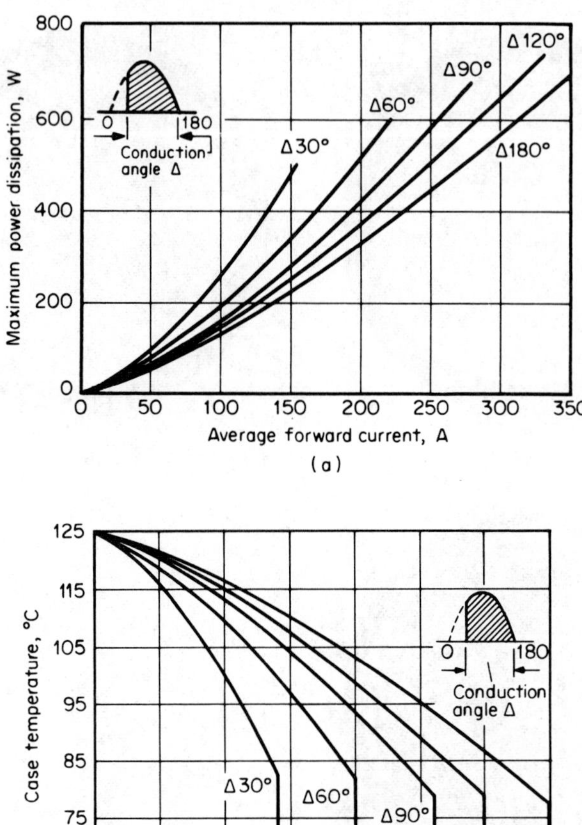


**FIG. 22-48** (*a*) Thyristor dissipation versus current and conduction angle; (*b*) permitted case temperatures. *(Westinghouse Electric Corp.)*

current versus losses and permissible case temperature. Sample curves are shown in Fig. 22-48. Maximum case temperature curves are generated from power dissipation curves with thermal impedance consideration, as for power rectifiers. Because SCRs possess the ability to control turn-on, the derating curves show a wider variety of conduction angles than those presented for power rectifiers. The curves show that as higher average currents are drawn for shorter conduction periods, losses rise and temperature derating is more severe. For stud-mounted devices, where connecting wires and solder are present, rms current limits should be carefully observed. For rectangular current waveforms, curves similar to Fig. 22-48 are also published. These curves are useful for rectifier systems in which "regular operation" is achieved as described in Ref. 10. Regular operation assumes current is heavily filtered by the use of a large choke.

The techniques of single-cycle surge current rating and multiple-cycle surge rating, and the overall relationship to $I^2t$ are identical to those for power rectifiers.

Devices operated at frequencies up to 20 kHz must turn on rapidly rising current wavefronts, conduct narrow-pulse-width currents at high peak levels, and support blocking volt-

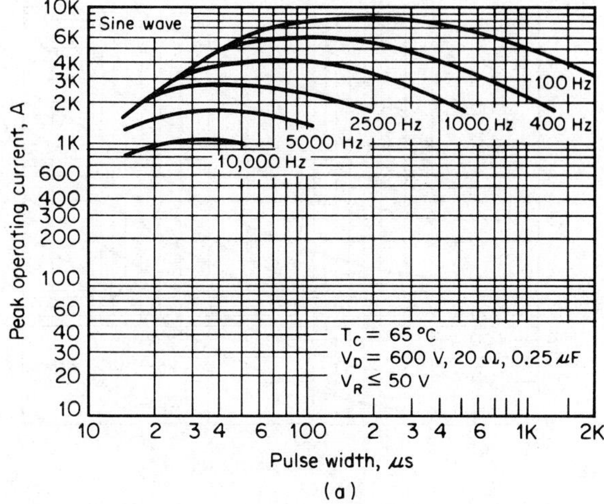


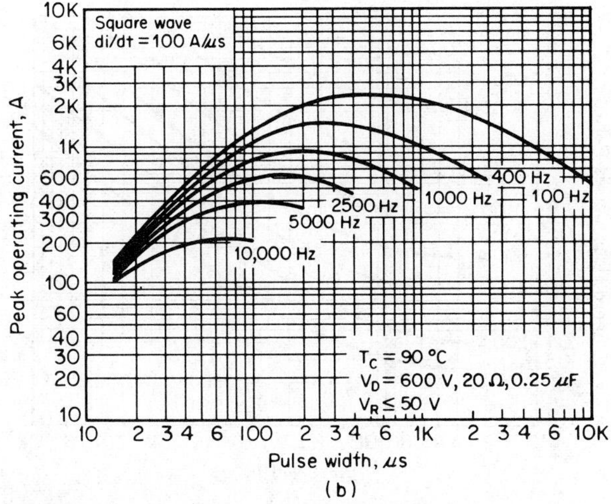


**FIG. 22-49** (*a*) Sine-wave pulse current ratings for thyristor; (*b*) Square-wave pulse current ratings for thyristor. *(Westinghouse Electric Corp.)*

ages with high rate of rise. Since many device parameters affect the ability to tolerate this type of duty, characterization requires that all the relevant parameters be specified concurrently.[11] Permissible operating point data are presented for half-sine and trapezoidal pulses. The characterizations indicate current pulse width, peak current, rate of rise of current, maximum pulse rate, switching voltage, forward blocking voltage recovery level and rate (*dv/dt*), case temperature, and the minimum turn-off time for which the ratings apply (Fig. 22-49). It should be noted that the specification of an *RC* snubber is also included in the data. The snubber repetitively discharges through the device and contributes losses.

Similar data are presented for other case temperatures, and extrapolations must be drawn for case temperatures in between. Also, for trapezoidal pulses, data are presented for different levels of *di/dt*. To determine heat-sink requirements for a device operating at one of the high-frequency points, energy loss curves are published. A sample is shown in Fig.

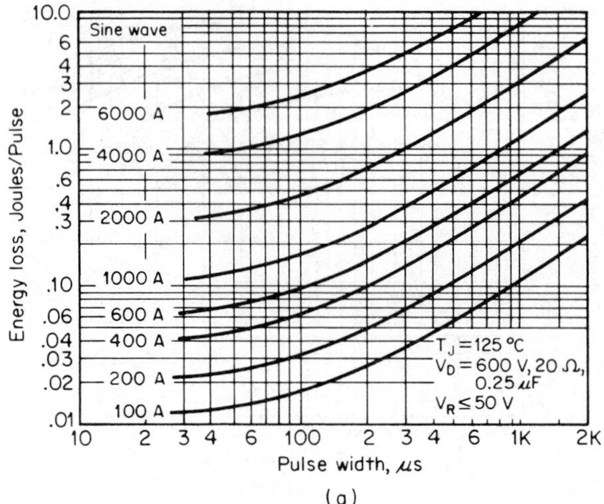


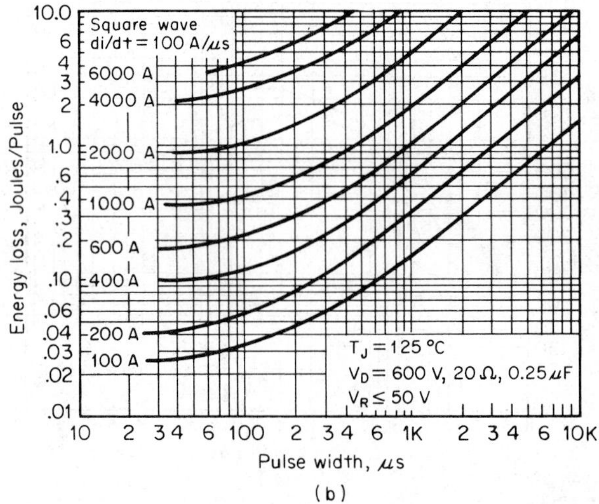


**FIG. 22-50** (*a*) Turn-on energy loss for sine-wave pulses; (*b*) turn-on energy loss for square-wave pulses. *(Westinghouse Electric Corp.)*

22-50. Losses are determined by multiplying the energy per pulse by the pulse rate. Then, a heat sink is selected with appropriate thermal impedance to limit case temperatures to desired levels. It is important that manufacturers' recommended gate drive for high-frequency rating not be diverged from.

***27. SCR Blocking Voltage.*** SCR voltage ratings are similar to power rectifier ratings. A reverse repetitive voltage rating $V_{RRM}$ is defined for which a rated leakage current $I_{RRM}$ is guaranteed not to be exceeded under repetitive application of 60- to 400-Hz power at maximum junction temperature. "At this voltage, reverse power dissipation is generally small and contributes little to the total dissipation in the thyristor."[12] As with power rectifiers, a transient voltage that causes higher than normal leakage can be tolerated if not applied repetitively. During the overvoltage period, "the instantaneous power dissipations may become significant, but still remain below the level which the manufacturer has found to be

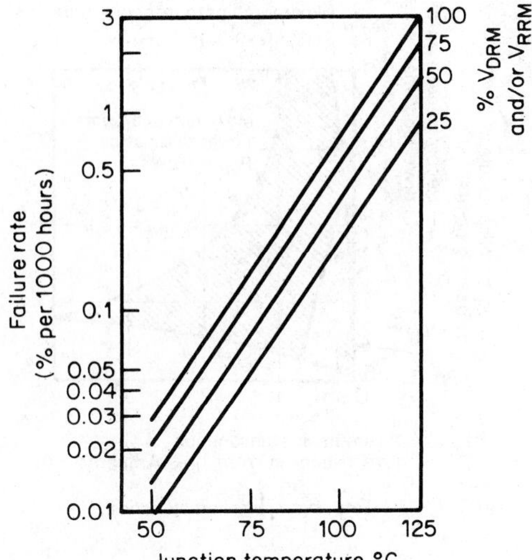


**FIG. 22-51** Thyristor reliability. *(Westinghouse Electric Corp.)*

destructive. While the energy dissipated during this time causes an increase in junction temperature, the level reached is not sufficient to cause thermal runaway, and removing the excess voltage within the time period specified will allow the junction temperature to rapidly drop back to the steady state operating level."[13]

The manufacturer also guarantees a repetitive forward blocking voltage $V_{DRM}$ at the maximum junction temperature, 125°C for thyristors. At rated leakage current $I_{DRM}$, power dissipation is insignificant and may be ignored. At temperatures higher than 125°C, leakage current and power dissipation rise, leading eventually to loss of forward blocking. For this reason, thyristors are limited to lower maximum junction temperature (125°C) than rectifiers (150 to 190°C).

Because of the possibility of turn-on by forward avalanche current, SCRs are not rated for a nonrepetitive transient forward blocking voltage.

Designs should always be such that blocking-voltage limitations are not violated; it is not reasonable to turn on a thyristor by forward breakover and expect survival. In applications, clippers and suppressors should be provided to guarantee this, as is done for rectifiers. Reliability benefits of blocking voltage safety margin are seen in Fig. 22-51.

***28. SCR Gate Drive.***[14] Because of the myriad SCR applications, a dc gate test condition (Fig. 22-52) with a resistive load is used to permit both the manufacturer and the user to ascertain basic gate parameters. It was not intended to reflect operational application requirements. These dc gate trigger requirements are normally given on SCR data sheets.

$I_{GT}$ is a dc gate current which causes the SCR to latch into conduction and remain on (hold); referred to as gate trigger current.

$V_{GT}$ is the dc gate cathode voltage which causes the SCR to latch into conduction and remain on (hold); referred to as gate trigger voltage.

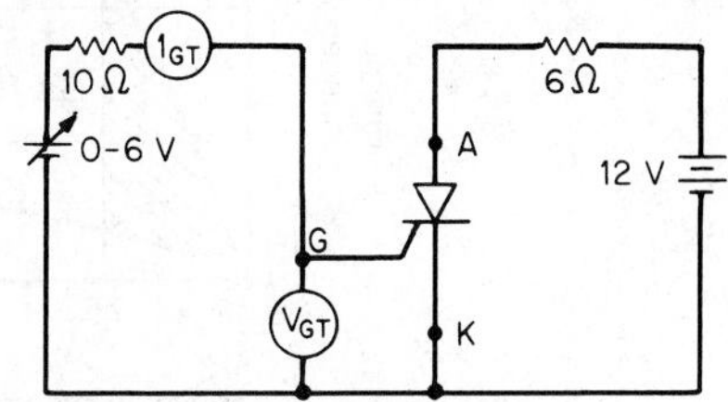


**FIG. 22-52** Gate-triggering test circuit.

$I_{GNT}$ is a dc gate current which, when applied to the gate cathode terminal, will still permit the SCR to block rated $V_{DRM}$; referred to as nontrigger gate current.

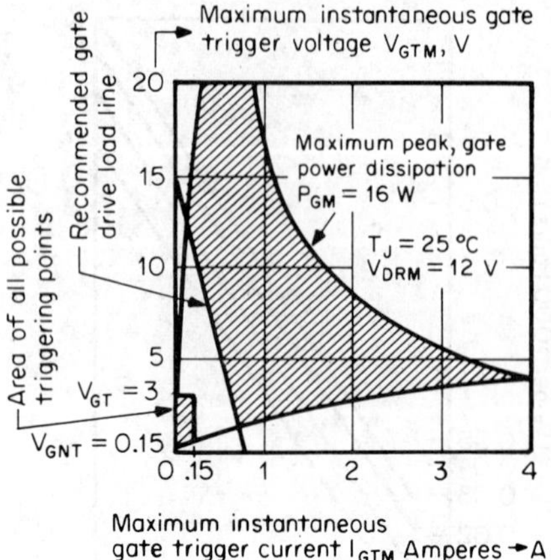


**FIG. 22-53** Gate-triggering characteristics.

$V_{GNT}$ is a dc gate voltage which, when applied to the gate cathode terminal, will still permit the SCR to block rated $V_{DRM}$; referred to as a nontrigger gate voltage.

Other published gate parameters include peak gate trigger current for pulse operation, $I_{GTM}$, peak and average gate power, $P_{GM}$ and $P_{G,av}$, and peak reverse gate voltage, $V_{GRM}$.

Maximum gate-triggering characteristics for typical dynamic gate SCRs are given in Fig. 22-53. Note that the ratings are at 25°C and that a recommended load line for moderate *di/dt* applications is given.

The characteristic effect of junction temperature on gate trigger current or voltage is depicted in Fig. 22-54, with an SCR for which $I_{GT}$ = 150 mA. Note that as an SCR is heated to 125°C, the required gate current to trigger typically is one-half the 25°C value, and that the −40°C value is approximately twice the 25°C value.

The information in Fig. 22-55 relates the measured value of dc gate trigger current to pulsed gate operation. It is necessary to increase gate drive amplitude for pulse widths less than 20 μs because of the charge turn-on concept ($q = \int I_{GT}\, dt_p$) described in the referenced literature.

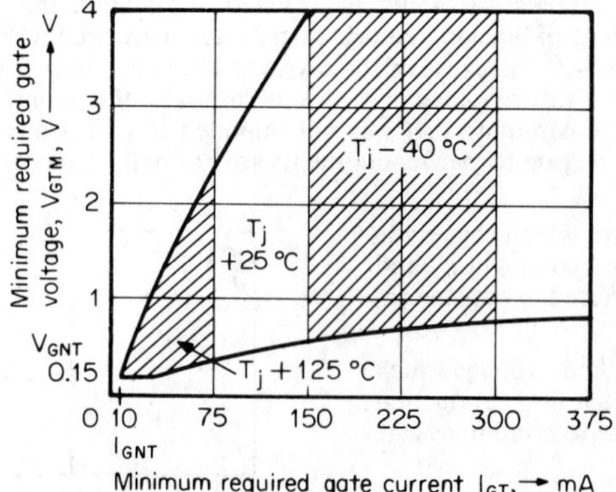


**FIG. 22-54** Effects of temperature on gate-triggering requirement.

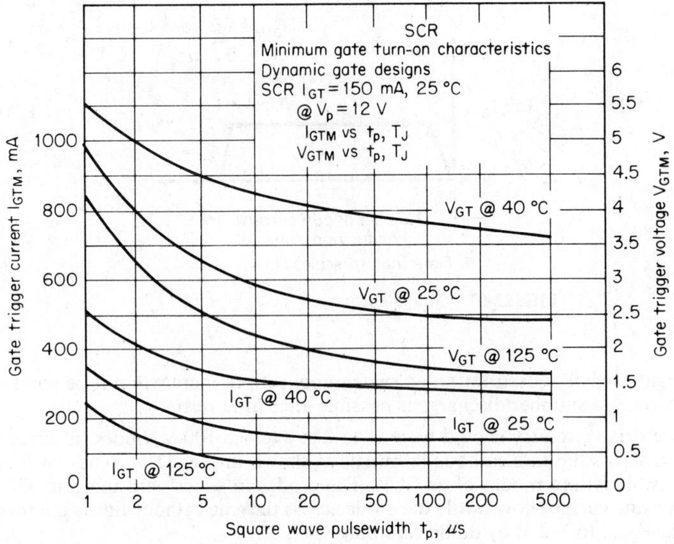


**FIG. 22-55** Pulsed turn-on gate-triggering requirements.

Dynamic gate SCRs may be triggered with either soft gate drive or hard gate drive. Devices with simple center-firing and edge-firing gates require hard gate drive.

Hard gate drive is needed where high repetitive *di/dt* is present. Figure 22-56 shows the suggested hard gate drive for an individual SCR. The cases of anode current conduction interval $\leq 20$ μs or picket-fence gate firing are not shown. Reference must then be made to the minimum pulsed gate trigger requirements to obtain the proper value of $I_{GT}$. This value of $I_{GT}$ is then used for the hard gate drive $I_{GTM}$ determination.

Soft gate drive (shown in Fig. 22-57) is adequate for moderate applications of dynamic gate type SCRs. If a snubber network is always available to discharge at turn-on, soft gate drive may be adequate even for some high *di/dt* applications, but further assistance is advised from an application engineer.

Dynamic gate SCRs are optimized to provide fast turn-on and low switching losses with a soft gate drive signal. Recognition of situations where dynamic gate action is limited, and a number of recommended design practices, are given below.

Gate the SCR when the anode voltage is positive. Allowing a positive gate while the SCR becomes reverse-biased limits device reliability.

Design the gate firing sequence such that the snubber network across the SCR is charged prior to gate triggering. This gives good dynamic gate action.

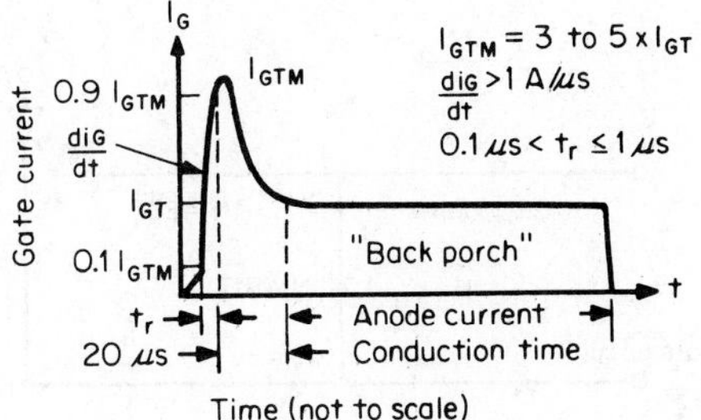


**FIG. 22-56** Hard gate drive.

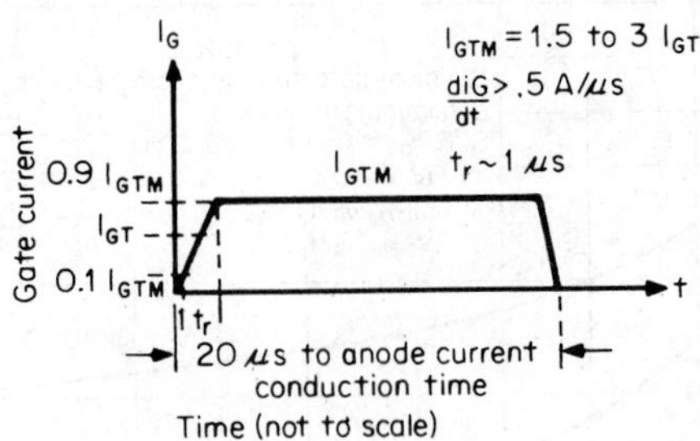


FIG. 22-57 Soft gate drive.

If a dc gate signal is used for an ac switch application, there will not be good dynamic gate action. No snubber discharge is possible after time zero.

The gate drive circuitry should have 1- to 2-A average 100-V diodes in series with the gate and across the gate cathode terminals as shown in Fig. 22-58. These will eliminate two possible failure modes of an SCR. The diodes in series with the gate will prevent negative gate current flow, while the diode across the gate cathode limits the reverse gate voltage, $V_{GRM}$, to ~2 V by diode clamping.

Provide open-circuit gate voltage over 20 V to prevent gate drive extinction. The instantaneous gate cathode voltage can exceed the source voltage in high *di/dt* applications.

Inductive loads can be troublesome if the gate drive is insufficient in amplitude or duration. Recommended practice is the use of the picket-fence or continuous drive. A picket fence is a high-frequency gate signal varying from 1 to 15 kHz, 200 to 50 μs wide, within a 60-Hz envelope such that the SCR is effectively continuously gated. The average gate current rating is maintained throughout conduction in continuous-drive circuits; the "back porch" anticipates worst-case power factor to ensure SCR latching and holding under all load conditions.

To prevent noise pickup in the gate connections, twist them together. Use either a twisted-wire pair or shielded cable from the gate pulse amplifier circuit. Locate the wires away from magnetic or high-current-carrying members in the circuit. Gate cathode leads should be as short as possible.

To minimize delay time variations between SCRs, use hard gate drive with as high a gate current rise time ($di_G/dt$) as possible; refer to Fig. 22-56.

Always use a resistor in series with each gate lead if triggering more than one SCR from the same source. Generally, 10 to 25 Ω is used to overcome gate cathode impedance variations.

***29. SCR di/dt Rating.*** The *di/dt* rating of an SCR defines the ability of a device to turn on a high rate of rise-current pulse. When *di/dt* rating of a thyristor is exceeded, excessive heat is created along the main cathode periphery during turn-on. This causes silicon-

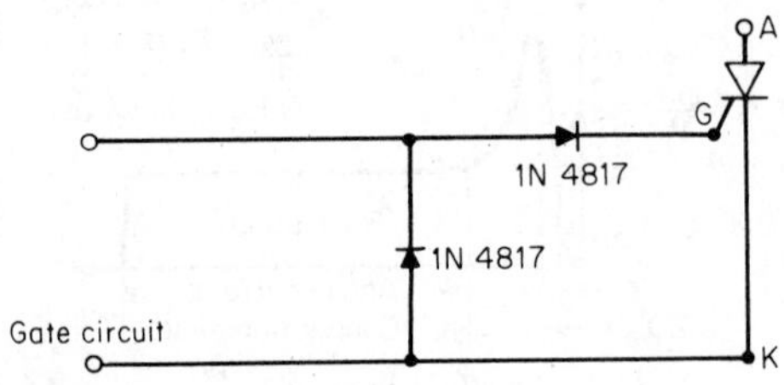


FIG. 22-58 Diodes in gate circuit.

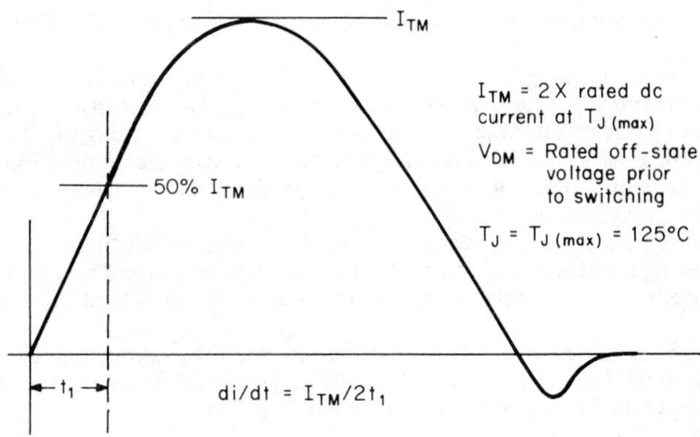


**FIG. 22-59** *di/dt* test pulse.

eutectic melt and catastrophic failure. For high-frequency devices, the initial hot-spot formation may increase turnoff time and decrease *dv/dt* capability, resulting in commutation failure.

The benchmark test pulse used to establish *di/dt* rating is shown in Fig. 22-59.[15] The rating test is run at 60 Hz and for 1000 h, with $t_1 \geq 1\ \mu s$, to establish the repetitive *di/dt* rating. For the test, the gate trigger pulse shall be specified as to pulse width, rise time, and gate source voltage and resistance.

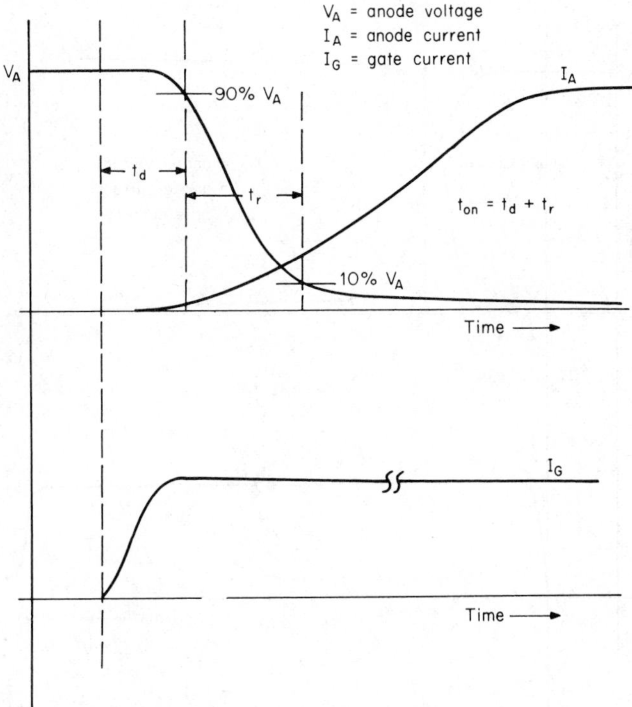


**FIG. 22-60** Thyristor turn-on waveforms, low *di/dt*.

For the nonrepetitive rating, the number of on-state current pulses is 300 at a repetition rate of 60 Hz.

Since the repetitive rating is defined only for a 60-Hz repetition rate, it is useful in low-frequency applications. For high-frequency applications, *di/dt* is intrinsically defined by the high-frequency characterization data. Current rate-of-rise failure in high-frequency systems usually occurs because of localized temperature rise followed by commutation failure.

***30. SCR Turn-on Time.*** SCR turn-on time is defined with reference to the switch anode-cathode voltage as shown in Fig. 22-60.

This definition does not address the true capability of a switch to turn on high-rate-of-rise currents. A better definition of turn-on time must specify current rate of rise. Thyristor turn-on time definition then offers three important areas for consideration. They are:

1. Turn-on current rate of rise is such that current rise time is much greater than anode voltage fall time. The definition of Fig. 22-60 is then valid, being the speed with which anode voltage falls for very low current density operation.
2. Turn-on current rate of rise is such that current rise time is comparable to voltage fall time. For these conditions, measurement of anode-cathode voltage fall time indicates switch turn-on speed; current density (and its variation) during turn-on is large.
3. Turn-on current rate of rise is such that current rise time attempts to be less than anode voltage fall time. Here, device limiting occurs and turn-on speed as measured by anode-cathode voltage has reached a limiting value. The observed current rate of rise for this condition is not what would exist as if the switch were ideal, but a reduced value allowed by the imperfect switch. Under a new turn-on time definition now adopted by the JEDEC JC-22 Committee on Thyristors, the JEDEC RS-397 (supplement 2) turn-on time stan-

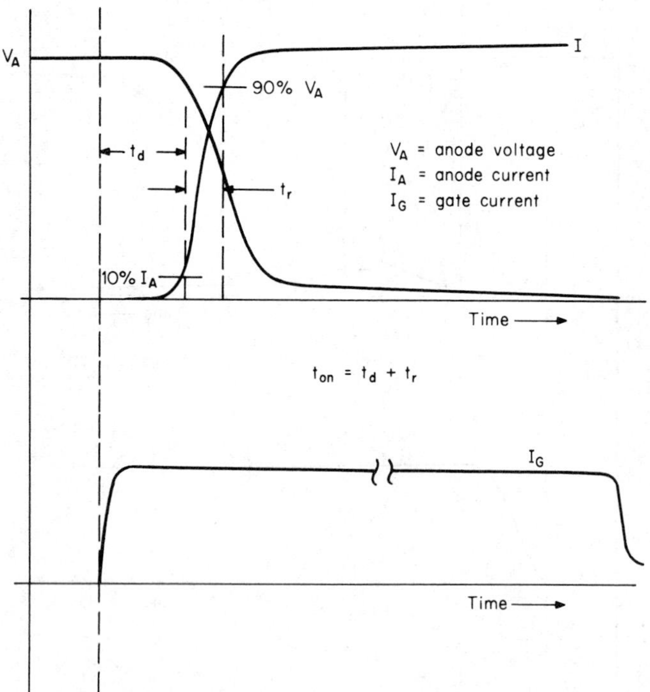


FIG. 22-61 Thyristor turn-on waveforms, high *di/dt*.

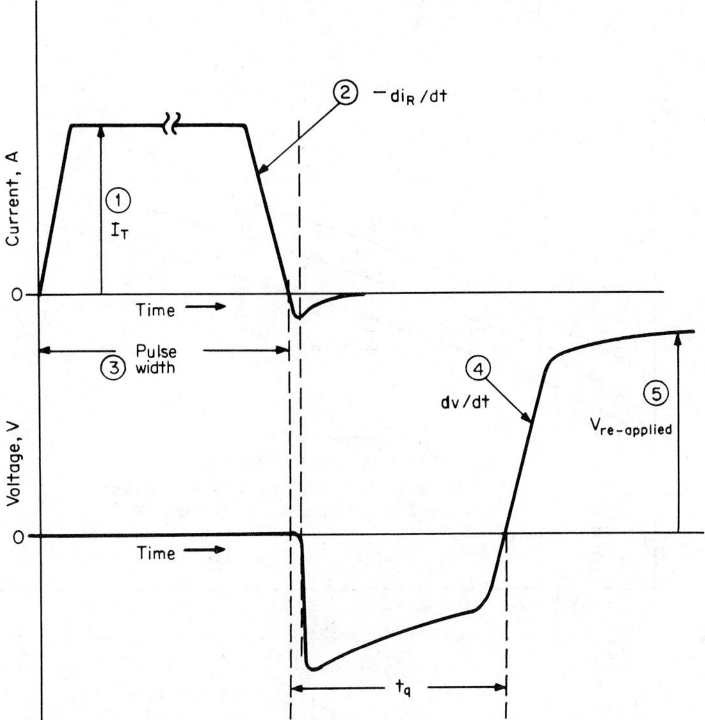


**FIG. 22-62** Turn-off parameters.

dard for thyristors is as indicated by Fig. 22-61. Although such operation causes high power loss, it is sometimes allowed in specially rated devices which switch high-rate-of-rise currents.

"Device limiting *di/dt* method of measurement of turn-on time should be specified for crowbar, pulse modulator, and laser diode applications."[16]

***31. SCR Turnoff Time.*** The parameters affecting SCR turnoff time are shown in Fig. 22-62.

The turnoff time is defined as the time from current zero crossing until anode-cathode voltage becomes zero while increasing toward forward bias. The relationships between turn-off time and other parameters are shown in normalized form in Fig. 22-63*a* to *f*. These results are typical measured $t_q$ values; the peak, midrange, and minimum values from a 25-sample lot are shown to indicate typical production spread. During $t_q$ testing, the $t_q$ interval is normally shortened until a collapse of reapplied *dv/dt* occurs. This is usually nondestructive if the collapse is followed by a limited current. For testing of some types of thyristors, such as high-voltage devices, $t_q$ testing even at a limited reapplied *dv/dt* source current can be destructive; in this case it may be necessary to test at a much lower level than the usual 0.8 $V_{DRM}$ value of most manufacturers' data sheets.

In some applications, SCR $V_{R,\text{peak}}$ is limited by a rectifier in reverse parallel with the commutated thyristor. Under these conditions, thyristor turnoff time is longer than if higher reverse voltages are applied.[17] Some curves showing turnoff times with inverse-parallel diode are presented in Fig. 22-64*a* and *b*.

***32. The Reverse Conducting Thyristor (RCT).*** A rectifier can be connected across a thyristor without inductance by fabricating the thyristor and rectifier together on a single silicon

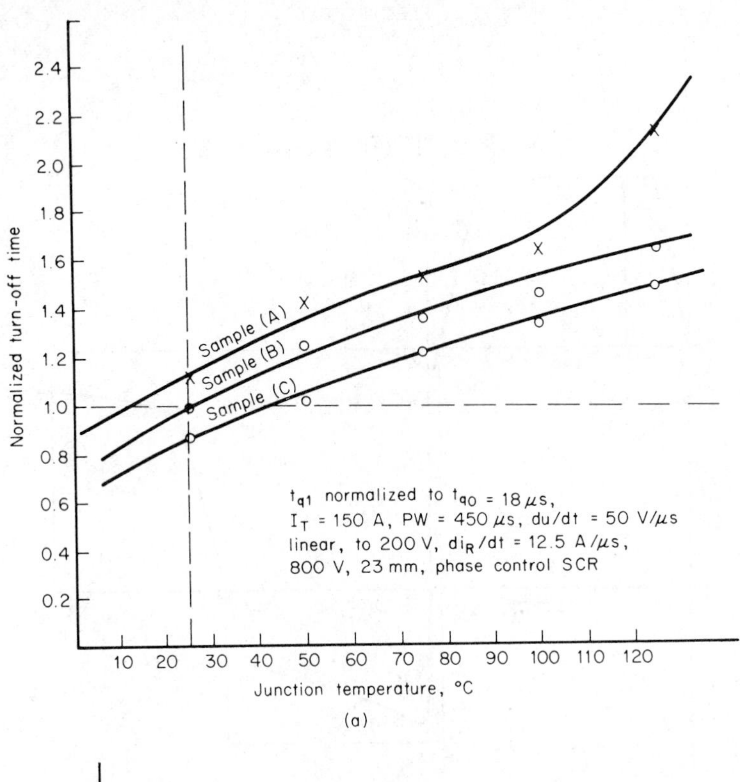


(a)

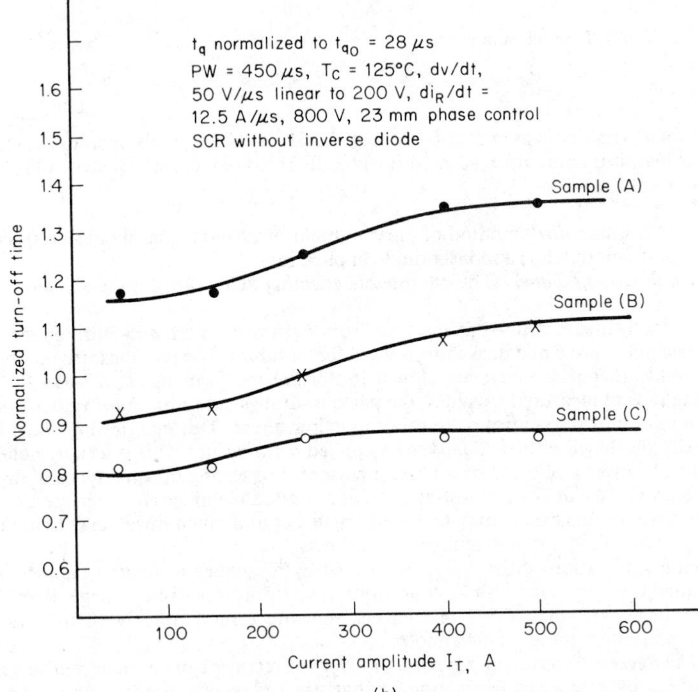


(b)

**FIG. 22-63** (*a*) Turn-off time versus temperature; (*b*) turn-off time versus forward current.

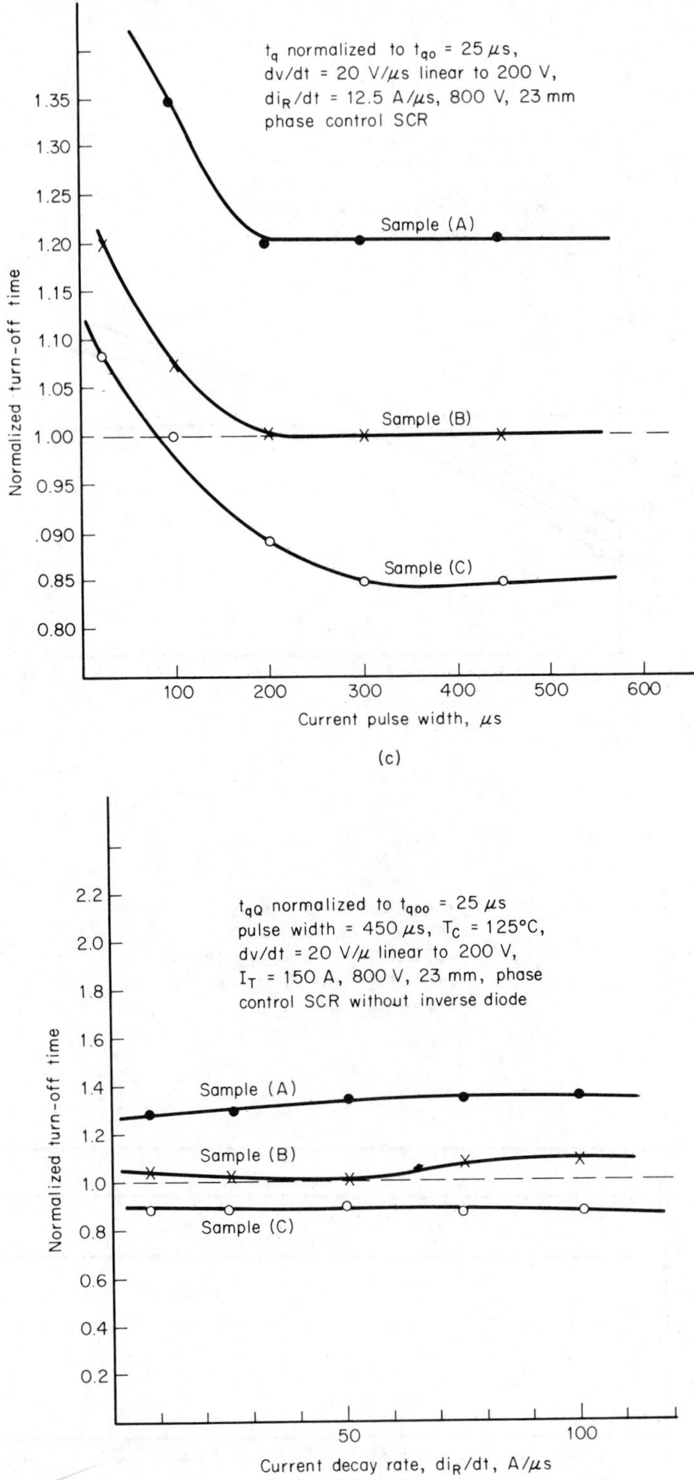


**FIG. 22-63** (*c*) turn-off time versus current pulse width; (*d*) turn-off time versus $di_R/dt$.

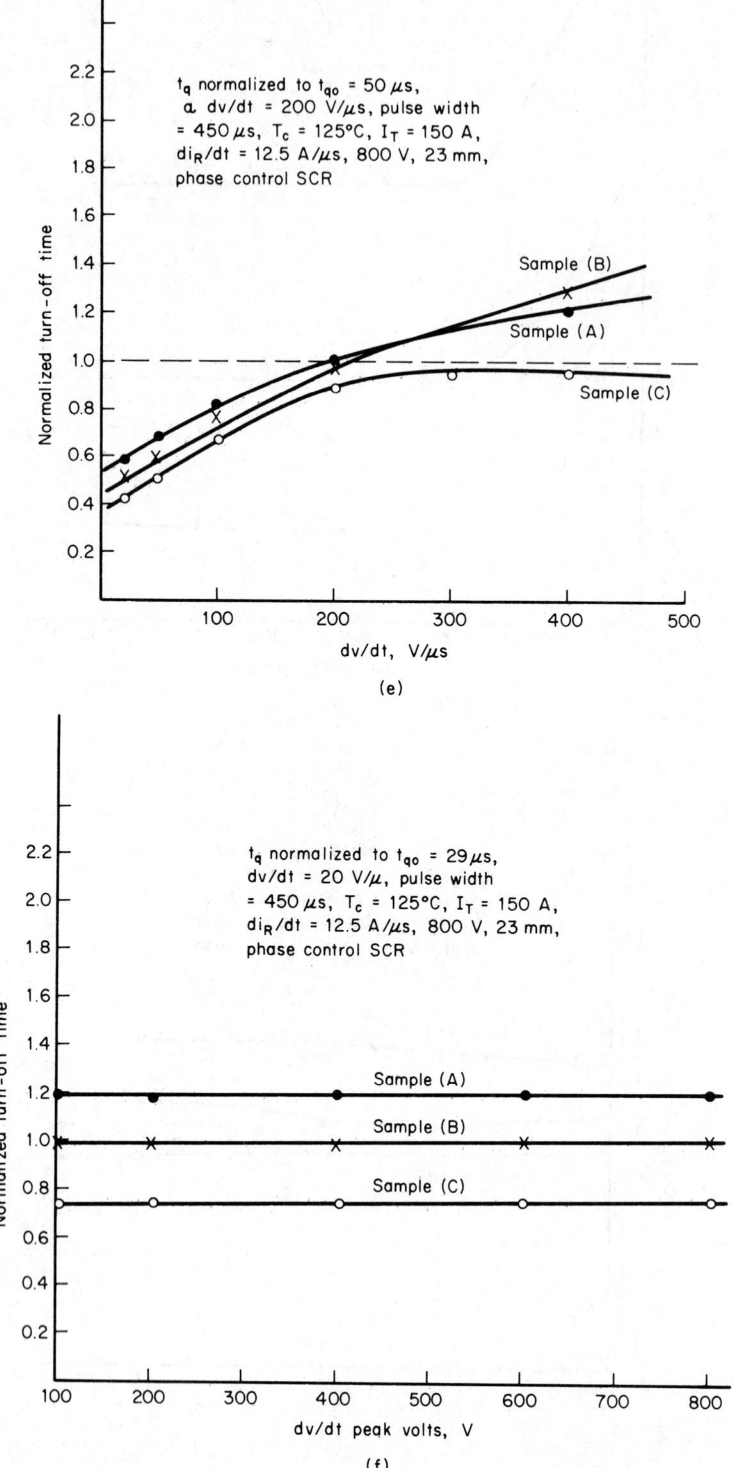


**FIG. 22-63** (*e*) turn-off time versus *dV/dt*; (*f*) turn-off time versus peak voltage.

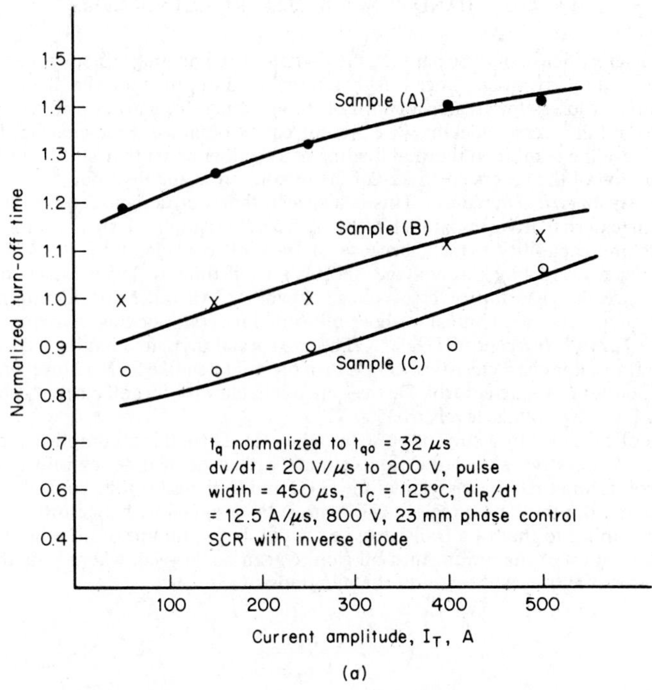


(a)

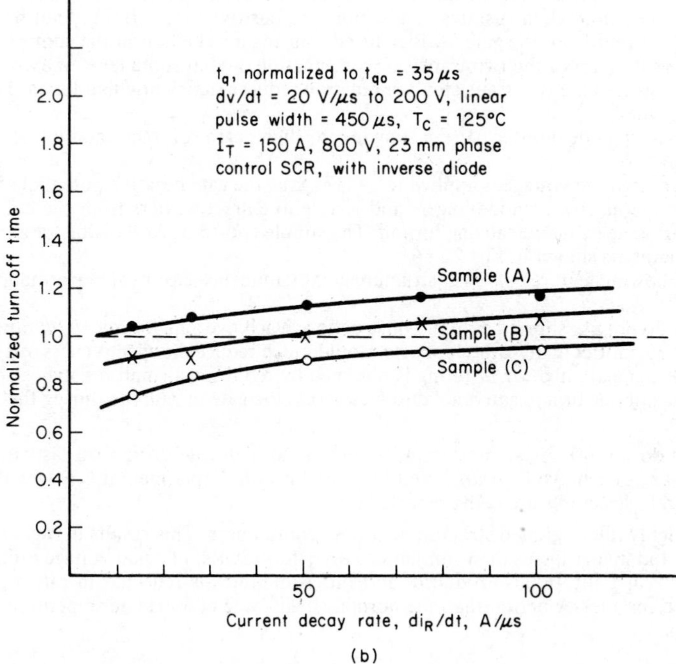


(b)

**FIG. 22-64** (*a*) Turn-off time versus current with inverse diode; (*b*) turn-off time versus $di_r/dt$ with inverse diode.

crystal. With a rectifier in inverse parallel, the thyristor need no longer block reverse voltage. This can be used to advantage, since it is then possible to improve the characteristics of the thyristor. Smaller losses for greater forward blocking voltage, or shorter recovery time than that achieved if full reverse blocking were present, can be obtained. Problems faced in applying the RCT are the combined thermal loading of a rectifier and a thyristor in a single crystal, and isolation of the reverse recovery of the rectifier from the thyristor.[18,19]

***33. The Asymmetric Thyristor.*** This is a special thyristor designed to be always operated with an external inverse-parallel rectifier. The thyristor is designed with very low reverse blocking capability, which, because of trade-off considerations, may be used to achieve higher forward blocking voltage, shorter turnoff time, or lower losses in a device identical in size that has full reverse voltage capability. All other considerations for the asymmetric thyristors are identical to those of normal reverse blocking thyristors.

***34. Gate Turnoff Switches (GTOs).*** GTOs are special thyristors having a combination of SCR and transistor characteristics. GTOs turn on and latch like SCRs; however, they can also turn off current by gate control. Devices are available with turnoff current capability to 1000 A and blocking voltage levels to 2500 V.

Turn-on of a GTO is by a current pulse into the gate. Turnoff is accomplished by switching 10 to 15 V negative across the gate cathode with a defined series impedance. This extracts current from the *p*-base region of the GTO and initiates turnoff. To facilitate extraction of carriers from the *p*-base region, GTOs must be designed with an emitter or cathode pattern very similar to that of a transistor, i.e., with a large amount of interdigitation. The following derivation of maximum turn-off gain, $G$, can be derived, where $\alpha_n$ is the gain of the *npn* transistor and $\alpha_p$ is the gain of the *pnp* transistor.

$$G = \frac{\alpha_p}{\alpha_n + \alpha_p - 1}$$

When a thyristor is on, the sum of $\alpha_n + \alpha_p$ approaches 1 and the $G$ becomes very large. Anode current alone then sustains conduction and positive gate current is not needed. By extracting current from the gate, $\alpha_n$ is reduced, causing a reduction in the sum of $\alpha_n + \alpha_p$. This lowers the gain of the two-transistor combination, and at some level of extracted gate current the gain of the two-transistor combination reduces sufficiently that forward blocking will be regained.

A typical cathode emitter pattern showing the high degree of interdigitation is shown in Fig. 22-65.

When a negative voltage is applied to a GTO gate, the gate negative potential effectively surrounds the narrow cathode fingers and is able to extract carriers from the emitter uniformly, reducing gain and causing turnoff. The simple two-transistor analog for a *pnpn* thyristor structure is shown in Fig. 22-66.

The following features of GTO structure greatly influence electrical performance:

GTOs do not use shorted emitter construction to achieve a high static *dv/dt* rating, since a shorted emitter (gate-cathode shorts) would much reduce the effectiveness of a negative turnoff signal. A high *dv/dt* rating is obtained by requiring a small resistance across the gate to cathode or a negative or zero bias across the gate to cathode during the static off period.

GTOs do not use regenerative gate structures, so high gate drive with fast rise time is necessary to achieve high *di/dt* and low loss turn-on. Experimental GTOs with amplifying gate structures are being reported.[20]

The highly interdigitated structure sacrifices emitter area. This results in higher forward drops and higher losses than similar-size thyristors (SCRs) of equal voltage rating. Surge current capability is also reduced, and this sometimes is reflected in ratings that are defined for a lesser period than the normal single-cycle of 60-Hz nonrepetitive thyristor ratings.

***35. GTO Voltage Ratings.*** Repetitive GTO voltage ratings are applied in the forward and reverse direction in a manner identical to SCRs. These voltage ratings are static and are not applicable when the GTO is turning on or off. Some data sheets show forward rep-

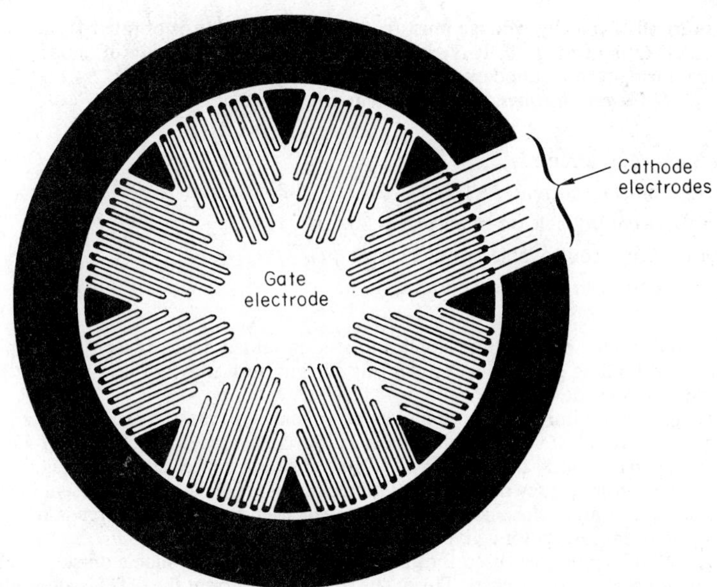


**FIG. 22-65** Interdigitated GTO structure. *(Ref. 27.)*

etitive voltage to be dependent upon gate bias conditions. In addition, there are variations in the maximum junction temperature at which the blocking level is guaranteed, with some manufacturers reducing $T_{J,\max}$ slightly below the normal SCR level of 125°C.

Some manufacturers do not specify the reverse blocking characteristics of the GTO. The structural reason for this is that these devices are fashioned with "anode shunts." These are very low impedance shunts in the *n* base of the anode equivalent *pnp* transistor, resulting in enhanced *dv*/*dt* and switching speed. When anode shunts are present, only very low reverse blocking voltage is possible; GTOs of this type resemble *npn* transistors in reverse blocking capability. A typical structure is shown in Fig. 22-67. GTOs without anode shunts will show full reverse voltage rating.

Most GTO ratings require a decrease in peak voltage level during turn-on and turnoff. Both restrictions are related to the ability of the GTO to absorb the peak energy during turn-

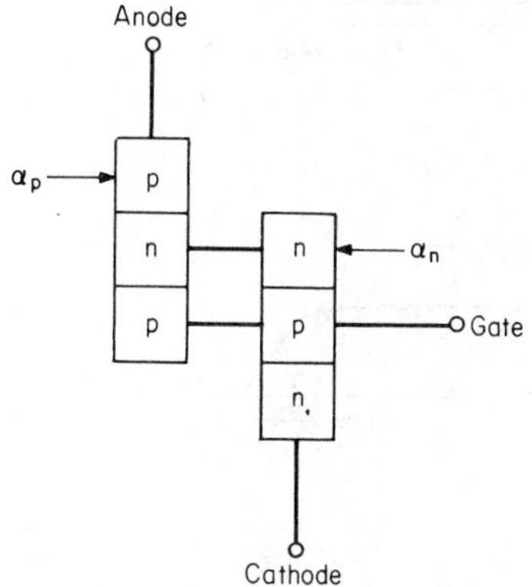


**FIG. 22-66** Two-transistor analog of GTO.

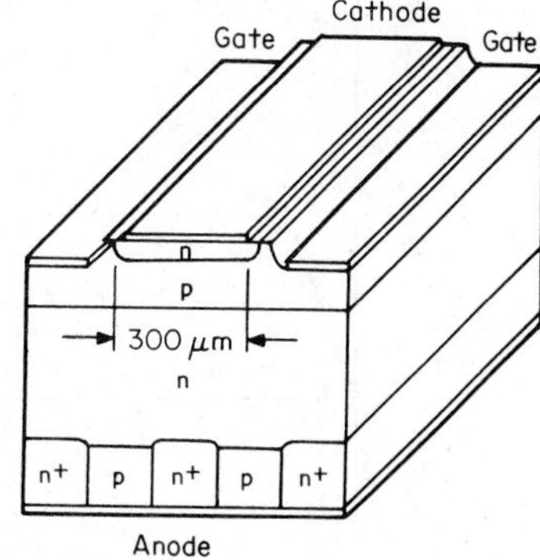


**FIG. 22-67** Anode shunts in GTO. *(Ref. 27.)*

on and turnoff. Typically, voltage must be reduced to two-thirds of rated $V_{DRM}$ during the time the GTO is turning off. It is also essential that the rate of rise of anode voltage be restricted during turnoff; anode voltage snubbers are always needed.

***36. GTO Current Ratings.*** Published current ratings of GTOs are

RMS on-state current, $I_{T,\mathrm{rms}}$

Average on-state current, $I_{T,\mathrm{av}}$, $I_T$

Repetitive controllable on-state current, $I_{TQ}$, $I_{TC}$, $I_{AC}$

Nonrepetitive, controllable on-state current, $I_{TCSM}$

Nonrepetitive, surge on-state current, $I_{TSM}$

The rms on-state current limit applies to stud-mounted devices where the chip is connected to the binding posts by wires. An rms current limit may not be given for "hockey-puck"-type package devices.

Average current limits may be found as dc on-state current, the average current of a 1-kHz square wave of 50% duty cycle, or the half-cycle average of a 50-Hz half sine wave. They are not particularly meaningful. GTOs need rating curves similar to Fig. 22-49, the typical SCR high-frequency rating system. Since much GTO data are at present incomplete in this important area, close cooperation between user and manufacturer is necessary to determine current ratings for a given application.

A variety of symbols are used for maximum repetitive controllable on-state current; as of yet no industry standard exists. This is the maximum current the GTO is allowed to turn off by gate control with manufacturer-specified maximum junction temperature, specified negative gate turnoff voltage, specified gate current rise time, and maximum specified rate

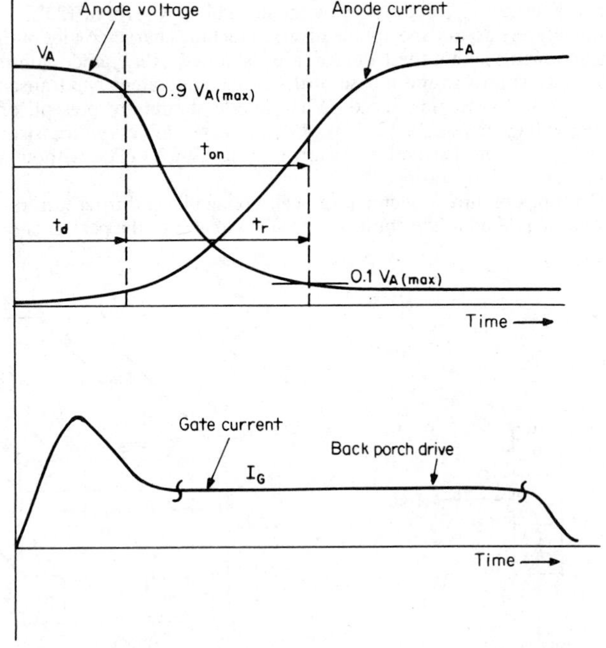


**FIG. 22-68** GTO turn-on.

of rise of anode voltage during anode current fall. The nonrepetitive controllable on-state current is, as implied, a current which may be turned off only occasionally. If applied in a repetitive manner, $T_{J,max}$ could be exceeded and permanent damage would result. Permanent damage to the GTO will result if an attempt is made to turn off more current than is within device capability. It is typical of present GTOs that average current ratings are much less than repetitive controllable current limits.

Nonrepetitive surge current ratings apply for manufacturer-specified gate drive; some early data sheets show this rating as one-half cycle of 50 Hz, but at least one manufacturer uses a 1- to 5-ms pulse for surge current specification.

***37. GTO Turn-on.*** Some GTO turn-on definitions are shown in Fig. 22-68. The definition of turn-on time may differ from one manufacturer to another. GTOs have good turn-on speed and *di*/*dt* capability, but most do not have a regenerative gate structure. Thus to achieve high turn-on speed from the fingered structure, high gate drive must be used with high rise time. Since holding current may be relatively high compared to SCRs, a back-porch drive may be required to forestall a drop out of conduction if currents fall below the holding level and then rise above it again during the desired conduction interval. Debate still exists

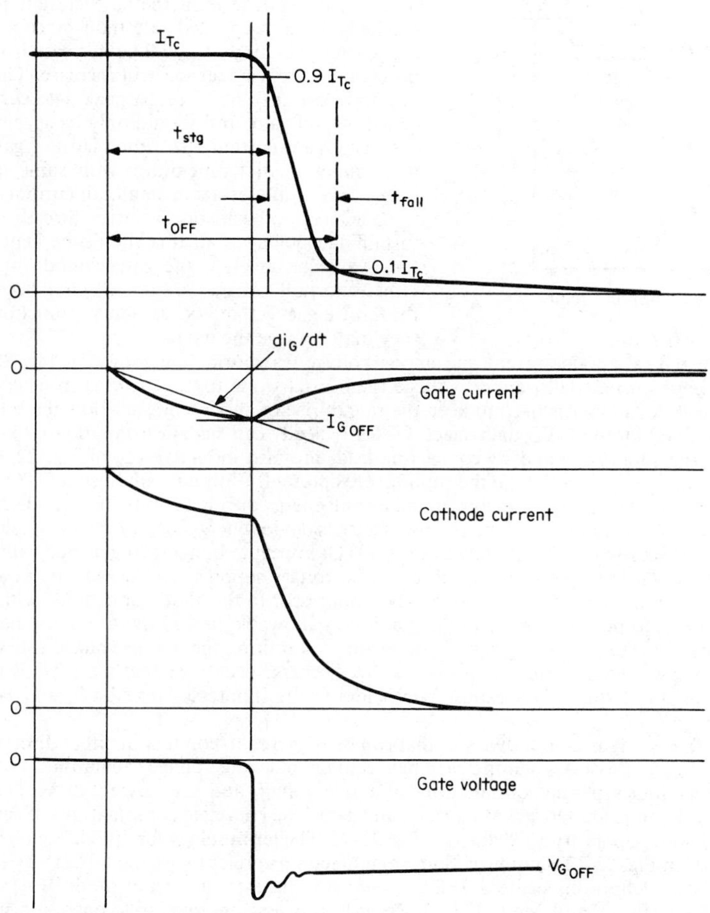


**FIG. 22-69** GTO turnoff.

as to whether this re-turn-on may be safely accomplished for a relatching without the use of a new spiked drive. It is thought that some fingers may drop out of conduction during current decrease below the holding level but not be turned back on by the low back-porch drive during relatching. If this is the case, few fingers may carry a large re-turn-on current. The problem may be overcome by a sophisticated drive system which senses GTO anode-cathode voltage and applies harder drive when anode voltage rises.

***38. GTO Turnoff.*** The important variables of GTO turnoff are shown in Fig. 22-69. Turnoff is initiated by applying the turn-off source voltage to the GTO gate cathode, as in Fig. 22-70. When S1 is closed, current from the gate is diverted from the cathode. Cathode current falls with the rise of extracted gate current. The rate of rise of gate current is a circuit function, to a first order given by $-V_{G,off}/L_s$. When sufficient gate current is extracted to meet the turnoff condition, anode current will abruptly fall. After anode current reaches 10% of maximum value, the fall time is considered to end and a GTO tail current period is considered to commence. The gate does not support reverse gate voltage until the gate current peak is reached. If the distributed gate inductance is small, gate current rise time, $t_{stg}$, will be decreased, and the peak gate current to turnoff will increase. The ratio of anode current turned off to peak gate current is called turnoff gain, and should only be specified for a specific gate current rise time. Turnoff gain is a weak function of temperature; with some devices long storage time results in small tail currents.

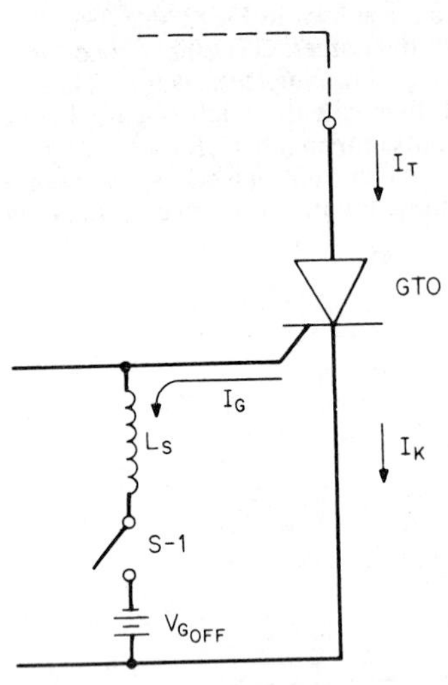

**FIG. 22-70** Turn-off circuit.

GTO turn-off capability varies directly as the turnoff gate voltage, but this should be kept below the avalanche level of the gate cathode junction. Greater turnoff voltages are not effective because of the limiting effect of the avalanche junction, and they may damage the device.

Some typical anode current and anode voltage waveforms are shown in Fig. 22-71*a*. When anode current falls, anode voltage rises in part due to $L \times (di/dt)$ from parasitic inductance. A snubber is used to keep the rate of rise of anode voltage below the tolerable level specified by the GTO data sheet. GTOs typically can sustain from 100 to 200 V/μs. Because the suppressor and its connecting leads are also inductive ($L_d$ of Fig. 22-71*b*), a small spike occurs even though the suppressor is present. Without a snubber, a GTO would fail as a result of its inability to absorb the turnoff energy during the time that anode current falls and anode voltage rises. The snubber restricts device energy loss during turnoff by limiting the rate of rise of anode voltage. Larger GTOs appear to be achieving turnoff with faster fall times, requiring even higher levels of anode voltage suppression. At some point, it may be physically impossible to connect a GTO suppressor to the anode or cathode with short enough leads to produce less than the maximum allowable turnoff *dv*/*dt*. Under these circumstances, GTO paralleling may be necessary.[21] Limiting the anode voltage rate of rise with $R_s$ and $C_s$ requires that $C_s$ of Fig. 22-71*b* discharge each time that the GTO is turned on. For higher-frequency operation, power loss in the suppressor resistor $R_s$ may be very high.

***39. Triacs.*** A triac is a thyristor that can be triggered to conduct in either direction, or if nontriggered, possesses symmetrical bidirectional blocking voltage capabilities. For these conditions, four separate combinations of anode voltage and gate current exist. The electrodes of the triac are termed $MT_2$, $MT_1$, and gate. The complete combinations of gate current and voltage polarity are defined in Fig. 22-72. The terminology for "mode" of operation as defined in Fig. 22-72 is not firm. Some manufacturers prefer to use the I, II, III, IV system of quadrant definition, while a JEDEC standard suggests the terms Mode I(+), Mode I(−), Mode III(−), and Mode III(+). Regardless of terminology, triac operation in each mode or quadrant is defined by Fig. 22-73. Operation in quadrant I is identical to SCR operation, quadrant II is defined as junction gate mode, and in quadrants III and IV the

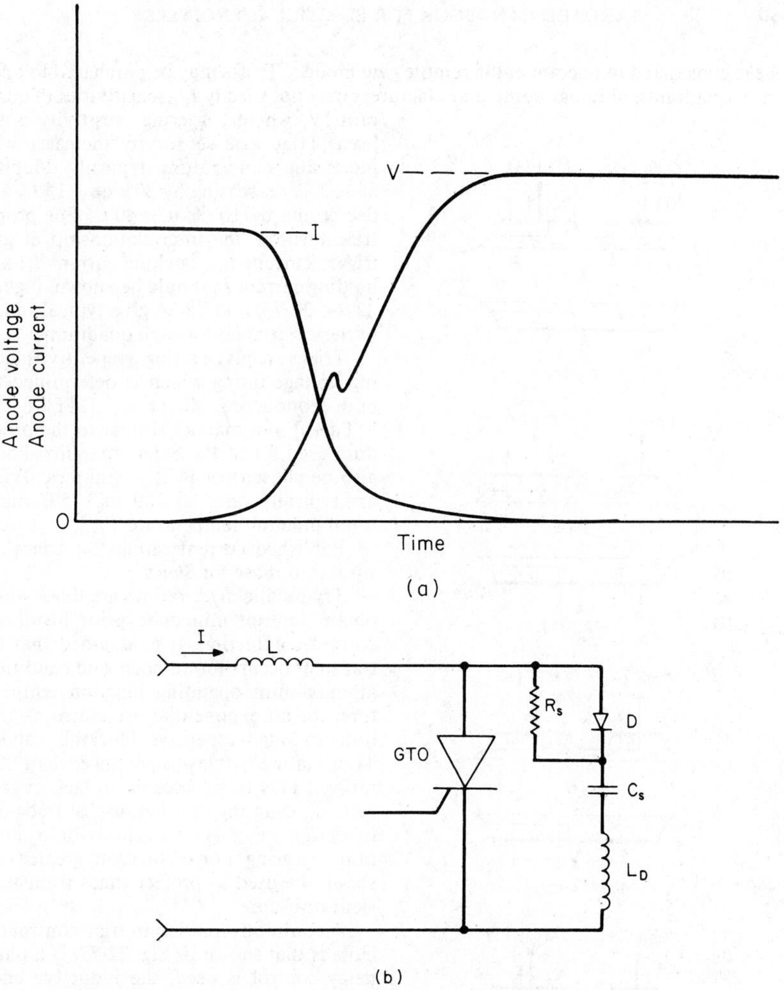

**FIG. 22-71** (*a*) Typical turn-off waveforms; (*b*) snubber showing parasitic inductance.

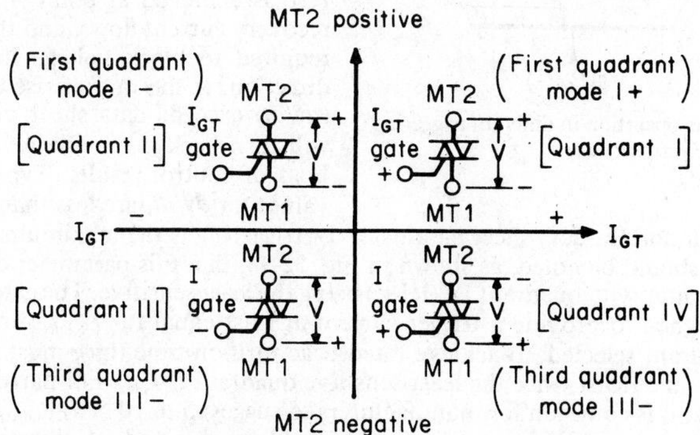


**FIG. 22-72** Triac triggering modes. *(Teccor Electronics Inc.)*

device is considered to operate in the remote gate mode.[22] Triacs may be purchased to operate in all quadrants, although some manufacturers may not specify $I_{GT}$ sensitivities in quadrant IV, where triggering sensitivity is the least. Triac gate sensitivity increases with increasing temperature, typically doubling at 100°C, decreasing by 30% at −15°C, and decreasing by 100% at −40°C. For proper triac turn-on, the interrelationship of gate trigger current $I_{GT}$, latching current $I_L$, and holding current $I_H$ should be known. Figures 22-74, 22-75, and 22-76 give typical ranges of these variables for each quadrant.

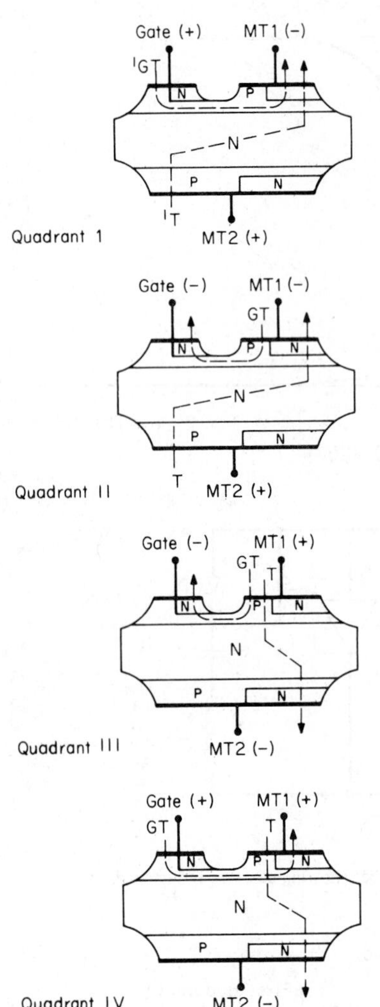


**FIG. 22-73** Triac operation in different triggering modes. *(Teccor Electronics Inc.)*

Triacs are given a single repetitive blocking voltage rating which is determined for both conducting directions [$MT_2(+)$ or $MT_1(+)$] in a manner similar to the procedure used for SCRs. Safety margins should also be chosen per SCR techniques. Triacs are typically rated at 100 to 115°C maximum junction temperature $T_{J,\max}$.

Published current ratings for triacs are similar to those for SCRs.

Triac static $dv/dt$ ratings are those which obtain for no immediate prior history of current conduction. It is assumed that the transient occurs for an open gate condition, at maximum operating junction temperature, for an exponential waveform asymptotic to rated repetitive blocking voltage. Triac static $dv/dt$ levels are lower than SCR ratings. This is not because of lack of gate-cathode shunting, but because of trade-offs in design necessary to achieve four-quadrant triggering. For this reason, greater care should be used to protect triacs from transient turn-on.

A common situation in triac control circuits is that shown in Fig. 22-77. If a phase delay control is used, the inductive effect causes triac current to carry over conduction beyond the point at which line voltage ($e_{in}$) reverses. The result is that as current zero is achieved at point $A$, some reverse recovery current flows, and then the triac is required to block voltage in the opposite direction. If the rate of rise of this voltage ($dv/dt$) exceeds data sheet $dv/dt_c$, the triac will turn back on in the reverse direction. Loss of control results. Typical data sheet values for $dv/dt_c$ are less than 10 V/μs.

The most important data sheet parameter for triggering is the maximum dc trigger current $I_{GT,\max}$. It should be noted, as shown in Fig. 22-74, that this parameter changes for the different quadrants, with quadrant IV (III+) being the least sensitive. The gate trigger circuit should be designed to provide a trigger current at least equal to $I_{GT,\max}$ for the particular operating quadrant selected; to achieve rated triac turn-on time (tgt), most manufacturers suggest a drive of at least twice the least-sensitive quadrant $I_{GT,\max}$. For narrow-pulse-width drives (≤10 μs) it is common for manufacturers to suggest 5 to 10 × $I_{GT,\max}$ overdrive. The particular manufacturer's recommendations should be obtained for this condition, since data sheets do not consider this need.

| Operating mode / Type | Quadrant I | II | III | IV |
|---|---|---|---|---|
| 4 A triac | 1 | 1.2 | 1.5 | 5 |
| 10 A triac | 1 | 1 | 1.5 | 3 |

Example: 4 A triac
if $I_{GT}$ (I) = 10 mA
then, $I_{GT}$ (II) = 12 mA
$I_{GT}$ (III) = 15 mA
$I_{GT}$ (IV) = 50 mA

**FIG. 22-74** Triac triggering currents. *(Teccor Electronics Inc.)*

Triacs are available in ratings to 80 A with voltages to 700 V. The standard areas of application are at power-line frequencies for light dimmers or temperature controls. Many triacs are presently now applied in solid-state relays (SSRs) in combination with a light isolator and a zero crossing detector. Units used for SSR duty usually are of the sensitive-gate variety and may be purchased upon special request. With the voltage ratings available, triacs are suitable for use with ac line voltages of 120 and 240 V ac. Although triacs are generally considered as 50- to 60-Hz control elements, it has been observed that triacs could successfully commutate a resistive load current in an ac circuit where the fundamental line frequency was 20 kHz.

***40. Reverse Blocking Diode Thyristor (RBDT).*** The RBDT is a two-terminal *pnpn* structure which is turned on by a fast-rising trigger directly applied across the anode/cathode. In the *dv/dt* mode of turn-on, a fast broad-area turn-on is achieved, yielding a switch suitable for very high *di/dt* duty. The RBDT structure is shown in Fig. 22-78. The small pinhole *p*-base emitter shunts used in the design are not shown. To achieve rated turn-on, the RBDT must be triggered with a rated trigger *dv/dt,* as defined in Fig. 22-79, of 5000 V/μs.[23] Presently, most devices are rated at 800 V blocking voltage and are recommended by the manufacturer to switch from no higher than 600 V. Loss characterizations are available for 3, 10, 20, and 30 μs conduction times. To achieve higher switch voltages, RBDTs are connected in series with no compensating elements. A typical five RBDT stack can switch from 3000 V and produce a 3-μs modulator pulse of 2500-A peak with a current rate of rise

| Operating mode / Type | Quadrant I | II | III | IV |
|---|---|---|---|---|
| 4 A triac | 1 | 5 | 0.7 | 1.2 |
| 10 A triac | 1 | 3 | 0.6 | 0.5 |

Example: 10 A triac
if $I_L$(I) = 20 mA
then, $I_L$(II) 60 mA
$I_L$(III) 12 mA
$I_L$(IV) 10 mA

**FIG. 22-75** Triac latching currents. *(Teccor Electronics Inc.)*

of 2500 A/$\mu$s. Losses may be improved by a factor of 20 to 25% by increased trigger *dv/dt*, but the 5000 V/$\mu$s required trigger has proved adequate through life testing. A typical trigger source, hold-off isolation diode, and load network are shown in Fig. 22-80.

| Type \ Operating mode | $I_H(+)$ | $I_H(-)$ |
|---|---|---|
| 4 A triac | 1 | 0.9 |
| 10 A triac | 1 | 0.7 |

Example: 10 A triac
if $I_H(+) = 20$ mA
then, $I_H(-) \approx 14$ mA

**FIG. 22-76** Triac holding currents. *(Teccor Electronics Inc.)*

RBDTs must be protected from excessive reverse recovery transients by suitable antiparallel rectifiers, not shown in Fig. 22-80. The reverse recovery capability of the 23-mm RBDT is no greater than that of a similar-sized thyristor, and although it may switch 2500 A/$\mu$s in the forward direction, high current decay rates require careful consideration to avoid excessive reverse recovery transients.

***41. Power MOSFETs (Metal-Oxide Semiconductor Field-Effect Transistors).*** Metal-oxide semiconductors differ from junction devices in that they do not possess barrier potentials or junctions, but depend on resistance modulation by the pinching effect of an electric field. Large devices of substantial power capability may now be found in the marketplace. These devices can feature currents of over 100 A and voltages approaching 1 kV (but not concurrently). Devices with these capabilities deserve consideration as power semiconductors. A basic field-effect transistor (FET) is shown in Fig. 22-81. Conductivity from drain to source is controlled by controlling the electric field developed from gate to source. At zero voltage from gate to source, conductivity from drain to source is very low. Figure 22-82 shows a typical drain-source blocking characteristic for zero gate-source voltage. At some voltage level determined by the resistivity of the silicon material used for the *n* channel, the drain source will avalanche and current can increase without bound. Before avalanche, the intrinsic leakage of the silicon channel is low and the device is considered off.

As gate-source voltage ($V_{GS}$) of the FET is increased, the characteristics shown in Fig. 22-83 develop. "As the drain-to-source voltage ($V_{DS}$) is increased, the current increases almost proportionally, though in practice drain-to-source resistance does increase at higher currents. At a certain current level, however, a channel pinchoff effect is reached within the device, and the operating characteristic moves into a constant current region."[24] Power MOSFETs are given the electrical symbols shown in Fig. 22-84, where the control variable (gate voltage) is compared with the control variable of a transistor (base current). Since the gate voltage controls drain current, the MOSFET closely resembles a vacuum tube where the output-input transfer function is best described by a transconductance (gm). Transconductance is the change of drain current caused by a 1-V change in gate voltage. A typical gate voltage-drain current transfer function for a power MOSFET is shown in Fig. 22-85. The transconductance is the slope of the transfer characteristic and approaches a constant at higher gate-source voltage.

A small-periphery MOSFET structure has a low current rating. Larger-current-capability devices are realized by paralleling a multiplicity of small source cells on one substrate, each of which contributes to the total current capability of the power MOSFET. A typical power MOSFET structure is shown in Fig. 22-86. The FET current is labeled "transistor current," and the heavy arrow shows the conduction path from source to drain of the built-in parasitic diode. The heavy banded lines are the *n* channel; the insulating metal oxide of the gate is also well illustrated. Because of this oxide insulation, a very high impedance exists between gate and source, with low dc leakage current.

The capacitance existing between gate and source ($C_{GS}$) and between drain and gate ($C_{DG}$) has significant impact upon dynamic switching performance. MOSFET capacitances are shown in Fig. 22-87. The impedance associated with the gate is important when dynamic changes are considered. If a MOSFET is blocking voltage and a voltage transient is induced across the drain-source terminals, the gate capacitance will divide the transient such that a portion of the drain transient will occur across the gate source. For this reason and also to guard against static transients when handling MOSFETs, an external Zener should be used

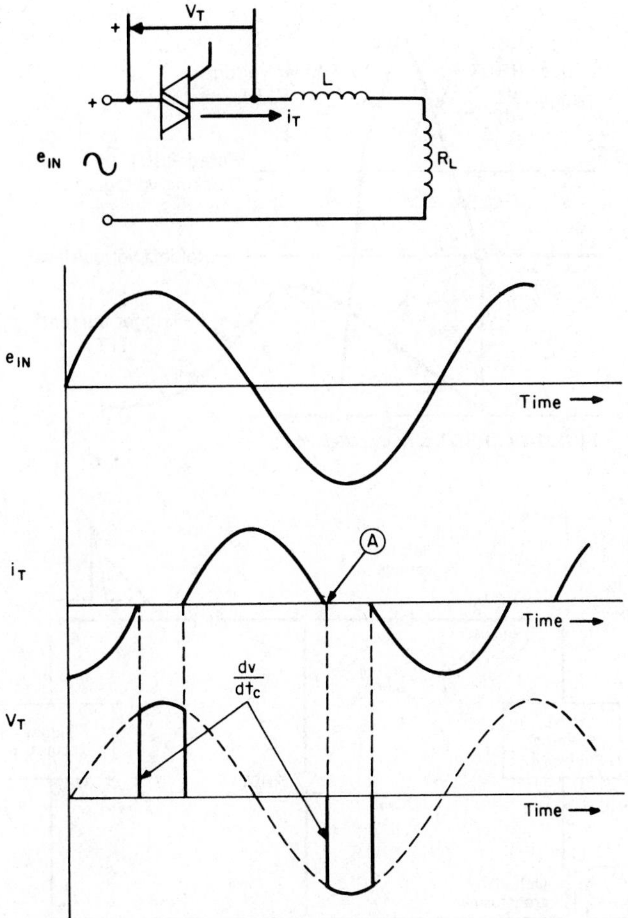


**FIG. 22-77** Triac waveforms with inductance load.

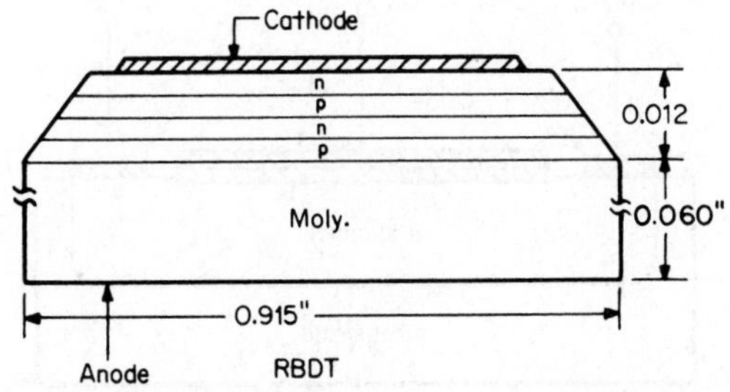


**FIG. 22-78** Reverse-blocking-diode thyristor structure. *(Ref. 26.)*

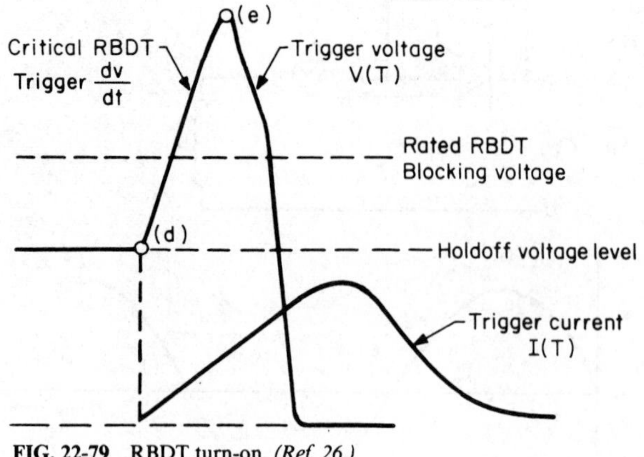


**FIG. 22-79** RBDT turn-on. *(Ref. 26.)*

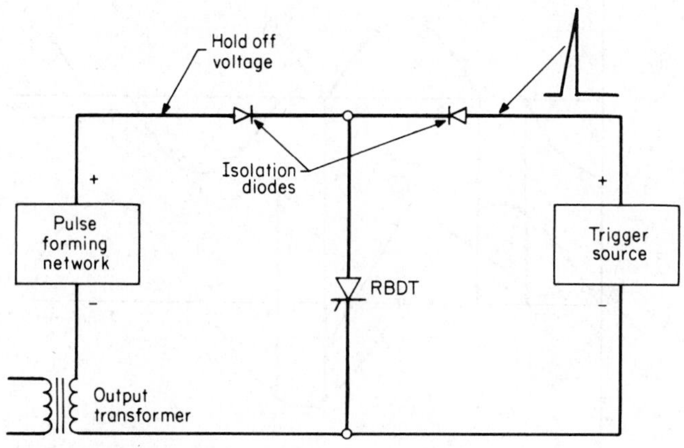


**FIG. 22-80** RBDT circuit. *(Ref. 26.)*

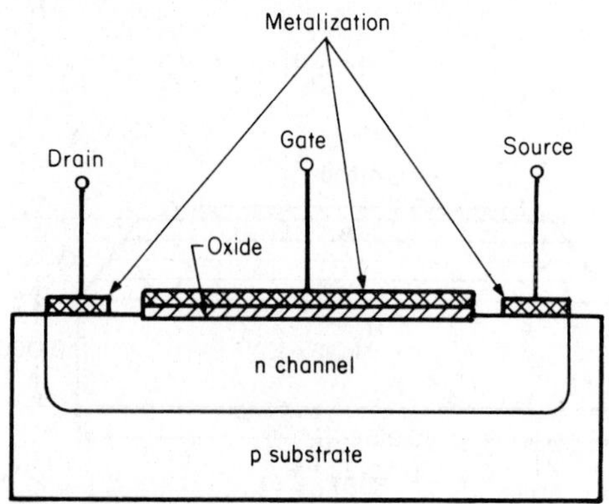


**FIG. 22-81** MOSFET structure.

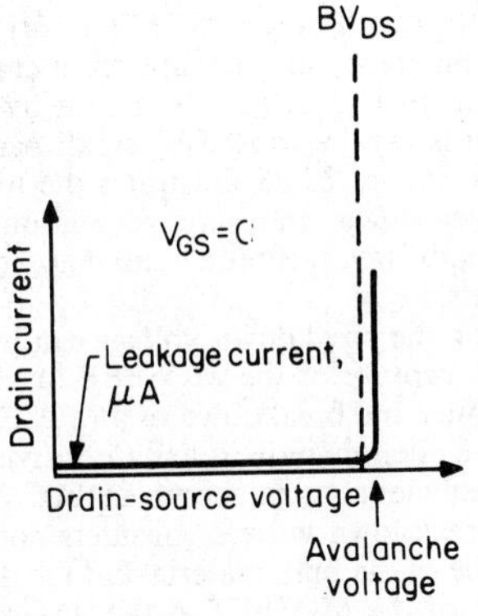


FIG. 22-82 MOSFET blocking characteristics.

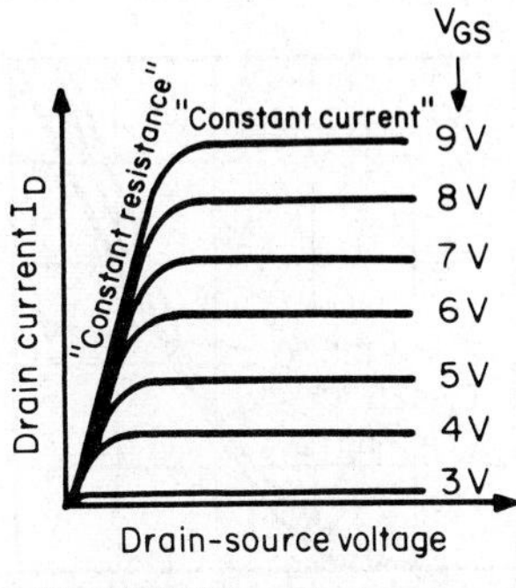


FIG. 22-83 MOSFET conduction characteristics. *(International Rectifier.)*

across the drain source to limit gate-source voltage to within the typical ±20-V maximum limits. If the gate-insulating oxide is punctured by voltage, the damage is permanent. In addition, before a MOSFET can be turned on, the gate-to-source and gate-to-drain capacitances must be charged. These capacitances, combined with the driving-source impedance, form the main time constants responsible for a MOSFETs turn-on and turnoff response. The gate-to-drain capacitance is actually smaller in value than the gate-to-source capacitance, but since the voltage excursion of the drain may be much larger than that of the gate, the gate-to-drain capacitance may require the most charge. In order to switch a MOSFET at high speed, the capacitances must be charged and discharged with transient currents substantially higher than the static gate leakage current. Most data sheets document the total gate charge $Q_g$ required to switch the transistor; peak gate transient current can be calculated once $Q_g$ is known.

MOSFET conduction is by "majority" carriers. This allows MOSFETs to conduct instantly, since there are no transit-time effects, and accounts for the positive temperature coefficient of MOSFET on-resistance, $R_{D,\text{on}}$. Positive-temperature-coefficient elements have inherent current ballasting or sharing properties making power MOSFETs second-breakdown-free since second breakdown is a current-crowding phenomenon. This same characteristic allows the direct paralleling of devices for higher current capability, with good current sharing between devices.

In the switching direction, MOS transistors are rated with a minimum breakdown voltage at $T_C$ = 25°C; this is designated drain-source breakdown voltage, $BV_{DSS}$, and voltage is defined as shown in Fig. 22-82, at some small drain leakage current. Typical leakage-current values range from 0.25 to 4.0 mA; by definition the breakdown voltage is established for $V_{GS}$ = 0.

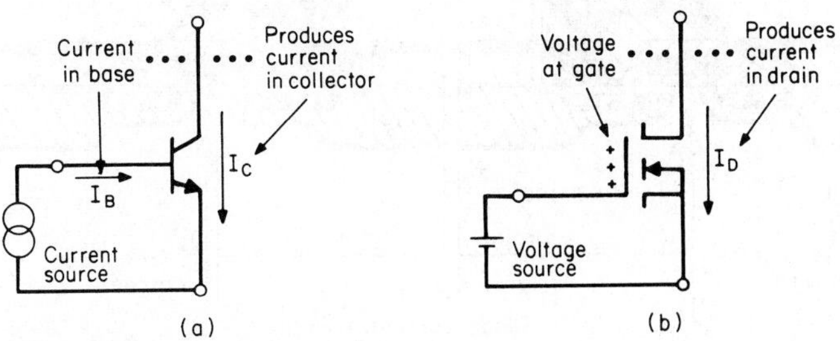


FIG. 22-84 Bipolar and MOSFET symbols: (*a*) bipolar transistor; (*b*) MOSFET.

The designer should be aware that voltage difficulties with power MOSFETs originate as a result of MOSFET turnoff conditions. With the capability to interrupt drain currents within 100 ns or less, large, destructive overvoltage transients can be caused by small parasitic inductances. Figure 22-88 illustrates the mechanism of overvoltage transient generation, and Fig. 22-89 indicates techniques used to control the transients.

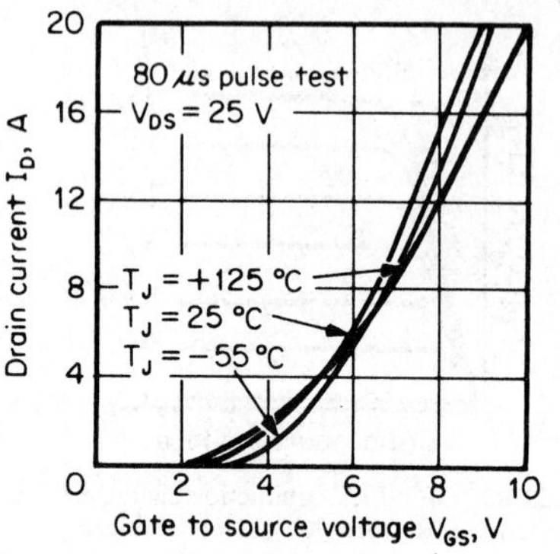


**FIG. 22-85** MOSFET transfer function. *(International Rectifier.)*

Exceeding the breakdown voltage can lead to catastrophic rupture of the MOSFET insulating dielectric. Since the breakdown of a MOSFET is an avalanche phenomenon, it has a positive temperature coefficient, as is shown in Fig. 22-90. The rated breakdown voltage considers not only the avalanche of the bulk material but the dielectric strength of the MOSFET at the surface and near the gate.

Power MOSFETs are thermally limited devices, and current ratings are determined by choosing the proper heat sink such that the bulk temperature $T_J$ is always less than $T_{J,\max}$ (150°C).

To determine the bulk temperature of the MOSFET, all losses must be known. These losses are as follows:

Static blocking losses due to $I_{CS}$, $P_L$ at $V_{GS} = 0$
Gate power losses, $P_G$
Switching losses, $P_S$
Conduction losses, $P_C$

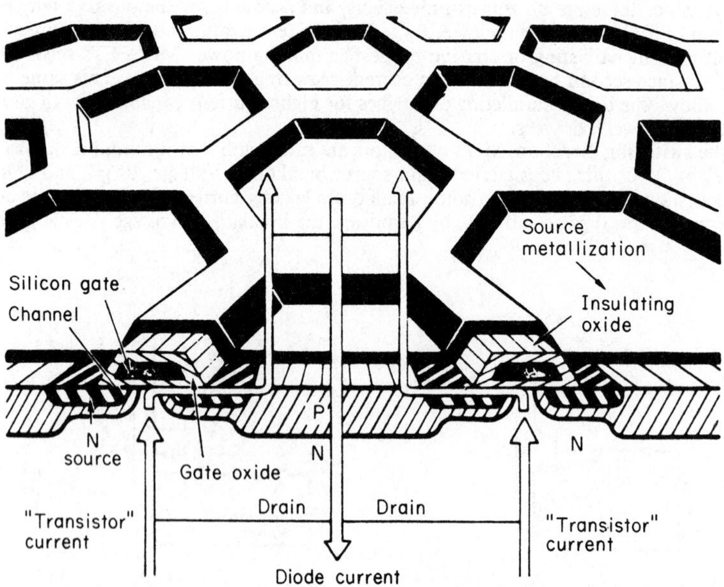


**FIG. 22-86** Power MOSFET structure. *(International Rectifier.)*

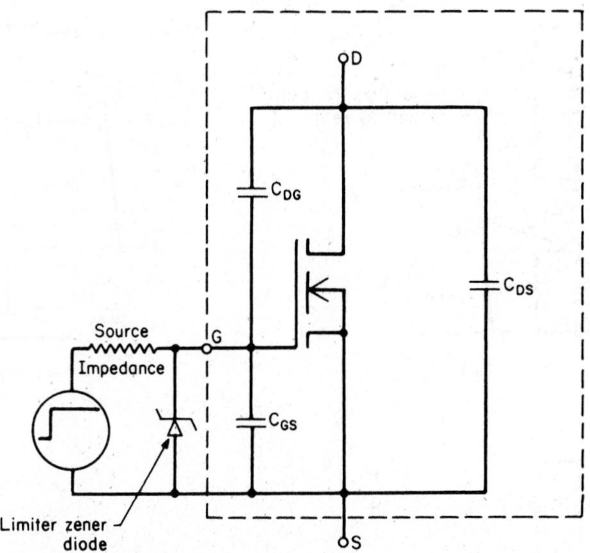


**FIG. 22-87** MOSFET capacitances. *(International Rectifier.)*

Static blocking losses $P_L$ are generally small and may be ignored, but may be troublesome if $V_{GS} \neq 0$. This can occur when the threshold of another switch is present from IC logic drive.

Gate power losses $P_G$ for low frequencies are small but can become important at frequencies >100 kHz. For frequencies >100 kHz, gate source impedance may have to be reduced to increase gate capacitive drive currents; then gate losses must be included in temperature calculations.

Switching losses $P_S$ may be determined as for any transistor.

To determine conduction losses $P_C$, the rms drain current, and the MOSFET on resistance, $R_{DS,on}$ must be known. $R_{DS,on}$ is a function of drain current $I_D$, gate-source voltage $V_{GS}$, and temperature $T_J$. The relationship of $R_{DS,on}$ to drain current $I_D$ and gate-source volt-

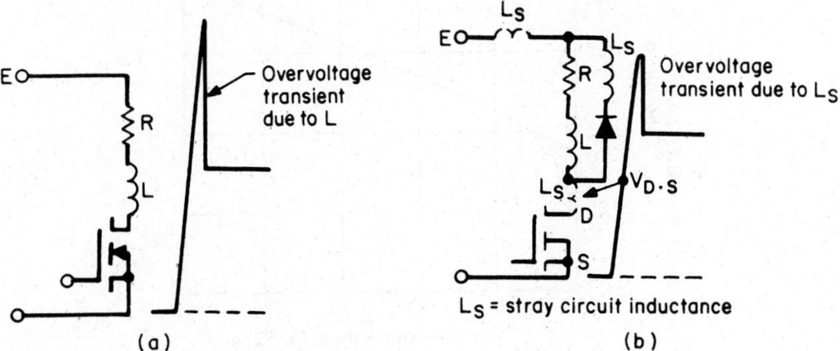


**FIG. 22-88** (*a*) MOSFET switching transient, unclamped inductive load; (*b*) MOSFET switching transient, clamped inductive load. *(International Rectifier.)*

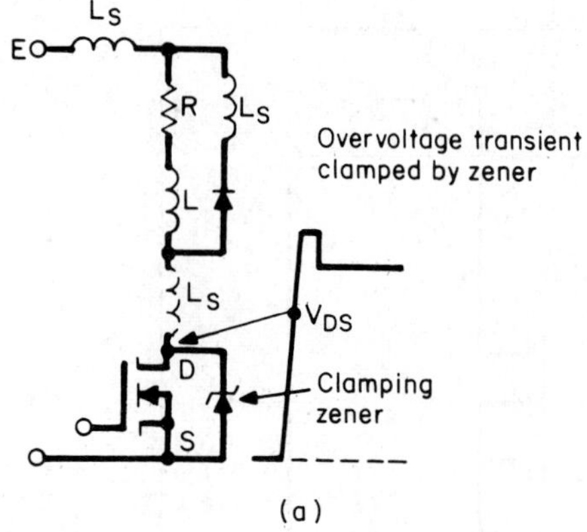


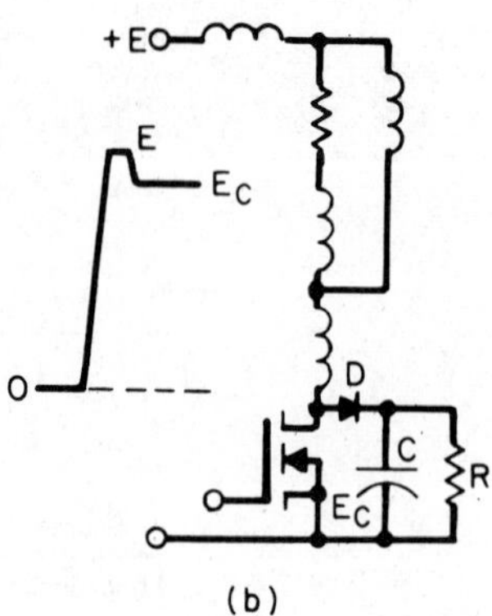


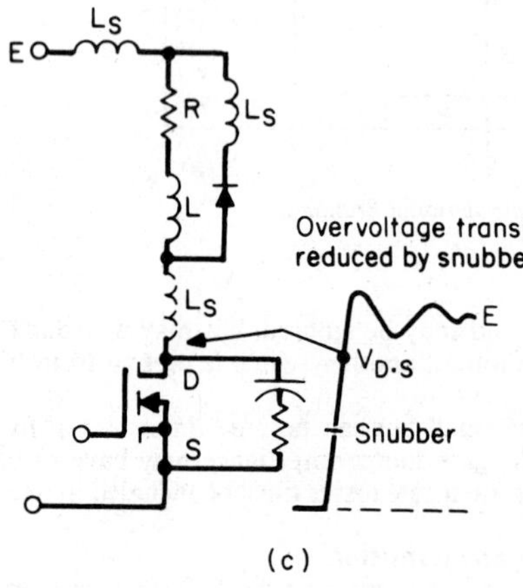


**FIG. 22-89** (*a*) MOSFET with shunt Zener: clamped inductive load with local drain-source Zener clamp; (*b*) MOSFET with snubber; (*c*) MOSFET with snubber and clamp: clamped inductive load with local drain-source snubber. *(International Rectifier.)*

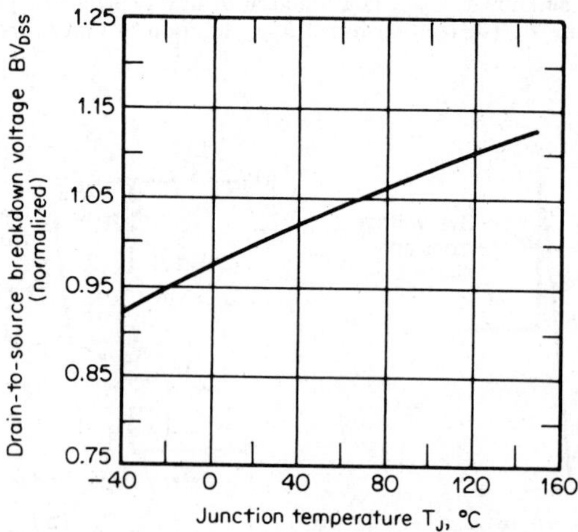


**FIG. 22-90** MOSFET avalanche versus temperature. *(International Rectifier.)*

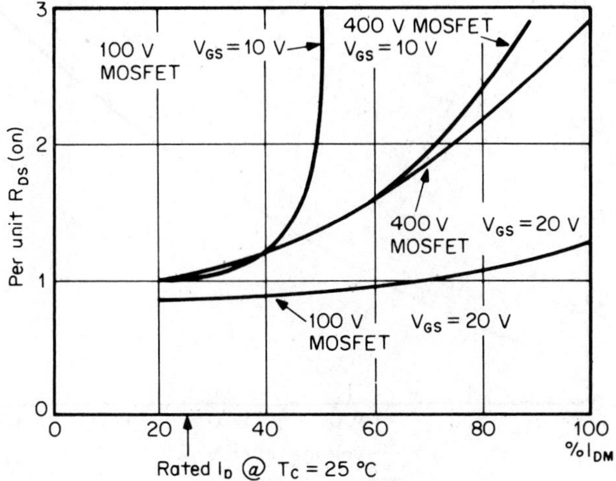


**FIG. 22-91** MOSFET on resistance. *(International Rectifier.)*

age $V_{GS}$ is shown in Fig. 22-91. For lowest loss switching, the drive for this MOSFET should be in the range 10 V < $V_{GS}$ < 20 V. It is presently uncertain how the gate drive should be specified for pulse drive $I_{DM}$, or even continuous drain current, inasmuch as the only guaranteed resistance levels on data sheets occur for a current of about $I_D/2$. At this current point and at $T_C$ = 25°C, a maximum $R_{DS,on}$ is specified on most data sheets. To determine $R_{DS,on}$ maximum at some other current level for a particular gate source voltage, check with the device manufacturer for specific recommendations.

Once $R_{DS,on}$ is known at 25°C for the worst-case drain current, then use a curve similar to Fig. 22-92 to determine $R_{DS,on}$ for the anticipated operating junction temperature and compute conduction losses from

$$P_C = I_D^2 \text{ rms } R_{DS,on} \text{ 25°C} \times R_{DS,on} \text{ normalized}$$

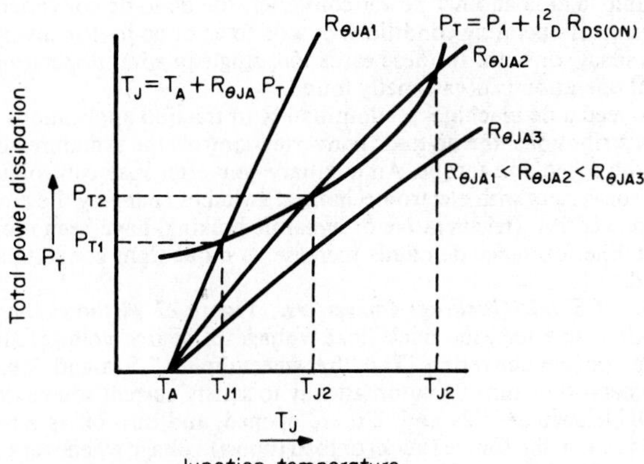


**FIG. 22-92** MOSFET thermal stability.

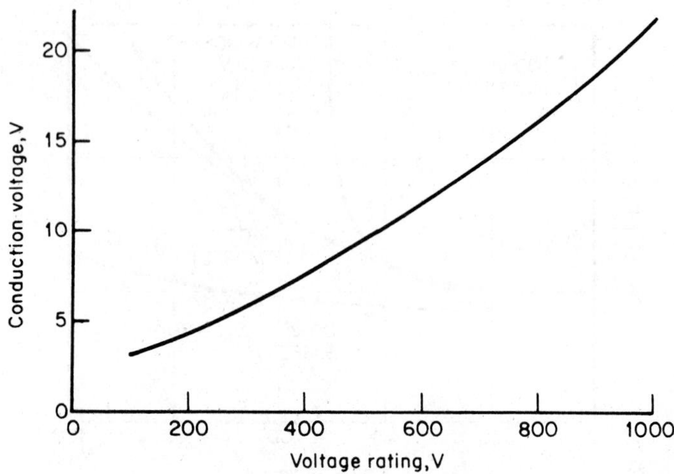


**FIG. 22-93** Conduction voltage versus blocking voltage. *(International Rectifier.)*

For a fixed-size MOSFET chip, the conduction voltage varies with design voltage rating as shown in Fig. 22-93. This shows $R_{DS,on}$ increasing about linearly with design breakdown voltage. High MOSFET current ratings with high voltage rating are difficult to achieve.

## DC-TO-DC CONVERTERS

*By P. WOOD*

***42. Applications.*** DC-to-dc converters are used primarily to serve electronic loads or electrical machines. In the former case, they are widely used to condition dc power from either the rectified utility supply or a dc source, such as a fuel cell, battery, or solar photovoltaic array for low-voltage information processing, and for instrumentation circuitry. In these cases only single-quadrant operation is required, but stringent specifications are usually imposed on unwanted component injection at both load and source terminals. Sometimes the electronic load is another power converter, the dc-to-dc converter being used to control the dc terminal operating conditions of a dc-to-ac or ac-to-dc converter for reasons of economy, efficiency, or both. In these cases too, single-quadrant operation is the norm, but two-quadrant operation is occasionally found.

When used to feed a dc machine, predominantly in traction applications with third rail or overhead dc distributions, the dc-to-dc converter controls the armature voltage and current, and hence the speed and torque. An auxiliary converter may control field excitation. Single-quadrant converters with electromechanical switches changing the circuit configuration for reverse power flow (regenerative or dynamic braking) have been most used, but as costs decrease and performance demands increase, two-quadrant converters may become more widely used.

***43. Topologies of Single-Quadrant Converters.*** Figure 22-94 shows the two basic single-quadrant dc-dc converters, the buck (load voltage $\leq$ source voltage) and boost (load voltage $\geq$ source voltage) converters. The "freewheel diodes," S1a and S1b, are autocomplementary switches—they turn on automatically to satisfy current source demands whenever the controllable switches S2a and S2b are opened, and turn off as a result of source commutations driven by the source (buck) or load (boost) voltage whenever the controllable switches are closed. The controllable switches may be implemented using transistors (bipolar or field effect), gate turn-off thyristors, or thyristors with impulse-commutating circuits.

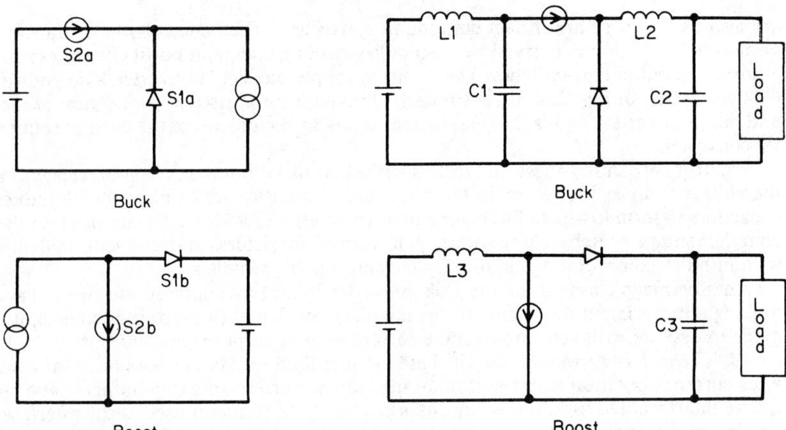


**FIG. 22-94** Idealized buck-and-boost single-quadrant dc-to-dc converters.

**FIG. 22-95** Buck-and-boost converters with interfaces.

Since there is no potential in the basic circuits acting to drive the commutation when a controllable switch opens, impulse commutations are needed. Thus one-half of the commutations are source commutations (controllable switches closing) and one-half are impulse commutations (controllable switches opening).

Figure 22-95 shows these converters with their usual practical interfaces. In the buck converter, *C*1 provides the voltage source, and in conjunction with *L*1 forms a low-pass filter limiting the ripple current burden on the actual dc source. *L*2 provides the current source, and in conjunction with *C*2 forms an *LC* filter limiting ripple voltage at the load. When the load is a machine, *C*2 is almost invariably omitted. If a machine field, then *L*2 also will not be needed since field inductances tend to be very large; if the converter drives a machine armature, *L*2 is sometimes still present, since the armature inductance may not be large enough to limit ripple current to tolerable levels.

Because the constraints on voltage ratio imposed by these topologies are sometimes objectionable and occasionally intolerable, the buck-boost converters shown in Fig. 22-96

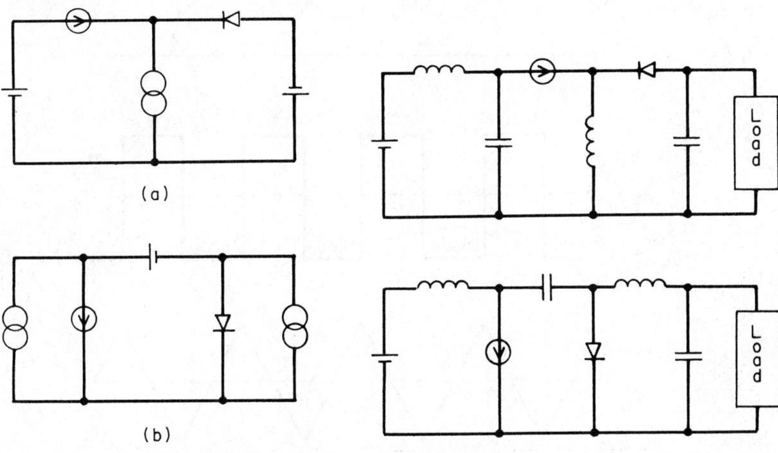


**FIG. 22-96** Idealized buck-boost converters.

**FIG. 22-97** Buck-boost converters with interfaces.

are also used. These are in fact compound converters. The "current-source transfer" converter of Fig. 22-96*a* is derived by cascade-connecting buck and boost converters and then eliminating redundant switches. The "voltage-source transfer" converter, also known as the Cúk converter, of Fig. 22-96*b*, is similarly derived from a cascade connection of the boost and buck converters and is the dual (planar topology reciprocal) of the current source transfer converter.

Figure 22-97 shows these converters with their usual interfaces. Again, capacitors form the voltage sources, inductors the current sources, and they are combined with inductors or capacitors to form low-pass filters for ripple attenuation. Diodes are again used for the autocomplementary switches, transistors, gate turnoff thyristors, and conventional thyristors with impulse commutating circuits for the controllable switches.

An advantage claimed for the Cúk converter is that by coupling the two reactors and making appropriate adjustments to the coupling coefficient (leakage inductance), it can be made to operate with zero input ripple current or zero output ripple current.

***44. Critical Inductance.*** In all these circuits there exists the possibility of discontinuous current operation as a result of an inadequate current-sourcing inductor at a current-source interface. Discontinuous current arises when the inductor used at the interface is too low in value to prevent the converter current from going to zero during a converter operating period. If the operating conditions depicted in Fig. 22-98 are examined, it can be seen that during diode conduction periods in the converters the source or load voltages (or both) act to reduce current-sourcing inductor currents. If the inductor is below a certain value for a given average current and diode conduction time, periods of zero current will result. The value of inductance which just allows the current to reach zero is called the critical induc-

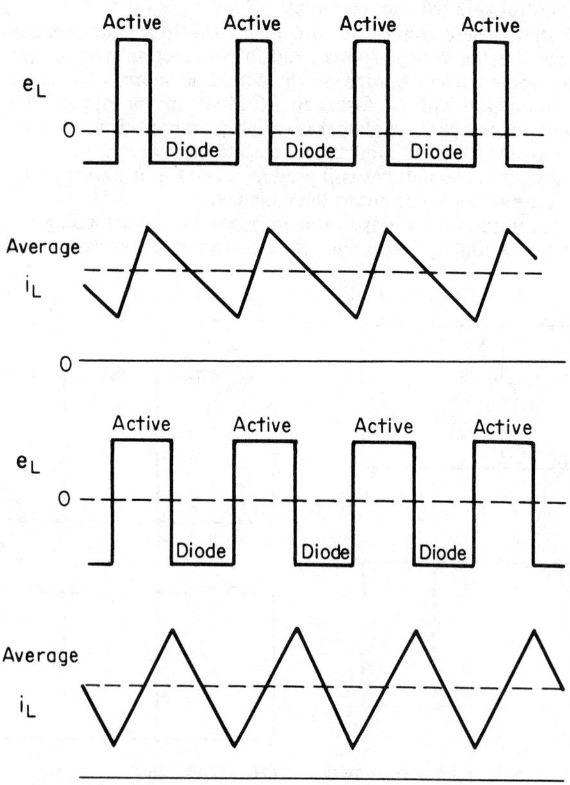

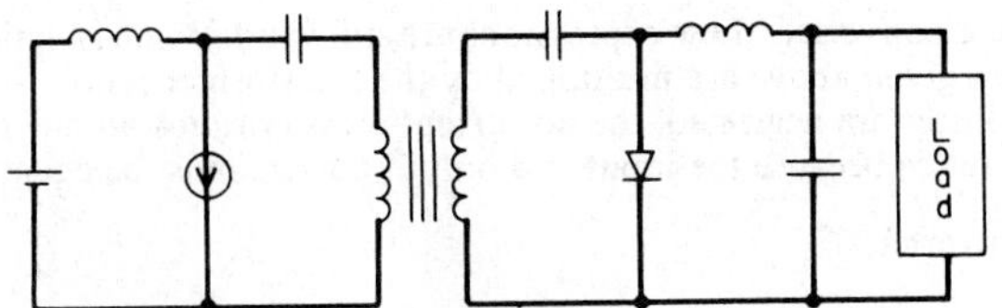


**FIG. 22-99** Cúk converter with isolation.

tance. It is a function of converter operating conditions—frequency, duty cycle (on time of controllable switch/total cycle time) and average dc current. Some converters are designed to operate in continuous current over their total anticipated operating range, some to operate partly in continuous current and partly in discontinuous current, and some in discontinuous current all the time. Apart from the obvious increase in ripple relative to average (wanted) components in the dependent current, discontinuous current is often considered undesirable because the converters' transfer functions are different for the continuous and discontinuous modes.

***45. Other Compound Topologies.*** In cases where the voltage and current transformation ratios needed are large, or when isolation is required between dc source and load, other topologies, or adaptations of the buck-boost converters, are used. Although called dc-to-dc converters in the literature, such converters are compounds consisting of a dc-to-ac converter, often single-ended, followed by a rectifier (single-quadrant ac-to-dc converter). The buck-boost converters are also, in fact, compound converters using this principle but not providing isolation.

The current-source transfer converter impresses no dc voltage on the current source it uses as the energy-transfer medium. Hence, the current-sourcing inductor can be replaced with a transformer. The resulting converter can have its transformation ratio for a given duty cycle modified by transformer turns ratio and provides isolation. When operated with discontinuous current, it is called a flyback converter; when operated with continuous current, it is called a forward converter.

Similarly, the voltage-source transfer converter impresses no dc current on its voltage-sourcing capacitor. Hence, if that component is split into two in series, a transformer may be interposed to provide isolation and, if desired, a transformation ratio. This configuration, shown in Fig. 22-99, has been called the optimum topology converter, but the virtue implied by that name is not truly ascribable to the circuit. Other compound converters of this kind, dc-to-ac followed by ac-to-dc, use more conventional half-bridge, midpoint, or bridge configuration dc-to-ac converters and full-wave rectifiers, midpoint or bridge.

***46. Transfer Ratios of Single-Quadrant Converters in Continuous Current Operation.*** All dc-to-dc converters are controlled by varying the duty cycle $D$ of their controllable switches. This is usually effected at constant frequency, simply by varying the ratio of unit value and zero value time periods in the switches' existence functions, but can also be done with varying frequency by maintaining constant unit- or zero-value periods while varying the other functions. Regardless of the technique used to vary $D$, the dc transfer ratios of all dc-to-dc converters are functions only of $D$.

For the buck converter,

$$\text{Output dc voltage} = D \times \text{input dc voltage}$$

For the boost converter,

$$\text{Output dc voltage} = \frac{1}{1 - D} \times \text{input dc voltage}$$

For the buck-boost converters,

$$\text{Output dc voltage} = -\frac{D}{1 - D} \times \text{input dc voltage}$$

For the compound converters these expressions are still valid. When isolation is introduced, the transfer ratios given above are multiplied by the transformer ratio.

If converter losses are neglected, the dc current transfer ratios are the reciprocals of the voltage transfer ratios because the input and output powers must be equal. Hence,

For the buck converter,

$$\text{Input dc current} = D \times \text{output dc current}$$

For the boost converter,

$$\text{Input dc current} = \frac{1}{1 - D} \times \text{output dc current}$$

For the buck-boost converter,

$$\text{Input dc current} = \frac{-D}{1 - D} \times \text{output dc current}$$

***47. Unwanted Components and Interfacing in Continuous Current Operation.*** The constant-frequency existence function for a controllable switch in a dc-to-dc converter operating at angular frequency $\omega (= 2\pi f)$ and duty cycle $D$ is given by

$$H = D + \frac{2}{\pi} \sum_{n=1}^{\infty} \frac{\sin (n\pi D)}{n} \cos (n\omega t)$$

The diode assumes the complementary existence function, $1 - H$. Thus the unwanted components and the dependent-quantity waveshapes are readily expressed.

Figure 22-100 shows the dependent voltage (output) and current (input) for an ideal buck converter. These waves are simply replicas of the switch existence function, and the unwanted components are simply

$$\text{Voltage: } \frac{2V_S}{\pi} \sum_{n=1}^{\infty} \frac{\sin (n\pi D)}{n} \cos (n\omega t)$$

$$\text{Current: } \frac{2I_L}{\pi} \sum_{n=1}^{\infty} \frac{\sin (n\pi D)}{n} \cos (n\omega t)$$

Figure 22-101 shows the dependent voltage (input) and current (output) waves for the ideal boost converter. These are simply replicas of the complementary existence function. The unwanted-component expressions are similar to those for the buck converter, with a change of sign and $V_L$ replacing $V_S$, $I_S$ replacing $I_L$.

To design a simple current-source interface for the buck converter, observe that

$$V_L = DV_S$$

and the inductor voltage during the on period, assuming a perfectly smooth load voltage, is

$$e_L = V_S - V_L = (1 - D)\, V_S = \frac{(1 - D)}{D} V_L$$

The on period is $D/f$ seconds, so the inductor is subjected to

$$\frac{D(1 - D)\, V_S}{f} = \frac{(1 - D)\, V_L}{f} \quad \text{V-s}$$

This, divided by $L$, gives the peak-to-peak ripple; since current excursions are linear, the peak ripple is one-half the peak-to-peak ripple, and the critical inductance is given by

$$L_C = \frac{D\,(1 - D)\, V_S}{2fI_L} = \frac{(1 - D)\, V_L}{2fI_L}$$

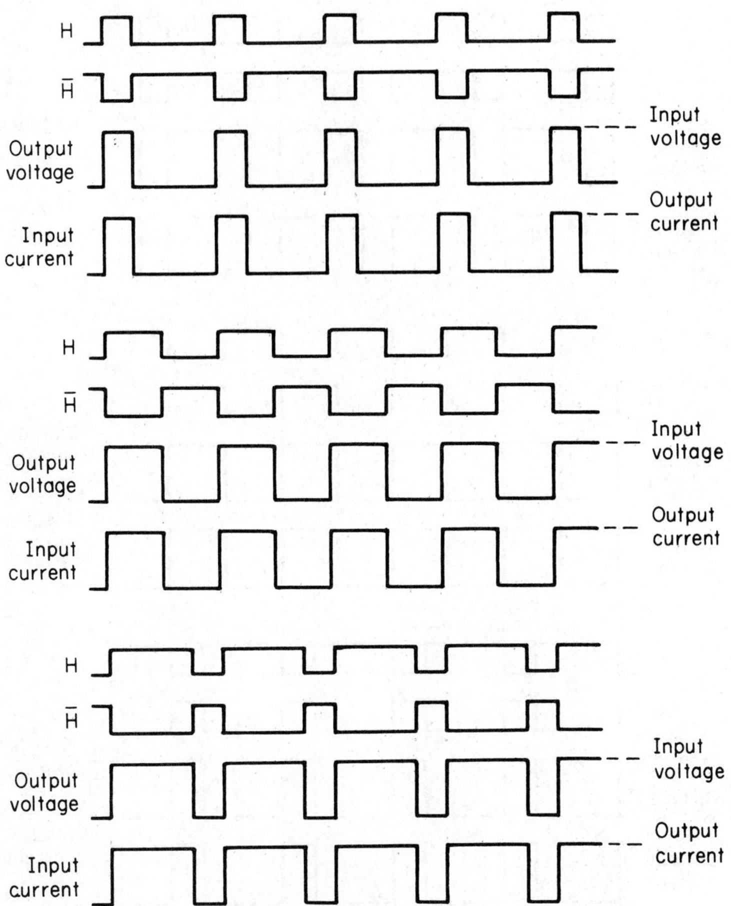


**FIG. 22-100** Idealized buck converter waveforms.

The inductance for a given peak-to-peak ripple stipulation is simply

$$L_{\Delta I} = \frac{D\,(1 - D)\,V_S}{f\,\Delta I} = \frac{(1 - D)\,V_L}{f\,\Delta I}$$

Since the rms ripple current is $\Delta I/2\sqrt{3}$, the inductance for a given rms ripple current is also easily found. For the boost converter, the expressions for volt-seconds, critical inductance, and inductance for given peak-to-peak or rms ripple are the same as those for the buck converter if $D$ is replaced by $1 - D$ and vice versa, $V_S$ replaced by $V_L$ and vice versa, and $I_L$ replaced by $I_S$.

To design an $LC$ filter for the buck converter for a given peak-to-peak ripple voltage at the load, observe that if this ripple is assumed to be too small to perturb the inductor behavior, then the peak-to-peak ripple voltage is given by

$$\Delta V = \Delta I/8fC$$

Substituting for $\Delta I$ and transposing gives

$$LC = \frac{(1 - D)\,V_L}{8f^2\,\Delta V} = \frac{D\,(1 - D)\,V_S}{8f^2\,\Delta V}$$

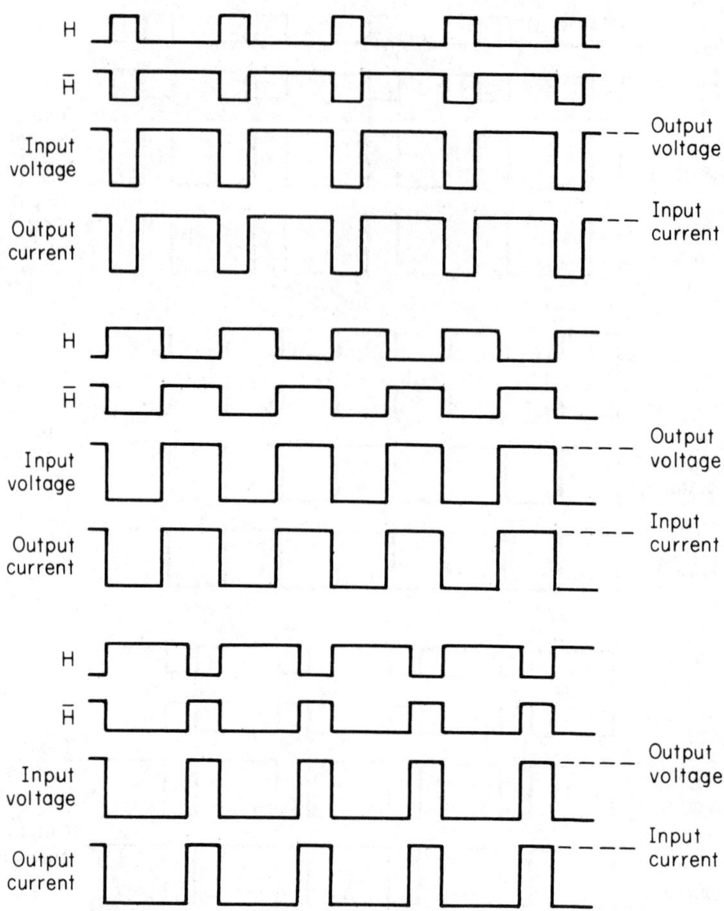


**FIG. 22-101** Idealized boost converter waveforms.

For the boost converter, the output capacitor needed to limit peak-to-peak ripple voltage to $\Delta V$ is given by

$$C = \frac{D(1-D)\,I_S}{f\,\Delta V} = \frac{DI_L}{f\,\Delta V}$$

An input capacitor to limit buck-converter input voltage ripple to $\Delta V$ can be found using the same expression, substituting $1 - D$ for $D$ and vice versa, and substituting $I_L$ for $I_S$ and vice versa. An $LC$ filter to limit buck-converter dc source peak-to-peak ripple current is defined by

$$LC = \frac{D(1-D)\,I_L}{8f^2\,\Delta I} = \frac{(1-D)\,I_S}{8f^2\,\Delta I}$$

An input capacitor to limit boost-converter voltage peak-to-peak ripple to $\Delta V$ is given by

$$C = \frac{D(1-D)\,V_L}{8f^2L\,\Delta V} = \frac{DV_S}{8f^2L\,\Delta V}$$

where $L$ is the current-sourcing inductor chosen. If the source peak-to-peak dc ripple current is to be limited to $\Delta I$, then the inductance required is given by

$$L' = \frac{D(1 - D^2)^{1.5}\, V_L}{36\sqrt{3f^3LC}\,\Delta I} \qquad \text{if } D < ½$$

and by

$$L' = \frac{(1 - D)\,[D(2 - D)]^{1.5}\, V_L}{36\sqrt{3f^3LC}\,\Delta I} \qquad \text{if } D > ½$$

where $L$, $C$ are the current-sourcing inductor and input filter capacitor.

For the current-source transfer buck-boost converter, the current-sourcing inductor carries a dc current given by

$$I_C = I_L + I_S = \frac{I_S}{D} = \frac{I_L}{1-D}$$

The volt-seconds applied are

$$\text{V}\cdot\text{s} = \frac{DV_S}{f} = \frac{(1 - D)\,V_L}{f}$$

Thus to limit the peak-to-peak ripple current in the inductor to $\Delta I$, the required inductance is

$$L = \frac{DV_S}{f\Delta I} = \frac{(1 - D)\,V_L}{f\Delta I}$$

and the critical inductance is

$$L_C = \frac{D^2V_S}{2fI_S} = \frac{(1 - D)^2V_L}{2fI_L}$$

The shunt capacitor needed to hold the peak-to-peak input ripple voltage to $\Delta V$ is

$$C_S = I_S/f\Delta V$$

The shunt capacitor needed to hold the peak-to-peak output ripple voltage to $\Delta V$ is

$$C_L = I_L/f\Delta V$$

Note that these values are independent of the duty cycle and the ripple current.

For the voltage-source transfer buck-boost converter, the voltage-sourcing capacitor supports a dc voltage

$$V_C = V_L + V_S = \frac{V_S}{1 - D} = \frac{V_L}{D}$$

To hold the peak-to-peak ripple voltage to $\Delta V$, the capacitance value needed is

$$C = \frac{DI_L}{f\Delta V} = \frac{(1 - D)\,I_S}{f\Delta V}$$

The critical capacitance (the capacitance that just maintains unipolar capacitor voltage) is

$$C_C = \frac{D^2I_L}{2fV_L} = \frac{(1 - D)^2I_S}{2fV_S}$$

The input inductance needed to limit the peak-to-peak input ripple current to $\Delta I$ is

$$L_S = DV_S/f\Delta I$$

and the critical inductance is thus

$$L_{CS} = DV_S/2fI_S$$

The output inductance needed to limit the peak-to-peak output ripple current to $\Delta I$ is

$$L_L = \frac{(1 - D)\, V_L}{f\,\Delta I}$$

and the critical inductance is

$$L_{CL} = \frac{(1 - D)\, V_L}{2fI_L}$$

***48. Transfer Ratios and Interfacing in Discontinuous Current Operation.*** Discontinuous current occurs when an inductor used as a current-sourcing interface is below the critical value. Operation in this region is characterized by the lack of a unique relationship between $D$ and the converter's voltage transfer ratio. The converter operating conditions need *two* specified variables, normally $V_L$ (or $V_L/V_S$) and $R_L$ ($= V_L/I_L$), and knowledge of the inductor value (or its ratio to the critical inductance).

For the buck converter, the equation relating $D$, $I_V$ ($= V_L/V_S$) and $R_L$ in discontinuous conduction is

$$D^2 = \frac{2fLK_V^2}{R_L\,(1 - K_v)}$$

where $L$ is the subcritical current-sourcing inductance.

The freewheel period, the fractional time for which the freewheel diode conducts, is defined by

$$F^2 = \frac{2fL\,(1 - K_V)}{R_L}$$

The output capacitor needed to limit peak-to-peak output ripple voltage to $\Delta V$ is given by

$$C_L = \frac{I_L\,[2 - (F + D)]^2}{4f}$$

The input capacitor needed to limit peak-to-peak input voltage ripple to $\Delta V$ is given by

$$C_S = \frac{K_V I_L\,(2 - D)^2}{4f} = \frac{I_S\,(2 - D)^2}{4f}$$

where $I_L$ and $I_S$ are the average output and input (load and source) currents, respectively. The peak current reached in each cycle is given by

$$I_P = \frac{FV_L}{fL} = \frac{FK_V V_S}{fL} = \frac{2I_L}{F + D} = \frac{2I_S}{D}$$

For the boost converter in discontinuous current, with $K_V = V_L/V_S$, $R_L = V_L/I_L$ and $L$ as the subcritical current-sourcing inductance

$$D^2 = \frac{2fLK_V\,(K_V - 1)}{R_L}$$

$$F^2 = \frac{2fLK_V}{R_L\,(K_V - 1)}$$

$$C_L = \frac{I_L\,(2 - F)^2}{4f\,\Delta V}$$

$$C_S = \frac{I_S\,[2 - (D + F)]^2}{4f\,\Delta V}$$

$$I_P = \frac{DV_S}{fL} = \frac{DV_L}{K_V fL} = \frac{2I_L}{F} = \frac{2I_S}{F + D}$$

For the current-source transfer buck-boost converter,

$$D^2 = \frac{2FLK_V^2}{R_L}$$

$$F^2 = \frac{2fL}{R_L}$$

$$C_L = \frac{I_L\,(2 - F)^2}{4f\,\Delta V}$$

$$C_S = \frac{I_S\,(2 - D)^2}{4f\,\Delta V}$$

$$I_P = \frac{DV_S}{fL} = \frac{FV_L}{fL} = \frac{2I_S}{D} = \frac{2I_L}{F}$$

and the average current in $L$ is given by

$$I_{CST} = (1 + K_V)\,I_L = \frac{(1 + K_V)\,I_S}{K_V}$$

The voltage-source transfer buck-boost converter has two current-sourcing inductors, $L_S$ and $L_L$. Discontinuous-current operation is defined not by zero current intervals in one or both of these, but by the diode having a fractional conduction period $F$ such that $F < 1 - D$. In discontinuous conduction, the converter will operate with *reverse* current in one of the inductances for part of the operating cycle. Which inductor suffers reverse current depends on $K_V$ and $L_L/L_S$. The pertinent design expressions are

$$D^2 = \frac{2fL_PK_V^2}{R_L}$$

where $L_P$ is the parallel combination of $L_L$ and $L_S$,

$$L_P = \frac{L_LL_S}{L_L + L_S}$$

$$F^2 = \frac{2fL_P}{R_L}$$

$$C_L = \frac{(1 + K_V)\,I_L}{4f\,\Delta V} \cdot \frac{L_P}{L_L} \cdot [2 - (F + D)]^2$$

$$C_S = \frac{1 + K_V}{K_V} \cdot \frac{I_L}{4fV} \cdot \frac{L_P}{L_S} \cdot [2 - (F + D)]^2$$

There are two peak currents, $I_{PS}$ and $I_{PL}$. Related to the average currents $I_S$ and $I_L$, they are given by

$$I_{PS} = I_S\left[\frac{L_P}{L_L} + \frac{L_P}{L_S}\left(\frac{2 - F}{D}\right)\right]$$

$$I_{PL} = I_L\left[\frac{L_P}{L_S} + \frac{L_P}{L_L}\left(\frac{2 - D}{F}\right)\right]$$

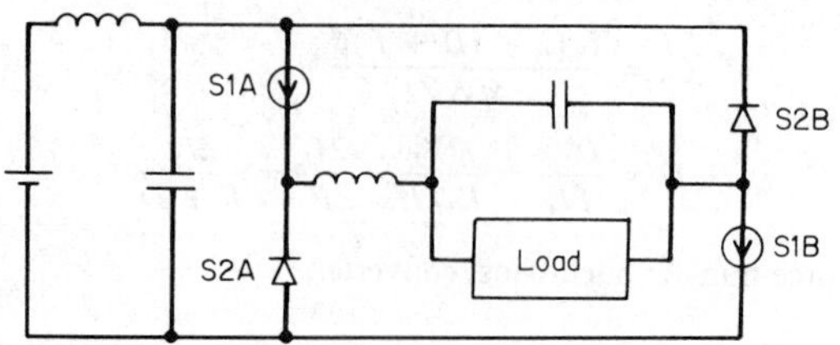


FIG. 22-102 Two-quadrant buck converter.

***49. Two-Quadrant DC-to-DC Converters.*** All the dc-to-dc converters can be designed to have two-quadrant capability, but rarely are. Figure 22-102 shows the two-quadrant buck converter topology. There are two possible modes of operation for this converter.

In mode 1, S1A and S1B are operated together with existence factors $H$, duty cycle $D$, and S2A and S2B assume existence functions $1 - H$ in continuous current operation. The voltage transfer function is then

$$K_V = V_L/V_S = 2D - 1$$

and interfacing component values are exactly twice those for the single-quadrant converter operating with the same $V_S$, $V_L$, $R_L$, and $D$.

In mode 2, S1B is kept closed ($H = 1$) for first-quadrant operation, with S2B open ($H = 0$) and S1A operated with duty cycle $D$. For fourth-quadrant operation, S1B is kept open, S2B closed, and S1A operated at duty cycle $D$. In this mode, the converter operates as a single-quadrant buck converter in the first quadrant, with the interfacing requirements previously defined therefore, and as a single-quadrant boost converter, with $V_L$ and $V_S$ reversing roles, in the fourth quadrant, with interfacing requirements previously defined.

In mode 2, it is possible to replace S1B and S2B by an electromechanical single-pole double-throw switch if the speed of transfer between first- and fourth-quadrant operation is not critical.

The two-quadrant boost converter is shown in Fig. 22-103. Again, two modes are possible. Mode 1 has S1A and S1B operated at duty cycle $D$ with S2A and S2B, assuming a duty cycle $1 - D$ in continuous current operation. The voltage transfer function is thus

$$K_V = V_L/V_S = \frac{1}{1 - 2D}$$

and interfacing requirements are as for the single-quadrant converter with all component values doubled.

In mode 2, S1B is left open, S2B kept closed for first-quadrant operation, and the converter becomes a single-quadrant boost converter. For second-quadrant operation, S1B is kept closed, S2B open, and the converter becomes a single-quadrant buck converter with $V_L$ and $V_S$ reversing roles. Again, S1B and S2B may be replaced by an electromechanical single-pole double-throw switch. The buck-boost converters have 2 two-quadrant versions

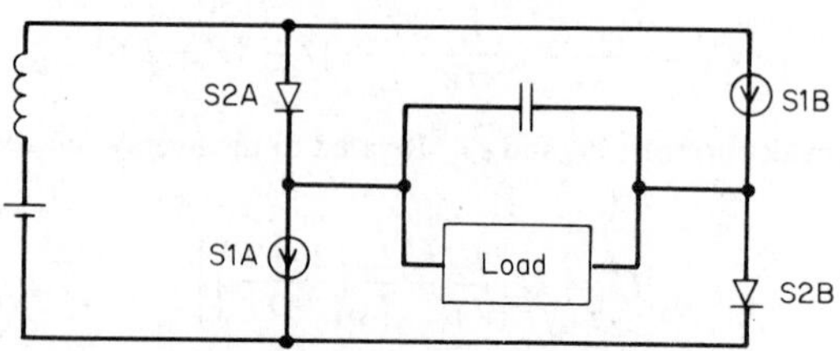


FIG. 22-103 Two-quadrant boost converter.

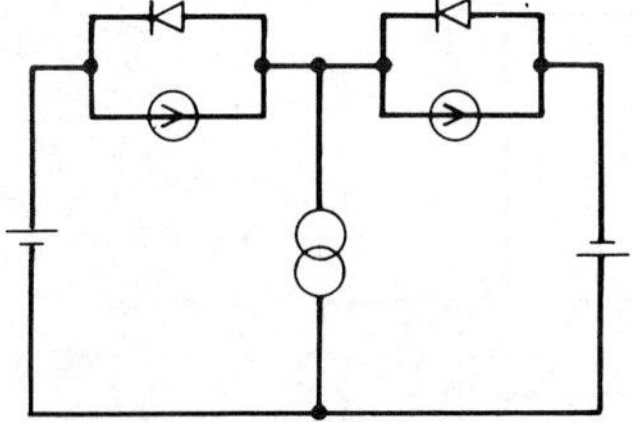

**FIG. 22-104** Two-quadrant CST buck-boost converter.

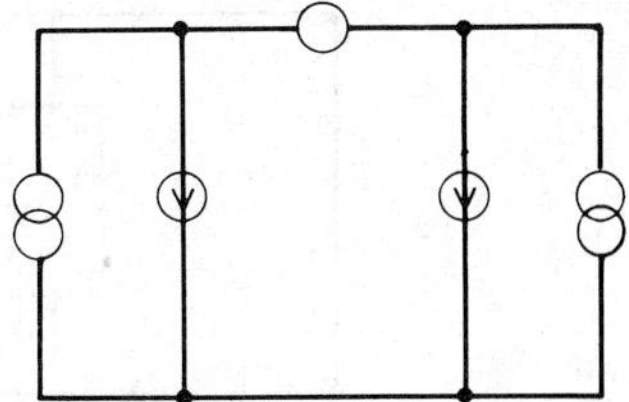

**FIG. 22-105** Two-quadrant VST buck-boost converter.

each. The natural versions, which assume that $V_S$ and $V_L$ are unipolar voltages, are shown in Figs. 22-104 and 22-105.

The two-quadrant current-source transfer buck-boost is obtained by giving each switch bidirectional current capability. Only mode 2 operation is possible, since the current in the current-sourcing inductor must reverse for operation to transfer from the first to the second quadrant. Interfacing requirements are the same as for the single-quadrant current-source transfer converter, with $V_S$ and $V_L$ reversing roles for second-quadrant operation.

The natural two-quadrant voltage-source transfer buck-boost converter of Fig. 22-105 is obtained by giving both switches bidirectional voltage-blocking capability. Again, only mode 2 operation is possible and reversal of power flow, involving a reversal of roles for $I_L$ and $I_S$, is accompanied by a reversal of voltage polarity on the transfer source; interfacing requirements are the same as for the single-quadrant converter.

The unnatural, or spurious, two-quadrant versions of the buck-boost converters are shown in Figs. 22-106 and 22-107. They assume polarity reversal in the defined quantities. Thus for the current-source transfer converter, reverse power flow is assumed to be accompanied by polarity reversals of both $V_S$ and $V_L$, the current in the transfer source (inductor) remains unidirectional, and the switches are given bidirectional voltage-blocking capability.

For the voltage-source transfer converter of Fig. 22-107, reverse power flow is assumed to be accompanied by polarity reversal of both $I_S$ and $I_L$, with the voltage transfer source retaining unipolar potential. Thus both switches are given bidirectional current flow capability, and only mode 2 operation is possible.

***50. Four-Quadrant DC-to-DC Converters.*** Four-quadrant buck-and-boost dc-to-dc converters have the same topologies as the dc to single-phase-ac voltage- and current-sourced converters, respectively. They may be operated in mode 1 or mode 2, and mode 1 operation always calls for double-value interfacing components, as compared to the single-quadrant converter or mode 2 operation.

Four-quadrant buck-boost converters are not possible. This is because in either the current-source transfer or voltage-source transfer converters, single-quadrant, natural two-quadrant, or spurious two-quadrant versions, the source and load quantities (voltage or current) must have opposite polarities. Power transfer between voltages or currents of the same polarity is not possible, and hence dc to four-quadrant-dc conversion is not possible. DC-to-ac conversion can be accomplished, however, by using two buck-boost converters, oppo-

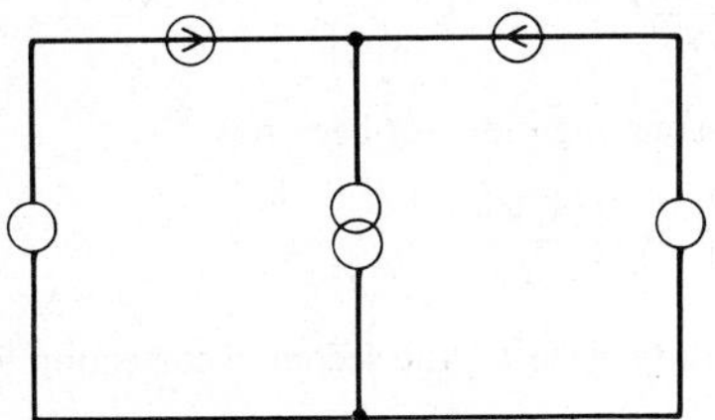

**FIG. 22-106** Spurious two-quadrant CST buck-boost converter.

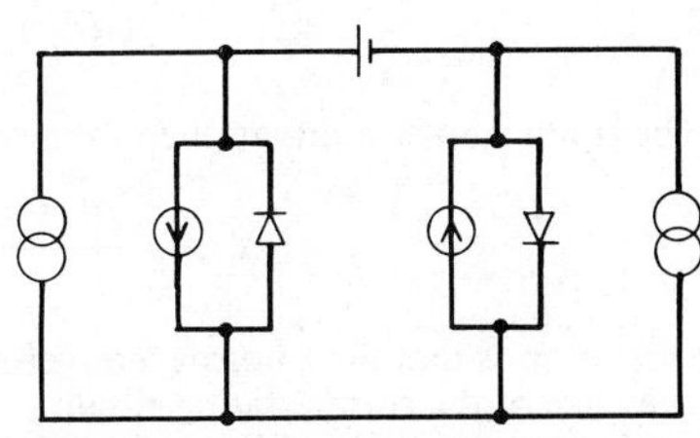

**FIG. 22-107** Spurious two-quadrant VST buck-boost converter.

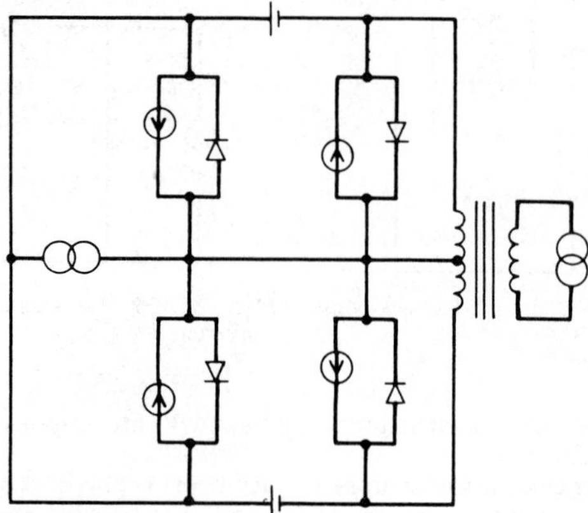

**FIG. 22-108** Buck-boost dc-to-ac converter.

sitely poled, and a center-tapped transformer so that $V_L$ or $I_L$ always has the appropriate polarity for one of these. Such a push-pull arrangement is shown in Fig. 22-108; it has little merit when compared to the more conventional dc-to-ac converters.

***51. Commutating Circuits for Thyristors in DC-to-DC Converters.*** If transistors (field-effect and bipolar) or gate turnoff thyristors are used as the controllable switch elements in dc-to-dc converters, no intrinsic difficulties arise. These devices can be made to assume the duty cycle desired simply by delivering to them appropriate base or gate drive signals. The impulse commutations required for converter operation are created when these switches turn off (open).

If thyristors are used, they need impulse commutating circuits to be effective. These circuits, viewed as part of the switch, can be used in any of the dc-to-dc converter topologies. The single-quadrant buck converter is used to examine their characteristics and design criteria.

Figure 22-109 shows the simplest commutating circuit, with synchrograms of the current and voltage waveforms. The key design criterion is

$$C_C = I_L\, t_q / V_S$$

where $tq$ is the reverse recovery time required by $Q_1$. Minimum off time for $Q_1$ is $2CV_S/I_L$; minimum on time is set by $L_C$, and is

$$t_{\text{on}} = \pi \sqrt{L_C C_C}$$

The commutating circuit modifies the dependent voltage and the apparent duty cycle. If $D$ is the duty cycle obtained by considering the time between triggering $Q_1$ and triggering $Q_2$, then the real duty cycle becomes

$$D' = D + \frac{2fCV_S}{I_L}$$

The ripple volt-seconds applied to the current-sourcing inductor becomes

$$\text{V}\cdot\text{s} = \frac{D'(1 - D')\, V_S}{f} + \frac{D'^2}{2}\frac{CV_S^2}{I_L}$$

The first term is that for a normal converter with duty cycle $D'$, the second a correction for the presence of the commutating circuit.

The current carried by $Q_1$ is also modified by the commutating circuit. Normally the peak current in $Q_1$ is $I_L$. It becomes

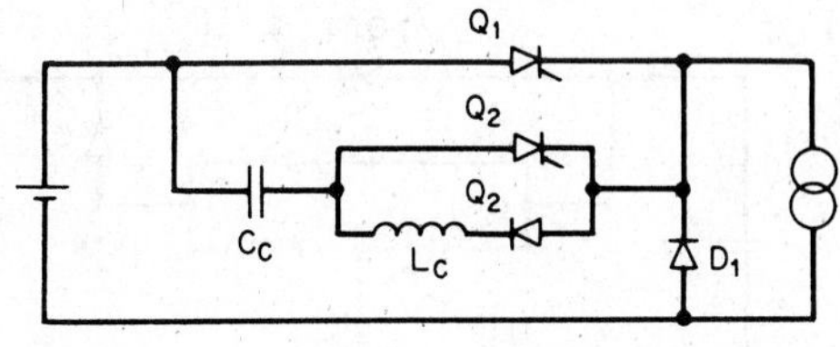


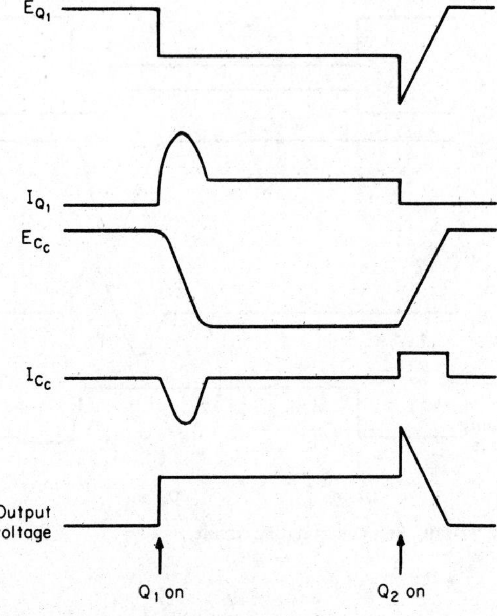


**FIG. 22-109** Hard commutating circuit.

$$I_{P1} = I_L + V_S \sqrt{C_C/L_C}$$

The average current in $Q_1$, normally $D'I_L$, becomes

$$I_{A1} = D'I_L + \pi f C_C V_S$$

These increases do not occur in $I_S$, so input interfacing can be designed as for a converter without a commutating circuit but having duty cycle $D'$.

An alternative commutating circuit is shown in Fig. 22-110, with synchrograms. A key design parameter for this circuit is

$$x = \frac{V_S}{I_L}\sqrt{\frac{C_C}{L_C}} \qquad x > 1$$

Then the design must have

$$C_C = \frac{xI_L t_q}{\pi - 2 \arcsin (1/x)}$$

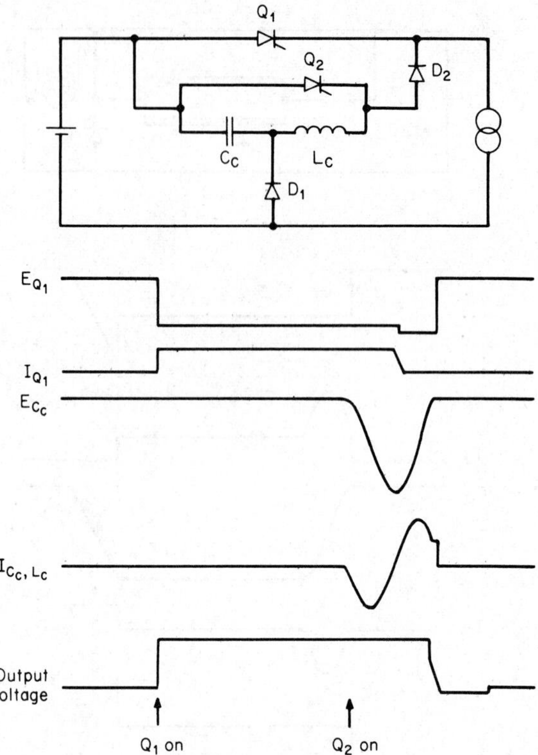


**FIG. 22-110** Soft commutating circuit.

and

$$L_C = \frac{V_S t_q}{x I_L \left[\pi - 2 \arcsin (1/x)\right]}$$

The duty cycle is modified from that obtained from the difference in triggering times for $Q_1$ and $Q_2$ as follows:

$$D' = \mathrm{D} + \left[2\pi - \arcsin\left(\frac{1}{x}\right)\right] f\sqrt{L_C C_C} + ½\frac{fCV_S}{I_L}\left(1 - \frac{1}{x}\sqrt{x^2 - 1}\right)^2$$

The perturbation in output ripple, as compared to a converter without the commutating circuit operating at duty cycle $D'$, is small and is usually ignored. It is, in fact, a slight

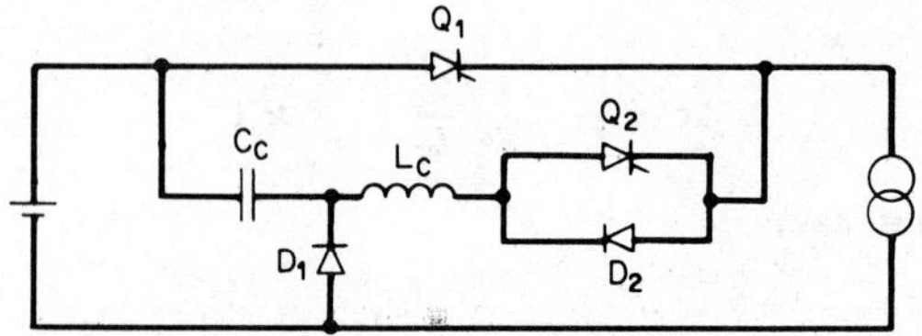


**FIG. 22-111** Alternative commutating circuit.

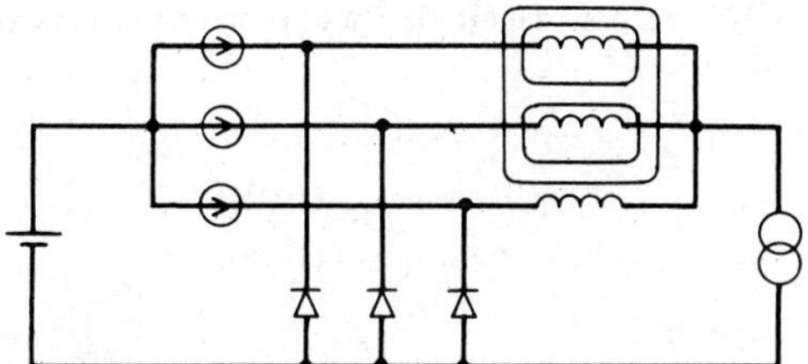

**FIG. 22-112** "Three-phase" buck converter.

*decrease* in the volt-second exposure of the current-sourcing inductor. There are no perturbations in $I_S$, or the peak current in $Q_1$, but the average current in $Q_1$ is slightly reduced because the duty cycle for $Q_1$ does not contain the term

$$\frac{1}{2}f\frac{CV_S}{I_L}\left(1-\frac{1}{x}\sqrt{x^2-1}\right)^2$$

This circuit has the disadvantage that the commutating path, $L_C - C_C$, must ring around when $Q_2$ is triggered before the commutation can begin. This problem is avoided in the variant shown in Fig. 22-111, which has similar defining equations except that the duty cycle is now

$$D' = D + \left[\pi - \arcsin\left(\frac{1}{x}\right)\right] f\sqrt{L_C C_C} + ½\frac{fCV_S}{I_L}\left(1-\frac{1}{x}\sqrt{x^2-1}\right)^2$$

A disadvantage of this circuit is that the peak current in $Q_1$ is increased to

$$I_{P1} = (1 + x)\, I_L$$

and the average current in $Q_1$ is increased by $f\pi x\sqrt{L_C C_C}\, I_L$.

***52. Polyphase DC-to-DC Converters.*** The interfacing (filtering) requirements of all the dc-to-dc converters are a function of converter operating frequency and load-source conditions. For a given load-source relationship, the only way to reduce interface component sizes in the simple converters is by increasing the frequency. Since this always increases switching losses, and frequency limitations exist for all converter power ratings due to the switching devices used, an alternative technique can be useful.

This technique is to use multiple converters with phase shifts between their switch existence functions, creating a "polyphase" converter. The converters share the load, having their phase-shifted dependent quantities combined. At voltage-source terminals, they can be directly parallel connected; at current-source terminals, an interphase reactor (sometimes called an interphase transformer) is used to support converter ripple voltage differences while parallel connecting the wanted components. The 3-phase single-quadrant buck converter of Fig. 22-112 exemplifies this type of converter.

The principle underlying this technique for improving the spectra (reducing the ripple effects) of converter-dependent quantities is often called *harmonic neutralization.* It is the fact that complete ac phasor sets sum to zero, and that when the angular displacements of such a set are multiplied by an integer other than an integer multiple of the number of phasors, a new set is formed that also sums to zero. When the multiplier is an integer multiple of the number of phasors, a zero sequence set is formed, a single phasor having amplitude $N$ times that of the original phase amplitudes, $N$ being the number of phasors in the original set.

Expressed mathematically, let

$$P_k = P\cos\left(\omega t - \frac{2k\pi}{N}\right) \qquad k = 0, 1, 2, 3, \ldots, N-1$$

define a complete set of $N$ phasors, magnitude $P$ and progressive angular displacement $2\pi/N$. The common 3-phase set,

$$P_A = P \cos \omega t$$
$$P_B = P \cos [\omega t - (2\pi/3)]$$
$$P_C = P \cos [\omega t - (4\pi/3)]$$

in the example most often met.

Then
$$\sum_{k=0}^{N-1} P_k = 0$$

Also, the sets

$$P_{k,M} = P \cos \left[ M \left( \omega t - \frac{2k\pi}{N} \right) \right] \qquad M \text{ any integer}$$

have

$$\sum_{k=0}^{N-1} P_{k,M} = 0$$

unless $M = JN$, $J$ an integer, in which case

$$P_k \sum_{k=1}^{N-1} P_{k,NJ} = NP \cos (JN\omega t)$$

The existence function for a switch in a dc-to-dc converter has the form

$$H = D + \frac{2}{\pi} \sum_{n=1}^{\infty} \frac{\sin (n\pi D)}{n} \cos (n\omega t)$$

If $N$ such switches are operated with appropriate phase displacements, then

$$H_k = D + \frac{2}{\pi} \Sigma \frac{\sin (n\omega D)}{n} \cos \left[ n \left( \omega t - \frac{2k\pi}{N} \right) \right]$$

It is clear that

$$\sum_{k=1}^{N-1} H_k = ND + \frac{2N}{\pi} \sum_{n=1}^{\infty} \frac{\sin (nN\pi D)}{nN} \cos (nN\omega t)$$

and that converter-dependent quantities, which are dc-defined quantities multiplied by the existence functions, will, when all $N$ are combined, exhibit an increase in ripple frequency to $Nf$ and a decrease of $1/N$ in the maximum amplitudes relative to wanted components. Hence interfacing components are reduced in value by the factor $N^2$. However, there are obviously penalties. $N$ complete sets of converter switches and controls must be used, and the interphase reactor must be added. The rating of this component is easy to establish, for each winding carries a dc current $I_L/N$ A and supports an absolute maximum ac excitation as follows:

$$\text{Volt-seconds on each winding} = \frac{V_S}{4f} \qquad \text{for } N \text{ even}$$

$$\text{Volt-seconds on each winding} = \frac{(N^2 - 1)\, V_S}{4N^2 f} \qquad \text{for } N \text{ odd}$$

This excitation can be equivalenced to a sinusoidal excitation voltage $V_e$ at frequency $f$, thus

$$V_e = \frac{\pi V_S}{4\sqrt{2}} \qquad N \text{ even}$$

$$V_e = \frac{\pi (N^2 - 1) V_S}{4N^2\sqrt{2}} \qquad N \text{ odd}$$

Since there are $N$ windings on the interphase reactor, its total winding rating requirement is

$$\text{VAW} = V_e I_L = \frac{\pi V_S I_L}{4\sqrt{2}} \qquad \text{volt-amperes, } N \text{ even}$$

$$= \frac{\pi (N^2 - 1) V_S I_L}{4N^2\sqrt{2}} \qquad \text{volt-amperes, } N \text{ odd}$$

The equivalent transformer rating is one-half the winding rating, i.e., the interphase reactor is equivalent in size to an $N$-phase transformer operating at $f$ hertz with a rating of

$$\text{VATE} = \frac{\pi V_S I_L}{8\sqrt{2}} \qquad \text{volt-amperes, } N \text{ even}$$

$$= \frac{\pi (N^2 - 1) V_S I_L}{8N^2\sqrt{2}} \qquad \text{volt-amperes, } N \text{ odd}$$

This component is basically traded for a reduction in current-sourcing reactor and filter capacitor size, both by a factor of $N^2$. The effectiveness of the technique in reducing passive component requirements is thus a function of the ripple current and voltage specifications. An equivalent inductor rating is obtained from

$$\text{VATE} = \pi f_e L I_L^2$$

where $f_e$ is the frequency at which size equivalence is sought, not necessarily the ripple frequency to which the inductor is subjected. An equivalent ac capacitor rating, if desired, is obtained from

$$\text{VACE} = \pi f_e C V_S^2 \qquad \text{for input capacitance}$$

$$= \pi f_e C V_L^2 \qquad \text{for output capacitance}$$

Generally, polyphase dc-to-dc converters are only effective if either or both of the following conditions are satisfied:

1. Parallel switching devices have to be used to carry the current.
2. Ripple limitations are very stringent.

If the first condition is not satisfied, increasing the converter operating frequency is normally a more effective design technique unless ripple constraints are extremely stringent.

The interphase reactor needed in practice will be somewhat larger than the simple equivalent transformer calculation indicates. This is because the combined effects of slight current imbalances among the $N$ branches of the polyphase converter and leakage flux from the core structure will result in a net dc magnetizing force on the core. This will have to be accommodated by gapping the core to avoid saturation and reducing the ac flux density allowed below that seen in normal transformer designs.

## *NATURALLY COMMUTATED CIRCUITS*

*By B. R. PELLY*

### AC-DC Converters

***53. General Considerations.*** The basic feature common to all the ac-dc converters described in the following paragraphs is that they are connected to a source of ac voltage which causes natural commutation. In most cases, power flow is from the ac terminals to a dc load, and the process is known as *rectification.* However, some members of this general family of converters can be controlled so that power flow occurs from the dc terminals back into the ac line. This process is known as *synchronous inversion* to distinguish it from inversion into a passive ac load. In the latter case, forced commutation is usually required.

The simplest member of this family of converters is the well-known *half-wave single-phase rectifier.* Although widely used for low-current dc power supplies, this type of rectifier is not used for higher power because of the large ripple voltage in its output and because the unidirectional current causes dc magnetization of transformer cores.

The number of different converter circuits is very large. Two basic configurations should first be distinguished, the *bridge circuit* (also known as double-way circuit) and the *midpoint circuit* (also known by the names center tap and single way).

Converter circuits are also distinguished according to whether the ac line is *single-phase* or *3-phase.* The effective number of phases of a 3-phase line can be further increased by connecting transformer windings to give intermediate phase shifts. Increasing the number of phases increases the *ripple frequency* of both the dc output voltage and the ac line current, making filtering easier. The ratio of the fundamental ripple frequency of the dc voltage to the ac line frequency is known as the *pulse number.*

If all the switching devices are diodes, the converter can operate only as an *uncontrolled rectifier* with the average dc output voltage fixed by the input ac voltage and by the circuit configuration. If half of the diodes in a bridge are replaced by thyristors, the average dc output voltage can be controlled by changing the phase angle at which the thyristors are fired, but the circuit is still capable only of rectifier action with power flow from the ac terminals to the dc terminals. Such circuits, known as *half-controlled converters* or *semiconverters,* belong to the category of *one-quadrant converters* since only one polarity of dc voltage and one polarity of dc current are possible.

Replacement of the diodes in a rectifier by thyristors produces a fully controlled converter, or *two-quadrant converter.* This type permits dc current to flow in only one direction, but the dc voltage may have either polarity. With one polarity, power flows from the ac to the dc terminals, and the converter acts as a *rectifier.* With the opposite polarity of dc voltage, net power flows from the dc terminals to the ac terminals, and the circuit acts as a *synchronous inverter.*

In some applications both polarities of both dc current and dc voltage must be permitted. This *four-quadrant action* can be achieved by interconnecting two similar two-quadrant converters, the combination being known as a *dual converter.*

The applications for this family of naturally commutated converters embrace a very wide range, including dc power supplies for electronic equipment, battery chargers, and speed controllers for fractional-horsepower motors, as well as dc supplies delivering many thousands of amperes for electrochemical and other industrial processes, high-performance reversing drives for dc machines rated at hundreds of horsepower, and high-voltage dc transmission in the megawatt power region.

Throughout the discussion which follows, unless otherwise stated, the following simplifying assumptions are made:

1. The voltage drop across switching devices is neglected while they are conducting, and the leakage current is neglected while they are blocking. Stray resistances are neglected.
2. Device turn-on and turnoff occur instantaneously.
3. The dc terminals are connected to an ideal filter (an infinite inductance), which suppresses all ripple current.

Ac-dc converters are treated in greater detail in Pelly[25] and Schaefer.[26]

***54. Two-Quadrant Converters.*** The circuit configurations, waveforms, and design relationships for various one- and two-quadrant converters are tabulated in subsequent paragraphs. In this paragraph, the operation of several typical circuits is discussed.

*Two-Pulse Midpoint Circuit.* Figure 22-113 shows a single-phase two-pulse midpoint converter and the associated source and load waveforms for various values of the firing delay angle $\alpha$. For $\alpha = 0$, the converter is equivalent to an uncontrolled rectifier, and the thyristors could be replaced by diodes. During the positive half cycle of the supply voltage, thyristor Th1 and transformer secondary S1 carry the load current, the voltage across the load is $v_{s1}$, and thyristor Th2 is reverse-biased. During the negative half cycle, Th2 and S2 carry the load current, the load voltage is $v_{s2}$, and Th1 is reverse-biased. The load-voltage waveform consists of a direct component $V_{d0}$ plus a superimposed ac ripple having a fundamental frequency which is twice the supply frequency (hence the name, two-pulse). The fundamental component of the supply current is in phase with the supply voltage.

As $\alpha$ is increased in the range $0 < \alpha < 90°$, the delay in firing causes the average load voltage to decrease, as shown in Fig. 22-113*b*. Note that the assumption of smooth load current, that is, constant throughout the cycle, implies a highly inductive load.* When the instantaneous load voltage goes negative, the reactive emf of the inductance forces power back into the source in order to maintain the current constant. However, over a half cycle, the net power flow is from the ac source to the dc load, and the supply current has a lagging power factor.

When $\alpha$ becomes equal to 90°, the instantaneous load voltage is negative for as long as it is positive, so that the average dc component of load voltage is zero (see Fig. 22-113*c*), and the supply current lags the supply voltage by 90°.

If $\alpha$ is increased beyond 90°, the continuous current flow can be maintained only if an external negative dc source is connected to the dc terminals. Net power flow is from the dc terminals to the ac terminals, and the converter is performing synchronous inversion (see Fig. 22-113*d*). Since the polarity of the average dc voltage has reversed, operation has shifted from quadrant I to quadrant IV. But because the current cannot reverse, quadrants II and III are forbidden. Hence this is a two-quadrant converter.

Finally, in Fig. 22-113*e*, $\alpha$ is nearly equal to 180°, and the dc voltage approaches its maximum negative value. In practice, $\alpha$ must be limited to about 160° or less to permit the thyristor which is being commutated off to regain its blocking ability before forward voltage is reapplied to it. Otherwise a commutation failure occurs. Operation in the inversion region is frequently described in terms of the advance angle $\beta = 180° - \alpha$. The margin of safety from commutation failure is described by the recovery angle $\delta$ between the completion of commutation and the next zero crossing at which forward voltage is reapplied.

*Three-Pulse Midpoint Circuit.* The simplest type of phase-controlled converter which operates from a 3-phase supply is the three-pulse midpoint circuit, shown in Fig. 22-114 with idealized waveforms. The zigzag connection of the transformer secondary windings prevents dc magnetization of the transformer core by permitting equal and opposite currents to flow in the two secondary windings in each phase.

The waveforms illustrate that this circuit has the same basic operating characteristics as the two-pulse circuit of Fig. 22-113. That is, continuous control of the mean dc terminal voltage from maximum positive to maximum negative is achieved by controlling the phase of the thyristor firing pulses through a theoretical range of 180°. This is accompanied by a continuously increasing shift in the phase of the input current from 0 to 180° lagging. In fact, these characteristics are common to all two-quadrant converters.

*Six-Pulse Midpoint Circuit.* The outputs of two three-pulse converters having mutually displaced input voltages can be combined in parallel through an interphase reactor as shown in Fig. 22-115. Each three-pulse converter operates independently of the other. In the ideal case, the load current is shared equally between the two groups, and there is no dc magnetization of the core of the interphase reactor. In practice, some relatively small unbalance of currents may occur.

---

*Under this assumption, Th1 cannot cease conduction until Th2 is fired. Therefore, each thyristor conducts for 180°. If, however, the load is purely resistive, each thyristor will cease conduction when its half of the supply voltage goes negative, and the current will pulsate.

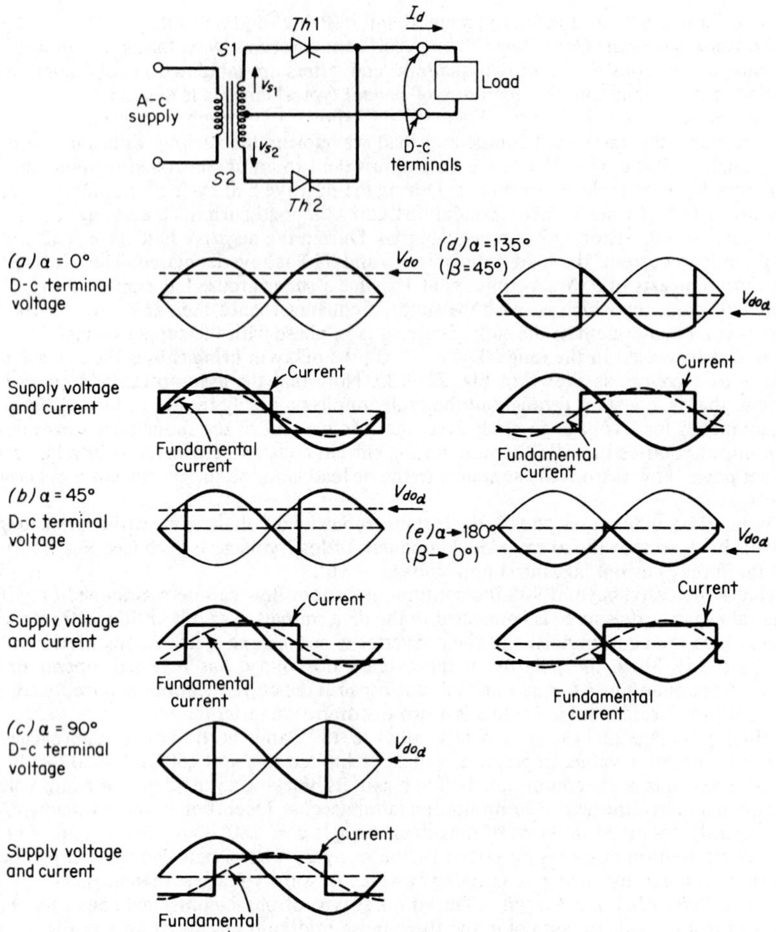


**FIG. 22-113** Two-pulse midpoint converter circuit and associated waveforms (smooth direct current assumed).

Because of the phase displacement between the ac ripple voltages at the dc terminals of the individual groups, the fundamental frequency of the ripple in the output voltage is six times the input frequency. The fundamental frequency of the ripple voltage across the interphase reactor is three times the input frequency.

If the interphase reactor is eliminated by making a solid connection between the dc terminals of the three-pulse groups, a six-pulse voltage waveform is still obtained at the output. However, the utilization factor of the circuit decreases because each thyristor then conducts for only 60° instead of the previous 120°.

*Six-Pulse Bridge Circuit.* The dc terminals of 2 three-pulse groups can be connected in series with one another to give an overall six-pulse operation. The resulting bridge, one of the most commonly used converter circuits, is shown in Fig. 22-116. So far as the ac lines are concerned, the bridge circuit contains two similar oppositely poled groups of rectifying devices; thus, it draws a balanced current from the line, ideally with no dc component.

*Higher Pulse Numbers.* Other converter-circuit configurations having higher pulse numbers can be constructed by connecting the dc terminals of individual groups, with suit-

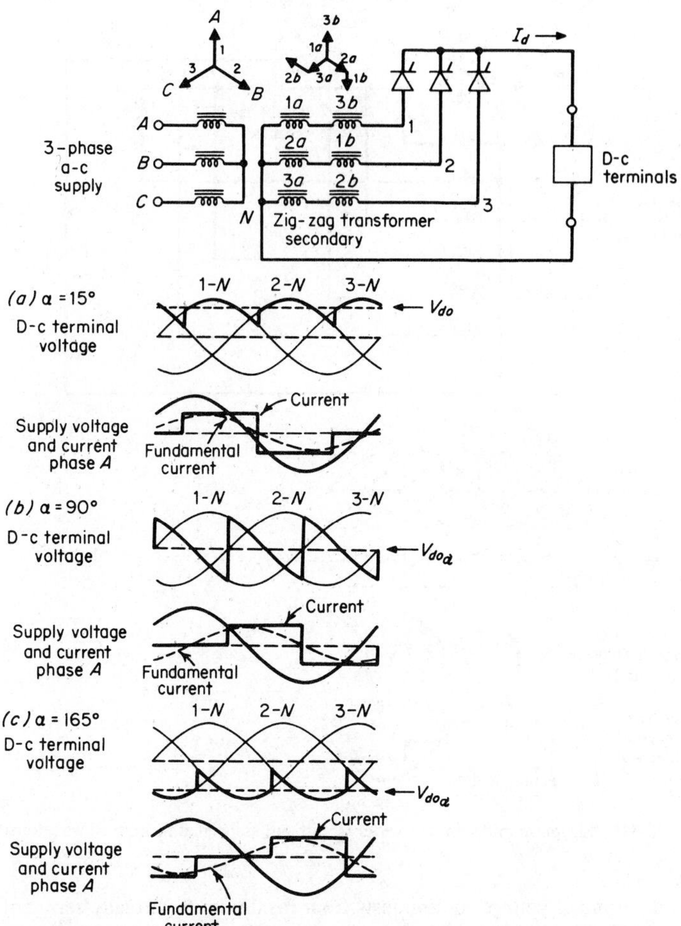


**FIG. 22-114** Three-pulse midpoint converter circuit and associated theoretical waveforms.

ably displaced ac voltages, in series or parallel with one another, or by combining series and parallel connections into one system.

In practice, a thyristor conduction angle of 120° is greatly preferred. Thus, almost all practical multipulse converter circuits comprise combinations of the basic three-pulse commutating groups. Each group within the system operates essentially independently of all the other groups. When the dc terminals of individual groups must be connected in parallel, the connections are made through interphase reactors to maintain independent operation of the groups. On the other hand, groups can be connected in series with "solid" connections at the dc terminals. Series connections of bridges, however, require isolation between the transformer secondaries connected to the individual bridges.

***55. One-Quadrant Converters.*** Many applications require operation with only one polarity of dc output voltage; that is, they operate only in the rectifying mode. In this case, it is generally advantageous to connect uncontrolled diodes into certain parts of the circuit.

In bridge-connected circuits (but not midpoint circuits) uncontrolled diodes can be used in place of half of the thyristors. With this *half-controlled converter,* it is possible to control

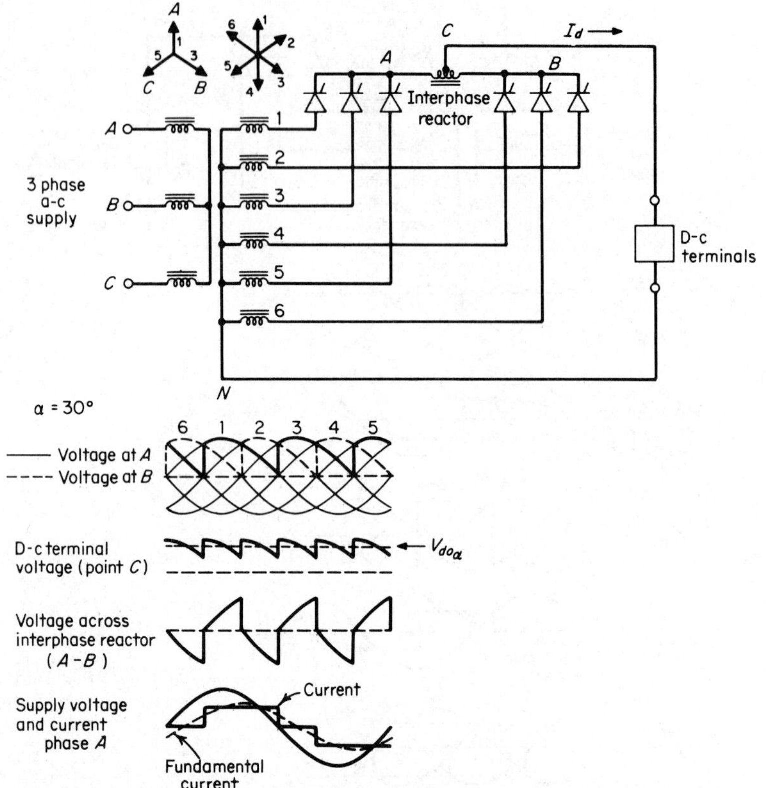


**FIG. 22-115** Six-pulse midpoint converter circuit and associated theoretical waveforms.

the mean dc terminal voltage continuously from maximum to virtually zero, but reversal of the mean voltage is not possible.

The half-controlled bridge has economic advantages over the fully controlled circuit because diodes are less expensive than equivalent thyristors. In addition, the input-power factor at relatively low levels of output voltage is improved over that of a fully controlled converter. Except for a single-phase bridge circuit, however, this advantage is obtained at the expense of a 2:1 reduction in ripple frequency at the dc terminals.

Either bridge or midpoint two-quadrant converters can be limited to one-quadrant operation by connecting a *freewheeling diode* across the dc terminals to conduct when the terminal voltage instantaneously tends to go negative. The diode reduces the ripple and improves the input-power factor for low dc output voltages. A further feature of the freewheel diode is that it provides a bypass path for inductive load currents if the supply lines become disconnected, thereby preventing reverse-voltage surges.

Both the half-controlled converter and the fully controlled converter with freewheeling diode have the advantage that the ratio of input current to dc output current decreases as the output voltage is reduced toward zero. In an ideal two-quadrant converter this ratio remains constant.

***56. Commutation Overlap.*** The preceding discussion and the idealized waveforms that have been shown assume instantaneous commutation of current from one thyristor which is turning off to the next which is being turned on. In practice, circuit inductance causes conduction in the two devices to overlap for a time that is usually not negligible. The process

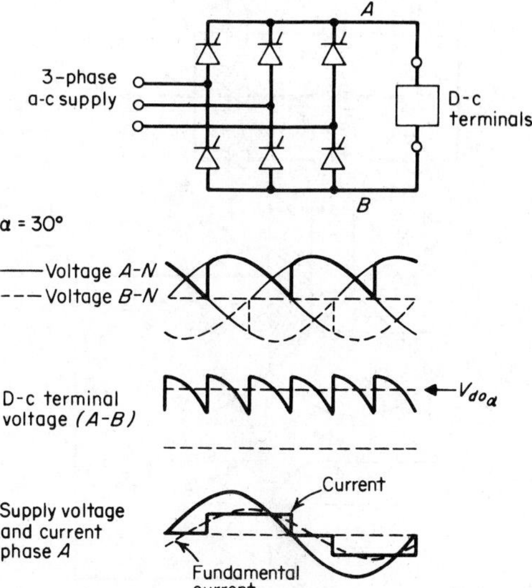


**FIG. 22-116** Six-pulse bridge converter and associated theoretical waveforms.

is known as commutation overlap, and its duration relative to the period of a cycle is expressed in terms of the overlap angle $u$.

The physical explanation of commutation overlap depends upon the fundamental voltage-current relationship of an inductor, $\Delta i = (1/L) \int v\, dt$, which states that the change in current is equal to the voltage-time area, that is, integral, divided by the inductance.* Hence transformer-leakage inductance and inductance in the ac line introduce a delay until the voltage-time area is sufficient to bring about the necessary redistribution of currents. During this delay, the current in the thyristor being turned on increases, and that in the thyristor turning off decreases at the same rate, since the total current is constant. If each thyristor has an equal series inductance, during the overlap the dc terminal voltage will be the average of the two source voltages to which the thyristors are connected.

The effects of commutation overlap are illustrated in Fig. 22-117 for a three-pulse group having inductance $L$ in series with each thyristor. The voltage at the dc terminal is shown as the current commutates from Th1 to Th2 after a phase delay $\alpha$. The voltage-time area required to change the current in $L_B$ from zero to $I_d$ is shown shaded and is subtracted from the ideal output-voltage waveform. The average value of voltage withheld from the dc terminals is directly proportional to the product of the direct current and the inductance, and it is independent of the firing-delay angle. Thus during the overlap, the output voltage follows the curve $(v_A + v_B)/2$.

In general, the relationship between the firing-delay angle $\alpha$ and the overlap angle $u$ for a three-pulse commutating group is

$$\cos \alpha - \cos(u + \alpha) = \sqrt{\frac{2}{3}}\,\frac{X_c I_d}{V_S}$$

For inverter operation, the corresponding relationship between the advance angle $\beta$ and the recovery angle $\delta = \beta - u$ is

*This relationship is widely useful in analyzing the smoothing action of an interphase or filter reactor and other aspects of converter operation.

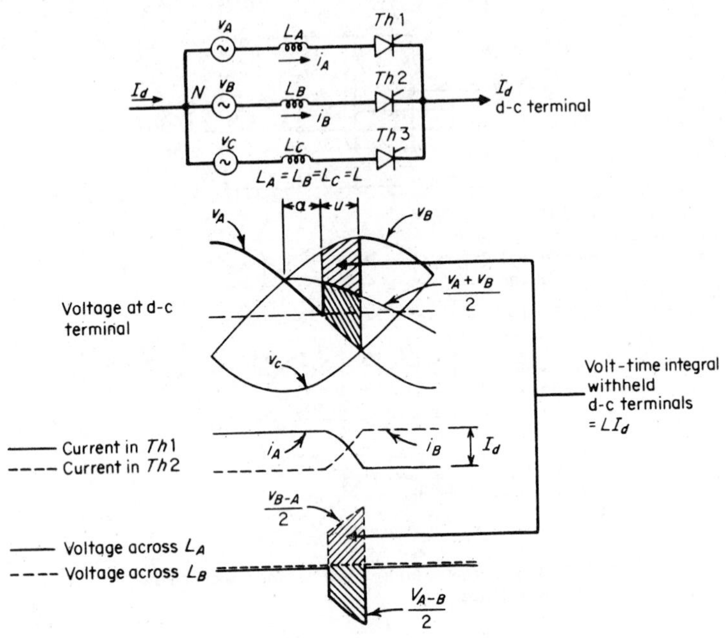


FIG. 22-117 The commutation process for a three-pulse group.

$$\cos \delta - \cos \beta = \sqrt{\frac{2}{3}} \frac{X_c I_d}{V_S}$$

For definitions of $X_c$, $I_d$, and $V_s$, see Table 22-4. These relationships are also valid for multipulse converters consisting of noninteracting three-pulse groups.

At the input side of a converter, the effect of commutation overlap is to cause rounding of the edges of the waveforms of line current. This means that the amplitudes of the higher-order harmonic terms are progressively reduced, compared with the theoretical amplitudes of these components with no overlap. In addition, the duration of each segment of the waveform of the input current is stretched by the overlap angle, resulting in a slight additional lagging phase shift of the fundamental component of current.

***57. Waveforms and Data for Converter Circuits.*** Table 22-4 lists the letter symbols most frequently used in the analysis of converter circuits. Tables 22-5 to 22-7 summarize the idealized waveforms and design relationships for the more common single-phase and 3-phase, one- and two-quadrant converters.

The relationships between firing angle and the principal harmonic components in the dc terminal voltage of these converters are shown in Figs. 22-118 and 22-119.

For all two-quadrant converters, the input-displacement factor is equal to the dc voltage ratio; and for all half-controlled converters, the input-displacement factor is equal to the square root of the dc voltage ratio. These relationships are indicated in the tables. The corresponding relationships for the converter circuits with freewheel diodes are illustrated in Fig. 22-120.

For all two-quadrant converters, the input-power factor and the dc voltage ratio are also directly proportional to one another, as indicated in the tables. The corresponding relationships for the one-quadrant converters are illustrated in Fig. 22-121.

**TABLE 22-4** Letter Symbols Used in the Analysis of Converter Circuits

| Symbol | Meaning |
|---|---|
| $V_S$ | Rms value of phase-to-neutral voltage at converter input terminals |
| $V_s$ | Peak value of $V_s$ |
| $V_n$ | Rms value of phase-to-neutral voltage at primary converter transformer |
| $h$ | Ratio of $V_n$ to $V_S$ |
| $V_d$ | Average value of voltage at dc terminals of converter under load, at any firing angle |
| $V_{d0\alpha}$ | Average value of voltage at dc terminals of converter at firing angle $\alpha$, with no commutation overlap |
| $V_{d_{max}}$ | Maximum possible average value of voltage at dc terminals of converter, obtained at $\alpha = 0°$, with no commutation overlap |
| $V_{FB}$ | Maximum instantaneous value of forward blocking voltage applied across thyristor |
| $V_{RB}$ | Maximum instantaneous value of reverse blocking voltage applied across thyristor |
| $V_D$ | Maximum instantaneous value of reverse blocking voltage applied across diode |
| $r$ | Ratio of $V_d$ to $V_{d,max}$ |
| $I_d$ | Direct current at output of converter |
| $I_1$ | Rms value of the fundamental component of converter input line current |
| $I_{1P}$ | Rms value of the "in-phase" or "power" component of $I_1$ |
| $I_{1Q}$ | Rms value of the "quadrature" or "reactive" component of $I_1$ |
| $I_{av,Th}$ ($I_{av,D}$) | Average value of thyristor (diode) current |
| $I_{rms,Th}$ ($I_{rms,D}$) | Rms value of thyristor (diode) current |
| $P_d$ | Average power at output of converter |
| $P_{d0}$ | Theoretical average power at output of converter, at $\alpha = 0°$ with no commutation overlap |
| $VA_0$ | Theoretical rms voltamperes of transformer windings at $\alpha = 0°$ with no commutation overlap |
| $L_s$ | Line-to-neutral commutating inductance at transformer secondary |
| $L_p$ | Line-to-neutral commutating inductance at transformer primary |
| $X_c$ | Commutating reactance at input frequency, referred to transformer secondary |
| $\alpha$ | Converter firing-delay angle, measured from the point at which the converter operates as if it were an uncontrolled rectifier circuit |
| $\beta$ | Inverter advance angle; angle in advance of the zero crossing of the line-to-line commutating voltage at which the commutation is initiated: $\beta = 180° - \alpha$ |
| $\delta$ | Inverter recovery angle; angle in advance of the zero crossing of the line-to-line commutating voltage at which the commutation is completed: $\delta = 180° - (\alpha + u)$ |
| $u$ | Commutation overlap angle |
| $u^*$ | Overlap angle for commutation of current into freewheeling path |
| $\phi$ | Displacement angle between fundamental component of converter input current and associated line-to-neutral voltage |
| $\cos \phi$ | Displacement factor of fundamental component of converter input current: $\cos \phi = I_{1P}/\sqrt{I_{1P}^2 + I_{1Q}^2}$ |
| $\lambda$ | Power factor at a given point in the converter input circuit; ratio of the average power to the rms voltamperes |
| $\mu$ | Distortion factor of the current at a given point in the converter input circuit; ratio of the rms value of the fundamental component to the total rms value $\mu = \lambda/(\cos \phi)$ |
| $p$ | Pulse number of converter = ratio of fundamental output ripple frequency to ac supply frequency (with steady delay angle) |
| $\omega$ | Angular frequency of input supply |

Figure 22-122 shows the principal-harmonic components present in the input line current of each of the converters shown in Tables 22-5 and 22-6.

All the above theoretical relationships assume ripple-free current at the dc terminals of the converter, with no commutation overlap.

***58. Four-Quadrant Converters.*** A four-quadrant converter, or *dual converter,* can operate with both polarities of both voltage and current at the dc terminals. Such converters permit, for example, dc motors to be driven and regeneratively braked in both forward and reverse directions. A six-pulse bridge four-quadrant converter formed by paralleling two oppositely polarized two-quadrant converters is shown in Fig. 22-123.

If both converters are active simultaneously, in principle one operates as a rectifier while the other operates as an inverter with the same average voltage. In practice, the instantaneous difference between the voltages of the two converters tends to cause a large circulating current. One solution is to parallel the two converters through a circulating-current reactor, as shown in Fig. 22-123. Another solution, which is usually preferable, involves deactivating the idle converter either by removing its firing pulses or by appropriately adjusting its relative delay angle. This control can be achieved automatically in various ways (Pelly[25]).

**TABLE 22-5** Waveforms and Data for Various Rectifier and Converter Circuits

| | 2-pulse midpoint rectifier | 2-pulse midpoint converter (2-quadrant) | 2-pulse midpoint converter with freewheel diode (1-quadrant) | 2-pulse bridge rectifier | 2-pulse bridge converter (2-quadrant) |
|---|---|---|---|---|---|
| Circuit | Circuit same as converter, with diodes instead of thyristors | $L_P$ $L_S$ | $L_P$ $L_S$ | Circuit same as converter, with diodes instead of thyristors | $L_P$ $L_S$ |
| D-c terminal voltage | | $V_d = \frac{2}{\pi}\hat{V}_S \cos\alpha - I_d \frac{X_c}{\pi}$<br>$X_c = \omega(\frac{2L_p}{h^2} + L_S)$<br>For harmonic distortion, see Fig. 9 | $V_d = \frac{2}{\pi}\hat{V}_S(\frac{1+\cos\alpha}{2}) - I_d \frac{X_c}{\pi}$<br>$X_c = \omega(\frac{2L_p}{h^2} + L_S)$<br>For harmonic distortion, see Fig. 10 | | $V_d = \frac{2}{\pi}\hat{V}_S \cos\alpha - I_d \frac{2X_c}{\pi}$<br>$X_c = \omega(\frac{L_p}{h^2} + L_S)$<br>For harmonic distortion, see Fig. 9 |
| Device voltage and current | | $V_{FB} = 2.0\hat{V}_S$ $I_{AV_{Th}} = 0.5I_d$<br>$V_{RB} = 2.0\hat{V}_S$ $I_{RMS_{Th}} = 0.707I_d$ | $V_{FB} = \hat{V}_S$ $I_{AV_{Th}} = 0.5I_d$, $I_{RMS_{Th}} = 0.707I_d$ $(\alpha = 0)$<br>$V_{RB} = 2\hat{V}_S$ $I_{AV_D} = I_d$, $I_{RMS_D} = I_d$ $(\alpha = \pi)$<br>$V_D = 2\hat{V}_S$ | | $V_{FB} = \hat{V}_S$ $I_{AV_{Th}} = 0.5I_d$<br>$V_{RB} = \hat{V}_S$ $I_{RMS_{Th}} = 0.707I_d$ |
| Transformer secondary voltage and current | | $I_{RMS} = 0.707I_d$ $\cos\phi = \cos\alpha = r$<br>$VA_o = 1.57P_{do}$ $\lambda = 0.637r$ | $I_{RMS} = 0.707I_d$ $(\alpha = 0)$<br>$VA_o = 1.57P_{do}$ $\cos\phi = \cos\frac{\alpha}{2} = \sqrt{r}$<br>$\lambda$ see Fig. 12 | | $I_{RMS} = I_d$ $\cos\phi = \cos\alpha = r$<br>$VA_o = 1.11P_{do}$ $\lambda = 0.9r$ |
| Transformer primary voltage and current | | $I_{RMS} = I_d/h$ $\cos\phi = \cos\alpha = r$<br>$VA_o = 1.11P_{do}$ $\lambda = 0.9r$<br>For harmonic distortion, see Fig. 13 | $I_{RMS} = I_d/h$ $(\alpha = 0)$<br>$VA_o = 1.11P_{do}$ $\cos\phi = \cos\frac{\alpha}{2} = \sqrt{r}$<br>$\lambda$-see Fig. 12<br>For harmonic distortion, see Fig. 13 | | $I_{RMS} = I_d/h$ $\cos\phi = \cos\alpha = r$<br>$VA_o = 1.11P_{do}$ $\lambda = 0.9r$<br>For harmonic distortion, see Fig. 13 |

| 2-pulse half-controlled bridge converter (1-quadrant) | 3-pulse midpoint rectifier | 3-pulse midpoint converter (2-quadrant) | 3-pulse midpoint converter with freewheel diode (1-quadrant) |
|---|---|---|---|
| 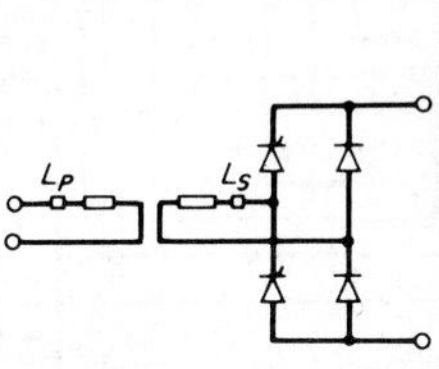 | Circuit same as converter, with diodes instead of thyristors | 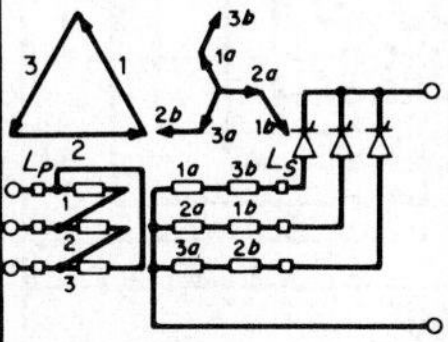 | 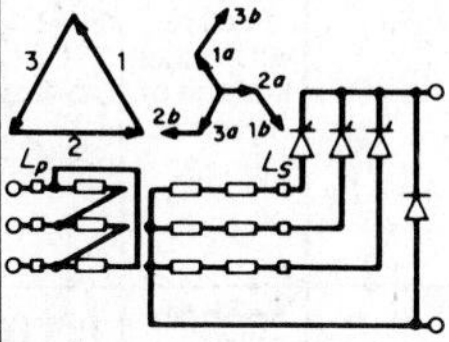 |
|  $V_d = \frac{2}{\pi}\hat{V}_S - I_d\,\frac{2X_c}{\pi}$ $(\alpha = 0)$<br>$= \frac{2}{\pi}\hat{V}_S\,\frac{(1+\cos\alpha)}{2} - I_d\frac{X_c}{\pi}$ $(\alpha \geqslant u^*)$<br>$X_C = \omega\left(\frac{L_p}{h^2} + L_S\right)$<br>For harmonic distortion, see Fig. 10 | 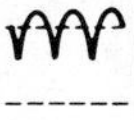 | 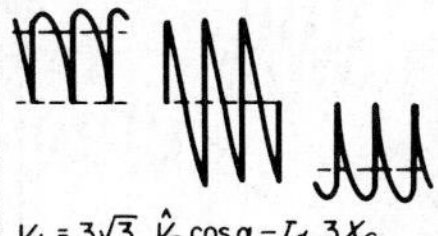 $V_d = \frac{3\sqrt{3}}{2\pi}\,\hat{V}_S\cos\alpha - I_d\,\frac{3X_C}{2\pi}$<br>$X_C = \omega\left(\frac{L_p}{h^2} + L_S\right)$<br>For harmonic distortion, see Fig. 9 |  $V_d = \frac{3\sqrt{3}}{2\pi}\hat{V}_S\cos\alpha - I_d\,\frac{3X_C}{2\pi}$ $(0 < \alpha \leqslant \frac{\pi}{6})$<br>$= \frac{3\sqrt{3}}{2\pi}\hat{V}_S\,\frac{\left[1+\cos(\alpha+\frac{\pi}{6})\right]}{\sqrt{3}} - I_d\,\frac{3X_C}{2\pi}$ $(\frac{\pi}{6} \leqslant \alpha \leqslant \frac{5\pi}{6})$<br>$X_C = \omega\left(\frac{L_p}{h^2} L_S\right)$<br>For harmonic distortion, see Fig. 10 |
| 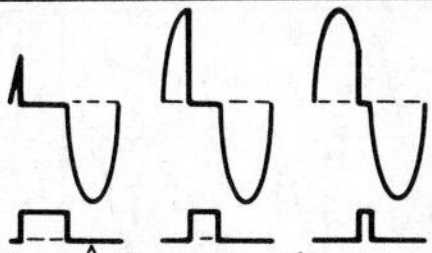 $V_{FB} = \hat{V}_S$<br>$V_{RB} = \hat{V}_S$<br>$V_D = \hat{V}_S$<br>$I_{AV_{Th}} = 0.5I_d$, $I_{RMS_{Th}} = 0.707I_d$ → $(\alpha = 0)$<br>$I_{AV} = I_d$, $I_{RMS} = I_d$ → $(\alpha = \pi)$ |  | 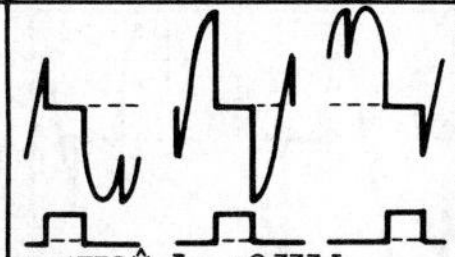 $V_{FB} = 1.732\hat{V}_S$ $I_{AV_{Th}} = 0.333I_d$<br>$V_{RB} = 1.732\hat{V}_S$ $I_{RMS_{Th}} = 0.577I_d$ | 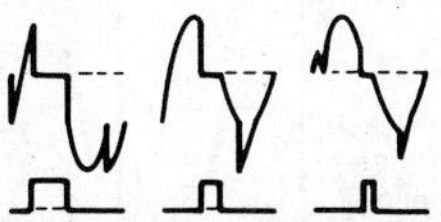 $V_{FB} = \hat{V}_S$<br>$V_{RB} = 1.732\hat{V}_S$<br>$V_D = 1.732\hat{V}_S$<br>$I_{AV_{Th}} = 0.333I_d$, $I_{RMS_{Th}} = 0.577I_d$ → $(0 \leqslant \alpha \leqslant \frac{\pi}{6})$<br>$I_{AV_D} = I_d$, $I_{RMS_D} = I_d$ → $(\alpha = \pi)$ |
| 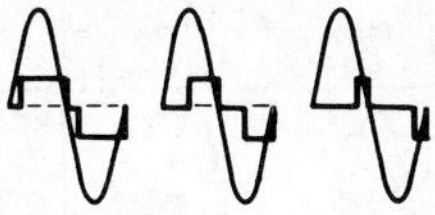 $I_{RMS} = I_d$ $(\alpha = 0)$<br>$VA_o = 1.11P_{do}$ $\cos\phi = \cos\frac{\alpha}{2} = \sqrt{r}$<br>$\lambda$ – see Fig. 12 |  | 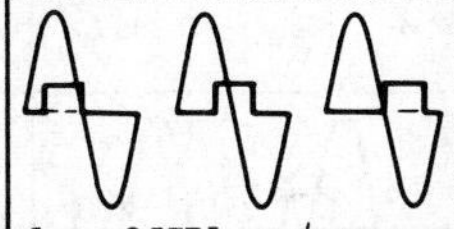 $I_{RMS} = 0.577I_d$ $\cos\phi = \cos\alpha = r$<br>$VA_o = 1.71P_{do}$ $\lambda = 0.585r$ | 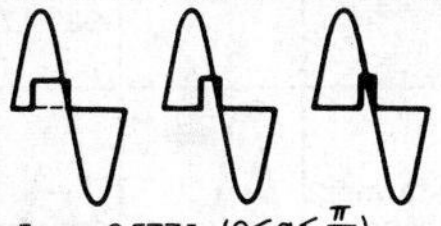 $I_{RMS} = 0.577I_d$ $(0 \leqslant \alpha \leqslant \frac{\pi}{6})$<br>$VA_o = 1.71P_{do}$ $\cos\phi, \lambda$, see Fig. 11, 12 |
| 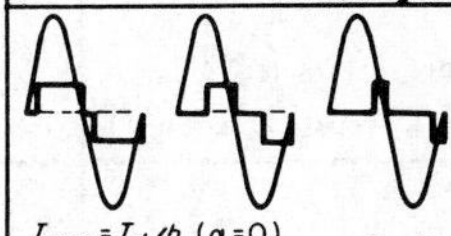 $I_{RMS} = I_d/h$ $(\alpha = 0)$<br>$VA_o = 1.11P_{do}$ $\cos\phi = \cos\frac{\alpha}{2} = \sqrt{r}$<br>$\lambda$ – see Fig. 12<br>For harmonic distortion, see Fig. 13 |  | 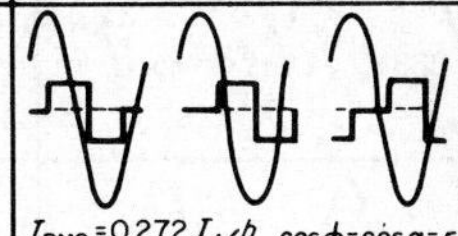 $I_{RMS} = 0.272\,I_d/h$ $\cos\phi = \cos\alpha = r$<br>$VA_o = 1.21P_{do}$ $\lambda = 0.827r$<br>For harmonic distortion, see Fig. 13 | 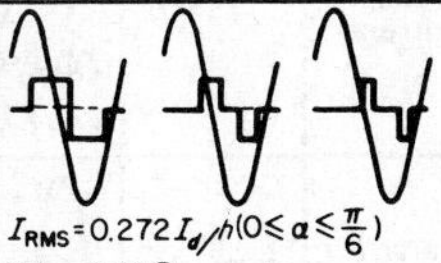 $I_{RMS} = 0.272\,I_d/h$ $(0 \leqslant \alpha \leqslant \frac{\pi}{6})$<br>$VA_o = 1.21P_{do}$ $\cos\phi, \lambda$ see Fig. 10, 11<br>For harmonic distortion, see Fig. 13 |

**TABLE 22-6** Waveforms and Data for Various Rectifier and Converter Circuits

| | 6-pulse midpoint rectifier | 6-pulse midpoint converter (2-quadrant) | 6-pulse midpoint converter with freewheel diode (1-quadrant) | 6-pulse bridge rectifier |
|---|---|---|---|---|
| Circuit | Circuit same as converter, with diodes instead of thyristors | 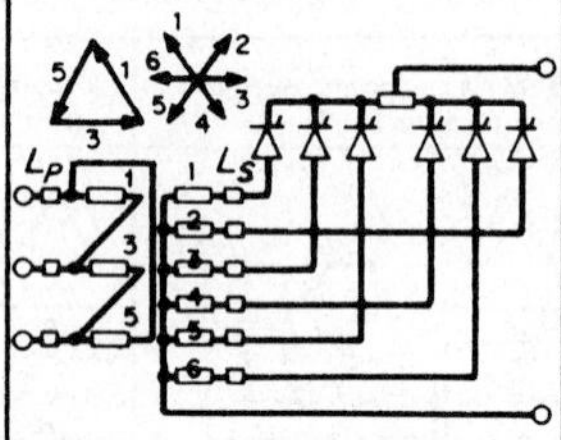 $L_P$, $L_S$ | 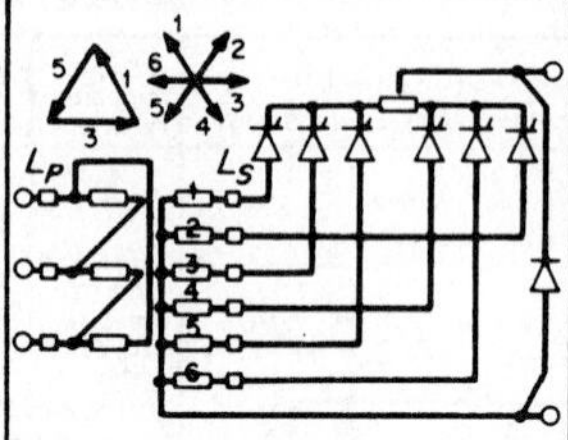 $L_P$, $L_S$ | Circuit same as converter, with diodes instead of thyristors |
| D-c terminal voltage | 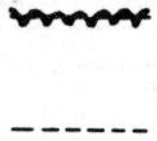 | 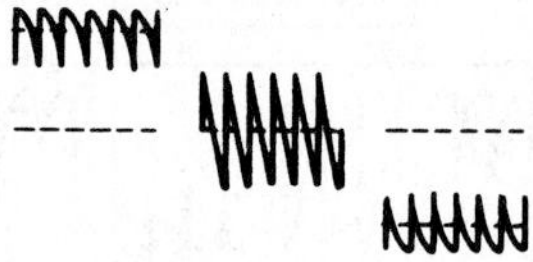 $V_d = \frac{3\sqrt{3}}{2\pi}\hat{V}_S \cos\alpha - I_d \frac{3X_c}{4\pi}$<br>$X_c = \omega\left(\frac{L_p}{h^2} + L_s\right)$<br>For harmonic distortion, see Fig. 9 |  $V_d = \frac{3\sqrt{3}}{2\pi}\hat{V}_S \cos\alpha - I_d \frac{3X_c}{4\pi}\ (0 \leqslant \alpha \leqslant \frac{\pi}{3})$<br>$= \frac{3\sqrt{3}}{2\pi}\hat{V}_S [1+\cos(\alpha + \frac{\pi}{3})] - I_d \frac{3X_c}{2\pi}$<br>$X_c = \omega\left(\frac{L_p}{h^2} + L_s\right)$ $\Big\vert\ (u^* + \frac{\pi}{3} \leqslant \alpha \leqslant 2\frac{\pi}{3})$<br>For harmonic distortion, see Fig. 10 |  |
| Device voltage and current |  | 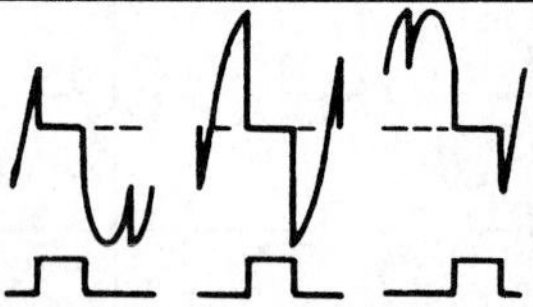 $V_{FB} = 1.732\ \hat{V}_S$ $\quad I_{AV_{Th}} = 0.167\ I_d$<br>$V_{RB} = 1.732\ \hat{V}_S$ $\quad I_{RMS_{Th}} = 0.288\ I_d$ | 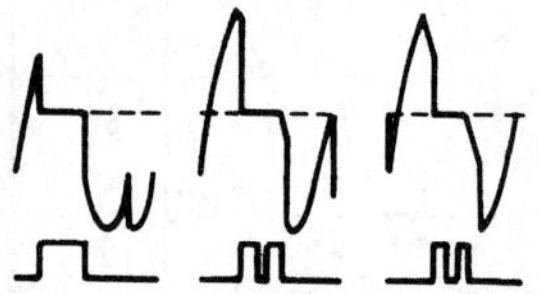 $V_{FB} = 1.5\ \hat{V}_S$ $\quad I_{AV_{Th}} = 0.167 I_d$ $\quad I_{AV_D} = I_d$<br>$V_{RB} = 1.732\hat{V}_S$ $\quad I_{RMS_{Th}} = 0.288 I_d$ $\quad I_{RMS_D} = I_d$<br>$(0 \leqslant \alpha \leqslant \frac{\pi}{3})$ $\quad (\alpha = \pi)$ |  |
| Transformer secondary voltage and current |  | 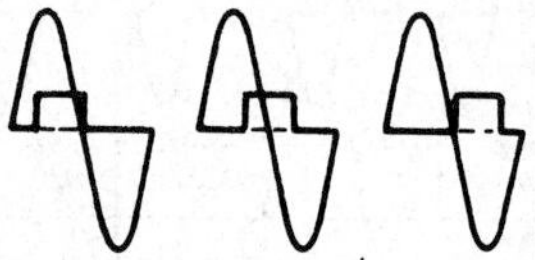 $I_{RMS} = 0.288\ I_d$ $\quad \cos\phi = \cos\alpha = r$<br>$VA_o = 1.48\ P_{do}$ $\quad \lambda = 0.675\ r$ | 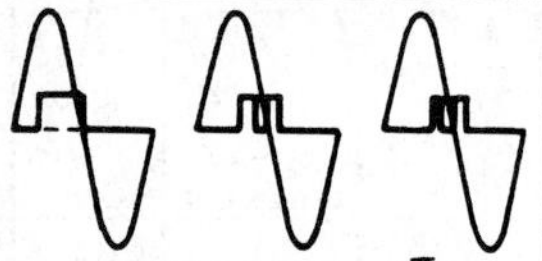 $I_{RMS} = 0.288 I_d$ $\quad (0 \leqslant \alpha \leqslant \frac{\pi}{3})$<br>$VA_o = 1.48\ P_{do}$ $\quad \cos\phi, \lambda$, see Fig. 11, 12 |  |
| Transformer primary voltage and current |  | 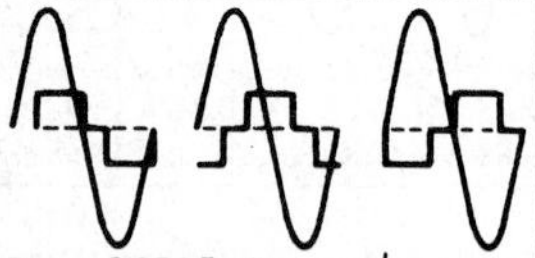 $I_{RMS} = 0.236 I_d/h$ $\quad \cos\phi = \cos\alpha = r$<br>$VA_o = 1.05 P_{do}$ $\quad \lambda = 0.955\ r$<br>For harmonic distortion, see Fig. 13 | 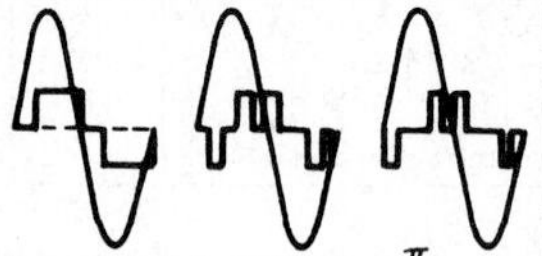 $I_{RMS} = 0.236 I_d/h\ (0 \leqslant \alpha \leqslant \frac{\pi}{3})$<br>$VA_o = 1.05\ P_{do}$ $\quad \cos\phi, \lambda$, see Fig. 11, 12<br>For harmonic distortion, see Fig 13. |  |

| 6-pulse bridge converter (2-quadrant) | 6-pulse bridge converter with freewheel diode (1-quadrant) | 3-pulse half-controlled bridge converter (1-quadrant) |
|---|---|---|
| 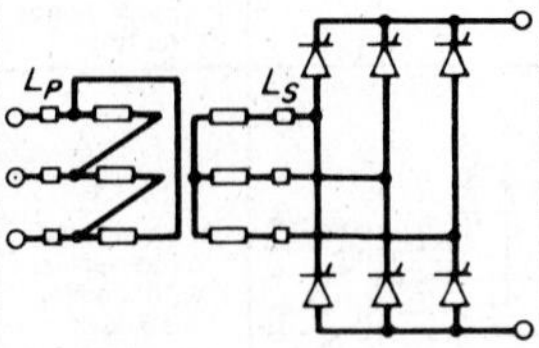 $L_p$, $L_s$ | 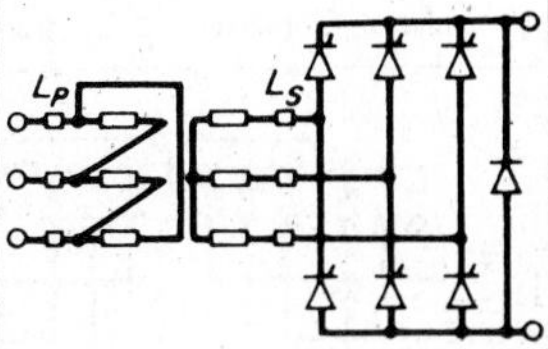 $L_p$, $L_s$ | 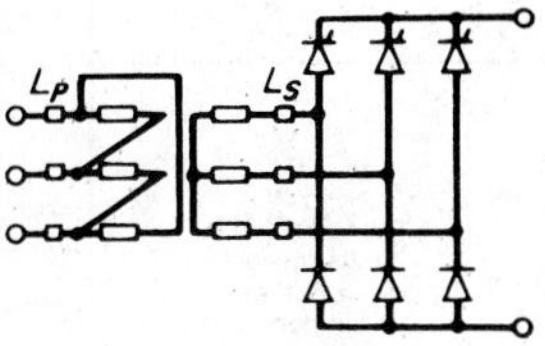 $L_p$, $L_s$ |
| 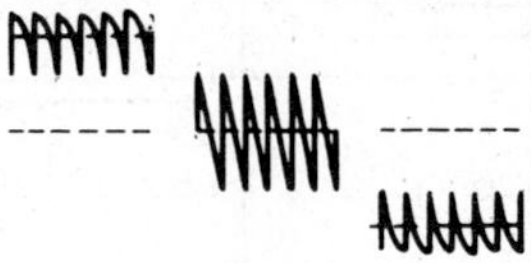 $V_d = 3\frac{\sqrt{3}}{\pi}\hat{V}_S \cos\alpha - I_d \frac{3X_c}{\pi}$<br>$X_c = \omega\left(\frac{L_p}{h^2} + L_s\right)$<br>For harmonic distortion, see Fig. 9 |  $V_d = \frac{3\sqrt{3}}{\pi}\hat{V}_S - I_d\frac{3X_c}{\pi}$ $\left(0 \leqslant \alpha \leqslant \frac{\pi}{3}\right)$<br>$= \frac{3\sqrt{3}}{\pi}\hat{V}_S\left[1+\cos\left(\alpha+\frac{\pi}{3}\right)\right] - I_d\frac{6X_c}{\pi}$<br>$X_c = \omega\left(\frac{L_p}{h^2} + L_s\right)$ $\left(u^* + \frac{\pi}{3} \leqslant \alpha \leqslant 2\frac{\pi}{3}\right)$<br>For harmonic distortion, see Fig. 10 | 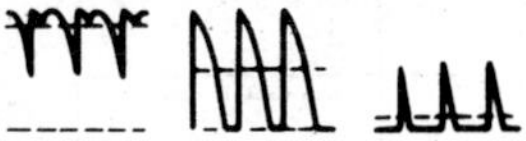 $V_d = \frac{3\sqrt{3}}{\pi}\hat{V}_S\left(\frac{1+\cos\alpha}{2}\right) - I_d\frac{3X_c}{\pi}$<br>$X_c = \omega\left(\frac{L_p}{h^2} + L_s\right)$<br>For harmonic distortion, see Fig. 10 |
| 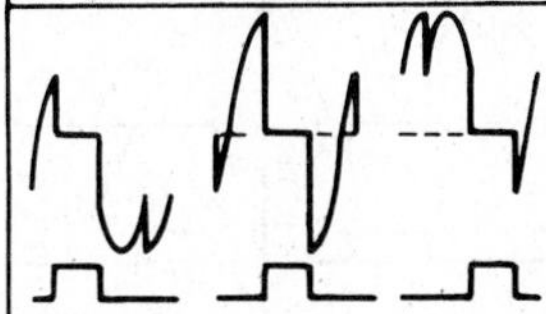 $V_{FB} = 1.732\,V_S$ $I_{AV_{Th}} = 0.333 I_d$<br>$V_{RB} = 1.732\,V_S$ $I_{RMS_{Th}} = 0.577 I_d$ | 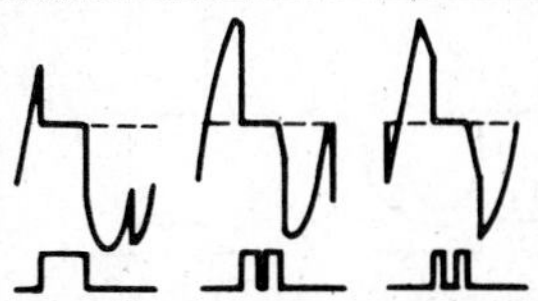 $V_{FB} = 1.5\hat{V}_S$ $I_{AV_{Th}} = 0.333 I_d$, $I_{RMS_{Th}} = 0.577 I_d$ $\left(0 \leqslant \alpha \leqslant \frac{\pi}{3}\right)$<br>$V_{RB} = 0.732\hat{V}_S$ $I_{AV_D} = I_d$, $I_{RMS_D} = I_d$ $(\alpha - \pi)$ | 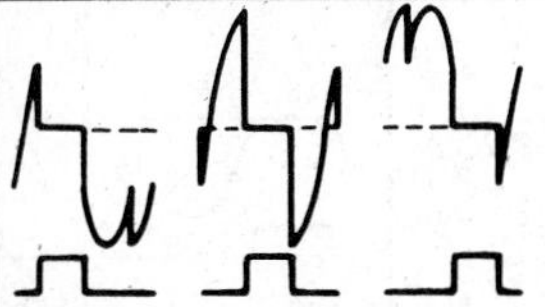 $V_{FB} = 1.732 V_S$ $I_{AV_{Th}} = 0.333 I_d$ $I_{AV_D} = 0.333 I_d$<br>$V_{RB} = 1.732 V_S$ $I_{RMS_{Th}} = 0.577 I_d$ $I_{RMS_D} = 0.577 I_d$ |
| 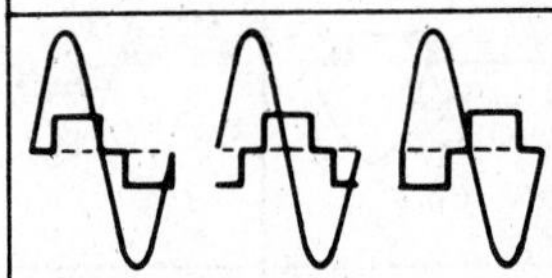 $I_{RMS} = 0.817 I_d$ $\cos\phi = \cos\alpha = r$<br>$VA_o = 1.05 P_{do}$ $\lambda = 0.955 r$ | 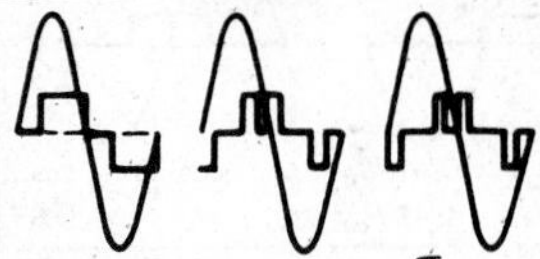 $I_{RMS} = 0.817 I_d$ $\left(0 \leqslant \alpha \leqslant \frac{\pi}{3}\right)$<br>$VA_o = 1.05 P_{do}$ $\cos\phi, \lambda$, see Fig. 11,12 | 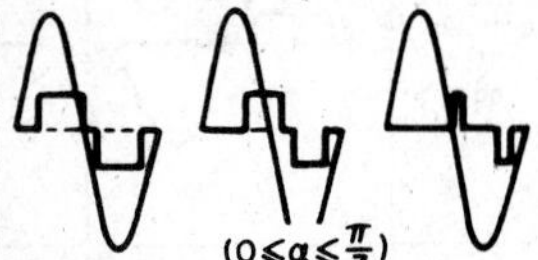 $I_{RMS} = 0.817 I_d$ $\left(0 \leqslant \alpha \leqslant \frac{\pi}{3}\right)$ $\cos\phi = \cos\frac{\alpha}{2} = \sqrt{r}$<br>$VA_o = 1.05 P_{do}$ $\lambda$-see Fig. 12 |
| 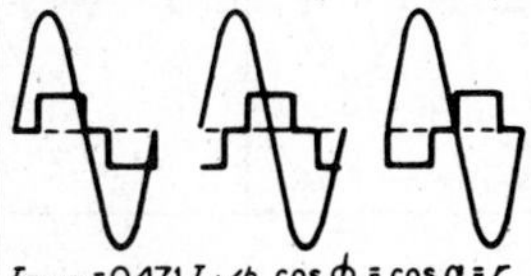 $I_{RMS} = 0.471 I_d/h$ $\cos\phi = \cos\alpha = r$<br>$VA_o = 1.05 P_{do}$ $\lambda = 0.955 r$<br>For harmonic distortion, see Fig. 13 | 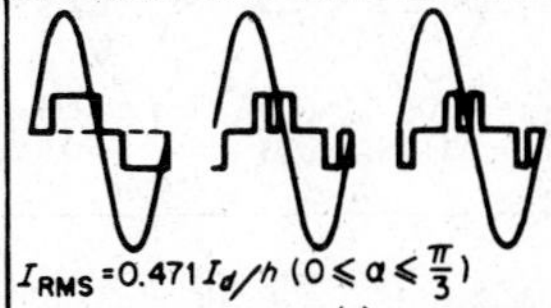 $I_{RMS} = 0.471 I_d/h$ $\left(0 \leqslant \alpha \leqslant \frac{\pi}{3}\right)$<br>$VA_o = 1.05 P_{do}$ $\cos\phi, \lambda$, see Fig. 11,12<br>For harmonic distortion, see Fig. 13 | 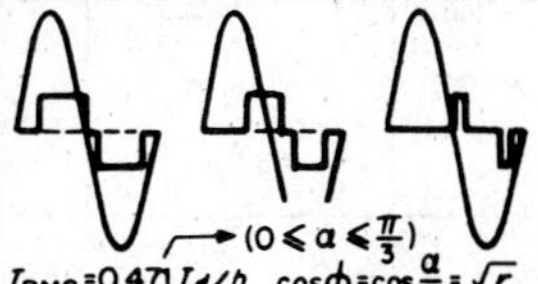 $I_{RMS} = 0.471 I_d/h$ $\left(0 \leqslant \alpha \leqslant \frac{\pi}{3}\right)$ $\cos\phi = \cos\frac{\alpha}{2} = \sqrt{r}$<br>$VA_o = 1.05 P_{do}$ $\lambda$, see Fig. 12<br>For harmonic distortion, see Fig. 13 |

**TABLE 22-7** Waveforms and Data for Various Rectifier and Converter Circuits

| | 12-pulse midpoint rectifier | 12-pulse midpoint converter (2-quadrant) | 12-pulse bridge rectifier |
|---|---|---|---|
| Circuit | Circuit same as converter, with diodes instead of thyristors | $L_P$ $T_1$ $L_S$ A B C; $L_S$ $T_2$ $L_P$ A B C | Circuit same as converter, with diodes instead of thyristors |
| D-c terminal voltage | | $V_d = \frac{3\sqrt{3}}{2\pi} \hat{V}_S \cos\alpha - I_d \frac{3X_C}{8\pi}$<br>$X_C = \omega\left(\frac{L_P}{h^2} + L_S\right)$<br>For harmonic distortion, see Fig. 9 | |
| Device voltage and current | | $V_{FB} = 1.732\ \hat{V}_S$ $\quad I_{AV\,Th} = 0.083\ I_d$<br>$V_{RB} = 1.732\ \hat{V}_S$ $\quad I_{RMS\,Th} = 0.144\ I_d$ | |
| Transformer secondary voltage and current | | $I_{RMS} = 0.144\ I_d$ $\quad \cos\phi = \cos\alpha = r$<br>$VA_o = 1.48\ P_{do}$ $\quad \lambda = 0.675\,r$ | |
| Transformer primary voltage and current | | $I_{RMS} = 0.204\ I_d/h\,(T1)$ $\quad \cos\phi = \cos\alpha = r$ $\quad VA = 1.05\ P_{do}$<br>$= 0.118\ I_d/h\,(T2)$ $\quad \lambda = 0.955 = r$<br>For harmonic distortion, see Fig. 13 | |

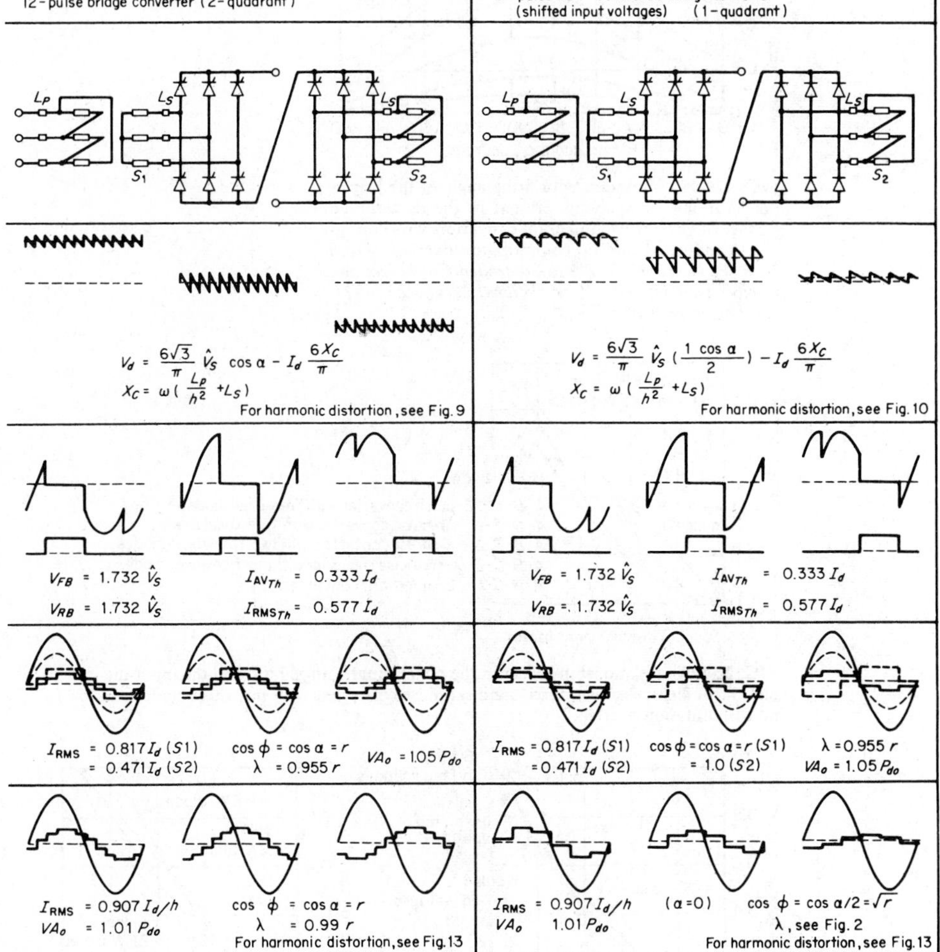

12-pulse bridge converter (2-quadrant)
6-pulse half-controlled bridge converter (shifted input voltages) (1-quadrant)
$L_P$
$L_S$
$S_1$
$S_2$
$V_d = \frac{6\sqrt{3}}{\pi}\hat{V}_S \cos\alpha - I_d\frac{6X_C}{\pi}$
$X_C = \omega(\frac{L_p}{h^2} + L_S)$
For harmonic distortion, see Fig. 9
$V_d = \frac{6\sqrt{3}}{\pi}\hat{V}_S (\frac{1 \cos\alpha}{2}) - I_d\frac{6X_C}{\pi}$
$X_C = \omega(\frac{L_p}{h^2} + L_S)$
For harmonic distortion, see Fig. 10
$V_{FB} = 1.732\,\hat{V}_S$
$V_{RB} = 1.732\,\hat{V}_S$
$I_{AV_{Th}} = 0.333\,I_d$
$I_{RMS_{Th}} = 0.577\,I_d$
$V_{FB} = 1.732\,\hat{V}_S$
$V_{RB} = 1.732\,\hat{V}_S$
$I_{AV_{Th}} = 0.333\,I_d$
$I_{RMS_{Th}} = 0.577\,I_d$
$I_{RMS} = 0.817\,I_d\,(S1)$
$= 0.471\,I_d\,(S2)$
$\cos\phi = \cos\alpha = r$
$\lambda = 0.955\,r$
$VA_o = 1.05\,P_{do}$
$I_{RMS} = 0.817\,I_d\,(S1)$
$= 0.471\,I_d\,(S2)$
$\cos\phi = \cos\alpha = r\,(S1)$
$= 1.0\,(S2)$
$\lambda = 0.955\,r$
$VA_o = 1.05\,P_{do}$
$I_{RMS} = 0.907\,I_d/h$
$VA_o = 1.01\,P_{do}$
$\cos\phi = \cos\alpha = r$
$\lambda = 0.99\,r$
For harmonic distortion, see Fig. 13
$I_{RMS} = 0.907\,I_d/h$
$VA_o$ $1.01\,P_{do}$
$(\alpha = 0)$
$\cos\phi = \cos\alpha/2 = \sqrt{r}$
$\lambda$, see Fig. 2
For harmonic distortion, see Fig. 13

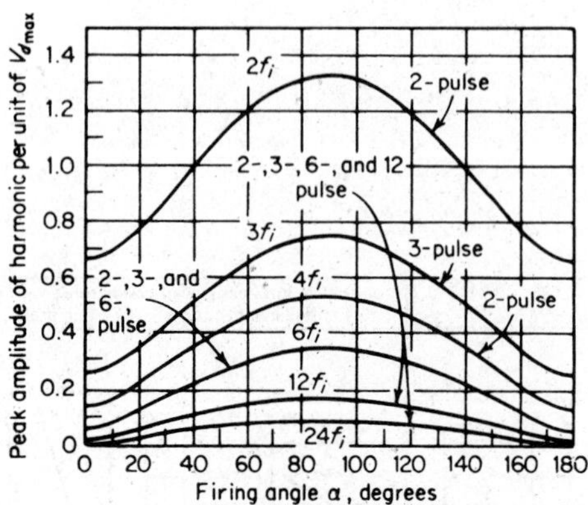


**FIG. 22-118** Variation with firing angle of the principal harmonic components present in the dc terminal voltage of various two-quadrant converters with continuous conduction and no commutation overlap. *(From B. R. Pelly, Thyristor Phase-Controlled Converters and Cycloconverters; New York, Wiley-Interscience, 1971. Used by permission.)*

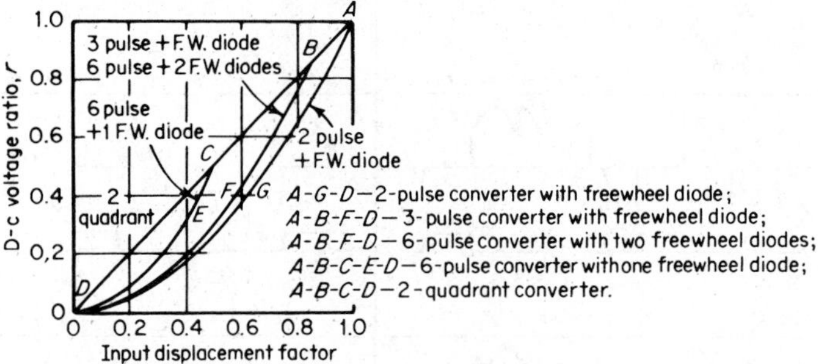


**FIG. 22-120** Relationships between the dc terminal voltage ratio and the input-displacement factor for two-, three-, and six-pulse converters with freewheel diodes and no commutation overlap.

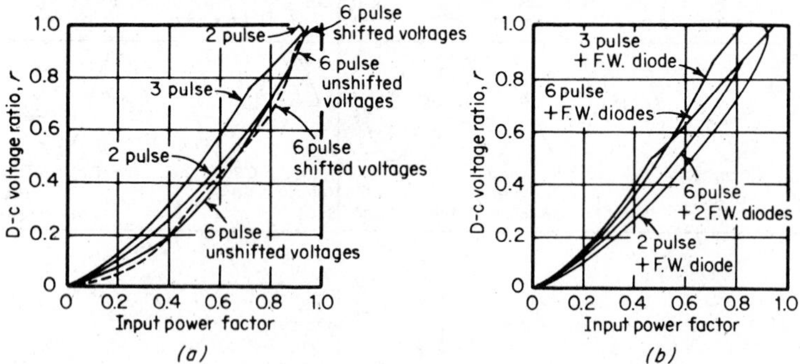


**FIG. 22-121** Relationships between the dc voltage ratio and the input power factor for various one-quadrant converters with no commutation overlap. These curves apply to transformer primary and secondary for bridge circuits and to primary for midpoint circuits. For transformer-secondary power factor of midpoint circuits, multiply by 0.707.

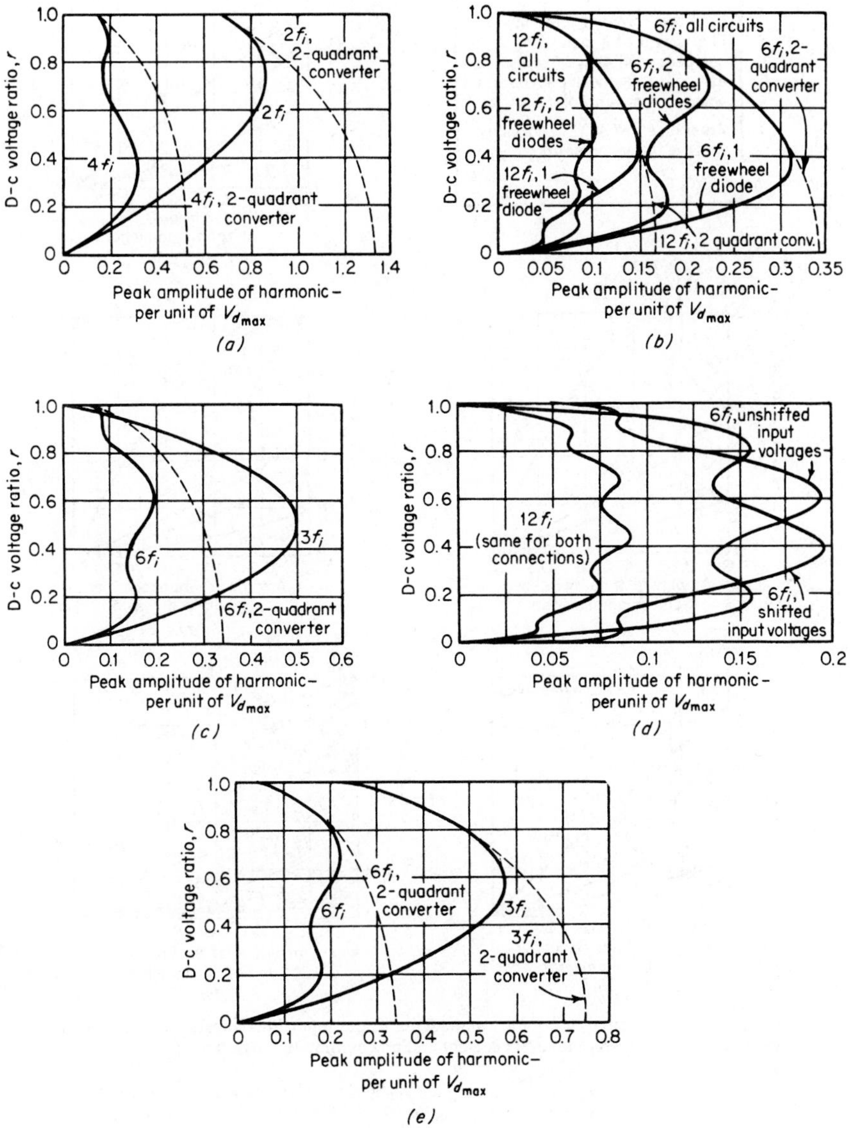


**FIG. 22-119** Principal harmonic components in the dc terminal voltage of various one-quadrant converters with no commutation overlap. Curves for corresponding two-quadrant converters are shown for comparison: (*a*) two-pulse half-controlled bridge circuit, and two-pulse midpoint circuit with freewheel diode; (*b*) six-pulse circuit with one and two freewheel diodes; (*c*) three-pulse half-controlled bridge circuit; (*d*) six-pulse half-controlled bridge circuit, with 30° "shifted" and "unshifted" input voltages for the two bridges; (*e*) 31-pulse circuit with freewheel diode. *(From B. R. Pelly, Thyristor Phase-Controlled Converters and Cycloconverters, New York, Wiley-Interscience, 1971. Used by permission.)*

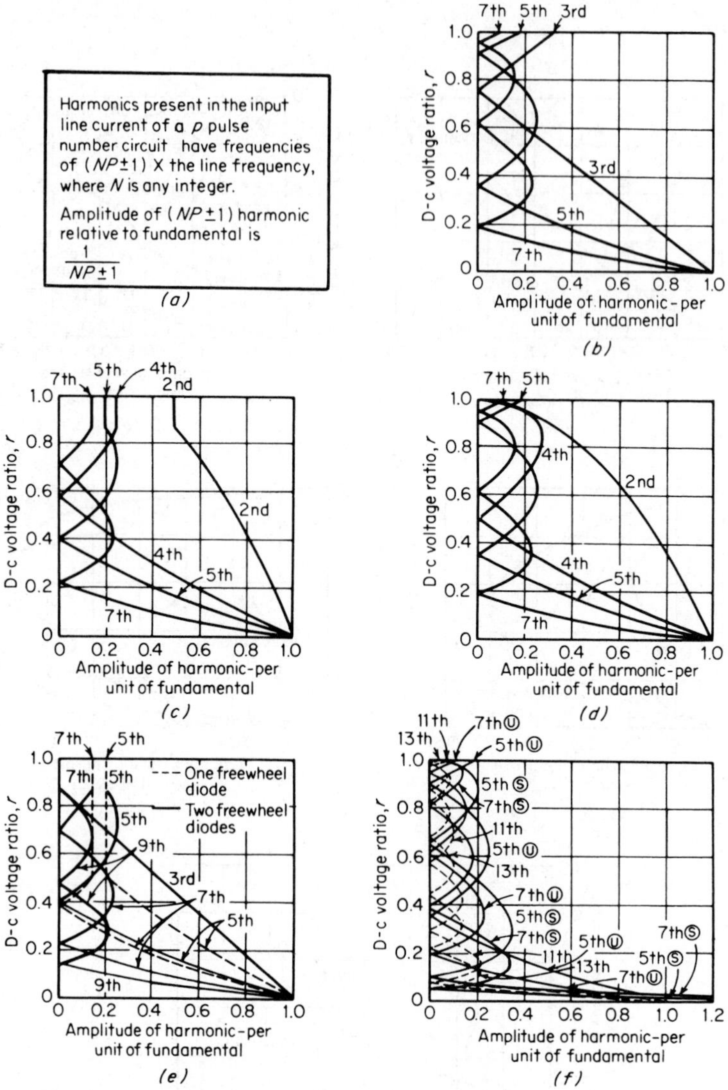

Harmonics present in the input line current of a $p$ pulse number circuit have frequencies of $(NP \pm 1) \times$ the line frequency, where $N$ is any integer.

Amplitude of $(NP \pm 1)$ harmonic relative to fundamental is $\frac{1}{NP \pm 1}$

(a)

(b)

(c)

(d)

(e)

(f)

**FIG. 22-122** Principal harmonic components in the input line current of various converter circuits, with smooth direct current and no commutation overlap.

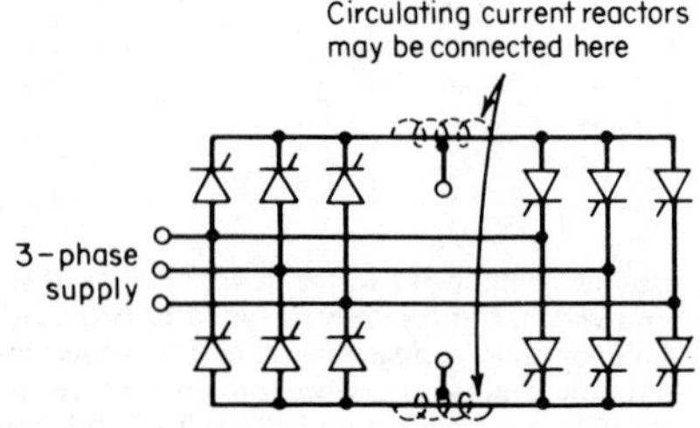


**FIG. 22-123** Six-pulse bridge dual-converter circuit.

## *INVERTERS*

*By T. M. HEINRICH and R. M. OATES*

***59. General Considerations.*** An inverter is a power converter in which the normal direction of power flow is from a dc source to an ac load. In contrast to naturally commutated converters, which may operate as synchronous inverters, as described previously (Pars. **53** to **58**), the thyristor inverters to be discussed here must be force-commutated unless the load happens to have a leading power factor. However, the power flow is still reversible. By properly phasing the control signals, the dc source can be made to absorb power from an active ac load, such as a synchronous motor which is being dynamically braked.

Table 22-8 shows the four basic inverter circuits, corresponding to a center-tapped load, a center-tapped source, and single- and 3-phase bridges. The relationships given were derived with the aid of the following simplifying assumptions: the switches operate instantaneously and have no voltage drop when closed or leakage current when open; the ideal filter removes all harmonics from the output voltage without attenuation or phase shift to the fundamental; and the load is resistive. As illustrated, the switches must block only one polarity of voltage, but they must be capable of conducting both polarities of current. In practice these switches are implemented by shunting a transistor or thyristor by a diode which carries the reverse current.

***60. Self-Commutated Inverters.*** The half-bridge thyristor inverter shown in Fig. 22-124 illustrates the principles of self-commutation. $R_0$ is the load resistor, $L_c$ and $C_c$ are the commutating components, and $L_F$ and $C_F$ constitute a simple filter to smooth the load-voltage waveform. When Th1 is turned on to initiate the first positive half cycle, $C_c$ charges to a voltage of $+E/2$. When the first negative half cycle is to be initiated, Th2 is turned on, causing the lower half of the source voltage to add to the voltage stored on $C_c$ across the lower half of $L_c$.

Because of the mutual coupling between the two halves of the commutating inductor, a voltage equal to $E$ is induced in the upper half. This voltage causes Th1 to become reverse-biased so that it begins to turn off. $C_c$ and $L_c$ are designed to hold a reverse bias on Th1 long enough for it to recover its blocking ability. Thus the firing of Th2 automatically transfers current from Th1 to Th2, and, following the commutation transient, the polarity of the load voltage reverses, causing $C_c$ to charge to $-E/2$. When Th1 is again fired, Th2 is turned off in a complementary manner, and the first full cycle of operation is complete.

During the time that $C_c$ is recharging toward $-E/2$, the current in the lower half of $L_c$ approximately doubles. Because of diode D2, the voltage across inductor $L_c$ cannot reverse by more than the combined forward drops of D2 and Th2. The excess current so trapped continues to circulate through D2 and Th2 until all the trapped inductor energy is dissipated. Aside from the extra dissipation in the devices, the trapped energy is bothersome in that it hinders the commutation process. If it is not removed, the inductor current for subsequent commutations will become progressively greater until $C_c$ can no longer supply sufficient commutating energy.

A resistor may be inserted in series with diodes D1 and D2 to dissipate the trapped energy, but for frequencies up to about 400 Hz, most of this energy can be recovered by employing a tapped transformer primary (McMurray and Shattuck[27]). A practical circuit with trapped-energy-recovery transformer is shown in Fig. 22-125. The tap at $n$ provides an additional voltage in the discharge loop to absorb energy from $L_c$. The energy absorbed by the transformer is passed along to the load or returned to the dc source if the load is unreceptive. The tap $n$ is generally placed at 10 to 20% of the primary turns, tending toward 20% if the dc input voltage is low and inverter frequency is high. Note that in Fig. 22-125 the commutating capacitor $C_c$ has been split between the +dc and −dc supplies. With this arrangement a center-tapped dc supply is unnecessary for commutation, and two half bridges can be combined into a full bridge, or three half bridges can be combined into a 3-phase bridge, as shown in Table 22-8.

The values of the commutating capacitor and commutating inductor are given by

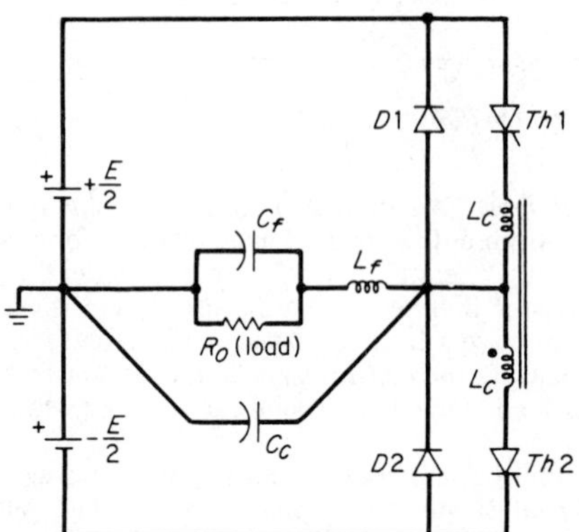


**FIG. 22-124** Half-bridge thyristor inverter with forced commutation.

$$C_c = \frac{t_r\hat{I}}{0.425E} \text{ farads} \quad \text{and} \quad L_c = \frac{t_rE}{0.425\hat{I}} \text{ henrys}$$

where $t_r$ is the turnoff time required by thyristor, $E$ is the total dc supply voltage, and $\hat{I}$ is the maximum thyristor anode current to be commutated.

***61. Auxiliary Commutation.*** Above 400 Hz, the trapped-energy problem of the McMurray-Shattuck circuit degrades circuit efficiency to such an extent that more complex circuits are justified. A half-bridge circuit which uses auxiliary thyristors for commutation is shown in Fig. 22-126. This circuit was suggested by W. McMurray. It has better voltage-regulation characteristics than the self-commutated circuit and can be used at frequencies up to about 5 kHz.

Operation of the circuit is initiated by firing Th1 and Th2A, thereby applying $+E/2$ to the load and charging the commutating capacitor $C$. When $C$ is fully charged, the current in Th2A goes to zero and it ceases conduction. To end the first half cycle, auxiliary thyristor Th1A is fired. Inductor $L$ limits the rate of current increase in D1 and Th1A. As the current in $L$ increases, the load current is diverted from Th1 to Th1A and $C$. After a delay of about 2.4 $\sqrt{LC}$ s, the forward drop across D1 reverse-biases Th1 and turns it off. Then Th2 is fired to begin the negative half cycle. In the meantime, $C$ charges to the opposite polarity for the next commutation before Th1A ceases conduction. Th2 is turned off by Th2A in the same way that Th1 was turned off by Th1A.

The values of the commutating components are given by

$$C = 0.893 \frac{\hat{I}t_r}{E} \text{ farads} \qquad L = 0.397 \frac{Et_r}{\hat{I}} \text{ henrys}$$

where $E$, $\hat{I}$, and $t_r$ are as defined previously.

The circuit as shown generates severe $dv/dt$ transients on all the thyristors, which require snubber circuits for protection (see Par. 77 on the protection of thyristors).

***62. Output-Voltage Waveform.*** For some applications, such as motor drives and dc-to-dc converters, a square-wave output from an inverter may be acceptable. Much of the time, however, sinusoidal voltage waveforms with limited total harmonic distortion are desired. A typical limit in equipment specifications would be 5% total harmonics relative to the magnitude of the fundamental frequency.

Various second- and third-order filter networks are commonly used to eliminate undesirable harmonics from the inverter output, but all tend to be large, heavy, costly, and, in general, highly load-dependent. For this reason, it is desirable to provide an inverter waveform which is inherently devoid of low-order harmonics. Higher-order harmonics can then be filtered with a relatively small network, producing an output waveform which is nearly sinusoidal. Common methods for producing such waveforms from square-wave inverters can be placed in two main categories: harmonic neutralization and pulse-width modulation.

**TABLE 22-8** Basic Inverter Circuits

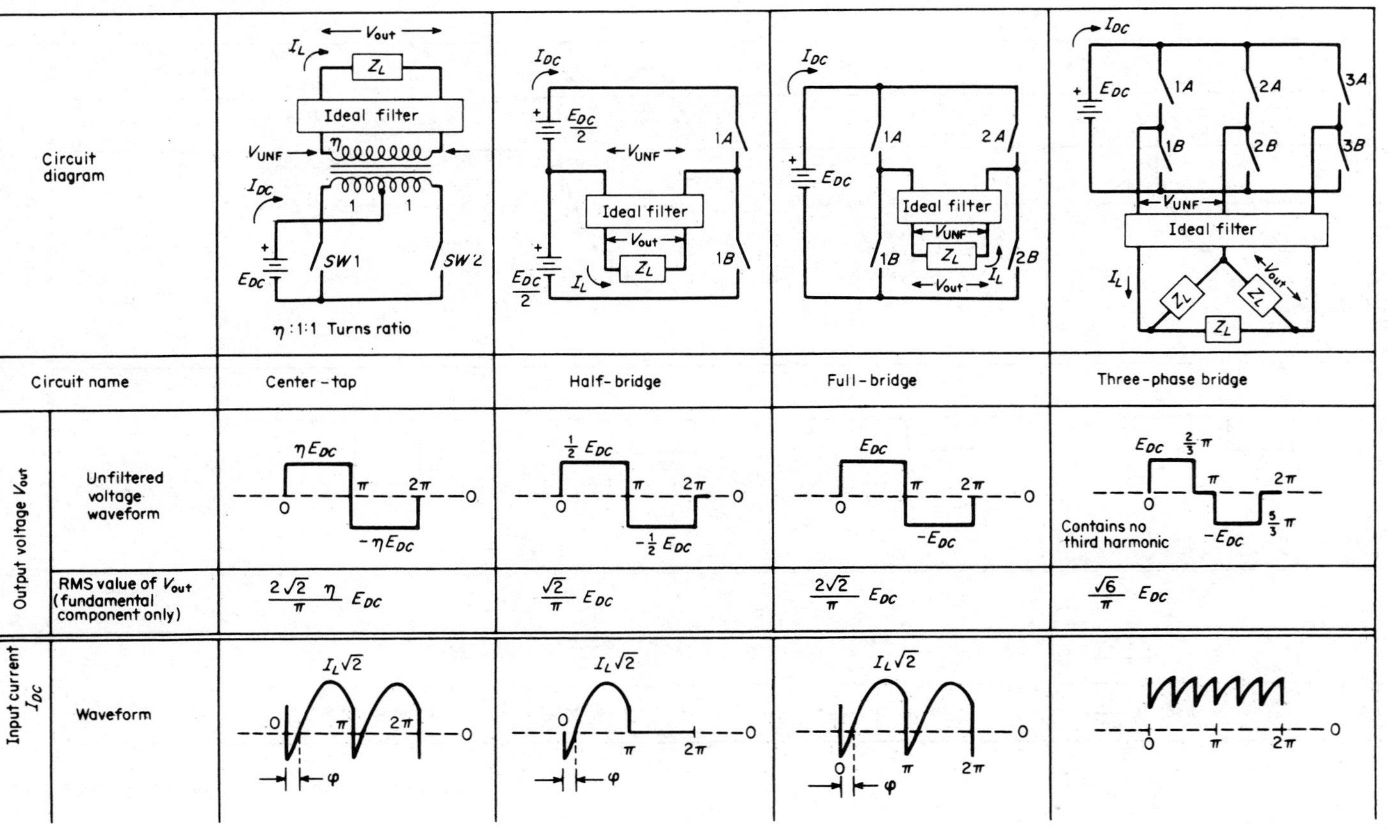

| | | Center-tap | Half-bridge | Full-bridge | Three-phase bridge |
|---|---|---|---|---|---|
| Circuit diagram | | $\eta$:1:1 Turns ratio | | | |
| Circuit name | | Center-tap | Half-bridge | Full-bridge | Three-phase bridge |
| Output voltage $V_{out}$ | Unfiltered voltage waveform | $\eta E_{DC}$, $-\eta E_{DC}$ | $\frac{1}{2} E_{DC}$, $-\frac{1}{2} E_{DC}$ | $E_{DC}$, $-E_{DC}$ | $E_{DC}$, $-E_{DC}$; Contains no third harmonic |
| | RMS value of $V_{out}$ (fundamental component only) | $\frac{2\sqrt{2}\,\eta}{\pi} E_{DC}$ | $\frac{\sqrt{2}}{\pi} E_{DC}$ | $\frac{2\sqrt{2}}{\pi} E_{DC}$ | $\frac{\sqrt{6}}{\pi} E_{DC}$ |
| Input current $I_{DC}$ | Waveform | $I_L\sqrt{2}$ | $I_L\sqrt{2}$ | $I_L\sqrt{2}$ | |

**TABLE 22-8** Basic Inverter Circuits (*Continued*)

| | | | | | |
|---|---|---|---|---|---|
| Input current $I_{DC}$ | $I_{DC}$ (avg value) | $\frac{2\sqrt{2}}{\pi}\ \eta\ I_L \cos\varphi$ | $\frac{\sqrt{2}}{\pi}\ I_L \cos\varphi$ | $\frac{2\sqrt{2}}{\pi}\ I_L \cos\varphi$ | $\frac{3\sqrt{2}}{\pi}\ I_L \cos\varphi$ |
| | $\frac{I_{PK}}{I_{DC}\text{ (avg)}}$ | $\frac{\pi}{2\cos\varphi}$ | $\frac{\pi}{\cos\varphi}$ | $\frac{\pi}{2\cos\varphi}$ | $\frac{\pi}{3\cos\varphi}\quad (0 \leqslant \omega \leqslant \frac{\pi}{6})$ |
| | $\frac{f_{\text{ripple}}}{f_{\text{inverter}}}$ | 2 | 1 | 2 | 6 |
| Switch stress | Voltage waveform | $2E_{DC}$; 0, $\pi$, $2\pi$; 0 | $E_{DC}$; 0, $\pi$, $2\pi$; 0 | $E_{DC}$; 0, $\pi$, $2\pi$; 0 | $E_{DC}$; 0, $\pi$, $2\pi$; 0 |
| | Current waveform | $\eta I_L\sqrt{2}$; $\eta I_L\sqrt{2}\cos\varphi$; 0; $\varphi$ | $I_L\sqrt{2}$; $I_L\sqrt{2}\cos\varphi$; 0; $\varphi$ | $I_L\sqrt{2}$; $I_L\sqrt{2}\cos\varphi$; 0; $\varphi$ | $I_L\sqrt{2}$; $I_L\sqrt{2}\cos\varphi$; 0; $\varphi$ |
| | RMS value of reverse current $I_{REV}$ | $\frac{1}{2}\eta I_L\ \sqrt{\frac{2\varphi - \sin 2\varphi}{\pi}}$ | $\frac{1}{2} I_L\ \sqrt{\frac{2\varphi - \sin 2\varphi}{\pi}}$ | $\frac{1}{2} I_L\ \sqrt{\frac{2\varphi - \sin 2\varphi}{\pi}}$ | $\frac{1}{2} I_L\ \sqrt{\frac{2\varphi - \sin 2\varphi}{\pi}}$ |
| | RMS value of forward current as a function of $I_{REV}$ | $\sqrt{\frac{\eta^2 I_L^2}{2} - (I_{REV})^2}$ | $\sqrt{\frac{I_L^2}{2} - (I_{REV})^2}$ | $\sqrt{\frac{I_L^2}{2} - (I_{REV})^2}$ | $\sqrt{\frac{I_L^2}{2} - (I_{REV})^2}$ |

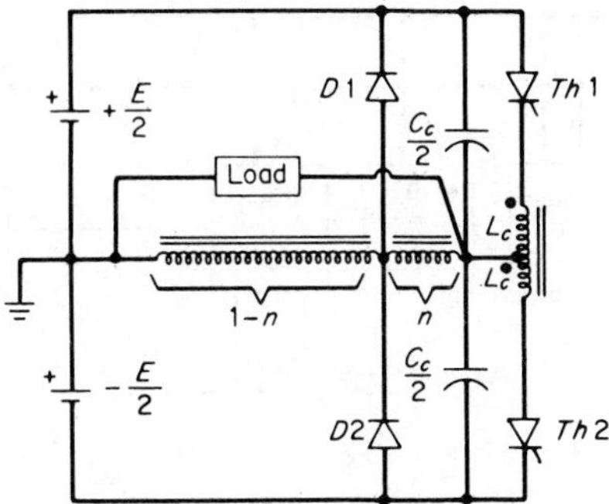


**FIG. 22-125** Half-bridge thyristor inverter with forced commutation and energy-recovery transformer.

Harmonic neutralization involves a combination of several phase-shifted square-wave inverters, each switching at the fundamental frequency, whereas pulse-width modulation involves switching a single inverter at a frequency higher than the fundamental (Kernick et al.[28]). Both schemes give satisfactory results, and actual selection of a method would depend on many factors such as the output-power level, the fundamental frequency, the speed of the switching devices, and the type of commutation circuit. Harmonic neutralization is especially suited for 3-phase outputs.

***63. Harmonic Neutralization.*** A harmonic-neutralized inverter consists of $N$ square-wave inverter stages which are sequentially phase shifted by $180/N$ electrical degrees (Kernick and Heinrich,[29] Heinrich[30]). In general, for a polyphase ac output, each inverter stage contributes to the output of each phase by means of a process of phasor addition performed by transformer windings. In place of an overall square-wave output containing all odd harmonics of the fundamental frequency, the output voltage is a stepped approximation to a sine wave in which most of the harmonics have been neutralized. The remaining harmonics occur in pairs and have frequencies of $2kN \pm 1$, where $k = 1, 2, 3, \ldots$. The amplitudes of the harmonics which remain, relative to the fundamental, are inversely proportional to their frequencies, as in a square wave.

Each stage of the inverter will share the total output power equally if the load is balanced, but the voltage contributed to each phase by each stage must be properly adjusted. In general, these voltages will not be equal but are given by $(\pi V_{rms} \cos \Psi_{MW})/\sqrt{2}N$, where $V_{rms}$ is

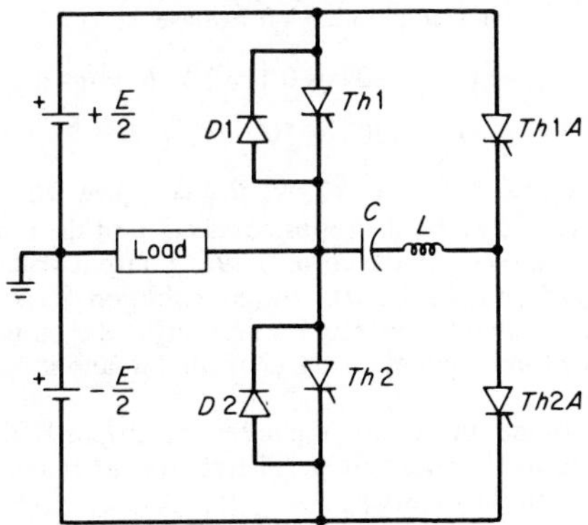


**FIG. 22-126** Auxiliary impulse-commutated inverter.

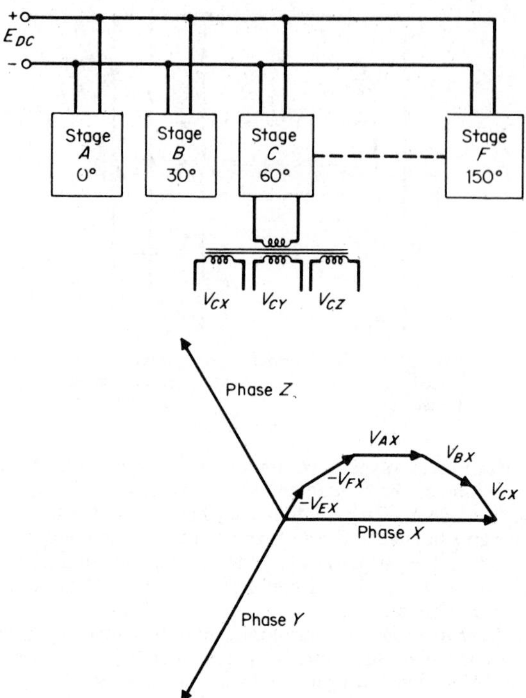


**FIG. 22-127** Six-stage harmonic-neutralized inverter.

the desired line-to-neutral output volage and $\Psi_{MW}$ is the phase angle between stage $M$ and phase $W$.

As an example, consider the 3-phase six-stage inverter shown in Fig. 22-127. The firing angles of the respective stages are separated by 180°/6 = 30°, giving the phasor diagrams shown for the fundamental components of the individual stages and phases. It is assumed that stage $A$ and phase $X$ are each arbitrarily assigned at 0° phase angle and the line-to-neutral voltage is to be 120 V. Using the relationship above, the transformer-turns ratio in the various stages should be chosen as follows for phase $X$:

$$V_{AX} = (\pi \times 120 \cos 0°)/\sqrt{2} \times 6 = 44 \text{ V}$$

$$V_{BX} = (\pi \times 120 \cos 30°)/\sqrt{2} \times 6 = 38 \text{ V}$$

Similarly, $V_{CX} = 22$ V, $V_{DX} = 0$, $V_{EX} = -22$ V, $V_{FX} = -38$ V. Since $V_{DX} = 0$, no winding is needed, and phase $X$ is formed by the series connection of the other five windings.

The individual square waves and the corresponding output-voltage waveform for phase $X$ are shown in Fig. 22-128. In a similar way, the contribution of each stage to the other two phases can be calculated. The only harmonics present in the output waveform and their amplitudes relative to the fundamental are the eleventh (1/11) and thirteenth (1/13), twenty-third (1/23) and twenty-fifth (1/25), etc.

The ripple frequency of the current into the inverter is $2N$, or 12 times the line frequency, thereby reducing the size of the input filter. The combined rating of the transformers is about 1.4 times the rating of the inverter.

Although the inverter described synthesizes the output from isolated single-phase stages, many variations are possible depending on the particular application. For instance, the same result could be achieved using a pair of 3-phase bridge inverters and 3-phase transformers (Oates[31]).

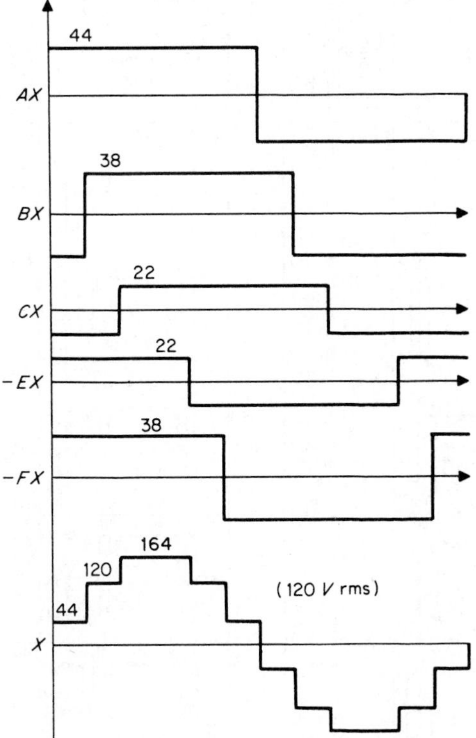


**FIG. 22-128** Waveforms from individual inverter stages summed to form phase *X*.

***64. Pulse-Width Modulation (PWM).*** Pulse-width-modulated inverters approximate sine-wave outputs by switching the power elements at a rate higher than the fundamental frequency. These inverters, categorized by the manner in which this switching takes place, fall into three basic groups: programmed waveform, modulated carrier, and optimum response (bang-bang) (Kernick and Haque[32]). Although the harmonic content of the output varies in the three methods, the distribution of the harmonics is always more favorable with higher switching rates.

In a *programmed-waveform PWM inverter,* the power stage or inverter is given a fixed switching pattern which is periodic. This pattern is designed to produce the best possible waveform for the number of switching operations permitted per cycle. Center-tap and half-bridge inverters are capable of providing positive or negative but not zero instantaneous output. The output from such an inverter is called a *noncommutated waveform.* The full-bridge inverter can also produce a zero instantaneous voltage, and its output is called *commutated.* In general, the commutated waveforms give lower distortion for the same number of switching operations per cycle. Figure 22-129 presents a summary of useful programmed waveforms of both types along with their harmonic content.

*Carrier-modulated PWM* is usually accomplished by comparing a reference sine wave at the fundamental frequency to a sawtooth signal having a fixed frequency higher than the fundamental (Ravas et al.[33]). The power elements are switched at the zero crossing of these two signals, as shown in Fig. 22-130. Distortion of the output waveform occurs at the carrier frequency and its sidebands and at multiples of the carrier frequency and their associated sidebands. This distortion may or may not be harmonic, depending on whether or not the carrier frequency is synchronized with the fundamental reference. The magnitude of the

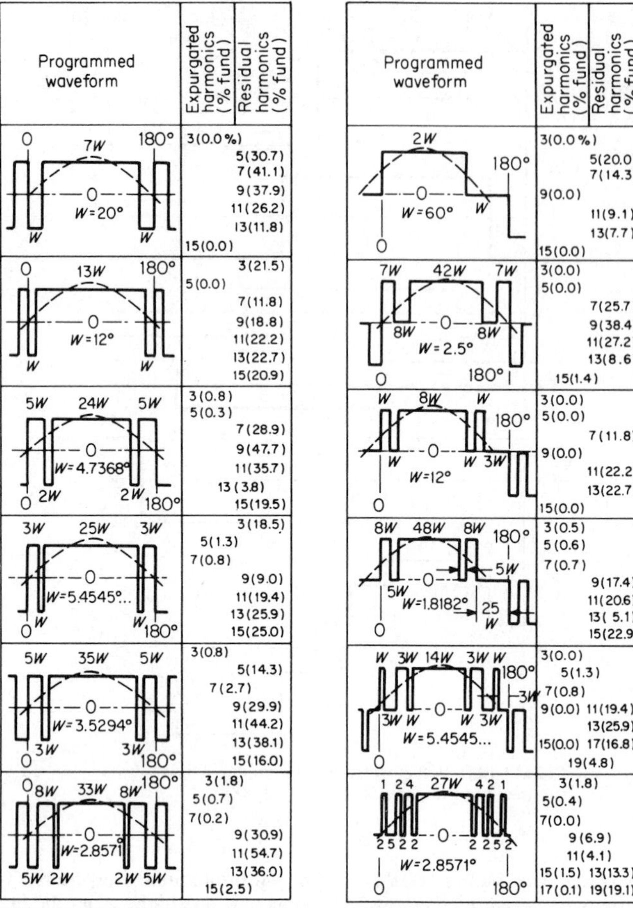


**FIG. 22-129** Summary of programmed waveforms and their harmonic content. $W$ is the unit increment of time for each waveform in degrees.

distortion depends on the degree of modulation (relative magnitude of the sine-wave peak compared to the carrier peak) and is lowest at 100% modulation.

Another type of PWM, known as *optimum-response switching,* is shown in Fig. 22-131 (Geyer and Kernick[34]). This scheme, unlike the others, must operate with an output filter, and it must have closed-loop control. Hysteresis in the feedback path sets the allowable deviation of the output from a sinusoidal reference. The switching rate varies throughout the cycle and is determined by the amount of hysteresis and by the characteristics of the load and filter. Very high switching rates are generally required to keep the error small.

The control for such an inverter is very simple, and voltage regulation is automatically accomplished. However, many applications will not permit optimum-response PWM because of the inherent voltage ripple and the asynchronous output waveform.

***65. Voltage Control.*** Most inverter applications require direct control of the output voltage. For motor-drive inverters, it must be continuously variable from zero to full value, depending upon torque and speed requirements. For ac power-supply inverters, the voltage must be held nearly constant over a certain load and input range. In addition, many inverters are required to provide a specified amount of current into a short circuit, making it necessary to cut back the output voltage to nearly zero. A typical load profile is shown in Fig. 12-132.

Varying the dc input voltage and internal pulse-width control are the most common methods of controlling the output voltage where this control is not inherent, as it is in carrier-modulated PWM and optimum-response inverters.

*DC-Input Control.* Control of the dc input is the most straightforward method. If an inverter's switching pattern remains invariant, the output voltage is directly proportional to the dc-input voltage for all types of inverters. If the power source is ac, a phase-controlled rectifier can be used to control the dc input to the inverter. If the power source is dc, it is necessary to use a dc regulator, that is, a chopper.

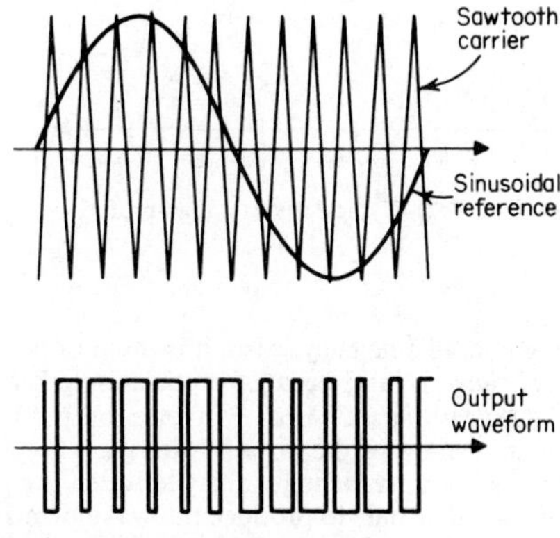


**FIG. 22-130** Carrier-modulated pulse-width-modulation waveforms.

The main advantages of using dc-input control are that the switching requirements and control complexity are not increased. In addition, the harmonic content of the output-voltage waveform does not vary with the input voltage. However, it has the disadvantage that the power must often pass through an extra stage of conditioning, thus reduc-

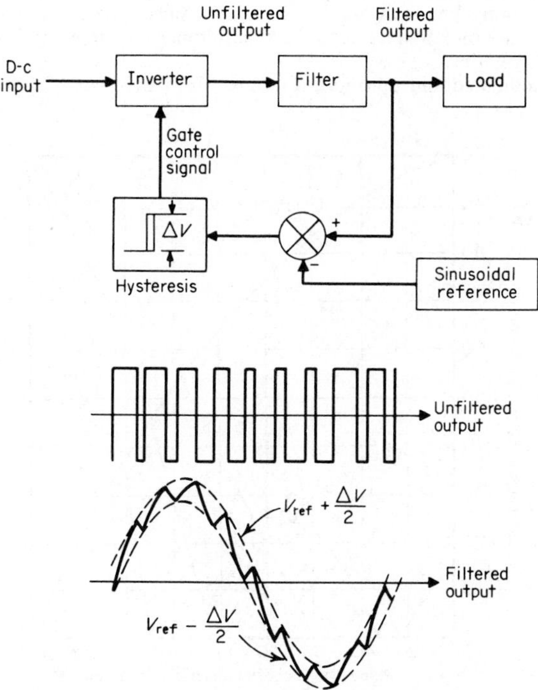


**FIG. 22-131** Optimum-response pulse-width modulation.

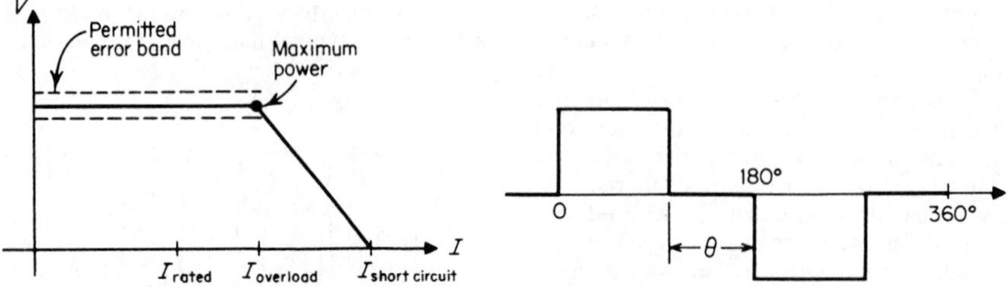


**FIG. 22-132** Typical inverter load profile.

**FIG. 22-133** Pulse-width voltage control of a full-bridge inverter.

ing overall efficiency. Also, it is often impossible to use input-voltage control when the control range is large because reliable forced commutation depends on input voltage.

*Pulse-Width Control.* Inverter output voltage can also be controlled by varying the conduction time of the power switches. The pulse width of a full-bridge inverter can be controlled by introducing a delay between the turnoff of each pair of switches and the turn-on of the other pair to produce the waveform shown in Fig. 22-133. The rms value of the fundamental component of this waveform varies as the cosine of half of the delay angle $\theta$. The fundamental frequency remains unchanged. All the odd harmonics are present, but their magnitudes change with $\theta$. Figure 22-134 shows the variation of the third, fifth, and seventh harmonics, expressed as a percentage of the fundamental voltage. Notice that as $\theta$ approaches 180°, the harmonics become as large as the fundamental.

Center-tap and half-bridge circuits cannot be modulated in this simple way but require more complex switching at a higher frequency.

Pulse-width techniques can be used to control the output voltage of harmonic-neutralized inverters by controlling each individual inverter stage. The output continues to obey the cosine dependence on $\theta$, and all neutralized harmonics remain neutralized. The remaining harmonics vary with $\theta$.

For carrier-modulated and optimum-response PWM inverters, pulse-width control is

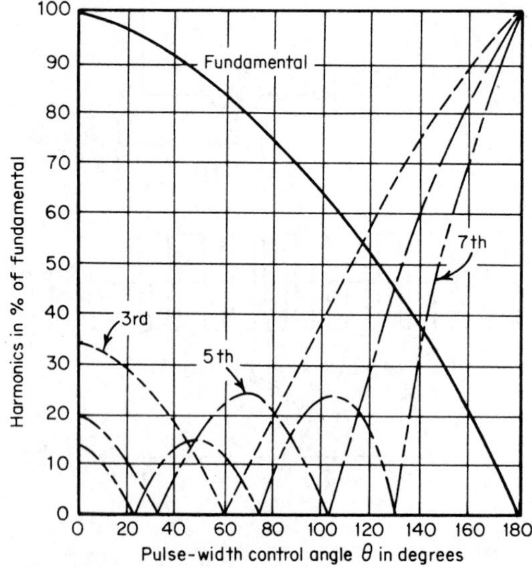


**FIG. 22-134** Variation of harmonics for a pulse-width-controlled square wave.

accomplished by simply reducing the width of each pulse. This reduction occurs automatically as the amplitude of the sinusoidal reference is decreased.

## POWER-FREQUENCY CHANGERS

*By L. GYUGYI*

***66. Basic Principles and Circuits.*** Power-frequency changers are static systems usually employing solid-state switching devices, capable of directly, that is, without an intermediate dc link, converting single or polyphase ac power of a given frequency to single or polyphase power of a chosen frequency. They may be used to link two ac power systems of different frequencies, to provide power at controllable frequency for variable-speed ac motor drives, or to convert the output of variable-speed ac generators to constant frequency.

Functionally, frequency changers are wave synthesizers. They fabricate the output-voltage wave(s) of desired amplitude and frequency by sequentially applying appropriate segments of the input-voltage wave(s) to the output. This is accomplished by arrays of static switches arranged to make *bilateral connections,* for controlled time intervals, between the input and output terminals, that is, between the supply voltages and loads.

Frequency changers generally require controllable power switches with intrinsic turn-on and turnoff ability (such as transistors and gate turn-off thyristors) or switches with controllable turn-on ability (such as thyristors and triacs) complemented by auxiliary forced-commutation circuitry to implement controllable turnoff. A notable exception is the naturally commutated cycloconverter (Par. **71**), which utilizes conventional controlled rectifiers.

The basic circuit configurations of static frequency changers are identical with polyphase converters characterized by their pulse number (see Tables 22-5 to 22-7) except that each unidirectional thyristor is replaced by a bidirectional ac switch. As in converters, increased pulse number leads to reduced distortion of the output-voltage and input-current waves. In practical applications, frequency changers are often required to produce 3-phase output; in this case three identical converter circuits, one for each output phase, are employed. Three-pulse frequency changers with single-phase and 3-phase output are shown in Fig. 22-135*a* and *b*, respectively. The bilateral-switch symbols represent any one of the previously described bidirectional solid-state switch arrangements.

Frequency changers fabricate the output-voltage wave with the desired (or "wanted") frequency and amplitude by sequentially connecting the input voltages to the output(s) for appropriate time intervals. The output-voltage wave(s) are thus composed of segments of the input-voltage waves. The length of each segment is determined by the duration of closure of the corresponding switch. However, an output-voltage wave of given frequency and amplitude can be obtained in several distinctly different ways (Gyugyi[1,35]) characterized by the control (modulation) of the repetition rate and/or duration of switch closures. The method of output-waveform fabrication uniquely determines the external performance characteristics of the frequency changer, the most important of which are the distortion of the output-voltage and input-current waves and the input-displacement and power factors.

***67. Fundamental Principles.*** Consider the simple three-pulse frequency-changer circuits shown in Fig. 22-135. These circuits convert 3-phase input power of frequency $f_I$ into a single- or 3-phase output power of frequency $f_O$. The relationship between the input and the generated output *voltage waves* can be described by the matrix equation

$$[v_O(t)] = [H(t)][v_I(t)]$$

or

$$\begin{bmatrix} v_{O1}(t) \\ v_{O2}(t) \\ v_{O3}(t) \end{bmatrix} = \begin{bmatrix} h_{11}(t) & h_{12}(t) & h_{13}(t) \\ h_{21}(t) & h_{22}(t) & h_{23}(t) \\ h_{31}(t) & h_{32}(t) & h_{33}(t) \end{bmatrix} \begin{bmatrix} v_{I1}(t) \\ v_{I2}(t) \\ v_{I3}(t) \end{bmatrix} \qquad (22\text{-}1)$$

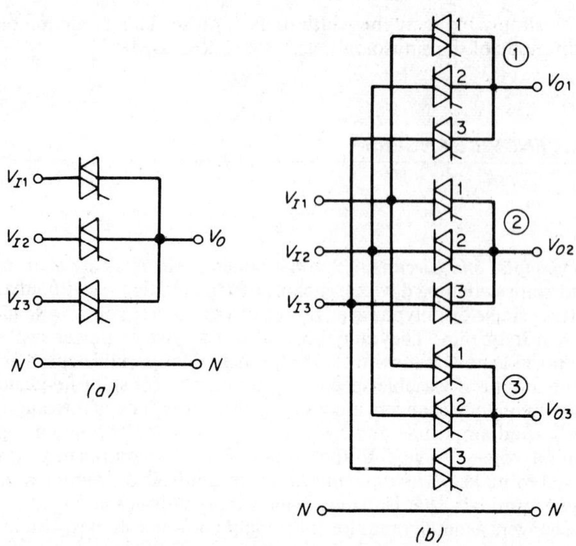


**FIG. 22-135** Three-pulse frequency changers (*a*) with single-phase output and (*b*) with 3-phase output.

where $v_{O1}$, $v_{O2}$, $v_{O3}$ are the time functions of generated voltage waves; $v_{I1}$, $v_{I2}$, $v_{I3}$ are the three input-voltage waves, which are usually sinusoids, that is,

$$v_{I1} = V_I \sin \omega_I t \qquad v_{I2} = V_I \sin\left(\omega_I t - \frac{2\pi}{3}\right) \qquad v_{I3} = V_I \sin\left(\omega_I t - \frac{4\pi}{3}\right)$$

and each $h_{ij}$ ($i = 1, 2, 3$; $j = 1, 2, 3$) is a time-varying existence function which defines whether a given switch $h_{ij}$, connecting output terminal $i$ to input terminal $j$, is open ($h_{ij} = 0$) or closed ($h_{ij} = 1$) at a given time $t$.

The input *current waves* drawn from the supply by a three-pulse frequency changer can be similarly expressed in terms of the output (load) currents:

$$[i_I(t)] = [H(t)]t[i_O(t)]$$

or

$$\begin{bmatrix} i_{I1}(t) \\ i_{I2}(t) \\ i_{I3}(t) \end{bmatrix} = \begin{bmatrix} h_{11}(t) & h_{21}(t) & h_{31}(t) \\ h_{12}(t) & h_{22}(t) & h_{32}(t) \\ h_{13}(t) & h_{23}(t) & h_{33}(t) \end{bmatrix} \begin{bmatrix} i_{O1}(t) \\ i_{O2}(t) \\ i_{O3}(t) \end{bmatrix} \tag{22-2}$$

where $i_{I1}$, $i_{I2}$, $i_{I3}$ are the three input current waves, $i_{O1}$, $i_{O2}$, $i_{O3}$ are the three output current waves, which for computations are usually assumed to be symmetrically displaced sinusoids, and each $h_{ij}$ is an appropriate existence function introduced in Eq. (22-1).

The basic sets of the three existence functions and related input- and output-voltage waveforms of the three-pulse power circuit shown in Fig. 22-135 are illustrated in Fig. 22-136*a* and *b* for the trivial case of zero output frequency and zero output voltage. Note that the output-voltage wave of Fig. 22-136*a* is identical with the output of a unidirectional naturally commutated ac-dc converter conducting continuous positive load current when the delay angle $\alpha$ is 90° (see Fig. 22-114). Similarly that of Fig. 22-136*b* is obtained from a converter conducting negative load current at $\alpha = 90°$. A bidirectional converter employing bilateral turnoff switches can produce either of these two waveforms, depending on which of the two sets of complementary existence functions ($h_{ij}$ or $h_{ij\pi}$) describes its operation. This

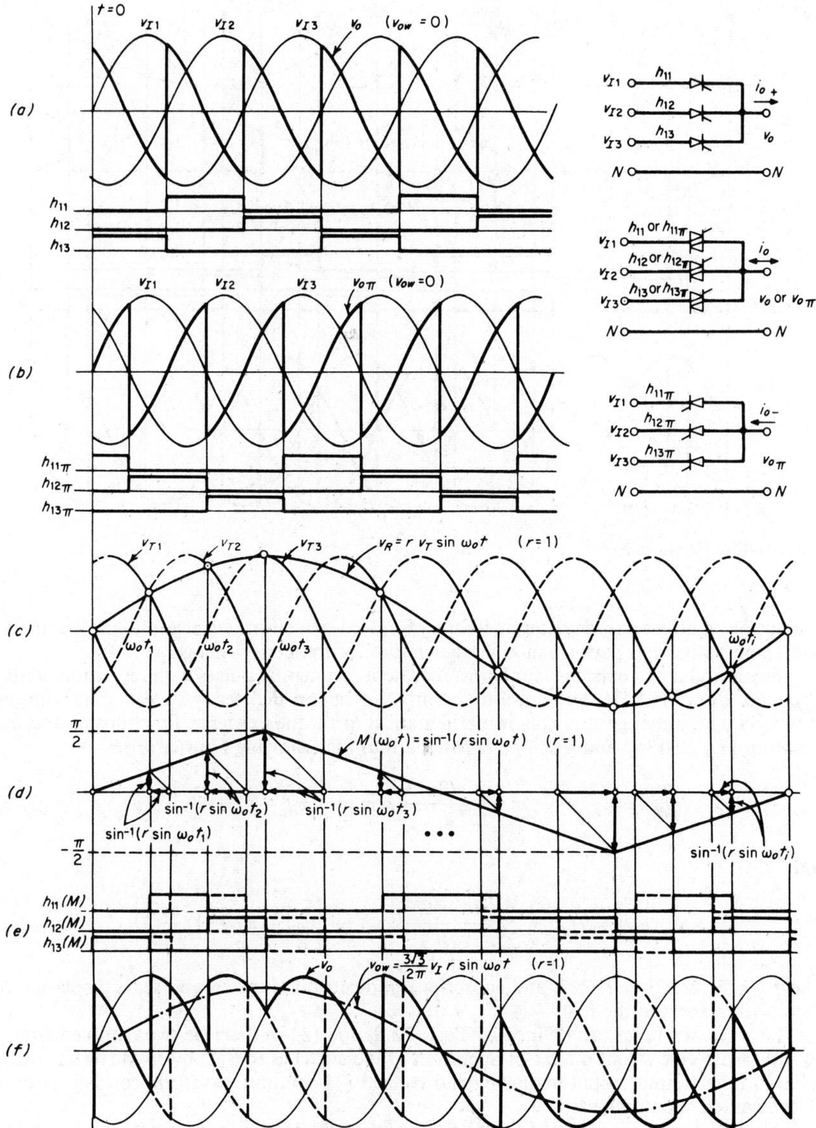


**FIG. 22-136** Waveforms illustrating the generation of two complementary waveforms $V_0$ and $V_{0\sigma}$.

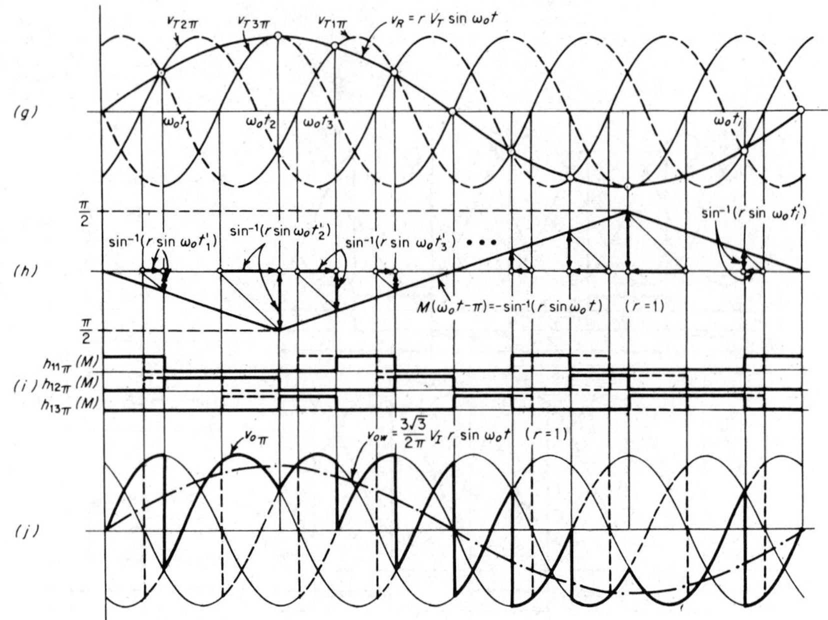


**FIG. 22-136** (*Continued*)

free option is utilized in devising methods of output-waveform fabrication which provide desired operating and performance characteristics for frequency changers.

The unmodulated existence functions represent rectangular pulses with repetition period $1/f_I$, pulse duration $1/3f_I$, and amplitude unity. To obtain the steady-state output voltages of the frequency changer in explicit mathematical form, the existence functions $h_{ij}$ and $h_{ij\pi}$ shown in Fig. 22-136$a$ and $b$ can be expanded into the following Fourier series:

$$h_{ij}(\omega_I t) = \frac{1}{3} - \frac{2}{\pi} \sum_{n=0}^{\infty} \frac{\sin n(2\pi/3)}{n} \cos \left\{ n \left[ \omega_I t - (j-1)\frac{2\pi}{3} \right] \right\} \tag{22-3}$$

and

$$h_{ij\pi}(\omega_I t) = \frac{1}{3} + \frac{2}{\pi} \sum_{n=0}^{\infty} \frac{\sin n(\pi/3)}{n} \cos \left\{ n \left[ \omega_I t - (j-1)\frac{2\pi}{2} \right] \right\} \tag{22-4}$$

where $i = 1, 2, 3$, $j = 1, 2, 3$, and subscript $\pi$ indicates that the second set is displaced by $\pi$ rad with respect to the first.

The existence functions defined by Eqs. (22-3) and (22-4) describe two complementary but otherwise equivalent modes of operation of the switches in the power converter which result in zero desired output frequency and voltage. (The output waveform consists entirely of "unwanted" components.)

A desired output frequency differing from zero is obtained by appropriately modulating the repetition frequency of the basic existence functions, which is equivalent to varying the commencement and/or duration of the conduction intervals of the corresponding switches. Mathematically this means that a modulating function $M(\omega_O t)$ ($\omega_O = 2\pi f_O$; $f_O$ is the "wanted" output frequency) is added to the arguments of the basic existence functions given by Eqs. (22-3) and (22-4). To keep the wanted (fundamental) component identical in the two output-voltage waves obtainable by the use of the two complementary sets of existence functions, the modulating functions must also be mutually displaced by $\pi$. Similarly, for balanced 3-phase output, the modulating functions used to generate the three output voltage waves must also be mutually displaced by $2\pi/3$. Thus

$$h_{ij}(\omega_I t,\omega_O t) = \frac{1}{3} - \frac{2}{\pi}\sum_{n=0}^{\infty}\frac{\sin n(2\pi/3)}{n}\cos\left(n\left\{\omega_I t + M\left[\omega_O t - (i-1)\frac{2\pi}{3}\right] - (j-1)\frac{2\pi}{3}\right\}\right) \tag{22-5}$$

and

$$h_{ij\pi}(\omega_i t,\omega_O t) = \frac{1}{3} + \frac{2}{\pi}\sum_{n=0}^{\infty}\frac{\sin n(\pi/3)}{n}\cos\left(n\left\{\omega_I t + M\left[\omega_O t - (i-1)\frac{2\pi}{3} - \pi\right] - (j-1)\frac{2\pi}{3}\right\}\right) \tag{22-6}$$

The modulating function in Eqs. (22-5) and (22-6) can be visualized as a time-varying angle which effectively changes (modulates) the repetition frequency of the existence functions from or about their quiescent frequency $f_I$.

Mathematical expressions for the resulting output-voltage waveforms of a general $P$-pulse frequency changer can be obtained from Eqs. (22-1) and (22-5) and (22-1) and (22-6) in terms of the modulating functions $M(\omega_O t)$ and $M(\omega_O t - \pi)$, respectively (Gyugyi[1,35]):

$$\begin{aligned} v_{Oi} = {} & \frac{3\sqrt{3}}{2\pi} V_I \sin\left[M\left(\omega_O t - (i-1)\frac{2\pi}{3}\right)\right] \\ & + \frac{3\sqrt{3}}{2\pi} V_I \sum_{k=1}^{\infty}\left(\frac{\sin\{Pk\omega_I t + (Pk-1)M[\omega_O t - (i-1)(2\pi/3)]\}}{Pk-1}\right. \\ & \left. + \frac{\sin\{Pk\omega_I t + (Pk+1)M[\omega_O t - (i-1)(2\pi/3)]\}}{Pk+1}\right) \end{aligned} \tag{22-7}$$

and

$$\begin{aligned} v_{Oi\pi} = {} & -\frac{3\sqrt{3}}{2\pi} V_I \sin\left[M\left(\omega_O t - (i-1)\frac{2\pi}{3} - \pi\right)\right] \\ & - \frac{3\sqrt{3}}{2\pi} V_I \sum_{k=1}^{\infty}(-1)^k\left(\frac{\sin\{Pk\omega_I t + (Pk-1)M[\omega_O t - (i-1)(2\pi/3) - \pi]\}}{Pk-1}\right. \\ & \left. + \frac{\sin\{Pk\omega_I t + (Pk+1)M[\omega_O t - (i-1)(2\pi/3) - \pi]\}}{Pk+1}\right) \end{aligned} \tag{22-8}$$

where $i$ = 1, 2, 3, $P$ is the pulse number, and $M(\omega_O t)$ specifies the modulation of the basic repetition frequency $f_I$ of the existence functions.

Similar equations can be written for the input-current wave, and after laborious computation the performance characteristics can be numerically obtained.

Equations (22-7) and (22-8) indicate that the modulating function entirely determines the operation and performance characteristics of a frequency-changer circuit defined by its pulse number $P$. The modulating function is actually a mathematical description of the control defining the operation of the power switches and thereby the method of output-waveform fabrication. Various control methods (modulating functions) can be applied to the same power circuit to generate output-voltage (or input-current) waveforms of widely differing characteristics to meet practical requirements. In the following, the five most important operation modes, resulting in practically desirable output waveforms and performance characteristics, will be summarized and illustrated for the case of the single-phase output, three-pulse circuit shown in Fig. 22-135*a*. All the output waveforms considered can be derived from the two complementary output-voltage waveforms obtained from Eqs. (22-7) and (22-8) by the substitution of the modulating function

$$M(\omega_O t) = \sin^{-1}(r\sin\omega_O t) \tag{22-9}$$

where $r$ is the output-voltage ratio, that is, $r = V_O/V_{O,\max}$ ($r \leq 1$). The practical derivation of this modulating function and the subsequent generation of the two complementary waveforms $v_O$ and $v_{O\pi}$ are graphically illustrated for $r = 1$ in Fig. 22-136*c* to *j*.

Figures 22-136*c* and *d* illustrate that the modulating function $M(\omega_O t)$ of Eq. (22-136*c*) is a mathematical expression for the well-known sine-wave crossing technique widely used to control the firing angle of thyristor converters (Pelly,[25] Chap. 9). Using this technique, the magnitude of $M(\omega_O t)$, and thus the modulation of each existence function $h_{ij}$ about its quiescent point (zero output), is determined by the crossing point of a corresponding timing wave $v_T$ with a reference sinusoid $v_R$. Derivation of the complementary sets of $M(\omega_O t)_\pi$ and $h_{ij\pi}$ are shown in Fig. 22-136*g* to *i*. The timing waves $v_{T1}$, $v_{T2}$, $v_{T3}$, $v_{T1\pi}$, $v_{T2\pi}$, and $v_{T3\pi}$ are opposite half-period sections of sine waves synchronized to the source voltages with a phase relationship such that at zero reference the mean of the output voltages $v_O$ and $v_{O\pi}$ is zero.

Figures 22-136*e*, *f*, *i*, and *j* show that the output waveforms $v_O$ and $v_{O\pi}$, at $r = 1.0$, are generated by the act of periodically stepping up and down ($v_O$) and down and up ($v_{O\pi}$) the original $f_I$ repetition frequency of the existence functions, and thus that of the power switches, to $f_I + f_O$ and $f_I - f_O$, respectively.

The spectral characteristics (frequency and amplitude) of the two complementary output-voltage waveforms $v_O$ and $v_{O\pi}$ shown in Fig. 22-136*f* and *j* are identical, as are those of the corresponding input-current waves. Since, however, the two complementary waveforms have a mutually complementing internal relationship (certain characteristics of the output cycle of $v_O$, observable during a given *half* period, are identical to those of $v_{O\pi}$, observable during the following output half cycle and vice versa), it is possible to fabricate new output waveshapes from the two complementary waveforms which satisfy given output- and/or input-performance requirements. In the following section, synthesis of the four important frequency-changer output waveforms, using the two basic complementary waves, is described, and the pertinent operating conditions and performance characteristics are summarized.

***68. Practical Frequency Changers.*** Utilizing the properties of the complementary output-voltage waveforms derived in the preceding paragraphs, frequency changers having the following special, *mutually exclusive* characteristics can be devised:

1. Unity or controllable-input displacement factor
2. Natural (input-line) commutation of the power switches
3. Unity input-power factor and minimum output-voltage distortion
4. Unrestricted output-to-input frequency ratio

To establish the necessary operating conditions for the above characteristics, consider Fig. 22-137*a* and *b* where the two three-pulse complementary output waveforms $v_O$ and $v_{O\pi}$, together with the voltage waves ($v_{I1}$, $v_{I2}$, and $v_{I3}$) of the 3-phase supply and an assumed sinusoidal load current $i_0$ having an arbitrary phase angle $\phi_0$, are shown. The input-current waveshapes, $i_{I1}$, $i_{I2}$, and $i_{I3}$, flowing in supply phases 1, 2, and 3, are shaded by vertical, horizontal, and crosshatched lines, respectively. The following observations may be made.

*Input-Phase-Angle Characteristics.* The input-current waves are composed of current "blocks" cut out of the load current. The phase position of the individual current blocks with respect to the corresponding input voltages over a complete output cycle determines the input phase angle $\phi_I$, that is, the angle between the fundamental component of the input-current wave and the corresponding phase voltage, and the displacement factor ($\cos \phi_I$). In the case of $v_O$ (Fig. 22-137*a*), the phase angles of the input-current blocks lag the corresponding phase voltages during the positive output-*current* half cycle, and then they lead those during the negative half cycle. Conversely, in the case of $v_{O\pi}$ (Fig. 22-137*b*), the phase position of the input-current blocks is opposite; that is, they lead the corresponding phase voltages during the positive output-current half cycle, and they lag those during the negative half cycle. The net input phase angle averaged over a complete output cycle is therefore zero, and the input displacement is unity for both $v_O$ and $v_{O\pi}$ regardless of the load phase angle.

*Commutation Characteristics.* At each switching point of waveform $v_O$ the "incoming" input voltage is more positive than the "outgoing" one. The opposite is true for waveform $v_{O\pi}$. Consequently, $v_O$ satisfies the conditions of natural commutation for a *positive* unidirectional converter, and $v_{O\pi}$ satisfies those for a *negative* converter. Therefore, only during the positive (negative) half cycle of the output *current* can $v_O(v_{O\pi})$ be produced by a naturally commutated converter.

*Spectral Characteristics.* The switches of the converter generating waveform $v_O$ are operated at a constant rate of $f_I + f_O$ during the half-cycle interval when the slope of the wanted (fundamental) component is positive, and they are operated at a constant rate of $f_I - f_O$ when the slope of the wanted component is negative. Conversely, the switches of the converter generating $v_{O\pi}$ are operated at the "slow" rate of $f_I - f_O$ to produce the half-cycle sections of the output waveform with positive slope, and they are operated at the "fast" rate of $f_I + f_O$ to produce the other half-cycle sections with negative slope. The half-cycle waveform intervals with "fast" switching rates can generally be characterized by unwanted components having frequencies consisting of *sums* of multiples of $f_I$ and $f_O$, whereas the intervals with "slow" switching rates can be characterized by unwanted components having frequencies consisting of *differences* of multiples of $f_I$ and $f_O$. The total waveform, $v_O$ as well as $v_{O\pi}$, therefore can be characterized by a frequency spectrum consisting of both sums and differences of multiples of the input and output frequencies.

Utilizing these complementary characteristics of waveforms $v_O$ and $v_{O\pi}$, the operating and performance characteristics of the following practically important frequency changers can be established.

***69. Unity-Displacement-Factor Frequency Changer (UDFFC).*** As was established under *input-phase-angle characteristics* in Par. **68**, the input-displacement factor of a bilateral converter generating either complementary output-voltage waveform $v_O$ or $v_{O\pi}$ is unity independently of the load-power factor. For this reason, a frequency changer controlled to fabricate either $v_O$ or $v_{O\pi}$ is termed a unity-displacement-factor frequency changer (UDFFC).

The pertinent characteristics of the UDFFC are summarized in the table adjoining the waveforms of Fig. 22-137*a* and *b*.

***70. Controllable-Displacement-Factor Frequency Changer (CDFFC).*** The CDFFC utilizes power switches with intrinsic turnoff ability (or with external forced commutation). The bidirectional converter is controlled so as to generate alternating half-period intervals of the complementary waveforms $v_O$ and $v_{O\pi}$. The phase position of the input current during successive input cycles used to fabricate $v_O$ ($v_{O\pi}$) varies from lagging (leading) to leading (lagging) as the ac output current goes through a full cycle. Thus, by appropriately choosing alternate half-period sections of $v_O$ and $v_{O\pi}$ to fabricate the final output waveform, the input-displacement factor may be made lagging, leading, or unity regardless of the phase angle of the output (load) current. The half-period sections constituting the output waveform can conveniently be defined for a given input phase angle (displacement factor) by an angle $\sigma$ (measured at $\omega_0 = 2\pi f_O$ angular frequency) specifying the point of changeover between $v_O$ and $v_{O\pi}$ with respect to the zero crossing point of the output current, as illustrated in Fig. 22-137*c*. The relationship between the input phase angle and angle $\sigma$ is shown in Fig. 22-138 for various $\phi_0$ load phase angles.

The intervals during which the converter switches are operated at high and low rates ($f_I + f_O$ and $f_I - f_O$) depend upon the angle $\sigma$, that is, the output-to-input displacement-factor transfer. The frequencies of the dominant unwanted components may thus be either the sums or differencies, or both, of multiples of $f_I$ and $f_O$. The frequency spectrum is therefore a function of $\sigma$.

The characteristics of the CDFFC are summarized in the table adjoining Fig. 22-137*c*. The unique input characteristic of the CDFFC offers a number of intriguing application possibilities. One of them is a power-generating system which utilizes a squirrel-cage induction machine and a CDFFC. In this system the reactive-excitation requirement of the induction generator and the reactive kilovoltampere demand of the load are provided by the static frequency changer itself. The frequency changer thus has two basic functions: (1) it converts the generally variable generator frequency to precisely regulated output frequency, and (2) it provides a controllable excitation for the induction machine.

***71. Naturally Commutated Cycloconverter (NCC).*** The naturally commutated cycloconverter, in compliance with the conditions outlined under *commutation characteristics* in Par. **68**, consists of two unidirectional inverse-parallel connected converter circuits (dual converter). The positive and negative converters are controlled to produce output waveforms $v_O$ and $v_{O\pi}$, respectively. However, the positive converter is gated on only during the positive half cycles of the output *current*, and the negative converter is operated only during the negative half cycles. The output waveform is therefore composed of half-period seg-

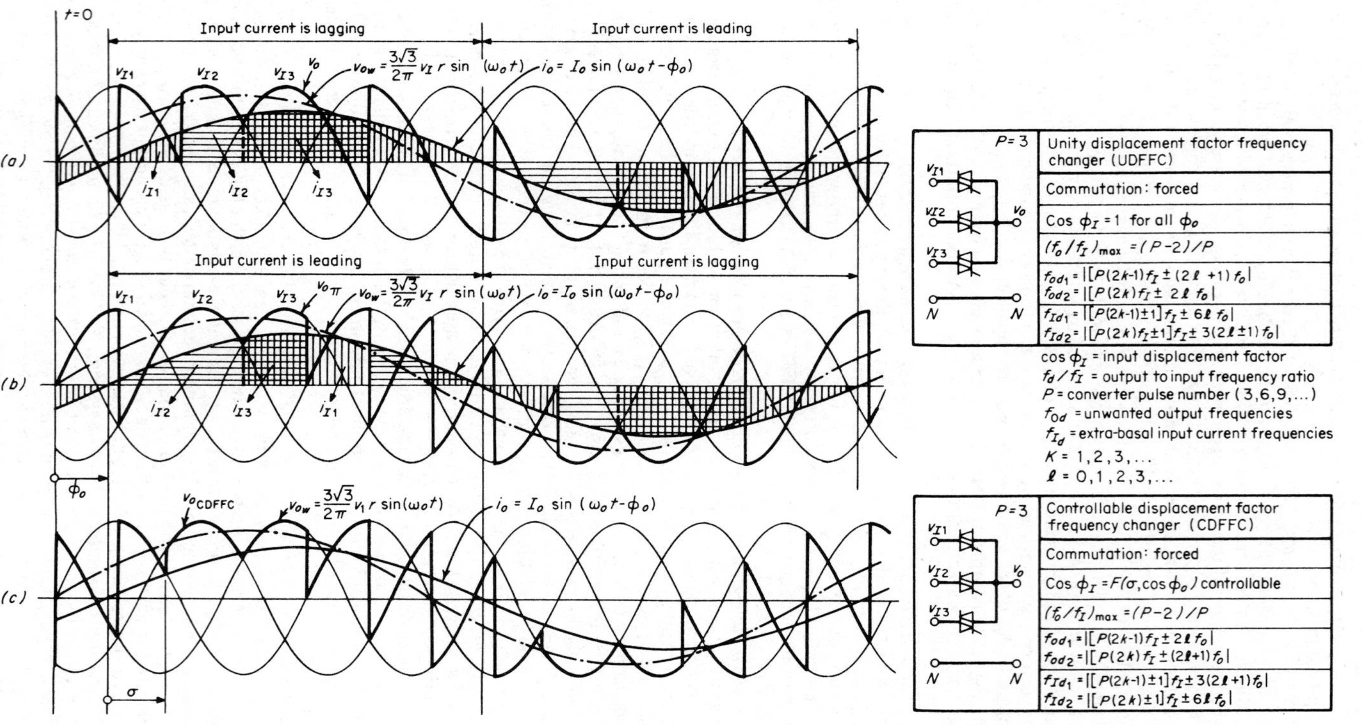
t=0
Input current is lagging
Input current is leading
$v_{I1}$ $v_{I2}$ $v_{I3}$ $v_o$
$v_{ow} = \frac{3\sqrt{3}}{2\pi} v_I r \sin(\omega_o t)$
$i_o = I_o \sin(\omega_o t - \phi_o)$
$i_{I1}$ $i_{I2}$ $i_{I3}$
(a)
Input current is leading
Input current is lagging
$v_{o\pi}$
(b)
$\phi_o$
$v_{o_{CDFFC}}$
$v_{ow} = \frac{3\sqrt{3}}{2\pi} v_1 r \sin(\omega_o t)$
$i_o = I_o \sin(\omega_o t - \phi_o)$
$\sigma$
(c)
P=3
N
Unity displacement factor frequency changer (UDFFC)
Commutation: forced
$\cos\phi_I = 1$ for all $\phi_o$
$(f_o/f_I)_{max} = (P-2)/P$
$f_{od_1} = |[P(2k-1)f_I \pm (2\ell+1) f_o|$
$f_{od_2} = |[P(2k)f_I \pm 2\ell f_o|$
$f_{Id_1} = |[P(2k-1) \pm 1] f_I \pm 6\ell f_o|$
$f_{Id_2} = |[P(2k) f_I \pm 1] f_I \pm 3(2\ell \pm 1) f_o|$
$\cos\phi_I$ = input displacement factor
$f_o/f_I$ = output to input frequency ratio
$P$ = converter pulse number (3,6,9,...)
$f_{od}$ = unwanted output frequencies
$f_{I_d}$ = extra-basal input current frequencies
$K$ = 1,2,3,...
$\ell$ = 0,1,2,3,...
Controllable displacement factor frequency changer (CDFFC)
Commutation: forced
$\cos\phi_I = F(\sigma, \cos\phi_o)$ controllable
$(f_o/f_I)_{max} = (P-2)/P$
$f_{od_1} = |[P(2k-1)f_I \pm 2\ell f_o|$
$f_{od_2} = |[P(2k) f_I \pm (2\ell+1) f_o|$
$f_{Id_1} = |[P(2k-1) \pm 1] f_I \pm 3(2\ell+1) f_o|$
$f_{Id_2} = |[P(2k) \pm 1] f_I \pm 6\ell f_o|$

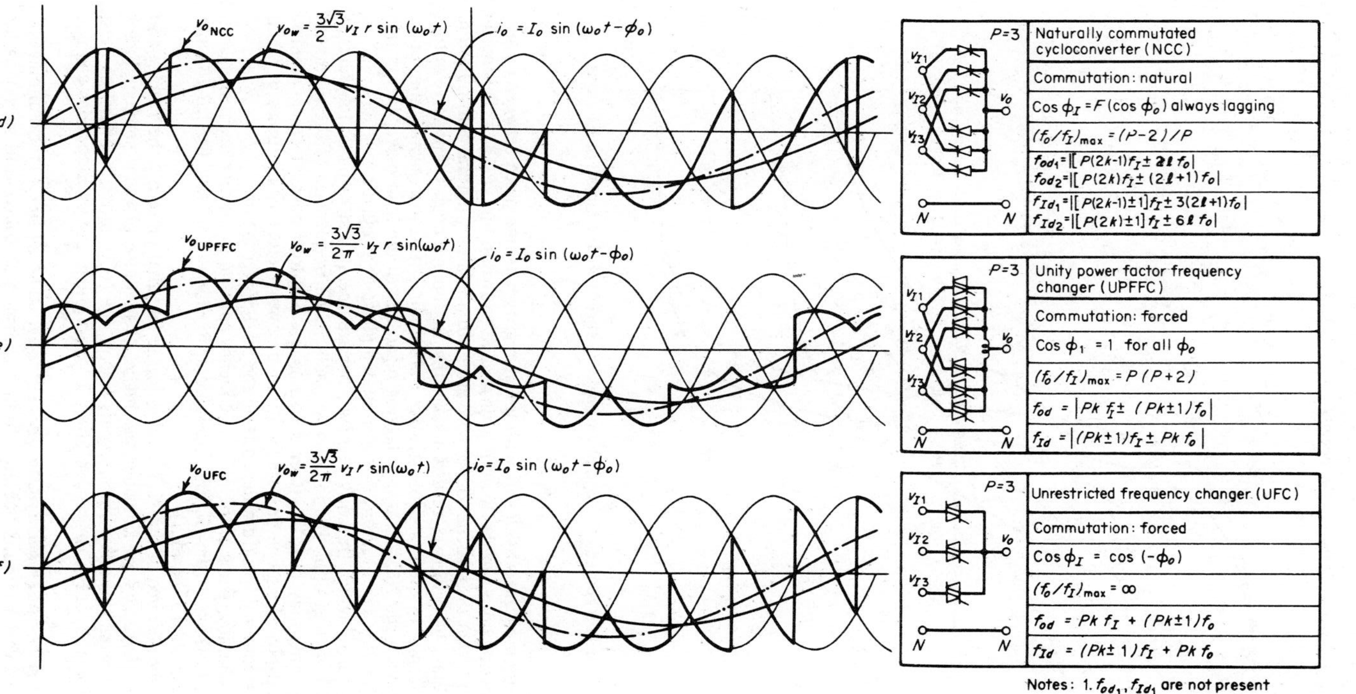

(d)
$v_{o_{NCC}}$
$v_{ow} = \frac{3\sqrt{3}}{2} v_I \, r \sin(\omega_o t)$
$i_o = I_o \sin(\omega_o t - \phi_o)$
(e)
$v_{o_{UPFFC}}$
$v_{ow} = \frac{3\sqrt{3}}{2\pi} v_I \, r \sin(\omega_o t)$
$i_o = I_o \sin(\omega_o t - \phi_o)$
(f)
$v_{o_{UFC}}$
$v_{ow} = \frac{3\sqrt{3}}{2\pi} v_I \, r \sin(\omega_o t)$
$i_o = I_o \sin(\omega_o t - \phi_o)$
P=3
$v_{I1}$ $v_{I2}$ $v_{I3}$ $v_o$ N N
Naturally commutated cycloconverter (NCC)
Commutation: natural
$\cos\phi_I = F(\cos\phi_o)$ always lagging
$(f_o/f_I)_{max} = (P-2)/P$
$f_{od_1} = |[P(2k-1)f_I \pm 2\ell f_o|$
$f_{od_2} = |[P(2k)f_I \pm (2\ell+1)f_o|$
$f_{Id_1} = |[P(2k-1)\pm 1]f_I \pm 3(2\ell+1)f_o|$
$f_{Id_2} = |[P(2k)\pm 1]f_I \pm 6\ell f_o|$
P=3
$v_{I1}$ $v_{I2}$ $v_{I3}$ $v_o$ N N
Unity power factor frequency changer (UPFFC)
Commutation: forced
$\cos\phi_1 = 1$ for all $\phi_o$
$(f_o/f_I)_{max} = P(P+2)$
$f_{od} = |Pk f_I \pm (Pk\pm 1)f_o|$
$f_{Id} = |(Pk\pm 1)f_I \pm Pk f_o|$
P=3
$v_{I1}$ $v_{I2}$ $v_{I3}$ $v_o$ N N
Unrestricted frequency changer (UFC)
Commutation: forced
$\cos\phi_I = \cos(-\phi_o)$
$(f_o/f_I)_{max} = \infty$
$f_{od} = Pk f_I + (Pk\pm 1)f_o$
$f_{Id} = (Pk\pm 1)f_I + Pk f_o$


Notes: 1. $f_{od_1}, f_{Id_1}$ are not present if $P$ is even ($P=6p$; $p=1,2,\ldots$)

2. Input current frequencies are given for balanced three-phase output.

3. $f_{od}$ and $f_{Id}$ may become harmonics (i.e. integral multiples of $f_o$ and $f_I$) at certain $f_o/f_I$ ratios.

**FIG. 22-137** Waveforms illustrating the derivation of practical frequency-changer output waveforms from the two complementary waves.

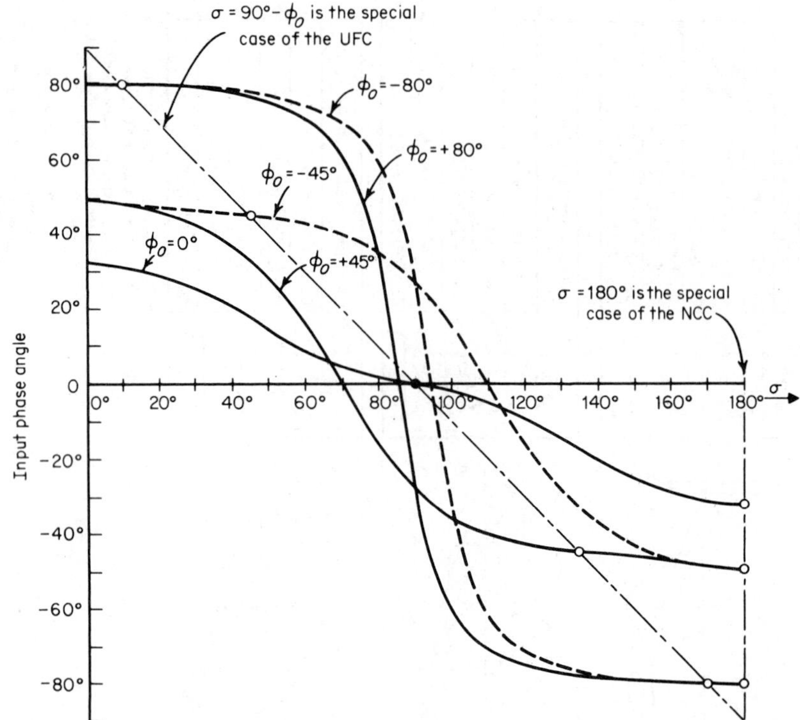


FIG. 22-138 Relationship between the load phase angle $\pi_\sigma$, angle $\sigma$, and the input phase angle.

ments of the complementary waveforms with the changeover taking place between $v_O$ and $v_{O\pi}$ at the zero crossing points of the ac output current, as shown in Fig. 22-137*d*. This mode of operation ensures that the switches of the converter can operate by natural commutation.

The output voltage waveform $v_O$ results in a lagging input phase angle for positive output current, and $v_{O\pi}$ results in the same for negative output current. Consequently, the input-displacement factor of the NCC is always lagging regardless of the load-power factor; this characteristic is inherent for all naturally commutated phase-controlled converters.

The intervals during which the converter switches are operated at high and low rates ($f_I + f_O$ and $f_I - f_O$) depend upon the load-power factor. The frequencies of the dominant unwanted components may thus be either the sums (leading load-power factor) or differences (lagging load-power factor), or both (unity load-power factor), of multiples of $f_I$ and $f_O$; therefore, the frequency spectrum depends on the load-power factor.

Note that the characteristics of the NCC (frequency spectra, input-displacement factor, etc.) are identical to those of the CDFFC operated at fixed $\sigma \equiv 180°$ (see Fig. 22-138 and tables adjoining Fig. 22-137*c* and *d*).

The practical significance of the NCC is due to the fact that presently the voltage and current ratings of thyristors are considerably higher than those of other semiconductor switches having internal turnoff ability. For this reason, the NCC currently offers the most economical if not the only feasible solution to very high power frequency-changer applications.

***72. Unity-Power Factor Frequency Changer (UPFFC).*** The UPFFC is the combination of two bilateral converter circuits operated from a common ac source and supplying the same load. One converter is controlled to generate $v_O$, and the other $v_{O\pi}$. The final output waveform is produced by summing or generating the arithmetic mean of the two complementary waveforms (see Fig. 22-137*e*). Note that the output rating of the combined system is the sum of the ratings of the constituent converters.

The input-displacement factor of each constituent converter, and thus that of the combined system, is unity. The advantage of this arrangement is that in addition to unity input-displacement factor (regardless of load-power factor), certain groups of unwanted components present in the output-voltage (input-current) waves of the constituent converters cancel out, resulting in greatly improved frequency spectra, increased $f_O/f_I$ ratio, and rms distortion decreased by a factor of $\sqrt{2}$ (see table adjoining Fig. 22-137*e*). The reduction in the distortion of the input-current wave results in a "near unity" input-power factor $\lambda$, which is the product of the input-displacement and current-distortion factors ($\lambda = 0.9$ for a 3-pulse, $\lambda = 0.977$ for a 6-pulse, and $\lambda = 0.995$ for a 12-pulse system).

The UPFFC is particularly advantageous in applications where the required output power is higher than that obtainable from a single converter, in which case multiple converters can be used advantageously to increase the power rating as well as to improve the performance of the system.

***73. Unrestricted Frequency Changer (UFC).*** The UFC utilizes a single bilateral converter whose switches are operated at the constant "fast" rate of $f_I + f_O$. Therefore, as discussed under *spectral characteristics* in Par. **68**, the output waveform of the UFC can be synthesized from half-period sections of the two complementary waveforms, $v_O$ providing the output when the slope of the wanted component is positive and $v_{O_\pi}$ when it is negative. The changeover points between $v_O$ and $v_{O_\pi}$ thus coincide with the peaks of the wanted voltage component (see Fig. 22-137*f*). Because of the constant switching rate of $f_I + f_O$, the frequencies of the unwanted components are only sums of multiples of $f_I$ and $f_O$ and therefore are always higher than $f_O$, regardless of the $f_O/f_I$ ratio (see table adjoining Fig. 22-137*f*). Consequently, the UFC can generate a high-quality output waveform having a frequency which may be lower or higher than the ac supply frequency or equal to it. The described operation of the converter switches also results in a unique output- to input-power-factor transfer characteristic; that is, the UFC reflects the negative of the load phase angle back to the source (Fig. 22-138) and therefore the input- and output-displacement factors are mirror images of each other (an inductive load is seen capacitive and vice versa).

The UFC is an ideal system to provide ac output power over a wide frequency range (which may extend from zero to well above the ac supply frequency) to control the speed of ac motors.

***74. Control of the Output Voltage.*** In the previously described operation modes, frequency changers supply ac power at the maximum output voltage obtainable from the given supply voltages. In many practical applications, for example, speed control of ac motors and regulated ac supplies, the effective value of the output-voltage waveform, that is, the amplitude of the wanted component, has to be controllable independently of the input source. This can be achieved by varying either the depth of modulation used to generate the output waveform or the conduction intervals of the switches while maintaining their repetition rate as required for maximum output voltage (pulse-width modulation). The first type of voltage control can be accomplished simply through sine-wave crossing control by varying the amplitude of the reference wave. This method is compatible, that is, does not significantly affect the output and input performance characteristics obtained at maximum voltage, with the UDFFC, NCC, CDFFC, and UPFFC (Gyugyi[1,35]). Typical three-pulse NCC and UPFFC output waveforms with a relative amplitude of 0.7 ($v_O/v_{O,\max} = 0.7$) are shown in Fig. 22-139*a* and *b*, respectively.

Pulse-width-modulation voltage control is the only type which is completely compatible with the UFC (Gyugyi[1,35]). Its essence is to subdivide the conduction periods into *active* and *passive* intervals. During the active interval, the switches are operated in the usual manner. During the passive interval, the load is reconnected to the input phase used for the preceding active interval, as illustrated by the three-pulse UFC waveform in Fig. 22-139*c*. [Note that for even-pulse-number converters (six, twelve, etc.), the output voltage is zero and the load is actually shorted during the passive intervals.] By controlling the relative duration of the active and passive intervals within the original conduction period, the mean output voltage can be continuously varied from maximum to zero at any given output frequency.

The amplitudes of the dominant unwanted components present in the output voltage waves of three-, six-, and twelve-pulse NCC, CDFFC, UPFFC, and UFC are given as per

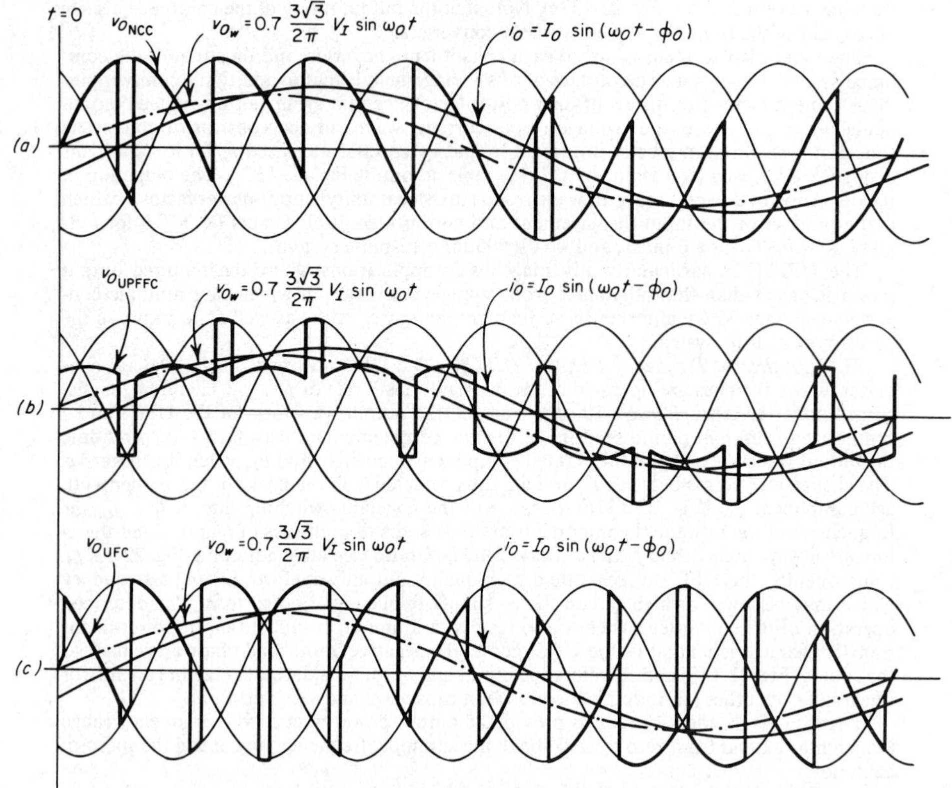


**FIG. 22-139** Waveforms illustrating amplitude control of the output voltage by reducing the depth of modulation (*a* and *b*, NCC and UPFFC) and employing pulse-width modulation (*c*, UFC).

unit values of $v_{O,\max}$ in Table 22-9 for five discrete values of the output-voltage ratio $r = v_O/v_{O,\max}$. The data presented are pertinent to the sine-wave crossing control for the NCC (at unity load-power factor, that is, $\phi_O = 0$ or 180°), the CDFFC (at $\sigma + \phi_O = 0$ or 180°), and the UPFFC. For the UFC, the data are relevant to PWM control.

## Power Filters

***75. General Considerations.*** Of necessity, high-power controls and converters utilize switching devices because these permit high power efficiency to be achieved. But the switching action generates transients and spurious frequencies (usually harmonics of the fundamental frequency) which may have intolerable effects on the power source, the load, and/or other nearby equipment, by way of electromagnetic interference (EMI) radiated or conducted through the supply line. Frequently, the internal design of a power converter, such as a harmonic-neutralized inverter, is chosen to minimize the most troublesome spurious frequencies. In addition it is often necessary to filter the input, the output, or both.

Because all power filters must handle substantial voltamperes, their component dissipation, cost, size, and weight are all important design factors, although these are usually not optimized by conventional small-signal filter-synthesis procedures. Basically, a low-loss filter operates by storing energy during an unwanted peak in a waveform and then discharging

**TALBE 22-9** Amplitudes of the Dominant Unwanted Frequency Components of the Output Voltage for Various Voltage Ratios

| Number of pulses | | $l$ | $r$ = 1.0 | 0.8 | 0.6 | 0.4 | 0.2 |
|---|---|---|---|---|---|---|---|
| **NCC at $\phi_0 = 0$ and CDFFC at $\sigma + \phi_0 = \{ {0 \atop 180^\circ}$** | | | | | | | |
| 3 | $3f_I \pm 2lf_o$ | 0 | 0.318 | 0.470 | 0.596 | 0.682 | 0.734 |
| | | 1 | 0.272 | 0.276 | 0.230 | 0.163 | 0.084 |
| | | 2 | 0.164 | 0.112 | 0.071 | 0.040 | 0.018 |
| | | 3 | 0.064 | 0.038 | 0.025 | 0.015 | 0.007 |
| | | 4 | 0.021 | 0.018 | 0.013 | 0.008 | 0.004 |
| 6 | $6f_I \pm (2l+1)f_o$ | 0 | 0.006 | 0.069 | 0.130 | 0.178 | 0.208 |
| | | 1 | 0.055 | 0.123 | 0.138 | 0.120 | 0.090 |
| | | 2 | 0.105 | 0.111 | 0.084 | 0.060 | 0.047 |
| | | 3 | 0.100 | 0.065 | 0.045 | 0.036 | 0.032 |
| | | 4 | 0.055 | 0.035 | 0.029 | 0.026 | 0.025 |
| | | 5 | 0.026 | 0.023 | 0.022 | 0.021 | 0.020 |
| | | 6 | 0.022 | 0.021 | 0.018 | 0.017 | 0.017 |
| 12 | $12f_I \pm (2l+1)f_o$ | 0 | 0.002 | 0.022 | 0.035 | 0.057 | 0.091 |
| | | 1 | 0.001 | 0.021 | 0.045 | 0.071 | 0.059 |
| | | 2 | 0.007 | 0.029 | 0.061 | 0.056 | 0.029 |
| | | 3 | 0.005 | 0.052 | 0.051 | 0.030 | 0.018 |
| | | 4 | 0.027 | 0.053 | 0.032 | 0.017 | 0.013 |
| | | 5 | 0.050 | 0.036 | 0.018 | 0.012 | 0.010 |
| | | 6 | 0.049 | 0.020 | 0.012 | 0.010 | 0.009 |
| | | 7 | 0.027 | 0.013 | 0.009 | 0.008 | 0.007 |
| | | 8 | 0.011 | 0.009 | 0.008 | 0.007 | 0.006 |
| **UPFCC** | | | | | | | |
| 3 | $3f_I$ | | 0.000 | 0.099 | 0.310 | 0.530 | 0.690 |
| | $3f_I \pm 2f_o$ | | 0.250 | 0.275 | 0.210 | 0.110 | 0.028 |
| | $3f_I \pm 4f_o$ | | 0.125 | 0.050 | 0.018 | 0.004 | 0.000 |
| 6 | $6f_I \pm f_o$ | | 0.000 | 0.061 | 0.025 | 0.169 | 0.163 |
| | $6f_I \pm 3f_o$ | | 0.000 | 0.098 | 0.133 | 0.068 | 0.012 |
| | $6f_I \pm 5f_o$ | | 0.100 | 0.089 | 0.031 | 0.006 | 0.000 |
| | $6f_I \pm 7f_o$ | | 0.071 | 0.013 | 0.003 | 0.000 | 0.000 |
| 12 | $12f_I \pm f_o$ | | 0.000 | 0.014 | 0.020 | 0.047 | 0.086 |
| | $12f_I \pm 3f_o$ | | 0.000 | 0.019 | 0.035 | 0.063 | 0.033 |
| | $12f_I \pm 5f_o$ | | 0.000 | 0.027 | 0.047 | 0.040 | 0.003 |
| | $12f_I \pm 7f_o$ | | 0.000 | 0.026 | 0.046 | 0.007 | 0.000 |
| | $12f_I \pm 9f_o$ | | 0.000 | 0.052 | 0.012 | 0.001 | 0.000 |
| | $12f_I \pm 11f_o$ | | 0.045 | 0.019 | 0.001 | 0.000 | 0.000 |
| | $12f_I \pm 13f_o$ | | 0.038 | 0.002 | 0.000 | 0.000 | 0.000 |
| **UFC** | | | | | | | |
| 3 | $3f_I + 2f_o$ | | 0.500 | 0.678 | 0.822 | 0.922 | 0.981 |
| | $3f_I + 4f_o$ | | 0.250 | 0.038 | 0.171 | 0.346 | 0.464 |
| 6 | $6f_I + 5f_o$ | | 0.200 | 0.350 | 0.399 | 0.335 | 0.196 |
| | $6f_I + 7f_o$ | | 0.143 | 0.071 | 0.235 | 0.229 | 0.179 |
| 12 | $12f_I + 11f_o$ | | 0.091 | 0.176 | 0.037 | 0.141 | 0.162 |
| | $12f_I + 13f_o$ | | 0.077 | 0.121 | 0.111 | 0.075 | 0.148 |

the energy back into the circuit during an unwanted trough. The cost of the filter obviously increases as the required energy storage increases.

Power filters may be classified according to whether their main purpose is to improve the power waveform or to remove EMI. Filters for waveform improvement usually deal with frequencies in the audio range. EMI filters are usually concerned with frequencies of 455 kHz or higher, although coupling to telephone lines or interference with low and very low frequency communications can be a problem at much lower frequencies.

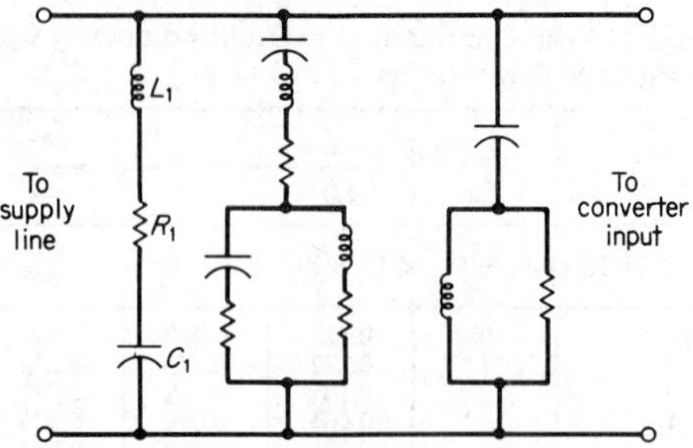


**FIG. 22-140** Typical input filter for waveform improvement.

***76. Input Filters.*** Input filters for waveform improvement normally consist of three to five series-resonant traps across the input power lines. These traps provide a low-impedance path in which the dominant low-order harmonic currents required by the converter can circulate. Without a filter, these currents would have to circulate through the source impedance of the supply line, thereby deteriorating the voltage waveform.

$L_1C_1R_1$ in Fig. 22-140 is a single-tuned trap of this type. The $Q$ of these traps is relatively high, with only enough damping to accommodate variations in line frequency and changes in component values due to initial tolerances, aging, and temperature variation. Although the ease of adjustment is decreased, the cost of filter components can be reduced by combining two single-tuned traps into a double-tuned trap, as shown in the middle of Fig. 22-140.

In contrast to the low-order harmonics which must be individually suppressed by high-$Q$ traps, higher-order harmonics can usually be adequately suppressed by a single damped filter section like that shown at the right of Fig. 22-140.

It should be noted that all these input filters draw leading current at the fundamental frequency, a property which is often useful for power-factor correction. It should also be observed that poles of impedance interleave the zeros which suppress the dominant harmonics. Care must be taken to assure that residual harmonics at other frequencies do not excite undesired resonance at these poles.

If the output frequency is variable, either intentionally (as in a variable-frequency cycloconverter) or because of appreciable variation in the supply frequency, high-$Q$ traps are unsatisfactory and broadband filters must be used.

Design procedures for input-power filters are given in Cory (Ref. 36, Chap. 7) and Kimbark (Ref. 37, Chap. 8.)

Input filters for EMI suppression usually consist of one or more low-pass *LCL* sections between the converter and the supply lines. The input and output terminals of the filter are transposed from the usual low-pass section, as shown in Fig. 22-141, because it is desired to minimize the current-transfer ratio rather than the voltage-transfer ratio. With this transposition, treating the converter as a current source and the supply line as a zero-impedance load, conventional design tables can be used (Geffe,[38] App. 3). Second-order filters are usually critically damped, that is, Butterworth response, while filters of higher order can be made to have a steeper edge to the stop band by using the Chebyshev design criterion.

***77. Output Filters.*** Output filters for waveform improvement may be divided into two categories: dc filters for rectifiers and choppers and ac filters for inverters and cycloconverters.

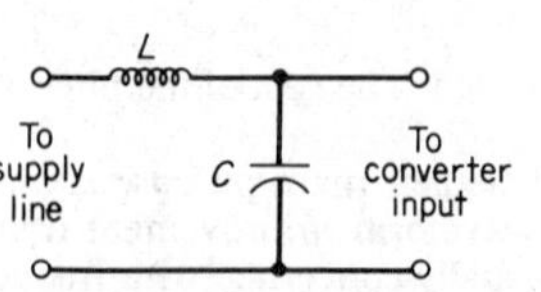


**FIG. 22-141** Transposed low-pass input filter to prevent EMI from entering the supply line.

Conventional single-section low-pass *LC* filters are used almost universally for dc applications. The inductor is usually chosen to be larger than the critical value which will maintain continuous current for the worst-case ripple-load combination (see Distler and Munshi). The capacitor is then chosen to obtain the desired reduction in ripple voltage. However, the resonant frequency must also be chosen so that it does not coincide with a residual harmonic below the fundamental ripple frequency. Although in theory these harmonics of the supply frequency are canceled, the cancellation is never perfect in practice.

The amplitudes of the unfiltered harmonics are summarized in Figs. 22-118 and 22-119, and design data for rectifier filters are presented in Terman (Ref. 39, Sec. 8) and Langford-Smith (Ref. 40, Chap. 31).

Simple second-order low-pass sections are ordinarily not used to filter the ac output of inverters and cycloconverters because of their insertion loss at the fundamental frequency. However, the series arm can be resonated to minimize this loss in fixed-frequency equipment, as in the Ott filter, which also supplies commutating capacitance (Ott, Rice). The shunt arms can be series-resonated at the dominant low-order harmonics and/or parallel-resonated at the fundamental frequency.

Passive filters are not suitable for use with variable-frequency converters unless the frequency range is restricted so that the lowest harmonic of the lowest fundamental frequency is considerably higher than the highest fundamental frequency.

***78. Filter Components.*** The capacitors used in power filters are required to pass high currents. Therefore, they should always be of extended-foil construction with a dielectric having a low loss over the required frequency range. Paper-oil, plastic-film, polycarbonate, or mica capacitors are generally suitable but must be chosen to have adequate transient ratings as well as steady-state ratings. Filters for dc applications nearly always use electrolytic capacitors because of their lower cost and smaller volume, even if their limited ripple-current capacity necessitates overdesign of the filter to stay within their ratings. The filter designer must also remember that in certain applications, for example, a rectifier feeding dc power to an inverter, the load may cause significant additional ripple currents.

The design of inductors for dc filters is well established (Terman,[39] Sec. 2; Langford-Smith,[40] Chap. 5.6). Chokes for ac filters may be air- or iron-cored, depending on the required inductance, kilovoltampere rating, and frequency range. Even iron-cored inductors are designed with an air gap which determines the inductance. The design of low-loss inductors is complicated by skin effect and winding capacitance, which often dictate the use of strip, tubular, or Litz conductors, and by fringing flux at the gap, which may require using powdered iron to distribute the gap.

***79. Active Filters.*** Conventional passive filters frequently represent a substantial part of the total cost, weight, and size of power electronics equipment. For this reason, improved types of filters are constantly being sought. Active filters represent a new approach still in its infancy about which practically nothing has been published. The approach is related to the theory of low-level active-feedback filters but requires considerable adaptation to achieve power efficiency within the capability of available devices. In effect, a power operational amplifier in a feedback loop with a single energy-storage element (an inductor or capacitor) serves to minimize the difference between the actual waveform and the desired waveform. The "filter" may become an integral part of the converter itself (see the optimum-response inverter described in Fig. 22-131). Although the cost of active filters tends to be high at present, their adaptability permits problems to be solved that could not be solved otherwise, as in a variable-frequency inverter, for instance.

## Bibliography

***80. Bibliography for Power Electronics***


1. Gyugyi, L., and Pelly, R. B.: *Static Power Frequency Changers;* New York, Wiley, 1975.
2. Wood, P.: *Switching Power Converters;* New York, Van Nostrand Reinhold, 1981.
3. Gentry, F. E.: Forward Current Surge Failure in Semiconductor Rectifiers; *AIEE Trans.* vol. 77, part I, November 1958, pp. 476–750.
4. Motorola Silicon Rectifier Manual, p. 8.
5. General Electric SCR Manual, p. 38.
6. Ghandi: *Semiconductor Power Devices;* New York, Wiley, 1977.
7. Hower, P. L. et al.: A New Method of Characterizing the Switching Performance of Power Transistors; IEEE-IAS Annual Meeting, 1978, pp. 1044–1049.

8. Hower, P. L., Blackburn, D. L., et al.: Stable Hot Spots and Second Breakdown in Power Transistors; *IEEE-PESC,* 1976, pp. 234–246.
9. General Electric Co. Transistor Manual, pp. 2-5–2-7.
10. Schaefer, J.: *Rectifier Circuits: Theory and Design;* New York, Wiley, 1965.
11. Dyer, R. F.: The Rating and Application of SCRs Designed for Switching at High Frequencies; *IEEE-IGA Trans.* vol. IGA-2, no. 1, pp. 5–15.
12. Gutzwiller, F. W., and Sylvan, T. D.: Power Semiconductor Ratings Under Transient and Intermittent Loads; *AIEE Trans.* part I, *Communications and Electronics,* 1960, pp. 699–706.
13. Recommended Standards for Thyristors, EIA-NEMA, Standard RS-397, June 1972, par. 7.3.1.2.
14. Westinghouse Applications Data 54-540, SCR Gate Turn-On Characteristics, December 1976, pp. 1–3.
15. See Ref. 13.
16. Borst, D. W., and Blatt, F. M.: Measurement of Gate-Controlled Turn-On Time of High-Current or Pulse-Modulated Thyristors; IEEE-IAS Annual Meeting, 1982, pp. 747–749.
17. Ericksson, F. W., and Carroll, E.: Establishing the Required Turn-Off Time for Thyristors with Companion Bypass Diode; Power Conversion International Proceedings, 1981.
18. De Bryune, P., and Jacklin, A. A.: The Reverse Conducting Thyristor and Its Application; *Brown Boveri Review,* 1979.
19. De Bryune, P., and Jacklin, A. A.: Why the Reverse Conducting Thyristor Can Improve the Design and Competitiveness of your Circuits; IEEE-IAS Annual Meeting, 1979, pp. 1109–1114.
20. Suzuki, T., et al.: Switching Characteristics of High Power Buried Gate Turn-Off Thyristor; IEEE-IEDM Annual Meeting, 1982, pp. 492–495.
21. Fukui, H., Hisao, A., and Miga, H.: Paralleling of Gate Turn-Off Thyristors; IEEE-IAS Annual Meeting, 1982, pp. 741.
22. Teccor Electronics Inc.: Gating, Latching, and Holding of SCRs and Triacs; product catalog, pp. 99–102.
23. Brewster, J. B., and Sherbondy, G. F.: Complete Characterisation Studies Verify RBDT-RSR Reliability; *IEEE Proc. Pulse Power Modulators* (special issue), October 1979, pp. 1462–1468.
24. Internal Rectifier Corp., Hexfet Databook, HDB-2, 1982–83, p. A10.
25. Pelly, B. R.: *Thyristor Phase-Controlled Converters and Cycloconverters;* New York, Wiley-Interscience.
26. Schaefer, J.: *Rectifier Circuits: Theory and Design;* New York, Wiley, 1965.
27. McMurray, W., and Shattuck, D. P.: A Silicon Controlled Rectifier Inverter with Improved Commutation, *Trans. AIEE,* 1961, vol. 80, Pt. 1, pp. 531–542.
28. Kernick, A., et al.: Static Inverter with Neutralization of Harmonics, *Trans. AIEE,* 1962, vol. 81, Pt. 2, pp. 59–68.
29. Kernick, A., and Heinrich, T. M.: Controlled Current Feedback in a Static Inverter with Neutralization of Harmonics, *IEEE Trans. Aerosp.,* 1964, vol. 2, pp. 985–992.
30. Heinrich, T. M.: Static Inverter with Neutralization of Harmonics, M.S. dissertation, University of Pittsburgh, Pa., 1967.
31. Oates, R. M.: Inverter Harmonic Neutralization Using Interphase Reactors, M.S. dissertation, University of Pittsburgh, Pa., 1970.
32. Kernick, A., and Haque, I. U.: Programmed Waveform Static Inverter, *Proc. 23d Ann. Power Sources Conf.,* 1969, pp. 59–63, sponsored by U.S. Army Electronics Command, Fort Monmouth, N.J.
33. Ravas, R. J., et al.: Staggered Phase Carrier Cancellation: A New Circuit Concept for Lightweight Static Inverters, *EASTCON Tech. Conv. Rec., Suppl. IEEE Trans.,* 1967, vol. AES-3, no. 6, pp. 432-444; *IEEE Pub.* 10-C-57.
34. Geyer, M. A., and Kernick, A.: Time Optimal Response Control of a Two-Pole Single-Phase Inverter, *Power Cond. Spec. Conf. Rec.,* 1939, pp. 101–109, *IEEE Pub.* 71C15-AES.
35. Gyugyi, L.: Generalized Theory of Static Power Frequency Changers, Ph.D. thesis, University of Salford, England, 1970.
36. Cory, B. J. (ed.): *High Voltage Direct Current Convertors and Systems;* London, Macdonald, 1965.
37. Kimbark, E. W.: *Direct Current Transmission,* vol. I; New York, Wiley-Interscience, 1971.
38. Geffe, P.: *Simplified Modern Filter Design;* Rider, New York, 1964.

39. Terman, F. E.: *Radio Engineers' Handbook;* New York, McGraw-Hill, 1943.

40. Langford-Smith, F. *Radiotron Designer's Handbook;* Harrison, N.J., RCA, 1953.

## *INDUSTRIAL ELECTRONICS*

*By DONALD GALLER*

### Introduction

***81. General Applications.*** The use of electronic equipment in industry has been increasing at a very rapid rate. Electronics has made it possible to automate many industrial processes, thus reducing costs below the cost of human labor to do the same work. A complete summary of the use of electronics in industry is beyond the scope of this handbook; however, the use of electronic apparatus can influence the design and operation of the electric (power) equipment in an industrial plant, and is discussed. The engineer responsible for design and operation of electric equipment will need to consider these influences, some of which are described below.

Electronics has made possible new and more effective safety devices. In addition, with electronics it has become possible to achieve greater accuracy in measurements, to use more precise methods of control, and to provide improved supervision of industrial operations. In most cases, these improved operations have required that the performance of the plant electric system be more reliable, have better regulation, and have improved protective devices.

Electronic controls are used in industry to monitor and control physical quantities within industrial processes. The simplest form of an electronic control system is shown in Fig. 22-142*a* It consists of three elements: (1) the process to be monitored or controlled, (2) a transducer, and (3) an indicator. The transducer converts a physical quantity into an electrical signal, usually a variable voltage or variable current, which is used to operate an indicator. This is called an *open-loop system* because the control action is fixed and the value of the physical quantity is free to vary due to changes in the process, changes in temperature, and other effects.

A *closed-loop* system is constructed to maintain the monitored quantity at a prescribed value. A simplified closed-loop system is shown in Fig. 22-142*b*. It contains two additional elements not shown in Fig. 22-142*a*: (1) an electronic control circuit, and (2) an actuator. The electronic control circuit compares the reference value of the controlled quantity with the actual value and produces an electrical control signal. The control signal causes the actuator to physically modify the control action to compensate for changes in the controlled quantity, thereby maintaining the reference value.[1,2*]

The National Electrical Manufacturer's Association standards ICS 3-100, General Standards for Industrial Systems; ICS 3-104, Classification of Systems and Subsystems; and ICS 3-106A, Performance of Feedback Control Systems contain definitions and general information on closed-loop control systems.

Modern closed-loop control systems use digital computers in place of the electronic control circuit.[3] A digital control system is shown in Fig. 22-142*c*. A digital computer performs the comparison between the reference and actual values of the controlled quantity and produces a digital control value. The digital control value is converted into an electrical signal by a digital-to-analog (D/A) converter. The signal representing the physical quantity is converted into digital form by an analog-to-digital (A/D) converter. D/A and A/D converters are generally part of the digital computer equipment.

***82. Impact of Integrated Circuits.*** In recent years the application of solid-state and integrated circuit devices in industry has become widespread. The focus of the development

---

*Superior numbers refer to bibliography, Par. **120**.

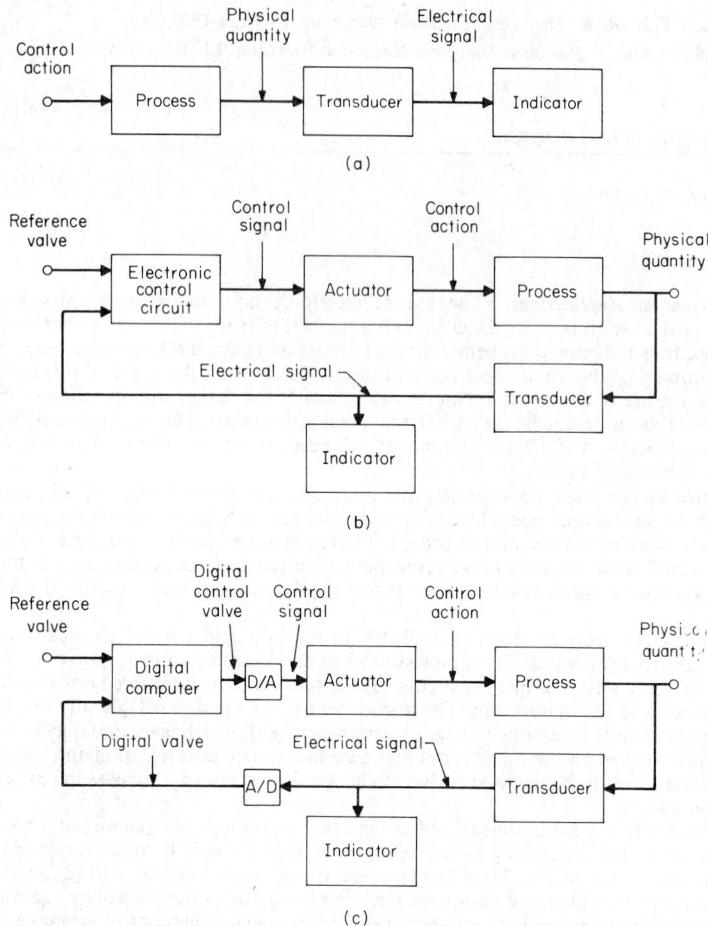


**FIG. 22-142** Electronic control systems. (*a*) Open-loop; (*b*) closed-loop analog control; (*c*) closed-loop digital control.

of integrated circuits has historically been on digital logic circuits. Currently available *large-scale integrated* (LSI) circuits may contain several hundred to several thousand digital logic gates in a single package. In the early 1970s the first *microprocessors* became available, and allowed a functional computer to be constructed from just two or three LSI circuits. The trend toward increasing levels of circuit integration has led to electronic equipment that is smaller and more reliable as a result of the reduced number of components.

The introduction of microprocessors into industrial electronics has had a dramatic impact on components and equipment. The use of LSI and microprocessors in devices such as transducers, display and monitoring devices, and actuators has increased the performance and reliability of these components both as independent devices and when used with control computer equipment. Microprocessors have been incorporated into control computer equipment, reducing their size and cost and increasing their performance. The impact has been that microprocessor-based computers have invaded the applications previously assigned to large central computers used for industrial process control while microprocessor-based sequencers and programmable controllers have seen increased use in applications previously implemented using relay logic or discrete transistors.

Recent developments in computer hardware and software have led to the development of *distributed control systems* for industrial process control.[4] These systems consist of several microprocessor-based control computers connected by a data communication network. The computers share the tasks of controlling an industrial application by sending data and instructions to each other along the network. Some network designs allow the computers in the network to automatically reallocate tasks if an individual computer fails. Distributed process control networks have the potential to be much more flexible and reliable than single-computer systems as a result of this property. Distributed computer process control will have an increasing impact on the industrial electronic environment in the future and is a direct result of the development of integrated circuitry and microprocessors.

## Solid-State Devices

***83. Solid-State Devices for Power Conversion.*** Discrete solid-state devices are used widely in industrial equipment to amplify electrical signals and for power conversion. The symbols and characteristics for the major solid-state devices used for power conversion are shown in Table 22-10. These devices fall into three categories: (1) diodes, (2) transistors, and (3) thyristors.

Diodes are two-terminal semiconductor junction devices which block the flow of current in one direction and conduct in the other direction. Their main application is in the rectification of 60-Hz ac power for dc power sources within industrial equipment.

Transistors are three-terminal semiconductor devices which are used for amplification and power switching. The bipolar transistors are current-controlled. Current flowing in the base-emitter circuit modulates a much larger current flowing in the collector-emitter circuit.[5] The MOSFET (metal-oxide semiconductor field-effect transistor) is a voltage-controlled device.[6] The gate-to-source voltage controls the current flowing in the drain-source circuit. The current and voltage ratings of MOSFET devices have only increased to the levels suitable for industrial applications since the early 1980s.

Thyristors are typically three-terminal devices and are constructed of four layers of semiconductor material. The most common thyristor devices are the reverse blocking triode thyristor or semiconductor-controlled rectifier (SCR) and the bidirectional triode thyristor or triac.[7] These devices are similar in that they remain in a blocking (nonconducting) state until triggered into a conducting state by the application of gate current. The thyristor will then remain in the conducting state without gate current until the anode current is reduced to zero. The device then returns to the blocking state until retriggered. The main difference between the SCR and the triac is that the SCR can only be triggered into conduction in one direction while the triac can be triggered into conduction in either direction.

Symbols for solid-state devices are found in the American National Standards Institute Standard ANSI Y32.2-1975, Graphic Symbols for Electrical and Electronics Diagrams.

***84. Optoelectronic Solid-State Devices.*** Optoelectronic solid-state devices are an important group of solid-state devices widely used in industrial electronic equipment.[8] The symbols and characteristics of the major optoelectronic solid-state devices are shown in Table 22-11. These devices fall into three categories: (1) light-emitting devices, (2) light-sensitive devices, (3) optically coupled isolators.

The most common light-emitting device is the light-emitting diode or LED. These are diodes that emit visible light or infrared radiation when conducting. The visible-light LEDs are available in red, yellow, and green light-emitting types. Single LEDs are used to replace filament lamps and for status indicators in low-voltage circuits. Several types of multisegment displays are also commercially available. These consist of several LEDs mounted in a special arrangement in a single package so that numbers or letters can be displayed by selective activation of the LEDs. These *alphanumeric* displays are widely used in calculating equipment and test instruments.

Photovoltaic cells generate a current proportional to the amount of light (intensity × area) falling on the cell. These are also called solar cells since they are most commonly used for solar-powered portable equipment and for solar energy (solar to electric) conversion. Large arrays are usually used for solar energy conversion. Single devices are sometimes used in photoelectric controls and for light-metering equipment used in photographic applica-

**TABLE 22-10** Solid-state Devices for Power Conversion

| Applications | Device | Symbol | Characteristics |
|---|---|---|---|
| **Diode** | | | |
| • Rectification<br>• Signal detection<br>• Limiting | Diode | (A) (K) $I_A$ | Conducts current in one direction, blocks current in the other direction. Anode (A) must be positive with respect to cathode (K) for conduction. |
| **Transistors** | | | |
| • Amplification<br>• Power switching<br>• Power control<br>• Inverters | Bipolar transistor (NPN) | (C) (B) (E) $I_C$ $I_B$ | Base current $I_B$ modulates larger collector current $I_C$. Base must be positive with respect to emitter to establish $I_B$. |
| | Bipolor transistor (PNP) | (C) (B) (E) $I_C$ $I_B$ | Complement to NPN transistor. |
| | Darlington transistor (NPN) | (C) (B) (E) | Two stage transistor amplifier integrated into one package. Operates as one transistor with very high gain. |
| | Power MOSFET* (Nchannel) (Enchancement) | (D) (G) (S) $I_D$ | Gate-source voltage $V_{GS}$ modulates drain current $I_D$. $V_{GS}$ must be positive to establish $I_D$. |
| | Power MOSFET (P channel) (Enhancement) | (D) (G) (S) $I_D$ | Complement to N channel MOSFET. |
| **Thyristors** | | | |
| • Power switching<br>• AC phase control<br>• Solid state relays<br>• Inverters (SCR) | Silicon controlled rectifier (SCR) | (A) (G) (K) $I_A$ $I_G$ | Gate current $I_G$ triggers device into conducting state, allowing larger $I_A$ to flow. conducting state persists until $I_A$ is removed. |
| | Bidirectional triode thyristor (triac) | (A1) (G) (A2) $I_G$ | Similar to SCR but can be triggered into conduction in either direction. |

*Metal oxide semiconductor field effect transistor.

Note: (A) = anode; (B) = base; (C) = collector; (D) = drain; (E) = emitter; (G) = gate; (K) = cathode; (S) = source.

**TABLE 22-11** Optoelectronic Solid-state Devices

| Applications | Device | Symbol | Characteristics |
|---|---|---|---|
| **Light-emitting devices** | | | |
| • Indicators<br>• Light sources | Light-emitting diode (LED) | 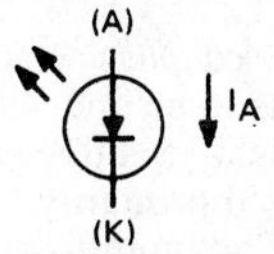 (A) (K) $I_A$ | Similar to conventional diode but emits visible or infrared radiation when conducting. |
| **Photovoltaic devices** | | | |
| • Solar power conversion<br>• Photoelectric controls | Photovoltaic cell (solar cell) | 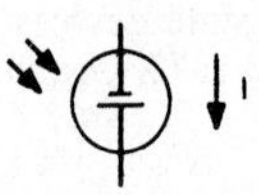 I | Semiconductor junction device that generates a current proportional to the amount of light falling on the junction. |
| **Photosensitive devices** | | | |
| • Photoelectric controls<br>• Position transducers<br>• Optical communication<br>• Photographic equipment | Photodiode | 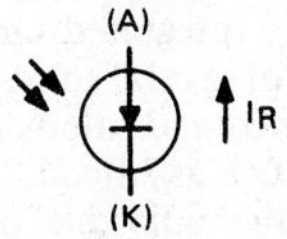 (A) (K) $I_R$ | Reverse leakage current increases with amount of light falling on the junction. Visible or infrared types available. |
| | Phototransistor | 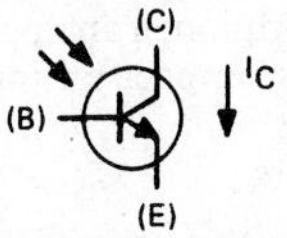 (C) (B) (E) $I_C$ | Similar to conventional bipolar transistor but base current generated by incident light. Visible or infrared types available. |
| | Light activated silicon controlled rectifier (LASCR) | 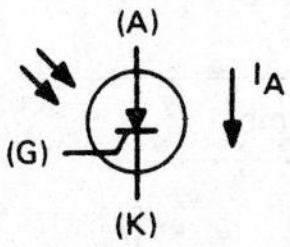 (A) (G) (K) $I_A$ | Similar to conventional SCR but can be triggered by light. |
| **Optically coupled isolator** | | | |
| • Signal isolation<br>• Relay replacement<br>• SCR trigger circuits | Opto-coupler transistor (output) | 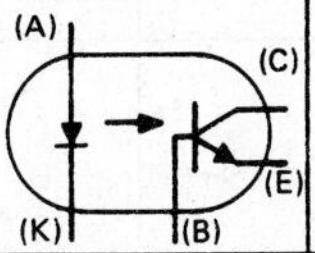 (A) (K) (C) (E) (B) | LED and phototransistor integrated into one package. Devices are optically coupled but electrically isolated. LED optically induces base current |
| | Opto-coupler (Darlington output) | 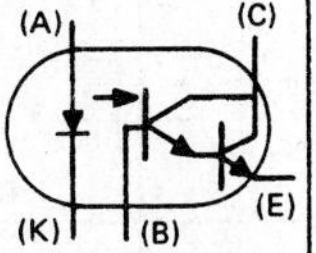 (A) (K) (C) (E) (B) | LED and Darlington in one package. Darlington provides higher sensitivity to LED current. |
| | Opto-coupler (LASCR output) | 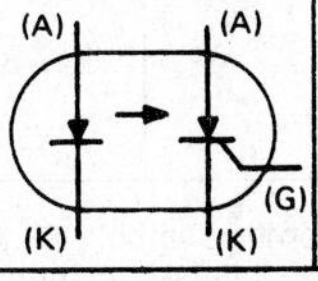 (A) (K) (A) (G) (K) | LED and LASCR in one package. LED triggers LASCR into conduction. |

Note: (A) = anode; (B) = base; (C) = collector; (E) = emitter; (G) = gate; (K) = cathode.

tions. Photodiodes, phototransistors, and light-activated SCRs (LASCRs) are light-controlled versions of the conventional devices. These devices are light-sensitive but do not convert the light directly into electric energy as does the photovoltaic cell. They are widely used in photoelectric controls, position transducers, optical communication systems, and in photographic equipment.

Optically coupled isolators or *optocouplers* consist of an LED and a light-sensitive device integrated into a single package so as to be optically coupled but electrically isolated. They are available with several types of light-sensitive output circuits, some of which are shown in Table 22-11. Optocouplers are used primarily for signal isolation. Devices are currently available that can withstand 3750-V ac input to output. They are widely used for isolating and transmitting digital data, for triggering SCRs and transistors in power conversion equipment, and as replacements for electromechanical relays.

Symbols for optoelectronic solid-state devices are found in American National Standards Institute Standard ANSI Y32.2-1975, Graphic Symbols for Electrical and Electronic Diagrams.

## Integrated Circuits

***85. Digital Integrated Circuits.*** Integrated circuits (ICs) first became commercially available around 1960. They consist of several solid-state devices combined in a single package. The individual devices are formed and interconnected by specially processing a single substrate, or chip, of base material such as silicon. Over the past two decades the trend has been to improve ICs by: (1) making the individual devices smaller in order to provide more functions in a single package, (2) increasing their maximum frequency of operation, and (3) reducing their power requirements. Although analog ICs have developed as well, the greatest advancement and by far the greatest impact on industrial electronic equipment has been that of digital IC technology.

**TABLE 22-12** Functions and Symbols of Basic Logic Gates

| Function | Symbol | Boolean expression | Truth table |
|---|---|---|---|
| Buffer | A B | B = A | A B<br>0 0<br>1 1 |
| Inverter (NOT) | A B * | $B = \bar{A}$ | A B<br>0 1<br>1 0 |
| AND | A B C | C = A • B | A B C<br>0 0 0<br>0 1 0<br>1 0 0<br>1 1 1 |
| OR | A B C | C = A + B | A B C<br>0 0 0<br>0 1 1<br>1 0 1<br>1 1 1 |

* The negation indicator symbol, O, can be added to any logic symbol to indicate that a signal is internally inverted. When added to the output of the AND and OR symbols, the NAND (NOT-AND) and NOR (NOT-OR) symbols are formed respectively.

Digital logic circuits perform operations on binary information. Each signal in a digital circuit can take on one of two values representing a binary 1 or 0. A single binary digit is called a *bit*. Bits may be assembled into groups called *words* representing multidigit binary numbers. Commonly used word sizes are frequently referred to by special names such as a *nibble* (4 bits) and a *byte* (8 bits).

***86. Digital Logic.*** The most basic digital logic operations are performed by circuits called *gates*. These operations are described by truth tables or in mathematical notation using boolean algebra.[9] The functions and circuit symbols of some basic logic gates are shown in Table 22-12. These circuits can be used to construct much more complex functions such as binary adders, memory circuits, and counters. More than several hundred logic functions are presently commercially available. These circuits are widely used in industrial equipment to replace relay logic, to replace analog control circuits, in data collection and display equipment, and in computers.

Digital ICs can be classified as to complexity by the number of gates in each package as shown in the following table:[10]

| Level of integration | Logic gates/ package | Development |
|---|---|---|
| Small-scale integration (SSI) | 3–30 | Early 1960s |
| Medium-scale integration (MSI) | 30–300 | Middle to late 1960s |
| Large-scale integration (LSI) | 300–3000 | Early to mid-1970 |
| Very large scale integration (VLSI) | 3000 | Beyond late 1970s |

SSI, MSI, and LSI ICs used in industrial and commercial equipment are commonly packaged in dual-in-line or DIP packages. VLSI circuits are presently under development for military applications and for use in special-purpose computers.

There are several methods of fabrication currently being used to manufacture digital ICs. Each fabrication method yields most of the same logic functions but with circuitry that differs in characteristics such as operating voltage, power consumption, and maximum operating frequency. Table 22-13 shows the characteristics of several logic families.[12–14] The two logic families most commonly used in commercial and industrial electronics are transistor-transistor logic (TTL) and complementary-metal-oxide-semiconductor (CMOS) logic.

TTL operates with supply voltages of 5 V dc and has a maximum frequency of about 35 MHz. The power dissipation of standard TTL is about 10 mW per gate. Most TTL logic functions are also available in versions with higher speed and/or lower power than the standard components. TTL is used in applications requiring high-speed operation. CMOS operates with supply voltages of 3 to 18 V dc and has a maximum frequency of about 4 MHz. Power dissipation is about 4 mW per gate at 4 MHz. CMOS power dissipation decreases almost linearly with frequency and static power requirements are extremely low, typically about 5 nW per gate. CMOS is used in applications requiring extremely low power operation, in portable equipment, and in applications where the speed of TTL is not required.

***87. Flip-Flops.*** The basic digital logic memory element used in counting and sequencing circuits is a *flip-flop*.[9] Flip-flops have two stable states and can be put in one state or the other by an input pulse. Once the input pulse is removed, the circuit will remain in a given state until a pulse forcing it into the other state is applied. This property allows the flip-flop to be used as a memory element.

Several types of flip-flops are presently commercially available as standard components within logic families such as TTL and CMOS.[11,12] Two of these, the *JK* flip-flop and the *D*-type flip-flop, are shown in Fig. 22-143. Both types have $Q$ and $\overline{Q}$ complementary outputs which indicate the state of the flip-flop. The flip-flop changes state either by clocked operation, using the clock and data or $J$ and $K$ inputs, or by the preset and clear inputs. In clocked operation, the flip-flop changes state after the leading edge (logic 0 to logic 1 transition) of a clock pulse. The state after the clock pulse is determined by the condition of the data inputs as shown in the truth tables of Fig. 22-143. The preset and clear inputs override the clocked operation and force the flip-flop into the $Q = 1$ ($\overline{Q} = 0$) and $Q = 0$ ($\overline{Q} = 1$) state, respectively.

**Table 22-13** Characteristics of Integrated-Circuit Digital Logic Families

| Logic family | Series | Supply voltage, V | Logic 0 voltage, V | Logic 1 voltage, V | Gate propagation delay, ns | Gate power dissipation | Maximum frequency, MHz |
|---|---|---|---|---|---|---|---|
| TTL | Low power | 5 | 0–0.2 | 3.5–5 | 33 | 1 mW | 3 |
| | Standard | 5 | 0–0.2 | 3.5–5 | 10 | 10 mW | 35 |
| | Low power Schottky | 5 | 0–0.2 | 3.5–5 | 9.5 | 2 mW | 45 |
| | High speed | 5 | 0–0.2 | 3.5–5 | 6 | 22 mW | 50 |
| | Advanced low-power Schottky | 5 | 0–0.2 | 3.5–5 | 4 | 2 mW | 50 |
| | Schottky | 5 | 0–0.2 | 3.5–5 | 3 | 19 mW | 125 |
| CMOS | Standard | 3–18 | 0–0.05 | 4.95–5 (5-V supply) | 125 | 5 nW (static) 4 mW @ 4 MHz | 4 |
| | High speed | 2–6 | 0–0.1 | 4.9–5 (5-V supply) | 15 | 2.5 nW (static) 2 mW @ 1 MHz | 40 |
| ECL | Standard | −5.2 | −1.7 | −0.9 | 2 | 25 mW | 165 |
| | High speed | −5.2 | −1.7 | −0.9 | 1 | 25 mW | 250 |

NOTE: TTL = Transistor-transistor logic; CMOS = complementary metal-oxide semiconductor; ECL = emitter-coupled logic.

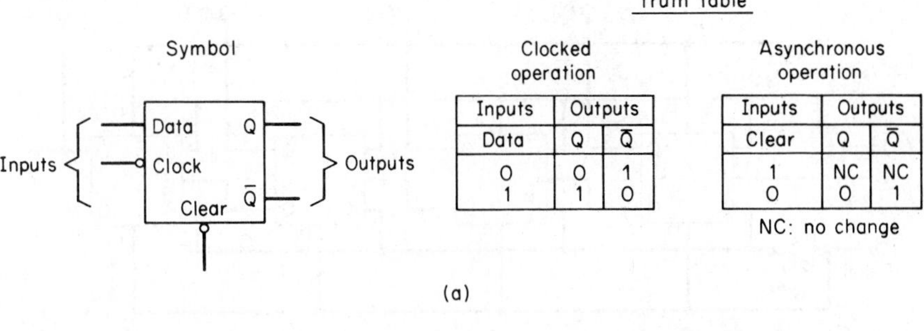


| Inputs | Outputs | |
|---|---|---|
| Data | Q | $\bar{Q}$ |
| 0 | 0 | 1 |
| 1 | 1 | 0 |

| Inputs | Outputs | |
|---|---|---|
| Clear | Q | $\bar{Q}$ |
| 1 | NC | NC |
| 0 | 0 | 1 |

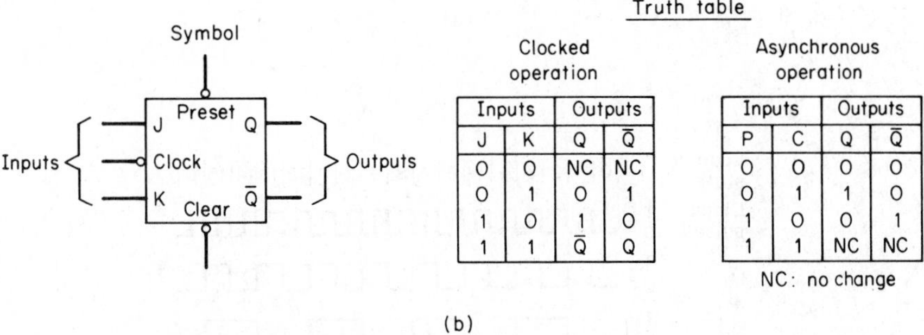


| Inputs | | Outputs | |
|---|---|---|---|
| J | K | Q | $\bar{Q}$ |
| 0 | 0 | NC | NC |
| 0 | 1 | 0 | 1 |
| 1 | 0 | 1 | 0 |
| 1 | 1 | $\bar{Q}$ | Q |

| Inputs | | Outputs | |
|---|---|---|---|
| P | C | Q | $\bar{Q}$ |
| 0 | 0 | 0 | 0 |
| 0 | 1 | 1 | 0 |
| 1 | 0 | 0 | 1 |
| 1 | 1 | NC | NC |

**FIG. 22-143** Operation of flip-flop circuits. (*a*) *D*-type flip-flop; (*b*) *JK* flip-flop.

***88. Counting Circuits.*** Counting circuits or *counters* are used for timing, totaling events, and in sequencing applications. Counters can be constructed from flip-flops as shown in Fig. 22-144 which illustrates a *binary counter.*[9] The circuit consists of four *JK* flip-flops as shown in Fig. 22-144*a*. The flip-flops change state after each clock (input) pulse as shown in Fig. 22-144*b*. The state of the flip-flops identifies the state of the counter. The state of the counter can be thought of as a four-bit binary number in which the *Q*4 output is the most significant bit (MSB). The state-to-state advancement of the counter is shown in the *state diagram* of Fig. 22-144*c*. This counter is a binary counter because it counts in the natural binary counting sequence and because it has maximum count length. The maximum count length for an *N*-stage counter is $2^N$.

Counters can be designed to have virtually any count length and sequence required by the application. Decade counters are frequently used in applications where decimal information is to be counted and displayed, such as digital timers. A decade counter consisting of four *D*-type flip-flops is shown in Fig. 22-145. The circuit in Fig. 22-145*a* shows the additional gating circuits required to shorten the count length from 16 to 10. The counting sequence is shown in Fig. 22-145*b*. The counter can be forced into states outside the normal sequence when power is applied to the circuit. Proper design assures that the counter will not *lock up* in state sequences outside the normal sequence. The state diagram of Fig. 22-145*c* shows how the counter will sequence out of these states and return to the normal counting sequence.

The counter circuit shown in Fig. 22-144 uses the output of each flip-flop to provide the clock input to the next flip-flop. Because of small delays called *propagation delays,* the output of the first flip-flop does not change state at the same time the clock pulse is applied. Because the flip-flops are cascaded, the delays in each stage add up and the total delay from the first clock input to the last flip-flop output may be significant. The counter is said to be an *asynchronous counter* because all the flip-flops do not change state at the same time. The circuit shown in Fig. 22-145 is a *synchronous counter* because the clock inputs to all the flip-flops are applied at the same time, causing all flip-flops to change state at the same time.

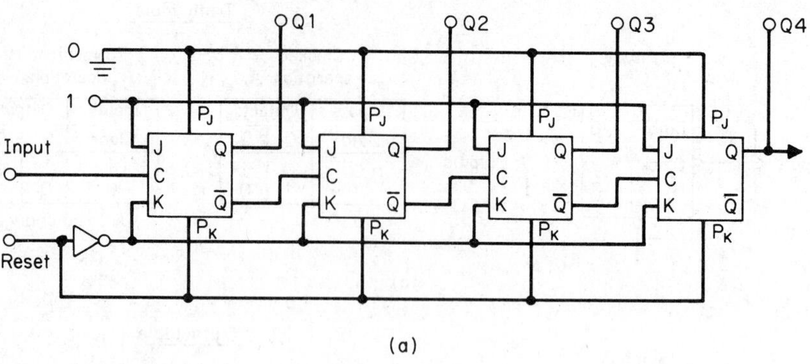


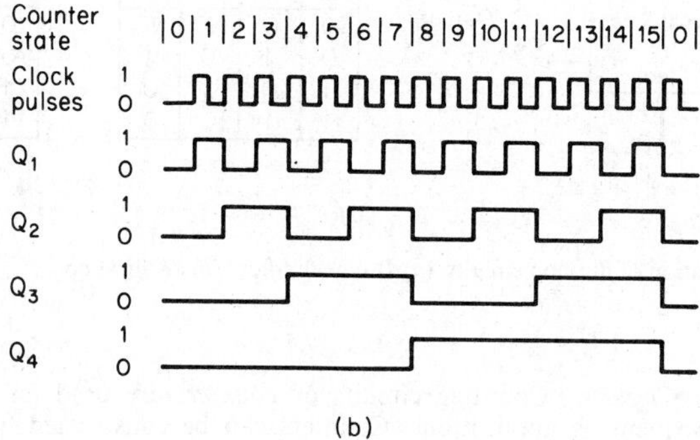


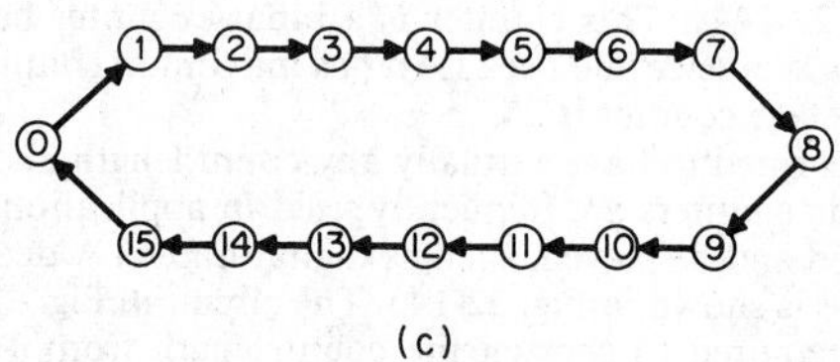


**FIG. 22-144** Operation of four-stage binary counter. (*a*) Flip-flop circuit; (*b*) output sequence; (*c*) state diagram.

Synchronous counters are used in event counting and timing circuits where it is necessary to test all the outputs at once to determine the state of the counter. Errors can result if the state of an asynchronous counter is tested during the time the clock pulse is propagating through the counter. For this reason asynchronous counters are used in applications such as frequency division where only the output of the last stage is used. Many types of asynchronous and synchronous counters, including those shown in Figs. 22-144 and 22-145 are presently commercially available in single IC packages.

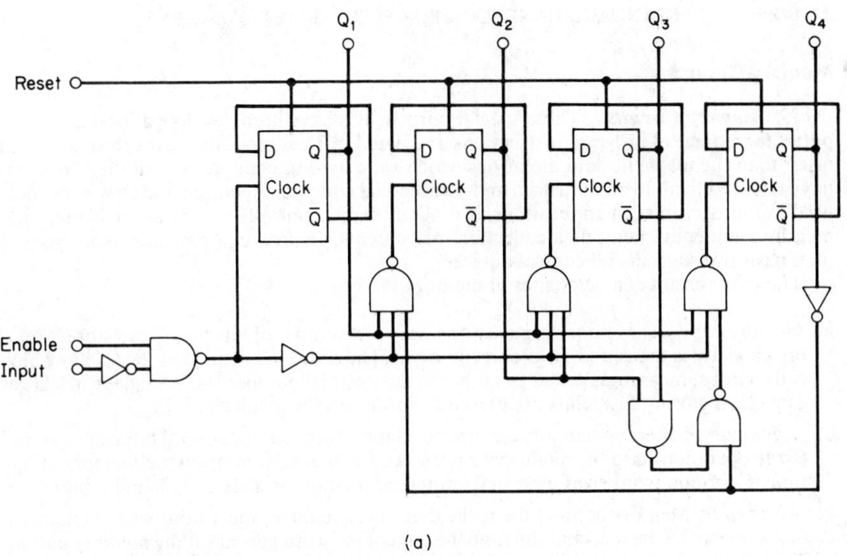


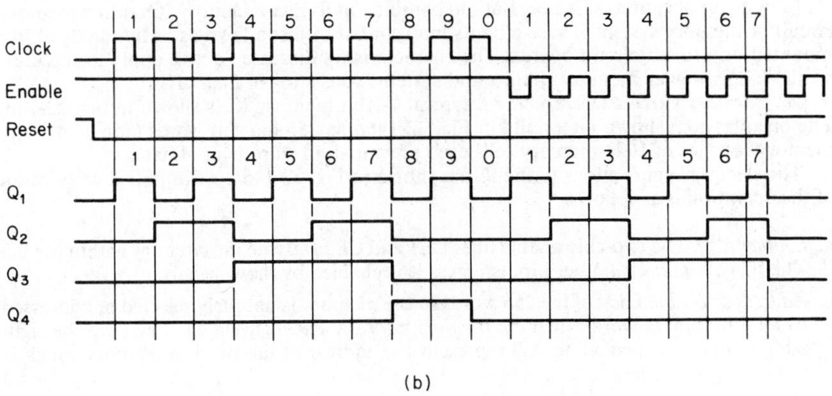


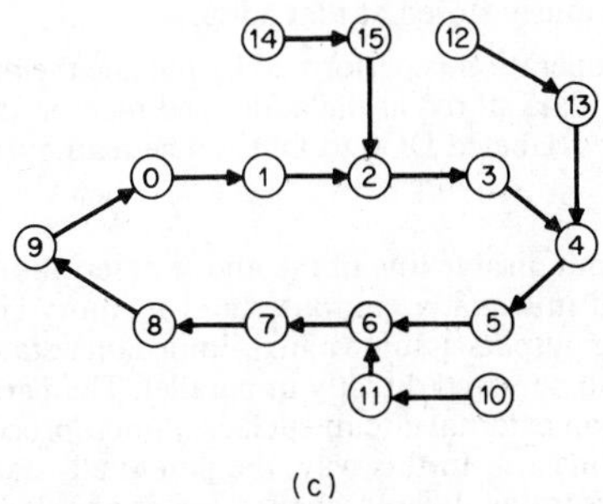


**FIG. 22-145** Operation of a decade counter. (*a*) Flip-flop circuit; (*b*) output sequence; (*c*) state diagram.

## Memory Circuits

***89. Memory Circuits.*** These special-purpose ICs have been developed for use in computers for storage of program instructions and data.[15,16] These circuits are much smaller and faster than the magnetic core memories which were used in computers until the 1960s and have replaced them in all industrial and commercial computing equipment. They are widely used in industry in such applications as programmable controllers, programmable test and measurement equipment, data collection instruments, security equipment, and all types of intelligent transducers and control devices.

Three important characteristics of memory ICs are

**a.** *Memory capacity.* Memory capacity or size is the number of bits the IC can store. Memory capacity is measured in kilobits or *Kbits.* The unit K is defined as $2^{10}$ (1024) which is the closest meaningful binary number to the decimal number 1000. Single IC packages capable of storing 512 Kbits are presently commercially available.

**b.** *Organization.* Memory circuits are organized into words so that several bits can be stored (*write* operation) into the memory or retrieved (*read* operation) from the memory at one time. Common word sizes used in commercial memory ICs are 1, 2, 4, and 8 bits.

**c.** *Access time.* Memory access time is the delay between the time a word within the memory is selected to be read and the time the data of the word appears at the memory output. This delay is a result of delays in the logic within the IC and is a measure of how fast it can operate. Commercially available memory ICs have access times from 20 ns to 1000 ns (1 $\mu$s).

In many applications it is necessary to combine individual memory ICs into a memory circuit or memory system. These circuits may consist of just a few ICs or hundreds of ICs mounted on printed circuit boards. The characteristics discussed above apply to the individual ICs as well as memory circuits and systems consisting of many ICs.

***90. Memory Circuit Operation.*** A typical 1-Kbit memory IC is shown in Fig. 22-146. The organization, shown pictorially in Fig. 22-146*a*, is 256 four-bit words (256 $\times$ 4). The memory consists of 1024 memory cells each of which can store 1 bit of data.

The electrical connections to the IC are shown in Fig. 22-146*b*. A simplified description of the operation is as follows:

**a.** *Chip enable.* The two-chip enable lines CE1 and CE2 activate the circuitry within the IC. The IC will ignore all other inputs unless it is enabled by these signals.

**b.** *Address selection.* Each of the 256 words in the memory is uniquely selected or addressed by an 8-bit binary word called the *memory address.* The logic state (1 and 0) of the eight address lines (labeled A0 to A7) represent the address of the word in memory which is to be read or written.

**c.** *Write operation.* A write operation is performed by presenting data on the four data input lines (labeled DI0 to DI3) and then pulsing the read-write control line (labeled R/W). During the write pulse, the data on the input lines are stored at the selected address, destroying the data previously stored at that address.

**d.** *Read operation.* A read operation is performed by pulsing the output enable line. During the output enable pulse, data stored at the addressed memory location are presented on the four data output lines (labeled DO0 to DO3). The read operation does not affect the contents of memory.

The function of the output disable line in the above description illustrates an important design feature incorporated into many computer and memory circuits. When the output enable line is not active, the outputs assume a high-impedance state. This feature allows the outputs of several circuits to be wired directly in parallel. The parallel signal lines, called a *data bus,* are controlled by an external circuit such as a microprocessor so that only one set of outputs is active at any instant. In this way, the potentially damaging condition of two outputs on the same line assuming different logic states is avoided.

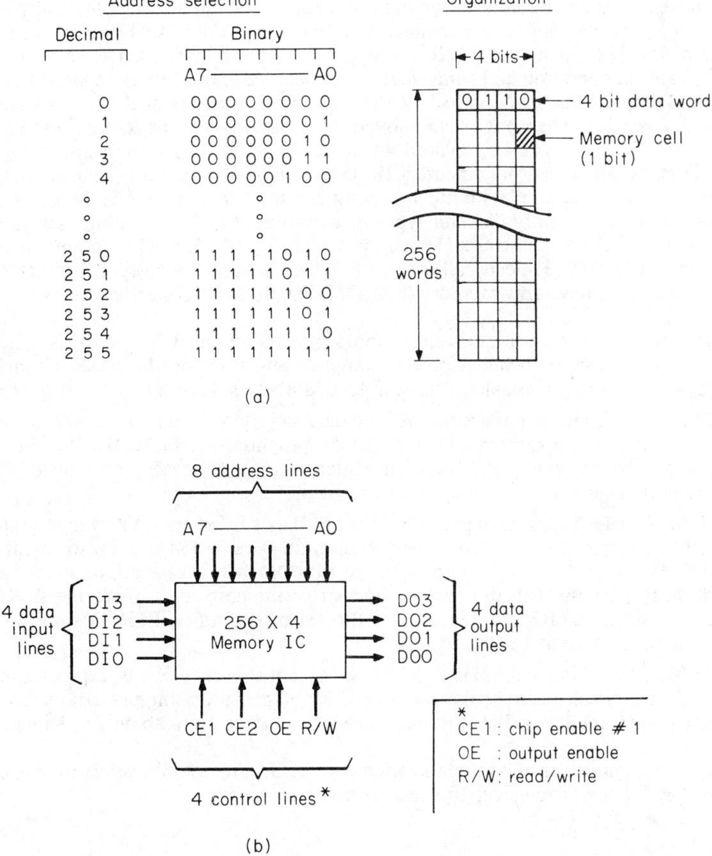


**FIG. 22-146** Typical 1-Kbit-memory IC operation. (*a*) Organization; (*b*) electrical configuration.

The sharing of the bus at different times is called *multiplexing*. It has the advantage of simplifying interconnect wiring in digital circuits. Because the outputs can assume logic 1, logic 0, and high-impedance states these circuits are said to have *three-state* outputs.

***91. Random Access Memory.*** For the memory circuit shown in Fig. 22-146, a delay exists between the time an address is presented on the address line and the time the output data are valid for that address. This delay is called the *memory access time* and is a result of delays in the logic within the ICs. The memory circuit (and the IC used to construct it) is called a random access memory (RAM) because the memory access time is independent of the memory address being selected. Memory addresses can be selected at random and the same memory access delay will result. In a sequential access memory (SAM), the access time depends heavily on the particular memory address being selected.

Two types of RAM that are presently being used are *static* RAM and *dynamic* RAM. These are sometimes referred to as SRAM and DRAM, respectively. The individual memory cells in SRAM circuits consist of a flip-flop memory element. The contents of each cell are valid as long as power remains applied. The individual memory cells in DRAM circuits consist of a capacitive-charge storage circuit. Since charge leaks off of the storage circuit, external refresh circuits are used to maintain the memory contents by periodically reading and rewriting the data. This is the primary disadvantage of DRAM. However, because the structure of the DRAM memory cell is simpler than that of SRAM, DRAM circuits are generally faster in operation and somewhat less expensive than equivalent SRAM circuits.

***92. Nonvolatile Memory Circuits.*** RAM memory circuits are said to be *volatile* since their contents are lost when power is removed. In computer systems RAM is used to store programs and data that can be replaced when the power is turned on again. *Nonvolatile* memory circuits called read-only memory (ROM) circuits are used for permanent data storage. ROMs are manufactured with the data permanently stored in them and they cannot be erased or altered electrically. Several types of nonvolatile memory circuits are presently commercially available. These are: (1) read-only memory (ROM), (2) programmable read-only memory (PROM), (3) electrically programmable read-only memory (EPROM), and (4) electrically alterable read-only memory (EAROM). These are described as follows:

**ROM:** The contents are permanently stored into the circuit when it is manufactured. The data are present whenever power is applied and it cannot be erased or modified. ROMs are presently available in sizes up to 1 Mbits (1024 Kbits).

**PROM:** PROMs are manufactured with no data stored in them and are *programmed* by the user with an instrument called a PROM programmer. Once the PROM is programmed, it behaves like a ROM and its contents cannot be erased or modified. PROMs are presently available in sizes up to 512 Kbits.

**EPROM:** EPROMs are similar to PROMs but they can be erased and reprogrammed. EPROMs are erased by exposing them to ultraviolet radiation in an instrument called an EPROM eraser. A quartz window in the EPROM IC package allows the ultraviolet radiation to pass through the package and strike the chip itself. Once the EPROM is erased, it can be reprogrammed using a PROM programmer. EPROMs are presently available in sizes up to 1 Mbit.

**EAROM:** EAROMs are similar to EPROMs except that they can be reprogrammed in circuit by electrical means without the need for a special erasing procedure. EAROMs became commercially available in the mid 1980s in sizes up to about 256 Kbits.

Although all four types of nonvolatile memory circuits are random access memories, the acronym RAM is used exclusively for read-write devices.

## Microprocessors and Microcomputers

***93. Microprocessors.*** In the early 1970s large-scale integrated (LSI) circuits were developed that performed the basic functions of the central processing unit (CPU) found in large data processing computers. These LSI circuits, called *microprocessors,* could be interconnected with as few as two other LSI circuits to form a working computer. The LSI computers are called *microcomputers* because they are dramatically smaller in size than their predecessors. When compared to the larger computers, microcomputers have reduced performance but their basic operation is similar. Microprocessors and microcomputers are used in all types of industrial applications including machine tool control, medical electronics, automobiles, process control, and robots.[17]

A microprocessor, like the CPU of the larger computer, executes *instructions* stored in memory circuits. Instructions are executed one at a time, each causing the microprocessor to perform a specific operation, such as adding two numbers. A *program* is a sequence of instructions that performs a useful task, such as adding up a list of numbers.

As the microprocessor executes an instruction, the data it operates on are a group of bits called a *word.* The number of bits per word is the *word size.* The word size of a microprocessor gives a relative indication of its data processing capability. Microprocessors with larger word sizes generally have more processing capability because more bits are processed with each instruction.

Processing requirements, compatibility with other circuitry, and ease of programming are among the primary considerations when selecting the word size needed for a particular application; 4-, 8-, 16- and 32-bit microprocessors are now commercially available. Eight-

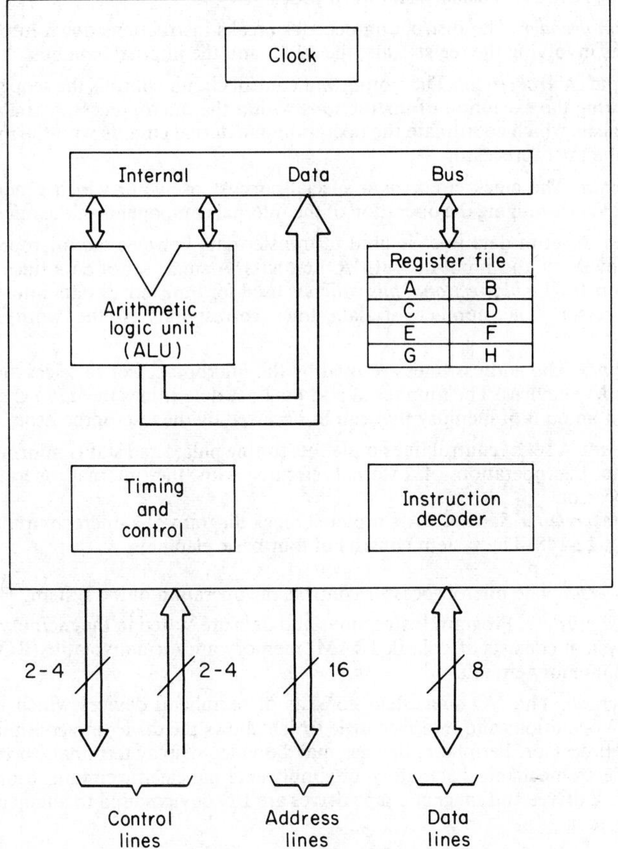


**FIG. 22-147** Simplified block diagram of a typical microprocessor.

bit and 16-bit microprocessors are being used in most industrial applications. The 32-bit microprocessors became commercially available in the early 1980s and are being used in some new equipment.[18]

A simplified block diagram of a microprocessor is shown in Fig. 22-147. The arrangement of the internal components of the microprocessor is called the *architecture.* The architecture affects the microprocessor's ability to perform certain kinds of operations. Most microprocessors have an architecture similar to the one shown and have similar internal components.[19] Wide arrows in Fig. 22-147 depict parallel signal lines or *buses* which connect the microprocessor to other circuits. Arrowheads indicate whether signals are transmitted to or from the microprocessor.

The internal components are described as follows:

**a.** *Register file.* The register file is a set of memory elements, each capable of storing one word of data.

**b.** *ALU.* The arithmetic and logic unit (ALU) is the circuit within the microprocessor that performs the data processing operations. Operations are usually performed on data stored in the register file.

**c.** *Internal data bus.* An internal data bus is used to provide a data path between the register file and the ALU. It is usually one word wide.

**d.** *Instruction decoder.* The instruction decoder breaks instructions down into more basic operations involving the register file, the ALU, and the internal data bus.

**e.** *Timing and control circuit.* The timing and control circuit controls the sequence of operations during the execution of instructions within the microprocessor. It also generates timing signals which coordinate the operation of external circuits with the internal operation of the microprocessor.

**f.** *Clock circuit.* The clock circuit is a special-purpose oscillator which generates timing pulses for synchronizing the operation of the internal components of the microprocessor.

**g.** *Data lines.* A set of data lines is used to transfer data between the microprocessor and external memory and input-output (I/O) circuits. A single set of data lines is generally used to construct a *bidirectional bus* which is used for transferring data into or out of the microprocessor. The number of data lines corresponds to the word size of the microprocessor.

**h.** *Address lines.* The address lines are used by the microprocessor to select memory locations and I/O devices. The number of address lines determines the *address capacity,* the maximum amount of memory that can be accessed by the microprocessor.

**i.** *Control lines.* A set of control lines transmits timing pulses and status information which coordinate the operation of external circuits with the internal operation of the microprocessor.

***94. Microcomputer System.*** A simplified block diagram of a microcomputer system is shown in Fig. 22-148. The system consists of four basic elements:

**a.** *Microprocessor.* The microprocessor controls the operation of the system.

**b.** *Memory subsystem.* Program instructions and data are stored in the memory subsystem. The subsystem consists of volatile (RAM) memory and/or nonvolatile (ROM, PROM, EPROM) memory circuits.

**c.** *I/O subsystem.* The I/O subsystem consists of peripheral devices which perform the actual I/O operations and interface logic which allows the devices to communicate with the microprocessor. Peripheral devices may be video display terminals, printers, annunciators, electromechanical actuators, or simply external electric circuits. Storage devices such as disk drives and magnetic tape drives are I/O devices used to augment the memory subsystem.

**d.** *Bus structure.* The bus structure is an arrangement of data, address, and control signals which allow the microprocessor to control and communicate with the memory and I/O subsystems.

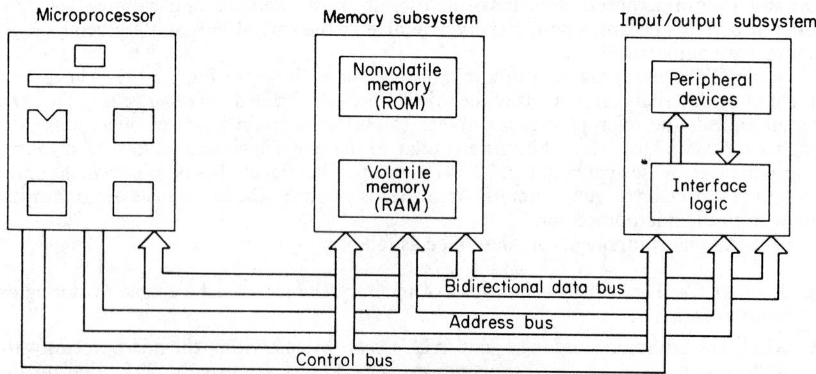


**FIG. 22-148** Simplified block diagram of a microcomputer system.

***95. Features of Microprocessors.*** The following features should be considered when selecting a microprocessor for a specific application:

1. Word size
2. Address capacity
3. Number of registers
4. Instruction set flexibility
5. Clock frequency (speed of operation)
6. Supply voltages required
7. Power dissipation
8. Compatibility with other logic families

***96. Programming.*** The *instruction set* of a microprocessor is the collection of all the instructions it can execute. Microprocessor instructions can be broken down by function into six general categories.

1. Arithmetic
2. Logic
3. Data transfer
4. Program control
5. Input-output
6. Processor control

A program is constructed from a sequence of individual instructions. The most basic form of instruction is a binary word with a 1 and 0 bit pattern that the microprocessor is designed to recognize. These instructions are called *machine-code* or *machine-language* instructions. They are the only form of instruction that the microprocessor can execute directly.

Programs are written using programming *languages.* The most fundamental programming language is called *assembly language.* Assembly language uses an abbreviated set of symbols called *mnemonics* to refer to the machine-code instructions of the microprocessor. A special-purpose program called an *assembler* is used to translate the assembly language program into machine code before it is executed.[20]

High-level programming languages allow programs to be written in a notation that resembles algebra. Many such programming languages are available for use on large computer systems. Among these BASIC (acronym for Beginners' All-Purpose Symbolic Instruction Code) and FORTRAN (acronym for Formula Translator) are also used on microcomputer systems. A brief description of programming features found in BASIC is contained in the American National Standards Institute Standard ANSI X3.60-1978, American National Standard for Minimal BASIC. A description of FORTRAN is contained in ANSI X3.9-1978.

## Transducers

***97. Hall-Effect Devices.*** When a current is passed between opposite edges of a piece of foil, a magnetic field through the foil will induce a voltage between the other two edges of the foil, as shown in Fig. 22-149. The magnitude of the voltage is proportional to the flux density of the magnetic field and to the current flowing through the foil. This phenomenon is known as the *Hall effect* and was first recorded in 1879. There have been practical applications of the Hall effect only in the past few years.

Combining a Hall-effect device and solid-state amplifiers on the same chip produces a very economical assembly. By maintaining a constant current, a voltage is obtained proportional to the magnetic field, permitting the design of magnetic probes. When used with a magnetic structure surrounding a conductor, the device becomes effectively an ammeter.

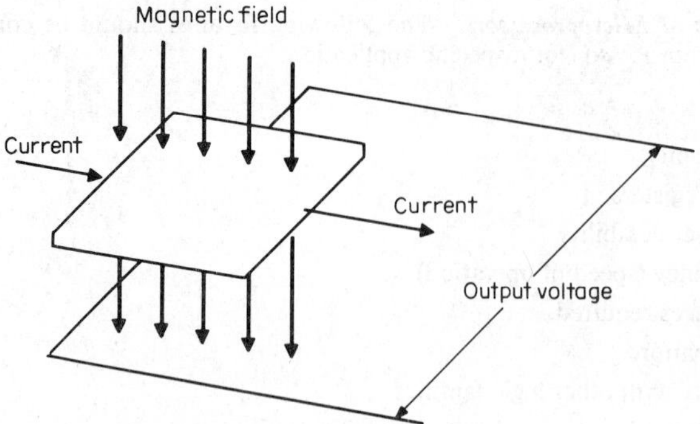


**FIG. 22-149** Operation of a Hall-effect device.

Conversely, maintaining a constant magnetic field while varying the current produces a current-to-voltage transducer. If the same current produces the magnetic field and passes through the foil, the output voltage is proportional to the square of the current. Devices such as wattmeters and varmeters may be produced by appropriate connections to voltage and current sources in the circuit to be measured.

For binary-type circuits, Hall-effect devices may be used as logic gates or as binary sensors (switches) to detect various parameters without requiring mechanical or electrical contact with the device being detected. Switching speeds of 100,000 operations per second with output signal rise and fall times under 0.5 $\mu$s have been reported.

***98. Input transducers*** are the basic devices which make measurements and control systems possible. In Standard S37.1-1969, Electrical Transducer Nomenclature and Terminology, the Instrument Society of America defines a transducer as "a device which provides a usable output in response to a specified measurand"; a measurand as "a physical quantity, property or condition which is measured"; and the output as "the electrical quantity, produced by a transducer, which is a function of the applied measurand."

Standard S37.1 establishes many standards on nomenclature, definitions, and methods of measurement for transducers. Other standards published by the American National Standards Institute and by the United States Nuclear Regulatory Commission Agency govern the design and application of transducers for specific measurands.

The chemical industry has developed a large variety of transducers capable of determining the physical or chemical properties of the gases, liquids, or solids in industrial processes. These devices use many of the electrochemical or physical properties of the material being measured. Some are capable of continuous measurements; others sample the material periodically, and may use some reagent to obtain a property that can be measured or to isolate the element or compound to be measured.

There has been a trend toward standardization of the output voltage or output current from transducers. Low-level signals are generally in the $\pm 50$ to $\pm 500$ mV range; higher levels are usually 0 to 1 or 0 to 10 V. Commonly used current ranges are 1 to 5, 4 to 20, or 10 to 50 mA.

The Instrument Society of America Standard S50.1-1975, Compatibility of Analog Signals for Electronic Industrial Process Instruments, describes the requirements for the compatibility of dc signals transmitted between systems of equipment.

Where the signal must be sent over a longer distance, variable frequencies are commonly used, such as 0 to 100 Hz or 0 to 4000 Hz. The variable-frequency signal can be amplified as necessary for the distance involved, and converted to a variable voltage or variable current where necessary. Limiting in the converters makes the output independent of changes in the amplifiers.

Where data are desired in digital form, it is practical to design transducers for such things as shaft position or linear position to operate directly in a binary code. If "gray codes" are used to eliminate uncertainty at critical positions, a conversion from gray code to binary code will be needed. This may be built into the transducers.

Many transducers produce very small electrical outputs. For the electrical engineer this may create problems of providing adequate power circuits or problems from transients in the power source of the transducer.

## Relays

***99. Protective Relaying and Coordination.*** The electric system in an industrial installation must have protection to isolate faults which may occur in the power distribution circuits or in the power-using equipment. These protective devices usually are fuses at the lower-voltage, lower-power levels, and circuit breakers at the higher-voltage and/or higher-power levels. The larger circuit breakers are often controlled by protective relays which monitor some parameter in order to detect the existence of a fault. The fault detection is usually a built-in feature of smaller circuit breakers.

In a properly designed system these protective devices are "coordinated"; i.e., the device nearest the fault should trip, limiting the power outage to those circuits which must be deenergized to isolate the fault. Each protective device must be capable of interrupting the maximum fault currents which could occur.

Protective relays, or the tripping devices in smaller circuit breakers, have generally used one of five electromechanical devices: (1) solenoids with plungers, (2) balance beams, (3) polar elements, (4) armatures or clappers, and (5) induction disks or cups. The development of solid-state components has led to new methods of fault detection having features not practical in the electromechanical devices. Power for operation of these solid-state fault detection circuits is usually obtained from the circuit being protected. The solid-state circuits are generally quite compact, frequently being housed within the body of the circuit breaker. Adjustments can be provided in the circuits so that the settings are flexible, making closer coordination possible between the breakers.

***100. Solid-State Relays.*** Solid-state relays (SSRs) are packaged electronic circuit assemblies that are used to replace electromechanical relays in power control applications. Parameter definitions, test methods, and application notes for ac SSRs are presented in Electronic Industries Association Standard EIA RS-443 (ANSI/EIA RS-443-1979).

An SSR consists of a low-voltage input circuit coupled to a power control circuit by means of a transformer or an optoelectronic isolator (optocoupler). The input circuit is electrically isolated from the power control circuit. The power control circuit uses solid-state power devices (transistors or thyristors) instead of electromechanical contacts to switch the load current. For this reason, SSRs offer increased reliability, absence of electromagnetic interference (EMI) due to arcing, and, in many cases, reduced cost when compared to electromechanical relays.[21]

SSRs have four terminals, two for the input signal and two for the load circuit. The input terminals are electrically isolated from the load circuit terminals which conduct load current when a low-voltage signal is applied to the input terminals. Input-output isolation voltage ratings can be as high as 3000 V ac. SSRs are commercially available which can sense either ac or dc input signals, and many types can be activated directly from 5-V logic signals.

AC and dc output SSRs are available. AC SSRs have current ratings from below 1 A to about 40 A and voltage ratings of either 120 or 240 V ac. DC SSRs have current ratings from below 1 A to about 5 A and voltage ratings up to 100 V dc.

Package styles and mounting requirements of SSRs depend on the output current rating. They are usually encapsulated or molded into a plastic block and either have screw terminals or pins for direct soldering. Depending on the current rating, SSRs are either mounted directly on a heat sink, in a relay rack, or soldered directly onto a printed circuit board.

Most SSRs are ac output types which use thyristors as the current-switching devices. The circuitry within the SSR derives the thyristor gate current from the load circuit itself. As a result of this operation and off-state leakage currents in the thyristors, a small leakage current flows in the SSR when it is in the off state. This is called *off-state leakage current* and is on the order of 1 mA. The forward conduction voltage of the thyristor causes an *on-state voltage drop* which is on the order of 1 V. This is analogous to contact voltage drop in electromechanical relays and is the primary source of power dissipation in SSRs.

Once triggered into the conducting state, the thyristors in the output circuit of an SSR will continue to conduct as long as the load current remains above their specified holding current. Normally the thyristors are triggered at the beginning of each half-cycle of ac line voltage and continue to conduct until the load current drops below the holding current parameter of the thyristors. Most SSRs have a *minimum load current* requirement that assures that the thyristors will conduct normally.

Thyristors are susceptible to failure caused by $dV/dt$ and by overvoltages. Most SSRs include an *RC* snubbing network in parallel with the thyristors to reduce the high $dV/dt$ that can result from switching certain kinds of loads. Transient-suppression devices such as Zener diodes or varistors are used to protect the thyristors against line voltage transients when the SSR is in the off state.

Most ac SSRs have an internal *zero voltage turn-on* feature which delays the triggering of the thyristors until the ac line voltage is near zero. Since the load voltage is zero when the SSR begins conducting, inrush current and the resultant EMI are substantially reduced. Some manufacturers offer special versions of their SSRs without the zero voltage feature. These are called *random turn-on* SSRs.

The following parameters should be considered when selecting an SSR for a particular application:

1. Output voltage rating
2. Output current rating
3. Input signal voltage
4. Input impedance or current requirement
5. Cooling requirements
6. Input-output isolation voltage
7. Off-state leakage current
8. On-state voltage drop
9. Minimum load current
10. Internal $dV/dt$ limiting (*RC* snubber)
11. Internal transient voltage protection
12. Zero voltage turn-on or random turn-on

## Uninterruptible Power Supplies

***101. Uninterruptible Power Supplies.*** Uninterruptible power supplies (UPS) incorporate complex equipment that must be planned and specified carefully before purchase. These systems, consisting of solid-state rectifier-inverters (usually backed up by engine-generators), are being used to supply power to computers, on-line data processors, process controllers, and other critical loads, to prevent costly power interruptions.[22,23]

The heart of the UPS is the rectifier-inverter unit, or module, which accepts ac line power and delivers transient-free ac power to the critical load. A battery supplies power to the inverter when the ac line power source is interrupted up to several minutes. The configuration of a typical UPS in shown in Fig. 22-150. Rectifier-inverter units are employed in many combinations to supply a critical load—singly, in parallel, with a bypass switch, backed up by engine-generator sets, etc. The particular combination selected is determined by the magnitude of the critical load power, the pattern of anticipated ac line interruption, and the sensitivity and critical nature of the load.

Commercial rectifier-inverter units are built from about 1 to 250 kVA and have been arranged in systems up to 2000 kVA. Units are designed to supply critical loads at typical commercial single-phase and 3-phase voltages and frequency, including 400 Hz. Batteries typically use lead-calcium cells and usually are selected to provide full-load power for 10 min. Battery ampere-hour ratings drop rapidly for short discharge times at high current, so that the cost saving resulting from selecting 5 min of capacity instead of 10 is minor. Except in the integral KVA ratings, rectifier-inverter units are usually not "off-the-shelf" items; they are custom-made to variations of a standard design.

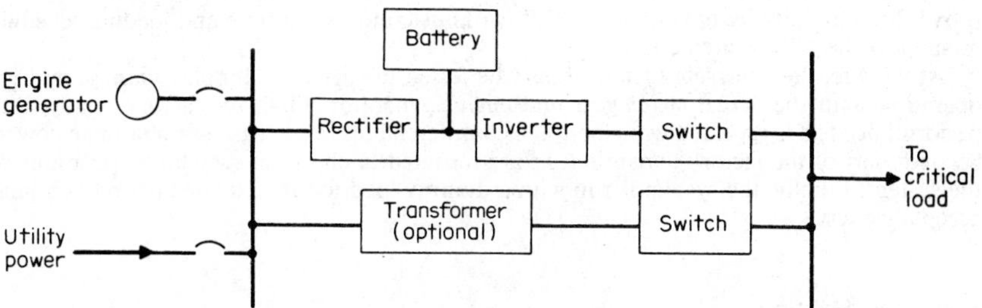


**FIG. 22-150** Block diagram of a typical UPS configuration.

The cost of a rectifier-inverter unit with 10 min of full-load battery power ranges from about \$5000/kVA at 1 kVA to \$500/kVA at 250 kVA. Costs of engine-generator sets lie in the \$125 to \$1200/kVA range. In large, several-hundred kVA redundant-unit systems, the cost for units and engine-generators approaches \$1000 per load kW. To the component costs must be added the costs of installation of the UPS, the electrical distribution and supply system, and additional building space—all of which can double the original component costs.

Equipment specs represent the translation of the customer's requirements and design into hardware. In preparation, the customer can be guided by the vendor's specifications, or modify an existing set, or retain an experienced consultant to prepare specifications. The Federal Aviation Administration Specification FAA-E-2473 is an excellent basic document that can serve as a model and be modified for a customer's particular application.

Specification must contain the following items:

1. *The steady-state and transient electric power requirements of the critical-load equipment to be supplied by the UPS.* Most computer manufacturers will make this information available. The conditions under which the UPS must meet these load requirements must be given (commercial power failure, battery condition, engine-generator operation, and internal UPS faults).
2. *The required reliability and maintainability of the UPS.* This is usually expressed as the mean time between failure (MTBF), in hours, of the UPS to meet the electric power requirements of the critical load, and the mean time to repair (MTTR), in hours, any failed rectifier-inverter unit of the system. The specification must describe how these requirements are to be measured and how they are to be demonstrated by the vendor.
3. *The terms of acceptance of the UPS by the customer.* These terms must include the tests to be conducted within the vendor's plant, the tests of the complete installation at the site, and the equipment performance for acceptance.

The most important step in purchasing a UPS is the testing. The vendor and customer must agree upon and carry out a meaningful set of tests at the factory and at the site. The purpose of testing is twofold: to see that the equipment meets the requirements of the specifications, and to expose any design or manufacturing flaws.

At the factory, the complete UPS, if possible, should be assembled and tested—including the rectifier-inverter units, the control console, bypass switches, and batteries. The switchgear and engine-generators may have to be tested independently and shipped to the site.

The first set of UPS tests is for steady-state performance; these tests should be conducted for variation of load, line voltage, power factor, frequency, and temperature and relative humidity (if adverse environment is expected). The second set of tests is for dynamic performance, such as for tripping the input power line, isolating and inserting of parallel units, transferring to and from bypass, switching load, short circuiting the output terminals, and fuse clearing. The third set of tests is to demonstrate the ability of a failed rectifier-inverter unit to isolate itself from the system while the UPS continues to supply critical load, under such conditions as loss of SCR gate drive, failure of inverter commutation, blower failure, and fault in the common frequency and load-division circuits. The final set of tests is to prove the reliability by operating the UPS without failure for a time and loading schedule to support the MTBF predictions.

At the site, the complete system should be tested to check on shipping damage and for operation with the actual switchgear and engine-generator. First, the factory steady-state performance test is repeated using both the commercial line and engine-generator ac power. Second, part of the factory dynamic tests are repeated to check on switching operations of the system. Finally, the system is run with a dummy load for an extended period as a final acceptance test.

## Electronic Heating

***102. Electronic Heating.*** Basically there are two types of high-frequency heating: induction heating and dielectric heating. Both methods share a common principle: power in the form of alternating current is transferred from an electric circuit, where it originates, into the material, where it is converted into heat by the resistance of the material and raises the temperature of all or part of that material. The heat actually originates inside the body of the material being heated, but the energy which causes this heating is furnished externally.

There are two basic differences between these methods. Only conductors of electricity can be heated by induction heating, whereas dielectric heating applies only to nonconducting materials. The second difference lies in the apparatus used to transmit the energy from the electric circuit into the material. In the case of induction heating, the heating element is a coil of copper wire or tubing through which the alternating current flows. Energy is transferred by induction—thus the term *induction heating.* In dielectric heating, on the other hand, the material to be heated is placed between two metallic plates, which are alternately charged to positive and negative voltages with respect to each other. The entire apparatus resembles a capacitor, with the material acting as the dielectric—thus, the term *dielectric heating.*

The electrical engineer responsible for plant operation may find that the electric load produced by electronic heating apparatus is cyclic as each workpiece is heated. Where many similar machines are in use, it may become desirable to stagger the times each machine requires peak energy. In addition, there may be a need for periodic tests to be sure the radiated energy and conducted energy on supply lines is within safe limits. Failures in shielding or filters may not otherwise be detected. The possibility that cooling lines may conduct electrical energy should not be overlooked.

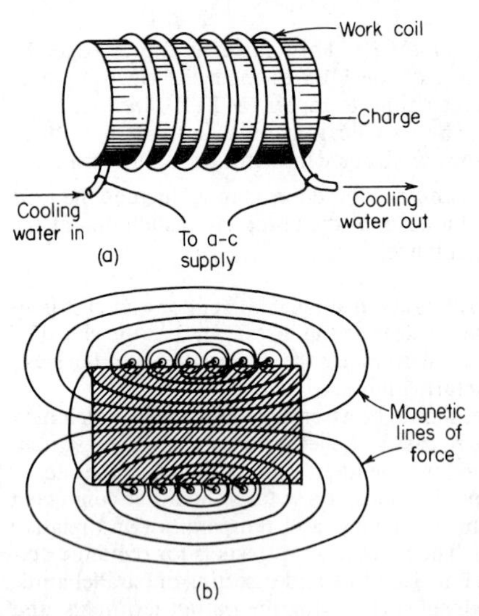

Work coil
Charge
Cooling water in
Cooling water out
To a-c supply
(a)
Magnetic lines of force
(b)

**FIG. 22-151** Basic induction-heating setup. A high-frequency alternating current passed through the coil heats the metallic charge. The depth of heating depends on the frequency of the heater current.

***103. Induction Heating.*** In induction-heating apparatus, the material to be heated is placed in an alternating electromagnetic field. The nature of the field required depends on the size, shape, and material of the work.[24]

Figure 22-151*a* shows a typical case. A metallic bar is to be heated for purposes of hardening. The coil surrounding the bar is referred to as the work coil. Its function is analogous to that of the primary winding of a transformer. The work, or charge, being a conductor, acts as a short-circuited secondary. When the primary, or work coil, is energized, a current is induced in the secondary, or work. The magnitude of this induced current will depend on the amplitude and frequency of the alternating current flowing in the work coil. It is the induced current that causes the heating of the material.

The lines of magnetic flux which are set up in the charge by the current-carrying coil are parallel to the charge axis in the case cited, as shown by the cutaway drawing (Fig. 22-151*b*). The distribution of this flux (and thus the distribution of heating) with the work will depend on several factors: (1) the electric and magnetic properties of the material, (2) the dimensional cross section of the work itself, and (3) the frequency of the applied alternating current. The first two of these factors are usually fixed by the job at hand. The remaining variable is the frequency of the alternating current which is made to flow in the work coil and which determines the rate at which the magnetic field within the charge builds up and collapses and then builds up in the other direction.

The selection of frequency for a particular job depends on the application. If, for instance, a piece of metal is to be melted, it may be placed in a suitable crucible, which is surrounded by a fairly low frequency work coil. As the charge experiences the regularly reversing magnetic field, it begins to heat uniformly and continues until the desired temperature is attained.

If the frequency is raised, skin effect (in which the current tends to flow only in a thin shell on the charge surface) comes into play and the secondary currents induced in the charge tend to remain on the outside. Since more heating occurs in the vicinity of the high-flux density, only the outside is heated by induction. This sort of arrangement is advantageous where the outside of the material is to be heat-treated for high surface hardness and the interior is to remain soft. Uniform hardening would cause the piece to be brittle; no hardening at all would cause the piece to be easily damaged. Such a piece might be used in the bearing of a crankshaft. By the same means, the cutting tip of a lathe tool may be hardened, while the main part is left soft and strong.

The induction-heating effect may be noticed when frequencies as low as 60 Hz (as in a standard power line) are used, but higher frequencies almost always are employed in actual practice. The production of relatively large amounts of power at frequencies of hundreds of kilohertz is accomplished by special-purpose transistor or thyristor inverter circuits. Above these frequencies, up to several megahertz, vacuum tubes are used. In some processes where only a few hundred hertz are required, frequency-multiplying rotating machinery operating from regular 60-Hz input can be used to better advantage.

Induction-heating work coils may take on a number of shapes. The example shown in Fig. 22-151 is typical but by no means a standard form. The technique of designing a work coil that will produce just the right amount of heat in just the right portion or portions of

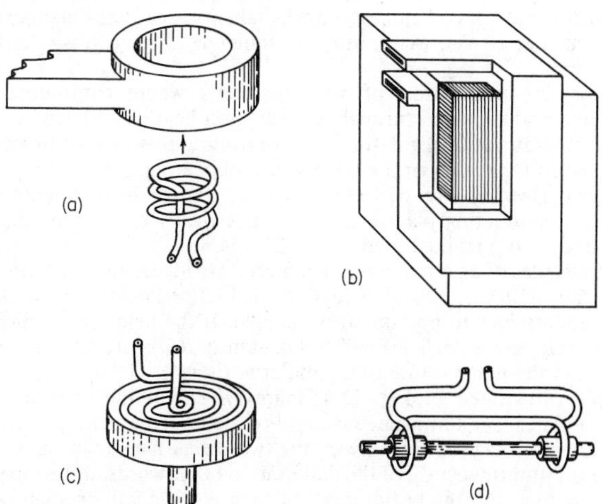


**FIG. 22-152** Many special induction-heating coil forms are used. In *a*, the inside of the connecting-rod bearing is to be heated. Drawings *b*, *c*, and *d* show other special cases for localized heating.

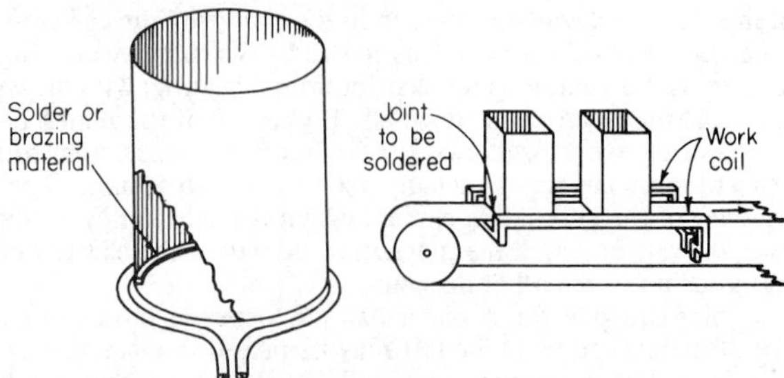


**FIG. 22-153** When induction heating is used for soldering or brazing, the heat produced must be concentrated in a small area surrounding the proposed joint to prevent distortion of the work.

the charge is a highly specialized science. A few typical configurations are illustrated in Fig. 22-152. In *a,* the induction-heating coil passes inside the bearing shown. Thus the inner surface is heated by the flow of high-frequency current within the coil. In *b,* the coil is concentrated around the corner of the piece of tool steel. Thus only that corner is heated. In *c,* the surface of the piece immediately under the spiral-wound work coil is heated. A system for applying heat in two different portions of the piece simultaneously is shown in *d.* Here the induction-heating current flows through two coils, and each coil imparts a certain amount of heating in the desired locations.

Very special work coils are sometimes required. In applications where induction heating is used in soldering or brazing (Fig. 22-153), it is usually important that only the portions which actually come in contact with the solder or brazing material be heated. Otherwise distortion of the assembly may occur. In cases like these the work coil may take the form of a knife-edged single-turn loop, which fits snugly around but does not touch the immediate vicinity of the proposed joint. Since the work coil is also part of an electrical circuit, its inductance and resistance must be considered in choosing coil shape and material. Low resistance is advantageous, since less power will be wasted in the work coil and the overall operation will be more efficient. Where moderately high frequencies are involved, the diameter of the wire or tubing used in the work coil should be large because of the skin effect. In most high-powered induction-heating equipment, provisions are made for circulating cooling water through the work-coil tubing. Care is taken to see that this does not hinder the operation of the device by short-circuiting out the ends of the coil where the water enters and leaves.

Frequently in the production of strip materials where continuous processes are employed, the material is passed through an induction-heating coil, where it is brought up to the proper temperature. In regulating and controlling this type of process the speed of travel through the coil may determine the amount of heating.

***104. Dielectric Heating.*** In dielectric heating, nonconducting materials are placed between a pair of conducting plates which are electrically charged by high-frequency ac potential. The basic setup is illustrated in Fig. 22-154.

When the molecules of a nonconducting material are placed in an electric field, they tend to line up physically with the lines of force of that field. If the field is reversed, these particles tend to reverse themselves to line up with the field. If the field is periodically reversed at some fairly high rate, these particles will be constantly realigning themselves. In doing so, heat is created, and the temperature of the material rises.

The examples illustrated in Fig. 22-154 are typical of the dielectric heating work. Between the two plates is a homogeneous piece of the nonconducting material. At any one time, when a certain voltage exists between the two plates, a share of the voltage difference is assumed by each minute section of the material. In other words, if the top plate is positive with respect to the bottom, the bottom half of each particle will be slightly more positive than the top half.

When the material being heated is not homogeneous, a different situation exists. Depending on the electrical characteristics of the different materials, a larger or smaller

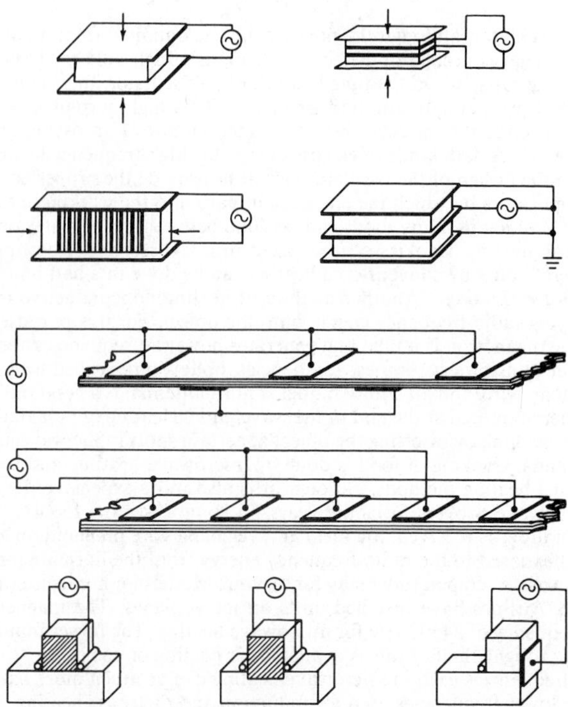

**FIG. 22-154** Dielectric heating apparatus may take many forms. Basically, all these contain two metallic plates that are alternately charged at a high rate. Here glued joints are heated.

share of the voltage difference will be assumed by each material. The voltage that each material is responsible for determines the amount of heating that will occur in that material.

Figure 22-154 shows a system for applying even heat to speed the drying of a glued joint. The voltage change is gradual throughout the thickness of the material, since there is no change in the type of material between any two opposite points on the electrodes.

Where other than rectangular work is to be heated by the dielectric method, air gaps can be permitted, as long as their share of the voltage drop is taken into consideration and enough voltage is applied between the plates to allow sufficient drop across the material. Air acts just like another material (which it is) and must be treated as such.

The first commercial use of dielectric heating had to do with introducing artificial fevers into various parts of the human body (diathermy). The most desirable characteristic of the artificial fevers thus produced was that the fever spread uniformly and extended well below the surface of the body, which was not the case with simple heat-radiation devices. Dielectric heating is applicable to almost all nonconducting materials, such as wood, plastic, certain liquids, etc. One of the most common applications of dielectric heating is the preheating of plastic materials. By this means much time is saved in the plastics industry. Originally the material had to be heated very slowly in order that it be heated uniformly. By dielectric heating, only a few moments are required, instead of hours.

Dielectric heating is also widely used in the processing of both natural and synthetic rubber products. Again, even heating is required—and obtained—by this method. Where heat is required for drying, such as in the printing and dyeing industries, dielectric heating serves very well. In the lumber industry and its associated fields dielectric heating is often used to process lumber and to speed and perfect the drying of glued joints.

The electric power required for both types of high-frequency heating is usually provided by some sort of oscillator, although the power could be obtained from an amplifier driven by an oscillator. What is needed in dielectric heating is a high voltage. This is provided at the terminals of the capacitor in the tuned circuit of an oscillator. In inductive heating, the need is for high current at some desired frequency. This high current is provided by the current flowing through the inductance of the tuned circuit of an oscillator. Therefore, a tube oscillator provides both kinds of electric energy for high-frequency heating. The major engineering is in the design of the oscillator circuit to provide the proper energy and in the design of the applicators by which the energy is imparted to the workpiece to be heated.

Heating and cooking of many foods can be done between metal plates on a production basis by dielectric heating. Cookies can be baked in a few seconds. Roasting of meat that would take several hours by conventional heating can be done in a half hour.

***105. Microwave Heating.*** Another method of heating nonconductive material is that of literally spraying radio-frequency energy onto the object. For this purpose, a magnetron tube is used. The magnetron is a tube that generates power at frequencies above 1000 MHz for modern radar transmitters. At these frequencies, hollow pipes called *waveguides* are better conductors than wire, and an empty funnel, a horn antenna, is a good radiator of microwave energy when mounted at the end of the waveguide. Heating of nonconductive objects or materials can be done by inserting the object at certain spots in the waveguide or in front of the horn antenna. Cooking of food is done by microwave heating, and it is much faster than conventional heating methods, although it is also more expensive. Restaurants have been using microwave cookers for some years. In some cases the food is precooked and stored. When an order is received, the meat and vegetables are prepared on a dish, inserted in the oven and exposed to the radio-frequency energy from the horn antenna, and heated in a minute or two to a temperature ready for serving. Models for home use were introduced a few years ago. Airlines have installed units on jet airplanes. The magnetron usually is operated at a frequency of 2450 MHz for microwave heating. The power input is about 2000 W from the line; roughly half of this is available for heating or cooking.

The higher frequencies used in microwave heating can be much more hazardous to personnel than the lower frequencies used for induction and dielectric heating. Stringent standards have been issued by the federal and state governments, establishing safe limits for human exposure to microwave energy.

## Lasers

***106. Laser Measuring Devices.*** The usefulness of a laser for measurements depends on several properties of light produced by a laser which are not present in light produced by other sources. The main property of light from a laser is coherence; that is, the corresponding points on a wavefront are in phase, and the spatial and time properties can be predicted. The intensity of laser light can be very high; the light is highly monochromatic and highly collimated.

Interferometer measurements of distance are made using the principle shown in Fig. 22-155. Because the light is coherent and monochromatic, various positions of the moving reflector will produce reinforcement or interference in the light reaching the detector. By counting the number of wavelengths between two positions of the reflector, an accurate measurement of the distance between the two positions can be maintained. This technique can be adapted to measure velocity by determining wavelengths in a predetermined time interval.

*Q* switching in a laser source makes it possible to produce a pulse of light having a very short duration. Measurement of the time required for a pulse to make a round trip, similar to methods used in radar, provides accurate range measurements.

Collimated light from lasers has many applications in surveying and alignment previously performed with transits and theodolites. Because the light is monochromatic, light filters can be used to reduce interference from other light sources. Accuracies as much as six times greater than standard optical techniques have been claimed.

Measurements of part size by determining how much of a collimated beam is blocked have proved useful for such processes as wire drawing or extruding. The workpiece need not stop, and no mechanical contact is required. This technique has been adapted to mea-

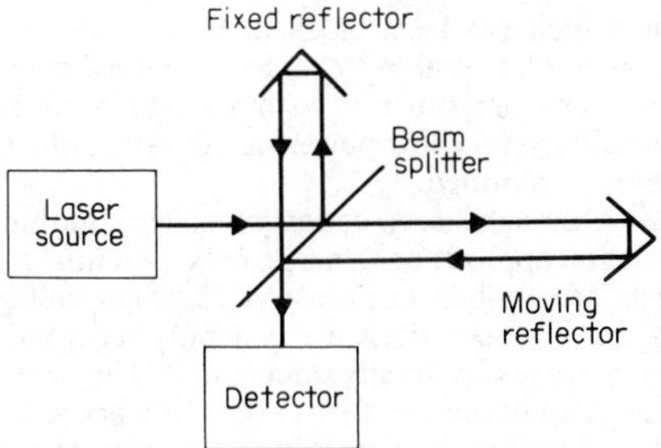


**FIG. 22-155** Principle of laser interferometer.

surements on discrete pieces. The pieces move at known speed, and the time interval during which light is blocked is measured.

***107. Lasers for Welding.*** Very high power densities can be achieved by focusing the beam of a high-power laser. Continuous power densities as high as $2.5 \times 10^5$ W/mm$^2$ and pulsed power densities over $10^{12}$ W/mm$^2$ have been reported. These power densities are higher than those achieved in flame welding, but are not as great as those possible with electron-beam welding.

The quality of the weld depends on many variables, including the frequency of the laser, the materials to be welded, the type of weld, etc. Extreme caution is necessary to provide adequate protection for personnel doing the welding and to protect those nearby.

The laser may be used for welds in thermoplastics where the materials selected will absorb rather than reflect the laser energy. When welding sheet plastics, beam splitting may be used so that the weld is heated from both sides.

Laser applications are generally limited by the continuous power available from the laser. This will depend on the type of laser but is generally less than 20 kW.

***108. Lasers for Cutting.*** Materials which are not readily cut by other means can be cut by the high power densities possible in the focused beam of a laser. The pattern-following processes frequently seen in flame cutting have been adapted by installing the laser in the position formerly occupied by the torch.

The laser cutting process has been modified to provide scribing for some applications, as with silicon wafers. Materials such as fabrics can also be profile cut or cut with numerically controlled systems. Lasers for this application operate at a power level of about 200 W.

***109. Lasers for Drilling.*** Lasers may be used for drilling by vaporization of the material at the focus of the beam. Energy levels must be high enough for vaporization to occur before the heat is diffused into the material. Techniques must be employed to remove the vapor above the workpiece so the light beam is not absorbed in the vapor. It has been found possible to drill holes where the thickness of the material is as high as 20 diameters of the hole being drilled. Multiple reflections from the sides of the hole confine the energy. The resulting hole may be slightly tapered with either the entrance or the exit diameter being greater depending on several variables in the machine setting. Some form of photoelectric device is commonly used to detect when the material has been completely penetrated.

***110. Lasers for Inspection Devices.*** Lasers can be used to sort, to measure size and shape, and to detect the presence of components. As inspection devices, lasers have the advantage of all optical instruments—no contact, no gage wear, no part damage, and fast operation. Since the light is collimated, lasers do not have the focus problems commonly encountered in other methods of optical inspection. Surface scratches, breaks, burrs, slivers, and dents can be readily detected. Where there is a surface pattern, such as on a printed circuit board, scanning with the laser beam may be used to compare the workpiece with a preprogrammed sample to detect errors in the workpiece.

Other inspection methods use the collimated beam to determine the location of the edge of the workpiece. For small pieces the width or diameter may be determined by determining the portion of the beam which has been blocked. In the graphic arts industry lasers are finding many uses, such as in plate making, character reading, color registration, etc., since their collimation and focusing are superior to other light sources. Lasers used in these inspection processes are usually very low power devices since the energy density must not damage the workpiece being examined.

***111. Laser Hazards.*** Although lasers, under controlled conditions, have been used beneficially in medical and dental applications, there can be potential safety hazards which cannot be ignored when they are applied in industry. The high voltages used and the fumes released in cutting and drilling operations are generally recognized. The hazards due to exposure to the laser beam are less generally understood. The principal hazard is normally to the eyes, but burns may also occur on other parts of the body. Because of the collimated beam, the hazard may be present at considerable distances from the laser source.

ANSI Standard Z136.1-1980, Safe Use of Lasers, contains recommendations for safe exposure limits. Other recommendations or mandatory requirements have been established by agencies of the federal and state governments responsible for employee safety. These standards are revised frequently as new experience is gained.

## Security Equipment

***112. Intrusion Prevention.*** The most common method of intrusion protection involves wiring that forms one or more circuits connected to all doors, windows, ventilation openings, and other means by which the intruder may gain access. Opening a door or window, or breaking a pane of glass, interrupts the circuit, causing an alarm either locally or at some location which is continuously staffed. This type of system can be employed only where there is structural support for the conductor.

Newer types of systems may use beams of light, either visible or invisible, or narrow radar beams. Interruption of a beam triggers an alarm. These beams effectively create a fence. If the intruder succeeds in penetrating the fence without triggering the alarm, detection is unlikely.

Area coverage can be provided by systems which use radar or ultrasonic energy throughout the protected area. Movements within the covered area produce a doppler shift in one or more receiving units, causing an alarm. False alarms may be a problem if the system can be triggered by movements of small animals or by objects which may be shifted by air movements.

*Intrusion alarm systems* must be supplied from a reliable power source. Standby alternate power sources may be desired.

***113. Alarm Systems.*** The most common alarm system is a fire alarm. In its simplest form, the fire-alarm system consists of call boxes which must be used by persons who discover the fire. For fire detection in areas which are unattended some or all of the time, some form of automatic fire alarm is indicated. The automatic fire alarm must detect two types of fires. One is the normal fire which produces both heat and smoke, the other is a smoldering fire which produces little heat, but a great deal of smoke and toxic gases. Much of the smoke and gases may be invisible. A complete fire alarm system includes both heat and smoke detectors. Two types of smoke detectors are available commercially: one type employs a light beam and photocell to detect the presence of smoke in the light beam; the other type senses the ionization of gases produced in the combustion, and will respond to the invisible products of combustion.

The automatic fire alarm is wired to alert personnel and may also be arranged to turn on automatic extinguishing equipment in the area of the fire. Considerable caution is needed to be certain that people who may be present when the automatic extinguishing system is triggered have adequate warning and time to vacate the premises.

Many other types of alarm systems are used in industry to alert personnel to the presence of an abnormal condition requiring corrective action. Any of the many types of input transducers may be used with appropriate circuitry to construct alarms.

An alarm system is useful only if it will perform reliably when needed, perhaps months since it last operated. This suggests the need for a warning device that announces failures in the alarm system. Many commercially available alarm devices include such warning devices.

The National Electrical Code includes requirements for alarm systems to ensure greater reliability in the completed installation. Publications of the National Fire Protection Association show several methods of employing redundant power sources which have been found effective to ensure continuity of power for the alarm system.

***114. Closed-Circuit Television.*** The use of closed-circuit television in industry has increased greatly as new applications have been developed. Probably the most common application of television has been by security forces, for whom television made it possible to observe activities in several areas simultaneously from a single location. The use of cameras which are sensitive to invisible parts of the light spectrum provides excellent anti-intrusion protection.

In industrial processes television has enabled observation of activities in locations where conditions are unsafe for personnel. Typical locations are the inside of boilers, where flame action can be observed, and areas where toxic fumes, temperature, or radiation may be beyond safe limits.

Television has also been applied for training where the process might otherwise be impossible to demonstrate. A typical application is micromanipulative work under microscopes. With television the instructor can show the process to a number of students simultaneously and can be sure that each student observes the specific items intended.

***115. Infrared Detection.*** Two types of systems have evolved which use infrared detectors to sense the presence or movement of an intruder in a protected area. These systems are referred to as *passive* and *active* infrared systems.

In the passive systems a sensor containing an infrared-sensitive phototransistor is placed in the area to be protected. Circuitry within the sensor detects the infrared radiation emitted by the intruder's body and triggers an alarm. Originally these systems were unreliable because they would be falsely triggered by warm air movement, pets, or other disturbances that altered the infrared radiation levels in the area. Newer sensors use two infrared-sensitive devices which are focused to monitor different zones within the protected area. Logic within the sensor circuitry triggers an alarm only when the two zones are activated in sequence, as would occur if an intruder walked through the protected area.

In the active systems each sensor consists of two housings. One housing contains an infrared-emitting diode and an infrared-sensitive phototransistor. The other housing contains an infrared reflector. When positioned in front of an entrance to the protected area, the two housings establish an invisible infrared beam. An intruder entering the area interrupts the beam and triggers an alarm. The active sensor systems are more reliable than the passive systems, but they require careful alignment when installed and cause false triggering if there is any mechanical shifting of the structure on which they are mounted.

***116. Microprocessor-Based Alarm Systems.*** To a large extent, digital integrated circuits have replaced relay logic in commercial and residential alarm systems. Many alarm systems are connected to a central monitoring station or police or fire station by telephone lines. In newer equipment, electronic circuits in the alarm control panel automatically dial a preprogrammed telephone number for the monitoring station and activate a display device which indicates that an alarm has been triggered. In addition, a digitally coded identification number may be transmitted to the monitoring equipment to identify the source of the alarm and the alarm condition (fire, burglary, etc.). The monitoring equipment usually transmits an acknowledgment signal back to the originating alarm equipment.

Much of the newer alarm equipment is microprocessor-based. Microprocessor-based autodialing equipment may have provisions for dialing a secondary telephone number if an acknowledgment is not received from the primary number. In addition, microprocessor-based equipment may also transmit code numbers identifying the type of alarm condition. Sensors for the various alarm conditions are wired to separate sets of terminals on the alarm panel and the appropriate alarm code is transmitted when the various terminals are activated.

The correspondence between alarm conditions and the transmitted alarm codes may be fixed or programmable. In either case the telephone numbers of the monitoring station(s) and identification codes must be programmed into the alarm control panel.

One programming method involves a matrix of connection points on a printed circuit board in the dialing equipment chassis. Information is programmed into the equipment by making connections on the matrix with jumpers or small screws. The information is read out by sequentially pulsing the rows of the matrix while testing the column connections.

Another method of programming uses programmable read-only memory (PROM) ICs to store the information. A PROM programmer is used to store data into the PROM representing the telephone numbers of the primary and secondary reporting stations and the codes to be transmitted for each alarm condition.

### Computers and Control Equipment

***117. Computers.*** Industrial applications of computers include process control, machine tool control, industrial robot control, data-gathering systems, and computer-aided design and manufacturing (CAD/CAM) systems. The principles of operation of computers are similar to those of the microprocessors and microcomputers as discussed in Pars. **93** to **96**. Three computer size classifications have evolved. These are main-frame computers, minicomputers, and microcomputers. For some number of years it was possible to distinguish these clearly by physical size or memory capacity, but advances in LSI, in particular microprocessor technology, have made the historical classifications nearly meaningless. Currently it is not uncommon to find microprocessor-based dedicated applications in which 64 Kbytes of memory are used. Some small desktop computers are being offered with up to 256 Kbytes of memory. Office computers or industrial process computers which were historically classified as minicomputers can be found with up to about 1 Mbyte. Main-frame computers commonly have more than 1 Mbyte.

***118. Programmable Controllers.*** NEMA Standard ICS 3-1978 contains definitions and general information on programmable controllers (PCs). Part ICS 3-304 of that standard includes the following general definition of a programmable controller.

> A digitally operating electronic apparatus which uses a programmable memory for the internal storage of instructions for implementing specific functions such as logic, sequencing, timing, counting and arithmetic to control, through digital or analog input/output modules, various types of machines or processes. A digital computer which is used to perform the functions of a programmable controller is considered to be within this scope. Excluded are drum and similar mechanical-type sequencing controllers.

The integration of MSI and LSI digital logic circuits into PCs has led to the commercial availability of a wide range of controller types with varying degrees of sophistication.

The general organization of a PC is shown in Fig. 22-156. It consists of four basic elements: (1) a processor, (2) memory circuits, (3) input and output (I/O) circuits and, (4) a control panel. These elements operate as follows:

*Processor.* The processor executes instructions stored in memory. Instructions may be of the following types:

**a.** Timing functions

**b.** Counting functions

**c.** Latch functions

**d.** Stepping-switch and shift-register functions

**e.** Arithmetic functions

**f.** Data transfer functions

*Memory.* Instructions are stored in volatile (RAM) or nonvolatile (ROM) memory circuits. Some programmable controllers use RAM with battery backup to provide data retention when ac input power is lost. The size of memory determines the number of program steps. Many PCs have memory that is expandable in fixed increments. PCs with up to 1000 program steps and up to 256 Kbytes are currently commercially available.

*I/O Circuits.* Solid-state power modules similar to solid-state relays provide the power control functions instead of electromechanical relays. Analog output modules are available in addition to ac and dc output modules. Analog output modules convert the digital values stored in the PC to varying voltage or current signals. AC, dc, and analog input modules are available as well. Both input and output modules provide electrical isolation from the internal circuitry of the PC. PCs with capability to support as many as 2048 separate I/O func-

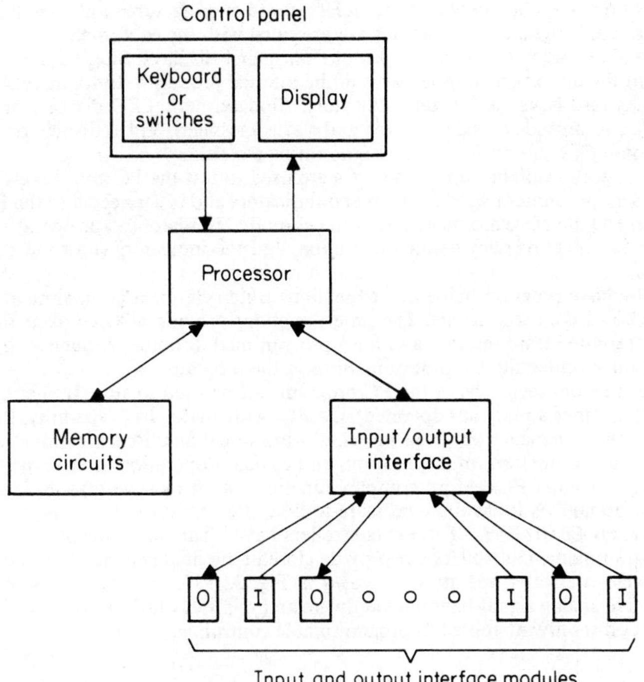


**FIG. 22-156** Organization of programmable controller.

tions are currently commercially available. I/O interface modules are installed as needed to meet application requirements.

*Control Panel.* The control panel provides a means of starting, stopping, and sequencing the operation of the PC. The status of relay and counter functions may be observed and changed from the front panel on some PCs.

The PC is programmed by storing instructions into its internal memory circuits. Some PCs can be programmed using buttons or a keyboard on the control panel. Most of the more complex equipment is programmed using separate programming panels or special-purpose computer systems. These computer-based PC programming systems often include printers to provide program listings, video displays, and special software for testing the PC program.

Programs are loaded into the PC in one of two ways: (1) the program is transferred into the PC by means of a cable connection to the programming equipment, or (2) the program is loaded into nonvolatile memory circuits (e.g., PROMs), which are then removed from the programming equipment and installed in the PC.

Programming methods will depend on the application. Front-panel programming is suitable for applications where the program will be changed frequently in the field and the programs are simple (e.g., using less than 20 relays). If the PCs are to be preprogrammed in large lots at a central facility it may not be necessary to have front-panel programmability. In this case the cost of a computer PC programming system may be justified. Field changes can be made by replacing PROMs or by using telephone-line interconnect features available on some PCs.

Programs indicate the interconnection of relay contacts and coils, timers, counters, as well as their connection with external equipment. This information is expressed in the program in one of three ways: (1) relay ladder diagrams, (2) boolean algebra expressions, and (3) symbolic representation. Relay ladder diagrams are most commonly used because they are easy to understand and are similar in appearance to conventional relay logic wiring diagrams.

Programming methods vary from manufacturer to manufacturer, although there is some standardization. Programming manuals are provided with the equipment.

PCs are available with various types of front-panel displays. Single LED status lights indicate that the unit is on and the status of the internal processor. Individual I/O interface modules may also have LED status indicators. Alphanumeric LED displays are available on some PCs to provide special warning and status messages. Video display units are provided on some PCs and on some PC programming panels.

I/O simulators available with some PCs are used to test the PC and the program. The simulator is an instrument with switches and indicators and is connected to the PC in place of the actual I/O interface modules. The switches on the simulator can be operated manually to simulate the effect of relay contact operation, and the indicators show the status of the I/O signals.

Most PCs have programmable timer functions which can be used to simulate the operation of motor-driven sequencers. The timer functions provide relay contacts that operate after a programmed time interval and for a programmed duration. Sequencer operation is effected by interconnecting the timer functions in the program.

Internal memory locations in the PC program can be used to store numeric data (e.g., the number of times a relay has operated). On PCs with analog I/O capability, the numeric data can represent signals from transducers. Mathematical functions are provided on some PCs. Addition, subtraction, multiplication, and division operations can be performed on numeric data in most PCs. More complex functions, such as square root extraction, are provided on some PCs to facilitate pressure-to-flow rate conversions.

***119. Process Controllers.*** Process controllers have a hardware organization similar to that of programmable controllers but provide closed-loop analog control of industrial processes. Process-controller operation is shown in Fig. 22-157. Process controllers generally have extensive analog signal-handling hardware and relatively little relay control capability of the type commonly attributed to programmable controllers.

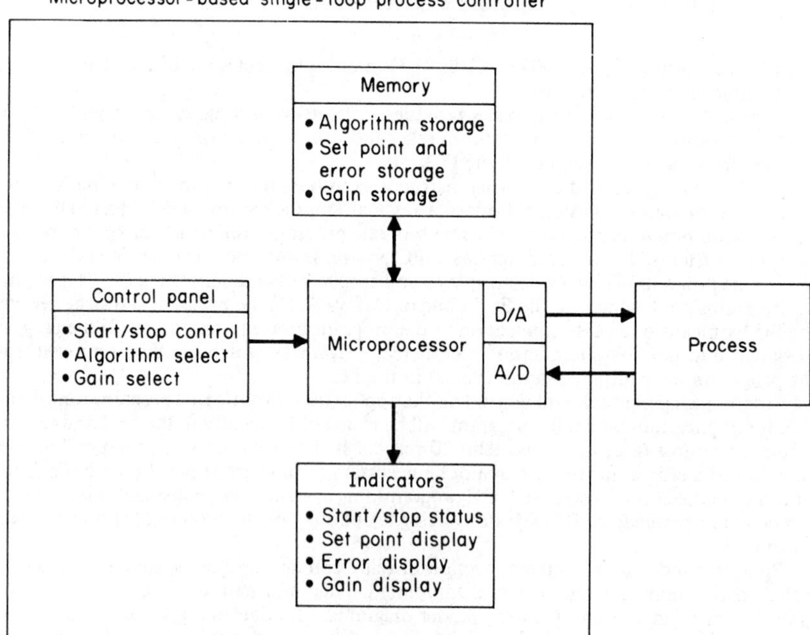


**FIG. 22-157** Organization of a single-loop process controller.

The configuration shown in Fig. 22-157 is called a *single-loop* process controller because there is only one feedback loop. *Multiloop* controllers have a single processor and control panel but have separate analog inputs and outputs for each loop.

Historically, process controllers were constructed from discrete components or individual analog ICs. These units did not need analog/digital (A/D) or digital/analog (D/A) conversion circuits. Microprocessor-based controllers became available in the mid-1970s and have substantially replaced the older analog controllers.

The main advantage that the microprocessor-based controllers have when compared to the older analog controllers is programmability. A variety of control programs are usually preprogrammed by the manufacturer. The user selects the particular control program, usually called a control *algorithm*, by means of front-panel controls. Control algorithms, and analog controllers, develop an error value which is the difference between the set-point value and the measured value of the process variable being controlled. The error value is the basis for a control value which is used as an input to the process. Control algorithms differ in the way the control value is developed from the error value.

Many controllers are supplied with algorithms based on PID (proportional-integral-derivative) control which was commonly used with the older analog controllers as well. In a PID algorithm the control value is developed as the sum of three terms related to the error (P term), the time-integral of the error (I term) and the time-derivative of the error (D term). The individual terms are multiplied by scale factors or gains which are used to adjust the time response of the controller or process system for certain desired characteristics. This adjusting procedure is called *tuning* and is usually done manually.

*Self-tuning* controllers automatically adjust the individual PID gains to maintain good performance, even if the process characteristics change during operation. Self-tuning control features were not available in analog process controllers and are a result of microprocessor-based technology.[25]

## Bibliography

### *120. Bibliography for Industrial Electronics*

1. Saucedo, R., and Schiring, E. E.: *Introduction to Continuous and Digital Control Systems;* New York, Macmillan Company, 1968.
2. Smith, C. L.: *Digital Computer Process Control;* Scranton, Pa., International Textbook Company, Haddon Craftsmen, Inc., 1972.
3. Franklin, G. F., and Powell, J. D.: *Digital Control of Dynamic Systems;* Reading, Mass.: Addison-Wesley Publishing Company, 1980.
4. Pluhar, K.: Distributed Control Manufacturers Offer a Variety of Systems to Worldwide Market; *Control Eng.,* vol. 30, no. 4, April, 1983, pp. 80–83.
5. *GE Transistor Manual,* 7th ed.; Syracuse, N.Y., General Electric Company, 1964.
6. *IR Hexfet Data Book;* El Segundo, Calif., International Rectifier Corporation, Semiconductor Division, 1982.
7. *GE SCR Manual,* 6th ed.; Auburn, N.Y., General Electric Co., 1979.
8. *Hewlett-Packard Optoelectronics Designer's Catalog;* Palo Alto, Calif., Hewlett-Packard Components, 1983.
9. Sifferlen, T. D., and Vartanian, V.: *Digital Electronics with Engineering Applications;* Englewood Cliffs, N.J.: Prentice-Hall, Inc., 1970.
10. VHSIC Systems and Technology; *Computer Magazine,* IEEE Computer Society, February, 1981, p. 13.
11. *The TTL Data Book for Design Engineers,* 2nd ed.; Dallas, Texas, Texas Instruments Inc., 1981.
12. *Motorola CMOS Integrated Circuits;* Austin, Texas: Motorola Inc., 1978.
13. *Advanced Low-Power Schottky and Advanced Schottky Logic Circuits Data Book;* Dallas, Texas, Texas Instruments Inc., 1982.
14. *Motorola MECL Integrated Circuits;* Austin, Texas, Motorola, Inc., 1982.
15. *The MOS Memory Data Book for Design Engineers;* Dallas, Texas, Texas Instruments Inc., 1978.
16. *Memory Data Book;* Santa Clara, Calif., National Semiconductor Corporation, 1976.

17. Bibbero, R. J.: *Microprocessors in Instruments and Control;* Process Control Division, Honeywell, Inc., and New York, John Wiley & Sons, Inc., 1977.

18. Gupta, A., and Toong, H. D. (eds.): *Advanced Microprocessors;* New York, The Institute of Electrical and Electronics Engineers, Inc., 1983.

19. *MCS-85 User's Manual;* Santa Clara, Calif., Intel Corporation, 1978.

20. Webster, J. G., and Simpson, W. D.: *Software Design for Microprocessors:* Dallas, Texas, Texas Instruments Inc., 1976.

21. Electromechanical vs. Solid State Relays; *Electronic Products,* Hearst Business Communications, Inc., March 28, 1983, p. 101.

22. Kusko, A., and Knutrud, T.: Specifying Uninterruptible Power Supplies; *Electrical Construction and Maintenance,* February, 1974.

23. Kusko, A., and Gilmore, F. E.: Application of Static Uninterruptible Power Systems to Computer Loads; *IEEE Trans. Ind. Gen. Appl.,* July/August 1970, vol. IGA-6, no. 4, pp. 330–336.

24. Davies, J., and Simpson, P.: *Induction Heating Handbook,* London, McGraw Hill Book Co., Ltd., 1979.

25. Cleaveland, P.: Programmable Controllers: A Technology Update; *Instruments and Control Systems,* 1983, vol. 56, no. 5, pp. 34–43.

26. Brewster, J. B., and Sherbondy, G. F.: Complete Characterization Studies Verify RBDT-RSR Reliability; Special Issue of IEEE Proceedings on Pulse Power Modulators, Oct. 1979, pp. 1462–1468.

27. Nagano, T., et al.: Characteristics of a 3000-V, 1000 A Gate Turn-Off Thyristor; 1981 IAS-IEEE Meeting Record, pp. 750–753.

28. Hower, P. L.: Power Switching Transistors—A Prognosis; International Electron Devices Meeting Record, IEEE-IDM, 1979, Washington, D.C., Electron Devices Society.

29. Brewster, J. B.: A 7 kW Inverter Using High Power Transistors; 1980 IAS-IEEE Meeting Record, Cincinnati, Ohio.

30. Hower, P. L., Brewster, J. B., and Morozovich, M.: A New Method of Characterizing the Switching Performance of Power Transistors; 1978 IAS-IEEE Meeting Record, Toronto, Canada.

31. Chu, C. K., et al.: Design Considerations of High Power Soft Recovery Rectifiers; 1982 IAS-IEEE Annual Meeting Record, San Francisco, Calif.

# SECTION 23
# ELECTRICITY IN TRANSPORTATION

**W. E. Murray**
*Principal Staff Engineer, Douglas Aircraft Company; Senior Member, IEEE*

**J. S. Young**
*Retired, McDonnell Douglas Astronautics Company*

**Edward F. Eaton**
*Manager, Marine Applications Engineering, General Electric Company; Member, IEEE*

**F. N. Houser**
*Consultant; formerly, Manager, Railroad Research Information Service, National Research Council; formerly Mechanical Editor, Railway Age; Managing Editor, Railway Locomotives and Cars; Member, ASME, IEEE*

**Robert W. McKnight**
*Manager, Communication and Signaling Engineering, Association of American Railroads*

**Enrico Levi**
*Professor of Electrophysics, Polytechnic University, New York; President, Enrico Levi, Inc.; Senior Member, IEEE*

**W. F. Hamilton**
*Consultant, Santa Barbara, Calif.*

## CONTENTS

*Numbers refer to paragraphs*

## ELECTRICITY IN AEROSPACE

By J. S. YOUNG and W. E. MURRAY

### Aerospace Electric Subsystems

*1. Electric Power in Aerospace Vehicles.* From the first aircraft, electric power has been in use in some form and with the accelerated development of both commercial and military aircraft, such use has expanded greatly. Then, with the development of the missile and its extension into space vehicles, the use of electric power has increased rapidly. Basically, the same criteria apply for aircraft, missile, and space-vehicle electric power generation and utilization; the difference occurs in the means of power generation and its application. In aircraft, a ready means of power generation has been available—the aircraft engine; in the missile and spacecraft, no such means generally exist. Again, on aircraft, the use of electric power was quickly applied to passenger or crew comfort items. This was not a requirement on the missile but has again become a prime item on manned spacecraft.

In this section, aircraft uses and development will be treated first, and then the divergencies due to missiles and spacecraft will be given, plus new concepts in the latter field.

### Aircraft Electric Subsystems

*2. Uses of Electric Power.* Electric power is used in an airplane to increase the effectiveness or utility of the airplane over that provided by the basic airfoil, airframe, and propulsion means. Most of the functions performed by electricity which contribute to improved effectiveness of aircraft may be included in the following major categories:

1. Lighting.
2. Instruments (flight, engine, position, etc.).
3. Actuators (flaps, landing gear, bomb doors, etc.).
4. Control systems (engine, propeller, warning, etc.).
5. Pumps and blowers.
6. Heating (deicing, cooking, special equipment, etc.).
7. Communication and navigation.
8. Offensive and defensive armament (turrets, radar, computers, etc.).
9. Flight control.

The many uses of electricity in present-day aircraft have required a capacity of 360 kW (480 kVA) on large bombers; 150-kVA generators are available for special applications.

*3. Types of Electric Power Subsystems.* There are three principal types of power subsystems in current use:

1. For many years, the most extensively used system for both military and commercial aircraft was 28 V dc. While inadequate for primary power on larger aircraft, 28-V dc power is still required in current designs, particularly in executive and other small aircraft.
2. The 400-Hz 3-phase 200-Y/115-V system is that preferred, with some applications of 400-Y/230-V now in development. This system usually operates with paralleled generators.
3. Variable-frequency (usually 400- to 800-Hz) single-phase 115-V or 3-phase 200-Y/115-V systems are often used to supplement a primary 28-V dc system. They have been considered for primary power, but the usage is not general. Load switching and lack of continuity of power make this system less desirable than the constant-frequency system.

Among other systems which have been used to a more limited extent are (1) 120-V dc (115-V dc is used widely by the British), (2) 12-V dc for small private aircraft, (3) 400-Hz Δ-connected 115-V, and (4) higher frequencies such as 800, 1600, 2000 or 3200 Hz single-phase or 3-phase.

Since practically all modern aircraft have metallic structures, one side of dc and single-phase ac and 1 phase of Δ-connected 115-V systems are grounded to save cable weight. The 200-Y/115-V system has 3-phase conductors, with the neutral grounded to permit application of single-phase loads between any phase conductor and ground.

***4. Voltage and Frequency Ratings.*** The voltage ratings of aircraft electric apparatus are shown in Table 23-1. These ratings are taken from the latest revision of IEEE Standard 127 and represent general usage. Frequency ratings are shown in Table 23-2 and are taken from the same source.

***5. Operating Conditions.*** Aircraft equipment must operate satisfactorily under the following conditions: (1) wide temperature range, (2) wide altitude range, (3) exposure to sand and dust, (4) wide range in humidity, (5) exposure to conditions promoting corrosion and fungus growth, (6) exposure to oil and other vapors, (7) explosion hazard, (8) vibration, (9) acceleration from landing shocks and maneuvers, and (10) changes in flight altitude.

The high degree of airplane mobility makes the full range of operating conditions possible on a single flight. Furthermore, this range is continually increasing, particularly for temperature, altitude, and vibration. The problem of equipment cooling has been aggravated by the ram-air-temperature rise encountered in supersonic aircraft and the need for more compact designs. Generators and other equipment have been developed to accept 120°C cooling air, and liquid cooling in the form of an oil mist is now being used in new generator designs.

***6. Importance of Weight.*** Weight is important for all air-borne equipment. Each additional pound of equipment requires 5 to 10 lb additional in airplane structure, propulsion

**TABLE 23-1** Aircraft-Apparatus Voltage Ratings

| | Direct current‡ | | | Alternating current | | |
|---|---|---|---|---|---|---|
| Nominal system designation (volts) | 14 | 28 | 120 | 115 | 115/200 | 230/400 |
| Generators: | | | | | | |
| Rated voltages | 15 | 30 | 125 | 120 | 120/208 | 240/416 |
| Voltage adjustment range, % | +0 to −15 | +0 to −15 | +0 to −15 | 115 ± 5 | 115/200 ± 5 | 230/400 ± 5 |
| Continuous-duty devices:* | | | | | | |
| Rated voltages | 13 | 27 | 115 | 115 | 115/200 | 230/400 |
| Voltage range, % | ±10 | ±10 | ±10 | ±5 | ±5 | ±5 |
| Intermittent-duty devices:* | | | | | | |
| Rated voltages | 12.5 | 26 | 115† | 115 | 115/200 | 230/400 |
| Voltage range, % | ±10 | ±10 | ±10 | +5 to −10† | +5 to −10 | +5 to −10 |
| Battery-operated devices (devices that must operate while generators are not in use): | | | | | | |
| Rated voltages | 11.5 | 23 | | | | |
| Voltage range, % | ±25 | 25 | | | | |
| Emergency-duty devices: | | | | | | |
| Rated voltages | .......... | .......... | .......... | 115 | 115/200 | 230/400 |
| Voltage range, % | .......... | .......... | .......... | +5 to −15 | +5 to −15 | +5 to −15 |
| Dielectric tests (rms volts for 1 min at 60 c or 120% of value for 5 s):§ | | | | | | |
| Factory-test volts | 1,500 | 1,500 | 1,500 | 1,500 | 1,500 | 1,800 |
| Field test or retest before and after use (clean and dry only):§ (75% of factory-test volts) | 1,125 | 1,125 | 1,125 | 1,125 | 1,125 | 1,260 |

*For operation from a dc voltage-regulated system. If operation is required from battery alone, use values for "battery-operated" devices.

†It is assumed that most 115-V wiring will be applied on the basis of thermal rating and that provisions for wider tolerances for voltage regulation on intermittent loads have not been made.

‡Direct-current devices such as starters, etc., and items operated during starting must operate under conditions of much wider voltage ranges. These devices are not covered by these standards.

§In accordance with IEEE Standard 135-1969, Aircraft, Missile and Space Equipment Electrical Insulation Tests, Recommended Practice for (1969).

NOTE: Apparatus is to function satisfactorily over the voltage and/or frequency ranges given (simultaneously), but with performance not necessarily in accordance with guarantees at rated voltage.

**TABLE 23-2** Aerospace Equipment Steady-State Frequency Ratings (Alternating Current)

| Nominal system frequency, Hz | 400 | 2,000 | 3,200 |
|---|---|---|---|
| Type I, standard | ±5% | ±5% | ±5% |
| Type II, optimum | ±1% | ±1% | ±1% |
| Type III, precision | ±0.01% | ±0.01% | ±0.01% |
| Type IV, broad | ±25% | ±25% | ±25% |

Type I. Standard frequency is in accordance with MIL-STD-704B, Military Standard, Electric Power, Aircraft, Characteristics and Utilization of, November 1975.

Type II. Optimum frequency respresents current state-of-the-art capability.

Type III. Precision frequency represents close tolerance power for systems with high accuracy requirements which cannot be met by using power with wider frequency tolerances.

Type IV. Broad frequency represents an ac generating system with broad frequency tolerances such as results from direct, or simple, coupling of an ac generator to a prime mover which is not regulated for frequency control.

Unless otherwise stipulated, type I, standard frequency, is assumed to be applicable.

engines, and fuel. Stated otherwise, decreased equipment weight increases pay load. Authorities differ on the exact value of weight saving in equipment, but it appears that an increased investment of $50 can be justified for each pound of aircraft weight saved. Thus, each pound of equipment weight may be worth as much as $500.

The high value of weight may justify a sacrifice in equipment life for achievement of lighter weight provided that reliability is not sacrificed. Typical required life is 1000 h for military aircraft and 5000 h or longer for commercial applications. For engine-mounted equipment, the no-maintenance interval may be matched with the engine-overhaul interval.

***7. Parts of an Aircraft Electric Subsystem.*** The aircraft electric system consists of three well-defined parts: (1) generation (or power supply); (2) distribution; (3) utilization.

## Power-Supply System

***8. Generator Prime Movers.*** Practically all aircraft generators which make up the main power-supply system receive their energy from the main engines either through mechanical shaft power or through compressor bleed air taken from gas-turbine-type engines. Figure 23-1 shows the various methods employed to extract accessory power from a typical jet engine. Auxiliary generators for ground or emergency operation may be driven by ram air or an auxiliary engine.

*a. Direct Engine Drive.* Generators mounted directly on pads on the accessory gear case of reciprocating engines are usually limited in power because of limitations on overhung moment imposed by vibration. Larger units may be mounted directly on turbine engine pads or separately mounted and shaft-driven. Direct-driven generators often operate over the entire engine-speed range from idle to military power. This range averages 3 or 3.5 to 1 in reciprocating and older single-rotor jet engines. However, some newer jet engines and the high-pressure rotor of dual rotor engines have speed ranges as low as 1.5 to 1. Turboprop engines often have only 10% variation for all flight conditions, although ground idle may be as low as 50% of maximum speed.

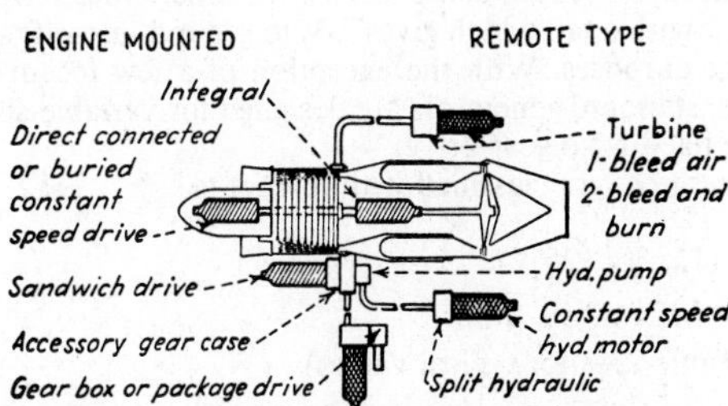


**FIG. 23-1** Possible methods of extracting power from a jet engine.

*b. Variable-Ratio Transmission.* In order to obtain constant speed from the main engines for constant-frequency ac systems, continuously variable-ratio hydraulic transmissions are used. A differential-type transmission has been widely adopted, since only an amount of power proportional to the difference between input and output speeds is handled hydraulically. Thus, space and weight are saved. Transmissions are operated either on engine oil or on a separate oil system including an air-oil or fuel-oil cooler. Governors hold steady-state speed within ±¼ to ½% and are provided with a means of sensing difference in real load between paralleled generators and of dividing it within 10%. Transient recovery times of the order of 0.5 to 2.0 s are obtained. The constant-speed hydraulic system now predominates in all jet-engine aircraft with all new installations utilizing the integrated generator design wherein the generator is a single-bearing machine bolted directly to the constant-speed drive.

*c. Engine-Compressor-Bleed Turbine Drive.* The pneumatic turbine is used where sufficient compressor bleed air can be obtained and where insufficient engine pads or space is available. Among its advantages are (1) suitability for constant-frequency systems and parallel operation, (2) flexibility as to location and number of units, and (3) ability to operate all generators from a common duct system after an engine failure. Turbines generally operate at much higher speeds than the generators; thus a speed-reducing gearbox is required. Automatic controls maintain constant speed and divide real load among paralleled generators. Methods of speed control employed are (1) throttling, (2) partial arc admission, and (3) variable-area nozzle, in order of increasing complexity and improved performance.

It is generally necessary to make a rather detailed application study before using air turbines for generator drives. This system is no longer being used in new designs.

*d. Auxiliary Engine Drive.* Small reciprocating-engine generator sets have been used for many years to (1) make aircraft self-sufficient, that is, provide starting power in absence of ground equipment; (2) provide power for servicing, lighting, etc., without operating the main engines (particularly in water-based aircraft); and (3) provide an emergency source of power under low-altitude flight conditions, including take-off and landing. Recent developments in small turbine engines have resulted in predominant use of this type in larger ratings with specific weight reduced to about 4 to 5 lb/kW for the prime mover.

*e. Ram-Air Turbine Drive.* The wind-driven generator of early aviation days is again finding application, this time as an emergency source of power which is particularly useful in single-engine aircraft to permit restarting after a jet-engine flame-out or to supply instruments and controls for an emergency landing. It also has an advantage as an emergency source when the main generating system fails in that a battery often has only sufficient capacity for a few minutes of operation whereas the turbine within its load rating is unlimited in this respect. Although such units are relatively heavy, they are more than competitive with a battery plus a small dc to ac inverter as an emergency source.

***9. Direct-Current Generator Characteristics.*** Multiengined aircraft designed before the advent of the jet engine generally featured 28-V dc systems paralleled to provide reliability. Many such aircraft are still in service and probably will be so for some years to come. In the meanwhile, a new series of dc generators have been developed for the small airplane market, and newly designed large aircraft feature ac systems. This section will deal briefly with the older dc systems, with some detail on the newer developments.

*1. Conventional DC Generators and Controls.* The conventional aircraft dc generators are usually self-excited by a shunt-field winding. Commutating and compensating series windings are nearly always used, in some cases with differential series fields. Usually, the series field is provided with a tap which gives 2-V to ground (aircraft structure) at full load. This is for paralleling purposes. With the exception of a few recent models for use with constant-speed drives, standard generators are designed for variable-speed operation as dictated by main-engine fixed-ratio gear drive.

Direct-current generators are classified with respect to

**1.** Operating voltage (15, 30, 125 V).

**2.** Current (50 to 400 A).

**3.** Speed range (see Table 23-3 for typical values).

**4.** Cooling air (40, 80, or 120°C at sea level with a pressure drop of 6 in of water).

In Table 23-3 are listed a number of standard generators in the 30-V 40°C class.

**TABLE 23-3** Characteristics of Typical Aircraft 30-V DC Generators (Conventional)

(Data from General Electric Co.)

| | | | | | | | | |
|---|---|---|---|---|---|---|---|---|
| Continuous rated current, A | 200 | 200 | 200 | 300 | 300 | 400 | 400 | 400* |
| Base speed, r/min | 2200 | 4400 | 3000 | 4550 | 3000 | 6000 | 4500 | 3000 |
| Maximum rated speed, r/min | 4500 | 8000 | 8000 | 8000 | 8000 | 8500 | 8000 | 8000 |
| Rated air flow, sea level, ft³/min | | 60 | 63 | 70 | 120 | 120 | 120 | 130 |
| Weight, lb | 48 | 34 | 44 | 50 | 60 | 51 | 61 | 73 |
| Lb per kW | 8.0 | 5.7 | 7.3 | 5.5 | 6.7 | 4.2 | 5.1 | 6.1 |
| Frame diam., in | | 6 | 6 | 6 | 6½ | 6½ | 6½ | 7 3/16 |
| Length including radial blast cap, in | | 10 13/16 | 12 3/16 | 13¾ | 15 1/16 | 13 9/16 | 15 1/16 | 17⅜ |

*This machine is a jet-engine starter-generator capable of delivering 93 lb·ft of torque with 1000-A current.

†Based on pressure drop of 6 in of water across the machine and inlet air temperature of 40°C at sea level. Some recently developed machines are designed for 120°C inlet air temperature.

NOTE: 1 ft³ = 0.0283 m³; 1 lb = 0.4536 kg; 1 in = 25.4 mm.

Generated armature voltage and field-resistance curves resemble those of a commercial generator at the base operating speed. As operating speed increases, the generated voltage is increased. The available output likewise increases (Fig. 23-2), as does also the electrical rate of response. The high available output at high speed provides a higher order of power-stability margin but entails possible thermal destruction of the generator in a short time if the high load is maintained. The high rate of electrical response is advantageous in quickly recovering from a shock-load addition but decreases the voltage-regulation stability margin. Generators usually have overload capacity of 150% of rated load for 2 min at 120% of base speed, starting from 40°C average winding temperature.

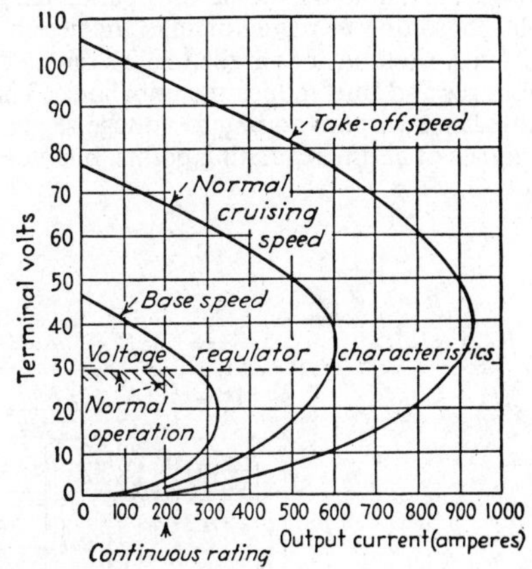


**FIG. 23-2** Aircraft dc generator external voltampere characteristics at various speeds.

2. *Brushless DC Generators and Controls* With the advent of the brushless ac generator, we have also the brushless dc generator. Both were made necessary by the brush-wear problems of commutators or slip rings at altitude. A pronounced benefit has been the lengthening of the time between overhauls. Essentially, the brushless dc generator is an ac machine with the commutator replaced by high-reliability silicon rectifiers. Operating voltages and currents are the same as those for conventional machines. The present trend is to make the brushless machine a direct replacement for the conventional brush machine.

The brushless dc machine development was accompanied by a more radical one, that which replaced the not-too-reliable carbon-pile voltage regulator with a transistorized version and the heavy control panel with a lighter, solid-state version. These, too, are designed as direct replacements of the conventional designs.

Some of the features of the brushless dc generator lend themselves to a simplification of the system. For example, since the brushless machine is an ac generator with rectifiers, there is no reason to use a cutout relay between the generator and the bus. The rectifiers serve to isolate it except when the machine is up to speed and is correctly excited.

The machine parameters resemble those of the brush machine in most cases. Table 23-3 may be used as a guide for these machines also.

***10. Direct-Current Generator Controls.*** Variable-speed aircraft generators do not inherently produce a sufficiently uniform voltage to serve the constant-potential power-supply system, and various failures must be protected against. Thus, controls are provided to accomplish the following:

1. Maintain voltage at the bus within specification limits.
2. Divide load current among the several generators in parallel.
3. Switch on the generator when its voltage is suitable.
4. Switch off the generator when its voltage is abnormal.
5. Provide short-circuit protection.
6. Provide means of manual supervision:
   *a.* Indication of performance.
   *b.* Manual switching.

A typical 28-V generating system is shown in Fig. 23-3, using an older style carbon-pile regulator, and related components for protection. A brushless type 28-V generator, transistor regulator, and solid-state protection and control system are shown in Fig. 23-4.

*a. Voltage Regulators.* The generator voltage regulator serves the dual purpose of automatically maintaining system voltage constant and balancing load current among paralleled generators. The older style voltage regulator consists of a solenoid operating against spring action on a carbon pile in series with the shunt field of the generator. New designs all use brushless dc generators with solid-state regulators. Regulators may be of the series-pass, pulse-width, or phase-delay type, all of which have a faster response than the carbon pile unit.

Where several generators are operated in parallel, an equalizer circuit of some type is provided on the voltage regulator for obtaining equal division of current among the several units. Current unbalance among paralleled generators is caused by error voltages, for example, those due to regulator inaccuracies in setting or droop or nonsymmetrical generator circuits. Methods of paralleling are directed at creating voltages to oppose these difference voltages and thus reduce the unbalance. The regulator-equalizer method described corrects unbalance without adding to voltage regulation at the bus. In fact, since bus voltage is the average of all the regulating points, bus-voltage regulation tends to be reduced by equalizer action.

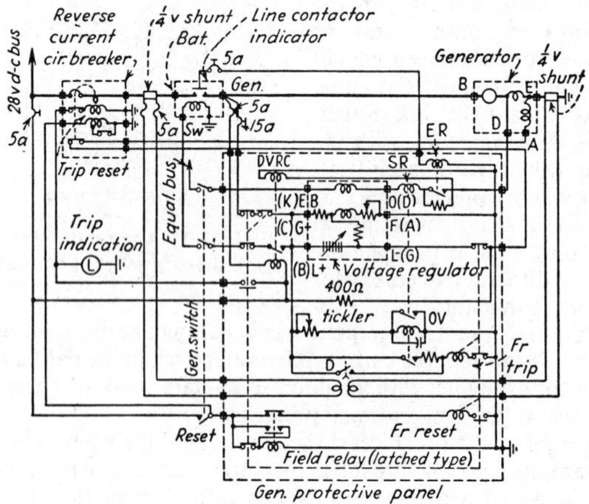


**FIG. 23-3** Typical generator circuit arrangement for 28-V dc system.

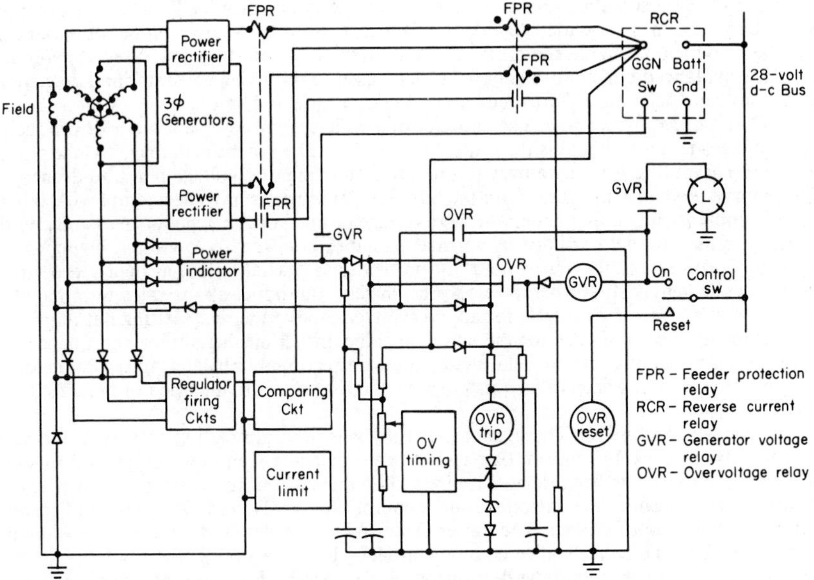


**FIG. 23-4** Brushless dc generator with transistor regulator and solid-state control system.

*b. Generator Switching.* Generator switching is accomplished by means of a main power-switching contactor in combination with appropriate pilot relays and controlled by them to accomplish automatically the following:

1. Close the contactor when the associated generator voltage reaches normal value and is of normal polarity.
2. Open the contactor when the associated generator voltage becomes subnormal.

Provision is made also to permit opening the contactor manually from a remote position. Where the generator is used also as a starter, provision is made for bypassing the pilot relay to close the contactor and operate the machine as a motor.

One type of pilot relay is polarized with a permanent magnet and has two coils. One coil senses the difference between generator and bus voltage, and the other senses generator current. When generator voltage exceeds bus voltage, the pilot relay closes and completes the circuit to the contactor. When the generator voltage drops below normal, reverse current flows through the current coil and opens the pilot relay removing the generator from the bus.

The "generator reverse-current cutout" consists of the line contactor and pilot relay mounted in one enclosure. In other system arrangements, the pilot relay is mounted in a separate control box (which often includes the voltage regulator) and senses reverse current through voltage drop across the generator series windings. The contactors used may either have a nominal 28-V interrupting rating or be designed with sufficient capacity to interrupt the maximum output voltage and current of a wide-speed-range generator under full field, as could occur in case of regulator failure.

*c. Short-Circuit Protection.* The generator relay provides some measure of protection to the system for generator and generator-feeder faults. However, a circuit breaker with a reverse-current trip having sufficient time delay to ride through normal switching transients is often used to provide (1) higher interrupting capacity (up to 12,000 A), (2) backup protection for a welded contactor or pilot relay failure, (3) auxiliary contacts to deenergize the generator (open the field) and prevent its continuing to feed the fault, and (4) interrupting capacity for overvoltage conditions using a shunt trip coil actuated by an overvoltage relay (see the following section). Forward current tripping is not provided, since it is accepted practice to leave an overloaded generator on the line, although an overtemperature warning may be provided to enable the crew to take corrective action. In a few cases, differential-current or balanced-current protection has been used to remove a generator and deenergize it without time delay in order to reduce the hazard of fire and smoke.

*d. Overvoltage Protection.* The capacity of a wide-speed-range generator for producing high voltages under full field is illustrated in Fig. 23-2. The potential damage to load equipment is so great that it is customary to provide fast-acting overvoltage relays to deenergize and remove the faulty machine from the bus. Several schemes have been employed, but all depend upon sensing either generator field voltage or bus voltage. The former is inherently selective, since the field voltage in normal machines is always below rated owing to the regulator series resistance, but it does not protect for a partial failure at high speed where the field voltage may still be below rated yet sufficient to produce above-rated terminal voltage. Bus-voltage sensing protects for all overvoltages above the relay setting but requires a more complex means of selecting the faulty machine. Either biasing of the overvoltage relay or the use of a separate selector relay acting on the directional unbalance in regulator equalizer circuit current is effective for this purpose. Such circuits are included in Figs. 23-3 and 23-4.

***11. Storage Batteries.*** The lead-acid battery was used almost exclusively in the past, but recently the nickel-cadmium battery has been replacing it in new design, and in some aircraft the battery has been deleted entirely. The nickel-cadmium battery is much lighter than the lead-acid but does introduce some complications. There is no method of reliably indicating the state of charge. The battery should be completely discharged occasionally. Battery capacity is adversely affected by incomplete discharge cycles. Twenty-four-volt batteries operating in conjunction with transformer rectifiers in flight and providing power for lights, control, and possible starting on the ground as well as for emergency flight operation are still found on most ac aircraft.

***12. Alternating-Current Generators.*** Alternating-current generators found on aircraft are almost exclusively of the synchronous type. They usually have an integral exciter which is so designed that it will build up on self-excitation, making the ac system independent of other power sources. Most designs have a small permanent-magnet pilot exciter for buildup and exciter control power which may also be used for other control functions. All new designs are the brushless type, which eliminates the brushes, commutator, and slip rings. This type of machine rectifies the output of the exciter and applies the result directly to the rotating field, eliminating the need for the commutator and slip rings and enhancing machine reliability.

Another important development is the variable-speed constant-frequency generator (VSCF). This machine delivers its constant-frequency output without the necessity for a hydromechanical constant-speed drive. The usual method of accomplishing this result is to use a 3-phase generator delivering 1300 to 2600 Hz depending upon engine speed. 1600 to 3200 Hz is being considered for future applications. This is synthesized into a 400-Hz 200-Y/115-V output voltage from the high-frequency supply, which is then filtered to produce a sine-wave (cycloconverter). The frequency synthesizer is a solid-state device operating silicon-controlled recifiers in the proper sequence to provide the desired output, utilizing an accurate frequency reference to maintain the output frequency constant. A block diagram of the VSCF system is shown in Fig. 23-5 (cycloconverter). Another approach is the dc link design wherein the generator output is rectified and transmitted to centrally located power conditioning equipment. Here is it converted back to ac with a conventional inverter. As an alternative, the rectifier and the inverter may be packaged in a single module. Design trade-offs are centered around the electromagnetic interference considerations. A cycloconverter switches at zero crossover points and minimizes switching noise but requires transmission of the high-frequency ac. The inverter operates in a switching mode that generates higher-frequency interference but eliminates the ac transmission problems. The interference question must be solved in either design approach.

Alternating-current generator characteristics are established by the requirements of Specification MIL-G-21480A/AS (September 1970). VSCF system requirements are established by Specification MIL-E-23001B (June 1976). Typical characteristics of ac generators of the conventional and brushless types are listed in Tables 23-4 and 23-5. Typical saturation and

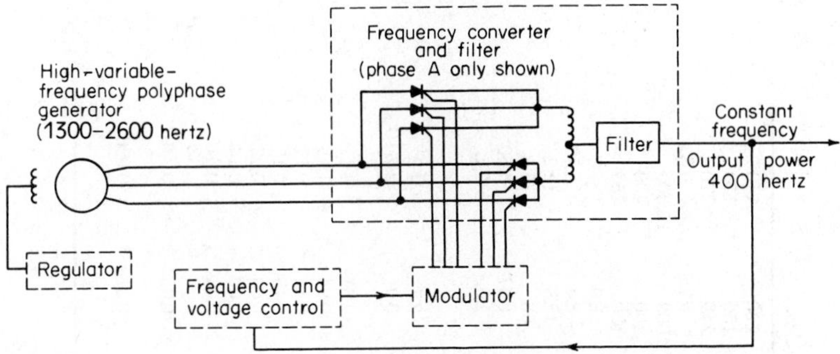


**FIG. 23-5** Block diagram of variable-speed constant frequency generator (cycloconverter).

short-circuit characteristics are shown in Fig. 23-6. Alternating-current generators are classified with respect to

1. Voltage (120- or 208-Y/120).
2. kVA (10 to 150).
3. Power factor (0.9 or 0.75).
4. Speed range and frequency (see Tables 23-4 and 23-5). New designs are 12,000 and 24,000 r/min.
5. Cooling air (40, 80, or 120°C at sea level with 6 or 11 in water-pressure drop).
6. Cooling oil, either sprayed oil mist or fluid flow designs (spray mist is becoming the standard).

**TABLE 23-4** Characteristics of Typical Aircraft 3-Phase AC Generators with Integral Exciters

(Data from General Electric Company)

| | | | | | | | | |
|---|---|---|---|---|---|---|---|---|
| Continuous rating, kva | 10 | 20* | 20 | 30 | 30 | 30 | 40 | 60 |
| Rated power factor | 0.75 | 0.75 | 0.75 | 0.9 | 0.75 | 0.75 | 0.75 | 0.75 |
| Min. rated speed, rpm | 5700 | 7600 | 5700 | 4000 | 4800 | 5700 | 5700 | 5700 |
| Max. rated speed, rpm | 6300 | 8400 | 6300 | 8000 | 7200 | 6300 | 6300 | 6300 |
| Frequency | 380/420 | 380/420 | 380/420 | 400/800 | 320/480 | 380/420 | 380/420 | 380/420 |
| Cooling air, S.L. temp., deg C | 80 | 120 | 80 | 40 | 40 | 80 | 80 | 80 |
| Cooling air, rated cfm | 140 | 220 | | 300 | 330 | 330 | 330 | 330 |
| Frame diam., in | 7 | $8\frac{1}{4}$ | 11 | 11 | 11 | 11 | 11 | 11 |
| Length with axial blast cap, in | $16\frac{15}{32}$ | $14\frac{5}{16}$ | $15\frac{5}{8}$ | $17\frac{5}{8}$ | $16\frac{7}{8}$ | $16\frac{7}{8}$ | $17\frac{5}{8}$ | $19\frac{1}{32}$ |
| Weight, lb | 45 | 38* | 62 | 99 | 87 | 75 | 87 | 115 |
| Lb per kva | 4.5 | 1.9* | 3.0 | 3.5 | 3.1 | 2.7 | 2.3 | 2.0 |
| Max. harmonic and % of fundamental | 5–1.90 | | 5–1.45 | 7–1.40 | 5–0.75 | 5–0.75 | 5–1.4 | 5–0.44 |
| Phase voltage unbalance with ⅓ current unbalance, max. % | 3.7 | | 3.6 | 3.9 | 3.8 | 3.0 | 2.9 | 2.9 |
| 3-phase S.C. current at base speed, p.u. | 5.0 | | 3.6 | 3.3 | 3.5 | 3.0 | 3.3 | 3.5 |
| Reactances at 400 cycles, p.u.:† | | | | | | | | |
| Synchronous, direct axis ($xd$) | 120 | 2.02 | 2.13 | 1.12 | 2.51 | 2.14 | 1.99 | 1.61 |
| Synchronous, quad. axis ($xq$) | 0.68 | 1.16 | 1.25 | 0.65 | 1.35 | 1.20 | 1.12 | 0.79 |
| Transient, direct axis ($x'd$) | 0.20 | 0.38 | 0.42 | 0.22 | 0.37 | 0.37 | 0.34 | 0.23 |
| Subtransient, direct axis ($x''d$) | 0.14 | 0.28 | 0.24 | 0.14 | 0.21 | 0.20 | 0.19 | 0.12 |
| Negative sequence ($x_2$) | 0.18 | 0.36 | 0.34 | 0.16 | 0.23 | 0.21 | 0.20 | 0.12 |
| Zero sequence ($x_0$) | 0.28 | 0.34 | 0.012 | | | 0.17 | 0.011 | 0.15 |
| Negative sequence resistance ($r_2$) | 0.035 | | 0.057 | | | 0.042 | 0.039 | 0.025 |
| Zero sequence resistance ($r_0$) | 0.026 | 0.32 | 0.039 | | | 0.026 | 0.024 | 0.020 |
| Efficiency, full load, 6000 rpm | 77 | 85 | 77 | 80 | 78 | 83 | 83 | |

*This machine is designed for static excitation and has no integral exciter.
†Calculated values.
NOTE: 1 in = 25.4 mm; 1 lb = 0.4536 kg; 1 $ft^3$ = 0.0283 $m^3$.

**TABLE 23-5** Characteristics of Typical Business Spray-Oil Cooled Aircraft 3-Phase AC Generators with Integral Exciters and Permanent-Magnet Generators

(Data from Westinghouse Co.)

| | | | | | | | |
|---|---|---|---|---|---|---|---|
| Basic rating, kVA | 30 | 40 | 60 | 90 | 105* | 120 | 150* |
| Continuous rating, kVA | 40 | 50 | 75 | 113 | 115 | 150 | 150 |
| Overload rating, kVA for 5 s | 60 | 80 | 120 | 180 | 210 | 240 | 270 |
| Min. rated power factor | 0.75 | 0.75 | 0.75 | 0.75 | 0.75 | 0.75 | 0.83 |
| Rated speed, r/min | 12,000 | 12,000 | 12,000 | 12,000 | 12,000 | 12,000 | 12,000 |
| Frequency, Hz | 400 | 400 | 400 | 400 | 400 | 400 | 400 |
| Cooling oil temperature, °C | 150 | 150 | 150 | 150 | 150 | 150 | 150 |
| Cooling oil flow, gal/min | 2.5 | 3.0 | 3.8 | 5.0 | 5.5 | 6.0 | 7.3 |
| Frame diameter, in | 7.20 | 7.20 | 7.20 | 8.4 | 8.40 | 8.4 | 8.50 |
| Length, in | 9.10 | 8.71 | 9.70 | 12.2 | 10.0 | 12.8 | 11.2 |
| Weight, lb | 29.0 | 34.5 | 42.9 | 55 | 65 | 68 | 75 |
| Lb/kVA (basic rating) | 0.97 | 0.86 | 0.72 | 0.61 | 0.62 | 0.57 | 0.50 |
| Max. harmonic and % of fundamental | 5-0.9 | 7-0.8 | 5-1.0 | 5-0.8 | 7-0.5 | 5-1.0 | 5-0.7 |
| Phase voltage unbalance with 2/3 basic rated current unbalance, max. %† | 4.6 | 3.5 | 3.4 | 3.5 | 3.7 | 3.5 | 4.2 |
| 3-phase S.C. current for 5 s, min. p.u. | 3.0 | 3.0 | 3.0 | 3.0 | 3.0 | 3.0 | 2.2(10 s) |
| Reactances at 400 Hz, p.u.† | | | | | | | |
| Synchronous, direct axis ($X_d$) | 2.38 | 2.42 | 2.70 | 2.83 | 2.73 | 2.61 | 4.11 |
| Synchronous, quad axis ($X_q$) | 0.87 | 0.88 | 0.98 | 1.03 | 0.99 | 0.94 | 1.48 |
| Transient, direct axis ($X'_d$) | 0.14 | 0.13 | 0.13 | 0.13 | 0.13 | 0.11 | 0.17 |
| Subtransient, direct axis ($X''_d$) | 0.12 | 0.11 | 0.12 | 0.11 | 0.11 | 0.10 | 0.15 |
| Subtransient, quad axis ($X''_q$) | 0.10 | 0.09 | 0.10 | 0.10 | 0.10 | 0.09 | 0.13 |
| Negative sequence ($X_2$) | 0.11 | 0.10 | 0.11 | 0.10 | 0.11 | 0.09 | 0.14 |
| Zero sequence ($X_0$) | 0.014 | 0.012 | 0.012 | 0.014 | 0.014 | 0.012 | 0.016 |
| Negative sequence resistance 25°C ($R_2$) p.u. | 0.065 | 0.044 | 0.037 | 0.036 | 0.040 | 0.032 | 0.049 |
| Zero sequence resistance, 25°C ($R_0$) p.u.† | 0.049 | 0.030 | 0.021 | 0.022 | 0.023 | 0.019 | 0.026 |
| Efficiency basic kVA min., p.f. | ‡ | ‡ | ‡ | ‡ | ‡ | ‡ | ‡ |

*Special application.

†Calculated values.

‡Typically 85%. Specific values are related to application parameters.

NOTE: 1 gal/min = 0.00063 $m^3/s$; 1 in = 25.4 mm; 1 lb = 0.4536 kg.

Generators fall into two classes, the conventional air-cooled brushless two-bearing designs which flange mount to the accessory pad and the newer spray oil-mist design. The new machines are designed as an integral part of the constant-speed drive and designated as integrated-drive generators (IDG) and integrated-drive generator systems (IDGS). They are one-bearing designs utilizing the drive unit output bearing as the second bearing. The stator housing then bolts directly to the drive unit housing. This is a key feature in achieving weight reduction. To reduce the overhung moment, the generator field is placed at the mounting end followed by the excitation generator and rotating rectifiers. Heavy amortisseur circuits are used in both direct and quadrature axes to reduce transient disturbances, to reduce phase-voltage unbalance, and to provide greater stability in parallel operation. A flexible drive shaft is normally used to reduce torsional oscillations from the engine drive. The shaft is provided with a friction damper to prevent breakage at resonant frequencies. The flexible shaft also helps in reducing the transient peak torques imposed on the drive during short circuits and synchronizing.

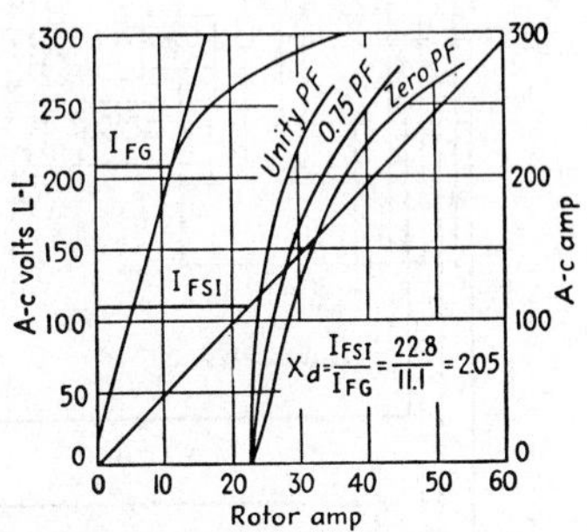


**FIG. 23-6** Typical aircraft ac generator, saturation and short-circuit characteristics.

***13. Alternating-Current Generator Control.*** The majority of newer ac systems have been constant-voltage and constant-frequency, and most are parallel systems. Controls are supplied to accomplish the following:

**a.** Maintain voltage at the bus within specification limits (Par. **14**).

**b.** Divide real and reactive load among the several generators in a parallel system (Par. **14**).

**c.** Switch the generator on the bus when frequency and phase voltages (and rotor angle in parallel systems) are suitable (Par. **15**).

**d.** Switch the generator off when it is the cause of abnormal frequency or any abnormal phase voltage (Pars. **16** and **17**).

**e.** Switch the generator off the bus when it is the cause of unequal generator current distribution in a parallel system (Par. **18**).

**f.** Provide protection against short circuits in the generating system (Par. **19**).

**g.** Provide manual and/or automatic means of supervision.

In many modern installations, equipment is provided to accomplish the foregoing automatically to simplify aircraft operation. In some systems, no manual supervision is required or provided for other than a reset switch to permit placing the automatic start-up sequence in effect after a generator has been tripped.

***14. Voltage Regulators.*** Voltage regulators are generally designed to sense and respond to the approximate average of the magnitudes of the three line-to-line voltages on 3-phase systems furnished with high-phase takeover if one phase exceeds the average. Reactive division has been obtained by feeding differential current signals from current transformers in one generator line into a discriminator circuit which biases the sensing circuit of the regulator according to the magnitude and sense of the difference between the reactive flow of the particular machine and the average flow of all machines. This produces a system in which bus voltage is the average of all the regulator operating points and no added regulation is caused by the reactive division circuitry.

Regulators in general use are the magnetic-amplifier type and the transistorized type. The transistor regulator has made possible more accurate voltage control and better transient control. Virtually all new design is transistorized. A schematic diagram of a magnetic-amplifier regulator is shown in Fig. 23-7, that of a transistorized regulator in Fig. 23-8 using a phase-delay technique. Pulse-width-modulation-type regulators are also in use.

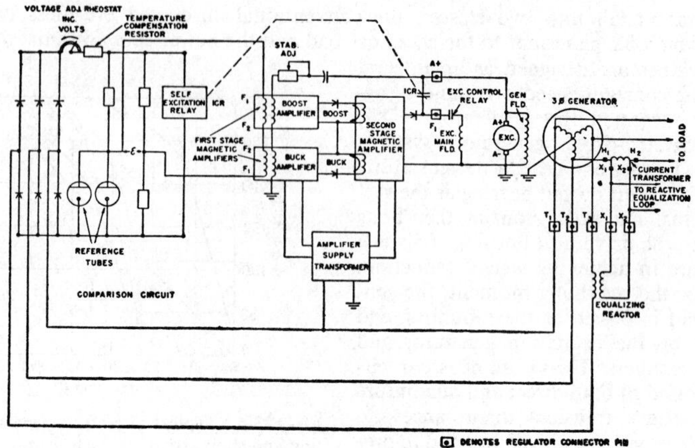


**FIG. 23-7** Diagram of a magnetic amplifier type of ac generator voltage regulator.

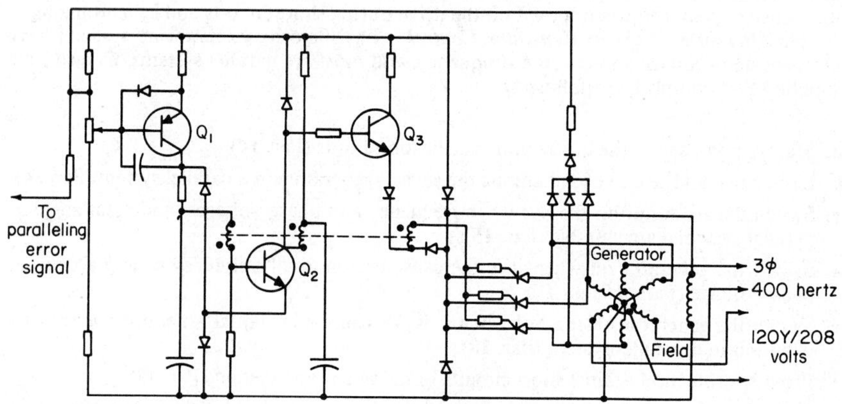


**FIG. 23-8** Brushless ac generator with transistor regulator using phase delay.

## Switching and Protection

***15. Switching Devices.*** A latch-type generator control relay is used to open and close the exciter field for energizing and deenergizing the machine. The protective equipment shuts the machine down by tripping this relay.

Both latch-type and magnetic hold-in contactors are used to connect the machine to the bus. They switch all 3 phases simultaneously and must be capable of interrupting the maximum fault current that may flow through them. Protective equipment isolates a faulty machine by tripping this contactor.

Generator switches are used for resetting, line-contactor switching, and, in some cases, deenergizing the machine. One type has three positions: "off," "on," and a spring-loaded "reset" position. The "off" position opens the line contactor; "on" closes the line contactor or initiates automatic paralleling; the "reset" position will close the generator control relay. New systems employ push-button type switch indicators of either the momentary or latching type depending on function need.

***16. Frequency-Sensing Devices.*** Induction-motor and transformer loads may overheat, or load-protective equipment may be operated owing to underfrequency operation. For this reason, frequency-sensing devices are used to open the line contactor when frequency is below permissible values.

Underfrequency relays operating off bus voltage have been used widely. A tuned circuit or a high- or low-pass filter is used as the sensing element. Many recent applications have used a mechanical underspeed switch on the drive to open the line contactor.

If a drive overspeeds for any reason, both the drive and the machine may be destroyed. For this reason an overspeed device is used to shift into full underdrive in hydraulic transmissions or to remove the source of energy from turbine drives. Selectivity in a parallel system is obtained by an overrunning clutch which prevents large speed increases in normal drives; thus, only the faulty drive control is actuated by the protection. The machine may be removed from the bus by underspeed and/or difference current protection.

Very accurate frequency control and real load division are incorporated as a part of the hydromechanical constant-speed drive. Some type of frequency standard is used, and all drives are compared against this standard and the speed regulated accordingly. Also, current transformers in one phase of each paralleled machine are connected in series. A sensing circuit coupled with a discriminator controls the individual drives to divide the real (kilowatts) load.

***17. Voltage Sensing.*** An inverse type of relay is used for overvoltage protection. Its voltage-time trip curve is designed to ride through normal system transients without tripping, but overvoltage transients in excess of normal will cause a trip and remove the machine from the system. These relays usually sense the same quantity as the regulator, that is, the approximate average magnitude of the three line-to-line voltages.

If the bus voltage falls below specifications, rotating loads may overheat, and static load performance is affected. Undervoltage relays sensing the approximate average of the three line-to-line voltages have been used to trip the line contactor under these conditions after a suitable delay.

Unbalanced phase voltages resulting from large unbalanced loads, unbalanced faults, and open conductors may seriously affect system performance owing to possible overvoltages in some phases and unbalanced currents that may be extremely large in some instances. Phase-voltage unbalanced relays are used to remove a machine with large voltage unbalance after a suitable delay to permit fault-protection coordination. These relays may be operated from sequence filters such as shown in Fig. 23-9. The negative-sequence relay has been used also to hold a generator off the line when it has reversed phase sequence due to maintenance error.

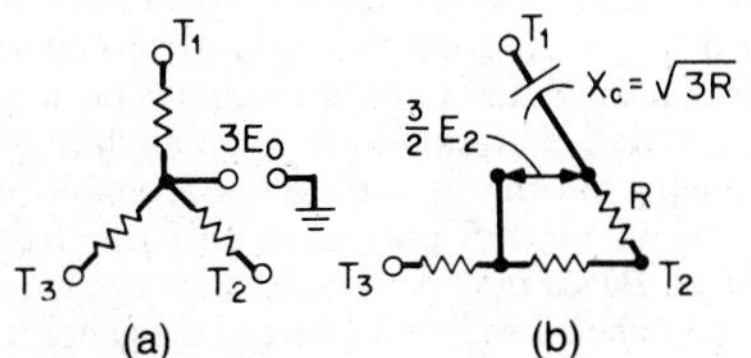


**FIG. 23-9** Sequence filters used for unbalanced voltage sensing: (*a*) zero sequence; (*b*) negative sequence.

***18. Unbalanced Generator Currents.*** If sizable differences in generator currents exist in a parallel system, total generating capacity is reduced. Furthermore, since load-division circuitry is designed to keep such unbalance within specifications for any normal conditions, excessive current differences are a direct indication of a failure within the generating system. A relay connected to a differential current-transformer loop senses the difference between one generator current and the average of all generator currents. It may be used to trip bus tie contactors and split a parallel system, thereby removing the unbalance. Time delay sufficient to ride through normal transient unbalances must be incorporated.

If a machine in a parallel system is over- or underexcited for a given load condition, differential reactive current will flow, bus voltage will be changed but not always sufficiently to actuate abnormal voltage relays, and, in some cases, out-of-step operation may result. One general approach has been to open bus tie contactors on difference current as described and allow over- and undervoltage protection to trip the faulty machine. The more commonly accepted method is to use differential reactive current to bias the over- and undervoltage relays. Thus, the abnormal machine voltage relays see a greater abnormality in voltage than do normal machine relays and trip the abnormal machine only.

If a feeder or machine phase is opened on a parallel system, phase voltages will be kept in good balance by the normal machines and loads are little affected. If, however, the phase in which load-division current transformers are located is opened, very severe unbalanced currents and neutral currents will flow in all machines and overheating in both normal and abnormal machines will result. Relays sensing neutral current have been used to trip bus tie contactors and split the system so that a larger unbalance voltage sufficient to operate unbalance voltage relays will exist in the abnormal machine and it will be tripped. Time delay sufficient to clear the worst fault involving neutral must be incorporated to avoid splitting the system on load- or distribution-system faults. This arrangement has been used in some cases to isolate a bus with a ground fault and to trip the associated machine.

***19. Short-Circuit Protection.*** As noted in the preceding, in case of a short circuit in a parallel system, unbalanced voltage protection may be used to isolate a machine and its associated bus and to trip the machine after a time delay of the order of 5 s.

The most commonly used method is to protect the machine and feeders with differential fault protection and rely on unbalanced and undervoltage relays for bus protection. This offers the advantages of high speed and fully selective operation, and the bus associated with the faulted machine or feeders need not be isolated from the system. The usual arrangement is to use one current transformer in each phase located at or near the feeder connection to the bus and one located on the neutral side of each machine phase. A difference in excess of 20 to 40 A between the two transformers on any phase is used to operate a relay which trips the machine.

***20. Generator-Circuit Arrangement.*** Figure 23-10 is a logic diagram of a generator system incorporating the above protective features. This is for a single generator of a multigenerator system and includes automatic paralleling, as well as features discussed previously. Refer to Fig. 23-11*c* for the distribution circuit configuration to which this logic circuit would be applicable. The logic circuit may be implemented either by relays or by solid-state logic. Both are in use, but new design is all solid-state.

## Distribution

***21. The distribution system*** is comprised of a number of buses, interconnecting cables, and associated protective devices.

***22. Buses.*** A bus is a junction of three or more circuits. Since damage of this point can make more than one circuit inoperative, the objective is to keep the area as small as possible and to give it good mechanical protection. The bus is usually a copper bar mounted on insulating material (aluminum has been used in a few installations). The bus and associated protective equipment are often enclosed in a metal housing or, in some cases, a Fiberglas housing in order to reduce the probability of short circuits.

Small aircraft may have only one bus, but some large aircraft have had more than 15 buses. Buses may be classified as to service: (*a*) source bus, (*b*) load bus. Source buses often need to be considered a part of the generating system in that generator protective equipment is designed to provide source-bus protection as well in many applications.

***23. Distribution Circuits.*** Figure 23-11 shows several system configurations in use today in both ac and dc systems. Each of these has advantages and disadvantages; the choice is dictated by the application and requirements of the particular airplane. Generally, larger airplanes tend toward a split or synchronizing source-bus arrangement with alternate feeds to vital load buses, whereas smaller, high-performance aircraft tend toward the simpler arrangements. However, a trend toward high-speed, fully selective protective equipment is discernible, since, for some loads, transient power disturbances of long duration are as bad as (or even worse than) complete loss of power to the load. As this trend continues, emphasis is expected to shift more toward the split-bus or ring-bus configurations at the expense of more complex protective equipment.

Load-transfer contactors as shown in Fig. 23-11 have been used widely to transfer loads from a dead bus over to a live one. These devices will transfer on loss of voltage or upon the presence of sizable voltage unbalance. Time delay sufficient to ride through fault clearing beyond the contactor must be incorporated to prevent chattering with a fault on the load side. Recently, consideration has been given to deletion of these in favor of automatic reclosing of bus tie contactors and/or fully selective generating-system protection. If a source bus is faulted or lost through sequential protection, the loads connected to that bus are lost.

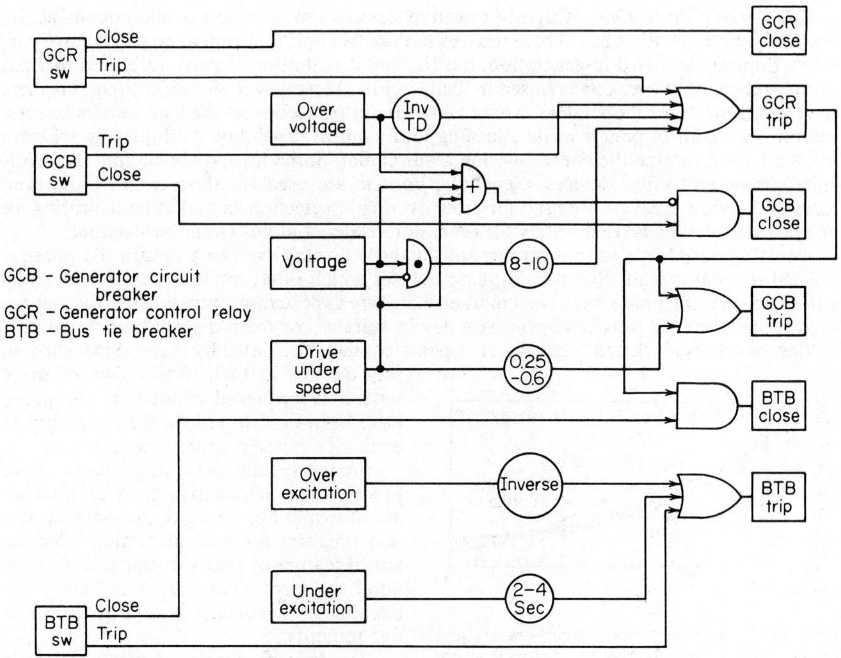


**FIG. 23-10** Block diagram of ac generator control system.

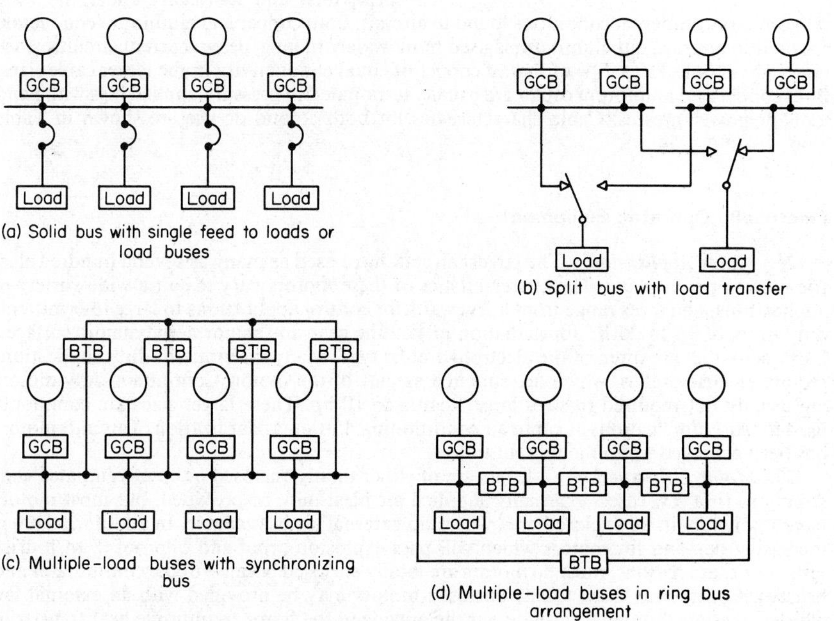


**FIG. 23-11** Distribution-circuit configuration commonly used in aircraft.

***24. Circuit Protection.*** Circuit-protective devices are provided in the circuits at the point of junction with a bus. These devices both detect and interrupt abnormally large currents. Limiters are used in generation, bus tie, and distribution circuits; fuses and thermal or magnetic circuit breakers are used in load circuits. Most new designs use circuit breakers having remote-control operators. These allow circuit protection at the load centers or even remote load control panels while affording local control capability. Multiplexing schemes are used to minimize the control-wiring requirement. Some low-power circuits use solid-state circuit protection devices. Generally, limiters are used for short-circuit protection; fuses and circuit breakers are used for cable overload protection as well as fault clearing. In ac systems, relaying is used widely for generator, feeder, and bus circuit protection.

Most 3-phase loads cannot operate satisfactorily or may even be a hazard if 1 phase is opened to clear a fault. Recently, 3-phase devices which interrupt all 3 phases with overcurrent in any one phase have been marketed and are expected to come into use for 3-phase load protection. No satisfactory 3-phase device suitable for main-distribution-circuit protection has been marketed other than a 3-phase contactor actuated by thermal-current-sensing devices in each phase. Two or more separately protected conductors per phase have been used to reduce the possibility of opening a phase because of a fault.

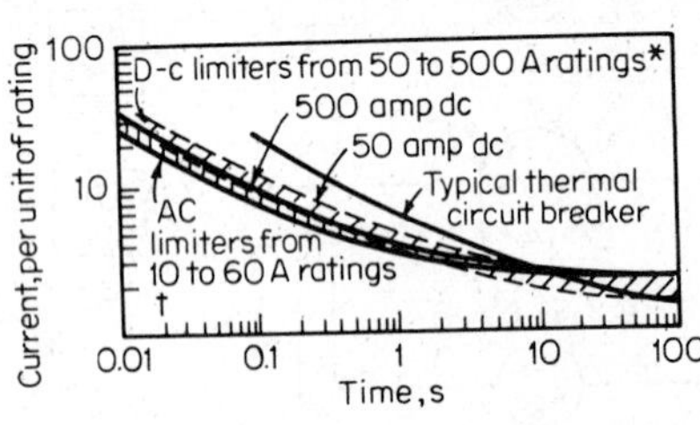


**FIG. 23-12** Typical aircraft circuit-breaker and current-limiter characteristics. (*Plotted from curves supplied by Bussman Manufacturing Company; †plotted from curves supplied by Burndy Engineering Company.)

Breakers, limiters, and fuses have inverse-time characteristics, some of which are shown in Fig. 23-12. Some magnetic circuit breakers for load protection offer the added feature of nearly instantaneous operation on large fault currents. This is an advantage in reducing system disturbances due to faults.

***25. Aircraft Cable.*** Aircraft cable is stranded of fine wire to obtain flexibility. The insulation withstands abrasion, has flame-retarding characteristics, and resists heat, fuel, and lubricants under the wide range of environmental conditions found in aircraft. Both copper and aluminum conductors are used in aircraft, but aluminum is used more widely in long, heavy-current circuits, since its weight is only 55 to 70% of that of copper of equal conductivity in the larger cable sizes. Both copper and aluminum cables are usually terminated in pressure-applied lugs with hand tools or power presses. Cable characteristics for both ac and dc use are shown in Table 23-6.

## Electrically Operated Equipment

***26. Motor Application.*** The largest aircraft have used as many as several hundred electric motors per airplane. The characteristics of these motors vary to suit a wide variety of applications, and sizes range from a few watts for control applications to large intermittent-duty units of 15 to 25 hp for actuation of landing gear and major aerodynamic surfaces. Large actuators are often of the electrohydraulic type. Many intermittent-duty applications require electric brakes, which are supplied as part of the motor. Continuous-duty motors are usually not required in sizes larger than 5 to 10 hp. These larger sizes are commonly used for such applications as cabin air conditioning. Little standardization of aircraft motors has been achieved up to the present time.

***27. Motor Construction.*** Motors are of either totally enclosed or open-ventilated construction. In a few cases, externally supplied air blast may be provided, but most motors have to operate in an enclosed area where no external air is available. In addition, there is increasing demand for motors which will pass explosion-proof and oilproof (½ pt/h drip) tests. For these reasons, most dc motors are totally enclosed because of the commutator and brushes. Continuous duty, totally enclosed motors may be provided with an external fan which forces air through a shroud over the outside of the frame to improve heat transfer to the surrounding air. Open-ventilated ac induction motors can often pass the required envi-

**TABLE 23-6** Aircraft-Cable Characteristics

| Wire or cable size | | | | | | Continuous-duty current A* | |
|---|---|---|---|---|---|---|---|
| Aluminum | Copper | Max.* dc resistance, Ω per 1000 ft at 20°C | 400-Hz positive† (and negative) sequence impedance, Ω per 1000 ft | 400-Hz zero† sequence impedance, Ω per 1000 ft | Nominal weight* of finished wire, lb per 1000 ft | Single wire in free air | Wires and cables in conduit or bundles |
| | AN-22 | 16.2 | | | 6.3 | | 5 |
| | AN-20 | 9.88 | 10.25 + j0.304 | 10.44 + j2.20 | 8.4 | 11 | 7.5 |
| | AN-18 | 6.23 | 6.44 + j0.289 | 6.33 + j1.93 | 11.6 | 16 | 10 |
| | AN-16 | 4.81 | 4.76 + j0.278 | 4.95 + j1.88 | 14.3 | 22 | 13 |
| | AN-14 | 3.06 | 2.99 + j0.264 | 3.18 + j1.80 | 21.0 | 32 | 17 |
| | AN-12 | 2.02 | 1.88 + j0.247 | 2.07 + j1.72 | 32.0 | 41 | 23 |
| | AN-10 | 1.26 | 1.10 + j0.228 | 1.29 + j1.64 | 47.5 | 55 | 33 |
| | AN-8 | 0.701 | 0.700 + j0.226 | 0.792 + j1.53 | 76.5 | 73 | 46 |
| | AN-6 | 0.445 | 0.440 + j0.221 | 0.632 + j1.42 | 115 | 101 | 60 |
| | AN-4 | 0.280 | 0.282 + j0.212 | 0.474 + j1.33 | 176 | 135 | 80 |
| | AN-2 | 0.183 | 0.188 + j0.206 | 0.380 + j1.24 | 278 | 181 | 100 |
| | AN-1 | 0.149 | | | 322 | 211 | 125 |
| | AN-0 | 0.116 | 0.127 + j0.204 | 0.319 + j1.15 | 415 | 245 | 150 |
| | AN-00 | 0.096 | | | 525 | 283 | 175 |
| | AN-000 | 0.071 | | | 657 | 328 | 200 |
| | AN-0000 | 0.056 | | | 820 | 380 | 225 |
| AL-8 | | 1.093 | 1.093 + j0.226 | 1.28 + j1.53 | 32 | 60 | 36 |
| AL-6 | | 0.641 | 0.648 + j0.221 | 0.840 + j1.42 | 55 | 83 | 50 |
| AL-4 | | 0.427 | 0.439 + j0.212 | 0.631 + j1.33 | 74 | 108 | 66 |
| AL-2 | | 0.268 | 0.282 + j0.206 | 0.474 + j1.24 | 107 | 152 | 82 |
| AL-1 | | 0.214 | | | 133 | 174 | 105 |
| AL-0 | | 0.169 | 0.188 + j0.204 | 0.370 + j1.15 | 167 | 202 | 123 |
| AL-00 | | 0.133 | | | 203 | 235 | 145 |
| AL-000 | | 0.109 | | | 242 | 266 | 162 |
| AL-0000 | | 0.085 | | | 296 | 303 | 190 |

*Data from Military Specification MIL-W-5086C, Wire, Electrical, 600-volt, Copper, Aircraft, December 1974; Military Specification MIL-W-5088B (ASG), Wiring, Aircraft, Installation of, March 1972; Military Specification MIL-W-7072, Wire, Electrical, 600-Volt, Aluminum, Aircraft, September 1976.

†Data from Impedance Data for 400-Cycle Aircraft Distribution Systems, by D. W. Exner and G. H. Singer, Jr., *AIEE Tech. Paper* 52-322, October 1952. Data for copper taken directly from this paper; data for aluminum extrapolated by the authors from this paper. Extrapolation method was spot-checked with data in Experimental Determination of 400-Cycle Impedance of Wire in Aircraft Power Distribution Circuits, by J. D. Andrew, *Trans. AIEEE Paper* 54-374, June 14, 1954. Positive sequence calculations checked within 5% and zero sequence within 17% of the data for the two aluminum conductors in this paper. New thin-walled insulation systems such as kapton and kynar will lower these values slightly.

NOTE 1: 1 ft = 0.3048 m; 1 lb = 0.4536 kg.

NOTE 2: Impedance data are based upon tight, equilateral bundle. Positive sequence data are nearly independent of skin distance†; zero sequence data are based upon 2-in skin distance.

ronmental tests and thus are used in the larger sizes for continuous duty because of substantial weight savings.

Motors are flange-mounted with the exception of a few cases in which they are built into an actuator or gear unit and then may be of single-bearing construction. Standard flange sizes are used. Square flanges with four boltholes are used in smaller sizes, and round flanges with additional boltholes to prevent distortion are used in larger sizes. Keyed shafts or splines are used for coupling to the load. Where leads are AN-10 cable or larger, terminal boards are generally used, whereas the leads are brought out directly or through connectors in the smaller sizes. Three-phase ac motors are 4-wire with four terminals, and continuous-duty dc motors require only two terminals. Intermittent-duty dc motors for reversing service have three terminals for split-series and four terminals for compound-wound types.

Brush problems encountered on dc motors at altitude have accelerated the development of brushless dc motors. A short treatise on these motors is included in the section on spacecraft electric subsystems.

***28. Motor Characteristics.*** Motor-design objective is the attainment of adequate performance with minimum weight and acceptable efficiency. To accomplish this, speed is usually high to reduce size and weight. Commonly used speeds for 400-Hz ac motors are 7300 and 11,000 r/min, corresponding to six-pole and four-pole designs, respectively. Direct-current motor speeds vary widely, but the trend is toward use of the ac motor speeds stated to increase the degree of interchangeability. Reduction gears are used frequently to obtain an acceptable speed for the driven equipment.

Aircraft motors are rated for either continuous or intermittent duty. The intermittent-duty cycles vary for the application, although 1 min on and 9 min off is typical. Temperature and altitude limit the rating and must be specified. In addition, where requirements for high-performance aircraft are especially severe, it is necessary to specify the maximum mounting-pad temperature and the velocity of the air surrounding the motor. Since an important consideration in these applications is also the total heat rejection which must be disposed of in the aircraft, it is desirable to achieve high efficiency in continuous-rated motors. Over 75% efficiency is usual for units larger than 1 hp, with somewhat lower values for intermittent-duty motors.

Direct-current motors may be shunt-, series-, or compound-wound, depending on the application. The split-series motor has been used extensively for reversing service because of the simplicity of the reversing control, which needs only to switch the input from one end of a tapped series winding to the other to reverse direction of rotation. Recently developed permanent-magnet motors are now commonly used. These offer improved performance, high efficiency, and reversing capability. Reversing occurs merely by reversing the input power polarity. New high-energy-density permanent magnets using rare earths such as samarium cobalt are used in some units to produce greater dynamic response for servo applications and to prevent inadvertent demagnetization. Speed-torque curves for typical aircraft motors fall within the limits shown in Fig. 23-13. Starting currents are in the order of 400% for series motors and 1000% for shunt motors.

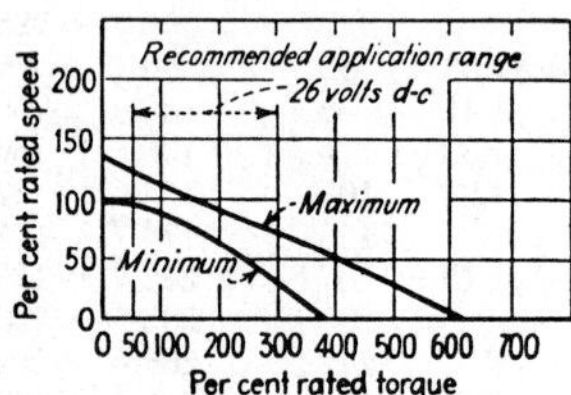


**FIG. 23-13** Range of aircraft dc motor speed-torque characteristics.

Alternating-current motors are usually of the 3-phase squirrel-cage induction type, although single-phase capacitor motors are used where no 3-phase power is available. Peak torque, locked-rotor torque, and power factor vary over wide limits, depending on the application requirements, although starting current is limited to 800% by military specifications.

***29. Conversion Equipment***

*a. Application.* Inverters have been used on practically all aircraft with dc generating systems to supply ac loads such as instruments, fluorescent lighting, radio and radar, and control systems such as automatic pilots and supercharger regulators. With increasing use of electronic equipment and other ac loads, and because rotary-inverter efficiencies are in the neighborhood of 50%, it is not unusual for 20 to 25% of the dc power to be used for inverters. The poor weight economy resulting from this situation has been a principal factor in the increasing use of ac power generation and the development of the static inverter, which has full-load efficiencies in the region of 75 to 80%. In aircraft having only ac power generation, it is often necessary to convert 10% or more to 28 V dc by means of transformer-rectifiers. However, the penalty is not so severe because of the higher efficiency (80 to 90%) and lower specific weight of the conversion units and the lower specific weight of the ac generating system. Of course, transformers are used for the conversion of voltage in aircraft as elsewhere. Step-up transformers are used principally in electronics for plate supplies, and step-down transformers are required in various load equipments such as lighting, certain low-voltage instruments, very small motors and, transistorized circuitry.

*b. Aircraft Inverters.* Inverters have been built to supply frequencies of 60 to 3200 Hz, although 400 Hz is most generally used. The rotary inverter is the most prevalent type in older aircraft but may be found as the emergency ac supply in more modern aircraft. Basically, it is a motor-generator set with control of frequency by dc motor speed control, of voltage by ac generator field control. Both single- and 3-phase inverters are found, and output voltages are usually 26, 115, and 200-Y/115. Automatic speed control is required to meet the close frequency tolerances required by today's aircraft equipment. The older inverters have carbon-pile regulators, but in order to meet both frequency and voltage requirements, new units all use solid-state regulators.

The trend to solid-state apparatus found in regulators is also found in the inverter field. In many cases, individual apparatus will incorporate a small inverter sufficient to supply the single load at high efficiency. The static inverter to replace the rotary inverter is characterized by lighter weight, higher efficiency, and freedom from mechanical faults, since it has no rotary armature, commutator, or slip rings and does away with the mechanical regulator. There are significant differences between the rotary and static inverters which must be taken into account when replacement is contemplated. Some are

1. The rotary contains its own fan as a matter of course; the static may require special cooling.
2. Voltage regulation of the rotary inverter is by regulation of the field for all 3 phases at once; the static regulation is commonly for each phase.
3. Usually, almost full inverter output may be taken from a single-phase on a 3-phase rotary inverter; only normal load may be so taken in the static inverter.
4. A static inverter is much more easily designed for a fixed load, and many are so designed. Care should be taken that such an inverter is not installed where varying loads occur. A possible arrangement for a static inverter is shown in Fig. 23-14. A series regulator controls the frequency of a 3-phase square-wave oscillator; the series regulator is controlled from the master frequency control. The output of the square-wave oscillator is fed into the bases of a transistor amplifier, and the output, which is a quasi-square wave, is transformed to the proper voltage and filtered to a sine wave. A feedback controls another series regulator to give the correct voltage output. This is only one method of producing an acceptable output; many others are in use. This one has the merit of suppressing the third harmonics. Additional steps in the waveform may be used to eliminate additional harmonics. The first harmonic will be 2*N*-1, where *N* is the number of discrete steps in the phase-to-phase voltage waveform. Here also pulse-width-modulation regulators are finding wide application because of their inherently higher efficiency.

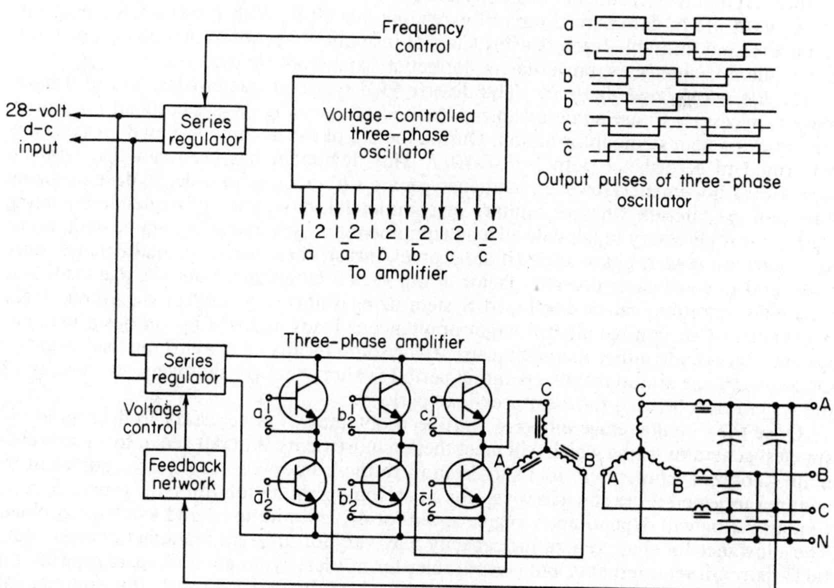


FIG. 23-14 Static inverter, 3-phase block diagram.

**5.** The number of reactive voltamperes supplied by the static inverter is fixed by the filter capacitors installed. This influences the types of loads to be supplied, the voltage regulation on surge loads, and the current that may be supplied on faults.

*c. Transformer-Rectifiers.* Ratings of aircraft transformer-rectifiers to supply 28-V dc power vary from 1 to 200 A. Three-phase units are customary because of the excessive ripple in the output of the single-phase type. They are built in both regulated and unregulated types. Voltage variation of an unregulated unit over the input voltage range is shown in Fig. 23-15. Such units can be operated in parallel from the same input power supply if consideration is given to the increased rectifier voltage drop with age. That is, a new unit may not divide load with an older unit within acceptable limits. Principal elements of the unregulated type are a transformer to reduce the voltage to about 35 V ac, a silicon rectifier stack, and a suitable enclosure. A fan is often used for cooling the stack, and a filter for smoothing the output ripple. The transformer usually has a Y-Δ-secondary to obtain 6-phase output for reducing ripple.

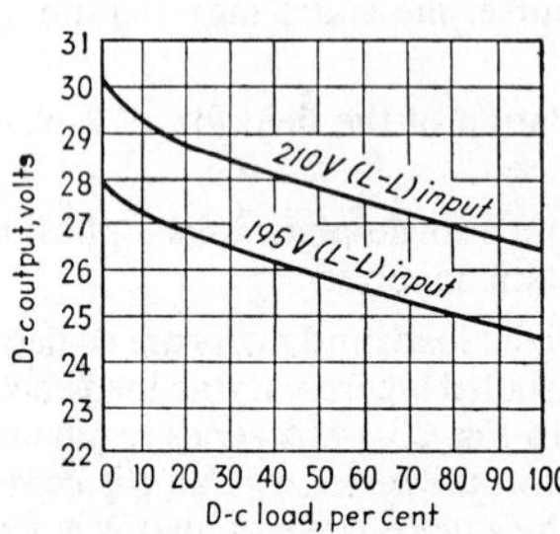


**FIG. 23-15** Unregulated transformer-rectifier voltampere characteristics. *(Data from General Electric Company.)*

While unregulated transformer-rectifiers have been used widely, there are many applications in which the loads require the narrower voltage limits obtained with a standard dc generator system. Saturable reactor and magnetic amplifier circuits have been used for this purpose. New designs are all solid-state and most use silicon controlled rectifiers (SCR) in a phase-delay regulating scheme. In this circuit a timing signal which is controlled by an error-detection amplifier turns the SCR on at the correct point in the waveform to maintain the desired output voltage. Pulse-width-modulation regulators are also used. By using the same equalizing voltage per ampere of load, it is possible to parallel and divide load with dc generators.

Improvements in rectifiers in recent years permit operation at plate temperatures up to 140°C, with life in commercial aircraft exceeding 50,000 h. With transformers, magnetic amplifiers, and saturable reactors using Class H insulation, operation in ambients of 120°C is possible. In some cases, power-factor-correction capacitors are used.

***30. Electrical Load Analysis.*** The diverse load types, characteristics, and utilization make it necessary to use some organized method of assessing the system load for power-generation-system sizing information. During the era of the dc aircraft a load analysis was performed in accordance with MIL-E-7017. This document has been canceled with no superseding document. However, load analyses are still required in order to develop sizing information. Aircraft, whether military or commercial, have a set of definable operating regimes. It is necessary to tabulate all load equipment for each operating regime with regard to its current, power, phase angle (if applicable), inrush characteristics, load balance, duty cycle, and in some cases diversity factor. From such a tabulation, both average loads and peak-load conditions can be developed. System sizing is normally based on the average load, but frequently on combat aircraft large simultaneous loads imposed by various armament systems necessitate either increased capacity or some means of establishing load priority. Of course, if the simultaneous operating period is short enough, the overload capacity of the system may be used for these peak-load periods.

Once the "total average amperes" has been obtained, it is necessary to determine the smallest generator rating which will meet these requirements. Aircraft generators reach ultimate stabilized temperature in 15 to 30 min; so these intervals usually are equivalent to continuous generator rating. However, for any given operating condition the generator continuous rating will depend upon engine speed (or drive limitations) and available cooling. The allowance for spare generating capacity may vary for different manufacturers and particular aircraft requirements, but general rules for military types are 50% spare capacity on single-engine aircraft and 100% spare capacity on multiengine aircraft. For commercial types, all essential loads must be supplied with one-half the engines inoperative. Of course, it is necessary to match the short-time load requirements as well as the continuous values with the corresponding generator ratings.

Space limitations permit brief discussion of only a few of the more important electric loads included in the load-analysis chart.

***31. Lighting.*** Requirements for some lighting affecting safety are rigidly prescribed and standardized by FAA regulations, as are position and navigation lights. Application of lighting for instruments and cockpits requires careful consideration in order to provide adequate illumination without distraction, glare, or eye fatigue. Lighting for passenger comfort and convenience is dictating the adoption of higher levels of illumination. Both incandescent and fluorescent lighting are used.

Landing lights should be so installed as to reduce to a minimum the veiling glare resulting when light in the beam is reflected back from fog, moisture, and dust particles in the air.

***32. Gyroscopic Instruments.*** Gyroscopes provide a reference axis. The gyroscope is usually a 3-phase or single-phase 400-Hz motor rated either 115 or 26 V. The rotor of the motor comprises the gyroscopic mass. Electrical gyroscopes are used in gyro-horizon instruments, drift meters (gyrostabilized), electric turn-and-bank indicators, remote-indicating earth-inductor compasses (gyrostabilized), automatic pilots, gunsights, and radar.

***33. Remote Indication.*** Self-synchronous devices provide one means for lightweight, reliable, and accurate indication from remote stations. Both dc and ac types are used in aircraft, sometimes with as many as four indications incorporated on one dial weighing less than 1 lb. Individual transmitters weigh approximately 0.3 lb. Liquid-level indication has also been provided by means of variation in the capacitance of a tank unit as the dielectric changes with the level of the fluid in the tank. The signal is amplified to operate the visual indicator. Indication of engine speed is accomplished with a permanent-magnet ac tachometer generator driving a magnetic drag indicator. Fuel flow rate is measured by a turbine-type flowmeter which generates an output similar to the tachometer, or by a hot-wire instrument. Multiplexing and digital data bases are being used in recent designs.

***34. Remote and Automatic Controls.*** Remote controls are very common and include solenoid valves, interlocks, control surface positioning, and crew and passenger comfort controls. Much of this is accomplished by servomechanisms on closed-circuit controllers. These vary from a simple on-off heater to a highly complex attitude control system. In military aircraft, gun or rocket pointing at speeds of one or more Mach numbers involves pointing the whole airframe, in many cases with complex computers on board both the aircraft and the projectile. Development of the inertial platform for nuclear submarines has led to parallel development of such systems for both aircraft and missiles, relieving the pilot of many of his duties on long-range flights. Large commercial transports today are capable of fully automatic take-off, enroute vectoring, approach, and landing in category III weather (no ceiling and 700-ft visibility), including flare, rollout, and automatic reverser and brake operation, without manual pilot operation. The advent of integrated circuits and the resulting microprocessors and special-purpose and general-purpose computers has greatly expanded the role of electronics in avionic systems. The expanded use of computers and microprocessors has had an impact on both the design and operation of most onboard electrical systems. Some of the new military aircraft have fly-by-wire systems wherein the only link between the control surface actuators and the pilot's control stick is electrical. Multiple channels are used for redundancy with voting logic to determine the correct set of inputs and outputs.

***35. Communications and Radar.*** Communication systems for modern aircraft have become integrated with position-sensing and display devices; Loran, Tacan, and various others have been developed and often serve for the dual purpose of communication and distance measuring. The complete communication system has in the past taken a large portion of the generated electric power of an airplane, but the introduction of solid-state equipment has tended to cause a decrease in total load. On the other hand, the miniaturization possible with transistors and integrated circuits has made feasible the use of more equipment so that the total load has continued to increase.

A very marked change has been brought about by the use of radar. Virtually all commercial aircraft are equipped with weather-avoidance radar and many with collision avoidance. A popular version of radar is height-measuring equipment, as well as modifications of the Doppler effect for distance measurement.

In military aircraft, an additional requirement is for jamming enemy radar and other electronic counter measures. These require large amounts of power.

## Spacecraft Electric Subsystems

***36. General.*** The whole subject of electric subsystems for spacecraft is in a continual state of change, and an authoritative statement today will be outmoded tomorrow. There are, however, some basics which can be explored, such as energy sources and possible means of conversion, typical distribution schemes, and exotic generation schemes that are being pursued.

***37. Electric Power in Spacecraft.*** In general, the usages specified in Par. **42** on electric systems in aircraft serve here, but with some modification.

**a.** *Unmanned Spacecraft*
  1. Control systems (attitude, orbital readjustment, computers, etc.).
  2. Communication (telemetry, radar, etc.).
  3. Pyrotechnics (separation squibs, retro and ullage rockets, engine starting, etc.).
  4. Pumps (fuel and oxidizer pumps; cooling, hydraulic).
  5. Precision timing devices (programmers).
  6. Navigation (gyros, inertial platforms, etc.).
  7. Camera controls.
  8. Actuators.
  9. Instrumentation.

**b.** *Manned Spacecraft*
  1. More complex communications requirements.
  2. Life-support systems (air regeneration, waste disposal or conversion).
  3. Body-temperature control. (The temperature of an unmanned spacecraft may be rigidly controlled, as many components will not operate at either low or high temperature. Thus, body-temperature control is only an extension of the normal control systems which may be installed on unmanned spacecraft.)

***38. Types of Electric-power Subsystems.*** Early spacecraft electric-power subsystems were at first extensions of aircraft practices, and this influence still lingers.

1. The most extensive use is found for the 28-V dc systems. The exact voltage used will vary considerably according to local need. The supply voltage for telemetry transducers is 5 V, and there are many instances of 56-V systems (two 28-V systems in series) in use for power. There is no real standardization; the missile designer has tended to consider each case as unique. Additionally, there are many voltage levels for the logic systems in use; usually these are internally generated in the subsystem itself.
2. The ac subsystem is a follow-on from the aircraft 200-Y/115-V system, but the general trend now is to use a dc to ac converter at the place of need rather than installing a central supply. This may change in the future when ion-propulsion systems are sufficiently developed.

***39. Operating Conditions.*** The operating conditions experienced in spacecraft are much more severe than those in aircraft, as follows:

**a.** Extreme temperature ranges, extending to very near absolute zero.
**b.** Pressure differential from atmospheric to that of space.
**c.** Possible high humidity in the spacecraft interior on the ground.
**d.** High vibration levels while being boosted into space.
**e.** High acceleration levels.
**f.** Zero-gravity (0-g) conditions in orbit.
**g.** In space, no protective atmospheric coating over materials, and oils, greases, etc., sublime off. Two bare surfaces may form a "weld" upon contact.

Any or all of the above may be operative at the same time. The spacecraft-electric-subsystem designer is required constantly to evaluate combinations of new conditions for his designs.

At this point, it should be emphasized that every watt of power used in a spacecraft must be dissipated to space as heat. That is, in a spacecraft no infinite heat sink is available such as that provided by the earth and its atmosphere. The watts loss due to inefficiency and the watts put into a device both remain aboard the spacecraft. This fact may complicate the spacecraft design immensely. The generation of large quantities of power must be associated with large radiators to radiate the generator inefficiency to space at a relatively high temperature. The generated wattage used on board must be radiated to space at a much lower temperature. Various methods to accomplish these purposes have been designed. Heat dissipation from the spacecraft is not a subject for this section. It is mentioned because it is a serious restraint upon and a primary consideration of the electrical-system design.

***40. The Total Complex.*** The spacecraft is only a part of a system and may well be only a small part. The spacecraft must be put into orbit by a booster rocket. This, in turn, must be supported by a large ground-support installation. The components of this system are shown in Fig. 23-16, giving the electrical interfaces.

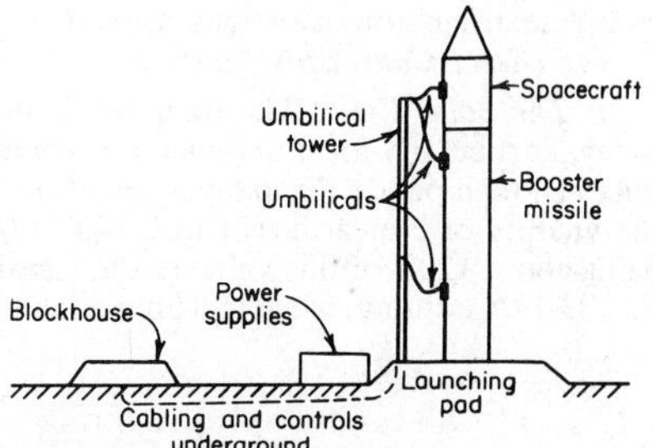


**FIG. 23-16** Booster rocket and spacecraft on launch pad. Not shown is the gantry, which surrounds the booster and spacecraft prior to iaunch, and which serves both for protection and as a work platform. It is mobile and is moved away a short while before launch.

1. The spacecraft sits on top of the booster, and there are interstage connectors between the two. Guidance and control components are commonly located in the spacecraft, and separation functions upon burnout of the booster may be controlled from there.
2. The umbilical connectors bring power to the spacecraft and booster prior to lift-off to prevent depletion of their power supplies. A number of functions are monitored through the umbilicals. These connections are severed at lift-off by flyaway disconnects, solenoid-release disconnects, lanyard-release disconnector, or explosive-charge disconnects.
3. The power supplies for both the booster and the spacecraft are located close to the pad for minimization of voltage drop.
4. All signals monitored through the umbilicals are brought to the control center. In addition, telemetry transmitters aboard the booster and spacecraft send intelligence which is monitored. The items monitored may include various critical voltages and currents, performance of the propulsion system, vibration about the spacecraft during launch, and, in manned spacecraft, the physiological condition of the astronauts. Firing controls, destruct controls, etc., are located in the control center.

Since the ground-support items are not flight hardware, there is no need for the miniaturization that exists in items installed in space vehicles. However, ground-support circuitry becomes quite sophisticated in most cases, since automated check-out of the vehicle systems is incorporated.

***41. Miniaturization and Weight Saving.*** Every ounce of weight in the spacecraft must be accelerated to orbital or earth-escape velocity, and this requires the expenditure of propellant. Thus, there has been an increasing push for weight savings. A 5-A switch in aerospace would probably be rated at less than 1 A in commercial use. At the same time, such equipment must operate under environmental conditions that a commercial component will never experience. The result has been an increased cost, but much more reliable components. The limiting criterion in almost all cases is the generation of heat and its dissipation. A parallel development with miniaturization has been the reduction of wattage dissipation. This has been quite successful. Relays have more efficient magnetic circuits; contacts are specially designed to be low-resistance; transistors have replaced vacuum tubes; and at present writing, integrated circuitry has replaced discrete components. The state of the art is changing so rapidly that it is difficult to predict the trend of the future except to say that extreme miniaturization will continue, solid-state logic will replace relaying, and all functions possible will be made automatic, but the automatic systems will incorporate manual backup where manpower is available.

***42. The Spacecraft Electric System.*** In common with the aircraft electric system, the spacecraft electric system consists of three well-defined parts: (1) generation (or power supply), (2) distribution, and (3) utilization.

***43. Spacecraft Electric-Power-Supply Methods.*** The spacecraft must carry its total power supply with it; refueling stations do not exist. However, orbiting refueling or resupply stations are being proposed and may be used in the near future. Power-supply methods are generally divided into two classifications: (1) direct conversion, for example, converting the photon energy of the sun directly into electric energy, and (2) indirect conversion, converting nuclear fission or chemical energy to heat and then using a heat-conversion apparatus for conversion to electric energy. Greater efficiencies are usually attainable with the first class, since Carnot efficiencies are not involved, but at this writing, larger amounts of power are available by the second. Each class will be treated separately. Refer to Sec. **11** for a more complete discussion of various conversion techniques.

***44. Direct-Conversion Methods***

*1. The Solar Cell.* The solar cell is the device most often used in energy conversion today, particularly for unmanned spacecraft. It is reliable and in the proper quantities produces enough power for many applications. It operates on the solar photon flux, which, in the vicinity of the earth, is about 1400 W/m$^2$. This value increases as one nears the sun, being about 120% of this value in the vicinity of Venus and 80% in the vicinity of Mars. In Fig. 23-17*a*, a thin piece of silicon is mounted on a lightweight backing system. This silicon

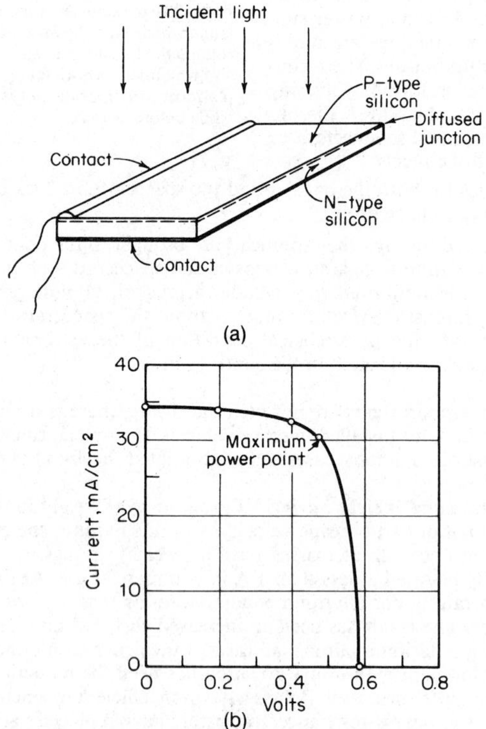


**FIG. 23-17** (*a*) Solar-cell construction; (*b*) characteristic curve at 1400 W/m$^2$.

wafer is made with an *n-p* junction near its surface. This may be either *n* on *p* or *p* on *n*; *p* on *n* is more radiation-resistant. The *n* layer contains an excess of electrons, the *p* layer an excess of holes; at the junction thermal diffusion has caused the electrons to migrate to the *p* layer and the holes to migrate to the *n* layer, leaving a thin layer of hole and electron-free silicon. A photon with the requisite energy, that is, of a wavelength less than 12,000 Å, will be absorbed by one of the silicon atoms in the equilibrium area, creating an electron-hole pair. The field at the junction will carry the electron into the *n* layer and the hole into the *p* layer, causing a voltage to form between the *n* and *p* layers.

Other types of solar cells are being developed, such as thin-film or dendritic, and semiconductors such as gallium arsenide which look promising. Present efficiencies may be surpassed by such developments.

Only a portion of the solar photons are of the energy necessary to create an electron-hole pair, and some of the pairs recombine before they can migrate to the proper side of the junction to be effective. The individual solar cell is small, usually 2 × 2 cm or 8 × 8 cm, and thus cannot supply much power. It is necessary to join thousands of the individual cells in series-parallel array or combination to obtain the voltage and current necessary to supply spacecraft loads. A typical voltampere curve of a solar cell is given in Fig. 23-17*b*.

Owing to the vary large number of cells required for supplying power the solar cell is best suited to loads of under 10 kW. The Skylab manned spacecraft mission used two large arrays. The main array was rated for 10.5 kW at the end of life. Larger systems are now feasible. Overall cell efficiency approaches 12%, with a practical limit of 16%. The cell is readily damaged by space radiation, and very thin covers are supplied for protection against such damage.

Maximum efficiency is not obtained unless the cell face is normal to the sun's rays, and various cell configurations are employed to accomplish this end, such as large paddles with the solar cells upon them and means for orienting the vehicle for maximum effect. The greatest use, however, is in vehicles which are spinning at a low rate. The outside of the vehicle is covered with cells so that at least a portion is exposed at all times. Such an arrangement gives an efficiency of 36% of the maximum possible but does have the virtue of not having to be oriented. Solar concentrators are used, but these require proper orientation to attain their potentialities.

*2. The Fuel Cell.* Fuel cells are devices by which fuels and oxidants are chemically combined to produce electrical energy. In this respect, the fuel cell is the exact opposite of the ordinary electrochemical reaction, where a current is passed through a chemical solution and hydrogen and oxygen are liberated. There are three general areas of fuel-cell development:

**a.** Hydrogen fuel cells.

**b.** Hydrocarbon fuel cells.

**c.** Biochemical fuel cells.

The hydrocarbon fuel cell is an attempt to use some hydrocarbon, such as hydrazine, directly as the fuel element, and it has attained some success. The biochemical fuel cell uses the fundamental electrochemistry of enzymes and bacteria in a life situation. Possibilities here include the conversion of waste products to useful products and the simultaneous generation of electricity. However, the hydrogen-oxygen fuel cell has seen the greatest development and is at present in use.

The hydrogen-oxygen fuel cell depends for its operation upon the ionization of gaseous hydrogen and oxygen at the cathode, with electron flow as a consequence. The chemical formulas underlying one such reaction are

Anode: $$2H_2 \rightarrow 4H^+ + 4e^-$$

Cathode: $$4e^- + 4H^+ + O_2 \rightarrow 2H_2O$$

The electrolyte here is acid, and the water is produced at the cathode. In the so-called Bacon cell, the electrolyte is a base, and the water is formed at the anode. It is mechanization of this process which has required development. The electrodes must be nonconsumable, and there must be a means of ionizing the hydrogen. The ion-exchange membrane is one such development, but other catalysts are also used.

The theoretical efficiency of the fuel cell is high compared with heat engines, running from 80 to 90%. This efficiency is not obtained in practice, since there is some heat of reaction generated, and there are circuit losses; around 55% is claimed. In addition, there is the voltage-regulation problem; the cell must be supplied with gaseous hydrogen and oxygen as a function of load, and the water formed must be carried away. There are various solutions to this problem, depending upon the particular development.

There have been several difficulties associated with the development of the fuel cell. These are listed below:

**a.** The pure bulk of the fuel and oxidizer in the gaseous form shows that it must be stored in the liquid form, and this, of course, at cryogenic temperatures. Further, there must be means of bringing the fuel and oxidizer to gaseous form before use. This storage and means for usage consume considerable space and weight and account for some of the complexity that accompanies a fuel-cell installation.

**b.** The fuel and oxidizer must be supplied in accordance with the load on the cell. Since hydrogen and oxygen are stored at cryogenic temperatures, there must be a means for conversion to usable pressures and volumes. To supply the necessary amperes capacity, it may be necessary to parallel several fuel-cell series banks.

The advantages of the fuel cell are so great that their development will undoubtedly be very rapid, and the advances in spacecraft power supplies will be reflected in the cells developed for industrial use. Here, fuels will be drawn from the hydrocarbons, while the oxidizer will be air. The utilization means are more complex and the efficiencies lower. The block diagram of a fuel-cell system is shown in Fig. 23-18.

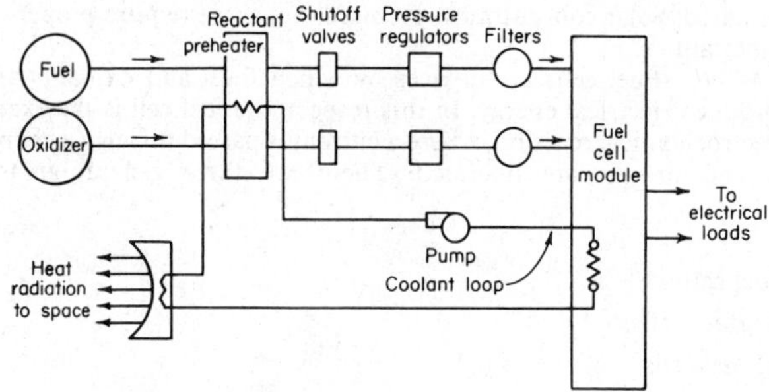


**FIG. 23-18** Fuel-cell system.

*3. The Magnetohydrodynamic (MHD) Generator.* The conventional dc generator consists of a conductor forced through a magnetic field. In the MHD generator, the conductor becomes a plasma, forced through a magnetic field. The plasma is contacted by conductors and the power drawn off. A simplified diagram of an MHD generator is shown in Fig. 23-19.

The conductive plasma flow, the electron flow, and the field are at right angles to one another; the voltage and power density per unit of volume flow are related thus:

$$E = (U \times B)n$$

$$PD = \sigma U^2 B^2 n(1 - n)$$

where $U$ is the plasma velocity, $B$ is the field strength, $n$ is the ratio of the load resistance to the sum of the internal and load resistances, and $\sigma$ is the conductivity.

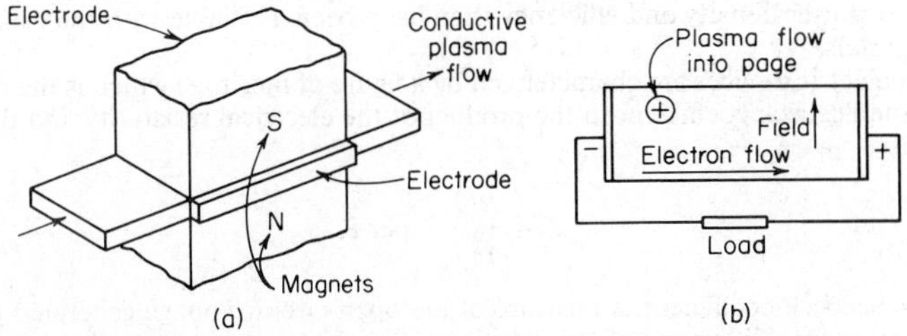


**FIG. 23-19** Schematic diagram of magnetohydrodynamic generator.

Several major problems have appeared which have been only partly solved:

**a.** Conductivity of the plasma. Unless the plasma is at an elevated temperature, it may not be ionized and conductive. Further, at such elevated temperatures, the plasma is likely to erode the magnet faces and the electrodes. Methods have been employed to make the plasma conductive, such as by seeding with readily dissociated molecules.

**b.** High field strength is exceptionally difficult, for to obtain a large amount of power, the "channel" must also be large. A solution here is the development of superconductive materials. In outer space, with cryogenic temperatures, it will be possible to obtain a field strength of 100 kG (10 tons) over a large area.

**c.** The high temperatures involved have made the development of new materials necessary. Research is currently advancing the state of the art here.

The MHD generator would seem to be well suited to the generation of electricity where a continuous plasma source is available, as in a rocket engine, and large amounts of power are to be extracted. The reverse process, where power is supplied and motor action results, is in use for pumping liquid metals.

*4. Atomic Fission.* When there is fission of a $^{235}$U atom, the result is two or more highly charged decay products, neutrons, and several beta particles. Fissionable material has been used to make an "atomic battery" by capture of the beta particles on an auxiliary electrode, insulated from that carrying the fissionable material. Over a period of time, a potential difference will build up as a result of beta-particle capture. The voltage may become high, but the internal impedance of the battery is very high, and no appreciable power output is realized.

Since the fission fragments are highly charged, some attempts are being made to isolate the charge particles and thus have direct conversion to electrical power.

***45. Indirect-Conversion Methods.*** All the methods to be described utilize the conversion of heat to electricity, the heat being obtained in a number of ways (see also Sec. **11**). These include

**a.** Solar energy.

**b.** Nuclear fission.
   **1.** Nuclear reactors.
   **2.** Radioisotopes.

The methods of obtaining heat are not a part of this write-up. It will suffice to say that a variety of types of solar-ray concentrators are used to obtain the necessary high temperatures, and the reactor power-conversion schemes are too numerous to cover in a short paragraph. There are many types of radioisotopes with different lengths of lives and different heating-generating capabilities. Which one is used will be governed by the individual requirements.

*1. Thermoelectricity.* This generation device has been in use for many years, and the chief contribution of aerospace technology has been the materials improvement, giving a much higher power density and efficiency than have been available with the usual thermocouple materials.

Thermoelectric devices are characterized by a figure of merit, $Z$, which is the ratio of the square of the Seebeck coefficient to the product of the electrical resistivity and the thermal conductivity.

$$Z = \frac{S^2}{\gamma k} \qquad \text{per K}$$

where $S$ = Seebeck coefficient, a measure of the open-circuit voltage generated per degree Celsius temperature difference of the junction, $\gamma$ = electrical conductivity, mhos per centimeter, and $k$ = thermoelectric conductivity, watt-centimeters per square centimeter per degree Celsius.

This formula says that the Seebeck coefficient should be high and the product of the electrical and thermoelectric conductivities should be low. In an electric generator, the conductivity should be high to prevent ohmic losses, so that thermoelectric conductivity must be very low. These requirements eliminate metals and insulators and leave only the semiconductors, and it is in this area that the greatest development has occurred.

Two materials in particular have been developed which are useful as thermoelectric generators, lead telluride (PbTe) and germanium silicon (GeSi). The peak Carnot efficiency with these materials is around 30%, and the peak temperature of the hot junction is 800°C, but the realizable efficiency is no more than 3 to 4% owing to heat conduction, contact resistance, etc.

Thermoelectric-development problems have centered on (1) making successful contact between the materials, (2) the fragility of the material, (3) and heat flow across the materials. Successful thermoelectric generators have been constructed, but in general they are less efficient than the solar cell, and since high temperatures are required, some sort of focusing collector is required for sunlight conversion.

*2. The Thermionic Generator.* Another device requiring high temperatures for operation is the thermionic generator shown schematically in Fig. 23-20. Cesium is extruded through the pores of a coarse cathode material and, having a low work function, is ionized

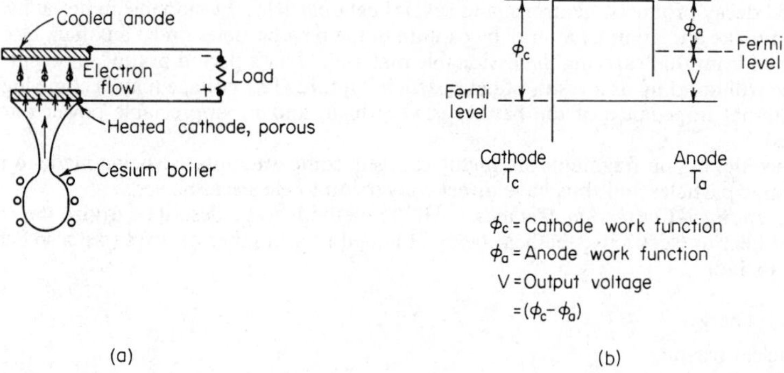


**FIG. 23-20** Thermionic generator: schematic (*a*); work function (*b*).

at the surface. Both cesium positive ions and electrons form about the cathodes, and dependent upon the elimination of the space charge provided by the cesium positive ions, the electrons travel to the anode. If a load is connected between cathode and anode, current will flow through it. The power generated is proportional to the difference of the Fermi levels at the cathode and anode. With cathode temperatures of 2200 K, a cesium thermionic-converter cell can provide nearly 1 V with 5 to 8 W/cm$^2$ of cathode surface and at an efficiency of over 15%. Theoretical efficiencies are above 40% but have not been realized in practice.

No thermionic converters have been installed in spacecraft at this writing. It appears probable that their most successful use will be when nuclear reactors are successfully adopted for space use. The thermionic converter is, of course, limited by Carnot efficiency. Its virtue is that the operating temperature is higher than that of other heat devices, and so the Carnot efficiency is higher.

*3. Rotary Devices.* Rotary devices are subject to the same ills that plague earth-bound motors and generators, bearings, seals, and the like. In addition, a rotating drive acts as a gyroscope and may inhibit maneuvering of a spacecraft. Counter-rotating designs could be employed to negate some of the gyroscopic effects. Nevertheless, the rotary device is proposed for use in cases where the loads are very high, and the state of the art in other conversion devices cannot supply the demand. Generally, the heat source is a nuclear reactor, and the problem is to transfer the heat to a generator located some distance from the reactor, so as not to be affected by nuclear radiation. In most cases, two series heat-exchange loops are used. Reference is made to the literature (Par. **50**) for further information.

Two general types of heat engines are used:

**a.** The turbine.

**b.** The piston engine—Brayton cycle, Stirling cycle, and Rankine cycle.

*4. The Electrostatic Generator.* Proposed, but not built, this generator holds promise for a high-power generator in space. It would operate in the space environment, which is a far better vacuum than can be produced on earth and which is necessary for the production of very high voltages at very low currents. This technology is an extension of the old "electrical-influence" machines used in early electrical experimentation. A well-designed machine here has at least as high an efficiency as the electromagnetic machine.

*5. Radioisotopes.* The use of radioisotopes is for heat sources only and is discussed only because their characteristics affect the design of the conversion means. The salient features of radioisotopes are

1. Their half-life, that is, the period in which they will decay to one-half their original potency. If a generator is to deliver 10 W for 0.38 year with the radioisotope polonium 210, the initial design must supply at least 20 W, for 0.38 year is the half-life of polonium 210 (see Table 23-7 for the characteristics of some representative radioisotopes).

**TABLE 23-7** Characteristics of Radioisotopes

| Radioisotope | Major radiation | Half-life, years | Power density W/cm³ |
|---|---|---|---|
| $Cs^{137}$ | $\alpha$, $\beta$, and $\gamma$ | 30 | 0.21 |
| $Sr^{90}$ | $\beta$ and $\gamma$ | 28 | 1.0 |
| $Ce^{144}$ | $\beta$ and $\gamma$ | 0.78 | 24.5 |
| $Po^{210}$ | $\alpha$ | 0.38 | 1210.0 |
| $Pu^{238}$ | $\alpha$ | 89 | 3.9 |

2. Their radiation characteristics, which influence the shielding requirements, since much spacecraft electronic equipment will not withstand nuclear radiation.
3. Safety, first, on handling prior to launch; and, second, the degree of burnup and atmospheric contamination to be expected upon reentry.
4. Reactions with conversion apparatus. Since good thermoelectric devices use semiconductors and semiconductors tend to have their lattice structure destroyed by some types of radiation, the radioisotope and the semiconductor must be compatible.

***46. Primary and Secondary Batteries.*** Batteries have been the primary source of power in all the early satellites and are the usual source for the spacecraft booster rocket. In the case of the booster rocket, the requirement is only for sufficient capacity to supply the loads until it burns out, and for this source a class of small, dense, high-energy batteries has been developed. Satellite or spacecraft service requires a battery which may be charged by solar cells or thermoelectric or thermionic devices to replace the normal usage of the satellite or spacecraft. The primary battery is one which is not rechargeable and, in aerospace use, is activated by addition of the electrolyte, either manually or automatically. The secondary battery is rechargeable; that is, after discharge, the electrochemical process may be reversed by application of a voltage.

The lead-acid type of secondary battery once was widely used in aircraft. This battery is not applicable in satellites or spacecraft, for a number of reasons:

**a.** Very corrosive electrolyte.

**b.** Cannot be hermetically sealed.

**c.** Weight.

**d.** Poor voltage regulation.

Both the primary and the secondary battery are employed in aerospace applications. The primary battery is particularly useful in booster-rocket applications, where the service is for a single shot; recharging and long life are not important. One variety of the primary battery is the quick-energized battery. The battery is charged and dry until a special mechanism forces the electrolyte into the plates. Such a mechanism may be a lanyard or an electrical signal activating an explosive device that shatters a diaphragm and releases gas to force the electrolyte into the proper area.

A feature of almost all the high-energy batteries is "peroxide voltage." That is, the battery when charged will have an open-circuit voltage somewhat higher than normal owing to the formation of a peroxide rather than an oxide. This voltage disappears after a very short interval of use and is only a small percentage of the battery capacity. A 28-V battery may read around 35 V on open circuit; the current transient occurring when the load is first connected may be damaging to some electronic equipment. Special precautions are often taken to get rid of this peroxide voltage by a small preload applied prior to the application of the actual load. This is necessary since much aerospace equipment is designed to operate within narrow limits and might well be damaged.

Special charging means must be provided where secondary batteries are used, as the special spacecraft batteries are much less tolerant of incorrect methods of charging than are the familiar lead-acid batteries.

Table 23-8 provides a comparison of the energy densities to be expected from the most common primary and secondary batteries. This is by no means complete; as various manufacturers claim special advantages for their particular batteries, only a reasonable estimate is given.

**TABLE 23-8** Power Densities

| Battery type | Power density, Wh/lb | |
|---|---|---|
| | Primary | Secondary |
| Zinc-silver oxide (An/KOH/AgO) | 80 | 40 |
| Silver-cadmium (AgO/KOH/Cd) | 33 | 30 |
| Nickel-cadmium (NiOOH/KOH/Cd) | 17 | 10 |

NOTE: 1 lb = 0.4536 kg.

Where very long life is expected, the two batteries showing the poorest density, the silver-cadmium and the nickel-cadmium, are the only ones which have been able to operate in space for a year or more and then only when seriously derated. In any event, every battery installation in a rocket or spacecraft is so special that very little standardization has occurred. It is not possible to present tables of dimensions, weights, etc., as with the airplane lead-acid battery. Among the items that make such a table impossible for spacecraft are

**1.** Quick-energized batteries are heavier and bulkier by reason of their actuation mechanisms.

2. The differences in life requirements in various missions change the battery size. A battery designed for low drain and charging rates will be very different from one that is to be quickly expended.
3. Good voltage regulation requires a heavier battery.
4. Sealed batteries designed for a long period of usage may have a stainless-steel noncorrosive case; a lighter battery without such requirements may have a magnesium case.

A very special type of battery is the thermal battery. Activation is by thermal means. When the internal temperature rises sufficiently, the chemical ingredients liquefy and the battery is usable. Life of the battery is very short, often being in seconds; longer-life units capable of operating for several minutes are now available. These batteries are used when large amounts of energy are needed for very short periods.

***47. Special Electric Utilization Equipment***

*1. Missile and spacecraft environmental and functional requirements* have created some entirely new equipment which will be briefly described here. No treatment will be given to special types of usual equipment such as radars, etc.

*2. Brushless DC Motors.* Altitude has always been a problem in dc generators and motors because of the commutator problem. This is accentuated in vehicles that operate above the atmosphere permanently, and the brushless dc motor has been developed to eliminate it. The ac motor is not directly applicable, since the majority of the power-generating devices in spacecraft are dc and conversion to ac power involves another inefficiency, that of the inverter. Two general types of brushless motors are in use, the inverter-induction motor and a dc motor with an electronic commutator.

**a.** The inverter-induction motor uses an inverter which is specially designed to work with a specific induction motor and which uses the motor windings as the usual filter. The operation is square-wave, and the combined efficiencies of the inverter and induction motor are at least as high as for a dc motor alone. The best of such combinations operate 3-phase, but 2-phase is also successful. In both cases, either the motor must be designed for low starting current or the inverter must be designed to saturate so that starting current is limited; otherwise the transistors or silicon-controlled rectifiers in the inverter will be overloaded.

**b.** The dc motor employs some position-sensing device to synchronize the electronic commutator with the armature of the motor. This device controls the operation of the transistors feeding the armature, assuring that the magnitude and phase of the current supplied are correct.

*3. The Electroexplosive Device (EED).* The EED is simply a resistance wire embedded in an explosive substance; it is used to initiate various sequences in the spacecraft, such as stage separation, automatically activated batteries, or the ignition of the main engines of a booster or sustainer rocket. Modern ordnance devices are all rated at 1 A, 1 W "no-fire" to prevent ignition from extraneous rf signals. Typical firing currents for these devices are on the order of 5A, so when several devices must be fired simultaneously large current pulses are imposed on the power source. Besides the rf threat, they are also sensitive to electrostatic discharge in both the pin-to-pin and pin-to-case mode. Recent designs now being used are tested to human body levels of electrostatic discharge in the pin-to-case mode. Typically this is a 500-pF capacitor charged to 25 kV, then discharged through a 5-kΩ resistor.

Another ordnance system used is the exploding-bridge wire (EBW). Here, a capacitor is charged to approximately 2 kV, then discharged through a fine wire. The wire fuses explosively, developing a high-energy pulse which detonates an explosive mixture.

*4. The Ion Engine.* After the spacecraft is free from the earth and is in space, very little acceleration is required to bring it to extremely high velocities. Small accelerations are not obtained efficiently with normal thrust-generating devices, and so the ion engine has been proposed for this usage. This engine accelerates relatively heavy ions to very high velocities and directs them so as to accelerate the spacecraft.

Figure 23-21 gives a schematic diagram of a simple ion engine. Cesium is ionized at the cathode, and the positive ions are accelerated toward a negatively charged grid. The grid being open-meshed, most of the ions pass through and, leaving the spacecraft, act to accel-

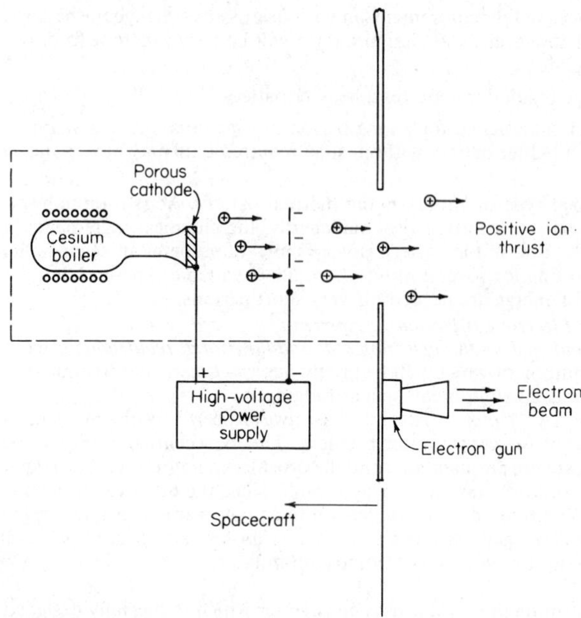


**FIG. 23-21** Ion-propulsion engine for spacecraft.

erate it. There are other grids in the engine designed to control the exit velocity of the ions. The ions, being positive, leave the vehicle with a net negative charge. This could build up to such proportions that no positive ions would escape and thus no thrust would be provided. The electron gun shown is intended to keep the spacecraft at or near zero potential.

*5. Guidance Systems.* Early missile guidance systems used an attitude reference unit (gyro package) to provide vehicle attitude information. Guidance then was effected through a ground-based radar and computer system which tracked the vehicle, compared the trajectory with the required trajectory, and issued correction commands as necessary. Intercontinental missiles were required to fly over the horizon, so on-board guidance systems were developed using an attitude reference system coupled to a complex analog-computer system.

Modern guidance systems now employ digital computers for this purpose which have increased the system capability, while significantly reducing power consumption and weight. These trends will continue as increased circuit integration is employed. Several missile and space booster systems now use a digital-computer guidance system.

Small tactical weapons employ a variety of guidance schemes. These range from the previously discussed command guidance through a multitude of seeker systems. Seeker systems use infrared, laser, video, radar, etc.

***48. Distribution and Control Systems***

*1. The booster* or ballistic-missile electric distribution and control system is quite simple, generally, being only a battery or batteries feeding loads, no battery charging or regulation systems being needed. The life of such systems is measured in minutes, at the most, and the emphasis is on reliability for the severe environment to be encountered.

When the spacecraft, manned or unmanned, is considered, the problem is compounded by the need for voltage regulation. A schematic of a power distribution system for a typical unmanned earth satellite is shown in Figure 23-22, and the salient features will be discussed briefly:

**a.** The power source is generalized and may be any of the devices discussed. Two are shown for reliability; blocking rectifiers prevent reverse current flow.

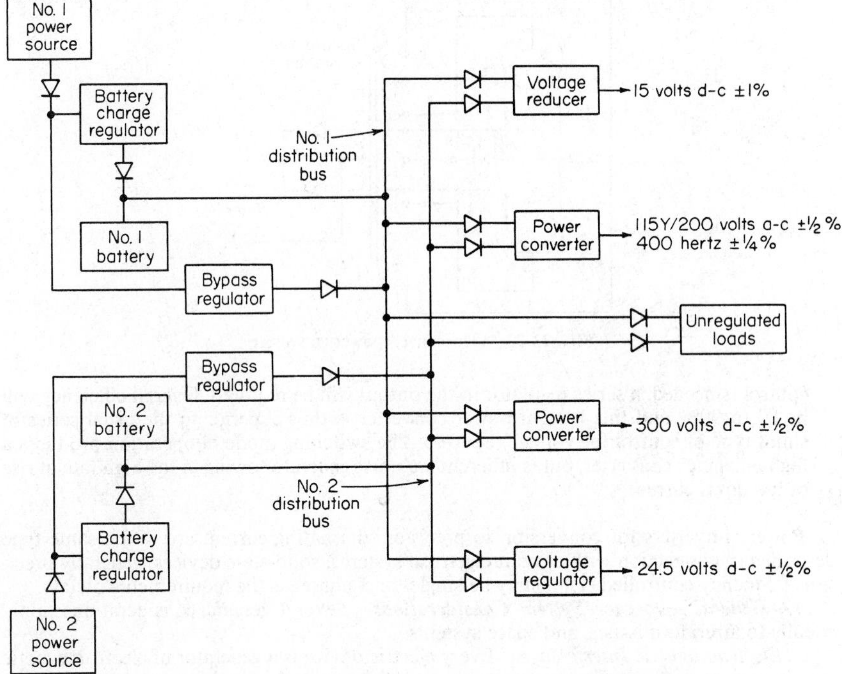


**FIG. 23-22** Distribution system for spacecraft.

**b.** Each power source charges a battery through a regulator, since spacecraft batteries require careful charging to prevent damage. Again, a blocking rectifier is used to permit battery charging but prevent the battery from feeding back into the regulator.

**c.** The bypass regulator is used to divert the power source current directly to the load bus to avoid charging rates in excess of those which the battery charge regulator can handle. The unregulated bus is kept at a maximum of 32 V dc; nominal is 28 V dc. Newer systems containing large solar array sources are being designed at 120 V dc on the primary bus.

**d.** All loads may be fed from either No. 1 or No. 2 distribution bus. Blocking rectifiers are used to decouple the distribution buses from one another.

*2. Conversion Devices.* Conversion devices used in missiles and spacecraft are generally similar to those used elsewhere but differ in some respects. The differences are noted briefly:

**a.** The voltage regulator may be complex, depending upon the regulation requirements. They are invariably solid-state.

**b.** The voltage reducer may be some form of lossy device or may be a dc to dc converter. This will be treated in more detail in the next paragraph.

**c.** The generalized power converter, or dc to dc converter, is shown in the schematic of Fig. 23-23. A dc supply powers an oscillator, the output of which is a square wave. The duty on the transistors is very light, since they are always operating in the switching mode. The voltage output is dependent upon the transformer ratio, and since the output is a square wave at a relatively high frequency, rectification and filtering are not difficult. Semiconductors and magnetic alloys are now available which allow operation up to 100 kHz at power levels from a few watts to several kilowatts. Where voltage regulation is required, a control of the base drive of the transistors will suffice in most cases. If more

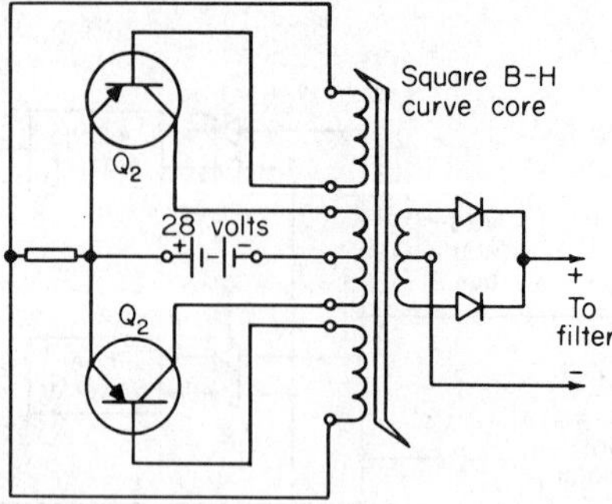


FIG. 23-23 Generalized power converter.

control is needed, a series regulator in the output will be required. Overall efficiency will be 80 to 90%, and this type of power converter is thus superior to the usual series of shunt type of control for voltage reducers. The switching mode of operation produces a high-efficiency converter, but is inherently a noise generator owing to the high rate of rise of the input current.

Power converters for conversion to precision alternating current are of the same type described in connection with aircraft electrical systems, solid-state devices, generally precision-frequency-controlled. They may be single- or 3-phase, as the requirements dictate.

***49. General Aerospace System Considerations.*** Several general considerations apply equally to aircraft, missiles, and space systems.

*1. Electromagnetic Interference.* Every electric device is a generator of electromagnetic interference at some amplitude and over some frequency spectrum. Correspondingly it is also susceptible to some amplitude over some frequency spectrum. Technical literature is full of detailed information on this subject. System and assembly design requirements are specified in a number of military specifications, standards, and handbooks as noted; all are good references.

MIL-E-6051D (July 1968).

MIL-STD-461B (April 1980), 462 (April 1980), 463A (June 1977).

MIL-STD-1541 (October 1973).

MIL-Handbook-DH1-4.

One of the important issues covered is with regard to power- and control-system grounding. Generally speaking, aircraft use airframe return wherein both the ac neutral and dc return are connected to structure. The dc return current and ac unbalanced load neutral current flow in the airframe. Missiles and space systems on the other hand have tended to use a single-point-ground concept wherein no intentional current flows in the airframe. This approach is in the interest of interference control and is now specified in some of the references. When properly designed, both systems will work, and in some cases, a mixture of the two has been successfully used. In either case it is important to avoid common impedances in either the airframe, the ground-attach points, or power-system conductors which can create interference among systems and subsystems.

*2. Bonding.* Airframe bonding is accomplished in order to create an equipotential ground plane out of the structure. This is important for providing the current path in an airframe power return system, to afford the maximum degree of shielding against external electromagnetic interference sources, and to provide a low-impedance path for lightning-stroke current. MIL-B-5087B defines airframe and aerospace ground-support facilities bonding requirements.

Modern airframes use many different high-strength aluminum alloys which are susceptible to corrosion. It has become standard practice to anodize such parts as a corrosion control measure. Where such structural elements are assembled with a number of driven rivets, a good bond results. However, where such elements are bolted together, high-resistance joints can result. MIL-C-5541, Class 3 finish, is a conductive conversion coat for aluminum and should be used when bonding such structures together. The conversion coat has poorer wear properties and corrosion control but offers sufficient protection in most cases. Corrosion can be further controlled by sealing the joint to exclude moisture.

*3. Electrostatics.* Static electrification is a natural phenomenon occurring in many places. Aircraft and missile and space boosters also have many sources of electrification such as engine plume charging, dust-particle and ice-crystal impingement, and vibration. Bonding as delineated above effectively controls the charge distribution on the stage, and static precipitators can be used to bleed it off to the atmosphere. However, modern systems use nonmetallics for both structural application and thermal insulation. Most of these are excellent dielectrics, creating isolated bodies which can accumulate a differential charge with respect to structure. Subsequent arc-over can occur which is a source of electromagnetic interference. These discharges have been known to upset digital computers used for guidance and to detonate electro-explosive devices. Methods must be incorporated for bleeding off any charge accumulation on these materials. Some of the uses of nonmetallics are fluid transfer lines, structural composites for solid rocket motor cases and aircraft control surfaces, silicone and epoxy insulations on payload pairings, insulating blankets comprised of many layers of aluminum foil and Mylar, Teflon, kapton, etc. All these can accumulate high levels of electrostatic charge.

Another source of electrostatics that has recently been identified is *space plasma charging.* It has been found that spacecraft in geosynchronous orbit can be differentially charged by substorm plasma. Here again, proper bonding procedures and controls on nonmetallic materials will preclude the problem. Materials with volume resistivities lower than $10^9$ $\Omega \cdot$cm are acceptable. Materials with higher volume resistivities must have some design feature incorporated to bleed off the charge. Woven-in conductive stranding, sprayed-on conductive coatings, and metallized surfaces which can be bonded have all been successfully used for this purpose.

*4. Corona.* The dielectric behavior of gases (corona) is discussed in Sec. **4,** which also provides an excellent bibliography of reference material. Dielectric behavior is governed by Paschen's law, which states that in a uniform field, the critical voltage at which breakdown will occur is a function of the absolute pressure $p$ and the spacing $d$. For air, the minimum voltage is approximately 340 V and occurs at a $pd$ product of approximately 0.6 mmHg/mm spacing.

Where missiles or spacecraft contain voltages near or above this voltage and must fly through the minimum $pd$ product region, special design precautions are required. The related altitudes for typical aerospace hardware are 70,000 to 250,000 ft. Any electrical system containing voltages close to the Paschen's law minimum which must operate through the critical pressure (altitude) region requires special design considerations:

**a.** Voltage levels involved.

**b.** The absolute pressure profile (combined flight profile plus the equipment bleed-down profile).

**c.** Ionizing radiation exposure.

**d.** Localized contaminants present.

**e.** Hardware design configuration.

A complete treatment of each of these is impractical. However, the Marshall Space Flight Center for NASA has published a design guide on this subject specifically for spacecraft applications. This document is MSFC-Std-531, titled *High-Voltage Design Criteria,* dated September 1978. It is a comprehensive document covering designs for four different voltage classifications as follows:

**a.** 0 to 50 V

**b.** 50 to 250 V

**c.** 250 to 5000 V

**d.** Above 5000 V

Designing according to these criteria will yield a trouble-free design.

***50. Bibliography.*** The flow of new material for both aircraft and spacecraft that is pertinent to this handbook is too extensive for inclusion. A brief bibliography of some of the key references has been included. Users are referred to the transactions of various society and industry meetings and workshops for current and more detailed information on the various subjects discussed.

*1. Aircraft*

Anderson, H. C., Crary, S. B., and Schultz, N. R.: Present D-C Aircraft Electric Supply Systems; *Trans. AIEE,* 1944, vol. 63, pp. 265–272.

Carlson, K. W., and Sherrard, G. S.: Distribution System Reliability of 28 Volt D-C Aircraft Electronic Systems; *AIEE Paper* 52-132; *Trans. AIEE,* vol. 71, pt. II, pp. 113–117.

Finison, H. J., and Kaufmann, R. H.: "DC Power Systems for Aircraft"; New York, John Wiley & Sons, Inc., 1952.

Caldwell, S. C., and Wood, A. J.: The Effects of Abnormal Conditions on Aircraft, Parallel A-C Power Systems; *Trans. AIEE,* 1953, vol. 72, pt. II, pp. 379–384.

Andrew, J. D.: Aircraft Power Distribution Circuitry, Experimental Determination of 400 Cycle Impedance of Wire On; *AIEE Paper* 54-374, *Trans. AIEE,* vol. 73, pt. II, pp. 469–478.

Kahle, H. A., and McConnell, H. M.: Analogue Computer Methods Applied to Steady-State A-C System Problems; *AIEE Paper* 56-766, *Trans. AIEE,* vol. 75, pt. II, pp. 279–282.

Smith, R. E.: A-C Generator: A Brushless Air-Cooled Aircraft; *AIEE Paper* 57-468; *Trans. AIEE,* vol. 76, pt. II, pp. 189–191.

Riaz: Energy Conversion Properties of Induction Machines on Variable Speed Constant-Frequency Generating Systems; *AIEE Paper* 58-917, *Trans. AIEE,* vol. 78, pt. II, pp. 25–30.

Hamer, W. J.: Aircraft Storage Batteries, *AIEE Paper* 60-849, *Trans. AIEE,* vol. 79, pt. II, pp. 277–287.

Gayek, H. W.: Transfer Characteristics of Brushless Aircraft Generator Systems, *IEEE Trans. Aerosp.,* April 1964, vol. AS-2, No. 2, pp. 913–928.

Plette, D. L., and Carlson, H. G.: Performance of a Variable Speed Constant Frequency Electrical System; *IEEE Trans. Aerosp.,* April 1964, vol. AS-2, pp. 957–970.

IEEE: Aircraft electrical systems; IEEE Std 128-1976 (formerly AIEE 750-1960, IEEE Std 128-1960 and IEEE Std 131-1960).

General Electric Co.: 150 KVA samarium colbalt VSCF starter/generator electrical system. Final technical report AFAPL-TR-78-104, December 1978.

Finke, R. C., and Sundberg, G. R.: Advanced electrical power system technology for the all electric aircraft; NASA technical memorandum 83390, May 17, 1983.

*2. Spacecraft*

Abrahamson, L. T.: Silver-Zinc Batteries or Source-Primary Electric Power for Pilotless Aircraft; *AIEE Paper* 54-491, *Trans. AIEE,* vol. 76, pt. II, pp. 297–300.

Schuh, N. F., and Tallent, R. J.: Solar Powered Thermoelectric General Design Considerations; *AIEE Paper* 59-847, *Trans. AIEE,* vol. 78, pt. II, pp. 345–352.

Beckman, C., Bedrosian, S. D., Bekowitz, R. S., and Chen, T. C.: Research Guidelines for Digital Computer—Controlled Launch Control and Checkout of Operational Satellite System, *IEEE Trans. Aeros.,* August 1963, vol. AS-1, no. 2, pp. 1065, 1073.

Levin, H., and Asam, A. R.: MHD Generator Relationships; *J. AIAA,* Feb. 13, 1963.

Klavan, L. S., and Yocheley, S. B.: Computer Controlled Launch Control and Checkout of Operational Satellite System; *IEEE Trans. Aerosp.,* August 1963, vol. AS-1, no. 2, pp. 1249–1261.

Szego, G. C., and Cohn, G. M.: Fuel Cells for Aerospace Applications; *Astronaut., and Aeronaut.,* May 1963.

Kahn, B., and Gourdine, M. C.: Electrodynamic Power Generation; *J. AIAA,* August 1964.

Kotnik, J. T., and Sater, B. L.: Power Conditioning Requirements of In Rockets; *IEEE Trans. Aerosp.,* April 1964, vol. AS-2, no. 2, pp. 497–504.

Gingrich, J. E.: Radioisotope Fueled Thermionic Space Power System; *IEEE Trans. Aerosp.,* April 1964, vol. AS-2, no. 2, pp. 669–674.

Gould, C. L.: Solar Cell Power Systems for Space Stations; *IEEE Trans. Aerosp.,* April 1964, vol. AS-2, no. 2, pp. 759–768.

Pierro, J. J., and Phillips, J. E.: Investigation of High Frequency Power Conversion and Generator Techniques; *IEEE Trans. Aerosp.*, June 1965, vol. AS-3, no. 2, pp. 411–422.

Rosen, A.: Large Discharges and Arcs on Spacecraft; *Astronaut., and Aeronaut.*, June 1975.

## MARINE POWER APPLICATIONS

*By EDWARD F. EATON*

### General Marine Considerations

***51. Introduction.*** Marine service conditions differ from industrial and land service primarily because of (1) the physical environmental conditions under which the equipment is operated and (2) the great importance of reliability in shipboard service. Equipment must be suitable for service under rolling and pitching of the vessel and for list of the vessel caused by hull damage or unbalanced loading.

In addition, the equipment must be able to withstand the adverse effects of moist, salt- or oil-laden atmospheres, as well as the vibrations incident to operation of other machinery and the pounding of the vessel in a heavy sea.

A 5° pitch, 22.5° momentary roll, and 15° permanent list are the usual values for merchant ships, and these parameters have a significant impact on the design of the lubricating systems on large rotating machinery and on other electrical equipment such as transformers and batteries. Internal transformer mounting arrangements often include a nonrigid sound-absorbing support system, and standard designs may require changes to meet the roll and pitch requirements. Space heaters are commonly required not only with the electrical equipment for deck machinery but with many other items to avoid moisture damage during periods when the ship may be inactivated.

***52. Safety.*** The safety of a ship at sea is dependent on the maintenance of propulsion, steering, lighting, communication, navigational, and other auxiliary services, which are, in turn, dependent upon the ship's-service power system. Vital services are safeguarded by duplicate units, alternate circuits, and emergency sources of supply.

The Government of the United States of America is a signatory to the International Convention for the Safety of Life at Sea, 1974, and compliance is causing changes in both American Bureau of Shipping (ABS) rules and U.S. Coast Guard regulations.

***53. United States Coast Guard.*** The U.S. Coast Guard is responsible by law for the enforcement of rules and regulations intended to safeguard life and property on American ships. In 1982, the U.S. Coast Guard issued Navigation Inspection Circular 10-82 which detailed planned review and inspection functions to be performed by the American Bureau of Shipping and accepted by the U.S. Coast Guard. Not included in these new procedures are "automated or centrally controlled systems for propulsion or auxiliary machinery, with associated logic, fault analysis, and piping or electrical equipment." Electrical engineering regulations issued by the Coast Guard cover the requirements for electrical equipment for the entire range of vessels coming under its cognizance. Compliance with these regulations, as applicable, is mandatory.

***54. American Bureau of Shipping.*** The American Bureau of Shipping is a private, nonprofit marine classification society which establishes rules and requirements as to design, construction, and inspection of ships and ships' machinery. This bureau classifies ships in accordance with their rules and thus provides assurance that industry-recognized good practices are incorporated in the vessel and its equipment.

***55. IEEE Marine Standards.*** The Institute of Electrical and Electronics Engineers' Committee on Marine Transportation is the author of IEEE Standard 45, Recommended Practice for Electric Installations on Shipboard. This publication, generally referred to as IEEE Marine Standards, is the basis for marine electrical engineering practices in this country. This widely accepted standard, in itself, is not a mandatory marine requirement, but both the U.S. Coast Guard and the American Bureau of Shipping use it as the basis of their rules and regulations.

***56. Marine Usage.*** The term "marine" is commonly associated with merchant marine ships and not with United States naval combat vessels. The following practices are applicable to merchant marine vessels and naval auxiliary vessels.

## Ship's-Service Power Systems

***57. Use and Capacity.*** The ship's service generation capacity now employed on new vessels ranges from 1500 to 3000 kW on cargo vessels and tankers to about 7500 kW on large refrigerated containerships. The usual practice is to employ two generating sets of equal capacity on cargo vessels, for normal loads. In addition a small emergency generating set is required for each ship.

***58. Standard Voltages.*** The voltages indicated in the following table have been accepted as standard on large merchant vessels in the United States:

| | Alternating current | Direct current |
|---|---|---|
| Lighting | 115 | 115 |
| Power utilization | 440-460-2300-4000 | 115 and 230 |
| Generation | 450-480-2400-4160 | 120 and 240 |

An ac generated voltage of 480, with 460 V for utilization, is a newly adopted marine standard which is in agreement with the present industrial standards and should prevail over the older standards of 450 and 440 V. For small ships, ac generation at 120, 208, 230, or 240 is acceptable. U.S. Navy ships generate at 450 V or 600 V and use 440-V motors.

***59. Standard Frequency.*** A frequency of 60 Hz is recognized as standard for ac lighting and power systems.

***60. Selection of Voltage and Distribution System, Direct Current.*** For vessels having little power apparatus, 120-V generators are recommended with 115-V light and power-distribution systems.

Where an appreciable amount of power apparatus is to be considered, 240/120-V 3-wire generators, 230-V power-distribution system, and 230/115-V 3-conductor lighting feeders should be selected. Branch circuits from lighting panelboards should be 115-V 2-wire.

***61. Selection of Voltage and Distribution System, Alternating Current.*** For small vessels having little power apparatus, 120-V 3-phase generators may be used, with 115-V 3-phase distribution for power and lighting. As an alternative, lighting feeders may be single-phase, balanced at the generator and distribution switchboard, to provide approximately equal load on the 3 phases.

For vessels with considerable power apparatus (particularly for intermediate-sized vessels), generators may be 230-V, 3-phase or 240-V, 3-phase; the power utilization at 220-V or 230-V, 3-phase, respectively; and lighting distribution 115-V, 3-phase, derived from transformers. As an alternative, generators may be 3-phase, 4-wire, 208/120-V. Power and lighting should be at 200/115-V, 3-phase.

For all vessels of a size and type to require a dual voltage system, first consideration should be given to the application of 480-V generators with power utilization at 460-V, 3-phase and lighting distribution at 115-V, 3-phase, 3-wire or 120/208-V, 3-phase, 4-wire from transformers.

For vessels having very large electrical systems requiring higher-voltage power generation, consideration should be given to generating first at 4160 V or at 2400 V with some power utilization at 4000 or 2300 V, 3-phase, respectively, with lower utilization voltages from transformers.

Either grounded or ungrounded ac electrical systems may be used for the above listed standard voltages. However, in the case of the 450- or 480-V, 3-phase systems, first consideration should be given to the use of an ungrounded system which is preferred.

***62. Advantages of Alternating Current.*** Alternating-current ship's-service power systems are now selected for virtually all new American merchant ships. In fact almost all oceangoing vessels built in this country since World War II have used ac ship's-service power. This type of electric power has been found advantageous on tankers, bulk carriers, general cargo vessels, and numerous special-purpose vessels.

The main advantage of the ac system comes through the use of squirrel-cage induction motors and across-the-line motor starters. This rugged and reliable motor, requiring extremely low maintenance, is used for literally hundreds of drives in sizes ranging from fractional horsepower to several hundred horsepower. The absence of commutators and brushes and the ability to start the motors directly across the line in all except the largest sizes make them exceedingly attractive for shipboard application. Most applications employ single-speed motors. Where the service requires it, such as for forced or induced fans on steam plants and certain pumps, multispeed motors can be supplied.

Another major advantage of ac power is the ability conveniently to employ a higher-voltage 3-phase distribution system. Generation is mostly at 480 V, with only a relatively small percentage of the power transformed to 115 V for lights and small devices.

Weight and first cost are lower for ac systems, with the possible exception of those ships in which dc power demands for cargo handling are an extremely large part of the total load.

For ships requiring direct current for various drives, the development of solid-state rectifiers (SCRs) permits easy conversion to dc from high-speed ac generating sources, thus eliminating much of the demand for dc generators, which are heavy and costly. For these SCR drives, there is no abrupt system voltage drop when starting the motors, and the ac source is not dedicated to any one load, as might well be the case with a dc generator.

Starting-voltage drops for motors should be limited as follows:

1. Several times per hour—10%.
2. Less frequently—18 to 23%.

It is desirable to keep the voltage drop to less than 15% for most conditions.

***63. Classification of Auxiliaries.*** Shipboard electric auxiliaries (Table 23-9) may be grouped and described as follows:

*a. Turbine, Engine, and Boiler-Room Auxiliaries.* These include pumps, fans, air compressors, etc., of the same general type as used on land. Special features provided for these auxiliaries are those having to do with the motion of the vessel and the corrosive atmosphere.

*b. Deck Auxiliaries.* These auxiliaries include steering gear, anchor windlass, capstans, boom topping, and lifeboat, deck, cargo, and mooring winches. These are specialized ship applications and, with the exception of the steering gear, are deck-mounted and directly exposed to seawater.

The steering gear has been classed as a deck auxiliary because it has been placed under the jurisdiction of deck officers. As the steering gear is very vital to the safe operation of the ship, much attention has been given to the rugged character of both motor and control units, to provide maximum reliability.

With the exception of the main cargo winches, practically all deck machinery is powered by ac motors, which are frequently of the two-speed type. Cargo winches, almost without exception, employ dc motors. The winches are normally used in pairs in the *burtoning method,* and the power for each pair of winch motors is derived from a single associated motor-generator set. Mooring winches may be of the constant-tension type for which dc motors and control are more appropriate. However, the explosion-proof enclosures required on the decks of tankers preclude economical application of dc motors for these drives, and ac solid-state variable-frequency drives are not yet fully developed. As a result, hydraulic systems powered by ac motors are often used for this application.

*c. Navigation and Communication Equipment.* This includes gyrocompass and autopilot, Loran and RDF equipment, radio, radar, sonar, intercommunication equipment, fire-alarm, and computerized collision-avoidance systems.

*d. General Auxiliaries.* Items such as ventilating blowers, fans, and air-conditioning equipment are employed in great numbers on modern ships.

*e. Hotel Load.* Electrical load so characterized is associated with personal comfort and includes electric cooking and heating units of all kinds, air conditioning, elevators, laundry equipment, special lighting, refrigeration, etc.

**TABLE 23-9** Ship's-Service Load for Typical Ships

| Gas turbine electric and steam | | | | |
|---|---|---|---|---|
| Type of ship | Product carrier | | LNG* carrier | |
| DWT | 35,000 | | 63,000 | |
| Shaft hp | 12,500 | | 40,000 | |
| Propulsion plant | Gas turbine electric | | Steam | |
| Total installed load, kW | 5,500 | | 10,000 | |
| Normal sea load, kW | 500 | | 1,400 | |
| Normal port load, kW | 300 | | 700 | |
| Cargo pump load, kW | 1,800 | | 3,800 | |
| | No. of motors | hp | No. of motors | hp |
| Representative motor applications: | | | | |
| Forced draft fan | | | 2 | 300/133 |
| Main circulating pump | | | 2 | 200 |
| Main condensate pump | | | 2 | 75 |
| Fuel-oil service pump | | | 2 | 40/20 |
| Fuel-oil transfer pump | | | 1 | 50 |
| Lubricating-oil service | | | 2 | 50 |
| Turning gear | 2 | 50 | 1 | 10 |
| Fire pump | 2 | 150 | 2 | 200 |
| Bilge pump | 1 | 7.5 | | |
| Bow thruster | 1 | 1000 | 1 | 1500 |
| Turbine starting pump | 1 | 250 | | |

| Diesel | | | | |
|---|---|---|---|---|
| Type of ship | Tanker | | Containership | |
| DWT | 44,000 | | 40,490 | |
| Shaft hp | 12,000 | | 43,200 | |
| Propulsion plant | Diesel | | Diesel | |
| Total installed load, kW | 4,400 | | 7,500 | |
| Normal sea load, kW | 600 | | 5,000 | |
| Normal port load, kW | 2,000 | | 5,000 | |
| Cargo pump load, kW | 1,300 | | — | |
| | No. of motors | hp | No. of motors | hp |
| Representative motor applications: | | | | |
| Main engine service pump | 2 | 40(SW) | 2 | 150(FW) |
| Main engine jacket water cooling | 2 | 25 | 2 | 100 |
| S.W. service pump | 2 | 25 | 2 | 400 |
| F.W. service pump | 2 | 40 | 2 | 200 |
| Main engine lubricating-oil pump | 2 | 60 | 2 | 125 |
| Main engine crosshead lubricating-oil pump | 2 | 25 | 2 | 100 |
| Fuel-oil booster pump | 2 | 10 | 2 | 25 |
| Fuel-oil transfer pump | 1 | 20 | 1 | 40/20 |

| | | | | |
|---|---|---|---|---|
| Main lubricating-oil cooling pump | 2 | 60 | | |
| F.W. cooling circulation pump | 2 | 40 | | |
| Fuel-oil service booster pump | 2 | 3 | | |
| Fuel-oil standby pump | 1 | 5 | | |
| Lubricating-oil standby pump | 1 | 30 | | |
| Deck auxiliaries: | | | | |
| Deck machinery hydraulic pump | | | 4 | 200 |
| Capstan/mooring | 7 | 75 | | |
| Boat winches | 2 | 10 | 2 | 15 |
| Steering gear pump | 2 | 50 | 2 | 100 |
| Anchor windlass | 2 | 75 | 2 | 150 |
| General auxiliaries: | | | | |
| Engine room supply | 2 | 50 | 4 | 40 |
| Engine room exhaust | 2 | 30 | 3 | 30 |
| S.S. refrigeration | 2 | 10 | 2 | 15 |
| Air compressor | 2 | 15 | 2 | 30 |
| Air conditioning | 2 | 40 | 2 | 40 |
| Cargo pumps | 8 | 300 | 8, 3 | 300, 200 |
| Cargo stripping pump | | | 7 | 100 |

| | | | | |
|---|---|---|---|---|
| Bilge pump | 2 | 10 | | |
| Main engine piston-cooling pump | 2 | 15 | 2 | 75 |
| Main engine fuel valve cooling | 2 | 3 | 2 | 7.5 |
| Cylinder oil transfer | 1 | 1 | 1 | 3 |
| D.O. transfer | 1 | 1.5 | 1 | 7.5 |
| Lubricating-oil transfer | 1 | 1.5 | 1 | 2 |
| Diesel gen. SW cooling | 3 | 10 | | |
| Main engine leakage transfer | 1 | 0.75 | 1 | 2/0.5 |
| Emergency fire pump | 1 | 125 | 1 | 150 |
| Deck auxiliaries: | | | | |
| Capstan/mooring | 6 | 75 | | |
| Boat winches | 2 | 15 | 2 | 10 |
| Steering gear pump | 2 | 50 | 2 | 150 |
| Anchor windlass | 2 | 100 | 2 | 100 |
| General auxiliaries: | | | | |
| Engine room supply | 3 | 35/8.75 | 2 | 125/31.25 |
| Engine room exhaust | 2 | 20/5 | 2 | 75/18.75 |
| S.S. refrigeration | 2 | 10 | 2 | 20 |
| S.S. air compressor | 1 | 20 | 1 | 50 |
| A/C compressor | 2 | 75 | 2 | 150 |
| Control air compressor | | | 1 | 60 |
| Cargo pumps | 6 | 300 | | |
| Ballast pumps | 4 | 75 | 1 | 200/113 |
| Fire and Butterworth pump | 1 | 300 | 1 | 250/150 |
| Potable water pump | 2 | 10 | 1 | 7.5 |
| Bow thruster | | | 1 | 2200 |
| Purifiers | | | 6 | 20 |
| Cargo manifold motor operating valve | 8 | 0.33 | | |

*LNG = liquified natural gas; DWT = dead weight tonnage; S.W. = salt water; F.W. = fresh water; S.S. = ship's service.

*f. Special auxiliaries* for particular services are to be found on such vessels as oil tankers having large pumps installed for handling cargo; self-unloading bulk carriers, for example, the taconite, coal, and limestone carriers on the Great Lakes; and vessels equipped to carry railroad cars. During the last 10 years bow thrusters with controllable and reversible pitch (C and RP) propellers have become popular for ease in maneuvering the long ships in use. Vertical squirrel-cage induction motors with autotransformer starters in the 500- to 1500-hp range are often used for this purpose. Thrusters with fixed-pitch propellers driven by dc motors and SCRs are also in service. Starting voltage drop for the ac drives is often a problem and must be carefully considered.

***64. Ship's-Service Load and Motor-Application Data.*** Data for a number of ships are shown in Table 23-9 for converted kilowatts and estimated demand kilowatts at sea and in port. The Society of Naval Architects and Marine Engineers publishes "Technical & Research" bulletins for various types of ships to establish service factors for connected loads and to assist in estimating refrigeration and most other electrical loads.

***65. Generating Sets.*** Two or more generating sets are required for ship's-service power. Rules require that the number and rating of sets be such as to provide one spare set at all times. Thus, most vessels use two sets of equal rating, each capable of supplying the full operating load requirements of the vessel. On steam-driven ships, the ship's-service generating sets are almost always turbine-driven. High-speed turbines, operating at the full-steam conditions of the propulsion plant, drive generators at a 60-Hz speed through reduction gears. Many foreign-flag ships built abroad have been outfitted with shaft-driven generating systems which can be electrically paralleled with conventional ship's-service systems. The usual components are a mechanical means of coupling to the propulsion shaft, a step-up gear and generator with excitation system, and a solid-state frequency converter to produce the same frequency (50 or 60 Hz) as the conventional system. Most of these configurations operate from between one-third and one-half speed to full speed on the main propulsion shaft and must include a source of reactive power such as capacitors or a synchronous condenser. Although expensive and fairly large, they often pay for themselves in a few years of usage by reduced fuel and maintenance costs on ships with a high percentage of underway time. Another alternate is a diesel set to supplement the steam-driven source, and small gas-turbine sets are being considered as well. Usually the emergency set (150 kW or smaller) is diesel-driven, but small gas turbines are available.

***66. Switchboards.*** Switchboards of the dead-front type are provided for the control, metering, and handling of the ship's-service power system. Generators are paralleled and voltages are load-adjusted from devices located at the main switchboard. Power feeders emanate from this board to the various major loads or to feeder panels and distribution boards arranged for supplying and switching many individual load circuits.

Automatic synchronizing equipment can be supplied if desired. Some systems are fully automatic, requiring only manual initiation, and others are semiautomatic. The latter type often include operator-monitoring equipment, which prevents breaker closing if voltage, frequency, and phase angle are not within acceptable limits.

***67. Continuity of Service.*** Continuity of service and adequacy of interrupting capacity are of paramount importance in ship's-service electric power systems. Figure 23-24 is a simplified diagram of an electric power system for a typical modern cargo vessel. The emergency set is programmed to start automatically in the event of a failure of the normal 480-V generation. Only services that are essential for safety in an emergency, such as lighting, fire pumps, and steering, are fed from the emergency bus. It is located above the uppermost continuous deck and outside the machinery space.

All vital auxiliaries are connected to the main 480-V bus through individual circuit breakers, each of which is capable of interrupting the maximum fault current available at that particular point. The majority of vital services is accommodated with molded-case circuit breakers either fused or unfused—depending upon the system short-circuit capacity. The relatively few very large auxiliaries are provided with open-frame air circuit breakers. Prospective short-circuit currents are calculated at the main buses and at the ends of each of the major feeder cables such as those connecting the main switchboard bus to the group controls. Appropriately rated circuit breakers are then selected for each location. The selection is based upon (*a*) the maximum prospective short-circuit current at the point of installation, (*b*) the degree of asymmetry of the circuit up to that point, and (*c*) the continuous current-carrying ability required by the load circuit.

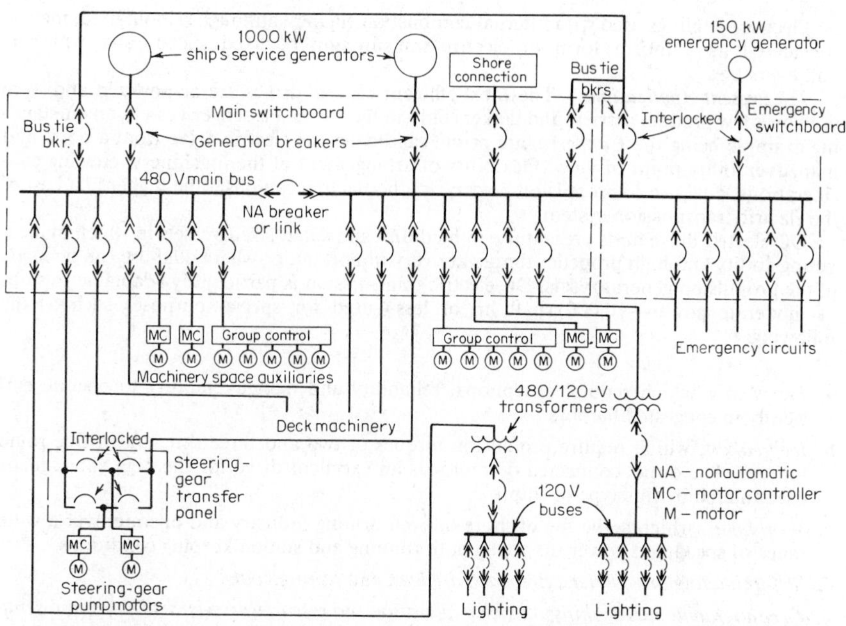


**FIG. 23-24** Ship's-service electric power system of a modern cargo ship.

***68. Group Control.*** The combining of groups of related motor controllers into one or more free-standing mechanical assemblies, known as *group controls,* has found increased favor with ship designers in view of the present trends toward centralization of control functions. Appearance is improved, servicing is made easier, and shipyard installation costs are usually diminished by the use of group controls. They have been successfully applied in the case of engine-room auxiliaries and deck-machinery auxiliaries. Generally the individual drawout motor controllers include a circuit breaker for short-circuit protection; a large number of optional accessories is available.

***69. High-Voltage AC Switchgear and Control.*** If the electrical system total load is large, or a number of single motors with high (over 300) horsepower ratings is involved, the electrical-system-design engineer may elect to use 2400-, 4160-, or even 6600-V equipment. Both engineering and economic considerations will heavily influence the voltage selection. For these higher-voltage systems, modified industrial-type metal-clad switchgear is offered. For ac motors up to 3500 hp at 6600 V, combination fused starters are available for full-voltage or reduced voltage closed-transition autotransformer starting duty. A circuit breaker may be used for this duty, but is more expensive in most cases. The marine industry has been most reluctant to go beyond 600 V ac on merchant ships or navy auxiliary ships.

## Electric Propulsion

***70. Electric Drives.*** The term "electric drive," when used in conjunction with ship propulsion, refers to the interposing of an electric generator and motor between the prime mover and either a fixed-pitch or controllable-and-reversible-pitch (CRP) propeller. The loading from a fixed-pitch wheel is proportional to speed cubed, and that from a CRP wheel is proportional to pitch angle. However, at reduced vessel speed on a ship with CRP propeller, the shaft speed often is reduced for more efficient operation.

Electric transmission systems can be classified in three basic types: alternating current, direct current, and ac-dc with power converters. The ac transmission is broadly a fixed-speed-ratio reversible drive; the dc transmission, a variable-speed-ratio reversible drive; the ac-dc system, a single ac power plant supplying a variety of ac and dc loads.

Electric couplings used with internal-combustion engines, although seldom used, may be considered as a fourth form of electric transmission of fixed speed ratio which is nonreversible.

*71.* ***Present Applications.*** The most efficient electric drives are ac-powered and were used on a few product carriers and tankers built in the 1970s. Passenger vessels, an outstanding example being the *Canberra,* are prime candidates for electric drive from a noise and maneuverability point of view. Flexibility of arrangement of the machinery, crewing considerations, dual use of propulsion power, or other factors can offset the higher first cost of the electric transmission system.

DC electric drive finds present use in medium- and small-sized vessels in which maneuvering ability and high propeller torque are very important, or where dual use can be made of the propulsion generator sets. DC electric transmission is particularly adaptable to ships of moderate power—10,000 shaft hp or less—used for special purposes such as the following:

**a.** *Ferryboats,* which require exceptional reliability and maneuverability to operate efficiently in congested harbors.

**b.** *Icebreakers,* which require propulsion torques of two and three times rated free route torques. The direct-connected dc motor is an excellent drive for the high shock loads encountered by this type of ship.

**c.** *Workboats,* which serve the offshore oil-well drilling industry and operate over a wide range of speed and power involving both running and station keeping conditions.

**d.** *Self-unloading ore carriers, fireboats, dredges,* and *fishing boats.*

**e.** *Oceanographic survey ships, oil-well exploration* and *research vessels, drill ships,* and *semi-submersibles, pipe-laying barges, cable layers* and *surveillance ships.*

The combination ac-dc system is used on ships where a single power plant, often with multiple prime movers, is used to supply power to both large alternating-current drives and direct-current drives. The recent trend is to this type of system, to supersede the dc-dc system, since it allows high-speed prime movers and ac generators to be used and the same generating plant can serve ac or dc propulsion, ship's service, and special function loads. Mining and drilling vessels, as well as the types listed above under dc electric drives, are often designed with combination ac-dc systems. Considerable advancement has been made during the last few years in the availability of high-current silicon-controlled-rectifier power converters, and this has led to their use as an efficient link between ac generation and variable-voltage reversible dc propulsion and thruster motors.

*72.* ***Advantages of Electric Drive.*** Electric drives offer a number of advantages, which include the following:

*a. Flexibility of Arrangement.* Electric drive increases the flexibility of vessel design because the generating sets may be located remotely from the propulsion motor. Likewise, the motor may be located so as to result in a short propeller shaft.

*b. Economy of Operation and Maintenance.* In vessels having multiple generators, part of the generating plant may be shut down when operating at sufficiently reduced power, resulting in better reduced-speed economy. Electric drive has demonstrated that its maintenance costs are low.

*c. Maneuvering.* With electric drive, it is possible to provide full power for reversing by electrical reversal of the motor while allowing the prime mover to continue rotation in one direction. Manipulation of controls for electric drive is exceedingly simple, and it also lends itself to multiple control stations where desired.

*d. Auxiliary Power from Propulsion Generator.* On certain vessels, advantage can be taken of the propulsion generator as a source of power for cargo handling or other special power service with consequent reduction in size of the ships'-service power-system generating sets.

*e. Nonreversing High-Speed Prime Mover.* Electric drive permits a satisfactory speed reduction between a high-speed prime mover and a low-speed, efficient propeller without the use of a large bull gear. Both ac and dc systems inherently allow easy reversing without any reverse reduction gearing.

*f. Minimum Noise.* For oceanographic research and surveillance ships, electric drive can obviate the need for reduction gearing. This is of invaluable assistance in any missions which use sonar equipment.

***73. Economic Considerations.*** To justify an electric drive system economically, the unique advantages and characteristics of such a system must be evaluated to offset the inherently higher first cost and lower transmission efficiencies. For example, a 20,000-hp ac system with a synchronous propulsion motor would have total transmission and excitation losses 3 to 3.5% greater than a direct-geared drive system. Smaller ac systems, induction-motor drives, and dc systems would have a greater differential. Factors to be considered include fuel and maintenance costs, crew size, and duty cycle as well as the value to the user of the advantages listed in Par. **72**.

## AC Electric Transmission Systems

***74. AC Propulsion Systems.*** The basic ac electric transmission system, illustrated in Fig. 23-25, consists of: one or more prime movers directly connected or geared to synchronous generators; one synchronous motor directly connected to each propeller; propulsion switchgear; and a propulsion control and excitation system.

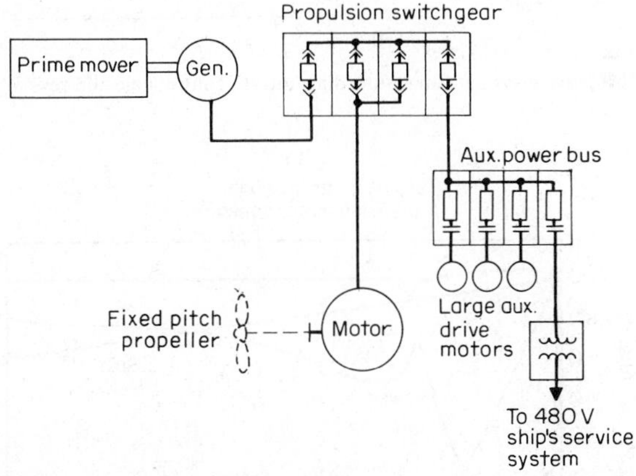


**FIG. 23-25** Basic ac electric propulsion system.

AC electric ship-propulsion systems can be broadly divided in two concepts: (a) those that use controllable-and-reversible-pitch propellers (CRPs) to control the magnitude and direction of thrust; and (b) those that use fixed-pitch propellers where the magnitude and direction of thrust are controlled by the speed and direction of rotation.

***75. AC System with CRP Propeller.*** The CRP system is best suited to ships on which economics dictate that one large power plant be used to provide all power for propulsion, large at-sea auxiliary loads, and ship's service.

This type of power plant, shown schematically in Fig. 23-26, is designed on the basis that the propulsion generator set or sets will provide 100% of the power used on the ship, 100% of the time. The generator, propulsion motor, and motor-starting system are designed to permit starting of the propulsion motor without exceeding an acceptable deviation in the voltage and frequency of the ship's power system.

In the event that the main generator set is secured, an auxiliary set can be used to provide power to the ship's-service load and cargo pump motors. In an emergency this unit can be used to provide propulsion power with the standby generator supplying power for the necessary ship's auxiliary loads and excitation power for the propulsion motor and auxiliary generator. In this emergency propulsion mode, the power demand must be limited by pitch and speed control of the propeller to prevent overloading the auxiliary set.

Typical crash-reversal chracteristics of this type of plant are shown in Fig. 23-27. As the reversal is executed, the motor power goes to zero and may reverse slightly. This sudden

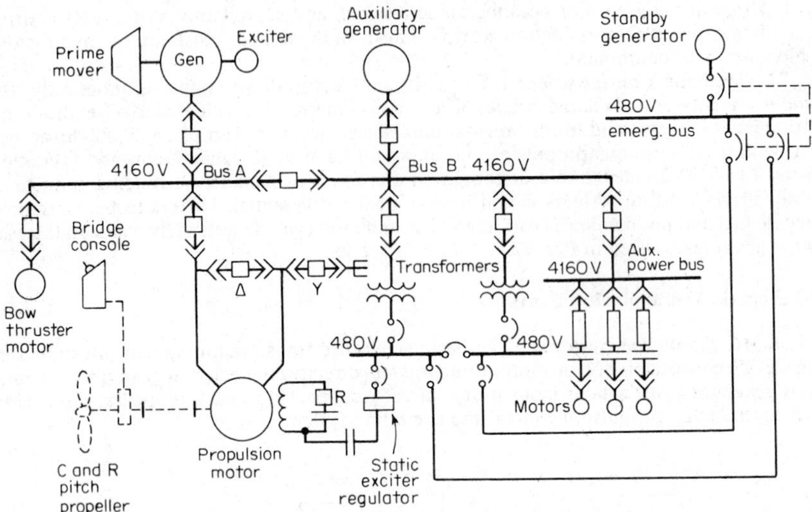


**FIG. 23-26** AC propulsion system (single-line diagram) for controllable-and-reversible-pitch (CRP) propeller.

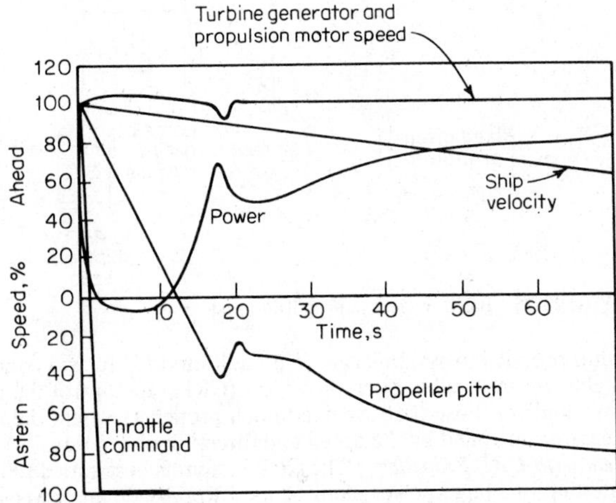


**FIG. 23-27** Typical time characteristics for full-speed reversal with CRP propeller.

decrease in power demand causes an increase in the generator-set speed. The overspeed should be limited to a maximum of about 106%. In this example, the propeller pitch changes at the rate of 8% per second, that is 25 s from full ahead to full astern pitch. After the pitch goes through zero, the power demand increases rapidly, causing a reduction in system speed. As the speed decreases below rated, the control system functions to decrease the propeller pitch and load until the speed recovers to rated. From this point on, the propeller pitch increases so that the maximum power available is delivered by the power plant.

***76. AC System with Fixed-Pitch Propeller.*** The fixed-pitch propeller plant (Fig. 23-28) is best suited to ships that have large in-port loads where dual use can be made of the propulsion power plant. Ship's-service power and other auxiliary loads cannot usually be sup-

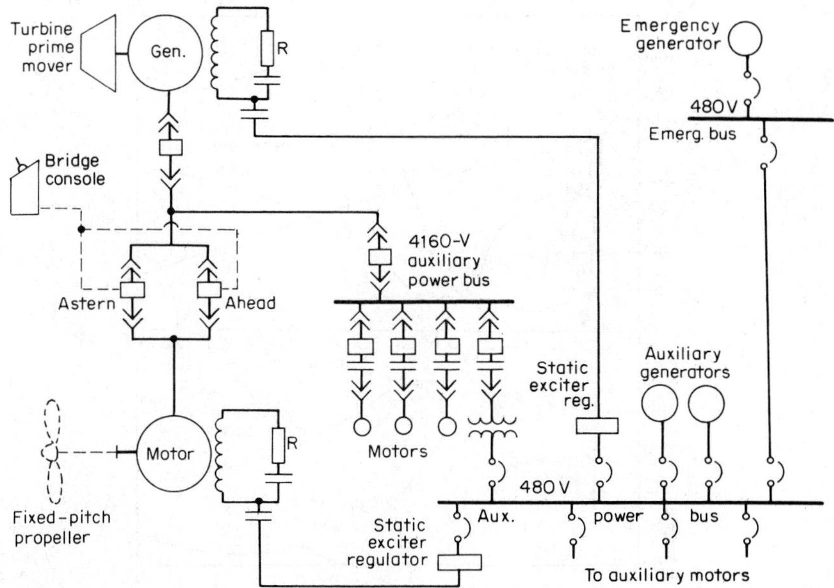


**FIG. 23-28** AC propulsion system (single-line diagram) for fixed-pitch (FP) propeller.

plied from the propulsion plant during maneuvering, reversing, or cruising at reduced speed.

The fixed-pitch electric transmission system performs the same function as the double reduction gear and reverse clutches do in a mechanical drive. The "gear ratio" between the generator speed and the propulsion motor is the ratio of poles on each rotor; that is, a 1200-r/min, 6-pole generator and 72-pole motor have a gear ratio of 12, while a 3600-r/min, 2-pole generator and a 72-pole motor have a gear ratio of 36.

Starting and reversing of the propeller is accomplished by operating the synchronous motor asynchronously, or as an induction motor. The electrical phase rotation applied to the motor through the "ahead" or "astern" breaker determines the direction of motor rotation.

The induction motor or slip mode of operation performs the same function as the clutches in a mechanical drive. Synchronization of the propulsion motor by an electrical field is synonymous with full clutch engagement. The excitation subsystems for the generator and motor automatically maintain a field power level that results in the highest power-transmission efficiency between the prime mover and propeller.

The electrical transmission system will operate in the synchronous mode over the maximum speed range of the prime mover. The propeller speed range can be extended below the minimum prime-mover speed by operating in the continuous slip or subsynchronous mode. This method of obtaining low propeller speeds can be employed with prime movers that have a minimum power level or speed that is above an acceptable maneuvering propeller speed. The subsynchronous operation increases the transmission losses, and most of these appear as $I^2R$ losses in the motor rotor. The propulsion motor must be designed for continuous dissipation of this slip energy.

Typical crash-reversal characteristics of fixed-pitch ac electric plants are shown in Fig. 23-29. At the reversal command, the fuel is reduced and the system speed drops to 70% rated, or slightly above the windmilling propeller speed, with full headway on the ship. At this time (5 s), the generator excitation is removed, the motor phase rotation reversed, and the motor excitation increased to produce approximately 200% armature current.

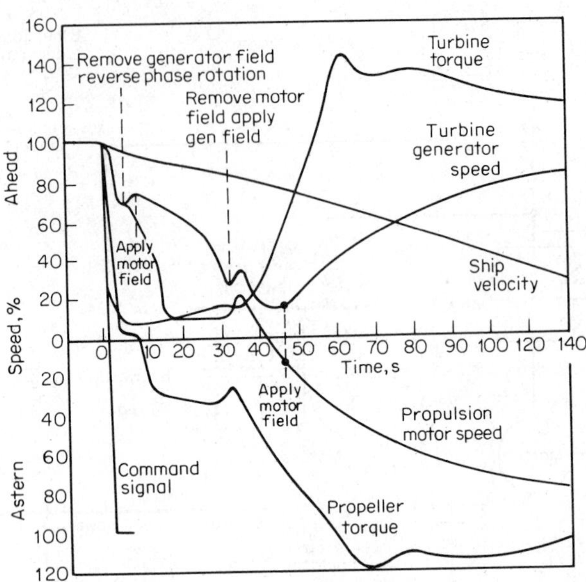


**FIG. 23-29** Typical time characteristics for full-speed reversal with fixed-pitch propeller.

Both the generator and motor speed decrease because of the plugging action on the generator and braking action on the motor. When the motor speed decreases to a preset value limit (in this example, 8% of ahead), the motor field is removed and the generator excitation applied. The resultant torque pulls the motor, operating as an induction machine, through zero and near to synchronous astern speed. Automatic field application then locks the system together so that it responds in the same manner as a geared system. The system accelerates at a maximum rate limited by the prime-mover power.

A dynamic braking resistor can be used in fixed-pitch systems during the reversing operation, to absorb the power generated by the motor owing to the action of the water passing through the propeller. This slows down the motor significantly, and when the ship's speed drops to about 65%, the dynamic brake can be disconnected and the propulsion motor connected to the propulsion bus in reverse-phase rotation.

***77. Generators and Motors.*** Generators are 3-phase, usually Y-connected, with a voltage rating between 2300 and 7000 V. They are direct-connected or geared to the prime movers and run at speeds ranging from 1200 to 5500 r/min. High-speed units have cylindrical rotors, and lower-speed units, 1800 r/min and below, are of salient-pole construction.

Propulsion motors are of the salient-pole synchronous type and are normally direct-connected to the propeller shaft. They must be equipped with heavy-duty amortisseur windings for starting, reversing, and low-speed maneuvering operations. Propeller speeds on most merchant ships are about 100 r/min.

Motors and generators are normally totally enclosed with surface air coolers using seawater for cooling. Ventilating air may be circulated by rotor fans or by seperate motor-driven blowers.

***78. Propulsion Switchgear.*** The propulsion switchgear employs air circuit breakers or air break contactors. The propulsion bus and circuit-interrupting device should be designed to handle the maximum fault current. Where circuit breakers are used for fault protection, they should be designed to interrupt the maximum fault current including the contribution from the propulsion motor and any other connected loads. Protection should include machine differential, and fault protection. In addition, ground-fault protection or alarms plus phase unbalance should be considered on the larger ratings.

***79. Propulsion Control.*** For the CRP propeller, control will involve starting of the propulsion motor at zero pitch and controlling the pitch to adjust the ahead and astern thrust. In addition, the prime-mover power and speed control may be combined with the pitch variation so as to optimize the propeller and prime-mover speed power characteristics.

For the fixed-pitch reversible system, the control must include means for starting and reversing the propulsion motor and varying the speed of the propulsion generator set(s). The actual control scheme is quite simple and involves the following steps for manual or automatic systems: (*a*) Close the ahead or astern breaker. (*b*) With the prime mover operating at maneuvering speed (20 to 40%), increase the generator field to start and accelerate the propulsion motor to a low value of slip. (*c*) Apply motor field to pull the motor into synchronism. (*d*) Increase the prime-mover speed to obtain the desired propulsion speed.

With certain prime movers, it is not possible to operate at the 15 to 30% speed required for maneuvering in channels or congested areas. In these cases it may be possible to operate the propulsion motor as an induction motor, and vary the slip by varying the generated voltage.

Both the CRP and fixed-pitch plants can be arranged for control at the bridge, engine room, or both and designed for limited or unmanned engine-room operation.

The CRP plant can be designed so that the propulsion power plant supplies all the ship's power at all times. The fixed-pitch plant, if designed for 60 Hz, can be arranged to supply ship's power in port where propulsion is not needed, and to operate at full power at sea.

***80. Excitation System.*** The excitation system must be capable of supplying excitation to the generator, the motor, or both, in proper sequence and magnitude during maneuvering or steady-state operation. Under steady-state operation, sufficient excitation must be provided to the motor and generator to preserve synchronism. The necessary value varies with the roughness of the sea, hull cleanliness, ship loading, and other factors which dictate the actual motor-torque demand.

For the fixed-pitch reversing propeller, a single propulsion exciter can be used to supply excitation power to both the propulsion generator and motor. Shaft-driven exciters are generally not suitable for this system.

For nonreversing propellers (CRPs), an exciter with its own automatic regulator and manual backup should be supplied for each propulsion generator, and a separate exciter with controls should be supplied for each propulsion motor. Generator and motor exciters can be of the shaft-driven, separate static, or motor-generator type. In all cases, the regulator should be of the static type to obtain fast response to fluctuations in propulsion loads.

## DC Electric Transmission Systems

***81. DC Propulsion Systems.*** The dc system consists of the following elements: (*a*) One or more prime movers directly connected or geared to direct-current generators. (*b*) Separately excited propulsion motor or motors directly coupled or geared to each propeller shaft. (*c*) An excitation system that may obtain its power from an exciter directly connected to the propulsion generator or the ship's-service electric power system or both. Alternatively, a motor-generator set may provide excitation power. (*d*) A control cubicle in the engine room, selective with one or more remote deck stations (optional but usually required) through which engine speed, generator excitation, and sometimes propulsion motor excitation are controlled.

***82. Armature Connections.*** Most dc electric propulsion plants utilize multiple prime movers, and the propulsion generators and motors can be connected either in series (Fig. 23-30) or in parallel (Fig. 23-31).

***83. Series System.*** The series system offers several advantages and for this reason is utilized to a much greater extent than the parallel system. The chief advantage of the series system is the fact that load division between engines is satisfactory regardless of small differences in engine speeds and generator voltages because the load will divide in proportion to the speed and voltage. Another advantage of the series system lies in the fact that it is possible to utilize full available engine horsepower when less than 100% rated horsepower capacity is available by employing motor field weakening up to some limiting value. As a general case it should be noted that a well-compensated machine can be expected to give entirely satisfactory performance with relatively low field strength and that in almost every

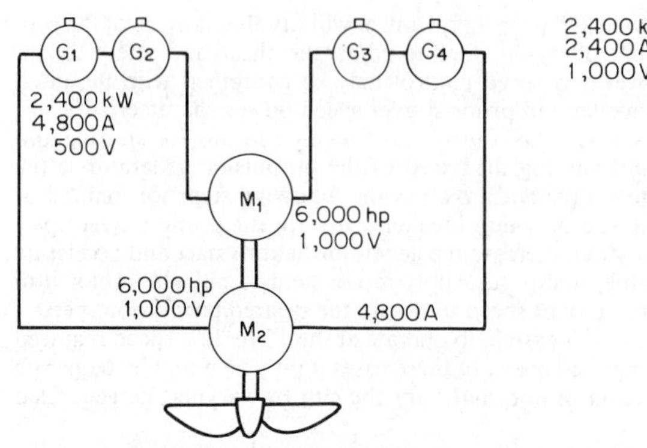


**FIG. 23-30** DC series propulsion system.

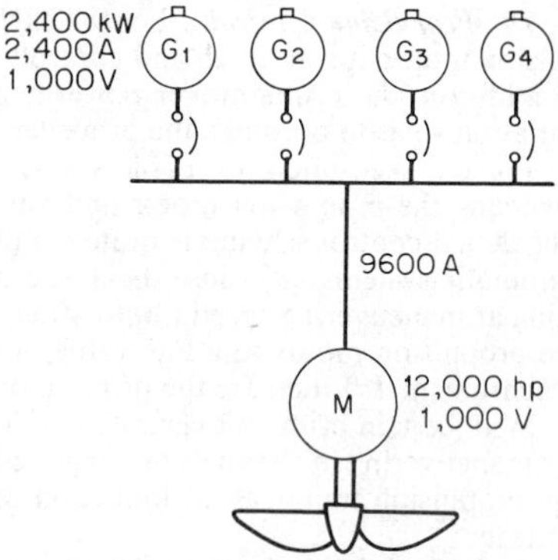

**FIG. 23-31** DC parallel propulsion system.

practical case complete utilization of available engine horsepower is possible with a series system.

***84. Parallel System.*** The use of parallel-connected dc generators requires careful matching of generator characteristics, precise speed-governing equipment for the engines, and accurate control of generator voltage to ensure equal division of load among the several engines. If less than 100% rated horsepower capacity is available, as for example, when an engine is secured, full utilization of the available horsepower cannot be realized without exceeding the normal generator armature current rating or the motor field current rating for most installations. Overdesign is one answer, of course, but this is costly.

Parallel armature connections are normally used only where the ratio of the number of generators to motor armatures is high and where relatively high shaft horsepowers are being considered. In such a case, the current in a series system could be sufficiently high to necessitate excessively large generators because of the number of brushes and the commutator surface area required to carry the high current. The use of such machines would not be desirable, owing to high centrifugal forces on the commutator and other design limitations.

***85. Generators.*** Generators are of the separately excited type rated 500 to 1000 V. They operate as variable-voltage units when connected to the propulsion system and at constant voltage when supplying the auxiliary power bus. The units are available in ratings as high as 2000 kW at speeds up to 1200 r/min and 5000 kW at lower speeds. Bearings are generally of the self-aligning sleeve type, forced-feed or flood-lubricated. Higher-speed generators (600 r/min and above) are usually ventilated by separate motor-driven blowers. Depending upon the application, machines can be either drip-proof-protected or totally enclosed with a sea-water-to-air heat exchanger. In the latter case, the ventilating air is filtered to remove the brush dust and then recirculated through the generator.

***86. Motors.*** Propulsion motors are similar to generators with regard to construction and enclosure. They may be of single- or double-armature types, the double-armature motor being longer but of smaller diameter for the same power. The horsepower rating and location within the hull influence the selection. High-speed motors are connected to the propeller shaft through gears; low-speed motors are rigidly coupled. The propeller-shaft thrust bearing may be incorporated in one of the motor-bearing housings if the motor is direct-connected. The design voltage of a single-armature motor should not exceed 1000 V. Low-speed direct-drive motors are available in ratings up to 15,000 hp per armature.

***87. Propulsion Control.*** A variable-voltage system is used with both the series loop and parallel armature circuits to control the speed and the direction of rotation of the propeller. The generated voltage is varied by controlling the magnitude and polarity of the generator excitation. A combination of variable engine speed and variable generator excitation can be used where it is advantageous to reduce fuel consumption and/or prime-mover maintenance. Constant horsepower at the full prime-mover speed rating can be obtained over the bollard (stalled ship) to free-route propeller-speed range for such applications as icebreakers and various workboats. The series system is easily protected against faults and severe overloads by field removal, and simple bypass switches or contactors can be used to add or remove a generator from the propulsion loop without removing power. The parallel system usually requires circuit breakers for fault protection and a means of paralleling the generators if they are to be added or removed from the propulsion circuit without shutting down the complete plant.

The dc electric transmission system is generally arranged for control from the bridge and engine room. Additional remote-control stations can easily be added, and the control system is designed for limited or unmanned engine-room operation.

***88. Excitation System.*** In a series loop system, a single exciter of the static or motor-generator type can be used to provide excitations to all propulsion generators. This exciter can be automatically regulated with a manual backup. An exciter and automatic regulator should be supplied for each generator in a parallel system.

The motor exciter can be of the static or motor-generator type and arranged to supply fixed field power with manual control or automatic regulation.

## AC-DC Systems with Power Conversion

***89. General.*** This type of system has superseded the dc-dc system for most large installations. It lends itself to supplying normal ac ship's-service power and any other loads, either ac or dc, simultaneously. For ships with widely varying loads, a number of medium-sized generating sets can be used as the demand varies, thus permitting maintenance on the secured units and both equalizing and minimizing the hours of operation on all the prime movers.

The navigation and other highly sensitive equipment on vessels with this type of system often have a small dedicated motor-generator set and a backup assigned to the critical loads involved. This isolates them from the power bus spikes, dips, and harmonics. Instead of a motor-generator set, electronic line conditioners or isolators can be installed to assure electrical separation. The use of uninterruptible power supplies or motor-generator sets provides the capability to ride through long- or short-term power interruptions, respectively. Care must be exercised in selecting and installing the ship's wiring to avoid picking up cross-talk from the power circuits and introducing spurious signals into the electronic and control circuits.

***90. Generators.*** AC generators with brushless exciters are normally utilized at either 600 or 4160 V depending on the size of the system and the generator. They must be designed specifically for use with SCR loads, since the nonlinear nature of the power-conversion process imposes more heating in the generator due to harmonics than the apparent kVA load. Several generating sets are usually provided to take advantage of the widely different loading cycles expected.

***91. Switchgear.*** Air circuit breakers are used to connect the generators to a common bus. Standard techniques for synchronizing and load sharing among units are similar to those used for any ship's-service systems. Inclusion of bus tie breakers or disconnects provides for continued operation on a split plant basis even in the event of a major casualty. Feeders to power converters and other loads are arranged so that the current in the switchgear bus will be well diversified. For practical limitations which occur in low-voltage circuit breakers, generator line currents exceeding 4000 A are to be avoided, and interrupting requirements should be kept below 85 kA symmetrical to avoid the need for fuses. Likewise on multiple generator installations, low-voltage switchgear main bus ratings in excess of 6600 A are not readily available. Figure 23-32 shows an electrical system typical of a drilling ship or a self-propelled semisubmersible offshore-drilling unit. In addition to the equipment shown, a small emergency generating set would normally be supplied.

***92. Drive Transformers.*** Modern techniques in the design of power converters have eliminated the need for transformers except on installations where the generated power is greater than 600 V ac. Thus, on 2300- or 4160-V systems either self-cooled or forced-air-cooled dry-type units are required. Under certain circumstances, the phase displacement of the primary and secondary windings can be used to cancel lower-order harmonic currents appearing on the ac power system. However, transformers cannot be justified on the basis of affording a reduction of harmonic content except in special cases.

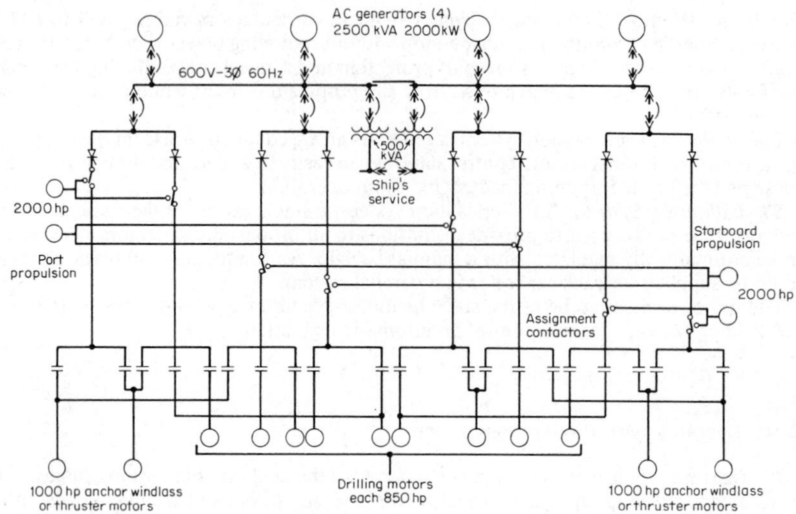


**FIG. 23-32** Single-line diagram of ac-dc electrical system for propulsion, positioning, and drilling loads.

***93. Power Converters.*** Typically, marine power converters are arranged in lineups which contain several individual converters fed from a common 600-V ac bus running the length of the lineup. Each individual converter bridge or group of bridges is connected to the ac source by a circuit breaker. The output of the converter, nominally 750 V dc, is either directly cable-connected to the motor armature or bussed to dc assignment contactors for those converters which can alternately be assigned to more than one motor. Section **22** of this handbook discusses the theory and circuitry of power converters in considerable depth. Six air-cooled silicon controlled rectifiers per bridge can be used with several bridges being paralleled for current capability up to the present practical limit of 5000 to 6000 hp. For greater horsepower a 12-pulse converter can be chosen using phase-shifting power transformers. Alternatively, multiple motors mechanically coupled can be used to achieve the desired shaft horsepower.

The firing circuits of the silicon controlled rectifiers are controlled by solid-state regulator circuits including microprocessors to provide the desired voltampere characteristics. Typically such regulator features would include: (*a*) voltage or speed regulation, (*b*) current limit, (*c*) current rate, (*d*) acceleration rate, (*e*) load balance, (*f*) voltage drop, (*g*) kVA limit, (*h*) kW limit.

Inverting capability is provided by the regulator circuits in conjunction with dc motor field reversal, thus enabling the system to slow down and reverse the propellers quickly by pumping inertial and hydraulic energy back into the power source.

Several icebreakers have been built using cycloconverters that are more costly than dc systems, but have the advantages of dc and also permit operation at low speeds at high torque and allow the use of very large ac motors.

***94. DC Drive Motors.*** Shunt-wound separately ventilated dc motors designed for SCR duty meet the needs of most propulsion. Vertical motors predominate for thrusters and thrusting drives. Occasionally, nonreversible series motors are applied to propeller drives where a nozzle or shroud is used to change the direction of thrust.

***95. Status of Ship Automation.*** Automation of shipboard systems, particularly propulsion and ship's-service power systems, has made very rapid progress in recent years. Practically all major new-construction ships are now automated to some degree, and programs are under way for modifying existing vessels by the addition of automatic equipment. The Department of Transportation, United States Coast Guard, has issued a Navigation and Vessel Inspection circular covering automated main and auxiliary machinery. This document was prepared as a guide to Coast Guard inspection personnel, shipbuilders, ship designers, and operators. IEEE and ABS documentation also discusses this subject in considerable depth, and spells out requirements to comply with the standards and rules involved.

ABS certifies automatic and remote-control systems by one of the following three classifications:

ACC—Automatic Control System Certified (constant machinery surveillance by at least one attendant)

ACCU—Automatic Control System for Unattended Engine Room Certified

ABCU—Automatic Bridge Control System for Unattended Engine Room

The rules identify the functions and capability of the equipment in considerable detail. In addition to reviewing the plans, ABS surveyors inspect the consoles at the manufacturer's plant as well as after installation on the ship.

***96. Purpose.*** In the marine industry, *automation* is used to describe systems of sensing, monitoring, indicating, and actuating equipment that get work done more economically, safely, and with improved reliability using automatic and remote-control equipment. Data logging of certain parameters is provided, although this should be confined to a minimum of essential items. Automation systems are aimed at reducing the work done by people and making the necessary work more easily accomplished. These automated systems are known by various names, including centralized control station, engine-room control station, central operations system, and machinery-control center. Dynamic displays of ship subsystems and trend displays are among the many special features available. Built in Test Equipment (BITE) systems facilitate servicing by technicians. An elementary block diagram of such a system is shown in Fig. 23-33.

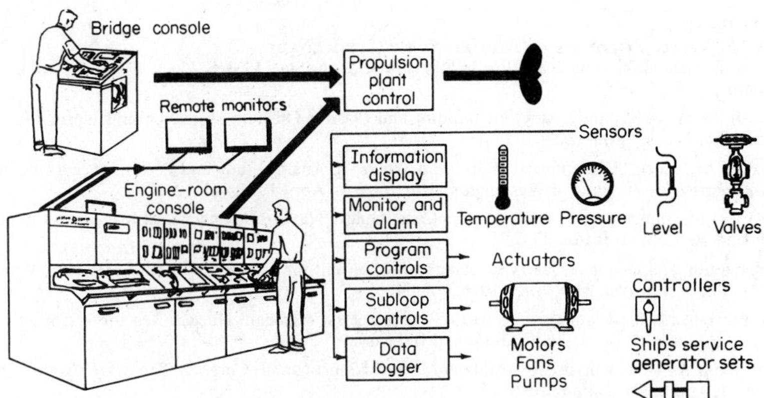


**FIG. 23-33** Block diagram of a large automated system.

The trend is to reduce the number of engineering-watch standers, for example, elimination of three-man watch on a steam plant. This can result in unattended machinery operation, and requires an extensive monitoring, control, and alarm system. Engineering personnel must be readily available to respond to the alarms as they may occur. All ships currently at sea require a manned bridge. Development work on the completely unmanned ship with satellite data communication is underway. Eventually this may be a feasible mode of operation, but some years remain before this can be anticipated as the usual method of navigating a ship at sea. Single-man engineering-watch standing is considerably easier to arrange for than completely unattended operation. Very high reliability is demanded from the equipment involved in the various automation systems. Unattended systems require remote monitor and alarm stations that normally consist of a color TV display with audible and visual alarms.

*97.* ***Advantages of Automation.*** Shipowners who install automation equipment are endeavoring to reduce operating costs by the following:

*a. Reduction in Operating Personnel.* Ships with well-automated power plants can be operated with fewer personnel. Direct propeller speed and direction control normally are accomplished from the bridge, and maintenance and engineering personnel are always on duty or on call.

*b. Improved Operating Efficiencies.* Improved and readily available data, as well as self-regulating control subsystems, reduce fuel consumption and help reduce maintenance requirements. Automatic programs are available for safely optimizing the loading and unloading of certain ships, large crude-oil tankers, for example. By proper loading, unloading, ballasting, or deballasting, very large vessels can be loaded or unloaded in many cases in less than 12 h. The programs involved will automatically fill or empty the proper fresh-water, ballast, fuel-oil, and cargo tanks in the proper order to avoid any safety problems which would be associated with unbalanced loading. With the cost of vessels running into the millions of dollars, and power plants from 15,000 to 50,000 shaft horsepower, time and fuel costs assume tremendous importance in the overall economics of operating these vessels. Displays alert personnel to impending failures.

*c. Better Vessel Utilization.* Reduced operating personnel reduces the required crew living spaces and support services, leaving more space for cargo. Crew sizes for recently built large tankers are generally under 20, with 10 or so a target after a year or two at sea. One recently built ship with a gas-turbine electric propulsion plant will go to sea initially with under 20 men. Ultimately it is expected that a crew of only 13 will be required. Some new ships have a crew of 8 and the machinery spaces are staffed only for service calls.

## Bibliography

***98. Reference Literature on Marine-Power Applications***

American Bureau of Shipping, Rules for Building and Classing Steel Vessels, 1985, 45 Eisenhower Drive, Paramus, N. J.

American Bureau of Shipping, Rules for Building and Classing Offshore Mobile Drilling Units, 1985, 45 Eisenhower Drive, Paramus, N.J.

Beverley, J. A., Koch, R. L., Stewart, E. C., and Weiks, J.: Analog Study and Performance of the New 440-Ft Ferries for the State of Washington; *Mar. Techol.*, April 1975, vol. 12, no. 2.

Department of Transportation, United States Coast Guard, Navigation and Vessel Inspection Circulars 1-69 and 10-82, Washington, D.C.

Fox, Benjamin, and Coleman, Harry C.: Alternating Current for Auxiliary Plants of Merchant Vessels; *Trans. Soc. Nav. Archit. Mar. Eng.*, 1946, vol. 54.

Geary, Elmer A.: N.S. Savannah—Electrical System in First Merchant Nuclear Reactor System; *Trans. AIEE,* 1962, vol. 81, pt. II, Applications and Industry.

Hall, William A.: Determination of Shipboard Electric Short-Circuit Currents; *Trans. AIEE,* 1962, vol. 81, pt. II, Applications and Industry.

Hansen, H. H.: Voltage Regulation of D-C Power Supply with Suddenly Applied Load; *Trans. AIEE,* 1962, vol. 81, pt. II, Applications and Industry.

IEEE Committee on Marine Transportation, Recommended Practice for Electric Installations on Shipboard, IEEE Std., 45-1983, New York, Institute of Electrical and Electronics Engineers.

Jacobsen, William E.: *Marine Engineering;* New York, Society of Naval Architects and Marine Engineers, 1971, chap. X, Electric Propulsion Drives.

Jacobsen, William E., and Koch, Richard L.: Diesel-Electric Propulsion for Polaris Submarine Tender; *Nav. Eng. J.,* August 1962, vol. 75, no. 3.

Koch, Richard L.: An Analysis of AC Electric Transmission Systems; *Trade Winds,* November 1973, General Electric Company, Schenectady, N.Y.

Melvin, Burr: *Marine Engineering;* New York, Society of Naval Architects and Marine Engineers, 1971, chap. XVII, Electrical Plants.

Smith, Frank V.: Electric Ship Propulsion, in *Marine Engineers' Handbook;* New York, McGraw-Hill Book Company, 1945.

Society of Naval Architects and Marine Engineers, "Marine Diesel Power Plant Performance Characteristics," T & R Bulletin 3-27, New York.

Thornbury, J. W.: Current Electrical Systems and Equipment for Warships; *Trans. Soc. Nav. Arch. Mar. Engs.*, 1961, vol. 69.

## ELECTRICAL APPLICATIONS IN RAILROADS

*By FREDERICK N. HOUSER and ROBERT W. MCKNIGHT*

### General Advantages

***99. History.*** During a 40-year period, beginning about 1895, the engineering and economics of classical railway electrification were developed and refined. In this period electrification overcame a number of major railroad problems, making possible numerous improvements and economies in operation. During its first 30 years railroad electrification was thought of almost solely in terms of power generated at central stations and distributed to trains by third rail or overhead trolley wire.

By 1925, the internal-combustion engine had been refined so that its size, rating, and reliability allowed it to be considered as a prime mover for railway service. After early attempts at producing mechanical transmissions had proved generally unsuccessful, electrical transmissions were found satisfactory as torque converters for gasoline and diesel engines. During the fourth decade of the twentieth century the internal-combustion engine became a significant factor in railway motive power. The following 25 years saw the diesel locomotive with electric transmission eliminate completely the steam locomotive in many parts of the world. The diesel-electric at times also displaced short stretches of conventional electrification which had been installed to overcome specific operating problems on otherwise all-steam lines.

By 1975, after 80 years of electrified U.S. mainline operation, and despite the fact that advanced technology was available domestically and from Europe and Japan, much of the first-generation American electrification had been scrapped.

***100. Reasons for Electrification.*** Electric traction has been applied for a variety of reasons since its initial use in 1895. The successful utilization in locomotives of diesel engines with electric transmissions has tended to cloud what were once clear-cut distinctions between the characteristics of operations based on central-station electrification and operations based on steam locomotives.

Smoke abatement was once a major advantage of electrification, especially in thickly settled districts near terminals and in lengthy tunnels.

Mountain grades often presented major operating problems with steam-locomotive operation which could be alleviated by electrifying. Because the slow-speed traction characteristics of electric and diesel-electric locomotives are practically identical, the diesel has readily supplanted full electrifications installed solely because of the high torque which can be developed in the series traction motor at low speeds.

Dynamic or regenerative braking, possible with locomotives and cars fitted with traction motors, has many operating advantages on grades and even in level territories where decelerations are frequent. Dynamic braking involves dissipation of energy in resistor banks mounted on the vehicle where the kinetic energy of the moving train is converted into electrical energy by operating the traction motors as generators. In regenerative braking, possible only with full electrification, the current produced by the traction motors is fed to the trolley wire so that a train descending a grade may actually be producing the electrical energy to power a train ascending a hill elsewhere on the same line. While regeneration is relatively simple in dc electrifications, the complications of phase matching and other problems have made it much less frequently used on ac electrifications. The advantage of concentrating the retarding effort on the motive-power unit so that the air-brake system need not be utilized is so great that many fully electrified lines are completely satisfied with dynamic braking where there is absolutely no feedback into the power system. In all cases there must be an air-brake system to take over in the event of a failure of the electric braking system and to bring the vehicles to a complete stop and hold them; the effect of electric braking decreases as the armatures rotate at low speeds and stops completely when they cease to rotate.

Another method of electric braking which received attention in the mid-1970s was flywheel energy storage in which a decelerating car or locomotive would transform the power produced into kinetic energy which would then be utilized to aid in acceleration when it was needed.

Suburban service in metropolitan areas places demands of service frequency and reliability on railroad plants which at one time could be met only by electrified operation. The rapid acceleration made possible by motors distributed throughout the train permits high schedule speeds without increasing the maximum speeds. The result of such operating characteristics is maximum capacity of crowded tracks during rush hours. Diesel-powered railcar trains with individually powered units can approach the accelerations which are possible on an electrified line. The possibility of using lightweight, high-performance gas turbines with short-time characteristics even more comparable with those of electric traction motors has been investigated. In the case of transit lines with subway operation, the possibility of utilizing anything but full electrification appears remote.

Transit and suburban operations impose requirements of operating flexibility which once could be achieved only with electric multiple-unit cars. The ability to change train consists rapidly and to reverse direction with no switching was once possible only with electric cars. Individually powered rail diesel cars have the same flexibility as electric multiple-unit cars. A recent development is push-pull operation, where conventional unpowered passenger cars are fitted with the multiconductor train line which is used for diesel-locomotive control and where certain unpowered cars are fitted with operator's compartments from which a trailing diesel locomotive can be controlled. Trains operating into stub-end terminals under push-pull operation have the diesel locomotive on the outbound end. In the outbound, or "pull," operation, the train is operated in the conventional manner from the locomotive cab. On the inbound, or "push," operation, the train is controlled from the cab in the leading passenger car, with the locomotive operated remotely at the far end of the train. By fitting all, or certain, intermediate coaches with cab controls, train consists can be varied almost as readily as with regular electric multiple-unit or rail-diesel-car operation. All types of multiple-unit operation involving cars, locomotive units, or cars and locomotive units are based on the pioneering work of the traction pioneer Frank J. Sprague, who developed the concept before the turn of the century (see Par. **187**).

***101. Fuel Saving.*** Full electrification is most popular in regular line-haul operations in countries where hydroelectric power is cheap and plentiful or where fuel oil is scarce or expensive or must be imported. Many early electrifications were justified economically on fuel-cost savings as compared with steam locomotives. Because the thermal efficiency of the diesel locomotive is four to five times greater than that of its steam predecessor, its thermal efficiency from fuel source to power output at the rail is approximately the same as the efficiency of electrified operation when all losses from the power plant through the transmission system to the locomotive are considered. In an era of declining liquid-fuel supply, however, the electrified railroad does enjoy the benefit of being able to utilize directly power generated from coal or nuclear sources, a characteristic it shares with none of the other land-transport modes except pipelines.

***102. Capital Charges.*** Installation of the power-transmission and -distribution system is a heavy charge against electrification. The maintenance of these facilities also is costly, and in some circumstances taxes are increased. If density of traffic is high, these costs may be divided by large train mileage, and the burden will not be great, but if density of traffic is low, this may be a serious obstacle to economic electrification.

In general, it may be said that straight electrification is most advantageous if there are

**a.** Dense traffic.

**b.** Traffic evenly distributed throughout the day and year.

***103. Maintenance Costs.*** Because full electric locomotives and cars have no prime movers, their maintenance costs are lower than for diesel or turbine-powered units. This must be countered, of course, by the maintenance costs of the transmission, conversion, and distribution facilities that are involved in full electrification. Experience has shown that, despite the higher maintenance costs of diesel-powered equipment, there is about the same availability, due largely to the standardization of servicing procedures, preventive maintenance practices, and standardized parts, which make rapid repair possible.

Ownership costs of electrically powered motive power must obviously be lower than those of independently powered units if complete electrification is to be justified and if there is no significant difference in operating costs. Experience has shown that electric locomotives are capable of economical operation for 30 years or more. The economic life of diesel-electric units has been approximately 10 to 15 years for road power and 20 to 25 years for switching locomotives.

A major advantage of electric motive power is its intrinsic overload capacity for short periods. This serves to increase greatly the effective rating of diesel-electric locomotives at low speeds and of electric locomotives throughout their entire speed range. As an example, a short grade may require capacity for, say, 15 min, which may not be required for longer periods and for which the overload capacity of traction motors and transformers may well be suited. Starting from stops or accelerating after slowdowns may also utilize overload capacity to advantage in maintaining fast schedules with heavy and high-speed trains. In preparing for full electrification, it is vital to know the characteristics of the traffic to be handled.

## General Principles of Train Operation

***104. Capacity of motive power*** to perform a given service depends upon the profile of the line, curves, weights of trains, frequency of stops, and schedule speed required.

***105. Frequency of stops*** is especially important, since the more stops for the same schedule speed, the more time is consumed and the greater the need for acceleration, hence the more horsepower required.

***106. Power of the locomotive*** is applied at the drawbar, and the weight of the locomotive for a given drawbar pull must be such that friction to prevent slipping will exist between the driving wheels and the rail.

***107. Train Resistance.*** The tractive effort is used in overcoming certain factors, that is, normal train resistance, grade resistance, curve resistance, wind resistance, air resistance, and acceleration, etc.

Basic train-resistance figures may be used to advantage. Values calculated will provide a basis for determining locomotive characteristics to meet particular speed requirements, designing for distribution system and substations, and determining energy consumption and power demand.

***108. Speed-Time Curve.*** Movement of a train can be represented graphically by what is known as a speed-time curve. After each stop the train must accelerate—at first perhaps rapidly, then more slowly until a balanced speed is reached. This balanced speed is maintained until a point is reached when the power is cut off and the train coasts until the brakes are applied to bring it to a standstill. The cycle can be indicated graphically (see Fig. 23-34). Acceleration is shown by the line *AB;* constant speed by the line *BC;* coasting, *CD;* braking, *DE;* and stop, *EF.*

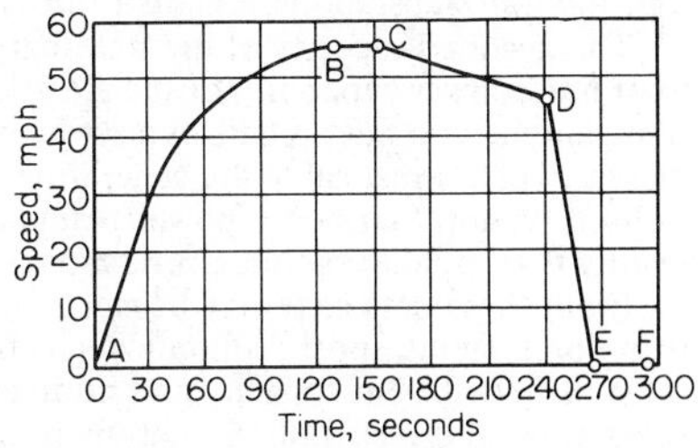


**FIG. 23-34** Typical speed-time curve.

***109. Distance-Time Curve.*** Distance is the first integral of speed wtih respect to time; hence the area included beneath the speed-time curve is a measure of the distance covered up to any point. The distance-time data may be constructed from the speed-time curve by a process of integration or by using one of the computerized train-performance programs which are widely available.

***110. Current-time curve*** can then be plotted for each run. From the motor-characteristic curve, the current for each value of tractive effort and speed can be transferred to the speed-

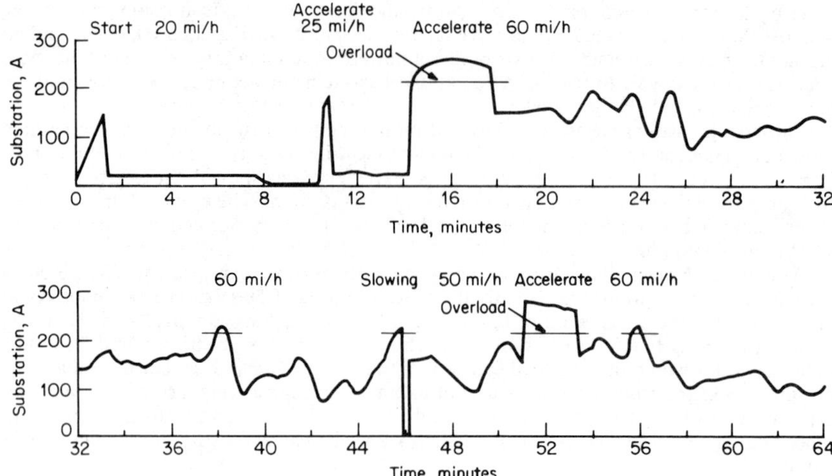


**FIG. 23-35** Typical current-time curve.

time curve, and the current requirements obtained (Fig. 23-35). The power input at any point is measured by the current times the voltage, which may be assumed constant for this analysis. (In ac power systems in figuring current requirements this must be divided by the power factor.) A *power-time curve* may be thus plotted.

***111. A graphic train diagram*** is built up, usually plotted with time as a base on a 24-h day and geographical distances as ordinates. This will show at a glance how many trains are on the line at any given instant and therefore, roughly, the power demand at that instant.

***112. Total power requirements*** in an electrified section for the system may be determined by adding the simultaneous values of all the power-time curves of trains on the system at one time. Average power per train can be determined by integrating the power-time curve over the total time and dividing by the same period. For ordinary purposes the average power may be multiplied by the total number of trains in order to arrive at the demand. The maximum possible demand will occur if all trains are starting simultaneously, which may be the case if schedules are disarranged for any reason. The energy input to the train is determined by integration of the power-time curve, and the summation of the energy input to all the trains gives the total energy. Such analysis of the demands and energy requirements is essential in designing a power plant or in negotiating for purchase of power. In figuring power-supply requirements, losses in the motive power and losses in power transmission, substations, and distribution to the moving trains must be taken into account.

The installed capacity of the stationary power plants does not have to be as great as the total horsepower capability of the electric locomotives they will supply, since all locomotives supplied do not operate at anywhere near full capacity at one time. This is not true of diesels, which must carry the generating capacity for full output at all times, whether it is utilized or not. Stationary power plants also lend themselves to ecological controls more readily than smaller mobile prime movers on locomotives.

From these data also may be calculated the characteristics of the transmission and distribution systems, most economical locations and sizes of substations, etc. The substations must be of sufficient capacity and number to supply the loads imposed. The transmission, substation, and distribution analysis is similar to that for any other type of power distribution, except that in the railroad the load center moves about. A balance must be maintained between cost of power losses and the capital and operating charges in the system, just as in any other power network.

Substations are located naturally in the vicinity of crossover, or interlocking, facilities, where trains may change from one track to another. At these points it is logical to sectionalize the distribution circuits in order to segregate electric troubles which may occur.

***113. Momentum Grade Operation.*** When a train is operated over a rolling profile, it will normally lose some speed while ascending grades. This loss of speed represents a portion of the kinetic energy of the moving train which is converted into some of the work of moving the train up the grade. This kinetic energy was produced by the work of the locomotive as the train approached the ascending grade, assisted possibly by a descending approach grade. In railroad circles this is commonly known as momentum operation.

In many cases, advantage may be taken of this form of operation to increase the maximum tonnage rating of a locomotive over a given profile to a value greater than that which would obtain if the locomotive were required to start the train on any grade or to maintain some specified minimum speed on any grade without benefit of momentum. The amount of increase depends upon a number of factors, including the type of locomotive, the length and steepness of the ascending grades, the speed at which the train can approach the grades, the average weight of cars in the train, and maximum permissible value of tractive effort. The latter may be determined by adhesion limits or, in some cases where electric propulsion is used, by the necessity of protecting the electrical apparatus from overheating. The minimum speed corresponding to this maximum tractive effort may be determined from the characteristic curves of the locomotive.

The relationships of speed, tractive effort, train resistance, and variations in profile are too involved to permit the development of a simple formula for taking account of this momentum effect. Computer programs offer ready solutions to these problems.

***114. Tonnage Rating.*** If the longest grade on the profile is also the steepest, it will determine the maximum tonnage rating. This will be the tonnage which the locomotive can haul up the grade at the minimum speed corresponding to the maximum permissible tractive effort or motor heating limits.

A greater gain in tonnage rating by this method can be expected for electric locomotives than for any type of prime mover locomotive. This is because of the short-time overload capacity of electric locomotives and their ability to draw power substantially greater than their continuous rating from the contact system.

## Systems of Electric Traction

***115. Systems.*** Electrification of railroads has been accomplished by means of a number of different methods mainly characterized by the system of distributing the power to the locomotives or motor cars. The principal systems may be designated as follows:

***116. Alternating-Current Single-Phase Systems.*** These systems utilize an overhead contact wire which formerly was always fed at lower-than-standard frequencies in order that the power could be used by commutator-type ac traction motors. In the United States this has been 11 kV, 25 Hz, and in Europe often 15 or 16⅔ Hz. Such a system requires frequency changers or a special generating network.

Commercial frequency (50 Hz in Europe and 60 Hz in North America) is now used exclusively in new ac electrifications. The change has come about as a result of the perfection of high-power solid-state rectifiers which make possible the use of dc traction motors on ac electric locomotives. All recent electrifications that are not constrained as extensions of existing systems are 25 kV, 50 Hz, in Europe and Asia. Both 25- and 50-kV 60-Hz electrifications have been successfully operated on U.S. industrial railroads. In studies of U.S. mainline electrification, it appears that 12.5, 25, and 50 kV would be used advantageously, depending on the electrical clearances existing in each application. The electrical power conditioning systems of locomotives and multiple-unit cars can be designed to operate successfully from two or more ac or dc supply voltages.

Commercial-frequency (50- or 60-Hz) distribution seems destined to be used on all future main-line electrifications. The light, single-phase contact system with its necessary circuit breakers represents the lowest-cost fixed installation which has ever been possible in the railway industry. Step-down transformers, supplied by the utility and covered in the power-rate structure, can be fed from the central power stations that serve commercial and residential loads. In all today's highly industrialized nations the load represented by even the heaviest-traffic rail lines is relatively small. Where once there was concern about imposing a main-line railroad or rapid-transit load on a public-utility network, today's rail loads are being handled readily.

Locomotives taking power from the single-phase distribution systems are, in general, of four types: (1) step-down transformer, with single-phase series-commutator traction motors; (2) phase converter, either synchronous or asynchronous, taking in single-phase power and delivering 3-phase power to induction-type traction motors; (3) motor-generator sets, where single-phase power in a synchronous motor drives a dc generator furnishing power to series-commutator traction motors; and (4) static rectifiers to change the alternating current to direct current for delivery to standard dc traction motors.

***117. Alternating-Current, 3-Phase System.*** This system originally utilized *two* overhead contact wires over each track, the third leg of the 3-phase circuit being connected to the rail, at ground potential. While this arrangement has disappeared completely, some work has been done in applying a 3-conductor 3-phase third rail for rapid-transit power supply.

***118. Direct-Current System.*** This system comprises distribution by means of overhead contact wire or third rail, depending upon the impressed voltage, which is usually nominally 600, 1200, 1500, 2400, or 3000 V, the power return being through the running rails, as in the single-phase system.

As much as 1000 V dc may be satisfactorily carried in a third rail, but higher voltages are usually carried in overhead-trolley wires.

*Rapid-Transit Systems.* The distribution is usually at a nominal voltage of 600 to 750 V. Rapid-transit systems are nearly always operated with a third rail. The traction motors are usually series-commutator type.

***119. System Comparison.*** The commercial-frequency (50- or 60-Hz) high-voltage (25- or 50-kV) ac system of railway electrification seems destined to be used for all major extensions. Only relatively short extensions of existing systems will utilize direct current or lower-voltage lower-frequency alternating current. Modern electronic techniques make practicable dual-, triple-, or even quadruple-voltage motive power, which can operate efficiently over routes where current must be collected from distribution systems at different voltages and frequencies and where even alternating and direct current must be utilized.

Commercial-frequency high-voltage ac electrification makes possible smaller support structures, lighter-weight and simplifed overhead wiring, and fewer substations. When such installations are properly proportioned, the cost of changes to the communication circuits will not be large and will produce a modern plant with greatly reduced maintenance and interference from the weather. Signal changes are necessary with ac electrification, but recent developments also make it possible for the costs of such changes to be minimized. Only in areas of very low overhead clearances is the cost of securing adequate electrical clearance for high-voltage systems an important factor. Commercial-frequency ac systems eliminate many problems of electrolytic corrosion.

## Power Supply

***120. The high-power controlled semiconductor rectifiers*** have opened a new era in railway electrification. The potential for the thyristor in power engineering can be compared with the revolution brought about by the transistor in communication electronics in the 1950s. Wherever electric power is used and the associated devices and machines must be switched and controlled, electronic components prove to be inertia-free, require no maintenance, and can be operated in any position; being unaffected by vibration, they are superior in every way to predecessor devices employing contacts. In some cases, as is true with railway electrification, they make solutions possible to problems which could not previously be overcome technically or economically. Semiconductors have altered many of the traditional approaches to railway electrification, and even to the electrical systems which diesel-electric locomotives carry.

***121. The single-phase system of distribution*** standard in the United States has been 25 Hz, with nominal trolley voltage of 11,000 V. In Europe the frequency was generally at 16⅔ Hz, though on an occasional railroad 25 Hz had been adopted. The trolley voltage in Europe was generally 15 kV. Since 1960, European railway systems have almost universally adopted 50-Hz, 25-kV distribution; the United States has chosen 60 Hz.

***122. Direct-current-system distribution*** is at voltage ranging from 600 V nominal potential to 1200, 1500, 2400, and 3000 V. The higher voltages are always carried in overhead trolley wires. It is obvious that the higher voltages will require relatively less copper or equivalent in the distribution circuits and that substations may be more widely spaced with higher-voltage circuits, all else being equal.

***123. Relatively greater voltage fluctuations*** may be tolerated in railroad distribution circuits than in commercial circuits, where dips may be troublesome to the users. The chief consideration, within limits of proper operations, lies, as in the design of commercial transmission lines, in the economic balance between capital charges of the power distribution system and fuel charges in heating the conducting materials. It is true that a serious drop in voltage will result in slowing the trains, and a relatively long period may affect the schedule; but if the voltage dip does not last long, the schedule is not seriously slowed. Voltage drops on the order of 10 to 15%, with occasional dips to 20%, are generally not excessive, depending upon the type of service operated.

***124. The voltage drop*** may be computed as in the case of commercial circuits from the load data assumed and the design of the distribution system which gives resistance (in dc systems) or impedance (in ac systems). The voltage is maintained at the desired point by means of feeders operated in connection with the trolley conductors.

***125. Substations.*** Railroad substations must be fed from a high-voltage line that normally extends from a utility substation. The cost of such lines is normally charged against the electrification project and thus an economic evaluation of potential substation sites must be made. The substation (Fig. 23-36) has protective switches on the high-voltage side of the transformers; metering for the utility may be either on the high- or low-voltage side of the transformer. The railway transformer must be mechanically strong and capable of withstanding many short circuits. Railway service is comparable to that of an arc-furnace supply. The single-phase substation transformer is normally connected to one phase of the three-phase system and Scott-connected transformers are not required. A 10- or 15-MVA transformer would have an impedance of 8 or 9%. Overload capability is 50% for 15 min and 100% for 5 min. The transformer must also be capable of high fault frequency (300 to 400 per year on average). Fixed taps 5% above and below nominal voltage can be provided.

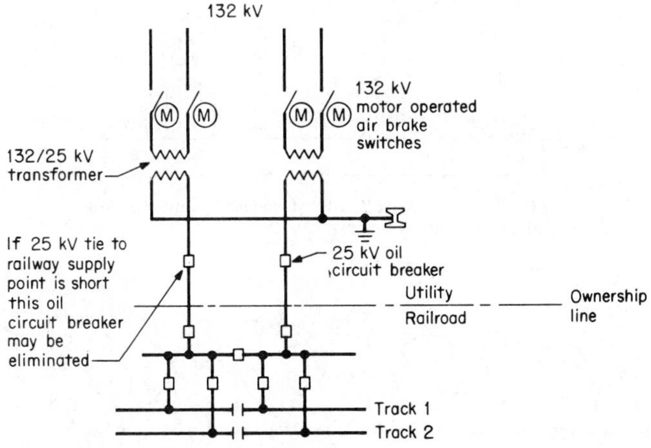


**FIG. 23-36** Typical substation switching diagram.

***126. Circuit Breakers.*** Low-side switching provides the protection from catenary faults and allows for isolation and/or interconnection, as required. Either air or oil circuit breakers, operating within four cycles, are used as the primary protection for each catenary section. On one major electrification, the second actuation of a breaker after a 15-s cycle operation will cause an interrupter switch to open, prior to breaker reclosure, thus isolating the fault. On another system, two circuit breakers are used on every catenary section. The main breaker provides fault protection. The second is a test breaker which has a series reactor to limit fault currents to 25% of normal values. The test breaker recloses up to three times to permit isolation of the particular catenary section experiencing faults. The main breaker then restores service. This low side switching must protect the system, clear temporary faults automatically (birds on lines at points of restricted clearance or insulator flashover), and isolate system elements.

***127. Grounding.*** A ground mat and rod system is required at each substation. The ground mat must be very well tied into the track system. Grounding is even more important in railroad work than in a normal power system. The current drawn by locomotives must return to the supply transformer through ground return; an additional ground wire may be part of the catenary. The proportion of the return current flowing through the earth depends on the relative conductivities of the rails, ground wire, and earth. There will be a voltage gradient along the track system. Such currents can cause corrosion problems.

***128. Trolley Sectionalizing.*** In ac or dc systems, the sections into which the contact conductor is divided may vary from a few hundred yards to several miles, depending upon the physical conditions and the importance of the service. The sectionalizing points are generally located at crossover points in order to enable trains to cross over from one track to another and run around a defective section.

***129. Phase Breaks.*** Phase breaks are necessary where asynchronous single-phase ac must be fed into different catenary sections. This results in momentary interruption of power which affects the equipment on the locomotive, produces current inrush, and may result in undesirable slack train action.

***130. Catenary Construction.*** If the power is distributed to high-speed trains by means of overhead conductors, catenary construction is utilized. The *contact wire* is supported from a so-called messenger by means of hangers of varying lengths (Fig. 23-37) so designed that the contact wire normally lies nearly parallel to the rails, the sag being taken in the supporting messenger. The contact wire, or trolley wire, is the conductor contacted by the pantograph or trolley. Typical cross sections are shown in Fig. 23-38.

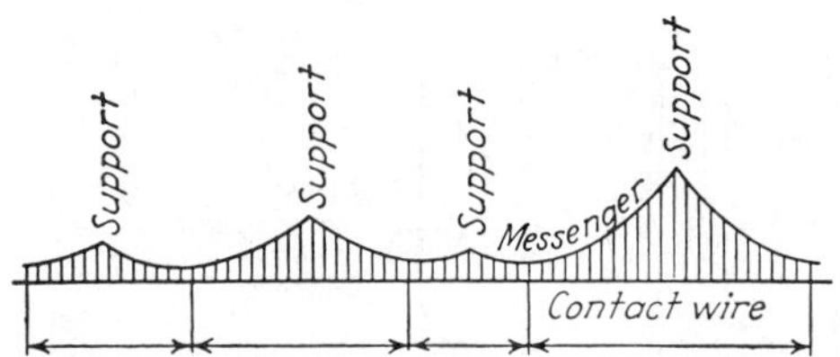


**FIG. 23-37** Varying lengths of catenary span symmetrical about points of support.

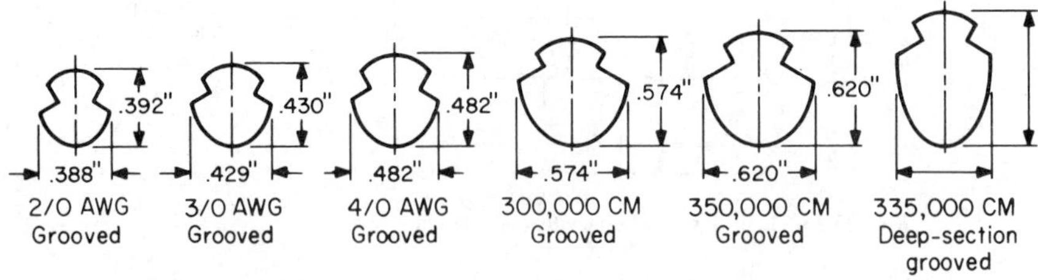


**FIG. 23-38** Cross sections of representative trolley wire.

***131. Catenary and Parabolic Curves.*** The term *catenary* is derived from the Latin *catena,* meaning "chain," and is the curve assumed by a completely flexible material hanging freely between two supports and loaded *uniformly throughout its length.* The formula for a catenary curve is somewhat complicated, so all design of overhead construction of the so-called catenary type is based upon the *parabola,* the formula for which is relatively simple. This procedure is justified because a flexible cable supported at two points and loaded *uniformly throughout the horizontal projection of its length* assumes a parabolic curve. The actual curve of the messenger supporting a contact wire lies between a parabola and a catenary, and errors are relatively small within the limits of the sags and stresses adopted in this type of work, with the accuracy that is economically possible in installing the construction also considered.

***132. Catenary Design.*** The principle of the track catenary design is the same basically as that of designs of transmission lines, but certain additional factors must be considered.

The parabola formula may be expressed thus:

$$S = \frac{w \times d^2}{8 \times T}$$

where $T$ = horizontal component of tension in messenger in pounds; $S$ = sag of messenger in feet; $d$ = length of span in feet; $w$ = weight of the system messenger, trolley wires, hangers, etc., in pounds per foot of span.

From this formula, with the weight per foot of the system (including messenger strand, trolley wire, and clips) and the permissible stress in the messenger known, the sag can be determined for any given span. The total tension in the messenger is greater than the horizontal component as expressed in the formula, the vertical component being only the dead weight of the system. Nevertheless, the vertical component is relatively so small as compared with the horizontal component (generally less than 2%) that the error may be safely ignored. A not uncommon sag is about 5 ft in a 300-ft span.

***133. Contact Wire Registration.*** To even wear on locomotive pantographs, contact wire is staggered from the centerline of the track at each support. Typically this may be 9 in either side of center on straight track. On curves the catenary approximates the track centers through a series of chords with support masts spaced to provide an acceptable level of stagger. Wind forces can displace the contact wire from its nominal position; with shorter spans wind displacement is reduced.

***134. Storm Loading.*** Edison Electric Institute has offered four separate design conditions: (a) storm loading: 80°F with 20 lb/ft$^2$ wind on wires and 30 lb/ft$^2$ wind on flat surfaces; (b) heavy loading: 0°F with 0.5 in of ice and 8 lb/ft$^2$ wind on wires plus 12 lb/ft$^2$ wind on flat surfaces; (c) heavy ice: 32°F with 1 in of ice and no wind; and (d) maximum operating wind: 65 mi/h wind with no ice.

Deflection due to wind across the track is of importance because if this is too great, the locomotive pantograph will ride off the wire in a bad storm. Deflection is calculated in the same way as messenger sag:

$$D = \frac{P \times d^2}{8T}$$

where $D$ = deflection due to wind, $d$ = span length, $T$ = combined tension of messenger and contact wires, $P$ = wind pressure per linear foot.

***135. Temperature Effects.*** An additional consideration is the temperature effect upon the messenger. Obviously, with the span fixed, the stress in the messenger will depend upon the temperature, all other things being equal, and the sag in any given distance must be so chosen that at the lowest possible temperature, as well as with the maximum assumed wind and ice loads, the stresses will not rise above a safe point.

***136. Messenger Sag.*** After the weights and stresses are determined, the sags are computed on the parabola formula—usually based upon the average temperature encountered (60°F in temperate latitudes)—and the hanger lengths determined on the basis that each hanger length is equivalent to the sag of the messenger at a span equal to twice the distance of that hanger from the low point of the span, plus a constant representing the shortest hanger in the span.

***137. Span Lengths and Hangers.*** The catenary spans usually cannot be of constant lengths but must be chosen to conform to local requirements, for example, track curves and overhead-highway-bridge structures with limited clearance. As will be noted in the formula, the sag of any span is a function of the square of the span. The stress in the messenger is thus (at an assumed temperature) approximately constant in adjacent spans regardless of the span length. It is thus convenient in laying out the catenary construction so to design it that the hanger sets or groups and messenger sags are symmetrical about each point of support. Since the messenger is horizontal at the low point in the sag or span, it is immaterial whether the span is symmetrical about the low point or is composed of two half spans of varying lengths (see Fig. 23-37). This method permits all hangers to be cut in bundles of standard lengths, varying only in number of uniformly spaced hangers chosen per span.

The foregoing analysis is based upon tangent-track construction where all hangers are normally vertical.

***138. Chord Construction on Curves.*** On curves it is sometimes the procedure to install the catenary around the curve in a series of chords, pulled off at intervals so spaced that the middle ordinate in any chord is not too great to permit safe passage of the collecting pantograph shoe (Fig. 23-39). On this basis the profile of the catenary is the same as that in tangent track. The spacing of pull-offs is dependent upon the sharpness of the curve. In general, the midordinate should not be greater than one-quarter the length either side of the center of the collecting shoe, which is usually about 4 ft long.

***139. Pull-off Chord Construction.*** The maximum length between pull-off points of a chord for a curve of a given radius, if at the point of pull-off the contact wire is 1 ft toward this to the outside of the curve and at the center 1 ft inside the centerline, is represented by

$$d = 4\sqrt{r}$$

where $d$ = span and $r$ = radius of curve.

***140. Inclined Catenary.*** It is common practice to install the catenary on curves with inclined hangers, each hanger taking a position that is a resultant of the vertical load of the contact wire and the horizontal load due to curve pull. This is known as *inclined-catenary* construction. The point of support of the messenger is offset toward the outside of the curve to a point that permits a horizontal as well as a vertical sag in the messenger (Fig. 23-40). This construction is very satisfactory for curves up to 4°. For sharper curves the offset of the messenger support is too great, and the longest hangers are too long to be satisfactory. Furthermore, not much distance is saved in pull-off spacing; so the inclined catenary construction is not much used in curves sharper than about 4°.

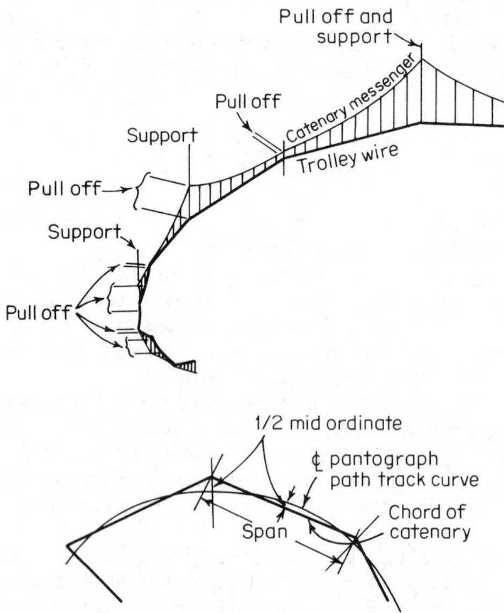


**FIG. 23-39** Catenary chord construction on curves. Perspective view, *above;* plan view, *below.*

***141. Calculating Inclined Catenary.*** In inclined-catenary construction the computations are, in general, of the same type as with tangent construction, the horizontal component of the sag being computed from the curve pull per foot, which is based on the tension of the trolley wire. With the degree of curvature of the track known, the radius and the required pull per foot of the curve are readily found:

$$F = \frac{T \times d}{R}$$

where $F$ = horizontal force normal to the wire in pounds, $T$ = tension in wires in pounds, $d$ = span in feet, $R$ = radius of curve in feet; $w$ = horizontal force per foot normal to wire, $w = T/R$.

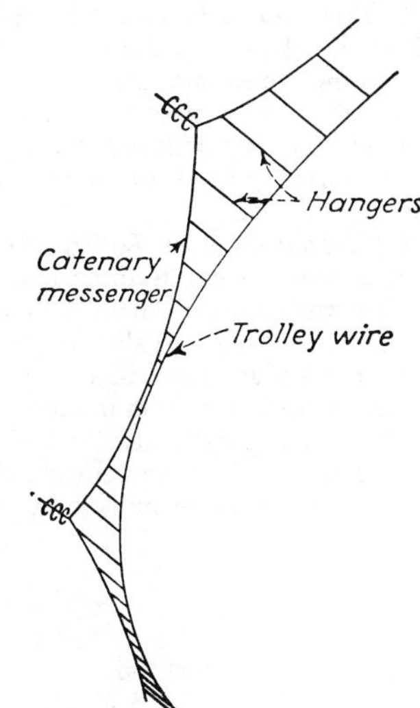


**FIG. 23-40** Inclined-catenary construction on curves, perspective view.

***142. Horizontal and Vertical Sags.*** From these data the horizontal sag may be obtained just as the vertical sag was. The resultant sag follows automatically, as does the length of the inclined hangers.

***143. Span Lengths and Sags.*** The span length most popular in this country is on the order of 300 ft between supports; recent studies have used 230 ft. The limitations are the practicable length of hangers and sags of the messenger, which, as has been indicated, varies as the square of the span. Large sags mean supports higher above the rail and therefore higher structures and greater deflection due to transverse wind loads.

***144. Contact Wire below Trolley.*** On many railroads a contact wire is installed immediately below the auxiliary or trolley wire, which thus becomes an auxiliary messenger. This design performs a double duty, in that the contact wire takes the wear of the sliding pantograph shoes and the supporting clips between hangers provide a flexible support without "hard spots," particularly undesirable in high-speed operation (see Fig. 23-41).

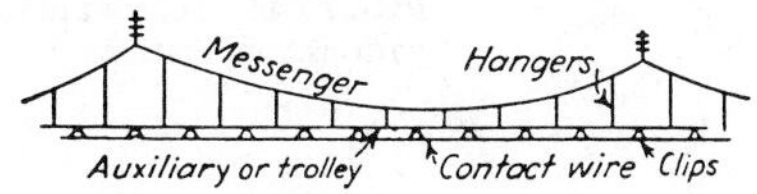


**FIG. 23-41** Typical arrangement for double trolley wire with one serving as contact wire, supported by clips between hangers from the other wire, which serves as an auxiliary messenger.

***145. Materials, Messenger.*** The material of the catenary construction depends upon local conditions. If electric conductivity is of importance, as in low-voltage distribution systems, the messenger may be of copper. Copper, however, has relatively low tensile strength. Bronze strand is sometimes used as a messenger on account of relatively higher tensile strength and satisfactory noncorrodible characteristics. A bronze alloy (silicon or cadmium) with conductivity on the order of 40 to 60% and occasionally 90% of that of copper is commonly used. Copper-covered steel strand is popular as a messenger material on account of its relatively high tensile strength and satifactory corrosion resistance. It has the advantage of a high melting point, so it is not so likely as some other materials to burn upon arcing contact. Galvanized heat-treated alloy steel of high tensile strength is sometimes employed for messenger satisfactorily, though it is subject to serious corrosion under certain atmospheric conditions. Stainless steel, which resists corrosion and has high tensile strength, is occasionally used but is costly.

***146. Materials, Trolley and Contact Wires.*** *Hard-drawn* copper trolley wire, of sizes between Nos. 2/0 and 4/0 and sometimes 6/0, is usually employed for distribution at all voltages. It is sometimes used as a contact wire but with steel pantograph shoes wears relatively rapidly. The contact wire installed below the copper wire is usually of bronze of an alloy possessing about 40% conductivity.

***147. Materials, Hangers.*** The hangers may be of stainless steel, cold-rolled galvanized steel, copper-covered steel rods, or bronze rods. Steel is relatively less expensive than other material, but copper offers advantages despite higher cost.

***148. Anchoring Trolley and Contact Wires.*** Trolley wire or auxiliary messenger and the contact wire in the United States have been anchored rigidly at each end of a section perhaps 4 or 5 mi long. Tension at installation must be so chosen that the material neither is overstressed at very low temperatures nor hangs too slackly at high temperatures. The temperature-stress curve is practically a straight line; so if the two limits, maximum and minimum stress, are correctly fixed, the wire will perform satisfactorily at all temperatures (see Fig. 23-42).

***149. Constant-Tension Trolley Wire.*** Where there is high-speed train operation, the constant-tension system minimizes the sag of the trolley wire so that there will be a minimum of intermittent current collection by the pantograph head as it passes under the wire. This system is widely used abroad.

At each end of an approximate 1-mi section, there are assemblies of weights and sheaves to maintain constant tension in messenger and contact wires under all temperature conditions. The center of each such section is anchored so that all wire movement is from the center to both ends. As temperature changes and the conductors contract or expand, the balance weight lets out or pulls in the wire to achieve a flat wire profile for the pantograph (Fig. 23-43).

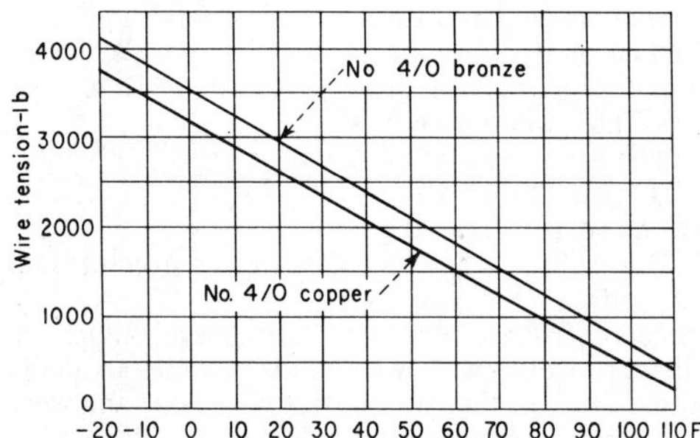


**FIG. 23-42** Tension in bronze and copper trolley wire at various temperatures.

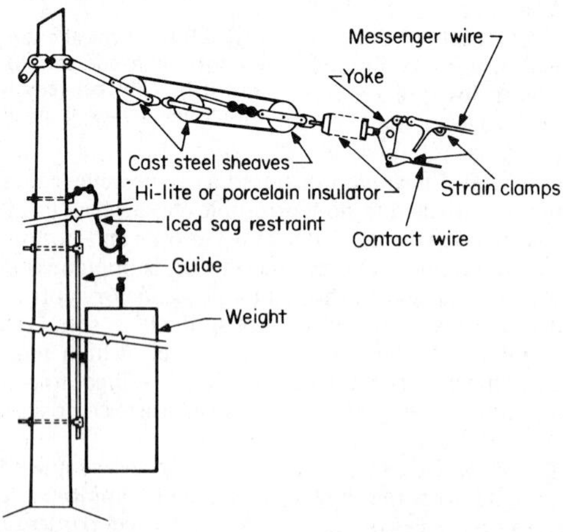


**FIG. 23-43** Constant-tension assembly.

***150. Types of Catenary Design.*** There are a number of different designs ranging from the simple catenary discussed above, in which the contact wire is suspended from a single messenger, to the compound catenary, where the contact wire is suspended from a secondary messenger (or auxiliary wire) which is in turn suspended from the primary messenger. For 25-kV applications, the simple catenary would be suitable for speeds to 95 km/h without tensioning and up to 120 when tensioned. For compound and tensioned catenary, speeds of up to 240 km/h are possible. Typical catenary designs used throughout the world are shown in Fig. 23-44.

On Japan's high-speed line, for instance, trains operate up to 240 km/h under a catenary consisting of a copper trolley wire with a cross section of 110 $mm^2$ and an auxiliary messenger of stranded cadmium copper wires of 60 $mm^2$; the messenger, also stranded cadmium copper, is 80 $mm^2$. The total area of this overhead is 250 $mm^2$. Other overhead contact-line facilities are a negative feeder of aluminum and booster transformers for prevention of induction on communication lines. Section breaks in this 25-kV 60-Hz distribution system consist of a dead section with a 10-Ω resistor at each end to prevent wear of the pantograph

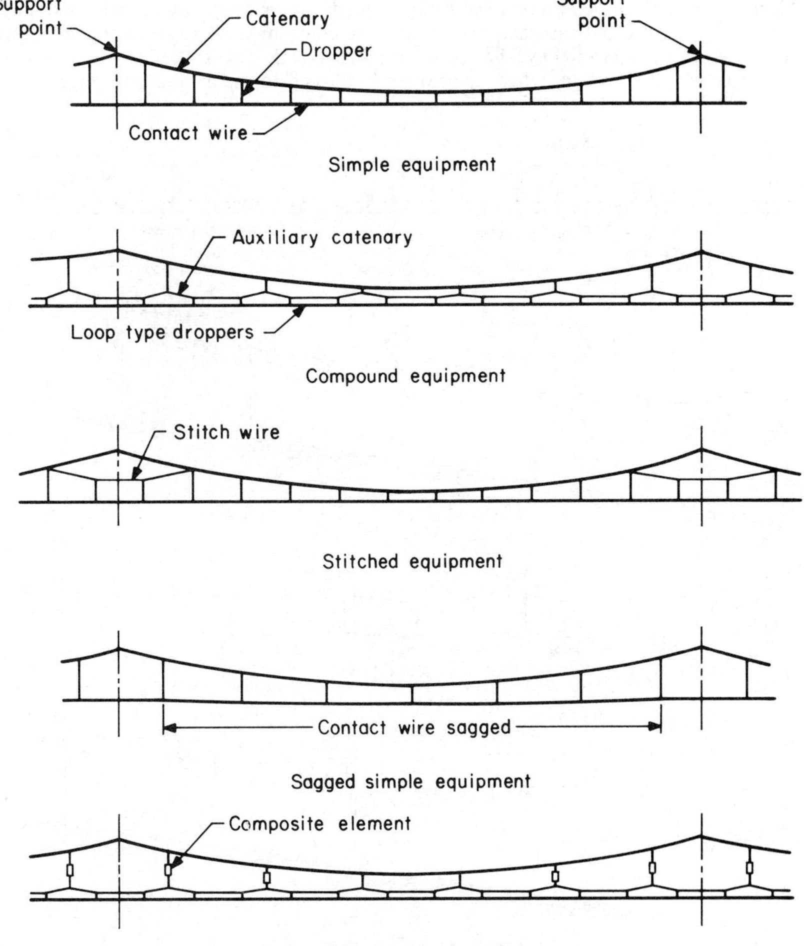


**FIG. 23-44** Types of catenary.

slider and trolley wire due to the breaking of large currents being drawn by accelerating trains.

***151. Low-Inertia Problems.*** The light overhead installation possible with commercial-frequency electrification does simplify construction and maintenance and makes possible low-cost supporting structures. At higher speeds, however, there has developed the problem of maintaining continuous contact between the trolley and pantograph because of the low inertia of the overhead. Springs and dampers are used in the Tokaido Line's contact system, adding considerably to initial and maintenance costs. Without proper damping, in either the contact wire or the pantograph, or both, serious resonant oscillation can develop, with the possibility of the pantograph shoe separating from the contact wire while drawing currents of 8 to 10 mW. A solution more economical than the constant-tension arrangement used in Japan has been the introduction of dampers in the pantographs and control of the level and stiffness of the contact wire to give the smoothest possible pantograph trajectory. Problems of satisfactory pantograph operation can be complicated when locomotives or cars with pantographs are operated in multiple units and follow each other closely under the contact wire. Because the contact wire cannot be kept at a constant height above the rail, particularly when existing routes are being electrified, its gradients must be carefully designed to avoid pantograph bounce. At high speeds a maximum contact-wire gradient of 1 in 500 has been found satisfactory. Actually the constant inertia, which the compound catenary was designed to achieve, has been reproduced by the British Railways with a lower-cost simple catenary system where the trolley is allowed to sag at midspan. Figure 23-45

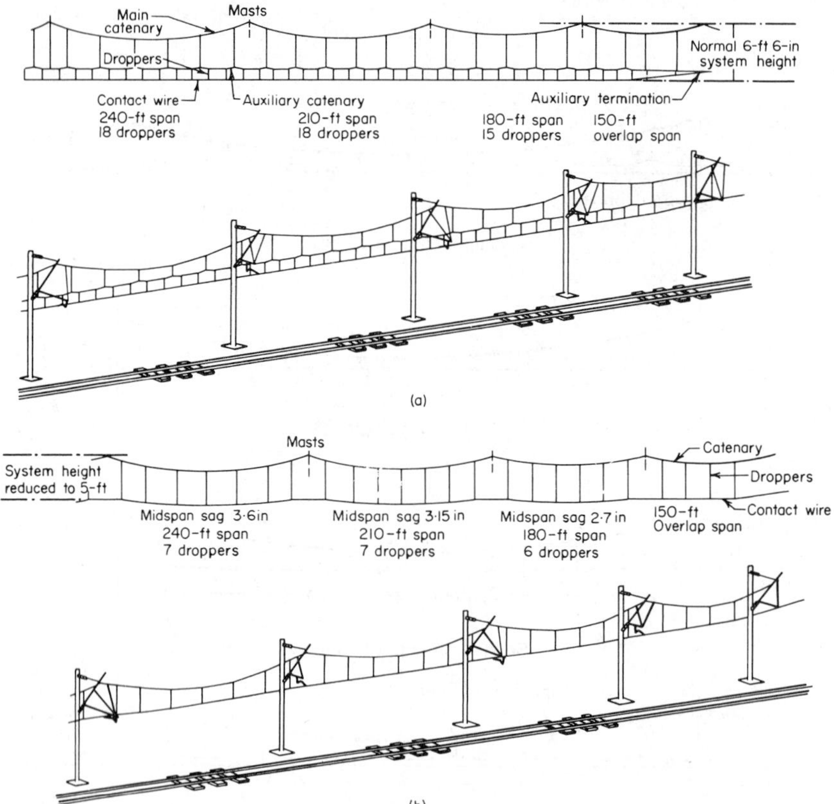


**FIG. 23-45** Lightweight commercial overload simplifies supporting structures. (*a*) Compound catenary; (*b*) sagged simple catenary.

shows not only typical catenary construction for commercial-frequency systems but the contemporary supporting structures for such a system.

***152. Supporting Structures.*** Catenary supporting structures are of varying designs. In many cases, especially when multiple tracks are involved, light, fabricated trusses or rolled sections, channels, or H beams are installed over the tracks to support the messengers. This construction has the advantage of also supporting track signals as required and also possessing a certain degree of stability along tracks in the event of messenger failure. The cross member, or bridge, is supported by posts, which may be tubular or of a rolled section. The posts may be self-supporting, or the entire unit may be of the so-called portal type, the overturning moment being taken care of by the corners, so that shear is imposed only upon the top of the foundation.

***153. Cross Catenary.*** One of the forms of construction is the cross catenary, supported usually by guyed poles, when there is sufficient width of right of way. From this cross span is hung the track-messenger construction. This type of construction is economically adapted to multiple-track electrification.

***154. Pole-and-Bracket Construction.*** For single- or double-track electrification, so-called pole-and-bracket construction is often employed (see Fig. 23-46). Sometimes the poles are located between tracks with double brackets. In some cases independent poles are used for each track. This has the advantage of complete independence of each track construction from the others—an advantage especially in the event of derailment of a train. Portal structures can also support brackets.

***155. Feeder Supports.*** The catenary supporting structures are available for carrying the traction-power feeder and auxiliary circuits, for example, signal-power-supply circuits.

***156. Clearance*** of the overhead trolley varies from a normal minimum of 16 ft above the top of the rail to a maximum of 24 to 25 ft on certain railroads.

***157. Third-rail construction*** is of two general types: (1) overrunning and (2) underrunning.

In the overrunning design the rail of low-carbon steel or steel/aluminum composite-construction is supported on porcelain ceramic or plastic insulators mounted on long track ties, generally every sixth tie. Rail joints are loosely installed to permit temperature expansion and contraction, and the joints are bonded to provide conductivity for large currents. Anchors are provided at intervals to prevent creepage. The rail is sometimes protected by a wood or composition shield to protect it from accidental contact (see Fig. 23-47). In these cases the shoe is horizontal and slips under the protective covering. In some instances when this arrangement is not practicable, the rail is protected by a strip of wood at one or both sides, the top being left open, directly under the third-rail shoe.

*Underrunning* third rails have the contact on the lower surface and are supported by gooseneck, or curved, brackets. The rail is protected by wood or composition trunking on three sides, the bottom only being left bare for the contact shoe which projects horizontally from the truck frame of the equipment. This arrangement is free from difficulty from ice and relatively free from snow trouble.

*Location of the third rail* is determined by the rolling stock passing along it. Sufficient clearance must be allowed to permit cars and locomotives of maximum dimensions to clear the rails; the Association of American Railroads clearance diagrams make provision for such location. On American railroads the distance from the track gage to the centerline of the third rail varies usually from 20 to 28 in, and the height above the top of the track rail, from zero to 7½ in.

***158. Rail Bonds.*** Rail bonds are usually stranded copper that has been swaged or welded into cylindrical terminals at right angles to the length. Gas or electric welding may be used for securing bonds to the head of the rail. Bonds may also be secured by mechanically forcing the terminals into holes in the rail by hydraulic compression or by expansion of the terminal through a drive pin in the unit.

*Welded bonds* are attached by gas or electric welding usually to the head of the rail.

*Impedance bonds* consist of a flat iron core surrounded by two coils, with the center point fed out for a terminal. This provides high impedance between rails and low impedance for the power-return current, which passes through each coil in balanced amount and, through the neutral, to the neighboring impedance bond or to cross bonding. Multiple tracks are cross-bonded at intervals to permit full advantage to be taken of all track conductivity possible. Since most electrified sections of railroads are also equipped with signal track circuits,

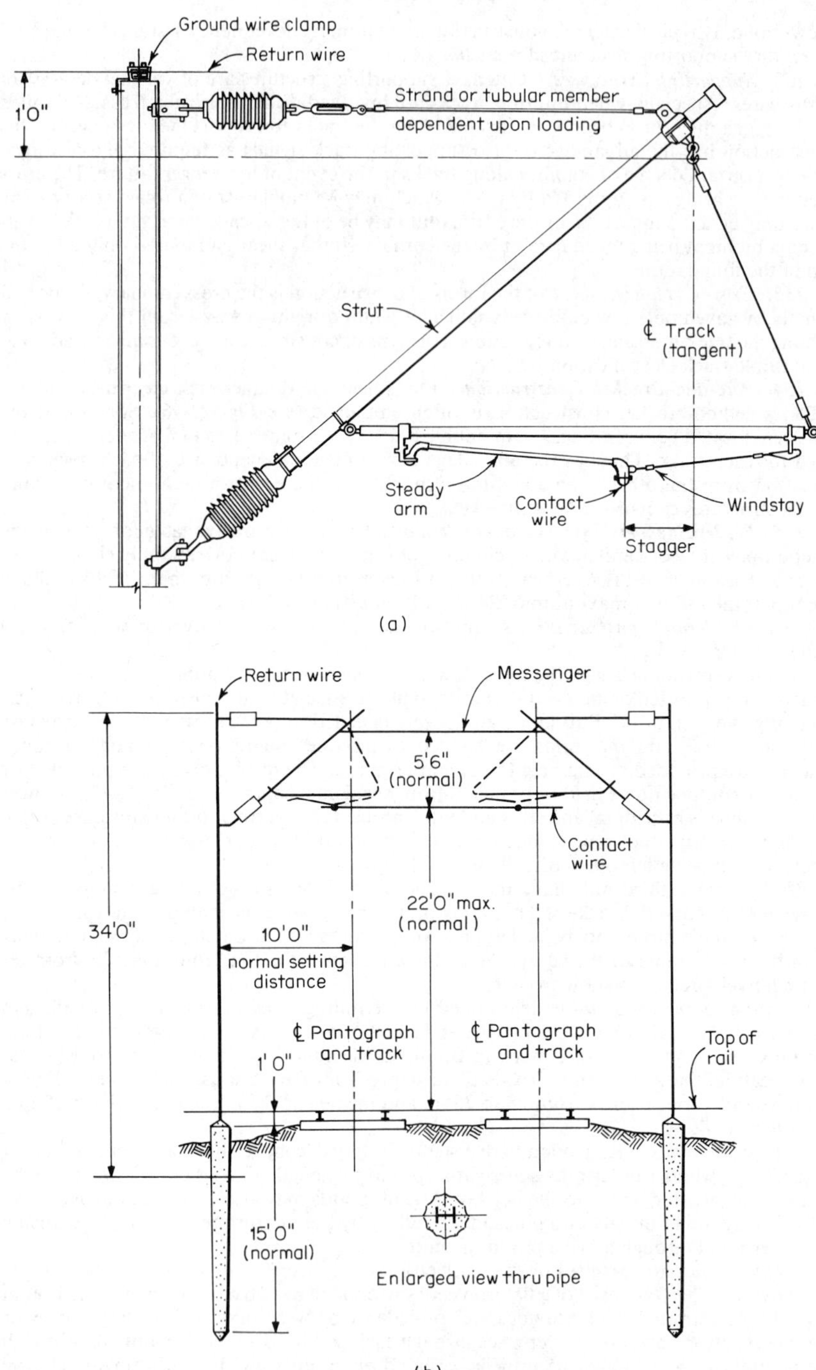


**FIG. 23-46** (*a*) Catenary support bracket assembly; (*b*) catenary structure for two tracks.

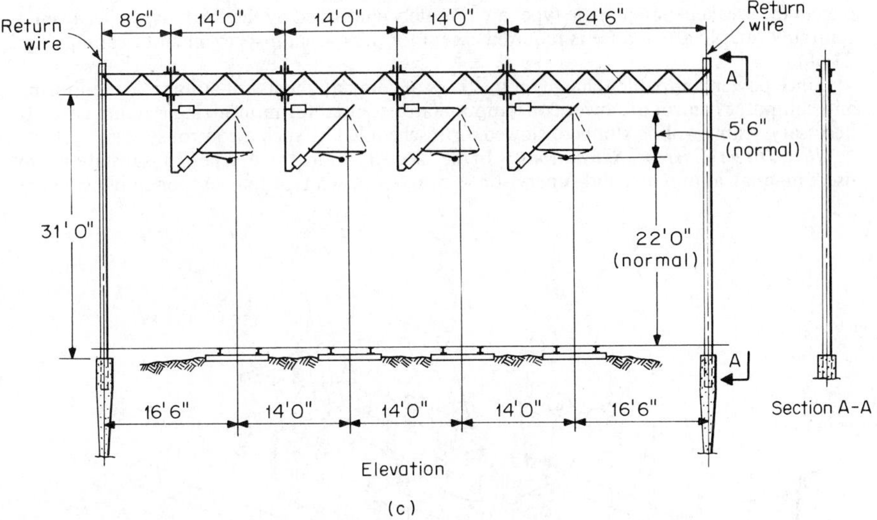


**FIG. 23-46** (*c*) Typical catenary structure for multiple tracks.

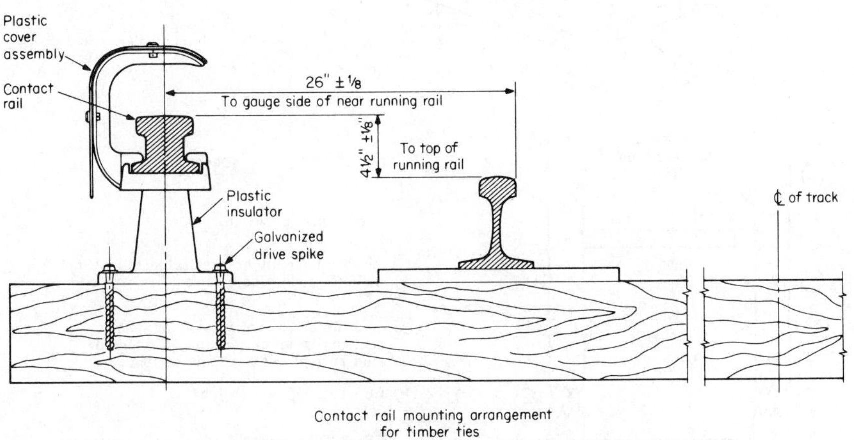


**FIG. 23-47** Overrunning third rail with top protection.

the cross bonding must be taken from the neutral points of impedance bonds, as the running rails must be insulated from each other by the ties in order to preserve the integrity of the track circuits (see Par. **220**).

***159. Substations.*** In single-phase electrification the substations are usually of two general types: (1) supply substation and (2) distribution substation.

Unless the power is to be used at commercial frequencies (50 or 60 Hz), the supply substation usually has facilities for changing the frequency from the commercial frequency to that of the railroad system. In some instances the power is generated directly at the required frequency, so the frequency changers are not necessary.

***160. Distribution substations*** for single-phase railroads are generally transformer stations with oil or air circuit breakers to control the trolley and feeder circuits, which are usually sectionalized at the substation. All this apparatus is usually outdoors. In many cases the circuit breakers (single-pole type) are remotely controlled by the load dispatcher from a central point. No attendance is required except for occasional inspection and cleaning (Fig. 23-48).

Since power companies charge on the basis of demand, the use of automatic monitoring and computer control of power consumption are essential for minimizing traction costs. If necessary, trains will be stopped, slowed, or rescheduled by such a control system.

***161. Direct-Current Substations.*** In dc electrification three types of substation are used: manual, automatic, and supervisory-controlled. Each type may use for conversion the

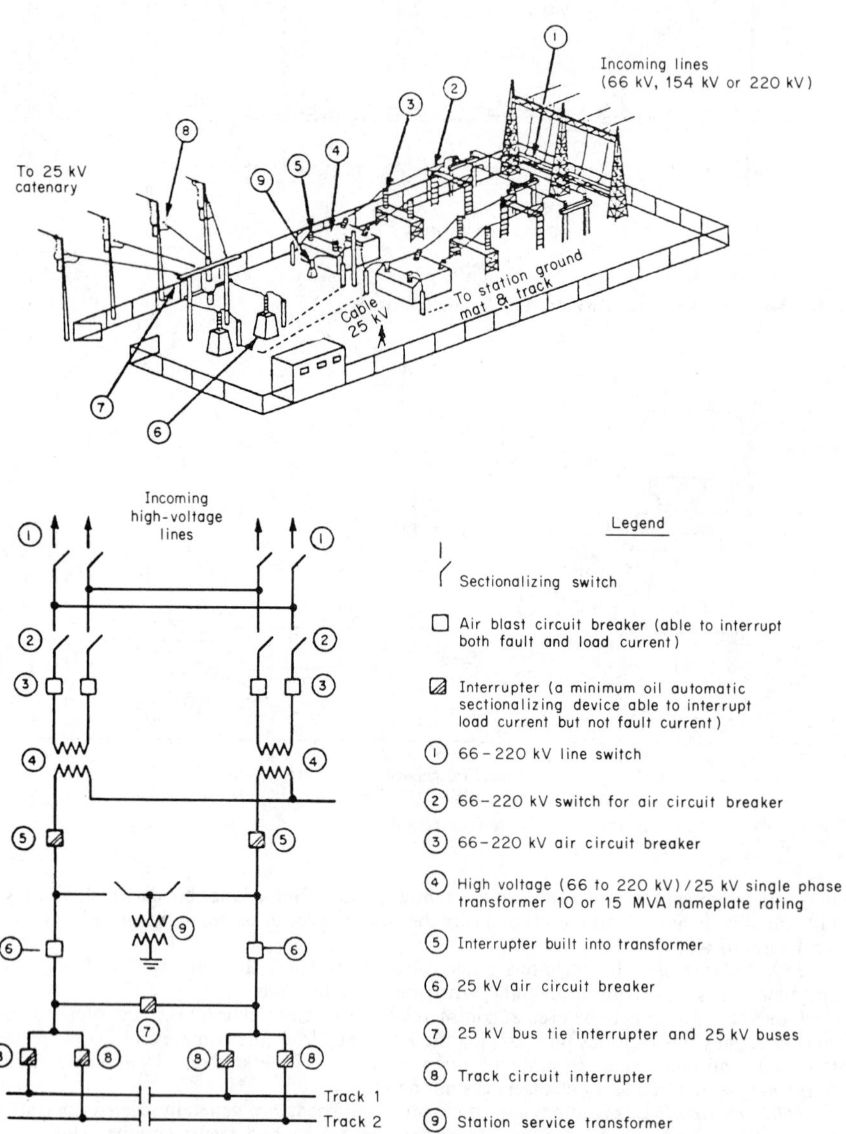


**FIG. 23-48** Typical railroad substation.

synchronous converter, the motor-generator set, the mercury arc, or the solid rectifier. *Manually operated substations* require attendants to start and stop the apparatus, open and close switches, and supervise other operations. *Automatic substations* replace the attendants with relays which automatically control the various operations in response to the load on the system or the voltage. The character and functioning of the automatic devices depend on the type of conversion apparatus and the conditions to be met. Design of the automatic apparatus is such that manual control is possible under abnormal conditions when desired.

*Supervisory control* substitutes for the manual operations relay-actuated movements, controlled from a central point. Indications of the condition of load, etc., are made at the load dispatcher's control board, who can energize circuits which perform the necessary actions at the substation. By this means the load dispatching for an entire system may be concentrated at one point, under the supervision of a load dispatcher who is thus at all times informed of the status of the load. This is similar to such operations in commercial power practice.

*Apparatus of dc substations* usually includes the following, which are common to all types: (1) incoming ac 3-phase feeders with lightning arresters; (2) high-tension circuit breakers; (3) step-down transformers; (4) switchboard, hand, automatic, or remotely operated, for control of incoming ac feeders and dc outgoing feeders; (5) conversion apparatus, details of which will vary with the types of converting apparatus; (6) outgoing feeders and switches.

***162. Synchronous converters*** for changing 3-phase alternating current to 600-V direct current were formerly standard equipment. The *compound-wound converter* is used especially with widely fluctuating output, for example, suburban or rapid-transit systems in which heavy units accelerate rapidly. The series field winding is usually adjusted for flat compounding at 600 V throughout the range of the load.

***163. Motor-generator sets*** have been used on some high-voltage dc systems instead of synchronous converters as it is difficult to design a converter running on a 60-Hz circuit for more than 750 V, whereas a generator may be built for 1500 V. Two 750-V machines may be run in series for a 1500-V electrification, but it is not considered practicable to use four in series for a 3000-V line. The efficiency of motor-generators is less than that of synchronous converters, but the former are better adpated for power-factor correction, and in some instances where the voltage of the primary feeder is low, step-down tranformers can be omitted by winding the synchronous motor for the voltage of the power supply.

***164. Electronic rectifiers*** have replaced other conversion equipment for the supply of direct current, both for 600-V city systems and for high-voltage electrifications. The connections resemble those for synchronous converters, because step-down transformers are required to obtain the correct ratio of voltages between alternating and direct current. Three-phase supply is suitable at any frequency, and the rectifier itself can be designed for any desirable number of phases, connections of the transformers changing the 3-phase power accordingly. Rectifiers can be built to have shunt or compound regulation in a manner somewhat similar to that with synchronous converters and are adaptable for regeneration of power.

***165. Advantages of rectifiers*** are high efficiency, particularly at high voltage, light weight; small floor space; and ease of automatic control.

*Operation of Automatic Substations.* Substations without attendants may be made to start automatically in various ways, but for the most part they function through remote control or automatically by means of a voltage relay, which closes its contacts at a particular value of reduced contact-line voltage. The machine remains in operation until cut out of service by a light-load control. It is necessary to provide a sequence of circuits to start the converter, bring it to synchronous speed, connect it to the dc circuit, and transfer to the several machines in the station the proper proportions of the total load. Operation of rectifier automatic substations is even simpler.

## Electrolytic and Inductive Coordination

***166. Earth Currents.*** Practically all railroad-electrification power distribution systems utilize the track rails to complete the traction circuit. Since the track rails are normally supported by wood ties on the ground, there is relatively low resistance between the rails and the ground, and much of the return current is likely to leak from the rails into the earth. In the case of dc systems this gives rise to electrolysis problems and in ac systems to inductive problems. Coordination is necessary with other agencies in either case to prevent interference from the grounded railroad circuits.

***167. Conditions Contributing to Electrolysis.*** The stray currents from the rail-return circuit may flow into the earth and into underground structures, for example, water pipes, gas pipes, and electric cable sheaths, returning to the rails or negative feeder taps in the vicinity of the substations or power plant. There is a neutral point where the rails are at earth potential. Between the substation and that point, the rails generally are negative with respect to the earth, and current tends to flow from earth or underground structures to the rail, whereas from the neutral point to points farther removed from the substation, the rails are positive with respect to the earth and current tends to flow from the rails into the earth.

***168. Electrolytic Damage.*** Electrolysis occurs where the current leaves metal underground structures to flow in the earth, that is, where the structures are positive to the earth. The extent of electrolytic damage is a function of the character of the metal and the amount of current and of time and soil conditions. One ampere flowing continuously for a year may dissolve 13 to 20 lb of iron or 75 lb of lead at the points where the current leaves the metal.

***169. Electrolysis-preventive methods*** may be divided into four classes:

1. *Insulating rails* from the ground so far as possible and installing insulating joints in pipelines or cable sheaths to increase the resistance of the current-return path through the earth, though care must be taken to ensure that stray currents shunting into the earth around the joints do not produce electrolytic corrosion in the positive side of the joint.
2. Ensuring that *rail bonding* is efficiently installed and maintained.
3. Installing *drainage* by bonding the metallic structures directly to the rails at points where the current tends to leave the structures to enter the earth and return to the rails.
4. Separate *return feeders,* insulated from the ground, which reduce the voltage drop in the rails, such feeders to be connected to the rails at various points, so chosen that the voltage at each location is about the same under normal conditions of load.

Electrolysis sometimes occurs in reinforced concrete. The oxides of iron thus formed may occupy more space than the original reinforcement and cause cracking of the concrete. It is a safe precaution to install insulating joints in all pipes and cable sheaths leading to or from reinforced concrete structures in the earth.

***170. Electrolysis effects due to alternating currents*** are less than 1% of those produced by direct current and thus have been found to be of little importance in practice.

***171. Electrolytic self-corrosion*** is often noticed when metallic structures are embedded in soil by varying properties, owing to local galvanic action and not to stray currents. For example, a piece of coke in contact with a pipe may cause a current to flow from the pipe to the coke and return elsewhere from the earth to the pipe, causing rapid local corrosion. It is very desirable on this account that pipes or metallic sheath cable not be buried in cinders, even though no stray direct current is present in the vicinity.

***172. Inductive coordination*** problems between traction power circuits and neighboring communication facilities are similar in many respects to those of commercial power and communication circuits. The principles are the same, and insofar as the traction power-transmission circuits are concerned, the problems are practically identical, as these are metallic circuits and may be transposed as desired and there is no earth current.

In the *traction power distribution systems,* the detail problems are somewhat different, because the ground must be used to some extent by the power circuit which returns to the substation or power plant through the rails, part of it unavoidably passing into the earth. This constitutes a loop between the trolley wire and the rail which may be as much as 24 ft apart and which it is, of course, impossible to transpose. Insofar as the ground-return circuit is concerned, the loop is roughly considered as twice the distance between the trolley and the rail, the earth current being considered as concentrated about an imaginary plane a distance below the ground equal to that of the trolley wire above ground, although this distance depends much on the nature of the ground at any point.

***173. Special Measures in Coordination.*** Since it is impossible to transpose the traction-distribution circuits, other means are necessary to avoid inductive disturbance in paralleling communication circuits. It is often not possible to separate geographically the power and communication facilities, as the communication facilities are usually necessary for railroad operation and the right of way is usually restricted in width.

The growing use of thyristors for phase-angle control of locomotive power has introduced special problems of electrical interference with both communication and signal circuits. While the problems can be overcome, the introduction of solid-state power conditioning systems on motive power, including diesel-electric locomotives, has produced appreciable signal-system problems.

## Motive Power

### General Classifications

***174. Electric locomotives*** may be grouped as follows:

1. Locomotives receiving all their *power from outside sources* by either overhead wire or third rail.
2. *Internal-combustion* locomotives: diesel, gasoline, or gas-turbine engines with electric power transmission to traction motors.
3. *Combination* of internal-combustion engine with collecting device and control to take power from an outside source. These last two classes are not, strictly speaking, "electric locomotives" but contain much electric equipment.

The electric locomotives in group 1 may be classified by the type of current—dc or single-phase ac—which they collect from the distribution system.

Direct-current locomotives are characterized by their method of control, a system that limits voltage from the overhead or third rail to the dc series traction motors from starting to full speed:

1. Resistance control involves a series of power switches progressively actuated during acceleration to remove resistance from the traction circuits as motor speed increases so that ultimately the motors operate at full line voltage.
2. Thyristor control in which silicon controlled rectifiers (choppers) serve to limit line voltage.

Alternating-current locomotives may be classified as follows:

1. Straight single-phase with step-down transformer supplying power to ac series-commutator traction motors.
2. Motor-generator in which single-phase power drives a large synchronous motor–dc generator set to furnish voltage-regulated power to dc series traction motors.
3. Rectifier delivering power through mercury-arc rectifiers or solid-state devices (diodes or silicon controlled rectifiers) for dc traction motors.
4. Inverter or cycloconverter power conditioning that produces 3 phases for ac traction motors of either the induction (asynchronous) or synchronous type.

Locomotives have been divided generally as follows:

1. Road passenger.
2. Road freight.
3. Road freight and passenger.
4. Road switcher.
5. Yard switcher.

***175. Classification of Driving and Trailer Axle or Wheel Arrangement.*** Starting at the front end of the locomotive designed for single-end operation or at either end of locomotives built for double-end operation, the axles and truck connections are designated in their consecutive order. Letters A, B, C, D represent the number of driving axles in a rigid wheel base; numerals 1, 2, 3, the number of guiding or trailing axles; and plus or minus signs, the absence or presence of connection between trucks.

*176. **Rating of locomotives*** is now usually expressed in terms of horsepower. This is more important than maximum tractive effort in hauling a train. Diesel-electric locomotives are usually rated in terms of horsepower at the diesel-engine shaft driving the main generator. To convert this figure to horsepower at the driving wheels, it is customary to multiply the rated horsepower by 82%. A great advantage of electric locomotives is their ability to draw on central power plants for short periods of time and accelerate rapidly or maintain speed in short grades. The diesel engine itself has little overload capacity. Gas turbines are somewhat more flexible in this regard.

The continuous rating is commonly based on temperature rise of transformer or traction motors.

The 1-h rating is based on similar temperature rise reached in 60 min, starting at normal temperature; overloads of shorter periods are similarly figured and are indicative of loads the locomotive can haul at given speeds for the periods indicated without overheating.

The starting rating is the sustained tractive effort that can be exerted during acceleration and may correspond to the 25% adhesion coefficient for a short time in high-speed passenger service and up to 10 min in freight service.

*177. **Similarity of Electric and Diesel-Electric Systems.*** Electrical and electronic developments since 1950 have served to cause electric and diesel-electric locomotives to have more and more in common. The successful application of ignitron rectifiers, followed by selenium and then silicon rectifiers, on ac electric locomotives has made it possible to drive them with the same dc traction motors used on diesel locomotives. The series dc motor has torque characteristics well suited for rail traction applications. The same rectifier developments have also made it possible, since 1965, to have the engine on diesel locomotives drive an alternator rather than the dc main generator, which had always been used prior to that time. Alternator output is rectified for delivery to conventional dc traction motors. The drive for higher diesel-electric locomotive ratings was, until then, being restricted by the physical dimensions and maintenance requirements of the dc main generator necessary to transmit the horsepower which improved diesel engines were capable of producing. A suitable dc generator for engine ratings of 3000 hp, or more, was of such size that it could not readily be fitted within railway clearance restrictions. The smaller-diameter lighter alternator without brushes fitted readily in a high-horsepower diesel locomotive and simultaneously cut maintenance requirements.

Developments in electronics are making possible new traction concepts. Along with more sophisticated control, the static converter, now feasible because of thyristors, can supply multiphase variable-frequency current. This ultimately has made possible the use of the asynchronous traction motor. The ability to supply power at variable frequency overcomes shortcomings that formerly ruled out the asynchronous motor for traction: lack of flexibility for speed regulation and low starting torque. The asynchronous motor offers the advantages of simplicity and ruggedness; it costs less than the commutator-type motor and requires little maintenance. Because higher rotating speeds are possible, it makes possible small, light units with high ratings.

While thyristor, or cycloconverter, design and operation are relatively simple with ac feed, the operation with a dc supply is relatively complicated because there is no following half wave to turn the static unit off. This is overcome with oscillating circuits incorporating capacitance, but as circuit complexity increases, there is a sacrifice of efficiency. The thyristor, however, can make possible the elimination of the conventional speed-control rheostat used on dc locomotives, using instead a converter with smooth output and no loss of power in speed-regulating resistances.

*178. **Power Sources.*** Electric locomotives, with no prime movers, collect current from a trolley or third rail. Diesel-electric and gas-turbine-electric locomotives generate current in a main generator or alternator.

*179. **Pantographs*** are mounted on the roof of the equipment and usually carry a sliding shoe for contact with the overhead trolley wire. They are always used for high-speed operation and when large amounts of current are to be collected. They consist of a jointed frame usually of steel tubing. The contact shoes are usually about 4 ft long, and these may be a single shoe or two shoes on each pantograph. The shoes may be straight throughout their length or cambered slightly or (especially abroad) may be in the form of an auxiliary bow. In this country the material is often steel with, sometimes, wearing plates of copper or bronze inserted, whereas aluminum or an alloy is often used abroad. The pressure varies from 20 to 35 lb in this country, but pressures from 10 to 20 lb are common abroad. The pantograph is usually held up by springs and lowered by air pressure, but this may be reversed, especially abroad, where the pantographs are often raised by air and dropped by gravity.

Since the trolley wire is placed above the tracks a distance that may vary between 15 ft 3 in and 25 ft, the pantographs must be so constructed that they will operate satisfactorily between these limits, maintaining continuous contact with the overhead wire at reasonably constant pressure at any height. They must be light enough so that they will follow the wire as it changes height above the rails to pass under overhead structures, etc., with minimum inertia effects even at high speeds. They must be sufficiently strong and rigid to resist air pressures, both head on, due to speed of the train, and transverse, due to wind; also they must resist stresses due to the sway of the locomotive at high speeds and, to some extent, stresses due to blows from the overhead system at deflectors or turnouts, etc., and from striking birds. If, however, there is serious trouble on the overhead system, the pantograph must be sufficiently pliable so that it will be damaged rather than pulling down the overhead structure.

Matching the development of the lightweight overhead construction in high-voltage electrifications has been the development of lightweight pantographs. The Faivlie (French) and the AMBR (British) designs consist essentially of an elbow-shaped frame which is really half the traditional diamond design. The low inertia of this assembly enables it to have a smooth trajectory with low shoe wear.

***180. Third-Rail Shoes.*** There are two general types of third rail: the *overrunning,* in which the contact is made on the upper surface, and the *underrunning,* where the contact is made on the under surface. In electrified main-line railroads in this country it is considered necessary to protect the rail effectively from accidental contact from trespassers, so the collecting shoes are in these cases horizontal whether the type is overrunning or underrunning.

In rapid-transit systems, where the right of way is fenced or is underground or elevated, more simple protection is adequate, and here often the shoes are merely supported directly over the rail and slide upon it.

## Transformers

***181. Single-phase transformers*** in electric locomotives are similar to those in stationary practice.

In *air-cooled transformers* the air is taken generally from the cab, as it is for the motor cooling. The cab forms a sort of settling chamber for air from outside. Careful screening through a filter is necessary in winter to prevent snow from getting into the cooling system and wetting the coils of transformer or motor. This is especially important with multiple-unit cars when the transformers are generally mounted under the car in a position to pick up dust and moisture, unless care is used in the design of air intake to provide screen and baffles.

***182. Voltage control*** is generally on the secondary side through switch groups, as described. It is sometimes foreign practice to make the voltage taps on the high-voltage side of the transformers. This has the advantage of less copper in the taps and smaller switches, which also may be all immersed in the oil used for cooling the transformers, but has the limitation of requiring greater kilovoltampere capacity in the transformer windings.

On equipment using dc supply the circuit is obviously necessarily grounded, because the rail is used for the negative power return to the substation. In ac systems the ground on the secondary winding of the main transformers is usually effected by grounding the midpoint of a reactor or an autotransformer connected to two of the secondary taps. This connection limits fault currents in case of ground faults.

Diesel-electric locomotives, through their first four decades, were propelled by direct current developed primarily in commutating-pole shunt-wound generators coupled to the engines. Output in such a system is regulated by varying engine speed and controlling generator excitation (Fig. 23-49). The load regulator and engine governor serve to maintain a constant-horsepower output at any throttle setting. If the engine demands more fuel than called for at that throttle setting, the load regulator reduces field excitation to avoid overloading the engine. In addition to the shunt and battery (separately excited) fields, the main generator is usually fitted with a starting field which is energized from the battery to motorize the generator and start the diesel engine. There are actually several types of diesel engine control, but all conform to this general principle.

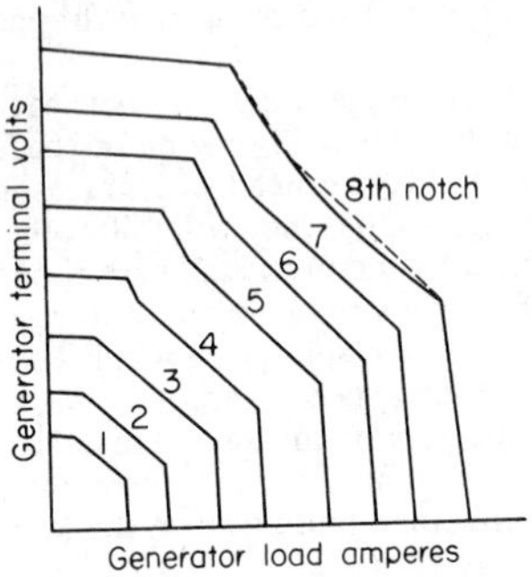

**FIG. 23-49** Generator demand for eight throttle positions.

When about 1965 diesel-engine ratings went above 2500 hp, the trend to alternators began. The small, lightweight alternator used on United States diesel-electrics for production of propulsion current is a rotating-field machine which develops 3-phase power in the stator windings. Because of the high ratings possible with alternators, it was practicable to eliminate some steps of transition (alternation of the series and parallel relation of the traction motors on a unit) and to reduce the amount of motor shunting over a locomotive's speed range. A separate starting motor is necessary.

***183. Locomotive Rectifiers.*** When current is not collected or generated as direct current on units which have dc traction motors, it is necessary that they be fitted with suitable converter or rectifier equipment. In the case of diesel-electric locomotives, semiconductor rectifiers were already perfected before the alternator came into use. Because the alternator is connected to a variable-speed diesel engine, the input current is of widely varying frequency. Rectifier banks on electric locomotives are fed with constant-frequency power. The same rectifiers are capable of being used in both types of locomotives. Typical silicon diodes have an average rated current of 250 A, a peak current of 500 A during each rectified half wave, and a maximum peak reverse voltage of 1200 V, corresponding to a reverse current of less than 6 mA under cold conditions and less than 20 mA under hot conditions. The forward drop is always in the range of 1 to 1.2 V/diode at the rated current.

***184. Diode Arrangement.*** Rated values lead to a theoretical number of diodes mounted in series and parallel, according to the circuit design. In this arrangement, however, the following must be considered:

1. Permissible overload current, which affects the number of diodes in parallel.
2. Possible surge voltage, which affects the number of diodes in series.
3. Cooling or forced cooling, which determines the physical arrangement of the rectifier assembly.
4. Protective or fault-detection devices and indicating devices for diode failure.

Prior to the adoption of "dry" rectifiers, electric locomotives and multiple-unit cars were fitted with ignitron (single-anode-tube) rectifiers. These units are considerably less rugged than selenium and silicon rectifiers and also require liquid cooling and antifreeze protection. Silicon diodes have been substituted for ignitrons in some electric locomotives (Fig. 23-50).

The growing similarities between electric and diesel-electric locomotives has led to consideration of ways that existing diesel-electrics might be converted for electric operation as a way of minimizing the capital investment that long deterred American electrification. Motive-power planning must be thorough when such a step is taken. Since the high-horsepower diesel-electric would be displaced by all-electric operation, new diesel-locomotive acquisitions might be concentrated in the areas of switching and road-switcher locomotives that would continue to operate the yards, sidings, and branches where electrification could not be justified. The displaced high-horsepower locomotives might be converted through installation of a transformer and drive motor in place of the diesel engine. This could diminish the financial impact of acquiring an entire fleet of new electric locomotives simultaneous with the costly installation of the catenary and power-supply system. The converted units could be used in service not requiring high horsepower, reserving such service for newly acquired electric locomotives.

***185. Thyristor Locomotives.*** A typical contemporary thyristor electric locomotive has the armature of each traction motor fed from two nonuniform converter bridges in series (Fig. 23-51). The thyristor branch in the bridge has four thyristors; the diode branch has five diodes. There are 16 thyristors and 20 diodes in each traction-motor converter for armature current. The two bridges are controlled in sequence. During the first period (*A*) the traction-motor voltage is continuously increased from zero to half voltage by controlling the firing angle of the thyristors in the first bridge; the armature current also passes through the diodes in the second bridge. During the next interval (*B*), motor voltage is controlled in the same manner by varying the firing angle of the thyristors in the second bridge. The thyristors in the first bridge then work as diodes. After the second bridge has been switched in, the motor voltage comes to its full value. During this control period, the current in the motor fields is at its full value to attain the highest power factor. After that, the field weakening to the desired speed occurs with the armature converter as a diode rectifier. Armature converters are controlled jointly while the field converters are individually controlled. Load sharing between the traction motors is accomplished in the motor fields which are automatically controlled so that the armature currents will be equal. Separate excitation of the motors achieves effective wheel slip control; load sharing involves a time lag so it does not intervene during a sudden slipping.

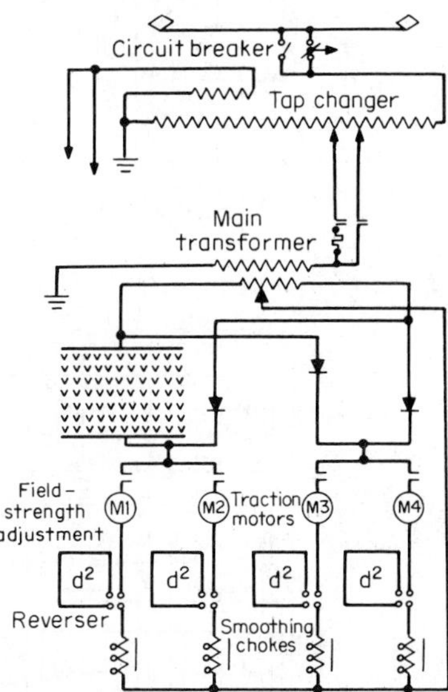


**FIG. 23-50** Silicon-diode rectifier installation in electric locomotive.

## Control Circuits

***186. Control-Battery Circuits.*** The switch control of electric and diesel-electric locomotives and multiple-unit cars is usually by means of a storage-battery supply (64 V or other convenient voltage) actuating electromagnetic or electropneumatic switches or contactors which control the power circuits. The controller is usually of the drum type, where a cylinder which may be revolved through any desired arc is equipped with insulated segments which press upon stationary fingers. Any desired circuit or combination of circuits can be set up by properly connecting the segments for contact as the controller handle is manipulated (see Fig. 23-52).

***187. Multiple-Unit Control.*** As the individual control circuits are set up by the controller contacts, they may be carried to one set of contactors, or switches, or by parallel arrangement may be carried by means of multiple-conductor cable to another motive-power unit, which is thus controlled in a manner identical with the lead unit in the train. A number of locomotive units or motor cars through a train can thus be simultaneously controlled by one man. This is known as multiple-unit control. It is desirable to incorporate interlocks or relays to ensure that the connections will take place in the proper sequence and to protect against overloads, failures of line voltage or motor overspeed due to slipping of drivers, etc.

***188. Automatic Control.*** In certain types of service, such as rapid-transit trains or suburban motor-car trains, automatic control is used. This is accomplished by inserting cur-

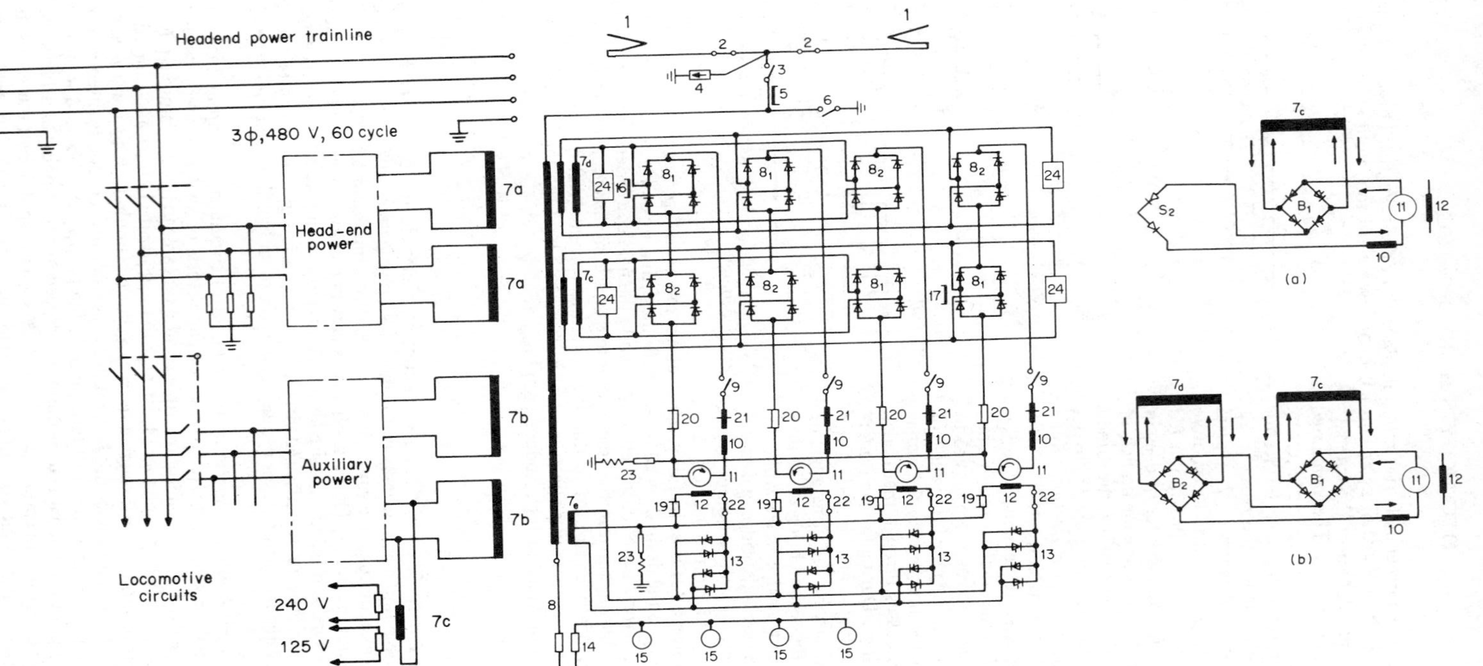


**FIG. 23-51** Thyristor control system. (1) Pantograph; (2) isolator; (3) main circuit breaker of ASEA design; (4) lightning arrester; (5) roof bushing with current transformer; (6) earthing switch; (7) main transformer; (7a) winding for train heating, 1000 V; (7b) winding of auxiliary-machine converter, 200 V; (7c,d) winding for thyristor bridges; (7e) winding for field supply; ($8_1$) nonuniform thyristor bridge 1; ($8_2$) nonuniform thyristor bridge 2; (9) traction-motor isolator; (10) smoothing reactor; (11) traction motor, armature; (12) traction motor, field winding; (13) thyristor converter for field supply; (14) earthing transformer; (15) earthing brush; (16–18) current transformers; (19–20) shunts; (21) measuring transductor; (22) isolator; (23) earthing current resistor and relay; (24) capacitor for telefilter and power factor correction.

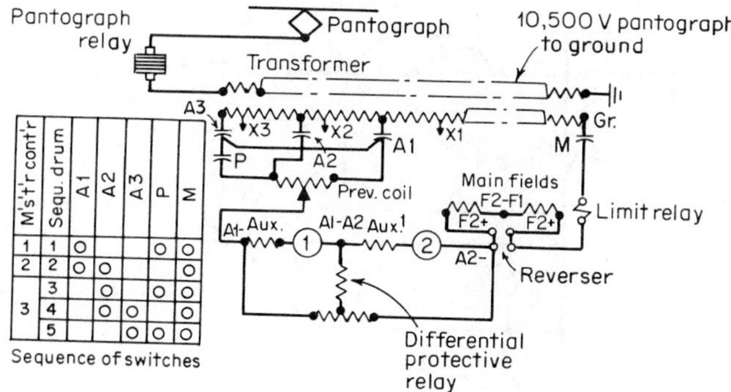


| M's't'r cont'r | Sequ. drum | A1 | A2 | A3 | P | M |
|---|---|---|---|---|---|---|
| 1 | 1 | O | | | O | O |
| 2 | 2 | O | O | | | O |
| 3 | 3 | | O | | O | O |
| | 4 | | O | O | | O |
| | 5 | | | O | O | O |

**FIG. 23-52** Control diagram, main circuit, two single-phase motors on 11 kV.

rent-limiting relays, with current coils, in the motor circuits which operate the control circuits when the current reaches a predetermined value and permit the next sequence to operate. By using this scheme, the engineer merely closes the main control, and the acceleration is automatically accomplished with no further manual operation.

***189. Switch Groups.*** The 64-V circuits from the controller operate contactors, or switches, which may be electromagnetic, electropneumatic, or semiconductors.

Rapid-transit cars have been fitted with cam-controller equipment in which cam switches function to regulate the voltage and current, and thus the speed and torque, of dc series motors by switching voltage dropping resistors in and out of the circuit (Fig. 23-53*a*). Here again the advent of solid-state thyristors provides several alternatives to cam-controller equipment. One is to supply dc power to series motors from a thyristor chopper which controls acceleration by very rapidly turning the motor current on and off in a controlled pattern (Fig. 23-53*b*). The chopper also provides for either regenerative or dynamic braking. A possible second approach is use of the thyristor chopper for control of voltage and current to shunt-wound dc motors instead of traditional series motors. A third approach is conversion of dc supply power to ac by a thyristor inverter with the 3-phase output (variable in voltage and frequency) utilized by standard ac induction motors. The inverter can also allow for regenerative or dynamic braking. Thyristors overcome problems of cam-controller contact maintenance and power losses incurred when voltage is controlled by resistors. The smooth control of thyristor systems also makes them more adaptable for use with automatic train control.

***190. Semiconductor Technology.*** The 1948 invention of the transistor and the 1957 development of the thyristor (silicon controlled rectifier, or SCR) have been the basis for effective power conversion equipment for ac, as well as dc, traction motors.

While French and British electrical engineers had established the 25-kW, commercial-frequency overhead power system as the electrification of the future, initially the alternating current had to be converted for dc traction motors, as already described. The alternator-equipped diesel-electric locomotive presented a similar challenge. The thyristor then began to have its impact; it is sometimes credited with making high-voltage ac electrification the world standard for mainlines.

While this was happening, the existing dc electrification of many mainline railroads and of virtually all existing and new rapid transit and streetcar lines did not seem to benefit from the new technology. Then the thyristor dc chopper was introduced. It differs from the original SCR, which requires only that it be "gated" to initiate operation and then switch off automatically. The chopper requires that there be a means of turning the device off, as well as on. This process is known as *forced commutation*. The application of choppers served to replace the classic resistance control used since the earliest electric locomotives and streetcars. It was possible to control starting voltage on traction motors and then gradually increase it to full line value without the energy losses associated with resistance control.

Even as all types of rolling stock could be fitted with dc traction motors supplied through solid-state power conditioning systems, the dc motor itself came under scrutiny. The dc traction motor involves a commutator and brush gear which limit rotational speed and

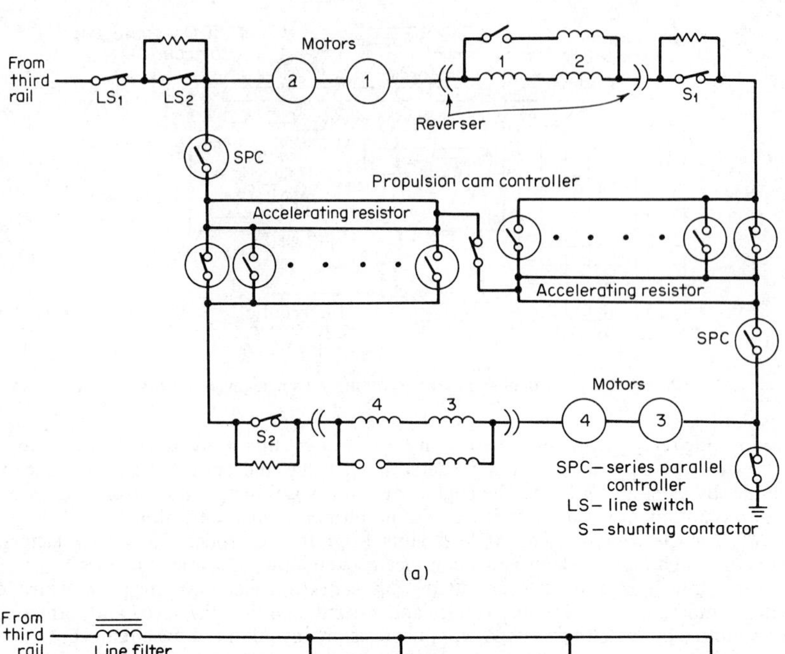


(a)

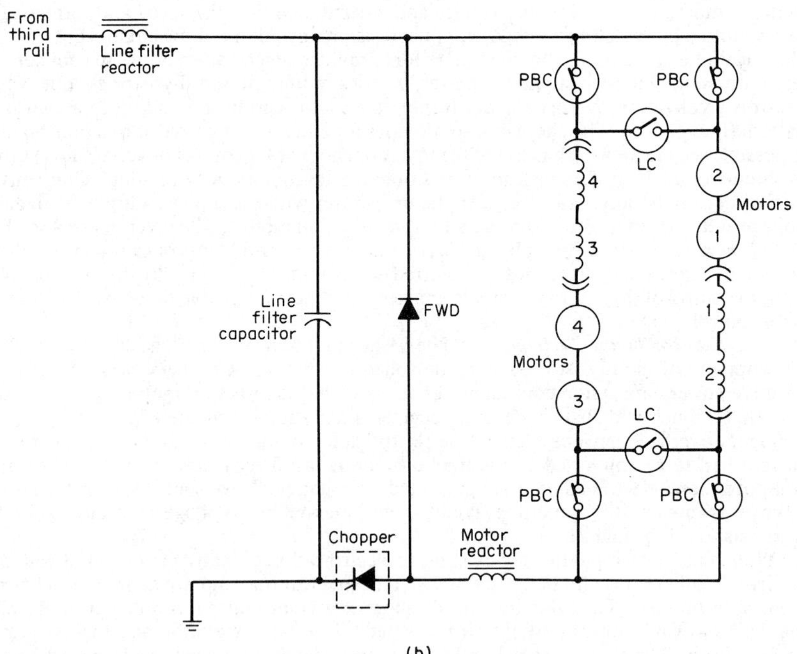


(b)

**FIG. 23-53** Rapid-transit-car power conditioning systems. (*a*) Cam-controller switches voltage-chopping resistors in and out of circuit (circles indicate cam-operated switches); (*b*) series dc motors can be supplied by thyristor chopper with circuit complete when chopper is switched on and energy stored in motor reactor and in inductance of motor when chopper is off (LC—loop controller, PBC—power brake changeover).

require routine maintenance. Even though the characteristics of the series or shunt-field dc motor are ideally suited to railroad requirements, it became apparent that modern controls might offer the potential for using ac motors with comparable performance. While the ac motor permits higher rotating speed, thus reducing dimensions and weight at equivalent power, it does require a power supply capable of variable-frequency output.

***191. AC Traction Motors.*** Two basic concepts for utilization of modern comutatorless ac traction motors have now evolved:

1. *Asynchronous squirrel-cage induction motors* with a power supply in which frequency and amplitude of supply voltage may be varied over a wide range by thyristor-controlled converters.
2. *Synchronous slip-ring motors* in which current inverters supply dc excitation to the rotor and armature windings are in the stator.

*Asynchronous System.* Figure 23-54 shows the 3-phase asynchronous concept as applied to a diesel-electric locomotive. The heart of this transmission is the set of four forced-commutated inverters operating in parallel that produce 3-phase voltage with variable amplitude and frequency. All motors are connected to the common 3-phase busbars.

In these inverters the phase modules are connected in parallel with the positive or negative terminals of the dc power supply. By triggering thyristors alternately, the positive or negative polarity of the dc voltage may be switched to the output. As a thyristor can only be switched on by a trigger impulse, but not switched off, additional thyristors, capacitors, and reactors are required to achieve the forced commutation. Motor rotation may readily be reversed by changing the triggering sequence, eliminating all reverser switchgear. The converter system with a dc voltage link was designed to be modular, feeding the same motor inverters from different prime movers such as the diesel engine or gas turbine, or in the case of electric locomotives, from the dc or ac catenary system.

*Synchronous System.* Figure 23-55 illustrates the concept of using a self-commutating traction motor identical to the traditional alternator. The rotor is supplied with dc excitation through slip rings. Armature windings are in the stator. The back emf developed within these windings may be used to turn off the inverter thyristors at the right moment. Electronic circuits actuate the thyristors in synchronism with motor rotation so as to switch current into the right winding at the proper instant. EMF generated at low speeds is too

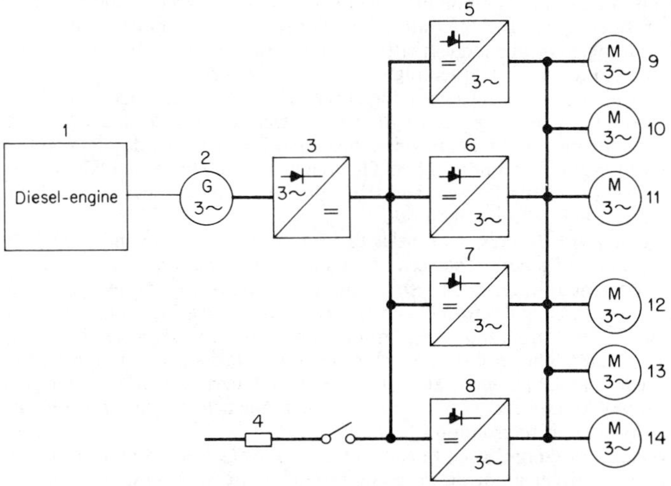


**FIG. 23-54** Three-phase transmission system of diesel-electric locomotive: (1) diesel engine, (2) three-phase generator, (3) diode rectifier, (4) resistor for electrified braking, (5–8) inverter units, and (9–14) traction motors (squirrel-cage).

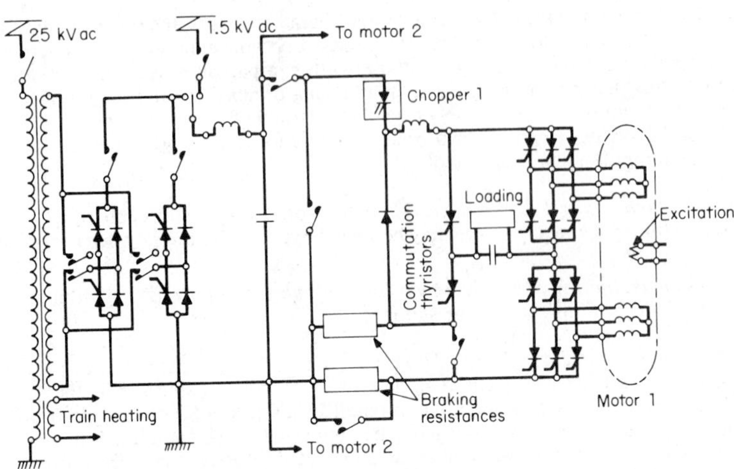


FIG. 23-55 A dual-voltage 5600-kW electric locomotive with monomotor trucks can operate under 25 kVac or 1500 V dc. The synchronous concept involves a second set of power circuits because motor 2 has its own chopper, commutation, and rheostatic braking circuits.

weak to turn off the thyristors. Sensors on the rotor transmit a signal that initiates firing of the commutation thyristors associated with an auxiliary commutation capacitor and smoothing coils. While each synchronous motor needs to be connected to its own inverter, it is possible to connect more than one inverter to a single power supply. Inverter-motor combinations may be energized either in parallel or in series from the same dc supply.

***192. Electric Braking.*** One of the important advantages in electrification or diesel-electric operation is in electric braking or in using the traction motors as generators and either feeding power back into the power distribution system or absorbing it in heating resistance on the locomotive or in some other manner. Since the energy of a moving train is measured by the mass and the square of the speed, braking at high speeds must absorb a very considerable amount of energy. The potential energy of a heavy train descending a long grade is also very considerable and, if the train is to be operated safely, must be absorbed in some manner. If this mechanical energy in either form is absorbed through brake shoes, great care must be exercised to avoid overheating the shoes and wheels.

Two other methods of electric braking are the magnetic track brake, which utilizes current produced by decelerating motors to energize electromagnetic shoes which act on the rails, and the eddy-current brake, in which high-speed vehicles are decelerated by eddy currents produced in the rails under them. The magnetic track brake tends to cause rail wear at points where it is regularly used, and the eddy-current brake can produce high rail temperatures which might affect metallurgy.

***193. Regenerative Braking.*** In order to feed power back into the supply system, the traction motors must generate power at a voltage higher than the supply voltage and at a reasonably constant voltage. In dc systems this condition is met by the traction motors acting as generators which will return the energy to the line with a slight increase above normal no-load speed. With compound-wound motors the increase in speed to obtain regeneration is somewhat greater. The speeds at which regeneration takes place are reduced if it is possible to increase the field somewhat during the braking period. Since braking is effected by the absorption of kinetic energy of the moving train, the field strength must be increased as the train slows down if retardation is to be uniform.

***194. Dynamic braking*** is used in some instances instead of regenerative braking. When this is done with either ac or dc series motors, they are short-circuited through suitable resistances. The residual magnetism in the poles will cause the machine to generate an emf which sends current through the local circuit. This current, passing through the field coils, builds up magnetism which in turn increases the voltage, causing a current to flow in amount limited by the resistance. The heating of the resistance absorbs the energy generated by the motor, and the train slows down. As the motors reduce speed, the voltage drops, and to obtain uniform retardation, the resistance must be reduced. The starting resistance grids are ordinarily also used in the braking. The chief limitation to dynamic braking lies in the relatively small amount of energy which can be dissipated by the amount of resistance it is possible conveniently to install in the locomotive.

### Traction-Motor Characteristics

***195. Direct-Current Series Motors.*** The railroad types are usually insulated for operation on 600- to 3000-V circuits, depending on distribution voltage. On 3000 V usually either two or three motors are operated permanently in series, and on some cars it is found desirable on account of improved commutation to wind motors of the small sizes for 300 V, placing two permanently in series for use with 600-V supply.

Since the same current in a series motor flows through armature and field circuits, the field strength varies with the load on the motor. The speed is reduced as the current increases on account of the higher flux density in the magnetic circuit.

***196. Field Excitation.*** Flexibility in performance of the series motor may be obtained by a change in the relation between the armature current and the field strength. The field current may be reduced by shunting the field coils with resistance, or there may be field coil taps. Either method will weaken the field flux and produce higher armature speed for any given armature current. This provides to some extent a method of controlling the speed.

***197. Terminal Voltage.*** Reduction in the voltage of the motor terminals of a series motor will cause it to run slower at any given current value, since the armature does not have to develop so much counter emf. The speed is thus very roughly proportional to the terminal voltage. Reduction in terminal voltage is obtained by placing resistance in series with the motor or by reducing the supply voltage by transformer taps in ac circuits or by placing the traction motors two or more in series.

***198. The compound-wound traction motor*** has an additional set of field coils fed directly from the line, forming a shunt circuit. This shunt winding supplies a considerably smaller proportion of the field ampere-turns than is customary in stationary practice and therefore does not give so flat a motor-speed characteristic as it does in its application to the industrial compound-wound motor. The compound winding tends to limit excessive speeds in descending grades, a characteristic of value in electric braking.

***199. Single-phase commutator series motors*** of several types have been used in railroad service. Their general construction is similar in many respects to that of dc series motors. Since the change in direction of the field and armature currents takes place simultaneously, the torque remains in the same direction. There are necessary, however, certain modifications in the single-phase motor as compared with the dc motor. The field coils can no longer be simple windings around a pole piece, because such windings would offer too high an impedance to the alternating current, so that fewer turns and more poles with a lower flux density are used. Likewise, the impedance of the armature, of rotor windings, must be designed with particular care to make the reactance low, and special compensating windings may be introduced in the field to effect this. In the voltage equation of the series ac motor, the resistance drop of the dc equation is displaced by a term for the impedance drop which may be sufficiently large to play a part in the output characteristics and thus make them somewhat different from those of the dc motor. This type of motor operates at 15, 16⅔, or 25 Hz.

***200. Single-Phase Motor Design.*** The current-heating limitations may be more rigid in ac motors than with dc motors because of the transformer action in the armature coils, which are short-circuited by the commutator brushes. Heavy short-circuit currents are induced in these coils, which act as secondary windings of a transformer of which the field coils are the primary. The magnitude of these currents can be limited by insertion of resistances in the armature leads or by keeping the induced voltage low by having a small number of turns in the primary or field coils. *Commutating interpoles* can be used to offset this short-circuit current during running if their flux is such as to induce a counter emf in the windings, opposite to that induced by the field flux in this transformer action. The conditions for overheating due to this action may be especially serious at starting, when a single armature coil may be short-circuited for some time.

***201. Asynchronous Motors.*** Polyphase induction motors have been operated from 3-phase supply and from a single-phase trolley circuit by use of a phase converter mounted on the locomotive. Various other means have been utilized for achieving variable speed from a traction system which is inherently single-speed.

The use of solid-state inverters had revived interest in the induction motor for traction purposes. With this system there would be no commutator maintenance and no other power contact devices; the induction motor is also the lowest-weight unit for its output when compared with dc and other types of ac motors. For a rapid-transit application with dc from a third rail or with a locomotive capable of producing dc from a diesel-driven generator or from a rectifier bank, the pulse-width-modulation (PWM) inverter has been studied. This inverter uses its solid-state devices to convert the dc to variable-voltage, variable-frequency ac which is then delivered to the traction motors. By simultaneously varying the voltage and frequency, motor torque can be closely regulated for precise speed control. The logic to produce the desired frequency-voltage relationships is performed with miniaturized printed circuits, while actual power is provided by heavy-duty thyristors. The 3-phase ac output first produces constant motor torque for acceleration. When a region of rated horsepower is reached, frequency is then increased further only to produce greater speed. Transition from motoring to braking mode requires only a change in inverter frequency by the control logic. There is no reconnection of power circuits.

***202. Induction Motors.*** The stator of an ac motor is a series of coils which produce a rotating magnetic field when excited by 3-phase ac. Speed of field rotation depends on the excitation frequency and number of motor poles. The rotor turns at a speed slightly different from that of field rotation, the phenomenon known as slip. The magnetic field resulting from the induced rotor current interacts with the rotating stator field to produce torque. If the rotor turns at a slower speed than the stator field (positive slip), electrical energy is produced. If the rotor turns at a faster speed than the stator field (negative slip), mechanical energy is converted to electrical energy and the motor regenerates. When the inverter controls the applied frequency to maintain a constant motor slip, and simultaneously controls the applied voltage to maintain a constant voltage/hertz ratio, the induction motor will produce constant torque. If inverter output voltage is held constant, the motor will produce constant power.

## Control of Traction Motors

***203. Motor Control.*** If a motor is connected across the supply circuit when the vehicle is standing still, the current it would take would be limited only by the resistance or inductance of the circuit and would far exceed the safe carrying capacity of the windings. The torque would also probably cause slipping of the driving wheels. Some means, therefore, is necessary to limit the starting current. It is also desirable to provide more than one economical running speed.

***204. DC Series Motor.*** Current and torque produced at standstill may be reduced by strengthening the field or lowering the terminal voltage or both. Motors may be placed in series, reducing the terminal voltage of each without loss in external resistance. External resistance may be placed in series with the motors to limit the starting current to any desired amount, and by varying the resistance the current may be kept constant during acceleration as desired, as the counter emf is being built up. Since maximum torque while starting demands full field strength, any shunts or reduced field connections are usually thrown out of action in starting.

***205. DC Compound-Wound Motor.*** Starting may be made with full armature current in the series-field coils and maximum current in the shunt-field coils. A starting resistance inserted in the armature circuit is reduced in steps until the armature and series field are connected across the line. Further speed increase is effected by reducing the shunt-field current in steps to the point where the shunt winding is disconnected, and the action is then identical with that of a plain series motor.

***206. Terminal voltage*** with single-phase motors is usually varied, not by inserting external resistance, but by voltage taps brought out from the supply transformer; so there is no rheostatic loss with ac motors as there is with dc motors.

Some ac electric locomotives are equipped with *transformer voltage taps* which can raise the motor terminal supply voltage at high speeds to counteract the speed-horsepower drooping characteristic inherent in the series motor. This gives the locomotive a great horsepower at high speeds, a valuable characteristic in hauling heavy trains on fast schedules.

*207. **Series-parallel operations*** of traction motors is usually adopted to economize on weight and produce good efficiency at low speeds and high tractive efforts.

The motors of a locomotive or car may be grouped in various combinations depending upon the number of motors in the vehicle.

With two motors, only the simple series and parallel positions of the motors are possible (see Fig. 23-56). With four or six motors or motors groups, series, series-parallel, and full-parallel positions give considerable flexibility. With supply voltages of 600 V dc, it is customary to use each motor individually in these combinations, though often two motors are connected permanently in series. With higher line voltages (1500 to 3000) it is customary to couple two motors permanently in series. With ac supply the transformer secondary voltage may be varied as desired, and these limitations do not exist.

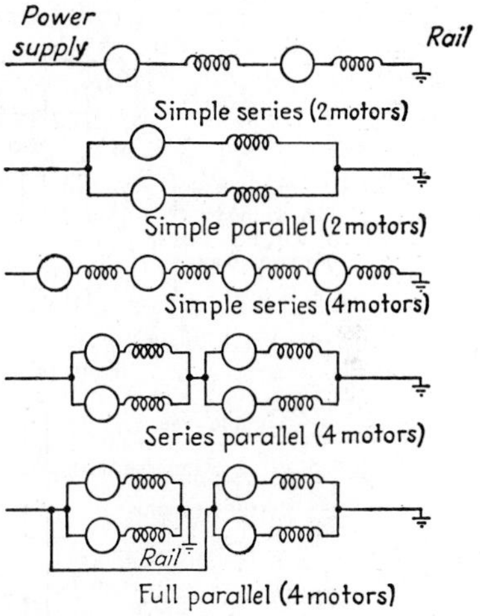

**FIG. 23-56** Arrangements of traction motors.

*208. **Transition from one connection combination*** to another must be made with as little variation in torque as possible. One scheme commonly used permits one of the motors at a time to change with resistance temporarily in its circuit.

*209. **Resistance grids*** are usually of cast iron, assembled in frames, usually mounted on insulated steel tie rods, and supported at each end by steel frames bolted securely to the locomotive or car structure. Edgewise-wound resistors may be used, wound from a noncorrodible, unbreakable strip of special resistance alloy with low temperature coefficient.

*210. **Preventive Coils.*** With alternating current, since the voltage supply is taken from transformer taps, the supply voltage is flexible. In starting, the voltage applied is low, and all points may be continuous running points. The problem of maintaining continuous torque in changing taps is an important one, just as in changing motor combinations. Two transformer taps cannot be tied together during the instant of change without some method of limiting short-circuit currents in the transformer windings involved. This is done by the use of preventive coils or impedance shunted across the taps during transition.

*211. **Pulsating-Current Motors.*** While the pulsation factor of dc traction motors has generally been suppressed to about 25%, motors capable of successful operation at a current pulsation factor of 50% have been developed. Such motors require lamination of certain internal components in a manner duplicating the construction of ac motors. Communications interference must be carefully controlled when current with such pulsation is involved. Motors capable of operating on pulsating dc current simplify considerably the circuitry associated with rectifiers. They also offer the possibility of another semiconductor system of control—the dc "chopper" concept.

*212. **Chopper Control.*** The replacement of the conventional rheostat of dc locomotives by a static converter incorporating thyristors is attractive even though oscillating dc circuits must be designed to turn the thyristors off after they have been fired. Rather than regulate voltage by rheostat control, the "chopper" can vary voltage or current by impulses set up in thyristors in series in the circuit.

The mean voltage supplied to the motor is then determined by the time of opening of the circuit, or its frequency. Two methods of varying the mean voltage are then possible (Fig. 23-57).

1. At a fixed frequency, the value of opening impulses is varied.
2. With impulses fixed, their frequency is varied.

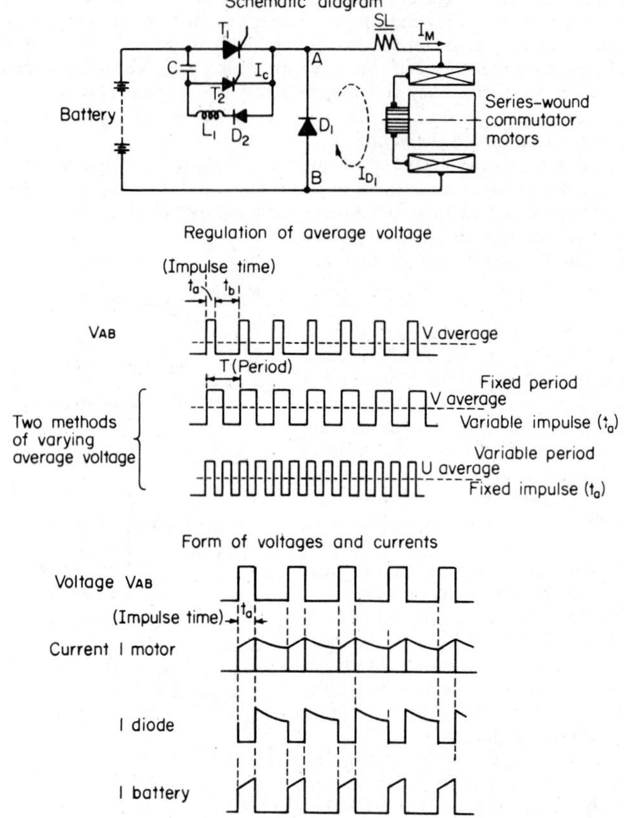


**FIG. 23-57** Feed for a commutator motor, impulse control.

## Method of Drive

***213. Single-Reduction Gearing.*** Diesel-electric locomotives and multiple-unit cars, along with recent electric locomotives also having dc traction motors, are usually powered with one motor per axle. The dc motor is normally "nose-suspended" with a single-reduction spur gear drive. This installation places the motor axis parallel with the axle. The motor frame incorporates bearings that are supported by the axle; the other end (nose) of the motor is supported resiliently by the truck frame. The pinion mounted on the armature shaft engages the drive gear, which is mounted by press fit on the axle just inboard of one of the wheels. This nose suspension maintains a fixed relation between the armature shaft and the axle. The motor is free to move through only a small arc, with the axle as the center, when track irregularities are encountered and when torque is being exerted. Such a motor, with most of its weight carried on the unsprung axle, must be extremely rugged because it is subjected to extreme shock and vibration as rail irregularities occur.

***214. Double-Reduction Gearing.*** Direct-current traction motors on some locomotives and cars are body-hung at right angles to the axles, with the torque from each transmitted to one or more axles by Cardan shafts which drive through right-angle gearboxes. Such a drive may incorporate double-reduction gearing. Double-reduction gearing, regardless of the motor location, permits smaller, higher-speed motors and has found wide application in transit cars. On many transit cars these light, economical motors are truck-mounted, usually supported entirely from the truck frame rather than being axle-suspended. The size can be reduced to such an extent that the motor is part of the gear unit, rather than the gear unit being part of the motor. Small, high-speed motors may safely be self-ventilating.

The driving of both axles in a truck by a single truck-mounted motor incorporating double-reduction gearing is growing in popularity. A single motor having the same capacity as two conventional traction motors is lighter and less costly than the two. The coupling of the two axles through the gear drive significantly reduces wheel slipping. The ability to incorporate gear shifting in a double-reduction drive makes possible the building of a "universal" type of locomotive which can exert its horsepower over a much wider speed range.

Conventional ac locomotives have been fitted with a variety of drives, including bipolar and quill types. Recent developments indicate that the costly, complicated drives once necessary with ac motors can be avoided in all future models.

## Signaling

***215. Railway Signaling.*** The initial purpose of railway signaling, safety, has been expanded to include train control which makes possible more efficient utilization of the railway plant, equipment, and manpower. It also speeds the movement of freight and passengers, enhancing the competitiveness of the rail mode. As train control has become more sophisticated, communications facilities have been increasingly integrated with the originally dispersed signal system. Automatic train control (ATC), centralized traffic control (CTC), and automatic train oepration (ATO) are among the technologies requiring more advanced data-transmission facilities.

By definition, a signal conveys an indication (*proceed,* for instance) by displaying an aspect (*green,* for example). Ideally each aspect should correspond to a single indication, so that the signal itself would convey complete information. This goal is met only in part because of the fundamentally different operating rules which apply in manual block, automatic block, interlockings, and in centralized traffic-control territories. One principle is that any likely defect in display or reading of an aspect should be of a nature to give a more restrictive indication. This fail-safe principle is illustrated by the fact that any imperfectly displayed signal must be interpreted as being at its most restrictive aspect. Signals normally fed by wayside line power are provided with automatically connected standby battery supply.

***216. Track Circuits.*** The basic track circuit was invented in 1872 and has been the basis for most of the automatic operation of signals by the trains themselves. This system eliminates the human element by automatically setting signals to stop behind a train, but the fail-safe nature of this basic arrangement also eliminates most of the hazards that could result from malfunction of the automatic portions of the system.

Because of the extremely variable and often considerable leakage of the current between the rails, numerous conflicting requirements are put upon the designer in selecting and adjusting circuit components to assure safety and not produce false stop indications each time that it rains. The basic track circuit has been dc, powered from low-voltage signal battery which is fed through a limiting resistor to keep the current flow within acceptable limits when the circuit is shunted by a train. Each segment of the railroad protected by such a circuit is separated electrically from adjacent segments by insulated joints in the rails. The rails themselves, with conventional joints, have a resistance of 0.015 to 0.050 Ω per 1000 ft of track; the leakage resistance from rail to rail varies from 1 Ω to several hundred ohms per thousand feet. The normal maximum length for steady current dc track circuits is about 1 mile. If the block to be protected is longer, then coded direct-current pulses or bursts can be used which will operate from one feed point (battery) to about 16,000 ft. Should insulated joints fail, the current from the adjacent circuit might augment the energizing current to cause false indications; this is usually guarded against by reversing the polarity of alternate circuits.

***217. Electrified Railroads.*** Electrified railroads using dc propulsion are signaled by alternating-current track circuits. AC electric railroads are signaled with ac circuits of high frequency, using frequency-sensitive relays. Key to this system is use of impedance bonds which readily pass propulsion currents in both rails around the insulated joints, but offer high resistance to the track circuit's ac current trying to flow from one rail to the other.

In electrifying existing railroads with signal systems, the proximity of high-voltage catenary to open-wire signal and communication lines can expose these circuits to the effects of electromagnetic and electrostatic induction. High voltages induced could produce personnel hazards as well as introducing the possibility of malfunction of the apparatus. Harmonics of the 60-Hz propulsion energy, as well as noise, would be induced into communication circuits. The following changes must all be considered: (1) Double-rail dc track circuits would be replaced with either single-rail dc circuits using ac immune track relays or with double-rail ac track circuits operating at a frequency not harmonically related to 60 Hz; (2) open-wire line circuits must be eliminated or placed in grounded shielded cable, preferably buried along the right of way; (3) protective devices must be installed to protect personnel and equipment during traction-system fault conditions; (4) open-wire signal power lines carried on catenary structures must be high-voltage and suitably insulated to avoid any inductive effects from catenary; (5) signal heads may have to be relocated for proper siting due to catenary supports.

With the rapidly growing use of welded rail, insulated joints require intentional introduction of discontinuities in what would otherwise be a continuous metal beam. There is the problem of joint maintenance, a compelling reason for installation of welded rail in the first place. The result has been introduction of the electronic, or audiofrequency, track circuit which can be limited in range without the use of insulated joints. Audiofrequency circuits are also used to impose additional sophisticated safety controls on track which also has conventional track circuits. It has been widely applied in grade-crossing warning systems.

***218. Phase-selected Track Circuits.*** Improvements in track circuits for electrified railroads include the phase-selective coded track circuit. This is a refinement of the standard direct-current coded track circuit. The circuit uses either mechanical or solid-state code transmitters to apply either coded 100- or 200-Hz energy to tracks. Phase-selective circuits can be up to 7000 ft in length for 100-Hz operation or up to 4500 ft for 200-Hz operation. The ac ballast resistance is 3 Ω per 1000 ft.

The current in the rails can be coded at rates of 75, 120, 180, or 270 pulses per minute. Also, the coded energy in the rails can be inductively coupled to passing trains for the control of cab signals or train control.

A simplified diagram of the phase-selective track circuit is shown in Fig. 23-58. At the sending end of the circuit (on the right), the code transmitter contact interrupts the 110 V ac feeding the autotransformer. The signal is coupled to the rail by the track transformer. For train detection only, a single code rate is used; for cab signaling, multiple rates are used in conjunction with the required speed limits. Combinations of inductor L1 and resistor R1 are used to obtain the proper phase relationship for normal operation.

At the receiving end, the coded track signal voltage is stepped up by the track transformer to the level required for the operation of the phase-selective detector. The detector input contains a tuned filter which minimizes the propulsion interference that would otherwise interrupt normal coding action.

With no signal in the rails, the reference voltage from the phase reference transformer is converted by the detector to positive dc, which moves the code-responsive magnetic-stick relay armature to the normal position. The correct phase signal from the rails reduces the positive dc voltage on one relay coil and applies positive dc voltage to the other coil, causing the armature to move to its reverse position. Thus a coded ac track signal of the correct phase causes the relay to follow the code rate.

The code-responsive relay can be used to drive various combinations of decoding equipment.

***219. Signal Operation.*** Signals are actuated automatically by a train in the block. This type is common on busy railroads. Automatic signals may employ either battery-current or ac supply. The track circuits are energized by power between 2 and 8 V, either ac or dc. The rails at each end of the block are cut and connected by insulated joints. There is sufficient insulation in the wood ties so that the rails constitute conductors in a closed circuit with a signal power supply at one end and a relay at the other to actuate the signal (see Figs. 23-59 and 23-60). In the diagrams of both Figs. 23-59 and 23-60, the arrows indicate the direction of current flow. In Fig. 23-59, starting from the positive post of the battery, current flows through the limiting resistance, the one rail, through the relay winding, the relay series resistance, and back through the other rail to the negative post of the battery. With the relay

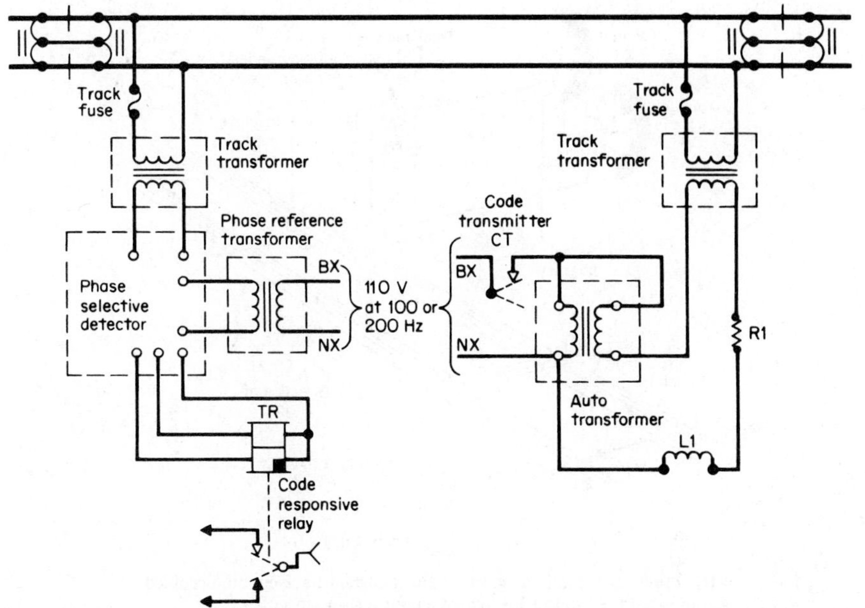


**FIG. 23-58** Simplified diagram of the phase-selective track circuit.

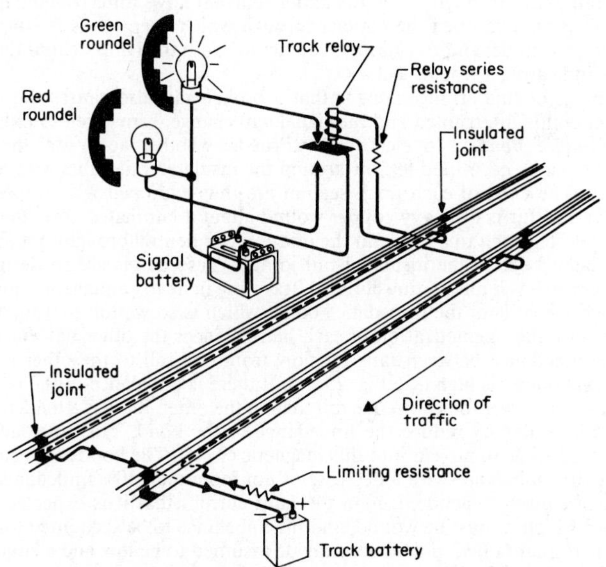


**FIG. 23-59** Schematic diagram of direct-current track circuit, unoccupied. Lamp behind green roundel is lighted, giving a proceed aspect.

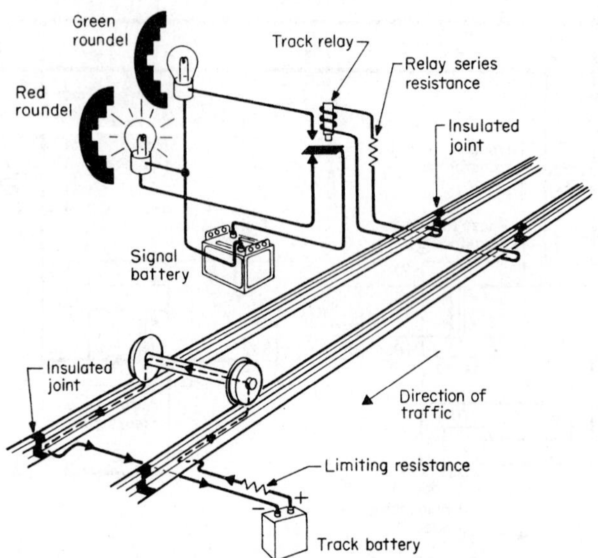


**FIG. 23-60** Schematic diagram of direct-current track circuit, occupied. Lamp behind red roundel is lighted, giving a stop aspect.

thus energized, it closes a contact to light the lamp (or to control a signal mechanism to its proceed aspect).

As the wheels and axles of a train move onto the track circuit, Fig. 23-60, they provide a path from rail to rail through which the battery current flows, thus robbing the relay of its current and causing it to open the contact through which energy was feeding to the lamp behind the green roundel and to close the contact to cause the lamp behind the red roundel to light, thus indicating a stop aspect.

An advantage of this arrangement is that a broken rail also causes the relay to drop because the circuit is interrupted and the signal will change from green to red.

***220. Impedance Bonds.*** In electrified railroads, whether ac or dc, the traction-rail return power must be permitted to flow around the insulated rail joints with low resistance or impedance. The ends of each track section are thus equipped with impedance bonds, consisting of a few turns of heavy copper wound about a laminated core, the terminals of which are connected each to a rail, and the midpoint or neutral brought out for connection to a similar bond beyond the insulated rail joints. These bonds are so designed that the traction power (direct or alternating current), traveling in nearly equal amounts in each rail, passes through the coils of the impedance bond, which is so wound that the impedance is very low, because the magnetization of each half balances the other half for currents thus flowing. The impedance between rails for flow from one rail to the other for the higher-frequency signal power is high (see Fig. 23-61). If there is unbalance in the rails which permits more traction power to flow in one rail than in the other, there is a tendency to saturate the iron core and thereby reduce the impedance of the bond. This unbalancing effect is limited by introducing an air gap into this magnetic circuit. The bonds are usually designed to withstand 20% unbalance with a decrease of not more than 10% impedance. The size of the bond is, of course, dependent upon the fault current that it is expected to carry, the impedance for which it must be wound, and the unbalance to be accounted for. When good track bonding is maintained, unbalance may be assumed to be low and a smaller bond can be employed (see Fig. 23-62).

***221. Cross Bonding.*** In multiple tracks, cross bonding is installed between the neutrals of the impedance bonds in adjacent tracks to permit advantage to be taken of power return

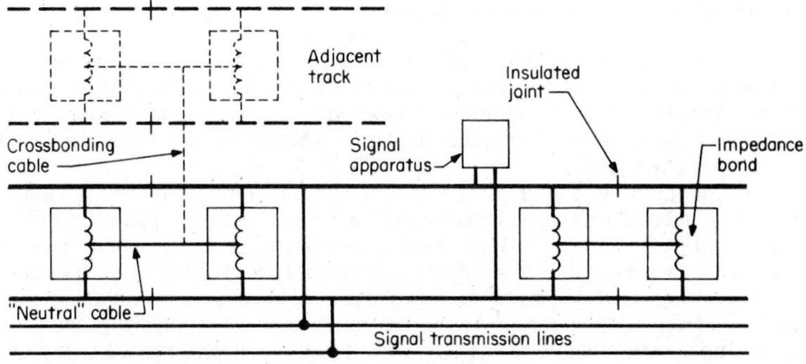


**FIG. 23-61** Impedance bonds used in ac or dc propulsion areas are installed in pairs and connected on both sides of insulated joints. The center tap arrangement of the bond coils permits propulsion current to flow around insulated rail joints while blocking the signal currents.

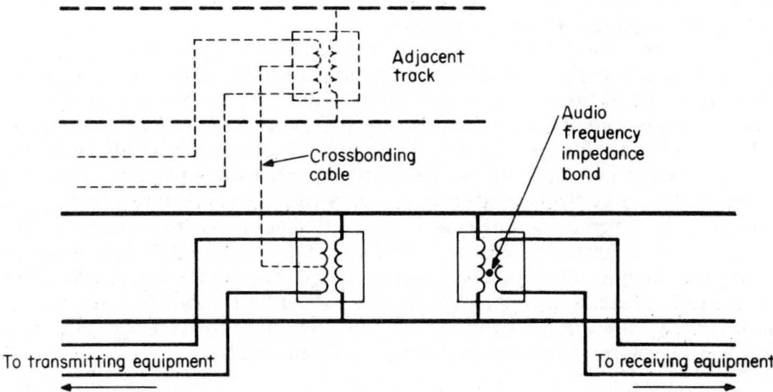


**FIG. 23-62** Audiofrequency minibonds are used in dc propulsion areas for transmitting specialized signals and control data through the rails. These bonds contain inductively coupled coils and are used with audiofrequency track circuits for the transmission of audiofrequencies for cab signals, speed command, and train detection signals.

using the conductivity in all tracks. Care must be taken in this connection not to have too frequent cross bonding, with possible danger of permitting, through leakage, faulty signal operation.

***222. Block Signals.*** The block on which the basic track circuit is applied is minimally the length required for the stopping of the fastest and/or heaviest train which is to be operated over the route. When problems of track capacity were encountered, multiple aspect signaling was introduced. The basic system would indicate the occupancy of the next block ahead of the train; automatic block signaling has at least provided a two-block indication assuring that the first restrictive indication will always be received in time to permit a stop short of the train ahead. This two-block system is an approximation of the train-spacing theory in which a train should carry along behind it a zone of protection exactly braking distance in length at all times. This was soon expanded to use of three-indication signals where the zone of protection is twice braking distance. When track capacity became acute, there came four-aspect, three-block signaling which can accommodate higher-speed trains with less sacrifice of capacity for slower trains, or it can handle denser traffic without delay while maintaining the same braking distance.

### Automatic Cab Signals and Speed Control

***223. Cab Signals.*** Although treated separately here, *cab signals* are often used in conjunction with wayside signals or incorporated as elements of a train control system. Cab signals are adaptable to all types of motive power and can provide the conventional signal aspects. They may be used with wayside signals or without wayside signals, in which case speed control is often used.

The information needed for control of the cab signal is transferred from the track to the locomotive by inductive coupling between the track and circuits on the locomotive. Since inductive coupling is effective only when an ac component is present in the track, steady-energy dc track circuits cannot be used alone. In electrified territory, ac track circuits may be used. But for nonelectrified railroads, practice is to superimpose an ac component on the dc used for wayside signal control. A system in which ac is continuously on the track for a clear block, but is not present for a restrictive block (train in block), provides two-indication cab signals.

For more than two indications, the ac track current can be coded at various rates, such as 75, 120, and 180 times a minute. The locomotive is then provided with decoding equipment with the cab signal controlled accordingly.

A typical three-indication cab signal system is shown in Fig. 23-63. For a clear indication, the coded ac (100 Hz) is coded at 180 times per minute. The approach indication uses a code rate of 75, and the absence of code, resulting from a train occupying the block or section of track, produces a restricting indication.

The code being fed to the track rails is picked up inductively by two receiver coils, one of which is mounted on each side of the locomotive (or the head end of multiple-unit passenger cars) in front of the leading wheels, directly over the rails, about 6 in above the rail head. The energy picked up by the receiver coils passes through the filter where frequencies other than 100 Hz are filtered out. The filter output is coupled to a solid-state amplifier which amplifies and processes the low-level coded input to produce two outputs. One of these outputs is dc, regardless of code rate; the other is a coded output applied to a 180-code rate decoder unit. When a 180-rate code is received, dc outputs are produced by both the decoder unit and the amplifier. With a 75 rate only, the amplifier dc output is present, but not the decoder output. With no code, neither output is present. Relays driven by these outputs control circuits to give the clear (green) indication for the 180-code rate, and the approach (yellow) indication for the 75-code rate. With no code received, both relays are released, and the restricting (red-over-yellow) indication results.

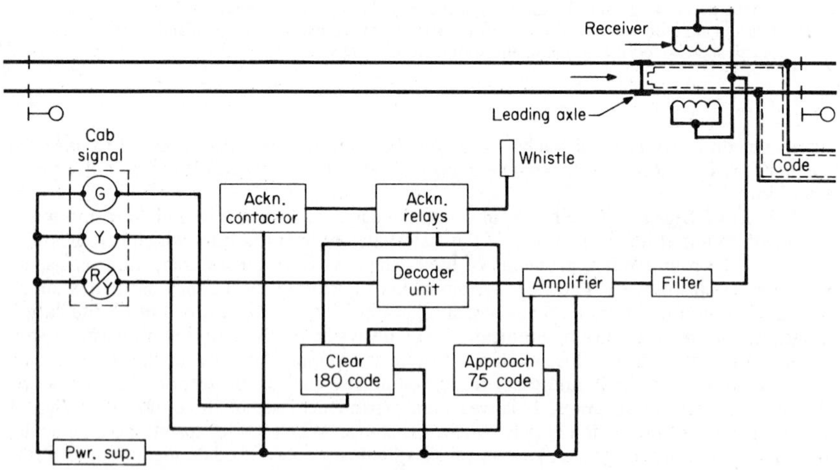


**FIG. 23-63** Schematic diagram of three-aspect cab signal system.

Associated with the signal control relays are acknowledging relays which cause a whistle to sound each time an aspect becomes more restrictive. The whistle is silenced by the operation of the acknowledging contactor, an action which proves the engineer is aware of the warning.

Figure 23-64 shows a typical locomotive circuit for a three-indication cab signal system.

When the 180-rate code is received, relays 180R and L are picked up. With relay 180R up, the green lamp in the cab signal display unit is lighted through a front contact on relay 180R and a back contact on acknowledging relay SP.

With relay 180R released and relay L still up, corresponding to receipt of the 75-rate code, the yellow lamp in the cab signal display unit is lighted through a back contact of 180R, a front contact of L, and a back contact of SP. With both relays 180R and L down (released), the red-over-yellow lamp is lighted through back contacts of relays 180R and L.

Acknowledgment operates as follows. When the cab signal is clear, whistle valve WV is energized through a front contact of relay 180R, and through back contacts of relays LP and

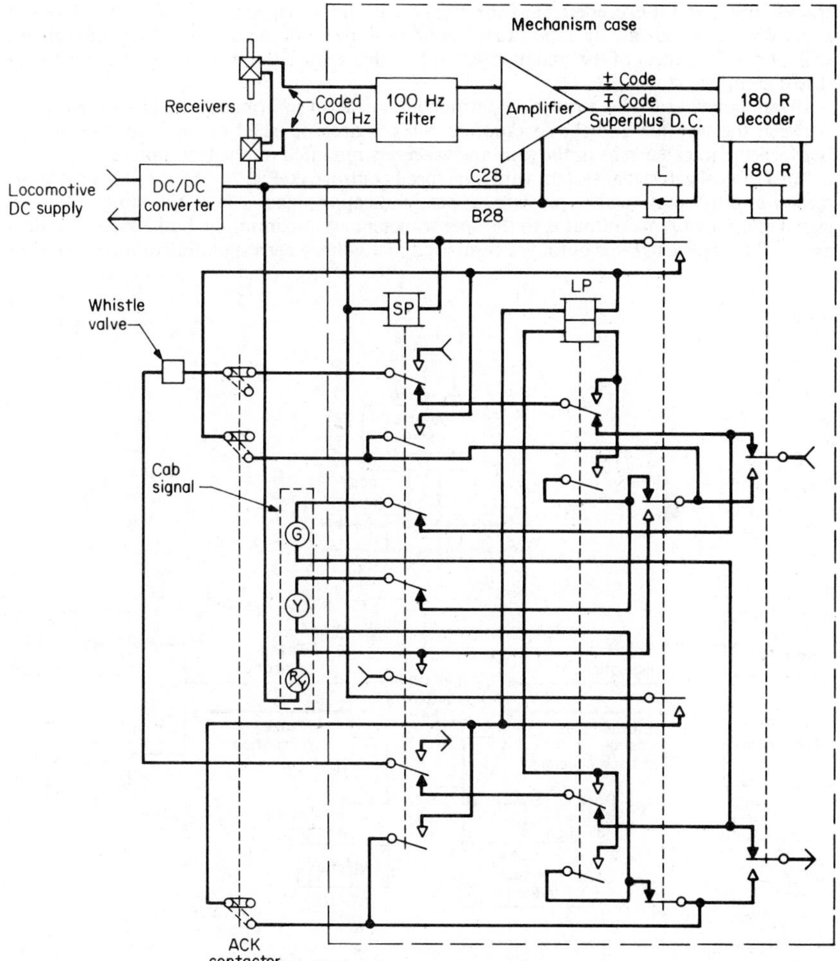


**FIG. 23-64** Simplified locomotive circuits for three-aspect cab signal system. (Relays shown in *clear* position.)

SP. If the signal changes from clear to approach, the relay 180R releases, deenergizing WV, causing the whistle to sound. Operating the acknowledging contactor energizes LP through a back contact of relay 180R and a normally open contact of the acknowledging contactor. When relay LP comes up, it sticks up on the circuit through relay 180R back contact, relay L front contact, and relay LP front contact. Release of the acknowledging contactor then reenergizes WV (stopping the whistle from sounding) through the relay 180R back contact, relay L front contact, two relay LP front contacts, relay SP back contact, and the normally closed acknowledging contact.

When a less restricting signal change occurs, a momentary "peep" of the whistle sounds. For example, when the signal changes from restricting to approach, relay L picks up and opens the stick circuit on relay SP. But relay SP is slow-release, so that it remains up for a period of about 3 s after relay LP picks up. When relay SP releases, relay LP remains up because the stick circuit through the relay L front contact has been reestablished. The whistle gives a momentary peep when SP crosses over from its front-to-back contact, but no acknowledgment is required since WV is energized when SP makes its back contact.

Train control systems are often added to cab signal systems to provide monitoring of a train's speed, and if exceeded, alert the engineer to the overspeed. If the engineer does not acknowledge or indicate by action that she or he is aware of the overspeed, the equipment will take over control of the braking system to either reduce the speed of the train or bring it to a complete stop.

Train control systems require a continuous transfer of information on block conditions between the wayside and the locomotive. This is accomplished by an inductive linkage between the track current in the rails and receivers mounted on the locomotive.

In a typical cab signal system with overspeed control (see Fig. 23-65), the axle generator, driven by a locomotive axle, provides a speed indication signal. After processing in the cab signal equipment box, output is to the speedometer and to control the brake system, if necessary. The axle generator output is compared to a voltage corresponding to the maximum

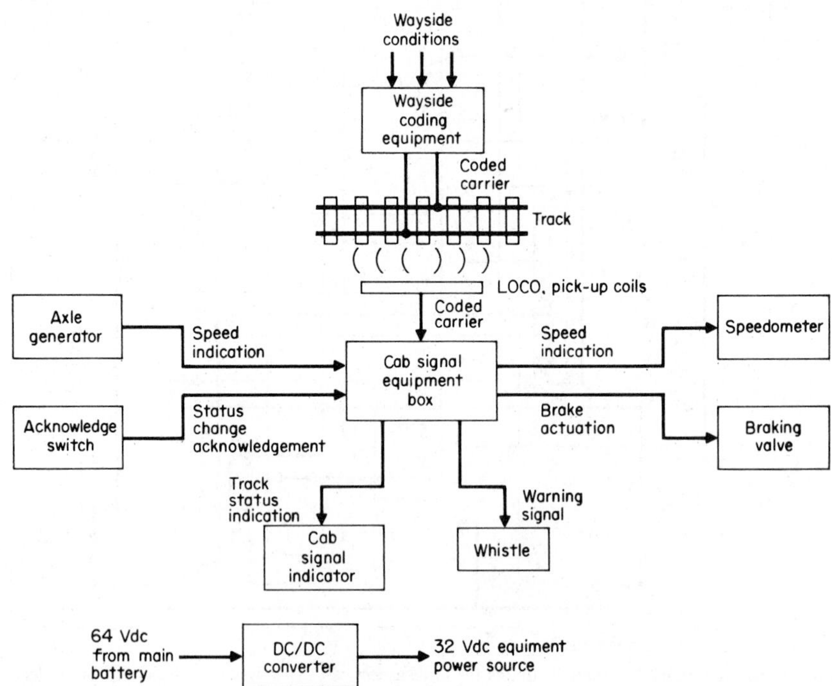


**FIG. 23-65** Block diagram of typical cab signal system with speed control.

allowable speed for the particular code rate. If the locomotive speed exceeds that which is allowable, a warning signal is sounded. If the engineer does not acknowledge and reduce speed, the speed control system takes over control of the brake system. Depending upon the type of control desired, the brakes may bring the train to a complete stop, or the brakes may only be applied until the train is below the permissible speed, at which time the brakes are released and the engineer may regain control.

***224. Centralized traffic control (CTC)*** is a development in railroad operation that has made possible important operating economies. Under this system all signals and switches at crossovers, passing sidings, or junctions on a given section of railroad are controlled from a central point. The trains are operated entirely by signal indications, and so no train orders are issued. All track switches thus installed in the section are remotely controlled, and the signals are either automatic or remotely controlled.

Modern CTC systems now use computers including microprocessors to handle specific functions, such as enabling one control machine to interface with several types of code systems that transmit codes and indications between the control machine and the field locations. Computer-aided dispatching is now considered a standard type of control for CTC.

The dispatcher or control machine operator has a keyboard unit to issue commands for controlling switches, signals, and auxiliary functions such as operating snow melters at remote track switches. The display unit showing the track diagram and signal positions may be on a video [cathode-ray tube (CRT)] unit, or it may be on a large panel of sufficient size to be seen from across a room.

The computer keeps track of trains moving over the territory, but it also records entry and leaving times for all trains at a location, such as ends of passing sidings. Other information recorded includes when switches are operated and locked in position, when signals are at clear or at stop, and when the dispatcher blocks sections of track for work by maintenance of way forces.

Communication links include line wires on poles along the right of way, microwave radio, buried cable, and the newest weatherproof, inductive-interference-proof system using fiber-optic cables.

One of the modern coding systems, for example, uses a minicomputer for message formatting, indication scanning, code validity check, and information processing. Transmission speeds can be up to 9600 bits per second. Such a system has a capacity of 62 stations (ends of sidings) with up to 256 controls per station. Indications (conditions in the field) for this 62-station system can be as high as 320 per station.

At the field locations, coding interface equipment is located to decode the control messages into controls for operating switches and signals. However, vital safety relay logic at these field locations only allows controls to be executed if it is safe to do so. On the reverse, conditions of switch positions and signal aspects displayed are sent as indications back to the control machine via the coding system. Here, in modern systems, a microprocessor in the field handles both the decoding of controls and the encoding of indications.

***225. Highway Grade Crossing Warning Devices.*** Highway grade crossing warning devices such as flashing-light signals and automatic gates are controlled automatically by track circuits in a manner similar to block signal operation. Current railroad practice is to use audiofrequency or frequency shift overlay track circuits to control highway grade crossing warning devices. These track circuits employ alternating current at various frequencies (500 Hz to 20 kHz) and are usually operated over conventional dc track circuits used for control of the automatic block signaling.

In a typical installation (see Fig. 23-66), the frequency shift overlay circuits are overlapped to provide operation of the warning devices while a train is occupying the roadway or street section of the crossing. Note that the f1 transmitter (f1 XMTR) sends its current to the f1 receiver (f1 RCVR) to the far side of the crossing. Similarly, an f2 transmitter (f2 XMTR) sends its signal over the crossing to an f2 receiver (f2 RCVR) overlapping the signal of f1. Thus two track circuits cover the entire approaches and the crossing. This compares with two approach circuits and an island circuit over the roadway, used in earlier installations prior to the development of the audiofrequency or frequency shift overlay circuits. An advantage of these audiofrequency or frequency shift overlay circuits is that insulated joints are not needed. Hence they are used extensively in welded rail territory.

In Fig. 23-66, when an eastbound train approaches, transmission from the f1 transmitter is shunted away from the f1 receiver and f1 OTR (f1 overlay track relay) drops, opening the circuit of XR (crossing relay) and activating the warning devices.

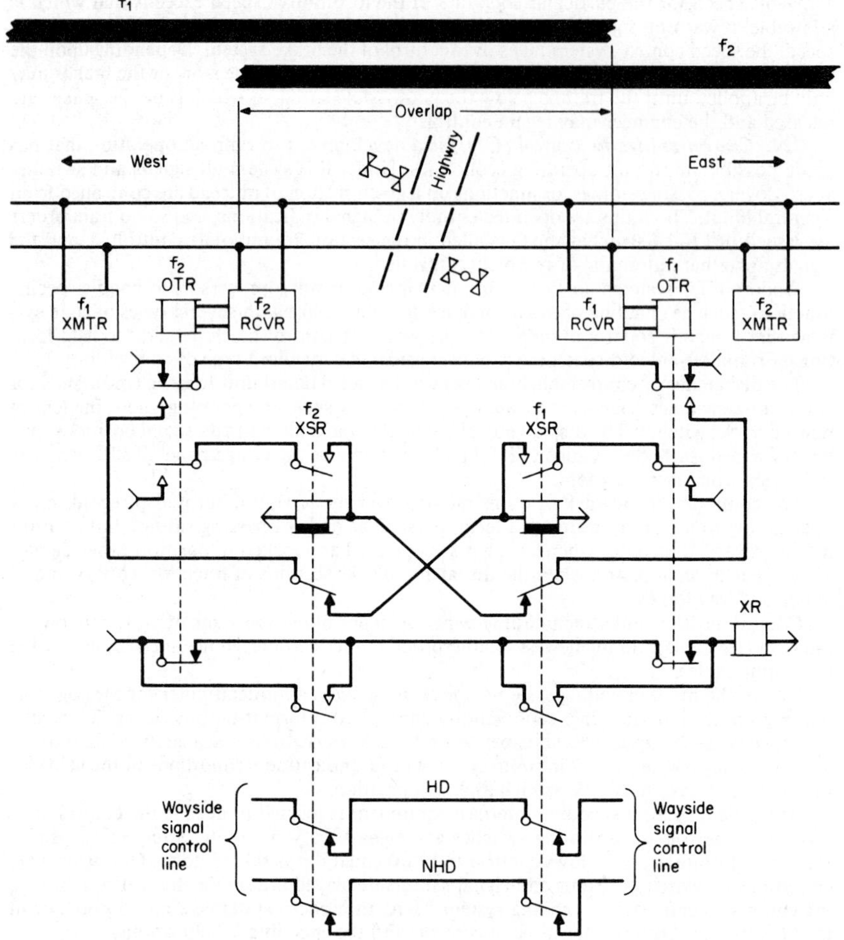


**FIG. 23-66** Highway grade crossing warning system controls showing overlapped frequency shift overlay track circuits unoccupied.

With the train across the highway, both transmitters are blocked from their receivers and both track relays (f1 OTR and f2 OTR) are down. When the train clears the crossing, the f1 RCVR again receives energy from the f1 XMTR, thus picking up the f1 OTR and completing the circuit to pick up the XR (crossing relay), which stops the flashing lights from operating and the gates rise, clearing the roadway.

## Automatic Train Control

*226. Application of automatic train control (ATC)* in the United States has been used in rapid transit systems within the last 20 years. ATC systems usually include three major subsystems: automatic train protection, automatic train operation, and automatic train supervision. An overall block diagram of a rapid transit ATC system is shown in Fig. 23-

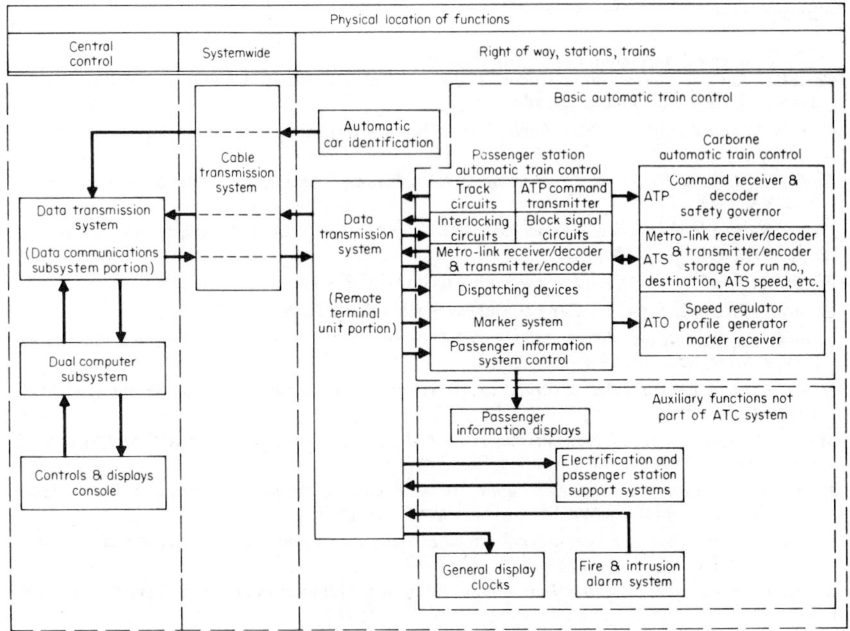


**FIG. 23-67** Block diagram of a rapid transit automatic train control system.

67. These subsystems are coordinated through a dual computer (active and backup) installation at a central control or operations headquarters.

The automatic train protection subsystem maintains safe stopping distances, between trains, ensures safe door operation, and provides control at interlockings.

The automatic train operation subsystem starts trains out of stations, regulates train speeds between stations, and brings trains to a properly positioned stop at stations.

The automatic train supervision subsystem automatically selects routes through interlockings (or switching locations), dispatches trains automatically, and furnishes the means to make trains responsive to supervisory commands from the central control.

Central control monitors the total transit system and initiates correction commands to smooth traffic flow. Correction commands may be generated automatically by computer; but if conditions warrant, the computer alerts a control center operator, who can take corrective action.

Three additional subsystems are (1) *data transmission* to transfer commands and data between wayside locations and central control; (2) *cable transmission,* a systemwide network to provide voice-band circuits for data transmission and other system communications; and (3) *automatic car identification,* which optically scans coded markers on each car to identify the car numbers and provide data for computer processing for car operating and maintenance purposes.

A key link in the automatic train control system is train-to-wayside communications. In one major system, this is accomplished by sending digital messages through the rails, using a carrier frequency higher than the channel frequencies used for track circuits and speed commands to the trains. At the wayside, the impedance bonds used for the track circuits and speed command functions also serve as coupling transformers for the train-to-wayside communications. The rail-to-train link is via separate coils assembled in the impedance bond housing. The train-to-rail link is via a horizontal loop (3 by 4 ft) encased in a plastic housing mounted under the lead car of the train.

## Bibliography

### 227. Selected Reference Books and Reports

*Canadian Railway Electrification Studies: Phase 1* (3 vols.); Canadian Institute for Guided Ground Transport, Kingston, Ontario, Canada, 1976.

*Corrosion Control Manual for Rail Rapid Transit;* National Technical Information Service, Springfield, VA. 1980.

*Dual-Mode Locomotive System Engineering* (2 vols.); National Technical Information Service, Springfield, VA, 1981.

*Electric Railroads and Trolley Systems: Past, Present, and Future;* National Technical Information Service, Springfield, VA, 1984.

*Electrification of Steam Railroads;* New York, McGraw-Hill Book Company, 1929.

*Elements of Railway Signaling;* General Railway Signal Co., Rochester, NY, 1979.

*Evaluation of Signal/Control System Equipment & Technology* (7 vols.); National Technical Information Service, Springfield, VA, 1978–1981.

*Joint Trackwork-Electrical Design Guidelines;* American Public Transit Association, Washington, D.C., 1980.

*Mutual Design of Overhead Transmission Lines and Railroad Communications & Signal Systems* (2 vols.); Electric Power Research Institute, Palo Alto, CA, 1983.

*Railraod Electrification in America's Future: An Assessment of Prospects and Impacts. Final Report;* National Technical Information Service, Springfield, VA, 1980.

*Railroad Electrification: A Report to the Railroad Electrification Committee of the Edison Electric Institute;* Edison Electric Institute, New York, NY, 1970.

*Railroad Electrification: A System Design Project;* National Technical Information Service, Springfield, VA, 1974.

*Railroad Electrification: The Issues;* Special Report 180. Transportation Research Board, Washington, D.C., 1977.

*Railroad Electromagnetic Compatibility (EMC). Proceedings of a Symposium Sponsored by the Railroad EMC Working Group;* National Technical Information Service, Springfield, VA, 1981.

*Railroad Electromagnetic Compatibility* (5 vols.); National Technical Information Service, Springfield, VA, 1978–1982.

*Railroad Operation & Railway Signaling;* Simmons-Boardman Publishing Corp., New York, NY, reprinted 1983.

*Railroad Track Structure, Electrification and Operations Management;* TR Record 802. Transportation Research Board, Washington, D.C., 1981.

*Railways and Energy;* The World Bank, Washington, D.C., 1984.

*Reduction of Peak Power Demand for Electric Rail Transit Systems;* NCTR&DP Report 3. Transportation Research Board, Washington, D.C., 1983.

*Report on the Visit to the USSR by the US Electrification Delegation;* National Technical Information Service, Springfield, VA, 1975.

*The Search for Safety: A History of Railroad Signals and the People Who Made Them;* Union Switch & Signal Division, American Standard, Pittsburgh, PA, 1981.

*Task Force on Railroad Electrification: A Review of Factors Influencing Railroad Electrification;* National Technical Information Service, Springfield, VA, 1973.

*Track Design and Electrification;* TR Record 939. Transportation Research Board, Washington, D.C., 1983.

*Wayside Energy Storage Study* (3 vols.); National Technical Information Service, Springfield, VA, 1979.

*When the Steam Roads Electrified;* Kambach Books, Milwaukee, WI, 1975.

### 228. Current Publications

*Proceedings,* American Railway Engineering Association, Washington, D.C. Reports of Communication & Signal Division, Engineering Division, and Mechanical Division, Association of American Railroads, Washington, D.C.

*International Railway Journal,* Falmouth, England.

*Mechanical Engineering* and *ASME Transactions,* New York, NY.

*Modern Railways,* Shepperton, England.

*Progressive Railroading,* Chicago, IL.

*Quarterly Reports, Railway Technical Research Institute,* Tokyo, Japan.

*Rail International,* Brussels, Belgium.

*Railway Age,* New York, NY.

*Railway Gazette International,* Sutton, England.

*Spectrum* and *IEEE Transactions,* New York, NY.

***229. Specific Articles***

Arai, K., and Watanabe, H.: Rail Potential and Countermeasures against Potential Rise; *Q. Rep. Railw. Tech. Res. Instit.,* no. 4, 1983, pp. 143–148.

Cooper, B.: Chopper Control; *Mod. Railw.,* April 1983, pp. 204–205.

Cooper, B.: Squirrel Cage or Static Commutator? *Mod. Railw.,* August 1983, pp. 426–427.

Cossie, A.: French Hail Synchronous Motor as Traction Panacea; *Railw. Gaz. Int.,* May 1983, pp. 349–351.

DuPuy, J.: Investing in Electrification; *Railw. Gaz. Int.,* January 1979, pp. 47–50.

Kendall, H. C.: Electrification: A Status Report; *Railw. Gaz. Int.,* June 1975.

Leander, P.: Thyristor Locomotives with Three-Phase Drives—Development for the Future; *ASEA J.,* no. 1, 1984, pp. 24–29.

Nouvion, F. F.: Three-Phase Motors in Electric Rail Traction, *IEEE Trans. Ind. Appl.,* September/October 1984, pp. 1152–1170.

Motive Power: Solid-State Technology Takes Over; *Int. Railw. J.,* August 1980, pp. 19–23.

Scott, P. R., and Rothman, M.: Computerized Evaluation of Overhead Equipment for Electric Railroad Traction; *IEEE Trans. Ind. Appl.,* May 1974.

Powers, W. F.: Asynchronous Traction Motors Will Meet the Synchronous Challenge; *Railw. Gaz. Int.,* March 1984, pp. 179–183.

Siemens, W. H.: Will 50 kV Become a World Standard? *Railw. Gaz. Int.,* April 1978, pp. 201–203.

Serdinov, S. M.: Electrification of Railways in the USSR; *Rail Int.,* January 1981, pp. 8–21.

## *LINEAR PROPULSION*

*By ENRICO LEVI*

### Introduction

***230. Definition.*** Linear propulsion relates to a novel type of electric drive for ground transportation systems. In such a scheme, the motor is split into two parts, of which one is carried by the vehicle and the other lies straight along the track. The force of interaction between these two structures is utilized directly as tractive effort, without the need for intermediate transmission or gear.

***231. History.*** The idea of associating power conversion with linear motion dates back to the early 1830s, when Faraday attempted to generate electricity by utilizing the streaming water of the Thames River as an active conductor, the geomagnetic field as the field excitation, and dangling wires as the electrodes. Other significant dates were 1841, when Wheatstone built the first linear motor; 1905, when linear propulsion was first proposed for ground transportation; and 1946, when a 1400-ft-long wound-rotor induction motor was constructed at Westinghouse for the purpose of catapulting small seaplanes from the deck of a ship (Ref. 33).* Soon thereafter, Prof. E. R. Laithwaite of the Imperial College in London initiated his pioneering work on the *linear induction motor* (LIM), and the recent popularity of this machine is largely due to his promotional effort (Refs. 1, 2, 3, 28).

***232. Wheel-on-Rail Adhesion.*** The major appeal of linear propulsion is the promise to overcome the limitations of the wheel in high-speed rail transportation. The wheel relies on friction in order to provide the required *tractive effort.* If slipping between the driving wheels

*References appear in Par. **269**.

and the rail is to be prevented, the following inequality must be satisfied between the tractive effort, or *thrust* $T$ (see Table 23-10) and the *weight* of the locomotive $W$:

$$T \leq W\alpha \tag{23-1}$$

where $\alpha$ is the *coefficient of adhesion* or *friction.* The coefficient of adhesion, which is a function of the weather and rail conditions, decreases with speed. According to Mueller, the

**TABLE 23-10** Symbols and Notation

| | | | |
|---|---|---|---|
| | **Capital letters** | | |
| $B$ | magnetic flux density | $P$ | power |
| DLIM | double-sided induction motor | $R$ | resistance |
| | | Re | real part of |
| $E$ | electric field intensity | | |
| | | $S$ | slip |
| $F$ | force | | |
| | | SLIM | single-sided induction motor |
| $H$ | magnetic field intensity | | |
| $I$ | current | SLM | synchronous linear motor |
| $J$ | current density | | |
| | | $T$ | thrust |
| $K$ | surface current density | | |
| | | $V$ | voltage |
| $L$ | motor length | | |
| | | $X$ | reactance |
| LCM | linear commutator motor | $W$ | weight |
| LIM | linear induction motor | | |
| | **Lowercase letters** | | |
| $a$ | thickness of reaction rail | $l$ | length |
| | | $t$ | time |
| $f$ | frequency | | |
| | | $v$ | velocity |
| $g$ | gap length | | |
| | | $x$ | direction of motion |
| $j$ | $\sqrt{-1}$ | | |
| | | $y$ | direction of current |
| $k$ | wave number | | |
| | | $z$ | direction of flux density |
| | **Greek letters** | | |
| $\alpha$ | coefficient of adhesion or friction | $\mu$ | permeability |
| $\beta$ | angle between $\dot{I}_f$ and $\dot{I}_a$ | $\tau$ | pole pitch |
| $\gamma$ | conductivity | $\varphi$ | angle between $\dot{V}_a$ and $\dot{I}_a$ |
| $\eta$ | efficiency | $\omega$ | radian frequency |
| | **Subscripts** | | |
| $a$ | armature | $p$ | pole face |
| $d$ | direct-axis | $ph$ | phase |
| $dif$ | diffusion | $q$ | quadrature |
| $e$ | electric | $s$ | synchronous |
| $f$ | field | 0 | free space |
| $l$ | lateral | 1 | primary, first root |
| $m$ | mechanical, magnetizing | 2 | secondary, second root |
| $n$ | normal | | |

Vectors are denoted by boldface Roman letters. The same letter in lightface type denotes either the magnitude or a component of the vector.

Phasors are indicated by a dot over the letter.

following relations hold:

$$\alpha = \frac{0.22}{1 + 0.036v} \quad \text{for acceleration} \tag{23-2}$$

and

$$\alpha = \frac{0.33}{1 + 0.072v} \quad \text{for deceleration} \tag{23-3}$$

where $v$ is the speed in m/s. For a set of operating conditions often used as a design basis [$v$ = 139 m/s (500 km/h) and $T = 10^5$ N] an individually powered car would require a weight in excess of 330 metric tons, or one order of magnitude larger than its expected payload.

***233. Suspension and Guidance.*** Linear motors also develop force components perpendicular to the direction of motion. These components can be utilized to fulfill the two remaining functions of the wheel: *suspension* and *guidance.* The resulting scheme, called *magnetic levitation* or *maglev,* when combined with on-ground installation of the energized side of the motor, completely eliminates the need for physical contact between the vehicle and the track (Ref. 25).

## General Configurations

***234. Open Ends.*** The basic difference between rotating and linear motors is that in the former the air gap and the magnetic structures are *endless,* whereas in the latter the air gap and one of the magnetic structures (short side) are *open-ended* (Fig. 23-68 and Refs. 34, 38–45).

***235. Machine Types.*** All types of existing electric motors can be realized in the linear version. In addition, the peculiar requirements of linear propulsion foster the development of novel topologies. As a result a great variety of schemes is being considered, and only the test of time will decide on their practicality. Table 23-11 lists the major lines of development.

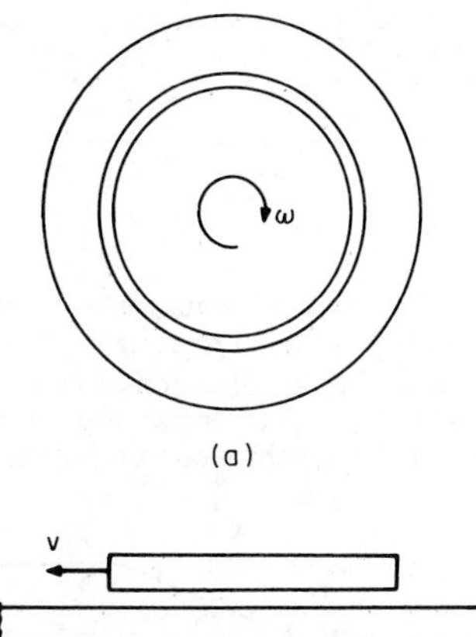


**FIG. 23-68** Magnetic structures in electrical machines. (*a*) Rotating; (*b*) linear.

***236. Energization.*** In general two alternatives exist with regard to which side of the motor is energized. Energization of the vehicle side (short) implies power collection through sliding contacts. The existing methods involving third rail or catenary do not perform satisfactorily at speeds in excess of 350 km/h. Energization by the wayside (long) is very expensive, especially since it seems likely that power supply at variable frequency will be required. Except for special applications, such as railroad switchyards, the cost of the power conditioning apparatus distributed along the track justifies the choice of this alternative only in the case of very high speeds, or of very dense traffic corridors.

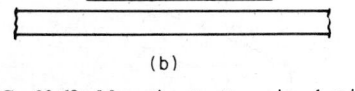

***237. Magnetic Circuit.*** Two alternatives also exist with regard to the magnetic circuit, depending on whether ferromagnetic cores or superconducting magnets are employed. The latter are feasible only in synchronous-motor drives energized by the wayside. They afford large suspension gaps (up to 30 cm); hence this propulsion system is contemplated in association with levitation by induced currents, that is, *repulsive maglev* (Refs. 25, 37, 50, 57, 62, 66). Iron-cored motors, instead, are better suited for smaller gaps (up to 3 cm) and energization by the vehicle side. They are compatible with ferromagnetic levitation, that is, *attractive maglev.* Because of the small gaps and the reliance on sliding contacts for power collection, propulsion systems employing iron-cored motors are likely to find application in the lower-speed range (up to 350 km/h) (Refs. 12, 25, 65, 78, 81).

**TABLE 23-11** Characteristic Features of Linear Propulsion Schemes

| Motor type | Windings | Magnetic circuit | Track | Supply frequency | Speed range, km/h |
|---|---|---|---|---|---|
| Induction (Fig. 23-72) | Reaction rail | Single-sided | Passive | Variable | 0–350 |
| | | Double-sided | | | 0–150 |
| | | Longitudinal flux | | | 0–350 |
| | | Transverse flux | | Industrial | 0–350 |
| | Wound secondary | Single-sided | Energized | | 0–150 |
| Romag (Fig. 23-73, Ref. 59 | Gramme | Variable reluctance | Passive | Variable | 0–150 |
| Synchronous (Fig. 23-75, Refs. 37, 50, 68) | Superconducting magnets | Air-cored | Energized | Variable | 350 or over |
| | Rubber-insulated windings | Iron-cored | | | 0–350 |
| Claw-pole (Fig. 23-70) | Field and armature on-board vehicle | Longitudinal and transverse | Passive, salient poles | Variable | 0–350 |
| Homopolar inductor (Fig. 23-71) | | | | | |
| Heteropolar inductor (Ref. 24) | | Longitudinal flux | | | 0–150 |
| Traklec (Ref. 15) (M. Barthalon, U.S Patent 3,707,924) | Armature only | Transverse flux | Passive, salient poles | Variable | 0–150 |

***238. Topology.*** The magnetic circuit also offers a number of alternatives with regard to its topology. In most rotating machines the lines of force of the main magnetic flux lie in planes parallel to the direction of motion (longitudinal or axial flux, Fig. 23-69). A problem arises with high-speed linear motors supplied at industrial frequencies (50 or 60 Hz), because of the synchronous tie between the speed of the traveling wave $v_s$ and the frequency $f$:

$$v_s = 2\tau f \tag{23-4}$$

where $\tau$ is the pole pitch. $\tau$, and hence the length of the end winding, becomes disproportionately large when compared with the length of the active portion of the conductor which is limited by the width of the track. The flux per pole and, hence, the height of the iron core behind the slots also become large. It is then more economical to have the magnetic-flux lines lie in planes perpendicular to the direction of motion (transverse flux) (Fig. 23-69 and Refs. 49, 63).

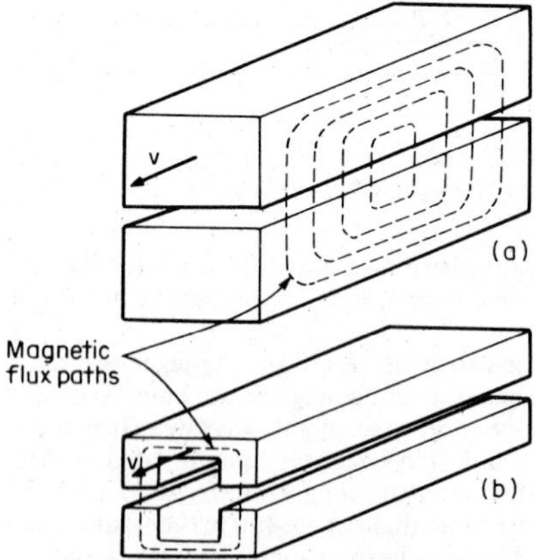


**FIG. 23-69** Topology of magnetic circuit. (*a*) Longitudinal or axial flux; (*b*) transverse flux.

A combination of longitudinal and transverse fluxes occurs in synchronously operating motors, as a result of the desire to have both the field excitation and the armature windings located on the vehicle side of the motor, so that the track is completely

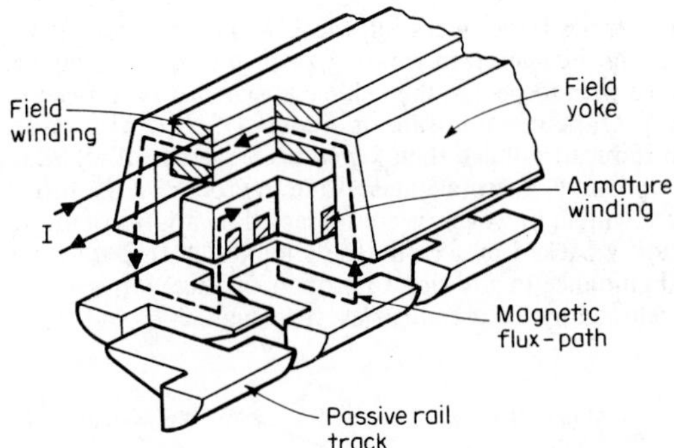


**FIG. 23-70** Sketch of claw-pole motor (Nadyne, J. J. Pierro, U.S. Patent 3,456,136, July 15, 1969).

passive. In this case the field winding is interlinked with the transverse flux and the armature winding with the longitudinal flux (Figs. 23-70, 23-71, and Refs. 15, 24, 26, 35, 44).

### Linear Induction Motor (LIM)

***239. Construction.*** The LIM is the motor type which has reached the most advanced stage of development. In its simplest form it consists of (*a*) a ferromagnetic structure located on the underside of the vehicle and carrying a winding energized by a polyphase power supply, and (*b*) a passive *reaction rail* in the form of a bar of conducting material, most commonly aluminum, embedded in the guideway. To reduce the reluctance of the magnetic-flux path, the reaction rail is either sandwiched between two energized structures (double-sided LIM or DLIM) or backed by a passive ferromagnetic structure (single-sided LIM or SLIM) (Fig. 23-72). The SLIM may require a more expensive rail but alleviates dynamic stability problems encountered with the DLIM.

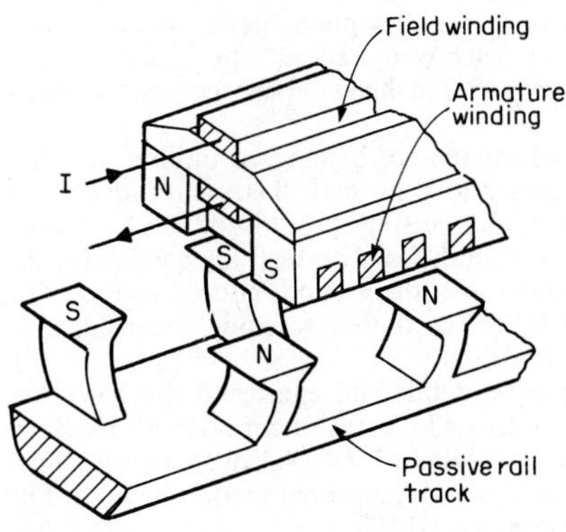


**FIG. 23-71** Sketch of homopolar inductor motor.

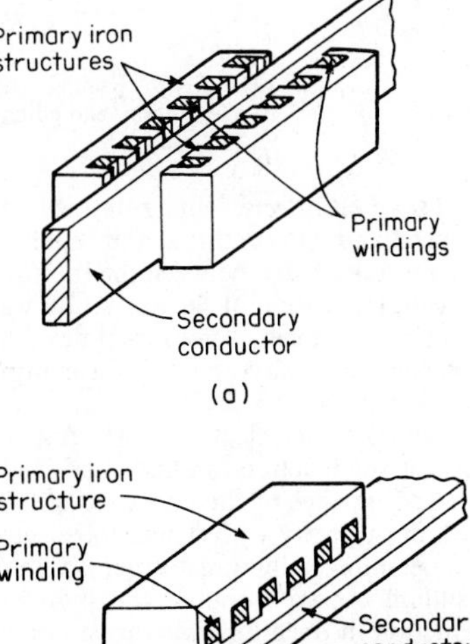


**FIG. 23-72** Linear induction motor configurations. (*a*) Double-sided (DLIM); (*b*) single-sided (SLIM).

***240. Operation.*** In its basic operation the LIM is similar to its rotating counterpart. The currents flowing in the energized winding, or *primary,* set up an electromagnetic wave that travels backward with respect to the vehicle at synchronous speed $v_s$. When the vehicle speed $v$ differs from $v_s$, the reaction rail, or *secondary,* slips with respect to this traveling wave. The motion-induced voltage then generates a system of alternating currents which interact with the impressed electromagnetic wave to produce a net thrust. There also develops a repulsive force which, however, is counteracted by a force of attraction, whenever the secondary conductor is backed by ferromagnetic material. In both cases the LIM can provide levitation and guidance in addition to propulsion. See, for instance, the attractive levitation scheme developed by Rohr Industries, Inc. (Fig. 23-73 and Refs. 12, 25, 59).

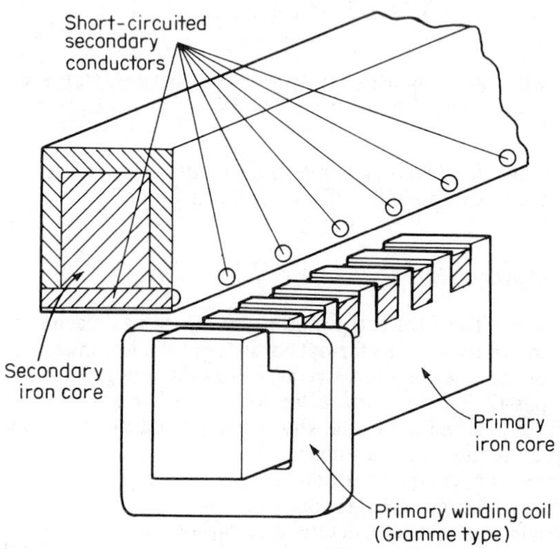


**FIG. 23-73** Romag scheme for combined propulsion, attractive levitation, and guidance.

***241. End Effects.*** In a rotating (endless) machine operating in steady state, currents are induced and, hence, thrust and repulsive force are produced, only when there exists a relative motion of *slip,* between the impressed traveling wave and the reaction rail. In the LIM, instead, truncation of the excitation wave introduces complex phenomena, and under certain circumstances, a net thrust develops, even at synchronous speed. The reason is that a truncated sine wave contains a continuum spectrum and hence gives rise also to faster waves (see also Par. **247**).

Another end effect seriously impairs the performance of high-speed motors. It arises because the reaction rail finds itself suddenly exposed to a magnetic field, as it slides under the leading edge of the energized structure. When the cruising speed of the train is higher than the velocity with which the electromagnetic field diffuses through the conductor, the voltage induced by transformer action predominates over the motion-induced voltage. The resultant currents oppose the inducing magnetic fields, thereby causing a considerable reduction in the attainable thrust (see also Par. **248**).

***242. One-Dimensional Model.*** The seriousness of this end effect and the hope that measures could be devised to alleviate it have motivated studies of considerable sophistication (Refs. 4, 12, 38, 39–41, 45, 52, 55, 56, 58, 60, 61, 64, 67, 69). Adequate insight, however, can be gained from the analysis of the simple one-dimensional model shown in Fig. 23-74, which represents a longitudinal cross section of a DLIM.

To a first approximation, one can assume that no variations occur in the directions perpendicular to the direction of motion ($\partial/\partial y = \partial/\partial z = 0$). Furthermore the slotted ferromagnetic structures are replaced by smooth structures of infinite permeability ($\mu = \infty$). The

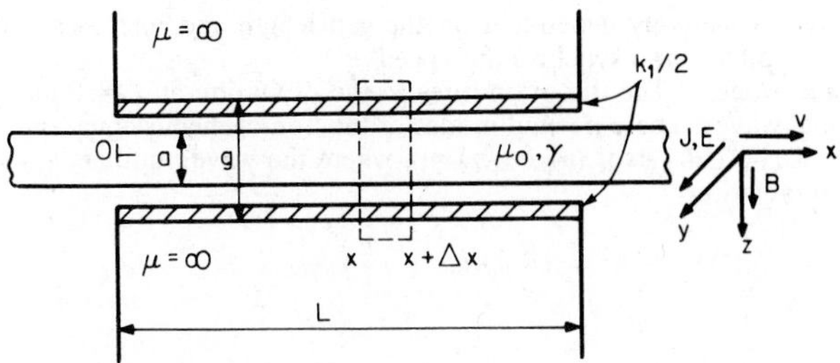


**FIG. 23-74** Model of a linear induction motor for one-dimensional analysis.

traveling wave impressed on the primary windings is idealized and described by two current sheets, each having a surface current density

$$\frac{\mathbf{K}_1}{2} = \frac{K_{1m}}{2}\cos\left(\omega t - \frac{\pi}{\tau}x\right)[\mathbf{M}(x) - \mathbf{M}(x - L)]\mathbf{y}_0 \tag{23-5}$$

where $\omega$ is the radian frequency of the supply, $t$ is the time, $\tau$ is the pole pitch, **M** is the unit step function, and boldface type denotes vectors.

***243. Basic Relations.*** A relation between **J**, the density of the current induced in the reaction rail, and the prevailing magnetic flux density **B** can be obtained by applying Ampere's law to an arbitrary loop of width $\Delta x$. Neglecting the displacement current, one gets

$$-\frac{\partial B}{\partial x} = \mu_0\frac{a}{g}J + \frac{\mu_0}{g}K_1 \tag{23-6}$$

$J$ can be related to the electric field intensity $E$ by using Ohm's law for a slow-moving medium:

$$J = \gamma(E - vB) \tag{23-7}$$

where $\gamma$ is the conductivity of the reaction rail. $E$, in turn, is related to $B$ by Faraday's law, which gives

$$\frac{\partial E}{\partial x} = -\frac{\partial B}{\partial t} \tag{23-8}$$

Introducing the last two equations into the first yields

$$\frac{\partial^2 B}{\partial x^2} - \mu_0\gamma v\frac{a}{g}\frac{\partial B}{\partial x} - \mu_0\gamma\frac{a}{g}\frac{\partial B}{\partial t} = -\frac{\mu_0}{g}\frac{\partial K_1}{\partial x} \tag{23-9}$$

***244. Synchronous Wave.*** A complete solution of this linear equation can be obtained by algebraization through Fourier transforms (Refs. 4, 55). The dominant waves, however, can be recognized immediately. By introducing a phasor $\dot{B}_s$, such that $B = \mathrm{Re}(\sqrt{2}\,\dot{B}_s \exp\{j[\omega t - (\pi/\tau)x]\})$, by making use of the synchronous tie relation $\omega = (\pi/\tau)v_s$, and by defining the slip as $S = (v_s - v)/v_s$, one obtains the synchronous wave for an endless machine:

$$B_s = \frac{K_1}{\sqrt{(\pi g/\mu_0\tau)^2 + (\gamma a S v_s)^2}} \tag{23-10}$$

and

$$\varphi_s = \tan^{-1}\frac{\pi g}{\mu_0\tau\gamma a S v_s} \tag{23-11}$$

where, neglecting the impedance of the primary, $\cos\varphi_s$ can be identified with the power factor of the machine. It appears that, since the ratio $g/a$ is practically constant, the power factor is not very sensitively dependent on the gap length and increases with increasing length of the pole pitch $\tau$ and synchronous speed $v_s$.

***245. Normal Modes.*** The discontinuities in the excitation at $x = 0$ and $x = L$ give rise to additional waves. These normal modes of the homogeneous equation can be found by letting $B = \text{Re}\,\{\sqrt{2}\,B_{1,2}\exp[j(\omega t - k_{1,2}x)]\}$, where the wave numbers $k_{1,2}$ are the roots of the dispersion relation:

$$k^2 - j\frac{a}{g}\mu_0\gamma v k + j\frac{a}{g}\mu_0\gamma\omega = 0 \tag{23-12}$$

One thus obtains

$$k_{1,2} = j\frac{a}{g}\frac{\mu_0\gamma v}{2}\left(1 \pm \sqrt{1 + j\frac{g}{a}\frac{4\omega}{\mu_0\gamma v^2}}\right) \tag{23-13}$$

***246. Thrust, Power, and Efficiency.*** The thrust is given, according to Biot-Savart's law, by

$$T = \int_0^L K_1 B\,dx \tag{23-14}$$

The electric power transferred through the gap into the secondary is

$$P_2 = \int_0^L K_1 E\,dx \tag{23-15}$$

where $E$ can be found from Eq. (23-8) as $\dot{E} = (\omega/k)\,\dot{B}$. The major contributions come from the waves which are not (or only slightly) attenuated [$k \simeq \text{Re}(k)$]. If one defines the phase velocity as $v_{ph} = \omega/\text{Re}(k)$, the secondary power becomes $P_2 = v_{ph}T$ and the secondary efficiency

$$\eta_2 = \frac{vT}{P_2} = \frac{v}{v_{ph}} \tag{23-16}$$

For the synchronous wave with $v_{ph} = v_s$, the efficiency reduces to $\eta_2 = 1 - S$, so that induction motors are designed to operate near synchronism.

***247. Low-Velocity Machines.*** Physical insight can be gained by considering the limiting cases of low and high velocities. In the low-velocity limit $\mu_0\gamma a v^2/(4\omega g) << 1$, the wave numbers are

$$k_1 \simeq (1 - j)\sqrt{\frac{a}{g}\frac{\mu_0\gamma\omega}{2}} \qquad k_2 \simeq -(1 - j)\sqrt{\frac{a}{g}\frac{\mu_0\gamma\omega}{2}} \tag{23-17}$$

It follows that $k_1$ is associated with waves diffusing in the direction of motion of the reaction rail with $v_{ph} = \sqrt{2\omega g/\mu_0\gamma a} = v_{dif}$. The excitation of these $B$ waves does not depend on the cruising speed $v$. Therefore, they are present also at the synchronous speed $v_s$. When $v_{dif} \approx v_s$, the $B_1$ wave excited by the leading edge of the current sheets at $x = 0$ interacts with the primary excitation $K_1$ to produce a thrust. It now becomes clear why, in the LIM, the tractive effort does not necessarily vanish for $S = 0$.

Furthermore one can now rewrite:

$$\frac{\mu_0\gamma a v^2}{4\omega g} = \frac{1}{2}\left(\frac{v}{v_{dif}}\right)^2 \tag{23-18}$$

and conclude, as predicted, that the natural scaling parameter for the cruising speed is the diffusion velocity.

***248. High-Velocity Machines.*** In the high-velocity limit $\tfrac{1}{2}(v/v_{dif})^2 >> 1$, the wave numbers become

$$k_1 \simeq \frac{\omega}{v} - j\frac{g}{a}\frac{\omega^2}{\mu_0\gamma v^3} \qquad k_2 \simeq -\frac{\omega}{v} + j\frac{a}{g}\frac{\mu_0\gamma v}{2} \tag{23-19}$$

In this case $k_l$ is associated with a practically unattenuated wave which is "frozen" in the reaction rail, since its phase velocity is $v$. Under rated conditions of operations $v$ is practically $v_s$. The $B_l$ wave, excited by the leading edge of the current sheets, then travels together with the synchronous $B_s$ wave. In fact $B_l$ practically cancels out $B_s$. This can be seen by realizing that, at high velocity, very little field diffuses out ahead of the leading edge. It follows that the boundary condition at $x = 0$ is $B \approx B_l + B_s \approx 0$. The small attenuation of the $B_l$ wave prevents a rapid buildup of the resultant $B$ under the primary and thus explains the poor performance of the LIM, when compared with a rotating machine.

A $B_l$ wave is also excited at the trailing edge, but its effect is minor. So is the effect of the $B_2$ waves, since they are strongly attenuated.

***249. Compensation of End Effect.*** Significant improvements in the performance of the LIM have been demonstrated by Laithwaite and Kuznetsov using a novel arrangement of primary winding (Ref. 79). Their approach is based on the realization that the $k_l$ natural mode which is excited at the leading edge and which is "frozen" in the reaction rail can be viewed as a dc field excitation. If this mode is made to interact with a special section of the primary winding having appropriate spatial displacement, pole pitch, and number of turns, this section may operate as an overexcited synchronous machine and provide power factor compensation. Moreover, the braking force resulting from trailing-edge effects is reduced. An efficiency and a power factor as high as 0.75 and 0.8, respectively, have been achieved.

***250. Skin and Edge Effects.*** The one-dimensional model, on which the preceding analysis was based, does not account for other effects which degrade the performance of the LIM. One is the skin effect, which becomes significant when the inequality $g/\tau << 1$ is not satisfied (Refs. 4, 7). Another is edge effects due to the limited width of the motor (Ref. 44). These are particularly significant in linear propulsion, because of the desire to keep the width of the track as small as possible.

***251. Wound-Secondary Machine.*** The end effects which occur when the reaction rail consists of a continuous sheet can be eliminated, at least theoretically, by using a polyphase wound secondary. With this scheme the series connection of the conductors forces the phase currents to be uniformly distributed along the length of the motor, thus preventing the concentration of out-of-phase current under the leading edge of the primary, and its deleterious effects. Unfortunately it would be very inefficient to extend the secondary circuits much beyond the length of the primary. A wound track, then, cannot be endless, but it must consist of a sequence of independent sections. This causes an undesirable transient to occur, whenever the leading edge of a secondary section is exposed to the primary (Ref. 4).

Because of this new end effect and because of the cost of the insulated windings, there seems to be little advantage in having the wound secondary serve as the passive rail. The future of a wound secondary seems to lie, instead, with energization from the wayside. A wound secondary on the vehicle side would provide on-board terminals from which acceleration and deceleration rates could be controlled. Moreover it would allow recovery of a portion of the secondary power (Ref. 4).

***252. Utilization of Slip Power.*** As was mentioned in Par. **246,** the fraction of the secondary electric power $P_2$ which the induction motor converts into mechanical form is $P_{conv}/P_2 = v/v = 1 - S$. If the secondary is short-circuited, the slip power $SP_2$ is dissipated into heat. Operation at reduced speed is then not merely uneconomical, but also imposes unacceptable thermal requirements on the insulation. The energy dissipated during acceleration from standstill exceeds the amount stored in kinetic form. During plugging, as may be required for emergency braking, it is three times as much (Ref. 7), hence the need for varying $f$ and, therefore, $v_s$, so that $S$ remains within tolerable bounds. With a wound secondary, instead, part of the slip power can be removed from the winding and utilized to energize auxiliary services. Thus power can be supplied at constant frequency and contactless ground transportation is brought within the realm of economic feasibility.

## Linear Synchronous Motor (LSM)

***253. Principle of Operation.*** Synchronous motors are so called because they develop a thrust when operating at the speed of the *synchronous* wave [$v = v_s = (\tau/\pi)\omega$]. The nonenergized structure is thus phase-locked into the traveling wave ($S = 0$) and the required system of interacting currents must be excited by a separate dc supply. The power involved, however, is small, since it covers only the ohmic loss; it can be drastically reduced by cryogenic cooling, or totally eliminated by using superconducting, or permanent magnets. Even with no excitation, a small thrust can be obtained, simply by relying on the variable-reluctance effect of a nonenergized structure having salient poles.

Operation at variable speed is achieved by varying the frequency of the supply through a power conditioner, such as a thyristor inverter.

***254. Characteristic Features.*** Since the induction of currents in the passive structure is not essential for the development of the thrust, the end effects which limit the efficiency of the LIM can be eliminated by preventing the flow of eddy currents. Moreover, in contrast with the LIM, the LSM can draw its magnetizing current requirements from the dc power supply. This allows operation at unity and even leading power factor, and results in smaller size for the motor and the power conditioner, even with larger clearance between the vehicle and the track. A larger clearance is advantageous in that it increases the ride quality and reduces the need for costly maintenance of the roadbed.

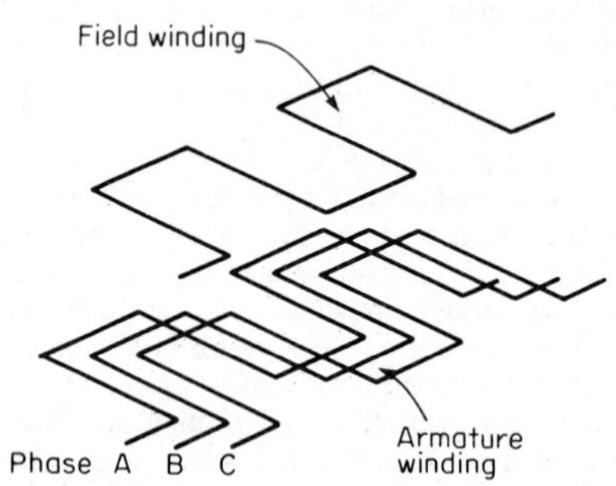


**FIG. 23-75** Schematic arrangement of winding in linear synchronous motor (LSM).

***255. Air-Cored Motors.*** In its simplest form an LSM consists of two arrays of conductors (see Fig. 23-75). The array carrying the polyphase systems of currents which set up the traveling wave is called the *armature* winding; the other carrying the dc excitation current is called the *field* winding.

Such a scheme is practical only when the field is superconducting and is located on the vehicle. It was first proposed, in conjunction with repulsive levitation, by Powell and Danby (Refs. 37, 50).

***256. Power Output.*** The mathematical model for an air-cored LSM is the same as for a round-rotor rotating machine (see Secs. **7** and **20**). In steady-state, the describing equation is

$$\dot{V}_a = jX_m(\dot{I}_a + \dot{I}_f) + (R_a + jX_a)\,\dot{I}_a \tag{23-20}$$

where $V$ stands for phase voltage, $\dot{I}$ for current, $R$ for resistance, $X$ for reactance, $a$ for armature, $f$ for field, and $m$ for magnetizing.

The electrical and mechanical powers are given respectively by

$$P_e = 3\text{Re}(\dot{V}_a\dot{I}_a{}^*) = 3V_a I_a \cos\varphi \tag{23-21}$$

and

$$P_m = 3\text{Re}[(\dot{V}_a - R_a\dot{I}_a)\dot{I}_a{}^*] = 3X_m I_a I_f \sin\beta \tag{23-22}$$

where $\varphi$ is the angle between $V_a$ and $\dot{I}_a$ and $\beta$ between $\dot{I}_f$ and $I_a$. Operation with constant armature and field currents presents special interest in view of the advantages of current-source inverter power supplies. Under these conditions both $P_m$ and the efficiency

$$\eta = \frac{1}{1 + R_a I_a^2/P_m} \tag{23-23}$$

peak for $\beta = \pi/2$.

***257. Linear Commutator Motor.*** At maximum power output, the power factor is likely to be low, particularly when $X_a$ is large, as is the case with energization by the wayside with sections extending over several kilometers. A simple relation between $\varphi$ and $\beta$ can be obtained by neglecting $R_a$, so that $P_m = P_e$, or

$$\left|\dot{I}_f + \frac{X_m + X_a}{X_m}\dot{I}_a\right| \cos\varphi = I_f \sin\beta \tag{23-24}$$

The phasor diagram of Fig. 23-76 shows that the power factor reaches a maximum when the phasor $\dot{I}_f + [(X_m + X_a)/X_m]\,\dot{I}_a$ is tangent to the circular locus of $\dot{I}_f$. This occurs for $\beta >$

$\pi/2$, an inherently unstable region. Operation with $\beta > \pi/2$ can be achieved by letting the firing of the inverter thyristors be controlled by the position of the field magnets. This scheme duplicates the action of the brushes on the commutator of a dc machine and endows linear propulsion with the excellent thrust speed and starting-performance characteristics of a dc motor drive. In this case the motor is more appropriately called *linear commutator motor* (LCM).

As can be seen from Fig. 23-76, when $\beta > \pi/2$, the field and armature excitation partially oppose one another, giving rise to a repulsive force which can be utilized for levitation purposes.

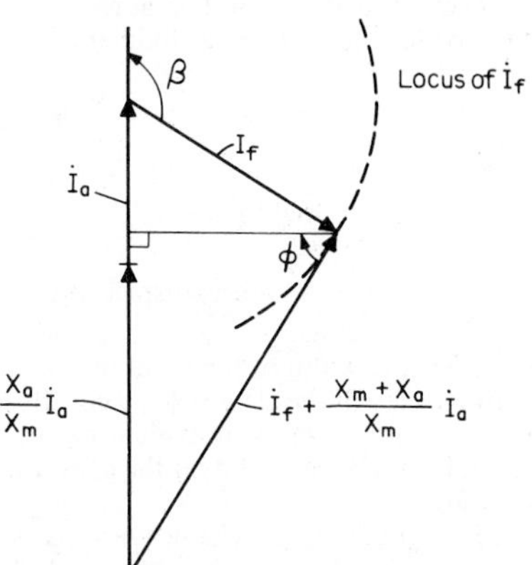


**FIG. 23-76** Phasor diagram of LSM operating with constant currents.

***258. Iron-Cored Motors.*** The thrust obtained from a given system of currents flowing in normal conductors can be enhanced by at least one order of magnitude by embedding both field and armature windings in iron structures. Moreover, significant savings in the cost of the track can be achieved by placing the field, as well as the armature winding, on board the vehicle, and by employing a nonenergized structure having salient poles. Of particular interest are those machine types in which the track does not carry an alternating flux and, hence, need not be thinly laminated: the *claw-pole* or *Nadyne* (Fig. 23-70) and the *homopolar inductor* (Fig. 23-71) (Refs. 15, 24, 26, 78, 81).

***259. Claw-Pole Type.*** The claw-pole type (Fig. 23-70) is so called because of the interdigital arrangement of the salient poles in the track. These poles are interlinked with the field excitation winding, which is carried by the vehicle, by a special ferromagnetic structure which closes the magnetic circuit behind the armature. Hence the portion of the magnetic circuit which carries a time-invariant flux is *transverse* to the direction of motion. In contrast, the portion of the magnetic circuit which interlinks with the armature winding, also carried by the vehicle, is *longitudinal,* as it would be in a conventional synchronous machine (see also Par. **238**).

One consequence of this unconventional arrangement is an increase of the field excitation requirements, as a result of the doubling of the air-gap length in the magnetic-flux path. The other is a considerable lengthening of the iron portion of the magnetic circuit with consequent increase in the weight of the motor.

***260. Homopolar Inductor Type.*** The homopolar inductor type (Fig. 23-71) is the linear version of a machine used at the turn of the century to generate high frequencies. Here, as in the previous machine, the field winding is interlinked with a *transverse* flux. The armature winding, instead, is divided into two sections, each interlinked with a *longitudinal* flux. In addition, the two armature structures carry the transverse flux, so that the magnetic flux density $B$ oscillates around an average value, instead of alternating between a positive and a negative peak, as in a conventional synchronous machine. The reduction in flux swing results in a proportionate increase in the size and weight of the motor.

***261. Theory of Operation.*** The iron-cored LSM does not differ in its basic mode of operation from a rotating salient-pole synchronous motor. Its steady-state performance, then, can be predicted according to the Blondel two-reaction theory (see Secs. **7** and **20**). The transient performance is best described by Park's equations (Ref. 32). It should be noted, however, that the typical values of the parameters reflect the complexity of the magnetic circuit. In general, the leakage fluxes are larger. In the claw-pole-type motor, the reactances include the effect of the additional gaps under the feet of the field yoke. In the homopolar inductor motor, $\tau$ corresponds to half the distance between two adjacent salient poles.

***262. Force Equations.*** Iron-cored motors develop strong forces of attraction, in addition to the thrust. These forces may exceed the thrust by one order of magnitude, so that they may provide the vehicle with full suspension and guidance. They also affect the motor dynamics in its six degrees of freedom (three translational and three rotational). The main force components can be determined by taking the partial derivative of the magnetic energy with respect to the appropriate coordinate. If additional subscripts $d$ and $q$ are introduced to denote the direct and quadrature axes, respectively, the following expressions are obtained for the homopolar inductor motor:

$$\text{Thrust} = T = \frac{3}{v}\left(X_{af}I_f I_a \sin\beta + \frac{X_{ad} - X_{aq}}{2} I_a^2 \sin 2\beta\right) \quad (23\text{-}25)$$

$$\text{Normal force} = F_n = \frac{3}{2\omega g}\left(X_{ff}I_f^2 + X_{ad}I_{ad}^2 + X_{aq}I_{aq}^2 + 2X_{af}I_f I_{ad}\right) \quad (23\text{-}26)$$

$$\text{Lateral-displacement restraining force} = F_l = \frac{g}{b} F_n \quad (23\text{-}27)$$

where $b$ is the width of the armature.

In the case of the claw-pole motor the energy stored under the feet of the yoke must also be taken into account in evaluating the reactances. Moreover, Eqs. (23-26) and (23-27) should be modified slightly if the gaps under the armature and the yoke feet differ in length and width.

***263. Rail Design.*** The heaviest parts of the motor are the yokes which carry the transverse flux. Their weight decreases with decreasing permeance in the interpolar space. For this purpose it is advantageous to bevel the pole-face tips and to limit the extent of the pole face $l_p$.

If the ferromagnetic rail of an iron-cored LSM were constructed of massive material, eddy currents would be induced, as in the case of the LIM, under the leading edge of the motor and would effectively prevent the penetration of the magnetic field. To overcome this end effect the rail structure must be laminated (Ref. 48). Technical and economical considerations militate against the thin laminations and special alloys used in the armature. Higher strength and lower cost of the track must be achieved at the expense of lower efficiency and larger size of the motor. It appears that structural-steel laminations up to a quarter of an inch in thickness could be employed without paying an intolerable penalty.

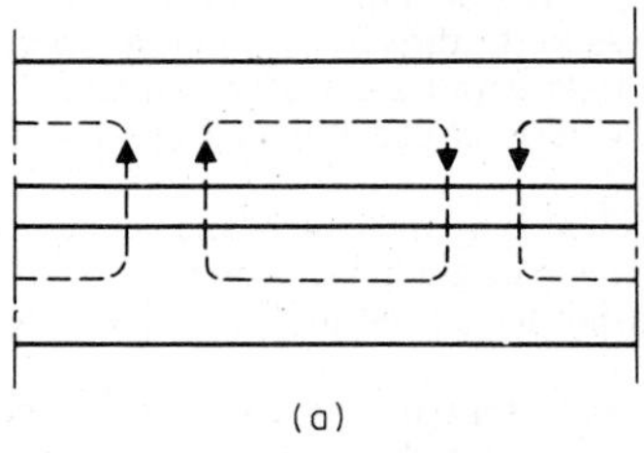


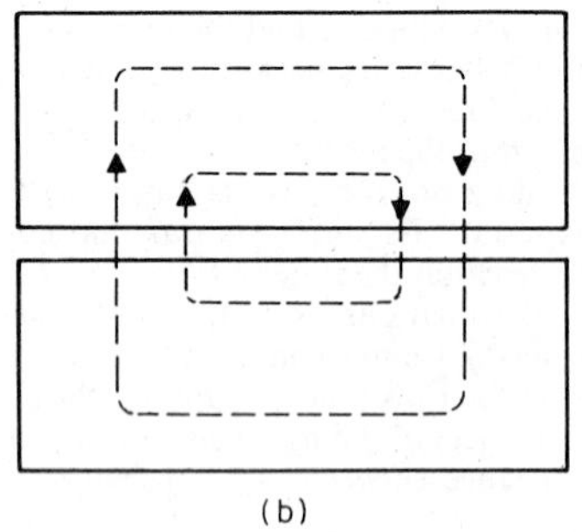


**FIG. 23-77** Comparison of flux patterns in the cores of (*a*) rotating (endless) motor, (*b*) linear motor.

***264. Parasitic Forces and Losses.*** Even with thinly laminated rails, end effects may give rise to eddy currents. They are caused by fluctuations in the permeance of the transverse flux path. Although the ohmic losses are generally small, the force fluctuations may be significant and in the case of the homopolar inductor motor may cause dangerous roll motions.

***265. Primary Winding.*** These end effects can be alleviated by making $L$, the length of the armature, equal to an even number of pole pitches $\tau$. The ideal choice would be $L = 2p\tau$. In this case, however, the armature core should be thicker than the core of a corresponding rotating motor, in order to carry the flux that is prevented from closing at the ends (Fig. 23-77 and Ref. 4). Additional length is required to accommodate either the flux or special extended windings (Refs. 14, 15, 47).

### Experimental Status

***266. Low-Speed Motors.*** These motors are being developed mostly for *personal* *rapid transit* (PRT) systems to propel vehicles in inner cities, or airports, at speeds ranging up to 80 km/h.

In the United States, TTI's system features air cushions for suspension and 1600 N thrust SLIMs for propulsion and regenerative braking. The operating clearance varies between 1 and 2 cm (Ref. 27). Rohr Industries' "Romag" (Fig. 23-73 and Ref. 59) features four SLIMs placed under the corners of the vehicle. They provide attractive maglev and guidance, as well as propulsion. Two types of ferromagnetic rails have been tried: one is notched to yield a variable-reluctance thrust; the other contains short-circuited conductors and is similar in operation to a squirrel-cage motor. A scheme developed by PRT Systems Inc. uses a SLIM interacting directly with a structural-steel I beam to supply both propulsion and partial levitation and guidance (Ref. 29). The Romag system has been adopted and further developed by Boeing.

In Germany a homopolar inductor propulsion system was developed by A. G. Siemens (Refs. 27, 65). MBB-Demag's system, the "cabinentaxi," carried three passengers at speeds up to 36 km/h. A maximum acceleration of 2.5 m/s$^2$ was provided by a SLIM which develops 23 N of thrust per kilogram of vehicle weight. Suspension and guidance were by means of solid-rubber-tired wheels. In contrast, KM's scheme, the "Transurban," employed servocontrolled electromagnets for support and guidance.

These early schemes have been superseded by Germany's M-Bahn rapid transit and downtown people-mover system (Ref. 74). Propulsion is provided by a long-stator iron-core synchronous linear motor whose armature winding is installed inside the guideway and is supplied at adjustable frequency. Permanent magnets carried by the vehicle fulfill the double function of field excitation and attractive levitation. Small wheels serve to guide the vehicle laterally and adjust the air gap to the length required by the weight of the vehicle and payload. A 1-km demonstration link using this scheme was opened for public service at the International Exhibition (IVA) in Hamburg, Germany, in June 1979. At the time of this writing, the system is ready for deployment in Berlin.

British Rail has developed an attractive maglev system propelled by a SLIM. It was due to be put into operation on a 620-m elevated guideway link to the Birmingham airport (Ref. 82).

In France a joint effort by Prof. M. Barthalon, Compagnie d'Energétique Linéaire, and Merlin Gerin produced a mass-transit system, "L'Urba." Suspension was by means of a dynamic vacuum cushion (Dynavac). Propulsion was provided by Gramme-ring-wound DLIMs, which developed thrusts of 1150 N at stalling and 300 N at rated speed of 84 km/h (Refs. 19, 22, 27).

Low-speed motors have also found practical application in drives for material-handling conveyors (Refs. 15, 18, 22, 28).

***267. High-Speed Motors.*** These motors are being developed to propel contact-free, surface-guided vehicles at speeds in excess of 250 km/h.

In the United States, the Garrett Corporation developed two test vehicles for the U.S. Department of Transportation (Refs. 12, 15). Both used DLIMs for propulsion. The "linear induction motor research vehicle" (LIMRV) was supported and guided by conventional steel wheels on steel rails. It reached a maximum speed of 410 km/h (Ref. 29). The 1860-kW motor was driven by an on-board generator and had the performance characteristics shown in Fig. 23-78. The same vehicle was later tested with a SLIM. A "track levitated vehicle" (TLV) was built but never tested on a high-speed track. It was powered by a 7460-kW motor fed through a thyristor inverter. A synchronous condenser was used for power factor compensation and commutation purposes. The motor design characteristics are summarized in Table 23-12 and Fig. 23-79. The U.S. Department of Transportation also sponsored the development of SLIMs of the homopolar inductor type (Ref. 17).

A repulsive suspension system using superconducting magnets and air-cored LSMs (Magneplane) has been jointly developed by MIT, Raytheon, and United Engineers. Typical design parameters are $T = 40$ kN, $v = 99$ m/s (356 km/h), with 0.25 m clearance between vehicle coils and guideway. The guideway armature would be excited in sections of 4 km length delivering 4 MW output at 36 Hz and 85% efficiency. Each vehicle would carry 140 passengers (Refs. 15, 25, 29).

Such a system is also been considered as a mass driver for outerspace applications (Ref. 80).

Significant effort toward the development of similar air-cored LSM–repulsive maglev

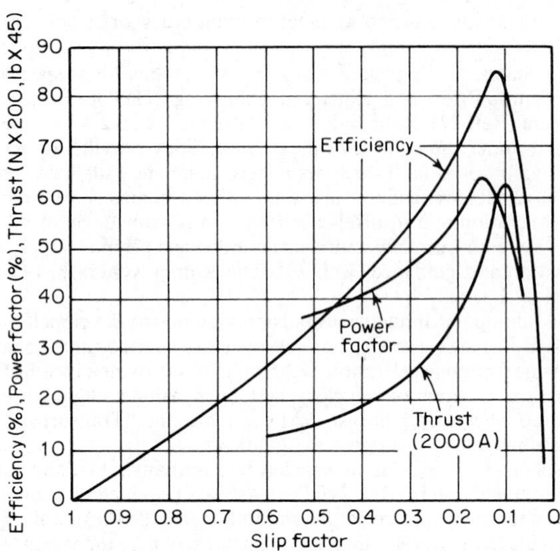


**FIG. 23-78** Performance characteristics of the Research Vehicle LIM at 157.1 Hz. *(U.S. Department of Transportation.)*

**TABLE 23-12** Design Characteristics of TLV LIM (Ref. 12)

| Specifications | | Data | |
|---|---|---|---|
| Weight | $1.33 \times 10^3$ kg | Pole pitch | 17.75 in |
| Thrust | $2.22 \times 10^4$ N | Stack height | 7.5 in |
| Voltage | 4120 $V_{1-n}$ 3-phase | Lamination length | 101.4 in |
| Frequency | 0–165 Hz | Lamination width | 3.625 in |
| Primary current | 530 A | Slots per pole | 15 |
| Power factor | 0.57 | Slot pitch | 1.183 in |
| Efficiency | 0.8 | Turns per coil | 4 |
| Reaction rail | 6061-T6 Al alloy | Coils | 75 per side |
| | ¼ in effective thickness | Slots | 85 |
| | 21.4 in high | Coil pitch | ⅔ |
| Air gap (electrical) | 1.5 in primary to primary | Conductor size | 0.258 in square with 0.143-in-dia. hole |
| Rail clearance | 0.3125 in per primary | Winding connection | Two halves in series wye-connected |
| Water cooling load | 249 kW | | $R_1 = 0.288$ Ω/phase at 150°C |
| Water flow | 45 gal/min | Equivalent circuit parameters at 165 Hz | $X_1 = 2.24$ Ω/phase |
| Water-pressure drop | 65 lb/in² | | $X_m = 18.9$ Ω/phase<br>$R_2 = 0.602$ Ω/phase at 20°C<br>$X_2 = 0.828$ Ω/phase |

NOTE: 1 in = 25.4 mm, 1 gal/min = 0.00063 m³s; 1 lb/in² = 6.895 kPa.

systems has been invested in Canada, Germany, and Japan (Refs. 15, 16, 29, 66, 70, 77). The Japanese National Railroad built a 7-km track in the Miyasaki prefecture. In 1979 the ML 500 R, 10-ton test vehicle reached a speed of 517 km/h (Ref. 77). In 1982 the track was converted into a U-shaped guideway and manned tests on an MLU001, three-car formation were completed.

In addition, Germany has demonstrated the feasibility of attractive maglev. The "sec-

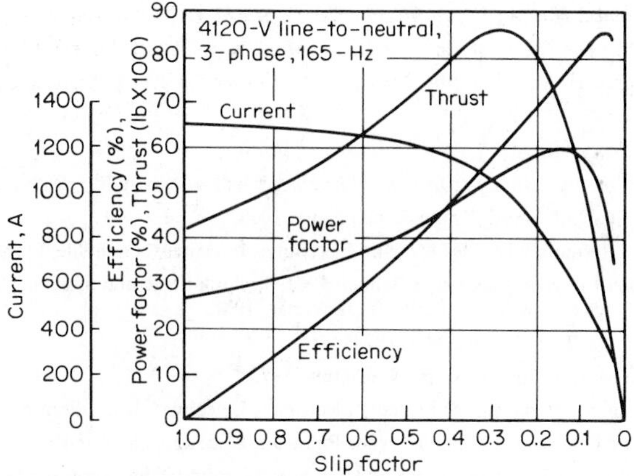


**FIG. 23-79** Projected performance characteristics of TLV LIM. *(From Ref. 12.)*

ond-generation" high-speed vehicles which included MBB's Komet and KM's "Transrapid 04" (Refs. 27, 65) have been followed by the TVE Transrapid 06 system (Ref. 75). It is based on the same long-stator concept described for the M-Bahn, except that control over the suspension and guidance is provided by electromagnets, instead of small wheels.

In France the "Aerotrain" developed by Société Bertin was powered by a 2-MVA DLIM and used air-cushion suspension (Refs. 22, 27).

Also using air cushions was the "Hovertrain" (RTV31) developed in the United Kingdom by Tracked Hovercraft, Ltd. It was propelled by a SLIM (Ref. 27).

To conclude this brief survey on the experimental status of linear propulsion, mention should be made of the "magnetic river" conceived by Prof. E. R. Laithwaite (Refs. 14, 15). It consists of a transverse flux, SLIM, so designed to provide the vehicle with full repulsive suspension and guidance. A small-scale model has demonstrated remarkably stable operation.

***268. Outlook.*** In low-speed applications, linear motors must compete with their rotating counterparts. Weighed against their inability to use gears for thrust multiplication and their need for larger air gaps are the following advantages: faster acceleration and deceleration and, hence, greater safety in case of emergency braking, improved ride dynamics and comfort, noise abatement, as a result of the absence of gears and flat wheels and the reduction in rail wear and tear, lower maintenance costs for propulsion and braking systems and tracks. Moreover only half of the motor need be carried by the vehicle, an advantage which becomes more pronounced with increasing density of traffic.

In the intermediate speed range iron-cored LSMs appear to have the edge over the LIMs, because they are less affected by end effects and can operate with larger clearances (3 cm), leading power factor, and better efficiency. Moreover their strong attractive forces can be controlled independently of the thrust.

At speeds of 350 km/h and over, the air-cored LSM seems to have no competitors, since it requires no power collectors and can operate with clearances up to 30 cm. Its practical realization, however, hinges on major technological advances in superconducting magnets and power electronics.

### *269. References on Linear Propulsion*

*Books*

1. Laithwaite, E. R.: *Induction Machines for Special Purposes;* George Newnes, London, 1966.
2. Laithwaite, E. R.: *Propulsion Without Wheels;* English University Press, New York, Hart Publishing Co., 1968.
3. Laithwaite, E. R.: *Linear Electric Motors,* Mills and Boon, London, 1971.
4. Yamamura, S.: *Theory of Linear Induction Motors;* Wiley, New York, 1972.

5. Nasar, S. A., and Boldea, I.: *Linear Motion Electric Machines;* Wiley, New York, 1976.
6. Poloujadoff, M.: *The Theory of Linear Induction Machinery;* Oxford University Press, Oxford, 1980.
7. Levi, E., and Panzer, M.: *Electromechanical Power Conversion;* Krieger Publishing Co., Melbourne, FL, 1982.

*Conference Proceedings*

8. *Proceedings of the Institution of Mechanical Engineers,* vol. 181, part 3G, 1966–1967.
9. Joint IRCA-UIC High-Speed Symposium, Vienna, 1968.
10. "Symposium International sur les Moteurs Electriques Linéaires," Grenoble, France, April 1970.
11. "Internationaler Congress Elektrische Bahnen 1971, Technik heute und morgen;" Munich, Oct. 11–15, 1971. Vortrags Sammelband, Berlin, VDE-Verlag, 1971.
12. *Proceedings of the IEEE,* vol. 61, May 1973.
13. "Simposio sui Motori Lineari;" Capri, Italy, June 1973.
14. "Symposium International sur les Moteurs Electriques Linéaires"; Lyon, France, May 1974.
15. "Linear Electric Machines," IEE Conference Publication no. 120, October 1974.
16. "Proceedings-International Conference on High Speed Ground Transportation;" January 1975.
17. "Statusseminar VII Magnetbahnentwicklung," Willingen, 1978.

*Review Articles*

18. Laithwaite, E. R.: Linear Induction Motors; *Proc. IEE,* vol. 104 A, pp. 461–470, 1957.
19. Remy, E.: Le moteur linéaire; and Remy, E., and Victorri, M.: Applications du moteur linéaire; *Rev. Gen. Electr.,* Vol. 78, no. 4, pp. 357–368, April 1969.
20. Laithwaite, E. R., and Barwell, F. T.: Application of Linear Induction Motor to High-Speed Transport Systems; *Proc. IEEE,* vol. 116, no. 3, pp. 713–724, 1969.
21. Laithwaite, E. R., and Nasar, S. A.: Linear-Motion Electrical Machines; *Proc. IEEE,* vol. 58, pp. 531–542, April 1970.
22. Poloujadoff, M.: Linear Induction Machines; *IEEE Spectrum,* vol. 8, pp. 72–80; February 1971, pp. 79–86, March 1971.
23. Wagner, A.: Bibliography on the Linear Induction Motor in High-Speed Ground Transportation, paper C72 448-9, IEEE Summer Power Meeting, San Francisco, 1972.
24. Levi, E.: Linear Synchronous Motors for High-Speed Ground Transportation; *IEEE Trans.,* Mag-9, no. 3, pp. 242–248, September 1973.
25. Kolm, H. H., and Thornton, R. D.: Electromagnetic Flight; *Sci. Am.,* vol. 229, no. 4, pp. 17–25, October 1973.
26. Guarino, M., Jr.: "Integrated Linear Electric Propulsion Systems for High Speed Transportation;" Symposium International sur les Moteurs Electriques Linéaires, Lyon, France, May 1974.
27. Ellison, A. J., and Bahmanyar, H.: Surface-Guided Transport Systems of the Future; *Proc. IEEE,* vol. 121, no. 11R, pp. 1224–1248, November 1974.
28. Laithwaite, E. R.: Linear Electric Machines—A Personal View; *Proc. IEEE,* vol. 63, no. 2, pp. 250–290, February 1975.
29. Thornton, R. D.: Magnetic Levitation and Propulsion, 1975; *IEEE Trans.,* vol. Mag-11, no. 4, pp. 981–995, July 1975.
30. Teodorescu, D.: Linear Motors, Present State and Development of Induction and Synchronous Linear Motors; *Elektr. Masch.,* vol. 59, pp. 94–101, 1980.
31. Petlenko, B. I.: Linear Electric Drives and Developmental Trends; *Elektrischestvo,* no. 9, pp. 43–47, 1981.

*Journal and Conference Papers*

32. Park, R. H.: Two-Reaction Theory of Synchronous Machines, Generalized Method of Analysis; part I, *AIEE Trans.,* vol. 48, pp. 716–730, July 1929.
33. A Wound Rotor 1400 ft. long; *Westinghouse Eng.,* vol. 6, p. 160, 1946.
34. Shturman, G. I., and Aronov, R. L.: End Effects in Induction Motors with Open Magnetic Circuit, *Elektrichestvo,* vol. 2, p. 54, 1947.
35. Kemper, H.: Elektrisch Angetriebene Eisenbahnfahrzeuge mit Elektromagnetischer Schwebefuehrung; *ETZ,* pp. 11–14, 1953.

36. Laithwaite, E. R., and Mamak, S.: Oscillating Synchronous Linear Machines, *Proc. IEE,* vol. 109A, pp. 415–429, 1962.

37. Powell, J. R., and Danby, G. R.: "High-Speed Transport by Magnetically Suspended Trains;" ASME Winter Annual Meeting, New York, Railroad Div. paper 66-WA/RR-5, 1966.

38. Brunelli, B.: Studio del motore ad induzione con statore ad arco; *L'energia Elettrica,* vol. 43, no. 10, pp. 1–13, 1966.

39. Poloujadoff, M.: Perfectionnemant a la théorie des moteurs d'induction linéaires destinés a la traction; *C. R. Acad. Sci. B (France),* vol. 263, pp. 605–607, 1966.

40. Laithwaite, E. R.: Some Aspects of Electrical Machines with Open Magnetic Circuits; *Proc. IEE,* vol. 115, no. 9, pp. 1275–1283, 1968.

41. Yamamura, S., Ito, H., and Ahmed, F.: End Effect of Linear Induction Motor; *J. Inst. Elec. Eng. (Tokyo),* vol. 89, pp. 459–467, 1968.

42. Poloujadoff, M., and Sabonnadiere, J. C.: Utilization d'une méthode de partition dans le resolution des certaines equations aux derivée partialles donc le domaine comparte une bande infinie; *C. R. Acad. Sci. B (France),* vol. 262, pp. 1412–1415, 1968.

43. Kalman, G. P., et al.: "Electrical Propulsion System for Linear Induction Motor Test Vehicle;" presented at the 4th Intersoc. Energy Conversion Engineering Conf., Washington, D.C., 1969.

44. Bolton, H.: Transverse Edge Effect in Sheet-Rotor Induction Motors; *Proc. IEE,* vol. 116, pp. 725–731, 1969.

45. Weh, H.: *Linearmotoren;* VDE Fachberichte, pp. 37–43, 1970.

46. Ooi, B. T., and White, D. C.: Traction and Normal Forces in the Linear Induction Motor; *IEEE Trans.,* PAS-90, pp. 638–645, 1970.

47. Kant, M., Mouillet, A., and Scheuer, J.- M.: Etude théoretique et expérimental des enroulements des moteurs linéaires à induction; *Rev. Gen. Elect.,* vol. 80, no. 1, pp. 13–19, 1971.

48. Appun, P., and Weh, H.: Wirbelstroeme im feststehenden Teil von Zugmagneten zur magnetischen Aufhaengung von Fahrzeugen; *ETZ-A,* vol. 92, no. 11, pp. 623–627, 1971.

49. Laithwaite, E. R., Eastham, J. F., Bolton, H., and Fellows, T. G.: Linear Motors with Transverse Flux; *Proc. IEE,* vol. 118, pp. 1761–1767, Dec. 1971.

50. Powell, J. R., and Danby, G. T.: "The Linear Synchronous Motor and High-Speed Ground Transport," presented at the 6th Intersoc. Energy Conversion Engineering Conf., Boston, 1971.

51. Wang, T. C.: Linear Induction Motor for High-Speed Ground Transportation; *IEEE Trans.,* vol. IGA-7, no. 5, pp. 632–642, 1971.

52. Weh, H., Braess, H., and Mosebach, H.: Die rechnerische Behandlung asynchroner linear Wandler; *Energy Conversion,* vol. 11, pp. 25–37, 1971.

53. Leitgeb, W.: Linearmotoren fuer Fahrzeugantriebe, *Siemens Z.,* vol. 45, pp. 177–180, 1971.

54. Rummich, E.: Machines Linéaires Synchrones, théorie et réalisation pratiques; *Bull. ASE,* vol. 63, no. 23, pp. 1338–1344, November 1972.

55. Yamamura, S., Ito, H., and Ishikawa, Y.: "Theories of the Linear Induction Motor and Compensated Linear Induction Motor;" IEEE Winter Power Meeting, New York, N.Y., Jan. 30–Feb. 4, 1972.

56. Iwamoto, M., Ohno, E., and Shinryo, Y.: "End Effect of High-Speed Linear Induction Motor;" Proc. 1972 7th Annual Meeting of IEEE Industry Application Society, Philadelphia, pp. 323–330, 1972.

57. Richards, P. L., and Tinkham, M.: Magnetic Suspension and Propulsion Systems for High-Speed Transportation; *J. Appl. Phys.,* vol. 43, no. 6, pp. 2680–2691, 1972.

58. Elliott, D. G.: "Numerical Analysis Method for Linear Induction Machines;" 12th Symposium on the Engineering Aspects of Magnetohydrodynamics, 1972 (publisher, Prof. J. A. Fox, University of Mississippi).

59. Ross, J. A.: Transportation System with Integrated Magnetic Suspension and Propulsion, *Proc. NEREM Transportation Conference,* pp. 266–269, 1972.

60. Oberretl, L.: Driedimentionale Berechnung des Linearmotors mit Beruecksichtigung der Endeffekte und der Wicklungsverteilung, *Arch. Elektrotech.* (Berlin), vol. 55, no. 4, pp. 181–190, 1973.

61. Kliman, G. B., and Elliott, D. G.: "Linear Induction Motor Experiments in Comparison with Mesh/Matrix Analysis," Paper C73 129-4 presented at the IEEE PES Winter Meeting, New York, Jan. 28–Feb. 2, 1973.

62. Davis, L. C., and Borcherts, R. H.: Superconducting Paddle Wheels, Screws, and Other Propulsion Units for High-Speed Ground Transportation; *J. Appl. Phys.,* vol. 44, no. 7, pp. 3294–3299, July 1973.

63. Eastham, J. F., and Laithwaite, E. R.: Linear Motor Topology; *Proc. IEE,* vol. 120, no. 3, pp. 337–343, 1973.

64. Dukowitz, J. K.: "Analysis of Linear Induction Machines with Discrete Windings and Finite Iron Length;" IEEE Industry Application Society Meeting, Milwaukee, Oct. 8–11, 1973.

65. Gutberlet, H. G.: The German Magnetic Transportation Program; *IEEE Trans.,* vol. Mag-10, no. 3, pp. 417–420, September 1974.

66. Slemon, G. R.: The Canadian Maglev Project on High-Speed Interurban Transportation; *IEEE Trans.,* vol. Mag-11, no. 5, pp. 1478–1483, September 1975.

67. Carpetis, C., and Peschka, W.: "Numerical Two- and Three-Dimensional Calculation of Linear Induction Machines and Comparison to Analytical and Experimental Results;" Sixth International Conference Magnetohydrodynamic Electrical Power Generation, Washington, D.C., June 9–13, 1975, pp. 231–246.

68. Weh, H.: Die Integration der Funktionen magnetisches Schweben and elektrischer Vortrieb; *ETZ-A,* vol. 96, no. 3, pp. 131–135, 1975.

69. Sabonnadiere, J. C., and Nicolas, A.: "A Three-Dimensional Analysis Method for Linear Induction Machines;" Paper A76-214-7, IEEE PES Winter Meeting, New York, Jan. 25–31, 1976.

70. Atherton, D. L., Belanger, P. R., Burke, P. E., Dawson, G. E., Eastham, A. R., Haynes, W. F., Ooi, B. T., Silvester, P., and Slemon, G. R.: The Canadian High-Speed Magnetically Levitated Vehicle System; *Can. Electr. Eng. J.,* vol. 3, pp. 3–26, 1978.

71. Haller, T. R., and Mischler, W. R.: A Comparison of Linear Induction and Linear Synchronous Motors for High-Speed Ground Transportation. *IEEE Trans.,* vol. Mag-14, pp. 924–926, 1978.

72. Klocker, P., and Porsch, C. P.: Rosy Rotating Test Rig for Air Cored Linear Synchronous Motors; *Electr. Eng. (Australia),* vol. 56, pp. 10–12, 1979.

73. Slemon, G. R.: An Experimental Study of a Homopolar Linear Synchronous Motor; *Electr. Mach. Electromech.,* vol. 4, pp. 57–70, 1979.

74. Weh, H.: Linear Synchronous Motor Development for Urban and Rapid Transit Systems. *IEEE Trans.,* vol. Mag-15, pp. 1422–1427, 1979.

75. Weh, H.: Linear Synchronous Propulsion with Permanent Magnet Excitation; *Proc. Int. Conf. Cybernetics Soc.,* pp. 1042–1049, 1980.

76. Baudon, Y., El Zawawi, A., and Ivanes, M.: Homopolar Synchronous Motor: Steady-State Analysis and Experimental Results; *Electr. Mach. Electromech.,* vol. 6, pp. 23–33, 1981.

77. Kyotani, Y.: Present Status of Research and Development of Superconducting Magnetic Levitation Railway. Research on Linear Motor Magnetic Levitation System in the Japanese National Railways; *Rail Int. (Belgium),* vol. 12, pp. 155–163, 1981.

78. Atherton, D. L.: A Magnetic Circuit Analysis of the Iron-Cored Homopolar Linear Synchronous Motor; *IEEE Trans.,* vol. Mag-17, pp. 1305–1310, 1981.

79. Laithwaite, E. R., and Kuznetsov, S. B.: Power Factor Improvement in Linear Induction Motors; *IEEE Proc.,* vol. 128, pp. 190–194, 1981.

80. Snow, W. R., Scott, Dunbar R., Kubby, J. A., and O'Neill, G. K.: Mass Driver Two: A Status Report; *IEEE Trans.,* vol. 18, pp. 127–133, 1982.

81. Atherton, D. L: Design and Control Studies of a Linear Synchronous Motor for Urban Transit Application; *IEEE Trans.,* vol. Mag-18, pp. 1847–1855, 1982.

82. Johnson, W., and Nenadovic, V.: World's First Maglev Operation Moves into the Test Phase; *Railw. Gaz. Int.,* pp. 260–262, 1983.

## ELECTRIC VEHICLES

*By W. F. HAMILTON*

***270. Introduction.*** An *electric vehicle* is propelled by an electric motor which is supplied with electric power from a rechargeable battery. A *hybrid electric vehicle* has an additional internal-combustion engine which may propel the vehicle mechanically, generate electricity to recharge the battery, or both. Its advantage is extending the range limitation of the pure electric vehicle. Owing primarily to the weight, cost, and limited capability of storage batteries, electric and hybrid vehicles for on-road use have generally been more expensive and less acceptable than comparable internal-combustion vehicles. Technological developments, however, promise substantial improvements in both performance and cost. An increased use of electric and hybrid vehicles, at least in specialized applications, appears likely within a few years.

***271. Historical Perspective.*** Early in the era of the automobile, electric vehicles (EVs) were competitive with steam and gasoline vehicles. They were relatively smooth, silent, dependable, simple to operate, and easy to maintain. In 1900, registrations of EVs in the United States exceeded those of either steam or gasoline vehicles, and the world speed record for automobiles was held by an electric car.

The driving range of EVs, however, was limited to perhaps 50 mi between recharges. Recharging the battery required hours, if not all night, and few recharging stations were available. Internal-combustion engine (ICE) vehicles, in contrast, could be refueled in a few minutes, and the spread of gasoline stations gave them a limitless radius of action. The technology of ICE vehicles advanced rapidly, and they soon captured most of the motor vehicle market.

Early developers built a variety of hybrid vehicles, attempting to enhance electric propulsion with the endurance and quick refuelability of the ICE. The complexity and weight of the hybrids, however, more than offset their added capability.

In 1912, the introduction of the self-starter overcame the greatest remaining obstacle to general use of ICE cars. Like the hybrids, this development brought together in one vehicle an ICE, an electric motor, and a storage battery. Here, however, the role of the electric components was reduced to a practical minimum: cranking the ICE.

By the 1920s, EVs had become little more than a curiosity on U.S. roads, as they remain today. In specialized off-road applications, however, EVs have successfully displaced many competing ICE vehicles. Over a third of the half-million U.S. industrial trucks are electric, as are roughly half of the half-million U.S. golf cars, much mine machinery, and an increasing number of service vehicles on airport ramps.

In the 1960s, U.S. interest in on-road EVs was reawakened by growing recognition of air pollution problems. By 1967, several acts to subsidize EV development had been introduced in the U.S. Congress. None, however, became law, and a government study of EVs eventually concluded that other means were preferable for reducing pollution due to automobiles.

In 1973, interest in EVs was redoubled by the Arab oil embargo and by the sharp rise in world oil prices subsequently brought about by the OPEC cartel. In the United States, a fledgling manufacturer was able to build and sell some 2000 small electric cars during 1974 to 1976 despite unfavorable reviews, and in 1976 an act to develop and demonstrate electric and hybrid vehicles was passed—over a presidential veto—by the U.S. Congress.

Under the 1976 act, the U.S. Department of Energy conducts a continuing program of research, development, and demonstration. It has made substantial progress in battery and power-train technology and has placed over 1500 EVs in a variety of applications, mostly commercial. Originally, the program's main goal was to demonstrate up to 10,000 EVs. It soon became clear, however, that available EVs were uneconomic due largely to short battery life, the high cost of low-volume production, and high maintenance expenses associated with immature power-train technology. Program emphasis has been redirected over the years to address these problems.

After the Iranian revolution and the gasoline shortages of 1979, expert opinion held that rising oil prices and repeated interruptions in the production of OPEC oil were likely. In 1980, General Motors (GM) announced a major project to develop and produce an electric passenger car. The Delco-Remy Division of GM devised a much-improved, high-performance nickel-zinc battery; a fleet of electric test vehicles was built; and extensive tests were conducted. Contrary to expectations, however, the price of gasoline fell rather than rose. When it became evident that the GM electric car would be economically competitive only at gasoline prices above $3 per gallon, over twice the then-current level, the project was dropped.

In Europe and Japan, the oil shortages and price increases of the 1970s also stimulated efforts to develop competitive EVs. Major automakers, encouraged and supported by government and electric utilities, expanded research on propulsion batteries and tested a variety of innovative EVs. Notable programs include those of Japan's Ministry of International

Trade and Industry (MITI), and West Germany's Gesellschaft Electrischen Strassenverkehr (GES), which is testing a fleet of 70 electric cars based on the Volkswagen Golf automobile. Only in the United Kingdom, however, have EVs entered regular production and commercial service.

The United Kingdom has long been unique in that it supports a significant fleet of on-road EVs (over 45,000 in 1980). The majority of these EVs are "milk floats," small trucks used for door-to-door delivery of dairy products. In this grueling slow-speed application, total annual costs of competing diesel vehicles are over 50% higher than those of EVs. Similar EVs are also used in other U.K. fleet applications for which their very limited performance is sufficient. But such applications are scarce because available long-life batteries have typically limited these economical EVs to a range of 25 to 50 mi and a top speed of only 15 to 25 mi/h.

Though incompatible with general road traffic, the U.K. milk floats have convincingly demonstrated the potential economy of electric propulsion. Accordingly, development effort in the United Kingdom has sought to expand applicability of EVs by raising performance to traffic-compatible levels without sacrificing their entire economic advantage. In 1984, this culminated in assembly-line production of the 50-mi, 50-mi/h Bedford electric van. This van was developed by an industry consortium, aided by a modest government subsidy which is to be eliminated by 1987.

***272. Current Electric Vehicles.*** The Bedford CF electric van is the leading EV in the world today. It is manufactured in the United Kingdom by Bedford Commercial Vehicles, a subsidiary of General Motors. The modular design of the electric propulsion system allows its installation instead of an ICE propulsion system on the assembly line used to build conventionally powered Bedford CF vans. The main characteristics of the electric version are shown in Table 23-13.

**TABLE 23-13** Characteristics of the Bedford CF Electric Van

| | |
|---|---|
| Range (city driving) | 50–60 mi (80–100 km) |
| Maximum speed | 50 mi/h (80 km/h) |
| Acceleration, 0–30 mi/h | 11–13 s |
| Gradability (startup) | 18 % min. |
| Recharge energy | 0.85 kWh/mi (0.53 kWh/km) |
| Payload | 2200 lb (1000 kg) |
| Curb weight | 5500 lb (2500 kg) |
| Gross weight | 7700 lb (3500 kg) |
| Battery weight | 2600 lb (1200 kg) |
| Motor weight | 300 lb (137 kg) |
| Motor rating, 30 min | 54 hp (40 kW) |

The Bedford's electric propulsion system is supplied by Lucas-Chloride EV Systems, a joint venture between major U.K. manufacturers of electrical equipment and storage batteries. Development of the propulsion system—motor, controller, battery, charger, and auxiliaries—began over 10 years ago and has included extensive field testing to assure reliable operation with low maintenance. The lead-acid battery is especially noteworthy because it is warranted for 4 years, over twice the typical lifetime of comparable EV batteries in use elsewhere. The battery is an assemblage of thirty-six 6-V modules. For reliability, connections between modules are welded, and modules are also interconnected by a single-point watering and venting system which greatly reduces the labor required for battery maintenance. The motor is a conventional separately excited dc machine about 12 in in diameter (300 mm). The controller uses efficient semiconductors to regulate motor current and voltage in response to driver demands for power; it is about 21 in (540 mm) in length and weighs about 60 lb (27 kg).

The ETV-1 (Electric Test Vehicle 1) was built from the ground up for the U.S. Department of Energy by General Electric and Chrysler in 1979. The battery pack is housed in a central "tunnel" which is a major structural element of the car. The separately excited dc motor of the ETV-1 drives the front wheels through a fixed gear reduction. As in the Bedford CF electric, there is no clutch, and reverse movement is accomplished by electrically reversing the motor. The controller is completely transistorized, includes a battery charger, and operates under the supervision of a microcomputer. The main characteristics of the ETV-1 are shown in Table 23-14.

**TABLE 23-14** Characteristics of the ETV-1 Electric Car

| | |
|---|---|
| Range: | |
| City driving | 45 mi (75 km) |
| Constant 35 mi/h | 120 mi (200 km) |
| Maximum speed | 70 mi/h (110 km/h) |
| Acceleration, 0–30 mi/h | 9 s |
| Payload | 600 lb (275 kg) |
| Curb weight | 3350 lb (1520 kg) |
| Battery weight | 1100 lb (500 kg) |
| Motor weight | 220 lb (100 kg) |
| Motor rating (continuous) | 20 hp (15 kW) |

***273. Hybrid Electric Vehicles.*** The Lucas hybrid car, built in 1983, is an adaptable test bed rather than a prototype for production. The power train is of the "parallel" configuration: the ICE and the motor can simultaneously supply mechanical power to the drive shaft and rear wheels. While the clutch is disengaged, the car may be driven as an all-electric vehicle, without any use of the ICE. Placement of the battery pack beneath the floor allows it to be relatively large without intruding into the seating and luggage space, which accommodates five persons. In consequence, battery capacity is large enough so that the range of the car on electricity alone would be sufficient on most days. For extended cruising, however, the clutch can be engaged and the car propelled entirely by the ICE. In this mode, it provides a freedom of action like that of conventional ICE cars.

The battery pack of the Lucas hybrid car comprises eighteen 12-V lead-acid modules similar in construction to those of the Bedford electric van. The motor and controller are uprated versions of those used in the Bedford, providing a maximum output of about 50 kW. The ICE is a Reliant 848-mL, 4-stroke, water-cooled unit rated at 30 kW. Maximum speed is about 70 mi/h (110 km/h) on the ICE alone, and about 75 mi/h on the electric motor alone. Typical range between recharges, without any use of the ICE, is 40 to 50 mi (65 to 80 km).

The *hybrid test vehicle* (HTV) was built for the U.S. Department of Energy by a team headed by General Electric. Like the Lucas hybrid, the HTV includes both an ICE and an electric motor in parallel configuration, so that either or both can provide mechanical drive power. The HTV, however, is considerably more complex and ambitious.

In the HTV, the ICE and the motor each drive a 3-speed transaxle through independent clutches. The motor is relatively small, and ICE power is used not only for long-distance cruising but also for high acceleration and speed in stop-start driving. Whenever electric power alone is inadequate to meet driver demands, the ICE must be started almost instantly and its output smoothly blended with that of the motor. To conserve fuel, the ICE is stopped whenever the electric propulsion subsystem alone suffices. A computer system senses driver inputs from accelerator and brake pedals and in response selects the appropriate gear in the transmission, smoothly modulates both clutches during gearshifts, starts and stops the ICE when appropriate, and adjusts the ICE throttle and the power controller for the motor.

The HTV is a conversion of a conventional five-passenger ICE car. To avoid any intrusion into the passenger or luggage space, the entire propulsion system was placed under the hood of the car, along with an air conditioner. This required very dense packaging. It also required a relatively small battery, limiting the capability of the HTV to operate on electricity alone. The conversion added 1272 lb (577 kg) to the weight of the unconverted vehicle. It included a 1.7-L Audi fuel-injected ICE producing 74 hp (55 kW), the dc motor of the ETV-1 uprated to 27 hp continuous or 46 hp peak (20 or 34 kW), and ten 12-V lead-acid battery modules weighing 760 lb.

Maximum speed of the HTV is about 60 m/h (100 km/h) on the motor alone, and exceeds 75 mi/h (120 km/h) on the ICE alone. Acceleration from 0 to 30 mi/h requires 8.4 s. Annual petroleum savings over a conventional 1985 car were projected at 50 to 70%, assuming use of an improved battery. With the battery actually developed for the HTV, which met all design requirements yet afforded unsatisfactory capacity in actual use, projected annual petroleum savings were 23 to 35%.

***274. The main components*** of an electric propulsion system are shown schematically in Fig. 23-80. They include

The propulsion battery, which stores electric energy for propelling the vehicle.

The controller, which adjusts the flow of power to an electric motor in accord with the wishes of the driver.

The motor, which converts electric power into mechanical power at its output shaft.

The transaxle (or transmission, drive shaft, differential, and axle), which reduces motor shaft speed to a suitable range for the driven wheels and balances the torques applied to them.

The electric power train is usually defined as the components between the propulsion battery and the driven wheels, i.e., the controller, motor, and transaxle. In addition to these main components, most EV power trains include various auxiliary components, such as

An on-board battery charger, which recharges the propulsion battery using the ac power supplied by electric utilities.

A battery state-of-charge meter (fuel gage) or range-meter (showing remaining range capability).

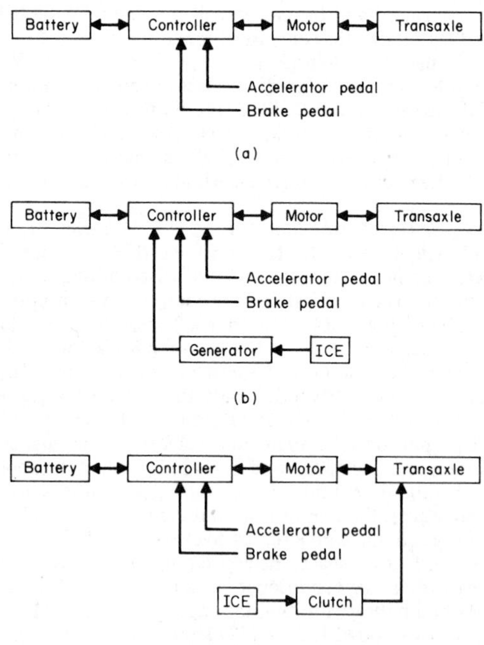


**FIG. 23-80** Main components of an electric propulsion system.

A 12-V accessory battery, used to power conventional automotive headlights, tail lights, signal lights, blowers, etc.

A dc-to-dc converter, which converts the high-voltage output of the propulsion battery to 12 V for operating accessories and maintaining the charge of the accessory battery.

In most recent EVs, all power-train components operate under the control and supervision of a microcomputer.

A hybrid power train (or, more specifically, hybrid-electric power train) generally includes all of the above plus an internal combustion engine and sometimes an electric generator as illustrated in Fig. 23-80.

In power-train design there are many possible trade-offs and compromises. AC motors are lighter and far cheaper than dc motors, for example, but are not necessarily preferable because they require a larger, more expensive controller. Similarly, multispeed transmissions are more expensive and less efficient than fixed reduction gears, but they may yet be desirable because they reduce the size and cost of both the motor and the controller required for a given level of vehicle performance.

At present, the choice between dc and ac systems, or between systems with and without multispeed transmissions, is not clearly defined. Eventually, however, ac systems with fixed gearing seem almost certain to be superior. They are mechanically the simplest, cheapest, and most reliable, and the cost of the complex controller they require falls with each new advance in semiconductor technology.

***275. Weight and Cost.*** The weight, cost, and performance of eight recent EV motors and controllers are shown in Table 23-15, along with comparable figures estimated for a widely used automotive ICE. With the exception of the Eaton permanent-magnet (pm) motor system, all the motor-controller systems were developed for the U.S. Department of Energy, beginning with the ETV-1 system in 1977 to 1979.

Combined motor-controller efficiencies typically peak at 85 to 88% and average 75 to 80% in stop-start driving cycles. Transaxle efficiencies are typically 90 to 95%, with the multispeed units toward the low end of the range.

The weights and costs of these systems show substantial progress since completion of the ETV-1 benchmark in 1979. The weight of the brushless (ac and pm) motors per unit output is roughly half that of the dc motor in the ETV-1, and their cost per unit output is perhaps a quarter that of dc motors. The weight and cost per unit output of complete brushless motor-controller systems are only about half those of dc systems.

***276. Design Choices.*** When the benchmark ETV-1 vehicle was designed in 1977, a careful optimization study led to the choice of a dc system including a chopper controller, separately excited motor, and fixed-ratio transaxle.

Each chopper is basically a high-speed transistor switch which interrupts the flow of current from the battery thousands of times per second. The ratio of "on" time to "off" time for each chopper is set by the microcomputer, which thereby controls the average current and voltage applied to the motor windings. At vehicle speeds below 30 mi/h (motor speed 2500 r/min), the field chopper supplies maximum field voltage and the armature chopper alone is used to control motor power. At higher speeds, the armature chopper applies full battery voltage and the field chopper alone controls motor speed, by field weakening.

The more recent Chrysler dc system simultaneously adjusts armature and field currents in order to control motor power and maximize motor efficiency at the same time. It also uses newer, faster transistors which allow chopping at 20 kHz. This eliminates acoustical noise from the electrical system and improves filtering so that both battery and motor currents are nearly pure dc, with little ripple.

After the ETV-1, other power-train design choices were explored at Eaton Corporation. To minimize motor-controller power ratings and consequent costs, the Eaton systems were designed around multispeed automatic transaxles. The Eaton dc system uses a single chopper to minimize costs, controlling armature current up to 9 mi/h and field current at all higher vehicle speeds. This scheme keeps chopper ratings relatively low but requires a large contactor to switch the chopper between field and armature circuits. The motor-transaxle package of the dc system is similar to that of the Eaton ac system. This package places the motor and transmission on a separate shaft adjacent to the differential and axle. The 3-phase inverter is both bulky and expensive, which led Eaton to devise a lower-cost alternative. Based on a novel single-phase motor using inexpensive ferrite pm's, this pm system requires

**TABLE 23-15** Characteristics of Recent EV Motors and Their Controllers

| System developer and type | Motor | | | | | | | Controller | | | |
|---|---|---|---|---|---|---|---|---|---|---|---|
| | Weight, kg | r/min, max | Output, kW Cont. | Output, kW Max | Cost, $ | $/kW, Cont. | kg/kW, Cont. | Weight, kg | Cost, $ | $/kW, Cont. | kg/kW, Cont. |
| GE ETV-1 dc | 99 | 5000 | 15 | 30 | 1230 | 82 | 6.6 | 41 | 1718 | 115 | 2.7 |
| Chrysler dc | 82 | 8000 | 26 | 45 | 2250 | 86 | 3.1 | 41 | 1500 | 57 | 1.6 |
| Eaton: | | | | | | | | | | | |
| dc | 82 | 5000 | 15 | 30 | 1913 | 125 | 5.3 | 30 | 1195 | 78 | 2.0 |
| ac | 55 | 12,500 | 19 | 34 | 375 | 20 | 3.0 | 43 | 1770 | 95 | 2.3 |
| pm | 61 | 11,000 | 19 | 34 | 675 | 36 | 3.3 | 23 | 1513 | 81 | 1.2 |
| Garret pm | 16 | 26,000 | 15 | 26 | 378 | 25 | 1.1 | 34 | 1020 | 68 | 2.3 |
| GE ac | 45 | 14,615 | 15 | 30 | 320 | 21 | 3.0 | 66 | 3338 | 223 | 4.4 |
| Ford ETX ac | 43 | 9000 | 20 | 40 | | | 2.1 | 40 | | | 2.0 |
| JPL ac | 74 | 12,000 | 25 | 52 | 330 | 13 | 3.0 | 17 | 2250 | 64 | 0.4 |
| GM 4-cycle ICE | 230 | 4500 | | 67 | 1500 | 22 | 3.4 | | | | |

NOTE: pm = permanent magnet; Cont. = Continuous.

an inverter using only two high-power transistor switches, rather than the six required in the conventional ac system.

Another pm motor system was developed at Garret AiResearch using aerospace technology. Weighing only 1.1 kg/kW, it is significantly lighter than others developed in the U.S. Department of Energy program. Its 3-phase inverter, however, remained very large. The high specific power of the motor reflects its use of rare earth-cobalt pm's and its operation at 26,000 r/min, considered to be the upper limit for grease-packed bearings.

At General Electric an experimental ac motor-controller system was developed after completion of the ETV-1. Expertise from this development was then brought to the Ford-GE ETX electric transaxle project. In the ETX the ac induction motor, two-speed planetary transmission, and axle are coaxial, one axle shaft running through the hollow rotor shaft of the motor. The resultant assembly is both light in weight and extraordinarily compact. Its overall length is under 24 in (60 cm), and its lowest point is less than 6 in (15 cm) below the axle centerline. Yet in an Escort-type test vehicle, it is expected to give performance rivaling that of the standard Escort with 1.6-L engine and to require a net energy output from the battery of under 250 Wh/mi (150 Wh/km).

The ETX system includes two digital computers. One serves as vehicle controller, directing and monitoring all the components of the propulsion system. The other controls the 3-phase current-controlled inverter, which synthesizes variable-frequency, variable-voltage alternating current for the induction motor. At low motor speeds, the inverter employs pulse-width modulation to provide ac power with voltage and frequency proportional to speed. At high motor speeds, the inverter operates in a simple six-step switching mode, providing constant-voltage ac power with frequency proportional to speed.

The ETX motor is built from inexpensive components already in volume production for commercial refrigeration. The inverter, however, is built around six specially developed 400-A Darlington transistor modules which are mounted on a liquid-cooled heat sink. It is also relatively large: 25 × 15 × 9 in (64 × 38 × 20 cm).

In the future, advances in power semiconductors should substantially reduce the size, weight, and cost of ac controllers. The controller being developed at the Cal. Tech. Jet Propulsion Laboratory (JPL), for example, uses transistorized power switching modules which recently become available from Japan. Manufactured for the large industrial market in variable-speed ac drives, the modules are both compact and inexpensive. The JPL controller is about half the size and weight of the ETX controller, yet offers higher output capability.

Another possibility for reducing controller size and cost is the use of a variable-reluctance motor (VRM), another type of brushless motor. A 3-phase VRM can be built using only three large semiconductor switches, half the number required in the typical controller for three-phase ac induction motors. The voltage rating of each switch, however, must be twice that required in the ac controller. The VRM itself may be slightly lighter and less expensive than ac induction motors and may also be amenable to use as a wheel motor in EVs (i.e., integrated into the wheel itself and driving it directly, without gearing).

VRM designs for vehicular propulsion have been developed in the United Kingdom, in France, and in the United States (at Ford). A current project at Massachusetts Institute of Technology (MIT), sponsored by the U.S. Department of Energy, produced a motor that weighs 60 kg and produces 60 kW of output from 5000 to 10,000 r/min at a 30% duty cycle. The end of the motor carries a shaft position sensor so that switching the current to the field windings of the motor can be properly timed. The rotor of the motor has no windings; made of soft iron, it consists simply of a stack of laminations on the motor shaft. The stator consists of a stack of laminations, with field windings for each phase placed on opposite pairs of teeth. When current is supplied to one such winding, the rotor moves toward the position of least magnetic reluctance, i.e., adjacent rotor teeth are pulled into alignment with the excited stator teeth. Sequential switching of currents to the windings for each phase causes continuous rotation of the rotor.

*277.* ***Commercial fleets*** of light trucks and vans are widely considered to be the most promising opportunity for introducing EVs into regular service. In some of these fleets, many vehicles already operate in limited territories or on fixed routes for which the limited range of EVs would be adequate. Furthermore, many are also parked overnight and maintained at a central garage, so recharging and servicing of EVs could easily be arranged. Finally, many fleet vehicles frequently start, stop, and idle, which leads to relatively high per-mile costs for fuel and maintenance. EV costs, in contrast, are little affected by such usage.

**TABLE 23-16** Daily Travel of Light-Duty Commercial Trucks and Vans

| Vehicle category | Typical daily travel, mi | | | | |
|---|---|---|---|---|---|
| | 0–30 | 30–60 | 60–90 | Over 90 | All |
| Total vehicles: | | | | | |
| Millions | 1.5 | 1.8 | 1.1 | 2.7 | 7.0 |
| Percent | 21 | 25 | 16 | 38 | 100 |
| Percent vehicles occasionally driven over: | | | | | |
| 30 mi/day | 12 | 25 | 16 | 38 | 91 |
| 60 mi/day | 8 | 12 | 16 | 38 | 74 |
| 90 mi/day | 4 | 7 | 9 | 38 | 58 |

Table 23-16, drawn from a 1983 survey of commercial fleet operators, shows that of 7 million commerical light trucks and vans in the United States,

3.3 million (46%) are typically driven less than 60 mi per day (100 km per day).

1.8 million (26%) are *never* driven over 60 mi per day.

1.4 million (20%) are driven on fixed routes of less than 60 mi per day.

The survey also found that of those trucks and vans typically driven under 60 mi per day, 66% were left on company premises overnight, and 18% were stopped with engines left running more than 100 times per day. Finally, the survey found that almost 90% of all light trucks and vans operate with typical payloads under 1000 lb (450 kg).

Accordingly, it seems clear that EVs which could be built today have many potential applications within commercial fleets. The extent to which they could be successfully marketed, however, is uncertain. Commerical enterprises usually seek to maximize profits by fulfilling their transportation requirements at minimum cost. Calculations suggest that EVs with working ranges of 50 mi and more would still be somewhat more expensive than competing ICE vehicles, even on a life-cycle cost basis with the full benefit of high-volume production.

Technological advances, however, should eliminate this cost disadvantage of EVs within a few years. And in the long run, rising prices for petroleum could bring EVs a substantial cost advantage relative to ICE vehicles. (It should be noted, however, that the ratio of gasoline to electricity prices in 1985 was virtually unchanged from that in 1975 or 1965.)

***278. Personal or household vehicles,*** unlike commercial light trucks and vans, are seldom if ever operated on fixed routes or in limited territories. Even secondary cars at multicar households are occasionally driven long distances. It is, however, at such households that electric automobiles are first expected to find application. Unlike single-car households, multicar households could own an electric vehicle without sacrificing the capability to make occasional long trips by automobile.

Table 23-17, based on a nationwide survey of travel at 25,000 representative households, shows that the average multicar household could accomplish 45% of all its annual travel with an EV of 60-mi useful range. Because many multicar households have more than two cars, this implies that the 60-mi EV would provide about as much annual travel as the average conventional car, about 10,000 mi (16,000 km).

A hybrid car with an electric range only two-thirds as large (40 mi, or 65 km) could electrify as much or more travel at these multicar households. The hybrid can make excursions of any length, so its full electric range is always used on long trips. The electric car, in contrast, cannot be used at all if its remaining range is insufficient to complete an excursion. As a result, some or all of its capability may go unused when a household makes a long excursion.

Unlike the electric car, the hybrid could reasonably be purchased by single-car as well as multicar households. Table 23-17 shows that if substituted for all household cars, hybrids with a 60-mi electric range would electrify 61% of annual travel. If 60-mi electric cars were

**TABLE 23-17** Potential Electrification of Household Travel

| EV or HV electric range, mi | Electrified travel, %: An EV or HV replaces 1 vehicle at each multicar household* | | An EV or HV replaces every vehicle at every household† | |
|---|---|---|---|---|
| | EV | HV | EV | HV |
| — | — | — | — | — |
| 40 | 34 | 48 | | |
| 50 | 40 | 53 | | |
| 60 | 45 | 57 | 49 | 61 |
| 80 | 52 | 63 | 56 | 67 |
| 100 | 57 | 66 | 61 | 70 |
| 125 | 61 | 69 | | |
| 150 | 64 | 70 | | |

*EV or HV is used in preference to other household vehicles.
†EV or HV is assigned to trips of vehicle replaced.
NOTE: HV = hybrid vehicle.
1 mi = 1.61 km.

somehow substituted for all household cars, only 49% of existing travel would be electrified, and the remainder would be impossible.

With daytime recharging between trips, the electric car could accomplish substantially more household travel. Daytime recharging at home only would increase by about 5% the amount of travel a 60-mi electric car could accomplish at the average multicar household. Daytime recharging at home, at the work place, and at the off-street parking used for shopping would increase by 20% the amount of travel the 60-mi car could accomplish. This, however, would require an extensive network of recharging stations which does not now exist. The advantage of daytime recharging is greater for electric cars with shorter range: the percentage increases above would be twice as large for electric cars with a useful range of 40 mi (65 km).

An electric car with range of 60 mi or more is clearly applicable to a reasonable share of household travel. Its market penetration, however, would probably be quite low. Surveys and studies of auto buyers suggest that even if offered at the low prices associated with production in very large quantities, electric cars of this sort might capture perhaps 1 to 2% of the U.S. auto market (which totals roughly 10 million units per year).

Despite their utility for much household travel, electric vehicles remain less capable than conventional vehicles. With a battery, they are also substantially more expensive to purchase. An investigation of consumer preferences in 1978 found that the average multivehicle household would have paid substantial premiums, if necessary, to purchase automobiles without the typical limitations of the electric car:

$6500 to avoid a 50-mi range as opposed to a 200-mi range.

$2500 or more to avoid a 7-h recharge time.

$2000 for high acceleration as opposed to slow or average.

Put another way, the average multicar household would be unlikely to choose an electric car even if its purchase price were much lower than that of competitive ICE cars.

The few households which would be likely to purchase an electric car are those attaching little importance to its limitations and its higher price, and much importance to its advantages: lower operating cost, higher reliability, low pollution, independence of the gasoline pump—and possibly its novelty.

The aversion of motorists to limited-range vehicles is the major reason for developing hybrids. Because they add to the advantages of electric propulsion a capability for long trips rivaling that of conventional cars, hybrids are projected to command a market share three to five times larger than that of pure electric vehicles.

## Bibliography

### *279. References for Electric Vehicles*

Hamilton, W. F.: *Electric Automobiles;* McGraw-Hill Book Company, New York, 1980.

Unnewehr, L. E., and Nasar, S. A.: *Electric Vehicle Technology;* Wiley, New York, 1982.

# SECTION 24

# ELECTROCHEMISTRY AND ELECTROMETALLURGY

**Keith F. Blurton**

*Vice President, PCK Technology Division, Kollmorgen Corporation. Member, American Chemical Society, Electrochemical Society, and American Association for the Advancement of Science*

**CONTENTS**

*Numbers refer to paragraphs*

## *PRINCIPLES*

### Introduction

***1. Definition.*** *Electrochemistry* is the study of the interconversion of chemical and electrical energy.

***2. Scope.*** This chapter discusses the fundamental principles of electrochemistry and the application of these principles to commercial operations. Detailed discussions are given of the conversion of electrical to chemical energy (electrodeposition, electrolytic refining, electrowinning, and electrosynthesis); only brief discussions, with particular reference to basic principles, are given for the devices converting chemical to electrical energy (batteries and fuel cells) because these are discussed elsewhere in this handbook.

***3. Electrolytes.*** *Electrolytic solutions* contain charged particles called ions which are *atoms* or groups of atoms which have lost or gained *electrons* and, as a result, are positively or negatively charged, respectively. Negatively charged ions are called *anions* and positively charged ions are called *cations.* The electrolyte may comprise as a solvent water, an organic liquid, a fused (or molten) salt, or a solid. Since the electrolyte is electrically neutral, the number of anions and cations must be equal.

In aqueous solutions (the most common solvent used in practice), electrolytic solutions are formed by dissolving a salt, an acid, or a base in water. These compounds may then be *completely* or *partially dissociated* into ions. Salts, such as copper sulfate ($CuSO_4$), silver nitrate ($AgNO_3$), and sodium chloride (NaCl); acids, such as sulfuric acid ($H_2SO_4$), hydrochloric acid (HCl), and nitric acid ($HNO_3$); and bases, such as sodium hydroxide (NaOH) and potassium hydroxide (KOH), are completely dissociated.

$$H_2SO_4 + H_2O \rightarrow H_3O^+ + SO_4^{2-} \qquad (24\text{-}1)$$

Weak acids, such as acetic acid ($CH_3$ COOH) and many other organic acids, and weak bases, such as ammonia ($NH_3$) and organic amines, are only partially dissociated.

$$CH_3COOH + H_2O \rightleftarrows H_3O^+ + CH_3COO^- \qquad (24\text{-}2)$$

and there is an equilibrium setup between the dissociated and undissociated species.

***4. Conductivity.*** Metals immersed in an electrolyte and connected to an external circuit are called *electrodes.* The passage of electricity from one electrode to the other results in the motion of the ions, the anions moving toward the *anode* and the cations to the *cathode.* A chemical reaction also occurs at the electrode-electrolyte interface.

The resistance of the electrolyte is similar to that of a metal, and the specific conductivity ($\kappa$) is given by

$$\kappa = L/AR \qquad (24\text{-}3)$$

where $L$ is the distance between electrodes (i.e., current path), $A$ is the electrode cross-sectional area, and $R$ is the experimentally measured solution resistance.

The value of the electrolyte-specific conductivity depends markedly on the nature of the electrolyte. For example, strong electrolytes have relatively high electrical conductivity because of the complete dissociation of the compound into ions, whereas weak electrolytes which are only partially ionized chemical compounds have very poor conductivity.

Electrolytic conductivity is also markedly dependent on the electrolyte concentration and, in general, increases with increasing electrolyte concentration. (See Table 24-1.)

The passage of current through an electrolyte differs in two major respects from that through a metallic conductor. First, the current is carried by ionic migration rather than by electrons, and second, electrolytic conductivity increases markedly with temperature.

**TABLE 24-1** Specific Conductivity of Aqueous Solutions at 25°C*

| Electrolyte, at concentration of 0.1 g · mol/L | Specific conductance, $\Omega^{-1}\ cm^{-1}$ |
|---|---|
| Sodium hydroxide | $2.21 \times 10^{-2}$ |
| Potassium chloride | $1.29 \times 10^{-2}$ |
| Silver nitrate | $1.09 \times 10^{-2}$ |
| Magnesium chloride | $9.71 \times 10^{-3}$ |
| Copper sulfate | $5.06 \times 10^{-3}$ |
| Acetic acid | $5.20 \times 10^{-4}$ |
| Ammonia | $3.54 \times 10^{-4}$ |

* Compare values with that of metallic copper, $5.80 \times 10^{5}\ \Omega^{-1}\ cm^{-1}$

*5. Electrode Reactions.* The process whereby chemical reactions occur at the electrode-electrolyte interfaces is by the gain or loss of electrons, i.e., change in *valence* of the ions. Electrons are gained at the cathode and lost at the anode, i.e., *reduction* occurs at the cathode and *oxidation* at the anode. For example, in the electrolysis of copper sulfate with a copper anode and inert (steel) cathode, the cathode (reduction) reaction is

$$Cu^{2+} + 2e \rightarrow Cu \tag{24-4}$$

and the anode (oxidation) reaction is

$$Cu \rightarrow Cu^{2+} + 2e \tag{24-5}$$

## Current Relationships

*6. Faraday's Laws.* Faraday's laws state: (1) the amount of product produced during electrolysis is directly proportional to the quantity of electricity flowing, and (2) the amounts of different chemical substances produced by a given quantity of electricity are proportional to their equivalent weights, where *equivalent weight* ($E$) is defined as

$$E = A/z \tag{24-6}$$

where $A$ is the atomic weight of an element and $z$ is the valence change. Therefore, for the deposition of copper described previously, the equivalent weight of copper is equal to 63.5/2 or 31.75.

*7. The Faraday.* Faraday's two laws can be summarized as

$$M = iEt/F \tag{24-7}$$

where $M$ is the weight in grams of substance reacting, $i$ is the current in amperes, $t$ is the time in seconds, and $F$ is the faraday. From Eq. (24-7), the faraday is defined as the quantity of electricity required to produce 1 gram-equivalent (i.e., the equivalent weight in grams) of product. From dimensional considerations, $F$ has the dimensions of ampere-seconds per gram-equivalent. Experimentally, $F$ has been found to be equal to 96,500 coulombs per gram-equivalent, or 23.06 kcal/volt · gram-equivalent.

The practical unit for the amount of electricity is the **coulomb** (C) defined as one ampere flowing for one second; 3600 C = 1 ampere hour (Ah), which is the practical unit used in battery technology.

***8. Coulometer.*** The coulometer is used for determining the quantity of electricity flowing through a cell, particularly when there are fluctuating currents. One of the most accurate coulometers is the silver coulometer, in which the amount of silver deposited from a silver nitrate solution is determined, and from it the quantity of electricity (*it*) is calculated using Eq. (24-7).

***9. Electrochemical Equivalents.*** The electrochemical equivalent ($e$) of an element is the weight of element deposited by 1 C of electricity. From Eq. (24-7), when $it = 1$ C,

$$e = E/F \tag{24-8}$$

The units of the electrochemical equivalents depend on the units of current and time and of the faraday, and values for selected elements are given in Table 24-2.

If the voltage of the cell reaction is known, values for energy consumption (or generation) may also be calculated since ampere-hours × volts = watt-hours.

***10. Current Efficiency.*** It is often found that the amount of chemical change occurring is less than that calculated from the values of the electrochemical equivalents. This is not due to any breakdown in Faraday's laws, but to a failure to take into account all reactions. For example, in electrodeposition side reactions, such as oxygen and hydrogen evolution, often occur and result in lower current efficiencies for metal deposition than calculated. If these side reactions are taken into account, however, strict adherence to Faraday's laws will be found.

Since one often wishes to know the deposition of a single species independent of side reactions, current efficiency is defined as

$$\text{Current efficiency} = \frac{\text{amount of desired chemical change}}{\text{theoretical amount from Faraday's law}} \times 100\% \tag{24-9}$$

***11. Current Density.*** The total current obtained from or applied to an electrochemical cell depends on the area of the electrodes. Therefore, current density is used to normalize the current-to-electrode area. This is the current divided by the geometric electrode area. Most reactions exhibit a range of current densities over which a particular electrochemical reaction will occur optimally, and these are dependent upon the nature of the reaction as well as the physical conditions such as temperature and stirring.

***12. Current Distribution.*** Faraday's laws only permit calculation of the total electrochemical reactions occurring at an electrode, but they do not give information on the reaction distribution across the electrode face. In many practical cells, the reaction is often hindered (or accelerated) on parts of the electrode due to electrode shape, uneven geometries of the anodes and cathodes, depletion of the reactants at the electrode surface due to insufficient stirring, or fortuitous resistances in the electrical circuit. As a result, metal may dissolve or be deposited at greater rates along the electrode surface. This is associated with uneven current values at various points on the electrode surface. Care is often taken in cell design to minimize uneven current distribution effects and to ensure uniform reaction rates along the electrodes. However, the vast majority of practical electrochemical cells have uneven current distribution which is manifested as, for example, uneven metal deposits, electrode shape change in secondary batteries, and low energy efficiency.

## Voltage Relationships for Reversible Cells

***13. Reversible Potentiometric Cells.*** There is a distinction between potentiometric cells, set up for studying cell reactions and calculating thermodynamic data, and practical *galvanic*

**TABLE 24-2** Electrochemical Equivalent of Elements

| Element (1) | Symbol (2) | Atomic weight (3) | Valence or valence change (4) | mg/C (5) | C/mg (6) | g/Ah or kg/kAh (7) | Ah/g or kAh/kg (8) |
|---|---|---|---|---|---|---|---|
| Any | | $A$ | $z$ | $0.0103631A/z$ | $96.50z/A$ | $0.037307A/z$ | $26.8045z/A$ |
| Aluminum | Al | 26.98 | 3 | 0.09320 | 10.729 | 0.33551 | 2.9804 |
| Antimony | Sb | 121.75 | 3 | 0.42059 | 2.3778 | 1.5140 | 0.66049 |
| Arsenic | As | 74.92 | 3 | 0.25880 | 3.8640 | 0.93168 | 1.0733 |
| Beryllium | Be | 9.012 | 2 | 0.04670 | 21.415 | 0.16811 | 5.9487 |
| Bismuth | Bi | 208.98 | 3 | 0.72189 | 1.3852 | 2.5988 | 0.38478 |
| Cadmium | Cd | 112.40 | 2 | 0.58241 | 1.7170 | 2.0967 | 0.47696 |
| Calcium | Ca | 40.08 | 2 | 0.20768 | 4.8152 | 0.74763 | 1.3375 |
| Chlorine | Cl | 35.453 | 1 | 0.36740 | 2.7218 | 1.3226 | 0.75605 |
| Chromium | Cr | 52.00 | 3 | 0.17962 | 5.5671 | 0.64664 | 1.5464 |
| | | | 6 | 0.089814 | 11.134 | 0.32332 | 3.0927 |
| Cobalt | Co | 58.93 | 2 | 0.30535 | 3.2750 | 1.0993 | 0.90972 |
| Copper | Cu | 63.54 | 1 | 0.65847 | 1.5187 | 2.3705 | 0.42185 |
| | | | 2 | 0.32924 | 3.0373 | 1.1852 | 0.84370 |
| Fluorine | F | 19.00 | 1 | 0.19690 | 5.0788 | 0.70883 | 1.4108 |
| Gold | Au | 196.97 | 1 | 2.0412 | 0.48990 | 7.3484 | 0.13608 |
| | | | 3 | 0.68041 | 1.4697 | 2.4495 | 0.40826 |
| Hydrogen | H | 1.008 | 1 | 0.010446 | 95.730 | 0.037605 | 26.592 |
| Iron | Fe | 55.85 | 1 | 0.57878 | 1.7278 | 2.0836 | 0.47993 |
| | | | 2 | 0.28939 | 3.4555 | 1.0418 | 0.95987 |
| | | | 3 | 0.19293 | 5.1833 | 0.69454 | 1.4398 |
| Lead | Pb | 207.2 | 2 | 1.0736 | 0.93143 | 3.8650 | 0.25873 |
| Magnesium | Mg | 24.31 | 2 | 0.12596 | 7.9388 | 0.45347 | 2.2052 |
| Manganese | Mn | 54.938 | 2 | 0.28466 | 3.5129 | 1.0248 | 0.97582 |
| Mercury | Hg | 200.59 | 1 | 2.0787 | 0.48105 | 7.4834 | 0.13363 |
| | | | 2 | 1.0394 | 0.96210 | 3.7417 | 0.26725 |
| Nickel | Ni | 58.71 | 2 | 0.30421 | 3.2872 | 1.0951 | 0.91312 |
| Oxygen | O | 16.00 | 2 | 0.082905 | 12.062 | 0.29846 | 3.3506 |
| Palladium | Pd | 106.4 | 2 | 0.55132 | 1.8138 | 1.9847 | 0.50384 |
| Platinum | Pt | 195.09 | 2 | 1.0109 | 0.98928 | 3.6391 | 0.27480 |
| | | | 4 | 0.50544 | 1.9786 | 1.8196 | 0.54957 |
| Potassium | K | 39.102 | 1 | 0.40522 | 2.4678 | 1.4588 | 0.68550 |
| Rhenium | Re | 186.2 | 7 | 0.27566 | 3.6277 | 0.99237 | 1.0077 |
| Rhodium | Rh | 102.91 | 2 | 0.53323 | 1.8753 | 1.9196 | 0.52092 |
| | | | 3 | 0.35549 | 2.8131 | 1.2797 | 0.78140 |
| | | | 4 | 0.26662 | 3.7507 | 0.95983 | 1.0419 |
| Selenium | Se | 78.96 | 4 | 0.20457 | 4.8884 | 0.73644 | 1.3579 |
| Silver | Ag | 107.868 | 1 | 1.1178 | 0.89457 | 4.0242 | 0.24849 |
| Sodium | Na | 22.99 | 1 | 0.23825 | 4.1973 | 0.85769 | 1.1659 |
| Tellurium | Te | 127.60 | 4 | 0.33058 | 3.0250 | 1.1901 | 0.84027 |
| Tin | Sn | 118.69 | 2 | 0.61500 | 1.6261 | 2.2140 | 0.45168 |
| | | | 4 | 0.30750 | 3.2520 | 1.1070 | 0.90333 |
| Titanium | Ti | 47.90 | 4 | 0.12410 | 8.0581 | 0.44675 | 2.2384 |
| Tungsten | W | 183.85 | 6 | 0.31755 | 3.1491 | 1.1432 | 0.87476 |
| Zinc | Zn | 65.37 | 2 | 0.33872 | 2.9523 | 1.2194 | 0.82008 |
| Zirconium | Zr | 91.22 | 4 | 0.23633 | 4.2313 | 0.85079 | 1.1754 |

cells, designed to produce useful energy where a current flows between the electrodes in an external circuit. When a metal is immersed in an electrolyte, an equilibrium is established across the metal-solution interface and this generates an electric potential difference across this region. When two dissimilar metals are immersed in the electrolyte, a *potential difference* exists between these electrodes which can be measured. For example, if copper and zinc are immersed in an electrolyte solution containing $Cu^{2+}$ and $Zn^{2+}$ cations, respectively, a potential difference will be generated across these two electrodes. (We assume for the sake of simplicity that the solutions do not mix and there is no potential difference across the junction. The former condition is often approximated in practice by placing a *membrane* between the two electrodes in the cell.)

Consider the case mentioned above of copper immersed in a solution containing $Cu^{2+}$ and zinc immersed in a solution containing $Zn^{2+}$, the two solutions being in intimate contact without mixing. The reaction at the copper electrode is

$$Cu^{2+} + 2e \rightarrow Cu \tag{24-10}$$

and the reaction at the zinc electrode is

$$Zn^{2+} + 2e \rightarrow Zn \tag{24-11}$$

The reactions as represented by Eqs. (24-10) and (24-11) will occur in a *thermodynamically reversible* manner when no net reaction takes place. This requires that in the measuring system, the potential difference between the two electrodes is exactly balanced by an externally applied potential difference so that no net reaction takes place and no current flows in the external circuit. This is achieved in practice by using a *potentiometer.* Note that the electrode equilibria attained are dynamic and they contain opposing electrode reactions taking place at the same rate. That is, the reaction represented by Eq. (24-10) is equal to that of the reaction represented by Eq. (24-11). Because these reactions are taking place under equilibrium conditions, the equations of thermodynamics are applicable.

***14. Thermodynamic Basis of Potentiometric Cells.*** An electrochemical cell is a method for carrying out chemical reactions, and a reaction carried out in this way is accompanied by the same energy change as when it occurs by other means. Under thermodynamically reversible conditions, the process occurs with maximum efficiency without production of heat and, therefore, without the loss of electrical energy. Therefore, the expenditure of electrical energy results in the performance of maximum work which, under constant temperature and pressure conditions, is equal to the *free energy change* of the system. For a cell reaction, the quantity of electricity required is $zF$ coulombs where $F$ is the faraday and $z$ is the number of electrons required in either electrode reaction during the completion of the stoichiometrical quantity of the overall cell reaction.[1] The electrical work of this cell is $zFE$ joules where $E$ is the potential difference of the cell. The work done on the system at constant pressure is $\Delta G$ (the increase of free energy of the system). Therefore, for a reversible cell at constant temperature and pressure,

$$\Delta G = -zFE \tag{24-12}$$

From thermodynamics

$$\Delta G = \Delta H - T\,\Delta S \tag{24-13}$$

where $\Delta H$ is the change of heat content (or enthalpy) of the system at constant pressure and temperature, and $\Delta S$ is the corresponding entropy change. Since $\Delta S$ is given by

$$\Delta S = -\left(\frac{\delta(\Delta G)}{\delta T}\right)_p \tag{24-14}$$

where the suffix $p$ denotes constant-pressure conditions. Therefore, from Eqs. (24-12), (24-13),

[1] The stoichiometry of an electrochemical reaction is obtained by multiplying each constituent in either the anodic or cathodic reactions by the smallest integer to ensure each equation has the same number of electrons and then adding the two equations.

and (24-14), the heat content change in a cell reaction at constant pressure is given by

$$\Delta H = -zF\left[E - T\left(\frac{\delta E}{\delta T}\right)_p\right] \qquad (24\text{-}15)$$

These series of equations show that: (1) the cell potential difference is related to the free energy change of the system, (2) the entropy change in a chemical reaction can be calculated from the temperature dependence of the cell potential difference, and (3) heat content changes in chemical reactions can be measured from electrochemical cells.

***15. Temperature Coefficient of Cell Reactions.*** From Eq. (24-15), a negative value of $\Delta H$ means that the heat content of the product is less than that of the reactants, so that heat is evolved during an exothermic reaction. And conversely, heat is absorbed during an endothermic reaction.

$\Delta H$ is calculated from the cell electromotive force (emf) and the temperature coefficient of cell emf. Typical values of $(\delta E/\delta T)_p$ are $10^{-3.5}$ V/°C. The difference between $\Delta G$ and $\Delta H$ is Eq. (24-15)

$$TzF(\delta E/\delta T)_p$$

which at 25°C (or 298 K) is $298 \times 2 \times 96{,}500 \times 10^{-3.5}$ J or 5 kcal

***16. Electromotive Force.*** The emf of the electrochemical cell is the maximum potential difference measured between the electrodes immersed in the cell electrolyte and is measured under conditions of zero net current. The term $E$ in Eqs. (24-12) and (24-15) is the cell emf.

***17. Calculation of Cell EMFs.*** For the cell reaction represented by the equation

$$A + B^{z+} \rightarrow A^{z+} + B \qquad (24\text{-}16)$$

$\Delta G$ can be calculated from the *Van't Hoff isotherm,* where

$$\Delta G = -RT \ln K + RT/zF \ln a_{A^{z+}} \times a_B/a_A \times a_{B^{z+}} \qquad (24\text{-}17)$$

where $R$ is the gas constant, $T$ is the absolute temperature, the $a$ terms represent the *activity* of the subscripted substances, and $K$ is the **equilibrium constant** of the reaction. The activity of an ionic species in solution is

$$a = Cj \qquad (24\text{-}18)$$

where $C$ is the concentration of the ionic species and $j$ is its *activity coefficient.* In dilute solution, activity coefficients equal 1, and throughout this section we will use concentration terms.

Substituting for $\Delta G$ from Eq. (24-12) in Eq. (24-17),

$$E = RT/zF \ln K + RT/zF \ln C_A \times C_{B^{z+}}/C_{A^{z+}} \times C_B \qquad (24\text{-}19)$$

When the concentrations of all species are unity, the second term on the right-hand side of the Eq. (24-19) becomes zero and

$$E = E^0 = RT/zF \ln K \qquad (24\text{-}20)$$

where $E^0$ is the *standard electromotive force* of the cell (defined as the cell emf when the concentration of reactants and products is unity). Equation (24-19), therefore, is written as

$$E = E^0 + RT/zF \ln C_A \times C_{B^{z+}}/C_{A^{z+}} \times C_B \qquad (24\text{-}21)$$

Equation (24-21) relates the cell emf to the value of the standard emf and the concentrations of the reactants and products.

***18. Expression of Electrode Potential.*** To derive the contributions of the separate electrodes to the cell emf, the cell reaction represented by Eq. (24-16) may be split into the individual electrode reactions. To carry out this derivation, the individual electrode reactions are written as reduction reactions (loss of electrons); thus,

$$A \rightarrow A^{z+} + Ze \qquad (24\text{-}22)$$

and

$$B \rightarrow B^{z+} + Ze \qquad (24\text{-}23)$$

Then, in an analogous manner to the derivation of Eq. (24-21), the *reversible electrode potentials* for half-cell reactions (24-22) and (24-23), respectively, may be calculated from

$$e_1 = e_1^0 + RT/zF \ln C_A/C_{A^{z+}} \qquad (24\text{-}24)$$

$$e_2 = e_2^0 + RT/zF \ln C_B/C_{B^{z+}} \qquad (24\text{-}25)$$

Where $e_1$ and $e_2$ are the reversible electrode potentials for half-cell reactions (24-22) and (24-23), respectively, and $e_1^0$ and $e_2^0$ are *standard electrode potentials.* The cell emf is given by

$$E = e_1 - e_2 \qquad (24\text{-}26)$$

and the standard cell emf is given by

$$E^0 = e_1^0 - e_2^0 \qquad (24\text{-}27)$$

Equation (24-21) can then also be obtained from Eqs. (24-24) through (24-27).

For solid electrodes, the activity of the metal is taken as unity, such that the electrode potentials for half-cell reactions (24-22) and (24-23), respectively, are

$$e_1 = e_2^0 - RT/zF \ln C_{A^{z+}} \qquad (24\text{-}28)$$

$$e_2 = e_2^0 - RT/zF \ln C_{B^{z+}} \qquad (24\text{-}29)$$

***19. The Zero of Potential.*** Only the cell emf is able to be experimentally measured, and the electrode potentials must be calculated from these measurements. Therefore, for comparison, we adopt an arbitrary zero of potential.

The zero of potential at any temperature is the potential corresponding to the reversible equilibrium between hydrogen gas at 1 atm pressure and hydrogen ions at unit activity, i.e., for the half-cell reaction

$$H_2 \rightarrow 2H^+ + 2e \qquad (24\text{-}30)$$

The potential of the reaction represented by Eq. (24-30) is the standard reversible hydrogen electrode potential and is arbitrarily set to the value of zero.

***20. Standard Electrode Potentials.*** Table 24-3 lists selected standard electrode potentials. The table contains values for elements and for redox couples where a valence change occurs.

***21. Electromotive Series.*** The elements in Table 24-3 embody the electromotive series of the elements, and this lists the order of reactivity predicted from thermodynamic considerations of the elements. For example, the reactive metals (i.e., those with a strong tendency to form ions) are on the negative end of the series and have increasingly large negative values of the electrode potential. Conversely, the noble metals such as platinum and gold have high positive electrode potentials. The electromotive series represents the ionizing tendency of metals, and a more reactive metal can be expected to displace a noble metal from a solution of its ions. The series also indicate the spontaneous direction of electrochemical reactions.

For example, for the reactions represented by Eqs. (24-10) and (24-11), Table 24-3 shows

**TABLE 24-3** Standard Reversible Electrode Potentials at 25°C

| Species in equilibrium | $e^0$, V | Species in equilibrium | $e^0$, V |
|---|---|---|---|
| $Li^+/Li$ | −3.05 | $Pb^{2+}/Pb$ | −0.130 |
| $K^+/K$ | −2.92 | $H^+/H$ | 0 |
| $Ba^{2+}/Ba$ | −2.90 | $Cu^{2+}/Cu$ | +0.34 |
| $Ca^{2+}/Ca$ | −2.87 | $O_2/OH^-$ | +0.4 |
| $Na^+/Na$ | −2.71 | $Cu^+/Cu$ | +0.52 |
| $Mg^{2+}/Mg$ | −2.37 | $I_2/I^-$ | +0.54 |
| $Be^{2+}/Be$ | −1.85 | $Fe^{3+}/Fe^{2+}$ | +0.77 |
| $Al^{3+}/Al$ | −1.66 | $Hg^{2+}/Hg$ | +0.79 |
| $Ti^{2+}/Ti$ | −1.63 | $Pd^{2+}/Pd$ | +0.98 |
| $Mn^{2+}/Mn$ | −1.18 | $Br_2(l)/Br^-$ | +1.073 |
| $Zn^{2+}/Zn$ | −0.76 | $Cr_2O_7^-/Cr^{3+}$ | +1.33 |
| $Cr^{3+}/Cr$ | −0.74 | $Cl_2(g)/Cl^-$ | +1.36 |
| $Ga^{3+}/Ga$ | −0.52 | $Ce^{4+}/Ce^{3+}$ | +1.61 |
| $Fe^{2+}/Fe$ | −0.44 | $Au^+/Au$ | +1.68 |
| $Cr^{3+}/Cr^{2+}$ | −0.41 | $F_2/F^-$ | +2.9 |
| $Cd^{2+}/Cd$ | −0.40 | | |
| $Ti^{3+}/Ti^{2+}$ | −0.37 | | |
| $Co^{2+}/Co$ | −0.28 | | |
| $Ni^{2+}/Ni$ | −0.25 | | |
| $Cu^{2+}/Cu^+$ | −0.15 | | |
| $Sn^{2+}/Sn$ | −0.14 | | |

that zinc will preferentially dissolve and form zinc cations, while cupric cations in the solution will deposit on zinc metal. Therefore, the overall cell reaction is

$$Zn + Cu^{2+} \rightarrow Zn^{2+} + Cu \qquad (24\text{-}31)$$

It should be noted that there are several limitations on the usefulness of Table 24-3. First, it indicates the order of reactivity as calculated from thermodynamic considerations; kinetic effects often modify the expected reaction. Complicating side reactions may also occur which confuse the situation and which are not readily understood. In addition, the potentials in this table apply exclusively to solutions at unit activity, whereas practical solutions seldom correspond to these conditions; under practical conditions the electrode potentials must be calculated from Eq. (24-28).

However, the electromotive series is extremely valuable for predicting electrochemical reactions occurring in practical electrochemical cells. Note that Table 24-3 relates to aqueous solutions; different values and indeed different potential scales are used for organic solvents, molten salts, and solid electrolytes.

***22. Calculation of Standard Cell EMF.*** $E^0$ can be calculated from the standard free energy. For example, the standard free energy for the reaction

$$H_2 + \tfrac{1}{2}O_2 = H_2O \qquad (24\text{-}32)$$

is 54.6 kcal/g·mol. The standard emf is then

$$E^0 = 56.6/2 \times 23.06 = 1.23 \text{ V}$$

By definition, $e^0$ for the hydrogen electrode is zero and, therefore, $e^0$ for the oxygen electrode is 1.23 V.

Similarly, from Table 24-3, $e^0$ for the zinc electrode is $-0.762$ V and for the Cu electrode is $+0.34$ V. Therefore, the emf of the cell from Eq. (24-27) is

$$E = 0.34 - (-0.76) = 1.1 \text{ V}$$

***23. Acidity and pH.*** Water dissociates slightly into ions by the equation

$$2H_2O \rightleftarrows H_3O^+ + OH^- \tag{24-33}$$

An equilibrium constant $K$ can be written for reaction (24-33) such that

$$K = a_{H_3O^+} \times a_{OH^-} / a_{H_2O} \tag{24-34}$$

where the $a$ terms are the activities of the species as defined in Eq. (24-18). Since $a_{H_2O}$ is essentially constant, the *ionic product of water* ($K_w$) is defined as

$$K_w = a_{H_3O^+} \times a_{OH^-} \tag{24-35}$$

At 25°C

$$K = 1 \times 10^{-14} \tag{24-36}$$

and, since from Eq. (24-33)

$$a_{H_3O^+} = a_{OH^-} \tag{24-37}$$

then

$$a_{H_3O^+} = 10^{-7} \text{ g-ions/L} \tag{24-38}$$

This is an extremely small number, and, therefore, for convenience, we define the term pH, which is the logarithm to the base 10 of the reciprocal of the hydrogen ion concentration in gram-ions per liter. From Eq. (24-38), the pH of water is 7. At this pH, the solution is said to be neutral; solutions having pH values less than 7 are acidic, and they are alkaline if the pH is more than 7.

***24. Buffer Solutions.*** It is often necessary in electrochemical measurements to use a solution of known and constant pH. Such a solution is called a buffer solution. They are commonly prepared from a weak acid and one of its salts or a weak base and one of its salts, and the effective solution pH can be adjusted by changing the concentration ratio of the components. Common buffer solutions are sodium acetate and acetic acid (pH 3.7 to 5.6), borax and sodium hydroxide (pH 9.2 to 11.0), and sodium citrate and sodium hydroxide (pH 5.0 to 6.7).

## Voltage Relationships for Galvanic Cells

***25. Galvanic Cells.*** Galvanic cells differ from potentiometric cells in that a net current is produced (or used) in the electrochemical reaction. *Batteries* and *fuel cells* generate current, while current is used in electrolytic cells to carry out the processes of *electrodeposition, electrorefining, electrowinning,* and *electrosynthesis.*

In energy-producing cells, the cell voltage during current flow is less than the thermodynamic emf. In contrast, in an electrolytic cell, the applied voltage is greater than the thermodynamic emf.

There are at least four contributions to the working voltage of a galvanic cell: (1) the reversible cell emf calculated from thermodynamic data ($E$), (2) the *activation overpotential* ($\eta_{act}$), (3) *concentration polarization* ($\eta_{conc}$), and (4) *ohmic polarization* ($\eta_{ohm}$).

The voltage ($V$) of an electricity-producing cell is

$$V = E - (\eta_{act} + \eta_{conc} + \eta_{ohm}) \tag{24-39}$$

and the voltage necessary to produce a chemical reaction in an electrolytic cell is

$$V = E + (\eta_{act} + \eta_{conc} + \eta_{ohm}) \tag{24-40}$$

***26. Activation Polarization.*** Chemical reactions involve an energy barrier that must be overcome by the reacting species. This activation energy results in activation polarization $\eta_{act}$ which is related to the current $i$ by the *Tafel equation*

$$\eta_{act} = a + b \log i \tag{24-41}$$

where $a$ and $b$ are constants obtained from the $\eta$ vs. log $i$ plot. Activation polarization is related to the slow step in the electrochemical reaction sequence. As temperature increases, activation polarization decreases.

***27. Concentration Polarization.*** Concentration polarization occurs when the electrode reaction is hindered by transport of the reacting ions through the electrolyte to the electrode surface. When the rate of the electrode reaction is completely governed by transport processes, the maximum (or *limiting*) current $i_L$ is reached.

The limiting current $i_L$ can be calculated from the diffusion coefficient $D$ of the reacting ions, the concentration of the reacting ions $C_B$ and the thickness of the diffusion layer $X$ by applying Fick's law:

$$i_L = zDFC_B/X \tag{24-42}$$

For an electrode free from activation polarization, the concentration overpotential is

$$\eta_{conc} = RT/zF \ln (1 - i/i_L) \tag{24-43}$$

***28. Total Electrode Polarization.*** The total electrode polarization is the sum of the activation and concentration polarizations

$$\eta = \eta_{act} + \eta_{conc} \tag{24-44}$$

***29. Ohmic Polarization.*** When an anode and cathode are connected to an external load, the circuit is complete and current flows. Resistance to conduction of ions through the electrolyte or electrode, and by contact resistances (between components), causes ohmic or $iR$ polarization. Because both electrodes and the electrolyte obey Ohm's law, the ohmic polarization is

$$\eta_{ohm} = iR \tag{24-45}$$

***30. Exchange Current ($i_0$).*** Current is related to the electrochemical reaction rate $r$ by

$$i = zFr \tag{24-46}$$

At equilibrium, the forward and reverse reaction rates of an electrochemical reaction are equal. They produce currents equal in magnitude but opposite in sign. The magnitude of the currents is called the equilibrium *exchange current.* High exchange currents indicate that the electrochemical reaction rate is high. Note that there is no net current, but the exchange current recognizes that equilibrium is a dynamic process.

The exchange current is measured from the Tafel equation, Eq. (24-35), by extrapolating the $\eta_{act}$ vs. log $i$ plot to $\eta = 0$.

***31. Total Cell Polarization.*** Cell polarization is the sum of the activation and concentration polarizations at the two electrodes and of the ohmic polarization.

Polarization cannot be eliminated, but it can be minimized by cell design. The *iR* loss can be reduced by minimizing electrode spacing and by efficient contacts between electrode and leads. Temperature and electrolyte composition also influence the cell resistance.

Concentration polarization is dependent on the mass transport properties of the system. Mass transfer is a function of temperature, pressure, concentration, and the physical characteristics of the system (i.e., stirring, etc.).

Activation polarization depends upon the fundamental electrochemical reaction rate, which is also a function of temperature, pressure, and concentration.

***32. Bibliography on Fundamental Principles***


Bockris, J. O'M., and Reddy, A.K.N. *Modern Electrochemistry,* New York, Plenum Press, 1970.

MacInnes, D.A. *The Principles of Electrochemistry,* New York, Dover Publications, Inc., 1961.

Potter, E.C. *Electrochemistry,* New York, The Macmillan Company, 1961.

Selley, N.J. *Experimental Approach to Electrochemistry,* New York, John Wiley & Sons, Inc., 1977.

Yeager, E., and Salkind, A.J. (eds.). *Techniques of Electrochemistry,* New York, John Wiley & Sons, Inc., 1973.


## USE OF ELECTRICAL ENERGY

***33. Electrolytic Reactions.*** Technical electrolytic processes may be either cathodic or anodic depending on which electrode contributes to the important reaction. Included in the former class are electroplating (including electroforming and electrotyping), electrowinning, electrorefining, and water electrolysis. Anodic processes include inorganic and organic synthesis and electropolishing.

### Electrodeposition

***34. Objectives.*** The aim of electroplating is to deposit metal on a part in order to provide a decorative or protective coating. It is also used to modify the surface properties of the substrate to improve solderability, reduce contact resistance, increase surface conductivity and reflectivity, and improve hardness and wear resistance. The selection of the electrodeposited metal depends on the application, and several are listed in Table 24-4.

***35. General Principles of Electroplating.*** The primary reaction is the deposition of metal

**TABLE 24-4** Selection of Plated Coatings

| Application | Platings Generally Useful |
|---|---|
| Corrosion prevention | Zinc, cadmium, lead |
| Decoration-protection: general. | Copper + nickel + chromium |
| Decoration-protection: special colors and effects | Bronze, brass, precious metals, tin-nickel alloy |
| Light reflectance | Rhodium, silver |
| Solderability | Tin, tin-lead alloy, cadmium, silver, gold |
| Bearing surfaces | Tin, tin-lead alloy, chromium, indium |
| Building up worn parts | Nickel, iron (with or without copper) |
| Hard surfacing | Chromium |
| Electrical contacts | Gold, silver, platinum group |
| Foods and beverages | Tin |
| Rubber bonding to metals | Brass |
| Stop-off in nitriding | Bronze, copper, tin |

from ions in solution. The part that is to receive a coating is made cathodic. The anode consists of either the same metal as the material being plated or it is inert. In the former case, the same amount of metal dissolves at the anode as is electroplated onto the cathode (at 100% current efficiency), and there is no need to add additional metal salt to the electrolyte. When an inert anode is used, additional salt is added to maintain the plating bath concentration constant.

In a practical electroplating solution, a constant current is applied across the anode and cathode with a regulated dc power supply such that the deposition occurs at a known rate. The voltage at the electrodes in the plating cell is relatively low and is generally between 1 and 6 V.

The voltage applied must be greater than the decomposition voltage of the electrolyte, which is equal to the sum of the thermodynamic reversible voltage and the cell polarization.

Since electroplating is used both for decoration and for protection of the base metal, it is important to plate a coating which is adherent, coherent, continuous, uniform, and decorative. With noble metal coatings, it is also necessary to achieve these properties within economic (i.e., thin) coating thicknesses. The practical operating conditions are selected to achieve the deposit with these characteristics.

***36. Preparation for Plating.*** The preparation of an object for plating involves: (1) the removal of oil, grease, or other organic matter; (2) the removal of rust, scale, oxides, or other inorganic coatings adhering to the metal; and (3) the mechanical preparation of the surface of the metal by polishing.

Organic material is usually removed by soaps, hot alkali solutions, or organic solvents such as chlorinated hydrocarbons (vapor degreasing to maintain contact between part and fresh solvent). The removal of inorganic material is carried out by immersion in acids and/or alkalis, or by using electrolytic cleaning in hot, alkaline solutions where the chemical action of the solution is supplemented by the mechanical effect of vigorous gas evolution at the surface. In general, the cleaning process involves proprietary formulations.

***37. The Plating Solution.*** The various constituents of the plating solution are selected to: (1) provide the metal ion in sufficient concentration, (2) increase the solution conductivity, (3) increase anode corrosion, (4) control the solution pH, and (5) modify the character of the deposit.

Plating solutions having insufficient ionic content will result in burned dendritic or powdery deposits at the usual operating current density. As a result, monitoring of the electroactive ion is an important procedure.

The free (i.e., noncomplexed) metal ion is often controlled at low values for specific plating bath operations. For example, when plating on zinc die castings from a copper sulfate solution, the potentials of zinc and copper are sufficiently far apart such that copper deposits on the zinc without the application of external potential; the resulting coating is unsatisfactory. By use of a complexing agent, cyanide ion, the free copper ion concentration is reduced by several orders of magnitude, so that copper no longer deposits on zinc unless an external potential is applied; good copper deposits are then obtained.

It is necessary to use a solution of high conductivity in order to minimize the power consumption owed to ohmic polarization in the solution, and also to increase the throwing power of the bath and reduce its tendency to form dendrite and rough deposits.

Anions, such as chloride ions, are often added to the bath to assist in anode corrosion where the anode is the same metal as the metal being deposited, so that the cell voltage is minimized (i.e., the anode polarization is decreased).

Solution pH must often be kept within narrow limits in order to avoid, on one hand, the excessive evolution of hydrogen and, on the other, the precipitation of basic salts or hydroxides in the electrolyte. Close regulation of the pH is generally accomplished by the use of buffered solutions.

Addition agents, usually organic compounds, are added to the electrolyte in small amounts relative to the other constituents in the solution to modify the character of the deposit. They promote the formation of cohesive, fine-grained deposits and minimize the formation of dendrites and nodules. In addition, they also permit the formation of bright deposits which require no mechanical polishing after plating, and level deposits which are smoother than the basis metal and, therefore, allow the use of a fairly rough substrate. The mechanism of these additives and plating baths is not fully clear, but it is known that they are consumed during the plating process and must be replenished periodically. It is probable that they become selectively absorbed on the cathode during the plating process, thus inhibiting the growth of certain crystal faces and promoting the formation of new nuclei at the expense of the growth of those already formed.

***38. Variations in Operating Conditions.*** In addition to the solution composition, variations in operating conditions also affect the character of the metal deposit. Those of particular importance are current density, temperature, agitation, cell design and throwing power, and the structure of the basis metal.

***39. Current Density.*** In general, it is desirable to operate at the maximum current density in order to deposit metal in the shortest possible time. Most plating solutions exhibit a range of current density over which the deposit is satisfactory in brightness and adhesion, and outside this range the deposit is unsatisfactory for a number of reasons. In a practical electroplating cell, this current density range must not be too narrow because it is not feasible to control current densities too rigidly, particularly with cathodes of irregular shapes. Deposits produced at too high a current density are dark, loosely adherent or powdery, and are said to be burned.

Current density is also important at the anode since too high a current density may cause passivation and result in gas evolution rather than metal dissolution as the anodic reaction.

***40. Temperature.*** A low electrolyte temperature favors formation of small crystals, and high temperature favors large crystal formation. In some cases, a temperature difference of only 15°C results in a 50% decrease in the metal deposit strength. On the other hand, high temperature often gives beneficial results because of: (a) increased salt solubility, thus permitting greater metal concentration and higher current densities; (b) increased conductivity, which also permits higher current density and reduces the tendency to form dendrites; and (c) decreased occlusion of hydrogen in a deposited metal. Since (a) and (b) tend to decrease crystal size, they may in some cases counteract the tendency of temperature alone to increase the crystal size.

***41. Agitation.*** Since the limiting current density depends on ionic transport to the cathode surface, agitation is important to permit higher plating rates. In practice, there is a limit to its usefulness in that it tends to stir up sediment and increases the adsorption of atmospheric carbon dioxide. Agitation is generally brought about by cathode agitation through the solution, particularly in automatic plating machines and continuous strip and wire plating. In still-tank plating, agitation may be accomplished by stirrers or by use of air. Agitation is often combined with continuous filtration.

***42. Cell Design and Throwing Power.*** In plating irregular-shaped objects, parts of the substrate are at higher current densities than other areas, and consequently more metal is deposited in these high-current-density areas. Since specifications for plating often call for a minimum thickness of metal, the excess metal plated upon the high-current-density areas represents a waste of metal, current, and time.

As a result, the cell and electrolyte composition and operating conditions are controlled to give good *throwing power.* This factor describes the ability of a cell system to deposit a uniform coating thickness over all the substrate area. There is no fully accepted measure of throwing power, and relative terms such as good, poor, etc., based on subjective judgment, are usually used to describe this property.

One way to achieve good throwing power is by modifying additives or by decreasing the ionic concentrations. Another is to shape the anodes to follow the main contours of the cathodes or to place the anodes further away from the cathodes so that the distance between an anode and the remotest areas of an adjacent cathode is proportionally little different from that between the anode and protruding areas of the same cathode.

***43. Structure of the Basis Metal.*** The substrate has a decided effect on the structure of the deposited metal, and defects in the former are reflected as defects in the plating. It is usually necessary, therefore, to give some mechanical or chemical pretreatment of the substrate to be plated prior to application of the deposited metal.

***44. Electroplating Applications.*** Table 24-5 lists a series of operating conditions for electroplating many of the common metals. It should be recognized that this is a condensed summary and almost all metals in this table can be plated by proprietary processes for which various advantages are claimed over these standard solutions.

***45. Alloy Plating.*** In order to plate two metals simultaneously, it is necessary that their deposition potentials be essentially equal. From Table 24-3, it can be seen that the standard electrode potential for the copper-copper(II) couple is over 1 V more anodic than that of the zinc-zinc(II) couple. Therefore, copper and zinc cannot be electrodeposited simultaneously from acid solution. As a result, brass (copper and zinc) alloy is electrodeposited by complexing the copper(II) ion with cyanide such that its concentration is low. Then by Eq. (24-28), the deposition potentials of copper and zinc will be essentially equal. For similar reasons, tin bronzes are electroplated from cyanide-stannate electrolytes, tin-nickel (65% tin) from chloride-fluoride solutions, and tin-lead from fluoborate solutions.

Many binary and ternary alloy systems have been investigated with the objective of producing alloys by electroplating. However, only a few have proven of commercial importance because the production of alloys by electroplating is complicated and requires careful control. However, in addition to the materials mentioned above, there is much interest in forming ferromagnetic films for the production of magnetic storage disks.

***46. Electroforming.*** Electroforming is the production of articles by electrodeposition upon a mandrel or mold. The mandrel is subsequently removed from the deposit, which then becomes the finished electroformed article.

The oldest and most used variety of electroforming is electrotyping, and the process is becoming of increasing importance as a means of manufacturing components of unusual shape or to close tolerances. Typical of such articles are venturi nozzles, waveguides, and reflectors. The production of master and stampers for phonograph records is another well-established use that illustrates the fine detail that electroforming is capable of.

Although any metal that can be electroplated can be electroformed, the principal emphasis is on nickel and copper. Molds or mandrels may be expendable (used once) or permanent and may be metallic or nonmetallic. The surface of a metallic mold is treated to prevent the adhesion of the deposited metal, so that the finished object can be removed; a nonmetallic mold is treated to render its surface conductive.

***47. Electrotyping.*** The objective of electrotyping is to reproduce printing, setup type, engravings, medals, etc. A mold of the object to be reproduced is first made, for example, of wax, by pressing the object in wax. If the mold is a nonconductor of electricity, as with wax, its surface is made conductive by: (1) coating with graphite, (2) pouring copper sulfate solution over the surface of the mold and dusting on finely divided iron filing to form a metallic copper coating, or (3) interposing the wax between a cathode and an anode of the metal to be plated and passing a high-tension discharge. By suitable electrical connections, the surface of the mold is then made a cathode in an acid copper electroplating bath. In the case of reproducing type matter, two cases containing prepared molds are always suspended back-to-back between two large copper anodes so that the conducting surfaces of the molds directly face the anodes. The copper shell is then separated from the mold on which it is deposited, and in order to give it the necessary strength for further use, it is backed with type metal.

***48. Brush Plating.*** Plating can sometimes be carried out in the field by using a brush wet with electrolyte; the technique is also useful for plating localized areas without the need to stop off the remainder of the piece.

***49. Continuous Plating.*** The tin plating of continuous coils of sheet steel in mechanized equipment was given impetus during World War II by the need to conserve tin; at present, all but a very minor proportion of tin plate is produced electrolytically, and electrotinning repre-

**TABLE 24-5** Summary of Electroplating Practice

| Metal (1) | Principal uses (2) | Type of solution (3) | Principal ingredients (4) | Temp, °C (5) | CD. A/m² (6) | Volts (7) | Cathode efficiency, % (8) | Throwing power (9) | Time to deposit 25 μm (10) |
|---|---|---|---|---|---|---|---|---|---|
| Cadmium | Protection | Cyanide | CdO, NaCN, brighteners | 20–35 | 150–450 | 1–4 | 90 | Good | 20 min |
| Chromium | Decorative<br>Engineering (hard)<br>Cylinder liners (porous) | Chromic acid | $CrO_3$, $H_2SO_4$ | 50 | 2500 | 6–8 | 15 | Poor | 2 h |
| Copper | Electroforming<br>Undercoat for other metals<br>Stop-off in casehardening, etc. | Acid<br>Cyanide<br>Rochelle<br>Many other types, e.g., fluoborate, pyrophosphate, all-potassium cyanide | $CuSO_4.5H_2O$, $H_2SO_4$<br>CuCN, NaCN, $Na_2CO_3$<br>Above + rochelle salts | 25–50<br>25–38<br>60–70 | 150–400<br>50–150<br>200–600 | 1–2<br>1.5–3<br>2–3 | 100<br>50<br>60 | Fair<br>Good<br>Good | 35 min<br>90 min<br>45 min |
| Gold | Decorative<br>Electronics | Cyanide<br>(Solutions vary considerably, depending on application) | $KAu(CN)_2$, $K_2CO_3$, KCN | 50–70 | 50–150 | 2–6 | 80 | Good | |
| Indium | Bearing surfaces | Cyanide<br>Sulfate<br>Fluoborate | $InCl_3$, NaCN, addition agent<br>$In_2(SO_4)_3$, $Na_2SO_4$<br>$In(BF_4)_3$, $H_3BO_3$, $NH_4BF_4$ | Room<br>Room<br>20–32 | 100–1500<br>200<br>500–1000 | | 40<br>75<br>50 | Good<br>Poor<br>Good | |
| Iron | Electroforming<br>Repair | Chloride<br>Sulfate | $FeCl_2$, $CaCl_2$<br>$FeSO_4(NH_4)_2SO_4$ | 85<br>Room | 600<br>200 | | 95<br>95 | | 20 min<br>1 h |
| Lead | Protection<br>Bearing surfaces | Fluoborate | $Pb(BF_4)_2$, $HBF_4$, glue | Room | 100–800 | 0.5 | 100 | Good | 40 min |
| Nickel | Protection<br>Decorative<br>Electroforming<br>Undercoat for Cr, etc. | Sulfate-chloride<br>Sulfamate<br>Fluoborate | $NiCl_2$, $NiSO_4$, $NH_4$ ion, $H_3BO_3$<br>(Formulations differ widely, depending on purpose)<br>Ni sulfamate, sulfamic acid<br>$Ni(BF_4)_2$, $HBF_4$, addition agents | 25–38 | Varies greatly | 0.5–3 | 95 | Fair | 30 min |
| Rhodium | Decorative<br>Optical | Sulfate<br>Phosphate | Prepared salts | 43–50 | 100–800 | 2.5–5 | 15 | | |

| | | | | | | | | | |
|---|---|---|---|---|---|---|---|---|---|
| Silver | Decorative<br>Protective<br>Bearing surfaces | Cyanide | AgCN, KCN, $K_2CO_3$, $CS_2$ (or Na in place of K) | 27 | 50–150 | 1 | 100 | Good | |
| Tin | Protection<br>Food and dairy<br>Bearings<br>Electrical<br>To enable easy soldering | Sulfate<br>Fluoborate<br>Other acid electrolytes<br>Stannate | $SnSO_4$, addition agents<br>$Sn(BF_4)_2$, $HBF_4$, addition agents<br><br>$Na_2$- or $K_2Sn(OH)_6$, Na- or KOH | Room<br>25–38<br><br>65–88 | 400<br>500<br><br>400 | 1–3<br><br><br>4–8 | 90<br>100<br><br>80 | Fair<br>Good<br><br>Excel-lent | 15 min<br>10 min<br><br>30 min |
| Zinc | Protection | Sulfate<br>Cyanide | $ZnSO_4$, $NH_4Cl$, addition agents<br>$Zn(CN)_2$, NaCN, NaOH, brighteners | 25–38<br>38 | 150–4000<br>100–500 | | 99<br>85 | Fair<br>Good | 10 min<br>40 min |
| | | | Alloys | | | | | | |
| Brass | Rubber-bonding<br>Decorative | Cyanide | CuCN, $Zn(CN)_2$, NaCN, $Na_2CO_3$ | 25–38 | 30–100 | 2–3 | 75 | Good | |
| Bronze | Decorative<br>Undercoat for chromium<br>Stop-off for steel | Cyanide-stannate | CuCN, KCN, KOH, $K_2Sn(OH)_6$, rochelle salt | 68 | 200–1000 | 3–6 | 70 | Excel-lent | 30 min |
| Lead-tin | Bearings<br>Solderability<br>Electrotyping | Fluoborate | $Sn(BF_4)_2$, $Pb(BF_4)_2$, $HBF_4$, addition agents | Room | 600 | 1–2 | 100 | Good | |
| Tin-zinc | Solderability<br>Substitute for cadmium | Cyanide-stannate | $Zn(CN)_2$, KCN, KOH, $K_2Sn(OH)_6$ | 65 | 100–750 | 4–5 | 80–95 | Excel-lent | 30 min |
| Tin-nickel | Printed circuits | Chloride-fluoride | $NiCl_2$, $SnCl_2$, $NH_4HF_2$, HF | 65 | 250 | 1–2 | 98 | Excel-lent | 30 min |

sents the largest single application of electroplating. On a somewhat smaller scale, other metals are continuously plated on steel strip, steel and copper wire, electrical terminals, and other substrates.

***50. Testing of Deposits.*** The specifications for the performance of electroplates is becoming more rigorous, and greater attention is directed toward testing methods. The principal properties subjected to test are

*Corrosion Resistance.* Although many accelerated weathering tests have been devised, correlation between the results of these tests and actual performance in the expected service must be determined before placing reliance on them. Thus, the neutral salt spray,[1] although appearing in many specifications, is often of questionable validity. For the specific case of automotive brightwork plated with copper-nickel-chromium, the CASS and Corrodkote[2] tests have wide acceptance and appear to predict service performance reasonably well. Outdoor exposure tests on stationary racks and observation of parts in actual service, although time consuming, are still required to confirm indications from accelerated tests.

*Porosity.* The presence of pores in a metallic coating affects its protective value, depending on whether the coating is cathodic or anodic to steel. Pores in a zinc coating do not greatly detract from its protective value, because the zinc protects the steel "sacrificially"; i.e., in the galvanic cell, which is set up under corrosive conditions, it is the zinc and not the steel that corrodes. If the metallic coating is cathodic to steel, the steel will preferentially corrode, and the steel may corrode more seriously about a pore than if it were not coated at all. The salt-spray test is also often a test of porosity. Pores may also be detected by the hot-water test and the ferroxyl test; each pore shows in the first test as a rust spot and in the second as a blue coloration.

*Thickness.* Other things being equal, the performance of an electroplate correlates fairly well with its thickness. Physical tests for thickness measurement involve x rays, beta rays, or other electromagnetic radiation reflected from the substrate, and several instruments and methods are based on this principle. Chemical tests rely on dissolving the coating in a standard reagent and either weighing the piece before and after stripping the coating or analyzing the stripping solution for the coating metal (tin plate is usually tested in this way). An alternative method is microscopic examination of a cross section, but this method is subject to many errors unless stringent precautions are taken. Several methods of thickness testing are detailed in ASTM specifications.

*Adhesion.* The quantitative measure of the force with which an electroplate adheres to the basis metal has been the subject of considerable study. However, most tests in actual use are rather qualitative and depend on the observer's experience. A coating is rated as either sufficiently adherent for the purpose or nonadherent, and no attempt is made to express the adhesion quantitatively.

*Other Properties.* Other tests are hardness, abrasion resistance, solderability, lacquer or paint adhesion, or such other characteristic as may be important in a particular application. Anodic coatings on aluminum are tested by electrical resistance measurements.

***51. Electroplating from Organic Solvents.*** Because certain technologically important coatings (e.g., aluminum, titanium) cannot be electrodeposited from aqueous solutions, research work has been proceeding to electroplate these metals from organic solvents. In this case, the solvent is an organic compound rather than water, and may be, for example, propylene carbonate, dimethyl sulfoxide, diethylether, alcohols, and nitrobenzene, to which a suitable salt is added to improve its conductivity. These processes are not common outside the research laboratories because of the need to totally restrict water from the system. However, pilot plant systems have been built for aluminum and are used in small-scale production situations.

---

[1] ASTM Designation B117.
[2] ASTM Designations B368 and B380.

***Bibliography on Electroplating***

Delahay, P.D., and Tobias, C.W. (eds.). *Advances in Electrochemistry and Electrochemical Engineering,* vol. 5, New York, John Wiley & Sons, Inc., 1967.

Graham, A.K. (ed.). *Electroplating Engineering Handbook,* 3d ed., New York, Van Nostrand-Reinhold Company, 1971.

Lowenheim, F.A. (ed.). *Modern Electroplating,* 3d ed., New York, John Wiley & Sons, Inc., 1974.

*Metal Finishing Guidebook Directory,* Hackensack, N.J., Metals & Plastics Publications, issued annually.

## Electrolytic Refining

***52. Fundamental Principles.*** Electrorefining is the purification of crude metals to commercially desirable levels. The crude metal is the anode in a suitable electrolyte, and the metal is simultaneously plated out as the cathode of the electrolytic cell. The principle of the process is that the impurities are either not dissolved and form a sludge in the electrolyte, or, if the impurities do dissolve, they are not plated out on the cathode.

Two types of cells are used in electrolytic refining: *monopolar* and *bipolar* systems. In the monopolar system, anodes and cathodes are placed alternately in the cell and electrically connected in parallel. With this arrangement, alternate electrodes in the bath dissolve, and refined metal is deposited on the others.

In the bipolar system, the arrangement of the electrodes in the cell is essentially the same, but only the two electrodes at opposite ends of the cell are directly placed in the electrical circuit. The intervening metal sheets immersed in the electrolyte act as bipolar electrodes, i.e., one side of each acts cathodically and the other acts anodically. Therefore, the cathodic side of each bipolar electrode receives a metal deposit, while the anodic side dissolves.

In general, electrolytic refining is more expensive than other refining processes. However, there are usually other factors to make this an economic process. For example, the values of precious metals recovered in copper refining more than offset the cost of the method.

***53. Copper Refining.*** The electrolyte is a copper sulfate solution containing free sulfuric acid. The copper content usually is between 3 and 3.75%, and the free sulfuric acid is between 15 and 18%. An anode of typical composition contains 99 to 99.3% copper, with impurities of silver, gold, and arsenic. The cathode, copper, is exceedingly pure, usually running about 99.98%, with hydrogen as the chief impurity. In order to have high electrical conductivity, copper must be essentially free from arsenic and antimony, while in order to prevent brittleness, copper must be free from tellurium and lead. Since the electrical conductivity of copper is very sensitive to impurities, this is commonly used as a measure of the purity.

The electrolyte is usually circulated, and this is more important as the current density increases. The current density is between 160 and 280 $A/m^2$, cell voltages are 100 to 200 V, and the electrolyte is heated. Soluble sulfates of impurities in the anode pass into the solution, which, therefore, needs purification at intervals. This is usually done by working up a certain quantity of electrolyte regularly into bluestone and adding fresh acid to the electrolyte.

***54. Silver Refining.*** The raw material for the production of pure silver consists of anode slimes from copper refining (mainly); anode slimes from the electrolysis of lead, nickel, and zinc; and concentrates obtained from the desilverization of lead.

The anode slimes are allowed to settle in order to remove the excess electrolyte; they are then washed, filtered, roasted, and leached in sulfuric acid to remove the copper. After again filtering and washing, the slimes are melted and refined in a small reverberatory furnace—a doré furnace. The doré metal is cast into anodes for electrolytic parting of the silver and gold by the *Balbach* or *Moebius* system.

In the Balbach process, the cells are mastic-lined concrete tanks with graphite slabs fitted to the bottom to form the cathode. The doré-metal anodes are supported in wooden frames above the cathode, and the frames are enclosed in a cloth case. During electrolysis, the silver is deposited on the cathode and the gold is collected as a slime in the cloth cases. A current density of 400 to 500 $A/m^2$ of anode is used. At this current density, the voltages range from 3 to 3.8 V per tank, and a current efficiency of 93 to 95% is obtained. The energy requirements are about 1 Wh/g of silver produced.

In the Moebius system, the anodes and cathodes are arranged vertically, and the cathodes are thin sheets of stainless steel. The doré-metal anodes are encased in a cloth bag to collect the gold slimes. Mechanically operated wooden scrapers brush the silver crystals from the cathodes into trays. A current density of about 500 $A/m^2$ of anode surface is employed. At this density, the voltage is about 2.7 V per tank, and a current efficiency of 98% is obtained.

***55. Gold Refining.*** Gold refining by the *Wohlwill* process with recovery of platinum and palladium employs a hot acid solution of gold chloride (7 to 8% gold, 10% hydrochloric acid) at a temperature of 65 to 70°C. The applicability of the Wohlwill process to alloys richer in silver has been achieved by employing pulsating current (obtained by superposing an alternating current on the direct current) instead of a purely direct current.

The gold cells are much smaller than the silver cells; approximately 300 by 400 mm and 300 mm deep. The anodes are cast slabs, suspended from the anode supports through a hole in the top of the anode. The current density is 500 to 700 $A/m^2$ of cathode surface. The cathodes are strips of fine gold rolled to a thickness of 0.25 mm and are removed for melting when they weigh 4 to 6 kg.

***56. Lead Refining.*** The *Betts* process employs a solution of lead fluosilicate, containing an excess of hydrofluosilicic acid (7 to 9% Pb and 5 to 8% free acid) with a very small addition of gelatin or glue, depositing lead in dense, coherent form and free from bismuth. The voltage per tank is 0.35 to 0.6 V, the temperature is 40°C, the energy consumption is from 105 to 120 kWh per metric ton of lead. The efficiency is about 90%, and the current density 160 to 220 $A/m^2$.

***57. Nickel Refining.*** Crude metal anodes contain about 95% nickel, 2.5% copper, and other impurities such as iron, cobalt, and sulfur. Since the impurities also dissolve in the electrolyte and would redeposit at the cathode, *diaphragm cells* are used and the anolyte and catholyte systems are kept separate, the anolyte being chemically purified before being returned to the system as catholyte.

## Electrowinning

***58. Fundamental Principles.*** Electrowinning is the extraction of metal from its ores in part by electrolysis. It is similar to electroplating in that the metal-containing ore is "dissolved" in a suitable electrolyte and the metal electroplated onto a cathode. In general, the prepared ore is treated with an aqueous electrolyte to leach out the compounds of the metal into an ionic solution. The electrolyte may then undergo further purification before being passed into the electrolytic cell where a considerable fraction of the metal is deposited on the cathode. The spent electrolyte is recirculated over fresh ore. This cyclic operation is capable of producing high-purity metal.

Although the cathodic reactions are the same in electrowinning and electrorefining, the processes differ in that the metal is introduced directly into the electrolyte in the former, while electrorefining consists of purifying an impure anode. A further difference is the energy requirements. Electrowinning commonly uses an inert anode, and the anode reaction is primarily oxygen evolution. Therefore, there is usually a considerable decomposition voltage.

The reason for using electrowinning rather than conventional chemical reduction methods for extracting a metal from its ores are either: (1) reduction with carbon is thermodynamically unfavorable at economically achievable temperatures, or (2) the ore is low-grade.

The first case applies to sodium, magnesium, calcium, and aluminum and, since these metals cannot be deposited by electrolysis from aqueous solutions, the ores are dissolved in the molten state. General features of fused-salt electrolyses are: (1) the use of mixtures of salts to lower the melting points, (2) the use of the heating effect of electric current to maintain the melt temperature, (3) the use of graphite anodes, at which the usual product is gas generation, (4) some means of protecting the molten metal from air and from the anode, and (5) applied potential differences are usually 6 to 8 V, and currents may be as high as $10^5$ A. The extraction of metals through electrolysis of aqueous solutions is less widely used, although it still is important in copper extraction and less so for lead, nickel, bismuth, indium, and tin.

***59. Aluminum.*** Aluminum production is the major electrochemical process in the United States and utilizes about 12% of the electricity production. This process occurs in the molten state and alumina (purified from bauxite ore) is dissolved in fused cryolite ($AlF_3 \cdot 3NaF$). This process was developed simultaneously and independently by Hall in the United States and Heroult in France and is generally known as the Hall process in the United States and the Heroult process in Europe. The fused bath contains about 5% alumina and is electrolyzed at a temperature of about 950°C. Carbon blocks are used as anodes, and the carbon lining of the furnace and the accumulated metal, as the cathodes. The liquid metal collects at the bottom of the bath and is periodically tapped off. As the alumina is used up, fresh material is added to the top of the bath. The heat required to maintain the bath in the liquid condition is furnished entirely by the current used in the electrolysis. A typical industrial cell operating at 40,000 A produces about 600 lb (130 kg) of 99.5% aluminum per day at 85% current efficiency. Energy consumption is about 15 kWh/kg of aluminum.

***60. Sodium.*** Sodium is produced by the electrolysis of a fused sodium chloride–calcium chloride mixture, and an equivalent quantity of chlorine is recovered as by-product.

The resulting sodium metal contains about 5% calcium, but this is removed by filtration outside the cell, leaving a final product containing less than 0.04% Ca.

The cell is composed of firebrick with an iron shell, a graphite anode, and iron cathodes. Chlorine gas is withdrawn from the anode compartment and liquid metallic sodium from the cathode compartment. Care must be exercised in handling the liquid sodium, because it is very fluid and finds its way through the smallest cracks, and, if exposed to oxygen, will burn very briskly.

Energy consumption is about 11 kWh/kg of sodium; 1.5 kg of chlorine is liberated at the same time. Cells are operated at about 300 V dc and from 20,000 to 40,000 A. Operating temperature is about 600°C, and current efficiency varies from 40 to 80% (for modern cells).

***61. Magnesium.*** Magnesium is made by electrolysis of a melt containing anhydrous magnesium chloride; seawater is the raw material. Magnesium hydrate is precipitated from the water by means of lime, and hydrate is converted to chloride by hydrochloric acid. The chloride is then dried and finally partly dehydrated.

Electrolytic decomposition is accomplished in rectangular cast-steel pots externally heated to maintain proper cell temperature and to reduce consumption of electric energy. The anode consists of large graphite bars, and the pot itself serves as cathode. The electrolyte is molten magnesium chloride, to which some sodium chloride is added to reduce the melting point and increase the conductivity. The cells operate on direct current at about 70,000 A with an energy consumption of 17.5 to 29 kWh/kg of magnesium. Chlorine gas is liberated at the anode as a by-product. Metallic magnesium formed at the cathode is lighter than the bath and, therefore, floats to the surface from which it is removed. The electrolyte is maintained at a nearly constant level by either a continuous or an intermittent feed of dehydrated magnesium chloride.

***62. Copper.*** About 10% of copper production results from the electrowinning process, the rest being produced by pyrometallurgy and refined electrolytically. The final products are identical.

In electrowinning of copper, the leaching solution is sulfuric acid, and since the electrolytic stripping of the leach solution regenerates the acid used in leaching, the process essentially consumes no acid. The feed electrolyte to the cells contains 3 to 5% copper and 2.5 to 3% free acid. It is electrolyzed until the copper content has been reduced to about 1% and the acid has increased to 7 or 8%, when it is withdrawn and used for leaching another batch of ore. Cell voltage is about 1.9 V, temperature 30 to 50°C, and current density 50 to 150 A/m$^2$.

***63. Zinc.*** Ores used for electrolytic reduction are complex zinc-lead, copper-iron sulfides with impurities of gold, silver, cadmium, and other metals. They are concentrated and roasted in preparation for electrolysis. The operation is cyclic, in that zinc is added to the circulating electrolyte in one department of the plant and deposited out in another.

Spent electrolyte from the tank house (high in acid and low in zinc) is used to leach calcined ore. The zinc oxides and sulfates dissolve in the solution, which becomes richer in zinc and lower in acid. The solution is purified and piped to the electrolytic tank house. Lead or lead with 1% silver is used for the insoluble anodes, and aluminum for the cathode starting sheets.

## Electrosynthesis

***64. Electrosynthesis.*** Electrosynthesis is the formation of compounds directly by electrochemical reactions or by subsequent chemical reaction. These reactions include the manufacture of elemental inorganics such as chlorine, fluorine, and hydrogen; of compounds such as chlorate, perchlorate, bromate, and hydrogen peroxide; and organic compounds.

***65. Chlor-Alkali Industry.*** Electrolysis of an aqueous sodium chloride solution results in a variety of products, depending on the cell design and operating conditions. Thus, if the anodic and cathodic products do not react to form other compounds, caustic soda (NaOH) and chlorine are the major products. (Although hydrogen is produced in the reaction, it is not commercially viable to collect and sell.) If the products react in a cold cell, sodium hypochlorite (bleaching powder, NaOCl) is produced, while sodium chlorate ($NaClO_3$) is produced when the cell is hot. If the electrolyte is potassium chloride rather than sodium hydroxide, caustic potash (KOH) is produced. This industry consumes about 2% of all electric energy produced in the United States.

Separation of the anodic product (chlorine) from the cathodic product is achieved either by using a mercury cell to prevent the formation of sodium hydroxide in the electrochemical cell, or by using a diaphragm to maintain separation of the anolyte and catholyte. Because of mercury pollution, new installations tend to be diaphragm cells.

***66. Chlorine and Caustic Soda.*** Chlorine and caustic soda are the two largest tonnage products produced by an electrochemical process. They are produced simultaneously at a ratio of 1.13 kg of sodium hydroxide for 1 kg of chlorine. Clearly, it is unusual for the demand of the products to be in the same ratio, and the need for chlorine is usually greater. (Chlorine is also produced as a by-product in the electrolysis of molten sodium chloride to form sodium and by the electrolysis of hydrochloric acid.)

In the diaphragm cell, nearly saturated sodium chloride brine is fed at a temperature of 60 to 70°C into the anolyte compartment and flows continuously through the diaphragm into the catholyte compartment, flow being maintained by a differential head. The cathode reaction is

$$2Cl^- \rightarrow Cl_2 + 2e$$

and the anode reaction is

$$2e + H_2O \rightarrow 2OH^- + H_2$$

Only a portion of the alkali chloride entering the cell is electrolyzed, the unreacted portion leaving with the hydroxide solution from the cathode compartment. The chlorine produced is evolved at the cathode, while the NaOH solution is passed to concentrators where it is concentrated by evaporation.

Historically, anodes have been graphite, but metallic anodes of tantalum clad with proprietary noble metal oxides are increasingly used. They do not erode as do graphite anodes (hence their name, *dimensionally stable anodes,* or DSA) and, thus, minimum ohmic polarization can more readily be achieved.

The principal anode reaction is the formation of chlorine from chloride ion which takes place at an efficiency of about 97%; side reactions include the formation of hypochlorite and the oxidation of the graphite anode to carbon dioxide. At the steel cathodes, the reaction is the evolution of hydrogen at practically 100% efficiency, leaving an equivalent amount of hydroxyl ions behind as caustic alkali.

Cell designs differ in detail, but all are similar in principle. They all incorporate vertical or horizontal graphite or metallic anodes, steel-screen cathodes, and deposited diaphragms. The early cells used asbestos diaphragms. However, there is also a tendency to replace these with membranes of *perfluorinated sulfonic acid polymers* (Nafion).

In the mercury cell, the anode process is the same as in the diaphragm cell—the evolution of chlorine—but the cathode reaction is the deposition of sodium into the mercury cathode to form sodium amalgam (or potassium amalgam in the electrolysis of potassium chloride). The amalgam is then transferred to another vessel, called a denuder or decomposer, where it reacts with water to form alkali hydroxide and regenerate the mercury.

The advantage of the mercury cell includes the greater purity of the caustic produced—or alternatively the greater ease and consequently cheaper plant needed to produce it—since it does not have to be separated from the original salt as in the diaphragm process, which produces a fairly dilute salt-caustic solution. On the other hand, energy requirements are somewhat higher, and the cells themselves are more expensive because of the investment in expensive mercury.

***67. Hyprochlorite (Bleaching Liquor).*** Whereas for the production of caustic and chlorine by electrolysis of sodium chloride the anodic and cathodic products are kept separate, the reverse requirement is fulfilled for the production of hypochlorites (bleaching liquors) by electrolysis of sodium chloride. Sodium hypochlorite is the result of the reaction of chlorine on caustic soda. To obtain the hypochlorite in the electrolytic cell itself, the electrodes are placed near each other, and the electrolyte is maintained in steady motion in order to mix the anodic and cathodic products.

***68. Chlorates.*** The production of sodium and potassium chlorates requires interaction between caustic soda or potash and chlorine under conditions of moderately high temperature (30 to 50°C) and absence of reducing conditions. For the latter purpose, the addition of chromate is used. No diaphragms are used.

***69. Fluorine.*** Fluorine has an anodic electrode potential of +2.65 V and, thus, it is not possible to produce fluorine by electrolysis of aqueous solutions. Therefore, fluorine is produced by electrolysis of a molten fluoride.

The electrolyte consists of potassium fluoride dissolved in liquid hydrofluoric acid at 75 to 105°C, the exact temperature depending on the ratio of KF to HF. The cathode, at which fluorine is produced, is mild steel, while the anode is graphite. The fluorine is liquefied and stored in steel cylinders.

***70. Organic Synthesis.*** Organic electrochemical reactions have been known for a long time. For example, in 1834, Faraday found that the acetate ion ($CH_3COO^-$) could be oxidized to carbon dioxide, and Kolbe in 1849 described the production of *n*-octane from valeric acid.

$$C_4H_9COO^- \rightarrow C_8H_{18} + 2CO_2 + 2e$$

Since that time, many electroorganic synthetic reactions have been discovered. However, there are few examples of industrial-scale electroorganic synthesis mainly because of the unsuitability of electrode and cell design.

Processes which are used in commercial-sized plants include the production of lead tetraethyl and lead tetramethyl (Nalco), the synthesis of adipodinitrile from acrylonitrile (Monsanto), and the oxidation of diacetone-L-sorbose to diacetone-2-keto-L-gulonic acid (Hoffman-LaRoche)—an intermediate in vitamin C manufacture. Each of these utilizes a unique cell design to permit intimate contact between the reactants and the electrode surface, thus minimizing electrode polarization and increasing system efficiency.

In the Nalco cell, the electrolytic cell is essentially a tube-and-shell heat exchanger. The heat exchanger tubes are used as cathodes which are separated from the anode by an insulating perforated polymer sleeve covering the inner side of the tubes. The tubes contain lead shot, which forms the sacrificial anode; these are continuously refilled from the top. The reactant is dissolved in an ether solution and is at sufficient concentration to supply enough solution conductivity. This ethereal solution is pumped through the tube and cooled to maintain a working temperature of 40 to 50°C.

The Monsanto cell uses a filter-press arrangement with hollow bipolar lead electrodes (with internal cooling). The divided cells use a cation-exchange membrane separating the anolyte from the catholyte. The product is separated from the catholyte by extraction with acrylonitrile.

The Hoffman-LaRoche process is carried out in a "Swiss-roll" electrolysis cell. Separator cloths, anodes, and cathodes are arranged as a sandwich, rolled up around an axis, and pressed into a cylindrical container with the necessary inlet and outlet for the electrolyte.

The characteristics which all these processes have in common are electrodes with high specific electrode area, a very high mass transfer rate, constructional simplicity, and ease of scale-up. In addition, the electrochemical cell is only part of the process, and product separation, reactant preparation, and temperature control are all important in plant design.

***71. Water Electrolysis.*** When a direct current is passed between two electrodes in contact with an aqueous solution to which a small quantity of acid or alkali has been added, water is decomposed to liberate hydrogen and oxygen by the equation

$$2H_2O \rightarrow 2H_2 + O_2$$

Hydrogen is generated at the cathode and oxygen is generated at the anode, and there is complete separation of these two gases. This reaction also occurs in, for example, battery overcharge situations where all the battery reactants have been reduced and the next favorable electrochemical reaction occurs.

From Table 24-3, it can be seen that at 25°C, the reversible cell voltage for the hydrogen-oxygen reaction is 1.23 V, and, therefore, the theoretical decomposition voltage for water is the same value. However, in practice, water cannot be decomposed at this voltage because of the cell overvoltage. Similar to other systems, overvoltages are reduced: (a) by increasing the operating temperature, (b) by optimum design of the electrode, and (c) by incorporating catalysts into the electrodes. Higher currents result in higher overvoltages, which result in lower efficiency of gas generation. However, the size of the cell, and thus its capital cost, is reduced if it is designed to operate at high current densities. There is, therefore, a trade-off between system efficiency and capital cost to achieve in the lowest hydrogen production cost.

***72. Water Electrolyzer Designs.*** Commercial water electrolyzers are available today. All operate using an aqueous alkaline electrolyte at relatively low temperatures. Alternative electrolyzers, of which the closest to commercialization uses an ion-exchange membrane, are also under development.

Commercial electrolyzers can be classified according to the construction techniques and are tank-type electrolyzers and filter-press-type electrolyzers. In the tank-type electrolyzer, a large vat holds the alkaline electrolyte. Electrodes consisting of mild-steel sheets welded to bus bars of steel and with alternate polarity are suspended in the vat. Anodes are nickel-plated, and the cathodes consist of the uncoated mild steel. An asbestos separator keeps the two electrodes apart in order to prevent mixing of the hydrogen and oxygen generated at adjacent electrodes. In this type of electrolyzer, each tank operates as one cell. Although it may carry thousands of amperes, only about 2 V is applied.

The filter-press-type electrolyzer was designed to minimize floor space and maintenance requirements associated with the tank-type system. A filter-press electrolyzer is constructed with alternating layers of electrodes and separators similar to the way a filter press is built. The electrodes are solid metal and are bipolar; one side of the electrode is the cathode of one cell, while the opposite side is the anode of the adjacent cell. With this type of construction, the individual cell voltages are additive within a stack. Filter-press electrolyzers usually cost more to construct than the tank type and they are, therefore, operated at higher current densities to generate more hydrogen per square foot. Therefore, the usual operating voltages of both tank-type and filter-press-type electrolyzers are about the same, and although the filter press is more expensive than the tank-type construction, benefits of floor space, maintenance, and electrical system costs favor this type of construction.

The electrolyte used is potassium hydroxide solution at about 28% concentration for maximum conductivity. Because of the absence of side reactions, the current efficiency in most types of electrolyzers is close to 100%, and the total electrical efficiency (measured in watt-hours) is about 70%. Since the performance of the electrolyzer is a function of both its operating current and voltage, there is a trade-off between the operating cost related to voltage and the capital cost related to the size of cell and, therefore, the operating current density. Electrolysis cells are normally tailor-made for their specific application and optimized to suit the cost of the available electric power.

Industrial plant capacities for hydrogen production range from as little as 500 standard cubic feet (scf) per day to over 40 million scf/day. The most common sizes in metallurgical and chemical processing applications are between 10,000 and 50,000 scf/day. The very large plants require up to 240 MW of electric energy and are installed where low-cost hydropower is available and are used to provide hydrogen for ammonia and, hence, fertilizer manufacture.

New approaches are also being made toward electrolyzer design. One of the more attractive uses a solid-polymer, acid-type ion-exchange electrolyte operating at temperatures in the 70 to 90°C range. This basic concept depends on the use of a perfluorinated, sulfonic acid polymer, ion-exchange electrolyte. When saturated with water, it is a good ionic conductor and is the only electrolyte required. Since the electrolyte is immobile and has a fixed concentration, it cannot leak or be carried over into corrodible parts of the system; therefore, the problems of water balance management are much simplified. The cells operate at very high current densities of 14,000 $A/m^2$ at about 2 V. However, they use expensive noble metal catalysts and, in order to reduce hydrogen cost, improved performance is required while using cheaper catalysts.

***73. Bibliography on Electrorefining, Electrowinning, and Electrosynthesis.***

Bockris, J. O'M., Conway, B.E., Yeager, E., and White, R.E. *Comprehensive Treatise of Electrochemistry,* vol. 2, New York, Plenum Press, 1981.

Hampell, C.A. *The Encyclopedia of Electrochemistry,* New York, Reinhold Publishing Corporation, 1964.

Mantell, C.L. *Electrochemical Engineering,* 4th ed., New York, McGraw-Hill Book Company, 1960.

## PRODUCTION OF ELECTRICAL ENERGY

***74. Scope.*** Chemical energy is converted into electrical energy in batteries and fuel cells. The former are energy-storage devices and store the chemical energy within the battery. Fuel cells are energy-conversion devices and produce energy all the time the required chemicals (fuel and oxidant) are supplied to the cell.

Similar to energy-using systems, there are overvoltages associated with energy-producing systems which result in lower energy and power than theoretical. Much of the system design is associated with improving these efficiencies together with other practical factors, such as shelf life, energy density, etc.

## Batteries

***75. Battery.*** A battery is an assembly of electrochemical cells which convert the energy produced by chemical reactions into low-voltage, direct-current electricity. Similar to other electrochemical cells, each cell contains three major components: a positive electrode or cathode, a negative electrode or anode, and the electrolyte.

***76. Primary Battery.*** Primary batteries convert chemical energy into electrical energy once only, and the original chemical state cannot be regained by passing electrical energy through the battery. They are discharged once and are discarded after use.

***77. Secondary Battery.*** After discharge of a secondary battery, their original chemical state can be regained by passing through the battery a quantity of electrical energy equivalent to that drawn during discharge. Secondary batteries are, therefore, distinguished from primary batteries by the feature of rechargeability.

***78. Definitions.*** *Ampere-hour (Ah) capacity* is the amount of electricity delivered per unit weight or per unit volume.

$$\text{Ah capacity} = \frac{\text{no. of Ah delivered}}{\text{battery weight}}$$

or

$$\text{Ah capacity} = \frac{\text{no. of Ah delivered}}{\text{battery volume}}$$

*Energy density* is the energy delivered per unit weight (or unit volume).

$$\text{Energy density} = \text{Ah capacity} \times \text{cell voltage}$$

Energy density is usually expressed in watt-hours (Wh) per pound (Wh/kg) or Wh/in$^3$ (Wh/cc).

In addition to the energy density of a cell, another important parameter is the rate at which the energy can be delivered, i.e., the power. *Power density* is the power delivered by the battery per unit weight or volume.

$$\text{Power density} = \frac{\text{current} \times \text{voltage}}{\text{battery weight}}$$

Units are watts/lb (watts/kg) or watts/in$^3$ (watts/cc).

Batteries rarely use 100% of the theoretical ampere-hour capacity. For primary cells, the reason is that mass transport and resistance effects limit the *utilization* of reactants, whereas for secondary cells, the energy used in one cycle is maintained below 100% in order to maximize battery life. For secondary battery applications, high energy density is only useful if the battery can be repetitively discharged, i.e., long cycle life at high reactant utilizations. Depth of discharge or utilization is defined as the ratio of the number of ampere-hours delivered to the theoretical number of ampere-hours calculated from Faraday's laws. Utilization depends on the battery discharge rate, and higher utilizations are achieved at slower discharge rates (i.e., lower currents).

The ampere-hour capacity available from the battery to a preselected voltage depends on the value of the current. At high load (i.e., high currents), the battery delivers far less capacity than at low load. A *capacity/time quotient* is, therefore, defined to normalize the discharge rate for cells of differing capacity. For example, C/5 is the current which completely discharges (or charges) the battery in 5 h; 4C is the current which discharges the cell in ¼ h. These values are independent of the battery size and weight but are characteristic only of the type of battery. For example, a radio battery has a discharge rate of about C/100, while an electric vehicle requires a C/3 discharge rate.

*Shelf life* is the loss of capacity of the cell during storage. The period beyond which it becomes uneconomic to store the battery is known as its shelf life.

*Cycle life* is the number of times a secondary battery can be discharged (or charged) before the battery voltage falls below a prescribed value.

Efficiency is an important parameter for secondary batteries. This is described by three parameters:

$$\textbf{Ah efficiency} = \frac{\text{Ah during discharge}}{\text{Ah required for charge}}$$

$$\textbf{Voltage efficiency} = \frac{\text{Average voltage during discharge}}{\text{Average voltage during charge}}$$

$$\textbf{Wh efficiency} = \text{Ah efficiency} \times \text{voltage efficiency}$$

Ampere-hour efficiency describes the fraction of electricity required to charge a cell which is recovered during discharge, and is, therefore, a measure of side reactions (usually gas evolution) occurring. Voltage efficiency is a measure of the polarizations during charge and discharge of the two electrode reactions.

***79. Discharge Voltage Characteristics.*** When a cell is discharged, i.e., a current is drawn from the cell, the voltage immediately decreases because of electrode polarization and ohmic polarization. There is then usually a period during which the cell voltage remains essentially constant. Subsequently, there is a sudden change in voltage with time. This point of inflection indicates that little useful energy can be obtained by continuing discharge beyond this point.

The shape of this curve depends on the battery and the discharge rate. There is a similar, though opposite, curve during charge at which further application of the charge current only results in gas evolution at the electrodes rather than reforming the chemical reactants.

***80. Battery Characteristics.*** The development of a commercial battery depends on the application which dictates the necessary battery characteristics. Table 24-6 lists some applications and the requirements for these applications in order of importance.

**TABLE 24-6** Battery Characteristics

| Auto SLI* | Electronic equipment | Emergency power | Military | Electric vehicles | Load leveling |
|---|---|---|---|---|---|
| Cost | Sealed operation | Calendar life | Specific energy | Cost | Cost |
| Specific power | Compactness | Maintenance | Specific power | Specific energy | Cycle life |
| Temperature range | Life | Cost | Compactness | Specific power | Energy efficiency |
| Ruggedness | Cost | | Ruggedness | Compactness | Temperature range |
| Life | | | Temperature range | Ruggedness | Maintenance |
| | | | Life | Cycle life | |
| | | | | Temperature range | |
| | | | | Energy efficiency | |
| | | | | Maintenance | |

* Starting, lighting, ignition.

***81. Battery Types.*** A large number of primary and secondary batteries are available or are being developed (Table 24-7) to satisfy the markets outlined in Table 24-6. There is no one universal battery which can satisfy all these requirements, and each has its particular market niche.

***82. Battery Design Features.*** As with other electrochemical cells, care is taken in the design of the battery to minimize cell polarization. This is achieved in several ways. First, the electrodes are fabricated to maximize their surface-area-to-volume ratio; this is achieved by using porous electrodes. Thus, a high specific area is formed at the interface between the active materials of the anodes and cathodes and of the electrolyte phase in order to obtain high currents during charge and discharge. At the same time, good electronic conduction must be achieved between the active battery material and the current collector which, in turn, must be stable and noncorrodible. Several types of electrode structures have been developed which satisfy this condition, depending on the nature of the battery.

Ohmic polarization is reduced by using electrolyte concentrations corresponding to maximum specific conductivity and by minimizing the anode-to-cathode distance. In order to have small anode-to-cathode separations, care must be taken that the electrodes are not in contact. This is particularly necessary for secondary batteries where electrode shape changes occur during cycling, with a resultant change in this distance. As a result, a separator is placed in the electrolyte between the two electrodes. This material is selected for maximum stability in the

**TABLE 24-7** Battery Types

| Type | Cathode | Anode | Electrolyte |
|---|---|---|---|
| | Commercial primary batteries | | |
| LeClanche | $MnO_2$ | Zn | $NH_4Cl-ZnCl_2$ |
| Alkaline manganese | $MnO_2$ | Zn | KOH |
| Rubens (or Duracell) | HgO | Zn | KOH |
| Panasonic high energy | $(CF_x)_n$ | Li | Propylene carbonate with $LiClO_4$ |
| Lithium (GTE) | $SOCl_2$ | Li | $SOCl_2$ with $LiCl-AlCl_3$ |
| Lithium (Mallory) | $SOCl_2$ | Li | Propylene carbonate with $LiClO_4$ |
| Lithium (Saft) | CuO | Li | Propylene carbonate with $LiClO_4$ |
| | Secondary batteries | | |
| Lead acid | $PbO_2$ | Pb | $H_2SO_4$ |
| NiCad | NiOOH | Cd | KOH |
| Nickel-iron | NiOOH | Fe | KOH |
| Silver-zinc | $Ag_2O_2$ | Zn | KOH |
| Zinc-halogen | $Br_2$ (or $Cl_2$) | Zn | $ZnBr_2$ (or $ZnCl_2$) |
| Zinc-air | $O_2$ | Zn | KOH |
| Nickel-hydrogen | NiOOH | $H_2$ | KOH |
| Sodium-sulfur | S | Na | $\beta$-$Al_2O_3$* |
| Lithium sulfide | Li | FeS (or $FeS_2$) | LiCl–KCl* |

* High-temperature operation.

electrolyte, low cost, and optimum porosity to minimize concentration gradients (and, hence, concentration polarization) in the electrolyte.

The remainder of the design considerations involve electrode configurations, low cost, stable case materials, methods of sealing, and low-cost manufacturing techniques, all of which are specific for a particular battery system. However, as a result of added materials to achieve a practical battery, practical energy densities are only about one-fourth to one-fifth of the theoretical value based on the weight of electroactive materials and electrolyte.

Battery design criteria are related to cost, performance, durability, safety, and environmental considerations. Initial battery costs are determined by materials and manufacturing costs, while manufacturing costs are determined by the ease of the manufacturing process and the number of individual components, hence the requirement for high cell voltages, reversible electrode reactions, high electrolyte conductivities, and minimal volumetric differences between reactants and products. Operating costs are influenced by cycle life, maintenance requirements, and energy efficiency of the charge-discharge cycle.

## Fuel Cells

***83. Definition.*** A fuel cell is an electrochemical device that continuously converts the chemical energy of the fuel and of the oxidant directly to electrical energy without an intermediate combustion process. Unlike a battery, a fuel cell does not run down or require recharging. It will continue to operate as long as the fuel (hydrogen), is fed to the anode, and the oxidant (air) is fed to the cathode. The electrodes, which are solid material of high internal area, act as reaction sites where the electrochemical transformation of the fuel and oxidant occurs. These reactions produce electrons which flow from one electrode to the other when the electrodes are connected together through an external circuit. Several types of electrolyte may be used, and the particular type is determined by the type of fuel and oxidant and by the cell operating temperature. The electrode reactions are

$$2H_2 \rightarrow 4H^+ + 4e \qquad \text{(anode)}$$

$$O_2 + 4H^+ + 4e \rightarrow 2H_2O \qquad \text{(cathode)}$$

giving as the overall cell reaction

$$2H_2 + O_2 \rightarrow 2H_2O$$

***84. Fuel Cell Systems.*** In order for a fuel cell power plant to be a useful part of an electricity generating system, it must be capable of using available fossil fuels and produce ac power. Therefore, a practical fuel cell power plant is composed of three major subsystems; fuel processor, fuel cell, and dc-to-ac inverter. Fuel, which may be basically any fossil fuel (coal, oil, natural gas), is fed into a fuel processor. The fuel processor converts the fossil fuel into a gas suitable for the fuel cell (i.e., mainly hydrogen); the gas entering the fuel cell also contains other constituents such as carbon monoxide and carbon dioxide, depending on the fuel processor system (coal gasifier, steam reformer). The third subsystem of the fuel-cell power plant is the inverter to convert the direct current of the fuel cell into alternating current.

Several types of fuel cells are being developed. These, characterized by the nature of the electrolyte, are alkaline fuel cells, acid fuel cells, and molten-carbonate fuel cells.

Alkaline fuel cells, using potassium hydroxide solution as the electrolyte, operate at temperatures of about 60 to 120°C. These cells operate on pure hydrogen and oxygen and are, therefore, of less use for terrestrial applications. They have demonstrated reliability in specific applications and were the fuel cell power plants for the Apollo and the Space Shuttle programs.

Many acid electrolytes have been considered for use in fuel cells because: (1) they are not contaminated by carbon dioxide in the fuel or air, and (2) complete purification of the fuel stream to pure hydrogen is not required. Acids which have been investigated are phosphoric acid, solid polymer electrolytes (based on perfluorinated sulfuric acid), and trifluoromethane sulfonic acid. The solid polymer electrolyte, similar to the acid water electrolysis electrolyte, was used in the Gemini space program but has not found application for commercial power plants because of the electrolyte cost. The most advanced acid power plant is based on phosphoric acid, but this system is limited because of (1) use of noble metal electrodes, (2) low cathode performance, (3) poisoning of the anode by carbon monoxide in the fuel, and (4) loss of acid electrolyte by evaporation, which causes corrosion of the cell components.

Molten-carbonate fuel cells operate at temperatures of 600 to 700°C using impure hydrogen and air, nonnoble metal electrodes, and an electrolyte of molten alkaline metal carbonates contained in a porous ceramic carrier. A characteristic of this cell is that high-quality waste heat is available for use in cogeneration or in bottoming cycles to increase overall system efficiency.

***85. Fuel Cell Characteristics.*** The fuel cell is an environmentally acceptable device since it is quiet and has no moving parts. Because it is not a combustion device, emissions such as $NO_x$, CO, and unburned hydrocarbons are not a problem. Such low-level emissions, coupled with the fuel cell's quiet, water-conserving operation, result in environmental acceptability and siting flexibility.

A single fuel cell normally generates power at approximately 0.5 to 1 V and can be connected in series stacked with other cells to obtain almost any desired voltage. The current produced is a function of the area of the single cells. The range of sizes, the modularity, and the load-following capabilities make the fuel cell system an attractive candidate for power generation in a variety of applications, including on-site and central plants for commercial, industrial, and residential use.

The major advantage of fuel cells is that they are highly efficient energy-conversion devices. They are inherently more efficient than conventional power generation devices because they do not suffer from the Carnot cycle limitation. In addition, because the system is located at the point of use, the waste heat generated by the irreversibilities of the electrode reactions is recoverable, thus further improving the system efficiency.

***86. Fuel-Cell Efficiency.*** The efficiency of converting gaseous fuels to dc power in a fuel cell power plant is a function of thermodynamic voltage and current efficiencies, and of the heating value of the composition of the fuel, which may contain nonelectrochemically active combustible species. The impact of system efficiency on fuel cell application is so important, and the calculation is such a good example of the equations developed earlier, that the calculation of system efficiency is presented here in some detail.

Only the free energy of a chemical reaction can be converted into electrical energy; therefore, $\Delta G$ is the maximum amount of energy available for conversion to electricity. Therefore, the maximum thermodynamic efficiency ($E_T$), i.e., the efficiency corresponding to no electrode polarization, is given as

$$E_T = \Delta G / \Delta H$$

A necessary corollary of this is that even a "perfect" fuel cell would have an efficiency less than 100%.

*Voltage Efficiency.* The voltage efficiency ($E_V$) is the ratio of the cell voltage on load ($V$) to the theoretical thermodynamic voltage at equilibrium ($E$) calculated from

$$E_V = V/E$$

*Current Efficiency.* The efficiency of an electrochemical cell is further reduced if all of the reactants are not converted to products, or if the electrodes are involved in an alternative reaction such as corrosion. The current efficiency ($E_I$) is the ratio of the actual current ($i$) to the current predicted by Faraday's laws ($i_F$):

$$E_I = i/i_F$$

*Heating Value Efficiency.* Product gases from most practical fuel processes will contain varying amounts of inert gases and hydrogen, carbon monoxide, and methane which can be burned in a conventional combustion device but which are electrochemically inactive (or poisons). There are, therefore, species in the fuel which are combustible and have heating values but which do not produce current. The heating value efficiency ($E_H$) is, therefore, defined as the ratio of the maximum amount of heat energy that can be produced by electrochemically active gases ($\Delta H$) to the amount of heat energy available from combustion of the fuel gas ($\Delta H_C$):

$$E_H = \Delta H / \Delta H_C$$

Only species involved in electricity production are included in $\Delta H$, but $\Delta H_C$ includes all combustible species and is determined from the thermodynamics of the mixtures. The lower heating value (LHV) of the fuel is used when the water is not condensed and the higher heating value (HHV) is used when it is.

The total fuel cell efficiency ($E_{FC}$) which describes the conversion of chemical energy in the feed gas to dc power is then defined as follows:

$$E_{FC} = E_T \cdot E_V \cdot E_I \cdot E_H$$

*Fuel Cell System Efficiency.* The previous equation describes only the fuel cell power section of the fuel cell system; the fuel processor, power conditioner, and waste-heat utilization are not included. The overall efficiency for a fuel cell system ($E_S$) without waste-heat utilization is given by

$$E_S = \frac{\text{ac power}}{\text{HHV or raw fuel into fuel processor}} = E_{FP} \cdot E_{FC} \cdot E_{PC}$$

where

$$E_{FP} = \frac{\text{LHV gaseous fuel from fuel processor to fuel cell}}{\text{HHV raw fuel into fuel processor}}$$

and

$$E_{PC} = \frac{\text{ac power}}{\text{dc power}}$$

Additional energy can be recovered in the system for heating or by integrating a bottoming cycle to produce additional electricity. For a system employing waste-heat utilization, the fuel cell system efficiency ($E_{SE}$) is given as

$$E_{SE} = \frac{\text{fuel cell + bottoming cycle ac + cogeneration BTUs}}{\text{HHV raw fuel into fuel processor}}$$

***87. Bibliography on Batteries and Fuel Cells.***

Falk, U., and Salkind, A.J. *Alkaline Storage Batteries,* New York, John Wiley & Sons, Inc., 1969.

Heise, G.W., and Cahoon, N.D. (eds.). *The Primary Battery,* New York, John Wiley & Sons, 1971.

Liebhafsky, H.A., and Cairns, E.J. *Fuel Cells and Fuel Batteries,* New York, John Wiley & Sons, Inc., 1968.

Williams, K.R. *An Introduction to Fuel Cells,* Amsterdam, Elsevier, 1966.

## OTHER ELECTROCHEMICAL PROCESSES

***88. Corrosion.*** Corrosion is the oxidation of a metal to form the metal salt, usually an oxide. Subsequent chemical attack may be retarded if a nonporous adherent film is formed on the metal surface, or it may continue if the film is porous with the resulting dissolution of the metal. In general, corrosion occurs as a result of electrochemical attack: *galvanic corrosion.*

One aspect of electrochemical corrosion is dissimilar-metal corrosion. This occurs when two dissimilar metals are in contact in the presence of an electrolyte. In this case, a galvanic couple is created and is short-circuited through the electrolyte. Clearly, there is a connection between the electromotive series of the elements (Table 24-3) and the susceptibility of an element to corrosion. The noncorrodible metals, platinum and gold, have the more anodic electrode potentials, while the more corrodible metals, zinc and tin, have the more cathodic electrode potentials. The metal of more negative electrode potential will tend to go into solution in the electrolyte, and, therefore, corrode.

A similar condition is obtained in an alloy which is not perfectly homogeneous or in a metal of which different parts have been subjected to different heat treatments or mechanical stresses. Under these conditions, certain parts will have a more cathodic potential than others, and in the presence of an electrolyte, a galvanic couple is formed. The electrolyte need only be rain water with impurities dissolved from the air or from the surface of the metal itself.

The electromotive series does not describe the complete picture of dissimilar-metal corrosion because, in many cases, a nonporous, adherent oxide layer is formed on the metal surface and inhibits further corrosion (for example, aluminum). As a result, the electromotive series only gives an indication of the tendency of metals to corrode.

An alternative mode of electrochemical corrosion is that caused by *differential aeration.* From Eq. (24-19), the potential of a metal in contact with oxygen is given by

$$e = e^0 + RT/4F \ln P_{O_2} + RT/F \ln a_{H^+}$$

where $P_{O_2}$ is the partial pressure of oxygen gas in equilibrium with dissolved oxygen in the electrolyte. This equation shows that if the oxygen concentration is greater at any part of a surface, this part is more positive (more noble) than at less-oxygenated surfaces. Therefore, corrosion may also be due to a differential supply of oxygen to a metal surface such as in nooks and crannies, and corrosion occurs at the area of the surface where there is restricted oxygen supply.

***89. Methods of Retarding Corrosion.*** The environment is of primary importance in determining methods of retarding corrosion. Metals which perform well in one type of environment may be entirely unsuited to another. The principal method of minimizing corrosion is to protect the metal from its environment by a coating of a more-corrosion-resistant material such as paint, organic coatings, or electrodeposit. Some metals, such as aluminum and chromium, form corrosion-resistant oxide coatings which retard further attack, but in any medium which reacts with these oxide films, the metals are not corrosion-resistant.

Paints and organic coatings must be free of pores and discontinuities in order to provide corrosion protection. This is also true of electrodeposits more noble than the basis metal, such as copper-nickel-chromium on steel. With zinc coating on steel, zinc becomes the anode and the steel, the cathode of the galvanic cell, and the steel does not corrode until all the zinc in its immediate neighborhood is consumed. Zinc plates are placed in marine boilers in order to make the iron cathodic and thus retard corrosion; zinc corrodes and must be replaced. Corrosion in boilers and condensers has also been retarded by application of an external voltage so as to make the iron a cathode.

***90. Corrosion by Stray Currents.*** Underground pipes and cables are corroded by stray electric currents in the ground. These currents may result from galvanic couples in the ground, track returns in electrified rails, or a variety of other causes.

The most widely used method of protection is cathodic protection, where the material to be protected is made more cathodic than the potential of the surrounding soil by the application of dc power obtained from small motor generators or batteries. The applied potential is small, being about 0.3 V between the protected part and the soil. The negative lead is attached to the metal structure to be protected and the positive lead to the ground some distance away. The number of units necessary to protect an underground structure cannot be calculated because this depends on the degree of corrosion protection afforded by coatings. General practice, therefore, is to install temporary units and determine the number and dc power rating required by checking the voltages between the metal and the ground.

***91. Electroanalysis.*** Electrochemical techniques are used in a variety of analysis methods to determine ionic concentrations.

***92. Measurement of pH.*** There are two important ways of measuring pH: colorimetric and potentiometric cells. The *colorimetric method* is based on the observation that the color of several organic dyes (indicators) changes over fairly narrow ranges of pH. This method is often used to give a rapid but approximate indication (to within 0.3 pH) of solution pH by use of indicator paper.

The *potentiometric method* measures the reversible potential difference between two electrodes, one of which is reversible to the hydrogen ion (its electrode potential depends on the hydrogen ion concentration) and a standard electrode. In principle, a variety of electrodes are available which are reversible to the hydrogen ion concentration. However, commercial pH meters use the glass electrode because of its stability and ease of use; a calomel electrode is used as the reference electrode.

***93. Ion-Selective Electrodes.*** Ion-selective electrodes are used to measure the concentration of the specified ion in a solution. The glass electrode used for measuring hydrogen ion concentration is one example. However, there are now specific electrodes commercially available for over 20 ions, including chloride, fluoride, bromide, iodide, cyanide, phosphate, calcium, magnesium, and the ammonium ions.

In these systems, the membrane is formed by a thin crystal of a solid electrolyte such as silver chloride or lanthanum fluoride, or by a liquid organic ion-exchange agent held on an inert, porous support. As with the glass electrode, the basic operating principle depends on the generation of a potential difference across the membrane due to the difference in ionic concentrations.

***94. Electrogravimetry.*** The solution containing suitable metal ions is electrolyzed at a controlled potential, sufficient to cause electrodeposition of one metal but not the other, and the weight of metal deposited measured.

***95. Conductimetric Titration.*** Conductimetric titrations are similar to ordinary burette titration, but the determination of the equivalence point is determined from the shape of the plot of conductance vs. volume of titrant.

***96. Potentiometric Titration.*** Since the cell potential depends on the concentration of the ionic species, potentiometric measurements may also be used to determine the equivalence point from the plot of cell potential vs. volume of titrant.

***97. Polarography.*** Polarography depends on the observation that when an electroactive ion is present in an ionic solution containing a large excess of other ions (supporting electrolyte), the limiting current is due only to the nature and concentration of the electroactive ion. Therefore, ionic concentrations can be determined from a standard calibration curve.

Although a variety of metal electrodes can be used, the most common is a dropping mercury electrode which consists of a flow of mercury through a fine capillary immersed in the electrolyte. The advantage of this technique is that the metal surface is constantly renewed and thus kept free of reaction products. Although known for over 60 years, this technique is not widely used because of the development of improved spectrophotometric techniques.

***98. Gas Analysis.*** The need to analyze gases in a variety of applications has led to the development of a variety of commercial electrochemical sensing techniques which are particularly valuable where only low power is available (portability). Examples of these analytical techniques are oxygen in biological systems and automobile exhausts, carbon monoxide, hydrogen sulfide, nitrogen oxide and sulfur dioxide in air, and alcohol on the breath.

The principle of operation of these systems is that a potential is applied to a sensing electrode and a signal (current) is generated when the gas is introduced in the cell.

***99. Electropolishing.*** Electropolishing is the formation of bright surfaces by anodically treating a metal in specially formulated electrolytes. The technique depends on the rapid dissolution of microprojections on the metal so that a smooth surface results.

***100. Electrochemical Machining.*** The process of electrochemical machining is similar to electropolishing in that selected portions of a metal are dissolved by applying an anodic potential to the metal. However, in this case the electrolyte is significantly more corrosive than in electropolishing, and the relationship between workpiece (anode) and cathode is selected to get maximum corrosion in the selected portions to form the desired shape. This technique is particularly useful to machine hard and strong alloys which are difficult to cut or grind mechanically. The electrolyte is corrosive, such as sodium chloride or sodium chlorate, and the cathode reaction is hydrogen evolution leading to the precipitation of hydroxides of the metal being machined. The cathode is shaped a little smaller than the required cavity in the anode workpiece, and the electrolyte is circulated and filtered to remove the precipitated metal hydroxide. Operation is at very high densities (1000 A/cm$^2$) to machine the metal and to prevent passivation.

***101. Immersion Coatings.*** Chemical replacement methods for applying coatings without the use of an electric current include immersion processes. The principle of operation is that the substrate will dissolve and the more noble metal ions in the solution be deposited (electromotive series) on a suitable substrate. As a result, the coatings are restricted to a thin layer. Immersion processes have specialized uses, but where applicable, they can be cheaper and simpler than standard electroplating techniques. Tin on copper and its alloys, tin-copper alloys on steel wire, and tin on aluminum alloy pistons are prominent among the major uses for immersion processes.

***102. Electroless Plating.*** Electroless plating is the chemical deposition of a metal on a nonconductor. Proprietary solutions are used that comprise essentially a salt of the metal to be deposited, a reducing agent, a complexing agent to prevent precipitation of the metal, and additives to stabilize the baths.

Most of electroless deposition is concerned with nickel, as a plastic pretreatment in a variety of applications, and with copper for the plated-through-hole in printed circuit boards. Electroless solutions are also commercially available for gold and nickel-cobalt alloys.

***103. Electrophoresis.*** Electrophoresis is a process for coating a metallic substrate with a polymeric coating. The polymer resin is suspended in an aqueous solution, but organic solvents may be used. The substrate is immersed in the tank containing the suspension and made the anode in a dc circuit. The substrate is coated for 1 to 3 min, withdrawn from the tank, and then heated to completely cure the film. Alternatively, some resin suspensions are used by making the part to be coated the cathode, and this depends on the nature of the resin.

The advantages of this process for covering metallic parts with polymeric coatings are: (1) a uniform coating is distributed on all parts of the substrate because the polymeric layer is insulating and builds to a self-limiting thickness; (2) the process may be conveyorized and is easily controlled; (3) there is practically no waste resin; and (4) unusual shapes, welds, and seams are all coated to the same thickness. This technology is used for coating (or painting) a wide variety of articles including auto bodies, refrigeration units, and beverage cans.

***104. Bibliography on Other Electrochemical Processes***

Duffy, J. I. (ed.). *Electroless and Other Non-electrolytic Plating Techniques. Recent Developments, Chemical Technology Review No. 171,* New Jersey, Noyes Data Corporation, 1980.

Duffy, J. I. (ed.). *Electrodeposition Processes, Equipment and Compositions. Chemical Technology Review No. 206,* New Jersey, Noyes Data Corporation, 1982.

Schweitzer, P. A. (ed.). *Corrosion and Corrosion Protection Handbook,* New York, Marcel Dekker, Inc., 1983.

Durst, R. A. (ed.). *Ion Selective Electrodes,* NBS Special Publication No. 314, Washington, D.C., 1969.

Crow, D. R. *Polarography of Metal Complexes,* New York, Academic Press, 1969.

# SECTION 25

# COMPUTER APPLICATIONS IN THE ELECTRIC POWER INDUSTRY*

**M. M. Adibi**
*President, IRD Corporation*

**J. D. Cypert**
*Manager, Energy Utilities Engineering Support Center, IBM Corporation*

**CONTENTS**

*Numbers refer to paragraphs*



* The authors acknowledge the assistance of the following who have reviewed the manuscript and provided helpful comments: T. E. Bishop (PP&L), Dr. F. I. Denny (EEI), Dr. R. D. Dunlop (DOE), P. M. Kushkowski (NUS), Dr. J. W. Lamont (EPRI), and D. W. Magee, Jr. (REMVEC).

## INTRODUCTION

***1. Growth of Computer Applications.*** The power industry is engaged in the generation, transmission, and distribution of electrical energy which is obtained by conversion from other forms of energy such as coal, gas, oil, nuclear, water, or other renewable energy. These activities often include mining, rail transport, shipping, slurry pipelines, and storage of energy in many forms. Many electric utilities are also engaged in the transmission and distribution of gas.

In its first 90 years of history, the industry expanded at a pace of nearly twice that of the overall economy, doubling roughly every 10 years. In the last decade the expansion has been at a much slower pace. The industry's annual rate of productivity improvement during the first period averaged approximately 5.5%, a rate more than three times that of increased productivity for the economy as a whole. The present productivity improvement rate is about 2%.

About 25% of primary energy is used in the generation of electricity. Almost half of the total requirement is satisfied with coal, a quarter with gas, and one-tenth with oil. Due to some recent setbacks, nuclear fuel has not met the industry's expectation, and at present it is not considered the future economic fuel. The use of coal is expected to rise in the coming decade.

Gas, oil, and coal emit nitrogen oxides, sulfur oxides, and particulates, all of which are objectionable. The costs of installing emission abatement equipment and expenditures for operating these devices are appreciable, increasing electric energy prices from 5 to 25%. Some of the emission standards, together with the high cost of cleaning fuels (precombustion) or stack gas (postcombustion), preclude the use of some of the available fuels.

In scheduling its day-to-day operation, and in planning for its future growth, the industry has made extensive use of analytical tools and mathematical models which, through optimization and simulation, help in the decision-making process. A schematic description of the ingredients of such models is shown in Fig. 25-1. As a consequence, the industry has long been one of the largest users of computers and among the most sophisticated in its modeling and computational techniques. This usage is quite understandable when one considers the high cost of power system equipment, the complexity of power systems, and the severe operational, reliability, and environmental requirements on the electricity supply.

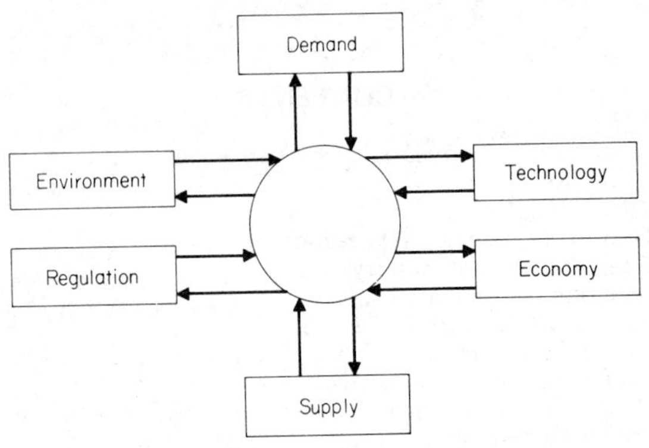


**FIG. 25-1** Model of the electric energy system.

Computer applications have assisted the industry in achieving its objectives: reducing the cost of energy delivered to consumers, improving the quality of service, enhancing the quality of the environment, and extending the life of existing equipment. These objectives have been achieved as follows:

**a.** Since the industry is one in which capital investment is usually high (over 10% of total spending by the nation's industries), unit costs have been reduced by operating facilities closer to their design limits, allowing better utilization of equipment.

**b.** Unit costs have also been reduced by automation, allowing operation with fewer personnel, and by optimization, lowering fuel consumption per kilowatthour delivered.

**c.** Electricity cannot readily be stored; therefore, production and consumption must be simultaneous. Hence, enough capacity is required to meet the maximum coincident demand or peak load of all customers. Interconnections between power systems provide important economies arising from different time patterns or diversity of use of the component systems in the network. They allow higher power system reliability at lower capital cost.

**d.** Quality of service has been improved by reducing the number, extent, and duration of service interruptions, thus providing a more reliable service.

**e.** Quality of environment has been maintained by operating facilities within acceptable bounds of emission, thermal discharge, waste disposal, and more effective land use.

Today the industry has reached a stage where computer systems are no longer merely an engineering tool. The effectiveness of computer applications is one of the key elements in achieving the basic functions associated with the planning, designing, construction, operation, and maintenance of the power system. In fact, engineering and computers have been integrated. This integration may be viewed as tending toward the construction of a utility industry information system. Such a system is shown in Fig. 25-2. It depicts a typical information system which may be viewed as a combination and integration of several functional information systems.

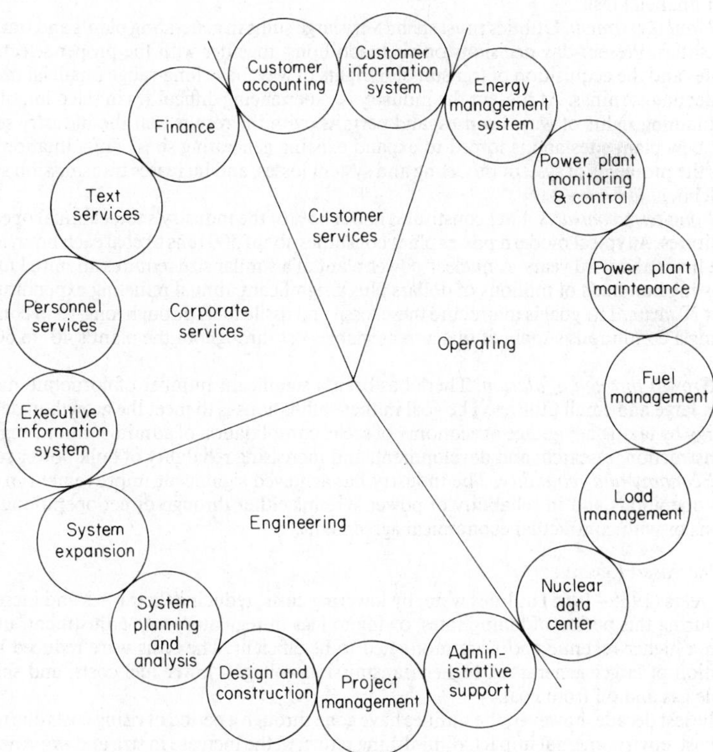


FIG. 25-2 Electric utility information systems.

Such an information system can extend management capabilities through information accessibility by economically maintaining complete central information files. These must be immediately accessible from any location in the company for information entry and retrieval by giving status information on any major maintenance or construction project by summarizing all revenue and expense for close cost control, and by furnishing daily technical, financial, and operating statements. The information system thus can provide meaningful data, at the proper time and location, to assist every level of management in making decisions which will result in close control of operations. Here, only the engineering and operating functions are covered. No attempt is made to deal with corporate and customer services which are very extensive fields of computer use. The division between business and technical computing has become less distinct as comprehensive data bases and data communication paths are used to serve both purposes.

***2. Goals of the Power Industry.*** The industry's purpose is to provide adequate, reliable, environmentally compatible electricity at reasonable cost with the ultimate goal of improving its productivity and net earnings. In spite of the differences between publicly and privately owned utilities, this goal is applicable to each, in different form. This goal is reached by pursuing a number of objectives as described below.

*a. Improved Financial Management*

- *Raising new capital.* Currently, the demand for electric power is growing at an annual rate of about 2 to 2.5%. The industry must meet the growth in size, but more important, it must replace old and obsolete equipment and retrofit plants for environmental and reliability considerations. To finance new plants, the industry has to raise more new capital over the coming 5 years than in the previous 10. The projected construction for the industry in the next decade runs into hundreds of billions of dollars. In view of interest rates and inflation, this is a most difficult financial task.
- *Plant investment.* Utilities must spend very large sums in generating plants and transmission facilities. Present-day decisions on such additions, together with the proper selection of plant sites and the acquisition of transmission rights of way, have long-range financial implications affecting earnings. At present the industry is experiencing difficulties in selection of plant sites, obtaining rights of way, licenses, and permits, with the results that the industry seldom obtains new plant sites and is forced to expand existing generating sites. This situation compounds the problems of system modeling and system losses, and increases transmission system dependency.
- *Long-term contracts.* Fuel constitutes about 35% of the industry's total annual operating expenditures. A typical modern power plant consumes about 500 tons of coal each hour, and its average life is about 30 years. A nuclear power plant of a similar size requires an initial nuclear core costing hundreds of millions of dollars plus a significant annual refueling expenditure for the next 30 years. The goal is to procure these fossil and fissile fuels through long-term contracts providing a continuous supply of fuel at reasonable cost throughout the plant's 40- to 60-year life.
- *Growth through affiliation.* There has been a significant number of corporate mergers between large and small utilities. The goal in these affiliations is to meet the growth in demand for energy by taking advantage of economy of scale; consolidation of administration, engineering, construction, research, and development; and increasing reliability of bulk power supply.
- *Economy and reliability.* The industry has achieved significant improvement in economy of operations and in reliability of power systems either through direct operational pool functions or with contractual economical agreements.

*b. Increased Revenue*

For 30 years (1935–1965) utilities were, by lowering costs, reducing their rates and increasing sales. During this period of falling rates, owing to lags in regulatory rate adjustment, utilities enjoyed a higher revenue and were motivated to be efficient. The costs were reduced by the installation of larger generators, higher transmission voltages, lower fuel costs, and shifts to available gas and oil from coal.

In the last decade, however, the utilities have gone through a period of rising costs due to rises in fuel cost, environmental impact, diminishing return in the increase in size and improvement in efficiency of units, and investing in new technologies such as nuclear plants.

During this rising-cost period, the regulatory lag in rate adjustments has had an adverse economic impact, and utility executives are seeking, and regulatory commissioners are just now beginning to project and apply, a regulatory lead. This approach demands detailed analysis of past and present operations and projection of future requirements by financial modeling, optimization schemes, and simulation techniques. For example, a recent regulatory decision calls for the inclusion of the cost of construction work-in-progress interest in the rate structure.

*c. Reduced Cost*

Cost reduction can be achieved by reducing investment per kilowatt of installation capacity for generation, transmission, and distribution, and reducing operating cost per kilowatthour of energy delivered.

*Reduced plant investment* can be achieved by installing larger generating plants (3000 MW), and by higher transmission (750 kV) and distribution (34.5 kV) voltages, by power pooling, interconnection planning, and coordination to gain further advantages of scale. Involved also are improved production and distribution facility utilization (that is, capacity factor and load factor) through peak shaving, reserve sharing, load diversity, and distribution load balancing. Other means of reducing costs include designing facilities with more precision and reducing the factor of safety, reducing construction and inventory costs, and operating the facilities closer to their design limits.

*Reduced operating expenditure* can be achieved by adopting new technology (such as breeders, biomass, and renewables) requiring lower fuel costs, by improving conventional and established methods of higher energy conversion efficiencies, by reducing energy losses in transmission and distribution facilities, and by interchanging energy with more economical resources and different time zones in different seasons to take advantage of diversity.

Other means are producing and distributing electricity with fewer personnel; minimizing the labor force and material inventory required for maintenance, repairs, and restoration of generation, transmission, and distribution facilities; and reducing customer accounting, general accounting, and administrative expenses.

*d. Improved Quality of Service*

Among the requirements in this category are reducing the frequency, duration, and extent of outages in the power supply, reducing voltage and frequency discontinuities and sudden excursions (power-line disturbances) to sensitive electronic loads and digital equipment, and improving customer services through prompt response to inquiries, requests, or complaints. It is also important to maintain the power supply within prescribed ranges and specifications, and to restore interruptions in service quickly.

*e. Enhanced Environment*

Means of improving environmental impact include reducing thermal discharge to natural bodies of water through the use of artificial lakes, cooling towers, and desalinization processes and advancing direct conversion of heat energy to electrical energy, as by magnetohydrodynamics, thermionics, and fuel cells. Also involved in conventional systems are reducing the release of combustion products (sulfur dioxide, nitrogen oxides, and particulate matter) in the atmosphere; reducing the frequency, duration, and intensity of pollution concentrates on urban areas; and providing more productive uses for fly ash. Safer storage of nuclear waste is of primary importance.

In the design of systems, selecting remote or underground sites for generating stations, improving aesthetics by the increased use of underground distribution facilities, and beautifying transmission towers and lines in harmony with the countryside are all being urged by environmentalists.

*f. Improved Employee Skill*

- *Labor.* In earlier years, the power industry had a labor force of about half a million employees, a small force when compared to its very high output. In the seventies, while the generating capacity doubled, the number of employees remained substantially the same. This was achieved through the operation of larger installations with fewer personnel, centralized control of generation and transmission, unattended substations, and minimizing maintenance and repair crews by automating dispatch procedures. This trend no longer holds.
- *Professional.* The design and construction of large installations such as generating sta-

tions and extra-high-voltage lines are often contracted out and are engineered and supervised by consulting firms. Thus, in effect, the consultants provide a common professional pool for all utilities. The electrical manufacturers have been primarily responsible for research and development of the industry; and the practice of accepting turnkey contracts is common. Thus manufacturers also provide a common pool of labor.

However, the advent of nuclear power, extra high voltage, and environmental limitations requires significant changes in utility systems and calls for an increase in both the quality and quantity of professional labor. The industry recognizes the need for this rapid increase in in-house skill. This can be provided by (1) improving the productivity and effectiveness of employees, (2) merging and affiliating with neighboring companies forming regional groups, and (3) maintaining aggressive in-house research and development as well as supporting institutions of higher learning and research organizations by sponsoring research and development efforts.

***3. Spectrum of Computer Usage.*** A review of engineering and operating computer applications indicates that they fall within several broad categories as shown in Fig. 25-3 and as described below.

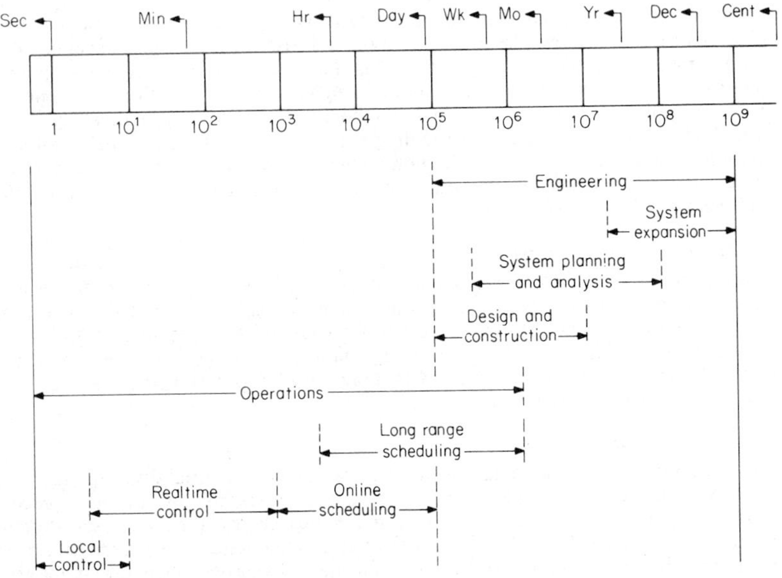


**FIG. 25-3** Spectrum of computer usage.

*a. System Expansion*

These applications are related to 20-, 10-, and 5-year construction programs and cover planning, design, and construction of new facilities. These functions are performed at least once a year and use long-range load forecasts and other predictions as input data. Batch data processing using large-scale computers, disk storage, and fast peripheral equipment such as printers and data entry equipment can effectively meet the computational requirements of this category. In recent years, however, the use of decision support systems (such as corporate models) requires interactive computing.

*b. System Planning and Analysis*

These applications deal with 3- and 1-year construction of new facilities and the economic and reliable operation of these additions in conjunction with other interconnected power systems. Nuclear fuel management, annual hydrothermal coordination, and coordination for firm transmission and generation planning are among these functions. Because these programs are more frequently called on, they normally reside on disk storage devices. Thus only changes in data and programs need be entered when using specific programs.

*c. Long-range Scheduling (Operating)*

These applications are related to annual, monthly, and daily operation of the power system. In this category are transmission and generation maintenance scheduling, unit commitment and withdrawal, and other functions dealing with both reliability and economy of operation. Remote data processing consisting of a central computer system, associated peripheral equipment, random-access files, and remote terminals meet the requirements of reliability and economy of operation. The user enters data and inquiries for new schedules through an appropriate person-machine terminal. These terminals can be located at all engineering and operating centers.

*d. On-line Scheduling*

These applications are related to security monitoring and determination of reserve indexes and hourly data recording. These schedules are performed at least once an hour, although some applications such as pumped storage scheduling are performed weekly and daily. They are based on historical data but also need current power system data such as facilities in and out of operation, generation outputs, and line flows. Therefore, they require direct data flow into the computer. The results, however, are presented to the user for consideration and execution. Because of the scheduling nature of these applications, very fast data acquisition is not a prerequisite; however, accuracy and timeliness of schedules are related to the extent that they include direct data acquisition.

*e. Real-time Control*

These regulating functions are carried out to meet the changing demands on the power system. Power system monitoring, security assessment, and display, rescheduling, and control of system frequency, tie-line flows, voltage conditions, and transmission flows are examples of this category. Other examples are closed-loop automatic control of generating units and interchange scheduling with neighboring companies and pool areas. These functions are performed in a time range of a few seconds to several minutes and therefore not only require direct data flows into the central computer but, in addition, require signals from the computer to the various remote controllers and actuators.

*f. Local Control*

These applications require a response speed beyond the capability of central computer control and related communication. Most of these functions are initiated immediately after a fault develops or a variable exceeds certain limits. Their objective is to react quickly and correct the situation or to isolate and contain a disturbance. These functions are performed in the few milliseconds to several seconds range and can best be handled by local computers: (1) by directly sensing variables and controlling through actuators (e.g., direct digital control of boilers or digital relaying of the substations), and (2) by superimposing the computer on the local controllers or protective relays in order to reset their operating positions. The latter applications are in the 1- to 10-s range.

The computational requirements shown in Fig. 25-3 cover both engineering and operating functions. These areas of computer activities are interrelated. From the above discussion it is clear that the power system operating functions do not all necessarily have to be performed in real time.

## ENGINEERING APPLICATIONS

As the electric utility industry has grown in size and complexity, modifications and additions to existing electric power networks have become increasingly costly. Therefore, it is vital that different design possibilities for additions and modifications to the network be studied in detail to determine their effect on the network, their effect during abnormal operating conditions, and their applicability as a flexible solution to current and future power demands.

The design and construction of planned facilities involves the efforts of a sizable engineering staff and a substantial investment in facilities. To provide support in these activities, computer programs have been developed for analysis of specified designs. The application of these programs contributes to the installation of reliable and economic facilities. The major engineering applications are shown in Fig. 25-2 and summarized below.

***4. System Expansion.*** The system expansion applications (Fig. 25-4) support the long-

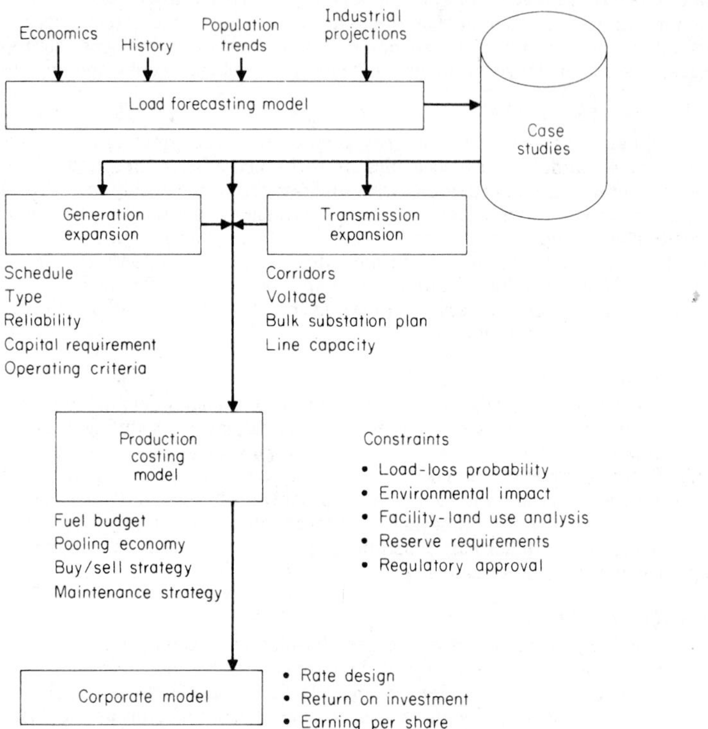


**FIG. 25-4** System expansion applications.

term (5 to 20 years) planning function for generation and transmission of power. The system expansion application area represents the typical decision support environment in that many cases are produced, and a variety of options and strategies are considered in the planning process. This area controls large common data sets from multiple sources. Lengthy reports are produced for internal documentation and regulation approval. In the past most of the processing was batch-oriented. Today there is a trend for on-line dialogue with the applications.

With the current economic outlook, the majority of emphasis in the industry will be to develop more efficient use of existing facilities rather than new construction. Load forecasting and production costing are becoming the most significant items in system expansion to predict load requirements and operating costs. Trade-offs between expansion and new facilities are increasingly important. The applications in this category are

*Load Forecasting.* This application is the basis for all planning functions. It utilizes historical data, trends, economic factors, and residential and industrial projections by geographic area to produce load requirements and load duration plots by area. It also predicts the load factor.

*Generation Mix Analysis.* This plans the optimum mix of peaking or base-loaded units, fuel type, and location of units to meet the future load requirements. It also provides a buy-and-sell analysis and accounts for reliability of generation.

*Production Costing.* This simulates the operation of the existing and planned generation facilities for several years in order to predict the fuel budget. It meets the load forecast and accounts for the generation availability and the hour-by-hour dispatch of generation. New techniques use statistical approaches versus detailed models.

*Loss of Load Probability.* This accounts for unit availability and the reserve requirement to produce a probability of loss of specific loads.

*Voltage Level Analysis.* This application is a tool to plan voltage levels of existing and planned transmission facilities. It provides trade-offs of network losses versus capital requirements.

*Environmental and Facility Land-Use Analysis.* This set of applications assists the planner in locating plants, substations, transmission towers, and lines. Trade-offs considered are expansion versus new facilities, right of way utilization, and environmental impact of planned facilities.

***5. System Planning and Analysis.*** This application area supports the short-term planning process and provides tools for analyzing incremental expansion. System planning and analysis applications are high in floating-point content and represent a significant computational requirement. Many cases are analyzed, and there is a rapid turnaround requirement. While there is on-line dialogue with the application and it is common to provide on-line display and edit of results, the trend is to use interactive graphics in all phases of this decision support process. The specific applications are shown in Fig. 25-5 and described below.

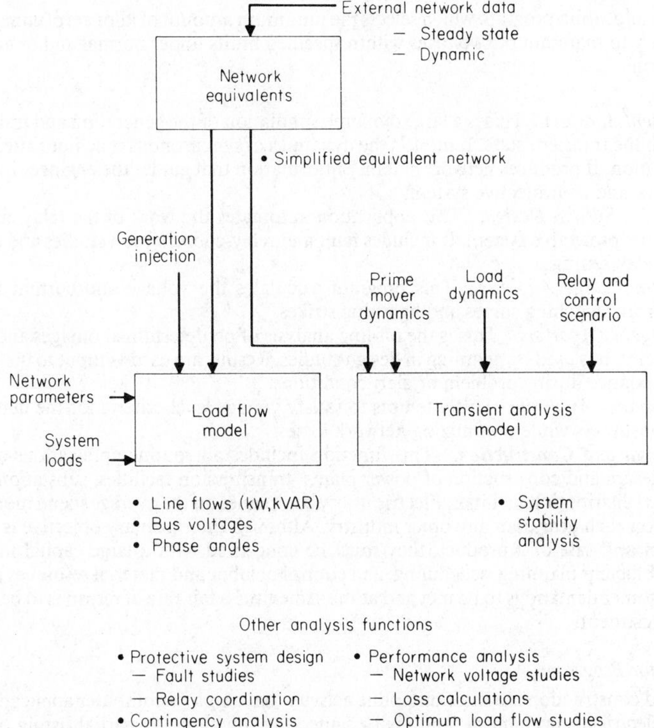


**FIG. 25-5** System planning and analysis applications.

*Load Flow.* The load flow program is one of the major tools of system planning and is utilized extensively. Important to the system planner are that input data errors be minimized and that there be an easy and rapid turnaround for answers when the frequency of program use is high. To accomplish this, interactive capability is provided with the ability to store base cases or numerous power system models on the computer's disks. The storage capability provides many different cases that the planning engineer can access for studying or varying a particular system condition. Load flow enables the power system planning engineer to simulate and solve various power system expansion alternatives in an interactive mode. It utilizes a limited graphics color terminal specified with a special set of graphic characters that presents results in the form of system one-line diagrams. The multicolor feature of the terminal is used to indicate heavily loaded lines, bus voltages outside normal limits, and open circuit breakers. The engineer working at such a terminal may, with a light pen and alphanumeric keyboard, remove, add, or change elements of the system being studied and request a solution from the host computer.

The simulation programs associated with the load flow program are a system of linked programs that have the following capabilities:

1. Basic programs involving calculation of voltages, power flows, angles, and interchanges between areas of a power system.
2. A network reduction program to represent large networks as equivalents in conjunction with the specified area to be studied.
3. A current-distribution program that indicates the sensitivity of response of the various circuits to outages of specified transmission lines.
4. The series of programs associated with the stored load flow files which permit accessing a particular case, deleting a case, adding a case, and changing a case.
5. The var allocation program which selects the minimum amount of kilovars of compensation necessary to maintain bus voltages within specified limits under normal and or emergency conditions.

*Transient Analysis.* This is a large dynamic simulation of the generation and transmission network in the transient state. It models the dynamics of synchronous machines after a system fault condition. It produces network stability information that guides the engineer in design of the network and its protective system.

*Protective System Design.* This application automates the work of the relay engineer in designing the protective system. It includes fault and relay-coordination studies and calculates complex relay settings.

*Switching Surge Analysis.* This program calculates the voltage and current transients resulting from switching surges and lightning strikes.

*Contingency Analysis.* This is the off-line analysis of predetermined outages and network contingencies. It is used by planning in design studies. Results are used as input to the operating area for guidance during problem or alert conditions.

*Performance Analysis.* This attempts to satisfy voltage level criteria for the network and specific consumers while minimizing network losses.

***6. Design and Construction.*** This function includes all routine applications associated with the design and construction of power plants, transmission facilities, substations, service centers, and distribution facilities. Electric utility companies on the average spend more for new construction each year than any other industry. Although their primary objective is the usual production and sale of a product, they must be concerned with a large capital investment program. Efficient planning, scheduling, and control of labor and material resources are necessary if customer demand is to be met and at the same time a fair rate of return is to be provided on the investment.

*a. Major Programs*

Design and construction is a multidiscipline activity that employs computer applications dealing with electrical, mechanical, and civil engineering functions. A partial listing of specific applications is described below.

The *tower analysis* program provides a summary of the maximum tension and maximum compression for each member of a three-dimensional structure over the entire load range specified. This program also spots structure locations; plots a profile of the transmission line; and calculates sag, insulator swing, and ground clearance.

*Line sag* calculates sag and tension of conductors under a given situation.

*Branch circuit design* uses the load, distance, number of cables in a raceway, wire temperature rating, motor starting and full-load amps to compute the voltage drops and sizes of breakers, cable, and conduit in a circuit.

*Structural design* programs are used to design concrete and steel structures using as input the structure configuration and loads such as floor, roof, and impact.

The *structural steel framing* program is used to design the beams, columns, girders, and baseplates of power plant structures.

The *foundation-slab analysis* program is used to design large, complex foundation mats. The results permit evaluation of various slab thicknesses, soil bearing pressures, shears and bending moments, and reinforcement areas.

The *concrete stack analysis* program is used to analyze proposed stacks by determining loadings, resulting stresses, and required steel reinforcement. This program is used extensively in the design of very tall concrete stacks selected for new power plants.

*Piping* programs are used to perform stress analysis of piping systems and determine hanger design information. The *power plant piping* program analyzes the flexibility of a piping system under the influence of temperature.

The *cable routing* program provides the shortest cable route between nodes, percent raceway fill, and number of cables in a tray or raceway.

*Interference analysis* resolves the interference between pipe, cable tray, and structures occupying the same space.

The *heating, ventilating, and air conditioning design* program uses thermal loads and the building configuration to calculate the size of refrigeration equipment and ductwork required.

*Fluid dynamics analysis* analyzes piping systems for pressure drop, flow distribution, and power requirements.

*Hydrologic analysis* is used to determine seepage flow networks, underground flow, and rainfall and runoff drainage for culvert and bridge size and design.

*Earthwork design* is used to design embankments and roadways and perform settlement and embankment stability analysis.

*Geotechnical evaluation* is used to evaluate soil testing results and determine the strength and swell of soil for dam and foundation design.

*Foundation design* programs are used to design foundation pile, pier, mat, and spread footing.

A *statistical analysis of equipment failures* means data collected on the frequency and cause of equipment failure is used to determine the likelihood of similar failures in the future based on various changes. The purpose of establishing an equipment operation data base is to record and summarize the specific causes of service interruption to generation, transmission, distribution, and communication system equipment as well as to customers. The data provide the basis for designing new systems to specific reliability levels, monitoring equipment and manufacturer adherence to desired availability standards, and carrying out maintenance scheduling activities.

The data base consists of a main file for each major equipment category and supplementary files which supply input to a family of programs designed to provide the engineers with periodic statistical reports. Engineers also have the ability to retrieve from this data base any combination of data of their own choosing.

*Transformer load management* consists of a series of programs to process manufacturing performance data for distribution transformers and derive an economic evaluation based on unit cost and expected loss contributions over the expected lifetime. Since distribution transformers represent such a substantial proportion of system investment, it is imperative that they be utilized to their fullest economic capability.

A large percentage of distribution transformers are nominally underloaded; that is, they are oversized for the load being served and, hence, waste money through overinvestment and excessive core losses. Overloaded transformers also waste money in terms of copper losses, loss of life, fuse and transformer burnouts and replacements, and the investigation of low-voltage complaints. In fact, records indicate that about 50% of customer voltage complaints are directly associated with heavily loaded transformers.

Studies indicate that if one could achieve a loading pattern which attempts to maintain transformers within their economic loading range of 80 to 160% of nameplate capacity, one could obtain an annual savings of millions of dollars.

*Drafting* includes engineering sketches and standard symbols used to lay out a drawing on a terminal and the results are printed or plotted. Included in this application are computer-aided design and drafting packages.

*Economic analysis* is used to make economic decisions between alternative sets of system designs or equipment.

Most of the above applications require common data. The trend is to treat the data as a corporate resource and to capture and maintain it in a common data base. This provides for consistency of data, avoids duplication, and minimizes errors due to entry of the same data in different programs. This common data base can then be used by many of the design and construction programs.

Resource management subsystems that support the design and construction applications are briefly described below. The purpose of resource management is to help utilities in more effective utilization, control, and management of their basic resource (people, equipment, and facilities) in the distribution system.

*b. Distribution Construction Information System*

This application supports the management of new investment and maintenance in the distribution area. It provides information for planning, scheduling, and controlling equipment, labor, and material resources, and becomes a tool to assist public utilities in providing consistent service and meeting customer demands while realizing fair rates of return on capital investment.[1*]

The term *distribution construction* refers to the entire process of work requesting, design, scheduling, reporting, and closing of that portion of the facilities closest in service to customers. New distribution work stems from three types of activity: system maintenance and improvement requirements, customer requests, and inspections or surveys.[2]

For many utilities the distribution system alone represents close to one-third of the total capital investment. Thus, it is not surprising to find continuing concern with the process by which facilities are constructed and maintained and by which costs are transferred to property accounts or charged to expense appropriations. This concern is often focused on improving the distribution work process to support the planning, design, scheduling, controlling, and tracking of jobs. Such improvements are undertaken to obtain a more effective and efficient work process leading to an earlier plant-in-service and to improving the utilization of the many resources devoted to distribution construction, maintenance, and operating tasks.[2]

Computerizing the distribution work process is desirable because of the significant capital investment and expense components of utility costs and because the work process has characteristics that lend themselves to a high degree of computerization and to improved productivity and control.[2]

*c. Distribution Facilities Information System*

This application provides the information required to plan, control, maintain, locate, account for, and manage the distribution facilities of an electric utility.[3] It is also referred to as an automated mapping and facilities management system. It is composed of a graphics system and a data base system.

The graphics system provides the interactive functions required to capture information needed to maintain a data base for facilities by locations. The user is required to define to the graphics system the facilities and the data elements which are associated with them. This includes its data fields, the pictures that are displayed on a map to represent it, and the connectivity requirements, if it is a network facility.

The graphics system employs a graphics work station composed of at least one high-resolution display, an alphanumeric display, a keyboard, a digitizing tablet, and various hard-copy devices. This work station is used to enter geographically related data, making it subsequently possible to display the data pictorially (maps), interact with the display (zoom in, window, edit, etc.), display facilities data and alter them, or to make additions or revisions. These are functions that formerly involved manual drafting and filing methods.

Maps or data generated at the work station may be stored in a common facilities data base

* Superior numbers refer to References, Par. **16**.

accessible to many users. The manner in which these data are stored varies. Some systems are able to retrieve only the map facets that were entered; the user must establish the relationship between adjoining facets. In other systems the data exist as a continuous network and the user requests only the portion and type of data needed. Storage techniques based on the common corporate data base management system are becoming prevalent because of the common requirements for facilities data by many departments throughout the company.

The production of maps, diagrams, and pictures is a by-product of this system. Of far greater importance is the network relationship of the facilities data. This allows such applications as feeder analysis, transformer load management, fuse coordination, branch circuit design, and fault current calculations to be executed using the common facilities data base.

*d. Materials Management Information System*

This system is used to plan for and control the flow of materials in and out of the company. A material items data base may be accessed by many departments for multiple purposes. The main subsystems are stores operations, materials planning, and purchasing.[3]

Stores operations relates to all day-to-day activities within the warehouse location. Included in these are functions such as stock inquiry handling, recording of stores transactions (receipts, issues), item location management, order and requisition initiation, and material reservation and allocation control.[3]

Materials planning refers to the control and management of an inventory, both repairable and expendable parts. The functions under this application are acquisition analysis, item forecasting, materials requirements planning, reporting, and stock taking control.[3]

The purchasing area includes the functions of ordering material from the suppliers and transferring to the inventory on receipt of the material. The functions within this are purchase order writing, quotation preparation, receiving, returns, implementation, quality assurance, vendor performance analysis, and invoice matching.[3]

*7.* ***Project Management.*** The objective of project management is to control project costs and schedules in the maintenance and construction of power system facilities. Power engineers have used manual project information systems for years. Now, there is rapid movement to automation of project control with provisions for on-line display and edit of results.

Project management in the engineering departments includes project control, project scheduling, and resource optimization tools. These tools are required to manage small procurement projects. However, they could also be major, long-term projects such as construction of a facility, installation of a major program, or daily tracking of activities within a department. Automated techniques, taking into consideration the control of time, resources, and costs, allow more productive utilization of project management personnel and stricter control of projects than manual methods. There are three major components of a project management system:

1. *Critical Path Method.* A network represents a project which consists of a mixture of serial and parallel activities and employs a combination of personnel resources, materials, and facilities. When time is associated with each activity within a network, critical path methodology can be used to analyze the network and determine the longest time path to completion of the project. All other time paths through the network will then have some slack in terms of the critical path.[3]
2. *Resource Management.* Project management and scheduling provides the means to plan and control a variety of projects. These systems permit tasks to be scheduled, resources assigned, costs allocated, and progress reported. Using this process, management can address identified problem areas and adjust their plans accordingly.[3]
3. *Project Costing and Estimating.* Cost control techniques involve the ability to estimate and assign costs for labor, material, facilities, test equipment, and other resources to all activities comprising the execution of all phases of a project. In addition, some application programs permit extending rates; accommodate matrix and other organization structures; compute general, administrative, and overhead expenses; and summarize and project costs over selected parts of projects as well as multiproject groups.[3]

*8.* ***Administrative Support.*** Administrative functions have been automated to serve the various requirements of engineering departments. Because these requirements are common throughout the company, integrated or common solutions are often used. Administrative support typically falls into the following categories.

*Text processing* is the preparation, output, and data entry and editing of text using an interactive host-based or stand-alone computer system. This service may be used by a secretary to compose a letter or modify an existing memo. It may also be used directly by an engineer for notes, lists, progress reports, and general documentation. Text processing in a power company is used to prepare and maintain operating standards and procedures, standard material lists, nuclear records, training manuals, maintenance and safety procedures, regulatory reports, and specifications.

*Administrative processing* allows the user to manage electronic document images. It includes activities such as copying and reproducing, document distribution, records file processing, mailing, and office correspondence.[3] Documents may be filed on disk, searched for, and retrieved. Document search may be by name or by complex search parameters as in nuclear records.

Other administrative services provide a convenient means for writing notes, reminders, messages, and appointment records. Typical functions include calendaring, tickler file (diary), meeting schedule, phone list, and to-do lists.[3]

*Text and data integration* applications provide the ability to include data created outside of text applications to form reports and letters. These may be used for the creation of manuals which include specification data to reduce redundant keystrokes and increase accuracy.[3]

*Communications* applications provide an informal and unstructured method of communication within an organization. This provides the means of handling messages that might otherwise require a phone call or memo. It also allows for distribution to multiple locations and receipt acknowledgment.[3]

*Personal computing* provides engineers with the tools, packages, and techniques that allow them to enter and manipulate data and accomplish in hours what might otherwise take weeks. The intelligent work station provides local processing, user-friendly interfaces, and access to host applications.[3]

*Presentation graphics* applications are used to present data in a pictorial form. The graphics may be displayed on a terminal or converted to a hard copy using a printer, plotter, terminal copier, or an attached camera device. Typical uses include the presentation of engineering or statistical data as line, bar, or pie charts, preparation of foils for a presentation, or drawing sketches or diagrams for inclusion in a publication.[3]

## OPERATING APPLICATIONS

The prime concern of the electric utility industry is to meet the consumer's power demand at all times and under all conditions. Electric utilities are continually seeking and have been most receptive to every available technique which would reduce the capital investment per kilowatt of installation, reduce the operating expenditures per kilowatthour of energy delivered, and improve the quality of service to the consumer.

Computer systems can be found in use at all levels of operation in power systems. At the generating-plant level, for example, they are used to control and monitor unit start-up and operating conditions. In bulk power substations they serve such functions as monitoring, event recording, and switching. And, at the system operating center level, they help to improve the economy of operation, improve the quality of service, and simplify system operation.

The major computer applications for the operating area of a utility are shown in Fig. 25-2 and briefly described below.

***9. Energy Management System (EMS).*** An EMS system provides automatic control of generation and monitoring and control of the transmission and distribution system. EMS applications assist the dispatcher in performing routine control functions, in maintaining the security of the power network, and in reducing operational costs. This application is critical to the operation of the power system and therefore has the highest executive visibility.

A typical EMS system (Fig. 25-6) consists of duplexed central computers and their associated peripherals, a color display person-machine subsystem, duplex front-end communications preprocessor equipment, and remote data acquisition and control equipment. An on-line

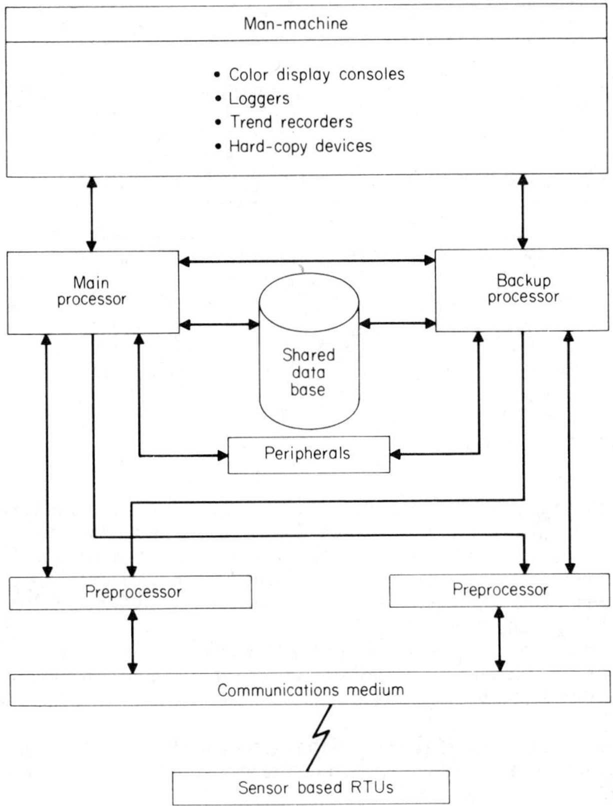


**FIG. 25-6** EMS architecture.

computer performs all the real-time functions, while a backup computer is in standby mode and also used for program development or for executing study applications. Real-time data are collected by remote terminal units (RTUs) and telemetered to the communications front end, which passes them to the central computer for processing. Resulting control action commands are routed over this same path to the appropriate power devices or generating units.

The dispatcher's interface to the EMS system is through color display terminals used to display one-line diagrams and tabular displays. These displays are updated cyclically or whenever a change is detected. Various data and operating conditions are highlighted with color. The dispatcher uses the displays to interact with the system such as acknowledging an alarm, changing operating limits, controlling devices, determining system status, or executing a program.

The major functions performed by an EMS system are shown in Fig. 25-7. The types of processing that these functions include are cyclic work that must be completed within a specific cycle; event-driven work that results from random alarms and events that occur on the network; real-time or demand work that has specific response-time requirements; and interactive work which is dispatcher-initiated that also has response-time requirements.

*a. Network Surveillance and Control*

This consists of continuous scanning of remote sensor-based units, acquiring all key network and generation data on a 2-s cycle, checking the data for problems, and presenting the status of

| Network surveillance and control | Generation commitment and control |
|---|---|
| • Network data acquisition<br>• Data conversion, limit checking<br>• Alarm processing<br>• Logging and reporting<br>• Load shedding power device control | • Automatic generation control<br>• Economic dispatch<br>• Load forecasting<br>• Unit commitment<br>• Reserve monitor |

| Interchange management | System security |
|---|---|
| • Production costing<br>• Transaction evaluation<br>• Interchange scheduling<br>• Interchange accounting | • Network status<br>• State estimation<br>• Contingency analysis<br>• Interactive load flow<br>• Training simulator |

**FIG. 25-7** EMS functions.

the network to the dispatcher through displays. The control function allows the dispatcher to take selective control action on power system devices such as circuit breakers.

*Network data acquisition* programs acquire power system data through periodic scanning of local and remote sensors attached to RTUs. The raw data are checked for missing or invalid data before being sent to the central processors by the communications front end. Retransmissions are requested for missing status data.

*Data conversion and limit checking* programs are used to process the raw data. The data are scaled and converted to engineering units and stored in a data base for use by application programs. Reasonability checks are made on the data and values not found to be within limits cause alarms.

*Alarm processing* programs generate alarms for scanned data and calculated results as changes of network status are discovered, limits are exceeded, or other invalid conditions are encountered.

*Logging and reporting* programs store selected data in historical log files, where they are available for analysis through displays and reports.

*Power device control* functions provide for the control of power system devices and for the placement or removal of device tags. Control actions made by the dispatcher to activate power devices are verified, then forwarded through the network to the proper destination.

*Load shedding* provides rapid access to data for guiding the action of the dispatcher during abnormal operating conditions. Typical of the data that are displayed is the amount of load relief that can be obtained through emergency output of all generating sources, generation requirements, interruptible loads, load curtailment, voltage reduction, and emergency backup through tie lines.

*b. Generation Commitment and Control*

This consists of a set of programs used to optimize the production and delivery of power for fuel savings.

*Automatic generation control* regulates generator output to match the load in order to maintain the frequency in an area and the sum of all active tie-line power exchanges with neighboring power systems. This program executes cyclically every 2 s.

*Economic dispatch* minimizes the cost of meeting the energy requirements of the system over a period of time and in a manner consistent with reliable service. Desired generator settings are computed and fed to the automatic generation control program. This cyclic program executes every 5 min and also on demand.

*Load forecasting* computes the total system hourly load for a specified number of days. It provides an adaptive forecasting system based on observed values of demand and estimated weather conditions. The program generally consists of three mathematical models. A load forecasting model uses past load data to compute hourly load forecasts. A weather forecasting model computes hourly weather forecasts based on past weather history. A weather correction model uses telemetered values of load and weather conditions to correct the forecast. The dispatcher may optionally enter these values through a display.

The historical load and weather data are stored and maintained in files which also contain special events that affect load such as holidays or unusual weather phenomena.

*Unit commitment* determines a schedule for optimum start-up and shut-down of thermal units which minimizes unit start-up costs subject to generation objectives, predicted area requirements (load forecast), security (spinning reserve requirements and off-system capacity), and operational constraints (unit minimum up and down times, limits, ramp rates, maintenance and derating schedule). It minimizes the operating costs of the dispatchable generating units. The total dispatchable generation is the sum of the load forecast and net scheduled interchange minus the total nondispatchable generation. The operating cost is defined as the sum of production, start-up, shut-down, and maintenance costs. Production cost is calculated by the use of input and output curves adjusted by fuel prices. Transmission losses are also considered using one or more sets of penalty factors.

The output of unit commitment is the hourly unit schedule. Several sets of schedules (or strategies) may be output for operator review and selection. The output is stored and made available to other applications such as transaction evaluation. This program is executed two times a day.

*Reserve monitor* calculates the available operating reserve necessary to meet company operating policies. Typically a spinning reserve, 30-min reserve, and 2-h reserve are computed. This program executes cyclically every 4 min or on demand. The dispatcher is alerted when there is inadequate reserve.

*Maintenance scheduling* assists operating personnel in scheduling generator maintenance in an optimum manner. It generates a maintenance schedule while taking into account constraints and limitations on available resources required to perform maintenance work. The maintenance schedule indicates when particular generating units will be out of service (unit outages). Maintenance is performed routinely at desired intervals or may be required because of unexpected forced outages.

During the maintenance of a given generator, the outage of this unit is compensated for by other capacities in the power system. This is done by maintaining a level of system reserves. The scheduling of generator outages is performed in a manner which maintains a flat megawatt reserve level or a consistent level of probabilistic level of risk, i.e., probability of emergency procedures.

### *c. Interchange Management*

Formal and informal interchange agreements may exist between neighboring companies. Operating as an interconnected system has economic and security advantages. While many power pools have control centers to manage the interconnection, most companies prefer to manage, or at least monitor, their own interchange of power. The interchange management programs provide the dispatcher with the ability to make good deals with the neighbors. These interactive routines allow the dispatcher to evaluate a buy or sell transaction before being committed to it. Once initiated the interchange scheduling program automatically schedules the transaction through the automatic generation control program.

*Production costing* provides the capability to compare a change in interchange with the current schedule. The program always starts by accessing a study file. The calculations are based on current or forecast system load and a proposed interchange schedule entered by the operator.

*Transaction evaluation* calculates the costs and savings associated with the sale and purchase of power with a selected interconnected company. The program can be used to evaluate past, present, and future operation costs. A production cost calculation is generally included as a part of this program or may be an independent routine.

For evaluating transactions, the transaction evaluation program computes the costs of savings of a proposed transaction by comparing the production costs computed from two economic dispatch calculations: A base economic dispatch calculation for the operating conditions with and without the proposed transaction.

The transaction evaluation program can usually be executed in two modes which specify the starting point for all calculations. These two modes are whether the unit commitment program will be called or not. Essentially, the two options determine how the generation schedule is to be determined. Mode A (also called economy A) is executed without unit commitment. Current system load, interchange schedules in study file, and existing unit commitment form the starting conditions. Mode B (also called economy B) is executed with unit commitment and is used for transactions in the future.

*Interchange scheduling* is used to process and display each scheduled transaction with interconnected utilities. It computes the total net scheduled interchange as a function of time for use by the automatic generation control program. This calculation considers the start and stop times, generator ramp rate, magnitude, and direction of each active interchange transaction. Interchange schedules and cost information may be logged and displayed.

*Interchange accounting* provides the capability to account for electric power in the system. This electric power, in the form of measured, calculated, or scheduled megawatts, includes megawatts which are generated, consumed, lost, passed through, sold, and purchased. It includes functions for logging, displaying, reporting, and updating data recorded by the realtime system.

*d. System Security*

These programs help to reduce the chance of a major outage or blackout condition. They operate on a study data base which has been cleansed of missing data and metering errors by the network status and state estimation programs.

*Network status* determines the current configuration of the network based on the status of circuit breakers and disconnect switches obtained from the real-time data and from manually entered status information for devices not telemetered. The program is executed periodically, whenever a status change is detected, or upon dispatcher request. In addition, the program develops the corresponding mathematical model of the network using the impedance data for the current base load flow or state estimation case. The model will be used in the subsequent calculation of real-time system conditions using the load flow or state estimation programs.

*State estimation* determines the current state of the power system, including voltage levels and power flows, and calculates loss factors for use by the economic dispatch program for generation scheduling. It filters real-time measurements to detect and eliminate known errors; estimates expected values of the next real-time measurements; and determines the current network configuration. It corrects the mathematical model of the current network configuration. Thus, the state estimation program acts as a filter between the raw real-time data and the real-time data requirements of other security applications. The state estimation program is executed whenever the network status program detects a change, periodically, or upon dispatcher request.

*Contingency analysis* computes the potential effect of contingencies involving the loss of generation and transmission facilities. A specific set of predefined contingencies is analyzed on a cyclic basis. It simulates a contingency and calculates the changes in bus voltages and power flows resulting from the contingency. The base conditions for this calculation are the bus voltages or power flows obtained from the load flow program.

*Interactive load flow (ILF)* allows the dispatcher to perform load flow studies for a scheduled outage or analyze corrective actions after an unexpected outage. ILF uses real-time data to project bus loads in a network. A bus is a connection where power lines change direction or voltage. A bus load is the load at this point. The ILF program stores mathematical models of the present and planned networks. A color display terminal is used to display one-line diagrams of these models. Colors are used to differentiate voltages, heavily loaded lines, bus voltages outside of limits, and open circuit lines. The dispatcher uses a light pen together with the terminal keyboard to retrieve previous load flow cases, make modifications, create displays, execute load flow cases, and store them.

*Training simulator* uses a model of the power network to produce realistic reactions to a dispatcher in training. This function is usually run on the backup central processing unit (CPU). The program can operate in either of two modes: monitor and simulation mode. In monitor or playback mode, the simulator reflects the changing status of the power system during the time period when a playback tape was written. The trainee can view displays, thus monitoring power system changes, but cannot take any control action.

In simulator mode, the trainer can input system changes, and the trainee is allowed to perform control actions. The simulator software will reflect the state of the power system based upon the changes input and control actions taken. In this manner the trainee can operate the control system without affecting the true state of the power system.

***10. Power Plant Monitoring and Control.*** This application approaches the traditional process control application. Several applications such as fuel monitoring and performance calculations increase the requirement for the data processing resource. The three major computer systems for fossil power plant monitoring and control are described below.

*a. Process Control System*

This is a closed-loop control system which takes its direction from EMS and automatically collects plant data by reading instruments. Physical and electrical parameters associated with the boiler, turbine, and generator are monitored on a continuous cyclic basis. Alarms and events are logged, and control of pumps, valves, and switches for routine functions and for start-up and shut-down are provided.

*b. Plant Monitoring System*

This is strictly a data collection system for fuel monitoring, performance calculations, and balance-of-plant calculations; no control actions are performed. Data are stored and retrieved as required to prepare reports and perform analysis. These reports include those required by the plant management, load dispatchers, and planning and engineering groups. Periodic reports are prepared to reflect plant operation. Unit incremental generation cost is determined periodically by collecting data that continuously reflect actual operating conditions. This information is transmitted to the dispatcher for use in load dispatching.

*c. Operational Monitoring System*

This is used by plant operators to enter manually collected operational data for record keeping, report writing, and engineering analysis.

In addition to these systems, the power plant may also use mini- or microcomputers for security systems, environmental systems, controlled access systems, and chemical analysis systems.

***11. Power plant maintenance*** stores pertinent plant maintenance information for analysis of maintenance costs and evaluation of equipment performance. The interactive portion of the system provides power plant personnel with the capability to enter problem data, planning data, and work execution data. Approval and verification functions are at each step of a work order's progress from problem description through work completion and commitment to history. Interactive functions also are provided for entry and maintenance of an equipment data base and for access to equipment history.[4]

The batch portion of the system provides for moving completed and rejected work from the active work data base to the equipment history data base. Batch functions are also used for work backlogs, scheduled work, and other reports.[4]

Because of its varied data requirements the maintenance information system also has interfaces to other computer systems in the power company. These are the materials information system for equipment stocking levels, the personnel information system for labor resources, and the general accounting system for cost tracking. Additionally, text processing services are frequently used.

***12. Fuel Management.*** Fuel is the single largest expenditure in power plant operations. Operating support personnel must plan for both short-term (1 year) and long-term (5+ years) fuel availability. Thus, control and improvement of fuel cost represent the most substantial contribution to the overall economy of power production. The system is used to administer the procurement of coal and oil. It facilitates the monitoring, reconciliation, and performance of fuel contracts. Accounting functions are used for fuel purchases, transportation costs, fuel usage, and inventory value. Also included are short- and long-term cash flow projections and managerial and regulatory reporting capabilities. The overall fuel management application is an integrated model of the load forecast, generation scheduling and dispatch, and fuel allocation with an objective of optimizing fuel contracts. This application is used routinely with growth and on demand as system perturbations occur.

***13. Load Management.*** The objective of the load management application is to improve the load factor (ratio of average load to peak load) and to be able to shed selected load during emergency conditions. Most utilities have pilot-tested some form of load management. The pilot solutions range from consumer guidance (voluntary) to complete control and metering of the consumer's load. While technical, and more important, institutional obstacles remain, several pilot solutions show promise as production load management systems.

As conceived at this time, the load management application has several conceptual solutions. The ones that appear to have promise are the following:

- *Load curtailment system.* This scheme employs one master computer at the power company and several computers at major industrial and commercial load centers. In this scheme there is a limited sensor-based data acquisition requirement and no closed-loop control. Reduction of load will be manual.
- *Automatic meter reading and load management.* This hierarchical approach employs the corporate customer information system and the EMS system to direct and receive the meter readings, an intermediate level to serve as a communication concentrator, and an intelligent data control unit to scan remote transponders at the meter locations.

***14. Nuclear Data Center.*** Functionally the nuclear data center (Fig. 25-8) consists of three systems, which are described below.

| Operational |
|---|
| Plant process computer system |
| Operator training simulator |
| Plant security system |
| Emergency response information system |
| Radiation monitoring system |

| Technical | |
|---|---|
| Fuel shuffling | Radioactive waste control |
| Fuel management | Meteorological/environment |
| Refueling outage scheduling | Chemical laboratory |
| Quality assurance | Health physics |
| Start-up testing | Technical services |

| Administrative | |
|---|---|
| Maintenance | Nuclear records management |
| Materials management | Financial control |
| Licensing | Plant operation |
| Document control | Personnel |
| | Training |

**FIG. 25-8** Nuclear data center functions.

*a. Operational System*

This consists of real-time and on-line monitoring of various subsystems in a nuclear plant. The plant process computer system in a nuclear plant is usually provided by the reactor vendor and integrated with the reactor. Additional monitoring facilities are now required by the Nuclear Regulatory Commission (NRC). One specific monitoring system is named the *emergency response information system,* which is now a firm requirement for all operating nuclear plants and a license requirement for plants under construction. It has the following components:

- *Safety parameter display system.* This provides additional monitoring and display facilities to the primary plant control room. The additional facilities provide continuous monitoring of critical parameters.

• *Technical support center.* This is a separate facility and staff within the plant to assist the primary plant control room during emergency conditions.

• *Emergency operations facility.* This is a separate facility and staff located off the plant site but in proximity to the plant. It will manage the overall emergency response and evaluate actual and potential radiological releases.

• *Nuclear data link.* This routes cyclic data from the plant to the NRC headquarters through a communication link. The critical data will be used to monitor an accident, to evaluate emergency conditions, to advise and assist the licensee, and to inform the general public.

*b. Administrative System*

This is used for maintenance and backup of the operational system. Storage and information retrieval systems are used for the storage and retrieval of large amounts of data as encountered in nuclear records management. These systems provide interactive direct access to, and fast scanning or searching of, large numbers of articles, reports, contracts, laws, general directives, or abstracts of publications.

*c. Technical System*

This is used for large simulation and nuclear fuel management programs. Computer programs simulating nuclear core physics in various levels of detail are presently being used for analysis, leading to continual improvement and enhancement of nuclear plants. These programs are used not only in the design of cores for nuclear reactors but also for the planning and decision making regarding fuel burnup and associated electrical generation for nuclear units. Core refueling schedules are of vital concern. Safety analysis for emergency situations is another area in which the simulation programs are used. Nuclear fuel management programs are used for the design of nuclear reactors and for nuclear fuel management calculations.

## ENGINEERING COMPUTING TRENDS

***15. Engineering computing trends are given here.***

*Costs.* Today the cost of the engineering user is 6 to 20 times the cost of the computer that the engineer uses. This range depends on the level of the user and the type of processing: machine-intensive (compiling, executing) or human-intensive (editing, scrolling). It is no longer cost-effective to economize on computing resources required by the engineering user.

*Interactive Processing.* This is replacing batch job processing. Interactive color terminals have replaced card-based systems. The planning engineer now uses a terminal to run an interactive load flow program instead of submitting trays of cards.

*Languages.* In addition to using FORTRAN for coding applications, the engineer now has the additional choice of programming languages such as APL, Basic, PL/1, and Pascal.

*Individual Terminals.* Productivity improvements are made when there is one terminal for one engineer. The terminal may be a stand-alone personal computer that is switched to the host computer only when required for data exchanges or additional computing capacity.

*On-Line Reporting.* This has replaced manual reporting of data. Although text processing systems are used to generate documents, text information can also be transmitted, distributed, and stored electronically to reduce the proliferation of paper documents.

*Results Selection and Editing.* This is now performed by scrolling a display instead of printing all results and sifting through reams of paper.

*Automated Data Flow.* The trend is to maintain one set of data that is available to all users. These data are protected by appropriate security features and procedures like any other corporate resource. However, the intent is to make common data available to all users and allow the appropriate users to update the common data. This common data base leads to accuracy, avoids duplication, and avoids loss of labor due to the reentry of data.

*Personal Computing.* While most of the computer applications that the engineer uses are complex programs, the productivity of the engineer is also increased by simple, direct tools for text processing, graphics, and project control. These tools are interactive and require a minimum of effort to use. These tools help the engineer make notes, drafts, lists, sketches, and establish personal meeting or project schedules.

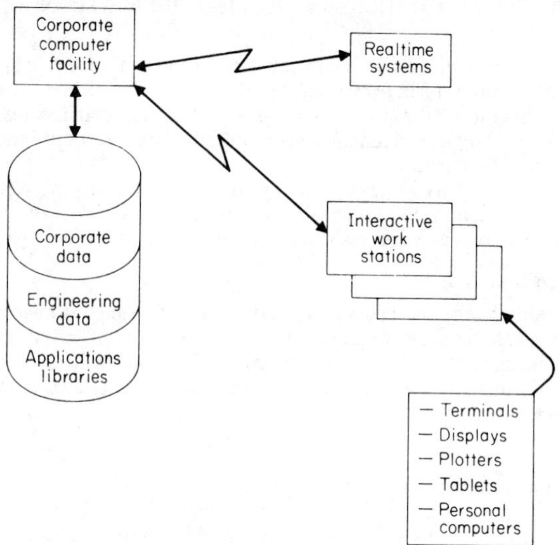


**FIG. 25-9** Power engineering centralized architecture.

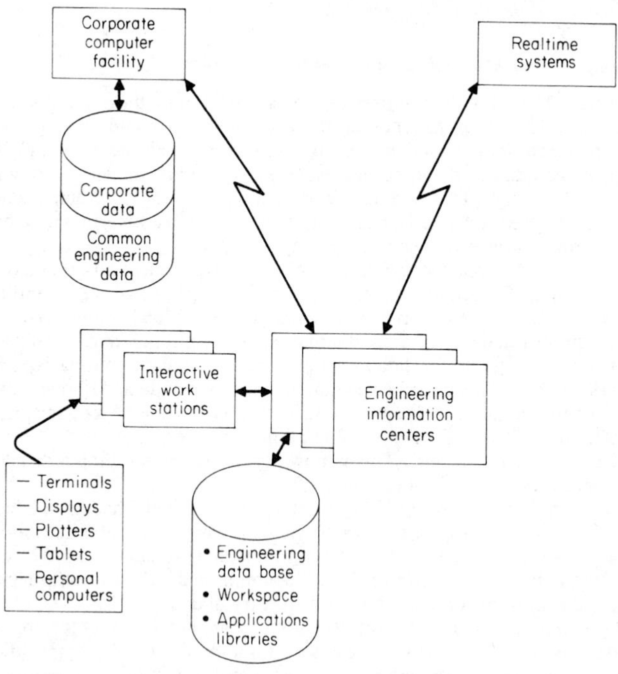


**FIG. 25-10** Power engineering distributed architecture.

*Computer Configurations.* Computer configurations such as shown in Figs. 25-9 and 25-10 are used to support engineering functions. Although the centralized computing approach has traditionally been used, the trend is to a distributed approach in which a separate engineering information center is assigned strictly to engineering work.

*Engineering User Service Objectives.* Engineering users require three categories of computing support:

1. *Performance requirements.* Less than 0.4-s response time is required for human-intensive work at a terminal. Longer response times reduce concentration and increase frustration. A predictable response time is required for machine-intensive work. Consistent timings for load flow are expected when operating under deadlines. High bandwidth terminals are required to support color graphics applications, which are used to reduce volumes of data to a presentation of results in a form that can be readily assimilated.
2. *Technical support requirements.* The engineering user requires a minimum of system programming knowledge to use the computing system. Included in this are user-friendly selection panels, menus, aids, and help functions; a library of programming tools, languages, and application packages, all of which are maintained by the central data processing department; and training facilities and technical support to assist engineers with the use of computing resources.
3. *Operational support requirements.* Security features and operational procedures are required to protect the user's data and access to the computing system. Included are security features such as password protection of user data files and sign-on's and operational procedures to provide backup of user data and prevent its destruction or loss.

### *16. References*

1. *Distribution Construction Information System, Description and Planning Guide,* GE20-0684, pp. 1–8, 1981 by IBM Corporation.
2. *Distribution Construction Information System, Overview,* GE20-0613, pp. 2, 4, 1981 by IBM Corporation.
3. *Applications and Abstracts,* G320-6131, pp. 60, 61, 263, 245, 246, 248, 270, 1983 by IBM Corporation.
4. *IBM 8100/DPPX Power Plant Maintenance System,* GB21-2729, p. 1, 1981 by IBM Corporation.

# SECTION 26
# ILLUMINATION

**Charles L. Amick, P.E.**
*Lighting Consultant, to Day-Brite Lighting Division, Emerson Electric Co.; Fellow, Illuminating Engineering Society of North America; Fellow, Chartered Institution of Building Services; Member, National Society of Professional Engineers*[1]

## CONTENTS

*Numbers refer to paragraphs*



[1] Includes some material from previous editions by Jack F. Parsons, Walter Sturrock, Karl A. Staley, and John E. Kaufman.

## RADIANT ENERGY AND LIGHT

*1.* For the principal purposes of illumination design, light is defined as *visually evaluated radiant energy.* The visible energy radiated by light sources is found in a narrow band in the electromagnetic spectrum (Fig. 26-1) approximately from 380 to 770 nanometers (nm). By

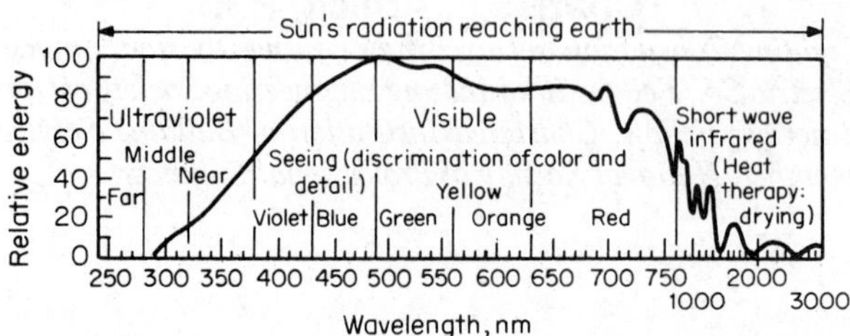


**FIG. 26-1** Ultraviolet, visible, and short-wave infrared are the three principal bands of the electromagnetic spectrum with which illuminating engineering is concerned.

extension, the art and science of illumination also include the applications of ultraviolet and infrared radiation. The principles of measurement, methods of control, and fundamentals of lighting system and equipment design in these fields closely parallel those long established in lighting practice.

## QUANTITIES, UNITS, AND CONVERSION FACTORS[1]

*2. **Luminous flux*** is the time rate of flow of light. See Table 26-1. Radiant energy in the visible region of the spectrum varies in its ability to produce visual sensation, the variation depending upon the wavelength. The ratio of the luminous flux to the corresponding radiant flux is known as *spectral luminous efficacy* (formerly luminosity factor) and is expressed in lumens per watt (lm/W). This varies with wavelength, having a maximum at approximately 555 nm. The data are plotted in Fig. 26-2. At very low levels of illumination the position of the maximum gradually shifts to 510 nm as a result of greater use of rod vision.

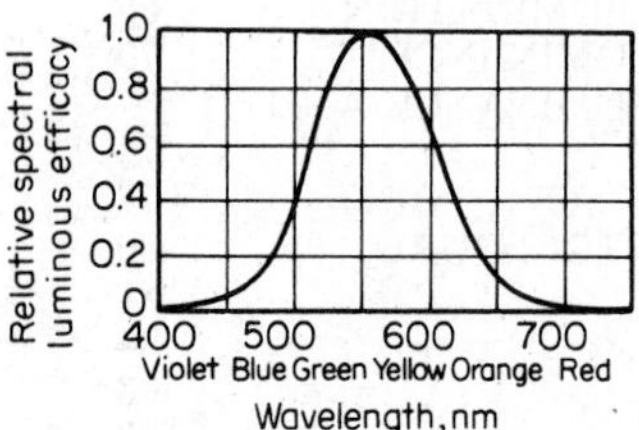


**FIG. 26-2** Spectral luminous efficacy (relative spectral luminous efficacy) of normal human eye.

From the foregoing it is apparent that two sources may radiate equal amounts of energy in the visible region of the spectrum but have different amounts of luminous flux emitted, depending upon the spectral distribution of the energy, the luminous flux being the integrated product of the energy per unit wavelength and the spectral luminous efficacies. *Lumen* is the unit of luminous flux. It is equal to the flux through a unit solid angle (steradian) from a point light source of one candela or to the flux on a square foot of surface, all points of which are one foot from a point source of one candela. Light sources are rated in lumens.

*3. **Luminous intensity*** is the luminous flux per unit solid angle in a specific direction. Hence, it is the luminous flux on a small surface normal to that direction, divided by the solid angle (in steradians) that the surface subtends at the source. (See Table 26-1.) The definition of luminous intensity applies strictly to a point light source. In practice, however, light emanating

[1] American National Standard Nomenclature and Definitions for Illuminating Engineering, RP-16-1980, Illuminating Engineering Society of North America.

**TABLE 26-1** Standard Units, Symbols, and Defining Equations for Fundamental Photometric Quantities

| Quantity* | Symbol* | Defining equation | Unit | Symbolic abbreviation |
|---|---|---|---|---|
| Luminous flux | $\Phi$ | $\Phi = dQ/dt$ | lumen† | lm |
| Luminous flux density at a surface | | | | |
| Luminous exitance | $M$ | $M = d\Phi/dA$ | lumen per square foot | $lm/ft^2$ |
| **Illuminance (Illumination)** | $E$ | $E = d\Phi/dA$ | footcandle (lumen per square foot)<br>lux ($lm/m^2$)†<br>phot($lm/cm^2$) | fc<br>lx<br>ph |
| Luminous intensity (candlepower) | $I$ | $I = d\Phi/d\omega$ ($\omega$ = solid angle through which flux from point source is radiated) | candela† (lumen per steradian) | cd |
| Luminance (photometric brightness) | $L$ | $L = d^2\Phi/d\omega(dA \cos \theta)$<br>$= dI/(dA \cos \theta)$<br>($\theta$ = angle between line of sight and normal to surface considered) | candela per unit area<br>stilb ($cd/cm^2$)<br>nit ($cd/m^2$†)<br>footlambert ($cd/\pi ft^2$)<br>lambert ($cd/\pi cm^2$)<br>apostilb ($cd/\pi m^2$) | $cd/in^2$, etc.<br>sb<br>$\mathbf{cd/m^2}$<br>fL<br>L<br>asb |
| Luminous efficacy | $K$ | $K = \Phi_r/\Phi_e$ | lumen per watt† | lm/W |

* Quantities may be restricted to a narrow wavelength band by adding the word spectral and indicating the wavelength. The corresponding symbols are changed by adding a subscript $\lambda$, e.g., $Q_\lambda$, for a spectral concentration or a $\lambda$ in parentheses, e.g., $K(\lambda)$, for a function of wavelength.

† International System (SI) unit.

from a source whose dimensions are negligible in comparison with the distance from which it is observed may be considered as coming from a point.

*Candlepower* is luminous intensity expressed in candelas.

*Candela* (formerly candle) is the unit of luminous intensity. One candela is defined as the luminous intensity of 1/600,000 square meter of projected area of blackbody radiator operating at the temperature of solidification of platinum under a pressure of 101,325 newtons per square meter. The original definition was in terms of the strength of a flame source, a standard candle.

***4. Illuminance*** is the density of the luminous flux incident on a surface; it is the quotient of the luminous flux by the area of the surface when the latter is uniformly illuminated. The term *illumination* is used to designate the act of illuminating or the state of being illuminated. Usually the context will indicate which meaning is intended, but the expression *level of illumination* is a term used to mean **illuminance** and should be discouraged.

*Illuminance* is the preferred term for the density of luminous flux incident on a point on a surface, but is subject to confusion with luminance and illuminants, especially when not clearly pronounced.

*Equivalent sphere illumination* is the level of illumination on a task from a source providing equal luminous intensity from all directions (sphere lighting) which would produce task visibility equivalent to that produced by a specific lighting environment.

*Footcandle* is the unit of illuminance when the foot is taken as the unit of length. It is the illumination on a surface one square foot in area on which there is a uniformly distributed flux of one lumen, or the illumination produced on a surface, all points of which are at a distance of one foot from a directionally uniform point source of one candela.

*Lux* is the International System (SI) unit of illuminance where the meter is taken as the unit of length (see Table 26-2).

***5. Luminance (photometric brightness)*** is the quotient of the luminous flux leaving or arriving at an element of a surface and propagated in directions defined by an elementary cone containing the given direction, by the product of the solid angle of the cone, and the area of the

**TABLE 26-2** Conversion Factors for Lighting Units

A. Illuminance

1 footcandle = 1 lumen per square foot
1 lumen-hour = 60 lumen-minutes
1 lux = 1 lumen per square meter = 1 meter-candela
1 phot = 1 lumen per square centimeter

| Number of → / Multiplied by ↘ / Equals number of ↓ | Footcandles | Lux | Phots | Milliphots |
|---|---|---|---|---|
| Footcandles | 1 | 0.0929 | 929 | 0.929 |
| Lux | 10.76 | 1 | 10,000 | 10 |
| Phots | 0.00108 | 0.0001 | 1 | 0.001 |
| Milliphots | 1.076 | 0.1 | 1,000 | 1 |

B. Luminance

1 footlambert = 1 lumen per square foot
1 millilambert = 0.001 lambert
1 apostilb (international) = 0.1 millilambert
1 lambert = 1 lumen per square centimeter
1 stilb = 1 candela per square centimeter
1 nit = 1 candela per square meter

| Number of → / Multiplied by ↘ / Equals number of ↓ | Foot-lamberts | Candelas per square meter | Milli-lamberts | Candelas per square inch | Candelas per square foot | Stilbs |
|---|---|---|---|---|---|---|
| Footlamberts | 1 | 0.2919 | 0.929 | 452 | 3.142 | 2919 |
| Candelas per sq m | 3.426 | 1 | 3.183 | 1550 | 10.76 | 10,000 |
| Millilamberts | 1.076 | 0.3142 | 1 | 487 | 3.382 | 3142 |
| Candelas per sq in | 0.00221 | 0.000645 | 0.00205 | 1 | 0.00694 | 6.45 |
| Candelas per sq ft | 0.3183 | 0.0929 | 0.2957 | 144 | 1 | 929 |
| Stilbs | 0.00034 | 0.0001 | 0.00032 | 0.155 | 0.00108 | 1 |

orthogonal projection of the element of the surface on a plane perpendicular to the given direction; or it is the luminous intensity of any surface in a given direction per unit of projected area of the surface as viewed from that direction. See Table 26-1.

*Footlambert* is a unit of luminance (photometric brightness) equal to $1/\pi$ candela per square foot, or to the uniform luminance of a perfectly diffusing surface emitting or reflecting light at the rate of one lumen per square foot, or to the average luminance of any surface emitting or reflecting light at that rate. The average luminance of any reflecting surface in footlamberts is, therefore, the product of the illuminance in footcandles by the reflectance of the surface.

*Candelas per square meter* is the SI unit of luminance when the meter is taken as the unit of length. Another term for this unit is *nit,* which is not commonly used in North America.

***6. Luminous efficacy*** is a quantity denoting the effectiveness of light sources. It is the ratio of the total luminous flux (lumens) to the total power input (watts). (The term *luminous efficiency* has in the past been extensively used for this quantity.) The maximum luminous efficacy of an "ideal" white source, defined as a radiator with constant output over the visible spectrum, is approximately 220 lm/W.

***7. Reflectance*** ($\rho$) is the ratio of reflected flux to incident flux. Measured values of reflectance depend upon the angles of incidence and view and on the spectral character of the incident flux. Because of the dependence, the angles of incidence and view and the spectral characteristics of the source should be specified.

***8. Transmittance*** ($\tau$) is the ratio of the transmitted flux to the incident flux. Measured values of transmittance depend upon the angle of incidence, the method of measurement of the transmitted flux, and the spectral character of the incident flux. Because of this dependence, complete information on the technique and conditions of measurement should be specified.

***9. Absorptance*** ($\alpha$) is the ratio of the flux absorbed by a medium to the incident flux. The sum of reflectance, transmittance, and absorptance is one.

***10. Brightness*** is the term that refers to the intensity of sensation resulting from viewing light sources and surfaces. This sensation is determined in part by the definitely measurable luminance defined above and in part by conditions of observation such as the state of adapta-

**TABLE 26-3** Color Temperature and Color Rendering Index of Some Common Light Sources

| Light source | Correlated color temperature, K | Color rendering index |
|---|---|---|
| Fluorescent: | | |
| Warm white | 3020 | 52 |
| Warm white deluxe | 2940 | 73 |
| Cool white | 4250 | 62 |
| Cool white deluxe | 4050 | 89 |
| Daylight | 6250 | 74 |
| High-intensity discharge: | | |
| Mercury | 5710 | 15 |
| Mercury improved color | 4430 | 32 |
| Metal halide | 3720 | 60 |
| Sodium, high-pressure | 2100 | 21 |
| Incandescent: | | |
| General service | 2600–3100 | 89–92 |
| Daylight: | | |
| Overcast sky | 6000–7000 | |
| Blue sky | 11,000–25,000 | |
| Sun, outside of earth's atmosphere | 6500 | |

tion of the eye. In much of the literature the term brightness, used alone, refers to both luminance and sensation. The context usually indicates which meaning is intended.

*11.* ***Color.*** Within the visible spectrum, wavelengths are distinguished one from another by their ability to excite in the human eye various color sensations. Thus the shorter wavelengths excite the color known as violet, and as the wavelengths increase, the color sensation gradually changes through blue, green, yellow, and orange and finally to red at the longer wavelengths of the visible spectrum. The color of the sensation produced by light of a composite character is determined by its spectral energy distribution. Color is defined as that quality of visual sensation which is associated with the spectral distribution of light.

*Color matching* is the process of adjusting the color of one area so that it is the same color as another.

*Color rendering* is a general expression for the effect of a light source on the color appearance of objects in conscious or subconscious comparison with their color appearance under a reference light source.

*Color rendering index* (of a light source) is the measure of the degree of color shift which objects undergo when illuminated by the light source, as compared with the color of those same objects when illuminated by a reference source of comparable color temperature. Values for common light sources vary from about 20 to 99. The higher the number, the better the color rendering (see Table 26-3).

*Correlated color temperature* (of a light source) is the absolute temperature (in kelvins) of a blackbody radiator whose chromaticity most nearly resembles that of the light source.

## INCANDESCENT LAMPS

*12.* ***Incandescent filament lamps*** are light sources in which light is produced by a filament heated to incandescence by an electric current. Of all commonly used light sources, incandescent lamps have the lowest initial cost, lowest luminous efficacy, and shortest life. As shown in Fig. 26-3, the major parts on an incandescent filament lamp are the filament, bulb, base, and fill gas.

*13.* ***Filament.*** The efficacy of light production by incandescent lamps depends on the temperature of the filament. Tungsten, because of its high melting point (3655 K), higher than

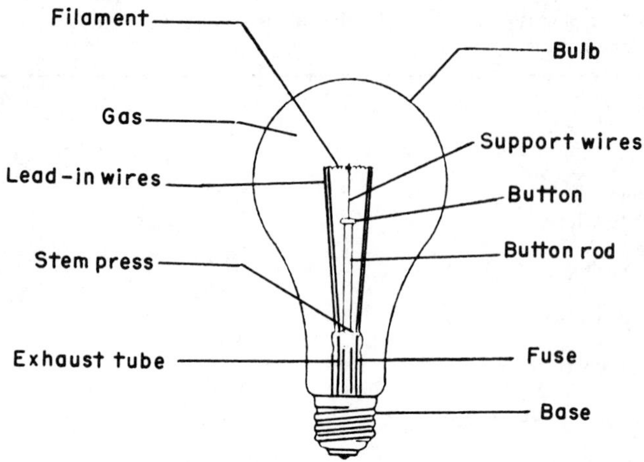


**FIG. 26-3** Incandescent filament lamp construction.

that of all other elements except carbon, is the most common filament material used today. Filament forms, sizes, and support constructions vary with different types of lamps as determined by lamp use. Filament forms are designated by a letter or group of letters followed by an arbitrary number. See Fig. 26-4. Most commonly used letters are C, for a helical coil; CC, for a coiled coil or a double helical coil; and S, for straight uncoiled wire. Coiling the filament increases the lamp's luminous efficacy. Coiling the coil further increases efficacy.

Mechanical problems associated with tungsten filaments make the incandescent lamp an inherently compact, somewhat spherical structure. The filament's length and diameter limit its

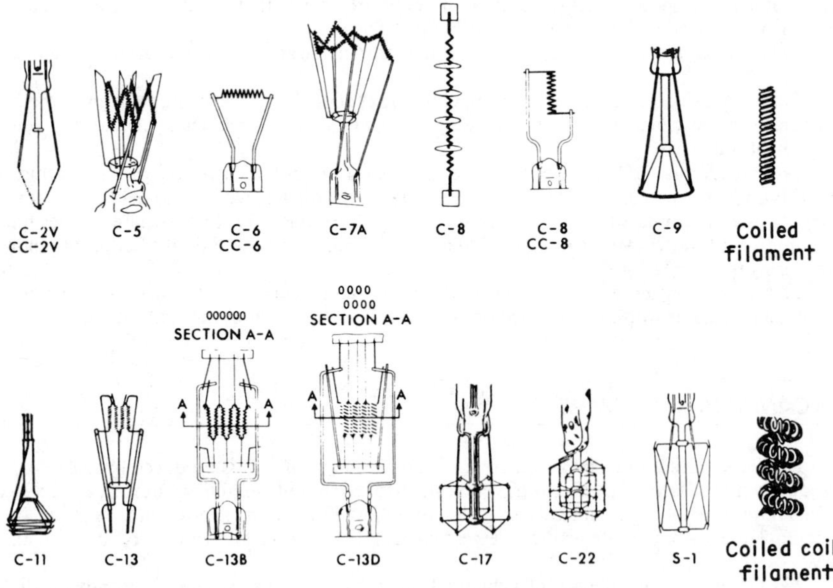


**FIG. 26-4** Typical incandescent lamp filament construction (not to scale).

range of operation between 1.5 and 300 V. At 1.5 V, the filament is very short and thick, and it becomes difficult to heat it without excessively heating its support wires. The lamps in the low-voltage (6- to 12-V) class, however, are relatively rugged and will withstand the shocks of motor-vehicle and similar applications. At voltages near 300, the filament is very long and slender; it is fragile and difficult to support.

*14. Bulbs.* Bulb shape, size, material, and finish vary according to application needs. Shapes range from tubular to spherical and from parabolic to flame form. Bulbs are designated by a letter referring to the shape (see Fig. 26-5) and by a number which is the maximum diameter

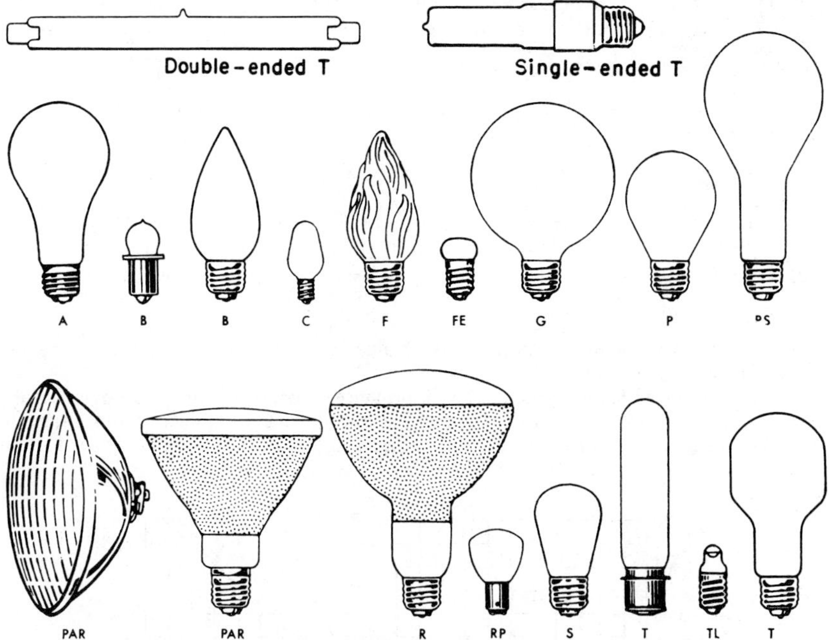


**FIG. 26-5** Typical bulb shapes and designations (not to scale).

in eighths of an inch; for example, A-19 designates an A shaped bulb with a diameter of 19/8 or 2⅜ in.

Most bulbs are made of lead or lime soft glass, although heat-resisting hard glass is used for high-temperature applications, and are frosted on the inside for moderate diffusion of the light without appreciably reducing light output. Clear, unfrosted lamps are used where accurate control of light is needed from a point or line source. Fused quartz and high-silica glass are used for other lamps.

*15. Base types* also vary according to application needs. They range from screw types for most general-service lamps to bipost and prefocus types where a high degree of accuracy in lamp positioning is important, such as in projection systems. Figure 26-6 shows some typical base shapes. Base size varies with lamp wattage, for heat dissipation, and voltage.

*16. Fill gas* is used in incandescent-filament lamps to reduce the rate of evaporation of the heated filament. Inert gases such as nitrogen, argon, and krypton are in common use today, with krypton used where its increased cost is justified by increased efficacy or increased lamp life. Halogen gases, for example, bromine and iodine, are also used in tungsten-halogen incandescent lamps to improve light output over the life of the lamp.

*17. Energy Characteristics.* Only a small percentage of the total radiation from incandescent lamps is in the visible spectrum, with the majority in the infrared spectrum. As the filament temperature is increased, the luminous efficacy increases with a maximum of 53 lm/W for an

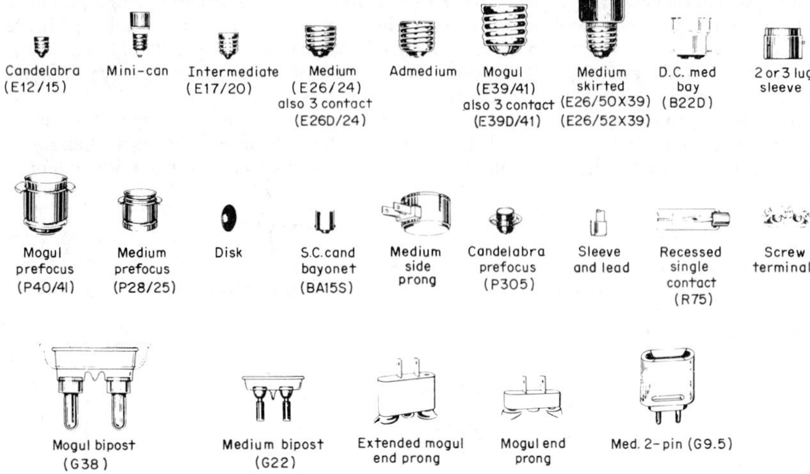


**FIG. 26-6** Common lamp bases (not to scale). IEC designations are shown where available.

uncoiled tungsten wire at its melting point. To obtain life, practical lamps operate at a temperature well below the melting point.

***18. Performance Characteristics.*** The performance of tungsten-filament lamps is affected by voltage, position of the bulb (if incorrect), size, construction, ambient temperature (if excessive), and quality of manufacture. The voltage characteristics[1] through a range of a few

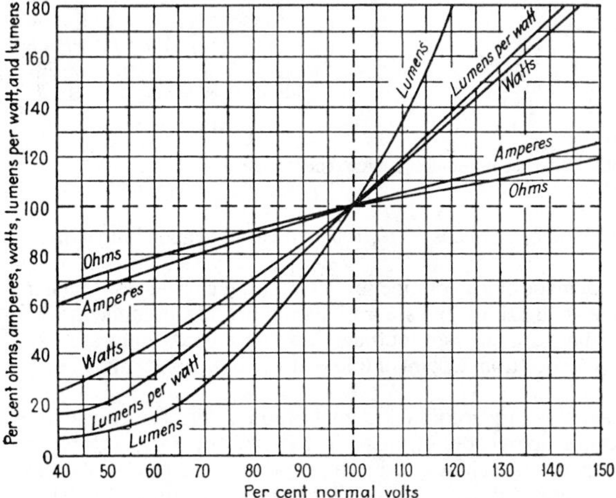


**FIG. 26-7** Characteristic curves for large gas-filled lamps showing the effect of operating a lamp at other than its rated voltage. These characteristics are averages of many lamps. They enable the user and designer to predetermine lamp performance under varying conditions.

[1] Weitz, C. E.: LD-1-Lamp Bulletin, General Electric Company, 1956.

volts above and below design volts may be expressed as simple exponential equations in the following relationships, where capitals represent normal rated values:

$$\frac{\text{Life}}{\text{LIFE}} = \left(\frac{\text{LUMENS}}{\text{lumens}}\right)^a = \left(\frac{\text{LUMENS/WATT}}{\text{lumens/watt}}\right)^b = \left(\frac{\text{VOLTS}}{\text{volts}}\right)^d = \left(\frac{\text{AMPS}}{\text{amps}}\right)^u$$

$$\frac{\text{lumens}}{\text{LUMENS}} = \left(\frac{\text{volts}}{\text{VOLTS}}\right)^k = \left(\frac{\text{lumens/watt}}{\text{LUMENS/WATT}}\right)^h = \left(\frac{\text{watts}}{\text{WATTS}}\right)^s$$

$$= \left(\frac{\text{amps}}{\text{AMPS}}\right)^y = \left(\frac{\text{ohms}}{\text{OHMS}}\right)^z$$

$$\frac{\text{LUMENS/WATT}}{\text{lumens/watt}} = \left(\frac{\text{LUMENS}}{\text{lumens}}\right)^f = \left(\frac{\text{VOLTS}}{\text{volts}}\right)^g = \left(\frac{\text{AMPS}}{\text{amps}}\right)^j$$

$$\frac{\text{amps}}{\text{AMPS}} = \left(\frac{\text{volts}}{\text{VOLTS}}\right)^t \text{ and } \left(\frac{\text{watts}}{\text{WATTS}}\right) = \left(\frac{\text{volts}}{\text{VOLTS}}\right)^n$$

Exponents $d$, $k$, and $t$ are taken as fundamentals, and other exponents are derived from them. Values given apply to lamps operated at 90 to 110% rated voltage. Outside that range, use the values from Fig. 26-7.

The theoretical life of lamps calculated by the exponential relationship of life and voltage is seldom realized in practical installations in the case of excessive "undervoltage" burning, since handling, cleaning, vibration, etc., introduce breakage factors which tend to reduce lamp life.

Exponents

| | Gas-filled | Vacuum |
|---|---|---|
| $a$ | 3.86 | 3.85 |
| $b$ | 7.1 | 7.0 |
| $d$ | 13.1 | 13.5 |
| $u$ | 24.1 | 23.3 |
| $k$ | 3.38 | 3.51 |
| $h$ | 1.84 | 1.82 |
| $s$ | 2.19 | 2.22 |
| $y$ | 6.25 | 6.05 |
| $z$ | 7.36 | 8.36 |
| $f$ | 0.544 | 0.550 |
| $g$ | 1.84 | 1.93 |
| $j$ | 3.40 | 3.33 |
| $t$ | 0.541 | 0.580 |
| $n$ | 1.54 | 1.58 |

***19. Lamp Output Depreciation.*** Because of filament evaporation throughout life, the filament of a lamp becomes thinner and thus consumes less power. The light output decreases as the lamp progresses through life because of lowering filament temperature and bulb blackening. Figure 26-8*a* shows the change in watts, amperes, lumens per watt, and lumens for a 200-W general-service lamp on constant-voltage service.

***20. Lamp Mortality and Renewal Rate.*** Lamp life is based on averages obtained from life-testing hundreds of thousands of lamps annually. A perfect mortality record would be one in which all lamps reached their rated life and then burned out. However, many factors inherent in lamp manufacture and lamp materials make it impossible for each individual lamp to operate for exactly the life for which it was designed. A typical mortality curve of a large group of lamps is illustrated in Fig. 26-8*b* where it is superimposed on a lumen depreciation curve from Fig. 26-8*a*.

The mortality curve influences the rate of lamp replacements for installations involving a large number of lamps. If individual lamps are replaced as they burn out, the replacement rate is

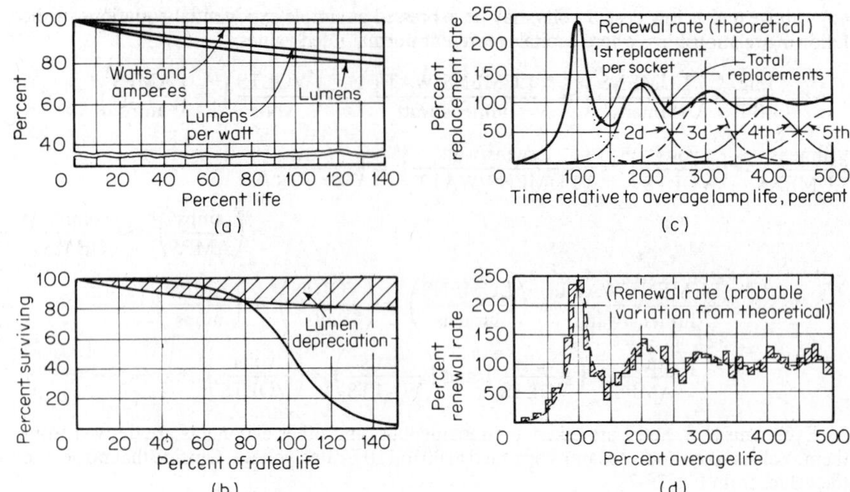


**FIG. 26-8** Life characteristics and renewal rate.

as shown in Fig. 26-8*c*. In a new installation relatively few burnouts would be expected during the first several hundred hours of operation, but as the design life is approached, the rate of burnout increases rapidly. After a burning period of four to five times the average lamp life, the renewal rate fluctuation finally reaches a steady or normal rate.

The dotted curve in Fig. 26-8*d* showing the theoretical rate of renewals holds only for an infinitely large installation. Departures from this curve in practical installations will, by the law of probability, more likely be represented by the solid block-shaped pattern. The larger the installation, the more closely the two curves tend to coincide. Complaints on life are occasionally encountered during those periods when chance dictates that renewals run higher than average, even though a record of the actual number of renewals over an extended period of time would show average rated life had been obtained.

***21. Average Life.*** The rated average laboratory life, defined as the average of a representative group of lamps burned under correct operating conditions on a 60-Hz circuit, ranges from 750 to 1500 h for the general-service types. As compared with life in the laboratory under controlled operating conditions, performance in service may differ widely. Lamp breakage and fluctuating line voltage tend to shorten life. Line-potential drop with resultant low-voltage operation often tends to lengthen life. Extended-service lamps with a rated life of 2500 h and longer are available in a range of sizes from 15 to 1500 W. They give less light than standard lamps under normal conditions but may be economically justified when labor costs to replace lamps are very high.

***22. Influence of Operating Conditions upon Lamp Performance.*** Tests show that ambient temperatures have little effect on performance characteristics. Very high temperatures, however, may cause mechanical difficulties.

On direct current, although the mortality rate is lower, the maintenance of light output is poorer than on alternating current.

Intermittent operation in general (not sign-flashing service) does not materially affect lamp performance. There is reason to believe that lamp life is shortened by voltage fluctuations, even though the voltage excess averaged over the life of the lamp is offset by an equal average voltage deficiency.

Except in the case of lamps designed for a particular position of operation, operating position has little effect on lamp performance. Shock and vibration are likely to impair the performance of lamps with filaments of small diameter to a greater extent than in the case of lamps with filaments of large diameter. Special types of lamps are available for use in installations where vibration is likely to be encountered and others, known as "rough-service" lamps, for use where they are likely to be subjected to shock. Neither of these two lamps will function properly in place of the other.

*23. Classes of Incandescent Lamps.* Incandescent lamps are divided and cataloged by manufacturers into three major groups: large lamps, miniature lamps, and photographic lamps. Large lamps are those normally used for interior and exterior general and task lighting. Miniature lamps are generally used in automotive, aircraft, and appliance applications. Photographic, as the name implies, are used in photography and projection service.

Some of the main classes of large lamps are as follows:

*General Service.* They are for general lighting on 120-V circuits (see Table 26-4). Sizes range from 10 to 1500 W with efficacies of 8 to nearly 24 lm/W.

*High Voltage* (220 to 300 V). They are for operation directly on circuits of 220 to 300 V. They are less rugged and have a lower efficacy than general-service lamps (see Table 26-5).

**TABLE 26-4** General-Service Lamps for 115-, 120-, and 125-V Circuits*

(Will operate in any position, but lumen maintenance is best for 40 to 1500 W when burned vertically base up)

| Watts | Bulb and other description | Base | Rated average life, h | Max overall length, in | Approx initial lumens | Rated initial lumens per watt | Depreciation factor, % output @ 70% rated life |
|---|---|---|---|---|---|---|---|
| 10 | S-14 inside frosted or clear | Med. | 1500 | 3½ | 80 | 8.0 | 89 |
| 15 | A-15 inside frosted | Med. | 2500 | 3½ | 126 | 8.4 | 83 |
| 25 | A-19 inside frosted | Med. | 2500 | 3⅞ | 230 | 9.2 | 79 |
| 40 | A-19 inside frosted | Med. | 1500 | 4¼ | 455 | 11.4 | 87.5 |
| 50 | A-19 inside frosted | Med. | 1000 | 4 7/16 | 680 | 13.6 | |
| 60 | A-19 inside frosted | Med. | 1000 | 4 7/16 | 860 | 14.3 | 93 |
| 75 | A-19 inside frosted | Med. | 750 | 4 7/16 | 1,180 | 15.7 | 92 |
| 100 | A-19 inside frosted | Med. | 750 | 4 7/16 | 1,740 | 17.4 | 90.5 |
| 150 | A-23 inside frosted or clear | Med. | 750 | 6 3/16 | 2,780 | 18.5 | 89 |
| 200 | A-23 inside frosted white or clear | Med. | 750 | 6 5/16 | 4,000 | 20.0 | 89.5 |
| 300 | PS-25 clear or inside frosted | Med. | 750 | 6 15/16 | 6,360 | 21.2 | 87.5 |
| 300 | PS-35 clear or inside frosted | Mog. | 1000 | 9⅜ | 5,860 | 19.6 | 86 |
| 500 | PS-35 clear or inside frosted | Mog. | 1000 | 9⅜ | 10,600 | 21.2 | 89 |
| 750 | PS-52 clear or inside frosted | Mog. | 1000 | 13 1/16 | 17,000 | 22.6 | 89 |
| 1000 | PS-52 clear or inside frosted | Mog. | 1000 | 13 1/16 | 23,600 | 23.6 | 89 |
| 1500 | PS-52 clear or inside frosted | Mog. | 1000 | 13 1/16 | 34,000 | 22.6 | 78 |

* Consult manufacturers' technical literature for current data, as values change frequently.

NOTE: 1 in = 25.4 mm.

**TABLE 26-5** Lamps for High-Voltage Service*

(Burned in any position)

| Watts | Bulb | Base | Rated average life, h | Max overall length, in | Depreciation factor, % output @ 70% rated life | Approx initial lumens† | Rated initial lumens per watt† |
|---|---|---|---|---|---|---|---|
| 25 | A-19 inside frosted | Medium | 1000 | 3⅞ | 86 | 220 | 8.8 |
| 50 | A-19 inside frosted | Medium | 1000 | 3⅞ | 79 | 490 | 9.8 |
| 100 | A-21 inside frosted or clear | Medium | 1000 | 5¼ | 90 | 1,280 | 12.8 |
| 200 | PS-30 inside frosted or clear | Medium | 1000 | 8 1/16 | 90 | 3,040 | 15.2 |
| 300 | PS-35 inside frosted or clear | Mogul | 1000 | 9⅜ | 89 | 4,890 | 16.3 |
| 500 | PS-40 inside frosted or clear | Mogul | 1000 | 9¾ | 87 | 9,270 | 18.5 |
| 750 | PS-52 inside frosted or clear | Mogul | 2000 | 13 | | 13,600 | 18.1 |
| 1000 | PS-52 inside frosted or clear | Mogul | 2000 | 13 | 82 | 18,000 | 18.0 |
| 1500 | PS-52 inside frosted or clear | Mogul | 2000 | 13 | | 27,000 | 18 0 |

* Consult manufacturers' technical literature for current data, as values change frequently.

† Data for 230 V.

NOTE: 1 in = 25.4 mm.

**TABLE 26-6** Extended Service (2500-h Rated Life) Incandescent Filament Lamps*
(For 120, 125, and 130 V)

| Watts | Bulb and Finish | Base | Max overall length, in | Approx initial lumens | Rated initial lumens per watt | Depreciation factor, % output @ 70% rated life |
|---|---|---|---|---|---|---|
| 15 | A-15 inside frosted | Med. | 3½ | 125 | 8.3 | 83 |
| 25 | A-19 inside frosted | Med. | 3¹⁵⁄₁₆ | 235 | 9.4 | 79 |
| 40 | A-19 inside frosted | Med. | 4¼ | 420 | 10.5 | 87.5 |
| 60 | A-19 inside frosted | Med. | 4⁷⁄₁₆ | 775 | 12.9 | 91.5 |
| 100 | A-19 inside frosted | Med. | 4⁷⁄₁₆ | 1,490 | 14.9 | 92.5 |
| 150 | A-23 inside frosted and clear | Med. | 6⁵⁄₁₆ | 2,350 | 15.7 | 89 |
| 150 | PS-25 inside frosted and clear | Med. | 6¹⁵⁄₁₆ | 2,300 | 15.3 | 85.5 |
| 200 | A-23 inside frosted and clear | Med. | 6³⁄₁₆ | 3,410 | 17.0 | 87.5 |
| 300 | PS-30 inside frosted and clear | Med. | 8¹⁄₁₆ | 5,190 | 17.3 | 79 |
| 300 | PS-35 inside frosted and clear | Mog. | 9⅜ | 5,100 | 17.0 | 84 |
| 500 | PS-40 inside frosted and clear | Mog. | 9¾ | 9,070 | 18.1 | 80 |
| 750 | PS-52 clear | Mog. | 13¹⁄₁₆ | 14,200 | 18.9 | 83 |
| 1000 | PS-52 clear | Mog. | 13¹⁄₁₆ | 19,800 | 19.8 | 78 |
| 1500 | PS-52 clear | Mog. | 13¹⁄₁₆ | 30,000 | 20.0 | |

* Consult manufacturers' technical literature for current data, as values change frequently.
NOTE: 1 in = 25.4 mm.

*Extended Service.* They have a life of 2500 or more hours and are intended for use in applications where a lamp failure causes an inconvenience, a nuisance, or a hazard to replace the lamp, or where replacement labor is expensive. They are less efficient than general-service lamps. See Table 26-6.

*General Lighting Tungsten-Halogen.* They are compact, have better lumen maintenance, and provide a whiter light and a longer life. Some typical lamps for general lighting are listed in Table 26-7.

**TABLE 26-7** Tungsten-Halogen Lamps for General Lighting*

| Watts | Volts | Bulb and finish | Base† | Max overall length, in | Rated life, h | Approx initial lumens | Approx initial lumens per watt | Depreciation factor, % output @ 70% rated life |
|---|---|---|---|---|---|---|---|---|
| | | | | Double-ended types | | | | |
| 200 | 120 | T-3 clear | RSC | 3⅛ | 1500 | 3,460 | 17.3 | |
| 300 | 120 | T-3 clear | RSC | 4¹¹⁄₁₆ | 2000 | 5,950 | 19.9 | 96 |
| 300 | 120 | T-3 clear | RSC | 3⅛ | 2000 | 5,650 | 18.5 | |
| 400 | 120 | T-4 clear | RSC | 3⅛ | 2000 | 7,750 | 19.4 | 96 |
| 500 | 120 | T-3 clear | RSC | 4¹¹⁄₁₆ | 2000 | 10,950 | 21.9 | 96 |
| 1000 | 120 | T-6 clear | RSC or RSC (rect.) | 5⅝ | 2000 | 23,400 | 23.4 | 96 |
| 1000 | 220 | T-3 clear | RSC | 10¹⁄₁₆ | 2000 | 21,400 | 21.4 | 96 |
| 1500 | 220 | T-3 clear | RSC | 10¹⁄₁₆ | 2000 | 35,800 | 23.2 | 96 |
| | | | | Single-ended types | | | | |
| 75 | 28 | T-3 clear | Min. | 2⅜ | 2000 | 1,400 | 21.4 | |
| 250 | 120 | T-4 clear | Minican | 3⅛ | 2000 | 4,850 | 19.4 | 96 |
| 250 | 120 | T-4 clear | D.C. Bay. | 2¹³⁄₁₆ | 2000 | 4,850 | 19.4 | 96 |
| 500 | 120 | T-4 frosted | Minican | 3¾ | 2000 | 10,000 | 20.0 | |

* Consult manufacturers' technical literature for current data, as values change frequently.
† RSC = recessed single contact, RSC (rect.) = rectangular recessed single contact.
NOTE: 1 in = 25.4 mm.

**TABLE 26-8** Basic Data on Standard Voltage Projector (PAR) and Reflector (R) Lamps†

| Watts | Bulb | Description | Base | Max. overall length, in | Approx. beam spread, deg[b] | Approx. beam lumens | Approx. total lumens | Approx. average candelas in central 10° cone[e] |
|---|---|---|---|---|---|---|---|---|
| Reflector lamps for spot lighting and floodlighting* | | | | | | | | |
| 30 | R-20 | Flood | Med. | 3 15/16 | 85 | 150 | 205 | 290 |
| 75 | R-30 | Spot | Med. | 5 3/8 | 50 | 400 | 850 | 1,730 |
| 75 | R-30 | Flood | Med. | 5 3/8 | 130 | 610 | 850 | 430 |
| 150 | R-40 | Spot | Med. | 6 1/2 | 37 | 835 | 1,825 | 7,000 |
| 150 | R-40 | Flood | Med. | 6 1/2 | 110 | 1,550 | 1,825 | 1,200 |
| 300 | R-40 | Spot | Med. | 6 1/2 | 35 | 1,800 | 3,600 | 13,500 |
| 300 | R-40 | Flood | Med. | 6 1/2 | 115 | 3,000 | 3,600 | 2,500 |
| 300 | R-40[a] | Spot | Med. | 6 7/8 | 35 | 1,800 | 3,600 | 13,500 |
| 300 | R-40[a] | Flood | Med. | 6 7/8 | 115 | 3,000 | 3,600 | 2,550 |
| 300 | R-40[a] | Spot | Mogul | 7 1/4 | 35 | 1,800 | 3,600 | 14,000 |
| 300 | R-40[a] | Flood | Mogul | 7 1/4 | 115 | 3,000 | 3,600 | 2,500 |
| 500 | R-40[a] | Spot | Mogul | 7 1/4 | 60 | 3,300 | 6,500 | 22,000 |
| 500 | R-40[a] | Flood | Mogul | 7 1/4 | 120 | 5,700 | 6,500 | 4,750 |
| 1000 | R-60[a,f] | Spot | Mogul | 10 1/8 | 32 | 11,500 | 18,300 | 135,000 |
| 1000 | R-60[a,f] | Flood | Mogul | 10 1/8 | 110 | 15,500 | 18,300 | 15,500 |
| ER lamps for spotlighting and floodlighting | | | | | | | | |
| 50 | ER-30 | Light I.F. | Med. | 6 3/8 | | | 525 | |
| 75 | ER-30 | Light I.F. | Med. | 6 3/8 | | | 850 | |
| 120 | ER-40 | Light I.F. | Med. | 7 3/4 | | | 1,475 | |
| Reflector lamps for general lighting[c] | | | | | | | | |
| 500 | R-52 | Wide beam | Mogul | 11 3/4 | 90 | | 7,550 | |
| 750 | R-52 | Wide beam | Mogul | 11 3/4 | 110 | | 13,000 | |
| 1000 | R-52[a] | Wide beam | Mogul | 11 3/4 | 110 | | 16,300 | |
| 1000 | RB-52 | Wide beam | Mogul | 12 11/16 | 130 | | 18,900 | |

**TABLE 26-8** *(Continued)*

| Watts | Bulb | Description | Base | Max. overall length, in | Approx. beam spread, deg[b] | Approx. beam lumens | Approx. total lumens | Approx. average candelas in central 10° cone[e] |
|---|---|---|---|---|---|---|---|---|
| | | | Projector lamps for spot lighting and floodlighting[d] | | | | | |
| 75 | PAR-38 | Spot | Med. skirted | 5⁵⁄₁₆ | 30 × 30 | 465 | 750 | 3,800 |
| 75 | PAR-38 | Flood | Med. skirted | 5⁵⁄₁₆ | 60 × 60 | 570 | 750 | 1,500 |
| 100 | PAR-38 | Spot | Med. skirted | 5⁵⁄₁₆ | 30 × 30 | | 1,250 | |
| 100 | PAR-38 | Flood | Med. skirted | 5⁵⁄₁₆ | 60 × 60 | | 1,250 | |
| 150 | PAR-38 | Spot | Med. skirted | 5⁵⁄₁₆ | 30 × 30 | 1,100 | 1,735 | 11,000 |
| 150 | PAR-38 | Flood | Med. skirted | 5⁵⁄₁₆ | 60 × 60 | 1,350 | 1,735 | 3,700 |
| 150 | PAR-38 | Spot | Med. side prong | 4⁵⁄₁₆ | 30 × 30 | 1,100 | 1,735 | 11,000 |
| 150 | PAR-38 | Flood | Med. side prong | 4⁵⁄₁₆ | 60 × 60 | 1,350 | 1,735 | 3,700 |
| 200 | PAR-46 | Narrow spot | Med. side prong | 4 | 17 × 23 | 1,200 | 2,325 | 32,500[g] |
| 200 | PAR-46 | Med. flood | Med. side prong | 4 | 20 × 40 | 1,300 | 2,325 | 11,200[g] |
| 250 | PAR-38[f] | Spot | Med. skirted | 5⁵⁄₁₆ | 26 × 26 | 1,600 | 3,180 | 25,000 |
| 250 | PAR-38[f] | Flood | Med. skirted | 5⁵⁄₁₆ | 60 × 60 | 2,400 | 3,180 | 6,500 |
| 300 | PAR-56 | Narrow spot | Mogul end prong | 5 | 15 × 20 | 1,800 | 3,750 | 70,000[g] |
| 300 | PAR-56 | Med. flood | Mogul end prong | 5 | 20 × 35 | 2,000 | 3,750 | 24,000[g] |
| 300 | PAR-56 | Wide flood | Mogul end prong | 5 | 30 × 60 | 2,100 | 3,750 | 10,000 |
| 500 | PAR-56[f] | Narrow spot | Mogul end prong | 5 | 15 × 32 | 4,900 | 7,650 | 96,000 |
| 500 | PAR-56[f] | Med. flood | Mogul end prong | 5 | 20 × 42 | 5,700 | 7,650 | 43,000 |
| 500 | PAR-56[f] | Wide flood | Mogul end prong | 5 | 34 × 66 | 5,725 | 7,650 | 19,000 |
| 500 | PAR-64 | Narrow spot | Extended Mogul end Prong | 6 | 13 × 20 | 3,000 | 6,000 | 110,000[g] |
| 500 | PAR-64 | Med. flood | | 6 | 20 × 35 | 3,400 | 6,000 | 35,000 |
| 500 | PAR-64 | Wide flood | | 6 | 35 × 65 | 3,500 | 6,000 | 12,000 |
| 1000 | PAR-64[f] | Narrow spot | Extended Mogul end Prong | 6 | 14 × 31 | 8,500 | 19,400 | 180,000 |
| 1000 | PAR-64[f] | Med. flood | | 6 | 22 × 45 | 10,000 | 19,400 | 80,000 |
| 1000 | PAR-64[f] | Wide flood | | 6 | 45 × 72 | 13,500 | 19,400 | 33,000 |

* The rated average life of reflector (R) lamps is 2000 h.

† Consult manufacturers' technical literature for current data, as values change frequently.

[a] Heat-resistant glass bulb.

[b] To 10% of maximum candlepower.

[c] Some of these types are also available for 230- to 260-V circuits.

[d] The rated average life of projector (PAR) lamps is 2000 h. All PAR lamps have bulbs of molded heat-resistant glass.

[e] Central cone defined as 5° cone for all spots and 10° cone for all floods.

[f] Halogen cycle.

[g] For horizontal burning; may be slightly lower for vertical operation.

**TABLE 26-9** Rough-Service Lamps for 115-, 120-, and 130-V Circuits and Vibration-Service Lamps for 120- and 130-V Circuits*

| Watts | Bulb and other description | Burning position | Base | Filament | Max overall length, in | Light center length, in | Approx initial lumens | Approx initial lumens per watt | Depreciation factor, % output @ 70% rated life |
|---|---|---|---|---|---|---|---|---|---|
| Vibration service (1000 h rated life) | | | | | | | | | |
| 25 | A-19 inside frosted or clear | Any | Med. | C-9 | 3 15/16 | 2½ | 260 | 10.4 | |
| 50 | A-19 inside frosted or clear | Not horizontal | Med. | C-9 | 3 15/16 | 2½ | 550 | 11.0 | 72 |
| 75 | A-21 inside frosted | Nor horizontal | Med. | C-9 | 5¼ | 3⅞ | 935 | 12.5 | |
| 100 | A-23 inside frosted or clear | Any | Med. | C-9 | 5 15/16 | 4 7/16 | 1340 | 13.4 | 83 |
| 100 | A-21 inside frosted or clear | Not horizontal | Med. | C-9 | 5 5/16 | 3⅞ | 1400 | 14.0 | |
| 150 | PS-25 inside frosted | Any | Med. | C-9 | 6 15/16 | 5¼ | 2390 | 15.9 | |
| Rough service (1000 h rated life) | | | | | | | | | |
| 25† | A-19 inside frosted | Any | Med. | C-17 | 3 15/16 | 2½ | 230 | 9.2 | |
| 50 | A-19 inside frosted | Any | Med. | C-22 | 3 15/16 | 2½ | 480 | 9.6 | 76 |
| 75† | A-21 inside frosted | Any | Med. | C-22 or C-17 | 4 7/16 | 2⅞ | 750 | 10.0 | |
| 100 | A-21 inside frosted | Any | Med. | C-17 | 5 5/16 | 3⅞ | 1250 | 12.5 | 79 |
| 150† | A-23 inside frosted | Any | Med. | C-17 | 6 1/16 | 4⅜ | 2130 | 14.2 | |
| 150 | PS-25 inside frosted | Any | Med. | C-17 | 6 15/16 | 5¼ | 2130 | 14.2 | 80 |
| 200† | PS-30 inside frosted or clear | Any | Med. | C-9 | 8 1/16 | 6 | 3380 | 16.9 | 82.5 |
| 200† | A-23 inside frosted or clear | Any | Med. | C-17 | 6 1/16 | 4⅜ | 3380 | 16.9 | |

* Consult manufacturers' technical literature for current data, as values change frequently.
† 115-V not available.
NOTE: 1 in = 25.4 mm.

**TABLE 26-10a.** Multiple Street-Lighting Lamps*
(Will operate in any position, but lumen maintenance best when burned vertically base up)

| Watts | Bulb | Base | Filament | Max. overall length, in | Light center length, in | Nominal lumens | Depreciation factor, % output @ 70% rated life |
|---|---|---|---|---|---|---|---|
| 58 | A-19 | Medium | C-9 | 4¼ | 2⅞ | 600 | |
| 92 | A-23 | Medium | C-9 | 6 | 4 7/16 | 1,000 | 89 |
| 189 | PS-25 | Mogul | C-9 | 7⅛ | 5⅜ | 2,500 | 86.5 |
| 189 | PS-25 | Medium | C-9 | 6 15/16 | 5¼ | 2,500 | 82 |
| 295 | PS-35 | Mogul | C-9 | 9⅜ | 7 | 4,000 | 82 |
| 405 | PS-40 | Mogul | C-9 | 9¾ | 7 | 6,000 | 80.5 |
| 620 | PS-40 | Mogul | C-7A | 9¾ | 7 | 10,000 | 79.5 |
| 860 | PS-52 | Mogul | C-7A | 13 1/16 | 9½ | 15,000 | 79.5 |

* 120 V; 3000 h life. Consult manufacturers' technical literature for current data.
NOTE: 1 in = 25.4 mm.

**TABLE 26-10b.** Series Street-Lighting Lamps*
(Except where noted, will operate in any position, but lumen maintenance best when burned vertically base up)

| Rated initial lumens | Initial volts | Amperes | Clear bulb | Burning position | Filament | Max. overall length, in | Light center length, in | Initial watts |
|---|---|---|---|---|---|---|---|---|
| 600 | 6.7 | 6.6 | PS-25 | Any | C-8 | 7⅛ | 5⅜ | 44 |
| 1,000 | 10.0 | 6.6 | PS-25 | Any | C-8 | 7⅛ | 5⅜ | 66 |
| 1,500 | 14.2 | 6.6 | PS-25 | Any | C-8 | 7⅛ | 5⅜ | 94 |
| 2,500 | 22.0 | 6.6 | PS-25 | Base up | C-2V | 7⅛ | 5⅜ | 146 |
| 2,500 | 22.4 | 6.6 | PS-35 | Any | C-2V | 9⅜ | 7 | 147 |
| 2,500 | 19.6 | 7.5 | PS-25 | Any | C-2V | 9⅜ | 7 | 147 |
| 4,000 | 34.2 | 6.6 | PS-35 | Any | C-2V | 9⅜ | 7 | 225 |
| 4,000 | 14.6 | 15 | PS-35 | Base up | C-2V | 9⅜ | 7 | 219 |
| 4,000 | 14.6 | 15 | PS-35 | Base down | C-2V | 9⅜ | 6¼ | 220 |
| 6,000 | 50.1 | 6.6 | PS-40 | Any | C-2V | 9¾ | 7 | 330 |
| 6,000 | 15.7 | 20 | PS-40 | Base up | C-2V | 9¾ | 7 | 315 |
| 10,000 | 86.2 | 6.6 | PS-40 | Any | C-7A | 9¾ | 7 | 570 |
| 10,000 | 26.1 | 20 | PS-40 | Base up | C-7 | 9¾ | 7 | 522 |
| 15,000 | 37.5 | 20 | PS-40 | Base up | C-7 | 9¾ | 7 | 750 |

* All Mogul base, 3000 h life. Consult manufacturers' technical literature for current data.
NOTE: 1 in = 25.4 mm.

*Reflectorized.* They are a group of lamps embodying integral reflecting surfaces. Bowl-silvered lamps are employed in direct-lighting equipment in which it is desired to shield the filament from view but principally in indirect equipment. Initial loss of light output due to the silvering is 6 to 10%; the rate of decline of light output is considerably greater than in clear-bulb lamps of corresponding sizes—60 to 80% greater in the case of 100- and 200-W lamps. However, a luminaire (of similar distribution) with an unprocessed lamp may produce less light because of poorer maintenance. In projector flood- and spotlight lamps (Table 26-8), the bulb is constructed of a molded bowl-shaped section of parabolic or other suitable profile, on the inner surface of which is a metal reflecting surface. This bowl is fused to a molded-glass cover plate, which may be clear or may consist of a pattern of lenses and prisms, depending upon the desired beam characteristics. Sealed-beam automobile lamps are of this construction. Reflector-type lamps are constructed with blown bulbs of suitable profiles (usually cylindrical for showcase lighting or parabolic for spotlighting) having parts of the inner surfaces covered with a reflecting metallic film. Their nominal life is usually 2000 h.

A number of projector lamps are available with dichroic filters (interference films) to control the spectral quality of the radiation in such a manner as to separate the heat from the light in the beam or to produce colored light without the usual losses due to absorption by filters. From 75 to 80% of the heat can be removed from the beam at a sacrifice of only 15 to 20% of the light. Colored dichroic lamps produce more deeply saturated colors with higher efficacy than is obtainable with color filters.

*Rough and Vibration Service.* They are for use where lamps are subjected to shock and vibration while in use. Filament construction differs. Rough-service lamps are available from 25 to 200 W, while those for vibration service range from 25 to 150 W. See Table 26-9.

*Street Lighting.* They are available for series burning and multiple burning. The former are rated according to lumen output and the latter according to watts. In general the lumen maintenance of series lamps is better than that of multiple lamps. The latter are rated on a basis such that their average lumens throughout life are approximately equivalent to those of series lamps of corresponding sizes. See Table 26-10.

## FLUORESCENT LAMPS

***24. Fluorescent lamps*** are low-pressure mercury electric-discharge lamps in which a phosphor coating transforms some of the ultraviolet energy generated by the discharge into light. The major parts of a fluorescent lamp (hot-cathode type) are the bulb (tube), electrodes, fill gas, phosphor coating, and bases, as shown in Fig. 26-9. When the proper voltage is applied across

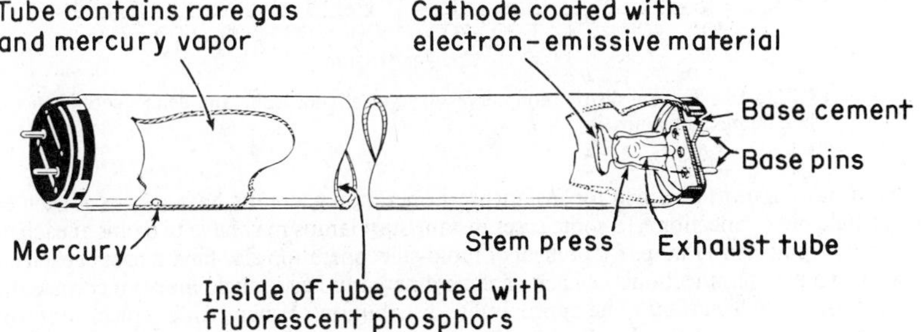


**FIG. 26-9** Cutaway view of fluorescent hot-cathode preheat-starting lamp.

the ends of the lamp, an arc is produced by current flowing between the electrodes through the fill gas (mercury vapor). This discharge generates some visible radiation, but mostly ultraviolet at 253.7 nm, which in turn excites the phosphor coating to emit light.

Fluorescent lamps are available commercially principally in four distinct types, depending upon their operating circuits: (1) hot-cathode, preheat-starting, (2) hot-cathode, instant-starting, (3) cold-cathode, and (4) rapid-start lamps.

***25. Bulb.*** Fluorescent lamp bulbs are basically tubular of small cross-sectional diameter. The bulb is available in straight, U-shaped, and circular configurations in bulb diameters from ⅝ to 2⅛ in. In straight lengths, they range from 6 to 96 in (nominal). Circular (circline) lamps have nominal overall diameters from 6½ to 16 in. U-shaped lamps are 25 in (hot-cathode) and 45 in (cold-cathode) in nominal overall length. Fluorescent lamps are designated by a letter indicating the tube cross section shape and a number indicating the diameter in eighths of an inch. A T-12 lamp has a tubular bulb of 1½ (12/8) in in diameter.

***26. Electrodes.*** There are two electrodes in each fluorescent lamp, one at each end, designed to operate as either "hot" or "cold" electrodes (or cathodes).

Hot-cathode lamps contain electrodes which are usually coiled-coil (or triple-coiled) tungsten filaments coated with one or more of the alkaline-earth oxides. By suitable circuit arrangements these cathodes can be heated to an electron-emitting temperature before the arc strikes, or they may be required to act momentarily as cold cathodes until they are heated by bombardment after the lamp has started. Lamps using these cathodes may be designed to carry currents of 1 to 2 A with low-voltage drop (10 to 12 V) at the electrodes. Unless the cathodes are kept at the proper temperature by means of an external heating circuit, they do not withstand frequent starting without correspondingly shorter life.

Cold-cathode lamps are those which use electrodes of tubular form of iron or nickel which may be coated on their inside surfaces with electron-emitting materials. These cathodes operate at temperatures which limit the lamps to low-current densities. The electrode drop in these lamps is relatively high (over 50 V), but they are not subject to short life as a result of frequent instant starting.

***27. Fill Gas.*** Droplets of liquid mercury are present in the fluorescent lamp and vaporize to a very low pressure during lamp operation. Small quantities of argon or other rare gases are included to aid starting and to improve efficacy.

***28. Phosphors.*** The chemical composition of the phosphor coating on the bulb interior surface determines the color of the light produced and, in part, lamp efficacy. Those lamps with phosphors producing good overall color rendering are generally of lower efficacy than those producing high output over limited wavelengths. Figure 26-10 shows typical spectral power distributions for the same size (wattage) lamp with two different phosphor compositions.

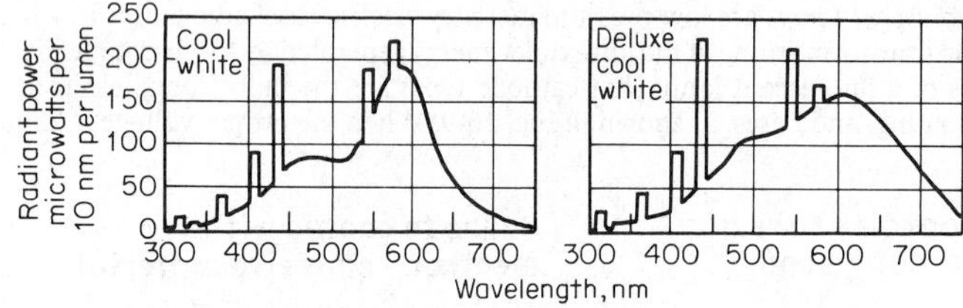


**FIG. 26-10** Spectral distribution curves for typical cool white and deluxe cool white fluorescent lamps.

***29. Bases.*** Lamps designed for instant-start operation generally have a base at each end with a single pin connection. (In some cases instant-start lamps may have two pins at each end electrically connected.) Lamps for preheat or rapid-start operation also have a base at each end, but with two pins (connections) in each. The circline lamp has a single four-pin connector.

***30. Energy Distribution.*** The approximate distribution of energy in a typical cool white fluorescent lamp is shown in Fig. 26-11.

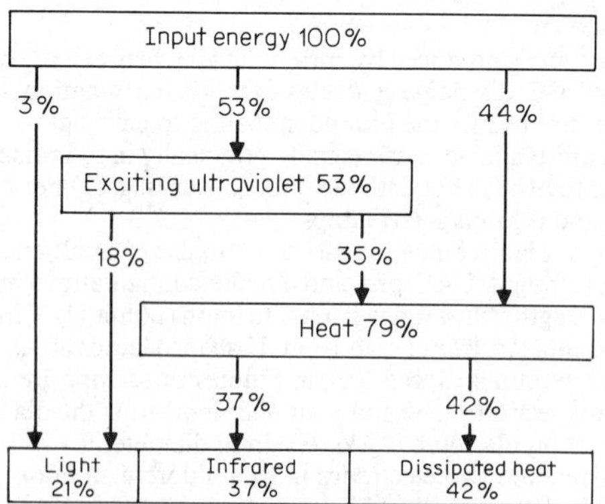


**FIG. 26-11** Energy distribution in a typical cool white fluorescent lamp.

**TABLE 26-11** Typical Fluorescent Lamps[a]

| Nominal length, in | Bulb | Base | Starting | Approx. lamp watts | Approx. lamp amps | Rated life, h[b] | Initial lumens cool white[b] | Lamp lumen depreciation factor, % | Luminance, footlamberts |
|---|---|---|---|---|---|---|---|---|---|
| 24 | T-12 | Med. bipin | Preheat | 20 | 0.380 | 9,000 | 1,270 | 85 | 2045 |
| 48 | T-12 | Med. bipin | Rapid | 41 | 0.430 | 20,000 | 3,150 | 84 | 2425 |
| 48 | T-12 | Med. bipin | Rapid | 34–35 | 0.450 | 20,000 | 2,770 | 84 | 2130 |
| 48 | T-12 | Med. bipin | Rapid | 34–35 | 0.450 | 20,000 | 2,925[d] | 84 | 2250 |
| 60 | T-8 | Med. bipin | Rapid | 32 | 0.265 | 20,000 | 2,900[e] | | |
| 48 | T-12 | Med. bipin | Rapid | 28 | | 15,000 | 2,550[f] | 85 | 1965 |
| 96 | T-12 | Single pin | Instant | 75 | 0.425 | 12,000 | 6,300 | 89 | 2270 |
| 96 | T-12 | Single pin | Instant | 60 | 0.440 | 12,000 | 5,500 | 89 | 1850 |
| 96 | T-12 | Recess D.C. | Rapid | 113 | 0.790 | 12,000 | 9,150 | 82 | 3295 |
| 96 | T-12 | Recess D.C. | Rapid | 95 | 0.810 | 12,000 | 8,300 | 82 | 2990 |
| 96 | T-12 | Recess D.C. | Rapid | 215 | 1.500 | 9,000 | 12,250 | 72 | 5780 |
| 96 | PG-17 | Recess D.C. | Rapid | 215 | 1.500 | 9,000 | 16,000 | 69 | 5065[c] |
| U-shaped 6-in spacing | T-12 | Med. bipin | Rapid | 40.5 | 0.430 | 12,000 | 2,935 | 84 | 2260 |
| Circline 16 in diam. | T-10 | 4-pin | Rapid | 41.5 | 0.415 | 12,000 | 2,580 | 77 | 2475 |

[a] Consult manufacturers' technical literature for current data, as values change frequently.

[b] For 40-W rapid-start lamps; deluxe cool white, 2200; daylight, 2615; white, 3185; warm white, 3175; deluxe warm white, 2165. Rated life based on 3 h per start.

[c] At 0° angle.

[d] Lite white.

[e] Rare-earth, higher CRI phosphor. Special ballast required.

[f] Lite white. Special ballast required.

NOTE: 1 in = 25.4 mm.

**TABLE 26-12** Fluorescent-Lamp Power Requirements (Bare Lamps in Still Air)

| | | | | Approx. watts consumed | | | |
|---|---|---|---|---|---|---|---|
| | | | | Single-lamp circuit | | Two-lamp circuit | |
| Lamp designation | Nominal length, in | Operating current, mA | Lamp | Ballast | Total | Ballast | Total |
| Preheat types: | | | | | | | |
| 20T12 | 24 | 380 | 20.5 | 5 | 25.5 | 10 | 41 |
| Instant-start: | | | | | | | |
| 96T12 | 96 | 425 | 75 | | | 22 | 172 |
| 96T12† | 96 | 425 | 75 | | | | 160 |
| 96T12‡ | 96 | 425 | 60 | | | | 128 |
| Rapid-start lightly loaded: | | | | | | | |
| 40T12 | 48 | 430 | 41 | 13* | 54* | 13* | 95* |
| 40T12† | 48 | 430 | 40 | 7* | 47* | 6* | 86* |
| 40T12‡ | 48 | 460 | 34–35 | | 41* | | 74* |
| Rapid-start medium loaded: | | | | | | | |
| 96T12 | 96 | 800 | 113 | | 140* | | 252* |
| 96T12† | 96 | 800 | 113 | | | | 237* |
| 96T12‡ | 96 | 840 | 95 | | | | 207* |
| Rapid start highly loaded: | | | | | | | |
| 96T12, 96PG17 | 96 | 1500 | 215 | | 260* | | 450* |

* Total watts consumed includes power required to heat lamp cathodes. Check for current data.
† With energy-saving full-output ballast.
‡ With energy-saving full-output ballast and energy-saving lamps.
NOTE: 1 in = 25.4 mm.

***31. Performance Characteristics.*** Table 26-11 lists some typical fluorescent lamps for general lighting along with their *light output* (initial lumens), *luminance, lamp lumen depreciation factor,* and *life.* Table 26-12 lists the power requirements for these lamps.

Light output is sensitive to surrounding (ambient) air temperature as shown in Fig. 26-12. Lamp wattage changes in a similar fashion, but not as drastically at ambients below normal. Lamps operated at ambient temperatures below 60°F should be enclosed to conserve their heat. Air movement over the lamp bulb has the effect of lowered ambient temperature.

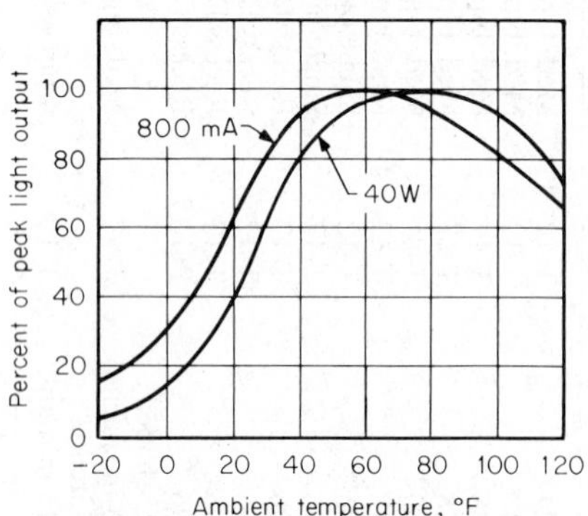


**FIG. 26-12** Effect of air temperature on the light output of 40-W and 800-mA fluorescent lamps.

Fluorescent lamps generally should be operated at voltages within ± 10% of their designed operating points for best performance. Decreased life and uncertain starting may result from operation at lower voltages, and at higher voltages there is danger of overheating of the ballast or transformer as well as decreased lamp life. One exception to this is found in the series operation of cold-cathode lamps, where an adjustable voltage supply makes possible operation over a wide range of illumination levels, that is, dimmer operation such as that used in stage lighting.

Failure of a hot-cathode fluorescent lamp usually results from loss of active material from the cathode or cathodes. This loss proceeds gradually throughout the life of the lamp and is accelerated by frequent starting. Depreciation of light output is caused principally by tube blackening and is rapid (as much as 10%) during the first 100 h but very gradual from that point on. For this reason the lamps are rated commercially on the basis of the lumen output after 100 h of operation.

***32. Fluorescent-Lamp Operation.*** Fluorescent lamps are best adapted to operate on ac circuits with reactance ballasts. Typical operating circuits are shown in Figs. 26-13 to 26-18.

Fluorescent-lamps are, to a considerable extent, dependent upon the characteristics of the ballast equipment. Typical of this is the effect of variations from rated line voltage on the conditions of lamp operation. Certified ballasts made in accordance with industry specifications and periodically field-checked by an independent laboratory are available for the more commonly used fluorescent lamps. They are to be distinguished by the letters CBM on the ballast case. Thermally protected "Class P" ballasts are required for fluorescent fixtures installed indoors, except fixtures with simple reactance ballasts.

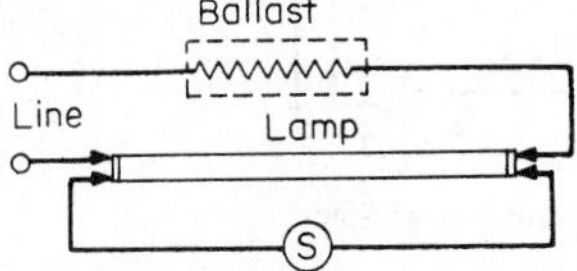


**FIG. 26-13** Single-lamp ballast for 4- to 40-W hot-cathode, preheat-starting fluorescent lamp. S = starting switch.

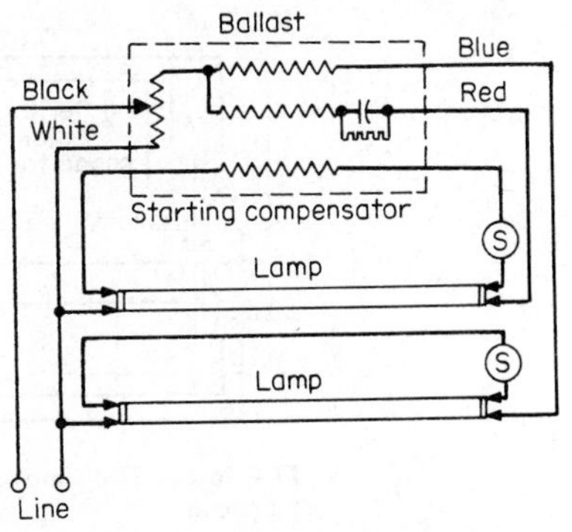


**FIG. 26-14** Two-lamp ballast circuit for 30- and 40-W hot-cathode, preheat-starting fluorescent lamps, showing built-in starting compensator.

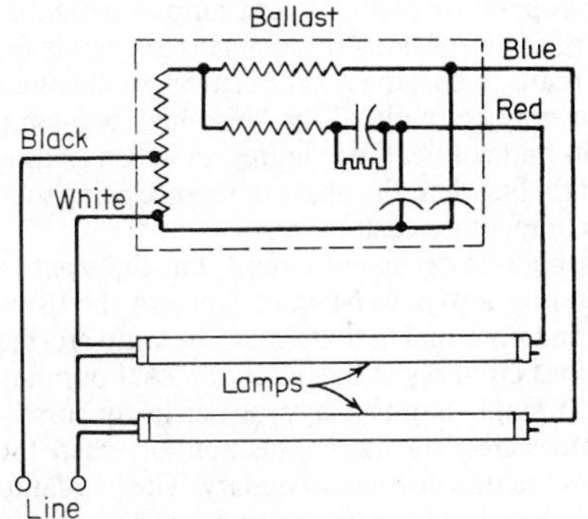


**FIG. 26-15** Two-lamp lead-lag ballast circuit for instant-starting hot-cathode lamps.

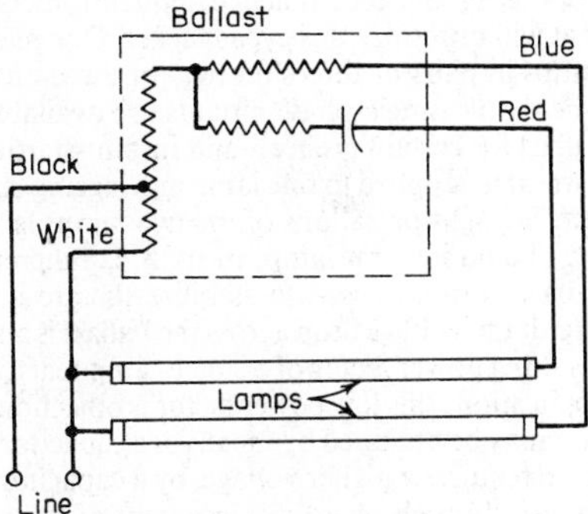


**FIG. 26-16** Two-lamp lead-lag ballast circuit for multiple operation of cold-cathode lamps.

When fluorescent lamps are operated on ac circuits, the light executes cyclic pulsations of considerably greater amplitude than those of incandescent lamps of equivalent lumen rating.

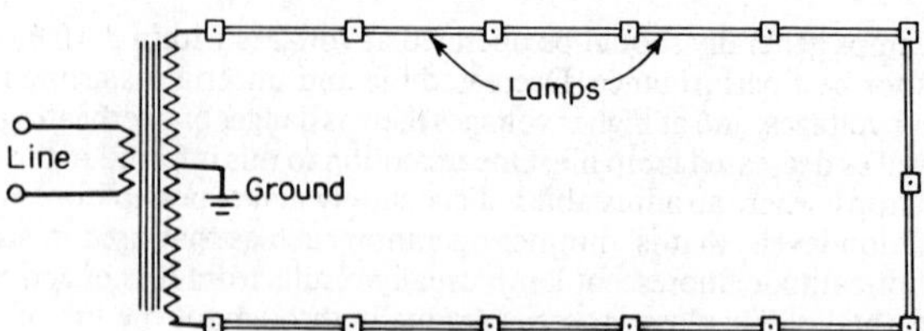


**FIG. 26-17** Circuit for the series operation of cold-cathode fluorescent lamps from a leakage reactance transformer.

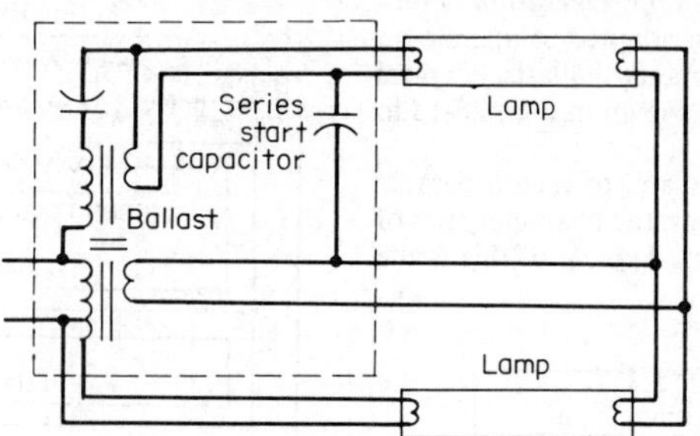


**FIG. 26-18** Two-lamp series rapid-start high-power-factor circuit.

Though at commercial frequencies (50 and 60 Hz) this cyclic flicker[1] is not usually noticeable, it may produce unpleasant stroboscopic effects when moving objects come into the field of view. For this reason, good practice frequently dictates the operation of fluorescent lamps in a manner that will minimize this cyclic flicker. One method of accomplishing this object is to operate the lamps in pairs or threes on two- or three-phase circuits, respectively. For critical applications where only single-phase circuits are available, use may be made of the "lead-lag" two-lamp ballast for certain preheat- and instant-starting hot- and cold-cathode lamps, in which leading current is supplied to one lamp and lagging current to the other, the phase difference being such that the light pulsations of the two lamps largely compensate each other.

The fluorescent lamp, in itself, is inherently a high-power-factor circuit, but the reactive ballast normally used to stabilize the arc is inherently low-power-factor. Since in the usual circuit the voltage drop across the ballast is approximately equal to that across the lamp arc, the resulting power factor of a single-lamp reactive-ballast circuit is of the order of 50%. For many applications this low power factor is objectionable. In single-lamp ballasts, power-factor correction may be obtained by means of a capacitor shunted across the line connections or, where the lamp requires a higher voltage, by a capacitor across the transformer secondary. The two-lamp ballast, through phase displacement of the lamp currents, or series capacitors, offers a ready means of power-factor correction and is usually designed to give a circuit power factor greater than 90%.

All inductive fluorescent ballasts emit a certain amount of noise; the noise increases with the lamp current. A sound rating for ballasts has been developed by some manufacturers from A (quietest) to F (noisiest). The amount of cumulative ballast noise which is tolerable depends on

[1] Brieger, Lawrence: Effect of Voltage Dip Duration on Cyclic Light Flicker; *Electr. Eng.*, August 1951, vol. 70, no. 8, p. 685.

Davidson, G. E.: Flicker in Lighting Systems; *Ont. Hydro Res. News,* October 1952, vol. 4, no. 4, p. 9.

Weise, W. R.: Cyclic Flicker of Fluorescent Lamps; *Electr. World,* Oct. 27, 1945, vol. 124, p. 80.

Eastman, A. A., and Campbell, J. H.: Stroboscopic and Flicker Effects from Fluorescent Lamps; *Illum. Eng.*, January 1952.

two sets of principal factors: (1) characteristics of the room and (2) characteristics of the luminaire.

Where direct current is available at circuit voltages comparable with the open-circuit voltages of the usual ac ballast circuits, fluorescent lamps may be operated from these sources. For such operation resistance must be added to the usual series reactance ballast (transformer ballasts are not applicable) to limit the operating current to the designed value. This causes a marked reduction in the overall efficacy of the lamp and circuit combination over that obtained in ac operation. Under dc operation, lamps more than a few feet in length will promptly develop a concentration of the mercury vapor at the negative end of the lamp, with the result that only a fraction of the bulb will give off light. This condition can be overcome through a periodic (about once in 4 h) reversal of the polarity of the lines feeding the lamps. The life of lamps is likely to be shorter on dc.

***33. Dimming.*** For dimming hot-cathode fluorescent lamps, a number of arrangements are available using autotransformers or SCR devices. Depending on the dimming system, 50 to 90% of full lamp brightness can be achieved, with dimming control down to one three-hundredth of full brightness. Autotransformer systems are available to dim two or more lamps, while up to 1200 lamps may be controlled on an SCR system. The lamps best suited to this application are the 40- and 30-W T-12 Rapid Start. The two wattages should not be used on the same circuit. For smooth operation, the lamps on any circuit should be made by the same manufacturer, at the same time, in the same color, and of the same age in use. Group replacement is the most satisfactory procedure. Lamps should be operated free from drafts at 50 to 80°F and should be seasoned 100 h at full brightness prior to dimming. Energy-saving 34-35 and 25-W lamps are not recommended. Special dimming ballasts are required.

***34. Electronic Ballasts.*** The use of solid-state electronic elements instead of conventional magnetic components can increase lamp efficacy by higher-frequency operation of lamps, and can reduce input power. Consult manufacturers for current data.

## HIGH-INTENSITY DISCHARGE LAMPS

***35. High-intensity discharge (HID)*** is a term denoting a general group of lamps consisting of mercury, metal halide, and high-pressure sodium lamps. A *mercury lamp* is an electric discharge lamp in which the major portion of the radiation is produced by the excitation of mercury atoms. A *metal halide lamp* is an electric discharge lamp in which the light is produced by the radiation from an excited mixture of a metallic vapor (mercury) and the products of the dissociation of halides (for example, halides of thallium, indium, sodium). A *high-pressure sodium lamp* is an electric discharge lamp in which the radiation is produced by the excitation of sodium vapor in which the partial pressure of the vapor during operation is of the order of $10^4$ N/m$^2$.

***36. Lamp Construction and Designation.*** HID lamps consist of a cylindrical transparent or translucent arc tube which confines the electric discharge and the associated gases. That tube is further enclosed in a glass bulb or outer jacket to exclude air to prevent oxidation of the metal parts and to stabilize operating temperatures and significantly reduce ultraviolet radiation emitted by the excitation of the vapors. The construction of a typical mercury lamp is shown in Fig. 26-19. The basic elements are the arc tube, fabricated from fused silica and filled with a drop of mercury and a rare gas at low pressure; the electrodes; and the outer envelope, which may or may not have a phosphor coating on the interior for improved color rendering. Metal halide lamps are very similar in construction to the mercury lamp, the major difference being the addition of a metal halide in the arc tube. The outer bulb may or may not have an inner phosphor coating.

The construction of a typical high-pressure sodium lamp is shown in Fig. 26-20. The basic components are the arc tube of translucent polycrystalline or single-crystal alumina, filled with sodium, mercury, and a rare gas (xenon); electrodes; and an outer borosilicate glass envelope.

A lamp designation system developed by the American National Standards Institute (ANSI) is currently in use.[1] It consists of five groups of letters or numbers: first a letter indicating the

[1] American National Standard Method for Designation of High Intensity Discharge Lamps, C78.380-1972, American National Standards Institute, New York.

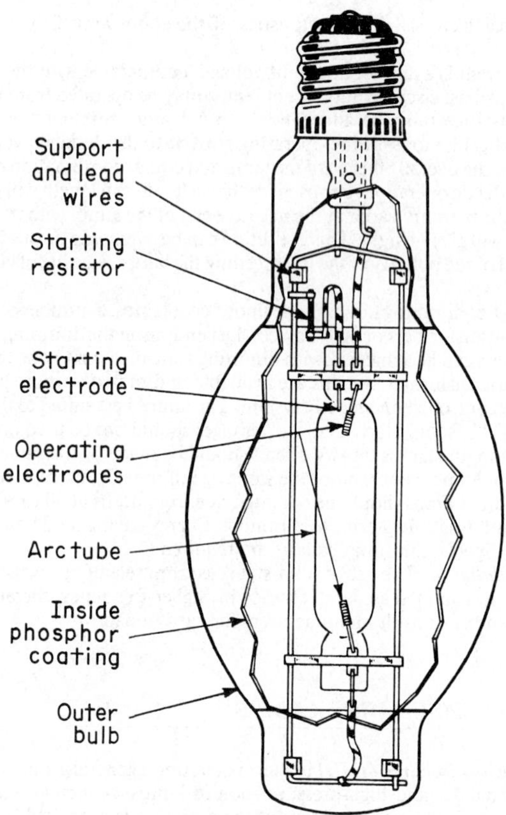


**FIG. 26-19** A 400-W phosphor-coated mercury lamp. Lamps of other sizes are constructed similarly.

type of lamp (H, mercury; M, metal halide; S, high-pressure sodium), followed by an arbitrary number designating electrical characteristics, followed by two arbitrary letters which describe the physical characteristics, then the lamp nominal wattage, and finally letters indicating the phosphor color. An example for a 400-W deluxe white mercury would be H33GL-400/DX.

***37. Lamp Characteristics.*** Light output from each of the three types of HID lamps has its own color appearance (chromaticity), and the spectral power distributions vary as shown in Fig. 26-21. Table 26-13 lists the radiated energy of typical 400-W HID lamps.

**TABLE 26-13** Energy Output for Some HID Lamps

| Type of energy | 400-W mercury | 400-W metal halide | 400-W high-pressure sodium |
|---|---|---|---|
| Light | 14.6% | 20.6% | 25.5% |
| Infrared | 46.4 | 31.9 | 37.2 |
| Ultraviolet | 1.9 | 2.7 | 0.2 |
| Conduction-convection | 27.0 | 31.1 | 22.2 |
| Ballast | 10.1 | 13.7 | 14.9 |

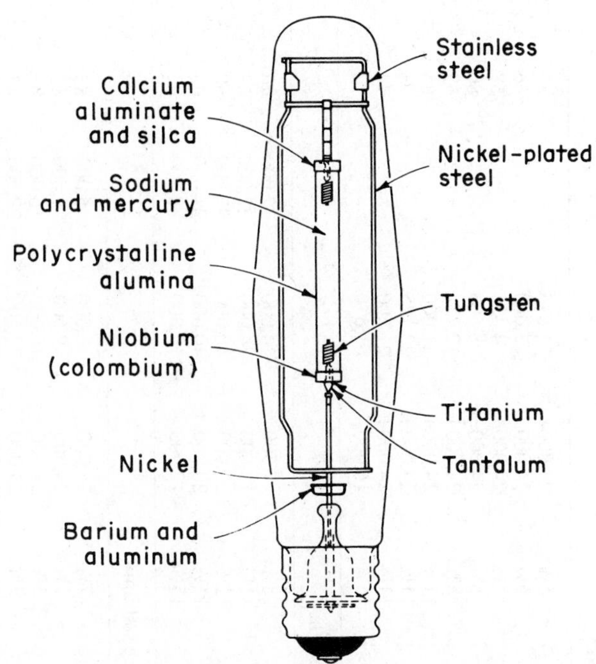


**FIG. 26-20** Construction of a typical high-pressure sodium lamp.

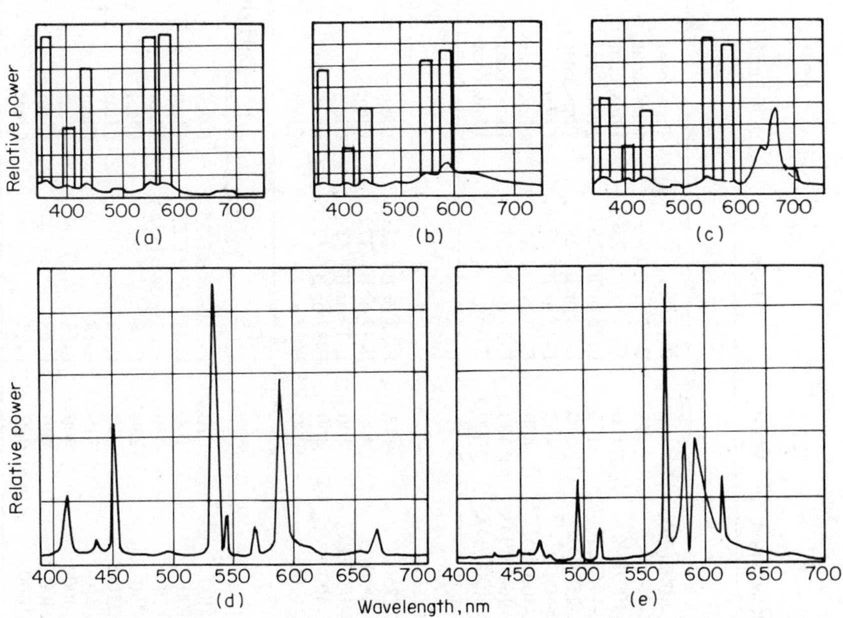


**FIG. 26-21** Spectral distribution curves for typical HID lamps. *(a)* Clear mercury, *(b)* phosphor-coated mercury, *(c)* improved-color phosphor-coated mercury, *(d)* sodium-thallium-indium iodide metal halide, and *(e)* high-pressure sodium.

**TABLE 26-14** Typical HID Lamps*

| ANSI designation | Watts | Bulb | Bulb finish | Base | Max. overall length, in | Life, h | Rated initial lumens Vert. | Rated initial lumens Horiz. | Lamp lumen depreciation factor† Vert. | Lamp lumen depreciation factor† Horiz. |
|---|---|---|---|---|---|---|---|---|---|---|
| **A. Mercury** | | | | | | | | | | |
| H46DL-40-50/DX | 50 | B-17 | Phos. coat | Med. | 5⅛ | 24,000 | 1,575 | | 65 | |
| H43AV-75/DX | 75 | B-17 or E-17 | Phos. coat | Med. | 6½ | 24,000 | 3,000 | | 66 | |
| H38JA-100/DX | 100 | D23½ or E-23½ | Phos. coat | Mog. | 7½ | 24,000+ | 4,425 | | 58 | |
| H39KC-175/DX | 175 | E-28 or BT-28 | Phos. coat | Mog. | 8 5/16 | 24,000+ | 8,600 | | 80 | |
| H37KC-250/DX | 250 | E-28 or BT-28 | Phos. coat | Mog. | 8 5/16 | 24,000+ | 12,775 | | 69 | |
| H33CD-400 | 400 | E-37 or BT-37 | Clear | Mog. | 11½ | 24,000+ | 21,000 | | 85 | |
| H33GL-400/DX | 400 | E-37 or BT-37 | Phos. coat | Mog. | 11½ | 24,000+ | 23,125 | | 74 | |
| H33GL-400/WDX | 400 | E-37 or BT-37 | Phos. coat | Mog. | 11½ | 24,000+ | 21,500 | | 74 | |
| H34GW-1000/DX | 1000 | BT-56 | Phos. coat | Mog. | 15⅜ | 16,000 | 61,670 | | 72 | |
| **B. Metal halide** | | | | | | | | | | |
| M57PE-175 | 175 | E-28 or BT-28 | Clear | Mog. | 8 5/16 | 7,500 | 14,000 | 12,000 | 70 | 62 |
| M59PJ-400/I/U | 400 | E-37 or BT-37 | Clear | Mog. | 11½ | 15,000 | 34,000 | 32,000 | 68 | |
| M59PJ-400/VBU | 400 | E-37 or BT-37 | Clear | Mog. | 11½ | 20,000 | 40,000 | 32,000 | 68 | |
| M59PK-400/BVU | 400 | E-37 or BT-37 | Phos. coat | Mog. | 11½ | 20,000 | 40,000 | 31,000 | 63 | |
| M47PA-1000 | 1000 | BT-56 | Clear | Mog. | 15 1/16 | 12,000 | 110,000 | 88,000 | 72 | |
| **C. High-pressure sodium** | | | | | | | | | | |
| S62ME-70 | 70 | E-23½ or BT-25 | Clear | Mog. | 7¾ | 24,000 | 6,000 | 6,000 | 83 | 83 |
| S54SB-100 | 100 | E-23½ or BT-25 | Clear | Mog. | 7¾ | 24,000 | 9,500 | 9,500 | 83 | 83 |
| S55SC-150 | 150 | E-23½ or BT-25 | Clear | Mog. | 7¾ | 24,000 | 16,000 | 16,000 | 83 | 83 |
| S50VA-250 | 250 | E-18 | Clear | Mog. | 9¾ | 24,000 | 27,500 | 27,500 | 83 | 83 |
| S50VA-250/S | 250 | E-18 | Clear | Mog. | 9¾ | 24,000 | 30,000 | 30,000 | 83 | 83 |
| S67MR-310 | 310 | E-18 | Clear | Mog. | 9¾ | 24,000 | 37,000 | 37,000 | 83 | 83 |
| S51WA-400 | 400 | E-18 | Clear | Mog. | 9¾ | 24,000 | 50,000 | 50,000 | 83 | 83 |
| S51WB-400 | 400 | E-18 | Diffuse | Mog. | 9¾ | 24,000 | 47,500 | 47,500 | 83 | 83 |
| S52XB-1000 | 1000 | E-25 | Clear | Mog. | 15 1/16 | 24,000 | 140,000 | 140,000 | 83 | 83 |

* Consult manufacturers' technical literature for current data, as values and types change frequently.
† Output at 70 percent of rated life.

***38. Performance Characteristics.*** Table 26-14 lists some typical HID lamps for general lighting along with their *light output* (initial lumens), *life,* and *lamp lumen depreciation factors.* For a qualitative comparison of HID lamps with incandescent and fluorescent lamps, see Table 26-15.

***39. HID Lamp Operation.*** The practical limit of an HID lamp's current-carrying capacity is how high a temperature its enclosing tube can withstand without rupturing. By connecting an impedance in series with the lamp, the current is controlled. In most lamps about one-half the supply voltage is absorbed by a series ballasting device. A variety of ballasts are available for operating lamps, singly or in pairs. Single lamp ballasts may have low (0.50 minimum) or high (0.90 minimum) power factor; two-lamp ballasts have inherently high power factor. The sim-

**TABLE 26-15** Qualitative Comparison of Incandescent, Fluorescent, and HID Light Sources

| Light source | Advantages | Disadvantages |
|---|---|---|
| Incandescent general-service | Compact size<br>No ballast<br>Good optical control<br>Good color rendering<br>Low cost<br>Dimmable<br>Good lumen maintenance | Short life<br>Low efficacy<br>Radiant-heat effects |
| Tungsten-halogen incandescent | Compact<br>No ballast<br>Good color rendering<br>Moderate life<br>Excellent optical control<br>Dimmable<br>Excellent lumen maintenance | Lamp handling is difficult during maintenance<br>High cost<br>Low efficacy<br>Radiant-heat effects<br>Operating temperature affects lamp life |
| Fluorescent | Linear shape<br>Moderate cost<br>High efficacy<br>Long life<br>Good lumen maintenance<br>Dimmable<br>Deluxe and special colors can give excellent color rendering | Ballast needed<br>Optical control limited<br>Ballasts may be noisy<br>Ambient temperature affects light output and color |
| Improved-color mercury (HID) | Moderate efficacy<br>Very long life<br>Good lumen maintenance<br>Burning position not critical<br>Dimmable to 25% | Starting takes 3–5 min<br>Does not restart immediately<br>Ballasts needed are large and may be noisy<br>Relatively high cost of lamp and ballast |
| Metal halide (HID) | High efficacy<br>Good color rendering<br>Medium life<br>Good optical control | Variations in color<br>Dimmable to 60%<br>Burn position very important<br>Ballasts large and may be noisy<br>High cost of lamp and ballast<br>Starting takes 2–10 min |
| High-pressure sodium (HID) | Very high efficacy<br>Long lamp life<br>Excellent lumen maintenance | Poor color rendering<br>Dimmable to 50–60%<br>Ballasts large and may be noisy<br>High cost of lamp and ballast<br>Starting takes 1–4 min<br>High luminance can cause control problems |

plest lamp ballast is the reactor type used in series with the lamp when line voltage is sufficient for reliable starting. This is not recommended where line-voltage fluctuations exceed 5%. A reactor-type ballast can be used when the line voltage is approximately twice the rated lamp voltage. The autotransformer-type ballast is used on circuits where the line voltage must be changed to suit the lamp requirements. Constant-wattage types of the autotransformer or isolated-secondary design are widely used because of better regulation, low line starting currents, and lower dropout voltage. They are also called "stabilized" or "regulated." Heavier wiring, oversized circuit breakers, and time-delay relays that may be required by the relatively high starting currents of non-CW and non-CWA ballasts are eliminated with stabilizing ballasts, as the starting current is less than the rated operating current.

One of the limitations of the HID lamp is the effect of power-supply interruptions. In the event of a power interruption or voltage dip lasting for more than 1 cycle, HID lamps extinguish and do not restart for several minutes. The exact magnitude of the voltage drop to cause this condition depends on the ballast design. Regulator ballasts withstand a greater drop than other types. The delay in lamp restarting is caused by the high pressure which develops in the arc tube during operation. The ballast open-circuit voltage is not sufficient to restart the lamp until the lamp cools and the pressure decreases. In installations where this characteristic might be a safety hazard, the use of a few incandescent or fluorescent luminaires along with the HID units assures emergency illumination until the HID lamps restart. Tungsten-halogen auxiliaries are available for HID industrial luminaires to provide standby illumination in event of momentary power failure.

A group of self-ballasted mercury lamps is available as replacements for incandescent sources in relighting applications. Self-extinguishing types should be used where there is the possibility of mercury and metal halide lamps continuing to operate after the outer bulb is broken or punctured.

## MISCELLANEOUS LAMPS

***40. Low-Pressure Sodium Lamps.*** These are sodium vapor lamps in which the partial pressure of the vapor during operation does not exceed a few newtons per square meter. Their light output is almost monochromatic, consisting of a double line in the yellow region of the spectrum at 589 and 589.6 nm. Table 26-16 lists some typical low-pressure sodium lamps along with their initial light output.

As with other electric discharge sources, a ballast is required. Starting time to full light output is 7 to 15 min, but the lamp will restart immediately after interruption of the power supply. The

**TABLE 26-16** Typical Low-Pressure Sodium Lamps for Street and Floodlighting Applications*

| Lamp designation | Type | Rated watts | Nominal volts | Amperes | Bulb diameter, in | Nominal length, in | Base | Initial lumens |
|---|---|---|---|---|---|---|---|---|
| SLI/H 60 | Linear | 60 | 80 | 0.83 | 1½ | 16¼ | G13/10 × 35 (Med. bipin) | 6,000 |
| SLI/H 200 | Linear | 200 | 135 | 1.6 | 1½ | 35¼ | G13/10 × 35 (Med. bipin) | 25,000 |
| SOX 18W | U tube | 18 | 57 | 0.35 | 2⅛ | 8½ | DC bay.† | 1,800 |
| SOX 35W | U tube | 35 | 70 | 0.60 | 2⅛ | 12¼ | DC bay.† | 4,800 |
| SOX 55W | U tube | 55 | 109 | 0.59 | 2⅛ | 16¾ | DC bay.† | 8,000 |
| SOX 90W | U tube | 90 | 112 | 0.94 | 2⅝ | 20¾ | DC bay.† | 13,500 |
| SOX 135W | U tube | 135 | 164 | 0.95 | 2⅝ | 30½ | DC bay.† | 22,500 |
| SOX 180W | U tube | 180 | 240 | 0.91 | 2⅝ | 44⅛ | DC bay.† | 33,000 |

* Consult manufacturers' technical literature for current data, as values change frequently.
† All have BY22d bases, a specific DC bayonet base.
NOTE: 1 in = 25.4 mm.

major application of these lamps is for floodlighting and streetlighting where monochromatic yellow light is acceptable and a high luminous efficacy is required.

***41. Sun Lamps.*** All mercury-vapor lamps generate ultraviolet rays which cause sunburn or sun tan on human skin. The erythemal rays are transmitted (280 to 320 nm) through special glass bulbs of quartz type. Arc-type generators and bulbs which look like conventional reflector lamps are the two most common kinds. The RS type is a self-contained device with a 275-W inner quartz tube in an outer bulb (R-40) and a resistance ballast incorporated within the bulb. It also contains a self-activating switch, similar in action to thermal-type starters used with fluorescent lamps.

***42. Glow Lamps.*** When sufficient voltage is applied to electrodes sealed within a bulb containing neon, argon, or helium, light is produced at the negative electrode. On direct current, one cathode glows; on alternating current, the reversal is so rapid, both electrodes appear to glow. The range of glow lamps is 1/25 to 3 W (Table 26-17). Their useful life varies approximately as the inverse of the cube of the current. A glow lamp has a negative voltampere characteristic; hence a limiting resistance is used in series with it. In conventional screw-base types, the resistor is concealed in the base. Average lamp life ranges between 7500 and 25,000 h.

Glow lamps have wide use in electronic circuitry, where their action is that of a practically instantaneous switch. At breakdown voltage, the lamp glows, and the switch is closed; at the extinguishing voltage, the lamp current drops to a fraction of its full value and may be considered as nonconducting, or open-circuit in certain circumstances. This on-off characteristic suits the glow lamp to the dichotomy of binary arithmetic as used in computers and logic circuitry in general. Other glow-lamp applications in electronic circuitry include oscillators, pulse generators, voltage regulators, and coupling networks.

***43. Electroluminescent Lamps.*** This type of lamp is a thin-area source in which light is produced by a phosphor excited by a pulsating electrical field. In essence, the lamp is a plate capacitor with a phosphor embedded in its dielectric and with one or both of its plates transparent. Green, blue, yellow, or white light may be produced by choice of phosphor. The green phosphor has the highest luminance. These lamps are available in ceramic and plastic form, flexible or with stiff backing, and are easily fabricated into simple or complex shapes. They have been used in decorative lighting, night lights, switchplates, instrument panels, clock faces, telephone dials, thermometers, and signs. Their application is limited to locations where the general illumination is low.

Luminance varies with applied voltage, frequency, and temperature, as well as with the type of phosphor. At 120 V, 60 Hz, the luminance of the ceramic form with the green phosphor is approximately 1 fL; the luminance of the plastic form may be as high as 8 fL under these conditions or up to 30 fL at 120 V, 400 Hz.

Life is long and power consumption low. There is no abrupt point at which the lamp fails; the time at which the luminance has fallen to 50% of initial is sometimes used as a measure of useful life. For the ceramic form, this is approximately 20,000 h at 120 V, 60 Hz. Approximate initial current and wattage values per square foot of lamp under these operating conditions are 60 mA and 3.5 W.

***44. Black-Light Lamps.***[1] Near-ultraviolet radiant energy (energy not visible to the human eye) causes certain materials to fluoresce or emit visible light. The normal human eye is sensitive only to radiant energy between 380 and 780 nm in wavelength. Thus, lamps which produce primarily near-ultraviolet radiant energy in the 320- and 380-nm range are popularly called "black" lights. This term is quite descriptive, since the ultraviolet energy from the "light" source cannot be seen by the human eye, but the effects of the radiation on special materials can be visually dramatic.

When black light is directed at a fluorescent material, an energy conversion takes place. The material or chemical sensitive to ultraviolet energy absorbs the energy, then reradiates it at longer wavelengths to which the eye is sensitive.

Mercury lamps with filters to absorb the visible light and transmit the near-ultraviolet are used for fluorescent effects. Called black-light lamps, they are generally enclosed in a red-purple filter glass bulb that looks black. Many materials fluoresce when irradiated by black-light lamps.

---

[1] Kraehenbuehl, J. O., and Chanon, H. J.: Technology of Brightness Production by Near-Ultraviolet Radiation; *Trans. IES,* February 1941.

**TABLE 26-17** Typical Glow Lamps*

| Gas used<br>Lamp no.<br>ANSI No. | Neon<br>A1B | Neon<br>B1A<br>NE-51 | Neon<br>A9A<br>NE-2E | Neon<br>A1C | Neon<br>B2A<br>NE-51H | Neon<br>C2A<br>NE-2H | Neon<br>C9A<br>NE-2J | Neon<br>B7A<br>NE-45 | Neon<br>B9A<br>NE-48 | Argon<br>J2A<br>AR-3 | Argon<br>J3A<br>AR-4 | Neon<br>F4A<br>NE-58 | Neon<br>J5A<br>NE-30 | Neon<br>L5A<br>NE-32 | Neon<br>R6A<br>NE-40 |
|---|---|---|---|---|---|---|---|---|---|---|---|---|---|---|---|
| Nominal watts | 1/25 | 1/25 | 1/12 | 1/7 | 1/7 | 1/4 | 1/4 | 1/4 | 1/4 | 1/4 | 1/4 | 1/2 | 1 | 1 | 3 |
| Circuit volts | | | | | | | | | | | | | | | |
| ac | 105–125 | 105–125 | 105–125 | 105–125 | 105–125 | 105–125 | 105–125 | 105–125 | 105–125 | 105–125 | 105–125 | 210–250 | 105–125 | 105–125 | 105–125 |
| dc | 105–125 | 105–125 | 105–125 | 150 | 150 | 150 | 150 | 105–125 | 105–125 | 135 | 135 | 210–250 | 105–125 | 105–125 | 105–125 |
| Bulb (clear) | T-2 | T-3¼ | T-2 | T-2 | T-3¼ | T-2 | T-2 | T-4½ | T-4½ | T-4½ | T-4½ | T-4½ | S-11 | G-10 | S-14 |
| Base† | Wire term. | Min. bay. | Wire term. | Wire term. | Min. bay. | Wire term. | S.C. mid. fl. | Cand. sc. | D.C. bay. | Cand. sc. | D.C. bay. | Cand. sc. | Med. sc. | D.C. bay. | Med. sc. |
| Max. overall length, in‡ | ½ | 1 3/16 | ¾ | ½ | 1 3/16 | ¾ | 15/16 | 1 17/32 | 1½ | 1 17/32 | 1½ | 1 17/32 | 2¼ | 2⅛ | 3½ |
| Max. initial breakdown volts | | | | | | | | | | | | | | | |
| ac | 65 | 65 | 65 | 95 | 95 | 95 | 95 | 65 | 65 | 85 | 85 | 75 | 65 | 75 | 60 |
| dc | 90 | 90 | 90 | 135 | 135 | 135 | 135 | 90 | 90 | 120 | 120 | 100 | 90 | 100 | 85 |
| Series resistance, Ω | 220,000 external | 220,000 external | 100,000 external | 47,000 external | 47,000 external | 30,000 external | 30,000 external | None | 30,000 external | None | 15,000 external | None | None | 7,500 external | None |
| Average ac life, h | 25,000 | 25,000 | 25,000 | 25,000 | 25,000 | 25,000 | 25,000 | 7,500 | 7,500 | 150 | 150 | 7,500 | 10,000 | 10,000 | 10,000 |

* Consult manufacturers' technical literature for current data.

† Wire term.—unbased wire terminals; Min. bay.—single-contact miniature bayonet; S.C. mid. fl.—single-contact midget flanged; Cand. sc.—candelabra screw; D.C. bay.—double-contact bayonet candelabra; Med. sc.—Medium screw.

‡ Length of glass only for unbased wire terminal lamps.

They are used for theatrical and advertising effects, industrial and food inspection, detection of counterfeits and forgeries, medical diagnosis, insect traps, crime and vermin detection, laundry marking, and copying equipment.

Tubular sources designated as BLB lamps, such as the 15-W T-8 and the 20- and 40-W T-12 lamps, have integral filters and may be operated with the same ballasts as corresponding fluorescent lamps.

The luminance of an irradiated fluorescent material is between 1 and 5 fL with printing inks and between 0.25 and 2.5 fL with interior paints, depending on the color. The apparent brightness increases considerably as the eyes become dark-adapted. Conversely, the effectiveness of black light is greatly reduced or entirely negated by a small amount of visible light.

***45. Carbon-Arc Sources.*** There are three basic types of carbon arcs: flame arc, low-intensity arc, and high-intensity arc. In the *flame arc* the light source is the entire arc stream made luminous by the addition of flame materials. This arc is used in photography and related industrial photochemical processes, where, with a special coring of rare-earth salts, it produces an essentially continuous radiation more closely approximating natural sunshine than any other artificial source. A special form of the flame arc is the so-called enclosed arc. It operates in a glass globe with limited access of air, the resulting nitrogen-rich atmosphere enhancing radiation in the violet and near-ultraviolet regions, which is quite effective in blueprinting and related copying processes. In the *low-intensity arc,* the principal light source is the incandescent tip of the positive carbon electrode, at or near its sublimation temperature. With a luminance of approximately 15,000 cd/cm$^2$ uniformly generated over a considerable area, this source found early application in motion-picture, searchlight, and other projection systems where a concentrated source capable of producing a well-defined narrow beam is required. The *high-intensity carbon arc* is distinguished primarily from the low-intensity one by the inclusion of rare-earth materials in the core of the positive electrode which volatilize into the arc stream as the electrode is consumed in burning. Thus, in addition to light from the incandescent carbon surface, there is a significant amount of light originating in the gaseous region immediately in front of the carbon as a result of the combination of a high current density and an atmosphere rich in flame materials. The color quality of the light from these sources is well adapted to color motion-picture photography and projection. Such arcs with luminances in the range from about 55,000 to 145,000 cd/cm$^2$ have largely replaced the earlier low-intensity carbon arcs in motion-picture and searchlight projection, while the practicability of operation at luminances in excess of 200,000 cd/cm$^2$ has been demonstrated in the laboratory.

## LUMINAIRES AND LIGHTING SYSTEMS

***46. Luminaires*** are complete lighting units consisting of a lamp or lamps together with the parts designed to distribute the light, to position and protect the lamps, and to connect the lamps to the power supply. They are classified by the CIE (International Commission on Illumination) according to the percentage of light output above and below the horizontal as follows:

| | |
|---|---|
| Direct | 0– 10% upward, 90–100% downward |
| Semidirect | 10– 40% upward, 60– 90% downward |
| General diffuse | 40– 60% upward, 40– 60% downward |
| Semi-indirect | 60– 90% upward, 10– 40% downward |
| Indirect | 90–100% upward, 0– 10% downward |

This classification system applies to all types of luminaires for general lighting in industrial, commercial, and residential applications. Luminaires are designed to redirect light and to increase the effective light-source area, thus decreasing brightness while absorbing no more of the light flux than necessary.

***47. Lighting systems*** are installations of one or more luminaires and are often classified in accordance with their layout or location with respect to the visual task or object lighted—general lighting, localized general lighting, and local (supplementary) lighting. They are also classified in accordance with the CIE type of luminaire used.

*General Lighting.* Lighting systems which provide an approximately uniform level of illumination on the work plane over the entire area are called general lighting systems. The luminaires are usually arranged in a symmetrical plan fitted into the physical characteristics of the area and blend well with the room architecture. They are relatively simple to install and require no coordination with furniture or machinery that may not be in place at the time of the installation. Perhaps the greatest advantage of general lighting systems is that they permit complete flexibility in task location.

*Localized General Lighting.* A localized general lighting system consists of a functional arrangement of luminaires with respect to the visual task or work areas. (It is sometimes called *task lighting.*) It also provides illumination for the entire room area. Such a lighting system requires special coordination in installation and careful consideration to ensure adequate general lighting for the room. This system has the advantages of better utilization of the light on the work area and the opportunity to locate the luminaires so that annoying shadows and direct and reflected glare are prevented.

*Local Lighting.* A local lighting system provides lighting only over a relatively small area occupied by the task and its immediate surround. The illumination may be from luminaires mounted near the task or from remote spotlights. (This system also is sometimes called *task lighting.*) It is an economical means of providing higher illumination levels over a small area, and it usually permits some adjustment of the lighting to suit the requirements of the individual. Improper adjustments may, however, cause annoying glare for nearby workers. Local lighting, by itself, is seldom desirable. To prevent excessive changes in adaptation, it should be used in conjunction with general lighting that is at least 20% of the local lighting level; it then becomes *supplementary lighting.*

*Direct Lighting.* When luminaires direct 90 to 100% of their output downward, they form a direct lighting system. The distribution may vary from widespread to highly concentrating, depending on the reflector material, finish, and contour and on the shielding or control media employed. *Troffers* and *downlights* are two forms of direct luminaires.

Direct lighting units can have the highest utilization of all types, but this utilization may be reduced in varying degrees by brightness-control media required to minimize direct glare. Veiling reflections may be excessive unless distribution of light is designed to reduce the effect.

Reflected glare and shadows may be problems with direct lighting unless close spacings are employed. Large-area units are also advantageous in this respect.

*Luminous ceilings, louverall ceilings,* and *large-area modular lighting elements* are forms of direct lighting having characteristics similar to those of indirect lighting discussed in paragraphs below.

*Semidirect Lighting.* The distribution from semidirect units is predominantly downward (60 to 90%) but with a small upward component to illuminate the ceiling and upper walls. The characteristics are essentially the same as for direct lighting except that the upward component will tend to soften shadows and improve room brightness relationships. Care should be exercised with close-to-ceiling mounting of some types to prevent overly bright ceilings directly above the luminaire. Utilization can approach, or even sometimes exceed, that of well-shielded direct units.

*General Diffuse Lighting.* When downward and upward components of light from luminaires are about equal (each 40 to 60% of total luminaire output), the system is classified as general diffuse. *Direct-indirect* is a special (non-CIE) category within the classification for luminaires which emit very little light at angles near the horizontal. Since this characteristic results in lower luminances in the direct-glare zone, direct-indirect luminaires are usually more suitable than general-diffuse luminaires which distribute the light about equally in all directions.

General-diffuse units combine the characteristics of direct lighting described above and those of indirect lighting described below. Utilization is somewhat lower than for direct or semidirect units, but it is still quite good in rooms with high reflectance surfaces. Brightness relationships throughout the room are generally good, and shadows from the direct component are softened by the upward light reflected from the ceiling.

Luminaires designed to provide a general-diffuse or direct-indirect distribution when pendant-mounted are frequently installed on or very close to the ceiling. Such mountings change the distribution to direct or semidirect since the ceiling acts as a top reflector redirecting the upward light back through the luminaire. Photometric data obtained with the luminaire equipped with top reflectors or installed on a simulated ceiling board should be employed to determine the luminaire characteristics for such application conditions.

*Semi-Indirect Lighting.* Lighting systems which emit 60 to 90% of their output upward are

**TABLE 26-18** Reflectances of New, Clean, Opaque Materials for Incidence of 20°

| Surface | Reflectance | Surface | Reflectance |
|---|---|---|---|
| Aluminum, polished | 0.67–0.73 | Nickel | 0.64 |
| Aluminum, anodic | 0.78–0.83 | Silver, polished | 0.92 |
| Aluminum paint | 0.55–0.70 | Silvered mirror | 0.85 |
| Brass, polished | 0.60 | Stainless steel | 0.55 |
| Chromium, polished | 0.65 | White painted enamel | 0.78–0.90 |
| Rhodium, polished | 0.75 | White paint | 0.75–0.90 |
| | | White porcelain enamel | 0.75–0.86 |

**TABLE 26-19** Reflectance, Transmittance, and Absorptance of Glasses for 20° Incidence

| Glass | Reflectance | Transmittance | Absorptance |
|---|---|---|---|
| Clear crystal | 0.08 | 0.91 | 0.01 |
| Rough or configurated crystal | 0.14 | 0.83 | 0.03 |
| White | 0.74 | 0.12 | 0.14 |
| Etched, single surface | 0.15 | 0.82 | 0.03 |
| Etched, double surface | 0.28 | 0.69 | 0.03 |

defined as semi-indirect. The characteristics of semi-indirect lighting are similar to those of indirect systems discussed below except that the downward component usually produces a luminaire luminance that closely matches that of the ceiling. However, if the downward component becomes too great and is not properly controlled, direct or reflected glare may result.

*Indirect Lighting.* Lighting systems classified as indirect are those which direct 90 to 100% of the light upward to the ceiling and upper sidewalls. In a well-designed installation the entire ceiling becomes the primary source of illumination, and shadows will be virtually eliminated. Also, since the luminaires direct very little light downward, both direct and reflected glare will be minimized if the installation is well planned. Luminaires whose luminance approximates that of the ceiling have some advantages in this respect. It is also important to suspend the luminaires a sufficient distance below the ceiling to obtain reasonable uniformity of ceiling luminance without excessive luminance immediately above the luminaires.

Since with indirect lighting the ceiling and upper walls must reflect light to the work plane, it is essential that these surfaces have high reflectances. Care is needed to prevent overall ceiling luminance from becoming too high and thus glaring.

***48. Control of light distribution*** is usually accomplished through reflection, refraction, transmission, absorption, and diffusion using glasses, plastics, metals, and woods, of various shapes, reflectance, transmittance, absorptance, polarization, and finish.

*Reflector* contour shapes including parabolic, ellipsoidal, hyperbolic, and spherical and *refractors* (lenses) utilizing prisms, cones, and spherical shapes are commonly used to produce a wide range of light-controlling devices. Flat or contoured *diffusers* are used to diffuse, color, or polarize the light according to the lighting needs. Tables 26-18 and 26-19 give reflectances, transmittances, and absorptances of certain typical light-control materials.

Polarizing multilayer diffusing panels for direct enclosed fluorescent luminaires are commercially available. These panels, when installed in suitable luminaires, vertically polarize a portion of the emitted light, the percentage of polarization depending on the angle of emission as shown in Fig. 26-22. Vertically polarized light, when reflected from nonmetallic surfaces, reduces veiling reflections and improves the contrast between the critical detail of the seeing task and its imme-

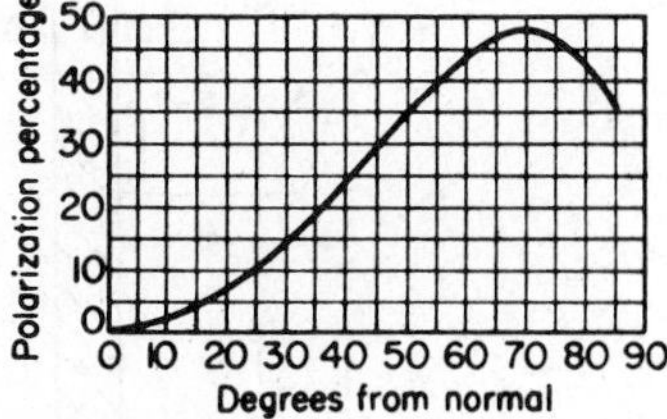


**FIG. 26-22** Polarization efficiency curve for a four-lamp 2-ft-wide troffer.

diate background. It is evident from Fig. 26-22 that polarization is most effective when the line of sight makes a large angle with the normal to the surface.

***49. Structural Lighting.*** Lighting systems which form a substantial part of the structure of a building, as distinguished from individual or groups of luminaires suspended from the ceiling or bracketed to walls, are examples of structural lighting.

The *luminous ceiling* of suspended plastic or other material is an outstanding form (Fig. 26-23). Other designs utilize horizontal sheets of corrugated or louvered plastic with vertical spacers forming acoustical baffles. In the sketch, recommended spacing of lamp rows and distance to diffusing media are shown. For a buildup of levels at desks, business machines, or other work surfaces, the combination of a *luminous ceiling with troffers* (Fig. 26-24) is illustrated. An acceptable design of a *soffit* using fluorescent luminaires is shown in Fig. 26-25. The illustration shows its application at a laboratory sink; however, other horizontal surfaces near walls where more critical seeing is done could be similarly lighted. The *open canopy* (Fig. 26-26) produces direct lighting, with an upward component for general space lighting. Other forms include additional downlights (incandescent) for highlighting furniture groups or merchandise displays. A *cornice-and-cove* plan is shown in Fig. 26-27. For a public dining room or a residence with a wall having ceiling-to-floor draperies, it provides light to the ceiling and

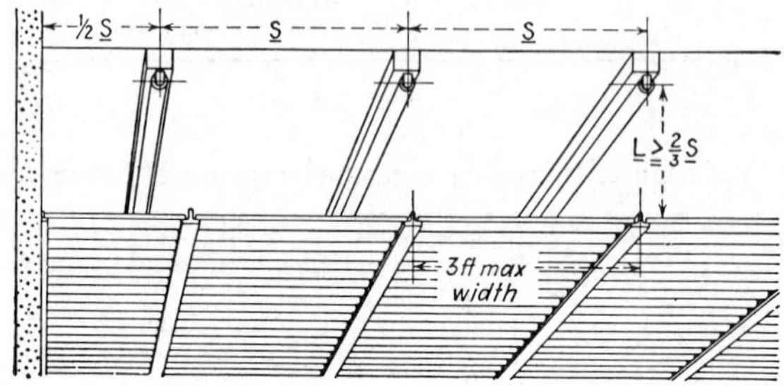


**FIG. 26-23** Luminous ceiling.

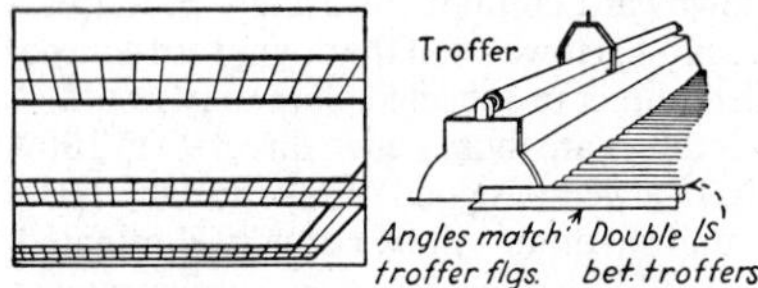


**FIG. 26-24** Luminous ceiling with troffers.

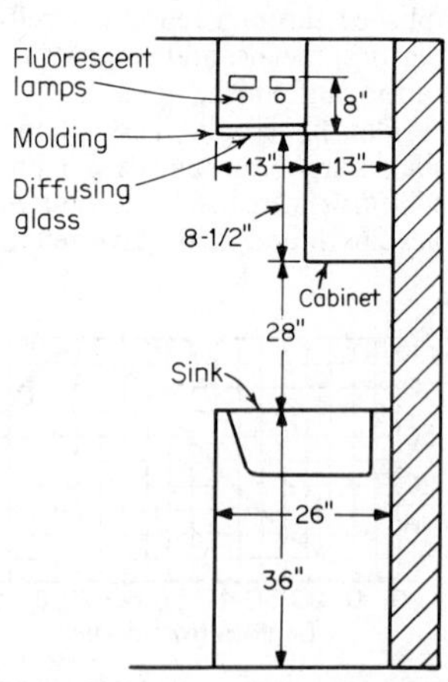


**FIG. 26-25** Soffit.

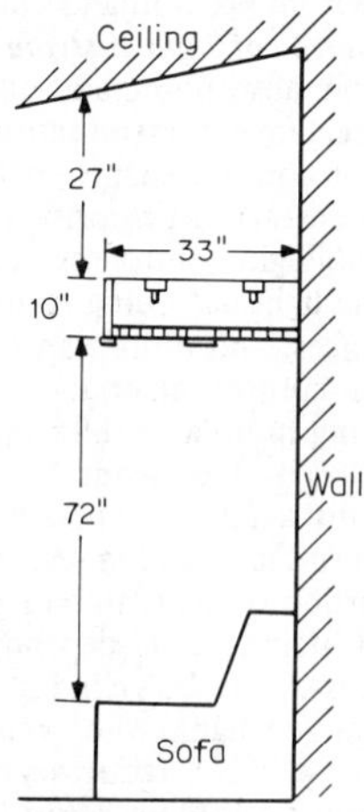


**FIG. 26-26** Open canopy.

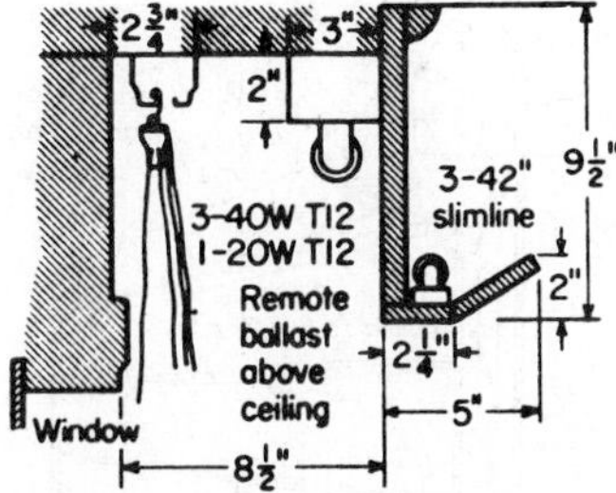

FIG. 26-27 Cornice and cove.

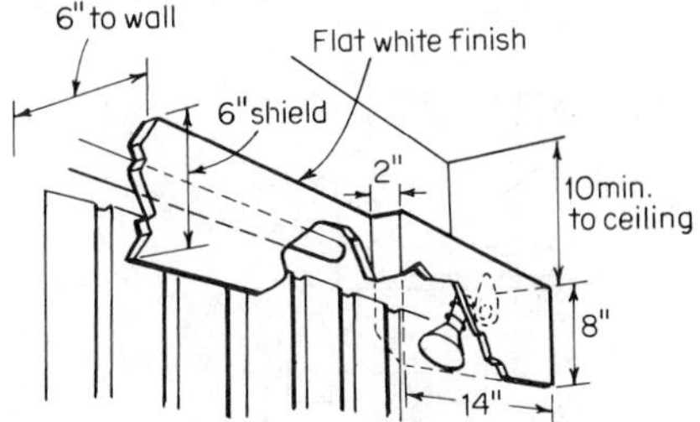


FIG. 26-28 Valance and spotlight.

down on the draperies in a pleasing manner. Reflecting surfaces within the cornice and cove are painted white. Other wall-lighting methods include the *valance* with *spotlight* as shown in Fig. 26-28. The spotlights in the end sections are separately controlled and are used to add supplementary light or highlighting. In the *closed coffer* (Fig. 26-29), the ceiling cavity is designed to fit the needs of the space to be lighted. The dimensions of the luminaire are 30 by 96 by 8 in for a room 7 by 11 ft. White corrugated vinyl plastic is the diffusing medium. Illuminated *open coffers* (Figs. 26-30 and 26-31) are equally acceptable in contemporary and classical interiors. Improved maintenance is effected by covering the concealed luminaire elements in the four sides with clear glass. Long, narrow coffers or free-form examples can be augmented by suspended incandescent luminaires for a combination of style, color effect, and sculptural distinction. The *shelf-valance* form (Fig. 26-32) is useful for display of books, merchandise, and art objects and for tasks not affected by veiling reflections. If the shelf were of transparent material, objects above it would be illuminated. By using white glass for the valance board (instead of plywood, as shown), the surface could serve as the background for a departmental sign in a shop or supermarket. The *room divider* (Fig. 26-33) is representative of residence use in the contemporary interior. It is also found in foyers in public buildings and many other interiors where unobtrusive divisions of space are desirable. It can serve as a storage cabinet and planter base; the shelves are adaptable to a wide variety of uses. The fluorescent ballasts are concealed below the bottom shelf.

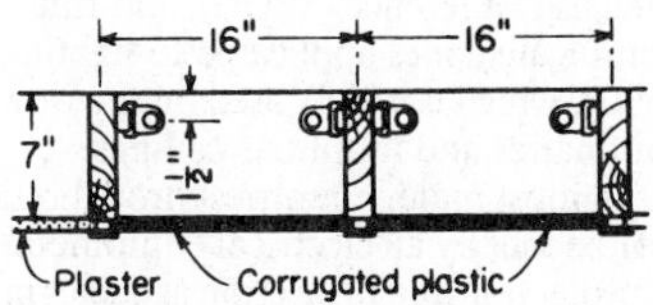


FIG. 26-29 Closed coffer.

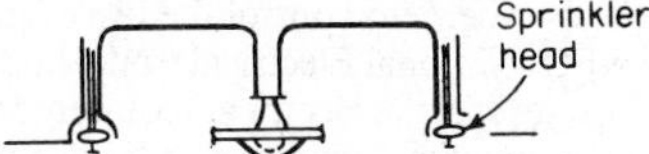


FIG. 26-30 Open coffer (incandescent).

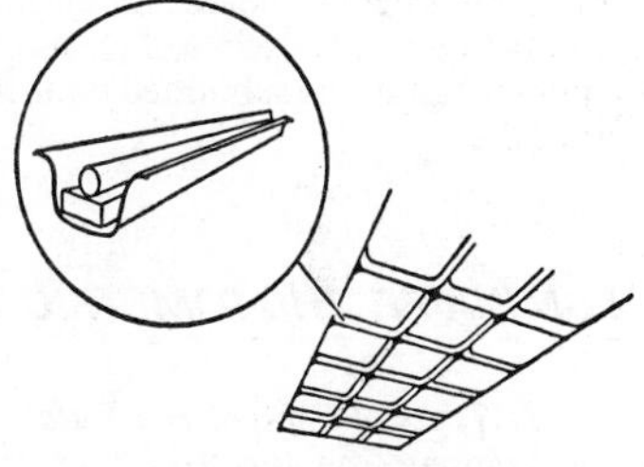

FIG. 26-31 Open coffer (fluorescent).

***50. Codes and Standards.*** Local codes, national codes, international codes, federal standards, professional standards, and manufacturers' standards relate to specific requirements which must be met in the construction and installation of a luminaire. Standards usually relate to minimum requirements (safety, construction, or performance) which may be exceeded to provide a better product.

Some codes and standards deal with fire and safety (electrical, mechanical, and thermal); others relate to performance (photometric) and construction (materials and finishes). They will vary to some extent depending on geographic location and end use of equipment. Conformance to the appropriate set of specifications is often determined by certified laboratory tests. Certification is often denoted by an identifying label. Local inspection agencies may or may not rely on conformance to national, federal, or industrial codes and standards.

Information regarding local codes may be obtained from electrical inspection departments.

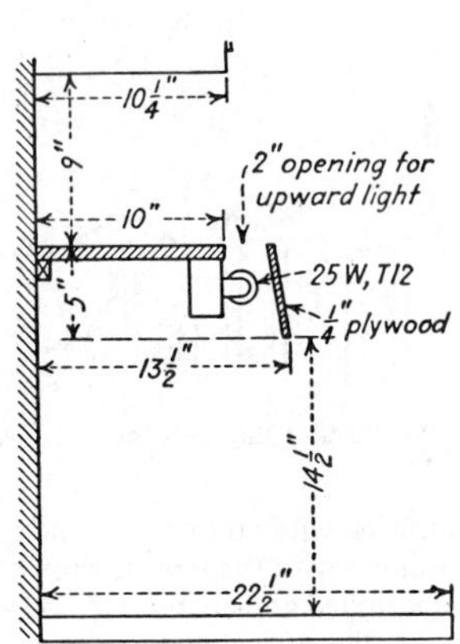


FIG. 26-32 Shelf valance.

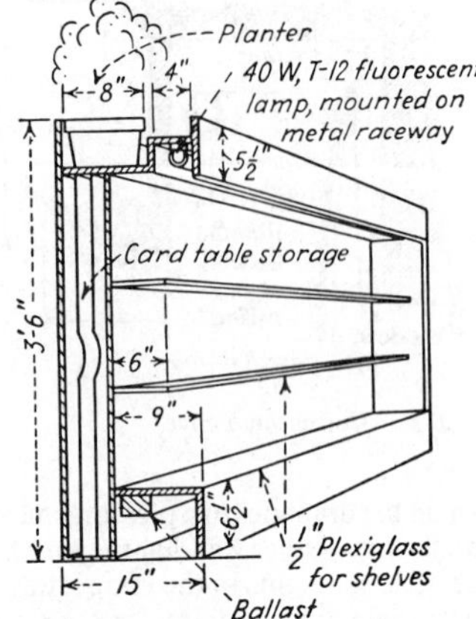


FIG. 26-33 Room divider.

Several other code jurisdictions may apply to regional and state codes. Building Officials Conference of American (BOCA), Southern Building Officials Conference (SBOC), and International Conference of Building Officials (ICBO) all promulgate codes applicable to lighting installations. ICBO issues the Uniform Building Code (UBC), which is widely used on the west coast and elsewhere, to control the use of materials in luminaires and luminous ceilings.

The NEC (National Electrical Code) and similar codes in most major countries throughout the world state specific electrical requirements which must be met by all electrical equipment, including luminaires. They have been developed by safety protection and inspection agencies in conjunction with fire-protection agencies.

The UL (Underwriters' Laboratories, Inc.) and similar groups in other countries publish minimum safety standards for electrical and associated products which are in conformance with the respective electrical codes of their country. They have testing laboratories to which equipment must be submitted for listing. Most manufacturers design luminaires to meet these standards.

## LUMINAIRE PHOTOMETRIC DATA

***51. Types of Photometric Data.*** Luminaire manufacturers provide laboratory-measured and -computed photometric and physical data required for a lighting design. The physical data include dimensions, finishes, light-control materials, number of lamps, types of lamps, etc. The photometric data include a luminous intensity (candlepower) distribution, zonal lumens, luminaire efficiency, coefficients of utilization, luminance coefficients, luminances, and visual comfort probability (VCP) values.

***52. Luminous Intensity (Candlepower) Distribution.*** A luminous intensity or candlepower distribution curve is a graphic presentation of the distribution of light about a lamp or luminaire. Such presentations contribute valuable information to guide the engineer in determining the suitability of lighting equipment for application in various fields. As a background for using distribution curves, it is first necessary to see how they are obtained.

The candlepower in any direction from a light source equals the illumination produced on a plane at right angles to the light rays times the square of the distance from the lamp to the point of measurement ($I = E \times D^2$). For accurate measurements the distance should be at least five times the largest dimension of the source. If in this way the average candlepower around the axis of a source is determined for any angle from the vertical, say, 25°, the average value becomes one point which can be plotted to a convenient scale on polar-coordinate paper. When the source is symmetric about its axis, as a bare lamp, taking several measurements around the axis at the 25° angle from the vertical usually shows them all to be about the same. However, in laboratory photometry any slight differences that might be present because of filament structure or other variations are compensated by rotating the lamp so that one reading represents an average value.

To get sufficient data for a lamp or luminaire, 20 readings are usually taken at angles of 0, 5, 15, 25, 35, 45°, etc., up to 180° (Fig. 26-34), and the candlepowers are computed and plotted on

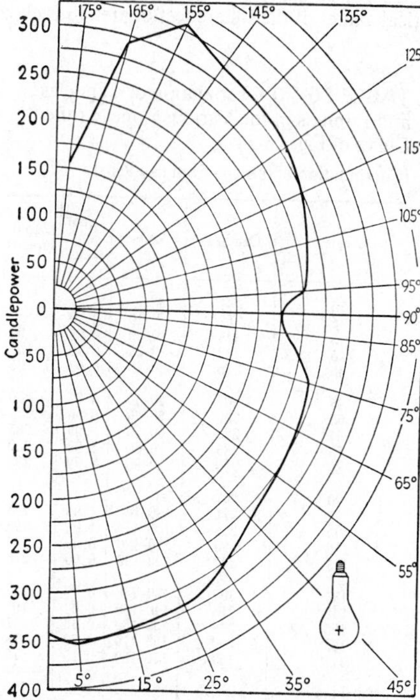


**FIG. 26-34** Distribution curve of a 3700-lm 200-W incandescent lamp. For convenience, the curves of symmetrical-type units show only for the 0 to 180° half. Actually the distribution of light includes 360°, that is, two symmetrical halves.

polar-coordinate paper. The line connecting a series of such points forms the candlepower distribution curve. The value at 90° is the candlepower straight out from the unit, while that at 0° is directly below. For concentrated light sources such as searchlights and spotlights, photometric readings are often required 1 or 2° apart, rather than at 10° intervals.

To determine the output of nonsymmetric or asymmetric luminaires, such as conventional fluorescent units, candlepower readings must be taken in a number of planes. For a fluorescent luminaire, candlepower readings are usually taken in five planes, at 0, 22½, 45, 67½, and 90° from a plane through the luminaire axis. Candlepower values are measured in each plane at 10° intervals (5, 15, 25°, etc.). On assuming that the five planes are at the angles 0, 22½, 45, 67½, and 90° and that the candlepower readings at such intervals are designated as *A, B, C, D,* and *E,* their weighted average is obtained by the formula

$$\text{cp} = \frac{A + 2B + 2C + 2D + E}{8}$$

In some laboratories, for similar tests, candlepower readings are taken in three planes only, crosswise (90°), lengthwise (0°), and at 45°. The candlepower values crosswise and lengthwise plus twice the values in 45° are added. The sum divided by 4 equals the average candlepower.

***53. Zonal and Total Lumens.*** The total lumens of a light source are obtained by adding the lumens in the various zones. The zonal lumen values are readily computed by multiplying the candlepower at the center of the zone by the zonal constant. For example, at 55°, which is the center of the 50 to 60° zone, the 300-cd reading multiplied by 0.90 gives 270 lm in that zone for the 200-W lamp (Fig. 26-34). Zonal constants for the usual 18 test zones as shown in Table 26-20 are simply numerical values of the square feet of area in each zone on the surface of a

**TABLE 26-20** Tabulation of Candelas and Lumens in the Various Zones in the Curve of Fig. 26-34

(Computation of lumens from candelas)

| Angle | Candelas | Zone | Zonal constant | Zonal lumens |
|---|---|---|---|---|
| 0 | 347 | | | |
| 5 | 352 | 0–10 | 0.10 | 35 |
| 15 | 348 | 10–20 | 0.28 | 97 |
| 25 | 342 | 20–30 | 0.46 | 157 |
| 35 | 326 | 30–40 | 0.63 | 205 |
| 45 | 307 | 40–50 | 0.77 | 236 |
| 55 | 300 | 50–60 | 0.90 | 270 |
| 65 | 288 | 60–70 | 0.99 | 285 |
| 75 | 285 | 70–80 | 1.06 | 302 |
| 85 | 259 | 80–90 | 1.09 | 282 |
| 90 | 257 | | | |
| 95 | 271 | 90–100 | 1.09 | 295 |
| 105 | 278 | 100–110 | 1.06 | 295 |
| 115 | 290 | 110–120 | 0.99 | 287 |
| 125 | 307 | 120–130 | 0.90 | 276 |
| 135 | 308 | 130–140 | 0.77 | 237 |
| 145 | 313 | 140–150 | 0.63 | 197 |
| 155 | 329 | 150–160 | 0.46 | 151 |
| 165 | 280 | 160–170 | 0.28 | 78 |
| 175 | 153 | 170–180 | 0.10 | 15 |
| Total lumens | ... | ....... | .... | 3,700 |

sphere having a radius of 1 ft. Since one candela generates one lumen of light per square foot of surface in the unit sphere, it follows that the lumens in each zone are equal to the square feet of sphere area in that zone multiplied by the candlepower at the center of the zone.

***54. Luminaire Efficiency.*** The efficiency of a luminaire is expressed in terms of its lumens output divided by the lumens generated by the lamp. For example, if the 200-W lamp in Fig. 26-34 is placed in a white-glass enclosing globe whose candlepower distribution shows an output of 2960 lm, the efficiency of the luminaire is 2960 divided by 3700, or 80%.

***55. Coefficients of Utilization.*** A coefficient of utilization (CU) is a number used in general lighting calculations and represents the ratio of the lumens from a luminaire received on the work plane to the lumens emitted by the luminaire's lamps alone. CUs are calculated using a uniform method[1] and are tabulated for various room shapes and reflectances. See Table 26-28.

***56. Luminance Coefficients.*** A luminance coefficient is a number used in discomfort glare evaluations and in determining the illumination at a point in an interior. It is a number similar to a coefficient of utilization and is used to determine room wall and ceiling luminances. See Table 26-28.

***57. Luminance and VCP.*** Maximum luminaire luminance is reported and average lumi-

[1] Recommended Procedure for Calculating Coefficients of Utilization, Wall Exitance Coefficients, and Ceiling-Cavity Exitance Coefficients; *Journal of IES,* October 1982, vol. 12, no. 1, p. 3.

nance calculated at various luminaire angles as part of direct-glare evaluations of visual comfort probability (VCP). Calculated VCP ratings of lighting installations using the specific luminaire photometered, for various size rooms of given reflectances and illumination levels, are reported. These ratings express the percent of people who, if seated in the most undesirable location within the installation, will be expected to find it acceptable.

## LIGHTING DESIGN

*58. General Process.* The lighting-design process can be broken down into four major parts: (1) determining the project goals, (2) collecting and selecting the nonillumination and illumination criteria, (3) making design decisions, and (4) evaluating the completed project. The design process may require changes in decisions (3) until the project goals have been satisfied.

*59. Project Goals.* These are the conceptual objectives of a whole project. The designer must have knowledge of the appearance requirements of the job, including the type of period of the architecture; whether a corporate image is to be projected; whether construction is for long life or of a speculative nature; whether young or old persons will be using the facility; and any other objectives of the project which will affect the lighting design.

*60. Nonillumination Criteria.* These are criteria which originate from outside the engineering discipline. They include identification of visual tasks and their locations and frequency, space dimensions, work-plane locations, reflectances, daylight availability, space temperature and dirt conditions, voltage, operating schedules, cost budgets, power and energy budgets, and codes. These criteria should represent the best available information which can be obtained from the owner, architect, or designer.

*61. Illuminance Criteria.* These are criteria which originate from within the illuminating engineering discipline. They include illuminance levels for task performance and safety, visual comfort data, luminance ratios, and light-loss factors. These criteria can be found in various American National Standards Institute practices (Office, School, Industrial, Roadway, Protective) and in IES recommended practices and committee reports and the *IES Lighting Handbook*.

*62. Design Decisions.* All the above criteria along with the project goals need to be evaluated when making the major design decisions. The major decisions include light-source selection, luminaire selection and mounting, maintenance-procedure determination, calculation-methods selection, luminaire-layout arrangement, and control (switching, dimming) planning. Decisions involving trade-offs and compromises are often necessary.

*63. Evaluation.* No design process is complete until one has evaluated the results to see that the project goals have been met. Furthermore, the lighting designer should see that the design is evaluated in terms of its impact on the total building energy utilization. Some of the evaluation procedures include illumination checks and life-cycle costs. In some cases, government or other energy limits may apply.

*64. Energy Utilization.* In any lighting design it is important to consider the following 12 recommendations prepared by the Illuminating Engineering Society for the better utilization of energy used for lighting. They are appropriate for designing new construction, in renovation work, and in operating and maintaining existing installations.

1. Design lighting for expected activity (light for seeing tasks with less light in surrounding nonworking areas).
2. Design with more effective luminaires and fenestration (use systems analysis based on life cycle).
3. Use efficient light sources (higher lumen per watt output).
4. Use more efficient luminaires.
5. Use thermal-controlled luminaires.
6. Use lighter finish on ceilings, walls, floor, and furnishings.
7. Use efficient incandescent lamps.

8. Turn off lights when not needed.
9. Control window brightness.
10. Utilize daylighting as practicable.
11. Keep lighting equipment clean and in good working condition.
12. Post instructions covering operation and maintenance.

***65. Lighting and the Thermal Environment.*** Effective building design requires a provision for efficient utilization or dissipation of lighting heat. The benefits of integrating building heat in lighting design are: (1) improved performance of the air-conditioning system, (2) more efficient handling of lighting heat, and (3) more efficient lamp performance if fluorescent lamps are the light source.

The control and removal of lighting heat before it enters the occupied space can reduce heat in that space, reduce air changes and fan horsepower, lower temperature differentials required in the space, enable a more economical cooling-coil selection because of the higher temperature differential across the coil, and reduce luminaire and ceiling temperature, thereby minimizing radiant effects.

The degree to which any of these benefits may be obtained depends on many variables such as the quantity of energy involved, the type of heat-transfer mechanism, the temperature difference between source and sink, and the velocity and quantity of fluids and/or air available for heat transfer. However, in most applications luminaire temperature will be higher than room temperature; so fluids at room temperature can be effective in heat transfer. Any unwanted heat that can be removed at room temperature or above can be removed much more economically than at lower temperatures. Various publications of ASHRAE[1] and IES should be consulted.

## QUANTITY AND QUALITY OF ILLUMINATION

***66. Levels of Illumination.*** Visual performance is a function of a number of fundamentally important factors. Some of these are the size of the object or detail to be seen, the contrast of the detail with its immediate background, the luminance of the object, the time available to see it, the luminance relation between the object and its surroundings, the visual capability of the human seeing machine, and the level of personal motivation.

Many lighting designers have used single-value illuminance recommendations published by IES since 1958.[2] In 1979, the Illuminating Engineering Society of North America (IESNA) adopted a range approach for the prescription of task-lighting levels. The new procedure involves illuminance ranges, plus a system using weighting factors for the selection of the top, middle, or lower value of the range. Designers can thus tailor their lighting proposals to the needs of the task and the persons performing those tasks. Such coordination became especially important during the 1970s, when the wisdom of conserving energy became generally recognized.

The designer must have several types of information about the visual tasks:

1. What are the task details that need to be seen?
2. How old are the persons who perform the tasks?
3. Is speed and/or accuracy a critical, important, or unimportant consideration?
4. Are the task details seen against a light, medium, or dark background?

These four types of information may require additional effort and investigation by the designer, but they must all be considered in order to effectively use the new IESNA illuminance selection system. Some typical ranges are given in Table 26-21. Weighting factors for areas or

[1] *ASHRAE Handbook.*

[2] Based on vision research sponsored by the Illuminating Engineering Research Institute. See Blackwell, H. R.: Development and Use of a Quantitative Method for Specification of Interior Illumination Levels; *Illum. Eng.*, June 1959, p. 317.

**TABLE 26-21** Typical Recommended Illuminance Ranges*

| Area or activity | Footcandles on task | Lux† |
|---|---|---|
| Industrial assembly or inspection | | |
| Simple | 20–30–50 | 200–300–500 |
| Moderately difficult | 50–75–100 | 500–750–1000 |
| Difficult | 100–150–200 | 1000–1500–2000 |
| Very difficult | 200–300–500‡ | 2000–3000–5000‡ |
| Exacting | 500–750–1000‡ | 5000–7500–10,000‡ |
| Machine shops | | |
| Rough bench or machine work | 20–30–50 | 200–300–500 |
| Medium bench or machine work, ordinary automatic machines, rough grinding, medium buffing and polishing | 50–75–100 | 500–750–1000 |
| Fine bench or machine work, fine automatic machines, medium grinding, fine buffing and polishing | 200–300–500‡ | 2000–3000–5000‡ |
| Extrafine bench or machine work, grinding, fine work | 500–750–1000‡ | 5000–7500–10,000‡ |
| Materials handling | | |
| Wrapping, packing, labeling | 20–30–50 | 200–300–500 |
| Picking stock, classifying | 20–30–50 | 200–300–500 |
| Loading, inside truck bodies and freight cars | 10–15–20 | 100–150–200 |
| Reading copied tasks | | |
| Ditto copy | 50–75–100§ | 500–750–1000§ |
| Mimeograph | 20–30–50 | 200–300–500 |
| Thermal copy, poor copy | 100–150–200§ | 1000–1500–2000§ |
| Xerograph | 20–30–50 | 200–300–500 |
| Xerography, third generation and greater | 50–75–100 | 500–750–1000 |
| Reading handwritten tasks | | |
| #3 pencil and softer leads | 50–75–100§ | 500–750–1000§ |
| #4 pencil and harder leads | 100–150–200§ | 1000–1500–2000§ |
| Ball-point pen | 20–30–50§ | 200–300–500§ |
| Felt-tip pen | 20–30–50 | 200–300–500 |
| Handwritten carbon copies | 50–75–100 | 500–750–1000 |
| Nonphotographically reproducible colors | 100–150–200 | 1000–1500–2000 |
| Chalkboards | 50–75–100§ | 500–750–1000§ |
| Reading printed tasks | | |
| 6-point type | 50–75–100§ | 500–750–1000§ |
| 8- and 10-point type | 20–30–50§ | 200–300–500§ |
| Glossy magazines | 20–30–50¶ | 200–300–500¶ |
| Maps | 50–75–100 | 500–750–1000 |
| Newsprint | 20–30–50 | 200–300–500 |
| Typed originals | 20–30–50 | 200–300–500 |
| Typed second carbon and later | 50–75–100 | 500–750–1000 |
| Telephone books | 50–75–100 | 500–750–1000 |

* From *IESNA Lighting Handbook*. For complete listing see handbook.

† Lux is SI unit equal to 0.0929 fc. For simplicity, the illuminance ranges above use a rounded value of 1 fc = 10 lux.

‡ Obtained by a combination of general and supplementary lighting.

§ Task subject to veiling reflections. Illuminance listed is not an ESI value.

¶ Especially subject to veiling reflections. It may be necessary to shield the task or to reorient it.

**TABLE 26-21a** Weighting Factors for Selecting Specific Values within Illuminance Ranges in Table 26-21*

| | Weighting factor† | | |
|---|---|---|---|
| Task and worker characteristics | −1 | 0 | +1 |
| Workers' ages | Under 40 | 40–55 | Over 55 |
| Speed and/or accuracy | Not important | Important | Critical |
| Reflectance of task background | More than 70% | 30–70% | Less than 30% |

* Adapted from the *IES Lighting Handbook,* for ranges 20–30–50 fc and above. See handbook for complete table and footnotes.

† Add the three factors algebraically. If the total is −2 or −3, use the lowest of the three illuminances in the range for the particular task. If the total is +2 or +3, use the highest of the three illuminances. Otherwise use the middle illuminance of the range.

activities where the recommended ranges are 20–30–50 fc or higher are given in Table 26-21a. Thus the user must be consulted before the appropriate illuminance level can be selected.

It is not unusual for building and plant engineers to target new or relighting projects for higher illuminance levels, based on experience with the activities involved. Too, since the IES recommendations are considered as minimums with respect to time, the user will be more familiar with probable maintenance procedures that will affect the design calculations.

In stores or sales areas, desirable patterns of brightness help capture the attention and interest of potential customers passing by. Once inside, the lighting system can help determine what they look at, what they buy, and how often they return. The illuminance selection procedure described here is not intended for spaces where merchandising is the principal activity and where lighting's function is the advantageous display of goods.

From 1972 to 1979 the IES recommended illuminances for many office and school tasks were given in terms of equivalent sphere illumination (ESI). ESI values are still useful as a tool in determining the effectiveness of measures to control veiling reflections and as a part of the evaluation of lighting systems. However, ESI values are not included in the illuminance ranges found in Table 26-21.

The illuminance levels for safety given in Table 26-22 are intended as target values considering all causes of light loss.

**TABLE 26-22** Illuminance Levels for Safety*

| Hazards requiring visual detection | Slight | | High | |
|---|---|---|---|---|
| Normal activity level† | Low | High | Low | High |
| Illumination level: | | | | |
| Footcandles | 0.5 | 1.0 | 2.0 | 5.0 |
| Lux | 5.4 | 11.0 | 22.0 | 54.0 |

* Minimum illumination for safety of personnel, absolute minimum at any time and at any location on any plane where safety is related to seeing conditions.

† Special conditions may require different levels of illumination. See "American National Standard Practice for Industrial Lighting," ANSI/IES RP-7-1983.

*67. **Veiling Reflections.*** Substantial losses in tasks contrast and hence in visibility and visual performance can result when light sources are reflected in such subtly specular (shiny) visual tasks as typing on bond paper. The apparent "veil" that is cast over a task when a light source is reflected in it may be so subtle as to be undetectable by the eye. Many factors contribute to veiling reflections and each of them, individually, has long been known. The problem is to integrate the effects of these interrelated factors. The factors are the visual task and its specularity; the worker's orientation, location, and viewing angles; and the lighting-system layout and luminaire light distribution and polarization.

The following are guidelines for reducing veiling reflections:

1. Where possible, written or printed tasks should be on matte paper using nongloss inks. The use of glossy paper stock and hard pencils should be minimized.
2. As long as the "geometry" of the situation is not changed, compensation for contrast losses can be made by increased illumination.
3. The use of luminaires with specific distributions and polarization characteristics for reducing reflections should be considered.
4. Side lighting such as from windows is effective in reducing veiling reflections. From the standpoint of visual comfort, workers should be positioned so their line of sight is parallel to or away from windows—rather than facing them.
5. Light-source positions on either side and behind the workers are preferred.
6. Where work positions can be determined, substantial gains can be made by not positioning lighting equipment in the general area above and forward of the position.
7. Any decision on a lighting installation should also include considerations of its efficiency and the visual comfort in the space.

***68. Visual Comfort.*** Visual comfort may occur when there are no overly high luminances within a worker's visual field. High luminances can also distract and can even reduce visibility. Luminaires and fenestrations which have luminances that are too high for the environment in which they are located will produce discomfort. Discomfort from direct glare is reduced by:

1. Decreasing the luminance of lighting equipment or other sources of objectionable glare, such as windows and overhead skylights.
2. Diminishing the area of uncomfortable luminances (with glare zone luminance constant).
3. Increasing the angle between the source and line of sight.
4. Increasing the general luminance in the room.

A rating system based on the degree of freedom from discomfort glare in a proposed lighting installation is called *visual comfort probability (VCP)*. Evaluation is based on the following factors which influence subjective judgments of visual comfort: room size and shape; room surface reflectances; illumination levels; luminaire type, size, luminance, maximum luminance, and light distribution; number of luminaires; luminance of the field of view; observer location and line of sight; and differences in individual glare sensitivity. Since each of these factors can vary considerably, a standard set of conditions has been established and used as a basis for VCP tables.

Direct discomfort glare should not be considered a problem in lighting installations if all three of the following conditions are satisfied:

1. The VCP is 70 or more.
2. The ratio of maximum to average luminaire luminance does not exceed 5 : 1 (preferably 3 : 1) at 45, 55, 65, 75, and 85° from nadir, crosswise and lengthwise.
3. Maximum luminaire luminances, crosswise and lengthwise, do not exceed the following values:

| Angle above nadir, deg | Max. luminance, fL (cd/m$^2$) |
|---|---|
| 45 | 2250 (7710) |
| 55 | 1605 (5500) |
| 65 | 1125 (3860) |
| 75 | 750 (2570) |
| 85 | 495 (1695) |

Wall reflectance: 50%; effective ceiling cavity reflectance 80%; effective floor cavity reflectance: 20%
Luminaire No. 000
**Work-plane illuminance: 100 footcandles**

| Room | | Luminaires lengthwise | | | | Luminaires crosswise | | | |
|---|---|---|---|---|---|---|---|---|---|
| W | L | 8.5* | 10.0* | 13.0* | 16.0* | 8.5* | 10.0* | 13.0* | 16.0* |
| 20 | 20 | 78 | 82 | 90 | 94 | 77 | 81 | 89 | 93 |
| 20 | 30 | 73 | 76 | 82 | 88 | 72 | 75 | 81 | 86 |
| 20 | 40 | 71 | 73 | 78 | 82 | 70 | 72 | 76 | 80 |
| 20 | 60 | 69 | 71 | 74 | 78 | 68 | 70 | 73 | 76 |
| 30 | 20 | 78 | 82 | 88 | 92 | 77 | 81 | 87 | 92 |
| 30 | 30 | 73 | 75 | 80 | 85 | 72 | 74 | 79 | 84 |
| 30 | 40 | 70 | 72 | 75 | 78 | 69 | 71 | 74 | 77 |
| 30 | 60 | 68 | 69 | 71 | 74 | 67 | 69 | 70 | 73 |
| 30 | 80 | 67 | 69 | 69 | 72 | 67 | 68 | 68 | 71 |
| 40 | 20 | 79 | 82 | 87 | 92 | 79 | 82 | 87 | 91 |
| 40 | 30 | 74 | 76 | 79 | 84 | 73 | 75 | 78 | 83 |
| 40 | 40 | 71 | 72 | 74 | 77 | 70 | 71 | 73 | 76 |
| 40 | 60 | 68 | 69 | 70 | 72 | 68 | 69 | 69 | 71 |
| 40 | 80 | 67 | 68 | 68 | 70 | 67 | 68 | 67 | 69 |
| 40 | 100 | 67 | 68 | 67 | 69 | 67 | 67 | 66 | 68 |
| 60 | 30 | 75 | 76 | 79 | 83 | 74 | 76 | 78 | 82 |
| 60 | 40 | 71 | 72 | 74 | 76 | 71 | 72 | 73 | 76 |
| 60 | 60 | 69 | 69 | 69 | 71 | 68 | 69 | 68 | 70 |
| 60 | 80 | 68 | 68 | 67 | 69 | 67 | 68 | 66 | 68 |
| 60 | 100 | 67 | 67 | 66 | 67 | 67 | 67 | 65 | 66 |
| 100 | 40 | 74 | 75 | 75 | 78 | 74 | 74 | 75 | 77 |
| 100 | 60 | 71 | 71 | 71 | 72 | 71 | 71 | 70 | 72 |
| 100 | 80 | 70 | 70 | 68 | 69 | 70 | 69 | 67 | 69 |
| 100 | 100 | 69 | 68 | 66 | 67 | 69 | 68 | 66 | 67 |

* Luminaire mounting heights above the floor, in feet.

**FIG. 26-35** Example of a typical tabulation of visual comfort probability (VCP) values.

Figure 26-35 is an example of a typical tabulation of calculated VCP values for a general lighting system providing 100 fc using fictitious luminaire no. "000."

***69. Luminance Distribution.*** The luminance relationship of the various surfaces in the visual field is important. When the eyes scan a task, an adaptation level is established consisting primarily of the task luminance. As the eyes leave the task and look at an area of a different luminance, there is a sudden loss of sensitivity to see the contrast of the detail in the new area until the eyes can readapt after an appreciable length of time. In order to see the detail of a visual task accurately and quickly, luminance ratios of appreciable areas in the visual environment should be kept low. Table 26-23 is an example of recommended luminance ratios for offices.[1]

**TABLE 26-23** Recommended Luminance Ratios for Offices*

To achieve a comfortable balance in the office and limit the effects of transient adaptations and disability glare, it is desirable and practical to limit luminance ratios between areas of appreciable size from normal viewpoints as follows:
1 – ⅓ between task and adjacent surroundings
1 – ⅕ between task and more remote darker surfaces
1 – 10 between task and more remote lighter surfaces

* These ratios are recommended as maximums; reductions are generally beneficial.

[1] American National Standard Practice for Office Lighting; ANSI/IES RP-1-1982.

To help achieve these ratios, room surfaces should be of high reflectance, as recommended for specific applications.

## CALCULATING MAINTAINED ILLUMINANCE

***70. Light-Loss Factor.*** In calculating maintained illuminance (the lowest level before maintenance procedures are instituted) a light-loss factor (LLF) is used to take into account losses in light output due to temperature and voltage variations, dirt accumulation on luminaire and room surfaces, lamp depreciation, maintenance procedures, and atmospheric conditions. The effect of these light losses can be seen in the example shown in Fig. 26-36. It is important

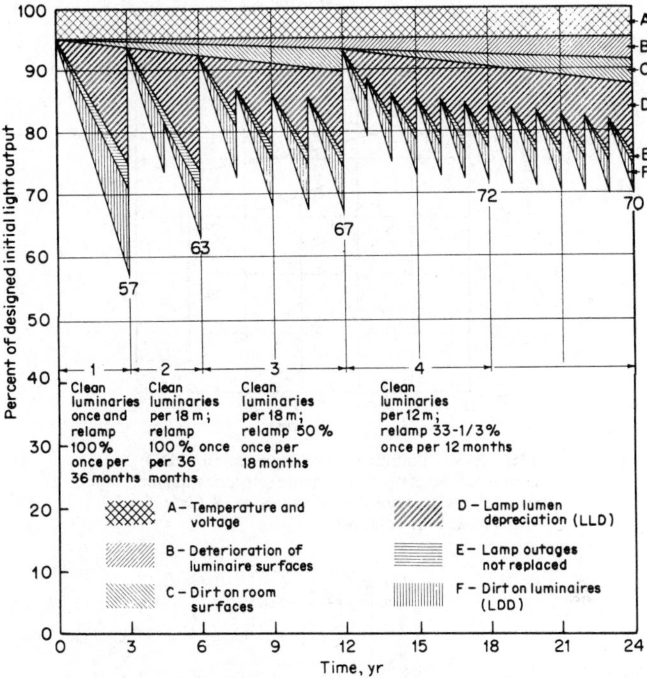


**FIG. 26-36** Effect of light loss on illumination level. Example shown is for enclosed surface-mounted luminaires with 40-W fluorescent lamps, operated 10 h a day, 5 days per week, 2600 h per year. Four maintenance systems are shown for comparison purposes. *(IES Lighting Handbook.)*

that all causes of light loss be investigated and maintenance procedures be established in the lighting-design stage to assure adequate illumination in an economical manner without wasting energy.

***71. LLF Determination.*** The LLF (formerly called maintenance factor) is mainly the product of the *lamp lumen depreciation factor (LLD),* the *luminaire dirt depreciation factor (LDD),* the *room-surface dirt depreciation factor (RSDD),* and *lamp burnouts (LBO).* Data on LLD are available from manufacturers. LBO is the ratio of lamps remaining lighted to the total in the original installation. LDD is obtained from curves, such as shown in Fig. 26-37, a knowledge of the amount of dirt in the room atmosphere, a knowledge of the length of time between luminaire cleanings, and the luminaire category (I to VI) from the manufacturer. RSDD is obtained from a curve and table as shown in Fig. 26-38, where room size, luminaire distribution, room atmosphere, and cleaning cycle are known.

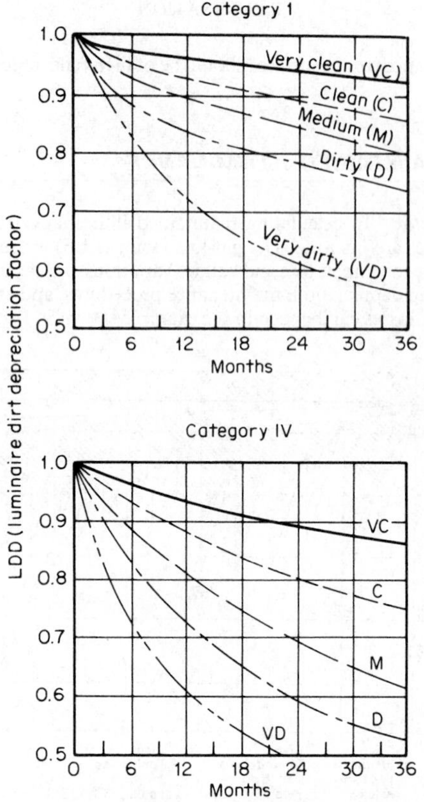


**FIG. 26-37** Luminaire dirt depreciation (LDD) factors for two of the six luminaire categories and for five degrees of atmosphere dirtiness. (*IES Lighting Handbook*, 1981 edition.)

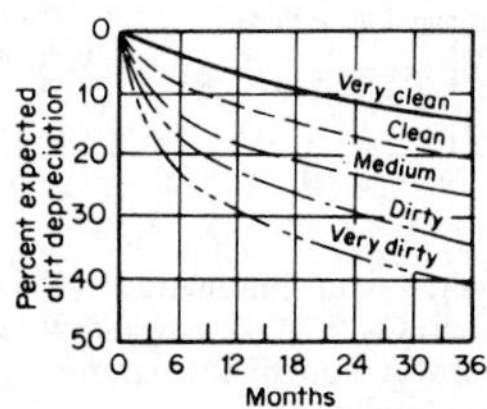


| Percent expected dirt depreciation | Luminaire distribution type | | | | | | | | | | | | | | | | | | | |
|---|---|---|---|---|---|---|---|---|---|---|---|---|---|---|---|---|---|---|---|---|
| | Direct | | | | Semidirect | | | | Direct-indirect | | | | Semi-indirect | | | | Indirect | | | |
| | 10 | 20 | 30 | 40 | 10 | 20 | 30 | 40 | 10 | 20 | 30 | 40 | 10 | 20 | 30 | 40 | 10 | 20 | 30 | 40 |
| Room-cavity ratio | | | | | | | | | | | | | | | | | | | | |
| 1 | .98 | .96 | .94 | .92 | .97 | .92 | .89 | .84 | .94 | .87 | .80 | .76 | .94 | .87 | .80 | .73 | .90 | .80 | .70 | .60 |
| 2 | .98 | .96 | .94 | .92 | .96 | .92 | .88 | .83 | .94 | .87 | .80 | .75 | .94 | .87 | .79 | .72 | .90 | .80 | .69 | .59 |
| 3 | .98 | .95 | .93 | .90 | .96 | .91 | .87 | .82 | .94 | .86 | .79 | .74 | .94 | .86 | .78 | .71 | .90 | .79 | .68 | .58 |
| 4 | .97 | .95 | .92 | .90 | .95 | .90 | .85 | .80 | .94 | .86 | .79 | .73 | .94 | .86 | .78 | .70 | .89 | .78 | .67 | .56 |
| 5 | .97 | .94 | .91 | .89 | .94 | .90 | .84 | .79 | .93 | .86 | .78 | .72 | .93 | .86 | .77 | .69 | .89 | .78 | .66 | .55 |
| 6 | .97 | .94 | .91 | .88 | .94 | .89 | .83 | .78 | .93 | .85 | .78 | .71 | .93 | .85 | .76 | .68 | .89 | .77 | .66 | .54 |
| 7 | .97 | .94 | .90 | .87 | .93 | .88 | .82 | .77 | .93 | .84 | .77 | .70 | .93 | .84 | .76 | .68 | .89 | .76 | .65 | .53 |
| 8 | .96 | .93 | .89 | .86 | .93 | .87 | .81 | .75 | .93 | .84 | .76 | .69 | .93 | .84 | .76 | .68 | .88 | .76 | .64 | .52 |
| 9 | .96 | .92 | .88 | .85 | .93 | .87 | .80 | .74 | .93 | .84 | .76 | .68 | .93 | .84 | .75 | .67 | .88 | .75 | .63 | .51 |
| 10 | .90 | .92 | .87 | .83 | .93 | .86 | .79 | .72 | .93 | .84 | .75 | .67 | .92 | .83 | .75 | .67 | .88 | .75 | .62 | .50 |

**FIG. 26-38** Room-surface dirt depreciation (RSDD) factors.

*72. Example of Calculation Procedure.* Given a direct type of luminaire of category IV, with 40-W fluorescent lamps (LLD = 0.84). Room atmosphere is very clean and cleaning is performed every 18 months. Room-cavity ratio is 2. Lamps are replaced at burnout.

LLD = 0.84 (from manufacturers' lamp tables)

LDD from Fig. 26-37 = 0.91

RSDD from Fig. 26-38 = 0.98

LBO = 1

$$LLF = 0.84 \times 0.91 \times 0.98 \times 1.0 = 0.75$$

## CALCULATION OF AVERAGE ILLUMINANCE AND LUMINANCE

*73. General Lighting.* The design of general lighting systems is governed by room dimensions, structural features, reflectances of room surfaces, mounting height of the luminaires, and the distribution and maintenance characteristics of the luminaire. The choice of the luminaire depends on the service to which it is to be put, which assumes a certain experience in selection, or other aids such as manufacturers' data, which assist the designer in making a selection appropriate from the standpoints of freedom from glare, efficiency, decorative value, and economy. The ultimate "brightness pattern" of the room is an important factor in the design.

The beginning concept of general lighting design is that of delivering a specified average footcandle level of illimination to a horizontal plane in a room. The light generated by the lamps in such a system is variously affected and considerably reduced by reflection, diffusion, and absorption as it impinges on reflectors and transmitting media in the luminaires and on ceilings, wall, floors, and on the objects in the room.

*74. Luminaire Spacing and Mounting Height.* The maximum permissible spacing for each type of luminaire is given in the Photometric Report provided by the manufacturer. These spacing limitations are related to the mounting height (usually above the work plane) of direct, semidirect, and general-diffuse luminaires and to the ceiling height for indirect and semidirect systems. Observance of such limitations usually will ensure satisfactory uniformity of illumination throughout the major portion of the room so that all parts of the area will be equally suitable for the intended use. Peripheral areas may require special treatment, as indicated below. Illuminance is usually considered uniform if the maximum and minimum values are within plus or minus one-sixth of the average illuminance in the area. Closer spacing than indicated by the spacing criterion improves uniformity and reduces shadows. The spacing-mounting-height relations apply not only to individual luminaires but to the spacing between continuous sections, luminous panels, troffers, or sections of coves.

The distance between luminaires and the wall should not exceed one-half the distance between luminaires. Where desks or benches might be located along the wall, the distance between luminaires and the wall should not exceed 2½ ft. Likewise, the ends of continuous rows of fluorescent luminaires should preferably be within 6 to 12 in of the wall. Additional luminaires or luminaires having a greater number of lamps may be required adjacent to the walls, particularly where walls have low reflectance. Where direct and semidirect luminaires are used under such conditions, the perimeter luminaires should be carefully located to avoid shadows on the work from the worker.

*75. Zonal-Cavity (Lumen) Method.*[1] The lumen method is used in calculating the illuminance that represents the average of all points on the work plane in an interior. It is based on the definition of illuminance as luminous flux per unit area, or

$$\text{Illuminance} = \frac{\text{luminous flux}}{\text{area}}$$

[1] Zonal Cavity Method of Calculating and using Coefficients of Utilization; *Illum. Eng.*, May 1964, vol. LIX, p. 309.

where luminous flux is expressed in lumens. If the area is in square feet, the illuminance is in footcandles (lumens per square foot); if the area is in square meters, the illuminance is in lux (lumens per square meter).

Because not all the lamp lumens will reach the work plane owing to losses in the luminaire and at the room surfaces, they must be multiplied by a coefficient of utilization which represents the portion that reaches the work plane. Thus,

$$\text{Initial illuminance} = \frac{\text{number of luminaires} \times \text{lamp lumens} \times \text{coefficient of utilization}}{\text{area}}$$

Since the design objective is usually the minimum maintained illuminance, factors must be applied to account for the estimated depreciation in lamp lumens, the estimated losses from dirt collection on the luminaire surfaces (including lamps), etc.

The formula thus becomes

$$\text{Maintained illuminance} = \frac{\text{number of luminaires} \times \text{lamp lumens} \times \text{CU} \times \text{LLF}}{\text{area}}$$

where CU = coefficient of utilization
LLF = light-loss factor
lamp lumens = total rated lamp lumens per luminaire

The zonal-cavity method considers the actual room as being made up of a ceiling cavity above the luminaires, a floor cavity beneath the work plane, and a room cavity located between the two (see Fig. 26-39).

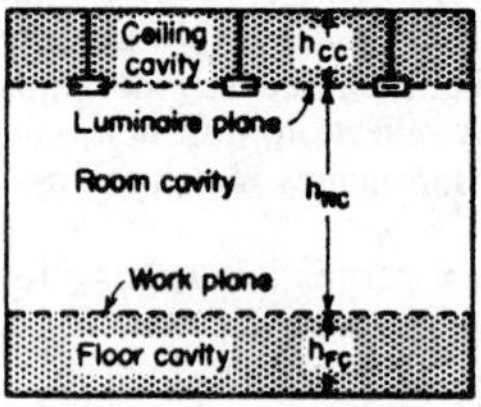

FIG. 26-39 Room cavities.

In the general case, all these cavities are present. In the case of recessed or surface-mounted luminaires, the ceiling cavity is simply the ceiling. When the illumination on the floor is to be determined, the floor cavity becomes the floor.

It is now possible to calculate numerical relationships called "cavity ratios" which may be used to determine effective reflectance of the floor and ceiling and then to find the coefficient of utilization.

The basic steps in the calculation of any average illuminance are as follows:

1. Determine *cavity ratios* for three cavities shown in Fig. 26-39 as follows:

$$\text{Room-cavity ratio, RCR} = \frac{5h_{RC}(L+W)}{LW}$$

$$\text{Ceiling-cavity ratio, CCR} = \frac{5h_{CC}(L+W)}{LW} = \text{RCR}\frac{h_{CC}}{h_{RC}}$$

$$\text{Floor-cavity ratio, FCR} = \frac{5h_{FC}(L+W)}{LW} = \text{RCR}\frac{h_{FC}}{h_{RC}}$$

   where $h_{RC}$ = height of room between luminaire plane and work plane; $h_{CC}$ = distance from luminaire plane to ceiling; $h_{FC}$ = height of work plane above floor; $L$ = room length; and $W$ = room width.

   Cavity ratios may also be found in Table 26-24 for typical-sized cavities which cover a wide range of room dimensions.

2. Obtain *effective ceiling-cavity reflectance* ($\rho_{cc}$) for combination of ceiling and wall reflectance to be employed, from Table 26-25. Note that, for surface-mounted or recessed luminaires, CCR = 0 and the ceiling reflectance may be used as the effective cavity reflectance.

3. Obtain *effective floor-cavity reflectance* ($\rho_{fc}$) for combination of floor and wall reflectances to be employed, from Table 26-25.

4. Obtain *coefficient of utilization* for the luminaire for 20% effective floor-cavity-reflectance condition (tables like 26-28), interpolating between tabulated values as required to match room size and ceiling- and wall-reflectance combinations.
5. If *effective floor-cavity reflectance* ($\rho_{fc}$) obtained in step 3 differs significantly from 20%, obtain multiplier from Table 26-26 or Table 26-27. Multiply the coefficient of utilization by this multiplier.
6. Determine *average maintained illuminance* by the following formula:

$$\text{fc} = \frac{(\text{rated lamp lumens/luminaire}) \times \text{CU} \times \text{LLF}}{\text{area/luminaire}}$$

If *initial illuminance* is desired, LLF is omitted from the formula.

When the desired average maintained illumination is known, the formula can more conveniently be expressed as

$$\text{Area/luminaire} = \frac{(\text{rated lamp lumens/luminaire}) \times \text{CU} \times \text{LLF}}{\text{fc}}$$

This area divided by the luminaire length gives the approximate spacing between continuous rows, or it may be divided into the total room area to determine the number of luminaires required.

*Example.* A room is 28 ft wide and 32 ft long and has a 10-ft ceiling height. Reflectances are: ceiling 80%, walls 50%, floor 10%. A recessed four-lamp fluorescent luminaire is to be used. Work plane is 2 ft 0 in. Find the coefficient of utilization.

1. Calculate *cavity ratios* as follows, or look up in table of cavity ratios (Table 26-24).

$$\text{CCR} = 0 \quad \text{(recessed units are used)}$$

$$\text{RCR} = \frac{(5)(8)(28 + 32)}{(28)(32)} = 2.7$$

$$\text{FCR} = \frac{(5)(2)(28 + 32)}{(28)(32)} = 0.67$$

2. In Table 26-25, look up *effective cavity reflectances* for ceiling and floor cavities. $\rho_{cc}$ for the ceiling cavity will be 80%, while $\rho_{fc}$ for the floor cavity will be 11%.
3. With the *room-cavity ratio* RCR known, it is now possible to find the coefficient of utilization for the luminaire in a room having an RCR of 2.7 and effective reflectances as follows (assume luminaire is type described in Table 26-28):

$$\rho_{cc} = 80\% \qquad \rho_w = 50\% \qquad \rho_{fc} = 20\%$$

This CU = 0.44. Note that this is for an effective floor reflectance of 20%, while the actual effective reflectance of the floor $\rho_{fc}$ is 11%. To correct for this, locate the appropriate multiplier in Table 26-26 for the RCR already calculated (2.7). It is 0.948 and is found by interpolating between the numbers for 80 $\rho_{cc}$ and between RCRs of 2.0 and 3.0. Then

$$\text{CU final} = 0.44 \times 0.948 = 0.42$$

4. *Illuminance level* can now be calculated if we know the area per luminaire, the lamp lumen rating, and the light-loss factor.

$$\text{fc in service} = \frac{(\text{rated lamp lumens/luminaire}) \times \text{CU} \times \text{LLF}}{\text{area/luminaire}}$$

***76. Luminance Calculations.*** The ability to predict the luminance is needed to (1) design lighting that promotes both visual comfort and good visual performance, (2) predict illumination at specific points within the environment, and (3) evaluate various criteria of the lighting system such as VCP. Luminance calculations are greatly simplified through the use of *lumi-*

**TABLE 26-24** Cavity Ratios

| Room dimensions | | Cavity depth | | | | | | | | | | | | | | | | | | | |
|---|---|---|---|---|---|---|---|---|---|---|---|---|---|---|---|---|---|---|---|---|---|
| Width | Length | 1.0 | 1.5 | 2.0 | 2.5 | 3.0 | 3.5 | 4.0 | 5.0 | 6.0 | 7.0 | 8 | 9 | 10 | 11 | 12 | 14 | 16 | 20 | 25 | 30 |
| 8 | 8 | 1.2 | 1.9 | 2.5 | 3.1 | 3.7 | 4.4 | 5.0 | 6.2 | 7.5 | 8.8 | 10.0 | 11.2 | 12.5 | | | | | | | |
| | 10 | 1.1 | 1.7 | 2.2 | 2.8 | 3.4 | 3.9 | 4.5 | 5.6 | 6.7 | 7.9 | 9.0 | 10.1 | 11.3 | 12.4 | | | | | | |
| | 14 | 1.0 | 1.5 | 2.0 | 2.5 | 3.0 | 3.4 | 3.9 | 4.9 | 5.9 | 6.9 | 7.8 | 8.8 | 9.7 | 10.7 | 11.7 | | | | | |
| | 20 | 0.9 | 1.3 | 1.7 | 2.2 | 2.6 | 3.1 | 3.5 | 4.4 | 5.2 | 6.1 | 7.0 | 7.9 | 8.8 | 9.6 | 10.5 | 12.2 | | | | |
| | 30 | 0.8 | 1.2 | 1.6 | 2.0 | 2.4 | 2.8 | 3.2 | 4.0 | 4.7 | 5.5 | 6.3 | 7.1 | 7.9 | 8.7 | 9.5 | 11.0 | | | | |
| | 40 | 0.7 | 1.1 | 1.5 | 1.9 | 2.3 | 2.6 | 3.0 | 3.7 | 4.5 | 5.3 | 5.9 | 6.5 | 7.4 | 8.1 | 8.8 | 10.3 | 11.8 | | | |
| 10 | 10 | 1.0 | 1.5 | 2.0 | 2.5 | 3.0 | 3.5 | 4.0 | 5.0 | 6.0 | 7.0 | 8.0 | 9.0 | 10.0 | 11.0 | 12.0 | | | | | |
| | 14 | 0.9 | 1.3 | 1.7 | 2.1 | 2.6 | 3.0 | 3.4 | 4.3 | 5.1 | 6.0 | 6.9 | 7.8 | 8.6 | 9.5 | 10.4 | 12.0 | | | | |
| | 20 | 0.7 | 1.1 | 1.5 | 1.9 | 2.3 | 2.6 | 3.0 | 3.7 | 4.5 | 5.3 | 6.0 | 6.8 | 7.5 | 8.3 | 9.0 | 10.5 | 12.0 | | | |
| | 30 | 0.7 | 1.0 | 1.3 | 1.7 | 2.0 | 2.3 | 2.7 | 3.3 | 4.0 | 4.7 | 5.3 | 6.0 | 6.6 | 7.3 | 8.0 | 9.4 | 10.6 | | | |
| | 40 | 0.6 | 0.9 | 1.2 | 1.6 | 1.9 | 2.2 | 2.5 | 3.1 | 3.7 | 4.4 | 5.0 | 5.6 | 6.2 | 6.9 | 7.5 | 8.7 | 10.0 | 12.5 | | |
| | 60 | 0.6 | 0.9 | 1.2 | 1.5 | 1.7 | 2.0 | 2.3 | 2.9 | 3.5 | 4.1 | 4.7 | 5.3 | 5.9 | 6.5 | 7.1 | 8.2 | 9.4 | 11.7 | | |
| 12 | 12 | 0.8 | 1.2 | 1.7 | 2.1 | 2.5 | 2.9 | 3.3 | 4.2 | 5.0 | 5.8 | 6.7 | 7.5 | 8.4 | 9.2 | 10.0 | 11.7 | | | | |
| | 16 | 0.7 | 1.1 | 1.5 | 1.8 | 2.2 | 2.5 | 2.9 | 3.6 | 4.4 | 5.1 | 5.8 | 6.5 | 7.2 | 8.0 | 8.7 | 10.2 | 11.6 | | | |
| | 24 | 0.6 | 0.9 | 1.2 | 1.6 | 1.9 | 2.2 | 2.5 | 3.1 | 3.7 | 4.4 | 5.0 | 5.6 | 6.2 | 6.9 | 7.5 | 8.7 | 10.0 | 12.5 | | |
| | 36 | 0.6 | 0.8 | 1.1 | 1.4 | 1.7 | 1.9 | 2.2 | 2.8 | 3.3 | 3.9 | 4.4 | 5.0 | 5.5 | 6.0 | 6.6 | 7.8 | 8.8 | 11.0 | | |
| | 50 | 0.5 | 0.8 | 1.0 | 1.3 | 1.5 | 1.8 | 2.1 | 2.6 | 3.1 | 3.6 | 4.1 | 4.6 | 5.1 | 5.6 | 6.2 | 7.2 | 8.2 | 10.2 | | |
| | 70 | 0.5 | 0.7 | 1.0 | 1.2 | 1.5 | 1.7 | 2.0 | 2.4 | 2.9 | 3.4 | 3.9 | 4.4 | 4.9 | 5.4 | 5.8 | 6.8 | 7.8 | 9.7 | 12.2 | |
| 14 | 14 | 0.7 | 1.1 | 1.4 | 1.8 | 2.1 | 2.5 | 2.9 | 3.6 | 4.3 | 5.0 | 5.7 | 6.4 | 7.1 | 7.8 | 8.5 | 10.0 | 11.4 | | | |
| | 20 | 0.6 | 0.9 | 1.2 | 1.5 | 1.8 | 2.1 | 2.4 | 3.0 | 3.6 | 4.2 | 4.9 | 5.5 | 6.1 | 6.7 | 7.3 | 8.6 | 9.8 | 12.3 | | |
| | 30 | 0.5 | 0.8 | 1.0 | 1.3 | 1.6 | 1.8 | 2.1 | 2.6 | 3.1 | 3.7 | 4.2 | 4.7 | 5.2 | 5.8 | 6.3 | 7.3 | 8.4 | 10.5 | | |
| | 42 | 0.5 | 0.7 | 1.0 | 1.2 | 1.4 | 1.7 | 1.9 | 2.4 | 2.9 | 3.3 | 3.8 | 4.3 | 4.7 | 5.2 | 5.7 | 6.7 | 7.6 | 9.5 | 11.9 | |
| | 60 | 0.4 | 0.7 | 0.9 | 1.1 | 1.3 | 1.5 | 1.8 | 2.2 | 2.6 | 3.1 | 3.5 | 3.9 | 4.4 | 4.8 | 5.2 | 6.1 | 7.0 | 8.8 | 10.9 | |
| | 90 | 0.4 | 0.6 | 0.8 | 1.0 | 1.2 | 1.4 | 1.6 | 2.0 | 2.5 | 2.9 | 3.3 | 3.7 | 4.1 | 4.5 | 5.0 | 5.8 | 6.6 | 8.3 | 10.3 | 12.4 |
| 17 | 17 | 0.6 | 0.9 | 1.2 | 1.5 | 1.8 | 2.1 | 2.3 | 2.9 | 3.5 | 4.1 | 4.7 | 5.3 | 5.9 | 6.5 | 7.0 | 8.2 | 9.4 | 11.7 | | |
| | 25 | 0.5 | 0.7 | 1.0 | 1.2 | 1.5 | 1.7 | 2.0 | 2.5 | 3.0 | 3.5 | 4.0 | 4.5 | 5.0 | 5.5 | 6.0 | 7.0 | 8.0 | 10.0 | 12.5 | |
| | 35 | 0.4 | 0.7 | 0.9 | 1.1 | 1.3 | 1.5 | 1.7 | 2.2 | 2.6 | 3.1 | 3.5 | 3.9 | 4.4 | 4.8 | 5.2 | 6.1 | 7.0 | 8.7 | 10.9 | |
| | 50 | 0.4 | 0.6 | 0.8 | 1.0 | 1.2 | 1.4 | 1.6 | 2.0 | 2.4 | 2.8 | 3.1 | 3.5 | 3.9 | 4.3 | 4.5 | 5.4 | 6.2 | 7.7 | 9.7 | 11.6 |
| | 80 | 0.4 | 0.5 | 0.7 | 0.9 | 1.1 | 1.2 | 1.4 | 1.8 | 2.1 | 2.5 | 2.9 | 3.3 | 3.6 | 4.0 | 4.3 | 5.1 | 5.8 | 7.2 | 9.0 | 10.9 |
| | 120 | 0.3 | 0.5 | 0.7 | 0.8 | 1.0 | 1.2 | 1.3 | 1.7 | 2.0 | 2.3 | 2.7 | 3.0 | 3.4 | 3.7 | 4.0 | 4.7 | 5.4 | 6.7 | 8.4 | 10.1 |
| 20 | 20 | 0.5 | 0.7 | 1.0 | 1.2 | 1.5 | 1.7 | 2.0 | 2.5 | 3.0 | 3.5 | 4.0 | 4.5 | 5.0 | 5.5 | 6.0 | 7.0 | 8.0 | 10.0 | 12.5 | |
| | 30 | 0.4 | 0.6 | 0.8 | 1.0 | 1.2 | 1.5 | 1.7 | 2.1 | 2.5 | 2.9 | 3.3 | 3.7 | 4.1 | 4.5 | 4.9 | 5.8 | 6.6 | 8.2 | 10.3 | 12.4 |
| | 45 | 0.4 | 0.5 | 0.7 | 0.9 | 1.1 | 1.3 | 1.4 | 1.8 | 2.2 | 2.5 | 2.9 | 3.3 | 3.6 | 4.0 | 4.3 | 5.1 | 5.8 | 7.2 | 9.1 | 10.9 |
| | 60 | 0.3 | 0.5 | 0.7 | 0.8 | 1.0 | 1.2 | 1.3 | 1.7 | 2.0 | 2.3 | 2.7 | 3.0 | 3.4 | 3.7 | 4.0 | 4.7 | 5.4 | 6.7 | 8.4 | 10.1 |
| | 90 | 0.3 | 0.5 | 0.6 | 0.8 | 0.9 | 1.1 | 1.2 | 1.5 | 1.8 | 2.1 | 2.4 | 2.7 | 3.0 | 3.3 | 3.6 | 4.2 | 4.8 | 6.0 | 7.5 | 9.0 |
| | 150 | 0.3 | 0.4 | 0.6 | 0.7 | 0.8 | 1.0 | 1.1 | 1.4 | 1.7 | 2.0 | 2.3 | 2.6 | 2.9 | 3.2 | 3.4 | 4.0 | 4.6 | 5.7 | 7.2 | 8.6 |

| | | | | | | | | | | | | | | | | | | | | | |
|---|---|---|---|---|---|---|---|---|---|---|---|---|---|---|---|---|---|---|---|---|---|
| 24 | 24 | 0.4 | 0.6 | 0.8 | 1.0 | 1.2 | 1.5 | 1.7 | 2.1 | 2.5 | 2.9 | 3.3 | 3.7 | 4.1 | 4.5 | 5.0 | 5.8 | 6.7 | 8.2 | 10.3 | 12.4 |
| | 32 | 0.4 | 0.5 | 0.7 | 0.9 | 1.1 | 1.3 | 1.5 | 1.8 | 2.2 | 2.6 | 2.9 | 3.3 | 3.6 | 4.0 | 4.3 | 5.1 | 5.8 | 7.2 | 9.0 | 11.0 |
| | 50 | 0.3 | 0.5 | 0.6 | 0.8 | 0.9 | 1.1 | 1.2 | 1.5 | 1.8 | 2.2 | 2.5 | 2.8 | 3.1 | 3.4 | 3.7 | 4.4 | 5.0 | 6.2 | 7.8 | 9.4 |
| | 70 | 0.3 | 0.4 | 0.6 | 0.7 | 0.8 | 1.0 | 1.1 | 1.4 | 1.7 | 2.0 | 2.2 | 2.5 | 2.8 | 3.0 | 3.3 | 3.8 | 4.4 | 5.5 | 6.9 | 8.2 |
| | 100 | 0.3 | 0.4 | 0.5 | 0.6 | 0.8 | 0.9 | 1.0 | 1.3 | 1.6 | 1.8 | 2.1 | 2.4 | 2.6 | 2.9 | 3.1 | 3.7 | 4.2 | 5.2 | 6.5 | 7.9 |
| | 160 | 0.2 | 0.4 | 0.5 | 0.6 | 0.7 | 0.8 | 1.0 | 1.2 | 1.4 | 1.7 | 1.9 | 2.1 | 2.4 | 2.6 | 2.8 | 3.3 | 3.8 | 4.7 | 5.9 | 7.1 |
| 30 | 30 | 0.3 | 0.5 | 0.7 | 0.8 | 1.0 | 1.2 | 1.3 | 1.7 | 2.0 | 2.3 | 2.7 | 3.0 | 3.3 | 3.7 | 4.0 | 4.7 | 5.4 | 6.7 | 8.4 | 10.0 |
| | 45 | 0.3 | 0.4 | 0.6 | 0.7 | 0.8 | 1.0 | 1.1 | 1.4 | 1.7 | 1.9 | 2.2 | 2.5 | 2.7 | 3.0 | 3.3 | 3.8 | 4.4 | 5.5 | 6.9 | 8.2 |
| | 60 | 0.3 | 0.4 | 0.5 | 0.6 | 0.7 | 0.9 | 1.0 | 1.2 | 1.5 | 1.7 | 2.0 | 2.2 | 2.5 | 2.7 | 3.0 | 3.5 | 4.0 | 5.0 | 6.2 | 7.4 |
| | 90 | 0.2 | 0.3 | 0.4 | 0.6 | 0.7 | 0.8 | 0.9 | 1.1 | 1.3 | 1.6 | 1.8 | 2.0 | 2.2 | 2.5 | 2.7 | 3.1 | 3.6 | 4.5 | 5.6 | 6.7 |
| | 150 | 0.2 | 0.3 | 0.4 | 0.5 | 0.6 | 0.7 | 0.8 | 1.0 | 1.2 | 1.4 | 1.6 | 1.8 | 2.0 | 2.2 | 2.4 | 2.8 | 3.2 | 4.0 | 5.0 | 5.9 |
| | 200 | 0.2 | 0.3 | 0.4 | 0.5 | 0.6 | 0.7 | 0.8 | 1.0 | 1.1 | 1.3 | 1.5 | 1.7 | 1.9 | 2.0 | 2.2 | 2.6 | 3.0 | 3.7 | 4.7 | 5.6 |
| 36 | 36 | 0.3 | 0.4 | 0.6 | 0.7 | 0.8 | 1.0 | 1.1 | 1.4 | 1.7 | 1.9 | 2.2 | 2.5 | 2.8 | 3.0 | 3.3 | 3.9 | 4.4 | 5.5 | 6.9 | 8.3 |
| | 50 | 0.2 | 0.4 | 0.5 | 0.6 | 0.7 | 0.8 | 1.0 | 1.2 | 1.4 | 1.7 | 1.9 | 2.1 | 2.5 | 2.6 | 2.9 | 3.3 | 3.8 | 4.8 | 5.9 | 7.2 |
| | 75 | 0.2 | 0.3 | 0.4 | 0.5 | 0.6 | 0.7 | 0.8 | 1.0 | 1.2 | 1.4 | 1.6 | 1.8 | 2.0 | 2.3 | 2.5 | 2.9 | 3.3 | 4.1 | 5.1 | 6.1 |
| | 100 | 0.2 | 0.3 | 0.4 | 0.5 | 0.6 | 0.7 | 0.8 | 0.9 | 1.1 | 1.3 | 1.5 | 1.7 | 1.9 | 2.1 | 2.3 | 2.6 | 3.0 | 3.8 | 4.7 | 5.7 |
| | 150 | 0.2 | 0.3 | 0.3 | 0.4 | 0.5 | 0.6 | 0.7 | 0.9 | 1.0 | 1.2 | 1.4 | 1.6 | 1.7 | 1.9 | 2.1 | 2.4 | 2.8 | 3.5 | 4.3 | 5.2 |
| | 200 | 0.2 | 0.2 | 0.3 | 0.4 | 0.5 | 0.6 | 0.7 | 0.8 | 1.0 | 1.1 | 1.3 | 1.5 | 1.6 | 1.8 | 2.0 | 2.3 | 2.6 | 3.3 | 4.1 | 4.9 |
| 42 | 42 | 0.2 | 0.4 | 0.5 | 0.6 | 0.7 | 0.8 | 1.0 | 1.2 | 1.4 | 1.6 | 1.9 | 2.1 | 2.4 | 2.6 | 2.8 | 3.3 | 3.8 | 4.7 | 5.9 | 7.1 |
| | 60 | 0.2 | 0.3 | 0.4 | 0.5 | 0.6 | 0.7 | 0.8 | 1.0 | 1.2 | 1.4 | 1.6 | 1.8 | 2.0 | 2.2 | 2.4 | 2.8 | 3.2 | 4.0 | 5.0 | 6.0 |
| | 90 | 0.2 | 0.3 | 0.3 | 0.4 | 0.5 | 0.6 | 0.7 | 0.9 | 1.0 | 1.2 | 1.4 | 1.6 | 1.7 | 1.9 | 2.1 | 2.4 | 2.8 | 3.5 | 4.4 | 5.2 |
| | 140 | 0.2 | 0.2 | 0.3 | 0.4 | 0.5 | 0.5 | 0.6 | 0.8 | 0.9 | 1.1 | 1.2 | 1.4 | 1.5 | 1.7 | 1.9 | 2.2 | 2.5 | 3.1 | 3.9 | 4.6 |
| | 200 | 0.1 | 0.2 | 0.3 | 0.4 | 0.4 | 0.5 | 0.6 | 0.7 | 0.9 | 1.0 | 1.1 | 1.3 | 1.4 | 1.6 | 1.7 | 2.0 | 2.3 | 2.9 | 3.6 | 4.3 |
| | 300 | 0.1 | 0.2 | 0.3 | 0.3 | 0.4 | 0.5 | 0.5 | 0.7 | 0.8 | 0.9 | 1.1 | 1.3 | 1.4 | 1.5 | 1.7 | 1.9 | 2.2 | 2.8 | 3.5 | 4.2 |
| 50 | 50 | 0.2 | 0.3 | 0.4 | 0.5 | 0.6 | 0.7 | 0.8 | 1.0 | 1.2 | 1.4 | 1.6 | 1.8 | 2.0 | 2.2 | 2.4 | 2.8 | 3.2 | 4.0 | 5.0 | 6.0 |
| | 70 | 0.2 | 0.3 | 0.3 | 0.4 | 0.5 | 0.6 | 0.7 | 0.9 | 1.0 | 1.2 | 1.4 | 1.5 | 1.7 | 1.9 | 2.0 | 2.4 | 2.7 | 3.4 | 4.3 | 5.1 |
| | 100 | 0.1 | 0.2 | 0.3 | 0.4 | 0.4 | 0.5 | 0.6 | 0.7 | 0.9 | 1.0 | 1.2 | 1.3 | 1.5 | 1.6 | 1.8 | 2.1 | 2.4 | 3.0 | 3.7 | 4.5 |
| | 150 | 0.1 | 0.2 | 0.3 | 0.3 | 0.4 | 0.5 | 0.5 | 0.7 | 0.8 | 0.9 | 1.1 | 1.2 | 1.3 | 1.5 | 1.6 | 1.9 | 2.1 | 2.7 | 3.3 | 4.0 |
| | 300 | 0.1 | 0.2 | 0.2 | 0.3 | 0.3 | 0.4 | 0.5 | 0.6 | 0.7 | 0.8 | 0.9 | 1.0 | 1.1 | 1.3 | 1.4 | 1.6 | 1.9 | 2.3 | 2.9 | 3.5 |
| 60 | 60 | 0.2 | 0.2 | 0.3 | 0.4 | 0.5 | 0.6 | 0.7 | 0.8 | 1.0 | 1.2 | 1.3 | 1.5 | 1.7 | 1.8 | 2.0 | 2.3 | 2.7 | 3.3 | 4.2 | 5.0 |
| | 100 | 0.1 | 0.2 | 0.3 | 0.3 | 0.4 | 0.5 | 0.5 | 0.7 | 0.8 | 0.9 | 1.1 | 1.2 | 1.3 | 1.5 | 1.6 | 1.9 | 2.1 | 2.7 | 3.3 | 4.0 |
| | 150 | 0.1 | 0.2 | 0.2 | 0.3 | 0.3 | 0.4 | 0.5 | 0.6 | 0.7 | 0.8 | 0.9 | 1.0 | 1.2 | 1.3 | 1.4 | 1.6 | 1.9 | 2.3 | 2.9 | 3.5 |
| | 300 | 0.1 | 0.1 | 0.2 | 0.2 | 0.3 | 0.3 | 0.4 | 0.5 | 0.6 | 0.7 | 0.8 | 0.9 | 1.0 | 1.1 | 1.2 | 1.4 | 1.6 | 2.0 | 2.5 | 3.0 |
| 75 | 75 | 0.1 | 0.2 | 0.3 | 0.3 | 0.4 | 0.5 | 0.5 | 0.7 | 0.8 | 0.9 | 1.1 | 1.2 | 1.3 | 1.5 | 1.6 | 1.9 | 2.1 | 2.7 | 3.3 | 4.0 |
| | 120 | 0.1 | 0.2 | 0.2 | 0.3 | 0.3 | 0.4 | 0.4 | 0.5 | 0.6 | 0.8 | 0.9 | 1.0 | 1.1 | 1.2 | 1.3 | 1.5 | 1.7 | 2.2 | 2.7 | 3.3 |
| | 200 | 0.1 | 0.1 | 0.2 | 0.2 | 0.3 | 0.3 | 0.4 | 0.5 | 0.5 | 0.6 | 0.7 | 0.8 | 0.9 | 1.0 | 1.1 | 1.3 | 1.5 | 1.8 | 2.3 | 2.7 |
| | 300 | 0.1 | 0.1 | 0.2 | 0.2 | 0.2 | 0.3 | 0.3 | 0.4 | 0.5 | 0.6 | 0.7 | 0.7 | 0.8 | 0.9 | 1.0 | 1.2 | 1.3 | 1.7 | 2.1 | 2.5 |
| 100 | 100 | 0.1 | 0.1 | 0.2 | 0.2 | 0.3 | 0.3 | 0.4 | 0.5 | 0.6 | 0.7 | 0.8 | 0.9 | 1.0 | 1.1 | 1.2 | 1.4 | 1.6 | 2.0 | 2.5 | 3.0 |
| | 200 | 0.1 | 0.1 | 0.1 | 0.2 | 0.2 | 0.3 | 0.3 | 0.4 | 0.4 | 0.5 | 0.6 | 0.7 | 0.7 | 0.8 | 0.9 | 1.0 | 1.2 | 1.5 | 1.9 | 2.2 |
| | 300 | 0.1 | 0.1 | 0.1 | 0.2 | 0.2 | 0.2 | 0.3 | 0.3 | 0.4 | 0.5 | 0.5 | 0.6 | 0.7 | 0.7 | 0.8 | 0.9 | 1.1 | 1.3 | 1.7 | 2.0 |
| 150 | 150 | 0.1 | 0.1 | 0.1 | 0.2 | 0.2 | 0.2 | 0.3 | 0.3 | 0.4 | 0.5 | 0.5 | 0.6 | 0.7 | 0.7 | 0.8 | 0.9 | 1.1 | 1.3 | 1.7 | 2.0 |
| | 300 | .... | 0.1 | 0.1 | 0.1 | 0.1 | 0.2 | 0.2 | 0.2 | 0.3 | 0.3 | 0.4 | 0.5 | 0.5 | 0.6 | 0.6 | 0.7 | 0.8 | 1.0 | 1.2 | 1.5 |
| 200 | 200 | .... | 0.1 | 0.1 | 0.1 | 0.1 | 0.2 | 0.2 | 0.2 | 0.3 | 0.3 | 0.4 | 0.5 | 0.5 | 0.6 | 0.6 | 0.7 | 0.8 | 1.0 | 1.2 | 1.5 |
| | 300 | .... | 0.1 | 0.1 | 0.1 | 0.1 | 0.1 | 0.2 | 0.2 | 0.2 | 0.3 | 0.3 | 0.4 | 0.4 | 0.5 | 0.5 | 0.6 | 0.7 | 0.8 | 1.0 | 1.2 |
| 300 | 300 | .... | .... | 0.1 | 0.1 | 0.1 | 0.1 | 0.1 | 0.2 | 0.2 | 0.2 | 0.3 | 0.3 | 0.3 | 0.4 | 0.4 | 0.5 | 0.5 | 0.6 | 0.7 | 0.8 |
| 500 | 500 | .... | .... | .... | .... | 0.1 | 0.1 | 0.1 | 0.1 | 0.1 | 0.1 | 0.2 | 0.2 | 0.2 | 0.2 | 0.2 | 0.3 | 0.3 | 0.4 | 0.5 | 0.6 |

NOTE: 1 ft = 0.3048 m.

**TABLE 26-25** Percent Effective Ceiling- or Floor-Cavity Reflectance for Various Reflectance Combinations†

| % base reflectance* | 90 | | | | 80 | | | | 70 | | | 50 | | | 30 | | | | 10 | | |
|---|---|---|---|---|---|---|---|---|---|---|---|---|---|---|---|---|---|---|---|---|---|
| **% wall reflectance** | **90** | **70** | **50** | **30** | **80** | **70** | **50** | **30** | **70** | **50** | **30** | **70** | **50** | **30** | **65** | **50** | **30** | **10** | **50** | **30** | **10** |
| Cavity ratio | | | | | | | | | | | | | | | | | | | | | |
| 0 | 90 | 90 | 90 | 90 | 80 | 80 | 80 | 80 | 70 | 70 | 70 | 50 | 50 | 50 | 30 | 30 | 30 | 30 | 10 | 10 | 10 |
| 0.1 | 90 | 89 | 88 | 87 | 79 | 79 | 78 | 78 | 69 | 69 | 68 | 59 | 49 | 48 | 30 | 30 | 29 | 29 | 10 | 10 | 10 |
| 0.2 | 89 | 88 | 86 | 85 | 79 | 78 | 77 | 76 | 68 | 67 | 66 | 49 | 48 | 47 | 30 | 29 | 29 | 28 | 10 | 10 | 9 |
| 0.3 | 89 | 87 | 85 | 83 | 78 | 77 | 75 | 74 | 68 | 66 | 64 | 49 | 47 | 46 | 30 | 29 | 28 | 27 | 10 | 10 | 9 |
| 0.4 | 88 | 86 | 83 | 81 | 78 | 76 | 74 | 72 | 67 | 65 | 63 | 48 | 46 | 45 | 30 | 29 | 27 | 26 | 11 | 10 | 9 |
| 0.5 | 88 | 85 | 81 | 78 | 77 | 75 | 73 | 70 | 66 | 64 | 61 | 48 | 46 | 44 | 29 | 28 | 27 | 25 | 11 | 10 | 9 |
| 0.6 | 88 | 84 | 80 | 76 | 77 | 75 | 71 | 68 | 65 | 62 | 59 | 47 | 45 | 43 | 29 | 28 | 26 | 25 | 11 | 10 | 9 |
| 0.7 | 88 | 83 | 78 | 74 | 76 | 74 | 70 | 66 | 65 | 61 | 58 | 47 | 44 | 42 | 29 | 28 | 26 | 24 | 11 | 10 | 8 |
| 0.8 | 87 | 82 | 77 | 73 | 75 | 73 | 69 | 65 | 64 | 60 | 56 | 47 | 43 | 41 | 29 | 27 | 25 | 23 | 11 | 10 | 8 |
| 0.9 | 87 | 81 | 76 | 71 | 75 | 72 | 68 | 63 | 63 | 59 | 55 | 46 | 43 | 40 | 29 | 27 | 25 | 22 | 11 | 9 | 8 |
| 1.0 | 86 | 80 | 74 | 69 | 74 | 71 | 66 | 61 | 63 | 58 | 53 | 46 | 42 | 39 | 29 | 27 | 24 | 22 | 11 | 9 | 8 |
| 1.1 | 86 | 79 | 73 | 67 | 74 | 71 | 65 | 60 | 62 | 57 | 52 | 46 | 41 | 38 | 29 | 26 | 24 | 21 | 11 | 9 | 8 |
| 1.2 | 86 | 78 | 72 | 65 | 73 | 70 | 64 | 58 | 61 | 56 | 50 | 45 | 41 | 37 | 29 | 26 | 23 | 20 | 12 | 9 | 7 |
| 1.3 | 85 | 78 | 70 | 64 | 73 | 69 | 63 | 57 | 61 | 55 | 49 | 45 | 40 | 36 | 29 | 26 | 23 | 20 | 12 | 9 | 7 |
| 1.4 | 85 | 77 | 69 | 62 | 72 | 68 | 62 | 55 | 60 | 54 | 48 | 45 | 40 | 35 | 28 | 26 | 22 | 19 | 12 | 9 | 7 |
| 1.5 | 85 | 76 | 68 | 61 | 72 | 68 | 61 | 54 | 59 | 53 | 47 | 44 | 39 | 34 | 28 | 25 | 22 | 18 | 12 | 9 | 7 |
| 1.6 | 85 | 75 | 66 | 59 | 71 | 67 | 60 | 53 | 59 | 53 | 45 | 44 | 39 | 33 | 28 | 25 | 21 | 18 | 12 | 9 | 7 |
| 1.7 | 84 | 74 | 65 | 58 | 71 | 66 | 59 | 52 | 58 | 51 | 44 | 44 | 38 | 32 | 28 | 25 | 21 | 17 | 12 | 9 | 7 |
| 1.8 | 84 | 73 | 64 | 56 | 70 | 65 | 58 | 50 | 57 | 50 | 43 | 43 | 37 | 32 | 28 | 25 | 21 | 17 | 12 | 9 | 6 |
| 1.9 | 84 | 73 | 63 | 55 | 70 | 65 | 57 | 49 | 57 | 49 | 42 | 43 | 37 | 31 | 28 | 25 | 20 | 16 | 12 | 9 | 6 |
| 2.0 | 83 | 72 | 62 | 53 | 69 | 64 | 56 | 48 | 56 | 48 | 41 | 43 | 37 | 30 | 28 | 24 | 20 | 16 | 12 | 9 | 6 |
| 2.1 | 83 | 71 | 61 | 52 | 69 | 63 | 55 | 47 | 56 | 47 | 40 | 43 | 36 | 29 | 28 | 24 | 20 | 16 | 13 | 9 | 6 |
| 2.2 | 83 | 70 | 60 | 51 | 68 | 63 | 54 | 45 | 55 | 46 | 39 | 42 | 36 | 29 | 28 | 24 | 19 | 15 | 13 | 9 | 6 |
| 2.3 | 83 | 69 | 59 | 50 | 68 | 62 | 53 | 44 | 54 | 46 | 38 | 42 | 35 | 28 | 28 | 24 | 19 | 15 | 13 | 9 | 6 |
| 2.4 | 82 | 68 | 58 | 48 | 67 | 61 | 52 | 43 | 54 | 45 | 37 | 42 | 35 | 27 | 28 | 24 | 19 | 14 | 13 | 9 | 6 |
| 2.5 | 82 | 68 | 57 | 47 | 67 | 61 | 51 | 42 | 53 | 44 | 36 | 41 | 34 | 27 | 27 | 23 | 18 | 14 | 13 | 9 | 6 |

| | | | | | | | | | | | | | | | | | | | | | |
|---|---|---|---|---|---|---|---|---|---|---|---|---|---|---|---|---|---|---|---|---|---|
| 2.6 | 82 | 67 | 56 | 46 | 66 | 60 | 50 | 41 | 53 | 43 | 35 | 41 | 34 | 26 | 27 | 23 | 18 | 13 | 13 | 9 | 5 |
| 2.7 | 82 | 66 | 55 | 45 | 66 | 60 | 49 | 40 | 52 | 43 | 34 | 41 | 33 | 26 | 27 | 23 | 18 | 13 | 13 | 9 | 5 |
| 2.8 | 81 | 66 | 54 | 44 | 66 | 59 | 48 | 39 | 52 | 42 | 33 | 41 | 33 | 25 | 27 | 23 | 18 | 13 | 13 | 9 | 5 |
| 2.9 | 81 | 65 | 53 | 43 | 65 | 58 | 48 | 38 | 51 | 41 | 33 | 40 | 33 | 25 | 27 | 23 | 17 | 12 | 13 | 9 | 5 |
| 3.0 | 81 | 64 | 52 | 42 | 65 | 58 | 47 | 38 | 51 | 40 | 32 | 40 | 32 | 24 | 27 | 22 | 17 | 12 | 13 | 8 | 5 |
| 3.1 | 80 | 64 | 51 | 41 | 64 | 57 | 46 | 37 | 50 | 40 | 31 | 40 | 32 | 24 | 27 | 22 | 17 | 12 | 13 | 8 | 5 |
| 3.2 | 80 | 63 | 50 | 40 | 64 | 57 | 45 | 36 | 50 | 39 | 30 | 40 | 31 | 23 | 27 | 22 | 16 | 11 | 13 | 8 | 5 |
| 3.3 | 80 | 62 | 49 | 39 | 64 | 56 | 44 | 35 | 49 | 39 | 30 | 39 | 31 | 23 | 27 | 22 | 16 | 11 | 13 | 8 | 5 |
| 3.4 | 80 | 62 | 48 | 38 | 63 | 56 | 44 | 34 | 49 | 38 | 29 | 39 | 31 | 22 | 27 | 22 | 16 | 11 | 13 | 8 | 5 |
| 3.5 | 79 | 61 | 48 | 37 | 63 | 55 | 43 | 33 | 48 | 38 | 29 | 39 | 30 | 22 | 26 | 22 | 16 | 11 | 13 | 8 | 5 |
| 3.6 | 79 | 60 | 47 | 36 | 62 | 54 | 42 | 33 | 48 | 37 | 28 | 39 | 30 | 21 | 26 | 21 | 15 | 10 | 13 | 8 | 5 |
| 3.7 | 79 | 60 | 46 | 35 | 62 | 54 | 42 | 32 | 48 | 37 | 27 | 38 | 30 | 21 | 26 | 21 | 15 | 10 | 13 | 8 | 4 |
| 3.8 | 79 | 59 | 45 | 35 | 62 | 53 | 41 | 31 | 47 | 36 | 27 | 38 | 29 | 21 | 26 | 21 | 15 | 10 | 13 | 8 | 4 |
| 3.9 | 78 | 59 | 45 | 34 | 61 | 53 | 40 | 30 | 47 | 36 | 26 | 38 | 29 | 20 | 26 | 21 | 15 | 10 | 13 | 8 | 4 |
| 4.0 | 78 | 58 | 44 | 33 | 61 | 52 | 40 | 30 | 46 | 35 | 26 | 38 | 29 | 20 | 26 | 21 | 15 | 9 | 13 | 8 | 4 |
| 4.1 | 78 | 57 | 43 | 32 | 60 | 52 | 39 | 29 | 46 | 35 | 25 | 37 | 28 | 20 | 26 | 21 | 14 | 9 | 13 | 8 | 4 |
| 4.2 | 78 | 57 | 43 | 32 | 60 | 51 | 39 | 29 | 46 | 34 | 25 | 37 | 28 | 19 | 26 | 20 | 14 | 9 | 13 | 8 | 4 |
| 4.3 | 78 | 56 | 42 | 31 | 60 | 51 | 38 | 28 | 45 | 34 | 25 | 37 | 28 | 19 | 26 | 20 | 14 | 9 | 13 | 8 | 4 |
| 4.4 | 77 | 56 | 41 | 30 | 59 | 51 | 38 | 28 | 45 | 34 | 24 | 37 | 27 | 19 | 26 | 20 | 14 | 8 | 13 | 8 | 4 |
| 4.5 | 77 | 55 | 41 | 30 | 59 | 50 | 37 | 27 | 45 | 33 | 24 | 37 | 27 | 19 | 25 | 20 | 14 | 8 | 14 | 8 | 4 |
| 4.6 | 77 | 55 | 40 | 29 | 59 | 50 | 37 | 26 | 44 | 33 | 24 | 36 | 27 | 18 | 25 | 20 | 14 | 8 | 14 | 8 | 4 |
| 4.7 | 77 | 54 | 40 | 29 | 58 | 49 | 36 | 26 | 44 | 33 | 23 | 36 | 26 | 18 | 25 | 20 | 13 | 8 | 14 | 8 | 4 |
| 4.8 | 76 | 54 | 39 | 28 | 58 | 49 | 36 | 25 | 44 | 32 | 23 | 36 | 26 | 18 | 25 | 19 | 13 | 8 | 14 | 8 | 4 |
| 4.9 | 76 | 53 | 38 | 28 | 58 | 49 | 35 | 25 | 44 | 32 | 23 | 36 | 26 | 18 | 25 | 19 | 13 | 7 | 14 | 8 | 4 |
| 5.0 | 76 | 53 | 38 | 27 | 57 | 48 | 35 | 25 | 43 | 32 | 22 | 36 | 26 | 17 | 25 | 19 | 13 | 7 | 14 | 8 | 4 |

* Ceiling, floor or floor of cavity.

† Tabular values based on 1.6 length-to-width ratio.

**TABLE 26-26** Multiplying Factors for 10% Effective Floor-Cavity Reflectance
(20% = 1.00)

| Effective ceiling cavity reflectance, $\rho_{cc}$ | 80 | | | | 70 | | | | 50 | | | 30 | | | 10 | | |
|---|---|---|---|---|---|---|---|---|---|---|---|---|---|---|---|---|---|
| % wall reflectance, $\rho_w$ | 70 | 50 | 30 | 10 | 70 | 50 | 30 | 10 | 50 | 30 | 10 | 50 | 30 | 10 | 50 | 30 | 10 |
| Room cavity ratio: | | | | | | | | | | | | | | | | | |
| 1 | 0.923 | 0.929 | 0.935 | 0.940 | 0.933 | 0.939 | 0.943 | 0.948 | 0.956 | 0.960 | 0.963 | 0.973 | 0.976 | 0.979 | 0.989 | 0.991 | 0.993 |
| 2 | 0.931 | 0.942 | 0.950 | 0.958 | 0.940 | 0.949 | 0.957 | 0.963 | 0.962 | 0.968 | 0.974 | 0.976 | 0.980 | 0.985 | 0.988 | 0.991 | 0.995 |
| 3 | 0.939 | 0.951 | 0.961 | 0.969 | 0.945 | 0.957 | 0.966 | 0.973 | 0.967 | 0.975 | 0.981 | 0.978 | 0.983 | 0.988 | 0.988 | 0.992 | 0.996 |
| 4 | 0.944 | 0.958 | 0.969 | 0.978 | 0.950 | 0.963 | 0.973 | 0.980 | 0.972 | 0.980 | 0.986 | 0.980 | 0.986 | 0.991 | 0.987 | 0.992 | 0.996 |
| 5 | 0.949 | 0.964 | 0.976 | 0.983 | 0.954 | 0.968 | 0.978 | 0.985 | 0.975 | 0.983 | 0.989 | 0.981 | 0.988 | 0.993 | 0.987 | 0.992 | 0.997 |
| 6 | 0.953 | 0.969 | 0.980 | 0.986 | 0.958 | 0.972 | 0.982 | 0.989 | 0.977 | 0.985 | 0.992 | 0.982 | 0.989 | 0.995 | 0.987 | 0.993 | 0.997 |
| 7 | 0.957 | 0.973 | 0.983 | 0.991 | 0.961 | 0.975 | 0.985 | 0.991 | 0.979 | 0.987 | 0.994 | 0.983 | 0.990 | 0.996 | 0.987 | 0.993 | 0.998 |
| 8 | 0.960 | 0.976 | 0.986 | 0.993 | 0.963 | 0.977 | 0.987 | 0.993 | 0.981 | 0.988 | 0.995 | 0.984 | 0.991 | 0.997 | 0.987 | 0.994 | 0.998 |
| 9 | 0.963 | 0.978 | 0.987 | 0.994 | 0.965 | 0.979 | 0.989 | 0.994 | 0.983 | 0.990 | 0.996 | 0.985 | 0.992 | 0.998 | 0.988 | 0.994 | 0.999 |
| 10 | 0.965 | 0.980 | 0.989 | 0.995 | 0.967 | 0.981 | 0.990 | 0.995 | 0.984 | 0.991 | 0.997 | 0.986 | 0.993 | 0.998 | 0.988 | 0.994 | 0.999 |

**TABLE 26-27** Multiplying Factors for 30% Effective Floor-Cavity Reflectance
(20% = 1.00)

| % effective ceiling cavity reflectance, $\rho_{cc}$ | 80 | | | | 70 | | | | 50 | | | 30 | | | 10 | | |
|---|---|---|---|---|---|---|---|---|---|---|---|---|---|---|---|---|---|
| % wall reflectance, $\rho_w$ | 70 | 50 | 30 | 10 | 70 | 50 | 30 | 10 | 50 | 30 | 10 | 50 | 30 | 10 | 50 | 30 | 10 |
| Room cavity ratio: | | | | | | | | | | | | | | | | | |
| 1 | 1.092 | 1.082 | 1.075 | 1.068 | 1.077 | 1.070 | 1.064 | 1.059 | 1.049 | 1.044 | 1.040 | 1.028 | 1.026 | 1.023 | 1.012 | 1.010 | 1.008 |
| 2 | 1.079 | 1.066 | 1.055 | 1.047 | 1.068 | 1.057 | 1.048 | 1.039 | 1.041 | 1.033 | 1.027 | 1.026 | 1.021 | 1.017 | 1.013 | 1.010 | 1.006 |
| 3 | 1.070 | 1.054 | 1.042 | 1.033 | 1.061 | 1.048 | 1.037 | 1.028 | 1.034 | 1.027 | 1.020 | 1.024 | 1.017 | 1.012 | 1.014 | 1.009 | 1.005 |
| 4 | 1.062 | 1.045 | 1.033 | 1.024 | 1.055 | 1.040 | 1.029 | 1.021 | 1.030 | 1.022 | 1.015 | 1.022 | 1.015 | 1.010 | 1.014 | 1.009 | 1.004 |
| 5 | 1.056 | 1.038 | 1.026 | 1.018 | 1.050 | 1.034 | 1.024 | 1.015 | 1.027 | 1.018 | 1.012 | 1.020 | 1.013 | 1.008 | 1.014 | 1.009 | 1.004 |
| 6 | 1.052 | 1.033 | 1.021 | 1.014 | 1.047 | 1.030 | 1.020 | 1.012 | 1.024 | 1.015 | 1.009 | 1.019 | 1.012 | 1.006 | 1.014 | 1.008 | 1.003 |
| 7 | 1.047 | 1.029 | 1.018 | 1.011 | 1.043 | 1.026 | 1.017 | 1.009 | 1.022 | 1.013 | 1.007 | 1.018 | 1.010 | 1.005 | 1.014 | 1.008 | 1.003 |
| 8 | 1.044 | 1.026 | 1.015 | 1.009 | 1.040 | 1.024 | 1.015 | 1.007 | 1.020 | 1.012 | 1.006 | 1.017 | 1.009 | 1.004 | 1.013 | 1.007 | 1.003 |
| 9 | 1.040 | 1.024 | 1.014 | 1.007 | 1.037 | 1.022 | 1.014 | 1.006 | 1.019 | 1.011 | 1.005 | 1.016 | 1.009 | 1.004 | 1.013 | 1.007 | 1.002 |
| 10 | 1.037 | 1.022 | 1.012 | 1.006 | 1.034 | 1.020 | 1.012 | 1.005 | 1.017 | 1.010 | 1.004 | 1.015 | 1.009 | 1.003 | 1.013 | 1.007 | 1.002 |

**TABLE 26-28** Coefficients of Utilization and Luminance Coefficients of a Generic Troffer

*a.* Coefficients of utilization

| Typical luminaire | Maint. Cat. | Spacing criterion[d] | $\rho_{CC}$[a] → | 80 | | | 70 | | | 50 | | | 30 | | | 10 | | | 0 |
|---|---|---|---|---|---|---|---|---|---|---|---|---|---|---|---|---|---|---|---|
| Typical distribution and % lamp lumens | | | $\rho_W$[b] → | 50 | 30 | 10 | 50 | 30 | 10 | 50 | 30 | 10 | 50 | 30 | 10 | 50 | 30 | 10 | 0 |
| | | | RCR[c] ↓ | Coefficients of utilization for 20% effective floor-cavity reflectance ($\rho_{FC}$ = 20) | | | | | | | | | | | | | | | |
| 4 lamp, 2-ft wide troffer with 45° plastic louver—multiply by 1.05 for 2 lamps and 0.95 for 6 lamps | IV | 1.0 | 0 | .59 | .59 | .59 | .58 | .58 | .58 | .55 | .55 | .55 | .53 | .53 | .53 | .51 | .51 | .51 | .50 |
| | | | 1 | .54 | .52 | .50 | .52 | .51 | .49 | .50 | .49 | .48 | .48 | .47 | .46 | .47 | .46 | .45 | .44 |
| | 0% ↑ | | 2 | .48 | .45 | .43 | .47 | .44 | .42 | .45 | .43 | .41 | .44 | .42 | .40 | .42 | .41 | .39 | .39 |
| | | | 3 | .43 | .40 | .37 | .42 | .39 | .37 | .41 | .38 | .36 | .40 | .37 | .36 | .39 | .37 | .35 | .34 |
| | | | 4 | .39 | .35 | .32 | .38 | .35 | .32 | .37 | .34 | .32 | .36 | .33 | .31 | .35 | .33 | .31 | .30 |
| | | | 5 | .35 | .31 | .28 | .35 | .31 | .28 | .34 | .30 | .28 | .33 | .30 | .28 | .32 | .29 | .27 | .26 |
| | 50% ↓ | | 6 | .32 | .28 | .25 | .32 | .28 | .25 | .31 | .27 | .25 | .30 | .27 | .25 | .29 | .26 | .24 | .23 |
| | | | 7 | .29 | .25 | .22 | .29 | .25 | .22 | .28 | .25 | .22 | .27 | .24 | .22 | .27 | .24 | .22 | .21 |
| | | | 8 | .26 | .22 | .20 | .26 | .22 | .20 | .25 | .22 | .20 | .25 | .22 | .19 | .24 | .21 | .19 | .18 |
| | | | 9 | .24 | .20 | .17 | .24 | .20 | .17 | .23 | .20 | .17 | .23 | .19 | .17 | .22 | .19 | .17 | .16 |
| | | | 10 | .22 | .18 | .16 | .22 | .18 | .16 | .21 | .18 | .16 | .21 | .18 | .15 | .20 | .17 | .15 | .15 |

*(continued)*

**TABLE 26-28** *(Continued)*

*b.* Luminance coefficients

Ceiling cavity luminance coefficients[e] for 20% effective floor-cavity reflectance ($\rho_{FC}$ = 20)

| $\rho_{CC}$→ | 80 | | | 70 | | | 50 | | | 30 | | | 10 | | |
|---|---|---|---|---|---|---|---|---|---|---|---|---|---|---|---|
| $\rho_W$→ | 50 | 30 | 10 | 50 | 30 | 10 | 50 | 30 | 10 | 50 | 30 | 10 | 50 | 30 | 10 |
| RCR ↓ | | | | | | | | | | | | | | | |
| 0 | .09 | .09 | .09 | .08 | .08 | .08 | .05 | .05 | .05 | .03 | .03 | .03 | .01 | .01 | .01 |
| 1 | .09 | .08 | .07 | .08 | .07 | .06 | .05 | .05 | .04 | .03 | .03 | .02 | .01 | .01 | .01 |
| 2 | .08 | .07 | .05 | .07 | .06 | .05 | .05 | .04 | .03 | .03 | .02 | .02 | .01 | .01 | .01 |
| 3 | .08 | .06 | .04 | .07 | .05 | .03 | .05 | .03 | .02 | .03 | .02 | .01 | .01 | .01 | .00 |
| 4 | .07 | .05 | .03 | .06 | .04 | .03 | .04 | .03 | .02 | .03 | .02 | .01 | .01 | .01 | .00 |
| 5 | .07 | .04 | .03 | .06 | .04 | .02 | .04 | .03 | .02 | .02 | .02 | .01 | .01 | .01 | .00 |
| 6 | .07 | .04 | .02 | .06 | .04 | .02 | .04 | .02 | .01 | .02 | .01 | .01 | .01 | .00 | .00 |
| 7 | .06 | .04 | .02 | .06 | .03 | .02 | .04 | .02 | .01 | .02 | .01 | .01 | .01 | .00 | .00 |
| 8 | .06 | .03 | .02 | .05 | .03 | .01 | .04 | .02 | .01 | .02 | .01 | .01 | .01 | .00 | .00 |
| 9 | .06 | .03 | .01 | .05 | .03 | .01 | .04 | .02 | .01 | .02 | .01 | .00 | .01 | .00 | .00 |
| 10 | .06 | .03 | .01 | .05 | .03 | .01 | .03 | .02 | .01 | .02 | .01 | .00 | .01 | .00 | .00 |

Wall luminance coefficients[e] for 20% effective floor-cavity reflectance ($\rho_{FC}$ = 20)

| $\rho_{CC}$→ | 80 | | | 70 | | | 50 | | | 30 | | | 10 | | | |
|---|---|---|---|---|---|---|---|---|---|---|---|---|---|---|---|---|
| $\rho_W$→ | 50 | 30 | 10 | 50 | 30 | 10 | 50 | 30 | 10 | 50 | 30 | 10 | 50 | 30 | 10 | WDRC[e] |
| RCR ↓ | | | | | | | | | | | | | | | | |
| 0 | | | | | | | | | | | | | | | | |
| 1 | .13 | .07 | .02 | .12 | .07 | .02 | .12 | .07 | .02 | .11 | .06 | .02 | .10 | .06 | .02 | .15 |
| 2 | .12 | .07 | .02 | .12 | .06 | .02 | .11 | .06 | .02 | .11 | .06 | .02 | .10 | .06 | .02 | .14 |
| 3 | .11 | .06 | .02 | .11 | .06 | .02 | .11 | .06 | .02 | .10 | .06 | .02 | .10 | .05 | .02 | .13 |
| 4 | .11 | .06 | .02 | .10 | .05 | .02 | .10 | .05 | .02 | .10 | .05 | .02 | .09 | .05 | .02 | .13 |
| 5 | .10 | .05 | .01 | .10 | .05 | .01 | .10 | .05 | .01 | .09 | .05 | .01 | .09 | .05 | .01 | .12 |
| 6 | .10 | .05 | .01 | .09 | .05 | .01 | .09 | .05 | .01 | .09 | .04 | .01 | .08 | .04 | .01 | .11 |
| 7 | .09 | .04 | .01 | .09 | .04 | .01 | .09 | .04 | .01 | .08 | .04 | .01 | .08 | .04 | .01 | .10 |
| 8 | .09 | .04 | .01 | .08 | .04 | .01 | .08 | .04 | .01 | .08 | .04 | .01 | .08 | .04 | .01 | .10 |
| 9 | .08 | .04 | .01 | .08 | .04 | .01 | .08 | .04 | .01 | .08 | .04 | .01 | .07 | .04 | .01 | .09 |
| 10 | .08 | .04 | .01 | .08 | .04 | .01 | .07 | .04 | .01 | .07 | .04 | .01 | .07 | .04 | .01 | .09 |

[a] $\rho_{CC}$ = percent effective ceiling-cavity reflectance.

[b] $\rho_W$ = percent wall reflectance.

[c] RCR = room-cavity ratio.

[d] Spacing criterion times the mounting height above the work plane gives the spacing between luminaires, which approximately is the dividing point between reasonably acceptable horizontal illuminance and a noticeably poorer uniformity.

[e] Alhough it is recommended that luminance coefficients and wall direct radiation coefficients be published to three decimal places, only two are shown here. Three-decimal-place data should be obtained from manufacturers of actual luminaires used.

NOTE: 1 ft = 0.3048 m.

*nance coefficients (LC)*. These coefficients, like coefficients of utilization, may be computed for any specific luminaire. The wall- and ceiling-cavity luminance coefficients for one generic luminaire are found in Table 26-28.

Luminance coefficients are similar to coefficients of utilization and may be substituted into a variation of the lumen method formula in place of the coefficient of utilization. The result obtained is either the average wall luminance or the average ceiling cavity luminance, rather than illumination on the work plane. Thus,

$$\text{Average initial wall luminance} = \frac{\text{total lamp lumens} \times \text{wall luminance coefficient}}{\pi \times \text{floor area}}$$

and

Average initial ceiling cavity luminance

$$= \frac{\text{total lamp lumens} \times \text{ceiling cavity luminance coefficient}}{\pi \times \text{floor area}}$$

If the area is expressed in square feet, the luminance is in candelas per square foot; if the area is in square meters, the luminance is in candelas per square meter. Luminance can be obtained in footlamberts by expressing area in feet and omitting the factor of $\pi$ in the denominators.

If the maintained average wall luminance or the maintained average ceiling cavity luminance is required, a light-loss factor is introduced.

## CALCULATION OF ILLUMINANCE AT A POINT

*77. **Point Calculation Methods.*** The calculation of the illuminance at a point, whether on a horizontal, a vertical, or an inclined plane consists of two parts: the direct component and the reflected component. The total of these two components is the illuminance at the point in question.

Table 26-29 is a matrix of the most commonly used systems of computing the direct component. Although each of the methods can be used to make calculations under all the conditions if sufficient data are available, the $x$ in the matrix indicates the conditions under which the method is easier to use.

*78. **Direct Component Illuminance.*** Most methods are based on the application of the *inverse-square law* (fc = cp/$D^2$) to data obtained from the candlepower distribution curve for

**TABLE 26-29** Methods of Determining the Direct Component of Illuminance at a Point*

| Source | Point | | | Linear | | | Area | | |
|---|---|---|---|---|---|---|---|---|---|
| Plane | Horiz. | Vert. | Inclined | Horiz. | Vert. | Inclined | Horiz. | Vert. | Inclined |
| 1. Inverse square | X | X | X | | | | | | |
| 2. Plan-scale method† | X | X | | | | | | | |
| 3. Angular coord.-DIC method‡ | | | | X | X | | | | |
| 4. IES-London-aspect factor method§ | | | | X | X | X | | | |
| 5. Illumination charts and tables | X | X | X | X | X | X | X | X | X |
| 6. Idealized source chart | | | | | | | X | X | |
| 7. Configuration factor | | | | | | | X | X | |

* The Determination of Illumination at a Point; *J. IES*, January 1974, vol. 3, p. 170.

† Goodbar, I.: New Methods for Point by Point Calculations; *Illum. Eng.*, January 1946, vol. 41, p. 39.

‡ Jones, J. R., LeVere, R. C., Ivanicki, N., and Chesebrough, P.: Angular Coordinate System for Computing Illumination at a Point; *Illum. Eng.*, April 1969, vol. 64, p. 296.

§ The Calculation of Direct Illumination from Linear Sources, *IES (London) Tech, Rep.* 11. The Illuminating Engineering Society, York House, Westminister Bridge Road, London, England.

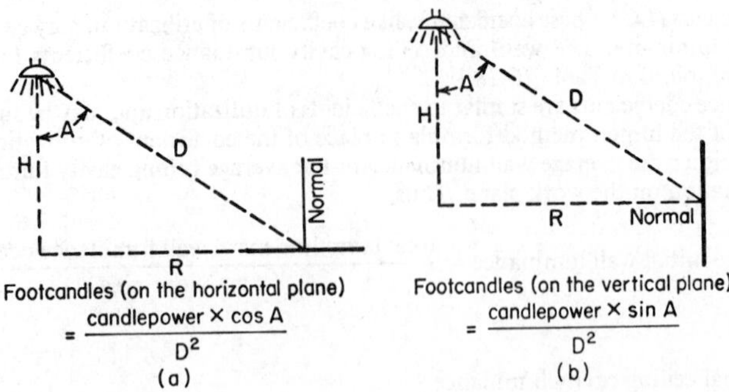


**FIG. 26-40** Fundamental relationships for point calculations where the inverse-square law applies.

the luminaire. In applying this formula to a horizontal surface, the footcandles at any point $P$ are equal to the candlepower directed toward $P$ multiplied by the cosine of the angle $A$ and divided by the square of the distance $D$ from the luminaire [fc = (cp × cos $A$)/$D^2$]. See Figs. 26-40 and 26-41. In this case $A$ is the angle between the axis of the luminaire and a line from the light center

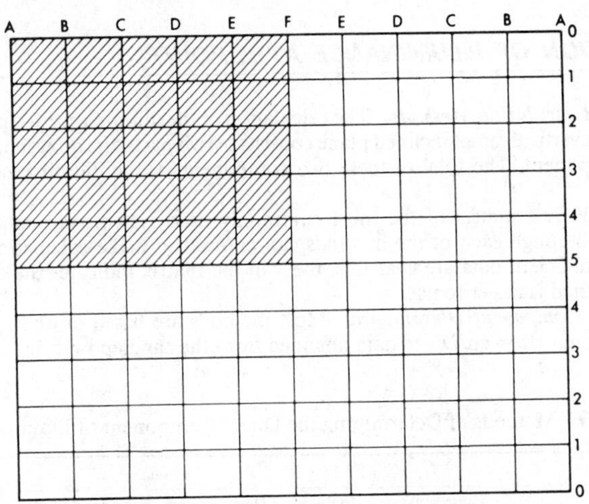


**FIG. 26-41** Chart for locating points on the work plane in a room. Letters are placed along length of room and numbers along width. Each block represents 10% of length or width. Position A0 is corner of a room and F5 is the center.

to point $P$. In many applications the computed footcandles of most interest are those obtained where angle $A$ is zero, and since its cosine then becomes unity, the formula is reduced to its simplest form: fc = (cp/$D^2$).

***79. Reflected Component Illuminance.*** The reflected illuminance component on horizontal surfaces is calculated in exactly the same manner as the average illuminance computed using the lumen method except that the RRC, the reflected radiation coefficient, is substituted for the CU, coefficient of utilization.

$$\text{Reflected illuminance (horizontal)} = \frac{\text{lamp lumens per luminaire} \times \text{RRC}}{\text{area per luminaire (on work plane)}}$$

**TABLE 26-30** Room Position Multipliers
(For all room-cavity ratios and for all points designated by a number and a letter as illustrated in Fig. 26-41)

| Points along width of room | RCR | Points along length of room A | B | C | D | E | F |
|---|---|---|---|---|---|---|---|
| 0 | 1 | .24 | .42 | .47 | .48 | .48 | .48 |
| | 2 | .24 | .36 | .42 | .44 | .46 | .46 |
| | 3 | .23 | .32 | .37 | .40 | .42 | .42 |
| | 4 | .22 | .28 | .32 | .35 | .37 | .37 |
| | 5 | .21 | .25 | .28 | .31 | .33 | .33 |
| | 6 | .20 | .23 | .26 | .28 | .29 | .30 |
| | 7 | .18 | .21 | .23 | .25 | .26 | .27 |
| | 8 | .17 | .18 | .21 | .22 | .22 | .23 |
| | 9 | .15 | .17 | .18 | .19 | .20 | .20 |
| | 10 | .14 | .16 | .16 | .17 | .18 | .18 |
| 1 | 1 | .42 | .74 | .81 | .83 | .84 | .84 |
| | 2 | .36 | .51 | .60 | .63 | .66 | .68 |
| | 3 | .32 | .40 | .48 | .51 | .53 | .57 |
| | 4 | .28 | .33 | .40 | .42 | .44 | .48 |
| | 5 | .25 | .29 | .33 | .36 | .38 | .42 |
| | 6 | .23 | .26 | .29 | .31 | .33 | .36 |
| | 7 | .21 | .23 | .26 | .28 | .29 | .30 |
| | 8 | .18 | .20 | .23 | .25 | .26 | .26 |
| | 9 | .17 | .18 | .20 | .21 | .22 | .23 |
| | 10 | .16 | .17 | .18 | .19 | .19 | .20 |
| 2 | 1 | .47 | .81 | .90 | .92 | .93 | .93 |
| | 2 | .42 | .60 | .68 | .72 | .78 | .83 |
| | 3 | .37 | .48 | .58 | .61 | .64 | .67 |
| | 4 | .32 | .40 | .48 | .50 | .52 | .57 |
| | 5 | .28 | .33 | .40 | .42 | .44 | .48 |
| | 6 | .26 | .29 | .35 | .37 | .38 | .40 |
| | 7 | .23 | .26 | .30 | .32 | .33 | .34 |
| | 8 | .21 | .23 | .26 | .27 | .28 | .29 |
| | 9 | .18 | .20 | .23 | .24 | .25 | .25 |
| | 10 | .16 | .18 | .19 | .21 | .22 | .22 |
| 3 | 1 | .48 | .83 | .92 | .94 | .95 | .95 |
| | 2 | .44 | .63 | .72 | .77 | .82 | .85 |
| | 3 | .40 | .51 | .61 | .65 | .69 | .71 |
| | 4 | .35 | .42 | .50 | .54 | .58 | .61 |
| | 5 | .31 | .36 | .42 | .46 | .49 | .52 |
| | 6 | .28 | .31 | .37 | .39 | .41 | .43 |
| | 7 | .25 | .28 | .32 | .34 | .35 | .36 |
| | 8 | .22 | .25 | .27 | .29 | .30 | .30 |
| | 9 | .19 | .21 | .24 | .25 | .26 | .26 |
| | 10 | .17 | .19 | .21 | .22 | .23 | .23 |
| 4 | 1 | .48 | .84 | .93 | .95 | .96 | .97 |
| | 2 | .46 | .66 | .78 | .82 | .85 | .86 |
| | 3 | .42 | .53 | .64 | .69 | .73 | .75 |
| | 4 | .37 | .44 | .52 | .58 | .62 | .64 |
| | 5 | .33 | .38 | .44 | .49 | .52 | .54 |
| | 6 | .29 | .33 | .38 | .41 | .43 | .45 |
| | 7 | .26 | .29 | .33 | .35 | .37 | .37 |
| | 8 | .22 | .26 | .28 | .30 | .31 | .32 |
| | 9 | .20 | .22 | .25 | .26 | .26 | .27 |
| | 10 | .18 | .19 | .22 | .23 | .23 | .24 |
| 5 | 1 | .48 | .84 | .93 | .95 | .97 | .97 |
| | 2 | .46 | .68 | .83 | .85 | .86 | .87 |
| | 3 | .42 | .57 | .67 | .71 | .75 | .77 |
| | 4 | .37 | .48 | .57 | .61 | .64 | .66 |
| | 5 | .33 | .42 | .48 | .52 | .54 | .56 |
| | 6 | .30 | .36 | .40 | .43 | .45 | .47 |
| | 7 | .27 | .30 | .34 | .36 | .37 | .38 |
| | 8 | .23 | .26 | .29 | .30 | .31 | .32 |
| | 9 | .20 | .23 | .25 | .26 | .27 | .27 |
| | 10 | .18 | .20 | .22 | .23 | .24 | .25 |

where $RRC = LC_w + RPM\,(LC_{cc} - LC_w)$
$LC_w$ = wall luminance coefficient (see Table 26-28)
$LC_{cc}$ = ceiling cavity luminance coefficient (see Table 26-28)
RPM = room position multiplier (see Table 26-30)

To determine the illuminance reflected to vertical surfaces, the approximate average value is determined using the same general formula, but substituting the WRRC, wall reflected radiation coefficient, for the CU.

$$\text{Reflected illuminance (vertical)} = \frac{\text{lamp lumens per luminaire} \times \text{WRRC}}{\text{area per luminaire (on work plane)}}$$

where $WRRC = LC_w/\rho_w - WDRC$
$\rho_w$ = average wall reflectance
WDRC = wall direct radiation coefficient (see Table 26-28)

## FLOODLIGHTING DESIGN AND PROCEDURE

***80. Beam-Lumen Method.*** All floodlighting-design methods include certain approximations, based on experience. In floodlighting systems containing a large number of luminaires, a detailed study of aiming diagrams and many calculations are usually required. A procedure which is useful for designing simpler systems is called the "beam-lumen" method. This method requires the solution of the two formulas (A and B), as discussed in the following paragraphs, and the coordination of the results.

In many locations in which floodlighting is proposed, there are some basic dimensions that can be assumed to be already fixed. For example, in ground-area floodlighting, the designer is usually able to locate points where the equipment should logically be placed, such as on nearby buildings, along high banks or fences, or on poles or towers. These locations establish the approximate perpendicular distance *D* from the floodlight to the plane of the surface to be lighted and the average aiming angles. They also guide the choice of floodlight type—narrow, medium, or broad-beam—as listed in Table 26-31. In like manner the choice of equipment for lighting vertical surfaces can be obtained by taking *D* as the horizontal distance from the luminaire to the plane in which the vertical surface is located.

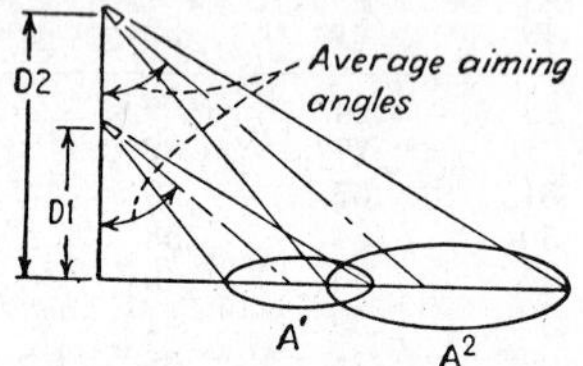


**FIG. 26-42** Spot areas (for the same beam spread and aiming angle) vary as the square of the distance *D*. Spot length *L* and spot width *W* vary as the distance *D*. The spot area may be determined from $\pi LW/4$.

The average aiming angle is measured from the perpendicular to the beam-axis line (Fig. 26-42). In a perimeter system in which the floodlights are mounted along or beyond the perimeter of an area, they will, of course,

**TABLE 26-31** Floodlight Beam Spread for Typical Areas

| Type of area | Approx. distance to area | Beam spread |
|---|---|---|
| Buildings 2-3 story lighted from curb posts | 10′–30′ | Broad, also Fresnel type (wide spread) |
| Buildings lighted from across street: | | |
| Areas 3,000 sq ft or less | 50′–100′ | Broad or medium |
| Areas 3,000 to 10,000 sq ft | 50′–100′ | Medium or narrow |
| Construction work: | | |
| Parking spaces | At perimeter | Broad or medium |
| Baseball, football | Behind bleachers | Broad and medium |

NOTE: 1 ft = 0.3048 m; 1 sq ft = 0.0929 m².

be aimed at various angles, but the average aiming angle used in computation is measured between the perpendicular and the centerline of the area to be lighted. When floodlights are on poles along the centerline of an area, the average aiming angle is measured between the pole (perpendicular) and a point halfway to the boundary (one-fourth of the width of the total area).

*Design Formulas and Procedure.* Formula A:

$$\text{Floodlights needed for coverage} = \frac{(\text{area lighted}) \times (\text{coverage factor})}{\text{beam-spot area}}$$

In this formula, *area* is the area to be lighted in square feet. This may be either a horizontal surface or a vertical one.

The *coverage factor* indicates the minimum number of directions from which each point in the area should be lighted, depending upon the use of the area. A coverage factor of 1 is acceptable in some applications, although in such systems one or two lamp burnouts might temporarily leave large, dark patches. Coverage factors greater than 1 therefore add desirable safety factors.

For example, a coverage factor of 2 is necessary for parking spaces and for protective lighting to reduce the effect of shadows between automobiles, rows of freight cars, piles of material, and similar bulky objects (see Table 26-32 for other recommended values).

The *beam-spot areas* at a 100-ft distance $D$ in formula A are given in Table 26-33 for various beam spreads and aiming angles of usual equipment having symmetrical candlepower distribution. In this table, $D$ is the perpendicular distance measured from the floodlight to the plane of the lighted surface. $L$ and $W$ are the lengths and widths of the ellipses formed when floodlights are aimed at an angle to the lighted surface. At 0° the area is assumed to be circular; at other angles, it is elliptical. At other distances and spreads and for similar beam spreads and aiming angles, the spot areas vary as the *square* of the distance $D$, while $L$ and $W$ values vary as the distance $D$. For example, if $D$ is 80 ft and a 30° beam-spread floodlighting unit is aimed at a 50° angle, the elliptical spot area as computed from Table 26-33 will be $(80)^2/(100)^2 \times 9978$, or 6386 ft$^2$. Likewise, the length $L$ of the ellipse will be $^{80}/_{100} \times 144.4$, or 115.5 ft, and the width $W$ will be $^{80}/_{100} \times 87.96 = 70.4$ ft.

Formula B:

$$\text{Floodlights needed for footcandles} = \frac{(\text{footcandles}) \times (\text{area lighted})}{(\text{light-loss factor}) \times (\text{utilization factor}) \times (\text{beam lumens})}$$

For this formula, typical *footcandle* recommendations are given in Table 26-32. The *light-loss factor* allows for dust and dirt and normal lamp depreciation. This is found under average conditions to be about 0.7. However, it may be as low as 0.3 for extremely dirty locations, where dust, dirt, and smoke are frequently suspended in the air.

The *utilization factor* or coefficient of beam utilization is the ratio of the lumens effectively lighting an area to the beam lumens, and it can be estimated from the following conditions:

1. If half or more than half of the floodlights are aimed so that all their beam lumens fall within an area, the overall utilization factor will be about 0.75.
2. If one-quarter to one-half of the floodlights are aimed so that all their beam lumens fall within an area, the overall utilization factor will be about 0.60.
3. If fewer than one-quarter of the floodlights can be aimed so that their beam lumens fall within an area, the overall utilization factor is likely to be not more than 0.40.

Most floodlights and projector- and reflector-type lamps as listed in the manufacturers' catalog are rated in beam lumens. These lumen ratings usually include only the light flux in that part of the beam in which the candlepower values are 10% or more of the maximum candlepower of the floodlight. A few typical beam-lumen values are given in Table 26-34.

As a general rule, it is wiser to design a system with a small number of floodlights with larger, more efficient lamps. This makes a simpler system to install, to control, and to maintain. Also from a control-of-light point of view it is desirable to choose a floodlighting unit having as narrow a beam spread as can be used and still maintain the coverage-factor requirements. It

**TABLE 26-32** Typical Recommended Illuminance Levels for Floodlighting, with Recommended Coverage Factors

| Area | Footcandles | Lux | Minimum coverage factor |
|---|---|---|---|
| Building: | | | 3–4 |
| General construction | 10 | 100 | |
| Excavation work | 2 | 20 | |
| Building exteriors and monuments: | | | 2 |
| Floodlighted | | | |
| Bright surroundings | | | |
| Light surfaces | 15 | 150 | |
| Medium-light surfaces | 20 | 200 | |
| Medium-dark surfaces | 30 | 300 | |
| Dark surfaces | 50 | 500 | |
| Dark surroundings | | | |
| Light surfaces | 5 | 50 | |
| Medium-light surfaces | 10 | 100 | |
| Medium-dark surfaces | 15 | 150 | |
| Dark surfaces | 20 | 200 | |
| Bulletin and poster boards: | | | 1–2 |
| Bright surroundings | | | |
| Light surfaces | 50 | 500 | |
| Dark surfaces | 100 | 1000 | |
| Dark surroundings | | | |
| Light surfaces | 20 | 200 | |
| Dark surfaces | 50 | 500 | |
| Open parking facilities* | | | |
| Low activity | 0.8† | 8† | |
| Medium activity | 2† | 20† | |
| High activity | 4† | 40† | |
| Covered parking facilities | | | |
| General parking and pedestrian areas | | | |
| Day | 5‡ | 50‡ | |
| Night | 5† | 50† | |
| Ramps and corners | | | |
| Day | 10‡ | 100‡ | |
| Night | 5† | 50† | |
| Entrance areas | | | |
| Day | 50‡ | 500‡ | |
| Night | 5† | 50† | |
| Service station (at grade) | | | 3–4 |
| Dark surrounding | | | |
| Approach | 1.5 | 15 | |
| Driveway | 1.5 | 15 | |
| Pump—island area | 20 | 200 | |
| Building faces (exclusive of glass) | 10§ | 100§ | |
| Service areas | 7 | 70 | |
| Light surrounding | | | |
| Approach | 3 | 30 | |
| Driveway | 5 | 50 | |
| Pump—island area | 30 | 300 | |
| Building faces (exclusive of glass) | 30§ | 300§ | |
| Service areas | 7 | 70 | |

* Based on recommended illuminance for pedestrian safety. Uniformity ratio 5 : 1.
† Average on pavement.
‡ Average on pavement, sum of electric lighting and daylight.
§ Vertical.

**TABLE 26-33** Spot Areas for Narrow-, Medium-, and Broad-Beam Floodlights at a 100-ft Distance*

| Aiming angle | 15-deg beam, narrow | | | 30-deg beam, medium | | | 50-deg beam, broad | | |
|---|---|---|---|---|---|---|---|---|---|
| | Spot area | *L* | *W* | Spot area | *L* | *W* | Spot area | *L* | *W* |
| 0 | 545 | 26.34 | 26.34 | 2,250 | 53.58 | 53.58 | 6,830 | 93.26 | 93.26 |
| 10 | 570 | 27.16 | 26.70 | 2,370 | 55.38 | 54.49 | 7,220 | 96.81 | 95.00 |
| 15 | 606 | 28.25 | 27.30 | 2,518 | 57.70 | 55.56 | 7,760 | 101.54 | 97.30 |
| 20 | 657 | 29.89 | 28.00 | 2,757 | 61.27 | 57.29 | 8,600 | 108.75 | 100.7 |
| 25 | 735 | 32.18 | 29.1 | 3,102 | 66.28 | 59.58 | 9,880 | 119.18 | 105.5 |
| 30 | 846 | 35.31 | 30.5 | 3,603 | 73.21 | 62.67 | 11,770 | 134.06 | 111.8 |
| 35 | 1,000 | 39.57 | 32.3 | 4,333 | 82.78 | 66.64 | 14,710 | 155.58 | 120.4 |
| 40 | 1,230 | 45.42 | 34.6 | 5,420 | 96.18 | 71.75 | 19,500 | 187.66 | 132.3 |
| 45 | 1,583 | 53.59 | 37.6 | 7,129 | 115.46 | 78.62 | 27,900 | 238.35 | 149.0 |
| 50 | 2,115 | 65.34 | 41.3 | 9,978 | 144.43 | 87.96 | 44,810 | 326.58 | 174.7 |
| 55 | 3,043 | 82.97 | 46.7 | 15,160 | 190.84 | 101.14 | 87,140 | 509.39 | 217.8 |
| 60 | 4,720 | 111.10 | 54.1 | 25,880 | 273.21 | 120.6 | 265,100 | 1072.99 | 314.6 |
| 65 | 8,165 | 160.19 | 64.9 | 54,480 | 447.94 | 154.85 | | | |
| 70 | 16,800 | 258.97 | 82.6 | 180,910 | 1000.2 | 230.3 | | | |

* The spot area for any other distance can be computed by multiplying the area in this table at the selected aiming angle by $D^2$ and dividing by 10,000.

**TABLE 26-34** Data on Typical Floodlighting Equipment*

| Floodlight watts | Lamp | Beam spread | Beam lumens |
|---|---|---|---|
| 500 | † | 25° | 3,900 |
| | | 60° | 5,300 |
| | | 150° × 50° | 5,300 |
| 1,000 | † | 15° | 9,500 |
| | | 55° | 9,200 |
| | | 120° × 25° | 9,200 |
| 1,500 | † | 15° | 12,300 |
| | | 65° | 17,200 |
| | | 148° | 19,600 |
| 400 | BT-37‡ | 142° | 13,900 |
| 1,000 | BT-56‡ | 150° × 114° | 28,300 |
| 500 | T-3§ | 75° × 20° | 4,500 |
| | T-3§ | 90° × 75° | 5,000 |
| 500 | PAR56NS¶ | 15° × 32° | 4,900 |
| | PAR56WF¶ | 34° × 66° | 5,725 |
| 1,000 | PAR64NS¶ | 14° × 31° | 8,500 |
| | PAR64WF¶ | 45° × 72° | 13,500 |

* See Table 26-8 for additional listing of reflector and projector lamps.
† Incandescent.
‡ Color-improved mercury.
§ Tungsten-halogen, tubular.
¶ Tungsten-halogen, projector.

should be remembered, however, that large floodlights are hard to conceal; this is important where their daytime appearance may be objectionable architecturally.

To solve formula B, after choosing the desired footcandles and determining the light-loss and utilization factors, a size of floodlight is chosen for trial calculation and its beam lumens (Table 26-34) are substituted in the equation.

When the dimensions or shape of an area points to the use of several types of floodlights, with different beam spreads, it is customary to divide the area into sections and plan a system for each of them. Buildings with setbacks are typical examples, also very tall structures such as towers or monuments. In setback buildings, you would design one setback at a time, selecting the type of floodlight most suitable for each. With towers or monuments, a similar approach is in order.

*Floodlighting Design Problem.* Assume that an area 200 by 200 ft is to be lighted to 5 fc. Also assume that it is located between two long buildings each 60 ft high, on top of which the floodlighting luminaires can be placed. By using a scale drawing to represent the luminaires placed 60 ft high and aimed toward the centerline of the 200-ft-wide work space, a protractor will show an aiming angle of about 60°. For trial-computation purposes it is best to start with the assumption that a narrow-beam-spread floodlighting unit will serve the coverage requirements and then change to a wider beam if found desirable. Hence, for this work space the lighting-design procedure is as follows:

1. From Table 26-33 the area which can be lighted for a 15° floodlight mounted 60 ft high at an aiming angle of 60° is

$$\frac{(60)^2}{(100)^2} \times 4720 = 1700 \text{ ft}^2$$

2. Solving formula A,

$$\text{No. of 15° units needed for coverage} = \frac{(\text{area}) \times (\text{coverage factor})}{\text{spot area}} = \frac{40{,}000 \times 3}{1700} = 70$$

3. Solving formula B. A light-loss factor of 0.3 is found for this dirty location, and a utilization factor of 0.7 may be assumed for narrow-beam floodlights in this area. If a 1000-W unit is selected for a trial computation, Table 26-34 shows that it may have as high as 9500 beam lm. Then the number of

$$\text{1000-W floodlights needed for footcandles} = \frac{(\text{footcandles}) \times (\text{area})}{\text{LLF} \times \text{UF} \times \text{beam lumens}} = \frac{5 \times 40{,}000}{0.3 \times 0.7 \times 9500} = 100$$

4. Since the number of floodlights (100) in formula B is greater than the number needed for adequate coverage, it could be concluded that, if 50 floodlight units are conveniently spaced on top of each building along the work area, a satisfactory 5-fc lighting installation would be provided. On the other hand, if a minimum number of floodlights are desired, formula B can be re-solved on the basis of using 1500-W luminaires having 12,300 beam lm (Table 26-34). In this case,

$$\text{1500-W floodlights needed} = \frac{5 \times 40{,}000}{0.3 \times 0.7 \times 12{,}300} = 77$$

Hence, this work area can be satisfactorily lighted with 5 fc maintained in service if thirty-nine 1500-W 15° floodlights are well distributed along each side of the work area and placed on top of the two 60-ft buildings.

As seen from the foregoing example, it is possible to use a fewer number of higher-wattage luminaires when the number of units (formula B) to provide the required footcandles is considerably greater than the number required for adequate coverage (formula A). On the other hand, when the trial computations show that the number of luminaires to provide the required footcandles is less than those for adequate coverage, it becomes necessary to recalculate formula B by using beam lumens from smaller-sized units, until one is found which brings the answer equal to or greater than that for formula A. In other words, the answer to formula B should preferably never be less than that for formula A in order to provide adequate illumination as well as satisfactory coverage. In all cases, consideration should be given to the use of lamps with the highest luminous efficacy of a color suitable for the application.

## ECONOMICS OF LIGHTING

***81. The Value of Lighting.*** The value of good lighting depends upon its use in the various fields of application.

Reports from a variety of industrial plants have indicated that better lighting created more satisfactory working conditions which in many instances meant an increase in production, reduced spoilage, fewer accidents, and less labor turnover. Likewise, in the store and other selling areas, good lighting properly applied has been recognized as a necessity, not only to permit the customer properly to inspect the merchandise but also to direct his or her attention to other items and thereby increase the volume of sales. Moreover, light and lighting are an integral

**TABLE 26-35** Cost-Analysis Outline for Lighting Systems

| General information | Lighting method 1 | Lighting method 2 |
|---|---|---|
| Installation data: | | |
| Type of installation | .... | .... |
| No. of rows | .... | .... |
| Luminaires per row | .... | .... |
| Lamps per luminaire | .... | .... |
| No. of lamps | .... | .... |
| Watts per lamp (including accessories) | .... | .... |
| Total watts | .... | .... |
| Maintained footcandles | .... | .... |
| **Calculation of complete expense** | | |
| Capital expense: | | |
| Estimated cost of each luminaire installed | .... | .... |
| Estimated wiring cost per luminaire | .... | .... |
| Cost per luminaire (luminaire plus wiring) | .... | .... |
| Number of luminaires | .... | .... |
| Total cost | .... | .... |
| Assumed years life | .... | .... |
| Total cost per year of life | .... | .... |
| Interest on investment (per year) | .... | .... |
| Taxes (per year) | .... | .... |
| Insurance (per year) | .... | .... |
| Total capital expense per year | .... | .... |
| Energy expense: | | |
| Total watts | .... | .... |
| Average hours used per year | .... | .... |
| kWh per year | .... | .... |
| Average rate per kWh | .... | .... |
| Total energy expense per year | .... | .... |
| Lamp-renewal expense: | | |
| No. of lamps | .... | .... |
| Avg. hours used per year | .... | .... |
| Total lamp hours per year | .... | .... |
| Rated lamp life, h | .... | .... |
| Avg. lamp renewals per year | .... | .... |
| Net price each | .... | .... |
| Replacement expense each (labor) | .... | .... |
| Net price plus replacement expense each | .... | .... |
| Total lamp-renewal expense per year | .... | .... |
| Cleaning expense: | | |
| No. of washings per year | .... | .... |
| Man-hours each (est.) | .... | .... |
| Man-hours for washing | .... | .... |
| No. of dustings per year | .... | .... |
| Man-hours each (est.) | .... | .... |
| Man-hours of dusting | .... | .... |
| Total man-hours | .... | .... |
| Expense per man-hour | .... | .... |
| Total cleaning expense per year | .... | .... |
| Repair expense: | | |
| Repairs (based on experience, allocation of repair man's time, etc.) | .... | .... |
| Est. total repair expense per year | .... | .... |
| **General information** | | |
| Recapitulation: | | |
| Total capital expense per year | .... | .... |
| Total energy expense per year | .... | .... |
| Total lamp renewal expense per year | .... | .... |
| Total cleaning expense per year | .... | .... |
| Est. total repair expense per year | .... | .... |
| Complete lighting expense for year | .... | .... |
| Complete lighting expense per footcandle per year | .... | .... |

part of many modern buildings where the architect has incorporated luminous elements for their functional use and aesthetic value.

For utilitarian installations which have resulted in an increase in factory production or an increase in sales in the store, a dollars-and-cents value can readily be given to better lighting. On the other hand, there is wide acceptance of light and lighting for its humanitarian and decorative aspects, which are of inestimable value.

***82. Cost of Lighting.*** The overall items for comparing different lighting systems must include both the initial and the operating cost. While one of these may be a dominant factor in the final selection, it is usually desirable to combine the two into some type of "total cost" indicator.

The computation of initial, operating and total annual cost for various systems considered for a given interior must be based on certain common assumptions, if the systems are to be fairly compared. Some of the important considerations are:

1. Equal illuminance results—since different systems may not produce equal illumination levels in service, all costs should be equated to an equal maintained footcandles basis.
2. Equal rates in amortizing the initial investment and allowing for interest, taxes, and insurance should be used.
3. Operating conditions, such as electrical-energy rate, burning hours per year, and starting frequency of the lamps, should be equal for the systems being considered.
4. Cleaning schedule should be appropriate to each type of system.
5. Uniform labor rates among systems should be used for estimating the cost of installations, cleaning, and relamping. Some users request life-cycle costs.

Table 26-35 tabulates a cost-analysis procedure for comparing the cost of two or more lighting systems and is self-explanatory. Other tabulations have also been proposed.[1]

***83. Group-Lamping Costs.*** The cost of lamp replacement is made up of the cost of the lamp and the cost of labor required to replace the lamp. When the sum of the costs is reduced, of course, the total annual cost of operating the lighting system is reduced. It is difficult to assess the exact overall cost reduction without having all the other facts about the installation, but it should be kept in mind that economical lamp replacement means better overall lighting economics.

With spot replacement, the total replacement cost per lamp is equal to the cost of the lamp plus the labor cost of replacement. Group-relamping cost, to compare with this, is equal to the lamp cost plus group-relamping labor cost plus the cost of any interim spot replacements, divided by the group-relamping interval to put both systems on an equal time basis. These costs can be expressed in the formulas which are presented and discussed in other publications. See also Par. **87**.

## LIGHTING MAINTENANCE

***84. Maintenance of Lighting.*** Good light maintenance is good economics; that is, good maintenance assures users that they actually get the light they pay for. With a well-maintained light system, the user also gets the better conditions that the system was designed to provide. These better seeing conditions contribute directly to higher productivity and improved morale in factories, offices, and schools. In stores, the better lighting that results from good maintenance helps to increase sales. In addition, all areas benefit from a better appearance and fewer work interruptions that come with good maintenance.

***85. Causes of Light Losses.*** Several factors contribute to light losses, and the effects of these factors vary with the kind of activity that takes place and the location of the establishment. For example, areas vary as to the amount and type of dirt in the air. Obviously, the amount of dirt in a foundry is greater than that in an air-conditioned office; and the amount and type of dirt in an office located near an industrial area are different from those in an office in the country; the

[1] Amick, C. L.: Economic Awareness in Design Planning; *Lighting Design and Application,* January, 1973, vol. 3, no. 1, p. 16.

black type of dirt characteristic of steel mills is certainly unlike the relatively light-colored dirt of a bakery or woodworking shop. It is important to recognize these variations in considering the light losses which result from dirt on the lamps and lighting units and dirt on room surfaces. These two, along with the unavoidable lamp-lumen depreciation, are the principal factors which cause light losses in every lighting installation.

***86. Benefits of Cleaning.*** A lighting system should not only be cleaned properly, it should be cleaned at the proper time. This combination produces a cleaning program which is a profitable investment because three principal benefits can be obtained.

1. *Better utilization of energy used for lighting.* Dirt absorbs light. Cleaning removes the dirt and thus helps to maintain the light level. As a result, visibility is maintained, which benefits the user.
2. *Reduced maintenance costs.* A good maintenance program calls for cleaning the lighting system at the most economical time. It makes use of the most efficient methods and equipment. In this way the time and materials required are reduced; thus maintenance costs are lowered.
3. *Better appearance.* Clean lighting systems improve the appearance of the working or selling area. This is conducive to improved morale, better housekeeping, and increased customer comfort and satisfaction.

To further improve the appearance of the area being lighted, walls and ceilings should be periodically cleaned and repainted as part of the cleaning program.

***87. Relamping Benefits.*** The lamps in a lighting system can be replaced individually as they burn out, or the entire installation can be replaced before the lamps reach their average life. Individual replacement is usually called "spot replacement"; mass replacement is called "group relamping." The labor costs saved by group relamping in large installations and in many small ones more than compensate for the value of the depreciated lamps that are thrown away before they burn out. Other advantages which always accompany group relamping are more light, fewer work interruptions, better appearance of the lighting system, and less maintenance of auxiliary equipment.

There are five principal advantages for group relamping. The first three apply to all lighting systems, the last two chiefly to fluorescent and HID systems.

1. *Reduced labor costs often mean net savings.* Group relamping saves on labor costs, largely because much of the travel time and setup time required to change lamps individually is eliminated.
2. *More light delivered.* All lamps depreciate in lumens continually as they burn. The earlier they are replaced, the higher the maintained illumination will be without adding to the use and cost of electric energy or the number of luminaires used.
3. *Fewer work interruptions.* Group relamping can be done at a convenient time—during vacation shutdowns or after working hours, for example, when there will be no interruption of operations. The number of interruptions to report burnouts or to replace them is greatly reduced.
4. *Better appearance of the lighting system.* Black ends, color variations, and differences in brightness between adjacent old and new fluorescent lamps are common when spot replacement is used. With group relamping, all the lamps are the same age, and appearance is far more uniform. Since most of the lamps are replaced before they burn out, the distraction of blinking, flashing, or swirling lamps is minimized.
5. *Less maintenance of auxiliary equipment.* Abnormal operating conditions that may occur at the end of lamp life can damage starters and ballasts. When most of the lamps are replaced before they reach the end of life, auxiliary equipment lasts longer.

## RADIANT-ENERGY DETECTORS

***88. Photovoltaic Cells.*** Selenium barrier-layer cells are of the self-generating type, with spectral sensitivity chiefly in the visible and near-ultraviolet regions, as indicated in Fig. 26-43.

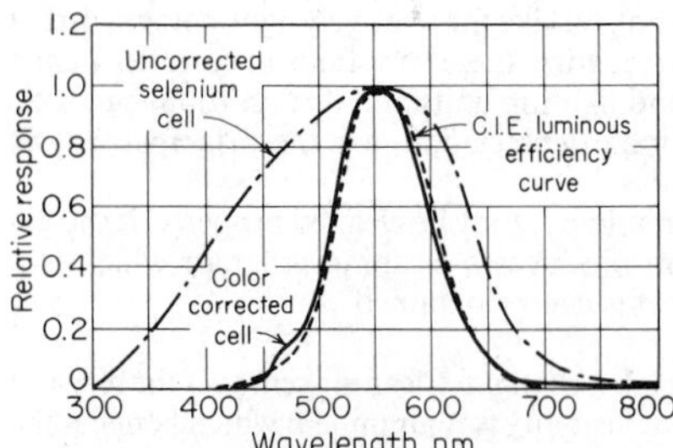


**FIG. 26-43** Average spectral sensitivity characteristics of selenium barrier-layer cells, compared with CIE spectral luminous efficiency curve.

These cells are available in a variety of physical shapes and sizes, hermetically sealed and color-corrected to match the CIE luminous efficiency curve. Photovoltaic cells generate a potential difference, when irradiated, which bears a relationship to the level of illumination. These cells are used in photometers, light meters, color-temperature meters, exposure meters, reflectometers, goniophotometers, glossmeters, fluorometers, colorimeters, refractometers, and pyrometers.

***89. Phototubes.*** Phototubes are photoemissive-type cells consisting of an anode and a cathode mounted in a transparent envelope which is either evacuated or gas-filled. When illuminated, the cathode emits electrons in proportion to the level of the illumination. An external voltage must be applied between the cathode and anode in order to sustain the electron flow from cathode to anode when the phototube is illuminated. Some of the materials used for the cathode in phototube construction are lithium, sodium, potassium, cesium, and rubidium.

The cells may be made in either the high-vacuum or the gas-filled type. The latter usually has greater sensitivity than the vacuum cells but is likely to give higher dark currents (current flow with the photocell shielded from all illumination).

Both high-vacuum and gas-filled photocells are used in photometric measurements, where a galvanometer and battery in series with the cell may suffice when illumination levels are high. For lower levels of illumination, the cells are used with vacuum-tube amplifier circuits in order to obtain high sensitivity. In extreme cases, where illumination levels are very low, it is sometimes necessary to use multiplier tubes, which then make it possible to detect currents as low as $5 \times 10^{-18}$ A, that is, 30 electrons per second.

The use of the multiplier phototube makes it possible to eliminate the vacuum-tube amplifier referred to above. These are high-vacuum photocells which incorporate, within the tube itself, a system for multiplying the photocurrent through secondary emission from successive electrodes (called "dynodes"). Such tubes usually require a total voltage of the order of 1000 V but are capable of delivering current up to 1 mA without further amplification. They are readily adaptable to measurements at extremely low light levels, when used with suitable amplifiers.

Phototubes are used in nephelometers, densitometers, recording microphotometers, color-transparency processing equipment, spectrophotometers, spectroradiometers, and the instruments listed under photovoltaic cells (Par. **88**).

Vacuum photocells usually give a linear response (constant microamperes per lumen) over a wide range of illumination levels, once saturation has been reached. Such is not the case with gas-filled cells, which normally are linear for only small portions of their characteristic curve.

Barrier-layer cells for photometric use should be selected on the basis of linearity of response. Certain cells will be linear over ranges of 100 : 1 with an accuracy of 1% or better. Such cells must be used in a low-resistance circuit in order to take full advantage of their linearity. Special circuits have been devised to compensate for nonlinearity of barrier-layer cells.

## VISUAL PHOTOMETERS

***90. Visual Photometers.*** Because of its power of adaptation to different brightnesses, the eye cannot measure photometric properties with any accuracy. However, it is possible to detect very small differences in photometric brightness (luminance) between two surfaces viewed simultaneously side by side, provided that they have nearly identical chromaticities (color). Visual photometers employ this principle and include a means for adjusting the photometric brightness of one or both fields, until the two appear equally bright.

***91. Lummer-Brodhun Photometer.*** The Lummer-Brodhun cube consists of two identical 45 by 90° prisms, with a pattern etched in the hypotenuse face of one. The two hypotenuse faces are assembled together to make optical contact. Where the surfaces are in optical contact, light is transmitted, thus presenting an opportunity to compare two fields.

*92. Flicker Photometer.* The flicker photometer was developed as a means of comparing two sources of different colors. A single field is alternately illuminated by two sources so that the observer sees a flicker, which may be due to color difference, difference in photometric brightness, or both. Above a certain rate of alternation the color sensations blend, and the observer adjusts the sources to eliminate the remaining flicker due to differences in field brightness. When the flicker disappears, a photometric balance has been reached. Quite accurate determinations of intensities from the light sources of different color characteristics are obtainable with this device.

*93. Macbeth Illuminometer.* The Macbeth illuminometer contains a Lummer-Brodhun cube, of which one field is a view of the test surface to be measured. The other field is lighted by a small comparison lamp mounted on a rack and pinion in a tube. The two fields are balanced by moving the comparison lamp back and forth in the tube. The inverse-square principle determines the result. The illuminometer may be equipped with a lens to restrict the test field and bring it into focus. The range can be extended over very wide limits by inserting calibrated neutral filters in either the test or the comparison lamp side. Colored filters may also be used to correct the color of the comparison lamp to that of a test surface.

*94. Luckiesh-Taylor Luminance Meter.* The Luckiesh-Taylor meter is a small, completely self-contained instrument for measurement of luminance or illumination. A lens in the eyepiece brings the external test field into focus in the same plane as the comparison field, which is seen as two small trapezoids against the circular test field. Light for the comparison field is supplied by a small lamp. The brightness of the comparison field is adjusted to match the test field by rotating a photographic-film gradient. An illuminated scale, calibrated in footlamberts and candles per square inch, is seen through a second eyepiece. Neutral filters greatly extend the range of values that can be measured.

## PHOTOELECTRIC PHOTOMETERS

*95. Photoelectric Photometers.* Photoelectric photometers are convenient to use, eliminate personal judgment and variations among individual observers, and under the best conditions, give very accurate results. Hence, these photometers have largely replaced visual devices for most routine measuring procedures. Photoelectric photometers may be divided into two classes: those employing photovoltaic and photoconductive cells, and those employing photoemissive tubes, which require additional equipment for operation.

*96. Photovoltaic Cell Meters.* Photovoltaic cells generate a voltage when radiant energy strikes the sensitive surface. However, the cells depart from linearity of response as the resistance of the circuit to which they are connected increases, and for precise results a current-balancing measuring instrument giving zero external resistance is necessary. The great majority of portable illuminance meters in use today consist of a barrier-layer cell, or cells, connected to a meter calibrated directly in footcandles.

Barrier-layer cells have a number of limitations which must be taken into consideration:

1. The spectral response is different from that of the human eye; consequently, color-correcting filters should be used.
2. Part of the light incident on the cell at wide angles from the normal is reflected from the cell surface and cover glass, causing the response to be as much as 25% low. Corrective cover glasses giving so-called cosine correction are used with these cells to measure illuminance.
3. The response of the cell varies with temperatures above and below 77°F. Prolonged temperatures above 120°F will permanently damage the cell.
4. The cell response when exposed to constant illuminance decreases slightly over a period of time before reaching its final value. The cell should be well fatigued before a series of measurements are begun.
5. The microammeters used in portable meters are subject to certain inherent limitations of accuracy. Manufacturing tolerances alone may result in an overall uncertainty of reading at

any point above one-quarter full scale of about ±8%. At lower scale readings, the percentage of error may be greater. The microammeter and cell should be checked frequently against a source of known illuminance.

6. When gaseous discharge sources are operated at frequencies above 60 Hz, precautions should be taken with regard to the effect of frequency on cell response.

***97. Freund Brightness Spot Meter.*** This is a photoelectric photometer for measuring the luminance of small areas. Models available cover ¼, ½, or 1½° field of view. A beam splitter allows a portion of the light from the objective lens to reach a reticule viewed by the eyepiece. The remainder of the light is reflected onto the field operative in front of the photomultiplier tube, and the output of the tube after amplification is read on a microammeter with a scale calibrated in footlamberts. One of the filters provided with the instrument approximately corrects the response of the photomultiplier to the standard spectral luminous efficiency curve. Full-scale deflection is produced by $10^{-2}$ to $10^{6}$ fL.

***98. Pritchard Photometer.*** This is a high-sensitivity precision photomultiplier photometer with interchangeable field apertures covering fields from 2 arc minutes to 3° in diameter. Full-scale sensitivity ranges are from $10^{-5}$ to $10^{7}$ fL. The readings of the light being measured are free from the effects of polarization, since there are no internal reflections of the beam. The spectral response of each photometer is individually measured and the filters to best match it to the standard spectral luminous efficiency curve are determined and inserted. Filters are also included to permit evaluation of polarization and color factors.

***99. Bibliography.*** References on illumination may be found in texts on the subject of illuminating engineering and in appropriate journals. A few are listed below.

*Texts*

Allphin, W.: *Primer of Lamps and Lighting,* 3d ed.; Reading, Mass., Addison-Wesley Publishing Company, 1973.

Amick, C. L.: *Fluorescent Lighting Manual,* 3d ed.; New York, McGraw-Hill Book Company, 1960.

Barrows, W. E.: *Light, Photometry, and Illuminating Engineering,* 3d ed.; New York, McGraw-Hill Book Company, 1951.

Boast, W. B.: *Illumination Engineering,* 2d ed.; New York, McGraw-Hill Book Company, 1953.

Flynn, J. E., and Mills, S. M.: *Architectural Lighting Graphics;* New York, Reinhold Publishing Corporation, 1962.

Hardy, A. C.: *Handbook of Colorimetry;* Cambridge, Mass., The Technology Press.

*IES Lighting Handbook;* New York, Illuminating Engineering Society of North America.

Jolly, L. B. W., Waldram, J. M., and Wilson, G. H.: *Theory and Design of Illuminating Engineering Equipment;* New York, John Wiley & Sons, Inc., 1931.

Kraehenbuehl, J. O.: *Electrical Illumination;* New York, John Wiley & Sons, Inc.

Moon, P.: *Scientific Basis of Illuminating Engineering;* New York, McGraw-Hill Book Company, 1936; Dover Publications, Inc., 1961.

Sharp, H. M.: *Introduction to Illumination;* Englewood Cliffs, N.J., Prentice-Hall, Inc., 1951.

Walsh, J. W. T.: *Photometry;* Princeton, N.J., D. Van Nostrand Company, Inc.; New York, Dover Publications, Inc., 1958.

Waymouth, J. F.: *Electric Discharge Lamps;* Cambridge, Mass., The MIT Press, 1971.

Wyszecki, G., and Stiles, W. S.: *Color Science;* New York, John Wiley & Sons, Inc., 1967.

*Journals*

*Lighting Design & Application.**

*Journal of the Illuminating Engineering Society.**

*Lighting Research & Technology.*†

---

* Formerly *Illuminating Engineering* and *Transactions of the Illuminating Engineering Society.*
† Formerly *Transactions of the IES* (Great Britain).

*J. Opt. Soc. Am.*
*Light Magazine.*
*Lighting Handbook* (Westinghouse Electric Corp., Bloomfield, N.J.).
*Light and Lighting* (Great Britain).
*Proc. Intern. Comm. on Illumination (Comm. Intern. de l'Eclairage).*
*International Lighting Review.*

# SECTION 27

# LIGHTNING AND SURGE PROTECTION

**Wendell Neugebauer**
*Project Manager, Electric Utility Systems Engineering Department, General Electric Company; Member, IEEE Overvoltage Protective Devices Subcommittee; Member, IEEE Working Groups on Metal Oxide Surge Arresters, Arrester Protection and Coordination of Transformer Insulation, High-Voltage System Arrester Application Guide, and Surge Protection of High-Voltage Cable-Connected Equipment.*

## CONTENTS

*Numbers refer to paragraphs*

## FUNDAMENTALS

***1. Surge arresters*** are applied on electric power systems to protect apparatus such as transformers, rotating machines, etc., against the effects of overvoltages resulting from lightning, switching surges, or other disturbances. Without such protection, flashovers and serious damage to equipment may occur with the result that the power supply to users may be jeopardized.

A surge arrester limits the magnitude of the voltage on the apparatus it is protecting by diverting current through itself. This current, in turn, causes a voltage drop across the source surge impedance, thus limiting the voltage at the equipment. If $E_S$ is the surge voltage, $V_A$ the voltage that the arrester will hold, and $Z$ the surge impedance, then the arrester current $I$ is given by

$$I = (E_S - V_A)/Z \tag{27-1}$$

The voltage $V_A$ is known as the surge arrester's protective level, while the current $I$ is the discharge current. The surge voltage $E_S$ is the voltage that would be present on the system when the arrester is not connected to the circuit.

***2. The protective level*** of a surge arrester is matched with the insulation withstand of the equipment it is protecting by allowing for a safety factor known as the protective ratio.[1]* The withstand levels for typical oil-filled transformers depend on the surge voltage time-to-crest and duration, as shown in Fig. 27-1. The protective level of arresters also depends on the surge

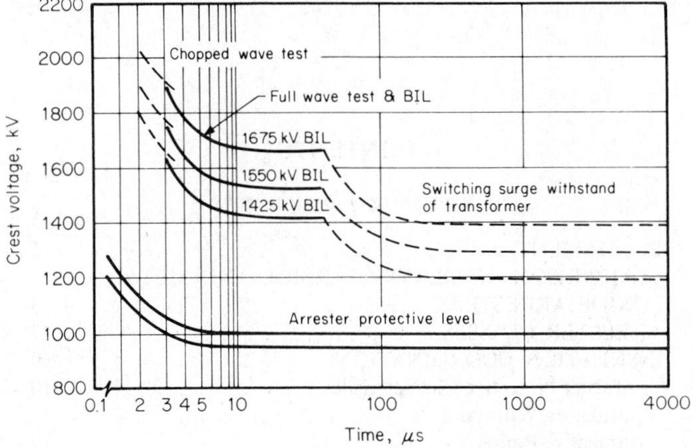


**FIG. 27-1** Transformer insulation withstand and arrester protective level.

waveform time-to-crest and is higher for the faster rising waves. The protective ratio is a measure of the difference between the insulation withstand and the arrester protective levels and is discussed further in the section on insulation coordination.

The energy $W$ that a surge arrester must dissipate as it performs its voltage-limiting function is

$$W = \int vi\,dt \tag{27-2}$$

where $v$ and $i$ are the instantaneous arrester voltage and current, respectively. The proper application of arresters depends on this energy, and different classes of arresters, having different energy capabilities, are manufactured to meet these needs.

* Superior numbers refer to References, Par. **35**.

## COMPARISON OF SILICON CARBIDE AND METAL OXIDE ARRESTERS

***3. Silicon Carbide Arresters.*** Before the advent of metal oxide arresters in the 1970s, the most common surge arresters were constructed with nonlinear resistors made of bonded silicon carbide placed in series with gaps.[2,3] The function of these series gaps was to isolate the resistors from the normal, steady-state system voltage. This was necessary to avoid overheating the resistors. Elaborate designs were developed for the gaps in order to ensure a consistent sparkover level and positive clearing (resealing) after the surge had passed. Such gaps were known as current-limiting gaps, an example of which is shown in Fig. 27-2. The sparkover of the main gap was initiated by either a trigger gap or a preionizer, neither of which was exposed to the main surge current. This prevented damage to the trigger mechanism so that the sparkover levels remained reasonably constant throughout the life of the arrester. Once the main gaps sparked over, a coil carrying the surge current produced a magnetic field which forced the arc into a serrated tooth chamber which cooled and stretched the arc so that it could easily be cleared after the surge had passed. The elongated arc also exhibited a significant voltage drop which, when added to the voltage drop across the nonlinear resistor, resulted in the switching discharge protective levels associated with such arresters. In the case of lightning surges, the discharge protective levels were essentially equal to the voltage drop across the nonlinear resistor because the arc elongation process took in the order of 400 $\mu$s, which is much longer than the typical lightning surge. For arresters with series gaps, the protective levels were the higher of the sparkover voltages or the subsequent discharge voltages.

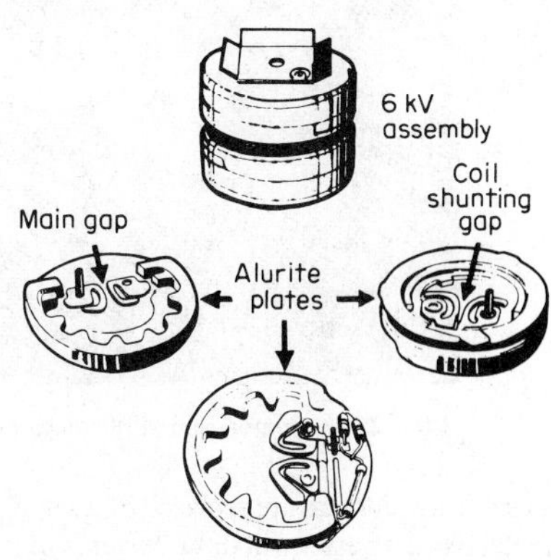


**FIG. 27-2** Current-limiting station arrester gap.

***4. Metal Oxide Arresters.*** Surge arresters made with metal oxide elements generally do not require series gaps[4] to isolate the elements from the steady-state voltages because the material, usually zinc oxide, is much more nonlinear than silicon carbide, as may be seen in Fig. 27-3.

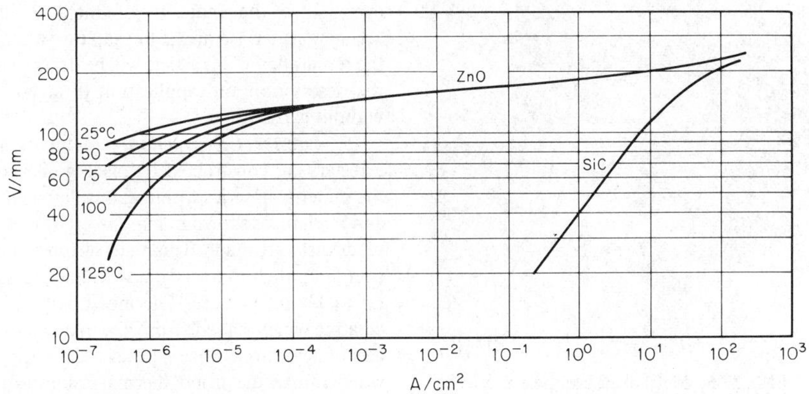


**FIG. 27-3** Normalized volt-ampere characteristic of zinc oxide and silicon carbide valve elements.

This characteristic results in negligible current through the elements when normal voltage is applied and leads to a much simpler arrester design. Figure 27-4 shows the parts of a silicon carbide arrester that are replaced by a set of zinc oxide disks. At the normal operating voltage, a metal oxide arrester will draw currents in the order of a milliampere from the system and will therefore not exhibit a significant temperature rise. A typical temperature rise is only a few degrees Celsius above ambient in the steady-state condition. Some metal oxide arresters are manufactured with series gaps for applications which require the special advantages of such gaps. Other arresters are manufactured with shunt gaps to further improve their nonlinearity. The function of a shunt gap is to short-circuit approximately 10% of an arrester's elements whenever the discharge current exceeds several hundred amperes. However, since the majority of metal oxide arresters in service are of the gapless variety, the remainder of this article will be devoted to the description and application of arresters without gaps.

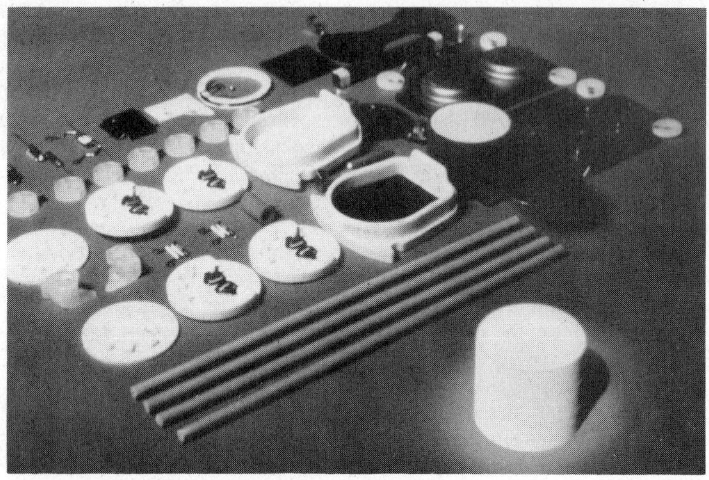

**FIG. 27-4** Comparison of silicon carbide and metal oxide arrester parts.

**FIG. 27-5** Metal oxide riser pole arrester.

***5. Arrester Construction.*** Metal oxide arresters are constructed by placing disks of the material in series within an insulator such as porcelain, as shown in Fig. 27-5. The metal oxide disks are made in diameters from a fraction of an inch to over 3 in, with thicknesses up to 1⅜ in or more. The metal oxide is a ceramic which is made from zinc oxide particles of a controlled size and several additives which impart the proper thermal stability and other electrical characteristics to the material. The sintering and heat-treatment processes to which the disks are subjected during manufacture are also critical in determining the final characteristics. A conducting layer, generally aluminum, is applied to the flat faces of each disk to ensure a proper contact and a uniform current distribution within the disk. On station, intermediate, and some distribution class arresters, the insulating housing is provided with end caps containing a pressure relief mechanism which can safely vent the arrester should an internal arc occur. An arrester is generally sealed to prevent external contaminants and moisture from entering the insulating housing and causing unwanted flashovers.

As in the case of silicon carbide arresters, metal oxide arresters are designed to have a nearly uniform voltage gradient along the length of the column.[5,6] For the higher ratings, above 150 kV approximately, a grading ring is mounted on the upper end of the arrester to achieve this uniform gradient. Such construction is shown for a 588-kV-rated arrester in Fig. 27-6. To avoid

**FIG. 27-6** Metal oxide arrester rated 588 kV showing design of grading rings.

disturbing the uniform gradient, the manufacturer's recommendation on minimum clearances to ground planes or other conductors should be observed when the arrester is installed. Figure 27-7 shows an arrester rated 588 kV which is constructed with five parallel columns of 3-in-diameter zinc oxide disks to discharge extremely high energy surges. In such a multicolumn design, the individual stacks of metal oxide must be matched to ensure proper sharing of the surge discharge energy.

***6. Arrester Field Testing.*** In general, it is impractical to fully test metal oxide arresters in the field. Surge counters and leakage current measurement devices are available so that some

**FIG. 27-7** 588-kV-rated, multicolumn arrester for application on an 800-kV system.

indication of an arrester's history can be obtained. Typical arrester leakage currents of station-class arresters operating at their continuous voltage capability are in the approximate range of 0.5 to 3 mA. A leakage current test may possibly detect the presence of moisture within an arrester. If this is the case, the arrester should be replaced.

## ARRESTER CLASSES

***7. In the United States, surge arresters are divided into three general classes[1]:*** (1) station arresters, which provide the best (lowest) available protective levels and which are capable of discharging the most energy; (2) intermediate arresters, which have somewhat higher protective characteristics and lower energy discharge capability than station arresters; and (3) distribution arresters, which provide the highest protective levels and the lowest energy dissipation capability. Table 27-1 indicates the range of rating, the protective levels, and the discharge capabilities of the three classes of arresters.

***8. Rating.*** Arrester rating is established by means of the duty cycle test[7] in which an arrester is subjected to an ac rms voltage equal to its rating for 24 min. During this time it is also exposed to lightning surges at 1-min intervals. The magnitude of these surges is 10 kA for station class arresters and 5 kA for intermediate or distribution class arresters. These lightning current pulses are timed to occur at points 15 electrical degrees apart on the power frequency wave, and have

**TABLE 27-1** Typical Metal Oxide Arrester Characteristics*

(Protective Levels in Per Unit of Crest of Rating)

| Arrester Class | Station | Station | Station | Intermediate | Distribution |
|---|---|---|---|---|---|
| Duty cycle rating, kV rms | 2.7–48 | 54–360 | 396 & above | 3–120 | 3–36 |
| MCOV† (per unit of rms rating) | 0.81–0.85 | 0.80–0.82 | 0.80–0.81 | 0.81–0.85 | 0.81–0.85 |
| Front-of-wave protective level | 2.1–2.5 | 1.74–2.03 | 1.91–1.98 | 2.2–2.5 | 2.54–2.64 |
| Front-of-wave test current—crest kA | 10 | 10 | 15–20 | 10 | 10 |
| Equivalent front-of-wave voltage time to crest, μs | 0.5 | 0.5 | 0.5 | 0.5 | 0.5–1.0 |
| Lightning protective level—Maximum discharge voltage using an 8 × 20 μs current wave of crest magnitude: 10 kA | 1.8–2.1 | 1.59–1.75 | 1.58–1.60 | 1.9–2.1 | 2.25–2.36 |
| 20 kA | 2.1–2.5 | 1.73–1.91 | 1.69–1.71 | 2.2–2.6 | 2.62–2.71 |
| 40 kA | 2.4–3.0 | 1.89–2.17 | 1.87–1.89 | 2.8–3.0 | 3.0–3.15 |
| Switching surge protective level | 1.4–1.7 | 1.39–1.54 | 1.37–1.42 | 1.4–1.5 | 1.6–1.9 |
| Current used to establish switching surge protective level crest amperes 45/90 | 500 | 3000 | 3000 | 500 | 500 |
| Energy capability, kJ/kV of rms rating | 4.0–4.2 | 7.2–7.8 | 13.0–13.1 | 2.7–2.9 | 1.75–1.9 |
| Transmission line discharge overhead line, mi | 150 | 150–200 | 200 | 100 | Not specified |
| High-current withstand crest, kA | 65 | 65 | 65 | 65 | 65 |
| Pressure relief, symmetrical, kA‡ | 25–65 | 25–63 | 40–63 | 16.1 | Not specified |

* From manufacturers' published data.
† MCOV = maximum continuous operating voltage.
‡ Higher levels by request.

an 8/20 waveshape. By 8/20 is meant a current wave which reaches crest in 8 $\mu$s and diminishes to half the crest value in 20 $\mu$s. Formerly such a wave was denoted as an $8 \times 20$ $\mu$s wave. The rating of an arrester should be clearly distinguished from its maximum continuous operating voltage (MCOV) level, which is generally between 80 and 90% of the rating.

## INSULATION COORDINATION

*9.* ***Insulation coordination*** is the process of bringing the insulation strengths of electrical equipment into the proper relationship with expected overvoltages and with the characteristics of surge protective devices.[8] After an arrester rating has been selected as described in "Selection of Arresters" below, the next step is to determine the insulation protection that the arrester will provide. This protection is dependent on the protective characteristics of the arrester, the lightning and switching surges expected on the system, and the insulation characteristics of the protected equipment. It is normally quantified in terms of the protective ratio, which is the ratio of the insulation withstand to the protective level:

$$\text{Protective ratio} = \frac{\text{insulation withstand}}{\text{arrester protective level}} \tag{27-3}$$

An alternative measure is the percent protective margin:

$$\text{Percent protective margin} = 100(\text{protective ratio} - 1) \tag{27-4}$$

### Arrester Protective Characteristics

*10.* ***The protective characteristics of gapless metal oxide arresters*** are defined by the discharge voltages appearing across the surge arrester for various discharge current waves as defined by industry standards.[7] For any particular metal oxide arrester being tested with a specific current wave, the discharge voltage is a function of the magnitude of the arrester current, and, in the impulse region, of the time to crest of the arrester current. The complete protective characteristics for some typical metal oxide arresters are given in Fig. 27-8, which

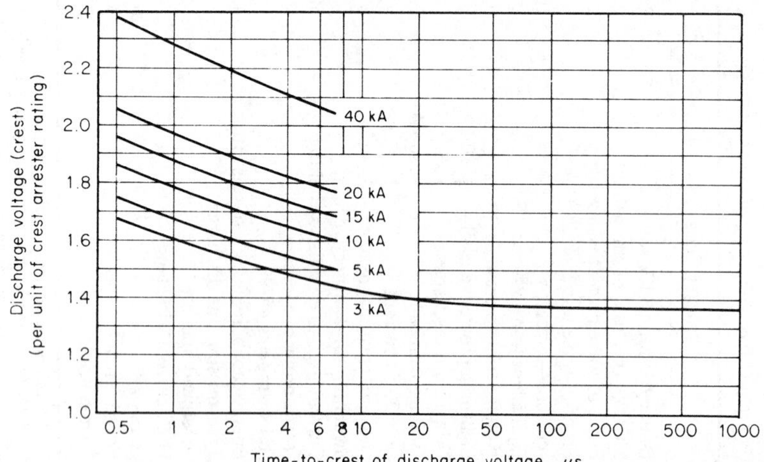


**FIG. 27-8** Typical protective characteristic for station-class arrester ratings 45 to 360 kV. The 3-kA curve is extended to show the switching surge protective level.

shows a family of crest discharge voltage curves. These curves are generated by applying current waves of various magnitudes and times-to-crest to the arrester.

## Insulation Withstand

*11. **The insulation withstand*** of equipment is usually defined at three points through the use of the standard switching surge and the full-wave [basic impulse insulation level (BIL)] and the chopped-wave tests.[9] To facilitate comparison with these three withstands, three corresponding points on the metal oxide arrester characteristic have been selected as indicated in Table 27-2.

**TABLE 27-2** Transformer Protective Characteristics

| Transformer insulation withstands | | Corresponding specific arrester protective characteristics |
|---|---|---|
| Switching surge | ⟷ | Switching surge |
| BIL | ⟷ | Impulse discharge voltage (8 × 20 μs current wave) |
| Chopped-wave | ⟷ | Equivalent front-of-wave |

The three protective levels are specific points on the total protective characteristic of an arrester and are defined as follows:

***12. Switching-Surge Protective Level.*** The switching-surge discharge voltage of a metal oxide arrester increases with increasing current. To define the arrester's switching-surge protective level, a "switching-surge coordination current" must be selected for the various voltage classes of arrester. Conservative estimates for such currents in modern power systems are

| System line-to-line voltage, kV | Switching-surge coordination current, A |
|---|---|
| 3–150 | 500 |
| 151–325 | 1000 |
| Above 325 | 2000 |

For the typical metal oxide arrester characteristics given in Table 27-1, crest currents of 500 and 3000 A are used to determine the indicated switching surge protective levels.

***13. Impulse Discharge Voltage (Lightning Protective Level).*** This is the crest discharge voltage that results when an 8/20 current impulse is forced through the arrester. The resultant crest voltages for a variety of crest currents are given in Table 27-1. These values are given on the continuous characteristic in Fig. 27-8 at a voltage time-to-crest of 7 μs, which is approximately the time to voltage crest for metal oxide valve elements when an 8/20 current wave is applied. The resulting voltage wave on the valve elements approximates the 1.2/50 wave used in insulation testing.

***14. To allow coordination with transformer insulation,*** a specific current impulse magnitude must be selected. The choice of discharge current to be used in insulation coordination depends on whether or not the installation is shielded against direct strokes. In shielded stations, the incoming lines must also be shielded for at least ½ mi from the station. With such shielding, the voltage magnitude of the surge at the start of the protected zone is usually assumed to be 1.2 times the negative critical flashover (CFO) of the line. Making the conservative assumption of zero attenuation during wave travel from the start of the shielding to the station, the maximum discharge current can be calculated as

$$\text{Arrester current} = \frac{2.4 \times \text{line CFO} - \text{arrester discharge voltage}}{\text{line surge impedance}} \tag{27-5}$$

Conservative estimates[1] of crest arrester currents in shielded stations are

| Maximum system voltage, kV | Surge current, A |
|---|---|
| 72.5 | 5000 |
| 121 | 5000 |
| 145 | 5000 |
| 242 | 10,000 |
| 362 | 10,000 |
| 550 | 15,000 |
| 800 | 20,000 |

***15. In nonshielded installations*** the lightning stroke current will follow several paths to ground according to the system impedances. Experience[10] indicates that arrester current estimates of 25 to 40 kA are very conservative, but 10 kA is commonly used for coordination, recognizing that there is some increased risk of equipment failure.

***16. Equivalent Front-of-Wave Protective Level.*** This is the arrester discharge voltage for current pulses having a time-to-crest of approximately 0.6 $\mu$s. This current wave results in a voltage wave across the arrester cresting in 0.5 $\mu$s. The resultant crest voltage is listed in Table 27-1 as the front-of-wave protective level for all the arrester ratings.

***17. Effect of Lead Length.*** In unshielded locations in particular, a significant voltage may be produced by the current rate of rise in the arrester lead. This is the length of lead, through which the discharge current must pass, that is in shunt with the line-to-ground insulation of the equipment. A normally accepted lead inductance is 1.0 $\mu$H/m. A rate of rise of current frequently used is 4000 A/$\mu$s. Because of the $L\ di/dt$ voltage drop in arrester leads, arrester connections should be made as short as possible; in fact, arresters are often mounted directly on equipment, distribution equipment in particular, to reduce the lead length to a minimum.

***18. Separation Effects.*** The maximum surge voltage that may appear at a location on the power system remote from an arrester is generally higher than the arrester's protective level. This rise in voltage is caused by the reflections of traveling waves on the conductors of the system and may be quite significant for lightning or fast-rising switching surges. Methods for determining the effects of separation between the arrester and the protected equipment, and the effects of multiple lines entering stations, are given in Ref. 1. Alternatively, digital simulation methods[11] have proved to be very useful in computing the voltage rise in complicated systems.

## Protective Ratios

***19. The insulation withstand*** of electrical equipment is generally quantified in terms of its basic impulse insulation level (BIL), which is based on tests[9] made with a voltage surge rising to crest in 1.2 μs and falling to half value in 50 μs, which is denoted as a 1.2/50 wave. Other equipment tests determine the chopped-wave withstand and the switching-surge withstand levels. Typically, for oil-filled equipment, the latter levels are related to the BIL as follows:

$$\text{Chopped-wave withstand/BIL} = 1.10 \text{ to } 1.15$$

$$\text{Switching-surge withstand/BIL} = 0.83$$

The manufacturer of the equipment should, however, be consulted for the exact insulation withstands.

The three-point method[1] is usually applied for insulation coordination. The minimum protective ratios recommended by ANSI and IEEE standards for satisfactory coordination are

$$\frac{\text{Switching-surge withstand}}{\text{Switching-surge protective level}} >= 1.15 \qquad (27\text{-}6)$$

$$\frac{\text{BIL}}{\text{Lightning protective level}} >= 1.20 \qquad (27\text{-}7)$$

$$\frac{\text{Chopped-wave withstand}}{\text{Front-of-wave protective level}} >= 1.20 \qquad (27\text{-}8)$$

The minimum protective ratios are usually exceeded by a considerable amount in actual power system applications. Utility system experience may dictate the use of higher minima for the protective ratios.

***20. Protection of Cable-Connected Equipment.*** Many of the problems associated with the protection of cable-connected equipment result from the practical difficulties of locating surge arresters at the equipment to be protected. If arresters can be installed at the individual equipment locations, application procedures similar to those already described may be used. If it is not possible to install arresters at the protected equipment, the protection available from an arrester at the overhead line-cable junction must be used to protect the cable-connected equipment as well as the cable.

Procedures[12,13,14] have been recommended to estimate the magnitudes of voltages at equipment locations as a function of cable length, velocity of surge propagation, cable surge impedance, the front time of the incoming surge, and the characteristics of the arrester at the overhead line-cable junction, including the effects of arrester lead length between the cable conductor and grounded sheath. Digital simulation methods[11] are useful for determining the protection afforded by arresters at the line-cable junction.

***21. Protection of Dry-type-insulated Equipment.*** The impulse-withstand strength of dry-type insulation for short-duration impulses is usually considered to be the same as the full-wave insulation strength (BIL). Procedures for protection are similar to those used for protection of oil-filled apparatus except that the front-of-wave protective level of the arrester is compared to the BIL of the equipment using a protective ratio of 1.20.

***22. Practices for the protection of rotating machines***[15–18] are similar to the practices used for the protection of other apparatus employing dry-type insulation with the additional requirement that the rate of rise of voltage applied to the machine must be reduced in some cases to reduce turn-to-turn stresses. Reduction in the rate of rise is accomplished by installing capacitors from phase to ground at the machine terminals. The magnitude of the voltage is limited by installing an arrester at the machine terminals. In addition, arresters are often installed on the overhead line 1500 to 2000 ft ahead of the machine location to limit the charging rate of the capacitors.

## SELECTION OF ARRESTERS

*23. The basic objective of arrester selection* is to determine the minimum rated arrester that will have a satisfactory service life on the power system. An arrester of the minimum practical rating is generally preferred because it provides the greatest margin of protection for the insulation. Use of a higher rating increases the capability of the arrester to survive on the power system.

*24. Table 27-3 lists arrester ratings* that would normally be applied on systems of various line-to-line voltages. To decide which rating is most appropriate for a particular application, consideration must be given to the following system stresses to which the arrester will be exposed:

1. Continuous system voltage
2. Temporary power frequency overvoltages
3. Switching surges
4. Lightning surges
5. Short-circuit current

The arrester finally selected should have sufficient capability to meet the anticipated service requirements for all these stresses as outlined below.

### Continuous System Voltage

*25. Arresters in service* are continuously exposed to a normal dc or power frequency voltage. For each arrester rating, there is a recommended limit to the magnitude of the voltage that may be continuously applied. This has been termed the "maximum continuous operating voltage" (MCOV) of the arrester. Table 27-1 shows typical MCOVs of metal oxide arresters. For these arresters, operation at voltages exceeding the continuous capability will cause the metal oxide elements to operate at higher than normal temperatures, which may lead to premature arrester failure or a decreased useful life.[19]

The arrester rating must be selected so that the continuous power system voltage applied to the arrester is less than, or equal to, its continuous voltage capability. Experience[20] over many years of arrester use indicates that the rating of an arrester should be at least 1.25 times the MCOV level in typical installations. It is for this reason that the MCOV levels as given in Table 27-1 are approximately 0.8 times the arrester rating.

### Temporary Power Frequency Overvoltages

*26. Temporary power frequency overvoltages* can be caused by line-to-ground faults, circuit backfeeding, load rejection, ferroresonance, and other system contingencies. These overvoltages depend strongly on the details of the system grounding.

*27. The permissible duration of such overvoltages* on typical arresters is given in Fig. 27-9. This duration is a function of the magnitude of the overvoltage, the ambient temperature, and the energy that may have been discharged in the arrester at the beginning of the overvoltage period. This initial energy would have raised the temperature of the metal oxide elements at the outset of the overvoltage period and would, therefore, allow a shorter permissible overvoltage duration. This initial energy may be due to the discharge associated with a switching surge. The heat generated within the disks when an electrical surge is discharged is transmitted to the

**TABLE 27-3** Arrester Ratings For Various System Voltages

| Nominal system L-L* voltage, kV | Arrester rating, kV: Grounded neutral circuits | Arrester rating, kV: High-impedance grounded, ungrounded, or temporarily ungrounded circuits |
|---|---|---|
| 2.4 | 2.7 | 2.7 |
| | — | 3.0 |
| 4.16 | 3.0 | — |
| | 4.5 | 4.5 |
| | — | 5.1 |
| 4.80 | 4.5 | — |
| | 5.1 | 5.1 |
| | — | 6.0 |
| 6.9 | 6.0 | — |
| | — | 7.5 |
| | — | 8.5 |
| 12.47 | 9.0 | — |
| | 10.0 | — |
| | — | 12 |
| | — | 15 |
| 13.2, 13.8 | 10.0 | — |
| | 12.0 | — |
| | — | 15 |
| | — | 18 |
| 23.0, 24.94 | 18 | — |
| | 21 | — |
| | 24 | 24 |
| | — | 27 |
| 34.5 | 27 | — |
| | 30 | — |
| | — | 36 |
| | — | 39 |
| 46 | 39 | — |
| | — | 48 |
| 69 | 54 | — |
| | 60 | — |
| | — | 66 |
| | — | 72 |

| Nominal system L-L voltage, kV | Arrester rating, kV: Grounded neutral circuits | Arrester rating, kV: High-impedance grounded, ungrounded, or temporarily ungrounded circuits |
|---|---|---|
| 115 | 90 | — |
| | 96 | — |
| | 108 | 108 |
| | — | 120 |
| 138 | 108 | — |
| | 120 | — |
| | — | 132 |
| | — | 144 |
| 161 | 120 | — |
| | 132 | — |
| | 144 | 144 |
| | — | 168 |
| 230 | 172 | — |
| | 180 | — |
| | 192 | — |
| | — | 228 |
| | — | 240 |
| 345 | 258 | — |
| | 264 | — |
| | 276 | — |
| | 288 | 288 |
| | 294 | 294 |
| | 300 | 300 |
| | 312 | 312 |
| 400 | 300 | — |
| | 312 | — |
| | 336 | — |
| | 360 | — |
| 525 | 396 | — |
| | 420 | — |
| | 444 | — |
| 765 | 588 | — |

* L-L = line-to-line voltages.

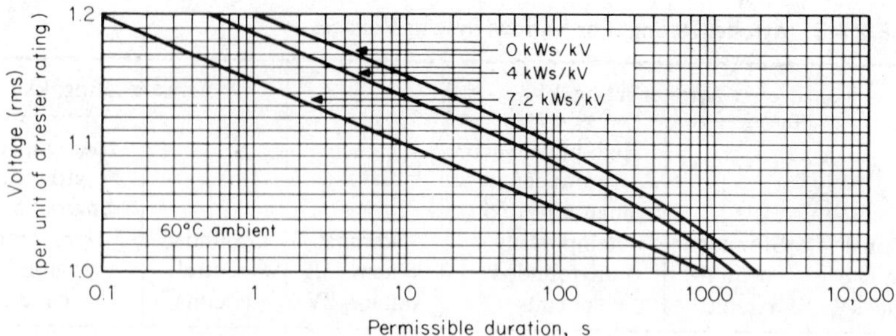


**FIG. 27-9** Typical temporary power frequency overvoltage capability of station-class arrester ratings 2.7 to 360 kV for various initial absorbed energies.

environment by means of conduction and convection. Some arresters depend on convection for transferring the heat from the disk to the porcelain, while in others, a deliberate heat-conducting path of silicon rubber is provided between the disks and the housing.[21] The time constant for cooling a station-class arrester is on the order of 90 min or more, depending on the method of heat transfer.

## Switching Surges

***28. The ability of a metal oxide arrester to dissipate switching surges*** can be quantified to a large degree in terms of energy.[4] The units used in quantifying the energy capability of metal oxide arresters are kilojoules per kilovolt of rating (kJ/kV).

***29. The maximum amount of energy*** that may be dissipated in metal oxide arresters of various types and ratings is given in Table 27-1. These energy capabilities apply for switching surges that occur in a system having surge impedances of several hundred ohms, as would be typical for overhead transmission circuits. The arrester current can be calculated by means of Eq. (27-1), and the corresponding energy is found by evaluating the integral of Eq. (27-2). Because of the increased magnitude of the currents in low-impedance circuits having cables or shunt capacitors as elements, the energy capability of metal oxide arresters used in such applications must be derated to some extent. The manufacturer of the arrester should be consulted for application guidance whenever low-surge impedance circuits are involved.

***30. The actual amount of energy discharged*** in a metal oxide arrester during a system switching surge is a complex function of both the arrester and the details of the system. The energy likely to be discharged can best be determined on a transient network analyzer (TNA) or with a digital circuit analysis program such as EMTP[11] (Electro-Magnetic Transients Program), where system and arrester details can be represented to a sufficient degree of accuracy.

## Lightning Surges

***31. Generally only a fraction of any lightning stroke current ever reaches an arrester*** because the lightning charge will travel in two directions if a transmission line is struck. Moreover, if the surge is great enough and there are transmission line towers intervening between the arrester and the struck point, the line insulators will probably flash over and limit the surge coming into the substation from the transmission system to at most 1.2 times the critical flashover (CFO-lightning) of the insulators. The energy dissipated in an arrester due to traveling lightning surges on transmission lines is generally small compared with the arrester's capability.

## Short-Circuit Current

*32. **Arresters equipped with a pressure relief mechanism*** are intended to fail nonviolently should an internal fault occur. An internal arc in the porcelain housing is transferred to the outside by venting the arrester through pressure-initiated openings in the end caps. The pressure relief capability of an arrester is a measure of the maximum short-circuit current that it can pass without violent failure.[7] Table 27-1 lists the pressure relief capability of the various classes of arresters. Once an arrester has safely vented, it may no longer possess its pressure relief capability and could shatter violently when subjected to fault current a second time. An arrester that has vented should be replaced immediately.

The arrester class to be selected for a given application should have a pressure relief capability greater than the maximum short-circuit current available at the intended arrester location.

## ARRESTER MODELING

*33. **Metal oxide arresters*** lend themselves readily to modeling for system studies. Typical voltage-current curves for metal oxide arresters are given in Fig. 27-10 for three different test

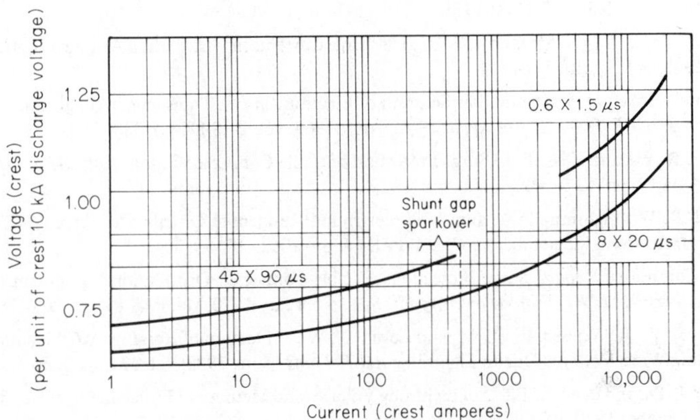


**FIG. 27-10** Typical maximum voltage-current characteristic for arresters constructed with 3-in-diameter metal oxide disks. For gapless arresters, do not use the upper 45/90 curve.

waves. The curves are normalized to the arrester's discharge voltage for an 8/20 current wave cresting at 10 kA. The upper 45/90 curve applies for metal oxide arresters made with shunt gaps where approximately 10 percent of the arrester's active length is shunted for currents exceeding several hundred amperes. For gapless arresters, the upper 45/90 μs curve does not apply, and only the lower curve should be used. These curves are generated by applying 0.6/1.5, 8/20, and 45/90 current waves to the arrester and measuring the resultant crest voltages.

*34. **In insulation coordination studies*** involving ordinary switching surges, the 45/90 characteristics should be used for modeling the arrester. For lightning surge or fast switching-surge studies, a conservative choice would be the use of the 0.6/1.5 characteristic because it yields the highest calculated voltages. Depending on the actual waveforms, however, the 8/20 characteristic may be an appropriate choice. When the purpose of the study is to determine discharge energy, a conservative choice would be a lower volt-ampere characteristic if the system surge impedance, as seen at the arrester location, is lower than the effective arrester impedance.

### *35. References*


1. *American National Standards Institute: Guide for the Application of Valve-Type Surge Arresters for Alternating Current Systems;* ANSI C62.2-1981.
2. Sakshaug, E. C.: Current Limiting Gap Arresters—Some Fundamental Considerations; *IEEE Trans. Power Appar. Syst.,* 1971, vol. PAS-90, no. 4, pp. 1563–1573.
3. Johnson, I. B., et al.: Surge Protection in Power Systems; *IEEE Tutorial Course,* Text 79 EH0144-6-PWR, 1978.
4. Sakshaug, E. C., Kresge, J. S., and Miske, S. A., Jr.: A New Concept in Station Arrester Design; *IEEE Trans. Power Appar. Syst.,* Mar/Apr 1977, vol. PAS-96, no. 2, pp. 647–656.
5. Oyama, M., Ohshima, I., Honda, M., Yamashita, M., and Kojima, S.: Analytical and Experimental Approach to the Voltage Distribution on Gapless Zinc-Oxide Surge Arresters; *IEEE Trans. Power Appar. Syst.,* Nov. 1981, vol. PAS-100, no. 11, pp. 4621–4627.
6. Csendes, Z. J., and Hamann, J. R.,: Surge Arrester Voltage Distribution Analysis by the Finite Element Method; *IEEE Trans. Power Appar. Syst.,* April 1981, vol. PAS-100, no. 4, pp. 1806–1813.
7. *American National Standards Institute: IEEE Standard for Surge Arresters for AC Power Circuits,* ANSI C62.1-1981.
8. *American National Standards Institute: Standard for Power Systems—Insulation Coordination,* ANSI C92.1-1982.
9. *American National Standards Institute: General Requirements for Distribution, Power, and Regulating Transformers;* ANSI C57.12.00-1973.
10. McEachron, K. B., and McMorris, W. A.: Discharge Currents in Distribution Arresters II; *AIEE Trans.,* June 1938, vol. 57, pp. 307–314.
11. Dommel, H. W.: Digital Computer Solution of Electromagnetic Transients in Single and Multiphase Networks; *IEEE Trans. Power Appar. Syst.,* April 1969, vol. PAS-88, pp. 388–399.
12. Witzke, R. W., and Bliss, T. J.: Surge Protection of Cable-Connected Equipment; *AIEE Trans.,* 1950, vol. 69, part I, pp. 527–542.
13. Powell, R. W.: Lightning Protection of Underground Residential Distribution Circuits; *IEEE Trans. Power Appar. Syst.,* September 1967, vol. PAS-86, pp. 1052–1056.
14. IEEE Committee Report: Surge Protection of Cable-Connected Distribution Equipment on Underground Systems; *IEEE Trans. Power Appar. Syst.,* February 1970, vol. PAS-89, pp. 263–267.
15. Rudge, W. J., Jr., Wieseman, R. W., and Lewis, W. W.: Protection of Rotating A-C Machines Against Traveling Wave Voltages Due to Lightning; *AIEE Trans.,* June 1933, vol. 52., no. 2, pp. 434–465.
16. Fielder, F. D., and Beck, E.: Effect of Lightning Voltages and Methods of Protecting Against Them; *AIEE Trans.,* October 1930, vol. 49, pp. 1577–1586.
17. Abetti, P. A., Johnson, I. B., and Schultz, A. J.: Surge Phenomena in Large Unit Connected Steam Turbine Generators; *AIEE Trans.,* Dec. 1952, vol. 71, part III, pp. 1035–1047.
18. Walsh, G. W.: A New Technology Station Class Arrester for Industrial and Commercial Power Systems; *Conference Record of the IEEE Industrial and Commercial Power System Technical Conference,* May 9–12, 1977, Catalogue No. 77CH1198-11A.
19. Tominaga, S., Shibuya, Y., Fujiwara, Y., Imataki, M., and Nitta, T.: Stability and Long-Term Degradation of Metal Oxide Surge Arresters; *IEEE Trans. Power Appar. Syst.,* July/Aug. 1980, vol. PAS-99, no. 4, pp. 1548–1556.
20. IEEE Working Group Report: Voltage Rating Investigation for Application of Lightning Arresters on Distribution Systems; *IEEE Trans. Power Appar. Syst.,* May/June 1972, vol. PAS-91, no. 3, pp. 1067–1074.
21. Lat, M. V.: Thermal Properties of Metal Oxide Surge Arresters; *IEEE Trans. Power Appar. Syst.,* July 1983, vol. PAS-102, pp. 2194–2202.

# SECTION 28

# STANDARDS IN ELECTROTECHNOLOGY

**Sava I. Sherr**

*Staff Director of Standards, Institute of Electrical and Electronics Engineers, Inc. (IEEE); formerly Managing Director, American National Standards Institute (ANSI); Vice President, U.S. National Committee of the International Electrotechnical Commission (USNC/IEC); member, Board of Directors, ANSI; member, Industry Functional Advisory Committee, U.S. Trade Representative, Department of Commerce; member, Executive Committee, Organizational Member Council, ANSI; Registered Professional Engineer, State of New York; Senior Member, IEEE*

## CONTENTS

*Numbers refer to paragraphs*

## HISTORY OF ELECTRICAL STANDARDS

***1. Early History.*** The early history of electrical standards stems from activities dominated by the American Institute of Electrical Engineers (AIEE). In 1884, the institute began actively to develop standard specifications for the growing electrical industry. In 1890, it proposed that the practical unit of self-induction be named the henry. At the same time, the institute appointed its first committee on standardization — the Committee on Units and Standards. The members of this committee were A. E. Kennelly, chairman, F. B. Crocker, W. E. Geyer, G. A. Hamilton, and G. B. Prescott, Jr. The institute also appointed a "Standard Wiring Table Committee" under the chairmanship of E. B. Crocker, to assign linear resistance of standard-conductivity copper wire and at standard temperatures.

A committee was also appointed to prepare a program for the delegates to the International Electrical Congress, held in Chicago in 1893, in regard to units, standards, and nomenclature. As a result of the congress, there were adopted units for magnetomotive force (gilbert), for flux (weber), for reluctance (oersted), and for flux density (gauss). Subsequently, as a result of correspondence with engineering organizations in England, France, and Germany, the term "inductance" was adopted to represent the coefficient of induction (with the symbol $L$) and the present definition of the term "reactance" was proposed by Steinmetz and adopted.

***2. First Electrical Standards.*** In 1896, a "National Conference of Standard Electrical Rules" was held. The institute's delegate, Professor F. B. Crocker, was made its president, and, in cooperation with other national organizations, the conference promulgated the "Underwriters Rules," which finally resulted in the National Electrical Code.

In 1897, the Units and Standards Committee recommended adoption of the standard of luminous intensity, or candlepower, as the output of the amyl acetate Hefner-Alteneck lamp. It also recommended that the Lummer-Brodhun photometer screen be adopted for measuring the mean horizontal intensity of incandescent lamps.

At the beginning of 1898, a discussion was organized on the subject of "Standardization of Generators, Motors, and Transformers." This resulted in the formation of the first AIEE product standards committee, which in 1899 published the first electrical standard under the unique title "Report of the Committee on Standardization."

***3. National Bureau of Standards.*** The institute was a prime mover in the endorsement of a bill before the U.S. Congress, in 1901, for establishing a national standardizing bureau in Washington, D.C., "for the construction, custody, and comparison of standards used in scientific and technical work." This bureau became known as the National Bureau of Standards and has had a marked influence on the growth of U.S. technology.

***4. International Electrical Standards.*** In 1904, an International Electrical Congress was held in St. Louis which set a precedent for international congresses related to electrical units and standards. The congress unanimously recommended the establishment of two committees. Committee 1 consisted of government representatives and was responsible for legal maintenance of units and standards. This committee has now evolved into the International Conference on Weights and Measures (GPMU). Committee 2, of which Lord Kelvin was elected president, was responsible for standards related to commercial products in the electrical industry and has now evolved into the International Electrotechnical Commission (IEC).

Another international body, the International Committee on Illumination (Commission International de l'Eclairage, CIE), had its first meeting in 1913. The CIE establishes international units, standards, and nomenclature, in the science and technology of light and illumination.

***5. American National Standards Institute (ANSI).*** The American Engineering Standards Committee (AESC) was organized in 1919 as a result of the action of five organizations spearheaded by AIEE. This organization has been aptly described as a "national clearinghouse for industrial standardization" and has now evolved into the American National Standards Institute. In its early years, this body was organized with 12 divisions, each based on its own area of technology. Few of these became active. The electrical engineering division actually became the strongest, even to the point of having its own bylaws. In 1926, under the auspices of this organization, then known as the American Standards Association (ASA), engineering abbreviations and symbols were standardized. The AIEE, in cooperation with ASA, sponsored in 1928 the development of a glossary of terms used in electrical engineering. This work was coordinated with the IEC.

It is interesting to note that, in the electrical industry, basic standardization was first in order of development, dating back before 1890. Technical standardization came next, with the formation of the Standards Committee of the AIEE in 1898. Manufacturing standardization came only as a result of World War I and did not take effect until 1920.

***6. Standardization through World War II.*** The needs for uniformity that became evident during World War I served as the impetus for the founding of the AESC and highlighted the advantages of standardization. This led to the establishment of the War Industries Board headed by Bernard M. Baruch. Postwar pressure developed for standards activities because of the serious economic problems in the building trades industry. This led to the development of various building codes and to transfer of responsibility for the National Electrical Code from the National Bureau of Standards to the AESC, which was then reorganized as the American Standards Association (ASA).

The military departments entered the standards field, originally through the War Industries Board. In 1921, a Federal Specifications Board was created to unify the specifications of government agencies. By 1942, a Joint Army-Navy Committee on Specifications had become operable and was responsible for military procurement documents known as "JAN Specs."

***7. Standardization in Current Times.*** International standardization activities were coordinated by the United Nations Organization. This activity resulted in 1947 in the founding of the ISO, the International Organization for Standards. This body was made responsible for standardization in all fields not already covered by the IEC. The two organizations, although separate and distinct, coordinate their activities, and share a common headquarters facility in Geneva, Switzerland. In the United States, the voluntary standards system is well developed, and most organizations coordinate their activities through ANSI, the American National Standards Institute, successor organization to ASA.

## STANDARDS AND THE LAW

***8. Legality of Standards.*** In the United States, the legality of standards activities is primarily affected by laws related to the fixing of prices and conspiracy in restraint of trade. The two key governmental agencies involved are the Federal Trade Commission (FTC) and the Department of Justice. Since standards activities involve meetings in which representatives of competing organizations make agreements that affect engineering and industrial practices (both of which have economic implications), such meetings must take place under conditions which are subject to carefully regulated procedures. Failing this, participants could be subject to charges of violation of antitrust or conspiracy statutes.

Trade associations are particularly vulnerable in this respect, as meetings restricted to their membership involve participants who tend to be exclusively competitive manufacturers, whereas meetings of committees of professional societies involve technical personnel who are more apt to be representative of the total industry (both manufacturers and users), independent consultants, government personnel, educators, and scientists. However, the degree of liability of participants in standards development activities is virtually negligible when these activities are conducted under the auspices of, and under the strict rules of, an organization experienced in standards development, that is, an organization whose procedures are designed to promote fair and unprejudiced participation by all eligible parties.

The basic legality of standards programs has never been questioned. The courts and the FTC have found the programs to be beneficial, except where participants have deliberately misused these programs to discriminate unfairly against competition, or have arbitrarily misused the standardization programs.

***9. Price Fixing.*** The setting of standards would not normally lend itself to charges of price fixing. Yet there has been a unique case in which a standard developed for the wheat used in macaroni called for a specific blend of scarce durum wheat with other wheats. The FTC held this to be an attempt to lower the total industry demand for durum wheat and ward off price competition. Such actions are not common, and normal standards activity operating under the well-defined rules of a recognized standards organization creates no problem.

***10. Antitrust Problems.*** Other problems in standards programs related to antitrust laws can arise from what may be interpreted as a boycott of nonstandard products and the limitation of production to standard products exclusively. The first problem is specifically related to certification requirements. Where such exist, criteria for certification of nonstandard products may be virtually nonexistent. Alternatively, it is possible that noncertified products may be excluded arbitrarily from the market as a result of industry practice. The second problem is more difficult, since some of the advantages of product standardization are simplification in manufacture and economies of scale associated with limitation in variety of product, where the differences between standard and nonstandard are found to be marginal. Thus the shifting of demand to standard sizes or values (as in electronic components, for example) has resulted in substantial price decreases as a result of increasing volume demand for fewer varieties of product.

***11. Voluntary Standards System.*** In general, standards within the United States are developed under a voluntary system. To the extent that their adoption is also voluntary, there is less vulnerability to legal liability. However, many standards are made mandatory either through reference in purchase specifications and contracts or through adoption by government bodies as regulatory documents. Under such circumstances compliance ceases to be voluntary and the effect of the document is to disqualify or limit the acceptability of certain products or services. This is another reason why such standards development and approval must take place under procedures that minimize possibility of discrimination against specific suppliers.

***12. Personal Liability.*** An area of legal concern for participants in the standards generation or approval process is the question of legal liability. A typical situation deals with the case where an accident occurs under circumstances where potentially negligent parties demonstrate that they faithfully complied with the provisions of the applicable safety standards. The question here is one of the extent of liability of those who participated in the generation or adoption of the standard. A somewhat equivalent situation arises in product liability cases. Any such claim in a legal action turns on allegations of negligence in writing the standard. The general conclusion held by counsel is that members of voluntary standards committees operating under procedures of such organizations as ANSI, IEEE, or ASTM are not likely to incur significant legal risks.

## THE VOLUNTARY STANDARDS SYSTEM

***13. Voluntary Development of Standards.*** In the United States, the development of engineering standards is the result of voluntary activities on the part of professional societies, trade associations, and other organizations that wholly or partially specialize in standards development. This voluntary system provides the broadest possible input of technology and points of view to the standards development process. Since many organizations independently develop documents which are approved as the standards of their organization, there often exist conflicting standards on the same subject. To minimize areas of such conflict, many documents are submitted by their sponsoring organizations to ANSI for adoption as American National Standards.

Standards approved under the voluntary system are, in effect, the recommendations of the approving organizations. They have no mandatory legal force or effect except as adopted by others. Thus such standards may be incorporated by reference in contracts between buyers and sellers, they may be referenced or cited as requirements by government bodies in regulatory documents, or they may be incorporated into laws and statutes by legislatures. Trade association standards are presumed to represent the technical requirements to which the association members warrant products. However, conformance to a standard cannot be presumed nor is it mandatory, failing conditions such as those enumerated.

The lack of a mandatory conformance requirement is, paradoxically, one of the advantages of the voluntary system, since it acts as an automatic screen to avoid enforcement of standards that are technically unsound or whose need is not established. Such standards, if developed and published, simply go unused. Table 28-1 provides a guide to the various fields of electrical standardization and the principal organizations participating in the relevant standards development programs.

***14. Problems of Voluntary Standardization.*** The major difficulties associated with the voluntary system relate to lack of coordination among independent standards developing organizations.

**TABLE 28-1** Major Organizations Involved in Electrical Standards

| | Appliances | Communication | Computer | Electronics | Industry applications | Instrumentation and measurement | Power | Wiring practice | Home entertainment | Other |
|---|---|---|---|---|---|---|---|---|---|---|
| Professional societies | | IEEE | IEEE | IEEE | IEEE | IEEE, ISA | IEEE | IEEE | IEEE, SMPTE | |
| Trade associations | AHAM | EIA, ECSA | CBEMA | EIA | | | NEMA | NEMA | EIA, NCTA | AAMI |
| Government and regulatory bodies | CPSC | FCC | NBS | | | NBS | NRC | OSHA | FCC, CPSC | |
| Miscellaneous | UL | | | | | | | NFPA, UL | | ASTM, FDA |

AAMI = Association for Advancement of Medical Instrumentation
AHAM = Association of Home Appliance Manufacturers
ASTM = American Society for Testing and Materials
CBEMA = Computer and Business Equipment Manufacturers Association
CPSC = Consumer Products Safety Commission
ECSA = Exchange Carriers Standards Association
EIA = Electronics Industries Association
FCC = Federal Communications Commission
FDA = Federal Department of Agriculture
IEEE = Institute of Electrical and Electronics Engineers
ISA = Instrument Society of America
NBS = National Bureau of Standards
NCTA = National Cable Television Association
NEMA = National Electrical Manufacturers Association
NFPA = National Fire Protection Association
NRC = Nuclear Regulatory Commission
OSHA = Occupational Safety and Health Administration
SMPTE = Society of Motion Picture and Television Engineers
UL = Underwriters' Laboratories

*Duplication of Effort.* With organizations operating through committees of volunteers working in similar fields, it is not unusual to find several organizations promulgating standards covering the same area or products.

*Conflicting Standards.* The duplication among the standards activities of different organizations often results in the promulgation of standards whose requirements are conflicting. This can often result from lack of information exchange among the responsible organizations. It also comes about because of the differing objectives of the organizations involved.

For example, associations of manufacturers promulgate standards that describe the characteristics of the products routinely produced by the industry. An organization of users, mindful of their specialized requirements, would develop a standard that takes into consideration their own specialized requirements. An organization concerned primarily with safety may develop a standard for the same product that neglects performance characteristics but delineates requirements aimed at protecting the consumer or the worker. It would indeed be fortuitous if a single product could meet the detailed requirements of all these documents at the same time, and be efficiently and economically produced.

*Copyright and Reproduction.* To maintain a consistent set of activities, and to assure that standards do not become obsolete, it is necessary that the organization promulgating a standard continue to discharge its responsibility for periodic review, reaffirmation, and revision of the document. To this end, it is imperative that the sponsoring organization retain control over its documents. Unfortunately, other organizations and even government regulatory agencies extract from or paraphrase requirements from existing documents. This undermines the demand for the original document and removes the incentive for the continued updating and maintenance of the standard by the sponsoring organization.

***15. Resolution of Problems.*** The viability of the voluntary standards system depends heavily on the existence of a coordinating organization that maintains oversight over the activities of the organizations involved in standards development. This function is discharged in the United States by ANSI.

## TERMINOLOGY IN STANDARDS

***16. Standards Terms.*** The following comprises a partial list of terms used by participants in standards activities. Many of these terms have unique and specialized meaning when used in the context of standardization, and a brief definition is given for each as applied in this context.

*Balance.* The characteristic of a standards approving unit (committee, subcommittee, or working group) which assures that all classifications of interests are represented and that no single classification has a representation sufficiently large to enable it to unduly influence the resulting output.

*Balanced Committee.* A committee so constituted as to maintain a balance among its members. Many committees are balanced among manufacturers, users, and general interest classifications.

*Basic Standard.* A standard common to all disciplines, or to an overall technology.

*Canvass.* A method used for approval of standards which is dependent on circulation of a draft document to a list of concerned organizations for review and ballot.

*Certification.* An attestation to the effect that a particular product or service meets the requirements of a relevant standard.

*Classification of Membership.* The classification assigned to a participant or member of a standards developing unit which identifies the member's functional relationship or interest in the subject to be standardized. Thus a participant may be a manufacturer of a product being standardized, a user or purchaser of the product, a technically qualified expert with no well-defined functional relationship (classified as general interest), a labor or insurance representative (in the case of safety standards), or a constructor (one who installs the product for use by others). A variety of other classifications is possible as dictated by the scope of the standards activity.

*Code.* *(a)* A body of recommendations of good practice to be followed during design, manufacture, construction, installation, operation, and maintenance to satisfy considerations of safety, quality, economy, or performance in a given application. *(b)* A particular form of identification marking or reference which serves the dual purpose of establishing in a systematic manner the complete identity of an individual product and of identifying its similarity with other products. It may consist of a brief, systematic combination of letters, numerals, and symbols.

*Consensus.* A substantial agreement of those concerned. It implies that no important interested parties are strongly opposed on substantive grounds, or alternatively, that any opposition is in a small minority and the changes required to effect agreement by this minority would lead to substantive disagreement by the majority. Consensus implies that all disagreements have been given careful consideration and all reasonable attempts have been made for their resolution.

*Designation.* A definite and distinguishing name or symbol given to a product or to a group of functionally similar products or to an abstract matter. It emphasizes the group similarity but does not bring out the differences among the various members of the group.

*Dimensional Interchangeability.* A condition in which the dimensions of two or more products are such that one can physically replace another in a given application.

*Dimensional Standard.* A standard whose main content is dimensions and sizes of a product or group of products.

*Functional Interchangeability.* A condition where the characteristics of two or more products are such that they are able to perform the same functions.

*Guide.* A standards document that provides alternative information which comprises good engineering practice. Guides may contain application information for use of products and may be tutorial in nature. The user should be cautioned that the use of the word "guide" in the title of a document does not guarantee that the document is in fact nonmandatory. There are many governmental regulatory guides which in fact set forth mandatory requirements. Conversely, many documents that are differently titled are in fact guides.

*Interface Standard.* A standard whose main purpose is to ensure coordination between systems.

*International Standard.* A standard that has been adopted by a recognized international standards body (such as IEC or ISO).

*Letter Ballot.* A ballot used in standards development to determine agreement on a draft standard, or to generate comments that will be instrumental in developing a document on which agreement can be achieved. Such ballots provide for affirmative and negative votes. Negative votes, however, must be accompanied by reasons in sufficient detail to enable the writers of the document to determine what steps need be taken in revision to change the vote from negative to affirmative. The primary advantage of a letter ballot is that it provides adequate time for the recipients to review thoroughly the document which is subject to ballot.

*Marking.* The action and the result of stamping, inscribing, printing, or labeling marks, symbols, letters, or numerals upon a product or its package for the purposes of identifying the product.

*May.* An operative verb used in a standards document which identifies a possible means for satisfying a requirement. For example, several alternative procedures may be indicated for measuring a particular characteristic or phenomenon, and the selection of the most suitable procedure is left to the user of the document.

*National Standard.* A standard that has been adopted by a recognized national standards body (such as ANSI), or a standard that is in effect recognized and used nationally in preference to other documents.

*Performance Characteristic.* A characteristic of a product which determines the product's suitability for a specific application.

*Product Standard.* A standard containing requirements to be met by a product or group of products, usually including, directly or by reference to other standards, all or some of the following elements: dimensions, performance characteristics, other characteristics, and test methods.

*Rating.* A characteristic of a product which is determined in an arbitrary, yet consistent, manner, based on the intended function of the product.

*Recommended Practice.* A standards document that provides information on good engineering practice. Such documents may contain application information for use of products.

*Safety Standard.* A standard whose primary purpose is to ensure the safety of people and property.

*Self-Certification.* An attestation by a manufacturer or supplier of a product or service that it meets the requirements of a relevant standard.

*Shall.* An operative verb used in a standards document which indicates a mandatory requirement that must be specifically complied with for conformance to the document.

*Should.* An operative verb used in a standards document which indicates a problem area that must be resolved and specifies a requirement, compliance with which resolves the problem. In this sense, the verb "should" can be read as "shall." Alternatively, it is allowable under the document to use some other method which can be proved to resolve adequately the condition or problem area addressed. In some cases, it is also possible to demonstrate clearly that the condition or problem area addressed does not in fact exist, or apply to the product or circumstance in a specific instance.

*Simplification.* A form of standardization consisting of the reduction of the number of types of products within a definite range to that number which is adequate to meet prevailing needs at a given time.

*Specification.* A standards document that specifies all the characteristics and conditions to be met by a product or service to be supplied to the purchaser. Such a document may refer to other standards, selecting among the specific allowable options. A specification is intended to be a complete purchasing document.

*Standard.* A document setting forth requirements normally dictated by customary practices in industry, science, or technology. Such documents may include and may standardize terms, definitions, symbols; methods of measurement, tests of parameters or performance of devices, apparatus, systems, or phenomena; characteristics, performance, and safety requirements; dimensions and ratings.

*Standardization.* An activity aimed at an increase of order, giving solutions for recurring problems in the spheres of scientific, technological, and economic activity. Generally it consists of the processes of formulating, issuing, and implementing standards.

*Terminology Standard.* A standard containing exclusively terms and their definitions.

*Test Standard.* A standard containing test methods which may be combined with other requirements related to testing, such as sampling, use of statistical methods, and sequence of tests.

*Third-Party Certification.* An attestation by a recognized, technically qualified, independent organization that a product or service supplied by others meets the requirements of a relevant standard. Such certification may be based on inspections and tests conducted by the certifying organization, or may be based on supervision, monitoring, or auditing by the organization of such tests which may be conducted by others. The tests may be performed by the manufacturer or supplier of the service or product while being witnessed or audited by the certifier.

*Unification.* A form of standardization in which two or more specifications are combined into one in such manner that the products obtained are interchangeable in use.

## MAJOR ORGANIZATIONS CONCERNED WITH ELECTRICAL STANDARDS

***17. General.*** The following sections give summary data covering the principal organizations that develop electrical standards under the voluntary system. A brief description of the nature, working practices, and standards responsibility, as well as a point of contact, is given for each organization.

***18. AAMI.*** Association for the Advancement of Medical Instrumentation, 5100 Wilson Blvd., Arlington, Va. 22209, Elizabeth Bridgeman.

AAMI is a professional association developing medical device standards. Its members are both individuals (physicians, engineers, and health professionals) and organizations (corporations and institutions). It develops standards in medical disciplines (such as anesthesia, neurology, pathology, and pediatrics) and in medical devices and instruments.

Standards subcommittees, comprised of volunteers, are responsible for development of standards which require approval of the AAMI Standards Committee. AAMI Standards may be submitted for adoption by ANSI, after public review, through notification in the *AAMI Journal* and canvass, followed by approval of the Board of Directors. AAMI works in ANSI through participation in the activities of the ANSI Medical Devices Standards Management Board.

*19. AHAM.* Association of Home Appliance Manufacturers, 20 North Wacker Drive, Chicago, Ill. 60606, John T. Weizorick.

AHAM's history as a manufacturers' trade association goes back over 50 years. It was formed in 1966 through the merger of the American Home Laundry Manufacturers' Association and the Consumer Appliances Division of the National Electrical Manufacturers Association. Its membership is comprised of companies who are manufacturers of major appliances and of portable appliances.

Working through its engineering committees, AHAM develops and updates appliance performance standards and submits such standards to the appropriate national standards organization for recognition. It also sponsors certification programs on room air conditioners, refrigerators and freezers, and humidifiers and dehumidifiers.

AHAM develops standards in a series of orderly steps which start with project authorization by concerned product boards and engineering committees, or by the Engineering, Standards and Safety Board (ESSB) of AHAM, or the board of directors. Standards are drafted by the AHAM professional staff working under the ESSB and the concerned engineering committee. Prior to approval by the ESSB, the document must be approved by the engineering committee and reviewed by the product board. Upon approval of the ESSB, the document becomes an AHAM Standard after 60 days, unless negative action is taken by the board of directors. AHAM routinely submits its standards for adoption by ANSI under the canvass method. AHAM participates in the activities of ANSI.

*20. ANSI.* American National Standards Institute, 1430 Broadway, New York, N.Y. 10018, Donald M. Peyton.

The American Engineering Standards Committee was organized in 1919 through the efforts of AIEE (now IEEE), ASTM, ASME, ASCE, and AIME to simplify and standardize production and construction. In the period since its founding, it has gone through several structural reorganizations and name changes, having been variously known as the American Standards Association, the United States of America Standards Institute, and currently the American National Standards Institute. The institute is a federation whose membership comprises organizations who develop or participate in development of standards, and who use or are concerned with the use of standards, and companies concerned with development of American National Standards. In addition, many governmental bodies and individual experts participate in the work of the institute.

The institute does not itself develop standards but functions, rather, as a coordinating body for the purpose of encouraging development and adoption of worthwhile standards as American National Standards. It looks to its organizational members, as well as to other concerned organizations, for accomplishing the task of standards development. These development activities may be performed wholly within one of these organizations, or in accredited committees organized and administered by one of these organizations and operating under a set of rules meeting the basic procedures of the institute.

A large number of standards that are processed for adoption by the institute are classified as proprietary, in that they are developed, approved, and published by a standards development organization. Such documents may be submitted to the institute for adoption under one of the three methods prescribed by ANSI procedures, namely, the accredited organization method, the accredited committee method, or the canvass method. These methods are described below.

The ANSI organization is shown in Fig. 28-1. Of the organizations shown, it is important to note the following functional responsibilities. The Executive Standards Council, together with its Standards Boards, is responsible for ANSI standards development and approval procedures, management and coordination of standards programs, and approval of the procedures of standards developing organizations who submit documents under the accredited organization method. The Board of Standards Review (BSR) has final authority for approval of American National Standards. The International Standards Council is responsible for policy coordination of activities of U.S. representatives to international standards bodies. The Consumer Council screens standards submitted to the institute to protect the interests of the consumer.

*Approval of American National Standards.* To be deserving of adoption as an American National Standard, a document must meet a series of requirements, among which are that it represents a consensus of all parties concerned or affected, is a needed standard, and is technically sound.

ANSI procedures are designed to verify that these requirements are met. As the final steps in the approval process, the document is made available for public comment and criticism (known

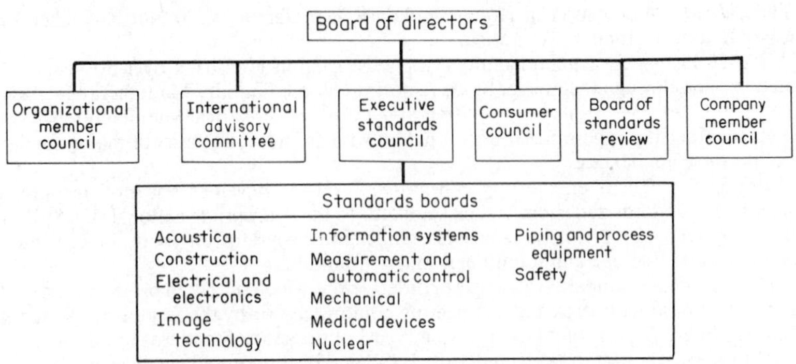


**FIG. 28-1** Organization chart—American National Standards Institute.

as public review) and is then submitted to the Board of Standards Review (BSR). The board reviews the history and record of the development of the standard and assures itself that adverse comments have been resolved or properly dealt with before granting its approval. Prior to submission to public review and to the BSR, the document must have received preliminary approval as a result of compliance with one of the three following methods.

**a.** *Accredited Organization.* An organization that develops standards under procedures that meet all the criteria for adopting such a standard as an American National Standard may request approval of its standards development procedures from the institute. The organization's procedures are examined by the BSR, which is empowered to designate the organization as an approved organization for development of American National Standards. Thereafter, the organization is empowered to submit documents it develops under these procedures directly to public review and the BSR.

**b.** *Accredited Standards Committee.* An Accredited Standards Committee is a committee whose membership is comprised of representatives of organizations having an interest in the development or adoption of standards within the committee's scope. The committee is administered by an organization which acts as the secretariat for the committee and which is fully responsible for its activities.

The committee's membership must represent a suitable balance of interested organizations so that a consensus of interested parties exists within the committee. Thus, there must not be a preponderance of manufacturers, or of users, or of labor representatives, for example, in the committee membership. Approval of documents is by letter ballot and, once consensus is achieved, documents approved by the committee are submitted to public review and the BSR.

**c.** *Canvass.* An organization whose standards development methods do not meet ANSI's criteria for consensus development may nevertheless achieve such a consensus by utilizing a canvass. Under this procedure, a list of organizations interested in the scope of the standard is developed by the sponsoring organization. This canvass list may be supplemented by additional organizations as a result of review and approval by the cognizant Standards Management Board or the Executive Standards Council. The draft standard is thereupon sent out for ballot by the sponsor to the organizations on the list. The sponsoring organization deals appropriately with all ballots submitted, and finally submits the document to public review and the BSR for verification of achievement of consensus and for final approval.

*Public Review.* Documents submitted for public review are listed in *Standards Action,* a biweekly publication of ANSI. This publication is available from ANSI to all concerned organizations and individuals, and it lists documents submitted for approval, together with instructions for procuring review copies. Public review is normally limited to 90 days.

***21. ASTM.*** American Society for Testing and Materials, 1916 Race Street, Philadelphia, Pa. 19103, William Cavanaugh.

Founded in 1898, ASTM is a scientific and technical organization whose charter purpose is "the development of standards on characteristics and performance of materials, products, systems and services; and the promotion of related knowledge." The society is a management system for the development of standards, operating through more than 126 main technical committees that function in prescribed fields under regulations designed to provide balanced representation among producers, users, consumers, and general-interest participants.

Standards are approved by ASTM in a four-stage process. After development, the sponsoring committee conducts a letter ballot. All negative votes resulting must be considered, and an affirmative vote of two-thirds of the committee membership is required to overcome a negative ballot. Upon approval by the committee, the standard is then submitted for ballot of the entire ASTM membership by means of an insert in the society's monthly publication, *ASTM Standardization News*. Affirmative votes of not less than 90% of those voting (and a return of not less than 50 ballots) is required for membership approval. Upon such approval, the document is submitted to the society's Standards Committee, which makes a determination that the procedures have been complied with, and thereupon grants final approval as an ASTM Standard.

*22. CBEMA.* Computer and Business Equipment Manufacturers Association, 1828 L Street NW, Washington, D.C. 20036, William Hanrahan.

CBEMA does not develop proprietary standards. Its standards activities take the form of acting as secretariat for accredited Committee X3, Computers and Information Processing.

CBEMA is heavily involved in international standardization through the work of ISO TC97, Information Systems, acting as secretariat for the US Technical Advisory Group, which is Committee X3. It also supports and participates, through the USNC, in IEC/TC74—Safety of Data Processing Equipment and Office Machines, and TC83, Information Technology Equipment. It also participates in other industry, government, and American National Standards committees related to environment, installation, and safety concerns of the computer and business equipment industry.

*23. ECSA.* Exchange Carriers Standards Association, 4 Century Drive, Parsippany, N.J. 07054, O. J. Gusella.

ECSA was founded in 1984 to bring together the telephone exchange carriers who took over the functions originally discharged by the Bell Telephone System. The association is concerned primarily with replacing the standardization function that was originally carried on within the Bell System. ECSA acts as the sponsor and secretariat for accredited Committee T1.

*24. EIA.* Electronic Industries Association, 2001 Eye Street NW, Washington, D.C. 20006, Allen Wilson.

Originally founded as the Radio Manufacturers Association in 1925, EIA is the trade association of the U.S. electronics industry. The association comprises seven divisions representing the various segments of the industry: Consumer Electronics, Communications, Industrial Electronics, Component Parts, Solid-State Products, Electron Tubes, Government Electronics. Standards are developed in committees which operate under divisional committee panels within the Engineering Department of EIA. Documents are first drafted in committee as Tentative Standards Proposals which, after review by the responsible panel chairman, the Engineering Department, and legal counsel, are issued as Standards Proposals. These proposals are available to the public during a 6-week review and comment period, and resulting comments are forwarded to the responsible committee chairman for resolution. After action with regard to comment resolution is completed, the proposal is subject to final approval by the Engineering Department Executive Committee and issued as an EIA Standard.

EIA participates in the work of ANSI, acting as secretariat for: C83, Components for Electronic Equipment and S4, Sound Recording. It also participates in the work of IEC, providing groups of technical advisers and experts for several IEC committees, and is represented on the Executive Committee of the USNC/IEC.

*25. IEEE.* Institute of Electrical and Electronics Engineers, 345 East 47 Street, New York, N.Y. 10017, Sava I. Sherr.

Formed in 1963 through merger of the American Institute of Electrical Engineers (founded in 1884) and Institute of Radio Engineers (founded in 1912), IEEE is now the largest engineering society in the world (over 250,000 members). It develops standards in the technical committees of its 31 professional groups and societies in such diverse subjects as broadcasting and communications, electrical practices for large industry (mining, textiles, shipbuilding, transportation, cement plants, and others), instrumentation and measurement, insulators and insulation, magnetics, motors and generators, nuclear power, power apparatus and systems, recording, symbols and units, and electrical transmission and distribution.

IEEE membership consists of qualified individuals in the engineering and scientific fields. There are no company memberships. The technical committees that generate IEEE Standards are comprised of qualified professional specialists (consultants, academicians, and engineers employed by manufacturers, utilities, government, and large industry users of electrical and electronic equipment and devices). Standards are adopted by consensus through ballots in technical committees, with full consideration given to all negative viewpoints. Final approval is by the IEEE Standards Board, consisting of 26 leading experts in diverse fields of electrotechnology. IEEE Standards are routinely submitted for adoption as American National Standards. Many have been adopted in government regulations, most are recognized internationally, and some have been adopted as national standards of other countries.

IEEE is a member and actively participates in the work of ANSI. In addition to representation on the administrative organizations of ANSI, it participates in the activities of over a hundred American National Standards committees. It administers the work of the following committees, acting as the secretariat organization:

C2, National Electrical Safety Code.

C16, Communications and Electronics Equipment.

C42, Definitions of Electrical Terms.

C61, Electric and Magnetic Quantities and Units.

C68, Techniques for Dielectric Tests.

C76, Apparatus Bushings.

C95, Radio Frequency Radiation Hazards.

N41, Controls, Instrumentation and Electrical Systems for Nuclear Power Generating Stations.

N42, Nuclear Instruments.

N44, Equipment and Materials for Medical Radiation Applications.

Y32, Graphic Symbols and Designations.

It also is a member of the U.S. National Committee of the IEC and is represented on its executive committee. It provides technical advisers to the USNC of IEC, and several of its committees function as the Committee of Experts to formulate the U.S. position for international meetings.

*26. ISA.* Instrument Society of America, 400 Stanwix St., Pittsburgh, Pa. 15222, Lois Ferson.

ISA generates proprietary standards in its technical committees. Its field is the technology of instrumentation and measurement, with particular attention to process instrumentation. Its standards are developed in the society's technical committees, and are approved by its Standards Board.

ISA participates in the work of ANSI. In the electrical field it provides the secretariat for accredited Committee C96, Temperature Measurement Thermocouples. In the process-control field it provides the secretariat for Subcommittee B16.0, Control Valves.

*27. NEMA.* National Electrical Manufacturers Association, 2101 L Street, NW, Washington, D.C. 20037, Bernard Falk.

NEMA is the largest trade organization for manufacturers of electrical products in the United States, and its 500 member companies are domestic firms varying in size from small companies to large diversified companies. It develops standards in the technical committees of its eight divisions covering products in such fields as building equipment, power electronics, industrial electrical equipment, insulation, lighting, power equipment, wire and cable, and radiation imaging products.

NEMA technical committees comprise engineers designated to represent member companies who are manufacturers of electrical equipment. Since manufacturers are most knowledgeable in the technology associated with their respective products, NEMA committees are highly competent in developing product standards that realistically take into consideration the economic trade-offs that are essential to practical standardization. Standards are adopted by consensus with final approval given by the NEMA Codes and Standards Committee.

NEMA Standards are generated in four classifications:

**a.** NEMA Standard—defines a commercially standardized product subject to repetitive manufacture.

**b.** Suggested Standard for Future Design—suggests an approach to future product improvement or development.

**c.** Authorized Engineering Information—included as part of other NEMA standards to explain data or information.

**d.** Official Standards Proposal—proposed draft for adoption by some other organization such as ANSI.

NEMA is a member of and actively participates in ANSI and is represented on over 100 American National Standards committees. It administers the work of the following committees, acting as the secretariat organization:

C8, Insulated Wire and Cable.

C9, Magnet Wire.

C18, Specifications for Dry Cells and Batteries.

C34, Static Power Converting Equipment.

C63, Radio-Electrical Coordination.

C64, Brushes for Electrical Machines.

C73, Attachment Plugs and Receptacles.

C87, Arc Welding Machines.

C89, Specialty Transformers.

C93, Carrier-Current Equipment and Potential Devices.

C97, Low Voltage Fuses, 600 Volts or Less.

C107, Disposal of Askarel Used in Electrical Equipment.

NEMA also acts (jointly with IEEE and/or EEI as cosecretariats) as the administrative secretariat of the following committees:

C19, Industrial Control Apparatus.

C29, Insulators for Electric Power Lines.

C37, Switchgear.

C50, Rotating Electrical Machinery.

C55, Capacitors.

C57, Transformers, Regulators, and Reactors.

C62, Surge Arresters.

C84, Preferred Voltage Ratings for AC Systems and Equipment.

C92, Insulation Coordination.

C119, Separable Insulated Connectors.

NEMA is a member of the USNC/IEC, is represented on its executive committee, provides technical advisers to the USNC, and participates in committees of experts to formulate U.S. positions for international meetings.

***28. NFPA.*** National Fire Protection Association, 470 Atlantic Avenue, Boston, Mass. 02210, Richard Stevens.

NFPA has been active in standards development since its founding in 1896. Working under the direction of its Standards Council, the technical committees of NFPA, composed of organization representatives, personal members, and liaison members from other technical committees, develop standards documents that are then subject to public review as a result of advance publication, and are finally approved at a semiannual meeting of the entire NFPA membership. Although dedicated to fire prevention, NFPA is responsible for a series of electrical standards, the most noted of which is the National Electrical Code.

***29. SAMA.*** Scientific Apparatus Makers Association, 1140 Connecticut Avenue NW, Washington, D.C. 20036.

Founded in 1918, SAMA is a trade association representing the scientific instrument industry. Its members are approximately 200 companies engaged in design, manufacture, and distribution of instruments, apparatus, and equipment used for measurement, analysis, testing, and control. SAMA is comprised of seven sections which cover analytical instruments, laboratory apparatus, measurement and test instruments, nuclear instruments, optical, process measurement and control, and scientific laboratory furniture and equipment. Within each section is a standards committee concerned with program operation. In addition, the SAMA Standardization Committee, which is responsible to the board of directors, manages the overall standards program.

SAMA is active within ANSI, acting as the secretariat for:

C39, Electrical Measuring Instruments.

C100, Electrical Reference Instruments and Devices.

It is also involved in international standards activity, providing representation to committees of both IEC and ISO.

***30. SMPTE.*** Society of Motion Picture and Television Engineers, 862 Scarsdale Avenue, Scarsdale, N.Y. 10583, Alex Alden.

SMPTE is a professional society of engineers working in the fields of motion pictures and television. It develops practices in its engineering committees on such subjects as color, photoinstrumentation, sound, television, and video-tape recording, and through the same committees initiates final development of American National Standards in similar fields. These are processed and adopted through the two American National Standards committees for which the SMPTE holds secretariats. These are: PH22, Motion Pictures and C98, Video-Tape Recording.

The process functions by consecutive approval of the pertinent engineering committees, the SMPTE Standards Committee, and the board of governors. Public review is achieved through publication in the *SMPTE Journal.* SMPTE also is active in the USNC/IEC and participates in ISO TC36, Motion-Picture Sound Recording.

***31. UL.*** Underwriters' Laboratories, Inc., 207 East Ohio Street, Chicago, Ill. 60611, Jack Bono.

Founded in 1894, UL is an independent, not-for-profit organization which maintains and operates laboratories for testing devices, systems, and materials with relation to public safety. Products so tested and meeting its requirements are eligible for UL "listing." UL maintains an inspection and follow-up program in factories where UL-listed devices are manufactured. UL representatives conduct in-factory and in-the-field inspections of manufacturers' procedures for assuring production compliance with UL requirements. Such requirements appear in appropriate UL Standards for Safety which are developed by UL under procedures that involve consultation with industry and government experts and consumers, among others.

Services available from UL include:

*UL's listing and follow-up service,* applicable to products evaluated with respect to hazards to life and property.

*UL's classifications and follow-up service,* applicable to products evaluated with respect to one or more of the following: specific hazards only, performance under specified conditions, regulatory codes, and other standards, including international standards.

*UL's certificate and follow-up service,* applicable to products comprising field-installed systems at a specific location, or to specific quantities of certain products where it is impractical to apply the listing mark or classification marking to the individual product.

*UL's recognition and follow-up program,* applicable to products evaluated only for use as components of end-product equipment.

The majority of insurance underwriters in the United States, and many federal, state, and municipal authorities either accept or require listing or classification by UL as a condition of their recognition of devices, systems, and materials having a bearing upon life and fire hazards.

UL is divided into several engineering departments, each dealing with distinct and separate subjects as follows: electrical; heating, air conditioning, and refrigeration; casualty and chemical hazards; burglary protection and signaling; fire protection; and marine. Each department has prepared standards for systems, materials, and appliances.

UL publishes (lists) the names of companies who have demonstrated the ability to provide products conforming to its requirements. Listing authorizes the manufacturer to use the laboratories' listing mark (classification marking, recognition marking, or certificate) on the listed products.

UL submits its standards to ANSI for adoption as American National Standards. It presently uses the canvass method almost exclusively to achieve national consensus. In addition, it is active in international standards, is represented on the Executive Committee of the USNC/IEC, and provides experts and technical advisers for developing U.S. positions for IEC and ISO committees.

## GOVERNMENTAL REGULATORY STANDARDS BODIES

***32. CPSC.*** Consumer Products Safety Commission, Washington, D.C. 20207.

The CPSC was established by congressional action in 1972. Its primary concern is the prevention of injuries associated with the use of consumer products. In this connection, it conducts an analysis of consumer products and has compiled a Consumer Product Hazard Index which ranks such products based on severity and frequency of accidents.

The commission is committed to the development of safety standards for various types of products including, in the electrical field, electric toys, TV sets, electrical appliances, and others. To assure compliance with developed standards, the commission can direct criminal actions against violators. Once a substantial product hazard is identified, the commission can force appropriate parties to give public notice of defects or failure to comply with standards. It can also order repair or replacement of hazardous products. The commission's work is based on a cause-effects analysis, starting with development of accident statistics and ending in a regulation which is subject to enforcement. Provision is made for input from interested parties in the form of inquiries, complaints, and petitions. The process is illustrated in Fig. 28-2.

*Standards Development.* Upon determination of the need for a standard by CPSC, a notice is published in the *Federal Register* explaining this need. The notice invites interested persons or organizations to submit an existing standard for consideration or an offer to develop a standard satisfying the need. Such offers are analyzed based on the technical competence of the offerer, the proposed time schedule, and the plan for development. The commission may contribute to the offerer's cost.

Thereafter, the commission may accept the proposal of an offerer and promulgate the resulting document as a Regulation, after publication in the *Federal Register.*

***33. EPA.*** Environmental Protection Agency, Washington, D.C. 20460.

The EPA was created by congressional action in 1970 for the purpose described by its title. Formulating standards and regulations for pollution abatement and control is an important responsibility of the agency. The enforcement of these regulations is another.

The EPA develops its regulations utilizing its own staff. It operates through the designation of a lead assistant administrator who is responsible for the preparation of all required materials and for ultimate recommendations to the administrator of the agency. The assistant administrator appoints a deputy as the line manager for the standard or regulation, who is responsible for its timely development. In turn, the deputy selects a Working Group chairperson who supervises the day-to-day operations leading to the assembly of a regulatory package.

The Working Group is convened with representation from interested offices and regions in EPA. The members are required to have technical competence in the subject and sufficient authority to present the views of their office.

A Steering Committee consists of high-level representatives from each assistant administrator's office, the administrator's office, the Office of Federal Activities, and the Office of Regional

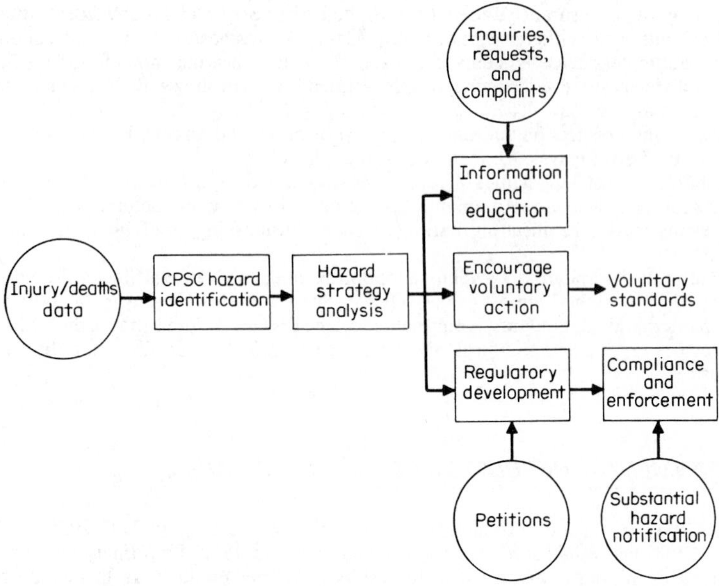


**FIG. 28-2** Flow chart—Consumer Products Safety Commission.

Liaison. Also participating are deputy administrators and Working Group chairpersons for specific packages being reviewed. The Steering Committee oversees development of all standards and regulations in the following manner:

**a.** *Working Group.* A decision by an assistant administrator that a need exists for a standard or regulation results in the designation of a Working Group chairperson. The chairperson prepares an outline of the standard to be developed and identifies the offices and specific individuals whose participation is desired. This material is circulated and the Working Group formed. Legal assistance is furnished by the Office of the General Counsel.

**b.** *Development Plan.* The Working Group formulates a development plan which includes the purpose of the standard, the methods to be followed in its development, the resources required, the schedule for development, and a decision regarding need for an Environmental Impact Statement. The lead office is responsible for the plan, and a consensus of the Working Group is not required.

**c.** *Review of Plan.* The development plan, together with comments from Working Group members, is reviewed by the lead administrator and the Steering Committee. Upon approval by the Steering Committee, implementation begins.

**d.** *Working Group Actions.* The Working Group may arrange for letting of contracts, researching literature, and conducting scientific and economic analyses. An Advance Notice of Proposed Rulemaking may be published in the *Federal Register,* formally soliciting information or comments. After all the information is assembled, the chairperson proceeds with drafting the standard. An Environmental Impact Statement is drafted in parallel, when required. The drafts are reviewed at subsequent group meetings, and each member keeps his or her office advised of developments. The completed draft is submitted to the Working Group for final review. However, consensus is not required, as the lead office has final responsibility.

**e.** *Proposed Regulatory Package.* The Working Group completes the package, which includes the draft regulation and Notice of Proposed Rulemaking; the Administrator's Action Memorandum covering the package, issues addressed, and consequences of the proposed action; and the Environmental Impact Statement, if required.

**f.** *Steering Committee Review.* The package is distributed to the Steering Committee for evaluation of its adequacy, to assure that appropriate alternatives were considered, to review major consequences of the proposed action, and to assure that the views of other offices are adequately reflected. After modifications resulting from this review are incorporated in the package, a briefing is held with the EPA administrator.

**g.** *External Review.* The package is distributed to other federal agencies for comment. Such comments may result in revision of the package, which after formal review and approval, is published as a Proposed Rulemaking in the *Federal Register.* Such publication provides for a public comment period of 30 to 60 days.

**h.** *Final Approval.* The Working Group is reconvened to consider the comments and make appropriate modifications. This may result in a reiteration of the process, with the ultimate result that the Final Rulemaking package is published in the *Federal Register* and becomes effective as a regulatory document.

***34. FCC.*** Federal Communications Commission, Washington, D.C. 20554.

The Federal Radio Commission was established in the late 1920s to deal with the problems caused by radio interference and frequency overlaps that were plaguing the broadcasting industry. The Federal Communication Act of 1934 converted this agency into the FCC.

Generation of standards by the FCC involves the rule-making process. Figure 28-3 shows the major steps involved in this process.

*Initiation.* Suggestions for changes in the FCC Rules and Regulations can come from sources outside the commission by either formal petition, legislation, court decision, or informal suggestion. In addition, a bureau or office within the FCC can initiate a Rule Making proceeding on its own.

*Evaluation.* When a petition of Rule Making is received, it is sent to the appropriate bureau or office for evaluation. If the bureau or office decides a particular petition is meritorious, it can request that Dockets assign a Rule Making (RM) number to the petition. A similar request is made when a bureau or office decides to initiate a Rule Making procedure on its own. A weekly notice is issued listing all accepted petitions for Rule Making; the public has 30 days to submit comments. The bureau or office then has the option of generating an agenda item requesting one of four actions by the commission. If a Notice of Inquiry (NOI) or Notice of Proposed Rule Making (NRPM) is issued, a Docket is instituted, and a Docket number is assigned.

*Actions.* Major changes to the rules are presented to the public either as an NOI or an NPRM. The commission issues an NOI when asking for information on a broad subject or to generate ideas on a given topic. An NPRM is issued when there is a specific change to the rules being proposed. If an NOI is issued, it must be followed by either an NPRM or a Memorandum Opinion and Order (MO&O) concluding the inquiry.

*Comment and Reply Evaluation.* When an NOI or NPRM has been issued, the public is given the opportunity to comment initially, and to respond to the comments that are made. It may be determined that an oral argument before the commission is needed to provide an opportunity for the public to testify before the commission, as well as for the bureaus or offices to present diverse opinions concerning the proposed rule change.

*Report and Order.* A Report and Order is issued by the commission stating the new or amended rule, or stating that the rules will not be changed. The proceeding may be terminated in whole or in part.

*Reconsideration.* Petitions for reconsideration may be filed by the public within 30 days; they are reviewed by the appropriate bureaus or offices and/or by the commission.

*Modification.* As a result of its review of a petition for reconsideration, the commission may issue an MO&O modifying its initial decision or denying the petition for reconsideration.

In addition to specific standards contained in FCC Rules, certain measurement methods have been developed by the Commission Laboratory, and these are described in Public Notices and Bulletins from the Office of the Chief Engineer. Such methods are specified in cases where specific data are required to be submitted to the commission. The data are dependent upon the measurement procedure used, and acceptable procedures are not available from other sources. In the event suitable measurement procedures are specified in other documents, such as IEEE Standards, reference to such procedures may be made in the rules.

The development of standards in the Commission's Rules is a joint undertaking between the operating bureaus and the Office of the Chief Engineer. A major effort of the Chief Engineer is to

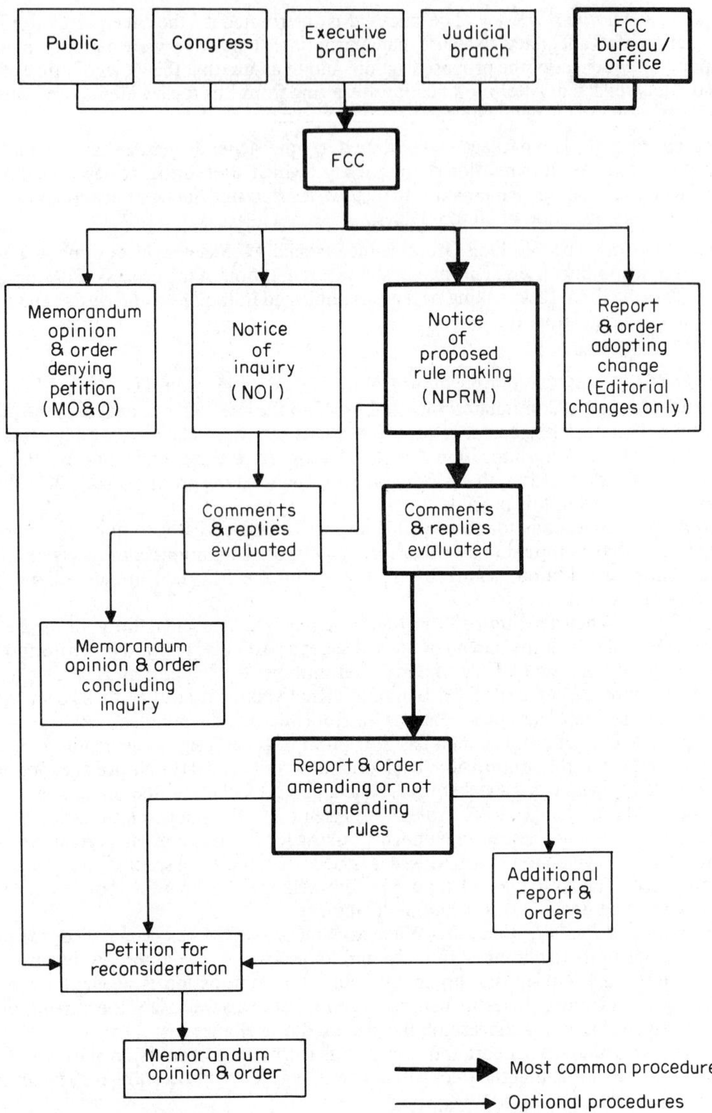


**FIG. 28-3** Flow chart—Federal Communications Commission.

maximize uniformity of the standards applicable to the various services regulated by the commission. The organization chart of the commission appears as Fig. 28-4.

In addition to standards development through the rule-making process, FCC actively participates with other bodies involved with standards. This allows the commission to see what standards efforts are being undertaken and permits both the standard body and the commission to become aware of each other's requirements. Some of the organizations with which the commission cooperates are:

The Institute of Electrical and Electronics Engineers (IEEE).

Electronics Industries Association (EIA).

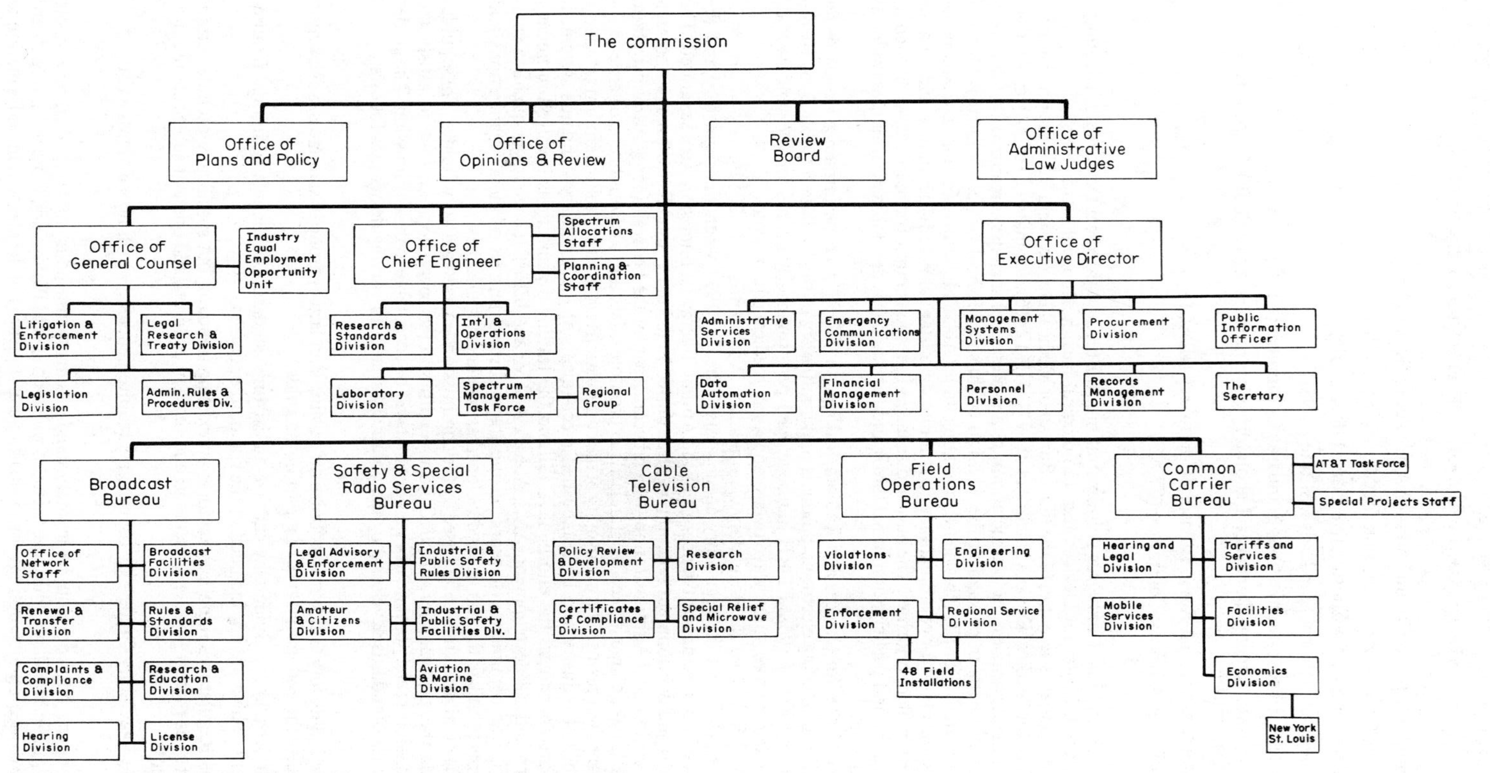


**FIG. 28-4** Organization chart — Federal Communications Commission.

Radio Technical Commission for Aeronautics (RTCA).

Radio Technical Commission for Maritime (RTCM).

Internation Radio Consultative Committee (CCIR).

American National Standards Institute (ANSI).

Federal Telecommunications Standards Committee (FTSC).

Law Enforcement Standards Laboratory Communications (LESL) Standards Review Committee.

Interdepartmental Radio Advisory Committee (IRAC).

Cable TV Advisory Committee (CTAC).

***35. FDA.*** Food and Drug Administration, A, Bureau of Medical Devices and Diagnostic Products; B, Center for Devices and Radiological Health, Rockville, Md. 20852.

The standards activities of the FDA Bureau of Medical Devices stem from the provisions of Title 1 of the Medical Devices Safety Act of 1971 (HR 12316). A conference of 20 organizations (government agencies, trade associations, and professional societies) organized by the bureau was instrumental in persuading the ANSI to establish a Medical Devices Standards Management Board (MDSMB). This board, operating under ANSI's Executive Standards Council, has the responsibility for coordination and stimulation of standards activities in this field, and the establishment of priorities. The bureau discharges its responsibilities primarily through encouragement of and participation in MDSMB and the standards development activities under its jurisdiction.

The Center for Devices and Radiological Health (CDRH) sets performance standards for radiation-producing electronic products under authority of the Radiation Control for Health and Safety Act of 1968. CDRH has developed performance standards for TV, cold-cathode gas-discharge tubes, microwave ovens, and x-ray medical equipment.

***36. NRC.*** Nuclear Regulatory Commission, Washington, D.C. 20555.

The NRC was the regulatory agency established in 1974 when the regulatory and the research and development functions of the Atomic Energy Commission were separated. The research and development function was transferred to the newly established Energy Research and Development Administration, which became part of the Federal Energy Administration.

NRC has an urgent need for standards to promote safety and to use as licensing criteria for nuclear power plants. The philosophy of the commission is to utilize to the maximum extent the expertise of industry and the established mechanisms of the voluntary standards system. It therefore encourages the efforts of ANSI and the leading organizations involved in nuclear power standards development (ANS, ASME, ASTM, and IEEE). It encourages qualified engineers in the commission to work on the standards committees of these organizations and attempts to utilize the standards developed as regulatory guidelines.

The documents most often used by the commission are its Regulatory Guides. These guides provide the applicant for an NRC license with specific suggestions on how to satisfy the commission's safety criteria in a manner that will be acceptable to the commission. The guides utilize the standards developed by the voluntary system to varying degrees, as follows:

**a.** The Regulatory Guide may endorse an existing standard or group of standards. Such action makes the endorsed standard a part of the guide.

**b.** The Regulatory Guide may endorse an existing standard, but include additional requirements which are not included in the standard endorsed.

**c.** The Regulatory Guide may endorse an existing standard, but specifically exclude certain portions thereof from endorsement; and may, in addition, add other or substitute requirements.

**d.** The Regulatory Guide may promulgate requirements for which no standards exist.

In-house, NRC develops basic safety standards that define policy on level of risk acceptable to the public, standards which are then promulgated as Regulations and Guides. Standards are also provided on assumptions for evaluating consequences of accidents. This information is then used as a basis for evaluating voluntary standards which are candidates for endorsement by NRC to the extent that such standards identify methodology for achieving the acceptable risk levels.

Ideally, from NRC's point of view, the voluntary standards system should be able to produce the necessary documents to achieve this objective. In point of fact, because of lack of knowledge and because of the state of the art, there are areas where industry cannot furnish valid answers to identified problem areas, and in such cases, NRC is forced to make some difficult judgments and develop requirements on its own recognizance.

The Regulatory Guide furnishes guidance but is not mandatory as a result of its promulgation. However, potential licensees must furnish the commission with a Preliminary Safety Analysis Report (PSAR) so that their applications may be considered. The PSAR spells out the methods to be used by the prospective licensee to assure compliance with policy requirements. Normally, a commitment to adhere to the letter of the applicable Regulatory Guides would be acceptable to the commission. Alternate methods of compliance, or exceptions to sections of the guides, would be acceptable if properly justified. However, such exceptions or alternate methods must be substantiated to the satisfaction of the commission and, in many cases, it is simpler for the applicant to comply with the guides in full.

A common and serious misunderstanding occurs when applicants commit themselves to compliance with Regulatory Guides that endorse standards utilizing the word "should" (see Par. **16** for the distinction between "should" and "shall"). The commission interprets a commitment to comply with the guide in such cases to mean that the applicant will treat every "should" as if it were written "shall." Therefore, if the applicant plans to comply in an area identified by "should" with a method different from the one described, the applicant must make this clear in the PSAR and not expect to resolve this at a later date. This caveat applies even if the identified problem does not exist.

NRC has indicated in many ways its desire to cooperate and work with the voluntary standards system. This has led to encouraging major productivity by the voluntary sector in the development of nuclear power standards, and the progress and achievements since 1973 have been striking. Other regulatory agencies may well profit from this experience.

***37. OSHA.*** Occupational Safety and Health Administration, Washington, D.C. 20210.

OSHA was established as a division of the U.S. Department of Labor through enactment in 1970 by Congress of the Occupational Safety and Health Act. This established, within the Department of Labor, a responsibility for occupational safety *in the workplace,* as distinguished from *public safety.* During its first year of operations, OSHA was empowered to adopt as regulations existing standards which had been developed under national consensus procedures. This authority did not extend beyond that time, and development of OSHA regulations was subsequently subject to the following procedures.

The director is responsible for establishing a project based on petitions, information, or data gathered from various sources, or requests from advisory committees. A project may also be initiated or authorized by the director. A project officer (OSPO) who is assigned to assume technical and administrative responsibility must consult with all interested OSHA personnel as well as appropriate people from other agencies. Normally, an Advance Notice of Proposed Rulemaking (ANPR) is issued and published for comment in the *Federal Register.*

Often, an advisory committee is formed which will make recommendations that are used, in conjunction with comments received as a result of publication of the ANPR, as well as with input from the National Institute of Occupational Safety and Health (NIOSH), to set up a schedule and to prepare a finalized draft proposal. This is processed through a series of reviews.

The proposal is first reviewed by the Regional Programs section of OSHA, and any changes deemed necessary are incorporated. This is followed by a Technical Review by a group consisting of the senior program manager, the OSPO, an OSHA senior scientist, a non-OSHA expert in the area (if available), and other concerned parties who may be invited to participate. Changes are recommended and may be incorporated. The director approves or disapproves the document based on these reviews. Approval is followed by a briefing conducted for the assistant secretary, who may recommend additional changes. The document is then submitted to the solicitor (SOL) for legal review.

SOL assigns a legal project officer who is responsible for legal review, assists the OSPO in solving legal problems, and represents OSHA with the OSPO in hearings and court actions. After approval by SOL, a revised draft is forwarded to the OSHA Information Office for further review. This review is followed by final approval by the assistant secretary, and the proposal is published in the *Federal Register* for comment.

A compilation of comments is published, together with a response from the OSPO, and an informal hearing may be held. The inputs from the public comments, the hearing, and internal comments are reviewed by the OSPO, and further changes may be made. The final standard is approved by the director, the Office of Information, and SOL, whereupon it is published in the *Federal Register* as an OSHA regulation.

## INTERNATIONAL ELECTROTECHNICAL COMMISSION (IEC)

***38. The International Electrotechnical Commission,*** located at 1 Rue de Varembe, Geneva, Switzerland, was an outgrowth of the International Electrical Congress held in St. Louis in 1904. The membership is comprised of the National Committees of the following 42 nations:

| | | |
|---|---|---|
| Argentina | Germany | Portugal |
| Australia | Hungary | Romania |
| Austria | India | South Africa |
| Belgium | Indonesia | South Korea |
| Brazil | Iran | Soviet Union |
| Bulgaria | Ireland | Spain |
| Canada | Israel | Sweden |
| China | Italy | Switzerland |
| Czechoslovakia | Japan | Turkey |
| Denmark | Mexico | United Kingdom |
| Egypt | Netherlands | United States |
| Finland | New Zealand | Yugoslavia |
| France | North Korea | |
| German Democratic Republic | Norway | |
| | Pakistan | |
| | Poland | |

A governing body, known as the IEC Council, comprised of the Presidents of the National Committees or their designees, meets annually in plenary session. The president, two vice-presidents, and a treasurer are each elected for 3-year terms. The headquarters organization is under the direction of a full-time director who acts as the General Secretary of the IEC.

Technical work is done in Technical Committees (currently numbering 78), many of which are further subdivided into subcommittees. The committees usually organize working groups that are responsible for the development of draft documents which are submitted for review, comment, and final approval as IEC Standards. Technical Committee work is, in turn, coordinated by a central body, known as the Committee of Action, consisting of nine council members, three of which are elected every 2 years to serve a 6-year term.

Each Technical Committee is originally organized through authorization of the Committee of Action. One of the member countries agrees to act as the secretariat, and a chairperson is appointed, usually from another country, for a 7-year term. The chairperson is eligible for reappointment for a supplemental term of 3 years. Membership on Technical Committees is by country, each having a single vote. Many committees meet annually, although other meeting schedules are common. At committee meetings, each country may be represented by a delegation, and at subsequent meetings the representatives of a given country may change. Thus delegates are not "members" of the committee and no continuity of individual participation can be assumed. The committees currently in existence are listed in Table 28-2.

***39. Standards Generating Procedures.*** To generate an IEC Standard, the committee frequently sets up a Working Group which is comprised of leading experts in the field of work whose services are offered by their National Committees. These experts participate on the Working Group as individual members, not as representatives of their countries. Initially, a Working Group develops a draft, which is forwarded by the Working Group Secretary to the Central Office in Geneva, and circulated as a Secretariat Document to all National Committees.

**TABLE 28-2** Committees of the International Electrotechnical Commission

| Committee | Title | Chair | Secretariat |
|---|---|---|---|
| 1 | Terminology | Switzerland | France |
| 2 | Rotating Machinery | UK | UK |
| 2A | Turbine-Type Generators | UK | USSR |
| 2B | Mounting Dimensions and Output Series | UK | France |
| 2F | Carbon Brushes, Holders Commutators, and Slip Rings | UK | USSR |
| 2G | Test Methods and Procedures | USA | USSR |
| 2H | Degrees of Protection, Methods of Cooling and Mounting | Germany | France |
| 2J | Classification of Insulation Systems for Rotating Machinery | USA | USA |
| 3 | Graphic Symbols | Switzerland | Switzerland |
| 3A | Graphic Symbols for Diagrams | Netherlands | Switzerland |
| 3B | Preparation of Diagrams, Charts and Tables. Item Designation | Norway | Sweden |
| 3C | Graphic Symbols for Use on Equipment | Germany | France |
| 4 | Hydraulic Turbines | USA | USA |
| 5 | Steam Turbines | UK | USA |
| 7 | Bare Aluminum Conductors | Italy | Brazil |
| 8 | Standard Voltages, Current Ratings and Frequencies | Italy | Italy |
| 9 | Electric Traction Equipment | Italy | France |
| 10 | Fluids for Electrotechnical Applications | Belgium | Italy |
| 10A | Hydrocarbon Insulating Oils | Italy | Belgium |
| 10B | Insulating Liquids Other than Hydrocarbon Oils | UK | Germany |
| 10C | Gaseous Insulants | Vacant | Italy |
| 11 | Recommendations for Overhead | Vacant | France |
| 12 | Radiocommunications | France | Netherlands |
| 12A | Receiving Equipment | Vacant | Japan |
| 12B | Safety | Sweden | Netherlands |
| 12C | Transmitting Equipment | UK | Netherlands |
| 12D | Antennas | Germany | Vacant |
| 12E | Microwave Systems | UK | Vacant |
| 12F | Equipment Used in Mobile Services | UK | USA |
| 12G | Cabled Distribution Systems | UK | Canada |
| 13 | Equipment for Electrical Energy Measurement and Load Control | Switzerland | Hungary |
| 14 | Power Transformers | UK | UK |
| 14B | On-Load Tap Changers | UK | UK |
| 14C | Reactors | Switzerland | Germany |
| 14D | Small Power Transformers and Special Transformers | Sweden | France |
| 15 | Insulating Materials | Germany | Italy |
| 15A | Short-Time Tests | Switzerland | Germany |
| 15B | Endurance Tests | Sweden | USA |
| 15C | Specifications | UK | Canada |
| 16 | Terminal Markings and Other Identifications | Sweden | Germany |
| 17 | Switchgear and Control Gear | Netherlands | Sweden |
| 17A | High-Voltage Switchgear and Control Gear | Netherlands | Sweden |
| 17B | Low-Voltage Switchgear and Control Gear | Sweden | France |
| 17C | High-Voltage Enclosed Switchgear and Control Gear | Netherlands | Germany |

**TABLE 28-2** *(Continued)*

| Committee | Title | Chair | Secretariat |
|---|---|---|---|
| 17D | Low-Voltage Switchgear and Control Gear Assemblies | Sweden | Germany |
| 18 | Electrical Installations in Ships | Netherlands | Netherlands |
| 18A | Cables and Cable Installations | Italy | Italy |
| 20 | Electric Cables | Italy | UK |
| 20A | High-Voltage Cables | Italy | UK |
| 20B | Low-Voltage Cables | UK | Netherlands |
| 21 | Secondary Cells and Batteries | Germany | Czech'vakia |
| 21A | Alkaline Accumulators | UK | Germany |
| 22 | Power Electronics | Germany | Switzerland |
| 22B | Semiconductor Converters | Germany | Sweden |
| 22D | Power Converters for Electric Rolling Stock | Vacant | France |
| 22E | Stabilized Power Supplies | UK | Vacant |
| 22F | Converters for High-Voltage DC Power Transmission | USA | USSR |
| 22G | Semiconductor Power Converters for Adjustable Speed Electric Drive Systems | France | USA |
| 23 | Electrical Accessories | France | Belgium |
| 23A | Electrical Conduits | France | UK |
| 23B | Plugs, Socket and Switches | Belgium | Italy |
| 23C | World-Wide Plug and Socket Systems | Sweden | South Africa |
| 23E | Circuit Breakers and Similar Equipment for Household Use | France | Italy |
| 23F | Connecting Devices | Switzerland | France |
| 23G | Appliance Couplers | Norway | Netherlands |
| 23H | Industrial Plugs and Sockets | UK | France |
| 23J | Switches for Appliances | Switzerland | Germany |
| 25 | Quantities and Units, and Their Letter Symbols | Switzerland | USA |
| 26 | Electric Welding | Vacant | Germany |
| 27 | Industrial Electroheating Equipment | France | Poland |
| 28 | Insulation Coordination | Canada | France |
| 28A | Insulation Coordination for Low-Voltage Equipment | USA | Germany |
| 29C | Measuring Devices | Germany | Denmark |
| 29D | Ultrasonics | Germany | USSR |
| 31 | Electrical Apparatus for Explosive Atmospheres | UK | UK |
| 31A | Flameproof Enclosures | Canada | USA |
| 31B | Sand-Filled Apparatus | Romania | France |
| 31C | Increased Safety Apparatus | Germany | UK |
| 31D | Pressurization and Associated Techniques | UK | France |
| 31G | Intrinsically Safe Apparatus | France | UK |
| 31H | Apparatus for Use in the Presence of Ignitable Dust | Belgium | USA |
| 31J | Classification of Hazardous Areas and Installation Requirements | UK | Yugoslavia |
| 32 | Fuses | UK | France |
| 32A | High-Voltage Fuses | UK | France |
| 32B | Low-Voltage Fuses | France | Germany |
| 32C | Miniature Fuses | USA | Germany |
| 33 | Power Capacitors | Switzerland | Italy |

**TABLE 28-2** *(Continued)*

| Committee | Title | Chair | Secretariat |
|---|---|---|---|
| 34 | Lamps and Related Equipment | Germany | UK |
| 34A | Lamps | Netherlands | UK |
| 34B | Lamp Caps and Holders | Sweden | Netherlands |
| 34C | Auxiliaries for Discharge Lamps | UK | UK |
| 34D | Luminaires | UK | UK |
| 35 | Primary Cells and Batteries | UK | Germany |
| 36 | Insulators | France | Italy |
| 36A | Insulated Bushings | France | Italy |
| 36B | Insulators for Overhead Lines | Italy | France |
| 36C | Insulators for Substations | USA | Sweden |
| 37 | Surge Arresters | France | USA |
| 38 | Instrument Transformers | Italy | Italy |
| 39 | Electronic Tubes | Germany | USSR |
| 40 | Capacitors and Resistors for Electronic Equipment | Sweden | Netherlands |
| 41 | Electric Relays | Switzerland | France |
| 41A | All-or-Nothing Relays | France | France |
| 41B | Measuring Relays and Protection Equipment | UK | France |
| 42 | High-Voltage Testing Techniques | Canada | Canada |
| 43 | Electric Fans for Household and Similar Purposes | India | India |
| 44 | Electrical Equipment for Industrial Machines | Switzerland | Vacant |
| 45 | Nuclear Instrumentation | France | Germany |
| 45A | Reactor Instrumentation | USA | France |
| 45B | Radiation Protection Instrumentation | UK | Italy |
| 46 | Cables, Wires, and Waveguides for Telecommunication Equipment | UK | USA |
| 46A | RF Cables | Finland | UK |
| 46B | Waveguides and Accessories | Sweden | USA |
| 46C | LF Cables and Wires | UK | France |
| 46D | Connectors for RF Cables | Switzerland | USA |
| 47 | Semiconductor Devices | USA | France |
| 47A | Integrated Circuits | Vacant | France |
| 47B | Microprocessor Systems | USA | Japan |
| 48 | Electromechanical Components for Electronic Equipment | Sweden | USA |
| 48B | Connectors | Switzerland | USA |
| 48C | Switches | UK | USA |
| 48D | Mechanical Structures for Electronic Equipment | UK | Germany |
| 49 | Piezoelectric Devices for Frequency Control | UK | USSR |
| 50 | Environmental Testing | France | UK |
| 50A | Shock and Vibration Tests | UK | UK |
| 50B | Climatic Tests | Sweden | Netherlands |
| 50C | Miscellaneous Environmental Tests | France | Germany |
| 50D | Fire Hazard Testing | UK | Germany |
| 51 | Magnetic Components and Ferrite Materials | Germany | UK |
| 52 | Printed Circuits | Belgium | Italy |
| 55 | Winding Wires | USA | France |
| 56 | Reliability and Maintainability | France | Sweden |
| 57 | Telecontrol, Teleprotection, and Associated Telecommunications for Electric Power Systems | Switzerland | Germany |

**TABLE 28-2** *(Continued)*

| Committee | Title | Chair | Secretariat |
|---|---|---|---|
| 58 | Methods of Measurement of Electrical Properties of Metallic Materials | Vacant | Vacant |
| 59 | Performance of Household Electrical Appliances | UK | France |
| 59A | Electric Dishwashers | Vacant | USA |
| 59B | Cooking Appliances | UK | Germany |
| 59C | Heating Appliances | UK | France |
| 59D | Home Laundry Appliances | Italy | France |
| 59E | Ironing and Pressing Appliances | Japan | Japan |
| 59F | Floor Treatment Appliances | UK | Sweden |
| 59G | Small Kitchen Machines | Vacant | Germany |
| 59H | Microwave Appliances | Sweden | USA |
| 60 | Recording | Germany | Netherlands |
| 60A | Sound Recording | USA | Belgium |
| 60B | Video Recording | USA | Germany |
| 61 | Safety of Household and Similar Appliances | France | USA |
| 61B | Safety of Household Microwave Ovens | Sweden | USA |
| 61C | Household Appliances for Refrigeration | Italy | France |
| 61D | Appliances for Air-Conditioning for Household and Similar Purposes | Denmark | USA |
| 61E | Safety of Electrical Commercial Catering Equipment | Italy | South Africa |
| 61F | Safety of Hand-Held Motor-Operated Electric Tools | Netherlands | Italy |
| 61G | Safety of Projectors | Germany | USA |
| 61H | Safety of Electrically-Operated Farm Appliances | Australia | New Zealand |
| 61J | Electrical Motor-Operated Cleaning Appliances | Switzerland | Germany |
| 62 | Electrical Equipment in Medical Practice | Finland | Germany |
| 62A | Common Aspects of Electrical Equipment Used in Medical Practice | France | Netherlands |
| 62B | X-Ray Equipment Operating up to 400 kV and Accessories | UK | Germany |
| 62C | High-Energy Radiation Equipment and Equipment for Nuclear Medicine | UK | Switzerland |
| 62D | Electromedical Equipment | Sweden | USA |
| 63 | Insulation Systems | USA | USSR |
| 64 | Electrical Installations in Buildings | Netherlands | Germany |
| 65 | Industrial-Process Measurement and Control | Germany | France |
| 65A | System Considerations | USA | UK |
| 65B | Elements of Systems | Germany | USA |
| 65C | Digital Data Communications for Measurement and Control Systems | France | UK |
| 66 | Measuring Equipment for Electronics | Vacant | Hungary |
| 66A | Generators | Vacant | Hungary |
| 66B | Oscilloscopes | Hungary | USA |
| 66C | Bridges and Meters | Vacant | Vacant |
| 66D | Analyzing Equipment | Netherlands | USA |
| 66E | Safety of Measuring, Control, and Associated Equipment | USA | UK |
| 68 | Magnetic Alloys and Steels | UK | Germany |

**TABLE 28-2** *(Continued)*

| Committee | Title | Chair | Secretariat |
|---|---|---|---|
| 69 | Electric Road Vehicles and Electric Industrial Trucks | UK | Sweden |
| 70 | Degrees of Protection by Enclosures | France | Germany |
| 71 | Electrical Installations for Outdoor Sites Under Heavy Conditions (Including Open-Cast Mines and Quarries) | USA | Australia |
| 72 | Automatic Controls for Household Use | Norway | USA |
| 73 | Short-Circuit Currents | Germany | Norway |
| 74 | Safety of Data Processing | UK | USA |
| 75 | Classification of Environmental Conditions | USA | Sweden |
| 76 | Laser Equipment | Sweden | USA |
| 77 | Electromagnetic Compatibility Between Electrical Equipment Including Networks | France | Germany |
| 77A | Equipment for Connection to the Public Low-Voltage Supply System | Belgium | France |
| 77B | Industrial and Other Non-Public Networks and Equipment Connected Thereto | France | Germany |
| 78 | Tools for Live Line Working | France | Canada |
| 79 | Alarm Systems | Belgium | Netherlands |
| 80 | Navigational Instruments | UK | France |
| 81 | Lightning Protection | Austria | Italy |
| 82 | Solar Photovoltaic Energy Systems | France | USA |
| 83 | Information Technology Equipment | Canada | Germany |
| 84 | Equipment and Systems in the Field of Audio, Video, and Audiovisual Engineering | USA | Netherlands |
| 85 | Measuring Equipment for Basic Electrical Quantities | Sweden | Hungary |
| 86 | Fiber Optics | France | USA |
| CMT | Electric Traction Equipment | | France |
| CISPR | Radio Interference | USA | Sweden |
| CISPR/A | Radio Interference Measurements | Sweden | USA |
| CISPR/B | Interference from Industrial, Scientific, and Medical RF Apparatus | UK | Canada |
| CISPR/C | Interference from Overhead Power Lines, High-Voltage Equipment, and Electric Traction Systems | France | France |
| CISPR/D | Interference Relating to Motor Vehicles and Internal Combustion Engines | Germany | Germany |
| CISPR/E | Interference Characteristics of Radio Receivers | Italy | Italy |
| CISPR/F | Interference from Motors, Household Appliances, Lighting Apparatus, and the Like | Netherlands | Netherlands |

Secretariat documents are reviewed by National Committees and comments returned to Geneva and in turn distributed by the Central Office. These comments serve as subjects of discussion at the next meeting of the Committee, and the document and comments may be referred back to the Working Group.

After one or more secretariat documents have been issued and discussed, the Committee may decide that a draft is ready for final ballot, and authorizes issuance of the draft standard under the six-month rule. Six-month rule documents are circulated by the Central Office for ballot, and a 20% negative vote of the national committees will defeat adoption. If in the judgment of the Technical Committee Chairman the comments submitted under the six-month rule are resolveable through minor editorial changes, such changes are made, and the document is reissued as a two-month rule document. If fewer than 20% negative votes are received, the document becomes an IEC Standard.

***40. Document Identification.*** The IEC Central Office uses a unique scheme to identify various sorts of documents and highlight those documents in particular that require various forms of attention. This scheme consists of both a document numbering system and a document color code.

*Numbering System.* A document number consists of three sections, the middle section of which is enclosed in parentheses. Section 1, or prefix, identifies the organization to which the document applies; Section 2, the parenthetical portion, identifies the issuing organization; and Section 3, the suffix, assigns a number to the document which is sequential with reference to the first two sections. Thus TC12(USA)16 would be the sixteenth document issued by the United States with reference to Technical Committee 12. In this scheme, council documents are identified with the prefix 01, and Committee of Action documents by the prefix 02. The parenthetical portion of all documents issued by a secretariat is identified as "(Secretariat)", not by the name of the country holding the secretariat. Thus TC1(Secretariat)23 would be a document of Technical Committee 1 that was approved for circulation by the committee. In fact, France is the secretariat of TC1. This scheme applies to all documents that are circulated by mail.

Since some documents are received too late for circulation prior to a meeting, or are brought to a meeting by the issuing delegation, these documents (known as "meeting documents") have a variation in the scheme. In meeting documents, the prefix includes the meeting place and the suffix is sequential with respect to the prefix based on time of distribution. Thus 01(The Hague/Denmark)5 is a document issued by Denmark for consideration of the council, and is the fifth council meeting document. With this system, delegates knowing the suffix of the last document issued would be aware of any documents they were missing.

*Color Code.* Documents are printed on different-colored paper stock which identifies the type of document being circulated. Table 28-3 illustrates the identification system.

**TABLE 28-3** Color Code for IEC Documents

| Color | Document type |
|---|---|
| Blue | Six-month rule documents |
| Buff | Two-month rule documents |
| Green | Meeting documents |
| Pink | Unapproved minutes of meetings |
| White | All other documents |

## U.S. NATIONAL COMMITTEE, IEC

***41.*** The U.S. National Committee (USNC) was founded by the American Institute of Electrical Engineers as a mechanism for U.S. participation in IEC work. It was supported solely by the institute until 1920, at which time its basis of support was broadened to include other organizations. In 1931, standards activities in the electrical field were consolidated under an Electrical Standards Committee (ESC). The USNC became affiliated with the ESC when the bylaws of the USNC were formally recognized in the constitution and bylaws of the ESC. ESC, in turn, became affiliated with the American Standards Association (ASA) when ASA formally recognized the constitution and bylaws of ESC, which eventually became the Electrical Standards Board (ESB) of ASA. These arrangements continued until 1966 when ASA reorganized, first as the United States of America Standards Institute (USASI), and later as the American National Standards Institute (ANSI).

The reorganization failed to deal with the status of USNC, but the USASI/ANSI Board of Directors agreed that the affiliation would continue. Subsequently, in 1973, ANSI and USNC signed a "Memorandum of Understanding" under which ANSI provided administrative support to USNC and pays its dues to IEC. In February, 1977, the USNC/IEC merged with and became a formal part of the ANSI organization, reporting to the Executive Standards Council on technical standards and to the International Standards Council on policy questions.

The USNC membership is comprised of all organizations involved or interested in international standards work in the field of electrotechnology. USNC is run by an Executive Committee whose voting members are elected officers (the president, four vice-presidents, and treasurer) and eight additional members. The secretary is assigned from ANSI staff. Each member of the Executive Committee acts also in the capacity of a Group Manager, supervising the U.S. activities for a selected number of IEC Technical Committees (TCs) through a group of individuals known as Technical Advisers (TA).

A TA is appointed for each TC in which the United States has an interest (76 of the present 78 TCs). The TA is, in turn, responsible for organizing the U.S. activity for his committee. They carry out their responsibilities by utilizing a Committee of Experts, usually an existing committee in the United States. This can be an existing American National Standards committee; or a technical committee of some other organization such as IEEE, NEMA, or EIA; or a special committee organized for the purpose. The TA is responsible for development of U.S. comments for his or her TC, for the designation of delegates to attend international meetings of his or her TC, and for ballots by the United States on six-month and two-month rule documents. Negative votes require concurrence of the Group Manager.

***42. Other International Organizations.*** In the field of electrotechnology, international standards are developed primarily in the IEC. However, other international standardization work takes place that impinges on electrotechnology because of overlap with other fields. The International Organization for Standardization (ISO) covers all fields of technology other than electrical. Among topics that are of interest in electrotechnology, it works in the areas of units, nuclear technology, computers, and information processing.

There are highly specialized organizations such as the Consultative Committee on Radio (CCIR) and the Consultative Committee on Telephony and Telegraphy (CCITT), both of which are associated with the International Telecommunications Union (ITU). The ITU develops binding agreements among its member countries, and representation is a governmental function, U.S. representation being controlled by the U.S. Department of State. In addition, the North Atlantic Treaty Organization (NATO) develops common military specifications, and U.S. representation is through the Department of Defense.

## ALPHABETICAL INDEX OF STANDARDS

***43. Introduction.*** The following list identifies available standards and their sponsors through subject headings. Many of the sponsors' standards are adopted as American National Standards and are available from ANSI. However, to avoid duplicate listings, all such documents appear under the publisher's identification (preceded by "ANSI/") given in parentheses. Documents published by ANSI are listed under ANSI identification.

Within each subject heading, standards are listed alphabetically, with introductory descriptive words disregarded for alphabetizing purposes. Numbers in parentheses are document numbers assigned by the sponsors. For brevity, closely related standards of a single organization are listed under abbreviated titles and the individual document numbers are separated by commas within the parenthetical material. Where documents have been adopted by more than one organization, the numbers assigned by both organizations are listed, separated by a semicolon. The identification "R" followed by four digits indicates that formal action has been taken by the organization reaffirming the document, the digits representing the year of last reaffirmation.

The following abbreviations are used in this section.

| | |
|---|---|
| ANSI | American National Standards Institute |
| ASTM | American Society for Testing and Materials |
| EIA | Electronics Industries Association |
| IEEE | The Institute of Electrical and Electronics Engineers |
| IPCEA | Insulated Power Cable Engineers Association |
| ISA | Instrument Society of America |

| | |
|---|---|
| NAB | National Association of Broadcasters |
| NEMA | National Electrical Manufacturers Association |
| NFPA | National Fire Protection Association |
| SAE | Society of Automotive Engineers |
| SMPTE | Society of Motion Picture and Television Engineers |
| UL | Underwriters' Laboratories |

***44. Abbreviations***

Abbreviations for Use on Drawings and in Text (ANSI Y1.1-1972)

***45. Accelerometers***

Specification Format Guide and Test Procedure for Linear, Single-Axis, Digital, Torque Balance Accelerometer (ANSI/IEEE 530-1978)

Specification Format Guide and Test Procedure for Linear, Single-Axis, Pendulous, Analog Torque Balance Accelerometer (IEEE 337-1972R1978)

***46. Access Control***

Access Control System Units (ANSI/UL 294-1983)

***47. Accidents and Accident Prevention***

Post Accident Monitoring Instrumentation for Nuclear Power Generating Stations (ANSI/IEEE 497-1981)

***48. Acoustics***

Absorption of Sound by the Atmosphere (ANSI S1.26-1978)

Acoustical Terminology (ANSI S1.1-1960R1976)

Measurement of Sound Pressure Levels (ANSI S1.13-1971R1976)

Preferred Reference Quantities for Acoustical Levels (ANSI S1.8-1969R1974)

Sound Level Meters (ANSI S1.4-1983)

***49. Actuators***

Qualification of Safety-Related Valve Actuators (IEEE 382-1980)

***50. Aerospace***

AC 400 Hz Aircraft Induction Motors (IEEE 137-1960)

Aerospace Equipment Voltage and Frequency Ratings (IEEE 127-1963)

Aircraft AC Generators (IEEE 138-1960)

Aircraft Circuit Breaker and Fuse Arrangement (ANSI/SAE AS3063A)

Aircraft Electric Systems (IEEE 128-1976)

Aircraft Generator and Regulator Characteristics (IEEE 136-1959)

Aircraft, Missile, and Space Equipment Electrical Insulation Tests (IEEE 135-1969)

ATLAS Test Language (ANSI/IEEE 416-1981)

DC Aircraft Rotating Machines (IEEE 132-1953)

DC Carbon Pile Voltage Regulators for Aircraft (IEEE 134-1955)

DC Tachometer Generators (IEEE 251-1963R1972)

Specification Format Guide and Test Procedure for Single-Axis Laser Gyros (ANSI/IEEE 647-1981)

Specification Format Guide and Test Procedure for Single-Degree-of-Freedom Rate-Integrating Gyros (ANSI/IEEE 517-1974R1980)

Specification Format for Single-Degree-of-Freedom Spring-Restrained Rate Gyros (IEEE 292-1969R1978)

Specification Format Guide and Test Procedure for Linear, Single-Axis, Digital, Torque Balance Accelerometer (ANSI/IEEE 530-1978)

Specification Format Guide and Test Procedure for Linear, Single-Axis, Pendulous, Analog Torque Balance Accelerometer (IEEE 337-1972R1978)

Strapdown Applications to Specification Format Guide and Test Procedure for Single-Degree-of-Freedom Rate-Integrating Gyros (ANSI/IEEE 529-1981)

Test Procedure for Single-Degree-of-Freedom Spring-Restrained Rate Gyros (IEEE 293-1969R1978)

***51. Air Conditioning*** (See Space Conditioning)

***52. Alarms***

Antitheft Alarms and Devices (UL1037)

Central Station Burglar Alarm Systems (UL611)

Design and Installation of Electric Pipe Heating Control and Alarm Systems for Power Generating Stations (ANSI/IEEE 622A/1984)

Measurement Procedure for Field Disturbance Sensor (RF Intrusion Alarm) (ANSI/IEEE 475-1983)
Power Supplies for Use with Burglar Alarm Systems (UL603)
Proprietary Burglar Alarm System Units (UL 1076)

***53. Ambient Conditions***

Reference Ambient Conditions for Test Measurements of Electrical Apparatus (ANSI/IEEE 3-1983)

***54. Ampacity***

Ampacities—Cables in Open-Top Cable Trays (NEMA WC51-1975R1980)
Ampacities, Including Effect of Shield Losses for Single-Conductor Solid-Dielectric Power Cable (NEMA WC50-1976R1982)

***55. Amplifiers***

Amplifiers and Preamplifiers for Semiconductor Radiation Detectors for Ionizing Radiation (ANSI/IEEE 301-1976R1982)
Magnetic Amplifiers (IEEE 107-1964R1979)

***56. Analyzers***

Spectrum Analyzers (IEEE 748-1979)

***57. Anodes***

Graphite Electrolytic Anodes (NEMA CG2-1969R1980)

***58. Antennas***

Antenna-Discharge Units (UL452)
Receivers Employing Ferrite Loop Core Antennas (ANSI/IEEE 189-1955R1972)
Land Mobile Communication Antennas (ANSI/EIA RS329A1978)
Terms for Antennas (ANSI/IEEE 145-1983)
Test Procedures for Antennas (ANSI/IEEE 149-1979)

***59. Armature***

Thyristor Converters and Motor Drives: Part 1—Converters for DC Motor Armature Supplies (ANSI/IEEE 444-1973)

***60. Armor***

Steel Armor and Associated Coverings for Impregnated-Paper-Insulated Wires and Cable (NEMA WC2-1980)

***61. Asbestos***

Asbestos, Asbestos-Varnished Cloth and Asbestos-Thermoplastic Insulated Wires and Cables (ANSI/NEMA WC1-1982)

***62. Askarel***

Chlorinated Aromatic Hydrocarbons (Askarels) for Capacitors (ANSI/ASTM D2233-80)
Test for Color of Chlorinated Aromatic Hydrocarbons (Askarels) (ANSI/ASTM D2129-79)
Test for Thermal Stability of Chlorinated Aromatic Hydrocarbons (Askarels) (ANSI/ASTM D1936-64R1980)
Transformer Askarel in Equipment (IEEE 76-1974)

***63. Aspect Ratio***

Measurement of Aspect Ratio and Geometric Distortion (ANSI/IEEE 202-1954R1978)

***64. Ate [Automatic Test Equipment]***

ATLAS Test Language (ANSI/IEEE 416-1981)

***65. ATLAS***

ATLAS Test Language (ANSI/IEEE 416-1981)
C/ATLAS Syntax (ANSI/IEEE 717-1982)
C/ATLAS Test Language (ANSI/IEEE 716-1982)
Guide to the Use of ATLAS (ANSI/IEEE 771-1980)

***66. Attenuators***

Fixed and Variable Attenuators (ANSI/IEEE 474-1973R1982)

***67. Audio and Audio Systems***

Protective Relay Applications of Audio Tones Over Telephone Channels (ANSI/IEEE C37.93-1976)
Terms for Audio and Electroacoustics (IEEE 151-1965R1971)
Volume Measurements of Electrical Speech and Program Waves (ANSI/IEEE 152-1953R1971)

***68. Automatic Control***

Automatic Generation Control on Electric Power Systems (IEEE 94-1970)

Definition, Specification, and Analysis of Manual, Automatic, and Supervisory Station Control and Data Acquisition (ANSI/IEEE C37.1-1979)
Terminology for Automatic Control (ANSI MC85.1M-1981)

***69. Availability***

Definitions for Use in Reporting Electric Generating Unit Reliability, Availability, and Productivity (ANSI/IEEE 762)

***70. Bandwidth***

Measurement of Impulse Strength and Impulse Bandwidth (ANSI/IEEE 376-1975)

***71. BASIC***

Real-Time BASIC for CAMAC (ANSI/IEEE 726-1982)

***72. Batteries***

Capacity Determination of Lead-Acid Industrial Storage Batteries for Motive Power Service (NEMA IB2-1974R1980)
Communications-Type Battery Chargers (NEMA PV7-1979)
Constant-Potential-Type Electric Utility (Semiconductor Static Converter) Battery Chargers (NEMA PE5-1983)
Cycle Life Testing of Lead-Acid Industrial Storage Batteries for Motive Power Service (NEMA IB3-1978)
Class 1E Battery Chargers and Inverters for Nuclear Power Generating Stations (ANSI/IEEE 650-1979)
Definitions for Lead-Acid Industrial Storage Batteries (NEMA IB1-1982)
Determination of Amperehour and Watthour Capacity of Lead-Acid Industrial Storage Batteries for Stationary Service (NEMA IB4-1979)
Dry Cells and Batteries (ANSIC18.1-1979)
Electric Battery Chargers (UL1236)
Emergency and Standby Power Systems for Industrial and Commercial Applications (ANSI/IEEE 446-1980)
Industrial Battery Chargers (UL1564)
Installation Design and Installation of Large Lead Storage Batteries for Generating Stations and Substations (IEEE 484-1981)
Lead-Acid Industrial Storage Batteries for Stationary Service (NEMA IB8-1981)
Life Testing of Lead-Acid Industrial Storage Batteries for Stationary Service (NEMA IB5-1979)
Maintenance, Testing, and Replacement of Large Lead Storage Batteries for Generating Stations and Substations (ANSI/IEEE 450-1980)
Mining Vehicle Battery Chargers (NEMA IB9-1981)
Qualification of Class 1E Lead Storage Batteries for Nuclear Power Generating Stations (ANSI/IEEE 535-1979)
Sealed Rechargeable Nickel-Cadmium Cylindrical Bare Cells (ANSI C18.2-1977)
Sizing Large Lead Storage Batteries for Generating Stations and Substations (IEEE 485-1983)
Standard Cells (ANSI C100.6-1976)
Testing Arrester Vents Used on Lead-Acid Industrial Storage Batteries for Stationary Service (NEMA IB7-1980)

***73. Boilers and Pressure Vessels***

Controls and Safety Devices for Automatically Fired Boilers (ANSI/ASME CSD-1-1982)
Electric Heating, Water Supply, and Power Boilers (ANSI/UL 834-1980)
Oil Fired Boiler Assemblies (ANSI/UL 726-1975)
Pressure Vessel Inspection Code (ANSI/API 510-1983)
Pressure Vessels for Human Occupancy (ANSI/ASME PVHO-1-1981)

***74. Boxes, Electric***

Electric Cabinets and Boxes (UL50)
Electrical Metallic Outlet Boxes (UL 514A)
Electrical Outlet Boxes and Fittings for Use in Hazardous Locations (ANSI/UL 886-1980)
Fittings for Conduit and Boxes (UL514B)
Nonmetallic Outlet Boxes and Covers (ANSI/UL 514C-1982)
Outlet Boxes and Fittings (ANSI/UL 514-1980)
Sheet-Steel Outlet Boxes, Device Boxes, Covers and Box Supports (ANSI/NEMA OS1-1973R1978)
Nonmetallic Outlet Boxes, Device Boxes, Covers and Box Supports (NEMA OS2-1980)

*75. Broadcast*

AM Broadcast Receivers (ANSI/IEEE 186-1948R1972)
FM Broadcast Receivers (ANSI/IEEE 185-1975)
Measurement of Conducted Interference Output to the Power Line from FM and TV Broadcast Receivers (ANSI/IEEE 213-1961R1974)
Monochrome TV Broadcast Receivers (ANSI/IEEE 190-1960)
Spurious Radiation from FM TV Broadcast Receivers (IEEE 187-1951)

*76. Brushes, Rotating Machinery*

Brushes for Electrical Machines (ANSI/NEMA CB1-1977)
Carbon Brushes (ANSI/IEEE 116-1975R1982)

*77. Buildings*

Electric Power Systems in Commercial Buildings (ANSI/IEEE 241-1983)
Graphic Symbols for Electrical Wiring and Layout Diagrams Used in Architecture and Building Construction (ANSI Y32.9-1972)

*78. Burst Measurements*

Burst Measurements in the Frequency Domain (IEEE 265-1966)
Burst Measurements in the Time Domain (IEEE 257-1964R1971)

*79. Buses*

Calculating Losses in Isolated-Phase Bus (ANSI/IEEE C37.23-1969R1977)
Microcomputer Systems Bus (ANSI/IEEE 796-1983)
Protective Relay Applications to Power System Buses (ANSI/IEEE C37.97-1979R1984)
Switchgear Assemblies Including Metal-Enclosed Bus (ANSI/IEEE C37.20-1969R1981)

*80. Bushings*

Loading Power Apparatus Bushings (ANSI/IEEE 757-1983)
Outdoor Apparatus Bushings, Rqm'ts & Tests (ANSI/IEEE 21-1976)
Outdoor Apparatus Bushings, Performance (ANSI/IEEE 24-1984)

*81. Busways*

Busways (NEMA BU1-1979)
Definitions and Requirements for High-Voltage Air Switches, Insulators, and Bus Supports (ANSI/IEEE C37.30-1971R1977)

*82. Busways and Fittings*

Electric Busways and Associated Fittings (ANSI/UL857-1981)
Electric Power Distribution for Industrial Plants (IEEE 141-1976)
Instructions for Safe Handling, Installation, Operation and Maintenance of Busways and Associated Fittings (NEMA BU1.1-1981)
Protection and Coordination of Industrial and Commercial Power Systems (IEEE 242-1975)

*83. Cabinets, Electric*

Electric Cabinets and Boxes (ANSI/UL50-1979)
Electric Wired Cabinets (ANSI/UL651981)
Test Blocks and Cabinets for Installation of Self-Contained "A" Base Watthour Meters (ANSI C12.8-1981)

*84. Cable Systems*

Electric Power Distribution for Industrial Plants (IEEE 141-1976)

*85. Cable Television*

Graphic Symbols for Grid and Mapping Diagrams Used in Cable Television Systems (ANSI/IEEE 623-1976)

*86. Cable Trays*

Ampacities—Cables in Open-Top Cable Trays (NEMA WC51-1975R1980)
Cable Tray Systems (NEMA VE1-1979)

*87. Cables*

Ampacities—Cables in Open-Top Cable Trays (NEMA WC51-1975R1980)
Ampacities, Including Effect of Shield Losses for Single-Conductor Solid-Dielectric Power Cable (NEMA WC50-1976R1982)
Armored Cable (UL4)
Asbestos, Asbestos-Varnished Cloth and Asbestos-Thermoplastic Insulated Wires and Cables (ANSI/NEMA WC1-1982)
Cable Penetration Fire Stop Qualification Test (ANSI/IEEE 634-1978)
Cable Plowing Guide (IEEE 590-1977)
Coaxial Communication Cable [CATV] (NEMA WC41-1975R1981)

Control and Low Voltage Cable Systems in Substations (IEEE 525-1978)
Cross-Linked Thermosetting-Polyethylene-Insulated Wire and Cable for the Transmission and Distribution of Electrical Energy (NEMA WC7-1982)
Design and Installation of Cable Systems in Power Generating Stations (IEEE 422-1977)
Ethylene-Propylene-Rubber-Insulated Wire and Cable for the Transmission and Distribution of Electrical Energy (NEMA WC8-1976)
Exposed Semiconducting Shields on Premolded High Voltage Cable Joints and Separable Insulated Connectors (IEEE 592-1977)
High Direct Voltage Tests on Power Cable Systems in the Field (ANSI/IEEE 400-1980)
High-Voltage AC Cable Terminations (IEEE 48-1975)
High-Voltage X-Ray Cables and Receptacles (NEMA XR7-1979)
Measuring Resistivity of Cable-Insulation Materials at High Direct Voltages (IEEE 402-1974R1982)
Metal-Clad Cables (UL1569)
Minimum Drum Diameter Reels for Cables (NEMA WC6-1975R1980)
Nonmetallic-Sheathed Cables (UL719)
Power Cable Joints (IEEE 404-1977)
Protective Coverings for Wire and Cable Reels (NEMA WC25-1981)
Radial Power Factor Tests on Insulating Tapes in Paper-Insulated Power Cable (IEEE 83-1963R1982)
Reference Standard for Electrical Wires, Cables, and Flexible Cords (UL 1581)
Rubber-Insulated Wires and Cables (UL 44)
Rubber-Insulated Wire and Cable for the Transmission and Distribution of Electrical Energy (NEMA WC3-1980)
Selecting and Testing Jackets for Cables (ANSI/IEEE 532-1982)
Selection and Design of Aluminum Sheaths for Cables (ANSI/IEEE 635-1980)
Service Entrance Cables (UL854)
Steel Armor and Associated Coverings for Impregnated-Paper-Insulated Wires and Cable (NEMA WC2-1980)
Test Method for Fire and Smoke Characteristics of Cables Used in Air Handling Spaces (UL910)
Thermoplastic-Insulated Wires and Cables (UL 83)
Thermoplastic-Insulated Underground Feeder and Branch Circuit Cables (UL 493)
Thermoplastic-Insulated Wire and Cable for the Transmission and Distribution of Electrical Energy (NEMA WC5-1973R1979)
Type Test of Class 1E Electric Cables, Field Splices and Connections for Nuclear Power Generating Stations (ANSI/IEEE 383-1974R1980)
Varnished-Cloth-Insulated Wire and Cable for the Transmission and Distribution of Electrical Energy (NEMA WC4-1976)
Wires and Cables with Varnished-Cloth Insulation (UL133)
Wires with Asbestos or Asbestos and Varnished Cloth or Tape Insulation (UL 115)

***88. Calculation***

Protection and Coordination of Industrial and Commercial Power Systems (IEEE 242-1975)
Simplified Method for Calculation of the Regression Line (IEEE 101A-1974R1980)

***89. CAMAC***

Block Transfers in CAMAC Systems (ANSI/IEEE 683-1976)
CAMAC Instrumentation and Interface Standards (ANSI/IEEE Camac-1982)
Modular Instrumentation and Digital Interface System (ANSI/IEEE 582-1982)
Multiple Controllers in a CAMAC Crate (ANSI/IEEE 675-1982)
Parallel Highway Interface System (ANSI/IEEE 596-1982)
Real-Time BASIC for CAMAC (ANSI/IEEE 726-1982)
Serial Highway Interface System (ANSI/IEEE 595-1982)
Subroutines for CAMAC (ANSI/IEEE 758-1979R1981)

***90. Cameras and Camera Systems***

Measurement and Characterization of Diode-Type Camera Tubes (IEEE 503-1978)
Measurement of Resolution of Camera Systems (ANSI/IEEE 211-1977)
Performance Measurement of Scintillation Cameras (NEMA NU1-1980)
Surveillance Cameras (UL983)

***91. Capacitors***

Capacitors (UL810)
Ceramic Dielectric Capacitors (ANSI/EIA RS198-1983)
Electrolytic Capacitors for AC Motor Starting (ANSI/EIA RS463-1979)
Electrolytic Capacitors for Long Life and General Purpose Application (ANSI/EIA RS395-1971R1983)
Encapsulated Dielectric Capacitors (ANSI/EIA RS198B3A-1981)
Film Dielectric Capacitors with Metallized Paper Electrodes for AC Applications (ANSI/EIA RS495-1982)
Fixed Composition Capacitors (ANSI/EIA RS335A-19872R1979)
Fixed Electrolytic Tantalum Capacitors (ANSI/EIA RS228B-1972R1983)
Fixed Film Dielectric Capacitors for DC Applications (ANSI/EIA RS376-1970R1983)
Fixed Paper and Film-Paper Dielectric Capacitors with Non-PCB Impregnants for AC Application (ANSI/EIA RS454-1978R1983)
Fixed Paper and Fixed Paper Polyester Film Capacitors in Non-Metallic Cases for DC Application (ANSI/EIA RS164A-1967)
Metallized Dielectric Capacitors for DC Application (ANSI/EIA RS377-1970R1983)
Metallized Film Dielectric Capacitors for AC Application (ANSI/EIA RS377-1970R1983)
Method of Test for Effective Series Resistance and Capacitance for Multilayer Ceramic Capacitors at High Frequencies (ANSI/EIA RS483-1981)
Molded and Dipped Mica Capacitors (ANSI/EIA RS153B-1972R1983)
Paper, Paper/Film Dielectric Capacitors for Power Semiconductor Applications (ANSI/EIA RS401-1973R1983)
Power Line Carrier Coupling Capacitors (ANSI C93.1-1981)
Protection of Shunt Capacitor Banks (ANSI/IEEE C37.99-1980)
Series Capacitors for Transmission and Distribution Line Compensation (ANSI C55.2-1973)
Shunt Power Capacitors (ANSI/IEEE 18-1980)

***92. Cement***

Cement Plant Electric Drives and Related Equipment (IEEE 499-1983)
Cement Plant Power Distribution (ANSI/IEEE 277-1983)
Improved Electrical Maintenance and Safety in the Cement Industry (IEEE 625-1979)

***93. Chargers***

Class 1E Battery Chargers and Inverters for Nuclear Power Generating Stations (ANSI/IEEE 650-1979)
Communications Type Battery Chargers (NEMA PV7-1979)
Constant-Potential-Type Electric Utility (Semiconductor Static Converter) Battery Chargers (NEMA PE5-1983)
Mining Vehicle Battery Chargers (NEMA IB9-1981)

***94. Charts***

Charts and Graphs (ANSI Y15.1, .2, .3-1979)
High Definition Facsimile Test Chart (IEEE 167A-1980)

***95. Circuit Breakers and Protectors***

AC High-Voltage Circuit Breakers (NEMA SG4-1975)
AC High-Voltage Circuit Breakers Rated on a Symmetrical Current Basis and a Total Current Basis (ANSI C37.12-1981)
Application Guide for AC High-Voltage Circuit Breakers Rated on a Symmetrical Current Basis (ANSI/IEEE C37.010-1979)
Calculation of Fault Current for Application of AC High-Voltage Circuit Breakers Rated on a Total Current Basis (ANSI/IEEE C37.5-1979)
Capacitance Current Switching for AC High-Voltage Circuit Breakers Rated on a Symmetrical Current Basis (ANSI/IEEE C37.012-1979)
Circuit Breakers and Circuit-Breaker Enclosures for Use in Hazardous Locations (UL877)
Conformance Testing of Metal-Enclosed Low-Voltage AC Power Circuit Breaker Switchgear Assemblies (ANSI C37.51-1979)
Definitions and Rating Structure for AC High-Voltage Circuit Breakers Rated on a Total Current Basis (ANSI/IEEE C37.3-1953R1982)
Electrical Control for AC High-Voltage Circuit Breakers Rated on a Symmetrical Current Basis and a Total Current Basis (ANSI C37.11-1979)

Field Discharge Circuit Breakers Used in Enclosures for Rotating Electrical Machinery (ANSI/IEEE C37.18-1979)

Interrupting Rating Factors for Reclosing Service for AC High-Voltage Circuit Breakers Rated on a Total Current Basis (ANSI C37.7-1960R1976)

Low-Voltage AC Non-Integrally Fused Power Circuit Breakers (Using Separately Mounted Current-Limiting Fuses) (ANSI/IEEE C37.27-1972)

Low-Voltage AC Power Circuit Breakers Used in Enclosures (ANSI/IEEE C37.13-1981)

Low-Voltage AC Power Circuit Protectors Used in Enclosures (ANSI/IEEE C37.27-1972)

Low-Voltage DC Power Circuit Breakers Used in Enclosures (ANSI/IEEE C37.14-1979)

Low-Voltage Power Circuit Breakers (NEMA SG3-1981)

Measurement of Sound Pressure Levels for AC Power Circuit Breakers (ANSI/IEEE C37.082-1982)

Metal-Enclosed Low-Voltage Power Circuit-Breaker Switchgear (UL1558)

Molded Case Circuit Breakers (NEMA AB1-1975R1981)

Molded Case Circuit Breakers and Circuit Breaker Enclosures (UL489)

Preferred Ratings for AC High-Voltage Circuit Breakers Rated on a Total Current Basis (ANSI C37.6-1971R1976)

Preferred Ratings, Related Requirements, and Application Recommendations for Low-Voltage Power Circuit Breakers and AC Power Circuit Protectors (ANSI C37.16-1980)

Procedures for Verifying the Performance of Molded Case Circuit Breakers (NEMA AB2-1980)

Protection and Coordination of Industrial and Commercial Power Systems (IEEE 242-1975)

Rating Structure for AC High-Voltage Circuit Breakers Rated on a Symmetrical Current Basis (ANSI/IEEE C37.04-1979)

Ratings and Related Required Capabilities for AC High-Voltage Circuit Breakers Rated on a Symmetrical Current Basis (ANSI C37.06-1979)

Synthetic Fault Testing of AC High-Voltage Circuit Breakers Rated on a Symmetrical Current Basis (ANSI/IEEE C37.081-1981)

Test Code for AC High-Voltage Circuit Breakers Rated on a Total Current Basis (ANSI C37.9-1953R1976)

Test Procedure for AC High-Voltage Circuit Breakers Rated on a Symmetrical Current Basis (ANSI/IEEE C37.09-1979)

Test Procedures for Low-Voltage AC Power Circuit Breakers Used in Enclosures (ANSI C37.50-1981)

Transient Recovery Voltage for AC High-Voltage Circuit Breakers Rated on a Symmetrical Current Basis (ANSI/IEEE C37.011-1979)

Trip Devices for AC and General Purpose DC Low-Voltage Power Circuit Breakers (IEEE C37.17-1972 & ANSI C37.17-1979)

***96. Circuit Protectors***

Low-Voltage AC Power Circuit Protectors Used in Enclosures (ANSI/IEEE C37.29-1981)

Protectors for Communication Circuits (ANSI/UL 497-1979)

Test Procedures for Low-Voltage AC Power Circuit Protectors Used in Enclosures (ANSI C37.52-1974R1980)

***97. Circuits***

Emergency and Standby Power Systems for Industrial and Commercial Applications (ANSI/IEEE 446-1980)

Fused-Power Circuit Devices (UL977)

Ground-Fault Circuit Interrupters (UL943)

Independence of Class 1E Equipment and Circuits (ANSI/IEEE 384-1981)

Measuring the Transmission Characteristics of Analog Voice Frequency Circuits (IEEE 743/1983)

Power Factor Measurement for Low-Voltage Inductive Test Circuits (ANSI/IEEE C37.26-1972R1976)

Surge Arresters for AC Power Circuits (ANSI/IEEE C62.1-1981)

Surge Voltages in Low-Voltage AC Power Circuits (ANSI/IEEE C62.41-1980)

***98. Cloth***

Asbestos, Asbestos-Varnished Cloth and Asbestos-Thermoplastic Insulated Wires and Cables (ANSI/NEMA WC1-1982)

***99. Codes***

DC Aircraft Rotating Machines (IEEE 132-1953)

DC Carbon Pile Voltage Regulators for Aircraft (IEEE 134-1955)
Industrial Control (IEEE 74-1958R1974)
Electrical Measurements in Power Circuits (IEEE 120-1955R1972)
National Board Inspection Code (A Manual for Boiler and Pressure Vessel Inspectors (ANSI/NB 23-1983)
National Electrical Code (ANSI/NFPA 70-1984)
National Electrical Safety Code (ANSI C2)
National Electrical Safety Code Interpretations (ANSI/IEEE C2 78-80)
Resistance Measurements (IEEE 118-1978)
Safety Code for Semiconductor Power Converters (NEMA PV3-1973R1979)
Terminology and Test Code for Shunt Reactors (ANSI/IEEE C57.21-1981)
Test Code for High-Voltage Air Switches (ANSI/IEEE C37.34-1971R1977)

***100. Coils***

Evaluation of Sealed Insulation Systems for AC Electrical Machinery Employing Form-Wound Stator Coils (IEEE 491-1972) COLOR CODING
Terms, Letter Symbols, and Color Code for Hall Effect Devices (IEEE 296-1969)

***101. Commercial Power***

Design of Reliable Industrial and Commercial Power Systems (ANSI/IEEE 493-1980)
Electric Power Systems in Commercial Buildings (ANSI/IEEE 241-1983)
Emergency and Standby Power Systems for Industrial and Commercial Applications (ANSI/IEEE 446-1980)
Grounding of Industrial and Commercial Power Systems (ANSI/IEEE 142-1982)
Protection and Coordination of Industrial and Commercial Power Systems (IEEE 242-1975)
Power System Analysis (ANSI/IEEE 399-1980)
Radio Noise Generated by Motor Vehicles and Affecting Mobile Communications Receivers (IEEE 263-1965)

***102. Communications and Communications Systems***

Coaxial Communication Cable [CATV] (NEMA WC41-1975R1981)
Electric Power Systems in Commercial Buildings (ANSI/IEEE 241-1983)
Emergency and Standby Power Systems for Industrial and Commercial Applications (ANSI/IEEE 446-1980)
FM Mobile Communications Receivers (ANSI/IEEE 184-1969)
Measurement of Spurious Emission from Land Mobile Communication Transmitters (ANSI/IEEE 377-1980)
Measuring Longitudinal Balance of Telephone Equipment Operating in the Voice Band (ANSI/IEEE 455-1976)
Measuring Transmission Performance of Telephone Sets (IEEE 269-1983)
Protection of Wire Line Communications Facilities Serving Electric Power Stations (ANSI/IEEE 487-1980)
Protectors for Communication Circuits (UL 497)
Service Conditions for Power Systems Communications Apparatus (281-1968)
Terms for Communication Switching (ANSI/IEEE 312-1977)

***103. Computer Languages***

ATLAS Test Language (ANSI/IEEE 416-1981)
C/ATLAS Syntax (ANSI/IEEE 717-1982)
C/ATLAS Test Language (ANSI/IEEE 716-1982)
Guide to the Use of ATLAS (ANSI/IEEE 771-1980)
Pascal Computer Programming Language (ANSI/IEEE 770X3.97-1983)
Real-Time BASIC for CAMAC (ANSI/IEEE 726-1982)

***104. Computers***

Computer-Type Pulse Transformers (IEEE272-1970R1976)
Dynamic Response Testing of Process Control Instrumentation (ISA S26)
Graphic Symbols for Distributed Control/Shared Display Instrumentation, Logic and Computer Systems (ISA S5.3)
Hardware Testing of Digital Process Computers (ANSI/ISA RP55.1-1975R1983)
Power System Analysis (ANSI/IEEE 399-1980)
Programmable Digital Computer Systems in Safety Systems of Nuclear Power Generating Stations (ANSI/IEEE/ANS 7432-1982)
Protection of Electronic Computer Data Processing Equipment (ANSI/NFPA 75-1981)

Safety for Electronic Data Processing Units and Systems (ANSI/UL 478-1979)
Terms for Analog Computers (IEEE 165-1977)
Terms for Electronic Digital Computers (IEEE 162-1963R1972)
Terms for Hybrid Computer Linkage Components (IEEE 166-1977)
Testing Bobbin Cores for Electronic Computers (IEEE 1964R1973)

***105. Conductivity***

Radio Methods for Measuring Earth Conductivity (IEEE 356-1974R1981)

***106. Conductors***

Conductor Self-Damping Measurements (IEEE 563-1978R1983)
Connectors for Use Between Aluminum or Aluminum-Copper Overhead Conductors (ANSI/NEMA CC3-1973R1978)
Impulse Voltage Tests on Insulated Conductors (IEEE 82-1963R1971)
Installation of Overhead Transmission Line Conductors (ANSI/IEEE 524-1980)
Insulated Wire Connectors for Use with Underground Conductors (UL 486D)
Performance of Aeolian Vibration Dampers for Single Conductors (ANSI/IEEE 664-1980)
Protection and Coordination of Industrial and Commercial Power Systems (IEEE 242-1975)
Test for Temperature Rise as a Function of Current in Printed Conductors (ANSI/EIA RS251A-1970)
Unplated Split-Bolt and Vise-Type Electrical Connectors for Copper Conductors (NEMA SG14-1958R1979)
Wire Connectors and Soldering Lugs for Use with Copper Conductors (UL 486A)
Wire Connectors for Use with Copper Aluminum (UL 486B)

***107. Conduit and Duct***

Communications Duct and Fittings for Underground Installations (ANSI/NEMA TC10-1978)
Corrugated Coilable Plastic Utilities Duct (ANSI/NEMA TC5-1978)
Electrical Impregnated Fiber Conduit (UL543)
Electrical Metallic Tubing (ANSI/UL 797-1983)
Electrical Plastic Tubing and Conduit (NEMA TC2-1978)
Extra-Strength Plastic Utilities Duct and Fittings for Underground Installations (ANSI/NEMA TC8-1978)
Fittings and Supports for Conduit and Cable Assemblies (ANSI/NEMA FB1-1977)
Fittings for Conduit and Boxes (UL514B)
Fittings for Plastic Utilities Duct for Underground Installations (ANSI/NEMA TC9-1978)
Flexible Metal Electrical Conduit (UL3-1979)
Flexible Nonmetallic Tubing (ANSI/UL3-1979)
Intermediate Metal Conduit (UL1242)
Liquid-Tight Flexible Steel Electrical Conduit (ANSI/UL360-1980)
Plastic Utilities Duct for Underground Installations (ANSI/NEMA TC6-1978)
Polyvinyl-Chloride Externally Coated Galvanized Rigid Steel Conduit and Electrical Metallic Tubing (NEMA RN1-1980)
PVC Fittings for Use with Rigid PVC Conduit and Tubing (NEMA TC3-1981)
Rigid Aluminum Conduit (ANSI C80.5-1983)
Rigid Metal Electrical Conduit (UL6)
Rigid PVC Conduit (ANSI/UL651&651A-1983)
Rigid Steel Conduit (ANSI C80.2-1983)
Smooth-Wall Coilable Electrical Plastic Utilities Duct (ANSI/NEMA TC5-1978)
Zinc-Coated Electrical Metallic Tubing (ANSI C80.3-1983)
Rigid Steel Conduit (ANSI C80.1-1983)

***108. Connectors and Connections, Electrical***

Application of Transformer Connections in Three-Phase Distribution Systems (ANSI/IEEE C57.105-1978)
Cable Clamping Test Procedure (ANSI/EIA RS364-43A-1983)
Connectors for Use Between Aluminum or Aluminum-Copper Overhead Conductors (ANSI/NEMA CC3-1973R1978)
Corona Testing (ANSI/EIA RS364-44-1983)
Electrical Flat Cable Type Connectors (ANSI/EIA RS429-1976R1981;ANSI/IPC FC218B)
Electrical Power Connectors for Substations (ANSI/NEMA CC1-1981)
Exposed Semiconducting Shields on Premolded High-Voltage Cable Joints and Separable Insulated Connectors (IEEE 592-1977)
Guide to Pin and Sleeve Plugs, Receptacles, and Connectors (NEMA PR3-1980)

High-Voltage Connectors for Nuclear Instruments (ANSI N42.4-1971R1978)
Humidity Test Procedure (ANSI/EIA RS364-31A-1983)
Ice Resistance of Mated Connectors (ANSI/EIA RS364-51-1983)
Insulated Wire Connectors for Use with Underground Conductors (UL 486D)
Insulation Resistance Test (ANSI/EIA RS364-21A-1983)
Plugs, Receptacles, and Connectors of the Pin and Sleeve Type for Hazardous Locations (NEMA FB11-1973R1978)
Precision Coaxial Connectors (ANSI/IEEE 287-1968)
Printed Wiring Board Electrical Connectors (ANSI/EIA RS406-1973R1979)
Qualifying Permanent Connections Used in Substation Grounding (IEEE 837-1984)
Sealed Insulated Underground Connector Systems (ANSI C119.1-1974)
Separable Insulated Connectors for Power Distribution Systems (ANSI/IEEE 386-1977)
Splicing Wire Connectors (UL 486C)
Unplated Split-Bolt and Vise-Type Electrical Connectors for Copper Conductors (NEMA SG14-1958R1979)
Wire Connectors and Soldering Lugs for Use with Copper Conductors (ANSI/UL 486A-1982)
Wire Connectors for Aluminum Conductors (ANSI/UL 486B-1982)
Withstanding Voltage Test Procedure (ANSI/EIA RS364-20A-1983)

***109. Construction***

Calibration and Control of Measuring and Test Equipment Used in the Construction and Maintenance of Nuclear Power Generating Stations (IEEE 498-1980)

***110. Containment***

Electric Penetration Assemblies in Containment Structures for Nuclear Power Generating Stations (IEEE 317-1983)

***111. Control Boards***

Qualification of Class 1E Control Boards, Panels, and Racks Used in Nuclear Power Generating Stations (IEEE 420-1982)

***112. Control Equipment and Systems***

Class 2 Transformers for Residential Controls (NEMA ST2-1973R1979)
Control and Low-Voltage Cable Systems in Substations (IEEE 525-1978)
Definition, Specification, and Analysis of Manual, Automatic, and Supervisory Station Control and Data Acquisition (ANSI/IEEE C37.1-1979)
Design of Display and Control Facilities for Central Control Rooms of Nuclear Power Generating Stations (IEEE 566-1977)
Design of the Control Room Complex for a Nuclear Power Generating Station (IEEE 567)
Dynamic Response Testing of Process Control Instrumentation (ISA S26)
Electric Control Apparatus for Land Transportation Vehicles (ANSI/IEEE 16-1955)
Electric Motor Control Centers (UL 845)
Electric Power Systems in Commercial Buildings (ANSI/IEEE 241-1983)
Electrical Guide for Control Centers (ISA RP60.8)
General Standards for Industrial Control and Systems (ANSI/NEMA ICS1-1978)
Harmonic Control and Reactive Compensation of Static Power Converters (ANSI/IEEE 519-1981)
Hot-Water Immersion Controls (NEMA DC12-1979)
Identification, Testing & Evaluation of the Dynamic Performance of Excitation Control Systems (ANSI/IEEE 421A-1978)
Industrial Control (IEEE 74-1958R1974)
Industrial Control Devices, Controllers and Assemblies (ANSI/NEMA ICS2-1978)
Installation of Electrical Equipment to Minimize Noise Inputs to Controllers from External Sources (ANSI/IEEE 518-1982)
Limit Controls (ANSI/UL353-1974)
Load Control for Use on Central Electric Heating Systems (NEMA DC22-1977)
Multiple Controllers in a CAMAC Crate (ANSI/IEEE 675-1982)
Oil Burner Primary Controls (NEMA DC9-1970R1980)
Procedures for Control of System Electromagnetic Compatibility (ANSI/IEEE C63.12-1983)
Qualifications and Certification of Instrumentation and Control Technicians in Nuclear Power Plants (ISA S67.14)
Qualifying Class 1E Motor Control Centers for Nuclear Power Generating Stations (ANSI/IEEE 649-1980)
Resistance Welding Control (ANSI/NEMA ICS5-1978)

Surface-Type Controls for Electric Water Heaters (NEMA DC5-1976)
Temperature Limit Controls for Electric Baseboard Heaters (NEMA DC10-1977)
Thyristor AC Power Controllers (ANSI/IEEE 428-1981)
Traffic Control Systems (NEMA TS1-1983)
Warm Air Limit and Fan Controls (NEMA DC4-1975R1980)

*113. Converters*

Constant-Potential-Type Electric Utility (Semiconductor Static Converter) Battery Chargers (NEMA PE5-1983)
Harmonic Control and Reactive Compensation of Static Power Converters (ANSI/IEEE 519-1981)
Safety Code for Semiconductor Power Converters (NEMA PV3-1973R1979)
Semiconductor Self-Commutated Converters (NEAM PV4-1973R1979)
Thyristor Converters and Motor Drives: Part 1—Converters for DC Motor Armature Supplies (ANSI/IEEE 444-1973)

*114. Cords, Electric*

Cord Reels (ANSI/UL 355-1980)
Cord Sets and Power Supply Cords (ANSI/UL 817-1979)

*115. Corona*

Corona (Partial Discharge) Measurements on Electronics Transformers (IEEE 436-1977)
Corona Testing (ANSI/EIA RS364-44-1983)
Partial Discharges (Corona) During Dielectric Tests (ANSI/IEEE 454-1973R1979)
Terms Relating to Overhead Power Lines Corona and Radio Noise (ANSI/IEEE 539-1979)

*116. Couplings, Electric*

Test Procedure for Electric Couplings (ANSI/IEEE 290-1980)

*117. Damping*

Conductor Self-Damping Measurements (IEEE 563-1978R1983)
Performance of Aeolian Vibration Dampers for Single Conductors (ANSI/IEEE 664-1980)

*118. Data*

Collection and Presentation of Electrical, Electronic, and Sensing Component and Mechanical Equipment Reliability Data for Nuclear Power Generating Stations (IEEE 500-1984)
Definition, Specification, and Analysis of Manual, Automatic, and Supervisory Station Control and Data Acquisition (ANSI/IEEE C37.1-1979)
Statistical Analysis of Thermal Life Test Data (IEEE 101-1972R1980)

*119. Data Processing*

Emergency and Standby Power Systems for Industrial and Commercial Applications (ANSI/IEEE 446-1980)

*120. Delay Lines*

Electromagnetic Delay Lines (ANSI/EIA RS242-1961)

*121. Designations*

Letter Designations for Radar Frequency Bands (IEEE 521-1976)
Reference Designations for Electronic Parts and Equipment (ANSI/IEEE 200-1975)

*122. Detectors*

Amplifiers and Preamplifiers for Semiconductor Radiation Detectors for Ionizing Radiation (ANSI/IEEE 301-1976R1982)
Calibration and Usage of Germanium Detectors for Measurement of Gamma-Ray Emission of Radionuclides (ANSI N42.14-1978)
Calibration and Usage of Sodium Iodide Detector Systems (ANSI N42.12-1980)
Germanium Gamma-Ray Detectors (ANSI/IEEE 325-1971R1982)
Germanium Semiconductor Detector Gamma-Ray Efficiency Using a Standard Marinelli (Reentrant) Beaker Geometry (ANSI/IEEE 680-1978)
High-Purity Germanium Detectors for Ionizing Radiation (ANSI/IEEE 645-1977)
Semiconductor Radiation Detectors (ANSI/IEEE 300-1982)

*123. Diagrams*

Electrical and Electronic Diagrams (ANSI Y14.15-1966R1973)
Graphic Symbols for Electrical Wiring and Layout Diagrams Used in Architecture and Building Construction (ANSI Y32.9-1972)
Graphic Symbols for Grid and Mapping Diagrams Used in Cable Television Systems (ANSI/IEEE 623-1976)
Graphic Symbols for Logic Diagrams (ANSI/IEEE 91-1984)

*124. Dictionaries*

Dictionary of Electrical and Electronics Terms (ANSI/IEEE 100-1984)
Dictionary of Terms for Computer-Aided Preparation of Product Definition Data (ANSI Y14.26.3-1975)

***125. Dielectric Tests***

Dielectric Test Requirements for Power Transformers for Operation at System Voltages from 115kV to 230kV (IEEE C57.12.14)
Dielectric Test Requirements for Power Transformers for Operation of Effectively Grounded Systems (IEEE 262B-1977)
Partial Discharges (Corona) During Dielectric Tests (ANSI/IEEE 454-1973R1979)
Techniques for High-Voltage Testing (ANSI/IEEE 4-1978)

***126. Diesel***

Diesel-Generator Units Applied as Standby Power Supplies for Nuclear Power Generating Stations (IEEE 387-1983)

***127. Digital***

Code and Format Conventions for Use with ANSI/IEEE Std 488-1978 (ANSI/IEEE 728-1982)
Digital Interface for Programmable Instrumentation (ANSI/IEEE 488-1978)
Digital Terms Relating to Television (ANSI/IEEE 847-1982)
Modular Instrumentation and Digital Interface System (ANSI/IEEE 582-1982)

***128. Dimmers***

Semiconductor Dimmers for Incandescent Lamps (NEMA WD2-1970R1980)

***129. Direction Finders***

Direction Finder Measurements (IEEE 173-1959)

***130. Distortion***

Measurement of Aspect Ratio and Geometric Distortion (ANSI/IEEE 202-1954R1978)
Video Signal Transmission Measurements of Linear Waveform Distortion (ANSI/IEEE 511-1979)

***131. Distribution***

Application of Transformer Connections in Three-Phase Distribution Systems (ANSI/IEEE C57.105-1978)
Cement Plant Power Distribution (ANSI/IEEE 277-1983)
Electric Power Distribution for Industrial Plants (IEEE 141-1976)
Electric Power Systems in Commercial Buildings (ANSI/IEEE 241-1983)
Power Operations Terminology Including Terms for Reporting and Analyzing Outages of Electrical Transmission and Distribution Facilities and Interruptions to Customer Service (ANSI/IEEE 346-1973)
Separable Insulated Connectors for Power Distribution Systems (ANSI/IEEE 386-1977)

***132. Documents and Documentation***

Engineering Drawing and Related Documentation Practices (ANSI Y14.26M-1981)
Identification of Documents Related to Class 1E Equipment and Systems for Nuclear Power Generating Stations (ANSI/IEEE 494-1974R1983)
Software Test Documentation (ANSI/IEEE 829-1983)

***133. Dosimeters***

Inspection and Test Specifications for Direct and Indirect Reading Quartz Fiber Pocker Dosimeters (ANSI N322-1977R1983)

***134. Drafting Practice***

Dictionary of Terms for Computer-Aided Preparation of Product Definition Data (ANSI Y14.26.3-1975)
Drawing Sheet Size and Format (ANSI Y14.1-1980)
Electrical and Electronic Diagrams (ANSI Y14.15-1966R1973)
Engineering Drawing and Related Documentation Practices (ANSI Y14.26M-1981)
Engineering Drawing and Related Documentation Practices—Parts Lists, Data Lists, and Index Lists (ANSI Y14.34M-1982)
Line Conventions and Lettering (ANSI Y14.2M-1979)

***135. Drawings***

Construction Drawings of Line Impedance Network for Measurement of Conducted Interference Output to the Power Line from FM and TV Broadcast Receivers (IEEE 214-1961)
Engineering Drawing and Related Documentation Practices (ANSI Y14.26M-1981)
Engineering Drawing and Related Documentation Practices—Parts Lists, Data Lists, and Index Lists (ANSI Y14.34M-1982)

***136. Drives***

Cement Plant Electric Drives and Related Equipment (IEEE 499-1983)
General Purpose Thyristor DC Drives (IEEE 597-1983)
Safety Standards for Construction and Guide for Selection, Installation and Operation of Adjustable-Speed Drive Systems (NEMA ICS3.1-1979)
Steam Turbines for Mechanical Drive Service (NEMA SM23-1979)
Thyristor Converters and Motor Drives: Part 1—Converters for DC Motor Armature Supplies (ANSI/IEEE 444-1973)

***137. Ducts***

Corrugated Polyolefin Coilable Plastic Utilities Duct (ANSI/NEMA TC5-1978)
Extra-Strength PVC Plastic Utilities Duct for Underground Installation (ANSI/NEMA TC8-1978)
Fittings for ABS and PVC Plastic Utilities Duct for Underground Installation (ANSI/NEMA TC9-1978)
PVC and ABS Plastic Communications Duct and Fittings for Underground Installation (ANSI/NEMA TC10-1983)
PVC and ABS Plastic Utilities Duct for Underground Installation (ANSI/NEMA TC6-1978)
Smooth Wall Coilable Polyethylene Electrical Plastic Duct (ANSI/NEMA TC7-1978)

***138. Duty and Duty Cycles***

Rating Electrical Apparatus for Short Time, Intermittent, or Varying Duty (IEEE 96-1969)

***139. Dynamic Performance***

Identification, Testing & Evaluation of the Dynamic Performance of Excitation Control Systems (ANSI/IEEE 421A-1978)

***140. Effluents***

Specification and Performance of On-Site Instrumentation for Continuously Monitoring Radioactivity in Effluents (ANSI N42.18-1980)

***141. Electric Equipment***

Quality Assurance Program Requirements for the Design and Manufacture of Class 1E Instrumentation and Electric Equipment for Nuclear Power Generating Stations (IEEE 467-1980)
Temperature Limits in Rating Electrical Equipment (IEEE 1-1969)
Thermal Evaluation of Insulation Systems for Electric Equipment (ANSI/IEEE 99-1980)
Transformer Askarel in Equipment (IEEE 76-1974)

***142. Electric Line Construction***

Anchor Rods and Nuts for Overhead Line Construction (ANSI C135.2-1979)
Bolt Type Insulator Pins with Lead Threads for Overhead Line Construction (ANSI C135.17-1979)
Bolts and Nuts for Overhead Line Construction (ANSI C135.1-1979)
Crossarm Gains (ANSI C135.33-1980)
Eye Bolts and Nuts for Overhead Line Construction (ANSI C135.4-1979)
Eyenuts and Eyelets for Overhead Line Construction (ANSI C135.5-1979)
Ground Rods for Overhead or Underground Line Construction (ANSI C135.30-1979)
Installation of Overhead Transmission Line Conductors (ANSI/IEEE 524-1980)
Pole-Top Insulator Pins with Lead Threads for Overhead Line Construction (ANSI C135.22-1979)
Spool Insulator Bolts (ANSI C135.31-1980)
Staples with Rolled Slash Points for Overhead Line Construction (ANSI C135.14-1979)
Underground Cable Racks and Cable Rack Hooks (ANSI C135.35-1980)

***143. Electric Power***

Electric Power Distribution for Industrial Plants (IEEE 141-1976)

***144. Electric Systems***

Aircraft Electric Systems (IEEE 128-1976)
Grounded 830V Three-Phase Electrical System for Oil Field Service (ANSI/IEEE 464-1981)

***145. Electrical Apparatus and Equipment***

Electrical Equipment for Use in Hazardous Locations (UL1064)
Rating Electrical Apparatus for Short Time, Intermittent, or Varying Duty (IEEE 96-1969)
Reference Ambient Conditions for Measurements of Electrical Apparatus (ANSI/IEEE 3-1983)
Relays and Relay Systems Associated with Electric Power Apparatus (ANSI/IEEE C37.90-1978R1982)
Temperature Measurement as Applied to Electrical Apparatus (IEEE 119-1974)

*146. Electroacoustics*

Terms for Audio and Electroacoustics (IEEE 151-1965R1971)
Volume Measurements of Electrical Speech and Program Waves (ANSI/IEEE 152-1953R1971)

*147. Electrodes*

Manufactured Graphite Electrodes (NEMA CG1-1980)

*148. Electromagnetism and Electromagnetic Radiation*

Electromagnetic Interference Filters (UL1283)
Electromagnetic Noise and Field Strength Instrumentation (ANSI C63.2-1980)
Electromagnetic Site Survey (IEEE 473-1983)
Measurement of Hazardous Electromagnetic Fields—RF and Microwave (ANSI C95.5-1981)
Measuring Electromagnetic Field Strength for Frequencies Below 1000 MHz in Radio Wave Propagation (IEEE 302-1969R1981)
Procedures for Control of System Electromagnetic Compatibility (ANSI/IEEE C63.12-1983)
Safety Level of Electromagnetic Radiation with Respect to Personnel (ANSI C95.1-1982)
Techniques and Instrumentation for the Measurement of Potentially Hazardous Electromagnetic Radiation at Microwave Frequencies (ANSI C95.3-1973R1979)

*149. Electrometers*

Interrelationship of Quartz-Fiber Electrometer Type Exposure Meters and Companion Exposure Meter Chargers (ANSI N42.6-1980)

*150. Elevators*

Electric Power Systems in Commercial Buildings (ANSI/IEEE 241-1983)
Elevators, Escalators, and Moving Walks (ANSI/ASME A17.1-1981)
Emergency and Standby Power Systems for Industrial and Commercial Applications (ANSI/IEEE 446-1980)
Inspection of Elevators, Escalators, and Moving Walks (ANSI/ASME A17.2-1982)

*151. Emergency*

Design of Reliable Industrial and Commercial Power Systems (ANSI/IEEE 493-1980)
Emergency and Standby Power Systems for Industrial and Commercial Applications (ANSI/IEEE 446-1980)
Emergency Lighting and Power Equipment (UL924)
Performance Specification for Reactor Emergency Radiological Monitoring Instrumentation (ANSI N320-1979)

*152. Emission*

Calibration and Usage of Germanium Detectors for Measurement of Gamma-Ray Emission of Radionuclides (ANSI N42.14-1978)
Measurement of Spurious Emission from Land Mobile Communication Transmitters (ANSI/IEEE 377-1980)
Radio Noise Emissions from Low-Voltage, Electrical and Electronic Equipment (ANSI C63.4-1981)

*153. Enclosures*

Circuit Breakers and Circuit-Breaker Enclosures for Use in Hazardous Locations (UL877)
Electric Power Systems in Commercial Buildings (ANSI/IEEE 241-1983)
Enclosures for Electrical Equipment (NEMA 250-1979)
Enclosures for Industrial Controls and Systems (ANSI/NEMA ICS6-1978)
Field Discharge Circuit Breakers Used in Enclosures for Rotating Electrical Machinery (ANSI/IEEE C37.18-1979)
Low Voltage AC Power Circuit Breakers Used in Enclosures (ANSI/IEEE C37.13-1981)
Low-Voltage AC Power Circuit Protectors Used in Enclosures (ANSI/IEEE C37.27-1972)
Low Voltage DC Power Circuit Breakers Used in Enclosures (ANSI/IEEE C37.14-1979)
Measurement of Shielding Effectiveness of Shielding Enclosures (IEEE 299-1969)
Molded Case Circuit Breakers and Circuit Breaker Enclosures (UL489)

*154. Excitation*

Excitation Systems for Synchronous Machines (IEEE 421-1972)
High Potential Test Requirements for Excitation Systems for Synchronous Machines (ANSI/IEEE 421B-1979)
Identification, Testing & Evaluation of the Dynamic Performance of Excitation Control Systems (ANSI/IEEE 421A-1978)

***155. Exposure Meters***

Interrelationship of Quartz-Fiber Electrometer Type Exposure Meters and Companion Exposure Meter Chargers (ANSI N42.6-1980)

***156. Facsimile***

High Definition Facsimile Test Chart (IEEE 167A-1980)

Terms for Facsimile (ANSI/IEEE 168-1956R1971)

Test Procedure for Facsimile (ANSI/IEEE 167-1966R1971)

***157. Failure***

Single-Failure Criterion to Nuclear Power Generating Station Class 1E Systems (ANSI/IEEE 379-1977)

***158. Faults and Fault Calculations***

Application Guide for Ground Fault Protective Devices for Equipment (NEMA PB1.2-1977)

Calculation of Fault Current for Application of AC High-Voltage Circuit Breakers Rated on a Total Current Basis (ANSI/IEEE C37.5-1979)

Electric Power Distribution for Industrial Plants (IEEE 141-1976)

Ground-Fault Circuit Interruptors (UL943)

Ground-Fault Sensing and Relaying Equipment (UL1053)

Maximum Electric Power Station Ground Potential Rise and Induced Voltage from a Power Fault (IEEE 367-1979)

Requirements for Overhead, Pad Mounted Dry Vault and Submersible Automatic Circuit Reclosers and Fault Interrupters for AC Systems (ANSI/IEEE C37.60-1981)

Synthetic Fault Testing of AC High-Voltage Circuit Breakers Rated on a Symmetrical Current Basis (ANSI/IEEE C37.081-1981)

***159. Fiber Optics***

Terms Relating to Fiber Optics (IEEE 812-1983)

***160. Fields and Field Intensity***

Electromagnetic Noise and Field Strength Instrumentation (ANSI C63.2-1980)

Field Intensity Above 300 MHz from RF Industrial, Scientific, and Medical Equipments (IEEE 139-1952)

Measurement of Electric and Magnetic Fields from AC Power Lines (ANSI/IEEE 644-1979)

Measurement of Hazardous Electromagnetic Fields—RF and Microwave (ANSI C95.5-1981)

Measurement Procedure for Field Disturbance Sensor (RF Intrustion Alarm) (ANSI/IEEE 475-1983)

Measuring Electromagnetic Field Strength for Frequencies Below 1000 MHz in Radio Wave Propagation (IEEE 302-1969R1981)

Measuring Field Strength, Continuous Wave, Sinusoidal (IEEE 284-1968)

Measuring Field Strength in Radio Wave Propagation (IEEE 291-1969R1981)

***161. Filters***

Electric Power Distribution for Industrial Plants (IEEE 141-1976)

Electromagnetic Interference Filters (UL1283)

Measurement of Electric Noise and Harmonic Filter Performance of High-Voltage DC Systems (ANSI/IEEE 368-1977)

Radio Interference Filters (ANSI/EIA RS416-1974R1981)

***162. Fire Protection***

Cable Penetration Fire Stop Qualification Test (ANSI/IEEE 634-1978)

Emergency and Standby Power Systems for Industrial and Commercial Applications (ANSI/IEEE 446-1980)

Life Safety Code (ANSI/NFPA 101-1981)

Uniform Coding for Fire Protection (ANSI/NFPA 901-1981)

***163. Fittings, Electrical***

Cellular Metal Floor Electrical Raceways and Fittings (UL 209)

Electrical Outlet Boxes and Fittings for Use in Hazardous Locations (UL 886)

Fittings for Conduit and Boxes (ANSI/UL514B-1983)

Porcelain Cleats, Knobs, and Tubes (ANSI/UL 511-1980)

PVC Fittings for Use with Rigid PVC Conduit and Tubing (NEMA TC3-1981)

Surface Metal Electrical Raceways and Fittings (UL 5)

Underfloor Electrical Raceways and Fittings (UL 884)

***164. Flashlights***

Electric Flashlights for Use in Hazardous Locations (ANSI/UL 783-1979)

*165. Flow Graphs*

Terms for Linear Signal Flow Graphs (IEEE 155-1960R1983)

*166. Flow Measurement*

Specification, Installation, and Calibration of Turbine Flow Meters (ANSI/ISA RP31.1-1977)

*167. Flux*

Measuring Recorded Flux of Magnetic Sound Records at Medium Wavelengths (ANSI/IEEE 347-1972)

*168. Fly Ash*

Laboratory Measurement and Reporting of Fly Ash Resistivity (ANSI/IEEE 548-1981)

*169. Frequency*

Aerospace Equipment Voltage and Frequency Ratings (IEEE 127-1963)

*170. Frequency Modulation*

Construction Drawing of Line Impedance Network for Measurement of Conducted Interference Output to the Power Line from FM and TV Broadcast Receivers (ANSI/IEEE 214-1961)
Measurement of Conducted Interference Output to the Power Line from FM and TV Broadcast Receivers (ANSI/IEEE 213-1961R1974)

*171. Fuses and Holders*

Conformance Test Procedure for High-Voltage Fuses (ANSI C37.53.1-1982)
Design Test for High-Voltage Fuses, Distribution Enclosed Single-Pole Air Switches, Fuse Disconnecting Switches and Accessories (ANSI/IEEE C37.49-1981)
Fuseholders (ANSI/UL512-1981)
Fuses (ANSI/UL198-1982)
High-Voltage Fuses (NEMA SG2-1981)
Low-Voltage Cartridge Fuses (ANSI C97.1-1972R1978)
Protection and Coordination of Industrial and Commercial Power Systems (IEEE 242-1975)
Service Conditions and Definitions for High-Voltage Fuses, Distribution Enclosed Single-Pole Air Switches, Fuse Disconnecting Switches and Accessories (ANSI/IEEE C37.40-1981)

*172. Gain*

Measurement of Differential Gain and Differential Phase (ANSI/IEEE 206-1960R1978)

*173. Geiger Counters*

Bases for Geiger-Muller Counter Tubes (ANSI N42.5-1965R1983)
Geiger-Muller Counters (ANSI/IEEE 309-1970)

*174. Generating Stations*

Maintenance, Testing, and Replacement of Large Lead Storage Batteries for Generating Stations and Substations (ANSI/IEEE 450-1980)

*175. Generation*

Automatic Generation Control on Electric Power Systems (IEEE 94-1970)

*176. Generators*

Aircraft AC Generators (IEEE 138-1960)
Aircraft Generator and Regulator Characteristics (IEEE 136-1959)
DC Machines (IEEE 113-1973)
DC Tachometer Generators (IEEE 251-1963R1972)
Diesel-Generator Units Applied as Standby Power Supplies for Nuclear Power Generating Stations (IEEE 387-1983)
Electric Motors and Generators for Use in Hazardous Locations (UL 674)
Emergency and Standby Power Systems for Industrial and Commercial Applications (ANSI/IEEE 446-1980)
Generator Ground Protection (ANSI/IEEE C37.101-1983)
Ground-Fault Neutralizers, Grounding of Synchronous Generator Systems, Neutral Grounding of Transmission Systems (IEEE 143-1954)
Marine Electric Motors and Generators (UL 1112)
Motors and Generators (ANSI/NEMA MG1-1978)
Operation and Maintenance of Hydro-Generators (ANSI/IEEE 491-1974R1981)
Operation and Maintenance of Turbine Generators (ANSI/IEEE 67-1972R1980)
Periodic Testing of Diesel-Generator Units Applied as Standby Power Supplies for Nuclear Power Generating Stations (IEEE 387-1983)
Polyphase Induction Motors and Generators (IEEE 112-1983)
Preparation of Equipment Specifications for Speed-Governing of Hydraulic Turbines Intended to Drive Electric Generators (ANSI/IEEE 125-1977)

Protection and Coordination of Industrial and Commercial Power Systems (IEEE 242-1975)
Renewal Parts for Motors and Generators (NEMA RP1-1981)
Requirements for Gas Turbine Driven Cylindrical-Rotor Synchronous Generators (ANSI C50.14-1977)
Safety Standard for Construction and Guide for Selection, Installation and Use of Electric Motors and Generators (ANSI/NEMA MG2-1977)
Specification for Speed-Governing of Internal Combustion Engine Generator Units (IEEE 126-1959R1983)
Speed-Governing of Steam Turbines Intended to Drive Electric Generators (IEEE 122-1959)
Synchronous Machines (ANSI/IEEE 115-1983)

***177. Graphic Symbols*** (See Symbols)

***178. Graphite***

Characterizing the Physical and Chemical Properties of EDM Graphite (NEMA CB2-1980)
Graphite Electrolytic Anodes (NEMA CG2-1969R1980)
Manufactured Graphite Electrodes (NEMA CG1-1980)

***179. Ground Faults and Ground-Fault Protection***

Application Guide for Ground-Fault Protective Devices for Equipment (NEMA PB1.2-1977)
Ground-Fault Neutralizers, Grounding of Synchronous Generator Systems, Neutral Grounding of Transmission Systems (IEEE 143-1954)
Ground Fault Sensing and Relaying Equipment (ANSI/UL 1053-1981)
Protection and Coordination of Industrial and Commercial Power Systems (IEEE 242-1975)

***180. Ground Potential***

Maximum Electric Power Station Ground Potential Rise and Induced Voltage from a Power Fault (IEEE 367-1979)

***181. Grounds and Grounding***

Electric Power Distribution for Industrial Plants (IEEE 141-1976)
Electric Power Systems in Commercial Buildings (ANSI/IEEE 241-1983)
Electrical Grounding and Bonding Equipment (UL467)
Generator Ground Protection (ANSI/IEEE C37.101-1983)
Ground-Fault Circuit Interrupters (UL943)
Ground-Fault Sensing and Relaying Equipment (UL1053)
Grounding and Bonding Equipment (ANSI/UL 467-1972)
Grounding of Industrial and Commercial Power Systems (ANSI/IEEE 142-1982)
Measuring Earth Resistivity (IEEE 81-1983)
Neutral Grounding Devices (IEEE 32-1972R1978)
Power System Analysis (ANSI/IEEE 399-1980)
Qualifying Permanent Connections Used in Substation Grounding (IEEE 837-1984)
Safety in AC Substation Grounding (IEEE 80-1976)

***182. Gyros***

Specification Format Guide and Test Procedure for Single-Axis Laser Gyros (ANSI/IEEE 647-1981)
Specification Format Guide and Test Procedure for Single-Degree-of-Freedom Rate Integrating Gyros (ANSI/IEEE 517-1974R1980)
Specification Format for Single-Degree-of-Freedom Spring-Restrained Rate Gyros (IEEE 292-1969R1978)
Strapdown Applications to Specification Format Guide and Test Procedure for Single-Degree-of-Freedom Rate Integrating Gyros (ANSI/IEEE 529-1981)
Test Procedure for Single-Degree-of-Freedom Spring-Restrained Rate Gyros (IEEE 293-1969R1978)

***183. Hall Effect***

Terms, Letter Symbols, and Color Code for Hall Effect Devices (IEEE 296-1969)

***184. Heaters and Heating Equipment***

Design and Installation of Electric Pipe Heating Control and Alarm Systems for Power Generating Stations (ANSI/IEEE 622A/1984)
Design and Installation of Electric Pipe Heating Systems for Nuclear Power Generating Stations (ANSI/IEEE 622/1979)
Electric Baseboard Heating Equipment (ANSI/UL 1042-1978)
Electric Heaters for Use in Hazardous Locations (ANSI/UL 823-1977)

Emergency and Standby Power Systems for Industrial and Commercial Applications (ANSI/IEEE 446-1980)
Induction and Dielectric Heating Equipment (IEEE 54-1955)
Field Intensity Above 300 MHz from RF Industrial, Scientific, and Medical Equipments (IEEE 139-1952)
Line-Voltage Integrally-Mounted Thermostats for Electric Heaters (NEMA DC13-1979)
Load Control for Use on Central Electric Heating Systems (NEMA DC22-1977)
Manual for Calculating Heat Loss and Heat Gain for Electric Comfort Conditioning (NEMA HE1-1980)
Surface-Type Controls for Electric Water Heaters (NEMA DC5-1976)
Temperature Limit Controls for Electric Baseboard Heaters (NEMA DC10-1977)

***185. High Potential***

High Potential Test Requirements for Excitation Systems for Synchronous Machines (ANSI/IEEE 421B-1979)

***186. High Voltage***

AC High-Voltage Circuit Breakers (NEMA SG4-1975)
Application Guide for AC High-Voltage Circuit Breakers Rated on a Symmetrical Current Basis (ANSI/IEEE C37.010-1979)
Application, Installation, Operation and Maintenance of High-Voltage Air Disconnecting and Load Interrupter Switches (ANSI/IEEE C37.35-1976R1981)
Calculation of Fault Current for Application of AC High-Voltage Circuit Breakers Rated on a Total Current Basis (ANSI/IEEE C37.5-1979)
Capacitance Current Switching for AC High-Voltage Circuit Breakers Rated on a Symmetrical Current Basis (ANSI/IEEE C37.012-1979)
Definitions and Requirements for High-Voltage Air Switches, Insulators, and Bus Supports (ANSI/IEEE C37.30-1971R1977)
Exposed Semiconducting Shields on Premolded High Voltage Cable Joints and Separable Insulated Connectors (IEEE 592-1977)
High Direct Voltage Tests on Power Cable Systems in the Field (ANSI/IEEE 400-1980)
High-Voltage Connectors for Nuclear Instruments (ANSI N42.4-1971R1978)
High-Voltage Fuses (NEMA SG2-1981)
High-Voltage Industrial Control Equipment (UL347)
High-Voltage Insulators (HV1-1978)
High-Voltage X-Ray Cables and Receptacles (NEMA XR7-1979)
Insulation Systems for Large High-Voltage Machines (ANSI/IEEE 434-1973R1979)
Insulation Testing of Large AC Rotating Machinery with High Direct Voltage (ANSI/IEEE 95-1977R1982)
Insulation Testing of Large AC Rotating Machinery with High Voltage at Very Low Frequencies (ANSI/IEEE 433-1974R1979)
Loading Guide for AC High-Voltage Switches (ANSI/IEEE C37.37-1979)
Measurement of Electric Noise and Harmonic Filter Performance of High-Voltage DC Systems (ANSI/IEEE 368-1977)
Measuring Resistivity of Cable-Insulation Materials at High Direct Voltages (IEEE 402-1974R1982)
Methods of Measurement of Radio Influence Voltage (RIV) of High-Voltage Apparatus (NEMA 107-1964R1981)
Rating Structure for AC High-Voltage Circuit Breakers Rated on a Symmetrical Current Basis (ANSI/IEEE C37.04-1979)
Safety in High-Voltage and High Power Testing (IEEE 510-1983)
Safety Requirements for X-Radiation Limits for AC High-Voltage Power Vacuum Interrupters Used in Power Switchgear (IEEE C37.85-1982R1978)
Switching Surge Testing of Extra-High-Voltage Switches (IEEE 271-1966R1977)
Synthetic Fault Testing of AC High-Voltage Circuit Breakers Rated on a Symmetrical Current Basis (ANSI/IEEE C37.081-1981)
Techniques for High Voltage Testing (ANSI/IEEE 4-1978)
Test Code for High-Voltage Air Switches (ANSI/IEEE C37.34-1971R1977)
Test Procedure for AC High-Voltage Circuit Breakers Rated on a Symmetrical Current Basis (ANSI/IEEE C37.09-1979)

Transient Recovery Voltage for AC High-Voltage Circuit Breakers Rated on a Symmetrical Current Basis (ANSI/IEEE C37.011-1979)

***187. Identification***

Implementation of Unique Identification in Power Plants and Related Facilities (IEEE 804-1983)
Principles and Definitions for Unique Identification in Power Plants and Related Facilities (IEEE 803-1983 & 803A-1983)
System Identification in Nuclear Power Plants and Related Facilities (IEEE 805-1984)

***188. Impedance***

Construction Drawings of Line Impedance Network for Measurement of Conducted Interference Output to the Power Line from FM and TV Broadcast Receivers (IEEE 214-1961)
Measuring Earth Resistivity (IEEE 81-1983)
Measuring Unbalanced Transmission-Line Impedance (IEEE 314-1971)

***189. Impulse Tests***

Impulse Voltage Tests on Insulated Conductors (IEEE 82-1963R1971)
Measurement of Impulse Strength and Impulse Bandwidth (ANSI/IEEE 376-1975)
Transformer Impulse Tests (ANSI/IEEE C57.98-1968)

***190. Inductors***

Charging Inductors (306-1969R1981)
High Reliability in Electronics Transformers and Inductors (ANSI/IEEE 392-1976R1982)
Power Filter Inductors for Electronic Equipment (ANSI/EIA RS197A-1973)
Testing Electronics Transformers and Inductors (IEEE 389-1979)

***191. Industrial Control***

Electric Industrial Control Equipment (UL508)
Enclosures for Industrial Controls and Systems (ANSI/NEMA ICS6-1978)
General Standards for Industrial Control and Systems (ANSI/NEMA ICS1-1978)
High-Voltage Industrial Control Equipment (ANSI/UL347-1978)
Industrial Control (IEEE 74-1958R1974)
Industrial Control Devices, Controllers and Assemblies (ANSI/NEMA ICS2-1978)
Industrial Control Equipment (ANSI/UL508-1983)
Industrial Control Equipment for Use in Hazardous Locations (ANSI/UL698-1973)
Industrial Systems (ANSI/NEMA ICS3-1978)
Resistance Welding Control (ANSI/NEMA ICS5-1978)
Terminal Blocks for Industrial Control Equipment and Systems (ANSI/NEMA ICS4-1977)

***192. Industrial Equipment***

Field Intensity Above 300 MHz from RF Industrial, Scientific, and Medical Equipment (IEEE 139-1952)

***193. Industrial Plants and Systems***

Cement Plant Power Distribution (ANSI/IEEE 277-1983)
Design of Reliable Industrial and Commercial Power Systems (ANSI/IEEE 493-1980)
Electric Power Distribution for Industrial Plants (IEEE 141-1976)
Emergency and Standby Power Systems for Industrial and Commercial Applications (ANSI/IEEE 446-1980)
Grounding of Industrial and Commercial Power Systems (ANSI/IEEE 142-1982)
Improved Electrical Maintenance and Safety in the Cement Industry (IEEE 625-1979)
Industrial Systems (ANSI/NEMA ICS3-1978)
Power System Analysis (ANSI/IEEE 399-1980)
Protection and Coordination of Industrial and Commercial Power Systems (IEEE 242-1975)

***194. Information Theory***

Terms for Information Theory (IEEE 171-1958)

***195. Installations***

Design and Installation of Cable Systems in Power Generating Stations (IEEE 422-1977)
Electrical Installations of Packaging Machinery and Associated Equipment (ANSI/IEEE 333-1980)
Electrical Installations on Shipboard (IEEE 45-1983)
Electric Installations on Textile Machinery (IEEE 77-1965R1972)
Installation Design and Installation of Large Lead Storage Batteries for Generating Stations and Substations (IEEE 484-1981)
Installation, Inspection, and Testing Requirements for Class 1E Instrumentation and Electric

Equipment at Nuclear Power Generating Stations (ANSI/IEEE 336-1980)
Installation of Electrical Equipment to Minimize Noise Inputs to Controllers from External Sources (ANSI/IEEE 518-1982)
Installation of Overhead Transmission Line Conductors (ANSI/IEEE 524-1980)
Test Blocks and Cabinets for Installation of Self-Contained "A" Base Watthour Meters (ANSI C12.8-1981)

***196. Instruments and Instrumentation***

AC-DC Transfer Instruments and Converters (ANSI C100.4-1973)
Analog Instruments—Panelboard Types, Electrical (UL1437)
ATLAS Test Language (ANSI/IEEE 416-1981)
Automatic Null-Balancing Electrical Measuring Instruments (ANSI C39.4-1966R1972)
Code and Format Conventions for Use with ANSI/IEEE Std 488-1978 (ANSI/IEEE 728-1982)
Compatability of Analog Signals for Electronic Industrial Process Instruments (ANSI/ISA S50.1-1975R1982)
DC Instrument Shunts (IEEE 316-1971)
Digital Interface for Programmable Instrumentation (ANSI/IEEE 488-1978)
Digital Measuring Instruments (ANSI C39.6-1983)
Digital Panel Instruments (NEMA II1-1976)
Direct-Acting Electrical Recording Instruments (ANSI C39.2-1964R1969)
Dynamic Response Testing of Process Control Instrumentation (ANSI/ISA S26-1968)
Electric Power Distribution for Industrial Plants (IEEE 141-1976)
Electrical Analog Indicating Instruments (ANSI C39.1-1981)
Electrical and Electronic Measuring and Controlling Instrumentation (ANSI C39.5-1974)
Electrical Indicating Instrument Relays (NEMA II2-1972R1977)
Electrical Instruments in Hazardous Atmospheres (ISA RP12.1)
Electrical Instruments in Hazardous Dust Locations (ISA S12.11)
Electromagnetic Noise and Field Strength Instrumentation (ANSI C63.2-1980)
Electronic Analog Voltmeters (ANSI C39.7-1975)
Graphic Symbols for Distributed Control/Shared Display Instrumentation, Logic and Computer Systems (ISA S5.3)
High-Voltage Connectors for Nuclear Instruments (ANSI N42.4-1971R1978)
Installation, Inspection, and Testing Requirements for Class 1E Instrumentation and Electric Equipment at Nuclear Power Generating Stations (ANSI/IEEE 336-1980)
Installation of Intrinsically Safe Instrument Systems in Hazardous Locations (ANSI/ISA RP12.6-1976)
Instrument Loop Diagrams (ISA S5.4)
Instrumentation Symbols and Identification (ISA S5.1)
Materials for Instruments in Radiation Service (ISA RP25.1)
Modular Instrumentation and Digital Interface System (ANSI/IEEE 582-1982)
Nuclear Safety-Related Instrument Sensing Line Piping and Tubing Standards for Use in Nuclear Power Plants (ISA S67.02)
Performance Criteria for Instrumentation Used for Inplant Plutonium Monitoring (ANSI N317-1980)
Performance Specification for Reactor Emergency Radiological Monitoring Instrumentation (ANSI N320-1979)
Post Accident Monitoring Instrumentation for Nuclear Power Generating Stations (ANSI/IEEE 497-1981)
Process Instrumentation Terminology (ANSI/ISA S51.1-1979)
Qualifications and Certification of Instrumentation and Control Technicians in Nuclear Power Plants (ISA S67.14)
Radiation Protection Instrumentation Test and Calibration (ANSI N322-1977R1983)
Requirements for Instrument Transformers (ANSI/IEEE C57.13-1978)
Setpoints for Nuclear Safety-Related Instrumentation Used in Nuclear Power Plants (ISA S67.04)
Shock Testing for Electrical Indicating Instruments (ANSI C39.3-1976)
Signal Connectors for Nuclear Instruments (ANSI N544-1968R1979)
Specification and Performance of On-Site Instrumentation for Continuously Monitoring Radioactivity in Effluents (ANSI N42.18-1980)
Specification of X- or Gamma-Radiation Survey Instruments (ANSI N13.4-1971R1983)

Techniques and Instrumentation for the Measurement of Potentially Hazardous Electromagnetic Radiation at Microwave Frequencies (ANSI C95.3-1973R1979)

***197. Insulating Materials, Systems, and Electric Insulation***

Acceptance and Maintenance of Insulating Oil in Equipment (ANSI/IEEE C57.106-1977)
Aircraft, Missile, and Space Equipment Electrical Insulation Tests (IEEE 135-1969)
Application Guide for Porcelain Suspension Insulators (NEMA HV 2-1974R1979)
Coated Electrical Sleeving (NEMA VS1-1962R1978)
Composite Slot and Phase Insulation (NEMA VF20-1978)
Double Insulation Systems for Use in Electrical Equipment (UL1097)
Evaluation and Classification of Insulation Systems for DC Machines (ANSI/IEEE 304-1977R1982)
Evaluation of Insulation Systems for Electronic Power Transformers (IEEE 266-1969R1981)
Evaluation of Sealed Insulation Systems for AC Electrical Machinery Employing Form-Wound Stator Coils (IEEE 491-1972)
Field Testing Power Apparatus Insulation (IEEE 62-1978)
General Systems of Insulating Materials (UL 1446)
High-Pressure Decorative Laminates (NEMA LD3-1980)
High-Voltage Insulators (HV1-1978)
Industrial Laminated Thermosetting Products (NEMA LI1-1971R1976)
Insulating Tape (UL 510)
Insulation Maintenance of Large AC Rotating Machinery (ANSI/IEEE 56-1977R1982)
Insulation Resistance Test (ANSI/EIA RS364-21A-1983)
Insulation Systems for Large High-Voltage Machines (ANSI/IEEE 434-1973R1979)
Insulation Testing of Large AC Rotating Machinery with High Direct Voltage (ANSI/IEEE 95-1977R1982)
Manufactured Electrical Mica (NEMA FI1-1977)
Measurement of Power Factor Tip-up of Rotating Machinery Stator Coil Insulation (ANSI/IEEE 286-1975R1981)
Measuring Resistivity of Cable-Insulation Materials at High Direct Voltages (IEEE 402-1974R1982)
Radial Power Factor Tests on Insulating Tapes in Paper-Insulated Power Cable (IEEE 83-1963R1982)
Systems of Insulating Materials—General (UL1446)
Systems of Insulating Materials for Random-Wound AC Electrical Machinery (ANSI/IEEE 117-1974R1979)
Systems of Insulation for Specialty Transformers (IEEE 259-1974R1980)
Temperature Indices of Industrial Thermosetting Laminates (NEMA LI5-1969R1979)
Testing Insulation Resistance of Electric Machinery (ANSI/IEEE 43-1974R1981)
Testing Turn-to Turn Insulation on Form-Wound Stator Coils for AC Rotating Electric Machines (IEEE 522-1977R1981)
Thermal Evaluation and Establishment of Temperature Indexes of Solid Electrical Insulating Materials (IEEE 98-1972)
Thermal Evaluation of Insulation Systems for AC Electric Machinery Employing Form-Wound Pre-Insulated Stator Coils (ANSI/IEEE 275-1981)
Thermal Evaluation of Insulation Systems for Electric Equipment (ANSI/IEEE 99-1980)

***198. Insulators***

Definitions and Requirements for High-Voltage Air Switches, Insulators, and Bus Supports (ANSI/IEEE C37.30-1971R1977)
Electrical Power Insulators (ANSI C29.1-1982)
Wet-Process Porcelain and Toughened Glass Insulators (Suspension Type) (ANSI C29.2-1983)
Wet-Process Porcelain Insulators (Apparatus, Cap, and Pin Type) (ANSI C29.8-1980)
Wet-Process Porcelain Insulators (Apparatus, Post Type) (ANSI C29.9-1983)
Wet-Process Porcelain Insulators (Spool Type) (ANSI C29.3-1980)
Wet-Process Porcelain Insulators (Strain Type) (ANSI C29.4-1977)
Wet-Process Porcelain Insulators, High-Voltage Pin Type (ANSI C29.6-1977)
Wet-Process Porcelain Insulators, Low and Medium Voltage Pin Type (ANSI C29.5-1977)

***199. Integrated Electronics***

Terms for Integrated Electronics (IEEE 274-1966R1980)

*200. Interconnections*

Protective Relaying of Utility-Consumer Interconnections (ANSI/IEEE C37.95.1973R1980)

*201. Interface*

Code and Format Conventions for Use with ANSI/IEEE Std 488-1978 (ANSI/IEEE 728-1982)
Digital Interface for Programmable Instrumentation (ANSI/IEEE 488-1978)

*202. Interference*

Construction Drawings of Line Impedance Network for Measurement of Conducted Interference Output to the Power Line from FM and TV Broadcast Receivers (IEEE 214-1961)
Electromagnetic Interference Filters (UL1283)
Field Intensity Above 300 MHz from RF Industrial, Scientific, and Medical Equipment (IEEE 139-1952)
Measurement of Conducted Interference Output to the Power Line from FM and TV Broadcast Receivers (ANSI/IEEE 213-1961R1974)

*203. Internal Combustion Engines*

Specification for Speed-Governing of Internal Combustion Engine Generator Units (IEEE 126-1959R1983)

*204. Interrupters and Interruptions*

Design of Reliable Industrial and Commercial Power Systems (ANSI/IEEE 493-1980)
Ground-Fault Circuit Interrupters (UL943)
Requirements for Overhead, Pad Mounted Dry Vault and Submersible Automatic Circuit Reclosers and Fault Interrupters for AC Systems (ANSI/IEEE C37.60-1981)
Safety Requirements for X-Radiation Limits for AC High-Voltage Power Vacuum Interrupters Used in Power Switchgear (IEEE C37.85-1982R1978)

*205. Inverters*

Class 1E Battery Chargers and Inverters for Nuclear Power Generating Stations (ANSI/IEEE 650-1979)

*206. Ionization Chambers*

Calibration and Usage of "Dose Calibrator" Ionization Chambers for the Assay of Radionuclides (N42.13-1978)

*207. Jackets*

Selecting and Testing Jackets for Cables (ANSI/IEEE 532-1982)

*208. Joints*

Exposed Semiconducting Shields on Premolded High Voltage Cable Joints and Separable Insulated Connectors (IEEE 592-1977)
Power Cable Joints (IEEE 404-1977)

*209. Labels*

Safety Labels for Padmounted Switchgear and Transformers Sited in Public Areas (NEMA 260-1982)

*210. Laminates*

High-Pressure Decorative Laminates (NEMA LD3-1980)
Industrial Laminated Thermosetting Products (NEMA LI1-1971R1976)
Temperature Indices of Industrial Thermosetting Laminates (NEMA LI5-1969R1979)

*211. Lamp Ballasts and Transformers*

Definitions for High-Intensity-Discharge Lamp Ballasts and Transformers (ANSI C82.9-1981)
Fluorescent Lamp Ballasts (ANSI/UL 935-1978)
High-Intensity-Discharge Lamp Ballasts (ANSI/UL 1029-1980)
High-Intensity-Discharge Lamp Reference Ballasts (ANSI C82.5-1983)
Incandescent Filament Lamp Transformers (ANSI C82.8-1963R1981)
Measurement of Fluorescent Lamp Ballasts (ANSI C82.2-1983)
Measurement of High-Intensity-Discharge Lamp Ballasts (ANSI C82.6-1980)
Mercury Lamp Transformers (ANSI C82.7-1983)
Semiconductor Dimmers for Incandescent Lamps (NEMA WD2-1970R1980)
Specifications for Fluorescent Lamp Ballasts (ANSI C82.1-1977R1982)
Specifications for Fluorescent Lamp Reference Ballasts (ANSI C82.3-1983)
Specifications for High-Intensity-Discharge Lamp Ballasts (ANSI C82.4-1978)

*212. Lasers*

Laser-Maser Terms (ANSI/IEEE 586-1980)
Safe Use of Lasers (ANSI Z136.1-1980)

Specification Format Guide and Test Procedure for Single-Axis Laser Gyros (ANSI/IEEE 647-1981)

*213. Letter Symbols*

Induction Motor Letter Symbols (ANSI/IEEE 58-1978)

*214. Life*

Statistical Analysis of Thermal Life Test Data (IEEE 101-1972R1980)

*215. Lighting*

Electric Lighting Fixtures (UL57)
Electric Lighting Fixtures for Use in Hazardous Locations (UL844)
Electric Power Systems in Commercial Buildings (ANSI/IEEE 241-1983)
Emergency and Standby Power Systems for Industrial and Commercial Applications (ANSI/IEEE 446-1980)
Emergency Lighting and Power Equipment (UL924)
Fluorescent Lighting Fixtures (UL1570)
Fluorescent Luminaires (NEMA LE1-1974R1980)
High Intensity Discharge Lighting Fixtures (UL1572)
Incandescent Lighting Fixtures (UL1571)
Lighting System Noise Criterion Ratings (NEMA LE2-1974R1980)
Outdoor Floodlighting Equipment (NEMA FA1-1973R1979)
Portable Electric Lighting Units for Use in Hazardous Locations (UL781)

*216. Lightning Arresters* (See Surge Arresters)

*217. Lightning Protection*

Grounding of Industrial and Commercial Power Systems (ANSI/IEEE 142-1982)
Installation Requirements for Lightning Protection Systems (UL96A)
Lightning Protection Code (ANSI/NFPA 78-1983)
Lightning Protection Components (ANSI/UL96-1981)

*218. Loads*

Electric Power Systems in Commercial Buildings (ANSI/IEEE 241-1983)
Load Control for Use on Central Electric Heating Systems (NEMA DC22-1977)
Power System Analysis (ANSI/IEEE 399-1980)

*219. Logic Diagrams*

Binary Logic Diagrams for Process Operations (ANSI/ISA S5.2-1976R1981)
Graphic Symbols for Logic Diagrams (ANSI/IEEE 91-1984)

*220. Losses*

Calculating Losses in Isolated-Phase Bus (ANSI/IEEE C37.23-1969R1977)

*221. Loudness*

Determining Objective Loudness Ratings of Telephone Connections (ANSI/IEEE 661-1979)

*222. Loudspeakers*

Loudspeaker Measurements (IEEE 219-1975)

*223. Luminance*

Measurement of Luminance Signal Levels (ANSI/IEEE 205-1958R1972)
Measurements of Luminance Signal-to-Noise Ratio in Video Magnetic Tape Recording Systems (IEEE 618-1984)

*224. Magnetic Amplifiers*

Magnetic Amplifiers (IEEE 107-1964R1979)

*225. Maintenance*

Calibration and Control of Measuring and Test Equipment Used in the Construction and Maintenance of Nuclear Power Generating Stations (IEEE 498-1980)
Design of Reliable Industrial and Commercial Power Systems (ANSI/IEEE 493-1980)
Electric Power Systems in Commercial Buildings (ANSI/IEEE 241-1983)
Improved Electrical Maintenance and Safety in the Cement Industry (IEEE 625-1979)
Maintenance, Testing, and Replacement of Large Lead Storage Batteries for Generating Stations and Substations (ANSI/IEEE 450-1980)
Operation and Maintenance of Hydro-Generators (ANSI/IEEE 491-1974R1981)
Protection and Coordination of Industrial and Commercial Power Systems (IEEE 242-1975)

*226. Masers*

Laser-Maser Terms (ANSI/IEEE 586-1980)

*227. Medical Equipment*

Field Intensity Above 300 MHz from RF Industrial, Scientific, and Medical Equipment (IEEE 139-1952)

***228. Meters and Metering***

Bolometric Power Meters (IEEE 470-1972)
Electric Power Distribution for Industrial Plants (IEEE 141-1976)
Electrical Meter Sockets (UL414)
Electricity Metering (ANSI C12.1-1982)
Electrothermic Power Meters (IEEE 544-1975)
Instrument Transformers for Metering Purposes (ANSI C12.11-1978)
Interrelationship of Quartz-Fiber Electrometer Type Exposure Meters and Companion Exposure Meter Chargers (ANSI N42.6-1980)
Magnetic Tape Pulse Recorders for Electricity Meters (ANSI C12.14-1982)
Mechanical Demand Registers (ANSI C12.4-1978)
Meter Sockets (ANSI/UL 414-1978)
Phase Shifting Devices Used in Metering (ANSI C12.6-1978R1983)
Test Blocks and Cabinets for Installation of Self-Contained "A" Base Watthour Meters (ANSI C12.8-1981)
Test Switches for Transformer-Rated Meters (ANSI C12.14-1982)
Watthour Meter Sockets (ANSI C12.7-1982)
Watthour Meters (ANSI C12.10-1978)

***229. Metric Practice***

Metric Practice Guide (ANSI/IEEE 268-1982)
Preferred Metric Units for Use in Electrical and Electronics Science and Technology (ANSI/IEEE 945-1982)

***230. Mica***

Manufactured Electrical Mica (NEMA FI1-1977)
Natural Muscovite Mica Splittings (ANSI/ASTM D2131-71)
Sampling and Testing Untreated Mica Paper Used for Electrical Insulation (ANSI/ASTM D-1677-80)

***231. Microcomputers***

Microcomputer Systems Bus (ANSI/IEEE 796-1983)

***232. Microphones***

Calibration of Microphones (ANSI S1.10-1966R1976)
Close Talking Pressure-Type Microphones (IEEE 258-1965R1971)
Laboratory Standard Microphones (ANSI S1.12-1967R1977)

***233. Microwave***

Measurement of Hazardous Electromagnetic Fields—RF and Microwave (ANSI C95.5-1981)
Techniques and Instrumentation for the Measurement of Potentially Hazardous Electromagnetic Radiation at Microwave Frequencies (ANSI C95.3-1973R1979)

***234. Mining***

Mining Vehicle Battery Chargers (NEMA IB9-1981)

***235. Modeling***

Power System Analysis (ANSI/IEEE 399-1980)

***236. Modulation Systems***

Terms for Modulation Systems (IEEE 170-1964)

***237. Modules***

Type Tests of Class 1E Modules Used in Nuclear Power Generating Stations (ANSI/IEEE 381-1977)

***238. Monitoring***

Performance Criteria for Instrumentation Used for Inplant Plutonium Monitoring (ANSI N317-1980)
Performance Specification for Reactor Emergency Radiological Monitoring Instrumentation (ANSI N320-1979)
Post Accident Monitoring Instrumentation for Nuclear Power Generating Stations (ANSI/IEEE 497-1981)
Specification and Performance of On-Site Instrumentation for Continuously Monitoring Radioactivity in Effluents (ANSI N42.18-1980)

***239. Motors***

AC 400 Hz Aircraft Induction Motors (IEEE 137-1960)
AC Motor Protection (ANSI/IEEE C37.76-1976R1981)
Auxiliary Devices for Motors in Class 1—Groups A, B, C, and D, Division 2 Locations (IEEE 303-1984)

Construction and Interpretation of Thermal Limit Curves for Squirrel-Cage Motors (ANSI/IEEE 620-1981)
DC Machines (IEEE 113-1973)
Electric Motor Control Centers (UL 845)
Electric Motors (UL 1004)
Electric Motors and Generators for Use in Hazardous Locations (UL 674)
Energy Management Guide for Selection and Use of Single-Phase Motors (NEMA MG10-1983)
Frame Assignments for AC Integral-Horsepower Induction Motors (NEMA MG13-1974R1979)
Impedance-Protected Motors (UL 519)
Induction Motor Letter Symbols (ANSI/IEEE 58-1978)
Marine Electric Motors and Generators (UL 1112)
Motors and Generators (ANSI/NEMA MG1-1978)
Polyphase Induction Motors and Generators (IEEE 112-1983)
Polyphase Induction Motors Having Liquid in the Magnetic Gap (ANSI/IEEE 252-1977)
Power System Analysis (ANSI/IEEE 399-1980)
Protection and Coordination of Industrial and Commercial Power Systems (IEEE 242-1975)
Qualifying Class 1E Motor Control Centers for Nuclear Power Generating Stations (ANSI/IEEE 649-1980)
Renewal Parts for Motors and Generators (NEMA RP1-1981)
Safety Standard for Construction and Guide for Selection, Installation and Use of Electric Motors and Generators (ANSI/NEMA MG2-1977)
Single Phase Induction Motors (ANSI/IEEE 114-1982)
Synchronous Machines (ANSI/IEEE 115-1983)
Thermal Protectors for Electric Motors (UL 547)
Thyristor Converters and Motor Drives: Part 1—Converters for DC Motor Armature Supplies (ANSI/IEEE 444-1973)
Type Tests of Continuous Duty Class 1E Motors for Nuclear Power Generating Stations (ANSI/IEEE 334-1974R1980)

***240. Motor Vehicles***

Radio Noise Generated by Motor Vehicles and Affecting Mobile Communications Receivers (IEEE 263-1965)

***241. Navigation***

Direction Finder Measurements (IEEE 173-1959)
Navigation Aid Terms (IEEE 172-1983)

***242. Networks***

Construction Drawings of Line Impedance Network for Measurement of Conducted Interference Output to the Power Line from FM and TV Broadcast Receivers (IEEE 214-1961)
Terms for Linear Passive Reciprocal Time Invariant Networks (IEEE 1960R1983)
Measurement of Conducted Interference Output to the Power Line from FM and TV Broadcast Receivers (ANSI/IEEE 213-1961R1974)

***243. Noise***

Airborne Sound Measurements on Rotating Electric Machinery (IEEE 85-1973R1980)
Electromagnetic Noise and Field Strength Instrumentation (ANSI C63.2-1980)
Installation of Electrical Equipment to Minimize Noise Inputs to Controllers from External Sources (ANSI/IEEE 518-1982)
Lighting System Noise Criterion Ratings (NEMA LE2-1974R1980)
Measurement of Electric Noise and Harmonic Filter Performance of High-Voltage DC Systems (ANSI/IEEE 368-1977)
Measurement of Radio Noise from Overhead Power Lines (ANSI/IEEE 430-1976)
Measurements of Luminance Signal-to-Noise Ratio in Video Magnetic Tape Recording Systems (IEEE 618-1984)
Radio Noise Emissions from Low-Voltage, Electrical and Electronic Equipment (ANSI C63.4-1981)
Radio Noise Generated by Motor Vehicles and Affecting Mobile Communications Receivers (IEEE 263-1965)
Terms Relating to Overhead Power Lines Corona and Radio Noise (ANSI/IEEE 539-1979)
Voice-Frequency Electrical-Noise Tests of Distribution Transformers (IEEE 467-1980)

***244. Nuclear Facilities***

Defining Safety-Related Features of Nuclear Fuel Cycle Facilities (ANSI N46.1-1980)
Facilities Handling Radioactive Materials (ANSI/NFPA 801-1980)
Nuclear Air-Cleaning Systems (ANSI/ASME N510-1980)
Protective Coatings for Light Water Nuclear Reactor Containment Facilities (ANSI N101.2-1972)
Protective Coatings for the Nuclear Industry (ANSI N512-1974)
Transducer and Transmitter Installation for Nuclear Safety Applications (ANSI/ISA S67.01-1979)

***245. Nuclear Power Plants***

Accident Monitoring Instrumentation for Nuclear Power Generating Stations (ANSI/IEEE 497-1981)
Application of the Single Failure Criterion to Nuclear Power Generating Stations (ANSI/IEEE 379-1977)
Boiling Water Reactor Plants (ANSI/ANS 52.1-1983)
Calibration and Control of Measuring and Test Equipment Used in the Construction and Maintenance of Nuclear Power Generating Stations (IEEE 498-1980)
Class 1E Battery Chargers and Inverters for Nuclear Power Generating Stations (ANSI/IEEE 650-1979)
Class 1E Power Systems for Nuclear Generating Stations (ANSI/IEEE 308-1980)
Collection and Presentation of Electrical, Electronic, and Sensing Component and Mechanical Equipment Reliability Data for Nuclear Power Generating Stations (IEEE 500-1984)
Concrete Radiation Shielding for Nuclear Power Plants (ANSI/ANS 6.4-1977)
Containment Isolation (ANSI/ANS N271-1976)
Cooling Water Systems in Nuclear Power Plants (ANSI/ANS 59.1-1979)
Design and Installation of Electric Pipe Heating Control and Alarm Systems for Power Generating Stations (ANSI/IEEE 622A/1984)
Design and Installation of Electric Pipe Heating Systems for Nuclear Power Generating Stations (ANSI/IEEE 622/1979)
Design of Display and Control Facilities for Central Control Rooms of Nuclear Power Generating Stations (IEEE 566-1977)
Design of the Control Room Complex for a Nuclear Power Generating Station (IEEE 567)
Design Qualification of Safety System Equipment Used in Nuclear Power Generating Stations (IEEE 627-1980)
Diesel-Generator Units Applied as Standby Power Supplies for Nuclear Power Generating Stations (IEEE 387-1983)
Earthquake Instrumentation Criteria for Nuclear Power Plants (ANSI/ANS 2.2-1978)
Electric Penetration Assemblies in Containment Structures for Nuclear Power Generating Stations (IEEE 317-1983)
Germanium Gamma-Ray Detectors (ANSI/IEEE 325-1971R1982)
High-Voltage Connectors for Nuclear Instruments (ANSI N42.4-1971R1978)
Identification of Documents Related to Class 1E Equipment and Systems for Nuclear Power Generating Stations (ANSI/IEEE 494-1974R1983)
Independence of Class 1E Equipment and Circuits (ANSI/IEEE 384-1981)
Installation, Inspection, and Testing Requirements for Class 1E Instrumentation and Electric Equipment at Nuclear Power Generating Stations (ANSI/IEEE 336-1980)
Light Water Reactor Coolant Pressure Boundary Leak Detection (ISA S67.03)
Materials for Instruments in Radiation Service (ISA RP25.1)
Nuclear Safety-Related Instrument Sensing Line Piping and Tubing Standards for Use in Nuclear Power Plants (ISA S67.02)
Organizations that Conduct Qualification Testing of Safety Systems Equipment for Use in Nuclear Power Generating Stations (IEEE 600-1983)
Performance Testing of Closed Cooling Water Systems (ANSI/ASME OM2-1982)
Periodic Testing of Nuclear Power Generating Station Safety Systems (ANSI/IEEE 338-1977R1984)
Post Accident Monitoring Instrumentation for Nuclear Power Generating Stations (ANSI/IEEE 497-1981)
Pre-Operational Testing Programs for Class 1E Power Systems for Nuclear Power Generating Stations (IEEE 415-1976)

Programmable Digital Computer Systems in Safety Systems of Nuclear Power Generating Stations (ANSI/IEEE/ANS 7432-1982)
Protection of Nuclear Power Plants Against Effects of Postulated Pipe Rupture (ANSI/ANS 58.2-1980)
Protection Systems for Nuclear Power Generating Stations (ANSI/IEEE 279-1971R1978)
Quality Assurance Requirements (ANSI/ASME NQA2-1983)
Qualification of Class 1E Control Boards, Panels, and Racks Used in Nuclear Power Generating Stations (IEEE 420-1982)
Qualification of Class 1E Lead Storage Batteries for Nuclear Power Generating Stations (ANSI/IEEE 535-1979)
Qualification of Safety-Related Valve Actuators (IEEE 382-1980)
Qualifying Class 1E Equipment for Nuclear Power Generating Stations (IEEE 323-1983)
Qualifying Class 1E Motor Control Centers for Nuclear Power Generating Stations (ANSI/IEEE 649-1980)
Quality Assurance Program Requirements for the Design and Manufacture of Class 1E Instrumentation and Electric Equipment for Nuclear Power Generating Stations (IEEE 467-1980)
Reliability Analysis in the Design and Operation of Safety Systems for Nuclear Power Generating Stations (ANSI/IEEE 577-1976)
Reliability Analysis of Nuclear Power Generating Station Protection Systems (ANSI/IEEE 352-1975R1980)
Safety-Related Control Air Systems (ANSI/ANS 59.3-1977)
Safety Systems for Nuclear Power Generating Stations (IEEE 630-1980)
Safety Systems in Nuclear Power Generating Stations (ANSI/ANS 4.1-1978)
Setpoints for Nuclear Safety-Related Instrumentation Used in Nuclear Power Plants (ISA S67.04)
Simulators for Use in Operator Training (ANSI/ANS 3.5-1981)
System Identification in Nuclear Power Plants and Related Facilities (IEEE 805-1984)
Systems and Components Important to Safety (ANSI/ANS 58.3-1977)
Technical Specifications for Nuclear Power Stations (ANSI/ANS 58.4-1979)
Transducer and Transmitter Installation for Nuclear Safety Applications (ISA S67.01)
Type Tests of Continuous Duty Class 1E Motors for Nuclear Power Generating Stations (ANSI/IEEE 334-1974R1980)
Seismic Qualification of Class 1E Equipment for Nuclear Power Generating Stations (IEEE 344-1975R1980)
Type Test of Class 1E Electric Cables, Field Splices and Connections for Nuclear Power Generating Stations (ANSI/IEEE 383-1974R1980)
Type Tests of Class 1E Modules Used in Nuclear Power Generating Stations (ANSI/IEEE 381-1977)

***246. Oil Field***

Grounded 830V Three-Phase Electrical System for Oil Field Service (ANSI/IEEE 464-1981)

***247. Oscilloscopes***

General-Purpose Laboratory Cathode-Ray Oscilloscopes (IEEE 311-1970)

***248. Outages***

Design of Reliable Industrial and Commercial Power Systems (ANSI/IEEE 493-1980)
Power Operations Terminology Including Terms for Reporting and Analyzing Outages of Electrical Transmission and Distribution Facilities and Interruptions to Customer Service (ANSI/IEEE 346-1973)

***249. Outlets*** (See Plugs, Outlets, and Receptacles)

***250. Packaging Machinery***

Electrical Installations of Packaging Machinery and Associated Equipment (ANSI/IEEE 333-1980)

***251. Panels and Panelboards***

Electric Panelboards (UL 67)
Panelboards (NEMA PB1-1977)
Qualification of Class 1E Control Boards, Panels, and Racks Used in Nuclear Power Generating Stations (IEEE 420-1982)

***252. Periodic Inspection and Testing***

Periodic Testing of Nuclear Power Generating Station Safety Systems (ANSI/IEEE 338-1977R1984)

Procedures for Periodic Inspection of Cobalt-60 and Cesium-137 Teletherapy Equipment (ANSI N449.1-1978)

*253. Phase*

Measurement of Differential Gain and Differential Phase (ANSI/IEEE 206-1960R1978)

*254. Photomultipliers*

Photomultipliers for Scintillation Counting (ANSI/IEEE 398-1972R1982)

*255. Piezoelectricity and Piezomagnetics*

Piezoelectricity (ANSI/IEEE 176-1978)

Piezomagnetic Nomenclature (IEEE 319-1971R1978)

*256. Plugs, Outlets, and Receptacles*

Attachment Plugs and Receptacles (ANSI C73-1973)

Electrical Attachment Plugs and Receptacles (ANSI/UL498-1980)

Electrical Power Outlets (UL 231)

Electrical Receptacle-Plug Combinations for Use in Hazardous Locations (UL 1010)

General Purpose Wiring Devices (NEMA WD1-1979)

Guide to Pin and Sleeve Plugs, Receptacles, and Connectors (NEMA PR3-1980)

High-Voltage X-Ray Cables and Receptacles (NEMA XR7-1979)

Plugs, Receptacles, and Connectors of the Pin and Sleeve Type for Hazardous Locations (NEMA FB11-1973R1978)

Power Outlets (ANSI/UL 231-1981)

Receptacle-Plug Combinations for Use in Hazardous Locations (ANSI/UL 1010-1977)

Receptacles and Switches for Use with Aluminum Wire (UL 1567)

*257. Potential*

Measuring Earth Resistivity (IEEE 81-1983)

*258. Power Apparatus*

Field Testing Power Apparatus Insulation (IEEE 62-1978)

*259. Power Circuits*

Electrical Measurements in Power Circuits (IEEE 120-1955R1972)

Fused Power-Circuit Devices (ANSI/UL 977-1976)

Safety Code for Semiconductor Power Converters (NEMA PV3-1973R1979)

Surge Voltages in Low-Voltage AC Power Circuits (ANSI/IEEE C62.41-1980)

*260. Power Factor*

Electric Power Distribution for Industrial Plants (IEEE 141-1976)

Measurement of Power Factor Tip-up of Rotating Machinery Stator Coil Insulation (ANSI/IEEE 286-1975R1981)

Power Factor Measurement for Low-Voltage Inductive Test Circuits (ANSI/IEEE C37.26-1972R1976)

Radial Power Factor Tests on Insulating Tapes in Paper-Insulated Power Cable (IEEE 83-1963R1982)

*261. Power Lines*

Measurement of Electric and Magnetic Fields from AC Power Lines (ANSI/IEEE 644-1979)

Measurement of Radio Noise from Overhead Power Lines (ANSI/IEEE 430-1976)

Power-Line Carrier Applications (ANSI/IEEE 643-1980)

Terms Relating to Overhead Power Lines Corona and Radio Noise (ANSI/IEEE 539-1979)

*262. Power Measurement*

Performance Test Code—Measurement of Indicated Power (ANSI/ASME PTC19.8-1970)

*263. Power Plants (Non-nuclear)*

Definitions for Use in Reporting Electric Generating Unit Reliability, Availability, and Productivity (ANSI/IEEE 762)

Design and Installation of Cable Systems in Power Generating Stations (IEEE 422-1977)

Implementation of Unique Identification in Power Plants and Related Facilities (IEEE 804-1983)

Installation Design and Installation of Large Lead Storage Batteries for Generating Stations and Substations (IEEE 484-1981)

Maintenance, Testing, and Replacement of Large Lead Storage Batteries for Generating Stations and Substations (ANSI/IEEE 450-1980)

Maximum Electric Power Station Ground Potential Rise and Induced Voltage from a Power Fault (IEEE 367-1979)

Nomenclature for Generating Station Electric Power Systems (ANSI/IEEE 505/1977)

Principles and Definitions for Unique Identification in Power Plants and Related Facilities (IEEE 803-1983 & 803A-1983)
Protection of Wire Line Communications Facilities Serving Electric Power Stations (ANSI/IEEE 487-1980)
Sizing Large Lead Storage Batteries for Generating Stations and Substations (IEEE 485-1983)
System Identification in Nuclear Power Plants and Related Facilities (IEEE 805-1984)

***264. Power Supplies***

Diesel-Generator Units Applied as Standby Power Supplies for Nuclear Power Generating Stations (IEEE 387-1983)
Power Supplies (ANSI/UL 1012-1981)
Power Supplies for Use with Burglar Alarm Systems (UL603)
Preferred Power Supply for Nuclear Power Generating Stations (ANSI/IEEE 765-1983)
Silicon Rectifier Units for Transportation Power Supplies (NEMA RI9-1968R1979)
Stabilized Power Supplies—Correct-Current Output (NEMA PY1-1972R1978)

***265. Power Systems***

Automatic Generation Control on Electric Power Systems (IEEE 94-1970)
Class 1E Power Systems for Nuclear Generating Stations (ANSI/IEEE 308-1980)
Communication Facilities Serving Electric Power Stations (ANSI/IEEE 487-1980)
Design of Reliable Industrial and Commercial Power Systems (ANSI/IEEE 493-1980)
Electric Power Systems in Commercial Buildings (ANSI/IEEE 241-1983)
Electrical Power System Device Function Numbers (ANSI/IEEE C37.2-1979)
Emergency and Standby Power Systems for Industrial and Commercial Applications (ANSI/IEEE 446-1980)
Grounding of Industrial and Commercial Power Systems (ANSI/IEEE 142-1982)
Harmonic Control and Reactive Compensation of Static Power Converters (ANSI/IEEE 519-1981)
Identification, Testing, and Evaluation of the Dynamic Performance of Excitation Control Systems (ANSI/IEEE 421A-1978)
Insulation Coordination (ANSI C92.1-1982)
Isolated Power Systems Equipment (UL 1047)
Maximum Electric Power Station Ground Potential Rise and Induced Voltage from a Power Fault (ANSI/IEEE 367-1979)
Nomenclature for Generating Station Electric Power Systems (ANSI/IEEE 505/1977)
Performance Test Code—Water and Steam in the Power Cycle (ANSI/ASME PTC19.11-1970)
Power Line Carrier Line Traps (ANSI C93.3-1981)
Power System Analysis (ANSI/IEEE 399-1980)
Preferred Power Supply for Nuclear Power Generating Stations (ANSI/IEEE 765-1983)
Preferred Voltage Ratings for AC Electrical Systems and Equipment (ANSI C92.2-1981)
Pre-Operational Testing Programs for Class 1E Power Systems for Nuclear Power Generating Stations (IEEE 415-1976)
Protection and Coordination of Industrial and Commercial Power Systems (IEEE 242-1975)
Protection of Shunt Capacitor Banks (ANSI/IEEE C37.99-1980)
Protective Relay Applications to Power System Buses (ANSI/IEEE C37.97-1979R1984)
Protective Relaying of Utility-Consumer Interconnections (ANSI/IEEE C37.95-1974R1980)
Reliable Industrial and Commercial Power Systems (ANSI/IEEE 493-1980)
Service Conditions for Power Systems Communications Apparatus (281-1968)
Voltage Ratings for Electric Power Systems and Equipment (ANSI C84.1-1982)

***266. Preamplifiers***

Amplifiers and Preamplifiers for Semiconductor Radiation Detectors for Ionizing Radiation (ANSI/IEEE 301-1976R1982)

***267. Printed Wiring***

Electrical Printed-Wiring Boards (UL 796)
Printed Wiring Board Electrical Connectors (ANSI/EIA RS406-1973R1979)

***268. Process Instrumentation***

Compatibility of Analog Signals for Electronic Industrial Process Instruments (ISA 50.1)
Dynamic Response Testing of Process Control Instrumentation (ISA S26)
Hardware Testing of Digital Process Computers (ISA RP55.1)

***269. Production and Productivity***

Definitions for Use in Reporting Electric Generating Unit Reliability, Availability, and Productivity (ANSI/IEEE 762)
Emergency and Standby Power Systems for Industrial and Commercial Applications (ANSI/IEEE 446-1980)

***270. Program Waves***

Volume Measurements of Electrical Speech and Program Waves (ANSI/IEEE 152-1953R1971)

***271. Propagation***

Measuring Electromagnetic Field Strength for Frequencies Below 1000 MHz in Radio Wave Propagation (IEEE 302-1969R1981)
Measuring Field Strength in Radio Wave Propagation (IEEE 291-1969R1981)
Radio Wave Propagation (ANSI/IEEE 211-1977)

***272. Protection***

AC Motor Protection (ANSI/IEEE C37.76-1976R1981)
Electric Power Distribution for Industrial Plants (IEEE 141-1976)
Electric Power Systems in Commercial Buildings (ANSI/IEEE 241-1983)
Emergency and Standby Power Systems for Industrial and Commercial Applications (ANSI/IEEE 446-1980)
Generator Ground Protection (ANSI/IEEE C37.101-1983)
Grounding of Industrial and Commercial Power Systems (ANSI/IEEE 142-1982)
Protection and Coordination of Industrial and Commercial Power Systems (IEEE 242-1975)
Protection of Wire Line Communications Facilities Serving Electric Power Stations (ANSI/IEEE 487-1980)
Protective Relay Applications of Audio Tones Over Telephone Channels (ANSI/IEEE C37.93.1976)
Protective Relay Applications of Power Transformers (ANSI/IEEE C37.91.1967R1980)
Protective Relay Applications to Power System Buses (ANSI/IEEE C37.97-1979R1984)
Protective Relaying of Utility-Consumer Interconnections (ANSI/IEEE C37.95.1973R1980)
Radiation Protection Instrumentation Test and Calibration (ANSI N322-1977R1983)
Reliability Analysis of Nuclear Power Generating Station Protection Systems (ANSI/IEEE 352-1975R1980)

***273. Pulse and Pulse Measurement***

Computer-Type Pulse Transformers (IEEE 272-1970R1976)
High-Power Pulse Transformers (IEEE 391-1976)
Low-Power Pulse Transformers (IEEE 390-1975)
Magnetic Tape Pulse Recorders for Electricity Meters (ANSI C12.14-1982)
Performance of Dial-Pulse (DP) Address Signaling Systems (ANSI/IEEE 753-1983)
Pulse Measurement and Analysis by Objective Techniques (ANSI/IEEE 181-1977)
Pulse Terms (IEEE 194-1977)

***274. Qualification***

Cable Penetration Fire Stop Qualification Test (ANSI/IEEE 634-1978)
Design Qualification of Safety System Equipment Used in Nuclear Power Generating Stations (IEEE 627-1980)
Organizations that Conduct Qualification Testing of Safety Systems Equipment for Use in Nuclear Power Generating Stations (IEEE 600-1983)
Qualification of Class 1E Control Boards, Panels, and Racks Used in Nuclear Power Generating Stations (IEEE 420-1982)
Qualification of Class 1E Lead Storage Batteries for Nuclear Power Generating Stations (ANSI/IEEE 535-1979)
Qualification of Safety-Related Valve Actuators (IEEE 382-1980)
Qualifying Class 1E Equipment for Nuclear Power Generating Stations (IEEE 323-1983)
Qualifying Class 1E Motor Control Centers for Nuclear Power Generating Stations (ANSI/IEEE 649-1980)
Qualifying Permanent Connections Used in Substation Grounding (IEEE 837-1984)

***275. Quality and Quality Assurance***

Quality Assurance Program Requirements for the Design and Manufacture of Class 1E Instrumentation and Electric Equipment for Nuclear Power Generating Stations (IEEE 467-1980)
Software Quality Assurance Plans (ANSI/IEEE 730-1981)

Speech Quality Measurements (IEEE 297-1969)

***276. Quantities***

Basic Per-Unit Quantities for AC Rotating Machines (ANSI/IEEE 86-1975)
Symbols for Quantities and Units Used in Electrical Science and Engineering (ANSI/IEEE 280-1968)

***277. Raceways and Fittings***

Cellular Metal Floor Electrical Raceways and Fittings (ANSI/UL 209-1981)
Surface Metal Electrical Raceways and Fittings (ANSI/UL 5-1979)
Underfloor Electrical Raceways and Fittings (ANSI/UL 884-1981)
Wireways, Auxiliary Gutters, and Associated Fittings (ANSI/UL 870-1981)

***278. Radar***

Letter Designations for Radar Frequency Bands (IEEE 521-1976)
Radar Definitions (ANSI/IEEE 686-1982)

***279. Radiation***

Amplifiers and Preamplifiers for Semiconductor Radiation Detectors for Ionizing Radiation (ANSI/IEEE 301-1976R1982)
Calibration and Usage of Germanium Detectors for Measurement of Gamma-Ray Emission of Radionuclides (ANSI N42.14-1978)
Geiger-Muller Counters (ANSI/IEEE 309-1970)
High-Purity Germanium Detectors for Ionizing Radiation (ANSI/IEEE 645-1977)
Performance Specification for Reactor Emergency Radiological Monitoring Instrumentation (ANSI N320-1979)
Radiation Protection Instrumentation Test and Calibration (ANSI N322-1977R1983)
Radio Frequency Radiation Hazard Warning Symbol (ANSI C95.2-1982)
Safety Requirements for X-Radiation Limits for AC High-Voltage Power Vacuum Interrupters Used in Power Switchgear (IEEE C37.85-1982R1978)
Semiconductor Radiation Detectors (ANSI/IEEE 300-1982)
Specification of X- or Gamma-Radiation Survey Instruments (ANSI N13.4-1971R1983)
Spurious Radiation from FM TV Broadcast Receivers (IEEE 187-1951)
Techniques and Instrumentation for the Measurement of Potentially Hazardous Electromagnetic Radiation at Microwave Frequencies (ANSI C95.3-1973R1979)

***280. Radio Frequency***

Field Intensity Above 300 MHz from RF Industrial, Scientific, and Medical Equipment (IEEE 139-1952)
Measurement of Hazardous Electromagnetic Fields—RF and Microwave (ANSI C95.5-1981)
Minimization of Interference from RF Heating Equipment (IEEE 140-1950)
Radio Frequency Radiation Hazard Warning Symbol (ANSI C95.2-1982)

***281. Radio Noise***

Electromagnetic Noise and Field-Strength Instrumentation (ANSI C63.2-1980)
Radio Noise Emissions from Low-Voltage, Electrical and Electronic Equipment (ANSI C63.4-1981)
Radio Noise from Overhead Power Lines (ANSI/IEEE 430-1976)
Radio Noise Generated by Motor Vehicles and Affecting Mobile Communications Receivers (IEEE 263-1965)
Terms Relating to Overhead Power Lines Corona and Radio Noise (ANSI/IEEE 539-1979)

***282. Radio Transmitters***

Supplement to Terms for Radio Transmitters (IEEE 182A-1961)
Terms for Radio Transmitters (IEEE 182-1961)

***283. Radioactivity and Radioactive Wastes***

Specification and Performance of On-Site Instrumentation for Continuously Monitoring Radioactivity in Effluents (ANSI N42.18-1980)

***284. Radiotelegraph***

Testing Radiotelegraph Transmitters (IEEE 183-1958)

***285. Railroad***

Electric Control Apparatus for Land Transportation Vehicles (IEEE 16-1955)
Rotating Electrical Machinery for Rail and Road Vehicles (ANSI/IEEE 11-1980)

***286. Rating***

Aerospace Equipment Voltage and Frequency Ratings (IEEE 127-1963)
Rating Electrical Apparatus for Short Time, Intermittent, or Varying Duty (IEEE 96-1969)

***287. Reactors***

Performance Specification for Reactor Emergency Radiological Monitoring Instrumentation (ANSI N320-1979)

Terminology and Test Code for Shunt Reactors (ANSI/IEEE C57.21-1981)

Transformers, Regulators and Reactors (NEMA TR1-1980)

***288. Receivers***

AM Broadcast Receivers (ANSI/IEEE 186-1948R1972)

FM Broadcast Receivers (ANSI/IEEE 185-1975)

FM Mobile Communications Receivers (ANSI/IEEE 184-1969)

Measurement of Conducted Interference Output to the Power Line from FM and TV Broadcast Receivers (ANSI/IEEE 213-1961R1974)

Monochrome TV Broadcast Receivers (ANSI/IEEE 190-1960)

Radio Noise Generated by Motor Vehicles and Affecting Mobile Communications Receivers (IEEE 263-1965)

Receivers Employing Ferrite Loop Core Antennas (ANSI/IEEE 189-1955R1972)

Spurious Radiation from FM TV Broadcast Receivers (IEEE 187-1951)

***289. Receptacles*** (See Plugs, Outlets, and Receptacles)

***290. Reclosers***

Application, Operation, and Maintenance of Automatic Circuit Reclosers (ANSI/IEEE C37.61-1973R1979)

Requirements for Overhead, Pad Mounted Dry Vault and Submersible Automatic Circuit Reclosers and Fault Interrupters for AC Systems (ANSI/IEEE C37.60-1981)

***291. Records and Recording***

Calibration of Mechanically Recorded Lateral Frequency Records (ANSI/IEEE 192-1958R1971)

Decoder for Reproducing Matrix Quadraphonic Disc Records (ANSI/EIA RS418-1975R1982 & RS430-1976)

Magnetic Tape Pulse Recorders for Electricity Meters (ANSI C12.14-1982)

Measurements of Luminance Signal-to-Noise Ratio in Video Magnetic Tape Recording Systems (IEEE 618-1984)

Measurement of Weighted Peak Flutter of Sound Recording and Reproducing Equipment (ANSI/IEEE 193-1971)

Measuring Recorded Flux of Magnetic Sound Records at Medium Wavelengths (ANSI/IEEE 347-1972)

Reproducing Discrete Four-Signal Disc Records (ANSI/EIA RS425-1975)

***292. Rectifiers***

Silicon Rectifier Units for Transportation Power Supplies (NEMA RI9-1968R1979)

***293. Reels***

Electric Cord Reels (UL355)

***294. Refrigeration***

Emergency and Standby Power Systems for Industrial and Commercial Applications (ANSI/IEEE 446-1980)

***295. Regression Line***

Simplified Method for Calculation of the Regression Line (IEEE 101A-1974R1980)

***296. Regulators***

Aircraft Generator and Regulator Characteristics (IEEE 136-1959)

Ferroresonant Voltage Regulators (IEEE 449-1984)

Loading Oil-Immersed Step-Voltage and Induction-Voltage Regulators (IEEE C57.95-1955)

Transformers, Regulators and Reactors (NEMA TR1-1980)

***297. Relays***

Electrical Indicating Instrument Relays (NEMA II2-1972R1977)

Field Testing of Relaying Current Transformers (ANSI/IEEE C57.13.1-1981)

Ground-Fault Sensing and Relaying Equipment (UL1053)

Protection and Coordination of Industrial and Commercial Power Systems (IEEE 242-1975)

Protective Relay Applications of Power Transformers (ANSI/IEEE C37.91.1967R1980)

Protective Relay Applications to Power System Buses (ANSI/IEEE C37.97-1979R1984)

Protective Relaying of Utility-Consumer Interconnections (ANSI/IEEE C37.95.1973R1980)

Relays and Relay Systems Associated with Electric Power Apparatus (ANSI/IEEE C37.90-1978R1982)

Relays for Electrical and Electronic Equipment (ANSI/EIA RS407A-1978)
Seismic Testing of Relays (ANSI/IEEE C37.98-1978)
Solid State Relays (ANSI/EIA RS443-1979)

***298. Reliability***

Collection and Presentation of Electrical, Electronic, and Sensing Component and Mechanical Equipment Reliability Data for Nuclear Power Generating Stations (IEEE 500-1984)
Definitions for Use in Reporting Electric Generating Unit Reliability, Availability, and Productivity (ANSI/IEEE 762)
Design of Reliable Industrial and Commercial Power Systems (ANSI/IEEE 493-1980)
High Reliability in Electronics Transformers and Inductors (ANSI/IEEE 392-1976R1982)
Power System Analysis (ANSI/IEEE 399-1980)
Reliability Analysis in the Design and Operation of Safety Systems for Nuclear Power Generating Stations (ANSI/IEEE 577-1976)
Reliability Analysis of Nuclear Power Generating Station Protection Systems (ANSI/IEEE 352-1975R1980)

***299. Resistance***

Electrical Resistance Heat Tracing (IEEE 515-1983)
Resistance Measurements (IEEE 118-1978)

***300. Resistivity***

Laboratory Measurement and Reporting of Fly Ash Resistivity (ANSI/IEEE 548-1981)
Measuring Earth Resistivity (IEEE 81-1983)
Measuring Resistivity of Cable-Insulation Materials at High Direct Voltages (IEEE 402-1974R1982)
Soil Thermal Resistivity Measurements (ANSI/IEEE 442-1981)

***301. Resistors***

Fixed Composition Resistors (ANSI/EIA RS172B-1969R1980)
Fixed Film Microelement Resistors (ANSI/EIA RS451-1978R1983)
Fixed Film Resistors—General Purpose (ANSI/EIA RS460-1980)
Fixed Film Resistors—High Resistance/High Voltage (ANSI/EIA RS452-1978R1983)
Fixed Film Resistors—Precision and Semiprecision (ANSI/EIA RS196A-1970R1983)
Fixed Wire-Wound Power Resistors (ANSI/EIA RS155B-1970R1981)
Fixed Wire-Wound Precision Resistors (ANSI/EIA RS299A-1965R1979)
Low-Power Fixed Wire-Wound Resistors (ANSI/EIA RS344-1968R1979)
Reference-Standard Electrical Resistors (ANSI/IEEE 310-1969R1975)
Resistor Networks—Fixed Film (ANSI/EIA RS451-1978R1983)
Variable Non-Wire-Wound Resistors—Lead Screw Actuated (ANSI/EIA RS360-1968R1981)
Variable Resistors, Commercial Non-Wire-Wound, User and Service Adjust (ANSI/EIA RS303A-1976R1981)
Variable Wire-Wound Precision Resistors (ANSI/EIA RS391-1971R1979)
Variable Wire-Wound Resistors (ANSI/EIA RS333-1967R1981)
Variable Wire-Wound Resistors—Lead Screw Actuated (ANSI/EIA RS345-1968R1981)

***302. Resolution***

Measurement of Resolution of Camera Systems (ANSI/IEEE 211-1977)

***303. Rheostats***

Wire-Wound Power Type Rheostats (ANSI/EIA RS322-1965R1979)

***304. Rotating Electrical Machinery***

AC Motor Protection (ANSI/IEEE C37.96-1976R1983)
Airborne Sound Measurements on Rotating Electric Machinery (IEEE 85-1973R1980)
Basic Per-Unit Quantities for AC Rotating Machines (ANSI/IEEE 86-1975)
Combustion Gas Turbine Driven Cylindrical Rotor Synchronous Generators (ANSI C50.14-1977)
Construction and Guide for Selection, Installation and Use of Electric Motors and Generators (ANSI/NEMA MG2-1977)
Construction and Interpretation of Thermal Limit Curves for Squirrel-Cage Motors (ANSI/IEEE 620-1981)
Cylindrical Rotor Synchronous Generators (ANSI C50.13-1977)
DC Aircraft Rotating Machines (IEEE 132-1953)
DC Machines (IEEE 113-1973)

Definitions for Use in Reporting Electric Generating Unit Reliability, Availability, and Productivity (ANSI/IEEE 762-1980)
Electric Motors and Generators for Use in Hazardous Locations (ANSI/UL 674-1979)
Energy Management Guide for Selection and Use of Polyphase Motors (NEMA MG10-1983)
Energy Management Guide for Selection and Use of Single-Phase Motors (NEMA MG10-1983)
Evaluation and Classification of Insulation Systems for DC Machines (ANSI/IEEE 304-1977R1982)
Evaluation of Sealed Insulation Systems for AC Electrical Machinery Employing Form-Wound Stator Coils (IEEE 491-1972)
Excitation Systems for Synchronous Machines (IEEE 421-1972)
Insulation Maintenance of Large AC Rotating Machinery (ANSI/IEEE 56-1977R1982)
Field Discharge Circuit Breakers Used in Enclosures for Rotating Electrical Machinery (ANSI/IEEE C37.18-1979)
Frame Assignments for AC Integral-Horsepower Induction Motors (NEMA MG13-1974R1979)
General Requirements for Synchronous Machines (ANSI C50.10-1977)
Generator Ground Protection Guide (ANSI/IEEE C37.101-1983)
Hermetic Refrigerant Motor-Compressors (ANSI/UL 984-1978)
High Potential Test Requirements for Excitation Systems for Synchronous Machines (ANSI/IEEE 421B-1979)
Insulation Maintenance for Rotating Electrical Machinery (ANSI/IEEE 432-1976)
Insulation Maintenance of Large AC Rotating Machinery (ANSI/IEEE 56-1977)
Insulation Systems for Large High-Voltage Machines (ANSI/IEEE 434-1973R1979)
Insulation Testing of Large AC Rotating Machinery with High Direct Voltage (ANSI/IEEE 95-1977R1982)
Insulation Testing of Large AC Rotating Machinery with High Voltage at Very Low Frequencies (ANSI/IEEE 433-1974R1979)
Measurement of Power Factor Tip-up of Rotating Machinery Stator Coil Insulation (ANSI/IEEE 286-1975R1981)
Motors and Generators (ANSI/NEMA MG1-1978)
Polyphase Induction Motors and Generators (IEEE 112-1983)
Polyphase Induction Motors for Power Generating Stations (ANSI C50.41-1982)
Polyphase Induction Motors Having Liquid in the Magnetic Gap (ANSI/IEEE 252-1977)
Requirements for Gas Turbine Driven Cylindrical-Rotor Synchronous Generators (ANSI C50.14-1977)
Requirements for Cylindrical-Rotor Synchronous Generators (ANSI C50.13-1977)
Rotating Electrical Machinery for Rail and Road Vehicles (ANSI/IEEE 11-1980)
Safety Standard for Construction and Guide for Selection, Installation and Use of Electric Motors and Generators (ANSI/NEMA MG2-1977)
Single Phase Induction Motors (ANSI/IEEE 114-1982)
Sound Level Prediction for Installed Rotating Electrical Machines (NEMA MG3-1974R1979)
Synchronous Generators and Generator/Motors for Hydraulic Turbine Applications (ANSI C50.12-1981)
Synchronous Machines (ANSI/IEEE 115-1983)
Synchronous Machines—General Requirements (ANSI C50.10-1977)
Systems of Insulating Materials for Random-Wound AC Electrical Machinery (ANSI/IEEE 117-1974R1979)
Testing Insulation Resistance of Rotating Machinery (ANSI/IEEE 43-1974R1981)
Testing Turn-to-Turn Insulation on Form-Wound Stator Coils for AC Rotating Electric Machines (IEEE 522-1977R1981)
Thermal Evaluation of Insulation Systems for AC Electric Machinery Employing Form-Wound Pre-Insulated Stator Coils (ANSI/IEEE 275-1981)
Thermal Protectors for Motors (ANSI/UL 547-1979)
Thyristor Converters for Motor Drives (ANSI/IEEE 444-1973)

***305. Safety***

Design Qualification of Safety System Equipment Used in Nuclear Power Generating Stations (IEEE 627-1980)

Electric Power Systems in Commercial Buildings (ANSI/IEEE 241-1983)
Electrical Safety Practices in Electrolytic Cell Line Working Zones (IEEE 463-1977)
Electricity Metering (ANSI C12.1-1982)
Emergency and Standby Power Systems for Industrial and Commercial Applications (ANSI/IEEE 446-1980)
Improved Electrical Maintenance and Safety in the Cement Industry (IEEE 625-1979)
Installation of Intrinsically Safe Instrument Systems in Hazardous Locations (ISA RP12.6)
Interrelationship of Quartz-Fiber Electrometer Type Exposure Meters and Companion Exposure Meter Chargers (ANSI N42.6-1980)
National Electrical Safety Code (ANSI C2)
Organizations that Conduct Qualification Testing of Safety Systems Equipment for Use in Nuclear Power Generating Stations (IEEE 600-1983)
Periodic Testing of Nuclear Power Generating Station Safety Systems (ANSI/IEEE 338-1977R1984)
Programmable Digital Computer Systems in Safety Systems of Nuclear Power Generating Stations (ANSI/IEEE/ANS 7432-1982)
Radio Frequency Radiation Hazard Warning Symbol (ANSI C95.2-1982)
Reliability Analysis in the Design and Operation of Safety Systems for Nuclear Power Generating Stations (ANSI/IEEE 577-1976)
Safety Code for Semiconductor Power Converters (NEMA PV3-1973R1979)
Safety in AC Substation Grounding (IEEE 80-1976)
Safety in High-Voltage and High Power Testing (IEEE 510-1983)
Safety Labels for Padmounted Switchgear and Transformers Sited in Public Areas (NEMA 260-1982)
Safety Level of Electromagnetic Radiation with Respect to Personnel (ANSI C95.1-1982)
Safety Requirements for X-Radiation Limits for AC High-Voltage Power Vacuum Interrupters Used in Power Switchgear (IEEE C37.85-1982R1978)
Safety Standard for Construction and Guide for Selection, Installation and Use of Electric Motors and Generators (ANSI/NEMA MG2-1977)
Safety Standard for Diagnostic Ultrasound Equipment (NEMA UL1-1981)
Safety Standards for Construction and Guide for Selection, Installation and Operation of Adjustable-Speed Drive Systems (NEMA ICS3.1-1979)
Safety Systems for Nuclear Power Generating Stations (IEEE 630-1980)
Techniques and Instrumentation for the Measurement of Potentially Hazardous Electromagnetic Radiation at Microwave Frequencies (ANSI C95.3-1973R1979)

***306. Scintillation and Scintillation Counting***

Performance Measurement of Scintillation Cameras (NEMA NU1-1980)
Performance Verification of Liquid Scintillation Counting Systems (ANSI N42.15-1980)
Photomultipliers for Scintillation Counting (ANSI/IEEE 398-1972R1982)

***307. Security Systems***

Electric Power Systems in Commercial Buildings (ANSI/IEEE 241-1983)

***308. Seismic***

Seismic Design of Substations (IEEE 693-1983)
Seismic Qualification of Class 1E Equipment for Nuclear Power Generating Stations (IEEE 344-1975R1980)
Seismic Testing of Relays (ANSI/IEEE C37.98-1978)

***309. Semiconductors***

Amplifiers and Preamplifiers for Semiconductor Radiation Detectors for Ionizing Radiation (ANSI/IEEE 301-1976R1982)
Germanium Semiconductor Detector Gamma-Ray Efficiency Using a Standard Marinelli (Reentrant) Beaker Geometry (ANSI/IEEE 680-1978)
Letter Symbols for Semiconductor Devices (IEEE 255-1963)
Power Switching Semiconductor Devices (UL 1557)
Semiconductor Power Rectifiers (ANSI C34.2-1968R1973)
Semiconductor Radiation Detectors (ANSI/IEEE 300-1982)
Semiconductor Terms (216-1960R1980)
Terms for Semiconductor Memory (ANSI/IEEE 662-1980)

***310. Sensors and Sensing Equipment***

Ground-Fault Sensing and Relaying Equipment (UL1053)

Inertial Sensor Terminology (IEEE 528-1983)
Measurement Procedure for Field Disturbance Sensor (RF Intrusion Alarm) (ANSI/IEEE 475-1983)

***311. Service and Service Conditions***

Detection and Determination of Generated Gases in Oil-Immersed Transformers and Their Relation to Serviceability of the Equipment (ANSI/IEEE C57.104-1978)
Electric Power Systems in Commercial Buildings (ANSI/IEEE 241-1983)
Electrical Service Equipment (UL 869)
Emergency and Standby Power Systems for Industrial and Commercial Applications (ANSI/IEEE 446-1980)
Power Operations Terminology Including Terms for Reporting and Analyzing Outages of Electrical Transmission and Distribution Facilities and Interruptions to Customer Service (ANSI/IEEE 346-1973)
Protection and Coordination of Industrial and Commercial Power Systems (IEEE 242-1975)
Service Conditions for Power Systems Communications Apparatus (281-1968)
Service-Entrance Cables (UL 854)
Specifying Service Conditions in Electrical Standards (IEEE 97-1969)

***312. Sheaths***

Selection and Design of Aluminum Sheaths for Cables (ANSI/IEEE 635-1980)

***313. Shields and Shielding***

Exposed Semiconducting Shields on Premolded High Voltage Cable Joints and Separable Insulated Connectors (IEEE 592-1977)
Magnetic Shield Efficiency in Attenuating Alternating Magnetic Fields (ANSI/ASTM A698-74R1980)
Measurement of Shielding Effectiveness of Shielding Enclosures (IEEE 299-1969)

***314. Ships***

Electric Installations on Shipboard (IEEE 45-1983)

***315. Short Circuits***

Liquid-Immersed Distribution, Power, and Regulating Transformers and Guide for Short-Circuit Testing of Distribution and Power Transformers (ANSI/IEEE C57.12.90-1980
Protection and Coordination of Industrial and Commercial Power Systems (IEEE 242-1975)
Power System Analysis (ANSI/IEEE 399-1980)

***316. Shunts***

DC Instrument Shunts (IEEE 316-1971)

***317. Signaling Systems***

Auxiliary Protective Signaling Systems (ANSI/NFPA 72B-1979)
Central Station Signaling Systems (ANSI/NFPA 71-1982)
Emergency and Standby Power Systems for Industrial and Commercial Applications (ANSI/IEEE 446-1980)
Heavy-Duty Electrically-Operated Audible Signaling Devices for General Use (NEMA SB27-1960R1978)
Interconnection Circuitry of Non-Coded Remote-Station Protective Signaling Systems (ANSI/NEMA SB3-1969R1978)
Local Protective Signaling Systems (ANSI/NFPA 72A-1979)
Performance of Dial-Pulse (DP) Address Signaling Systems (ANSI/IEEE 753-1983)
Proprietary Protective Signaling Systems (ANSI/NFPA 72D-1979)
Remote Station Protective Signaling Systems (ANSI/NFPA 72C-1982)

***318. Site***

Electromagnetic Site Survey (IEEE 473-1983)

***319. SI Units***

Letter Symbols for Units of Measurement (ANSI/IEEE 260-1978)
Metric Practice Guide (ANSI/IEEE 268-1982)

***320. Sockets***

Battery Socket Patterns (ANSI/EIA RS156B-1973R1983)
Receiver Type Sockets (ANSI/EIA RS367-1969R1982)
Watthour Meter Sockets (ANSI C12.7-1982)

***321. Software***

Software Configuration Management Plans (IEEE 828-1983)
Software Engineering Terminology (ANSI/IEEE 729-1983)

Software Quality Assurance Plans (ANSI/IEEE 730-1981)
Software Requirements Specifications (IEEE 830-1984)
Software Test Documentation (ANSI/IEEE 829-1983)

***322. Soil***

Soil Thermal Resistivity Measurements (ANSI/IEEE 442-1981)

***323. Solar***

Evaluating the Effect of Solar Radiation on Outdoor Metal-Clad Switchgear (ANSI/IEEE C37.24-1971R1984)
Terms for Solar Cells (IEEE 307-1969)

***324. Sound***

Airborne Sound Measurements on Rotating Electric Machinery (IEEE 85-1973R1980)
Measurement of Sound Pressure Levels for AC Power Circuit Breakers (ANSI/IEEE C37.082-1982)
Measurement of Weighted Peak Flutter of Sound Recording and Reproducing Equipment (ANSI/IEEE 193-1971)
Measuring Recorded Flux of Magnetic Sound Records at Medium Wavelengths (ANSI/IEEE 347-1972)
Sound Level Prediction for Installed Rotating Electrical Machines (NEMA MG3-1974R1979)

***325. Space Conditioning***

Air Conditioners, Central Cooling (UL465)
Electric Power Systems in Commercial Buildings (ANSI/IEEE 241-1983)
Emergency and Standby Power Systems for Industrial and Commercial Applications (ANSI/IEEE 446-1980)
Manual for Calculating Heat Loss and Heat Gain for Electric Comfort Conditioning (NEMA HE1-1980)

***326. Specialty Transformers***

Systems of Insulation for Specialty Transformers (IEEE 259-1974R1980)

***327. Spectrum***

Spectrum Analyzers (IEEE 748-1979)

***328. Speech***

Speech Quality Measurements (IEEE 297-1969)
Volume Measurements of Electrical Speech and Program Waves (ANSI/IEEE 152-1953R1971)

***329. Speed***

Preparation of Equipment Specifications for Speed-Governing of Hydraulic Turbines Intended to Drive Electric Generators (ANSI/IEEE 125-1977)
Specification for Speed-Governing of Internal Combustion Engine Generator Units (IEEE 126-1959R1983)
Speed-Governing of Steam Turbines Intended to Drive Electric Generators (IEEE 122-1959)

***330. Stability***

Power System Analysis (ANSI/IEEE 399-1980)

***331. Standby Power***

Design of Reliable Industrial and Commercial Power Systems (ANSI/IEEE 493-1980)
Electric Power Systems in Commercial Buildings (ANSI/IEEE 241-1983)
Emergency and Standby Power Systems for Industrial and Commercial Applications (ANSI/IEEE 446-1980)

***332. Startup Power***

Emergency and Standby Power Systems for Industrial and Commercial Applications (ANSI/IEEE 446-1980)

***333. Substations***

Control and Low Voltage Cable Systems in Substations (IEEE 525-1978)
Electrical Power Connectors for Substations (NEMA CC1-1981)
Grounding of Industrial and Commercial Power Systems (ANSI/IEEE 142-1982)
Installation Design and Installation of Large Lead Storage Batteries for Generating Stations and Substations (IEEE 484-1981)
Maintenance, Testing, and Replacement of Large Lead Storage Batteries for Generating Stations and Substations (ANSI/IEEE 450-1980)
Primary Unit Substations (NEMA 201-1970R1976)
Qualifying Permanent Connections Used in Substation Grounding (IEEE 837-1984)

Safety in AC Substation Grounding (IEEE 80-1976)
Secondary Unit Substations (NEMA 210-1970R1976)
Seismic Design of Substations (IEEE 693-1983)
Sizing Large Lead Storage Batteries for Generating Stations and Substations (IEEE 485-1983)
Unit Substations (UL 1062)

***334. Surge Arresters***

Air Gap Surge-Protective Devices (ANSI/IEEE C623.32-1981)
Gas Tube Surge-Protective Devices (ANSI/IEEE C62.31-1981)
Surge Arresters (NEMA LA1-1976R1980)
Surge Arresters for AC Power Circuits (ANSI/IEEE C62.1-1981)
Surge Voltages in Low-Voltage AC Power Circuits (ANSI/IEEE C62.41-1980)
Test Specifications for Gas Tube Surge Protective Devices (ANSI/IEEE C62.31-1981)
Test Specifications for Low-Voltage Air Gap Surge Protective Devices (ANSI/IEEE C62.32-1981)
Test Specifications for Varistor Surge Protective Devices (ANSI/IEEE C62.33-1982)
Valve Type Surge Arresters for AC Systems (ANSI C62.2-1981)
Varistor Surge-Protective Devices (ANSI/IEEE C62.33-1982)

***335. Surge Testing***

Surge Withstand Capability Tests (IEEE 472-1974R1979)
Switching Surge Testing of Extra-High-Voltage Switches (IEEE 271-1966R1977)

***336. Switchboards***

Deadfront Distribution Switchboards (NEMA PB2-1978)
Electrical Deadfront Switchboards (UL 891)

***337. Switches***

Application of Interruptor Switches to Capacitance Loads (IEEE 22A-1962R1977)
Application, Installation, Operation and Maintenance of High-Voltage Air Disconnecting and Load Interrupter Switches (ANSI/IEEE C37.35-1976R1981)
Automatic Transfer Switches (ANSI/UL 1008-1983)
Basic Sensitive Switches (ANSI/EIA RS437.1-1978R1983)
Clock-Operated Switches (ANSI/UL-917-1981)
Design Test for High-Voltage Fuses, Distribution Enclosed Single-Pole Air Switches, Fuse Disconnecting Switches and Accessories (ANSI/IEEE C37.49-1981)
Digital Switches (ANSI/EIA RS457-1982)
Dry Reed Switches (ANSI/EIA RS421A-1983)
Enclosed and Deadfront Switches for Use in Hazardous Locations (ANSI/UL 98-1981)
Enclosed Switches (NEMA KS1-1975R1981)
General Purpose Wiring Devices (NEMA WD1-1979)
Knife Switches (ANSI/UL 363-1980)
Loading Guide for AC High-Voltage Switches (ANSI/IEEE C37.37-1979)
Miniature Sensitive Switches (ANSI/EIA RS437-2-1978R1983)
Non-Sensitive Pushbutton Switches (ANSI/EIA RS446A-1981)
Receptacles and Switches for Use with Aluminum Wire (UL 1567)
Sensitive Switches (ANSI/EIA RS437-1980)
Service Conditions and Definitions for High-Voltage Fuses, Distribution Enclosed Single-Pole Air Switches, Fuse Disconnecting Switches and Accessories (ANSI/IEEE C37.40-1981)
Snap Switches (ANSI/UL 20-1979)
Special-Use Switches (ANSI/UL 1054-1980)
Subminiature Sensitive Switches (ANSI/EIA RS437-1978R1983)
Switches for Use in Hazardous Locations (ANSI/UL 894-1977)
Switching Surge Testing of Extra-High-Voltage Switches (IEEE 271-1966R1977)
Test Switches for Transformer-Rated Meters (ANSI C12.14-1982)
Toggle Switches (ANSI/EIA RS480-1981)

***338. Switchgear***

AC High-Voltage Circuit Breakers (NEMA SG4-1975)
AC Motor Protection (ANSI/IEEE C37.76-1976R1981)
Application Guide for AC High-Voltage Circuit Breakers Rated on a Symmetrical Current Basis (ANSI/IEEE C37.010-1979)
Application, Installation, Operation and Maintenance of High-Voltage Air Disconnecting and Load Interrupter Switches (ANSI/IEEE C37.35-1976R1981)

Application, Operation, and Maintenance of Automatic Circuit Reclosers (ANSI/IEEE C37.61-1973R1979)
Application, Operation, and Maintenance of Distribution Cutouts and Fuse Links, Secondary Fuses, Distribution Enclosed Single-Pole Air Switches, Power Fuses, Fuse Disconnecting Switches, and Accessories (ANSI C37.48-1969R1974)
Automatic Circuit Reclosers and Fault Interrupters for AC Systems (ANSI/IEEE C37.60-1981)
Calculating Losses in Isolated-Phase Bus (ANSI/IEEE C37.23-1969R1977)
Calculation of Fault Current for Application of AC High-Voltage Circuit Breakers Rated on a Total Current Basis (ANSI/IEEE C37.5-1979)
Capacitance Current Switching for AC High-Voltage Circuit Breakers Rated on a Symmetrical Current Basis (ANSI/IEEE C37.012-1979)
Circuit Breakers and Circuit-Breaker Enclosures for Use in Hazardous Locations (UL877)
Definition, Specification, and Analysis of Manual, Automatic, and Supervisory Station Control and Data Acquisition (ANSI/IEEE C37.1-1979)
Definitions and Requirements for High-Voltage Air Switches, Insulators, and Bus Supports (ANSI/IEEE C37.30-1971R1977)
Definitions for Power Switchgear (ANSI/IEEE C37.100-1981)
Design Test for High-Voltage Fuses, Distribution Enclosed Single-Pole Air Switches, Fuse Disconnecting Switches and Accessories (ANSI/IEEE C37.41-1981)
Distribution Cutouts and Fuse Links (ANSI C37.42-1981)
Distribution Enclosed Single-Pole Air Switches (ANSI C37.45-1981)
Distribution Fuse Disconnecting Switches, Fuse Supports, and Current-Limiting Fuses (ANSI C37.47-1981)
Distribution Oil Cutouts and Fuse Links (ANSI C37.44-1981)
Electrical Power System Device Function Numbers (ANSI/IEEE C37.2-1979)
Evaluating the Effect of Solar Radiation on Outdoor Metal-Clad Switchgear (ANSI/IEEE C37.24-1971R1984)
Field Discharge Circuit Breakers Used in Enclosures for Rotating Electrical Machinery (ANSI/IEEE C37.18-1979)
Generator Ground Protection (ANSI/IEEE C37.101-1983)
Interrupters Used in Power Switchgear (ANSI C37.85-1972R1978)
Loading Guide for AC High-Voltage Switches (ANSI/IEEE C37.37-1979)
Low-Voltage AC Non-Integrally Fused Power Circuit Breakers (Using Separately Mounted Current-Limiting Fuses) (ANSI/IEEE C37.27-1972)
Low-Voltage AC Power Circuit Breakers Used in Enclosures (ANSI/IEEE C37.13-1981)
Low-Voltage DC Power Circuit Breakers Used in Enclosures (ANSI/IEEE C37.14-1979)
Low-Voltage AC Power Circuit Protectors Used in Enclosures (ANSI/IEEE C37.27-1972)
Low-Voltage Power Circuit Breakers (NEMA SG3-1981)
Measurement of Sound Pressure Levels for AC Power Circuit Breakers (ANSI/IEEE C37.082-1982)
Metal-Enclosed Low-Voltage Power Circuit-Breaker Switchgear (UL1558)
Molded Case Circuit Breakers (NEMA AB1-1975R1981)
Molded Case Circuit Breakers and Circuit Breaker Enclosures (UL489)
Oil-Filled Capacitor Switches for AC Systems (ANSI/IEEE C37.66-1969R1982)
Power Factor Measurement for Low-Voltage Inductive Test Circuits (ANSI/IEEE C37.26-1972R1976)
Power Fuses and Fuse Disconnecting Switches (ANSI C37.46-1981)
Power Switchgear Assemblies (NEMA SG5-1981)
Power Switching Equipment (NEMA SG6-1974R1981)
Preferred Ratings, Manufacturing Specifications, and Application Guide for High-Voltage Air Switches, Bus Supports, and Switch Accessories (ANSI C37.32-1972)
Procedures for Verifying the Performance of Molded Case Circuit Breakers (NEMA AB2-1980)
Protection and Coordination of Industrial and Commercial Power Systems (IEEE 242-1975)
Protection of Shunt Capacitor Banks (ANSI/IEEE C37.99-1980)
Protective Relay Applications of Audio Tones Over Telephone Channels (ANSI/IEEE C37.93.1976)
Protective Relay Applications of Power Transformers (ANSI/IEEE C37.91.1967R1980)
Protective Relay Applications to Power System Buses (ANSI/IEEE C37.97-1979R1984)

Protective Relaying of Utility-Consumer Interconnections (ANSI/IEEE C37.95.1973R1980)
Rated Control Voltages and Their Ranges for High-Voltage Air Switches (ANSI C37.33-1970R1976)
Rating Structure for AC High-Voltage Circuit Breakers Rated on a Symmetrical Current Basis (ANSI/IEEE C37.04-1979)
Relays and Relay Systems Associated with Electric Power Apparatus (ANSI/IEEE C37.90-1978R1982)
Requirements for Overhead, Pad Mounted Dry Vault and Submersible Automatic Circuit Reclosers and Fault Interrupters for AC Systems (ANSI/IEEE C37.60-1981)
Requirements for Overhead, Pad Mounted Dry Vault and Submersible Line Sectionalizers for AC Systems (ANSI/IEEE C37.63-1983)
Safety Labels for Padmounted Switchgear and Transformers Sited in Public Areas (NEMA 260-1982)
Safety Requirements for X-Radiation Limits for AC High-Voltage Power Vacuum Interrupters Used in Power Switchgear (IEEE C37.85-1982R1978)
Seismic Testing of Relays (ANSI/IEEE C37.98-1978)
Service Conditions and Definitions for High-Voltage Fuses, Distribution Enclosed Single-Pole Air Switches, Fuse Disconnecting Switches and Accessories (ANSI/IEEE C37.40-1981)
Station Control and Data Acquisition (ANSI/IEEE C37.1-1979)
Switches, Insulators, and Bus Supports (ANSI/IEEE C37.30-1981)
Switchgear Assemblies Including Metal-Enclosed Bus (ANSI/IEEE C37.20-1969R1981)
Synthetic Fault Testing of AC High-Voltage Circuit Breakers Rated on a Symmetrical Current Basis (ANSI/IEEE C37.081-1981)
Test Code for High-Voltage Air Switches (ANSI/IEEE C37.34-1971R1977)
Test Procedure for AC High-Voltage Circuit Breakers Rated on a Symmetrical Current Basis (ANSI/IEEE C37.09-1979)
Transient Recovery Voltage for AC High-Voltage Circuit Breakers Rated on a Symmetrical Current Basis (ANSI/IEEE C37.011-1979)
Trip Devices for AC and General Purpose DC Low-Voltage Power Circuit Breakers (IEEE C37.17-1972)

***339. Switching***

Capacitance Current Switching for AC High-Voltage Circuit Breakers Rated on a Symmetrical Current Basis (ANSI/IEEE C37.012-1979)
Power Switching Equipment (NEMA SG6-1974R1981)
Power Switching Semiconductor Devices (UL 1557)
Power System Analysis (ANSI/IEEE 399-1980)
Switching Surge Testing of Extra-High-Voltage Switches (IEEE 271-1966R1977)
Terms for Communication Switching (ANSI/IEEE 312-1977)

***340. Symbols***

Definitions, Symbols, and Characterization of Metal-Nitride-Oxide Field-Effect Transistors (IEEE 581-1978)
Graphic Symbols for Distributed Control/Shared Display Instrumentation, Logic and Computer Systems (ISA S5.3)
Graphic Symbols for Electrical and Electronics Diagrams (ANSI/IEEE 315-1975)
Graphic Symbols for Electrical Wiring and Layout Diagrams Used in Architecture and Building Construction (ANSI Y32.9-1972)
Graphic Symbols for Grid and Mapping Used in CATV Systems (ANSI/IEEE 623-1976)
Graphic Symbols for Heat Power Apparatus (ANSI Z32.2.6-1950R1956)
Graphic Symbols for Heating, Ventilating and Air Conditioning (ANSI Z32.2.4-1949R1953)
Graphic Symbols for Logic Diagrams (ANSI/IEEE 91-1984)
Induction Motor Letter Symbols (ANSI/IEEE 58-1978)
Instrument Loop Diagrams (ISA S5.4)
Instrumentation Symbols and Identification (ANSI/ISA S5.1-1975R1981)
Letter Symbols for Illuminating Engineering (ANSI Y10.18-1976R1977)
Letter Symbols for Quantities Used in Electrical Science and Engineering (ANSI Y10.5-1968)
Letter Symbols for Semiconductor Devices (IEEE 255-1963)
Letter Symbols for SI Units and Certain Other Units of Measurement (IEEE 260-1978)
Letter Symbols for Thermoelectric Devices (IEEE 261-1965)

Letter Symbols for Units of Measurement (ANSI/IEEE 260-1978)
Preparation and Use of Symbols (IEEE 267-1966)
Radio Frequency Radiation Hazard Warning Symbol (ANSI C95.2-1982)
Reference Designations for Electrical and Electronic Parts and Equipments (ANSI/IEEE 200-1975)
Terms, Letter Symbols, and Color Code for Hall Effect Devices (IEEE 296-1969)
Warning Symbol (ANSI N2.1-1969)

***341. Tachometers***

DC Tachometer Generators (IEEE 251-1963R1972)

***342. Tape***

Flutter Measurement of Instrumentation Magnetic Tape Recorders/Reproducers (ANSI/EIA RS405-1972R1979)
Magnetic Tape Pulse Recorders for Electricity Meters (ANSI C12.14-1982)
Measurements of Luminance Signal-to-Noise Ratio in Video Magnetic Tape Recording Systems (IEEE 618-1984)
Timing Error Measurements of Instrumentation Magnetic Tape Recorders/Reproducers (ANSI/EIA RS413-1973R1979)

***343. Telephone***

Determining Objective Loudness Ratings of Telephone Connections (ANSI/IEEE 661-1979)
Measuring Longitudinal Balance of Telephone Equipment Operating in the Voice Band (ANSI/IEEE 455-1976)
Measuring Transmission Performance of Telephone Sets (IEEE 269-1983)
Phone Plugs and Jacks (ANSI/EIA RS453-1978R1983)
Protective Relay Applications of Audio Tones Over Telephone Channels (ANSI/IEEE C37.93.1976)

***344. Teletherapy***

Procedures for Periodic Inspection of Cobalt-60 and Cesium-137 Teletherapy Equipment (ANSI N449.1-1978)

***345. Television***

Construction Drawings of Line Impedance Network for Measurement of Conducted Interference Output to the Power Line from FM and TV Broadcast Receivers (IEEE 214-1961)
Digital Terms Relating to Television (ANSI/IEEE 847-1982)
Electrical Performance Standards for Monochrome Closed Circuit TV Cameras (ANSI/EIA RS420-1975)
Measurement of Aspect Ratio and Geometric Distortion (ANSI/IEEE 202-1954R1978)
Measurement of Conducted Interference Output to the Power Line from FM and TV Broadcast Receivers (ANSI/IEEE 213-1961R1974)
Measurement of Differential Gain and Differential Phase (ANSI/IEEE 206-1960R1978)
Measurement of Luminance Signal Levels (ANSI/IEEE 205-1958R1972)
Measurement of Resolution of Camera Systems (ANSI/IEEE 211-1977)
Monochrome TV Broadcast Receivers (ANSI/IEEE 190-1960)
Spurious Radiation from FM TV Broadcast Receivers (IEEE 187-1951)
Surveillance Camera Units (ANSI/UL 983-1980)
Terms Relating to TV (ANSI/IEEE 201-1979)

***346. Temperature***

Construction and Interpretation of Thermal Limit Curves for Squirrel-Cage Motors (ANSI/IEEE 620-1981)
Soil Thermal Resistivity Measurements (ANSI/IEEE 442-1981)
Temperature-Indicating and -Regulating Equipment (UL 873)
Temperature Limit Controls for Electric Baseboard Heaters (NEMA DC10-1977)
Temperature Limits in Rating Electrical Equipment (IEEE 1-1969)
Temperature Measurement as Applied to Electrical Apparatus (IEEE 119-1974)
Temperature Measurement Thermocouples (ANSI/ISA MC96.1)
Thermal Evaluation and Establishment of Temperature Indexes of Solid Electrical Insulating Materials (IEEE 98-1972)
Thermal Evaluation of Insulation Systems for AC Electric Machinery Employing Form-Wound Pre-Insulated Stator Coils (ANSI/IEEE 275-1981)

***347. Terminals and Terminal Blocks***

Electrical Terminal Blocks (UL 1059)

Quick-Connect Terminals (ANSI/NEMA DC2-1976)
Quick-Connect Terminals (ANSI/UL 310-1980)
Terminal Blocks for Industrial Control Equipment and Systems (ANSI/NEMA ICS4-1977)

***348. Test Charts***

High Definition Facsimile Test Chart (IEEE 167A-1980)

***349. Test Equipment and Testing***

ATLAS Test Language (ANSI/IEEE 416-1981)
C/ATLAS Syntax (ANSI/IEEE 717-1982)
C/ATLAS Test Language (ANSI/IEEE 716-1982)
Calibration and Control of Measuring and Test Equipment Used in the Construction and Maintenance of Nuclear Power Generating Stations (IEEE 498-1980)
Dielectric Test Requirements for Power Transformers for Operation at System Voltages from 115kV to 230kV (IEEE C57.12.14)
Electrical and Electronic Measuring and Testing Equipment (ANSI/IEEE C1244-1981)
Field Testing of Relaying Current Transformers (ANSI/IEEE C57.13.1-1981)
Impulse Dielectric Testing of High-Temperature Insulated Wire (NEMA HP1-1979)
Liquid-Immersed Distribution, Power, and Regulating Transformers and Guide for Short-Circuit Testing of Distribution and Power Transformers (ANSI/IEEE C57.12.90-1980)
Periodic Testing of Diesel-Generator Units Applied as Standby Power Supplies for Nuclear Power Generating Stations (IEEE 387-1983)
Safety in High-Voltage and High Power Testing (IEEE 510-1983)
Seismic Testing of Relays (ANSI/IEEE C37.98-1978)
Terminology and Test Code for Shunt Reactors (ANSI/IEEE C57.21-1981)
Test Code for High-Voltage Air Switches (ANSI/IEEE C37.34-1971R1977)
Test Procedure for Thermal Evaluation of Oil-Immersed Distribution Transformers (ANSI/IEEE C57.100-1974)
Transformer Impulse Tests (ANSI/IEEE C57.98-1968)

***350. Textile***

Electric Installations on Textile Machinery (IEEE 77-1965R1972)

***351. Thermoelectric Devices***

Letter Symbols for Thermoelectric Devices (IEEE 261-1965)

***352. Thermostats***

Line-Voltage Integrally-Mounted Thermostats for Electric Heaters (NEMA DC13-1979)
Line-Voltage Thermostats (NEMA DC15-1979)
Low-Voltage Room Thermostats (NEMA DC3-1978)

***353. Thyristors***

General Purpose Thyristor DC Drives (IEEE 597-1983)
Thyristor AC Power Controllers (IEEE 428-1981)
Thyristor Converters and Motor Drives: Part 1—Converters for DC Motor Armature Supplies (ANSI/IEEE 444-1973)

***354. Transducers***

Electrical Transducer Nomenclature and Terminology (ANSI/ISA S37.1-1969-R1982)
Potentiometric Displacement Transducers (ANSI/ISA S37.12-1977R1982)
Potentiometric Pressure Transducers (ANSI/ISA S37.6-1975R1982)
Terms for Transducers (IEEE 196-1051)
Transducer and Transmitter Installation for Nuclear Safety Applications (ISA S67.01)

***355. Transformers***

Acceptance and Maintenance of Insulating Oil in Equipment (ANSI/IEEE C57.106-1977)
Application of Transformer Connections in Three-Phase Distribution Systems (ANSI/IEEE C57.105-1978)
Arc Furnace Transformers (ANSI C57.17-1965R1978)
Computer-Type Pulse Transformers (IEEE272-1970R1976)
Corona (Partial Discharge) Measurements on Electronics Transformers (IEEE 436-1977)
Detection and Determination of Generated Gases in Oil-Immersed Transformers and Their Relation to Serviceability of the Equipment (ANSI/IEEE C57.104-1978)
Dielectric Test Requirements for Power Transformers for Operation at System Voltages from 115kV to 230kV (IEEE C57.12.14)
Dielectric Test Requirements for Power Transformers for Operation of Effectively Grounded Systems (IEEE 262B-1977)

Distribution and Power Regulating Transformers (ANSI C57.12.00-1980)
Dry-Type Distribution and Power Transformers (ANSI C57.12.91-1979)
Dry-Type Transformers for General Applications (ANSI/NEMA ST20-1972R1978)
Electronics Power Transformers (IEEE 295-1969R1981)
Evaluation of Insulation Systems for Electronics Power Transformers (IEEE 266-1969R1981)
Field Testing of Relaying Current Transformers (ANSI/IEEE C57.13.1-1981)
Guide for Loading Oil-Immersed Power Transformers with 65C Average Winding Rise (NEMA TR98-1978)
High-Power Pulse Transformers (IEEE 391-1976)
High-Power Wide-Band Transformers (IEEE 264-1977)
High Reliability in Electronics Transformers and Inductors (ANSI/IEEE 392-1976R1982)
Installation, Application, Operation, and Maintenance of Dry-Type General Purpose Distribution Transformers (ANSI/IEEE C57.94-1982)
Liquid-Filled Distribution Transformers Used in Pad-Mounted Installations Including Unit Substations (ANSI C57.12.27-1982)
Liquid-Filled Transformers Used in Unit Installations Including Unit Substations (ANSI C57.12.13-1982)
Liquid-Immersed Distribution, Power, and Regulating Transformers and Guide for Short-Circuit Testing of Distribution and Power Transformers (ANSI/IEEE C57.12.90-1980)
Load-Tap Changing Transformers (ANSI C57.12.30-1977)
Loading Mineral Oil-Immersed Overhead and Pad-Mounted Distribution Transformers (ANSI/IEEE C57.91-1981)
Loading Mineral Oil-Immersed Power Transformers (ANSI/IEEE C57.92-1981)
Loading Mineral Oil-Immersed Power Transformers (IEEE 756)
Loading Oil-Immersed Step-Voltage and Induction-Voltage Regulators (IEEE C57.95-1955)
Low-Power Pulse Transformers (IEEE 390-1975)
Oil-Immersed EHV Transformers (ANSI/IEEE C57.12.12-1980)
Oil-Immersed Transformers (ANSI/IEEE C57.12.11-1980)
Overhead Type Distribution Transformers (ANSI C57.12.20-1981)
Pad-Mounted Compartmental-Type Self-Cooled Single-Phase Distribution Transformers with High-Voltage Bushings (ANSI C57.12.21-1980)
Pad-Mounted Compartmental-Type Self-Cooled Single-Phase Distribution Transformers with Separable Insulated High-Voltage Connectors (ANSI C57.12.25-1981)
Pad-Mounted Compartmental-Type Self-Cooled Three-Phase Distribution Transformers with High-Voltage Bushings (ANSI C57.12.22-1980)
Pad-Mounted Compartmental-Type Self-Cooled Three-Phase Distribution Transformers with Separable Insulated High-Voltage Connectors (ANSI C57.12.26-1975)
Power Line Coupling Capacitor Voltage Transformers (ANSI C93.2-1976)
Protection and Coordination of Industrial and Commercial Power Systems (IEEE 242-1975)
Protective Relay Applications of Power Transformers (ANSI/IEEE C37.91.1967R1980)
Requirements for Instrument Transformers (ANSI/IEEE C57.13-1978)
Safety Labels for Padmounted Switchgear and Transformers Sited in Public Areas (NEMA 260-1982)
Sealed Dry-Type Power Transformers (ANSI C57.12.52-1981)
Secondary Network Transformers (ANSI C57.12.40-1982)
Specialty Transformers (ANSI/NEMA ST1-1978)
Specialty Transformers (ANSI/UL 506-1979)
Systems of Insulation for Specialty Transformers (IEEE 259-1974R1980)
Terminal Markings and Connections for Distribution and Power Transformers (ANSI C57.12.70-1978)
Terminology and Test Code for Shunt Reactors (ANSI/IEEE C57.21-1981)
Test Procedure for Thermal Evaluation of Oil-Immersed Distribution Transformers (ANSI/IEEE C57.100-1974)
Testing Electronics Transformers and Inductors (IEEE 389-1979)
Terminology for Power and Distribution Transformers (ANSI/IEEE C57.12.80-1978)
Transformer Askarel in Equipment (IEEE 76-1974)
Transformer Impulse Tests (ANSI/IEEE C57.98-1968)
Transformers for Residential Controls (NEMA ST2-1973R1979)
Transformers, Regulators and Reactors (NEMA TR1-1980)

Transformers, 230 000 Volts and Below (ANSI C57.12.10-1978)
Underground-Type Single-Phase Distribution Transformers with Separable Insulated High-Voltage Connectors (ANSI C57.12.23-1978)
Underground-Type Three-Phase Distribution Transformers (ANSI C57.12.24-1982)
Ventilated Dry-Type Distribution Transformers (ANSI C57.12.50-1981)
Ventilated Dry-Type Power Transformers (ANSI C57.12.51-1981)
Wide Band Transformers (IEEE 111-1983)

***356. Transistors***

Definitions, Symbols, and Characterization of Metal-Nitride-Oxide Field-Effect Transistors (IEEE 581-1978)
Testing Transistors (IEEE 218-1956R1980)

***357. Transmission***

Installation of Overhead Transmission Line Conductors (ANSI/IEEE 524-1980)
Measuring the Transmission Characteristics of Analog Voice Frequency Circuits (IEEE 743/1983)
Measuring Transmission Performance of Telephone Sets (IEEE 269-1983)
Power Operations Terminology Including Terms for Reporting and Analyzing Outages of Electrical Transmission and Distribution Facilities and Interruptions to Customer Service (ANSI/IEEE 346-1973)
Tapered Tubular Steel Structures (NEMA TT1-1983)
Video Signal Transmission Measurements of Linear Waveform Distortion (ANSI/IEEE 511-1979)

***358. Transmission Lines and Associated Construction Material***

Cross-Linked Thermosetting-Polyethylene-Insulated Wire and Cable for the Transmission and Distribution of Electrical Energy (NEMA WC7-1982)
Ethylene-Propylene-Rubber-Insulated Wire and Cable for the Transmission and Distribution of Electrical Energy (NEMA WC8-1976)
Galvanized Ferrous Guy Attachments, Wrap and Formed Guy Hooks, Guy Strain Plates, and Pole Eye Plates (NEMA PH11-1979)
Galvanized Ferrous Insulator Clevices (NEMA PH20-1979)
Galvanized Ferrous Washers (ANSI/NEMA PH10-1977)
Measuring Unbalanced Transmission-Line Impedance (IEEE 314-1971)
Plastic Riser U-Guards (NEMA PH41-1981)
Plastic Tree Guards (ANSI/NEMA PH44-1981)
Rigid Coaxial Transmission Lines (ANSI/EIA RS225-1959)
Rubber-Insulated Wire and Cable for the Transmission and Distribution of Electrical Energy (NEMA WC3-1980)
Solid and Semi-Solid Dielectric Transmission Lines (ANSI/EIA RS1991-1972R1983)
Thermoplastic-Insulated Wire and Cable for the Transmission and Distribution of Electrical Energy (NEMA WC5-1973R1979)
Varnished-Cloth-Insulated Wire and Cable for the Transmission and Distribution of Electrical Energy (NEMA WC4-1976)

***359. Transmission Systems***

Ground-Fault Neutralizers, Grounding of Synchronous Generator Systems, Neutral Grounding of Transmission Systems (IEEE 143-1954)

***360. Transmitters***

Electrically Actuated Transmitters (ANSI/UL 632-1980)
Measurement of Spurious Emission from Land Mobile Communication Transmitters (ANSI/IEEE 377-1980)
Supplement to Terms for Radio Transmitters (IEEE 182A-1961)
Terms for Radio Transmitters (IEEE 182-1961)
Testing Radiotelegraph Transmitters (IEEE 183-1958)
Transducer and Transmitter Installation for Nuclear Safety Applications (ISA S67.01)

***361. Trip and Trip Devices***

Trip Devices for AC and General Purpose DC Low-Voltage Power Circuit Breakers (IEEE C37.17-1972)

***362. Tubes, Electron***

Bases for Geiger-Muller Counter Tubes (ANSI N42.5-1965R1983)
Definitions on Electron Tubes (ANSI/IEEE 161-1971R1980)

Direct Interelectrode Capacitances of Electron Tubes (ANSI/EIA RS191C-1981)
Electron Tubes—Bases, Caps, and Terminals (ANSI/EIA RS209A-1963)
Measurement and Characterization of Diode-Type Camera Tubes (IEEE 503-1978)
Measurement of Dimensions of Focal Spots of Diagnostic X-Ray Tubes (NEMA XR5-1974R1979)
Methods of Testing Electron Tubes (ANSI/IEEE 158-1962R1971)
Test Specifications for Gas Tube Surge Protective Devices (ANSI/IEEE C62.31-1981)

***363. Tubing***

Electrical Metallic Tubing (UL797)
Electrical Plastic Tubing and Conduit (NEMA TC2-1978)
Extruded Insulating Tubing (UL 224)
Flexible Nonmetallic Tubing for Electrical Wiring (UL 3)
PVC Externally Coated Galvanized Rigid Steel Conduit and Electrical Metallic Tubing (NEMA RN1-1980)
PVC Fittings for Use with Rigid PVC Conduit and Tubing (NEMA TC3-1981)

***364. Turbine Generators***

Operation & Maintenance of Turbine Generators (ANSI/IEEE 67-1972R1980)
Requirements for Gas Turbine Driven Cylindrical-Rotor Synchronous Generators (ANSI C50.14-1977)
Speed Governing Systems for Hydraulic Turbine-Generator Units (ANSI/ASME PTC29-1980)

***365. Turbines***

Preparation of Equipment Specifications for Speed-Governing of Hydraulic Turbines Intended to Drive Electric Generators (ANSI/IEEE 125-1977)
Pumping Mode of Pump/Turbines (ANSI/ASME PTC18.1-1978)
Speed-Governing of Steam Turbines Intended to Drive Electric Generators (IEEE 122-1959)
Steam Turbines for Mechanical Drive Service (NEMA SM23-1979)

***366. Type Tests***

Type Test of Class 1E Electric Cables, Field Splices and Connections for Nuclear Power Generating Stations (ANSI/IEEE 383-1974R1980)
Type Tests of Class 1E Modules Used in Nuclear Power Generating Stations (ANSI/IEEE 381-1977)
Type Tests of Continuous Duty Class 1E Motors for Nuclear Power Generating Stations (ANSI/IEEE 334-1974R1980)

***367. Ultrasonics***

Safety Standard for Diagnostic Ultrasound Equipment (NEMA UL1-1981)

***368. Units***

Letter Symbols for Units of Measurement (ANSI/IEEE 260-1978)

***369. Valves***

Electrically Operated Valves (UL 429)
Electrically Operated Valves for Use in Hazardous Locations (UL 1002)
Qualification of Safety-Related Valve Actuators (IEEE 382-1980)

***370. Varistors***

Non-Linear Symmetrical Varistors (ANSI/EIA RS350-1968R1981)
Varistor Definitions and Test Methods (ANSI/EIA RS349-1968R1981)

***371. Vaults***

Electric Power Systems in Commercial Buildings (ANSI/IEEE 241-1983)

***372. Vibration***

Performance of Aeolian Vibration Dampers for Single Conductors (ANSI/IEEE 664-1980)

***373. Voltage***

Aerospace Equipment Voltage and Frequency Ratings (IEEE 127-1963)
Electric Power Distribution for Industrial Plants (IEEE 141-1976)
Electric Power Systems in Commercial Buildings (ANSI/IEEE 241-1983)
Impulse Voltage Tests on Insulated Conductors (IEEE 82-1963R1971)
Maximum Electric Power Station Ground Potential Rise and Induced Voltage from a Power Fault (IEEE 367-1979)
Methods of Measurement of Radio Influence Voltage (RIV) of High-Voltage Apparatus (NEMA 107-1964R1981)

Transient Recovery Voltage for AC High-Voltage Circuit Breakers Rated on a Symmetrical Current Basis (ANSI/IEEE C37.011-1979)

*374. Voltage Dividers*

Decade Resistive Voltage Dividers (ANSI C100.5-1975)
Decade Transformer Dividers (ANSI C100.1-1972)

*375. Voltage Regulators and Reactors*

DC Carbon Pile Voltage Regulators for Aircraft (IEEE 134-1955)
Ferroresonant Voltage Regulators (IEEE 449-1984)
Loading Dry-Type and Oil-Immersed Current-Limiting Reactors (ANSI C57.99-1965)
Requirements, Terminology and Test Code for Current-Limiting Reactors (ANSI C57.16-1958R1971)
Loading Oil-Immersed Step-Voltage and Induction-Voltage Regulators (ANSI C57.95-1955)
Requirements, Terminology, and Test Code for Shunt Reactors (ANSI C57.21-1981)
Requirements, Terminology, and Test Code for Step-Voltage and Induction-Voltage Regulators (ANSI C57.15-1968)

*376. Voltmeters*

Electronic Voltmeters (IEEE 108-1955)

*377. Waveguides*

Circular Waveguides (ANSI/EIA RS200A-1965)
Fundamental Waveguide Terms (ANSI/IEEE 146-1980)
Nonlinear, Active, and Nonreciprocal Waveguide Component Terms (IEEE 457-1982)
Rectangular Waveguides (ANSI/EIA RS261A-1965)
Terms for Waveguide Components (ANSI/IEEE 147-1979)
Waveguide and Waveguide Component Measurements (148-1959R1971)

*378. Welding*

Arc-Welding Machines, Transformer Type (UL551)
Electric Arc-Welding Apparatus (ANSI/NEMA EW1-1983)
Resistance Welding Control (ANSI/NEMA ICS5-1978)
Semiautomatic Wire Feed Systems for Arc Welding (NEMA EW3-1983)

*379. Wire and Wiring*

Armored Cable (ANSI/UL 4-1980)
Asbestos, Asbestos-Varnished Cloth and Asbestos-Thermoplastic Insulated Wires and Cables (ANSI/NEMA WC1-1982)
Asbestos, Asbestos-Varnished Cloth or Tape Insulation (ANSI/UL 115-1981)
Communication Cables (ANSI C8.47-1977)
Cross-Linked Thermosetting-Polyethylene-Insulated Wire and Cable for the Transmission and Distribution of Electrical Energy (NEMA WC7-1982)
Electric Power Systems in Commercial Buildings (ANSI/IEEE 241-1983)
Ethylene-Propylene-Rubber-Insulated Wire and Cable for the Transmission and Distribution of Electrical Energy (NEMA WC8-1976)
Flat Cable (ANSI/IPC FC222-1981)
Flexible Cord and Fixture Wire (ANSI/UL62-1979)
General Purpose Wiring Devices (NEMA WD1-1979)
Graphic Symbols for Electrical Wiring and Layout Diagrams Used in Architecture and Building Construction (ANSI Y32.9-1972)
Impulse Dielectric Testing of High-Temperature Insulated Wire (NEMA HP1-1979)
Magnet Wire (ANSI/NEMA MW1000-1981)
Reference Standard for Electrical Wires, Cables, and Flexible Cords (UL 1581)
Rubber-Insulated Wire and Cable for the Transmission and Distribution of Electrical Energy (NEMA WC3-1980)
Rubber-Insulated Wires and Cables (ANSI/UL 44-1983)
Thermoplastic-Insulated Underground Feeder and Branch Circuit Cables (ANSI/UL 493-1983)
Thermoplastic-Insulated Wire and Cable for the Transmission and Distribution of Electrical Energy (NEMA WC5-1973R1979)
Thermoplastic-Insulated Wires and Cables (ANSI/UL 83-1979)
Varnished-Cloth-Insulated Wire and Cable for the Transmission and Distribution of Electrical Energy (NEMA WC4-1976)
Varnished-Cloth-Insulated Wires and Cables ANSI/UL 133-1981)

Weather-Resistant Polyolefin-Covered Wire and Cable (ANSI C8.35-1975)
Wires with Asbestos or Asbestos and Varnished Cloth or Tape Insulation (UL 115)

### *380. Wireways*

Electrical Wireways, Auxiliary Gutters and Associated Fittings (UL 870)

### *381. X Ray*

High-Voltage X-Ray Cables and Receptacles (NEMA XR7-1979)
Location of Primary Operating Controls for Spot-Film Devices (NEMA XR6-1979)
Measurement of Dimensions of Focal Spots of Diagnostic X-Ray Tubes (NEMA XR5-1974R1979)
Radiation Safety for X-Ray Diffraction and Fluorescence Analysis Equipment (ANSI N43.2-1977)
Radiological Safety of Radiographic and Fluoroscopic Industrial X-Ray Equipment (ANSI/NBS 123-1976)
Test Methods for Diagnostic X-Ray Machines for Use During Initial Installation (NEMA XR8-1979)
X-Ray Cassettes, Fluoroscopic Diaphragms, and Minimum Power Supply Requirements (NEMA XR2-1974R1979)
X-Ray Equipment (ABSI/UL 187-1974)

### *382. Bibliography*

1. Pearce Trade Association Survey 310, TNEC Monograph no. 18, 1941.
2. Scott The Institute's First Half Century. *Elec. Eng.*, May 1934.
3. Kennelly The Work of the Institute in Standardization, *Elec. Eng.*, May 1934.
4. Lamb and Shields *Trade Association Laws and Practice;* Boston, Little, Brown, 1971.
5. *United States v. Southern Pine Association,* Civ. 275 Filed Feb. 21, 1941, E.D.La.
6. National Macaroni Manufacturers Association, FTC Docket 8524, 65FTC, 1964.
7. Products Liability of Members of Standards Writing Committees, American National Standards Institute, 1975.
8. Annex to ISO/STACO WG3 Definitions/28, International Organization for Standards, 1975.
9. 1984 Catalog, American National Standards Institute.
10. 1984 Summer Listing of IEEE Standards, Institute of Electrical and Electronics Engineers.
11. Standards for Safety, Underwriters Laboratories, July 1984.
12. NEMA Publications and Materials, National Electrical Manufacturers Association, 1983.
13. Testing for Public Safety, Underwriters Laboratories, 1975.
14. Procedures Manual for Standards Development, American Nuclear Society, 1974.
15. ISA Publications and Education Aids, Instrument Society of America, 1984.
16. Manual of Organization and Procedure, EP-1-F, Electronics Industries Association, 1970.
17. AAMI Procedural Guidelines for the Development of Medical Device Standards, American Association for Medical Instrumentation.
18. Alden, A. The SMPTE and the American National Standards Program, Society of Motion Picture and Television Engineers.
19. Statutes and Rules of Procedure, International Electrotechnical Commission.
20. Procedures for the Standards and Regulation Development Process, U.S. Environmental Protection Agency.
21. Fact Sheet No. 30: Standards Development under the Consumer Product Safety Act, Consumer Product Safety Commission, 1974.
22. Nobel, J. J. and Cangelosi, R. J. Standards and Specifications for Medical Devices, *Adv. Biomed. Eng.*, vol. 4, 1974.
23. Procedure for Promulgation, Modification, or Revocation of Standards, U. S. Department of Labor, Occupational Safety and Health Administration, 1974.

# INDEX

## ABOUT THE EDITORS

DONALD G. FINK was formerly Executive Director and General Manager of the Institute of Electrical and Electronics Engineers. He holds a BS degree in Electrical Communications from the Massachusetts Institute of Technology and an MS degree in Electrical Engineering from Columbia University.

Following his graduation, Mr. Fink was a research assistant at M.I.T. In 1939 he joined the staff of the journal *Electronics*, and served as its Editor in Chief from 1946 to 1952, when he joined the staff of the Philco Corporation. In 1956 and 1957 he was Editor of the *Proceedings of the IRE* and in 1958 served as President of the Institute of Radio Engineers. In 1961 he was appointed Vice President, Research, of the Philco Corporation, and later became Director of the Philco Scientific Laboratory.

Mr. Fink is a Fellow of the IEEE and IEE (London), and an Eminent Member of the Eta Kappa Nu. He is also the recipient of a number of distinguished awards for service to the U.S. Government. Mr. Fink holds two patents on stereophonic systems, and is the author of numerous books, including *Engineering Electronics*, *Television Engineering Handbook*, and *Radar Engineering*, and is the Editor in Chief of the *Electronics Engineers' Handbook*, now in its second edition.

H. WAYNE BEATY graduated from the University of Houston with a BS degree in Electrical Engineering and is a Senior Member of the IEEE. During his career he was a Senior Editor with McGraw-Hill's *Electrical World* magazine, Manager of Member Services for the Electric Power Research Institute, and is now Vice President of Loadmaster Systems, Inc. Mr. Beaty is also a consultant in the electrical engineering field.